PROTEINS, PEPTIDES AND AMINO ACIDS SOURCEBOOK

JOHN STEPHEN WHITE
DOROTHY CHONG WHITE

WHITE Technical Research GROUP, Argenta, IL

HUMANA PRESS
TOTOWA, NEW JERSEY

© 2002 Humana Press Inc.
999 Riverview Drive, Suite 208
Totowa, New Jersey 07512

For additional copies, pricing for bulk purchases, and/or information about other Humana titles, contact Humana at the above address or at any of the following numbers: Tel.: 973-256-1699; Fax: 973-256-8341; E-mail:humana@humanapr.com; Website: http://humanapress.com

Cover design by Patricia F. Cleary.

This publication is printed on acid-free paper. ⬭∞
ANSI Z39.48-1984 (American National Standards Institute) Permanence of Paper for Printed Library Materials.

Printed in the United States of America.   10 9 8 7 6 5 4 3 2 1
Library of Congress Cataloging-in-Publication Data

White, John Stephen, 1958–
        Proteins, peptides and amino acid sourcebook / John Stephen White, Dorothy Chong White.
              p. cm.
        Includes index.
        ISBN 0-89603-613-8 (alk. paper)
        1. Proteins–Handbooks, manuals, etc. 2. Peptides–Handbooks, manuals, etc. 3. Amino acids–Handbooks, manuals, etc. 4. Proteins–Directories. 5. Peptides–Directories. 6. Amino acids–Directories. I. White, Dorothy Chong. II. Title.
        QP551.W53 2002
        572′.6–dc21
                                                                                                2002022200

It is to our children—
Paul, Elise, Diana, Jessamyn, and Samuel
—that we dedicate this book.

Their understanding and encouragement were given without reservation
throughout a process that lasted longer than any of us bargained for.

PREFACE

Proteins, Peptides and Amino Acids SourceBook is the second in a series of reference books conceived to cover the explosive growth in commercially available biological reagents. The success of our first reference work, *Source Book of Enzymes* published in 1997, encouraged us to continue this series. Choosing proteins, peptides, and amino acids as the subject matter for the second volume was simple, given their preeminence in regulating biochemical processes and their importance to modern molecular biology.

The *SourceBook* series was inspired by our difficulty in locating a suitable replacement for a depleted reagent in the midst of an urgent research project. To our dismay, we found the reagent supplier out of business and the product line no longer available. Other reagent catalogs on our library bookshelf offered a narrow selection and incomplete functional information. We were ultimately able to locate a satisfactory alternative only by making countless inquiries and paging through innumerable product catalogs and technical data sheets. We needed—but could not find—a single resource that cataloged available compounds, organized them in a logical and accessible format, provided critical technical information to distinguish one from another, and told us where we could buy them.

We conceived the *SourceBook* series as a unique reference tool to satisfy this need. It has been carefully designed to inform users of the broad variety of reagent products available from commercial suppliers worldwide. *Proteins, Peptides and Amino Acids SourceBook* facilitates the selection process by providing systematic and comparative functional information about each product. The task of maintaining and searching scores of supplier catalogs and product data bulletins is eliminated.

Proteins, Peptides and Amino Acids SourceBook will be a practical tool for researchers in academia, industry, and government. Students and educators will regularly refer to its concise information and helpful indexes. Reagent suppliers and marketers will find the *SourceBook* an invaluable resource in conducting competitive assessments and identifying new product trends and opportunities. And to speed their identification and ordering of specific products, purchasing agents will rely on *Proteins, Peptides and Amino Acids SourceBook*'s distillation of critical product information into thorough reference tables and a comprehensive index of worldwide suppliers, brokers and distributors.

Proteins, Peptides and Amino Acids SourceBook is organized in a user-friendly format that combines the following unique features in one expansive volume:

- An encyclopedic comparison of more than 26,000 commercially available proteins, peptides, and amino acids.

- A worldwide listing of nearly 500 reagent suppliers, brokers and distributors.

- Critical functional information, proven applications and literature references.

- Comparative sequence, biological source, molecular weight, purity, biochemical and physical data, and physical form information to ensure the selection of suitable products.

- An informative User's Guide that takes the reader through the unique features of *Proteins, Peptides and Amino Acids SourceBook*.

- Helpful listing of abbreviations and acronyms used throughout the world in describing proteins, peptides and amino acids.

- Complete indexes of peptide sequences and names, synonyms and derivatives.

It is our hope that *Proteins, Peptides and Amino Acids SourceBook* finds a place on your bookshelf as a valued research tool, and that it eases the process of locating the reagent and supplier best suited to your needs and geographic location.

We made a concerted effort to make *Proteins, Peptides and Amino Acids SourceBook* an exhaustive and global compendium of all commercially available proteins, peptides and amino acids. That we fell short of this goal is due in part to our ignorance of additional suppliers and our inability to keep pace with the development and commercialization of new products. If we overlooked your products, please accept our apology and invitation to contribute to the next edition.

Because of the continuing discovery, development, and commercialization of new reagents and sources, the information in this book will quite naturally become dated. To address this shortcoming, the authors and publishers are committed to updating *Proteins, Peptides and Amino Acids SourceBook* and the *SourceBook* series on a regular basis. We welcome your suggestions for corrections and additions to the books. Particularly valued are your suggestions for improving the content, readability, accessibility and organization.

John Stephen White
Dorothy Chong White
WHITE Technical Research GROUP
8895 Hickory Hills Drive
Argenta, Illinois 62501 USA
Phone: 217-795-4437
E-mail: wtrg@argenta.net

About the Authors

Dorothy Chong White has worked for twenty years in the corn wet-milling industry, holding varied positions in basic research and product development for the A.E. Staley Manufacturing Company in Decatur, Illinois (a subsidiary of Tate & Lyle, North America). She has developed considerable expertise in manufacturing process troubleshooting and product characterization. She has the distinction of being a member of the research team which won technical approval for 100% use of high fructose corn syrup-55 as a replacement for sucrose in the carbonated beverage industry.

Ms. White holds two US patents and has been actively engaged in new product development from biotechnology for the past five years. Her scientific training was taken at the Universities of Illinois at Champaign-Urbana and Springfield.

John Stephen White developed an enduring fascination for the use of enzymes and other biological reagents through practical experience in the isolation and purification of enzymes for academic research, development of a multienzyme in vitro enzyme system to predict the in vivo digestibility of carbohydrate polymers, and large-scale applications of enzymes for food and industrial ingredient manufacture.

His training in biochemistry was taken at the University of California at San Diego and the University of Utah, Salt Lake City. Dr. White held joint appointments at the University of Illinois, Champaign-Urbana, as postdoctoral fellow and visiting assistant professor of biochemistry. He spent thirteen years in research and management in the corn wet-milling industry, one of the largest consumers of commercial bioreagents, and is the author of more than 25 journal articles, book chapters and reference books.

Dr. White has worked extensively with scientists from the food and beverage industries, academia and government to investigate nutritional issues surrounding traditional and novel foods and ingredients. He has enjoyed professional affiliations with the Calorie Control Council, the Corn Refiners Association, the Institute of Food Technologists, and the International Life Sciences Institute, which he has served as trustee and scientific advisor. Dr. White is founder and principal in the consulting firm of WHITE Technical Research GROUP, established in 1994 to serve scientific research and management.

Acknowledgments

The authors express their gratitude to the many marketing, technical service, customer service, research, and public affairs personnel from reagent suppliers who gave so generously of their time to provide us with technical information about their protein, peptide and amino acid products.

Finally, we thank our long-suffering editors and publishers at Humana Press for their support, encouragement and prodding during development of this book. We hope you agree that it was worth the wait.

Contents

Part 1. User's Guide

Like other volumes in this series, *Proteins, Peptides & Amino Acids SourceBook* was created with two primary purposes in mind: to increase accessibility to globally-available products and provide technical information to aid the product selection process. In Part 1, we will guide you through the product selection process using information found in *Proteins, Peptides & Amino Acids SourceBook*.

If you know the protein or amino acid name...

locating the compound is easy, since information in *Proteins, Peptides & Amino Acids SourceBook* is grouped alphabetically within these natural categories. Proteins are listed in Part 2, beginning on page 5. Amino acids and their derivatives are listed in Part 5, beginning on page 715.

If you cannot readily locate a protein or amino acid...

consult the General Index, beginning on page 1021. Compounds and page locations are listed alphabetically within this index by name, synonym and derivative formula.

If you are looking for a specific peptide...

the compound will be located in either of two sections, depending on whether sequence information was provided by the supplier. Peptides with known sequences are listed alphabetically by basic sequence (no modification) in Part 3, beginning on page 207. Peptides lacking supplier sequence information are listed alphabetically by name in Part 4, beginning on page 671.

If you cannot readily locate a peptide...

consult the Sequences Index, beginning on page 975. Basic peptide sequences, without chemical modifications, are arranged alphabetically within this index. If you still cannot locate the product, consult the General Index, beginning on page 1021. It may be that the compound you seek is a fragment of a larger protein, or that the name you seek is a synonym for a compound with a more common name. Compounds and page locations are listed alphabetically in this index by name, synonym and derivative formula.

If you only know the acronym...

consult the Glossary of Abbreviations & Acronyms (Appendix B), beginning on page 963. The product you seek likely appears in *Proteins, Peptides & Amino Acids SourceBook* under its full name.

Browsing can be beneficial...

even if you don't have a specific purpose in mind. You may browse the General and Sequences Indexes for names, synonyms, derivative formulas or basic sequences that complement your application. Alternatively, novel applications may occur to you while browsing information within the broad protein, peptide or amino acid sections of Parts 2-5.

Selecting the right product for your application...

is the final step once the compound is located. *Proteins, Peptides & Amino Acids SourceBook* simplifies this task by providing product-specific technical data in an easy-to-use format that encourages product comparisons.

Use the technical data as selection criteria for comparing and contrasting available products from different suppliers. Select that product which most nearly matches your requirements for sequence, sequence modification, chemical derivative, preparation form, purity, composition, activity, functionality and temperature/pH optima.

- Biological source—responsible for the wondrous variety in sequence, stability, reactivity, temperature and pH optima, molecular weight and activity.
- Molecular weight—given in units of molar mass for small molecules or Daltons for macromolecules.
- Preparation form—describes the physical form of the product (crystallized, lyophilized, etc.) and any added buffers, salts, stabilizers or preservatives. Discloses product contaminants.
- Product notes—lists compound importance, known applications and literature references.

Confused about...

a term or abbreviation used in the product data tables? Appendix B, beginning on page 963, contains an extensive Glossary of terms used in *Proteins, Peptides & Amino Acids SourceBook* that will clarify those with which you are unfamiliar.

To order the selected product...

simply record the supplier's catalog number from the product listing and locate their office, broker or distributor closest to you from the Reagent Supplier Appendix (Appendix A), beginning on page 941.

Conventions...

Verbal descriptions of product attributes have been condensed and standardized in order to simplify reader comparisons. The authors have endeavored to do this without compromising supplier product claims.

The abbreviations and acronyms used in this work are those in common use; they are listed in Appendix B of Part 6. Since they differ somewhat around the world, we adopted for use those that seemed most descriptive and straightforward. New abbreviations and acronyms were coined only where necessary to avoid ambiguity.

The authors have made every attempt to ensure that the information contained in this work is accurate. Since the field of proteins, peptides and amino acids is changing so rapidly, readers are encouraged to verify product information with suppliers before ordering.

Information for this book was gathered from supplier catalogs, data sheets and electronic data bases. Obvious discrepancies between suppliers were resolved where possible. The sheer volume of data in this work, however, made it impossible to verify the veracity of each supplier's claims.

Proteins, Peptides & Amino Acids SourceBook specifications...

- 5,300 protein products based on 2,065 unique molecular compounds are listed alphabetically in Part 2.
- 12,300 peptide products based on 4,700 unique sequences are listed alphabetically by basic sequence (Part 3) or name (Part 4).
- 8,400 amino acid products based on 6,030 unique molecular compounds are listed alphabetically in Part 5.
- 475 reagent suppliers, brokers and distributors based in 70 countries in Appendix A of Part 6.

4

Part 2. Proteins

4-IBBL

PeproTech 310-11   Human recombinant, expressed in *E. coli*
MW 19.5k   >98% (SDS-PAGE) & HPLC; <0.1 ng endotoxin per
mg (1EU/mg); lyophilized | Member of the emerging family of
ligands with structural homology to tumor necrosis factor; 185 AA;
$ED_{50} \leq 10$ng/mL; SA determined by the dose-dependent
stimulation of IL-8 production by human PBMC

6CKINE

Synonyms: Exodus II; SLC; T-Cell Activation Gene IV

Biodesign A52013M   *E. coli*   Purified | Species specificity:
mouse

Biodesign A52035H   *E. coli*   Purified

Chemicon GF086   Human   ≥95%

Biogenesis 4396-0150   Human r-DNA   Lyophilized

BioSource International PHC1474   Human recombinant

PeproTech 300-35   Human recombinant, expressed in *E. coli*
MW 12.2k   >98% (SDS-PAGE) & HPLC; <0.1 ng endotoxin per
mg (1EU/mg); lyophilized | In the CC chemokine family; similar to
Exodus-1 (aka MIP-3α & LARC); inhibits hemopoiesis & stimulating
chemotaxis; 111 AA; SA determined by its ability to chemoattract
total lymphocyte population

Sigma C 0720   Human recombinant, expressed in *E. coli*
Lyophilized from 30% acetonitrile/0.1% TFA containing 1.25 BSA

Biogenesis 4396-0350   Mouse r-DNA   Lyophilized

Sigma C 0845   Mouse recombinant, expressed in E. *coli*
Lyophilized from 30% acetonitrile/0.1% TFA containing 1.25 BSA |
Endotoxin tested; cell culture tested; long-term growth &
metabolism of most mammalian cells in culture require that growth
factors, hormones & other factors be present in the basal medium;
these factors are provided by fetal bovine serum &/or other sera
routinely used in cell culture protocols; endotoxin tested; cell
culture tested

PeproTech 250-13   Murine recombinant, expressed in *E. coli*
MW 12.0k   >98% (SDS-PAGE) & HPLC; <0.1 ng endotoxin per
mg (1EU/mg); lyophilized | In the CC chemokine family; similar to
Exodus-1 (MIP-3a/LARC); inhibits hemopoiesis & stimulating
chemotaxis 110 AA; SA determined by its ability to chemoattract
total murine T cell population

Abrin Toxin A

Sigma L 2017   *Abrus precatorius* (Jequirity bean)   Highly
purified by affinity & ion exchange chromatography; solution in
0.01 *M* potassium phosphate, pH 7.2, containing 0.15 *M* NaCl &
0.1% sodium azide; activity: >50 µg/mL (the lowest concentration
to agglutinate a 2% suspension of human erythrocytes after 1 hr at
25°C) | Lectin

Abrin Toxin A & C

Sigma L 9633   *Abrus precatorius* (Jequirity bean)   Highly
purified by affinity chromatography according to the method of
Wei, CH et al, *J Biol Chem*, 249: 3061, 1974; solution in 0.01 *M*
potassium phosphate, pH 7.2, containing 0.15 *M* NaCl & 0.1%
sodium azide; activity: <60 µg/mL (the lowest concentration to
agglutinate a 2% suspension of human erythrocytes after 1 hr at
25°C) | Lectin

Abrin Toxin C

Sigma L 1892   *Abrus precatorius* (Jequirity bean)   Highly
purified by affinity & ion exchange chromatography; solution in
0.01 *M* potassium phosphate, pH 7.2, containing 0.15 *M* NaCl &
0.1% sodium azide; activity: <60 µg/mL (the lowest concentration
to agglutinate a 2% suspension of human erythrocytes after 1 hr at
25°C) | Lectin; this is a further purification of Sigma L 9633 to
eliminate the abrin A content

Acetylcholine Transporter Protein

Biotrend 0030-5059   Rat, synthetic   >80% | Antigen

Acid Glycoprotein VI, α_1-

Synonyms: Orosomucoid

Fluka 50646   Human plasma   MW 45k   ≥98% (GE); 60%
protein; 6% water | Arnaud, P et al, *Meth Enzymol*, 163: 418,
1988; Eap, CB & Baumann, P, *Electrophoresis*, 9: 650, 1988

Acid Glycoprotein, α_1-

Synonyms: Orosomucoid; Seromucoid, α_1-

Fluka 50647   Bovine serum   ≥99% (gel electrophoresis); ≤5%
water; ~70% protein

Biodesign A50106H   Human   Purified

Fitzgerald 30-AA01   Human plasma   High purity |
Orosomucoid

ICN 153905   Human plasma   MW 44.1k   Lyophilized, salt
free, >99%

ICN 191348   Human plasma   MW 44.1k   Lyophilized, salt
free, >99% (SDS-PAGE) | Negative for HBs Ag & HIV antibodies;
Schmid, K etal, *Biochemistry*, 12:2711, 1973

Scipac P153-0   Orosomucoid, from serum/plasma   >99%;
lyophilized | Acute phase serum protein

Scipac P153-1   Orosomucoid, from serum/plasma   >96%;
lyophilized | Acute phase serum protein

Scipac P153-2   Serum/plasma   40-90%; lyophilized | Serum
protein

Actibind™ K

ICN 797091/797092   *Streptococcus*   Protein from a variant
of *Streptococcal* protein L that binds κ light chains &/or the F$_{ab}$
portion of immunoglobulins of all types; useful in the purification of
monoclonal Ab & F$_{ab}$ Ab fragments

Actin

ICN 771012   MW 43k   Purified | Suitable for use as a high
MW marker or as a standard; may be used as an antigen for Ab
development

Fluka 01813   Bovine muscle   ≥90% (gel electrophoresis);
≥99% protein; white powder

Cortex CP1110   Rabbit   >95%

Biodesign A08001R   Rabbit muscle   Purified

Biogenesis 0070-2504   Rabbit muscle   MW 46k   Purified;
50% glycerol; liquid | Stimulates myosin ATPase activity if first
polymerized

Fluka 01812   Rabbit muscle   MW ~43k   ~96% (gel
electrophoresis) | One of the major proteins of muscle & an
important component of all eukaryotic cells; useful as a reference
standard for SDS-PAGE & 2D-electrophoresis; role in muscle
contraction & relaxation, also involved in a variety of cellular
events: cell movement, cytokinesis, chromosome movement,
phagocytosis & exocytosis; Pudich, J, Watt, S, *JBC*, 246: 4866,
1971; Vandekerckhove, J & Weber, K, *Eur J Biochem*, 90: 451,
1978; Korn, ED, *Physiol Rev*, 62: 672, 1982; Korn, ED, *PNAS*, 75:
588, 1978

ICN 159848   Rabbit muscle   3µg/g inhibits 1 µg of DNAse I;
>90% | Prepared by a slightly different procedure (reference
below); can be used in cell transport & motion studies; inhibits
DNAse I & possibly controls DNAse I nucleolytic activity during the
cell cycle; Spudich, JA & S Watt, J Biol Chem, 246:4866, 1971

Actinin

Biodesign A08002C   Chicken gizzard   Purified

Actinin, α-

Sigma A 9776   Chicken Gizzard   MW ~100k   ~80% (SDS-
PAGE); suspension in 2 *M* (NH$_4$)$_2$SO$_4$ containing 20 *mM* Tris
acetate, pH 7.6, 20 *mM* sodium chloride, 0.1 *mM* EDTA, 15 *mM* β-
mercaptoethanol & 1 *mM* phenylmethylsulfonyl fluoride | Prepared
using a modification of the procedure of Neidel, JE & Cuatrecasas,
P, *Biochem Biophys Res Commun*, 91: 152, 1979

Actinomycin C

Synonyms: Actinomycin C_1; Actinomycin C_2; Actinomycin C_3

Sigma A 4639 *Streptomyces chrysomallus* ~95% (HPLC) | Mixture of actinomycins C_1, C_2, & C_3

Actinomycin D

Synonyms: Actinomycin IV; Actinomycin C_1

Sigma A 1410 *Streptomyces* species FW 1255.4 ~98% (HPLC); | Protein antibiotic that inhibits cell proliferation by forming a stable complex with double-stranded DNA & inhibiting RNA synthesis

Sigma A 4262 *Streptomyces* species FW 1255.4 ~95% (HPLC);

Actinomycin D, Mannitol

Sigma A 5156 Lyophilized powder; each vial contains 1 mg actinomycin D & 49 mg of mannitol | Soluble at 20 mg/mL in water

Actinomycin I

Sigma A 3012 *Streptomyces antibioticus* FW 1263.5 ~80% (HPLC)

Actinomycin V

Sigma A 2887 Streptomyces *antibioticus* FW 1269.4 ~95% (HPLC)

Activin A

Sigma A 4839 Bovine recombinant, expressed in *E. coli* Lyophilized from 40% acetonitrile/0.1% TFA | Endotoxin tested; cell culture tested; see Sigma C 0720

Acyl Carrier Protein

Sigma A 7303 *Escherichia coli* ≥60% pure by electrophoresis in 1% 2-β-MSH; lyophilized

Adenovirus Type VI

USBio A0880-02 Single peak by HPLC; 5 mg/mL protein; in MEM buffer | Suitable for antigenic applications in immunological protocols

Adenylate Cyclase Toxin

Alexis 630-088 *Bordetella pertussis* recombinant, expressed in *E. coli* MW ~177k ≥90%; lyophilized powder; soluble in water; reconstitute in 50 µL of sterile distilled water | Important virulence factor of *Bordetella pertussis* & a novel research tool for manipulation of cAMP levels in mammalian cells; ability to interact with target cells, insert into the cytoplasmic membrane & deliver its adenylate cyclase enzymatic domain to the cell interior; a calcium-binding protein & its ability to intoxicate target cells is calcium dependent; the cell entry process, unlike that of many other toxins, does not involve receptor-mediated endocytosis; Wolff, J et al, *PNAS*, 77: 3841, 1980; Confer, DL & Eaton, JW, *Science*, 217: 948, 1982; Weiss, A & Hewlett, EL, *Ann Rev Microbiol*, 40: 661, 1986; Glaser, P et al, *Mol Microbiol*, 2: 19, 1988; Sakamoto, H et al, *J Biol Chem*, 267: 13598, 1992; Mock, M & Ullmann, A, *Trends in Microbiol*, 1: 187, 1993; Hewlett, EL et al, *J Biol Chem*, 268: 7842, 1993; Szabo, G et al, *J Biol Chem*, 269: 22496, 1994; Benz, R et al, *J Biol Chem*, 269: 27231, 1994; Hackett, M et al, *Science*, 266: 433, 1994; Hackett, M et al, *J Biol Chem*, 270: 20250, 1995

ICN 195877 Recombinant, expressed in *E. coli* MW 177k 400 µmols cAMP/min/mg protein

Adhesive Protein

Synonyms: BioGlue

Sigma A 2707 Mytilus *edulis* (blue mussel) ~1 mg/mL in 5% acetic acid | On neutralization, forms very strong bonds to any surface

Aequorin

ICN 194084 Jellyfish (*Aequorea* sp.) Purified powder | A bioluminescent protein used in immunoassay procedures; reported to measure calcium serum & subcellular organelle levels <10 µM; Izutsu, KT & SP Felton, *Clin Chem*, 18:77, 1972

Aequorin Type III

Sigma A 4140 *Aequorea* species (jellyfish) ~0.5% protein (Bradford) in buffer salts consisting mainly of sodium EDTA, pH 7.5 | A bioluminescent protein; used to measure physiological levels of calcium in serum & subcellular organelles down to 10 µM; 1 mg solid will yield ~200 calcium assays; Izutsu, KT & Felton, SP, *Clin Chem*, 18: 77 (1972); Azzi, A & Chance, B, *Biochim Biophys Acta*, 189: 141 (1969)

Aflatoxin B_1, $(^3H(G))$-

ARC ART-247 MW 312.3 15-30 Ci/mmol; 0.55-1.11 TBq/mmol; in methanol | Radiochemical

Aflatoxin B_2, $(8,9-^3H)$-

ARC ART-617 MW 314.3 20-50 Ci/mmol; 0.74-1.85 TBq/mmol; in methanol | Radiochemical

Aflatoxin M_1, $(^3H(G))$-

ARC ART-616 MW 328.3 1-5 Ci/mmol; 37-185 GBq/mmol; in methanol | Radiochemical

Agglutinin, *Abrus precatorius*

Synonyms: Abrin; Lectin

Sigma L 9758 *Abrus precatorius* (Jequirity bean) Highly purified by affinity chromatography according to the method of Lin, J-Y et al, *Toxicon*, 19: 41, 1981; solution in 0.01 *M* potassium phosphate, pH 7.2, containing 0.15 *M* NaCl & 0.1% sodium azide; activity: <2 µg/mL (the lowest concentration to agglutinate a 2% suspension of human erythrocytes after 1 hr at 25°C) | Extremely hazardous! Be aware of the risks & familiar with safety procedures before you ruse this product; not blood group specific but has an affinity for D-galactose; seeds of *Abrus precatorius* contain a nontoxic agglutinin together with 2 major toxic proteins, Abrin A^2 also referred to as Abrin b+c & Abrin C^2 also referred to as Abrin a; conjugates are prepared from affinity purified lectin; activity is expressed in µg/mL & is determined from serial dilutions in phosphate buffered saline, pH 7.2 of a 1 mg/mL solution; Tomita, M et al, *Experientia*, 28: 84, 1972; Wei, CH et al, *J Biol Chem*, 249: 3061, 1974; Lin, J-Y et al, *Toxicon*, 19: 41, 1981; Olsnes, S et al, *J Biol Chem*, 249: 803, 1974

Agglutinin, *Abrus precatorius*, Biotin Conjugated

Synonyms: Abrin; Lectin

Sigma L 2266 *Abrus precatorius* (Jequirity bean) Solution in 0.01 *M* potassium phosphate, pH 7.3, containing 0.15 *M* NaCl & 0.1% sodium azide; contains ~3 moles biotin/mole protein | Extremely hazardous! Be aware of the risks & familiar with safety procedures before you ruse this product; not blood group specific but has an affinity for D-galactose; seeds of *Abrus precatorius* contain a nontoxic agglutinin together with 2 major toxic proteins, Abrin A^2 also referred to as Abrin b+c & Abrin C^2 also referred to as Abrin a; conjugates are prepared from affinity purified lectin; activity is expressed in µg/mL & is determined from serial dilutions in phosphate buffered saline, pH 7.2 of a 1 mg/mL solution; Tomita, M et al, *Experientia*, 28: 84, 1972; Wei, CH et al, *J Biol Chem*, 249: 3061, 1974; Lin, J-Y et al, *Toxicon*, 19: 41, 1981; Olsnes, S et al, *J Biol Chem*, 249: 803, 1974

Agglutinin, *Abrus precatorius*, FITC Conjugated

Synonyms: Abrin; Lectin

Sigma L 9883   *Abrus precatorius* (Jequirity bean)   Solution in 0.01 *M* potassium phosphate, pH 7.2, containing 0.15 *M* NaCl & 0.1% sodium azide; contains 1-2 moles FITC/mole protein | Extremely hazardous! Be aware of the risks & familiar with safety procedures before you ruse this product; not blood group specific but has an affinity for D-galactose; seeds of *Abrus precatorius* contain a nontoxic agglutinin together with 2 major toxic proteins, Abrin A[2] also referred to as Abrin b+c & Abrin C[2] also referred to as Abrin a; conjugates are prepared from affinity purified lectin; activity is expressed in µg/mL & is determined from serial dilutions in phosphate buffered saline, pH 7.2 of a 1 mg/mL solution; Tomita, M et al, *Experientia*, 28: 84, 1972; Wei, CH et al, *J Biol Chem*, 249: 3061, 1974; Lin, J-Y et al, *Toxicon*, 19: 41, 1981; Olsnes, S et al, *J Biol Chem*, 249: 803, 1974

Agglutinin, *Abrus precatorius*, Immobilized

Synonyms: Abrin; Lectin

Sigma L 0518   *Abrus precatorius* (Jequirity bean) Immobilized on 4% cross-linked beaded agarose; suspension in 1.0 *M* NaCl containing 0.01% thimerosal; contains 2-4 mg lectin/mL packed gel | Extremely hazardous! Be aware of the risks & familiar with safety procedures before you ruse this product; not blood group specific but has an affinity for D-galactose; seeds of *Abrus precatorius* contain a nontoxic agglutinin together with 2 major toxic proteins, Abrin A[2] also referred to as Abrin b+c & Abrin C[2] also referred to as Abrin a; conjugates are prepared from affinity purified lectin; activity is expressed in µg/mL & is determined from serial dilutions in phosphate buffered saline, pH 7.2 of a 1 mg/mL solution; Tomita, M et al, *Experientia*, 28: 84, 1972; Wei, CH et al, *J Biol Chem*, 249: 3061, 1974; Lin, J-Y et al, *Toxicon*, 19: 41, 1981; Olsnes, S et al, *J Biol Chem*, 249: 803, 1974

Agglutinin, *Agaricus bisporus*

Synonyms: Lectin

Sigma L 5640   *Agaricus bisporus* (Mushroom)   Highly purified by affinity chromatography; lyophilized powder containing ~10% protein (Biuret), balance primarily NaCl; activity: <16 µg protein/mL (the lowest concentration to agglutinate a 2% suspension of human blood group O erythrocytes after 1 hr at 25°C) | Lectin; not blood group specific but has an affinity for fetuin; ABA is a mixture of 2 phytohemagglutinins with similar specificities for carbohydrate receptors; activity is expressed in µg/mL & is determined from serial dilutions in phosphate buffered saline, pH 6.8 of a 1 mg/mL solution; Presant, CA & Kornfeld, S, *J Biol Chem*, 247: 6937, 1972

Agglutinin, *Anguilla anguilla*

Synonyms: Lectin

Sigma L 4141   *Anguilla anguilla* (Fresh water eel)   Highly purified by affinity chromatography; lyophilized powder containing ~30% protein (Biuret), balance primarily NaCl; activity: <20 µg/mL (the lowest concentration to agglutinate a 2% suspension of human blood group O erythrocytes after 1 hr at 25°C) | Anti-H blood group specific & has an affinity for α-L-fucosyl residues; conjugates are prepared from affinity purified lectin; activity is expressed in µg/mL & is determined from serial dilutions in phosphate buffered saline, pH 7.3 of a 1 mg/mL solution; Horejsi, V & Kocourek, J, *J Biochim Biophys Acta*, 538: 299, 1978; Kelly, C, *Biochem J*, 220: 221, 1984

Agglutinin, *Anguilla anguilla*, Biotin Conjugated

Synonyms: Lectin

ICN 153235   *Anguilla anguilla*   Purified by affinity chromatography | Specificity for α-L-Fuc

Agglutinin, *Anguilla anguilla*, FITC Conjugated

Synonyms: Lectin

ICN 153236   *Anguilla anguilla*   Purified by affinity chromatography | Specificity for α-L-Fuc

Agglutinin, *Arachis hypogaea*

Synonyms: Peanut Agglutinin; Lectin

Sigma L 0881   *Arachis hypogaea* (Peanut)   Affinity-purified; salt-free, lyophilized powder; activity: <0.1 µg/mL (the lowest concentration to agglutinate a 2% suspension of neuraminidase treated human blood group O erythrocytes after 1 hr at 25°C) | Does not agglutinate normal human erythrocytes but strongly agglutinates neuraminidase treated erythrocytes; potent anti-T activity similar to anti-T Ab in human sera; can be used to distinguish between human lymphocyte subsets; conjugates are prepared from affinity purified lectin; activity is expressed in µg/mL & is determined from serial dilutions in phosphate buffered saline, pH 6.8 of a 1 mg/mL solution; Bird, GWG, *Vox Sang*, 9: 748, 1964; Lotan, R et al, *J Biol Chem*, 250: 8518, 1975; London, J et al, *J Immunol*, 121: 438, 1978

Agglutinin, *Arachis hypogaea*, 10 nm Colloidal Gold Conjugated

Synonyms: PEANUT AGGLUTININ; LECTIN

Sigma L 1644   *Arachis hypogaea* (Peanut)   Mean particle size 8-12 nm; monodisperse; suspension in ~50% glycerol containing 0.15 *M* NaCl, 0.01 *M* sodium phosphate, pH 7.2, 0.02% PEG 20 & 0.02% sodium azide; concentration: A_{520} ~5.0 | Does not agglutinate normal human erythrocytes but strongly agglutinates neuraminidase treated erythrocytes; potent anti-T activity similar to anti-T Ab in human sera; can be used to distinguish between human lymphocyte subsets; conjugates are prepared from affinity purified lectin; activity is expressed in µg/mL & is determined from serial dilutions in phosphate buffered saline, pH 6.8 of a 1 mg/mL solution; Bird, GWG, *Vox Sang*, 9: 748, 1964; Lotan, R et al, *J Biol Chem*, 250: 8518, 1975; London, J et al, *J Immunol*, 121: 438, 1978

Agglutinin, *Arachis hypogaea*, Biotin Conjugated

Synonyms: Peanut Agglutinin; Lectin

Sigma L 6135   *Arachis hypogaea* (Peanut)   Lyophilized powder containing ~85% protein (Lowry), balance sodium citrate; contains ~3 moles biotin/mole protein | Does not agglutinate normal human erythrocytes but strongly agglutinates neuraminidase treated erythrocytes; potent anti-T activity similar to anti-T Ab in human sera; can be used to distinguish between human lymphocyte subsets; conjugates are prepared from affinity purified lectin; activity is expressed in µg/mL & is determined from serial dilutions in phosphate buffered saline, pH 6.8 of a 1 mg/mL solution; Bird, GWG, *Vox Sang*, 9: 748, 1964; Lotan, R et al, *J Biol Chem*, 250: 8518, 1975; London, J et al, *J Immunol*, 121: 438, 1978

Agglutinin, *Arachis hypogaea*, Ferritin Conjugated

Synonyms: Peanut Agglutinin; Lectin

Sigma L 1513   *Arachis hypogaea* (Peanut)   Sterile-filtered solution in 0.01 *M* potassium phosphate buffered saline, pH 7.0; contains 10-20 moles lectin bound/mole ferritin | Does not agglutinate normal human erythrocytes but strongly agglutinates neuraminidase treated erythrocytes; potent anti-T activity similar to anti-T Ab in human sera; can be used to distinguish between human lymphocyte subsets; conjugates are prepared from affinity purified lectin; activity is expressed in µg/mL & is determined from serial dilutions in phosphate buffered saline, pH 6.8 of a 1 mg/mL solution; Bird, GWG, *Vox Sang*, 9: 748, 1964; Lotan, R et al, *J Biol Chem*, 250: 8518, 1975; London, J et al, *J Immunol*, 121: 438, 1978

Agglutinin, *Arachis hypogaea*, FITC Conjugated

Synonyms: Peanut Agglutinin; Lectin

Sigma L 7381 *Arachis hypogaea* (Peanut) Lyophilized powder containing ~10% protein (Lowry), balance phosphate buffer salts & NaCl; contains 4-8 moles FITC/mole protein | Does not agglutinate normal human erythrocytes but strongly agglutinates neuraminidase treated erythrocytes; potent anti-T activity similar to anti-T Ab in human sera; can be used to distinguish between human lymphocyte subsets; conjugates are prepared from affinity purified lectin; activity is expressed in μg/mL & is determined from serial dilutions in phosphate buffered saline, pH 6.8 of a 1 mg/mL solution; Bird, GWG, *Vox Sang*, 9: 748, 1964; Lotan, R et al, *J Biol Chem*, 250: 8518, 1975; London, J et al, *J Immunol*, 121: 438, 1978

Agglutinin, *Arachis hypogaea*, Immobilized

Synonyms: Peanut Agglutinin; Lectin

Sigma L 2507 *Arachis hypogaea* (Peanut) Immobilized on cross-linked 4% beaded agarose; suspension in 1.0 *M* NaCl containing 0.01% thimerosal; contains 2-4 mg lectin/mL packed gel; 1 mL suspension yields ~0.5 mL packed gel | Does not agglutinate normal human erythrocytes but strongly agglutinates neuraminidase treated erythrocytes; potent anti-T activity similar to anti-T Ab in human sera; can be used to distinguish between human lymphocyte subsets; conjugates are prepared from affinity purified lectin; activity is expressed in μg/mL & is determined from serial dilutions in phosphate buffered saline, pH 6.8 of a 1 mg/mL solution; Bird, GWG, *Vox Sang*, 9: 748, 1964; Lotan, R et al, *J Biol Chem*, 250: 8518, 1975; London, J et al, *J Immunol*, 121: 438, 1978

Sigma L 6646 *Arachis hypogaea* (Peanut) Immobilized on 6% agarose macrobeads, particle size 200-300 μm; suspension in 1.0 *M* NaCl containing 0.01% thimerosal; contains 2-4 mg lectin/mL packed gel | Does not agglutinate normal human erythrocytes but strongly agglutinates neuraminidase treated erythrocytes; potent anti-T activity similar to anti-T Ab in human sera; can be used to distinguish between human lymphocyte subsets; conjugates are prepared from affinity purified lectin; activity is expressed in μg/mL & is determined from serial dilutions in phosphate buffered saline, pH 6.8 of a 1 mg/mL solution; Bird, GWG, *Vox Sang*, 9: 748, 1964; Lotan, R et al, *J Biol Chem*, 250: 8518, 1975; London, J et al, *J Immunol*, 121: 438, 1978

Agglutinin, *Arachis hypogaea*, Peroxidase Conjugated

Synonyms: Peanut Agglutinin; Lectin

Sigma L 7759 *Arachis hypogaea* (Peanut) Lyophilized powder containing ~90% protein (Modified Warburg-Christian), balance primarily sodium citrate; repurified after conjugation by affinity chromatography; peroxidase activity: 20-40 purpurogallin units/mg protein; unit definition: 1 unit forms 1 mg purpurogallin in 20 sec from pyrogallol at pH 6.0 at 20°C | Prepared from peroxidase type VI using a modification of the method of O'Sullivan, MJ et al, *FEBS Lett*, 95: 311, 1978, which favors low MW conjugates; does not agglutinate normal human erythrocytes but strongly agglutinates neuraminidase treated erythrocytes; potent anti-T activity similar to anti-T Ab in human sera; can be used to distinguish between human lymphocyte subsets; conjugates are prepared from affinity purified lectin; activity is expressed in μg/mL & is determined from serial dilutions in phosphate buffered saline, pH 6.8 of a 1 mg/mL solution; Bird, GWG, *Vox Sang*, 9: 748, 1964; Lotan, R et al, *J Biol Chem*, 250: 8518, 1975; London, J et al, *J Immunol*, 121: 438, 1978

Agglutinin, *Arachis hypogaea*, TRITC Conjugated

Synonyms: Peanut Agglutinin; Lectin

Sigma L 3766 *Arachis hypogaea* (Peanut) Lyophilized powder containing ~10% protein (Lowry), balance phosphate buffer salts & NaCl; contains 1.5-3 moles TRITC/mole protein | Does not agglutinate normal human erythrocytes but strongly agglutinates neuraminidase treated erythrocytes; potent anti-T activity similar to anti-T Ab in human sera; can be used to distinguish between human lymphocyte subsets; conjugates are prepared from affinity purified lectin; activity is expressed in μg/mL & is determined from serial dilutions in phosphate buffered saline, pH 6.8 of a 1 mg/mL solution; Bird, GWG, *Vox Sang*, 9: 748, 1964; Lotan, R et al, *J Biol Chem*, 250: 8518, 1975; London, J et al, *J Immunol*, 121: 438, 1978

Agglutinin, *Artocarpus integrifolia*

Synonyms: Lectin

Sigma L 3515 *Artocarpus integrifolia* (Jacalin, Jack fruit) Highly purified by affinity chromatography; lyophilized powder containing ~70% protein (Biuret), balance primarily phosphate buffer salts; activity: <10 μg/mL (the lowest concentration to agglutinate a 2% suspension of human blood group A erythrocytes after 1 hr at 25°C) | Extracted from Jack fruit seeds; glycoprotein reported to bind human serum & secretory IgA; not blood group specific; the most effective inhibitor of hemagglutination is methyl α-D-galactopyranoside; specific toward the Thomsen-Friedenreich (T) antigen but has subtle differences in its combining site when compared to the site of Arachis *hypogaea* agglutinin; also demonstrated to be a mitogen of T-lymphocytes; activity is expressed in μg/mL & is determined from serial dilutions in phosphate buffered saline, pH 6.8 of a 1 mg/mL solution; Roque-Barreira, MC & Campo-Neto, A, *J Immunol*, 134: 1740, 1985; Ahmed, H & Chatterjee, BP, *in Lectins, Biology, Biochemistry, Clinical Biochemistry*, Vol 5, Bog-Hansen, TC & van Driessche, E, eds, p125, de Gruyter & Co, Berlin, NY, 1986; Krishna Sastry, MV et al, *J Biol Chem*, 261: 11726, 1986

Agglutinin, *Artocarpus integrifolia*, Immobilized

Synonyms: Lectin

Sigma L 5147 *Artocarpus integrifolia* (Jacalin, Jack fruit) Immobilized on cross-linked 4% beaded agarose; suspension in 1.0 *M* NaCl containing 0.01% thimerosal; contains ~5 mg lectin/mL packed gel; binding capacity: 1 mL gel binds 1-2 mg human IgA | Extracted from Jack fruit seeds; glycoprotein reported to bind human serum & secretory IgA; not blood group specific; the most effective inhibitor of hemagglutination is methyl α-D-galactopyranoside; specific toward the Thomsen-Friedenreich (T) antigen but has subtle differences in its combining site when compared to the site of Arachis *hypogaea* agglutinin; also demonstrated to be a mitogen of T-lymphocytes; activity is expressed in μg/mL & is determined from serial dilutions in phosphate buffered saline, pH 6.8 of a 1 mg/mL solution; Roque-Barreira, MC & Campo-Neto, A, *J Immunol*, 134: 1740, 1985; Ahmed, H & Chatterjee, BP, *in Lectins, Biology, Biochemistry, Clinical Biochemistry*, Vol 5, Bog-Hansen, TC & van Driessche, E, eds, p125, de Gruyter & Co, Berlin, NY, 1986; Krishna Sastry, MV et al, *J Biol Chem*, 261: 11726, 1986

Agglutinin, *Artocarpus integrifolia*, Peroxidase Conjugated

Synonyms: Lectin

Sigma L 4650 *Artocarpus integrifolia* (Jacalin, Jack fruit) Lyophilized powder containing ~75% protein (Modified Warburg-Christian), balance primarily sodium citrate; repurified after conjugation by affinity chromatography; peroxidase activity: 50-100 purpurogallin units/mg protein; unit definition: 1 unit forms 1 mg purpurogallin in 20 sec from pyrogallol at pH 6.0 at 20°C | Prepared from peroxidase type VI using a modification of the method of O'Sullivan, MJ et al, *FEBS Lett*, 95: 311, 1978, which favors low molecular weight conjugates

Agglutinin, *Bandeiraea simplicifolia* I

Synonyms: Lectin

Sigma L 2380 *Bandeiraea simplicifolia* Salt-free; lyophilized; purified by affinity chromatography using the method of Hayes, CE & Goldstein, IJ, *J Biol Chem*, 249: 1904, 1974; activity: <20 μg/mL using human blood group A or B erythrocytes | Major affinity for terminal α-D-galactosyl residues with a secondary affinity for terminal *N*-acetyl-α-D-galactosaminyl residues; BS-I is a tetrameric lectin consisting of 2 types of subunits designated A & B; five BS-I isolectins with different subunit composition: BSI-B$_4$, BSI-AB$_3$, BSI-A$_2$B$_2$, BSI-A$_3$B & BSI-A$_4$; BSI-B4 is blood group B specific & has an exclusive affinity for terminal α-D-galactosyl residues, whereas BSI-A4 has blood group A specificity & has a major affinity for terminal *N*-acetyl-α-D-galactosaminyl residues; conjugates are prepared from the corresponding affinity purified lectin; agglutination activity expressed in μg/mL & is determined from serial dilutions of a 1 mg/mL solution using phosphate buffered saline, pH 6.8, containing calcium, magnesium & manganese as indicated on the data sheet accompanying the product; activity is the lowest concentration to agglutinate a 2% suspension of appropriate erythrocytes after 1 hr at 25°C; Hayes CE & Goldstein, IJ, *J Biol Chem*, 249: 1904, 1974; Wood, C et al, *Arch Biochem Biophys*, 198: 1, 1979; Murphy, LA & Goldstein, IJ, *J Biol Chem*, 252: 4739, 1977

Agglutinin, *Bandeiraea simplicifolia* I, Biotin Conjugated

Synonyms: Lectin

Sigma L 3759 *Bandeiraea simplicifolia* Lyophilized containing ~85% protein (E$^{1\%}_{280}$); balance primarily sodium citrate; contains 5 moles biotin/mole protein | See Sigma L 2380

Agglutinin, *Bandeiraea simplicifolia* I, FITC Conjugated

Synonyms: Lectin

Sigma L 9381 *Bandeiraea simplicifolia* Lyophilized containing ~10% protein (Lowry); balance phosphate buffer salts & NaCl; contains 1-2 moles FITC/mole protein | See Sigma L 2380

Agglutinin, *Bandeiraea simplicifolia* I, Peroxidase Conjugated

Synonyms: Lectin

Sigma L 3383 *Bandeiraea simplicifolia* Lyophilized containing ~90% protein (Modified Warburg-Christian); balance primarily sodium citrate; peroxidase activity: 10-30 purpurogallin units/mg protein; unit definition: 1 unit forms 1 mg purpurogallin in 20 sec from pyrogallol at pH 6.0 at 20°C | Prepared from peroxidase type VI using a modification of the method of Wilson, MB & Nakane, PK, *immunofluorescence & Related Staining Techniques*, ed, Knapp, P et al, Elsevier, North Holland, Biomedical Press 215, 1978 See Sigma L 2380

Agglutinin, *Bandeiraea simplicifolia* I, TRITC Conjugated

Synonyms: Lectin

Sigma L 5264 *Bandeiraea simplicifolia* Lyophilized containing ~10% protein (Lowry); balance phosphate buffer salts & NaCl; contains 3 moles TRITC/mole protein | See Sigma L 2380

Agglutinin, *Bandeiraea simplicifolia* II

Synonyms: Lectin

Sigma L 7508 *Bandeiraea simplicifolia* Lyophilized containing ~95% protein (E$^{1\%}_{280}$); balance primarily sodium citrate plus trace calcium & magnesium chloride; activity: <20 μg/mL; highly purified by affinity chromatography using a modification of the method of Iyer, PN et al, *Arch Biochem Biophys*, 177: 330, 1976; | BS-II does not agglutinate human erythrocytes of blood groups A, B or O but will agglutinate acquired B, T-activated or Tk polyagglutinable cells; has an affinity for *N*-acetyl-D-glucosamine; conjugates are prepared from affinity purified lectin; agglutination activity expressed in μg/mL & is determined from serial dilutions of a 1 mg/mL solution using phosphate buffered saline, pH 6.8, containing calcium, magnesium & manganese; activity is the lowest concentration to agglutinate a 2% suspension of *Bacteroides fragilisficin* treated human blood group O erythrocytes after 1 hr at 25°C; Judd, WJ et al, *Vox Sang*, 33: 246, 1977; Ebisu S et al, *carbohydrate Res*, 61: 129, 1978; Wood, C et al, *Arch Biochem Biophys*, 198: 1, 1979; Ebisu, S & Goldstein, IJ, *Meth Enzymol*, 50: 350, 1978

Agglutinin, *Bandeiraea simplicifolia* II, FITC Conjugated

Synonyms: Lectin

Sigma L 1259 *Bandeiraea simplicifolia* Lyophilized containing ~95% protein (E$^{1\%}_{280}$); balance primarily sodium citrate; contains 1-2 moles FITC/mole protein | See Sigma L 7508

Agglutinin, *Bandeiraea simplicifolia* II, Immobilized

Synonyms: Lectin

Sigma L 1889 *Bandeiraea simplicifolia* Immobilized on 4% beaded agarose; suspension in 50% glycerol containing 0.15 *M* NaCl, 0.01 *M* sodium phosphate & 0.02% sodium azide; 1 mL suspension yields ~0.5 mL packed gel; spacer: 16 atoms; contains 3-5 mg protein (Biuret)/mL packed gel; binding capacity: 1 mL packed gel binds 0.5-1.5 mg albumin, bovine p-aminophenyl-*N*-acetyl-β-D-glucosaminide; attached through the carbohydrate portion of the glycoprotein by reductive amination to aminoethylamino epoxy-activated 4% beaded agarose | See Sigma L 2380

Agglutinin, *Bandeiraea simplicifolia* II, Peroxidase Conjugated

Synonyms: Lectin

Sigma L 6650 *Bandeiraea simplicifolia* Lyophilized containing ~90% protein (Modified Warburg-Christian); balance primarily sodium citrate buffer salts & CaCl$_2$; peroxidase activity: ~70 units purpurogallin/mg protein; unit definition: 1 unit forms 1 mg purpurogallin in 20 sec from pyrogallol at pH 6.0 at 20°C | prepared from peroxidase type VI using a modification of the method of O'Sullivan, MJ et al, *FEBS Lett*, 95: 311, 1978; repurified by affinity chromatography after conjugation; see Sigma L 2380

Agglutinin, *Bandeiraea simplicifolia* Isolectin A$_4$

Synonyms: Lectin

Sigma L 1509 *Bandeiraea simplicifolia* Highly purified; lyophilized containing ~95% protein (E$^{1\%}_{280}$); balance primarily sodium citrate; activity: <16 μg/mL with human blood group A erythrocytes; >32 μg/mL with blood human blood group B erythrocytes | See Sigma L 2380

Agglutinin, *Bandeiraea simplicifolia* Isolectin A$_4$, FITC Conjugated

Synonyms: Lectin

Sigma L 0890 *Bandeiraea simplicifolia* Lyophilized containing ~95% protein (E$^{1\%}_{280}$); balance citrate buffer salts & CaCl$_2$; contains 2-4 moles FITC/mole protein isolectin A$_4$ | See Sigma L 2380

Agglutinin, *Bandeiraea simplicifolia* Isolectin B₄

Synonyms: Lectin

Sigma L 2140 *Bandeiraea simplicifolia* Lyophilized containing ~95% protein ($E^{1\%}_{280}$); balance citrate buffer salts & CaCl₂; contains 2-4 moles biotin/mole isolectin B₄ | See Sigma L 2380

Sigma L 3019 *Bandeiraea simplicifolia* Highly purified; lyophilized containing ~95% protein ($E^{1\%}_{280}$); balance primarily sodium citrate; activity: <8 μg/mL using human blood group B erythrocytes; does not agglutinate blood group A erythrocytes at 125 μg/mL | See Sigma L 2380

Agglutinin, *Bandeiraea simplicifolia* Isolectin B₄, FITC Conjugated

Synonyms: Lectin

Sigma L 2895 *Bandeiraea simplicifolia* Lyophilized containing ≥70% protein (Lowry); balance primarily sodium citrate; contains ~2 moles FITC/mole protein | See Sigma L 2380

Agglutinin, *Bandeiraea simplicifolia* Isolectin B₄, Peroxidase Conjugated

Synonyms: Lectin

Sigma L 5391 *Bandeiraea simplicifolia* Lyophilized containing ~95% protein (Modified Warburg-Christian); balance primarily sodium citrate buffer salts & CaCl₂; peroxidase activity: 40-160 units/mg protein; unit definition: 1 unit forms 1 mg purpurogallin in 20 sec from pyrogallol at pH 6.0 at 20°C | prepared from peroxidase type VI using a modification of the method of O'Sullivan, MJ et al, *FEBS Lett*, 95: 311, 1978; repurified by affinity chromatography after conjugation See Sigma L 2380

Agglutinin, *Bauhinia purpurea*

Synonyms: Lectin

ICN 153240 *Bauhinia purpurea* (Camels Foot Tree) Purified by affinity chromatography | Extracts of seeds are reported to contain an anti-N blood group specific lectin but after purification, BPA is not blood group specific; specificity for β-D-gal(1→3)-D-galNAc

Sigma L 6013 *Bauhinia purpurea* (Camels foot tree) Salt-free; lyophilized; highly purified by affinity chromatography using the method of Osawa, T et al, *Meth Enzymology*, 50: 367, 1978; activity: <8 μg/mL | Has an affinity for *N*-acetyl-D-galactosamine & D-galactose; extracts of *B. purpurea* seeds are reported to contain an anti-N blood group specific lectin but after purification BPA is not blood group specific; agglutination activity expressed in μg/mL & is determined from serial dilutions of a 1 mg/mL solution using phosphate buffered saline, pH 7.2; activity is the lowest concentration to agglutinate a 2% suspension of human erythrocytes after 1 hr at 25°C; Makela O & Makela, P, *Ann Med Exp Fenn*, 84: 402, 1956; Osawa, T et al, *Meth Enzymology*, 50: 367, 1978

Agglutinin, *Bauhinia purpurea*, Biotin Conjugated

Synonyms: Lectin

ICN 153241 *Bauhinia purpurea* (Camels Foot Tree) Purified by affinity chromatography | Specificity for β-D-gal(1→3)-D-galNAc

Sigma L 0768 *Bauhinia purpurea* (Camels foot tree) Lyophilized containing ~10% protein (Lowry); balance potassium phosphate buffer salts & NaCl; contains ~4 moles biotin/mole protein | See Sigma L 6013

Agglutinin, *Bauhinia purpurea*, FITC Conjugated

Synonyms: LECTIN

ICN 153242 *Bauhinia purpurea* (Camels Foot Tree) Purified by affinity chromatography | Specificity for β-D-gal(1→3)-D-galNAc

Agglutinin, *Bauhinia purpurea*, Horse Radish Peroxidase Conjugated

Synonyms: Lectin

ICN 153243 *Bauhinia purpurea* (Camels Foot Tree) Purified by affinity chromatography | Specificity for β-D-gal(1→3)-D-galNAc

Agglutinin, *Canavalia ensiformis*

Synonyms: Lectin; Concanavalin A

ICN 150710 *Canavalia ensiformis* (jack bean) Lyophilized; essentially salt- & CHO-free; isolated by affinity chromatography; mitogenic properties | Specificity for terminal α-D-man & α-D-glc residues; agglutinates red blood cells; Bittiges, H & HP Schnebli, Eds, Concanavalin A as a tool, J Wiley & Sons, London, 1976

ICN 194069 *Canavalia ensiformis* (jack bean) Lyophilized powder of 15% protein & the balance primarily NaCl | Specificity for terminal α-D-man & α-D-glc residues; agglutinates red blood cells; caution: mitogenic properties; Bittiges, H & HP Schnebli, Eds, Concanavalin A as a tool, J Wiley & Sons, London, 1976

ICN 195283 *Canavalia ensiformis* (jack bean) Highly purified; essentially salt free; mitogenic properties | Specificity for terminal α-D-man & α-D-glc residues; agglutinates red blood cells; Bittiges, H & HP Schnebli, Eds, Concanavalin A as a tool, J Wiley & Sons, London, 1976

Agglutinin, *Canavalia ensiformis* (¹⁴C-Me)-

Synonyms: Lectin; Concanavalin A

Sigma C 8665 *Canavalia ensiformis* MW ~102k 5-50 μCi/mg protein; solution in 10 *mM* sodium phosphate, pH 7.0, containing 2% NaCl in serum bottle | Radiochemical prepared from Sigma C 2010

Agglutinin, *Canavalia ensiformis* 10 nm Colloidal Gold Conjugated

Synonyms: Lectin; Concanavalin A

ICN 154017 Purified by affinity chromatography | Specificity for terminal α-D-man & α-D-glc residues

Agglutinin, *Canavalia ensiformis* 20 nm Colloidal Gold Conjugated

Synonyms: Lectin; Concanavalin A

ICN 154018 Purified by affinity chromatography | Specificity for terminal α-D-man & α-D-glc residues

Agglutinin, *Canavalia ensiformis* Agarose

Synonyms: Lectin; Concanavalin A

ICN 191474 *Canavalia ensiformis* 5 atoms hydrophilic spacer arm; 10 mg lectin/mL gel; suspension in 0.5 *M* acetate buffer, pH 6.0, 1 *M* NaCl, 4 *mM* MnCl₂, 0.02% NaN₃ | Specificity for terminal α-D-man & α-D-glc residues; used for purification of membrane glycoproteins hormones & receptors, IgM, α-fetoprotein, glycoprotein enzymes

Agglutinin, *Canavalia ensiformis* Biotin Conjugated

Synonyms: Lectin; Concanavalin A

ICN 154016 Purified by affinity chromatography | Specificity for terminal α-D-man & α-D-glc residues

Biogenesis 2250-1404 Canavalia ensiformis MW 104k 25 *mM* sodium borate-bicarbonate buffer, 1.0 *M* NaCl, 0.05% NaN3, pH 8.0; liquid | Useful in chromosome analysis, cell separation, cell toxicity and migration, blood grouping; most purified Con A preparations contain a small percentage of aggregates; high Con A concentrations are not stable in buffer solutions for long term

Agglutinin, *Canavalia ensiformis* FITC Conjugated

Synonyms: Lectin; Concanavalin A

ICN 153245 Purified by affinity chromatography | Specificity for terminal α-D-man & α-D-glc residues

Biogenesis 2250-1304   Canavalia ensiformis   MW 425.8
>95%; FITC isomer I; 25 mM sodium Borate, 0.9% NaCl, 0.01%
NaN_3, pH 7.8; liquid

Agglutinin, *Canavalia ensiformis* Horse Radish Peroxidase Conjugated

Synonyms: Lectin; Concanavalin A

ICN 153246   Purified by affinity chromatography | Specificity
for terminal α-D-man & α-D-glc residues

Agglutinin, *Canavalia ensiformis* Phycoerythrin Conjugated

Synonyms: Lectin; Concanavalin A

Biogenesis 2250-1204   Suspension

Agglutinin, *Canavalia ensiformis* Sodium Salt

Synonyms: Lectin; Concanavalin A

ICN 790016   2X crystallized | Not blood group specific, but
demonstrates an affinity for terminal α-D-man & α-D-glc residues

Agglutinin, *Canavalia ensiformis* Type III

Synonyms: Lectin; Concanavalin A

Sigma C 2631   *Canavalia ensiformis* (Jack bean)   Purified by
affinity chromatography; lyophilized containing ~15% protein
(Biuret); balance primarily NaCl; substantially free of carbohydrate;
activity: <500 µg solid/mL | Not blood group specific, but has
affinity for terminal α-D-glucosyl & α-D-mannosyl residues; Ca^{2+} &
Mn^{2+} ions are required for activity; Con A dissociates into dimers at
pH 5.6 or below; between pH 5.8 & pH 7.0 it exists as a tetramer;
above pH 7.0 higher aggregates are formed; exhibits mitogenic
activity which is dependent on its degree of aggregation;
succinylation results in an active dimer form which remains a dimer
above pH 5.6; conjugates are prepared from affinity purified lectin
type IV; immobilized Con A is prepared from affinity purified lectin
type III; agglutination activity expressed in µg/mL & is determined
from serial dilutions of a 1 mg/mL solution using phosphate
buffered saline, pH 6.8 containing Ca^{2+} & Mn^{2+}; activity is the
lowest concentration to agglutinate a 2% suspension of human
erythrocytes after 1 hr at 25°C; Reeke, GN et al, *Ann NY Acad Sci*,
234: 369, 1974; Kalb, AJ & Lustig, A, *Biochim Biophys Acta*, 168:
366, 1968; Gunther, GR et al, *Proc Natl Acad Sci USA*, 70: 1012,
1973; *Concanavalin A as a Tool*, ed Bittiges, H & Schnebli, HP, J
Wiley & Sons, London, 1976

Agglutinin, *Canavalia ensiformis* Type III-ASCL, Immobilized

Synonyms: Lectin; Concanavalin A

Sigma C 7911   *Canavalia ensiformis* (Jack bean)   Immobilized
on 4% cross-linked beaded agarose; contains ~10 mg lectin/mL
gel; binding capacity: 1 mL gel binds ~6 mg yeast mannan at pH
7.2 at 25°C; suspension in 1.0 M NaCl, containing 1 mM each $CaCl_2$,
$MgCl_2$ & $MnCl_2$ & 0.02% thimerosal; 1 mL suspension yields ~0.5
mL gel | See Sigma C 2631

Agglutinin, *Canavalia ensiformis* Type IV

Synonyms: Lectin; Concanavalin A

Sigma C 2010   *Canavalia ensiformis* (Jack bean)   Lyophilized;
essentially salt-free; highly purified; produced from Type III using
acetic acid by the method of Olson, MOJ & Liener, IE, *Biochemistry*,
6: 105, 1967; activity: <64 µg/mL | See Sigma C 2631

Agglutinin, *Canavalia ensiformis* Type IV Salt Free

Synonyms: Lectin; Concanavalin A

Sigma C 0412   *Canavalia ensiformis*   Purified; lyophilized;
sterilized by γ-irradiation; cell culture tested | Lectin; highly
specific polyvalent carbohydrate-binding proteins; useful in
polysaccharide studies, glycoprotein studies, enzyme tagging & cell
membrane studies, cell agglutination & cell typing; in tissue culture
certain lectins used to induce mitogenic activity; tested in a tissue
culture system using 3H-thymidine incorporation as a measure of
mitogenic activity; Goldstein, I & Hayes, C, *Adv Carbo Chem
Biochem*, 35: 127, 1978; Rosenberg, SA & Lipsky, PE, *J Immunol*,
122: 926, 1979

Sigma C 5275   *Canavalia ensiformis*   Purified; lyophilized;
filter-sterilized; cell culture tested | Lectin; highly specific
polyvalent carbohydrate-binding proteins; useful in polysaccharide
studies, glycoprotein studies, enzyme tagging & cell membrane
studies, cell agglutination & cell typing; in tissue culture certain
lectins used to induce mitogenic activity; tested in a tissue culture
system using 3H-thymidine incorporation as a measure of
mitogenic activity; Goldstein, I & Hayes, C, *Adv Carbo Chem
Biochem*, 35: 127, 1978; Rosenberg, SA & Lipsky, PE, *J Immunol*,
122: 926, 1979

Agglutinin, *Canavalia ensiformis* Type IV, Biotin Conjugated

Synonyms: Lectin; Concanavalin A

Sigma C 2272   Lyophilized containing ~85% protein ($E^{1\%}_{280}$);
balance primarily sodium citrate; contains 4-8 moles biotin/mole
protein

Agglutinin, *Canavalia ensiformis* Type IV, Ferritin Conjugated

Synonyms: Lectin; Concanavalin A

Sigma C 7898   *Canavalia ensiformis* (Jack bean)   Solution in
50% glycerol containing 0.3 M NaCl, 0.01 M sodium phosphate,
0.06 M $CaCl_2$ & 0.06 mM $MnCl_2$ & 0.02% sodium azide as
preservative; protein: 5-10 mg/mL (Biuret); contains ~1 mole
ferritin/mole lectin | See Sigma C 2631

Agglutinin, *Canavalia ensiformis* Type IV, FITC Conjugated

Synonyms: Lectin; Concanavalin A

Sigma C 7642   *Canavalia ensiformis* (Jack bean)   Lyophilized
containing ~10% protein (Biuret); balance phosphate buffer salts &
NaCl; contains 3-6 moles FITC/mole protein | See Sigma C 2631

Agglutinin, *Canavalia ensiformis* Type IV, TRITC Conjugated

Synonyms: Lectin; Concanavalin A

Sigma C 3636   *Canavalia ensiformis* (Jack bean)   Lyophilized
containing ~10% protein (Lowry); balance phosphate buffer salts &
NaCl; contains 0.5 mole TRITC/mole protein | See Sigma C 2631

Agglutinin, *Canavalia ensiformis* Type V

Synonyms: Lectin; Concanavalin A

Sigma C 7275   *Canavalia ensiformis* (Jack bean)   Lyophilized;
essentially salt-free; highly purified; contains trace $CaCl_2$ & $MnCl_2$
(<0.1 wt%); yields a substantially clear solution in water at 1%
(w/v); produced from Type III using the method of
Sophianopoulos, AJ & Sophianopoulos, JA, *Prep Biochem*, 11: 413,
1981; activity: <64 µg/mL | See Sigma C 2631

Agglutinin, *Canavalia ensiformis* Type V-A, Immobilized

Synonyms: Lectin; Concanavalin A

Sigma C 6904 *Canavalia ensiformis* (Jack bean) Immobilized on beaded agarose; spacer: hydrophilic 6 atoms (5 carbon atoms); contains ~10 mg lectin/mL gel; binding capacity: 1 mL gel binds 3-6 mg yeast mannan; suspension in 0.1 M acetate buffer, pH 6.0, containing 1 M NaCl, 1 M Mg^{2+}, 1 mM Ca^{2+}, 1 mM Mn^{2+}, 0.01% thimerosal | prepared by activation with p-nitrophenyl chloroformate which reduces ligand leakage & charge problems associated with cyanogen bromide activation; Wilchek, M & Miron, T, *Biochem Int*, 4: 629, 1982; see Sigma C 2631

Agglutinin, *Canavalia ensiformis* Type V-B, Immobilized

Synonyms: Lectin; Concanavalin A

Sigma C 6170 *Canavalia ensiformis* (Jack bean) Immobilized on 4% beaded agarose; contains ~15 mg lectin/mL gel; binding capacity: 1 mL gel binds ~6 mg yeast mannan; suspension in 0.1 M acetate buffer, pH 6.0, containing 1 M NaCl, 1 M Mg^{2+}, 1 mM Ca^{2+}, 1 mM Mn^{2+}, 0.02% thimerosal | prepared by activation with p-nitrophenyl chloroformate which reduces ligand leakage & charge problems associated with cyanogen bromide activation; Wilchek, M & Miron, T, *Biochem Int*, 4: 629, 1982; see Sigma C 2631

Agglutinin, *Canavalia ensiformis* Type V-B, Sepharose 4B

Synonyms: Lectin; Concanavalin A

Sigma C 9017 *Canavalia ensiformis* (Jack bean) Attached to Sepharose 4B; contains 10-16 mg lectin/mL gel; binding capacity: 1 mL gel binds ~6 mg yeast mannan; suspension in 0.1 M acetate buffer, pH 6.0, containing 1 M NaCl, & 1 mM each CaCl$_2$, MgCl$_2$ & MnCl$_2$$^+$ & 20% ethanol | See Sigma C 2631

Agglutinin, *Canavalia ensiformis* Type VI

Synonyms: Lectin; Concanavalin A

Sigma C 7647 *Canavalia ensiformis* (Jack bean) Lyophilized; essentially salt-free; highly purified by affinity chromatography; soluble in PBS at 4% (w/v); purified by modification of the method of Matsumoto et al, *Anal Biochem*, 116: 103, 1981; activity: <20 µg/mL | See Sigma C 2631

Agglutinin, *Canavalia ensiformis* Type VI, Immobilized

Synonyms: Lectin; Concanavalin A

Sigma C 7555 *Canavalia ensiformis* (Jack bean) Immobilized on 4% cross-linked beaded agarose; contains 15-30 mg lectin/mL gel; binding capacity: 1 mL gel binds 5-15 mg yeast mannan & 15-30 mg thyroglobulin protein/mL in 0.01 M phosphate buffered saline, pH 6.8, containing 1 mM each CaCl$_2$, MgCl$_2$ & MnCl$_2$; suspension in 0.1 M potassium phosphate, 1.0 M NaCl, pH 6.0, containing 1 mM each CaCl$_2$, MgCl$_2$ & MnCl$_2$ & 0.02% thimerosal | See Sigma C 2631

Agglutinin, *Canavalia ensiformis*, 10 nm Colloidal Gold Conjugated

Synonyms: Lectin; Concanavalin A

Sigma L 5021 *Canavalia ensiformis* (Jack bean) Lectin (Sigma C 7275) adsorbed to colloidal gold; mean particle size 8-12 nm; monodisperse; suspension in 0.02 M Tris buffered saline, pH 8.0, containing 20% glycerol, 1% PEG & 0.05% sodium azide; concentration: A$_{520}$ ~5.0; Benhamou, N & Ouellette, GB, *J Histochem Cytochem*, 34: 855, 1986 | See Sigma C 2631

Agglutinin, *Canavalia ensiformis*, 20 nm Colloidal Gold Conjugated

Synonyms: Lectin; Concanavalin A

Sigma L 3642 *Canavalia ensiformis* (Jack bean) Lectin (Sigma C 7275) adsorbed to colloidal gold; mean particle size 17-23 nm; monodisperse; suspension in 50% glycerol containing 0.15 M NaCl, 0.01 M sodium phosphate, pH 6.8, 0.1 mM Ca^{2+}, 0.1 mM Mn^{2+}, 0.02% PEG & 0.02% sodium azide; concentration: A$_{520}$ ~5.0; Benhamou, N & Ouellette, GB, *J Histochem Cytochem*, 34: 855, 1986 | See Sigma C 2631

Agglutinin, *Canavalia ensiformis*, 5 nm Colloidal Gold Conjugated

Synonyms: Lectin; Concanavalin A

Sigma L 8529 *Canavalia ensiformis* (Jack bean) Lectin (Sigma C 7275) adsorbed to colloidal gold; mean particle size 3.5-6.5 nm; monodisperse; solution in 20% glycerol containing 0.15 M NaCl, 0.02 M Tris-HCl, pH 6.8, 1% PEG & 0.05% sodium azide; concentration: A$_{520}$ ~5.0; Benhamou, N & Ouellette, GB, *J Histochem Cytochem*, 34: 855, 1986 | See Sigma C 2631

Agglutinin, *Canavalia ensiformis*, Acrylic Beads Immobilized

Synonyms: Lectin; Concanavalin A

Sigma C 6160 *Canavalia ensiformis* (Jack bean) Attached through a covalent bond to macroporous acrylic beads (oxirane acrylic beads, Sigma O 7628); spacer: hydrophilic 6 atoms (5 carbon atoms); contains ~10 mg lectin (Lowry)/g beads; binding capacity: 1 mL gel binds 1-2 mg yeast mannan; 1 g beads will swell to ~4 mL gel | See Sigma C 2631

Agglutinin, *Canavalia ensiformis*, Macroagarose Immobilized

Synonyms: Lectin; Concanavalin A

Sigma C 7044 *Canavalia ensiformis* (Jack bean) Immobilized on 6% agarose macrobeads; contains ~10 mg lectin/mL packed gel; binding capacity: 1 mL gel binds ~3.5 mg yeast mannan at pH 7.2 at 25°C; suspension in 1.0 M NaCl, containing 1 mM each CaCl$_2$, MgCl$_2$ & MnCl$_2$ & 0.02% thimerosal | See Sigma C 2631

Agglutinin, *Canavalia ensiformis*, Peroxidase Conjugated

Synonyms: Lectin; Concanavalin A

Sigma C 6397 *Canavalia ensiformis* (Jack bean) Lyophilized containing ~85% protein (Modified Warburg-Christian); balance primarily Tris-citrate buffer salts & trace calcium & manganese; peroxidase activity: 30-60 units purpurogallin/mg protein; unit definition: 1 unit forms 1 mg purpurogallin in 20 sec from pyrogallol at pH 6.0 at 20°C; prepared from peroxidase type VI using the method of Wilson, MB & Nakane, PK, *Immunofluorescence & Related Staining Techniques*, Knapp, P, ed Elsevier, North Holland Biomedical Press, p. 215, 1978, which promotes conjugation but prevents the interaction between Con A & peroxidase sugar residues | See Sigma C 2631

Agglutinin, *Canavalia ensiformis*, Succinyl

Synonyms: Lectin; Concanavalin A

ICN 153350 Purified by affinity chromatography | Specificity for α-D-Man,α-D-Glc residues

Sigma L 3885 *Canavalia ensiformis* (Jack bean) Highly purified; lyophilized containing ~95% protein (E$^{1\%}_{280}$); balance primarily sodium phosphate | Active dimeric (divalent) form which does not re-aggregate to form tetramers above pH 5.6; agglutination activity is greatly reduced by the succinylation procedure but reactivity with mannose or glucose residues is retained; conjugates are prepared from affinity purified lectin before succinylation; Gunther, GR et al, *Proc Natl Acad Sci USA*, 70: 1012, 1973; see Sigma C 2631

Agglutinin, *Canavalia ensiformis*, Succinyl Biotin Conjugated

Synonyms: Lectin; Concanavalin A

ICN 153351 Purified by affinity chromatography | Specificity for α-D-Man,α-D-Glc residues

Sigma L 0767 *Canavalia ensiformis* (Jack bean) Lyophilized containing ~95% protein (Biuret); balance primarily sodium phosphate; biotinylated utilizing an aminocaproyl spacer; contains 1-3 moles biotin/mole protein | See Sigma L 3885 & Sigma C 2631

Agglutinin, *Canavalia ensiformis*, Succinyl FITC Conjugated

Synonyms: Lectin; Concanavalin A

ICN 153352 Purified by affinity chromatography | Specificity for α-D-Man,α-D-Glc residues

Sigma L 9385 *Canavalia ensiformis* (Jack bean) Lyophilized containing ~95% protein ($E^{1\%}_{280}$); balance primarily sodium phosphate; contains 1.5 moles FITC/mole protein | See Sigma L 3885 & Sigma C 2631

Agglutinin, *Canavalia ensiformis*, Succinyl Horse Radish Peroxidase Conjugated

Synonyms: Lectin; Concanavalin A

ICN 153353 Purified by affinity chromatography | Specificity for α-D-Man,α-D-Glc residues

Agglutinin, *Caragana arborescens*

Synonyms: Lectin

Sigma L 4503 *Caragana arborescens* (Siberian pea tree) Lyophilized; highly purified (affinity chromatography) containing ~15% protein; balance phosphate buffer salts & NaCl; activity: <10 µg/mL | Not blood group specific, but has an affinity for *N*-acetyl-D-galactosamine; conjugates are prepared from affinity purified lectin; agglutination activity expressed in µg/mL & is determined from serial dilutions of a 1 mg/mL solution using phosphate buffered saline, pH 6.8; activity is the lowest concentration to agglutinate a 2% suspension of human blood group A erythrocytes after 1 hr at 25°C; Bloch, R et al, *J Biol Chem*, 251: 5929, 1976

Agglutinin, *Caragana arborescens*, Biotin Conjugated

Synonyms: Lectin

Sigma L 9637 *Caragana arborescens* (Siberian pea tree) Lyophilized containing ~85% protein ($E^{1\%}_{280}$); balance primarily sodium citrate; contains 3-6 moles biotin/mole protein | See Sigma L 4503

Agglutinin, *Caragana arborescens*, FITC Conjugated

Synonyms: Lectin

Sigma L 9512 *Caragana arborescens* (Siberian pea tree) Lyophilized containing ~10% protein (Lowry); balance phosphate buffer salts & NaCl; contains 2-5 moles FITC/mole protein | See Sigma L 4503

Agglutinin, *Cicer arietinum*

Synonyms: Lectin

Sigma L 3141 *Cicer arietinum* (Chick pea) Purified; lyophilized containing ~80% protein (Biuret); balance primarily sodium acetate & NaCl; contains <20 µg/mL | Not blood group specific, but agglutinates human erythrocytes that have undergone treatment with papain; also it has an affinity for glycoproteins like fetuin & IgM; agglutination activity expressed in µg/mL & is determined from serial dilutions of a 1 mg/mL solution using phosphate buffered saline, pH 6.8; activity is the lowest concentration to agglutinate a 2% suspension of papain treated human erythrocytes after 1 hr at 25°C; Kolberg, J et al, *Hoppe-Seyler's Z Physiol Chem*, 364: 655, 1983

Agglutinin, *Codium fragile* (subspecies *tomentosoides*)

Synonyms: Lectin

Sigma L 2638 *Codium fragile* (Green marine aglae) Highly purified (affinity chromatography); lyophilized; salt-free: <2 µg/mL | Not blood group specific, but has an affinity for *N*-acetyl-D-galactosamine; agglutination activity expressed in µg/mL & is determined from serial dilutions of a 1 mg/mL solution using saline (0.9% NaCl), pH 6.8; activity is the lowest concentration to agglutinate a 2% suspension of human blood group A erythrocytes after 1 hr at 25°C; Rogers, DJ et al, *Lectins, Biology, Biochemistry, Clinical Biochemistry*, eds TC Bog-Hansen, E Van Driessche, Vol 5, p. 155, 1986

Sigma L 6510 *Codium fragile* (Green marine aglae) Partially purified; lyophilized; activity: <30 µg/mL | See Sigma L 2638

Agglutinin, *Cytisus scoparius*

Synonyms: Lectin

Sigma L 4891 *Cytisus scoparius* (Scotch broom) Highly purified (affinity chromatography); lyophilized containing ~20% protein (Biuret); balance NaCl & sodium phosphate buffer salts; activity: 6 µg/mL | Not blood group specific; 2 lectins have been reported in the seeds, both having an affinity for *N*-acetyl-D-galactosamine & D-galactose; agglutination activity expressed in µg/mL & is determined from serial dilutions of a 1 mg/mL solution using phosphate buffered saline, pH 7.2; activity is the lowest concentration to agglutinate a 2% suspension of neuraminidase-treated human type A RBC after 1 hr at 25°C; Young, NM et al, *Biochem J*, 222: 41, 1984

Agglutinin, *Datura stramonium*

Synonyms: Lectin

Sigma L 2766 *Datura stramonium* (Jimson weed, Thorn apple) Highly purified (affinity chromatography); essentially salt-free; lyophilized; activity: ≤10 µg/mL | Not blood group specific; 2 lectins but has an affinity for oligomers of *N*-acetyl-glucosamine & *N*-acetyllactosamine; a glycoprotein containing ~35% carbohydrate; agglutination activity expressed in µg/mL & is determined from serial dilutions of a 1 mg/mL solution using phosphate buffered saline, pH 6.8; activity is the lowest concentration to agglutinate a 2% suspension of human erythrocytes after 1 hr at 25°C; Young, NM et al, *Biochem J*, 222: 41, 1984; Crowley, JI et al, *Arch Biochem Biophys*, 231: 524, 1984

Agglutinin, *Dolichos biflorus*

Synonyms: Lectin

ICN 151015 *Dolichos biflorus* 1 mg/mL solution | Isolated by affinity chromatography; has affinity for terminal *N*-acetyl-α-D-galactosaminyl residues; also has anti-A₁ human blood group specificity & is useful in distinguishing between A₁ & A₂ blood types; Etzler, MW & EA Kabat, *Biochem*, 9:869, 1970; Bird, GWG, *Blut*, 21:366, 1970

Sigma L 1135 *Dolichos biflorus* (Horse gram) Purified (affinity chromatography); lyophilized containing ~90% protein (Biuret); balance primarily Tris-succinate buffer salts; activity: <10 µg/mL | Anti-A₁ human blood group specificity, useful for distinguishing between A₁ & A₂ blood types; strong anti-Cad activity & has an affinity for terminal *N*-acetyl-α-D-galactosaminyl residues; conjugates are prepared from affinity purified lectin; agglutination activity expressed in µg/mL & is determined from serial dilutions of a 1 mg/mL solution using phosphate buffered saline, pH 6.8; activity is the lowest concentration to agglutinate a 2% suspension of human blood group A erythrocytes after 1 hr at 25°C; Bird, GWG, *Blut*, 21: 366, 1970; Etzler, ME & Kabat, EA, *Biochemistry*, 9: 869, 1970; Etzler, ME, *Meth Enzymol*, 28: 340, 1972

Sigma L 6887 *Dolichos biflorus* (Horse gram) Partially purified; salt-free; lyophilized; activity: <100 µg/mL | See Sigma L 1135

Sigma L 7762 *Dolichos biflorus* (Horse gram) Lyophilized powder in 5 mL reagent vial; reconstitute with 5.0 mL of 0.9% saline solution or 0.01 *M* phosphate buffered saline for a working strength reagent for determining human Type A₁ blood in tube or tile tests | See Sigma L 1135

Agglutinin, *Dolichos biflorus*, 10 nm Colloidal Gold Conjugated

Synonyms: Lectin

ICN 154026   *Dolichos biflorus*   Purified by affinity chromatography | Specificity for α-D-galNAc

Sigma L 4643   *Dolichos biflorus* (Horse gram)   Mean particle size 8-12 nm; monodisperse; suspension in ~50% glycerol containing 0.15 M NaCl, 0.01 M sodium phosphate, pH 7.2, 0.02% PEG 20 & 0.02% sodium azide; concentration: A_{520} ~5.0; Lucocq, JM & Roth, J, *Techniques in Immunocytochemistry*, Bullock & Petrusz, eds, Academic Press, 3: 203, 1985 | See Sigma L 1135

Agglutinin, *Dolichos biflorus*, 20 nm Colloidal Gold Conjugated

Synonyms: LECTIN

ICN 154027   *Dolichos biflorus*   Purified by affinity chromatography | Specificity for α-D-galNAc

Agglutinin, *Dolichos biflorus*, 5 nm Colloidal Gold Conjugated

Synonyms: Lectin

ICN 154025   *Dolichos biflorus*   Purified by affinity chromatography | Specificity for α-D-galNAc

Agglutinin, *Dolichos biflorus*, Biotin conjugated

Synonyms: Lectin

ICN 153255   *Dolichos biflorus*   Purified by affinity chromatography | Specificity for α-D-galNAc

Sigma L 6533   *Dolichos biflorus* (Horse gram)   Lyophilized containing ~25% protein (Biuret); balance primarily HEPES buffer salts & NaCl; contains 6-10 moles biotin/mole protein | See Sigma L 1135

Agglutinin, *Dolichos biflorus*, FITC Conjugated

Synonyms: Lectin

ICN 153256   *Dolichos biflorus*   Purified by affinity chromatography | Specificity for α-D-galNAc

Agglutinin, *Dolichos biflorus*, FTIC Conjugated

Synonyms: LECTIN

Sigma L 9142   *Dolichos biflorus* (Horse gram)   Salt-free; lyophilized; may contain residual Tris-succinate buffer salts; contains 3-8 moles FITC/mole protein | See Sigma L 1135

Sigma L 9533   *Dolichos biflorus* (Horse gram)   Lyophilized containing ~10% protein (Lowry); balance phosphate buffer salts & NaCl; contains 3-6 moles FITC/mole protein | See Sigma L 1135

Agglutinin, *Dolichos biflorus*, Horse Radish Peroxidase Conjugated

Synonyms: Lectin

ICN 153257   *Dolichos biflorus*   Purified by affinity chromatography | Specificity for α-D-galNAc

Agglutinin, *Dolichos biflorus*, Immobilized

Synonyms: Lectin

Sigma L 9894   *Dolichos biflorus* (Horse gram)   Immobilized on 4% beaded agarose; contains 2-4 mg protein (Lowry)/mL gel; suspension in 0.01 M potassium phosphate, pH 6.8, containing 0.5 M NaCl & 0.02% sodium azide | See Sigma L 1135

Agglutinin, *Dolichos biflorus*, Peroxidase Conjugated

Synonyms: Lectin

Sigma L 4258   *Dolichos biflorus* (Horse gram)   Lyophilized containing ~90% protein (Biuret); balance primarily sodium citrate; peroxidase activity: ~60 units purpurogallin/mg protein; unit definition: 1 unit forms 1 mg purpurogallin in 20 sec from pyrogallol at pH 6.0 at 20°C; prepared from peroxidase type VI using a modification of the method of O'Sullivan, MJ et al, *FEBS Lett*, 95: 311, 1978, which favors low molecular weight conjugates | See Sigma L 1135

Agglutinin, *Dolichos biflorus*, TRITC Conjugated

Synonyms: Lectin

Sigma L 9658   *Dolichos biflorus* (Horse gram)   Lyophilized containing ~15% protein (Lowry); balance phosphate buffer salts & NaCl; contains 1-3 moles TRITC/mole protein | See Sigma L 1135

Agglutinin, *Erythrina christagalli*

Synonyms: Lectin

ICN 153258   *Erythrina christagalli*   Purified by affinity chromatography; | Specificity for β-D-gal(1→4)-D-GlcNAc

Agglutinin, *Erythrina christagalli*, 10 nm Colloidal Gold Conjugated

Synonyms: Lectin

ICN 154029   *Erythrina christagalli*   Purified by affinity chromatography | Specificity for β-D-gal(1→4)-D-GlcNAc

Agglutinin, *Erythrina christagalli*, 20 nm Colloidal Gold Conjugated

Synonyms: Lectin

ICN 154030   *Erythrina christagalli*   Purified by affinity chromatography | Specificity for β-D-gal(1→4)-D-GlcNAc

Agglutinin, *Erythrina christagalli*, 5 nm Colloidal Gold Conjugated

Synonyms: Lectin

ICN 154028   *Erythrina christagalli*   Purified by affinity chromatography | Specificity for β-D-gal(1→4)-D-GlcNAc

Agglutinin, *Erythrina christagalli*, Biotin Conjugated

Synonyms: Lectin

ICN 153259   *Erythrina christagalli*   Purified by affinity chromatography | Specificity for β-D-gal(1→4)-D-GlcNAc

Agglutinin, *Erythrina christagalli*, FITC Conjugated

Synonyms: Lectin

ICN 153260   *Erythrina christagalli*   Purified by affinity chromatography | Specificity for β-D-gal(1→4)-D-GlcNAc

Agglutinin, *Erythrina christagalli*, Horse Radish Peroxidase Conjugated

Synonyms: Lectin

ICN 153261   *Erythrina christagalli*   Purified by affinity chromatography | Specificity for β-D-gal(1→4)-D-GlcNAc

Agglutinin, *Erythrina corallodendron*

Synonyms: Lectin

Sigma L 2142 *Erythrina corallodendron* (Coral tree)
Lyophilized; highly purified by affinity chromatography; salt-free; activity: <10 µg/mL | Not blood group specific but has an affinity for *N*-acetyllactosamine, *N*-acetyl-D-galactosamine, lactose & D-galactose; mitogenic activity for human peripheral blood lymphocytes, predominantly T cells; agglutination activity expressed in µg/mL & is determined from serial dilutions of a 1 mg/mL solution using phosphate buffered saline, pH 7.2; activity is the lowest concentration to agglutinate a 2% suspension of human erythrocytes; Gilboa-Garber, N & Mizrahi, L, *Can J Biochem*, 59: 315, 1981; Lis, H et al, *Phytochemistry*, 24: 2803, 1985; Sharon, N et al, *Phytochemistry*, 24: 2803, 1985

Agglutinin, *Erythrina corallodendron*, Biotin Conjugated

Synonyms: Lectin

Sigma L 0893 *Erythrina corallodendron* (Coral tree)
Lyophilized containing ~10% protein (Lowry); balance potassium phosphate buffer salts & NaCl; contains ~5 moles biotin/mole protein | See Sigma L 2142

Agglutinin, *Erythrina cristagalli*

Synonyms: Lectin

Sigma L 5390 *Erythrina cristagalli* (Coral tree) Lyophilized; highly purified by affinity chromatography; salt-free; activity: <0.5 µg/mL | Not blood group specific but has an affinity D-galactose & D-galactosides; reported to be mitogenic for human peripheral blood T lymphocytes; agglutination activity expressed in µg/mL & is determined from serial dilutions of a 1 mg/mL solution using phosphate buffered saline, pH 6.8; activity is the lowest concentration to agglutinate a 2% suspension of trypsinized human blood group O erythrocytes after 1 hr at 25°C; Iglesias, JL, *Eur J Biochem*, 123: 247, 1982

Agglutinin, *Erythrina cristagalli*, Biotin Conjugated

Synonyms: Lectin

Sigma L 3266 *Erythrina cristagalli* (Coral tree) Lyophilized containing ~60% protein (Lowry); balance primarily citrate buffer salts; contains ~2 moles biotin/mole protein | See Sigma L 5390

Agglutinin, *Erythrina cristagalli*, FITC Conjugated

Synonyms: Lectin

Sigma L 3391 *Erythrina cristagalli* (Coral tree) Lyophilized containing ~5% protein (Lowry); balance potassium phosphate buffer salts & NaCl; contains ~5 moles FITC/mole protein | See Sigma L 5390

Agglutinin, *Erythrina cristagalli*, Peroxidase Conjugated

Synonyms: Lectin

Sigma L 9015 *Erythrina cristagalli* (Coral tree) Lyophilized containing ~60% protein (Lowry); balance phosphate buffer salts; peroxidase activity: ~15-60 units purpurogallin/mg protein; unit definition: 1 unit forms 1 mg purpurogallin in 20 sec from pyrogallol at pH 6.0 at 20°C | See Sigma L 5390

Agglutinin, *Euonymus europaeus*

Synonyms: Lectin

ICN 153262 *Euonymus europaeus* Purified by affinity chromatography

Sigma L 7400 *Euonymus europaeus* (Spindle tree)

Lyophilized containing ~5% protein (Lowry); balance primarily NaCl & phosphate buffer salts; activity: <1 µg/mL using neuraminidase-treated type B red blood cells | Has anti-B+H blood group specificity & is reported to be most specific for blood group B oligosaccharides having the structure α-D-Gal(1→3)(α-L-Fuc(1→20)-β-D-Gal(1→3/4)-β-D-GlcNac; the B+H specificity is an intrinsic property of a single lectin binding site; agglutination activity expressed in µg/mL & is determined from serial dilutions of a 1 mg/mL solution using phosphate buffered saline, pH 7.2; activity is the lowest concentration to agglutinate a 2% suspension of neuraminidase-treated human blood group type B erythrocytes after 1 hr at 25°C; Petryniak, J et al, *Arch Biochem Biophys*, 178: 118, 1977

Agglutinin, *Euonymus europaeus*, Biotin Conjugated

Synonyms: Lectin

ICN 153263 *Euonymus europaeus* Purified by affinity chromatography | Specificity for α-D-gal(1→3)-D-gal

Agglutinin, *Euonymus europaeus*, FITC Conjugated

Synonyms: Lectin

ICN 153264 *Euonymus europaeus* Purified by affinity chromatography | Specificity for α-D-gal(1→3)-D-gal

Agglutinin, *Euonymus europaeus*, Horse Radish Peroxidase Conjugated

Synonyms: Lectin

ICN 153265 *Euonymus europaeus* Purified by affinity chromatography | Specificity for α-D-gal(1→3)-D-gal

Agglutinin, *Galanthus nivalis*

Synonyms: Lectin

Sigma L 8275 *Galanthus nivalis* (Snowdrop) Lyophilized; salt-free; highly purified by affinity chromatography using the method of Kaku, H & Goldstein, IJ, *Meth Enzymol*, 179: 327, 1989; activity: <20 µg/mL | Agglutinates rabbit erythrocytes; human erythrocytes are not agglutinated; GNL sugar specificity is directed to the nonreducing end of the terminal α-D-mannosyl residue of glycoconjugates; agglutination activity expressed in µg/mL & is determined from serial dilutions of a 1 mg/mL solution using phosphate buffered saline, pH 7.2; activity is the lowest concentration to agglutinate a 2% suspension of rabbit erythrocytes after 1 hr at 25°C; Kaku, H & Goldstein, IJ, *Meth Enzymol*, 179: 327, 1989; Shibuya, M et al, *Arch Biochem Biophys*, 267: 676, 1988

Agglutinin, *Galanthus nivalis*, Immobilized

Synonyms: Lectin

Sigma L 8775 *Galanthus nivalis* (Snowdrop) Immobilized on cross-linked 4% beaded agarose; contains 2-4 mg protein/mL packed gel; suspension in 1.0 *M* NaCl containing 0.01% thimerosal; binding capactiy: 1 mL gel binds 5-10 mg yeast mannan at pH 7.2 at 25°C; 1 mL suspension yields ~0.5 mL packed gel | See Sigma L 8275

Agglutinin, *Glycine max*

Synonyms: Lectin

ICN 152066 *Glycine max* (Soybean) Purified by affinity chromatography; mitogenic properties | Affinity for *N*-acetyl-D-galactosamine; useful in separation of blood cell populations; agglutinates mouse B-cells but not T-cells; Reisner, Y, A Ravid & N Sharon, *BBRC*, 72:1585, 1976; Reisner, Y etal, *PNAS*, 75:2933; 1978

Sigma L 1395 *Glycine max* (Soybean)   Purified by affinity chromatography; essentially salt-free; lyophilized; activity: <80 µg/mL | Not blood group specific, but has an affinity for *N*-acetyl-D-galactosamine; conjugates are prepared from affinity purified lectin; agglutination activity expressed in µg/mL & is determined from serial dilutions of a 1 mg/mL solution using phosphate buffered saline, pH 6.8; activity is the lowest concentration to agglutinate a 2% suspension of human blood group A erythrocytes after 1 hr at 25°C; Lis, H et al, *Biochim Biophys Acta*, 211: 582, 1970; Lotan, R et al, *J Biol Chem*, 249: 1219, 1974

Agglutinin, *Glycine max*, 10 nm Colloidal Gold Conjugated

Synonyms: Lectin

ICN 154049 *Glycine max* (Soybean)   Purified by affinity chromatography | Specificity for D-galNAc

Sigma L 4768 *Glycine max* (Soybean)   Mean particle size 8-12 nm; monodisperse; suspension in ~50% glycerol containing 0.15 *M* NaCl, 0.01 *M* sodium phosphate, pH 7.2, 0.02% PEG 20 & 0.02% sodium azide; concentration: A_{520} ~5.0; Horisberger, *Techniques in Immunocytochemistry*, Bullock & Petrusz, eds, Academic Press, 3: 155, 1985 | See Sigma L 1395 & Sigma L 2650

Agglutinin, *Glycine max*, 20 nm Colloidal Gold Conjugated

Synonyms: Lectin

ICN 154050 *Glycine max* (Soybean)   Purified by affinity chromatography | Specificity for D-galNAc

Agglutinin, *Glycine max*, 5 nm Colloidal Gold Conjugated

Synonyms: Lectin

ICN 154048 *Glycine max* (Soybean)   Purified by affinity chromatography | Specificity for D-galNAc

Agglutinin, *Glycine max*, Biotin Conjugated

Synonyms: Lectin

ICN 153266 *Glycine max* (Soybean)   Purified by affinity chromatography | Specificity for D-galNAc

Sigma L 3395 *Glycine max* (Soybean)   Lyophilized containing ~75% protein (Biuret); balance sodium citrate; contains 2-4 moles biotin/mole protein | See Sigma L 1395

Agglutinin, *Glycine max*, FITC Conjugated

Synonyms: Lectin

ICN 153267 *Glycine max* (Soybean)   Purified by affinity chromatography | Specificity for D-galNAc

Sigma L 1020 *Glycine max* (Soybean)   Lyophilized containing ~10% protein (Lowry); balance phosphate buffer salts & NaCl; contains 1-2 moles FITC/mole protein | See Sigma L 1395

Agglutinin, *Glycine max*, Horse Radish Peroxidase Conjugated

Synonyms: Lectin

ICN 153268 *Glycine max* (Soybean)   Purified by affinity chromatography | Specificity for D-galNAc

Agglutinin, *Glycine max*, Immobilized

Synonyms: Lectin

Sigma L 1145 *Glycine max* (Soybean)   Immobilized on 4% cross-linked beaded agarose; spacer: ~20 atoms; contains 2-4 mg protein (Biuret)/mL gel; suspension in 0.01 *M* phosphate buffered saline containing 50% glycerol, 0.01 *M* *N*-acetyl-D-galactosamine & 0.02% sodium azide | See Sigma L 1395 & Sigma L 2650

Agglutinin, *Glycine max*, Peroxidase Conjugated

Synonyms: Lectin

Sigma L 1270 *Glycine max* (Soybean)   Packaged in microcone vials for ease of reconstitution & recovery at microliter volumes | See Sigma L 1395 & Sigma L 2650

Sigma L 2650 *Glycine max* (Soybean)   Lyophilized containing ~95% protein (Modified Warburg-Christian); balance primarily sodium citrate; peroxidase activity: ~50 units purpurogallin/mg protein; unit definition: 1 unit forms 1 mg purpurogallin in 20 sec from pyrogallol at pH 6.0 at 20°C; prepared from peroxidase type VI using a modification of the method of O'Sullivan, MJ et al, *FEBS Lett*, 95: 311, 1978, which favors low molecular weight conjugates; repurified after conjugation by affinity chromatography | See Sigma L 1395

Agglutinin, *Glycine max*, TRITC Conjugated

Synonyms: Lectin

Sigma L 4511 *Glycine max* (Soybean)   Lyophilized containing ~10% protein (Lowry); balance phosphate buffer salts & NaCl; contains ~1 mole TRITC/mole protein | See Sigma L 1395

Agglutinin, *Griffonia simplicifolia* GS-I

Synonyms: Lectin

ICN 153269 *Bandeiraea simplicifolia*   Purified by affinity chromatography | Specificity for D-GluNAc

ICN 153270 *Bandeiraea simplicifolia*   Purified by affinity chromatography | Specificity for D-GluNAc

Agglutinin, *Griffonia simplicifolia* GS-I B4

Synonyms: Lectin

ICN 153273 *Bandeiraea simplicifolia*   Purified by affinity chromatography | Specificity for D-GalNAc, α-D-Gal

Agglutinin, *Griffonia simplicifolia* GS-I B4, Biotin Conjugated

Synonyms: Lectin

ICN 153274 *Bandeiraea simplicifolia*   Purified by affinity chromatography | Specificity for D-GalNAc, α-D-Gal

Agglutinin, *Griffonia simplicifolia* GS-I B4, FITC Conjugated

Synonyms: Lectin

ICN 153275 *Bandeiraea simplicifolia*   Purified by affinity chromatography | Specificity for D-GalNAc, α-D-Gal

Agglutinin, *Griffonia simplicifolia* GS-I B4, Horse Radish Peroxidase Conjugated

Synonyms: Lectin

ICN 153276 *Bandeiraea simplicifolia*   Purified by affinity chromatography | Specificity for D-GalNAc, α-D-Gal

Agglutinin, *Griffonia simplicifolia* GS-I, 10 nm Colloidal Gold Conjugated

Synonyms: Lectin

ICN 154052 *Bandeiraea simplicifolia*   Purified by affinity chromatography | Specificity for D-GluNAc

Agglutinin, *Griffonia simplicifolia* GS-I, 20 nm Colloidal Gold Conjugated

Synonyms: Lectin

ICN 154053 *Bandeiraea simplicifolia*   Purified by affinity chromatography | Specificity for D-GluNAc

Agglutinin, *Griffonia simplicifolia* GS-I, 5 nm Colloidal Gold Conjugated

Synonyms: Lectin

ICN 154051 *Bandeiraea simplicifolia* Purified by affinity chromatography | Specificity for D-GluNAc

Agglutinin, *Griffonia simplicifolia* GS-I, FITC Conjugated

Synonyms: Lectin

ICN 153271 *Bandeiraea simplicifolia* Purified by affinity chromatography | Specificity for D-GluNAc

Agglutinin, *Griffonia simplicifolia* GS-I, Horse Radish Peroxidase Conjugated

Synonyms: Lectin

ICN 153272 *Bandeiraea simplicifolia* Purified by affinity chromatography | Specificity for D-GluNAc

Agglutinin, *Griffonia simplicifolia* GS-II

Synonyms: Lectin

ICN 150422 *Bandeiraea simplicifolia* 1 mg/mL solution; purified by affinity chromatography | The second *Bandeiraea simplicifolia* lectin GS-II; affinity for *N*-acetyl-D-glucosamine; agglutinates "acquired β-cells", activated T-cells & T_k polyagglutinable cells

Agglutinin, *Griffonia simplicifolia* GS-II, 20 nm Colloidal Gold Conjugated

Synonyms: Lectin

ICN 154055 *Bandeiraea simplicifolia* Purified by affinity chromatography | Specificity for α-D-Gal

Agglutinin, *Griffonia simplicifolia* GS-II, 5 nm Colloidal Gold Conjugated

Synonyms: Lectin

ICN 154054 *Bandeiraea simplicifolia* Purified by affinity chromatography | Specificity for α-D-Gal

Agglutinin, *Griffonia simplicifolia* GS-II, Biotin Conjugated

Synonyms: Lectin

ICN 153277 *Bandeiraea simplicifolia* Purified by affinity chromatography | Specificity for α-D-Gal

Agglutinin, *Griffonia simplicifolia* GS-II, FITC Conjugated

Synonyms: Lectin

ICN 153278 *Bandeiraea simplicifolia* Purified by affinity chromatography | Specificity for α-D-Gal

Agglutinin, *Griffonia simplicifolia* GS-II, Horse Radish Peroxidase Conjugated

Synonyms: Lectin

ICN 153279 *Bandeiraea simplicifolia* Purified by affinity chromatography | Specificity for α-D-Gal

Agglutinin, *Helix aspersa*

Synonyms: Lectin

ICN 153280 *Helix aspersa* Purified by affinity chromatography | Specificity for D-GalNAc

Sigma L 6635 *Helix aspersa* (Garden snail) Lyophilized; salt-free; highly purified by affinity chromatography using a modified method of Hammarstrom, S & Kabat, EA, *Biochemistry*, 8: 2969, 1969; activity: <2 μg/mL | Anti-A human blood group specificity & an affinity for terminal *N*-acetyl-α-D-galactosaminyl residues; conjugates are prepared from affinity purified lectins; agglutination activity expressed in μg/mL & is determined from serial dilutions of a 1 mg/mL solution using phosphate buffered saline, pH 6.8; activity is the lowest concentration to agglutinate a 2% suspension of human blood group A erythrocytes after 1 hr at 25°C; Pemberton, RT, *Vox Sang*, 16: 457, 1969; Uhlenbruck, G et al, *Anim Bld Gps Biochem Genet*, 3: 125, 1972; Hammarstrom, S et al, *Scand J Immuno*, 1: 295, 1972

Agglutinin, *Helix aspersa*, Biotin Conjugated

Synonyms: Lectin

ICN 153281 *Helix aspersa* Purified by affinity chromatography | Specificity for D-GalNAc

Sigma L 8764 *Helix aspersa* (Garden snail) Salt-free lyophilized powder; contains ~2 moles biotin/mole lectin | See Sigma L 6635

Agglutinin, *Helix aspersa*, FITC Conjugated

Synonyms: Lectin

ICN 153282 *Helix aspersa* Purified by affinity chromatography | Specificity for D-GalNAc

Sigma L 3764 *Helix aspersa* (Garden snail) Salt-free lyophilized powder; contains ~2 moles FITC/mole protein | See Sigma L 6635

Agglutinin, *Helix aspersa*, Horse Radish Peroxidase Conjugated

Synonyms: Lectin

ICN 153283 *Helix aspersa* Purified by affinity chromatography | Specificity for D-GalNAc

Agglutinin, *Helix aspersa*, Sulforhodamine 101 Acid Chloride (Texas Red) Conjugated

Synonyms: Lectin

Sigma L 3889 *Helix aspersa* (Garden snail) Salt-free lyophilized powder; contains ~1.5 moles Texas Red/mole protein | See Sigma L 6635

Agglutinin, *Helix aspersa*, TRITC Conjugated

Synonyms: Lectin

Sigma L 4014 *Helix aspersa* (Garden snail) Salt-free lyophilized powder; contains ~1 mole TRITC/mole protein | See Sigma L 6635

Agglutinin, *Helix pomatia*

Synonyms: Lectin

ICN 151229 *Helix pomatia* (Roman or edible snail) Purified by affinity chromatography; 1 mg/mL solution | Specific for the anti-A human blood group; affinity for terminal *N*-acetyl-β-D-galactosaminyl residues

Sigma L 3382 *Helix pomatia* (Roman or edible snail) Salt-free; lyophilized; highly purified by affinity chromatography by a modification of the method of Hammarstrom, S & Kabat, EA, *Biochemistry*, 8: 2696, 1969; activity: <4 μg/mL | Anti-A human blood group specificity & has an affinity for terminal *N*-acetyl-α-D-galactosaminyl residues; conjugates are prepared from affinity purified lectin; agglutination activity expressed in μg/mL & is determined from serial dilutions of a 1 mg/mL solution using phosphate buffered saline, pH 6.8; activity is the lowest concentration to agglutinate a 2% suspension of human blood group A erythrocytes after 1 hr at 25°C; Hammarstrom, S et al, *Scand J Immunol*, 1: 295, 1972; Hammarstrom, S & Kabat, EA, *Biochemistry*, 10: 1684, 1971

Sigma L 7760 *Helix pomatia* (Roman or edible snail) Salt-free; lyophilized; partially purified; activity: <10 μg/mL | See Sigma L 3382

Agglutinin, *Helix pomatia*, 10 nm Colloidal Gold Conjugated

Synonyms: Lectin

ICN 154057 *Helix pomatia* (Roman or edible snail)   Purified by affinity chromatography | Specificity for D-GalNAc

Sigma L 4770 *Helix pomatia* (Roman or edible snail)   Mean particle size 8-12 nm; monodisperse; suspension in ~50% glycerol containing 0.15 *M* NaCl, 0.01 *M* sodium phosphate, pH 6.8, 0.1 m*M* Ca^{2+}, 0.1 m*M* Mn^{2+}, 0.02% PEG 20 & 0.02% sodium azide; concentration: A_{520} ~5.0; Brown, D & Orci, L, *J Histochem Cytochem*, 34: 1057, 1986 | See Sigma L 3382

Agglutinin, *Helix pomatia*, 20 nm Colloidal Gold Conjugated

Synonyms: Lectin

ICN 154058 *Helix pomatia* (Roman or edible snail)   Purified by affinity chromatography | Specificity for D-GalNAc

Agglutinin, *Helix pomatia*, 5 nm Colloidal Gold Conjugated

Synonyms: Lectin

ICN 154056 *Helix pomatia* (Roman or edible snail)   Purified by affinity chromatography | Specificity for D-GalNAc

Sigma L 2765 *Helix pomatia* (Roman or edible snail)   Mean particle size 3.5-6.5 nm; monodisperse; suspension in ~50% glycerol containing 0.15 *M* NaCl, 0.01 *M* sodium phosphate, pH 6.8, 0.1 m*M* Ca^{2+}, 0.1 m*M* Mn^{2+}, 0.02% PEG 20 & 0.02% sodium azide; concentration: A_{520} ~5.0; Brown, D & Orci, L, *J Histochem Cytochem*, 34: 1057, 1986 | See Sigma L 3382

Agglutinin, *Helix pomatia*, Biotin Conjugated

Synonyms: Lectin

ICN 153284 *Helix pomatia* (Roman or edible snail)   Purified by affinity chromatography | Specificity for D-GalNAc

Sigma L 6512 *Helix pomatia* (Roman or edible snail)   Salt-free; lyophilized; contains 2-4 moles biotin/mole protein | See Sigma L 3382

Agglutinin, *Helix pomatia*, FITC Conjugated

Synonyms: Lectin

ICN 153285 *Helix pomatia* (Roman or edible snail)   Purified by affinity chromatography | Specificity for D-GalNAc

Sigma L 1034 *Helix pomatia* (Roman or edible snail)   Solution in 0.01 *M* Tris, pH 8.0, containing 0.15 *M* NaCl & 0.1% sodium azide; contains 5-8 moles FITC/mole protein & ≥0.5 mg protein/mL; activity: <1 µg/mL | See Sigma L 3382

Agglutinin, *Helix pomatia*, Horse Radish Peroxidase Conjugated

Synonyms: Lectin

ICN 153286 *Helix pomatia* (Roman or edible snail)   Purified by affinity chromatography | Specificity for D-GalNAc

Agglutinin, *Helix pomatia*, Immobilized

Synonyms: Lectin

Sigma L 8639 *Helix pomatia* (Roman or edible snail)   Immobilized on 4% cross-linked beaded agarose; contains 1.5 mg lectin/mL packed gel; suspension in 0.5 *M* NaCl, 10 m*M* phosphate buffer, pH 7.3, 0.02% sodium azide | See Sigma L 3382

Agglutinin, *Helix pomatia*, Peroxidase Conjugated

Synonyms: Lectin

Sigma L 6387 *Helix pomatia* (Roman or edible snail)   Lyophilized containing ~95% protein (Lowry); balance primarily sodium citrate; peroxidase activity: 50-120 units purpurogallin/mg protein; unit definition: 1 unit forms 1 mg purpurogallin in 20 sec from pyrogallol at pH 6.0 at 20°C; prepared from peroxidase type VI using a modification of the method of O'Sullivan, MJ et al, *FEBS Lett*, 95: 311, 1978, which favors low molecular weight conjugates; repurified after conjugation by affinity chromatography | See Sigma L 3382

Agglutinin, *Helix pomatia*, Sulforhodamine 101 Acid Chloride (Texas Red) Conjugated

Synonyms: Lectin

Sigma L 9136 *Helix pomatia* (Roman or edible snail)   Salt-free; lyophilized; contains ~1.5 moles Texas Red/mole protein | See Sigma L 3382

Agglutinin, *Helix pomatia*, TRITC Conjugated

Synonyms: Lectin

Sigma L 1261 *Helix pomatia* (Roman or edible snail)   Salt-free; lyophilized; contains ~1 mole TRITC/mole protein | See Sigma L 3382

Agglutinin, *Laburnum alpinum*

Synonyms: Lectin

ICN 153287 *Laburnum alpinum*   Purified by affinity chromatography

Agglutinin, *Laburnum alpinum*, Biotin Conjugated

Synonyms: Lectin

ICN 153288 *Laburnum alpinum*   Purified by affinity chromatography

Agglutinin, *Laburnum alpinum*, FITC Conjugated

Synonyms: Lectin

ICN 153289 *Laburnum alpinum*   Purified by affinity chromatography

Agglutinin, *Laburnum alpinum*, Horse Radish Peroxidase Conjugated

Synonyms: Lectin

ICN 153290 *Laburnum alpinum*   Purified by affinity chromatography

Agglutinin, *Lathyrus odoratus*

Synonyms: Lectin

Sigma L 4651 *Lathyrus odoratus* (Sweet pea)   Lyophilized containing ~50% protein (Biuret); balance NaCl; highly purified by affinity chromatography using the method of Kolberg, J et al, *FEBS Lett*, 117: 281, 1980; activity: <20 µg/mL | Not blood group specific; specific for α-mannosyl end-groups, D-glucose & *N*-acetyl glucosamine; agglutination activity expressed in µg/mL & is determined from serial dilutions of a 1 mg/mL solution using phosphate buffered saline, pH 7.2; activity is the lowest concentration to agglutinate a 2% suspension of human erythrocytes after 1 hr at 25°C; Ticha, M et al, *Acta Biol Med Germ*, 39: 649, 1980

Agglutinin, *Lens culinaris*

Synonyms: Lectin

ICN 151542 *Lens culinaris* (Lentil)   Purified by affinity chromatography; lyophilized; electrophoretically pure | Specificity for terminal α-D-Man & α-D-Gluc residues; stimulates human lymphocytes in culture to incorporate ^{3}H-thymidine; comprised of 2 isomers differing in AA content & electrophoretic mobility: hemagglutinins A (LCH-A) & B (LCH-B)

ICN 153294   *Lens culinaris* (Lentil)   Purified by affinity chromatography | Specificity for α-D-Man

Sigma L 9267   *Lens culinaris* (Lentil)   MW 49k (LcH-A), MW 49k (LcH-B)   Salt-free; lyophilized; highly purified by affinity chromatography; activity: <16 μg/mL; contains two major bands on electrophoresis corresponding to isolectins LcH-A & LcH-B | Not blood group specific but has an affinity for α-D-mannosyl & α-D-glucosyl residues; LcH is comprised of two isomers: LcH-A (MW 49,000) & LcH-B (MW 49,000); LcH-A is mitogenic; conjugates are prepared from purified lectin; agglutination activity expressed in μg/mL & is determined from serial dilutions of a 1 mg/mL solution using phosphate buffered saline, pH 6.8; activity is the lowest concentration to agglutinate a 2% suspension of human erythrocytes after 1 hr at 25°C; Howard, IK et al, *J Biol Chem*, 246: 1590, 1971

Agglutinin, *Lens culinaris*, 10 nm Colloidal Gold Conjugated

Synonyms: Lectin

ICN 154063   *Lens culinaris* (Lentil)   Purified by affinity chromatography | Specificity for α-D-Man

Sigma L 2144   *Lens culinaris* (Lentil)   MW 49k (LcH-A), MW 49k (LcH-B)   Mean particle size 8-12 nm; monodisperse; suspension in ~50% glycerol containing 0.15 M NaCl, 0.01 M sodium phosphate, pH 7.2, 0.1 mM Ca^{2+}, 0.1 mM Mn^{2+}, 0.02% PEG 20 & 0.02% sodium azide; concentration: A$_{520}$ ~5.0; Horisberger, M, *Techniques in Immunocytochemistry*, Bullock & Petrusz, eds, Academic Press, 3: 155, 1985 | See Sigma L 9267

Agglutinin, *Lens culinaris*, 20 nm Colloidal Gold Conjugated

Synonyms: Lectin

ICN 154064   *Lens culinaris* (Lentil)   Purified by affinity chromatography | Specificity for α-D-Man

Agglutinin, *Lens culinaris*, 5 nm Colloidal Gold Conjugated

Synonyms: Lectin

ICN 154062   *Lens culinaris* (Lentil)   Purified by affinity chromatography | Specificity for α-D-Man

Sigma L 1769   *Lens culinaris* (Lentil)   MW 49k (LcH-A), MW 49k (LcH-B)   Mean particle size 3.5-6.5 nm; monodisperse; suspension in ~50% glycerol containing 0.15 M NaCl, 0.01 M sodium phosphate, pH 7.2, 0.1 mM Ca^{2+}, 0.1 mM Mn^{2+}, 0.02% PEG 20 & 0.02% sodium azide; concentration: A$_{520}$ ~5.0; Horisberger, M, *Techniques in Immunocytochemistry*, Bullock & Petrusz, eds, Academic Press, 3: 155, 1985 | See Sigma L 9267

Agglutinin, *Lens culinaris*, Agarose

Synonyms: Lectin

ICN 191473   *Lens culinaris* (Lentil)   Prepacked column; 5 atoms hydrophilic spacer arm; 2 mg lectin/mL gel; suspension in PBS, 0.02% NaN$_3$ | Specificity for α-D-man; useful for purification of detergent solubilized membrane glycoproteins, cell surface Ag, glycoproteins, viral glycoproteins

Agglutinin, *Lens culinaris*, Biotin Conjugated

Synonyms: Lectin

ICN 153291   *Lens culinaris* (Lentil)   Purified by affinity chromatography | Specificity for α-D-Man

Sigma L 4143   *Lens culinaris* (Lentil)   MW 49k (LcH-A), MW 49k (LcH-B)   Lyophilized containing ~85% protein (Lowry); balance primarily sodium citrate; contains 2-5 moles biotin/mole lectin | See Sigma L 9267

Agglutinin, *Lens culinaris*, FITC Conjugated

Synonyms: Lectin

ICN 153293   *Lens culinaris* (Lentil)   Purified by affinity chromatography | Specificity for α-D-Man

Sigma L 9262   *Lens culinaris* (Lentil)   MW 49k (LcH-A), MW 49k (LcH-B)   Lyophilized containing ~40% protein (Lowry); balance primarily HEPES buffer salts & NaCl; contains ~2 moles FITC/mole protein | See Sigma L 9267

Agglutinin, *Lens culinaris*, Horse Radish Peroxidase Conjugated

Synonyms: Lectin

ICN 153292   *Lens culinaris* (Lentil)   Purified by affinity chromatography | Specificity for α-D-Man

Agglutinin, *Lens culinaris*, Immobilized

Synonyms: Lectin

Sigma L 4018   *Lens culinaris* (Lentil)   MW 49k (LcH-A), MW 49k (LcH-B)   Immobilized on 4% beaded agarose; contains 2-4 mg lentil lectin/mL gel; suspension in 0.9% NaCl, 1 mM CaCl$_2$, 1 mM MnCl$_2$ & 0.01% thimerosal; similar to Sigma L 0511 but produced by Sigma | See Sigma L 9267

Agglutinin, *Lens culinaris*, Sepharose 4B

Synonyms: Lectin

Sigma L 0511   *Lens culinaris* (Lentil)   MW 49k (LcH-A), MW 49k (LcH-B)   Attached to Sepharose 4B; contains ~2 mg lentil lectin/mL packed gel; suspension in 20% ethanol containing 0.9% NaCl, 1 mM CaCl$_2$, 1 mM MnCl$_2$ | See Sigma L 9267

Agglutinin, *Lens culinaris*, TRITC Conjugated

Synonyms: Lectin

Sigma L 5764   *Lens culinaris* (Lentil)   MW 49k (LcH-A), MW 49k (LcH-B)   Lyophilized containing ~10% protein (Lowry); balance primarily phosphate buffer salts & NaCl; contains ~0.5 mole TRITC/mole protein | See Sigma L 9267

Agglutinin, *Limax flavus*

Synonyms: Lectin

ICN 153295   *Limax flavus*   Purified by affinity chromatography

Agglutinin, *Limax flavus*, 10 nm Colloidal Gold Conjugated

Synonyms: Lectin

ICN 154066   *Limax flavus*   Purified by affinity chromatography

Agglutinin, *Limax flavus*, 20 nm Colloidal Gold Conjugated

Synonyms: Lectin

ICN 154067   *Limax flavus*   Purified by affinity chromatography

Agglutinin, *Limax flavus*, 5 nm Colloidal Gold Conjugated

Synonyms: Lectin

ICN 154065   *Limax flavus*   Purified by affinity chromatography

Agglutinin, *Limax flavus*, Biotin Conjugated

Synonyms: Lectin

ICN 153296   *Limax flavus*   Purified by affinity chromatography

Agglutinin, *Limax flavus*, FITC conjugated

Synonyms: Lectin

ICN 153297   *Limax flavus*   Purified by affinity chromatography

Agglutinin, *Limax flavus*, Horse Radish Peroxidase Conjugated

Synonyms: Lectin

ICN 153298   *Limax flavus*   Purified by affinity chromatography

Agglutinin, *Limulus polyphemus*

Synonyms: Bacterial Agglutinin; Lectin; Limulin III

ICN 153299   *Limulus polyphemus*   Purified by affinity chromatography | Specificity for NeuNAc

Sigma L 7908   *Limulus polyphemus* (Horseshoe crab)   Affinity purified; lyophilized containing ~50% protein (Bradford); balance primarily NaCl, Tris succinate & calcium acetate | Not human blood group specific; affinity for sialic acid (*N*-acetylneuraminic acid), glucuronic acid & phosphorylcholine analogs; horse erythrocytes are most strongly agglutinated; requires Ca^{2+} for activity; agglutination activity expressed in μg/mL & is determined from serial dilutions of a 1 mg/mL solution using 50 *mM* Tris buffered saline, pH 7.2, containing 0.1 *M* $CaCl_2$; activity is the lowest concentration to agglutinate a 1.5% suspension of horse erythrocytes after 1 hr at 25°C; Marchelonis, JJ & Edelman, GM, *J Mol Biol*, 32: 453, 1968; Fernandez-Moran, H et al, *J Mol Biol*, 32: 467, 1968

Sigma L 2263   *Limuius polyphemus* (Limulin)   Lyophilized containing ~25% protein (Modified Warburg-Christian); balance primarily Tris succinate, NaCl & calcium acetate; purified by affinity chromatography; salt-free; activity: <16 μg/mL with blood group B erythrocytes & <8 μg/mL with *S. aureus* cells | Agglutinates Staphylococcus *aureus* cells, as well as human & horse erythrocytes; major affinity is for *N*-acetylated D-hexosamines, but is reported to have anti-galactan specificity; agglutination activity expressed in μg/mL & is determined from serial dilutions of a 1 mg/mL solution using phosphate buffered saline, pH 7.4 containing 0.1 *mM* $CaCl_2$; activity is the lowest concentration to agglutinate a 2% suspension of human blood group B erythrocytes after 1 hr at 25°C; bacterial agglutination activity is determined from serial dilutions in 0.01 *M* Tris, 0.15 *M* NaCl, pH 7.4, containing 0.01 *M* $CaCl_2$ of a 1 mg/mL solution; activity is the lowest concentration to agglutinate a 0.083% suspension of *S. aureus* cells; Gilbride, KJ & Pistole, TG, *Prog Clin Biol Res*, 29: 525, 1979

Agglutinin, *Limulus polyphemus*, Biotin Conjugated

Synonyms: Lectin; Limulin III

ICN 153300   *Limulus polyphemus*   Purified by affinity chromatography | Specificity for NeuNAc

Agglutinin, *Limulus polyphemus*, FITC Conjugated

Synonyms: Lectin; Limulin III

ICN 153301   *Limulus polyphemus*   Purified by affinity chromatography | Specificity for NeuNAc

Sigma L 8520   *Limulus polyphemus* (Horseshoe crab)   Lyophilized containing ~90% protein (Biuret); balance Tris buffer salts & calcium acetate; contains >15 moles FITC/mole protein | See Sigma L 7908

Agglutinin, *Limulus polyphemus*, Horse Radish Peroxidase Conjugated

Synonyms: Lectin; Limulin III

ICN 153302   *Limulus polyphemus*   Purified by affinity chromatography | Specificity for NeuNAc

Agglutinin, *Lotus tetragonolobus*

Synonyms: Lectin

ICN 153303   *Lotus tetragonolobus*   Purified by affinity chromatography | Specificity for α-L-fucose

Agglutinin, *Lotus tetragonolobus*, Biotin Conjugated

Synonyms: Lectin

ICN 153304   *Lotus tetragonolobus*   Purified by affinity chromatography | Specificity for α-L-fucose

Agglutinin, *Lotus tetragonolobus*, FITC Conjugated

Synonyms: Lectin

ICN 153305   *Lotus tetragonolobus*   Purified by affinity chromatography | Specificity for α-L-fucose

Agglutinin, *Lotus tetragonolobus*, Horse Radish Peroxidase Conjugated

Synonyms: Lectin

ICN 153306   *Lotus tetragonolobus*   Purified by affinity chromatography | Specificity for α-L-fucose

Agglutinin, *Lycopersicon esculentum*

Synonyms: Lectin

Sigma L 2886   *Lycopersicon esculentum* (Tomato)   Highly purified by affinity chromatography; lyophilized containing ≥40% lectin; balance NaCl; activity: <2 μg/mL | Not blood group specific; affinity for *N*-acetyl-β-D-glucosamine oligomers; glycoprotein containing approx. equal amounts of protein & carbohydrate & is reported to inhibit the mitogenic activity of phytohemagglutinin from Phaseolus vulgaris; conjugates are prepared from affinity purified lectin; agglutination activity expressed in μg/mL & is determined from serial dilutions of a 1 mg/mL solution using phosphate buffered saline, pH 7.3; activity is the lowest concentration to agglutinate a 2% suspension of human erythrocytes after 1 hr at 25°C; Nachbar, MA et al, *J Biol Chem*, 255: 2056, 1980

Agglutinin, *Lycopersicon esculentum*, Biotin Conjugated

Synonyms: Lectin

Sigma L 0651   *Lycopersicon esculentum* (Tomato)   Lyophilized containing ~25% lectin; balance primarily sodium citrate; contains 3-5 moles biotin/mole lectin | See Sigma L 2886

Agglutinin, *Lycopersicon esculentum*, FITC Conjugated

Synonyms: Lectin

Sigma L 0401   *Lycopersicon esculentum* (Tomato)   Solution in 10 *mM* HEPES, 0.15 *M* NaCl, pH 7.5, containing 0.1 *mM* Ca^{2+} & 0.08% sodium azide; contains 3-6 moles FITC/mole lectin | See Sigma L 2886

Agglutinin, *Lycopersicon esculentum*, Sulforhodamine 101 Acid Chloride (Texas Red) Conjugated

Synonyms: Lectin

Sigma L 9139   *Lycopersicon esculentum* (Tomato)   Lyophilized containing ~50% lectin; balance primarily NaCl; contains ~1 mole Texas Red/mole lectin | See Sigma L 2886

Agglutinin, *Lycopersicon esculentum*, TRITC Conjugated

Synonyms: Lectin

Sigma L 9511   *Lycopersicon esculentum* (Tomato)   Lyophilized containing ~50% lectin; balance primarily NaCl; contains 1-2 moles TRITC/mole lectin | See Sigma L 2886

Agglutinin, *Maackia amurensis*

Synonyms: Lectin

Sigma L 8025   *Maackia amurensis*   Lyophilized; essentially salt-free; highly purified by affinity chromatography using a modification of the method of Wang, W-C & Cummings, RD, *J Biol Chem*, 263: 4576, 1988; activity: <10 μg/mL | Interacts with sialic acid-containing glycoconjugates; consists of 2 molecular species, a strongly hemagglutinating hemagglutinin (MAH) & a strongly mitogenic hemagglutinin (MAL); mitogenic activity of the latter is inhibited by α-sialyl-(2→3)-lactose; MAA could be useful for fractionation of sialylated oligosaccharides; agglutination activity expressed in μg/mL & is determined from serial dilutions of a 1 mg/mL solution using phosphate buffered saline, pH 7.2; activity is the lowest concentration to agglutinate a 2% suspension of fresh human erythrocytes, type O, after 1 hr incubation at 25°C; Knibbs, RN et al, *J Biol Chem*, 266: 83, 1991; Kawaguchi, T et al, *J Biol Chem*, 249: 2786, 1974

Agglutinin, *Maclura pomifera*

Synonyms: Lectin

ICN 153307   *Maclura pomifera*   Purified by affinity chromatography | Specificity for α-D-gal, α-D-galNAc

Sigma L 6141   *Maclura pomifera* (Osage orange)   Lyophilized; salt-free; highly purified by affinity chromatography using the method of Bausch, NJ & Poretz, RD, *Biochemistry*, 16: 5790, 1977; activity: <1 µg/mL | Not human blood group specific, but is reported to bind specifically to T lymphocytes; affinity for terminal α-D-galactosyl & *N*-acetyl-D-galactosaminyl residues; conjugates are prepared from affinity purified lectin; agglutination activity expressed in µg/mL & is determined from serial dilutions of a 1 mg/mL solution using phosphate buffered saline, pH 6.8; activity is the lowest concentration to agglutinate a 2% suspension of human erythrocytes after 1 hr at 25°C; Jones, JM & Feldman, JD, *J Immunol*, 3: 1765, 1973; Bausch, NJ & Poretz, RD, *Biochemistry*, 16: 5790, 1977

Agglutinin, *Maclura pomifera*, 10 nm Colloidal Gold Conjugated

Synonyms: Lectin

ICN 154069   *Maclura pomifera*   Purified by affinity chromatography | Specificity for α-D-gal, α-D-galNAc

Agglutinin, *Maclura pomifera*, 20 nm Colloidal Gold Conjugated

Synonyms: Lectin

ICN 154070   *Maclura pomifera*   Purified by affinity chromatography | Specificity for α-D-gal, α-D-galNAc

Agglutinin, *Maclura pomifera*, 5 nm Colloidal Gold Conjugated

Synonyms: Lectin

ICN 154068   *Maclura pomifera*   Purified by affinity chromatography | Specificity for α-D-gal, α-D-galNAc

Agglutinin, *Maclura pomifera*, Biotin Conjugated

Synonyms: Lectin

ICN 153308   *Maclura pomifera*   Purified by affinity chromatography | Specificity for α-D-gal, α-D-galNAc

Sigma L 2013   *Maclura pomifera* (Osage orange)   Lyophilized containing ~85% protein ($E^{1\%}_{280}$); balance primarily sodium citrate; contains ~2 moles biotin/mole protein | See Sigma L 6141

Agglutinin, *Maclura pomifera*, FITC Conjugated

Synonyms: Lectin

ICN 153309   *Maclura pomifera*   Purified by affinity chromatography | Specificity for α-D-gal, α-D-galNAc

Sigma L 4383   *Maclura pomifera* (Osage orange)   Lyophilized containing ~10% protein (Lowry); balance phosphate buffer salts & NaCl; contains ~2 moles FITC/mole protein | See Sigma L 6141

Agglutinin, *Maclura pomifera*, Horse Radish Peroxidase Conjugated

Synonyms: Lectin

ICN 153310   *Maclura pomifera*   Purified by affinity chromatography | Specificity for α-D-gal, α-D-galNAc

Agglutinin, *Maclura pomifera*, Peroxidase Conjugated

Synonyms: Lectin

Sigma L 4401   *Maclura pomifera* (Osage orange)   Lyophilized containing ~90% protein (Modified Warburg-Christian); balance primarily sodium citrate; peroxidase activity: 20-50 units purpurogallin/mg protein; unit definition: 1 unit forms 1 mg purpurogallin in 20 sec from pyrogallol at pH 6.0 at 20°C; prepared from peroxidase type VI using a modification of the method of O'Sullivan, MJ et al, *FEBS Lett*, 95: 311, 1978, which favors low molecular weight conjugates; repurified by affinity chromatography after conjugation | See Sigma L 6141

Agglutinin, *Maclura pomifera*, TRITC Conjugated

Synonyms: Lectin

Sigma L 5889   *Maclura pomifera* (Osage orange)   Lyophilized containing ~10% protein (Lowry); balance phosphate buffer salts & NaCl; contains 0.25-1 mole TRITC/mole protein | See Sigma L 6141

Agglutinin, *Momordica charantia*

Synonyms: Lectin

Sigma L 4644   *Momordica charantia* (Bitter pear melon)   Lyophilized containing ~25% protein (Biuret); balance primarily NaCl & sodium phosphate; highly purified by affinity chromatography using a modification of the method of Ng, TB et al, *Int J Peptide Prot Res*, 28: 163, 1986; activity: <1 µg/mL | The seeds of Momoridca *charantia* contain two proteins with inhibitory activity on protein synthesis in a cell-free system; one is a high potency protein synthesis inhibitor, momordin; the other is a low potency protein synthesis inhibitor which is also a hemagglutinating lectin; MCA is a glycoprotein which exerts insulinomimetic, antilipolytic & lipogenic activities on isolated rat adipocytes; also agglutinates adipocytes & shows affinities for Gal & GalNAc; Barbieri, L et al, *Biochem J*, 18: 443, 1980; Ng, TB et al, *Int J Peptide Prot Res*, 28: 163, 1986

Agglutinin, *Naja mossambica mossambica*

Synonyms: Lectin; Snake Venom Agglutinin; SVAM

Sigma L 4515   *Naja mossambica mossambica* (Mossambica cobra)   Lyophilized containing ~90% protein (Biuret); balance acetate buffer salts; highly purified; activity: <20 µg/mL using glutaraldehyde fixed erythrocytes; <100 µg/mL using trypsinized erythrocytes | Not blood group specific, but has an affinity for heparin; agglutination activity expressed in µg/mL & is determined from serial dilutions of a 1 mg/mL solution using phosphate buffered saline, pH 6.8; activity is the lowest concentration to agglutinate a 2% suspension of appropriate human erythrocytes after incubation at 25°C; Oglivie, ML & Gartner, TK, *J Herpetology*, 18: 285, 1984

Agglutinin, *Naja naja kaouthia*

Synonyms: Lectin; Snake Venom Agglutinin; SVAK

Sigma L 8648   *Naja naja kaouthia*   Lyophilized containing ~70% protein (Lowry); balance acetate buffer salts; highly purified; activity: <15 µg/mL using glutaraldehyde-fixed type A human red blood cells | Not blood group specific, but has a high affinity for heparin; agglutination activity expressed in µg/mL & is determined from serial dilutions of a 1 mg/mL solution using phosphate buffered saline, pH 6.8; activity is the lowest concentration to agglutinate a 2% suspension of appropriate human erythrocytes after 1 hr incubation at 25°C; Oglivie, ML & Gartner, TK, *J Herpetology*, 18: 285, 1984

Agglutinin, *Narcissus pseudonarcissus*

Synonyms: Lectin

Sigma L 5650   *Narcissus pseudonarcissus* (Daffodil)   Lyophilized; essentially salt-free; highly purified by affinity chromatography using a modification of the method of Kaku, H & Goldstein, IJ, *Meth Enzymol*, 179: 327, 1989; activity: <1 µg/mL | Agglutinates trypsinized rabbit erythrocytes with high efficiency & untreated rabbit erythrocytes less so; human erythrocytes are not agglutinated; NPA sugar specificity is directed to terminal & internal α-D-mannosyl residues of glycoconjugates; agglutination activity expressed in µg/mL & is determined from serial dilutions of a 1 mg/mL solution using phosphate buffered saline, pH 7.2; activity is the lowest concentration to agglutinate a 2% suspension of trypsinized rabbit erythrocytes after 1 hr incubation at 25°C; VanDamme, JM et al, *Physiol Plantarum*, 73: 52, 1988; Kaku, H & Goldstein, IJ, *Meth Enzymol*, 179: 327, 1989

Agglutinin, *Perseau americana*

Synonyms: Lectin

Sigma L 8513   *Perseau americana* (Avocado)   Partially purified; lyophilized; essentially salt-free; activity: <40 μg/mL | Not blood group specific; agglutination activity expressed in μg/mL & is determined from serial dilutions of a 1 mg/mL solution using phosphate buffered saline, pH 6.8; activity is the lowest concentration to agglutinate a 2% suspension of human erythrocytes after 1 hr incubation at 25°C; Meade, NA et al, *Carbo Res*, 78: 349, 1980

Agglutinin, *Phaseolus coccineus*

Synonyms: Lectin

ICN 193553   *Phaseolus coccineus* (Scarlet runner bean) Purified, lyophilized; 250-300 μg/mL; mix of 5 lectins | Not blood group specific

Sigma L 3138   *Phaseolus coccineus* (Scarlet runner bean) Affinity purified; lyophilized containing ~75% protein (Biuret); ~10% carbohydrate as glycoprotein; balance primarily NaCl; activity: <8 μg/mL; mitogenic at <10 μg/mL | Not blood group specific; agglutination is not inhibited by monosaccharides but is inhibited by fetuin; conjugates are prepared from affinity purified lectin; agglutination activity expressed in μg/mL & is determined from serial dilutions of a 1 mg/mL solution using phosphate buffered saline, pH 6.8; activity is the lowest concentration to agglutinate a 2% suspension of human erythrocytes after 1 hr at 25°C; mitogenic activity is determined by ^{3}H-thymidine incorporation in lymphocyte cultures; Ochoa, JL & Kristiansen, T, *Biochim Biophys Acta*, 705: 396, 1982; Immunology Series No. 8 "*Procedural Guide*", p. 11, 1978, US Dept HEW, PHS, CDC, Bureau of Laboratories, Atlanta, GA

Agglutinin, *Phaseolus coccineus*, Biotin Conjugated

Synonyms: Lectin

Sigma L 4514   *Phaseolus coccineus* (Scarlet runner bean) Lyophilized containing ~85% protein ($E^{1\%}_{280}$); balance primarily sodium citrate; contains 5-10 moles biotin/mole protein | See Sigma L 3138

Agglutinin, *Phaseolus coccineus*, TRITC Conjugated

Synonyms: Lectin

Sigma L 4389   *Phaseolus coccineus* (Scarlet runner bean) Lyophilized containing ~10% protein (Lowry); balance phosphate buffer salts & NaCl; contains 1-2 moles TRITC/mole protein | See Sigma L 3138

Agglutinin, *Phaseolus limensis*

Synonyms: Lectin

ICN 153314   *Phaseolus limensis*   Purified by affinity chromatography | Specificity for α-D-galNAc

Sigma L 3897   *Phaseolus limensis* (Lima bean)   Lyophilized; essentially salt-free; highly purified by modification of the method of Bessler, W & Goldstein, AJ, *Arch Biochem Biophys*, 165: 444, 1974; activity: <10 μg/mL | Consists of at least 3 molecular components: Component III (LBL₄), component II (LBL₈, a dimeric form of component III) & component I, a higher molecular weight form; components differ in their mitogenic activity toward human lymphocytes with component II (LBL₈) showing the highest activity; the lectin has an affinity for GalNAc & an α-GalNAc-containing oligosaccharide, α-D-GalNAc-(1→3)-(α-L-Fuc-(1→2))-D-Gal; it preferentially agglutinates type A erythrocytes; activity is the lowest concentration to agglutinate a 2% suspension of human blood group A erythrocytes, type O, after 1 hr incubation at 25°C; Roberts, DD et al, *J Biol Chem*, 257: 9198, 1982; Pandolfino, ER et al, *J Biol Chem*, 258: 9203, 1983

Agglutinin, *Phaseolus limensis*, Biotin Conjugated

Synonyms: Lectin

ICN 153315   *Phaseolus limensis*   Purified by affinity chromatography | Specificity for α-D-galNAc

Agglutinin, *Phaseolus limensis*, FITC Conjugated

Synonyms: Lectin

ICN 153316   *Phaseolus limensis*   Purified by affinity chromatography | Specificity for α-D-galNAc

Agglutinin, *Phaseolus limensis*, Horse Radish Peroxidase Conjugated

Synonyms: Lectin

ICN 153317   *Phaseolus limensis*   Purified by affinity chromatography | Specificity for α-D-galNAc

Agglutinin, *Phaseolus vulgaris*

Synonyms: Agglutinin, Wax Bean; Erythroagglutinin; Lectin; Leukoagglutinin; Phytohemagglutinin; Phytohemagglutinin E; Phytohemagglutinin L; Phytohemagglutinin M; Phytohemagglutinin P

ICN 153318   *Phaseolus vulgaris*   Purified by affinity chromatography | Specificity for D-galNAc

ICN 153322   *Phaseolus vulgaris*   Purified by affinity chromatography | Specificity for D-galNAc

ICN 153326   *Phaseolus vulgaris*   Purified by affinity chromatography | Specificity for D-galNAc

ICN 151884   *Phaseolus vulgaris* (Red kidney bean)   Partially purified mixture of isolectins; mitogenic properties | Agglutinates erythrocytes of all human blood groups & many mammalian blood groups

ICN 151885   *Phaseolus vulgaris* (Red kidney bean)   Purified by affinity chromatography; high erythroagglutinating activity, low mitogenic activity | Inhibited by certain oligosaccharides; Kornfeld, R & S Kornfeld, *JBC*, 245:2536, 1970; Yachnin, A & RH Svenson, *Immunology*, 22:871, 1972

ICN 151886   *Phaseolus vulgaris* (Red kidney bean)   Purified by affinity chromatography; high mitogenic activity, low erythroagglutinin activity; 1.0-2.0 mg/mL | Yachnin, A & RH Svenson, *Immunology*, 22:871, 1972

Sigma L 2646   *Phaseolus vulgaris* (Red kidney bean) Lyophilized; salt-free; activity: <40 μg/mL with erythrocytes; mitogenic at <10 μg/mL | Mucoprotein form See Sigma L 8629

Sigma L 2769   *Phaseolus vulgaris* (Red kidney bean)   Purified; lyophilized; salt-free; purified by a modification of the method of Leavitt, RD et al, *J Biol Chem*, 252: 2961, 1977; mitogenic at <5 μg/mL; does not agglutinate erythrocytes at 250 μg/mL; tested for leucoagglutination | Same comments as for Sigma L 8629

Sigma L 8629   *Phaseolus vulgaris* (Red kidney bean)   Purified; lyophilized; essentially salt-free; purified by a modification of the method of Leavitt, RD et al, *J Biol Chem*, 252: 2961, 1977; activity <10 μg/mL with erythrocytes; tested for leucoagglutination | PHA consists of 2 molecular species, an erythroagglutinin (PHA-E) which has low mitogenic activity & high erythroagglutinating activity & a leucoagglutinin (PHA-L) which has high mitogenic & leucoagglutinating activity, but very low erythroagglutinating activity; PHA-E is not blood group specific but agglutination can be inhibited by certain oligosaccharides; PHA-P is the protein form of PHA prior to separation & purification of erythroagglutinin & leucoagglutinin; PHA-M is the mucoprotein form; erythroagglutinin, leucoagglutinin & phytohemagglutinin conjugates are prepared from the corresponding purified lectins, Sigma L 8629, Sigma L 2769 & Sigma L 9017; agglutination activity expressed in μg/mL & is determined from serial dilutions of a 1 mg/mL solution using phosphate buffered saline, pH 6.8; activity is the lowest concentration to agglutinate a 2% suspension of either human erythrocytes or human leukocytes (10^7/mL in saline) after 1 hr at 25°C; mitogenic activity is determined by ^{3}H-thymidine incorporation in lymphocyte cultures; Yachnin, A & Svenson, RH, *Immunology*, 22: 871, 1972; Felsted et al, *J Biol Chem*, 252: 2967, 1977; Kornfeld, R & Kornfeld, S, *J Biol Chem*, 245: 2536, 1970; Rigas, DA & Osgood, EE, *J Biol Chem*, 212: 607, 1955; *Immunology Series No. 8* "*Procedural Guide*", p. 11, 1978, US Dept HEW, PHS, CDC, Bureau of Laboratories, Atlanta, GA

Sigma L 8754   *Phaseolus vulgaris* (Red kidney bean) Lyophilized; salt-free; activity: <16 μg/mL with erythrocytes; mitogenic at <10 μg/mL | See Sigma L 8629

Sigma L 8902   *Phaseolus vulgaris* (red kidney Bean)
Lyophilized; salt-free; mitogenic at ~10 μg/mL; cell culture tested
| Lectin

Sigma L 9017   *Phaseolus vulgaris* (Red kidney bean)   Affinity
purified; lyophilized; essentially salt-free; activity: <16 μg/mL with
erythrocytes; mitogenic at <10 μg/mL | See Sigma L 8629

Sigma L 9132   *Phaseolus vulgaris* (red kidney bean)
Essentially salt-free; lyophilized; activity: <16 μg/mL with
erythrocytes; mitogenic at <10 μg/mL; cell culture tested |
Lectin; highly specific polyvalent carbohydrate-binding proteins;
useful in polysaccharide studies, glycoprotein studies, enzyme
tagging & cell membrane studies, cell agglutination & cell typing; in
tissue culture certain lectins used to induce mitogenic activity;
tested in a tissue culture system using 3H-thymidine incorporation
as a measure of mitogenic activity; Goldstein, I & Hayes, C, *Adv
Carbo Chem Biochem*, 35: 127, 1978; Rosenberg, SA & Lipsky, PE,
J Immunol, 122: 926, 1979

ICN 152265   *Phaseolus vulgaris* (Wax bean)   Purified by
affinity chromatography | No defined CHO binding specificity;
agglutinates a wide variety of cells; shows selective & differential
agglutinability of tumor cells; Sela, BA etal, *BBA*, 310:273, 1973

Agglutinin, *Phaseolus vulgaris*, 10 nm Colloidal Gold Conjugated

Synonyms: Lectin; Phytohemagglutinin E; Phytohemagglutinin L

ICN 154084   *Phaseolus vulgaris*   Purified by affinity
chromatography | Specificity for D-galNAc

ICN 154087   *Phaseolus vulgaris*   Purified by affinity
chromatography | Specificity for D-galNAc

Agglutinin, *Phaseolus vulgaris*, 20 nm Colloidal Gold Conjugated

Synonyms: Lectin; Phytohemagglutinin E; Phytohemagglutinin L

ICN 154085   *Phaseolus vulgaris*   Purified by affinity
chromatography | Specificity for D-galNAc

ICN 154088   *Phaseolus vulgaris*   Purified by affinity
chromatography | Specificity for D-galNAc

Agglutinin, *Phaseolus vulgaris*, 5 nm Colloidal Gold Conjugated

Synonyms: Lectin; Phytohemagglutinin E; Phytohemagglutinin L

ICN 154083   *Phaseolus vulgaris*   Purified by affinity
chromatography | Specificity for D-galNAc

ICN 154086   *Phaseolus vulgaris*   Purified by affinity
chromatography | Specificity for D-galNAc

Agglutinin, *Phaseolus vulgaris*, Biotin Conjugated

Synonyms: Erythroagglutinin; Lectin; Leukoagglutinin;
Phytohemagglutinin; Phytohemagglutinin E; Phytohemagglutinin L;
Phytohemagglutinin P

ICN 153319   *Phaseolus vulgaris*   Purified by affinity
chromatography | Specificity for D-galNAc

ICN 153323   *Phaseolus vulgaris*   Purified by affinity
chromatography | Specificity for D-galNAc

Sigma L 3509   *Phaseolus vulgaris* (Red kidney bean)
Lyophilized containing ~85% protein (E$^{1\%}_{280}$); balance primarily
sodium citrate; contains ~15 moles biotin/mole protein | See
Sigma L 8629

Sigma L 7019   *Phaseolus vulgaris* (Red kidney bean)
Lyophilized containing ~70% protein (E$^{1\%}_{280}$); balance primarily
sodium citrate; contains 6-12 moles biotin/mole protein | Same
comments as for Sigma L 8629

Sigma L 8512   *Phaseolus vulgaris* (Red kidney bean)
Lyophilized containing ~85% protein (Lowry); balance primarily
sodium citrate; contains ~6 moles biotin/mole protein | See
Sigma L 8629

Agglutinin, *Phaseolus vulgaris*, FITC Conjugated

Synonyms: Lectin; Erythroagglutinin; Leukoagglutinin;
Phytohemagglutinin; Phytohemagglutinin E; Phytohemagglutinin L;
Phytohemagglutinin P

ICN 153320   *Phaseolus vulgaris*   Purified by affinity
chromatography | Specificity for D-galNAc

ICN 153324   *Phaseolus vulgaris*   Purified by affinity
chromatography | Specificity for D-galNAc

Sigma L 0895   *Phaseolus vulgaris* (Red kidney bean)
Lyophilized; salt-free; contains 2-5 moles FITC/mole protein | See
Sigma L 8629

Sigma L 1520   *Phaseolus vulgaris* (Red kidney bean)
Lyophilized; salt-free; contains ~2 moles FITC/mole protein | See
Sigma L 8629

Sigma L 8006   *Phaseolus vulgaris* (Red kidney bean)
Lyophilized containing ~10% protein (Lowry); balance potassium
phosphate buffer salts & NaCl; contains 2-5 moles FITC/mole
protein | Same comments as for Sigma L 8629

Agglutinin, *Phaseolus vulgaris*, Horse Radish Peroxidase Conjugated

Synonyms: Phytohemagglutinin E; Phytohemagglutinin L

ICN 153321   *Phaseolus vulgaris*   Purified by affinity
chromatography | Specificity for D-galNAc

ICN 153325   *Phaseolus vulgaris*   Purified by affinity
chromatography | Specificity for D-galNAc

Agglutinin, Phaseolus *vulgaris*, Immobilized

Synonyms: Erythroagglutinin; Lectin; Leukoagglutinin;
Phytohemagglutinin E; Phytohemagglutinin L

Sigma L 3007   *Phaseolus vulgaris* (Red kidney bean)
Immobilized on cross-linked 4% beaded agarose; contains 2-4 mg
lectin/mL packed gel; suspension in 1.0 *M* NaCl | Same comments
as for Sigma L 8629

Sigma L 3132   *Phaseolus vulgaris* (Red kidney bean)
Immobilized on cross-linked 4% beaded agarose; contains 2-4 mg
lectin/mL packed gel; suspension in 1.0 *M* NaCl containing 0.01%
thimerosal as preservative; 1 mL suspension yields ~0.5 mL
packed gel | See Sigma L 8629

Agglutinin, *Phaseolus vulgaris*, TRITC Conjugated

Synonyms: Lectin; Erythroagglutinin; Phytohemagglutinin E

Sigma L 6139   *Phaseolus vulgaris* (Red kidney bean)
Lyophilized containing ~10% protein (Lowry); balance phosphate
buffer salts & NaCl; contains ~1 mole TRITC/mole protein | See
Sigma L 8629

Agglutinin, *Phytolacca americana*

Synonyms: Lectin; Pokeweed Mitogen

Sigma L 8777   *Phytolacca americana*   Lyophilized; partially
purified TCA precipitate; essentially salt-free; aseptically filled;
activity: <20 μg/mL; mitogenic at ~2.5 μg/mL; cell culture tested
| Lectin

ICN 153327   *Phytolacca americana* (Pokeweed)   Purified by
affinity chromatography; salt & sugar free; <20 μg/mL;
agglutinates fresh human 2% type O erythrocytes in 0.01*M* PBS,
pH 7.5 | Specificity for (D-glcNAc)₃; mitogenic properties at ~2.5
μg/mL

Sigma L 9379 *Phytolacca americana* (Pokeweed) Partially purified; TCA precipitate; lyophilized; essentially salt-free; activity: <20 μg/mL; mitogenic at ~2.5 μg/mL | Extracted from pokeweed roots; has hemagglutinating, leukoagglutinating & mitogenic properties; not blood group specific, but has affinity for *N*-acetyl-β-D-glucosamine oligomers; conjugates are prepared from Sigma L 9379 product; agglutination activity expressed in μg/mL & is determined from serial dilutions of a 1 mg/mL solution using phosphate buffered saline, pH 6.8; activity is the lowest concentration to agglutinate a 2% suspension of human blood group A erythrocytes after 1 hr at 25°C; mitogenic activity was determined by ^{3}H-thymidine incorporation in lymphocyte cultures; Yokoyama, K et al, *Biochim Biophys Acta*, 538: 384, 1978; *Immunology Series No. 8*, Procedural guide p. 11, May 1978, US Dept HEW PHS CDC, Bureau of Laboratories, Atlanta, GA

Agglutinin, *Phytolacca americana*, 10 nm Colloidal Gold conjugated

Synonyms: Lectin; Pokeweed Mitogen

ICN 154090 *Phytolacca americana* (Pokeweed) Purified by affinity chromatography | Specificity for (D-glcNAc)₃

Agglutinin, *Phytolacca americana*, 20 nm Colloidal Gold conjugated

Synonyms: Lectin; Pokeweed Mitogen

ICN 154091 *Phytolacca americana* (Pokeweed) Purified by affinity chromatography | Specificity for (D-glcNAc)₃

Agglutinin, *Phytolacca americana*, 5 nm Colloidal Gold conjugated

Synonyms: Lectin; Pokeweed Mitogen

ICN 154089 *Phytolacca americana* (Pokeweed) Purified by affinity chromatography | Specificity for (D-glcNAc)₃

Agglutinin, *Phytolacca americana*, Biotin conjugated

Synonyms: Lectin; Pokeweed Mitogen

ICN 153328 *Phytolacca americana* (Pokeweed) Purified by affinity chromatography | Specificity for (D-glcNAc)₃

Sigma L 8387 *Phytolacca americana* (Pokeweed) Lyophilized containing ~85% protein ($E^{1\%}_{280}$); balance primarily sodium citrate; contains 2-4 moles biotin/mole protein | See Sigma L 9379

Agglutinin, *Phytolacca americana*, FITC conjugated

Synonyms: Lectin; Pokeweed Mitogen

ICN 153329 *Phytolacca americana* (Pokeweed) Purified by affinity chromatography | Specificity for (D-glcNAc)₃

Sigma L 8631 *Phytolacca americana* (Pokeweed) Lyophilized containing ~10% protein (Biuret); balance phosphate buffer salts & NaCl; contains ~2 moles FITC/mole lectin | See Sigma L 9379

Agglutinin, *Phytolacca americana*, Immobilized

Synonyms: Lectin; Pokeweed Mitogen

Sigma L 2882 *Phytolacca americana* (Pokeweed) Immobilized on cross-linked 4% beaded agarose; contains 1-3 mg lectin/mL packed gel; suspension in 1.0 *M* NaCl; 1 mL suspension yields ~0.5 mL packed gel | See Sigma L 9379

Agglutinin, *Pisum sativum*

Synonyms: Lectin

Sigma L 5380 *Pisum sativum* (Pea) Purified by affinity chromatography; lyophilized; salt-free; activity: <20 μg/mL | Not blood group specific, but has affinity for terminal α-D-glucosyl & α-D-mannosyl residues; PSA lectin is a mitogen similar to Con A; conjugates are prepared from affinity purified lectin; agglutination activity expressed in μg/mL & is determined from serial dilutions of a 1 mg/mL solution using phosphate buffered saline, pH 6.8; activity is the lowest concentration to agglutinate a 2% suspension of human erythrocytes after 1 hr at 25°C; Trowbridge, IS, *J Biol Chem*, 249: 6004, 1974

Agglutinin, *Pisum sativum*, Biotin Conjugated

Synonyms: Lectin

Sigma L 3884 *Pisum sativum* (Pea) Lyophilized containing ~70% protein ($E^{1\%}_{280}$); balance primarily sodium citrate; contains ~4 moles biotin/mole protein | See Sigma L 5380

Agglutinin, *Pisum sativum*, FITC Conjugated

Synonyms: Lectin

Sigma L 0770 *Pisum sativum* (Pea) Lyophilized; salt-free; contains 2-4 moles FITC/mole protein (Lowry) | See Sigma L 5380

Agglutinin, *Pisum sativum*, Immobilized

Synonyms: Lectin

Sigma L 2257 *Pisum sativum* (Pea) Immobilized on cross-linked 4% beaded agarose; contains 2-4 mg lectin/mL packed gel; suspension in 1.0 *M* NaCl containing 0.01% thimerosal; 1 mL suspension yields ~0.5 mL packed gel | See Sigma L 5380

Agglutinin, *Pisum sativum*, TRITC Conjugated

Synonyms: Lectin

Sigma L 6639 *Pisum sativum* (Pea) Lyophilized containing ~10% protein (Lowry); balance phosphate buffer salts & NaCl; contains 0.5-1 mole TRITC/mole protein | See Sigma L 5380

Agglutinin, *Pseudomonas aeruginosa*

Synonyms: Lectin

Sigma L 9895 *Pseudomonas aeruginosa* Lyophilized containing ~80% protein (Biuret); balance primarily sodium citrate plus traces of calcium chloride, magnesium chloride & manganese chloride; highly purified by affinity chromatography using a modification of the method of Gilboa-Garber, N, *Meth Enzymol*, 83: 378, 1982; activity: <15 μg/mL | Major affinity for D-galactose & its derivatives; extracts from Pseudomonas aeruginosa cells contain another lectin, PA-II which is specific for Fucose; PA-I agglutinates all types of human & animal erythrocytes & interacts also with unicellular protozoa, algae & bacteria; induces mitogenic stimulation in cultured human lymphocytes pretreated with neuraminidase; agglutination activity expressed in μg/mL & is determined from serial dilutions of a 1 mg/mL solution using phosphate buffered saline, pH 7.0; activity is the lowest concentration to agglutinate a 2% suspension of neuraminidase treated human blood group A erythrocytes after 1 hr at 25°C; Gilboa-Garber, N, *Meth Enzymol*, 83: 378, 1982

Agglutinin, *Psophocarpus tetragonolobus*

Synonyms: Lectin

Sigma L 2138 *Psophocarpus tetragonolobus* (Winged bean) Purified; lyophilized containing ~65% protein (Biuret); balance primarily phosphate buffers & NaCl; activity: <30 μg/mL | Not blood group specific but has an affinity for *N*-acetyl-D-galactosamine; conjugates are prepared from purified lectin; agglutination activity expressed in μg/mL & is determined from serial dilutions of a 1 mg/mL solution using phosphate buffered saline, pH 6.8; activity is the lowest concentration to agglutinate a 2% suspension of human erythrocytes after 1 hr at 25°C; Appukuttan, PS & Basu, D, *Anal Biochem*, 113: 253, 1981

Agglutinin, *Psophocarpus tetragonolobus*, Biotin Conjugated

Synonyms: Lectin

Sigma L 3014 *Psophocarpus tetragonolobus* (Winged bean) Lyophilized containing ~85% protein ($E^{1\%}_{280}$); balance primarily sodium citrate; contains ~3 moles biotin/mole protein | See Sigma L 2138

Agglutinin, *Psophocarpus tetragonolobus*, FITC Conjugated

Synonyms: Lectin

Sigma L 3264 *Psophocarpus tetragonolobus* (Winged bean) Lyophilized containing ~10% protein (Biuret); balance phosphate buffer salts & NaCl; contains ~3 moles FITC/mole protein | See Sigma L 2138

Agglutinin, *Psophocarpus tetragonolobus*, Peroxidase Conjugated

Synonyms: Lectin

Sigma L 3139 *Psophocarpus tetragonolobus* (Winged bean) Lyophilized containing ~10% protein (Modified Warburg-Christian); balance phosphate buffer salts & NaCl; peroxidase activity: 20-60 units purpurogallin/mg protein; unit definition: 1 unit forms 1 mg purpurogallin in 20 sec from pyrogallol at pH 6.0 at 20°C | See Sigma L 2138

Agglutinin, *Ptilota plumosa*

Synonyms: Lectin

Sigma L 9260 *Ptilota plumosa* (Red marine algae) Partially purified; lyophilized; salt-free; activity: <50 µg/mL | Has anti-B blood group specificity; has an affinity for terminal α-D-galactosyl residues; agglutination activity expressed in µg/mL & is determined from serial dilutions of a 1 mg/mL solution using phosphate buffered saline, pH 7.3; activity is the lowest concentration to agglutinate a 2% suspension of papain treated human blood group B erythrocytes after 1 hr incubation at 25°C; Rogers, DJ et al, *Med Lab Sci*, 34: 195, 1977

Agglutinin, *Ricinus communis* 120

Synonyms: Lectin

ICN 791451 *Ricinus communis*

Agglutinin, *Ricinus communis* 60

Synonyms: Lectin

ICN 791401 *Ricinus communis*

Agglutinin, *Ricinus communis* A

Synonyms: Lectin; Ricin A Chain

Sigma L 9514 *Ricinus communis* (Castor bean) MW 60k (RCA_{60}), MW 120k (RCA_{120}) Prepared from Toxin RCA_{60}; purified by affinity chromatography; electrophoretically pure (SDS-PAGE); solution in 40% glycerol containing 10 mM phosphate, pH 6.0, 0.15 M NaCl, 10 mM galactose & 0.5 mM DTT; extremely hazardous | Occurs in 2 forms designated RCA_{60} & RCA_{120} according to their MW of ~60,000 & 120,000 respectively; neither is blood group specific; RCA_{60} also referred to as RCA_{II}, Ricin D or RCL III is extremely toxic, inhibits protein synthesis & has an affinity for *N*-acetyl-D-galactosamine; RCA_{120} also referred to as RCA_I or RCL (I+II) is an agglutinin & has an affinity for terminal β-D-galactosyl residues; conjugates are prepared from affinity purified toxin, RCA_{60} or agglutinin RCA_{120}; agglutination activity expressed in µg/mL & is determined from serial dilutions of a 1 mg/mL solution using phosphate buffered saline, pH 7.2; activity is the lowest concentration to agglutinate a 2% suspension of either human erythrocytes after 1 hr at 25°C; Nicolson, GL & Blaustein, J, *Biochim Biophys Acta*, 266: 543, 1972; Nicolson, GL et al, *Biochemistry*, 13: 196, 1974; Wei, CH & Koh, C, *J Mol Biol*, 123: 707, 1978; Lin, TS & Li, SL, *Eur J Biochem*, 105: 453, 1980

Agglutinin, *Ricinus communis* A, Deglycosylated

Synonyms: Lectin; Ricin A Chain

Sigma L 4022 *Ricinus communis* (Castor bean) MW 60k (RCA_{60}), MW 120k (RCA_{120}) Solution in 40% glycerol containing 10 mM phosphate, pH 6.0, 0.15 M NaCl, 10 mM galactose & 0.5 mM DTT; extremely hazardous | The carbohydrate moiety of Ricin A chain has been modified using a metaperiodate-cyanoborohydride mixture; the products still inhibits protein synthesis in cell-free systems & its removal from the blood stream by the liver is markedly decreased; immunotoxins prepared from deglycosylated Ricin A chain will therefore be more effective in killing target cells; Thorpe, PE et al, *Eur J Biochem*, 147: 197, 1985; Skilleter, DN et al, *Biochim Biophys Acta*, 842: 12, 1985; Thorpe, PE et al, *Cancer Res*, 48: 6396, 1988; seeSigma L 9514

Agglutinin, *Ricinus communis* B

Synonyms: Lectin; Ricin B Chain

Sigma L 9639 *Ricinus communis* (Castor bean) MW 60k (RCA_{60}), MW 120k (RCA_{120}) Prepared from Toxin RCA_{60}; purified by affinity chromatography; electrophoretically pure (SDS-PAGE); solution in 10 mM phosphate, pH 6.5, containing 0.15 M NaCl, 10 mM galactose, 0.5 mM DTT & 0.02% sodium azide; extremely hazardous | See Sigma L 9514

Agglutinin, *Ricinus communis* RCA_{120}

Synonyms: Lectin

Sigma L 7886 *Ricinus communis* (Castor bean) Highly purified by affinity chromatography using the method of Lin, TS & Li, SL, *Eur J Biochem*, 105: 453, 1980; activity: <1 µg/mL; solution in 0.005 M sodium phosphate, pH 7.2, 0.2 M NaCl & 0.1% sodium azide; extremely hazardous | See Sigma L 9514

Agglutinin, *Ricinus communis* RCA_{120}, Biotin Conjugated

Synonyms: Lectin

Sigma L 2641 *Ricinus communis* (Castor bean) Agglutinin RCA_{120}; contains ~3 moles biotin/mole protein; solution in 0.01 M potassium phosphate, pH 7.3, 0.15 M NaCl & 0.1% sodium azide; extremely hazardous | See Sigma L 9514

Agglutinin, *Ricinus communis* RCA_{120}, FITC Conjugated

Synonyms: Lectin

Sigma L 4638 *Ricinus communis* (Castor bean) Contains 4-6 moles FITC/mole protein; solution in 0.01 M potassium phosphate, pH 7.2, 0.15 M NaCl & 0.1% sodium azide; extremely hazardous | See Sigma L 9514

Agglutinin, *Ricinus communis* RCA_{120}, Immobilized

Synonyms: Lectin

Sigma L 2390 *Ricinus communis* (Castor bean) Immobilized on cross-linked 4% beaded agarose; contains 2-4 mg protein/mL packed gel; suspension in 1.0 M NaCl containing 0.02% sodium azide; 1 mL suspension yields ~0.5 mL packed gel; prepared by activation with p-nitrophenyl chloroformate; Wilchek, M & Miron, T, *Biochem Int*, 4: 629, 1982; extremely hazardous | See Sigma L 9514

Agglutinin, *Ricinus communis* RCA_{120}, Peroxidase Conjugated

Synonyms: Lectin

Sigma L 2758 *Ricinus communis* (Castor bean) Lyophilized containing ~75% protein (Modified Warburg-Christian); balance primarily sodium citrate; peroxidase activity: 5-20 units/mg protein; unit definition: 1 unit forms 1 mg purpurogallin in 20 sec from pyrogallol at pH 6.0 at 20°C; prepared from peroxidase type VI using two-step glutaraldehyde method of Avrameas, S & Ternyck, T, *Immunochemistry*, 8: 1175, 1971; repurified after conjugation by affinity chromatography; extremely hazardous | See Sigma L 9514

Agglutinin, *Ricinus communis* RCA₆₀

Synonyms: Lectin; Toxin RCA₆₀

Sigma L 8508   *Ricinus communis* (Castor bean)   MW 60k (RCA₆₀), MW 120k (RCA₁₂₀)   Highly purified by affinity chromatography using the method of Lin, TS & Li, SL, *Eur J Biochem*, 105: 453, 1980; activity: <20 μg/mL; solution in 0.01 M potassium phosphate, pH 7.0, 0.15 M NaCl & 0.1% sodium azide; extremely hazardous | See Sigma L 9514

Agglutinin, *Ricinus communis* RCA₆₀, FITC Conjugated

Synonyms: Lectin; Toxin RCA₆₀

Sigma L 8633   *Ricinus communis* (Castor bean)   MW 60k (RCA₆₀), MW 120k (RCA₁₂₀)   Contains ~1 mole FITC/mole protein; solution in 0.01 M potassium phosphate, pH 7.2, 0.15 M NaCl & 0.1% sodium azide; extremely hazardous | See Sigma L 9514

Agglutinin, *Ricinus communis* RCA₆₀, Immobilized

Synonyms: Lectin; Toxin RCA₆₀

Sigma L 1265   *Ricinus communis* (Castor bean)   MW 60k (RCA₆₀), MW 120k (RCA₁₂₀)   Immobilized on cross-linked 4% beaded agarose; contains 1.5-3 mg protein/mL packed gel; suspension in 1.0 M NaCl containing 0.02% sodium azide; 1 mL suspension yields ~0.5 mL packed gel; prepared by activation with p-nitrophenyl chloroformate; Wilchek, M & Miron, T, *Biochem Int*, 4: 629, 1982; extremely hazardous | See Sigma L 9514

Sigma L 2632   *Ricinus communis* (Castor bean)   MW 60k (RCA₆₀), MW 120k (RCA₁₂₀)   Immobilized on cross-linked 4% beaded agarose; contains 2-4 mg protein/mL packed gel; suspension in 1.0 M NaCl containing 0.02% thimerosal; 1 mL suspension yields ~0.5 mL packed gel; extremely hazardous | See Sigma L 9514

Agglutinin, *Ricinus communis* RCA₆₀, Peroxidase Conjugated

Synonyms: Lectin; Toxin RCA₆₀

Sigma L 2633   *Ricinus communis* (Castor bean)   MW 60k (RCA₆₀), MW 120k (RCA₁₂₀)   Lyophilized containing ~75% protein (Modified Warburg-Christian); balance primarily sodium citrate; peroxidase activity: 5-20 units/mg protein; unit definition: 1 unit forms 1 mg purpurogallin in 20 sec from pyrogallol at pH 6.0 at 20°C; prepared from peroxidase type VI using two-step glutaraldehyde method of Avrameas, S & Ternyck, T, *Immunochemistry*, 8: 1175, 1971; repurified after conjugation by affinity chromatography; extremely hazardous! be aware of the risks & familiar with safety procedures before you use this product | See Sigma L 9514

Agglutinin, *Robinia pseudoacacia*

Synonyms: Lectin

ICN 153336   *Robinia pseudoacacia*   Purified by affinity chromatography | Not blood group specific

Agglutinin, *Robinia pseudoacacia*, Biotin Conjugated

Synonyms: Lectin

ICN 153337   *Robinia pseudoacacia*   Purified by affinity chromatography

Agglutinin, *Robinia pseudoacacia*, FITC Conjugated

Synonyms: Lectin

ICN 153338   *Robinia pseudoacacia*   Purified by affinity chromatography

Agglutinin, *Robinia pseudoacacia*, Horse Radish Peroxidase Conjugated

Synonyms: Lectin

ICN 153339   *Robinia pseudoacacia*   Purified by affinity chromatography

Agglutinin, *Salvia horminum*

Synonyms: Lectin

ICN 153340   *Salvia horminum*   Purified by affinity chromatography

Agglutinin, *Salvia sclarea*

Synonyms: Lectin

ICN 153341   *Salvia sclarea*   Purified by affinity chromatography

Agglutinin, *Sambucus nigra*

Synonyms: Lectin

Sigma L 6890   *Sambucus nigra* (Elder)   Lyophilized containing ~30% protein (Lowry); balance primarily NaCl; highly purified by modification of the method of Broekaert, WF et al, *Biochem J*, 221: 163, 1984 | Isolated from the inner bark (bast tissue) of elder stems & branches; not blood group specific but has an affinity for α-NeuNAc-(2→6)-Gal, α-NeuNAc-(2→6)-GalNAc & to a lesser extent, α-NeuNAc-(2→3)-Gal residues; agglutination activity expressed in μg/mL & is determined from serial dilutions of a 1 mg/mL solution using phosphate buffered saline, pH 7.3; activity is the lowest concentration to agglutinate a 2% suspension of human blood group A erythrocytes after 1 hr incubation at 25°C; Broakaert, WF et al, *Biochem J*, 221: 163, 1984; Shibuya, N et al, *J Biol Chem*, 262: 1596, 1987

Agglutinin, *Solanum tuberosum*

Synonyms: Lectin

ICN 152063   *Solanum tuberosum* (Potato)   Purified by affinity chromatography | Not blood group specific, but does have affinity for *N*-acetyl-β-D-glucosamine oligomers

Sigma L 4266   *Solanum tuberosum* (Potato)   Lyophilized containing ~20% protein (Lowry); ~20% carbohydrate present as glycoprotein; balance primarily NaCl; highly purified by affinity chromatography using a modification of the method of Matsumoto, I et al, *J Biochem*, 258: 2886, 1983; activity: <2 μg/mL | Not blood group specific; affinity for *N*-acetyl-β-D-glucosamine oligomers; glycoprotein containing approx. equal amounts of protein & carbohydrate; conjugates are prepared from affinity purified lectin; agglutination activity expressed in μg/mL & is determined from serial dilutions of a 1 mg/mL solution using phosphate buffered saline, pH 6.8; activity is the lowest concentration to agglutinate a 2% suspension of human erythrocytes after 1 hr at 25°C; Allen, AK & Neuberger, A, *Biochem J*, 135: 307, 1973

Sigma L 7885   *Solanum tuberosum* (Potato)   Lyophilized containing ~80% protein (Lowry); balance carbohydrate present as glycoprotein; partially purified; salt-free; activity: <75 μg/mL | See Sigma L 4266

Agglutinin, *Solanum tuberosum*, Biotin Conjugated

Synonyms: Lectin

ICN 153342   *Solanum tuberosum* (Potato)   Purified by affinity chromatography | Specificity for (D-glc NAc)₃

Agglutinin, *Solanum tuberosum*, FITC Conjugated

Synonyms: Lectin

ICN 153344   *Solanum tuberosum* (Potato)   Purified by affinity chromatography | Specificity for (D-glc NAc)₃

Sigma L 9408   *Solanum tuberosum* (Potato)   Lyophilized containing ~30% lectin; balance primarily HEPES buffer salts & NaCl; contains ~1 mole FITC/mole lectin | See Sigma L 4266

Agglutinin, *Solanum tuberosum*, Horse Radish Peroxidase Conjugated

Synonyms: Lectin

ICN 153345   *Solanum tuberosum* (Potato)   Purified by affinity chromatography | Specificity for (D-glc NAc)₃

Agglutinin, *Sophora japonica*

Synonyms: Lectin

ICN 153346 *Sophora japonica* Purified by affinity chromatography | Specificity for β-D-galNAc

Sigma L 6138 *Sophora japonica* (Japanese pagoda tree) Lyophilized; salt-free; highly purified by affinity chromatography using the method of Poretz, RD, *Methods in Enzymology*, 28: 349, 1972; activity: <20 μg/mL | Has anti-A & anti-B human blood group specificity; has an affinity for *N*-acetyl-D-galactosamine & D-galactose; agglutination activity expressed in μg/mL & is determined from serial dilutions of a 1 mg/mL solution using phosphate buffered saline, pH 8.0; activity is the lowest concentration to agglutinate a 2% suspension of human blood group B erythrocytes after 1 hr at 25°C; Irimura, T et al, *Carbo Res*, 39: 317, 1975

Agglutinin, *Sophora japonica*, Biotin Conjugated

Synonyms: Lectin

ICN 153347 *Sophora japonica* Purified by affinity chromatography | Specificity for β-D-galNAc

Sigma L 3016 *Sophora japonica* (Japanese pagoda tree) Lyophilized containing ~10% protein (Lowry); balance potassium phosphate buffer salts & NaCl; contains ~2 moles biotin/mole protein | See Sigma L 6138

Agglutinin, *Sophora japonica*, FITC Conjugated

Synonyms: Lectin

ICN 153348 *Sophora japonica* Purified by affinity chromatography | Specificity for β-D-galNAc

Agglutinin, *Sophora japonica*, Horse Radish Peroxidase Conjugated

Synonyms: Lectin

ICN 153349 *Sophora japonica* Purified by affinity chromatography | Specificity for β-D-galNAc

Agglutinin, *Tetragonolobus purpureas*

Synonyms: Lectin; Agglutinin, *Lotus*

Sigma L 9254 *Tetragonolobus purpureas* (Winged or asparagus pea) Salt-free; lyophilized; highly purified by affinity chromatography; activity: <32 μg/mL | Anti-H blood group specificity & an affinity for α-L-fucosyl residues & an unusually high affinity for α-L-fucose residues on type II chain blood group oligosaccharides; conjugates are prepared from purified lectin; agglutination activity expressed in μg/mL & is determined from serial dilutions of a 1 mg/mL solution using phosphate buffered saline, pH 6.8; activity is the lowest concentration to agglutinate a 2% suspension of human blood group O erythrocytes after 1 hr at 25°C; Pereira, MEA & Kabat, EA, *Ann NY Acad Sci*, 234: 301, 1974

Agglutinin, *Tetragonolobus purpureas*, Biotin Conjugated

Synonyms: Lectin; Agglutinin, *Lotus*

Sigma L 3134 *Tetragonolobus purpureas* (Winged or asparagus pea) Lyophilized containing ~85% protein (E$^{1\%}_{280}$); balance primarily sodium citrate; contains ~6 moles biotin/mole protein | See Sigma L 9254

Agglutinin, *Tetragonolobus purpureas*, FITC Conjugated

Synonyms: Lectin; Agglutinin, *Lotus*

Sigma L 5644 *Tetragonolobus purpureas* (Winged or asparagus pea) Salt-free; lyophilized; contains 2-6 moles FITC/mole protein | See Sigma L 9254

Agglutinin, *Tetragonolobus purpureas*, Immobilized

Synonyms: Lectin; Agglutinin, *Lotus*

Sigma L 3257 *Tetragonolobus purpureas* (Winged or asparagus pea) Immobilized on 4% beaded agarose; contains 1-2 mg protein/mL packed gel; spacer: 6 carbon; suspension in 0.01 M phosphate buffered saline containing 50% glycerol, 0.1 mM CaCl$_2$, 0.02% sodium azide; 1 mL suspension yields ~0.5 mL packed gel | See Sigma L 9254

Agglutinin, *Tetragonolobus purpureas*, Peroxidase Conjugated

Synonyms: Lectin; Agglutinin, *Lotus*

Sigma L 1508 *Tetragonolobus purpureas* (Winged or asparagus pea) Lyophilized containing ~90% protein (Modified Warburg-Christian); balance phosphate buffer salts; peroxidase activity: 10-30 units purpurogallin/mg protein; unit definition: 1 unit forms 1 mg purpurogallin in 20 sec from pyrogallol at pH 6.0 at 20°C; prepared from peroxidase type II using the two step glutaraldehyde method of Avrameas, S & Ternyck, T, *Immunochemistry*, 8: 1175, 1971; repurified after conjugation by affinity chromatography | See Sigma L 9254

Sigma L 5759 *Tetragonolobus purpureas* (Winged or asparagus pea) Lyophilized containing ~90% protein (Modified Warburg-Christian); balance sodium citrate; peroxidase activity: 40-100 units purpurogallin/mg protein; unit definition: 1 unit forms 1 mg purpurogallin in 20 sec from pyrogallol at pH 6.0 at 20°C; prepared from peroxidase type VI using a modification of the method of O'Sullivan, MJ et al, *FEBS Lett*, 95: 311, 1978, which favors low molecular weight conjugates; repurified after conjugation by affinity chromatography | See Sigma L 9254

Agglutinin, *Tetragonolobus purpureas*, TRITC Conjugated

Synonyms: Lectin; Agglutinin, *Lotus*

Sigma L 6764 *Tetragonolobus purpureas* (Winged or asparagus pea) Lyophilized containing ~75% protein (Lowry); balance potassium phosphate buffer salts; contains ~1 mole TRITC/mole protein | See Sigma L 9254

Agglutinin, *Trichosanthes kinlowii*

Synonyms: LECTIN

ICN 153358 *Trichosanthes kinlowii* Purified by affinity chromatography

Agglutinin, *Trichosanthes kinlowii*, Biotin Conjugated

Synonyms: Lectin

ICN 153359 *Trichosanthes kinlowii* Purified by affinity chromatography

Agglutinin, *Trichosanthes kinlowii*, FITC Conjugated

Synonyms: Lectin

ICN 153360 *Trichosanthes kinlowii* Purified by affinity chromatography

Agglutinin, *Trichosanthes kinlowii*, Horse Radish Peroxidase Conjugated

Synonyms: Lectin

ICN 153361 *Trichosanthes kinlowii* Purified by affinity chromatography

Agglutinin, *Triticum vulgaris*

Synonyms: Agglutinin, Wheat Germ; Lectin

Sigma L 0636 *Triticum vulgaris* (wheat germ) Highly purified; essentially salt-free; lyophilized; aseptically processed; cell culture tested | Lectin; highly specific polyvalent carbohydrate-binding proteins; useful in polysaccharide studies, glycoprotein studies, enzyme tagging & cell membrane studies, cell agglutination & cell typing; in tissue culture certain lectins used to induce mitogenic activity; tested in a tissue culture system using 3H-thymidine incorporation as a measure of mitogenic activity; Goldstein, I & Hayes, C, *Adv Carbo Chem Biochem*, 35: 127, 1978; Rosenberg, SA & Lipsky, PE, *J Immunol*, 122: 926, 1979

Sigma L 1263 *Triticum vulgaris* (wheat germ) Sterile-filtered solution of 0.01 M potassium phosphate buffered saline, pH 7.0; contains ~20-40 moles lectin bound/mole ferritin; tested by electron microscopy for ability to label cells | See Sigma L 9640

Sigma L 3143 *Triticum vulgaris* (wheat germ) Highly purified; essentially salt-free; lyophilized; sterilized by γ-irradiation; cell culture tested | Lectin; highly specific polyvalent carbohydrate-binding proteins; useful in polysaccharide studies, glycoprotein studies, enzyme tagging & cell membrane studies, cell agglutination & cell typing; in tissue culture certain lectins used to induce mitogenic activity; tested in a tissue culture system using 3H-thymidine incorporation as a measure of mitogenic activity; Goldstein, I & Hayes, C, *Adv Carbo Chem Biochem*, 35: 127, 1978; Rosenberg, SA & Lipsky, PE, *J Immunol*, 122: 926, 1979

Sigma L 4895 *Triticum vulgaris* (wheat germ) Lyophilized containing ~50% protein (Lowry); balance phosphate buffer salts; contains ~2 moles FITC/mole protein | See Sigma L 9640

Sigma L 5142 *Triticum vulgaris* (wheat germ) Lyophilized containing ~85% protein ($E^{1\%}_{280}$); balance primarily sodium citrate; contains 2-4 moles biotin/mole protein | See Sigma L 9640

Sigma L 5266 *Triticum vulgaris* (wheat germ) Lyophilized containing ~10% protein (Lowry); balance phosphate buffer salts & NaCl; contains ~1 mole TRITC/mole protein | See Sigma L 9640

Sigma L 9640 *Triticum vulgaris* (wheat germ) Salt-free; lyophilized; highly purified; activity: <20 μg/mL | Not blood group specific but has an affinity for N-acetyl-β-D-glucosaminyl residues & N-acetyl-β-D-glucosamine oligomers; contains no protein-bound carbohydrate; conjugates are prepared from purified lectin; agglutination activity expressed in μg/mL & is determined from serial dilutions of a 1 mg/mL solution using phosphate buffered saline, pH 6.8; activity is the lowest concentration to agglutinate a 2% suspension of human erythrocytes after 1 hr at 25°C; Nagata, Y & Burger, MM, *J Biol Chem*, 249: 3116, 1974

ICN 152266 *Triticum vulgaris* (Wheat) Purified by affinity chromatography; lyophilized; essentially salt- & CHO-free; mitogenic properties | Specificity for N-acetyl-β-D-glucosaminyl residues & N-acetyl-β-D-glucamine oligomers; specifically agglutinates erythrocytes & lymphocytes

Agglutinin, *Triticum vulgaris*, 10 nm Colloidal Gold Conjugated

Synonyms: Lectin; Agglutinin, Wheat Germ

Sigma L 1894 *Triticum vulgaris* (wheat germ) Mean particle size 8-12 nm; monodisperse; suspension in ~50% glycerol containing 0.15 M NaCl, 0.01 M sodium phosphate, pH 7.2, 0.02% PEG 20 & 0.02% sodium azide; concentration: A_{520} ~5.0; Horisberger, M, *Techniques in Immunocytochemistry*, Bullock & Petrusz, eds, Academic Press, 3: 155, 1985 | See Sigma L 9640

ICN 154123 *Triticum vulgaris* (Wheat) Purified by affinity chromatography | Specificity for (β(1→4)-D-glcNAc)₂

Agglutinin, *Triticum vulgaris*, 20 nm Colloidal Gold Conjugated

Synonyms: Lectin; Agglutinin, Wheat Germ

ICN 154124 *Triticum vulgaris* (Wheat) Purified by affinity chromatography | Specificity for (β(1→4)-D-glcNAc)₂

Agglutinin, *Triticum vulgaris*, 5 nm Colloidal Gold Conjugated

Synonyms: Lectin; Agglutinin, Wheat Germ

ICN 154122 *Triticum vulgaris* (Wheat) Purified by affinity chromatography | Specificity for (β(1→4)-D-glcNAc)₂

Agglutinin, *Triticum vulgaris*, Agarose

Synonyms: Lectin; Agglutinin, Wheat Germ

ICN 191472 *Triticum vulgaris* (Wheat) Supplied in a pre-packed column; 5 atoms hydrophilic spacer arm; 5 mg lectin/mL gel; suspension in PBS, 0.02% NaN₃ | Specificity for (β(1→4)-D-glcNAc)₂; useful for glycoproteins, polysaccharides, major sialoglycoprotein (glycophorin A) from human erythrocyte membrane, subcellular particles, cells (esp. T-lymphocytes)

Agglutinin, *Triticum vulgaris*, Biotin Conjugated

Synonyms: Lectin; Agglutinin, Wheat Germ

ICN 153362 *Triticum vulgaris* (Wheat) Purified by affinity chromatography | Specificity for (β(1→4)-D-glcNAc)₂

Agglutinin, *Triticum vulgaris*, Evans Blue Conjugated

Synonyms: Lectin; Agglutinin, Wheat Germ

Sigma L 9884 *Triticum vulgaris* (wheat germ) Lyophilized containing ~85% protein; balance citrate buffer salts; contains ~2 moles Evans Blue/mole protein | See Sigma L 9640

Agglutinin, *Triticum vulgaris*, FITC Conjugated

Synonyms: Lectin; Agglutinin, Wheat Germ

ICN 153363 *Triticum vulgaris* (Wheat) Purified by affinity chromatography | Specificity for (β(1→4)-D-glcNAc)₂

Agglutinin, *Triticum vulgaris*, Horse Radish Peroxidase Conjugated

Synonyms: Lectin; Agglutinin, Wheat Germ

ICN 153364 *Triticum vulgaris* (Wheat) Purified by affinity chromatography | Specificity for (β(1→4)-D-glcNAc)₂

Agglutinin, *Triticum vulgaris*, Immobilized

Synonyms: Lectin; Agglutinin, Wheat Germ

Sigma L 1882 *Triticum vulgaris* (wheat germ) Immobilized on 4% cross-linked beaded agarose; contains 5-10 mg lectin/mL packed gel; suspension in 1.0 M NaCl & 0.02% thimerosal; 1 mL suspension yields ~0.5 mL packed gel | See Sigma L 9640

Agglutinin, *Triticum vulgaris*, Immobilized on Macrobeads

Synonyms: Lectin; Agglutinin, Wheat Germ

Sigma L 1394 *Triticum vulgaris* (wheat germ) Immobilized on 6% agarose macrobeads; particle size: 200-300 μm; contains ~6 mg lectin/mL gel; suspension in 0.9% NaCl & 0.01% thimerosal; similar to Sigma L 6257 but produced by Sigma | See Sigma L 9640

Agglutinin, *Triticum vulgaris*, Peroxidase Conjugated

Synonyms: Lectin; Agglutinin, Wheat Germ

Sigma L 3892 *Triticum vulgaris* (wheat germ) Lyophilized containing ~90% protein (Modified Warburg-Christian); balance primarily citrate buffer; peroxidase activity: 50-200 units purpurogallin/mg protein; unit definition: 1 unit forms 1 mg purpurogallin in 20 sec from pyrogallol at pH 6.0 at 20°C; prepared from peroxidase type VI using a modification of the method of O'Sullivan, MJ et al, *FEBS Lett*, 95: 311, 1978, which favors low molecular weight conjugates; repurified after conjugation by affinity chromatography | See Sigma L 9640

Sigma L 7017 *Triticum vulgaris* (wheat germ) Lyophilized containing ~90% protein (Modified Warburg-Christian); balance primarily citrate buffer; peroxidase activity: 50-200 units purpurogallin/mg protein; unit definition: 1 unit forms 1 mg purpurogallin in 20 sec from pyrogallol at pH 6.0 at 20°C; prepared from peroxidase type VI using a modification of the method of O'Sullivan, MJ et al, *FEBS Lett*, 95: 311, 1978, which favors low molecular weight conjugates; repurified after conjugation by affinity chromatography; packaged in microcone vials for ease of reconstitution & recovery of microliter volumes | Same as Sigma L 3892 but packaged in microcone vials; see Sigma L 9640

Agglutinin, *Triticum vulgaris*, Peroxidase Inactivated Conjugated

Synonyms: Lectin; Agglutinin, Wheat Germ

Sigma L 0390 *Triticum vulgaris* (wheat germ) Lyophilized containing ~90% protein (Modified Warburg-Christian); balance primarily sodium citrate; peroxidase activity: <0.01 unit purpurogallin/mg protein; unit definition: 1 unit forms 1 mg purpurogallin in 20 sec from pyrogallol at pH 6.0 at 20°C; prepared from inactivated peroxidase, Sigma P 6278, using a modification of the method of O'Sullivan, MJ et al, *FEBS Lett*, 95: 311, 1978 | Potentially useful as an alternate probe in dual neuronal transport studies when conjugated with an appropriate marker; see Sigma L 9640

Agglutinin, *Triticum vulgaris*, Sepharose 6MB

Synonyms: Lectin; Agglutinin, Wheat Germ

Sigma L 6257 *Triticum vulgaris* (wheat germ) Immobilized on Sepharose 6MB; particle size: 200-300 μm; contains ~5 mg lectin/mL gel; suspension in 0.9% NaCl & 0.01% thimerosal; similar to Sigma L 1394 but not produced by Sigma | See Sigma L 9640

Agglutinin, *Triticum vulgaris*, Succinyl

Synonyms: Lectin; Agglutinin, Wheat Germ

ICN 153354 Purified by affinity chromatography

Agglutinin, *Triticum vulgaris*, Succinyl Biotin Conjugated

Synonyms: Lectin; Agglutinin, Wheat Germ

ICN 153355 Purified by affinity chromatography

Agglutinin, *Triticum vulgaris*, Succinyl FITC Conjugated

Synonyms: Lectin; Agglutinin, Wheat Germ

ICN 153356 Purified by affinity chromatography

Agglutinin, *Triticum vulgaris*, Succinyl Horse Radish Peroxidase Conjugated

Synonyms: Lectin; Agglutinin, Wheat Germ

ICN 153357 Purified by affinity chromatography

Agglutinin, *Ulex europaeus* I

Synonyms: Lectin

Sigma L 5505 *Ulex europaeus* (Gorse or Furze) MW 68k Essentially salt-free; lyophilized; purified by affinity chromatography using the method of Pereira, MEA et al, *Arch Biochem Biophys*, 185: 108, 1978; activity: <8 μg/mL | Has anti-H blood group specificity; 2 types of lectin: UEA I has an affinity for L-fucose & UEA II has an affinity for *N,N'*-diacetylchitobiose; UEA I MW was initally found to be 170,000; later reports indicate that UEA I may form aggregates at neutral & basic pH & that the correct MW is 68,000; conjugates are prepared from affinity purified lectin; agglutination activity expressed in μg/mL & is determined from serial dilutions of a 1 mg/mL solution using phosphate buffered saline, pH 6.8; activity is the lowest concentration to agglutinate a 2% suspension of human blood group O erythrocytes after 1 hr at 25°C; Matsumoto, I & Osawa, T, *Biochim Biophys Acta*, 194: 180, 1969; Horejsi, V & Kocourek, J, *Biochim Biophys Acta*, 336: 329, 1974; Allen, HJ & Johnson, EAZ, *Carbo Res*, 58: 253, 1977

ICN 152179 *Ulex europaeus* (Gorse seeds) Purified by affinity chromatography; 1-2 mg/mL solution | Specificity for L-fucose & H-antigen bearing RBC; useful for identification of A & B blood groups, & secretors; Boyd, WC & E Sharpleigh, *J Lab Clin Med*, 44:235, 1985; Pereira, MEA etal, *Arch Biochem Biophys*, 185:108, 1978; Boyd, WC & E Sharpleigh, *Blood*, 9:1195, 1954

Agglutinin, *Ulex europaeus* I + II

Synonyms: Lectin

Sigma L 6762 *Ulex europaeus* (Gorse or Furze) Salt-free; lyophilized; partially purified; activity: <500 μg/mL | See Sigma L 5505

Agglutinin, *Ulex europaeus* I, 10 nm Colloidal Gold Conjugated

Synonyms: Lectin

Sigma L 4893 *Ulex europaeus* (Gorse or Furze) MW 68k Mean particle size 8-12 nm; monodisperse; suspension in ~50% glycerol containing 0.15 M NaCl, 0.01 M sodium phosphate, pH 7.2, 0.02% PEG 20 & 0.02% sodium azide; concentration: A_{520} ~5.0; Horisberger, M, *Techniques in Immunocytochemistry*, Bullock & Petrusz, eds, Academic Press, 3: 155, 1985 | See Sigma L 5505

ICN 154126 *Ulex europaeus* (Gorse seeds) Purified by affinity chromatography | Specificity for α-L-fucose

Agglutinin, *Ulex europaeus* I, 20 nm Colloidal Gold Conjugated

Synonyms: Lectin

ICN 154127 *Ulex europaeus* (Gorse seeds) Purified by affinity chromatography | Specificity for α-L-fucose

Agglutinin, *Ulex europaeus* I, 5 nm Colloidal Gold Conjugated

Synonyms: Lectin

ICN 154125 *Ulex europaeus* (Gorse seeds) Purified by affinity chromatography | Specificity for α-L-fucose

Agglutinin, *Ulex europaeus* I, Biotin Conjugated

Synonyms: Lectin

Sigma L 8262 *Ulex europaeus* (Gorse or Furze) MW 68k Lyophilized containing ~85% protein ($E^{1\%}_{280}$); balance primarily sodium citrate; contains ~3 moles biotin/mole protein | See Sigma L 5505

ICN 153365 *Ulex europaeus* (Gorse seeds) Purified by affinity chromatography | Specificity for α-L-fucose

Agglutinin, *Ulex europaeus* I, Ferritin Conjugated

Synonyms: Lectin

Sigma L 4764 *Ulex europaeus* (Gorse or Furze) MW 68k Sterile-filtered solution of 0.01 M potassium phosphate buffered saline, pH 7.0; contains 1-5 moles lectin/mole ferritin | Tested by electron microscopy for ability to label cells; see Sigma L 5505

Agglutinin, *Ulex europaeus* I, FITC Conjugated

Synonyms: Lectin

Sigma L 9006 *Ulex europaeus* (Gorse or Furze)   MW 68k
Lyophilized containing ~10% protein (Lowry); balance potassium phosphate buffer salts & NaCl; contains 2-5 moles FITC/mole protein | See Sigma L 5505

ICN 153366 *Ulex europaeus* (Gorse seeds)   Purified by affinity chromatography | Specificity for α-L-fucose

ICN 153368 *Ulex europaeus* (Gorse seeds)   Purified by affinity chromatography | Specificity for α-L-fucose

Agglutinin, *Ulex europaeus* I, Horse Radish Peroxidase Conjugated

Synonyms: Lectin

ICN 153367 *Ulex europaeus*   Purified by affinity chromatography | Specificity for α-L-fucose

Agglutinin, *Ulex europaeus* I, Immobilized

Synonyms: Lectin

Sigma L 2382 *Ulex europaeus* (Gorse or Furze)   MW 68k
Immobilized on 4% cross-linked beaded agarose; contains 2-4 mg lectin/mL packed gel; suspension in 1.0 *M* NaCl with 0.01% thimerosal; binding capacity: ~3 mg BSA-fucosylamide/mL gel | See Sigma L 5505

Agglutinin, *Ulex europaeus* I, Peroxidase Conjugated

Synonyms: Lectin

Sigma L 8146 *Ulex europaeus* (Gorse or Furze)   MW 68k
Lyophilized containing ~80% protein (Lowry); balance primarily sodium citrate buffer salts; peroxidase activity: 15-60 units purpurogallin/mg protein; unit definition: 1 unit forms 1 mg purpurogallin in 20 sec from pyrogallol at pH 6.0 at 20°C; prepared from peroxidase type VI using the two step glutaraldehyde method of Avrameas, S & Ternyck, T, *Immunochemistry*, 8: 1175, 1971; repurified after conjugation by affinity chromatography | See Sigma L 5505

Agglutinin, *Ulex europaeus* I, TRITC Conjugated

Synonyms: Lectin

Sigma L 4889 *Ulex europaeus* (Gorse or Furze)   MW 68k
Lyophilized containing ~10% protein (Lowry); balance potassium phosphate buffer salts & NaCl; contains 4-8 moles TRITC/mole lectin | See Sigma L 5505

Agglutinin, *Ulex europaeus* II

Synonyms: Lectin

ICN 780211 *Ulex europaeus*

Sigma L 6391 *Ulex europaeus* (Gorse or Furze)   Lyophilized containing ~25% protein (Biuret); balance primarily phosphate buffer salts & NaCl; highly purified by affinity chromatography; activity: <50 µg/mL; Horejsi, V, *Biochim Biophys Acta*, 577: 389, 1979 | See Sigma L 5505

ICN 152180 *Ulex europaeus* (Gorse seeds)   Purified by affinity chromatography | Specificity for *N,N*'-diacetylchitobiose; Matsumoto, I & T Osawa, *BBA*, 194:180, 1969

Agglutinin, *Ulex europaeus* II, Horse Radish Peroxidase Conjugated

Synonyms: Lectin

ICN 153369 *Ulex europaeus*   Purified by affinity chromatography | Specificity for (D-glcNAc)$_2$

Agglutinin, *Vicia ervilla*

Synonyms: Lectin

ICN 152261 *Vicia ervilla* seeds   Purified by affinity chromatography; lyophilized; essentially free of salt & CHO

Agglutinin, *Vicia faba*

Synonyms: Lectin

ICN 152262 *Vicia faba* (Broad bean)   Purified by affinity chromatography; lyophilized; essentially free of salt & CHO; mitogenic properties | Specificity for D-man, D-gluc; Matsumoto, I etal, *J Biochem*, 93:763, 1983;

Sigma L 6263 *Vicia faba* (Fava bean, Broad bean)
Lyophilized containing ~20% protein (Biuret); balance primarily NaCl, phosphate buffer salts & stabilizer; highly purified by affinity chromatography using the method of Matsumoto, I, *J Biochem*, 93: 763, 1983; activity: <20 µg/mL | Not blood group specific; affinity for D-mannose & D-glucose; agglutination activity expressed in µg/mL & is determined from serial dilutions of a 1 mg/mL solution using phosphate buffered saline, pH 6.8; activity is the lowest concentration to agglutinate a 2% suspension of human erythrocytes after 1 hr at 25°C; Matsumoto, I et al, *J Biochem*, 93: 763, 1983

Agglutinin, *Vicia faba*, Biotin Conjugated

Synonyms: Lectin

ICN 153371 *Vicia faba* (Broad bean)   Purified by affinity chromatography | Specificity for D-man, D-gluc

Agglutinin, *Vicia faba*, FITC Conjugated

Synonyms: Lectin

Sigma L 4265 *Vicia faba* (Fava bean, Broad bean)
Lyophilized containing ~60% protein (Lowry); balance potassium phosphate buffer salts; contains ~4 moles FITC/mole protein | See Sigma L 6263

Agglutinin, *Vicia faba*, Horse Radish Peroxidase Conjugated

Synonyms: Lectin

ICN 153372 *Vicia faba* (Broad bean)   Purified by affinity chromatography | Specificity for D-man, D-gluc

Agglutinin, *Vicia sativa*

Synonyms: Lectin

Sigma L 2770 *Vicia sativa*   Lyophilized containing ~35% protein (Biuret); balance primarily NaCl, sodium phosphate salts & 20% sucrose; highly purified by affinity chromatography using the method of Falasca, A et al, *Biochim Biophys Acta*, 577: 71, 1979; activity: <32 µg/mL | Not blood group specific; affinity for D-mannose & D-glucose; mitogenic activity on human peripheral blood lymphocytes & mouse lymphocytes; agglutination activity expressed in µg/mL & is determined from serial dilutions of a 1 mg/mL solution using phosphate buffered saline, pH 6.8; activity is the lowest concentration to agglutinate a 2% suspension of human erythrocytes after 1 hr at 25°C; Gebaver, G et al, *Hoppe-Seyler's Z Physiol Chem*, 360: 1727, 1979; Falasca, A et al, *Biochim Biophys Acta*, 577: 71, 1979

ICN 152263 *Vicia sativa* seeds   Purified by affinity chromatography; lyophilized; essentially free of salt & CHO

Agglutinin, *Vicia villosa*

Synonyms: Lectin

ICN 153373 *Vicia villosa*   Purified by affinity chromatography | Specificity for D-galNAc

Sigma L 4011   *Vicia villosa* (Hairy vetch)   Lyophilized containing 95% protein (Biuret); balance primarily sodium citrate; affinity purified by a modification of the method of Grubhoffer, L et al, *Biochem J*, 195: 626, 1981; two major bands on SDS-PAGE; may contain at least two isolectins; activity <65 μg/mL using blood group A₁ erythrocytes; does not agglutinate blood group B or O erythrocytes at 250 μg/mL | Has an affinity for *N*-acetyl-D-galactosamine; there are at least 3 tetrameric VVA isolectins composed of subunits A & B, designated B₄, A₂B₂ & A₄; isolectin B₄ is specific for Tn activated erythrocytes; A₄ has anti-A₁ blood group specificity & A₂B₂ has a specificity intermediate between A₄ & B₄; there is an additional non-blood group specific lectin in seed extracts; conjugates are prepared from affinity purified lectin; agglutination activity expressed in μg/mL & is determined from serial dilutions of a 1 mg/mL solution using phosphate buffered saline, pH 6.8; activity is the lowest concentration to agglutinate a 2% suspension of appropriate erythrocytes after 1 hr at 25°C; Tollefsen, CE & Kornfeld, R, *J Biol Chem*, 258: 5165, 1983; Lee, LT et al, *Blood*, 58: 1228, 1981

Sigma L 7513   *Vicia villosa* (Hairy vetch)   Highly purified; lyophilized containing ~95% protein (Biuret); balance primarily sodium phosphate buffer salts; affinity purified by a modification of the method of Tollefsen, CE & Kornfeld, R, *J Biol Chem*, 258: 5165, 1983; single major band on SDS electrophoresis; may contain at least two isolectins; activity <0.5 μg/mL using neuraminidase & β-galactosidase treated blood group O erythrocytes (Tn-activated); does not agglutinate untreated blood group A, B or O erythrocytes at 250 μg/mL | See Sigma L 4011

Agglutinin, *Vicia villosa*, 10 nm Colloidal Gold Conjugated

Synonyms: Lectin

ICN 154129   *Vicia villosa*   Purified by affinity chromatography | Specificity for D-galNAc

Agglutinin, *Vicia villosa*, 20 nm Colloidal Gold Conjugated

Synonyms: Lectin

ICN 154130   *Vicia villosa*   Purified by affinity chromatography | Specificity for D-galNAc

Agglutinin, *Vicia villosa*, 5 nm Colloidal Gold Conjugated

Synonyms: Lectin

ICN 154128   *Vicia villosa*   Purified by affinity chromatography | Specificity for D-galNAc

Agglutinin, *Vicia villosa*, Biotin Conjugated

Synonyms: Lectin

ICN 153374   *Vicia villosa*   Purified by affinity chromatography | Specificity for D-galNAc

Sigma L 6012   *Vicia villosa* (Hairy vetch)   Lyophilized containing ~90% protein (Biuret); balance primarily sodium citrate; contains ~3 moles biotin/mole protein | See Sigma L 4011

Sigma L 7638   *Vicia villosa* (Hairy vetch)   Lyophilized containing ~90% protein (Biuret); balance primarily sodium citrate; contains 3-6 moles biotin/mole protein | See Sigma L 4011

Agglutinin, *Vicia villosa*, FITC Conjugated

Synonyms: Lectin

ICN 153375   *Vicia villosa*   Purified by affinity chromatography | Specificity for D-galNAc

Sigma L 5887   *Vicia villosa* (Hairy vetch)   Lyophilized containing ~90% protein (Biuret); balance primarily sodium citrate; contains 3-6 moles FITC/mole protein | See Sigma L 4011

Sigma L 7763   *Vicia villosa* (Hairy vetch)   Lyophilized containing ~90% protein (BCA); balance primarily sodium citrate; contains 3-6 moles FITC/mole protein | See Sigma L 4011

Agglutinin, *Vicia villosa*, Horse Radish Peroxidase Conjugated

Synonyms: Lectin

ICN 153376   *Vicia villosa*   Purified by affinity chromatography | Specificity for D-galNAc

Agglutinin, *Vicia villosa*, Immobilized

Synonyms: Lectin

Sigma L 7888   *Vicia villosa* (Hairy vetch)   Immobilized on 4% beaded agarose; spacer: 28 atom; contains 3-6 mg protein (E$^{1\%}_{280}$)/mL; suspension in 0.01 *M* phosphate, 0.15 *M* NaCl containing 50% glycerol, 0.02% sodium azide & trace metals | See Sigma L 4011

Sigma L 9388   *Vicia villosa* (Hairy vetch)   Immobilized on 4% beaded agarose; spacer: 12 atom; contains 2-4 mg protein (Biuret)/mL packed gel; suspension in 0.01 *M* phosphate buffered saline containing 50% glycerol & 0.02% sodium azide; 1 mL suspension yields ~0.5 mL packed gel | See Sigma L 4011

Agglutinin, *Vicia villosa*, Peroxidase Conjugated

Synonyms: Lectin

Sigma L 5641   *Vicia villosa* (Hairy vetch)   Lyophilized containing ~95% protein (E$^{1\%}_{280}$); balance primarily sodium citrate buffer salts; peroxidase activity: ~50 units/mg protein; unit definition: 1 unit forms 1 mg purpurogallin in 20 sec from pyrogallol at pH 6.0 at 20°C; prepared from peroxidase type VI using a modification of the method of O'Sullivan, MJ et al, *FEBS Lett*, 95: 311, 1978; repurified after conjugation by affinity chromatography | See Sigma L 4011

Agglutinin, *Viscum album*

Synonyms: Lectin

ICN 152264   *Viscum album* (European mistletoe)   Purified by affinity chromatography | Specificity for terminal β-D-gal residues; inhibits allergen induced histamine release *in vitro* from human leukocytes; inhibits protein synthesis similar to *Ricinus communis* agglutinin-II

Sigma L 0261   *Viscum album* (European mistletoe)   Lyophilized; salt-free; partially purified; activity: <50 μg/mL using papain treated human blood group O erythrocytes; extremely hazardous | See Sigma L 2511

Sigma L 2511   *Viscum album* (European mistletoe)   Lyophilized; salt-free; highly purified by affinity chromatography using a modification of the method of Ziska, P et al, *Experientia*, 34: 123, 1978; activity: <4 μg/mL using human blood group O erythrocytes; extremely hazardous | Not blood group specific; affinity for β-D-galactosyl residues; inhibits protein synthesis similarly to Ricin (RCA₆₀) & inhibits allergen induced histamine release *in vitro* from human leukocytes; agglutination activity expressed in μg/mL & is determined from serial dilutions of a 1 mg/mL solution using phosphate buffered saline, pH 6.8; activity is the lowest concentration to agglutinate a 2% suspension of appropriate human erythrocytes after 1 hr at 25°C; Stirpe, F et al, *Biochem J*, 190: 843, 1980; Sehrt, I & Luther, P, *Lectins: Biology Biochemistry & Clinical Biochemistry*, 2: 45, 1981, ed Bog-Hansen, DeGruyter, Berlin

Agglutinin, *Viscum album*, Biotin Conjugated

Synonyms: Lectin

ICN 153377   *Viscum album* (European mistletoe)   Purified by affinity chromatography | Specificity for β-D-gal

Agglutinin, *Viscum album*, FITC Conjugated

Synonyms: Lectin

ICN 153378   *Viscum album* (European mistletoe)   Purified by affinity chromatography | Specificity for β-D-gal

Agglutinin, *Wisteria floribunda*

Synonyms: Lectin

Sigma L 2016 *Wisteria floribunda* (Japanese wisteria) Lyophilized containing ~95% protein ($E^{1\%}_{280}$); balance primarily sodium citrate; activity: No agglutination of Type A RBC's at 250 µg/mL | Native lectin which has been disulfide-reduced; free sulfhydryls (one per subunit) are blocked with *N*-ethylmaleimide to prevent reoxidation; MW & valency are reduced to ~1/2 of the native lectin; agglutination activity is greatly reduced but reactivity with *N*-acetyl-D-galactosamine is retained; conjugates are prepared from affinity purified lectin following reduction & blocking; seeSigma L 8258

Sigma L 8258 *Wisteria floribunda* (Japanese wisteria) Lyophilized containing ~95% protein ($E^{1\%}_{280}$); balance primarily sodium citrate; highly purified (affinity chromatography); activity: <16 µg/mL | Not blood group specific but has an affinity for *N*-acetyl-D-galactosamine; agglutination activity expressed in µg/mL & is determined from serial dilutions of a 1 mg/mL solution using phosphate buffered saline, pH 6.8; activity is the lowest concentration to agglutinate a 2% suspension of human blood group A erythrocytes after 1 hr at 25°C; Kurokawa et al, *J Biol Chem*, 251: 5686, 1976

Agglutinin, *Wisteria floribunda*, Biotin Conjugated

Synonyms: Lectin

Sigma L 1516 *Wisteria floribunda* (Japanese wisteria) Lyophilized containing ~95% protein (Biuret); balance primarily sodium citrate; protein is biotinylated utilizing aminocaproyl spacer; contains ~4 moles biotin/mole protein | See Sigma L 8258

Sigma L 1766 *Wisteria floribunda* (Japanese wisteria) Lyophilized containing ~95% protein ($E^{1\%}_{280}$); balance primarily sodium citrate; biotinylated via aminocaproyl spacer; contains ~4 moles biotin/mole dimer | See Sigma L 2016 & Sigma L 8258

Agglutinin, *Wisteria floribunda*, FITC Conjugated

Synonyms: Lectin

Sigma L 1641 *Wisteria floribunda* (Japanese wisteria) Lyophilized containing ~95% protein (Biuret); balance primarily sodium citrate; contains ~3 moles FITC/mole protein | See Sigma L 8258

Aggrecan

Sigma A 1960 Bovine articular cartilage MW >2.5x10[6] Lyophilized, essentially salt-free; sterile-filtered | Proteoglycan of high molecular weight containing a core protein of 210-250k substituted with 100-150k glycosaminoglycan chains; the majority of the chains are chondroitin/dermatan sulfate, the minority being keratan sulfate glycosaminoglycans; will form a macro-molecular complex in the presence of hyaluronic acid causing an increase of relative viscosity of ~30%; Roughley, PJ & Lee, ER, *Micros Res Tech*, 28: 385, 1994; Ruoslahti, E, *J Biol Chem*, 264: 13369, 1989; Hardingham, TE & Muir, H, *Biochim Biophys Acta*, 279: 401, 1972

Agrostin

Sigma A 7928 *Agrostemma githago* seeds Lyophilized powder containing ~30% protein (Lowry); balance primarily glucose & sodium phosphate buffer salts | A ribosome inactivating protein; Stirpe, F et al, *Biochem J*, 216: 617, 1983; Stirpe, F & Barbieri, L, *FEBS Lett*, 195: 1, 1986

Akt1/PKB α, Active

USBio A1125-06 Human recombinant, expressed in Sf9 cells MW ~60k ≥95%; purified using Ni-NTA agarose, activated with PDK1 & repurified; frozen solution in 50 µL of 50 *mM* Tris-HCl, pH 7.5, 150 *mM* NaCl, 150 *mM* imidazole, 0.1% β-MSH, 0.1 *mM* EGTA, 270 *mM* sucrose, 1 *mM* benzamidine, 0.2 *mM* PMSF, 0.03% Brij-35 | N-terminus His-tagged fusion protein corresponding to human Akt1; SA:: 154 nmol phosphate/min/mg enzyme transferred to Akt/SGK substrate peptide (RPRAATF); active kinase detected by comparing incorporation of (^{32}P) into Akt/SGK substrate peptide

Akt1/PKB α, Inactive

USBio A1125-08 Human recombinant, N-terminus His-tagged fusion protein expressed in Sf9 cells MW ~60k ≥80%; purified using Ni-NTA agarose; liquid in 50 *mM* Tris, pH 7.5, 0.1% (v/v) β-MSH, 0.1 *mM* EGTA, 1 *mM* benzamidine, 0.2 *mM* PMSF, 1 mg/mL BSA, 50% glycerol | Corresponds to the human sequence

Akt1/PKB α-GST, Inactive Agarose

USBio A1125-07 Rat fusion protein, expressed in *E. coli* MW ~83k Purified using glutathione sepharose; 50 µg of GST-Akt1/PKBa bound to 100 µL packed gel beads (200 µL suspension); suspension of 50% agarose gel slurry in 0.05 *M* PBS, pH 7.2, 50% glycerol | Full-length bovine cDNA was fused in frame with GST; recognized by anti-peptide Ab directed against the sequence corresponding to residues 466-480 of Rat Akt1; cellular homolog of the viral oncogene v-akt; phosphorylates & inactivates GSK-3; growth factor or insulin stimulation of mouse NIH 3T3 fibroblasts or Rat-1 cells activates Akt1; activation is concomitant with phosphorylation of Thr[308] & Ser[473]; it is prevented by inhibitors of phosphatidylinositol 3-kinase

Albumin

Synonyms: Bovine Serum Albumin; Ovalbumin; Human Serum Albumin; Rat Serum Albumin

Amersham US10867 Bovine Protease free | Suitable as an ELISA blocking agent

ICN 105033 Bovine Clinical reagent grade, RIA grade; 98-99%; fatty acid-free powder; low levels of metabolites & enzymes; <1 micro U/g insulin RIA, T3 & T4 ≤ detectable levels, pH 7.0 ± 0.2, <2% sulfated ash, <5% moisture | Designed for the most sensitive research & diagnostic applications; especially suitable for RIA & enzyme kits as a protein base

ICN 150269 Bovine 30% Serological solution isotonic with respect to human erythrocytes; stabilized with ~10-15 mg octanoic acid/g protein; 0.1% NaN₃ as a preservative

ICN 150270 Bovine 30% Serological solution isotonic with respect to human erythrocytes; <1 mg FA/g protein; stabilizer free; 0.1% NaN₃ as a preservative

ICN 194120 Bovine Nuclease free; ≥90%; no detectable exonuclease, endonuclease, ribonuclease or protease; some degradation products may exist; 50 mg/ml in 50% aqueous glycerol, neutral pH

ICN 194773 Bovine Cell culture reagent, γ-irradiated; crystalline, 98-99%; essentially globulin-free

ICN 194774 Bovine Cell culture reagent; lyophilized; essentially globulin-free

ICN 194775 Bovine Cell culture reagent; powder; <0.1 ng/mg endotoxin

ICN 194776 Bovine Cell culture reagent; powder; <0.1 ng/mg endotoxin | Prepared using a salt fractionation procedure with ion exchange chromatography

ICN 55918 Bovine Lyophilized, purified antigen

ICN 810012/810013/810014/810015 Bovine Crystalline powder; the purest bovine albumin available

ICN 810032 ICN 810033 ICN 810034 ICN 810035 ICN 810036 Bovine Lyophilized, low salt; RIA insulin grade (<10 µU insulin/g powder; ≥98%

ICN 810061 Bovine 35% Solution; sterile

ICN 810101 Bovine Path-o-Cyte® 4 solution | Used as a medium in density gradient centrifugation or as an addition to tissue culture growth medium; stable ≥2 yr @ 5°C

ICN 810111 Bovine Path-o-Cyte® 5 solution | Used as a medium in density gradient centrifugation or as an addition to tissue culture growth medium; stable ≥2 yr @ 5°C

ICN 810133 Bovine 30% Solution; cap free, sterile

ICN 810706 Bovine 30% Solution; non-sterile

ICN 810783 ICN 810784 Bovine 30% Polymer enhanced solution | Prepared from high quality Fraction V powder intended for *in vitro* diagnostic use in immunohematological procedures requiring high protein diluent; can be used for serological procedures; potentiates the expression of weak Ag-Ab reactions & avidity of RBC

ICN 820471 ICN 820472   Bovine   Fatty acid free; 0.2 mg endotoxin

ICN 840052 ICN 840053 ICN 840054 ICN 840055   Bovine Microbiological grade

Sigma 850-100   Bovine   For use in the determination of fibrin/fibrinogen degradation products in serum in the *Staphylococcal* Clumping Test per Sigma Procedure No. 850 or other usage; crystallized; 100 mg/vial; initial fractionation by cold alcohol precipitation

Sigma A 0336   Bovine   Globulin free; 30% solution; essentially IgG free; ≤25 ng IgG/mg protein (HPLC); aseptically filled; ~0.85% NaCl; no stabilizer added

Sigma A 1662   Bovine   30% solution; aseptically filled; stabilized with 10-15 mg octanoic acid/g protein; contains ~0.85% NaCl & 0.1% sodium azide as preservative

Sigma A 2153   Bovine   Fraction V powder; prepared by a modification of Cohn, using cold ethanol, pH, & low temperature precipitation, followed by additional pH adjustment step prior to final drying; pH of a 1% (w/v) aqueous solution is ~7; ≥96% (electrophoresis); initial fractionation by cold alcohol precipitation; remainder mostly globulins; Cohn, EJ et al, *J Am Chem Soc*, 68: 459, 1946

Sigma A 3174   Bovine   30% solution; high avidity; aseptically filled; stabilized with 10-15 octanoic acid/gram protein; contains ~0.85% NaCl & 0.1% sodium azide as preservative; tested for avidity with known incomplete antibody (not saline agglutinable)

Sigma A 3299   Bovine   30% solution; high avidity; aseptically filled; no stabilizer added; contains ~0.85% NaCl & 0.1% sodium azide as preservative; tested for avidity with known incomplete antibody (not saline agglutinable)

Sigma A 3424   Bovine   30% solution; ultra-high avidity; aseptically filled; no stabilizer added; contains ~0.85% NaCl & 0.1% sodium azide as preservative; tested for avidity with known incomplete antibody (not saline agglutinable)

Sigma A 3675   Bovine   Fraction V powder; ≥98% (electrophoresis); initial fractionation by salt fractionation; low endotoxin: ≤0.1 ng/mg of endotoxin detection

Sigma A 3902   Bovine   Vitamin B_{12} & B_{12} binding factor deficient; initial fractionation by cold alcohol precipitation; 1 g contains < 2.0 ng vitamin B_{12} & will bind < 2.0 ng of vitamin B_{12} | For use in vitamin B_{12} assays

Sigma A 4378   Bovine   1x crystallized; ≥97% albumin (agarose electrophoresis); initial fractionation by cold alcohol precipitation; prepared using method IV of Cohn | Cohn, EJ et al, *J Am Chem Soc*, 69: 1753, 1947

Sigma A 4628   Bovine   5% solution; aseptically filled; contains ~0.70% NaCl; no stabilizer or preservative added

Sigma A 6003   Bovine   ≥96% albumin (agarose electrophoresis); essentially fatty acid free (~0.005%); initial fractionation by cold alcohol precipitation; prepared from Fraction V albumin (A 4503) | Chen, RF, *J Biol Chem*, 242: 173, 1967

Sigma A 7034   Bovine   22% solution; aseptically filled; contains ~0.85% NaCl; 0.1% sodium azide as preservative

Sigma A 7159   Bovine   25% solution in Tyrode's Buffer; aseptically filled; no preservative added

Sigma A 7284   Bovine   30% solution; aseptically filled; contains ~0.85% NaCl; no stabilizer added; 0.1% sodium azide as preservative

Sigma A 7409   Bovine   35% solution; aseptically filled; contains ~0.85% NaCl; no preservative added

Sigma A 7511   Bovine   Crystallized; ≥97% albumin (agarose electrophoresis); essentially fatty acid free (~0.005%); initial fractionation by cold alcohol precipitation; prepared from albumin (A 4378) | Chen, RF, *J Biol Chem*, 242: 173, 1967

Sigma A 7534   Bovine   35% solution; aseptically filled; contains ~0.85% NaCl; 0.1% sodium azide as preservative

Sigma A 7638   Bovine   Lyophilized powder ≥99% (agarose electrophoresis); essentially globulin free; initial fractionation by cold alcohol precipitation; prepared from bovine albumin Fraction V (A 4503)

Sigma A 8327   Bovine   30% solution; aseptically filled; contains ~0.85% NaCl; no stabilizer or preservative added

Sigma A 8577   Bovine   30% solution; protease: <0.0001 unit/mg protein; alkaline phosphatase: <0.001 unit/mg protein; peroxidase: <0.001 unit/mg protein; aseptically filled; contains ~0.85% NaCl; no stabilizer or preservative added

Sigma A 9085   Bovine   Lyophilized powder ≥98% (agarose electrophoresis); essentially IgG free; <25 ng IgG/mg protein (HPLC)

Sigma A 9205   Bovine   30% solution; ≥96% albumin (agarose electrophoresis); essentially fatty acid free (~0.005%); initial fractionation by cold alcohol precipitation; aseptically filled; contains ~0.85% NaCl; no preservative added

Sigma A 0281   Bovine albumin   Prepared from bovine albumin (A 7638); lyophilized powder ≥99% (electrophoresis); essentially fatty acid free & globulin free; initial fractionation by cold alcohol precipitation; Chen, RF, *J Biol Chem*, 242: 173, 1967

Biogenesis 0220-1404   Bovine blood   MW 68k   Purified; from 0.05 M NH_4HCO_3; lyophilized

ICN 103700   Bovine plasma   Crystalline, 98-99%; pH 5.2 ± 0.2 (1% solution), <0.5% sulfated ash, <0.1% carbohydrate, <5.2% ± 0.2% moisture | Prepared fresh by the Cohn cold ethanol fractionation method, followed by crystallization at low temperature from an alcohol-containing solution; not heated at any stage in the process; can be utilized as a nutrient for tissue culture, preparation of protein standards & as an antigen in immunological studies in sensitive research applications

ICN 194771   Bovine plasma   Cell culture reagent, 96-99%; see ICN 103700, <5.2% moisture | See ICN 103700

Fitzgerald 30-AB73   Bovine serum   Reagent grade; high purity; pH 7.0

Fitzgerald 30-AB74   Bovine serum   Protease free; high purity; pH 7.0

Fitzgerald 30-AB79   Bovine serum   High purity; fatty acid free; pH 7.0

Fluka 05468   Bovine serum   MW ~67k   Fatty acid free; ≥97% (gel electrophoresis); ≥95% protein; ≤5% water; ≤3% (GE) globulin; ~0.005% fatty acids | From crystallized & lyophilized albumin isolated by Cohn method IV; initial fractionation by cold alcohol precipitation; stimulated the release of lysophosphatidylcholine from cultured rat hepatocytes; Cohn, EJ et al, *J Am Chem Soc*, 68: 1753, 1947; Baisted, DJ et al, *Biochem J*, 253: 693, 1988

Fluka 05470   Bovine serum   MW ~67k   ≥98% (gel electrophoresis); ≥95% protein; ≤5% water; ≤0.5% residue on ignition, Na; ≤0.1% SO₄; ≤0.2% Cl; ≤0.01% Ca; ≤0.005% K; ≤0.001% Cu, Fe, Mg; ≤0.0005% Cd, Co, Cr, Mn, Ni, Pb, Zn | Peter, T Jr, *Adv Protein Chem*, 37: 161, 1985

Fluka 05471   Bovine serum   MW ~67k   ≥97% (gel electrophoresis); ≥95% protein; ≤5% water; ≤2% α-globulin; ≤1% β-globulin | Prepared by Cohn method IV; initial fractionation by cold alcohol precipitation; S Anders Sevall, J, *Biochemistry*, 27: 5038, 1988; Cohn, EJ et al, *J Am Chem Soc*, 69: 1753, 1947

Fluka 05478   Bovine serum   MW ~67k   For vitamin B_{12} assay; ≥97% (gel electrophoresis); ≥90% protein; ≤8% water; vitamin B_{12} capacity ≤2 ng/g BSA

Fluka 05489   Bovine serum   MW ~67k   ≥98% (gel electrophoresis); ~50 mg BSA/mL water

Fluka 05490   Bovine serum   MW ~67k   Low fatty acid; ~98% (gel electrophoresis); ≤1% residue on ignition; ≤3% water; ≤1% fatty acid

Fluka 05491   Bovine serum   MW ~67k   Molecular biology grade; ≥99% protein; acetylated; ≤5% water; no detectable DNases, RNases, proteases, phosphatases | Prepared from BSA by acetylation in order to inactivate nucleases; Ternynck, T et al, *Meth Enzymol*, 150: 117, 1987

Fluka 05496   Bovine serum   MW ~67k   Molecular biology grade; solution in 50 *mM* Tris HCl, 0.1 *M* NaCl, 0.25 *mM* EDTA-disodium salt, 1 *mM* mercaptoethanol, 50% glycerol; ~20 mg/mL; no detectable DNases, RNases, proteases, phosphatases | Acetylation of BSA to inactivate the nucleases; suitable for the stabilization of restriction enzymes

Sigma A 2065 Bovine serum MW 66k Vial contains enough FITC-conjugated protein to run 50 mini-gels or 25 standard size gels; protein band be visualized by using UV light or Brilliant Blue stain | Fluorescent marker for SDS-page & protein transfer; suitable for use as molecular weight standards in both SDS-PAGE & transfer membranes

Biogenesis 0220-1604 Canine serum Purified; no preservatives; lyophilized

Sigma A 3686 Chicken Lyophilized powder; essentially globulin free; ~99% (agarose electrophoresis)

Biogenesis 0220-1724 Chicken egg white MW 44,287 Tested negative for *Salmonella, Staphylococcus, Coliform, Listeria* and *E. coli*; purified; no preservatives; lyophilized

Fluka 05438 Chicken egg white MW ~45k ≥98% (gel electrophoresis); ≥99% protein; 6 mol mannose/mol ovalbumin carbohydrate content | Prepared from crystallized, lyophilized & desalted ovalbumin

ICN 191224 Chicken egg white Lyophilized, ≥90% | Ovalbumin

ICN 55925 Dog Purified Ag, lyophilized

Sigma A 3184 Dog Essentially fatty acid free (~0.005%); prepared from Fraction V albumin (A 9263) | Chen, RF, *J Biol Chem*, 242: 173, 1967

Sigma A 3185 Goat Essentially fatty acid free (~0.005%); prepared from Fraction V albumin (A 2514) | Chen, RF, *J Biol Chem*, 242: 173, 1967

Sigma A 4164 Goat Lyophilized powder; essentially globulin free; ~99% (agarose electrophoresis); prepared from goat albumin (A 2514)

Sigma A 3060 Guinea pig Essentially fatty acid free (~0.005%); prepared from Fraction V albumin (A 2639) | Chen, RF, *J Biol Chem*, 242: 173, 1967

Sigma A 6539 Guinea pig Lyophilized powder; essentially globulin free; ~99% (agarose electrophoresis); prepared from guinea pig albumin (A 2639)

Biogenesis 0220-1804 Guinea pig serum 15.7% nitrogen; purified; no preservatives; lyophilized

ICN 55934 Hamster Purified Ag, lyophilized

Sigma A 3434 Horse Lyophilized powder; essentially globulin free; ~99% (agarose electrophoresis); prepared from horse albumin (A 9888)

Sigma A 9888 Horse Fraction V powder

Cortex CP0925U Human >98%; liquid; 25%

Cortex CS1052 Human >98%; powder

ICN 103712 Human 30% Solution, aseptically filled; 0.85% NaCl, 0.1% NaN$_3$

ICN 55912 Human Purified Ag, lyophilized

ICN 823011 ICN 823012 Human Crystalline, lyophilized

ICN 823030 ICN 823031 ICN 823032 ICN 823033 ICN 823034 ICN 823036 Human 1X Crystalline, lyophilized; 97-99%; low B$_{12}$ & folate | Prepared using Cohn method IV; source material is tested for HbsAg, anti-HCV, anti-HIV-1, anti-HIV-2 & syphilis

ICN 823471 ICN 823474 Human Low endotoxin | Highly suitable for serum-free culture methods

Sigma A 1887 Human Source material has tested negative for HIV & HBsAg; ≥96% albumin; essentially fatty acid free (~0.005%); Chen, RF, *J Biol Chem*, 242: 173, 1967

Sigma A 3173 Human Low endotoxin (≤0.1 ng/mg solid) | Source material has tested negative for HIV & HBsAg

Sigma A 3782 Human Lyophilized powder; essentially globulin free; ~99% (agarose electrophoresis); essentially fatty acid free (~0.005%); prepared from essentially globulin free human albumin (A 8763) | Source material has tested negative for HIV & HBsAg

Sigma A 4327 Human ≥96%; protease: <0.0001 units/mg protein; alkaline phosphatase: <0.001 units/mg protein; peroxidase: <0.001 units/mg protein; prepared from Fraction V powder (A 1653) | Source material has tested negative for HIV & HBsAg

Sigma A 6784 Human 10% Solution; aseptically filled; contains ~0.85% NaCl & 0.05% sodium azide | Source material has tested negative for HIV & HBsAg

Sigma A 6909 Human 30% Solution; aseptically filled; contains ~0.85% NaCl; 0.1% sodium azide as preservative | Source material has tested negative for HIV & HBsAg

Sigma A 8763 Human Lyophilized powder; essentially globulin free; ~99% (agarose electrophoresis); prepared from human albumin (A 1653) | Source material has tested negative for HIV & HBsAg

Sigma A 9080 Human 30% Solution; from human plasma; aseptically filled; contains ~0.85% NaCl | Source material has tested negative for HIV & HBsAg

Sigma A 9511 Human 1x crystallized & lyophilized; 97-99%; prepared from human albumin (A 1653) per Cohn | Source material has tested negative for HIV & HbsAgCohn, EJ et al, *J Am Chem Soc*, 69: 1753, 1947

Biogenesis 0220-0806 Human serum MW 65k pI: 4.7; affinity purified; from 0.02 *M* sodium phosphate, 0.14 *M* NaCl, pH 7.6; lyophilized

Biogenesis 0220-0809 Human serum MW 28k Tested negative for HCV, HIV-1 and HIV-2 antibodies and HBsAg; affinity purified; from 0.05 *M* NH4HC03, pH 8.0; lyophilized

Fluka 05418 Human serum MW ~68k Fatty acid free, globulin free; ≥99% (gel electrophoresis); ≤10% water; ≤0.6% α-globulin; ≤0.4% β-globulin; ≤0.01% fatty acid | Prepared from crystallized & lyophilized Cohn fraction V; Lightner, DA et al, *JBC*, 263: 16669, 1988

Fluka 05420 Human serum MW ~68k ≥98% (gel electrophoresis); ≤10% water; ≤1 residue on ignition | Peters, T Jr, *Adv Protein Chem*, 37: 161, 1985

Scipac P199-1 Human serum/plasma >96%; lyophilized; suitable for bulk requirements | Serum protein

ICN 55941 Mouse Purified Ag, lyophilized

Sigma A 3559 Mouse Lyophilized powder; essentially globulin free; ~99% (agarose electrophoresis); prepared from mouse albumin (A 3139)

Sigma A 1056 Mouse Fraction V albumin Essentially fatty acid free (~0.005%); Chen, RF, *J Biol Chem*, 242: 173, 1967 | Prepared from Fraction V albumin (A 3139)

Biogenesis 0220-1904 Mouse serum MW 67k Purified; essentially salt free; lyophilized

ICN 823514 Mouse serum Lyophilized

Sigma A 4414 Pig Lyophilized powder; essentially globulin free; ~99% (agarose electrophoresis; prepared from pig albumin (A 2764)

Sigma A 9422 Pig Essentially fatty acid free (~0.005%); prepared from Fraction V albumin (A 2764) | Chen, RF, *J Biol Chem*, 242: 173, 1967

Sigma A 1173 Pig albumin Lyophilized powder; essentially globulin free; ~99% (agarose electrophoresis); essentially fatty acid free (~0.005%); | Prepared from essentially globulin free albumin (A 4414)

Sigma A 1830 Porcine serum ≥98% (agarose electrophoresis); initial fractionation by heat shock

ICN 55948 Rabbit Purified Ag, lyophilized

ICN 824514 Rabbit Crystalline | Used in RIA & EIA systems requiring a homologous protein diluent

Sigma A 9438 Rabbit Essentially fatty acid free (~0.005%); prepared from Fraction V | Chen, RF, *J Biol Chem*, 242: 173, 1967

Sigma A 0764 Rabbit albumin Globulin free; lyophilized powder; essentially globulin free; ~99% (agarose electrophoresis) | Prepared from rabbit albumin (Sigma A 0639)

Biogenesis 0220-2004 Rabbit serum MW 66k >98% albumin; no preservatives; lyophilized | Used as a protein standard or stabilizer for dilute protein samples

Fluka 05456 Rabbit serum MW ~68k Globulin free; ≥98% (gel electrophoresis); ≥95% protein; ≤5% water

ICN 55952 Rat Purified Ag, lyophilized

Sigma A 4538 Rat Lyophilized powder; essentially globulin free; ~99% (agarose electrophoresis); prepared from rat albumin (A 6272)

Sigma A 6414 Rat Lyophilized powder; essentially globulin free; ~99% (agarose electrophoresis); essentially fatty acid free (~0.005%); prepared from essentially globulin free albumin (A 4538)

Sigma A 2018   Rat Fraction V   Prepared from Fraction V (A 6272); essentially fatty acid free (~0.005%); Chen, RF, *J Biol Chem*, 242: 173, 1967

Biogenesis 0220-2439   Rat serum   Purified; from 0.05 *M* NH$_4$HCO$_3$, pH 8.0; lyophilized

Fluka 05464   Rat serum   MW ~67k   ≥98% (gel electrophoresis); ≥98% protein; ≤10% water

Fluka 05466   Rat serum   MW ~67k   Fatty acid free; ≥98% (gel electrophoresis); ≤10% water; ≤0.01% fatty acid

ICN 825014   Rat serum   Lyophilized

Scipac P166-1   Serum/plasma   >96%; lyophilized; suitable for research and the most demanding applications | Serum protein

Sigma A 6289   Sheep   Essentially fatty acid free (~0.005%); prepared from Fraction V (A 3264) | Chen, RF, *J Biol Chem*, 242: 173, 1967

Scipac P140-0   Urine of patients with chronic renal tubular proteinuria   >99%; lyophilized; sterile filtered through 0.2µm membrane; available on request | Urine protein

Scipac P140-1   Urine of patients with chronic renal tubular proteinuria   >98%; lyophilized | Urine protein

Albumin, (^{125}I)-
Synonyms: Bovine Serum Albumin

ICN 68031   Bovine Serum   ~1mCi/mg, ~37 MBq/mg; 0.1 *M* KPO$_4$ buffer, pH 7.5

ICN 68113   Human serum   ~1 µCi/µg, ~37 kBq/µg; 0.1 M KPO$_4$, pH 7.5

ICN 68120   Rabbit serum   ~1 µCi/µg; 0.1 *M* KPO$_4$ buffer, pH 7.5

Albumin, (^{14}C-Me)-
Synonyms: Bovine Serum Albumin; Ovalbumin

ARC ARC-422   Bovine serum   MW 69k   3-30 µCi/mmol; 111-1111 KBq/mg; in 0.01 *M* sodium phosphate, pH 7.2, buffer | Radiochemical

Sigma A 7417   Bovine serum   MW ~66k (monomer), ~132k (dimer)   5-50 Ci/mg protein; solution in 40 *mM* potassium phosphate, pH 7.0, in Combi-vial | RadiochemicalPrepared from Sigma A 7517

Sigma A 6418   Chicken egg white   MW ~45k   5-50 Ci/mg protein; solution in 40 *mM* potassium phosphate, pH 7.0, in Combi-vial | Radiochemical prepared from Sigma A 7642

Albumin, 10 nm Colloidal Gold Conjugated
Sigma A 4292   Bovine   Mean particle diameter 8-12 nm; monodisperse; bovine-biotinamidocaproyl (A 6043) adsorbed to colloidal gold; suspension in 50% glycerol containing 0.15 *M* NaCl, 0.01 *M* sodium phosphate, pH 7.4, 0.02% PEG 20, & 0.02% sodium azide; concentration: A$_{520}$ ~5.0; | Useful in the detection of biotinylated compounds using streptavidin as bridging proteinBonnard, °C et al, *Immunolabeling for Electron Microscopy*, Polak & Varndell, eds, Elsevier Science Publishers, p 95, 1984

Sigma A 5179   Bovine   Mean particle size 8-12 nm; monodisperse; albumin bovine (A 0281) adsorbed to colloidal gold; suspension in ~50% glycerol containing 0.15 *M* NaCl, 0.01 *M* Tris, 0.02% PEG 20, & 0.02% sodium azide, pH 7.6; concentration: A$_{520}$ ~5.0; | Suitable for use as a control

Albumin, 20 nm Colloidal Gold Conjugated
Sigma A 4417   Bovine   Mean particle diameter 17-23 nm; monodisperse; bovine-biotinamidocaproyl (A 6043) adsorbed to colloidal gold; suspension in 50% glycerol containing 0.15 *M* NaCl, 0.01 *M* sodium phosphate, pH 7.4, 0.02% PEG 20, & 0.02% sodium azide; concentration: A$_{520}$ ~5.0; | Useful in the detection of biotinylated compounds using streptavidin as bridging proteinBonnard, °C et al, *Immunolabeling for Electron Microscopy*, Polak & Varndell, eds, Elsevier Science Publishers, p 95, 1984

Albumin, 2X Crystallized
Synonyms: Ovalbumin

Fluka 05450   Chicken egg white   MW ~45k   Lyophilized powder; salt-free; ≥80% (gel electrophoresis); ≤3% loss on drying; ≤1% residue on ignition

Albumin, 5 nm Colloidal Gold Conjugated
Sigma A 5547   Bovine   Mean particle diameter 3.5-6.5 nm; monodisperse; bovine-biotinamidocaproyl (A 6043) adsorbed to colloidal gold; suspension in 0.15 *M* NaCl, 0.01 *M* sodium phosphate, pH 7.4, 0.02% PEG 20, & 0.02% sodium azide; concentration: A$_{520}$ ~5.0; | Useful in the detection of biotinylated compounds using streptavidin as bridging proteinBonnard, °C et al, *Immunolabeling for Electron Microscopy*, Polak & Varndell, eds, Elsevier Science Publishers, p 95, 1984

Albumin, 5X Crystallized
Synonyms: Ovalbumin

Fluka 05440   Chicken egg white   MW ~45k   Lyophilized powder; ≥95% (gel electrophoresis); ≤6% loss on drying; ≤1% residue on ignition | Marshall, RD & Neuberger, A, *Glycoproteins*, BBA Library, vol 5: (A Gottschalk, ed), 2nd, 732, 1972, Elsevier. Amsterdam

Albumin, Ac-
Synonyms: Bovine Serum Albumin

Amersham US10848   Bovine   Nuclease free | Tested as an enzyme stabilizer in PCR

Promega R3961   Bovine serum   Acetylated by the method of Gonzalez to ensure the inactivation of any contaminating nuclease activities & is dialyzed extensively with DI water to remove impurities | Used as an enzyme stabilizer or as a carrier protein; Gonzalez, N et al, *Arch Biochem Biophys*, 182: 404, 1977

Albumin, Azo-
Synonyms: Sulfanilic Acid-Azoalbumin; Bovine Serum Albumin

ICN 100308   Bovine   Prepared from bovine albumin Fraction V; a soluble chromogenic substrate for proteolytic enzymes

Sigma A 2382   Bovine albumin Fraction V   Prepared from bovine albumin Fraction V (A 4503); soluble chromogenic substrate for proteolytic enzymes; Tomarelli, RM et al, *J Lab Clin Med*, 34: 428, 1949

Albumin, Biotin Conjugated
Sigma A 8549   Bovine   Lyophilized powder containing ~90% protein (Biuret); balance sodium citrate buffer salts; 8-12 moles biotin/mole albumin

Albumin, Biotinamido-Caproyl Conjugated
Sigma A 6043   Bovine   Lyophilized powder containing ~90% labeled protein (Biuret); balance sodium citrate buffer salts; 8-12 moles biotin/mole albumin; prepared from albumin (A7638) | Costello, SM et al, *Clin Chem*, 25: 1572, 1979

Albumin, Carboxymethyl-
Sigma A 6285   Bovine   Lyophilize d powder; essentially salt-free; <0.02 mole sulfhydryl/mole albumin & ≤1.5 moles *S*-carboxymethyl-cysteine/mole albumin

Sigma A 9437   Bovine   Dialyzed & lyophilized powder; prepared from Fraction V (A 0281) reduced & carboxymethylated; contains ≥30 moles *S*-carboxymethylcysteine/mole albumin

Albumin, Cohn Fraction V
Synonyms: Bovine Serum Albumin

ICN 100117   Bovine   35% Solution; aseptically filled & sterilized

ICN 103703 Bovine Powder; 95-98% | Manufactured under non-denaturing conditions to yield a product with native fatty acid profile—exogenous short chain fatty acid stabilizers are never added; suitable as a nutrient in growth media, for increasing viscosity &/or protein in certain antibiotic assay media, as a binding inhibitor in growth media & *in vitro* biochemical or structural studies, & in the formulation of veterinary vaccines

ICN 840532 ICN 840533 ICN 840534 ICN 840535 Bovine Powder; pH 5.2

ICN 841032 ICN 841033 ICN 841034 ICN 841035 Bovine Powder; pH 7.0

Fitzgerald 30-AB75 Bovine serum High purity; pH 7.0

Fitzgerald 30-AB76 Bovine serum High purity; pH 5.2

Albumin, Con-
Synonyms: Ovotransferrin; Ovalbumin

Sigma C 0755 Chicken egg white Substantially Iron free

Sigma C 0880 Chicken egg white Iron complex

Sigma C 7786 Chicken egg white Substantially Iron free; cell culture tested

Albumin, Cross-Linked Molecular Weight Standards for SDS-PAGE

Sigma A 9392 Bovine MW 66k (monomer), 132k (dimer), 198k (trimer), 264k (tetramer) Lyophilized; ~7 mg/vial | See Sigma MW-SDS-280

Albumin, Crude Powder
Synonyms: Egg Albumin; Ovalbumin

Fluka 05461 Chicken egg white MW ~45k ≤8% loss on drying; ≤4% residue on ignition; ≤2% Na; ≤1% K; ≤0.1% Ca, Mg; ≤0.005% K; ≤0.001% Fe; ≤0.0005% Cd, Co, Cr, Cu, Mn, Ni, Pb, Zn

Albumin, Cysteinyl

Sigma A 0161 Bovine Lyophilized powder; essentially salt-free; <0.02 mole sulfhydryl/mole albumin; Isles, TE & Jocelyn, PC, *Biochem J*, 88: 84, 1963

Albumin, Dimer

Sigma A 9039 Bovine Lyophilized powder; ~90% (electrophoresis); "Native" dimer chromatographically purified from A 4503

Albumin, Dinitrophenyl

Sigma A 6661 Human Lyophilized powder; contains 30-40 moles DNP/mole albumin | Source material has tested negative for HIV & HbsAgDNP-albuminLittle, SR & Eisen, HN, *Methods in Immunol & Immunochem*, Vol 1: 128, 1967

Albumin, FITC Conjugated
Synonyms: Bovine Serum Albumin

Sigma A 9771 Bovine Prepared from crystallized & lyophilized bovine albumin (A 4378)

Fluka 05493 Bovine serum MW ~67k ≥97% (gel electrophoresis); ≥98% protein; ≤10% water; ~10 moles/mole BSA FITC content | Prepared from crystallized & lyophilized BSA obtained by Cohn method IV; Lemke, H et al, *J Immuno Meth*, 121: 175, 1989; Cortese, JD et al, *J Cell Biol*, 113: 1331, 1991

Fluka 05494 Canine serum MW ~67k ≥95% (gel electrophoresis); ~95% protein; ≤5% water; FITC content: ~10 moles/mole BSA

Sigma A 4407 Dog ~10 moles FITC/mole albumin; prepared from dog albumin (A 9263)

ICN 55883 Human Purified Ag, lyophilized

Sigma A 7016 Human ~10 moles FITC/mole albumin; prepared form essentially globulin free albumin | Source material has tested negative for HIV & HbsAgFITC-albumin

Sigma A 2889 Rat ~10 moles FITC/mole albumin; prepared from essentially globulin free rat albumin | FITC-albumin

Albumin, Fraction II
Synonyms: Ovalbumin

Sigma A 5253 Chicken egg Crude; dried egg white | Ovalbumin

Albumin, Fraction III
Synonyms: Ovalbumin

Sigma A 5378 Chicken egg ≥90% (agarose electrophoresis); crystallized & lyophilized

Albumin, Fraction V
Synonyms: Bovine Serum Albumin; Human Serum Albumin; Ovalbumin

Amersham US70195 Bovine ≥98%, pH (1%) 7.0; ≥96% protein

Amersham US70244 Bovine RIA grade; ≥98%, pH (1%) 7.0

ICN 100152 Bovine Fatty acid poor (<0.2 mg/g protein); 98-99%

ICN 100153 Bovine Microbiological grade | Intended for use in growth of *Leptospires* & other fastidious organisms; tested for ability to support rapid initiation of growth & high cell yield of *Leptospires*

ICN 152401 Bovine Fatty acid free (<0.005%); 98-99%

ICN 160069 Bovine 98-99%; 1-2 mg Fatty acids/g protein, pH 7.02 (1%, isotonic), <2% sulfated ash, <6% moisture, <20ppm heavy metals | More highly purified than Cohn Fraction V powder

ICN 160069/100153/152401 Bovine Powder; pH ~7.0

ICN 194772 Bovine Cell culture reagent; fatty acid free (<0.05 mg/g protein); low endotoxin

ICN 810531 ICN 810532 ICN 810533 ICN 810534 ICN 810535 Bovine Powder; pH ~5.2

ICN 820012 ICN 820013 ICN 820015 ICN 820016 Bovine Fatty acid poor

ICN 820022 ICN 820024 ICN 820025 ICN 820026 Bovine Lyophilized; fatty acid free

ICN 820451 ICN 820452 Bovine Protease, HRP & alkaline phosphatase free

Sigma A 1933 Bovine Prepared by salt fractionation, ion exchange & gel filtration chromatography; powder; low endotoxin (endotoxin <0.1 ng/mg); cell culture tested

Sigma A 2058 Bovine Prepared by salt fractionation, ion exchange & gel filtration chromatography; powder; low endotoxin (endotoxin <0.1 ng/mg); IgG free (IgG <0.05% by agarose gel electrophoresis); cell culture tested

Sigma A 2934 Bovine Powder; ~99%; low endotoxin: ≤1 ng/mg; essentially γ-globulin free; prepared from pasteurized serum; purified by heat treatment & organic solvent precipitation

Sigma A 3059 Bovine Powder; ~99%; protease free; essentially γ-globulin free; prepared from pasteurized serum; purified by heat treatment & organic solvent precipitation

Sigma A 3156 Bovine Crystalline; lyophilized; essentially globulin free; sterilized by γ-irradiation; cell culture tested

Sigma A 3294 Bovine Powder; ≥98% (electrophoresis); protease ≤0.005% units/mg solid; initial fractionation by heat shock; remainder mostly globulins

Sigma A 3912 Bovine Powder; ≥96% (electrophoresis); remainder mostly globulins; prepared with charcoal treatment & extensive dialysis to reduce low molecular weight substances; pH of a 1% (w/v) aqueous solution is ~5.2

Sigma A 4161 Bovine Lyophilized; essentially globulin free; cell culture tested

Sigma A 4503 Bovine Powder; ≥96% (electrophoresis); initial fractionation by cold alcohol precipitation; remainder mostly globulins; prepared by a modification of Cohn, using cold ethanol, pH, & low temperature precipitation; pH of a 1% (w/v) aqueous solution is ~5.2 | Cohn, EJ et al, *J Am Chem Soc*, 68: 459, 1946

Sigma A 4919 Bovine Powder; low endotoxin (endotoxin <0.1 ng/mg); cell culture tested

Sigma A 6793 Bovine Powder; ≥98% (electrophoresis); remainder mostly globulins; prepared by multi-temperature ethanol fractionation with extensive charcoal treatment, deionization & dialysis; pH of a 1% (w/v) aqueous solution is ~7

Sigma A 6918 Bovine Powder; ≥98% (electrophoresis); remainder mostly globulins; prepared by multi-temperature ethanol fractionation with extensive charcoal treatment, deionization & dialysis; pH of a 1% (w/v) aqueous solution is ~5.2

Sigma A 7030 Bovine Powder; ≥98% (electrophoresis); prepared from pasteurized bovine serum & further processed to be essentially fatty acid free (<0.02%); essentially γ-globulin free; pH of a 1% solution in 0.15 M NaCl is ~7 | Processed to reduce interference of T_3, T_4 & insulin in RIA assays; suitable as diluent in ELISA applications

Sigma A 7888 Bovine RIA grade; powder; ≥96% (electrophoresis); initial fractionation by cold alcohol precipitation; pH of a 1% aqueous solution is ~5.2 | Suitable for insulin RIA procedures; may yield an insulin blank of up to 0.1 μunit/mg in certain procedures

Sigma A 7906 Bovine Powder; ≥98% (electrophoresis); remainder mostly globulins; prepared with charcoal treatment & extensive dialysis to reduce low molecular weight substances; pH of a 1% solution in 0.15 M NaCl is ~7

Sigma A 8022 Bovine Powder; ≥96% (electrophoresis); remainder mostly globulins; prepared from pasteurized bovine serum; purified by heat treatment & organic solvent precipitation; pH of a 1% (w/v) aqueous solution is ~5.4

Sigma A 8412 Bovine 7.5% Solution; prepared in DPBS; sterile-filtered; endotoxin tested; cell culture tested

Sigma A 8806 Bovine Fatty acid free (FFA <0.005%); low endotoxin (endotoxin <0.1 ng/mg); cell culture tested

Sigma A 8918 Bovine 35% Solution; prepared in DPBS; sterile-filtered; endotoxin tested; cell culture tested

Sigma A 9306 Bovine Powder; ≥97%; essentially γ-globulin free; initial fractionation by salt fractionation; low endotoxin: ≤0.1 ng/mg of endotoxin detection; from bovine plasma produced in New Zeal &

Sigma A 9418 Bovine 96-99%; remainder mostly globulins; cell culture tested

Sigma A 9430 Bovine Powder; ≥98% (electrophoresis); initial fractionation by heat shock; low endotoxin: ≤1 ng endotoxin/mg

Sigma A 9543 Bovine Powder; ≥98% (electrophoresis); low endotoxin: ≤0.1 ng/mg; purified by heat treatment & organic solvent precipitation

Sigma A 9576 Bovine 30% Solution; prepared in DPBS; sterile-filtered; endotoxin tested; cell culture tested

Sigma A 9647 Bovine Powder; ≥96% (electrophoresis); remainder mostly globulins; purified by heat treatment & organic solvent precipitation; pH of a 1% (w/v) aqueous solution is ~7

Fluka 05473 Bovine serum MW ~67k ≥96% (gel electrophoresis); ≥95% protein; ≤3% water | Prepared by multi-temperature ethanol fractionation with extensive charcoal treatment, deionization & dialysis

Fluka 05475 Bovine serum MW ~67k RIA grade; ≥96% (gel electrophoresis); ≥90% protein; ≤3 water | Cohn, EJ et al, *J Am Chem Soc*, 68: 459, 1946

Fluka 05476 Bovine serum MW ~67k ≥96% (gel electrophoresis); ≥95% protein; ≤5% water | Prepared by a modified Cohn procedure using cold ethanol, pH & low temperature precipitation followed by pH adjustment prior to final drying

Fluka 05480 Bovine serum MW ~67k Lyophilized; ≥97% (gel electrophoresis); ≤3% water; ≤1% residue on ignition

Fluka 05481 Bovine serum MW ~67k ≥96% (gel electrophoresis); ≤3% water | Prepared from pasteurized bovine serum; initial fractionation by heat shock; purified by heat treatment & organic solvent precipitation

Fluka 05482 Bovine serum MW ~67k ≥96% (gel electrophoresis); ≥95% protein; ≤5% water | Prepared by a modified Cohn procedure using cold ethanol, pH & low temperature precipitation; Lepri, L et al, *Chromatographis*, 36: 297, 1993; Cohn, EJ et al, *J Am Chem Soc*, 68: 459, 1946

Fluka 05484 Bovine serum MW ~67k ≥96% (gel electrophoresis); ≤3% water | Prepared by a modified Cohn procedure using cold ethanol, pH & low temperature precipitation followed by pH adjustment prior to final drying; Sozuki, N, *JBC*, 263: 5037, 1988; Cohn, EJ et al, *J Am Chem Soc*, 68: 459, 1946

Fluka 05492 Bovine serum MW ~67k Low endotoxin; ≥75% (gel electrophoresis); ≥95% protein; ≤5% water; ≤1 U/mg endotoxin

ICN 151429 Bovine serum Protease free; >97% by electrophoresis

Fluka 05445 Canine serum ≥95% (gel electrophoresis); ≥95% protein; ≤5% water | *Merck*, 12: 8613

Sigma A 4662 Cat Powder

Sigma A 3014 Chicken Powder

Sigma A 5503 Chicken egg ≥98% (agarose electrophoresis); crystallized & lyophilized; essentially salt-free

Sigma A 9263 Dog Powder

Sigma A 5287 Donkey Powder

Sigma A 2514 Goat Powder

Fluka 05483 Guinea pig serum MW ~62k ≥96% (gel electrophoresis); ≤5% water | *Merck*, 12: 8613

Sigma A 5409 Hamster Powder

Amersham US10878 Human 96-99%

ICN 810171 Human 10% Solution; pH 7.1 ± 0.3, 0.05% NaN_3

ICN 823022 Human Lyophilized | Prepared by a modification of the Cohn procedure

ICN 823051 Human 30% Solution; pH 7.1 ± 0.3, 0.1% NaN_3

ICN 823234 Human Lyophilized; FA free | Prepared by a modification of the Chen low pH charcoal method; ~0.2 mg/g FA

Sigma A 1653 Human 96-99% albumin; powder; remainder mostly globulins; Cohn, EJ, *J Am Chem Soc*, 68: 459, 1946 | Source material has tested negative for HIV & HbsAg

Sigma A 2817 Human Powder; 96-99% albumin; similar to A 1653, but specifically for bulk usage | Source material has tested negative for HIV & HBsAg

ICN 191349 Human plasma Non-denatured; lyophilized, salt-free, homogeneous; >97% monomer purity; <3% α- or β-globulins (electrophoresis) | Prepared under proprietary non-denaturing conditions; negative for HBsAg & HIV Ab

ICN 823001 ICN 823002 Human plasma Protease, HRP & AP free | Used to eliminate background interference in ELISA or other enzyme assay systems; prepared from plasma so there is no discernable proteolytic, peroxidase or alkaline phosphatase activity by common assay methods

Fluka 05430 Human serum MW ~68k ≥85% (gel electrophoresis); ≤8% water; ≤3% residue on ignition | Domenici, E et al, *Chromatographia*, 29: 170, 1990; noctor, TAG & Wainer, IW, *Liq Chromat*, 16: 783, 1993

ICN 55858 Mouse Purified, lyophilized

Sigma A 3139 Mouse Powder

Fluka 05487 Ovine serum ≥95% (gel electrophoresis); ≥95% protein; ≤5% water

Sigma A 2764 Pig Powder; 96-99%; initial fractionation by cold alcohol precipitation

ICN 55864 Rabbit Purified, lyophilized

Sigma A 0639 Rabbit Powder

Fluka 05455 Rabbit serum MW ~68k ≥95% (gel electrophoresis); ≥90% protein; ≤10% water

ICN 824522 Rabbit serum Used in RIA & EIA systems requiring a homologous protein diluent

ICN 55869 Rat Purified, lyophilized

Sigma A 6272 Rat Powder

Fluka 05465 Rat serum MW ~67k ≥95% (gel electrophoresis); ≥90% protein; ≤10% water; ≤3% α-globulin; ≤2% β-globulin | Schnitzer, JE et al, *PNAS*, 85: 6773, 1988

Sigma A 4297 Rhesus monkey Powder

Sigma A 3264 Sheep Powder

Sigma A 4650 Turkey Powder

Sigma A 7269   Turkey egg   Crystallized & lyophilized; essentially salt-free; ~1% extraneous protein detected by electrophoresis on cellulose acetate in barbital buffer, pH 8.6, ionic strength 0.075; contains 8-12 moles mannose/mole albumin

Albumin, Fraction V Heat Shock Fraction

Synonyms: Bovine Serum Albumin

Fluka 05479   Bovine serum   MW ~67k   Protease free; ≥96% (gel electrophoresis); ≤5% water; ≤0.001% alkaline phosphatase, peroxidase; ≤0.00001% protease | Initial preparation by heat shock

Fluka 05488   Bovine serum   MW ~67k   ≥98% (gel electrophoresis); ≥95% protein; ≤3% water | Initial fractionation by heat shock; charcoal treatment & extensive dialysis to reduce low MW substances & a pH-adjustment prior to final drying; Pauly, DF & McMillin, JB, *JBC*, 263: 18160, 1988

Albumin, Fraction VI

Synonyms: Ovalbumin

Sigma A 2512   Chicken egg   ~99% (agarose electrophoresis); crystallized & lyophilized; essentially salt-free; a further purification of Sigma A 5503 to reduce mannose content; each mole of ovalbumin protein contains 5-6 moles of mannose as part of its native structure per Huang, CC et al, *Carbohyd Res*, 13: 127, 1970

Sigma A 5015   Turkey egg   Crystallized & lyophilized; essentially salt-free; ~1% extraneous protein detected by electrophoresis on cellulose acetate in barbital buffer, pH 8.6, ionic strength 0.075; contains 4-6 moles mannose/mole albumin | Ovalbumin

Albumin, Fraction VII

Synonyms: Ovalbumin

Sigma A 7641   Chicken egg   Crystallized & lyophilized; essentially salt-free; essentially free of *S*-Ovalbumin

Albumin, Fucosylamide

Synonyms: Albumin-1-Amido-1-Deoxy-L-Fucose

Sigma A 6033   Bovine   Lyophilized powder containing ~85% glycoprotein (Biuret); balance primarily citrate buffer salts; 15-25 moles monosaccharide/mole albumin; prepared from bovine albumin (A 7638) blocked by reductive amination with glyceraldehyde, coupled to fucosylamine by amidation

Albumin, Fucosylamide Biotin Conjugated

Sigma A 4042   Bovine   Biotin-labeled; lyophilized powder containing ~85% glycoprotein (Biuret); balance primarily citrate buffer salts; 3-6 moles biotin & 15-25 moles fucosylamide/mole BSA; prepared from bovine albumin (A 7638) by biotinylating using aminocaproyl spacer, blocking by reductive amination with glyceraldehyde, & coupling to fucosylamine by amidation

Albumin, Fucosylated 10 nm Colloidal Gold Conjugated

Synonyms: Bovine Serum Albumin

ICN 154035   Bovine serum

Albumin, Fucosylated 20 nm Colloidal Gold Conjugated

Synonyms: Bovine Serum Albumin

ICN 154036   Bovine serum

Albumin, Fucosylated 5 nm Colloidal Gold Conjugated

Synonyms: Bovine Serum Albumin

ICN 154034   Bovine serum

Albumin, Galactosamide

Synonyms: Albumin-2-Amido-2-Deoxy-D-Galactose

Sigma A 5908   Bovine   Lyophilized powder containing ~85% glycoprotein (Biuret); balance primarily citrate buffer salts; contains 15-25 moles monosaccharide/mole albumin; prepared from bovine albumin (A 7638) blocked by reductive amination with glyceraldehyde, coupled to galactosamine by amidation

Albumin, Galactosylated 10 nm Colloidal Gold Conjugated

Synonyms: Bovine Serum Albumin

ICN 154038   Bovine serum

Albumin, Galactosylated 20 nm Colloidal Gold Conjugated

Synonyms: Bovine Serum Albumin

ICN 154039   Bovine serum

Albumin, Galactosylated 5 nm Colloidal Gold Conjugated

Synonyms: Bovine Serum Albumin

ICN 154037   Bovine serum

Albumin, Galactosylated-β-o- 10 nm Colloidal Gold Conjugated

Synonyms: Bovine Serum Albumin

ICN 154041   Bovine serum

Albumin, Galactosylated-β-o- 20 nm Colloidal Gold Conjugated

Synonyms: Bovine Serum Albumin

ICN 154042   Bovine serum

Albumin, Galactosylated-β-o- 5 nm Colloidal Gold Conjugated

Synonyms: Bovine Serum Albumin

ICN 154040   Bovine serum

Albumin, Glucosamide

Synonyms: Albumin-2-Amido-2-Deoxy-D-Glucose

Sigma A 6158   Bovine   Lyophilized powder containing ~85% glycoprotein (Biuret); balance primarily citrate buffer salts; contains 10-20 moles monosaccharide/mole albumin; prepared from bovine albumin (A 7638) blocked by reductive amination with glyceraldehyde, coupled to glucosamine by amidation

Albumin, Glucosylated 10 nm Colloidal Gold Conjugated

Synonyms: Bovine Serum Albumin

ICN 154044   Bovine serum

Albumin, Glucosylated 15 nm Colloidal Gold Conjugated

Synonyms: Bovine Serum Albumin

ICN 154043   Bovine serum

Albumin, Glucosylated 20 nm Colloidal Gold Conjugated

Synonyms: Bovine Serum Albumin

ICN 154045   Bovine serum

Albumin, Glucosylated-β-o- 10 nm Colloidal Gold Conjugated

Synonyms: Bovine Serum Albumin

ICN 154047   Bovine serum

Albumin, Glycated

Synonyms: Human Serum Albumin

Sigma A 8426   Bovine   Lyophilized powder containing ~95% protein (Biuret); balance primarily citrate buffer salts; 1-2 moles hexose (as fructosamine)/mole albumin; glycated *in vitro* | Armbruster, DA, *Clin Chem*, 33: 2513, 1987; Furth, AJ, *Anal Biochem*, 175: 347, 1988

Sigma A 8301   Human   Lyophilized powder containing ~95% protein (Biuret); balance primarily citrate buffer salts; glycated *in vitro*; 1-5 moles hexose (as fructosamine)/mole albumin | Source material has tested negative for HIV & HbsAg;Armbruster, DA, *Clin Chem*, 33: 2513, 1987; Furth, AJ *Anal Biochem*, 175: 347, 1988

Fitzgerald 30-AA72   Human plasma   Immunogen grade

USBio A1327-53   Human serum   ≥97% (SDS-PAGE); free of nonglycated albumin; 1 mg/mL; lyophilized; 1 glyco group/mole albumin | Suitable for antigenic applications in immunological protocols

Albumin, Glycosylated

Synonyms: Human Serum Albumin

Biogenesis 0220-3056   Human serum   Purified; no preservatives; lyophilized | Free of un-reacted albumin; elevated levels is a marker for diabetes; Garlick & Mazer, *JBC*, 258:6142, 1983; Dolhofer & Wieland, , *FEBS Letts*, 103:282, 1983; Williams et al, *PNAS*, 78:2393, 1981

Albumin, Heat Shock Fraction

Synonyms: Bovine Serum Albumin

ICN 810661 ICN 810662 ICN 810663 ICN 810667   Bovine   Reagent grade; pH ~7.0

ICN 810682 ICN 810683 ICN 810684 ICN 810685   Bovine   Heat-shocked fractionate; low endotoxin

Sigma A 3803   Bovine   ≥98% (electrophoresis); initial fractionation by heat shock; processed to reduce fatty acid content to ~0.005%; pH of a 1% (w/v) aqueous solution is ~7 | Suitable as diluent in ELISA applications

Fluka 05477   Bovine serum   MW ~67k   Enzyme immunoassay grade; ≥98% (gel electrophoresis); ≥95% protein; ≤5% water; ≤0.02% fatty acid | Initial fractionation by heat shock; further purified to reduce fatty acids & interfering impurities for immunoassays

Albumin, Lactosyl

Sigma A 5783   Bovine   Lyophilized powder containing ~85% glycoprotein (Biuret); balance primarily citrate buffer salts; 10-20 moles disaccharide/mole albumin; prepared from bovine albumin (A 7638) & α-lactose coupled via reductive amination | Schwartz, BA & Gray, GR, *Arch Biochem Biophys*, 181: 542, 1977Albumin-*N*-1-(deoxylactitol)

Albumin, Lewis x-Conjugated

Synonyms: Bovine Serum Albumin

Calbiochem 434632   Bovine Serum   MW 66k (BSA), 529.5 (Lewis x)   Lyophilized solid averaging 10-12 carbohydrate moieties/molecule of protein | Neoglycoprotein containing 14-atom spacer; Lewis x is a human cancer marker & embryonic antigen

Albumin, Low Folate Vitamin B₁₂

Scipac P198-1   Serum/plasma   >99%; lyophilized; suitable for research and the most demanding applications | Serum protein

Albumin, Maltosyl

Synonyms: Albumin-*N*-1-(Deoxymaltitol)

Sigma A 5283   Bovine   Lyophilized powder containing ~85% glycoprotein (Biuret); balance primarily citrate buffer salts; 10-20 moles disaccharide/mole albumin; prepared from bovine albumin (A 7638) & maltose coupled via reductive amination | Schwartz, BA & Gray, GR, *Arch Biochem Biophys*, 181: 542, 1977

Albumin, Mannosylated 10 nm Colloidal Gold Conjugated

Synonyms: Bovine Serum Albumin

ICN 154072   Bovine serum

Albumin, Mannosylated 20 nm Colloidal Gold Conjugated

Synonyms: Bovine Serum Albumin

ICN 154073   Bovine serum

Albumin, Mannosylated 5 nm Colloidal Gold Conjugated

Synonyms: Bovine Serum Albumin

ICN 154071   Bovine serum

Albumin, Mannosylated-α-o- 10 nm Colloidal Gold Conjugated

Synonyms: Bovine Serum Albumin

ICN 154075   Bovine serum

Albumin, Mannosylated-α-o- 20 nm Colloidal Gold Conjugated

Synonyms: Bovine Serum Albumin

ICN 154076   Bovine serum

Albumin, Mannosylated-α-o- 5 nm Colloidal Gold Conjugated

Synonyms: Bovine Serum Albumin

ICN 154074   Bovine serum

Albumin, Methylated

Synonyms: Bovine Serum Albumin

Sigma A 1009   Bovine   Used with diatomaceous earth for the separation & identification of natural & derived nucleic acids using column chromatography; M Andell, JD & Hershey, AD, *Anal Biochem*, 1: 66, 1960

Fluka 05485   Bovine serum   ≥96% (gel electrophoresis); ≥90% protein; ≤10% water | Prepared from Fluka 05482 by esterification with methanol; M Andell, JD & Hershey, AD, *Anal Biochem*, 1: 66, 1960

Albumin, Modified Cohn Fraction V Heat Shock Fraction

ICN 840042 ICN 840043 ICN 840044 ICN 840045   Bovine   Powder; 96-99%

Albumin, Monomer

Synonyms: Bovine Serum Albumin

Sigma A 1900   Bovine albumin   Purified from bovine albumin (A 4503) as described for mercaptalbumin per Janatova, J et al, *J Biol Chem*, 243: 3612, 1968; lyophilized powder containing ~98% monomer

40

Albumin, Monomer Standard

ICN 810282 Bovine Meets specifications of the National Committee for Clinical Laboratory Standards, ASC-1 | Treated to remove high MW contaminants; recommended for use as a standard for Biuret, Folin-Lowry, gel permeation chromatography, amino acid analysis, PAGE & other test systems

ICN 810291 Bovine Meets specifications of the National Committee for Clinical Laboratory Standards, ASC-1; 7% sterile solution without preservatives, >98% monomer (HPLC); 2 mL/ampoule | Treated to remove high MW contaminants; recommended for use as a standard for Biuret, Folin-Lowry, gel permeation chromatography, amino acid analysis, PAGE & other test systems

ICN 810301 Bovine 7% sterile solution without preservatives, >98% monomer (HPLC) | Treated to remove high MW contaminants; recommended for use as a reference standard

Albumin, N-Acetylglucosaminylated 10 nm Colloidal Gold Conjugated
Synonyms: Bovine Serum Albumin

ICN 153990 Bovine Serum

Albumin, N-Acetylglucosaminylated 20 nm Colloidal Gold Conjugated
Synonyms: Bovine Serum Albumin

ICN 153991 Bovine Serum

Albumin, N-Acetylglucosaminylated- 5 nm Colloidal Gold Conjugated
Synonyms: Bovine Serum Albumin

ICN 153989 Bovine Serum

Albumin, N-Acetylglucosaminylated-β-o- 10 nm Colloidal Gold Conjugated
Synonyms: Bovine Serum Albumin

ICN 153993 Bovine Serum

Albumin, N-Acetylglucosaminylated-β-o- 20 nm Colloidal Gold Conjugated
Synonyms: Bovine Serum Albumin

ICN 153994 Bovine Serum

Albumin, N-Acetylglucosaminylated-β-o- 5 nm Colloidal Gold Conjugated
Synonyms: Bovine Serum Albumin

ICN 153992 Bovine Serum

Albumin, N-Acetyllactosaminylated-β-o- 10 nm Colloidal Gold Conjugated
Synonyms: Bovine Serum Albumin

ICN 153996 Bovine Serum

Albumin, N-Acetyllactosaminylated-β-o- 20 nm Colloidal Gold Conjugated
Synonyms: Bovine Serum Albumin

ICN 153997 Bovine Serum

Albumin, N-Acetyllactosaminylated-β-o- 5 nm Colloidal Gold Conjugated
Synonyms: Bovine Serum Albumin

ICN 153995 Bovine Serum

Albumin, N-Ac-β-D-Glucosamide Resorufin Labeled

Sigma A 6052 Bovine Resorufin labeled; lyophilized powder; essentially salt-free; contains ~30 moles N-acetylglucosamine/mole BSA | Gabius, H-J et al, *Anal Biochem*, 165: 349, 1987

Albumin, Naphthol Blue Black
Synonyms: Black Albumin

Sigma A 1777 Bovine BSA impregnated with naphthol blue-black dye; Amido Black 10B stained BSA; useful as a sedimentation marker for density gradient centrifugation; Kouvonen, I et al, *Anal Biochem*, 89: 306, 1978

Albumin, Partially Denatured
Synonyms: Human Serum Albumin

Biogenesis 0220-0804 Human serum Tested negative for HIV and HBsAg; <20 ppm heavy metals; affinity purified; from 0.02 *M* sodium phosphate, 0.14 *M* NaCl, pH 7.6; lyophilized

Albumin, Pre-
Synonyms: Transthyretin

Cortex CP1074 >95%

USBio P6000-10 Human ≥98% (SDS-PAGE); no contaminants detected; single band by SDS-PAGE, IEP, &/or RID; lyophilized in 50 *mM* Na$_3$PO$_4$, pH 7.5, 150 *mM* NaCl | Suitable for antigenic applications in immunological protocols

Biodesign A50185H Human plasma Purified

Fitzgerald 30-AP36 Human plasma High purity

Sigma P 7528 Human plasma Lyophilized powder; salt-free; ~95% by non-reducing polyacrylamide electrophoresis; E$_{280}$ at 1%=14.0 | Thyroxine binding prealbumin; Raz, A et al, *J Biol Chem*, 244: 12, 1969

Biogenesis 7600-0604 Human serum Lyophilized

Scipac P171-0 Serum/plasma >99%; lyophilized; from Scipac P171-1; recommended for use as an immunogen | Nutritional protein

Scipac P171-1 Serum/plasma >96%; lyophilized | Nutritional protein

Scipac P171-2 Serum/plasma 40-90%; lyophilized | Nutritional protein

Albumin, Rhodamine Conjugated

ICN 55897 Bovine Lyophilized, purified antigen

Albumin, RITC Conjugated

ICN 655011 Bovine Liquid

Albumin, Sialyl Lewis x Conjugated
Synonyms: Bovine Serum Albumin

Calbiochem 565951 MW 66k (BSA), 820.8 (Sialyl Lewis x) Lyophilized solid averaging 10-12 carbohydrate moieties/molecule of protein | Neoglycoprotein containing 3-atom spacer; for the study of inflammation & cell adhesion processes involving sialyl Lewis x & related analogs

Calbiochem 565952 MW 66k (BSA), 820.8 (Sialyl Lewis x) Lyophilized solid averaging 10-12 carbohydrate moieties/molecule of protein | Neoglycoprotein containing 14-atom spacer; for the study of inflammation & cell adhesion processes involving sialyl Lewis x & related analogs

Albumin, Standard
Synonyms: Bovine Serum Albumin

Fitzgerald 30-AB70 Bovine serum High purity; pH 7.0

Fitzgerald 30-AB71 Bovine serum High purity; pH 5.2

Sigma A 1533 Human 5 mL each of the following concentrations: 2, 4, 6, 8 & 10 g/dL albumin; albumin in 0.85% NaCl; 0.05% sodium azide added as preservative | Available as a set only & cannot be ordered separately

Albumin, Sulforhodamine 101 Acid Chloride

Synonyms: Texas Red-Albumin

Sigma A 2164   Bovine albumin   Prepared from crystallized & lyophilized bovine albumin (A 4378); lyophilized powder; salt-free; contains ~2-3 moles sulforhodamine 101 acid chloride/mole protein

Albumin, TRITC Conjugated

Synonyms: Bovine Serum Albumin

Sigma A 2289   Bovine albumin   Prepared from crystallized & lyophilized bovine albumin (A 4378); lyophilized powder; salt-free; contains 1 mole TRITC/mole protein

Albumin, α-D-Galactopyranosylphenyl Isothiocyanate FITC Conjugated

Sigma A 2420   Bovine   Lyophilized powder; salt-free; contains 15-20 moles α-D-galactopyranose & 1-3 moles fluorescein/mole albumin; Monsigny, M et al, *Biol Cell*, 51: 187, 1984

Albumin, α-D-Galactopyranosylphenyl Isothiocyanate TRITC Conjugated

Sigma A 6544   Bovine   Lyophilized powder; essentially salt-free; contains 15-25 moles α-D-galactopyranose & ~1 mole TRITC/mole protein | Monsigny, M et al, *Biol Cell*, 51: 187, 1984

Albumin, α-D-Glucopyranosylphenyl Isothiocyanate FITC Conjugated

Sigma A 5543   Bovine   Lyophilized powder; essentially salt-free; contains 15-20 moles α-D-glucopyranose & 2-3 moles FITC /mole protein | Monsigny, M et al, *Biol Cell*, 51: 187, 1984

Albumin, α-D-Mannopyranosylphenyl Isothiocyanate

Sigma A 8303   Bovine   Lyophilized powder; salt-free; contains 15-25 moles α-D-mannopyranose/mole albumin | Monsigny, M et al, *Biol Cell*, 51: 187, 1984

Albumin, α-D-Mannopyranosylphenyl Isothiocyanate Biotin Conjugated

Sigma A 7924   Bovine   Lyophilized powder; salt-free; contains 15-20 moles α-D-mannopyranose & ~3 moles biotin/mole albumin | Monsigny, M et al, *Biol Cell*, 51: 187, 1984

Albumin, α-D-Mannopyranosylphenyl Isothiocyanate FITC Conjugated

Sigma A 7790   Bovine   Lyophilized powder; salt-free; contains 15-20 moles α-D-mannopyranose & 1.5-3.0 moles fluorescein/mole albumin | Monsigny, M et al, *Biol Cell*, 51: 187, 1984

Albumin, α-D-Mannopyranosylphenyl Isothiocyanate TRITC Conjugated

Sigma A 7915   Bovine   Lyophilized powder containing ~95% glycoprotein (Lowry); balance primarily sodium carbonate buffer salts; 15-20 moles α-D-mannopyranose & ~1 mole TRITC/mole protein | Monsigny, M et al, *Biol Cell*, 51: 187, 1984

Albumin, α-L-Fucopyranosylphenyl Isothiocyanate FITC Conjugated

Sigma A 5793   Bovine   Lyophilized powder; essentially salt-free; contains 15-20 moles α-L-fucopyranose & ~2-3 moles FITC/mole protein | Monsigny, M et al, *Biol Cell*, 51: 187, 1984

Albumin, α-L-Fucopyranosylphenyl Isothiocyanate TRITC Conjugated

Sigma A 5918   Bovine   Lyophilized powder containing ~95% glycoprotein (Lowry); balance primarily sodium carbonate buffer salts; contains 15-25 moles α-L-fucopyranose & ~1 mole TRITC/mole protein | Monsigny, M et al, *Biol Cell*, 51: 187, 1984

Albumin, β-D-Galactopyranosylphenyl Isothiocyanate FITC Conjugated

Sigma A 8165   Bovine   Lyophilized powder; essentially salt-free; contains 15-20 moles β-D-galactopyranose & 2-3 moles FITC/mole protein | Monsigny, M et al, *Biol Cell*, 51: 187, 198

Albumin, β-D-Galactopyranosylphenyl Resorufin Conjugated

Sigma A 6177   Bovine   Resorufin labeled; lyophilized powder; essentially salt-free; contains ~30 moles α- galactose/mole BSA | Gabius, H-J et al, *Anal Chem*, 165: 349, 1987

Albumin, β-D-Glucopyranosylphenyl Isothiocyanate FITC Conjugated

Sigma A 7172   Bovine   Lyophilized powder; essentially salt-free; contains 15-20 moles β-D-glucopyranose & 2-3 moles FITC/mole protein | Monsigny, M et al, *Biol Cell*, 51: 187, 1984

Albumin, β-D-Xylopyranosylphenyl Isothiocyanate FITC Conjugated

Sigma A 3955   Bovine   Lyophilized powder containing ~95% glycoprotein (Lowry); balance primarily sodium carbonate buffer salts; 15-20 moles β-D-xylopyranose & 2-3 moles fluorescein/mole albumin | Monsigny, M et al, *Biol Cell*, 51: 187, 1984

Albumin, β-Lactosylphenyl Isothiocyanate Biotin Conjugated

Sigma A 7799   Bovine   Lyophilized powder containing ~15-20 moles disaccharide & 2-3 moles biotin/mole albumin | Monsigny, M et al, *Biol Cell*, 51: 187, 1984

Albumin, β-Lactosylphenyl Isothiocyanate FITEC Conjugated

Sigma A 8040   Bovine   Lyophilized powder; salt-free; contains 15-20 moles β-lactopyranose & 1.5-2 moles fluorescein/mole albumin | Monsigny, M et al, *Biol Cell*, 51: 187, 1984

Albumin, β-Lactosylphenyl Isothiocyanate TRITC Conjugated

Sigma A 7665   Bovine   Lyophilized powder; essentially salt-free; contains 15-25 moles β-D- lactopyranose & ~1 mole TRITC/mole protein | Monsigny, M et al, *Biol Cell*, 51: 187, 1984

Albumin-*p*-Aminophenyl-*N*-Ac-β-D-Galactosaminide

Synonyms: Bovine Serum Albumin

Sigma A 1159   Bovine albumin   Lyophilized powder containing ~90% glycoprotein (Biuret); balance primarily Tris-citrate buffer salts; 15-25 moles monosaccharide/mole albumin | Prepared from bovine albumin (A7638) blocked by reductive amination with glyceraldehyde, coupled to *p*-aminophenyl-*N*-acetyl-β-D-galactosaminide by amidation

Albumin-*p*-Aminophenyl-*N*-Ac-β-D-Glucosaminide

Synonyms: Bovine Serum Albumin

Sigma A 1034   Bovine albumin   Lyophilized powder containing ~90% glycoprotein (Biuret); balance primarily Tris-citrate buffer salts; 15-25 moles monosaccharide/mole albumin | Prepared from bovine albumin (A7638) blocked by reductive amination with glyceraldehyde, coupled to *p*-aminophenyl-*N*-acetyl-β-D-glucosaminide by amidation

Albumin-*p*-Aminophenyl-α-D-Mannopyranoside

Sigma A 4664   Bovine   Lyophilized powder containing ~90% glycoprotein (Biuret); balance primarily citrate buffer salts; 20-30 moles monosaccharide/mole albumin; prepared from bovine albumin (A7638) blocked by reductive amination with glyceraldehyde, coupled to *p*-aminophenyl-α-D-mannopyranoside by amidation

Alcohol Dehydrogenase

Sigma A 2190　Horse liver　MW 39.8k　Vial contains enough FITC-conjugated protein to run 50 mini-gels or 25 standard size gels; protein band be visualized by using UV light or Brilliant Blue stain | Fluorescent marker for SDS-page & protein transfer; suitable for use as molecular weight standards in both SDS-PAGE & transfer membranes

Alcohol Dehydrogenase, Carbonic Anhydrase, Catalase, Lysozyme, Phosphorylase B, Trypsin Inhibitor

Sigma SDS-6B　Contains six biotinylated proteins (~0.1 mg total): lysozyme, phosphorylase b, catalase, carbonic anhydrase, alcohol dehydrogenase & trypsin inhibitor; ~33% Protein, 33% NaCl & 33% sucrose | Molecular weight standard mixture; sDS molecular weight standard mixture

Allophycocyanin

ICN 150274　Algae　Purified; 100 mM NaPO₄, 60% (NH₄)₂SO₄, 0.02% NaN₃ | Bousiba, S & A Richmond, *Arch Microbiol*, 120:155-159, 1979

Amino Acid Oxidase, Apo-D-

Biogenesis 0380-0206　Porcine kidney　MW 38-39k (monomer)　Purified; lyophilized | Tu et al, *Arch Biochem Biophys*, 159:889 1973; Parkin & Hultin, *Biotech & Bioeng XXI*, 939, 1979

Amino Acid Oxidase, D-

Biogenesis 0380-0104　Porcine kidney　MW 38-39k Purified; 3.2M ammonium sulphate, pH 7.0; suspension | *Arch Biochem Biophys*, 159:889, 1973

Amphiregulin

Oncogene PF047　Human recombinant　MW 11k　>97% (SDS-PAGE); lyophilized; biological activity: EC₅₀ of 5-15 ng/mL as determined by the ability to stimulate the proliferation of Balb/3T3 cells | Species reactivity: human; for proliferation studies; 98 amino acid protein

ICN 195724　Human recombinant, expressed in *E. coli* Lyophilized; ≥97% | BSA used as a carrier protein

Amphiregulin Epidermal Growth Factor

Synonyms: Keratinocyte Autocrine Factor

Sigma A 7080　Human recombinant, expressed in *E. coli*　MW 18k　≥97% (SDS-PAGE); 0.2 μm filtered & lyophilized from PBS containing 500 μg BSA; proliferative activity is tested in culture by using a mouse fibroblast cell line BALB/3T3; endotoxin tested; see Sigma E 1264 | Glycosylated polypeptide originally isolated from the media of phorbol ester treated MCF-7 human breast carcinoma cells; stimulates the growth of normal epithelial cells, fibroblasts & keratinocytes though inhibits the growth of several aggressive tumor cell lines such as HTB-10 neuroblastoma cells, A431 epidermoid & HTB-132 breast carcinoma cells; Marquardt, H et al, *Science*, 223: 1079, 1984

Amyloglucosidase

Sigma A 2910　*Aspergillus niger*　pI 3.6; vial contains ~2 mg | IEF Marker

Amyloid A

Biogenesis 0490-1802　Mouse acute phase serum　MW 12k Contaminants: mainly Apolipoprotein A1; semi-pure; PBS buffer, pH 7.3, 0.1% NaN₃; liquid

Amyloid A, Apo-Serum

Synonyms: Serum Amyloid A, Apo-

Biogenesis 0490-1739　Human r-DNA *E. coli*　MW 11.7k Purified; PBS buffer, pH 7.2, 0.1% NaN₃; liquid

PeproTech 300-13　Human recombinant, expressed in *E. coli* MW 11.7k　>98%; 104 aa; lyophilized from 3 mM Tris pH 7.6; exerts its biological activity in the concentration range of 10.0-50.0 μg/mL

Amyloid P

Alexis 200-007　Human plasma　MW ~25-28k　>95% (SDS-PAGE & cellulose acetate electrophoresis); 500 μg protein (Lowry & Pierce); liquid in PBS containing 10 mM sodium azide; purified by affinity chromatography & ion exchange chromatography | From citrated human plasma converted to serum by kaolin

Amyloid P, Serum

Biogenesis 0490-2752　Human serum　MW 25-28k (monomeric)　Tested negative for HIV and HBsAg antibodies; purified from PBS, with 10 mM NaN₃; lyophilized | Pepys et al, *Ann NY Acad Sci*, 389:286, 1982

Angiogenin

Synonyms: Growth Factor

Oncogene PF075　Human recombinant　MW 14k (non-glycosylated form)　>97% (SDS-PAGE); lyophilized; biological activity: 1 μg of recombinant human angiogenin produces an absorbance change at 260 nm of ~2.0-3.0 based on a ribonucleolytic assay using yeast tRNA | Species reactivity: yeast, human

ICN 195723　Human recombinant, expressed in *E. coli* Lyophilized, ≥97%; activity based on ribonucleolytic activity vs. yeast tRNA | ANG

Sigma A 6955　Human recombinant, expressed in E. *coli*　MW 14.4k　≥97% (SDS-PAGE); 0.2 μm filtered & lyophilized from phosphate saline containing 2.5 mg BSA; endotoxin tested; cell culture tested | A single chain polypeptide with an isoelectric point of pI>9.5angiogenesis is the formation of blood vessels or capillaries from existing blood vessels that occurs in response to specific signal; angiogenin stimulates capillary & umbilical vein endothelial cells to produce diacylglycerol; member of the ribonuclease superfamily; ribonucleolytic activity of angiogenin toward most RNase A substrates is lower than that of RNase A; however, the ribonucleolytic activity of angiogenin is essential to its angiogenic activity since inhibition of the angiogenin RNase activity inhibits the angiogenic activity; also important in tumor growth, where the capillary network formed by the tumor increases the blood-born nutrients to the tumor that allow it to growFett, J et al, *Biochemistry*, 24: 5480, 1985; Folkman, J et al; in: *International Review of Experimental Pathology*, Richter, G (ed), Academic Press, New York, p. 207, 1976

Angiostatin

Synonyms: Plasminogen, Human; Human plasminogen

Biodesign A86885H　>98%

Oncogene 176700　MW 50k　Lyophilized from 20 mM NaCl; ≥98% (SDS-PAGE); no plasmin, plasminogen detected; soluble in water; prepared from fluid shown by certified tests to be negative for HBsAg & HIV & HCV antibodies | Proteolytic fragment of plasminogen; a specific inhibitor of endothelial cell proliferation & one of the most potent & specific natural inhibitors of angiogenesis & metastatic tumor growth; significantly inhibits bFGF-induced endothelial cell proliferation & migration at concentrations ranging from 300 nM-1.0 μM; Sim, BK et al, *Cancer Res*, 57: 1329, 1997; Wu, Z e al, *BBRC*, 236: 651, 1997; Cao, Y et al, *JBC*, 271: 29461, 1996; O'Reilly, MS et al, *Cell*, 79: 315, 1994

Calbiochem 176700　Human　MW 50k　Lyophilized from 20 mM NaCl; ≥98% (SDS-PAGE); plasmin, plasminogen: none detected; soluble in water; prepared from fluid has been shown to be negative for HBsAg & for antibodies to HIV & HCV | A proteolytic fragment of plasminogen; a specific inhibitor of endothelial cell proliferation & one of the most potent & specific natural inhibitors of angiogenesis & metastatic tumor growth; human angiostatin significantly inhibits bFGF-induced endothelial cell proliferation & migration at concentrations ranging from 300 nM-1.0 μM; Sim, BK et al, *Cancer Res*, 57: 1329, 1997; Wu, Z et al, *Biochem Biophys Res Com*, 236: 651, 1997; Cao, Y et al, *J Biol Chem*, 271: 29461, 1996; O'Reilly, MS et al, *Cell*, 79: 315, 1994

Biogenesis 0559-5050   Human plasma   MW ~40k   Tested negative for antibodies to HIV I, HIV-II, HCV, and *T. pallidum*, and HBsAg; purified; liquid

Angiotensin Converting Enzyme

Chemicon AG761   Human   ≥95% | Cellular biochemistry/regulatory protein used in immunoblotting (Western)

Chemicon AG782   Rat   ≥95% | Cellular biochemistry/regulatory protein used in immunoblotting (Western)

Angiotensinogen

Synonyms: Renin Substrate

Biodesign A50108H   Human plasma   Purified

Sigma A 2562   Human plasma   >95% (SDS-PAGE); lyophilized powder containing sodium phosphate buffer salts, pH 6.8; protein determined by ($E_{280}^{1\%}$) | Tewksbury, DA, *Fed Proc*, 42: 2724, 1983

Sigma A 2283   Porcine plasma   Lyophilized powder containing ~65% protein (Modified Warburg-Christian); balance primarily sodium citrate; activity: 1,500-3,000 units/g angiotensinogen protein; unit definition: 1 unit yields 1.0 nmole angiotensin I in the presence of renin at pH 6.0 at 37°C measured by RIA | Suitable for measuring renin by RIA

Annexin

Sigma A 2699   Bovine liver   MW 32.5k & 35k   Lyophilized from solution containing dithiothreitol; affinity purified | Fauvel, J et al, *FEBS Lett*, 216: 45, 1987

Sigma A 2824   Bovine liver   MW 67k   Lyophilized from solution containing dithiothreitol; affinity purified | Fauvel, J et al, *FEBS Lett*, 216: 45, 1987

Sigma A 2449   Bovine lung   MW 36k   Lyophilized from solution containing dithiothreitol; affinity purified | Gerke, V & Weber K, *EMBO J*, 3: 227, 1984; Glenney, JB Jr et al, *J Cell Biol*, 104: 503, 1987; Pepinsky, RB et al, *J Biol Chem*, 263: 10799, 1988

Sigma A 2574   Bovine lung   MW 32.5k & 35k   Lyophilized from solution containing dithiothreitol; affinity purified | Gerke, V & Weber K, *EMBO J*, 3: 227, 1984; Glenney, JB Jr et al, *J Cell Biol*, 104: 503, 1987; Pepinsky, RB et al, *J Biol Chem*, 263: 10799, 1988

Annexin I

Biodesign A80108B   Bovine lung   Purified

Annexin II

Biodesign A80109B   Bovine lung   Purified

Annexin III

Biodesign A80110B   Bovine lung   Purified

Annexin IV

Biodesign A80111B   Bovine liver   Purified

Annexin V

Synonyms: PAP-1; Calphosbindin I; Lipocortin V; Annexin V, rh

Sigma A 9460   Human placenta   MW 33k   ≥90% (SDS-PAGE); solution in 40 *mM* Tris-HCl, pH 7.5, containing 150 *mM* NaCl, 1 *mM* DTT & 0.05% sodium azide | Buhl, W-J et al, *Eur J Cell Biol*, 56: 381, 1991; Schlaepfer, DD et al, *Biochemistry*, 31: 1886, 1992

Alexis 201-018   Human recombinant, expressed in *E. coli*   MW 35.8k   Lyophilized; >98% (SDS-PAGE & HPLC); reconsitituted solution contains 20 *mM* phosphate buffer, pH 7.0, 0.02% Tween 80, 130 *mM* arginine HCl

Alexis BMS306/a   Recombinant expressed in *E. coli*   MW 35.8k   >98% (SDS-gel electrophoresis & reverse phase HPLC); 30 μg lyophilized powder; reconsitituted Solution contains 20 *mM* phosphate buffer, 0.02% Tween 80, 130 *mM* arginine hydrochloride, pH 7.0 | Exhibits anti-phospholipase activity & binds to phosphatidylserine

Annexin V, APC Conjugated

Synonyms: Annexin V, APC Conjugated rh-

Alexis 209-252   Human recombinant, expressed in *E. coli*   MW 35.8k   ≥98% (SDS-PAGE & HPLC); liquid containing 50 *mM* TRIS, 100 *mM* NaCl, 1% BSA & 0.02% sodium azide, pH 7.4 | APC has an excitation min of 650 nm & an emission max of 660 nm; suitable for dual staining experiments to detect phosphatidylserine whilst cell-surface protein can be detected with a suitable MAb (FITC-labeled or using an appropriately labeled secondary antibody)

Alexis BMS306APC/a   Recombinant expressed in *E. coli*   MW 35.8k   >98% (SDS-gel electrophoresis & reverse phase HPLC); 30 μg lyophilized powder; reconsitituted solution contains 20 *mM* phosphate buffer, 0.02% Tween 80, 130 *mM* arginine hydrochloride, pH 7.0 | APC shows excitation maxima of 564 (495) nm & 650 nm & emission maxima of 576 nm & 660 nm, respectively; useful detection of phosphatidylserine; allows simple secondary labeling by staining, eg, the membrane surface proteins with a monoclonal antibody for further cellular characterization

Annexin V, Biotin Conjugated

Synonyms: Annexin V, Biotin Conjugated rh-

Sigma A 7810   3-6 moles biotin/mole annexin V; solution in 50 *mM* Tris-HCl, pH 7.5, containing 100 *mM* NaCl | Useful for detection of apoptotic cells in conjunction with streptavidin coupled to alkaline phosphatase, peroxidase, or FITC

Alexis 209-002   Human recombinant, expressed in *E. coli*   >98% (SDS-PAGE & HPLC); liquid containing 50 *mM* TRIS, 100 *mM* NaCl, 1% BSA & 0.02% sodium azide, pH 7.4 | Flow cytometry applications; Andree, HA et al, *J Biol Chem*, 265: 4923, 1990; Fadok, VA et al, *J Immuno*, 148: 2207, 1992; Koopman, G et al, *Blood*, 84: 1415, 1994; Homburg, CH et al, *Blood*, 85: 532, 1995; Vermes, I et al, *J Immunol Meth*, 184: 39, 1995; Martin, S et al, *J Exp Med*, 182: 1545, 1995; Rovere, P et al, *J Immunol*, 156: 4631, 1996; Boersma, AWM et al, *Cytometry*, 24: 123, 1996

Alexis BMS306BT/a   Recombinant expressed in *E. coli*   MW 35.8k   >98% (SDS-gel electrophoresis & reverse phase HPLC); 30 μg lyophilized powder; reconsitituted solution contains 20 *mM* phosphate buffer, 0.02% Tween 80, 130 *mM* arginine hydrochloride, pH 7.0 | See Alexis BMS306FI/a

Annexin V, Cy3.18 Conjugated

Sigma A 4963   Cy3.18 conjugate; contains 1-3 moles Cy3/mole annexin V; solution in 50 *mM* Tris-HCl, pH 7.5, containing 100 *mM* NaCl | Useful for detection of apoptotic cells in conjunction with streptavidin coupled to alkaline phosphatase, peroxidase, or FITC

Annexin V, FITC Conjugated

Synonyms: Annexin V, FITC Conjugated rh-

Sigma A 9210   1-2 moles FITC/mole annexin V; ~50 μg/mL in 50 *mM* Tris-HCl, pH 7.5, containing 100 *mM* NaCl | Can be used for detection of apoptotic cells by flow cytometry

Alexis 209-250   Human recombinant, expressed in E. *coli*   MW 35.8k   >98% (SDS-gel electrophoresis & HPLC); liquid containing 50 *mM* TRIS, 100 *mM* NaCl, 1% BSA & 0.02% sodium azide, pH 7.4 | Flow cytometry applications; member of a family of proteins that are structurally related & exhibit Ca^{2+}-dependent phospholipid-binding properties; binds to various phospholipid species & shows its highest specificity for phosphatidylserine; situated on the inner leaflet of the plasma membrane; when cell death occurs, phosphatidylserine is translocated in the outer layer of the membrane, ie. the external surface of the cell; this occurs in the early phases of apoptosis during which the cell membrane itself remains intact; FITC-labeled annexin V can easily be used for the quantification of apoptotic cells; in contrast to apoptosis, necrosis is accompanied by the loss of cell membrane integrity & leakage of cellular constituents into the environment; therefore, the measurement of annexin V binding, performed simultaneously with a dye exclusion test using propidium iodide is a perfect assay to detect apoptotic cells & to discriminate between apoptosis & necrosis; thus, annexin V FITC represents a fast, simple & reliable method for the detection & quantification of apoptotic cells on a single cell basis; loss of translocation of phosphatidylserine seems to be a universal phenomenon of apoptosis; thus likely that apoptotic cells of all cell types can be quantified by staining with annexin V FITC & propidium iodide; see Alexis 209-002

Alexis BMS306FI/a   Recombinant expressed in E. *coli*   MW 35.8k   >98% (SDS-gel electrophoresis & reverse phase HPLC); 30 μg lyophilized powder; reconstituted solution contains 20 *mM* phosphate buffer, 0.02% Tween 80, 130 *mM* arginine hydrochloride, pH 7.0 | Exhibits anti-phospholipase activity & binds to phosphatidylserine; useful for detecting apoptotic cells by flow cytometry; Boersma, AWM et al, *Cytometry*, 24: 123, 1996; Homburg, CHE et al, *Blood*, 85: 532, 1995; Koopman, G et al, *Blood*, 84: 1415, 1995; Martin, SJ et al, *J Exp Med*, 182: 1545, 1995; Rovere P et al, *J Immunol*, 156: 4631, 1996; Vermes, I et al, *J Immunol Methods*, 184: 39, 1995

Annexin V, PE Conjugated

Synonyms: Annexin V, PE Conjugated rh-

Alexis BMS306PE/a   Recombinant expressed in E. *coli*   MW 35.8k   >98% (SDS-gel electrophoresis & reverse phase HPLC); 30 μg lyophilized powder; reconstituted solution contains 20 *mM* phosphate buffer, 0.02% Tween 80, 130 *mM* arginine hydrochloride, pH 7.0 | PE shows excitation maxima of 564 (495) nm & 650 nm & emission maxima of 576 nm & 660 nm, respectively; useful detection of phosphatidylserine; allows simple secondary labeling by staining, eg, the membrane surface proteins with a monoclonal antibody for further cellular characterization

Annexin V, R-PE Conjugated

Alexis 209-251   Human recombinant, expressed in E. coli   MW 35.8k   ≥98% (SDS-gel electrophoresis & HPLC); liquid containing 50 *mM* TRIS, 100 *mM* NaCl, 1% BSA & 0.02% sodium azide, pH 7.4 | R-PE has an excitation min of 564 nm & an emission max of 576 nm; suitable for dual staining experiments to detect phosphatidylserine whilst cell-surface protein can be detected with a suitable MAb (FITC-labeled or using an appropriately labeled secondary antibody)

Annexin VI

Biodesign A80113B   Bovine liver   Purified

Anthopleura Toxin A

Sigma T 3529   *Anthopleura xanthogrammica*

Antichymotrypsin, αI-

Biodesign A50104H   Human   Purified

USBio A2298-03   Human   ≥98%; no contaminants detected; single band by SDS-PAGE, IEP, &/or RID; lyophilized from 0.02 *M* Tris, 0.2 *M* NaCl, pH 7.5 | Suitable for antigenic applications in immunological protocols

USBio A2298-05   Human   ≥98%; no contaminants detected; single band by SDS-PAGE, IEP, &/or RID; lyophilized from 0.02 *M* Tris, 0.2 *M* NaCl, pH 7.5 | Suitable for antigenic applications in immunological protocols

Biogenesis 0600-8004   Human plasma   MW 68k   Tested negative for HBsAg, HCV and HIV antibodies; purified; from 37.5 μl of 150 *mM* NaCl, 20 *mM* TRIS-HCl, pH 8.0; lyophilized | Inhibitor of Chymotrypsin and Cathepsin G SA; Travis et al, *Biochem*, 17:5647, 1979

ICN 191347   Human plasma   MW 68k   Salt-free, lyophilized; negative for HBsAg & HIV Ab; 95-100% inhibitory activity; >99% (SDS-PAGE), single arc by IEP | Travis, J, D Garner & J Bowen, *Biochemistry*, 17:5647, 1978; Travis, J & GS Salvesen, *Annu Rev Biochem*, 42:655, 1983

ICN 770941 ICN 770942 ICN 770943   Human plasma   98% | Excellent for immunization

Cortex CP3002   Plasma   >95%

Scipac P159-1   Pooled serum/plasma   >96%; lyophilized; available on request | Acute phase serum protein

Scipac P159-5   Pooled serum/plasma   ~90%; frozen in TRIS buffer | Acute phase serum protein; binds PSA

Antigen II, Non-Specific Cross Reacting

Biogenesis 6882-5107   Human meconium   Tested negative for HBsAg, antibodies to HCV and HIV-1/2; purified; 1 *mM* Tris-HCl, pH 7.0; liquid

Antigen S-100

Cortex CP1079U   Human brain   >98%

Antiplasmin, αII-

Biogenesis 0620-4802   Human serum   MW 70k   SA: 5 IU/mg; tested negative for HBsAg and HIV antibodies; 1 mg Tris-Sodium citrate and 1.5 mg NaCl, pH 7.5; lyophilized; purified | Saito et al, Circ Res, 34:641, 1974

Cortex CP3004   Plasma   >95%

Antithrombin III

USBio A2298-25   Human   ≥98%; no contaminants detected; single band by SDS-PAGE, IEP, &/or RID; 5 mg/vial supplied in 20 *mM* Tris HCl, 0.1 *M* sodium citrate, 0.15 *M* NaCl, pH 8.3 | Suitable for antigenic applications in immunological protocols

Biodesign A50109H   Human plasma   Purified | Platelets & hemostasis reagents

ICN 153572   Human plasma   MW 65k   >95%; lyophilized, 50 *mM* Tris-HCl, pH 8, 0.15M NaCl | Wide-spectrum inhibitor; Rosenburg, RD & PS Damus, *JBC*, 248:6490, 1973

ICN 194187   Human plasma   Highly purified; <0.02 PEU antigenic heparin cofactor II | Wide-spectrum inhibitor; Rosenburg, RD & PS Damus, *JBC*, 248:6490, 1973

ICN 194936   Human plasma   50% Gycerol/H_2O | Important serine protease inhibitor in the coagulation cascade; Rosenburg, RD et al, *J Clin Invest*, 74:1, 1984

Biogenesis 0620-7002   Human serum   MW 65k   Tested negative for HBsAg, HCV and HIV antibodies; 20 *mM* Tris/HCl, 100 *mM* sodium citrate, 150 *mM* NaCl, pH 8.3; liquid | Single band by SDS-PAGE

Cortex CP3007U   Plasma   >98%

Scipac P214-1   Plasma   >96%; lyophilized | Hemostasis protein

Antitrypsin, αI-

ICN 191346   Human plasma   MW 52k   >95% (SDS-PAGE); lyophilized, 1 mg protein, 30 *mM* $NaPO_4$ buffer, 300 *mM* NaCl | Prepared from plasma negative for HBsAg & HIV Ab; Travis, J & Salvesen, GS, *Annu Rev Biochem*, 42:655, 1983

USBio A2298-32   Human plasma   ≥95%; lyophilized in 644 μL 30 *mM* sodium phosphate, 300 *mM* NaCl | Suitable for antigenic applications in immunological protocols

Biogenesis 0640-5604   Human serum   MW 54k   Tested negative for HBsAg, HIV I and II antibodies and HCV antibodies; purified; from 20 *mM* ammonium bicarbonate buffer; lyophilized

Cortex CP3003   Plasma   >96%

APO-1/Fas

Synonyms: sAPO-1/Fas; APO-1/Fas:Fc-IgG Fusion Protein

Alexis BMS314   Human recombinant, expressed in human embryo kidney cells   MW 60k   >95% (SDS-PAGE); 50 μg lyophilized powder | The extracellular domain of human APO-1/Fas (AA 1-154) is fused to the Fc portion of human IgG1; inhibits the activity of APO-1/Fas ligand of human & mouse; inhibits soluble APO-1/Fas ligand-mediated lysis of APO-1 sensitive cells; MW measured under reducing conditions

APO-1/Fas Ligand

Synonyms: sAPO-1/Fas Ligand

Alexis BMS309   Human recombinant, produced in human embryo kidney cells   >95% (SDS-PAGE); 5 μg lyophilized powder at 0.1 mg/mL | MW 32k (nonglycosylated), 35k (glycosylated) under reducing conditions; the extracellular domain of human APO-1/Fas ligand (AA 103-281 is fused at the *N*-terminus to a 26 AA linker protein & tag; glycosylation of recombinant form is similar or identical to natural form; recombinant human soluble APO-1/Fas ligand recognizes the APO-1Fas receptor of human, mouse & rat; a 40k type II transmembrane protein belonging to the TNF family; interaction between APO-1/Fas ligand & APO-1/Fas induces apoptosis of APO-1/Fas sensitive cells at a concentration ≥50 ng/mL, e.g. on A20 B lymphoma cells; implicated in CTL-mediated–killing, activation-induced cell death, creation of immune-privileged sites & tissue homeostasis; the extracellular part of APO-1/Fas ligand can be cleaved off by a metalloprotease, generating soluble APO-1/Fas ligand

Apoferritin

ICN 100260   Horse spleen   ≥40 mg/mL 0.15*M* NaCl | Prepared from crystalline, Iron free, cadmium-free ferritin; Criehton, RR, *Structure & Bonding*, 17:67, 1974

Apolactoferrin

Synonyms: Lactoferrin, Iron Depleted

Sigma A 1835   Human milk   ~90% (SDS-PAGE); lyophilized powder; salt-free

Apolipoprotein AI

Cortex CP1061   >95%

Cortex CP2012   >95%

Sigma A 4422   Bovine plasma   Lyophilized from 0.01 M ammonium bicarbonate; ~95% (SDS-PAGE) | Apolipoprotein A-I & A-II are directly purified from HDL; apolipoprotein B is prepared from purified LDL using SDS or sodium deoxycholate to remove lipids; Burstein, M, *J Lipid Res*, 11: 583, 1970; Rudel, LL, *Biochem J*, 139: 89, 1974

ICN 59414   Human   Purified Ag; lyophilized

Biogenesis 0650-0311   Human plasma   MW 28.4k   Tested negative for HIV-1, HIV-2, HCV antibodies and HBsAg; purified; 0.1 M Na-hydrocarbonate; lyophilized | Segret & Albers , *Meth Enzymol*, 128, Academic Press, 1986; Fielding et al, *BBRC*, 46:1493, 1972; Tall & Small, *Adv Lipid Res*, 17:1, 1980

Fitzgerald 30-AA15   Human plasma   High purity

Fitzgerald 30-AC15   Human plasma   Control/calibrator

Fluka 10817   Human plasma   MW ~280k   ≥90% (GE); ≥97% protein content; lyophilized from 0.01 *M* ammonium hydrogen carbonate | A principle protein component of high-density lipoproteins; Gennis, RB & Jonas, A, *Ann Rev Biophys Bioeng*, 6: 195, 1977

ICN 153906   Human plasma   MW 28,016   10*mM* Tris, 0.1% NaN₃, pH 8.0 buffer; 0.8-3.0 mg/mL protein; >98%

Sigma A 9284   Human plasma   Lyophilized from 0.01 M ammonium bicarbonate; ~85% (SDS-PAGE) | Apolipoprotein A-I & A-II are directly purified from HDL; apolipoprotein B is prepared from purified LDL using SDS or sodium deoxycholate to remove lipids; Burstein, M, *J Lipid Res*, 11: 583, 1970; Rudel, LL, *Biochem J*, 139: 89, 1974

USBio A2299-10   Human plasma   ≥98%; no contaminants detected; single band by SDS-PAGE, IEP, &/or RID; ~1 mg/mL supplied in 0.01 *M* ammonium bicarbonate, pH 7.4 | Suitable for antigenic applications in immunological protocols

Biodesign A95120H   Human plasma HDL   Purified | Species specificity: human

Biodesign A23100M   Murine plasma   Purified | Species specificity: mouse

Scipac P188-3   Pooled serum/plasma   >96%; frozen in TRIS buffer; 1-5 mg/mL; available on request | Apolipoprotein

Biogenesis 0650-0329   Rabbit plasma   MW 27.6k theoretical, confirmed by SDS-PAGE   pI: 5.2 +/- 0.05; purified; 10 *mM* triethylamine bicarbonate, pH 7.5; purified | Segret & Albers , *Meth Enzymol*, 128, Academic Press, 1986; Fielding et al, *BBRC*, 46:1493, 1972; Pan et al, *Eur J Biochem*, 170:99, 1987; Chao et al, *JBC*, 259:5306, 1984

Scipac P188-2   Serum/plasma   30-80%; frozen in TRIS buffer; | Apolipoprotein

Apolipoprotein AII

ICN 59415   Human   Purified Ag; lyophilized

USBio A2299-34   Human   ≥98%; no contaminants detected; single band by SDS-PAGE, IEP, &/or RID; 1 mg/mL; lyophilized | Suitable for antigenic applications in immunological protocols

Biogenesis 0650-0604   Human plasma   MW 17,380   Tested negative for HBsAg and for antibodies to HIV and HCV; purified; 10 *mM* NH₄HCO₃, pH 7.4; liquid | Binds phospholipids during lipoprotein metabolism; displaces lecithin-cholesterol acyltransferase bound to lipoprotein; influences HDL functional states and contributes to arteriosclerosis; Breslow, *PNAS USA*, 90:8314, 1993

ICN 153907   Human plasma   MW 17.4k   10*mM* Tris, 50*mM* NaCl, 1.0*mM* EDTA, 1.0*mM* TSF, pH 8.0; 0.8-1.5 mg/mL protein; ≥98%

Sigma A 8909   Human plasma   Lyophilized from 0.01 M ammonium bicarbonate; ~97% (SDS-PAGE) | Apolipoprotein A-I & A-II are directly purified from HDL; apolipoprotein B is prepared from purified LDL using SDS or sodium deoxycholate to remove lipids; Burstein, M, *J Lipid Res*, 11: 583, 1970; Rudel, LL, *Biochem J*, 139: 89, 1974

Fitzgerald 30-AA20   Human plasma (HDL)   High purity

Biodesign A95122H   Human plasma HDL   Purified | Species specificity: human

Apolipoprotein AV

Cortex CP8101   >95%

Apolipoprotein B

Synonyms: Lipoprotein, Low Density

Cortex CP1060   >95%

ICN 59416   Human   Purified Ag; lyophilized

USBio A2299-48   Human   ≥96%; no contaminants detected; single band by SDS-PAGE, IEP, &/or RID; 9.4 mg/mL supplied in 0.15 *M* NaCl, pH 7.5, 0.01% EDTA | Suitable for antigenic applications in immunological protocols

Fitzgerald 30-AA25   Human plasma   High purity

Sigma A 9159   Human plasma   Lyophilized from buffer containing 10 mM sodium deoxycholate, 0.05 M sodium carbonate & 0.05 M NaCl, pH 10.0; ~97%; delipidated with sodium deoxycholate | Apolipoprotein A-I & A-II are directly purified from HDL; apolipoprotein B is prepared from purified LDL using SDS or sodium deoxycholate to remove lipids; Burstein, M, *J Lipid Res*, 11: 583, 1970; Rudel, LL, *Biochem J*, 139: 89, 1974

Sigma A 9937   Human plasma   Lyophilized from 117 mM sodium phosphate containing 3.1 mM SDS, pH 7.4; ~97%; delipidated with SDS | Apolipoprotein A-I & A-II are directly purified from HDL; apolipoprotein B is prepared from purified LDL using SDS or sodium deoxycholate to remove lipids; Burstein, M, *J Lipid Res*, 11: 583, 1970; Rudel, LL, *Biochem J*, 139: 89, 1974

Biogenesis 0650-1004   Human plasma LDL   MW 550
Tested negative for antibodies to HBsAg, HCV and HIV 1 and 2;
purified; 10 *mM* sodium deoxycholate, 0.1 *M* sodium bicarbonate,
0.5 *M* NaCl, pH 11; liquid | Helenius et al, *Biochem*, 10:2542,
1971; LaemLli, *Nature*, 227:680, 1970; Ginsbury, et al, *JBC*,
259:6667, 1984

Apolipoprotein B100

Biodesign A34300H   Human plasma   Purified | Species
specificity: human

ICN 153908   Human plasma   MW 549k   10*mM* Na
deoxycholate, 50*mM* NaCl, 10*mM* NH_4HCO_3, pH 9.0; 0.8-1.5
mg/mL protein; ≥95%

USBio A2299-49   Human plasma   ~2 mg/mL in 100*mM* Tris,
2*mM* SDS, pH 8.0 | Delipidated & purified by gel filtration; useful
in Lowry protein determination assay; none detected: Apo(a),
ApoC, ApoE; cross reacts with anti-LDL

Scipac P193-2   Plasma   40-90%; lyophilized | Apolipoprotein

Apolipoprotein CI

Synonyms: Lipoprotein, Very Low Density

Cortex CP2013   >95%

ICN 59417   Human   Purified Ag; lyophilized

USBio A2299-60   Human   ≥98%; no contaminants detected;
single band by SDS-PAGE, IEP, &/or RID; 2.0 mg/mL supplied in
0.01 *M* NH_4HCO_3, pH 7.4 | Suitable for antigenic applications in
immunological protocols

Biogenesis 0650-1204   Human fresh plasma   MW 6,613
≥95% (SDS-PAGE); tested negative for HIV and HBsAg; from 10
mM ammonium bicarbonate; lyophilized

Biodesign A95126H   Human plasma   Purified | Species
specificity: human

Fitzgerald 30-AA32   Human plasma   High purity

ICN 194957   Human plasma   MW 6613   Lyophilized, >95%
| Partially activates lecithin-cholesterol acyltransferase & inhibits
lipase activity; Chen, CH, *Circulation*, 76:IV-117, 1987

Apolipoprotein CII

Synonyms: Lipoprotein, Very Low Density; Lipoprotein Lipase
Cofactor

ICN 59418   Human   Purified Ag; lyophilized

USBio A2299-64   Human   ≥98%; no contaminants detected;
single band by SDS-PAGE, IEP, &/or RID; 1 mg/mL; lyophilized in
0.01 *M* NH_4HCO_3, pH 7.4 | Suitable for antigenic applications in
immunological protocols

Biodesign A95128H   Human plasma   Purified | Species
specificity: human

Fitzgerald 30-AA27   Human plasma   High purity

ICN 194958   Human plasma   MW 8.8k   Lyophilized, >95%
| Low serum levels indicate Nephrotic Syndrome (Tangier disease);
Jackson, RL & G Holdsworth, *Methods Enzymol*, 128, Chap 14,
1986

Biogenesis 0650-1404   Human serum   Tested negative for
HIV antibodies and HBsAg; purified; 0.01 *M* ammonium
bicarbonate, non-sterile; lyophilized | Jonas et al, *Biochem*,
20:3802, 1981; Herbert et al, *JBC*, 248:4941, 1972; Schonfeld et
al, *J Lipid Res*, 19:645, 1977; Formisano et al, *JBC*, 253:354, 1978

Apolipoprotein CIII

Synonyms: Lipoprotein, Very Low Density

Cortex CP2015   >95%

ICN 59419   Human   Purified Ag; lyophilized

USBio A2299-72   Human   ≥98%; no contaminants detected;
single band by SDS-PAGE, IEP, &/or RID; 4.0 mg/mL supplied in
0.01 *M* NH_4HCO_3, pH 7.4 | Suitable for antigenic applications in
immunological protocols

Biodesign A95129H   Human plasma   Purified | Species
specificity: human

Biogenesis 0650-1804   Human plasma   Tested negative for
HIV antibodies and HBsAg; purified; 0.01 *M* ammonium bicarbonate
with 5 *M* guanidine-HCl; liquid | Catapano et al, *J Lipid Res*,
19:1047, 1978; Kane, *Anal Biochem*, 53:350, 1973; Jonas et al,
Biochem, 20:3802, 1981; Herbert et al, *JBC*, 248:4941, 1972;
Schonfeld et al, *J Lipid Res*, 19:645, 1977; Formisano et al, *JBC*,
253:354, 1978

Fitzgerald 30-AA28   Human plasma   High purity

ICN 194959   Human plasma   MW 8750   10*mM* ammonium
carbonate, pH 7.5; >95% | May inhibit lipoprotein lipase activation
by Apo C-II; Catapano, AL etal, *J Lipid Res*, 19:1047, 1978

Apolipoprotein E

Synonyms: Lipoprotein, Very Low Density

ICN 59432   Human   Purified Ag; lyophilized

USBio A2299-73   Human   ≥98%; no contaminants detected;
single band by SDS-PAGE, IEP, &/or RID; 1 mg/mL in 0.1 *M* Tris,
0.1 *M* ammonium bicarbonate, pH 7.5 | Suitable for antigenic
applications in immunological protocols

Biodesign A95130H   Human plasma   Purified | Species
Specificity: human

ICN 194960   Human plasma   MW 34.2k   10*mM* ammonium
carbonate, pH 7.5; >95% | A VLDL component & subclass of HDL;
LDL receptor ligand & participant in immunoregulation & cell growth
differentiation; Mahley, RW, *Science*, 240:622, 1988

Fitzgerald 30-AA30   Human plasma (VLDL)   High purity

Biogenesis 0650-2104   Human plasma VLDL   MW 34.2k
Tested negative for HBsAg, and antibodies to HCV and HIV;
purified; 50 *mM* NH_4HCO_3, pH 7.4; liquid | No free monomer at
concentrations >1 mg/mL; exists as monomer in buffers containing
5 *M* guanidine HCL or 6 *M* urea

ICN 195016   Human recombinant, expressed in *E. coli*   MW
34k   Lyophilized, >95%

Cortex CP2016   Plasma   >95%

Biodesign A95199H   Rabbit plasma   Purified | Species
Specificity: human

Apolipoprotein E IV

USBio A2299-77   Expressed in baculovirus insect cell culture
system   ≥95% (SDS-PAGE); liquid in 0.7 *M* ammonium
carbonate, pH 7.5; pI: 6.7 (for the primary isoform of rApo E4)

Biogenesis 0650-2409   Human r-DNA baculovirus system
MW 34k   Purified; pI: 6.7; 0.7 *M* NH_4HCO_3, pH 7.5; liquid |
Soluble in aqueous solutions; may be no free monomer at
concentrations >1 mg/mL; apolipoproteins exist as monomers in
buffers containing 5 *M* guanidine HCl or 6 *M* urea; Corder et al,
Science, 261:921, 1993; Strittmatter et al, *PNAS USA*, 90:8098,
1993; Gretsch et al, *PNAS USA*, 88:8530, 1991; Rall et al, *Meth
Enzymol*, 128:273, 1986

ICN 194962   Human recombinant, isoform E4   MW 34k
Lyophilized, >98% | Gretch, DG etal, *PNAS*, 88:8530, 1991

Apolipoprotein EII

Biogenesis 0650-2209   Human r-DNA expressed in baculovirus
insect cell culture system   MW 34k   Purified; pI: 6.25 (for the
primary isoform of rApo E2); 700 *mM* NH_4HCO_3; liquid |
Strittmatter et al, *PNAS USA*, 90:8098, 1993; Gretsch et al, *PNAS
USA*, 88:8530, 1991

Apolipoprotein EIII

Biogenesis 0650-2309   Human r-DNA   MW 34k   Purified;
pI: 6.35 (for the primary isoform of rApo E3); 0.7 *M* NH_4HCO_3, pH
7.5; liquid | No free monomer at concentrations >1 mg/mL; exists
as monomer in buffers containing 5 *M* guanidine HCL or 6 *M* urea;
Strittmatter et al, *PNAS USA*, 90:8098, 1993; Gretsch et al, *PNAS
USA*, 88:8530, 1991; Rall et al, *Meth Enzymol*, 128:273, 1986

ICN 194961   Human recombinant, isoform E3   MW 34k
Lyophilized, >98% | Gretch, DG etal, *PNAS*, 88:8530, 1991

Apolipoprotein H

Synonyms: Glycoprotein I, β_2-

ICN 59653   Human   Purified Ag; lyophilized

Scipac P195-1   Serum/plasma   >96%; lyophilized |
Apolipoprotein

Apolipoprotein H/βII

Synonyms: Glycoprotein IIβ_2-

Biodesign A11083H   Human plasma   Purified | Species
Specificity: human

Apolipoprotein SAA

Biodesign A52313H   *E. coli*   Purified | Species Specificity:
human

ICN 195013   Human recombinant, expressed in *E. coli*   MW
11.5k   Lyophilized, >98%; 5-50 µg/mL

Aprotinin

Sigma A 2315   Bovine lung   MW 6.5k   Vial contains enough
FITC-conjugated protein to run 50 mini-gels or 25 standard size
gels; protein band be visualized by using UV light or Brilliant Blue
stain | Fluorescent marker for SDS-page & protein transfer;
suitable for use as molecular weight standards in both SDS-PAGE &
transfer membranes

Arf2, His-Tagged

Calbiochem 181321   Rat brain recombinant, produced by
overexpression of a full-length Arf2 cDNA clone in *E. coli*   MW
21.9k   ≥90% (SDS-PAGE); liquid in 200 *mM* BME, 50 *mM* Tris-
HCl, 10% glycerol, 2% SDS, 0.005% bromophenol blue, pH 6.8 |
GTP binding protein; ADP-ribosylation factors (Arfs) are GTP-
binding proteins that act as allosteric activators of NAD: arginine
ADP-ribosyltransferase activity of cholera toxin; they regulate
intracellular vesicular traffic & stimulate the activity of
phospholipase D; localized to the perinuclear Golgi structure in
cells; suitable for use as a positive control or to assay for cross-
reactivity by Western blotting; Liang, JO & Kornfield, S, *J Biol
Chem*, 272: 4141, 1997; Hosaka, M et al, *J Biochem*, 120: 813,
1996; Price, SR et al, *Mol Cell Biochem*, 159: 15, 1996; Moss, J et
al, *J Biol Chem*, 270: 12327, 1995; Price, SR et al, *J Biol Chem*,
267: 17766, 1992; Steams, T et al, *Mol Cell Biol*, 10: 6690, 1990

Arf3, His-Tagged

Calbiochem 181323   Human brain recombinant, produced by
overexpression of a full-length Arf3 cDNA clone in *E. coli*   MW
21.7k   ≥90% (SDS-PAGE); liquid in 200 *mM* BME, 50 *mM* Tris-
HCl, 10% glycerol, 2% SDS, 0.005% bromophenol blue, pH 6.8 |
GTP binding protein; ADP-ribosylation factors (Arfs) are GTP-
binding proteins that act as allosteric activators of NAD: arginine
ADP-ribosyltransferase activity of cholera toxin; they regulate
intracellular vesicular traffic & stimulate the activity of
phospholipase D; associated with the Golgi & endoplasmic
reticulum; expression of Arf3 is developmentally regulated; suitable
for use as a positive control or to assay for cross-reactivity by
Western blotting; Tsai, SC et al, *J Biol Chem*, 266: 8213, 1991;
Price, SR et al, *Mol Cell Biochem*, 159: 15, 1996; Moss, J et al, *J
Biol Chem*, 270: 12327, 1995; Price, SR et al, *J Biol Chem*, 267:
17766, 1992

Arf4, His-Tagged

Calbiochem 181325   Human brain recombinant, produced by
overexpression of a full-length Arf4 cDNA clone in *E. coli*   MW
21.6k   ≥90% (SDS-PAGE); liquid in 200 *mM* BME, 50 *mM* Tris-
HCl, 10% glycerol, 2% SDS, 0.005% bromophenol blue, pH 6.8 |
GTP binding protein; ADP-ribosylation factors (Arfs) are GTP-
binding proteins that act as allosteric activators of NAD: arginine
ADP-ribosyltransferase activity of cholera toxin; they regulate
intracellular vesicular traffic & stimulate the activity of
phospholipase D; human Arf4 exhibits ~96% homology with rat
Arf4; plays a role in the recruitment of vesicle coat proteins in the
secretory pathway; suitable for use as a positive control or to assay
for cross-reactivity by Western blotting; Deitz, SB et al, *Mol Cell
Biol*, 16: 3275, 1996; Price, SR et al, *Mol Cell Biochem*, 159: 15,
1996; Moss, J et al, *J Biol Chem*, 270: 12327, 1995; Monaco, L et
al, *PNAS*, 87: 2206, 1990

Arf5, His-Tagged

Calbiochem 181327   Human Neuroblastoma (NIE115)
recombinant, produced by overexpression of a full-length Arf5
cDNA clone in *E. coli*   MW 21.6k   ≥90% (SDS-PAGE); liquid in
200 *mM* BME, 50 *mM* Tris-HCl, 10% glycerol, 2% SDS, 0.005%
bromophenol blue, pH 6.8 | GTP binding protein; ADP-ribosylation
factors (Arfs) are GTP-binding proteins that act as allosteric
activators of NAD: arginine ADP-ribosyltransferase activity of
cholera toxin; they regulate intracellular vesicular traffic &
stimulate the activity of phospholipase D; predominantly cytosolic
but can be recruited to the Golgi membranes upon binding to a
non-hydrolyzable analog of GTP; suitable for use as a positive
control or to assay for cross-reactivity by Western blotting; Liang,
JO & Kornfield, S, *J Biol Chem*, 272: 4141, 1997; Cavenagh, MM et
al, *J Biol Chem*, 271: 21767, 1996; Tsuchiya, M et al, *J Biol Chem*,
266: 2772, 1991; Moss, J et al, *J Biol Chem*, 270: 12327, 1995;
Price, SR et al, *J Biol Chem*, 267: 17766, 1992

Arf6, His-Tagged

Calbiochem 181329   Human Neuroblastoma (NIE115)
recombinant, produced by overexpression of a full-length Arf6
cDNA clone in *E. coli*   MW 21.2k   ≥90% (SDS-PAGE); liquid in
200 *mM* BME, 50 *mM* Tris-HCl, 10% glycerol, 2% SDS, 0.005%
bromophenol blue, pH 6.8 | GTP binding protein; ADP-ribosylation
factors (Arfs) are GTP-binding proteins that act as allosteric
activators of NAD: arginine ADP-ribosyltransferase activity of
cholera toxin; they regulate intracellular vesicular traffic &
stimulate the activity of phospholipase D; localized in cell plasma
membranes & membranes of secretory chromaffin granules;
thought to play a role in receptor-mediated endocytosis &
exocytosis; suitable for use as a positive control or to assay for
cross-reactivity by Western blotting; D'Souza-Schorey, C et al,
Science, 267: 1175, 1995; Cavenagh, MM et al, *J Biol Chem*, 271:
21767, 1996; Tsuchiya, M et al, *J Biol Chem*, 266: 2772, 1991;
Moss, J et al, *J Biol Chem*, 270: 12327, 1995

Asialoglycophorin

Sigma A 9791   Human blood type MN   Lyophilized powder |
Predominantly asialoglycophorin A

Atroporin

ICN 159838   *Crotalus atrox* (Western diamondback rattlesnake)
MW ~35k   One band (SDS-PAGE) | An anti-cancer protein; 0.5
µg kills various cancer cells *in vitro* (10⁵ cells/culture); no effect on
normal mouse kidney, spleen & liver cells up to 5 µg; prevents &
causes regression of ascitic tumors formed by mouse myeloma
cells; show enhanced cytolytic activity when used in combination
with Kaotree

Avidin

Synonyms: Biotin Enzymes Inhibitor

ICN 100303   MW 70k   10 U/mg, 1 U binds 1 µg D-biotin, pH
8.9 | Glycoprotein with tetrameric structure; high Specificity for
biotin; a highly specific inhibitor of biotin enzymes

ICN 55827   Lyophilized from water with 0.01% thimerosal

ICN 150047   Chicken egg white   MW 70k   10-15 U/mg solid, 1 U binds 1 µg D-biotin, pH 8.9; lyophilized | Useful in a biotin-avidin system for labeling biomolecules for receptor studies, immunoassays & immunohistological methods

ICN 150407   Chicken egg white   MW 70k   10-15 U/mg protein, 1 U binds 1 µg D-biotin, pH 8.9; affinity purified

Sigma A 9390   Chicken egg white   Lyophilized powder; chromatographically purified; 1 unit binds 1.0 µg of D–biotin; 10-15 units/mg protein

Sigma A 8706   Recombinant expressed in corn   Affinity purified; 1 unit binds 1.0 µg of D–biotin

Avidin, 4-(2-Aminoethylamido)-Succinyl-

Sigma A 7909   Lyophilized powder containing ~90% protein (Biuret); balance sodium citrate

Avidin, Ac-

ICN 153844   10-13 U/mg solid, crystalline; isoelectric point 8.5 (IEF) | A chemically modified avidin with lower isoelectric point than native avidin; reacted with acetic anhydride

Avidin, Agarose

ICN 191323   Chicken egg white   1 mg avidin/mL gel; capacity ~60 nmoles d(+) biotin/mL packed gel; suspension in PBS, 0.02% NaN_3 | 5 atoms hydrophilic spacer arm; used for isolation of biotinyl-peptides, proteins, ligands & immobilization of biotinylated enzymes

Avidin, Alkaline Phosphatase Conjugated

ICN 55963   Liquid in 0.03 M Tris buffered saline, pH 8.0, 1% BSA, 10% glycerol, 0.05% NaN_3

Avidin, Ferritin Conjugated

Sigma A 4030   Solution in 50% glycerol, 0.25 M NaCl, 0.01 M sodium phosphate, pH 6.8, containing 0.02% sodium azide; ~10 mg/mL protein; contains ~1 mole ferritin/mole of avidin; prepared by reductive alkylation by a modification of the procedure of Bayer | Labeled with equine spleen ferritin; Bayer, EA et al, *J Histochem Cytochem*, 24: 933, 1976

Sigma A 5405   Lyophilized powder; ~30% protein; balance primarily Tris buffer salt; prepared by modification of the procedure of Sullivan | Labeled with equine spleen ferritin; actual ferritin to avidin ratio given on label; Sullivan, MJ et al, *FEBS Lett*, 95: 311, 1978

Avidin, FITC Conjugated

ICN 55880   Liquid in 0.02 M PBS, pH 7.3, 10% glycerol, 0.05% NaN_3

Sigma A 2901   Lyophilized powder containing ~80% protein (A_{280}); balance primarily sodium citrate; contains 2-4 moles fluorescein isothiocyanate/mole avidin

Avidin, Horse Radish Peroxidase Conjugated

ICN 55898   Liquid in 0.02 M PBS, pH 7.3, 10% glycerol, 0.01% thimerosal

Avidin, Peroxidase Conjugated

Sigma A 3151   Lyophilized powder containing ~80% protein ($E^{1\%}_{280}$); balance primarily citrate buffer; contains 1-2 moles peroxidase/mole avidin; peroxidase activity: 80-160 units/mg protein; avidin activity: 5-10 units/mg protein; 1 unit forms 1 mg of purpurogallin/20 seconds from pyrogallol at pH 6.0, 20°C; coupled by a modification of Sullivan | Sullivan, MJ et al, *FEBS Lett*, 95: 311, 1978

Avidin, Rhodamine Isothiocyanate Conjugated

Sigma A 3026   Lyophilized powder containing ~90% protein (Biuret); balance primarily citrate buffer salts; contains 1-2 moles rhodamine isothiocyanate/mole avidin;

Avidin, Streptavidin Labeled

Amersham US11681   *Streptomyces avidinii*   A tetrameric protein containing four high specificity binding sites for biotin

Avidin, Succinyl Labeled

Sigma A 3907   Lyophilized powder containing ~90% protein ($E^{1\%}_{280}$); balance sodium citrate; prepared by succinylation by a modification of the procedure of Klapper & Klotz | Klapper, MH & Klotz, IM, *Meth Enzymol*, 25: 531, 1972

Avidin, Texas Red Labeled

ICN 55894   Liquid in 0.02 M PBS, pH 7.3, 10% glycerol, 0.05% NaN_3

Avidin, β-Galactosidase Labeled

Sigma A 2930   Lyophilized powder containing ~60% protein ($E^{1\%}_{280}$); balance primarily Tris-succinate with a trace of dithiothreitol; galactosidase activity: 200-600 units/mg protein; avidin activity: 2-4 units/mg protein; 1 unit hydrolyzes 1.0 µmole o-nitrophenyl β-D-galactoside/min at pH 7.3, 37°C; prepared from avidin (A 9275) partially acetylated, & β-galactosidase (G 5635) by modification of Sullivan | Sullivan, MJ et al, *FEBS Lett*, 95: 311, 1978

Azurin

Sigma A 3672   Pseudomonas *aeruginosa*   Lyophilized powder containing ~70% protein (Lowry); balance primarily ammonium acetate buffer salts

B Cell Activating Factor

PeproTech 310-13   Human recombinant, expressed in *E. coli*   MW 17.0k   >98% (SDS-PAGE) & HPLC; <0.1 ng endotoxin per mg (1EU/mg); lyophilized | Newly discovered novel ligand of the TNF family; important role as costimulator of B cell proliferation & function; soluble; 153 AA; ED_{50} < 10 ng/mL; SA determined by the dose-dependent stimulation of IL-8 production by Human PBMC

B Cell Attracting Chemokine I

PeproTech 300-47   Human recombinant, expressed in *E. coli*   MW 10.1k   >98% (SDS-PAGE) & HPLC; <0.1 ng endotoxin per mg (1EU/mg); lyophilized | Recently discovered chemokine;85 AA; SA determined by its ability to chemoattract human B cells

Bacitracin

Synonyms: Polypeptide Antibiotic

Amersham US11805   *Bacillus subtilis*   38 U/mg (as is) | An antibiotic polypeptide

Bacitracin, Zinc

Synonyms: Polypeptide Antibiotic

Amersham US11810   *Bacillus subtilis*   ≥40 U/mg (as is); 2-10% Zn | An antibiotic polypeptide

BAD Control Proteins

Oncogene PF086   MW 76k   Contains a positive control phosphorylated by PKA *in vitro* & the non-phosphorylated Bad fusion protein | Detects a ~76k mouseBAD peptide including residues Se^{112}, Ser^{136}, and Ser^{155}, fused to paramyosin

BAD, Agarose

USBio B0003-50   Murine full-length, His-tagged Bad fusion protein expressed in *E. coli*   ~90% | Bound with ProBond™ nickel-chelating resin to agarose

BAD, Soluble

USBio B0003-60   Murine recombinant, full-length Bad fused with an N-terminal His^6-tag, expressed in *E. coli*   Applications: kinase assay

Bafilomycin AI

USBio B0003-70   >96% by HPLC; soluble in DMSO, EtOH or acetone; lyophilized | A macrolide antibiotic that acts as a specific potent inhibitor of vacuolar-type ATPases; blocks lysosomal cholesterol transport in macrophages

B-Cell Stimulation Factor, Pre-

Synonyms: Stromal Cell Derived Factor β

Calbiochem 512777   Human recombinant, expressed in *E. coli* MW 8.5k   ≥97% (SDS-PAGE); lyophilized from filter-sterilized 30% acetonitrile, 0.1% TFA containing 50 μg BSA/μg PBSF/SDF-1β; biological activity: ED_{50}=50-100 ng/mL as measured by its ability to chemoattract human T lymphocytes cultured in the presence of IL-2; endotoxin: ≤100 pg/μg PBSF/SDF-1β | Supports the proliferation of a stromal cell-dependent pre-B cell line; found to be a chemoattractant for T lymphocytes & monocytes but not neutrophils; recently shown to be a ligand for the CSCR4 (fusin/LESTR) receptor that functions as a coreceptor for lymphocyte-tropic HIV-1 strains; Bleul, C et al, *Nature*, 382: 829, 1996; Oberlin, E et al, *Nature*, 382: 833, 1996; Tashiro, K et al, *Science*, 261: 600, 1993; Nagasaqa, T et al, *PNAS*, 91: 2305, 1994

ICN 195794   Human recombinant, expressed in *E. coli*   ≥97%; lyophilized; max chemotaxis = 1 μg/mL, by chemoattractant ability to Fusin (CXCR-4)

Bcl-2, Agarose

USBio B0807-06   Murine full-length Bcl-2, GST fusion protein expressed in *E. coli*   5 μg of fusion protein, 54kD, bound to 50 μL of packed glutathione-agarose beads in PBS with10% glycerol; provided as a 50% slurry for a total volume of 100 μL; frozen solution | Bound to glutathione-agarose; overexpression of Bcl-2 blocks apoptosis in response to a number of inductive stimuli; forms heterodimers with Bax & other family members; phosphorylation of Bcl-2 may decrease its anti-apoptotic activity

B-DNF

BioSource International PHC7014   Human recombinant

Bence Jones Protein κ

Cortex CP8105U   >98%

Cortex CP8106   >95%

Betacellulin

IBT ARU020, ARU100   Bovine recombinant, expressed in *E. coli* MW 8995   Lyophilized from 50 *mM* acetic acid, 0.05% (v/v) TFA | In the EGF family; 80 AA heparin-binding protein; synthesized as a transmembrane precursor; the soluble cytokine, containing 1 EGF structural motif, is released by proteolytic cleavage

IBT ASU010   Rat recombinant, expressed in *E. coli*   MW 9039 Lyophilized from 50 *mM* acetic acid, 0.05% (v/v) TFA | Binds the EGF receptor (ErbB-1) which then dimerizes with an EGF receptor family one (ErbB-1, ErbB-2, ErbB-3 or ErbB-4) to signal through the tyrosine kinase pathway; at high concentrations EGF & BTC can induce cell growth & differentiation in the absence of ErbB-1 (TGF-a cannot); potent mitogen for Balb/c 3T3 fibroblasts, retinal pigment epithelial cells & vascular smooth muscle cells

Betacellulin Epidermal Growth Factor

Sigma B 3670   Human recombinant, expressed in *E. coli*   MW 9.5k   ≥97% (SDS-PAGE); 0.2 μm filtered & lyophilized from PBS containing 500 μg BSA; bioactivity is measured in cell proliferation assay using BALB/3T3 fibroblasts; endotoxin tested; see Sigma E 1264 | New member of the EGF family of cytokines

Biglycan

Sigma B 8041   Bovine articular cartilage   MW >2.5x10^6 Lyophilized, essentially salt-free; sterile-filtered | 200-350k proteoglycan consisting of a 45k core protein & 2 chrondroitin/dermatan sulfate glycosaminoglycan chains; interacts with collagen type I as well as with fibronectin & TGF-β; Roughley, PJ et al, *Matrix Biol*, 14: 51, 1994; Schonherr, E et al, *J Biol Chem*, 270: 2776, 1995; Pogany, G et al, *Arch Biochem Biophys*, 313: 102, 1994

Bone Morphogenic Protein II

Kamiya   Recombinant   MW 12k   >85% & <95% (SDS-PAGE)

Bone Morphogenic Protein, Natural Cocktail

Kamiya

Botulinum Toxin A

Calbiochem 203674   *Clostridium botulinum*   MW 500k Liquid in 200 *mM* NaCl, 50 *mM* NaOAc, pH 6.0, stabilized with hemagglutinin; single band purity (disc gel electrophoresis); highly toxic: LD_{50}≤50 mg/kg | Presynaptic toxin that acts at neuromuscular junctions & inhibits acetylcholine release; inhibits catecholamine secretion from bovine adrenal medullary cells; does not affect the entry of Ca^{2+} into cells; Schiavo, G et al, *Ann NY Acad Sci*, 710: 65, 1994; in *Botulinum Neurotoxins & Tetanus Toxin* (Simpson, LL, ed), Academic Press, San Diego, CA

Botulinum Toxin A Heavy Chain

Calbiochem 203652   *Clostridium botulinum*   MW 100k Lyophilized solid; single band purity (SDS-PAGE); toxic: LD_{50}≤200 mg/kg but >50 mg/kg | Subunit of botulinum toxin A responsible for membrane binding & internalization; useful in binding studies; Gill, DM, *Microbiol Rev*, 46: 86, 1982

Botulinum Toxin A Light Chain

Calbiochem 203650   *Clostridium botulinum*   MW 50k Lyophilized solid; single band purity (SDS-PAGE); toxic: LD_{50}≤200 mg/kg but >50 mg/kg | Enzymatic subunit of botulinum toxin A responsible for blocking acetylcholine release from synaptic vesicles; mimics the intact toxin if injected into cells; Gill, DM, *Microbiol Rev*, 46: 86, 1982; Simpson, LL, in *Methods in Neuroscience* (Conn, PM, ed), Academic Press, 8: 56, 1992

Botulinum Toxin B

Calbiochem 203672   *Clostridium botulinum*   MW 500k Liquid in 200 *mM* NaCl, 50 *mM* NaOAc, pH 6.0, stabilized with hemagglutinin; single band purity (disc gel electrophoresis); highly toxic: LD_{50}≤50 mg/kg | Neurotoxin that inhibits catecholamine secretion from bovine adrenal medullary cells & acts at neuromuscular junctions, repressing the release of acetylcholine

Botulinum Toxin B Heavy Chain

Calbiochem 203656   *Clostridium botulinum*   MW 100k Lyophilized solid; single band purity (SDS-PAGE); toxic: LD_{50}≤200 mg/kg but >50 mg/kg | Subunit of botulinum toxin B responsible for membrane binding & internalization; useful in binding studies; Gill, DM, *Microbiol Rev*, 46: 86, 1982

Botulinum Toxin B Light Chain

Calbiochem 203654   *Clostridium botulinum*   MW 50k Lyophilized solid; single band purity (SDS-PAGE); toxic: LD_{50}≤200 mg/kg but >50 mg/kg | Enzymatic subunit of botulinum toxin B responsible for blocking acetylcholine release from synaptic vesicles; mimics the intact toxin if injected into cells; Gill, DM, *Microbiol Rev*, 46: 86, 1982; Simpson, LL, in *Methods in Neuroscience*(Conn, PM, ed), Academic Press, 8: 56, 1992

Botulinum Toxin C

Calbiochem 203676 *Clostridium botulinum* MW 500k
Liquid in 200 *mM* NaCl, 50 *mM* NaOAc, pH 6.0, stabilized with hemagglutinin; single band purity (disc gel electrophoresis); highly toxic: $LD_{50} \leq 50$ mg/kg | Neurotoxin that inhibits catecholamine secretion from bovine adrenal medullary cells & acts at neuromuscular junctions, repressing the release of acetylcholine; exhibits ADP-ribosyltransferase activity

Botulinum Toxin D

Calbiochem 203677 *Clostridium botulinum* MW 300k
Single band purity (disc gel electrophoresis); liquid in 200 *mM* NaCl, 50 *mM* NaOAc, pH 6.0; highly toxic: $LD_{50} \leq 50$ mg/kg | Potent neurotoxin that acts at presynaptic terminals of cholinergic neurons, blocking the release of acetylcholine; inhibits catecholamine secretion from bovine adrenal medullary cells & human neutrophils; causes GTP-stimulated ADP-ribosylation of a 22 kDa protein isolated from mouse brain membranes; Nath, J et al, *J Immunol*, 152: 1370, 1994

Botulinum Toxin E

Calbiochem 203673 *Clostridium botulinum* MW 300k
Liquid in 200 *mM* NaCl, 50 *mM* NaOAc, pH 6.0; highly toxic: $LD_{50} \leq 50$ mg/kg | Neurotoxin that acts at neuromuscular junctions, repressing the release of acetylcholine

Botulinum Toxin F

Calbiochem 203679 *Clostridium botulinum* MW 235k
Liquid in 200 *mM* NaCl, 50 *mM* NaOAc, pH 6.0; single band purity (disc gel electrophoresis); highly toxic: $LD_{50} \leq 50$ mg/kg | Neurotoxin that acts at neuromuscular junctions, repressing the release of acetylcholine; in *Botulinum Neurotoxins & Tetanus Toxin* (Simpson, LL, ed), Academic Press, San Diego, CA

Botulinum Toxoid Type A

Calbiochem 203653 *Clostridium botulinum* Lyophilized solid from 10 *mM* sodium phosphate buffer, pH 7.5; non-toxic as determined by LD_{50} assay in mice | Prepared by formaldehyde inactivation of botulinum toxin type A

Botulinum Toxoid Type B

Calbiochem 203658 *Clostridium botulinum* Lyophilized solid from 10 *mM* sodium phosphate buffer, pH 7.5; non-toxic as determined by LD_{50} assay in mice | Prepared by formaldehyde inactivation of botulinum toxin type A

Bradykinin

ARC ART-701 MW 1060.24 2,3-(Prolyl-3,4-H(N)) 90-120 Ci/mmol; 3.33-4.44 TBq/mmol; in 0.2% TFA: CH₃CN (80:20) | Radiochemical

Breast Tumor Marker Antigen

Synonyms: Ca 15-3®

ICN 771001 ICN 771002 ICN 771003 Human ascites >96%; iodination grade; >50,000 U/mL | for immunization or labeling

Bungarotoxin, FITC-α-

Synonyms: Snake Toxin

Sigma T 9641 *Bungarus multicinctus* ~1 mole FITC/mole α-bungarotoxin | Fluorescent label for nicotinic receptors on the motor endplate

Sigma T 3783 *Bungarus multicinctus* 0.5-1 mole FITC/mole β-bungarotoxin | Fluorescent label for cholinergic terminals

Bungarotoxin, α-

Synonyms: Snake Toxin; Phospholipase A₂ Presynaptic, Neurotoxic

Sigma T 3019 *Bungarus multicinctus* Binds irreversibly to motor endplate acetylcholine receptors; prevents opening of nicotinic receptor-associated ion channels

Sigma B-137 *Bungarus multicinctus* (Elapidae snake) White solid; peptide content & salt form information are provided with each lot | Neurotoxin which binds irreversibly with post-synaptic cholinergic receptors to produce neuromuscular blockage; potent neurotoxin; Mebs et al, *Biochem Biophys Res Commun*, 44: 711, 1971; Kalash et al, *Neuroendocrinology*, 49: 462, 1989; Kamiya et al, *Brain Res Bull*, 8: 431, 1982

Alexis 630-050 *Bungarus multicinctus* MW 6979.0 ≥98%; lyophilized powder; soluble in water or aqueous buffers; potent neurotoxin | Halliwell, JV, *J Neurochem*, 39: 543, 1982; Kondo, K et al, *J Biochem*, 91: 1519 & 1531, 1982; Rehm, H & Betz, H, *J Biol Chem*, 259: 6865, 1984; Petersen, M et al, *Neurosci Lett*, 68: 141, 1986; Schmidt, RR & Betz, H, *Biochemistry*, 28: 8346, 1989; Danse, JM et al, *Nucl Acids Res*, 18: 4609, 1990; Strong, PN, *Pharmacol ther*, 46: 137, 1990

Sigma T 5644 *Bungarus multicinctus* Toxic phospholipase A₂ | Destroys synaptic vesicles & inhibits ACh release

C Reactive Protein

ICN 194983 Human ascites >95%; supplied in 2 *mM* CaCl₂, 150 *mM* NaCl, 20 mM Tris, pH 7.5 | Useful in immunological studies & in rheumatoid arthritis diagnosis

Sigma C 4063 Human plasma Solution in 0.02 M Tris, 0.25 M NaCl, pH 8.0, containing 0.1% sodium azide; protein determined by Lowry

ICN 150713 Human pleural & ascites fluid & plasma >99%; 1.0-3.0 mg protein/mL in 220 *mM* NaCl, 0.1% NaN₃, 20 mM Tris, pH 8.0 | Recommended for immunological applications in the preparation of antisera

ICN 152315 Human pleural & ascites fluid & plasma >95%; 1.5-3.0 mg protein/mL in 220 *mM* NaCl, 0.1% NaN₃, 20 mM Tris, pH 8.0 | Suitable for use as a standard in CRP assays &/or kit calibration for diagnostic purposes

Biodesign A15200H Human pleural fluid 60-80% | Cardiac markers

Biodesign A97201H Human pleural fluid >99% | Cardiac markers

ICN 194982 Human serum >99%; supplied in 2 *mM* CaCl₂, 140 *mM* NaCl, 0.1% NaN₃, 10 mM Tris, pH 8.0 | For use as a calibration standard, structural & functional studies, & an Ag for antisera production

Sigma C 8898 *Limulus* polyphemus Hemagglutinin free; lyophilized powder containing ~60% protein (modified Warburg-Christian); balance primarily NaCl, Tris succinate, calcium acetate; prepared from C 7023 by removal of hemagglutinin

C1

Cortex CP1093 >95%

C1 Esterase Inhibitor

Cortex CP2041U >98%

C-10

Synonyms: MRP-1

Biodesign A52006M *E. coli* MW 10.7k Purified | Species specificity: mouse

Chemicon GF076 Murine ≥95%

PeproTech 250-06 Murine recombinant, expressed in *E. coli* MW 10.7k >98% (SDS-PAGE) & HPLC; <0.1 ng endotoxin per mg (1EU/mg); lyophilized with no additives | Member of the CC chemokine family; related to macrophage inflammatory protein-1 alpha (MIP-1a); 95 AA; SA determined by its ability to chemoattract Balb/c mouse spleen MNCs

C-10 Chemokine

Sigma C 0835   Mouse recombinant, expressed in *E. coli* ≥97% (SDS-PAGE); 0.2 µm filtered & lyophilized from PBS containing 0.5 mg BSA; endotoxin tested | Member of the CC or β chemokine class which act primarily as chemoattractants & activate monocytes, dendritic cells, T lymphocytes, natural killer cells, B lymphocytes, basophils & eosinophils; originally identified as a transcript that is induced in bone marrow cells upon stimulation with GM-CSF; the precursor form of C-10 consists of 116 AA; to generate the mature C-10 (95 AA) the precursor cleaves its hydrophobic signal peptide; a combination of chemoattractant & cytokine, describing proteins structurally defined as chemoattractants for leukocytes; Schall, T, in: *The Cytokine Handbook*, Thomson, A (ed), Academic Press, San Diego, p. 419, 1994; Murphy, PM, *Ann Rev Immunol*, 12: 593, 1994; Miller, MD et al, *Crit Rev Immunol*, 12: 17, 1992; Oppenheim, JJ et al, *Ann Rev Immunol*, 9: 617, 1991; Rot, A et al, *J Exp Med*, 176: 1489, 1992; Rollins, BJ et al, *Mol Cell Biol*, 11: 3215, 1991; Walter, S et al, *Int J Cancer*, 49: 431, 1991; Tanaka, Y et al, *Nature*, 361: 79; 1993

C1Q

Cortex CP1002U   >98%

USBio C0010-10   Human   ≥98%; no contaminants detected; single band by SDS-PAGE, IEP, &/or RID; 1 mg/mL (≥500,000 U/mg) supplied in PBS buffer, 40% glycerol, pH 7.2, no preservative | Suitable for antigenic applications in immunological protocols

C2

Cortex CP2021   >98%

C3

Cortex CP1036U   >98%

C3b

Cortex CP1044   >95%

C3c

Cortex CP1040   >95%

C3d

Cortex CP1043   >95%

C4

Cortex CP1037   >95%

C4a

Cortex CP1091U   >98%

C4b

Cortex CP2011U   >98%

C4b Binding Protein

ICN 194188   Human plasma   Purified; <0.25 µg protein S (ELISA)

C5

Cortex CP2026U   >98%

C5b,6

Cortex CP2031U   >98%; complex

C6

Cortex CP2032U   >98%

C7

Cortex CP2033U   >98%

C8

Cortex CP2034U   >98%

C9

Cortex CP1050   >95%

CA72-4 Cancer Antigen

USBio C0100   Human   ≥98%; no contaminants detected; single band by SDS-PAGE, IEP, &/or RID; 100K U/mL (Centocor RIA) supplied in 0.1 *M* PBS, pH 7.4, 0.05% NaN₃, 2.5% sucrose | Suitable for antigenic applications in immunological protocols, calibrators & controls; some background levels of CA-125 & CA-19-9

Cag Antigen, *Helicobacter pylori*

IBT HPA-5000-4, HPA-5000-5   *E. coli*   MW 30k   >90%; 0.50 mg/mL solution in PBS, 0.1% SDS, 10 *mM* EDTA, pH 7.4 | Covers Glu^{748} to Glu^{1015} of *H. pylori* cytotoxin-associated gene A

Calbindin

Synonyms: CaBP-9k

Biogenesis 0100-0078   Porcine synthetic intestine   Purified; lyophilized | Schroder et al, *J Physiol*, 429:715, 1996

Calbindin-D₉ₖ

Synonyms: CaBP-9k

Calbiochem 206500   Porcine intestine   MW 8840   ≥98% (HPLC); white lyophilized solid; soluble in water | Intestinal vitamin D-dependent calcium-binding protein; functions as a cytosolic buffer for Ca^{2+} ions in duodenal enterocytes & facilitates transepithelial active transport of Ca^{2+} ions; also present in pig uterus & placenta; binds two equivalents of Ca^{2+}; Hitchman, AJW & Harrison, JE, *Can J Biochem*, 50: 758, 1972; Schroder, B et al, *J Physiol*, 429: 715, 1996

Calcineurin

Synonyms: Calmodulin Binding Protein; Modulator Binding Protein; Protein Phosphatase 2B; Ca^{++}/Calmodulin-Dependent Serine/Threonine Phosphatase

Fluka 21044   Bovine brain   ~30 U/mg; ~1% protein content; 1 U corresponds to the amount of protein causing 50% inhibition of the activated 3',5'-cyclic nucleotide activity, when assayed with 2 U calmodulin & 0.1 *mM* Ca^{2+} in an enzyme coupled system at pH 7.5, 30°C | Aitken, A et al, *Dev Biochem*, 25: 113, 1983

ICN 150545   Bovine brain   Lyophilized; ~1% protein, with buffers, salts & stabilizers; 1U causes 50% inhibition of activated phosphodiesterase 3'-5' cyclic nucleotide activity when assayed in 2 U activator & 0.1 *mM* Ca^{2+} in an enzyme coupled system

Sigma C 1907   Bovine brain   Lyophilized powder containing ~1% protein (Lowry); balance 0.5% EGTA, buffer salts & stabilizers; activity: 2,500-5,000 units/mg protein; unit definition: 1 unit causes a 50% inhibitions of the activated phosphodiesterase, 3':5'-cyclic nucleotide (P 9529) activity when assayed with 2 units of activator (P 2277) & 0.1 mM Ca^{++} in an enzyme coupled system at pH 7.5, 30°C | Involved in T-lymphocyte activation; may be involved in hyperphosphorylation of tau in Alzheimer's disease; inhibits calmodulin-induced activation of cyclic nucleotide-dependent phosphodiesterase

Biogenesis 0100-0122   Human r-DNA *E. coli*   MW 60k (A), 15k (B)   Purified; 25 *mM* TRIS, 3 *mM* MgCl₂, 1 *mM* DTT, 1 *mM* EGTA, pH 7.4; liquid | Recombinant human calcineurin A and B co-expressed with yeast myristoyl-Co-A:protein N-myristoyltransferase; the resulting highly active calcineurin is N-myristoylated on the CaNB-subunit, as is the native protein; Mondragon et al, *Biochem*, 36:4934, 1997

Calcineurin, Bα-

Chemicon AG640   Bovine brain   ≥95% | Purified protein for apoptosis & signal transduction

Chemicon AG655   Recombinant   ≥95% | Purified protein for apoptosis & signal transduction

Calcineurin, Bα- His Tag

Chemicon AG656   Recombinant   ≥95% | Purified protein for apoptosis & signal transduction

Calciseptine

ICN 193934   *Dendroaspis p. polyepis*   MW 7036   A polypeptide toxin consisting of 60 residues that selectively blocks L-type voltage-dependent Ca^{2+} channels in various cells; De Weille etal, *PNAS*, 88:2437, 1991

Calcium Binding Protein

Sigma C 1540   Bovine mucosa   Vitamin-induced; lyophilized powder containing ~60% protein (Lowry); balance salts; prepared by the modification of the method of Bryant & Andrews | Bryant, DW & Andrews, P, *Biochem J*, 211: 709, 1983

Calmodulin

Synonyms: Phosphodiesterase 3:5'-Cyclic Nucleotide Activator; Phosphodiesterase 3':5'-Cyclic Nucleotide Activator; Phosphodiesterase 3':5'-Cyclic Nucleotide Activator, Gold Conjugated; Phosphodiesterase 3':5'-Cyclic Nucleotide Activator, FITC Conjugated; Nitric Oxide Synthase Activator

ICN 195697   Lyophilized; virtually salt free; 15,000-30,000 U/mg; 1 U activates 0.016 U 3',5'-cyclic-nucleotide phosphodiesterase to 50% V_{max} when saturated with activator in the presence of 0.01 m*M* Ca^{2+} | Used to activate CaM-dependent phosphodiesterase, calcineurin, CaM kinases, etc

Alexis 202-024   Bovine brain   ≥98% (SDS gel electrophoresis); essentially salt-free; lyophilized powder; specific activity: >20,000 U/mg protein (Lowry); 1 unit stimulated 0.016 activated units of phosphodiesterase 3':5'-cyclic nucleotide to 50% of the max activity of the enzyme when saturated with activator in the presence of 0.01 m*M* Ca^{2+} at 30°C, pH 7.5

Biogenesis 1740-2384   Bovine brain   >98% (SDS-PAGE); essentially salt free; SA: >4000 U/mg protein; purified material; no preservatives; lyophilized

Calbiochem 208690   Bovine brain   MW 16,723   ≥95% (SDS-PAGE); solid lyophilized from 50 m*M* Tris-HCl, 150 m*M* NaCl, 2 m*M* EDTA, pH 7.6; specific activity: ≥12,500 units/mg protein; 1 unit gives rise to 50% of the maximal enzyme activation of a standard level of activator-deficient 3',5'-cyclic nucleotide phosphodiesterase | *Merck Index*, 12: 1767

Calbiochem 208694   Bovine brain   MW 16,723   High Purity ≥99% (SDS-PAGE); solid lyophilized from 1.7 m*M* HEPES, 30 μ*M* CaCl₂, pH 7.0; specific activity: ≥13,000 units/mg protein; 1 unit gives rise to 50% of the maximal enzyme activation of a standard level of activator-deficient 3',5'-cyclic nucleotide phosphodiesterase | *Merck Index*, 12: 1767

ICN 195691   Bovine brain   MW 16.7k   Lyophilized; 1.7 m*M* HEPES, pH 7, 30 m*M* CaCl₂; 13,000 U/mg; 1 U activates 3',5'-cyclic-nucleotide phosphodiesterase to 50% V_{max}; ≥99% | Used to activate CaM-dependent phosphodiesterase, calcineurin, CaM kinases, etc

ICN 195692   Bovine brain   Lyophilized; virtually salt free; >40,000 U/mg; 1 U activates 0.016 U 3',5'-cyclic-nucleotide phosphodiesterase to 50% V_{max} when saturated with activator in the presence of 0.01 m*M* Ca^{2+}; ≥98% | Used to activate CaM-dependent phosphodiesterase, calcineurin, CaM kinases, etc

Sigma P 0809   Bovine brain   MW acidic proteins (<30k) Crude; lyophilized powder containing calmodulin & *S*-100 proteins (BCA) | Convenient source of low actual percentages of protein activator & *S*-100 proteins as determined by HPLC provided on label

Sigma P 1062   Bovine brain   MW 3350 (spacer to albumin-coated colloidal gold)   5 nm colloidal Gold labeled; mean particle diameter 3.5-6.5 nm, monodisperse; calmodulin from Sigma P 2277 coupled through PEG; suspension in ~50% glycerol containing 0.15 M NaCl, 0.01 M BES, pH 7.0, 0.02% PEG 20, 0.02% sodium azide; concentration: A_{520} ~5.0 | For the enhanced detection of calmodulin binding compounds; Fujimoto, K et al, *J Histochem Cytochem*, 37: 249, 1989

Sigma P 1187   Bovine brain   MW 3350 (spacer to albumin-coated colloidal gold)   10 nm colloidal Gold labeled; mean particle diameter 8-10 nm, monodisperse; calmodulin from Sigma P 2277 coupled through PEG; suspension in ~50% glycerol containing 0.15 M NaCl, 0.01 M BES, pH 7.0, 0.02% PEG 20, 0.02% sodium azide; concentration: A_{520} ~5.0 | For the enhanced detection of calmodulin binding compounds; Fujimoto, K et al, *J Histochem Cytochem*, 37: 249, 1989

Sigma P 2277   Bovine brain   ~95% (SDS-PAGE); essentially salt-free; lyophilized powder; activity: >40,000 U/mg protein (Lowry); unit definition: 1 U stimulates 0.016 activated U of phosphodiesterase 3':5'-cyclic nucleotide, Sigma P 0520, in a 3 mL reaction volume at pH 7.5 & 30°C to 50% of maximum activity of the enzyme when saturated with activator in the presence of 0.01 m*M* Ca^{2+} | Ca^{2+} binding protein that is required for activation of cyclic nucleotide-dependent phosphodiesterase; involved in intracellular Ca^{2+} homeostasis; O'Neil, KT & DeGrado, WF, *Trends Pharmacol Sci*, 15: 59, 1990

Sigma P 3922   Bovine brain   Essentially salt-free; lyophilized powder; activity: >40,000 U/mg protein (Lowry); prepared by labeling Sigma P 2277 with tetramethylrhodamine isothiocyanate; contains 0.5-1.0 mole TRITC/mole of protein | Useful for the direct localization of calmodulin binding inside the cell, i.e. binding on mitochondria, & as a probe in studying mitosis *in vivo*; Zavortink et al, *Exp Cell Res*, 149: 375, 1983; Pardue, RL et al, *Cell*, 23: 533, 1981

Sigma P 4046   Bovine brain   Essentially salt-free; lyophilized powder; activity: >40,000 U/mg protein (Lowry); prepared by labeling Sigma P 2277 with fluorescein isothiocyanate; contains 0.5-1.0 mole FITC/mole of protein | Useful as a probe in studying mitosis *in vivo*; Zavortink et al, *Exp Cell Res*, 149: 375, 1983

ICN 195693   Bovine heart   Lyophilized; 2500-10,000 U/mg; 1 U activates 0.016 U 3',5'-cyclic-nucleotide phosphodiesterase to 50% V_{max} when saturated with activator in the presence of 0.01 m*M* Ca^{2+}; ≥90% | Used to activate CaM-dependent phosphodiesterase, calcineurin, CaM kinases, etc

Sigma P 0270   Bovine heart   Lyophilized powder containing ~90% protein (Lowry); balance primarily buffer salts as imidazole & magnesium sulfate; activity: 2,500-10,000 U/mg protein; purified by modification of Teo | Protein activator required for the Ca^{2+} dependent activation of phosphodiesterase 3':5'-cyclic nucleotide, Sigma P 0520; Teo, TS et al, *J Biol Chem*, 248: 588, 1973

Fluka 21275   Bovine testes   MW ~17k   Lyophilized calcium complex; ≥70,000 U/mg protein; ~3.5% Ca; 1 U is the amount required to get 50% of the max activity of calcium-dependent, calmodulin-free phosphodiesterase

ICN 195694   Bovine testes   Lyophilized; virtually salt free; >40,000 U/mg; 1 U activates 0.016 U 3',5'-cyclic-nucleotide phosphodiesterase to 50% V_{max} when saturated with activator in the presence of 0.01 m*M* Ca^{2+}; ≥98% | Used to activate CaM-dependent phosphodiesterase, calcineurin, CaM kinases, etc

Sigma P 1431   Bovine testes   >98% (SDS-PAGE); essentially salt-free; lyophilized powder; activity: >40,000 U/mg protein (Lowry) | Ca^{2+} binding protein that is required for activation of cyclic nucleotide-dependent phosphodiesterase; involved in intracellular Ca^{2+} homeostasis; O'Neil, KT & DeGrado, WF, *Trends Pharmacol Sci*, 15: 59, 1990

Calbiochem 208695   Chicken recombinant, expressed in *E. coli*   MW 16.7k   ≥95% (SDS-PAGE); liquid in 100 m*M* KCl, 10 m*M* Tris-HCl, 1 m*M* DTT, 1 m*M* EDTA, pH 7.5; biological activity: Will activate nitric oxide synthase

ICN 195894   Chicken recombinant, expressed in *E. coli*   MW 16.7k   ≥95%

Biogenesis 1740-2656   Human brain   Lyophilized

Calbiochem 208698   Human brain   MW 16.7k   ≥98% (SDS-PAGE); salt-free lyophilized solid; specific activity: ≥40,000 units/mg protein; 1 unit is gives rise to 50% of the maximal enzyme activation of a standard level of activator-deficient 3',5'-cyclic nucleotide phosphodiesterase | Prepared from tissue of individuals that have been shown by certified tests to be negative for HBsAg & for antibodies to HIV & HCV

USBio C1035   Human brain   ≥98%; no contaminants detected; single band by SDS PAGE, IEP, &/or RID; 40K U/mg; salt-free lyophilized | Suitable for antigenic applications in immunological protocols; 1 U increases the maximal enzyme activation of activator-deficient 3', 5'-cyclic nucleotide phosphodiesterase by 50%

Biodesign A86810H   Human brain tissue   >98% | Apoptosis & signal transduction

Fluka 21272   Human erythrocytes   MW ~17k   ≥99.0% (HPLC); lyophilized calcium complex stabilized with $CaCl_2$ & imidazole hydrochloride; ~50,000 U/mg protein; ~1% Ca; 1 U is the amount required to get 50% of the max activity of calcium-dependent, calmodulin-free phosphodiesterase | CaM

ICN 195695   Human erythrocytes   Lyophilized; virtually salt free; 30,000-40,000 U/mg; 1 U activates 0.016 U 3',5'-cyclic-nucleotide phosphodiesterase to 50% V_{max} when saturated with activator in the presence of 0.01 mM Ca^{2+}; ≥90% | Used to activate CaM-dependent phosphodiesterase, calcineurin, CaM kinases, etc

Biogenesis 1740-2699   Ovine testis   Purified; lyophilized

ICN 195696   Porcine brain   Lyophilized; virtually salt free; >40,000 U/mg; 1 U activates 0.016 U 3',5'-cyclic-nucleotide phosphodiesterase to 50% V_{max} when saturated with activator in the presence of 0.01 mM Ca^{2+}; ≥95% | Used to activate CaM-dependent phosphodiesterase, calcineurin, CaM kinases, etc

Sigma P 1915   Porcine brain   >95% (SDS-PAGE); essentially salt-free; lyophilized powder; activity: >40,000 U/mg protein (Lowry) | Ca^{2+} binding protein that is required for activation of cyclic nucleotide-dependent phosphodiesterase; involved in intracellular Ca^{2+} homeostasis; O'Neil, KT & DeGrado, WF, *Trends Pharmacol Sci*, 15: 59, 1990

Sigma P 5779   Spinach   Essentially salt-free; lyophilized powder; activity: 15,000-30,000 U/mg protein (Lowry) | Ca^{2+} binding protein that is required for activation of cyclic nucleotide-dependent phosphodiesterase; involved in intracellular Ca^{2+} homeostasis; O'Neil, KT & DeGrado, WF, *Trends Pharmacol Sci*, 15: 59, 1990

Calmodulin Bα His Tag

Chemicon AG10P   ≥95% | Purified protein for apoptosis & signal transduction

Calmodulin Kinase II Inhibitor

USBio C1036-80   Recombinant, expressed in *E. coli*   MW 70k   >70% | Expressed mainly in brain & testis

Calmodulin, 10 nm Colloidal Gold Conjugated

Synonyms: Phosphodiesterase 3:5'-Cyclic Nucleotide Activator, Gold Conjugated

ICN 195703   Bovine brain   Lyophilized

Calmodulin, 20 nm Colloidal Gold Conjugated

Synonyms: Phosphodiesterase 3:5'-Cyclic Nucleotide Activator, Gold Conjugated

ICN 195704   Bovine brain   Lyophilized

Calmodulin, 5 nm Colloidal Gold Conjugated

Synonyms: Phosphodiesterase 3:5'-Cyclic Nucleotide Activator, Gold Conjugated

ICN 195702   Bovine brain   Lyophilized

Calmodulin, AEDANS Conjugated

Synonyms: Phosphodiesterase 3:5'-Cyclic Nucleotide Activator, AEDANS Conjugated

ICN 195698   Bovine brain   Lyophilized; >30,000 U/mg; 1 U activates 0.016 U 3',5'-cyclic-nucleotide phosphodiesterase to 50% V_{max} when saturated with activator in the presence of 0.01 mM Ca^{2+}

Calmodulin, Agarose

ICN 191303   5 atoms hydrophilic spacer arm; 1 mg Calmodulin per mL gel; suspension in PBS, 0.02% NaN3 | Applications: ATPases, protein kinases, phosphodiesterases, proteins in neurotransmission

Calmodulin, Biotin Conjugated

Calbiochem 208697   Bovine brain   Liquid in PBS 1 mg/mL BSA, 0.02% NaN_3; may be carcinogenic/teratogenic | Biotin-CaM; useful for the study of calmodulin-binding proteins; Hern Andex, EO et al, *Tissue Cell*, 26: 849, 1994; Kincaid, RL et al, *Methods Enzymol*, 159: 605, 1988

Calmodulin, Dansyl-

Synonyms: Phosphodiesterase 3:5'-Cyclic Nucleotide Activator, Dansyl-

ICN 195699   Bovine brain   Lyophilized; dansyl chloride labeled; ~30,000 U/mg; 1 U activates 0.016 U 3',5'-cyclic-nucleotide phosphodiesterase to 50% V_{max} when saturated with activator in the presence of 0.01 mM Ca^{2+}

Calmodulin, FITC Conjugated

Synonyms: Phosphodiesterase 3:5'-Cyclic Nucleotide Activator, FITC Conjugated

ICN 195701   Bovine brain   Lyophilized; >40,000 U/mg; 1 U activates 0.016 U 3',5'-cyclic-nucleotide phosphodiesterase to 50% V_{max} when saturated with activator in the presence of 0.01 mM Ca^{2+}

Calmodulin, Immobilized

Calbiochem 208702   Bovine brain   Cross-linked agarose matrix coupled to 1 mg/mL calmodulin in PBS buffer containing 0.02% NaN_3 | Tariq Khan, M et al, *J Biol Chem*, 269: 10016, 1994; Lukas, TJ & Watterson, DM, *Methods Enzymol*, 157: 328, 1988

Calmodulin, TRITC Conjugated

Synonyms: Phosphodiesterase 3:5'-Cyclic Nucleotide Activator, TRITC Conjugated

ICN 195705   Bovine brain   Lyophilized; >40,000 U/mg; 1 U activates 0.016 U 3',5'-cyclic-nucleotide phosphodiesterase to 50% V_{max} when saturated with activator in the presence of 0.01 mM Ca^{2+}

Calmodulin-Dependent Phosphodiesterase

Chemicon AG641   Bovine brain   ≥95% | Purified protein for apoptosis & signal transduction

Calpain Inhibitor I

Biogenesis 1740-6110   Non-species synthetic   Lyophilized

Calpain Inhibitor II

Biogenesis 1740-6120   Non-species synthetic   Lyophilized

Calpain Inhibitor III

Biogenesis 1740-6130   Non-species synthetic   Lyophilized

Calpain Inhibitor IV

Biogenesis 1740-6140   Non-species synthetic   Lyophilized

Calpastatin

Synonyms: Calcium Activated Neutral Proteinase Inhibitor; Protease Inhibitor

Calbiochem 208901 Human erythrocytes MW 250k Liquid in 30 mM Tris-HCl, 1 mM EDTA, 1 mM EGTA, 600 μM DTT, 40% glycerol, pH 7.4; specific activity: ≥1000 units/mg protein; one unit is the amount of protein that inhibits the increase in absorbance at 750 nm by 1.0 induced by one unit of calpain I at 30°C, pH 7.5 using casein as a substrate | Tetrameric protein that is a specific, endogenous inhibitor of the calcium-activated neutral proteinases; prepared from blood that has been shown to be negative for HBsAg & for antibodies to HIV & HCV; Melloni, E et al, *Biochem Biophys Res Comm*, 106: 731, 1981; Shigeta, K et al, *Biochem Int*, 9: 327, 1994; Parkes, C, *Proteinase Inhibitors* (Barrett & Salvesen, eds), 19: 571, Elsevier Science Publishers BV, 1986

Calbiochem 208900 Human recombinant MW 14k Homogeneous purity (SDS-PAGE); lyophilized solid; soluble in water | Domain I of human calpastatin; endogenous protease inhibitor that acts specifically on calpain I; exhibits greater inhibitory capacity than calpain inhibitors I & Ii; not available for sale in Japan; Salamino, F et al, *Biochem Biophys Res Comm*, 199: 1326, 1994; Vemori, T et al, *Biochem Biophys Res Comm*, 166: 1485, 1990; Asada, K et al, *J Enzyme Inhib*, 3: 39, 1989

Calreticulin

Sigma C 4714 Bovine liver MW 60k Lyophilized powder; ≥90% (SDS-PAGE); essentially salt-free | A high-specificity calcium binding protein which plays a dynamic role in calcium homeostasis; binds directly to the DNA domain of hormone receptors as well as to the regulatory cytoplasmic domain of proteins of the integrin family; possesses a chaperone function; Mery, L et al, *J Biol Chem*, 271: 9332, 1996; Burns, K et al, *Nature*, 367: 476, 1994; Dedhar, S, *TIBS*, 19: 269, 1994; Wada, I et al, *J Biol Chem*, 270: 20298, 1995

Calyculin A

USBio C1036-70 >98% by HPLC; lyophilized; soluble in DMSO & EtOH, insoluble in H_2O | Potent inhibitor (Ki ~ 0.1 nM) with high specificity for the PP-1 & PP-2 classes of protein Ser/Thr phosphatase

Cancer Antigen 125

Cortex CP1062 Low cross grade purity

Cortex CP1062P Standard grade purity

Biodesign A32303H Human fluids Ovarian cancer calibrator grade; low cross reactivity | Tumor markers, cancer antigens & oncogenes

Biodesign A97182H Human fluids Ovarian cancer calibrator grade | Tumor markers, cancer antigens & oncogenes

Cancer Antigen 15-3

USBio C0050-20 Human ≥99%; no contaminants detected; single band by SDS-PAGE, IEP, &/or RID; CA125 ≤1%, CA 19-9 ~6%, CA 72-4 ≤0.05%; 130K U/mL supplied in 0.01 M phosphate, 0.1 M NaCl, pH 7.2, 0.1% NaN₃ | Suitable for antigenic applications in immunological protocols

USBio C0050-24 Human ≥98%; no contaminants detected; single band by SDS-PAGE, IEP, &/or RID; 6,100 U/mL (CIS RIA) supplied in 0.1 M NaCl, 0.01 M phosphate, pH 7.4, & 0.1% NaN₃ | Suitable for antigenic applications in immunological protocols

USBio C0050-21 Human fluids; ≥90% (HPLC); ≤10% background cancer antigens, mainly CA 125; ~20kI U/mL (CIS CA 15-3 ELISA RIA assay); liquid in PBS, pH 7.4, 0.1% NaN₃ | Suitable for antigenic applications in immunological protocols.

Cancer Antigen 15-3™

Cortex CP1064 Low cross purity

Cortex CP1064P Standard grade purity

Biodesign A32231H Cell culture Breast cancer calibrator grade; low cross reactivity | Tumor markers, cancer antigens & oncogenes; CA 15-3™ is a trademark of Centocor Inc, Malvern, PA

Biodesign A32618H Cell culture Breast cancer calibrator grade | Tumor markers, cancer antigens & oncogenes; CA 15-3™ is a trademark of Centocor Inc, Malvern, PA

Biodesign A37211H Cell culture Breast cancer calibrator grade | Tumor markers, cancer antigens & oncogenes; CA 15-3™ is a trademark of Centocor Inc, Malvern, PA

Biodesign A33130H Human fluids Breast cancer calibrator grade; low cross reactivity | Tumor markers, cancer antigens & oncogenes; CA 15-3™ is a trademark of Centocor Inc, Malvern, PA

Biodesign A33187H Human fluids Breast cancer calibrator grade | Tumor markers, cancer antigens & oncogenes; CA 15-3™ is a trademark of Centocor Inc, Malvern, PA

Biodesign A97183H Human fluids Breast cancer antigen grade | Tumor markers, cancer antigens & oncogenes; CA 15-3™ is a trademark of Centocor Inc, Malvern, PA

Cancer Antigen 19-9

Cortex CP1063 Low cross grade purity

Cortex CP1063P Standard grade purity

Biodesign A37116H Cell culture Gastrointestinal cancer calibrator grade | Tumor markers, cancer antigens & oncogenes

USBio C0075-13 Human ≥70%; no contaminants detected; single band by SDS-PAGE, IEP, &/or RID; 210K U/mL (CIS RIA) supplied in 20 mM phosphate buffer, 150 mM NaCl, pH 5.2, 0.1% NaN₃, 0.2m filtered | Suitable for antigenic applications in immunological protocols

USBio C0075-14 Human ≥99%; no contaminants detected; single band by SDS-PAGE, IEP, &/or RID; CA 125: 0%, CA 15-3: 0%, CA 72-4 ≤ 1%; ~1 M U/mL (CIS RIA) supplied in PBS buffer, 0.1% NaN₃ | Suitable for antigenic applications in immunological protocols

Biodesign A32302H Human fluids Gastrointestinal cancer calibrator grade; low cross reactivity | Tumor markers, cancer antigens & oncogenes

Biodesign A97184H Human fluids Gastrointestinal cancer antigen grade | Tumor markers, cancer antigens & oncogenes

Biodesign A97185H Human fluids Gastrointestinal cancer antigen grade | Tumor markers, cancer antigens & oncogenes

Cancer Antigen 19-9®

Synonyms: GI-Pancreatic Tumor Marker Antigen

ICN 770961 ICN 770962 ICN 770963 Human ascites >65-85%; standard grade; >50,000 U/mL | For controls & calibrators

ICN 770971 ICN 770972 ICN 770973 Human ascites >96%; iodination grade; >50,000 U/mL | For immunization or labeling

Cancer Antigen 242™

Biodesign A37705H Cell culture Calibrator grade | Tumor markers, cancer antigens & oncogenes; CA 242™ is a trademark of Kabi Pharmacia

Cancer Antigen 27-29

Cortex CP1096 Low cross grade purity

Cancer Antigen 50

Cortex CP1066 Low cross grade purity

Cortex CP1066P Standard grade purity

Cancer Antigen 50

Biodesign A37115H Cell culture Calibrator grade | Tumor markers, cancer antigens & oncogenes

Cancer Antigen 50

Biogenesis 1695-0356 Human tissue fluids Liquid

Cancer Antigen 72-4

Cortex CP1065 Low cross grade purity

Cortex CP1065P   Standard grade purity

USBio C0100-10   Human   ≥98%; no contaminants detected; single band by SDS-PAGE, IEP, &/or RID; 100K U/mL (Centocor RIA) supplied in 0.1 M PBS, pH 7.4, 0.05% NaN_3, 2.5% sucrose | Suitable for antigenic applications in immunological protocols, calibrators & controls, & other applications where background contaminants need to be minimized; very low background levels (<1%) of CA-125 & CA-19-9

ICN 770981 ICN 770982 ICN 770983   Human ascites >96%; iodination grade; >50,000 U/mL | Human antigen for immunization or labeling

Biodesign A32200H   Human fluids   Calibrator grade | Tumor markers, cancer antigens & oncogenes

Calbiochem 209930   Human fluids, TAG-72   Liquid in PBS, pH 7.4 with 0.1% NaN_3; activity: 5000-50,000 units/mL; contaminants: traces of α_1-acid glycoprotein, CA 15-3, CA 19-9 & CA 125, CEA & HAS; prepared from fluids of individuals shown to be negative for HBsAg & for antibodies to HIV & HCV | A glycoprotein that is one of the most sensitive & specific markers for monitoring gastric cancers; elevated in serum of individuals with breast, colon, lung, ovary, rectum & stomach cancers; suitable for use in immunoassays & as a standard; Guadagni, F et al, *Cancer Invest*, 13: 227, 1995; Alles, AJ et al, *Ann Surg*, 219: 131, 1994; Johnson, VG et al, *Cancer Res*, 46: 850, 1986; Klug, TL et al, *Int J Cancer*, 38: 661, 1986

Cancer Antigen, Breast

Synonyms: Cancer Antigen 15-3

Biogenesis 1695-0056   Human tissue fluids   SA: 34,850 U/mL; tested negative for HIV 1 and 2, HCV and HbsAg; purified; PBS buffer, pH 7.4, 0.1% NaN_3; liquid

Biogenesis 1695-0066   Human tissue fluids   Semi-pure; liquid

Cancer Antigen, Gastrointestinal

Synonyms: Cancer Antigen 19-9

Biogenesis 1695-0156   Human tissue fluids   Purified; liquid

Biogenesis 1695-0166   Human tissue fluids   Semi-pure; liquid

Cancer Antigen, Ovarian

Synonyms: Cancer Antigen 125

USBio C0050-10   Human fluids, ovarian cancer   ≥95%; no contaminants detected; single band by SDS-PAGE, IEP, &/or RID; CA 19-9 ≤1.0%, CA72-4 ≤1.0%, CA15-3 ≤0.1%; ≥120K U/mL (ROCHE) supplied in 0.03 M PBS, pH 7.2, 0.05% NaN_3, sucrose | Suitable for antigenic applications in immunological protocols.

Biogenesis 1695-0556   Human pleural fluids   Tested negative for HIV 1, 2, HCV and HBsAg; purified; PBS buffer, pH 7.4, 0.1% NaN_3; liquid

Biogenesis 1695-0566   Human tissue fluids   Semi-pure; liquid

USBio C0050   Human, ovarian cancer   ≥70% with trace amounts of CEA, CA19-9, CA15-3, CA 72-4; 200K U/mL (ROCHE) supplied in 0.02 M PBS, pH 7.2, 0.05% NaN_3 | Suitable for antigenic applications in immunological protocols

Cancer Associated Antigen B

Synonyms: Cancer Antigen 15-3

Calbiochem 209915   Breast tumor, Human fluids   MW 400k Liquid in PBS, pH 7.4 with 0.1% NaN_3; concentration: 10,000-200,000 units/mL (RIA); contaminants: CA 19-9 & CA 125: <25%; prepared from fluids of individuals shown to be negative for HBsAg & for antibodies to HIV & HCV | A glycoprotein normally present at levels < 30 units/mL, found in significantly higher levels in serum of women with metastatic breast cancer; exists in serum & fluid as a high molecular weight complex of >1,000,000; suitable for use in immunoassays & as an immunogen; Barros, AC et al, *Eur J Surg Oncol*, 20: 130, 1994; Reddish, M et al, *J Tumor Marker Oncol*, 7: 1, 1993; Hilkens, J et al, *Int J Cancer*, 34: 197, 1984; Kufe, D et al, *Hybridoma*, 3: 223, 1984

Cancer Associated Antigen GI

Synonyms: Cancer Antigen 19-9

Calbiochem 209920   Gastrointestinal tumor, Human fluids MW 200k   Liquid in PBS, pH 7.4 with 0.1% NaN_3; activity: 50,000-200,000 units/mL (CIS-EIA); contaminants: CA 15-3 & CA 125: <25%; prepared from fluids of individuals shown to be negative for HBsAg & for antibodies to HIV & HCV | A glycoprotein that is elevated in serum of individuals with colorectal, gastric or pancreatic cancers; the antigen is defined by a monoclonal antibody that recognizes a carbohydrate determinant, sialylated lacto-*N*-fucopentaose II, a Lewis a blood group antigen; suitable for use in immunoassays as an immunogen & in enzyme/radiolabeling; Deugnier, YM et al, *Gut*, 35: 1107, 1994; Diez, M et al, *Anticancer Res*, 14: 2819, 1994; Klug, TL et al, *Cancer Res*, 48: 1505, 1988; Kufe, D et al, *Hybridoma*, 3: 223, 1984; Steinberg, W, *Am J Gastroenterology*, 4: 350, 1985; Koprowski, H et al, *Science*, 212: 53, 1981

Calbiochem 209922   Gastrointestinal tumor, Human, Cell culture-derived   MW 200k   Liquid in PBS, pH 7.4 with 0.1% NaN_3; ≥95% (SDS-PAGE); activity: 10,000-2,000,000 units/mL; contaminants: CA 15-3 & CA 125: <25%; prepared from cell culture supernatants of a human colon adenocarcinoma cell line | A cell surface glycoprotein found at elevated levels in serum of individuals with colorectal, gastric or pancreatic cancers; mainly secreted by pancreatic & bile duct cells; identified by a monoclonal antibody that recognizes a carbohydrate determinant, sialylated lacto-*N*-fucopentaose II, a Lewis a blood group antigen; suitable for use in immunoassays as an immunogen; Deugnier, YM et al, *Gut*, 35: 1107, 1994; Diez, M et al, *Anticancer Res*, 14: 2819, 1994; Klug, TL et al, *Cancer Res*, 48: 1505, 1988; Steinberg, W, *Am J Gastroenterology*, 4: 350, 1985; Koprowski, H et al, *Science*, 212: 53, 1981

Cancer Associated Antigen O

Synonyms: Cancer Antigen 125

Calbiochem 209925   Ovarian tumor, Human fluids   MW 200k Liquid in PBS, pH 7.4 with 0.1% NaN_3; activity: 50,000-200,000 units/mL; contaminants: CA 19-9 & CA 27.29: <25%; prepared from fluids of individuals shown to be negative for HBsAg & for antibodies to HIV & HCV | A cell surface glycoprotein that is widely used as a marker for ovarian epithelial cancers; exists in serum & fluid as a high molecular weight complex of >100k; suitable for use in immunoassays, as an immunogen & in enzyme/radiolabeling; O'Riordan, DK et al, *Gut*, 36: 303, 1995; Rustin, GJ, *Ann Oncol*, 4: 571, 1993; Davis, HM et al, *Cancer Res*, 46: 6143, 1986; Canney, PA et al, *Br J Cancer*, 50: 765, 1984

Carbonic Anhydrase

Sigma C 1311   Bovine erythrocyte   MW 29k   Vial contains enough FITC-conjugated protein to run 50 mini-gels or 25 standard size gels; protein band be visualized by using UV light or Brilliant Blue stain | Fluorescent marker for SDS-page & protein transfer; suitable for use as molecular weight standards in both SDS-PAGE & transfer membranes

Carbonic Anhydrase I

Sigma C 6653   Human erythrocytes   pI 6.6; vial contains ~2 mg | IEF Marker

Carbonic Anhydrase II

Sigma C 3666   Bovine erythrocytes   pI 5.4; vial contains ~1 mg | IEF Marker

Sigma C 6403   Bovine erythrocytes   pI 5.9; vial contains ~2 mg | IEF Marker

Carcinoembryonic Antigen

Synonyms: CD66

Dako X0556   >95% | Antigen useful as an immunogen

Cortex CP1001   Cell line   >95%

USBio C1300-14   Human   ≥98%; no contaminants detected; single band by SDS-PAGE, IEP, &/or RID; 1.55 mg/mL (EIA) supplied in 0.15 M phosphate buffer, pH 7.3, 0.07% NaN_3 | Suitable for antigenic applications in immunological protocols

Biodesign A32100H Human ascites ≥95% | Tumor markers, cancer antigens & oncogenes

Biodesign A32137H Human ascites ≥95% | Tumor markers, cancer antigens & oncogenes

Calbiochem 219369 Human colon adenocarcinoma cell line MW 180k Lyophilized solid; single band purity (SDS-PAGE) | Found in tumors in adults; presence indicates carcinogenic activity; purified CEA has been the most widely used general marker in determining & monitoring the status of malignant disease; derived from an established cell line

Fitzgerald 30-AC25 Human colon carcinoma Standard purity >70%

Fitzgerald 30-AC30 Human colon carcinoma High purity >96%

USBio C1300-16 Human colon carcinoma liver metastases No contaminants detected by SDS-PAGE or Rocket IEP; ~1.0 mg/mL supplied in 0.15 M phosphate buffer, pH 7.3, 0.1% NaN_3 | Suitable for antigenic applications in immunological protocols

Biogenesis 1820-5904 Human liver metastases Semi-pure; liquid

Biogenesis 1820-5804 Human liver metastases of colon adenocarcinoma MW 180k by SDS-PAGE (single subunit) Tested negative for HBsAg, HIV 1 and 2 and HCV antibodies; pI: 3.7; purified; 10 mM TRIS, pH 8.0, 0.1% NaN_3; liquid

Biodesign A81125H Human metastatic liver 98% | Tumor markers, cancer antigens & oncogenes

Calbiochem 219368 Human tumor MW 180k Liquid in 150 mM NaCl, 20 mM phosphate buffer, pH 7.3; preservative- & reductant-free; no perchloric acid or detergents were used; ≥98% (IEP); single major band purity (SDS-PAGE); prepared from tissue of individuals shown to be negative for HBsAg & for antibodies to HIV & HCV | Found in tumors in adults; a useful marker for breast cancer & adenocarcinomas of the esophagus & stomach; derived from liver metastases of colon adenocarcinoma via saline extraction, ion exchange & gel filtration; Kim, YH et al, *Cancer*, 75: 451, 1995; Esteben, JM et al, *Cancer*, 74: 1575, 1994; Jacobs, EL & Haskell, CM, *Curr Probl Cancer*, 15: 299, 1991; Robertson, JF et al, *Br J Cancer*, 64: 757, 1991

Cortex CP3012 Liver >98%

Cortex CP3012P Liver >40%

Fitzgerald 30-AC26 Recombinant High purity >60%

Fitzgerald 30-AC27 Recombinant High purity >95%

Cardiotoxin

Synonyms: Snake Toxin; *Naja nigricollis* Toxin γ; Protein Kinase C Inhibitor

Sigma C 9759 *Naja mossambica mossambica* Mixture of cardiotoxins

Sigma C 1777 *Naja naja kaouthia* Mixture of cardiotoxins

Calbiochem 217504 *Naja nigricollis* MW 6827.4 $C_{298}H_{493}N_{81}O_{77}S_{12}$ ≥95% (IEF); lyophilized solid; contaminants: phospholipase A_2 activity <2%; LD_{50}≤2000 mg/kg; not available for sale outside of the United States | A cytolytic toxin that causes depolarization of skeletal muscle fibers *in vitro*; stimulates Ca^{2+} transport & ATP hydrolysis by the sarcolemmal Ca^{2+}/Mg^{2+} ATPase; action is strongly potentiated by phospholipase A2; Raynor, RL et al, *J Biol Chem*, 266: 2753, 1991; Grognet, JM et al, *Mol Immunol*, 23: 132, 1986; Huang, JL & Trumble, WR, *Toxicon*, 29: 31, 1991; Jang, JY et al, *Biochemistry*, 36: 4635, 1997; *Merck Index*, 12: 1884

Cardiotrophin I

Biodesign A52332H *E. coli* MW 21.5k Purified

Chemicon GF058 Human ≥95%

Biogenesis 1828-5030 Human r-DNA Lyophilized

BioSource International PHC1594 Human recombinant

Calbiochem 218200 Human recombinant, expressed in *E. coli* MW 21.5k ≥99% (SDS-PAGE & HPLC); lyophilized solid; biological activity: ED_{50}≤1 ng/mL as measured by the ability of CT-1 to induce proliferation of TF-1 cells in a dose-dependent manner | A cytokine with structural similarities to interleukin-6; promotes the survival of neonatal cardiomyocytes via the activation of an anti-apoptotic signaling pathway requiring MAP kinases; induces a distinct hypertrophic response to endothelin-1 in cultured neonatal cardiomyocytes; induces heat shock protein accumulation in cultured cardiac cells & protects them from stressful stimuli; Ch Andrasekar, B et al, *Immunol Lett*, 61: 89, 1998; Kuwahara, K et al, *J Cardiovasc Pharmacol*, 31: 5354, 1998; Stephanou, A et al, *J Mol Cell Cardiol*, 30: 849, 1998; Sheng, Z et al, *J Biol Chem*, 272: 5783, 1997

PeproTech 300-32 Human recombinant, expressed in *E. coli* MW 21.5k >98% (SDS-PAGE) & HPLC; <0.1 ng endotoxin per mg (1EU/mg); lyophilized | Member of the leukemia inhibitory/ciliary neurotrophic factor/Oncostatin M/interleukin-11 family of cytokines; affects cells through interaction with the gp 130 receptor subunit; 201 AA; ED_{50} < 1.0 ng/mL; SA > 1 x 10^6 U/mg; SA determined by the dose-dependent proliferation of TF-1 cells

Casein

ICN 101289 Hammerstein grade; typical analysis: 13.9% N_2, 6.0% moisture, 1.1% ash; 2.0% ash (phosphate free), pH 5.5 (2% suspension)

ICN 104815 A standard casein; ANRC reference protein

ICN 904520 Vitamin free; precipitated casein, extracted with ethyl alcohol; from New Zealand lactic acid

ICN 904798 Vitamin free; micropulverized

Fluka 22078 Bovine milk Low vitamin; ~5% water; ≤0.1% Ca, K, Na; ≤0.01% Zn; ≤0.005% Ca, Co, Cu, Ni, Pb; ≤0.2% lactic acid; ~0.2 ppm thiamine hydrochloride; ~0.5 ppm riboflavin; ~0.4 ppm pyridoxine hydrochloride, folic acid; ~4 ppm pantothenic acid; ~0.01 ppm biotin, vitamin B_{12}

Fluka 22080 Bovine milk ≤5% residue on ignition; ≤15% water | Can be cross-linked with peroxidase; Matheis, G & Whitaker, JR, *J Prot Chem*, 3: 35, 1984; Hinterwaldner, *Coating*, 17: 174, 1984

Sigma C 3400 Bovine milk Vitamin free; yellow-tan powder; essentially "vitamin free"; from lactic acid precipitated New Zeal & casein extracted with ethyl alcohol

Sigma C 5679 Bovine milk Essentially vitamin free; insect cell culture tested

Sigma C 5890 Bovine milk Purified powder | After dephosphorylation, a suitable substrate for protein kinase; mayer, SE et al, *Meth Enzymol*, 38: 66, 1974

Sigma C 6554 Bovine milk High protein; insect cell culture tested

Sigma C 7078 Bovine milk Technical

Sigma C 7906 Bovine milk Dietary fiber control

Sigma C 0536 Goat milk Lyophilized powder; ~90% protein (Biuret); lactose content <3%

Sigma C 3335 Human milk Lyophilized powder; ≥90% (electrophoresis); containing ~60% protein (Biuret);

Sigma C 7164 Sheep milk Lyophilized powder containing ~85% protein (Biuret); lactose content <3%

Casein Sodium Salt

Amersham US12865

ICN 102896 ICN 902896 Sodium Caseinate

Sigma C 8654 Bovine milk

Casein Type I, FITC Conjugated

Synonyms: Protease Substrate

Sigma C 0403 Bovine milk Lyophilized powder; essentially salt-free; 5-20 μg FITC/mg solid | Twining, SS, *Fed Proc*, 42: 1951, 1983; Twining, SS, *Anal Biochem*, 143: 30, 1984

Proteins

Casein Type II, FITC Conjugated

Synonyms: Protease Substrate

Sigma C 3777 Bovine milk Lyophilized powder; essentially salt-free; 20-50 μg FITC/mg solid | Twining, SS, *Fed Proc*, 42: 1951, 1983; Twining, SS, *Anal Biochem*, 143: 30, 1984

Casein Type III, FITC Conjugated

Synonyms: Protease Substrate

Sigma C 0528 Bovine milk Lyophilized powder; essentially salt-free; 50-100 μg FITC/mg solid | Twining, SS, *Fed Proc*, 42: 1951, 1983; Twining, SS, *Anal Biochem*, 143: 30, 1984

Casein Vitafree

Amersham US12866 Free of vitamins | Hot alcohol extracted New Zealand casein

Casein, (^{14}C-Me)-

ARC ARC-426 MW 23.6k 0.5-5 μCi/mg; 18.5-185 KBq/mg; in 0.01 *M* sodium phosphate, pH 7.2 | Radiochemical

Sigma C 5784 5-50 μCi/mg protein; solution in 40 *mM* potassium phosphate, pH 7.0, in serum bottle | Radiochemicalprepared from Sigma C 5890

Casein, (^{14}C-Me)-α-

Sigma C 5909 5-50 μCi/mg protein; solution in 40 *mM* potassium phosphate, pH 7.0, in serum bottle | Radiochemicalprepared from Sigma C 6780

ARC ART-758 MW 23.6k 5-10 μCi/mg; 185-370 KBq/mg; in 0.01 *M* sodium phosphate, pH 7.2 | Radiochemical

Sigma C 6034 5-50 μCi/mg protein; solution in 40 *mM* potassium phosphate, pH 7.0, in serum bottle | Radiochemicalprepared from Sigma C 6905

Sigma C 6159 5-50 μCi/mg protein; solution in 40 *mM* potassium phosphate, pH 7.0, in serum bottle | Radiochemicalprepared from Sigma C 0406

Casein, Agarose

ICN 191282 5 atoms hydrophilic spacer arm; 10-15 mg ligand per mL gel; suspension in PBS, 0.02% NaN3 | Iodinatable reagent for protease activity measurement; protein kinase substrate

Casein, Azo-

Synonyms: Protease Substrate

Fluka 11610 $E^{1\%}_{440}$ in 0.1 *M* NaOH: 32-38 | Suitable as substrate for proteolytic enzymes; Rowan, AD & Buttle, DJ, *Meth Enzymol*, 244: 555, 1994; Peyronel, DV & Cantera, AMB, *Electrophoresis*, 16: 1894, 1995

ICN 100863

Sigma A 2765 $E^{1\%}_{440}$ in 0.1 N NaOH: 32-38 | Sulfanilamide-azocasein; Tomarelli, RM et al, *J Lab Clin Med*, 34: 428, 1949

Casein, Azo- Sodium Salt

Fluka 11615 Lyophilized | See Fluka 11610

Casein, Dephosphorylated

Sigma C 4032 Bovine milk Lyophilized powder; ≥80% dephosphorylated

Sigma C 4765 Bovine milk 5% solution; aseptically filled; hydrolyzed & partially dephosphorylated | Prepared for the assay of cyclic-AMP dependent protein kinase; other substrates such as protamine & histone might be evaluated for this purpose

Casein, Hammersten

Amersham US12840 Blocking agent used in Western blotting

Casein, High Nitrogen

Amersham US12845 ≥95% Protein; ≤10% moisture; ≤1.5% fat | Prepared exclusively from New Zealand casein

ICN 901293 Purified

Casein, Low Trace Element

ICN 960128 Bovine milk 30 mesh; precipitated by HCl rather than lactic acid to remove most of the trace metals usually present in casein

Casein, *N,N*-Dimethylated Salt Free

Synonyms: Protease Substrate

Sigma C 9801 Bovine milk Lyophilized powder; <10% reactivity with 2,4,6-trinitrobenzenesulfonic acid (TNBS) compared to non-methylated casein; prepared by reductive methylation of C 5890 by the method of Cabacungan | Cabacungan, JC et al, *Anal Biochem*, 124: 272, 1982

Casein, Resorufin Labeled

Synonyms: Casein, 1-(Resorufin-4-Carbonyl)Piperidine-4-Carboxylic Acid; Protease Substrate

Fluka 22097 ~90 μg resorufin/mg casein | Substrate for protease activity in beer; Mochaba, F et al, *Proc Congr-Eur Brew Conv 24th*, 597, 1993

Casein, α-

ICN 100251 Prepared by urea fractionation

Fluka 22084 Bovine milk ≥90% (GE); off-white powder; ≥80% protein content; ≤5% water

Sigma C 6780 Bovine milk Lyophilized powder; chromatographically purified; ≥70% protein (Biuret); ~85% α$_s$-casein by electrophoresis

Casein, α- Dephosphorylated

Sigma C 8032 Bovine milk Lyophilized powder; ≥80% enzymatically prepared from C 6780

Casein, α- Low Protease

Synonyms: Plasmin Substrate; Fibrinolysin Substrate

ICN 195096 Bovine milk Processed to minimize contaminating protease | Suitable substrate for the assay of plasmin

Sigma C 7891 Bovine milk Lyophilized powder containing 10-20% NaCl; ~60% α$_s$-casein by electrophoresis; balance primarily β-casein

Casein, α- No Protease

Fluka 22085 Bovine milk ≥60% (GE); ≥70% protein content; ≤40% β-casein; no detectable proteases

Casein, β-

ICN 100321 Lyophilized; ≥90%; <2% α-casein

Fluka 22086 Bovine milk ≥80% (GE); lyophilized white powder; ≥80% protein content

Casein, β- Dephosphorylated

Sigma C 8157 Bovine milk Lyophilized powder; ≥80% dephosphorylated; enzymatically prepared from C 6905

Casein, β- Salt Free

Sigma C 6905 Bovine milk Lyophilized powder; essentially salt-free; ≥90% β-casein by electrophoresis

Casein, γ-

ICN 100653 Bovine milk

Fluka 22087   Bovine milk   ≥70% (GE); beige powder; ≥90% protein content

Casein, κ-
Sigma C 0406   Bovine milk   Lyophilized powder; ≥80% κ-casein by electrophoresis

Caspase I
Chemicon CC126   Human active recombinant   Purified protein for apoptosis & signal transduction

Caspase II
Chemicon CC127   Human active recombinant   Purified protein for apoptosis & signal transduction

Caspase III
Synonyms: CPP-32
Kamiya
Chemicon CC119   Human active recombinant   Purified protein for apoptosis & signal transduction

Caspase III Fluorometric Substrate
USBio C2087-19   ≥95% by HPLC; lyophilized | Useful in Apoapain activity assay *in vitro* (5–50 μM); solubilize lyophilized in 1 mL of DMSO

Caspase III, Active
BioSource International PHZ0014   Human recombinant

Caspase IX
Chemicon CC120   Human active recombinant   Purified protein for apoptosis & signal transduction

Caspase VI
Chemicon CC122   Human active recombinant   Purified protein for apoptosis & signal transduction
Kamiya   Human recombinant

Caspase VI, Active
BioSource International PHZ0034   Human recombinant

Caspase VII
Kamiya
Chemicon CC125   Human active recombinant   Purified protein for apoptosis & signal transduction

Caspase VIII
Kamiya
Chemicon CC123   Human active recombinant   Purified protein for apoptosis & signal transduction

Caspase X
Chemicon CC128   Human active recombinant   Purified protein for apoptosis & signal transduction

Cathepsin B
Cortex CP3009   Liver   >95%

Cathepsin C
Biogenesis 1910-8709   Bovine spleen

Cathepsin D
Cortex CP3090   >95%

Biogenesis 1910-9004   Bovine spleen   Lyophilized
Biogenesis 1910-9037   Human kidney   Liquid
Biodesign A50161H   Human liver   Purified | Tumor markers, cancer antigens & oncogenes
Biogenesis 1910-9016   Human liver   Lyophilized
USBio C2097-26   Human liver tissue   ≥300 U/mg; 1 mg/mL; lyophilized from 2 *mM* sodium phosphate, pH 6.5 | Suitable for antigenic applications in immunological protocols

Cathepsin G
Biogenesis 1910-9847   Human   Liquid
USBio C2097-50   Human   ≥98%; no contaminants detected; single band by SDS-PAGE, IEP, &/or RID; 2-4 U/mg; lyophilized, salt free | Suitable for antigenic applications in immunological protocols
Cortex CP3010   Neutrophil   >95%

Cathepsin H
Biogenesis 1910-9957   Human   Liquid
USBio C2097-60   Human liver tissue   ≥80% (SDS-PAGE); trace human enzyme contaminants; ~0.5 mg/mL in 30 *mM* sodium acetate buffer, pH 5.0, 1 *mM* EDTA, 0.08 *M* NaCl | Suitable for antigenic applications in immunological protocols

CC Chemokine
Synonyms: C-10
ICN 195770   Murine recombinant, expressed in *E. coli*   ≥97%; lyophilized; ED$_{50}$ = 0.05-0.2 μg/mL activity

CC Chemokine Receptor I
Chemicon GF090   Human   ≥95% | Blocks function

CC Chemokine Receptor VIII, 2nd Extracellular Loop
Chemicon GF096   Human   ≥95% | Blocks function

CC Chemokine Receptor VIII, N-Terminal
Chemicon GF095   Human   ≥95% | Blocks function

CD137 Ligand-muCD8 Fusion Protein
Synonyms: CD137L-muCD8 Fusion Protein
Alexis ANC-503-020   Human   Purified liquid; free of azide & carrier protein & stabilized with 0.5 mg/mL gentamycin sulfate | A soluble fusion protein consisting of the extracellular (184 AA) domain of human CD137L fused to the extracellular domain (167 AA) of mouse CD8α; transfectant cell line: mouse myeloma cell line P3x63Ag8.653; applications in flow cytometry & immunohistochemistry (frozen sections); binds to CD137 & blocks binding of anti-CD137 monoclonal antibody; human CD137L (4-1BB Ligand) is a type II transmembrane protein constitutively expressed by monocytes, B cells & neuroblastoma cells; binding of CD137 (4-1BB) to CD137L induces monocyte activationAlderson, MR, *Eur J Immunol*, 24: 2219, 1994; Langstein, J et al, *J Immunol*, 160: 2488, 1998

CD137 Ligand-muCD8 Fusion Protein, Biotin Conjugated
Synonyms: CD137L-muCD8 Fusion Protein, Biotin Conjugated
Alexis ANC-503-030   Human   Liquid; stabilized with 0.04% sodium azide | A soluble fusion protein consisting of the extracellular (184 AA) domain of human CD137L fused to the extracellular domain (167 AA) of mouse CD8α; transfectant cell line: mouse myeloma cell line P3x63Ag8.653; applications in flow cytometry & immunohistochemistry (frozen sections); binds to CD137 & blocks binding of anti-CD137 monoclonal antibody; human CD137L (4-1BB Ligand) is a type II transmembrane protein constitutively expressed by monocytes, B cells & neuroblastoma cells; binding of CD137 (4-1BB) to CD137L induces monocyte activationAlderson, MR, *Eur J Immunol*, 24: 2219, 1994; Langstein, J et al, *J Immunol*, 160: 2488, 1998

CD137 Ligand-muCD8 Fusion Protein, FITC Conjugated

Synonyms: CD137L-muCD8 Fusion Protein, FITC Conjugated

Alexis ANC-503-040 Human Liquid; stabilized with 0.04% sodium azide | A soluble fusion protein consisting of the extracellular (184 AA) domain of human CD137L fused to the extracellular domain (167 AA) of mouse CD8α; transfectant cell line: mouse myeloma cell line P3x63Ag8.653; applications in flow cytometry & immunohistochemistry (frozen sections); binds to CD137 & blocks binding of anti-CD137 monoclonal antibody; human CD137L (4-1BB Ligand) is a type II transmembrane protein constitutively expressed by monocytes, B cells & neuroblastoma cells; binding of CD137 (4-1BB) to CD137L induces monocyte activationAlderson, MR, *Eur J Immunol*, 24: 2219, 1994; Langstein, J et al, *J Immunol*, 160: 2488, 1998

CD137 Ligand-muCD8 Fusion Protein, Purified

Synonyms: CD137L-muCD8 Fusion Protein, Purified

Alexis ANC-503-820 Human Purified liquid; preservative free | A soluble fusion protein consisting of the extracellular (184 AA) domain of human CD137L fused to the extracellular domain (167 AA) of mouse CD8α; transfectant cell line: mouse myeloma cell line P3x63Ag8.653; applications in flow cytometry & immunohistochemistry (frozen sections); binds to CD137 & blocks binding of anti-CD137 monoclonal antibody; human CD137L (4-1BB Ligand) is a type II transmembrane protein constitutively expressed by monocytes, B cells & neuroblastoma cells; binding of CD137 (4-1BB) to CD137L induces monocyte activationAlderson, MR, *Eur J Immunol*, 24: 2219, 1994; Langstein, J et al, *J Immunol*, 160: 2488, 1998

CD137 Ligand-muCD8 Fusion Protein, R-PE

Synonyms: CD137L-muCD8 Fusion Protein, R-PE

Alexis ANC-503-050 Human Liquid; stabilized with 0.04% sodium azide | A soluble fusion protein consisting of the extracellular (184 AA) domain of human CD137L fused to the extracellular domain (167 AA) of mouse CD8α; transfectant cell line: mouse myeloma cell line P3x63Ag8.653; applications in flow cytometry & immunohistochemistry (frozen sections); binds to CD137 & blocks binding of anti-CD137 monoclonal antibody; human CD137L (4-1BB Ligand) is a type II transmembrane protein constitutively expressed by monocytes, B cells & neuroblastoma cells; binding of CD137 (4-1BB) to CD137L induces monocyte activationAlderson, MR, *Eur J Immunol*, 24: 2219, 1994; Langstein, J et al, *J Immunol*, 160: 2488, 1998

CD137-huIg Fusion Protein

Synonyms: 4-1BB-huIg Fusion Protein

Alexis ANC-502-020 Human MW 55k Purified liquid; free of azide & carrier protein & stabilized with 0.5 mg/mL gentamycin sulfate | A soluble fusion protein consisting of the extracellular (186 AA) domain of human CD137 fused to human IgG1 Fc; transfectant cell line: CHO; applications in flow cytometry & immunohistochemistry (frozen sections); CD137 Ig fusion protein blocks binding of anti-human CD137 to activated CEM human tumor cells & also binds to CD137 Ligand on Raji cells; human CD137 (4-1BB) is expressed on activated T cells within 24-48 hours of activation; CD137 is a type I membrane protein & a member of the tumor necrosis factor (TNF) receptor superfamily; CD137 appears to be important for T cell proliferation & survival & induces monocyte activationGiarni-Wagner, BA, *Cellular Immunol*, 169: 91, 1996; Schwarz, H et al, *Blood*, 87: 2839, 1996; Alderson, MR, *Eur J Immunol*, 24: 2219, 1994; Langstein, J et al, *J Immunol*, 160: 2488, 1998

Alexis ANC-502-820 Human Purified liquid; preservative free | Comments & references are the same as Alexis ANC-502-020

CD137-huIg Fusion Protein, Biotin Conjugated

Alexis ANC-502-030 Human Liquid; stabilized with 0.04% sodium azide | Comments & references are the same as Alexis ANC-502-020

CD137-huIg Fusion Protein, FITC Conjugated

Alexis ANC-502-040 Human Liquid; stabilized with 0.04% sodium azide | Comments & references are the same as Alexis ANC-502-020

CD137-huIg Fusion Protein, R-PE

Alexis ANC-502-050 Human Liquid; stabilized with 0.04% sodium azide | Comments & references are the same as Alexis ANC-502-020

CD152 (CTLA-4).Ig:Fc Fusion Protein

Alexis ANC-501-020A, -820, -030A, -040, -050 Human MW 110k Purified liquid; free of azide & carrier protein & stabilized with 0.5 mg/mL gentamycin sulfate | A soluble dimeric fusion protein consisting of the extracellular (125 AA) domain of human CTLA-4 fused to mouse IgG2a Fc; transfectant cell line: BHK; applications in flow cytometry & immunohistochemistry (frozen sections); CD152 Ig binds with high Specificity to human or mouse CD80 (B7-1) & CD86 (B7-2) & blocks binding of anti-CD80 (B7-1) & anti-CD86 (B7-2) Mabs; also binds to mouse CD80/CD86; human CD152 (CTLA-4) muIg is a cell surface glycoprotein expressed at low levels on activated T cells; CD152 is a high Specificity receptor for the costimulatory molecules CD80 (B7-1) & CD86 (B7-2) & appears to function as a negative regulator of T cell activation; a soluble fusion protein combining the extracellular (125 AA) domain of human CD152 & mouse IgG2a Fc (CTLA-4) was developedLindsten, T, *J Immunol*, 151: 3489, 1993; Morton, PA et al, *J Immunol*, 156: 1047, 1996; Walunas, TL et al, *Immunity*, 1: 405, 1994; Kar Andikar, NJ et al, *J Exp Med*, 184: 783, 1996; Cross, AH et al, *J Clin Invest*, 95: 2783, 1995

Alexis 202-038 Mouse MW ~110k ≥99% (SDS-PAGE); 50 μg in 50 *mM* sodium phosphate buffer, pH 7.5, 100 *mM* KCl, 150 *mM* NaCl; 50 μg protein | A soluble dimeric fusion protein consisting of the extracellular (160 AA) domain of mouse CTLA-4 fused to mouse IgG2a Fc; transfectant cell line: NS.1; applications in flow cytometry, functional studies & immunohistochemistry (frozen sections); mouse CD152 Ig has biological activity & binds with high specificity to mouse CD80 (B7-1) & CD86 (B7-2) & also binds to human, rat, pig & monkey CD80 (B7-1) & CD86 (B7-2) proteins & blocks the binding of anti-CD80 (B7-1) & anti-CD86 (B7-2) Mabs as well as block the interaction of CD80 &/or CD86 with cell surface CD28 &/or CD152; is a potent immunosupressive agentSteurer, W et al, *J Immunol*, 155: 1165, 1995; Borriello, F et al, *Immunity*, 6: 303, 1997; Perez, VL et al, *Immunity*, 6: 411, 1997

CD154 Fusion Protein

Synonyms: CD40L Soluble

Alexis 201-036 Human recombinant, expressed in E. *coli* MW 48k Soluble; lyophilized powder containing 50 μg protein & purified from E. *coli* by ion exchange & gel filtration chromatography; dissolve in 0.5 mL water to give a solution with a concentration of 20 μg/mL containing 12.5 *mM* HEPES, 0.05 *mM* EDTA, 2.5% glycerol, 0.5% BSA, 150 *mM* NaCl, 1.5 *mM* KCl, 4 *mM* Na_2HPO_4, 0.8 *mM* KH_2PO_4, pH 7.2 | Monomer MW 15k; noncovalent trimer consisting of residues Gly[116] to Leu[261] of human CD40L; identical to the final construct made for determination of the crystal structure; biologically active; capable of co-stimulating human B cell proliferation in the presence of hIL-4, anti-IgM (or both); in the presence of a cytokine mixture, it can induce differentiation of human B cells to secrete IgM & IgG; Karpusas et al, *Structure*, 3: 1031 & 1426, 1995; Mazzei, GJ et al, *J Biol Chem*, 270: 7025, 1995

CD154 Ligand-muCD8 Fusion Protein

Synonyms: CD40L-muCD8 Fusion Protein

Alexis ANC-505-020, -820, -030, -040, -050   Human
Purified liquid; free of azide & carrier protein & stabilized with 0.5 mg/mL gentamycin sulfate | Soluble molecule consisting of the extracellular (213 AA) domain of human CD154 fused to the extracellular domain (167 AA) of mouse CD8α; transfectant cell line: Mouse myeloma cell line P3x63Ag8.653; applications in flow cytometry & immunohistochemistry (frozen sections); CD154-muCD8 binds to CD40 & blocks binding of anti-CD40 monoclonal antibody; human CD154 (CD40 Ligand) is a member of the tumor necrosis factor (TNF) family & is expressed on the surface of activated T cells; interaction of CD154 & CD40 is essential for isotype switching in B cells; known genetic defects that alter this interaction lead to impaired immune system function; CD154 has been shown to be hyperexpressed by B & T cells in SLE patients; CD154 has been reported to be expressed on vascular endothelial cells, smooth muscle cells & macrophages indicating a possible role for the CD40-CD154 immunoregulatory signaling mechanism during inflammation & immunity in atherogenesisGray, D, *Seminares in Immunol*, 6: 303, 1994; Pietravalle, F et al, *J Biol Chem*, 271: 5965, 1996; noelle, RJ, *Immunity*, 4: 415, 1996; Desai-Mehta, A et al, *J Clin Invest*, 97: 2063, 1996; Grewal, S & Flavell, RA, *Immunol Today*, 17: 410, 1996; Mach, F et al, *PNAS*, 94: 1931, 1997

CD4 Antigen

Biodesign A49162B   Baculovirus   95%

CD40 Ligand, Soluble

Synonyms: rhsCD40 Ligand

Alexis BMS308/a   Human recombinant, expressed in *E. coli*
MW 48k   50 μg lyophilized powder; reconstituted solution at 20 μg/mL in 0.5 mL water contains 12.5 *mM* HEPES, 0.05 *mM* EDTA, 2.5% glycerol, 0.5% BSA, 150 *mM* NaCl, 1.5 *mM* KCl, 4 *mM* Na$_2$HPO$_4$, 0.8 *mM* KH$_2$PO$_4$, pH 7.2 | Monomer MW 15k; noncovalent trimer starting at Gly^{116} in the human DS40L sequence; purification from E. coli by ion exchange & gel filtration chromatography

CD40 Ligand/Thrombin Receptor Activator Peptide

Chemicon GF101   Human   ≥95%

PeproTech 310-02   Human recombinant, expressed in *E. coli*
MW 16.3k   >98% (SDS-PAGE & HPLC); <0.1 ng endotoxin per mg (1EU/mg); lyophilized | Type II membrane protein; effectuates the helper function of T cells on resting B cells; 149 AA residues comprising the receptor binding TNF-like domain of CD40L; stimulates IL-8 induction by human PBMC; SA determined by the stimulation of IL-12 induction by human peripheral blood mononuclear cells (PBMC)

CD40-mulg Fusion Protein

Alexis ANC-504-020   Human   Purified liquid; free of azide & carrier protein & stabilized with 0.5 mg/mL gentamycin sulfate | A soluble fusion protein consisting of the extracellular (193 AA) domain of human CD40 fused to mouse IgG2a Fc; transfectant cell line: mouse myeloma cell line P3x63Ag8.653; applications in flow cytometry & immunohistochemistry (frozen sections); CD40-mulg fusion protein blocks binding of anti-human CD40 to Raji human tumor cells; human CD40 is a member of the tumor necrosis factor (TNF) receptor family & is present on all B cells except plasma cells; CD40 is also found on some epithelial cells, carcinomas & lymphoid dendritic cells; plays an important role in B cell activation & the interaction with its ligand CD154 (CD40 ligand) is essential for isotype switchingFoy, TA, *Ann Rev Immunol*, 14: 591, 1996; Ozaki, ME et al, *J Immunol*, 159: 214, 1997

Alexis ANC-504-820   Human   Purified liquid; preservative free; see Alexis ANC-504-020

CD40-mulg Fusion Protein, Biotin Conjugated

Alexis ANC-504-030   Human   Liquid; stabilized with 0.04% sodium azide; see Alexis ANC-504-020

CD40-mulg Fusion Protein, FITC Conjugated

Alexis ANC-504-040   Human   Liquid; stabilized with 0.04% sodium azide; see Alexis ANC-504-020

CD40-mulg Fusion Protein, R-PE

Alexis ANC-504-050   Human   Liquid; stabilized with 0.04% sodium azide; see Alexis ANC-504-020

Cdc42 Protein

ICN 195960   Rat brain recombinant, expressed in *E. coli*   MW 23,980   A basic component of morphogenesis & cell cycle progression in eukaryotic cells; suitable for a positive control or western blotting cross-reactivity assay

Cdc42, His-Tagged

Calbiochem 219430   Rat brain recombinant, produced by overexpression of a full-length Cdc42 cDNA clone in *E. coli*   MW 23,980   ≥90% (SDS-PAGE); liquid in 200 *mM* BME, 50 *mM* Tris-HCl, 10% glycerol, 2% SDS, 0.005% bromophenol blue, pH 6.8 | GTP binding protein; fundamental component of morphogenesis & cell cycle progression in eukaryotic cells; stimulates the formation of filopodia; suitable for use as a positive control or to assay for cross-reactivity by Western blotting; Olson, MF et al, *Science*, 269: 1270, 1995; Simon, MN et al, *Nature*, 376: 702, 1995

Cdc42Hs, GST-Tagged

Calbiochem 219432   Human recombinant, expressed in *E. coli*
MW 48,260   ≥90% (SDS-PAGE); liquid in 20 *mM* HEPES, 5 *mM* MgCl$_2$, 1 *mM* NaN$_3$, 40% glycerol, pH 8.0; GTPγS binding: ≥400 mmol/mol; use 100 ng/blot probed with Anti-Cdc42 | GTP binding protein; small Rho GTPase that belongs to the Ras superfamily that regulates cytoskeletal rearrangements in response to growth factor stimulation; acts on targets in the Golgi & directs polarized growth at the plasma membrane; homolog of the yeast Cdc42 cell division cycle gene product; useful as a positive control in Western blotting & as an affinity precipitation reagent using immobilized glutathione resin; Erickson, JW et al, *J Biol Chem*, 271: 26850, 1996; Ridley, A et al, *Curr Biol*, 6: 1256, 1996; Shinjo, K et al, *PNAS*, 87: 9853, 1990

Cdc42Hs, His-Tagged

Calbiochem 219434   Human recombinant, expressed in *E. coli*
MW 21,960   ≥90% (SDS-PAGE); liquid in 150 *mM* NaCl, 20 *mM* Tris-HCl, 5 *mM* imidazole, 40% glycerol, pH 7.9; GTPγS binding: ≥600 mmol/mol; use 100 ng/blot probed with Anti-Cdc42 | GTP binding protein; small Rho GTPase that belongs to the Ras superfamily that regulates cytoskeletal rearrangements in response to growth factor stimulation; homolog of the yeast Cdc42 cell division cycle gene product; useful as a positive control in Western blotting & as an affinity precipitation reagent using Ni^{2+}-charged metal chelate resin; Ridley, A et al, *Curr Biol*, 6: 1256, 1996; Shinjo, K et al, *PNAS*, 87: 9853, 1990

Ceruloplasmin

Cortex CP1010   >95%

Biodesign A50143H   Human plasma   Purified

Calbiochem 239799   Human plasma   MW 134k   Lyophilized from 100 *mM* KCl, 50 *mM* potassium phosphate buffer, 20 *mM* ε-aminocaproic acid, pH 6.8; ≥95% (SDS-PAGE); soluble in water; prepared from plasma shown to be negative for HBsAg & for antibodies to HIV & HCV | Serum copper transport & Iron oxidizing protein; plays an important role in antioxidant protection against organic & inorganic oxygen radicals generated by iron & ascorbate; Gutteridge, JM & Quinlan, GJ, *Biochim Biophys Acta*, 1156: 144, 1993; Harris, ED et al, *Proc Soc Exp Biol Med*, 196: 130, 1991; Krsek-Staples, JA & Webster, RO, *Free Radic Biol Med*, 14: 115, 1993; *Merck Index*, 12: 2049

Biogenesis 1940-0404   Human serum   Lyophilized

ICN 194974   Human serum   MW 134k   Lyophilized; >95% | Serum copper transport & iron oxidizing protein; plays an important role in antioxidant protection against organic & inorganic O$_2$ radicals from iron & ascorbate; Gutteridge, JM & GJ Quinlan, *BBA*, 1156:144, 1993

Proteins

Chaperonin 10

Synonyms: Growth Related Oncogene ES

Sigma C 7438 *E. coli* recombinant, overexpressed in *E. coli* ≥90% (SDS-PAGE); lyophilized powder containing Tris buffer salts, KCl, DTT & trehalose as stabilizer; inhibition of GroEL ATPase ~50% at a molar ratio of 2:1 (GroES:GroEL) | Belongs to the family of heat-shock molecular chaperones found in prokaryotes & in eukaryotic organelles; assist the folding of nascent, organelle-imported or stress-destabilized polypeptides; *in vitro*, purified GroEL together with purified GroES in presence of Mg-ATP facilitate refolding & reactivation of denatured proteins, eg, the photosynthetic enzyme rubisco & the mitochondrial enzyme rhodanese; the folding activity of a 1:1 molar mixture of GroEL:GroES was tested using urea-denatured rhodanese; at least 2-fold reactivation was obtained; Goloubinoff, P et al, *Nature*, 342: 884, 1989; Mendoza, JA et al, *J Biol Chem*, 266: 13044, 1991

Chaperonin 60

Synonyms: Growth Related Oncogene EL

Sigma C 7688 *E. coli* recombinant, overexpressed in *E. coli* >95% (SDS-PAGE); lyophilized powder containing Tris buffer salts, KCl, DTT & trehalose as stabilizer; ATPase activity >80 nmol/min/mg protein at 30°C

Chaperonin 60+10

Synonyms: Growth Related Oncogene EL+ES

Sigma C 7563 *E. coli* recombinant, overexpressed in *E. coli* 1:1 Mixture; >95% (SDS-PAGE); lyophilized powder containing Tris buffer salts, KCl, DTT & trehalose as stabilizer

Chimera

Synonyms: Flt-1/Fc

ICN 195784 Human recombinant, expressed in Sf21 ≥97%; lyophilized; 15-30 ng/mL will typically inhibit ½ the biological response to 4 ng/mL VEGF

Cholera Toxin

Sigma C 3012 *Vibrio cholerae* When reconstituted to 1 mL, solution will contain 0.05 M Tris buffer salts, pH 7.5, 0.2 M NaCl, 0.003 M NaN$_3$, 0.001 M EDTA • Na^2; prepared by a modification of the methods of Rappaport & Mekalanos | Toxin that consists of an A subunit surrounded by five B subunits, which attach the toxin to ganglioside G$_{M1}$ on the cell surface; the A subunit catalyzes ADP-ribosylation of the α-subunit of G proteins, reducing GTPase activity & activating the a-subunit; catalyzes ADP-ribosylation of cell membrane adenylate cyclase; Rappaport, RS et al, *Infection & Immunity*, 9: 294, 1974; Mekalanos, JJ et al, ibid, 20: 552, 1978

Sigma C 8052 *Vibrio cholerae* ~95% (SDS-PAGE); lyophilized powder containing ~5% protein (Lowry-TCA); balance primarily Tris buffer salts, NaCl, sodium azide, EDTA·Na^2; activity: 10^5-10^6 units/mg protein; unit definition: 1 unit of cholera toxin increases the adenylate cyclase activity by a factor of 6, measured as (^{3}H)-cAMP present, using ~70,000 CHO cells after incubation for 90 minutes at 37°C with (^{3}H)-adenine followed by 15 minute incubation with 1 mM 3-isobutyl-1-methylxanthine | Composed of one A chain (28 kD) subunit that catalyses the ADP-ribosylation activity of the cell membrane adenylate cyclase, & five B chain (11.8 kD) subunits that bind with high affinity to cell membrane ganglioside-GM Salomon, Y, *Meth Enzymol*, 195: 22, 1991; Mekalanos, JJ, *Meth Enzymol*, 165: 169, 1988 Toxin that consists of an A subunit surrounded by five B subunits, which attach the toxin to ganglioside G$_{M1}$ on the cell surface; the A subunit catalyzes ADP-ribosylation of the α-subunit of G proteins, reducing GTPase activity & activating the a-subunit; catalyzes ADP-ribosylation of cell membrane adenylate cyclase

Calbiochem 227035 *Vibrio cholerae*, Type Inaba 569B MW 84k Single band purity (disc gel electrophoresis); solid lyophilized from 200 mM NaCl, 50 mM Tris, 3 mM NaN$_3$, 1 mM EDTA, pH 7.5; SA: ≥a standard cholera toxin preparation determined in an ADP-ribosylation assay; highly toxic: LD$_{50}$ ≤50 mg/kg; may be carcinogenic/teratogenic | Useful for studies involving adenylate cyclase, AMP & related membrane-transport phenomena;

Calbiochem 227036 *Vibrio cholerae*, Type Inaba 569B MW 84k Identical to Calbiochem 227035, but azide free; single band purity (disc gel electrophoresis); solid lyophilized from 200 mM NaCl, 50 mM Tris, 1 mM EDTA, pH 7.5; highly toxic: LD$_{50}$ ≤50 mg/kg; may be carcinogenic/teratogenic | Useful for tissue culture applications; consists of a single A subunit surrounded by five B subunits; the A subunit catalyzes the ADP-ribosylation of the α-subunit of G-proteins, reducing intrinsic GTPase activity & activating the α-subunit; the B subunits are responsible for the attachment of the native toxin on the cell surface; DiRita, VJ et al, *PNAS*, 88: 5403, 1991; Gilman, AG, *Ann Rev Biochem*, 56: 615, 1987; Moss, J & Vaughan, M, *Curr Top Cell Regul*, 32: 49, 1992

Cholera Toxin A Subunit

Sigma C 2398 *Vibrio cholerae* Lyophilized powder; when reconstituted to 1 mL, solution will contain 0.05 M Tris buffer salts, pH 7.5, 0.2 M NaCl, 0.003 M NaN$_3$, 0.001 M EDTA·Na^2; prepared by a modification of the method of Lai | Catalyzes ADP-ribosylation of the α-subunit of G proteins, reducing GTPase activity & activating the α-subunit; catalyzes ADP-ribosylation of cell membrane adenylate cyclase; Lai, CY et al, *J Infect Dis*, 133: S23, 1976; Mekalanos, JJ, *Meth Enzymol*, 165: 169, 1988

Sigma C 8180 *Vibrio cholerae* Lyophilized powder containing ~5% protein (Lowry); balance Tris buffer salts, NaCl, sodium azide, sodium EDTA; ADP-ribosylation activity: measured by ADP-ribosylation of poly-L-arginine following the incorporation of ADP-ribose from NAD-(adenine-UL-^{14}C) to TCA-precipitable material; 1 μg of cholera toxin A subunit will cause the incorporation of at least 1 picomoles of ADP-ribose in 30 min at 30°C; cholera toxin B subunit: ≤0.5% (SDS-PAGE); prepared by a modification of the method of Lai | Catalyzes ADP-ribosylation of the a-subunit of G proteins, reducing GTPase activity & activating the α-subunit; catalyzes ADP-ribosylation of cell membrane adenylate cyclase; Lai, CY et al, *J Infect Dis*, 133: S23, 1976; Mekalanos, JJ, *Meth Enzymol*, 165: 169, 1988

Calbiochem 227037 *Vibrio cholerae*, Type Inaba 569B MW 28k Solid lyophilized from 200 mM NaCl, 50 mM Tris, 3 mM NaN$_3$, 1 mM EDTA, pH 7.5; in an ADP-ribosylation assay, the specific enzymatic activity of a typical lot is ~2.0 times that of cholera toxin on a weight basis; contaminants B subunit & intact toxin: none detected by native PAGE; highly toxic: LD$_{50}$ ≤50 mg/kg; may be carcinogenic/teratogenic | The subunit responsible for activation of adenylate cyclase; catalyzes the ADP-ribosylation of the α-subunit of G-proteins, reducing intrinsic GTPase activity & activating the α-subunit

Cholera Toxin B Subunit

Synonyms: Choleragenoid

Sigma C 7771 *Vibrio cholerae* Lyophilized powder containing Tris buffer salts, NaCl, sodium azide, sodium EDTA; antitoxin combining power activity: >30 toxoid units/μg protein (Lowry); permeability factor (PF) activity: <0.05%; bioassay not run by Sigma | Cholera toxin subunit that binds ganglioside G$_{M1}$ on the cell surface; Lai, CY et al, *J Infect Dis*, 133: S23, 1976; Craig, JP, in *Microbial Toxins*, Vol 2A, 189, Kadis, S et al, eds Academic Press Inc, New York

Sigma C 9903 *Vibrio cholerae* ≥95% (SDS-PAGE); lyophilized powder containing ~5% protein (Lowry); balance Tris buffer salts, NaCl, sodium azide, sodium EDTA; activity: measured by ELISA using ganglioside G$_{M1}$-coated multiwell plates, rabbit anti-cholera toxin B subunit antibodies & peroxidase-labeled goat anti-rabbit IgG as the second antibody; 50% saturation of binding was achieved with 0.05-1 μg of cholera toxin B subunit/mL; cholera toxin A subunit: ≤0.5% (SDS-PAGE); prepared by a modification of the method of Lai | Cholera toxin subunit that binds ganglioside G$_{M1}$ on the cell surface; Lai, CY et al, *J Infect Dis*, 133: S23, 1976; Craig, JP, in *Microbial Toxins*, Vol 2A, 189, Kadis, S et al, eds Academic Press Inc, New York

Calbiochem 227039 *Vibrio cholerae*, Type Inaba 569B MW 55k Solid lyophilized from 200 mM NaCl, 50 mM Tris, 3 mM NaN$_3$, 1 mM EDTA, pH 7.5; highly toxic: LD$_{50}$ ≤50 mg/kg; may be carcinogenic/teratogenic | Portion of cholera toxin responsible for binding to GM1 ganglioside receptors in membranes; blocks action of toxin on intact cells; Mulhein, SA et al, *J Membr Biol*, 109: 21, 1989

Cholera Toxin B Subunit Type Inaba 569B, Peroxidase Conjugated

Calbiochem 227041 Lyophilized solid from 10 mM sodium phosphate buffer, pH 7.5; specific activity: 125 purpurogallin units/mg HRP; toxic: $LD_{50} \leq 50$ mg/kg | Suitable for the demonstration of dendritic branching in retrogradely labeled neurons; 50 times more sensitive than free peroxidase as both an orthogradely & retrogradely transported marker in the rat visual system; Raappana, P & Arvidsson, J, *J Comp Neurol*, 328: 103, 1993; Trojanowski, JQ, *J Neurosci Methods*, 9: 185, 1983

Cholera Toxin B Subunit, Biotin Conjugated

Synonyms: Choleragenoid

Sigma C 9972 *Vibrio cholerae* Lyophilized powder containing ~40% protein (Lowry); balance sodium phosphate buffer salts, sodium azide, sodium EDTA; biotin content: ~1.0 mole/mole protein; activity: measured by ELISA using ganglioside GM_1-coated multiwell plates, rabbit anti-cholera toxin B subunit antibodies & peroxidase-labeled goat anti-rabbit IgG as the second antibody; 50% saturation of binding was achieved with 0.1-1 µg of cholera toxin B subunit-biotin conjugate/mL; conjugated B subunit gives a similar value for 50% binding to that of unconjugated B subunit from which it is prepared | Cholera toxin subunit that binds ganglioside G_{M1} on the cell surfaceLai, CY et al, *J Infect Dis*, 133: S23, 1976; Craig, JP, in *Microbial Toxins*, Vol 2A, 189, Kadis, S et al, eds Academic Press Inc, New York

Cholera Toxin B Subunit, FITC Conjugated

Synonyms: Choleragenoid

Sigma C 1655 *Vibrio cholerae* Lyophilized powder containing ~20% protein (Lowry); balance Tris buffer salts, NaCl, sodium azide, sodium EDTA; FITC content: ~1.0 mole/mole protein; activity measured by ELISA using ganglioside GM_1-coated multiwell plates, rabbit anti-cholera toxin B subunit antibodies & peroxidase-labeled goat anti-rabbit IgG as the second antibody; 50% saturation of binding was achieved with 0.01-1 µg of cholera toxin B subunit-FITC conjugate/mL | Cholera toxin subunit that binds ganglioside G_{M1} on the cell surface; Lai, CY et al, *J Infect Dis*, 133: S23, 1976; Craig, JP, in *Microbial Toxins*, Vol 2A, 189, Kadis, S et al, eds Academic Press Inc, New York

Cholera Toxin B Subunit, Peroxidase Conjugated

Synonyms: Choleragenoid

Sigma C 4672 *Vibrio cholerae* Lyophilized powder; vial contains ~42 µg cholera toxin B subunit, 100 µg (10-30 units) horseradish peroxidase & buffer salts from 0.1 mL of 0.01 M sodium phosphate, pH 7.5; unit definition: 1 unit forms 1.0 mg of purpurogallin from pyrogallol/20 sec at pH 6.0 at 20°C | Cholera toxin subunit that binds ganglioside G_{M1} on the cell surface; Lai, CY et al, *J Infect Dis*, 133: S23, 1976; Craig, JP, in *Microbial Toxins*, Vol 2A, 189, Kadis, S et al, eds Academic Press Inc, New York

Cholesterol Esterase

Biogenesis 2070-0004 Porcine pancreas

Chorionic Gonadotropin

Cortex CP1011P	Human	>40%
Cortex CP1011U	Human	>98%
Biodesign A81351M	Human pregnancy urine	50%
Biodesign A81355M	Human pregnancy urine	98%

Calbiochem 230734 Human urine MW 36.7k Standard grade; lyophilized solid; immunopotency: ≥3000 IU/mg (WHO 1st IRP 75/551); soluble in water; shown by certified tests to be negative for HBsAg & for antibodies to HIV & HCV; may be carcinogenic/teratogenic | Glycoprotein hormone synthesized by chorionic tissue of the placenta & found in urine during pregnancy; present in body fluids of patients with trophoblastic disease & ovarian tumors; *Merck Index*, 12: 2273

Calbiochem 869031 Human urine MW 36.7k Iodination grade; lyophilized solid; ≥95% (SDS-PAGE); immunopotency: ≥16,000 IU/mg (WHO 1st IRP 75/551); soluble in water; shown by certified tests to be negative for HBsAg & for antibodies to HIV & HCV; may be carcinogenic/teratogenic | *Merck Index*, 12: 2273

Chorionic Gonadotropin, Intact

USBio C5069-21 Human ≥98% (SDS-PAGE); lyophilized from ammonium bicarbonate buffer; SA ~15000 IU/mg by 2nd (3rd IS for hCG 75/537, WHO); hLH <0.2%, hFSH <0.05%, hTSH <0.05% | Suitable for antigenic applications in immunological protocols

Biogenesis 2090-0454 Human urine MW 36.7k SA: 14.1 KIU/mg; 2 subunits; pI: 4.9; hLH/hFSH/hTSH <0.2%; purified; from 10 mM phosphate buffer, 150 mM NaCl, 0.1% NaN₃, pH 7.4; lyophilized

Biogenesis 2090-0459 Human urine MW 38.4k <2% hCG alpha & hCG beta, hPL undetectable; tested negative for HBsAg, HCV, HIV-1 & HIV-2 antibodies, HIV-Ag, HTLV I & II antibodies; 10,000 IU/mg; purified; 0.05 M NH₄HCO₃, pH 8.0; lyophilized | A potent LH agonist; displays a weak FSH-like SA; Bahl, *JBC*, 244:567, 1969; Birken, *Ann Endocrinol*, 45:297, 1984

Biogenesis 2090-0489 Human urine MW 38.4k <0.1% hPL, no other hormones detected; tested negative for HCV, HIV-1, HIV-2 and HBsAg; semi-pure; 0.05 M NH₄HCO₃, pH 8.0, essentially salt free; lyophilized

Biogenesis 2090-0504 Human urine 0.004% hLH, 0.04% hFSH; tested negative for HBsAg, HCV, HIV I, HIV II and syphilis; SA: 8,000 IU/mg; semi-pure; lyophilized

Chorionic Gonadotropin, α-

Cortex CP1038U	Human	>98%

USBio C5069-52 Human ≥98%; hLH ≤0.1%, hFSH ≤0.05%, hTSH ≤0.05%, hCGb ≤0.1%, hCG ≤1%; lyophilized from 50 mM ammonium bicarbonate buffer | Suitable for antigenic applications in immunological protocols

Biogenesis 2090-1204 Human pituitary Tested negative for antibodies to HIV and HBsAg; <1% beta subunit; purified; PBS buffer, pH 7.2, no preservatives; liquid

Biodesign A81251M	Human pregnancy urine	50%
Biodesign A81255M	Human pregnancy urine	98%

Biogenesis 2090-1209 Human urine MW 14,930 <0.2% hCGß; SA: 1000 IU/mg; tested negative for Hepatitis B surface antigen, Hepatitis C (HCV antibody), HIV-1, HIV-2 antibody; purified; 0.05 M NH₄HCO₃, pH 8.0; lyophilized | Bahl, *BBRC*, 40:422, 1970

USBio C5069-86 ≥98%; hCG ≤0.2%; FSH, TSH & LH: none detected; no contaminants detected; single band by SDS-PAGE, IEP, &/or RID; lyophilized from 0.05 M ammonium bicarbonate | Suitable for antigenic applications in immunological protocols

Cortex CP1012U	Human	>98%

Biogenesis 2090-1809 Human pregacy urine MW 23,470 <0.1% hCG-alpha; tested negative for HBsAg, HCV and HIV-1 and 2; SA: 1,000 IU/mg; purified; 50 mM ammonium bicarbonate buffer, pH 8.0; lyophilized

Biodesign A81451M	Human pregnancy urine	50%
Biodesign A81455M	Human pregnancy urine	98%
Biogenesis 2090-1804	Human urine	Lyophilized

Calbiochem 969126 Human urine MW 22.2k Iodination grade; lyophilized solid; >98% (SDS-PAGE); immunopotency: ≥1000 IU/mg (WHO 1st IRP 75/551); soluble in water; α-Subunit: <2.0% by RIA; shown by certified tests to be negative for HBsAg & for antibodies to HIV & HCV; may be carcinogenic/teratogenic | Glycoprotein hormone produced by the trophoblastic cells of the placenta; determines the specificity & activity of the hormone; inhibits the growth of Kaposi sarcoma-derived cells; *Merck Index*, 12: 2273; Albini, A et al, *AIDS*, 11: 713, 1997

c-H-Ras, GST-Tagged

Calbiochem 553329 Human recombinant, expressed in *E. coli* MW 48,299 ≥90% (SDS-PAGE); liquid in 20 mM HEPES, 5 mM MgCl₂, 40% glycerol, 1 mM NaN₃, pH 8.0; GTPγS binding: ≥400 mmol/mol | GTP binding protein; small GTPase that regulates growth & proliferation in response to growth factor stimulation; mutants defective in GTP hydrolysis are potent oncogenes; useful as a positive control in Western blotting & as an affinity precipitation reagent using immobilized glutathione resin; Boguski, M & McCormick, F, *Nature*, 366: 643, 1993

c-H-Ras, His-Tagged

Calbiochem 553330 Human recombinant, expressed in *E. coli* MW 21,999 ≥90% (SDS-PAGE); liquid in 150 m*M* NaCl, 20 m*M* Tris-HCl, 40% glycerol, 5 m*M* imidazole, pH 7.9; GTPγS binding: ≥500 mmol/mol | GTP binding protein; small GTPase that regulates growth & proliferation in response to growth factor stimulation; mutants defective in GTP hydrolysis are potent oncogenes; useful as a positive control in Western blotting & as an affinity precipitation reagent using Ni²⁺-charged metal chelate resin; Boguski, M & McCormick, F, *Nature*, 366: 643, 1993

Chromogranin A

Sigma C 9335 Bovine adrenal medulla MW 48k ≥90% (SDS-PAGE); lyophilized powder | High-capacity low-affinity Ca⁺⁺ binding protein; major protein in secretory vesicles; precursor of biologically active peptide (Pancreasatin) & is involved in the sorting of secretory proteins & granules biogenesis; interacts with membrane proteins, including the inositol 1,4,5-triphosphate receptor/Ca⁺⁺ channel; a very hydrophilic protein that migrates on SDS-PAGE with an apparent MW of 75 kD; Yoo SH & Albanesi, JP, *J Biol Chem*, 265: 14414, 1990; Winkler, H & Fischer-Colbrie, R, *Neuroscience*, 49: 497, 1992; Sigafoos, J et al, *J Anat*, 183: 253, 1993; Huttner, WB et al, *TIBS*, 16: 27, 1991; Yoo, SH & Lewis, MS, *Biochemistry*, 34: 632, 1995

Chymotrypsinogen A Type II, α-

Sigma C 4879 Bovine pancreas 6X Crystallized; lyophilized powder; essentially salt-free; activity: 40-60 units/mg solid after activation to α-chymotrypsin; unit definition: 1 U hydrolyzes 1.0 μmole BTEE/min at pH 7.8 at 25°C; may contain up to 1 U α-chymotrypsin/mg prior to activation by trypsin

Clathrin

Sigma C 5823 Bovine brain Lyophilized powder containing ~10% protein; balance Tris & MES buffers; clathrin heavy chains represent more than 60% of the total protein; contains clathrin assembly proteins | The major protein of coated vesicles; involved in receptor-mediated endocytosis & recycling of synaptic vesicles; clathrin triskelion binds to adaptins on the cytoplasmic domain of membrane receptors forming coated pits that develop into endocytotic vesicles; Nandi, Pk et al, *Biochem*, 19: 5917, 1980; Woodman, PG & Warren, G, *J Cell Biol*, 112: 1133, 1991

Cleavage Control Protein

Calbiochem 69069-3 Cleaved into two proteolytic fragments of 35k & 13k, (Thrombin) or 32k & 16k (Enterokinase) which are easily visualized by SDS-PAGE | Used to monitor the performance of either thrombin or enterokinase cleavage conditions

Cobra Venom Factor

Synonyms: Cobra Venom Anti-Complementary Protein

Calbiochem 233550 *Naja naja* MW 75k, 51k & 29-31k >99% (SDS-PAGE); liquid in 140 m*M* NaCl, 10 m*M* Tris, pH 8.0; no phospholipase activity detected; 1.24 μg protein=1.0 unit of functional activity when *in vitro* anticomplementary activity is measured by the method of Ballow & Cochrane; LD₅₀≤2000 mg/kg; not available for sale outside of the United States | 3 subunits; mediates specific cytotoxicity via the alternative pathway of human complement activation; Bogers, WM et al, *Eur J Immunol*, 23: 433, 1993; Muller, B & Muller-Ruchholtz,W, *Leuk Res*, 11: 461, 1987; Ballow, M & Cochrane, CG, *J Immunol*, 103: 944, 1969

Calbiochem 233552 *Naja naja kaouthia* MW 68k (α), 48k (β), 30k (γ) >99% (SDS-PAGE); liquid in PBS, pH 7.2; no phospholipase activity detected; 4-6 μg purified CVF=1.0 unit of functional activity when *in vitro* anticomplementary activity is measured by the method of Cochrane, CG et al; LD₅₀≤2000 mg/kg; not available for sale outside of the United States | 3 S-S subunits; structural & functional analog of cobra C3; in the presence of Factor B, Factor D & Mg²⁺, CVF forms a stable CVF,Bb complex with is a C3/C5 convertase enzyme, however the CVF,Bb complex is not susceptible to regulation by Factors H & I; Fritzinger, DC et al, *J Immunol*, 149: 3554, 1992; Vogel, CW & Muller-Eberhard, HJ, *J Immunol Methods*, 73: 203, 1984; Cochrane, CG, *J Immunol*, 105: 55, 1970

Sigma C 8406 *Naja naja kaouthia* Lyophilized powder containing ~35% protein (Biuret); balance primarily NaCl & ammonium acetate; major band on gel electrophoresis >80%

Cobratoxin, α-

Synonyms: Snake Toxin

Sigma C 6903 *Naja naja kaouthia* Chromatographically purified | Binds nicotinic receptors & blocks cholinergic neurotransmission at the neuromuscular junction; binding is irreversible

Collagen

Fluka 27662 Bovine Achilles' tendon MW ~80k Lyophilized | Miller, EJ & Gay, S, *Meth Enzymol*, 144: 3, 1987

ICN 150703 Bovine dermal Aqueous, 3 mg/mL; 99.9% | suitable for cell culture & many biochemical applications

ICN 151458 Bovine dermal Tissue culture grade; sterile; enzyme-solubilized, 3 mg/mL; 99%

ICN 160084 Calf skin Soluble | Suitable for gel formation, platelet aggregation & assay of collagenase by viscometry; Gallop & Seifert, *Methods Enzymol*, VI:635, 1963

Sigma 885-1 Calf skin Freeze-dried vial containing ~2 mg collagen with buffer salts

Collagen Mixture, Crude

Calbiochem 234112 Calf skin Lyophilized solid; soluble in acetic acid | A crude mixture of various collagens from calf skin; *Merck Index*, 12: 2543; Gallop, PM & Siefert, S, *Methods Enzymol*, 6: 635, 1963

Collagen Mixture, Types I/II

Biogenesis 2150-2556 Porcine

Biogenesis 2150-2656 Rat tail tendon MW ~300k 60-65% collagen type I, 30-35% collagen type III; <0.5% non-collagenous proteins, ~10% cross-linked collagen type I dimers and trimers, <5% other collagen types; purified; essentially salt free; lyophilized | Molecular composition: *M*[a1(I)1a2(I)2], native triple-helical structure is preserved (characteristic optical rotation spectrum and ability to form microfibrils); dissolved collagen (0.5 *M* acetic acid, pH 2.5) retains immunogenic properties; thermal degradation converts dissolved collagen to gelatin

Collagen Mixture, Types I/III

Biogenesis 2150-2306 Bovine

ICN 193492 Bovine 95% Type I, 5% Type III; sterile; aqueous, 0.3% pure bovine collagen buffered with 0.1% acetate, pH 3.2-3.8 (20°C);

Biogenesis 2150-2356 Canine

Biogenesis 2150-2506 Mouse tail tendons MW 300k 45% collagen type I & III, 10% collagen type IV, <1% collagen type V, <0.5% non-collagenous proteins; purified; essentially salt free; lyophilized | Dissolved collagen (0.05-0.5 *M* acetic acid, pH 2.5 at 4°C) retains immunologic properties of native collagen types I+III; thermal denaturation converts dissolved collagen to gelatin; molecular composition: *M*[a1(I)1a2(I)2], native triple helix; Rhodes & Miller, *Biochem*, 17:3442, 1979

Biogenesis 2150-2606 Rabbit

Collagen Type I

Synonyms: Collagenase Substrate

ICN 160083 Bovine Achilles tendon Insoluble; Type I (predominantly) | Einbinder, J & M Schubert, *JBC*, 188:335, 1951

Sigma C 8886 Bovine Achilles tendon Sigma Type II; insoluble; prepared by modification of the method of Neuman | Neuman, RE, *Arch of Biochem*, 24: 289, 1949; Bornstein, P & Traub, W, *The Proteins*, IV: 412, 1979

Sigma C 9879 Bovine Achilles tendon Sigma Type I; insoluble; prepared by method of Einbinder & Schubert | Not suitable for coating glasswareEinbinder, J & Schubert, M, *J Biol Chem*, 188: 335, 1951; Bornstein, P & Traub, W, *The Proteins*, IV: 412, 1979

Chemicon CC072 Bovine fetal skin Purified | Extracellular matrix protein

Biodesign A33120B Bovine placenta Purified

Biogenesis 2150-0515 Bovine skin MW ~300k ~10% cross-linked collagen type I dimers & trimers, <5% other collagen types, <0.5% non-collagenous proteins; purified; essentially salt free; lyophilized | Native, triple-helical structure is preserved (characteristic optical rotation spectrum and ability to form microfibrils); molecular composition: [a1(I)1a2(I)2], native triple helix; dissolved collagen (0.5 M acetic acid, pH 2.5) retains immunogenic properties; thermal degradation converts dissolved collagen to gelatin

Fluka 27664 Calf skin Soluble; prepared by a modification of the procedure of Gallop, PM & Seifter, S, *Meth Enzymol*, 6: 635, 1963 | Electrophoresis test: consistent with calf skin collagen; Yannas, IV et al, *PNAS*, 86: 933, 1989

ICN 150026 Calf skin Soluble, lyophilized | For attachment of primary cultures of epithelioid cells & many other cell types; Gallop & Seifert, *Methods Enzymol*, VI:635, 1963

Sigma C 3511 Calf skin Sigma Type III; acid soluble; prepared by modification of the method of Gallop | Gallop, PM & Seifter, S, *Meth Enzymol*, VI, 635, 1963, 1949; Bornstein, P & Traub, W, *The Proteins*, IV: 412, 1979

Sigma C 8919 Calf skin 0.1% solution in 0.1 N acetic acid; sterile-filtered; cell culture tested

Sigma C 9791 Calf skin Acid soluble; cell culture tested | suitable for tissue culture & cell biology studies; prepared by a modification of Gallop, PM & Seifter, S, *Meth Enzymol*, VI: 635, 1963

Chemicon CC090 Chicken fetal tissue Purified | Extracellular matrix protein

Biogenesis 2150-1308 Goat skin MW 300k <0.5% non-collagenous proteins, typically cross-linked collagen type I dimers and timers (approx 10%); purified; essentially salt free; lyophilized | Molecular composition: a1(I)1a2(I)2, native triple helix; dissolved collagen (0.5 M acetic acid, pH 2.5) retains immunogenic properties; thermal degradation converts dissolved collagen to gelatin

Biodesign A33704H Human placenta Purified

Biogenesis 2150-0030 Human placenta MW 300k <0.5% non-collagenous proteins; typical contaminants are cross-linked collagen type I dimers and trimers (approx 10%); other collagen types constitute <5% according to immunoassay and analytical chromatography; tested negative for HBsAg and HIV-1/2 antibodies; purified; contains 10% NaCl; lyophilized

Calbiochem 234149 Human placenta Liquid in 10 mM acetic acid; ≥95% (SDS-PAGE); prepared from tissue of individuals shown by certified tests to be negative for HBsAg & antibodies to HIV & HCV | Highly purified preparation; useful as a gel or thin coating for the attachment of cells; *Merck Index*, 12: 2543; Karsenty, G et al, *J Biol Chem*, 266: 24842, 1991; Klasson, SC et al, *Coll Relat Res*, 6: 397, 1986

Chemicon CC050 Human placenta Purified | Extracellular matrix protein

Sigma C 7774 Human placenta Sigma Type VIII; acid soluble; prepared by modification of the pepsin extraction & salt fractionation method of Niyibizi | Niyibizi, C et al, *J Biol Chem*, 259: 14170, 1984; Bornstein, P & Traub, W, *The Proteins*, IV: 412, 1979

Sigma C 1809 Kangaroo tail Lyophilized from 10 mM glacial acetic acid; acid soluble; cell culture tested | suitable for tissue culture & cell biology studies; prepared by a modification of the pepsin extraction method of Niyibizi, et al

Sigma C 3929 Kangaroo tail Acid soluble; prepared by modification of the pepsin extraction & salt precipitation method of Niyibizi | Niyibizi, C et al, *J Biol Chem*, 259: 14170, 1984; Bornstein, P & Traub, W, *The Proteins*, IV: 412, 1979

Biogenesis 2150-1425 Mouse embryos MW ~300k Cross linked collagen type I dimers and trimers represent approximately 10%; <3% collagen type III, <0.5% non-collagenous proteins; purified; essentially salt free; lyophilized | Extracted from washed dissected tissue into dilute acetic acid after mild pepsin treatment; collagen type I was purified by using differential salt precipitation; dissolved collagen (0.1 M acetic acid, pH 3.0) retains immunogenic properties; thermal denaturation converts the collagen to gelatin; Rhodes & Miller, *Biochem*, 17:3442, 1979

Fluka 27666 Rat tail Soluble; for cell biology; prepared by a modification of the procedure of Bernstein, MB, *Lab Invest*, 7: 134, 1958 | Electrophoresis test: consistent with rat tail collagen

Sigma C 7661 Rat tail Acid soluble; cell culture tested | suitable for tissue culture & cell biology studies; prepared by a modification of the extraction method of Bornstein, MB, *Lab Invest*, 7: 134, 1958

Sigma C 8897 Rat tail Sigma Type VII; acid soluble; prepared by modification of the method of Bornstein | Preparation method from Bornstein, MB, *Lab Invest*, 7: 134, 1958; Bornstein, P & Traub, W, *The Proteins*, IV: 412, 1979

USBio C7510-18 Rat tail Tested for uniform gelation & attachment/spreading of PC-12 rat cells; negative for the presence of mycoplasma, bacteria & fungi; ~4 mg/mL in 0.02 N HOAc; do not freeze

Biogenesis 2150-1915 Rat tail tendon <0.5% non-collagenous proteins, typically cross-linked collagen type I dimers and trimers (approx 10%); purified; essentially salt free; lyophilized | Molecular composition: a1(I) 1a2(I)2 native triple helix; dissolved collagen (0.1 M acetic acid, pH 3.0) retains immunogenic properties; thermal denaturation (100ºC for 5 min) converts collagen to gelatin

Collagen Type I, FITC Conjugated

Chemicon CC111F Bovine inner skin Purified | Extracellular matrix protein; not for sale in Japan

Collagen Type II

Calbiochem 234184 Bovine MW 100k Salt-free lyophilized solid; ≥95% (SDS-PAGE) | Highly purified preparation isolated from the triple helical domain of Gn-HCl-extracted & pepsin-digested bovine joint cartilage collagen Type II; subunit composition: $\alpha1(II)_3$; does not react with anti-collagen Type I & anti-collagen Type III by Western Blotting; *Merck Index*, 12: 2543; Sieper, J et al, *Arthritis Rheum*, 39: 41, 1996; Boissier, M-C et al, *Arthritis Rheum*, 33: 1, 1990

Biogenesis 2150-0535 Bovine articular cartilage Essentially homogenous by SDS-PAGE; purified; from dilute acetic acid as acid-soluble monomers; lyophilized | Useful in the study of the cartilage-collagen induced model of arthritis in rats and mice, autoimmunity studies (eg tests for autoimmune responses to cartilage collagen in rheumatoid patients) & collagen-chemistry comparison studies

USBio C7510-21 Bovine articular cartilage MW 30k Highly purified ≥99% (SDS-PAGE); lyophilized; soluble in acidic buffer (max 4 mg/mL), but difficult to dissolve in neutral buffer | Purified for immunization; suitable for antigenic applications in immunological protocols; used for collagen-induced arthritis (CIA) in experimental animals; to avoid fibril formation under neutral conditions, use a buffer containing 0.15-0.2 M NaCl & keep on ice; to prepare a collagen gel, dissolve in neutral buffer containing 0.15-0.2 M NaCl & incubate at 37°C for 1-2 hr

Biodesign A33122B Bovine cartilage Purified

Chemicon CC110 Bovine hyaline cartilage Purified | Extracellular matrix protein; not for sale in Japan

Sigma C 7806 Bovine nasal septum Acid soluble; prepared by modification of the pepsin extraction method of Niyibizi | Niyibizi, C et al, *J Biol Chem*, 259: 14170, 1984; Bornstein, P & Traub, W, *The Proteins*, IV: 412, 1979

USBio C7510-26 Bovine sternal cartilage MW 30k Highly purified ≥99% (SDS-PAGE); solution in 0.05 M acetic acid; free of pepsin & proteoglycans | Highly purified for ELISA; used for assaying Ab in sera from collagen-induced arthritis (CIA) models & humans by ELISA; polymeric or fibrilar collagen can bind nonspecifically to Ig in serum samples, secondary Ab & avidin-peroxidase, creating high, false-positive reactions; this highly purified, polymeric-free, ELISA grade collagen prevents false positives

Sigma C 1188 Bovine tracheal cartilage Acid soluble; prepared by modification of the pepsin extraction method of Trentham | Trentham, et al, *J Exp Med*, 146: 857, 1977; Bornstein, P & Traub, W, *The Proteins*, IV: 412, 1979

USBio C7510-22 Chick sternal cartilage   MW 30k   Highly purified ≥99% (SDS-PAGE); lyophilized; soluble in acidic buffer (max 4 mg/mL), but difficult to dissolve in neutral buffer | Purified for immunization; suitable for antigenic applications in immunological protocols; used for collagen-induced arthritis (CIA) in experimental animals; to avoid fibril formation under neutral conditions, use a buffer containing 0.15-0.2 M NaCl & keep on ice; to prepare a collagen gel, dissolve in neutral buffer containing 0.15-0.2 M NaCl & incubate at 37°C for 1-2 hr

USBio C7510-27 Chick sternal cartilage   MW 30k   Highly purified ≥99% (SDS-PAGE); solution in 0.05 M acetic acid; free of pepsin & proteoglycans | Highly purified for ELISA; used for assaying Ab in sera from collagen-induced arthritis (CIA) models & humans by ELISA; polymeric or fibrilar collagen can bind nonspecifically to Ig in serum samples, secondary Ab & avidin-peroxidase, creating high, false-positive reactions; this highly purified, polymeric-free, ELISA grade collagen prevents false positives

Chemicon CC092 Chicken sternal cartilage   Purified | Extracellular matrix protein

Sigma C 9301 Chicken sternal cartilage   suitable for tissue culture & cell biology studies; prepared by a modification of Trentham, DE et al, *J Exp Med*, 146: 857, 1977

Calbiochem 234185 Human   MW 100k   Salt-free lyophilized solid; ≥95% (SDS-PAGE); prepared from tissue of individuals shown by certified tests to be negative for HBsAg & antibodies to HIV & HCV | Highly purified preparation isolated from the triple helical domain of Gn-HCl-extracted & pepsin-digested human joint cartilage collagen Type II; subunit composition: $\alpha1(II)_3$; does not react with anti-collagen Type I & anti-collagen Type III by Western Blotting; *Merck Index*, 12: 2543; Ricard-Blum, S et al, *J Cell Biochem*, 27: 347, 1985; Boissier, M-C et al, *Arthritis Rheum*, 33: 1, 1990

Biodesign A22314H Human cartilage   Purified

Chemicon CC052 Human cartilage   Purified | Extracellular matrix protein

Biogenesis 2150-0070 Human hip/knee joint cartilage   Purified; dilute acetic acid; lyophilized | Useful in the study of the cartilage-collagen induced model of arthritis in rats and mice, autoimmunity studies & collagen-chemistry comparison studies

USBio C7510-23 Human sternal cartilage   MW 300k   Highly purified ≥99% (SDS-PAGE); lyophilized; soluble in acidic buffer (max 4 mg/mL), but difficult to dissolve in neutral buffer | Purified for immunization; suitable for antigenic applications in immunological protocols; used for collagen-induced arthritis (CIA) in experimental animals; to avoid fibril formation under neutral conditions, use a buffer containing 0.15-0.2 M NaCl & keep on ice; to prepare a collagen gel, dissolve in neutral buffer containing 0.15-0.2 M NaCl & incubate at 37°C for 1-2 hr

USBio C7510-28 Human sternal cartilage   MW 30k   Highly purified ≥99% (SDS-PAGE); solution in 0.05 M acetic acid; free of pepsin & proteoglycans | Highly purified for ELISA; used for assaying Ab in sera from collagen-induced arthritis (CIA) models & humans by ELISA; polymeric or fibrilar collagen can bind nonspecifically to Ig in serum samples, secondary Ab & avidin-peroxidase, creating high, false-positive reactions; this highly purified, polymeric-free, ELISA grade collagen prevents false positives

USBio C7510-29 Monkey sternal cartilage   Highly purified ≥99% (SDS-PAGE); solution in 0.05 M acetic acid; free of pepsin & proteoglycans | Highly purified for ELISA; used for assaying Ab in sera from collagen-induced arthritis (CIA) models & humans by ELISA; polymeric or fibrilar collagen can bind nonspecifically to Ig in serum samples, secondary Ab & avidin-peroxidase, creating high, false-positive reactions; this highly purified, polymeric-free, ELISA grade collagen prevents false positives

USBio C7510-30 Mouse sternal cartilage   MW 30k   Highly purified ≥99% (SDS-PAGE); solution in 0.05 M acetic acid; free of pepsin & proteoglycans | Highly purified for ELISA; used for assaying Ab in sera from collagen-induced arthritis (CIA) models & humans by ELISA; polymeric or fibrilar collagen can bind nonspecifically to Ig in serum samples, secondary Ab & avidin-peroxidase, creating high, false-positive reactions; this highly purified, polymeric-free, ELISA grade collagen prevents false positives

USBio C7510-24 Porcine sternal cartilage   MW 30k   Highly purified ≥99% (SDS-PAGE); lyophilized; soluble in acidic buffer (max 4 mg/mL), but difficult to dissolve in neutral buffer | Purified for immunization; suitable for antigenic applications in immunological protocols; used for collagen-induced arthritis (CIA) in experimental animals; to avoid fibril formation under neutral conditions, use a buffer containing 0.15-0.2 M NaCl & keep on ice; to prepare a collagen gel, dissolve in neutral buffer containing 0.15-0.2 M NaCl & incubate at 37°C for 1-2 hr

USBio C7510-31 Porcine sternal cartilage   MW 30k   Highly purified ≥99% (SDS-PAGE); solution in 0.05 M acetic acid; free of pepsin & proteoglycans | Highly purified for ELISA; used for assaying Ab in sera from collagen-induced arthritis (CIA) models & humans by ELISA; polymeric or fibrilar collagen can bind nonspecifically to Ig in serum samples, secondary Ab & avidin-peroxidase, creating high, false-positive reactions; this highly purified, polymeric-free, ELISA grade collagen prevents false positives

USBio C7510-25 Rat sternal cartilage   MW 30k   Highly purified ≥99% (SDS-PAGE); lyophilized; soluble in acidic buffer (max 4 mg/mL), but difficult to dissolve in neutral buffer | Purified for immunization; suitable for antigenic applications in immunological protocols; used for collagen-induced arthritis (CIA) in experimental animals; to avoid fibril formation under neutral conditions, use a buffer containing 0.15-0.2 M NaCl & keep on ice; to prepare a collagen gel, dissolve in neutral buffer containing 0.15-0.2 M NaCl & incubate at 37°C for 1-2 hr

USBio C7510-32 Rat sternal cartilage   MW 30k   Highly purified ≥99% (SDS-PAGE); solution in 0.05 M acetic acid; free of pepsin & proteoglycans | Highly purified for ELISA; used for assaying Ab in sera from collagen-induced arthritis (CIA) models & humans by ELISA; polymeric or fibrilar collagen can bind nonspecifically to Ig in serum samples, secondary Ab & avidin-peroxidase, creating high, false-positive reactions; this highly purified, polymeric-free, ELISA grade collagen prevents false positives

Collagen Type II, FITC Conjugated

Chemicon CC110F Bovine hyaline cartilage   Purified | Extracellular matrix protein

Collagen Type III

Biodesign A33124B Bovine placenta   Purified

Chemicon CC081 Bovine placenta   Purified | Extracellular matrix protein

Biogenesis 2150-0555 Bovine skin   MW 300k   <0.5% non-collagenous proteins; typical contaminants are cross-linked collagen type II dimers and trimers (approximately 10%); other collagen types constitute <5% by immunoassay and analytical chromatography; purified; lyophilized | a1(III)3 native triple helix; dissolved preparations (0.1 M acetic acid, pH 3.0) retain immunochemical properties; thermal degradation to gelatin occurs at 100°C for 5 minutes

Chemicon CC078 Bovine skin   Purified | Extracellular matrix protein

Biodesign A33123H Human placenta   Purified

Biogenesis 2150-0110 Human placenta   MW ~300k   <0.5% non-collagenous proteins, typically cross-linked collagen type III dimers and trimers (approx 10%); tested negative for HBsAg and HIV-1/2 antibodies; molecular composition: a1(III)3; purified; essentially salt free; lyophilized

Chemicon CC054 Human placenta   Purified | Extracellular matrix protein

Sigma C 4407 Human placenta   Sigma Type X; acid soluble; prepared by modification of the pepsin extraction & salt fractionation method of Hill & Harper | Hill, RJ & Harper, E, *Anal Biochem*, 141: 83, 1984; Bornstein, P & Traub, W, *The Proteins*, IV: 412, 1979

Biogenesis 2150-1955 Rat healthy tendons   MW 300k   <0.5% non-collagenous proteins, typically cross-linked collagen type III dimers and trimers (approx 10%); purified; essentially salt free; lyophilized | Molecular composition: a1(III)3, native triple helix; dissolved collagen (0.5 M acetic acid, pH2.5) retains immunogenic properties; thermal denaturation (100°C for 5 min) converts collagen to gelatin

Collagen Type III, FITC Conjugated

Chemicon CC112F   Bovine skin   Purified | Extracellular matrix protein; not for sale in Japan

Collagen Type IV

Synonyms: Collagenase Substrate

Biodesign A33126B   Bovine placenta   Purified

Chemicon CC083   Bovine placenta   Purified | Extracellular matrix protein

Biodesign A33125H   Human placenta   Purified

Biogenesis 2150-0150   Human placenta   MW 340k   <2% Collagen type I & type V, <1% collagen type III, <0.5% non-collagenous proteins; molecular composition: alpha 1(V)2, alpha 2(V) (native triple helix); tested negative for HBsAg and HIV antibodies; purified; essentially salt free; lyophilized | Dissolved collagen (0.5 *M* acetic acid, pH 2.5) retains immunogenic properties; thermal degradation converts dissolved collagen to gelatin; Glanville et al, *Eur J Biochem*, 95:383, 1979; Sage et al, *Biochem*, 18:3815, 1979; Klasson et al, Coll Rel Res, 6:397, 1986

Calbiochem 234154   Human placenta   MW 130k Aseptically filled liquid in 10 *mM* acetic acid; homogeneous purity (SDS-PAGE); prepared from tissue of individuals shown by certified tests to be negative for HBsAg & antibodies to HIV & HCV | Highly purified preparation; useful as a reference standard for collagen Type IV; typically used as a thin coating on tissue culture surfaces; *Merck Index*, 12: 2543; Klasson, SC et al, *Coll Relat Res*, 6: 397, 1986

Chemicon CC076   Human placenta   Purified | Extracellular matrix protein

Fluka 27663   Human placenta   MW ~125k   Prepared by a modified pepsin extraction method of Glanville, RW et al, *Eur J Biochem*, 95: 383, 1979 | Electrophoresis test: consistent with basement membrane collagen; Bailey, AJ et al, *FEBS Lett*, 99: 361, 1979; Konigsberg, IR, *Meth Enzymol*, 58: 525, 1979

Sigma C 5533   Human placenta   Lyophilized from 10 *mM* glacial acetic acid; acid soluble; cell culture tested | PCR negative for HIV & Hepatitis B; prepared by a modification of the pepsin extraction method of Niyibizi, et al

Sigma C 7521   Human placenta   Sigma Type VI; acid soluble; prepared by modification of the pepsin extraction method of Glanville; 3 bands following SDS polyacrylamide gel electrophoresis under reducing conditions consistent with basement membrane collagen | Glanville, RW et al, *Eur J Biochem*, 95: 383, 1979; Bailey AJ et al, *FEBS Lett*, 99: 361, 1979; Bornstein, P & Traub, W, *The Proteins*, IV: 412, 1979

Sigma C 0543   Mouse sarcoma   Lyophilized; sterile-filtered; chloroform treated; pepsin is not used in the preparation; 0.75 mg/vial; cell culture tested | Isolated form basement membrane of Engelbreth-Holm-Swarm mouse sarcoma; suitable for attachment of epithelial, endothelial, muscle & nerve cells in culture

Collagen Type IV, FITC Conjugated

Chemicon CC113F   Bovine placenta   Purified | Extracellular matrix protein; not for sale in Japan

Collagen Type V

Biodesign A33128B   Bovine placenta   Purified

Chemicon CC084   Bovine placenta   Purified | Extracellular matrix protein

Biodesign A33127H   Human placenta   Purified

Biogenesis 2150-0190   Human placenta   <2% collagen type I & IV, <1% collagen type III, <0.5% non-collageneous proteins; tested negative for HBsAg and HIV antibodies; purified; essentially salt free; lyophilized | Molecular composition: [alpha1(V)2, alpha2(V)]; native triple helix; dissolved collagen (0.5 *M* acetic acid, pH 2.5) retains immunogenic properties; thermal degradation converts dissolved collagen to gelatin; Glanville et al, *Eur J Biochem*, 95:383, 1979; Sage et al, *Biochem*, 18:3815, 1979; Klasson et al, Coll Rel Res, 6:397, 1986

Calbiochem 234161   Human placenta   Consists of three subunits: α1(V), α2(V) & α3(V) of MW 110,000, 115,000 & 125,000, respectively; liquid in 10 *mM* acetic acid; homogeneous purity (SDS-PAGE); prepared from tissue of individuals shown by certified tests to be negative for HBsAg & antibodies to HIV & HCV | *Merck Index*, 12: 2543

Chemicon CC077   Human placenta   Purified | Extracellular matrix protein

Sigma C 3657   Human placenta   Sigma Type IX; acid soluble; prepared by modification of the pepsin extraction & salt fractionation method of Niyibizi | Niyibizi, C et al, *J Biol Chem*, 259: 14170, 1984; Bornstein, P & Traub, W, *The Proteins*, IV: 412, 1979

Collagen Type V, FITC Conjugated

Chemicon CC084F   Bovine placenta   Purified; FITC conjugated | Extracellular matrix protein; not for sale in Japan

Collagen Type VI

Biodesign A33130B   Bovine placenta   Purified

Chemicon CC086   Bovine placenta   Purified | Extracellular matrix protein

Biodesign A33129H   Human placenta   Purified

Biogenesis 2150-0230   Human placenta   Tested negative for HIV-1 and 2 antibodies and for HBsAg; <1% collagen types III, IV & V, <0.5% non-collageneous protein; purified; essentially salt free; lyophilized | Dissolved collagen (0.1 *M* acetic acid, pH 3.0) retains immunogenic properties; thermal denaturation converts collagen to gelatin; Miller & Rhodes, *Meth Enzymol*, 82, 1982

Collagen Type XI

Biogenesis 2150-0715   Bovine articular cartilage   Purified; dilute acetic acid; lyophilized | Supplied as acid-soluble monomers; useful in the study of the cartilage-collagen induced model of arthritis in rats and mice, autoimmunity studies & collagen chemistry comparison studies

Collagen, Azo Dye Impregnated

Sigma A 4341   Dye-impregnated hide powder | Azocollsuitable as a non-specific protease substrate; the rate of dye release may vary from lot to lotChavira, R Jr et al, *Anal Biochem*, 136: 446, 1984

Collagen, Carrier Insoluble Bone

Kamiya

Colony Stimulating Factor Tpo

Synonyms: Thrombopoietin

Sigma T 4309   Human recombinant, expressed in mouse myeloma cell line NSO   ≥97% (SDS-PAGE); 0.2 μm filtered & lyophilized in PBS containing 250 μg BSA; activity measured in a cell proliferation assay using MO7e cells; endotoxin tested | 335 AA; due to glycoylation the protein has an apparent MW of 75 kD in SDS-PAGE; precursor form consists of 356 AA; to generate the mature Tpo (335 AA), the precursor cleaves a 21 AA signal peptide; human, mouse & dog Tpo shows 69-75% AA homology; Tpo is the ligand for the receptor encoded by the *c-Mpl* proto-oncogene, acts as a stimulator of the development of megakaryocyte precursors of platelets; similar to erythropoietin, Tpo leads to an increase in the number of circulating platelets; affects the entire thrombopoietic process with stronger effects in the later stages; other thrombopoietic cytokines include stem cell factor (SCF), IL-3, IL-6 & IL-11

Sigma T 4184   Mouse recombinant, expressed in mouse myeloma cell line NSO   ≥97% (SDS-PAGE); 0.2 μm filtered & lyophilized in PBS containing 250 μg BSA; activity measured in a cell proliferation assay using MO7e cells; endotoxin tested; see Sigma T 4309

Colony Stimulating Factor, Granulocyte

Synonyms: Colony Stimulating Factor G

Biodesign A52023H   *E. coli*   MW 18.5k   Purified

Biodesign A52255H   *E. coli*   MW 19k   Purified | Species specificity: mouse

Chemicon GF051   Human   ≥95%

BioSource International PHC2034   Human recombinant

Calbiochem 234370   Human recombinant, expressed in *E. coli* MW 18k   ≥97% (SDS-PAGE); lyophilized from filter-sterilized 20 *mM* acetic acid containing 50 µg HSA/µg G-CSF; biological activity: ED_{50}=20-60 pg/mL as measured in a cell proliferation assay with the murine myeloblastic cell line NFS-60; endotoxin: ≤100 pg/µg G-CSF; may be carcinogenic/teratogenic | Pleiotropic cytokine best known for its specific effects on proliferation, differentiation & activation of hematopoietic cells of neutrophilic granulocyte lineage; causes neutrophil sequestration in rabbit lungs; inano, H et al, *Am J Respir Cell Mol Biol*, 19: 167, 1998; Nicola, NA et al, *Ann Rev Biochem*, 58: 45, 1989

ICN 154138   Human recombinant, expressed in *E. coli* Lyophilized from a sterile filtered solution containing 50 µg human serum albumin/µg cytokine in 10 *mM* acetic acid; ≥95%; activity: ED_{50} = 0.02-0.06 ng/mL measured in a cell proliferation assay with murine myeloblastic cell line NFS-60 | Best known for its proliferation, differentiation & activation effects on neutrophilic granulocyte hematopoietic cells

PeproTech 300-23   Human recombinant, expressed in *E. coli* MW 18.7k   >98% (SDS-PAGE) & HPLC; <0.1 ng endotoxin per mg (1EU/mg); lyophilized | Potent stimulator of bone marrow cells, especially those of neutrophil lineage; can enhance the survival & activate the immunological functions of mature neutrophils; 174 AA; $ED_{50} \leq$ 0.1 ng/mL; SA $\geq$ 1 x 10^7 U/mg; SA determined by the dose-dependent stimulation of the proliferation of murine M-NSF-60 cells

Sigma G 0407   Human recombinant, expressed in *E. coli*   MW 19.6k   ≥97% (SDS-PAGE); 0.2 µm filtered & lyophilized from 10 *mM* acetic acid containing 100 µg BSA; activity measured in a cell proliferation assay using NFS-60 cells; endotoxin tested | Produced by monocytes & fibroblasts; stimulates granulocyte colony formation, activates neutrophils, differentiates certain myeloid leukemic cell lines & is a potent activator of mature granulocytes; natural human G-CSF is a glycoprotein having 177 AA; shares ~75% AA sequence homology & has biological cross-reactivity with murine G-CSF; Shirafuji, N et al, *Exp Hematol*, 17: 116, 1989; Metcalf, D, *Cell*, 43: 5, 1985; Groopman, JE, *Cell*, 50: 5, 1987; Souza, LM et al, *Science*, 232: 61, 1986; Morstyn, G & Burgess, A, *Cancer Res*, 48: 5624, 1988

BioSource International PMC2034   Mouse recombinant

Calbiochem 234371   Mouse recombinant, expressed in *E. coli* MW 19k   >97% (SDS-PAGE); lyophilized from filter-sterilized 20 *mM* acetic acid containing 50 µg BSA/µg G-CSF; biological activity: ED_{50}=10-30 pg/mL as measured in a cell proliferation assay using a mouse myeloblastic cell line NFS-60; endotoxin: ≤100 pg/µg G-CSF; may be carcinogenic/teratogenic | Pleiotropic cytokine best known for its specific effects on proliferation, differentiation & activation of hematopoietic cells of neutrophilic granulocyte lineage; induces binding of STAT1 & STAT3 to IFN-γ response region in human neutrophils; Borolenta, C et al, *FEBS Lett*, 386: 239, 1996; Moore, MAS, *Ann Rev Immunol*, 9: 159, 1991; Nillson, SK et al, *Blood*, 86: 66, 1995; Gabrilove, JL, *Growth Factors*, 6: 187, 1992; Tsuchiya, M et al, *PNAS*, 83: 7633, 1986; *Merck Index*, 12: 4558

Sigma G 8160   Mouse recombinant, expressed in *E. coli*   MW 19k   ≥97% (SDS-PAGE); 0.2 µm filtered & lyophilized from 20 *mM* acetic acid containing 250 µg BSA; activity measured in a cell proliferation assay using NFS-60 cells; endotoxin tested; see Sigma G 0407

ICN 195761   Murine recombinant, expressed in *E. coli*   ≥95%; activity: ED_{50} = 0.01-0.03 ng/mL | G-CSF

PeproTech 250-05   Murine recombinant, expressed in *E. coli* MW 19.0k   >98% (SDS-PAGE) & HPLC; <0.1 ng endotoxin per mg (1EU/mg); lyophilized from 10 *mM* NaCitrate pH 4.0 | Potent stimulator of bone marrow cells, especially those of neutrophil lineage; 179 AA; $ED_{50} \leq$ 1.0 ng/mL; SA $\geq$ 10^6 U/mg; SA determined by the dose-dependent stimulation of the proliferation of murine M-NFS-60 cells

Colony Stimulating Factor, Granulocyte-Macrophage

Synonyms: Colony Stimulating Factor GM

Biodesign A52303H   *E. coli*   MW 14k   Purified

Biodesign A52503H   *E. coli*   MW 14k   Purified | Species specificity: mouse

Chemicon GF004   Human   ≥95%

BioSource International PHC2014   Human recombinant

Oncogene PF014   Human recombinant   MW 14k   >98% (SDS-PAGE); lyophilized; biological activity: half maximal stimulation of granulocyte & macrophage colony formation from human bone cells 0.1 ng/mL | Species reactivity: human; for proliferation studies & Western blot

Calbiochem 234373   Human recombinant, expressed in *E. coli* MW 14k   ≥97% (SDS-PAGE); lyophilized from filter-sterilized PBS containing 50 µg BSA/µg GM-CSF; biological activity: ED_{50}=20-80 pg/mL as measured in a cell proliferation assay with TF-1 cells; endotoxin: ≤100 pg/µg GM-CSF; may be carcinogenic/teratogenic | Stimulates proliferation, maturation & function of hematopoietic cells; target cells include macrophages, granulocytes & eosinophils; mediates host defense & inflammation & is associated with tumor growth & tumor growth & metastasis; Moore, MAS, *Ann Rev Immunol*, 9: 159, 1991; Tsuruta, N et al, *Cancer*, 82: 2173, 1998; *Merck Index*, 12: 4559

Fitzgerald 30-AG10   Human recombinant, expressed in *E. coli*

Harlan BT-3006   Human recombinant, expressed in *E. coli* Lyophilized; 0.002 mg

Harlan BT-3007   Human recombinant, expressed in *E. coli* Lyophilized; 0.01 mg

ICN 154137   Human recombinant, expressed in *E. coli* Frozen; ≥95%; activity: ED_{50} = 0.02-0.08 ng/mL | Stimulates the production of superoxide anion by eosinophils; stimulates granulocyte & macrophage production in human & murine bone marrow cell cultures

PeproTech 300-03   Human recombinant, expressed in *E. coli* MW 14.0k   >98% (SDS-PAGE) & HPLC; <0.1 ng endotoxin per mg (1EU/mg); lyophilized | Potent species-specific stimulator of bone marrow cells; stimulates precursor cells of granulocytes, macrophages, & eosinophils; 123 AA; $ED_{50} \leq$ 0.1 ng/mL; SA $\geq$ 10^7 U/mg; SA determined by the dose-dependent stimulation of the proliferation of human TF-1 cells

Sigma G 5035   Human recombinant, expressed in *E. coli*   MW 18-22k   ≥97% (SDS-PAGE); 0.2 µm filtered & lyophilized in PBS containing 100 µg BSA; activity tested in culture using human TF-1 cells; endotoxin tested | Induces myeloid progenitor cells from bone marrow to form colonies containing macrophages & granulocytes in a semisolid media; acts upon mature macrophages, eosinophils & neutrophils to stimulate various functional activities; acidic glycoprotein which binds to high affinity receptors on GM-CSF sensitive cells; although it shares 54% AA sequence homology with mouse GM-CSF, their biological actions are species-specific; other growth factors & CSFs modulate receptor binding or actions of GM-CSF; Kitamura, T et al, *J Cell Physiol*, 140: 323, 1989; Metcalf, D, *Blood*, 67: 257, 1986; Nicola, N, *Immunol Today*, 8: 134, 1987; Wong, G et al, *Science*, 228: 810, 1985; Morstyn, G & Burgess, A, *Cancer Res*, 48: 5624, 1988; Mazur, e & Cohen, J, *Clin Pharmacol Ther*, 46: 250, 1989

ICN 150708   Human recombinant, expressed in yeast containing the GM-CSF gene originally cloned from peripheral blood t-cell DNA & purified by HPLC   Frozen; ≥95%; ≥1.25 x 10^7 U/mg protein; 1 U induces half-maximal ^{3}H-TdR incorporation by TF-1 cells in a 96 hr bioassay

Biogenesis 4740-1004   Mouse r-DNA   Lyophilized

Biogenesis 4740-1115   Mouse r-DNA

BioSource International PMC2014   Mouse recombinant

Calbiochem 234374   Mouse recombinant, expressed in *E. coli*
MW 14.8k   >97% (SDS-PAGE); lyophilized from filter-sterilized
PBS containing 50 μg BSA/μg GM-CSF; biological activity:
ED_{50}=100-300 pg/mL as measured in a cell proliferation assay
using a factor-dependent cell line, DA-3; endotoxin: ≤100 pg/μg
GM-CSF; may be carcinogenic/teratogenic | Pleiotropic cytokine
that can stimulate proliferation, maturation & function of
hematopoietic cells; improves the response rate to antibiotic
therapy in cancer subjects; Moore, MAS, *Ann Rev Immunol*, 9: 159,
1991; Anaissie, EJ et al, *Am J Med*, 100: 15, 1996; *Merck Index*,
12: 4559; Watanabe, S et al, *J Biol Chem*, 271: 12681, 1996;
Nicola, NA, *Ann Rev Biochem*, 58: 45, 1989; Gough, NM et al,
EMBO J, 4: 645, 1985

Sigma G 0282   Mouse recombinant, expressed in *E. coli*   MW
23k   ≥97% (SDS-PAGE); 0.2 μm filtered & lyophilized in PBS
containing 250 μg BSA; proliferative activity tested in culture using
mouse DA-3 cells; endotoxin tested; see Sigma G 5035 | Acidic
glycoproteinIhle, JN et al, *Advances in Viral Oncology*, G Klein (ed),
Raven Press, New York, NY, 4: 95, 1984

Chemicon GF026   Murine   ≥95%

Fitzgerald 30-AG15   Murine recombinant, expressed in *E. coli*

Harlan BT-5112   Murine recombinant, expressed in *E. coli*
Lyophilized; 0.005 mg

Harlan BT-5113   Murine recombinant, expressed in *E. coli*
Lyophilized; 0.025 mg

ICN 195007   Murine recombinant, expressed in *E. coli*   MW
14k   Lyophilized | GM-CSF

PeproTech 315-03   Murine recombinant, expressed in *E. coli*
MW 14.0k   >98% (SDS-PAGE) & HPLC; <0.1 ng endotoxin per
mg (1EU/mg); lyophilized | Potent, species-specific stimulator of
bone marrow cells; stimulates precursor cells of granulocytes,
macrophages, & eosinophils; 124 AA; $ED_{50} \leq$ 0.2 ng/mL; SA ≥ 5 x
10^6 U/mg; SA determined by the dose-dependent stimulation of the
proliferation of murine FDC-P1 cells

ICN 150709   Murine recombinant, expressed in yeast & purified
by HPLC   Frozen; ≥90%; ≥5.0 x 10^6 U/mg protein; 1 U stimulates
half-maximal proliferation of FDCP2-1D cells; optimal colony
formation is typically obtained using 10-50 ng/mL | GM-CSF

BioSource International PRC2014   Rat recombinant

Colony Stimulating Factor, Macrophage

Synonyms: Colony Stimulating Factor I; Colony Stimulating Factor
M

Biodesign A52025H   *E. coli*   MW 18.4k   Purified

BioSource International PHC2024   Human

Chemicon GF053   Human   ≥95%

Biogenesis 4740-1255   Human r-DNA   Liquid

Calbiochem 234376   Human recombinant, expressed in *E. coli*
MW 37k   ≥97% (SDS-PAGE); lyophilized from sterile-filtered PBS
containing 50 μg BSA/μg M-CSF; biological activity: ED_{50}=0.5-1.5
ng/mL as measured in a cell proliferation assay using the M-CSF-
dependent mouse monocytic cell line M-NFS-60; endotoxin: ≤100
pg/μg M-CSF; may be carcinogenic/teratogenic | Stimulates
formation of macrophage colonies from bone marrow hematopoietic
progenitor cells; increases osteoclastic bone resorption in adults;
induces cell death in HIV-1 infected monocytes; Edwards, M et al,
Bone, 22: 325, 1998; Bergamini, A et al, *Immunol Lett*, 42: 35,
1994; Hattersley, G et al, *Biochem Biophys Res Comm*, 177: 526,
1991

PeproTech 300-25   Human recombinant, expressed in *E. coli*
MW 36.8k   >98%; 316 AA; lyophilized with no additives; ED_{50}:
1.0-5.0 ng/mL as determined by a cell proliferation assay using
murine M-NFS-60 cells

Sigma M 6518   Human recombinant, expressed in *E. coli*   MW
18.5k   ≥97% (SDS-PAGE); 0.2 μm filtered & lyophilized from PBS
containing 100 μg BSA; proliferative activity tested in culture using
mouse M-NFS-60 cells; endotoxin tested | Glycoprotein containing
an *N*-terminal Met; produced by monocytes, fibroblasts &
endothelial cells; stimulates the formation of macrophage colonies,
enhances antibody-dependent cell mediated cytotoxicity by
monocytes & macrophages & inhibits bone resorption by
osteoclasts; appears in a few different MW forms due to variations
in glycosylation; 159 AA; Halenbeck, R et al, *Biotechnology*, 7:
710, 1989; Metcalf, D, *Blood*, 67: 257, 1986; Mufson, RA et al,
Cellular Immunol, 119: 182, 1989; Kawasaki, ES et al, *Science*,
230: 291, 1985; Hattersley, G et al, *J Cell Physiol*, 137: 199, 1988

ICN 152368   Human recombinant, expressed in yeast
Frozen; ≥90%; ≥2 x 10^5 proliferation U/mg protein; 1 U stimulates
half-maximal proliferation of non-adherent mouse bone marrow
cells | Supports proliferation & differentiation of macrophage
colonies in bone marrow cultures; ironically, human *M*-CSF is a
strong inducer of mouse macrophage colony formation, but is
significantly less active on human cells; only effective on mouse
cells & gives rise to macrophage colonies in murine bone marrow
cultures

ICN 160008   Human urine   MW ~100k   Sterile filtered with
1% BSA in H_2O; ≥1 x 10^7 U/mg protein; 1 U stimulates colony
formation of mouse bone marrow cells in a soft agar assay

Calbiochem 234378   Mouse recombinant, expressed in *E. coli*
MW 26k   ≥97% (SDS-PAGE); lyophilized from sterile-filtered PBS
containing 50 μg BSA/μg M-CSF; biological activity: ED_{50}=0.5-1.5
ng/mL as measured in a cell proliferation assay using the M-CSF-
dependent mouse monocytic cell line M-NFS-60; endotoxin: ≤100
pg/μg M-CSF; may be carcinogenic/teratogenic | Stimulates
formation of macrophage colonies from bone marrow hematopoietic
progenitor cells; Ladner, MB et al, *PNAS*, 85: 6706, 1988;
Bergamini, A et al, *Immunol Lett*, 42: 35, 1994; Hattersley, G et al,
Biochem Biophys Res Comm, 177: 526, 1991

Sigma M 9170   Mouse recombinant, expressed in *E. coli*
≥97% (SDS-PAGE); 0.2 μm filtered & lyophilized from PBS
containing 500 μg BSA; activity tested in culture by cell
proliferation assay using mouse monocytic cell line M-NFS-60;
endotoxin tested; see Sigma M 6518 | Halenbeck, R et al,
Biotechnology, 7: 710, 1989

ICN 160007   Murine recombinant, expressed in *E. coli*
Lyophilized with BSA carrier; >97%; activity: ED_{50} = 0.5-1.5
ng/mL; 1 U stimulates half maximal cell proliferation using a M-CSF
dependent murine monocytic cell line, M-NFS-60 | Supports the
proliferation & differentiation of macrophage colonies in bone
marrow cultures

Colony Stimulating Factor, Macrophage Human

IBT GF-390-3, GF-390-4   *E. coli*   >92%; 1 mg/mL solution in
0.1% mannitol, 0.1 M citrate, pH 6.5 | SA = 6x10^7 U/mg
(measured by the mouse bone marrow colony-forming assay)

Colony Stimulating Factor, Murine Granulocyte

Chemicon GF059   Murine   ≥95%

Complement

USBio C7849-10   Goat   Lyophilized complete with buffered
saline ionic diluent

ICN 55852 ICN 55854   Guinea pig   Purified

USBio C7849-15   Guinea pig   Lyophilized complete with
buffered saline ionic diluent

USBio C7849-20   Hamster   Lyophilized complete with
buffered saline ionic diluent

ICN 55860   Mouse   Purified

USBio C7849-25   Mouse   Lyophilized complete with buffered
saline ionic diluent; ~85 mg protein /mL; immunoelectrophoresis:
to pass standard

ICN 55866   Rabbit   Purified

USBio C7849-35   Rabbit   Lyophilized complete with buffered
saline ionic diluent

ICN 55870   Rat   Purified

Complement B

Synonyms: Properdin Factor B

ICN 191383   Human serum   Functionally & biochemically pure (PAGE); 1 mg protein/mL; in PBS, 40% glycerol

Sigma C 4909   Human serum   Lyophilized from 0.05 M Tris buffer, pH 7.5, containing 0.02 M NaCl, 0.01 M EDTA, 0.01 M ε-aminocaproic acid; single band by immunoelectrophoresis at 20 μg protein/gel against both anti-B & anti-whole serum | Suitable for radioiodination

Complement C1

Sigma C 2660   Human serum   ~90% by nonreducing SDS-urea gel; frozen solution containing 0.1 mg/mL protein in 20 mM Tris-HCl, 154 mM NaCl, 1 mM CaCl₂, 0.03 mM p-nitrophenyl-p'-guanidinobenzoate, pH 7.4 | Functionally active by a hemolytic assay; suitable for radioiodination

Complement C1 Esterase Inhibitor

Sigma C 2412   Human   MW ~105k   ~95% (SDS-PAGE); frozen solution containing 0.4 mg/mL protein in 20 mM Tris-HCl, 240 mM NaCl, 1 mM EDTA, pH 7.4 | Functionally active; suitable for radioiodination

Complement C1q

ICN 191391   Human serum   Functionally & biochemically pure (PAGE); 1 mg protein/mL; in PBS, 40% glycerol

Sigma C 0660   Human serum   Lyophilized from 0.05 M Tris buffer, pH 7.3, containing 0.5 M NaCl; single band by immunoelelctrophoresis at 20 μg protein/gel against both anti-C1q & anti-whole serum; no visible reaction against anti-IgG, anti-IgM, anti-IgA sera at 20 μg protein/gel | Suitable for radioiodination

Scipac P102-1   Pooled serum/plasma   >96%; lyophilized | Complement protein

Complement C3

ICN 191390   Human serum   Functionally & biochemically pure (PAGE); 1 mg protein/mL; in PBS, 40% glycerol

Sigma C 0651   Human serum   ~95% (SDS-PAGE); lyophilized from 10 mM sodium phosphate buffer; pH 7.2, containing 150 mM NaCl | Partial aggregation of C3 may be observed when product is electrophoresed; suitable for radioiodination

Sigma C 2910   Human serum   Frozen solution containing 1 mg/mL in 15 mM sodium phosphate buffer, 150 mM NaCl, pH 7.2; activity: ≥50 C3H50 units/mg using C3 deficient serum | Functionally pure by a hemolytic assay using deficient sera; suitable for radioiodination

USBio C7850-12   Human serum   ≥97%; no contaminants detected; single band by SDS-PAGE, IEP, &/or RID; ~5 mg/mL RID (CAP 4) supplied in PBS buffer, pH 7.2, 0.1% NaN₃ | Suitable for antigenic applications in immunological protocols

Scipac P150-0   Pooled serum/plasma   >99%; frozen in sodium phosphate buffer; immunoaffinity absorbed; <1% IgG, IgM, IgG and C4 combined totals by RID | Complement protein

Scipac P150-1   Pooled serum/plasma   >96%; lyophilized | Complement protein

Scipac P150-2   Serum/plasma   40-90%; lyophilized; available on request | Complement protein

Complement C4

USBio C7850-17   Human   Purity: Trace amounts of C3; 3.2 mg/mL supplied in PBS, pH 7.2, & 0.1% NaN₃ | Suitable for antigenic applications in immunological protocols

ICN 191389   Human serum   Functionally & biochemically pure (PAGE); 1 mg protein/mL; in PBS, 40% glycerol

Sigma C 8195   Human serum   Frozen solution containing 1 mg/mL in 25 mM sodium phosphate buffer, 100 mM NaCl, pH 7.2, containing 0.02% sodium azide; activity: ≥300,000 C4H50 U/mg protein (not determined by Sigma) | Suitable for radioiodination

Scipac P151-0   Pooled serum/plasma   >99%; frozen in sodium phosphate buffer; immunoaffinity absorbed; <1% IgG, IgM, IgG and C3 combined totals by RID | Complement protein

Scipac P151-3   Pooled serum/plasma   >96%; frozen in sodium phosphate buffer | Complement protein

Scipac P151-2   Serum/plasma   40-90%; lyophilized | Complement protein

Complement C4b Binding Protein

Sigma C 2537   Human plasma   MW 500k   >70% (SDS-PAGE); lyophilized from ~20 μg protein in 100 μL of 20 mM Tris buffer, 0.1 M NaCl, pH 7.4; a macromolecular glycoprotein of & is composed of 7 identical subunit chains of 70 kD | Regulatory factor for both complement system & blood coagulation; suitable for radioiodination; Nagasawa, S et al, Immunochemistry, 14: 749, 1977; Dahlback, B, J Biochem, 209: 847, 1983

Complement C5

ICN 191388   Human serum   Functionally & biochemically pure (PAGE); 1 mg protein/mL; in PBS, 40% glycerol

Sigma C 3160   Human serum   ≥75% by (SDS-PAGE); frozen solution containing 1 mg/mL in 10 mM sodium phosphate, 150 mM NaCl, pH 7.2; hemolytic activity: >300,000 C5H50 U/mg protein | Functionally active by a sensitive hemolytic assay; suitable for radioiodination

Complement C5a

Fluka 60897   Human recombinant, expressed in E. coli   MW ~8.6k   ≥95% (GE); protein modification: non-glycosylated form with glutathione bound to Cys-27, ~65% of C5a carry a methionyl residue at Thr-1 (NH₂-terminal end) | Wilkinson, PC, Meth Enzymol, 162: 127, 1988; Pike, MC & Snyderman, R, ibid, 162: 236, 1988; Janatova, J, ibid, 162: 579, 1988; Daumy, GO et al, Biochim Biophys Acta, 967: 326, 1988

Sigma C 5788   Human recombinant, expressed in E. coli   MW ~8.6k (non-glycosylated, with glutathione attached to Cys²⁷) ~95% (SDS-PAGE; HPLC); lyophilized; mixture of C5a (~35%) & C5a having an added methionyl residue at the amino terminus (~65%) | Exhibits biological activities similar to serum-derived C5a; suitable for radioiodination

Complement C6

ICN 191387   Human serum   Functionally & biochemically pure (PAGE); 1 mg protein/mL; in PBS, 40% glycerol

Sigma C 3285   Human serum   ≥80% by (SDS-PAGE); frozen solution containing 1 mg/mL in 10 mM sodium phosphate, 150 mM NaCl, pH 7.2; hemolytic activity: >300,000 C6H50 U/mg protein | Functionally active by a sensitive hemolytic assay; suitable for radioiodination

Complement C7

ICN 191386   Human serum   Functionally & biochemically pure (PAGE); 1 mg protein/mL; in PBS, 40% glycerol

Sigma C 2787   Human serum   ≥60% by (SDS-PAGE); frozen solution containing 1 mg/mL in 10 mM sodium phosphate, 150 mM NaCl, pH 7.2; hemolytic activity: >200,000 C7H50 U/mg protein | Functionally active by a sensitive hemolytic assay; suitable for radioiodination

Complement C8

ICN 191385   Human serum   Functionally & biochemically pure (PAGE); 1 mg protein/mL; in PBS, 40% glycerol

Sigma C 3535   Human serum   ≥85% by (SDS-PAGE); frozen solution containing 1 mg/mL in 15 mM sodium phosphate, 150 mM NaCl, pH 7.2; activity: ≥125,000 C8H50 U/mg protein using C8 deficient serum | Functionally pure by a sensitive hemolytic assay using deficient sera; suitable for radioiodination

Complement C9

ICN 191384   Human serum   Functionally & biochemically pure (PAGE); 1 mg protein/mL; in PBS, 40% glycerol

Sigma C 3660 Human serum Frozen solution at 1 mg/mL containing 15 mM sodium phosphate, 135 mM NaCl, pH 7.4; activity: ≥70,000 C9H50 U/mg protein using C9 deficient serum | Functionally pure by a sensitive hemolytic assay using deficient sera; suitable for radioiodination

Complement D

Sigma C 5688 Human plasma >90% (SDS-PAGE); frozen solution containing 100 µg/mL in PBS, pH 7.2 | Suitable for radioiodination; Niemann, MA et al, *J Immunol*, 132: 809, 1984

Complement H

Sigma C 5813 Human plasma >90% (SDS-PAGE); solution containing 1 mg/mL in PBS, pH 7.2 | C3b-binding protein which regulates the formation & function of complement C3 & C5 convertases; suitable for radioiodination; Fearon, DT & Austen KF, *Proc Natl Acad Sci USA*, 74: 1683, 1977; Pangburn, MK & Muller-Eberhard, HJ, *Springer Semin Immunopath*, 7: 63, 1984

Complement III

Biogenesis 2222-5704 Human plasma MW ~180k Tested negative for antibodies to HBsAg, HCV and HIV-1 and 2; purified; from 0.01 *M* PBS, pH 7.2; lyophilized | Cellulose acetate electrophoresis shows one band only with beta2-electrophoretic mobility

Complement IIIb

Biogenesis 2222-5909 Human Liquid

Complement Iq

Biogenesis 2221-5504 Human serum MW 410k Tested negative for HIV1/2 antibodies, HBsAg and HCV antibodies; purified; 0.01M EDTA, 0.3M NaCl, pH 7.5; lyophilized | Single band in PAGE

Complement IV

Biogenesis 2222-7704 Human serum Liquid

Complement IX

Biogenesis 2222-9054 Human serum Liquid

Complement P

Sigma C 6063 Human plasma ≥95% (SDS-PAGE); frozen solution containing 1 mg/mL in PBS, pH 7.2 | Suitable for radioiodination; Pangburn MK, *Meth Enzymol*, 162: 639, 1988

Complement V

Biogenesis 2222-8454 Human serum Liquid

Complement VI

Biogenesis 2222-8654 Human serum Liquid

Complement VII

Biogenesis 2222-8854 Human serum Liquid

Complement VIII

Biogenesis 2222-8954 Human serum Liquid

Component LS III

Synonyms: Snake Toxin

Sigma T 0409 *Laticauda semifasciata* FW 6837.6 Lyophilized powder containing ~80% protein (Lowry); balance potassium phosphate buffer salts | Postsynaptic neurotoxin; Maeda, N & Tamiya, N, *Biochem J*, 141: 389, 1974

Conalbumin

Synonyms: Ovotransferrin

ICN 194981 Chicken egg MW 76k >95% | Binding protein which can transport metal ions such as Cu^{2+}, Fe^{2+}, Mn^{2+} & Zn^{2+}; Szekacs, A etal, *Anal Biochem*, 207:291, 1992

Fluka 27695 Chicken egg white MW ~77k ≥89% (GE); lyophilized; ≤0.005% Fe; ≤6% loss on drying; ≤0.05% residue on ignition | Crichton, RR et al, *Eur J Biochem*, 164: 485, 1987; Mano, N et al, *J Chromatog*, 603: 105, 1992

Connexin, 32 Control Peptide

Chemicon AG632 ≥95% | Purified protein for apoptosis & signal transduction; for use with Chemicon AB1721

Connexin, 40 Control Peptide

Chemicon AG634 ≥95% | Purified protein for apoptosis & signal transduction; for use with Chemicon AB1726

Connexin, 43 Control Peptide

Chemicon AG633 ≥95% | Purified protein for apoptosis & signal transduction; for use with Chemicon AB1727

Chemicon AG678 Purified protein for apoptosis & signal transduction; for use with Chemicon MAB3067/AB1721

Corticosteroid Binding Globulin

Synonyms: Transcortin; Steroid Binding Protein

USBio C7902 Human ≥60%; no contaminants detected; single band by SDS-PAGE, IEP, &/or RID; 1 mg/mL supplied in PBS buffer, pH 7.5 | Suitable for antigenic applications in immunological protocols

USBio C7902-10 Human ≥99%; no contaminants detected; single band by SDS-PAGE, IEP, &/or RID; 1 mg/mL supplied in PBS buffer, pH 7.5 | Suitable for antigenic applications in immunological protocols

Biogenesis 2319-3004 Human serum MW 55.7k Tested negative for HBsAg and HIV-1 and HCV antibodies; affinity purified; 0.05 *M* NH_4HCO_3, pH 8.0; lyophilized

Calbiochem 235200 Human serum MW 55.7k Solid lyophilized from 50 *mM* NH_4HCO_3, pH 8.0; ≥99% (SDS-PAGE); soluble in aqueous buffers & water; prepared from serum shown by certified tests to be negative for HBsAg & antibodies to HIV & HCV | Regulates the concentration of free cortisol & progesterone; Ghose-Dastidar, J, *PNAS*, 88: 6408, 1991; Smith, CL & Hammond, GL, *Endocrinology*, 128: 983, 1991

Cortisol Binding Globulin

Biogenesis 2330-6809 Human serum MW 55.7k Tested negative for HBsAg and HIV-1 and HIV-2 antibodies; affinity purified; 0.05 *M* NH_4HCO_3, pH 8.0, salt free; lyophilized

C-Reactive Protein

Cortex CP1000U >98%

Biogenesis 1707-2004 Human plasma MW 114k Tested negative for HIV I and II antibody, HBsAg and HCV antibody; purified; 20 *mM* Tris, 0.28 *M* NaCl, 0.09% NaN3 and 5 *mM* CaCl2, pH 8.0, 0.2 *µM* filtered; liquid

Biogenesis 1707-2029 Human pleural ascites Purified; liquid

Fitzgerald 30-AC05 Human serum High purity

Fitzgerald 30-AC10 Human serum Standard purity

Fitzgerald 30-CC30 Human serum Control/calibrator

Fitzgerald 30-AC07 Recombinant expressed in *E. coli* >99%

Scipac P100-0 Serum/plasma >99%; liquid in TRIS buffer; immunoaffinity absorbed; 1-5 mg/mL | Cardiac marker protein

Scipac P100-7 Serum/plasma >96%; liquid in TRIS buffer; 1-5 mg/mL | Cardiac marker protein

C-Reactive Protein, 60-80%

USBio C7907-24   Sterile filtered (0.2 μm) | Suitable for antigenic applications in immunological protocols

C-Reactive Protein, 95-98%

USBio C7907-26   Human serum ≥80% (RID Behring, SDS-PAGE); clear to light amber, colorless; sterile filtered (0.2 μm); 2-4 mg/mL supplied in 0.1 M Tris-HCl, 0.2 M NaCl, pH 7.5, 2 mM CaCl₂, 0.1% NaN₃ | Suitable for antigenic applications in immunological protocols

Creatine Kinase BB

USBio C7910-10   Recombinant   ≥99%; no contaminants detected; single band by SDS-PAGE, IEP, &/or RID; 1 mg/mL supplied in Tris buffered saline, pH7.2, 10 mM β-MSH, 50% glycerol, 0.1% NaN₃ | Suitable for antigenic applications in immunological protocols

Creatine Kinase BB Isoenzyme

USBio C7910-14   Human brain   ≥99%; no contaminants were detected to CK-MB & CK-mM; 0.69 mg/mL protein, 340 U/mL enzyme activity at 30°C, pH 6.5; 500 U/mg; purified preparation in 5 mM succinate, 1 mM EDTA, 5 mM β-MSH, 0.01 M NaCl, 50% glycerol, pH 7.2 | Suitable for antigenic applications in immunological protocols

Creatine Kinase MB

USBio C7910-11   Recombinant   ≥99%; no contaminants detected; single band by SDS-PAGE, IEP, &/or RID; 1 mg/mL, 950 IU/mg supplied in Tris buffered saline, pH7.2, 10 mM β-MSH, 50 % glycerol, 0.1% NaN₃ | Suitable for antigenic applications in immunological protocols

Creatine Kinase MB Isoenzyme

Scipac P190-8   Heart tissue   1-10%; liquid in TRIS buffer; no contamination from other isoforms | Cardiac marker protein

USBio C7910-18   Human heart   No contaminants detected; single band by SDS-PAGE, IEP, &/or RID; 300 U/mL total CK; 0.14 mg/mL CKMB (mass) supplied in 0.05 M Tris, 10mm β-MSH, 50% glycerol, 0.15 M NaCl, pH 7.5 | Suitable for antigenic applications in immunological protocols

USBio C7910-19   Human heart   ≥99%; no contaminants detected; single band by SDS-PAGE, IEP, &/or RID; supplied in 0.01 M Tris, 5 mM β-MSH, 50% glycerol, pH 7.2, no preservatives added | Suitable for antigenic applications in immunological protocols

Creatine Kinase MM Isoenzyme

USBio C7910-25   Human skeletal muscle   ≥98%; no contaminants detected; single band by SDS-PAGE, IEP, &/or RID; ~2 mg/mL protein, ~430 U/mg protein; purified preparation in 10 mM Tris HCl, 150 mM NaCl, 1 mM Na EDTA, 10 mM β Mercaptoethanol, 0.15 M NaCl, 50% glycerol, pH 7.4 | Suitable for antigenic applications in immunological protocols

USBio C7910-12   Recombinant   ≥99%; no contaminants detected; single band by SDS-PAGE, IEP, &/or RID; supplied in Tris buffered saline, pH7.2, 10 mM β-MSH, 50% glycerol, 1 mM EDTA | Suitable for antigenic applications in immunological protocols

Crosstide

Synonyms: Akt Substrate

USBio C7935   ≥95% after Sephadex G-10 chromatography in 10% HOAc; frozen solution in sterile deionized H₂O | Suitable as a substrate for Akt/PKB when used in conjunction with immunoprecipitated Akt/PKB; may also be phosphorylated by other protein kinases, including Rsk-2

Crotoxin, A Subunit

Calbiochem 238476   *Crotalus durissus terrificus*   MW 9160 >99% (SDS-PAGE); lyophilized solid; soluble in water; no phospholipase A₂ detected (titrimetric assay); not available for sale outside of the United States | Acidic non-toxic, non-enzymatic subunit; Faure, G et al, *Eur J Biochem*, 223: 161, 1994; Aird, SD et al, *Biochim Biophys Acta*, 1040: 217, 1990

Crotoxin, B Subunit

Calbiochem 238477   *Crotalus durissus terrificus*   MW 14,040 >99% (SDS-PAGE); lyophilized solid; soluble in water; activity: 60 units/mg; one unit is the amount of enzyme that releases 1.0 μmol fatty acid/minute at 25°C, pH 8.0 using egg yolk emulsion as the substrate; not available for sale outside of the United States | Basic phospholipase A₂ subunit; Faure, G et al, *Eur J Biochem*, 223: 161, 1994; Aird, SD et al, *Arch Biochem Biophys*, 249: 296, 1986; Mascarenhas, YP et al, *Eur Biophys J*, 21: 199, 1992

Cyclin H

IBT TA-350-1   Human recombinant, expressed in *E. coli*

Cyclophilin A

Alexis 201-023   Human recombinant, expressed in *E. coli* ≥90%; 0.5 mg/mL in 20 mM Tris HCl, pH 7.8 | May be used as a control in Western Blot experiments with Alexis 210-124; used to catalyze the *cis-trans*-isomeration of X-Pro-peptide bonds; protein folding reactions that are limited by the isomeration of X-Pro-peptide bonds are accelerated by cyclophilin A

Cyfra 21-1

Synonyms: Cytokeratin 19

USBio C8910   Human   TPA: ~ 2,000 K U/mL, 3,000 K U/mL (Boehringer Mannheim EIA); liquid tissue culture supernatant in Tris buffer, 8 M urea, 10 mM EDTA, pH 7.0 | Suitable for antigenic applications in immunological protocols

Fitzgerald 30-AC69   Human cell line   Standard purity

Cortex CP1067r   Recombinant   >95%

Fitzgerald 30-AC68   Recombinant expressed in *E. coli*   >96%

Cystatin

Synonyms: Cysteine Protease Inhibitor; Ficin Inhibitor; Papain Inhibitor; Thiol Protease Inhibitor

Biogenesis 2409-8107   Chicken egg white   Liquid

Fluka 30065   Chicken egg white   MW ~12.5k   ≥95% (GE) | Barrett, AJ, *Meth Enzymol*, 80: 771, 1981; Barrett, AJ, *Trends Biochem Sci*, 12: 193, 1987; Bode, W et al, *EMBO J*, 7: 2593, 1988

Fluka 30066   Chicken egg white   Solution; ≥85% (GE); in 0.01 M Tris buffer, pH 8.0, with 50% glycerol; 1 mg protein/mL | Anastasi, A et al, *Biochem J*, 211: 129, 1983; Turk, V & Bode, W, *FEBS Lett*, 285: 213, 1991

ICN 194984   Chicken egg white   MW 12.7k   Activity: 10-15 BAEE U/mg protein; 1 mg inhibits ~1.5 mg of papain | Competitive & reversible inhibitor; Anastasi, A etal, *Biochem J*, 211:129, 1983

Sigma C 0408   Chicken egg white   Solution in 10 mM Tris buffer, pH 8.0, containing 50% glycerol; 1 mg protein inhibits 50% of the activity of 40-80 BAEE units of papain, Sigma P 4762, at pH 6.8, 40°C; 1 BAEE unit of papain hydrolyzes 1.0 μmole Nα-benzoyl-L-arginine ethyl ester/min at pH 6.2, 25°°C; protein by E₂₈₀ at 1% | Barret, AJ, *Meth Enzymol*, 80: 771, 1981; Anastasi, A et al, *Biochem J*, 211: 129, 1983

Sigma C 8917   Chicken egg white   Lyophilized powder; 1 mg protein inhibits 50% of the activity of 40-80 BAEE units of papain, Sigma P 4762, at pH 6.8, 40°C; 1 BAEE unit of papain hydrolyzes 1.0 μmole Nα-benzoyl-L-arginine ethyl ester/min at pH 6.2, 25°°C; protein by E₂₈₀ at 1% | Barret, AJ, *Meth Enzymol*, 80: 771, 1981; Anastasi, A et al, *Biochem J*, 211: 129, 1983

Cystatin A

Synonyms: Steffin A; Thiol Protease Inhibitor

ICN 194985 Human placenta MW 12k >99%; 50 µg/mL in 10 mM Tris-HCl, pH 7.8 | Competitive & reversible thiol-protease inhibitor; Brain, J etal, Hoppe Seyler's *Z Physiol Chem*, 364:1475, 1983

Biogenesis 2409-8257 Human plasma MW ~12k Purified; 10 *mM* Tris/HCl, pH 7.8; liquid

Cystatin B

Synonyms: Steffin B; Thiol Protease Inhibitor

ICN 194986 Human plasma MW 12k >99%; 50 µg/mL in 10 mM NaOAc, pH 5.5 | Competitive & reversible thiol-protease inhibitor; Brain, J etal, Hoppe Seyler's *Z Physiol Chem*, 364:1475, 1983

Cystatin C

Biogenesis 2409-8457 Human plasma MW 13.5k Purified; 10 *mM* sodium acetate buffer, pH 5.5; liquid | Brizin et al, *BBRC*, 118:103, 1984

ICN 194987 Human plasma Lyophilized; 20 inhibitory U/mg protein; 1 U inhibits 1.0 U of papain in 20 min at 25°C | Bazin, J etal, *BBRC*, 118:103, 1984

Cytochrome c

Sigma C 9197 Bison heart MW 12,327 ≥95%; prepared without TCA

Biogenesis 2450-0004 Bovine heart Purified; essentially salt free; lyophilized | Supplied predominantly in the oxidized form; manufactured without TCA

Sigma C 2037 Bovine heart MW 12,327 ≥95%; prepared without TCA

Sigma C 3006 Bovine heart MW 12,327 Practical grade; ≥60% based on containing variable amounts of reduced cytochrome c; prepared without TCA

Sigma C 3131 Bovine heart MW 12,327 ≥95%; prepared without TCA

Sigma C 0761 Chicken heart MW 12,222 ≥95%; prepared without using TCA

Sigma C 4013 Dog heart MW 12,241 ≥95%; prepared without TCA

Biogenesis 2450-0104 Equine heart Lyophilized

ICN 101467 Horse heart MW 12,384 >90%

Sigma C 2506 Horse heart MW 12,384 ≥95%; prepared without TCA

Sigma C 7752 Horse heart MW 12,384 ≥95%; prepared without TCA | Formerly listed as Type VI

Sigma C 8266 Horse heart MW 12,384 Practical grade; ≥95% based on containing variable amounts of reduced cytochrome c; prepared using TCA

R&D Systems 709-CC-010 Human placenta 95%; lyophilized | Species specificity: human cytochrome C; used in *in vitro* apoptosis assays

Sigma C 4011 Pigeon breast muscle MW 12,173 ≥95%; prepared without TCA

Sigma C 9261 Pigeon heart MW 12,173 ≥95%; prepared without TCA

Sigma C 0886 Porcine heart MW 12,327 ≥95%; prepared without using TCA

Sigma C 9136 Rabbit heart MW 12,220 ≥95%; prepared without TCA

Sigma C 7892 Rat heart MW 12,132 ≥95%; prepared without TCA

Sigma C 2436 *Saccharomyces cerevisiae* MW 12,588 ≥85%; prepared without TCA | Care has been taken to maintain the native form of the protein; it has not been artificially oxidized or reduced during purification

Sigma C 2136 Sheep heart MW 12,327 ≥95%; prepared without TCA

Sigma C 2011 Tuna heart MW 12,170 ≥95%; prepared without TCA

Cytochrome c, (^{14}C-Me)-

Sigma C 7664 Horse heart MW ~12.3k 5-50 µCi/mg protein; solution in 10 *mM* sodium phosphate, pH 7.0, in serum bottle | Radiochemicalprepared from Sigma C 7752

Cytochrome c, Acid Modified

Sigma C 3256 Horse heart MW 12,384 Acid modified; purity ~90% based on millimolar extinction coefficient at 550 nm of 27.8 and essentially "Fraction II, pH 7" of Margoliash prepared using TCA; contains inactive cytochrome polymers which can be converted back to native cytochrome | High rate of ascorbic acid oxidation & low enzymatic activity in a cytochrome oxidase system; Margoliash, E, *Biochem J*, 56: 535, 1954; Schejter, A et al; *Biochim Biophys Acta*, 73: 641, 1963

Cytochrome c, Biotin Conjugated

Sigma C 2022 Horse heart Lyophilized powder containing ~80% protein; balance sodium citrate buffer salts; contains 4-6 moles biotin/mole cytochrome c

Cytochrome c, C551

Sigma C 9533 *Pseudomonas aeruginosa* Solution in 0.05 M ammonium azide, pH 4.5, containing 0.02% sodium azide | Rosen, R & Pecht, I, *Biochemistry*, 15: 775, 1976

Cytochrome c, DITC Glass Coupled

Sigma C 1155 *Candida krusei* Coupled to DITC glass via the α- & ε-amino groups; 1-2 nmoles/mg glass as determined by AA analysis | Use-tested as a standard for solid phase protein sequencing analysis

Cytochrome c, Partially Acetylated

Sigma C 4186 Horse heart Partially acetylated; lyophilized powder containing ~90% protein; balance potassium phosphate buffer salts; ~60% of the lysine residues are acetylated (ninhydrin) | Suitable for detection of superoxide radicals in biological systems containing cytochrome c reductases or oxidases; Azzi, A et al; *BBRC*, 65: 597, 1975

Cytochrome F

Sigma C 2285 Spinach Lyophilized powder; essentially salt-free; prepared by a modification of the method of Ho & Krogman; A_{554}/A_{280} ~0.5 | Ho & Krogman, *J Biol Chem*, 255: 3855, 1980

Sigma C 0168 Turnip Lyophilized powder; essentially salt-free; prepared by the method of Gray; A_{554}/A_{280} ~0.8 | Gray, JC, *Eur J Biochem*, 82: 133, 1978

Cytochrome H

Sigma C 7523 *Helix pomatia* digestive glands

Cytochrome P450

Synonyms: Pentoxyresorufin O-De-Ethylase; P450 2B4

Sigma C 7552 Rabbit liver microsomes Induced with phenobarbital; lyophilized powder containing ~10% protein (Bradford); balance potassium phosphate buffer, pH 7.5, EDTA & stabilizer; activity: 150-450 units/mg protein; unit definition: 1 unit releases 1.0 pmole of resorufin from pentoxyresorufin/min at pH 7.6 at 37°C; contains 0.1-0.4 unit cytochrome-P450 reductase/mg protein | Ubiquitous heme-containing enzymes found in prokaryotes & eukaryotes; part of a super-family of enzymes found in mammalian liver whose role is removal of xenobiotic compounds from the body; Schenkman, JB, *Handbook of Experimental Pharmacology*, Vol 105: pp 3-14, Springer-Verlag, Berlin Heidelberg, 1993; Haugon, DA & Coon, MJ, *J Biol Chem*, 251: 7929, 1976

Cytokeratin

Biogenesis 5550-0404 Human epidermis Liquid

Cytokeratin 18

ICN 771032   MW 45k   Highly purified protein for use as a high MW marker, a standard, or an Ag for various purposes

Biodesign A08004B   Bovine liver   >98%

Biogenesis 5553-1804   Bovine liver   Lyophilized

Cytokeratin 19

Biodesign A08014H   *E. coli*   >95%

Cytokeratin 8

ICN 771022   MW 52.5k   Highly purified protein for use as a high MW marker, a standard, or an Ag for various purposes

Biodesign A08003B   Bovine liver   >95%

Biogenesis 5553-0804   Bovine liver   MW 52k   pI: 6.4; purified; 30 *mM* Tris/HCl, pH 8.0, 9 *M* urea, 2 *mM* DTT; lyophilized | Franke et al, *Exp Cell Res*, 131:299ff, 1981; Quinian et al, *J Mol Biol*, 178:365ff, 1984

Cytokine Induced Neutrophil Chemoattractant I Chemokine

Sigma C 9709   Rat recombinant, expressed in *E. coli*   MW ~7.8k   ≥97% (SDS-PAGE); 0.2 μm filtered & lyophilized in 30% acetonitrile/0.1% trifluoroacetic acid containing 0.5 mg BSA; biological activity measured by its ability to induce myeloperoxidase release from cytochalasin B-treated neutrophils; endotoxin tested | Member of the C-X-C or α chemokine class which act primarily on neutrophils as a chemoattractant & activator & plays an important role in the infiltration of neutrophils into inflammatory sites; originally purified from media conditioned by IL-1β stimulated rat kidney epithelioid cells (NRK-52E); CINCs may be the rat equivalent of human GROs; polypeptide of 72 AA

Cytokine Induced Neutrophil Chemoattractant IIb

BioSource International PRC1564   Rat recombinant

BioSource International PRC1565   Rat recombinant

Cytolysin, *Stoichactis*

Calbiochem 569415   *Stoichactis helianthus* (sea anemone), synthetic   MW 16,977   ≥90% (HPLC); lyophilized solid; soluble in water; LD$_{50}$≤2000 mg/kg | Toxin belonging to the group of channel-forming polypeptides; one of the most potent hemolysins known; this toxin contains a binding site specific for sphingomyelin; Blumenthal, KM & Kem, WR, *J Biol Chem*, 258: 5574, 1983

Cytomegalovirus Glycoprotein B

IBT CMA-1400-3, CMA-1400-4   CHO cells   MW 140k   >90%; 0.50 mg/mL solution in 90 *mM* sodium citrate, 200 *mM* NaCl, pH 6.0 | Covers Met[1] to Val[907] of human CMV glycoprotein B; glycosylated

Cytomegalovirus Glycoprotein P50

IBT CMA-1410-3, CMA-1410-4   *E. coli*, as a fusion protein with human superoxide dismutase   MW 62k   >85%; 0.50 mg/mL solution in PBS, 0.1% SDS, 1 *mM* EDTA, pH 7.4 | Covers Met[1] to Gly[433] of glycoprotein p50 of the CMV AD169 strain

Cytomegalovirus Glycoprotein P65

IBT CMA-1420-3, CMA-1420-4   Yeast cells, as a fusion with human superoxide dismutase   MW 79k   >90%; 0.50 mg/mL solution in PBS, 0.1% SDS, 1 *mM* EDTA, pH 7.5 | Covers Met 1 to Gly 561 of the glycoprotein p65 of the CMV AD169 strain

DARPP-32

Chemicon AG657   Recombinant   ≥95% | Purified protein for apoptosis & signal transduction

D-Dimer

Biodesign A86870H   Human plasma   >95% | Platelets & hemostasis reagents

Scipac P202-4   Plasma   Extract; frozen in sodium phosphate buffer | Hemostasis protein

Defensin Iα (NP-1)

BioSource International PHC1615   Human recombinant

Defensin α

Synonyms: NP-1; BD-2

Chemicon GF099   Human   ≥95%

Chemicon GF100   Human   ≥95%

Dengue 2 Antigen, Strain 16681

Biodesign R02220   Purified | Infectious disease antigen

Desmin

ICN 771042   MW 53k   Highly purified protein for use as a high MW marker, a standard, or an Ag for various purposes

Biodesign A08005C   Chicken stomach   >98%

Desmosine

ICN 191378   Bovine neck ligament   Crystalline; 99% (AA analysis); hygroscopic

DHEA Sulfate

USBio D3228   Iodination grade; ≥99%; no contaminants detected; single band by SDS-PAGE, IEP, &/or RID; white crystalline powder | Suitable for antigenic applications in immunological protocols

USBio D3228-05   ≥98%; white crystalline powder; MP: 150-190°C

Diphtheria Toxin

Sigma D 2918   *Corynebacterium diphtheriae*   Lyophilized powder containing sodium phosphate buffer & lactose | Inhibits protein synthesis by catalyzing ADP-ribosylation of eukaryotic aminoacyltransferase II; not assayed by Sigma; Grollman, AP & Huang, M-T, in Protein Synthesis, A Series of Advances, 1976

Diphtheria Toxin, (Glu[52])-

Sigma D 7544   *Corynebacterium diphtheriae*   Lyophilized powder containing sodium phosphate buffer & lactose | Not assayed by Sigma; CRM 197; sold by weight of (Glu[52])-diphtheria toxin

Diphtheria Toxin, Un-nicked

Calbiochem 322326   *Corynebacterium diphtheriae*   MW 63k   Single band purity (disc gel electrophoresis); solid lyophilized from sterile 10 *mM* Tris, 1 *mM* EDTA, pH 7.5; soluble in aqueous buffers; toxic: LD$_{50}$ ≤200 mg/kg but >50 mg/kg; may be carcinogenic/teratogenic | Catalyzes ADP-ribosylation of eukaryotic aminoacyltransferase II (EF2) using NAD as substrate, thereby inhibiting protein synthesis; also induces internucleosomal breakdown; causes DNA fragmentation & cytolysis in U937 cells; activation requires nicking with a protease followed by reduction with DTT; Kochi, SK & Collier, RJ, *Exp Cell Res*, 208: 296, 1993; Chang, MP et al, *J Biol Chem*, 264: 15261, 1989; Pappenheimer, AM Jr, *Ann Rev Biochem*, 46: 69, 1977

DNA Fragmentation Factor 45/ICAD

USBio D3224-41 Human recombinant, expressed in *E. coli* MW ~48k ≥85% (SDS-PAGE, Coomassie blue staining); purified using Ni-NTA agarose; supplied as 50 µg of His-tagged DFF45/ICAD in 500 µL PBS, 50% glycerol | Recombinant human full length His-tagged fusion protein; a heterodimer of 40kD & 45kD subunits; caspase 3 cleaves the 45kD subunit (DFF45) to generate an active factor that causes DNA fragmentation without further requirement of caspase 3 or other cytosolic factors; an N-terminus His-tagged fusion protein expressed in E. coli corresponding to the human sequence

DNA/Protein A, Agarose

USBio D3956 Salmon sperm recombinant Suitable for use in immunoprecipitation; recombinant Protein A covalently bound to agarose by alkylamine linkage

DnaJ Protein

Calbiochem 323100 *E. coli* MW 41k Liquid in 100 *mM* NaCl, 50 *mM* Tris, 1 *mM* DTT, 50% glycerol, pH 7.5; DNases, proteases, RNases: none detected | Molecular chaperone that is essential for the activation of substrate binding properties of the DnaK chaperone; plays an integral role in protein folding & in mediating protein-protein interactions in both normal & stressed cells; enhances the ATPase activity of DnaK *in vivo*; useful for *in vitro* protein folding studies; Wall, D et al, *J Biol Chem*, 270: 2139, 1995; Gething, MJ & Sambrook, J, *Nature*, 355: 33, 1992; Landry, SF et al, *Nature*, 355: 455, 1992; Liberek, K et al, *PNAS*, 88: 2874, 1991

DnaK Protein

Calbiochem 323105 *E. coli* MW 72k >95% (SDS-PAGE); liquid in 100 *mM* NaCl, 50 *mM* Tris, 1 *mM* DTT, 50% glycerol, pH 7.5; DNases, proteases, RNases: none detected | Molecular chaperone & a member of the HSP70 family; plays an integral role in protein folding & in mediating protein-protein interactions in both normal & stressed cells; functions as a monomer with a single peptide-binding site; possesses ATPase activity which facilitates the release of bound proteins & the disaggregation of protein complexes *in vivo*; useful for *in vitro* protein folding studies; Wall, D et al, *J Biol Chem*, 270: 2139, 1995; Gething, MJ & Sambrook, J, *Nature*, 355: 33, 1992; Landry, SF et al, *Nature*, 355: 455, 1992; Langer, T et al, *Nature*, 356: 683, 1992

Drap 1/P28

IBT TA-300-1 Human recombinant, expressed in *E. coli*

DsbA Protein

Calbiochem 324500 *E. coli* MW 21k >95% (SDS-PAGE); liquid in 100 *mM* NaCl, 50 *mM* Tris, 50% glycerol, pH 7.5; DNases, proteases, RNases: none detected | A monomeric periplasmic *E. coli* protein that appears to be necessary for correct formation of disulfide bonds in exported proteins *in vivo*; catalyzes the exchange of disulfide bonds & the oxidation of free sulfhydryl groups *in vitro*; useful for facilitating the refolding of inactive proteins containing reduced or misformed disulfide bonds; Akiyama, Y et al, *J Biol Chem*, 267: 22440, 1992; Schirra, HJ et al, *Biochemistry*, 37: 6263, 1998; Bardwell, JCA et al, *Cell*, 67: 581, 1992

Ecarin

Synonyms: Prothrombin Activator

Sigma E 0504 *Echis carinatus* venom ~50 units/vial; contains thimerosal & lactose; unit definition: 1 unit activates prothrombin to produce 1 unit of amidolytic activity at pH 8.4 at 37°C; 1 amidolytic unit hydrolyzes 1.0 µmole of *N-p-*tosyl-Gly-Pro-Arg-*p*-nitroanilide/min at pH 8.4, 37°C

Echistatin Disintegrin

Synonyms: Integrin Inhibitor

Sigma E 1518 *Echis carinatus* >95% (SDS-PAGE); lyophilized; sterilized by γ-irradiation | Disintegrins represent a novel family of integrin β1 & β3 inhibitor proteins isolated from viper venoms; low molecular weight, cysteine-rich peptides containing the Arg-Gly-Asp (RGD) sequence; the most potent known inhibitors of integrin function; they interfere with cell adhesion to the extracellular matrix including adhesion of melanoma cells & fibroblasts to fibronectin & are potent inhibitors of platelet aggregation

E-C-L Cell Attachment Matrix

USBio E0275 Engelbreth-Holm-Swarm (EHS) mouse tumor Protein determined by Bradford dye binding assay using gamma globulin as the standard; 1 mg/mL frozen liquid in 0.05 *M* Tris-HCl, pH 7.4, 0.15 *M* NaCl

Elastin

ICN 101636 Bovine neck ligament Powder

Sigma E 1625 Bovine neck ligament Powder

Calbiochem 324695 Human lung Solid; prepared from tissue of individuals shown by certified tests to be negative for HBsAg & antibodies to HIV & HCV | Insoluble protein prepared form normal human lung by alkaline hydrolysis according to Lansing procedure; useful for measurement of elastolytic enzymes in tissues; *Merck Index*, 12: 3577; Reilly, CF & Travis, J, *Biochim Biophys Acta*, 621: 147, 1980; Lansing, AI et al, *Anat Rec*, 114: 555, 1952

ICN 191169 Human lung powder Used for measurement of elastolytic enzymes in tissue; Reilly, CF & J Travis, *BBA*, 621:147, 1980

Elastin, Congo Red

Synonyms: Elastase Substrate

ICN 101637 Bovine neck ligament Impregnated with Congo Red dye, ground powder; ε495, 1% susp, blank value 0.02, digested value >10

Sigma E 0502 Bovine neck ligament Elastin impregnated with Congo red dye; prepared from Sigma 1625 | Substrate for the estimation of elastase

Elastin, Fluorescein

Synonyms: Elastase Substrate

ICN 100620 Elastin covalently labeled with Fluorescein Isothiocyanate

Elastin, Orcein

Synonyms: Elastase Substrate

ICN 100618 Impregnated with orcein

Sigma E 1500 Bovine neck ligament Elastin impregnated with orcein; prepared from Sigma 1625 | Substrate for the estimation of elastase

Endostatin, rh-

Calbiochem 324742 Human recombinant, expressed in *Spodoptera frugiperda* MW 21,231 Lyophilized solid; ≥95% (SDS-PAGE); soluble in water | C-terminal proteolytic fragment of collagen XVIII that specifically inhibits endothelial cell proliferation & potently inhibits antiogenesis & tumor growth; Oh, SP et al, *Genomics*, 19: 494, 1994; O'Reilly, MS et al, *Cell*, 88: 277, 1997

Endostatin, rm-

Calbiochem 324743 Mouse recombinant, expressed in *Spodoptera frugiperda* MW 21,397 Lyophilized solid; ≥95% (SDS-PAGE); soluble in water | C-terminal proteolytic fragment of collagen XVIII that specifically inhibits endothelial cell proliferation & potently inhibits antiogenesis & tumor growth; Hohenester, E et al, *EMBO J*, 17: 1656, 1998; O'Reilly, MS et al, *Cell*, 88: 277, 1997

Endothelial Cell Growth Supplement

USBio E3010

USBio E3010-05 Mitogenic for many cell types under reduced- or serum-free conditions, such as mammalian, avian, & human endothelial cells, smooth muscle cells, keratinocytes, melanocytes & hybridomas; often fully substitutes for feeder layers in culture of fastidious cells

Endothelial Mitogen

Biogenesis 4110-5004 Bovine hypothalamus Microorganisms & mycoplasma: not detected, endotoxins: 0.5 Eu/mg protein; 0.1 M NaCl, 5 mM, NaH2PO4, pH 7.4; lyophilized | Useful in vascular endothelial cells; hybridoma cell cloning, Balb/C-3T3 cells, keratinocytes

Enterotoxin B

Synonyms: Staphylococcal Enterotoxin B

Calbiochem 324798 *Staphylococcus aureus* Lyophilized solid from 5 mM potassium phosphate buffer, pH 6.8; single major band purity (SDS-PAGE); highly toxic: LD$_{50}$≤50 mg/kg | A heat-stable bacterial superantigen that activates the immune system to produce a burst of anti-inflammatory cytokines; Hasko, G et al, *Eur J Immunol*, 28: 1417, 1998

Fluka 45182 *Staphylococcus aureus* MW 29k ≥95% (GE); 25% protein content; contains sodium phosphate | Iandolo, JJ & Tweten, RK, *Meth Enzymol*, 165: 43, 1988

Eotaxin CC Chemokine

Biodesign A52250H *E. coli* MW 8.4k Purified | Species specificity: mouse

Biodesign A52321H *E. coli* MW 8.3k Purified

Chemicon GF042 Human ≥95%

Biogenesis 4182-4050 Human r-DNA Purified; lyophilized

BioSource International PHC1434 Human recombinant

PeproTech 300-21 Human recombinant, expressed in *E. coli* MW 8.3k >98% (SDS-PAGE) & HPLC; <0.1 ng endotoxin per mg (1EU/mg); lyophilized | Chemokine characterized by its high chemotactic selectivity for eosinophils; the location of the four Cys residues in Eotaxin places it in the β-chemokine family (CC) of cytokines (along with RANTES, MCP-3 & MIP-1α); 74 AA; SA determined by its ability to chemoattract human peripheral blood eosinophils

Sigma E 7127 Human recombinant, expressed in *E. coli* MW ~8.4k ≥97% (SDS-PAGE); 0.2 µm filtered & lyophilized in 30% acetonitrile, 0.1% trifluoroacetic acid containing 1 mg BSA; biological activity measured by its human eosinophil chemotactic activity; endotoxin tested | Member of the CC or β chemokine class which act primarily as chemoattractants & activate monocytes, dendritic cells, T lymphocytes, natural killer cells, B lymphocytes, basophils & eosinophils; originally isolated from bronchoalveolar lavage fluid of guinea pigs sensitized by aerosol challenge with ovalbumin; polypeptide of 74 AA; precursor form is 97 AA; to generate the mature Eot (74 AA), the precursor cleaves a 23 amino-terminal AA signal peptide; Eot is chemotactic for eosinophils, but not mononuclear cells or neutrophils; human Eot shows ~60% AA homology to mouse & guinea pig eotaxin; shows high identity with MCP-1, 2 & 3

Biogenesis 4182-4250 Mouse r-DNA Purified; lyophilized

BioSource International PMC1434 Mouse recombinant

Sigma E 9008 Mouse recombinant, expressed in *E. coli* ≥97% (SDS-PAGE); 0.2 µm filtered & lyophilized in 30% acetonitrile, 0.1% trifluoroacetic acid containing 1 mg BSA; biological activity measured by its human eosinophil chemotactic activity; endotoxin tested; see Sigma E 7127

Chemicon GF043 Murine ≥95%

Eotaxin CC Chemokine, Murine

PeproTech 250-01 Murine recombinant, expressed in *E. coli* MW 8.4k >98% (SDS-PAGE) & HPLC; <0.1 ng endotoxin per mg (1EU/mg); lyophilized with no additives | Shares many biological & physical characteristics of human Eotaxin; chemokine exhibiting a high chemotactic selectivity for eosinophils; 74 AA; SA determined by its ability to chemoattract purified eosinophils

Eotaxin II CC Chemokine

Synonyms: MPIF-2

Biodesign A52333H *E. coli* MW 8.8k Purified

Chemicon GF062 Human ≥95%

Biogenesis 4182-4650 Human r-DNA Lyophilized

PeproTech 300-33 Human recombinant, expressed in *E. coli* MW 8.8k >98% (SDS-PAGE) & HPLC; <0.1 ng endotoxin per mg (1EU/mg); lyophilized | Recently discovered CC chemokine; characterized by its high chemotactic selectivity for eosinophils; 78 AA; SA determined by its ability to chemoattract human peripheral blood eosinophils

PeproTech 250-22 Murine recombinant, expressed in *E. coli* MW 10.3k >98% (SDS-PAGE) & HPLC; <0.1 ng endotoxin per mg (1EU/mg); lyophilized | Recently discovered CC chemokine characterized by its high chemotactic selectivity for eosinophils; 93 AA; SA determined by its ability to chemoattract murine lymphocytes cells

Eotaxin III CC Chemokine

Synonyms: TSC

Biogenesis 0100-0126 Human r-DNA Lyophilized

PeproTech 300-48 Human recombinant, expressed in *E. coli* MW 8.4k >98% (SDS-PAGE) & HPLC; <0.1 ng endotoxin per mg (1EU/mg); lyophilized | Recently discovered CC chemokine; binds to the CCR3 receptor & is highly chemotactic for eosinophils;71 AA; SA determined by its ability to chemoattract human CCR3/HEK 293 cells

Epidermal Growth Factor

Synonyms: Urogastrone, β-

Harlan BT-4016	Lyophilized; 0.1 mg; with carrier protein
Harlan BT-4017	Lyophilized; 0.1 mg; without carrier protein
Harlan BT-5014	Receptor Grade; lyophilized; 1 mg; receptor grade
Harlan BT-5016	Tissue culture grade; lyophilized; 1 mg; tissue culture grade

Biodesign A52115H *E. coli* MW 6k Purified

Chemicon GF001 Human ≥95%

Biogenesis 4220-1004 Human r-DNA ^{125}I-conjugated

Biogenesis 4220-0704 Human r-DNA *E. coli* Purified; contains ~ 100 µg phosphate salts, aseptic; lyophilized | Kelly & Hunter, *Clin Sci*, 79:425, 1990

BioSource International PHG0062 Human recombinant

Oncogene PF011 Human recombinant MW 6348 (AA analysis) >92% (HPLC); lyophilized with 100 µg BSA; reconstitute in 10mM acetic acid; biological activity: fully active in EGF receptor binding & mitogenesis assays | Species reactivity: human; for proliferation studies & binding studies; exerts its biological effects in the concentration range of 10-100 pM

Fitzgerald 30-AE40 Human recombinant, expressed in *E. coli*

Harlan BT-3000 Human recombinant, expressed in *E. coli* Lyophilized; 0.1 mg

Harlan BT-3001 Human recombinant, expressed in *E. coli* Lyophilized; 1 mg

PeproTech 100-15 Human recombinant, expressed in *E. coli* MW 6.2k >97%; 53 AA; lyophilized with no additives; ED$_{50}$: 0.2-1.0 ng/mL as determined by the stimulation of thymidine uptake by Balb/c 3T3 cells

Sigma E 9644 Human recombinant, expressed in *E. coli* ≥97% (SDS-PAGE); 0.2 µm filtered & lyophilized from PBS, pH 7.4; proliferative activity is measured in culture using the BALB/3T3 cell line; endotoxin tested; see Sigma E 1264 | Rubin, JS et al, *Proc Natl Acad Sci USA*, 88: 415, 1991

Sigma E 1264   Human recombinant, expressed in *S. cerevisiae*   MW 6k   ≥98% (SDS-PAGE); 0.2 μm filtered & lyophilized from an acetic acid solution; endotoxin tested | A polypeptide originally discovered & purified from mouse submaxillary glands by Cohen & Levi-Montalcini; isolated from human urine; initially named β-urogastrone; structurally homologous to Transforming Growth Factor α; mitogenic for a variety of epidermal & epithelial cells including fibroblasts, glial cells, mammary epithelial cells, vascular & corneal endothelial cells, bovine granulosa, rabbit chondrocytes, HeLa & SV40-3T3 cells; shown to accelerate wound healing; EGF receptor is a 170 kD glycoprotein having EGF-activated protein tyrosine kinase activity; platelet derived Growth Factor (PDGF) transmodulates the EGF receptor by reducing both its EGF affinity & its kinase activity probably via the activation of a separate cellular kinase; Levi-Montalcini, R & Cohen, S, *Ann NY Acad Sci*, 85: 324, 1960; Cohen, S, *J Biol Chem*, 237: 1555, 1962; Cohen, S & Carpenter, G, *Proc Natl Acad Sci USA*, 72: 1317, 1975; Carpenter, G & Cohen, S, *Ann Rev Biochem*, 48: 193, 1979; Brown, G et al, *N Engl J Med*, 321: 76, 1989; Schlessinger, J, *Biochemistry*, 27: 3119, 1988; Davis, R & Czech, M, *J Biol Chem*, 262: 6832, 1988; Shoyab, M et al, *Proc Natl Acad Sci USA*, 85: 6528, 1988; Carpenter, G & Zendegui, J, *Anal Biochem*, 153: 279, 1985

Biogenesis 4220-1104   Mouse

BioSource International PMG0062   Mouse recombinant

Biogenesis 4220-1404   Mouse submaxillaries   MW 6.1k   Purified; 0.01 *M* sodium acetate, sterile; lyophilized | Cohen, *JBC*, 237:1555, 1962; Taylor et al, *JBC*, 247:5928, 1972; Taylor et al, *PNAS,* 67:164, 1970

Chemicon EA135   Murine   ≥95%

Chemicon EA140   Murine   Semi-pure

Biodesign A3B808H   *Pichia pastoris*   Pure; entire native sequence

Biogenesis 4220-1454   Rat adult male submandibular glands   MW 5377   Purified; from 0.01 *M* sodium acetate solution; lyophilized

Chemicon EA144   Recombinant   ≥95%

Sigma E 1257   Submaxillary glands of adult male mice   Receptor grade; 0.2 μm filtered & lyophilized from an ammonium acetate solution; isolated by gel filtration & ion-exchange chromatography; purity determined by SDS-PAGE; endotoxin tested; see Sigma E 1264

Sigma E 4127   Submaxillary glands of adult male mice   Tissue culture grade; 0.2 μm filtered & lyophilized from an ammonium acetate solution; purified by gel filtration; endotoxin tested; this product is less extensively purified than product Sigma E 1257 as determined by SDS-PAGE; see Sigma E 1264

Sigma E 6135   Submaxillary glands of adult male mice   Lyophilized; isolated by reverse phase chromatography; purity determined by AA analysis & HPLC; tested by receptor binding radioimmunoassay; see Sigma E 1264

Epidermal Growth Factor Receptor

Promega V5551   A-431 Tumor cell line   MW 170k   1 unit is the amount of EGF receptor required to catalyze the transfer of 1 pmol of phosphate onto angiotensin II/min at 30°C | Glycoprotein; comprises an extracellular domain that binds EGF, a single membrane-spanning domain & a cytoplasmic domain that has intrinsic protein tyrosine kinase activity; ligands that bind the receptor are EGF, transforming growth factor-α, vaccinia virus growth factor & amphiregulin; substrates include the pp60src-derived peptides, angiotensin II, ras GTPase activating protein, c-erb B2, lipocortin I & phospholipase C-γ; Todderud, G & Carpenter, G, *BioFactors*, 2: 11, 1989; Carpenter, G & Cohen, S, *JBC*, 265: 7709, 1990; Weber, W et al, *JBC*, 259: 14631, 1984

Sigma E 2645   Human carcinoma A431 cells   Affinity purified; lyophilized powder with trehalose as cryoprotectant; reconstitution with 100 μL 10% glycerol in distilled water gives 50 mM HEPES, pH 7.6, 150 mM NaCl, 0.05% Triton X-100, 1 mM DTT & 10% trehalose; activity: ≥15,000 units/mg protein (Bradford) | A receptor protein tyrosine kinase that mediates the activity of epidermal growth factorPanayotou, G et al, *Receptor Purification*, 1: 289, 1990Unit definition: 1 unit catalyzes the incorporation of 1 pmol of phosphate from γ-^{32}P-ATP into poly(Glu, Tyr), 4:1 at 30°C/min

Sigma E 3641   Human carcinoma A431 cells   Affinity purified; solution in 50% glycerol, containing 50 mM HEPES, pH 7.6, 150 mM NaCl, 0.1% Triton, 1 mM DTT & 10% trehalose; activity: ≥15,000 units/mg protein (Lowry) | See Sigma E 2645

Epidermal Growth Factor Receptor, Extracellular Domain

IBT GR-010-3, GR-010-5   Human recombinant, expressed in *Spodoptera rugiperda* insect cells (Sf9)   >85%; 0.50 mg prot/mL in 10 *mM* Tris-HCl, pH 7.0 | Binds radioiodinated EGF; inhibits EGF-mediated proliferation of fibroblasts

Epidermal Growth Factor, Human

IBT GF-010-5, GF-010-8   Yeast   >97%; lyophilized powder | Mitogenic activity measured by stimulation of ^{3}H-thymidine incorporation into human foreskin fibroblasts; EGF activity determined by receptor binding assay (RBA) using A-431 cells

Epidermal Growth Factor, Long

ICN 198785   Human recombinant, expressed in *E. coli*   MW 12,297   >95%; lyophilized | Potent analog of EGF; useful as a growth factor supplement for serum-free or low serum cell cultures

Epidermal Growth Factor, Long Human

IBT NU200, NM001, NM005   MW 12,298   Dried from 0.1 *M* acetic acid | Analog of EGF with a 53 AA N-terminal extension; suited for cell culture in serum free & low serum media

Epinephrine, (-)-L-(*N*-Me-^{3}H)-

Synonyms: Adrenoceptor Agonist

ARC ART-809   MW 183.2   70-87 Ci/mmol; 2.59-3.22 TBq/mmol; in 0.2 *N* HOAc: EtOH (9:1) | Radiochemical

Epithelial Neutrophil Activating Peptide 78

Biodesign A52322H   *E. coli*   Purified | ENA-78

PeproTech 300-22   Human recombinant, expressed in *E. coli*   MW 8.0k   >98% (SDS-PAGE) & HPLC; <0.1 ng endotoxin per mg (1EU/mg); lyophilized | Chemoattractant & activator for neutrophils; belongs to the IL-8 subgroup of the CXC family of chemokines; 74 AA; SA determined by its ability to chemoattract CXCR2 transfected HEK cells

Epithelial Neutrophil Activating Peptide 78 Chemokine

Sigma E 9769   Human recombinant, expressed in *E. coli*   MW 8.3k   ≥97% (SDS-PAGE); 0.2 μm filtered & lyophilized from PBS containing 500 μg BSA; activity measured in culture by its ability to induce myeloperoxidase release from neutrophils; endotoxin tested | Member of the C-X-C or α supergene family; originally discovered from the conditioned medium of human pulmonary epithelial cells (A549) stimulated with TNF-α or IL-1β; protein with 78 AA containing 4 cysteines positioned identically to those of IL-8; shares 53% sequence homology with NAP-2 & 52% sequence homology with GROα; shares several properties of neutrophil activation with NAP-2 & IL-8; induces chemotactic activity in neutrophils as well as release of elastase from cytochalasin-B-pretreated neutrophils & the induction of cytosolic calcium release; neutrophils migrate in response to ENA-78 into inflamed joins of patients with rheumatoid arthritis; Schroder, J et al, *J Immunol* 139: 3474, 1987General references: Walz, A et al, *J Exp Med*, 174: 1355, 1991; Walz, A et al, in: *Chemotactic Cytokines*, Plenum Publishing Corp, 1991; Koch, A et al, *Journal of Clin Invest*, 94: 1012, 1994

Epstein Barr Virus Antigen

Biodesign R70100   Lysate | Infectious disease antigen

Epstein Barr Virus Early Antigen

USBio E3440-10   EBV early diffuse   ≥96%; no contaminants detected; single band by SDS-PAGE, IEP, &/or RID; 1 mg/mL supplied in PBS | Suitable for antigenic applications in immunological protocols; useful marker of chronic or acute infection; suitable applications include EIA at concentrations of <1µg/mL for coating plates & Western blot analysis

USBio E3440   EBV infected cells   ≥98%; no contaminants detected; single band by SDS-PAGE, IEP, &/or RID; 1 mg/mL supplied in TNE buffer, pH 8.0 | Suitable for antigenic applications in immunological protocols

Epstein Barr Virus Early Antigen D

Biodesign R93110   Affinity purified | Infectious disease antigen

Epstein Barr Virus Early Antigen R Complex p17

Biodesign R93120   Affinity purified | Infectious disease antigen

Epstein Barr Virus Membrane Protein gp350/250

Biodesign R93125   Affinity purified | Infectious disease antigen

Epstein Barr Virus Nuclear Antigen

USBio E3440-21   E. coli   MW 27k   ≥99%; no contaminants detected; single band by SDS-PAGE, IEP, &/or RID; 1 mg/mL supplied in PBS | Suitable for antigenic applications in immunological protocols; useful marker of chronic infection & useful in IgA & IgM studies; suitable applications include EIA at concentrations of ≤1 µg/mL for coating plates & Western blot analysis

USBio E3440-11   RAJI strain   ≥98%; no contaminants detected; single band by SDS-PAGE, IEP, &/or RID; 1 mg/mL supplied in TNE buffer, pH 8.0 | Suitable for antigenic applications in immunological protocols

Epstein Barr Virus Nuclear Antigen 1

Biodesign R57523   Purified | Infectious disease antigen

Biodesign R65916   Recombinant   MW 78k   Purified | Infectious disease antigen

Epstein Barr Virus Viral Capsid Antigen

USBio E3440-22   Suitable for use in IFA, ELISA & antigenic applications in immunological protocols

USBio E3440-23   EBV-p18   ≥99%; no contaminants detected; single band by SDS-PAGE, IEP, &/or RID; 1 mg/mL supplied in PBS | Suitable for antigenic applications in immunological protocols

Epstein Barr Virus Viral Capsid Antigen 125

Biodesign R70150   Affinity purified | Infectious disease antigen

Epstein Barr Virus Viral Capsid Antigen 160

Biodesign R93105   Affinity purified | Infectious disease antigen

Erabutoxin A

Synonyms: Snake Toxin

Sigma E 6763   *Laticauda semifasciata*   Lyophilized powder containing ~80% protein (Lowry); balance potassium phosphate buffer salts | Postsynaptic neurotoxin; Sato, S & Tamiya, N, *Biochem J*, 122: 453, 1971

Erabutoxin B

Synonyms: Snake Toxin

Sigma E 4888   *Laticauda semifasciata*   FW 6860.7 Lyophilized powder containing ~65% protein (Lowry); balance potassium phosphate buffer salts | Postsynaptic neurotoxin; Sato, S & Tamiya, N, *Biochem J*, 122: 453, 1971

Erythropoietin

Fluka 45678   Human recombinant, expressed in CHO cells Important hormone in erythrocyte production; induces cytosolic protein phosphorylation & dephosphorylation in erythroid cells; Bailey, SC et al, *JBC*, 266: 24121, 1991

USBio E3455-06   Human recombinant, expressed in CHO cells SA ≥100 KIU/mg; 100-120 µg/mL in 20 *mM* sodium citrate, 150 *mM* NaCl, pH 7.4, 0.1% NaN$_3$ | Principal hormone involved in the regulation of erythrocyte differentiation & the maintenance of a physiological level of circulating erythrocyte mass; indicated for treatment of neutropenia or anemia secondary to ZDV- or Ganciclovir-induced bone marrow suppression

Fitzgerald 30-AE25   Human recombinant, from cDNA expressed in CHO   High purity

ICN 151073   Human recombinant, from Epo cDNA expressed in transfected cell lines   >97% (SDS PAGE & HPLC); sterile filtered solution, 50% glycerol, 25 *mM* HEPES, pH 7.2; ultra pure grade | Graber, SE & SB Krantz, *Ann Rev Med*, 29:51, 1978; Cotes, PM etal, *J Haematol*, 50:427, 1982

Sigma E 5627   Human recombinant, produced by cDNA expressed in CHO cells   Activity: ~100,000 units/mg protein; reconstitution with 1 mL distilled water contains 0.1 M NaCl, 0.01 M NaH$_2$PO$_4$, pH 7.0, & 0.1 mg/mL lactose | Glycoprotein that is the principal regulator of red blood cell growth & differentiation; bioassay not run by SigmaMiyake, T et al, *J Biol Chem*, 252: 5558, 1977; Dordal, MS et al, *Endocrinol*, 116: 2293, 1985; Davis, JM et al, *Biochem*, 26: 2633, 1987; Bailey, SC et al, *J Biol Chem*, 266: 24121, 1991; Hanspal, M et al, *J Biol Chem*, 266: 15626, 1991

ICN 152301   Human urine   Lyophilized; ultra pure grade; purified by immunoadsorbent column chromatography using monoclonal Ab; ~80,000 U/mg protein | Sialoglycoprotein which stimulates differentiation & proliferation of cells in the relatively late stages of erythropoiesis

Sigma E 2514   Human urine   Activity: ~100 units/mg solid | See Sigma E 5627

Sigma E 2639   Human urine   Activity: ~500 units/mg solid | See Sigma E 5627

Cortex CP3039r   Recombinant   >95%

Erythropoietin, Natural

USBio E3455-05   Human urine   ≥80%; lyophilized; SA: 30-40 IU/mg | Principal hormone involved in the regulation of erythrocyte differentiation & the maintenance of a physiological level of circulating erythrocyte mass; indicated for treatment of neutropenia or anemia secondary to ZDV- or Ganciclovir-induced bone marrow suppression

Erythropoietin, Soluble Receptor

ICN 195782   Human recombinant, NSO-expressed   Typically 30-60 ng/mL will inhibit half the biological response in the presence of 0.2 U/mL Epo

Estrogen Receptor-α

Calbiochem 330655 Human recombinant MW 66k ≥80% (SDS-PAGE); liquid in 500 mM KCl, 50 mM Tris, 2 mM DTT, 1 mM EDTA, 1 mM sodium vanadate, 10% glycerol, 0.02% NaN$_3$, pH 7.5; specific activity: ≥5000 units/mg protein; one unit is the amount of enzyme that will bind 1.0 pmol of ^{3}H-estradiol in 2 hours at 22°C, pH 7.5 | Hormone-inducible transcription factor that can positively or negatively regulate expression of many genes involved in tissue growth & differentiation; essential for induction of the oxytocin receptor by estrogen; binds to estradiol (K$_d$=300 pM) & to a fluorescein-labeled estrogen response element (ERE; K$_d$=10 nM) suitable for *in vitro* transcription assays; Young, LT et al, *NeuroReport*, 9: 933, 1998; Tzukeman, MT et al, *Mol Endocrinol*, 8: 21, 1994; Beekman, JM et al, *Mol Endocrinol*, 7: 1266, 1993; Oboum, JD et al, *Biochemistry*, 32: 6229, 1993

Alexis 201-015 Recombinant, *Baculovirus*-infected Sf9 cells 50 µg purified active protein in 50 mM Tris, pH 7.5, 10% glycerol, 0.5 M KCl, 1 mM EDTA, 2 mM DTT & 1 mM sodium vanadate containing 0.02% sodium azide | This protein is hormone-binding & estrogen response element binding competent; may be used as a control in Western Blot or for Gel Supershift Assays with an appropriate estrogen response element & an antibody; can be used to screen chemical compounds for agonist or antagonists activity; Tzukerman, MT et al, *Mol Endocrinol*, 8: 21, 1994

Calbiochem 330657 Human recombinant MW 53k ≥80% (SDS-PAGE); liquid in 400 mM KCl, 50 mM bis-Tris-propane, 2 mM DTT, 1 mM EDTA, 10% glycerol, pH 9.0; specific activity: ≥5000 units/mg protein; one unit is the amount of estrogen receptor that will bind 1.0 pmol of ^{3}H-estradiol overnight at 4°C as measured by quantitation of estradiol:receptor complexes in a hydroxylapatite assay | Purified, soluble & functionally active protein with post-translational modifications similar to those found in human cells; a hormone-inducible transcription factor that shares high homology with ER-α, especially in the DNA binding protein domain, suggesting that both receptors interact with related DNA response elements; Kuiper, GG et al, *Endocrinology*, 138: 683, 1997; Paech, K et al, *Science*, 277: 1508, 1997; Mosselman, S et al, *FEBS Lett*, 392: 49, 1996; Oboum, JD et al, *Biochemistry*, 32: 6229, 1993

Alexis 201-033 Human recombinant, from *Baculovirus*-infected Sf9 cells ≥80%; ~40 µg at ~0.36 mg/mL protein; liquid; specific activity: ≥5000 pmole (^{3}H)-estradiol bound/mg protein | Kuiper, GGJM et al, *Endocrinology*, 138: 863, 1997; Mosselman, S et al, *FEBS Lett*, 392: 49, 1996; Paech, K et al, *Science*, 277: 1508, 1997

Excision Repair Cross Complement Group III

IBT TA-030-1 Human recombinant, expressed in *E. coli*

Excision Repair Cross Complement Group VI

IBT TA-060-1 Human recombinant, expressed in *E. coli*

Factor B

Cortex CP2036U >98%

Biogenesis 4400-9254 Human serum Liquid

Factor D

Cortex CP2037U >98%

Biogenesis 4400-9365 Human serum MW 24k Tested negative for HBsAg and antibodies to HIV and HCV; free of related complement & factor proteins; purified; PBS buffer, pH 7.2; liquid

Factor H

Cortex CP2038U >98%

Biogenesis 4400-9554 Human serum MW 150k Tested negative for antibodies to HBsAg, HCV, HIV 1 and 2; purified; PBS buffer, pH 7.2; liquid

Factor I

Cortex CP6028 >95% | Inactivator

Biogenesis 4400-9754 Human serum Liquid

Factor II

Sigma F 5132 Human plasma Lyophilized powder; reconstitution with 1 mL DI water contains the indicated activity in 0.5 M NaCl, 0.05 M Tris-HCl, pH 8.0; contains <0.01 unit of Factor X/unit of Factor II; no detectable thrombin; activity: 5-20 units/mg protein (E$_{280}$ at 1% = 15.5); unit definition: 1 unit is equal to the amount contained in 1 mL normal human plasma | Source material negative for HIV & HBsAgProthrombin

Factor IX

Synonyms: Christmas Factor; Plasma Thromboplastin Component

Cortex CP2054U Bovine >98%

ICN 194089 Bovine Plasma A zymogen precursor to serine protease IXa; Fujikawa, K etal, *Biochem*, 12:4938, 1973

Cortex CP2053U Human >98%

ICN 194193 Human Plasma Highly purified

Factor P

Synonyms: Properdin

Cortex CP2044U >98%

Biogenesis 4400-9954 Human serum Liquid

Factor V

Synonyms: Proaccelerin

Cortex CP3105U Bovine >98%

ICN 194928 Bovine Plasma 50% glycerol, CaCl$_2$ | A glycoprotein procofactor activated by thrombin to form the active cofactor, Factor Va; Nesheim, M etal, *Methods Enzymol*, 80:249, 1981

Cortex CP3104 Human >98%

ICN 194190 Human Plasma Purified | Activates the conversion of prothrombin to thrombin

ICN 194927 Human Plasma 50% glycerol, CaCl$_2$ | A glycoprotein procofactor activated by thrombin to form the active cofactor, Factor Va; Nesheim, M etal, *JBC*, 254:508, 1979

Factor VII

Cortex CP2052U Human >98%

Sigma F 6509 Human plasma Frozen solution in 0.02 M sodium citrate, pH 6.0 with 1 mM benzamidine HCl; contains <0.01 unit of Factors II & X/unit of Factor VII; ratio of clotting activity to amidolytic activity (VII activity ratio) is 0.9-1.5; activity: 1000-2000 units/mg protein (E$_{280}$ at 1% = 13.9); unit definition: 1 unit is equal to the amount contained in 1 mL normal human plasma | Source material negative for HIV & HBsAgProconvertinSeligsohn, U et al, *Blood*, 52: 978, 1978; Bajaj, SP et al, *J Biol Chem*, 256: 253, 1981

ICN 194191 Human recombinant, expressed in yeast >95% (reversed phase HPLC); 200 µg powder/vial, lyophilized from 100 µL buffer; 3 clot U/µg using one-stage clotting assay; <3% FVIIA (reduced SDS-PAGE) | A vitamin K-dependent glycoprotein synthesized *in vivo* by the liver; participates in the extrinsic coagulation pathway

Factor X

Synonyms: Stuart Prower Factor

Cortex CP2100U Bovine >98%

ICN 194090 Bovine Plasma A precursor to serine protease Xa; Discipio, RG etal, *Biochem*, 16:698, 1977

Sigma F 4003 Bovine plasma In 0.14 M NaCl, 0.04 M citrate buffer, pH 5.8; contains <0.01 unit of Factors II, V, VII & IX/unit of Factor X; Thrombin- & Activated Factor X-free; activity: ~80 U/mg protein (E$_{280}$ at 1% = 12.4); unit definition: 1 unit consists of that amount contained in 1 mL normal human plasma

ICN 153578 Bovine Serum >95% (reversed phase HPLC); <3% Factor Xa | A circulating precursor that is activated to a serine protease (Factor Xa) during coagulation

Cortex CP2048U Human >98%

ICN 194195 Human Plasma Purified

Sigma F 4634 Human plasma Lyophilized powder; reconstitution with 1 mL water, vial contains the indicated activity in 0.5 M NaCl, 0.05 M Tris-HCl, pH 8.0; activated Factor X-free; <0.01 unit of Factor II & VII/unit of Factor X; activity: ~75 units/mg protein (E_{280} at 1% = 11.6); unit definition: 1 unit consists of that amount contained in 1 mL normal human plasma | Source material negative for HIV & HBsAg

Factor X, Gla-Domainless

Cortex CP2055U Human >98%

Factor XI

Cortex CP2101U Human >98%

Factor XII

Cortex CP2102U Human >98%

Factor XIII

Cortex CP2103U Human >98%

FADD, Agarose

USBio F0019-53 Human recombinant, expressed in *E. coli* 200 µg of FADD-agarose in 66 µL of a 50% slurry of PBS/50% glycerol | Full-length human FADD his-tagged fusion protein expressed in E. coli & bound with nickel-chelating resin to agarose; originally isolated as a protein that bound to the cytoplasmic domain of Fas in the yeast two-hybrid system; sequence analysis revealed a region homologous to the death domain of Fas & TNFR-1; subsequent studies show that FADD associates with Fas through interaction of the death domains; when overexpressed in several cell lines, FADD induces apoptosis (can be blocked by CrmA, an inhibitor of Caspase-1); may play a role in Fas-mediated apoptosis

Fas

Kamiya Recombinant intracellular fragment >95% (SDS-PAGE)

Fas Ligand

Synonyms: TNFRSF6

USBio F0019-65

Kamiya Human MW 32k >95% (SDS-PAGE); soluble, with cross linker

Oncogene PF033 Human recombinant MW 35k (dimer) >95% (SDS-PAGE); lyophilized | Bacterial recombinant protein corresponding to the extra-cellular domain of human Fas ligand; species reactivity: mouse, rat, human; used in cytotoxicity assays

R&D Systems 526-SA-050 Recombinant NSO-expressed 95%; lyophilized; ED_{50}: 0.4-1.2 µg/mL | Species specificity: mouse Fas ligand

Fas Ligand Control Peptide

Chemicon AG626 ≥95% | Purified protein for apoptosis & signal transduction

Fas Ligand Inhibitor

Synonyms: CD95 Ligand Inhibitor

Kamiya MW 60k

Fas Ligand Protein

Kamiya MW 35k Soluble

Fas Ligand, Soluble

Synonyms: CD95 Ligand, Soluble

Alexis 522-001 Human recombinant MW 35k (reducing conditions) Lyophilized containing PBS; affinity purified; >95% (SDS-PAGE); 5 µg protein | The extracellular domain of human Fas ligand (AA 103-281) is fused at the *N*-terminus to a linker peptide (26 AA) & a tag; glycosylation of recombinant, human sFas ligand is homologous to natural human Fas ligand; recombinant human sFas ligand is produced in the human cell line HEK 293; rhsFas ligand recognizes human, mouse & rat Fas receptor; Kills Fas-sensitive cells at concentrations <50 ng/mL; Russell, JH et al, *PNAS*, 90: 4409, 1993; Krammer, PH et al, *Curr Opin Immunol*, 6: 279, 1994; Kagi, D et al, *Science*, 265: 528, 1994; Iowin, B et al, *Science*, 267: 1449, 1995; Bellgrau, D et al, *Nature*, 377: 630, 1995; Griffith, TS et al, *Science*, 270: 1189, 1995; Tanaka, M et al, *EMBO J*, 14: 1129, 1995; Mariani, SM et al, *Eur J Immunol*, 25: 2303, 1995; Tanaka, M et al, *Nature Med*, 2: 317, 1996; Hahne, M et al, *Science*, 274: 1363, 1996; Hahne, M et al, *Science*, 274: 1363, 1996; French, LE et al, *J Cell Biol*, 133: 335, 1996; Bodmer, JL et al, *Immunity*, 6: 79, 1997; Irmler, M et al, *Nature*, 388: 190, 1997; Thome, M et al, *Nature*, 386: 517, 1997; Pitti, RM et al, *Nature*, 396: 699, 1998

Fas Protein

Synonyms: TNFRSF6; Apolipoprotein I/CD95

ICN 198749 Human recombinant, expressed in *E. coli* >95% | Encompasses the entire Fas sequence

R&D Systems 326-FS-050 Human recombinant, expressed in NSO >97%; lyophilized with a carrier protein; ED_{50}: 10-25 ng/mL | Member of the TNF receptor superfamily; ligation of Fas by FasL or anti-Fas antibody can induce apoptotic cell death in cells expressing Fas; the extracellular domain of human or mouse Fas is fused to the Fc region of human IgG1

R&D Systems 435-FA-050 Murine recombinant, expressed in *Sf*21 >97%; lyophilized with a carrier protein | See R&D Systems 326-FS-050

Fas Protein, Intracellular Death Domain

Calbiochem 341288 Human recombinant, expressed in *E. coli* MW 45k Liquid in 200 *mM* reduced glutathione, 200 *mM* NaCl, 50 *mM* Tris, 10% glycerol, pH 8.0; single band purity (SDS-PAGE) | A transmembrane protein of the TNF/NGF receptor family; induces apoptosis in Fas-bearing cells; involved in down-regulation of immune responses & T cell mediated cytotoxicity; produced as a fusion protein of human Fas linked to glutathione-*S*-transferase; Burke, G, *Cell*, 81: 9, 1995; Nagata, S & Goldstein, P, *Science*, 267: 1449, 1995

ICN 195861 Human recombinant, expressed in *E. coli* MW 45k Transmembrane protein of the TNF/NGF receptor family that induces apoptosis in Fas-bearing cells; participant in down-regulation of the immune response & in T-cell mediated cytotoxicity

Fas Protein, Intracellular Fragment

Synonyms: Fas Antigen

Oncogene PF072 >80% (SDS-PAGE); 100 mg in 1 mL of 50 *mM* HEPES, pH 7.5 containing 0.5 *M* NaCl, 0.05% TWEEN® 20, 0.01 *M* β-mercaptoethanol & 50% ammonium sulfate | Used in western blot or in an ELISA format; reported that Fas antibodies raised against the intracellular domain of Fas will recognize #PF072; Western blot & ELISA

Fasciculin I

Sigma F 3918 *Dendroaspis angusticeps* (Eastern Green Mamba) Protein components of the venom of the eastern green mamba snakeCholinesterase inhibitors that inhibit acetylcholinesterase ~10,000 times more efficiently than butyrylcholinesteraseKarlsson, E et al, *J Physiol* (Paris), 79: 232, 1984

Fasciculin II

Sigma F 4293 *Dendroaspis angusticeps* (Eastern Green Mamba) Believed to differ from fasciculin I only by a single AA substitution (Tyr for Asx); see Sigma F 3918

Fas-huIg Fusion Protein

Synonyms: CD95-huIg Fusion Protein, Human

Alexis ANC-506-020 Human Purified; liquid; free of azide & carrier protein; stabilized with 0.5 mg/mL gentamycin sulfate; transfectant cell line: CHO | Soluble fusion protein consisting of the extracellular (175AA) domain of human Fas (CD95) fused to human IgG1 Fc (234AA); useful in flow cytometry, immunohistochemistry (frozen sections); Fas-huIg fusion protein blocks binding of anti-human Fas antibody to cells expressing Fas; human Fas is a type 1 cell surface glycoprotein that is strongly unregulated on activated T cells, B cells, NK cells & thymocytes*Leukocyte Typing V* (Schlossman, SF et al, eds) Oxford University Press, Oxford, p. 1142, 1995

Fas-huIg Fusion Protein R-PE

Synonyms: CD95-huIg Fusion Protein, Human

Alexis ANC-506-050 Human Liquid; stabilized with 0.04% sodium azide; transfectant cell line: CHO | See Alexis ANC-506-020

Fas-huIg Fusion Protein, Biotin Conjugated

Synonyms: CD95-huIg Fusion Protein, Human

Alexis ANC-506-030 Human Liquid; stabilized with 0.04% sodium azide; transfectant cell line: CHO | See Alexis ANC-506-020

Fas-huIg Fusion Protein, FITC Conjugated

Synonyms: CD95-huIg Fusion Protein, Human

Alexis ANC-506-040 Human Liquid; stabilized with 0.04% sodium azide; transfectant cell line: CHO | See Alexis ANC-506-020

Fatty Acid Binding Protein

Scipac P196-1 Heart tissue >96%; lyophilized | Cardiac marker protein

Cortex CP2049 Human >95%

Fitzgerald 30-AF14 Human cardiac High purity

Biodesign A86865H Human heart >98% | Cardiac markers

USBio F0019-76 Human heart tissue ≥98%; no contaminants detected; single band by SDS-PAGE, IEP, &/or RID; 1 mg/mL; lyophilized | Suitable for antigenic applications in immunological protocols

Ferredoxin

Sigma F 6671 *Clostridium pasteurianum* Lyophilized powder containing ~30% ferredoxin by weight; balance Trizma buffer; isolated by modified procedure of Rabinowitz | Iron containing proteins that serve as electron-transfer catalysts in photosynthesis; sold on the basis of mg ferredoxin based on E_{390} at 1% = 34; package sizes are based on ferredoxin content; Rabinowitz, J, *Methods in Enzymology*, 24: 431, 1972

Sigma F 4029 *Porphyra umbilicalis* (Red marine algae) Chromatographically purified, lyophilized powder | Iron containing proteins that serve as electron-transfer catalysts in photosynthesis; package sizes are based on ferredoxin; Andrew, PW et al, *Eur J Biochem*, 69: 243, 1976

Sigma F 3013 Spinach Lyophilized powder containing ~25% ferredoxin by weight; balance is Trizma buffer; partially purified per method by Tagawa & Arnon | Iron containing proteins that serve as electron-transfer catalysts in photosynthesis; package sizes are based on ferredoxin; Tagawa & Arnon, *Nature*, 195: 537, 1962

Sigma F 5875 Spinach Frozen solution containing 1-3 mg ferredoxin/mL of 0.15 M Trizma buffer, pH 7.5; contains NaCl; partially purified per method by Tagawa & Arnon | Iron containing proteins that serve as electron-transfer catalysts in photosynthesis; package sizes are based on ferredoxin; Tagawa & Arnon, *Nature*, 195: 537, 1962

Sigma F 2513 Spirulina species Chromatographically purified, lyophilized powder | Iron containing proteins that serve as electron-transfer catalysts in photosynthesis; package sizes are based on ferredoxin

Ferritin

Biogenesis 4420-5409 Equine spleen MW 500k 30% iron (bipridyl method); pI: 4.1-5.6; purified; 150 mM NaCl with 0.002% NaN3; liquid

Calbiochem 341475 Equine spleen MW 500k Highly purified; liquid in 150 mM NaCl, 0.02% NaN3; >90% (size exclusion chromatography) | Major iron storage protein found in spleen, liver & intestinal mucosa

Calbiochem 341476 Equine spleen MW 500k Chromatographically purified; cadmium-free; iquid in 150 mM NaCl, 0.1% NaN3; >90% (size exclusion chromatography); purified chromatographically without using cadmium precipitation

USBio F4015-19 Heart ≥98%; no contaminants detected; single band by SDS-PAGE, IEP, &/or RID; 2-3 mg/mL supplied in 150 mM NaCl, 0.1% NaN3, 10 mM Tris, pH 7.0 | Suitable for antigenic applications in immunological protocols

Fluka 46230 Horse spleen MW 900k ≥75% (GE); ≥36 mg/mL protein (Lowry); ~16% Fe; ≤0.05% Cd | Fe-saturated ferritin; Crichton, RR & Charl-Wauters, M, *Eur J Biochem*, 164: 485, 1987

ICN 100646 Horse spleen 6X re-crystallized; Cd removed; solution in 0.15 M NaCl; 100 mg ferritin/mL

ICN 151119 Horse spleen 2X crystallized; may contain ≤1% Cd (as % of ferritin)

ICN 960272 Horse spleen 96%; >48 mg/mL; <% Cd | Marker in electron microscopy; species specific but not organ specific

Biogenesis 4420-4304 Human heart pI: 4.5-4.8; H subunit content is high (50-60%); purified; liquid

Fitzgerald 30-AF05 Human heart High purity

ICN 151120 Human heart 0.15 M NaCl, 0.1% NaN3; >99% (PAGE); 0.4 mg/mL (Lowry)

Biodesign A10152H Human liver Purified

Biogenesis 4420-4804 Human liver Tested negative for HBsAg, HCV, HIV 1 and 2 and syphilis; purified; 0.05 M Tris, pH 7.5 with 0.1% NaN3; liquid | Addison et al, *FEBS Letts*, 164:139, 1983

Calbiochem 341482 Human liver MW 450k Sterile-filtered; liquid in 150 mM NaCl, 0.1% NaN3, pH 7.0; ≥95% (SDS-PAGE); prepared from tissue of individuals shown to be negative for HBsAg & for antibodies to HIV & HCV | Major iron storage protein; suitable for use in immunoassays as an immunogen & in enzyme/radiolabeling; Addison, JM et al, *FEBS Lett*, 164: 139, 1983

Fitzgerald 30-AF10 Human liver High purity

ICN 151121 Human liver 0.15 M NaCl, 0.1% NaN3; >98% (PAGE single peak on sepharose CL-6B); 2.25 mg/mL (Lowry)

USBio F4015-21 Human liver No contaminants detected; ~1 mg/mL supplied in 0.9% NaCl, 0.1% NaN3 | Suitable for antigenic applications in immunological protocols

Biogenesis 4420-5004 Human placenta 0.15 M NaCl and 0.02% NaN3; liquid | Placental ferritin has a high content of the H-subunit (10-20%); its sequence differs in a small number of positions from the major component of spleen ferritin; some of these differences are also found in the liver ferritin; tested and found negative for the HBsAg and HTLV III antibody; Wustefeld & Crichton, *FEBS Letts*, 150:43, 1982; Addison et al, *FEBS Letts*, 164:139, 1983

USBio F4015-24 Human placenta >95% by PAGE (PI ~ 4.9); chromatographically purifiied by DEAE; ethanol/thermal denatured; liquid, sterile filtered in 0.15 M NaCl & 0.02% NaN3 | High content of the H-subunit (10-20%); sequence differs in a small number of positions from the major component of spleen ferritin; some of these differences are also found in liver ferritin; suitable for antigenic applications in immunological protocols

Fitzgerald 30-AF13 Human placental High purity

Biogenesis 4420-5204 Human spleen

Fitzgerald 30-AF15 Human spleen High purity

USBio F4015-22 Human spleen No contaminants detected (SDS-PAGE); ~3 mg/mL (Lowry) supplied in 0.01 M sodium phosphate, 0.15 M NaCl, pH 7.2, 0.1% Kathon (preservative) | Suitable for antigenic applications in immunological protocols

Cortex CP1003 Liver >95%

Scipac P103-9 Liver >96%; in TRIS buffer; sterile filtered through 0.2µm membrane; 3-10 mg/mL | Nutritional protein; tumor marker

Biogenesis 4420-5519	Rat liver	Liquid

Biogenesis 4420-5539　Rat liver　Purified; 0.9% NaCl with 0.02% NaN$_3$; liquid

Cortex CP1004	Spleen	>95%

Ferritin H Chain

Biogenesis 4420-6009	Human r-DNA	Liquid

Calbiochem 341490　Human recombinant　MW 507k　Liquid in 150 mM NaCl, 20 mM Tris-HCl, 0.02% NaN$_3$; >95% (SDS-PAGE) | Contains the metal binding site of ferritin which confers ferroxidase activity to the protein; Levi, S et al, *J Biol Chem*, 269: 30334, 1994; Levi, S et al, *J Mol Biol*, 238: 649, 1994

Ferritin L Chain

Biogenesis 4420-7009	Equine r-DNA	Liquid
Biogenesis 4420-7109	Human r-DNA	Liquid

Calbiochem 341491　Human recombinant　MW 478k　Liquid in 150 mM NaCl, 20 mM Tris-HCl, 0.02% NaN$_3$; >95% (SDS-PAGE) | Lacks detectable ferroxidase activity; Levi, S et al, *J Mol Biol*, 238: 649, 1994

Ferritin, Apo-

Calbiochem 178440　Equine spleen　MW 460k　Lyophilized; ≥90% (SDS-PAGE); iron: <0.01%; soluble in dilute buffers & water | Protein shell of ferritin molecule lacking iron; large amounts are present in pancreatic β-cells where it acts as an antioxidant; Sun, S et al, *J Biol Chem*, 267: 25160, 1992; McDonald, MJ et al, *FASEB J*, 8: 777, 1994; de Silva, D et al, *Arch Biochem Biophys*, 298: 259, 1992; Crichton, RR, *Structure Bonding*, 17: 67, 1974

Ferritin, Cationized

ICN 911141　9.0-12.0 mg/mL in sterile 0.15M NaCl | This polycationic derivative of ferritin is useful for labeling negative charges on cell surfaces; lacks the inherent disadvantages of other cationic dyes such as ruthenium red, alcian blue, thorium hydroxide, & colloidal iron; has the advantage of quantitating the surface charges with greater precision due to its smaller diameter size; effective on a variety of cells, & its geometry permits easy membrane surface counting

Fes/Fps

USBio F4050-05　Human recombinant, produced by Baculovirus expression in Sf9 cells　~90% (DEAE chromatography, Mono-Q, FPLC); 250 ng/25 μL; packaged in 4 vials, each vial containing 250 ng in 25 mlLof 10 mM Tris-HCl, pH 7.5, containing 0.25 mM EGTA, 0.25 mM EDTA, 0.125 M NaCl with 40% glycerol, & 5 mM β-MSH | Transfers ~16nmole phosphate/min/mgkinase (1 U defined as 1nmole/min); useful in proteinkinase assay; undergoes autophosphorylation & catalyzes the phosphorylation of poly-Glu-Tyr (4:1) *in vitro*

Fetoprotein Cell Line Antigen, α-

USBio F4100-24　AFP cell line　≥98%; no contaminants detected; single band by SDS-PAGE, IEP, &/or RID; 1 mg/mL supplied in 0.1 M Tris buffer, 0.2 M NaCl, pH 7.4, 0.1% NaN$_3$ | Suitable for antigenic applications in immunological protocols

USBio F4100-27　AFP cell line　98%, 50% protein (SDS-PAGE); 0.5-5 mg/mL; 0.2 μm filtered | Suitable for antigenic applications in immunological protocols

Fetoprotein Receptor, α-

Biodesign A86886H	Fetal tissue	Purified

Fetoprotein, α-

Cortex CP1007	>95%
Cortex CP1007U	>98%
Fitzgerald 30-AA06	AFP cell line	High purity >98%

Scipac P107-1　Amniotic fluid　>96%; lyophilized | Tumor marker

Scipac P107-2　Amniotic fluid　3-12%; lyophilized | Tumor marker

Biodesign A32260H	Cell culture	Antigen grade

USBio F4100-18　Human　95%; contains trace amounts of albumin; 1.1 mg/mL supplied in 0.1 M Tris buffer, 0.2 M NaCl, 0.1% NaN$_3$, pH 7.5 | Suitable for antigenic applications in immunological protocols

USBio F4100-19　Human　≥98%; single band by SDS-PAGE, IEP, &/or RID; 1 mg/mL liquid in 0.1 M Phosphate buffer, pH 7.5, 0.15 M NaCl | Suitable for antigenic applications in immunological protocols as well as calibrators & controls

Fitzgerald 30-AA05	Human cord plasma	High purity >98%

Fitzgerald 30-AA10　Human cord plasma　Standard purity >60%

Biodesign A15108H　Human cord serum　60-80% | Tumor markers, cancer antigens & oncogenes

Calbiochem 341498　Human cord serum　Liquid in 200 mM NaCl, 100 mM Tris-HCl, 0.1% NaN$_3$, pH 7.5; >95% (SDS-PAGE); prepared from serum shown to be negative for HBsAg & for antibodies to HIV & HCV | Major fetal serum glycoprotein, classified as an oncofetal protein, synthesized in the liver, yolk sac & developing fetal gastrointestinal tract; serves as a modulator of various cell growth regulatory pathways during embryonic development in vertebrates; a useful tumor marker; higher serum levels aid in the diagnosis, classification & monitoring of non-seminomatous testicular cancer & primary hepatocellular (liver) carcinoma; elevated maternal α-fetoprotein levels are common in chorioangioma; suitable for use in immunoassays & as an immunogen; Jeng, LB et al, *Am J Obstet Gynecol*, 172: 219, 1995; Ruoslahti, E et al, *Methods Enzymol*, 84: 3, 1982; Anderson, T et al, *Ann Int Med*, 90: 373, 1979; Khong, TY & George, K, *Am J Perinatol*, 11: 245, 1994; Mizejewski, GJ, *Life Sci*, 56: 1, 1994

Sigma F 1510　Human cord serum　Solution in 0.1 M phosphate buffer, pH 7.2, 15 mM sodium azide & 2.5% sucrose

USBio F4100-26　Human cord serum　60-80% (SDS-PAGE); some major serum proteins; ~0.5 mg/mL | Suitable for antigenic applications in immunological protocols

ICN 195002　Human fetal cord serum　>95%; lyophilized | Suitable for use as an immunogen, iodinated tracer & as a clinical calibrator

ICN 770931 ICN 770932 ICN 770933　Human fetal cord serum　>97%; iodination grade | Ideal for labeling or immunization purposes

Fetoprotein, αI-

Biogenesis 4520-5704　Human amniotic fluid　MW ~66k Tested negative for HIV 1 and 2 antibodies, HBsAg and HCV antibodies; purified; PBS buffer, pH 7.2, 0.1% NaN$_3$; liquid

Biogenesis 4520-5804　Human amniotic fluid　MW 66k　3-12% of total protein is alpha fetoprotein by radial immunodiffusion; may contain traces of buffer salts; tested negative for HIV antibodies, HBsAg and for HCV antibodies; semi-pure; PBS buffer, pH 7.2; liquid

Biogenesis 4520-5849	Human fetal serum	Lyophilized
Biogenesis 4520-5869	Human fetal serum	

Fetuin

Biogenesis 4430-2204	Bovine serum	Lyophilized

Fetuin I

ICN 104874　Fetal bovine serum　A glycoprotein recovered from the globulin fraction of fresh calf serum by ammonium sulfate fractionation; Pederson, KO J Phys & Colloid Chem, 51:164 1947

Fetuin II

ICN 152410　Neonatal calf serum　A glycoprotein recovered from the globulin fraction of calf serum by ammonium sulfate fractionation; Pederson, KO J Phys & Colloid Chem, 51:164 1947

Fetuin, 10 nm Colloidal Gold Conjugated

ICN 154032

Fetuin, 20 nm Colloidal Gold Conjugated
ICN 154033

Fetuin, 5 nm Colloidal Gold Conjugated
ICN 154031

Fibrin
ICN 901687   Bovine blood   Thrombin treated fibrinogen; washed 6 times

Fibrin Degradation Product D-Dimer
Biogenesis 4440-0256   Human serum   Tested negative for HBsAg, HIV-1 and HIV-2 antibodies, HCV and syphilis; purified; essentially salt free; liquid | A specific degradation product of cross-linked fibrin: used as a marker of venous thrombo embolism for the diagnosis of deep venous thrombosis of the lower limbs and pulmonary embolism

Fibrin Degradation Product D-Monomer
Biogenesis 4440-0276   Human serum   MW 45k under reducing conditions, 94k under non-reducing conditions   Tested negative for HBsAg and HIV-1 antibodies; purified; 50mM TBS, pH 8.0, with 1 *mM* Tranexamic Acid, 0.05% NaN3; liquid | Homogenous by electrophoresis

Fibrin Degradation Product E
Biogenesis 4440-0456   Human serum   MW 50k   Tested negative for HBsAg and HIV antibodies; E_{280} nm (1%): 0.9; purified; 0.15 *M* NaCl, 0.01 *M* sodium phosphate, pH 7.5, 0.1% mannitol and residual urea; lyophilized | Marder & Francis, *Ann NY Acad Sci*, 408:397

Fibrin Degradation Product X
Biogenesis 4440-0626   Human serum   MW 240k   Non-reactive for HBsAg, HIV I/II antibodies and HCV antibody; purified; 50 *mM* Tris-HCl, pH 8.0, 0.1M NaCl, 10 *mM* e-amino caproic acid; liquid

Fibrin Degradation Product Y
Biogenesis 4440-0826   Human serum   MW 175k   Non-reactive for HBsAg and anti HIV I/II and anti-HCV antibodies; purified; 20 *mM* Tris-HCl, pH 8.0, 0.5M NaCl, 1 *mM* tranexamic acid, 0.05% NaN3; liquid

Fibrin, Blue
Synonyms: Pepsin Substrate
ICN 158051   Pepsin substrate at low pH; high blanks result at high pH; Nelson, WL etal, *Anal Biochem*, 2:39, 1961

Fibrinogen
Synonyms: Coagulation Factor I
ICN 154165   Bovine   >80% clottable protein; >70% protein; <3% moisture
Fluka 46312   Bovine plasma   MW 341k   ≥70% protein; clottable protein: ≥80%; 15% NaCl; 10% sodium citrate; ≤10% water; white powder
ICN 820212 ICN 820215   Bovine Plasma   75%; 75% clottable; lyophilized | Used for preparation of fibrin plates for analysis of fibrinolytic enzymes, as a substrate for clotting assays & for study of fibrinogen degradation products
ICN 820224 ICN 820225   Bovine Plasma   95%; 95% clottable; lyophilized | Used for preparation of fibrin plates for analysis of fibrinolytic enzymes, as a substrate for clotting assays & for study of fibrinogen degradation products
ICN 820244   Bovine Plasma   Plasminogen free
USBio F4200   Human   Homogeneous by 6% SDS-PAGE, ≥95% clottable; plasminogen has been depleted; ~24 mg/mL supplied in 20 *mM* citric acid-HCl/glycine, pH 7.4 | Suitable for antigenic applications in immunological protocols

Fluka 46313   Human plasma   MW 341k   ~50% protein; contains sodium citrate & NaCl; clottability: ~95% of protein; ≤0.05% plasminogen; ≤0.01% plasmin; ≤6% water | Fuller, GM et al, *Meth Enzymol*, 163: 474, 1988
Biogenesis 4440-8604   Human plasma (Cohn fraction I) Protein ~45% by Biuret; virtually free from plasminogen and plasmin; tested negative for antibodies to HIV-1, HIV-2, antibodies to HCV, HBsAg, and for HIV-1 antigen; purified; contains ~20% sodium citrate & 30% NaCl; lyophilized | 90-93% of protein is clottable

Fibrinogen Degradation Product
Sigma F 9036   Human plasma   Frozen solution in 10 mM Tris-HCl, pH 7.4, containing 150 mM NaCl & 2 mM $CaCl_2$ | Presence of fibrinogen degradation products is a clinical marker for a wide variety of disease states; protein determined by LowrySoria, J & Soria, °C, *Gaz Med Fr*, 86: 1099-1106, 1979

Fibrinogen Degradation Product D
ICN 194073   Human Plasma   Purified protein; glycine & NaCl stabilizers | Thermolabile protein fragment used as a clinical marker for various states of disease; Amiral, J etal, *Fibrinogen & its Derivatives*, 285, 1986

Fibrinogen Degradation Product E
ICN 194074   Human Plasma   Purified protein; glycine & NaCl stabilizers | Thermolabile protein fragment used as a clinical marker for various states of disease

Fibrinogen Fraction I
ICN 151122   Bovine Plasma   ~75% protein
Sigma F 4753   Bovine plasma   Type IV; ~60% protein (~95% of protein clottable); contains ~15% sodium citrate & ~25% NaCl | Protein determined by Biuret method
Sigma F 8630   Bovine plasma   Type I-S; ~75% protein (>75% of protein clottable); contains ~10% sodium citrate & ~15% NaCl | Protein determined by Biuret method
Sigma F 8513   Cat plasma   ~50% protein (>60% of protein clottable); contains ~20% sodium citrate & ~30% NaCl | Protein determined by Biuret method
Sigma F 7128   Dog plasma   ~55% protein (>60% of protein clottable); contains ~20% sodium citrate & ~25% NaCl | Protein determined by Biuret method
Sigma F 9631   Guinea pig plasma   ~50% protein (>90% of protein clottable); contains ~20% sodium citrate & ~30% NaCl | Protein determined by Biuret method
ICN 151123   Human Plasma   ~65% protein
Sigma F 3879   Human plasma   Type I; ~60% protein (>90% of protein clottable); contains ~15% sodium citrate & ~15% NaCl | Protein determined by Biuret method;source material negative for HIV & HBsAg
Sigma F 4129   Human plasma   Type III; ~65% protein (>60% of protein clottable); contains ~15% sodium citrate & ~20% NaCl | Protein determined by Biuret method;source material negative for HIV & HBsAg
Sigma F 4883   Human plasma   Essentially plasmin(ogen) free; ~50% protein (~95% of protein clottable); contains ~20% sodium citrate & ~30% NaCl | Protein determined by Biuret method;source material negative for HIV & HBsAg
Sigma F 4385   Mouse plasma   ~50% protein (>80% of protein clottable); contains ~20% sodium citrate & ~30% NaCl | Protein determined by Biuret method
Sigma F 2629   Pig plasma   ~70% protein (>70% of protein clottable); contains ~12% sodium citrate & ~18% NaCl | Protein determined by Biuret method
Sigma F 6755   Rat plasma   ~70% protein (>60% of protein clottable); contains ~12% sodium citrate & ~18% NaCl | Protein determined by Biuret method
Sigma F 9754   Sheep plasma   ~75% protein (>80% of protein clottable); contains ~10% sodium citrate & ~15% NaCl | Protein determined by Biuret method

Fibrinogen Reference

Sigma 880-10   Human plasma   ~250 mg/dL; lyophilized citrated human plasma containing buffer & preservative | Fibrinogen value listed on label is determined by clottable protein method; for use in the determination of fibrinogen per Sigma Procedure No. 880

Fibrinogen, (^{125}I)-

ICN 68033   Human   ~1 μCi/μg, ~37kBq/μg; 0.1 M KPO$_4$, pH 7.5, 0.5% BSA

Fibroblast Growth Factor

Synonyms: Endothelial Cell Growth Factor, β-

ICN 160020   Bovine brain   Prepared according to Gospodurowicz; membrane filtered before lyophilization; optimal conc: 50-100 ng/mL for Balb/c 3T3 cells | Gospodurowicz, D, *JBC*, 253:3736, 1978

ICN 160037   Bovine Pituitary Gland   Prepared according to Gospodurowicz; membrane filtered before lyophilization; optimal conc: 10-50 ng/mL for Balb/c 3T3 cells | Gospodurowicz, D, *JBC*, 250:2515, 1975

Sigma F 3133   Bovine pituitary glands   Tissue culture grade; 0.2 μm filtered & lyophilized from a solution of sodium phosphate & sodium chloride containing chicken egg albumin as a stabilizer; activity/unit weight of protein is less than that of purified bFGF; endotoxin tested; see Sigma F 5542 | Purified by a modification of the method of Gospodarowicz, D, *J Biol Chem*, 250: 2515, 1975

Sigma E 1388   Human recombinant, expressed in *E. coli* $\geq$97% (SDS-PAGE); 0.2 μm filtered & lyophilized from PBS containing 1.25 mg BSA; proliferative activity is tested in culture using quiescent NR6R-3T3 fibroblasts; endotoxin tested; see Sigma F 5542 | Rizzino, A et al, *Cancer Res*, 48: 4266, 1988

Fibroblast Growth Factor I

Sigma F 2897   Human recombinant, expressed in 293T cells MW ~110-120k   Provided as a transfected cell extract & is in SDS-PAGE loading buffer; ready-to-use in immunoblotting techniques; see Sigma F 5542 | A predominant doublet detected by specific antibodies to FGFR-1; Recommended use in SDS-PAGE as a positive control is 5-10 μL of undiluted product/gel lane

Fibroblast Growth Factor II

Chemicon GF085   Murine $\geq$95%

Fibroblast Growth Factor IV

Chemicon GF098   Human $\geq$95%

Biogenesis 0100-0130   Human r-DNA   Lyophilized

Oncogene PF076   Human recombinant   MW 14 & 16k >97% (SDS-PAGE); lyophilized; biological activity: EC$_{50}$ of 0.05-0.15 ng/mL as determined in a mitogenic assay measuring FGF dependent ^{3}H-thymidine incorporation in quiescent NR6R-by 3T3 fibroblasts | Species reactivity: human; for proliferation studies

ICN 160071   Human recombinant, expressed in *E. coli* $\geq$97%; lyophilized; ED$_{50}$ = 0.05-0.15 ng/ml (by stimulation of ^{3}H-thymidine uptake by quiescent NR6R 3T3 fibroblasts)

PeproTech 100-31   Human recombinant, expressed in *E. coli* MW 19.0k   >98% (SDS-PAGE) & HPLC; <0.1 ng endotoxin per mg (1EU/mg); lyophilized | Heparin binding growth factor; stimulates proliferation & activation of cells that express the FGF receptors; 182 AA; ED$_{50}$ $\leq$ 0.5 ng/mL; SA $\geq$ 10^6 U/mg; SA determined by dose-dependent stimulation of thymidine uptake by BaF3 cells expressing FGF receptors

Sigma F 2278   Human recombinant, expressed in *E. coli* $\geq$97% (SDS-PAGE); 0.2 μm filtered & lyophilized from 20 mM sodium phosphate buffer, pH 7.0, containing 1.25 mg human serum albumin; proliferative activity is tested in culture using quiescent NR6R-3T3 fibroblasts; endotoxin tested; see Sigma F 5542 | Rizzino, A et al, *Cancer Res*, 48: 4266, 1988

Fibroblast Growth Factor IX

Synonyms: Glial Activating Factor

Biodesign A52030H   *E. coli*   MW 23.3k   Purified | Species specificity: mouse

Chemicon GF097   Human $\geq$95%

Biogenesis 0100-0133   Human r-DNA   Lyophilized

PeproTech 100-23   *Human recombinant, expressed in* Human recombinant, expressed in *E. coli*   MW 23.4k   >98% (SDS-PAGE) & HPLC; <0.1 ng endotoxin per mg (1EU/mg); lyophilized | Heparin binding growth factor; stimulates proliferation & activation of cells that express the FGF receptors; 207 AA; ED$_{50}$ $\leq$ 0.5 ng/mL; SA $\geq$ 10^6 U/mg; SA determined by dose-dependent stimulation of thymidine uptake by BaF3 cells expressing FGF receptors

Sigma F 1168   Human recombinant, expressed in *Sf* 21 insect cells   MW 25k $\geq$97% (SDS-PAGE); 0.2 μm filtered & lyophilized from PBS containing 1.25 mg BSA; bioactivity is measured in a bioassay using BALB/3T3 cells; endotoxin tested; see Sigma F 5542

ICN 195735   Human recombinant, *Sf*21 expressed $\geq$97%; lyophilized; activity: ED$_{50}$ = 1-2 ng/mL

Biogenesis 0100-0134   Mouse r-DNA   Lyophilized

BioSource International PMG0014   Mouse recombinant

Chemicon GF040   Murine $\geq$95%

ICN 195006   Murine recombinant, expressed in *E. coli*   MW 22k   Lyophilized

PeproTech 450-30   Murine recombinant, expressed in *E. coli* MW 23.3k   >98% (SDS-PAGE) & HPLC; <0.1 ng endotoxin per mg (1EU/mg); lyophilized | Heparin binding growth factor; stimulates proliferation & activation of glial cells & other cells that express FGF receptors; 205 AA; ED$_{50}$ $\leq$ 0.5 ng/mL; SA $\geq$ 2 x 10^6 U/mg; SA determined by the dose-dependent stimulation of thymidine uptake by BaF3 expressing FGF receptors

Fibroblast Growth Factor Receptor (flg-5) Extracellular Domain

IBT GR-030-3, GR-030-5   Human recombinant, expressed in *Spodoptera frugiperda* insect cells (Sf9)   >85%; 1.0 mg prot/mL, 10 mM Tris-HCl, pH 7.0 | Binds radioiodinated basic FGF; inhibits basic FGF-mediated proliferation of adrenocortically capillary endothelial cells

Fibroblast Growth Factor V

ICN 160172   Human recombinant, expressed in *E. coli* $\geq$97%; lyophilized; ED$_{50}$ = 0.05-0.15 ng/ml (by stimulation of ^{3}H-thymidine uptake by quiescent NR6R 3T3 fibroblasts)

Oncogene PF077   Human recombinant, expressed in *E. coli* MW 27k   >97% (SDS-PAGE); lyophilized; biological activity: EC$_{50}$ of 0.05-0.1 μg/mL in the presence of 0.1 μg/mL heparin as determined by FGF dependent ^{3}H-thymidine incorporation in quiescent NR6R-by 3T3 fibroblasts | Recombinant human protein based on a DNA sequence encoding mature FGF-5 expressed in *E. coli*; species reactivity: human; for proliferation studies

PeproTech 100-34   Human recombinant, expressed in *E. coli* MW 27.6k   >98% (SDS-PAGE) & HPLC; <0.1 ng endotoxin per mg (1EU/mg); lyophilized | Heparin binding growth factor; stimulates proliferation & activation of cells that express the FGF receptors; 252 AA; ED$_{50}$ $\leq$ 0.5 ng/mL; SA $\geq$ 10^6 U/mg; SA determined by dose-dependent stimulation of thymidine uptake by BaF3 cells expressing FGF receptors

Sigma F 4537   Human recombinant, expressed in *E. coli* $\geq$97% (SDS-PAGE); purified by sequential chromatography; 0.2 μm filtered & lyophilized from phosphate buffered solution containing 2.5 mg BSA; bioactivity is determined by measuring the FGF-5 dependent ^{3}H-thymidine incorporation in quiescent NR6R-3T3 fibroblasts; endotoxin tested; see Sigma F 5542 | Rizzino, A et al, *Cancer Res*, 48: 4266, 1988

Fibroblast Growth Factor VI

Biogenesis 0100-0131   Human r-DNA   Lyophilized

ICN 160073   Human recombinant, expressed in *E. coli* $\geq$97%; lyophilized; ED$_{50}$ = 0.05-0.15 ng/ml (by stimulation of ^{3}H-thymidine uptake by quiescent NR6R 3T3 fibroblasts)

PeproTech 100-30   Human recombinant, expressed in *E. coli*
MW 18.7k   >98% (SDS-PAGE) & HPLC; <0.1 ng endotoxin per
mg (1EU/mg); lyophilized | Newly discovered heparin binding
growth factor; stimulates proliferation & activation of cells that
express the FGF receptors; 168 AA; $ED_{50} \leq 0.5$ ng/mL; $SA \geq 10^6$
U/mg; SA determined by dose-dependent stimulation of thymidine
uptake by BaF3 cells expressing FGF receptors

Sigma F 4662   Human recombinant, expressed in *E. coli*
≥97% (SDS-PAGE); purified by sequential chromatography; 0.2 μm
filtered & lyophilized from PBS containing 0.05% CHAPS & 2.5 mg
BSA; bioactivity is determined by measuring the FGF-5 dependent
^{3}H-thymidine incorporation in quiescent NR6R-3T3 fibroblasts;
endotoxin tested; see Sigma F 5542 | Rizzino, A et al, *Cancer
Res*, 48: 4266, 1988

Fibroblast Growth Factor VII

USBio F4299   Recombinant (FGF-7/KGF)   Sterilized through a
0.2m membrane filter & packaged aseptically | Contains 163 AA
residues; a potent mitogen forkeratinocytes & epithelial cells; 1–
10ng/mL produced a 11-fold maximal stimulation of (^{3} H)-
thymidine incorporation in mousekeratinocytes; ED_{50} = 0.2-0.8
ng/mL; optimal KGF concentration in cell culture varies from 0.01-
10 ng/mL, depending on cell type

Fibroblast Growth Factor VIII

Synonyms: Androgen Induced Growth Factor

Biogenesis 0100-0132   Human r-DNA   Lyophilized

PeproTech 100-25   *Human recombinant, expressed in* Human
recombinant, expressed in *E. coli*   MW 22.4k   >98% (SDS-
PAGE) & HPLC; <0.1 ng endotoxin per mg (1EU/mg); lyophilized |
Heparin binding growth factor; stimulates proliferation & activation
of cells that express the FGF receptors; 193 AA; $ED_{50} \leq 0.5$ ng/mL;
$SA \geq 10^6$ U/mg; SA determined by dose-dependent stimulation of
thymidine uptake by BaF3 cells expressing FGF receptors

Fibroblast Growth Factor VIIIβ

ICN 195783   Murine recombinant, expressed in *E. coli*   ≥97%;
lyophilized; ED_{50} = typically 1-3 ng/mL in the presence of 0.1
μg/mL heparin

Fibroblast Growth Factor X

Biogenesis 0100-0135   Human r-DNA   Lyophilized

PeproTech 100-26   Human recombinant, expressed in *E. coli*
MW 19.3k   >98% (SDS-PAGE) & HPLC; <0.1 ng endotoxin per
mg (1EU/mg); lyophilized | Heparin binding growth factor;
stimulates proliferation & activation of cells that express the FGF
receptors; most related to KGF/FGF-7; expressed during
development & preferentially in adult lungs 170 AA; $ED_{50} \leq 0.5$
ng/mL; $SA \geq 1 \times 10^6$ U/mg; SA determined by the dose-dependent
stimulation of thymidine uptake by BaF3 cells expressing FGF
receptors

Fibroblast Growth Factor XVI

Biogenesis 0100-0136   Human r-DNA   Lyophilized

PeproTech 100-29   Human recombinant, expressed in *E. coli*
MW 23.7k   >98% (SDS-PAGE) & HPLC; <0.1 ng endotoxin per
mg (1EU/mg); lyophilized | Newly discovered heparin binding
growth factor; stimulates proliferation & activation of cells that
express the FGF receptors; 207 AA; $ED_{50} \leq 0.5$ ng/mL; $SA \geq 10^6$
U/mg; SA determined by dose-dependent stimulation of thymidine
uptake by BaF3 cells expressing FGF receptors

Fibroblast Growth Factor XVII

Biogenesis 0100-0137   Human r-DNA   Lyophilized

PeproTech 100-27   Human recombinant, expressed in *E. coli*
MW 22.7k   >98% (SDS-PAGE) & HPLC; <0.1 ng endotoxin per
mg (1EU/mg); lyophilized | Heparin binding growth factor;
stimulates proliferation & activation of cells that express the FGF
receptors; 195 AA; $ED_{50} \leq 0.5$ ng/mL; $SA \geq 10^6$ U/mg; SA
determined by dose-dependent stimulation of thymidine uptake by
BaF3 cells expressing FGF receptors

Fibroblast Growth Factor XVIII

Biogenesis 0100-0138   Human r-DNA   Lyophilized

PeproTech 100-28   Human recombinant, expressed in *E. coli*
MW 21.2k   >98% (SDS-PAGE) & HPLC; <0.1 ng endotoxin per
mg (1EU/mg); lyophilized | Newly discovered heparin binding
growth factor; stimulates proliferation & activation of cells that
express the FGF receptors; 182 AA; $ED_{50} \leq 0.5$ ng/mL; $SA \geq 10^6$
U/mg; SA determined by dose-dependent stimulation of thymidine
uptake by BaF3 cells expressing FGF receptors

Fibroblast Growth Factor, Acidic

Synonyms: Endothelial Cell Growth Factor; Heparin Binding
Growth Factor (I or α); Retina Derived Growth Factor; Astroglial
Growth Factor I; Endothelial Cell Growth Factor, β-

Biogenesis 4460-4104   Bovine brain   MW 17k & 20k
Purified; from 0.5 mL 0.001 *M* sodium phosphate, pH 7.0;
lyophilized | Stimulates growth of bovine capillary endothelial cells
by 3-5 fold over 5% calf serum at 10-25 ng/mL FGF and 10 μg/mL
heparin; the 17K peptide is derived from the 20K peptide by
restricted proteolysis; Burgess et al, *JBC*, 260:11389, 1985; Jaye
et al, *Science*, 233:541, 1986; Lobb et al, *JBC*, 261:1924, 1986

ICN 154132   Bovine brain   >95% (SDS-PAGE visualized by
silver stain & *N*-terminus analysis); lyophilized with BSA carrier;
ED_{50} = 0.1-0.5 ng/mL (by dose-dependent stimulation of ^{3}H-
thymidine incorporation by NR6R 3T3 fibroblasts)

Sigma F 5267   Bovine brain   MW 14-16k   ≥90% (SDS-
PAGE); 0.2 μm filtered & lyophilized from a solution of sodium
phosphate & sodium chloride containing 100 μg BSA; endotoxin
tested; see Sigma F 5542 | Protein isolated by a modification of
the method of Gospodarowicz, involving heparin affinity
chromatography; Gospodarowicz, D et al, *Proc Natl Acad Sci USA*,
81: 6963, 1984

ICN 152306   Bovine hypothalmus   ~95%; lyophilized;
reconstitute with 1 ml sterile solution of 0.05 *M* Na_2HPO_4, pH 7;
suggested conc in cell culture = 1-20 ng/mL | Stimulates the
growth of bovine capillary endothelial cells by 3- to 5-fold over 5%
FBS at 10-25 ng FGF & 10 μg/ml heparin; structurally &
functionally nearly identical to ECGF (endothelial cell growth
factor), eye-derived growth factor II, & RDGF (retina-derived
growth factor)

Biodesign A52117H   *E. coli*   MW 15.5k   Purified

Chemicon GF002   Human   ≥95%

Biogenesis 4460-4236   Human r-DNA   Lyophilized

BioSource International PHG0014   Human recombinant

Oncogene PF002   Human recombinant   MW 15.9k   >98%
(SDS-PAGE); lyophilized; biological activity: half maximal
stimulation of ^{3}H-thymidine uptake by 3T3 cells ~1 ng/mL |
Species reactivity: human; for proliferation studies & Western blot

Fitzgerald 30-AF16   Human recombinant, expressed in *E. coli*

Harlan BT-3002   Human recombinant, expressed in *E. coli*
Lyophilized; 0.025 mg | Cross-reactive with human, mouse, rat,
canine, feline, rabbit, bovine, equine, swine, primate, guinea pig,
ovine & avian

Harlan BT-3003   Human recombinant, expressed in *E. coli*
Lyophilized; 0.1 mg | Cross-reactive with human, mouse, rat,
canine, feline, rabbit, bovine, equine, swine, primate, guinea pig,
ovine & avian

ICN 153482   Human recombinant, expressed in *E. coli*   ≥98%
(SDS-PAGE); lyophilized; ED_{50} = 1.0-2.0 ng/mL (by ^{3}H-thymidine
uptake by Balb/c 3T3 cells) | Heparin binding protein which
stimulates the proliferation of a wide variety of cells

ICN 153509   Human recombinant, expressed in *E. coli*
Receptor grade; ≥96%; lyophilized carrier free; unspecified activity,
but measured by stimulation of ^{3}H-thymidine uptake by Balb/c 3T3
cells

ICN 154570   Human recombinant, expressed in *E. coli*   MW
~18k   ≥95%; lyophilized; ED_{50} = 0.5 ng/mL (by stimulation of
^{3}H-thymidine uptake by Balb/c 3T3 cells; <0.1 ng endotoxin/μg
protein | Goustin, AS etal, *Cancer Res*, 46:1015, 1986;
Gospodurowicz, D etal, *PNAS*, 81:6963, 1984; Ross, R etal, *Cell*,
46:155, 1986

PeproTech 100-17A   Human recombinant, expressed in *E. coli*
MW 15.8k   >95%; 140 AA; lyophilized from 0.1 *M* NaCl, 10 *mM*
Tris pH 7.6; ED_{50}: 5.0-10.0 ng/mL as determined by the
stimulation of thymidine uptake by BaF3 cells expressing the FGF
receptors | Heparin binding growth factor; stimulates proliferation
of a wide variety of cells including mesenchymal, neuroectodermal
& endothelial; 140 AA; $ED_{50} \le 10$ ng/mL; SA $\ge 10^5$ U/mg; SA
determined by stimulation of the proliferation of thymidine uptake
by BaF3 cells expressing FGF receptors

Sigma F 5542   Human recombinant, expressed in *E. coli*
$\ge$97% (SDS-PAGE); 0.2 µm filtered & lyophilized in PBS containing
1.25 mg BSA; proliferative activity is tested in culture using
quiescent NR6R-3T3 fibroblasts; endotoxin tested | Rizzino, A et
al, *Cancer Res*, 48: 4266, 1988; Fibroblast Growth Factor Acidic & –
Basic exert similar mitogenic actions on a variety of mesoderm-
derived cells, including BALB/3T3 fibroblasts, capillary endothelial
cells, myoblasts, vascular smooth muscle cells, mesothelial cells,
glial & astroglial cells & adrenal cortex cells; they aFGF & bFGF
share common cellular receptors, but differ in their specific
activities, depending on the individual cell type under study; unless
otherwise listed, the mitogenic activities of all FGF preparations are
tested in cell culture using fetal bovine heart endothelial cells;
Gospodarowicz, D et al, *Endocrine Rev*, 8: 95, 1987; Neufeld, G &
Gospodarowicz, D, *J Biol Chem*, 261: 5631, 1986; Lobb, R et al,
Anal Biochem, 154: 1, 1986

ICN 195781   Human recombinant, *N*-terminal extended form
expressed in *E. coli* $\ge$97%; lyophilized; ED_{50} = 0.1-0.3 ng/mL in
the presence of 10 µg/mL heparin

Fibroblast Growth Factor, Basic

Synonyms: Heparin Binding Growth Factor (II or β); Eye Derived
Growth Factor I; Cartilage Derived Growth Factor; Astroglial
Growth Factor II; β Heparin Binding Growth Factor (II or β)

Chemicon FA009-1 $\ge$95%

ICN 154147   Bovine Brain   >95% (by *N*-terminus analysis &
SDS-PAGE visualized by silver stain); lyophilized with BSA carrier;
ED_{50} = 0.05 - 0.3 ng/mL (by dose dependent stimulation of ^{3}H-
thymidine incorporation by NR6R 3T3 fibroblasts)

ICN 152323   Bovine Pituitary Glands   >95%; sterile frozen
solution | Heparin binding growth factor which exhibits mitogenic
activity for fibroblasts, glial cells, & some endothelial cells; may be
used to decrease or replace serum in cell cultures; facilitates early
passage of fibroblasts & aids in cell proliferation, differentiation &
development studies

Sigma F 5392   Bovine pituitary glands   MW 16-18k $\ge$90%
(SDS-PAGE); 0.2 µm filtered & lyophilized from a solution of
sodium phosphate & sodium chloride containing 100 µg BSA;
endotoxin tested; see Sigma F 5542 | Isolated by a modification
of the method of Gospodarowicz involving heparin affinity
chromatographyGospodarowicz, D et al, *Proc Natl Acad Sci USA*,
81: 6963, 1984

Biodesign A52118H   *E. coli*   MW 17k   Purified

Chemicon GF003   Human $\ge$95%

Biogenesis 4460-4252   Human r-DNA   Lyophilized

BioSource International PHG0024   Human recombinant

Fitzgerald 30-AF17   Human recombinant, expressed in *E. coli*

Harlan BT-3004   Human recombinant, expressed in *E. coli*
Lyophilized; 0.025 mg | Cross-reactive with human, mouse, rat,
canine, feline, rabbit, bovine, equine, swine, primate, guinea pig,
ovine & avian

Harlan BT-3005   Human recombinant, expressed in *E. coli*
Lyophilized; 0.1 mg | Cross-reactive with human, mouse, rat,
canine, feline, rabbit, bovine, equine, swine, primate, guinea pig,
ovine & avian

PeproTech 100-18B   Human recombinant, expressed in *E. coli*
MW 17.2k   >95%; 155 AA; lyophilized from 150 *mM* NaCl, 5 *mM*
Tris pH 7.6; ED_{50}: 0.5-1.0 ng/mL as determined by the stimulation
of thymidine uptake by BaF3 cells expressing the FGF receptors |
Heparin binding growth factor; stimulates proliferation of a wide
variety of cells including mesenchymal, neuroectodermal &
endothelial; 154 AA; $ED_{50} \le 0.5$ ng/mL; SA $\ge 2 \times 10^6$ U/mg; SA
determined by dose-dependent stimulation of the proliferation of
thymidine uptake by BaF3 cells expressing the FGF receptors

Sigma F 0291   Human recombinant, expressed in *E. coli*
$\ge$97% (SDS-PAGE); 0.2 µm filtered & lyophilized from 10 *mM* Tris,
pH 6.0, containing 1.25 mg BSA; proliferative activity is tested in
culture using fetal bovine heart endothelial cells; endotoxin tested;
see Sigma F 5542 | Gospodarowicz, D et al, *Endocrine Rev*, 8: 95,
1987

Sigma F 9786   Human recombinant, expressed in *E. coli*
Heparin stabilized; 0.67 µg/mL in high salt PBS; cell culture tested;
endotoxin tested; see Sigma F 5542 | Gospodarowicz, D et al,
Endocrine Rev, 8: 95, 1987

Oncogene PF003   Recombinant   MW 18k   >98% (SDS-
PAGE); lyophilized with 100 µg BSA; reconstitute with sterile dH_2O;
biological activity: half maximal stimulation of ^{3}H-thymidine uptake
by 3T3 cells ~0.3 ng/mL | Species reactivity: human; for
proliferation studies & Western blot

Fibroblast Growth Factor, Basic (^{125}I)-

Synonyms: Fibroblast Growth Factor Cytokine Induced Neutrophil
Chemoattractant

ICN 68104   Human recombinant   >50 µCi/µg, >18.5 MBq/µg;
0.1 *M* KPB, pH 7.5, 75 *mM* NaCl, 0.5% BSA

Fibroblast Growth Factor, Basic Human

IBT GF-030-3, GFR-030-5   Yeast   MW 17.5k   >97%;
lyophilized powder | Mitogen which stimulates cell growth for
fibroblasts, endothelial cells, myoblasts, glial cells & smooth muscle
cells; 154 AA residues; 55% homologous with acidic FGF, including
2 conserved Cys residues

Fibronectin

Synonyms: Globulin Protein, Cold Insoluble; Fibronectin, Cellular

BioSource International PHE0023

Biogenesis 4470-3729   Bovine plasma   Purified; from 0.05 *M*
Tris, 0.001 *M* $CaCl_2$, pH 7.4; lyophilized

Calbiochem 341631   Bovine plasma   MW 440k   Liquid in
150 *mM* NaCl, 20 *mM* sodium phosphate buffer, pH 7.3; single
band purity (SDS-PAGE) | Purified from pooled bovine plasma;
effective agent for promoting attachment of cells to commonly-
used culture substrates; *Merck Index*, 12: 4119

Chemicon FC014   Bovine plasma   Purified | Extracellular
matrix protein

ICN 150025   BOVINE PLASMA   Lyophilized | Used as an
attachment factor in cell culture work

Sigma F 1141   Bovine plasma   0.1% solution; 1 mg
protein/mL in 0.5 *M* NaCl, 0.05 *M* Tris, pH 7.5; cell culture tested |
Homogeneity is evaluated by immunoelectrophoresis

Sigma F 4759   Bovine plasma   Lyophilized from 0.05 *M* Tris
buffered saline, pH 7.5; cell culture tested | Homogeneity is
evaluated by immunoelectrophoresis

Biogenesis 4470-3724   Bovine serum   Purified; 30 *mM*
TRIS/Cl, pH 7.8 with 30% glycerol; liquid

Biogenesis 4470-4104   Equine serum

Cortex CP3108U   Human   >98%

ICN 55913   Human   Lyophilized | Purified antigen

ICN 158220   Human cellular   MW 271k (reduced SDS-PAGE)
or 542k   Highly purified; aseptically filled & lyophilized; free of
tenascin & other large proteins | A component of the extracellular
matrix; involved in attachment of cells to their substrate, long-term
studies of wound healing & prevention of spread of metastatic
tumor cells; unlike plasma fibronectin, this product is excreted into
the medium by human cellular fibroblasts

ICN 771111 ICN 771112   HUMAN CELLULAR   MW 256k
Highly purified | Cytoskeletal protein used as a high MW marker,
as a standard or an antigen for various research procedures

Calbiochem 341633   Human fibroblast   MW 550k   Solid
lyophilized from 100 *mM* NaCl, 50 *mM* sodium phosphate buffer, pH
7.5; $\ge$95% (SDS-PAGE); prepared from tissue of individuals shown
by certified tests to be negative for HBsAg & antibodies to HIV &
HCV | *Merck Index*, 12: 4119; Krejci, K & Fritz, H, *FEBS Lett*, 64:
152, 1976

Sigma F 2518   Human foreskin fibroblasts   Lyophilized in CAPS buffered saline; sterilized by chloroform dialysis | Homogeneity is evaluated by immunoelectrophoresis; differ from plasma fibronectin in the presence of additional polypeptide segments & in altering morphology of transformed cells & hemagglutination

Sigma F 6277   Human foreskin fibroblasts   Lyophilized from 0.05 M phosphate buffer, pH 7.5, containing 0.1 M NaCl; aseptically processed | Homogeneity is evaluated by immunoelectrophoresis

Biogenesis 4470-2809   Human plasma   Lyophilized

Calbiochem 341635   Human plasma   MW 440k   Liquid in PBS; single major band purity (SDS-PAGE); no contaminants detected by IEP with anti-fibronectin & anti-human serum; prepared from plasma shown by certified tests to be negative for HBsAg & antibodies to HIV & HCV | Effective agent for promoting attachment of cells to commonly-used culture substrates; *Merck Index*, 12: 4119

Chemicon FC010   Human plasma   Purified | Extracellular matrix protein

Chemicon FC010-10mg   Human plasma   Purified | Extracellular matrix protein

Chemicon FC010-5mg   Human plasma   Purified | Extracellular matrix protein

ICN 151126   Human plasma   Tissue culture grade; >95% (SDS-PAGE); lyophilized | Purified from the Cohn Fraction I of human plasma by a modification of the procedure of Engvall & Ruoslahti, *Int J Cancer*, 20:1, 1977

ICN 194072   Human plasma   Purified; <1 μg factor VIII & fibrinogen | Useful as a substrate for the promotion of attachment & replication of culture cells; Mosesson, MW, *Blood*, 56:145, 1980

ICN 194931   Human plasma   Lyophilized | Mosesson, MW & RA Umfleet, *JBC*, 245:5728, 1970

Promega G5291   Human plasma   Plasma-derived protein; plasma tested & found negative for HIV-1 antibody & Hepatitis B antigen | Used as an attachment for the culture of many cell types under reduced or serum-free conditions; promotes the attachment of fibroblasts & other mesenchymally derived cell types such as human umbilical vein endothelial, capillary endothelial, aortic endothelial & kidney epithelial cells

Sigma F 0895   Human plasma   0.1% solution; 1 mg protein/mL in 0.05 M Tris buffered saline, pH 7.5 | Homogeneity is evaluated by immunoelectrophoresis; source material tested negative for HBsAg, HCV & HIV antibody

Sigma F 2006   Human plasma   Lyophilized from 0.05 M Tris buffered saline, pH 7.5 | Homogeneity is evaluated by immunoelectrophoresis; source material tested negative for HBsAg & HIV antibody

Biogenesis 4470-2807   Human serum   Liquid

Biogenesis 4470-4504   Mouse plasma   Purified; 50 mM TRIS buffer, pH 7.4, 1 mM CaCl2; lyophilized | Engvall & Ruoslahti, *J Cancer*, 20:1, 1977

Calbiochem 341655   Mouse plasma   MW 220k   Liquid in 150 mM NaCl, 20 mM sodium phosphate buffer, pH 7.3; single major band purity (SDS-PAGE); no contaminants detected by IEP with anti-fibronectin & anti-mouse serum | Effective agent for promoting attachment of cells to commonly-used culture substrates; *Merck Index*, 12: 4119

Chemicon FC015   Murine plasma   Purified | Extracellular matrix protein

Calbiochem 341650   Rabbit plasma   MW 220k   Liquid in 150 mM NaCl, 20 mM sodium phosphate buffer, pH 7.3; single major band purity (SDS-PAGE) | Purified from pooled rabbit plasma; effective agent for promoting attachment of cells to commonly-used culture substrates; *Merck Index*, 12: 4119

Biogenesis 4470-4904   Rat plasma   Purified; 50 mM Tris buffer, pH 7.4 with 1 mM CaCl$_2$; aseptic; lyophilized | Engvall & Ruoslahti, *J Cancer*, 20:1, 1977

Calbiochem 341668   Rat plasma   MW 220k   Liquid in 100 mM NaCl, 20 mM sodium phosphate buffer, pH 7.3, preservative & reductant-free; single major band purity (SDS-PAGE); no contaminants detected by IEP with anti-fibronectin & anti-rat serum | Purified from pooled rat plasma; effective agent for promoting attachment of cells to commonly-used culture substrates; *Merck Index*, 12: 4119

Sigma F 0635   Rat plasma   Lyophilized from 0.05 M Tris buffered saline, pH 7.5 | Homogeneity is evaluated by immunoelectrophoresis

Fibronectin Fragment III$_1$-C

Sigma F 3542   Human recombinant, expressed in *E. coli*   MW ~7k   >95% (SDS-PAGE); lyophilized, essentially salt-free; sterile-filtered | Promotes cross-linking of fibronectin to from matrix fibrile-like multimers

Fibronectin Proteolytic Fragment

Synonyms: Gelatin Binding Fragment; Heparin & Gelatin Binding Fragment; Fibronectin, 70 k Fragment; Heparin Binding Fragment; Fibronectin, 30 k Fragment

Sigma F 0162   Human plasma fibronectin   >90% (SDS-PAGE); lyophilized from phosphate buffered saline with sucrose as a cryoprotectant | Used for mapping regions, functions & activities of fibronectin; all fragments are purified by affinity chromatography; small proteolytic fragments may be present; source material tested negative for HBsAg & HIV

Sigma F 0287   Human plasma fibronectin   >90% (SDS-PAGE); lyophilized from phosphate buffered saline with sucrose as a cryoprotectant | Used for mapping regions, functions & activities of fibronectin; all fragments are purified by affinity chromatography; small proteolytic fragments may be present; source material tested negative for HBsAg & HIV

Sigma F 9911   Human plasma fibronectin   MW 30k   >90% (SDS-PAGE); lyophilized from phosphate buffered saline with sucrose as a cryoprotectant | Used for mapping regions, functions & activities of fibronectin; all fragments are purified by affinity chromatography; small proteolytic fragments may be present; source material tested negative for HBsAg & HIV

Fibronectin, (^{125}I)-

ICN 68066   ~1 μCi/μg, ~37 kBq/μg; 0.1 M KPO$_4$, pH 7.5, 0.5% BSA

Fibronectin, Cell Attachment

Chemicon F1904   MW 120k   Purified | Extracellular matrix protein

Fibronectin, Cell Binding Fragment

USBio F4310   Human serum   MW ~110k   Single band on SDS-PAGE; lyophilized from 0.02 M Tris, pH 7.4, 0.15 M NaCl | Exists in 3 forms: a soluble dimeric form (plasma fibronectin), oligomers which are transiently attached to the cell surface (cell-surface fibronectin) & highly insoluble fibrils in the extracellular matrix (matrix fibronectin); the 110kD cell-binding fragment serves as a good substrate for cells that attach via the a5b1 integrin receptor & other integrin receptors that recognize the Arg-Gly-Asp sequence; does not contain the Hep-2/CS1 region recognized by a4b1 integrin

Fibronectin, Cellular

Calbiochem 341658   Mouse fibroblast   MW 550k   Solid lyophilized from 100 mM NaCl, 50 mM sodium phosphate buffer, pH 7.5 | *Merck Index*, 12: 4119; Krejci, K & Fritz, H, *FEBS Lett*, 64: 152, 1976

Fibronectin, Heparin Binding

Chemicon F1903   MW 40k   Purified | Extracellular matrix protein

Fibronectin-Like Engineered Protein Polymer

Sigma F 5022   MW 72,738 (gene sequence), ~110k (SDS-PAGE)   Lyophilized polymer supplied with diluent in separate vial; diluent contains 4.5 M LiClO$_4$; sterilized by autoclaving; cell culture tested | Recombinant polymer that incorporates multiple copies of the RGD attachment ligand of human fibronectin interspaced between repeated structural peptide units; U.S. Patent No. 5,514,581

Fibronectin-Like Engineered Protein Polymer Plus

Sigma F 8141 Lyophilized; sterilized by autoclaving; cell culture tested | A positively charged water soluble recombinant polymer similar to Sigma F 5022; U.S. Patent No. 5,514,581

Fibronogen

Calbiochem 341573 Bovine plasma MW 330k Solid lyophilized from 150 mM NaCl, 20 mM citrate, 10 mM phosphate buffer, pH 7.4; clottable proteins: >95%; purity homogenous (SDS-PAGE); soluble in water | Plasma glycoprotein synthesized & secreted by hepatic parenchymal cells; Merck Index, 12: 4116

Calbiochem 341576 Human plasma MW 341k Solid lyophilized from 150 mM NaCl, 20 mM citrate, 10 mM phosphate buffer, pH 7.4; clottable proteins: >90%; purity homogenous (SDS-PAGE); soluble in water; prepared from plasma shown by certified tests to be negative for HBsAg & antibodies to HIV & HCV | Plasma glycoprotein, essential for clotting of blood, synthesized & secreted by hepatic parenchymal cells; Merck Index, 12: 4116

Fibronogen Fragment D

Calbiochem 341600 Human plasma MW 85k Lyophilized from 150 mM NaCl, 400 mM glycine; single-band purity (SDS-PAGE); ≤1% Fragment E; soluble in water; prepared from plasma shown by certified tests to be negative for HBsAg & antibodies to HIV & HCV | Thermolabile fragment; complete proteolysis of single fibrinogen molecule produces two D & one E fragments; D & E fragments have no common antigenic determinants; therefore, no cross-reaction occurs

Fibronogen Fragment E

Calbiochem 341605 Human plasma MW 50k Lyophilized from 0.9% NaCl, 3% glycine; single-band purity (SDS-PAGE); ≤2% Fragment D (antigen determination); soluble in water; prepared from plasma shown by certified tests to be negative for HBsAg & antibodies to HIV & HCV | Thermostable fragment

Fibronogen, Plasminogen Depleted

Calbiochem 341578 Human plasma MW 341k Solid lyophilized from 150 mM NaCl, 20 mM citrate, 10 mM phosphate buffer, pH 7.4; clottable proteins: >95%; purity homogenous (SDS-PAGE); soluble in water; prepared from plasma shown by certified tests to be negative for HBsAg & antibodies to HIV & HCV

Filamin

Biodesign A08006C Chicken gizzard >90%

FK Binding Protein

Sigma F 5398 Human recombinant, expressed in E. coli Enzyme which catalyzes cis-trans isomerization of X-Pro peptide bonds (ie a peptidyl prolyl isomerase) in synthetic substrates; FK binding protein characterized by binding to, & inhibition by, the immunosuppressant, FK-506; Fischer, G et al, Biomed Biochim Acta, 43: 1101, 1984; Handschumacher, RE et al, Science, 226: 544, 1984

Flagellin Antigen, *Helicobacter pylori*

IBT HPA-5040-4, HPA-5040-5 E. coli MW 50k >92%; .50 mg/mL solution 0 in PBS, 0.1% SDS, 1 mM EDTA, pH 7.4 | Covers Met[1] to Thr[511] of H. pylori flagellin antigen

Flavoridin Disintegrin

Sigma F 0412 Trimeresurus flavoviridis >95% (SDS-PAGE); lyophilized; sterilized by γ-irradiation | Disintegrins represent a novel family of integrin β1 & β3 inhibitor proteins isolated from viper venoms; low molecular weight, cysteine-rich peptides containing the Arg-Gly-Asp (RGD) sequence; the most potent known inhibitors of integrin function; they interfere with cell adhesion to the extracellular matrix including adhesion of melanoma cells & fibroblasts to fibronectin & are potent inhibitors of platelet aggregation

Flt-3 Ligand

Synonyms: Flk-2 Ligand

Biodesign A52019H E. coli MW 17.6k Purified

Chemicon GF038 Human ≥95%

BioSource International PHC1124 Human recombinant

ICN 195004 Human recombinant, expressed in E. coli MW 17k Lyophilized

PeproTech 300-19 Human recombinant, expressed in E. coli MW 17.6k >98%; 155 AA; lyophilized with no additives; ED50: 1.0-5.0 ng/mL as determined by the stimulation of Balb/c bone marrow cells that have been depleted of T cells & B cells

Flt-3 Ligand 78

Synonyms: Tyrosine II Ligand, Fms-Like

Biogenesis 0100-0140 Human r-DNA Lyophilized

Flt-3/Flk-2 Ligand

Sigma F 3422 Human recombinant, expressed in NSO cells Lyophilized from 40% acetonitrile/0.1% TFA containing 0.25 mg BSA; cell culture tested; endotoxin tested

Folate Binding Protein

Fitzgerald 30-AF20 Bovine High purity

USBio F5750 Bovine ≥98%; no contaminants detected; single band by SDS-PAGE, IEP, &/or RID; ~1-2 mg/mL supplied in PBS buffer containing 0.1 M sodium phosphate, 0.03 M NaCl, pH7.5 | Suitable for antigenic applications in immunological protocols

Biogenesis 4550-1004 Bovine milk Liquid

Fluka 47605 Bovine milk ~30% protein; binding capacity: 1 μg protein binds ~2 mg folic acid at pH 7.5, 25°C

Sigma F 0504 Bovine milk Purified over 1000 fold; lyophilized powder containing ~30% protein (modified Warburg-Christian); balance primarily buffer salt as sodium citrate; may contain trace BME; binding capacity: 1 mg protein binds 1.0-3.0 μg of folic acid at 25°C at pH 7.5; 1 mg protein sufficient for ~4000 folate assays | May also bind N-methyltetrahydrofolic acid & other folate analogues; this protein may be used to replace the crude preparations previously used as the source for this binding ligand in radioassays for serum folate; Waxman, S et al, Blood, 38: 219, 1971; Rothenberg, SP et al, New England J Med, 286: 1335, 1972; Dunn, RT & Foster, LB, Clin Chem, 19: 1101, 1973

Cortex CP8104 Human >95%

Follicle Stimulating Hormone

Cortex CP1027 >95%

Cortex CP1027P >40%

Cortex CP1027U >98%

USBio F5900-32 Bovine ≥98% (SDS-PAGE); bovine TSH: <0%, bovine LH: <0%, bovine GH: <0%, bovine Prl: <0%; ≥ 2 IU/mg (25 mg/vial); lyophilized | Suitable for antigenic applications in immunological protocols

Biogenesis 4561-6604 Bovine pituitary MW 34k <1% bLH; purified; 0.05 M NH4HCO3, pH 8.0, essentially salt free; lyophilized

Fitzgerald 30-AF28 Bovine pituitary Immunization grade

Biogenesis 4561-6804 Equine pituitary MW 34k ≤1% eLH/eTSH; purified; 0.05 M NH4HCO3, pH 8.0; lyophilized

Biogenesis 4560-6104 Human pituitary Lyophilized

Biogenesis 4560-6139 Human pituitary Tested negative for HCV, HIV-1 and HIV-2 antibodies and HBsAg; semi-pure; 210 IU/vial hLH, 230 mIU/vial hTSH; 0.05 M NH4HCO3, pH 8.0; lyophilized

Biogenesis 4560-6204 Human pituitary MW 34k 3000 IU/mg; <1% hLH/hTSH/hCG, <0.1% hGH, <0.01% hPRL; tested negative for HBsAg, HCV, HIV 1 and HIV 2; purified; 0.05 M NH4HCO3, pH 8.0; lyophilized

Biogenesis 4560-6404 Human pituitary

Calbiochem 869001   Human pituitary   MW 35.5k   Iodination grade; lyophilized solid; immunopotency: ≥2000 IU/mg (WHO 1st IRP 69/104); soluble in water; hLH, hTSH: ≤0.5%; hCG: ≤0.1%; negative for HBsAg & for antibodies to HIV & HCV; may be carcinogenic/teratogenic | Two-chain glycoprotein gonadotropic hormone that induces maturation of the Graafian follicles of the ovary; Promotes the development of germinal cells in males; FSH regulates Sertoli cells by acting on G-protein-linked cell surface FSH receptors; activates cytosolic soluble protein tyrosine kinase (CyPTk); *Merck Index*, 12: 4299; Costrici, N et al, *Endocrinology*, 136: 4705, 1995

Sigma F 4021   Human pituitary   RIA activity: ~7000 IU/mg based on IRP 68/140; also contains 0.1 mg salts from 0.05 M phosphate buffer, pH 7.4 | Two-chain glycoprotein hormone; α-chain isn't active, biological specificity is attributed to the β-chain; induces maturation of Graafian follicles of the ovary; promotes development of germinal cells in males; activates cytosolic tyrosine kinase; bioassay not run by Sigma

Biogenesis 4561-7004   Ovine pituitary   MW 34k   <2% ovine LH, <1% ovine PRL; purified; essentially salt free; lyophilized

Biogenesis 4561-7204   Porcine pituitary   MW 34k   <0.3% pPRL, <0.2% pLH/pTSH, <0.1% pGH; purified; from 0.05 M NH_4HCO_3, pH 8.0, essentially salt free; lyophilized | Closset & Hennen, *Eur J Biochem,* 86:105, 1978

Calbiochem 344115   Porcine pituitary   MW 36k   Lyophilized solid; FSH/LH ratio: 500:1 (RIA); 1 U = activity in 1 U of the NIH FSH S1 standard; harmful: LD_{50}≤2000 mg/kg; may be carcinogenic/teratogenic | *Merck Index*, 12: 4299

Sigma F 2293   Porcine pituitary   50 units/vial by the Steelman-Pohley assay using the Armour FSH Standard G-94 | Steelman, SL & Pohley, FM, *Endocrinology*, 53: 604, 1953

Biogenesis 4561-7404   Rat pituitary   <1.0% rLH/rTSH, <0.1% rPRL/rGH; purified; 0.05 M NH_4HCO_3, pH 8.0; lyophilized | Boujon et al, *J Comp Path*, 109:163, 1993

Sigma F 4520   Sheep pituitary   50 units/vial by the Steelman-Pohley assay using NIH-FSH-S17 reference standard; Steelman, SL & Pohley, FM, *Endocrinology*, 53: 604, 1953 | Two-chain glycoprotein hormone; α-chain isn't active, biological specificity is attributed to the β-chain; induces maturation of Graafian follicles of the ovary; promotes development of germinal cells in males; activates cytosolic tyrosine kinase; bioassay not run by Sigma

Follicle Stimulating Hormone, Intact

USBio F5900-25   Human   ≥98% (SDS-PAGE); hLH <1.0%, hTSH <0.2%, hCG, hGH, hPrl <0.1%; ≥ 3800 IU/mg (2nd IRP 78/549, WHO); lyophilized from human pituitary glands; highly purified | Suitable for antigenic applications in immunological protocols

Fitzgerald 30-AF25   Human pituitary   Standard purity

USBio F5900-24   Human pituitary   ~95% (SDS-PAGE); hLH ≤1.0%, hTSH ≤1.0%, hCG/hGH/hPrl ≤0.01%; ≥1000 IU/mg (2nd IRP 78/549, WHO); lyophilized in 50 mM ammonium bicarbonate | Suitable for antigenic applications in immunological protocols

USBio F5900-23   Human pituitary glands   ≥99% (SDS-PAGE); hLH <1.0%, hTSH <0.5%, hCG/hGH/hPrl <0.1%; 4400 IU/mg; lyophilized from 0.05 M ammonium bicarbonate; affinity purified | Suitable for antigenic applications in immunological protocols

Fitzgerald 30-AF31   Postmenopausal urine   High purity

Follicle Stimulating Hormone, α-

Biogenesis 4561-6629   Bovine pituitary   Purified; lyophilized

USBio F5900-27   Human   ≥98% (SDS-PAGE); hFSH beta: <2%; lyophilized | Suitable for antigenic applications in immunological protocols

Fitzgerald 30-AF40   Human pituitary   Affinity purity

Biogenesis 4561-7054   Ovine pituitary   Lyophilized

Biogenesis 4561-7229   Porcine pituitary   Purified; lyophilized

Biogenesis 4561-6649   Bovine pituitary   Purified; lyophilized

Biogenesis 4560-8204   Human pituitary   MW 18.5k   Tested negative for HBsAg and HTLV III antibody; *Endocrinology*, (1984) 114, 2223-2227; purified; from 0.05 M NH_4HCO_3, pH 8.0, essentially salt free; lyophilized

Fitzgerald 30-AF45   Human pituitary   Affinity purity

USBio F5900-29   Human pituitary glands   ≥98% by HPLC; hFSH Alpha: <0.13%; hLH <0.2%; lyophilized in 50 mM ammonium bicarbonate | Suitable for antigenic applications in immunological protocols

Biogenesis 4561-7084   Ovine pituitary   Lyophilized

Biogenesis 4561-7249   Porcine pituitary   Purified; lyophilized

Forskolin

USBio F6025   >99% by HPLC; lyophilized; soluble in DMSO (5 mg/mL) & ethanol (6 mg/mL), insoluble in water | Forskolin is an activator of adenylate cyclase, EC50=4mM, leading to an increase in the intracellular concentration of cAMP

Fractalkine

Synonyms: CX3C

Biodesign A52331H   *E. coli*   MW 8.5k   Purified | Chemokine

PeproTech 300-31   Human recombinant, expressed in *E. coli*   MW 8.5k   >98% (SDS-PAGE) & HPLC; <0.1 ng endotoxin per mg (1EU/mg); lyophilized | Member of the newly discovered CX3C family of chemokines which contains both chemokine & mucin domain; 76 AA, comprising only the chemokine domain of Fractalkine; SA determined by its ability to chemoattract human T cells

Fyn

USBio F9500-06   Bovine thymus membrane   SA: 1 U/μL; essentially free of other proteinkinase contamination; frozen solution in 0.1% NP-40 with 10% glycerol | Transfers PO-2kinase to cdc2 (6–20) peptide; useful in the Protein Kinase Assay or to prepare phosphotyrosyl peptides; partially purified by DEAE-sepharose, hydroxyapatite & phenyl sepharose columns, followed by Sephacryl S-200 gel filtration

F-β1 Cytokine

Synonyms: Transforming Growth Factor βI

Alexis BMS307   Human recombinant   >98% (SDS-PAGE & HPLC); 1 μg lyophilized from 5 mM HCl containing 10 μg BSA; biological activity: stimulates the anchorage independent colony formation of NRK49F cells in soft agar at 0.1-2.0 ng/mL, in the presence of 2 ng/mL of EGF or TGF-β | Produced by culturing 293 (transformed primary human embryonal kidney) cells transfected with a recombinant plasmid containing the hTGF-β1 cDNA; inhibits the growth of epithelial cells (e.g. hepatocytes or mink lung cells)

G Protein, Gαγ

Chemicon GPR002   Recombinant   ≥95% | Purified protein for apoptosis & signal transduction

G Protein, Gβγ Dimer

Chemicon AG600   Bovine brain   ≥95% | Purified protein for apoptosis & signal transduction

Galactosidase, β-

Sigma G 7279   *E. coli*   MW 116k   Vial contains enough FITC-conjugated protein to run 50 mini-gels or 25 standard size gels; protein band be visualized by using UV light or Brilliant Blue stain | Fluorescent marker for SDS-page & protein transfer; suitable for use as molecular weight standards in both SDS-PAGE & transfer membranes

Galaptin

Sigma G 8777   Bovine spleen   Lyophilized containing ~10% protein (Bradford); balance NaCl, Tris buffer salts & DTT | Lectin; endogenous mammalian galactoside-binding lectin; allen, HJ et al, *Arch Biochem Biophys*, 256: 523, 1987

Gamma Globulin

USBio G2000 Bovine Total protein >95%; heavy metals: <500 ppm; moisture: 3.0 %; pH (1.0%, 0.15 M NaCl): 7.1; lyophilized | Suitable for antigenic applications in immunological protocols;

Gamma Globulin Fraction II

USBio G2000-10 Human Manufactured from human plasma collected & tested in the US.

GCP-2

Biodesign A52041H *E. coli* Purified

Chemicon GF063 Human ≥95%

PeproTech 300-41 Human recombinant, expressed in *E. coli* >98% (SDS-PAGE) & HPLC; <0.1 ng endotoxin per mg (1EU/mg); lyophilized | Member of the CXC family of chemokines; promotes neutrophil chemotaxis & degranulation; 73 AA; SA determined by its ability to chemoattract human neutrophils

Gelatin

Fluka 04055 Porcine skin Powder; ≤15% loss on drying; ≤2% residue on ignition; ≤0.005% heavy metals; ≤0.02% SO₂; ≤0.01% peroxides; ≤0.0001% As

Gelatin, Agarose

ICN 191300 5 atoms hydrophilic spacer arm; 3-6 mg gelatin/mL gel; suspension in PBS, 0.02% NaN₃ | Fibronectin applications

Gelatin, Amplification Grade

Amersham US70086 Porcine skin Enzyme stabilizer; functionally tested in PCR

Gelatin, European Pharmacopoeia

Fluka 18808 White, foil, "Gold" | Meets analytical specification of European Pharmacopoeia

Fluka 33223 Foil; ≤15% loss on drying; ≤2% residue on ignition; ≤0.005% heavy metals; ≤0.02% SO₂ | Meets European Pharmacopoeia for bacteriology

Fluka 48723 Gelatina | Conforms to European Pharmacopoeia

Gelatin, Glycerol-

Fluka 49927 Porcine skin 88.7 g/L gelatin from porcine skin, 554 mL/L glycerol, 10.8 g/L phenol | For mounting histochemical slides

Gelatin, High Gel Strength

Fluka 48724 Porcine skin 250 g Bloom; 240-270 g Bloom gel strength; ≤15% loss on drying; ≤2% residue on ignition; ≤0.005% Fe, Cu; ≤0.2% Cl, Ca, Na; ≤0.001% Cr, Zn; ≤0.0005% Cd, Co, Mn, Ni, Pb; ≤0.05% K, Mg | For microbiology; Busk Jr, GC, *Food Tech*, 38: 59, 1984; Kozlov, PV & Burdygina, GI, *Polymer*, 24: 651, 1983

Gelatin, Low Gel Strength

Fluka 48719 Porcine skin 80 g Bloom; 70-90 g Bloom gel strength; ≤15% loss on drying; ≤2% residue on ignition; ≤0.005% Fe, Cu; ≤0.2% Cl, Ca, Na; ≤0.001% Cr, Zn; ≤0.0005% Cd, Co, Mn, Ni, Pb; ≤0.05% K, Mg | For microbiology

Fluka 48720 Porcine skin 60 g Bloom; 60-80 g Bloom gel strength; ≤15% loss on drying; ≤2% residue on ignition; ≤0.005% Fe, Cu; ≤0.2% Cl, Ca, Na; ≤0.001% Cr, Zn; ≤0.0005% Cd, Co, Mn, Ni, Pb; ≤0.05% K, Mg | For microbiology

Gelatin, Medium Gel Strength

Fluka 48722 Porcine skin 180 g Bloom; 170-190 g Bloom gel strength; ≤15% loss on drying; ≤2% residue on ignition; ≤0.005% Fe, Cu; ≤0.2% Cl, Ca, Na; ≤0.001% Cr, Zn; ≤0.0005% Cd, Co, Mn, Ni, Pb; ≤0.05% K, Mg | For microbiology; Levine, M & Carpenter, DC, *J Bact*, 8: 297, 1923

Gelatin, Medium Inositol

Fluka 17155 Composition: 120 g/L gelatin, 5 g/L yeast extract, 5 g/L disodium hydrogen phosphate, 10 g/L inositol, 0.05 g/L phenol red | For cultivation of *Plesiomonas shigelloides* from foods

Gelatin, Phosphate Buffer

Fluka 48726 4 g/L Sodium dihydrogen phosphate & 2 g/L gelatine, pH 6.2 | For microbiology; for qualitative toxin detection in food products eg. when *Cl botulinum* is suspected; Rose, SA et al, *J Appl Bact*, 65: 223, 1988

Gelatin, Phosphate Salt/Agar

Fluka 17149 10 g/L Gelatin, 10 g/L NaCl, 5 g/L dipotassium phosphate & 15 g/L agar, pH 7.2 | GPS Agar; for microbiology; for characterization of *Vibrio cholerae* from food

Gelatin, Salt/Agar

Fluka 17150 15 g/L Gelatin, 30 g/L NaCl, 4 g/L peptic digest of animal tissue, 1 g/L yeast extract & 15 g/L agar, pH 7.2 | For microbiology; for the cultivation & differentiation of *Vibrio* species from food

Gelatin, Teleostean

Sigma G 7765 Cold water fish skin ~45% aqueous solution with 0.15% propyl *p*-hydroxybenzoate & 0.2% methyl *p*-hydroxybenzoate as preservatives

Gelatin, Type A

Synonyms: Prionex

ICN 960102 Food grade; flaked 50 bloom

ICN 960317 Food grade; 100 Bloom

Sigma G 0411 Highly purified; aqueous solution containing ~10% protein (Biuret); aseptically processed | Protein stabilizer offered as an alternative to BSA & HSA

ICN 901771 Bovine skin Food grade; 225 bloom

Sigma G 1890 Porcine skin ~300 Bloom; derived from acid-cured tissue; cell culture tested | The higher the Bloom number the stronger the gel; type A is derived from acid-cured tissue, Type B from lime-cured tissue

Sigma G 2500 Porcine skin ~300 Bloom | The higher the Bloom number the stronger the gel; Type A is derived from acid-cured tissue, Type B from lime-cured tissue

Sigma G 2625 Porcine skin ~175 Bloom | The higher the Bloom number the stronger the gel; Type A is derived from acid-cured tissue, Type B from lime-cured tissue

Sigma G 6144 Porcine skin ~75-100 Bloom | The higher the Bloom number the stronger the gel; Type A is derived from acid-cured tissue, Type B from lime-cured tissue

Sigma G 8150 Porcine skin ~300 Bloom; protease: None detected | Useful as a blocking reagent for Western blots

Sigma G 9136 Porcine skin ~300 Bloom; lyophilized; sterilized by γ-irradiation; derived from acid-cured tissue; cell culture tested | The higher the Bloom number the stronger the gel; Type A is derived from acid-cured tissue, Type B from lime-cured tissue

Gelatin, Type B

Sigma G 1393 Bovine skin 2% Solution; prepared in tissue culture grade water; derived from lime-cured tissue; sterilized by autoclaving; endotoxin tested; cell culture tested | The higher the Bloom number the stronger the gel; Type A is derived from acid-cured tissue, Type B from lime-cured tissue

Sigma G 6650 Bovine skin ~75 Bloom | The higher the Bloom number the stronger the gel; Type A is derived from acid-cured tissue, Type B from lime-cured tissue

Sigma G 9382 Bovine skin ~225 Bloom | The higher the Bloom number the stronger the gel; Type A is derived from acid-cured tissue, Type B from lime-cured tissue

Sigma G 9391   Bovine skin   ~225 Bloom; derived from lime-cured tissue; cell culture tested | The higher the Bloom number the stronger the gel; Type A is derived from acid-cured tissue, Type B from lime-cured tissue

Gelonin

Sigma G 2394   *Gelonium multiflorum*   Lyophilized powder containing ~90% protein (E_{280} at 1%=6.7); balance primarily NaCl & sodium phosphate buffer salts | Ribosome inactivating protein; Stirpe, F et al, *J Biol Chem*, 255: 6947, 1980; Stirpe F & Barbieri, L, *FEBS Lett*, 195: 1, 1986

Gelsolin

Biogenesis 4628-4030   Bovine plasma   5% protein, remainder buffer salts; purified; NaCl, Tris buffer salt & EGTA; lyophilized

Sigma G 8032   Bovine plasma   >95% (SDS-PAGE); lyophilized powder containing ~5% protein; balance NaCl, Tris buffer salt & EGTA; activity: 20-100 U/mg protein; 1 U reduces viscosity difference between actin solution & buffer by 50% in a 1 mL mixture containing 1-2 mg F-actin, 0.15 M KCl, 20 mM Tris, pH 7.6, 0.2 mM $CaCl_2$, 0.2 mM ATP & 1 mM DTT at 28°C | Actin-severing protein found in mammalian cells & blood plasma; reported to decrease the viscosity of cystic fibrosis sputum samples *in vitro* & to suppress human bladder cancer; Wen, D et al, *Biochemistry*, 35: 9700, 1996; Vasconcellos, CA et al, *Science*, 263: 969, 1994; Tanaka, M et al, *Cancer Res*, 55: 3228, 1995

Sigma G 1538   Human plasma   ~90% (SDS-PAGE); lyophilized powder containing ~10% protein (TCA-Lowry); balance Tris buffer salts, pH 7.6, NaCl, DTT & EGTA; activity: 100-300 U/mg protein; 1 U reduces viscosity difference between actin solution & buffer by 50% in a 1 mL mixture containing 1-2 mg F-actin, 0.15 M KCl, 20 mM Tris, pH 7.6, 0.2 mM $CaCl_2$, 0.2 mM ATP & 1 mM DTT at 28°C | Actin-severing protein found in mammalian cells & blood plasma; reported to decrease the viscosity of cystic fibrosis sputum samples *in vitro* & to suppress human bladder cancer; Wen, D et al, *Biochemistry*, 35: 9700, 1996; Vasconcellos, CA et al, *Science*, 263: 969, 1994; Tanaka, M et al, *Cancer Res*, 55: 3228, 1995

Gelsolin, FITC Conjugated

Sigma G 1158   Bovine plasma   >95% (SDS-PAGE); lyophilized powder containing ~5% protein (Lowry); balance NaCl, Tris buffer salt & EGTA; prepared form gelsolin purified from bovine plasma; 2-6 moles FITC/mole protein

GITR Ligand

Synonyms: TNFRSF18

R&D Systems 694-GL-025   *SF*21-Expressed   97%; lyophilized | Species specificity: human GITR ligand

Gliadin

ICN 101778   Wheat gluten   A prolamin

Sigma G 3375   Wheat gluten   Crude

Glial Fibrillary Acidic Protein

ICN 771062   MW 52k   Highly purified | High MW marker, standard or antigen for various research purposes

Biodesign A08007B   Bovine spinal cord   Purified

Biodesign A86823H   Human brain   Purified

Calbiochem 345996   Human brain   MW 55k   ≥90% (SDS-PAGE); lyophilized solid; soluble in water; prepared from tissue of individuals shown to be negative for HBsAg & for antibodies to HIV & HCV | Marker of sclerotic plaques in brain & gliosed brain; component of astroglial intermediate filaments; Liem, RK & Hutchison, SB, *Biochemistry*, 21: 3221, 1982; Massod, K et al, *J Neurochem*, 61: 160, 1993; Dahl, D, *Brain Res*, 57: 343, 1973

Biogenesis 4650-0717   Human normal brain   MW 43-49k (several bands)   Tested negative for antibodies to HIV 1 and 2, and HBsAg; purified; sodium phosphate 2.0 *mM*, 1.33 *M* urea, 27 *mM* sodium bicarbonate pH 8.5; liquid | Dahl, *Brain Res*, 57:343, 1973; Liem, *Biochem*, 21:3221, 1982

Globin

Sigma G 3633   Human hemoglobin   Dialyzed & lyophilized; A_{405}/A_{280} ~0.1; recombines with hematin in the presence of 0.1 M Tris, pH 11.0 | Ascoli, F et al, *Meth Enzymol*, 76: 75, 1981

Globin, Gc-

ICN 153573   Human plasma   MW 52k   >98%; mixed type; lyophilized

Globulin

Sigma EG   Chicken egg white   Contains several globulins as demonstrated by strip electrophoresis; substantially free of albumin

Sigma G 3884   *Cucurbita pepo* (pumpkin)   2X Crystallized & lyophilized | Possible substitute for edestin (a globulin from hemp seed)

Globulin Cohn Fraction II&III

Sigma G 2263   Bovine   Lyophilized powder containing ~75% protein (Biuret); balance primarily NaCl; primarily β- & γ-globulins

Sigma G 6765   Dog   Lyophilized powder containing ~75% protein (Biuret); balance primarily NaCl; primarily γ-globulins

Sigma G 5640   Goat   Lyophilized powder containing ~75% protein (Biuret); balance primarily NaCl; primarily γ-globulins

Sigma G 6015   Horse   Lyophilized powder containing ~75% protein (Biuret); balance primarily NaCl; primarily γ-globulins; remainder mostly β-globulins

Sigma G 2388   Human   Lyophilized powder containing ~75% protein (Biuret); balance primarily NaCl; primarily γ- & β-globulins | Source material negative for HIV & HBsAg

Sigma G 4390   Pig   Lyophilized powder containing ~75% protein (Biuret); balance primarily NaCl; primarily γ-globulins

Sigma G 4765   Rabbit   Lyophilized powder containing ~75% protein (Biuret); balance primarily NaCl; primarily γ-globulins

Sigma G 4890   Rat   Lyophilized powder containing ~75% protein (Biuret); balance primarily NaCl; primarily β- & γ-globulins

Sigma G 4265   Sheep   Lyophilized powder containing ~75% protein (Biuret); balance primarily NaCl; primarily γ-globulins

Globulin Cohn Fraction II, γ-

Fluka 49030   Bovine blood   MW 150k   ≥95% (GE); ≤5% loss on drying; ≤2% residue on ignition

Globulin Cohn Fraction III

Sigma G 4633   Bovine   Lyophilized powder containing ~75% protein (Biuret); balance primarily NaCl; predominantly β- & γ-globulins; plasminogen activity

Sigma G 6763   Human   Predominantly β- & γ-globulins; plasminogen activity | Source material negative for HIV & HBsAg

Globulin Cohn Fraction IV

Sigma G 9762   Bovine   Coprecipitation of Cohn Fractions IV-1 & IV-4; predominantly α- & β-globulins; remainder ~30% albumin

Globulin Cohn Fraction IV-1

Sigma G 8512   Bovine   Predominantly α-globulins

Sigma G 7015   Dog   Predominantly α–globulins

Sigma G 5390   Goat   Predominantly α–globulins

Sigma G 5765   Horse   Predominantly α–globulins

Sigma G 2011   Human   Predominantly α-globulins | Source material negative for HIV & HBsAg

Sigma G 6140   Pig   Predominantly α–globulins

Sigma G 5140   Rat   Predominantly α–globulins

Globulin Cohn Fraction IV-4

Sigma G 8637   Bovine   Predominantly α- & β-globulins; remainder ~30% albumin

Sigma G 6890   Dog   Predominantly α- & β-globulins

Sigma G 5890	Horse	Predominantly α- & β-globulins
Sigma G 3387	Human	Coprecipitation of Cohn Fractions IV-1 & IV-4; predominantly α- & β-globulins
Sigma G 3637	Human	Predominantly α- & β-globulins | Source material negative for HIV & HBsAg
Sigma G 6265	Pig	Predominantly α-globulins
Sigma G 4640	Rabbit	~50% α- & β-globulins; remainder mostly albumin
Sigma G 5015	Rat	Predominantly α-globulins
Sigma G 4140	Sheep	Predominantly β-globulins

Globulin Fraction II, γ-

Fluka 49000 Human blood MW 150k ≥98% (GE); stabilized with 10% glycine; ≤1% residue on ignition

Globulin, (Me-^{14}C)-

ARC ARC-428 MW 150k Methylated; 30-60 µCi/mg; 0.11-1.11 MBq/mmol; in 0.01 M sodium phosphate, pH 7.2 \| Radiochemical; dissociates into subunits of ~22.5 & 53 k Da under reducing conditions

Globulin, Corticosteroid Binding

Sigma G 2653 Human plasma MW 56k Lyophilized from 1.25 mM sodium phosphate, pH 7.4, containing 0.05% sodium azide \| Glycosylated β-globulin present in serum that regulates the concentration of glucocorticoids & progesterone; Heyns, W, *Adv Steroid Biochem*, 6: 59, 1977; Ghose-Dastidar, J et al, *Proc Natl Acad Sci (USA)*, 88: 6408, 1991

Globulin, Cortisol Binding

Cortex CP3040U >98%

Globulin, Gc-

Synonyms: Vitamin D Binding Protein

Sigma G 8764 Human plasma ~90% (SDS-PAGE); salt-free lyophilized powder \| Group specific component

Globulin, Sex Hormone Binding

Cortex CP2020	Human	>95%
Cortex CP2020P	Human	>40%

Sigma G 2778 Human plasma MW 94k Solution in 50 mM Tris, pH 7.4, containing 50 mM calcium, 10% glycerol & 0.01% sodium azide \| Produced in the liver that has a high affinity binding site for androgens & estrogens; Heyns, W, *Adv Steroid Biochem*, 6: 59, 1977

Biodesign A86849H Human pregnancy serum >90%

Globulin, Thyro-

Cortex CP1028	>95%
Cortex CP1028U	>98%

Fluka 89385 Bovine thyroid glands MW 660k ≥98% (GE); ≤2% γ-globulin; 0.7% iodine \| Iodinated precursor protein of the thyroid hormones; Rawitch, AB et al, *JBC*, 258: 2079, 1983; Mercken, L et al, *Nature*, 316: 647, 1985

Fluka 89387 Porcine thyroid glands ≤4% Residue on ignition; 1% iodine

Globulin, Thyroxine Binding

Cortex CP1025		>95%
Cortex CP1025P		>40%
Fitzgerald 30-AT30	Human serum	High purity
Fitzgerald 30-AT35	Human serum	Standard purity

Globulin, α-

ICN 823064 Human plasma 80% protein (nitrogen analyzer); prepared as Cohn Fraction IV

Globulin, αI Micro

Cortex CP1021	>95%
Cortex CP1021P	>40%

Globulin, αII Micro

Cortex CP3005		>95%
Cortex CP1022	Human	>95%
Cortex CP1022P	Human	>40%
Cortex CP1022U	Human	>98%

Globulin, β-Thrombo

Cortex CP3097U	Human	>98%

Globulin, γ-

ICN 55847	Bovine	Purified

Sigma G 5009 Bovine Prepared from Cohn Fraction II, III; electrophoretic purity ~99%; <4% NaCl

ICN 191478 Bovine plasma 98%; white crystalline; lyophilized

ICN 820412 ICN 820412 Bovine plasma >98%; analyzed by cellulose acetate electrophoresis; prepared as Cohn Fraction II; \| Used as a co-precipitant in RIA methods employing PEG or $(NH_4)_2SO_4$ precipitation procedures; also ideal as a starting material for isolation of bovine IgG subclasses

ICN 820423 ICN 820421 Bovine plasma >98%; analyzed by cellulose acetate electrophoresis; prepared as Cohn Fraction II; labile enzyme free \| Used as a co-precipitant in RIA methods employing PEG or $(NH_4)_2SO_4$ precipitation procedures; also ideal as a starting material for isolation of bovine IgG subclasses

Sigma G 7516 Bovine plasma Electrophoretic purity ~99%; <5% NaCl

Sigma G 4904 Bovine serum 16% Solution; ~0.85% NaCl & 0.1% sodium azide as preservative; no stabilizers added

Sigma G 7515 Cat Prepared from Cohn Fraction II, III; electrophoretic purity ~99%; 2-4% NaCl

ICN 55851 Goat Purified

Sigma G 9513 Goat Prepared from Cohn Fraction II, III; electrophoretic purity ~99%; 2-4% NaCl

ICN 191479 Goat plasma 95%; white crystalline; lyophilized

Sigma G 2638 Guinea pig Prepared from Cohn Fraction II, III; electrophoretic purity ~99%; may contains ≤6% NaCl

ICN 822041 Guinea pig plasma >90%; analyzed by cellulose acetate electrophoresis; prepared as Cohn Fraction II; lyophilized, low salt

Sigma G 2387 Horse Prepared from Cohn Fraction II, III; electrophoretic purity ~99%; 2-4% NaCl

ICN 55838 Human Purified antigen; lyophilized

Sigma G 4386 Human Prepared from Cohn Fraction II, III; electrophoretic purity ~99%; <4% NaCl \| Source material negative for HIV & HBsAg

ICN 823101 ICN 823102 Human plasma >95%; analyzed by cellulose acetate electrophoresis; prepared as Cohn Fraction II

ICN 55861 Mouse Purified antigen; lyophilized

Sigma G 9894 Mouse Prepared from Cohn Fraction II, III; electrophoretic purity ≥90%; ≤4% NaCl

ICN 191480 Mouse plasma ≥90%; white crystalline; lyophilized

ICN 823541 Mouse plasma >90% analyzed by cellulose acetate electrophoresis; prepared as Cohn Fraction II

Sigma G 2512 Pig Prepared from Cohn Fraction II, III; electrophoretic purity ~99%; 2-4% NaCl

ICN 55867 Rabbit Purified

Sigma G 0261 Rabbit Prepared from Cohn Fraction II, III; electrophoretic purity ~99%; <4% NaCl

ICN 191481 Rabbit plasma ≥98%; white crystalline; lyophilized

ICN 824551 Rabbit plasma >90% analyzed by cellulose acetate electrophoresis; prepared as Cohn Fraction II

Sigma G 2018   Rabbit plasma   Electrophoretic purity ~99%; <5% NaCl

ICN 55871   Rat   Purified

Sigma G 2885   Rat   Prepared from Cohn Fraction II, III; electrophoretic purity ~98%; 2-4% NaCl

ICN 825041   Rat plasma   >90% analyzed by cellulose acetate electrophoresis; prepared as Cohn Fraction II

Sigma G 9887   Sheep   Prepared from Cohn Fraction II, III; electrophoretic purity ~99%; <4% NaCl

Globulin, γ-(^{14}C-Me)-

Sigma G 0647   Bovine plasma   MW ~150k (subunit MW 23k & 50k)   5-50 μCi/mg protein; solution in 10 mM sodium phosphate, pH 7.0, containing 0.1% NaCl, in serum bottle | Radiochemical; prepared from Sigma G 7516

Glucagon

Sigma G 3157   Porcine pancreas   FW 3482.8   Sterilized by γ-irradiation; cell culture tested

Sigma G 9154   Porcine pancreas   FW 3482.8   Cell culture tested

Glucose Dependent Insulinotropic Hormone

Biogenesis 4665-6404   Porcine synthetic   Purified; lyophilized

Glucose Oxidase

Sigma G 7146   *Aspergillus niger*   pI 4.2; vial contains ~2 mg | IEF Marker

Glucose Regulating Protein 78

Sigma G 1285   Hamster recombinant, expressed in *E. coli*   Solution in 37 mM Tris-HCl, pH 7.5, containing 37 mM NaCl

Glutathione *S*-Transferase Tag

Synonyms: GST Tag

Upstate 12-350   *E. coli*   MW 26k   Frozen solution | Fusion protein tag; purified from *E. coli* lysate by glutathione-agarose chromatography

Gluten

ICN 101815   Wheat   12.3% total N; ~80% protein

Sigma G 5004   Wheat   Crude; ~80% protein 7% fat

Glycogen Phosphorylase Isoenzyme BB

USBio G8170-05   Human   ≥95%; no contaminants detected; 1 mg/mL liquid in glycerol buffer, 1 mM β-glycerophosphate, 1 mM EDTA, 0.5 M NaCl, pH 7.8 | Suitable for antigenic applications in immunological protocols; single band by SDS-PAGE, IEP, &/or RID

Glycohemoglobin Control-E

Sigma G 1012   Human blood   Elevated; assayed, freeze-dried preparations of stabilized human blood containing glycosylated hemoglobin, Hb A$_1$ | For use as controls in the determination of Hb A$_1$ by Sigma Procedure No. 440

Glycohemoglobin Control-N

Sigma G 2012   Human blood   Normal; assayed, freeze-dried preparations of stabilized human blood containing glycosylated hemoglobin, Hb A$_1$ | For use as controls in the determination of Hb A$_1$ by Sigma Procedure No. 440

Glycophorin, Blood Type B Negative

Sigma G 9511   Human blood   Lyophilized powder; predominantly glycophorin A; may produce a hazy solution in water

Glycophorin, Blood Type MM

Sigma G 7903   Human blood   Lyophilized powder; predominantly glycophorin A

Glycophorin, Blood Type MN

Sigma G 5017   Human blood   Lyophilized powder; predominantly glycophorin A; may produce a hazy solution in water

Glycoprotein

ICN 820611   Bovine albumin   Isolated from albumin supernatants & prepared as Cohn Fraction VI; lyophilized; >90% by cellulose acetate electrophoresis; ~10% carbohydrate, ≥50% protein by nitrogen analyzer

Glycoprotein 130

ICN 195739   Human recombinant, expressed in *Sf*21   Soluble, ≥97%; lyophilized; ED$_{50}$ = 0.5-2.0 μg/mL in the presence of 10 ng/mL of IL-6 sR & 20 ng/mL of IL-6

Glycoprotein I, β$_2$-

Cortex CP3094U   >98%

Fitzgerald 30-AB23   Calf thymus   High purity

ICN 194075   Human plasma   Purified, containing glycine & NaCl | Possible target of antiphospholipid Ab & a component involved in phospholipid immunogenicity in the APA syndrome; McNeal, HP etal, *PNAS*, 87:4120, 1990

Glycoprotein, His-Rich

ICN 194077   Human plasma   Purified, containing glycine & NaCl; <0.01 PEU plasminogen & α2-antiplasmin | Non-enzymatic single-chain glycoprotein that regulates plasminogen binding to fibrin by interacting at the high-affinity Lys binding site; demonstrates an anti-inflammatory effect; Lijwen, HR etal, *JBC*, 258:3803, 1983

Glycoprotein, Specific β-1

USBio S5370   Human   ≥95%; no contaminants detected; single band by SDS-PAGE, IEP, &/or RID; 1 mg/mL; lyophilized in PBS buffer | Suitable for antigenic applications in immunological protocols

Biogenesis 4732-0307   Human placental serum   Lyophilized

Glycoprotein, Tamm-Horsfall

Cortex CP1029P   >95%

Glycoprotein, α$_1$-Acid

Synonyms: Orosomucoid

Sigma G 3643   Bovine   Purified from serum; 99%

Sigma G 9885   Human   Purified from Cohn Fraction VI; 99% | Hao, YL & Wickerhauser, M, *Biochim Biophys Acta*, 322: 99, 1973

USBio A0550   Human   ≥98%; no contaminants detected; single band by SDS-PAGE, IEP, RID; 1 mg/mL specific protein; lyophilized | Suitable for antigenic applications in immunological protocols

Biogenesis 4730-2004   Human plasma   Lyophilized

Cortex CP3001U   Plasma   >98%

Sigma 0 0514   Rat   Purified from Cohn Fraction VI; 99%

Sigma G 3394   Rat   Purified from serum; 99%

Sigma G 6401   Sheep   Purified from Cohn Fraction VI; 99% (agarose electrophoresis)

Glycoprotein, α$_2$-HS

Biodesign A50107H   Human plasma   Purified

Glycoprotein, α₂-HS-

Sigma G 0516   Human plasma   ≥95% (SDS-PAGE); lyophilized from 20 mM Tris-HCl, pH 8.0 with 200 mM NaCl | Lebreton, JP et al, *J Clin Invest*, 64: 1118, 1979

Glycoprotein, β₁-

Fitzgerald 30-AS43   Human retroplacental serum   High purity

Gonadotropin, Menopausal

Biogenesis 6062-1009   Human urine   Crude; lyophilized

Gonadotropin, Pregnant Mare Serum

Biogenesis 7646-0604   Equine pregnant mare serum   MW 43-63k   Generally 10-30%, crude preparation; essentially salt free in 50 mM bicarbonate buffer, pH 8.0; lyophilized | Contaminants are typically equine serum proteins

Calbiochem 367222   Equine pregnant mare serum Lyophilized solid; biopotency: ≥2000 units/mg dry weight; soluble in water; may be carcinogenic/teratogenic | Complex glycoprotein with combined follicle stimulating hormone & interstitial cell-stimulating hormone action; Lapolt, PS et al, *Endocrinology*, 130: 1289, 1992; Sato, EF et al, *FEBS Lett*, 303: 121, 1992

Biogenesis 7646-0504   Equine serum   MW 43-63k Purified; from 0.05 M NH₄HCO₃, essentially salt free; lyophilized | Moore et al, *JBC*, 255:6923, 1980

Gonadotropin, α-Pregnant Mare Serum

Biogenesis 7646-0704   Equine serum   Lyophilized

Biogenesis 7646-0804   Equine serum   Lyophilized

G-Protein G₁₃α-Subunit, His-Tagged

Calbiochem 371718   BHK21 cells recombinant, produced by overexpression of a full-length G₁₃α cDNA clone in *E. coli*   MW 45.2k   Immunoblot standard; ≥90% (SDS-PAGE); liquid in 200 mM BME, 50 mM Tris-HCl, 10% glycerol, 2% SDS, 0.005% bromophenol blue, pH 6.8 | Participates in the regulation of cell movement as well as in development angiogenesis; useful as a positive control or to assay for cross-reactivity by Western blotting; Offermanns, S et al, *Science*, 275: 533, 1997; Strathmann, MP & Simon, MI, *PNAS*, 88: 5582, 1991

G-Protein G₁α-III-Subunit

Calbiochem 371761   Recombinant   MW 41k   Immunoblot standard; liquid; partially purified by DEAE fractionation of bacterial lysate; use 10 µL/blot | Useful as a positive control or to assay for cross-reactivity with the G₁α-3-subunit by Western blot

G-Protein G₁α-III-Subunit, His-Tagged

Synonyms: Pertussis Toxin Substrate

Calbiochem 371762   Rat brain recombinant, produced by overexpression of a full-length G₁α–3 cDNA clone in *E. coli*   MW 43.2k   Immunoblot standard; ≥85% (SDS-PAGE); liquid in 200 mM BME, 50 mM Tris-HCl, 10% glycerol, 2% SDS, 0.005% bromophenol blue, pH 6.8 | Functions to couple the activation of receptors to the inhibition of adenylyl cyclase; ubiquitous tissue distribution; useful as a positive control or to assay for cross-reactivity by Western blotting; Simon, MI et al, *Science*, 252: 802, 1991; Jones, DT & Reed, RR, *J Biol Chem*, 262: 14241, 1987; Ioth, H et al, *PNAS*, 83: 3776, 1986

G-Protein G₁α-III-Subunit, Myristoylated

Calbiochem 371799   Rat recombinant   MW 40,528   ≥95% (SDS-PAGE); biologically active & myristoylated preparation of G₁α-3-subunit; liquid in 3 mM MgCl₂, 20 mM HEPES, 100 mM NaCl, 1 mM EDTA, pH 8.0 | Functionally interacts with a variety of G₁-coupled receptors including the adenosine A₁, angiotensin II, 5-HT₁ₐ & 5-HT₁Dβ receptors

G-Protein G₁α-II-Subunit

Calbiochem 371759   Recombinant   MW 40k   Immunoblot standard; liquid; partially purified by DEAE fractionation of bacterial lysate; use 10 µL/blot | Useful as a positive control or to assay for cross-reactivity with the G₁α-2-subunit by Western blot

G-Protein G₁α-II-Subunit, His-Tagged

Synonyms: Pertussis Toxin Substrate

Calbiochem 371760   Human brain recombinant, produced by overexpression of a full-length G₁α–2 cDNA clone in *E. coli*   MW 41.7k   Immunoblot standard; ≥90% (SDS-PAGE); liquid in 200 mM BME, 50 mM Tris-HCl, 10% glycerol, 2% SDS, 0.005% bromophenol blue, pH 6.8 | Functions in transducing extracellular signals leading to the inhibition of adenylyl cyclase & may also stimulate certain potassium channels; found in the plasma membranes of all tissues; useful as a positive control or to assay for cross-reactivity by Western blotting; Simon, MI et al, *Science*, 252: 802, 1991; Ioth, H et al, *PNAS*, 83: 3776, 1986

G-Protein G₁α-II-Subunit, Myristoylated

Calbiochem 371796   Rat recombinant   MW 40,505   ≥90% (SDS-PAGE); biologically active & myristoylated preparation of G₁α-2-subunit; liquid in 3 mM MgCl₂, 20 mM HEPES, 100 mM NaCl, 1 mM EDTA, pH 8.0 | Functionally interacts with a variety of G₁-coupled receptors including the adenosine A₁, angiotensin II, 5-HT₁ₐ & 5-HT₁Dβ receptors

G-Protein G₁α-I-Subunit

Calbiochem 371756   Recombinant   MW 41k   Immunoblot standard; liquid; partially purified by DEAE fractionation of bacterial lysate; use 10 µL/blot | Useful as a positive control or to assay for cross-reactivity with the G₁α-1-subunit by Western blot

G-Protein G₁α-I-Subunit, His-Tagged

Calbiochem 371758   Rat brain recombinant, produced by overexpression of a full-length G₁α–1 cDNA clone in *E. coli*   MW 43.1k   Immunoblot standard; ≥90% (SDS-PAGE); liquid in 200 mM BME, 50 mM Tris-HCl, 10% glycerol, 2% SDS, 0.005% bromophenol blue, pH 6.8 | Functions in transducing extracellular signals leading to the inhibition of adenylyl cyclase & may also stimulate certain potassium channels; found mainly in the plasma membranes of neuronal tissues; useful as a positive control or to assay for cross-reactivity by Western blotting; Simon, MI et al, *Science*, 252: 802, 1991; Jones, DT & Reed, RR, *J Biol Chem*, 262: 14241, 1987

G-Protein G₁α-I-Subunit, Myristoylated

Calbiochem 371793   Rat recombinant   MW 40,351   ≥95% (SDS-PAGE); biologically active & myristoylated preparation of G₁α-1-subunit; liquid in 3 mM MgCl₂, 20 mM HEPES, 100 mM NaCl, 1 mM EDTA, pH 8.0 | Functionally interacts with a variety of G₁-coupled receptors including the adenosine A₁, angiotensin II, 5-HT₁ₐ & 5-HT₁Dβ receptors

G-Protein Gₒα-Subunit

Calbiochem 371767   Recombinant   MW 39k-40k Immunoblot standard; liquid; partially purified by DEAE fractionation of bacterial lysate; use 10 µL/blot | Useful as a positive control or to assay for cross-reactivity with the Gₒα-subunit by Western blot

G-Protein G$_o$α-Subunit, His-Tagged

Synonyms: Pertussis Toxin Substrate

Calbiochem 371774 Rat brain recombinant, produced by overexpression of a full-length G$_o$α cDNA clone in *E. coli* MW 41.2k Immunoblot standard; ≥85% (SDS-PAGE); liquid in 200 *mM* BME, 50 *mM* Tris-HCl, 10% glycerol, 2% SDS, 0.005% bromophenol blue, pH 6.8 | A GTP-binding α-subunit of Go that is most abundant in endocrine & neuronal tissues; thought to play a pivotal role in growth cone function, coordinating the effects of both extracellular signals & intracellular growth protein; useful as a positive control or to assay for cross-reactivity by Western blotting; Strathmann, NP et al, *PNAS*, 87: 6477, 1990; Jones, DT & Reed, RR, *J Biol Chem*, 262: 14241, 1987; Ioth, H et al, *PNAS*, 83: 3776, 1986

G-Protein G$_o$α-Subunit, Myristoylated

Calbiochem 371790 Rat recombinant MW 40,074 ≥95% (SDS-PAGE); biologically active & myristoylated preparation of G$_o$α-subunit; liquid in 3 *mM* MgCl$_2$, 20 *mM* HEPES, 100 *mM* NaCl, 1 *mM* EDTA, pH 8.0 | Functionally interacts with a variety of G$_i$-coupled receptors including the adenosine A$_1$, angiotensin II, 5-HT$_{1A}$ & 5-HT$_{1Dβ}$ receptors

G-Protein G$_{q/11}$α-Subunit, His-Tagged

Calbiochem 371781 Human retina recombinant, produced by overexpression of a full-length G$_{q/11}$α cDNA clone in *E. coli* MW 43.1k Immunoblot standard; ≥90% (SDS-PAGE); liquid in 200 *mM* BME, 50 *mM* Tris-HCl, 10% glycerol, 2% SDS, 0.005% bromophenol blue, pH 6.8 | Involved in the stimulation of phospholipase C; distributed ubiquitously in tissues; provided as a denatured protein; useful as a positive control or to assay for cross-reactivity by Western blotting; Macrez-Lepretre, N et al, *J Biol Chem*, 272: 5261, 1997; Dippel, E et al, *PNAS*, 93: 1391, 1996; Strathmann, NP et al, *PNAS*, 87: 9113, 1990

G-Protein G$_q$α-Subunit, His-Tagged

Calbiochem 371765 Rat brain recombinant, produced by overexpression of a full-length G$_q$α cDNA clone in *E. coli* MW 44.9k Immunoblot standard; ≥90% (SDS-PAGE); liquid in 200 *mM* BME, 50 *mM* Tris-HCl, 10% glycerol, 2% SDS, 0.005% bromophenol blue, pH 6.8 | Participates in the stimulation of phospholipase C; useful as a positive control or to assay for cross-reactivity by Western blotting; Ku, CY et al, *Endocrinology*, 136: 1509, 1995

G-Protein G$_s$α-Subunit

Calbiochem 371764 Recombinant MW 45k Immunoblot standard; liquid; partially purified by DEAE fractionation of bacterial lysate; use 10 μL/blot | Useful as a positive control or to assay for cross-reactivity with the G$_s$α–subunit by Western blot

G-Protein G$_s$α-Subunit, His-Tagged

Calbiochem 371766 Rat brain recombinant, produced by overexpression of a full-length G$_s$α cDNA clone in *E. coli* MW 48.5k Immunoblot standard; ≥90% (SDS-PAGE); liquid in 200 *mM* BME, 50 *mM* Tris-HCl, 10% glycerol, 2% SDS, 0.005% bromophenol blue, pH 6.8 | Involved in the regulation of several cellular processes including calcium channel modulation & oocyte maturation; useful as a positive control or to assay for cross-reactivity by Western blotting; Gallo, CJ et al, *J Cell Biol*, 130: 275, 1995; Fong, HKW et al, *J Physiol*, 487: 291, 1987

G-Protein G$_t$α-I-Subunit, His-Tagged

Calbiochem 371785 Bovine retina recombinant, produced by overexpression of a full-length G$_t$α–1 cDNA clone in *E. coli* MW 41.1k Immunoblot standard; ≥85% (SDS-PAGE); liquid in 200 *mM* BME, 50 *mM* Tris-HCl, 10% glycerol, 2% SDS, 0.005% bromophenol blue, pH 6.8 | Is the 39 kDa GTP-binding α-subunit of rod transducin that functions in coupling rhodopsin to cGMP phosphodiesterase; useful as a positive control or to assay for cross-reactivity by Western blotting; Medynski, DS et al, *PNAS*, 82: 4311, 1985; Fung, BKK et al, *PNAS*, 78: 152, 1981

G-Protein Gβ-III-Subunit, His-Tagged

Calbiochem 371772 Human retina recombinant, produced by overexpression of a full-length Gβ–3 cDNA clone in *E. coli* MW 39.9k Immunoblot standard; ≥90% (SDS-PAGE); liquid in 200 *mM* BME, 50 *mM* Tris-HCl, 10% SDS, 0.005% bromophenol blue, pH 6.8 | Useful as a positive control or to assay for cross-reactivity by Western blotting; Levine, MA et al, *PNAS*, 87: 2329, 1990

G-Protein Gβ-II-Subunit, His-Tagged

Calbiochem 371771 Human retina recombinant, produced by overexpression of a full-length Gβ–2 cDNA clone in *E. coli* MW 41.1k Immunoblot standard; ≥90% (SDS-PAGE); liquid in 200 *mM* BME, 50 *mM* Tris-HCl, 10% glycerol, 2% SDS, 0.005% bromophenol blue, pH 6.8 | Shares 90% AA identity with bovine transducin Gβ-1-subunit; useful as a positive control or to assay for cross-reactivity by Western blotting; Fong, HKW et al, *PNAS*, 84: 3792, 1987

G-Protein Gβ-V-Subunit, His-Tagged

Calbiochem 371775 Rat brain recombinant, produced by overexpression of a full-length Gβ–5 cDNA clone in *E. coli* MW 41.4k β Immunoblot standard; ≥90% (SDS-PAGE); liquid in 200 *mM* BME, 50 *mM* Tris-HCl, 10% glycerol, 2% SDS, 0.005% bromophenol blue, pH 6.8 | Shown to stimulate the activity of the β2 isotype of phospholipase C; useful as a positive control or to assay for cross-reactivity by Western blotting; Watson, AJ et al, *J Biol Chem*, 269: 22150, 1994

G-Protein Gγ-C-Subunit, His-Tagged

Calbiochem 371794 Human retina recombinant, produced by overexpression of a full-length Gγ–c cDNA clone in *E. coli* MW 10.2k Immunoblot standard; ≥90% (SDS-PAGE); liquid in 200 *mM* BME, 50 *mM* Tris-HCl, 10% glycerol, 2% SDS, 0.005% bromophenol blue, pH 6.8 | Gγ-8-Subunit; γ-subunit of cone transducin that has been localized exclusively to con photoreceptors of human retinas by immunostaining; C-terminal CLIS motif for post-translational farnesylation; useful as a positive control or to assay for cross-reactivity by Western blotting; Ong, OC et al, *J Biol Chem*, 270: 8495, 1995

G-Protein Gγ-III-Subunit, His-Tagged

Calbiochem 371789 Rat brain recombinant, produced by overexpression of a full-length Gγ–3 cDNA clone in *E. coli* MW 9.4k Immunoblot standard; ≥90% (SDS-PAGE); liquid in 200 *mM* BME, 50 *mM* Tris-HCl, 10% glycerol, 2% SDS, 0.005% bromophenol blue, pH 6.8 | γ-Subunit of the heterotrimeric G-proteins that is expressed exclusively in brain; C-terminal CALL motif for post-translational geranylgeranylation; useful as a positive control or to assay for cross-reactivity by Western blotting; Cali, JJ et al, *J Biol Chem*, 267: 24023, 1992; Gautam, N et al, *PNAS*, 87: 7973, 1990

G-Protein Gγ-II-Subunit, His-Tagged

Calbiochem 371788 Human retina recombinant, produced by overexpression of a full-length Gγ–2 cDNA clone in *E. coli* MW 10.3k Immunoblot standard; ≥90% (SDS-PAGE); liquid in 200 *mM* BME, 50 *mM* Tris-HCl, 10% glycerol, 2% SDS, 0.005% bromophenol blue, pH 6.8 | Dimerizes with the Gβ-1- & Gβ-2-subunits; useful as a positive control or to assay for cross-reactivity by Western blotting; Meister, M et al, *Eur J Biochem*, 234: 171, 1995; Ray, K & Ganguly, R, *J Biol Chem*, 267: 6086, 1992; Wall, MA et al, *Cell*, 83: 1047, 1995

G-Protein Gγ-VII-Subunit, His-Tagged

Calbiochem 371792   Rat brain recombinant, produced by overexpression of a full-length Gγ–7 cDNA clone in *E. coli*   MW 8640   Immunoblot standard; ≥90% (SDS-PAGE); liquid in 200 *mM* BME, 50 *mM* Tris-HCl, 10% glycerol, 2% SDS, 0.005% bromophenol blue, pH 6.8 | γ-Subunit of the heterotrimeric G-proteins that is expressed in brain & to a lesser extent in heart, kidney, spleen & lung; C-terminal CIIL motif for post-translational geranylgeranylation; useful as a positive control or to assay for cross-reactivity by Western blotting; Cali, JJ et al, *J Biol Chem*, 267: 24023, 1992

G-Protein Gγ-V-Subunit, His-Tagged

Calbiochem 371791   Rat brain recombinant, produced by overexpression of a full-length Gγ–5 cDNA clone in *E. coli*   MW 8430   Immunoblot standard; ≥90% (SDS-PAGE); liquid in 200 *mM* BME, 50 *mM* Tris-HCl, 10% glycerol, 2% SDS, 0.005% bromophenol blue, pH 6.8 | γ-Subunit of the heterotrimeric G-proteins that is expressed in a number of tissues, including kidney, liver & lung; C-terminal CSFL motif for post-translational geranylgeranylation; useful as a positive control or to assay for cross-reactivity by Western blotting; Cali, JJ et al, *J Biol Chem*, 267: 24023, 1992; Fisher, KJ & Aronson, NN, *Mol Cell Biol*, 12: 1585, 1992

G-Protein βγ-Subunit

Calbiochem 371768   Bovine brain   Liquid; purified G-protein βγ-subunit in 1 *mM* DTT, 50 *mM* HEPES, 1 *mM* EDTA, 0.1% LUBROL, pH 7.6 | Diluted form gives sufficient G-protein βγ-subunit for six standard curves; βγ-subunit is biologically active & can be used in reconstitution experiments

G-Protein, Functional

Calbiochem 371739   Bovine brain   Mixture of purified pertussis toxin-sensitive G-proteins from bovine brain; the following heterotrimeric γ-proteins: $G_{o}\alpha$ (4-5 μM), $G_{i}\alpha$-1 (1-2 μM), $G_{i}\alpha$-2 (1-2 μM) & $G_{i}\alpha$-3 (<1 μM); liquid in 50 *mM* HEPES, 1 *mM* DTT, 1 *mM* EDTA, 0.1% LUBROL, pH 7.6 | Suitable for functional studies of G-proteins; βγ-subunit complex associated with these α-subunits represents the predominant form found in bovine brain; characterized by GTPγS binding activity & by immunoblotting using subunit-specific antisera

G-Protein, Immunoblot Standard

Calbiochem 371736   Bovine brain   Purified liquid; subunits: $G_{i}\alpha$-1, $G_{i}\alpha$-2, $G_{i}\alpha$-3, $G_{o}\alpha$, $G_{s}\alpha$, $G_{z}\alpha$, Gβ-1, Gβ-2 & Gγ; amount loaded for detection depends on the antisera used & the blotting protocol followed; generally 0.2-1.0 μL/lane will suffice | G-protein useful as an immunoblot standard or positive control; characterized by immunoblotting with subunit-specific antisera

Granzyme B

Kamiya   Purified enzyme

GRB2 SH2-Domain-Protein A Fusion Protein

Sigma G 5650   Recombinant, expressed in *E. coli*   MW 37k   ≥90% (SDS-PAGE); lyophilized powder containing 3-5% protein (Bradford); balance potassium phosphate & NaCl | Fusion protein containing the SH2 domain of mouse GRB2 (AA 50-161) fused to protein A; key protein in signal transduction, GRB2 transfers the signal from tyrosine kinases such as EGF receptor to the small G protein Ras. GRB2 contains an SH2 (Src homology 2) domain enabling binding to tyrosine-phosphorylated receptors; fusion protein enables detection & purification of tyrosine-phosphorylated proteins (ie tyrosine-phosphorylated EGF receptor) utilizing the affinity of protein A for the Fc domain of IgG; Olivier, JP et al, *Cell*, 73: 179, 1993; Margolis, B, *Progr Biophys Molec Biol*, 62: 223, 1994

GRB2, Agarose

USBio G8960-05   Recombinant   MW ~52k   50μg bound to 25μL of agarose beads & suspended to a final volume of 50μl (50% gel slurry) in PBS containing 2 *mM* DTT and 10% glycerol; frozen suspension | Used to precipitate mouse Sos 1 (175kD) from a RIPA lysate (500 μg) of NIH 3T3 cells; binds EGFR, SHC & Sos proteins

Growth Factor

Fitzgerald 30-AH05	Human pituitary	High purity >98%
Fitzgerald 30-AH06	Human pituitary	Standard purity >60%

Growth Factor, Hepatocyte

Synonyms: Scatter Factor; Hepatopoietin A

Sigma H 1404   Human recombinant, expressed in *Sf* 21 insect cell lines   ≥95% (SDS-PAGE); 0.2 μm filtered & lyophilized from 20 mM sodium phosphate & 0.35 M NaCl, pH 7.0, containing 250 μg BSA; endotoxin tested | Potent mitogen for epithelial cells; stimulates the growth of hepatocytes, renal tubular epithelial cells, epidermal keratinocytes, epidermal melanocytes, Mv1Lu (mink lung epithelial cells) & BALB/MK (mouse keratinocytes); inhibits the growth of B6/F1 (mouse melanoma) cells, KB (human squamous carcinoma cells) & HepG2 (human hepatoma) cells; has molecular mass of 82-85 kD; HGF gene spans ~70 kb & consists of 18 exons, interrupted by 17 introns; organization of the human HGF gene highly homologous to that of human plasminogen; HGF maps to the long arm of chromosome 7, 7q21.1; mitogenic activity is tested in culture using the monkey epithelial cell line 4MBr-5; Furlong, RA et al, *BioEssays*, 14: 613, 1992; Nakamura, T et al, *Progress in Growth Factor Research*, 3: 67, 1991; Petersen, TE et al, *J Biol Chem*, 265: 6104, 1990; Weidner, KM et al, *Proc Natl Acad Sci USA*, 88: 7001, 1991; Fukuyama, R et al, *Genomics*, 11: 410, 1991; Rubin JS et al, *Proc Natl Acad Sci USA*, 88: 415, 1991

PeproTech 100-39   Human recombinante expressed in Baculovirus-infected High-5 cells   MW 80k (disulfide linked heterodimer)   >98% (SDS-PAGE) & HPLC; <0.1 ng endotoxin per mg (1EU/mg); lyophilized | Potent mitogen for mature parenchymal hepatocyte cells; growth factor for a broad spectrum of tissues & cell types; α-chain (463 AA) + β-chain (234 AA); ED_{50} ~ 20-40 ng/mL; SA determined by dose-dependant stimulation of monkey 4MBr-5 cell proliferation

Growth Hormone

Synonyms: Somatotropin

Cortex CP2042	Bovine	>95%	
Biogenesis 4750-1504	Bovine pituitary	MW 22k	Purified; from 0.05 *M* NH₄HCO₃, pH 8.0; lyophilized	Closset et al, *Biochem*, 214:885, 1983
USBio G8999-05	Bovine pituitary	≥95% (SDS-PAGE); lyophilized	
Biogenesis 4750-2059	Equine pituitary	Lyophilized	
Chemicon GC065	Human	≥95%	
Cortex CP1042	Human	>95%	
Cortex CP1042P	Human	>40%	
USBio G9000-12	Human	≥98% (SDS-PAGE); hPrl ≤ 1.1%, hLH ≤ 0.1%, hFSH ≤ 0.1%, hTSH ≤ 0.1%; 2.6 IU/mg; lyophilized from 0.05 *M* ammonium bicarbonate	Suitable for antigenic applications in immunological protocols	
USBio G9000-14	Human	≥70% (SDS-PAGE); hPrl <1.1%, hLH <0.1%, hFSH <0.1%, hTSH <0.1%; 1.4 IU/mg specific activity (1st IRP, WHO); lyophilized from 0.05 *M* ammonium bicarbonate;	Suitable for antigenic applications in immunological protocols	
Biodesign A81555M	Human pituitary	97%	
Biogenesis 4750-0504	Human pituitary	Lyophilized	
Biogenesis 4750-0509	Human pituitary	MW 22k	0.02% hLH/hFSH/hTSH, 0.01% hPRL; purified; 0.05 *M* NH₄HCO₃, pH 8.0; lyophilized	Closset et al, *Biochem*, 214:885, 1983
Biogenesis 4750-0704	Human pituitary	0.4% hPRL, 0.004% hLH; tested negative for HBsAg, HCV, Syphilis and HIV I and II; semi-pure; from 50 *mM* ammonium bicarbonate; lyophilized	

Biogenesis 4750-0759 Human pituitary MW 22k 20 K <0.5%; tested negative for HBsAg, HCV and HIV 1 and 2; purified; 0.05 M NH$_4$HCO$_3$, pH 8.0, essentially salt free; lyophilized | Closset et al, *Biochem*, 214:885, 1983

Calbiochem 869008 Human pituitary MW 21.7k Iodination grade; lyophilized solid; immunopotency: ≥5 IU/mg (WHO 1st IRP); soluble in sodium bicarbonate; hLH, hTSH, hFSH, hPRL: ≤0.5%; shown by certified tests to be negative for HBsAg & for antibodies to HIV & HCV); may be carcinogenic/teratogenic | Single chain polypeptide hormone essential for growth of all tissues; increases fat mobilization; *Merck Index*, 12: 8864

Biogenesis 4750-3004 Ovine pituitary Semi-pure; lyophilized

Biogenesis 4750-4104 Porcine pituitary MW 22k Purified; 0.05 M NH$_4$HCO$_3$, pH 8.0; lyophilized | Closset et al, *Biochem*, 214:885, 1983

Biogenesis 4750-5204 Rat pituitary <0.1% rLH/rTHS/rPRL/rFSH; 2 IU/mg immunological potency; purified; 0.05 M NH4HC03, pH 8.0. Essentially salt free; lyophilized

Biogenesis 4750-5204-50ug Rat pituitary <0.1% rLH/rTHS/rPRL/rFSH; 2 IU/mg immunological potency; purified; 0.05 M NH$_4$HCO$_3$, pH 8.0. Essentially salt free; lyophilized

Growth Hormone, *Acanthopagrus butcheri* (Bream)

IBT GHAU020 Recombinant, expressed in *E. coli* MW 21,410 Dried under dry nitrogen at a slight vacuum (-25kPa) from 5 mM sodium phosphate buffer, pH 8.5, containing m-brGH/mannitol/glycine (1:5:1 w/w) | Shares 97% AA sequence identity with gilthead sea bream (*Sparus aurata*), 94% with red sea bream (*Pagrus major*), 93% with tuna (*Thunnus thynnus*), 91% with barramundi (*Lates calcarifer*), 63% with salmon (*Oncorhynchuseta*) & 32% with human GH

Growth Hormone, Rat

IBT GHCU020 Recombinant, expressed in *E. coli* MW 22k 97%; dried under dry nitrogen at a slight vacuum (-25kPa) from 10 mM sodium phosphate buffer, pH 8.8, containing m-rGH/mannitol/glycine (1:5:1 w/w)

Growth Hormone, Salmon/Trout

IBT GHBU020 Recombinant, expressed in *E. coli* Dried under dry nitrogen at a slight vacuum (-25kPa) from a solution containing salmon-trout GH/mannitol/sodium bicarbonate (1:2:1 w/w) | Stimulates growth in salmon/trout & enhances the adaptability of salmon parr moving from fresh to sea H$_2$O; GH levels increase naturally during the parr-smolt transformation in Atlantic salmon (*Salmo salar*)

Growth Related Oncogene

BioSource International PRC1065 Rat recombinant

Growth Related Oncogene ES Protein

Calbiochem 368610 *E. coli* MW 41k Liquid in 100 mM NaCl, 50 mM Tris, 1 mM DTT, 50% glycerol, pH 7.5; DNases, proteases, RNases: none detected | Heat shock protein that is a member of the chaperonin family; consists of a single ring of six to eight identical subunits of MW 10,000; plays an integral role in protein folding & in mediating protein-protein interactions in both normal & stressed cells; in the presence of ATP binds GroEL inhibiting its ATPase activity; mayhew, M et al, *Nature*, 379: 420, 1996; Gething, MJ & Sambrook, J, *Nature*, 355: 33, 1992; Chen, S et al, *Nature*, 371: 261, 1994; Hemmingsen, SM et al, *Nature*, 333: 330, 1988

Growth Related Oncogene α/KC

Biodesign A52011M *E. coli* MW 7.8k Purified | Species specificity: mouse

PeproTech 250-11 Murine recombinant, expressed in *E. coli* MW 7.8k >98% (SDS-PAGE) & HPLC; <0.1 ng endotoxin per mg (1EU/mg); lyophilized | Belongs to the C-X-C family of chemokines; 72 AA; SA determined by its ability to chemoattract human neutrophil population

Growth Related Oncogene α/Melanoma Growth Stimulating Activity

Sigma G 0657 Human recombinant, expressed in *E. coli* MW 7.9k ≥97% (SDS-PAGE); 0.2 µm filtered & lyophilized in 10% acetonitrile, 0.1% trifluoroacetic acid containing 500 µg BSA; activity tested in culture using human neutrophils; endotoxin tested | GROα gene initially cloned from T24 cells & the gene in melanoma cells encoding melanoma growth stimulating protein (MGSA) are identical; Human cells contain 3 closely related but distinct GRO genes: GROα, GROβ & GROγ; GROβ & GROγ share 93% & 82% identity respectively, with GROα at the nucleotide level; GROs are members of the chemokine α subfamily characterized by the separations with 1 AA of the first 2 cysteine residues: C-X-C in the AA sequence; in normal cells, human mRNA GRO expression found in foreskin fibroblasts, in synovial fibroblasts, chrondocytes & bone cells; characterization of the GROα receptor indicates the presence of low & high affinity receptors on human neutrophils; Schroder, J et al, *J Immunol* 139: 3474, 1987 General References for GRO: Anisowicz, A et al, *Proc Natl Acad Sci USA*, 84: 7188, 1987; Richmond, A et al, *EMBO J*, 7: 2025, 1988; Haskill, S et al, *Proc Natl Acad Sci USA*, 87: 7732, 1990; Sager, R et al, *Cytokines*, 4: 96, 1992; Goldring, M et al, *J Bone Miner Res*, 4: 402, 1989

Growth Related Oncogene α/KC

Synonyms: Cytokine Induced Neutrophil Chemoattractant

PeproTech 400-10 Rat recombinant, expressed in *E. coli* MW 7.8k >98% (SDS-PAGE) & HPLC; <0.1 ng endotoxin per mg (1EU/mg); lyophilized | Promotes neutrophil chemotaxis & degranulation; 72 AA; SA determined by its ability to chemoattract rat neutrophils

Growth Related Oncogene α/Melanoma Growth Stimulating Activity

BioSource International PHC1065 Human recombinant

PeproTech 300-11 Human recombinant, expressed in *E. coli* MW 7.8k >98% (SDS-PAGE) & HPLC; <0.1 ng endotoxin per mg (1EU/mg); lyophilized | Promotes neutrophil chemotaxis & degranulation; stimulates mitogenesis in certain human melanoma cells; 73 AA; SA determined by its ability to chemoattract human neutrophils

Growth Related Oncogene β

Biodesign A52039H *E. coli* MW 7.9k Purified

Chemicon GF064 Human ≥95%

PeproTech 300-39 Human recombinant, expressed in *E. coli* MW 7.9k >98% (SDS-PAGE) & HPLC; <0.1 ng endotoxin per mg (1EU/mg); lyophilized | Member of the CXC family of chemokines; promotes neutrophil & basophil chemotaxis & degranulation; specifically inhibits growth factor-stimulated proliferation of capillary endothelial cells in a dose-dependent manner; 73 AA; SA determined by its ability to chemoattract CXCR2 transfected 293 cells

Sigma G 7909 Human recombinant, expressed in *E. coli* ≥97% (SDS-PAGE); 0.2 µm filtered & lyophilized in 30% acetonitrile, 0.1% trifluoroacetic acid containing 500 µg BSA; activity tested in culture by its ability to induce myeloperoxidase release from cytochalasin B treated neutrophils; endotoxin tested; see Sigma G 0657 | Schroder, J et al, *J Immunol* 139: 3474, 1987

Growth Related Oncogene β/Macrophage Inflammatory Peptide II

BioSource International PRC1074 Rat recombinant

PeproTech 400-11 Rat recombinant, expressed in *E. coli* MW 7.9k >98% (SDS-PAGE) & HPLC; <0.1 ng endotoxin per mg (1EU/mg); lyophilized | Promotes neutrophil chemotaxis & degranulation; 73 AA; SA determined by its ability to chemoattract rat neutrophils

Growth Related Oncogene β/Macrophage Inflammatory Protein II

Biodesign A52410H *E. coli* MW 8k Purified | Species specificity: rat

Chemicon GF036   Rat   ≥95%

Growth Related Oncogene γ

Biodesign A52040H   *E. coli*   MW 7.9k   Purified

Chemicon GF065   Human   ≥95%

PeproTech 300-40   Human recombinant, expressed in *E. coli*
MW 7.9k   >98% (SDS-PAGE) & HPLC; <0.1 ng endotoxin per mg
(1EU/mg); lyophilized | Member of the CXC family of chemokines;
promotes neutrophil & basophil chemotaxis & degranulation; 73
AA; SA determined by its ability to chemoattract CXCR2 transfected
293 cells

Sigma G 7784   Human recombinant, expressed in *E. coli*
≥97% (SDS-PAGE); 0.2 µm filtered & lyophilized in 30%
acetonitrile, 0.1% trifluoroacetic acid containing 500 µg BSA;
activity tested in culture by its ability to induce myeloperoxidase
release from cytochalasin B treated neutrophils; endotoxin tested;
see Sigma G 0657 | Schroder, J et al, *J Immunol* 139: 3474, 1987

Growth Related Oncogene/KC

Biodesign A52401H   *E. coli*   MW 8k   Purified | Species
specificity: rat

Chemicon GF034   Rat   ≥95%

Growth Related Oncogene/Melanoma Growth Stimulating Activity

Biodesign A52311H   *E. coli*   MW 8.5k   Purified

Chemicon GF005   Human   ≥95%

Fitzgerald 30-AG19   Human recombinant, expressed in *E. coli*

GrpE Protein

Calbiochem 368650   *E. coli*   MW 41k   Liquid in 100 *mM*
NaCl, 50 *mM* Tris, 1 *mM* DTT, 50% glycerol, pH 7.5; DNases,
proteases, RNases: none detected | Molecular chaperone; plays
an integral role in protein folding & in mediating protein-protein
interactions in both normal & stressed cells; enhances the ATPase
activity of DnaK in the presence of DnaJ *in vivo*; useful for *in vitro*
protein folding studies; Gething, MJ & Sambrook, J, *Nature*, 355:
33, 1992; Liberek, K et al, *PNAS*, 88: 2874, 1991

Haptoglobin

USBio H1820-10   Human   ≥90%; no contaminants detected;
single band by SDS-PAGE, IEP, &/or RID; 1 mg; lyophilized, salt
free preparation | Suitable for antigenic applications in
immunological protocols; phenotypes: 1-1, 2-1, 2-2

Biodesign A50150H   Human plasma   Purified

Fitzgerald 30-AH02   Human plasma   High purity

Sigma H 1511   Human plasma   From pooled plasma; 98-
100%; essentially salt-free lyophilized powder; biological activity: 1
mg haptoglobin binds 0.5-0.9 mg hemoglobin | Occurs as 3 major
phenotypes: type 1-1, type 2-1, type 2-2; Javid & Liang, *J Lab Clin
Med*, 82: 991, 1973

Biogenesis 4890-0504   Human serum   Lyophilized

ICN 823351   Human serum   Sterile filtered; isolated from
Cohn Fraction IV proteins by selective precipitation & then purified
by ion-exchange; a low-salt, lyophilized powder

Cortex CP3018   Plasma/Mixed Type   >95%

Scipac P119-1   Serum/plasma   >96%; lyophilized; standard
mix of phenotypes | Acute phase protein

Scipac P119-2   Serum/plasma   40-90%; lyophilized; standard
mix of phenotypes; available on request | Acute phase protein

Haptoglobin, Phenotype 1-1

Sigma H 0138   Human plasma   Phenotype 1-1; 98-100%;
essentially salt-free lyophilized powder; biological activity: 1 mg
haptoglobin binds 0.5-0.9 mg hemoglobin | Occurs as 3 major
phenotypes: type 1-1, type 2-1, type 2-2; Javid & Liang, *J Lab Clin
Med*, 82: 991, 1973

Haptoglobin, Phenotype 2-1

Sigma H 9887   Human plasma   Phenotype 2-1; 98-100%;
essentially salt-free lyophilized powder; biological activity: 1 mg
haptoglobin binds 0.5-0.9 mg hemoglobin | Occurs as 3 major
phenotypes: type 1-1, type 2-1, type 2-2; Javid & Liang, *J Lab Clin
Med*, 82: 991, 1973

Haptoglobin, Phenotype 2-2

Sigma H 9762   Human plasma   Phenotype 2-2; 98-100%;
essentially salt-free lyophilized powder; biological activity: 1 mg
haptoglobin binds 0.5-0.9 mg hemoglobin | Occurs as 3 major
phenotypes: type 1-1, type 2-1, type 2-2; Javid & Liang, *J Lab Clin
Med*, 82: 991, 1973

Haptoglobulin, Types 1 & 2 Mixture

ICN 191334   Human plasma   MW Hp-1: 86k, Hp-2: 400k, Hp-
2-1: 200k   98% (PAGE); salt-free, lyophilized; negative for
HBsAg & HIV Ab | Polymorphic & identical the to untreated form
in human serum; Smithies, O, *Biochem J*, 61:629, 1955; Bowman,
BH & A Kurosky, *Adv Hum Genet*, 12:189, 1982

HBeAg

USBio H1826   Recombinant expressed in *E. coli*   ≥90%; no
contaminants detected; single band by SDS-PAGE, IEP, &/or RID; 1
mg/vial; lyophilized; reconstitute with dH₂O | Suitable for
antigenic applications in immunological protocols

Heat Shock Factor I

Alexis 201-024   Human recombinant, expressed in *E. coli*
extracts transformed with an expression plasmid   150 µg
sonicated *E. coli* extract at a total protein concentration of 6 mg/mL
in 25 *mM* HEPES, pH 7.9, 12.5 *mM* MgCl, 0.1 *mM* EDTA, 10%
glycerol, 1 *mM* DTT, 0.1% NP-40 & 300 *mM* KCl & protease
inhibitors | Positive control for 25 Gel Shift Assays or 50 Western
Blots with antiserum to heat shock factor 1

Heat Shock Protein 25

Sigma H 1154   Mouse recombinant, expressed in *E. coli*
Lyophilized powder | Homologous to crystallin, an abundant
protein of eye lens; Engel, K et al, *Biomed Biochem Acta*, 50: 1065,
1991

Heat Shock Protein 27

Sigma H 1273   Human recombinant, expressed in *E. coli*
Lyophilized powder; composition of buffer salts given on data sheet
| Engel, K et al, *Biomed Biochem Acta*, 50: 1065, 1991

Heat Shock Protein 32

Synonyms: Heme Oxygenase I

Sigma H 9028   Rat recombinant, expressed in *E. coli*
Lyophilized from 50 mM ammonium bicarbonate, pH 8.0 |
Shibahara, S et al, *Proc Natl Acad Sci Usa*, 82: 7865, 1985

Heat Shock Protein 60

Sigma H 8903   Human recombinant, expressed in *E. coli*
Lyophilized; Tris buffer salts, NaCl, DTT | Jindal, S et al, *Mol Cell
Biol*, 9: 2279, 1989; assists in synthesis, translocation, correct
folding & subunit assembly of proteins while consuming ATP. Shows
some ATPase activity even in the absence of other peptides.;
homologous to GroEL of *E. coli* (also called chaperonin)

Heat Shock Protein 70

Sigma H 1648   Bovine brain   Biotin labeled; lyophilized; actual
composition of buffer salts on data sheet | Schlossman, DM, *J Cell
Biology*, 99: 723, 1984; assists in synthesis, translocation, correct
folding & subunit assembly of proteins while consuming ATP. Shows
some ATPase activity even in the absence of other peptides.;
homologous to DnaK of *E. coli* (also called chaperonin)

Sigma H 9776   Bovine brain   Lyophilized powder; >95% (SDS-PAGE); containing ~10% protein (Lowry); balance Tris buffer salts | Welch, WJ, *Physiol Rev*, 72: 1063, 1992; assists in synthesis, translocation, correct folding & subunit assembly of proteins while consuming ATP. Shows some ATPase activity even in the absence of other peptides.; homologous to DnaK of *E. coli* (also called chaperonin)

Sigma H 8778   Human recombinant, expressed in *E. coli* >90% (SDS-PAGE); lyophilized from 50 mM Tris-HCl, pH 7.5, 100 mM NaCl, 0.1 mM PMSF & 1 mM DTT | Gething, M-J & Sambrook, J, *Cell*, 355: 33, 1992; assists in synthesis, translocation, correct folding & subunit assembly of proteins while consuming ATP. Shows some ATPase activity even in the absence of other peptides.; homologous to DnaK of *E. coli* (also called chaperonin)

Sigma H 9651   *Mycobacterium smegmatis*   ~85% (SDS-PAGE); lyophilized; Tris buffer salts | Strong affinity for ATP & a low ATPase activity; in many pathogens, HSP 70 considered & important antigen; Welch, WJ, *Physiol Rev*, 72: 1063, 1992; Palleros, DR et al, *FEBS Lett*, 336: 124, 1993; Young, D et al, *Proc Natl Acad Sci USA*, 85: 4267, 1988; assists in synthesis, translocation, correct folding & subunit assembly of proteins while consuming ATP; shows some ATPase activity even in the absence of other peptides; homologous to DnaK of *E. coli* (also called chaperonin)

Heat Shock Protein 90

Sigma H 6774   Bovine brain   Lyophilized; ≥95% (SDS-PAGE); Tris buffer salt | Plays a pivotal role in regulating steroid receptors

Heat Shock Protein DnaJ

Sigma D 4419   *E. coli*   ≥85% (SDS-PAGE); solution in 40 mM Tris, pH 7.5, 80 mM NaCl, 0.8 mM DTT & 20% (v/v) glycerol | Bardwell, JCA & Craig, EA, *Proc Natl Acad Sci*, 81: 848, 1984

Sigma D 4928   *E. coli*   MW 41k   ≥80% (SDS-PAGE); solution in 25 mM HEPES, pH 7.2, containing 100 mM KCl, 25 mM NaCl, 1 mM DTT & 10% (v/v) glycerol | Basic heat shock protein referred to as a "co-chaperone" because it is known to assist DnaK-dependent chaperone activities

Hemocyanin

Synonyms: Hemolymph Plasma Powder

Sigma H 4506   From horseshoe crab blood after the amoebocyte lysate (E-Toxate®) has been removed   Crude; lyophilized; ~60% protein (Biuret) which is predominantly hemocyanin; balance is NaCl & other naturally occurring components, including lectins | Oxygen-exchange protein of crustaceans

Biogenesis 4860-0602   Keyhole limpet   Lyophilized

USBio K0300   Keyhole limpet (*Megathura crenulata*)   ≥90%, ≥60% protein; light gray to blue; lyophilized in buffer; slightly soluble in H₂O, readily soluble in saline | Protein composed of five subunits; used as an immunological carrier for mammalian Ab production, specifically rabbits & mice; since KLH is Lys rich, primary amines will promote peptide attachment after dissociation

ICN 193550   *Limulus polyphemus* hemolymph (horseshoe crab)   Lyophilized

Sigma H 3009   *Limulus polyphemus* hemolymph (horseshoe crab)   Biotin-labeled; solution in 50% glycerol containing ~ 0.08 M NaCl, 1 mM calcium chloride, 0.01% sodium azide; 3-6 mg protein (E₂₈₀ at 1%)/mL; 40-100 moles biotin/mole protein (12-30 nmoles biotin/mg protein)

ICN 151233   *Megathura crenulata* (keyhole limpet)   >99%; lyophilized

Sigma H 2133   *Megathura crenulata* (keyhole limpets)   Lyophilized powder containing ~90% protein (Biuret); ~0.2% copper; prepared by ultracentrifugation

Sigma H 5654   *Megathura crenulata* (keyhole limpets)   Succinylated; hghly water soluble; lyophilized with stabilizing buffer; reconstitution with 2 mL DI water yields an opalescent solution of KLH at 5 mg/mL in sodium phosphate buffer, pH 7.4, containing 0.32 M NaCl & 20 mM sucrose; no preservatives added | Provides a large number of available carboxyl groups for conjugation

Sigma H 7017   *Megathura crenulata* (keyhole limpets)   Lyophilized with stabilizing buffer; reconsitution with 2 mL DI water yields an opalescent solution of KLH at 10 mg/mL in 31 mM sodium phosphate buffer, pH 7.4, containing 0.46 M NaCl & 41 mM sucrose; no preservatives added

Hemocyanin Type VIII

Sigma H 1757   *Limulus polyphemus* hemolymph (horseshoe crab)   Lyophilized powder containing ~95% protein (Biuret); ~0.15% copper; a peak eluted from a DEAE Sephadex column with a buffer containing EDTA

Hemocyanin, Cross-Linked Molecular Weight Standard for SDS-PAGE

Sigma H 2757   *Limulus polyphemus*   MW 70k (monomer), 140k (dimer), 210k (trimer), 280k (tetramer)   Lyophilized; ~2.5 mg/vial | See Sigma MW-SDS-280

Hemocyanin, Keyhole Limpet

Fluka 51522   *Megathura crenulata* (keyhole limpet)   ≥90% (GE); ≥85% protein; 0.2% Cu; prepared by ultracentrifugation | Carrier protein, used as antigen; Kyewski, BA et al, *Nature*, 308: 196, 1984; Bennett, AP et al, *Ann Clin Biochem*, 24: 374, 1987

Hemofiltrate CC Chemokine

Synonyms: Colony Inhibition Factor, Macrophage

Sigma H 0656   Human recombinant, expressed in *E. coli*   MW 8.7k   ≥97% (SDS-PAGE); 0.2 μm filtered & lyophilized in PBS containing 500 μg BSA; activity tested in culture by its ability to chemoattract cultured human monocytes; endotoxin tested | Non-glycosylated polypeptide of 74 AA; member of the CC or β chemokine class; originally isolated from the hemofiltrate of patients with chronic renal failure; the precursor form of HCC-1 consists of 93 AA; to generate the mature HCC-1 (74 AA), the precursor cleaves a 19 AA signal peptide; expressed constitutively in normal tissues & is present in high concentrations in human plasma; shows ~46% AA identity with MIP-1α & MIP-1β & 29-37% sequence identity with other CC chemokines

Hemofiltrate CC Chemokine I

Chemicon GF066   Human   ≥95%

PeproTech 300-38   Human recombinant, expressed in *E. coli* MW 8.4k   >98% (SDS-PAGE) & HPLC; <0.1 ng endotoxin per mg (1EU/mg); lyophilized | First isolated from the hemofiltrate of patients with chronic renal failure; 46% AA sequence homology with MIP-1a/MIP-1b & 29%-37% homology with other CC chemokines; 72 AA; SA determined by its ability to chemoattract human monocytes

Hemoglobin

Synonyms: Methemoglobin; Ferrohemoglobin; Oxyhemoglobin

ICN 10049J7   2% solution | Used with GC Agar Base

ICN 1004817   Beef blood   Autoclavable preparation | Used with GC Agar Base

Amersham US16891   Bovine   2x crystallized; ≥99.0%

ICN 100714   Bovine   Standardized for protease assay

Sigma H 2500   Bovine   Lyophilized powder | Oxygen transporter, NO scavenger; since native hemoglobin is readily oxidized in air, these preparations may be predominantly methemoglobin

Fluka 51290   Bovine blood   MW 64.5k   ≥94% (GE); ≤2% loss on drying; ≤5% residue on ignition; 2X crystallized; lyophilized; salt-free powder | Oxygen carrier protein; Brunori, M et al, *Top Mol Struct Biol*, 7: 263, 1985

Calbiochem 3745   Bovine erythrocytes   MW 64.5k   Lyophilized; ≥95% (SDS-PAGE); soluble in water; primarily ferric-hemoglobin & must be reduced to the ferrous form to bind molecular oxygen | Major oxygen-transporting component of red blood cells; a nitric oxide scavenger; blocks carbachol-stimulated cGMP production; Bredt, DS & Snyder, SH, *PNAS*, 86: 9030, 1989; Castoldi, AF et al, *Brain Res*, 610: 57, 1993; *Merck Index*, 12: 4682

Sigma H 2625   Bovine erythrocytes   Substrate powder; prepared from washed, lysed & dialyzed erythrocytes | Oxygen transporter, NO scavenger; since native hemoglobin is readily oxidized in air, these preparations may be predominantly methemoglobin; useful as a protease substrate

Sigma H 3760   Bovine erythrocytes   Dried erythrocytes; methemoglobin & oxyhemoglobin content is not determined | Oxygen transporter, NO scavenger; since native hemoglobin is readily oxidized in air, these preparations may be predominantly methemoglobin

Biogenesis 4870-2002   Bovine RBCs   Lyophilized

ICN 151234   Bovine red blood cells   MW 64.5k   >98% (electrophoresis); VAT mixed with ether; lyophilized | *Ann Rev Biochem*, 12:327, 1979

ICN 151235   Bovine red blood cells   MW 64.5k   >98% (electrophoresis); VAT mixed with ether; spray dried; 40% iron by weight | *Ann Rev Biochem*, 12:327, 1979

Sigma H 4632   Horse   Lyophilized powder | Oxygen transporter, NO scavenger; since native hemoglobin is readily oxidized in air, these preparations may be predominantly methemoglobin

Sigma H 7379   Human   Lyophilized powder | Negative for HIV & Hepatitis B antigen; oxygen transporter, NO scavenger; since native hemoglobin is readily oxidized in air, these preparations may be predominantly methemoglobin

Biogenesis 4870-7056   Human RBCs   Lyophilized

Sigma H 5633   Mouse   Lyophilized powder | Oxygen transporter, NO scavenger; since native hemoglobin is readily oxidized in air, these preparations may be predominantly methemoglobin

Biogenesis 4870-5004   Ovine RBCs   Lyophilized

Sigma H 4131   Pig   Lyophilized powder | Oxygen transporter, NO scavenger; since native hemoglobin is readily oxidized in air, these preparations may be predominantly methemoglobin

Sigma H 0256   Pigeon   Lyophilized powder | Oxygen transporter, NO scavenger; since native hemoglobin is readily oxidized in air, these preparations may be predominantly methemoglobin

Sigma H 7255   Rabbit   Lyophilized powder | Oxygen transporter, NO scavenger; since native hemoglobin is readily oxidized in air, these preparations may be predominantly methemoglobin

Sigma H 3883   Rat   Lyophilized powder | Oxygen transporter, NO scavenger; since native hemoglobin is readily oxidized in air, these preparations may be predominantly methemoglobin

Sigma H 2750   Sheep   Lyophilized powder | Oxygen transporter, NO scavenger; since native hemoglobin is readily oxidized in air, these preparations may be predominantly methemoglobin

Sigma H 0142   Turkey   Lyophilized powder | Oxygen transporter, NO scavenger; since native hemoglobin is readily oxidized in air, these preparations may be predominantly methemoglobin

Hemoglobin A$_0$, Ferrous

Sigma H 0267   Human, stabilized   Chromatographically purified & lyophilized from ammonium bicarbonate containing ~50% nonionic stabilizers; ~98% by agarose electrophoresis; reconsititution with buffer gives >90% ferrous hemoglobin | Suitable for electrophoresis & chromatography standard, but has not been tested for functional equivalence against native preparations (unlyophilized ferrous hemoglobins); Package size indicates the amount of hemoglobin as determined by the procedure of Drabkin, DL, *J Biol Chem*, 164: 703, 1946

Hemoglobin A1c

Cortex CP1048U   >98%

Fitzgerald 30-AH31   Erythrocyte lysates   Immunogen grade

Scipac P186-2   Erythrocytes   >98% hemoglobin; 25-50% HbA1c; lyophilized; available on request | Diabetes protein

Scipac P186-4   Erythrocytes   >98% hemoglobin; 25-50% HbA1c; frozen in sodium phosphate buffer; 150-200 mg Hb/mL | Diabetes protein

Scipac P186-7   Erythrocytes   >90%; frozen in sodium phosphate buffer | Diabetes protein

USBio H1850-15   Human erythrocyte lysates   No reaction against any glycohemoglobins, including Hb A1a, HbA1b, HbF, & HbAo; 5-10 mg/mL liquid in 1 *mM* KCN, pH 7.5 | Suitable for antigenic applications in immunological protocols; 1 glyco group per β chain monomer N-terminal Val

Hemoglobin A$_2$, Ferrous

Sigma H 0266   Human, stabilized   Chromatographically purified & lyophilized from ammonium bicarbonate containing ~50% nonionic stabilizers; ~98% by agarose electrophoresis; reconsititution with buffer gives >90% ferrous hemoglobin | Suitable for electrophoresis & chromatography standard, but has not been tested for functional equivalence against native preparations (unlyophilized ferrous hemoglobins); Package size indicates the amount of hemoglobin as determined by the procedure of Drabkin, DL, *J Biol Chem*, 164: 703, 1946

Hemoglobin Ao

Fitzgerald 30-AH32   Erythrocyte lysates   Immunogen grade

USBio H1850-18   Human   No reaction against any glycohemoglobins, including Hb A1a, HbA1b, HbF, & HbA1c; 1 mg/mL; lyophilized preparation | Suitable for antigenic applications in immunological protocols; <0.05 glyco group per monomer hemoglobin

Hemoglobin Controls, High Level

Sigma H 4268   Human plasma   Assayed values for hemoglobin, data sheet included

Hemoglobin Controls, Low Level

Sigma H 3268   Human plasma   Assayed values for hemoglobin, data sheet included

Hemoglobin F

Cortex CP1089   >95%

Hemoglobin S, Ferrous

Sigma H 0392   Human, stabilized   Chromatographically purified & lyophilized from ammonium bicarbonate containing ~50% nonionic stabilizers; ~98% by agarose electrophoresis; reconsititution with buffer gives >90% ferrous hemoglobin | Sickle cell hemoglobin; all other comments same as for Sigma H 0267

Hemoglobin Standards

Sigma 525-18   Reconstitutes to 50 mL | For the colorimetric determination of total hemoglobin in blood per Sigma Procedure No. 525

Sigma 527-11   Set contains 3 mL each of standards with hemoglobin levels of 15, 30 & 45 mg/dL in human plasma

Sigma 527-30   30 mg/dL | For the colorimetric determination of hemoglobin in plasma per Sigma Procedure No. 527

Hemoglobin, (^{14}C-Me)-

Sigma H 4390   Bovine   MW ~64.5k   5-50 μCi/mg protein; solution in 40 *mM* potassium phosphate, pH 7.0, in Combi-vial | Radiochemical; prepared from Sigma H 2500

Hemoglobin, (^{3}H(G))-

ARC ART-707   Human   1-5 Ci/mmol; 37-185 GBq/mmol; in sterile water | Radiochemical

Hemoglobin, Cross-Linked Molecular Weight Standard for SDS-PAGE

Sigma H 2507   Bovine   MW 16k (monomer), 32k (dimer), 48k (trimer), 64k (tetramer)   Lyophilized; ~3 mg/vial | See Sigma MW-SDS-280

Hemoglobin, Glycated

Fitzgerald 30-AH33   Erythrocyte lysates   Immunogen grade; glycated

USBio H1850-22   Human erythrocyte lysates   No reaction against Hb A1a, HbA1b, HbF, HbAo & HbA1c; 1 mg/mL; lyophilized preparation | Suitable for antigenic applications in immunological protocols; ~1 glyco group per hemoglobin tetramer; glycated in non A1c positions at E-amino groups of lysines residues

Hemoglobin, HbA1c Free

Scipac P211-1   Erythrocytes   >98%; lyophilized; no HbA1c visible by FPLC | Diabetes protein

Scipac P211-3   Erythrocytes   >98%; frozen in sodium phosphate buffer; no HbA1c visible by FPLC | Diabetes protein

Hemoglobin, N,N-Dimethylated

Sigma H 9891   Bovine blood   Lyophilized, essentially salt-free powder; retains ~20% reactivity with TNBS when compared with nonmethylated hemoglobin; prepared by reductive methylation of Sigma H 2500 by method of Cabacungan | Cabacungan, JC et al, *Anal Biochem*, 124: 272, 1982

Hemoglobin, Oxy-

Biogenesis 4870-4056   Human erythrocytes   MW 66k (including 0.34% iron)   62% oxyhemoglobin, 7% methemoglobin; tested negative for Hepatitis and HIV; semi-pure; lyophilized

Hemolysin, Kanagawa

Sigma H 3142   *Vibrio parahaemolyticus*   Lyophilized powder containing ~50% protein (Lowry); balance Tris-HCl, EDTA, phenylmethylsulfonyl fluoride & sodium azide; activity: ≥400 U/mg protein; 1 hemolytic U causes 50% lysis of a 1% suspension of human red blood cells in phosphate buffered saline, pH 7.0, after 2 hrs incubation at 37°C followed by refrigeration for 12-24 hrs at 4°C | Cherwonogrodzky, JW & Clark, AG, *FEMS Microbiol Lett*, 15: 175, 1982

Hemolysin, α-

Synonyms: Toxin, α-

Sigma H 9395   *Staphylococcus aureus*   ~60% protein (Lowry); balance primarily sodium citrate buffer; activity: ≥10,000 U/mg protein; 1 hemolytic U causes 50% lysis of a 1% suspension of rabbit red blood cells in phosphate buffered saline, pH 7.0, containing 1% bovine serum albumin after 30 min at 37°C followed by refrigeration for 30 min at 4°C | Channel-forming protein similar to complement & perforin, penetrating the cell membrane & creating a defined size pore; stimulates cellular phospholipase activity; Thelestam, M & Blomqvist, L, *Toxicon*, 26: 51, 1988; Fink, D et al, *Cellular Signalling*, 1: 387, 1989

Hemopexin

P180-5   Serum/plasma   >90%; lyophilized; available on request | Serum protein

Heparan Sulfate

Synonyms: Heparitin Sulfate; Heparin Monosulfate

Sigma H 5393   Bovine intestinal mucosa   MW$_{Ave}$ ~7.5k Fast-moving fraction; sodium salt; 90+% (electrophoresis)

Sigma H 7640   Bovine kidney   Sodium salt; similar to Sigma H 9637

Sigma H 9637   Bovine kidney   Sodium salt

Sigma H 9902   Porcine intestinal mucosa   Fast-moving fraction: sodium salt; 90+% (electrophoresis)

Heparan Sulfate Proteoglycan

Sigma H 4777   Engelbreth-Holm-Swarm mouse sarcoma   In 50 *mM* Tris HCl, 150 *mM* NaCl, 1 *mM* EDTA, 0.1 *mM* PMSF, pH 7.4; protein: ≥400 µg/mL; glycosaminoglycan: ≥400 µg/mL; uronic acid: ≥100 µg/mL; sterile-filtered | Isolated from basement membrane of Engelbreth-Holm-Swarm mouse sarcoma; composed of a core protein covalently bound to heparan sulfate chains; sold on the basis of µg protein; for cell culture use

Heparin

Sigma 210-6   Sodium salt; endotoxin-free; preweighed vial: 300 USP units; no preservatives | Bioassay not run by Sigma; sufficient anti-coagulant for 5 mL of blood; not for injection; suitable for use in gram negative endotoxin detection per Sigma Technical Bulletin No. 210

Sigma 840-20   Sodium salt; siliconized glass vials each containing 20 USP units; no preservatives | For use in the histochemical demonstration of nitro blue tetrazolium reduction in neutrophils by procedure in Sigma Procedure No. 840; bioassay not run by Sigma

Sigma H 0777   Bovine intestinal mucosa   Sodium salt; no preservatives; activity: ≥140 USP units/mg | Bioassay not run by Sigma

Sigma H 4898   Bovine lung   Sodium salt; contains no preservatives; activity: ≥140 USP units/mg | Bioassay not run by Sigma

Sigma H 9266   Ovine intestinal mucosa   Sodium salt; no preservatives; activity: ~160 USP units/mg | Bioassay not run by Sigma

Sigma H 0878   Porcine intestinal mucosa   Lithium salt; activity: ≥150 USP U/mg; no preservatives | Bioassay not run by Sigma

Sigma H 1636   Porcine intestinal mucosa   Sodium salt; low calcium content; prepared from Sigma H 3393; activity: ~170 USP U/mg; no preservatives | Bioassay not run by Sigma

Sigma H 2149   Porcine intestinal mucosa   MW$_{ave}$ ~6k Sodium salt; low molecular weight; no preservatives; prepared by enzymatic depolymerization | Bioassay not run by Sigma

Sigma H 3393   Porcine intestinal mucosa   Sodium salt; grade I-A Typical activity: ~170 USP U/mg; no preservatives | Bioassay not run by Sigma

Sigma H 3400   Porcine intestinal mucosa   MW$_{AVE}$ ~3k Sodium salt; low MW; no preservatives; activity: Anti-X$_a$ 75-125 IU/mg; anticlotting 30-50 USP units/mg | Bioassay not run by Sigma; depolymerized by peroxidolysis (free-radical induced cleavage)

Sigma H 5152   Porcine intestinal mucosa   MW$_{AVE}$ ~4k Sodium salt;, Tyramine & FITC labeled; low MW; no preservatives; solution: 1 µg/mL in water; activity: Anti-X$_a$ ~100 U/mg; Anti II$_a$ ~45 U/mg | Bioassay not run by Sigma

Sigma H 5277   Porcine intestinal mucosa   MW$_{AVE}$ ~4k Sodium salt;, tyramine labeled low MW; no preservatives; solution: 1 µg/mL in water; activity: Anti-X$_a$ ~100 U/mg; Anti II$_a$ ~45 U/mg | Bioassay not run by Sigma

Sigma H 5515   Porcine intestinal mucosa   Sodium salt; crude; unbleached; no preservatives; activity: ≥160 IU/mg | Bioassay not run by Sigma

Sigma H 6279   Porcine intestinal mucosa   Ammonium salt; activity: ~140 USP U/mg; no preservatives | Bioassay not run by Sigma

Sigma H 7155   Porcine intestinal mucosa   Zinc salt; no preservatives; activity: ≥140 USP units/mg | Bioassay not run by Sigma

Sigma H 7405   Porcine intestinal mucosa   Sodium salt; deaminated; low molecular weight (mean >5 kD) mono-aldehyde; heparin activity: >100 USP units/mg; antithrombin activity: >100 U/mg | Prepared by nitrous acid deamination of porcine mucosal heparin by a modification of the method of Kosakai, M et al, *J Biochem*, 83: 1567, 1978

Sigma H 8398   Porcine intestinal mucosa   Calcium salt; activity: ≥140 USP U/mg; no preservatives | Bioassay not run by Sigma

Sigma H 9399   Porcine intestinal mucosa   Sodium salt; activity: ≥140 USP U/mg; no preservatives | Bioassay not run by Sigma

Heparin Binding Epidermal Growth Factor-Like Growth Factor

Oncogene PF078   Human recombinant   MW 9.5k (SDS-PAGE) >97% (SDS-PAGE); lyophilized in a sterile-filtered PBS (pH 7.4) solution containing 50 mg of BSA per 1 mg of cytokine | Heterogeneously O-glycosylated; migrates as an approximately 12k protein in SDS-PAGE; activity measured as ability to stimulate (^{3}H)thymidine incorporation in the EGF-responsive mouse fibroblast cell line, Balb/3T3; the ED_{50} for this effect is typically 2.0-5.0 ng/mL

Sigma E 4643   Human recombinant, expressed in *Spodoptera frugiperda* 21   ≥97% (SDS-PAGE); 0.2 µm filtered & lyophilized in PBS containing 2.5 mg BSA; activity is measured by its ability to stimulate ^{3}H-thymidine incorporation in the EGF-responsive BALB/3T3 cell line; endotoxin tested; see Sigma E 1264 | originally purified from the conditioned medium of human U-937 histiocytic lymphoma cells based on its ability to bind heparin; member of the EGF family & accordingly named heparin binding EGF-like growth factor; produced in the insect cell line *Sf 21* by infection with a recombinant baculovirus containing a DNA sequence which encodes the first 148 of the human HB-EGF precursor; purified HB-EGF is mitogenic for the BALB/3T3 cell line & is a very potent mitogen for smooth muscle cells; Higashiyama, S et al, *Science*, 251: 936, 1991

Heparin, Benzalkonium

Sigma H 7280   Prepared from porcine mucosal heparin & alkyl (C_{12}-C_{14}) dimethylbenzylammonium chloride; activity: ~60 USP units/mg; benzalkonium content: ~60%

Heparin-Albumin

Sigma H 0403   Lyophilized powder containing ~50% protein (Biuret); balance is primarily heparin; 3-6 moles heparin/mole albumin | Albumin may be further derivatized through available primary amines for labeling in different detection systems; heparin, ~170 USP units/mg, coupled through terminal formyl by reductive amination to BSA

Heparin-Albumin, Biotin Conjugated

Sigma H 4016   Lyophilized powder containing ~50% protein (Biuret); balance is primarily heparin & ~5% citrate buffer salts; contains 4-6 moles biotin/mole protein | Heparin-albumin coupled to biotin by amide bond through aminocaproyl spacer; Grulich-Henn, J et al, *Thrombosis & Haemostasis*, 64: 420, 1990; Zou, S et al, *Comp Biochem Physiol*, 1038: 889, 1992

Heparin-Albumin, Gold Conjugated

Sigma H 3278   20 nm Colloidal Gold Labeled; mean particle diameter 17-23 nm; monodisperse; suspension in ~50% glycerol containing 0.15 M NaCl, 0.01 M BES, pH 7.2, 0.02% PEG 20, 0.02% sodium azide; concentration: A_{520} ~5.0 | Heparin covalently linked to albumin as carrier, adsorbed to colloidal gold for detection of heparin-binding compounds

Sigma H 5641   10 nm Colloidal Gold Labeled; MEAN particle diameter 8-12 nm; monosdisperse; suspension in ~50% glycerol containing 0.15 M NaCl, 0.01 M sodium phosphate, pH 7.0, 0.02% PEG 20, 0.02% sodium azide; concentration: A_{520} ~5.0 | Heparin covalently linked to albumin as carrier, adsorbed to colloidal gold for detection of heparin-binding compounds

Sigma H 9516   5 nm Colloidal Gold Labeled; mean particle diameter 3.5-6.5 nm; monosdisperse; suspension in ~50% glycerol containing 0.15 M NaCl, 0.01 M sodium phosphate, pH 7.0, 0.02% PEG 20, 0.02% sodium azide; concentration: A_{520} ~5.0 | Heparin covalently linked to albumin as carrier, adsorbed to colloidal gold for detection of heparin-binding compounds

Hepatitis A Virus Antigen

Biodesign R9A001   Antigen Strain pHM175   Purified | Infectious disease antigen

Hepatitis B Virus Antigen e Epitope

Biodesign R65915   Recombinant   MW 17k   Purified; AA 160 | Infectious disease antigen

Hepatitis B Virus Core Antigen

IBT HBA-020-4, HBA-020-5, HBA-020-7   E. coli   >90%; 0.50 mg/mL in 100 *mM* NaCO$_3$, pH 9.3

Biodesign R3B601   Recombinant   Purified; AA 183 | Infectious disease antigen

Biodesign R65914   Recombinant   MW 2500k   Purified; AA 180 | Infectious disease antigen

USBio H1905-02   Recombinant expressed in *E. coli*   ≥95% (SDS-PAGE, 280 nm, Bradford); 10 mg/mL liquid in 7.5 *mM* phosphate buffer, pH 7.2, 75 *mM* NaCl, 50% glycerol | Suitable for antigenic applications in immunological protocolsHBV core antigen HBcAg (recombinant) 1 to 183 of HBV core antigen (18kD, ayw); cloned from HBV 320 genome; Reacts strongly with human HBV positive serum

Fitzgerald 30-AH39   Recombinant, Ecto-Domain (modified yeast)

USBio H1905   Yeast (*Pichia pastoris*) recombinant   MW 24k (protein), 27k (glycoprotein)   ≥95% (SDS-PAGE); 1 mg/mL frozen liquid in 0.4 M NaCl, 0.05 M sodium acetate, pH 4.5 | Suitable for antigenic applications in immunological protocols; reacts with HB core Ag sera & human monoclonals in ELISA & Western blot; a capsid structure in excess of 2 million Daltons

Hepatitis B Virus e Antigen

Fitzgerald 30-AH18   Recombinant, expressed in *E. coli*

Hepatitis B Virus Pre-S1 Antigen

Biodesign R65913   Recombinant   MW 13.5k   Purified; AA 108 | Infectious disease antigen

Hepatitis B Virus Surface Antigen

IBT HBA-010-4, HBA-010-5, HBA-010-7   Genetically engineered CHO cells   >95%; 1.0 mg/mL solution in 10*mM* Tris-HCl, pH 7.5 | Contains the preS$_2$ region

Cortex CP2019r   Recombinant   >95%

Fitzgerald 30-AH38   Recombinant, expressed in *E. coli*

Hepatitis B Virus Surface Antigen, Subtype ad

Biodesign R36001   Purified | Infectious disease antigen

USBio H1910-27   Human plasma   MW 24k (protein), 27k (glycoprotein)   ≥99% (HPLC); complete virions are not detected; heat inactivated (10 hours @ 60°C); 2.0 mg/mL (OD 280 nm) liquid in PBS buffer, 5% sucrose, 0.01% NaN$_3$ | Suitable for antigenic applications in immunological protocols

Fitzgerald 30-AH16   Human plasma human-ad   Affinity purified

USBio H1910-20   Human recombinant, from modified yeast   MW 24k (protein), 27k (glycoprotein)   ≥98% (SDS-PAGE); ~1 mg/mL (OD 280 nm) liquid in PBS, 0.1% NaN$_3$ | Suitable for antigenic applications in immunological protocols; reacts with HBsAg sera & human monoclonals in ELISA & Western blot; 226 AA

Hepatitis B Virus Surface Antigen, Subtype adw

Biodesign R3B602   Recombinant   Purified; AA 226 | Infectious disease antigen

Hepatitis B Virus Surface Antigen, Subtype ay

Biodesign R36002   Purified | Infectious disease antigen

USBio H1910-28   Human plasma   MW 24k (protein), 27k (glycoprotein)   ≥99%; complete virions are not detected; heat inactivated (10 hours @ 60°C); 2.0 mg/mL (OD 280 nm) supplied in PBS buffer, 5% sucrose, 0.01% NaN$_3$ | Suitable for antigenic applications in immunological protocols

Fitzgerald 30-AH17   Human plasma human-ay   Affinity purified

Hepatitis B Virus Surface Antigen, Subtype ayw

USBio H1910-21 Human recombinant, expressed in *Saccharomyces cerevisae* containing the plasmid pCGA7   Purity: ≥98% (SDS-PAGE) supplied in 0.05 M phosphate buffer, 0.2 M NaCl, pH 7.2 | Suitable for use with pRc/CMV-HBs(S) or other plasmids to validate DNA-based immunization methods; can be used in ELISA & other immunological assays; purified by clarification, microfiltration, ultrafiltration, adsorption chromatography, ion-exchange chromatography, ultracentrifugation; tested in ELISA with anti-HBsAg antibodies; MW: 2 million Daltons; morphologically HBsAg possesses a subunit diameter of 18-22nm

Biodesign R86870 Recombinant   Purified | Infectious disease antigen

Hepatitis C Virus Core Antigen

IBT HCA-070-3, HCA-070-4 Yeast cells   MW 36k   >90%; 0.7 mg/mL solution in sodium borate, 10 mM thiocyanate, 10 mM DTT, 5 mM EDTA, pH 9.5 | Covers Met[1] to Gly[120] of HCV polyprotein; fusion protein with human Superoxide Dismutase; reacts in ELISA with serum from HCV positive individuals

Hepatitis C Virus eII Antigen

IBT HCA-090-2 CHO cells   MW 55k   >90%; 0.2 mg/mL solution in 10 mM sodium phosphate, 0.1 M NaCl, 1% Trition X-100 | Covers Ala[384] to Lys[715] of HCV polyprotein; runs anomalously on PAGE

Hepatitis C Virus NS3 Antigen

IBT HCA-110-3, HCA-110-4 *E. coli*, as a fusion protein with human superoxide dismutase   MW 45k   >95%; 1 mg/mL solution in 50 mM Tris-HCl, 50 mM NaCl, 10 mM DTT, 1 mM EDTA, pH 7.4 | Covers Ala[1192] to Cys[1457] of HCV polyprotein; equivalent to c33c antigen

Hepatitis C Virus NS3/NS4 Antigen

IBT HCA-100-3, HCA-100-4 Yeast cells, as a fusion with human superoxide dismutase   MW 53k   >95%; 0.50 mg/mL solution in 20 mM Tris-HCl, 0.1% SDS, 10 mM DTT, pH 7.0 | Covers Asp[1569] to Pro[1931] of HCV polyprotein; equivalent to C-100 antigen

Hepatitis C Virus NS4 Antigen

IBT HCA-120-3, HCA-120-4 *E. coli*, as a fusion protein with human superoxide dismutase   MW 29k   >95%; 1 mg/mL solution in 50 mM sodium borate, 0.5 M NaCl, 20 mM DTT, 2 mM EDTA, pH 8.4 | Covers Ile[1694] to Leu[1735] of HCV polyprotein; equivalent to 5-1-1 antigen

Hepatitis C Virus NS5 Antigen

IBT HCA-130-3, HCA-130-4 Yeast cells, as a fusion with human superoxide dismutase   MW 150k   >95%; 0.8mg/mL solution in 20 mM MES, 100 mM NaCl, 10 mM DTT, 1 mM EDTA, 0.1% SDS, pH 6.0). | Covers Asn[2054] to Cys [2995] of HCV polyprotein

Hepatitis C Virus Nucleocapsid p22

USBio H1920-20 Recombinant expressed in *Pichia pastoris* ≥97%; no contaminants detected; single band by SDS-PAGE, IEP, &/or RID; 1 mg/mL supplied in Tris, saline, pH 8.0 | Suitable for antigenic applications in immunological protocols; 173 AA; mature nucleocapsid core protein

Hepatitis D Virus p24

Biodesign R65892 Recombinant   Purified | Infectious disease antigen

Heregulin

Oncogene PF048 Human recombinant   MW 7k   >97% (SDS-PAGE); lyophilized in 15% acetonitrile & 0.1% TFA containing 50 mg of BSA per 1 mg of cytokine; biological activity: EC_{50} of 20-40 μg/mL as determined by its inhibitory effect on human breast cancer cell line SK-BR-3 or in a serum-free proliferation assay using human MCF7 cells | Mitogenic for Schwann & various epithelial cells; species reactivity: human; for proliferation studies; shown to be mitogenic for Schwann cells & various epithelial cells

Heregulin α Epidermal Growth Factor

Sigma H 5529 Human recombinant, expressed in *E. coli*   MW 7k   ≥97% (SDS-PAGE); 0.2 μm filtered & lyophilized from 15% acetonitrile & 0.1% trifluoroacetic acid solution containing 2.5 mg BSA; bioactivity is measured in a bioassay using SK-BR-3 cells; endotoxin tested; see Sigma E 1264 | Has an EGF-like domain

Heregulin β, Epidermal Growth Factor Domain

USBio H2030-50

Herpes Simplex Virus I HF

Biodesign R86871 Purified | Infectious disease antigen

Herpes Simplex Virus I McIntyre

Biodesign R57145 Purified | Infectious disease antigen

Biodesign R70002 Lysate | Infectious disease antigen

Herpes Simplex Virus I N-Terminal Glycoprotein D

Biodesign R3B501 Recombinant   Purified; glycosylated | Infectious disease antigen

Herpes Simplex Virus II Glycoprotein G

IBT HS2A-470-3, HS2A-470-4 Yeast cells, as a fusion with human superoxide dismutase   MW 50k   >90%; 1 mg/mL solution in PBS, 0.1% SDS, 1 mM EDTA, pH 7.4 | Comprises Met[1] to Asp[190]

High Density Lipoprotein

USBio H2038 Human   ≥98%; no contaminants detected; single band by SDS-PAGE, IEP, &/or RID; ~1-5 mg/mL supplied in borate/saline buffer | Suitable for antigenic applications in immunological protocols

Fitzgerald 30-AH08 Human plasma   High purity

Hirudin

Fluka 53287 Leech   MW 7027   ~100 U/mg; ≥90% protein; 1 U corresponds to the amount of inhibitor which reduces the thrombin activity by 1 U at pH 7.5, 37°C, 0.274 mM Tos-Gly-Pro-Arg-4NA•AcOH as substrate, 1.348 U/L thrombin | Thrombin specific inhibitor; Chang, JY, *FEBS Lett*, 164: 307, 1983

Sigma H 9022 Leech   Lyophilized; activity: ≥6000 U/mg protein; 1 U neutralizes 1 NIH U of thrombin using a fibrinogen assay at 37°C | Markwardt, F, *Meth Enz*, 19: 924, 1970; Bagdy, D et al, *Meth Enz*, 45: 669, 1976

Sigma H 0393 Leech recombinant (Lys[47])-rHV2 variant, produced by cDNA expressed in Saccharomyces *cerevisiae* Activity: 7000-14,000 U/mg protein; 1 U neutralizes 1 NIH U of thrombin at 37°C based on direct comparison to an NIH thrombin reference standard (Lot J) | Markwardt, F, *Meth Enz*, 19: 924, 1970; Bagdy, D et al, *Meth Enz*, 45: 669, 1976; Loison, G et al, *Bio/Technology*, 6: 72, 1988

Sigma H 7016 Leeches   Activity: ≥1000 U/mg protein; 1 U neutralizes 1 NIH U of thrombin at 37°C based on direct comparison to an NIH thrombin reference standard (Lot J) | Markwardt, F, *Meth Enz*, 19: 924, 1970; Bagdy, D et al, *Meth Enz*, 45: 669, 1976

Sigma H 7380 Leeches   Lyophilized from 0.045 M NaCl-0.02 M Tris buffer, pH 7.5; activity: 300-1000 U/mg protein; 1 U neutralizes 1 NIH U of thrombin at 37°C based on direct comparison to an NIH thrombin reference standard (Lot J) | Markwardt, F, *Meth Enz*, 19: 924, 1970; Bagdy, D et al, *Meth Enz*, 45: 669, 1976

Hirudin, (Lys⁴⁷)-

Fluka 53288   Leech recombinant, from *S. cerevisiae*   ~600 U/mg; ≥90% protein; 1 U corresponds to the amount of inhibitor which reduces the thrombin activity by 1 U at pH 7.5, 37°C, 0.274 *mM* Tos-Gly-Pro-Arg-4NA•AcOH as substrate, 1.348 U/L thrombin | Loison, G et al, *Biotechnology*, 6: 72, 1988

Hirudin, (Tyr⁶³)-

Synonyms: Hirudin, Desulfato-; Hirudin, Desulfo-

Fluka 53289   Leech recombinant, from yeast C₂₈₇H₄₄₀N₈₀O₁₁₀S₆   ~600 U/mg; ≥95% (HPLC); ≥90% protein; 1 U corresponds to the amount of inhibitor which reduces the thrombin activity by 1 U at pH 7.5, 37°C, 0.274 *mM* Tos-Gly-Pro-Arg-4NA•AcOH as substrate, 1.348 U/L thrombin; ≤10% water; ≤0.05 U/mg endotoxin | Most potent, specific inhibitor of thrombin; Meyhack, B et al, *Thromb Res Suppl*, 7: 33, 1987; Grossenbacher, H et al, *ibid*, 7: 34, 1987; Markwardt, F et al, *Pharmazie*, 43: 202, 1988

Histone Core

USBio H5110-10A   Chicken erythrocytes   ≥90% (SDS PAGE & coomassie blue staining) lyophilized; aseptically reconstitute to 1 mg/mL in sterile water & aliquot to avoid repeated freezing & thawing | Purified by acid extraction & TCA precipitation from chicken erythrocytes for use as a substrate in histone acetyl-transferase (HAT); an effective substrate for a number of lysine acetyl-transferases

Histone H1

USBio H5110-02   ≥95%; frozen solution in 1ml sterile water, sterilized through a 0.2um membrane filter & packaged aseptically | An effective substrate for a number of serine/threoninekinases; tested by using PKCa to phosphorylate 10 µg histone H1; purified as a Lys-rich fraction

Histone Type II-A

Sigma H 9250   Calf thymus   Lyophilized powder; unfractionated whole histone | Luck, JM et al, *J Biol Chem*, 233: 1407, 1958; Satake, K et al, *J Biol Chem*, 235: 2801, 1960

Histone Type II-AS

Sigma H 7755   Calf thymus   Prepared by extraction in 1 M NaCl solution, precipitation in water, acid extraction & dialysis & lyophilization | Do not confuse Sigma "Type" designations with the "Fraction" designations used by Luck & co-workers; Luck, JM et al, *J Biol Chem*, 233: 1407, 1958; Satake, K et al, *J Biol Chem*, 235: 2801, 1960

Histone Type III-S

Sigma H 5505   Calf thymus   Lysine-rich fraction as isolated & described by de Nooij; characterized as mainly subgroup f₁ by SDS-PAGE | Luck, JM et al, *J Biol Chem*, 233: 1407, 1958; Satake, K et al, *J Biol Chem*, 235: 2801, 1960; nooij, EH & Westenbrink, HGK, *Biochim Biophys Acta*, 62: 608, 1962

Histone Type III-SS

Sigma H 4524   Calf thymus   Isolated as a lysine-rich fraction; characterized as mainly subgroup f₁ by SDS-PAGE | Tested & found suitable as a substrate for protein kinase C; phosphorylation of this histone may also be suitable in other protein kinase systems; commonly used for chromosome-reconstitution studies; Luck, JM et al, *J Biol Chem*, 233: 1407, 1958; Satake, K et al, *J Biol Chem*, 235: 2801, 1960; nooij, EH & Westenbrink, HGK, *Biochim Biophys Acta*, 62: 608, 1968; Cicirelli, MF et al, *J Biol Chem*, 263: 2009, 1988; Cole, RD, *Int J Peptide Protein Res*, 30: 433, 1987

Histone Type II-S

Sigma H 6005   Calf thymus   Prepared by extraction in 1 M NaCl solution, precipitation in water, acid extraction & reprecipitation with alcohol | Luck, JM et al, *J Biol Chem*, 233: 1407, 1958; Satake, K et al, *J Biol Chem*, 235: 2801, 1960

Histone Type VIII-S

Sigma H 4380   Calf thymus   Arginine-rich subgroup f₃ isolated by a modification of the method of Johns; other subgroups | Luck, JM et al, *J Biol Chem*, 233: 1407, 1958; Satake, K et al, *J Biol Chem*, 235: 2801, 1960; Johns, EW, *Biochem J*, 92: 55, 1964

Histone Type VII-S

Sigma H 4255   Calf thymus   Slightly Lysine-rich subgroup f₂b as isolated & described by Johns; other subgroups | Luck, JM et al, *J Biol Chem*, 233: 1407, 1958; Satake, K et al, *J Biol Chem*, 235: 2801, 1960; Johns, EW, *Biochem J*, 92: 55, 1964

Histone Type VI-S

Sigma H 6881   Calf thymus   Mixture of arginine-rich subgroup f₂a1 & slightly Lysine-rich subgroup f₂a2 as isolated & described in Johns; other subgroups | Luck, JM et al, *J Biol Chem*, 233: 1407, 1958; Satake, K et al, *J Biol Chem*, 235: 2801, 1960; Johns, EW, *Biochem J*, 105: 611, 1967

HIV I & II C-Terminal gp120 + gp41/gp36

Biodesign R49132   Recombinant   Purified | Infectious disease antigen

HIV I C-Terminal gp120 + gp41

Biodesign R49550   Recombinant   Purified | Infectious disease antigen

HIV I Envelope Protein 101, SF-2 Isolate

IBT HI1A-610-4, HI1A610-5   Yeast cells, as a fusion with human superoxide dismutase   MW 41.7k   >90%; 1 mg/mL solution in PBS, 20 *mM* DTT, 0.1% SDS, 2 *mM* EDTA, pH 7.4 | Covers Ala⁵⁴⁸ to Leu⁷⁶⁷ of HIV-1 (SF2 isolate) envelope protein

HIV I Envelope Protein 131, SF-2 Isolate

IBT HI1A-620-4, HI1A-620-5   Yeast cells, as a fusion with human superoxide dismutase   MW 35.6k   >90%; 1 mg/mL solution in PBS, 10 *mM* EDTA, 0.1% SDS, 1 *mM* EDTA, pH 7.5 | Covers Val⁴⁷³ to Leu⁶⁶⁸ (D Ala⁵¹⁸ to Val⁵⁴⁶) of HIV-1 (SF-2 isolate) envelope protein

HIV I Envelope Protein gag p17, SF-2 Isolate

IBT HI1A-710-4, HI1A-710-5   Yeast cells   MW 14.9k >90%; 1 mg/mL solution in PBS, 1 *mM* EDTA, pH 7.5 | Covers Met⁵ to Tyr¹³⁸ of HIV-1 gag protein (SF-2 isolate)

HIV I Envelope Protein gp101, SF-2 Isolate

USBio H6003-05   Recombinant from CHO cells   ≥95% purified by ion-exchange chromatography & gel filtration; ~1 mg/mL liquid in PBS, 1 *mM* EDTA, 1 *mM* EGTA | Suitable for antigenic applications in immunological protocols; contains the SF2 isolate AA 39-517

HIV I Envelope Protein gp120

Biodesign R65906   Recombinant   Purified | Infectious disease antigen

Biodesign R65919   Recombinant   Purified | Infectious disease antigen

USBio H6003-08   Recombinant from CHO cells   MW 120k (Glycosylated)   ≥95% purified by ion-exchange chromatography & gel filtration; ~1 mg/mL liquid in 2X PBS, 1 *mM* EDTA, 1 *mM* EGTA (pH 7.4) | Suitable for antigenic applications in immunological; contains the SF2 isolate AA 39-517

HIV I Envelope Protein gp120, SF-2 Isolate

IBT HI1A-600-4, HI1A-600-5   CHO cells   MW 120k >95%; 1.0 mg/mL solution in 2X PBS, 1 *mM* EDTA, 1 *mM* EGTA, pH 7.4 | Covers Glu³⁹ to aa Arg⁵¹⁷ of HIV 1 (SF2 isolate) envelope protein

HIV I Envelope Protein gp131

USBio H6003-10 Recombinant by the fusion between yeast cells & human superoxide dismutase from an isolate of HIV-1 (SF-2) envelope protein MW 35.6k ≥90% purified by ion-exchange chromatography & gel filtration; 1 mg/mL liquid in PBS, 10 mM EDTA, 0.1%SDS, 1 mM EGTA | Suitable for antigenic applications in immunological protocols; corresponds to Val^{473}-Leu^{668} (D-Ala^{518}-Val^{546})

HIV I Envelope Protein gp31, SF-2 Isolate

IBT HI1A-700-4, HI1A-700-5 Yeast cells, as a fusion with human superoxide dismutase MW 46k >85%; 1 mg/mL solution in PBS, 10 mM DTT, 0.1% SDS, 2 mM EDTA, pH 7.3 | Covers Met^{737} to Asp^{1003} of HIV-1 envelope protein (SF-2 isolate)

HIV I Envelope Protein gp41

Biodesign R65908 MW 146k Purified | Infectious disease antigen

Biodesign R65907 Recombinant Purified | Infectious disease antigen

USBio H6003-16 Recombinant ≥95% (SDS-PAGE, 280 nm, Bradford); 1 mg/mL liquid in 8 M urea, 20 mM Tris-HCl, pH 8.0, 10 mM β-MSH | Suitable for antigenic applications in immunological protocols; HIV I Envelope region AA 466-753; β-galactosidase (114kD) fused at the N-terminus; reacts strongly with human HIV positive serum

HIV I Envelope Protein p31, SF-2 Isolate

USBio H6003-15 Recombinant in genetically engineered yeast cells as a fusion protein with human superoxide dismutase ≥85%; ~1 mg/mL liquid in PBS, 10 mM DTT, 0.1% SDS, 2 mM EDTA (pH7.3) | Suitable for antigenic applications in immunological protocols; covers Met^{737}-Asp^{1003} of the HIV-1 envelope protein (SF-2 isolate); purified by ion-exchange chromatography & gel filtration

HIV I gag gp24

Biodesign R65909 Recombinant Purified | Infectious disease antigen

Biodesign R65910 Recombinant Purified | Infectious disease antigen

HIV I gag p17, SF-2 Isolate

USBio H6003-20 Recombinant yeast MW 14.9k ≥90%; 0.4 mg/mL liquid in PBS at 0.4 mg/mL in PBS, 1 mM EDTA (7.5) | Covers Met^5-Tyr^{138} of the HIV-1 gag protein (SF-2 isolate); the p17 protein is from Met^5-Tyr^{138} of the gag region of the SF2 isolate of HIV1; suitable for antigenic applications in immunological protocols; purified by ion-exchange chromatography & gel filtration

HIV I gag p24, SF-2 Isolate

USBio H6005-11 Recombinant yeast (*Pichia pastoris*) ≥90%; 1 mg/mL liquid in PBS, 1 mg/mL in 30 mM Tris-HCl, 1 mM EDTA, (pH 9.0) | Suitable for antigenic applications in immunological protocols; covers Pro^{139}-Leu^{369} of the HIV-1 gag protein (SF-2 isolate); 231 AA; purified by ion-exchange chromatography & gel filtration

IBT HI1A-720-4, HI1A-720-5 Yeast cells MW 27k >90%; 1 mg/mL solution in 30 mM Tris-HCl, 1mM EDTA, pH 9.0 | Covers Pro^{139} to Leu^{369} of HIV-1 gag protein (SF-2 isolate)

HIV I gag p24, Strain IIIB

USBio H6003-25 Recombinant ≥95% (SDS-PAGE, 280 nm, Bradford); 1 mg/mL; liquid in 8 M urea, 20 mM Tris-HCl, pH 8.0, 10 mM β-MSH | Suitable for antigenic applications in immunological protocols; HIV-1 gag region AA 77-436; β-galactosidase (114kD) fused at the N-terminus; reacts strongly with human HIV positive serum

HIV I gag Protein

ICN 158377 Recombinant, expressed in *E. coli* >95% | Reacts with HIV positive serum; used to elicit reactive Ab

HIV I gp160 Envelope Protein

ICN 198752 Recombinant, expressed in *E. coli* >95% | Reacts strongly with HIV positive serum; used to elicit reactive Ab

HIV I gp36 Envelope Protein

ICN 198753 Recombinant, expressed in *E. coli* >95% | Reacts strongly with HIV positive serum; used to elicit reactive Ab

HIV I gp41-1

USBio H6004 Recombinant ≥95%; no contaminants detected; single band by SDS-PAGE, IEP, &/or RID; 1 mg/mL supplied in Tris, PBS, pH 8.0 | Suitable for antigenic applications in immunological protocols; glycosylated recombinant ecto-domain of gp41 (546-682 of HxB2); heterogeneously glycosylated & is seen predominately as a series of bands from 16,000 to 28,000 Daltons on Western blot; recommended that the product be incubated at 40°C for 30 minutes for SDS PAGE

HIV I nef

USBio H6004-14 Recombinant ≥95% (SDS-PAGE, 280 nm, Bradford); 1 mg/mL liquid in 0.01 M sodium carbonate, 0.01 M EDTA, 0.014 M β-MSH | Suitable for antigenic applications in immunological protocols; HIV-1 nef region AA 3-190; β-galactosidase (114kD) fused at the N-terminus; reacts strongly with human HIV positive serum

HIV I nef Protein

ICN 198754 Recombinant, expressed in *E. coli* >95% | Reacts strongly with HIV positive serum; used to elicit reactive Ab

HIV I N-Terminal gp24

Biodesign R3B304 Recombinant Purified | Infectious disease antigen

HIV I N-Terminal gp41-1

Biodesign R3B301 Recombinant Purified; glycosylated | Infectious disease antigen

HIV I p24

Biodesign R49301 Recombinant Purified | Infectious disease antigen

HIV II Envelope Protein (390-702)

Biodesign R8A114 Recombinant MW 34k Purified | Infectious disease antigen

HIV II Envelope Protein 201

USBio H6009-10 Recombinant ≥95%; ~1 mg/mL liquid in PBS, 1 mM EDTA, 1 mM EGTA | Suitable for antigenic applications in immunological protocols; purified by ion-exchange chromatography & gel filtration

USBio H6009-11 Recombinant ≥95%; ~1 mg/mL liquid in PBS, 1 mM EDTA, 1 mM EGTA | Purified by ion-exchange chromatography & gel filtration

IBT HI2A-800-4, HI2A-800-5 Yeast cells, as a fusion with human superoxide dismutase MW 33k >92%; 1 mg/mL solution in PBS, 0.1% SDS, 2 mM EDTA, pH 7.5 | Covers Ala^{548} to Lys^{708} (D Val^{686}-Ala^{704}) of the envelope antigen of HIV-2 isolate UC1.

HIV II Envelope Protein 300

USBio H6009-12 Recombinant ≥95%; ~1 mg/mL liquid in PBS, 1 mM EDTA, 1 mM EGTA | Suitable for antigenic applications in immunological protocols; purified by ion-exchange chromatography & gel filtration

IBT HI2A-810-4, HI2A-810-5 Yeast cells, as a fusion with human superoxide dismutase MW 37k Covers Tyr^{457} to Asp^{654} (D Gly^{509}-Ala^{530}) of the envelope protein of HIV-2 (ISYR isolate)

HIV II Envelope Protein 310

USBio H6009-14 Recombinant ≥95%; ~1 mg/mL liquid in PBS, 1 mM EDTA, 1 mM EGTA | Suitable for antigenic applications in immunological protocols; purified by ion-exchange chromatography & gel filtration

HIV II Envelope Protein 310, ISYR Isolate

IBT HI2A-820-4, HI2A-820-5 Yeast cells, as a fusion with human superoxide dismutase MW 37k >90%; 1.0 mg/mL solution in PBS, 0.10% SDS, 2 mM EDTA, pH 7.5 | Covers Tyr 457 to Asp 654 (D Gly 509-Ala 530) of the envelope protein of HIV-2 (ISYR isolate)

HIV II Envelope Protein 320

USBio H6009-16 Recombinant ≥95%; ~1 mg/mL liquid in PBS, 1 mM EDTA, 1 mM EGTA | Suitable for antigenic applications in immunological protocols; purified by ion-exchange chromatography & gel filtration

HIV II Envelope Protein 320, UC2 Isolate

IBT HI2A-830-4, HI2A-830-5 Yeast cells, as a fusion with human superoxide dismutase MW 37k >90% | Covers Tyr^{469} to Asp^{666} (D Gly^{521}-Ala^{542}) of the envelope protein of HIV-2 (UC2 isolate)

HIV II Envelope Protein gp36

Biodesign R65911 Recombinant MW 148k Purified | Infectious disease antigen

HIV II gag p26

USBio H6009-20 Recombinant ≥95%; ~1 mg/mL liquid in PBS, 1 mM EDTA, 1 mM EGTA | Suitable for antigenic applications in immunological protocols; purified by ion-exchange chromatography & gel filtration

HIV II gag p26, UC-1 Isolate

IBT HI2A-850-4, HI2A-850-5 Yeast cells, as a fusion with human superoxide dismutase MW 26k >90% | Covers Val^{132} to Leu^{360} of HIV-2 gag protein (UC-1 isolate)

HIV II gp36

USBio H6009-40 Recombinant ≥95%; no contaminants detected; single band by SDS-PAGE, IEP, &/or RID; 1 mg/mL supplied in Tris, PBS, pH 8.0 | Suitable for antigenic applications in immunological protocols; glycosylated recombinant ecto-domain of gp 36 (534-654 of ST)

HIV II N-Terminal gp36

Biodesign R3B306 Recombinant Purified | Infectious disease antigen

HIV Inhibitory Protein

ICN 159836 *Oxyuranus scutellatus* (Australian taipan) venom MW 13k Purified (single band by SDS-PAGE) | Inhibits replication of HIV-1 & -2 viruses in cell culture

HTLV I Envelope Protein 701

USBio H7950-10 Recombinant Suitable for antigenic applications in immunological protocols

HTLV I Envelope Protein 701, ATK-1 Isolate

IBT HTIA-900-4, HTIA-900-5 Yeast cells, as a fusion with human superoxide dismutase MW 46k >90%; 0.5 mg/mL solution in PBS, 0.1% SDS, 1 mM EDTA, pH 7.5 | Covers Asp^{166} to Arg^{440} of HTLV-I envelope protein (isolate ATK-1)

HTLV I Envelope Protein 702

USBio H7950-20 Recombinant Suitable for antigenic applications in immunological protocols

HTLV I Envelope Protein 702, ATK-1 Isolate

IBT HTIA-910-4, HTIA-910-5 Yeast cells, as a fusion with human superoxide dismutase MW 44.5k >90%; 0.5 mg/mL solution in PBS, 0.1% SDS, 1 mM EDTA, pH 7.5 | Covers Asp^{166} to Arg^{440} (D Ala^{313} to Ala^{331}) of HTLV-I envelope protein (isolate ATK-1)

HTLV I Envelope Protein RE1

ICN 158369 MT-2 isolate, expressed in *E. coli* Immunoreactive with ATL(+) patient sera | Spans the envelope region; includes regions unique to the transmembrane & 175 AA from the highly immunogenic external portions of the HIV envelope, as well as the junction portion of the envelope

HTLV I Envelope Protein RE3

ICN 158370 MT-2 isolate, expressed in *E. coli* Immunoreactive with ATL(+) patient sera | Spans the envelope region; includes regions unique to the transmembrane & 142 AA from the highly immunogenic external portions of the HIV envelope, as well as the junction portion of the envelope

HTLV I Envelope Protein RE5

ICN 158371 MT-2 isolate, expressed in *E. coli* Immunoreactive with ATL(+) patient sera | Spans the envelope region; includes 128 AA from the immunogenic regions unique to the transmembrane & 6 AA from the highly immunogenic external portions of the HIV envelope, as well as the junction portion of the envelope

HTLV I Envelope Protein RE6

ICN 158372 MT-2 isolate, expressed in *E. coli* Immunoreactive with ATL(+) patient sera | Spans the envelope region; includes 128 AA from the immunogenic regions unique to the transmembrane & 6 AA from the highly immunogenic external portions of the HIV envelope, as well as the junction portion of the envelope (RE3 + RE5)

HTLV I gag p19

USBio H7950-30 Recombinant Suitable for antigenic applications in immunological protocols

HTLV I gag p19, ATK-1 Isolate

IBT HTIA-980-4, HTIA-980-5 Yeast cells, as a fusion with human superoxide dismutase MW 30.7k >90%; 0.50 mg/mL solution in PBS, 1 mM EDTA, pH 7.4 | Covers Met^1 to Leu^{130} of HTLV-I gag protein (isolate ATK-1)

HTLV I gag p24

USBio H7950-40 Recombinant Suitable for antigenic applications in immunological protocols

HTLV I gag p24, ATK-1 Isolate

IBT HTIA-990-4, HTIA-990-5 Yeast cells MW >95%; 1 mg/mL solution in PBS, 0.1% SDS, 2 mM EDTA, pH 7.5 | Covers Pro^{131} to Leu^{344} of HTLV-I gag protein (isolate ATK-1)

HTLV I gp46

Biodesign R49142 Recombinant Purified | Infectious disease antigen

HTLV I p24

Biodesign R49152 Recombinant Purified | Infectious disease antigen

HTLV II Envelope Protein 801

USBio H7951-10   Suitable for antigenic applications in immunological protocols

HTLV II Envelope Protein 801, H6.0/MO Isolate

IBT HTIIA-1160-4, HTIIA-1160-5   Yeast cells, as a fusion with human superoxide dismutase   MW   >90%; 1 mg/mL solution in PBS, 0.1% SDS, 1 mM EDTA, pH 7.5 | Covers Ala[163] to Arg[436] of HTLV-II envelope protein (H6.0/MO isolate)

HTLV II Envelope Protein 802

USBio H7951-20   Suitable for antigenic applications in immunological protocols

IBT HTIIA-1170-4, HTIIA-1170-5   Yeast cells, as a fusion with human superoxide dismutase   MW 30k   >90%; 1 mg/mL solution in PBS, 0.1% SDS, 1 mM EDTA, pH 7.4 | Covers Ala[309] to Gln[440] of HTLV-II envelope protein (lambda H6.0 isolate)

HTLV II gag p19

USBio H7951-30   Suitable for antigenic applications in immunological protocols

HTLV II gag p19, Lambda H6.0 Isolate

IBT HTIIA-1200-4, HTIIA-1200-5   Yeast cells, as a fusion with human superoxide dismutase   MW 15k   >90%; 1 mg/mL solution in PBS, 0.1% SDS, 1 mM EDTA, pH 7.5 | Covers Met[1] to Phe[136] of HTLV-II gag protein (lambda H6.0 isolate)

HTLV II gag p24

USBio H7951-40   Suitable for antigenic applications in immunological protocols

HTLV II gag p24, Lambda H6.0 Isolate

IBT HTIIA-1190-4, HTIIA-1190-5   Yeast cells, as a fusion with human superoxide dismutase   MW 24k   >90%; 1 mg/mL solution in PBS, 0.1% SDS, 1 mM EDTA, pH 7.5 | Covers Pro[137] to Leu[350] of HTLV-II gag protein (lambda H6.0 isolate)

HVEM/Fc Chimera

Synonyms: TNFRSF14

R&D Systems 356-HV-100   Human recombinant, NSO-expressed   >95%; lyophilized; ED$_{50}$: 0.5-2 μg/mL | Species specificity: human HVEM; LIGHT, a member of the TNF family can trigger apoptosis of cells expressing both HVEM & LTβR & its cytotoxicity can be blocked by HVEM- or LTβR-Fc fusion proteins; Zhai, Y et al, *J Clin Invest*, 63: 1142, 1998

I-309

Biodesign A52037H   *E. coli*   >99%

Chemicon GF067   Human   ≥95%

PeproTech 300-37   Human recombinant, expressed in *E. coli*   MW 8.5k   >98% (SDS-PAGE) & HPLC; <0.1 ng endotoxin per mg (1EU/mg); lyophilized | Small glycoprotein secreted by activated T lymphocytes; structurally related to other CC chemokines; 74 AA; SA determined by its ability to chemoattract total human T cell population

Immunoglobulin A

Sigma I 0633   Human   Reagent grade; lyophilized; purified from pooled colostrum; ≥95%; essentially salt-free preparation may be reconstituted with 150 mM NaCl | Human & animal proteins may be used as antigens, standards, coating proteins & blocking agents; also used as starting materials for the preparation of immunogens & solid-phase immunoadsorbents

Sigma I 1010   Human   Reagent grade; lyophilized; purified from pooled colostrum; ≥95%; essentially salt-free preparation may be reconstituted with 150 mM NaCl | Human & animal proteins may be used as antigens, standards, coating proteins & blocking agents; also used as starting materials for the preparation of immunogens & solid-phase immunoadsorbents

Sigma I 2636   Human   Reagent grade liquid; purified from pooled colostrum; ≥95%; supplied in buffered solution with 15 mM sodium azide as preservative | Human & animal proteins may be used as antigens, standards, coating proteins & blocking agents; also used as starting materials for the preparation of immunogens & solid-phase immunoadsorbents

USBio I1890-10   Human   ≥50%; no contaminants detected; single band by SDS-PAGE, IEP, &/or RID; 3.4 mg/mL protein; supplied in 0.1 M NaCl, 0.1 M phosphate, 0.1% NaN$_3$, pH 7.5 | Suitable for antigenic applications in immunological protocols

USBio I1890-12   Human   ≥98%; no contaminants detected; single band by SDS-PAGE, IEP, &/or RID; 1.65 mg/mL supplied in 0.1 M Tris, pH 8.0, 0.1 M NaCl & 0.1% NaN$_3$ | Suitable for antigenic applications in immunological protocols

Biogenesis 5111-5504   Human colostrum   Tested negative for Ab to HIV and HBsAg; <10% moisture; purified; essentially salt free; lyophilized | Useful as a antigen, standard, blocking agent or coating protein in a variety of immunoassays including ELISA, dot immunobinding, Western immunoblotting, immunodiffusion and immunoelectrophoresis

Biogenesis 5104-6017   Human myeloma serum   Lyophilized

Biogenesis 5104-6004   Human serum   Purified; 0.02 M potassium phosphate, 0.15 M NaCl, pH 7.2 with 0.01% NaN$_3$; sterile filtered; liquid

Biogenesis 5107-1004   Human serum   Liquid

USBio I1890-13   Human, secretory (colostrum)   Suitable for antigenic applications in immunological protocols; single band in immunoelectrophoresis with anti-IgA antibodies; reconstitute to 1 mg/mL in 150mM NaCl

Biogenesis 5105-1004   Mouse ascites   Purified; phosphate buffer, pH 7.0; liquid

Biogenesis 5107-5004   Rat myeloma   Liquid

Immunoglobulin A, α-

Dako X0594   >95% | Antigen useful as an immunogen

Immunoglobulin A, κ- (TEPC 15)

Synonyms: Myeloma Protein, Mouse

Sigma M 1421   Mouse   Purified immunoglobulin prepared from ascites fluid by fractionation & ion-exchange or affinity chromatography; tested by immunoelectrophoresis & SDS-PAGE; each vial contains 1 mg protein (by extinction); liquid in 0.02 M Tris buffered saline, pH 8.0 with 0.02% sodium azide as preservative | Potter, M, *Phy Rev*, 52: 631, 1972

Sigma M 7269   Mouse   Clarified ascites fluid prepared by centrifugation & filtration; characterized by immunoelectrophoresis; each vial contains ≥5 mg specific myeloma immunoglobulin as determined by immunoelectrophoresis & quantitative densitometry; lyophilized from 0.01 phosphate buffered saline, pH 7.2 | Potter, M, *Phy Rev*, 52: 631, 1972

Immunoglobulin A, λ- (MOPC 315)

Synonyms: Myeloma Protein, Mouse

Sigma M 2046   Mouse   Purified immunoglobulin prepared from ascites fluid by fractionation & ion-exchange or affinity chromatography; tested by immunoelectrophoresis & SDS-PAGE; each vial contains 1 mg protein (by extinction); liquid in 0.02 M Tris buffered saline, pH 8.0 with 0.02% sodium azide as preservative | Potter, M, *Phy Rev*, 52: 631, 1972

Sigma M 2396   Mouse   Clarified ascites fluid prepared by centrifugation & filtration; characterized by immunoelectrophoresis; each vial contains ≥5 mg specific myeloma immunoglobulin as determined by immunoelectrophoresis & quantitative densitometry; lyophilized from 0.01 phosphate buffered saline, pH 7.2 | Potter, M, *Phy Rev*, 52: 631, 1972

Immunoglobulin D

Biogenesis 5112-0506   Human serum   Lyophilized

Immunoglobulin E

Biogenesis 5118-6316   Human   Serum; liquid

Biogenesis 5118-6004   Human myeloma serum   No detectable IgG, IgM or IgA; may contain ≤2% IgE fragments; SA: 1.24×10^6 IU/mL; tested negative for HIV 1 and 2, HBV and HCV; purified; 10 mM potassium phosphate, 100 mM NaCl, 0.1 % NaN₃, pH 8.0; liquid

Biogenesis 5118-6104   Human myeloma serum   Tested negative for HIV 1 and 2, HBsAg and Hepatitis C; serum; with 0.1% NaN₃; liquid

USBio I1900-51   Human myeloma serum   ≥95% (SDS-PAGE); ~1 mg/mL supplied in PBS, pH 7.5, 0.1% NaN₃ | Suitable for antigenic applications in immunological protocols

Biogenesis 5118-8704   Rat   >80% by agarose electrophoresis; purified; PBS buffer, pH 7.2, 0.1% NaN₃; liquid | Bazin et al, *J Immunol Meth*, 71:9, 1984

Biogenesis 5118-8710   Rat myeloma   Purified; PBS buffer, pH 7.2, 0.1% NaN₃; liquid | *Eur J Immunol*, 4:44, 1974; *Immunology*, 26:713, 1974

Biogenesis 5118-7204   Rat myeloma serum   Purified Ig

Immunoglobulin E, κ-

USBio I1900-50   Human myeloma serum (Thomas)   ≥99%; no contaminants detected; single band by SDS-PAGE, IEP, &/or RID; ~1 mg/mL supplied in 0.1 M Tris, 0.2 M NaCl, pH 7.5, 0.1% NaN₃ | Suitable for antigenic applications in immunological protocols

Immunoglobulin G

Sigma I 5506   Bovine   Reagent grade; lyophilized; purified immunoglobulin from pooled normal serum; ≥95%; essentially salt-free preparation may be reconstituted with 150 mM NaCl | Human & animal proteins may be used as antigens, standards, coating proteins & blocking agents; also used as starting materials for the preparation of immunogens & solid-phase immunoadsorbents

Sigma I 9640   Bovine   Technical grade; liquid; purified immunoglobulin from pooled normal serum; ≥80%; supplied in 0.01 M phosphate buffered saline, pH 7.2, with 15 mM sodium azide as preservative | Human & animal proteins may be used as antigens, standards, coating proteins & blocking agents; also used as starting materials for the preparation of immunogens & solid-phase immunoadsorbents

USBio I1903-07   Bovine   Protease: none detected; 98% protein purity; lyophilized, pH 7.0 | Suitable for antigenic applications in immunological protocols;

USBio I1903-09   Bovine   Protease: none detected; 98% protein purity; lyophilized, pH 7.0 | Suitable for antigenic applications in immunological protocols;

USBio I1903   Bovine normal sera   Highly purified; white; total protein: 5.0 ± 1.0g/dL; OD: 0.01 to 0.03; γ-globulin: ≥99.0%; Na: ≤20 mEq/L; K: ≤0.5 mEq/L; Cl: ≤1 mEq/L; pH: 7.0 ± 0.3; microbial: 0 cf U/mL; moisture: ≤2.0%; ash: ≤2.0%; protease: none detected | Suitable for antigenic applications in immunological protocols; isolated by an exclusive fractionation process & further processed by ion exchange or gel filtration chromatography; used for *in vitro* diagnostics as a blocking agent, reference standard, antigen or coating protein & in biological systems for immunogen preparation

Biogenesis 5124-5004   Bovine serum   Purified; 0.01 M phosphate buffer with 15 mM NaCl, pH 7.2; lyophilized

Sigma I 4256   Cat   Reagent grade; lyophilized; purified from pooled normal serum; ≥95%; essentially salt-free preparation may be reconstituted with 150 mM NaCl | Human & animal proteins may be used as antigens, standards, coating proteins & blocking agents; also used as starting materials for the preparation of immunogens & solid-phase immunoadsorbents

Sigma I 4881   Chicken   Reagent grade; lyophilized from phosphate buffer (~80% protein); purified from pooled normal serum; ≥95%; preparation may be reconstituted with 150 mM NaCl | Human & animal proteins may be used as antigens, standards, coating proteins & blocking agents; also used as starting materials for the preparation of immunogens & solid-phase immunoadsorbents

USBio I1903-17   Chicken   ≥98%; no contaminants detected; single band by SDS-PAGE, IEP, &/or RID; 10 mg/mL supplied in PBS, pH 7.2, preservative-free | Suitable for antigenic applications in immunological protocols

USBio I1903-15   Chicken pooled normal sera   Highly purified; lyophilized; purity (SDS PAGE): 98 ± 2%; pH: 7.0 ± 0.2 after reconstitution; essentially salt free (≤ 1.0%) | Suitable for antigenic applications in immunological protocols; isolated by an exclusive fractionation process & further processed by ion exchange or gel filtration chromatography; used for *in vitro* diagnostics as a blocking agent, reference standard, antigen or coating protein & in biological systems for immunogen preparation; purified by ion exchange or gel filtration chromatography

Sigma I 4006   Dog   Reagent grade; lyophilized; purified from pooled normal serum; ≥95%; essentially salt-free preparation may be reconstituted with 150 mM NaCl | Human & animal proteins may be used as antigens, standards, coating proteins & blocking agents; also used as starting materials for the preparation of immunogens & solid-phase immunoadsorbents

USBio I1903-20   Equine pooled normal sera   Purity (SDS PAGE): 98 ± 2%; pH: 7.0 ± 0.2 after reconstitution; lyophilized; essentially salt free (≤ 1.0%) | Suitable for antigenic applications in immunological protocols; isolated by an exclusive fractionation process & further processed by ion exchange or gel filtration chromatography; used for *in vitro* diagnostics as a blocking agent, reference standard, antigen or coating protein & in biological systems for immunogen preparation; purified by ion exchange or gel filtration chromatography

Sigma I 5256   Goat   Reagent grade; lyophilized; purified from pooled normal serum; ≥95%; essentially salt-free preparation may be reconstituted with 150 mM NaCl | Human & animal proteins may be used as antigens, standards, coating proteins & blocking agents; also used as starting materials for the preparation of immunogens & solid-phase immunoadsorbents

Sigma I 9140   Goat   Technical grade; liquid; purified immunoglobulin from pooled normal serum; ≥80%; supplied in 0.01 M phosphate buffered saline, pH 7.2, with 15 mM sodium azide as preservative | Human & animal proteins may be used as antigens, standards, coating proteins & blocking agents; also used as starting materials for the preparation of immunogens & solid-phase immunoadsorbents

USBio I1903-31   Goat   ≥98%; no contaminants detected; single band by SDS-PAGE, IEP, &/or RID; 20 mg/mL supplied in 0.05 M PBS | Suitable for antigenic applications in immunological protocols

Biogenesis 5160-5004   Goat n/a serum   Purified; 10 mM sodium phosphate, 15 mM NaCl, pH 7.2; lyophilized

USBio I1903-25   Goat pooled normal sera   Purity (SDS PAGE): 98 ± 2%; pH: 7.0 ± 0.2 after reconstitution; lyophilized; essentially salt free (≤ 1.0%) | Suitable for antigenic applications in immunological protocols; isolated by an exclusive fractionation process & further processed by ion exchange or gel filtration chromatography; used for *in vitro* diagnostics as a blocking agent, reference standard, antigen or coating protein & in biological systems for immunogen preparation; purified by ion exchange or gel filtration chromatography

Sigma I 4756   Guinea pig   Reagent grade; lyophilized; purified from pooled normal serum; ≥95%; essentially salt-free preparation may be reconstituted with 150 mM NaCl | Human & animal proteins may be used as antigens, standards, coating proteins & blocking agents; also used as starting materials for the preparation of immunogens & solid-phase immunoadsorbents

USBio I1903-44   Guinea pig   ≥98%; no contaminants detected; single band by SDS-PAGE, IEP, &/or RID; 20 mg/mL supplied in 0.05 M PBS | Suitable for antigenic applications in immunological protocols; highly purified solution used for *in vitro* diagnostics as a blocking agent, reference standard, antigen or coating protein & in biological systems for immunogen preparation

USBio I1903-40   Guinea pig pooled normal sera   Purity (SDS PAGE): ≥90%; pH: 7.0 ± 0.2 after reconstitution; lyophilized; essentially salt free | Suitable for antigenic applications in immunological protocols; isolated by an exclusive fractionation process & further processed by ion exchange or gel filtration chromatography; used for *in vitro* diagnostics as a blocking agent, reference standard, antigen or coating protein & in biological systems for immunogen preparation; purified by ion exchange or gel filtration chromatography

Biogenesis 5166-5029   Guinea pig serum   Lyophilized

USBio I1903-50   Hamster   ≥98%; no contaminants detected; single band by SDS-PAGE, IEP, &/or RID; 25 mgs/mL; lyophilized from 0.01 M sodium phosphate, 0.14 M NaCl, pH 7.4; no preservative added | Suitable for antigenic applications in immunological protocols

Sigma I 4631   Horse   Reagent grade; lyophilized; purified from pooled normal serum; ≥95%; essentially salt-free preparation may be reconstituted with 150 mM NaCl | Human & animal proteins may be used as antigens, standards, coating proteins & blocking agents; also used as starting materials for the preparation of immunogens & solid-phase immunoadsorbents

Sigma I 2511   Human   Reagent grade; liquid; purified from pooled normal serum; ≥95%; supplied in buffered solution with 15 mM sodium azide as preservative | Human & animal proteins may be used as antigens, standards, coating proteins & blocking agents; also used as starting materials for the preparation of immunogens & solid-phase immunoadsorbents

Sigma I 4506   Human   Reagent grade; lyophilized; purified from pooled normal serum; ≥95%; essentially salt-free preparation may be reconstituted with 150 mM NaCl | Human & animal proteins may be used as antigens, standards, coating proteins & blocking agents; also used as starting materials for the preparation of immunogens & solid-phase immunoadsorbents

Sigma I 8640   Human   Technical grade; liquid; purified immunoglobulin from pooled normal serum; ≥80%; supplied in 0.01 M phosphate buffered saline, pH 7.2, with 15 mM sodium azide as preservative | Human & animal proteins may be used as antigens, standards, coating proteins & blocking agents; also used as starting materials for the preparation of immunogens & solid-phase immunoadsorbents

Biogenesis 5172-9017   Human myeloma serum   Tested negative for HBsAg and HIV antibody; purified; lyophilized

Biogenesis 5172-9007   Human serum   Tested negative for HBsAg, HCV, HIV-1, HIV-2; purified; essentially salt free, from 50 mM NH$_4$HCO$_3$, pH 8.0; lyophilized

Biogenesis 5175-5004   Human serum   Liquid

Biogenesis 5212-3004   Human serum   Purified; 0.02 M potassium phosphate, 0.15 M NaCl, pH 7.2; liquid

Fluka 56834   Human serum   ≥95% (GE); ≥95% protein | Standard in solid-phase radioimmunoassay specific for human IgG; Gorevic, PD et al, *Meth Enzymol*, 116: 3, 1985; Frade, R et al, *Meth Enzymol*, 93: 155, 1983; Creswick, P, *Meth Enzymol*, 108: 254, 1984; Lindstrom, J et al, *Meth Enzymol*, 74: 432, 1981

Sigma I 5381   Mouse   Reagent grade; lyophilized from phosphate buffer (~80% protein); purified from pooled normal serum; ≥95%; preparation may be reconstituted with 150 mM NaCl | Human & animal proteins may be used as antigens, standards, coating proteins & blocking agents; also used as starting materials for the preparation of immunogens & solid-phase immunoadsorbents

Sigma I 8765   Mouse   Technical grade; liquid; purified immunoglobulin from pooled normal serum; ≥80%; supplied in 0.01 M phosphate buffered saline, pH 7.2, with 15 mM sodium azide as preservative | Human & animal proteins may be used as antigens, standards, coating proteins & blocking agents; also used as starting materials for the preparation of immunogens & solid-phase immunoadsorbents

USBio I1904-24   Mouse pooled normal sera   Single band (SDS-PAGE); supplied in 10 mg/mL, 10 mM sodium phosphate, 0.15 M NaCl, no preservative, pH 7.2 | Suitable for antigenic applications in immunological protocols; highly purified & isolated by an exclusive fractionation process & further processed by ion exchange or gel filtration chromatography; used for *in vitro* diagnostics as a blocking agent, reference standard, antigen or coating protein & in biological systems for immunogen preparation

USBio I1904-26   Mouse pooled normal sera   Purity (SDS PAGE): 98 ± 2%; pH: 7.0 ± 0.2 after reconstitution; lyophilized; essentially salt free | Suitable for antigenic applications in immunological protocols; highly purified & isolated by an exclusive fractionation process & further processed by ion exchange or gel filtration chromatography; used for *in vitro* diagnostics as a blocking agent, reference standard, antigen or coating protein & in biological systems for immunogen preparation; purified by ion exchange or gel filtration chromatography

Biogenesis 5183-1504   Mouse serum   Affinity purified Ig; PBS buffer, pH 7.2; liquid

USBio I1904-12   Murine ascites   10 mg/mL supplied in 10 mM sodium phosphate, 0.15 M NaCl, pH 7.2, no preservative added | Suitable for antigenic applications in immunological protocols

Fluka 56832   Murine serum   ≥95% (GE); ≥80% protein; lyophilized from 10 mM sodium phosphate & 15 mM NaCl solution, pH 7.2; ≤5% water | Used in preparation of anti-mouse Ig for bridging primary monoclonal antibodies to PAP or APAAP complexes; Lansdorp, PM et al, *Meth Enzymol*, 121: 855, 1986; Soloski, MJ & Vitetta, ES, *Meth Enzymol*, 108: 549, 1984

Biogenesis 5184-4006   Ovine serum   Lyophilized

Sigma I 4382   Pig   Reagent grade; lyophilized; purified from pooled normal serum; ≥95%; essentially salt-free preparation may be reconstituted with 150 mM NaCl | Human & animal proteins may be used as antigens, standards, coating proteins & blocking agents; also used as starting materials for the preparation of immunogens & solid-phase immunoadsorbents

Biogenesis 5190-5004   Porcine serum   Affinity purified Ig; 0.05 M NH$_4$HCO$_3$, pH 8.0, salt free; lyophilized

Sigma I 5006   Rabbit   Reagent grade; lyophilized; purified from pooled normal serum; ≥95%; essentially salt-free preparation may be reconstituted with 150 mM NaCl | Human & animal proteins may be used as antigens, standards, coating proteins & blocking agents; also used as starting materials for the preparation of immunogens & solid-phase immunoadsorbents

Sigma I 8140   Rabbit   Technical grade; liquid; purified immunoglobulin from pooled normal serum; ≥80%; supplied in 0.01 M phosphate buffered saline, pH 7.2, with 15 mM sodium azide as preservative | Human & animal proteins may be used as antigens, standards, coating proteins & blocking agents; also used as starting materials for the preparation of immunogens & solid-phase immunoadsorbents

USBio I1904-30   Rabbit   Purity (SDS PAGE): 98 ± 2%; pH: 7.0 ± 0.2 after reconstitution; lyophilized; essentially salt free | Suitable for antigenic applications in immunological protocols; highly purified & isolated by an exclusive fractionation process & further processed by ion exchange or gel filtration chromatography; used for *in vitro* diagnostics as a blocking agent, reference standard, antigen or coating protein & in biological systems for immunogen preparation; purified by ion exchange or gel filtration chromatography

Biogenesis 5196-5004   Rabbit serum   Affinity purified; 0.05 M NH$_4$HCO$_3$, pH 8.0; lyophilized

Fluka 56830   Rabbit serum   ≥95% (GE); ≥90% protein; ≤10% water | Used in production of anti-rabbit immunoglobulin G; white, ME, *J Animal Sci*, 67: 3144, 1989

Sigma I 4131   Rat   Reagent grade; lyophilized; purified from pooled normal serum; ≥95%; essentially salt-free preparation may be reconstituted with 150 mM NaCl | Human & animal proteins may be used as antigens, standards, coating proteins & blocking agents; also used as starting materials for the preparation of immunogens & solid-phase immunoadsorbents

Sigma I 8015   Rat   Technical grade; liquid; purified immunoglobulin from pooled normal serum; ≥80%; supplied in 0.01 M phosphate buffered saline, pH 7.2, with 15 mM sodium azide as preservative | Human & animal proteins may be used as antigens, standards, coating proteins & blocking agents; also used as starting materials for the preparation of immunogens & solid-phase immunoadsorbents

USBio I1904-52   Rat   ≥98%; no contaminants detected; single band by SDS-PAGE, IEP, &/or RID; 10 mg/mL supplied in PBS with NaCl, pH 7.2 | Suitable for antigenic applications in immunological protocols

USBio I1904-50 Rat pooled normal sera Highly purified; ≥90% (SDS PAGE): pH: 7.0 ± 0.2 after reconstitution, stable; lyophilized; essentially salt free lyophilized | Suitable for antigenic applications in immunological protocols; highly purified lyophilized isolated from by an exclusive fractionation process & further processed by ion exchange or gel filtration chromatography; used for in vitro diagnostics as a blocking agent, reference standard, antigen or coating protein & in biological systems for immunogen preparation

Biogenesis 5207-3004 Rat serum Purified; PBS buffer, pH 7.2; lyophilized

Sigma I 4385 Rhesus (monkey) Reagent grade; lyophilized; purified from pooled normal serum; ≥95%; essentially salt-free preparation may be reconstituted with 150 mM NaCl | Human & animal proteins may be used as antigens, standards, coating proteins & blocking agents; also used as starting materials for the preparation of immunogens & solid-phase immunoadsorbents

Sigma I 8265 Sheep Technical grade; liquid; purified immunoglobulin from pooled normal serum; ≥80%; supplied in 0.01 M phosphate buffered saline, pH 7.2, with 15 mM sodium azide as preservative | Human & animal proteins may be used as antigens, standards, coating proteins & blocking agents; also used as starting materials for the preparation of immunogens & solid-phase immunoadsorbents

USBio I1904-64 Sheep No contaminants detected by IEP; 70 mg/mL; lyophilized; 20 mM sodium phosphate, 150 mM NaCl, pH 7.3; no preservative | Suitable for antigenic applications in immunological protocols

USBio I1904-56 Sheep pooled normal sera 98 ± 2% (SDS PAGE); pH: 7.0 ± 0.2 after reconstitution, stable; lyophilized; essentially salt free | Suitable for antigenic applications in immunological protocols; highly purified lyophilized isolated from by an exclusive fractionation process & further processed by ion exchange or gel filtration chromatography; used for in vitro diagnostics as a blocking agent, reference standard, antigen or coating protein & in biological systems for immunogen preparation

Immunoglobulin G, Affinity Purified

USBio I1904-38 Rabbit Filtration: 0.2um | Suitable for antigenic applications in immunological protocols

Immunoglobulin G, Fc

USBio I1903-64 Human ≥60%; no contaminants detected; single band by SDS-PAGE, IEP, &/or RID; 7.5 mg/mL supplied in 0.1 M buffer, 0.2 M NaCl, 0.1% NaN₃ as preservative, pH 7.6 | Suitable for antigenic applications in immunological protocols

USBio I1903-74 Human ≥98%; no contaminants detected; single band by SDS-PAGE, IEP, &/or RID; ~2 mg/mL supplied in 0.05 M Tris, 0.2 M NaCl, pH 8.0, 0.1% NaN₃ | Suitable for antigenic applications in immunological protocols

USBio I1904-16 Mouse ≥98%; no contaminants detected; single band by SDS-PAGE, IEP, &/or RID; 0.7 mg/mL supplied in PBS, pH 7.5, 0.01% NaN₃ used as a preservative | Suitable for antigenic applications in immunological protocols; prepared from highly purified mouse serum- delipidated & fractionated, purified ion-exchange chromatography & papain digestion

Immunoglobulin G, FITC Conjugated

Sigma F 7381 Goat FITC conjugated to goat IgG; purified; liquid at ~20 mg/mL protein concentration in 0.01 M phosphate buffered saline, pH 7.4, containing 15 mM sodium azide as preservative | Non-reactive with human IgG, IgA, IgM, Bence Jones Kappa & Lambda, normal human serum & the appropriate animal serum proteins by immunoassay

Sigma F 9636 Human FITC conjugated to human IgG; purified; liquid at ~20 mg/mL protein concentration in 0.01 M phosphate buffered saline, pH 7.4, containing 15 mM sodium azide as preservative | Non-reactive with human IgG, IgA, IgM, Bence Jones Kappa & Lambda, normal human serum & the appropriate animal serum proteins by immunoassay

Sigma F 7256 Rabbit FITC conjugated to rabbit IgG; purified; liquid at ~20 mg/mL protein concentration in 0.01 M phosphate buffered saline, pH 7.4, containing 15 mM sodium azide as preservative | Non-reactive with human IgG, IgA, IgM, Bence Jones Kappa & Lambda, normal human serum & the appropriate animal serum proteins by immunoassay

Immunoglobulin G, H&L

USBio I1903-90 Human ≥96%; no contaminants detected; single band by SDS-PAGE, IEP, &/or RID; 80 mg/mL (OD 280 nm) supplied in 0.1 M sodium phosphate buffer, 0.2 M NaCl, pH 7.4 | Suitable for antigenic applications in immunological protocols

Immunoglobulin G, γ-

Dako X0593 >95% | Antigen useful as an immunogen

Immunoglobulin G1

Biogenesis 5219-3004 Human myeloma serum or urine Tested negative for HBsAg, and antibodies to HIV and HCV; purified serum; Tris buffer (see s); liquid

Biogenesis 5220-3059 Mouse Affinity purified Ig; PBS buffer, pH 7.2, 10 mM NaN₃, 1 mg/mL BSA; liquid | Negligible cross reactivity with human cell surface antigens on tissue sections or in cellular preparations; useful for estimating non-specific binding of mouse monoclonals to cell surface antigens; useful as an isotype control for indirect immunofluoresence when using mouse monoclonal antibodies; suitable for whole blood, Ficoll-separated preparations, frozen and paraffin embedded tissue sections

Biogenesis 5220-3062 Mouse FITC conjugated; PBS buffer, pH 7.2, 10 mM NaN₃, 1 mg/mL BSA; liquid | Negligible cross reactivity with human cell surface antigens on tissue sections or in cellular preparations; useful for estimating non-specific binding of mouse monoclonals to cell surface antigens; useful as an isotype control for direct immunofluoresence when using mouse monoclonal antibodies. Suitable for whole blood, Ficoll-separated preparations, frozen and paraffin embedded tissue sections

Biogenesis 5220-3067 Mouse PE conjugated

Biogenesis 5220-2959 Mouse clone MOPC-21 Liquid

Biogenesis 5221-3059 Rat Affinity purified Ig; PBS buffer, pH 7.2, 1% BSA, 0.1% NaN₃; liquid | Negligible cross-reactivity with human cell surface antigens on tissue sections or in cellular preparations; useful as an isotype control for indirect immunofluoresence when using rat monoclonal antibodies; suitable for whole blood, Ficoll-separated preparations, frozen and paraffin embedded sections

Immunoglobulin G1, FITC Conjugated

Biogenesis 5221-3062 Rat

Immunoglobulin G1, PE Conjugated

Biogenesis 5221-3067 Rat

Immunoglobulin G1, κ-

USBio I1904-77 Human myeloma ≥96%; no contaminants detected; single band by SDS-PAGE, IEP, &/or RID; 3-4 mg/mL supplied in 0.02 M Phosphate buffer, 0.15 M NaCl, pH 7.5 & 0.05% NaN₃ | Suitable for antigenic applications in immunological protocols

Sigma I 3889 Human myeloma plasma Purified & isolated by fractionation, ion-exchange &/or affinity chromatography; purity & identity determined by immunoelectrophoresis, indirect ELISA & SDS-PAGE; lyophilized from phosphate buffer, pH 7.2 | IgG subclass; each purified immunoglobulin represents a single subclass & light chain type

Immunoglobulin G1, κ- (MOPC 21)
Synonyms: Myeloma Protein, Mouse

Sigma M 7894 Mouse Clarified ascites fluid prepared by centrifugation & filtration; characterized by immunoelectrophoresis; each vial contains ≥5 mg specific myeloma immunoglobulin as determined by immunoelectrophoresis & quantitative densitometry; lyophilized from 0.01 phosphate buffered saline, pH 7.2 | Potter, M, *Phy Rev*, 52: 631, 1972

Sigma M 9269 Mouse Purified immunoglobulin prepared from ascites fluid by fractionation & ion-exchange or affinity chromatography; tested by immunoelectrophoresis & SDS-PAGE; each vial contains 1 mg protein (by extinction); liquid in 0.02 *M* Tris buffered saline, pH 8.0 with 0.02% sodium azide as preservative | Potter, M, *Phy Rev*, 52: 631, 1972

Immunoglobulin G1, κ- (MOPC 31C)
Synonyms: Myeloma Protein, Mouse

Sigma M 1398 Mouse Clarified ascites fluid prepared by centrifugation & filtration; characterized by immunoelectrophoresis; each vial contains ≥5 mg specific myeloma immunoglobulin as determined by immunoelectrophoresis & quantitative densitometry; lyophilized from 0.01 phosphate buffered saline, pH 7.2 | Potter, M, *Phy Rev*, 52: 631, 1972

Sigma M 9035 Mouse Purified immunoglobulin prepared from ascites fluid by fractionation & ion-exchange or affinity chromatography; tested by immunoelectrophoresis & SDS-PAGE; each vial contains 1 mg protein (by extinction); liquid in 0.02 *M* Tris buffered saline, pH 8.0 with 0.02% sodium azide as preservative | Potter, M, *Phy Rev*, 52: 631, 1972

Immunoglobulin G1, λ-

Sigma I 4014 Human myeloma plasma Purified & isolated by fractionation, ion-exchange &/or affinity chromatography; purity & identity determined by immunoelectrophoresis, indirect ELISA & SDS-PAGE; lyophilized from phosphate buffer, pH 7.2 | IgG subclass; each purified immunoglobulin represents a single subclass & light chain type

Immunoglobulin G2

Biogenesis 5225-3004 Human myeloma serum

Immunoglobulin G2, α-

USBio I1904-17 Murine ascites 1 mg/mL sterile-filtered liquid in PBS, pH 7.2, 0.01% NaN₃ | Suitable for use as a control or standard in flow cytometry & immunohistochemistry; prepared from immunodeficient murine ascites by protein A chromatography; exhibits <1% purity cross reactivity against other murine & human heavy or light chains isotypes by ELISA

Immunoglobulin G2, α- FITC Conjugated

USBio I1904-17A Murine ascites MW 390k 1 mg/mL; lyophilized in PBS, pH 7.2, 10 mg/mL BSA, 0.01% Thimerosal, conjugated to FITC | Suitable for use as a control or standard in flow cytometry; fluorochrome/protein ratio: 2.1 moles FITC per mole of murine IgG2a; prepared from immunodeficient murine ascites by Protein A chromatography; exhibits less than 1% purity cross reactivity against other murine & human heavy or light chains isotypes by ELISA

Immunoglobulin G2, κ-

USBio I1904-79 Human myeloma ≥96%; no contaminants detected; single band by SDS-PAGE, IEP, &/or RID; 2 mg/mL supplied in 0.02 *M* Phosphate buffer, 0.15 *M* NaCl, pH 7.4 & 0.05% NaN₃ | Suitable for antigenic applications in immunological protocols; the heavy chain of myeloma IgG may appear as either a single or double band on gel electrophoresis

Sigma I 4139 Human myeloma plasma Purified & isolated by fractionation, ion-exchange &/or affinity chromatography; purity & identity determined by immunoelectrophoresis, indirect ELISA & SDS-PAGE; lyophilized from phosphate buffer, pH 7.2 | IgG subclass; each purified immunoglobulin represents a single subclass & light chain type

Sigma I 4264 Human myeloma plasma Purified & isolated by fractionation, ion-exchange &/or affinity chromatography; purity & identity determined by immunoelectrophoresis, indirect ELISA & SDS-PAGE; lyophilized from phosphate buffer, pH 7.2 | IgG subclass; each purified immunoglobulin represents a single subclass & light chain type

Immunoglobulin G2a

Biogenesis 5230-3062 Mouse FITC conjugated; PBS buffer, pH 7.2, 1% BSA, 0.1% NaN₃; liquid | Negligible cross-reactivity with human cell surface antigens on tissue sections or in cellular preparations; useful as isotype control (non-specific binding) for direct and indirect immunofluorescence when using mouse monoclonal antibodies; suitable for whole blood, Ficol separated preparations, frozen and paraffin embedded sections

Biogenesis 5230-3029 Mouse clone RPC-5 Liquid

Biogenesis 5230-3004 Mouse monoclonal myeloma (2031/13) Purified; PBS buffer, pH 7.2, 0.1% NaN₃; liquid | IgG2a is the only subclass detected in ELISA subclass assay

Biogenesis 5231-2559 Rat monoclonal

Biogenesis 5230-3059 Unconjugated monoclonal Affinity purified Ig

Immunoglobulin G2a, FITC Conjugated

Biogenesis 5231-2562 Rat

Immunoglobulin G2a, PE Conjugated

Biogenesis 5230-3067 Mouse Affinity purified Ig

Biogenesis 5231-2567 Rat

Immunoglobulin G2a, κ- (UPC 10)
Synonyms: Myeloma Protein, Mouse

Sigma M 7769 Mouse Clarified ascites fluid prepared by centrifugation & filtration; characterized by immunoelectrophoresis; each vial contains ≥5 mg specific myeloma immunoglobulin as determined by immunoelectrophoresis & quantitative densitometry; lyophilized from 0.01 phosphate buffered saline, pH 7.2 | Potter, M, *Phy Rev*, 52: 631, 1972

Sigma M 9144 Mouse Purified immunoglobulin prepared from ascites fluid by fractionation & ion-exchange or affinity chromatography; tested by immunoelectrophoresis & SDS-PAGE; each vial contains 1 mg protein (by extinction); liquid in 0.02 *M* Tris buffered saline, pH 8.0 with 0.02% sodium azide as preservative | Potter, M, *Phy Rev*, 52: 631, 1972

Immunoglobulin G2a, λ- (HOPC 1)
Synonyms: Myeloma Protein, Mouse

Sigma M 6034 Mouse Purified immunoglobulin prepared from ascites fluid by fractionation & ion-exchange or affinity chromatography; tested by immunoelectrophoresis & SDS-PAGE; each vial contains 1 mg protein (by extinction); liquid in 0.02 *M* Tris buffered saline, pH 8.0 with 0.02% sodium azide as preservative | Potter, M, *Phy Rev*, 52: 631, 1972

Immunoglobulin G2b

Biogenesis 5236-3059 Mouse Affinity purified Ig; PBS buffer, pH 7.2, 1 mg/mL BSA, 10 *mM* NaN₃; liquid | Negligible cross-reactivity with human cell surface antigens on tissue sections or in cellular preparations; useful for estimating non-specific binding of mouse monoclonals to cell surface antigens; useful as an isotype control for indirect IF when using mouse monoclonal antibodies; suitable for whole blood, Ficoll-separated preparations and frozen and paraffin embedded tissue sections

Biogenesis 5236-3062 Mouse FITC conjugated; PBS buffer, pH 7.2, 1 mg/mL BSA, 10 *mM* NaN₃; liquid | Negligible cross-reactivity with human cell surface antigens on tissue sections or in cellular preparations; useful as an isotype control for direct IF when using mouse monoclonal antibodies; suitable for whole blood, Ficoll-separated preparations, frozen and paraffin embedded tissue sections

Biogenesis 5236-3067 Mouse PE conjugated

Biogenesis 5236-3004	Mouse clone MOPC-195	
Biogenesis 5237-2562	Rat	FITC conjugated
Biogenesis 5237-2567	Rat	PE conjugated

Immunoglobulin G2b, κ- (MOPC 141)

Synonyms: Myeloma Protein, Mouse

Sigma M 7644   Mouse   Clarified ascites fluid prepared by centrifugation & filtration; characterized by immunoelectrophoresis; each vial contains ≥5 mg specific myeloma immunoglobulin as determined by immunoelectrophoresis & quantitative densitometry; lyophilized from 0.01 phosphate buffered saline, pH 7.2 | Potter, M, *Phy Rev*, 52: 631, 1972

Sigma M 8894   Mouse   Purified immunoglobulin prepared from ascites fluid by fractionation & ion-exchange or affinity chromatography; tested by immunoelectrophoresis & SDS-PAGE; each vial contains 1 mg protein (by extinction); liquid in 0.02 M Tris buffered saline, pH 8.0 with 0.02% sodium azide as preservative | Potter, M, *Phy Rev*, 52: 631, 1972

Immunoglobulin G2b, κ- (MOPC 195)

Synonyms: Myeloma Protein, Mouse

Sigma M 1395   Mouse   Clarified ascites fluid prepared by centrifugation & filtration; characterized by immunoelectrophoresis; each vial contains ≥5 mg specific myeloma immunoglobulin as determined by immunoelectrophoresis & quantitative densitometry; lyophilized from 0.01 phosphate buffered saline, pH 7.2 | Potter, M, *Phy Rev*, 52: 631, 1972

Immunoglobulin G3

Biogenesis 5248-3004	Human myeloma serum	Liquid
Biogenesis 5249-3004	Mouse myeloma serum	Liquid

Immunoglobulin G3, κ- (FLOPC 21)

Synonyms: Myeloma Protein, Mouse

Sigma M 1645   Mouse   Clarified ascites fluid prepared by centrifugation & filtration; characterized by immunoelectrophoresis; each vial contains ≥5 mg specific myeloma immunoglobulin as determined by immunoelectrophoresis & quantitative densitometry; lyophilized from 0.01 phosphate buffered saline, pH 7.2 | Potter, M, *Phy Rev*, 52: 631, 1972

Sigma M 3645   Mouse   Purified immunoglobulin prepared from ascites fluid by fractionation & ion-exchange or affinity chromatography; tested by immunoelectrophoresis & SDS-PAGE; each vial contains 1 mg protein (by extinction); liquid in 0.02 M Tris buffered saline, pH 8.0 with 0.02% sodium azide as preservative | Potter, M, *Phy Rev*, 52: 631, 1972

Immunoglobulin G3, λ-

USBio I1904-85   Human myeloma   ≥96%; no contaminants detected; single band by SDS-PAGE, IEP, &/or RID; ~2.0 mg/mL supplied in 0.02 M Phosphate buffer, 0.15 M NaCl, pH 7.5 | Suitable for antigenic applications in immunological protocols

Sigma I 4514   Human myeloma plasma   Purified & isolated by fractionation, ion-exchange &/or affinity chromatography; purity & identity determined by immunoelectrophoresis, indirect ELISA & SDS-PAGE; lyophilized from phosphate buffer, pH 7.2 | IgG subclass; each purified immunoglobulin represents a single subclass & light chain type

Immunoglobulin G3, λ- (Y5606)

Synonyms: Myeloma Protein, Mouse

Sigma M 7519   Mouse   Clarified ascites fluid prepared by centrifugation & filtration; characterized by immunoelectrophoresis; each vial contains ≥5 mg specific myeloma immunoglobulin as determined by immunoelectrophoresis & quantitative densitometry; lyophilized from 0.01 phosphate buffered saline, pH 7.2 | Potter, M, *Phy Rev*, 52: 631, 1972

Sigma M 9019   Mouse   Purified immunoglobulin prepared from ascites fluid by fractionation & ion-exchange or affinity chromatography; tested by immunoelectrophoresis & SDS-PAGE; each vial contains 1 mg protein (by extinction); liquid in 0.02 M Tris buffered saline, pH 8.0 with 0.02% sodium azide as preservative | Potter, M, *Phy Rev*, 52: 631, 1972

Immunoglobulin G4

Biogenesis 5254-3004   Human myeloma plasma   Tested negative for HBsAg, anti-HCV, anti-HBc and anti HIV; purified; 20 mM sodium phosphate buffer, pH 7.4, with 150 mM NaCl and 0.05% NaN$_3$; liquid

Immunoglobulin G4, κ-

USBio I1904-91   Human myeloma   ≥96%; no contaminants detected; single band by SDS-PAGE, IEP, &/or RID; 3.2 mg/mL supplied in 0.02 M Phosphate buffer, 0.15 M NaCl, pH 7.5 & 0.05% NaN$_3$ as preservative | Suitable for antigenic applications in immunological protocols

Sigma I 4389   Human myeloma plasma   Purified & isolated by fractionation, ion-exchange &/or affinity chromatography; purity & identity determined by immunoelectrophoresis & SDS-PAGE; lyophilized from phosphate buffer, pH 7.2 | IgG subclass; each purified immunoglobulin represents a single subclass & light chain type

Sigma I 4639   Human myeloma plasma   Purified & isolated by fractionation, ion-exchange &/or affinity chromatography; purity & identity determined by immunoelectrophoresis, indirect ELISA & SDS-PAGE; frozen liquid in 0.02 M Tris buffered saline, pH 8.0 | IgG subclass; each purified immunoglobulin represents a single subclass & light chain type

Sigma I 4764   Human myeloma plasma   Purified & isolated by fractionation, ion-exchange &/or affinity chromatography; purity & identity determined by immunoelectrophoresis, indirect ELISA & SDS-PAGE; frozen liquid in 0.02 M Tris buffered saline, pH 8.0 | IgG subclass; each purified immunoglobulin represents a single subclass & light chain type

Immunoglobulin Heavy Chain Binding Protein

Synonyms: GRP78

Sigma B 1174   Calf liver   MW 78k   Lyophilized powder containing 20 mM HEPES, pH 7.5, 25 mM KCl, 5 mM MgCl$_2$ & 100 mg/mL trehalose | Molecular chaperone found in endoplasmic reticulum lumen; involved in protein folding & translocation through the ER membranes; low basal level of ATPase activity which is stimulated by various peptides; Gething, M-J & Sambrook, J, *Nature*, 355: 33, 1992; Wei, J et al, *J Biol Chem*, 270: 26677, 1995; Blond-Elguindi, S et al, *J Biol Chem*, 268: 12730, 1993; Shirai, N et al, *J Biochem*, 121: 787, 1997

Immunoglobulin M

Sigma I 8135   Bovine   Reagent grade; liquid; purified from pooled normal serum; ≥95%; supplied in buffered solution with 15 mM sodium azide as preservative | Human & animal proteins may be used as antigens, standards, coating proteins & blocking agents; also used as starting materials for the preparation of immunogens & solid-phase immunoadsorbents

Biogenesis 5272-3004   Bovine serum   Purified; 0.1 M Tris chloride, 0.5 M NaCl, pH 8.0, 0.1% NaN$_3$; liquid (sterile)

Biogenesis 5273-3004   Canine   Purified; 0.1 M TRIS Chloride, 0.5 M NaCl and 0.1% NaN$_3$, pH 8.0; liquid

Biogenesis 5273-5509   Feline serum   Liquid

Sigma I 8260   Human   Reagent grade; liquid; purified from pooled normal serum; ≥95%; supplied in buffered solution with 15 mM sodium azide as preservative | Human & animal proteins may be used as antigens, standards, coating proteins & blocking agents; also used as starting materials for the preparation of immunogens & solid-phase immunoadsorbents

USBio I1905-25   Human myeloma   ≥96%; no contaminants detected; single band by SDS-PAGE, IEP, &/or RID; hImmunoglobulin M: 100%, hIgG: 0%, hIgA: 0%, hIgE: 0%, hIgD: 0%; 10 mg/mL; lyophilized from 0.05 M Tris, pH 8.0, 0.2 M NaCl, 0.1% NaN$_3$ | Suitable for antigenic applications in immunological protocols

Biogenesis 5275-5504 Human myeloma serum Tested negative for HBsAg and HIV antibody; purified; lyophilized

Biogenesis 5275-5004 Human plasma Tested negative for HIV-1, HBsAg and anti HCV; purified; PBS buffer, pH 7.2, 0.01% NaN₃; liquid

Biogenesis 5278-8704 Human serum Liquid

Biogenesis 5276-5059 Mouse Unconjugated

Biogenesis 5276-5062 Mouse FITC conjugated

Biogenesis 5276-5067 Mouse PE conjugated

Biogenesis 5276-4930 Mouse clone ABPC-22

Biogenesis 5276-4950 Mouse clone MOPC-104E

Biogenesis 5276-5004 Mouse serum Purified Ig; PBS buffer, pH 7.2, 0.1% NaN₃; liquid

Biogenesis 5276-6504 Porcine serum Purified; 0.1 M Tris chloride, 0.5 M NaCl, pH 8.0 with 0.1% NaN₃; sterile filtered; liquid

Biogenesis 5277-5059 Rat Purified Ig; PBS buffer, pH 7.2, 10 mM NaN₃; liquid | Negligible cross-reactivity with human cell surface antigens on tissue sections or in cellular preparations; useful for estimation of non-specific binding of rat monoclonals to cell surface antigens; useful as an isotype control for indirect immunofluoresence when using rat monoclonal antibodies. Suitable for whole blood, Ficoll-separated preparations, frozen and paraffin embedded sections

Biogenesis 5277-5062 Rat FITC conjugated

Biogenesis 5277-5067 Rat PE conjugated

Biogenesis 5277-5004 Rat serum Purified; 0.1 M Tris Chloride, 0.5 M NaCl, pH 8.0, 0.1% NaN3; sterile filtered; liquid

Immunoglobulin M, κ- (ABPC 22)

Synonyms: Myeloma Protein, Mouse

Sigma M 7394 Mouse Clarified ascites fluid prepared by centrifugation & filtration; characterized by immunoelectrophoresis; each vial contains ≥5 mg specific myeloma immunoglobulin as determined by immunoelectrophoresis & quantitative densitometry; lyophilized from 0.01 phosphate buffered saline, pH 7.2 | Potter, M, *Phy Rev*, 52: 631, 1972

Immunoglobulin M, κ- (TEPC 183)

Synonyms: Myeloma Protein, Mouse

Sigma M 1520 Mouse Clarified ascites fluid prepared by centrifugation & filtration; characterized by immunoelectrophoresis; each vial contains ≥5 mg specific myeloma immunoglobulin as determined by immunoelectrophoresis & quantitative densitometry; lyophilized from 0.01 phosphate buffered saline, pH 7.2 | Potter, M, *Phy Rev*, 52: 631, 1972

Sigma M 3795 Mouse Purified immunoglobulin prepared from ascites fluid by fractionation & ion-exchange or affinity chromatography; tested by immunoelectrophoresis & SDS-PAGE; each vial contains 1 mg protein (by extinction); liquid in 0.05 M Tris & 0.5 M NaCl, pH 8.0 with 0.02% sodium azide as preservative | Potter, M, *Phy Rev*, 52: 631, 1972

Immunoglobulin M, λ- (MOPC 104E)

Synonyms: Myeloma Protein, Mouse

Sigma M 2521 Mouse Clarified ascites fluid prepared by centrifugation & filtration; characterized by immunoelectrophoresis; each vial contains ≥5 mg specific myeloma immunoglobulin as determined by immunoelectrophoresis & quantitative densitometry; lyophilized from 0.01 phosphate buffered saline, pH 7.2 | Potter, M, *Phy Rev*, 52: 631, 1972

Sigma M 5170 Mouse Purified immunoglobulin prepared from ascites fluid by fractionation & ion-exchange or affinity chromatography; tested by immunoelectrophoresis & SDS-PAGE; each vial contains 1 mg protein (by extinction); liquid in 0.05 M Tris & 0.5 M NaCl, pH 8.0 with 0.02% sodium azide as preservative | Potter, M, *Phy Rev*, 52: 631, 1972

Immunoglobulin M, μ-

Dako X0595 >95% | Antigen useful as an immunogen

Immunoglobulin κ Light Chain

Biogenesis 5268-6307 Human myeloma urine Tested negative for HBsAg and HIV antibody; purified; lyophilized

Biogenesis 5269-6207 Human myeloma urine Tested negative for HBsAg and antibody to HIV; purified; lyophilized

Immunoglobulin μ

USBio I1905-10 Human normal plasma ≥98%; no contaminants detected; single band by SDS-PAGE, IEP, &/or RID; 10.0 mg/mL supplied in 0.05 M Tris, 0.2 M NaCl, pH 8.0 & 0.1% NaN₃ ∣ Suitable for antigenic applications in immunological protocols

Inhibin

Synonyms: Follicle Stimulating Hormone Suppressing Protein

Sigma I 9149 Porcine ovaries 2 k IU/vial | Bioassay not run by Sigma

Inhibitor I

Chemicon AG649 Recombinant ≥95% | Purified protein for apoptosis & signal transduction

Insulin

Fluka 57590 Bovine pancreas $C_{254}H_{377}N_{65}O_{75}S_6$ ≥90% (GE); 25 U/mg; ≤10% loss on drying; ≤0.003% proinsulin-like immunoreactivity; 0.5% Zn; 1 U corresponds to the efficiency of 0.04167 mg international standard substance | Standard for MALDI-MS; Homan, JDH & Terpstra, J, *Discoveries Pharmacol*, 2: 429, 1984

Sigma I 1882 Bovine pancreas FW 5733.5 Lyophilized; sterilized by γ-irradiation; cell culture tested

Sigma I 5500 Bovine pancreas ≥27 USP U/mg; anhydrous (HPLC); zinc content: ~0.5%

Sigma I 6634 Bovine pancreas FW 5733.5 Crystalline; cell culture tested

Sigma I 0259 Human recombinant, expressed in *E. coli* FW 5807.6 ~28 USP U/mg | Insulin regulates the cellular uptake, utilization, & storage of glucose, AA & fatty acids & inhibits the breakdown of glycogen, protein & fat; Smith, *Amer J Med*, 40: 662, 1966; Kono, T, *Vitamins & Hormones*, 7: 1003, 1988

Sigma I 2767 Human recombinant, expressed in *E. coli* FW 5733.5 Sodium salt; crystalline; cell culture tested

Fitzgerald 30-AI51 Human recombinant, expressed in yeast High purity

Sigma I 1507 Human synthetic, from porcine insulin FW 5807.6 ~24 IU/mg protein; 95-98% (HPLC); crystalline | Morihara, I et al, *Nature*, 280: 412, 1979

Fitzgerald 30-AI49 Ovine pancreas High purity

Biogenesis 5330-1002 Porcine pancreas Lyophilized

Fluka 57595 Porcine pancreas $C_{256}H_{381}N_{65}O_{76}S_6$ ≥85% (GE); 25 U/mg; ≤10% loss on drying; ≤0.003% proinsulin-like immunoreactivity; 0.4% Zn; 1 U corresponds to the efficiency of 0.04167 mg international standard substance | Kaarsholm, NC et al, *Biochemistry*, 28: 4427, 1989; Markussen, J, *Human Insulin by Tryptic Transpeptidation of Porcine Insulin & Biosynthetic Precursors*, MTP Press Ltd, 240, 1987, Lancaster UK

Sigma I 5523 Porcine pancreas FW 5777.6 ~24 IU/mg; crystalline; zinc content: ~0.5%

Cortex CP1041r Recombinant >98%

USBio I7660-29 Sheep ≥95%; no contaminants detected; single band by SDS-PAGE, IEP, &/or RID; ~20 IU/mg | Suitable for antigenic applications in immunological protocols

Insulin Like Growth Factor I

Oncogene PF015 Human recombinant MW 7.5k >97% (SDS-PAGE); lyophilized; biological activity: half maximal stimulation of ³H-thymidine uptake by Balb/C 3T3 cells ~1.0 ng/mL; reconstitute in 10mM acetic acid | Species reactivity: human; for proliferation studies & Western blot

PeproTech 100-11   Human recombinant, expressed in *E. coli* MW 7.6k   >97%; 70 AA; lyophilized with no additives; ED$_{50}$: 0.5-1.0 ng/mL as determined by the stimulation of thymidine uptake by Balb/c 3T3 cells

Insulin Like Growth Factor I Soluble Receptor

Oncogene PF061   Human recombinant   MW 100k (non-glycosylated), >200k (following glycosylation & tetramerization) >95% (SDS-PAGE); lyophilized; biological activity: binds IGF-I in solution but is not an effective antagonist | Species reactivity: human; for binding studies; the IGF-I sR is highly expressed in all cell types & tissues

Insulin Receptor

Sigma I 9266   Rat liver   Purified by affinity chromatography on wheat germ agglutinin; solution in 50% glycerol containing 50 mM HEPES, pH 7.6, 150 mM NaCl, 0.1% Triton; vial contains 250 U; 1 U catalyzes the incorporation of 1 pmol/min of phosphate from γ-^{32}P-ATP into poly(Glu,Tyr), 4:1 at 30°C | Zick, Y et al, *J Biol Chem*, 258: 75, 1983

Insulin Receptor Substrate I

Synonyms: Phosphatidylinositol-3-Kinase Activator

Upstate 12-335   Recombinant, Produced in Sf9 insect cells MW 170k   Frozen solution

Insulin, Agarose

Sigma I 2508   Bovine   Bovine insulin, Sigma I 5500, coupled preferentially through the α-amino via active ester of diaminodipropylaminosuccinyl linked 4% beaded agarose at pH 6.0; contains 2-5 mg insulin (E$_{280}$ at 1%)/mL gel; suspension in 0.1 *M* phosphate buffered saline containing 0.02% sodium azide | Sold on the basis of packed gel volume

Sigma I 9635   Porcine   Porcine insulin, Sigma I 3505, coupled preferentially through the α-amino via active ester to epichlorohydrin-activated diaminodipropylaminosuccinyl; contains 1-2 mg insulin by BCA/mL gel; suspension in 0.15 M NaCl, 0.01 M Tris, 0.4 mM ZnCl$_2$, 0.02% sodium azide, pH 7.4, 50% glycerol | Sold on the basis of packed gel volume; Cuatrecasas, P & Parikh, I, *Meth Enz*, 34: 653, 1974; Fujita-Yamaguchi, Yoko, et al, *J Biol Chem*, 258: 5045, 1983

Insulin, Arg-

Sigma I 5389   Human recombinant, expressed in *E. coli*   FW 5963.8   ≥97% (HPCL) | Can replace bovine or human insulin in mammalian cell culture; ~1/3rd the binding capacity of human or bovine insulin to human placental insulin receptors

Insulin, Biotin Conjugated

Sigma I 2258   Bovine   Lyophilized powder containing ~50% insulin (E$_{280}$ at 1%); balance primarily sodium phosphate buffer salts; ~1 mole biotin/mole insulin; prepared from bovine insulin, Sigma I 5500, coupled preferentially through 1 or both α-amino groups via active ester to biotin at pH 6.0 | Cuatrecasas, P & Parikh, I, *Biochemistry*, 11: 2291, 1972

Insulin, Biotinamidocaproyl Conjugated

Sigma I 5636   Bovine pancreas   Lyophilized powder containing ~80% insulin (E$_{280}$ at 1%); balance primarily sodium phosphate buffer salts; ~1 mole biotin/mole insulin; prepared from bovine insulin, Sigma I 5500, coupled preferentially through 1 or both α-amino groups via active ester to biotinamidocaproic acid at pH 6.0 | Cuatrecasas, P & Parikh, I, *Biochemistry*, 11: 2291, 1972; Kohanski, AR & Lane, MD, *J Biol Chem*, 260: 5014, 1985

Insulin, FITC Conjugated

Sigma I 2383   Bovine pancreas   Lyophilized powder containing ~90% insulin (Biuret); balance primarily sodium phosphate buffer salts; ~1 mole FITC/mole insulin; prepared from bovine insulin, Sigma I 5500, coupled preferentially through the α-amino to FITC | Tietz, F et al, *Biochim Biophys Acta*, 59: 336, 1962

Insulin, Gold Conjugated

Sigma I 0391   Porcine pancreas   10 nm colloidal gold labeled; mean particle diameter 8-12 nm; monodisperse; from porcine insulin, Sigma I 3505, coupled through spacer to albumin coated colloidal gold; suspension in ~50% glycerol containing 0.15 *M* NaCl, 0.01 *M* BES, 0.02% PEG 20 & 0.02% sodium azide; concentration: A$_{520}$ ~5.0 | Moll, UM et al, *Histochemistry*, 86: 83, 1986

Insulin, Hybri-Max®

Sigma I 4011   Bovine pancreas   ≥27 USP U/mg; anhydrous (HPLC); zinc content: ~0.5%; endotoxin tested; hybridoma tested

Insulin, Peroxidase Conjugated

Sigma I 2133   Bovine pancreas   Lyophilized powder containing ~90% insulin (E$_{280}$ at 1%); balance primarily Tris-citrate buffer salts; 0.7-1.4 moles peroxidase/mole insulin; prepared from bovine insulin, Sigma I 5500, coupled preferentially through the α- or ε-amino via active ester to Type VI peroxidase, Sigma P 8375, at pH 7.0; affinity purified to substantially remove unlabeled insulin; peroxidase activity: 200-300 U/mg protein; unit definition: 1 U forms 1 mg purpurogallin/20 sec from pyrogallol at pH 6.0 at 20°C | Cuatrecasas, P & Parikh, I, *Biochemistry*, 11: 2291, 1972

Insulin, Zinc

Calbiochem 40769   Bovine   MW 5733.5   White solid; activity: ≥27 USP U/mg dry weight; soluble in 0.01 N HCl; bioburden: ≤300 organisms/g; endotoxin: ≤20 EU/mg; may be carcinogenic/teratogenic | *Merck Index*, 12: 5011

Calbiochem 407694   Human recombinant   MW 5807.7 ≤95% (HPLC); white solid; activity: ≥24 USP U/mg dry weight; soluble in 0.01 N HCl; may be carcinogenic/teratogenic | *Merck Index*, 12: 5011

Calbiochem 407693   Porcine   MW 6k   White solid; activity: ≥25 USP U/mg dry weight; soluble in 0.01 N HCl; may be carcinogenic/teratogenic | *Merck Index*, 12: 5011

Insulin-Like Growth Factor Binding Protein I

Biogenesis 5345-5029   Amniotic fluid   Lyophilized

Sigma I 0524   Human hepatoma, cell line HepG2   MW 30k ~90% (SDS-PAGE); solution in Tris buffered saline, pH 7.4; unit definition: 1 U binds 1 ng IGF at IGF concentrations of 10 ng/mL at pH 6.0 at 25°C; phosphorylated activity: ≥250 U/vial | Binds IGF-I & IGF-II with very high affinity; *in vivo*, IGFBP-1 is phosphorylated on Ser residues; affinity towards IGF increases with the degree of phosphorylation; has both stimulatory & inhibitory effects on IGF-induced DNA synthesis; effects appear to correlate with the degree of phosphorylation; Baxter, RC & Martin, JL, *Prog in Growth Factor Res*, 1: 49, 1989; Clemmons, DR, *Growth Regul*, 2: 80, 1992; Frost, RA & Cheng, L, *J Biol Chem*, 266: 18082, 1991; Jones, JI et al, *Proc Natl Acad Sci USA*, 88: 7481, 1991; Koistinen, R et al, *Clin Chim Acta*, 215: 189, 1993

Sigma I 1649   Human hepatoma, cell line HepG2   MW 30k ~90% (SDS-PAGE); solution in Tris buffered saline, pH 7.4; unit definition: 1 U binds 1 ng IGF at IGF concentrations of 10 ng/mL at pH 6.0 at 25°C; dephosphorylated activity: ≥200 U/vial | Made by dephosphorylation of IGFBP-1; binds IGF-I & IGF-II with very high affinity; *in vivo*, IGFBP-1 is phosphorylated on Ser residues; affinity towards IGF increases with the degree of phosphorylation; has both stimulatory & inhibitory effects on IGF-induced DNA synthesis; effects appear to correlate with the degree of phosphorylation; Baxter, RC & Martin, JL, *Prog in Growth Factor Res*, 1: 49, 1989; Clemmons, DR, *Growth Regul*, 2: 80, 1992; Frost, RA & Cheng, L, *J Biol Chem*, 266: 18082, 1991; Jones, JI et al, *Proc Natl Acad Sci USA*, 88: 7481, 1991; Koistinen, R et al, *Clin Chim Acta*, 215: 189, 1993

USBio I7661-16B   Human recombinant   >98%; lyophilized; rehydrate in 10 *mM* acetic acid because IGF adheres to container surfaces | Purified by Phenyl-Sepharose chromatography, IGF-BP1 affinity chromatography, Reverse Phase HPLC C-4 column

Upstate 12-129   Human recombinant, expressed in CHO cells MW 25k   Lyophilized powder > 98%

Insulin-Like Growth Factor Binding Protein I, Human

Synonyms: Placental Protein 12

IBT BP1BU020 Human amniotic fluid MW 25.3k >95% (FPLC, SDS-PAGE) | 218 AA protein; one of 6 circulating proteins that bind Insulin-like Growth Factors (IGF-I & IGF-II) with high affinity, modulating their metabolic & mitogenic effects; produced by the liver, ovarian granulosa cells, decidualised endometrium & other cell types; binds both IGF-I & IGF-II; serum levels regulated by insulin-induced inhibition of IGFBP-1 production; inhibits smooth muscle migration in response to IGF-I or IGF-II; stimulates wound healing in response to IGF-II

Insulin-Like Growth Factor Binding Protein II, Bovine

IBT BP2A2U100 >95% (animal/media grade) | Contains genuine recombinant bIGFBP-2 as the major species, along with a 4 AA N-terminal extension & a 3 AA C-terminal deletion

IBT BP2A1U020 Recombinant, receptor grade, expressed in CHO cells MW 30,776 >95% (HPLC, N-terminal sequencing, MS) | 284 AA protein; one of 6 high affinity IGF binding proteins; important role in stabilizing & regulating IGFs *in vivo*; binds IGF-I & -II; ED_{50} for IGF-I < 10 *nM*

Insulin-Like Growth Factor Binding Protein II, Gly-Ala-Arg-Ala-

IBT BP2AG1U020 Recombinant, expressed in CHO cells MW 31,131 >95% (HPLC) | 288 AA variant of bovine IGFBP-2; generated by alternative processing of the signal sequence by host mammalian cells; 4 AA longer than bIGFBP-2 at the N-terminus; binds IGF-II with a similar affinity to bIGFBP-2; 4-times the affinity for IGF-I

Insulin-Like Growth Factor Binding Protein II, Human

IBT BP2BU020 Human recombinant receptor grade, expressed in CHO cells MW 31k >95% | 289 AA protein; one of 6 'classical' IGF binding proteins; important role in stabilizing & regulating activity of IGFs *in vivo*; binds IGF-I & -II & modifies their biological activity & bioavailability

IBT rIGFBP2-5 Recombinant, expressed in insect cells (*Spodoptera frugiperda*, Sf9) infected with recombinant viruses MW 31k >90% (SDS-PAGE) | Purified by affinity chromatography & HPLC

Insulin-Like Growth Factor Binding Protein III

Upstate 12-131 Human recombinant, expressed in CHO cells MW 47k Lyophilized

USBio I7661-16J Human recombinant, expressed in CHO cells Lyophilized from 0.04% TFA containing a trace of acetonitrile | IGFBP-3 will adhere to container surfaces: dissolve with low volumes of 10*mM* acetic acid; purified by phenyl-sepharose chromatography, gel filtration, IGF-BP3 affinity chromatography, reverse phase HPLC

IBT BP3BU015 Human recombinant, expressed in human cells MW 43-45k >95% | Glycoprotein; most abundant IGFBP in serum & milk; produced by non-parenchymal hepatic cells; circulates in serum; binds IGF-I or IGF-II with an acid labile subunit (ALS) to form a 150k circulating complex at a serum concentration of ~100 *nM*; the half-life of IFG-I in serum is increased from ~10-25 min if bound to IGFBP-3 & to ~15 h if complexed with IGFBP-3 & ALS

IBT IGFBP3-5 Human recombinant, expressed in mouse myeloma cells MW 43-45k >98%; contains BSA as carrier protein for stabilisation

Insulin-Like Growth Factor Binding Protein III (N109D)

USBio I7661-16K DsbA(mut) recombinant protein expressed in E. coli & cleaved with 3C protease; the protein was renatured & further purified using ion exchange & hydrophobic interaction chromatography; binding Activity: IGFBP-3 (N109D) binds IGF-I like wild type; the N109D substitution makes the protein more soluble when expressed in E. coli

Insulin-Like Growth Factor Binding Protein III (N109D, K228E, R230G)

USBio I7661-16L DsbA(mut) recombinant protein expressed in E. coli & cleaved with 3C protease; the N109D substitution makes the protein more soluble when expressed in E. coli; binding Activity: This IGFBP-3 mutant binds IGF-I like the wild type; the K228E & R230G mutations cause the protein to lose its ability to bind collagen

Insulin-Like Growth Factor Binding Protein IV

IBT BP4BU020 Human recombinant, expressed in CHO cells MW 24-28k (glycosylated) >95% | 237 AA glycosylated protein; one of 6 'classical' IGF binding proteins; important role in stabilizing & regulating IGF *in vivo*

IBT rIGFBP4-5 Human recombinant, expressed in insect cells (*Spodoptera frugiperda*, Sf9) infected with recombinant viruses MW 24k >90% (SDS-PAGE) | Purified by affinity chromatography & HPLC

IBT BP4CU015 Rat recombinant, expressed in *E. coli* MW 25.7k (non-glycosylated) >95%; non-glycosylated | 233 AA protein

Insulin-Like Growth Factor Binding Protein V

USBio I7661-16T Human recombinant, expressed in CHO cells Lyophilized | Purified by phenyl-sepharose, IGF-I affinity & reverse phase HPLC C-4 column chromatography

IBT rIGFBP5-5 Human recombinant, expressed in insect cells (*Spodoptera frugiperda*, Sf9) infected with recombinant viruses MW 30k >90% (SDS-PAGE); forms a double band in PAGE, indicating a possible carbohydrate heterogenity; ~50% of the product lacks the first 2 AA | Purified by affinity chromatography & HPLC

IBT BP5BU10 Human, expressed in mammalian cells MW 31-32k >95%; glycosylated | 252 AA protein; one of 6 'classical' IGF binding proteins; stabilizes & regulates IGF *in vivo*; like IGFBP-3, forms a ternary complex with acid-labile subunit (ALS); preferentially binds IGF-II; also binds with high affinity to extracellular matrix components which protect it from proteolysis; has a nuclear targeting sequence; independent of the presence of IGF-I, fluorescently labelled IGFBP-5 is translocated to the nucleus of actively dividing cells

Upstate 12-333 Recombinant, expressed in CHO cells MW 30k Lyophilized

Insulin-Like Growth Factor Binding Protein VI

IBT 1 BP6BU015 Human recombinant, expressed in CHO cells MW 28-30k (glycosylated) >95% (HPLC) | Major glycosylated protein species contains 216 AA; minor isoform with a 2 AA N-terminal extension is produced as a result of alternative processing by CHO cells used for rhIGFBP-6 production

IBT rIGFBP6-5 Human recombinant, expressed in insect cells cells (*Spodoptera frugiperda*, Sf9) infected with recombinant viruses MW 28k >95% (SDS-PAGE) | Purified by affinity chromatography & HPLC

Insulin-Like Growth Factor I

Biogenesis 5345-0654 Human r-DNA *E. coli* MW 7.6k Endotoxin: <0.1 ng/µg of IGF-1; ED_{50} ≤1.0 ng/mL; purified; no preservatives; lyophilized

BioSource International PHG0074 Human recombinant

BioSource International PMG0071 Mouse recombinant

Insulin-Like Growth Factor I (Y60L), Null

USBio N8000 DsbA(mut) recombinant protein expressed in *E. coli* Cleaved with 3C protease; renatured & further purified using ion exchange & hydrophobic interaction chromatography; binds IGF binding proteins like wild type IGF-I, however, the Y60L mutation results in an approximate 20-fold loss in affinity for the IGF-I receptor & little affinity for the IGF-2 receptor

Insulin-Like Growth Factor I Binding Protein I

Sigma I 2024   A complex of IGF-I & dephosphorylated IGFBP-1; solution in Tris buffered saline, pH 7.4 | Essential for the coordination & regulation of the biological activities of the IGFs; could be a powerful tool for studying those activities both *in vitro* & *in vivo*; Jones, JI & Clemmons, DR, *Endocrine Reviews*; 16: 3, 1995

Insulin-Like Growth Factor II

Biogenesis 5345-3004   Human r-DNA   Lyophilized

BioSource International PHG0084   Human recombinant

Insulin-Transferrin, Sodium Selenite Media Supplement

Sigma I 1884   Lyophilized; each vial contains: 25 mg insulin from bovine pancreas, 25 mg human transferrin (substantially iron free) & 25 μg sodium selenite; each vial prepares 5 L medium; sterilized by γ-irradiation; cell culture tested

Integrin $\alpha 1\beta 1$

Synonyms: VLA-1

Chemicon CC1012   Human   Highly purified using collagen-sepharose, fibronectin-sepharose & immunoaffinity chromatography; formulated with Triton X-100 | For ligand binding studies & as a positive control for electrophoresis or immunoblotting

Chemicon CC1015   Human   Highly purified using collagen-sepharose, fibronectin-sepharose & immunoaffinity chromatography; formulated with Triton X-100 | For ligand binding studies & as a positive control for electrophoresis or immunoblotting

Integrin $\alpha 5\beta 1$

Synonyms: VLA-5

Chemicon CC1026   Human   Highly purified using collagen-sepharose, fibronectin-sepharose & immunoaffinity chromatography; formulated with octyl-β-D-glucopyranoside | For ligand binding studies & as a positive control for electrophoresis or immunoblotting

Chemicon CC1027   Human   Highly purified using collagen-sepharose, fibronectin-sepharose & immunoaffinity chromatography; formulated with octyl-β-D-glucopyranoside | For ligand binding studies & as a positive control for electrophoresis or immunoblotting

Chemicon CC1052   Human   Highly purified using collagen-sepharose, fibronectin-sepharose & immunoaffinity chromatography; formulated with Triton X-100 | For ligand binding studies & as a positive control for electrophoresis or immunoblotting

Chemicon CC1055   Human   Highly purified using collagen-sepharose, fibronectin-sepharose & immunoaffinity chromatography; formulated with Triton X-100 | For ligand binding studies & as a positive control for electrophoresis or immunoblotting

Integrin $\alpha V\beta 3$

Chemicon CC1018   Human   Highly purified using collagen-sepharose, fibronectin-sepharose & immunoaffinity chromatography; formulated with Triton X-100 | For ligand binding studies & as a positive control for electrophoresis or immunoblotting

Chemicon CC1019   Human   Highly purified using collagen-sepharose, fibronectin-sepharose & immunoaffinity chromatography; formulated with Triton X-100 | For ligand binding studies & as a positive control for electrophoresis or immunoblotting

Chemicon CC1020   Human   Highly purified using collagen-sepharose, fibronectin-sepharose & immunoaffinity chromatography; formulated with octyl-β-D-glucopyranoside | For ligand binding studies & as a positive control for electrophoresis or immunoblotting

Chemicon CC1021   Human   Highly purified using collagen-sepharose, fibronectin-sepharose & immunoaffinity chromatography; formulated with octyl-β-D-glucopyranoside | For ligand binding studies & as a positive control for electrophoresis or immunoblotting

Integrin $\alpha V\beta 5$

Chemicon CC1022   Human   Purified; formulated with octyl-β-D-glucopyranoside | For ligand binding studies & as a positive control for electrophoresis or immunoblotting

Chemicon CC1023   Human   Purified; formulated with octyl-β-D-glucopyranoside | For ligand binding studies & as a positive control for electrophoresis or immunoblotting

Chemicon CC1024   Human   Purified; formulated with Triton X-100 | For ligand binding studies & as a positive control for electrophoresis or immunoblotting

Chemicon CC1025   Human   Purified; formulated with Triton X-100 | For ligand binding studies & as a positive control for electrophoresis or immunoblotting

Interferon

Sigma I 2396   Human leukocyte   Aseptically filled; solution in phosphate buffered saline, pH ~7.3; unit definition: 1 U is that amount of interferon which protects 50% of the indicator cells from viral cytopathology | Potency determined by viral plaque reduction method using human epithelial cells & vesicular *stomatitus* virus; not assayed by Sigma; αIFH, Le

Sigma I 4268   Rabbit   Produced in RK13 Rabbit kidney cell cultures by stimulation with Parainfluenza 1 virus; lyophilized, composition of buffer salts given in accompanying data sheet; reconstitute to 1 mL; activity: 1-3X10^5 International Reference Units/vial; unit definition: see that for Sigma I 2396; potency: established by multiple comparative assays with the interferon reference reagents provided by the Antiviral Substances Program, NIAID, NIH, using a modified vital dye-uptake assay | Finter, NB, *J Gen Virol*, 5: 419, 1969

Sigma I 4023   Rat   Produced in rat kidney cell cultures by stimulation with Poly I:C; lyophilized from 0.4 M glycine HCl, pH 3.5; non-irradiated, sterile-filtered; reconstitute to 1 mL; activity: $\geq$1X10^5 International Reference Units/mg protein; unit definition: see that for Sigma I 2396; potency: established by multiple comparative assays with the interferon reference reagents provided by the Antiviral Substances Program, NIAID, NIH, using a modified vital dye-uptake assay; protein concentration determined by Coomassie blue binding assay with bovine albumin as standard | Finter, NB, *J Gen Virol*, 5: 419, 1969; Bradford, M, *Anal Biochem*, 72: 248, 1976

Interferon Inducible T-Cell α-Chemoattractant

Biodesign A52046H   *E. coli*   MW 15.6k   Purified

Chemicon GF088   Human $\geq$95%

PeproTech 300-46   Human recombinant, expressed in *E. coli* MW 8.3k   >98% (SDS-PAGE) & HPLC; <0.1 ng endotoxin per mg (1EU/mg); lyophilized | Novel non-ELR CXC chemokine; regulated by interferon; potent chemoattractant activity for IL-2 activated T cells, but not for freshly isolated unstimulated T cells, neutrophils, or monocytes; 73 AA; SA determined by its ability to chemoattract IL-2 activated human T cells

Interferon-α

Fitzgerald 30-AI90   Human recombinant, expressed in *E. coli*

Calbiochem 407293   Mouse recombinant, expressed in *E. coli* MW 19,323 $\geq$95% (SDS-PAGE); liquid in PBS containing 0.1% BSA; activity: $\geq$3.5x10^4 U/mL; 1 U of interferon/mL of medium causes a 50% inhibition of EMCV-induced cytopathic effect on mouse L929 cells; units are determined with respect to the NIH international reference standard for mouse IFN-α | Antiviral & anti-neoplastic agent; suppresses bone marrow function & induces apoptotic cell death in non-melanoma skin cancer; IFN-α signaling is mediated by JAK1 & TYK2 tyrosine kinases; *Merck Index*, 12: 5016; Krishnan, K et al, *Eur J Biochem*, 247: 298, 1997; Rodriquez-Villanueva, S & McDonnel, TJ, *Int J Cancer*, 61: 110, 1995; Familletti, PC et al, *Methods Enzymol*, 78: 387, 1981

BioSource International PRC4014   Rat recombinant

Interferon-α & -β

Sigma I 1258   Mouse   Produced in mouse L929 cell cultures by stimulation with Poly I:C; lyophilized, composition of buffer salts given in accompanying data sheet; reconstitute to 1 mL; non-irradiated; sterile-filtered; activity: ≥1X10[5] International Reference Units/vial; potency: established by multiple comparative assays with the interferon reference reagents provided by the Antiviral Substances Program, NIAID, NIH, using a modified vital dye-uptake assay | Finter, NB, *J Gen Virol*, 5: 419, 1969

Interferon-α A

BioSource International PHC4014   Human recombinant Pure

Sigma I 4276   Human recombinant, expressed in *E. coli* Solution in phosphate buffered saline; endotoxin tested; cell culture tested

BioSource International PMC4016   Mouse recombinant Pure

Fitzgerald 30-AI96   Murine recombinant, expressed in *E. coli*

Interferon-α A Subtype

BioSource International PHC4814   Human recombinant

Interferon-α A/D

BioSource International PHC4044   Human recombinant Liquid

BioSource International PHC4045   Human recombinant

Sigma I 4401   Human recombinant, expressed in *E. coli* Solution in phosphate buffered saline containing 0.1% BSA; endotoxin tested; cell culture tested

Interferon-α A-p1

BioSource International PHC4114   Human recombinant

Interferon-α B-p1

BioSource International PHC4124   Human recombinant

Interferon-α D

BioSource International PHC4054   Human recombinant

Interferon-α2c

**Alexis BMS305   Human recombinant, produced in *E. coli*   MW 19.3k   >98% (SDS-gel electrophoresis prior to addition of human serum albumin); 16.5 μg lyophilized powder at 15 μg/mL; lyophilized in phosphate-buffered saline containing 20 mg/mL BSA; specific activity: 3.2x10[8] NIH G0 23-901-527 units/mg; 2.3x10[8] NIH Gxa 01-901-535 units/mg

Interferon-β

BioSource International PHC4024   Human recombinant

Calbiochem 407297   Human recombinant, expressed in *E. coli* MW 20,027   ≥95% (SDS-PAGE); liquid in PBS containing 0.1% BSA; activity: ≥1x10[5] U/mL; SA: ≥2.0x10[7] U/mg protein; 1 U of interferon/mL of medium causes a 50% inhibition of VSV-induced cytopathic effect on WISH cells; units are determined with respect to the NIH international reference standard for human IFN-β | Antiviral & anti-neoplastic agent; used as a therapeutic agent in multiple sclerosis; has higher anti-proliferative effects on breast cancer cells than α- & γ-interferon; inhibits mitogen-induced astrocyte proliferation; expressed during the triggering stage of macrophage cytocidal activation; *Merck Index*, 12: 5017; Malik, O et al, *Neuroimmunology*, 86: 155, 1998; Runkel, L et al, *J Biol Chem*, 273: 8003, 1988; Familletti, PC et al, *Methods Enzymol*, 78: 387, 1981; Dhib-Jalbut, S, *Mult Scler*, 3: 397, 1997

Fitzgerald 30-AI72   Human recombinant, expressed in *E. coli*

Sigma I 4151   Human recombinant, expressed in *E. coli* Solution in phosphate buffered saline containing 0.1% BSA; endotoxin tested; cell culture tested

BioSource International PMC4024   Mouse recombinant

Calbiochem 407298   Mouse recombinant, expressed in *E. coli* MW 19,735   ≥90% (SDS-PAGE); liquid in PBS containing 0.1% BSA; activity: ≥2x10[5] U/mg; 1 U of interferon/mL of medium causes a 50% inhibition of EMCV-induced cytopathic effect on mouse L929 cells; units are determined with respect to the NIH international reference standard for mouse IFN-β | Antiviral & anti-neoplastic agent; has higher anti-proliferative effects on breast cancer cells than α- & γ-interferon; *Merck Index*, 12: 5017; Dianzani, F, *Gut*, 34: 574, 1993; Familletti, PC et al, *Methods Enzymol*, 78: 387, 1981

**Sigma I 5143   Natural human, produced in MG 63 osteosarcoma cells   Produced by superinduction; ≥90% (SDS-PAGE); 0.2 μm filtered in phosphate buffered saline containing 0.1% human serum albumin; activity ≥1x10[5] U/vial; units are evaluated by a bioassay system comparing the inhibition of cytopathic effects caused by a challenge virus (vesicular stomatitis virus, VSV) in indicator fibroblast cells against the NIH/NIAID International Standard, Gb 23-902-531; endotoxin tested | Van Damme, J & Billiau, A, *Meth Enzymol*, 78: 101, 1981

BioSource International PRC4024   Rat recombinant

Biodesign A52302H   *E. coli*   MW 17k   Purified

Biodesign A52420H   *E. coli*   MW 15.6k   Purified | Species specificity: rat

Chemicon IF002   Human   ≥95%

Sigma I 6507   Human   Produced from human buffy coats by induction with A23187 & mezerein; solution in phosphate buffered saline adjusted to 2-4mg/mL total protein with human serum albumin; non-irradiated; sterile-filtered; activity: ≥1X10[6] International Reference Units/mg protein; unit definition: see that for Sigma I 2396; activity determined by inhibitions of Sindbis virus-induced cytopathic effect in FL cells; titer is reciprocal of dilution at which 50% cells exhibit cytopathic effect; WHO reference material used as standard; each vial contains ~1X10[6] U; protein concentration determined by Coomassie blue method | Sedmak & Grossberg, *Anal Biochem*, 79: 544, 1977

Chemicon IF3   Human leukocyte   Purified

Chemicon IF6   Human leukocyte   Purified

Biogenesis 5362-5204   Human r-DNA   Lyophilized

BioSource International PHC4031   Human recombinant

BioSource International PHC4834   Human recombinant

Calbiochem 407306   Human recombinant, expressed in *E. coli* MW 17k   ≥97% (SDS-PAGE); lyophilized from sterile-filtered 100 *mM* NH₄OAc, pH 7.0 containing 50 μg BSA/μg IFN-γ; biological activity: ED₅₀=8-15 U/mL as measured by inhibition of cytopathic effects in HeLa cells infected with EMC virus; SA: ≥1x10[7] U/mg protein; endotoxin: ≤100 pg/μg IFN-γ | A multifunctional protein consisting of 144 AA; exhibits antiviral & antitumor properties; regulates the development of specific immune responses; *Merck Index*, 12: 5018; Kohji, K et al, *J Int J Cancer*, 58: 380, 1994; Meager, A, *Lymphokines & Interferons, A Practical Approach* (Clemens, MJ et al, eds), IRL Press, Oxford, p. 129, 1987

Fitzgerald 30-AI71   Human recombinant, expressed in *E. coli*

Harlan BT-3008   Human recombinant, expressed in *E. coli* Lyophilized; 0.025 mg

Harlan BT-3009   Human recombinant, expressed in *E. coli* Lyophilized; 0.1 mg

PeproTech 300-02   Human recombinant, expressed in *E. coli* MW 16.7k   >98% (SDS-PAGE) & HPLC; <0.1 ng endotoxin per mg (1EU/mg); lyophilized | Lymphoid factor which possesses potent anti-viral activity; stimulates macrophages & NK cells; 143 AA; ED₅₀ ≤ 0.05 ng/mL; SA ≥ 2 x 10[7] U/mg; SA determined in a viral resistance assay

Sigma I 1520   Human recombinant, expressed in *E. coli* ≥97% (SDS-PAGE); 0.2 μm filtered & lyophilized from 0.1 *M* ammonium acetate, pH 7.0; activity measured in an anti-viral assay using HeLa cells infected with EMC virus; endotoxin tested | Produced by activated T cells & natural killer cells stimulated by alloantigens, tumors & mitogens; exerts a variety of biological effects including antiviral activity, inhibition of cell or tumor growth & promotion of terminal differentiation of B cells into immunoglobulin-producing cells; activates macrophages, boosts cytotoxicity of natural killer cells & stimulates T cell cytotoxicity; synergistic with TNF-α in its cytotoxicity, but acts on cells via specific cell surface receptors; Hibino, Y et al, *J Biol Chem*, 266: 6948, 1991; Vilcek, J et al, *Lymphokines*, 11: 1, 1985; Gresser, I et al, *Proc Natl Acad Sci USA*, 66: 1052, 1970; Knight, E Jr, *Nature*, 262: 302, 1976; Perussia, B et al, *J Exp Med*, 158: 1092, 1983; Opdenakker, G et al, *Experimenta (Basel)*, 45: 513, 1989; Friedman, RM et al, *Adv Immunol*, 34: 97, 1983; Vilcek, J et al, *Interferon & the Immune System*, Elsevier, North Holland, Amsterdam, 1984; Fransen, L et al, *Cell Immunol*, 100: 260, 1986; Pestka, S et al, *Ann Rev Biochem*, 56: 727, 1987; Meager, A, in: *Lymphokines & Interferons, A Practical Approach*, Clemens, MJ et al, eds, IRL Press

Interferon-γ

Sigma I 3265   Human recombinant, expressed in *E. coli*   MW 15.5k   ≥97% (SDS-PAGE); lyophilized from phosphate buffered saline; proliferative activity tested in culture using WiDr human colon adenocarcinoma cells in a MTT dye assay which is a modification of the method of Pfizenmaier, K et al, *Cancer Research*, 45: 3503, 1985; endotoxin tested | Produced by activated T cells & natural killer cells stimulated by alloantigens, tumors & mitogens; exerts a variety of biological effects including antiviral activity, inhibition of cell or tumor growth & promotion of terminal differentiation of B cells into immunoglobulin-producing cells; activates macrophages, boosts cytotoxicity of natural killer cells & stimulates T cell cytotoxicity; synergistic with TNF-α in its cytotoxicity, but acts on cells via specific cell surface receptors; Hibino, Y et al, *J Biol Chem*, 266: 6948, 1991; Vilcek, J et al, *Lymphokines*, 11: 1, 1985; Gresser, I et al, *Proc Natl Acad Sci USA*, 66: 1052, 1970; Knight, E Jr, *Nature*, 262: 302, 1976; Perussia, B et al, *J Exp Med*, 158: 1092, 1983; Opdenakker, G et al, *Experimenta (Basel)*, 45: 513, 1989; Friedman, RM et al, *Adv Immunol*, 34: 97, 1983; Vilcek, J et al, *Interferon & the Immune System*, Elsevier, North Holland, Amsterdam, 1984; Fransen, L et al, *Cell Immunol*, 100: 260, 1986; Pestka, S et al, *Ann Rev Biochem*, 56: 727, 1987

Alexis BMS303   Human recombinant, produced in *E. coli*   MW 33k (dimer composed of two identical 16.5k monomers)   >98% (SDS-gel electrophoresis prior to addition of human serum albumin); 110 μg lyophilized powder at 100 μg/mL; lyophilized in phosphate-buffered saline containing 10 mg/mL BSA; specific activity: 2x10⁷ NIH Gg 23-901-530 units/mg; biological activity: assessed in a bioassay using inhibition of the cytopathic encephalomyocarditis virus effect on human lung carcinoma cell line A549 as test parameter

Biogenesis 5362-5226   Mouse r-DNA   Lyophilized

BioSource International PMC4034   Mouse recombinant

Calbiochem 407303   Mouse recombinant, expressed in *E. coli* MW 15,521   ≥95% (SDS-PAGE); liquid in PBS containing 0.1% BSA; activity: ≥2x10⁵ U/mL; SA: ≥5x10⁷ U/mg protein; 1 U of interferon/mL of medium causes a 50% inhibition of EMCV-induced cytopathic effect on mouse L929 cells; units are determined with respect to the NIH international reference standard for mouse IFN-γ | A multifunctional protein exhibiting antiviral & antitumor properties; regulates the development of specific immune responses; protects against bacterial sepsis in murine models; *Merck Index*, 12: 5018; Zantl, N et al, *Infect Immun*, 66: 2300, 1998; Familletti, PC et al, *Methods Enzymol*, 78: 387, 1981

Sigma I 5517   Mouse recombinant, expressed in *E. coli*   ≥95% (SDS-PAGE); 0.2 μm filtered solution containing 0.1% BSA; 1 U is the amount of IFN-γ required to induce half-maximal inhibition of proliferation of WEHI-279 cells; tested in culture using a modification of the method of Reynolds, DS et al, *J Immunol*, 139: 767, 1987; endotoxin tested | Produced by activated T cells & natural killer cells stimulated by alloantigens, tumors & mitogens; exerts a variety of biological effects including antiviral activity, inhibition of cell or tumor growth & promotion of terminal differentiation of B cells into immunoglobulin-producing cells; activates macrophages, boosts cytotoxicity of natural killer cells & stimulates T cell cytotoxicity; synergistic with TNF-α in its cytotoxicity, but acts on cells via specific cell surface receptors; Hibino, Y et al, *J Biol Chem*, 266: 6948, 1991; Vilcek, J et al, *Lymphokines*, 11: 1, 1985; Gresser, I et al, *Proc Natl Acad Sci USA*, 66: 1052, 1970; Knight, E Jr, *Nature*, 262: 302, 1976; Perussia, B et al, *J Exp Med*, 158: 1092, 1983; Opdenakker, G et al, *Experimenta (Basel)*, 45: 513, 1989; Friedman, RM et al, *Adv Immunol*, 34: 97, 1983; Vilcek, J et al, *Interferon & the Immune System*, Elsevier, North Holland, Amsterdam, 1984; Fransen, L et al, *Cell Immunol*, 100: 260, 1986; Pestka, S et al, *Ann Rev Biochem*, 56: 727, 1987

Chemicon IF005   Murine   ≥95%

Alexis BMS312   Murine recombinant, expressed in *E. coli*   MW 15.6k   >98% (SDS-gel electrophoresis, before addition of human serum albumin); lyophilized in bicarbonate buffer containing human serum albumin; bioactivity: 10⁷ U/mg

Fitzgerald 30-AI74   Murine recombinant, expressed in *E. coli*

PeproTech 315-05   Murine recombinant, expressed in *E. coli* MW 15.6k   >98% (SDS-PAGE) & HPLC; <0.1 ng endotoxin per mg (1EU/mg); lyophilized | Regulatory protein produced by activated NK cells & CD4⁺TCRab⁺, CD8⁺TCRab⁺, & TCRgd+ T cells; specifically binds to a single class of high affinity receptors; 134 AA; ED$_{50}$ ≤ 0.1 ng/mL; SA ≥ 1 x 10⁷ U/mg; SA determined by a cytopathic effect inhibition assay with murine L929 cells challenged with EMC virus

Biogenesis 0100-0105   Rat r-DNA   Lyophilized

BioSource International PRC4034   Rat recombinant, expressed in *E. coli*

BioSource International PRC4035   Rat recombinant, expressed in *E. coli*

Calbiochem 407304   Rat recombinant, expressed in *E. coli* MW 15,635   ≥95% (SDS-PAGE); liquid in PBS containing 0.1% BSA; activity: ≥2x10⁵ U/mL; SA: ≥5x10⁶ U/mg protein; 1 U of interferon/mL of medium causes a 50% inhibition of EMCV-induced cytopathic effect on mouse L929 cells; units are determined with respect to the NIH international reference standard for mouse IFN-γ | A multifunctional protein exhibiting antiviral & antitumor properties; regulates the development of specific immune responses; inhibits rat Leydig cell steroidogenesis; *Merck Index*, 12: 5018; Lin, T et al, *Endocrinology*, 139: 2217, 1998; Familletti, PC et al, *Methods Enzymol*, 78: 387, 1981

Fitzgerald 30-AI78   Rat recombinant, expressed in *E. coli*

PeproTech 400-20   Rat recombinant, expressed in *E. coli* MW 15.6k   >98% (SDS-PAGE) & HPLC; <0.1 ng endotoxin per mg (1EU/mg); lyophilized | Lymphoid factor with potent anti-viral activity; stimulates macrophages & NK cells; 135 AA; ED$_{50}$ ≤ 0.1 ng/mL; SA ≥ 1 x 10⁷ U/mg; SA determined by a cytopathic effect inhibition assay with murine L929 cells challenged with EMC virus

Sigma I 2651 Rat recombinant, expressed in *E. coli* ≥97% (SDS-PAGE); 0.2 µm filtered in 30% acetonitrile, 0.1% trifluoroacetic acid containing 5 mg BSA; activity measured in an anti-viral assay using L-929 cells infected with EMC virus; endotoxin tested | Produced by activated T cells & natural killer cells stimulated by alloantigens, tumors & mitogens; exerts a variety of biological effects including antiviral activity, inhibition of cell or tumor growth & promotion of terminal differentiation of B cells into immunoglobulin-producing cells; activates macrophages, boosts cytotoxicity of natural killer cells & stimulates T cell cytotoxicity; synergistic with TNF-α in its cytotoxicity, but acts on cells via specific cell surface receptors; Hibino, Y et al, *J Biol Chem*, 266: 6948, 1991; Vilcek, J et al, *Lymphokines*, 11: 1, 1985; Gresser, I et al, *Proc Natl Acad Sci USA*, 66: 1052, 1970; Knight, E Jr, *Nature*, 262: 302, 1976; Perussia, B et al, *J Exp Med*, 158: 1092, 1983; Opdenakker, G et al, *Experimenta (Basel)*, 45: 513, 1989; Friedman, RM et al, *Adv Immunol*, 34: 97, 1983; Vilcek, J et al, *Interferon & the Immune System*, Elsevier, North Holland, Amsterdam, 1984; Fransen, L et al, *Cell Immunol*, 100: 260, 1986; Pestka, S et al, *Ann Rev Biochem*, 56: 727, 1987

BioSource International PSC4034	Swine recombinant

Biodesign A52355H T.ni cells MW 35k Purified | Species specificity: mouse

Interferon-γ Inducible Protein 10

BioSource International PHC1084	Human recombinant
BioSource International PMC1084	Mouse recombinant

Interferon-γ Inducible Protein-10

Biodesign A52016M *E. coli* MW 8.7k Purified	Species specificity: mouse
Biodesign A52312H *E. coli* MW 9k Purified	
Chemicon GF033 Human ≥95%	
Fitzgerald 30-AI95 Human recombinant, expressed in *E. coli*	

PeproTech 300-12 Human recombinant, expressed in *E. coli* MW 8.5k >98%; 78 AA; lyophilized with no additives; activity determined by its ability to chemoattract human T-lymphocytes using a concentration range of 10.0-50.0 ng/mL

Sigma I 3400 Human recombinant, expressed in *E. coli* MW ~8.7k ≥97% (SDS-PAGE); 0.2 µm filtered & lyophilized in PBS containing 500 µg BSA; activity measured in a cell proliferation assay using the factor-dependent human erythroleukemic cell line, TF-1; endotoxin tested | Member of the C-X-C or α chemokine class; doesn't contain the ELR domain immediately preceding the first cysteine residue near the amino terminus; other chemokines in this group include mouse CRG, Mig, PBSF/SDF-1 & PF4; originally identified as an IFN-γ-inducible gene in monocytes, fibroblasts & endothelial cells; chemoattractant for activated T lymphocytes; potent inhibitor of angiogenesis & displays a thymus-dependent anti-tumor effect; polypeptide of 78 AA, precursor form of human IP-10 is 98 AA; to generate the mature IP-10, the precursor cleaves its 21 AA signal peptide; 67% AA homology to mouse CRG-2

Chemicon GF093	Murine	≥95%

PeproTech 250-16 Murine recombinant, expressed in *E. coli* MW 8.7k >98% (SDS-PAGE) & HPLC; <0.1 ng endotoxin per mg (1EU/mg); lyophilized | Produced by several cell types during the delayed-type hypersensitivity response; acts as a chemoattractant towards monocytes, lymphocytes & certain T cells; 77 AA; SA determined by its ability to chemoattract IL-2 activated T cells

Interferon-γ Type I (Universal)

Fitzgerald 30-AI73	Human recombinant, expressed in *E. coli*

Interferon-γ, Monokine Induced

Biodesign A52026H *E. coli* MW 11.7k Purified

Interferon-ω

Alexis BMS304 Human recombinant, produced in *E. coli* MW 20k >95% (SDS-gel electrophoresis prior to addition of human serum albumin); 25 µg lyophilized powder at 23 µg/mL; lyophilized in phosphate-buffered saline containing 20 mg/mL BSA; specific activity: 1x10⁸ NIH Gxa 01-901-535 units/mg; biological activity: assessed in a bioassay using inhibition of the cytopathic encephalomyocarditis virus effect on human lung carcinoma cell line A549 as test parameter

Interleukin gp130 Soluble Fragment

Sigma G 7534 Human recombinant, expressed in *Sf* 21 insect cells ≥97% (SDS-PAGE); 0.2 µm filtered & lyophilized from phosphate buffered saline containing 500 µg BSA; activity is measured in culture by its ability to inhibit IL-6 soluble receptor enhancement of IL-6 activity with the mouse myeloid leukemia cell line, M1; endotoxin tested | IL-6 receptor (IL-6R) consists of 2 chains: IL-6R & gp130; interaction of IL-6 with the IL-6R initiates the association of gp130 with the IL-6R & a signal is transduced through gp130; Saito, T et al, *J Immuno*, 147: 168, 1991; multifunctional protein originally discovered in the media of cells stimulated with double stranded RNA; appears to be directly involved in the responses that occur after infection & injury & may prove to be as important as IL-1 & TNF-α in regulating the acute phase response; reported to be produced by fibroblasts, activated T cells, activated monocytes or macrophages & endothelial cells; acts upon a variety of cells including fibroblasts, myeloid progenitor cells, T cells, B cells & hepatocytes; appears to interact with IL-2 in the proliferation of T lymphocytes; also potentiates the proliferative effect of IL-3 on multipotential hematopoietic progenitors; Billiau, A, *Immunol Today*, 8: 84, 1987; Gauldie, J et al, *Proc Natl Acad Sci USA*, 84: 7251, 1987; Van Snick, J, *Ann Rev Immunol*, 8: 253, 1990; nordan, R & Potter, M, *Science*, 233: 566, 1986; Van Snick, J et al, *Proc Natl Acad Sci USA*, 83: 9679, 1986; Van Damme, J et al, *Eur J Immunol*, 17: 1, 1987

Interleukin I Receptor

Biogenesis 5375-5175	r-DNA

Interleukin I Soluble Receptor Type II

Sigma I 8148 Human recombinant, expressed in *Sf* 21 insect cells ≥97% (SDS-PAGE); 0.2 µm filtered & lyophilized from phosphate buffered saline containing 1.25 mg BSA; receptor mediated activity measured by its ability of D10.G4.1 cells; endotoxin tested | Activates T cell lymphocytes which then proliferate & secrete Interleukin-2; released primarily from stimulated macrophages & monocytes but has been shown to be released from several other cell types & plays a key role in inflammatory & immune responses; closely related agents IL-1α & IL-1β share 62% homology in AA sequence & elicit nearly identical biological responses; both are ~17k with some heterogeneity in the amount of glycosylation; Symons, JA et al, in: *Lymphokines & Interferons, A Practical Approach*, Clemens, MJ et al, eds, IRL Press, Oxford, p. 272, 1987; Gery, I et al, *J Exp Med*, 136: 128, 1972; Oppenheim, J et al, *Immunol Today*, 7: 45, 1986; Durum, S et al, *Ann Rev Immunol*, 3: 263, 1985; Aarden, L et al, *J Immunol*, 123: 2928, 1979

Interleukin Iα

Synonyms: Endogenous Pyrogen; Mitogenic Protein Helper Peak I; T-Cell Replacing Factor III; B-Cell Activating Factor; B-Cell Differentiation Factor; Lymphocyte Activating Factor; IL-1α

Biodesign A52011H *E. coli* MW 17.9k Purified	Species specificity: mouse		
Biodesign A52201H *E. coli* MW 18k Purified			
Chemicon IL001 Human ≥95%			
Biogenesis 5375-0005 Human r-DNA Lyophilized			
BioSource International PHC0014 Human recombinant			

Oncogene PF012 Human recombinant MW 17k >98% (SDS-PAGE); with 100 µg BSA; reconstitute in sterile PBS; biological activity: half maximal stimulation of ³H-thymidine uptake by murine C3H/HeJ thymocytes ~0.1 ng/mL | Species reactivity: human; for proliferation studies & Western blot

Calbiochem 407611 Human recombinant, expressed in *E. coli* MW 17.5k ≥97% (SDS-PAGE); lyophilized from sterile-filtered PBS containing 50 μg BSA/μg IL-1α; biological activity: ED_{50}=3-10 pg/mL as measured in a cell proliferation assay; endotoxin: ≤100 pg/μg IL-1α | Potent stimulator of bone resorption & stimulator of nuclear phospholipase C; induces nitric oxide synthase; *Merck Index*, 12: 5019; Kilbourn, RG et al, *J Natl Cancer Inst*, 84: 1008, 1992; Tsan, MF & White, JE, *Am J Physiol*, 266: L316, 1994

Fitzgerald 30-AI56 Human recombinant, expressed in *E. coli*

Harlan BT-2 k Human recombinant, expressed in *E. coli* Lyophilized; 0.002 mg | Cytokine

Harlan BT-2001 Human recombinant, expressed in *E. coli* Lyophilized; 0.01 mg | Cytokine

PeproTech 200-01A Human recombinant, expressed in *E. coli* MW 18.0k >98% (SDS-PAGE) & HPLC; <0.1 ng endotoxin per mg (1EU/mg); lyophilized | Potent immuno-modulator; mediates a wide range of immune & inflammatory responses; 159 AA; ED_{50} ≤ 0.001 ng/mL; SA ≥ 10^9 U/mg; SA determined by the dose-dependent stimulation of murine D10S cells

Sigma I 3894 Human recombinant, expressed in *E. coli* MW ~17k ≥97% (SDS-PAGE); 0.2 μm filtered & lyophilized from phosphate buffered saline solution containing 100 μg human serum albumin; proliferative activity tested in culture using the T cell line D10.G4.1; endotoxin tested | Activates T cell lymphocytes which then proliferate & secrete Interleukin-2; released primarily from stimulated macrophages & monocytes but has been shown to be released from several other cell types & plays a key role in inflammatory & immune responses; closely related agents IL-1α & IL-1β share 62% homology in AA sequence & elicit nearly identical biological responses; both are ~17k with some heterogeneity in the amount of glycosylation; Symons, JA et al, in: *Lymphokines & Interferons, A Practical Approach*, Clemens, MJ et al, eds, IRL Press, Oxford, p. 272, 1987; Gery, I et al, *J Exp Med*, 136: 128, 1972; Oppenheim, J et al, *Immunol Today*, 7: 45, 1986; Durum, S et al, *Ann Rev Immunol*, 3: 263, 1985; Aarden, L et al, *J Immunol*, 123: 2928, 1979

Biogenesis 5375-0025 Mouse r-DNA Lyophilized

BioSource International PMC0014 Mouse recombinant

Calbiochem 407613 Mouse recombinant, expressed in *E. coli* MW 18k ≥97% (SDS-PAGE); lyophilized from sterile-filtered PBS containing 50 μg BSA/μg IL-1α; biological activity: ED_{50}=3-7 pg/mL as measured in a cell proliferation assay; endotoxin: ≤100 pg/μg IL-1α | Affects the differentiation & function of cells involved in inflammatory & immune responses; potent stimulator of bone resorption; *Merck Index*, 12: 5019; Weidmann, B et al, *Nature*, 370: 434, 1994; Dinarello, CA & Wolff, SM, *N Engl J Med*, 328: 106, 1993

Sigma I 5396 Mouse recombinant, expressed in *E. coli* ≥97% (SDS-PAGE); 0.2 μm filtered & lyophilized from phosphate buffered saline containing 250 μg BSA; proliferative activity tested in culture using the mouse helper T cell line D10.G4.1; endotoxin tested | Activates T cell lymphocytes which then proliferate & secrete Interleukin-2; released primarily from stimulated macrophages & monocytes but has been shown to be released from several other cell types & plays a key role in inflammatory & immune responses; closely related agents IL-1α & IL-1β share 62% homology in AA sequence & elicit nearly identical biological responses; both are ~17k with some heterogeneity in the amount of glycosylation; Symons, JA et al, in: *Lymphokines & Interferons, A Practical Approach*, Clemens, MJ et al, eds, IRL Press, Oxford, p. 272, 1987; Gery, I et al, *J Exp Med*, 136: 128, 1972; Oppenheim, J et al, *Immunol Today*, 7: 45, 1986; Durum, S et al, *Ann Rev Immunol*, 3: 263, 1985; Aarden, L et al, *J Immunol*, 123: 2928, 1979

Chemicon IL023 Murine ≥95%

Fitzgerald 30-AI41 Murine recombinant, expressed in *E. coli*

PeproTech 211-11A Murine recombinant, expressed in *E. coli* MW 17.9k >98% (SDS-PAGE) & HPLC; <0.1 ng endotoxin per mg (1EU/mg); lyophilized with no additives | Potent immuno-modulator; mediates a wide range of immune & inflammatory responses; 156 AA; ED_{50} ≤ 0.002 ng/mL; SA ≥ 5 x 10^9 U/mg; SA determined by the dose-dependent stimulation of murine D10S cells

PeproTech 400-01A Rat recombinant, expressed in *E. coli* MW 17.7k >98% (SDS-PAGE) & HPLC; <0.1 ng endotoxin per mg (1EU/mg); lyophilized | Potent immuno-modulator; mediates a wide range of immune & inflammatory responses; 155 AA; the ED_{50} ≤ 0.005 ng/m; SA ≥ 2 x 10^9 U/mg; SA determined by the dose-dependent proliferation of murine D10S

Sigma I 3901 Rat recombinant, expressed in *E. coli* Lyophilized from phosphate buffered saline containing 250 μg BSA; endotoxin tested; cell culture tested

BioSource International PSC0014 Swine recombinant

BioSource International PSC0015 Swine recombinant

Biodesign A52121H *E. coli* MW 17k Purified | Species specificity: mouse

Biodesign A52400H *E. coli* MW 17.3k Purified | Species specificity: rat

Biogenesis 5375-5005 Human r-DNA *E. coli* Purified; PBS buffer, pH 7.0; lyophilized

BioSource International PHC0814 Human recombinant

Oncogene PF013 Human recombinant MW 17k >98% (SDS-PAGE); lyophilized with 100 μg BSA.; biological activity: half maximal stimulation of ^{3}H-thymidine uptake by murine C3H/HeJ thymocytes ~0.1 ng/mL | Species reactivity: human & mouse; for proliferation studies & Western blot; will act on both human & mouse cells

Alexis 520-001 Human recombinant, expressed in *E. coli* MW ~17k Cell culture grade; ≥98% (SDS-PAGE); lyophilized powder; salt-free; soluble in water, PBS or most aqueous buffers; activity: $2x10^8$ U/mg protein; biological activity: the ED_{50} as the dose-dependent stimulation of thymidine uptake by murine C3H/HeJ thymocytes is 0.1 ng/mL; lyophilized powder can be reconstituted in sterile water to a concentration of 10 μg/100 μL; for further dilution carrier protein should be added to avoid loss of bioactivity | For most *in vitro* applications, exerts its biological activity in the concentration range of 0.1 to 10 ng/mL

Calbiochem 407615 Human recombinant, expressed in *E. coli* MW 17k ≥97% (SDS-PAGE); lyophilized from sterile-filtered PBS containing 50 μg BSA/μg IL-1β; biological activity: ED_{50}=3-10 pg/mL as measured in a cell proliferation assay; endotoxin: ≤100 pg/μg IL-1β | Major form of interleukin-1 secreted by monocytes & macrophages; induces nitric oxide synthase in pancreatic & smooth muscle cells & suppresses apoptosis in rat ovarian follicles; involved in inflammatory & immune responses & promotes wound healing; *Merck Index*, 12: 5019; Chun, SY et al, *Endocrinology*, 136: 3120, 1995; Kunz, D et al, *Biochem J*, 304: 337, 1994; Eizirik, DL et al, *FEBS Lett*, 317: 62, 1993

Fitzgerald 30-AI57 Human recombinant, expressed in *E. coli*

PeproTech 200-01B Human recombinant, expressed in *E. coli* MW 17.0k >98% (SDS-PAGE) & HPLC; <0.1 ng endotoxin per mg (1EU/mg); lyophilized | Potent immuno-modulator; mediates a wide range of immune & inflammatory responses including activation of B & T cells; 153 AA; ED_{50} ≤ 0.1 ng/mL; SA ≥ 10^7 U/mg; SA determined by the dose-dependent stimulation of thymidine uptake by murine C3H/HeJ thymocytes

Sigma I 4019 Human recombinant, expressed in *E. coli* ≥97% (SDS-PAGE & N-terminal sequence analysis); lyophilized from phosphate buffered saline containing 100 μg human serum albumin; endotoxin tested; cell culture tested

Biogenesis 5375-5026 Mouse r-DNA Lyophilized

BioSource International PMC0814 Mouse recombinant

Calbiochem 407617 Mouse recombinant, expressed in *E. coli* MW 17k ≥97% (SDS-PAGE); lyophilized from sterile-filtered PBS containing 50 μg BSA/μg IL-1β; biological activity: ED_{50}=5-10 pg/mL as measured in a cell proliferation assay; endotoxin: ≤100 pg/μg IL-1β | Suppresses apoptosis in rat ovarian follicles; involved in inflammatory & immune responses & promotes wound healing; *Merck Index*, 12: 5019; Chun, SY et al, *Endocrinology*, 136: 3120, 1995; Weidmann, B et al, *Nature*, 370: 434, 1994; Dinarello, CA & Wolff, SM, *N Engl J Med*, 328: 106, 1993

Sigma I 5271   Mouse recombinant, expressed in *E. coli*   ≥97% (SDS-PAGE); 0.2 µm filtered & lyophilized from phosphate buffered saline containing 250 µg BSA; proliferative activity measured in culture using the mouse helper T cell line D10.G4.1; endotoxin tested | Activates T cell lymphocytes which then proliferate & secrete Interleukin-2; released primarily from stimulated macrophages & monocytes but has been shown to be released from several other cell types & plays a key role in inflammatory & immune responses; closely related agents IL-1α & IL-1β share 62% homology in AA sequence & elicit nearly identical biological responses; both are ~17k with some heterogeneity in the amount of glycosylation; Symons, JA et al, in: *Lymphokines & Interferons, A Practical Approach*, Clemens, MJ et al, eds, IRL Press, Oxford, p. 272, 1987; Gery, I et al, *J Exp Med*, 136: 128, 1972; Oppenheim, J et al, *Immunol Today*, 7: 45, 1986; Durum, S et al, *Ann Rev Immunol*, 3: 263, 1985; Aarden, L et al, *J Immunol*, 123: 2928, 1979

Chemicon IL014	Murine	≥95%

Fitzgerald 30-AI58   Murine recombinant, expressed in *E. coli*

Harlan BT-5100   Murine recombinant, expressed in *E. coli* Lyophilized; 0.005 mg | Cytokine

Harlan BT-5101   Murine recombinant, expressed in *E. coli* Lyophilized; 0.025 mg | Cytokine

PeproTech 211-11B   Murine recombinant, expressed in *E. coli* MW 17.5k   >98% (SDS-PAGE) & HPLC; <0.1 ng endotoxin per mg (1EU/mg); lyophilized with no additives | Potent immuno-modulator; mediates a wide range of immune & inflammatory responses; 153 AA; $ED_{50} \leq 0.002$ ng/mL; $SA \geq 5 \times 10^9$ U/mg; SA determined by the dose-dependent stimulation of murine D10S cells

Chemicon IL024	Rat	≥95%
Biogenesis 0100-0096	Rat r-DNA	Lyophilized

BioSource International PRC0814   Rat recombinant

Fitzgerald 30-AI42   Rat recombinant, expressed in *E. coli*

PeproTech 400-01B   Rat recombinant, expressed in *E. coli* MW 17.3k   >98% (SDS-PAGE) & HPLC; <0.1 ng endotoxin per mg (1EU/mg); lyophilized | Potent immuno-modulator; mediates a wide range of immune & inflammatory responses, including the activation of B & T cells; 153 AA; $ED_{50} \leq 0.1$ ng/mL; $SA \geq 1 \times 10^7$ U/mg; SA determined by the dose-dependent stimulation of murine D10S cells

Sigma I 2393   Rat recombinant, expressed in *E. coli*   ≥97% (SDS-PAGE); 0.2 µm filtered & lyophilized from phosphate buffered saline containing 500 µg BSA; biological activity measured by its ability to stimulate proliferation in the mouse T cell line D10.G4.1; endotoxin tested | Activates T cell lymphocytes which then proliferate & secrete Interleukin-2; released primarily from stimulated macrophages & monocytes but has been shown to be released from several other cell types & plays a key role in inflammatory & immune responses; closely related agents IL-1α & IL-1β share 62% homology in AA sequence & elicit nearly identical biological responses; both are ~17k with some heterogeneity in the amount of glycosylation; Symons, JA et al, in: *Lymphokines & Interferons, A Practical Approach*, Clemens, MJ et al, eds, IRL Press, Oxford, p. 272, 1987; Gery, I et al, *J Exp Med*, 136: 128, 1972; Oppenheim, J et al, *Immunol Today*, 7: 45, 1986; Durum, S et al, *Ann Rev Immunol*, 3: 263, 1985; Aarden, L et al, *J Immunol*, 123: 2928, 1979

Interleukin Iβ

Synonyms: IL-1β

BioSource International PSC0814   Swine recombinant

Interleukin II

Synonyms: IL-2; T-Cell Growth Factor

Biodesign A52202H	*E. coli*	MW 5k	Purified

Biodesign A52222M   *E. coli*   Purified | Species specificity: mouse

Chemicon IL002	Human	≥95%
Biogenesis 5376-1127	Human r-DNA	Lyophilized

BioSource International PHC0024   Human recombinant

Oncogene PF004   Human recombinant   MW 15k   >98% (SDS-PAGE); lyophilized with 100 µg BSA; biological activity: 4×10^5 U/mg as measured in the CTLL-2 cell proliferation assay | Species reactivity: human; for proliferation studies & Western blot; 4×10^5 U/mg protein measured in the CTLL-2 cell proliferation assay

Calbiochem 407623   Human recombinant, expressed in *E. coli* MW 14.7k   >97% (SDS-PAGE); lyophilized from sterile-filtered PBS containing 50 µg BSA/µg IL-2; biological activity: ED_{50}=250-500 pg/mL as measured in a cell proliferation assay with an IL-2-dependent murine cytotoxic T cell line; endotoxin: ≤100 pg/µg IL-2 | Variety of immunological functions; acts as an autocrine agent to promote proliferation & maturation of activated T cells; active in murine cell lines; exhibits ~60% AA homology with murine IL-2; inhibits dexamethasone-induced apoptosis in T lymphocytes; induces nitric oxide synthase; *Merck Index*, 12: 5020; Migliorati, G et al, *Pharmacol Res*, 30: 43, 1994; Nakamura, Y et al, *Cancer Res*, 54: 5757, 1994; Karanth, S et al, *PNAS*, 90: 3383, 1993

Fitzgerald 30-AI59   Human recombinant, expressed in *E. coli*

Fluka 57600   Human recombinant, expressed in *E. coli*   MW 15.5k   ≥98% (GE); 10 k U/mL; solution in Dulbecco-PBS containing 1 mg BSA/mL; 1 U corresponds to the U activity in the colorimetric MTT-assay with CTLL-2 cells; Gillis, S et al, *J Immunol*, 120: 2027, 1978 | For tissue culture; promotes the growth of IL-2 dependent lymphocytes; Smith, KA, *Ann Rev Immunol*, 2: 319, 1984; Robb, RJ et al, *PNAS*, 81: 6486, 1984

Harlan BT-2004   Human recombinant, expressed in *E. coli* Lyophilized; 0.01 mg | Cytokine

Harlan BT-2005   Human recombinant, expressed in *E. coli* Lyophilized; 0.05 mg | Cytokine

PeproTech 200-02   Human recombinant, expressed in *E. coli* MW 15.4k   >98% (SDS-PAGE) & HPLC; <0.1 ng endotoxin per mg (1EU/mg); lyophilized | Potent immuno-modulator; mediates a wide range of immune & inflammatory responses; 134 AA; $ED_{50} \leq 0.1$ ng/mL; $SA \geq 10^7$ U/mg; SA determined by dose-dependent stimulation of murine CTLL-2 cells

Sigma I 2644   Human recombinant, expressed in *E. coli*   MW 15.5k   0.2 µm filtered & lyophilized from 50 *mM* sodium acetate, pH 4.0, containing 500 µg BSA; endotoxin tested | Glycoprotein purified from conditioned media of a human T leukemic cell line; an immunomodulatory factor produced by certain subsets of T lymphocytes; has been isolated from a variety of cell cultures & recombinant systems; has been shown to promote long term growth of activated T cells, activation & proliferation of NK cells & induction of γ-interferon & B cell growth factor secretion; IL-2 proliferative activity is tested in culture using a modification of a biological assay, whereby 1 U is the amount required to induce half-maximal proliferation of CTLL-2 cells as measured in a MTT dye assay or in a ³H-thymidine biological assay; Smith, K, *Science*, 240: 1169, 1988; Morgan, D et al, *Science*, 193: 1007, 1976; Ortaldo, J et al, *J Immuno*, 133: 779, 1984; Farrar, J et al, *J Immunol Res*, 63: 129, 1982; inaba, K et al, *J Exp Med*, 158: 2040, 1983; Coligan, J et al, ed, *Current Protocols in Immunology*, 1: 6.3.1, 1991; Gearing A et al, in: *Lymphokines & Interferons, A Practical Approach*, Clemens, MJ et al, eds, IRL Press, Oxford, p. 296, 1987

Sigma T 3267   Human recombinant, expressed in *E. coli*   MW 15.5k   0.2 µm filtered solution containing 0.1% BSA in 1 mL phosphate buffered saline; endotoxin tested | Glycoprotein purified from conditioned media of a human T leukemic cell line; an immunomodulatory factor produced by certain subsets of T lymphocytes; has been isolated from a variety of cell cultures & recombinant systems; has been shown to promote long term growth of activated T cells, activation & proliferation of NK cells & induction of γ-interferon & B cell growth factor secretion; IL-2 proliferative activity is tested in culture using a modification of a biological assay, whereby 1 U is the amount required to induce half-maximal proliferation of CTLL-2 cells as measured in a MTT dye assay or in a ³H-thymidine biological assay; Smith, K, *Science*, 240: 1169, 1988; Morgan, D et al, *Science*, 193: 1007, 1976; Ortaldo, J et al, *J Immuno*, 133: 779, 1984; Farrar, J et al, *J Immunol Res*, 63: 129, 1982; inaba, K et al, *J Exp Med*, 158: 2040, 1983; Coligan, J et al, *Current Protocols in Immunology*, 1: 6.3.1, 1991; Gearing A et al, in: *Lymphokines & Interferons, A Practical Approach*, Clemens, MJ et al, eds, IRL Press, Oxford, p. 296, 1987

Biogenesis 5376-1205	Mouse r-DNA	Lyophilized

BioSource International PMC0024   Mouse recombinant

Calbiochem 407627 Mouse recombinant, expressed in *E. coli* MW 17.2k >97% (SDS-PAGE); lyophilized from sterile-filtered 50 *mM* NH₄OAc, 1 *mM* DTT, pH 4.0 containing 50 μg BSA/μg IL-2; biological activity: ED$_{50}$=100-400 pg/mL as measured in a cell proliferation assay using an IL-2-dependent mouse cytotoxic T cell line, CTLL-2 | Lymphokine that functions as an autocrine factor, driving the expansion of antigen-specific cells; also acts as a paracrine factor, influencing the activity of other cells, both within & outside of the immune system; mediates the regression of metastatic cancers & increases phagocytic activity of kupffer cells; *Merck Index*, 12: 5020; Umlauf, SW et al, *Mol Cell Biol*, 15: 3197, 1995; Kashima, K et al, *Nature*, 313: 402, 1985

Sigma I 0523 Mouse recombinant, expressed in *E. coli* ≥97% (SDS-PAGE); 0.2 μm filtered & lyophilized from 50 *mM* ammonium acetate & 1 *mM* DTT, pH 4.0, containing 1 mg BSA; activity is tested in culture by a cell proliferation assay using the mouse helper T cell line CTLL-2; endotoxin tested | Glycoprotein purified from conditioned media of a human T leukemic cell line; an immunomodulatory factor produced by certain subsets of T lymphocytes; has been isolated from a variety of cell cultures & recombinant systems; has been shown to promote long term growth of activated T cells, activation & proliferation of NK cells & induction of γ-interferon & B cell growth factor secretion; IL-2 proliferative activity is tested in culture using a modification of a biological assay, whereby 1 U is the amount required to induce half-maximal proliferation of CTLL-2 cells as measured in a MTT dye assay or in a ³H-thymidine biological assay; Smith, K, *Science*, 240: 1169, 1988; Morgan, D et al, *Science*, 193: 1007, 1976; Ortaldo, J et al, *J Immuno*, 133: 779, 1984; Farrar, J et al, *J Immunol Res*, 63: 129, 1982; inaba, K et al, *J Exp Med*, 158: 2040, 1983; Coligan, J et al, ed, *Current Protocols in Immunology*, 1: 6.3.1, 1991; Gearing A et al, in: *Lymphokines & Interferons, A Practical Approach*, Clemens, MJ et al, eds, IRL Press, Oxford, p. 296, 1987

Chemicon IL031 Murine ≥95%

PeproTech 212-12 Murine recombinant, expressed in *E. coli* MW 17.2k >98% (SDS-PAGE) & HPLC; <0.1 ng endotoxin per mg (1EU/mg); lyophilized from 10 *mM* Na Citrate pH 4.0 | Potent lymphoid cell growth factor; exerts its biological activity primarily on T cells; 149 AA; ED$_{50}$ ≤ 1.5 ng/mL; SA ≥ 6.6 x 10⁵ U/mg; SA determined by the dose-dependent stimulation of murine CTLL-2 cells

Sigma I 6013 Natural human MW 15-17k 0.2 μm filtered solution containing 0.1% BSA in 1 mL phosphate buffered saline; endotoxin tested | Glycoprotein purified from conditioned media of a human T leukemic cell line; an immunomodulatory factor produced by certain subsets of T lymphocytes; has been isolated from a variety of cell cultures & recombinant systems; has been shown to promote long term growth of activated T cells, activation & proliferation of NK cells & induction of γ-interferon & B cell growth factor secretion; IL-2 proliferative activity is tested in culture using a modification of a biological assay, whereby 1 U is the amount required to induce half-maximal proliferation of CTLL-2 cells as measured in a MTT dye assay or in a ³H-thymidine biological assay; Smith, K, *Science*, 240: 1169, 1988; Morgan, D et al, *Science*, 193: 1007, 1976; Ortaldo, J et al, *J Immuno*, 133: 779, 1984; Farrar, J et al, *J Immunol Res*, 63: 129, 1982; inaba, K et al, *J Exp Med*, 158: 2040, 1983; Coligan, J et al, ed, *Current Protocols in Immunology*, 1: 6.3.1, 1991; Gearing A et al, in: *Lymphokines & Interferons, A Practical Approach*, Clemens, MJ et al, eds, IRL Press, Oxford, p. 296, 1987

Sigma T 0892 Natural rat MW 15-17k 0.2 μm filtered solution & lyophilized from a solution of phosphate buffered saline containing 1 mg BSA; endotoxin tested | Glycoprotein purified from conditioned media of a rat splenocyte culture stimulated by Concanavalin A; immunomodulatory factor produced by certain subsets of T lymphocytes; has been isolated from a variety of cell cultures & recombinant systems; has been shown to promote long term growth of activated T cells, activation & proliferation of NK cells & induction of γ-interferon & B cell growth factor secretion; IL-2 proliferative activity is tested in culture using a modification of a biological assay, whereby 1 U is the amount required to induce half-maximal proliferation of CTLL-2 cells as measured in a MTT dye assay or in a ³H-thymidine biological assay; Smith, K, *Science*, 240: 1169, 1988; Morgan, D et al, *Science*, 193: 1007, 1976; Ortaldo, J et al, *J Immuno*, 133: 779, 1984; Farrar, J et al, *J Immunol Res*, 63: 129, 1982; inaba, K et al, *J Exp Med*, 158: 2040, 1983; Coligan, J et al, ed, *Current Protocols in Immunology*, 1: 6.3.1, 1991; Gearing A et al, in: *Lymphokines & Interferons, A Practical Approach*, Clemens, MJ et al, eds, IRL Press, Oxford, p. 296, 1987

BioSource International PRC0024 Rat recombinant

BioSource International PSC0024 Swine recombinant

Interleukin II, Human

IBT CI-420-3 Recombinant, expressed in *E. coli* MW 15k >95%; lyophilized in the presence of stabilizer (mannitol), low levels of SDS (preventing aggregation), & sodium phosphate | Lymphokine; activates & proliferates T cells, B cells & other lymphokine activatedkiller cells; 133 AA, including 3 Cys residues, 2 of which are involved in a disulfide bond essential for biological activity; 62% homology with murine IL-2; active on murine cell lines

Interleukin III

Synonyms: Hemopoietin, Pan-Specific; Multicolony Stimulating Factor; Mast Cell Growth Factor Burst Promoting Activity; Histamine Producing Cell Stimulating Factor; P-Cell Stimulating Factor; WEHI-3 Factor; Colony Forming U-Stimulating Activity; IL-3

Biodesign A52131H *E. coli* MW 15k Purified | Species specificity: mouse

Biodesign A52203H *E. coli* MW 15k Purified

Chemicon IL003 Human ≥95%

Biogenesis 5377-3002 Human r-DNA Lyophilized

BioSource International PHC0034 Human recombinant

Calbiochem 407629 Human recombinant, expressed in *E. coli* MW 17.5k ≥97% (SDS-PAGE); lyophilized from sterile-filtered PBS containing 50 μg BSA/μg IL-3; biological activity: ED$_{50}$=100-400 pg/mL as measured in a cell proliferation assay with a human IL-3-dependent cell line; endotoxin: ≤100 pg/μg IL-3 | Stimulates the production & differentiation of macrophages, neutrophils, basophils, mast cells & eosinophils; as hematopoietic colony-stimulating factor, IL-3 interacts with very early multipotent hematopoietic progenitor cells; supports colony formation in soft agar of bone marrow cells of the granulocyte, macrophage, megakaryocyte & erythrocyte lineages; *Merck Index*, 12: 5021; Dercksen, MW et al, *Br J Cancer*, 68: 996, 1993; Theodossiou, C et al, *Cancer*, 74: 2808, 1994

Fitzgerald 30-AI60 Human recombinant, expressed in *E. coli*

Harlan BT-2007 Human recombinant, expressed in *E. coli* Lyophilized; 0.01 mg | Cytokine

Harlan BT-2008 Human recombinant, expressed in *E. coli* Lyophilized; 0.05 mg | Cytokine

PeproTech 200-03 Human recombinant, expressed in *E. coli* MW 15.0k >98% (SDS-PAGE) & HPLC; <0.1 ng endotoxin per mg (1EU/mg); lyophilized | Species-specific colony-stimulating factor; stimulates colony formation of megakaryocytes, neutrophils, & macrophages from bone marrow cultures; 133 AA; ED$_{50}$ ≤ 0.1 ng/mL; SA ≥ 10⁷ U/mg; SA determined by dose-dependant stimulation of human TF-1 cell proliferation

Sigma I 1646   Human recombinant, expressed in *E. coli*   MW 15k   ≥97% (SDS-PAGE); 0.2 µm filtered & lyophilized from phosphate buffered saline containing 500 µg BSA; proliferative activity is tested in culture using human TF-1 cells; endotoxin tested | Multifunctional protein, originally called colony forming U-stimulating activity (CFU-SA) & is produced by activated T lymphocytes; supports the formation of multilineage colonies in the early development of multipotent hematopoietic progenitor cells; shown to induce colony formation of macrophages, neutrophils, mast cells & megakaryocytes from agar-suspended bone marrow cells; also interacts with IL-2 to stimulate growth of T lymphocytes & to induce IgG secretion from activated B cells; Cerny, J et al, *Nature*, 249: 63, 1974; Luger, T et al, *J Immunol*, 134: 915, 1985; Schrader, J et al, *Immunol Rev*, 76: 79, 1983; Santoli, D et al, *J Immunol*, 141: 519, 1988; Tadmori, W et al, *J Immunol*, 142: 1950, 1989; Schrader, J et al, *Ann Rev Immunol*, 4: 205, 1986; Kitamura, T et al, *J Cell Physiol*, 140: 323, 1989; Kuwaki, T et al, *Biochem Biophys Res Commun*, 161: 16, 1989

Sigma I 7389   Human recombinant, expressed in *E. coli*   MW 15k   ≥95% (SDS-PAGE); 0.2 µm filtered & lyophilized; proliferative activity is tested in culture using human TF-1 cells whereby 1 U is the amount required to induce half-maximal incorporation of 3H-thymidine; activity expressed in Reference U (NIBSC reference preparation for IL-3 code 88/87); SA is given in the lot specific data sheet; endotoxin tested | Multifunctional protein, originally called colony forming U-stimulating activity (CFU-SA) & is produced by activated T lymphocytes; supports the formation of multilineage colonies in the early development of multipotent hematopoietic progenitor cells; shown to induce colony formation of macrophages, neutrophils, mast cells & megakaryocytes from agar-suspended bone marrow cells; also interacts with IL-2 to stimulate growth of T lymphocytes & to induce IgG secretion from activated B cells; Cerny, J et al, *Nature*, 249: 63, 1974; Luger, T et al, *J Immunol*, 134: 915, 1985; Schrader, J et al, *Immunol Rev*, 76: 79, 1983; Santoli, D et al, *J Immunol*, 141: 519, 1988; Tadmori, W et al, *J Immunol*, 142: 1950, 1989; Schrader, J et al, *Ann Rev Immunol*, 4: 205, 1986; Kitamura, T et al, *J Cell Physiol*, 140: 323, 1989; Kuwaki, T et al, *Biochem Biophys Res Commun*, 161: 16, 1989

Biogenesis 5377-3055   Mouse r-DNA   Lyophilized

BioSource International PMC0034   Mouse recombinant

Calbiochem 407631   Mouse recombinant, expressed in *E. coli*   MW 15k   ≥97% (SDS-PAGE); lyophilized from sterile-filtered PBS containing 50 µg BSA/µg IL-3; biological activity: ED_{50}=50-100 pg/mL as measured in a cell proliferation assay with a factor-dependent murine myeloblastic cell line; endotoxin: ≤100 pg/µg IL-3 | Stimulates production & differentiation of macrophages, neutrophils, basophils, mast cells & eosinophils; *Merck Index*, 12: 5021; Muther, H et al, *Growth Factors*, 10: 17, 1994

Sigma I 4144   Mouse recombinant, expressed in *E. coli*   MW 28k   ≥97% (SDS-PAGE); 0.2 µm filtered & lyophilized from phosphate buffered saline containing 500 µg BSA; activity is measured in cell proliferation assay using the NFS-60 cell line; endotoxin tested | Glycoprotein; Holmes, K et al, *Proc Natl Acad Sci USA*, 82: 6687, 1985; general comments & references are the same as for Sigma I 7389

Chemicon IL015   Murine   ≥95%

Fitzgerald 30-AI61   Murine recombinant, expressed in *E. coli*

Harlan BT-5102   Murine recombinant, expressed in *E. coli*   Lyophilized; 0.005 mg | Cytokine

Harlan BT-5103   Murine recombinant, expressed in *E. coli*   Lyophilized; 0.025 mg | Cytokine

PeproTech 213-13   Murine recombinant, expressed in *E. coli*   MW 15.1k   >98% (SDS-PAGE) & HPLC; <0.1 ng endotoxin per mg (1EU/mg); lyophilized with no additives | Species-specific; stimulates colony formation of megakaryocytes, neutrophils, & macrophages from bone marrow cultures; 135 AA; $ED_{50} \leq 0.1$ ng/mL; SA ≥ 10^7 U/mg; SA determined by the dose-dependent stimulation of the proliferation of murine IL-3-dependent FDC-P1 cells

Interleukin IIIβ

PeproTech 400-03   Rat recombinant, expressed in *E. coli*   MW 16.3k   >98% (SDS-PAGE) & HPLC; <0.1 ng endotoxin per mg (1EU/mg); lyophilized | Species-specific colony-stimulating factor; stimulates colony formation of megakaryocytes, neutrophils & macrophages from bone marrow cultures; 144 AA; $ED_{50} \leq 10$ ng/mL; SA ≥ 1.0 x 10^5 U/mg; SA determined by the dose-dependent stimulation of the proliferation of thymidine uptake by murine MC-9 cells

Interleukin IV

Synonyms: B-Cell Stimulatory Factor I; T-Cell Growth Factor II; Mast Cell Growth Factor II; IL-4

Biodesign A52141H   *E. coli*   MW 13k   Purified | Species specificity: mouse

Biodesign A52204H   *E. coli*   MW 14k   Purified

Biodesign A52404R   *E. coli*   MW 14k   Purified | Species specificity: rat

Chemicon IL004   Human   ≥95%

Biogenesis 5377-6002   Human r-DNA *E. coli*   MW 14.9k   SA: ≥5 x 10(e)6 U/mg; ED_{50} ≤0.2 ng/mL; endotoxin <0.1 ng/µg (1 EU/µg); purified; lyophilized

BioSource International PHC0044   Human recombinant

Calbiochem 407635   Human recombinant, expressed in *E. coli*   MW 14k   >97% (SDS-PAGE); lyophilized from sterile-filtered PBS containing 50 µg BSA/µg IL-4; biological activity: ED_{50}=50-200 pg/mL as measured in a TF-1 proliferation assay with PHA-activated human peripheral blood lymphocytes; endotoxin: ≤100 pg/µg IL-4 | Has profound effects on proliferation & differentiation of T & B lymphocytes; also affects a number of other cell types, including immature erythroid precursors, bone marrow & induced macrophages, myelomonocytic precursors, megakaryocyte precursors & mast cells; activity is species-specific; inhibits dexamethasone-induced apoptosis in thymocytes & peripheral T lymphocytes; not active on murine cells; Migliorati, G et al, *Pharmacol Res*, 30: 43, 1994

Fitzgerald 30-AI62   Human recombinant, expressed in *E. coli*

Harlan BT-2009   Human recombinant, expressed in *E. coli*   Lyophilized; 0.005 mg | Cytokine

Harlan BT-2010   Human recombinant, expressed in *E. coli*   Lyophilized; 0.025 mg | Cytokine

PeproTech 200-04   Human recombinant, expressed in *E. coli*   MW 14.9k   >98% (SDS-PAGE) & HPLC; <0.1 ng endotoxin per mg (1EU/mg); lyophilized | Potent lymphoid cell growth factor; stimulates growth & survivability of certain B & T cells; 129 AA; $ED_{50} \leq 0.5$ ng/mL; SA ≥ 2 x 10^6 U/mg; SA determined by dose-dependant stimulation of human TF-1 cell proliferation

Sigma I 4269   Human recombinant, expressed in *E. coli*   MW 14k   ≥97% (SDS-PAGE); 0.2 µm filtered & lyophilized from phosphate buffered saline containing 250 µg BSA; proliferative activity is tested in culture using human PHA activated human peripheral blood lymphocytes; endotoxin tested | Lymphokine with profound effects on the growth & differentiation of immunologically competent cells; complex glycoprotein released by a subset of activated T cells; treatment of IL-4 with specific glycosidases yields an active 15-16kD polypeptide; human & mouse IL-4 share a 50% AA sequence homology but their biological actions are species-specific; Howard, M et al, *J Exp Med*, 155: 914, 1982; Mosmann, T et al, *Proc Natl Acad Sci USA*, 83: 5654, 1986; Howard, M et al, *Immunol Rev*, 78: 185, 1984; Park, L et al, *J Exp Med*, 166: 476, 1987; Paul W & Ohara, J, *Ann Rev Immunol*, 5: 429, 1987

Biogenesis 5377-6025   Mouse r-DNA   Lyophilized

BioSource International PMC0044   Mouse recombinant

Calbiochem 407637 Mouse recombinant, expressed in *E. coli* MW 14k ≥97% (SDS-PAGE); lyophilized from sterile-filtered PBS containing 50 μg BSA/μg IL-4; biological activity: ED$_{50}$=1.0-2.0 ng/mL as measured in a cell proliferation assay with a factor-dependent murine cell line; endotoxin: ≤100 pg/μg IL-4 | Induces the differentiation of CD4+ T cells into T-helper-2 like cells; exhibits anti-neoplastic & anti-inflammatory properties; overproduction of IL-4 is liked to osteoporosis in murine models; induces MAP-kinase activation & Shc phosphorylation inkeratinocytes; Wery, S et al, *J Biol Chem*, 271: 8529, 1996; Lacey, DL et al, *J Cell Biochem*, 53: 122, 1993; Lewis, DB et al, *PNAS*, 90: 11618, 1993

Sigma I 1020 Mouse recombinant, expressed in *E. coli* ≥97% (SDS-PAGE); 0.2 μm filtered & lyophilized from phosphate buffered saline containing 250 μg BSA; proliferative activity is tested in culture using mouse HT-2 cells; endotoxin tested | Lymphokine with profound effects on the growth & differentiation of immunologically competent cells; complex glycoprotein released by a subset of activated T cells; treatment of IL-4 with specific glycosidases yields an active 15-16kD polypeptide; human & mouse IL-4 share a 50% AA sequence homology but their biological actions are species-specific; Howard, M et al, *J Exp Med*, 155: 914, 1982; Mosmann, T et al, *Proc Natl Acad Sci USA*, 83: 5654, 1986; Howard, M et al, *Immunol Rev*, 78: 185, 1984; Park, L et al, *J Exp Med*, 166: 476, 1987; Paul W & Ohara, J, *Ann Rev Immunol*, 5: 429, 1987

Chemicon IL016 Murine ≥95%

Fitzgerald 30-AI63 Murine recombinant, expressed in *E. coli*

Harlan BT-5104 Murine recombinant, expressed in *E. coli* Lyophilized; 0.01 mg | Cytokine

Harlan BT-5105 Murine recombinant, expressed in *E. coli* Lyophilized; 0.05 mg | Cytokine

PeproTech 214-14 Murine recombinant, expressed in *E. coli* MW 13.5k >98% (SDS-PAGE) & HPLC; <0.1 ng endotoxin per mg (1EU/mg); lyophilized with no additives | Potent lymphoid cell growth factor; stimulates growth & survivability of certain B & T cells; 120 AA; ED$_{50}$ ≤ 0.1 ng/mL; SA ≥ 10^7 U/mg; dose-dependent proliferation of the IL-4-dependent murine CT.4S cells

Chemicon IL037 Rat ≥95%

Biogenesis 0100-0098 Rat r-DNA Lyophilized

BioSource International PRC0044 Rat recombinant

PeproTech 400-04 Rat recombinant, expressed in *E. coli* MW 14k >98% (SDS-PAGE) & HPLC; <0.1 ng endotoxin per mg (1EU/mg) | Potent lymphoid cell growth factor; growth & survivability of certain B & T cells; 121 AA; ED$_{50}$ ≤ 1.5 ng/mL; SA ≥ 6.6 x 10^5 U/mg; SA determined by the dose-dependent stimulation of the proliferation of con A activated rat spleen cells

Sigma I 3650 Rat recombinant, expressed in *E. coli* ≥97% (SDS-PAGE); 0.2 μm filtered & lyophilized from phosphate buffered saline containing 250 μg BSA; biological activity is measured by its ability to stimulate proliferation of rat splenocytes; endotoxin tested | Lymphokine with profound effects on the growth & differentiation of immunologically competent cells; complex glycoprotein released by a subset of activated T cells; treatment of IL-4 with specific glycosidases yields an active 15-16kD polypeptide; human & mouse IL-4 share a 50% AA sequence homology but their biological actions are species-specific; Howard, M et al, *J Exp Med*, 155: 914, 1982; Mosmann, T et al, *Proc Natl Acad Sci USA*, 83: 5654, 1986; Howard, M et al, *Immunol Rev*, 78: 185, 1984; Park, L et al, *J Exp Med*, 166: 476, 1987; Paul W & Ohara, J, *Ann Rev Immunol*, 5: 429, 1987

BioSource International PSC0044 Swine recombinant

Interleukin IV Receptor Soluble Fragment

Sigma I 6021 Human recombinant, expressed in *Sf* 21 insect cells ≥97% (SDS-PAGE); 0.2 μm filtered & lyophilized from phosphate buffered saline containing 1.25 mg BSA; receptor mediated activity is measured by its ability to inhibit the IL-4 proliferation of TF-1 cells | Lymphokine with profound effects on the growth & differentiation of immunologically competent cells; complex glycoprotein released by a subset of activated T cells; treatment of IL-4 with specific glycosidases yields an active 15-16kD polypeptide; human & mouse IL-4 share a 50% AA sequence homology but their biological actions are species-specific; Howard, M et al, *J Exp Med*, 155: 914, 1982; Mosmann, T et al, *Proc Natl Acad Sci USA*, 83: 5654, 1986; Howard, M et al, *Immunol Rev*, 78: 185, 1984; Park, L et al, *J Exp Med*, 166: 476, 1987; Paul W & Ohara, J, *Ann Rev Immunol*, 5: 429, 1987; Kitamura, T et al, *J Cell Physiol*, 140: 323, 1989

Interleukin V

Synonyms: B-Cell Growth Factor II; T-Cell Replacing Factor; Eosinophil Differentiating Factor; Immunoglobulin A Enhancing Factor; Eosinophil Colony Stimulating Factor; IL-5

Chemicon IL005 Human ≥95%

Biogenesis 5377-9002 Human r-DNA Lyophilized

BioSource International PHC0054 Human recombinant

Fitzgerald 30-AI91 Human recombinant, expressed in *E. coli*

PeproTech 200-05 Human recombinant, expressed in *E. coli* MW 26.0k >98% (SDS-PAGE) & HPLC; <0.1 ng endotoxin per mg (1EU/mg); lyophilized | Potent T cell derived factor; stimulates growth of certain B cells & the differentiation of eosinophils; 232 AA; ED$_{50}$ ≤ 0.15 ng/mL; SA ≥ 6 x 10^6 U/mg; SA determined by dose-dependant stimulation of human TF-1 cell proliferation

Sigma I 5273 Human recombinant, expressed in *Sf* 21 cells MW 45-50k ≥97% (SDS-PAGE); 0.2 μm filtered & lyophilized from phosphate buffered saline containing 100 μg BSA; proliferative activity is tested in culture using human TF-1 cells; endotoxin tested | Product of activated T lymphocytes & exhibits activity on eosinophils, B cells & thymocytes; dimeric glycoprotein although glycosylation is not required for activity; human & mouse IL-5 exhibit homology at the nucleotide & AA levels & show species cross-reactivity; Sanderson, CJ et al, *J Exp Med*, 162: 60, 1985; Yokota, T et al, *Proc Natl Acad Sci USA*, 84: 7388, 1987; Takatsu, K et al, *Immunol Rev*, 102: 107, 1988; Clutterbuck, E et al, *Blood*, 73: 1504, 1989; Ramos, T, *Immunol Lett*, 21: 277, 1989; McKenzie, DT et al, *J Immunol*, 139: 2661, 1987; tominaga, A et al, *J Immunol*, 140: 1175, 1988; Kitamura, T et al, *J Cell Physiol*, 140: 323, 1989; Kuwaki, T et al, *Biochem Biophys Res Commun*, 161: 16, 1989;

Calbiochem 407641 Human recombinant, expressed in *Spodoptera frugiperda* MW 32-34k >97% (SDS-PAGE); lyophilized from sterile-filtered PBS containing 50 μg BSA/μg IL-5; biological activity: ED$_{50}$=100-200 pg/mL as measured in a cell proliferation assay with the IL-5-dependent human cell line TF-1; endotoxin: ≤100 pg/μg IL-5 | Stimulates growth & differentiation of eosinophils; a potent primer of eosinophil migration; activates a 45-k MAP kinase & jak2 tyrosine kinase; Bates, ME et al, *J Immunol*, 156: 711, 1996; Bozza, PT et al, *Immunopharmacol*, 27: 131, 1994

Biogenesis 5377-9025 Mouse r-DNA Liquid

Sigma I 1145 Mouse recombinant, expressed in *Sf* 21 cells using a recombinant *baculovirus* expression vector containing a mIL-5 cDNA ≥97% (SDS-PAGE); 0.2 μm filtered & lyophilized from phosphate buffered saline containing 250 μg BSA; proliferative activity is tested in culture using TF-1 cells; endotoxin tested | Product of activated T lymphocytes & exhibits activity on eosinophils, B cells & thymocytes; dimeric glycoprotein although glycosylation is not required for activity; human & mouse IL-5 exhibit homology at the nucleotide & AA levels & show species cross-reactivity; Sanderson, CJ et al, *J Exp Med*, 162: 60, 1985; Yokota, T et al, *Proc Natl Acad Sci USA*, 84: 7388, 1987; Takatsu, K et al, *Immunol Rev*, 102: 107, 1988; Clutterbuck, E et al, *Blood*, 73: 1504, 1989; Ramos, T, *Immunol Lett*, 21: 277, 1989; McKenzie, DT et al, *J Immunol*, 139: 2661, 1987; tominaga, A et al, *J Immunol*, 140: 1175, 1988; Kitamura, T et al, *J Cell Physiol*, 140: 323, 1989; Kuwaki, T et al, *Biochem Biophys Res Commun*, 161: 16, 1989

Calbiochem 407647   Mouse recombinant, expressed in *Spodoptera frugiperda*   MW 32-34k   >97% (SDS-PAGE); lyophilized from sterile-filtered PBS containing 50 μg BSA/μg IL-5; biological activity: ED$_{50}$=200-600 pg/mL as measured in a cell proliferation assay with the IL-5-dependent human cell line TF-1; endotoxin: ≤100 pg/μg IL-5

Interleukin V Receptor α-Chain Soluble Fragment

Sigma I 5646   Human recombinant, expressed in *Sf 21* insect cells   ≥97% (SDS-PAGE); 0.2 μm filtered & lyophilized from phosphate buffered saline containing 1.25 mg BSA; receptor mediated activity is measured by its ability to inhibit the IL-5 proliferation of TF-1 cells | Lymphokine with profound effects on the growth & differentiation of immunologically competent cells; complex glycoprotein released by a subset of activated T cells; treatment of IL-4 with specific glycosidases yields an active 15-16kD polypeptide; human & mouse IL-4 share a 50% AA sequence homology but their biological actions are species-specific; Howard, M et al, *J Exp Med*, 155: 914, 1982; Mosmann, T et al, *Proc Natl Acad Sci USA*, 83: 5654, 1986; Howard, M et al, *Immunol Rev*, 78: 185, 1984; Park, L et al, *J Exp Med*, 166: 476, 1987; Paul W & Ohara, J, *Ann Rev Immunol*, 5: 429, 1987; Ktamura, T et al, *J Cell Physiol*, 140: 323, 1989

Interleukin VI

Synonyms: Plasmacytoma Growth Factor; Interferon-β2; Monocyte Derived Human B-Cell Growth Factor; B-Cell Stimulating Factor; Hepatocyte Stimulating Factor; IL-6; Interleukin Hybridoma/Plasmacytoma I

Biodesign A52006H   *E. coli*   MW 21.7k   Purified | Species specificity: rat

Biodesign A52161H   *E. coli*   MW 21k   Purified | Species specificity: mouse

Biodesign A52206H   *E. coli*   MW 20.5k   Purified

Chemicon IL006   Human   ≥95%

Biogenesis 5378-2002   Human r-DNA   Lyophilized

BioSource International PHC0064   Human recombinant

Fitzgerald 30-AI64   Human recombinant, expressed in *E. coli*

Harlan BT-2012   Human recombinant, expressed in *E. coli*   Lyophilized; 0.005 mg | Cytokine

Harlan BT-2013   Human recombinant, expressed in *E. coli*   Lyophilized; 0.025 mg | Cytokine

PeproTech 200-06   Human recombinant, expressed in *E. coli*   MW 20.9k   >98% (SDS-PAGE) & HPLC; <0.1 ng endotoxin per mg (1EU/mg); lyophilized | Potent lymphoid cell growth factor; affects B-lymphocytes, T lymphocytes, & hybridoma cells; also affects cytotoxic T cells in combination with other factors such as IL-2 & γ-interferon; 184 AA; ED$_{50}$ ≤ 0.15 ng/mL; SA ≥ 6 x 10^6 U/mg; SA determined by the dose-dependent stimulation of the proliferation of IL-6-dependent murine 7TD1 cells

Sigma I 1395   Human recombinant, expressed in *E. coli*   MW 26k   ≥97% (SDS-PAGE); 0.2 μm filtered solution & lyophilized from phosphate buffered saline containing 500 μg BSA; biological activity measured in cell proliferation assay using the T1165.85.2.1 cell line; endotoxin tested | Multifunctional protein originally discovered in the media of cells stimulated with double stranded RNA; appears to be directly involved in the responses that occur after infection & injury & may prove to be as important as IL-1 & TNF-α in regulating the acute phase response; reported to be produced by fibroblasts, activated T cells, activated monocytes or macrophages & endothelial cells; acts upon a variety of cells including fibroblasts, myeloid progenitor cells, T cells, B cells & hepatocytes; appears to interact with IL-2 in the proliferation of T lymphocytes; also potentiates the proliferative effect of IL-3 on multipotential hematopoientic progenitors; Billiau, A, *Immunol Today*, 8: 84, 1987; Gauldie, J et al, *Proc Natl Acad Sci USA*, 84: 7251, 1987; Van Snick, J, *Ann Rev Immunol*, 8: 253, 1990; nordan, R & Potter, M, *Science*, 233: 566, 1986; Van Snick, J et al, *Proc Natl Acad Sci USA*, 83: 9679, 1986; Van Damme, J et al, *Eur J Immunol*, 17: 1, 1987

Sigma I 7764   Human recombinant, expressed in *E. coli*   MW 26k   ≥98% (SDS-PAGE); 0.2 μm filtered solution in 1 mL phosphate buffered saline containing 0.1% BSA; endotoxin tested | Multifunctional protein discovered in the media of cells stimulated with double stranded RNA; appears to be directly involved in the responses that occur after infection & injury & may prove to be as important as IL-1 & TNF-α in regulating the acute phase response; reported to be produced by fibroblasts, activated T cells, activated monocytes or macrophages & endothelial cells; acts upon a variety of cells including fibroblasts, myeloid progenitor cells, T cells, B cells & hepatocytes; appears to interact with IL-2 in the proliferation of T lymphocytes; also potentiates the proliferative effect of IL-3 on multipotential hematopoientic progenitors; Billiau, A, *Immunol Today*, 8: 84, 1987; Gauldie, J et al, *Proc Natl Acad Sci USA*, 84: 7251, 1987; Van Snick, J, *Ann Rev Immunol*, 8: 253, 1990; nordan, R & Potter, M, *Science*, 233: 566, 1986; Van Snick, J et al, *Proc Natl Acad Sci USA*, 83: 9679, 1986; Van Damme, J et al, *Eur J Immunol*, 17: 1, 1987

Biogenesis 5378-2105   Mouse r-DNA   Lyophilized

BioSource International PMC0064   Mouse recombinant, expressed in *E. coli*

Calbiochem 407652   Mouse recombinant, expressed in *E. coli*   MW 20.3k   ≥97% (SDS-PAGE); lyophilized from sterile-filtered PBS containing 50 μg BSA/μg IL-6; biological activity: ED$_{50}$=200-800 pg/mL as measured in a cell proliferation assay with a factor-dependent murine plasmacytoma cell line; endotoxin: ≤100 pg/μg IL-6 | Stimulates production of acute phase proteins by hepatocytes; also known to induce skeletal muscle protein breakdown; induces growth & differentiation of B cells, T cells & hepatocytes; stimulates growth & inhibits constitutive, protein synthesis-independent apoptosis of murine B cell hybridoma 7TD1; active on murine cells; Goodman, MN et al, *Proc Soc Exp Biol Med*, 205: 182, 1994; Liu, J et al, *Cell Immunol*, 155: 229, 1994

Calbiochem 407654   Mouse recombinant, expressed in *E. coli*   MW 20.6k   >97% (SDS-PAGE); lyophilized from sterile-filtered NaOAc & NaCl, pH 4.0 containing 50 μg BSA/μg IL-6; biological activity: ED$_{50}$=50-200 pg/mL as measured in a cell proliferation assay using a factor-dependent plasmacytoma cell line; endotoxin: ≤100 pg/μg IL-6 | Stimulates the production of acute phase proteins by hepatocytes; induces growth & differentiation of B cells, hepatocytes, keratinocytes & nerve cells; Matsuda, T et al, *Biochem Biophys Res Comm*, 202: 637, 1994

Sigma I 9646   Mouse recombinant, expressed in *E. coli*   ≥97% (SDS-PAGE); 0.2 μm filtered solution & lyophilized from phosphate buffered saline containing 250 μg BSA; proliferative activity is measured in culture using the mouse plasmacytoma cell line T1165.85.2.1; endotoxin tested | Multifunctional protein originally discovered in the media of cells stimulated with double stranded RNA; appears to be directly involved in the responses that occur after infection & injury & may prove to be as important as IL-1 & TNF-α in regulating the acute phase response; reported to be produced by fibroblasts, activated T cells, activated monocytes or macrophages & endothelial cells; acts upon a variety of cells including fibroblasts, myeloid progenitor cells, T cells, B cells & hepatocytes; appears to interact with IL-2 in the proliferation of T lymphocytes; also potentiates the proliferative effect of IL-3 on multipotential hematopoientic progenitors; Billiau, A, *Immunol Today*, 8: 84, 1987; Gauldie, J et al, *Proc Natl Acad Sci USA*, 84: 7251, 1987; Van Snick, J, *Ann Rev Immunol*, 8: 253, 1990; nordan, R & Potter, M, *Science*, 233: 566, 1986; Van Snick, J et al, *Proc Natl Acad Sci USA*, 83: 9679, 1986; Van Damme, J et al, *Eur J Immunol*, 17: 1, 1987

Chemicon IL017   Murine   ≥95%

Fitzgerald 30-AI65   Murine recombinant, expressed in *E. coli*

Harlan BT-5106   Murine recombinant, expressed in *E. coli*   Lyophilized; 0.005 mg | Cytokine

Harlan BT-5107   Murine recombinant, expressed in *E. coli*   Lyophilized; 0.025 mg | Cytokine

PeproTech 216-16   Murine recombinant, expressed in *E. coli*   MW 21.7k   >98% (SDS-PAGE) & HPLC; <0.1 ng endotoxin per mg (1EU/mg); lyophilized with no additives | Potent lymphoid cell growth factor; stimulates growth & survivability of certain B & T cells; 187 AA; ED$_{50}$ ≤ 0.02 ng/mL; SA ≥ 5 x 10^8 U/mg; SA determined by the dose-dependent stimulation of the proliferation of IL-6-dependent murine 7TD1 cells

Sigma I 3268   Natural human, produced in MG-63 osteosarcoma cells induced with IL-1β   MW 26k   ≥90% (SDS-PAGE); 0.2 μm filtered solution in 1 mL phosphate buffered saline containing 0.1% BSA; activity is ≥1x10⁴ U/vial; endotoxin tested | Multifunctional protein originally discovered in the media of cells stimulated with double stranded RNA; appears to be directly involved in the responses that occur after infection & injury & may prove to be as important as IL-1 & TNF-α in regulating the acute phase response; reported to be produced by fibroblasts, activated T cells, activated monocytes or macrophages & endothelial cells; acts upon a variety of cells including fibroblasts, myeloid progenitor cells, T cells, B cells & hepatocytes; appears to interact with IL-2 in the proliferation of T lymphocytes; also potentiates the proliferative effect of IL-3 on multipotential hematopoientic progenitors; Billiau, A, *Immunol Today*, 8: 84, 1987; Gauldie, J et al, *Proc Natl Acad Sci USA*, 84: 7251, 1987; Van Snick, J, *Ann Rev Immunol*, 8: 253, 1990; nordan, R & Potter, M, *Science*, 233: 566, 1986; Van Snick, J et al, *Proc Natl Acad Sci USA*, 83: 9679, 1986; Van Damme, J et al, *Eur J Immunol*, 17: 1, 1987

Chemicon IL025	Rat	≥95%

Biogenesis 0100-0103	Rat r-DNA	Lyophilized

Biogenesis 0100-0104	Rat r-DNA	Lyophilized

BioSource International PRC0064   Rat recombinant

PeproTech 400-06   Rat recombinant, expressed in *E. coli*   MW 21.7k   >98% (SDS-PAGE) & HPLC; <0.1 ng endotoxin per mg (1EU/mg); lyophilized | Potent lymphoid cell growth factor; growth & survivability of certain B & T cells; 187 AA; ED₅₀ ≤ 0.01 ng/mL; SA ≥ 10⁸ U/mg; SA determined by the dose-dependent stimulation of the proliferation of IL-6-dependent murine 7TD1 cells

BioSource International PSC0064   Swine recombinant

Interleukin VI Receptor Soluble Fragment

Sigma I 5771   Human recombinant, expressed in *Sf* 21 insect cells   ≥97% (SDS-PAGE); 0.2 μm filtered & lyophilized from phosphate buffered saline containing 250 μg BSA; receptor mediated activity is measured by its ability to increase the IL-6 inhibition of M1 cells | Multifunctional protein originally discovered in the media of cells stimulated with double stranded RNA; appears to be directly involved in the responses that occur after infection & injury & may prove to be as important as IL-1 & TNF-α in regulating the acute phase response; reported to be produced by fibroblasts, activated T cells, activated monocytes or macrophages & endothelial cells; acts upon a variety of cells including fibroblasts, myeloid progenitor cells, T cells, B cells & hepatocytes; appears to interact with IL-2 in the proliferation of T lymphocytes; also potentiates the proliferative effect of IL-3 on multipotential hematopoientic progenitors; Billiau, A, *Immunol Today*, 8: 84, 1987; Gauldie, J et al, *Proc Natl Acad Sci USA*, 84: 7251, 1987; Van Snick, J, *Ann Rev Immunol*, 8: 253, 1990; nordan, R & Potter, M, *Science*, 233: 566, 1986; Van Snick, J et al, *Proc Natl Acad Sci USA*, 83: 9679, 1986; Van Damme, J et al, *Eur J Immunol*, 17: 1, 1987

Interleukin VI Receptor, Soluble

Calbiochem 407653   Human recombinant, expressed in *Spodoptera frugiperda*   MW 38k   ≥97% (SDS-PAGE); lyophilized from sterile-filtered PBS containing 50 μg BSA/μg IL-6 receptor; biological activity: ED₅₀=5-15 ng/mL as measured by its ability to enhance IL-6-induced growth inhibition of mouse M1 myeloid leukemic cells; endotoxin: ≤100 pg/μg IL-6 receptor | Soluble form of the IL-6 receptor that binds IL-6 & mediates IL-6 signaling through interaction with gp 130; elevated serum levels of soluble IO-6 receptors have been shown to be associated with a number of pathological states including multiple myelomas, adult T cell leukemia, interstitial lung infection & HIV infection; Schobitz, B et al, *FASEB J*, 9: 659, 1995; Yokoyama, A et al, *Clin Exp Immunol*, 100: 325, 1995; Suzuki, H et al, *Eur J Immunol*, 23: 1078, 1993; Lust, JA et al, *Cytokine*, 4: 96, 1992; Sugita, T et al, *J Exp Med*, 171: 2001, 1990; Yamasaki, K et al, *Science*, 241: 825, 1988

Interleukin VI, Human

IBT CI-460-33   Recombinant, expressed in *Spodoptera frugiperda* insect cells (Sf9)   MW 20-25k (glycoprotein)   >85%; frozen 0.5 mg/mL solution in PBS; 50 μL/vial | Role in mediating inflammatory & immune responses initiated by infection or injury; acts on B cells by stimulating differentiation & Ab secretion; exhibits growth factor activity for mature thymic or peripheral T cells; enhances differentiation of cytotoxic T cells in the presence of IL-2 or IFNτ; glycoprotein composed of 184 AA; 42% homology with murine IL-6

Interleukin VII

Synonyms: IL-7

Biodesign A52171H   *E. coli*   MW 14.5k   Purified | Species specificity: mouse

Biodesign A52207H   *E. coli*   MW 17k   Purified

Chemicon IL007	Human	≥95%

Biogenesis 5378-4005	Human r-DNA	Lyophilized

BioSource International PHC0074   Human recombinant

Calbiochem 407658   Human recombinant, expressed in *E. coli*   MW 17k   >97% (SDS-PAGE); lyophilized from sterile-filtered PBS containing 50 μg HSA/μg IL-7; biological activity: ED₅₀=200-500 pg/mL measured as T cell growth factor activity in a cell proliferation assay with PHA-activated human peripheral blood lymphocytes; endotoxin: ≤100 pg/μg IL-7 | Growth & differentiation factor for human & murine T cells & an important factor in regulation of B cell & T cell development; acts on human CD8+ T cells to augment toxicity; Rich, BE & Leder, P, *J Exp Med*, 181: 223, 1995

Fitzgerald 30-AI66   Human recombinant, expressed in *E. coli*

Harlan BT-2014   Human recombinant, expressed in *E. coli* Lyophilized; 0.005 mg | Cytokine

Harlan BT-2015   Human recombinant, expressed in *E. coli* Lyophilized; 0.025 mg | Cytokine

PeproTech 200-07   Human recombinant, expressed in *E. coli*   MW 17.4k   >98% (SDS-PAGE) & HPLC; <0.1 ng endotoxin per mg (1EU/mg); lyophilized | Potent lymphoid cell growth factor; affects Pre-B, Pro-B & early T cells; also affects mature T cells in combination with other factors such as IL-2; 152 AA; ED₅₀ ≤ 0.5 ng/mL; SA ≥ 2 x 10⁶ U/mg; SA determined by the dose-dependant stimulation of the proliferation of murine IXN/2B cells

Sigma I 5896   Human recombinant, expressed in *E. coli*   MW25k   ≥97% (SDS-PAGE); 0.2 μm filtered solution & lyophilized from phosphate buffered saline containing 250 μg BSA; activity measured in cell proliferation assay using PHA activated human peripheral blood lymphocytes | Lymphoid cell growth factor which affects pre-B, pro-B & early T cells; first isolated by Namen in 1988; supports the growth of early B cells from long-term lymphoid bone marrow cultures; mitogenic for thymocytes & co-mitogenic with PHA & Con A; also stimulates the proliferation of CD4/CD8 cells; proliferative response of thymocytes to IL-7 is not affected by antibodies to the T cell growth factors such as IL-2, IL-4 & IL-6; mature T cells respond to IL-7 & Con A but not to IL-7 alone; Ab against IL-2 affect its activity suggesting that it functions through IL-2 production; glycoprotein that has 6 Cys residues which are important for biological activity; human & mouse IL-7 have 60% AA sequence homology; Henney, CS, *Immunol Today*, 10: 170, 1989; Namen, AE et al, *Nature*, 333: 57, 1988; Namen, AE et al, *J Exp Med*, 167: 988, 1988; Conlon, PJ et al, *Blood*, 74: 1368, 1989; Suda, T et al, *J Immunol*, 144: 3039, 1990; Morrissey, PJ et al, *J Exp Med*, 169: 707, 1989; Goodmin, AG et al, *Proc Natl Acad Sci USA*, 86: 302, 1989

Biogenesis 5378-4025	Mouse r-DNA	Lyophilized

BioSource International PMC0074   Mouse recombinant

Sigma I 4892   Mouse recombinant, expressed in *E. coli*   ≥97% (SDS-PAGE); 0.2 μm filtered solution & lyophilized from phosphate buffered saline containing 250 μg BSA; activity measured in cell proliferation assay using PHA activated human peripheral blood lymphocytes; endotoxin tested | Lymphoid cell growth factor which affects pre-B, pro-B & early T cells; first isolated by Namen in 1988; supports the growth of early B cells from long-term lymphoid bone marrow cultures; mitogenic for thymocytes & co-mitogenic with PHA & Con A; also stimulates the proliferation of CD4/CD8 cells; proliferative response of thymocytes to IL-7 is not affected by antibodies to the T cell growth factors such as IL-2, IL-4 & IL-6; mature T cells respond to IL-7 & Con A but not to IL-7 alone; Ab against IL-2 affect its activity suggesting that it functions through IL-2 production; glycoprotein that has 6 Cys residues which are important for biological activity; human & mouse IL-7 have 60% AA sequence homology; Henney, CS, *Immunol Today*, 10: 170, 1989; Namen, AE et al, *Nature*, 333: 57, 1988; Namen, AE et al, *J Exp Med*, 167: 988, 1988; Conlon, PJ et al, *Blood*, 74: 1368, 1989; Suda, T et al, *J Immunol*, 144: 3039, 1990; Morrissey, PJ et al, *J Exp Med*, 169: 707, 1989; Goodmin, AG et al, *Proc Natl Acad Sci USA*, 86: 302, 1989

Chemicon IL018   Murine   ≥95%

Fitzgerald 30-AI67   Murine recombinant, expressed in *E. coli*

Harlan BT-5108   Murine recombinant, expressed in *E. coli*
Lyophilized; 0.005 mg | Cytokine

Harlan BT-5109   Murine recombinant, expressed in *E. coli*
Lyophilized; 0.025 mg | Cytokine

PeproTech 217-17   Murine recombinant, expressed in *E. coli*
MW 15.0k   >98% (SDS-PAGE) & HPLC; <0.1 ng endotoxin per mg (1EU/mg); lyophilized with no additives | Potent lymphoid cell growth factor; affects Pre-B, Pro-B & early T cells; also affects mature T cells in combination with other factors such as IL-2; 129 AA; ED$_{50}$ ≤ 0.2 ng/mL; SA ≥ 5 x 10^6 U/mg; SA determined by the dose-dependant stimulation of the proliferation of murine IXN/2B cells

Interleukin VIII

Synonyms: rNAP-1; IL-8

Biodesign A52208H   *E. coli*   MW 9k   Purified | IL-8; species specificity: endothelial

Biodesign A52280H   *E. coli*   MW 8.5k   Purified | IL-8

Chemicon IL008   Human   ≥95%

Oncogene PF006   Human recombinant   MW 8919 (AA analysis)   Lyophilized with 100 μg BSA; reconstitute with sterile PBS; biological activity: half maximal activity as determined by elastase release from cytochalasin B treated neutrophils; 10 ng/mL | Species reactivity: human; for proliferation studies & Western blot

BioSource International PHC0084   Human recombinant endothelial

BioSource International PHC0884   Human recombinant monocyte

Alexis 520-003   Human recombinant, expressed in *E. coli* Cell culture grade; >98% (HPLC & SDS-PAGE); lyophilized from 200 μL in PBS containing 137 *mM* NaCl, 2.7 *mM* KCl, 4.3 *mM* Na$_2$HPO$_4$, 1.4 *mM* KH$_2$PO$_4$, pH 7.18 & 4 mg/mL D-mannitol; soluble in water; biological & SA: 1.9-2.5 *nM* measured by induction of myeloperoxidase release from neutrophils | 8-80 ng/mL (tissue culture); suitable for cell culture application; neutrophil chemotactic factor produced by LPS-stimulated human blood mononuclear leukocytes & other cell types, including fibroblasts, epithelial cells, chondrocytes & endothelial cells; Matsushima, K & Oppenheim, JJ, *Cytokine*, 1: 2, 1989; Holmes, WE et al, *Science*, 253: 1278, 1991; Murphy, PM & Tiffany, HL, *Science*, 253: 1280, 1991; Kishikava, K et al, *Prostaglandins*, 44: 261, 1992; Chwalisz, K et al, *Human Reproduction*, 9: 2173, 1994

Calbiochem 407673   Human recombinant, expressed in *E. coli* MW 8k   ≥97% (SDS-PAGE); lyophilized from sterile-filtered PBS containing 50 μg BSA/μg IL-8; biological activity: ED$_{50}$=150-300 ng/mL as measured by induction of myeloperoxidase release from human neutrophils; endotoxin: ≤100 pg/μg IL-8 | Potent proinflammatory cytokine is produced by monocytes, fibroblasts and keratinocytes in response to stimulation by LPS, IL-1 or TNF-α & by T lymphocytes in response to PHA stimulation; induces adhesion of neutrophils to endothelial cells; may be involved in chronic inflammation; Mazzucchelli, L et al, *Am J Pathol*, 144: 997, 1994; Harada, A et al, *J Leukoc Biol*, 56: 559, 1994

Fitzgerald 30-AI92   Human recombinant, expressed in *E. coli*

Harlan BT-2016   Human recombinant, expressed in *E. coli*
Lyophilized; 0.01 mg | Cytokine

Harlan BT-2017   Human recombinant, expressed in *E. coli*
Lyophilized; 0.05 mg | Cytokine

Sigma I 1645   Human recombinant, expressed in *E. coli*   MW 8k   ≥97% (SDS-PAGE); 0.2 μm filtered solution & lyophilized from phosphate buffered saline containing 500 μg BSA; biological activity is tested in culture using human neutrophils; endotoxin tested | formerly called monocyte-derived neutrophil chemotactic factor; belongs to the chemokine α or C-X-C family; mature form has 4 Cys residues as do the other members of the chemokine family & the first 2 Cys residues are separated by Gln; mature human IL-8 consists of 72 AA with a molecular mass of 8kD; exhibits chemotactic activity *in vitro* for T cells, basophils, as measured by enzymes including myeloperoxidase, α-mannosidase & β-glucuronidase; white, M et al, *Immunol Lett*, 22: 151, 1989; Larsen, CG et al, *Science*, 243: 1464, 1989; Mukaida, N et al, *Microbiol Immunol*, 36(8): 773, 1992; Yoshimura, T et al, *Proc Natl Acad Sci USA*, 84: 9233, 1987

Biogenesis 5378-6045   r-DNA   Lyophilized

Biogenesis 5378-6055   r-DNA   Lyophilized

BioSource International PSC0884   Swine recombinant

Interleukin VIII (72a.a)

Synonyms: IL-8 (72a.a)

PeproTech 200-08M   Human recombinant, expressed in *E. coli* MW 8.4k   >98% (SDS-PAGE) & HPLC; <0.1 ng endotoxin per mg (1EU/mg); lyophilized with no additives | Promotes neutrophil chemotaxis & degranulation; 72 AA IL-8 is the predominant form secreted by monocytes & lymphocytes; SA determined by ability to chemoattract human peripheral blood neutrophils

Interleukin VIII (77a.a)

PeproTech 200-08   Human recombinant, expressed in *E. coli* MW 8.9k   >98% (SDS-PAGE) & HPLC; <0.1 ng endotoxin per mg (1EU/mg); lyophilized with no additives | Promotes neutrophil chemotaxis & degranulation; 77 AA IL-8 is the predominant form secreted by endothelial cells; SA determined by ability to chemoattract human peripheral blood neutrophils

Interleukin IX

Synonyms: IL-9

Chemicon IL009   Human   ≥95%

Biogenesis 0100-0102   Human r-DNA   Lyophilized

BioSource International PHC0094   Human recombinant

Fitzgerald 30-AI69   Human recombinant, expressed in *E. coli*

Harlan BT-2018   Human recombinant, expressed in *E. coli*
Lyophilized; 0.01 mg | Cytokine

Harlan BT-2019   Human recombinant, expressed in *E. coli*
Lyophilized; 0.05 mg | Cytokine

PeproTech 200-09   Human recombinant, expressed in *E. coli* MW 14k   >98% (SDS-PAGE) & HPLC; <0.1 ng endotoxin per mg (1EU/mg); lyophilized | T-cell derived growth factor produced preferentially by CD4+ helper cells; a pleiotropic cytokine with multiple functions on cells of lymphoid, myeloid, & mast cell lineages; 114 AA; ED$_{50}$ ≤ 0.2 ng/mL; SA ≥ 5 x 10^6 U/mg; SA determined by the dose-dependent stimulation of the proliferation of human MO7e cells

Sigma I 3394   Human recombinant, expressed in *Sf* 21 cells using a recombinant *baculovirus* expression vector   MW 8k ≥97% (SDS-PAGE); 0.2 μm filtered solution & lyophilized from phosphate buffered saline containing 500 μg human serum albumin; proliferative activity is tested in culture using MO7e cells; endotoxin tested | First identified as a T cell derived growth factor; high affinity receptors for IL-9 on a variety of hematopoietic cells including T cells, mast cells & macrophages; high sequence homology between mouse & human IL-9; overall sequence homology between human & mouse IL-9 cDNA's is 56% & 67% identity at the AA & nucleotide levels, respectively; MO7e cells are responsive to both mouse & human IL-9 while only mouse IL-9 can stimulate mouse P40-responsive cell lines; mouse IL-9 enhances erythroid burst formation by normal mouse bone marrow cells; mouse IL-9 induces day 15 fetal thymocyte proliferation in the presence of IL-2 & enhances the mast cell growth elicited by IL-3 or IL-4; mouse IL-9 supports the growth of certain helper T cell clones; human IL-9 supports erythroid colony formation & synergizes with IL-4 in the production of IgE & IgG; Yang, Y-C, *Leukemia & Lymphoma*, 8: 441, 1992; Van Snick, J et al, *J Exp Med*, 169: 363, 1989; Suda, T et al, *J Immunol*, 144: 1783, 1990; Hultner, L et al, *J Immunol*, 142: 3440, 1989; Donahue, RE et al, *Blood*, 75: 2271, 1990; Petit-Frere, C et al, *Cytokine*, 3: 466, 1991; Uyttenhove, C et al, *Proc Natl Acad Sci USA*, 85: 6934, 1988

Calbiochem 407681   Human recombinant, expressed in *Spodoptera frugiperda*   MW 16-25k   >97% (SDS-PAGE); lyophilized from sterile-filtered PBS containing 50 μg HSA/μg IL-9; biological activity: ED_{50}=500-1 k pg/mL measured as cell proliferation assay with an IL-9-dependent human megakaryocytic leukemic cell line; endotoxin: ≤100 pg/μg IL-9 | Cytokine with pleiotropic effects on mast cells & T cell lines; can enhance the survival of human T cell lines & in synergy with EPO supports erythroid colony formation; exhibits 56% & 67% homology with murine IL-9 at the AA & nucleotide levels, respectively; not active on murine cells; Bauer, JH et al, *J Biol Chem*, 273: 9255, 1998; Houssiau, FA et al, *J Immunol*, 154: 2624, 1995

Biogenesis 0100-0093   Mouse r-DNA   Lyophilized

BioSource International PMC0094   Mouse recombinant

Sigma I 3269   Mouse recombinant, expressed in *Sf* 21 cells using a recombinant *baculovirus* expression vector   ≥97% (SDS-PAGE); 0.2 μm filtered solution & lyophilized from phosphate buffered saline containing 500 μg BSA; proliferative activity is tested in culture using TS-1 cells; endotoxin tested | First identified as a T cell derived growth factor; high affinity receptors for IL-9 on a variety of hematopoietic cells including T cells, mast cells & macrophages; high sequence homology between mouse & human IL-9; overall sequence homology between human & mouse IL-9 cDNA's is 56% & 67% identity at the AA & nucleotide levels, respectively; MO7e cells are responsive to both mouse & human IL-9 while only mouse IL-9 can stimulate mouse P40-responsive cell lines; mouse IL-9 enhances erythroid burst formation by normal mouse bone marrow cells; mouse IL-9 induces day 15 fetal thymocyte proliferation in the presence of IL-2 & enhances the mast cell growth elicited by IL-3 or IL-4; mouse IL-9 supports the growth of certain helper T cell clones; human IL-9 supports erythroid colony formation & synergizes with IL-4 in the production of IgE & IgG; Yang, Y-C, *Leukemia & Lymphoma*, 8: 441, 1992; Van Snick, J et al, *J Exp Med*, 169: 363, 1989; Suda, T et al, *J Immunol*, 144: 1783, 1990; Hultner, L et al, *J Immunol*, 142: 3440, 1989; Donahue, RE et al, *Blood*, 75: 2271, 1990; Petit-Frere, C et al, *Cytokine*, 3: 466, 1991; Uyttenhove, C et al, *Proc Natl Acad Sci USA*, 85: 6934, 1988

Chemicon IL019   Murine   ≥95%

Fitzgerald 30-AI70   Murine recombinant, expressed in *E. coli*

PeproTech 219-19   Murine recombinant, expressed in *E. coli* MW 14.3k   >98% (SDS-PAGE) & HPLC; <0.1 ng endotoxin per mg (1EU/mg); lyophilized with no additives | Potent lymphoid cell growth factor; stimulates growth of certain T cells, mast cells, & megakaryoblastic cells; 127 AA; $ED_{50} \leq$ 0.1 ng/mL; SA $\geq 10^{7}$ U/mg; SA determined by the dose-dependent stimulation of the proliferation of murine TS1.C3 cells

Interleukin X

Synonyms: Cytokine Synthesis Inhibitory Factor; IL-10

Biodesign A52019R   *E. coli*   MW 18.7k   Purified | Species specificity: rat

Biodesign A52101H   *E. coli*   MW 18.5k   Purified | Species specificity: mouse

Biodesign A52210H   *E. coli*   MW 18.5k   Purified

Chemicon IL010   Human   ≥95%

Biogenesis 5378-7550   Human r-DNA   Lyophilized

BioSource International PHC0104   Human recombinant

Fitzgerald 30-AI53   Human recombinant, expressed in *E. coli*

Harlan BT-2020   Human recombinant, expressed in *E. coli* Lyophilized; 0.005 mg | Cytokine

Harlan BT-2021   Human recombinant, expressed in *E. coli* Lyophilized; 0.025 mg | Cytokine

PeproTech 200-10   Human recombinant, expressed in *E. coli* MW 18.6k   >98% (SDS-PAGE) & HPLC; <0.1 ng endotoxin per mg (1EU/mg); lyophilized with no additives | 160 AA; > 80% sequence homology with Epstein-Barr Virus protein BCRFI; inhibits macrophage-mediated cytokine synthesis; suppresses the delayed-type hyper-sensitivity response; stimulates the Th2 cell response which results in elevated antibody production; $ED_{50} \leq$ 2.0 ng/mL; SA $\geq$ 5 x 10^{5} U/mL; SA determined by the dose-dependent co-stimulation (with murine IL-4) of MC/9 cells

USBio I8432   Human recombinant, expressed in *E. coli*   MW 18.6k   Sterile filtered, lyophilized, additive-free; soluble in water & most aqueous buffers | Suitable for antigenic applications in immunological protocols; equivalent to native Interleukin-10; 160 AA residues; shares >80% purity sequence homology with the Epstein-Barr Virus protein BCRFI; biological activities include inhibition of macrophage-mediated cytokine synthesis, suppression of the delayed-type hypersensitivity response & stimulation of the Th2 cell response (which results in elevated Ab production); exerts its biological activity in the concentration range of 0.2 to 20 ng/mL

Sigma I 3519   Human recombinant, expressed in *Sf* 21 cells using a recombinant *baculovirus* expression vector   ≥97% (SDS-PAGE); 0.2 μm filtered solution & lyophilized from phosphate buffered saline containing 250 μg human serum albumin; proliferative activity is tested in culture using MC/9 cells; endotoxin tested | Important regulator of the functions of lymphoid & myeloid cells; blocks the activation of cytokine synthesis & several accessory functions of macrophages; human & mouse IL-10 share a 73% sequence homology, however human acts on both human & mouse target cells while mouse has species-SA; in the mouse, the cellular sources of IL-10 consist of Th0 & Th2 T cell clones, thymocytes, B cells, B cell lymphomas, macrophages, mast cell lines and keratinocytes; in the human, the cellular sources of IL-10 consist of CD4+ T cells & T cell clones, thymocytes, B cells, B cell lymphomas, macrophages, mast cell lines and keratinocytes; stimulates the growth of stem cells, mast cells & thymocytes; enhances cytotoxic T cell development & co-stimulates B cell differentiation & Ig secretion; biological activity is determined in a cell proliferation assay using MC/9 cells; Rousset, F et al, *Proc Natl Acad Sci USA*, 89: 1890, 1992; Thompson-Snipes, L et al, *J Exp Med*, 173: 507, 1991; Chen, W-F et al, *J Immunol*, 147: 528, 1991; Moore, K et al, *Ann Rev Immunol*, 11: 165, 1993; Rennick, D et al, *Progress in Growth Factor Research*, 4: 207, 1992; Vieira, P et al, *Proc Natl Acad Sci USA*, 88: 1172, 1991

Calbiochem 407700   Human recombinant, expressed in *Spodoptera frugiperda*   MW 18.6k   >97% (SDS-PAGE); lyophilized from sterile-filtered PBS containing 50 μg BSA/μg IL-10; biological activity: ED_{50}=0.5-1.0 ng/mL measured as cell proliferation assay using a mouse mast cell line MC/9; endotoxin: <100 pg/μg IL-10 | Pleiotropic cytokine that exerts either immunosuppressive or immunostimulatory effects on a variety of cell types; exhibits anti-inflammatory effects by suppressing macrophage proliferation; potent modulator of monocyte/macrophage function; Wang, CQ et al, *J Cell Physiol*, 166: 305, 1996; Fleming, SD et al, *J Immunol*, 156: 1143, 1996; Niro, H et al, *Int Immunol*, 6: 661, 1994; Moore, K et al, *Ann Rev Immunol*, 11: 165, 1993; Vieira, P et al, *PNAS*, 88: 1172, 1991

Biogenesis 5378-8050   Mouse r-DNA   Lyophilized

BioSource International PMC0104   Mouse recombinant

Sigma I 3019 Mouse recombinant, expressed in *E. coli* ≥97% (SDS-PAGE); 0.2 μm filtered solution & lyophilized from phosphate buffered saline containing 250 μg BSA; proliferative activity is tested in culture using MC/9 cells; endotoxin tested | Important regulator of the functions of lymphoid & myeloid cells; blocks the activation of cytokine synthesis & several accessory functions of macrophages; human & mouse IL-10 share a 73% sequence homology, however human acts on both human & mouse target cells while mouse has species-SA; in the mouse, the cellular sources of IL-10 consist of Th0 & Th2 T cell clones, thymocytes, B cells, B cell lymphomas, macrophages, mast cell lines and keratinocytes; in the human, the cellular sources of IL-10 consist of CD4+ T cells & T cell clones, thymocytes, B cells, B cell lymphomas, macrophages, mast cell lines and keratinocytes; stimulates the growth of stem cells, mast cells & thymocytes; enhances cytotoxic T cell development & co-stimulates B cell differentiation & Ig secretion; biological activity is determined in a cell proliferation assay using MC/9 cells; Rousset, F et al, *Proc Natl Acad Sci USA*, 89: 1890, 1992; Thompson-Snipes, L et al, *J Exp Med*, 173: 507, 1991; Chen, W-F et al, *J Immunol*, 147: 528, 1991; Moore, K et al, *Ann Rev Immunol*, 11: 165, 1993; Rennick, D et al, *Progress in Growth Factor Research*, 4: 207, 1992; Vieira, P et al, *Proc Natl Acad Sci USA*, 88: 1172, 1991

Calbiochem 407702 Mouse recombinant, expressed in *Spodoptera frugiperda* MW 18k >97% (SDS-PAGE); lyophilized from sterile-filtered PBS containing 50 μg BSA/μg IL-10; biological activity: ED50=0.3-0.6 ng/mL measured as cell proliferation assay using a mouse mast cell line MC/9; endotoxin: <100 pg/μg IL-10 | Pleiotropic cytokine that exerts either immunosuppressive or immunostimulatory effects on a variety of cell types; potent modulator of monocyte/macrophage function; Wang, CQ et al, *J Cell Physiol*, 166: 305, 1996; Fleming, SD et al, *J Immunol*, 156: 1143, 1996; Niro, H et al, *Int Immunol*, 6: 661, 1994; Moore, K et al, *Ann Rev Immunol*, 11: 165, 1993; Vieira, P et al, *PNAS*, 88: 1172, 1991; Moore, K et al, *Science*, 248: 1230, 1990

Chemicon IL020 Murine ≥95%

Fitzgerald 30-AI54 Murine recombinant, expressed in *E. coli*

Harlan BT-5110 Murine recombinant, expressed in *E. coli* Lyophilized; 0.005 mg | Cytokine

Harlan BT-5111 Murine recombinant, expressed in *E. coli* Lyophilized; 0.025 mg | Cytokine

PeproTech 210-10 Murine recombinant, expressed in *E. coli* MW 18.7k >98% (SDS-PAGE) & HPLC; <0.1 ng endotoxin per mg (1EU/mg); lyophilized with no additives | 160 AA; > 80% sequence homology with the Epstein-Barr Virus protein BCRFI; ED50 ≤ 2.0 ng/mL; SA ≥ 5 x 10⁵ U/mg; SA determined by the dose-dependent co-stimulation (with IL-4) of the proliferation of murine MC/9 cells

Chemicon IL035 Rat ≥95%

Biogenesis 0100-0100 Rat r-DNA Lyophilized

BioSource International PRC0104 Rat recombinant

PeproTech 400-19 Rat recombinant, expressed in *E. coli* MW 18.7k >98% (SDS-PAGE) & HPLC; <0.1 ng endotoxin per mg (1EU/mg); lyophilized | 160 AA; > 80% sequence homology with the Epstein-Barr Virus protein BCRFI; the ED50 ≤ 10 ng/mL; SA determined by the dose-dependent inhibition of antigen-specific T cell proliferation

BioSource International PSC0104 Swine recombinant

Interleukin XI

Synonyms: Adipogenesis Inhibitory Factor; IL-11

Biodesign A52211H *E. coli* MW 19.5k Purified

Chemicon IL011 Human ≥95%

Biogenesis 5378-8150 Human r-DNA Lyophilized

BioSource International PHC0114 Human recombinant

Fitzgerald 30-AI55 Human recombinant, expressed in *E. coli*

Harlan BT-2022 Human recombinant, expressed in *E. coli* Lyophilized; 0.01 mg | Cytokine; cross-reactive with primate cells

Harlan BT-2023 Human recombinant, expressed in *E. coli* Lyophilized; 0.05 mg | Cytokine; cross-reactive with primate cells

PeproTech 200-11 Human recombinant, expressed in *E. coli* MW 19.1k >98% (SDS-PAGE) & HPLC; <0.1 ng endotoxin per mg (1EU/mg); lyophilized with no additives | Potent lymphoid cell growth factor; stimulates growth & survivability of certain B & T cells; 178 AA; ED50 ≤ 10 ng/mL; SA ≥ 10⁵ U/mg; SA determined by the dose-dependent stimulation of the proliferation of murine 7TD1 cells

Sigma I 3644 Human recombinant, expressed in *Sf 21* cells using a recombinant *baculovirus* expression vector containing the cDNA sequence for human IL-11 MW 19k, migrates as a 23k band in SDS-PAGE ≥97% (SDS-PAGE); 0.2 μm filtered solution & lyophilized from phosphate buffered saline containing 250 μg human serum albumin; biological activity is measured in a cell proliferation assay using T11 cells, a subline of the IL-6 dependent murine plasmacytoma cell line T1165.85.1 | Acts on hematopoietic progenitor cells & stromal cells; human IL-11 gene consists of 5 exons & 4 introns & was mapped on chromosome 19 at band 19q13.3-q13.4; enhances the proliferation of IL-6 dependent plasmacytoma cells; stimulates the production of erythrocytes, megakaryocytes & stimulates T cell development of antibody producing B cells; biological activity is measured in a cell proliferation assay using T11 cells, a subline of the IL-6 dependent murine plasmacytoma cell line T1165.85.1; Paul, SR et al, *Proc Natl Acad Sci USA*, 87: 7512, 1990; Kawashima, I et al, *Progress in Growth Factor Research*, 4: 191, 1992; nordan, RP et al, *J Immunol*, 139: 813, 1987; Paul, SR et al, *Leukemia Research*, 16: 247, 1992; Quesniaux, VFJ et al, *Blood*, 80: 1218, 1992

Calbiochem 407705 Human recombinant, expressed in *Spodoptera frugiperda* MW 18.6k >97% (SDS-PAGE); lyophilized from sterile-filtered PBS containing 50 μg BSA/μg IL-11; biological activity: ED50=60-240 pg/mL measured as cell proliferation assay using T11 a subline of the IL-6-dependent mouse plasmacytoma cell line T1165.85.2.1 that has been adapted to grow in IL-11; endotoxin: <100 pg/μg IL-11 | Pleiotropic cytokine thought to be involved in hematopoiesis, lymphopoiesis, acute phase responses & in the development of adipocytes, neurons & osteoclasts; acts as a synergistic factor for the proliferation of human myeloid leukemia cells; activates MAP kinases, JAK tyrosine kinases & pp90^rsk; Lemoli, RM et al, *Br J Haematol*, 91: 319, 1995; Yang, YC & Yin, T, *Ann NY Acad Sci*, 762: 40, 1995; Du, XX & Williams, DA, *Blood*, 83: 2023, 1994; Yang, YC, *Stem Cells*, 11: 474, 1993; Paul, SR et al, *PNAS*, 87: 7512, 1990

Interleukin XII

Synonyms: Natural Killer Cell Stimulatory Factor; Cytotoxic Lymphocyte Maturation Factor; IL-12

Chemicon IL029 Human ≥95%

Biogenesis 5378-8350 Human r-DNA sf21 insect cells MW 70k ED50 ≤0.1 ng/mL; endotoxin: <0.1 ng/μg; purified; from 20 mM Tris pH 7.6, 0.12 M NaCl, HSA; lyophilized | 70k disulfide linked heterodimeric protein comprised of disulfide-bonded 35k (p35) and 40k (p40) subunits

BioSource International PHC0124 Human recombinant

Fitzgerald 30-AI86 Human recombinant, expressed in CHO cells

Harlan BT-3036 Human recombinant, expressed in *E. coli* Lyophilized; 0.01 mg | Cytokine

Sigma I 2276 Human recombinant, expressed in *Sf 21* insect cells Lyophilized from phosphate buffered saline containing 250 μg BSA; endotoxin tested; cell culture tested | Identified as a factor secreted by human Epstein-Barr (EBV)-transformed B cell lines; 75kD disulfide-linked heterodimer of a 35kD subunit & 40kD subunit; produced predominantly by monocytes & NK cells & induces T cells & NK cells to produce IFN-γ; human IL-12 is not active on mouse cells but murine IL-12 is active on both murine & human lymphocytes; Stern, A et al, *Proc Natl Acad Sci USA*, 87: 6808, 1990; Trinchieri, G et al, *Progress in Growth Factor Research*, 4: 355, 1992; Schoenhaut, DS et al, *J Immunol*, 148: 3433, 1992; Kobayashi, M et al, *J Exp Med*, 170: 827, 1989

PeproTech 200-12 Human recombinant, expressed in *sf21* insect cells MW 75k >95%; lyophilized from 20 mM Tris pH 7.6, 0.12 N NaCl & 50 μg human serum albumin (protease-free)/mg of IL-12; ED50: 0.1-0.2 ng/mL as determined by a cell proliferation assay using PHA-activated human lymphoblast

Calbiochem 407711   Human recombinant, expressed in *Spodoptera frugiperda*   MW 75k   ≥95% (SDS-PAGE); lyophilized from sterile-filtered PBS containing 50 µg BSA/µg IL-12; biological activity: ED_{50}=50-200 pg/mL measured by its ability to stimulate the proliferation of PHA-activated human lymphoblasts; endotoxin: ≤100 pg/µg IL-12 | Pleiotropic cytokine produced by monocytes/macrophages, B cells & connective tissue-type mast cells; potent inducer of IFN-γ production & T cell differentiation & function; Kobayashi, M et al, *J Exp Med*, 170: 827, 1989; Kang, K et al, *J Immunol*, 156: 1402, 1996; Riemann, H et al, *J Immunol*, 156: 1799, 1996; Brunda, MJ et al, *J Leukoc Biol*, 55: 280, 1994; Smith, TJ et al, *Eur J Immunol*, 24: 822, 1994; Gubler, U et al, *PNAS*, 88: 4143, 1991; Stern, AS et al, *PNAS*, 87: 6808, 1990

Biodesign A52212H   Insect cells   Purified

Biogenesis 5378-8450   Mouse r-DNA   Lyophilized

BioSource International PMC0124   Mouse recombinant

Sigma I 8523   Mouse recombinant, expressed in *Sf* 21 insect cells   MW 1254k   ≥97% (SDS-PAGE); 0.2 µm filtered solution & lyophilized from phosphate buffered saline containing 50 µg BSA per 1 µg of the cytokine; bioactivity is measured in a bioassay using PHA-activated human lymphoblasts; endotoxin tested | Identified as a factor secreted by human Epstein-Barr (EBV)-transformed B cell lines; 75kD disulfide-linked heterodimer of a 35kD subunit & 40kD subunit; produced predominantly by monocytes & NK cells & induces T cells & NK cells to produce IFN-γ; human IL-12 is not active on mouse cells but murine IL-12 is active on both murine & human lymphocytes; Stern, A et al, *Proc Natl Acad Sci USA*, 87: 6808, 1990; Trinchieri, G et al, *Progress in Growth Factor Research*, 4: 355, 1992; Schoenhaut, DS et al, *J Immunol*, 148: 3433, 1992; Kobayashi, M et al, *J Exp Med*, 170: 827, 1989

Calbiochem 407713   Mouse recombinant, expressed in *Spodoptera frugiperda*   MW 54k   ≥97% (SDS-PAGE); lyophilized from sterile-filtered PBS containing 50 µg BSA/µg IL-12; biological activity: ED_{50}=50-200 pg/mL measured by its ability to stimulate the proliferation of PHA-activated human lymphoblasts; endotoxin: ≤100 pg/µg IL-12 | Pleiotropic cytokine produced by monocytes/macrophages, B cells & connective tissue-type mast cells; Zou, JJ et al, *J Biol Chem*, 270: 5864, 1995; Schoenhaut, DS et al, *J Immunol*, 148: 3433, 1992; Gillessen, S et al, *Eur J Immunol*, 25: 200, 1995; Brunda, MJ et al, *J Leukoc Biol*, 55: 280, 1994; Smith, TJ et al, *Eur J Immunol*, 24: 822, 1994; Wolf, SF et al, *Stem Cells*, 12: 154, 1994

Chemicon IL032   Murine   ≥95%

PeproTech 210-12   Murine recombinant, expressed in CHO cells (Chinese Hamster Ovarian cells)   MW 70k disulfide-linked heterodimer with 35k (p35) & 40k (p40) subunits   >98% (SDS-PAGE) & HPLC; <0.1 ng endotoxin per mg (1EU/mg); lyophilized%; 0.5xPBS+1.5 mg/mL BSA pH 6.0; ED_{50}: 0.05-0.1 ng/mL as determined by the stimulation of IFN-γ production by murine splenocytes co-stimulated with IL-12 | Regulatory protein produced by activated B-lymphocytes & macrophages; ED_{50} ≤ 0.1 ng/mL; SA ≥ 10^7 U/mg; SA determined by the stimulation of IFN-g production by murine splenocytes co-stimulated with IL-12

Interleukin XIII

Synonyms: IL-13

Biodesign A52213H   *E. coli*   MW 13k   Purified

Chemicon IL012   Human   ≥95%

Biogenesis 5378-8550   Human r-DNA   Lyophilized

BioSource International PHC0134   Human recombinant

Calbiochem 407715   Human recombinant, expressed in *E. coli*   MW 12k   >97% (SDS-PAGE); lyophilized from sterile-filtered PBS containing 50 µg BSA/µg IL-13; biological activity: ED_{50}=3.0-6.0 ng/mL measured as cell proliferation assay using a human factor-dependent cell line TF-1; endotoxin: <100 pg/µg IL-13 | Pleiotropic cytokine that shares many of the properties of IL-4; exhibits IL-4 like activities on monocytes/macrophages & human B cells but has no effect on T cells; potent regulator of STAT6 & JAK3 in NK & T cells; regulates inflammatory & immune responses; inhibits inducible nitric oxide synthase in human mesangial cells; Y, CR et al, *J Immunol*, 161: 218, 1998; Saura, M et al, *Biochem J*, 313: 641, 1996; Xi, X et al, *Br J Haematol*, 90: 921, 1995; Zurawski, G & DeVries, JE, *Stem Cells*, 12: 169, 1994; McKenzie, ANJ et al, *PNAS*, 90: 3735, 1993; Minty, A et al, *Nature*, 362: 248, 1993

Fitzgerald 30-AI93   Human recombinant, expressed in *E. coli*

Harlan BT-2026   Human recombinant, expressed in *E. coli* Lyophilized; 0.005 mg | Cytokine

Harlan BT-2027   Human recombinant, expressed in *E. coli* Lyophilized; 0.025 mg | Cytokine

PeproTech 200-13   Human recombinant, expressed in *E. coli* MW 12.5k   >98% (SDS-PAGE) & HPLC; <0.1 ng endotoxin per mg (1EU/mg); lyophilized with no additives | Immunoregulatory protein produced by activated T-lymphocyte; stimulates B cell proliferation & immunoglobulin production; 114 AA; ED_{50} ≤ 1.0 ng/mL; SA ≥ 10^6 U/mg; SA determined by the dose-dependant stimulation of the proliferation of human TF-1 cells

Sigma I 1771   Human recombinant, expressed in *E. coli*   MW 10k   ≥97% (SDS-PAGE); 0.2 µm filtered solution & lyophilized from phosphate buffered saline containing 250 µg BSA; proliferative activity is tested in culture using human TF-1 cells; endotoxin tested | Pleiotropic cytokine produced by activated Th2 cells in the mouse & human; secreted mainly as an unglycosylated protein of 132 AA; 4.3-kb DNA fragment of the mouse IL-13 gene was sequenced & occurs as a single copy mapping to chromosome 11; in the human, a 4.6-kb DNA segment of the IL-13 gene occurs as a single copy & maps to chromosome 5; induces B cell proliferation & also induces IgE switching; both mouse & human IL-13 induce proliferation of the human erythroleukemic cell line TF-1; McKenzie, A et al, *Proc Natl Acad Sci USA*, 90: 3735, 1993; Defrance, T et al, *J Exp Med*, 179: 135, 1994; McKenzie, A et al, *J Immunol*, 150: 5436, 1993; Cock, B et al, *International Immunology*, 5: 657, 1993; Thomson, AW et al, in: *The Cytokine Handbook*, Thomson, A, ed, Academic Press, London, p. 257, 1994

Calbiochem 407717   Mouse recombinant, expressed in *E. coli* MW 11.5k   >97% (SDS-PAGE); lyophilized from sterile-filtered PBS containing 50 µg BSA/µg IL-13; biological activity: ED_{50}=3.0-6.0 ng/mL measured as cell proliferation assay using a human factor-dependent cell line TF-1; endotoxin: ≤100 pg/µg IL-13 | Pleiotropic cytokine that shares many of the properties of IL-4; exhibits IL-4 like activities on monocytes/macrophages & human B cells but has no effect on T cells; regulates inflammatory & immune responses; inhibits bone resorption by suppressing cyclooxygenase-2-dependent prostaglandin synthesis in osteoblasts; potent inhibitor of inducible nitric oxide synthase in smooth muscle cells; Onoe, Y et al, *J Immunol*, 156: 758, 1996; Ruetten, H & Thiemermann, C, *Shock*, 8: 409, 1997; Label-Binay, S et al, *Eur J Immunol*, 25: 2340, 1995; Zurawski, G & DeVries, JE, *Stem Cells*, 12: 169, 1994; Brown, KD et al, *J Immunol*, 142: 679, 1989; Minty, A et al, *Nature*, 362: 248, 1993; Cherwinski, HM et al, *J Exp Med*, 166: 1229, 1987

Sigma I 1896   Mouse recombinant, expressed in *E. coli*   ≥97% (SDS-PAGE); 0.2 µm filtered solution & lyophilized from phosphate buffered saline containing 250 µg BSA; proliferative activity is tested in culture using TF-1 cells; endotoxin tested | Pleiotropic cytokine produced by activated Th2 cells in the mouse & human; secreted mainly as an unglycosylated protein of 132 AA; 4.3-kb DNA fragment of the mouse IL-13 gene was sequenced & occurs as a single copy mapping to chromosome 11; in the human, a 4.6-kb DNA segment of the IL-13 gene occurs as a single copy & maps to chromosome 5; induces B cell proliferation & also induces IgE switching; both mouse & human IL-13 induce proliferation of the human erythroleukemic cell line TF-1; McKenzie, A et al, *Proc Natl Acad Sci USA*, 90: 3735, 1993; Defrance, T et al, *J Exp Med*, 179: 135, 1994; McKenzie, A et al, *J Immunol*, 150: 5436, 1993; Cock, B et al, *International Immunology*, 5: 657, 1993; Thomson, AW et al, in: *The Cytokine Handbook*, Thomson, A, ed, Academic Press, London, p. 257, 1994

PeproTech 210-13   Murine recombinant, expressed in *E. coli* MW 12.3k   >98% (SDS-PAGE) & HPLC; <0.1 ng endotoxin per mg (1EU/mg); lyophilized | Immunoregulatory protein produced by activated T-lymphocytes; stimulates B cell proliferation & immunoglobulins production; 111 AA; ED_{50} ≤ 4.0 ng/mL; SA ≥ 2.5 x 10^5 U/mg; SA determined by the dose-dependant stimulation of the proliferation of human TF-1 cells

BioSource International PRC0134   Rat recombinant

Interleukin XV

Synonyms: IL-15

Biodesign A52215H   *E. coli*   MW 13k   Purified

Chemicon IL013   Human   ≥95%

Biogenesis 5378-8950 Human r-DNA MW 12.9k Purified; no preservatives; lyophilized; SA: ≥2 x 10^6 U/mg; ED$_{50}$ ≤0.5 ng/mL

BioSource International PHC0154 Human recombinant

Fitzgerald 30-AI94 Human recombinant, expressed in *E. coli*

Harlan BT-2024 Human recombinant, expressed in *E. coli* Lyophilized; 0.005 mg | Cytokine

PeproTech 200-15 Human recombinant, expressed in *E. coli* MW 12.9k >98% (SDS-PAGE) & HPLC; <0.1 ng endotoxin per mg (1EU/mg); lyophilized with no additives | Potent lymphoid cell growth factor; exerts its biological activities primarily on T cells; 114 AA; ED$_{50}$ ≤ 0.5 ng/mL; SA ≥ 2 x 10^6 U/mg; SA determined by the dose-dependent stimulation of the proliferation of murine CTLL-2 cells

Sigma I 8648 Human recombinant, expressed in *E. coli* MW 12.5k ≥97% (SDS-PAGE); 0.2 μm filtered & lyophilized from an acetonitrile & trifluoroacetic acid solution containing 500 μg BSA; bioactivity is tested with M07e cells; endotoxin tested | First isolated from the supernatant of a cultured simian kidney epithelial cell line; cDNA encodes a 162 AA peptide with a 48 AA leader sequence; no sequence homology with any other known cytokine; competes for binding sites with IL-2 as both IL-2 & IL-15 stimulate the growth of cells through the IL-2 receptor; Grabstein, K et al, *Science*, 264: 965, 1944

BioSource International PMC0154 Mouse recombinant

PeproTech 210-15 Murine recombinant, expressed in *E. coli* MW 13.3k >98% (SDS-PAGE) & HPLC; <0.1 ng endotoxin per mg (1EU/mg); lyophilized | Potent lymphoid cell growth factor; exerts biological activities primarily on T cells; 115 AA; ED$_{50}$ ≤ 20 ng/mL; SA ≥ 5 x 10^4 U/mg; SA determined by the stimulation of the proliferation of murine CTLL-2 cells

BioSource International PSC0154 Swine recombinant

Interleukin XVI

Synonyms: Lymphocyte Chemoattractant Factor; IL-16

Biodesign A52216H *E. coli* MW 13.5k Purified

Chemicon IL021 Human ≥95%

Biogenesis 5378-9150 Human r-DNA Lyophilized

BioSource International PHC0164 Human recombinant

Fitzgerald 30-AI76 Human recombinant, expressed in *E. coli*

PeproTech 200-16 Human recombinant, expressed in *E. coli* MW 13.5k >98% (SDS-PAGE) & HPLC; <0.1 ng endotoxin per mg (1EU/mg); lyophilized from PBS | Stimulates a migratory response in CD4$^+$ T cells, CD4$^+$ monocytes & eosinophils; 130 AA; SA determined by its ability to chemoattract human CD4+ T-lymphocytes

Interleukin XVII

Synonyms: CTLA-8; IL-17

Biodesign A52217H *E. coli* MW 15.5k Purified

Chemicon IL022 Human ≥95%

Biogenesis 5378-9350 Human r-DNA Lyophilized

BioSource International PHC0174 Human recombinant

Fitzgerald 30-AI77 Human recombinant, expressed in *E. coli*

PeproTech 200-17 Human recombinant, expressed in *E. coli* MW 15.5k >98% (SDS-PAGE) & HPLC; <0.1 ng endotoxin per mg (1EU/mg); lyophilized with no additives | Identified from a CD4$^+$ T cell DNA library; can be induced from primary peripheral blood CD4$^+$ T cells upon stimulation; cytokine exhibiting a high degree of AA identity with HVS13, an open reading frame from a T lymphotropic *Herpesvirus saimiri* & with murine CTLA8; a disulfide-linked homodimer of two subunits, each containing 136 AA; ED$_{50}$ ≤ 2.0 ng/mL; SA ≥ 5 x 10^5 U/mg; SA determined by the dose-dependent induction of IL-6 in primary human foreskin fibroblasts

Sigma I 3525 Human recombinant, expressed in *E. coli* MW ~ 16k ≥97% (SDS-PAGE); 0.2 μm filtered & lyophilized in phosphate buffered saline containing 2.5 mg BSA; bioactivity is measured by its ability to induce IL-6 production by NHDF cells; endotoxin tested | T cell-derived hematopoietic cytokine originally cloned from a T cell hybridoma produced by the fusion of a mouse cytotoxic T cell clone & a rat T lymphoma; exhibits multiple biological activities on a variety of cells including: the induction of IL-6, IL-8 & G-CSF production in fibroblasts, the enhancement of surface expression of ICAM-1 in fibroblasts, activation of NF-κB & co-stimulation of T cell proliferation; polypeptide of 136 AA; precursor form consists of 155 AA; to generate the mature IL-17 (136 AA) the precursor cleaves a 19 AA signal peptide; human IL-17 shows ~62.5% AA homology to mouse IL-17 & 58% AA homology to rat IL-17

Sigma I 4026 Mouse recombinant, expressed in *E. coli* Lyophilized from 30% acetonitrile/0.1% TFA containing 1.25 mg BSA; endotoxin tested; cell culture tested | T cell-derived hematopoietic cytokine originally cloned from a T cell hybridoma produced by the fusion of a mouse cytotoxic T cell clone & a rat T lymphoma; exhibits multiple biological activities on a variety of cells including: the induction of IL-6, IL-8 & G-CSF production in fibroblasts, the enhancement of surface expression of ICAM-1 in fibroblasts, activation of NF-κB & co-stimulation of T cell proliferation; polypeptide of 136 AA; precursor form consists of 155 AA; to generate the mature IL-17 (136 AA) the precursor cleaves a 19 AA signal peptide; human IL-17 shows ~62.5% AA homology to mouse IL-17 & 58% AA homology to rat IL-17

Interleukin XVIII

Synonyms: Interferon-γ Inducing Factor; IL-18

Biodesign A52218H *E. coli* MW 18.3k Purified

Biodesign A52504M *E. coli* MW 18.1k Purified | Species specificity: mouse

Chemicon IL030 Human ≥95%

Biogenesis 5378-9550 Human r-DNA Lyophilized

PeproTech 200-18 Human recombinant, expressed in *E. coli* MW 18.3k >98% (SDS-PAGE) & HPLC; <0.1 ng endotoxin per mg (1EU/mg); lyophilized | Pleiotropic cytokine produced by monocyte/macrophage cells; like IL-12, plays an important role in cell-mediated immune responses; ED$_{50}$ ≤ 5.0 ng/mL; SA ≥ 2 x 10^4 U/mg; SA determined by the dose-dependent stimulation of IFN-γ production by human PBMC co-stimulated with human IL-12

PeproTech 210-18 Human recombinant, expressed in *E. coli* MW 18.3k >98%; 158 AA; lyophilized from 5 m*M* Tris pH 8.0, 75 m*M* NaCl; ED$_{50}$: 5.0-10.0 ng/mL as determined by the stimulation of IFN-γ production using human PBMC co-stimulated with IL-12

Biogenesis 5378-9650 Mouse r-DNA Lyophilized

BioSource International PMC0184 Mouse recombinant

Chemicon IL033 Murine ≥95%

PeproTech 315-04 Murine recombinant, expressed in *E. coli* MW 18.1k >98% (SDS-PAGE) & HPLC; <0.1 ng endotoxin per mg (1EU/mg); lyophilized | Pleiotropic cytokine produced by monocyte/macrophage cells; like IL-12, plays an important role in cell-mediated immune responses; 158 AA; exhibits biological activities in its monomeric form; ED$_{50}$ = 12ng/mL; SA determined by the dose-dependent stimulation of IFN-g production by murine lymph node cells

Interleukin XVIII Binding Protein

Biogenesis 0100-0106 Mouse r-DNA Lyophilized

Intracellular Adhesion Molecule I Protein

Synonyms: Intracellular Adhesion Molecule I, rh Soluble; Intracellular Adhesion Molecule I, rhs-

Alexis BMS313 Human recombinant MW 82k >99% (SDS-gel electrophoresis); 50 μg in phosphate-buffered saline | Adhesion molecule produced in CHO cells; corresponds in structure with natural human circulating ICAM-1, truncated at the ectodomain side of the beginning of the transmembrane region (AA 472)

Intrinsic Factor

Cortex CP8103   Porcine   >95%

Fitzgerald 30-AI15   Porcine gastric mucosa   High purity

Sigma I 6006   Porcine gastric mucosa   Contains <5% non-intrinsic factor (R-protein); activity: 10 k-25 k U/mg protein; lyophilized powder containing 20-40% protein (Lowry); balance primarily potassium phosphate; unit definition: 1 U binds 1 ng Vitamin B_{12} at pH 7.5 at 25°C using a modification of the procedure of Allen; | Allen, RH & Mehlman, CS, *J Biol Chem*, 248: 3670, 1973

USBio I8445   Porcine gastric mucosa   ≥98% (SDS-PAGE); ≤2% R protein | Suitable for antigenic applications in immunological protocols

Biogenesis 5390-0004   Porcine stomach   >98% pure based on a B12-binding assay using cobinamide and anti-intrinsic factor blocking antibodies; purified; distilled H_2O; liquid

Sigma I 7140   Rat stomach   Contains <5% non-intrinsic factor (R-protein); activity: 10 k-25 k U/mg protein; lyophilized powder containing 20-40% protein (Lowry); balance primarily potassium phosphate; unit definition: 1 U binds 1 ng Vitamin B_{12} at pH 7.5 at 25°C using a modification of the procedure of Allen | Allen, RH & Mehlman, CS, *J Biol Chem*, 248: 3670, 1973

IRS-1

USBio I8700-05   Rat recombinant produced in Sf9 insect cells MW ~170k   Frozen solution in 50 *mM* Tris-HCl, pH 7.8, containing 1 *mM* NaCl | Not phosphorylated; immunoblot positive control; chromatographically purified (Sephacryl S-200HR);

jak2 Immune Complex, Agarose

USBio J0901-06   Murine recombinant, expressed by baculovirus in Sf9 insect cells   MW 130k   Frozen suspension in 150 *mM* NaCl, 50 *mM* Tris-HCl, pH 8.0, 10% glycerol, 0.1 *mM* EDTA, 0.1 *mM* sodium orthovanadate, 50 *mM* NaF, 0.5% NP-40 | Purified by immunoprecipitation using jak2 antibody bound to Protein A agarose

JE/MCP-1

Chemicon GF077   Murine   ≥95%

Jo-1

Fitzgerald 30-AJ75   Calf thymus   High purity

Jo-1 Antigen

Biodesign A07302B   Bovine/Rabbit mixture   Purified | Autoimmune reagent

Biodesign A2A450R   Rabbit thymus   Purified | Autoimmune reagent

Kallikrein

Biodesign A50190H   Human plasma   Purified

Fitzgerald 30-AK10   Human plasma   High purity

USBio K0005   Human plasmakallikrein   ≥95%; no contaminants detected; single band by SDS-PAGE, IEP, &/or RID; 0.5-1 mg/mL; liquid in 20 *mM* Tris HCl, 100 *mM* NaCl, pH 7.8 | Suitable for antigenic applications in immunological protocols

Biogenesis 5543-6207   Human urine   MW 34k/41k   >98% (SDS-PAGE); SA: 4.9 U/mg; 50 *mM* TRIS/HCl, pH 7.75; liquid

Cortex CP3024   Plasma   >95%

Biogenesis 5543-6257   Porcine pancreas   Lyophilized

Biogenesis 5543-6307   Rat urine   Liquid

KC

Chemicon GF078   Murine   ≥95%

Keratan Sulfate Proteoglycan

Sigmak 3009   Bovine cornea   Lyophilized from a sterile-filtered, essentially salt-free solution | KSPG together with dermatan/chondroitin sulfate PG (decorin) represents the major class of proteoglycans in the corneal stroma; both are thought to play important roles in corneal transparency by modulating collagen fibril formation; Lumican also acts as a filler in the extracellular matrix of the cornea; bovine KSPG consists of 3 isoforms each with a different core protein & pattern of glycosylation; two 37k & one 25k core proteins (designated 37A, 37B & 25); the 37B isoform is homologous to chicken, mouse & human lumican & is referred to as lumican; Axelsson, I & Heingard, D, *Biochem J*, 145: 491, 1975; Rada, JA et al, *Exp Eye Res*, 56: 635, 1993; Uma, L et al, *Biochim Biophys Acta*, 1294: 8, 1996; Funderburgh, JL et al, *Biochem Soc Trans*, 19: 871, 1991; Funderburgh, JL et al, *Invest Ophthalmol Vis Sci*, 36: 2296, 1995; Chakravarti, S et al, *Genomics*, 27: 481, 1995

Keratin

ICN 902111   Hooves, horns   Purified protein powder

ICN 151390   Human epidermis   ~94%; 1-5 mg/mL in 8 *M* urea, 25 *mM* Tris, 0.1 *M* BME, 1 *mM* EDTA, pH 7.4

ICN 151391   Human epidermis   ~96%; 3-4 mg protein/mL in 8 *M* urea, 50 *mM* Tris, 0.1 *M* BME, 0.1% NaN_3, pH 7.4

Sigma K 0253   Human epidermis   ~8 mg protein (Biuret TCA)/mL; solution in 8 M urea, 50 mM Tris, 0.1 M BME & 0.1% sodium azide as preservative

Keratin Azure

Sigma K 8500

Keratin Sulfate

ICN 190220   Bovine cornea   Sodium salt

Keratinocyte Growth Factor

Synonyms: Fibroblast Growth Factor VII

Biodesign A52119H   E. coli   MW 19k   Purified

Chemicon GF008   Human   ≥95%

BioSource International PHG0094   Human recombinant

Fitzgerald 30-AK20   Human recombinant, expressed in *E. coli*

Harlan BT-3014   Human recombinant, expressed in *E. coli* Lyophilized; 0.01 mg

Harlan BT-3015   Human recombinant, expressed in *E. coli* Lyophilized; 0.05 mg

PeproTech 100-19   Human recombinant, expressed in *E. coli* MW 18.9k   >98% (SDS-PAGE) & HPLC; <0.1 ng endotoxin per mg (1EU/mg); lyophilized | Potent mitogen forkeratinocytes & epithelial cells; 163 AA; ED_{50} ≤ 10 ng/mL; SA ≥ 10^7 U/mg; SA determined by the dose-dependent stimulation of thymidine uptake by KGF-responsive BaF3 cells

Sigma K 1757   Human recombinant, expressed in *E. coli*   MW 19k   ≥97% (SDS-PAGE); 0.2 μm filtered solution & lyophilized from phosphate buffered saline containing 500 μg BSA; proliferative activity is tested in culture using monkey epithelial cell line 4MBr-5; endotoxin tested | Epithelial cell specific mitogen, responsible for the normal proliferation & differentiation of human epithelial cells; member of the family of fibroblast growth factors; secreted by stromal fibroblasts, derived from major epithelial organs including skin & gastrointestinal tract, in culture; & is expressed *in vivo* in dermis, but not epidermis; KGF transcripts are found in dermal fibroblasts, epidermal melanocytes & malignant melanoma cells; acts as a potent mitogen for human keratinocytes in culture, equivalent to EGF; Particularly active as a mitogen for BALB/MK cells, a continuous mouse keratinocyte line; has a molecular mass of 19kD; has been tested in culture using a modification of the biological assay of Rubin; Rubin, JS et al, *Proc Natl Acad Sci USA*, 86: 802, 1989; Marchese, C et al, *J Cellular Physiology*, 144: 326, 1990; Finch, PW et al, *Science*, 245: 752, 1989; Albino, AP et al, *Cancer Research*, 51: 4815, 1991; Weissman, BE et al, *Cell*, 32: 599, 1983

Keratinocyte Growth Supplement

Biogenesis 5560-5004 Bovine pituitary Lyophilized

Kininogen

Sigma K 1632 Human plasma High Lyophilized from 4 mM sodium acetate, pH 5.3, 0.15 M NaCl; prepared by kallikrein digestion of human kininogen; purified to remove kallikrein and kinin

Kininogen, HMW-

Biogenesis 5575-5559 Human Liquid

Biogenesis 5575-5539 Human plasma MW 110k Tested negative for HBsAg and antibodies against HCV, HBV, HIV 1, HIV 2 and HTLV I/II; purified; 4 mM Sodium acetate-HCl, 0.15 M NaCl, pH 5.3; liquid

Kistrin Disintegrin

Synonyms: ADAM/Disintegrins Protein

Chemicon CC1032 ≥95%

Sigma K 4755 *Agkistrodon rhodostoma* >95% (SDS-PAGE); lyophilized; sterilized by γ-irradiation | Disintegrins represent a novel family of integrin β1 & β3 inhibitor proteins isolated from viper venoms; low molecular weight, cysteine-rich peptides containing the Arg-Gly-Asp (RGD) sequence; the most potent known inhibitors of integrin function; they interfere with cell adhesion to the extracellular matrix including adhesion of melanoma cells & fibroblasts to fibronectin & are potent inhibitors of platelet aggregation

Lactalbumin

Synonyms: Agglutinin, *Laburnum alpinum*

Amersham US18035 Bovine

ICN 102128 Milk Denatured, non-soluble

Sigma L 7252 Milk ~80% protein, 4% lactose | Non-soluble denatured protein fraction from milk

Lactalbumin, α-

Fluka 61289 Bovine milk Calcium depleted; ≥90% (GE); 85% protein content; 0.1% calcium | McKenzie, HA & White, FA, *Adv Prot Chem*, 41: 173, 1991

Sigma L 7263 Bovine milk (Tetraethyl)rhodamine B isothiocyanate-labeled α-lactalbumin; lyophilized powder; essentially salt-free; ~0.15-0.55 μmole RITC/mg solid | Useful reagent for fluorescent labeling of hormones, surface antigens, lectins & other biologically active molecules; Shecter, Y et al, *Proc Natl Acad Sci USA*, 75: 2135, 1978

Sigma L 8151 Bovine milk MW 14.2k Vial contains enough FITC-conjugated protein to run 50 mini-gels or 25 standard size gels; protein band be visualized by using UV light or Brilliant Blue stain | Fluorescent marker for SDS-page & protein transfer; suitable for use as molecular weight standards in both SDS-PAGE & transfer membranes

Fluka 61288 Human milk Salt-free; ≥90% (GE) | Stimulator for lactose synthase

Sigma L 7269 Human milk ≥90% (electrophoresis); lyophilized | Alters substrate specificity of galactosyltransferase to increase the rate of lactose formation; complex of galactosyltransferase & α-lactalbumin is called lactose synthase; Brodbeck, U et al, *J Biol Chem*, 242: 1391, 1967

Lactalbumin, α-Carboxymethyl

Sigma L 5888 Bovine Dialyzed & lyophilized powder; contains ≥6 moles of S-carboxymethylcysteine/mole α-lactalbumin | <1% active as native α-lactalbumin in stimulating galactosyltransferase to produce lactose

Lactalbumin, α-Type I

Sigma L 5385 Bovine milk ~85% by polyacrylamide gel electrophoresis; calcium saturated; lyophilized powder which may contain traces of ammonium sulfate & sodium phosphate; literature reports the existence of both a high & a low affinity site for binding calcium | Alters substrate specificity of galactosyltransferase to increase the rate of lactose formation; complex of galactosyltransferase & α-lactalbumin is called lactose synthase; Brodbeck, U et al, *J Biol Chem*, 242: 1391, 1967; Hiraoka, Y et al, *Biochem Biophys Res Commun*, 95: 1098, 1980; Permyakov, EA et al, *Biochem Biophys Res Commun*, 100: 191, 1981; Kronman, MJ et al, *J Biol Chem*, 256: 8582, 1981

Lactalbumin, α-Type III

Sigma L 6010 Bovine milk ~85% (PAGE); calcium depleted; lyophilized powder which may contain traces of ammonium sulfate & sodium phosphate; contains <0.3 moles of calcium/mole α-lactalbumin | Alters substrate specificity of galactosyltransferase to increase the rate of lactose formation; complex of galactosyltransferase & α-lactalbumin is called lactose synthase; Brodbeck, U et al, *J Biol Chem*, 242: 1391, 1967

Lactoferrin

Sigma L 4765 Bovine colostrum ~90%; essentially salt-free; purity by SDS gel electrophoresis

ICN 151535 Bovine milk ~98% | Growth factor for cell & tissue culture studies; source of iron for certain cell lines

ICN 1522333 Bovine milk Purified; free-flowing powder; 96-98% protein; 17% iron saturation | Antimicrobial activity against a variety of microorganisms; useful cell culture growth factor; source of iron to certain cell lines

Sigma L 9507 Bovine milk ~90%; purity by SDS gel electrophoresis

ICN 55839 Human Purified Ag; lyophilized

ICN 150203 Human colostrum MW ~77k >99% (sequence HPLC); free of lysozyme & secretory immunoglobulin | Antimicrobial activity against a variety of microorganisms; useful cell culture growth factor; source of iron to certain cell lines

Biogenesis 5605-2011 Human milk Purified; essentially salt free; lyophilized

ICN 160046 Human milk ≥98%; lyophilized | Source of iron for certain cell lines

ICN 194692 Human milk Cell culture reagent; ≥98%; lyophilized | Growth factor for cell & tissue culture studies; a source of iron for certain cell lines

Sigma L 0520 Human milk ~90%; purity by SDS gel electrophoresis; chromatographically purified; lyophilized powder containing ~90% protein (Biuret); balance NaCl

Sigma L 3770 Human milk ~90%; purity by SDS gel electrophoresis; iron saturated; iron content ≥0.15% (w/w)

Sigma L 4894 Human milk ~90% (SDS gel electrophoresis); cell culture tested

Lactoferrin, Gold Conjugated

Sigma L 3647 Bovine colostrum 10 nm colloidal gold labeled; mean particle diameter 8-12 nm; monodisperse; lactoferrin from bovine colostrum (Sigma L 4765) coupled through spacer to albumin coated colloidal gold; suspension in ~50% glycerol containing 0.15 M NaCl, 0.01 M BES, pH 7.0, 0.25% BSA & 0.02% sodium azide; concentration: A_{520}~5.0 | Particularly useful in detection of DNA; Benhamou, N, *J Electron Micr Tech*, 12: 1, 1989

Lactogen, Placental

Biogenesis 7400-0509 Human placenta MW 22k <0.1% hGH/hCG, <0.01% human placental GH; tested negative for HBsAg, HCV, HIV-1 and 2; purified; 0.05 M NH_4HCO_3; lyophilized

Cortex CP1015 Human placenta >95%

Cortex CP1015P Human placenta >40%

Lactoglobulin A, β-

Sigma L 5137   Bovine milk   pI 5.1; vial contains ~2 mg | IEF Marker

Sigma L 8005   Bovine milk   ≥90% (PAGE)

Lactoglobulin A, β-(^{14}C-Me)-

ARC ARC-429   MW 18,367   3-30 µCi/mg; 111-1111kBq/mg; in 0.01 M sodium phosphate, pH 7.2 | Radiochemical

Lactoglobulin, β-

ICN 100363   Bovine   3X crystallized

Fluka 61329   Bovine milk   MW 17.5k   ~90% (GE) | Mediates the asymmetric self-condensation of β-ionylideneacetaldehyde, retinal & related compounds; Asato, AE et al, *Tetrahedron Lett*, 33: 3105, 1992

ICN 151536   Bovine milk   Lyophilized powder; 1 mg solid binds 10-15 ng folate, Dunn method | Dunn & foster, *Clin Chem*, 19:1101, 1973

Sigma L 0130   Bovine milk   ~90% (PAGE); 3X crystallized & lyophilized; a further purification of Sigma L 6879; contains β-lactoglobulins A & B which can be isolated chromatographically | May not contain folate binding protein; not recommended for folate analysis; Piez, KA et al, *J Biol Chem*, 236: 2912, 1961

Sigma L 2506   Bovine milk   ~80% (PAGE); lyophilized powder; 1 mg solid binds 3-15 ng folate; contains β-lactoglobulins A & B which can be isolated chromatographically | Suitable for analysis of folate by method of Dunn & Foster, *Clin Chem*, 19: 1101, 1973; Piez, KA et al, *J Biol Chem*, 236: 2912, 1961

Sigma L 3908   Bovine milk   ~90% (PAGE); chromatographically purified & lyophilized; contains β-lactoglobulins A & B which can be isolated chromatographically | May not contain folate binding protein; not recommended for folate analysis; Piez, KA et al, *J Biol Chem*, 236: 2912, 1961

Sigma L 6879   Bovine milk   ~85% (PAGE); crystallized & lyophilized; contains β-lactoglobulins A & B which can be isolated chromatographically | May not contain folate binding protein; not recommended for folate analysis; Piez, KA et al, *J Biol Chem*, 236: 2912, 1961

Laminin

Biogenesis 5620-0604   Human placenta   MW 170k-190k Purified; tested negative for HBsAg and HIV and HCV antibodies; 50 mM Tris-HCl, pH 8.2, 300 mM NaCl; liquid | Useful in quantitative and qualitative immunochemical methods (eg ELISA), for production of antibodies, neurile stimulation assay and as an attachment factor in cell culture; Wever et al, *JBC*, 258:12654, 1983; Engvall et al, *J Cell Biol*, 103:2457, 1986

Calbiochem 428012   Human placenta   MW 170k & 190k Liquid in 300 mM NaCl, 50 mM Tris-HCl, pH 8.2; ≥95% (SDS-PAGE); prepared from tissue of individuals shown by certified tests to be negative for HBsAg & antibodies to HIV & HCV | Multi-functional non-collagenous glycoprotein found in extracellular matrix; involved in promotion of cellular adhesion, platelet adhesion & neurite regeneration & binding to Type IV collagen, glycoaminoglycan & heparin; useful for antiserum production, ELISA, Ouchterlony, cell-attachment & neurite-stimulation assays; biologically & immunologically identical to intact laminin; *Merck Index*, 12: 5364; Timpl, R & Brown, JC, *Matrix Biol*, 14: 274, 1994; Engvall, E et al, *J Cell Biol*, 103: 2457, 1986

BioSource International PHE0033   Mouse

Biogenesis 5620-2004   Mouse EHS sarcoma   Purified; 0.05 M Tris, 0.15 M NaCl, pH 7.4 with 50 µg/mL gentamycin; liquid | Can be used in cell culture, especially in the attachment and/or spreading of epithelial, endothelial and neuronal cells

USBio L1225   Mouse, purified from basement membrane of the Engelbreth-Holm-Swarm (EHS) mouse tumor   Purified using salt precipitation & IX chromatography; tested negative for bacteria, fungi & mycoplasma; frozen liquid in 0.05 M Tris-HCl, pH 7.4, 0.15 M NaCl

Chemicon CC095   Murine   Purified | Extracellular matrix protein

Laminin, Pepsinized

Chemicon AG56P   Human placenta   Purified; pepsinized | Extracellular matrix protein

Laminin-Like Engineered Protein Polymer

Sigma L 6515   Recombinant   MW 75,639 (gene sequence), ~110k (SDS-PAGE)   Lyophilized polymer supplied with diluent in separate vial; diluent contains 4.5 M LiClO$_4$; sterilized by autoclaving; cell culture tested | Incorporates multiple copies of the IKVAV ligand from the laminin alpha chain interspaced between repeated structural peptide U; U.S. Patent No. 5,211,657

Latrotoxin, α-

Synonyms: Ltx, α-

Alexis 630-027   *Latrodectus tredecemguttatus*   MW 130k ≥97%; white lyophilized powder; soluble in water; potent neurotoxin | Causes massive neurotransmitter release from a wide variety of central & peripheral synaptic junctions of vertebrates using Ca^{2+}-dependent & Ca^{2+}-independent pathways; useful pharmacological tool in the studies of synaptic vesicles exocytosis of different neutrotransmitters; Frontali, N et al, *J Cell Biol*, 68: 462, 1976; Stahl, B et al, *J Biol Chem*, 269: 24770, 1994; Grasso, A, *Biochim Biophys Acta*, 439: 406, 1976; Valtorta, F et al, *J Cell Biol*, 107: 2717, 1988; Osipenko, ON et al, *Toxicon*, 31: 1123, 1993; Parpura, V et al, *FEBS Lett*, 360: 266, 1995

Calbiochem 428025   *Latrodectus tredecimguttatus*   MW 130k Lyophilized solid; one distinct band purity (SDS-PAGE); biological activity: stimulates neurotransmitter release in both Ca^{2+}-free & Ca^{2+}-containing media; soluble in water; LD$_{50}$≤2 k mg/kg | Causes massive release of exocytotic neurotransmitter synaptic vesicles from a wide variety of central & peripheral synaptic junctions of vertebrates via Ca^{2+}-dependent & –independent parallel mechanisms at concentrations as low as 100 pM; stimulates Ca^{2+}-independent GABA & glutamate release from cortical astrocytes in culture; Parpura, V et al, *FEBS Lett*, 360: 266, 1995; Storchak, LG et al, *FEBS Lett*, 351: 267, 1994; Osipenko, ON et al, *Toxicon*, 31: 1123, 1993; Valtorta, F et al, *J Cell Biol*, 107: 2717, 1988

Lck (p56)

USBio L1565-06   Bovine thymus   Purified by DEAE-Sepharose, hydroxyapatite, & phenyl-Sepharose columns, followed by Sephacryl S-200 gel filtration; essentially free of other proteinkinase contamination; frozen solution in 75 mL of 25 mM HEPES, pH 7.0,10% glycerol, 0.1% IGEPAL CA-630

Lectin

Sigma L 1277   *Lens culinaris* (lentil)   pI 8.2, 8.6, 8.8; vial contains ~1 mg | IEF Marker

Lectin Mixture

Sigma L-7S   7 Lectins: Arachis *hypogaea* (Peanut, Sigma L 0881, affinity: galactose), Canavalia *ensiformis* (Jack bean, Sigma C 2010, affinity: glucose, mannose), Lens *culinaris* (Lentil, Sigma L 9267, affinity: glucose, mannose), Lotus *tetragonolobus* (Winged pea, Sigma L 9254, affinity: L-fucose), Ricinus *communis* (Castor bean, Sigma L 8508, Toxin RCA$_{60}$, affinity: GalNAc), Ricinus *communis* (Castor bean, Sigma L 7886, Agglutinin RCA$_{120}$, affinity: galactose), Triticum *vulgaris* (Wheat germ, Sigma L 9640; affinity: GlcNAc); extremely hazardous! Be aware of the risks & familiar with safety procedures before you use this product

Lens Proteins

Sigma L 2394   Bovine eye lens   Water soluble; lyophilized preparation of a water extract

Leptin

Synonyms: Anti-Obesity Protein

Biodesign A52327H   *E. coli*   Purified

Chemicon GF052   Human   ≥95% | Cellular biochemistry/regulatory protein

Biogenesis 5633-1439   Human r-DNA *E. coli*   MW 16k
Endotoxin <0.1 ng/µg; purified; 10 *mM* Sodium citrate, pH 4.0;
lyophilized | Biological activity in the ob/ob and NZO mouse
obesity models

BioSource International PHP0013   Human recombinant

Alexis 201-034   Human recombinant, expressed in *E. coli*
MW 16k   ≥95% (SDS-PAGE & HPLC); lyophilized from 0.5 mg
leptin/mL in 10 *mM* sodium citrate, pH 4.0; endotoxin <0.1 ng/µg
leptin | Leptin; product of the *ob* (obese) gene; a protein
consisting of 146 AA residues; produced in the adipose tissue &
considered to play an important role in appetite control, fat
metabolism & regulation of body weight; targets the central
nervous system, particularly hypothalamus, affecting food intake;
leptin levels are high in most obese individuals; Zhang, Y et al,
Nature, 372: 425, 1994; Considine, RV et al, *J Clin Invest*, 95:
2986, 1995; Pelleymounter, MA, *Science*, 269: 540, 1995;
Lonnqvist, F et al, *Nature Med*, 1: 950, 1995; Maffei, M et al,
Nature Med, 1: 1155, 1995; Considine, RV et al, *New Engl J Med*,
334: 292, 1996; Considine, RV et al, *BBRC*, 220: 735, 1996

Calbiochem 429700   Human recombinant, expressed in *E. coli*
MW 16k   ≥97% (SDS-PAGE); lyophilized from sterile-filtered PBS,
carrier-free; biological activity: ED$_{50}$=0.4-2 ng/mL as measured by
its ability to induce proliferation of leptin-dependent rOB-R
transfected murine BAF3 cells; endotoxin: ≤100 pg/µg leptin |
Product of the obese (ob) gene; a ligand for the OB receptor (OB-
R); mice with mutations of the ob gene have been found to be
obese & diabetic & to have reduced activity, metabolism & body
temperature; *Merck Index*, 12: 5466; Ookuma, M et al, *Diabetes*,
47: 219, 1998; Campfield, LA et al, *Science*, 269: 546, 1995;
Halaas, JL et al, *Science* 269: 543, 1995; Pelleymounter, MA et al,
Science, 269: 540, 1995; Zhang, Y et al, *Nature*, 372: 425, 1994

Fitzgerald 30-AL12   Human recombinant, expressed in *E. coli*

ICN 195807   Human recombinant, expressed in *E. coli*
>95%; lyophilized

PeproTech 300-27   Human recombinant, expressed in *E. coli*
MW 16k   >98% (SDS-PAGE) & HPLC; <0.1 ng endotoxin per mg
(1EU/mg); lyophilized | Protein product of the *ob* (obese) gene in
mice; involved in appetite control; the *ob* gene is expressed in
adipose tissue & is thought to regulate the body's fat stores; mice
with the *ob/ob* genotype also develop a form of diabetes similar to
type II (non-insulin dependent);. 146 AA; biologically active in two
different mouse obesity models, *ob/ob* and NZO

Sigma L 4146   Human recombinant, expressed in *E. coli*   MW
~16k   ≥97% (SDS-PAGE); 0.2 µm filtered solution & lyophilized
from phosphate buffered saline; biological activity is measured by
its ability to lower body weight & food intake in the mutant obese
C57BL/6J mice (ob/ob) following daily intraperitoneal injection;
endotoxin tested | Product of the ob gene; suppresses feeding &
leads to body weight reduction; the only cell currently reported to
secrete leptin is the mature adipocyte; factors reported to induce
leptin secretion include insulin & inflammatory mediators such as
LPS, IL-1β & TNF-α; feedback loop may exist where insulin
stimulates OB secretion & circulating OB inhibits insulin production;
insulin-producing pancreatic β-cells have OB receptors & it is
suspected that adipose cell size is a major determinant of OB
mRNA expression; non-glycosylated polypeptide of 146 AA;
precursor form of leptin consists of 167 AA; to generate the mature
leptin (146 AA) the precursor cleaves a 21 AA signal peptide;
human leptin shows ~85% AA homology to mouse leptin & 84% AA
homology to rat leptin; mouse & rat leptin show ~96% AA
homology with each other; Stephens, TW et al, *Nature*, 377: 530,
1995; Zhang, Y et al, *Nature*, 372: 425, 1994; Leroy, P et al, *J Biol
Chem*, 271: 2365, 1996; Gettys, TW et al, *Endocrinology*, 137:
4054, 1996; Malmstrom, R et al, *Diabetologia*, 39: 993, 1996;
Grunfeld, C et al, *J Clin Invest*, 97: 2152, 1996; Mizuno, TM et al,
Proc Natl Acad Sci USA, 93: 3434, 1996; Kieffer, TJ et al, *Biochem
Biophys Res Commun*, 224: 522, 1996; Collins, S & Surwit, RS, *J
Biol Chem*, 271: 9437, 1996; Cohen, SL et al, *Nature*, 382: 589,
1996; Ogawa, Y et al, *J Clin Invest*, 96: 1647, 1996

Biogenesis 5633-1479   Mouse r-DNA   MW 16k   Endotoxin:
<0.1 ng/µg of leptin; purified; no preservatives; lyophilized | 147
AA

BioSource International PMP0013   Mouse recombinant

Alexis 201-035   Mouse recombinant, expressed in *E. coli*   MW
16k   ≥95% (SDS-PAGE & HPLC); lyophilized from 0.5 mg
leptin/mL in 10 *mM* sodium citrate, pH 4.0; endotoxin <0.1 ng/µg
leptin | Product of the *ob* (obese) gene, consisting of 146 AA
residues; produced in the adipose tissue & considered to play an
important role in appetite control, fat metabolism & regulation of
body weight; studies have shown that it may also influence
reproductive function; mice with *ob/ob* genotype develop a form of
diabetes similar to type II in humans; Zhang, Y et al, *Nature*, 372:
425, 1994; Pelleymounter, MA, *Science*, 269: 540, 1995; Halaas,
JL et al, *Science*, 269: 543, 1995; Maffei, M et al, *Nature Med*, 1:
1155, 1995

Calbiochem 429705   Mouse recombinant, expressed in *E. coli*
MW 16k   ≥97% (SDS-PAGE); lyophilized from sterile-filtered PBS,
carrier-free; biological activity: ED$_{50}$=0.2-1 ng/mL as measured by
its ability to induce proliferation of leptin-dependent rOB-R
transfected murine BAF3 cells; endotoxin: ≤100 pg/µg leptin |
Product of the obese (ob) gene; a ligand for the OB receptor (OB-
R); mice with mutations of the ob gene have been found to be
obese & diabetic & to have reduced activity, metabolism & body
temperature; suppresses insulin secretion by inhibiting activities of
Ca^{2+}-dependent PKC isoforms; *Merck Index*, 12: 5466; Ookuma, M
et al, *Diabetes*, 47: 219, 1998; Campfield, LA et al, *Science*, 269:
546, 1995; Halaas, JL et al, *Science* 269: 543, 1995;
Pelleymounter, MA et al, *Science*, 269: 540, 1995; Zhang, Y et al,
Nature, 372: 425, 1994

Sigma L 3772   Mouse recombinant, expressed in *E. coli*   ≥97%
(SDS-PAGE); 0.2 µm filtered solution & lyophilized from phosphate
buffered saline; biological activity is measured by its ability to lower
body weight & food intake in the mutant obese C57BL/6J mice
(ob/ob) following daily intraperitoneal injection; endotoxin tested |
Product of the ob gene; suppresses feeding & leads to body weight
reduction; the only cell currently reported to secrete leptin is the
mature adipocyte; factors reported to induce leptin secretion
include insulin & inflammatory mediators such as LPS, IL-1β & TNF-
α; feedback loop may exist where insulin stimulates OB secretion &
circulating OB inhibits insulin production; insulin-producing
pancreatic β-cells have OB receptors & it is suspected that adipose
cell size is a major determinant of OB mRNA expression; non-
glycosylated polypeptide of 146 AA; precursor form of leptin
consists of 167 AA; to generate the mature leptin (146 AA) the
precursor cleaves a 21 AA signal peptide; human leptin shows
~85% AA homology to mouse leptin & 84% AA homology to rat
leptin; mouse & rat leptin show ~96% AA homology with each
other; Stephens, TW et al, *Nature*, 377: 530, 1995; Zhang, Y et al,
Nature, 372: 425, 1994; Leroy, P et al, *J Biol Chem*, 271: 2365,
1996; Gettys, TW et al, *Endocrinology*, 137: 4054, 1996;
Malmstrom, R et al, *Diabetologia*, 39: 993, 1996; Grunfeld, C et al,
J Clin Invest, 97: 2152, 1996; Mizuno, TM et al, *Proc Natl Acad Sci
USA*, 93: 3434, 1996; Kieffer, TJ et al, *Biochem Biophys Res
Commun*, 224: 522, 1996; Collins, S & Surwit, RS, *J Biol Chem*,
271: 9437, 1996; Cohen, SL et al, *Nature*, 382: 589, 1996; Ogawa,
Y et al, *J Clin Invest*, 96: 1647, 1996

Chemicon GF050   Murine   ≥95% | Cellular
biochemistry/regulatory protein

Fitzgerald 30-AL13   Murine recombinant, expressed in *E. coli*

ICN 195015   Murine recombinant, expressed in *E. coli*   MW
16k   >95%; lyophilized

PeproTech 450-31   Murine recombinant, expressed in *E. coli*
MW 16k   >98% (SDS-PAGE) & HPLC; <0.1 ng endotoxin per mg
(1EU/mg); lyophilized | Protein product of the *ob* (obese) gene in
mice; involved in appetite control; expressed in adipose tissue & is
thought to regulate the body's fat stores; mice with the *ob/ob*
genotype also develop a form of diabetes similar to type II (non-
insulin dependent; 147 AA; biologically active in two different
mouse obesity models, *ob/ob* and NZO

Lethal Toxin Inhibiting Factor

ICN 159835   *Didelphis virginiana* (opossum) serum   MW 66k
Purified, single band by SDS-PAGE | Inhibits the lethality of
various venoms & toxins; histamine blocker when tested on mast
cells; ≥0.5 µg LTIF mixed with lethal doses of venoms or toxins
inhibits the lethality of the venom injected IP in mice

Leukemia Inhibitory Factor

Synonyms: Differentiation-Inhibition Activity For Murine Embryonic Stem Cells; Human Interleukin for DA Cells; Hepatocyte Stimulating Factor III; Cholinergic Neuronal Differentiation Factor; Lipoprotein Lipase Inhibitor; Differentiation-Inhibition Activity For Murine Embryonic Stem Cells; Interleukin for DA Cells, Human; Hepatocyte Stimulating Factor III; Cholinergic Neuronal Differentiation Factor; Lipoprotein Lipase Inhibitor

Chemicon LIF1005   Human recombinant   ≥95%

Chemicon LIF1010   Human recombinant   ≥95%

Oncogene PF044   Human recombinant   MW 20k   Liquid; >95% (SDS-PAGE); biological activity: EC_{50} of 0.5 ng/mL as determined by its ability to induce differentiation in murine M1 myeloid leukemia cells; 5 mg (5×10^5 units) supplied in 0.5 mL of PBS, pH 7.4, 0.02% Tween 20; reconstitution buffer should contain 0. | Species reactivity: human; for proliferation studies; sequence analysis predicts a mature protein of 180 AA

Fitzgerald 30-AL50   Human recombinant, expressed in *E. coli*

Harlan BT-3016   Human recombinant, expressed in *E. coli*   Lyophilized; 0.01 mg

Harlan BT-3017   Human recombinant, expressed in *E. coli*   Lyophilized; 0.05 mg

ICN 158412   Human recombinant, expressed in *E. coli*   ≥98%; ED_{50} = 5.0-10.0 ng/mL, 1 U inhibits M1 mouse myelomonocytic leukemia cell proliferation | Active in human & mouse systems

PeproTech 300-05   Human recombinant, expressed in *E. coli*   MW 19.7k   >98% (SDS-PAGE) & HPLC; <0.1 ng endotoxin per mg (1EU/mg); lyophilized | Lymphoid factor which promotes long-term maintenance of embryonic stem cells by suppressing spontaneous differentiation; 180 AA; $ED_{50} \leq 0.01$ ng/mL; SA $\geq 10^8$ U/mg; SA determined by the M1 cell differentiation assay

Sigma L 5283   Human recombinant, expressed in *E. coli*   ≥95% (SDS-PAGE); 0.2 μm filtered in phosphate buffered saline, pH 7.4 containing 0.02% Tween 20; proliferative activity is measured in culture using the human leukemic cell line TF-1 | Multifunctional glycoprotein that induces macrophage differentiation & supresses the proliferation of the murine M1 myeloid leukemia cell line; LIF plays an important role along with IL-6 & G-CSF in the regulation of early hematopoietic stem cells; also important in the release of calcium from bone tissue; Smith, AG et al, *Nature*, 336: 688, 1988; Moreau, JF et al, *Nature*, 336: 690, 1988; Baumann, H et al, *J Immunol*, 143: 1163, 1989; Gearing, DP et al, *EMBO J*, 6: 3995, 1987; Smith, AG et al, *Dev Biol*, 121: 1, 1987; Yamamoni, T et al, *Science*, 246: 1412, 1989; Abe, E et al, *Proc Natl Acad Sci USA*, 83: 5958, 1986; Mori, M et al, *Physchem Biophys Pres Commun*, 160: 1085, 1989; Leary, AG et al, *Blood*, 75: 1960, 1990; Gough, NM et al, *Proc Natl Acad Sci USA*, 85: 2623, 1988

Sigma L 5158   Mouse recombinant, expressed in *E. coli*   ≥95% (SDS-PAGE); 0.2 μm filtered in phosphate buffered saline, pH 7.4 containing 0.02% Tween 20; proliferative activity is measured in culture using the M1 murine cell line | Multifunctional glycoprotein that induces macrophage differentiation & supresses the proliferation of the murine M1 myeloid leukemia cell line; LIF plays an important role along with IL-6 & G-CSF in the regulation of early hematopoietic stem cells; also important in the release of calcium from bone tissue; Smith, AG et al, *Nature*, 336: 688, 1988; Moreau, JF et al, *Nature*, 336: 690, 1988; Baumann, H et al, *J Immunol*, 143: 1163, 1989; Gearing, DP et al, *EMBO J*, 6: 3995, 1987; Smith, AG et al, *Dev Biol*, 121: 1, 1987; Yamamoni, T et al, *Science*, 246: 1412, 1989; Abe, E et al, *Proc Natl Acad Sci USA*, 83: 5958, 1986; Mori, M et al, *Physchem Biophys Pres Commun*, 160: 1085, 1989; Leary, AG et al, *Blood*, 75: 1960, 1990; Gearing, DP et al, *EMBO J*, 6: 3995, 1987

Chemicon LIF2005   Murine recombinant   ≥95%

Chemicon LIF2010   Murine recombinant   ≥95%

Oncogene PF045   Murine recombinant   MW 20k   Liquid; >97% (SDS-PAGE); biological activity: EC_{50} of 0.5 ng/mL as determined by its ability to induce differentiation in murine M1 myeloid leukemia cells | Species reactivity: mouse; for proliferation studies; sequence analysis predicts a mature protein of 181 AA

Fitzgerald 30-AL51   Murine recombinant, expressed in *E. coli*

ICN 160052   Murine recombinant, expressed in *E. coli*   ≥97%; ED_{50} = 0.03-0.1 ng/mL, 1 U induces murine leukemic DA-1a cell proliferation | ~1000X decreased activity on human cells vs human LIF

PeproTech 250-02   Murine recombinant, expressed in *E. coli*   MW 20k   >98% (SDS-PAGE) & HPLC; <0.1 ng endotoxin per mg (1EU/mg); lyophilized from PBS | Lymphoid factor, which promotes long-term maintenance of embryonic stem cells by suppressing spontaneous differentiation; 180 AA; $ED_{50} \leq 0.01$ ng/mL; SA $\geq 10^8$ U/mg; SA determined by the M1 cell differentiation assay

USBio L2024-08   Recombinant

Leukemia Inhibitory Factor Soluble Receptor-α

Oncogene PF046   Human recombinant   MW 100-110k (glycosylated)   Lyophilized in PBS containing 50 mg of BSA per 1 mg of cytokine; >97% (SDS-PAGE); biological activity: EC_{50} of 3-6 ng/mL as determined by its ability to inhibit proliferation of TF-1 cells in the presence of 0.3 ng/mL of recombinant human LIF | Species reactivity: human; has a predicted MW of 89k, but as a result of heterogeneous glycosylation has a MW of 100-110 k; shown to bind LIF & has antagonistic activity

ICN 195726   Human recombinant, expressed in S*f*21   ≥97%; ~3-6 μg/mL inhibits half the biological response of 0.3 ng/mL LIF

Leukemia Virus gp70, Feline

USBio L2041-12   Feline   ≥96%; no contaminants detected; single band by SDS-PAGE, IEP, &/or RID; no trace of p27 or transmembrane visible; ~1 mg/mL supplied in PBS containing 0.1% NaN_3 | Suitable for antigenic applications in immunological protocols

Leukemia Virus p27, Feline

USBio L2041-16   Tissue culture   ≥96%; no contaminants detected; single band by SDS-PAGE, IEP, &/or RID; ≤0.01% Gp70 with a trace of precursor material visible by Western blot; ~1 mg/mL (OD 280 nm) supplied in PBS containing 0.1% NaN_3 | Suitable for antigenic applications in immunological protocols

Leukoagglutinin

Sigma L 4144   *Phaseolus vulgaris* (red kidney bean)   Purified; salt-free; lyophilized; tested for leucocyte agglutination activity; mitogenic at <5 μg/mL; cell culture tested; endotoxin tested | Lectin; highly specific polyvalent carbohydrate-binding proteins; useful in polysaccharide studies, glycoprotein studies, enzyme tagging & cell membrane studies, cell agglutination & cell typing; in tissue culture certain lectins used to induce mitogenic activity; tested in a tissue culture system using 3H-thymidine incorporation as a measure of mitogenic activity; Goldstein, I & Hayes, C, *Adv Carbo Chem Biochem*, 35: 127, 1978; Rosenberg, SA & Lipsky, PE, *J Immunol*, 122: 926, 1979

Levetin, γ-

ICN 194993   Egg yolk   MW 15k   >98%; white solid | Immunologically equal to chicken serum γ-globulin

Light Chains, κ

Scipac P163-1   Pooled urine of Bence-Jones patients   >96%; lyophilized | Urine protein

Scipac P164-1   Pooled urine of Bence-Jones patients   >96%; lyophilized | Urine protein

Lipopolysaccharide Induced CXC Chemokine

Biodesign A52017M   *E. coli*   >98% | Species specificity: mouse

BioSource International PMC1595   Mouse recombinant

PeproTech 250-17   Murine recombinant, expressed in *E. coli*   MW 9.9k   >98% (SDS-PAGE) & HPLC; <0.1 ng endotoxin per mg (1EU/mg); lyophilized | Novel murine neutrophil-chemoattractant CXC chemokine; 93 AA; SA determined by its ability to chemoattract human neutrophils

Lipoprotein

Sigma L 3626 Bovine plasma Solution in 10 mM sodium bicarbonate, pH 7.4; ~20 mg protein/mL; aseptically filtered | Predominantly HDL Apolipoprotein A-I & A-II are directly purified from HDL; apolipoprotein B is prepared from purified LDL using SDS or sodium deoxycholate to remove lipids; Burstein, M, *J Lipid Res*, 11: 583, 1970; Rudel, LL, *Biochem J*, 139: 89, 1974

Lipoprotein Concentrate

Sigma L 9906 Human Low endotoxin; cholesterol: 8-11 g/L; protein: (Biuret): 9-20 g/L; pH: 7.0-8.6; aseptically filtered | Apolipoprotein A-I & A-II are directly purified from HDL; apolipoprotein B is prepared from purified LDL using SDS or sodium deoxycholate to remove lipids; endotoxin level given on the label; IgG: undetectable; Burstein, M, *J Lipid Res*, 11: 583, 1970; Rudel, LL, *Biochem J*, 139: 89, 1974

Lipoprotein, High Density

Synonyms: Lipoprotein, α-; Lipoprotein, α-; Lipoprotein, α-

Cortex CP2107 >95%

ICN 59433 Human Purified; 1.063-1.210 g/mL

Biodesign A95124H Human plasma Purified | Species specificity: human

Biodesign A95332H Human plasma Purified; 3 subfraction; d: 1.120-1.210 | Species specificity: human

Biogenesis 5685-2004 Human plasma Purified; 0.05 *M* Tris-HCl, 0.15 *M* NaCl, 0.3 *mM* EDTA, pH 7.4; liquid

Fluka 62332 Human plasma 10 mg protein/mL; free from contamination by other lipoprotein classes; aseptically filtered solution in 0.15 *M* NaCl, 0.01% EDTA, pH 7.4 | Useful for cellulose acetate electrophoresis

Sigma L 2014 Human plasma Solution in 0.15 M NaCl with 0.01% EDTA, pH 7.4; aseptically filtered | Apolipoprotein A-I & A-II are directly purified from HDL; apolipoprotein B is prepared from purified LDL using SDS or sodium deoxycholate to remove lipids; freezing may cause structural change & denature lipoproteins; Burstein, M, *J Lipid Res*, 11: 583, 1970; Rudel, LL, *Biochem J*, 139: 89, 1974

Sigma L 5277 Human plasma ~10 mg protein/vial (modified Lowry); cholesterol: >25 µg/mg protein; aseptically filtrated then lyophilized from 1 mL HDL solution in 0.15 M NaCl & 0.01% EDTA at pH 7.4; heat treated at 60°C for 10 hours | Apolipoprotein A-I & A-II are directly purified from HDL; apolipoprotein B is prepared from purified LDL using SDS or sodium deoxycholate to remove lipids; suitable for use as a nutritional source of fatty acids & lipids; Burstein, M, *J Lipid Res*, 11: 583, 1970; Rudel, LL, *Biochem J*, 139: 89, 1974

Biodesign A94005H Human serum Purified | Species specificity: human

Biogenesis 5685-2025 Human serum DiI conjugated

Lipoprotein, Low Density

Synonyms: Lipoprotein, β-

Cortex CP2017 >95%

ICN 59392 Human Purified; 1.019-1.063 g/mL

Fitzgerald 30-AL88 Human fluids High purity

Biogenesis 5685-3204 Human plasma Tested negative for HIV-1 and HIV-2, HBsAg, the antibody to Human T-Lymphotropic Virus Type I, HCV and HBcAg; purified; 0.05 *M* Tris-HCl, 0.15 *M* NaCl, 0.3 *mM* EDTA, pH 7.4, 0.22 micron filtered; liquid

Biogenesis 5685-3557 Human plasma Purified; 0.05 *M* Tris-HCl, 0.15 *M* NaCl & 0.3 *mM* EDTA, pH 7.4; sterile; liquid | Binds LDL receptor in peritoneal macrophages

Fluka 62331 Human plasma 5 mg protein/mL; free from contamination by other lipoprotein classes; aseptically filtered solution in 0.15 *M* NaCl, 0.01% EDTA, pH 7.4 | Useful for cellulose acetate electrophoresis

Sigma L 2139 Human plasma Solution in 0.15 M NaCl with 0.01% EDTA, pH 7.4; aseptically filtered | Apolipoprotein A-I & A-II are directly purified from HDL; apolipoprotein B is prepared from purified LDL using SDS or sodium deoxycholate to remove lipids; Burstein, M, *J Lipid Res*, 11: 583, 1970; Rudel, LL, *Biochem J*, 139: 89, 1974

Sigma L 5402 Human plasma ~5 mg protein/vial (modified Lowry); cholesterol: >500 µg/mg protein; aseptically filtrated then lyophilized from 1 mL LDL solution in 0.15 M NaCl & 0.01% EDTA at pH 7.4; heat treated at 60°C for 10 hours | Apolipoprotein A-I & A-II are directly purified from HDL; apolipoprotein B is prepared from purified LDL using SDS or sodium deoxycholate to remove lipids; Burstein, M, *J Lipid Res*, 11: 583, 1970; Rudel, LL, *Biochem J*, 139: 89, 1974

Biogenesis 5685-3255 Human serum DiO conjugated

Biogenesis 5685-3577 Human serum ^{125}I conjugated

Lipoprotein, Low Density Ac-

Biogenesis 5685-3404 Human serum Tested negative for HIV 1 and 2 antibodies, HBsAg, HTLV I antibodies, HCV and HBcAg; purified; 0.05 *M* Tris-HCl, 0.15 *M* NaCl and 0.3 *mM* EDTA, pH 7.4; sterile filtered; liquid

Biogenesis 5685-3502 Human serum ^{125}I conjugated

Lipoprotein, Very Low Density

Synonyms: Lipoprotein, Pre-β-; Lipoprotein, Pre-β-

Cortex CP2018 >95%

ICN 59393 Human Purified; 1.006-1.063 g/mL

Biodesign A34013H Human plasma Purified | Species specificity: human

Fluka 62329 Human plasma 1 mg protein/mL; free from contamination by other lipoprotein classes; aseptically filtered solution in 0.15 *M* NaCl, 0.01% EDTA, pH 7.4 | Useful for cellulose acetate electrophoresis

Sigma L 2264 Human plasma Solution in 0.15 M NaCl with 0.01% EDTA, pH 7.4; filtered through a 0.45 µm membrane | Apolipoprotein A-I & A-II are directly purified from HDL; apolipoprotein B is prepared from purified LDL using SDS or sodium deoxycholate to remove lipids; Burstein, M, *J Lipid Res*, 11: 583, 1970; Rudel, LL, *Biochem J*, 139: 89, 1974

Biogenesis 5685-4004 Human serum Liquid

Biogenesis 5685-4104 Rabbit serum Liquid

Lipoprotein, α-

Cortex CP8102 >95%

ICN 59602 ICN 59603 Human Purified

Fitzgerald 30-AL37 Human fluids High purity

Biodesign A19132H Human plasma Purified | Species specificity: human

Biogenesis 5684-9804 Human plasma Purified; 10 *mM* NaCl, 1 *mM* EDTA, 0.01% NaN₃, pH 7.2; liquid | A protein concentration >3 mg/mL will lead to extensive aggregation; tested negative for HBsAg and HIV

ICN 59604 Human plasma Purified; 1.063-1.210 g/mL; isolated by sequential isopycnic untracentrifugation, followed by lysine affinity chromatography

Sigma L 0526 Human plasma Solution in 10 mM Tris-HCl containing 1 mg/mL Na₂EDTA, pH 7.2; filtered through a 0.45 µm membrane | Apolipoprotein A-I & A-II are directly purified from HDL; apolipoprotein B is prepared from purified LDL using SDS or sodium deoxycholate to remove lipids; Burstein, M, *J Lipid Res*, 11: 583, 1970; Rudel, LL, *Biochem J*, 139: 89, 1974

Sigma L 2532 Human plasma Lyophilized powder | Apolipoprotein A-I & A-II are directly purified from HDL; apolipoprotein B is prepared from purified LDL using SDS or sodium deoxycholate to remove lipids; Burstein, M, *J Lipid Res*, 11: 583, 1970; Rudel, LL, *Biochem J*, 139: 89, 1974

USBio L2600   Human plasma   ≥98%; no contaminants detected; single band by SDS-PAGE; purified by Lys-Sepharose affinity chromatography; ~0.5 mg/mL; liquid in 10 mM NaCl, 1 mM EDTA, pH 7.2, 0.01% NaN$_3$ | Predominant polymorph 4, minor polymorph 8; suitable for antigenic applications in immunological protocols

USBio L2600-12   Human plasma   no contaminants detected; single band by SDS-PAGE, IEP, &/or RID; ~0.5 mg/mL; supplied in 0.1 M Tris buffer containing stabilizers, pH 7.6, 0.1% NaN$_3$ added as preservative | Suitable for antigenic applications in immunological protocols

Liver Cell Growth Factor

Chemicon LC010   ≥95%

Liver Expressed Chemokine

Synonyms: NCC-4

Biodesign A52044H   *E. coli*   Purified

Liver-Expressed Chemokine

Synonyms: NCC-4

PeproTech 300-44   Human recombinant, expressed in *E. coli*   MW 11.2k   >98% (SDS-PAGE) & HPLC; <0.1 ng endotoxin per mg (1EU/mg); lyophilized | Recently discovered CC chemokine; chemotactic on monocytes; 97 AA; SA determined by its ability to chemoattract total human monocytes

LTβR/Fc Chimera

Synonyms: TNFRSF3

R&D Systems 629-LR-100   Human NSO-expressed   >95%; lyophilized | Species specificity: human lymphotoxin β receptor

Luffin

Synonyms: Ribosome Inactivating Protein

Sigma L 7146   *Luffa aegyptiaca* seeds   Lyophilized powder containing ~15% protein (Lowry); balance primarily sodium phosphate, pH 7.2 | Kishida, K et al, *FEBS Lett*, 153: 209, 1983

Luteinizing Hormone

Synonyms: ICSH; PLH

Cortex CP1018   >95%

Cortex CP1018P   >40%

Biogenesis 5720-5104   Bovine pituitary   MW 30k   ≤1.0% bTSH, <0.1% bGH/bFSH; purified; 0.05 M NH$_4$HCO$_3$, pH 8.0, essentially salt free; lyophilized

Biogenesis 5720-5304   Bovine pituitary   Lyophilized

Fitzgerald 30-AL27   Bovine pituitary gland   Immunization grade

USBio L7500-30   Bovine pituitary glands   ≥98% (SDS-PAGE); 0.65 IU/mg; ≥ 10 mg/vial; NIH Reference Standard; lyophilized | Suitable for antigenic applications in immunological protocols

USBio L7500-31   Bovine pituitary glands   ≥80% (SDS-PAGE); lyophilized | Suitable for antigenic applications in immunological protocols; SA ~1 IU/mg

Biogenesis 5720-6104   Equine pituitary   Lyophilized

Sigma L 9773   Equine pituitary   ~15 U/vial based on NIH standard LH-S1 using the rat Leydig cell assay; also contains sodium phosphate buffer salts; 1 NIH LH-S1 U=~2000 IU | Not assayed by Sigma; 2 chain glycoprotein hormone, the α-chain no active, biological specificity is attributed to the β-chain; induces ovulation, spermatogenesis & synthesis of sex steroids

Biogenesis 5720-2159   Human pituitary   Essentially salt free; 32.5 mIU/vial hTSH, 370 mIU/vial hFSH; tested negative for HCV, HIV-1 and HIV-2 antibodies and HBsAg; crude; from 0.05 M NH$_4$HCO$_3$, pH 8.0; lyophilized

Biogenesis 5720-2204   Human pituitary   Lyophilized

Biogenesis 5720-2209   Human pituitary   MW 28k   <1% hTSH, <0.3% hFSH, <0.2% hGH, <0.05% hPRL; tested negative for HBV, HCV, HIV-1, HIV-2, HTLV 1 and 2, HIV-1 Ag; purified; 0.05 M NH$_4$HCO$_3$, pH 8.0; lyophilized | Closset et al, *Eur J Biochem*, 57:325, 1975

Biogenesis 5720-2304   Human pituitary   SA: 10,500 IU/mg; <0.1% hTSH, <0.06% hFSH, <0.01% hGH/hPRL/hCG; purified; 50 mM ammonium bicarbonate; lyophilized

Biogenesis 5720-2404   Human pituitary   Lyophilized

Biogenesis 5720-4304   Human pituitary   Lyophilized

Calbiochem 869003   Human pituitary   MW 28.5k   Lyophilized solid; potency: ≥5000 IU/mg (WHO IRP 68/40); soluble in water; hFSH, hTSH: ≤0.5%; hCG: ≤0.05%; shown by certified tests to be negative for HBsAg & for antibodies to HIV & HCV; may be carcinogenic/teratogenic | Two-chain glycoprotein hormone required for ovulation, spermatogenesis & the biosynthesis of sex steroids; biological specificity has been attributed to β-chain; *Merck Index*, 12: 5499

Sigma L 5259   Human pituitary   ≥5000 IU/mg; sold on basis of weight of RIA-active LH | Not assayed by Sigma; 2 chain glycoprotein hormone, the α-chain no active, biological specificity is attributed to the β-chain; induces ovulation, spermatogenesis & synthesis of sex steroids

Sigma L 8650   Human pituitary   Lyophilized from 0.1 mL of 0.05 M sodium phosphate buffer, pH 7.4; <1% α-subunit | Subunit responsible for the biological specificity of LH

Biodesign A81755M   Human pituitary glands   98% (affinity purified)

USBio L7500-28   Human pituitary glands   ≥98%; affinity purified; hLH Alpha: ≤2.0%; 1 mg/mL | Suitable for antigenic applications in immunological protocols; SA: 11,000 IU/mg (1st IRP 68/40, WHO)

Biogenesis 5720-7004   Ovine pituitary   ≤1% oTSH/oFSH; purified; liquid

Biogenesis 5720-7304   Ovine pituitary   Lyophilized

Sigma L 5269   Ovine pituitary   Lyophilized from 5 mM sodium phosphate buffer; 25 U/vial based on HIH standard LH-525 using the ascorbic acid depletion assay | Not assayed by Sigma; 2 chain glycoprotein hormone, the α-chain no active, biological specificity is attributed to the β-chain; induces ovulation, spermatogenesis & synthesis of sex steroids

Biogenesis 5720-8104   Porcine pituitary   MW 30k   <1.0% pTHS, 0.1% pPRL, <0.1% pGH/pFSH; purified; 0.05 M NH$_4$HCO$_3$, pH 8.0; lyophilized

Biogenesis 5720-8204   Porcine pituitary   Lyophilized

Biogenesis 5720-9104   Rat pituitary   <0.3% rTSH, <0.1% rPRL/rGH/rFSH; purified; 0.05 M NH$_4$HCO$_3$, pH 8.0; lyophilized

Luteinizing Hormone, Intact

USBio L7500-24   Human   ≥97% (SDS-PAGE), hTSH <2.0%, hGH <0.1%, hFSH <0.3%, hPrl <0.1%, hCG <0.1%; 12,000 IU/mg; lyophilized in 50 mM ammonium bicarbonate buffer | Suitable for antigenic applications in immunological protocols

USBio L7500-25   Human   ≥60% (SDS-PAGE); hTSH <1.0%, hGH <0.1%, hFSH <0.3%, hPrl <0.1%, hCG <0.1%; 5, 500 IU/mg; lyophilized in 50 mM ammonium bicarbonate buffer | Suitable for antigenic applications in immunological protocols

Fitzgerald 30-AL15   Human pituitary gland   Affinity purity

Fitzgerald 30-AL20   Human pituitary gland   High purity

Fitzgerald 30-AL25   Human pituitary gland   Standard purity

USBio L7500-23   Human pituitary glands   ≥98%; hFSH <0.05%; hPrl <0.005%, hTSH <0.0001%; 14,400 IU/mg; lyophilized | Suitable for antigenic applications in immunological protocols.

Luteinizing Hormone, α-

Biogenesis 5720-5254   Bovine pituitary   Lyophilized

USBio L7500-11   Human   <2.0% hLH-β; lyophilized | Suitable for antigenic applications in immunological protocols

Biogenesis 5720-3204   Human pituitary   MW 14.5k   0.02% hLH; tested negative for HBsAg, HCV, HIV-1 and 2; purified; 0.05 M NH$_4$HCO$_3$, pH 8.0. Essentially salt free; lyophilized | Closset et al, *Eur J Biochem*, 57:325, 1975

Fitzgerald 30-AL30	Human pituitary gland	High purity
Biogenesis 5720-7204	Ovine pituitary	Lyophilized
Biogenesis 5720-8154	Porcine pituitary	Lyophilized
Fitzgerald 30-AL35	Human pituitary gland	High purity

Luteotropic Hormone

Synonyms: Prolactin; Lactogenic Hormone

Sigma L 7009   Human pituitary gland   ~30 IU/mg; SA by RIA per an international reference preparation of LTH for immunoassay | Induces lactation; inhibits secretion of gonadotropins; release is inhibited by dopamine; bioassay not run by Sigma

ICN 155277   Human pituitary glands   30 IU/mg

Sigma L 6520   Sheep pituitary gland   20-50 IU/mg | Induces lactation; inhibits secretion of gonadotropins; release is inhibited by dopamine; bioassay not run by Sigma

ICN 155278   Sheep pituitary glands   20-50 IU/mg

Lymphotactin

Biodesign A52020H	E. coli	MW 10.2k	Purified
Chemicon GF039	Human	≥95%	
BioSource International PHC1134	Human recombinant		

ICN 193969   Human recombinant, expressed in *E. coli* Lyophilized; chemotaxis typically >20 ng/mL

PeproTech 300-20   Human recombinant, expressed in *E. coli* MW 10.0k   >98% (SDS-PAGE) & HPLC; <0.1 ng endotoxin per mg (1EU/mg); lyophilized | Part of the C family of chemokines; exhibits chemotactic activity toward lymphocytes, but not towards monocytes or neutrophils; 92 AA; SA determined by its ability to chemoattract human peripheral blood lymphocytes

Lyn (p56)

USBio L8050-06   Bovine spleen membranes   Essentially free of other proteinkinase contamination; frozen solution in 75 mL of 25 *mM* HEPES, pH 7.0, 0.1% NP-40 with 10% glycerol; 1unit/mL | Partially purified by DEAE-Sepharose, hydroxyapatite, phenyl Sepharose columns, & Sephacryl S-200 gel filtration

Lysenin

Synonyms: Sphingomyelin-Specific Binding Protein

Peptides International PLN-4802-v   Coelomic fluid of earthworm (*Eisenia foetida*); natural product   MW 33k   Salt free lyophilized powder

Maceration Stimulating Factor

Sigma M 3793   *Aspergillus japonicus*   Lyophilized containing ~20% protein (Lowry); balance acetate buffer salts; sterile-filtered; activity: 1 μg/mL protein causes a 2-6 fold activation in the maceration activity of a 1 U/mL solution of polygalacturonase (Sigma P 5079) on potato slices at pH 4.5, 40°C in a 3 hr reaction; contaminants: <1.0 U/mg polygalacturonase; <0.25 U/mg protein pectin lyase | Ishii, S, *Phytopathology*, 67: 994, 1977

Macroglobulin, αI

Biogenesis 6220-1004   Human urine from patients with chronic renal tubular proteinuria   MW ~30k (SDS-PAGE)   RBP, b-2-microglobulin and albumin were undetectable by RID; yellow-brown color; tested negative for HIV I and II antibodies, HBsAg and HCV antibodies; purified; 0.02 *M* ammonium bicarbonate. May contain traces of buffer salts; lyophilized | Under certain conditions a dimer may appear

Macroglobulin, αII

| Biodesign A50114H | Human | Purified |

Fluka 63013   Human plasma   MW 725k   ≥90% (GE); ≥20% protein content; lyophilized from 0.025 *M* Tris-HCl, pH 8.0 containing 0.1 *M* NaCl | Ishibashi, H, et al, *Meth Enzymol*, 163: 485, 1988; Roche, PA et al, *Biochemistry*, 28: 7629, 1989

Biogenesis 5850-2004   Human serum   MW 725k   Protein E [280 nm/1 cm, 0.1%] = 0.81; tested negative for HBsAg and HCV antibodies, and HIV-1 and 2 antibodies; purified; 0.02 *M* TRIS, 0.15 *M* NaCl, 0.1 *M* sucrose, may contain traces of buffer salts; lyophilized | Barret et al, *Meth Enzymol*, 80:737, 1981

| Biodesign A17131H | Human | 95% |
| Biodesign A1B001H | Human | 78% |

Macrophage Chemotactic & Activating Factor

Harlan BT-3018   Human recombinant, expressed in *E. coli* Lyophilized; 0.01 mg

Harlan BT-3019   Human recombinant, expressed in *E. coli* Lyophilized; 0.05 mg

Macrophage Derived Chemokine

Synonyms: STDP-1

Biodesign A52036H	E. coli	Purified; 67 AA
Biodesign A52336H	E. coli	Purified; 69 AA
BioSource International PHC1204	Human recombinant	

PeproTech 300-36   Human recombinant, expressed in *E. coli* MW 8.0k   >98% (SDS-PAGE) & HPLC; <0.1 ng endotoxin per mg (1EU/mg); lyophilized | 67AA; unique member of the CC chemokine family; highly expressed in macrophages & in monocyte-derived dendritic cells; but not in monocytes, naturalkiller cells, or several cell lines of epithelial, endothelial, or fibroblast origin; 67 AA; SA determined by its ability to chemoattract human T cells

PeproTech 300-36A   Human recombinant, expressed in *E. coli* MW 8.1k   >98% (SDS-PAGE) & HPLC; <0.1 ng endotoxin per mg (1EU/mg); lyophilized | unique member of the CC chemokine family; highly expressed in macrophages & monocyte-derived dendritic cells; but not in monocytes, naturalkiller cells, or several cell lines of epithelial, endothelial, or fibroblast origin; 69 AA (Gly & Pro at the N-terminus); SA determined by its ability to chemoattract human T

| BioSource International PMC1575 | Mouse recombinant | |

PeproTech 250-23   Murine recombinant, expressed in *E. coli* MW 7.8k   >98% (SDS-PAGE) & HPLC; <0.1 ng endotoxin per mg (1EU/mg); lyophilized | Unique member of the CC chemokine family; highly expressed in macrophages & monocyte-derived dendritic cells; not expressed in monocytes, naturalkiller cells or several cell lines of epithelial, endothelial or fibroblast origin; 68 AA; SA determined by its ability to chemoattract total murine T cell population

Macrophage Inflammatory Peptide IIIα

| BioSource International PHC1234 | Human recombinant |
| BioSource International PHC1244 | Human recombinant |

Macrophage Inflammatory Peptide IV

| BioSource International PHC1254 | Human recombinant |

Macrophage Inflammatory Peptide Iα

BioSource International PHC1104	Human recombinant
BioSource International PMC1024	Mouse recombinant
BioSource International PRC1024	Rat recombinant
BioSource International PHC1034	Human recombinant

Macrophage Inflammatory Peptide V

| BioSource International PHC1544 | Human recombinant |

Macrophage Inflammatory Protein 3β Chemokine

Synonyms: ELC

Sigma M 3552   Human recombinant, expressed in *E. coli* Lyophilized from 30% acetonitrile/0.1% TFA containing 1.25 mg BSA; endotoxin tested; cell culture tested; see Sigma M 8668

Macrophage Inflammatory Protein II

Biodesign A52015M *E. coli* MW 7.8k Purified | Species specificity: mouse

Sigma M 8668 Mouse recombinant, expressed in *E. coli* MW ~8k ≥97% (SDS-PAGE); 0.2 μm filtered & lyophilized in 30% acetonitrile, 0.1% trifluoroacetic acid containing 500 μg BSA; activity measured by its ability to induce myeloperoxidase release from human neutrophils; endotoxin tested | Member of the C-X-C or α chemokine class; contains the ELR domain immediately preceding the first cysteine residue near the amino terminus; act primarily on neutrophils as chemoattractants & activators including neutrophil degradation with release of myeloperoxidase & other enzymes; originally identified as a heparin-binding protein secreted from a murine macrophage cell line in response to endotoxin stimulation; polypeptide of 73 AA; precursor form consists of 100 AA; to generate the mature MIP-2, the precursor cleaves it amino-terminal 27 AA; show 60% AA homology to human GROβ & GROγ

IBT MI-260-3, MI-260-4 Mouse recombinant, expressed in yeast >95 %; 1.0 mg prot/mL in 0.1 *M* ammonium acetate

Chemicon GF089 Murine ≥95%

ICN 195764 Murine recombinant, expressed in *E. coli* ≥97%; lyophilized; ED_{50} = 0.15-0.30 ng/mL

PeproTech 250-15 Murine recombinant, expressed in *E. coli* MW 7.8k >98% (SDS-PAGE) & HPLC; <0.1 ng endotoxin per mg (1EU/mg); lyophilized | Promotes neutrophil chemotaxis & degranulation; 73 AA; SA determined by its ability to chemoattract total human neutrophils

PeproTech 350-03 Viral recombinant, expressed in *E. coli* MW 7.9k >98%; 70 AA; lyophilized with no additives; activity determined by the inhibitory effect on monocyte migration response to human MIP-1α; using a concentration range of 1.0 μg-10 μg/mL of viral MIP-2 will inhibit 25 ng/mL of human MIP-1α

Macrophage Inflammatory Protein II, Viral

Biodesign A52353H *E. coli* Purified

Macrophage Inflammatory Protein III

Synonyms: MPIF-1

Biodesign A52029H *E. coli* MW 11.3k Purified

PeproTech 300-29 Human recombinant, expressed in *E. coli* MW 11.3k >98% (SDS-PAGE) & HPLC; <0.1 ng endotoxin per mg (1EU/mg); lyophilized | Member of the CC chemokine family of cytokines; chemotactic activity on resting T lymphocytes & monocytes; 99 AA; SA determined by its ability to chemoattract human T cell population

Macrophage Inflammatory Protein IIIα

Synonyms: LARC; Exodus; Exodus; LARC; LARC; Exodus; Exodus III/ELC; Exodus III/ELC

Biodesign A52329H *E. coli* MW 8k Purified

Chemicon GF069 Human ≥95%

PeproTech 300-29A Human recombinant, expressed in *E. coli* MW 8.0k >98% (SDS-PAGE) & HPLC; <0.1 ng endotoxin per mg (1EU/mg); lyophilized | Member of the b-chemokine (CC) family of cytokines; 70 AA; SA determined by its ability to chemoattract human T cells

Sigma M 3677 Human recombinant, expressed in *E. coli* Lyophilized from 30% acetonitrile/0.1% TFA containing 1.25 mg BSA; endotoxin tested; cell culture tested;

Biodesign A52923H *E. coli* MW 8k Purified

Chemicon GF070 Human ≥95%

PeproTech 300-29B Human recombinant, expressed in *E. coli* MW 8.8k >98% (SDS-PAGE) & HPLC; <0.1 ng endotoxin per mg (1EU/mg); lyophilized | Member of the b-chemokine (CC) family of cytokines; important role in the inflammatory response of human T cells & tissue macrophages; 77 AA; SA determined by its ability to chemoattract human T cells

Macrophage Inflammatory Protein IIα

IBT MI-310-3, MI-310-4 Human recombinant, expressed in yeast >95 %; 1.0 mg prot/mL in 0.1 *M* ammonium acetate

IBT MI-320-3, MI-320-4 Human recombinant, expressed in yeast >95 %; 1.0 mg prot/mL in 0.1 *M* ammonium acetate | Biological activity determined by its use as a cofactor for CSF-dependent myelopoiesis

Macrophage Inflammatory Protein IV

Synonyms: Pulmonary & Activation Regulated Chemokine

Biodesign A52034H *E. coli* MW 7.8k Purified

Chemicon GF071 Human ≥95%

PeproTech 300-34 Human recombinant, expressed in *E. coli* MW 7.8k >98% (SDS-PAGE) & HPLC; <0.1 ng endotoxin per mg (1EU/mg); lyophilized | Chemotactic for T-lymphocytes but not for monocytes or granulocytes; 69 AA; SA determined by its ability to chemoattract human T lymphocytes

Macrophage Inflammatory Protein Iα

Synonyms: MRP2; CCF18

Biodesign A52009M *E. coli* MW 7.8k Purified | Species specificity: mouse

Biodesign A52015H *E. coli* MW 8k Purified | Species specificity: rat

Biodesign A52308H *E. coli* MW 8k Purified

Chemicon GF010 Human ≥95%

Calbiochem 441523 Human recombinant, expressed in *E. coli* MW 7.5k ≥97% (SDS-PAGE); lyophilized from filter-sterilized 30% acetonitrile, 0.1% TFA containing 50 μg BSA/μg MIP-1α; biological activity: ED_{50}=2.0-5.0 ng/mL as measured by its ability to inhibit mouse hematopoietic stem cell proliferation in an *in vitro* colony assay that detects primitive cells; endotoxin: ≤100 pg/μg MIP-1α | Low MW chemokine involved in inflammation, hematopoiesis & immunoregulation; exhibits chemoattractant activity on monocytes & chemoattractant & proadhesive effects on lymphocytes; inhibits primitive hematopoietic stem cells proliferation; Hunter, MG et al, *Blood*, 86: 4400, 1995; Ritter, LM et al, *Mol Cell Biol*, 15: 3110, 1995; Taub, DD et al, *Science*, 260: 355, 1993; Nirsimloo, N & Gordon, MY, *Leuk Res*, 19: 319, 1995; Oppenheim, J et al, *Ann Rev Immunol*, 9: 617, 1991; Wolpe, SD & Cerami, A, *FASEB J*, 3: 2565, 1989

Fitzgerald 30-AM50 Human recombinant, expressed in *E. coli*

ICN 160261 Human recombinant, expressed in *E. coli* ≥99%; lyophilized; ED_{50} = 50.0 ng/mL, 1 U gives maximal chemotactic activity on human blood monocytes | Chemoattractant & pro-inflammatory activities targeted to monocytes & tissue macrophages

PeproTech 300-08 Human recombinant, expressed in *E. coli* MW 7.8k >98% (SDS-PAGE) & HPLC; <0.1 ng endotoxin per mg (1EU/mg); lyophilized | Important role in the inflammatory response of blood monocytes & tissue macrophages; 70 AA; SA determined by its ability to chemoattract human monocytes

Sigma M 6292 Human recombinant, expressed in *E. coli* ≥97% (SDS-PAGE); 0.2 μm filtered & lyophilized from 30% acetonitrile, 0.1% trifluoroacetic acid containing 500 μg BSA; activity tested in culture using an *in vitro* colony assay; endotoxin tested | Member of the β chemokine subfamily characterized by a CC configuration at the first 2 cysteines; originally copurified from endotoxin-stimulated mouse macrophages; further analysis showed that MIP-1 is composed of 2 distinct, but highly related proteins, MIP-1α & MIP-1β; although other cytokines, such as IL-1α, IL-1β & TNF have endogenous pyrogen activity & the pyrogenic effects of these cytokines can be inhibited by cyclooxygenase blockers, the pyrogenicity of MIP-1α & MIP-1β is unaffected by these agents; Graham, GJ et al, *Nature*, 344: 442, 1990General References for MIPs: Wolpe, S et al, J Exp Med, 167: 570, 1988; Schall, T, *Cytokine*, 3: 165, 1991; Miller, M et al, *Critical Review in Immunology*, 12 (1,2): 17, 1992

IBT MI-290-3, MI-290-4 Human recombinant, expressed in yeast >95 %; 1.0 mg prot/mL in 0.1 *M* ammonium acetate

Calbiochem 441526 Mouse recombinant, expressed in *E. coli*
MW 7.8k ≥97% (SDS-PAGE); lyophilized from filter-sterilized
30% acetonitrile, 0.1% TFA containing 50 μg BSA/μg MIP-1α;
biological activity: ED$_{50}$=4.0-6.0 ng/mL as measured by its ability
to inhibit mouse hematopoietic stem cell proliferation in an *in vitro*
colony assay that detects primitive cells; endotoxin: ≤100 pg/μg
MIP-1α | Low MW chemokine involved in inflammation,
hematopoiesis & immunoregulation; exhibits chemoattractant
activity on monocytes & chemoattractant & proadhesive effects on
lymphocytes; inhibits primitive hematopoietic stem cells
proliferation; Hunter, MG et al, *Blood*, 86: 4400, 1995; Ritter, LM
et al, *Mol Cell Biol*, 15: 3110, 1995; Taub, DD et al, *Science*, 260:
355, 1993; Nirsimloo, N & Gordon, MY, *Leuk Res*, 19: 319, 1995;
Oppenheim, J et al, *Ann Rev Immunol*, 9: 617, 1991; Wolpe, SD &
Cerami, A, *FASEB J*, 3: 2565, 1989

Sigma M 6167 Mouse recombinant, expressed in *E. coli*
≥97% (SDS-PAGE); 0.2 μm filtered & lyophilized from 30%
acetonitrile, 0.1% trifluoroacetic acid containing 500 μg BSA;
activity tested in culture using an *in vitro* colony assay; endotoxin
tested; see Sigma M 6292 | Graham, GJ et al, *Nature*, 344: 442,
1990

IBT MI-270-3, MI-270-4 Mouse recombinant, expressed in
yeast >95 %; 1.0 mg prot/mL in 0.1 *M* ammonium acetate

ICN 158413 Murine recombinant, expressed in *E. coli* ≥97%;
lyophilized; ED$_{50}$ = 2.0-5.0 ng/mL, 1 U gives dose-dependent
inhibition of murine hematopoietic stem cell proliferation in an *in
vitro* colony assay (CFU-A)

PeproTech 250-09 Murine recombinant, expressed in *E. coli*
MW 7.8k >98% (SDS-PAGE) & HPLC; <0.1 ng endotoxin per mg
(1EU/mg); lyophilized with no additives | Member of the CC
chemokine family; plays an important role in the inflammatory
response; 69 AA; SA determined by its ability to chemoattract
mouse Balb/c peripheral blood MNCs & mouse Balb/c splenocytes

Chemicon GF048 Rat ≥95%

ICN 195805 Rat recombinant, expressed in *E. coli* >95%;
lyophilized; significant chemotaxis at 200 ng/mL

PeproTech 400-15 Rat recombinant, expressed in *E. coli*
MW 7.8k >98% (SDS-PAGE) & HPLC; <0.1 ng endotoxin per mg
(1EU/mg); lyophilized | Important role in the inflammatory
response of blood monocytes & tissue macrophages; 69 AA; SA
determined by its ability to chemoattract rat peritoneal
macrophages

Biodesign A52309H *E. coli* MW 8k Purified

Chemicon GF011 Human ≥95%

Fitzgerald 30-AM51 Human recombinant, expressed in *E. coli*

ICN 159351 Human recombinant, expressed in *E. coli* ≥99%;
lyophilized; ED$_{50}$ = 50.0 ng/mL, 1 U gives maximal chemotactic
activity on human blood monocytes | Chemoattractant & pro-
inflammatory activities targeted to monocytes; may reverse MIP-1α
inhibitory effects on hematopoietic stem cell proliferation

PeproTech 300-09 Human recombinant, expressed in *E. coli*
MW 7.6k >98% (SDS-PAGE) & HPLC; <0.1 ng endotoxin per mg
(1EU/mg); lyophilized | Important role in the inflammatory
response of blood monocytes & tissue macrophages; 69 AA; SA
determined by its ability to chemoattract human monocytes

Sigma M 6417 Human recombinant, expressed in *Sf 21* cells
≥97% (SDS-PAGE); 0.2 μm filtered & lyophilized from 30%
acetonitrile, 0.1% trifluoroacetic acid containing 500 μg BSA;
activity tested in culture using an *in vitro* colony assay; endotoxin
tested; see Sigma M 6292 | Graham, GJ et al, *Nature*, 344: 442,
1990

Calbiochem 441529 Human recombinant, expressed in
Spodoptera frugiperda MW 7.8k ≥97% (SDS-PAGE);
lyophilized from filter-sterilized 30% acetonitrile, 0.1% TFA
containing 50 μg BSA/μg MIP-1β; biological activity: ED$_{50}$=40-60
ng/mL as measured by its ability to inhibit mouse hematopoietic
stem cell proliferation in an *in vitro* colony assay that detects
primitive cells; endotoxin: ≤100 pg/μg MIP-1β | Low MW chemokine
involved in inflammation, hematopoiesis & immunoregulation;
exhibits chemoattractant activity on monocytes & chemoattractant
& proadhesive effects on lymphocytes; Napolitano, M et al, *J Biol
Chem*, 266: 17531, 1991; Schall, TJ, *Cytokine*, 3: 165, 1991;
Taub, DD et al, *Science*, 260: 355, 1993; Lipes, MA et al, *PNAS*,
85: 9704, 1988; Oppenheim, J et al, *Ann Rev Immunol*, 9: 617,
1991

IBT MI-300-3, MI-300-4 Human recombinant, expressed in
yeast >95 %; 1.0 mg prot/mL in 0.1 *M* ammonium acetate

Calbiochem 441532 Mouse recombinant, expressed in *E. coli*
MW 7.8k >97% (SDS-PAGE); lyophilized from filter-sterilized
30% acetonitrile, 0.1% TFA containing 50 μg BSA/μg MIP-1β;
biological activity: ED$_{50}$=80-100 ng/mL as measured by its ability
to inhibit mouse hematopoietic stem cell proliferation in an *in vitro*
colony assay that detects primitive cells; endotoxin: ≤100 pg/μg
MIP-1β | Low MW chemokine involved in inflammation,
hematopoiesis & immunoregulation; exhibits chemoattractant
activity on monocytes & chemoattractant & proadhesive effects on
lymphocytes; inhibits primitive hematopoietic stem cells
proliferation; Hunter, MG et al, *Blood*, 86: 4400, 1995; Ritter, LM
et al, *Mol Cell Biol*, 15: 3110, 1995; Taub, DD et al, *Science*, 260:
355, 1993; Nirsimloo, N & Gordon, MY, *Leuk Res*, 19: 319, 1995;
Oppenheim, J et al, *Ann Rev Immunol*, 9: 617, 1991; Wolpe, SD &
Cerami, A, *FASEB J*, 3: 2565, 1989

Sigma M 6542 Mouse recombinant, expressed in *E. coli*
≥97% (SDS-PAGE); 0.2 μm filtered & lyophilized from 30%
acetonitrile, 0.1% trifluoroacetic acid containing 500 μg BSA;
activity tested in culture using an *in vitro* colony assay; endotoxin
tested; see Sigma M 6292 | Graham, GJ et al, *Nature*, 344: 442,
1990

ICN 160256 Murine recombinant, expressed in insect cell line
Sf21 ≥97%; lyophilized; ED$_{50}$ = 40.0-60.0 ng/mL, 1 U gives
dose-dependent inhibition of murine hematopoietic stem cell
proliferation in an *in vitro* colony assay (CFU-A)

Biodesign A52512M *E. coli* MW 11.6k Purified | Species
specificity: mouse

Chemicon GF084 Murine ≥95%

PeproTech 250-12 Murine recombinant, expressed in *E. coli*
MW 11.6k >98% (SDS-PAGE) & HPLC; <0.1 ng endotoxin per
mg (1EU/mg); lyophilized | Monokine with inflammatory,
pyrogenic & chemokinetic CC properties; when bound to a high-
affinity receptor, it activates CC calcium release in neutrophils; 101
AA; SA determined by its ability to chemoattract human neutrophils
and human T cells

Macrophage Inflammatory Protein V
Synonyms: Lkn-1

Biodesign A52043H *E. coli* MW 10.1k Purified

Chemicon GF083 Human ≥95%

PeproTech 300-43 Human recombinant, expressed in *E. coli*
MW 10.1k >98% (SDS-PAGE) & HPLC; <0.1 ng endotoxin per
mg (1EU/mg); lyophilized | Chemotactic for T-lymphocytes but
not for monocytes or granulocytes; 92 AA; SA determined by its
ability to chemoattract human T lymphocytes

Macrophage Migration Inhibitory Factor

ICN 195792 Human recombinant, expressed in *E. coli* ≥97%;
lyophilized | Useful as a ELISA calibrator

Macrophage/Monocyte Chemoattractant Protein I
Synonyms: Macrophage Chemotatic & Activating Factor

PeproTech 400-12 Rat recombinant, expressed in *E. coli*
MW 14.1k >98% (SDS-PAGE) & HPLC; <0.1 ng endotoxin per
mg (1EU/mg); lyophilized | Important role in the inflammatory
response of blood monocytes & tissue macrophages; 125 AA; SA
determined by its ability to chemoattract human monocytes

Macrophage/Monocyte Chemotactic & Activating Factor
Synonyms: Macrophage/Monocyte Chemoattractant Protein I

Biodesign A52304H *E. coli* MW 8.5k Purified

USBio M1203-05 E. coli ≥99% (SDS-PAGE); no
contaminants detected; single band by SDS-PAGE, IEP, &/or RID;
~0.1 mg/mL; lyophilized | Suitable for antigenic applications in
immunological protocols

Chemicon GF012 Human ≥95%

ICN 195008 Human recombinant, expressed in *E. coli* MW
8.5k Lyophilized; 20 ng/mL activity

Sigma M 5662   Human recombinant, expressed in *E. coli*
Maximal chemotactic activity is achieved at 10-100 ng/mL medium | Matsushima, *J Exp Med*, 169, 1485, 1989

Macrophage/Monocyte Chemotactic & Activating Factor I

Biodesign A52012H   *E. coli*   MW 14.1k   Purified | Species specificity: rat

Fitzgerald 30-AM52   Human recombinant, expressed in *E. coli*

Macrophage/Monocyte Chemotactic & Activating Factor I/JE

Biodesign A52010M   *E. coli*   MW 13.8k   Purified | Species specificity: mouse

Macrophage/Monocyte Chemotactic Protein I

Synonyms: Macrophage Chemotatic & Activating Factor; ; Macrophage Chemotatic & Activating Factor

BioSource International PHC1014   Human recombinant

PeproTech 250-10   Murine recombinant, expressed in *E. coli* MW 13.8k   >98% (SDS-PAGE) & HPLC; <0.1 ng endotoxin per mg (1EU/mg); lyophilized | Important role in the inflammatory response of blood monocytes & tissue macrophages; 125 AA; SA determined by its ability to chemoattract Balb/c mouse spleen MNCs & total human neutrophils

BioSource International PRC1014   Rat recombinant

Macrophage/Monocyte Chemotactic Protein II

Biodesign A52315H   *E. coli*   MW 8k   Purified

BioSource International PHC1114   Human recombinant

PeproTech 300-15   Human recombinant, expressed in *E. coli* MW 8.9k   >98% (SDS-PAGE) & HPLC; <0.1 ng endotoxin per mg (1EU/mg); lyophilized | 76 AA; important role in the inflammatory response of blood monocytes & tissue macrophages; SA determined by its ability to chemoattract human peripheral blood monocytes

Macrophage/Monocyte Chemotactic Protein III

Biodesign A52008M   *E. coli*   MW 8.5k   Purified | Species specificity: mouse

BioSource International PHC1574   Human recombinant

PeproTech 300-17   Human recombinant, expressed in *E. coli* MW 9.0k   >98% (SDS-PAGE) & HPLC; <0.1 ng endotoxin per mg (1EU/mg); lyophilized | 76 AA; important role in the inflammatory response of blood monocytes, tissue macrophages & eosinophils; SA determined by its ability to chemoattract human peripheral blood monocytes

BioSource International PMC1574   Mouse recombinant

PeproTech 250-08   Murine recombinant, expressed in *E. coli* MW 8.5k   >98% (SDS-PAGE) & HPLC; <0.1 ng endotoxin per mg (1EU/mg); lyophilized with no additives | Plays an important role in the inflammatory response of blood monocytes & T-cells; 74 AA; SA determined by its ability to chemoattract Balb/c mouse spleen MNCs

Macrophage/Monocyte Chemotactic Protein IV

Biodesign A52024H   *E. coli*   MW 8.5k   Purified

BioSource International PHC1154   Human recombinant

PeproTech 300-24   Human recombinant, expressed in *E. coli* MW 8.6k   >98% (SDS-PAGE) & HPLC; <0.1 ng endotoxin per mg (1EU/mg); lyophilized | Member of the β-chemokine family (CC) of cytokines; powerful eosinophil chemoattractant (like Eotaxin); elicits chemoattraction of monocytes & T lymphocytes (like MCP-3); 75 AA; SA determined by its ability to chemoattract human eosinophils

Macrophage/Monocyte Chemotactic Protein V

Biodesign A52004H   *E. coli*   MW 9.2k   Purified | Species specificity: mouse

BioSource International PMC1164   Mouse recombinant

PeproTech 250-04   Murine recombinant, expressed in *E. coli* MW 9.2k   >98% (SDS-PAGE) & HPLC; <0.1 ng endotoxin per mg (1EU/mg); lyophilized with no additives | Member of the b-chemokine family (CC) of cytokines; strongly chemotactic for human monocytes & murine peritoneal macrophages; not active on mouse neutrophils; 82 AA; SA determined by its ability to chemoattract human peripheral blood monocytes

Mammary Carcinoma

Fitzgerald 30-AM42   Human fluid   Standard purity

Mannan Binding Protein

Biogenesis 5937-1100   Human serum   Purified; PBS buffer, pH 7.2, 0.1% NaN$_3$; liquid

Matrix Metalloproteinase II, Proenzyme

Synonyms: Collagen Substrate (Type IV, V, VII, X); Elastin Substrate; Gelatin Substrate (Type I)

Oncogene PF037   >95% (SDS-PAGE); in 5mM TRIS, pH 7.5, 0.1mM CaCl$_2$, 0.005% Brij-35, 20% glycerol; 1 mM 1,10-phenanthroline added to prevent auto-activation & -degradation; liquid | A simple activation step is included in protocol; species reactivity: human; for substrate cleavage assay, Western blot & zymography

Matrix Metalloproteinase III, Proenzyme

Oncogene PF063   >95% (SDS-PAGE); lyophilized in 50 mM HEPES, pH 7.3 containing 0.18 M NaCl; reconstitute in dH$_2$O | Used as a positive control or standard, for zymographic analysis or substrate assay

Matrix Metalloproteinase IX, Proenzyme

Synonyms: Collagen Substrate (Type IV, V, VII, X); Elastin Substrate; Gelatin Substrate (Type I, V)

Oncogene PF038   >95% (SDS-PAGE); in 5mM TRIS, pH 7.5, 0.1mM CaCl$_2$, 0.005% Brij-35, 20% glycerol; 1 mM 1,10-phenanthroline added to prevent auto-activation & -degradation; liquid | A simple activation step is included in protocol; species reactivity: human; for substrate cleavage assay, Western blot & zymography

Matrix Metalloproteinase VII, Proenzyme

Synonyms: Promatrilysin

Oncogene 538540   Human recombinant, expressed in *E. coli* MW 28k   Liquid in 150 mM NaCl, 25 mM Tris-HCl, 5 mM CaCl$_2$, 0.02% NaN$_3$, 0.01% BRIJ®35, pH 7.5; SA: ≥1400 U/mg protein; 1 U is the amount of enzyme that digests 1.0 µg AZOCOLL® substrate/min at 37°C, pH 7.5, in the presence of 500 µM p-aminophenylmercuric acetate | Latent form believed to be a key factor in the regulation of metastasis; activatable by organomercurials; Kihira, Y et al, *Urol Oncol*, 2: 20, 1996; Woessner, JF, *Meth Enzymol*, 248: 485, 1995; Crabbe, T et al, *Biochemistry*, 31: 8500, 1992

Melanocyte Stimulating Hormone, Iγ

Biogenesis 6045-2502   Human synthetic melanocytes Purified; lyophilized

Melanocyte Stimulating Hormone, β-

Biogenesis 6045-1504   Human synthetic melanocytes Purified; lyophilized

Melanostatin

Synonyms: Melanocyte-Stimulating Hormone-Release Inhibiting Factor; Macrophage Migration Inhibitory Factor I

ICN 153167 MW 284.4 $C_{13}H_{24}N_4O_3$ Inhibits the release of melanocyte-stimulating hormone from the pituitary gland; Bjoekman, S & Sievertsson, Naunyn-Schmiedeberg's *Arch Pharmac*, 298:79, 1977; Chiu, S & RK Mishra, *Eur J Pharmac*, 53:119, 1979

Meromyosin, Heavy

Sigma M 8141 Chicken muscle Solution in 50% glycerol containing 0.5 M KCl & 0.025 M potassium phosphate, pH 6.2; ATPase activity: 1-3 U/mg protein (Biuret); unit definition: 1 U liberates 1.0 µmole inorganic phosphorus from ATP/min at pH 9.0 at 25°C in the presence of calcium | Tryptic fragments of myosin, prepared by the method of Lowey, S & Cohen, C, *J Mol Biol*, 4: 293, 1962; light meromyosin forms filaments but does not bind to actin; heavy meromyosin binds to actin but does not form filaments; only heavy meromyosin exhibits ATPase activity

Sigma M 9014 Rabbit muscle Solution in 50% glycerol containing 0.5 M KCl & 0.025 M potassium phosphate, pH 6.2; ATPase activity: 1-3 U/mg protein (Biuret); unit definition: 1 U liberates 1.0 µmole inorganic phosphorus from ATP/min at pH 9.0 at 25°C in the presence of calcium

Merosin

Chemicon CC085 Human placenta Purified | Extracellular matrix protein

Mesoglycan

Sigma H 0519 Bovine intestine Sodium salt; anti-clotting activity: <50 IU/mg; dermatan sulfate: 25-60%; chondroitin sulfate: 3-15%; remainder is mostly a fast-moving electrophoretic fraction with ~10% slow-moving electrophoretic fraction | Heparin-like substance

Metallothionein

Sigma M 4766 Horse kidney Essentially salt-free; contains 4-7% metal as Cd + Zn; may contain both metallothionein-1A & metallothionein-1B | Kojima, Y et al, *Proc Natl Acad Sci USA*, 73: 3413, 1976

Sigma M 7641 Rabbit liver Essentially salt-free; contains ~7% metal as Cd + Zn; contains both form I & form II | Nordberg, GF et al, *Biochem J*, 126: 491, 1972

Metallothionein I

Sigma M 5267 Rabbit liver Essentially salt-free; contains ~7% metal as Cd + Zn | Form that elutes first from an anion exchange column & is probably identical to Form II of Nordberg, GF et al, *Biochem J*, 126: 491, 1972

Metallothionein II

Sigma M 5392 Rabbit liver Essentially salt-free; contains ~7% metal as Cd + Zn | Form that elutes first from an anion exchange column & is probably identical to Form II of Nordberg, GF et al, *Biochem J*, 126: 491, 1972

Metallothionein II, Zinc

Sigma M 9542 Contains 5-8% metal as Zn & ≤0.5% as Cd; prepared from metallothionein II | Comeau, RD et al, *Prep Biochem*, 22: 151, 1992

Methemoglobin

Sigma M 9250	Bovine	Crystallized; dialyzed; lyophilized
Sigma M 5882	Dog	Crystallized; dialyzed; lyophilized
Sigma M 4257	Human	Crystallized; dialyzed; lyophilized
Sigma M 8383	Pig	Crystallized; dialyzed; lyophilized
Sigma M 3759	Pigeon	Crystallized; dialyzed; lyophilized
Sigma M 6007	Rabbit	Crystallized; dialyzed; lyophilized

Microcystin LR

Synonyms: Hepatotoxin Cyclic Heptapeptide; Protein Phosphatase I & 2A Inhibitor

Sigma M 2912 *Microcystis aeruginosa* FW 995.2 $C_{49}H_{74}N_{10}O_{12}$ ~95% | No effect on protein kinase; Carmichael, WE, *Handbook of Natural Toxins*, Vol 3: Marine Toxins & Venoms (Tu, A, ed) pp 121-147, 1988, Marcel Dekker, New York; Honkanen, RE et al, *J Biol Chem*, 265: 19401, 1990

Microcystin RR

Synonyms: Protein Phosphatase 2A Inhibitor

Sigma M 1537 *Microcystis aeruginosa* FW 1038.2 ≥95% (HPLC) | Lower toxicity than microcystin LR; Matsushima, R et al, *Biochem Biophys Res Commun*, 171: 867, 1990; Shirai, M et al, *Appl Environ Microbiol*, 57: 1241, 1991

Microcystin, Sepharose

USBio M3889-05 *Microcystis aeruginosa* ≥99% | Useful for purification of Ser/Thr protein phosphatases & associated regulatory subunits; recommended for purification of the catalytic & regulatory subunits of novel protein phosphatases from all eukaryotic cells; binding capacity: 5 mg of PP1g catalytic subunit/mg of microcystin agarose

Microgloblin, βII-

Fitzgerald 30-AM11 Human urine Standard purity >60%

Microglobulin, α-

USBio M3890-18 Human ≥60%; SDS-PAGE analysis detects a major band of the αI Microgloblin protein at 30kD; 3.25 mg/mL (Lowry protein & RID) supplied in PBS buffer, pH 7.5, containing 0.05% NaN₃; standard grade | Suitable for antigenic applications in immunological protocols

Microglobulin, αI-

Fitzgerald 30-AM05 Human urine High purity

Fitzgerald 30-AM06 Human urine Standard purity

ICN 153904 Human urine MW ~27,150 >95%; lyophilized, yellow-brown color

USBio M3890-12 Human urine Highly purified, ≥98% (SDS-PAGE); no contaminants detected; single band by SDS-PAGE, IEP, &/or RID; liquid in 0.1 M ammonium bicarbonate, 0.05% NaN₃ | Suitable for antigenic applications in immunological protocols; purified from human urine of patients with chronic renal tubular proteinuria

Scipac P121-1 Urine of patients with chronic renal tubular proteinuria >96%; lyophilized | Urine protein

Scipac P121-2 Urine of patients with chronic renal tubular proteinuria 10-50%; lyophilized; sterile filtered through 0.2µm membrane | Urine protein

Scipac P121-6 Urine of patients with chronic renal tubular proteinuria 40-90%; lyophilized; sterile filtered through 0.2µm membrane | Urine protein

Microglobulin, αII-

Synonyms: Globulin, βII-

ICN 55833 Human Purified Ag; lyophilized

ICN 191345 Human plasma MW 725k >95% (SDS-PAGE); lyophilized; protein:glycine stabilizer = 1:1; negative for HBsAg & HIV Ab | Barrett, AJ, *Methods Enzymol*, 80:737E, 1981

Calbiochem 475823 Human urine MW 11.8k Liquid in PBS, pH 7.3; ≥98% (SDS-PAGE); activity: 10,000-2,000,000 U/mL; contaminants: CA 15-3 & CA 125: <25%; prepared from urine shown to negative for HBsAg & for antibodies to HIV & HCV | Useful marker for the determination of AIDS progression in HIV-infected subjects; increased levels indicate increased proliferation of tumorous masses; Zabay, JM et al, *J Acquired Immune Defic Syndr Hum Retrovirol*, 8: 266, 1995; Mady, BJ et al, *J Immunol*, 147: 3139, 1991; Lampson, LA et al, *J Immunol*, 144: 512, 1990; Ninomiya, Y & Arakawa, M, *Diabetes Res*, 10: 129, 1989; Calzia, R et al, *Adv Exp Med Biol*, 257: 225, 1989

Fitzgerald 30-AM10   Human urine   High purity >98%

Fluka 69767   Human urine   MW 11.6k   ≥90% (GE); lyophilized from phosphate buffered NaCl solution, pH 7.3 | Low MW; prepared by the method of Berggard, I & Bearn, AG, *JBC*, 243: 4095, 1968; Appella, E & Sawicki, JA, *Meth Enzymol*, 108: 494, 1984

ICN 153903   Human urine   MW 11.8k   30-60% total protein; lyophilized

ICN 770951, 770952, 770953   Human urine   Iodination grade; >96% (SDS-PAGE) | Excellent for controls or calibrators in enzyme assays

Sigma M 4890   Human urine   Lyophilized powder containing ~20% protein (Lowry); balance primarily NaCl & phosphate buffer salts, pH 7.3 | Prepared by the method of Berggard, I & Bearn, AG, *J Biol Chem*, 243: 4095, 1968, with a reported MW of 11.6 kD

USBio M3890-13   Human urine   Highly purified, ≥98% (SDS-PAGE); no contaminants detected; single band by SDS-PAGE, IEP, &/or RID; 1 mg/vial; lyophilized from 20 mM NH₄HCO₃ | Suitable for antigenic applications in immunological protocols; purified from human urine of patients with chronic renal tubular proteinuria

USBio M3890-17   Human urine   Highly purified, ≥98% (SDS-PAGE); no contaminants detected; single band by SDS-PAGE, IEP, &/or RID; 5.5 mg/mL (RID); liquid in PBS buffer, pH7.2, 0.1% NaN₃ | Suitable for antigenic applications in immunological protocols; purified from human urine of patients with chronic renal tubular proteinuria

USBio M3890-19   Human urine   MW   60-80%; ~3 mg/mL supplied in 0.1 M ammonium carbonate, 0.1% NaN₃; standard grade | Suitable for antigenic applications in immunological protocols; purified from human urine of patients with chronic renal tubular proteinuria

Biogenesis 6240-0804   Human urine from patients with chronic renal tubular proteinuria   MW ~12k (electrophoresis & chromatography)   Tested negative for HIV I and II antibodies, Hepatitis B surface antigen and Hepatitis C antibodies; purified; from 0.02 M NH₄HCO₃, may contain traces of buffer salts; lyophilized

Biogenesis 6240-0824   Human urine from patients with chronic renal tubular proteinuria   MW ~12k   Tested negative for antibodies to HIV I and II, HBsAg and HCV antibodies; semi-pure; from 0.02 M NH₄HCO₃, may contain traces of buffer salts; lyophilized

P122-2   Urine of patients with chronic renal tubular proteinuria 40-90%; | Tumor marker

Scipac P122-1   Urine of patients with chronic renal tubular proteinuria   >98%; lyophilized | Tumor marker

Microtubule Associating Protein II

Synonyms: Epidermal Growth Factor Receptor Kinase Substrate

Sigma M 4914   Bovine brain, heat stable fraction   Lyophilized from PIPES buffer containing NaCl, EGTA, magnesium sulfate, DTT, trehalose, protease inhibitors; 200 μg protein (Lowry-Peterson)/vial | High MW MAP found in dendrites; Vallee, RB, *J Cell Biol*, 92: 435, 1982; Wiche, G, *Biochem J*, 259: 1, 1989

Midkine

BioSource International PHC7035   Human recombinant

Oncogene PF049   Human recombinant   MW ~13.3k   >97% (SDS-PAGE); lyophilized in PBS | Identified in senile plaques of Alzheimer's disease patients; activity measured by its ability to enhance neurite growth of cerebral cortical neurons of E₁₀ chick embryos; optimal neurite outgrowth observed when neurons were plated on 96 well culture plates pre-coated with 100 mL/mL of 3.0-8.0 mg/mL of human recombinant midkine; protein titration recommended; species reactivity: human; for proliferation studies;

ICN 195729   Human recombinant, expressed in *E. coli*   ≥ 97%; lyophilized | Enhances neurite outgrowth of cerebral cortical neurons

PeproTech 450-16   Human recombinant, expressed in *E. coli*   MW 13.4k   >98% (SDS-PAGE) & HPLC; <0.1 ng endotoxin per mg (1EU/mg); lyophilized | New member of the heparin-binding neurotrophic factor family; structural homologs of Pleiotrophin—highly conserved among species; important roles in development & carcinogenesis; several important biological effects, including promotion of neurite extension & neuronal survival; 123 AA; SA determined by its ability to enhance neurite outgrowth of cerebral cortical neurons of E10 chick embryos

Mixtures/Standards
Albumin

Sigma A 7517   Bovine   MW ~66k   Albumin; 25 mg/vial | For SDS-PAGE

Fluka 82516   Bovine serum   Ampoule; 1.0 mg BSA/mL 0.15 M NaCl | Protein sequencing standard

Sigma A 8654   Bovine serum   MW ~66k (monomer), 132k (dimer)   Albumin | MW markers for nondenaturing PAGE systems

Sigma P 0834   Bovine serum   2 mg BSA/mL in 0.9% NaCl containing 0.05% sodium azide; sealed ampoules | Protein sequencing standard

Sigma P 0914   Bovine serum   1 mg BSA/mL in 0.15 M NaCl containing 0.05% sodium azide as preservative; sealed ampoules containing 1 mL | Protein sequencing standard; micro standard set

Sigma P 5304   Bovine serum   20 g BSA/dL in 0.85% NaCl containing 0.1% sodium azide | Protein sequencing standard

Sigma P 6529   Bovine serum   Set contains 10 mL each of BSA protein standards with concentration of 2, 4, 6, 8, 10 g/dL in 0.85% NaCl containing 0.1% sodium azide | Protein sequencing standard

Sigma P 7656   Bovine serum   2 mg BSA/vial; freeze-dried; assayed value on vial label | Protein sequencing standard; micro standard also available as part of a kit

Sigma A 8529   Chicken egg   MW ~45k   Albumin | MW markers for nondenaturing PAGE systems

Sigma A 7642   Egg   MW ~45k   Albumin; 25 mg/vial | For SDS-PAGE

Sigma 610-11   Human serum   5 mL each of human serum albumin protein standards with concentrations of 15, 30 & 50 mg/dL in 0.85% NaCl containing 0.1% sodium azide | Protein sequencing standard; micro standard set

Mixtures/Standards
Albumin, Alcohol Dehydrogenase, Carbonic Anhydrase, β-Galactosidase, Myosin, Trypsin Inhibitor

Sigma F 3526   MW 20.1-205k   Lyophilized blue powder; six proteins conjugated with FITC: albumin (bovine serum), β-galactosidase (*E. coli*), myosin (rabbit muscle), carbonic anhydrase (bovine erythrocytes), alcohol dehydrogenase (horse liver) & trypsin inhibitor (soybean) | High molecular weight standard; fluorescent marker for SDS-page & protein transfer; suitable for use as molecular weight marker; visualized with UV light

Mixtures/Standards
Albumin, Aprotinin, Bradykinin, Insulin Chain B, α-Lactalbumin, Myoglobin, Triosephosphate Isomerase

Sigma M 3546   Each vial contains 200 μL protein markers in 0.1 M Tris-HCl, pH ~8.5, 4 mM EDTA, 3 mM sodium azide & 40% glycerol: BSA, Myoglobin (horse heart), triosephosphate isomerase, aprotinin (bovine lung), α-lactalbumin (bovine milk), insulin chain b (bovine), bradykinin; supplied with 10 mL Tris-Tricine Sample Buffer (Sigma S 3047) | Protein MW markers; ultra low MW range for SDS-PAGE; designed for MW determinations in Tris-Tricine SDS-PAGE systems; formulated to yield 6 bands with about equal intensity when stained with Brilliant Blue G

Mixtures/Standards
Albumin, Aprotinin, Carbonic Anhydrase, Fructose-6-Phosphate Kinase, β-Galactosidase, Glutamic Dehydrogenase, Glyceraldehyde-3-Phosphate Dehydrogenase, α-Lactalbumin, Myosin, Ovalbumin, Phosphorylase B, Trypsin Inhibitor, Trypsinogen

Sigma M 4038   MW 6.5-205k   Lyophilized; ready for use after reconstitution with deionized water; yields 13 bands of equal intensity when stained with Brilliant Blue; contains: glutamic dehydrogenase, albumin, fructose-6-phosphate kinase, phosphorylase b, β-galactosidase, glyceraldehyde-3-phosphate dehydrogenase, ovalbumin & myosin, aprotinin, α-lactalbumin, trypsin inhibitor, trypsinogen, carbonic anhydrase; each vial sufficient for 100 applications on PhastGels, 30 applications on mini (10x10 cm) gels & 20 applications for large (16x18 cm) gels | SigmaMarkers protein standards designed for use on PhastGel media & SDS-PAGE (Laemmli) gels; wide molecular weight range

Mixtures/Standards
Albumin, Aprotinin, Carbonic Anhydrase, Fumarase, β-Galactosidase, α-Lactalbumin, β-Lactoglobulin, Phosphorylase B

Sigma M 6539   Vial contains 200 µL protein marker in 300 mM NaCl, 100 mM DTT, 3 mM sodium azide & 50% glycerol: BSA, fumarase (porcine heart), carbonic anhydrase (bovine erythrocytes), β-galactosidase (E. coli), phosphorylase b (rabbit muscle); β-lactoglobulin (bovine milk), α-lactalbumin (bovine milk), aprotinin bovine lung) | Silver Stain SDS-PAGE wide MW standard mixture; designed for MW determinations on silver stained SDS-PAGE (Laemmli) gels; formulated to yield bands with about equal intensity & no background when stained using Rapid Silver Stain (RSK-1) or AG-5 or AG-25 silver stain kits; supplied with 1 vial Laemmli sample buffer (Sigma S 3401) containing 6 mL 4% SDS, 20% glycerol, 10% BME & 0.004% bromphenol blue in 0.125 M Trizma base, pH 6.8; supplied with 1 vial Laemmli sample buffer (Sigma S 3401) containing 6 mL 4% SDS, 20% glycerol, 10% BME & 0.004% bromphenol blue in 0.125 M Trizma base, pH 6.8

Mixtures/Standards
Albumin, Aprotinin, Carbonic Anhydrase, β-Galactosidase, α-Lactalbumin, Myosin, Ovalbumin, Trypsin Inhibitor

Sigma C 3437   MW 6.5-205k   Solution contains 2.5 mg of 8 proteins in 62 mM Tris-HCl, pH 7.5, 2% SDS, 0.01 mM EDTA, 100 mM DTT, 4 M urea, 0.005% bromphenol blue & 30% glycerol: aprotinin (bovine lung, blue), α-lactalbumin (bovine milk, purple), trypsin inhibitor (soybean, green), carbonic anhydrase (bovine erythrocytes, orange), ovalbumin (chicken egg, yellow), albumin (bovine serum, pink), β-galactosidase (E. coli, turquoise) & myosin (rabbit muscle, blue); ready for use | Color markers for SDS-PAGE & protein transfer; wide molecular weight range; recommended for use on a 4-20% SDS-gradient gel; see Sigma C 6210

Mixtures/Standards
Albumin, Aprotinin, Carbonic Anhydrase, Glyceraldehyde-3-Phosphate Dehydrogenase, α-Lactalbumin, Ovalbumin, Trypsin Inhibitor, Trypsinogen

Sigma M 3913   MW 6.5-66k   Lyophilized; ready for use after reconstitution with deionized water; yields 8 bands of equal intensity when stained with Brilliant Blue; contains: aprotinin, α-lactalbumin, trypsin inhibitor, trypsinogen, carbonic anhydrase, glyceraldehyde-3-phosphate dehydrogenase, ovalbumin & albumin; each vial sufficient for 100 applications on PhastGels, 30 applications on mini (10x10 cm) gels & 20 applications for large (16x18 cm) gels | SigmaMarkers protein standards designed for use on PhastGel media & SDS-PAGE (Laemmli) gels; low Molecular Weight Range

Mixtures/Standards
Albumin, Carbonic Anhydrase, β-Casein, Glyceraldehyde-3-Phosphate Dehydrogenase, α-Lactalbumin, Trypsin Inhibitor

Sigma M 4399   MW 14-70k   Solution in 10 mM sodium phosphate, pH 7.0, containing 0.1% SDS & 0.1% BME in Combi-vial; mixture of seven ¹⁴C-methylated proteins: α-lactalbumin (14,200), trypsin inhibitor (20,100), β-casein (23,600), carbonic anhydrase (29,000), glyceraldehyde-3-phosphate dehydrogenase (36,000), chicken egg albumin (45,000), bovine serum albumin (66,000) | Protein MW markers, ¹⁴C-Me-; radiochemical

Mixtures/Standards
Albumin, Carbonic Anhydrase, Conalbumin, Galactosidase, Ovalbumin, Phosphorylase A, Soybean Trypsin Inhibitor

Alexis 850-051-KI01   MW 21-114k   Kit contains phosphotyrosine-modified proteins: galactosidase (114 k), phosphorylase A (94 k), conalbumin (78 k), BSA (67 k), ovalbumin (47 k), carbonic anhydrase (30 k), soybean trypsin inhibitor (21 k); pre-blended as lyophilisate; malachite green added as tracking dye | Malachite green, tracking dye additive, used to visualize individual lanes after blotting onto NC or PVDF membranes; sufficient for 50-100 immunoblots with phosphotyrosine-specific MAbs

Mixtures/Standards
Albumin, Carbonic Anhydrase, Cytochrome C, Globulins, Lactoglobulin A, Myosin, Ovalbumin, Phosphorylase B,

ARC ARC-430   MW 12.3-200k   Choice of 3 or 5 individual markers in 0.01 M sodium phosphate, pH 7.2, 3-30 µCi/mg; 111-1111 KBq/mg: Myosin (MW 200,000, ARC-433), Globulins (MW 150,000, ARC-428), (MW 97,400, ARC-432), Bovine Albumin (MW 69,000, ARC-422), Ovalbumin (MW 46,000, ARC-431), Carbonic anhydrase (MW 30,000, ARC-425), Lactoglobulin A (MW 18,367, ARC-429), Cytochrome c (MW 12,300, ARC-427) | Protein MW Markers, ¹⁴C-Me-; radiochemical

Mixtures/Standards
Albumin, Carbonic Anhydrase, Fumarase, β-Galactosidase, Phosphorylase B

Sigma M 5505   Vial contains 200 µL protein marker in 300 mM NaCl, 100 mM DTT, 3 mM sodium azide & 50% glycerol: BSA, fumarase (porcine heart), carbonic anhydrase (bovine erythrocytes), β-galactosidase (E. coli), phosphorylase b (rabbit muscle) | Silver Stain SDS-PAGE high MW standard mixture; designed for MW determinations on silver stained SDS-PAGE (Laemmli) gels; formulated to yield bands with about equal intensity & no background when stained using Rapid Silver Stain (RSK-1) or AG-5 or AG-25 silver stain kits; supplied with 1 vial Laemmli sample buffer (Sigma S 3401) containing 6 mL 4% SDS, 20% glycerol, 10% BME & 0.004% bromphenol blue in 0.125 M Trizma base, pH 6.8; supplied with 1 vial Laemmli sample buffer (Sigma S 3401) containing 6 mL 4% SDS, 20% glycerol, 10% BME & 0.004% bromphenol blue in 0.125 M Trizma base, pH 6.8

Mixtures/Standards
Albumin, Carbonic Anhydrase, Fumarase, α-Lactalbumin, β-Lactoglobulin

Sigma M 5630   Vial contains 200 µL protein marker in 300 mM NaCl, 100 mM DTT, 3 mM sodium azide & 50% glycerol: BSA, fumarase (porcine heart), carbonic anhydrase (bovine erythrocytes), β-lactoglobulin (bovine milk), α-lactalbumin (bovine milk) | Silver stain SDS-PAGE low MW standard; designed for MW determinations on silver stained SDS-PAGE (Laemmli) gels; formulated to yield bands with about equal intensity & no background when stained using Rapid Silver Stain (RSK-1) or AG-5 or AG-25 silver stain kits; supplied with 1 vial Laemmli sample buffer (Sigma S 3401) containing 6 mL 4% SDS, 20% glycerol, 10% BME & 0.004% bromphenol blue in 0.125 M Trizma base, pH 6.8

Mixtures/Standards
Albumin, Carbonic Anhydrase, β-Galactosidase, α-Lactalbumin, Myosin

Sigma M 3797   MW 14.2-205k   Solution in 10 *mM* sodium phosphate, pH 7.2, containing 0.1% SDS, 0.1% BME & 1 *mM* EDTA in Combi-vial; mixture of six ^{14}C-methylated proteins: α-lactalbumin (14,200), carbonic anhydrase (29,000), chicken egg albumin (45,000), bovine serum albumin (66,000), β-galactosidase (116,000), myosin (205,000) | Protein MW markers, ^{14}C-Me-; radiochemical

Mixtures/Standards
Albumin, Carbonic Anhydrase, β-Galactosidase, α-Lactalbumin, Myosin, Ovalbumin, Phosphorylase B, Trypsin Inhibitor

Sigma M 2789   Lyophilized solid; contains ~3.5 mg total of 8 proteins: α-lactalbumin (bovine milk), trypsin inhibitor (soybean), carbonic anhydrase (bovine erythrocytes), ovalbumin (egg), albumin (bovine), phosphorylase b (rabbit muscle), β-galactosidase (*E. coli*), myosin (rabbit muscle); must be reconstituted with SDS-2X Sample Buffer (Sigma S 9788) | SDS protein standards; suitable for determination of protein MWs on coated protein CE columns

Mixtures/Standards
Albumin, Carbonic Anhydrase, β-Galactosidase, Myosin, Ovalbumin

Sigma C 3312   MW 29k-205k   Solution contains 1.5 mg total of five proteins in 62 *mM* Tris-HCl, pH 7.5, 2% SDS, 0.01 *mM* EDTA, 100 *mM* DTT, 4 M urea, 0.005% bromphenol blue & 30% glycerol: albumin (bovine serum, pink), β-galactosidase (E. coli, Turquoise), myosin (rabbit muscle, blue), carbonic anhydrase (bovine erythrocytes, orange) & ovalbumin (chicken egg, yellow); ready for use | Color markers for SDS-PAGE & protein transfer; high molecular weight range; see Sigma C 6210

Mixtures/Standards
Albumin, Carbonic Anhydrase, β-Galactosidase, Myosin, Ovalbumin, Phosphorylase B

Sigma SDS-6H   Contains 3 mg of a lyophilized mixture of the 6 proteins: carbonic anhydrase (bovine erythrocytes), albumin (bovine), ovalbumin, phosphorylase b (rabbit muscle), β-galactosidase (*E. coli*), myosin (rabbit muscle) | Protein MW markers; high MW range mixture

Mixtures/Standards
Albumin, Carbonic Anhydrase, β-Galactosidase, Myosin, Phosphorylase B

Fluka 69811   MW 30-200k   Composition: carbonic anhydrase (bovine erythrocytes, 29,000); albumin (chicken egg white, 45,000); albumin (bovine serum, 66,000); phosphorylase b (rabbit muscle, 97,400); β-galactosidase (*E. coli*, 116,000); myosin (rabbit muscle, 205,000) | Protein MW markers

Mixtures/Standards
Albumin, Carbonic Anhydrase, Glyceraldehyde-3-Phosphate Dehydrogenase, α-Lactalbumin, Ovalbumin, Trypsin Inhibitor, Trypsinogen

Synonyms: Dalton Mark VII-L™

Sigma SDS-7   Contains 3.5 mg of a lyophilized mixture of the 7 proteins: trypsin inhibitor (soybean), albumin (bovine), ovalbumin, glyceraldehyde-3-phosphate dehydrogenase (rabbit muscle), α-lactalbumin (bovine milk), trypsinogen, PMSF treated (bovine pancreas), carbonic anhydrase (bovine erythrocytes) | See Sigma MW-SDS-70L

Mixtures/Standards
Albumin, Carbonic Anhydrase, Glyceraldehyde-3-Phosphate Dehydrogenase, α-Lactalbumin, Trypsin Inhibitor, Trypsinogen

Synonyms: Dalton Mark VII-L

Fluka 69810   MW 14.2-67k   Composition: trypsin inhibitor (soybean, 20,000); α-lactalbumin (bovine milk, 14,200); trypsinogen (PMSF-treated, bovine pancreas, 24,000); carbonic anhydrase (bovine erythrocytes, 29,000); albumin (chicken egg white, 45,000); albumin (bovine serum, 67,000); (rabbit muscle, 36,000) | Protein MW markers for electrophoresis; Laemmli, UK et al, *Nature*, 227: 680, 1970

Sigma MW-SDS-70L   MW 14-70k   Contains 1 vial each of 7 proteins, 1 vial of Sigma SDS-7; proteins are: carbonic anhydrase (bovine erythrocytes), albumin (egg), albumin (bovine), trypsin inhibitor (soybean), α-lactalbumin (bovine milk), trypsinogen, PMSF treated (bovine pancreas), glyceraldehyde-3-phosphate dehydrogenase (rabbit muscle) | Molecular weight standards for SDS-PAGE; useful in the procedure of Laemmli as described in Technical Bulletin No. MWS-877L or the procedure of Weber & Osborn as described in Bulletin No. MWS-877

Mixtures/Standards
Albumin, Carbonic Anhydrase, α-Lactalbumin

Sigma M 4774   MW 14-132k   Solution in 40 *mM* potassium phosphate, pH 7.0, in Combi-vial; mixture of five ^{14}C-methylated proteins: α-lactalbumin (14,200), carbonic anhydrase (29,000), chicken egg albumin (45,000), bovine serum albumin (66,000 monomer & 132,000 dimer), | Protein MW markers, ^{14}C-Me-; radiochemical

Mixtures/Standards
Albumin, Carbonic Anhydrase, α-Lactalbumin, Urease

Sigma MW-ND-500   MW 14-500k   Contains 1 vial each of the following 5 proteins (1 mg each) & Technical Bulletin No. MKR-137: α-lactalbumin (bovine milk), carbonic anhydrase (bovine erythrocytes), albumin (chicken egg), albumin (bovine), urease (Jack bean) | Protein molecular weight standards for nondenaturing PAGE; allows retention of the characteristics of the "native" protein, whereas SDS gel electrophoresis causes denaturation of the proteins, leading to losses of most enzymatic or biological properties; proteins can be examined by electrophoresis in nondenaturing systems to determine several characteristics: MW of homogenous proteins; &, for nonhomogenous proteins, differences in charge (charge isomers) or MW (MW isomers); procedure for determining MW in a nondenaturing system is a modification of the methods of Bryan & Davis; Hedrick, JL & Smith, AJ, *Arch Biochem Biophys*, 126: 155, 1968; Bryan, JK, *Anal Biochem*, 78: 513, 1977; Davis, BJ, *Ann NY Acad Sci*, 121: 404, 1964

Mixtures/Standards
Albumin, Carbonic Anhydrase, β-Lactoglobulin, Lysozyme

Sigma M 4524   MW 14-70k   Solution in 10 *mM* sodium phosphate, pH 7.0, containing 0.1% SDS & 0.1% BME in Combi-vial; mixture of five ^{14}C-methylated proteins: egg white lysozyme (14,300), β-lactoglobulin (18,400), carbonic anhydrase (29,000), chicken egg albumin (45,000), bovine serum albumin (66,000) | Protein MW markers, ^{14}C-Me-; radiochemical

Mixtures/Standards
Albumin, Fructose-6-Phosphate Kinase, β-Galactosidase, Glutamic Dehydrogenase, Glyceraldehyde-3-Phosphate Dehydrogenase, Myosin, Ovalbumin, Phosphorylase B

Sigma M 3788 MW 36-205k Lyophilized; ready for use after reconstitution with deionized water; yields 8 bands of equal intensity when stained with Brilliant Blue; contains: glutamic dehydrogenase, albumin, fructose-6-phosphate kinase, phosphorylase b, β-galactosidase, glyceraldehyde-3-phosphate dehydrogenase, ovalbumin & myosin; each vial sufficient for 100 applications on PhastGels, 30 applications on mini (10x10 cm) gels & 20 applications for large (16x18 cm) gels | SigmaMarkers protein standards designed for use on PhastGel media & SDS-PAGE (Laemmli) gels; high molecular weight range

Mixtures/Standards
Albumin, Globulin

Sigma 540-10 5 g/dL albumin, 3.0 g/dL globulin (8.0 g/dL total protein) in 0.85% NaCl containing 0.1% sodium azide | Protein sequencing standard

Mixtures/Standards
Albumin, Hemocyanin, Hemoglobin

Sigma MW-SDS-280 MW 16-280k Contains 1 vial each of the following 3 cross-linked proteins & Technical Bulletin No. MWS-877: cross-linked albumin (bovine), cross-linked hemocyanin (*Limulus polyphemus*), cross-linked hemoglobin (bovine) | Cross-linked protein molecular weight standards for SDS-PAGE; recommended for use in a modified system by Weber, K & Osborn, M, *J Biol Chem*, 244: 4406, 1969; anomalous migration is seen in the SDS system of Laemmli, UK, *Nature*, 227: 680, 1970, & therefore the markers are not recommended for use in this system

Mixtures/Standards
Albumin, β-Lactoglobulin, Lysozyme, Pepsin, Trypsinogen

Synonyms: Dalton Mark VI-L; Dalton Mark VI™; Dalton Mark VI™ Mixture

Fluka 69814 MW 14-67k Composition: lysozyme (chicken egg white, 14,000); β-lactoglobulin (bovine milk, 18,400); trypsinogen (PMSF-treated, bovine pancreas, 24,000); pepsin (porcine stomach, 36,000); albumin (chicken egg white, 45,000); albumin (bovine serum, 67,000); bromophenol blue as tracking dye | Protein MW markers for electrophoresis; Weber, K & Osborn, MJ, *JBC*, 244: 4406, 1969

Sigma MW-SDS-70 MW 10-70k Contains 1 vial each of 6 proteins (25 mg each), 1 vial of Sigma SDS-6; proteins are: β-lactoglobulin (bovine milk), albumin (egg), albumin (bovine), lysozyme (chicken egg white), pepsin (porcine stomach mucosa), trypsinogen, PMSF treated (bovine pancreas) | Molecular weight standards for SDS-PAGE; five proteins are characterized by a single band; β-lactoglobulin contains two subunits which migrate as two closely-spaced bands; pepsin migrates anomalously in the Laemmli gel system; Laemmli, UK, *Nature*, 227: 680, 1970

Sigma SDS-6 Contains 13.5 mg of a lyophilized mixture of six proteins plus brophenol blue tracking dye: lysozyme, β-lactoglobulin, pepsin, trypsinogen (PMSF treated), albumin (egg), albumin (bovine) | For SDS-PAGE

Mixtures/Standards
Albumin, Myosin, Superoxide Dismutase

Upstate 12-354 MW 16k, 32k, 66k, 215k Myosin/Superoxide Dismutase/BSA/Muscle; frozen solution; nitrated | Nitrotyrosine immunoblotting control

Mixtures/Standards
Alcohol Dehydrogenase, Aprotinin, Carbonic Anhydrase, Catalase, β-Galactosidase, α₂-Macroglobulin, Phosphorylase B, Trypsin Inhibitor

Sigma B 2787 MW 6.5k-180k Contains nine biotinylated proteins (~0.1 mg total): aprotinin (bovine lung), α₂-macroglobulin (human plasma), β-galactosidase (E. *coli*), phosphorylase b (rabbit muscle), catalase (bovine liver), carbonic anhydrase (bovine erythrocytes), alcohol dehydrogenase (horse liver) & trypsin inhibitor (soybean); ~33% protein, 33% NaCl & 33% sucrose | Wide molecular weight standard mixture; molecular weight markers for SDS-PAGE & protein transfer

Mixtures/Standards
Alcohol Dehydrogenase, Aprotinin, Carbonic Anhydrase, α-Lactalbumin, Trypsin Inhibitor

Sigma F 3401 MW 6.5-39.8k Lyophilized blue powder; five proteins conjugated with FITC: aprotinin (bovine lung), α-lactalbumin (bovine milk), carbonic anhydrase (bovine erythrocytes), alcohol dehydrogenase (horse liver) & trypsin inhibitor (soybean) | Low molecular weight standard; fluorescent marker for SDS-page & protein transfer; suitable for use as molecular weight marker; visualized with UV light

Mixtures/Standards
Amyloglucosidase

Sigma A 8437 *Aspergillus niger* MW ~89k, 70k pI ~3.8; vial contains ~200 □L in 8 *M* urea & 2% BME | Marker for 2D electrophoresis; sufficient for 20-40 applications on gels that will be stained with Brilliant Blue or 200-400 applications on gels that will be silver stained

Mixtures/Standards
Amyloglucosidase, Carbonic Anhydrase, Myoglobin, Ovalbumin

Sigma M 3411 pI 7.6-3.8 Vial contains ~200 µL of a mixture of 4 proteins in 8 *M* urea & 2% BME: amyloglucosidase (A. *niger*), carbonic anhydrase (human erythrocytes), myoglobin (horse heart), ovalbumin; markers provide a diagonal line across a 2D gel (as the pI increases, the MW decreases) | Marker for 2D electrophoresis; sufficient for 20-40 applications on gels that will be stained with Brilliant Blue or 200-400 applications on gels that will be silver stained

Mixtures/Standards
Apomyoglobin

Synonyms: Apomyoglobin

Sigma A 8548 Horse skeletal muscle Apomyoglobin coupled via the α- & ε-amino groups to DITC glass (Sigma G 9764); 1-2 nmoles/mg glass as determined by AA analysis | Protein sequencing standard; use -tested

Sigma A 8673 Horse skeletal muscle Lyophilized; ~60 nmoles/vial | Protein sequencing standard; use-tested & prepared by the method of Rothgeb, TM & Gurd, FRN, *Meth Enzymol*, 52: 473, 1978

Mixtures/Standards
Aprotinin, Bradykinin, Insulin Chain B, α-Lactalbumin, Myoglobin, Triosephosphate Isomerase

Sigma C 6210 MW 1.06-26.6 Each vial contains 200 µL of a solution of six polypeptides in 10 m*M* Tris-HCl, pH 7.0, 0.5% SDS, 2 m*M* EDTA, 10 m*M* DTT, 0.01% sodium azide & 33% glycerol: myoglobin (horse heart, violet), triose-phosphate isomerase (rabbit muscle, orange), aprotinin (bovine lung, blue), α-lactalbumin (bovine milk, red), insulin chain b (bovine, blue), bradykinin (blue); total protein content: ~1.5 mg; ready for use | Color markers for SDS-PAGE & protein transfer; ultra low molecular weight range; MW of pre-stained proteins are altered by the attachment of dyes; apparent molecular weights for each lot of color markers are determined using SigmaMarkers as standards & are supplied on the label

Mixtures/Standards
Aprotinin, Carbonic Anhydrase, Cytochrome c

Fluka 69883   MW 6.5-66k   Composition: aprotinin (bovine lung, 6500); cytochrome c (equine heart, 12,400); carbonic anhydrase (bovine erythrocytes, 29,000); albumin (bovine serum, 66,200); dextran blue | Protein MW markers for gel filtration

Mixtures/Standards
Aprotinin, Carbonic Anhydrase, α-Lactalbumin, Ovalbumin, Trypsin Inhibitor

Sigma C 3187   MW 6.5-45k   Solution contains 1.5 mg total of five proteins in 62 mM Tris-HCl, pH 7.5, 2% SDS, 0.01 mM EDTA, 100 mM DTT, 4 M urea, 0.005% bromphenol blue & 30% glycerol: aprotinin (bovine lung, blue), α-lactalbumin (bovine milk, purple), trypsin inhibitor (soybean, green), carbonic anhydrase (bovine erythrocytes, orange) & ovalbumin (chicken egg, yellow); ready for use | Color markers for SDS-PAGE & protein transfer; low molecular weight range; see Sigma C 6210

Mixtures/Standards
Carbonic Anhydrase

Sigma C 2273   Bovine erythrocytes   MW ~29k   Carbonic anhydrase; 5 mg/vial | For SDS-PAGE

Sigma C 5024   Bovine erythrocytes   MW ~29k   MW markers for nondenaturing PAGE systems

Sigma C 4806   Human erythrocytes   MW ~29k   pI ~7.0; vial contains ~200 µL in 8 M urea & 2% BME | Marker for 2D electrophoresis; sufficient for 20-40 applications on gels that will be stained with Brilliant Blue or 200-400 applications on gels that will be silver stained

Mixtures/Standards
Carbonic Anhydrase, α-Lactalbumin, β-Lactoglobulin A, β-Lactoglobulin B

Synonyms: Protein Calibration Standard A

Sigma P 2818   Lyophilized solid; vial contains ~240 µg total of 5 proteins: α-lactalbumin, β-lactoglobulin A, β-lactoglobulin B, human carbonic anhydrase & bovine carbonic anhydrase | Suitable for use as a calibration standard in CE

Mixtures/Standards
Cross-Linked Peptides

Fluka 69827   MW 56-280k   Lyophilized; mixture of oligomeric peptides which have been chemically cross-linked with MW: 56,000, 112,000, 168,000, 224,000, 280,000 | Protein MW markers for electrophoresis; Weber, K et al, *Meth Enzymol*, (CHW Hirs & Timosheff, SM, eds)26: 44306, 1972, Academic Press, New York; Steele, J Ch & Nielsen, TB, *Anal Biochem*, 84: 218, 1978

Mixtures/Standards
Cytochrome c

Sigma C 7337   Horse heart   Lyophilized; vial contains 100 µg | Protein sequencing standard; peptide map control; use -tested; suitable as a control for proteolytic digestions; tested by trypsin digestions; data sheet accompanies each order

Mixtures/Standards
Fructose-6-Phosphate Kinase

Sigma F 0387   Rabbit muscle   MW 84k (unstained)   Pre-stained MW marker for SDS-PAGE; produces highly visible blue protein bands on SDS-PAGE gels when used with either the procedure of Laemmli, UK, *Nature*, 227: 680, 1970 or that of Weber, L & Osborn, M, *J Biol Chem*, 244: 4406, 1969; pre-stained marker can be transferred from gels to solid supports such as nitro-cellulose, nylon or PVDF membranes; MWs of pre-stained proteins are somewhat altered by the attachment of dye

Mixtures/Standards
Fructose-6-Phosphate Kinase, Fumarase, β-Galactosidase, Lactate Dehydrogenase, α₂-Macroglobulin, Pyruvate Kinase, Triosephosphate Isomerase

Fluka 69813   MW 27-180k   Pre-stained; composition: triosephosphate isomerase (rabbit muscle, 26,600); β-galactosidase (*E. coli*, 116,000); lactate dehydrogenase (rabbit muscle, 36,500); fumarase (porcine heart, 48,500); pyruvate kinase (chicken muscle, 58,000); fructose-6-phosphate kinase (rabbit muscle, 84,000); α₂-macroglobulin (human plasma, 180,000) | Protein MW markers for electrophoresis; for precise MW determinations on Western Blots, Fluka biotinylated MW Standard Mixture (Fluka #69881) is recommended; Laemmli, UK et al, *Nature*, 227: 680, 1970; Tsang, VCW et al, *Anal Biochem*, 143: 304, 1984

Sigma SDS-7B   Contains a lyophilized mixture of the 7 pre-stained proteins are: triosephosphate isomerase, lactic dehydrogenase, fumarase, pyruvate kinase, fructose-6-phosphate kinase, β-galactosidase, α₂-macroglobulin | Pre-stained MW marker for SDS-PAGE; produces highly visible blue protein bands on SDS-PAGE gels when used with either the procedure of Laemmli, UK, *Nature*, 227: 680, 1970 or that of Weber, L & Osborn, M, *J Biol Chem*, 244: 4406, 1969; pre-stained marker can be transferred from gels to solid supports such as nitro-cellulose, nylon or PVDF membranes; MWs of pre-stained proteins are somewhat altered by the attachment of dye

Mixtures/Standards
Fructose-6-Phosphate Kinase, β-Galactosidase, Lactic Dehydrogenase, Ovalbumin, Pyruvate Kinase, Triosephosphate Isomerase

Sigma P 1677   MW 30-120k   Solution contains six pre-stained proteins (2 mg total protein) in 4 M urea, 2% SDS, 100 mM DTT, 0.01 mM EDTA, 1 mM sodium azide & 33% glycerol: triosephosphate isomerase, lactic dehydrogenase, ovalbumin, pyruvate kinase, fructose-6-phosphate kinase, β-galactosidase; ready to use | Pre-stained MW marker for SDS-PAGE; produces highly visible blue protein bands on SDS-PAGE gels when used with either the procedure of Laemmli, UK, *Nature*, 227: 680, 1970 or that of Weber, L & Osborn, M, *J Biol Chem*, 244: 4406, 1969; pre-stained marker can be transferred from gels to solid supports such as nitro-cellulose, nylon or PVDF membranes; MWs of pre-stained proteins are somewhat altered by the attachment of dye

Mixtures/Standards
Fumarase

Sigma F 0262   Porcine heart   MW 48.5k (unstained)   Pre-stained MW marker for SDS-PAGE; produces highly visible blue protein bands on SDS-PAGE gels when used with either the procedure of Laemmli, UK, *Nature*, 227: 680, 1970 or that of Weber, L & Osborn, M, *J Biol Chem*, 244: 4406, 1969; pre-stained marker can be transferred from gels to solid supports such as nitro-cellulose, nylon or PVDF membranes; MWs of pre-stained proteins are somewhat altered by the attachment of dye

Mixtures/Standards
β-Galactosidase

Sigma G 6017   *E. coli*   MW 116k (unstained)   Pre-stained MW marker for SDS-PAGE; produces highly visible blue protein bands on SDS-PAGE gels when used with either the procedure of Laemmli, UK, *Nature*, 227: 680, 1970 or that of Weber, L & Osborn, M, *J Biol Chem*, 244: 4406, 1969; pre-stained marker can be transferred from gels to solid supports such as nitro-cellulose, nylon or PVDF membranes; MWs of pre-stained proteins are somewhat altered by the attachment of dye

Sigma G 8511   *E. coli*   MW ~116k   Contains 0.5 mg/vial | For SDS-PAGE

Mixtures/Standards
Gelatin

Sigma P 0959   2 mg gelatin/mL in 0.9% NaCl containing 0.05% sodium azide; sealed ampoules | Protein sequencing standard

Mixtures/Standards
Glucagon

Sigma G 7774 Lyophilized; vial contains 100 µg; HPLC purified | Protein sequencing standard; use -tested as a protease substrate; data sheet accompanies each order

Mixtures/Standards
Glyceraldehyde-3-Phosphate Dehydrogenase

Sigma G 5262 Rabbit muscle MW ~36k Contains 5 mg/vial | For SDS-PAGE

Mixtures/Standards
Insulin Chain B, Oxidized

Sigma I 1764 Bovine Lyophilized; vial contains 100 µg; HPLC purified | Protein sequencing standard; use -tested as a protease substrate; data sheet accompanies each order

Mixtures/Standards
α-Lactalbumin

Sigma L 4385 Bovine milk MW ~14.2k MW markers for nondenaturing PAGE systems

Sigma L 6385 Bovine milk MW ~14.2k Contains 5 mg/vial | For SDS-PAGE

Mixtures/Standards
Lactic Dehydrogenase

Sigma L 3891 Rabbit muscle MW 36.5k (unstained) Pre-stained MW marker for SDS-PAGE; produces highly visible blue protein bands on SDS-PAGE gels when used with either the procedure of Laemmli, UK, *Nature*, 227: 680, 1970 or that of Weber, L & Osborn, M, *J Biol Chem*, 244: 4406, 1969; pre-stained marker can be transferred from gels to solid supports such as nitro-cellulose, nylon or PVDF membranes; MWs of pre-stained proteins are somewhat altered by the attachment of dye

Mixtures/Standards
β-Lactoglobulin

Sigma L 4756 Bovine milk MW ~18.4k For SDS-PAGE

Mixtures/Standards
Lysozyme

Sigma P 1084 2 mg lysozyme/mL in 0.9% NaCl containing 0.05% sodium azide; sealed ampoules | Protein sequencing standard

Sigma L 4631 Chicken egg white MW ~14.3k For SDS-PAGE

Mixtures/Standards
α₂-Macroglobulin

Sigma M 3398 Human plasma MW 180k (unstained) Pre-stained MW marker for SDS-PAGE; produces highly visible blue protein bands on SDS-PAGE gels when used with either the procedure of Laemmli, UK, *Nature*, 227: 680, 1970 or that of Weber, L & Osborn, M, *J Biol Chem*, 244: 4406, 1969; pre-stained marker can be transferred from gels to solid supports such as nitro-cellulose, nylon or PVDF membranes; MWs of pre-stained proteins are somewhat altered by the attachment of dye

Mixtures/Standards
Melittin

Sigma M 1407 Bee venom Lyophilized; vial contains 100 ☐g; HPLC purified | Protein sequencing standard; use -tested as a protease substrate; data sheet accompanies each order

Mixtures/Standards
Myoglobin

Fluka 69825 MW 2.5-17k Cyanogen bromide cleavage of horse heart myoglobin yields 6 peptide fragments (fragment #/positions/Daltons): 1/1-153/17,000; 2/1-131/14,500; 3/56-153/10,700; 4/56-131/8200; 5/1-55/6300; 6/132-153/2500 | Protein MW markers for electrophoresis; Kratzin, HD et al, *Anal Biochem*, 183: 1, 1989

Sigma M 3286 Horse heart Contains ~250 µg of lyophilized carbamylated myoglobin & 40 µg of methyl red; marker displays ~20 spots on a line parallel to the IEF axis on IEF-urea or 2D gel | 2D marker; suitable for use as an internal or external calibration standard

Sigma M 7911 Horse heart MW ~17k pI ~7.6; vial contains ~200 µL in 8 *M* urea & 2% BME | Marker for 2D electrophoresis; sufficient for 20-40 applications on gels that will be stained with Brilliant Blue or 200-400 applications on gels that will be silver stained

Mixtures/Standards
Myoglobins

Sigma MW-SDS-17S MW 2.5-17k Contains 7 polypeptides: myoglobin (1-153), myoglobin I+II (1-131), myoglobin I+III (56-153), myoglobin I (1-55), glucagon, myoglobin III (132-153) | Molecular weight standards for SDS-PAGE; recommended for use in modifications of the systems of Schagger, H & von Jagow, G, *Anal Biochem*, 166: 368, 1987 & Swank, RT & Munkres, KD, *Anal Biochem*, 39: 462, 1971

Mixtures/Standards
Myosin

Sigma M 3889 Rabbit muscle MW ~205k Contains 0.25 mg/vial | For SDS-PAGE

Mixtures/Standards
Ovalbumin

Sigma O 4757 MW ~45k pI ~5.1; vial contains ~200 µL in 8 *M* urea & 2% BME | Marker for 2D electrophoresis; sufficient for 20-40 applications on gels that will be stained with Brilliant Blue or 200-400 applications on gels that will be silver stained

Mixtures/Standards
Pepsin

Sigma P 1143 Porcine stomach mucosa MW ~34.7k For SDS-PAGE

Mixtures/Standards
Phosphorylase B

Sigma P 8906 Rabbit MW: 97.4k (monomer), 194.8k (dimer), 292k (trimer), 389.6k (tetramer), 487k (pentamer), 584.4k (hexamer) Lyophilized; ~3 mg/vial | Cross-linked proteins for high molecular weights for SDS-PAGE; NOT recommended for use in the Laemmli system: Laemmli, UK, *Nature*, 227: 680, 1970; not included in kit MW-SDS-280; Relative band intensity decreases as molecular weight increases; trace bands (heptamer through nonomer) may also be detected; comments are the same as Sigma MW-SDS-280

Sigma P 4649 Rabbit muscle MW ~97.4k (subunit) Contains 0.5 mg/vial | For SDS-PAGE

Mixtures/Standards
Pyruvate Kinase

Sigma P 5788 Chicken muscle MW 58k (unstained) Pre-stained MW marker for SDS-PAGE; produces highly visible blue protein bands on SDS-PAGE gels when used with either the procedure of Laemmli, UK, *Nature*, 227: 680, 1970 or that of Weber, L & Osborn, M, *J Biol Chem*, 244: 4406, 1969; pre-stained marker can be transferred from gels to solid supports such as nitro-cellulose, nylon or PVDF membranes; MWs of pre-stained proteins are somewhat altered by the attachment of dye

Mixtures/Standards
Triosephosphate Isomerase

Sigma T 9400   Rabbit muscle   MW 26.6k (unstained)   Pre-stained MW marker for SDS-PAGE; produces highly visible blue protein bands on SDS-PAGE gels when used with either the procedure of Laemmli, UK, *Nature*, 227: 680, 1970 or that of Weber, L & Osborn, M, *J Biol Chem*, 244: 4406, 1969; pre-stained marker can be transferred from gels to solid supports such as nitrocellulose, nylon or PVDF membranes; MWs of pre-stained proteins are somewhat altered by the attachment of dye

Mixtures/Standards
Trypsin Inhibitor

Sigma T 9767   Soybean   MW ~20.1k   Contains 5 mg/vial |
For SDS-PAGE

Mixtures/Standards
Trypsinogen

Sigma T 9011   Bovine pancreas   MW ~24k   25 mg/vial,
PMSF treated | For SDS-PAGE

Mixtures/Standards
Urease

Sigma U 7752   Jack bean   MW ~272k (monomer), 545k (dimer)   Contains ~20% DTT | MW markers for nondenaturing PAGE systems

Mixtures/Standards
Unspecified

Sigma M 0671   Recombinant   MW 15, 25, 35, 50, 75, 100 & 150k   Solution in 125 *mm* Tris-HCl, pH 6.8, 2% SDS, 10% glycerol, 200 m*M* BME, 0.007% bromophenol blue; mixture contains seven precisely sized proteins: Total protein concentration: ~800 µg/mL | Protein MW markers for SDS-PAGE & Western Blotting recombinant proteins; have not been glycosylated so they produce sharp bands & allow precise size determination; each marker carries a 15-AA sequence that binds to *S*-Protein–Alkaline Phosphatase conjugate, allowing enzyme-linked visualization on Western Blots; ~5 µL yields visible bands on mini-gel with Coomassie blue stain

Sigma SDS-PRO-CE   Contains SDS protein standards, 2X SDS sample buffer, orange G solution, SDS protein separation medium, SDS washing solution & technical bulletin | SDS protein calibration kit; suitable for determination of MWs of protein subunits; a coated capillary is recommended

Sigma Silver-3   Contains one each of the three silver stain SDS-PAGE standards (Sigma M 5630, Sigma M 5505, Sigma M 6539), three vials of sample buffer & product information sheet | Silver Stain SDS-PAGE MW standard mixture; designed for MW determinations on silver stained SDS-PAGE (Laemmli) gels; formulated to yield bands with about equal intensity & no background when stained using Rapid Silver Stain (RSK-1) or AG-5 or AG-25 silver stain kits; supplied with 1 vial Laemmli sample buffer (Sigma S 3401) containing 6 mL 4% SDS, 20% glycerol, 10% BME & 0.004% bromphenol blue in 0.125 *M* Trizma base, pH 6.8

Sigma SMARKER-3   Contains one each of the three SigmaMarkers protein standard kits: Sigma M 3913, Sigma M 3788 & Sigma M 4038 | SigmaMarkers protein standards designed for use on PhastGel media & SDS-PAGE (Laemmli) gels

MKK7 β, Active

USBio M4100   Human recombinant, expressed in *E. coli*   Full-length MKK7β with an N-terminal GST-tag; SA: 550 U/mg when maximally activated

Monellin

ICN 155720   *Dioscorephyllum cumminsii* (serendipity berry)   Intensely sweet protein

Sigma M 7755   *Dioscorephyllum cumminsii* (serendipity berry)   Partially purified | Intensely sweet protein

Monocyte Chemotactic Protein I

Synonyms: Macrophage/Monocyte Chemoattractant Protein I; Macrophage/Monocyte Chemotactic & Activating Factor; Monocyte Chemotactic & Activating Factor; Macrophage/Monocyte Chemoattractant Protein I

PeproTech 300-04   Human recombinant, expressed in *E. coli*   MW 8.6k   >98% (SDS-PAGE) & HPLC; <0.1 ng endotoxin per mg (1EU/mg); lyophilized | Important role in the inflammatory response of blood monocytes & tissue macrophages; 76 AA; SA determined by its ability to chemoattract human monocytes

Sigma M 6667   Human recombinant, expressed in *E. coli*   ≥97% (SDS-PAGE); 0.2 µm filtered & lyophilized from PBS containing 500 µg BSA; activity tested in culture by measuring monocyte chemotactic activity; endotoxin tested | Product of the human JE gene; precursor form consists of 99 AA with a signal peptide sequence consisting of 23 amino-terminal AA; mature form has 4 cysteine residues; a member of the β chemokine subfamily characterized by the first 2 cysteine residues in an adjacent position CC; in *vitro* MCP-1 will act on monocytes to initiate chemotaxis, induce superoxide anion release, induce the release of lysosomal enzymes & augment cytostatic activity; in *vivo*, will induce macrophage infiltration; Matsushima, K et al, *J Exp Med*, 169: 1485, 1989General References for MCP1: Furutani, Y et al, *Biochem Biophys Res Commun*, 159: 249, 1989; Mukaida, N et al, *Microbiol Immunol*, 36(8): 773, 1992; Miller, M et al, *Critical Review in Immunology*, 12 (1,2): 17, 1992

Fitzgerald 30-AM46   Murine recombinant, expressed in *E. coli*

Chemicon GF041   Rat   ≥95%

Fitzgerald 30-AM45   Rat recombinant, expressed in *E. coli*

ICN 195804   Rat recombinant, expressed in *E. coli*   >95%; lyophilized; max chemotactic activity = 100 ng/mL on human monocytes & 10 ng/mL on eosinophils

Monocyte Chemotactic Protein II

Synonyms: Macrophage/Monocyte Chemoattractant Protein II

Chemicon GF013   Human   ≥95%

Fitzgerald 30-AM53   Human recombinant, expressed in *E. coli*

ICN 195009   Human recombinant, expressed in *E. coli*   MW 8k   >99%; lyophilized; 50 ng/mL activity

Sigma M 4292   Human recombinant, expressed in *E. coli*   MW ~9k   ≥97% (SDS-PAGE); 0.2 µm filtered & lyophilized in PBS containing 500 µg BSA; activity measured by its human monocyte chemotactic activity; endotoxin tested | Member of the CC or β chemokine class; acts primarily as chemoattractants & activate monocytes, dendritic cells, T lymphocytes, natural killer cells, B lymphocytes, basophils & eosinophils; originally identified as monocyte chemotactic proteins produced by human MG-63 osteosarcoma cells; 76 AA; shares 62% AA sequence homology with MCP-1 & 58% with MCP-3

USBio M4500   Human recombinant, expressed in *E. coli*   Highly purified, ≥99% (SDS-PAGE); no contaminants detected; single band by SDS-PAGE, IEP, &/or RID; 0.1 mg/mL; lyophilized | Suitable for antigenic applications in immunological protocols

Monocyte Chemotactic Protein III

Synonyms: Macrophage/Monocyte Chemoattractant Protein III

Biodesign A52317H   *E. coli*   MW 8.5k   Purified

Chemicon GF014   Human   ≥95%

Fitzgerald 30-AM54   Human recombinant, expressed in *E. coli*

ICN 195010   Human recombinant, expressed in *E. coli*   MW 8.5k   >99%; lyophilized; 50 ng/mL activity

Sigma M 8543   Human recombinant, expressed in *E. coli*   MW ~9k   ≥97% (SDS-PAGE); 0.2 µm filtered & lyophilized in 30% acetonitrile, 0.1% trifluoroacetic acid containing 500 µg BSA; activity measured by its human monocyte chemotactic activity; endotoxin tested | Member of the CC or β chemokine class; acts primarily as chemoattractants & activate monocytes, dendritic cells, T lymphocytes, natural killer cells, B lymphocytes, basophils & eosinophils; originally identified as monocyte chemotactic proteins produced by human MG-63 osteosarcoma cells; 76 AA; shares 71% AA sequence homology with MCP-1 & 58% with MCP-2

USBio M4500-10   Human recombinant, expressed in *E. coli*
Highly purified, ≥99% (SDS-PAGE); no contaminants detected; single band by SDS-PAGE, IEP, &/or RID; 0.1 mg/mL; lyophilized | Suitable for antigenic applications in immunological protocols

Monocyte Chemotactic Protein III-II
Synonyms: Macrophage/Monocyte Chemoattractant Protein III-II

Chemicon GF079   Murine   ≥95%

Monocyte Chemotactic Protein IV
Synonyms: Macrophage/Monocyte Chemoattractant Protein IV

Chemicon GF054   Human   ≥95%

Fitzgerald 30-AM55   Human recombinant, expressed in *E. coli*

ICN 195800   Human recombinant, expressed in *E. coli*
>95%; lyophilized; max chemotactic activity = 100 ng/mL on human monocytes & 10 ng/mL on eosinophils

Monocyte Chemotactic Protein V
Synonyms: Macrophage/Monocyte Chemoattractant Protein V

Chemicon GF080   Murine   ≥95%

ICN 195806   Murine recombinant, expressed in *E. coli*   >95%; lyophilized

Monokine, Interferon-γ Induced
Synonyms: C-X-C Chemokine; C-X-C Chemokine

Chemicon GF055   Human   ≥95%

BioSource International PHC1374   Human recombinant

ICN 195799   Human recombinant, expressed in *E. coli*   ≥95%; lyophilized; max chemotactic activity = 100 ng/mL on peripheral blood T-lymphocytes

PeproTech 300-26   Human recombinant, expressed in *E. coli*
MW 11.7k   >98% (SDS-PAGE) & HPLC; <0.1 ng endotoxin per mg (1EU/mg); lyophilized | Produced by macrophages & other cells; member of the a-chemokine family (C-X-C) of cytokines; acts as a chemoattractant toward monocytes, lymphocytes, & certain T cells; 103 AA; SA determined by its ability to chemoattract human peripheral blood T lymphocytes

BioSource International PMC1375   Mouse recombinant

PeproTech 250-18   Murine recombinant, expressed in *E. coli*
MW 12.2k   >98% (SDS-PAGE) & HPLC; <0.1 ng endotoxin per mg (1EU/mg); lyophilized | Produced by macrophages & other cells; belongs to the α-chemokine family (C-X-C) of cytokines; 105 AA; SA determined by its ability to chemoattract human lymphocytes

MSK1, Active
USBio M4692-51

Myelin
Biogenesis 6418-0896   Human brain tissue   MW ~70k (glycoprotein)   Tested negative for HBsAg, HTLV, and antibodies to HIV-1; purified; lyophilized | Contains glycoproteins, pricipally cholesterol phosphatides and cerebrosides in a high degree of unsaturation, characteristic of myelin; norton & Poduslo, *J Neurochem*, 21:749, 1973; Salzer et al, *J Cell Biol*, 104:957, 1987

Myelin Basic Protein
USBio M9758   ≥95% (SDS-PAGE & Coomassie blue staining); frozen solution in 10 *mM* MOPS, pH 7.0, 0.05% NaN₃ | A substrate for phosphorylation by several different proteinkinases including MAPK, PKA, calmodulin-dependent proteinkinase, PKC & phosphorylasekinase; even highly specific proteinkinases such as Raf1, MEK & MEKK can utilize MBP as an alternative substrate

Biogenesis 6420-0100   Bovine brain   78.5% protein by Biuret; purified; essentially salt free; lyophilized | Useful as iodination or immunoassay standard, substrate for phosphorylation for various proteinkinases (including Raf1, Mek, etc); Addison, *Horm Metabol Res*, 16:311, 1984

Fluka 70019   Bovine brain   ≥50% (GE); ≥90% protein

Sigma M 1891   Bovine brain   Lyophilized powder; may contain traces of urea-glycine buffer salts; ~50% (SDS-PAGE) | Major structural protein of CNS myelin; mutation of the myelin basic protein gene induces dysmyelination; used to induce experimental allergic encephalomyelitis

Sigma M 2295   Guinea pig brain   Lyophilized powder; may contain traces of urea-glycine buffer salts; ~50% (SDS-PAGE) | Major structural protein of CNS myelin; mutation of the myelin basic protein gene induces dysmyelination; used to induce experimental allergic encephalomyelitis; *J Immunol*, 129(3): 1209, 1982

Biodesign A86879H   Human brain   >90%

Biogenesis 6420-3006   Human nerve tissue   Tested negative for HBsAg and HIV-1; purified; 10 *mM* HCl; liquid | Addison, *Horm Metabol Res*, 16:311, 1984

Biogenesis 6420-3310   Mouse brain   Purified; 10 *mM* HCl, 0.1% NaN3; liquid

Sigma M 2016   Rabbit brain   Lyophilized powder; may contain traces of urea-glycine buffer salts; ~50% (SDS-PAGE) | Major structural protein of CNS myelin; mutation of the myelin basic protein gene induces dysmyelination; used to induce experimental allergic encephalomyelitis

Myelin Basic Protein, Dephosphorylated
USBio M9758-06   Bovine brain   95% (SDS-PAGE) & Coomassie blue staining; provided in 500 µL of 10 *mM* MOPS, pH 7.0, 0.3 *mM* MnCl₂, 1.56 *mM* EDTA, 156 ng inactive lambda phosphatase, 0.05% NaN₃; 5 mg/mL | Developed for use in radioactive & non-radioactivekinase assays; purified using SP-Sepharose TM HPLC & de-phosphorylated using Lambda protein phosphatase

Myeloma Proteins Immunoglobulin A (κ)
Synonyms: TEPC 15

ICN 50326   Ascites   Purified by salt precipitation, ion exchange & bioaffinity chromatography; dialyzed into 0.02 *M* Tris, 0.14 *M* NaCl, pH 8.1; adjusted to 1.0 mg/mL, filtered, vialed, stored frozen at −70°C

Myeloma Proteins Immunoglobulin A (λ₂)
Synonyms: MOPC 315

ICN 50325   Ascites   Purified by salt precipitation, ion exchange & bioaffinity chromatography; dialyzed into 0.02 *M* Tris, 0.14 *M* NaCl, pH 8.1; adjusted to 1.0 mg/mL, filtered, vialed, stored frozen at −70°C

Myeloma Proteins Immunoglobulin G₁ (κ)
Synonyms: MOPC 21

ICN 50327   Ascites   Purified by salt precipitation, ion exchange & bioaffinity chromatography; dialyzed into 0.02 *M* Tris, 0.14 *M* NaCl, pH 8.1; adjusted to 1.0 mg/mL, filtered, vialed, stored frozen at −70°C

Myeloma Proteins Immunoglobulin G₂ₐ (κ)
Synonyms: UPC 10; RPC 5

ICN 50328   Ascites   Purified by salt precipitation, ion exchange & bioaffinity chromatography; dialyzed into 0.02 *M* Tris, 0.14 *M* NaCl, pH 8.1; adjusted to 1.0 mg/mL, filtered, vialed, stored frozen at −70°C

ICN 50329   Ascites   Purified by salt precipitation, ion exchange & bioaffinity chromatography; dialyzed into 0.02 *M* Tris, 0.14 *M* NaCl, pH 8.1; adjusted to 1.0 mg/mL, filtered, vialed, stored frozen at −70°C

Myeloma Proteins Immunoglobulin G₂ᵦ (κ)
Synonyms: MOPC 195; MOPC 141

ICN 50330   Ascites   Purified by salt precipitation, ion exchange & bioaffinity chromatography; dialyzed into 0.02 *M* Tris, 0.14 *M* NaCl, pH 8.1; adjusted to 1.0 mg/mL, filtered, vialed, stored frozen at −70°C

ICN 50331 Ascites Purified by salt precipitation, ion exchange & bioaffinity chromatography; dialyzed into 0.02 *M* Tris, 0.14 *M* NaCl, pH 8.1; adjusted to 1.0 mg/mL, filtered, vialed, stored frozen at –70°C

Myeloma Proteins Immunoglobulin G₃ (κ)

Synonyms: FLOPC 21;; J 606

ICN 50332 Ascites Purified by salt precipitation, ion exchange & bioaffinity chromatography; dialyzed into 0.02 *M* Tris, 0.14 *M* NaCl, pH 8.1; adjusted to 1.0 mg/mL, filtered, vialed, stored frozen at –70°C

ICN 50333 Ascites Purified by salt precipitation, ion exchange & bioaffinity chromatography; dialyzed into 0.02 *M* Tris, 0.14 *M* NaCl, pH 8.1; adjusted to 1.0 mg/mL, filtered, vialed, stored frozen at –70°C

Myeloma Proteins Immunoglobulin M (κ)

Synonyms: TEPC 183

ICN 50336 Ascites Purified by salt precipitation, ion exchange & bioaffinity chromatography; dialyzed into 0.02 *M* Tris, 0.14 *M* NaCl, pH 8.1; adjusted to 1.0 mg/mL, filtered, vialed, stored frozen at –70°C

Myeloma Proteins Immunoglobulin M (λ₁)

Synonyms: MOPC 104E

ICN 50335 Ascites Purified by salt precipitation, ion exchange & bioaffinity chromatography; dialyzed into 0.02 *M* Tris, 0.14 *M* NaCl, pH 8.1; adjusted to 1.0 mg/mL, filtered, vialed, stored frozen at –70°C

Myoglobin

Sigma H 3029 Lyophilized; each vial contains 20 nmoles myoglobin | Hydrolysis standard for AA analysis

Cortex CP1030U Cardiac >98%

Sigma M 8007 Dog heart 95-100%; crystallized & lyophilized; essentially salt-free; iron content: ~0.20%

Sigma M 7382 Dog skeletal muscle 95-100%; crystallized & lyophilized; essentially salt-free; iron content: ~0.20%

Fluka 70030 Equine heart MW 17.8k ≥90% (GE); 2X crystallized; lyophilized; essentially salt-free | Oxygen carrier protein; Brunori, M et al, Top Mol Struct Biol, 7: 263, 1985; Kleparnik, K et al, *Electrophoresis*, 14: 475, 1993

Fluka 70025 Equine skeletal muscle MW 18.8k ≥80% (GE); ≤10% water; ≤0.3% iron

Scipac P136-3 Heart tissue >96%; frozen in sodium phosphate buffer | Cardiac marker protein

Scipac P136-4 Heart tissue 40-90%; frozen in sodium phosphate buffer | Cardiac marker protein

Sigma M 1882 Horse heart ≥90% (PhastGel); lyophilized; essentially salt-free; iron content: ≥0.20%

Sigma M 9267 Horse heart pI 6.8, 7.2; vial contains ~2 mg | IEF Marker

ICN 100862 Horse skeletal muscle ≥98%; salt-free; lyophilized; 0.3% iron

Sigma M 0630 Horse skeletal muscle 95-100%; crystallized & lyophilized; essentially salt-free; iron content: ~0.30%

ICN 55840 Human Control; 1 mg/mL

USBio M9800 Human ≥97% (SDS-PAGE); no contaminants detected; single band by SDS-PAGE, IEP, &/or RID; 1.3 mg/mL protein; supplied in 0.02 *M* Tris HCl & 1 *mM* EDTA, pH 8.4 | Suitable for antigenic applications in immunological protocols

USBio M9800-10 Human ≥60% (SDS-PAGE); no contaminants were detected by IEP; 0.7 mg/mL protein; supplied in 0.02 *M* Tris HCl & 1 *mM* EDTA, pH 8.4 | Suitable for antigenic applications in immunological protocols

Biogenesis 6450-1089 Human cardiac muscle <3% HAS, <1% other proteins, <0.5% hemoglobin, 0% parvalbumin; tested negative for HBsAg and HTLV II antibodies; purified; 0.05 *M* NH₄HCO₃, pH 8.0; lyophilized | Totally soluble from 0-100 µg/mL; at higher concentrations, 25-35% of the product does not solubilize because of the tendancy of myoglobin to clot; molar extinction coefficient: 3.4 x 10,000 (at 280 nm); *Res Exp Med*, 171:71, 1977

Biodesign A31210H Human heart Antigen grade | Cardiac markers

Biodesign A32215H Human heart Calibrator grade | Cardiac markers

Biogenesis 6450-1104 Human heart Lyophilized

Fitzgerald 30-AM20 Human heart High purity

Fitzgerald 30-AM21 Human heart Standard purity

Sigma M 6036 Human heart ≥95% (SDS-PAGE); 2 mg protein (Lowry)/mL in 20 mM Tris-HCl, 1 mM EDTA & 50% glycerol, pH 8.5

USBio M9800-18 Human heart ≥97% (SDS-PAGE); no contaminants detected; single band by SDS-PAGE, IEP, &/or RID; 0.4 mg/mL specific protein RIA supplied in 0.02 *M* Tris HCl & 1 *mM* EDTA, pH 8.4 | Suitable for antigenic applications in immunological protocols

Cortex CP3031P Human skeletal muscle Whole molecule >40%

Sigma M 1277 Sheep skeletal muscle ~70%; lyophilized; essentially salt-free; iron content: ~0.25%

Scipac P210-3 Skeletal muscle >96%; in sodium phosphate buffer | Urine protein

Scipac P210-4 Skeletal muscle 40-90%; in sodium phosphate buffer | Urine protein

Fluka 70035 Sperm whale recombinant, expressed *E. coli* MW 17k ≥96% (GE); 2 mg/mL protein | Springer, BA & Sliger, SG, *PNAS*, 84: 8961, 1987

Sigma M 7527 Sperm whale recombinant, expressed in *E. coli* 95-100%; solution in 0.02 M Tris-HCl, pH 8.0; iron content: ~0.30% | Produced by a synthetic gene expressed in *E. coli*; contains an *N*-terminal methionine not present in the natural product; AA sequence not confirmed by Sigma; Springer, BA & Sligar, SG, *Proc Natl Acad Sci USA*, 84: 8961, 1987

Myosin

Sigma M 6643 Bovine muscle Calcium activated; solution in 50% glycerol containing 0.6 M KCl & 10 mM potassium phosphate buffer, pH 6.8; ATPase U definition: 1 U liberates 1.0 µmole of inorganic phosphorus from ATP/min at pH 9.0 at 25°C in the presence of calcium; activity: 0.3-1.0 U/mg protein (Biuret)

Sigma M 1270 Chicken gizzard Calcium activated; solution in 50% glycerol containing 0.6 M KCl & 10 mM potassium phosphate buffer, pH 6.8; ATPase U definition: 1 U liberates 1.0 µmole of inorganic phosphorus from ATP/min at pH 9.0 at 25°C in the presence of calcium; activity: 0.1-0.3 U/mg protein (Biuret)

Sigma M 7266 Chicken muscle Calcium activated; solution in 50% glycerol containing 0.6 M KCl & 10 mM potassium phosphate buffer, pH 6.8; ATPase U definition: 1 U liberates 1.0 µmole of inorganic phosphorus from ATP/min at pH 9.0 at 25°C in the presence of calcium; activity: 0.5-1.5 U/mg protein (Biuret)

Biogenesis 6490-2956 Human heart MW 500k Tested negative for HBsAg and HIV-1; purified; 50% Glycerol; liquid

Sigma M 0531 Porcine heart Calcium activated; solution in 50% glycerol containing 0.6 M KCl & 10 mM potassium phosphate buffer, pH 6.8; ATPase U definition: 1 U liberates 1.0 µmole of inorganic phosphorus from ATP/min at pH 9.0 at 25°C in the presence of calcium; activity: 0.1-0.3 U/mg protein (Biuret)

Sigma M 0273 Porcine muscle Calcium activated; solution in 50% glycerol containing 0.6 M KCl & 10 mM potassium phosphate buffer, pH 6.8; ATPase U definition: 1 U liberates 1.0 µmole of inorganic phosphorus from ATP/min at pH 9.0 at 25°C in the presence of calcium; activity: 0.1-0.5 U/mg protein (Biuret)

Biogenesis 6490-3004 Rabbit muscle MW 500k Purified; 50% Glycerol; liquid

Fluka 70045   Rabbit muscle   MW 500k   ≥50% (GE); 0.5-2.5 U/mg protein; solution in 50% glycerol & 0.6 *M* KCl, pH 6.8; 1 U corresponds to the amount of enzyme which liberates 1 μmol inorganic phosphate from ATP/min at pH 9.0, 25°C, in the presence of calcium | Frederiksen, DW & Cunningham, LW, *Meth Enzymol*, 85: 55, 1982

ICN 153887   Rabbit muscle   ≥90% (4% PAGE); 5% glycerol, 0.6 *M* KCl, pH 7; soluble in 0.5 *M* KCl; stable for 6 months when stored at −20°C; 5 mg/mL protein (Biuret)

Sigma M 0163   Rabbit muscle   MW 205k   Vial contains enough FITC-conjugated protein to run 50 mini-gels or 25 standard size gels; protein band be visualized by using UV light or Brilliant Blue stain | Fluorescent marker for SDS-page & protein transfer; suitable for use as molecular weight standards in both SDS-PAGE & transfer membranes

Sigma M 1636   Rabbit muscle   Calcium activated; solution in 50% glycerol containing 0.6 M KCl & 10 mM potassium phosphate buffer, pH 6.8; ATPase U definition: 1 U liberates 1.0 μmole of inorganic phosphorus from ATP/min at pH 9.0 at 25°C in the presence of calcium; activity: 0.5-1.5 U/mg protein (Biuret)

Myosin Heavy Chain

Synonyms: Myosin, ATPase Inactive Whole Chain

Sigma M 7659   Rabbit muscle   Solution in 50% glycerol containing 0.6 M KCl & 0.005 M potassium phosphate buffer, pH 6.5; ATPase activity: <0.01 U/mg protein

Myosin Light Chain

Fluka 70048   Bovine muscle   Powder; ≥98% protein; composition: 15% 27,000; 40% 24,000; 45% 18,000

Sigma M 6648   Bovine muscle   Lyophilized powder containing ~85% protein (Biuret); balance primarily KCl, Tris & EDTA; three major bands on SDS electrophoresis | Prepared by the method of Holt, JC & Lowey, S, *Biochemistry*, 14: 4600, 1975

Sigma M 7518   Chicken muscle   Lyophilized powder containing ~85% protein (Biuret); balance primarily KCl, Tris & EDTA; three major bands on SDS electrophoresis | Prepared by the method of Holt, JC & Lowey, S, *Biochemistry*, 14: 4600, 1975

Fitzgerald 30-AM25   Human cardiac, left ventricle   High purity

USBio M9850-10   Human heart left ventricle   ≥98% (SDS-PAGE); no contaminants detected; single band by SDS-PAGE, IEP, &/or RID; 1 mg/mL; lyophilized | Suitable for antigenic applications in immunological protocols

Sigma M 9891   Rabbit muscle   Lyophilized powder containing ~85% protein (Biuret); balance primarily KCl, Tris & EDTA; three major bands on SDS electrophoresis | Prepared by the method of Holt, JC & Lowey, S, *Biochemistry*, 14: 4600, 1975

Myosin Light Chain I

Cortex CP3030U   Human ventricular   >98%

Myosin Subfragment I

Sigma M 5772   Rabbit muscle   Solution in 50% glycerol containing 0.5 M KCl & 0.025 M potassium phosphate, pH 6.2; ATPase activity: 1-4 U/mg protein (Biuret); unit definition: 1 U liberates 1.0 μmole of inorganic phosphorus from ATP/min at pH 9.0 at 25°C in the presence of calcium | Produced from a chymotryptic digest of myosin in the presence of EDTA

Myosin Subfragment II

Synonyms: Myosin Long Subfragment II

Sigma M 5897   Rabbit muscle   Solution in 50% glycerol containing 0.5 M KCl & 0.025 M potassium phosphate, pH 6.2; significantly free of ATPase activity | Produced from a chymotryptic digest of heavy meromyosin

Myosin, (¹⁴C-Me)-

ARC ARC-433   MW 200k   3-30 μCi/mg; 111-1111 KBq/mg; in 0.01 *M* sodium phosphate, pH 7.2 | Radiochemical

Sigma M 8922   Bovine muscle   5-50 μCi/mg protein; solution in 10 *mM* sodium phosphate, pH 7.0, with 1% SDS, 1% BME & 1 *mM* EDTA in Combi-vial | Radiochemical; prepared from Sigma M 6643

Myotoxin I

Sigma M 9047   *Crotalus viridis concolor* (midget faded rattlesnake) venom   Components of rattlesnake venom causing instantaneous paralysis of bitten prey; some sequence homology with crotamine; presumed to have similar modes of action; not assayed by Sigma; Volpe, P et al, *Arch Biochem Biophys*, 246: 90, 1986; Ohkura, M et al, *Eur J Pharmacol*, 268: R1, 1994

Myotoxin II

Sigma M 9172   *Crotalus viridis concolor* (midget faded rattlesnake) venom   Components of rattlesnake venom causing instantaneous paralysis of bitten prey; some sequence homology with crotamine; presumed to have similar modes of action; not assayed by Sigma; Volpe, P et al, *Arch Biochem Biophys*, 246: 90, 1986; Ohkura, M et al, *Eur J Pharmacol*, 268: R1, 1994

Nerve Growth Factor

ICN 191130   *Echis multisguamatus* venom   Enriched NGF fraction

IBT GF-022-5, GF-022-8   Mouse submaxillary glands   >95 %; lyophilized powder | Fully active as measured by the receptor binding assay using recombinant NGF receptor extracellular domain expressed in transformed CHO cells

ICN 191132   *Naja oxiana* venom   Electrophoretically homogeneous

Sigma N 8133   *Vipera lebetina* (snake) venom   MW 32.5k   Purified by HPLC; ≥95% (SDS-PAGE); 0.2 μm filtered & lyophilized from sodium acetate buffer solution; endotoxin tested | Siigur, E et al, *Comp Biochem Physiol*, 81B: 211, 1985; general comments & references are the same as for Sigma B 3795

Nerve Growth Factor 2.5S

Chemicon NC011   ≥95%

Harlan BT-5017   Lyophilized; 1 mg | Cross-reactive with human, mouse, rat, canine, feline, rabbit, bovine, equine, swine, primate, guinea pig, ovine & avian

USBio N2050-05   The minimum concentration needed for neurite outgrowth of PC-12 rat pheochromocytoma cells is 10ng/mL

ICN 150022   Male mouse submaxillary gland   ≥95%; lyophilized; ED₅₀ = 10.0-25.0 ng/mL, 1 U stimulates chick dorsal root ganglia neurite outgrowth dose-dependently | Each lot is sterile filtered & tested for bacteria, fungi & mycoplasma

Sigma N 6009   Male mouse submaxillary glands   0.2 μm filtered & lyophilized from sodium acetate buffer solution; endotoxin tested | Essentially the β-subunit of NGF-7S when isolated under initially dissociative conditions using a modification of the method of Bocchini, V & Angeletti, P, *Proc Natl Acad Sci USA*, 64: 787, 1969

Biogenesis 6620-1004   Mouse   Lyophilized

Biogenesis 6620-1015   Mouse   ¹²⁵I conjugated

ICN 160040   Mouse submaxillary gland   >95%; lyophilized; bioassayed in a rat pheochromocytoma cell line | Bocchini, V & PU Angeletti, *PNAS*, 64:787, 1969; Greene, LA, *Brain Res*, 133:350, 1977

Calbiochem 480352   Mouse submaxillary glands   MW 26k   ≥95% (SDS-PAGE); lyophilized from PBS | Enhances survival, phagocytosis & superoxide production in murine neutrophils; reported to stimulate extracellular matrix invasion by human myeloma cells; *Merck Index*, 12: 6562; Hermann, JL et al, *Mol Cell Biol*, 4: 1205, 1993; Raffioni, S & Bradshaw, RA, *PNAS*, 89: 9121, 1992; Traverse, S et al, *Biochem J*, 288: 351, 1992; Levi-Montalcini, R, *In Vitro Cell Dev Biol*, 23: 227, 1987

Nerve Growth Factor 2.5S, Grade I

Alexis 521-006   Male mouse submaxillary glands   MW 26k ≥98% (SDS-PAGE); lyophilized; mNGF 2.5S concentration is estimated spectroscopically from its extinction coefficient & AA analysis; biological activity: bioassayed for neurotrophic activity using the rat pheochromocytoma cell PC12 cultures over a period of 7-14 days; activity is in the range of 0.1-10 ng/mL; recommended concentration to be used *in vitro* for maintenance of sympathetic & sensory nerve cultures is 50 ng/mL medium | Dimer protein consisting of two identical subunits; Thoenen, H, *TINS*, 14: 165, 1991; Thoenen, H et al, *CR Acad Sci III*, 316: 1158, 1993; Maness, LM et al, *Neurosci & Biobehav Rev*, 18: 143, 1994; Klein, R, *FASEB J*, 8: 738, 1994; Barinaga, M, *Science*, 264: 772, 1994; Nishi, R, *Science*, 265: 1052, 1994; Davies, AM, *Nature*, 368: 193, 1994; Bradshaw, RA, *Ann Rev Biochem*, 47: 191, 1978

Nerve Growth Factor 2.5S, Grade II

Alexis 521-007   Male mouse submaxillary glands   MW 26k ≥90% (SDS-PAGE); lyophilized; mNGF 2.5S concentration is estimated spectroscopically from its extinction coefficient & AA analysis; biological activity: bioassayed for neurotrophic activity using the rat pheochromocytoma cell PC12 cultures over a period of 7-14 days; activity is in the range of 0.5-50 ng/mL; recommended concentration to be used *in vitro* for maintenance of sympathetic & sensory nerve cultures is 50 ng/mL medium | Dimer protein consisting of two identical subunits; Thoenen, H, *TINS*, 14: 165, 1991; Thoenen, H et al, *CR Acad Sci III*, 316: 1158, 1993; Maness, LM et al, *Neurosci & Biobehav Rev*, 18: 143, 1994; Klein, R, *FASEB J*, 8: 738, 1994; Barinaga, M, *Science*, 264: 772, 1994; Nishi, R, *Science*, 265: 1052, 1994; Davies, AM, *Nature*, 368: 193, 1994; Bradshaw, RA, *Ann Rev Biochem*, 47: 191, 1978

Nerve Growth Factor 7S

Chemicon NC010   ≥95%

Harlan BT-5023   Lyophilized; 1 mg | Cross-reactive with human, mouse, rat, canine, feline, rabbit, bovine, equine, swine, primate, guinea pig, ovine & avian

USBio N2050-06   Dose for significant outgrowth of neurite using PC-12 rat pheochromocytoma cells: 10ng/mL; 7S NGF has a variety of effects on sensory & sympathetic neuron growth & development; 7S NGF is also needed for sympathetic nerve cells development & maintenance

ICN 150174   Male mouse submaxillary gland   ≥90%; lyophilized; ED$_{50}$ = 5.0-10.0 ng/mL; 1 U stimulates chick dorsal root ganglia neurite outgrowth dose-dependently | Each lot is sterile filtered & tested for bacteria, fungi & mycoplasma

Alexis 521-008   Male mouse submaxillary glands   MW 130k >97% (SDS-PAGE); lyophilized; 100 μg determined by Lowry & Pierce method; biological activity: bioassayed for neurotrophic activity using the rat pheochromocytoma cell PC12 cultures over a period of 7-14 days; activity is in the range of 0.5-50 ng/mL; recommended concentration to be used *in vitro* for maintenance of sympathetic & immature sensory nerve cells is 50-100 ng/mL medium | Dimer protein consisting of two identical subunits; Thoenen, H, *TINS*, 14: 165, 1991; Thoenen, H et al, *CR Acad Sci III*, 316: 1158, 1993; Maness, LM et al, *Neurosci & Biobehav Rev*, 18: 143, 1994; Klein, R, *FASEB J*, 8: 738, 1994; Barinaga, M, *Science*, 264: 772, 1994; Nishi, R, *Science*, 265: 1052, 1994; Davies, AM, *Nature*, 368: 193, 1994; Varon, S et al, *Biochemistry*, 6: 2202, 1967

Sigma N 0513   Male mouse submaxillary glands   MW 130k 0.2 μm filtered & lyophilized from sodium phosphate buffered solution containing 500 μg BSA; endotoxin tested | Nerve growth factor; protein isolated using a modification of the method of Varon, S et al, *Biochem*, 6: 2202, 1967; generally thought that NGF-7S consists of 5 subunits (2a, 1b, 2g); only the b-subunit has neurotrophic activity

ICN 160072   Mouse submaxillary gland   MW 140k lyophilized | Stimulates neurite-like fiber outgrowth *in vitro* using a pheochromocytoma cell line; useful for *in vitro* maintenance of sympathetic & immature sensory nerve cells; Varon, S etal, *Biochem*, 6:2202, 1987

Calbiochem 480354   Mouse submaxillary glands   MW 130k Lyophilized solid; EGF: none detected | Promotes neuron survival & neurite outgrowth in newborn rat brain; composed of three subunits: (α:β:γ):2:1:2; the α-subunit is an inactive serine proteinase & the γ-subunit is an active serine proteinase capable of processing the precursor form of β-NGF; *Merck Index*, 12: 6562; Bax, B et al, *Structure*, 5: 1275, 1997; Shao, N et al, *Brain Res*, 609: 338, 1993; Varon, S et al, *Methods Neurochem*, 3: 203, 1972

Fluka 72183   Murine submaxillary gland   MW 130k Prepared by gel filtration, ion-exchange chromatography, sterilize-filtered & lyophilized from 5 *mM* sodium phosphate buffer, pH 6.8 | Neurofilament outgrowth observed at 30 ng/mL; Yanker, BA & Shooter, EM, *Ann Rev Biochem*, 51: 845, 1982; Vale, RD & Shooter, EM, *Meth Enzymol*, 109: 21, 1985

Nerve Growth Factor R/Fc Chimera

Synonyms: Neurotrophin R, p75

R&D Systems 367-NR-050   Human recombinant, expressed in *SF21*   >95%; lyophilized; ED$_{50}$: 0.2-0.6 μg/mL | Species specificity: human NGF R; member of the TNF receptor superfamily; ligands for NGF R include NGF, BDNF, NT-3 & NT-4; shown to regulate cell migration, gene expression & to mediate apoptosis; recombinant NGF R binds NGF with high affinity & is a potent NGF antagonist; Barker, PA & Murphy, RA, *Mol Cell Biochem*, 110: 1, 1992; Bamji, AX et al, *J Cell Biol*, 140: 911, 1998; Feinstein, E et al, *Trends Biochem Sci*, 20: 342, 1995

Nerve Growth Factor Receptor, Human Extracellular Domain

IBT GR-020-3, GR-020-5   Chinese hamster ovary cells (CHO cells)   >85%; 1.0 mg prot/mL in 20 *mM* HEPES, 0.1 M NaCl, pH 7.3 | Binds NGF

Nerve Growth Factor β

Biodesign A52451H   *E. coli*   MW 28k   Purified

Chemicon GF028   Human   ≥95%

Biogenesis 6620-2030   Human r-DNA   Lyophilized

BioSource International PHG0124   Human recombinant

Alexis 521-005   Human recombinant, expressed in *E. coli* MW 26k   ≥98% (SDS-PAGE); lyophilized; 5 μg determined by AA analysis; biological activity: rhβ-NGF preparation bioassayed for neurotrophic activity using the rat pheochromocytoma cell PC12 cultures & the effective dose 50% was found as 2 ng/mL medium | Dimer protein consisting of two identical 119 AA subunits associated through strong hydrophobic interactions; Thoenen, H, *TINS*, 14: 165, 1991; Thoenen, H et al, *CR Acad Sci III*, 316: 1158, 1993; Maness, LM et al, *Neurosci & Biobehav Rev*, 18: 143, 1994; Klein, R, *FASEB J*, 8: 738, 1994; Barinaga, M, *Science*, 264: 772, 1994; Nishi, R, *Science*, 265: 1052, 1994; Davies, AM, *Nature*, 368: 193, 1994

Fitzgerald 30-AN15   Human recombinant, expressed in *E. coli* High purity

PeproTech 450-01   Human recombinant, expressed in *E. coli* MW 27.0k (dimer)   >98% (SDS-PAGE) & HPLC; <0.1 ng endotoxin per mg (1EU/mg); lyophilized | Potent neurotrophic factor that supports the growth & survivability of nerve and/or glial cells; active form is formed by two identical 119 AA subunits held together by strong hydrophobic interactions; ED$_{50}$ = 2.0- 5.0 ng/mL; SA determined by the dose-dependent induction of choline acetyl transferase activity in rat basal forebrain primary septal cultures

Calbiochem 480275   Human recombinant, expressed in mouse myeloma cell line NS0   MW 26k   ≥97% (SDS-PAGE); lyophilized from filter-sterilized 0.2% acetic acid containing 50 μg BSA/μg β-NGF; biological activity: ED$_{50}$=800-1500 pg/mL as measured in a cell proliferation assay using a factor-dependent human erythroleukemic cell line; endotoxin: ≤100 pg/μg β-NGF | Involved in neuroimmune interactions & inflammation; *Merck Index*, 12: 6562; Leon, A et al, *PNAS*, 91: 3739, 1994

ICN 160062   Human recombinant, expressed in NS0   ≥97%; lyophilized with carrier; ED$_{50}$ >0.3 ng/mL; 1 U proliferates human erythroleukemic TF-1 cells dose-dependently | Measured by ability to support survival & to stimulate neurite outgrowth of embryonic chick dorsal root ganglia

Sigma N 1408   Human recombinant, expressed in NSO murine myeloma cells   ≥97% (SDS-PAGE); lyophilized from 0.2% acetic acid containing 5 mg BSA; endotoxin tested; cell culture tested

Sigma N 2393   Male mouse submaxillary glands   Purified by HPLC; ≥95% (SDS-PAGE); 0.2 μm filtered & lyophilized from sodium phosphate buffered solution; endotoxin tested | Purified using a modification of the method of Varon, S et al, *Biochem*, 7: 1296, 1968

ICN 152303   Mouse submaxillary gland   MW ~13k   Isolated from 7S-NGF; ≥98% | Responsible for survival & fiber outgrowth of sympathetic neurons & embryonic sensor neurons from dorsal root ganglia

ICN 159840   *Naja naja kaouthia* (Thailand cobra) venom   MW ~13k   Purified (single band by SDS-PAGE); suggested use = 1-5 ng/mL media | Produces neurite outgrowth on rat pheochromocytoma (PC-12) cells within 24 hrs

R&D Systems 256-GF-100   NSO-expressed   >97%; lyophilized; ED_{50}: 0.5-1 ng/mL | Species specificity: human

Oncogene PF043   Rat recombinant   MW 13.2k   >97% (SDS-PAGE); lyophilized in 0.2% acetic acid containing 50 mg BSA per 1 mg of cytokine; reconstitution buffer should contain 0.1% serum; biological activity: EC_{50} of 0.5-1.0 ng/mL as determined by a cell proliferation assay using factor-dependent TF-1 cells | Species reactivity: rat; for proliferation studies; once reconstituted, exists as a non-disulfide linked homodimer; activity measured in a cell proliferation assay using a factor-dependent human eythroleukemic cell line, TF-1

Sigma N 2513   Rat recombinant, expressed in *E. coli*   ≥97% (SDS-PAGE); 0.2 μm filtered & lyophilized in 0.2% acetic acid containing 5 mg BSA; biological activity was measured in a cell proliferation assay using the factor-dependent human erythroleukemic cell line, TF-1; endotoxin tested

ICN 195793   Rat recombinant, expressed in *Sf*21   ≥97%; lyophilized; ED_{50} = 0.5-1.0 ng/mL

R&D Systems 556-NG-100   *SF*21-expressed   >97%; lyophilized; ED_{50} = 0.5-1 ng/mL | Species specificity: rat

Nerve Growth Factor, (^{125}I)-

ICN 68118   Human recombinant   ~40 μCi/μg, ~1.5 MBq/μg; 0.1 *M* KPB, ph7.5, 0.25% BSA

Netropsin

Synonyms: Sinanomycin; Congocidin

Sigma N 9653   *Streptomyces netropsis*   FW 503.4 $C_{18}H_{26}N_{10}O_3 \cdot 2HCl$ Dihydrochloride; ≥98% | Unusual *N*-methylpyrrole-containing oligopeptide that binds to AT-rich sequences of dsDNA, especially in the minor groove; protects such regions from DNase I & other endonucleases, also inhibits topoisomerases; disrupts the cell cycle, prolonging G1 & arresting in G2; Zimmer, C et al, *Nucl Acid Res*, 8: 2999, 1980; Beerman, TA et al, *Biochim Biophys Acta*, 1090: 52, 1991; Poot, M et al, *Exp Cell Res*, 218: 326, 1995

Neural Cell Adhesion Molecule

Chemicon AG265   Chicken   ≥95% | Immunoglobulin superfamily adhesion molecule

Neurofilament

Biodesign A08008B	Bovine spinal cord	MW 68k	>98%
Biodesign A08009B	Bovine spinal cord	>98%	
Biodesign A08010B	Bovine spinal cord	MW 200k	>98%

Sigma N 1022   Bovine spinal cord   Lyophilized from a solution containing 6 M urea, 10 mM sodium phosphate, 5 mM EDTA, 1% BME, pH 7.5 | Intermediate filaments found in axons of large myelinated fibers, most neurons, astrocytes & Schwann cells; prepared using a modification of Dahl, D et al, *Anal Biochem*, 126: 165, 1982

Neurofilament 160

ICN 771082   MW 160k   Highly purified protein useful as a high MW marker, standard or Ag

Neurofilament 200

ICN 771072   MW 200k   Highly purified protein useful as a high MW marker, standard or Ag

Neurofilament 68

ICN 771092   MW 68k   Highly purified protein useful as a high MW marker, standard or Ag

Neurophysin I

ICN 155820   Bovine pituitary   MW 9330.5   ≥95%; ~80% protein

Sigma N 2404   Bovine pituitary   A protein found in vasopressin- & oxitocin-containing neurons in the hypothalamus that is associated with the transport of these hormones to the posterior pituitary

Biogenesis 6740-0807   Porcine pituitary   Liquid

Neurophysin I+II

Biogenesis 6740-0917   Porcine pituitary   Liquid

Neuroprotective Factor II, Activity Dependent

BioSource International PHC1094   Human recombinant

Neurotactin

Synonyms: Fractalkine

BioSource International PHC1174   Human recombinant

Neurotoxin I

Sigma T 3643   *Naja naja oxiana*   Lyophilized powder purified by gel filtration & ion exchange chromatography | Postsynaptic neurotoxin; not assayed by Sigma; Grishin, EV et al, *FEBS Lett*, 45: 118, 1974

Neurotoxin II

Sigma T 3768   *Naja naja oxiana*   Lyophilized powder purified by gel filtration & ion exchange chromatography | Postsynaptic neurotoxin; not assayed by Sigma; Grishin, EV et al, *FEBS Lett*, 36: 77, 1973

Neurotrophic Factor Soluble Receptor α, Ciliary

Synonyms: Neurotrophic Factor, Ciliary; Neurotrophic Factor, Ciliary

BioSource International PHC7015   Human recombinant

Oncogene PF041   Human recombinant   MW 43k (glycosylated)   >97% (SDS-PAGE); lyophilized; biological activity: EC_{50} of 0.2-0.4 μg/mL as determined by the ability to enhance TF-1 cell proliferation in the presence of 20 ng/mL recombinant human CNTF protein | Species reactivity: human; for proliferation studies; binds CNTF in solution; complex acts on cells that express only LIF Rβ & gp130 but not CNTF Rα; contains 324 AA with a predicted MW ~36k; as a result of glycosylation, migrates as a 43k band in SDS-PAGE

BioSource International PRC7014   Rat recombinant

Oncogene PF042   Rat recombinant   MW 43k (glycosylated)   >97% (SDS-PAGE); lyophilized; biological activity: EC_{50} of 0.05-0.15 μg/mL as determined by the ability to enhance TF-1 cell proliferation in the presence of 10 ng/mL recombinant rat CNTF protein | Species reactivity: rat; for proliferation studies; binds CNTF in solution; complex acts on cells that express only LIF Rβ & gp130 but not CNTF Rα

Neurotrophic Factor, Brain Derived

Biodesign A52452H	*E. coli*	MW 28k	Purified
Chemicon GF029	Human	≥95%	
Fitzgerald 30-AN16	Human brain recombinant, expressed in *E. coli* High purity		
Biogenesis 1504-1030	Human r-DNA	Lyophilized	

Sigma B-147   Human recombinant synthetic   >96%; lyophilized; carrier & additive free; activity levels are provided with each lot; reconsituted in water to a concentration of 100 μg/ml; may be diluted with buffered solution | Nerve growth factor (NGF)-related protein that increases neuronal survival *in vitro*Hughes et al, *Neuroscience*, 57: 319, 1993; Murphy et al, *J Neurosci*, 13: 2853, 1993; Weiss et al, *Science*, 260: 1072, 1993

Alexis 521-009   Human recombinant, expressed in *E. coli* >97% (SDS-PAGE); lyophilized; 1 μg/5 μg (AA analysis); biological activity: activity was evaluated for the ability to support the survival of rat basal forebrain primary septal cultured neurons; effective concentration of 50% (50 ng/mL) was determined measuring the induction of choline acetyltransferase enzymatic activity | Dimer protein consisting of two identical 119 AA subunits associated through strong hydrophobic interactions; Thoenen, H, *TINS*, 14: 165, 1991; Thoenen, H et al, *CR Acad Sci III*, 316: 1158, 1993; Maness, LM et al, *Neurosci & Biobehav Rev*, 18: 143, 1994; Klein, R, *FASEB J*, 8: 738, 1994; Barinaga, M, *Science*, 264: 772, 1994; Nishi, R, *Science*, 265: 1052, 1994; Davies, AM, *Nature*, 368: 193, 1994

Calbiochem 203702   Human recombinant, expressed in *E. coli* MW 28k   >97% (SDS-PAGE); lyophilized solid; soluble in most aqueous buffers & water; biological activity: Active in rat basal forebrain primary septal cultured neurons; induces choline acetyltransferase (EC_{50}=50 ng/mL) | Neuronal survival & growth promoting factor; its actions are mediated via the TrkB receptor; *Merck Index*, 12: 6570; Erickson, JT et al, *J Neurosci*, 16: 5361, 1996; Leibrock, J et al, *Nature*, 341: 149, 1989

PeproTech 450-02   Human recombinant, expressed in *E. coli* MW 27.0k (dimer)   >98% (SDS-PAGE) & HPLC; <0.1 ng endotoxin per mg (1EU/mg); lyophilized | Potent neurotrophic factor that supports the growth & survivability of nerve and/or glial cells; active form is formed by two identical 119 AA subunits held together by strong hydrophobic interactions; ED_{50} = 50 ng/mL; SA determined by the dose-dependent induction of choline acetyl transferase activity in rat basal forebrain primary septal cultures

Sigma B 3795   Human recombinant, expressed in *Sf* 21 cells MW 13.6k   ≥97% (SDS-PAGE); 0.2 μm filtered & lyophilized from acetonitrile & trifluoroacetic solution containing 250 μg BSA; bioactivity is measured by its ability to support the survival & stimulate neurite outgrowth of cultured embryonic chick dorsal root ganglia (DRG); endotoxin tested | Member of the Nerve growth factor (NGF) family of neurotrophic factors also named neurotrophins; specific factor for sensory & peripheral sympathetic neurons; induces formation of neurite filaments on chick embryo dorsal root ganglia & has been used in the culture & study of adrenergic neurons of sympathetic ganglia, pheochromocytoma cells (PC12) & some neoplastic cells of neural crest origin; all NGF products are biologically tested in cell culture for their ability to induce neurite-like filament growth from chick dorsal root ganglia or to induce proliferation of TF-1 or PC12 cells; Levi-Montalcini, R et al, *Cancer Res*, 14: 49, 1954; Server, A et al, *Adv Protein Chem*, 31: 339, 1977; Kitamura, T et al, *J Cell Physiol*, 140: 323, 1989; Greene, L, *J Cell Biol*, 78: 747, 1978

Neurotrophic Factor, Ciliary

Biodesign A52455H   *E. coli*   MW 23k   Purified | Species specificity: rat

Oncogene PF030   Human recombinant   MW 22.8k   >97% (SDS-PAGE); lyophilized; biological activity: EC_{50} of 50-150 ng/mL as determined by a cell proliferation assay of factor-dependent TF-1 cells; EC_{50} of 1.0-3.0 ng/mL as determined by the ability to support survival & stimulate neurite outgrowth of cultured embryonic chick dorsal root ganglia | Species reactivity: human; for proliferation studies; the expressed recombinant protein lacks the N-terminal methionine residue

Fitzgerald 30-AC36   Human recombinant, expressed in *E. coli*

PeproTech 450-13   Human recombinant, expressed in *E. coli* MW 22.9k   >98% (SDS-PAGE) & HPLC; <0.1 ng endotoxin per mg (1EU/mg); lyophilized | Potent neural factor originally characterized as a survivability factor for chick ciliary neurons *in vitro*; promotes survivability & differentiation of other neural cell types; 200 AA; $ED_{50} \leq$ 2.0 ng/mL; SA $\geq$ 5 x 10^5 U/mg; SA determined by the dose-dependent proliferation of human TF-a cells

Sigma C 3710   Human recombinant, expressed in *E. coli*   MW 22.8k   ≥97% (SDS-PAGE); 0.2 μm filtered & lyophilized from phosphate buffered saline containing 500 μg BSA; bioactivity is measured in a cell proliferation assay using the TF-1 cell line; endotoxin tested | Survival factor for neuronal cell types; member of the nerve growth factor (NGF) family of neurotrophic factors also named neurotrophins; specific factor for sensory & peripheral sympathetic neurons; induces formation of neurite filaments on chick embryo dorsal root ganglia & has been used in the culture & study of adrenergic neurons of sympathetic ganglia, pheochromocytoma cells (PC12) & some neoplastic cells of neural crest origin; all NGF products are biologically tested in cell culture for their ability to induce neurite-like filament growth from chick dorsal root ganglia or to induce proliferation of TF-1 or PC12 cells; Levi-Montalcini, R et al, *Cancer Res*, 14: 49, 1954; Server, A & Shooter, E, *Adv Protein Chem*, 31: 339, 1977; Kitamura, T et al, *J Cell Physiol*, 140: 323, 1989; Greene, L, *J Cell Biol*, 78: 747, 1978

Chemicon GF035   Rat   ≥95%

Biogenesis 2104-9030   Rat r-DNA   Lyophilized

Alexis 521-002   Rat recombinant, expressed in *E. coli*   MW 23k   >98% (SDS-PAGE); lyophilized; 5 μg (AA analysis); biological activity: activity was originally characterized as a neurotrophic growth factor promoting survival of chick ciliary neurons *in vitro*; other neuronal cell lines such as IMR 32 cells are also dependent on rrCNTF for survival; effective dose of 50% (50 ng/mL) is sufficient for the induction of choline acetyltransferase activity; exerts *in vitro*, biological effects in a concentration range between 0.01-10.0 ng/mL medium | Protein consisting of 200 AA; Sendtner, M et al, *Nature*, 345: 440, 1990; Korsching, S, *J Neurosci*, 13: 2739, 1993; Thoenen, H, *TINS*, 14: 165, 1991; Thoenen, H et al, *CR Acad Sci III*, 316: 1158, 1993; Maness, LM et al, *Neurosci & Biobehav Rev*, 18: 143, 1994; Klein, R, *FASEB J*, 8: 738, 1994; Barinaga, M, *Science*, 264: 772, 1994; Nishi, R, *Science*, 265: 1052, 1994; Davies, AM, *Nature*, 368: 193, 1994

Calbiochem 231000   Rat recombinant, expressed in *E. coli* MW 23k   >98% (SDS-PAGE); lyophilized solid; soluble in most aqueous buffers & water; biological activity: Induces choline acetyltransferase (EC_{50}=50 ng/mL) | Neuronal survival that is known to promote the survival & differentiation of a variety of neuronal cell lines; prevents lesion-induced death of motor neurons in adults; reported to increase K^+ currents in SK-*N*-SH neuroblastoma cells; *Merck Index*, 12: 2333; Lesser, SS & Lo, DC, *J Neurosci*, 15: 253, 1995; Sendtner, M et al, *J Neurobiol*, 25: 1436, 1994; Sendtner, M et al, *Nature*, 345: 440, 1990

Fitzgerald 30-AC37   Rat recombinant, expressed in *E. coli*

PeproTech 450-50   Rat recombinant, expressed in *E. coli* MW 22.7k   >98% (SDS-PAGE) & HPLC; <0.1 ng endotoxin per mg (1EU/mg); lyophilized | Potent neural factor originally characterized as a survivability factor for chick ciliary neurons *in vitro*; promotes survivability & differentiation of other neural cell types; 199 AA; $ED_{50} \leq$ 0.05 ng/mL; SA $\geq$ 2 x 10^7 U/mg; SA determined by the dose-dependent induction of choline acetyl transferase activity in IMR32 cells

Sigma C 3835   Rat recombinant, expressed in *E. coli*   MW 22.8k   ≥97% (SDS-PAGE); 0.2 μm filtered & lyophilized from phosphate buffered saline solution containing 500 μg BSA; bioactivity is measured in a bioassay using the TF-1 cell line; endotoxin tested | Survival factor for neuronal cell types; member of the Nerve growth factor (NGF) family of neurotrophic factors also named neurotrophins; specific factor for sensory & peripheral sympathetic neurons; induces formation of neurite filaments on chick embryo dorsal root ganglia & has been used in the culture & study of adrenergic neurons of sympathetic ganglia, pheochromocytoma cells (PC12) & some neoplastic cells of neural crest origin; all NGF products are biologically tested in cell culture for their ability to induce neurite-like filament growth from chick dorsal root ganglia or to induce proliferation of TF-1 or PC12 cells; Levi-Montalcini, R et al, *Cancer Res*, 14: 49, 1954; Server, A & Shooter, E, *Adv Protein Chem*, 31: 339, 1977; Kitamura, T et al, *J Cell Physiol*, 140: 323, 1989; Greene, L, *J Cell Biol*, 78: 747, 1978

Neurotrophic Factor, Glial Derived

Biodesign A52450H   *E. coli*   MW 30k   Purified

Chemicon GF030   Human   ≥95%

USBio N2180-11   Human   ≥98% (SDS-PAGE); no contaminants detected; single band by SDS-PAGE, IEP, &/or RID; 1 mg/mL; lyophilized | Suitable for antigenic applications in immunological protocols

Fitzgerald 30-AN17   Human glial recombinant, expressed in *E. coli*   High purity

Biogenesis 4649-5030   Human r-DNA   Lyophilized

BioSource International PHC7044   Human recombinant

Alexis 521-001   Human recombinant, expressed in *E. coli* MW 30k   >98% (SDS-PAGE); lyophilized; 1 μg/5 μg (AA analysis); biological activity: promotes survival of dopaminergic midbrain neurons; fully biologically active when compared to the native factor by the dose-dependent ^{3}H-dopamine up-take assay measured with rat midbrain primary cultures | Disulfide linked homodimeric protein consisting of two 134 AA chains; Thoenen, H, *TINS*, 14: 165, 1991; Thoenen, H et al, *CR Acad Sci III*, 316: 1158, 1993; Maness, LM et al, *Neurosci & Biobehav Rev*, 18: 143, 1994; Klein, R, *FASEB J*, 8: 738, 1994; Barinaga, M, *Science*, 264: 772, 1994; Nishi, R, *Science*, 265: 1052, 1994; Davies, AM, *Nature*, 368: 193, 1994; Lin, LF et al, *Science*, 260: 1130, 1993

Calbiochem 345872   Human recombinant, expressed in *E. coli* MW 30k   ≥97% (SDS-PAGE); lyophilized solid; biological activity: Fully biologically active when compared to native GDNF by does-dependent ^{3}H-dopamine uptake assay; soluble in most aqueous buffers & water | Growth promoting factor that enhances neuronal survival of dopaminergic midbrain neurons; activates receptor tyrosine kinase Ret & promotes kidney morphogenesis; also an activator of MAP kinase; its biological activity is destroyed by reduction of disulfide bonds; Vega, QC et al, *PNAS*, 93: 10657, 1996; Worby, CA et al, *J Biol Chem*, 271: 23619, 1996; Lin, LF et al, *J Neurochem*, 63: 758, 1994; Lev-Fen, H et al, *Science*, 260: 1130, 1993

ICN 193955   Human recombinant, expressed in *E. coli*   MW 30k   Promotes growth & survival of dopaminergic neurons & motor neurons; Lev-Fen etal, *Science*, 260:1130, 1993

ICN 193956   Human recombinant, expressed in *E. coli*   MW ~30k   ≥95% | Promotes growth & survival of dopaminergic neurons & motor neurons

PeproTech 450-10   Human recombinant, expressed in *E. coli* MW 30.1k   >98% (SDS-PAGE) & HPLC; <0.1 ng endotoxin per mg (1EU/mg); lyophilized | Specifically promotes dopamine uptake & survival of midbrain neurons; disulfide-linked homodimer, formed by two identical 134 AA subunits; ED$_{50}$ = 5-10 ng/mL; SA determined by the dose-dependent dopamine uptake by rat ventral mesencephalic cultures

Sigma G 1777   Human recombinant, expressed in the mouse myeloma cell line NSO   ≥97% (SDS-PAGE); 0.2 μm filtered & lyophilized from phosphate buffered saline containing 500 μg BSA; bioactivity is measured by its ability to support the survival & stimulate neurite growth of cultured embryonic chick dorsal root ganglia; endotoxin tested | Nerve growth factor promotes neuron survival in both the central & peripheral nervous systems; motor neurons, midbrain dopaminergic neurons, Purkinje cells & sympathetic neurons are among the neuronal subpopulations affected by GDNF; shows significant homology to members of the TGF-β superfamily; cells known to express GDNF include: Sertoli cells, type 1 astrocytes, Schwann cells, neurons, pinealocytes & skeletal muscle cells; is a disulfide-linked dimeric protein consisting of two 134 AA peptides with a predicted mass of 15 kD each; human GDNF shows ~93% AA homology to rat GDNF

Oncogene PF039   Rat recombinant   MW ~15k (monomer)   >97% (SDS-PAGE); lyophilized in PBS containing 50 mg of BSA per 1 mg of cytokine; biological activity: EC$_{50}$ of 1.0-3.0 ng/mL as determined by the ability to support the survival & stimulate neurite outgrowth of cultured embryonic chick dorsal root ganglia | Species reactivity: rat; native GDNF exists as a disulfide homodimeric glycoprotein; a novel member of the TGF-β superfamily; activity measured by its ability to support survival & stimulate neurite outgrowth of cultured embryonic chick dorsal root ganglia; the EC$_{50}$ for this effect is typically 1-3 ng/mL; protein titration recommended

Sigma G 1401   Rat recombinant, expressed in *Sf* 21 insect cells   ≥97% (SDS-PAGE); 0.2 μm filtered & lyophilized from phosphate buffered saline containing 500 μg BSA; bioactivity is measured by its ability to support the survival & stimulate neurite growth of cultured embryonic chick dorsal root ganglia; endotoxin tested | Nerve growth factor promotes neuron survival in both the central & peripheral nervous systems; motor neurons, midbrain dopaminergic neurons, Purkinje cells & sympathetic neurons are among the neuronal subpopulations affected by GDNF; shows significant homology to members of the TGF-β superfamily; cells known to express GDNF include: Sertoli cells, type 1 astrocytes, Schwann cells, neurons, pinealocytes & skeletal muscle cells; is a disulfide-linked dimeric protein consisting of two 134 AA peptides with a predicted mass of 15 kD each; human GDNF shows ~93% AA homology to rat GDNF

ICN 195777   Rat recombinant, expressed in *Sf*21   ≥97%; ED$_{50}$ = 1-3 ng/mL

Neurotrophin III

Biodesign A52453H   *E. coli*   MW 28k   Purified

Chemicon GF031   Human   ≥95%

USBio N2200   Human   ≥98% (SDS-PAGE); no contaminants detected; single band by SDS-PAGE, IEP, &/or RID; 1 mg/mL; lyophilized | Suitable for antigenic applications in immunological protocols

Biogenesis 6751-5330   Human r-DNA   Lyophilized

BioSource International PHC7034   Human recombinant

Alexis 521-003   Human recombinant, expressed in *E. coli* MW 28k   ≥98% (SDS-PAGE); lyophilized; 1 μg/5 μg (AA analysis); biological activity: supports survival of certain CNS neurons; fully biologically active when compared to the native factor; effective dose 50% enables survival of rat basal forebrain primary septal cultures as measured from the induction of the choline acetyltransferase enzymatic activity of the cells | Dimer protein consisting of two identical 119 AA subunits which are associated through strong hydrophobic interactions; Thoenen, H, *TINS*, 14: 165, 1991; Thoenen, H et al, *CR Acad Sci III*, 316: 1158, 1993; Maness, LM et al, *Neurosci & Biobehav Rev*, 18: 143, 1994; Klein, R, *FASEB J*, 8: 738, 1994; Barinaga, M, *Science*, 264: 772, 1994; Nishi, R, *Science*, 265: 1052, 1994; Davies, AM, *Nature*, 368: 193, 1994; Maisonpierre, PC et al, *Science*, 247: 1446, 1990; Korsching, *J Neurosci*, 13: 2739, 1993

Calbiochem 480875   Human recombinant, expressed in *E. coli* MW 28k   ≥98% (SDS-PAGE); lyophilized solid; biological activity: fully biologically active when compared to native NT-3; ED$_{50}$=50 ng/mL for induction of choline acetyltransferase activity in rat basal forebrain primary septal culture cells | Potent neurotrophic factor that supports the survival of certain CNS neurons; increases DNA-binding of several transcription factors; *Merck Index*, 12: 6570; Iwata, E et al, *Biochim Biophys Acta*, 1311: 85, 1996; Zhou, XF & Rush, RA, *Brain Res*, 643: 162, 1994; Hohn, A et al, *Nature*, 344: 339, 1990; Mainsonpierre, PC et al, *Science*, 247: 1447, 1990

Fitzgerald 30-AN18   Human recombinant, expressed in *E. coli* High purity

PeproTech 450-03   Human recombinant, expressed in *E. coli* MW 27.2k   >98% (SDS-PAGE) & HPLC; <0.1 ng endotoxin per mg (1EU/mg); lyophilized | Potent neurotrophic factor that supports the growth & survivability of nerve and/or glial cells; active form is formed by two identical 119 AA subunits held together by strong hydrophobic interactions; ED$_{50}$ = 20-50 ng/mL; SA determined by the dose-dependent induction of choline acetyl transferase activity in rat basal forebrain primary septal cultures

Sigma N 1905   Human recombinant, expressed in *Sf* 21 insect cells   MW 13.6k   ≥97% (SDS-PAGE); 0.2 μm filtered & lyophilized from 30% acetonitrile & 0.1% TFA solution containing 250 μg BSA per 1 μg of the cytokine; bioactivity is measured in a neurite outgrowth bioassay using chick dorsal root ganglia; endotoxin tested | Neurotrophic factor that is required for the differentiation & survival of specific neuronal subpopulations in the central & the peripheral nervous system

Neurotrophin IV

Biodesign A52454H   *E. coli*   MW 28k   Purified

Chemicon GF032   Human   ≥95%

USBio N2200-12   Human   ≥98% (SDS-PAGE); no contaminants detected; single band by SDS-PAGE, IEP, &/or RID; 1 mg/mL; lyophilized | Suitable for antigenic applications in immunological protocols

Biogenesis 6751-5430   Human r-DNA   Lyophilized

BioSource International PHC7024   Human recombinant

Alexis 521-004   Human recombinant, expressed in *E. coli* ≥98% (SDS-PAGE) with no additives; lyophilized; 1 μg/5 μg (AA analysis); biological activity: a potent neurotrophic factor promoting survival of certain CNS neurons; fully biologically active when compared to the native factor; effective dose 50% enables survival of rat basal forebrain primary septal cultures as measured from the induction of the choline acetyltransferase enzymatic activity of the cells | Dimer protein consisting of two identical 130 AA subunits which are associated through strong hydrophobic interactions; Thoenen, H, *TINS*, 14: 165, 1991; Thoenen, H et al, *CR Acad Sci III*, 316: 1158, 1993; Maness, LM et al, *Neurosci & Biobehav Rev*, 18: 143, 1994; Klein, R, *FASEB J*, 8: 738, 1994; Barinaga, M, *Science*, 264: 772, 1994; Nishi, R, *Science*, 265: 1052, 1994; Davies, AM, *Nature*, 368: 193, 1994; Hohn, A et al, *Nature*, 344: 339, 1990; Fandl, NJ et al, *J Biol Chem*, 269: 755, 1994

Calbiochem 480877   Human recombinant, expressed in *E. coli* MW 28k   >98% (SDS-PAGE); lyophilized solid; biological activity: fully biologically active when compared to native NT-4; ED₅₀=50 ng/mL for induction of choline acetyltransferase activity in rat basal forebrain primary septal culture cells | Potent neurotrophic factor that supports the survival of certain CNS neurons; a neuronal-signaling molecule involved in the compensatory adjustments of muscle fibers to oxidative dysfunction; Walker, UA & Schon, EA, *Ann Neurol*, 43: 536, 1998

Fitzgerald 30-AN19   Human recombinant, expressed in *E. coli* High purity

Sigma N 1780   Human recombinant, expressed in *Sf* 21 insect cells   MW 14k   ≥97% (SDS-PAGE); 0.2 μm filtered & lyophilized from 30% acetonitrile & 0.1% TFA solution containing 250 μg BSA; bioactivity is measured in a neurite outgrowth bioassay using chick dorsal root ganglia; endotoxin tested | Neurotrophic factor that is required for the differentiation & survival of specific neuronal subpopulations in the central & the peripheral nervous system

Neurotrophin IV/V

PeproTech 450-04   Human recombinant, expressed in *E. coli* MW 28.1k   >98% (SDS-PAGE) & HPLC; <0.1 ng endotoxin per mg (1EU/mg); lyophilized | Potent neurotrophic factor that supports the growth & survivability of nerve and/or glial cells; active form is formed by two identical 130 AA subunits held together by strong hydrophobic interactions; ED₅₀ = 20-50 ng/mL; SA determined by the dose-dependent induction of choline acetyl transferase activity in rat basal forebrain primary septal cultures

Neurturin

Biodesign A52511H   *E. coli*   Purified

Chemicon GF081   Human   ≥95%

BioSource International PHC7064   Human recombinant

Calbiochem 480890   Human recombinant, expressed in *E. coli* MW 23.6k   ≥98% (SDS-PAGE); ≥98% (HPLC); lyophilized solid; biological activity: shown to support the survival of 65% of newborn rat sympathetic neurons at 100 ng/ L | Belongs to the transforming growth factor-β (TGF-β)-related neurotrophic factors known collectively as glial cell-line derived neurotrophic factor (GDNF) family; NTN is expressed in the nigrostriatal system & exerts potent effects on survival & function of midbrain dopaminergic (DA) neurons; Hishiki, T et al, *Cancer Res*, 58: 2158, 1998; Horger, BA et al, *J Neurosci*, 18: 4929, 1998

PeproTech 450-11   Human recombinant, expressed in *E. coli* MW 23.6k (homodimer)   >98% (SDS-PAGE) & HPLC; <0.1 ng endotoxin per mg (1EU/mg); lyophilized | With Human GDNF, comprises a family of TGF-beta related neurotrophic factor that has trophic influences on a variety of neuronal populations; promotes the survivability of certain sympathetic & sensory neurons through interaction with distinct set of GDNF-like receptors; a protein consisting of two identical subunits of 103 AA each; supports the survival of 65% of newborn rat sympathetic neurons at 100 ng/mL

Neutrophil Activating Protein II

Biodesign A52314H   *E. coli*   MW 8k   Purified

Chemicon GF015   Human   ≥95%

Fitzgerald 30-AN20   Human recombinant, expressed in *E. coli* 70 AA

Harlan BT-2028   Human recombinant, expressed in *E. coli* Lyophilized; 0.01 mg

Harlan BT-2029   Human recombinant, expressed in *E. coli* Lyophilized; 0.05 mg

ICN 195012   Human recombinant, expressed in *E. coli*   MW 8.5k   >98%; lyophilized; 50 ng/mL

PeproTech 300-14   Human recombinant, expressed in *E. coli* MW 7.6k   >98% (SDS-PAGE) & HPLC; <0.1 ng endotoxin per mg (1EU/mg); lyophilized | Promotes neutrophil chemotaxis & degranulation; 70 AA; SA determined by its ability to chemoattract human neutrophils

USBio N2250   Human recombinant, expressed in *E. Coli* Highly purified, ≥98% (SDS-PAGE); lyophilized; no contaminants detected; single band by SDS-PAGE, IEP, &/or RID; 0.1 mg/mL | Suitable for antigenic applications in immunological protocols

NF-κB

Synonyms: p50

Alexis 201-026   Human recombinant, expressed in *E. coli* ≥90% (SDS-PAGE & Western Blot analysis) | Kleran, M et al, *Cell*, 62: 1007, 1990

NF-κB Receptor Activator, Soluble

Biodesign A52001H   *E. coli*   Purified

Nuclear Inhibitor of Protein Phosphatase I

Calbiochem 482250   Bovine thymus recombinant   MW 38.5k ≥95% (SDS-PAGE); liquid in 20 mM Tris-HCl, 500 μM benzamidine, 500 μM DTT, 500 μM PMSF, 5 μM leupeptin, 60% glycerol, pH 7.4; activity: stoichiometric amounts of NIPP-1 completely inhibit the catalytic subunit of PP1 using various substrates | Potent & specific inhibitor of protein phosphatase used to distinguish PP1 from other major serine/threonine protein phosphatases including PP2A, PP2B & PP2C; a model substrate for phosphorylation by protein kinase A & casein kinase II; involved in the targeting of PP1 to RNA-associated substrates & in the dephosphorylation of transcription factors like CREB & the tumor suppressor Rb; Van Eynde, A et al, *J Biol Chem*, 270: 28068, 1995; Jagiello, I et al, *J Biol Chem*, 272: 22067, 1997; Jagiello, I et al, *J Biol Chem*, 270: 17257, 1995; Van Eynde, A et al, *Biochem J*, 297: 447, 1994; Beullens, M et al, *J Biol Chem*, 268: 13172, 1993; Beullens, M et al, *J Biol Chem*, 267: 16538, 1992

Obese Protein Control Peptide I

Chemicon AG766   Purified | Cellular biochemistry/regulatory protein used in enzyme immunoassay

Obese Protein Control Peptide II

Chemicon AG768   Purified | Cellular biochemistry/regulatory protein used in enzyme immunoassay

Oncostatin M

Biodesign A52130H   *E. coli*   MW 26k   Purified

Chemicon GF016   Human   ≥95%

BioSource International PHC5014   Human recombinant

Fitzgerald 30-AO10   Human recombinant, expressed in *E. coli*

PeproTech 300-10   Human recombinant, expressed in *E. coli* MW 26k   >98% (SDS-PAGE) & HPLC; <0.1 ng endotoxin per mg (1EU/mg); lyophilized | Important growth regulating cytokine; variably affects a number of tumor & normal cells; exerts inhibitory effects on the growth of A375 melanoma & other cancer cells; but augments the growth of normal fibroblasts, AIDS-related Kaposi sarcoma cells, & certain other cells; 227 AA; ED₅₀ ≤ 2 ng/mL; SA ≥ 5 x 10⁵ U/mg; SA determined by the dose-dependent stimulation of the proliferation of human TF-1 cells

Sigma O 9635 Human recombinant, expressed in *E. coli* ≥97% (SDS-PAGE); 0.2 µm filtered & lyophilized from 35% acetonitrile & 0.1% TFA solution containing 500 µg BSA; proliferative activity is tested in culture by using a human erythroleukemic cell line TF-1; endotoxin tested | Growth-regulating cytokine affecting a number of tumor & normal cells; first identified by its ability to inhibit the growth of A375 melanoma cells & other human tumor cells but not inhibit the growth of normal human fibroblasts; acts synergistically with TGF β1 to inhibit the proliferation of tumor cells like A375 melanoma cells; secreted by macrophages & activated T lymphocytes; affects a wide variety of normal & tumor cells; induces an increase in LDL receptor expression & LDL uptake by hepatoma cells; cultured human endothelial cells are induced to increase IL-6 production; activates synovial fibroblast-like cells to produce urokinase type plasminogen activator; OSM, LIF, G-CSF, IL-6 & CNTF are structurally related members of the same cytokine family sharing similarities in their primary AA sequences predicted secondary structure & receptor components; Brown, TJ et al, *J Immunol*, 139: 2977, 1987; Grove, RI et al, *J Biol Chem*, 266: 18194, 1991; Brown, TJ et al, *J Immunol*, 147: 2175, 1991; Hamilton, JA et al, *Biochem Biophys Res Commun*, 180: 652, 1991; Bazan, JF et al, *Neuron*, 7: 197, 1991; Kitamura, T et al, *J Cell Physiol*, 140: 323, 1989

Sigma O 1637 Mouse recombinant, expressed in *E. coli* ≥97% (SDS-PAGE); 0.2 µm filtered & lyophilized from phosphate buffered saline containing 1.25 mg BSA; biological activity was measured by its ability to stimulate ^{3}H-thymidine incorporation in quiescent NIH/3T3 cells; endotoxin tested | Growth-regulating cytokine affecting a number of tumor & normal cells; first identified by its ability to inhibit the growth of A375 melanoma cells & other human tumor cells but not inhibit the growth of normal human fibroblasts; acts synergistically with TGF β1 to inhibit the proliferation of tumor cells like A375 melanoma cells; secreted by macrophages & activated T lymphocytes; affects a wide variety of normal & tumor cells; induces an increase in LDL receptor expression & LDL uptake by hepatoma cells; cultured human endothelial cells are induced to increase IL-6 production; activates synovial fibroblast-like cells to produce urokinase type plasminogen activator; OSM, LIF, G-CSF, IL-6 & CNTF are structurally related members of the same cytokine family sharing similarities in their primary AA sequences predicted secondary structure & receptor components; Brown, TJ et al, *J Immunol*, 139: 2977, 1987; Grove, RI et al, *J Biol Chem*, 266: 18194, 1991; Brown, TJ et al, *J Immunol*, 147: 2175, 1991; Hamilton, JA et al, *Biochem Biophys Res Commun*, 180: 652, 1991; Bazan, JF et al, *Neuron*, 7: 197, 1991; Kitamura, T et al, *J Cell Physiol*, 140: 323, 1989

Osteocalcin

Synonyms: Bone Gla Protein; Vitamin K-Dependent Protein

Cortex CP4062U	Bovine	>98%
Biodesign A95020B	Bovine bone	Purified
Biogenesis 7060-1054	Bovine bone	Lyophilized
Biogenesis 7060-1104	Bovine bone	Liquid
Biogenesis 7060-1204	Bovine bone	^{125}I conjugated

Calbiochem 499050 Bovine bone MW 5834.4 $C_{263}H_{372}N_{66}O_{82}S_2$ ≥98% (SDS-PAGE); liquid in 75 mM NaCl, 10 mM sodium phosphate buffer, pH 7.4; pI 4.0-4.5 | Single chain vitamin K-dependent protein produced by osteoblasts & present at high concentrations in bone; binds to phospholipid vesicles in the presence of calcium ions (K_d=6 µM); also binds hydroxylapatite; good marker of bone turnover; may play a role in bone mineralization, bone resorption & bone formation; Carter, SD et al, *J Anim Sci*, 74: 2719, 1996; Watson, KE et al, *J Clin Invest*, 93: 2106, 1994; Tracy, RP et al, *J Bone Min Res*, 5: 451, 1990; Gendreau, MA et al, *J Biol Chem*, 264: 6972, 1989

ICN 194940 Bovine bone 50% glycerol, 0.01 M Tris, 0.075 M NaCl, pH 7.4 | Produced in osteoblasts; Hauschka, PV etal, *PNAS*, 73:1447, 1975

Biogenesis 7060-1277	Canine bone	Liquid
Biogenesis 7060-1297	Canine bone	^{125}I conjugated
Cortex CP3015U	Human	>98%
Biogenesis 7060-1855	Mouse bone	Purified; RIA buffer containing BSA; lyophilized
Biogenesis 7060-1865	Mouse bone	10 µCi/mL, SA: 148 µC/µg; RIA buffer; liquid

Biogenesis 7060-2504 Rat bone MW 5.734k Purified; from 10 µl solution of 30 mM sodium phosphate buffer pH 7.5; lyophilized | Suitable for use as RIA standard or for iodination; adsorbed non-specifically to glass and plastic services unless protected by RIA buffer

Biogenesis 7060-2604 Rat bone 225 µC/µg, 10 µC/mL; ^{125}I conjugated; in RIA buffer (see s); liquid

Osteonectin

Synonyms: BM-40; SPARC; BM-40; SPARC

Cortex CP3091U	Bovine	>98%
Biodesign A95010B	Bovine bone	Purified

ICN 194943 Bovine bone Kelm, RJ etal, *Blood*, 80:3112, 1992

Calbiochem 499240 Bovine brain MW 32k ≥95% (SDS-PAGE); liquid in 150 mM NaCl, 20 mM Tris-HCl, 2 mM CaCl$_2$, pH 7.4 | Acidic, Ca^{2+}-binding glycoprotein found in a variety of embryonic & adult tissues containing actively proliferating & remodeling cells; one of the most abundant glycoproteins secreted by osteoblasts; reported to play a role in the differentiation & maintenance of dermis; involved in the regulation of bone mineralization; may also play a role in the disengagement of cells from the extracellular matrix; Hunzelman, N et al, *J Invest Dermatol*, 110: 122, 1998; Kelm, RJ et al, *J Biol Chem*, 269: 30147, 1994; Yost, JC & Sage, EH, *J Biol Chem*, 268: 25790, 1993; Sage, EH & Borstein, PJ, *J Biol Chem*, 266: 14831, 1991

Cortex CP3095U	Human	>98%

Calbiochem 499250 Human platelets MW 32.7k ≥98% (SDS-PAGE); liquid in 150 mM NaCl, 20 mM Tris-HCl, 2 mM CaCl$_2$, pH 7.4 | Acidic, Ca^{2+}-binding glycoprotein found in a variety of embryonic & adult tissues containing actively proliferating & remodeling cells; differs from the bone-derived osteonectin in its glycosylation pattern & its collagen binding specificity; prepared from serum that has been shown by certified tests to be negative for HBsAg & for antibodies to HIV & HCV; Villarreal, XC et al, *Biochemistry*, 28: 6483, 1989; Kelm, RJ et al, *J Biol Chem*, 269: 30147, 1994; Kelm, RJ & Mann, KG, *J Biol Chem*, 266: 9632, 1991; Sage, EH & Borstein, PJ, *J Biol Chem*, 266: 14831, 1991

ICN 194942 Human platelets Non-collagenous glycoprotein; potently inhibits hydroxyapatite-seeded crystal growth; Fisher, LW etal, *JBC*, 262:9702, 1987

Osteoprotegerin

Synonyms: Osteoprotegerin:Fc, rh-; Osteoprotegerin Ig; Osteoprotegerin:Fc, rh-; Osteoclastogenesis Inhibitory Factor

Kamiya	>95% (SDS-PAGE)

Alexis 522-007 Human embryo kidney cells recombinant >95% (SDS-PAGE); lyophilized powder containing PBS; 25 µg protein | Interacts with human & mouse TRAIL & RANKL/TRANCE; the Cys-rich region of human osteoprotegerin (AA 22-202) is fused to the Fc portion of human IgG1; inhibits soluble TRAIL (sTRAIL)-mediated lysis of TRAIL sensitive cells (conc range: 5-20 ng/mL) & blocks RANKL-induced osteoclastogenesis & stimulation of dendritic cells; Simonet, WS et al, *Cell*, 89: 309, 1997; Emery, JG et al, *J Biol Chem*, 273: 14363, 1998; Yasuda, H et al, *PNAS*, 95: 3597, 1998; Yasuda, H et al, *Endocrinology*, 139: 1329, 1998; Lacey, DL et al, *Cell*, 93: 165, 1998

PeproTech 450-14 Human recombinant, expressed in *E. coli* MW 19.9k >98% (SDS-PAGE) & HPLC; <0.1 ng endotoxin per mg (1EU/mg); lyophilized | Member of the TNF receptor superfamily; specifically acts on bone tissues; increases bone mineral density & bone volume associated with a decrease of active osteoclast number; 175 AA; SA determined by its ability to neutralize the stimulation of U937 cells

Osteoprotegerin/Fc Chimera

Synonyms: TNFRSF11B

R&D Systems 459-MO-100 NSO-expressed >95%; lyophilized; ED$_{50}$: 8-15 ng/mL | Species specificity: mouse OPG; member of the TNF receptor superfamily & exists as a soluble secreted protein; TRANCE & TRAIL shown to be ligands for OPG

R&D Systems 805-OS-100 *SF*21-expressed >90%; lyophilized; ED$_{50}$: 8-24 ng/mL | Species specificity: human OPG; member of the TNF receptor superfamily & exists as a soluble secreted protein; TRANCE & TRAIL shown to be ligands for OPG

Ovalbumin

ICN 950512 Chicken egg Lyophilized; 5X crystallized; no lysozyme; prepared from fresh chicken egg whites by (NH$_4$)$_2$SO$_4$ fractionation & repeated crystallization at pH 4.5

ICN 825051 Duck egg 95%; lyophilized

Ovalbumin, (Me-^{14}C)-

ARC ARC-431 MW 46k 3-30 μCi/mg; 111-1111 KBq/mg; in 0.01 *M* sodium phosphate, pH 7.2 | Radiochemical

Ovarian Tumor Marker Antigen

Synonyms: CA 125®

ICN 770991 ICN 770992 ICN 770993 Human ascites >96%; iodination grade; >50,000 U/mL | For immunization or labeling

Oxynor

ICN 159839 *Oxyuranus scutellatus* (Australian taipan) venom MW ~13.5k Single band by SDS-PAGE; 0.1 μg/mL in media promotes neurotrophic growth equivalent to 10% FBS | Cell growth factor similar to epidermal growth factor; promotes growth of various eukaryotic cells & keratinocytes; produces neurite outgrowth on rat adrenal pheochromocytoma (PC-12) cells; useful for accelerating wound healing

p13

Amersham VPF001 Recombinant Cell proliferation & cell cycle signals

Oncogene PF001 Recombinant MW 13k (dimer, non-denaturing) 99% (SDS-PAGE); lyophilized; reconstitute in PBS | Use to elute with p13 agarose columns

p13 suc1, Agarose

USBio P0999 Recombinant GST-p13 suc1 MW 39k ≥95% (SDS-PAGE) & Coomassie Stain; liquid suspension in PBS, 50% glycerol, 0.05% NaN$_3$ | Non-covalently bound to glutathione-agarose; useful for isolation of p34 cdc2 & histonekinase assay of bound p34 cdc2

p13, Agarose

Oncogene PF001A Recombinant MW 13k (dimer, non-denaturing) 99% (SDS-PAGE); highly purified protein coupled to agarose; 1.2 mg yeast p13 coupled to 0.5 mL agarose in PBS containing 0.1% sodium azide & 30% glycerol; reconstitute in PBS | Used to precipitate 13-mitotickinase complex where non-ionic detergents (ie NP40, DOC) are used

p21 H-ras^{Gly12}

Oncogene WA01 MW 21k (may form dimer of 42k) >95% (SDS-PAGE); lyophilized | Western Blot standard; protein is denatured in SDS

p21 K-ras^{Asp12}

Oncogene WA03 MW 21k >95% (SDS-PAGE); lyophilized | Western Blot standard; protein is denatured in SDS

p21 K-ras^{Gly12}

Oncogene WA02 MW 21k (may form dimer of 42k) >95% (SDS-PAGE); lyophilized | Western Blot standard; protein is denatured in SDS

Calbiochem WA02 Human recombinant ≥95% (SDS-PAGE); denatured human recombinant p21 K-Ras Gly^{12} protein; denatured in SDS; reconstitute in 100 μL SDS-PAGE buffer containing DTT; use 10 μL/lane; heat to 100°C prior to use | Immunoblot standard suitable as a positive control in Western blotting

p21 K-ras^{Val12}

Oncogene WA04 MW 21k (may form dimer of 42k) >95% (SDS-PAGE); lyophilized | Western Blot standard; protein is denatured in SDS

p21 N-ras^{Gly12}

Oncogene WA05 MW 21k (may form dimer of 42k) >95% (SDS-PAGE); lyophilized | Western Blot standard; protein is denatured in SDS

Pancreozymin

Synonyms: Cholecystokinin

ICN 190265 Porcine intestine ~2-4 Crick U/mg solid | Crick, J etal, *J Physiol*, 110:367, 1950

Sigma P 4429 Porcine intestine 4-6 Crick U/mg solid; also contains 0.5-1 Crick U of secretin/mg solid | Not assayed by Sigma; Crick, J et al, *J Physiol*, 110: 367, 1950

Parvalbumin

Biogenesis 7179-8009 Rat muscle Purified; 0.05 *M* NH$_4$HCO$_3$, pH 8.0, essentially salt free; lyophilized

PCAF, Active

USBio P3114-75 Recombinant, expressed in *E. coli* ~30%; supplied as 50 μg free enzyme in 500 μL of TBS, pH 8.0, 25 *mM* glutathione, 50% glycerol | GST fusion protein corresponding to AA 352-832; a direct link exists between hyper-acetylation of chromatin & transcriptional activation; PCAF possesses intrinsic histone acetylase activity; primarily acetylates Lys^{14} of H3 but also less efficiently acetylates Lys^8 of H4; also acetylates p53 in response to DNA damage; the distinct patterns of acetylation by PCAF may contribute to transcriptional regulation of important genes

Peripheral Type Benzodiazepine Receptor

R&D Systems 6360-025-01 Human placenta MW 18k Control protein; found in most steroidogenic tissues in the outer mitochondrial membrane in association with a 34 k voltage-dependent anion channel protein (VDAC); thought to be part of the mitochondrial permeability transition pore;

Persephin

PeproTech 450-12 Human recombinant, expressed in *E. coli* MW 20.6k (homodimer) >98% (SDS-PAGE) & HPLC; <0.1 ng endotoxin per mg (1EU/mg); lyophilized | Novel neurotrophic factor ~40% identical to GDNF & NTN; promotes the survival of ventral midbrain dopaminergic neurons in culture; supports the survival of motor neurons in culture; a protein consisting of two identical subunits of 96 AA each; induces RET phosphorylation at 0.1-1.0 ng/mL; binds to mammalian GFRa4 at Kd = 100*pM*; other members of the GDNF family (Artemin, GDNF, Neurturin) do not bind to mammalian GFRa4

Pertussis Toxin

Synonyms: Islet Activating Protein

Alexis 630-003 *Bordetella pertussis* MW 94k Lyophilized powder; when reconstituted with 0.5 mL distilled water, each vial contains 50 μg protein in 0.01 *M* sodium phosphate buffer, pH 7.0, with 0.05 *M* NaCl; potent toxin | Major protein toxin produced by virulent strains of *Bordetella pertussis*; the purified protein consists of 5 dissimilar subunits: *S*-1 (MW 28 k), *S*-2 (MW 23 k), *S*-3 (MW 22 k), *S*-4 (MW 11.7 k) & *S*-5 (MW 9.3 k), in a molar ratio of 1:1:1:2:1; *S*-1 (A protomer) is responsible for the enzymatic activity of the toxin; together, *S*-2, *S*-3, *S*-4 & *S*-5 comprise the B oligomer, responsible for binding the toxin to the cell surface

Calbiochem 516560 *Bordetella pertussis* Purity: five distinct bands (SDS-PAGE); solid lyophilized from 50 *mM* NaCl, 10 *mM* phosphate buffer, pH 7.0; adenylate cyclase activity: ≤2.5 pmol/min/µg in the presence of calmodulin; harmful: LD₅₀ ≤2000 mg/kg | A protein endotoxin that catalyzes ADP-ribosylation of guanine nucleotide-binding regulatory protein G_I, G_O & G_T; used in the study of adenylate cyclase regulation & the role of G_I proteins; holotoxin activity is determined in a CHO cell assay; Hewlett, EL et al, *Infect Immuno*, 40: 1198, 1983

Sigma P 0317 *Bordetella pertussis* Lyophilized powder containing phosphate buffered saline & lactose | Catalyzes the ADP-ribosylation of G_I, G_O & G_T guanine nucleotide-binding regulatory proteins; potentiates insulin secretion from mammalian pancreatic islet cells; Sumi, T & Ui, M, *Endocrinology*, 97: 352, 1975

Sigma P 9452 *Bordetella pertussis* Solution in 50% glycerol containing 50 mM Tris, 10 mM glycine, 0.5 M NaCl, pH 7.5; sterile-filtered | Catalyzes the ADP-ribosylation of G_I, G_O & G_T guanine nucleotide-binding regulatory proteins; potentiates insulin secretion from mammalian pancreatic islet cells; Sumi, T & Ui, M, *Endocrinology*, 97: 352, 1975

Pertussis Toxin, A Protomer

Calbiochem 516854 *Bordetella pertussis* MW 28k Purity: single major band (SDS-PAGE); solid lyophilized from 10 *mM* Tris-HCl, 100 *µM* EDTA, 0.04% CHAPS, pH 8.0; ≤0.1% holotoxin by CHO cell assay; harmful: LD₅₀ ≤2000 mg/kg | Enzymatic component of the holotoxin; both NAD-glycohydrolase & ADP-ribosyltransferase activities; unable to penetrate cells in the absence of the B oligomer; Moss, J et al, *J Biol Chem*, 258: 11879, 1983

Pertussis Toxin, B Protomer

Calbiochem 516852 *Bordetella pertussis* Purity: four distinct bands (SDS-PAGE); lyophilized solid; ≤0.1% holotoxin by CHO cell assay | Pentameric cell-binding component responsible for binding of the holotoxin to eukaryotic cell surfaces, facilitating entry of the A protomer into receptive cells; also elicits a variety of physiological responses, such as mitogenesis in human T cells, enhancement of aggregation of human platelets, elevation of cytosolic Ca^{2+} levels & neutralization of antibody response in mice; Banga, S et al, *J Biol Chem*, 262: 14871, 1987; Hazes, B et al, *J Mol Biol*, 258: 661, 1996

PF4

BioSource International PHC7054 Human recombinant

Phosphatidylinositol-3-Kinase p85 CT-SH2 Domain, Agarose

USBio P4185-28 Human fusion protein, expressed in *E. coli* Purified by glutathione-agarose chromatography; 2 mg/mL; frozen liquid in 25 µL total volume (20% bead slurry) of PBS, 10% glycerol, 2 *mM* DTT, & 0.02% NaN₃ | Provided as a fusion protein partner with glutathione-S-transferase derived from pGEX vector; recognizes various phosphotyrosine containing proteins in cell lysates; Western Blot: the fusion protein is strongly recognized by PI3-Kinase Ab & is not recognized by PI3-Kinase N-SH3 Ab

Phosphatidylinositol-3-Kinase p85 CT-SH2 Domain, Soluble

USBio P4185-29 Human, fusion protein expressed in *E. coli* Purified by glutathione-agarose chromatography & eluted with glutathione; 1 mg/mL; frozen solution in 100 µL total volume of TBS, pH 8.0,10% glycerol, 25 *mM* glutathione & 2 *mM* DTT | AA 624–718; provided as fusion protein partner with glutathione-S-transferase derived from pGEX vector; phosphatidylinositol (PI) metabolism has been associated with the intracellular signaling of many different transmembrane receptors & may play an important role in mitogenesis induced by growth factors; after treatment of cells with growth factors such as EGF or PDGF, PI 3-kinase phosphorylates PI & phosphorylated forms of PI at the D3 position of its inositol ring & PI(3) phosphate, PI(3,4) bisphosphate, & PI(3,4, 5) triphosphate rapidly accumulate; PI 3-kinase is a dimer composed of an 85kD subunit & a 110kD subunit; at least 3 genes (α, β & γ) areknown to encode the 85kD subunit, which has no catalytic activity & is thought to regulate thekinase activity of the 110kD catalytic subunit

Phosphatidylinositol-3-Kinase p85 NT-SH2 Domain, Agarose

USBio P4185-33 Human, fusion protein expressed in *E. coli* Purified by glutathione-agarose chromatography; 50 µg of GST-(PI 3-Kinase) N-SH2 domain provided in 25 µL total volume (20% bead slurry) of PBS,10% glycerol, 2 *mM* DTT, 0.02% NaN₃ | AA 333–428; provided as fusion protein partner with glutathione-S-transferase derived from pGEX; recognizes various phosphotyrosine containing proteins in cell lysates; Western Blot: the fusion protein is strongly recognized by PI 3-Kinase N-SH2 Ab & is not recognized by PI 3-Kinase N-SH3 Ab

Phosphatidylinositol-P3 Dependent Kinase, Active

USBio P3123 Human recombinant, expressed in Sf9 cells MW ~67k ≥70%; liquid in 50 µL of 50 *mM* Tris, pH 7.5, 0.1 *mM* EGTA, 0.1% β-MSH, 0.15 *M* NaCl, 0.27 *M* sucrose, 1 *mM* benzamidine, 200 µM PMSF, 1 mg/mL BSA | N-terminus His-tagged fusion protein, corresponding to the human sequence; has a kinase domain that is distantly related to Akt/PKB; PDK1, like Akt/PKB, contains a PH domain that tightly binds to PtdIns P3; activation of Akt/PKB is concomitant with phosphorylation of Thr³⁰⁸ & Ser⁴⁷³ & is prevented by inhibitors of phosphatidylinositol 3-Kinase; P3-dependent Kinase-1 (PDK1) phosphorylates Akt1 on Thr³⁰⁸ in the activation loop of thekinase domain of Akt/PKB; PDK2 phosphorylates Akt/PKB on Ser⁴⁷³ near the carboxyl-terminal; phosphorylation of these two sites is sufficient to fully activate Akt/PKB; SA: 625 U/mg

Phosvitin

Sigma P 1253 Egg yolk A phosphoprotein containing 8-10% phosphorus; Molar N/P ratio ~2.7; completely soluble in water

Phycocyanin C

ICN 151879 Algae PB, pH 7.0, 60% saturated (NH₄)₂SO₄, 2 mg/mL | Fluorescent phycobiliprotein easily coupled to biological molecules; coupled to proteins, enzymes, nucleic acids; superior labeling compared to fluorescein & rhodamine; used in fluorescence immunoassays & fluorescent labeling of DNA probes

Sigma P 0796 *Aphanotheca halophytica* Suspension in 50% ammonium sulfate containing 0.15 M Tris, pH 7.4; A₆₂₀/A₂₈₀>4.4; sold as mg protein based on E₆₂₀ at 1%=76.5 | US Patent No. 4,859,582; Teale, FWJ & Dale, RE, *Biochem J*, 116: 161, 1970; Oi, VT et al, *J Cell Biol*, 93: 981, 1982

Sigma P 7165 *Porphyra tenera* ("Nori") Lyophilized powder containing ~30% protein (Lowry); balance primarily sucrose, dithioerythritol & sodium azide as preservatives | US Patent No. 4,859,582

Sigma P 2172 *Spirulina sp.* Partially purified lyophilized powder containing ~40% protein (Lowry); balance primarily sucrose, dithioerythritol & sodium azide as preservatives | US Patent No. 4,859,582

Sigma P 6161 *Spirulina sp.* Highly purified lyophilized powder containing ~30% protein; balance primarily sucrose, dithioerythritol & sodium azide as preservatives | US Patent No. 4,859,582

Phycocyanin C, Pyridyldisulfide Derivative

Sigma P 0664 *Spirulina sp.* Lyophilized powder containing ~20% protein (Lowry); balance primarily sucrose & sodium azide as preservatives | US Patent No. 4,859,582

Phycocyanin R

Sigma P 1536 *Porphyridium cruentum* Lyophilized powder containing ~30% protein; balance primarily sucrose, dithioerythritol & sodium azide as preservatives | US Patent No. 4,859,582

Phycoerythrin B

ICN 151880 Algae PB, pH 7.0, 60% saturated $(NH_4)_2SO_4$, 2 mg/mL | Fluorescent phycobiliprotein easily coupled to biological molecules; coupled to proteins, enzymes, nucleic acids, polypeptide hormones, drug & vitamins; superior labeling compared to fluorescein & rhodamine; used in enumeration of T & B-lymphocytes by fluorescence microscopy, fluorescence-labeling of DNA-probes & fluorescence immunoassays

Sigma P 1286 *Porphyridium cruentum* Lyophilized powder containing ~30% protein (Lowry); balance primarily sucrose, dithioerythritol & sodium azide as preservatives | US Patent No. 4,859,582; Gantt, E & Lipshultz, CA, *Biochemistry*, 13: 2960, 1974

Phycoerythrin B, Biotin Conjugated

Sigma P 0788 *Porphyridium cruentum* Lyophilized powder containing ~30% protein (Lowry); balance primarily sucrose, dithioerythritol & sodium azide as preservatives; contains ~5 moles biotin/mole protein | US Patent No. 4,859,582

Phycoerythrin C

ICN 151881 Algae Fluorescent phycobiliprotein easily coupled to biological molecules; superior labeling compared to fluorescein & rhodamine; similar uses as phycoerythrin B

Phycoerythrin R

ICN 151882 Algae Fluorescent phycobiliprotein easily coupled to biological molecules; superior labeling compared to fluorescein & rhodamine

Sigma P 0159 *Corallina officinalis* Suspension in 50% ammonium sulfate, 50 mM sodium phosphate buffer, pH 7.0; sold on the basis of mg protein, based on E_{565} at 1%=81.7 | Oi, VT et al, *J Cell Biol*, 93: 981, 1982; US Patent No. 4,859,582

Sigma P 3663 *Porphyra tenera* ("Nori") Lyophilized powder containing ~30% protein (Lowry); balance primarily sucrose, dithioerythritol & sodium azide as preservatives | US Patent No. 4,859,582

Sigma P 8912 *Porphyra tenera* ("Nori") Lyophilized powder containing ~30% protein (Lowry); balance primarily sucrose, dithioerythritol & sodium azide as preservatives | Oi, VT et al, *J Cell Biol*, 93: 981, 1982; US Patent No. 4,859,582

Phycoerythrin R, Biotin Conjugated

Sigma P 7540 *Porphyra tenera* ("Nori") Lyophilized powder containing ~30% protein (Lowry); balance primarily sucrose, dithioerythritol & sodium azide as preservatives; contains ~5 moles biotin/mole protein | Oi, VT et al, *J Cell Biol*, 93: 981, 1982; US Patent No. 4,859,582

Phycoerythrin R, Pyridyldisulfide Derivative

Sigma P 7415 *Porphyra tenera* ("Nori") Lyophilized powder containing ~30% protein (Lowry); balance primarily sucrose & sodium azide as preservatives | US Patent No. 4,859,582

Pituitary Acetone Powder

Sigma P 3034 Carp

Sigma P 3909 Salmon

Pituitary Extract

Sigma P 1476 Bovine 0.2 μm filtered solution in phosphate buffered saline at a concentration of ~14 mg protein/mL; endotoxin tested; cell culture tested

Placenta Growth Factor

ICN 195732 Human recombinant, expressed in *E. coli* ≥97%; lyophilized | Binds with high activity to Flt-1 but not KDR/Flk-1

Placental Lactogen, Human

USBio P4220-20 Human ≥90% (SDS-PAGE); no contaminants detected; single band by SDS-PAGE, IEP, &/or RID; 1 mg/mL Lowry & DPC EIAkit supplied in PBS, pH 7.4, no preservative | Suitable for antigenic applications in immunological protocols

Plasmin

Synonyms: Fibrinolysin

Cortex CP1071 >95%

USBio P4256 Human ≥98% (SDS-PAGE); no contaminants detected; single band by SDS-PAGE, IEP, &/or RID; 1 mg/mL supplied in 100 *mM* sodium phosphate | Suitable for antigenic applications in immunological protocols

Biodesign A50192H Human plasma Purified

Fluka 80955 Human plasma MW 85k Lyophilized; 0.1-0.3 U/mg; ≤1 U corresponds to the amount of enzyme which hydrolyzes 1 μmol of Tos-Gly-Pro-Lys-4-NA•AcOH/min at pH 8.2, 25°C | Serine protease with trypsin-like specificity; Robbins, KC et al, *Meth Enzymol*, 80: 379, 1981; Saksela, O & Rifkin, DB, *Ann Rev Cell Biol*, 4: 93, 1988; Wiman, B & Collen, D, *Nature*, 272: 549, 1978

Plasminogen

Synonyms: Profibrinolysin; Plasma Trypsinogen

Cortex CP1072 >95%

Cortex CP2109U Bovine >98%

ICN 194097 Bovine plasma Precursor to the serine protease, plasmin

Sigma P 9156 Bovine plasma Lyophilized powder containing ~5% protein (Biuret); balance primarily NaCl, EDTA, lysine, Tris buffer salts; ε-aminocaproic acid free; activity: 3-5 U/mg protein; unit definition: 1 U produces a ΔA_{275} of 1.0 from α-casein/20 min at pH 7.5, 37°C, when measuring perchloric acid soluble products in a volume of 5.0 mL; activity determined after activation to plasmin with urokinase | Not suitable for clot formation procedure for streptokinase; 1 Sigma U = 3 WHO U (1st British standard-78/646)

Sigma P 3281 Horse plasma Lyophilized powder containing ~5% protein (Biuret); balance primarily NaCl, EDTA, lysine, Tris buffer salts; ε-aminocaproic acid free; activity: 3-6 U/mg protein; <0.01 U plasmin/U plasminogen; unit definition same as for Sigma P 9156 | Not suitable for clot formation procedure for streptokinase; 1 Sigma U = 3 WHO U (1st British standard-78/646)

USBio P4256-25 Human ≥98% (SDS-PAGE); no contaminants detected; single band by SDS-PAGE, IEP, &/or RID; 1 mg/mL; lyophilized from 0.02 *M* Tris HCl buffer, pH 7.5 with 2 *mM* EDTA | Suitable for antigenic applications in immunological protocols

Biodesign A50182H Human plasma Purified | Platelets & hemostasis reagents

Biogenesis 7440-1004 Human plasma MW 90k Free of Lys and EACA; tested negative for HBsAg, HIV and HCV antibodies; purified; 50 *mM* sodium phosphate, 2 *mM* EDTA, 2 mg D-mannitol, 2 mg NaCl, pH 7.5; liquid

Calbiochem 528175 Human plasma MW 90k Solid lyophilized from 20 *mM* Tris-HCl, 2 *mM* EDTA, pH 7.5; ≥95% (SDS-PAGE); SA: ≥120 U/mg protein; one U is the amount of enzyme that will hydrolyze 1.0 mmol *N*-Tosyl-Arg ethyl ester in 30 min at 37°C, pH 8.0; soluble in water; prepared from plasma shown by certified tests to be negative for HBsAg & antibodies to HIV & HCV | Single chain glycoprotein containing 790 AA residues; found in normal plasma at ~12 mg/100 mL; inactive precursor of the serine protease plasmin; converted to the active protease by cleavage at Arg[560]; *Merck Index*, 12: 7679; Korner, G et al, *J Cell Physiol*, 154: 456, 1993

Calbiochem 528178 Human plasma MW 90k EACA-free; Lys-free; solid lyophilized from 10 mg D-mannitol, 10 mg NaCl & phosphate buffer, pH 7.5; SA: ≥10 U/mg protein; one U is the amount of enzyme that will hydrolyze 1.0 mmol *N*-Tosyl-Gly-Pro-Lys-pNA/min at 25°C, pH 7.8; soluble in water; prepared from plasma shown by certified tests to be negative for HBsAg & antibodies to HIV & HCV | *Merck Index*, 12: 7679

Fluka 80959 Human plasma MW 90k Lyophilized; 0.1-0.3 U/mg after activation with urokinase; ≤1% plasmin; 1 U corresponds to the amount of enzyme which hydrolyzes 1 μmol of Tos-Gly-Pro-Lys-4-NA•AcOH/min at pH 8.2, 25°C

ICN 191342 Human plasma MW 90-94k >98% (SDS-PAGE); lyophilized; 20 *mM* Tris-HCl buffer, 2 *mM* EDTA, pH 7.5; 1 U hydrolyzes 1 μmole *N*-Tosyl-L-Arginine Ester/30 min at 37°C, pH 8.0 | Sottrup-Jensen, L etal, *Prog Chem Fibrinolysis Thrombolysis*, 3:191, 1978

ICN 194079 Human plasma Purified protein; glycine & NaCl stabilizers | Rijken, DC etal, *Thromb Haemostas*, 60(5):867, 1993

ICN 194094 Human plasma Precursor to the serine protease, plasmin

Sigma P 5661 Human plasma Lyophilized powder containing ~5% protein (Biuret); balance primarily NaCl, EDTA, lysine, Tris buffer salts; ε-aminocaproic acid free; activity: 6-9 U/mg protein; <0.0001 U plasmin/U plasminogen; unit definition same as for Sigma P 9156 | Source material negative for HIV & HBsAg; suitable for clot formation procedure for streptokinase; 1 Sigma U = 3 WHO U (1st British standard-78/646)

Sigma P 7397 Human plasma Lyophilized powder containing ~2% protein (Lowry); balance primarily NaCl, EDTA, lysine, Tris buffer salts; ε-aminocaproic acid free; activity: 6-9 U/mg protein; <0.01 U plasmin/U plasminogen; unit definition same as for Sigma P 9156 | Source material negative for HIV & HbsAg; not suitable for clot formation procedure for streptokinase; 1 Sigma U = 3 WHO U (1st British standard-78/646)

Sigma P 2284 Rabbit plasma Lyophilized powder containing ~5% protein (Biuret); balance primarily NaCl, EDTA, lysine, Tris buffer salts; ε-aminocaproic acid free; activity: 4-10 U/mg protein; <0.01 U plasmin/U plasminogen; unit definition same as for Sigma P 9156 | Not suitable for clot formation procedure for streptokinase; 1 Sigma U = 3 WHO U (1st British standard-78/646)

Biogenesis 7440-6059 Rat serum Lyophilized

Plasminogen Activator Inhibitor I

Cortex CP1123 >95%

Calbiochem 528208 Human recombinant, expressed in mutant MW 43k Liquid in 150 *mM* NaCl, 50 *mM* sodium phosphate buffer, 1 *mM* EDTA, pH 6.6; biological activity: >99% (uPA assay); >95% (SDS-PAGE); highly purified | Highly purified preparation of an altered form of human PAI-1 containing four mutated AA; virtually unable to go latent & is stable at elevated temperature & pH for extended periods of time; inhibits uPA & tPA; Berkenpas, MB et al, *EMBO J*, 14: 2969, 1995

Calbiochem 528213 Human recombinant, expressed in mutant MW 43k Liquid in 150 *mM* NaCl, 50 *mM* sodium phosphate buffer, 1 *mM* EDTA, pH 6.6; biological activity: ≥90% (uPA assay); ≥90% (SDS-PAGE); highly purified | Contains a single minor conservative AA substitution Ile[91]→Leu[91] which gives the inhibitor increased half life (~4 fold increase over the native recombinant form)

Calbiochem 528214 Rat recombinant MW 43k Liquid in 200 *mM* NaCl, 50 *mM* sodium phosphate buffer, 1 *mM* EDTA, pH 6.6; biological activity: >90% (uPA assay); >95% (SDS-PAGE) | Inhibits human uPA

Plasminogen Affinity Form I, Glu-

Synonyms: Profibrinolysin

Cortex CP2105U Human >98%

ICN 194095 Human plasma Isolated by gradient elution | Carbohydrate variant; precursor to the serine protease, plasmin

Plasminogen Affinity Form II, Glu-

Synonyms: Profibrinolysin

Cortex CP2106U Human >98%

ICN 194096 Human plasma Isolated by gradient elution | Carbohydrate variant; precursor to the serine protease, plasmin

Plasminogen Lysine Binding Site I

Synonyms: Plasminogen LBS I

Sigma P 1667 Human plasminogen Lyophilized from 25 mM ammonium bicarbonate; protein by Biuret; obtained from purified human plasminogen after digestion with elastase | Contains the 1st 3 triple-loop structures (number 1-3) in the plasmin A-chain (Kringle 1+2+3); this part of the plasmin(ogen) molecule shown to bind α2-antiplasmin; source material negative for HIV & HBsAg; Sottrup-Jensen, L et al, in Progress in Chemical Fibrinolysis & Thrombolysis, Vol 3: 191, Davidson, JF et al, eds, Raven Press, New York, 1978; Wiman, B et al, *Biochim Biophys Acta*, 579: 142, 1979

Plasminogen, Glu-

Synonyms: Profibrinolysin

Cortex CP2104U Human >98%

Calbiochem 528180 Human plasma MW 90k Liquid in 50 *mM* Tris-HCl, 100 *mM* NaCl, pH 7.5; single band purity (SDS-PAGE); no plasmin activity detected; SA: 23-27 U/mg protein; one U is the amount of enzyme that changes absorbance by 1.0 U at 275 nm/20 min at 37°C, pH 7.5 using casein as substrate; prepared from plasma shown by certified tests to be negative for HBsAg & antibodies to HIV & HCV | Can be activated to the serine protease plasmin via action of streptokinase, tissue plasminogen activator or urokinase; *Merck Index*, 12: 7679

Sigma P 2422 Human plasma Lyophilized from 0.02 M phosphate buffer, pH 7.3; contains 0.1 M NaCl; vial contains ~0.5 mg protein (E[280] at 1%); unit definition same as for Sigma P 9156 | Source material negative for HIV & HBsAg; not suitable for clot formation procedure for streptokinase; 1 Sigma U = 3 WHO U (1st British standard-78/646)

Plasminogen, Lys-

Synonyms: Profibrinolysin

Cortex CP2108U >98%

Calbiochem 528185 Human plasma MW 83k Liquid in 50 *mM* Tris-HCl, 100 *mM* NaCl, pH 7.4; homogeneous purity (SDS-PAGE); prepared from plasma shown by certified tests to be negative for HBsAg & antibodies to HIV & HCV | Purified from homogeneous Glu-plasminogen by activation with plasmin; activation results in the release of a 76 residue peptide (Glu[1]-Lys[76]); Lys[77]-plasminogen readily converted to Lys[77]-plasmin by any of the common plasminogen activators; *Merck Index*, 12: 7679

Platelet Activating Factor, Lyso *N*-(Me-^{14}C)-

ARC ARC-405 50-60 mCi/mmol; 1.85-2.22 GBq/mmol; in toluene:ethanol (1:1) | Radiochemical

Platelet Derived Endothelial Cell Growth Factor

Sigma P 5208   Human recombinant, expressed in Sf 21 insect cells   MW 45k   ≥97% (SDS-PAGE); 0.2 μm filtered & lyophilized from phosphate buffered saline containing 500 μg BSA; proliferative activity is measured in culture by using a human umbilical vein endothelial cells; endotoxin tested | Endothelial cell mitogen originally purified from human platelets; doesn't bind to heparin & doesn't stimulate the proliferation of fibroblasts; this is in contrast with the effect of PDGF which stimulates the growth of human foreskin fibroblasts, but was inactive on endothelial cells; stimulates endothelial cells *in vitro* & *in vivo*; chemotactic for bovine aortic endothelial cells however doesn't induce smooth muscle cell migration; involved in angiogenesis & has potent angiogenic activity both in the developing vascular system of the chick chorioalantoic membrane & vascularization of tumor cells in nude mice; has a pI of 4.0-4.8; Miyazano, K et al, *J Biol Chem*, 262: 4098, 1987; Lobb, RR et al, *Analyt Biochem*, 154: 1, 1986; Ishikawa, F et al, *Nature*, 338: 557, 1989; Usuki, K et al, *Cell Regulation*, 1: 577, 1990

ICN 195740   Human recombinant, expressed in Sf21   ≥97%; lyophilized; ED$_{50}$ = 20-40 ng/mL

Platelet Derived Growth Factor

ICN 154131   Human platelet   >97%; lyophilized carrier free; ED$_{50}$ = 1.0-3.0 ng/mL, 1 U stimulates dose-dependent uptake of ^{3}H-thymidine by NR6R-3T3 fibroblasts

Biogenesis 7460-0504	Human platelets	Lyophilized

Calbiochem 521200   Human platelets   MW 28k, 32k   ≥97% (SDS-PAGE); lyophilized from filter-sterilized 25% acetonitrile, 0.1% TFA; biological activity: ED$_{50}$=1.0-3.0 ng/mL as measured in a mitogenic assay using quiescent NR6-3T3 fibroblasts; endotoxin: ≤100 pg/μg PDGF | Consists of ~70% PDGF-AB & may contain ~30% PDGF-BB; action is mediated by cell surface α- & β-receptors; *Merck Index*, 12: 7683; Abboud, HE et al, *J Cell Physiol*, 158: 140, 1994; Soma, Y et al, *Exp Cell Res*, 212: 274, 1994; Hammacher, A et al, *J Biol Chem*, 263: 16493, 1988

Sigma P 8147   Human platelets   MW 28-31k   ≥95% (SDS-PAGE); lyophilized & carrier-free; 0.25 μg/vial; endotoxin tested | Principal mitogen found in mammalian serum & is released from platelets during clot formation; elicits multifunctional actions with a variety of cells, including mitogenesis of mesoderm-derived cells, increased extracellular matrix synthesis, & chemotaxis & activation of neutrophils, monocytes & fibroblasts; mitogenic for dermal & tendon fibroblasts vascular smooth muscle cells, glial cells & chondrocytes; appears to interact with Transforming Growth Factor-1 in accelerating wound healing; pathogenic in arteriosclerosis & neoplasia; the mitogenic activities of all PDGF products are tested in culture using Swiss 3T3 cells or NR6-3T3 fibroblasts; Pierce, G et al, *J Cell Biol*, 109: 429, 1989; Ross, R et al, *Proc Natl Acad Sci USA*, 71: 1207, 1974; Ross, R, *Arteriosclerosis*, 1: 293, 1981; Raines, E et al, *Meth Enzymol*, 109: 749, 1985

ICN 150204   Human platelets (outdated)   Lyophilized; ≥5x10^4 U/mg protein, 1 U stimulates DNA synthesis in 50% of the 3%3 cells in a confluent monolayer using a microtiter plate (0.2 mL/well) assay system; partially purified via cation-exchange, size-exclusion & hydrophobic systems

ICN 150020   Porcine platelet   Partially purified, sterile solution in 400 mL 1 *M* NaCl, 10 m*M* NaPO$_4$, pH 7.4, 50% (v/v) ethanediol; 1 U gives 50% maximal stimulation of untransformed Swiss 3T3 cells in a DNA synthesis assay in 2 mL serum-free medium in a 30 mm tissue culture dish | Stroobant, P & MD Waterfield, *EMBO J*, 3:2963, 1984

ICN 153503   Porcine platelet   ≥95%; frozen liquid, 30% acetonitrile, 0.1% TFA; ≥30,000 U/mg protein, 1 U gives 50% maximal stimulation of untransformed Swiss 3T3 cells in a DNA synthesis assay in 2 mL serum-free medium in a 30 mm tissue culture dish

Biogenesis 7460-2002	Porcine platelets	Liquid

Calbiochem 521300   Porcine platelets   MW 38k   >97% (SDS-PAGE); lyophilized from filter-sterilized 30% acetonitrile, 0.1% TFA; endotoxin: ≤100 pg/μg PDGF | Consists primarily of PDGF-BB homodimers; *Merck Index*, 12: 7683; Fretto, LJ et al, *J Biol Chem*, 268: 3625, 1993; Stroobant, P & Waterfield, MD, *EMBO J*, 3: 2963, 1984

Sigma P 8953   Porcine platelets   ≥97% (SDS-PAGE); lyophilized & carrier-free; biological activity was determined by measuring the PDGF dependent ^{3}H-thymidine incorporation in quiescent NR6R-3T3 fibroblasts; endotoxin tested | Principal mitogen found in mammalian serum & is released from platelets during clot formation; elicits multifunctional actions with a variety of cells, including mitogenesis of mesoderm-derived cells, increased extracellular matrix synthesis, & chemotaxis & activation of neutrophils, monocytes & fibroblasts; mitogenic for dermal & tendon fibroblasts vascular smooth muscle cells, glial cells & chondrocytes; appears to interact with Transforming Growth Factor-1 in accelerating wound healing; pathogenic in arteriosclerosis & neoplasia; the mitogenic activities of all PDGF products are tested in culture using Swiss 3T3 cells or NR6-3T3 fibroblasts; Pierce, G et al, *J Cell Biol*, 109: 429, 1989; Ross, R et al, *Proc Natl Acad Sci USA*, 71: 1207, 1974; Ross, R, *Arteriosclerosis*, 1: 293, 1981; Raines, E et al, *Meth Enzymol*, 109: 749, 1985

Biogenesis 7460-0604	r-DNA	Lyophilized
Biogenesis 7460-0625	r-DNA	^{125}I conjugated; liquid
Biogenesis 7460-0654	r-DNA	Lyophilized
Biogenesis 7460-0704	r-DNA	Liquid
Biogenesis 7460-1004	r-DNA	^{125}I conjugated; liquid

Platelet Derived Growth Factor AA

Biodesign A52113H	*E. coli*	MW 26.5k	Purified
Chemicon GF017	Human	≥95%	
BioSource International PHG0034	Human recombinant		

Oncogene PF009   Human recombinant   MW 30k (dimer, SDS-PAGE)   >97% (SDS-PAGE); lyophilized; biological activity: half maximal stimulation of ^{3}H-thymidine uptake by NIH/3T3 fibroblasts ~2 ng/mL | Species reactivity: human & mouse; for proliferation studies & Western blot; will act on both human & mouse cells

Sigma P 3076   Human recombinant   ≥97% (SDS-PAGE); 0.2 μm filtered & lyophilized from 30% acetonitrile & 0.1% TFA without stabilizer proteins; endotoxin tested | Recombinant human PDGF-AA is the dimer of the A chain of human PDGF expressed in *E. coli*; principal mitogen found in mammalian serum & is released from platelets during clot formation; elicits multifunctional actions with a variety of cells, including mitogenesis of mesoderm-derived cells, increased extracellular matrix synthesis, & chemotaxis & activation of neutrophils, monocytes & fibroblasts; mitogenic for dermal & tendon fibroblasts vascular smooth muscle cells, glial cells & chondrocytes; appears to interact with Transforming Growth Factor-1 in accelerating wound healing; pathogenic in arteriosclerosis & neoplasia; the mitogenic activities of all PDGF products are tested in culture using Swiss 3T3 cells or NR6-3T3 fibroblasts; Pierce, G et al, *J Cell Biol*, 109: 429, 1989; Ross, R et al, *Proc Natl Acad Sci USA*, 71: 1207, 1974; Ross, R, *Arteriosclerosis*, 1: 293, 1981; Raines, E et al, *Meth Enzymol*, 109: 749, 1985

Calbiochem 521215   Human recombinant, expressed in *E. coli*   MW 29k   ≥97% (SDS-PAGE); lyophilized from filter-sterilized 30% acetonitrile, 0.1% TFA; endotoxin: ≤100 pg/μg PDGF-AA | Disulfide-linked dimer of two a-chain monomers; AA sequence is the long form of mature human PDGF-A chain & is identical to that deduced from the native human nucleotide sequence except for the addition of an N-terminal methionine group; implicated in the differentiation of cells of the oligodendrocyte lineage; *Merck Index*, 12: 7683; Butt, AM et al, *J Neurosci Res*, 48: 588, 1997; Westermark, B & Heldin, C-H, *Cancer Res*, 51: 5087, 1991; Betsholtz, C et al, *Nature*, 320: 695, 1986

Fitzgerald 30-AP28	Human recombinant, expressed in *E. coli*

ICN 153476   Human recombinant, expressed in *E. coli*   MW ~28k   >97%; lyophilized; ED$_{50}$ = 1.0-5.0 ng/mL; 1 U stimulates ^{3}H-thymidine incorporation by NR6R-3T3 fibroblasts

PeproTech 100-13A   Human recombinant, expressed in *E. coli*   MW 28.5k (A chain homodimer)   >98% (SDS-PAGE) & HPLC; <0.1 ng endotoxin per mg (1EU/mg); lyophilized | Potent mitogen for a wide range of cell types including fibroblasts, smooth muscle & connective tissue; composed of a dimer of two chains, A & B; present as AA or BB homodimers or AB heterodimer; 250 amino acids; ED$_{50}$ ≤ 1.0 ng/mL; SA ≥ 10^6 U/mg; SA determined by dose-dependent stimulation of thymidine uptake by BALB/c 3T3 cells

IBT GF-080-3, GF-080-5 Human recombinant, expressed in yeast >95 %; lyophilized powder | Composed of 2 identical polypeptide chains attached by disulfide bonds; fully active as measured by mitogenic assay involving stimulation of ^{3}H-thymidine incorporation into NIH-3T3 cells

Amersham VPF009 Recombinant Human & rat cross-reactivity | Useful for proliferation assay; growth/death factor interactions

Platelet Derived Growth Factor AA, α-

Harlan BT-3020 Human recombinant, expressed in *E. coli* Lyophilized; 0.01 mg

Harlan BT-3021 Human recombinant, expressed in *E. coli* Lyophilized; 0.025 mg

Platelet Derived Growth Factor AB

Biodesign A52000H *E. coli* MW 25.5k Purified

BioSource International PHG0134 Human recombinant

Sigma P 3326 Human recombinant ≥97% (SDS-PAGE); 0.2 μm filtered & lyophilized from 30% acetonitrile & 0.1% TFA without carrier proteins; endotoxin tested | Recombinant human PDGF-AB is the heterodimer of the A & B chains of PDGF expressed in *E. coli* & disulfide linked; principal mitogen found in mammalian serum & is released from platelets during clot formation; elicits multifunctional actions with a variety of cells, including mitogenesis of mesoderm-derived cells, increased extracellular matrix synthesis, & chemotaxis & activation of neutrophils, monocytes & fibroblasts; mitogenic for dermal & tendon fibroblasts vascular smooth muscle cells, glial cells & chondrocytes; appears to interact with Transforming Growth Factor-1 in accelerating wound healing; pathogenic in arteriosclerosis & neoplasia; the mitogenic activities of all PDGF products are tested in culture using Swiss 3T3 cells or NR6-3T3 fibroblasts; Pierce, G et al, *J Cell Biol*, 109: 429, 1989; Ross, R et al, *Proc Natl Acad Sci USA*, 71: 1207, 1974; Ross, R, *Arteriosclerosis*, 1: 293, 1981; Raines, E et al, *Meth Enzymol*, 109: 749, 1985

Sigma P 6684 Human recombinant ≥95% (SDS-PAGE); 0.2 μm filtered solution of 0.2 mL acetate buffered saline without stabilizer proteins; endotoxin tested | Recombinant human PDGF-AB is the heterodimer of the A & B chains of PDGF expressed in *E. coli* & linked by glutathione-facilitated dimerization; principal mitogen found in mammalian serum & is released from platelets during clot formation; elicits multifunctional actions with a variety of cells, including mitogenesis of mesoderm-derived cells, increased extracellular matrix synthesis, & chemotaxis & activation of neutrophils, monocytes & fibroblasts; mitogenic for dermal & tendon fibroblasts vascular smooth muscle cells, glial cells & chondrocytes; appears to interact with Transforming Growth Factor-1 in accelerating wound healing; pathogenic in arteriosclerosis & neoplasia; the mitogenic activities of all PDGF products are tested in culture using Swiss 3T3 cells or NR6-3T3 fibroblasts; Pierce, G et al, *J Cell Biol*, 109: 429, 1989; Ross, R et al, *Proc Natl Acad Sci USA*, 71: 1207, 1974; Ross, R, *Arteriosclerosis*, 1: 293, 1981; Raines, E et al, *Meth Enzymol*, 109: 749, 1985

Calbiochem 521220 Human recombinant, expressed in *E. coli* MW 27k ≥97% (SDS-PAGE); lyophilized from filter-sterilized 30% acetonitrile, 0.1% TFA; endotoxin: ≤100 pg/μg PDGF-AB | Produced by the *in vitro* dimerization of PDGF-A & PDGF-B monomers; *Merck Index*, 12: 7683; Hannink, M & Donoghue, DJ, *Biochim Biophys Acta*, 989: 1, 1989; Williams, LT, *Science*, 243: 1564, 1989; Betsholtz, C et al, *Nature*, 320: 695, 1986; Deuel, TF, *Ann Rev Cell Biol*, 3: 443, 1987; Johnson, A et al, *EMBO J*, 3: 921, 1984

ICN 160063 Human recombinant, expressed in *E. coli* >97%; lyophilized; ED$_{50}$ = 1.0-3.0 ng/mL; 1 U stimulates ^{3}H-thymidine incorporation by NR6R-3T3 fibroblasts | Homodimer

PeproTech 100-00AB Human recombinant, expressed in *E. coli* MW 25.5k (heterodimer, 13.3k A & 12.2k B chain) >98% (SDS-PAGE) & HPLC; <0.1 ng endotoxin per mg (1EU/mg); lyophilized | Potent mitogen for a wide range of cell types including fibroblasts, smooth muscle & connective tissue; composed of a dimer of two chains, A & B; present as AA or BB homodimers or AB heterodimer; ED$_{50}$ ≤ 1.0 ng/mL; SA ≥ 10^6 U/mg; SA determined by dose-dependent stimulation of thymidine uptake by BALB/c 3T3 cells

Platelet Derived Growth Factor BB

Biodesign A52114H *E. coli* MW 25k Purified

Chemicon GF018 Human ≥95%

BioSource International PHG0044 Human recombinant

Oncogene PF010 Human recombinant MW 32k (homodimer) >97% (SDS-PAGE); lyophilized with 100 μg BSA; biological activity: half maximal stimulation of ^{3}H-thymidine uptake by 3T3 fibroblasts ~2 ng/mL | Species reactivity: human; for proliferation studies & Western blot

Sigma P 3201 Human recombinant ≥97% (SDS-PAGE); 0.2 μm filtered & lyophilized from 30% acetonitrile & 0.1% TFA without stabilizer proteins; biological activity is tested in culture by measuring its ability to stimulate ^{3}H-thymidine incorporation in NR6R-3T3 fibroblasts; endotoxin tested | Recombinant human PDGF-BB is the dimer of the B chain of human PDGF expressed in *E. coli*; principal mitogen found in mammalian serum & is released from platelets during clot formation; elicits multifunctional actions with a variety of cells, including mitogenesis of mesoderm-derived cells, increased extracellular matrix synthesis, & chemotaxis & activation of neutrophils, monocytes & fibroblasts; mitogenic for dermal & tendon fibroblasts vascular smooth muscle cells, glial cells & chondrocytes; appears to interact with Transforming Growth Factor-1 in accelerating wound healing; pathogenic in arteriosclerosis & neoplasia; the mitogenic activities of all PDGF products are tested in culture using Swiss 3T3 cells or NR6-3T3 fibroblasts; Pierce, G et al, *J Cell Biol*, 109: 429, 1989; Ross, R et al, *Proc Natl Acad Sci USA*, 71: 1207, 1974; Ross, R, *Arteriosclerosis*, 1: 293, 1981; Raines, E et al, *Meth Enzymol*, 109: 749, 1985

Sigma P 4306 Human recombinant ≥95% (SDS-PAGE); 0.2 μm filtered & lyophilized without stabilizer proteins; endotoxin tested | Recombinant human PDGF-BB is the dimer of the B chain of human PDGF expressed in *E. coli*; principal mitogen found in mammalian serum & is released from platelets during clot formation; elicits multifunctional actions with a variety of cells, including mitogenesis of mesoderm-derived cells, increased extracellular matrix synthesis, & chemotaxis & activation of neutrophils, monocytes & fibroblasts; mitogenic for dermal & tendon fibroblasts vascular smooth muscle cells, glial cells & chondrocytes; appears to interact with Transforming Growth Factor-1 in accelerating wound healing; pathogenic in arteriosclerosis & neoplasia; the mitogenic activities of all PDGF products are tested in culture using Swiss 3T3 cells or NR6-3T3 fibroblasts; Pierce, G et al, *J Cell Biol*, 109: 429, 1989; Ross, R et al, *Proc Natl Acad Sci USA*, 71: 1207, 1974; Ross, R, *Arteriosclerosis*, 1: 293, 1981; Raines, E et al, *Meth Enzymol*, 109: 749, 1985

Calbiochem 521225 Human recombinant, expressed in *E. coli* MW 25k ≥97% (SDS-PAGE); lyophilized from filter-sterilized 30% acetonitrile, 0.1% TFA; endotoxin: ≤100 pg/μg PDGF-BB | Disulfide-linked dimer of two 109 AA B-chain monomers; potent chemoattractant for neutrophils, mesenchymal & mononuclear cells; same AA sequence as mature human PDGF-B; rapidly activates protein kinase D in vascular smooth muscle cells; *Merck Index*, 12: 7683; Westermark, B & Heldin, C-H, *Cancer Res*, 51: 5087, 1991; Abedi, H et al, *FEBS Lett*, 427: 209, 1998; Abboud, HE et al, *J Cell Physiol*, 158: 140, 1994; Johnson, A et al, *EMBO J*, 3: 921, 1984

Fitzgerald 30-AP29 Human recombinant, expressed in *E. coli*

ICN 153477 Human recombinant, expressed in *E. coli* >97%; lyophilized; ED$_{50}$ = 1.0-3.0 ng/mL; 1 U stimulates ^{3}H-thymidine incorporation by NR6R-3T3 fibroblasts | Homodimer; suitable for receptor binding applications & iodination procedures

PeproTech 100-14B Human recombinant, expressed in *E. coli* MW 24.3k (B chain homodimer) >98% (SDS-PAGE) & HPLC; <0.1 ng endotoxin per mg (1EU/mg); lyophilized | Potent mitogen for a wide range of cell types including fibroblasts, smooth muscle & connective tissue; composed of a dimer of two chains, A & B; present as AA or BB homodimers or AB heterodimer; 250 amino acids; ED$_{50}$ ≤ 1.0 ng/mL; SA ≥ 10^6 U/mg; SA determined by dose-dependent stimulation of thymidine uptake by BALB/c 3T3 cells

IBT GF-070-3, GF-070-5 Human recombinant, expressed in yeast >97 %; lyophilized powder | Fully active as determined by its mitogenic activity measured by stimulation of ^{3}H-thymidine incorporation into human foreskin fibroblasts

Sigma P 4056   Rat recombinant, expressed in *E. coli*   ≥97% (SDS-PAGE); 0.2 μm filtered & lyophilized from 30% acetonitrile & 0.1% TFA containing 2.5 mg BSA; biological activity is tested is measured by its ability to stimulate ^{3}H-thymidine incorporation in quiescent NR6R-3T3 fibroblasts; endotoxin tested | Principal mitogen found in mammalian serum & is released from platelets during clot formation; elicits multifunctional actions with a variety of cells, including mitogenesis of mesoderm-derived cells, increased extracellular matrix synthesis, & chemotaxis & activation of neutrophils, monocytes & fibroblasts; mitogenic for dermal & tendon fibroblasts vascular smooth muscle cells, glial cells & chondrocytes; appears to interact with Transforming Growth Factor-1 in accelerating wound healing; pathogenic in arteriosclerosis & neoplasia; the mitogenic activities of all PDGF products are tested in culture using Swiss 3T3 cells or NR6-3T3 fibroblasts; Pierce, G et al, *J Cell Biol*, 109: 429, 1989; Ross, R et al, *Proc Natl Acad Sci USA*, 71: 1207, 1974; Ross, R, *Arteriosclerosis*, 1: 293, 1981; Raines, E et al, *Meth Enzymol*, 109: 749, 1985

Amersham VPF010   Recombinant   Human cross-reactivity | Useful for proliferation assay; growth/death factor interactions

Platelet Derived Growth Factor BB, β-

Harlan BT-3022   Human recombinant, expressed in *E. coli* Lyophilized; 0.01 mg

Harlan BT-3023   Human recombinant, expressed in *E. coli* Lyophilized; 0.025 mg

Platelet Derived Growth Factor Receptor α, Human Extracellular Domain

IBT GR-080-3, GR-080-5   Human recombinant, expressed in *Spodoptera frugiperda* insect cells (Sf9)   >85%; 0.50 mg prot/mL in 10 mM Tris-HCl, pH 7.0 | Binds the ligand (PDGF), as demonstrated by immobilizing the PDGF receptor extracellular domain & establishing competition of binding with radioiodinated PDGF

Platelet Derived Growth Factor Receptor β (Fc Chimera)

Oncogene PF079   Human recombinant   MW 84k/150k (glycosylated)   >97% (SDS-PAGE); lyophilized; biological activity: EC$_{50}$ of 0.01-0.03 μg/mL in the presence of 4 ng/mL of recombinant human PDGF BB as measured by its ability to inhibit PDGF BB induced ^{3}H-thymidine incorporation in NR6R-3T3 fibroblasts | Species reactivity: human; for proliferation studies; as a result of glycosylation, the recombinant human PDGF Rβ/Fc migrates as a 150 k protein in SDS-PAGE

Platelet Derived Growth Factor Receptor β, Human Extracellular Domain

IBT GR-070-3, GR-070-5   Human recombinant, expressed in *Spodoptera frugiperda* insect cells (Sf9)   >85%; 0.50 mg prot/mL in 10 mM Tris-HCl, pH 7.0 | Binds the ligand (PDGF), as demonstrated by immobilizing the PDGF receptor extracellular domain & establishing competition of binding with radioiodinated PDGF

Platelet Derived Growth Factor Soluble Receptor α

Oncogene PF080   Human recombinant   MW 56k   >97% (SDS-PAGE); lyophilized; biological activity: EC$_{50}$ of 1-3 μg/mL in the presence of 10 ng/mL of recombinant human PDGF AA as measured by its ability to inhibit PDGF AB or recombinant human PDGF AA in quiescent NR6R-3T3 fibroblasts | Species reactivity: human; for competition studies

Platelet Factor IV

Biodesign A52316H	*E. coli*	MW 8k	Purified
Chemicon GF019	Human	≥95%	
Cortex CP3096U	Human	>98%	

Sigma F 1385   Human fresh platelet-rich plasma   Lyophilized from 1 mL protein solution containing 0.4 M NaCl & 10 mM Tris, pH 8.2 | In double diffusion assay, protein gives a single arc against anti-human platelet factor 4 & no arc against anti-human whole serum; Campbell, PJ, *J Biol Stand*, 2: 259, 1974

ICN 195808   Human platelets   ≥98%; lyophilized; negative for HBsAg & HIV | Heparin-binding protein from α-granules of activated platelets

Fitzgerald 30-AP50   Human recombinant, expressed in *E. coli*

ICN 160264   Human recombinant, expressed in *E. coli*   ≥98%; lyophilized; ED$_{50}$ = 50 ng/mL; 1 U exerts maximal chemotactic activity on human fibroblasts in a modified Boyden chamber | Heparin neutralizing protein which affects the immune response; may increase the normal immune response or restore suppressed immune systems

PeproTech 300-16   Human recombinant, expressed in *E. coli* MW 7.8k   >98% (SDS-PAGE) & HPLC; <0.1 ng endotoxin per mg (1EU/mg); lyophilized | 70 AA; structurally related to human IL-8 & GROa; chemoattractant for human fibroblasts & certain other cells; SA determined by its ability to chemoattract human fibroblasts

Pleiotrophin

Synonyms: Heparin Binding Brain Mitogen; Heparin Binding Growth Factor VIII; Heparin Binding Growth Associated Molecule; Osteoblast Specific Factor I

BioSource International PHC7075   Human recombinant

Oncogene PF050   Human recombinant   MW 15.3k   >97% (SDS-PAGE); lyophilized in PBS; biological activity: optimal neurite outgrowth of cerebral cortical neurons of E10 chick embryos was observed with 3-8 μg/mL of recombinant human pleiotrophin | Species reactivity: human; for proliferation studies; can be used as an attachment substrate to stimulate neurite outgrowth in mixed cultures of embryonic rat, mouse or chicken brain cells

Sigma P 5333   Human recombinant, expressed in *Sf* 21 insect cells   ≥97% (SDS-PAGE); 0.2 μm filtered & lyophilized from phosphate buffered saline; pleiotrophin is measured in culture by neurite outgrowth of chick embryos; endotoxin tested | May be the first member of a family of developmentally regulated cytokines; active in growth & development; pleiotrophin gene is highly expressed in brain, uterus, gut, muscle, lung & skin; pleiotrophin mRNA is expressed in osteoblasts, chondrocytes, fibroblasts, astrocytes, Schwann cells & tumor cells; mitogenic & neurite outgrowth activity; extraordinary conservation between AA sequences of bovine, human & rat species; Li, YS et al, *Science*, 250: 1690, 1990; Tezuka, KI et al, *Biochem Biophys Res Commun*, 173: 246, 1990; Milner, PG et al, *Biochem Biophys Res Commun*, 165: 1096, 1990; Merenmies, J et al, *J Biol Chem*, 265: 28, 1990; Hampton, BS et al, *Mol Biol Cell*, 3: 85, 1992

ICN 195727   Human recombinant, expressed in *Sf*21   ≥97%; lyophilized | Enhances neurite outgrowth of cerebral cortical neurons

Pokeweed Antiviral Toxin

ICN 158825   *Phytolacca americana* (pokeweed) leaves   MW 29k (PAP), MW 30k (PAP II)   PAP mitogen free (SDS-PAGE); 1 U inhibits protein synthesis by 50% (IC$_{50}$) in a cell-free translation system | Hemitoxin with A chain but no B chain activity; plant protein with antiviral & anticellular activity; specific site of action is the Ef-2 mediated translocation step to the elongation cycle during protein synthesis; exists as two forms: pokeweed antiviral protein (PAP, 29k MW) & PAP II (30k MW)

PRAK, Active

USBio P5600-05

PRAK, Inactive

USBio P5600-07   Recombinant, expressed in Sf9 cells Recombinant full-length human PRAK with an N-terminal His-tag; purified using Ni-NTA agarose

Pregnancy Specific βI-Glycoprotein

Synonyms: SP1; Schwangerschaft Protein

Calbiochem 529580   Human retroplacental serum   MW 100k
Lyophilized solid; ≥90% (SDS-PAGE); immunological activity: fully active in ELISA with anti-SP1 antibodies; prepared from serum shown to be negative for HBsAg & for antibodies to HIV & HCV | Produced by syncytiotrophoblastic cells; useful for screening normal & pathological pregnancy, screening & monitoring trophoblastic disease & differentiating between benign & malignant trophoblastic disease; lower levels are also indicative of Down's syndrome; Joe, TW et al, *Biochim Biophys Acta*, 1219: 195, 1994; Kalenga, MK et al, *Eur J Pharm Mol Pharmacol*, 16: 231, 1994; Wu, S-M & Chan, W-Y, *Am J Human Genetics*, 55: A290, 1994

Proinsulin

ICN 156374   Human recombinant, expressed in *E. coli*

Sigma P 4672   Human recombinant, expressed in *E. coli*
Lyophilized; may contain trace of NH₄HCO₃

Prolactin

Cortex CP1114	>95%
Cortex CP1114P	>40%

ICN 151951   Human   Iodination grade; >98%; lyophilized; 30 IU/mg; <0.3% hFSH, hTSH; <1.0% hLH | WHO/IRP 75/504

USBio P9009-21   Human   ≥80% (SDS-PAGE); hFSH/hTSH/hCG <0.01%, hLH <0.03%, hGH <0.1%; lyophilized from 0.5 M ammonium bicarbonate | Suitable for antigenic applications in immunological protocols

Biogenesis 7770-0959   Human pituitary   MW ~22.8k (multiple bands on SDS-PAGE)   Tested negative for antibodies to HBsAg, HCV, HIV-1 and HIV-2; SA: 100 mIU/vial; purified; from 0.05 M NH₄HCO₃, pH 8.0; lyophilized

Biogenesis 7770-1009   Human pituitary   MW 23k   <0.30% hLH, <0.1% hGH, <0.05% hFSH, <0.2% hTSH; essentially salt free; tested negative for HBsAg, HCV, HIV 1 and 2; purified; from 0.05 M NH₄HCO₃, pH 8.0; lyophilized

Biogenesis 7770-1009-1mg   Human pituitary   MW 23k   <0.30% hLH, <0.1% hGH, <0.05% hFSH, <0.2% hTSH; essentially salt free; tested negative for HBsAg, HCV, HIV 1 and 2; purified; from 0.05 M NH₄HCO₃, pH 8.0; lyophilized

Biogenesis 7770-1104   Human pituitary   Lyophilized

Calbiochem 869039   Human pituitary   MW 22.8k
Iodination grade; lyophilized solid; immunopotency: ≥30 IU/mg (WHO 1st IRP 75/504); soluble in ethanol & methanol; hLH, hTSH, hGH: ≤1.0%; hFSH: ≤0.5%; shown by certified tests to be negative for HBsAg & for antibodies to HIV & HCV; may be carcinogenic/teratogenic | Single chain peptide hormone required for lactation in mammals; inhibits secretion of gonadotropins; prolactin receptors are present in breast tissue, adrenals, ovary, testis, kidney & liver; *Merck Index*, 12: 7961

Fitzgerald 30-AP05	Human pituitary gland	Affinity purity
Fitzgerald 30-AP10	Human pituitary gland	Standard purity

USBio P9009-25   Human pituitary gland   ≥98% (SDS-PAGE); hFSH/hTSH/hCG <0.01%, hLH <0.03%, hGH <0.1%; lyophilized from 0.5 M ammonium bicarbonate | Suitable for antigenic applications in immunological protocols

Biodesign A86848H   Human pituitary glands   98%

USBio P9009-27   Human pituitary glands   ≥40% (SDS-PAGE); lyophilized from 50mM ammonium bicarbonate

Biogenesis 7770-3004   Ovine pituitary   MW 24k   Purified; from 0.05 M ammonium bicarbonate buffer, pH 8.0; lyophilized | A member of the growth-hormone placental lactogen family; has 2 disulfide bonds and one Asn-Thr-Ser glycosylation site; partially glycosylated representing 70% of the circulating prolactin

Biogenesis 7770-3509   Porcine pituitary   <0.5% pGH, <0.1% pLH/pFSH/pTSH; purified; 0.05 M NH₄HCO₃; lyophilized

Biogenesis 7770-5504   Rat pituitary   <0.5% rGH, <0.05% rLH, <0.03% rTSH/rFSH; purified; 0.05M ammonium bicarbonate, pH 8.0; lyophilized | A member of the growth hormone-placental lactogen family; 2 disulfide bonds and free amino and carboxy-terminal AA; 1 Asn-Thr-Ser glycosylation site; partially glycosylated and is thought to represent 70% of circulating prolactin

Fitzgerald 30-AP06   Recombinant expressed in *E. coli*

USBio P9009-26   Recombinant, expressed in *E. coli*   ≥99% (SDS-PAGE); no contaminants detected; single band by SDS-PAGE, IEP, &/or RID; lyophilized | Suitable for antigenic applications in immunological protocols; SA: 20-30 IU

Pronectin F

BioSource International PNE0013

Prostaglandin F2a

USBio P9054-05   ≥99%; lyophilized | Suitable for antigenic applications in immunological protocols

Prostate Specific Antigen

Cortex CP1017	>95%	
Cortex CP1017P	>40%	
Cortex CP1017U	>98%	
Fitzgerald 30-AP15E	Enzymatically active	High purity

USBio P9054-52   Human   Iodination grade, 99%; PSA content determined by Abbott ELISA, total protein by Lowry & Optical density; SDS-PAGE shows single band; 2.7 mg/mL (Abbott IMX) supplied in 50 mM Tris buffer, 150 mM NaCl, pH 8.0; sterile filtered; no azide added | Suitable for antigenic applications in immunological protocols

Biodesign A32310H   Human fluids   ≥50% | Tumor marker, cancer antigens & oncogenes

Biodesign A32874H   Human fluids   ≥95% | Tumor marker, cancer antigens & oncogenes

Biogenesis 7820-0504   Human seminal fluid   Tested negative for HIV I and II antibodies, Hepatitis B surface antigen and HCV antibody; purified; 0.05 M Phosphate buffer, pH 7.5, 0.15 M NaCl, 0.1% NaN₃; liquid

Biogenesis 7820-0604   Human seminal fluid   Liquid

Calbiochem 539832   Human seminal fluid   Liquid in 10 mM Tris, 0.1% NaN₃, pH 8.0; sterile-filtered; preservative- & reductant-free; no perchloric acid or detergents were used; ≥98% (SDS-PAGE); prepared from fluids of individuals shown to be negative for HBsAg & for antibodies to HIV & HCV | Prostate tumor marker; higher levels are reported in patients with prostate cancer; Culkin, DJ et al, *Prostate*, 26: 1, 1995

Calbiochem 539834   Human seminal fluid   MW 30k   Liquid in 150 mM NaCl, 50 mM Tris-HCl, 0.1% NaN₃, sterile-filtered; >95% (SDS-PAGE); enzymatic activity: fully biologically active as tested by its ability to hydrolyze the synthetic peptide substrate MeO-Suc-Arg-Pro-Tyr-pNA; prepared from fluids of individuals shown to be negative for HBsAg & for antibodies to HIV & HCV | Enzymatically active; single-chain glycoprotein, with one Asn-linked carbohydrate side chain; a preeminent clinical tumor marker in the management of patients with prostate cancer; Higashihara, E et al, *J Urol*, 156: 1964, 1996; Stamey, TA et al, *N Engl J Med*, 317: 909, 1987; Papsidero, LD et al, *Cancer Res*, 40: 2428, 1980; Kablin, JN, *Geriatrics*, 47: 23, 1992

Fitzgerald 30-AP15	Human seminal fluid	High purity
Fitzgerald 30-AP16	Human seminal fluid	Standard purity

Sigma P 3338   Human seminal fluid   ≥95% (SDS-PAGE); highly purified; solution in 0.15 M phosphate buffered saline, pH 7.4 containing 0.1% sodium azide | Not assayed by Sigma; Ambruster, DA, *Clin Chem*, 39: 181, 1993; Zhou, AM et al, *Clin Chem*, 39: 2483, 1993

USBio P9054-51   Human seminal fluid   ~75% (SDS-PAGE); ~1 mg/mL; supplied in PBS buffer, pH 7.5, 0.1% Proclin added as a preservative; 0.2 µm sterile filtered | Suitable for antigenic applications in immunological protocols; major band at ~30kD

Biogenesis 7820-0370   Human, monoclonal mouse   >90% (SDS-PAGE); affinity purified Ig; PBS buffer, pH 7.2, 0.1% NaN₃; liquid

Scipac P117-7   Seminal fluid   >96%; liquid in phosphate buffer; sterile filtered through 0.2µm membrane; 0.5-5 mg/mL | Tumor marker

Scipac P117-8   Seminal fluid   40-90%; liquid in phosphate buffer; sterile filtered through 0.2µm membrane; 0.5-5 mg/mL | Tumor marker

Prostate Specific Antigen, ACT Complex

Cortex CP1097   >95%

Biodesign A31029H   Human seminal fluid   >95% | Tumor marker, cancer antigens & oncogenes

Fitzgerald 30-AP13   Human seminal fluid   High purity

Fitzgerald 30-AP21   Human seminal fluid   Standard purity

USBio P9054-62   Human seminal fluid & ACT from human plasma   >96% (SDS-PAGE); major band at 95kD; ~1 mg/mL liquid in 50 mM PBS, pH 7.5, 0.05% NaN_3 | Suitable for antigenic applications in immobilised in immunological protocols

Scipac P192-3   Seminal fluid & serum   >96%; frozen in phosphate buffer; 0.75-3 mg Complex/mL | Tumor marker

Prostate Specific Antigen, Azide Free

Cortex CP1099   >95%; azide-free

Scipac P117-9E   Seminal fluid   >96%; liquid in phosphate buffer; azide free | Tumor marker

Prostate Specific Antigen, Enzymatically Active

Cortex CP1095   >95%

USBio P9054-66   Human seminal fluid   Iodination grade, 99%; enzymatically active PSA antigen; PSA content determined by Abbott ELISA, total protein by Lowry & optical density; SDS-PAGE shows single band; 1.2 mg/mL (Abbott IMX) sterile-filtered liquid in PBS, 0.1% NaN_3 | Suitable for antigenic applications in immunological protocols

Prostate Specific Antigen, Enzymatically Active Azide Free

Cortex CP1105   >95%; azide-free

Prostate Specific Antigen, Free

Cortex CP1118   >95%

Fitzgerald 30-AP14   Human seminal fluid   High purity

USBio P9054-60   Human seminal fluid   Iodination grade, ≥99%; SDS-PAGE shows a major band at ~30kD; no contamination with the complex Alpha-1 antichymotrypsin; ~1 mg/mL supplied in 100 mM PBS, pH 7.4, 2.5% sucrose, 0.02% NaN_3 | Suitable for antigenic applications in immunological protocols

Prostate Specific Protein 94

USBio P9054-64   Human   ≥99% (SDS-PAGE); no contaminants detected; single band by SDS-PAGE, IEP, &/or RID; 1 mg/mL; lyophilized | Suitable for antigenic applications in immunological protocols

Fitzgerald 30-AP30   Human seminal fluid   High purity

Proteasome α-Subunit 20S

ICN 193626   *Methanosarcina thermophila* recombinant, expressed in *E. coli*   MW ~24k

ICN 193627   *Methanosarcina thermophila* recombinant, expressed in *E. coli*   MW ~22k

Protein 14-3-3

Synonyms: Protein 130; Protein 131

Biogenesis 7835-1109   Human pathogen-free brain   MW 26k & 29k   98% (SDS-PAGE); 2 subunits; purified; 50 mM Tris-HCl, pH 7.4; liquid | Boston et al, *J Neurobiochem*, 38:1466, 1982

Protein A

Cortex CX4511   >95%

ICN 55832   Lyophilized; salt free | Tested for binding to human IgG using radial immunodiffusion

Fitzgerald 30-AP75   Recombinant, expressed in *E. coli*

Biogenesis 7840-0604   S. aureus   MW 42k   Purified; no preservatives; lyophilized | Good binding with: human IgG1/IgG2/IgG4, mouse IgG2a/IgG2b/IgG3, rat IgG1/IgG2c, Guinea pig IgG1/IgG2, rabbit IgG, dog IgGa/IgGb/IgGc/IgGd, cow IgG2; low binding with: human IgM/IgA2, mouse IgG1, sheep/Goat IgG2

Biogenesis 7840-2054   S. aureus   allophycocyanin conjugated

Biogenesis 7840-2074   S. aureus   ; biotin conjugated; 25 mM borate, 0.9% NaCl, pH 8.0, with BSA and 0.05% NaN_3; liquid | >85% activity vs immobilised human gamma globulins; >90% reactive against avidin matrix

Biogenesis 7840-2104   S. aureus   FITC conjugated

Biogenesis 7840-2154   S. aureus   FluoroBlue conjugated

Biogenesis 7840-2204   S. aureus   ; HRP conjugated; liquid | Homogenous band on SDS-Page at 50k

Biogenesis 7840-2264   S. aureus   phycocyanin conjugate

Biogenesis 7840-0704   S. aureus r-DNA   Lyophilized

ICN 797001/797002   S. aureus, Cowan I   Lyophilized; Ab binding achieved in <30 min at 4-37°C | Bacterial adsorbent; binds the F_c portion of immunoglobulin of many different animal species

ICN 797051   S. aureus, Cowan I   1 mg typically binds 8-12 mg human IgG | Used as a lymphocytic mitogen; stimulates polyclonal Ab secretion from human B cells

ICN 153891   S. aureus, Cowan I recombinant   Binding grade; lyophilized; 12-15 mg human IgG/mg solid binding capacity | Carboxy truncated with 301 AA residues; non-binding regions removed to minimize potential steric interference; F_c-binding domains exhibit affinity for IgG subclasses in most mammalian species

ICN 987051   S. aureus, Cowan I recombinant   Lyophilized; salt free; 98%; SA equal to the most active Protein A preparations from S. aureus

ICN 150050   S. aureus, Cowan strain   MW ~42k   Lyophilized, essentially salt free; 6-8 mg human IgG/mg solid binding capacity | Single polypeptide chain; 4 regions with binding activity; detects cell surface Ag & circulating immune complexes; useful in isolating IgG, as a lymphocyte mitogen, & to stimulate polyclonal Ab production from human B cells

ICN 150051   S. aureus, Cowan strain   Lyophilized, essentially salt free; 9-11 mg human IgG/mg solid binding capacity

ICN 150052   S. aureus, Cowan strain   Lyophilized, essentially salt free; 11-14 mg human IgG/mg solid binding capacity

Biogenesis 7840-2284   Staphylococcal   MW 240k   ); PE conjugated; 3.0 M ammonium sulphate, pH 7.0, 50 mM sodium phosphate & 0.02% NaN_3; liquid | Conjugate molar extinction coefficient: 1.9 x 106 at 565nm; emmission max: 575nm (using excitation at 500nm

Fitzgerald 30-AP76   *Streptococcus aureus*   High purity

Protein A, (^{125}I)-

ICN 68038   Immunological grade; >30 µCi/µg, >1.11 MBq/µg; 0.1 M KPB, pH 7.5, EtOH (1:1) with 0.5% BSA

ICN 68049   Immunological grade; 2-10 µCi/µg, 74-370 kBq/µg; 0.1 M KPB, pH 7.5, EtOH (1:1) with 0.5% BSA

ICN 68061   Immunological grade; 70-100 µCi/µg, 2.59-3.7 MBq/µg; 0.1 M KPB, pH 7.5, EtOH (1:1) with 0.5% BSA

Protein A, Actibind SF

ICN 684951

Protein A, Actibind-Ald

ICN 684921   Prepk Col

ICN 684941   Magnetic

ICN 685011

Protein A, Agarose

ICN 191284 ICN 797011 5 atoms hydrophilic spacer arm; 1.2-1.5 mg protein A/mL gel; ~12-15 mg human IgG/mL gel binding capacity; suspension in PBS, 0.02% NaN₃ ¦ Useful for purification of some IgG molecules

ICN 191314 ICN 678791 Binding reagent; preweighed solid buffer mixture | Facilitates binding of monoclonal mouse IgG to protein A

ICN 191315 ICN 678781 Fast flow grade; 5 atoms hydrophilic spacer arm; 1.2-1.5 mg Protein A/mL gel; ~12-15 mg human IgG/mL gel binding capacity; suspension in PBS, 0.02% NaN₃; 3000 cmh⁻¹ flow rate possible, medium pressure matrix | Useful for purification of some IgG molecules

ICN 191316 ICN 678891 Elution reagent; nondenaturing nontoxic neutral buffer solution, effective pH 4.0-7.0 | Elution of Ag from immunoadsorbents or immunoglobulins from Protein a

Oncogene IP02 Solution | Recombinant bacterial agarose conjugates for the immunoprecipitation of antibodies or antibody containing complexes

Oncogene IP06 Recombinant bacterial agarose conjugates for the immunoprecipitation & purification of antibodies or antibody containing complexes

Protein A, Alkaline Phosphatase Conjugated

Sigma P 9650 Lyophilized powder containing ~50% protein (Biuret); balance primarily tris-aspartate containing trace magnesium chloride & zinc sulfate; 1-2 moles Protein A (Sigma P 6650) conjugated/mole alkaline phosphatase (Sigma P 5521); alkaline phosphatase activity: 300-900 U/mg protein; 1 U hydrolyzes 1.0 µmole p-nitrophenyl phosphate/min at pH 10.4, 37°C; binding capacity: 2-5 mg of human IgG/mg protein | Coupled by a modification of the procedure of O'Sullivan, MS et al, *FEBS Lett*, 95: 311, 1978 which favors low MW conjugates

Fluka 82494 *S. aureus*, Bovine intestinal mucosa ≥300 U/mg protein; lyophilized; 50% protein content; binding capacity: 2 mg Human IgG/mg protein; contains TRIS-aspartate salts & traces of magnesium sulfate & zinc sulfate; 1 U corresponds to the amount of enzyme which hydrolyzes 1 µmol 4-nitrophenyl phosphate/min at pH 9.8, 25°C; 0.5 M alkaline phosphatase/M protein A | Sensitive reagent to immunoscreen an expression cDNA plasmid library; Tuan, RS & Fitzpatrick, DF, *Anal Biochem*, 159: 329, 1986

Protein A, Biotin Conjugated

ICN 622651 Typical dilution range = 1:200-1:5000, depending on procedure | May replace labeled second Ab in many immunological studies

ICN 678741

Sigma P 2065 *S. aureus* Lyophilized powder containing ~95% protein (Biuret); balance primarily sodium citrate; prepared from Sigma P 6031, coupled to biotin by an amide bond through an aminocaproyl spacer; contains 3-5 moles biotin/mole protein A

ICN 191367 *S. aureus*, Cowan I Conjugated with AH-biotin hydroxysuccinimide ester to minimize steric interaction between biotinylated protein & avidin or streptavidin

Protein A, DTAF Conjugated

Sigma P 9899 Lyophilized, essentially salt-free; 10-30 µg DTAF/mg solid; binding capacity: 4-9 mg of human IgG/mg solid | Blakeslee, D & Baines, MG, *J Immunological Methods*, 13: 320, 1977

Protein A, Extracellular

Fluka 82485 *Staphylococcus aureus* MW 41k ≥90% protein; lyophilized; ~90% (GE) | Nature's universal anti-antibody; Surolia, A et al, *Trends Biochem Sci*, 7: 74, 1982; Lindmark, R et al, *J Immunol Meth*, 62: 1, 1983

Fluka 82493 *Staphylococcus aureus* MW 41k Binding capacity: ~10 mg human IgG/mg solid

Protein A, Extracellular Agarose CL-4B

Fluka 82483 *Staphylococcus aureus* Immobilized on cross-linked CNBr-activated agarose 4B stabilized with lactose; binding capacity: 20 mg human IgG/mL gel; 1 g powder yields 4 mL swollen gel | Immobilized for affinity chromatography

Fluka 82486 *Staphylococcus aureus* Immobilized on cross-linked 4-nitrophenylcarbonate-activated agarose 4B; 50% suspension in 0.9% NaCl containing 0.02% thimerosal; binding capacity: 10 mg human IgG/mL gel | Immobilized for affinity chromatography; Wilchek, M & Miron, T, *Biochem Int*, 4: 629, 1982

Fluka 82487 *Staphylococcus aureus* Immobilized on hydrophilic macroporous oxirane acrylic beads (150 µm) with an electroneutral mechanism for binding proteins via a C₂-spacer; binding capacity: 5 mg human IgG/mL gel | Immobilized for affinity chromatography

Fluka 82491 *Staphylococcus aureus* Immobilized on cross-linked CNBr-activated agarose 4B; 50% suspension in 0.5 M NaCl containing 0.02% thimerosal; binding capacity: 30 mg human IgG/mL gel | Immobilized for affinity chromatography

Fluka 82492 *Staphylococcus aureus* Immobilized on cross-linked CNBr-activated agarose 4B stabilized with lactose; binding capacity: 20 mg human IgG/mL gel; 1 g powder yields 4 mL swollen gel | Immobilized for affinity chromatography

Protein A, Ferritin Conjugated

Sigma P 6530 Sterile filtered solution containing 0.025 M sodium phosphate buffer, pH 7.5, & 0.025 M NaCl; ~25:1 mole ratio of protein A (Sigma P 6650) & Ferritin (Type I, Sigma F 4503); unbound protein removed by gel filtration chromatography | Conjugated by the method of Carlsson, J et al, *Biochem J*, 173: 723, 1978, using SPDP; useful in electron mocroscopy; Wolf, P, *Anal Biochem*, 129: 143, 1983; Templeton, CL et at, *FEBS Lett*, 85: 95, 1978

Protein A, FITC Conjugated

ICN 55881 Liquid in 0.02 M PBS, pH 7.3, 1% BSA, 10% glycerol, 0.05% NaN₃; tested fro appropriate fluorochrome: protein ratio & immunofluorescence on purified rabbit IgG

ICN 622801 Typical dilution range = 1:50-1:200 with PBS; modified Goding procedure | May replace labeled second Ab in many immunological studies

ICN 797061

Sigma P 5145 Lyophilized, essentially salt-free; ≥50 µg FITC/mg solid; binding capacity: ≥5 mg of human IgG/mg solid | McKinney, RM et al, *Anal Biochem*, 14: 421, 1966

ICN 191371 *S. aureus*, Cowan I Lyophilized, salt free; FITC:Protein A = 2.5-3.0; Abs max = 495 nm, Emission max = 515 nm; recommended dilutions = 1:50 to 1:200 with PBS

Fluka 82484 *Staphylococcus aureus* ≥95% (GE); binding capacity: ~10 mg human IgG/mg solid; conjugate: 0.1 mg FITC/mg | For fluorescence; McKinney, RM et al, *Anal Biochem*, 14: 421, 1966

Protein A, FITC/Gold Conjugated

ICN 154095	5 nm colloidal gold
ICN 154096	10 nm colloidal gold
ICN 154097	15 nm colloidal gold

Protein A, Gold Conjugated

ICN 154092	5 nm colloidal gold
ICN 154093	10 nm colloidal gold
ICN 154094	15 nm colloidal gold
ICN 678621 ICN 678622	5 nm colloidal gold | Useful in electron microscopy
ICN 678631 ICN 678632	10 nm colloidal gold | Useful in electron microscopy
ICN 678641 ICN 678642	20 nm colloidal gold | Useful in electron microscopy

Sigma P 1039   10 nm Colloidal Gold labeled; mean particle diameter 8-12 nm; monodisperse; suspension in 50% glycerol containing 0.15 M NaCl, 0.01 M sodium phosphate, pH 7.4, 0.02% PEG 20, 0.02% sodium azide; concentration: A_{520} ~5.0 | Protein A, extracellular (Sigma P 6031) adsorbed to colloidal gold; Horisberger, M & Clerc, MF, *Histochem*, 82: 219, 1985

Sigma P 9660   5 nm Colloidal Gold labeled; mean particle diameter 3.5-6.5 nm; monodisperse; suspension in 50% glycerol containing 0.15 M NaCl, 0.01 M sodium phosphate, pH 7.4, 0.02% PEG 20, 0.02% sodium azide; concentration: A_{520} ~5.0 | Protein A, extracellular (Sigma P 6031) adsorbed to colloidal gold; Horisberger, M & Clerc, MF, *Histochem*, 82: 219, 1985

Sigma P 9785   20 nm Colloidal Gold labeled; mean particle diameter 17-23 nm; monodisperse; suspension in 50% glycerol containing 0.15 M NaCl, 0.01 M sodium phosphate, pH 7.4, 0.02% PEG 20, 0.02% sodium azide; concentration: A_{520} ~5.0 | Protein A, extracellular (Sigma P 6031) adsorbed to colloidal gold; Horisberger, M & Clerc, MF, *Histochem*, 82: 219, 1985

Fluka 82479   *Staphylococcus aureus*   Labeled with colloidal gold (5 nm); suspension in 50% glycerol containing 0.15 M NaCl, 0.01 M sodium phosphate, pH 7.4, 0.02% PEG 200 & 0.02% NaN₃; blotting test: ≤10 ng human IgG | Horisberger, M & Clerc, MF, *Histochem*, 82: 219, 1985

Fluka 82481   *Staphylococcus aureus*   Labeled with colloidal gold (10 nm); suspension in 50% glycerol containing 0.15 M NaCl, 0.01 M sodium phosphate, pH 7.4, 0.02% PEG 200 & 0.02% NaN₃; blotting test: ≤5 ng human IgG | Roth, J, *J Microsc*, 143: 125, 1986

Fluka 82482   *Staphylococcus aureus*   Labeled with colloidal gold (20 nm); suspension in 50% glycerol containing 0.15 M NaCl, 0.01 M sodium phosphate, pH 7.4, 0.02% PEG 200 & 0.02% NaN₃; blotting test: ≤2 ng human IgG | Bendayan, M, *J Elect Microsc Tech*, 1: 243, 1984; Taatjes, DJ et al, *Eur J Cell Biol*, 45: 151, 1987

Protein A, Horse Radish Peroxidase Conjugated

ICN 55901   Liquid in 0.01 M Na PB, 0.15 m NaCl, pH 7.4, 1% Ovalbumin, 40% glycerol, 0.1% proclin

ICN 622811   Typical dilution range = 1:1500-1:2500 for immunoblotting, 1:3000-1:5000 for ELISA | May replace labeled second Ab in many immunological studies

ICN 191374   *S. aureus*, Cowan I   1.5 mg conjugate/mL; recommended dilution = 1:3000 to 1:5000 for ELISA

Protein A, Insoluble

Sigma P 7155   *S. aureus* (Cowan strain) cells   Formalin treated; ~10% (wet wt/vol) of essentially non-viable cells in 0.04 M sodium phosphate buffer, pH 7.2, 0.15 M NaCl, 0.05% sodium azide; binding capacity: ≥1.2 mg human IgG/mL suspension; produced in pure culture | Immunoadsorbent prepared to ensure reproducible binding of IgG; may be used directly as an immunoadsorbent; not intended for use as a starter culture; not processed or packaged aseptically

Sigma P 9151   *S. aureus* (Cowan strain) cells   Lyophilized cell powder containing ~20% PVP & 0.5% sodium azide; binding capacity: ≥4.0 mg human IgG bound/100 mg solid; produced in pure culture | Immunoadsorbent; yields a superior cell suspension for use as an IgG adsorbent; not intended for use as a starter culture; not processed or packaged aseptically

Fluka 82496   *Staphylococcus aureus*   Crude, lyophilized cell powder of *Staphylococcus aureus*, containing 20% PVP & 0.5% NaN₃; binding capacity: 0.05 mg human IgG/mg solid | For immunoprecipitation & isolation of antigens from cells; Kessler, SW, *Meth Enzymol*, 73: 442, 1981

Fluka 82503   *Staphylococcus aureus*   Crude 10% suspension of non-viable *Staphylococcus aureus* cells in 0.05 M potassium phosphate buffer, pH 7.5, containing 0.2% NaN₃; binding capacity: 1 mg human IgG/mg solid

Protein A, MagaBeads™-Immobilized

Cortex CM3510   Uniform magnetizable particles | Used for protein & DNA separation techniques, cell isolation, enzyme immobilization & bacterial capture

Cortex CM3511   Uniform magnetizable particles | Used for protein & DNA separation techniques, cell isolation, enzyme immobilization & bacterial capture

Protein A, MagaCell™-Immobilized

Cortex CM2510   Magnetizable cellulose/iron oxide | Large porous surface which offers a high capacity for binding large quantities of biomolecules; vicinal hydroxyl groups are activated by employing surface chemistries including cyanogen bromide, carbodiimidazole, epoxide & periodate

Cortex CM2511   Magnetizable cellulose/iron oxide | Large porous surface which offers a high capacity for binding large quantities of biomolecules; vicinal hydroxyl groups are activated by employing surface chemistries including cyanogen bromide, carbodiimidazole, epoxide & periodate

Protein A, Peroxidase Conjugated

Sigma P 8651   Lyophilized powder containing ~90% protein (A_{205}); balance primarily sodium citrate; binding capacity: 2-5 mg of human IgG/mg solid; peroxidase activity: 100-200 U/mg protein; 1 U forms 1 mg purpurogallin/20 sec from pyrogallol at pH 6.0 at 20°C | Extracellular Protein A coupled to peroxidase, Type VI by a modification of the procedure of O'Sullivan, MJ et al, *FEBS Lett*, 95: 311, 1978, which favors low MW conjugates

Fluka 82504   *Staphylococcus aureus*, Horseradish   Binding capacity: 2-5 mg human IgG/mg solid; 150 U/mg; lyophilized; 10% sodium citrate; 1 mol peroxidase/mol protein A; 1 U corresponds to the amount of enzyme which oxidizes 1 μmol ABT/min at pH 5.0, 25°C

Protein A, Peroxidase/Gold Conjugated

ICN 154098   5 nm colloidal gold

ICN 154099   10 nm colloidal gold

ICN 154100   15 nm colloidal gold

Protein A, Pregnancy Associated Plasma

Biodesign A86864H   Pooled retroplacental blood   Purified

Protein A, Sepharose 6MB

Fluka 82507   Immobilized for affinity chromatography

Protein A, Sepharose CL-4B

Fluka 82506   Immobilized for affinity chromatography; Lindmark, R et al, *J Immunol Meth*, 62: 1, 1983

Protein A, Soluble

Sigma P 7837   Recombinant, expressed in *E. coli*   MW 45k ≥95% (HPLC); sterile-filtered solution in water; protein concentration: ≥50 mg/mL; endotoxin tested | US Patent No. 5,151,350

Sigma P 2164   Recombinant, produced by cDNA expressed in *E. coli*   MW ~15k   ≥90% (SDS-PAGE); IgG-binding fragment

Sigma P 4931   *S. aureus*   Prepared from Sigma P 6031; aseptically filled

Sigma P 6031   *S. aureus*   Purified from culture medium of a protein A-secreting *S. aureus* strain; lyophilized, essentially salt-free; binding capacity: 7-14 mg of human IgG/mg solid | Bacterial strain is a derivative of a strain from Cohen, S & Sweeney, HM, *J Bact*, 140: 1028, 1979

Sigma P 3838   *S. aureus* (Cowan strain) cell walls   Purified from cell walls; lyophilized, essentially salt-free; binding capacity: 7-14 mg of human IgG/mg solid

Sigma P 3963   *S. aureus* (Cowan strain) cell walls   Purified from cell walls; lyophilized, essentially salt-free; binding capacity: 4-9 mg of human IgG/mg solid

Sigma P 9267   *S. aureus* (Cowan strain) cell walls   Partially purified from cell walls; lyophilized, essentially salt-free; binding capacity: 3-6 mg of human IgG/mg solid | Prepared by a modification of Sjoquist, J et al, *Eur J Biochem*, 29: 572, 1972, using ion exchange chromatography

Fluka 82526 *Staphylococcus aureus* (Cowan strain) Binding capacity: 10 mg human IgG/mg; powder

Protein A, Sulforhodamine 101 Acid Chloride (Texas Red)

Sigma P 8162 Lyophilized, essentially salt-free; ~1:1 mole ratio protein A: Texas Red; binding capacity: 4-9 mg of human IgG/mg solid; purified by gel filtration

Protein A, TRITC Conjugated

Sigma P 1775 Lyophilized, essentially salt-free; ~6:1 mole ratio TRITC: protein A; binding capacity: 4-9 mg of human IgG/mg solid | Amante, L et al, *J Immunological Methods*, 1: 289, 1972

Protein A, β-Galactosidase Conjugated

Sigma P 7650 Solution in 45% glycerol, 0.01 M potassium phosphate, pH 7.3, containing 10^{-6} M magnesium chloride, 0.15 M NaCl, 1% BSA & 10 ppm 4-chloro-3,5-dimethylphenol; ~3:1 mole ratio conjugate of protein A & β-galactosidase | Used in immunosorbent detection of human IgG

Protein A/G

Cortex CX4513 >95%

ICN 154601 Recombinant, *S. aureus* Cowan I strain Protein A gene & binding domains of *Strept.* sp. Lancefield Group G Protein G strain MW 45-47k Lyophilized; crystalline; salt free; 14-18 gm human IgG/mg fusion protein binding capacity | Unique fusion protein consisting of the F_c-binding portions of Protein A & Protein G; will sufficiently bind F_c of IgG from: human, mouse, rabbit, goat, bovine, sheep, porcine, dog, cat, horse; will not bind mouse IgA or IgM, BSA or HSA

Protein B

Biogenesis 7845-5006 *S. aureus* group B ; 0.1 M PBS, pH 7.5; liquid | Specific for the Fc portion of both subclasses of serum and secretory human IgA; in*fect Immun*, 57:1573, 1989; *Biotechniques*, 10:748, 1991

Protein C

Cortex CP3047U Bovine >98%

ICN 194913 Bovine plasma 50% glycerol/H_2O | Precursor to activated protein C (APC); Kisiel, W et al, *Methods Enzymol*, 80:320, 1981

Cortex CP2110U Human >98%

ICN 194080 Human plasma Purified protein containing glycine & NaCl; <0.05 µg Factors II, VII, IX, X & Protein S | Inactivates factors V & VIII in the presence of phospholipids & Ca; Stenflo, J & RM Bertina (ed), *Churchill Livingstone*, 21, 1988

ICN 194912 Human plasma 50% glycerol/H_2O | Precursor to activated protein C (APC); Kisiel, W et al, *Methods Enzymol*, 80:320, 1981

Biogenesis 7850-0709 Plasma Liquid

Protein C, Activated

ICN 194185 Human plasma Purified; <0.01 µg Factors II, VII, X & Protein S

Protein G

Synonyms: IgG Fc Receptor Type III; IgG Fc Receptor Type III; IgG Fc Receptor Type III; Protein G'

Biogenesis 7860-0207	Alkaline phosphate conjugated
Biogenesis 7860-0504	Biotin conjugated
Cortex CX4512	>95%

Sigma P 2169 Group C *Streptococcus sp.* (derivative of strain 26RP66) Formalin fixed; ~10% (wet wt/vol) Cell suspension in 5 mM potassium phosphate, pH 7.2, 150 mM NaCl, 0.05% sodium azide; binding capacity: ≥100 µg human IgG/mL suspension | IgG Fc Receptor Type III; binds more efficiently to bovine, goat & sheep IgG than does Protein A; Bjorck, L & Kronvall, G, *J Immunol*, 133: 969, 1984; Taatjes, DJ et al, *Eur J Cell Biol*, 45: 151, 1987

Sigma P 9659 Group C *Streptococcus sp.* (derivative of strain 26RP66) Purified; lyophilized, salt-free; ~9 mg human IgG/mg solid (determined by radial immunodiffusion) | Binds more efficiently to bovine, goat & sheep IgG than does Protein A; Reis, K et al, *J Immunol*, 132: 3098, 1984; Becker, W, *Immunochem*, 6: 539, 1969; Bjorck, L & Kronvall, G, *J Immunol*, 133: 969, 1984; Taatjes, DJ et al, *Eur J Cell Biol*, 45: 151, 1987

Biogenesis 7860-0104 r-DNA *E. coli* MW 22k; pI: 6; purified; essentially salt free; lyophilized | Binds Fc-fragment of all subclasses of IgG from human, mouse, rat, rabbit, sheep and Goat; does not bind human IgM, IgA, IgD or albumin

Sigma P 5170 Recombinant Lyophilized, essentially salt-free; binding capacity: ~5 mg human IgG/mg solid | Engineered to eliminate non-specific binding with human serum albumin; binds more efficiently to bovine, goat & sheep IgG than does Protein A; Bjorck, L & Kronvall, G, *J Immunol*, 133: 969, 1984; Taatjes, DJ et al, *Eur J Cell Biol*, 45: 151, 1987

ICN 152342 ICN 672651 Recombinant, *Streptococcal* origin Lyophilized; salt free; freely soluble in H_2O & standard buffers; stable pH 2-10 | 2 IgG-binding B regions; binds both F_c & F_{ab} fragments of IgG; does not cross react with human albumin

Fluka 82489 *Staphylococcus* species (group C) Binding capacity: 9 mg human IgG/mg solid | Reis, K et al, *J Immunol*, 132: 3098, 1984; Bjorck, L & Kronvall, G, *J Immunol*, 133: 969, 1984; Akerstrom, B et al, *J Immunol*, 135: 2589, 1985

Fluka 82509 *Staphylococcus* species recombinant from *E. coli* MW 33.8k ~85% (GE); ≥90% protein content; binding capacity: 5 mg human IgG/mg solid | Protein G binds more efficiently to bovine, goat & sheep IgG than does protein A; no binding to human serum albumin; Bjorck, L & Kronvall, G, *J Immunol*, 133: 969, 1984

Sigma P 4689 *Streptococcus sp.* recombinant, expressed in *E. coli* Lyophilized from a Tris-HCl buffer solution | Genetically engineered truncated protein G which retains its affinity for IgG, but lacks albumin- & Fab binding sites & membrane-binding regions; Goward, CR et al, *Biochem J*, 267: 171, 1990

Fitzgerald 30-AP77	*Streptococcus* species	High purity
Biogenesis 7860-0056	Streptococcus spp	Suspension
Biogenesis 7860-0556	Streptococcus spp	FITC conjugated
Biogenesis 7860-1904 conjugated	Streptococcus spp	Allophycocyanin
Biogenesis 7860-2004	Streptococcus spp	PE conjugated

Protein G Plus, Agarose Conjugated

Oncogene IP04 Suspension | Recombinant bacterial agarose conjugates for the immunoprecipitation of antibodies or antibody containing complexes

Oncogene IP08 Recombinant bacterial agarose conjugates for the immunoprecipitation & purification of antibodies or antibody containing complexes

Protein G Plus/Protein A, Agarose Conjugated

Oncogene IP05 Suspension | Recombinant bacterial agarose conjugates for the immunoprecipitation of antibodies or antibody containing complexes

Oncogene IP10 Recombinant bacterial agarose conjugates for the immunoprecipitation & purification of antibodies or antibody containing complexes

Protein G, (^{125}I)-

ICN 68089 2-15 µCi/µg, 74-555 kBq/µg; 0.1 M KPB, pH 7.5, 0.5% BSA:EtOH (1:1)

Protein G, Biotin Conjugated

Synonyms: Protein G'

Sigma P 8045 Recombinant from Protein G' Lyophilized powder containing ~90% protein (Biuret); balance primarily sodium citrate; contains 2-4 moles d-biotin/mole protein; binding capacity: ~5 mg of human IgG/mg protein | Prepared using biotinamidocaproate *N*-hydroxysuccinimide which incorporates an aminocaproyl spacer

Protein G, FITC Conjugated

Sigma P 7670 Lyophilized, essentially salt-free; ≥20 µg FITC/mg solid; binding capacity: 6-10 mg of human IgG/mg solid | McKinney, RM et al, *Anal Biochem*, 14: 421, 1966

Protein G, Gold

Sigma P 1796 20 nm Colloidal Gold labeled; mean particle diameter 17-23 nm; monodisperse; suspension in 50% glycerol containing 0.15 M NaCl, 0.01 M Tris, pH 7.4, 0.02% PEG 20, 0.02% sodium azide; concentration: A_{520} ~5.0 | Protein G, Sigma P 9659, adsorbed to colloidal gold; Bendayan, M, *J Electron Microscopy Technique*, 6: 7, 1987; Bendayan, M & Gavzon, S, *J Histochem Cytochem*, 6: 597, 1988

Protein G, (^{125}I)-

ICN 68089 2-15 µCi/µg, 74-555 kBq/µg; 0.1 *M* KPB, pH 7.5, 0.5% BSA:EtOH (1:1)

Protein G, Biotin Conjugated

Synonyms: Protein G'

Sigma P 8045 Recombinant from Protein G' Lyophilized powder containing ~90% protein (Biuret); balance primarily sodium citrate; contains 2-4 moles d-biotin/mole protein; binding capacity: ~5 mg of human IgG/mg protein | Prepared using biotinamidocaproate *N*-hydroxysuccinimide which incorporates an aminocaproyl spacer

Protein G, FITC Conjugated

Sigma P 7670 Lyophilized, essentially salt-free; ≥20 µg FITC/mg solid; binding capacity: 6-10 mg of human IgG/mg solid | McKinney, RM et al, *Anal Biochem*, 14: 421, 1966

Protein G, Gold

Sigma P 1796 20 nm Colloidal Gold labeled; mean particle diameter 17-23 nm; monodisperse; suspension in 50% glycerol containing 0.15 M NaCl, 0.01 M Tris, pH 7.4, 0.02% PEG 20, 0.02% sodium azide; concentration: A_{520} ~5.0 | Protein G, Sigma P 9659, adsorbed to colloidal gold; Bendayan, M, *J Electron Microscopy Technique*, 6: 7, 1987; Bendayan, M & Gavzon, S, *J Histochem Cytochem*, 6: 597, 1988

Protein G, Gold Conjugated

ICN 154101	5 nm colloidal gold
ICN 154102	10 nm colloidal gold
ICN 154103	15 nm colloidal gold
ICN 678651 ICN 678652	5 nm colloidal gold	Useful for electron microscopy
ICN 678661 ICN 678662	10 nm colloidal gold	Useful for electron microscopy
ICN 678671 ICN 678672	20 nm colloidal gold	Useful for electron microscopy

Sigma P 1546 5 nm Colloidal Gold labeled; mean particle diameter 3.5-6.5 nm; monodisperse; suspension in 50% glycerol containing 0.15 M NaCl, 0.01 M Tris, pH 7.4, 0.02% PEG 20, 0.02% sodium azide; concentration: A_{520} ~5.0 | Protein G, Sigma P 9659, adsorbed to colloidal gold; Bendayan, M, *J Electron Microscopy Technique*, 6: 7, 1987; Bendayan, M & Gavzon, S, *J Histochem Cytochem*, 6: 597, 1988

Sigma P 1671 10 nm Colloidal Gold labeled; mean particle diameter 8-12 nm; monodisperse; suspension in 50% glycerol containing 0.15 M NaCl, 0.01 M Tris, pH 7.4, 0.02% PEG 20, 0.02% sodium azide; concentration: A_{520} ~5.0 | Protein G, Sigma P 9659, adsorbed to colloidal gold; Bendayan, M, *J Electron Microscopy Technique*, 6: 7, 1987; Bendayan, M & Gavzon, S, *J Histochem Cytochem*, 6: 597, 1988

Protein G, Horse Radish Peroxidase Conjugated

ICN 152343 ICN 672681 Recombinant Conjugated with highly purified HRP; stabilized with 0.05% methiolate preservative; recommended dilutions 1:1000 to 1:3000 for ELISA, 1:5000 for Western Blotting

Protein G, Immobilized

Fluka 82508 *Staphylococcus* species (group C) Attached to CNBr-activated 4% cross-linked beaded agarose stabilized with lactose; binding capacity: 15 mg human IgG/mL gel; 1 g powder yields 4 mL swollen gel | For affinity chromatography

Protein G, MagaBeads™-Immobilized

Cortex CM3520 Uniform magnetizable particles | Used for protein & DNA separation techniques, cell isolation, enzyme immobilization & bacterial capture

Cortex CM3521 Uniform magnetizable particles | Used for protein & DNA separation techniques, cell isolation, enzyme immobilization & bacterial capture

Protein G, MagaCell™-Immobilized

Cortex CM1520 Magnetizable cellulose/iron oxide | Large porous surface which offers a high capacity for binding large quantities of biomolecules; vicinal hydroxyl groups are activated by employing surface chemistries including cyanogen bromide, carbodiimidazole, epoxide & periodate

Cortex CM1521 Magnetizable cellulose/iron oxide | Large porous surface which offers a high capacity for binding large quantities of biomolecules; vicinal hydroxyl groups are activated by employing surface chemistries including cyanogen bromide, carbodiimidazole, epoxide & periodate

Protein G, Peroxidase Conjugated

Sigma P 8170 Recombinant Lyophilized powder containing ~80% protein ($E^{1\%}$); balance primarily sodium citrate; protein G activity: >95% of conjugate binds to IgG agarose; peroxidase activity: ~170 U/mg protein; 1 U forms 1 mg purpurogallin/20 sec from pyrogallol at pH 6.0 at 20°°C; conjugate purified by IgG affinity chromatography | Labeled with Peroxidase, Type VI, by a modification of the procedure of O'Sullivan, MJ et al, *FEBS Lett*, 95: 311, 1978, which favors low MW conjugates

Protein Gene Product 9.5

Biogenesis 7863-2108 Human brain (pathogen free) Purified; 50 *mM* TRIS/HCl, pH 7.4 with 1 *mM* beta mercaptoethanol; liquid | Thompson et al, *Brain Res*, 278:224, 1983

Protein Kinase A Heat Stable Inhibitor, Isoform α

Calbiochem 539488 Rabbit recombinant, expressed in *E. coli* MW 8k Liquid in 75 *mM* NaCl, 30 *mM* MES, 10 *mM* BME, 100 µM EDTA, 50% glycerol, pH 6.5; >90% (SDS-PAGE); phosphatases, proteases, RNases: none detected; activity: 9000 U/mL; one unit is the amount of PKI-α that inhibits 1.0 unit of PKA by >95% | 77 AA protein that is a highly specific inhibitor of the protein kinase A (PKA) catalytic subunit; not known to inhibit any other kinase; shown that PKI-α expression & intracellular localization vary as a function of cell cycle progression; Wen, W et al, *J Biol Chem*, 270: 2041, 1995; Baude, EJ et al, *J Biol Chem*, 269: 2316, 1994; Thomas, J et al, *J Biol Chem*, 266: 10906, 1991; Scott, JP et al, *PNAS*, 82: 5732, 1985

Protein Kinase A, cAMP-Dependent

Alexis 202-027 Bovine heart MW ~40k ≥95% (SDS-PAGE); 20 µg/mL solution in 25 *mM* MES, pH 6.5, 0.1 *mM* EDTA, 30 *mM* BME, 50% ethylene glycol & 100 *mM* NaCl; SA: ~1000 U/µg protein; 1 unit is the amount of enzyme that will transfer 1.0 pmole of phosphate to histone/minute at 30°C | Catalytic subunit

Protein Kinase A, Catalytic Subunit

USBio P9102-91E Bovine heart Frozen solution in 30 *mM* potassium phosphate, pH 7.0, 1 *mM* DTT, 1 *mM* EDTA, 0.15 *M* KCl | Tested & incorporated 3304 pmol phosphate/min/µg enzyme into the Kemptide substrate peptide; purified to electrophoretic homogeneity by anion-exchange chromatography (DEAE-Sephacel), ammonium sulfate fractionation, cation exchange chromatography (CM-Sephadex C-50) in the absence & presence of cAMP, & gel filtration (Sephacryl S200)

Protein Kinase C$_\alpha$, Standard

Calbiochem 539617 Human recombinant MW 76,799 Liquid; possesses no enzymatic activity | Denatured human recombinant protein kinase C$_\alpha$ expressed in baculoviral insect expression system; for use as a positive control & for discriminating PKC isozymes by Western blotting

Protein Kinase C$_{\beta I}$, Standard

Calbiochem 539619 Human recombinant MW 76,790 Liquid; possesses no enzymatic activity | Denatured human recombinant protein kinase C$_{\beta I}$ expressed in baculoviral insect expression system; for use as a positive control & for discriminating PKC isozymes by Western blotting

Protein Kinase C$_{\beta II}$, Standard

Calbiochem 539621 Human recombinant MW 76,933 Liquid; possesses no enzymatic activity | Denatured human recombinant protein kinase C$_{\beta II}$ expressed in baculoviral insect expression system; for use as a positive control & for discriminating PKC isozymes by Western blotting

Protein Kinase C$_\delta$, Standard

Calbiochem 539623 Human recombinant MW 77,517 Liquid; possesses no enzymatic activity | Denatured human recombinant protein kinase C$_\delta$ expressed in baculoviral insect expression system; for use as a positive control & for discriminating PKC isozymes by Western blotting

Calbiochem 539625 Human recombinant MW 84,474 Liquid; possesses no enzymatic activity | Denatured human recombinant protein kinase C$_\epsilon$ expressed in baculoviral insect expression system; for use as a positive control & for discriminating PKC isozymes by Western blotting

Calbiochem 539627 Human recombinant MW 78,366 Liquid; possesses no enzymatic activity | Denatured human recombinant protein kinase C$_\gamma$ expressed in baculoviral insect expression system; for use as a positive control & for discriminating PKC isozymes by Western blotting

Calbiochem 539629 Human recombinant MW 77.6k Liquid; possesses no enzymatic activity | Denatured human recombinant protein kinase C$_\eta$ expressed in baculoviral insect expression system; for use as a positive control & for discriminating PKC isozymes by Western blotting

Calbiochem 539632 Human recombinant MW 118k Liquid; possesses no enzymatic activity | Denatured human recombinant protein kinase C$_\mu$ expressed in baculoviral insect expression system; for use as a positive control & for discriminating PKC isozymes by Western blotting

Protein Kinase Inhibitor Type II

Sigma P 8140 Bovine heart Lyophilized powder containing ~50% protein (Biuret) & 30% glycerophosphate & 20% EDTA; activity: 1.0 µg inhibits 0.75-2.0 phosphorylating units of cAMP-dependent protein kinase | Fractionated essentially by procedure of Walsh, DA et al, *J Biol Chem*, 246: 1977, 1971; general comments for Sigma Protein kinase inhibitors apply

Protein Kinase Inhibitor Type III

Sigma P 0393 Porcine heart Lyophilized powder containing ~50% protein (Biuret) & 30% glycerophosphate & 20% EDTA; activity: 1.0 µg inhibits 1-3 units of cAMP-dependent protein kinase | Fractionated essentially by procedure of Walsh, DA et al, *J Biol Chem*, 246: 1977, 1971; general comments for Sigma Protein kinase inhibitors apply

Protein Kinase Inhibitor, Crude

Sigma P 5015 Rabbit muscle Lyophilized powder containing ~95% protein (Biuret) & 5% potassium phosphate; activity: 1.0 µg inhibits 0.2-1.0 unit of cAMP-dependent protein kinase | Fractionated essentially by procedure of Walsh, DA et al, *J Biol Chem*, 246: 1977, 1971, through TCA precipitation; general comments for Sigma Protein kinase inhibitors apply

Protein Kinase M

Chemicon AG610 Rat brain ≥95% | Catalytic subunit of PKC; purified protein for apoptosis & signal transduction

Protein L

Sigma P 3101 *Peptostreptococcus magnus* recombinant, expressed in *E. coli* MW 35,824 Lyophilized, essentially salt-free; >98% (SDS-PAGE); contains four Ig-binding domains | Binds immunoglobulins (Ig) primarily through kappa light chain interactions without interfering with the antigen binding sites of Igs; Bjorck, L, *Immunol*, 140: 1194, 1988; Kastern, W et al, *J Biol Chem*, 267: 12820, 1992

Protein L, Peroxidase Conjugated

Sigma P 3226 Recombinant Solution in 0.01 mM phosphate buffer with 50% glycerol

Protein Phosphatase 2A Inhibitor

Synonyms: I_1^{PP2A}; I_2^{PP2A}

Calbiochem 539612 Bovine kidney MW 30k >95% (SDS-PAGE); liquid in 25 *mM* Tris-HCl, 14 *mM* BME, 1 *mM* benzamidine; 100 µ*M* PMSF, 1 *mM* EDTA, 10% glycerol, pH 7.4 | Potent, heat-& specific inhibitor of protein phosphatase 2A (PP2A); inhibits PP2A with myelin basic protein, histone H1 or phosphorylase as substrates but not with casein as the substrate; inhibits PP2A in a noncompetitive manner; does not inhibit protein phosphatase 1, protein phosphatase 2B, protein phosphatase 2C or pyruvate dehydrogenase phosphatase; Li, M et al, *Biochemistry*, 35: 6998, 1996; Li, M et al, *Biochemistry*, 34: 1988, 1995

Calbiochem 539614 Bovine kidney MW 20k >95% (SDS-PAGE); liquid in 25 *mM* Tris-HCl, 14 *mM* BME, 1 *mM* benzamidine; 100 µ*M* PMSF, 1 *mM* EDTA, 10% glycerol, pH 7.4 | Potent, heat-& specific inhibitor of protein phosphatase 2A (PP2A); inhibits PP2A with myelin basic protein, histone H1 or phosphorylase as substrates but not with casein as the substrate; inhibits PP2A in a noncompetitive manner; does not inhibit protein phosphatase 1, protein phosphatase 2B, protein phosphatase 2C or pyruvate dehydrogenase phosphatase; Li, M et al, *Biochemistry*, 34: 1988, 1995

Protein Phosphatase 2A Inhibitor I$_1$

Synonyms: I_1^{PP2A}

ICN 195929 Bovine kidney MW 30k ≥95%; 150 ng/vial | Potent, specific, stable inhibitor of PP2A with myelin basic protein, Histone H1 or phosphorylase as substrates, but not with casein; will not inhibit PP1, PP2B, PP2C or pyruvate dehydrogenase phosphatase

Protein Phosphatase 2A Inhibitor I$_2$

Synonyms: I_2^{PP2A}

ICN 195928 Bovine kidney MW 20k ≥95%; 150 ng/vial | Potent, specific, stable inhibitor of PP2A with myelin basic protein, Histone H1 or phosphorylase as substrates, but not with casein; will not inhibit PP1, PP2B, PP2C or pyruvate dehydrogenase phosphatase

Protein Phosphatase 2BA

Chemicon AG645 *N. crassa* recombinant ≥95% | Purified protein for apoptosis & signal transduction

Protein Phosphatase I Catalytic Subunit, α-Isoform

Sigma P 7937 Rabbit recombinant, expressed in *E. coli* Lyophilized powder containing imidazole buffer, pH 7.4, NaCl, DTT, EDTA, MnCl$_2$, Tween 20 & trehalose as stabilizer; activity: 5000-15,000 U/mg protein; 1 U hydrolyzes 1 nmole of p-nitrophenyl phosphate/min at pH 7.0 at 30°C | Protein phosphatase 1 (PP1) is a heterodimeric enzyme with serine/threonine phosphatase activity; comprises a catalytic subunit & a targeting subunit or a specific protein inhibitor; targeting subunit determines the enzyme localization, e.g. to glycogen particles (subunit G) or to myofibrils (subunit M); PP1 is involved in glycogen metabolism & in regulation of muscle contractility; also implicated in cell cycle & transcriptional regulation; properties in common with the native rabbit muscle protein: size is 37.5 kDa, requires Mn^{2+} for activity, specific activity for phosphorylase a & inhibition by okadaic acid, microcystin LR & phosphatase inhibitor 2 (I-2); Faux, MC & Scott, JD, *Trends Biochem Sci*, 21: 312, 1996; Cohen, P, *Ann Rev Biochem*, 58: 453, 1989; Hunter, T, *Cell*, 80: 225, 1995; Zhang, Z et al, *J Biol Chem*, 267: 1484, 1992

Protein Phosphatase I Inhibitor

Synonyms: Inhibitor II

USBio I7653 Human ≥95% as determined by SDS PAGE; frozen solution in 20 *mM* ammonium acetate, pH 7.0, 0.1 *mM* EGTA & 1 *mM* DTT | A heat-stable protein that inhibits PP1 but not PP2, which makes it a useful tool for distinguishing between PP1 & PP2 activities in cells & during purification of PP; purified by ion exchange chromatography

Protein Phosphatase Inhibitor II

Calbiochem 539516 Rabbit muscle recombinant, expressed in *E. coli* MW 22.8k >95% (SDS-PAGE); liquid in 50 *mM* Tris-HCl, 1 *mM* EDTA, 50% glycerol, 0.01% BRIJ, pH 7.0; specific activity: 500,000 U/mg; 1 U inhibits 0.01 units of PP1 by 50%; 1 U of PP1 releases 1.0 nmol phosphate from phosphorylase/min at 30°C, pH 7.0; DNases, phosphatases, proteases, RNases: none detected | 204 AA, heat-stable protein that specifically inhibits the catalytic subunit of protein phosphatase 1; for distinguishing type 1 from type 2 protein phosphatases; a good substrate for CKI, CKII, GSK3β & PKA; Park, I et al, *J Biol Chem*, 269: 944, 1994; Cohen, P, *Methods Enzymol*, 201: 389, 1991

Sigma P 8218 Rabbit recombinant, expressed in *E. coli* Lyophilized powder containing ~15% protein (Bradford); balance Tris buffer, pH 7, EDTA & leupeptin; activity: 5000-25,000 U/mg protein; 1 U inhibits 1 unit of protein phosphatase 1 (PP1) catalytic subunit by 50%; 1 unit of PP1 releases 1.0 nmol phosphate/min at pH 7.5 at 30°C from phosphorylase a | Specific inhibitor of protein phosphatase 1 (PP1); constitutes the regulatory subunit of the cytosolic form of PP1 known also at ATP-Mg dependent phosphatase or protein phosphatase 1I (PP1I); activation of PP1I can be achieved by phosphorylation of I-2 on the Thr-72 by glycogen synthase kinase-3 (GSK-3); activation is increased by casein kinase phosphorylation in a synergistic mechanism; rabbit muscle I-2 is a 204 AA heat stable protein with MW of 31 kDa; inhibits PP1 at nanomolar concentrations; protein phosphatase inhibitor-2 cloned & expressed in *E. coli* to yield an active protein; Cohen, P, *Ann Rev Biochem*, 58: 453, 1989; Plyte, S et al, *Biochim Biophys Acta*, 1114: 147, 1992; DePaoli-Roach, A, *J Biol Chem*, 259: 12144, 1984; Cohen, P et al, *Meth Enzym*, 159: 427, 1988; Park, I-K et al, *J Biol Chem*, 269: 944, 1994

Protein S

Cortex CP3109U Human >98%

Biogenesis 7861-1079 Human plasma MW 69k Tested negative for all communicable diseases including HIV-1, HIV-2, HBsAg and HCV; purified; 20 *mM* Tris-HCl, 0.1 *M* NaCl, 1 *mM* Benzamidine, pH 7.4; liquid

ICN 194081 Human plasma Purified, containing glycine & NaCl | Vitamin K-dependent protein & cofactor of activated protein C (APC); increases affinity of APC for phospholipids leading to Factor V & VIII inhibition; plays an essential role in regulation of the coagulation system; Dahlback, B, *Biochem J*, 209:837, 1983

ICN 194932 Human plasma In 50% glycerol/H$_2$O | Vitamin K-dependent protein & cofactor of activated protein C (APC); functions in the coagulation & complement cascade; Walker, FJ, *Semin Thromb Haemostas*, 10:131, 1984

Protein Z

Synonyms: Vitamin K-Dependent Protein

ICN 194082 Human plasma Purified protein | Thrombin associates with phospholipid vesicles in the presence of Protein Z; Hogg, DJ & J Stenflo, *JBC*, 17:266, 1991

Protein, α-

ICN 900539

Proteoglycan Aggregate

ICN 191484 Bovine nasal cartilage ≥80% aggregate purity; lyophilized; <1% collagen (AA analysis); galactosamine:glucosamine ≥ 10:1; no preservatives, reconstitute with PBS | Isolated under dissociative conditions by finely mincing tissue at low T in the presence of protease inhibitors

ICN 971501 Bovine nasal cartilage Binds hyaluronic acid at the protein core near one end of the molecule, while most of the chondroitin sulfate (CS) chains are at the opposite end with a keratin sulfate (KS) rich region; CS chains are sulfated at 4 & 6 positions; substrate for proteoglycan digesting enzymes such as chondroitinase ABC or AC; standard & competitive labels in RIA & EIA for quantitating proteoglycan Ag recognized by ICN monoclonal Ab when labeled

ICN 971511 Bovine nasal cartilage Protein core has ~100 chondroitin sulfate chains & ~40 keratin sulfate chains attached; binding site for hyaluronic acid is located on the protein core near one end of the molecule; most of the CS chains are located at the opposite end, with a KD-rich region between; CS chains are sulfated in the 6- (80%) & 4- (20%) positions; substrate for digesting enzymes like chondroitinase ABC & AC; when labeled, is a competitive label in RIA & EIA for quantitating proteoglycan Ag recognized by monoclonal Ab

ICN 191485 Rat chondrosarcoma ≥80% aggregate purity; lyophilized; <1% collagen (AA analysis); galactosamine:glucosamine ≥ 10:1; no preservatives, reconstitute with PBS | Isolated under dissociative conditions by finely mincing tissue at low T in the presence of protease inhibitors

ICN 971521 Rat chondrosarcoma Similar to ICN 971501; lacks keratin sulfate; chondroitin sulfate sulfated almost entirely in the 4-position of the galactosamine residue

Proteoglycan Monomer

ICN 191486 Bovine nasal cartilage MW 250k Protein core ≥80% monomer purity; lyophilized; <1% collagen (AA analysis); galactosamine:glucosamine ≥ 10:1 | Isolated under dissociative conditions; protein core has chondroitin sulfate & keratin sulfate chains attached; prepared from proteoglycan aggregate by CsCl density gradient centrifugation in 4 *M* guanidine HCl

ICN 191487 Rat chondrosarcoma MW 250k Protein core ≥90% monomer purity; lyophilized; protein:uronic acid = 0.22-0.26; uronic acid:galactosamine = 1.00-1.05 | Isolated under dissociative conditions; protein core has chondroitin sulfate & keratin sulfate chains attached; prepared from proteoglycan aggregate by CsCl density gradient centrifugation in 4 *M* guanidine HCl

ICN 971531 Rat chondrosarcoma Lacking keratin sulfate; chondroitin sulfate chains are sulfated almost entirely in the 4-position of the galactosamine residue; standard or substrate for proteoglycan digesting enzymes like chondroitinase ABC or AC; when labeled, serves as a competitive label in RIA & EIA

Prothrombin

Synonyms: Factor II; Factor II

Cortex CP3049U   Bovine   >98%

ICN 154164   Bovine plasma   High purity grade; >125 NIH U/mg protein, 1 U produces 1 US unit of clotting activity in the presence of sufficient thromboplastin & Ca; <3% moisture;

ICN 194915   Bovine plasma   50% glycerol:H$_2$O | Precursor to α-thrombin; Mann, KG etal, *Methods Enzymol*, 45:156, 1976

ICN 101033   Bovine plasma fraction III-2   pI = 4.2

Cortex CP3048U   Human   >98%

Biodesign A86863H   Human plasma   >98% | Platelets & Hemostasis reagents

ICN 194189   Human plasma   Purified; <0.1 μg Factor X, proteins C & S; <0.5 μg Factor VII

ICN 194914   Human plasma   50% glycerol:H$_2$O | Precursor to α-thrombin; Kisiel, W etal, *Biochem Biophys ACTA*, 304:103, 1973

ICN 159841   *Oxyuranus scutellatus* (Australian taipan) venom   MW ~56k   Single band (SDS-PAGE) | Useful in blood Factor II assays; provides precise clotting time vs use of crude venom

Pseudomonas Exotoxin A

Sigma P 0184   *Pseudomonas aeruginosa*   Lyophilized powder containing ~30% protein (E$_{280}$ at 1%); balance primarily Tris buffer, EDTA, NaCl & lactose | Purified by a modification of a method by Kozak; shown to be toxic to animals & to cell lines, & to inhibit protein synthesis via ADP ribosylation of elongation factor 2; Kozak, KJ & Saelinger, CB, *Meth Enzymol*, 165: 147, 1988; Middlebrook, JL & Dorland, RB, *Can J Microbiol*, 23: 183, 1977; Iglewsky, BH et al, *Inf Immun*, 15: 138, 1977

PTEN

USBio P9182   Human recombinant GST-fusion protein purified from *E. coli*   ≥60% (SDS-PAGE & Coomassie blue staining); frozen solution in 100 μL of 20 mM Tris-HCl, pH 8.0, 1 mM EDTA, 10 mM glutathione, 50% glycerol | PTEN is a candidate tumor suppressor gene mapping to the homozygous deletion on human chromosome 10q23; appears to be mutated at considerable frequency in human cancers; the predicted PTEN protein has a protein tyrosine phosphatase domain & extensive homology to tensin; purified PTEN catalyzes dephosphorylation of PtdIns(3,4,5)P3, specifically at position 3 on the inositol ring; possibility that PTEN acts *in vivo* as a phosphoinositide 3-phosphatase by regulating PtdIns(3,4, 5)P3 levels; although PTEN has the consensus sequence of a protein tyrosine phosphatase, it dephosphorylates p-nitrophenylphosphate & other synthetic & protein substrates poorly; the greatest catalytic activity has been observed with the highly negatively charged, multiply phosphorylated polymer of (Glu-Tyr)n; immunoblotting with anti-GST antibody showed that the major contaminant at 27-29kD was a GST protein; SA: 1.12 pmol/min/ng using 25ng PTEN with PtdIns(3,4, 5)P3 (PIP3); purified by glutathione-sepharose chromatography

Rab 1A

Chemicon GPR055   Recombinant   ≥95% | Purified protein for apoptosis & signal transduction

Rab 5A

Chemicon GPR070   Recombinant   ≥95% | Purified protein for apoptosis & signal transduction

Rab 8A

Chemicon GPR074   Recombinant   ≥95% | Purified protein for apoptosis & signal transduction

Rab10, His-Tagged

Calbiochem 552114   Rat brain recombinant, produced by overexpression of a full-length Rab10 cDNA clone in *E. coli*   MW 25.6k   ≥85% (SDS-PAGE); liquid in 200 mM BME, 50 mM Tris-HCl, 10% glycerol, 2% SDS, 0.005% bromophenol blue, pH 6.8 | GTP binding protein; small GTPase that has been found to be associated with membranes in the perinuclear region & apical transport vesicles; C-terminal XXCC motif for post-translational geranylgeranylation; suitable for use as a positive control or to assay for cross-reactivity by Western blotting; Chen, YT et al, *PNAS*, 90: 6508, 1993; Huber, LA et al, *J Cell Biol*, 123: 35, 1993

Rab11B, His-Tagged

Calbiochem 552115   Human brain recombinant, produced by overexpression of a full-length Rab11B cDNA clone in *E. coli*   MW 27.2k   ≥90% (SDS-PAGE); liquid in 200 mM BME, 50 mM Tris-HCl, 10% glycerol, 2% SDS, 0.005% bromophenol blue, pH 6.8 | GTP binding protein; plays a role in the vesicular transport machinery; corresponds to yeast Ypt3; suitable for use as a positive control or to assay for cross-reactivity by Western blotting; Lai, F et al, *Genomics*, 22: 610, 1994

Rab13, His-Tagged

Calbiochem 552118   Human brain recombinant, produced by overexpression of a full-length Rab13 cDNA clone in *E. coli*   MW 25,495   ≥90% (SDS-PAGE); liquid in 200 mM BME, 50 mM Tris-HCl, 10% glycerol, 2% SDS, 0.005% bromophenol blue, pH 6.8 | GTP binding protein; localized in the junctional complex region of a variety of epithelia, including intestinal, renal & hepatic; also co-localizes with a tight junction marker protein, ZO-1; disruption of tight junctions by incubation in low Ca^{2+} medium induces the redistribution of Rab13; suitable for use as a positive control or to assay for cross-reactivity by Western blotting; Zahraoui, A et al, *J Cell Biol*, 124: 101, 1994

Rab14, His-Tagged

Calbiochem 552120   Rat brain recombinant, produced by overexpression of a full-length Rab14 cDNA clone in *E. coli*   MW 25k   ≥90% (SDS-PAGE); liquid in 200 mM BME, 50 mM Tris-HCl, 10% glycerol, 2% SDS, 0.005% bromophenol blue, pH 6.8 | GTP binding protein; small GTPase thought to function in the regulation of membrane trafficking; transcripts of Rab14 are detected in brain, heart, kidney, lung & other tissues; C-terminal XXCC motif for post-translational geranylgeranylation; suitable for use as a positive control or to assay for cross-reactivity by Western blotting; Elferink, LA et al, *J Biol Chem*, 267: 5768, 1992

Rab15, His-Tagged

Calbiochem 552122   Rat brain recombinant, produced by overexpression of a full-length Rab15 cDNA clone in *E. coli*   MW 27,008   ≥90% (SDS-PAGE); liquid in 200 mM BME, 50 mM Tris-HCl, 10% glycerol, 2% SDS, 0.005% bromophenol blue, pH 6.8 | GTP binding protein; expression patterns suggest that Rab15 may act in concert with Rab3A in regulating aspects of synaptic vesicle membrane flow within the nerve terminal; suitable for use as a positive control or to assay for cross-reactivity by Western blotting; Elferink, LA et al, *J Biol Chem*, 267: 5768 & 22693, 1992

Rab18, His-Tagged

Calbiochem 552125   Rat brain recombinant, produced by overexpression of a full-length Rab18 cDNA clone in *E. coli*   MW 25,757   ≥90% (SDS-PAGE); liquid in 200 mM BME, 50 mM Tris-HCl, 10% glycerol, 2% SDS, 0.005% bromophenol blue, pH 6.8 | GTP binding protein; small protein involved in vesicular transport; localized on the apical side of epithelial cells & may play a role in endocytosis & membrane recycling; suitable for use as a positive control or to assay for cross-reactivity by Western blotting; Yu H et al, *Gene*, 132: 273, 1993; Lang, V et al, *Plant Mol Biol*, 21: 581, 1992; Lutcke, A et al, *J Cell Sci*, 107: 3437, 1994

Rab1A, His-Tagged

Calbiochem 552080   Rat brain recombinant, produced by overexpression of a full-length Rab1A cDNA clone in *E. coli*   MW 25.3k   ≥90% (SDS-PAGE); liquid in 200 *mM* BME, 50 *mM* Tris-HCl, 10% glycerol, 2% SDS, 0.005% bromophenol blue, pH 6.8 | GTP binding protein; small GTPase that is located in the endoplasmic reticulum, pre-Golgi intermediates & Golgi stack; required for transport between the endoplasmic reticulum & the *cis*-Golgi compartment; a C-terminal XXCC motif for post-translational geranylgeranylation; suitable for use as a positive control or to assay for cross-reactivity by Western blotting; Plutner, H et al, *J Cell Biol*, 115: 31, 1991; Zahraoui, A et al, *J Biol Chem*, 264: 12394, 1989

Rab2, His-Tagged

Calbiochem 552090   Rat brain recombinant, produced by overexpression of a full-length Rab2 cDNA clone in *E. coli*   MW 24.5k   ≥90% (SDS-PAGE); liquid in 200 *mM* BME, 50 *mM* Tris-HCl, 10% glycerol, 2% SDS, 0.005% bromophenol blue, pH 6.8 | GTP binding protein; small GTP-binding protein associated with structures characteristic of the intermediate between the endoplasmic reticulum & the *cis*-Golgi compartment; a C-terminal XXCC motif for post-translational geranylgeranylation & membrane targeting; suitable for use as a positive control or to assay for cross-reactivity by Western blotting; Lotti, VR et al, *J Cell Biol*, 118: 43, 1992; Tisdale, EJ et al, *J Cell Biol*, 119: 749, 1992; Touchot, N et al, *PNAS*, 84: 8210, 1987

Rab24, His-Tagged

Calbiochem 552127   Rab24 cDNA clone in *E. coli*   MW 24.3k   ≥85% (SDS-PAGE); liquid in 200 *mM* BME, 50 *mM* Tris-HCl, 10% glycerol, 2% SDS, 0.005% bromophenol blue, pH 6.8 | GTP binding protein; small GTPase localized to the endoplasmic reticulum, Golgi apparatus & late endosomes by immunostaining; C-terminal CCXX motif that is a potential site for post-translational geranylgeranylation; suitable for use as a positive control or to assay for cross-reactivity by Western blotting; Olkkonen, VM et al, *J Cell Sci*, 106: 249, 1993

Rab3A

Calbiochem 552100   Human recombinant, expressed in *E. coli*   MW 21k   ≥90% (SDS-PAGE); liquid in 45 *mM* NaCl, 20 *mM* Tris-HCl, 5 *mM* MgCl$_2$, 50% glycerol, 1 *mM* DTT, pH 8.0; GDP binding: ≥1:4 stoichiometry (GDP:Rab3A) | GTP binding protein; member of the Rab family of small GTP-binding proteins; involved in the fusion of exocytotic vesicles with the plasma membrane in neural & pituitary cells; Physiological substrate for geranylgeranyltransferase II; Prenylation at the C-terminus modulates interaction with proteins that control the guanine nucleotide-bound state of the protein; Cox, AD & Der, CJ, *Curr Opin Cell Biol*, 4: 1008, 1992; Geppert, M et al, *Nature*, 369: 493, 1994; Bourne, HR et al, *Nature*, 348: 125, 1990

Rab3A, His-Tagged

Calbiochem 552105   Rat brain recombinant, produced by overexpression of a full-length Rab3A cDNA clone in *E. coli*   MW 27.6k   ≥90% (SDS-PAGE); liquid in 200 *mM* BME, 50 *mM* Tris-HCl, 10% glycerol, 2% SDS, 0.005% bromophenol blue, pH 6.8 | GTP binding protein; small GTP-binding protein associated with synaptic vesicles in neurons; has been shown to control late steps of exocytosis expressed in various regions of the rat brain; suitable for use as a positive control or to assay for cross-reactivity by Western blotting; Li, JY et al, *Eur J Cell Biol*, 67: 297, 1995; Singh, G et al, *Am J Physiol*, 269: G400, 1995; Stettler, O et al, *Eur J Neurosci*, 7: 720, 1995

Rab4A, His-Tagged

Calbiochem 552107   Human brain recombinant, produced by overexpression of a full-length Rab4A cDNA clone in *E. coli*   MW 25k   ≥90% (SDS-PAGE); liquid in 200 *mM* BME, 50 *mM* Tris-HCl, 10% glycerol, 2% SDS, 0.005% bromophenol blue, pH 6.8 | GTP binding protein; small GTPase localized in early endosomes that is involved in endocytosis & controls exit from early endosomes; also contains a C-terminal CXC motif for post-translational geranylgeranylation & a single site for phosphorylation by p34^{cdc2} kinase; suitable for use as a positive control or to assay for cross-reactivity by Western blotting; Van der Slujis, P et al, *Cell*, 70: 729, 1992; van der Slujis, P et al, *EMBO J*, 11: 4379, 1992; van der Slujis, P et al, *PNAS*, 88: 6313, 1991; Zahraoui, A et al, *J Biol Chem*, 264: 12394, 1989

Rab5A, His-Tagged

Calbiochem 552110   Human retina recombinant, produced by overexpression of a full-length Rab5A cDNA clone in *E. coli*   MW 25.9k   ≥90% (SDS-PAGE); liquid in 200 *mM* BME, 50 *mM* Tris-HCl, 10% glycerol, 2% SDS, 0.005% bromophenol blue, pH 6.8 | GTP binding protein; small GTPase that regulates the fusion of endocytic vesicles to early endosomes; also involved in axonal & dendritic endocytosis; corresponds to yeast Ypt51; suitable for use as a positive control or to assay for cross-reactivity by Western blotting; Li, G et al, *Arch Biochem Biophys*, 316: 529, 1995; Singer-Kruger, B et al, *J Cell Sci*, 108: 3509, 1995; deHoop, MJ et al, *Neuron*, 13: 11, 1994

Rab6A, His-Tagged

Calbiochem 552111   Human brain recombinant, produced by overexpression of a full-length Rab6A cDNA clone in *E. coli*   MW 26,314   ≥90% (SDS-PAGE); liquid in 200 *mM* BME, 50 *mM* Tris-HCl, 10% glycerol, 2% SDS, 0.005% bromophenol blue, pH 6.8 | GTP binding protein; small GTP-binding protein of the Ras superfamily that mediates intra-Golgi vesicular trafficking; C-terminal CXC motif that is geranylgeranylated on both cysteines; suitable for use as a positive control or to assay for cross-reactivity by Western blotting; Feldmann, G et al, *Biol Cell*, 83: 121, 1995; Beranger, F et al, *Mol Cell Biol*, 14: 744, 1994

Rab7, His-Tagged

Calbiochem 552112   Rat brain recombinant, produced by overexpression of a full-length Rab7 cDNA clone in *E. coli*   MW 24.7k   ≥90% (SDS-PAGE); liquid in 200 *mM* BME, 50 *mM* Tris-HCl, 10% glycerol, 2% SDS, 0.005% bromophenol blue, pH 6.8 | GTP binding protein; small 23.5 kDa GTPase localized in late endosomes; regulates late endocytic membrane traffic between endosomes & lysosomes; C-terminal CXC motif for post-translational geranylgeranylation; suitable for use as a positive control or to assay for cross-reactivity by Western blotting; Vitelli, R et al, *J Biol Chem*, 272: 4391, 1997; Chavrier, P et al, *Cell*, 62: 317, 1990; Bottger, G et al, *J Biol Chem*, 271: 29191, 1996

Rab8, His-Tagged

Calbiochem 552113   Human brain recombinant, produced by overexpression of a full-length Rab8 cDNA clone in *E. coli*   MW 26,389   ≥90% (SDS-PAGE); liquid in 200 *mM* BME, 50 *mM* Tris-HCl, 10% glycerol, 2% SDS, 0.005% bromophenol blue, pH 6.8 | GTP binding protein; small Ras-like GTPase that regulates polarized membrane transport in neurons & in epithelial cells; regulates transport process during morphological maturation in 95% of neurons; localized in the Golgi regions, in vesicular structures & in the basolateral plasma membrane; suitable for use as a positive control or to assay for cross-reactivity by Western blotting; Peranen, J et al, *J Cell Biol*, 135: 153, 1996; Huber, LA et al, *Mol Cell Biol*, 15: 918, 1993; Huber, LA et al, *J Cell Biol*, 123: 35, 1993

RabGDI-α, His-Tagged

Synonyms: RabGDP Dissociation Inhibitor

Calbiochem 552128 Rat brain recombinant, produced by overexpression of a full-length RabGDI-α cDNA clone in *E. coli* MW 53,240 ≥70% (SDS-PAGE); liquid in 200 *mM* BME, 50 *mM* Tris-HCl, 10% glycerol, 2% SDS, 0.005% bromophenol blue, pH 6.8 | GTP binding protein; forms complexes with cytoplasmic GDP-bound Rab proteins & releases Ras proteins for association with GTP; functions in vesicle-membrane transport to recycle & regulate Rab GTPases; in the cytoplasm, interaction between Rab6 protein & RabGDI is enhanced by the geranylgeranylation of Rab6; suitable for use as a positive control or to assay for cross-reactivity by Western blotting; Horiuchi, H et al, *J Biol Chem*, 270: 11257, 1995; Beranger, F et al, *J Biol Chem*, 269: 13637, 1994

Rac1 (G12V)-GST, Agarose Conjugated

Calbiochem 553501 Human Purified; in a 50% agarose slurry | GTP binding protein; Gly¹² to Val mutation was introduced to eliminate intrinsic GTPase activity from the recombinant Rac1-GST fusion protein; Rac1 (G12V)-GST is supplied bound to glutathione-agarose beads to facilitate adsorption of Rac-binding proteins such as PAK1 & PAK2 from cellular extracts; full-length soluble Rac1 (G12V) can be released from the beads by incubation with thrombin, or the Rac1 (G12V)-GST fusion protein can be eluted with 10 *mM* glutathione

Rac1, GST-Tagged

Calbiochem 552134 Human recombinant, expressed in *E. coli* MW 48,451 ≥90% (SDS-PAGE); liquid in 20 *mM* HEPES, 5 *mM* MgCl₂, 40% glycerol, 1 *mM* NaN₃, pH 8.0; GTPγS binding: ≥400 mmol/mol | GTP binding protein; small GTPase that regulates lamellipodia formation in response to growth factor stimulation, as well as activation of NADPH-oxidase in phagocytic lymphocytes & JNK-MAP kinase cascade; useful as a positive control in Western blotting & as an affinity precipitation reagent using immobilized glutathione resin; Kuroda, S et al, *J Biol Chem*, 271: 23363, 1996; Lamarche, N et al, *Cell*, 87: 519, 1996; Ridley, AJ et al, *Cell*, 70: 401, 1992

Rac1, His-Tagged

Calbiochem 552137 Human recombinant, expressed in *E. coli* MW 22,151 ≥90% (SDS-PAGE); liquid in 150 *mM* NaCl, 20 *mM* Tris-HCl, 40% glycerol, 5 *mM* imidazole, pH 7.9; GTPγS binding: ≥600 mmol/mol | GTP Binding Protein small GTPase that belongs to the Ras superfamily & regulates lamellipodia formation in response to growth factor stimulation as well as activation of NADPH-oxidase in phagocytic lymphocytes; useful as a positive control in Western blotting & as an affinity precipitation reagent using Ni²⁺-charged metal chelate resin; Lamarche, N et al, *Cell*, 87: 519, 1996; Ridley, AJ et al, *Cell*, 70: 401, 1992

Calbiochem 552132 Rat brain recombinant, produced by overexpression of a full-length Rac1 cDNA clone in *E. coli* MW 22,560 ≥90% (SDS-PAGE); liquid in 200 *mM* BME, 50 *mM* Tris-HCl, 10% glycerol, 2% SDS, 0.005% bromophenol blue, pH 6.8 | GTP binding protein; small protein of the Rho family that regulates the reorganization of the actin cytoskeleton in all eukaryotic cells; reported to stimulate rapid polymerization of actin & to induce membrane ruffling when injected into quiescent Swiss 3T3 cells; may be involved in controlling the c-Jun amino-terminal kinase (JNK) pathway; suitable for use as a positive control or to assay for cross-reactivity by Western blotting; Best, A et al, *J Biol Chem*, 271: 3756, 1996; Coso, OA et al, *Cell*, 81: 1137, 1995; Qui, RG et al, *Nature*, 374: 457, 1995; Ridley, AJ et al, *Cell*, 70: 401, 1992

Rac1-GST, Agarose Conjugated

Calbiochem 553503 Full-length human Rac1 recombinant, amplified by PCR & cloned into *E. coli* Purified; in a 50% agarose slurry | GTP binding protein; supplied bound to glutathione-agarose beads to facilitate adsorption of Rac-binding proteins such as PAK1 & PAK2 from cellular extracts; full-length soluble Rac1 can be released from the beads by incubation with thrombin, or the Rac1-GST fusion protein can be eluted with 10 *mM* glutathione

Rac2, His-Tagged

Calbiochem 552151 Human recombinant, produced by overexpression of a full-length Rac2 cDNA clone in *E. coli* MW 21,910 ≥90% (SDS-PAGE); liquid in 200 *mM* BME, 50 *mM* Tris-HCl, 10% glycerol, 2% SDS, 0.005% bromophenol blue, pH 6.8 | GTP binding protein; small GTPase that belongs to the Ras superfamily; regulates the activity of NADPH oxidase in phagocytic cells; suitable for use as a positive control or to assay for cross-reactivity by Western blotting; Dorseuil, O et al, *J Biol Chem*, 271: 83, 1996; el Benna, J et al, *J Biol Chem*, 269: 6729, 1994

Raf-1 Truncated, Active

USBio R0495-08

Raf-1, Active

USBio R0495-05 Human recombinant from Sf9 insect cell lysate In 50 μL of 20 *mM* Tris-acetate, pH 7.5, 0.27 *M* sucrose, 1 *mM* EDTA, 1 *mM* EGTA, 1 *mM* sodium orthovanadate, 10 *mM* β-glycerophosphate, 50 *mM* NaF, 5 *mM* sodium pyrophosphate, 1% Triton X-100, 0.1% β-MSH, 0.2 *mM* PMSF, 5 mg/mL leupeptin, 5 mg/mL aprotinin | Full length untagged recombinant human Raf-1; at the convergence of two lines of inquiry: the search for activators of MAP Kinase Kinase (MEK) & the search for effectors of the Ras oncoprotein; Raf-1 binds the effector loop of p21 Ras when Ras is in complex with GTP, by virtue of its amino-terminal regulatory domain; while the interaction itself is insufficient to activate Raf-1, recruitment of Raf-1 to the plasma membrane consequent to Ras binding results in Raf-1 activation & the requirement for Ras can be bypassed if Raf-1 is constitutively localized to the membrane; Raf-1 associates with members of the 14-3-3 family of proteins, which appear to protect active Raf-1 from phosphatase action; important regulatory phosphorylation events occur on Tyr³⁴⁰,³⁴¹ & Ser²⁵⁹,⁴⁹⁹ (activating), Ser⁴³ (prevents Ras:GTP binding) & Ser⁶²¹ (constitutive, required for activity)

RalB, His-Tagged

Calbiochem 553201 Rat brain recombinant, produced by overexpression of a full-length RalB cDNA clone in *E. coli* MW 24.4k ≥85% (SDS-PAGE); liquid in 200 *mM* BME, 50 *mM* Tris-HCl, 10% glycerol, 2% SDS, 0.005% bromophenol blue, pH 6.8 | GTP binding protein; small GTP-binding protein associated with the synaptic vesicles & human platelet dense granules; detected in adrenal glands, brain, kidney, spleen & testis; thought to mediate a downstream signaling pathway from Ras that facilitates cellular transformation; a C-terminal CCLL motif & is most likely geranylgeranylated; suitable for use as a positive control or to assay for cross-reactivity by Western blotting; Hermann, C et al, *J Biol Chem*, 271: 6794, 1996; Mark, BL et al, *Biochem Biophys Res Commun*, 225: 40, 1996; Wildey, GM et al, *Biochem Biophys Res Commun*, 194: 552, 1993

Ran, His-Tagged

Calbiochem 553203 Rat brain recombinant, produced by overexpression of a full-length Ran cDNA clone in *E. coli* MW 25.5k ≥90% (SDS-PAGE); liquid in 200 *mM* BME, 50 *mM* Tris-HCl, 10% glycerol, 2% SDS, 0.005% bromophenol blue, pH 6.8 | GTP binding protein; small GTP-binding protein located in the nucleus; regulates the check point between completion of DNA synthesis & initiation of mitosis; also required for import of proteins into the nucleus; suitable for use as a positive control or to assay for cross-reactivity by Western blotting; Ren, M et al, *J Cel Biol*, 120: 313, 1993; Moore, MS et al, *PNAS*, 91: 10212, 1994; Bischoff, FR et al, *PNAS*, 88: 10830, 1991

RANK

Synonyms: TNFRSF11A; NF-κB Receptor Activator

R&D Systems 683-RK-100 NSO-Expressed >95%; lyophilized; ED₅₀: 4-10 ng/mL | Species specificity: human; member of the TNF receptor superfamily; TRANCE (TNF-related activation-induced cytokine) is the ligand for RANK; a number of biological functions are mediated through RANK, including activation of NF-κB; soluble RANK can block TRANCE-induced biological activity; Anderson, DM et al, *Nature*, 390: 175, 1997; Nakagawa, N et al, *BBRC*, 245: 382, 1998

R&D Systems 692-RK-100   NSO-Expressed   >95%; lyophilized; ED_{50}: 5-15 ng/mL | Species specificity: mouse; member of the TNF receptor superfamily; TRANCE (TNF-related activation-induced cytokine) is the ligand for RANK; a number of biological functions are mediated through RANK, including activation of NF-κB; soluble RANK can block TRANCE-induced biological activity; Anderson, DM et al, *Nature*, 390: 175, 1997; Nakagawa, N et al, *BBRC*, 245: 382, 1998

RANK Ligand

BioSource International PHP0034   Human recombinant

RANK Ligand, Soluble

Synonyms: sRANKL; OPGL; TRANCE; ODF; TNFRSF11A; NF-κB Receptor Activator

Chemicon GF091   Human   ≥95% | Purified protein for apoptosis & signal transduction

Alexis 522-012   Human recombinant   Novel member of the TNF family expressed in activated T cells, lymph nodes & in stromal cell lines; interacts with its receptor RANK expressed on mature dendritic cells (DC) & mature osteoclasts, leading to the inhibition of apoptosis, probably through the upregulation of bcl-x; useful tool for enhancing DC & osteoclast survival & activity; Lacey, DL et al, *Cell*, 93: 165, 1998; Anderson, DA et al, *Nature*, 390: 175, 1997; Wong, BR et al, *J Exp Med*, 186: 2075, 1997

PeproTech 310-01   Human recombinant, expressed in *E. coli*   MW 20.0k   >98% (SDS-PAGE) & HPLC; <0.1 ng endotoxin per mg (1EU/mg); lyophilized | Receptor activator of NF-KappaB; newly discovered member of the TNFR family; a dendritic cell membrane protein comprising the full-length of the TNF-like extracellular domain of RANKL; ED_{50} < 10.0 ng/mL; SA determined by the dose-dependent stimulation of IL-8 production by human PBMC

RANKL

Kamiya   MW 27.8k   >98% (SDS-PAGE) & HPLC

RANTES

Synonyms: MCP2; Macrophage/Monocyte Chemoattractant Protein II; hSIS δ Protein

Biodesign A52007M   *E. coli*   MW 7.8k   Purified | Species specificity: mouse

Biodesign A52013H   *E. coli*   MW 7.9k   Purified | Species specificity: rat

Biodesign A52306H   *E. coli*   MW 8k   Purified

Chemicon GF020   Human   ≥95%

BioSource International PHC1054   Human recombinant

Calbiochem 553500   Human recombinant, expressed in *E. coli* MW 7.8k   ≥97% (SDS-PAGE); lyophilized from filter-sterilized 30% acetonitrile, 0.1% TFA containing 50 μg BSA/μg RANTES; biological activity: ED_{50}=50-200 ng/mL as measured by its monocyte chemotactic ability; endotoxin: ≤100 pg/μg RANTES | RANTES (regulated on activation, normal T cell expressed & secreted) is a member of the β (CC) chemokine family; chemoattractant for unstimulated CD4$^+$/CD45RO$^+$ memory T cells; induces Ca^{2+} mobilization in T cells; acts as an antigen-independent activator of T cells *in vitro*; Bacon, KB et al, *Science*, 269: 1727, 1995; Devergne, O et al, *J Exp Med*, 179: 1689, 1994

Fitzgerald 30-AR50   Human recombinant, expressed in *E. coli*

Harlan BT-3024   Human recombinant, expressed in *E. coli* Lyophilized; 0.01 mg

Harlan BT-3025   Human recombinant, expressed in *E. coli* Lyophilized; 0.05 mg

ICN 158419   Human recombinant, expressed in *E. coli*   ≥98%; lyophilized; ED_{50} = 50.0 ng/mL, 1 U is the amount of RANTES needed for maximal chemotactic activity on human blood monocytes | Important member of the PF4 superfamily of chemoattractant proteins shown to selectively attract T cells of the CD4+/CD45RO+ phenotype *in vitro*; useful with human IL-8 & MCAF/MCP1 in inflammatory response research

PeproTech 300-06   Human recombinant, expressed in *E. coli* MW 7.8k   >98% (SDS-PAGE) & HPLC; <0.1 ng endotoxin per mg (1EU/mg); lyophilized | Chemoattractant for peripheral blood monocytes; selectively attracts T cells of the CD4$^+$/CD45RO$^+$ phenotype *in vitro*; 68 AA; SA determined by its ability to chemoattract human blood monocytes

Sigma R 6267   Human recombinant, expressed in *E. coli*   MW 7.8k   ≥97% (SDS-PAGE); 0.2 μm filtered & lyophilized from 30% acetonitrile, 0.1% trifluoroacetic acid containing 500 μg BSA; activity tested in culture by measuring monocyte chemotactic activity; endotoxin tested | Member of the chemokine superfamily, platelet factor 4 (PF4) which is characterized by 4 positionally conserved cysteine residues & has been subdivided according to the position of the first 2 cysteines into 2 branches, the chemokine α family & the chemokine β subfamilies; member of the chemokine β family; chemoattractant for peripheral blood monocytes & will selectively attract T cells of the CD4+/CD45RO+ phenotype *in vitro*; Schall, T et al, *Nature*, 347: 669, 1990; Staeckle, M et al, *New Biol*, 2: 313, 1990; Brown, K et al, *J Immunol*, 142: 679, 1989; Schall, T et al, *Cytokine*, 3(3): 165, 1991; Schall, T et al, *Nature*, 347: 669, 1990

BioSource International PMC1054   Mouse recombinant

Chemicon GF082   Murine   ≥95%

PeproTech 250-07   Murine recombinant, expressed in *E. coli* MW 7.8k   >98% (SDS-PAGE) & HPLC; <0.1 ng endotoxin per mg (1EU/mg); lyophilized with no additives | Chemo-attractant for peripheral blood monocytes; selectively attracts T cells of the CD4$^+$/CD45RO$^+$ phenotype *in vitro*; 68 AA; SA determined by its ability to chemoattract total human lymphocyte population & total murine T cell population

Chemicon GF045   Rat   ≥95%

BioSource International PRC1054   Rat recombinant

Fitzgerald 30-AR51   Rat recombinant, expressed in *E. coli*

ICN 193968   Rat recombinant, expressed in *E. coli*   ≥95%; lyophilized | Homolog of human RANTES, a chemoattractant to monocytes & eosinophils

PeproTech 400-13   Rat recombinant, expressed in *E. coli* MW 7.9k   >98% (SDS-PAGE) & HPLC; <0.1 ng endotoxin per mg (1EU/mg); lyophilized | Chemoattractant for peripheral blood monocytes; selectively attracts T cells of the CD4$^+$/CD45RO$^+$ phenotype *in vitro*; 69 AA; SA determined by its ability to chemoattract rat peritoneal macrophages

Rap1a, His-Tagged

Calbiochem 553205   Human brain recombinant, produced by overexpression of a full-length Rap1a cDNA clone in *E. coli*   MW 22.1k   ≥90% (SDS-PAGE); liquid in 200 *mM* BME, 50 *mM* Tris-HCl, 10% glycerol, 2% SDS, 0.005% bromophenol blue, pH 6.8 | GTP binding protein; small GTP-binding protein that antagonizes the Ras transforming activity through tight binding to Ras-GAP; interferes with Ras-dependent Raf1 activation by inhibiting binding of Ras to the cysteine-rich region of Raf1; localized in the endoplasmic reticulum, late endosomes & lysosomes, indicating its potential role in the regulation of intracellular protein degradation; a C-terminal CLLL motif for post-translational geranylgeranylation & a single site for phosphorylation by protein kinase A; suitable for use as a positive control or to assay for cross-reactivity by Western blotting; Hu, CD et al, *J Biol Chem*, 272: 1702, 1997; Pizon, V et al, *J Cell Sci*, 107: 1661, 1994; Rubinfeld, B et al, *Cell*, 65: 1033, 1991; Kitayama, H et al, *Cell*, 56: 77, 1989

Rap1b, His-Tagged

Calbiochem 553207 Human brain recombinant, produced by overexpression of a full-length Rap1b cDNA clone in *E. coli* MW 21.9k ≥90% (SDS-PAGE); liquid in 200 *mM* BME, 50 *mM* Tris-HCl, 10% glycerol, 2% SDS, 0.005% bromophenol blue, pH 6.8 | GTP binding protein; small GTP-binding protein that belongs to the Rap subgroup of Ras-related proteins; reported to have a stimulating effect on B-Raf activity; distribution of Rap1b is restricted to late endosome/lysosomal structures, suggesting a potential role in the regulation of intracellular protein degradation; a C-terminal CQLL motif for post-translational geranylgeranylation & a single site for phosphorylation by protein kinase A; suitable for use as a positive control or to assay for cross-reactivity by Western blotting; Kuroda, S et al, *J Biol Chem*, 271: 14680, 1996; Pizon, V et al, *J Cell Sci*, 107: 1661, 1994; Lapetina, E et al, *PNAS*, 86: 3131, 1989; Pizon, V et al, *Nucl Acids Res*, 16: 7719, 1988

Rap2a, His-Tagged

Calbiochem 553209 Human brain recombinant, produced by overexpression of a full-length Rap2a cDNA clone in *E. coli* MW 21.7k ≥90% (SDS-PAGE); liquid in 200 *mM* BME, 50 *mM* Tris-HCl, 10% glycerol, 2% SDS, 0.005% bromophenol blue, pH 6.8 | GTP binding protein; small GTP-binding protein expressed mainly in brain, muscle, platelet & testis; thought to play a role in the regulation of granule organization; a C-terminal CNIQ motif for post-translational farnesylation; suitable for use as a positive control or to assay for cross-reactivity by Western blotting; Carnero, A et al, *Biochem Biophys Res Commun*, 216: 748, 1995; Farrell, FX et al, *Biochem J*, 289: 349, 1993; Pizon, V et al, *Oncogene*, 3: 201, 1988

Ras

Sigma R 9894 Human wild type recombinant, expressed in *E. coli* MW 21k ~95% (SDS-PAGE); solution in 50% glycerol containing 20 mM Tris, pH 7.6, 5 mM MgCl$_2$, 50 mM NaCl, 1 mM DTT; GDP binding: ≥0.25 mole GDP/mole H-Ras | Guanine-nucleotide binding protein that couples tyrosine kinase receptor signals to MAP kinase cascade; Ras is a substrate for farnesyltransferase (Ftase), when farnesylated it is anchored to the membrane; membrane localization of Ras is crucial for its cellular activity; inhibition of Ras farnesylation is a target for new anti-cancer drug development, since abnormally active Ras is present in >50% of human cancer; Tucker, J et al, *EMBO J*, 5: 1351, 1986; Egan, SE & Weinberg, RA, *Nature*, 365: 781, 1993; Tamanoi, F, *TIBS*, 18: 349, 1993

Ras (G12V)-GST, Agarose Conjugated

Calbiochem 553574 Human Purified; in a 50% agarose slurry | GTP binding protein; binds Raf-1, GAP & SOS in cellular extracts; a glycine 12 to valine mutation was introduced to eliminate intrinsic GTPase activity from the recombinant Ras-GST fusion protein; Ras (G12V)-GST is supplied bound to glutathione-agarose beads to facilitate adsorption of Ras-binding proteins such as Raf1, GAP & SOS from cellular extracts; full-length soluble Ras (G12V) can be released from the beads by incubation with factor Xa, or the Ras (G12V)-GST fusion protein can be eluted with 10 *mM* glutathione; Vojtek, AB et al, *Cell*, 74: 205, 1993

Ras Modified

Synonyms: Ras-Cys-Val-Leu-Leu

Sigma R 0145 Human recombinant, expressed in *E. coli* MW 21k ~95% (SDS-PAGE); solution in 50% glycerol containing 20 mM Tris HCl, pH 8.0, 5 mM MgCl$_2$, 45 mM NaCl, 1 mM DTT | Modified Ras protein which terminates in the tetrapeptide Cys-Val-Leu-Leu; substrate for geranylgeranyl transferase I; Cox, AD et al, *Mol Cell Biol*, 12: 2606, 1992

Ras, CVLL Type

Synonyms: Geranylgeranyltransferase Substrate

Calbiochem 553322 Human recombinant, expressed in *E. coli* MW 20k ≥90% (SDS-PAGE); liquid in 45 *mM* NaCl, 5 *mM* MgCl$_2$, 20 *mM* Tris-HCl, 1 *mM* DTT, 50% glycerol, pH 8.0; GDP binding: ≥1:4 stoichiometry of GDP:H-Ras | GTP binding protein; Marshall, MS et al, *FASEB J*, 9: 1311, 1995; Casey, P et al, *PNAS*, 88: 8631, 1991

Ras, Wild-Type

Synonyms: Farnesyltransferase Substrate

Calbiochem 553325 Human recombinant, expressed in *E. coli* MW 20k ≥90% (SDS-PAGE); liquid in 45 *mM* NaCl, 5 *mM* MgCl$_2$, 20 *mM* Tris-HCl, 1 *mM* DTT, 50% glycerol, pH 8.0; GDP binding: ≥1:4 stoichiometry of GDP:H-Ras | GTP binding protein; Marshall, MS et al, *FASEB J*, 9: 1311, 1995; Reiss, Y et al, *Cell*, 62: 81, 1990

Ras-GST, Agarose Conjugated

USBio R1198-20 Recombinant MW ~47k ≥95%; frozen suspension of 50% agarose gel slurry suspended in PBS for a total volume of 50 µL per vial; each vial containing 25 µg of GST-Ras is bound to 25 µL of packed agarose beads | Non-covalently bound to glutathione-agarose (50% bead slurry); although it is supplied linked to glutathione-agarose, GST-Ras protein can be eluted from the beads with 10*mM* glutathione; full length, soluble Ras can be released by incubation of the beads with factor Xa

RecA Protein

Fluka 83543 0.5 U/mg protein (≥1 mg/mL); 1 U corresponds to the amount of enzyme which liberates 1 µmol inorganic phosphate from ATP in the presence of single stranded calf thymus DNA/min at pH 7.5, 37°C | Rigas, B et al, *PNAS*, 83: 9591, 1986

Fluka 83548 *E. coli* KM 1842 ~0.5 U/mg protein (~5 mg/mL); 1 U corresponds to the amount of enzyme which liberates 1 µmol inorganic phosphate from ATP in the presence of single stranded calf thymus DNA/min at pH 7.5, 37°C | Koob, M, *Meth Enzymol*, 216: 321, 1992

Amersham E70028Y/E70028Z purified from an exonuclease I-deficient *E. coli* that carries the cloned *recA* gene from *E. coli*. MW 37.8k 1-5 mg/mL by A$_{280}$; >95% pure (PAGE); free of endonuclease, exonuclease, DNA & RNA; 20 mM Tris-HCl, pH 7.5, 1.0 mM DTT, 0.1 mM EDTA, 20% glycerol | Participates in general recombination, repair of DNA & regulation of repair mechanisms; has been observed *in vitro* to have DNA-dependent ATPase activity, promote proteolytic cleavage of repressors & to catalyze DNA strand-pairing & exchange; forms helical filaments with ss- & duplex DNA

Promega M1691 Recombinant, expressed in *E. coli* Binds cooperatively & stoichiometrically to ss-DNA & is active in strand exchange as a nucleoprotein filament containing 1 RecA monomer (38 kDa) per 3 bases of ss-DNA; can locate & pair a ss-DNA sequence to its homologous ds-DNA sequence in the presence of ATP-γ-S; Honigberg, SM et al, *PNAS*, 83: 9591, 1986; Ferrin, LJ & Camerini-Otero, RD, *Science*, 254: 1494, 1991; Taidi-Laskowski, B et al, *Nucl Acids Res*, 16: 8157, 1988; Rigas, B et al, *PNAS*, 83: 9591, 1986; Koob, M et al, *Nucl Acids Res*, 20: 5831, 1992; Krasnow, MA et al, *Nature*, 304: 559, 1983

Renin

Biogenesis 7930-0004 Human kidney Tested negative for HBsAg; purified; lyophilized

Respiratory Syncytial Virus Antigen

Biodesign R02712 Lysate | Infectious disease antigen

Restrictin

Synonyms: Tenascin-R; Janusin

Chemicon CC116 Chicken brain MW 180k Chromatographically purified | Extracellular matrix protein; a large glycoprotein that forms disulfide-linked trimers; structurally related to tenascin & contains EGF-like repeats, fibronectin type III repeats & a region homologous to fibrinogen; the expression of restrictin is limited to the nervous system; can function as an anti-adhesive molecule

Restrictocin

Synonyms: Ribosome Inactivating Protein

Sigma R 0389 *Aspergillus restrictus* Lyophilized from 0.3 mL of 0.5% NaCl; ~70% of protein is active restrictocin | Acts by specifically cleaving rRNA

Retinol Binding Protein

Cortex CP1024	>95%
Cortex CP1024P	>40%
Cortex CP1024U	>98%

USBio R1701 Human ≥98% (SDS-PAGE); no contaminants detected; single band by SDS-PAGE, IEP, &/or RID; 1.15 mg/mL (Lowry) supplied in 10 mM Tris-HCl; pH 7.4, 150 mM NaCl | Suitable for antigenic applications in immunological protocols

Fitzgerald 30-AR20 Human urine High purity

Biogenesis 7970-0504 Human urine from patients with chronic renal tubular proteinuria MW 21k Tested negative for HIV 1 and 2 antibodies, hepatitis B surface antigen and HCV antibodies; purified; from 0.02M NH_4HCO_3, may contain traces of buffer salts; lyophilized

Scipac P124-0 Urine of patients with chronic renal tubular proteinuria >99%; lyophilized; < 0.1% A1M, B2M and albumin | Nutritional protein

Scipac P124-1 Urine of patients with chronic renal tubular proteinuria >98%; lyophilized | Nutritional protein

Scipac P124-2 Urine of patients with chronic renal tubular proteinuria 40-90%; lyophilized; available on request | Nutritional protein

RhoA, GST-Tagged

Calbiochem 555466 Human recombinant, expressed in *E. coli* MW 48,769 ≥90% (SDS-PAGE); liquid in 20 mM HEPES, 5 mM $MgCl_2$, 40% glycerol, 0.06% NaN_3, pH 8.0; GTPγS binding: ≥400 mmol/mol | GTP binding protein; small GTPase that belongs to the Ras superfamily & regulates stress fiber formation in response to growth factor stimulation; useful as a positive control in Western blotting & as an affinity precipitation reagent using immobilized glutathione resin; Ridley, AJ et al, *Curr Biol*, 6: 1256, 1996; Ridley, AJ & Hall, A, *Cell*, 70: 389, 1992

RhoA, His-Tagged

Calbiochem 555470 Human recombinant, expressed in *E. coli* MW 22,469 ≥90% (SDS-PAGE); liquid in 150 mM NaCl, 20 mM Tris-HCl, 40% glycerol, 5 mM imidazole, pH 7.9; GTPγS binding: ≥600 mmol/mol | GTP binding protein; small GTPase that belongs to the Ras superfamily & regulates stress fiber formation in response to growth factor stimulation; useful as a positive control in Western blotting & as an affinity precipitation reagent using Ni^{2+}-charged metal chelate resin; Ridley, AJ et al, *Curr Biol*, 6: 1256, 1996; Ridley, AJ & Hall, A, *Cell*, 70: 389, 1992

RhoB

Chemicon GPR082 Recombinant ≥95% | Purified protein for apoptosis & signal transduction

RhoB, His-Tagged

Calbiochem 555475 Rat brain recombinant, produced by overexpression of a full-length RhoB cDNA clone in *E. coli* MW 24,844 ≥85% (SDS-PAGE); liquid in 200 mM BME, 50 mM Tris-HCl, 10% glycerol, 2% SDS, 0.005% bromophenol blue, pH 6.8 | GTP binding protein; small GTP-binding protein that is implicated in the regulation of the microfilamental network & in cell transformation; Perinuclear & vesicular localization of the endogenous RhoB protein is also inducible by growth factors; also plays a role in the G_1/S phase transition &/or in the S phase of the cell cycle; suitable for use as a positive control or to assay for cross-reactivity by Western blotting; Armstrong, SA et al, *J Biol Chem*, 270: 7864, 1995; Fritz, G et al, *J Biol Chem*, 270: 25172, 1995; Zalcman, G et al, *Oncogene*, 10: 1935, 1995; Hall, A et al, *Ann Rev Cell Biol*, 10: 31, 1994

Rhodopsin

Calbiochem 555520 Bovine retina MW 30,007 ≥95% (SDS-PAGE); dark-adapted membrane suspension in 20 mM Tris-HCl, 10 mM BME, 100 μM EDTA, pH 7.5; activity: activates equimolar amounts of transducin as measured by GTPγS binding; must open in complete darkness | G-protein-coupled receptor localized in retinal rod outer segment membranes; absorption of light by the chromophore 11-*cis*-retinal in the active site of rhodopsin is the initial event in the conversion of light energy to neural excitation; Ting, TD et al, *Methods Neurosci*, 15: 180, 1993; Papermaster, DS, *Biochemistry*, 13: 2438, 1982

Riboflavin Binding Protein

Fluka 83806 Chicken egg white MW 32.3k ≥98.0% (GE);. ~10 U/mg protein; 1 U corresponds to the binding of 1 μg riboflavin at pH 7.5, 25°C | Very heat-stable phosphoglycoprotein; dissociation constant for riboflavin: 1.3×10^{-9} M; occurs readily <pH 4.0; Becvar, J & Palmer, G, *JBC*, 257: 5607, 1982; Sanberlich, ME, *Ann Rev Nutr*, 4: 377, 1984; Tillotson, JA & Bashor, MM, *Anal Biochem*, 107: 214, 1980

Riboflavin Binding Protein, Apo Form

Sigma R 8628 Chicken egg white Lyophilized powder; binding capacity: 1 mg of protein (Biuret) binds 5-15 μg of riboflavin; essentially free of riboflavin & salts | Useful in the fluorometric determination of riboflavin; 1 mg sufficient for 20 riboflavin determinations; Farrell, HM Jr et al, *Biochim Biophys Acta*, 194: 433, 1969; Tillotson, JA & Bashor, MM, *Anal Biochem*, 107: 214, 1980

Riboflavin Binding Protein, Holo Form

Sigma R 3754 Chicken egg white Lyophilized powder; contains 5-15 μg of protein-bound riboflavin/mg protein | Not useful in the determination of riboflavin

Ribonuclease S Protein Grade XII-PR

Sigma R 6250 Bovine pancreas A protein component obtained from RNase S (protease-modified RNase A); when mixed with S-peptide, full enzymatic activity is restored | Richard, R & Vithayathil, P, *J Biol Chem*, 234: 1459, 1959

Ribosomal P Antigen

Biodesign A07305B Bovine/rabbit mixture Purified | Autoimmune reagent

Ristocetin

Sigma R 7752 Sulfate salt; >90% Ristocetin A; balance primarily Ristocetin B | Glycopeptide antibiotic complex; Jordan, DC in *Antibiotics*, Vol 1, D Gottlieb & P Shaw, eds, Springer-Verlag (New York: 1967), 84

Ristocetin Reagent

Sigma 885-7 Reconstitute with 1 mL water to obtain Ristocetin solution, ~15 mg/mL, with buffer

ROK α/ROCK-II, Active

USBio R2998

Rubella Core Protein

Biodesign R3B901   Recombinant   >95% | Infectious disease antigen

S-100 Antigen

USBio S0051   Human   ≥98% (SDS-PAGE); no contaminants detected; single band by SDS-PAGE, IEP, &/or RID; 40,000 U/mg; lyophilized | Suitable for antigenic applications in immunological protocols

S-100 Protein

Calbiochem 559284   Bovine brain   MW 21k   >95% (SDS-PAGE); lyophilized solid; solubilized in 100 *mM* Tris, pH 7.5 | Calcium-modulated protein with some structural similarity to calmodulin; interacts with glial fibrillary acidic protein & inhibits its polymerization in a Ca^{2+}-dependent manner; originally detected in neural tissue, principally in the cytoplasm of glial cells; also found in chondrocytes, T lymphocytes & skin; involved in microtubule dissociation & inhibition of microtubule assembly; Delvalle, ME et al, *Neurosci Lett*, 168: 247, 1994; Lackmann, M et al, *J Biol Chem*, 267: 7499, 1992; Marshak, DR et al, *PNAS*, 78: 6793, 1981

Sigma S 6552   Bovine brain   Lyophilized powder containing ≥50% (w/w) *S*-100 proteins; actual percentage determined by HPLC provided on label

Calbiochem 559291   Human brain   MW 21k   ≥98% (SDS-PAGE); salt-free lyophilized powder; solubilized in PBS, pH 7.2 | Calcium-binding protein found in neuronal tissue, principally in the cytoplasm of glial cells; also found in chodrocytes, T lymphocytes & skin; prepared from tissues of individuals that have been shown by certified tests to be negative for HBsAg & for antibodies to HIV & HCV

Fitzgerald 30-AS07   Human brain   High purity >98%

S-100A (α,α) Protein

Calbiochem 559287   Bovine brain   MW 21k   ≥90% (SDS-PAGE); ≥95% (HPLC); lyophilized solid | Homodimeric S-100 isoform composed of αα-subunits; Ogoma, Y et al, *J Biol Macromol*, 14: 279, 1992; Usui, A et al, *Clin Chem*, 36: 639, 1990

Sigma S 6927   Bovine brain   ~90% (HPLC); ≥85% (PAGE); lyophilized powder containing ~90% protein | Ca^{2+} binding protein found in glial cell cytoplasm in two major dimeric forms, *S*-100A (α,β) & *S*-100B (β,β) & one minor form *S*-100 (α,α); inhibits microtubule assembly

S-100A Protein

Sigma S 6802   Bovine brain   ~90% (HPLC); ≥70% (PAGE); lyophilized powder containing ~90% protein (Biuret) | Ca^{2+} binding protein found in glial cell cytoplasm in two major dimeric forms, *S*-100A (α,β) & *S*-100B (β,β) & one minor form *S*-100 (α,α); inhibits microtubule assembly

S-100B Protein

Biogenesis 8200-0990   Bovine brain   94% protein (Biuret); purified

Calbiochem 559290   Bovine brain   MW 21k   >98% (SDS-PAGE); ≤10% α-chain (HPLC); lyophilized solid; soluble in aqueous buffer, pH >6.0 | Homodimeric S-100 isoform composed of ββsubunits; Momotani, E et al, *J Comp Pathol*, 108: 291, 1993; Baudier, J et al, *PNAS*, 89: 11627, 1992; Donato, R et al, *Cell Calcium*, 12: 713, 1991; Isobe, T et al, *Biochim Biophys Acta*, 494: 222, 1977

Sigma S 6677   Bovine brain   ~90% (HPLC); ≥70% (PAGE); lyophilized powder containing ~90% protein (Biuret) | Ca^{2+} binding protein found in glial cell cytoplasm in two major dimeric forms, *S*-100A (α,β) & *S*-100B (β,β) & one minor form *S*-100 (α,α); inhibits microtubule assembly

SAA

BioSource International PHA0014   Human

SAPK2/p38 Inhibitor

Synonyms: SB 203580

USBio S0096-26   ≥98% by HPLC; soluble in DMSO at 50 mg/mL; pale yellow solid | A highly specific inhibitor (IC_{50}: 600 *nM*); suppresses the activation of MAPKAP Kinase-2 & MAPKAP Kinase-3; does not significantly inhibit SAPK/JNK or Erk/MAPkinases at 100*mM*; inhibits IL-1 & TNF-α production from LPS-stimulated human monocytes & the human monocyte cell line THP-1; inhibits activation of the HIV-1 long terminal repeat by IL-1 & TNF-α; an effective inhibitor of inflammatory cytokine production *in vivo* in mice & rats

Saporin

Synonyms: Ribosome Inactivating Protein

Sigma S 9896   *Saponaria officinalis* seeds   Lyophilized powder containing ~20% protein (Lowry); balance primarily glucose & sodium phosphate buffer salts | Stirpe, F et al, *Biochem J*, 216: 617, 1983; Stirpe, F & Barbieri, L, *FEBS Lett*, 195: 1, 1986

sCD40L/Trap

BioSource International PHP0024   Human recombinant

Scl-70 Antigen

Biodesign A07301B   Bovine/rabbit thymus mixture   Purified | Autoimmune reagent

Biodesign A2A550R   Rabbit thymus   Purified | Autoimmune reagent

SDF-1α

BioSource International PHC1354   Human recombinant
BioSource International PHC1364   Human recombinant

Secretory Leukocyte Protease Inhibitor

ICN 195731   Human recombinant, expressed in *E. coli*   ≥97%; inhibits trypsin activity at a 1:1 molar ratio to active trypsin

Selectin, E-

Synonyms: Selectin, rhsE-; CD62E; ELAM-1

Alexis BMS316   Human recombinant produced in an insect cell line   >95%; 170 µg/mL in PBS | Chimera of the CA21 epitope, a part of the L-selectin cytoplasmic domain & truncated E-selectin (lectin domain, EGF domain, & two consensus repeats)

Calbiochem 561300   Human recombinant, expressed in CHO cell line   MW 58.8k   Lyophilized from PBS containing calcium & magnesium with a stabilizer; ≥95% (SDS-PAGE); activity: shown to bind to E-selectin ligands on U937 cells | Recombinant form of E-selectin consisting of 535 AA minus the transmembrane & cytoplasmic domains; transiently expressed on vascular endothelial cell surfaces where it is able to bind to ligands such as sialyl Lewis x found on leukocytes; this initial binding event is followed by a second interaction involving ICAM & VCAM-1 leading to vascular penetration & leukocyte invasion into the extracellular matrix tissue

Selectin, L-

Synonyms: CD62L; LECAM-1; MEL-14

Calbiochem 561303   Human recombinant, expressed in CHO cell line   MW 33k   Lyophilized from PBS containing calcium & magnesium with a stabilizer; ≥95% (SDS-PAGE); activity: shown to bind to L-selectin ligands on LS180 cells | Recombinant form of E-selectin consisting of 294 AA minus the transmembrane & cytoplasmic domains; expressed on leukocytes & acts along with P-selectin & E-selectin in establishing an initial interaction between circulating leukocytes & the endothelium

Selectin, P-

Synonyms: CD62P; GMP-140; LECAM-3; PADGEM

Calbiochem 561306   Human recombinant, expressed in CHO cell line   MW 80k   Lyophilized from PBS containing calcium & magnesium with a stabilizer; ≥95% (SDS-PAGE); activity: shown to bind to P-selectin ligands on U937 cells | Recombinant form of E-selectin consisting of 730 AA minus the transmembrane & cytoplasmic domains; expressed on activated platelets & endothelial cells; a specific interaction between P-selectin & P-selelctin Glycoprotein ligand-1, present on numerous cells types, promotes adhesion & acts along with L-selectin in establishing an initial interaction between circulating leukocytes & the endothelium

Serum Amyloid A

Scipac P127-1   Serum/plasma   >96%; lyophilized | Acute phase protein

Scipac P127-2   Serum/plasma   10-50%; lyophilized | Acute phase protein

Serum Proteins

Sigma S 2396   Human   Lyophilized powder containing ~95% protein (Biuret); balance primarily citrate buffer salts; 1 mL serum yields ~ 45 mg lyophilized proteins; 1-3 moles hexose (as fructosamine)/mole albumin | Glycated *in vitro* & packaged as 'serum equivalents'; Armbruster, DA, *Clin Chem*, 33: 2513, 1987; Furth, AJ, *Anal Biochem*, 175: 347, 1988

Sex Hormone Binding Globulin

Synonyms: Testosterone-Estradiol Binding Globulin

USBio S1012-54   Human pregnancy plasma   ≥98% (SDS-PAGE); purified by affinity chromatography; no contaminants detected; single band by SDS-PAGE, IEP, &/or RID; lyophilized in 0.01 *M* PBS, 0.02% NaN$_3$ | Suitable for antigenic applications in immunological protocols.

Biogenesis 8280-1004   Human serum   MW 85k   Tested negative for HBsAg, HCV antibody and HIV 1 and 2 antibodies; purified; 0.05 *M* Tris HCl, pH 7.4, with 10% glycerol; liquid

Calbiochem 581228   Human serum   MW 94k   Liquid in 10 *mM* CaCl$_2$, 10 *mM* Tris, 50% glycerol, pH 7.4; ≥90% (SDS-PAGE); prepared from serum shown by certified tests to be negative for HBsAg & antibodies to HIV & HCV | Circulating transport glycoprotein originating from the liver & possessing a high affinity binding site for the gonadal steroids; Loukovaara, M et al, *J Clin Endocrinol Metab*, 80: 160, 1995

Scipac P145-1   Serum/plasma   >98%; lyophilized | Acute phase protein

Scipac P145-2   Serum/plasma   10-50%; lyophilized; available on request | Acute phase protein

SHP-1/SHPTP-1, Agarose Conjugated

USBio S1013-26   Recombinant human SHPTP1   Frozen preparation in 50 µg of SHP-1/GST fusion protein bound to glutathione-agarose beads & formulated as a 50% agarose bead slurry in 125 µL of 25 *mM* HEPES, 150 *mM* NaCl, 5 *mM* DTT, and 10% glycerol | Bound to glutathione-agarose beads for the easy removal of enzyme by centrifugation or filtration following dephosphorylation reactions

Single Strand Binding Protein

Amersham E70032Y/E70032Z   Cloned; *E. coli* strain M5248/pKAC27   1-5 mg/mL by A$_{280}$; 4 identical 18,900 DA subunits; >98% pure (PAGE); free of non-specific endonuclease, exonuclease & ribonuclease; 50 mM Tris-HCl, pH 7.5, 200 mM NaCl, 1 mM EDTA, 50% glycerol | Subunits bind with high affinity & cooperatively to ss-DNA; subunits don't bind well to ds-DNA; involved in DNA replication & recombination *in vivo*; DNA binding & recombination have been studied *in vitro*; used to visualize ss-DNA by electron microscopy; used with recA protein for carrying out site-directed mutagenesis & to select specific sequences from libraries of ds-DNA; may also stimulate specific DNA polymerases used in DNA sequencing reactions; in conjunction with appropriate oligonucleotides & restriction endonucleases, has been used to target restriction endonuclease digestion to specific restriction sites in ss-DNA for subsequent mutagenesis

Promega M3011   *E. coli*   MW 75.6k (4 identical 18.9k subunits)   Binds with high affinity in a cooperative manner to ss-DNA but not well to ds-DNA; involved in DNA replication & recombination *in vivo*; Sancer et al, *PNAS*, 78: 4274, 1981; Chase, JW & Williams, KR, *Ann Rev Biochem*, 55: 103, 1986; Krauss, G et al, *Biochemistry*, 20: 5346, 1981; Weiner, JH et al, *JBC*, 250: 1972, 1975

Sigma S 3917   *E. coli*   >95% (SDS-PAGE); solution (1.5 mg/mL) in 20 *mM* Tris-HCl, pH 8.0, 0.5 *M* NaCl, 0.1 *mM* EDTA, 0.1 *mM* DTT, 50% glycerol; DNase, RNase: none detected | Binds with high specificity to ss-DNA; useful in enhancing the specificity of PCR & in enabling the sequencing of problematic DNA templates; Schwarz, K et al, *Nucl Acids Res*, 18: 1079, 1990

Smith Antigen

Biodesign A07303B   Bovine spleen/thymus mixture   Purified | Autoimmune reagent

Biodesign A2A650R   Rabbit thymus   Purified | Autoimmune reagent

Smith/RNP Antigen

Biodesign A07304B   Bovine/Rabbit mixture   Purified | Autoimmune reagent

Biodesign A2A250R   Rabbit thymus   Purified | Autoimmune reagent

Snake Toxin

Sigma T 4307   *Crotalus vergrandis* (Uracoan rattlesnake)

snRNP

Biodesign A08050M   Hela cells   MW 70k   Purified | Autoimmune reagent

snRNP A Protein

Biodesign A08051M   Hela cells   Purified | Autoimmune reagent

snRNP B Protein

Biodesign A08052M   Hela cells   Purified | Autoimmune reagent

Sodium Glucose Transporter I Control Peptide

Chemicon AG661   ≥95% | Purified protein for apoptosis & signal transduction; for use with Chemicon AB1352

SPARC

Synonyms: Osteonectin; BM-40

Sigma S 5174   Mouse parietal yolk sac (PYS-2) cells   MW 43k   ~80% (SDS-PAGE); lyophilized from phosphate buffered saline | Secreted protein acidic & rich in cysteine; calcium binding glycoprotein; binds albumin, collagen & thrombospondin; inhibits spreading of endothelial & smooth muscle cells & fibroblasts; Lane, TF & Sage, HE, *FASEB J*, 8: 163, 1994; Sage, HE, *J Cell Biol*, 109: 341, 1989

sRANK Receptor

PeproTech 310-08   Human recombinant, expressed in *E. coli* MW 19.3k   >98% (SDS-PAGE) & HPLC; <0.1 ng endotoxin per mg (1EU/mg); lyophilized | Receptor activator of NF-KappaB; member of the TNFR family, derived from dendritic cells; soluble protein containing 175 AA residues comprising the full-length of the TNF receptor-like extracellular domain of RANK Receptor; SA determined by its ability to suppress the production of IFN-g from human PBMCs

src SH2 Protein, Agarose Conjugated

Oncogene SH01A   Liquid; negative control: BSA conjugated with biotin or to agarose; positive control: any cell line | Reacts with any phosphoproteins that bind src via the SH2 domain; recombinant SH2 domain of c-src coupled to agarose; species reactivity: broad range of species; for immunoprecipitation, antibody purification & affinity chromatography

src SH2 Protein, Biotin Conjugated

Oncogene SH01B   Liquid; negative control: BSA conjugated with biotin or to agarose; positive control: any cell line | Reacts with any phosphoproteins that bind src via the SH2 domain; conjugated recombinant SH2 domain of c-src; species reactivity: broad range of species; for immunoprecipitation, antibody purification & Western blot

SS-A (RO) Antigen

Biodesign A07300B Bovine Purified	Autoimmune reagent
Biodesign A2A750B Porcine spleen Purified	Autoimmune reagent

SS-A60 (RO) Antigen

Biodesign A43130H   Recombinant, expressed in *E. coli* Purified | Autoimmune reagent

SS-B (La) Antigen

Biodesign A41022H   Bovine/Rabbit   Purified | Autoimmune reagent

Biodesign A2A350R   Rabbit thymus   Purified | Autoimmune reagent

Biodesign A43315H   Recombinant, expressed in *E. coli* Purified | Autoimmune reagent

SSL1/p44

IBT TA-400-1   Human recombinant, expressed in *E. coli*

Stem Cell Factor

Synonyms: c-*kit* Ligand; Mast Cell Growth Factor; c-*kit* Ligand; Mast Cell Growth Factor

Biodesign A52307H   *E. coli*   MW 18.5k   Purified

Chemicon GF021   Human   ≥95%

Calbiochem 569600   Human recombinant, expressed in *E. coli* MW 18.5k   >97% (SDS-PAGE); lyophilized from sterile-filtered PBS containing 50 µg BSA/µg SCF; biological activity: ED_{50}=2.5-5.0 ng/mL as measured in a cell proliferation assay using a factor-dependent human erythroleukemic cell line; endotoxin: ≤100 pg/µg SCF | Hematopoietic growth factor that stimulates the growth of cells of multiple lineage; McNiece, IK et al, *Leuk Lymphoma*, 15: 405, 1994; Huang, E et al, *Cell*, 63: 225, 1990; Martin, FH et al, *Cell*, 63: 203, 1990

Fitzgerald 30-AS60   Human recombinant, expressed in *E. coli*

Harlan BT-3026   Human recombinant, expressed in *E. coli* Lyophilized; 0.01 mg

Harlan BT-3027   Human recombinant, expressed in *E. coli* Lyophilized; 0.05 mg

PeproTech 300-07   Human recombinant, expressed in *E. coli* MW 18.4k   >98% (SDS-PAGE) & HPLC; <0.1 ng endotoxin per mg (1EU/mg); lyophilized | Hematopoietic growth factor; exerts its activity at the early stages of hematopoiesis; stimulates proliferation of myeloid, erythroid, & lymphoid progenitors in bone marrow cultures; 164 AA; $ED_{50} \leq 0.01$ ng/mL; SA $\geq 10^5$ U/mg; SA determined by the dose-dependent stimulation of the proliferation of the human MO7e cells

Sigma S 7901   Human recombinant, expressed in *E. coli*   MW 18.5k   ≥97% (SDS-PAGE); 0.2 µm filtered & lyophilized from phosphate buffered saline containing 500 µg BSA; proliferative activity is tested using TF-1 cells; endotoxin tested | Single chain; Kitamura, T et al, *J Cell Physiol*, 140: 323, 1989; peptide growth factor/cytokine with broad activities, especially hematopoiesis; among SCF's many activities are the ability to act on early hematopoietic progenitor/stem cells & to stimulate the proliferation & survival of mast cells; also one of the most potent stimulators of multilineage progenitors (CFU-GEMM) in both human & murine bone marrow cells; acts synergistically with other growth factors including erythropoietin, G-CSF, M-CSF, GM-CSF, IL-3 & IL-6 to increase the number & size of colonies of hematopoietic progenitors; appears to play an important role in the survival, proliferation or migration of primordial germ cells & melanoblasts during development & maturation stages; Zsebo, KM et al, *Cell*, 63: 195, 1990; nocka, K et al, *EMBO J*, 9: 3287, 1990; Williams, DE et al, *Cell*, 63: 167, 1990; Broxmeyer, HE et al, *Blood*, 77: 2142, 1991; Martin, F et al, *Cell*, 63: 203, 1990; Orr-Urtreger, A et al, *Development*, 109: 911, 1990

Biogenesis 8407-5066   Mouse r-DNA   Liquid

Calbiochem 569610   Mouse recombinant, expressed in *E. coli* MW 18.6k   ≥97% (SDS-PAGE); lyophilized from sterile-filtered PBS containing 50 µg BSA/µg SCF; biological activity: ED_{50}=5.0-10.0 ng/mL as measured in a cell proliferation assay using a factor-dependent human erythroleukemic cell line; endotoxin: ≤100 pg/µg SCF | Hematopoietic growth factor; Morstyn, G et al, *Oncology*, 51: 205, 1994; Huang, E et al, *Cell*, 63: 225, 1990; Martin, FH et al, *Cell*, 63: 203, 1990

Sigma S 9915   Mouse recombinant, expressed in *E. coli*   MW 18.5k   ≥97% (SDS-PAGE); 0.2 µm filtered & lyophilized from phosphate buffered saline containing 250 µg BSA; proliferative activity is tested using TF-1 cells; endotoxin tested | Single chain; Kitamura, T et al, *J Cell Physiol*, 140: 323, 1989; peptide growth factor/cytokine with broad activities, especially hematopoiesis; among SCF's many activities are the ability to act on early hematopoietic progenitor/stem cells & to stimulate the proliferation & survival of mast cells; also one of the most potent stimulators of multilineage progenitors (CFU-GEMM) in both human & murine bone marrow cells; acts synergistically with other growth factors including erythropoietin, G-CSF, M-CSF, GM-CSF, IL-3 & IL-6 to increase the number & size of colonies of hematopoietic progenitors; appears to play an important role in the survival, proliferation or migration of primordial germ cells & melanoblasts during development & maturation stages; Zsebo, KM et al, *Cell*, 63: 195, 1990; nocka, K et al, *EMBO J*, 9: 3287, 1990; Williams, DE et al, *Cell*, 63: 167, 1990; Broxmeyer, HE et al, *Blood*, 77: 2142, 1991; Martin, F et al, *Cell*, 63: 203, 1990; Orr-Urtreger, A et al, *Development*, 109: 911, 1990

Chemicon GF049   Murine   ≥95%

PeproTech 250-03   Murine recombinant, expressed in *E. coli* MW 18.3k   >98% (SDS-PAGE) & HPLC; <0.1 ng endotoxin per mg (1EU/mg); lyophilized with no additives | Hematopoietic growth factor; exerts its activity at the early stages of hematopoiesis; stimulates proliferation of myeloid, erythroid, & lymphoid progenitors in bone marrow cultures; 164 AA; $ED_{50} \leq 20$ ng/mL; SA $\geq 5 \times 10^4$ U/mg; SA determined by the dose-dependent stimulation of the proliferation of the human MO7e cells

Biogenesis 8407-5056   r-DNA   Liquid

Stem Cell Growth Factor

Biodesign A52122H   *E. coli*   Purified

Chemicon GF072   Human   ≥95%

PeproTech 100-22   Human recombinant, expressed in *E. coli* MW 29.0k   >95%; 165 AA; lyophilized from 25 *mM* HEPES, pH 6.5, 0.25 NaCl, 0.25 *mM* DTT; activity determined by a cell proliferation assay using human MO7e cells in the absence of serum

Stem Cell Growth Factor α

PeproTech 100-22A Human recombinant, expressed in *E. coli* MW 33.9k >98% (SDS-PAGE) & HPLC; <0.1 ng endotoxin per mg (1EU/mg); lyophilized | Newly discovered hematopoietic growth factor; exerts its activity at the early stages of hematopoiesis; non-glycosylated species-specific cytokine that can support growth of primitive hematopoietic cells; in combination with EPO or GM-CSF, promotes proliferation of erthroid or myeloid progenitors, respectively; $ED_{50} \leq 7$ ng/mL; SA determined by its inhibitory effect on the proliferation of TF-1 cells previously grown in media containing GM-CSF

PeproTech 100-22B Human recombinant, expressed in *E. coli* MW 29k >98% (SDS-PAGE) & HPLC; <0.1 ng endotoxin per mg (1EU/mg); lyophilized | Newly discovered hematopoietic growth factor; exerts its activity at the early stages of hematopoiesis; non-glycosylated species-specific cytokine that can support growth of primitive hematopoietic cells; in combination with EPO or GM-CSF, promotes proliferation of erthroid or myeloid progenitors, respectively; $ED_{50} \leq 7$ ng/mL; SA determined by its inhibitory effect on the proliferation of TF-1 cells previously grown in media containing GM-CSF

Streptavidin

Fluka 85878 *Streptomyces avidinii* MW 60k ~14 U/mg protein; lyophilized from 10 *mM* potassium phosphate; affinity purified; 1 U corresponds to the amount of protein which binds 1 μg (+)-biotin at pH 7.5; binding capacity: streptavidin binds 1 molecule of biotin/subunit | Comparison of biotin binding & absorbance measurements indicates that ≤5% binding sites are occupied; Fuccillo, DA, *BioTechniques*, 3: 494, 1985; Haeuptle, M-T et al, *JBC*, 258: 305, 1983

Sigma S 4762 *Streptomyces avidinii* Affinity purified; lyophilized powder; essentially salt-free; activity: ~14 U/mg protein (E_{282} at 1%); 1 mole of streptavidin binds 4 moles of biotin; 1 U binds 1.0 μg biotin | Chalet, L & Wolf, F, *Arch Biochem Biophys*, 106: 1, 1964; Green, NM, *Meth Enzymol*, 18A: 418, 1970; Haeuptle, M-T et al, *J Biol Chem*, 258: 305, 1983

Sigma S 0677 *Streptomyces avidinii*; recombinant, expressed in *E. coli* Lyophilized from 0.02 M potassium phosphate buffer, pH 6.5, with no preservative added; 1 mole of streptavidin binds 4 moles of biotin; 1 U binds 1.0 μg biotin | Chalet, L & Wolf, F, *Arch Biochem Biophys*, 106: 1, 1964; Green, NM, *Meth Enzymol*, 18A: 418, 1970; Haeuptle, M-T et al, *J Biol Chem*, 258: 305, 1983

Streptavidin Albumin, Gold Conjugated

Sigma S 4275 *Streptomyces avidinii* 10 nm Colloidal Gold; mean particle size 8-12 nm; monodisperse; streptavidin, (Sigma S 4762) coupled through spacer to albumin-coated colloidal gold for enhanced detection of biotinylated compounds; suspension in ~50% glycerol containing 0.15 M NaCl, 0.01 M BES, pH 7.4, 0.25% BSA & 0.02% sodium azide | Liesi, P et al, *J Histochem Cytochem*, 34: 923, 1986; Bonnard, C et al, *Immunolabeling for Electron Microscopy*, JM Polak, IM Varndell, eds, Elsevier Science Publishers, New York, NY, p 95, 1984

Streptavidin, Agarose CL-4B

Fluka 85881 50% Suspension in 0.01 *M* sodium phosphate buffer, pH 7.2, containing 0.15 M NaCl & 0.02% NaN₃; streptavidin attached to 4% beaded cross-linked agarose via a C_6-spacer (1.2 mg streptavidin/mL packed gel); binding capacity: 4 mg biotinylated rabbit IgG/mL packed gel | Buckie, JW & Cook, MW, *Anal Biochem*, 156: 463, 1986

Streptavidin, Alkaline Phosphatase Conjugated

Sigma S 2890 Lyophilized powder containing ~50% protein (Biuret); balance primarily trehalose with EPPS, phosphate & traces of $MgCl_2$ & zinc chloride; streptavidin activity: 4-8 U/mg protein; alkaline phosphatase activity: 700-1400 DEA U/mg protein; 1 U hydrolyzes 1.0 μmole *p*-nitrophenyl phosphate/min at pH 9.8, 37°C

Sigma S 5795 Lyophilized powder containing ~50% protein (BCA); balance primarily trehalose & sodium citrate; streptavidin coupled to polymerized alkaline phosphatase, from calf intestine mucosa, by thioether linkage; purified by gel filtration chromatography; streptavidin activity: 1-4 U/mg protein; alkaline phosphatase activity: 800-1600 DEA U/mg protein; 1 U hydrolyzes 1.0 μmole *p*-nitrophenyl phosphate/min at pH 9.8, 37°C | For highly sensitive detection of biotin conjugates, at least 10-fold more sensitive than monomeric enzyme conjugate, Sigma S 2890

Streptavidin, FITC Conjugated

Sigma S 3762 Lyophilized powder; essentially salt-free; contains 4-8 moles FITC/mole streptavidin

Streptavidin, Gold Conjugated

Sigma S 1139 10 nm Colloidal Gold labeled; mean particle size 8-12 nm; monodisperse; streptavidin, (Sigma S 4762) adsorbed to colloidal gold for detection of biotinylated compounds; suspension in ~50% glycerol containing 0.15 M NaCl, 0.02% PEG, pH 7.4, 0.01 M phosphate buffer & 0.02% sodium azide; concentration: A_{520} ~5.0 | Liesi, P et al, *J Histochem Cytochem*, 34: 923, 1986

Sigma S 2390 5 nm Colloidal Gold labeled; mean particle size 3.6-6.5 nm; monodisperse; streptavidin, (Sigma S 4762) adsorbed to colloidal gold for detection of biotinylated compounds; suspension in ~50% glycerol containing 0.15 M NaCl, 0.02% PEG, pH 7.4, 0.01 M phosphate buffer & 0.02% sodium azide; concentration: A_{520} ~5.0 | Liesi, P et al, *J Histochem Cytochem*, 34: 923, 1986

Sigma S 6514 20 nm Colloidal Gold labeled; mean particle size 17-23 nm; monodisperse; streptavidin, (Sigma S 4762) adsorbed to colloidal gold for detection of biotinylated compounds; suspension in ~50% glycerol containing 0.15 M NaCl, 0.02% PEG, pH 7.4, 0.01 M phosphate buffer & 0.02% sodium azide; concentration: A_{520} ~5.0 | Liesi, P et al, *J Histochem Cytochem*, 34: 923, 1986

Streptavidin, Isoluminol Conjugated

Sigma S 8532 Lyophilized powder containing ~85% protein (Biuret); balance sodium phosphate buffer salts; prepared by the method of Brockelbank | Suitable as a chemiluminescent detection reagent in avidin/biotin systems; Brockelbank, JL et al, *Ann Clin Biochem*, 21: 284, 1984; Wood, WG & Missler, V, in *Luminescence, Immunoassay & Molecular Applications*, CRC Press, 141, 1990

Streptavidin, MagaBeads™-Immobilized

Cortex CM3450 Uniform magnetizable particles | Used for protein & DNA separation techniques, cell isolation, enzyme immobilization & bacterial capture

Cortex CM3455 Uniform magnetizable particles | Used for protein & DNA separation techniques, cell isolation, enzyme immobilization & bacterial capture

Streptavidin, MagaCell™-Immobilized

Cortex CM5450 Magnetizable cellulose/iron oxide | Large porous surface which offers a high capacity for binding large quantities of biomolecules; vicinal hydroxyl groups are activated by employing surface chemistries including cyanogen bromide, carbodiimidazole, epoxide & periodate

Cortex CM5455 Magnetizable cellulose/iron oxide | Large porous surface which offers a high capacity for binding large quantities of biomolecules; vicinal hydroxyl groups are activated by employing surface chemistries including cyanogen bromide, carbodiimidazole, epoxide & periodate

Streptavidin, Maleimide

Sigma S 9415 *Streptomyces avidinii* Lyophilized powder containing ~90% protein (E$_{280}$ at 1%); balance sodium citrate; streptavidin activated with maleimidocaproic acid *N*-hydroxysuccinimide ester (Sigma M 4650); streptavidin activity: 10-16 U/mg protein; contains 4-8 moles maleimide/mole protein | Suitable for direct conjugation to compounds containing free sulfhydryl groups; Liu, FT et al, *Biochem*, 18(4): 690, 1979; Kitagawa, T et al, *Chem Pharm Bull*, 29(4): 1131, 1981; Duncan, RJS et al, *Anal Biochem*, 132: 68, 1983

Streptavidin, Peroxidase Conjugated

Sigma S 5512 Lyophilized powder containing ~80% protein (E$_{280}$ at 1%); balance citrate buffer salts; purified by affinity chromatography; streptavidin activity: 5-9 U/mg protein; 1 U binds 1.0 μg biotin; peroxidase activity: 80-150 U/mg protein; 1 U forms 1 mg purpurogallin/20 sec from pyrogallol at pH 6.0, 20°C | Labeled with Type VI peroxidase by a modification of the method of O'Sullivan, MJ et al, *FEBS Lett*, 95: 311, 1978; sold on the basis of weight of protein

Sigma S 9420 Lyophilized powder containing ~50% protein (BCA); balance primarily trehalose & sodium citrate; purified by gel filtration chromatography; streptavidin activity: 1-4 U/mg protein; peroxidase activity: 100-200 U/mg protein; 1 U forms 1 mg purpurogallin/20 sec from pyrogallol at pH 6.0, 20°C | Streptavidin coupled to polymerized horseradish peroxidase by thioether linkage; for highly sensitive detection of biotin conjugates; at least 10-fold more sensitive than monomeric enzyme conjugate, Sigma S 5512

Fluka 85876 *Streptomyces avidinii*/ Horseradish ~5 U/mg activity; powder; ≥80 U/mg peroxidase activity; 1 U corresponds to the amount of enzyme which oxidizes 1 μmol ABTS/min at pH 6.0, 25°C; 1 U corresponds to the amount of protein which binds 1 μg (+)-biotin at pH 8.9 | Labeled by a modification of the method of Sullivan, MJ et al, *FEBS Lett*, 95: 311, 1978; purified by affinity chromatography

Streptavidin, Sulforhodamine 101 Acid Chloride (Texas Red) Conjugated

Sigma S 7261 Lyophilized powder; essentially salt-free; purified by affinity chromatography; streptavidin activity: 8-16 U/mg protein (E$_{280}$ at 1%); 1 U binds 1.0 μg biotin; streptavidin: Texas Red molar ratio: ~ 1:2 | Green, NM, *Meth Enzymol*, 18A: 418, 1970

Streptavidin, β-Galactosidase Conjugated

Sigma S 3887 Lyophilized powder containing ≥80% protein (Biuret); balance primarily Tris buffer salts; streptavidin activity: 1-2 U/mg protein; 1 U binds 1 μg biotin; β-galactosidase activity: 300-700 U/mg protein; 1 U hydrolyzes 1.0 μmole *o*-nitrophenyl-β-D-galactoside to *o*-nitrophenol/min at pH 7.3, 37°C | O'Sullivan, M et al, *FEBS Lett*, 95: 311, 1978; Green, NM, *Meth Enzymol*, 18A: 418, 1970; streptavidin & β-Galactosidase Grade VIII are conjugated by a modification of the method of O'Sullivan & purified by gel filtration; sold on basis of weight of active streptavidin

Stromal Cell Derived Factor Iα

Biodesign A52028H	*E. coli*	MW 8k	Purified
Chemicon GF073	Human	≥95%	

PeproTech 300-28A Human recombinant, expressed in *E. coli* MW 8.0k >98% (SDS-PAGE) & HPLC; <0.1 ng endotoxin per mg (1EU/mg); lyophilized | Recently discovered protein belonging to the a-chemokine (C-X-C) family of cytokines; 68 AA; SA determined by its ability to chemoattract human peripheral T cells activated with PHA and IL-2

PeproTech 250-20A Murine recombinant, expressed in *E. coli* MW 7.9k >98% (SDS-PAGE) & HPLC; <0.1 ng endotoxin per mg (1EU/mg); lyophilized | Recently discovered protein belonging to the α-chemokine (C-X-C) family of cytokines; 68 AA; SA determined by its ability to chemoattract human peripheral blood monocytes

Biodesign A52328H	*E. coli*	MW 8.5k	Purified
Chemicon GF074	Human	≥95%	

PeproTech 300-28B Human recombinant, expressed in *E. coli* MW 8.5k >98% (SDS-PAGE) & HPLC; <0.1 ng endotoxin per mg (1EU/mg); lyophilized | Recently discovered protein belonging to the a-chemokine (C-X-C) family of cytokines; 72 AA; SA determined by its ability to chemoattract human peripheral T cells activated with PHA and IL-2

PeproTech 250-20B Murine recombinant, expressed in *E. coli* MW 8.5k >98% (SDS-PAGE) & HPLC; <0.1 ng endotoxin per mg (1EU/mg); lyophilized | Recently discovered protein belonging to the a-chemokine (C-X-C) family of cytokines; 72 AA; SA determined by its ability to chemoattract human peripheral blood monocytes

Stromal Cell Derived Factor Iβ/Pre-B Cell Growth Stimulating Factor

Sigma S 8406 Human recombinant, expressed in *E. coli* Lyophilized from 30% acetonitrile/0.1% TFA containing 0.5 mg BSA; endotoxin tested; cell culture tested

Stromal Cell Derived Factor α

USBio S7975-55	
USBio S7975-60	

Sulodexide

Sigma H 1642 Bovine intestine Sodium salt; anti-clotting activity: 50-70 IU/mg (WHO STD); dermatan sulfate: 20-35%; chondroitin sulfate: 2-7%; remainder is mostly a fast-moving electrophoretic fraction with ~5% slow-moving electrophoretic fraction | Heparin-like substance; not assayed by Sigma

Superfibronectin

Sigma S 5171 Human Solution in 0.05 *M* Tris buffered saline; sterile-filtered | A complex of recombinant human FF III$_1$-C & human plasma fibronectin; resembles *in vivo* matrix form of fibronectin; source material tested for HBsAg & HIV antibody

T4 Gene 32 Protein

Amersham E70029Y/E70029Z Cloned from E. coli strain M5248/pYS6 MW 33.5k >95% pure (SDS-PAGE); free of contaminating, non-specific endonuclease, exonuclease & ribonuclease; 20 mM Tris-HCl, pH 8.0, 100 mM NaCl, 0.15 mM EDTA, 1mM β-MSH, 50% glycerol | ss-DNA binding protein required for T4 DNA replication, recombination & repair; binds co-operatively to ss-DNA & is required in stoichiometric rather than catalytic quantities; also binds ss-RNA (10-10^4 lower affinity than DNA), allowing it to control its own rate of synthesis at the level of translation; widely used in studies of DNA-protein interactions & for marking regions of ss-DNA in cytological preparations viewed by electron microscopy; on a primed ss-DNA template, its addition results in a 5-10 fold increase in the rate of synthesis by T4 DNA polymerase; eliminates pausing when sequencing through regions of ds- & ss-DNA with strong secondary structure

TAG-72 Carcinoma Marker

Synonyms: CA 72-4

Biogenesis 8580-0109 Human fluids Tested negative for HBsAg and HIV and HCV antibodies; contaminants: trace alpha1-acid glycoprotein, CA 15-3, CA 19-9, CA 125, CEA and HAS; purified; SA: 10,500 U/mL; PBS buffer, pH 7.4, 0.1% NaN$_3$; liquid | Useful in immunoassays, standards and controls; Guadagni et al, *Cancer Invest*, 13:227, 1995

Taicatoxin

Alexis 630-031 *Oxyuranus scutellatus scutellatus* MW 52k ≥97% (SDS-PAGE); lyophilized powder; soluble in water; potent neurotoxin | Reversible, selective & voltage-dependent blocker of L-type, voltage-gated Ca^{2+} channels in excitable membranes; Brown, AM et al, *Circul Res*, 61: Suppl. 1, I6, 1987; Possani, LD et al, *Toxicon*, 30: 1343, 1992

Calbiochem 574785 *Oxyuranus scutellatus scutellatus* MW 52k >97% (SDS-PAGE); lyophilized powder; soluble in water; biological activity: shown to block spontaneous or K$^+$ -induced contractions of cardiac cells; harmful; LD$_{50}$ ≤2000 mg/kg | Potent selective & reversible blocker of L-type Ca^{2+} channels in excitable cardiac membranes; also blocks apamin-sensitive after-hyperpolarizing slow tail K$^+$ currents in rat chromaffin cells; Doorty, KB et al, *J Biol Chem*, 272: 19925, 1997; Possani, LD et al, *Toxicon*, 30: 1343, 1992

Taipoxin

Alexis 630-029 *Oxyuranus scutellatus scutellatus* MW 46k ≥97% (SDS-PAGE); lyophilized powder; soluble in water; potent neurotoxin | Extremely potent glycoprotein; blocks irreversibly Ca^{2+}-dependent neuromuscular transmission; binds with high affinity to neuronal pentraxin; Fohlman, J et al, *Eur J Biochem*, 68: 457, 1976; Schlimgen, AK et al, *Neuron*, 14: 519, 1995

Tamm Horsfall Glycoprotein

Synonyms: Uromucoid

Biogenesis 8595-0210 Human normal urine Purified; in an aqueous solution with 0.02% NaN$_3$; liquid | Suitable for use in competitive binding assays and for coating microtiter plates; Tamm & Horsfall, *Proc Soc Exp Biol Med*, 74:108, 1950; *Kidney Int*, 16:279, 1979; *Science*, 236:83, 1987; *Science*, 237:1479, 1987

Biogenesis 8595-0204 Human urine MW 75k Purified; tested negative for HBsAg and HIV antibodies; aqueous solution containing 0.02% NaN$_3$; liquid | Useful in competitive binding assays and for coating to polystyrene plates for solid phase ELISA and RIA

Scipac P135-1 Normal urine >96%; lyophilized | Urine protein

TATA Binding Protein

Synonyms: TFIID

Promega E3081 Human recombinant, expressed in *E. coli* MW 38k General transcription factor involved in the formation of an active complex *in vitro* capable of specifically initiating RNA synthesis by RNA polymerases I, II & III; exhibits sequence-specific DNA binding; used in gel shift assays, footprinting assays & transcriptional activation *in vitro*; Sharp, PA, *Cell*, 68: 819, 1992; Patterson, MG et al, *Science*, 248: 1625, 1990; Gaston, K et al, *Nucl Acids Res*, 20: 3391, 1992; Wiley SR et al, *PNAS*, 89: 5814, 1992

Tau Proteins

Sigma T 7675 Bovine brain Purified by affinity chromatography; >90% (SDS-PAGE); lyophilized powder containing Tris buffer salt, NaCl, EGTA, DTT & sucrose as stabilizer; partially phosphorylated | Tau proteins are composed of several isoforms localized mainly in neuronal axons; stimulators of microtubule polymerization & are found in an abnormal state of phosphorylation in Alzheimer disease; tested as substrate for Protein Kinase C; Mercken, M et al, *J Neurochem*, 58: 548, 1992; Mandelkow, EM & Mandelkow, E, *Trends Biochem Sci*, 18: 480, 1993; Lindwall, G & Cole, RD, *J Biol Chem*, 259: 12241, 1984

T-Cell Attracting Chemokine, Cuteaneous

Synonyms: ALP; Skinkine; Eskine; MILC

PeproTech 250-26 Murine recombinant, expressed in *E. coli* MW 10.9k >98% (SDS-PAGE) & HPLC; <0.1 ng endotoxin per mg (1EU/mg); lyophilized | Predominantly expressed in the skin; selectively attracts skin-associated memory T-lymphocytes; 95 AA; SA determined by its ability to chemoattract CXCR3 transfected HEK/293 cells

T-Cell Growth Factor

Chemicon TG203 High purity

Tenascin

Chemicon CC115 Chicken brain Purified | Extracellular matrix protein

Chemicon CC066 Human Purified | Extracellular matrix protein

Biogenesis 8640-0502 Human tumour cell line Purified; 50 mM phosphate buffer, pH 7.4, 0.4 M NaCl, 0.025% NaN$_3$; liquid | Plays an active role in the development of the CNS and mesenchymal derived organs; present in adult tumor vasculature and has functions in cell adhesion

Testosterone-Estradiol Binding Globulin

Biogenesis 8681-5009 Human serum MW 85k Tested negative for HBsAg, and HCV, HIV-1 and HIV-2 antibodies; affinity purified; 0.05 M Tris HCl, pH 7.4 with 10% glycerol; liquid

Tetanus Toxin C-Fragment

Calbiochem 582235 *Clostridium tetani* MW 47k Single major band purity (SDS-PAGE); lyophilized solid from 10 mM sodium phosphate buffer, pH 7.5; soluble in water; may contain trace amounts of intact toxin; LD$_{50}$≤2000 mg/kg | C-terminal binding portion of the heavy chain of tetanus toxin; the intact toxin is known to block neurotransmitter release at inhibitory synapses in cultured spinal cord cells; shows trans-synaptic retrograde transport in the central nervous system; Helting, TB & Zwisler, O, *J Biol Chem*, 252: 187, 1977; Poulain, B, *Pathol Biol*, 42: 173, 1994; Williamson, LC et al, *Soc Neurosci*, 19: Abstract 770.10, 1993; Evinger, C & Erichsen, JT, *Brain Res*, 380: 383, 1986

Tetanus Toxin C-Fragment, FITC Conjugated

Calbiochem 582239 Lyophilized solid from 10 mM sodium phosphate buffer, pH 7.5; contains ~5 µg FITC bound/mg of C-fragment & shows no measurable toxicity; FITC binding ratio: >5 µg FITC/mg C-fragment | Robbins, N & Polak, J, *J Neurocytol*, 17: 545, 1988; Wood, BT et al, *J Immunol*, 95: 225, 1965

Tetanus Toxin C-Fragment, Horseradish Peroxidase Conjugated

Calbiochem 582241 Lyophilized solid from 10 mM sodium phosphate buffer, pH 7.5; soluble in water; non-toxic as determined by LD$_{50}$ assay in mice | Fishman, PS & Savitt, JM, *Exp Neurol*, 106: 197, 1989; Avrameas, S & Ternynck, T, *Immunochemistry*, 8: 1175, 1971

Tetanus Toxoid

Calbiochem 582231 *Clostridium tetani* Lyophilized solid from 10 mM sodium phosphate buffer, pH 7.5; soluble in water; non-toxic as determined by LD$_{50}$ assay in mice | Prepared by formaldehyde inactivation of tetanus toxin

TFN-α Cytokine

Alexis BMS311 Murine recombinant, expressed in *E. coli* MW 17.3k >98% (SDS-gel electrophoresis, before addition of human serum albumin); solution, 3.4 mg/mL in 50 mM NaH$_2$PO$_4$, 0.4 M NaCl, pH 7.0; bioactivity: 3x10^7 U/mg (cytotoxicity on LM-cells)

Thapsigargin

USBio T3700 >99%, by HPLC; lyophilized | Potent cell permeable IP3- independent intracellular calcium releaser; stimulates arachidonic acid metabolism in macrophages; inhibits microsomal Ca^{2+}-ATPase, IC$_{50}$: 2- 20 nM; mouse skin tumor promoter with potency somewhat weaker than teleocidin or PMA, but does not bind to proteinkinase C or induce ornithine decarboxylase activity

Thaumatin

Sigma T 7638 *Thaumatococcus daniellii* A mixture of Thaumatin I & Thaumatin II with traces of other sweet proteins | ~10,000 times sweeter than sucrose on a molar basis; *Eur J Biochem*, 31: 22, 1972

Thioredoxin

Fluka 89032 *Spirulina* species   Powder; ≥60% (GE); ≥60% protein content

Thrombin

Biogenesis 8810-0506   Bovine serum   Lyophilized

Biogenesis 8810-1006   Human plasma   Tested negative for all communicable diseases, including HIV-1, HIV-2, HBsAg and HCV; SA: 3265 NIH U/mg, 11656 NIH U/mL; 50 mM sodium citrate/0.2 M NaCl/0.1% PEG-8000, pH 6.5; liquid | Activated from homogenous thrombin with factor Xa, factor Va and phospholipid (all removed after activation)

Biogenesis 8810-1006-serum   Human plasma   Tested negative for all communicable diseases, including HIV-1, HIV-2, HBsAg and HCV; SA: 3261 NIH U/mg, 6392 NIH U/mL; 50 mM sodium citrate/0.2 M NaCl/0.1% PEG-8000, pH 6.5; liquid | Activated from homogenous thrombin with factor Xa, factor Va and phospholipid (all removed after activation)

Thromboglobulin, β-

Biogenesis 8830-0502   Human platelets   Lyophilized

ICN 153511   Human platelets   ≥95% (silver stained SDS-PAGE); 30% (v/w) acetonitrile, 0.1% trifluoroacetic acid | Moore, S etal, *BBA*, 379:360, 1975; Begg, GS etal, *Biochem*, 17:1739, 1978

ICN 194939   Human platelets   In 25 mM HEPES, 150 mM NaCl, pH 7.4 | Senior, RM etal, *J Cell Bio*, 96:382, 1983

ICN 194933   Rabbit lung   In 0.02% polidocanol, NaN₃ | Membrane glycoprotein; protein C cofactor in the anticoagulant pathway; Esmon, CT etal, *PNAS*, 78:2249, 1981

Thrombomodulin

Cortex CP1083   Human   >95%

Cortex CP4001U   Rabbit   >98%

Thrombomodulin Fragment (EGF 4-5-6)

Synonyms: M388L

Alexis 201-001   Human recombinant, from yeast   MW ~30k ≥95% (reversed-phase HPLC & *N*-terminal sequencing); purified by reversed-phase HPLC; 10 μg protein by AA analysis; lyophilized from Tris-buffered saline containing NaCl, pH 7.4; biological activity: Specific activity for activation of protein C is 1.1x10⁶ U/mg (1 U activates 1 nmole of activated protein C/minute at 25ºC); contains high mannose sugars at two *N*-linked glycosylation sites | Contains only the last 3 EGF-like domains from the extracellular domain; the mutation of M³⁸⁸ to L³⁸⁸ has been shown to increase the activity of this fragment by approx. a factor of two; fragment corresponds to AA E³⁴⁶-K⁴⁶⁶ of human thrombomodulin with methionine³⁸⁸ mutated to leucine; also active as an inhibitor of fibrinogen clotting; thrombomodulin is an endothelial cell surface protein that forms a 1:1 complex with thrombin; the resulting complex is inhibited for fibrinogen cleavage & is capable of activating protein C; full-length thrombomodulin is 70 kDa & contains a large extracellular domain, six EGF-like domains, a sulfated region, a transmembrane domain, & a short intracellular domain; Esmon, NL et al, *Thrombomodulin: Progr Hemostasis Thrombosis*, 9: 29, 1989; Esmon, CT, *J Biol Chem*, 264: 4743, 1989; Tsiang, M et al, *J Biol Chem*, 267: 6164, 1992; Glaser, CB et al, *J Clin Invest*, 90: 2565, 1992; Parkinson, JF et al, *BBRC*, 185: 567, 1992; Nagashima, M et al, *J Biol Chem*, 268: 2888, 1993

Thromboplastin

ICN 154162   Bovine lung extract   >12,000 US U/mg protein, 1 U converts 1 U prothrombin to thrombin in the presence of calcium; <3% moisture

Thrombopoietin

Biodesign A52018H   *E. coli*   MW 18.6k   Purified

Chemicon GF037   Human   ≥95%

BioSource International PHC1144   Human recombinant

Cortex CP9131r   Human recombinant   >95%

ICN 195745   Human recombinant, expressed in *Sf*21   ≥97%; lyophilized; ED_{50} = 1-3 ng/mL

Calbiochem 605218   Human recombinant, expressed in *Spodoptera frugiperda*   MW 35k   >97% (SDS-PAGE); lyophilized from sterile-filtered PBS containing 50 μg BSA/μg TPO; biological activity: ED_{50}=1.0-3.0 ng/mL as measured in cell proliferation assay using MO7e cells; endotoxin: ≤100 pg/μg TPO | Glycopeptide hormone that is a key regulator of megakaryocytopoiesis & thrombopoiesis *in vitro* & *in vivo*; ligand for the receptor encoded by the c-Mpl proto-oncogene; promotes maturation of megakaryocytes & increases platelet size & number; acts as both a proliferative & a maturation factor of megakaryocytes; *Merck Index*, 12: 9528; Foster, D et al, *PNAS*, 91: 13023, 1994; Lok, S & Foster, D, *Stem Cells*, 12: 586, 1994; McDonald, TP et al, *Am J Pediatr Hematol Oncol*, 14: 8, 1992; de Sauvage, FJ et al, *Nature*, 369: 533, 1994

ICN 195771   Mouse recombinant, expressed in NSO   ≥97%; lyophilized; ED_{50} = 0.2-0.6 ng/mL

Thrombopoietin, Mpl Ligand

Synonyms: MGDF

PeproTech 300-18   Human recombinant, expressed in *E. coli* MW 16.9k   >98% (SDS-PAGE) & HPLC; <0.1 ng endotoxin per mg (1EU/mg); lyophilized | Stimulates proliferation & maturation of megakaryocytes; promotes increased circulation levels of platelets *in vivo*; 158 AA residues comprise the receptor binding domain of the Mpl-ligand protein; $ED_{50} \leq 1.0$ ng/mL; SA $\geq 10^6$ U/mg; SA determined by the dose-dependent stimulation of the proliferation of human MO7e cells

Thrombospondin

Cortex CP3099U   Human   >98%

Biogenesis 8835-0056   Human platelets   Purified; PBS buffer with 0.6 M NaCl and 1 mM Ca2+; liquid

Calbiochem 605225   Human platelets   Lyophilized from 600 mM NaCl, 20 mM Tris-HCl, 1 mM CaCl₂, 20% sucrose, pH 8.0; ≥95% (SDS-PAGE); soluble in water; prepared from platelets shown to be negative for HBsAg & for antibodies to HIV & HCV | Adhesive Glycoprotein released in response to platelet activation by α-thrombin; inhibits FGF-induced neovascularization of the rat cornea; inducer of platelet aggregation; synthesized by a number of different cells & secreted into the extracellular matrix; involved in the regulation of cellular proliferation; Michaund, M & Poyet, O, *Anticancer Res*, 14: 1127, 1994; Koch, AE et al, *Pathobiology*, 61: 1, 1993; Castle, VP et al, *J Biol Chem*, 268: 2899, 1993; Tolsma, SS et al, *J Cell Biol*, 122: 497, 1993; Tuszynski, GP et al, *J Cell Biol*, 120: 513, 1993; Schon, P et al, *Eur J Cell Biol*, 59: 329, 1992; Good, DJ et al, *PNAS*, 87: 6624, 1990; *Merck Index*, 12: 9529

ICN 194085   Human platelets   Purified protein | Platelet protein found in alpha granules; binds to platelet membranes in a calcium-dependent mechanism; Agbanyo, FR & EF Plow, *Thromb Haemostas*, 69:563, 1993

ICN 194934   Human platelets   In 50% glycerol, Tris buffer | Heparin binding glycoprotein responsible for platelet aggregation & adherence; functions in cell-matrix interactions; Lawler, JW etal, *JBC*, 253:8609, 1978

Sigma T 7043   Human platelets   20 μg protein/vial; lyophilized from a solution containing sucrose, NaCl & Tris buffer salts | For cell culture use

Thymidine Phosphorylase

Oncogene PF081   Human recombinant   MW 49k   >97% (SDS-PAGE); lyophilized; biological activity: EC₅₀ of 20-40 ng/mL as determined by the ability to stimulate ³H-thymidine incorporation in human umbilical vein endothelial cells | Recombinant human protein based on a DNA sequence encoding AA 11-482 of PD-ECGF expressed in Sf21 cells; species reactivity: human; for proliferation studies

Thymus and Activation Regulated Chemokine

Biodesign A52330H   *E. coli*   MW 8k   Purified

Chemicon GF075   Human   ≥95%

BioSource International PHC1264   Human recombinant

PeproTech 300-30 Human recombinant, expressed in *E. coli*
MW 8.0k >98% (SDS-PAGE) & HPLC; <0.1 ng endotoxin per mg
(1EU/mg); lyophilized | Recently discovered protein belonging to
the b-chemokine (CC) family of cytokines; 71 AA; SA determined
by its ability to chemoattract human T cells

Thymus Expressed Chemokine

Biodesign A52045H *E. coli* Purified

Chemicon GF087 Human ≥95% | Purified protein for
apoptosis & signal transduction

BioSource International PHC1625 Human recombinant

PeproTech 300-45 Human recombinant, expressed in *E. coli*
MW 14.2k >98% (SDS-PAGE) & HPLC; <0.1 ng endotoxin per
mg (1EU/mg); lyophilized | Novel CC chemokine identified in the
thymus of a mouse & human; chemotactic activity for activated
macrophages, dendritic cells & thymocytes; 127 AA; SA determined
by its ability to chemoattract human monocytes

Thyroglobulin

Dako X0553 >95% | Antigen useful as an immunogen

Sigma T 1001 Bovine Iodine: ~1%; nitrogen: ~14.5%; ash:
≥4%; electrophoretically heterogeneous

Biogenesis 8900-1354 Bovine thyroid glands 95%; no
preservatives; lyophilized

Biogenesis 8900-1009 Human thyroid MW 600-900k
<0.4% IgG, <0.3% IgA, 0.3% IgM; tested negative for HCV and
HIV-1 and 2 antibodies and HBsAg; purified; 0.01M PO4 pH 7.8;
liquid

Fitzgerald 30-AT02 Human thyroid High purity >98%

Biodesign A86852H Human thyroid glands 98%

Biogenesis 8900-1004 Human thyroid glands MW 330k
subunit tested negative for HIV I and II antibodies, HBsAg, and
HCV antibodies; purified; 0.02 M NH4HCO3, may contain traces of
buffer salts; lyophilized

Biogenesis 8900-1039 Human thyroid glands MW 600-900k
0.4% IgG, 0.3% IgM, <0.3% IgA; tested negative for HBsAg and
HTLV III antibody; semi-pure; 2.25 mL 0.02 M PO4, pH 7.4 with
42% ammonium sulfate; suspension

USBio T5300-10 Human thyroid glands MW ~600k
≥98%; no contaminants detected; single band by SDS-PAGE;
lyophilized from 0.02 M NH4HCO3 | Suitable for antigenic
applications in immunological protocols; thyroglobulin levels are
indicated in the diagnostic analysis for thyroid carcinoma & Graves
disease; can be used to measure thyroid uptake levels

Sigma T 1126 Porcine Iodine: ~1%; nitrogen: ~14.5%; ash:
≥4%; electrophoretically heterogeneous

Scipac P128-0 Thyroid tissue >98%; lyophilized; protein A
treated (IgG low); 10 mg min pack size | Autoimmunity protein;
thyroid function protein

Scipac P128-1 Thyroid tissue >96%; lyophilized; 10 mg min
pack size | Autoimmunity protein; thyroid function protein

Thyroid Microsomal Antigen

Cortex CP1031 Crude Grade

USBio T5350 Human ≥50% (SDS-PAGE); 1.87 mg/mL
supplied in 0.05 M phosphate buffer, 0.15 M NaCl, pH 7.5, 0.1%
NaN3 | Suitable for antigenic applications in immunological
protocols

Scipac P137-0 Thyroid tissue 0.3-3 mg/mL; frozen in
phosphate buffer; isolated by chromatography; Triton X-100
treated | Thyroid function protein

Scipac P137-5 Thyroid tissue 1-5 mg/mL; frozen in
phosphate buffer; produced by differential centrifugation method;
solubilized by dispersion and ultrasonication | Thyroid function
protein

Scipac P137-6 Thyroid tissue 1-5 mg/mL; frozen in
phosphate buffer; protein A treated; produced by differential
centrifugation method; solubilized by dispersion and ultrasonication
| Thyroid function protein

Thyroid Stimulating Hormone

Synonyms: Thyrotropic Hormone; Thyrotropic Hormone;
Thyrotropic Hormone; Thyrotropic Hormone; Thyrotropic Hormone

Cortex CP9026U Bovine >98%

Biogenesis 8921-1004 Bovine pituitary Purified; lyophilized

Biogenesis 8921-1004-1mg Bovine pituitary Purified;
lyophilized

Biogenesis 8921-1204 Bovine pituitary MW 28k Semi-
pure; lyophilized

Calbiochem 609385 Bovine pituitary MW 28k Lyophilized
solid; potency: ≥0.7 IU/mg as measured against the WHO standard
or equivalent; 1 IU is the activity in 20 mg of the labeled
preparation of the USP reference substance; soluble in aqueous
buffers & water; harmful: LD50≤2000 mg/kg; may be
carcinogenic/teratogenic | Glycopeptide hormone that exerts mild,
continuous stimulation on the thyroid, resulting in maintenance of
activity; its secretion is inhibited by somatostatin; *Merck Index*, 12:
9931

Sigma T 8931 Bovine pituitary 10 IU/vial; activity: 2 IU/mg
protein; contains phosphate buffer salts; not assayed by Sigma |
2 chain glycoprotein hormone, the α-chain not active, biological
specificity attributed to the β-chain; activates adenylate cyclase in
the thyroid gland, thus stimulating iodine uptake, thyroxine
synthesis & release; goitrogenic

Fitzgerald 30-AT14 Bovine pituitary glands Standard grade

Biogenesis 8921-5157 Canine pituitary Contains trace
buffer salts; *J Clin Endocrin Metab* 31, 331 (1970); semi-pure; 50
mM NH4HCO3, pH 8.0; lyophilized

Cortex CP1026 Human >95%

Cortex CP1026P Human >40%

Biogenesis 8920-0989 Human pituitary 250 mIU/vial; 190
IU/vial hLH, 150 IU/vial hFSHhFSH; tested negative for antibodies
to HCV, HIV-1 and HIV-2 and HBsAg; semi-pure; 0.05 M NH4HCO3,
pH 8.0; lyophilized

Biogenesis 8920-1004 Human pituitary MW 28k 1.8%
hLH, <1.0% hFSH; tested negative for HBsAg, HCV and HIV-1 and
2; purified; 0.05 M NH4HCO3, pH 8.0; lyophilized | *Arch Int
Physiol Biochim*, 85:905, 1977

Biogenesis 8920-1204 Human pituitary Semi-pure;
lyophilized

Biogenesis 8920-1404 Human pituitary affinity purified

Calbiochem 869006 Human pituitary MW 25k Iodination
grade; lyophilized solid; immunopotency: ≥6 IU/mg (WHO 1st IRP
68/38); hLH: ≤0.3%; hGH, hFSH: ≤0.2%; hPRL: ≤0.1%; soluble in
iodination buffer; prepared from tissue shown by certified tests to
be negative for HBsAg & antibodies to HIV & HCV; harmful:
LD50≤2000 mg/kg; may be carcinogenic/teratogenic | *Merck
Index*, 12: 9931

Sigma T 9265 Human pituitary Lyophilized powder
containing hormone & salts from ~0.1 mL 0.05 M phosphate buffer,
pH 7.4; activity: ~7 IU/mg hormone by RIA; not assayed by Sigma
| Sold on the basis of mg hormone; 2 chain glycoprotein hormone,
the α-chain not active, biological specificity attributed to the β-
chain; activates adenylate cyclase in the thyroid gland, thus
stimulating iodine uptake, thyroxine synthesis & release;
goitrogenic

Biodesign A81159M Human pituitary glands 99%

Sigma T 4533 Human recombinant, expressed in CHO cells
Activity: ≥4 IU/mg protein; not assayed by Sigma | 2 chain
glycoprotein hormone, the α-chain not active, biological specificity
attributed to the β-chain; activates adenylate cyclase in the thyroid
gland, thus stimulating iodine uptake, thyroxine synthesis &
release; goitrogenic

Biogenesis 8923-1004 Porcine pituitary MW 32k <0.1%
pLH, <0.05% pFSH; virtually salt free; purified; 0.05M NH4HCO3,
pH 8.0; lyophilized

Sigma T 8785 Porcine pituitary 10 IU/vial; activity: 5 IU/mg
protein; not assayed by Sigma | 2 chain glycoprotein hormone,
the α-chain not active, biological specificity attributed to the β-
chain; activates adenylate cyclase in the thyroid gland, thus
stimulating iodine uptake, thyroxine synthesis & release;
goitrogenic

Biogenesis 8924-1954   Rat pituitary glands   rFSH/rLH <1.0%, rGH/rPRL <0.1%; purified; from 0.05 M NH$_4$HCO$_3$, pH 8.0; lyophilized

Biogenesis 8924-1954-50µg   Rat pituitary glands   <1.0% rFSH/rLH, <0.1% rGH/rPRL; purified; from 0.05 M NH$_4$HCO$_3$, pH 8.0; lyophilized

Thyroid Stimulating Hormone, Intact
Synonyms: Thyrotropic Hormone

USBio T5400-09   Bovine   SA ≥25 U/mg; lyophilized; FSH/GH ≤0.10%, LH ≤0.25%, Prl ≤0.01% | Suitable for antigenic applications in immunological protocols

USBio T5400-09A   Bovine   SA ≥2 U/mg; lyophilized; bovine FSH/GH ≥0.10%, LH ≥0.25%, Prl ≥0.01% | Suitable for antigenic applications in immunological protocols

Fitzgerald 30-AT11   Bovine pituitary gland   High purity

USBio T5400-14   Human   ~60% (SDS-PAGE); ~4.0 IU/mg; lyophilized; hLH ≤1.4%, hGH ≤0.1%, hFSH ≤0.2%, hPrl/hCG ≤0.1% | Suitable for antigenic applications in immunological protocols

USBio T5400-14A   Human   ~60% (SDS-PAGE); ~2 IU/mg; lyophilized from 50 mM ammonium bicarbonate; hLH ≤0.6%, hFSH ≤0.4%, hPrl ≤0.01%, hGH ≤0.1% | Suitable for antigenic applications in immunological protocols

Fitzgerald 30-AT05   Human pituitary gland   Affinity purity

Fitzgerald 30-AT10   Human pituitary gland   High purity

Fitzgerald 30-AT15   Human pituitary gland   Standard purity

USBio T5400-15   Human pituitary glands   ≥98% (SDS-PAGE); 8.0 IU/mg (2nd IRP 80/558, WHO); lyophilized; hLH <2.5%, hFSH <0.2%, hPrl/hCG <0.1% | Suitable for antigenic applications in immunological protocols

USBio T5400-07   Human recombinant prepared from mammalian cell culture   ≥98% (SDS-PAGE); ~4.0 IU/mg (WHO); lyophilized; hLH/hGH/hFSH ≤ 0.1%, hPrl ≤ 0.1%, hCG ≤ 0.1% | Suitable for antigenic applications in immunological protocols

Fitzgerald 30-AT09   Recombinant, expressed in mammalian cell culture

Thyroid Stimulating Hormone, α-
Synonyms: Thyrotropic Hormone; Thyrotropic Hormone; Thyrotropic Hormone

USBio T5400-20   Human   ≥98%; no contaminants detected; single band by SDS-PAGE, IEP, &/or RID; lyophilized; human TSH β <1.00%, hFSH/hLH/h GH/hPrl <0% | Suitable for antigenic applications in immunological protocols

Biogenesis 8925-1004   Human pituitary   Purified; lyophilized

Fitzgerald 30-AT20   Human pituitary gland   High purity

Biogenesis 8923-1054   Porcine pituitary   Purified; lyophilized

USBio T5400-25   Bovine   ≥98% (SDS-PAGE); ≥ 30 IU/mg; lyophilized | Suitable for antigenic applications in immunological protocols

Fitzgerald 30-AT26   Bovine pituitary gland   High purity

USBio T5400-24   Human   ≥98% (SDS-PAGE); lyophilized; hTSHa: <1.00%, hFSH/hLH/hGH/hPrl <0% | Suitable for antigenic applications in immunological protocols

Biogenesis 8926-1004   Human pituitary   Purified; lyophilized

Fitzgerald 30-AT25   Human pituitary gland   High purity

Biogenesis 8923-1104   Porcine pituitary   Purified; lyophilized

Thyroxine Binding Globulin

Sigma T 2022   Human   MW 54k   Solution in 0.05 M Tris, pH 8.6, containing 0.25 M NaCl, 0.3 M glycine & 0.1% sodium azide; T$_4$ content: <0.1 mole T$_4$/mole protein; T$_4$ binding capacity: ≥0.7 mole T$_4$/mole protein; protein determined by Lowry | A glycoprotein produced in the liver that is the primary carrier of thyroxine & triiodothyronine in serum; Janssen, OE & Refetoff, S, *J Biol Chem*, 267: 13998, 1992

USBio T5461-15   Human   MW 60k   ≥98%; single band by SDS-PAGE; 4.0 mg/mL (RID, Lowry) supplied in 0.01 M PBS, pH 7.5, 0.1 % NaN$_3$ | Suitable for antigenic applications in immunological protocols

ICN 153980   Human plasma   MW 58k   >95%; 150 mM NaCl, 0.1% NaN$_3$, pH 8.5

Biogenesis 8970-1004   Human serum   Lyophilized

Biogenesis 8970-1054   Human serum   Lyophilized

Calbiochem 612075   Human serum   MW 54-64k   Sterile-filtered liquid in 140 mM NaCl, 10 mM Tris, 0.1% sodium azide, pH 8.0; ≥99% (SDS-PAGE); molar T$_4$: <10% of the molar TBG; prepared from serum shown by certified tests to be negative for HBsAg & antibodies to HIV & HCV | Liver glycoprotein & a major thyroid hormone carrier in serum; exhibits high sequence homology with α$_1$-antitrypsin; affinity for the hormone is thought to be temperature-sensitive; Miura, Y et al, *Endocrinol J*, 40: 127, 1993; Janssen, OE et al, *J Biol Chem*, 267: 13998, 1992

Scipac P125-1   Pooled serum/plasma   >98 | Thyroid function protein

Scipac P125-3   Pooled serum/plasma   >96%; frozen in TRIS buffer | Thyroid function protein

Scipac P125-4   Pooled serum/plasma   40-90%; frozen in sodium phosphate buffer; suitable for bulk requirements | Thyroid function protein

Biogenesis 8970-3009   Rat serum   Affinity isolated; liquid

Scipac P125-2   Serum/plasma   40-90%; lyophilized; suitable for bulk requirements | Thyroid function protein

Thyroxine Binding Globulin, Azide Free

Scipac P125-0   Pooled serum/plasma   >98%; frozen in TRIS-HCl | Thyroid function protein

Thyroxine, Sodium Salt
Synonyms: T4

USBio T5460-17   ≥99% by HPLC; lyophilized | Suitable for antigenic applications in immunological protocols

Tissue Factor
Synonyms: CD xxx

Biogenesis 9010-6006   Human r-DNA   MW 44k   >95% (SDS-PAGE); promotes clotting in a 2-stage prothrombin time test after relipidation; 10 mM TRIS/HCl, pH 8.0, 150 mM NaCl, 0.01% CHAPS & 200 mM mannitol; lyophilized | Nemerson et al, *J Clin Invest*, 48:322, 1969; Carson et al, *Science*, 208:307, 1980

Tissue Inhibitor of Metalloproteinase I
Synonyms: TIMP-1; TIMP-1

Amersham VPF020   Bovine recombinant   Bovine cross-reactivity | Growth/death factor interactions

Calbiochem PF020   Bovine recombinant   Lyophilized solid containing 100 μg BSA; >98% (SDS-PAGE); manufactured by Fuji Chemical Industries, Ltd; not available for sale in Japan | Glycoprotein that is expressed in a variety of cell types; forms a non-covalent stoichiometric complex with latent & active MMPs; preferentially binds & inhibits MMP-9; alters the metastatic potential of cancer cells & inhibits invasion & metastasis in animal models; for use in SDS-PAGE as a Western blot standard & in competition studies; Johnson, MD et al, *Proc Am Assoc Cancer Res*, 32: 81, 1991; Freudenstein, J et al, *Biochem Biophys Res Comm*, 171: 250, 1990; Liotta, LA & Stetler-Stevenson, WG, *Seminars in Cancer Biology*, (Gottesman, MM, ed), 1(2): 99, 1990

Oncogene PF020   Bovine recombinant   >98% (SDS-PAGE); lyophilized with BSA; purified from transfected CHO cells; IC$_{50}$ against 6.7 x 10^{-9} M; native human MMP-1 is 1-5 x 10^{-9} M | Species reactivity: bovine; for Western blot; CHO cell derived recombinant protein; migrates as a 24 k protein under reducing conditions in SDS-PAGE; manufactured by Fuji Chemical Industries; not available for sale in Japan

Calbiochem 612080   Human neutrophil granulocyte   MW 28k   Liquid in 200 *mM* NaCl, 50 *mM* Tris-HCl, 5 *mM* CaCl$_2$, 1 μ*M* ZnCl$_2$, 0.05% BRIJ 35, 0.05% NaN$_3$, pH 7.0; ≥90% (SDS-PAGE); from stimulated neutrophils shown by certified tests to be negative for HBsAg & antibodies to HIV & HCV | Member of a family of inhibitors that participate in the activation & regulation of MMP activity; forms a non-covalent stoichiometric complex with latent & active MMPs; binds to pro-MMP-9 & MMP-9 via their C-terminal domains; Kolkenbrock, H et al, *Biol Chem*, 377: 529, 1996; Kolkenbrock, H et al, *Biol Chem*, 376: 495, 1995

Biogenesis 9013-1559   Human recombinant   Purified; acetate buffer; liquid | Inhibits all active forms of the MMP family

Calbiochem PF019   Human recombinant   Lyophilized solid containing 100 μg BSA; >98% (SDS-PAGE); manufactured by Fuji Chemical Industries, Ltd; not available for sale in Japan | For use in SDS-PAGE as a Western blot standard & in competition studies; Johnson, MD et al, *Proc Am Assoc Cancer Res*, 32: 81, 1991; Liotta, LA & Stetler-Stevenson, WG, *Seminars in Cancer Biology*, (Gottesman, MM, ed), 1(2): 99, 1990

Oncogene PF019   Human recombinant   MW 24k   >98% (SDS-PAGE); lyophilized with BSA; purified from transfected CHO cells; IC$_{50}$ against 6.7 x 10^{-9} *M*; native human MMP-1 is 1-5 x 10^{-9} *M* | Species reactivity: human; for Western blot; manufactured by Fuji Chemical Industries; not available for sale in Japan

Amersham VPF019   Human recombinant,   Human cross-reactivity | Growth/death factor interactions

Tissue Inhibitor of Metalloproteinase II

Synonyms: TIMP-2; TIMP-2

Biogenesis 9013-2559   Human r-DNA   Purified; 25 *mM* Sodium cacodylate, 1 *M* NaCl, 10 *mM* CaCl2, 0.02% NaN3, 0.05% brij, pH 7.5; liquid

Amersham VPF021   Human recombinant   Human cross-reactivity | Growth/death factor interactions

Calbiochem PF021   Human recombinant   Lyophilized solid containing 100 μg BSA; >98% (SDS-PAGE); manufactured by Fuji Chemical Industries, Ltd; not available for sale in Japan | Glycoprotein that is expressed in a variety of cell types; forms a non-covalent stoichiometric complex with latent & active MMPs; preferentially binds & inhibits MMP-2; alters the metastatic potential of cancer cells & inhibits invasion & metastasis in animal models; for use in SDS-PAGE as a Western blot standard & in competition studies; Johnson, MD et al, *Proc Am Assoc Cancer Res*, 32: 81, 1991; Boone, TC et al, *PNAS*, 87: 2800, 1990; Liotta, LA & Stetler-Stevenson, WG, *Seminars in Cancer Biology*, (Gottesman, MM, ed), 1(2): 99, 1990

Oncogene PF021   Human recombinant   MW 24k (reducing conditions)   >98% (SDS-PAGE); lyophilized with BSA; purified from transfected CHO cells | Species reactivity: human; for Western blot; CHO cell derived recombinant protein; migrates as a 24 k protein under reducing conditions in SDS-PAGE; manufactured by Fuji Chemical Industries; not available for sale in Japan

Calbiochem 612084   Human rheumatoid synovial fibroblast   MW 24k   Liquid in 200 *mM* NaCl, 50 *mM* Tris-HCl, 5 *mM* CaCl$_2$, 1 μ*M* ZnCl$_2$, 0.05% BRIJ 35, 0.05% NaN$_3$, pH 7.0; ≥90% (SDS-PAGE); from culture medium of human rheumatoid synovial fibroblasts shown by certified tests to be negative for HBsAg & antibodies to HIV & HCV | Member of a family of inhibitors that participate in the activation & regulation of MMP activity; forms a non-covalent stoichiometric complex with latent & active MMPs; inhibits the activities of MMP-1, MMP-2, MMP-12 & transin; Greene, J et al, *J Biol Chem*, 271: 30375, 1996; Miyazaki, K et al, *J Biol Chem*, 268: 14387, 1993

Tissue Necrosis Factor Receptor I

Kamiya   Human   >95% (SDS-PAGE)

BioSource International PHR3015   Human recombinant

Tissue Necrosis Factor Receptor II

Kamiya   Human   >95% (SDS-PAGE)

BioSource International PHR3025   Human recombinant

Tissue Necrosis Factor Related Activation Inducing Cytokine

Synonyms: TRANCE; RANK Ligand; TNFSF11

R&D Systems 462-TR-010   NSO-Expressed   >95%; lyophilized; ED$_{50}$: 5-15 ng/mL | Species specificity: mouse TRANCE; a member of the TNF family & involved in regulating the function of dendritic cells & osteoclasts; RANK is the cell surface signaling receptor for TRANCE; osteoprotegerin also binds TRANCE & serves as a decoy receptor that counterbalances the effects of TRANCE

Tissue Necrosis Factor Related Apoptosis Inducing Ligand

Synonyms: ApoII Ligand; TNFRSF10

Alexis 522-003   Human recombinant   ≥95% (SDS-PAGE); lyophilized containing 10 μg protein; contains PBS | Binds to human & mouse TRAIL-R1 (DR4) & TRAIL-R2 (DR5); does not induce apoptosis in the absence of the enhancer (Alexis Prod. No. 804-034); Kit (Alexis Prod. No. 850-018) contains rhsTRAIL & enhancer; Wiley, SR et al, *Immunity*, 3: 673, 1995; Marsters, SA et al, *Curr Biology*, 6: 750, 1996; Pitt, RM et al, *J Biol Chem*, 271: 12687, 1996; Pan, G et al, *Science*, 276: 111, 1997; Pan, G et al, *Science*, 277: 815, 1997; Sheridan, JP et al, *Science*, 277: 818, 1997; Thome, M et al, *Nature*, 386: 517, 1997; Irmler, M et al, *Nature*, 388: 190, 1997

Kamiya   Human recombinant (Apo-2L)   MW 28k   >95% (SDS-PAGE).

Alexis 522-004   Human recombinant embryo kidney cells   MW 54k (chimera protein) under reducing conditions   >95% (SDS-PAGE); lyophilized containing 25 μg protein; contains PBS | The extracellular domain of rhTRAIL-R1 (AA 24-239) is fused to the Fc portion of human IgG1 & inhibits soluble TRAIL (sTRAIL)-mediated lysis of TRAIL sensitive cells (concentration range: 2-10 ng/mL); for the detection of surface TRAIL by flow cytometry use rhTRAIL-R2.Ig:Fc-FITC (Alexis Prod. No. 522-005F); Golstein, P et al, *Curr Biol*, 7: 750, 1997; Marsters, SA et al, *Curr Biology*, 7: 1003, 1997; Schneider, P et al, *FEBS Lett*, 416: 329, 1997; Pan, G et al, *Science*, 276: 111, 1997; Schneider, P et al, *Immunity*, 7: 831, 1997; Walczak, H et al, *EMBO J*, 16: 5386, 1997

Alexis 522-005   Human recombinant embryo kidney cells   MW 46k (chimera protein) under reducing conditions   >95% (SDS-PAGE); lyophilized containing 50 μg protein; contains PBS | The extracellular domain of rhTRAIL-R2 (AA 52-212) is fused to the Fc portion of human IgG1 & inhibits soluble TRAIL (sTRAIL)-mediated lysis of TRAIL sensitive cells (concentration range: 0.1-0.5 ng/mL) & can be used for immunoprecipitation of TRAIL; References are the same as for Alexis 522-004

Alexis 522-005F   Human recombinant embryo kidney cells   MW 46k (chimera protein) under reducing conditions   >95% (SDS-PAGE); lyophilized containing 50 μg protein; contains PBS | The extracellular domain of rhTRAIL-R2 (AA 52-212) is fused to the Fc portion of human IgG1 & inhibits soluble TRAIL (sTRAIL)-mediated lysis of TRAIL sensitive cells (concentration range: 0.1-0.5 ng/mL) & can be used for immunoprecipitation of TRAIL; rhTRAIL-R2.Ig:Fc-FITC (Alexis Prod. No. 522-005F) is useful for the detection of surface TRAIL by flow cytometry; References are the same as for Alexis 522-004

Alexis 522-006   Human recombinant embryo kidney cells   MW 66k (chimera protein) under reducing conditions   >95% (SDS-PAGE); lyophilized containing 25 μg protein; contains PBS | The extracellular domain of rhTRAIL-R3 (AA 25-240) is fused to the Fc portion of human IgG1 & inhibits soluble TRAIL (sTRAIL)-mediated lysis of TRAIL sensitive cells (concentration range: 50-100 ng/mL); for the detection of surface TRAIL by flow cytometry use rhTRAIL-R2.Ig:Fc-FITC (Alexis Prod. No. 522-005F); References are the same as for Alexis 522-004

Alexis 522-011   Human recombinant embryo kidney cells   MW 54k (chimera protein) under reducing conditions   >95% (SDS-PAGE); lyophilized containing 25 μg protein; contains PBS | The extracellular domain of rhTRAIL-R4 (AA 56-212) is fused to the Fc portion of human IgG1 & inhibits soluble TRAIL (sTRAIL)-mediated lysis of TRAIL sensitive cells (concentration range: 2-10 ng/mL); for the detection of surface TRAIL by flow cytometry use rhTRAIL-R2.Ig:Fc-FITC (Alexis Prod. No. 522-005F); References are the same as for Alexis 522-004

PeproTech 310-04 Human recombinant, expressed in *E. coli* MW 19.6k >98% (SDS-PAGE) & HPLC; <0.1 ng endotoxin per mg (1EU/mg); lyophilized | Cytotoxic protein; activates rapid apoptosis in tumor cells but not in normal cells; comprises the full-length of the TNF-like extracellular domain of TRAIL; $ED_{50} \leq 10.0$ ng/mL; induction of apoptosis of LANCap (human prostate cancer cells) by TRAIL/Apo2L was typically 2-3 fold higher after 48 hours; SA determined by the dose-dependent stimulation of IL-8 production by human PBMC

R&D Systems 375-TL-010 NSO-Expressed >97%; lyophilized; ED_{50}: 4-12 ng/mL | Species specificity: human TRAIL; member of the TNF family of cytokines; several TRAIL receptors have been identified including decoy receptors that function to antagonize TRAIL-induced apoptosis

Tissue Necrosis Factor Related Apoptosis Inducing Ligand Receptor I

Kamiya Human recombinant >95% (SDS-PAGE)

Tissue Necrosis Factor Related Apoptosis Inducing Ligand Receptor II

Kamiya Human recombinant >95% (SDS-PAGE)

Tissue Necrosis Factor Related Apoptosis Inducing Ligand Receptor II, FITC Conjugated

Kamiya Human recombinant >95% (SDS-PAGE)

Tissue Necrosis Factor Related Apoptosis Inducing Ligand Receptor III

Kamiya Human recombinant >95% (SDS-PAGE)

Tissue Necrosis Factor Related Apoptosis Inducing Ligand, Soluble

Alexis BMS310 Human recombinant, produced in bacteria MW 28k >95% (SDS-PAGE); 10 μg lyophilized powder at 0.1 mg/mL | The extracellular domain of human TRAIL (AA 95-281) is fused at the *N*-terminus to a FLAG-tag & a 8 AA linker protein; the enhancer is an antibody reacting with rh sTRAIL thereby increasing its activity; in the presence of the enhancer, rh sTRAIL induces apoptosis of TRAIL sensitive cells at concentration ≥1 ng/mL e.g. if added with enhancer, rh sTRAIL induces apoptosis of Jurkat T lymphoma cells in a concentration of 1-100 ng/mL

Tissue Necrosis Factor Related Apoptosis Inducing Ligand/Apo2 Ligand

Biodesign A52104H *E. coli* Purified

Chemicon GF092 Human ≥95% | Purified protein for apoptosis & signal transduction

Tissue Necrosis Factor Related Apoptosis-Inducing Ligand II, N-Terminal Peptide

Chemicon AG620 ≥95% | Purified protein for apoptosis & signal transduction

Tissue Necrosis Factor Related Apoptosis-Inducing Ligand III, N-Terminal Peptide

Chemicon AG621 ≥95% | Purified protein for apoptosis & signal transduction

Tissue Necrosis Factor Related Apoptosis-Inducing Ligand Receptor III

R&D Systems 630-TR-100 NSO-Expressed >95%; lyophilized; ED_{50}: 2.5-7 ng/mL | Species specificity: human TRAIL Receptor 3/Fc Chimera

Tissue Necrosis Factor Related Apoptosis-Inducing Ligand Receptor II

R&D Systems 631-T2-100 NSO-Expressed >95%; lyophilized; ED_{50}: 0.7-2 ng/mL | Species specificity: human TRAIL Receptor 2/Fc Chimera

Tissue Necrosis Factor Related Apoptosis-Inducing Ligand Receptor IV

R&D Systems 633-TR-100 NSO-Expressed >95%; lyophilized; ED_{50}: 30-60 ng/mL | Species specificity: human TRAIL Receptor 4/Fc Chimera

Tissue Necrosis Factor Related Apoptosis-Inducing Ligand Receptor I

R&D Systems 347-DR-100 *SF*21-Expressed >97%; lyophilized; ED_{50}: 1-3 ng/mL | Species specificity: human TRAIL Receptor 1/Fc Chimera

Tissue Necrosis Factor α

Kamiya	Human	>95% (SDS-PAGE); soluble
BioSource International PHC3011		Human recombinant
Kamiya	Mouse	>95% (SDS-PAGE); soluble
BioSource International PMC3014		Mouse recombinant
BioSource International PRC3014		Rat recombinant
BioSource International PSC3014		Swine recombinant

Tissue Necrosis Factor α Receptor I, Soluble

Synonyms: Tumor Necrosis Factor Receptor Type I; p60

PeproTech 310-07 Human recombinant, expressed in *E. coli* MW 18.3k >98% (SDS-PAGE) & HPLC; <0.1 ng endotoxin per mg (1EU/mg); lyophilized | 162 AA residues comprising the extra cellular domain of TNF; ED_{50} = 0.05 mg/mL; SA determined by its inhibitory effect of the TNF-alpha mediated cytotoxicity in murine L929 cells

Tissue Necrosis Factor α Receptor II, Soluble

Synonyms: Tumor Necrosis Factor Receptor Type II

PeproTech 310-12 Human recombinant, expressed in *E. coli* MW 18.9k >98% (SDS-PAGE) & HPLC; <0.1 ng endotoxin per mg (1EU/mg); lyophilized | Inhibitor of TNF & can block TNF bioactivity; soluble; 174 AA residues comprising the extra cellular domain of TNF; ED_{50} = 0.125 mg/mL; SA determined by its inhibitory effect of the TNF-alpha mediated cytotoxicity in murine L929 cells

Tissue Necrosis Factor β

Synonyms: Lymphotoxin

BioSource International PHC3024 Human recombinant

Tissue Plasminogen Activator

USBio T5600-15 Human ≥95%; no contaminants detected; single band by SDS-PAGE, IEP, &/or RID; two-chain tPA; 50 mg salt-free, lyophilized preparation | Suitable for antigenic applications in immunological protocols

Fitzgerald 30-AT50 Human melanoma cell line Standard purity

Biogenesis 9020-1026 Human serum MW 65k Purified; from 100 m*M* PBS, 35 mg/mL L-arginine, 0.01% Tween 80; lyophilized | Bos et al, *Biotherapy*, 5:187, 1992

Tissue Plasminogen Activator, Single-Chain

Calbiochem 612200 Human MW 65k Lyophilized solid; ≥95% (SDS-PAGE); soluble in water | Thrombolytic agent isolated from a melanoma cell line; serine protease that contains a single chain of 527 AA; binds to fibrin via lysine binding sites at its amino terminus & activates bound plasminogen; cleaves plasminogen to form active plasmin, the proteolytic enzyme that dissolves clots; reported to be involved in neurite outgrowth, regeneration & migration; Plays an important role in excitotoxin-induced neuronal degeneration; dihydropyridine Ca^{2+} antagonists increase levels of tPA; Tsirka, SE et al, *Nature*, 377: 340, 1995; Winther, K et al, *J Cardiovasc Pharmacol*, 19: S21, 1992

Sigma T 7776 Human melanoma cell culture MW 68k Lyophilized with each vial containing 10 μg t-PA with BSA & d-mannitol; fibrinolytic activity: ~400,000 IU/mg t-PA | Fibrinolytic serine protease found in many tissues & body fluids; works by converting plasminogen to plasmin, which then dissolves fibrin, a major component of blood clots; amidolytic activity is determined by the method of Verheijen using the substrate D-Ile-Pro-Arg p-nitroanilide; Bachmann, F & Kruithof, E, *Seminars in Thrombosis & Hemostasis*, 10: 6, 1984; Klausner, A, *Bio/Technology*, 4: 706, 1986; Verheijen, JH et al, *Methods of Enzymatic Analysis*, 3rd ed. (Bergmeyer, J & Grassi, M, eds), Vol 5: 425, 1984

Tissue Plasminogen Activator, Two-Chain

Sigma T 4055 Human melanoma cell culture MW 30k & 40k (2 chains) Solution in 1 M ammonium bicarbonate with each vial containing 50 μg t-PA; fibrinolytic activity: ≥200,000 IU/mg t-PA | Fibrinolytic serine protease found in many tissues & body fluids; works by converting plasminogen to plasmin, which then dissolves fibrin, a major component of blood clots; amidolytic activity is determined by the method of Verheijen using the substrate D-Ile-Pro-Arg p-nitroanilide; Bachmann, F & Kruithof, E, *Seminars in Thrombosis & Hemostasis*, 10: 6, 1984; Klausner, A, *Bio/Technology*, 4: 706, 1986; Verheijen, JH et al, *Methods of Enzymatic Analysis*, 3rd ed. (Bergmeyer, J & Grassi, M, eds), Vol 5: 425, 1984

Sigma T 4432 Human melanoma cell culture MW 30k & 40k (2 chains) Lyophilized with each vial containing 10 μg t-PA with BSA & D-mannitol; fibrinolytic activity: ≥300,000 IU/mg t-PA | Fibrinolytic serine protease found in many tissues & body fluids; works by converting plasminogen to plasmin, which then dissolves fibrin, a major component of blood clots; amidolytic activity is determined by the method of Verheijen using the substrate D-Ile-Pro-Arg p-nitroanilide; Bachmann, F & Kruithof, E, *Seminars in Thrombosis & Hemostasis*, 10: 6, 1984; Klausner, A, *Bio/Technology*, 4: 706, 1986; Verheijen, JH et al, *Methods of Enzymatic Analysis*, 3rd ed. (Bergmeyer, J & Grassi, M, eds), Vol 5: 425, 1984

Tissue Polypeptide Antigen

Calbiochem 612312 Human breast epithelium carcinoma cell line Liquid in 130 *mM* NaCl, 80 *mM* sodium phosphate buffer, 20 *mM* potassium phosphate buffer, 20 *mM* KCl, 0.1% NaN_3, pH 7.4, sterile-filtered; activity: 100-500 units/mL (RIA); antigen derived from cell culture | Reported to be related to cytoplasmic intermediate filaments; serves as a tumor marker, particularly in breast cancer evaluation; elevated serum levels of hTPA are present in breast, gastrointestinal, gynecologic, lung & urologic cancers; suitable for use in immunoassays & as an immunogen; Gion, M et al, *Eur J Clin Chem Clin Biochem*, 32: 779, 1994; Giovagnoli, MR et al, *Anticancer Res*, 14: 635, 1994; Kuman S et al, *J Urol*, 53: 578, 1981; Luthgens, M & Schlegal, G, *J Tumor Marker Oncol*, 2: 261, 1987

Sigma T 9181 Human cell culture Partially purified solution in 0.15 M phosphate buffered saline, pH 7.4, containing 0.1% sodium azide; activity: ≥100 U/mL; 1 U is an arbitrary unit related to a reference antigen preparation using the Byk Sangtec-RIA method | Cancer-associated TPA; Kumar, S et al, *Br J Urology*, 53: 578, 1981; Luthgens, M & Schlegel, G, *J Tumor Marker Oncol*, 2: 261, 1987

Toxic Shock Syndrome Toxin I

Synonyms: Staphylococcal Enterotoxin F

Sigma T 5662 *Staphylococcus aureus* Contains ~50% protein (Lowry); balance primarily NaCl & sodium phosphate buffer | A superantigen for T-lymphocytes; Marrack, P & Kappler, J, *Science*, 248: 705, 1990; Misfeldt, ML, *Infection Immun*, 58: 2409, 1990; Blanco, L et al, *Infection Immun*, 58: 3020, 1990

Toxin, *Pasteurella multocida*

Calbiochem 512742 *Pasteurella multocida* recombinant MW 14.7k >95% (SDS-PAGE); liquid in 50 mM Tris-HCl, 10% glycerol, pH 7.5; harmful: LD_{50} ≤2000 mg/kg | Bacterial protein toxin that binds to & enters eukaryotic cells via receptor-mediated endocytosis; acts intracellularly to initiate the inositol trisphosphate signaling pathway, Ca^{2+} mobilization & DNA synthesis; stimulates tyrosine phosphorylation of multiple substrates including focal adhesion kinase; exerts its action on the α-subunit of the G_q family of G-proteins; used for the activation of G_q-protein-coupled phosphatidylinositol-specific phospholipase C (PLC) intact cells, making PMT useful as a tool for studying G_q-linked signal transduction; Wilson, BA et al, *J Biol Chem*, 272: 1268, 1997; Lacerda, HM et al, *J Biol Chem*, 271: 439, 1996; Higgins, TE et al, *PNAS*, 89: 4240, 1992; Murphy, AC et al, *J Biol Chem*, 267: 25296, 1992

Toxoplasma gondii Surface Antigen

Biogenesis 9070-2029 *T. gondii* strain RH Purified; 50 *mM* TRIS/HCl pH 8.0, 0.005% Merthiolate, 40 μM PMSF, 100 IU/mL aprotinin; liquid | Contains high concentration of *T. gondii* surface antigens, mainly P30 antigen; useful in EIA

Transcription Factor IIA

IBT TA-200-1 Human recombinant, expressed in *E. coli*

Transcription Factor IIB

IBT TA-210-1 Human recombinant, expressed in *E. coli*

Transcription Factor IIF

Synonyms: RAP-30

IBT TA-250-1 Human recombinant, expressed in *E. coli*

Transcription Factor IIF, Large Subunit

Synonyms: RAP-74

IBT TA-100-1 Human recombinant, expressed in *E. coli*

Transcription Factor YY1

Synonyms: NF1

IBT TA-150-1 Human recombinant, expressed in *E. coli*

Transducin

Calbiochem 616410 Bovine retina MW 40, 36 & 8k, (α-, β- & γ-subunits) ≥95% (SDS-PAGE); liquid in 20 *mM* Tris-HCl, 10 *mM* BME, 100 *mM* NaCl, 5 *mM* magnesium acetate, 1.5 *mM* NaN_3, 50% glycerol, pH 7.5; activity: activates equimolar amounts of transducin as measured by GTPγS binding; must open in complete darkness | The G-protein transducin couples the reception of a photon of light by a rhodopsin molecule to the activation of cyclic GMP phosphodiesterase; heterotrimer; biologically active preparation characterized by GTPγS binding to the α-subunit; Bubis, J & Khorana, HG, *J Biol Chem*, 265: 12995, 1990

Transferrin

Synonyms: Siderophilin; Siderophilin, Partially Saturated; Siderophilin, Partially Iron Loaded; Serotransferrin

Scipac P158-5 >98%; lyophilized; H_2O dialyzed (low NH_3); suitable for bulk requirements; 1g min pack size | Nutritional protein

Fluka 90192 Bovine serum MW 77k ≥80.0% (GE); ≤6% water; ≥98% protein; ≤0.01% iron

ICN 152154   Guinea pig   ≥98%; purified by affinity chromatography; concentrated; supplied sterile & frozen | Iron binding protein displaying bacteristatic & fungistatic characteristics

Cortex CP4060   Human   >95%

ICN 55915   Human   Purified

Sigma T 3309   Human   ≥98% (agarose electrophoresis); iron content 300-600 µg/g; endotoxin level <1 EU/mg protein

Sigma T 7559   Human   5% solution in water; aseptically filled; ≥98% (agarose electrophoresis); iron content 300-600 µg/g; endotoxin level <1 EU/mg protein

Sigma T 8158   Human   ~98%; partially iron saturated; source material tested for HBsAg & HIV antibodies; endotoxin tested; cell culture tested | Heat treated at 60°C for 10 hours

Fluka 90190   Human blood serum   MW 79,550   ~95.0% (GE); ≤0.1% iron | Crichton, RR & Charl-Wauters, M, Eur J Biochem, 164: 485, 1987; Gorinsky, B, Adv Red Blood Cell Biol (DJ Weatherall, et al, eds), 7, 1981, Raven, New York

Biodesign A50132H   Human plasma   Purified

Fitzgerald 30-AT31   Human plasma   High purity

ICN 160076   Human plasma   ≥95% total protein; per vial: 10 mg transferrin, 10 mg mannitol, 1 mg NaCl; readily soluble in 1 mL H2O | Conveys essential metabolites to cultured cells; binds & makes iron available to cells in a recognizable form; cellular iron is an enzymatic cofactor in key metabolic pathways such as ATP generation; strong promoter of cell growth in culture; frequent cell culture supplement in serum-free media

ICN 55943   Mouse   Purified

Biogenesis 9100-5504   Mouse serum   Lyophilized

ICN 152155   Murine   90%; purified by affinity chromatography; concentrated; sterile filtered

ICN 55953   Rat   Purified

Scipac T102-1   Serum/plasma   >96%; lyophilized; iron content 0.3-0.8 mg/g protein; endotoxin <1.0 EU/mg; available on request | Cell culture protein

Transferrin Receptor

Synonyms: CD71; CD71

Cortex CP1113   Human   >95%

Biodesign A86001H   Human placenta   Purified

Biogenesis 9110-0300   Human placenta   Tested negative for HIV I and II antibodies, Hepatitis B surface antigen and Hepatitis C antibodies; purified; 0.1 M HEPES, pH 7.5, with 0.15 M NaCl & 0.05% NaN3; liquid

Fitzgerald 30-AT60   Human placenta   High purity

USBio T8199-20   Human placenta   ≥95%; no contaminants were detected by SDS-PAGE analysis & IEP; lyophilized in PBS, pH 7.4, 0.01% NaN3 | Suitable for antigenic applications in immunological protocols

Biogenesis 9110-0320   Human serum   Liquid

Scipac P185-3   Placenta   >96%; frozen in HEPES buffer; 0.5-2 mg/L | Nutritional protein

Scipac P185-8   Serum/plasma   Extract; frozen in Phosphate buffer; >50 mg of sTfR /L | Nutritional protein

Transferrin, 30% Iron Saturated

ICN 820551   Bovine plasma   Highly purified | Suitable growth factor for primary, continuous 7 transformed cell lines

ICN 823421   Human plasma   Highly purified; trace IgG | Excellent growth factor for primary, continuous 7 transformed cell lines

Transferrin, 90% Iron Saturated

ICN 820571   Bovine plasma   Highly purified; trace IgG | Ideal growth factor for primary, continuous 7 transformed cell lines

ICN 823431   Human plasma   Highly purified; no detectable IgG | Excellent growth factor for primary, continuous 7 transformed cell lines

Transferrin, Apo-

Synonyms: Siderophilin, Iron Poor; Siderophilin, Iron Poor; Siderophilin, Iron Poor

Sigma T 0178   Bovine   MW 76-81k   ~98% (agarose electrophoresis) | Non-heme iron transport protein; Aisen, P & Listowsky, I, Ann Rev Biochem, 49: 357, 1980

Sigma T 1428   Bovine   ≥98%; substantially iron free; cell culture tested

ICN 152334   Bovine plasma   98% (FPLC); buffer-free, off-white powder | Useful transport factor in cell culture to convey essential metabolites to cells; important factor when culturing cells in defined medium

Sigma T 6011   Dog   MW 76-81k   ~98% (agarose electrophoresis) | Non-heme iron transport protein; Aisen, P & Listowsky, I, Ann Rev Biochem, 49: 357, 1980

Sigma T 0904   Guinea pig   MW 76-81k   ~98% (agarose electrophoresis) | Non-heme iron transport protein; Aisen, P & Listowsky, I, Ann Rev Biochem, 49: 357, 1980

Sigma T 1147   Human   >97% (agarose electrophoresis); substantially iron free; source material tested for HBsAg & HIV antibodies; cell culture tested

Sigma T 2036   Human   ~98%; substantially iron free; source material tested for HBsAg & HIV antibodies; endotoxin tested; cell culture tested

Sigma T 2252   Human   MW 76-81k   >98% (agarose electrophoresis) | Non-heme iron transport protein; human source material negative for HIV & HBsAg; Aisen, P & Listowsky, I, Ann Rev Biochem, 49: 357, 1980

Sigma T 4382   Human   MW 76-81k   ≥98% (agarose electrophoresis); iron content ≤30 µg/g | Non-heme iron transport protein; human source material negative for HIV & HBsAg; Aisen, P & Listowsky, I, Ann Rev Biochem, 49: 357, 1980

Sigma T 5391   Human   >97%; lyophilized; substantially iron free; source material tested for HBsAg & HIV antibodies; sterilized by γ-irradiation; endotoxin tested; cell culture tested

Sigma T 0523   Mouse   MW 76-81k   ~98% (agarose electrophoresis) | Non-heme iron transport protein; Aisen, P & Listowsky, I, Ann Rev Biochem, 49: 357, 1980

Sigma T 6136   Rabbit   MW 76-81k   ~98% (agarose electrophoresis) | Non-heme iron transport protein; Aisen, P & Listowsky, I, Ann Rev Biochem, 49: 357, 1980

Sigma T 6013   Rat   MW 76-81k   ~98% (agarose electrophoresis) | Non-heme iron transport protein; Aisen, P & Listowsky, I, Ann Rev Biochem, 49: 357, 1980; Gordon, AH & Louis, LN, Biochem J, 88: 409, 1963

Scipac P130-1   Serum/plasma   >96%; lyophilized; <30mg Fe/g protein | Nutritional protein

Scipac T100-1   Serum/plasma   >98%; lyophilized; <15 mg Cl/mL; <10 mg Na/mL; <0.03 mg Fe/g protein; <1.0 EU endotoxin/mg; <5% moisture; available on request | Cell culture protein

Sigma T 5037   Sheep   MW 76-81k   ~98% (agarose electrophoresis) | Non-heme iron transport protein; Aisen, P & Listowsky, I, Ann Rev Biochem, 49: 357, 1980

Transferrin, Biotin Conjugated

Sigma T 3915   Human   Prepared from human holotransferrin, coupled to biotin by an amide bond through an aminocaproyl spacer; lyophilized powder containing ~90% protein (Biuret); balance primarily sodium citrate; contains 4-8 moles d-biotin/mole transferrin

Transferrin, Holo-

Synonyms: Siderophilin, Iron Saturated

Sigma T 0156   Bovine   >95% (agarose electrophoresis); similar to Sigma T 1408

Sigma T 1283   Bovine   ~98%; endotoxin tested; cell culture tested

Sigma T 1408   Bovine   ~98% (agarose electrophoresis)

ICN 152335   Bovine plasma   98% (FPLC); buffer-free, off-white powder | Useful transport factor in cell culture to convey essential metabolites to cells; important factor when culturing cells in defined medium

Sigma T 0665   Human   ~98%; source material tested for HBsAg & HIV antibodies; endotoxin tested; cell culture tested

Sigma T 4132   Human   ≥98% (agarose electrophoresis); iron content 1200-1600 µg/g | Human source material negative for HIV & HBsAg

Sigma T 7434   Human   5% solution in water; aseptically filled (0.1 µm filter); ≥98% (agarose electrophoresis); iron content 1200-1600 µg/g; endotoxin level <1 EU/mg protein | Human source material negative for HIV & HBsAg

Scipac T101-1   Pooled serum/plasma   >96%; lyophilized; iron content 1.2-1.7 mg/g protein; endotoxin <1.25 EU/mg | Cell culture protein

Scipac P133-1   Serum/plasma   >98%; lyophilized | Nutritional protein

Transferrin, Iron Poor

ICN 820561   Bovine plasma   <1.0% iron, low endotoxin, trace IgG

ICN 823411   Human plasma   Highly purified; <1.0% iron, trace to non-detectable IgG | Ideal growth factor for primary, continuous 7 transformed cell lines

Transferrin, Iron Saturated Holo-

USBio T8199-12   Human   ≥99%; no contaminants were detected by SDS-PAGE analysis & IEP; <0.1% Fe; 10 mg/mL supplied in 50% mannitol & 5% NaCl as stabilizers | Suitable for antigenic applications in immunological protocols

Transferrin, Partially Iron Saturated

Scipac P158-0   Pooled serum/plasma   >99%; lyophilized; ultra-pure grade suitable for cell culture, research and the most demanding applications | Nutritional protein

Scipac P158-1   Serum/plasma   >96%; lyophilized; suitable for bulk requirements; 1g min pack size | Nutritional protein

Transferrin, Poly(Lys) FITC Conjugated

Sigma T 0288   Human   Prepared from apo-transferrin covalently liked to FITC-labeled polylysine; iron incorporated after purification of conjugate; lyophilized powder containing ~90% protein (Biuret); balance primarily HEPES buffer salts; contains 0.3-0.6 moles polylysine/mole transferrin & ~1 mole FITC/mole polylysine | Potentially useful for receptor-mediated transport of polyanions, e.g. DNA, & the detection of transferrin-binding compounds; source material negative for HIV & HBsAg; Wagner, E et al, *Bioconj Chem*, 2: 226, 1991; Kurrie, A et al, *Biochemistry*, 29: 8274, 1990

Transforming Growth Factor α

Oncogene WA32   MW 5546   >98% (HPLC); liquid | Western Blot standard; protein is denatured in SDS; species reactivity: human

Biodesign A52116H   *E. coli*   MW 5.5k   Purified

Chemicon GF022   Human   ≥95%

Chemicon TG011   Human   ≥95%

Biogenesis 9130-0254   Human r-DNA *E. coli*   MW 5547 (50 AA)   Purified; no preservatives; lyophilized | Equivalent to native TGF in receptor binding assay on cultured A431 cells and in mitogenesis assay on fibroblasts

Amersham VPF007   Human recombinant   Human cross-reactivity | Growth/death factor interactions

BioSource International PHG0051   Human recombinant

Oncogene PF008   Human recombinant   MW 5546 (AA analysis)   >98% (HPLC); lyophilized with 100 µg BSA; biological activity: fully active in EGF receptor binding & mitogenesis assays | Species reactivity: human; for proliferation studies & binding studies; exerts its biological effects in the concentration range of 10-100 p*M*; reconstitute in 10m*M* acetic acid

Calbiochem 616430   Human recombinant, expressed in *E. coli*   MW 6k   ≥97% (SDS-PAGE); lyophilized from 30% acetonitrile, 0.1% TFA; biological activity: ED_{50}=0.1-0.4 ng/mL as measured by the ability to stimulate ^{3}H-thymidine incorporation in the mouse fibroblast cell line Balb/3T3; endotoxin: ≤100 pg/µg TGF-α | Plays an important role in cell-cell adhesion & cell growth & differentiation; reported to play a role in the development of hormonal insensitivity in estrogen-receptor-positive breast cancers; Nicholson, RI et al, *Cancer Res*, 54: 1684, 1994; Massague, J, *J Biol Chem*, 265: 21393, 1990

Fitzgerald 30-AT62   Human recombinant, expressed in *E. coli*

Harlan BT-3028   Human recombinant, expressed in *E. coli*   Lyophilized; 0.025 mg

Harlan BT-3029   Human recombinant, expressed in *E. coli*   Lyophilized; 0.1 mg

IBT UU020, UM100   Human recombinant, expressed in *E. coli*   MW 5546   >95%; dried from 0.1 *M* acetic acid | Promotes proliferation & differentiation of cell types displaying the EGF receptor; biological effects of TGF-alpha are mediated by autocrine & paracrine mechanisms in adult & embryonic cells, as well as in a variety of tumours & retrovirally transformed cells

ICN 150217   Human recombinant, expressed in *E. coli*   >95%; lyophilized; ED_{50} = 0.1-10 ng/mL, measured as the ability to maximally stimulate human foreskin fibroblast proliferation | Growth factor released by cancer cells; structurally related to epidermal growth factor (EGF); binds EGF receptors; mediates cell growth

ICN 153507   Human recombinant, expressed in *E. coli*   ≥95%; lyophilized; ED_{50} = 2.0 ng/mL, measured via dose dependent ^{3}H-thymidine uptake by 3T3 cells | 50 residue protein with activity similar to EGF; stimulates a wide range of epithelial & epidermal cell types; related to autocrine growth of selected transformed cells

PeproTech 100-16A   Human recombinant, expressed in *E. coli*   MW 5.5k   >98%; 50 AA; lyophilized with no additives; ED_{50}: 0.2-0.7 ng/mL as determined by the stimulation of thymidine uptake by Balb/c 3T3 cells

Sigma T 7924   Human recombinant, expressed in *E. coli*   MW 5.5k   ≥97% (SDS-PAGE); 0.2 µm filtered & lyophilized from 30% acetonitrile & 0.1% trifluoroacetic acid; endotoxin tested | Reversibly confers a transformed phenotype upon normal non-neoplastic cells, such as normal rat kidney fibroblasts; this activity also requires the presence of TGF-β which potentiates the action of TGF-α via a separate receptor; TGF-α is structurally homologous to EGF & exerts its action through the EGF receptor; mitogenic activities of all TGF-α products are tested in culture using BALB/MK cells, 4MBr-5 cells or BALB/3T3 cells; Derynck, R, *Cell*, 54: 593, 1988; Anzano, M et al, *Cancer Res*, 42: 4776, 1982; Marquardt, H et al, *Proc Natl Acad Sci USA*, 80: 4684, 1983; Todaro, G et al, *Proc Natl Acad Sci USA*, 77: 5258, 1980

Sigma T 9533   Rat synthetic   ~90% (HPLC); lyophilized & γ-irradiated; endotoxin tested | Reversibly confers a transformed phenotype upon normal non-neoplastic cells, such as normal rat kidney fibroblasts; this activity also requires the presence of TGF-β which potentiates the action of TGF-α via a separate receptor; TGF-α is structurally homologous to EGF & exerts its action through the EGF receptor; mitogenic activities of all TGF-α products are tested in culture using BALB/MK cells, 4MBr-5 cells or BALB/3T3 cells; Derynck, R, *Cell*, 54: 593, 1988; Anzano, M et al, *Cancer Res*, 42: 4776, 1982; Marquardt, H et al, *Proc Natl Acad Sci USA*, 80: 4684, 1983; Todaro, G et al, *Proc Natl Acad Sci USA*, 77: 5258, 1980

Biogenesis 9130-0304   r-DNA   [125]I conjugated

Amersham V654205   Recombinant   Human cross-reactivity | Growth/death factor interactions

Transforming Growth Factor α, Long Human

IBT VU200, VM001   Recombinant, expressed in *E. coli*   MW 7113   >90%; dried from 0.1 *M* acetic acid | 64 AA analog of human TGF-α, comprising the complete human TGF-α sequence & a 14 AA extension peptide at the N-terminus; engineered as an inexpensive, high quality potent analog for use in serum-free or low serum cell culture

Transforming Growth Factor α, N-Ac-Ethyl-Amide

Chemicon TG013   ≥95%

Transforming Growth Factor α, *N*-Ac-OMe

Chemicon TG014   ≥95%

Transforming Growth Factor β

Chemicon TG010   Human   ≥95%

Biogenesis 9130-1504   Human platelets   Purified to homogeneity (SDS-PAGE); lyophilized; significant losses may result by with neutral buffers, use of glass implements or extensive manipulations; supplemented with 3.5 μg BSA to enhance recovery and stability | Assoian et al, *JBC*, 258:7155, 1983; Childs et al, *PNAS*, 79:5312, 1982

Biogenesis 9130-1704   Porcine platelets ^{125}I conjugated

Biogenesis 9130-1604   r-DNA   Lyophilized

Transforming Growth Factor βI

Synonyms: Latency Associated Peptide

Chemicon GF056   Human   ≥95%

Oncogene PF051   Human   MW 25k   >97% (SDS-PAGE); lyophilized in 30% acetonitrile, 0.1% TFA containing 50 mg of BSA per 1 mg of cytokine; reconstitute in sterile 4 mM HCl containing at least 0.1% HSA or BSA; biological activity: EC_{50} of 0.02-0.06 ng/mL as determined by the ability to inhibit IL-4-dependent ^{3}H-thymidine incorporation in mouse HT-2 cells | Species reactivity: human & mouse; stimulatory for cells of mesenchymal origin & inhibitory for cells of epithelial or neuroectodermal origin; for proliferation studies & growth inhibition assay

Biogenesis 9130-1506   Human platelets   Purified Ig; lyophilized

Calbiochem 616450   Human platelets   MW 25k   >97% (SDS-PAGE); lyophilized aseptically in the presence of 50 μg BSA/μg TGF-β1; biological activity: ED_{50}=20-60 pg/mL; endotoxin: ≤100 pg/μg TGF-β1 | Promotes apoptosis in resting human B lymphocytes, glioma cells & trigeminal neurinomal cells; prepared from blood shown by certificate to be negative for HBsAg & antibodies to HIV & HCV; *Merck Index*, 12: 9707; Lomo, J et al, *J Immunol*, 154: 1634, 1995; Marushige, K & Marushige, Y, *Anticancer Res*, 14: 2419, 1995

BioSource International PHG9104   Human recombinant

Oncogene PF016   Human recombinant   MW ~25k   >92% (SDS-PAGE); lyophilized from a volatile buffer; contains 100 μg lipid free BSA as carrier; reconstitute in 50 mL of 10*mM* HCl/10% EtOH in siliconized vial; biological activity: a concentration of 1-5 ng/mL for initial cell inhibition assays is recommended | Reactive on many cell types

Sigma T 7039   Human recombinant, expressed in a mammalian cell line   MW 25k   ≥97% (SDS-PAGE); 0.2 μm filtered & lyophilized from an acetonitrile/TFA solution containing 50 μg BSA per μg of cytokine; endotoxin tested | Multifunctional protein capable of influencing cell proliferation, differentiation & a variety of cellular functions; stimulates growth of cells of mesenchymal origin, but inhibits growth of hepatocytes, epithelial cells, T & B lymphocytes; known to interact with several other agents including EGF, FGF, TGF-α, PDGF & IL-2; bioactivity of TGF-β is tested in culture using Mv1Lu cells, a mink lung epithelial cell line, whereby 1 U is the amount of TGF-b required to induce half-maximal inhibition of Mv1Lu cells as measured in an MTT dye assay or by measuring ^{3}H-thymidine incorporation in mouse HT-2 cells; Cheifetz, S et al, *Cell*, 48: 409, 1987; Roberts, A & Sporn, M, *Adv Cancer Res*, 51: 107, 1988; Sporn, M et al, *Science*, 233: 532, 1986; Tsang, M et al, *Lymphokine Res*, 9: 607, 1990

PeproTech 100-21   Human recombinant, expressed in A293 cells   MW 25.0k   >98% (SDS-PAGE) & HPLC; <0.1 ng endotoxin per mg (1EU/mg); lyophilized | Member of a superfamily of homologous, disulfide-linked, homodimeric proteins; regulates proliferation & differentiation of normal & transformed cells; 112 AA/subunit; $ED_{50} \le 0.05$ ng/mL; SA ≥ 2 x 10^7 U/mg; SA determined by the ability to inhibit mouse IL-4-dependent proliferation of mouse HT-2 cells

PeproTech 100-21R   Human recombinant, expressed in Baculovirus High-5 infected cells   MW 25.0k   >98% (SDS-PAGE) & HPLC; <0.1 ng endotoxin per mg (1EU/mg); lyophilized | Member of a superfamily of homologous, disulfide-linked, homodimeric proteins; regulates proliferation & differentiation of normal & transformed cells; 112 AA/subunit; $ED_{50} \le 0.05$ ng/mL; SA ≥ 2 x 10^7 U/mg; SA determined by the ability to inhibit mouse IL-4-dependent proliferation of mouse HT-2 cells

Calbiochem 616455   Human recombinant, expressed in CHO cell line   MW 25k   ≥97% (SDS-PAGE); lyophilized from filter-sterilized solution of 30% acetonitrile, 0.1% TFA containing 50 μg BSA/μg TGF-β1; biological activity: ED_{50}=20-60 pg/mL; endotoxin: ≤100 pg/μg TGF-β1 | Promotes apoptosis in resting human B lymphocytes, glioma cells & trigeminal neurinomal cells; induces the accumulation of extracellular matrix in various diseases; *Merck Index*, 12: 9707; Yamazaki, M et al, *Am J Pathol*, 144: 221, 1994; Lomo, J et al, *J Immunol*, 154: 1634, 1995; Marushige, K & Marushige, Y, *Anticancer Res*, 14: 2419, 1995

Sigma T 3408   Human recombinant, expressed in *Sf* 21 insect cells   ≥97% (SDS-PAGE); 0.2 μm filtered & lyophilized from phosphate buffered saline containing 1.25 mg BSA; activity is tested in culture by using a mouse T cell line HT-2; endotoxin tested | Multifunctional protein capable of influencing cell proliferation, differentiation & a variety of cellular functions; stimulates growth of cells of mesenchymal origin, but inhibits growth of hepatocytes, epithelial cells, T & B lymphocytes; known to interact with several other agents including EGF, FGF, TGF-α, PDGF & IL-2; bioactivity of TGF-β is tested in culture using Mv1Lu cells, a mink lung epithelial cell line, whereby 1 U is the amount of TGF-b required to induce half-maximal inhibition of Mv1Lu cells as measured in an MTT dye assay or by measuring ^{3}H-thymidine incorporation in mouse HT-2 cells; Cheifetz, S et al, *Cell*, 48: 409, 1987; Roberts, A & Sporn, M, *Adv Cancer Res*, 51: 107, 1988; Sporn, M et al, *Science*, 233: 532, 1986; Tsang, M et al, *Lymphokine Res*, 9: 607, 1990

Sigma T 1654   Natural human platelets   MW 25k   ≥97% (SDS-PAGE); aseptically prepared & lyophilized from acetonitrile solution containing 50 μg BSA; endotoxin tested | Multifunctional protein capable of influencing cell proliferation, differentiation & a variety of cellular functions; stimulates growth of cells of mesenchymal origin, but inhibits growth of hepatocytes, epithelial cells, T & B lymphocytes; known to interact with several other agents including EGF, FGF, TGF-α, PDGF & IL-2; bioactivity of TGF-β is tested in culture using Mv1Lu cells, a mink lung epithelial cell line, whereby 1 U is the amount of TGF-b required to induce half-maximal inhibition of Mv1Lu cells as measured in an MTT dye assay or by measuring ^{3}H-thymidine incorporation in mouse HT-2 cells; Cheifetz, S et al, *Cell*, 48: 409, 1987; Roberts, A & Sporn, M, *Adv Cancer Res*, 51: 107, 1988; Sporn, M et al, *Science*, 233: 532, 1986; Tsang, M et al, *Lymphokine Res*, 9: 607, 1990

Oncogene PF052   Porcine   MW 25k   >97% (SDS-PAGE); lyophilized in 25% acetonitrile, 0.1% TFA containing 50 mg of BSA per 1 g of cytokine; reconstitute in sterile 4 mM HCl containing at least 0.1% HSA or BSA; biological activity: EC_{50} of 0.02-0.06 ng/mL | Species reactivity: pig & mouse; stimulatory for cells of mesenchymal origin & inhibitory for cells of epithelial or neuroectodermal origin; for proliferation studies & growth inhibition assay

Biogenesis 9131-2002   Porcine platelets   Purified Ig; lyophilized

Biogenesis 9131-3002   Porcine platelets   Purified Ig; lyophilized

Calbiochem 616460   Porcine platelets   MW 25k   >97% (SDS-PAGE); lyophilized aseptically in the presence of 50 μg BSA/μg TGF-β1; biological activity: ED_{50}=20-60 pg/mL; endotoxin: ≤100 pg/μg TGF-β1 | *Merck Index*, 12: 9707

Sigma T 5050   Porcine platelets   ≥97% (SDS-PAGE visualized by silver stain); 0.2 μm filtered & lyophilized in the presence of 50 μg BSA; bioactivity is measured by its inhibitory effect on ^{3}H-thymidine incorporation in the mouse IL-4 dependent HT-2 cells; endotoxin tested | Multifunctional protein capable of influencing cell proliferation, differentiation & a variety of cellular functions; stimulates growth of cells of mesenchymal origin, but inhibits growth of hepatocytes, epithelial cells, T & B lymphocytes; known to interact with several other agents including EGF, FGF, TGF-α, PDGF & IL-2; bioactivity of TGF-β is tested in culture using Mv1Lu cells, a mink lung epithelial cell line, whereby 1 U is the amount of TGF-b required to induce half-maximal inhibition of Mv1Lu cells as measured in an MTT dye assay or by measuring ^{3}H-thymidine incorporation in mouse HT-2 cells; Cheifetz, S et al, *Cell*, 48: 409, 1987; Roberts, A & Sporn, M, *Adv Cancer Res*, 51: 107, 1988; Sporn, M et al, *Science*, 233: 532, 1986; Tsang, M et al, *Lymphokine Res*, 9: 607, 1990

Amersham V616455   Recombinant   Human cross-reactivity | Growth/death factor interactions

---

Transforming Growth Factor βI Receptor II Soluble Fragment

Sigma T 3301   Human recombinant, expressed in NSO mouse myeloma cells   ≥97% (SDS-PAGE); 0.2 μm filtered & lyophilized from phosphate buffered saline containing 1.25 mg BSA; receptor mediated activity is measured by its ability to inhibit the TGF-β1 bioactivity in HT-2 cells | Multifunctional protein capable of influencing cell proliferation, differentiation & a variety of cellular functions; stimulates growth of cells of mesenchymal origin, but inhibits growth of hepatocytes, epithelial cells, T & B lymphocytes; known to interact with several other agents including EGF, FGF, TGF-α, PDGF & IL-2; bioactivity of TGF-β is tested in culture using Mv1Lu cells, a mink lung epithelial cell line, whereby 1 U is the amount of TGF-b required to induce half-maximal inhibition of Mv1Lu cells as measured in an MTT dye assay or by measuring ^{3}H-thymidine incorporation in mouse HT-2 cells; Cheifetz, S et al, *Cell*, 48: 409, 1987; Roberts, A & Sporn, M, *Adv Cancer Res*, 51: 107, 1988; Sporn, M et al, *Science*, 233: 532, 1986; Tsang, M et al, *Lymphokine Res*, 9: 607, 1990

Transforming Growth Factor βI&II

Biogenesis 9131-4002   Porcine platelets   Purified Ig; lyophilized

---

Transforming Growth Factor βII

Oncogene PF017   Human recombinant   MW ~25k   >92% (SDS-PAGE); lyophilized from a volatile buffer; contain 100 μg lipid free BSA as carrier; reconstitute in 50 mL of 10mM HCl/10% EtOH in siliconized vial; biological activity: a concentration of 1-5 ng/mL for initial cell inhibition assays is recommended | Reactive on many cell types

IBT GF-240-2   Human recombinant, expressed in *E. coli*   MW 25k (dimer)   >98%; 1 mg/mL solution, pH 2.5 | Stimulatory for cells of mesenchymal origin; inhibitory for cells of epithelial or neuroectodermal origin

Sigma T 7289   Human recombinant, expressed in *E. coli*   MW 25k (dimer)   ≥93% (SDS-PAGE); 0.2 μm filtered liquid containing 30% acetonitrile & 0.1% TFA; endotoxin tested | Multifunctional protein capable of influencing cell proliferation, differentiation & a variety of cellular functions; stimulates growth of cells of mesenchymal origin, but inhibits growth of hepatocytes, epithelial cells, T & B lymphocytes; known to interact with several other agents including EGF, FGF, TGF-α, PDGF & IL-2; bioactivity of TGF-β is tested in culture using Mv1Lu cells, a mink lung epithelial cell line, whereby 1 U is the amount of TGF-b required to induce half-maximal inhibition of Mv1Lu cells as measured in an MTT dye assay or by measuring ^{3}H-thymidine incorporation in mouse HT-2 cells; Cheifetz, S et al, *Cell*, 48: 409, 1987; Roberts, A & Sporn, M, *Adv Cancer Res*, 51: 107, 1988; Sporn, M et al, *Science*, 233: 532, 1986; Tsang, M et al, *Lymphokine Res*, 9: 607, 1990

Oncogene PF053   Porcine   MW 25k   >97% (SDS-PAGE); lyophilized in 30% acetonitrile, 0.1% TFA containing 50 mg of BSA per 1 mg of cytokine; reconstitute in sterile 4 mM HCl containing at least 0.1% HSA or BSA; biological activity: EC$_{50}$ of 0.1-0.2 ng/mL as determined by the ability to inhibit IL-4-dependent ^{3}H-thymidine incorporation in mouse HT-2 cells | Species reactivity: pig & mouse; stimulatory for cells of mesenchymal origin & inhibitory for cells of epithelial or neuroectodermal origin; for proliferation studies

Sigma T 5300   Porcine platelets   ≥97% (SDS-PAGE visualized by silver stain); 0.2 μm filtered & lyophilized in the presence of 50 μg BSA; bioactivity is measured by its inhibitory effect on ^{3}H-thymidine incorporation in the mouse IL-4 dependent HT-2 cells; endotoxin tested | Multifunctional protein capable of influencing cell proliferation, differentiation & a variety of cellular functions; stimulates growth of cells of mesenchymal origin, but inhibits growth of hepatocytes, epithelial cells, T & B lymphocytes; known to interact with several other agents including EGF, FGF, TGF-α, PDGF & IL-2; bioactivity of TGF-β is tested in culture using Mv1Lu cells, a mink lung epithelial cell line, whereby 1 U is the amount of TGF-b required to induce half-maximal inhibition of Mv1Lu cells as measured in an MTT dye assay or by measuring ^{3}H-thymidine incorporation in mouse HT-2 cells; Cheifetz, S et al, *Cell*, 48: 409, 1987; Roberts, A & Sporn, M, *Adv Cancer Res*, 51: 107, 1988; Sporn, M et al, *Science*, 233: 532, 1986; Tsang, M et al, *Lymphokine Res*, 9: 607, 1990

Biogenesis 9130-1616   r-DNA   Purified Ig; liquid

Transforming Growth Factor βIII

Calbiochem PF073   Human recombinant   MW 25k   >97% (SDS-PAGE); lyophilized; biological activity: ED$_{50}$=30-100 pg/mL | For use in growth inhibition assays & other assays designed to study cellular responses & receptor interactions involving TGF; expression of TGF-β3 is linked to progression osteosarcoma; Kloen, P et al, *Cancer*, 80: 2230, 1997

Oncogene PF073   Human recombinant   MW 25k (disulfide-linked homodimeric)   >97% (SDS-PAGE); lyophilized; biological activity: EC$_{50}$ of 0.01-0.03 ng/mL as determined by the ability to inhibit IL-4-dependent ^{3}H-thymidine incorporation by mouse HT-2 cells | Recombinant human protein based on a DNA sequence encoding TGF-β3 prepropeptide (containing chicken TGF-β signal & latency associated peptide sequence & human mature TGF-β3 sequence) expressed in Sf21 cells; species reactivity: human & mouse; for growth inhibition assay

PeproTech 100-36   Human recombinant, expressed in *E. coli*   MW 25.0k   >98% (SDS-PAGE) & HPLC; <0.1 ng endotoxin per mg (1EU/mg); lyophilized | Member of a superfamily of homologous, disulfide-linked, homodimeric proteins; regulate proliferation & differentiation of normal & transformed cells 112 AA; ED$_{50}$ ≤ 0.05 ng/mL; SA ≥ 2 x 10^8 U/mg; SA determined by ability to inhibit mouse IL-4-dependent proliferation of mouse HT-2 cells

Sigma T 5425   Human recombinant, expressed in *Sf* 21 insect cells   ≥97% (SDS-PAGE); purified by sequential chromatography; 0.2 μm filtered & lyophilized from 35% acetonitrile & 0.1% TFA containing 100 μg BSA; bioactivity is measured by its inhibitory effect on ^{3}H-thymidine incorporation in the mouse IL-4 dependent HT-2 cells; endotoxin tested | Multifunctional protein capable of influencing cell proliferation, differentiation & a variety of cellular functions; stimulates growth of cells of mesenchymal origin, but inhibits growth of hepatocytes, epithelial cells, T & B lymphocytes; known to interact with several other agents including EGF, FGF, TGF-α, PDGF & IL-2; bioactivity of TGF-β is tested in culture using Mv1Lu cells, a mink lung epithelial cell line, whereby 1 U is the amount of TGF-b required to induce half-maximal inhibition of Mv1Lu cells as measured in an MTT dye assay or by measuring ^{3}H-thymidine incorporation in mouse HT-2 cells; Cheifetz, S et al, *Cell*, 48: 409, 1987; Roberts, A & Sporn, M, *Adv Cancer Res*, 51: 107, 1988; Sporn, M et al, *Science*, 233: 532, 1986; Tsang, M et al, *Lymphokine Res*, 9: 607, 1990

Transforming Growth Factor βV

Sigma T 2926   Amphibian recombinant, expressed in Sf 21 cells ≥97% (SDS-PAGE); purified by sequential chromatography; 0.2 µm filtered & lyophilized from 30% acetonitrile & 0.1% TFA containing 100 µg BSA; proliferative activity is tested in culture by using a mouse cell line HT-2; endotoxin tested | Multifunctional protein capable of influencing cell proliferation, differentiation & a variety of cellular functions; stimulates growth of cells of mesenchymal origin, but inhibits growth of hepatocytes, epithelial cells, T & B lymphocytes; known to interact with several other agents including EGF, FGF, TGF-α, PDGF & IL-2; bioactivity of TGF-β is tested in culture using Mv1Lu cells, a mink lung epithelial cell line, whereby 1 U is the amount of TGF-b required to induce half-maximal inhibition of Mv1Lu cells as measured in an MTT dye assay or by measuring ^{3}H-thymidine incorporation in mouse HT-2 cells; Cheifetz, S et al, *Cell*, 48: 409, 1987; Roberts, A & Sporn, M, *Adv Cancer Res*, 51: 107, 1988; Sporn, M et al, *Science*, 233: 532, 1986; Tsang, M et al, *Lymphokine Res*, 9: 607, 1990

Trichosanthin

Sigma T 0794   *Trichosanthes kirilowii*   Solution in 0.05 M Tris, pH 7.4, containing 50% glycerol; >99% | Active principle of the traditional Chinese medicinal plant with abortifacient activity; potent protein synthesis inhibitor *in vitro* & an inhibitor of HIV replication; Maraganore, JM et al, *J Biol Chem*, 262: 11628, 1987; Barbieri, L et al, *J Chromatogr*, 408: 235, 1987; McGrath, MS et al, *Proc Natl Acad Sci* USA, 86: 2844, 1989

Sigma T 3797   *Trichosanthes kirilowii*   Lyophilized powder containing ~15% protein (Lowry); balance primarily sodium phosphate & preservatives; ~95% (SDS-PAGE) | Active principle of the traditional Chinese medicinal plant with abortifacient activity; potent protein synthesis inhibitor *in vitro* & an inhibitor of HIV replication; Maraganore, JM et al, *J Biol Chem*, 262: 11628, 1987; Barbieri, L et al, *J Chromatogr*, 408: 235, 1987; McGrath, MS et al, *Proc Natl Acad Sci* USA, 86: 2844, 1989

Trichostatin A

Synonyms: Histone Deacetylase Inhibitor

USBio T8375   Streptomyces   >98%; powder; soluble in DMSO, DMF, acetonitrile, & EtOH | Incubation of A431 cells with 50–100 ng/mL Trichostatin A for 24 hours inhibited deacetylation of histones *in vivo*

Tropomyosin

Sigma T 4770   Bovine muscle   Lyophilized powder; may contain Tris buffer salts, CaCl$_2$ & mercaptoethanol; SDS electrophoresis shows 2 major bands | In the absence of Ca^{2+}, inhibits muscle contractility by blocking the myosin binding sites on actin; prepared by a modification of the procedure of Greaser, ML & Gergely, J, *J Biol Chem*, 246: 4226, 1971

Sigma T 3026   Chicken gizzard   Lyophilized powder; may contain Tris buffer salts; SDS electrophoresis shows 2 major bands | Active principle of the traditional Chinese medicinal plant with abortifacient activity; potent protein synthesis inhibitor *in vitro* & an inhibitor of HIV replication; Maraganore, JM et al, *J Biol Chem*, 262: 11628, 1987; Barbieri, L et al, *J Chromatogr*, 408: 235, 1987; McGrath, MS et al, *Proc Natl Acad Sci* USA, 86: 2844, 1989

Sigma T 1646   Chicken muscle   Lyophilized powder; may contain Tris buffer salts; SDS electrophoresis shows 2 major bands | Active principle of the traditional Chinese medicinal plant with abortifacient activity; potent protein synthesis inhibitor *in vitro* & an inhibitor of HIV replication; Maraganore, JM et al, *J Biol Chem*, 262: 11628, 1987; Barbieri, L et al, *J Chromatogr*, 408: 235, 1987; McGrath, MS et al, *Proc Natl Acad Sci* USA, 86: 2844, 1989

Cortex CP3034   Human   >95%

Sigma T 2400   Porcine muscle   Lyophilized powder; may contain Tris buffer salts; SDS electrophoresis shows 2 major bands | Active principle of the traditional Chinese medicinal plant with abortifacient activity; potent protein synthesis inhibitor *in vitro* & an inhibitor of HIV replication; Maraganore, JM et al, *J Biol Chem*, 262: 11628, 1987; Barbieri, L et al, *J Chromatogr*, 408: 235, 1987; McGrath, MS et al, *Proc Natl Acad Sci* USA, 86: 2844, 1989

Sigma T 3640   Rabbit muscle   Lyophilized powder; may contain Tris buffer salts; SDS electrophoresis shows 2 major bands | Active principle of the traditional Chinese medicinal plant with abortifacient activity; potent protein synthesis inhibitor *in vitro* & an inhibitor of HIV replication; Maraganore, JM et al, *J Biol Chem*, 262: 11628, 1987; Barbieri, L et al, *J Chromatogr*, 408: 235, 1987; McGrath, MS et al, *Proc Natl Acad Sci* USA, 86: 2844, 1989

Troponin

Synonyms: Actin Associated Protein

Sigma T 4895   Bovine muscle   Lyophilized powder containing ~85% protein (Biuret); balance Tris buffer salts; SDS gel electrophoresis shows the 3 major bands for troponin | In the absence of Ca^{2+}, inhibits muscle contractility by causing tropomyosin to block the myosin binding sites on actin; prepared by a modification of the procedure of Greaser, ML & Gergely, J, *J Biol Chem*, 246: 4226, 1971

Sigma T 1771   Chicken muscle   Lyophilized powder containing ~90% protein (Biuret); balance Tris buffer salts; SDS gel electrophoresis shows the 3 major bands for troponin | In the absence of Ca^{2+}, inhibits muscle contractility by causing tropomyosin to block the myosin binding sites on actin; prepared by a modification of the procedure of Greaser, ML & Gergely, J, *J Biol Chem*, 246: 4226, 1971

Sigma T 2275   Porcine muscle   Lyophilized powder containing ~85% protein (Biuret); balance Tris buffer salts; SDS gel electrophoresis shows the 3 major bands for troponin | In the absence of Ca^{2+}, inhibits muscle contractility by causing tropomyosin to block the myosin binding sites on actin; prepared by a modification of the procedure of Greaser, ML & Gergely, J, *J Biol Chem*, 246: 4226, 1971

Sigma T 3515   Rabbit muscle   Lyophilized powder containing ~85% protein (Biuret); balance Tris buffer salts; SDS gel electrophoresis shows the 3 major bands for troponin | In the absence of Ca^{2+}, inhibits muscle contractility by causing tropomyosin to block the myosin binding sites on actin; prepared by a modification of the procedure of Greaser, ML & Gergely, J, *J Biol Chem*, 246: 4226, 1971

Troponin C

Synonyms: Actin Associated Protein

Biodesign A86857H   Human heart   >98% | Cardiac marker

Calbiochem 648475   Human heart   MW 18k   ≥95% (SDS-PAGE); liquid in phosphate buffer containing 250 mM NaCl, pH 7.2 | Regulates muscle contractions by Ca^{2+} binding, which results in the reorganization of the interactions between the troponin-tropomyosin complex; prepared from tissues of individuals that have been shown by certified tests to be negative for HBsAg & for antibodies to HIV & HCV; Malnic, B et al, *J Biol Chem*, 273: 10594, 1998

Chemicon AG751   Human heart   ≥95%

Fitzgerald 30-AT44   Human heart   High purity

Cortex CP3035   Rabbit   >95%

Fitzgerald 30-AT64   Recombinant, expressed in *E. coli*   98%

Troponin Complex

Synonyms: Troponin C, Cardiac I & T

USBio T8665-12   Human   ≥98%; no contaminants detected; single band by SDS-PAGE, IEP, &/or RID; 0.14 mg/mL supplied in 2 mM NaHCO$_3$, 5 mM CaCl$_2$, pH 7.5 | Suitable for antigenic applications in immunological protocols; Troponin Complex is native & corresponds to the presentation of troponin complex in serum of AMI patients

Biodesign A86862H   Human heart   Cardiac marker

Fitzgerald 30-AT49   Human heart   95% | Troponin C, I & T

Fitzgerald 30-AT54   Human heart   97% | Troponin C, I & T

Troponin I

Synonyms: Actin Associated Protein

Cortex CP3036   Human   >95%

USBio T8665-13 Human cardiac ≥95%; no contaminants were detected by SDS-PAGE analysis & IEP; <0.1% Fe; ~2 mg/mL; lyophilized | Suitable for antigenic applications in immunological protocols

USBio T8665-16 Human cardiac ≥98%; no contaminants were detected by SDS-PAGE analysis & IEP; ≤ 0.1% Fe; ~2 mg/mL; lyophilized | Suitable for antigenic applications in immunological protocols

Biodesign A86813H Human heart >98% | Cardiac marker

Biodesign A86853H Human heart >95% | Cardiac marker

Biogenesis 9202-0707 Human heart Tested negative for antibodies to HBsAg, HCV and HIV1; purified; from 10 *mM* HCl; lyophilized | Suitable for iodination, as standard or for antibody production

Chemicon AG750 Human heart ≥95%

Fitzgerald 30-AT43 Human heart High purity

Sigma T 9924 Human heart Affinity-purified, lyophilized powder | Inhibiting subunit of troponin, responsible for preventing actin-myosin binding & ATPase activity

Chemicon AG752 Human skeletal ≥95%

Biodesign A86824H Human skeletal muscle >95% | Cardiac marker

Biodesign A86825H Human skeletal muscle >95% | Cardiac marker

Fitzgerald 30-AT48 Human skeletal muscle High purity

USBio T8665-21 Human skeletal muscle ≥95%; no contaminants were detected by SDS-PAGE analysis & IEP; <0.1% Fe; 1 mg/mL; lyophilized from 10 *mM* HCl | Suitable for antigenic applications in immunological protocols

Fitzgerald 30-AT63 Recombinant, expressed in *E. coli* 96%

Biodesign A3B003H Recombinant, expressed in *Pichia pastoris* >95% | Cardiac marker

Troponin I, Cardiac Calibrator Set

Synonyms: Actin Associated Protein

Biodesign A86860H Human heart Calibrator set (NHS base) | Cardiac marker

Troponin T

Synonyms: Actin Associated Protein; Actin Associated Protein; Actin Associated Protein

Cortex CP3037 Human >95%

USBio T8665-23 Human ≥98%; no contaminants were detected by SDS-PAGE analysis & IEP; <0.1% Fe; 1.45 mg/mL; lyophilized | Suitable for antigenic applications in immunological protocols; suitable for Western Blot, Immunohistochemistry & ELISA Capture

Biogenesis 9202-1107 Human heart Tested negative for HBsAg, HIV-1 and HIV-2 antibodies and HCV; purified; 10 *mM* HCl; lyophilized

Chemicon AG754 Human heart ≥95%

Fitzgerald 30-AT38 Human heart High purity

Sigma T 0175 Human heart ≥60% (SDS-PAGE); affinity-purified, lyophilized powder | Tropomyosin-binding subunit of troponin

Chemicon AG756 Human skeletal ≥95%

Biogenesis 9202-1157 Human skeletal muscle Tested negative for HBsAg and HIV antibodies; purified; PBS buffer, pH 7.2; lyophilized

Biodesign A3B002H Recombinant, expressed in *Pichia pastoris* >95% | Cardiac marker

Troponin, Complex

Scipac P184-4 Heart tissue 1-10%; frozen in TRIS buffer; >10,000 ng/mL; contains Troponin I/T/C complex | Cardiac marker protein

Trypsin Inhibitor

Synonyms: Ovomucoid; PCI; Hageman Factor Inhibitor; Popcorn Inhibitor; Ovomucoid; SBTI; Kunitz Soybean Trypsin Inhibitor

Fluka 93616 Chicken egg white Powder; ~10,000 U/mg; 1 U corresponds to the amount of inhibitor which reduces the trypsin activity by 1 BAEE-U; 1 BAEE-U is the amount of enzyme which increases the absorbance at 253 nm by 0.001/min at pH 7.6, 25°C; no chymotrypsin inhibition detected | Prepared by method C of Tomimatsu Y et al, *Arch Biochem Biophys*, 115: 536, 1966

Fluka 93621 Chicken egg white Powder; ~10,000 U/mg; 1 U corresponds to the amount of inhibitor which reduces the trypsin activity by 1 BAEE-U; 1 BAEE-U is the amount of enzyme which increases the absorbance at 253 nm by 0.001/min at pH 7.6, 25°C; chymotrypsin inhibition 12 U/mg

Fluka 93622 Corn kernels MW 12,028 Solution in 20 mM TRIS HCl, 30 mM NaCl, pH 8.2; ≥30,000 U/mL; 1 U corresponds to the amount of inhibitor which reduces the trypsin activity by 1 BAEE-U; 1 BAEE-U is the amount of enzyme which increases the absorbance at 253 nm by 0.001/min at pH 7.6, 25°C

ICN 198905 Duck egg 95% (SDS-PAGE); lyophilized; 1 mg inhibits 1 mg trypsin with activity of 10,000 BAEE U/mg protein

Fluka 93609 Lima beans Powder; ~10,000 U/mg; 1 U corresponds to the amount of inhibitor which reduces the trypsin activity by 1 BAEE-U; 1 BAEE-U is the amount of enzyme which increases the absorbance at 253 nm by 0.001/min at pH 7.6, 25°C | Birk, Y, *Meth Enzymol*, 45: 707, 1976

Fluka 93618 Soybean Powder; ~5000 U/mg; 1 U corresponds to the amount of inhibitor which reduces the trypsin activity by 1 BAEE-U; 1 BAEE-U is the amount of enzyme which increases the absorbance at 253 nm by 0.001/min at pH 7.6, 25°C; chymotrypsin inhibition ≤30 U/mg

Fluka 93619 Soybean MW 20k Lyophilized; ~12,000 U/mg; 1 U corresponds to the amount of inhibitor which reduces the trypsin activity by 1 BAEE-U; 1 BAEE-U is the amount of enzyme which increases the absorbance at 253 nm by 0.001/min at pH 7.6, 25°C | Prepared by method of Kunitz, M, *J Gen Physiol*, 29: 149, 1946

Fluka 93620 Soybean Lyophilized; essentially salt-free; ~10,000 U/mg; 1 U corresponds to the amount of inhibitor which reduces the trypsin activity by 1 BAEE-U; 1 BAEE-U is the amount of enzyme which increases the absorbance at 253 nm by 0.001/min at pH 7.6, 25°C | Prepared by method of Kunitz, M, *J Gen Physiol*, 29: 149, 1946; Ozawa, K & Laskowski, M, *JBC*, 241: 3955, 1966; Birk, Y, *Meth Enzymol*, 45: 700, 1976; Katoh, S et al, *Polym Prep J Am Chem Soc, Div Polym Chem*, 21: 94, 1980

Sigma T 1021 Soybean pI 4.6; vial contains ~2 mg | IEF Marker

Sigma T 9416 Soybean MW 20.1k Vial contains enough FITC-conjugated protein to run 50 mini-gels or 25 standard size gels; protein band be visualized by using UV light or Brilliant Blue stain | Fluorescent marker for SDS-page & protein transfer; suitable for use as molecular weight standards in both SDS-PAGE & transfer membranes

Fluka 93623 Turkey egg white Powder; ~11,000 U/mg; 1 U corresponds to the amount of inhibitor which reduces the trypsin activity by 1 BAEE-U; 1 BAEE-U is the amount of enzyme which increases the absorbance at 253 nm by 0.001/min at pH 7.6, 25°C; chymotrypsin inhibition 28 U/mg

Trypsin Inhibitor Type III-O

Sigma T 2011 Chicken egg white Purified ovomucoid free of ovoinhibitor prepared from Type II-O using method C of Tomimatsu; 1 mg inhibits ~1 mg trypsin with activity of ~10,000 BAEE U/mg protein (Biuret) | Selective inhibitor; 1 Trypsin U = ΔA_{253} of 0.001/min with BAEE as substrate at pH 7.6, 25°C; reaction volume=3.2 mL (1 cm light path); 1 chymotrypsin U hydrolyzes 1.0 μmole BTEE/min at pH 7.8, 25°C; Tomimatsu, Y et al, *Arch Biochem Biophys*, 115: 536, 1966

Trypsin Inhibitor Type II-L

Sigma T 9378 Lima bean Crude powder; 1 mg inhibits ~1 mg trypsin with activity of ~10,000 BAEE U/mg protein (Biuret) | 1 Trypsin U = ΔA_{253} of 0.001/min with BAEE as substrate at pH 7.6, 25°C; reaction volume=3.2 mL (1 cm light path); 1 chymotrypsin U hydrolyzes 1.0 μmole BTEE/min at pH 7.8, 25°C; prepared by a modification of the method of Tauber, H et al, *J Biol Chem*, 179: 1155, 1949 (modified)

Trypsin Inhibitor Type II-O

Sigma T 9253 Chicken egg white Partially purified ovomucoid containing ovoinhibitor; 1 mg inhibits ~1 mg trypsin with activity of ~10,000 BAEE U/mg protein (Biuret); may inhibit ≤0.3 mg chymotrypsin with activity of ~40 BTEE U/mg protein | 1 Trypsin U = ΔA_{253} of 0.001/min with BAEE as substrate at pH 7.6, 25°C; reaction volume=3.2 mL (1 cm light path); 1 chymotrypsin U hydrolyzes 1.0 μmole BTEE/min at pH 7.8, 25°C; ovomucoid, from chicken egg white, does not itself inhibit chymotrypsin; chymotrypsin inhibition is a measure of ovoinhibitor contamination by method of Feeney, RE et al, *J Biol Chem*, 238: 1415, 1963; Lineweaver, H & Murray, CW, *J Biol Chem*, 171: 565, 1947

Trypsin Inhibitor Type II-S

Sigma T 9128 Soybean MW ~20k Crude; soluble powder; 1 mg inhibits 1.5-2.5 mg trypsin with activity of ~10,000 BAEE U/mg protein (Biuret) | 1 Trypsin U = ΔA_{253} of 0.001/min with BAEE as substrate at pH 7.6, 25°C; reaction volume=3.2 mL (1 cm light path); 1 chymotrypsin U hydrolyzes 1.0 μmole BTEE/min at pH 7.8, 25°C

Trypsin Inhibitor Type II-T

Sigma T 4385 Turkey egg white 1 mg inhibits 0.9-1.3 mg trypsin with activity of ~10,000 BAEE U/mg protein (Biuret); 1 mg inhibits 0.4-1.0 mg α-chymotrypsin with activity of ~40 BTEE U/mg protein | 1 Trypsin U = ΔA_{253} of 0.001/min with BAEE as substrate at pH 7.6, 25°C; reaction volume=3.2 mL (1 cm light path); 1 chymotrypsin U hydrolyzes 1.0 μmole BTEE/min at pH 7.8, 25°C; Lineweaver, H & Murray, CW, *J Biol Chem*, 171: 565, 1947

Trypsin Inhibitor Type I-P

Synonyms: BPTI

Sigma T 0256 Bovine pancreas Crystallized & lyophilized, essentially salt-free; 1 mg inhibits 1.5-3.0 mg trypsin with activity of ~10,000 BAEE U/mg protein (Biuret); 1 mg inhibits ~0.8 mg α-chymotrypsin with activity of ~40 BTEE U/mg protein | 1 Trypsin U = ΔA_{253} of 0.001/min with BAEE as substrate at pH 7.6, 25°C; reaction volume=3.2 mL (1 cm light path); 1 chymotrypsin U hydrolyzes 1.0 μmole BTEE/min at pH 7.8, 25°C; prepared by the method of Kunitz & Northrup, *J Gen Physiol*, 19: 991, 1936

Trypsin Inhibitor Type I-S

Sigma T 9003 Soybean MW ~20k Lyophilized; chromatographically prepared; 1 mg inhibits 1-3 mg trypsin with activity of ~10,000 BAEE U/mg protein (Biuret) | 1 Trypsin U = ΔA_{253} of 0.001/min with BAEE as substrate at pH 7.6, 25°C; reaction volume=3.2 mL (1 cm light path); 1 chymotrypsin U hydrolyzes 1.0 μmole BTEE/min at pH 7.8, 25°C

Sigma T 9008 Soybean MW ~20k 1% sterile filtered solution prepared form Type I-S | 1 Trypsin U = ΔA_{253} of 0.001/min with BAEE as substrate at pH 7.6, 25°C; reaction volume=3.2 mL (1 cm light path); 1 chymotrypsin U hydrolyzes 1.0 μmole BTEE/min at pH 7.8, 25°C; for use in the staphylococcal clumping & serial dilution-protamine sulfate tests for detecting fibrin/fibrinogen degradation products

Trypsin Inhibitor Type IV-O

Synonyms: Chymotrypsin Inhibitor

Sigma T 1886 Chicken egg white Purified ovoinhibitor prepared from Type II-O using method C of Tomimatsu; 1 mg inhibits ~1 mg trypsin with activity of ~10,000 BAEE U/mg protein (Biuret); 1 mg inhibits ~1 mg chymotrypsin with activity of ~40 BTEE U/mg protein | Selective inhibitor; 1 Trypsin U = ΔA_{253} of 0.001/min with BAEE as substrate at pH 7.6, 25°C; reaction volume=3.2 mL (1 cm light path); 1 chymotrypsin U hydrolyzes 1.0 μmole BTEE/min at pH 7.8, 25°C; Tomimatsu, Y et al, *Arch Biochem Biophys*, 115: 536, 1966

Trypsin Inhibitor, DITC Glass Conjugated

Sigma T 9024 Soybean MW ~20k Inhibitor Sigma T 9003 covalently attached to DITC controlled pore glass (80-120 mesh, 700 Angstrom average pore size); 1 g solid yields ~2 mL packed volume; 1 g solid inhibits 2-25 mg trypsin with activity of ~10,000 BAEE U/mg protein (Biuret) | 1 Trypsin U = ΔA_{253} of 0.001/min with BAEE as substrate at pH 7.6, 25°C; reaction volume=3.2 mL (1 cm light path); 1 chymotrypsin U hydrolyzes 1.0 μmole BTEE/min at pH 7.8, 25°C

Trypsin Inhibitor, Ovomucoid Gold Conjugated

Sigma T 2654 Chicken egg white 10 nm Colloidal Gold labeled; mean particle size 8-12 nm; monodisperse; suspension in ~40% glycerol containing 0.15 M NaCl, 0.01 M potassium phosphate, pH 7.2, 0.02% PEG 20; 0.02% sodium azide; concentration: A_{520} ~5.0; Sigma T 2011 adsorbed to colloidal gold | Useful in a 2-step (indirect) labeling of lectin binding sites with GlcNAc specificity; 1 Trypsin U = ΔA_{253} of 0.001/min with BAEE as substrate at pH 7.6, 25°C; reaction volume=3.2 mL (1 cm light path); 1 chymotrypsin U hydrolyzes 1.0 μmole BTEE/min at pH 7.8, 25°C; Geoghegan, WD & Ackerman, GA, *J Histochem Cytochem*, 31: 1394, 1983

Trypsin/Chymotrypsin Inhibitor

Synonyms: Bowman-Birk Inhibitor

Sigma T 9777 Soybean Lyophilized powder containing ~80% protein (Biuret); balance primarily phosphate buffer salts, pH 7.6; 1 mg inhibits 3-5 mg trypsin with activity of ~10,000 BAEE U/mg protein; unit definition: 1 trypsin U=ΔA_{253} of 0.001/min with BAEE as substrate at pH 7.6, 25°C; reaction volume=3.2 mL (1 cm light path); 1 chymotrypsin U hydrolyzes 1.0 μmole BTEE/min at pH 7.8, 25°C | Birk, Y, *Int J Peptide Protein Res*, 25: 113, 1985

Trypsinogen

Sigma T 1143 Bovine pancreas Dialyzed & lyophilized, essentially salt-free; activity: 10,000-15,000 BAEE U/mg protein (E_{280} at 1%) after activation to trypsin; contains <1000 BAEE U/mg protein prior to activation; unit definition: 1 BAEE U=ΔA_{253} of 0.001/min with BAEE as substrate at pH 7.6, 25°C & a reaction volume of 3.2 mL (1 cm light path) | Wilimowska-Pelc, A & Mejbaum-Katzenellenbogen, W, *Anal Biochem*, 90: 816, 1978

Sigma T 1146 Bovine pancreas pI 9.3; vial contains ~2 mg | IEF Marker

Tryptase

Sigma T 7063 Human lung Highly purified; solution in 10 mM MES, 300 mM NaCl, 0.02 mM heparin, pH 6.1, containing 0.02% sodium azide; activity: ≥5000 U/mg; unit definition: 1 U hydrolyzes 1.0 μmole *N*-benzyl-DL-arginine-pNA | *J Biol Chem*, 259: 11046, 1984; Addington, AK & Johnson, D, *Biochem*, 197: 13511, 1996; Schwartz, LB & Bradford, TR, *J Biol Chem*, 261: 7372, 1986

Tryptic Soy Broth

Sigma T 8261 Soybean Soybean casein digest broth

Tryptone

Synonyms: EZMix™

Sigma T 2559   Pancreatic digest of casein; same formulation as Sigma T 9410 with the added advantage of being dust-free, allowing easier weighing & handling | Used to supplement media in the cultivation of anaerobes

Sigma T 9410   ~13% total nitrogen | Pancreatic digest of casein; used to supplement media in the cultivation of anaerobes (clostridia & other fermenting organisms, e.g. Lactobacilli)

Tubulin

Sigma T 4925   Bovine brain   Lyophilized powder containing MES buffer salts, EGTA, EDTA, $MgCl_2$, DTT, GTP, leupeptin, aprotinin & sucrose as stabilizer; contains ~15% microtubule associated proteins; vial contains > 7.5 mg protein in assembled form which will give >5 mg soluble protein after disassembly | Primary protein of microtubules; Ringel, I & Horwitz, SB, *J Pharmacol Exp Ther*, 259: 855, 1991; purified by temperature-dependent assembly-disassembly cycles

Biogenesis 9280-3050   Porcine brain   Liquid

Tubulin, α- & β-

ICN 771121/771122   Bovine brain   MW 55k   10 mg/mL in 80 mM PIPES buffer, 1 mM EGTA, 1 mM GTP, 10% glycerol; 1 U = 5.0 mg purified protein; 1.0 U/mL increases A_{340} from 0-0.650 (35°C, 30 min), microtubule mass = 3.6 mg/mL; mean microtubule length ~10 mm (immunofluorescence microscopy); average microtubules/mL = 1.0×10^{12} | Heterodimer composed of α- & β-tubulin; polymerization forms microtubules 25 nm (diameter) X ~1650 heterodimers/µm length

Tumor Necrosis Factor Receptor I

R&D Systems 225-B1-025   *E. coli*-Expressed   >97%; lyophilized; ED_{50}: 0.03-0.06 µg/mL | Species specificity: human; soluble TNF RI

R&D Systems 425-R1-050   *E. coli*-Expressed   >97%; lyophilized; ED_{50}: 0.5-1.5 µg/mL | Species specificity: mouse; soluble TNF RI

Chemicon GF103   Human   ≥95%

R&D Systems 372-RI-050   NSO-Expressed   >95%; lyophilized; ED_{50}: 0.4-1 ng/mL | Species specificity: human; domain fused to human IgG1 Fc

R&D Systems 430-RI-050   NSO-Expressed   >90%; lyophilized; ED_{50}: 0.2-0.6 ng/mL | Species specificity: mouse; domain fused to human IgG1 Fc

Tumor Necrosis Factor Receptor I, Soluble

Oncogene PF055   Human recombinant   MW 18k (non-glycosylated)   >97% (SDS-PAGE); lyophilized; biological activity: EC_{50} of 0.03-0.06 µg/mL as determined by the ability to inhibit the cytotoxic effects of 0.25 ng/mL recombinant human TNF-α in mouse L929 cells | Species reactivity: mouse, human; neutralizes the biological activity of TNF-α at least 10-fold more efficiently than it neutralizes TNF-β

Oncogene PF056   Murine recombinant   MW 21k (non-glycosylated)   >97% (SDS-PAGE); lyophilized; biological activity: EC_{50} of 0.5-1.5 µg/mL as determined by the ability to inhibit the cytotoxic effects of 0.1 ng/mL recombinant mouse TNF-α in mouse L929 cells | Species reactivity: mouse; neutralizes the biological activity of TNF-α at least 10-fold more efficiently than it neutralizes TNF-β

Tumor Necrosis Factor Receptor II, Soluble

R&D Systems 226-B2-025   *E. coli*-Expressed   >97%; lyophilized; ED_{50}: 0.2-0.6 µg/mL | Species specificity: human

R&D Systems 426-R2-050   *E. coli*-Expressed   >97%; lyophilized; ED_{50}: 1-3 µg/mL | Species specificity: mouse

Oncogene PF057   Human recombinant   MW 20k (non-glycosylated)   >97% (SDS-PAGE); lyophilized; biological activity: EC_{50} of 0.2-0.06 µg/mL as determined by the ability to inhibit the cytotoxic effects of 0.25 ng/mL TNF-α in mouse L929 cells | Species reactivity: mouse, human; neutralizes the biological activity of both TNF-α & TNF-β with approximately equal efficiency

Oncogene PF058   Murine recombinant   MW 25k (non-glycosylated)   >97% (SDS-PAGE); lyophilized; biological activity: EC_{50} of 1-3 µg/mL as determined by the ability to inhibit the cytotoxic effects of 0.1 ng/mL TNF-α in mouse L929 cells | Species reactivity: mouse; neutralizes the biological activity of both TNF-α & TNF-β with approximately equal efficiency

R&D Systems 726-R2-050   NSO-Expressed   >95%; lyophilized; ED_{50}: 0.004-0.016 µg/mL | Species specificity: human

Tumor Necrosis Factor α

Synonyms: TNFSF2; Cachectin, rh-; Cachectin; Cachectin, rm-

Biodesign A52014H   *E. coli*   MW 17.9k   Purified | Species specificity: rat

Biodesign A52150H   *E. coli*   MW 18.5k   Purified | Species specificity: mouse

Biodesign A52301H   *E. coli*   MW 17.5k   Purified

R&D Systems 210-TA-010   *E. coli*-Expressed   >97%; lyophilized; ED_{50}: 0.02-0.05 ng/mL | Species specificity: human

R&D Systems 210-TA-050   *E. coli*-Expressed   >97%; lyophilized; ED_{50}: 0.02-0.05 ng/mL | Species specificity: human

R&D Systems 410-MT-010   *E. coli*-Expressed   >97%; lyophilized; ED_{50}: 0.02-0.05 ng/mL | Species specificity: mouse

R&D Systems 410-MT-050   *E. coli*-Expressed   >97%; lyophilized; ED_{50}: 0.02-0.05 ng/mL | Species specificity: mouse

R&D Systems 410-TRNC-010   *E. coli*-Expressed   >97%; lyophilized; ED_{50}: 5-10 pg/mL | Species specificity: mouse; amino terminal truncated form of mouse TNF-α

R&D Systems 410-TRNC-050   *E. coli*-Expressed   >97%; lyophilized; ED_{50}: 5-10 pg/mL | Species specificity: mouse; amino terminal truncated form of mouse TNF-α

R&D Systems 510-RT-010   *E. coli*-Expressed   >97%; lyophilized; ED_{50}: 10-20 pg/mL | Species specificity: rat

R&D Systems 510-RT-050   *E. coli*-Expressed   >97%; lyophilized; ED_{50}: 10-20 pg/mL | Species specificity: rat

R&D Systems 690-PT-025   *E. coli*-Expressed   >97%; lyophilized; ED_{50}: 0.003-0.018 ng/mL | Species specificity: porcine

Chemicon GF023   Human   ≥95%

Biogenesis 9295-1302   Human r-DNA expressed in *E. coli*   MW 17.5k   SA: ~2x10(e)7 U/mg; endotoxin <0.1 ng/µg; purified; additive and carrier free; lyophilized | 157 AA; Carswell et al, *PNAS USA*, 72:3666, 1975; Aggerwal et al, *JBC*, 260:2345, 1985; Rosenblum & Donato, *Crit Rev Immunol*, 9:21, 1989

Oncogene PF007   Human recombinant   MW 17k   >97% (SDS-PAGE); lyophilized; biological activity: half maximal cytolysis of murine L929 cells in the presence of Actinomycin D is 0.2 ng/mL | Species reactivity: mouse, human; used in cytotoxicity assays & Western blot; use 2 mg per gel lane, 68 for Western blot; acts on human & mouse cells

Calbiochem 654205   Human recombinant, expressed in *E. coli*   MW 17.5k   ≥97% (SDS-PAGE); lyophilized from sterile-filtered PBS containing 50 µg BSA/µg TNF-α; biological activity: ED_{50}=20-50 pg/mL as measured in a cytotoxicity assay with the TNF-α-susceptible murine L-929 cell line in the presence of Actinomycin D; endotoxin: ≤100 pg/µg TNF-α; May be carcinogenic/teratogenic | Activates a variety of immune defense mechanisms by interactions with polymorphonuclear leukocytes, T cells, antibody-producing B lymphocytes, fibroblasts & hematopoietic bone marrow cells; activity is not species-specific; induces apoptosis in human blood & bone marrow neutrophils & in endothelial cells; increases the iNOS levels in vascular smooth muscle cells; involved in pathophysiological processes of several chronic & acute diseases; stimulates stress activated protein (SAP) kinase; *Merck Index*, 12: 9943; Tsuchida, H et al, *J Immunol*, 154: 2403, 1995; Westwick, JK et al, *J Biol Chem*, 270: 22689, 1995; Polunovsky, VA et al, *Exp Cell Res*, 214: 584, 1994; Koide, M et al, *FEBS Lett*, 318: 213, 1993; Vilcek, J & Lee, TH, *J Biol Chem*, 266: 7313, 1991

Fitzgerald 30-AT70   Human recombinant, expressed in *E. coli*

Harlan BT-3030   Human recombinant, expressed in *E. coli* Lyophilized; 0.01 mg

Harlan BT-3031   Human recombinant, expressed in *E. coli* Lyophilized; 0.05 mg

PeproTech 300-01A   Human recombinant, expressed in *E. coli* MW 17.4k   >98% (SDS-PAGE) & HPLC; <0.1 ng endotoxin per mg (1EU/mg); lyophilized | Potent lymphoid factor; exerts cytotoxic effects on a wide range of tumor cells & certain other target cells; 157 AA; $ED_{50} \leq 0.05$ ng/mL; $SA \geq 2 \times 10^7$ U/mg; SA determined by the cytolysis of murine L929 cells in the presence of actinomycin D

Sigma T 6674   Human recombinant, expressed in *E. coli*   MW 17k   ≥97% (SDS-PAGE); 0.2 µm filtered & lyophilized from phosphate buffered saline containing 500 µg BSA; endotoxin tested | Two closely related agents, TNF-α & TNF-β elicit similar inflammation & anti-tumor activities by acting upon the same cellular receptor; they share a 30% homology in primary structure & both may exist in multimeric forms under certain conditions; TNF-α is secreted by lipopolysaccharide-stimulated macrophages while TNF-β is secreted by activated T-lymphocytes; TNF cause cytolysis of certain transformed cells & are directly toxic to vascular endothelial cells; they are mitogenic to certain other cell types; TNF-α is undergoing clinical trials for treatment of certain cancers; Sigma's growth factors, cytokines & receptor research reagents are for research only; TNF cytolytic activity is tested in cell culture using a mouse fibrosarcoma line, L929 cells or its derivatives, whereby 1 U is the amount of TNF required to induce half-maximal cytolysis in the presence of actinomycin D measured by crystal violet staining; Hass, P et al, *J Biol Chem*, 260: 12214, 1985; Carswell, E et al, *Proc Natl Acad Sci USA*, 72: 3666, 1975; Ruddle, N et al, *J Exp Med*, 128: 1267, 1968; Helson, L et al, *Nature*, 258: 731, 1975; Sato, N et al, *JNCI*, 76: 1113, 1986; Jones, A & Selby, P, *Prog Growth Factor Res*, 1: 107, 1989; Lejeune, F et al, in: *Tumor Necrosis Factor: Molecular & Cellular Biology & Clinical Relevance*, Fiers, W et al, eds, Karger, Basel p.1, 1993; Sheehan, K et al, *J Immunol*, 142: 3884, 1989

Alexis 520-002   Human recombinant, expressed in yeast   MW ~17k   Cell culture grade; ≥98% (SDS-PAGE); lyophilized containing 50 µg D-mannitol/µg rhTNF-α; soluble in water or most aqueous buffers; activity: 1×10^8 units/mg protein; biological activity: ED_{50} as determined by the cytolysis of murine L929 cells in the presence of mitomycin C is 0.2 ng/mL | For most *in vitro* applications, rhTNF-α exerts its biological activity in the concentration range of 0.05 to 20 ng/mL

IBT GF-090-3, GF-090-4   Human recombinant, expressed in yeast   MW 17k (monomer)   >95%; 1 mg/mL solution in 0.02 *M* Tris-HCl, 0.15 *M* NaCl, pH 8.0 | Antitumor activity *in vivo* & *in vitro*; activates polymorphonuclear leukocytes; antiviral activity; induces release of Interleukin-1 or colony stimulating factor from various sources

Sigma T 0157   Human recombinant, expressed in yeast   MW 17k   >95% (SDS-PAGE); 0.2 µm filtered solution in 1 mL phosphate buffered saline containing 500 µg BSA; activity: 2×10^7 U/mg; cell culture tested; endotoxin tested | Two closely related agents, TNF-α & TNF-β elicit similar inflammation & anti-tumor activities by acting upon the same cellular receptor; they share a 30% homology in primary structure & both may exist in multimeric forms under certain conditions; TNF-α is secreted by lipopolysaccharide-stimulated macrophages while TNF-β is secreted by activated T-lymphocytes; TNF cause cytolysis of certain transformed cells & are directly toxic to vascular endothelial cells; they are mitogenic to certain other cell types; TNF-α is undergoing clinical trials for treatment of certain cancers; Sigma's growth factors, cytokines & receptor research reagents are for research only; TNF cytolytic activity is tested in cell culture using a mouse fibrosarcoma line, L929 cells or its derivatives, whereby 1 U is the amount of TNF required to induce half-maximal cytolysis in the presence of actinomycin D measured by crystal violet staining; Hass, P et al, *J Biol Chem*, 260: 12214, 1985; Carswell, E et al, *Proc Natl Acad Sci USA*, 72: 3666, 1975; Ruddle, N et al, *J Exp Med*, 128: 1267, 1968; Helson, L et al, *Nature*, 258: 731, 1975; Sato, N et al, *JNCI*, 76: 1113, 1986; Jones, A & Selby, P, *Prog Growth Factor Res*, 1: 107, 1989; Lejeune, F et al, in: *Tumor Necrosis Factor: Molecular & Cellular Biology & Clinical Relevance*, Fiers, W et al, eds, Karger, Basel p.1, 1993; Sheehan, K et al, *J Immunol*, 142: 3884, 1989

Alexis BMS301   Human recombinant, produced in *E. coli*   MW 52k (trimer composed of 17.3k monomers)   >98% (SDS-gel electrophoresis prior to addition of human serum albumin); 10 µg lyophilized powder in phosphate-buffered saline containing 10 mg/mL BSA; SA: 5×10^7 units/mg; bioactivity: L-M cytotoxicity assay was performed in the presence of 1 µg/mL Actinomycin D

Biogenesis 9295-1355   Mouse r-DNA

Calbiochem 654245   Mouse recombinant, expressed in *E. coli* MW 17k   ≥97% (SDS-PAGE); lyophilized from sterile-filtered PBS containing 50 µg BSA/µg TNF-α; biological activity: ED_{50}=20-50 pg/mL as measured in a cytotoxic assay with the TNF-α-susceptible murine L-929 cell line in the presence of Actinomycin D; endotoxin: ≤100 pg/µg TNF-α; May be carcinogenic/teratogenic | Has the ability to kill certain tumor cells directly; referred to as an inflammatory cytokine as it initiates the cascade of other cytokines & factors that make up the immune system's response to infection & cancer; reported to mediate changes in bone metabolism during inflammation; *Merck Index*, 12: 9943; Scharla, SH et al, *Eur J Endocrinol*, 131: 293, 1994; Eck, MJ et al, *J Biol Chem*, 267: 2119, 1992

Sigma T 7539   Mouse recombinant, expressed in *E. coli*   MW 17k   ≥97% (SDS-PAGE); 0.2 µm filtered & lyophilized from phosphate buffered saline, pH 7.4 containing 500 µg BSA per µg of cytokine; endotoxin tested | Two closely related agents, TNF-α & TNF-β elicit similar inflammation & anti-tumor activities by acting upon the same cellular receptor; they share a 30% homology in primary structure & both may exist in multimeric forms under certain conditions; TNF-α is secreted by lipopolysaccharide-stimulated macrophages while TNF-β is secreted by activated T-lymphocytes; TNF cause cytolysis of certain transformed cells & are directly toxic to vascular endothelial cells; they are mitogenic to certain other cell types; TNF-α is undergoing clinical trials for treatment of certain cancers; Sigma's growth factors, cytokines & receptor research reagents are for research only; TNF cytolytic activity is tested in cell culture using a mouse fibrosarcoma line, L929 cells or its derivatives, whereby 1 U is the amount of TNF required to induce half-maximal cytolysis in the presence of actinomycin D measured by crystal violet staining; Hass, P et al, *J Biol Chem*, 260: 12214, 1985; Carswell, E et al, *Proc Natl Acad Sci USA*, 72: 3666, 1975; Ruddle, N et al, *J Exp Med*, 128: 1267, 1968; Helson, L et al, *Nature*, 258: 731, 1975; Sato, N et al, *JNCI*, 76: 1113, 1986; Jones, A & Selby, P, *Prog Growth Factor Res*, 1: 107, 1989; Lejeune, F et al, in: *Tumor Necrosis Factor: Molecular & Cellular Biology & Clinical Relevance*, Fiers, W et al, eds, Karger, Basel p.1, 1993; Sheehan, K et al, *J Immunol*, 142: 3884, 1989

Chemicon GF027   Murine   ≥95%

PeproTech 315-01A   Murine recombinant, expressed in *E. coli* MW 17.5k   >98% (SDS-PAGE) & HPLC; <0.1 ng endotoxin per mg (1EU/mg); lyophilized | Potent lymphoid factor; exerts cytotoxic effects on a wide range of tumor cells & certain other target cells; 156 AA; $ED_{50} \leq 0.1$ ng/mL; $SA \geq 1 \times 10^7$ U/mg; SA determined by the cytolysis of murine L929 cells in the presence of actinomycin D

Tumor Necrosis Factor α

Fitzgerald 30-AT72   Murine recombinant, expressed in in *E. coli*

Chemicon GF046   Rat   ≥95%

PeproTech 400-14   Rat recombinant, expressed in *E. coli* MW 17k   >98% (SDS-PAGE) & HPLC; <0.1 ng endotoxin per mg (1EU/mg); lyophilized | Potent lymphoid factor; exerts cytotoxic effects on a wide range of tumor cells & certain other target cells; 157 AA; $ED_{50} \leq 0.05$ ng/mL; $SA \geq 2 \times 10^7$ U/mg; SA determined by the cytolysis of murine L929 cells in the presence of actinomycin D

Fitzgerald 30-AT73   Rat recombinant, expressed in in *E. coli*

Tumor Necrosis Factor α, Soluble

Alexis 522-008   Human recombinant   MW 19k under reducing conditions   Cell culture grade; ≥95% (SDS-PAGE); lyophilized containing 50 µg protein | The extracellular domain is fused at the *N*-terminus to FLAG-tag & an 8 AA linker peptide; Recognizes human & mouse TNF-R1; combined with an enhancer (Alexis Kit Prod. No. 850-060-KI01), recognizes human & mouse TNF-R1, but only human TNF-R2; Schneider, P et al, *J Exp Med*, 187: 1205, 1998

Alexis 522-009   Mouse recombinant   MW 20k under reducing conditions   Cell culture grade; >95% (SDS-PAGE); lyophilized containing 50 μg protein | The extracellular domain of the mouse TNF-α (AA 77-235) is fused at the *N*-terminus to FLAG-tag & an 8 AA linker peptide; recognizes human, mouse & rat TNF-R1; combined with an enhancer (Alexis Kit Prod. No. 850-061-Kl01), recognizes human & mouse TNF-R1 & TNF-R2; Schneider, P et al, *J Exp Med*, 187: 1205, 1998

Tumor Necrosis Factor β

Synonyms: TNFSF2; Lymphotoxin; Lymphotoxin

Biodesign A52310H   *E. coli*   MW 18.5k   Purified

R&D Systems 211-TB-010   *E. coli*-Expressed   >95%; lyophilized; ED_{50}: 0.02-0.05 ng/mL | Species specificity: human

R&D Systems 211-TB-050   *E. coli*-Expressed   >95%; lyophilized; ED_{50}: 0.02-0.05 ng/mL | Species specificity: human

Chemicon GF024   Human   ≥95%

Biogenesis 9295-3305   Human r-DNA

Calbiochem 654215   Human recombinant, expressed in *E. coli*   MW 18.8k   ≥95% (SDS-PAGE); lyophilized from sterile-filtered PBS containing 50 μg BSA/μg TNF-β; biological activity: ED_{50}=20-50 pg/mL as measured in a cytotoxicity assay with the TNF-β-susceptible murine L-929 cell line in the presence of Actinomycin D; endotoxin: ≤100 pg/μg TNF-β; May be carcinogenic/teratogenic | Secreted by activated T lymphocytes; structurally & functionally related to TNF-α & is active on murine cells; shown to induce necrosis of tumors in mouse models; exerts inflammatory & cytotoxic effects; *Merck Index*, 12: 9943; Loetscher, H et al, *Cancer Cells*, 3: 221, 1991

Fitzgerald 30-AT71   Human recombinant, expressed in *E. coli*

Harlan BT-3032   Human recombinant, expressed in *E. coli* Lyophilized; 20 μg

Harlan BT-3033   Human recombinant, expressed in *E. coli* Lyophilized; 0.05 mg

PeproTech 300-01B   Human recombinant, expressed in *E. coli* MW 18.6k   >98% (SDS-PAGE) & HPLC; <0.1 ng endotoxin per mg (1EU/mg); lyophilized | Potent lymphoid factor; exerts cytotoxic effects on a wide range of tumor cells & certain other target cells; 172 AA; $ED_{50} \leq$ 0.05 ng/mL; SA $\geq 2 \times 10^7$ U/mg; SA determined by the cytolysis of murine L929 cells in the presence of actinomycin D

PeproTech 315-01B   Human recombinant, expressed in *E. coli* MW 18.6k   >98%; 171 AA; lyophilized from 10 *mM* NaPB pH 7.2, 20 *mM* NaCl; ED_{50}: 0.05-1.0 ng/mL as determined by the cytolysis of murine L929 cells in the presence of Actinomycin D

Sigma T 7799   Human recombinant, expressed in *E. coli*   MW 18.8k   ≥95% (SDS-PAGE); 0.2 μm filtered & lyophilized from phosphate buffered saline containing 500 μg BSA; endotoxin tested | Two closely related agents, TNF-α & TNF-β elicit similar inflammation & anti-tumor activities by acting upon the same cellular receptor; they share a 30% homology in primary structure & both may exist in multimeric forms under certain conditions; TNF-α is secreted by lipopolysaccharide-stimulated macrophages while TNF-β is secreted by activated T-lymphocytes; TNF cause cytolysis of certain transformed cells & are directly toxic to vascular endothelial cells; they are mitogenic to certain other cell types; TNF-α is undergoing clinical trials for treatment of certain cancers; Sigma's growth factors, cytokines & receptor research reagents are for research only; TNF cytolytic activity is tested in cell culture using a mouse fibrosarcoma line, L929 cells or its derivatives, whereby 1 U is the amount of TNF required to induce half-maximal cytolysis in the presence of actinomycin D measured by crystal violet staining; Hass, P et al, *J Biol Chem*, 260: 12214, 1985; Carswell, E et al, *Proc Natl Acad Sci USA*, 72: 3666, 1975; Ruddle, N et al, *J Exp Med*, 128: 1267, 1968; Helson, L et al, *Nature*, 258: 731, 1975; Sato, N et al, *JNCI*, 76: 1113, 1986; Jones, A & Selby, P, *Prog Growth Factor Res*, 1: 107, 1989; Lejeune, F et al, in: *Tumor Necrosis Factor: Molecular & Cellular Biology & Clinical Relevance*, Fiers, W et al, eds, Karger, Basel p.1, 1993; Sheehan, K et al, *J Immunol*, 142: 3884, 1989

Alexis BMS302   Human recombinant, produced in *E. coli*   MW 16.5k   >98% (SDS-gel electrophoresis prior to addition of human serum albumin); 10 μg lyophilized powder in phosphate-buffered saline containing 10 mg/mL BSA; SA: 3×10^8 units/mg; bioactivity: L-M cytotoxicity assay was performed in the presence of 1 μg/mL Actinomycin D | This product is the 16 kDa form of the full length 18 kDa human TNF-β protein & expresses identical bioactivity to that displayed by natural human TNF-β

TWEAK

Chemicon GF102   Human   ≥95%

PeproTech 310-06   Human recombinant, expressed in *E. coli* MW 17.0k   >98% (SDS-PAGE) & HPLC; <0.1 ng endotoxin per mg (1EU/mg); lyophilized | New secreted ligand in the tumor necrosis factor family; weakly induces apoptosis; induced interleukin-8 synthesis in a member of cell lines; comprises the TNF-like extracellular domain of TWEAK; ED_{50} < 10ng/mL; SA determined by the dose-dependent stimulation of IL-8 production by human PBMC

Tyrphostin A25

Synonyms: Protein Tyrosine Kinase Inhibitor

USBio T9250   >99%; lyophilized; soluble in DMSO or EtOH | IC_{50}: 3 *mM*

Tyrphostin A51

Synonyms: Protein Tyrosine Kinase Inhibitor

USBio T9250-05   >99%; lyophilized; soluble in DMSO or EtOH | IC_{50}: 800 *nM*

Tyrphostin AG 1288

Synonyms: Protein Tyrosine Kinase Inhibitor

USBio T9250-10   >99%; lyophilized; soluble in DMSO or DMF | Has activity similar to that of Tyrphostin A10; prevents lipopolysaccharide-induced lethal toxicity (septic shock) in mice

Tyrphostin AG 490

Synonyms: jak2 Protein Tyrosine Kinase Inhibitor

USBio T9250-15   >99% by HPLC; lyophilized; soluble in DMSO | Specific & potent inhibitor; also, inhibits EGF receptor autophosphorylation, IC_{50}: 100 *nM*; inhibits DNA synthesis & cell growth; induces apoptosis; blocks growth of leukemic cells *in vitro* & *in vivo*

Ubiquitin

Synonyms: ATP-Dependent Proteolytic Factor

Biogenesis 9400-1502   Bovine RBCs   Lyophilized

Fluka 93950   Bovine red blood cells   MW 181.19   ≥90.0% (GE); ≥90% protein | Stimulation of ATP-dependent proteolysis; Wilkinson, KD & Audhya, TK, *JBC*, 256: 9235, 1981

Sigma U 6253   Bovine red blood cells   MW 8.5k Lyophilized, essentially salt-free powder; ≥90% (SDS gel electrophoresis); contains ≥5.4 moles glycine/mole protein | 76-AA protein bound covalently by a conjugating enzyme to proteins that are targeted for degradation by 26S proteosome; occurs in virtually all eukaryotes, including plants (thus its name); sequence is highly conserved, identical in animal sources from insects to humans; Wilkinson, KD & Audhya, TK, *J Biol Chem*, 256: 9235, 1981; Coux, O et al, *Ann Rev Biochem*, 65: 801, 1996

Ubiquitin, (His[10])-

R&D Systems 701-UB-025   *E. coli*-Expressed   95%; frozen without a carrier protein | Species specificity: human His[10] ubiquitin; used in *in vitro* apoptosis assays

Ubiquitin, (Me)-

Sigma U 1632   Bovine   Lyophilized, essentially salt-free powder; ≥90% (SDS-PAGE); prepared from bovine ubiquitin by reductive methylation of primary amino groups | Can be ligated to proteins but cannot form polyubiquitin chains; inhibits the elongation of polyubiquitin chains & inhibits ubiquitin-dependent protein degradation; Hershko, A & Heller, H, *Biochem Biophys Res Commun*, 128: 1079, 1985; Zeigenhagen, R et al, *FEBS Lett*, 271: 71, 1990; Hershko, A et al, *J Biol Chem*, 266: 16476, 1990

Ubiquitin+1, (His[10])-

R&D Systems 703-UB-025   *E. coli*-Expressed   95%; frozen without a carrier protein | Species specificity: human His[10] ubiquitin+1; used in *in vitro* apoptosis assays

Urease Large Subunit Antigen, *Helicobacter pylori*

IBT HPA-5030-4, HPA-5030-5   Yeast cells   MW 63k   >90%; 0.50 mg/mL solution in PBS, 0.1% SDS, 10 m*M* EDTA, pH 7.4 | Covers Met[1] to Lys[559] of *H. pylori* urease large subunit

Urease Small Subunit Antigen, *Helicobacter pylori*

IBT HPA-5020-4, HPA-5020-5   Yeast cells   MW 27k   >92%; .50 mg/mL solution 0 in PBS, 0.1% SDS, 10 m*M* EDTA, pH 7.4 | Covers Met[1] to Glu[238] of *H. pylori* urease small subunit

Urinary Protein Lyophilizate

Sigma U 8126   Human male urine   Lyophilized preparation containing 30-70% protein (Lowry) & ~ 2% sodium phosphate; contains those components of urine >10,000 1 mg protein represents ~20-40 mL urine | Sold on basis of protein content

Urinary Trypsin Inhibitor

Scipac P205-1   Urine of patients with chronic renal tubular proteinuria   >96%; lyophilized; available on request | Urine protein

Urine Protein I

Synonyms: Clara Cell 16 Protein

Scipac P174-4   Urine of patients with chronic renal tubular proteinuria   30-70%; in Phosphate buffer; available on request | Urine protein

Urogastrone

Sigma U 9129   Human pregnancy urine   Crude extract containing ~ 40% protein (Biuret) | Potent stimulator of cell proliferation; structurally similar to & has the same intrinsic biological activity as EGF; Morimoto, T et al, *Yakugaku Zasshi*, 89: 215, 1969

Vac (Toxin) Antigen, *Helicobacter pylori*

IBT HPA-5010-4, HPA-5010-5   Yeast cells, as a fusion with human superoxide dismutase   MW 72k   >90%; 0.50 mg/mL solution in PBS, 0.1% SDS, 1 m*M* EDTA, 10 m*M* DTT, pH 7.5 | Covers Gly[311] to Ile[819] of *H. pylori* vacuolating protein

Vascular Endothelial Growth Factor

Synonyms: Vascular Endothelial Growth Factor 165; Placenta Growth Factor; Vasculotropin; Vasculotropin, rh-

Kamiya   MW 19-22k   Purified from culture medium by heparin-agarose chromatography

Biodesign A52120H   *E. coli*   MW 38k   purified

Biodesign A52532M   *E. coli*   Purified | Species specificity: mouse

Chemicon GF025   Human   ≥95%

Chemicon GF104   Human   ≥95%

Biogenesis 9532-7008   Human r-DNA   MW 38k   Purified; lyophilized | Whole dimeric protein is expressed; antigen is two 165 AA chains covalently linked by disulfide bonds; demonstrates mitogenic activity

BioSource International PHG0114   Human recombinant

Fitzgerald 30-AV15   Human recombinant, expressed in *E. coli*

Harlan BT-3034   Human recombinant, expressed in *E. coli*   Lyophilized; 0.01 mg

Harlan BT-3035   Human recombinant, expressed in *E. coli*   Lyophilized; 0.05 mg

ICN 159302   Human recombinant, expressed in *E. coli*   ≥98%; lyophilized; 1-100 ng/mL, determined by mitogenic activity on human dermal microvascular endothelial cells | Stimulates endothelial cell growth, angiogenesis & capillary permeability

PeproTech 100-20   Human recombinant, expressed in *E. coli*   MW 38.2k   >98% (SDS-PAGE) & HPLC; <0.1 ng endotoxin per mg (1EU/mg); lyophilized | Homodimeric protein secreted by a variety of vascularized tissues; reported activities include stimulation of endothelial cell growth, angiogenesis & capillary permeability; two 165 AA polypeptide chains; SA determined by its mitogenic activity on human dermal microvascular endothelial cells using a concentration range of 1-100 ng/mL

Sigma P 1588   Human recombinant, expressed in *E. coli*   MW 29k   ≥97% (SDS-PAGE); 0.2 μm filtered & lyophilized from phosphate buffered saline solution containing 500 μg BSA; bioactivity is measured in a cell-based bioassay using HUVEC cells; endotoxin tested | 165 AA; angiogenic growth factor which is heat & acid stable; dimeric heparin-binding glycoprotein; able to promote the growth of vascular endothelial cells isolated from bovine adrenal cortex, cerebral cortex, fetal & adult aorta & human umbilical vein; doesn't have a mitogenic effect on cultured corneal endothelial cells, vascular smooth muscle cells, BHK-12 fibroblasts, keratinocytes, human sarcoma cells or lens epithelial cells; biological activity is measured by its ability to stimulate ^{3}H-thymidine incorporation in human umbilical vein endothelial cells (HUVEC); member of the VEGF family of growth factors; Ferrara, N et al, *Biochem Biophy Res Commun*, 161: 851, 1989; Ferrara, N et al, *Endocrine Reviews*, 13: 18, 1992; Conn, G et al, *Proc Natl Acad Sci USA*, 87: 1323, 1990

Sigma V 7259   Human recombinant, expressed in *Sf* 21 cells   MW 42k   ≥97% (SDS-PAGE); 0.2 μm filtered & lyophilized from 30% acetonitrile & 0.1% TFA containing 250 μg BSA; endotoxin tested | 165 AA; angiogenic growth factor which is heat & acid stable; dimeric heparin-binding glycoprotein; able to promote the growth of vascular endothelial cells isolated from bovine adrenal cortex, cerebral cortex, fetal & adult aorta & human umbilical vein; doesn't have a mitogenic effect on cultured corneal endothelial cells, vascular smooth muscle cells, BHK-12 fibroblasts, keratinocytes, human sarcoma cells or lens epithelial cells; biological activity is measured by its ability to stimulate ^{3}H-thymidine incorporation in human umbilical vein endothelial cells (HUVEC); member of the VEGF family of growth factors; Ferrara, N et al, *Biochem Biophy Res Commun*, 161: 851, 1989; Ferrara, N et al, *Endocrine Reviews*, 13: 18, 1992; Conn, G et al, *Proc Natl Acad Sci USA*, 87: 1323, 1990

Calbiochem 676472   Human recombinant, expressed in *Spodoptera frugiperda*   MW 42k   ≥97% (SDS-PAGE); lyophilized from filter-sterilized 30% acetonitrile, 0.1% TFA containing 50 μg BSA/μg VEGF; biological activity: ED$_{50}$=2.0-6.0 ng/mL as measured by its ability to stimulate ^{3}H-thymidine incorporation into human umbilical vein endothelial cells; endotoxin: ≤100 pg/μg VEGF | Heparin-binding glycoprotein with potent antiogenic, endothelial cell-specific mitogenic & vascular permeability-enhancing activities; increases endothelial cell proliferation after vascular injury; Burke, PA et al, *Biochem Biophys Res Comm*, 207: 348, 1995; Neufeld, G et al, *Prog Growth Factor Res*, 5: 89, 1994

Biogenesis 0100-0107   Mouse r-DNA   Lyophilized

BioSource International PMG0114   Mouse recombinant

Sigma V 4512   Mouse recombinant, expressed in *Sf* 21 insect cells   ≥97% (SDS-PAGE); lyophilized from 30% acetonitrile & 0.1% TFA containing 250 µg BSA; endotoxin tested; cell culture tested | 165 AA; angiogenic growth factor which is heat & acid stable; dimeric heparin-binding glycoprotein; able to promote the growth of vascular endothelial cells isolated from bovine adrenal cortex, cerebral cortex, fetal & adult aorta & human umbilical vein; doesn't have a mitogenic effect on cultured corneal endothelial cells, vascular smooth muscle cells, BHK-12 fibroblasts, keratinocytes, human sarcoma cells or lens epithelial cells; biological activity is measured by its ability to stimulate ^{3}H-thymidine incorporation in human umbilical vein endothelial cells (HUVEC); member of the VEGF family of growth factors; Ferrara, N et al, *Biochem Biophy Res Commun*, 161: 851, 1989; Ferrara, N et al, *Endocrine Reviews*, 13: 18, 1992; Conn, G et al, *Proc Natl Acad Sci USA*, 87: 1323, 1990

Chemicon GF060   Murine   ≥95%

PeproTech 450-32   Murine recombinant, expressed in *E. coli*   MW 39k   >98% (SDS-PAGE) & HPLC; <0.1 ng endotoxin per mg (1EU/mg); lyophilized | Homodimeric protein secreted by a variety of vascularized tissues; stimulates endothelial cell growth, angiogenesis & capillary permeability; homodimeric protein with two two 165 AA polypeptide chains; mitogenic activity was obtained for stimulation of (^{3}H)thymidine incorporation & cell proliferation in a concentration range of 1.0-5.0 ng/mL; SA determined by the mitogenic activity on human umbilical vein endothelial cells and bovine aortic endothelial cells

ICN 195772   Murine recombinant, expressed in *Sf*21   ≥97%; lyophilized; ED_{50} = 2-4 ng/mL

BioSource International PRG0114   Rat recombinant

Vascular Endothelial Growth Factor 121

Oncogene PF083   Human recombinant   MW 28k (predicted homodimeric)   >97% (SDS-PAGE); lyophilized; biological activity: EC_{50} of 5-10 ng/mL as determined by the ability to stimulate ^{3}H-thymidine uptake in human umbilical vein endothelial cells | Species reactivity: human; for proliferation studies

ICN 195749   Human recombinant, expressed in *E. coli*   ≥97%; lyophilized; ED_{50} = 5-10 ng/mL

Vascular Endothelial Growth Factor 165

Oncogene PF074   Human recombinant   MW 42k (glycosylated) homodimeric   >97% (SDS-PAGE); lyophilized; biological activity: EC_{50} of 2-6 ng/mL as determined by the ability to stimulate ^{3}H-thymidine uptake in human umbilical vein endothelial cells | Species reactivity: human; for proliferation studies

Vascular Endothelial Growth Factor Receptor I

Oncogene PF082   Recombinant   MW ~100k/123k (glycosylated, SDS-PAGE)   >97% (SDS-PAGE); lyophilized in a sterile-filtered PBS solution containing 50 mg of BSA per 1 mg of cytokine; reconstitute in sterile PBS containing at least 0.1% serum albumin

Vascular Endothelial Growth Factor/Placenta Growth Factor Heterodimer

ICN 195748   Human recombinant, expressed in *E. coli*   ≥97%; lyophilized; ED_{50} = 100-200 ng/mL

Venom, Bee

Sigma V 3125   *Apis mellifera* (honey bee)   Dried whole venom; free of transglycosidase; contains numerous other enzymes & substances; entire non-volatile part of the venom obtained by stimulating bees to sting a sheet from which the venom is then gathered & dried | Contains hyaluronidase of the β-*N*-acetyl glucosaminidase type which hydrolyzes hyaluronic acid yielding a tetra & a hexasaccharide; Barker, SA et al, *Nature*, 199: 693, 1963; *Clin Chim Acta*, 8: 902, 1963; 9: 339, 1964

Sigma V 3250   *Apis mellifera* (honey bee)   Natural suspension; essentially whole venom including a major part of the volatile constituents; ~50% solids; free of transglycosidase; contains numerous other enzymes & substances | Contains hyaluronidase of the β-*N*-acetyl glucosaminidase type which hydrolyzes hyaluronic acid yielding a tetra & a hexasaccharide; Barker, SA et al, *Nature*, 199: 693, 1963; *Clin Chim Acta*, 8: 902, 1963; 9: 339, 1964

Sigma V 3375   *Apis mellifera* (honey bee)   Lyophilized whole venom; this preparation comes from a different supplier & might be similar to the higher priced Sigma V 3125, but not enough data to be certain | Contains hyaluronidase of the β-*N*-acetyl glucosaminidase type which hydrolyzes hyaluronic acid yielding a tetra & a hexasaccharide; Barker, SA et al, *Nature*, 199: 693, 1963; *Clin Chim Acta*, 8: 902, 1963; 9: 339, 1964

Sigma V 6878   *Megabombus fervidus* (bumble bee)

Sigma V 7003   *Megabombus fervidus* (bumble bee)

Vimentin

Biodesign A08011B   Bovine lens   >98%

Biogenesis 9550-1004   Bovine lens   MW 57k   pI: 5.3; purified; 10 *mM* Sodium phosphate, pH 7.5, 2 *mM* DDT, 6 *M* urea, 1mM EDTA; lyophilized | Bloemendal et al, *FEBS Letts*, 180:2191, 1985

Vinculin

Biodesign A08012C   Chicken gizzard   >90% | Platelets & hemostasis reagents

Vitamin B2 Binding Protein

Biogenesis 9579-9095   Chicken egg white   MW ~30-35k   1 mg of protein (Biuret) will bind 5-15 µg of riboflavin; pI: 3.9-4.5; essentially free of riboflavin and salts; lyophilized | Osuga et al, *Arch Biochem Biophys*, 124:560, 1968; Farrell et al, *Int J Biochem*, 1:168, 1975; Froehlich et al, *Comp Biochem Physiol*, 66B:397, 1980; Miller et al, *Comp Biochem Physiol*, 69B:681, 1981

Vitamin D Binding Protein

Synonyms: Globulin, Gc-

Biodesign A50674H   Human plasma   Purified

Biogenesis 9580-2750   Human plasma   MW 52k   Tested negative for HBsAg and for antibodies to HIV and HCV; purified; 150 *mM* NaCl, 20 *mM* sodium phosphate, pH 7.4; lyophilized

Vitronectin

Synonyms: Serum Spreading Factor; Serum Spreading Factor; Serum Spreading Factor

Biogenesis 9590-0102   Bovine plasma   Purified; PBS buffer, pH 7.2, 0.1% NaN$_3$; liquid | S-Protein of the SC5b-7 complex is vitronectin

Sigma V 9881   Bovine plasma   Non-sterile; may be filtered; lyophilized in buffered saline; cell culture tested | Antigenically unrelated to fibronectin

BioSource International PHE0011   Human

Biogenesis 9590-1504   Human outdated plasma   Tested negative for HBsAg and HIV1 antibodies; purified; 20 *mM* ammonium carbonate buffer; liquid

Alexis 200-082   Human plasma   >95% (SDS-PAGE); liquid containing 100 µg purified protein in 100 µL sterile 0.15 *M* PBS, pH 7.4; activity: cell adhesion activity to NRK-49F cells is observed at concentration >0.5 µg/mL; negative for HBs antigen & HIV antibody | Multifunctional adhesion glycoprotein with binding sites for e.g. collagen, heparin, perforin & integrins; ligand for the Ca^{2+}-dependent CD41/CD61 complex, the major integrin of platelets; Suzuki, S et al, *EMBO J*, 4: 2519, 1985; Preissner, KT et al, *Ann Rev Cell Biol*, 7: 275, 1991; Felding-Habermann, B & Cheresh, DA, *Curr Opin Cell Biol*, 5: 864, 1993

Calbiochem 681105   Human plasma   MW 65k & 75k Lyophilized solid; >95% (SDS-PAGE); soluble in water & aqueous buffers; prepared from plasma of individuals shown by certified tests to be negative for HBsAg & antibodies to HIV & HCV | Extracellular matrix protein that promotes cell adhesion; mixture of 75 kDa & 65 kDa polypeptides; biochemically & immunologically distinct from fibronectin; intact protein & SDS-PAGE bands react with monoclonal antibodies to vitronectin, bind to heparin & promote cell adhesion; *Merck Index*, 12: 10167; Hayman, EG et al, *PNAS*, 80: 4003, 1983

Chemicon CC080   Human plasma   Purified | Extracellular matrix protein

Promega G5381   Human plasma   MW 75k   A major plasma glycoprotein; circulates as a single-chain moiety of 75 kDa & a double-chain moiety of 65 kDa & 10 kDa; the plasma has been tested & found negative for HIV-1 antibody & Hepatitis B antigen | Contains the AA structural motif, Arg-Gly-Asp which is involved in cell attachment; inhibits the activity of antithrombin III on thrombin, binds β-endorphins & plasminogen activator inhibitor & prevents lysis *in vitro* of bystander cells by C5β-9 & perforin

Sigma V 8379   Human plasma   Non-sterile; may be filtered; lyophilized in buffered saline; cell culture tested | Antigenically unrelated to fibronectin

Biogenesis 9590-1755   Mouse serum   Purified; 20 *mM* ammonium carbonate buffer; liquid | Suitable for ELISA and promotion of cell adhesion; S-Protein of the SC5b-7 complex is vitronectin; Hayman et al, *PNAS* USA, 80:4003, 1983

Cortex CP3038   Plasma   >95%

Sigma V 0132   Rat plasma   Non-sterile; may be filtered; lyophilized in buffered saline; cell culture tested | Antigenically unrelated to fibronectin

Von Willebrand Factor
Cortex CP4002U   Human   >98%

Von Willebrand Factor, Factor VIII-Free
Cortex CP4003U   Human   >98%; factor VIII-free

Wortmannin
Synonyms: Phosphatidylinositol-3-Kinase Inhibitor

USBio W6000   Fungal   ≥95% by HPLC; lyophilized under inert gas; white to off-white powder; soluble in DMSO & EtOH | A fungal metabolite, cell-permeable, irreversible inhibitor of phosphatidylinositol 3-kinase (PI 3-Kinase), IC_{50}: 5 *nM*; blocks the catalytic activity of PI 3-Kinase without affecting upstream signaling events such as insulin receptor tyrosinekinase activity; also inhibits the activities of myosin light chainkinase & PI 4-kinase at concentrations a hundred times higher than those required to inhibit PI 3-kinase

Zein
Fluka 96095   Maize (corn)   MW 25-29k   ≤4% loss on drying; ≤1% residue on ignition; alcohol-soluble mixture of two polypeptides | Larkins, BA et al, *Trends Biochem Sci*, 9: 306, 1984

ICN 103306   Maize (corn)

Zymosan A
Fluka 97340   *Saccharomyces cerevisiae*   Major components are glucose & mannose after hydrolysis; ≤5% loss on drying | Complex cell wall polysaccharides & glycoproteins from yeast; Bar-Shavit, Z, & Goldman, R, *Meth Enzymol*, 132: 326, 1986; Absolom, DR, *Meth Enzymol*, 132: 138, 1986; Allen, RC, *Meth Enzymol*, 133: 458, 1986

Part 3. Peptides

Sequences and Modifications

(Cys)-Ala-Leu-Arg-Ile-Asp-Glu-Asp-Glu-Lys-Ala-Gly-Gln-Lys Trifluoroacetate Salt

Synonyms: Spastin (130-142), C~

Biogenesis 8350-1050   Human synthetic   MW 1574.47
>80% (HPLC); lyophilized

(Cys)-Gly-Ser-Leu-Gly-Ser-Gln-Pro-Leu-Leu-Lys-Pro-Ser-Pro-Tyr-Gly-Gln-Ser-Gly

Synonyms: Signal Transducer and Activator of Transcription VI Blocking Peptide; Signal Transducer and Activator of Transcription VI (806-823)

Calbiochem 575141   Mouse synthetic   MW 1876.1
$C_{81}H_{130}N_{22}O_{27}S$ Lyophilized | Based on the mouse STAT6 (806-823) with a Cys (C) residue added; coupled to KLH, used as the immunogen for the production of Anti-STAT6; suitable for use in immunoabsorption for immunoprecipitation, immunocytochemistry, Western blotting & dot blots

(Cys)-His-Asp-Asn-Leu-Lys-Gln-Leu-Met-Leu-Gln Acetate Salt

Synonyms: G Protein Alpha 13 (368-377)

Biogenesis 4737-2320   Human, mouse synthetic   MW 1332.7
>85% (HPLC); lyophilized

(Cys)-Ile-Arg-Glu-Leu-Lys-Pro-Glu-Gln-Val-Lys-Asn-Met-Ser Acetate Salt

Synonyms: Spastin (561-573), C~

Biogenesis 8350-1350   Human synthetic   MW 1671.5
>80% (HPLC); lyophilized

(Cys)-Leu-Glu-Asp-Asn-Glu-Glu-Arg-Met-Ser-Arg-Leu-Ser-Lys Acetate Salt

Synonyms: Glutamic Acid Decarboxylase (516-528)

Biogenesis 4670-6641   Human 65 k isoform synthetic   MW 1710.5   >80% (HPLC); liquid

(Cys)-Lys-Gln-Leu-Glu-Val-Ile-Arg-Ser-Gln-Gln-Lys-Arg-Gln-Gly-Thr-Ser

Synonyms: TAK1 C-Terminal Blocking Peptide (1-16)

Calbiochem 575369   Mouse synthetic   MW 2117.4
$C_{86}H_{153}N_{31}O_{29}S$ Lyophilized | Based on mouse TAK1 (563-579) with a Cys C residue added & coupled to KLH; this peptide coupled to KLH, was used as the immunogen for the production of Anti-TAK1; for use in immunoabsorption for immunoprecipitation, immunocytochemistry, Western blotting & dot blots

(Cys)-Phe-Pro-Met-Ile-Ser-Lys-Arg-Pro-Glu-His-Leu-Arg-Met-Asn-Leu

Synonyms: Ribosomal S6 Kinase C-Terminal Blocking Peptide (488-502)

Calbiochem 559283   Synthetic   MW 1972.5
$C_{86}H_{142}N_{26}O_{21}S_3$ Lyophilized | Based on human & rat S6 kinase (488-502) with a Cys C residue added; this peptide coupled to KLH, was used as the immunogen for the production of Anti-Ribosomal S6 Kinase; for use in immunoabsorption for immunoprecipitation, immunocytochemistry, Western blotting & dot blots

(Cys)-Phe-Ser-Lys-Ser-Gln-Thr-Asp-Val-Tyr-Asn-Asp-Ser Acetate Salt

Synonyms: Spastin (204-215), C~

Biogenesis 8350-1250   Human synthetic   MW 1492.9
>80% (HPLC); lyophilized

(Cys)-Pro-Gly-Lys-Lys-Pro-Thr-Pro-Ser-Leu-Leu-Ile

Synonyms: Anti-Nuclear Inhibitor of Protein Phosphatase I Blocking Peptide (341-351)

Calbiochem 482255   Bovine thymus   MW 1253.6
$C_{57}H_{100}N_{14}O_{15}S$ Lyophilized solid; soluble in water | Immunization & blocking peptide for Anti-NIPP-1

(Cys)-Ser-Asn-Pro-Ala-Ala-Thr-Gln-Ser-Glu-Ile-Asp-Phe-Leu-Ile-Glu Acetate Salt

Synonyms: Glutamic Acid Decarboxylase (571-585), (Glu^{579})-

Biogenesis 4670-6681   Human 67/65 k isoform synthetic
>80% (HPLC); liquid

Abu-Ala

Bachem G-1440.0250   MW 174.2 $C_7H_{14}N_2O_3$ Store at -15°C

Abu-Arg
D-Abu-CHA-Arg-pNA·2AcOH

Synonyms: Pefachrome®PK; Pefa-5019

Pentapharm 080-31, 080-03   MW 652.7   Highly sensitive chromogenic peptide substrate for plasmakallikrein (PK); v_{max}:7.48 μmol/min, K_M:0.175 *mM*

Abu-Asn-Arg-Leu-Glu-Ala-Ser-Ser-Arg-Ser-Ser-Lys
Mca-γ-Abu-Asn-Arg-Leu-Glu-Ala-Ser-Ser-Arg-Ser-Ser-Lys(Dnp) Amide

Bachem M-2260.0001   Store at -15°C

Abu-Gly

Bachem G-1445.0250   MW 160.17 $C_6H_{12}N_2O_3$ Store at -15°C

Abu-Ile-His-Pro-Phe-His-Leu-Val-Ile-His-Thr
DABCYL-γ-Abu-Ile-His-Pro-Phe-His-Leu-Val-Ile-His-Thr-EDANS

Bachem M-2050.0001   MW 1798.15 $C_{90}H_{120}N_{22}O_{16}S$ Store at -15°C

Abu-Ser-Gln-Asn-Tyr-Pro-Ile-Val-Gln
DABCYL-γ-Abu-Ser-Gln-Asn-Tyr-Pro-Ile-Val-Gln-EDANS

American Peptide 72-3-30   MW 1535.7 $C_{72}H_{96}N_{17}O_{19}S$ Substrate cleaved by the HIV-1 protease at the Tyr-Pro position resulting in a time-dependent increase in fluorescence intensity; Matayoshi, ED et al, *Science*, 247:954, 1990; Pennington, MW et al, *Peptides, Proc 22nd Eur Peptide Symp*, Interlaken, Switzerland, 936, 1992

Bachem M-1865.0001   Store at -15°C

Abz-Ala-Ala-Phe-Phe
Abz-Ala-Ala-Phe-Phe-pNA

Synonyms: Elastinolytic Metalloprotease Substrate

Neosystem SC1298   MW 708.75   Substrate for an elastinolytic metalloprotease from *Aspergillus fumigatus*; Markaryan, A et al, *Infection & Immunity*, 62:2149-2157, 1994

Abz-Ala-Arg-Val-Nle-Phe-Glu-Ala-Nle
Abz-Ala-Arg-Val-Nle-*p*-Nitro-Phe-Glu-Ala-Nle Amide

Synonyms: Anthranilyl-HIV Protease Substrate V

Bachem H-1168.0001   Store at -15°C

Abz-Ala-Gly-Leu-Ala
Abz-Ala-Gly-Leu-Ala-*p*-Nitrobenzylamide

Bachem H-6675.0025   MW 583.65 $C_{28}H_{37}N_7O_7$ Store at -15°C

Abz-Ala-Phe-Ala-Phe-Asp-Val-Phe-Tyr-Asp
Abz-Ala-Phe-Ala-Phe-Asp-Val-Phe-Tyr(NO₂)-Asp

Synonyms: Aspartate-Specific Protease Fluorescent Substrate

American Peptide 81-0-50   MW 1258.3 $C_{62}H_{71}N_{11}O_{18}$ *Eur J Biochem*, 206:103, 1992

Abz-Ala-Phe-Ala-Phe-Asp-Val-Phe-Tyr-Asp
Abz-Ala-Phe-Ala-Phe-Asp-Val-Phe-3-Nitro-Tyr-Asp

Bachem M-2475.0001   MW 1258.31 $C_{62}H_{71}N_{11}O_{18}$ Store at -15°C

Abz-Arg-Val-Lys-Arg-Gly-Leu-Ala-Tyr-Asp
Abz-Arg-Val-Lys-Arg-Gly-Leu-Ala-*m*-Nitro-Tyr-Asp

Bachem M-2115.0001   MW 1241.37 $C_{54}H_{84}N_{18}O_{16}$ Store at -15°C

Abz-Arg-Val-Nle-Phe-Glu-Ala-Nle
Abz-Arg-Val-Nle-*p*-Nitro-Phe-Glu-Ala-Nle Amide

Synonyms: Anthranilyl-HIV Protease Substrate VI

Bachem H-1204.0001   Store at -15°C

Abz-Gln-Val-Val-Ala-Gly-Ala
Abz-Gln-Val-Val-Ala-Gly-Ala-Ethylenediamine-Dnp

Bachem M-2100.0001   MW 870.92 $C_{38}H_{54}N_{12}O_{12}$ Store at -15°C

Abz-Glu-Thr-Leu-Phe-Gln-Gly-Pro-Val-Phe
Abz-Glu-Thr-Leu-Phe-Gln-Gly-Pro-Val-*p*-Nitro-Phe Amide

Bachem M-2075.0001   Store at -15°C

Abz-Gly Hydrochloride

Bachem E-2920.0001   MW 230.65 $C_9H_{10}N_2O_3 \cdot HCl$ Store at -15°C

Abz-Gly-Ala-Ala-Pro-Phe-Tyr-Asp
Abz-Gly-Ala-Ala-Pro-Phe-3-Nitro-Tyr-Asp

Bachem M-2480.0001   MW 903.9 $C_{42}H_{49}N_9O_{14}$ Store at -15°C

Abz-Gly-Gly-Ala-Ser-Ser-Arg-Leu-Tyr-Arg
Abz-tBu-Gly-tBu-Gly-Asn(Me)₂-Ala-Ser-Ser-Arg-Leu-3-Nitro-Tyr-Arg

Bachem M-2450.0001   MW 1384.56 $C_{61}H_{97}N_{19}O_{18}$ Store at -15°C

Abz-Gly-Phe-Pro
Abz-Gly-*p*-Nitro-Phe-Pro

Bachem M-1100.0050   MW 483.48 $C_{23}H_{25}N_5O_7$ Store at -15°C

Abz-Gly-Pro-Leu-Ala
4-Abz-Gly-Pro-D-Leu-D-Ala-NHOH

Bachem N-1405.0005   MW 490.56 $C_{23}H_{34}N_6O_6$ Store at -15°C

Abz-Gly-Pro-Leu-Ala
4-Abz-Gly-Pro-D-Leu-D-Ala-NH

Synonyms: Matrix Metalloproteinases Inhibitor I; Collagenase Inhibitor; Gelatinase Inhibitor; Stromelysin Inhibitor

Calbiochem 444250   MW 490.6 $C_{23}H_{34}N_6O_6$ Solid; ≥98% (HPLC); soluble in water | Inhibitor of human matrix metalloproteinases & interstitial & granulocyte collagenases, granulocyte gelatinase & skin fibroblast stromelysin; retains its activity even after prolonged incubation with PRONASE Protease or human granulocyte elastase; Odake, S et al, *Biochem Biophys Res Comm*, 199:1442, 1994

Abz-Gly-Pro-Leu-Ala
4-Abz-Gly-Pro-D-Leu-D-Ala-NHOH

Synonyms: Matrix Metalloproteinase Inhibitor I

ICN 196018   MW 490.6 $C_{23}H_{34}N_6O_6$ ≥98% | Human matrix metalloproteinase inhibitor

---

Abz-Gly-Trp-Thr-Leu-Asn-Ser-Ala-Gly-Tyr-Leu-Lys-Tyr
Abz-Gly-Trp-Thr-Leu-Asn-Ser-Ala-Gly-Tyr-Leu-Lys(retro-*m*-Nitro-Tyr-H) Amide

Synonyms: Galanin (1-10)-Lys(retro-*m*-Nitro-Tyr-H), (Abz-Gly[1])-

Bachem M-2365.0500   Human   Store at -15°C

Abz-Lys-Ala-Arg-Val-Nle-Phe-Glu-Ala-Nle
Abz-Lys-Ala-Arg-Val-Nle-*p*-Nitro-Phe-Glu-Ala-Nle Amide

Synonyms: Anthranilyl-HIV Protease Substrate IV

Bachem H-1052.0001   MW 1209.41 $C_{56}H_{88}N_{16}O_{14}$ Store at -15°C

Abz-Lys-Pro-Leu-Gly
***N*-Me-Abz-Lys-Pro-Leu-Gly**

Bachem H-3758.0005   Store at -15°C

Abz-Lys-Pro-Leu-Gly-Leu-Ala-Arg
Abz-Lys-Pro-Leu-Gly-Leu-Dap(Dnp)-Ala-Arg Amide

Bachem H-2638.0001   MW 1124.27 $C_{50}H_{77}N_{17}O_{13}$ Store at -15°C

Abz-Lys-Pro-Leu-Gly-Leu-Ala-Arg
***N*-Me-Abz-Lys-Pro-Leu-Gly-Leu-Dap(Dnp)-Ala-Arg Amide**

Bachem M-2145.0001   MW 1124.27 $C_{50}H_{77}N_{17}O_{13}$ Store at -15°C

Abz-Thr-Ile-Nle-Phe-Gln-Arg
Abz-Thr-Ile-Nle-*p*-Nitro-Phe-Gln-Arg Amide

Synonyms: Anthranilyl-HIV Protease Substrate

Bachem H-2992.0001   Store at -15°C

Ac-Ala-Thr-Gln-Arg-Leu-Ala-Asn-Phe-Leu-Val-Arg-Ser-Ser-Asn-Asn-Leu-Gly-Pro-Val-Leu-Pro-Pro-Thr-Asn-Val-Gly-Ser-Asn-Thr-Tyr Amide

Synonyms: Amylin (8-37), Ac-

Bachem H-8665.0500   Rat   MW 3242.64 $C_{142}H_{229}N_{43}O_{44}$ Store at -15°C

Ac-Ala-Thr-Gln-Arg-Leu-Ala-Asn-Phe-Leu-Val-His-Ser-Ser-Asn-Asn-Phe-Gly-Ala-Ile-Leu-Ser-Ser-Thr-Asn-Val-Gly-Ser-Asn-Thr-Tyr Amide

Synonyms: Amylin (8-37), Ac-

Bachem H-2744.0500   Human   MW 3225.53 $C_{140}H_{218}N_{42}O_{46}$ Store at -15°C

Ac-Arg-Arg-Pro-Tyr-Ile-Leu

Synonyms: Neurotensin (8-13), (N$^\alpha$-Ac)-

Sigma N 0511 FW 859.0 C$_{40}$H$_{66}$N$_{12}$O$_9$ ≥85% (HPLC) | Bioactive peptide; smallest active fragment of neurotensin; Granier, C et al, *Eur J Biochem*, 124:117, 1982

Ac-Asp-Arg-Val-Tyr-Ile-His-Pro-Phe-His-Leu

Synonyms: Angiotensin I, Ac-

Bachem H-1015.0005 MW 1338.53 C$_{64}$H$_{91}$N$_{17}$O$_{15}$ Store at -15°C

Ac-Asp-Arg-Val-Tyr-Ile-His-Pro-Phe-His-Leu-Leu-Val-Tyr-Ser

Synonyms: Angiotensinogen (1-14), (N-Ac)-; Renin Substrate

American Peptide 12-4-10 Porcine MW 1801.1 C$_{87}$H$_{125}$N$_{21}$O$_{21}$

Ac-Asp-Arg-Val-Tyr-Ile-His-Pro-Phe-His-Leu-Val-Ile-His-Asn

Synonyms: Angiotensinogen (1-14), (N-Ac)-; Pre-Angiotensinogen, (1-14), Ac-; Renin Substrate Tetradecapeptide, Ac-

American Peptide 12-1-79 Human MW 1802.1 C$_{85}$H$_{124}$N$_{24}$O$_{20}$

ICN 152753 Human MW 1802.1 Kageyama, R etal, *Biochemistry*, 23:3603, 1984

Ac-Ser-Phe-Pro-Trp-Met-Glu-Ser-Asp-Val-Thr

Synonyms: Prepro-Thyrotropin Releasing Hormone (160-169), Ac-

Neosystem SC485 MW 1240.35 Bioactive; thyrotropin releasing hormone-related peptide

Ac-Tyr-Gly-Gly-Phe-Met-Thr-Ser-Glu-Lys-Ser-Gln-Thr-Pro-Leu-Val-Thr-Leu

Synonyms: Endorphin, γ-; Lipotropin (61-77), β-

American Peptide 28-4-41 MW 1901.2 C$_{85}$H$_{133}$N$_{19}$O$_{28}$S

Ac-Tyr-Val-Ala-Asp
N-Ac-Tyr-Val-Ala-Asp-AMC

Synonyms: Interleukin Iβ Converting Enzyme Substrate III

ICN 195849 MW 665.7 C$_{33}$H$_{39}$N$_5$O$_{10}$ ≥97%; excitation = 380 nm, emission = 460 nm | Fluorogenic substrate

Ac-Val-Val-Sta-Ala-Sta

Synonyms: Acetylpepstatin

ICN 195663 MW 643.8 C$_{31}$H$_{57}$N$_5$O$_9$ ≥95% | Inhibits HIV-1 & HIV-2 proteinase

Adrenocorticotropic Hormone (1-39)

Biogenesis 0178-0399 Human pituitary Tested negative for Hepatitis B & C and HIV-1 and -2; purified; essentially salt free in 0.05 M ammonium bicarbonate, pH 8.0; lyophilized | RIA potency: 100 IU/mgH; Kappeler & Schwyzer, *Helv Chim Acta*, 44:1136, 1961

Aib-His-Nal-Nal-Phe-Lys
Aib-His-D-Nal$_2$-D-Phe-Lys Amide

Synonyms: Ipamorelin

Neosystem SC1329 MW 711.94 Bioactive; hypothalamic releasing hormone; newly discovered peptide with highly potent & selective growth hormone-releasing activity; Raun, K et al, *Mol Pharm Endocrinol*, 139:552-561, 1998

Akt1/PKB Alpha (1-149), PH Domain

USBio A1125-10 Human recombinant GST fusion protein fragment ~43kD ≥95% purity (SDS-PAGE); affinity purified by glutathione-agarose; supplied as 50 μg of Akt1/PKBa in 50 μL of 50 mM Tris-HCl, pH 7.5, 0.1 mM EGTA, 0.1% β-MSH, 0.03% Brij-35, 150 mM NaCl with 50% glycerol | Corresponds to the first 149 AA of human Akt1/PKBa containing the Pleckstrin Homology (PH) domain; used to identify Akt/PKB binding proteins; the PH domain can be used to block the antigen binding site on Anti-Akt1/PKBa, PH Domain antibody, which immunoprecipitates active Akt/PKB

Ala-Phe-Ala-Ser-Lys-Lys-Leu-Lys-Pro-Ala
N-Ac-β-(2-Naphthyl)-D-Ala-D-p-Chloro-Phe-β-(3-Pyridyl)-D-Ala-Ser-N$^\varepsilon$-(Nicotinoyl)-Lys-N$^\varepsilon$-(Nicotinoyl)-D-Lys-Leu-N$^\varepsilon$-D-(Isopropyl)-Lys-Pro-D-Ala Amide

Synonyms: Antide

ICN 152922 MW 1591.3 An LH-RH antagonist with high antiovulatory & negligible histamine release activity; Lungqvist, A etal, *BBRC*, 148:1849, 1987

Ala-Abu

Bachem G-1170.0250 MW 174.2 C$_7$H$_{14}$N$_2$O$_3$ Store at -15°C

Ala-Ala
D-Ala-D-Ala (1-^{14}C)

ARC ARC-1205 MW 160.2 CH$_3$CH(NH$_2$)CONHCH(CH$_3$)COOH 50-60 mCi/mmol; 1.85-2.22 GBq/mmol; EtOH:H$_2$O (95:5) | Radiochemical

Ala-Ala
D-Ala-D-Ala (2,3-^{3}H)

ARC ART-685 MW 160.2 CH$_3$CH(NH$_2$)CONHCH(CH$_3$)COOH 20-40 Ci/mmol; 0.74-1.48 TBq/mmol; in EtOH | Radiochemical

Ala-Ala
BOC-Ala-Ala

Bachem A-2680.0001 MW 260.29 C$_{11}$H$_{20}$N$_2$O$_5$ Store at -15°C

Ala-Ala
FMOC-Ala-Ala

Bachem B-1470.0001 MW 382.42 C$_{21}$H$_{22}$N$_2$O$_5$ Store at -15°C

Ala-Ala
Z-Ala-Ala

Bachem C-1045.0001 MW 294.31 C$_{14}$H$_{18}$N$_2$O$_5$ Store at -15°C

Ala-Ala
Z-Ala-Ala Amide

Bachem C-1050.0001 MW 293.32 C$_{14}$H$_{19}$N$_3$O$_4$ Store at -15°C

Ala-Ala
Z-Ala-Ala-OMe

Bachem C-1055.0001 MW 308.34 C$_{15}$H$_{20}$N$_2$O$_5$ Store at -15°C

Ala-Ala
Z-D-Ala-D-Ala-OMe

Bachem C-3420.0001 MW 308.34 C$_{15}$H$_{20}$N$_2$O$_5$ Store at -15°C

Ala-Ala
Ac-Ala-Ala

Bachem G-1000.0250   MW 202.21 $C_8H_{14}N_2O_4$ Store at -15°C

Ala-Ala
Ac-Ala-Ala-OMe

Bachem G-1005.0250   MW 216.24 $C_9H_{16}N_2O_4$ Store at -15°C

Ala-Ala

Bachem G-1120.0001   MW 160.17 $C_6H_{12}N_2O_3$ Store at -15°C

Ala-Ala
Ala-D-Ala

Bachem G-1125.0250   MW 160.17 $C_6H_{12}N_2O_3$ Store at -15°C

Ala-Ala
D-Ala-Ala

Bachem G-1130.0250   MW 160.17 $C_6H_{12}N_2O_3$ Store at -15°C

Ala-Ala
D-Ala-D-Ala

Bachem G-1135.0250   MW 160.17 $C_6H_{12}N_2O_3$ Store at -15°C

Ala-Ala
DL-Ala-DL-Ala

Bachem G-1140.0001   MW 160.17 $C_6H_{12}N_2O_3$ Store at -15°C

Ala-Ala
α-Ala-Ala

Bachem G-1145.0250   MW 160.17 $C_6H_{12}N_2O_3$ Store at -15°C

Ala-Ala
α-Ala-α-Ala

Bachem G-1150.0250   MW 160.17 $C_6H_{12}N_2O_3$ Store at -15°C

Ala-Ala
Ala-Ala-OMe Hydrochloride

Bachem G-1160.0001   MW 210.66 $C_7H_{14}N_2O_3 \cdot HCl$ Store at -15°C

Ala-Ala
D-Ala-D-Ala-OMe Hydrochloride

Bachem G-1165.0001   MW 210.66 $C_7H_{14}N_2O_3 \cdot HCl$ Store at -15°C

Ala-Ala
Ala-Ala-OtBu Hydrochloride

Bachem G-4060.0001   MW 252.74 $C_{10}H_{20}N_2O_3 \cdot HCl$ Store at -15°C

Ala-Ala
Ala-Ala-βNA

Bachem K-1040.0250   MW 285.35 $C_{16}H_{19}N_3O_2$ Store at -15°C

Ala-Ala
D-Ala-D-Ala-βNA Hydrochloride

Bachem K-1045.0250   MW 321.81 $C_{16}H_{19}N_3O_2 \cdot HCl$ Store at -15°C

Ala-Ala
Ala-Ala-βNA TFA

Bachem K-1255.0250   MW 399.37 $C_{16}H_{19}N_3O_2 \cdot C_2HF_3O_2$ Store at -15°C

Ala-Ala
Ala-Ala-pNA

Bachem L-1085.0250   MW 280.28 $C_{12}H_{16}N_4O_4$ Store at -15°C

Ala-Ala
BOC-Ala-Ala-pNA

Bachem L-1160.0250   MW 380.4 $C_{17}H_{24}N_4O_6$ Store at -15°C

Ala-Ala
Z-Ala-Ala-pNA

Bachem L-1495.0001   MW 414.42 $C_{20}H_{22}N_4O_6$ Store at -15°C

Ala-Ala
Suc-Ala-Ala-pNA

Bachem L-1505.0050   Store at -15°C

Ala-Ala
L-Ala-L-Ala

Fluka 05250   MW 160.17 $C_6H_{12}N_2O_3$ ≥99% (titration); mp:280-285°C

Ala-Ala
DL-Ala-DL-Ala

ICN 100057   MW 160.2 $C_6H_{12}N_2O_3$ Crystalline

Ala-Ala
Ala-β-Ala

ICN 100310   MW 160.2 $C_6H_{12}N_2O_3$ Crystalline

Ala-Ala

ICN 100324   MW 160.2 $C_6H_{12}N_2O_3$ Crystalline; 99%

Ala-Ala
β-Ala-β-Ala

ICN 100931   MW 160.2 $C_6H_{12}N_2O_3$

Ala-Ala
β-Ala-Ala

ICN 100939   MW 160.2 $C_6H_{12}N_2O_3$ Crystalline

Ala-Ala
BOC-Ala-Ala-NHO-Bz

Synonyms: Subtilisin Inhibitor I; Thermitase Inhibitor; Serine Protease Inhibitor

ICN 195920   MW 379.4 $C_{18}H_{25}N_3O_6$ >95%

Ala-Ala
(Z-Ala-Ala)₂ MR

Synonyms: Cathepsin B Subtrate

Kamiya   MW 1474   4 HBr   >99% (HPLC)

Ala-Ala
Suc-Ala-Ala-pNA

Peptides International SAA-3117   MW 380.36 $C_{16}H_{20}N_4O_7$
>99% (HPLC); amorphous powder

Ala-Ala
Ala-Ala-p-Nitroanilide Hydrochloride

Sigma A 2538   FW 316.7 $C_{12}H_{16}N_4O_4 \cdot HCl$ Bieth, J et al,
Biochem Med, 11:350, 1974

Ala-Ala
N-Ac-D-Ala-D-Ala

Sigma A 6441   FW 202.2 $C_8H_{14}N_2O_4$ ≥98%

Ala-Ala
β-Ala-Ala

Sigma A 9252   FW 160.2 $C_6H_{12}N_2O_3$

Ala-Ala

Sigma A 9502   FW 160.2 $C_6H_{12}N_2O_3$

Ala-Ala
N-CBZ-β-Ala-β-Ala

Sigma C 1876   FW 294.3 $C_{14}H_{18}N_2O_5$

Ala-Ala
N-CBZ-Ala-Ala

Sigma C 2126   FW 294.3 $C_{14}H_{18}N_2O_5$

Ala-Ala
(2R)-2-Mercaptomethyl-4-Methylpentanoyl-β-(2-Naphthyl)-Ala-Ala Amide

Synonyms: Collagenase Inhibitor

Sigma M 3531   FW 429.6 $C_{23}H_{31}N_3O_3S$ ≥95% (TLC) |
Bioactive peptide

Ala-Ala
CBZ-Ala-Ala

USBio C2098-18   MW 294.3 $C_{14}H_{18}N_2O_5$ ≥99%

Ala-Ala Amide Hydrochloride

Bachem G-1155.0250   MW 195.65 $C_6H_{13}N_3O_2 \cdot HCl$ Store
at -15°C

Ala-Ala-Abz
Ala-Ala-4-Abz

Bachem G-4130.0250   MW 279.3 $C_{13}H_{17}N_3O_4$ Store at
-15°C

Ala-Ala-Ala
Ala-Ala-Ala

Bachem H-1225.0250   MW 231.25 $C_9H_{17}N_3O_4$ Store at
-15°C

Ala-Ala-Ala
Ala-Ala-D-Ala

Bachem H-1230.0250   MW 231.25 $C_9H_{17}N_3O_4$ Store at
-15°C

Ala-Ala-Ala
Ala-D-Ala-Ala

Bachem H-1235.0250   MW 231.25 $C_9H_{17}N_3O_4$ Store at
-15°C

Ala-Ala-Ala
D-Ala-Ala-Ala

Bachem H-1240.0250   MW 231.25 $C_9H_{17}N_3O_4$ Store at
-15°C

Ala-Ala-Ala
D-Ala-D-Ala-D-Ala

Bachem H-1245.0250   MW 231.25 $C_9H_{17}N_3O_4$ Store at
-15°C

Ala-Ala-Ala
α-Ala-α-Ala-α-Ala

Bachem H-1250.0250   MW 231.25 $C_9H_{17}N_3O_4$ Store at
-15°C

Ala-Ala-Ala
For-Ala-Ala-Ala

Bachem H-3010.0050   MW 259.26 $C_{10}H_{17}N_3O_5$ Store at
-15°C

Ala-Ala-Ala
Suc-Ala-Ala-Ala-AMC

Bachem I-1310.0050   MW 488.5 $C_{23}H_{28}N_4O_8$ Store at
-15°C

Ala-Ala-Ala
Ala-Ala-Ala-4MβNA

Bachem J-1015.0050   MW 386.45 $C_{20}H_{26}N_4O_4$ Store at
-15°C

Ala-Ala-Ala
Glutaryl-Ala-Ala-Ala-4MβNA

Bachem J-1190.0050   MW 500.55 $C_{25}H_{32}N_4O_7$ Store at
-15°C

Ala-Ala-Ala
Ac-Ala-Ala-Ala-pNA

Bachem L-1000.0050   MW 393.4 $C_{17}H_{23}N_5O_6$ Store at
-15°C

Ala-Ala-Ala
Ala-Ala-Ala-pNA Hydrochloride

Bachem L-1090.0250   MW 351.36 $C_{15}H_{21}N_5O_5$ Store at
-15°C

Ala-Ala-Ala
Suc-Ala-Ala-Ala-pNA

Bachem L-1385.0050   MW 451.44 $C_{19}H_{25}N_5O_8$ Store at
-15°C

Ala-Ala-Ala
Ac-Ala-Ala-Ala

Bachem M-1010.0100   MW 273.29 $C_{11}H_{19}N_3O_5$ Store at
-15°C

Ala-Ala-Ala
Ac-Ala-Ala-Ala-OMe

Bachem M-1020.0250   MW 287.32 $C_{12}H_{21}N_3O_5$ Store at
-15°C

Ala-Ala-Ala
Ala-Ala-Ala-OMe

Bachem M-1090.0250   MW 245.28 $C_{10}H_{19}N_3O_4$ Store at
-15°C

Ala-Ala-Ala
N-Suc-L-Ala-L-Ala-L-Ala pNA

Synonyms: Pancreatic Elastase Substrate

Fluka 85975   MW 451.44 $C_{19}H_{25}N_5O_8$ ≥97.0% (enzymatic) | Chromogenic substrate; Geiger, R, *Methods of Enzymatic Analysis* (HU Bergmeyer, ed), 3rd vol, 5:170, 1984, Verlag Chemie Weinheim; Kasafirek, E et al, *FEBS Lett*, 40:353, 1974

Ala-Ala-Ala
BOC-Ala-Ala-Ala-NHO-Bz

Synonyms: Elastase Inhibitor; PPE Inhibitor

ICN 195978   MW 450.5 $C_{21}H_{30}N_4O_7$ Serine protease inhibitor of pancreatic elastase & thermitase

---

Ala-Ala-Ala
Suc-Ala-Ala-Ala-MCA

Synonyms: Elastase Substrate

Peptides International MAA-3133-v   MW 488.50 $C_{23}H_{28}N_4O_8$ >99% (HPLC); lyophilized amorphous powder | Mumford, RA et al, *JBC*, 255:2227, 1980

Ala-Ala-Ala
Suc-Ala-Ala-Ala-*p*NA

Synonyms: Elastase Substrate; STANA

Peptides International SAA-3071   MW 451.44 $C_{19}H_{25}N_5O_8$ >99% (HPLC); amorphous powder; bulk | Bieth, J et al, *Biochem Med*, 11:350, 1974

Peptides International SAA-3071-v   MW 451.44 $C_{19}H_{25}N_5O_8$ >99% (HPLC); lyophilized amorphous powder | Bieth, J et al, *Biochem Med*, 11:350, 1974

Ala-Ala-Ala
Ala-Ala-Ala-*p*-Nitroanilide Hydrochloride

Sigma A 2663   FW 387.8 $C_{15}H_{21}N_5O_5 \cdot$ HCl   Bieth, J et al, *Biochem Med*, 11:350, 1974

Ala-Ala-Ala
Ala-Ala-Ala-OMe Acetate Salt

Synonyms: Tri-L-Alanine-OMe

Sigma A 4775   FW 305.3 $C_{10}H_{19}N_3O_4 \cdot C_2H_4O_2$

Ala-Ala-Ala
N-Ac-Ala-Ala-Ala

Sigma A 5510   FW 273.3 $C_{11}H_{19}N_3O_5$

Ala-Ala-Ala
N-Ac-Ala-Ala-Ala-OMe

Sigma A 5635   FW 287.3 $C_{12}H_{21}N_3O_5$

Ala-Ala-Ala

Synonyms: Tri-L-Alanine

Sigma A 9627   FW 231.3 $C_9H_{17}N_3O_4$

Ala-Ala-Ala
N-Glutaryl-Ala-Ala-Ala 4-Methoxy-β-Naphthylamide

Synonyms: Elastase Substrate

Sigma G 7010   FW 500.6 $C_{25}H_{32}N_4O_7$

Ala-Ala-Ala
N-Suc-Ala-Ala-Ala-AMC

Sigma S 3261   FW 488.5 $C_{23}H_{28}N_4O_8$

Ala-Ala-Ala
N-Suc-Ala-Ala-Ala *p*-Nitroanilide

Synonyms: Elastase Substrate

Sigma S 4760   FW 451.4 $C_{19}H_{25}N_5O_8$ Bieth, B et al, *Biochem Med*, 11:350, 1974

Ala-Ala-Ala Amide Hydrochloride

Bachem H-1255.0250   MW 266.73 $C_9H_{18}N_4O_3 \cdot$ HCl   Store at -15°C

Ala-Ala-Ala-Ala
Ac-Ala-Ala-Ala-Ala

Bachem H-1005.0050   MW 344.37 $C_{14}H_{24}N_4O_6$ Store at -15°C

Ala-Ala-Ala-Ala
Ac-Ala-Ala-Ala-Ala-OMe

Bachem H-1010.0250   MW 358.4 $C_{15}H_{26}N_4O_6$ Store at -15°C

Ala-Ala-Ala-Ala

Bachem H-1260.0250   MW 302.33 $C_{12}H_{22}N_4O_5$ Store at -15°C

Ala-Ala-Ala-Ala
D-Ala-D-Ala-D-Ala-D-Ala

Bachem H-1265.0250   MW 302.33 $C_{12}H_{22}N_4O_5$ Store at -15°C

Ala-Ala-Ala-Ala

Synonyms: Tetra-L-Alanine

Sigma A 4900   FW 302.3 $C_{12}H_{22}N_4O_5$

Ala-Ala-Ala-Ala-Ala

Bachem H-1270.0250   MW 373.41 $C_{15}H_{27}N_5O_6$ Store at -15°C

Ala-Ala-Ala-Ala-Ala
D-Ala-D-Ala-D-Ala-D-Ala-D-Ala

Bachem H-1275.0250   MW 373.41 $C_{15}H_{27}N_5O_6$ Store at -15°C

Ala-Ala-Ala-Ala-Ala
Suc-Ala-Ala-Ala-Ala-Ala-pNA

Bachem L-1325.0025   Store at -15°C

Ala-Ala-Ala-Ala-Ala

Synonyms: Penta-L-Alanine

Sigma A 5025   FW 373.4 $C_{15}H_{27}N_5O_6$

Ala-Ala-Ala-Ala-Ala-Ala

Bachem H-1280.0250   MW 444.49 $C_{18}H_{32}N_6O_7$ Store at -15°C

Ala-Ala-Ala-Ala-Ala-Ala
D-Ala-D-Ala-D-Ala-D-Ala-D-Ala-D-Ala

Bachem H-1285.0250   MW 444.49 $C_{18}H_{32}N_6O_7$ Store at -15°C

Ala-Ala-Ala-Ala-Glu-Glu-Glu-Glu-Glu

Bachem H-1325.0050   MW 947.91 $C_{37}H_{57}N_9O_{20}$ Store at -15°C

Ala-Ala-Ala-Tyr

Bachem H-1390.0050   MW 394.43 $C_{18}H_{26}N_4O_6$ Store at -15°C

Ala-Ala-Ala-Tyr-Ala

Bachem H-1395.0050   MW 465.51 $C_{21}H_{31}N_5O_7$ Store at -15°C

Ala-Ala-Ala-Tyr-Ala-Ala

Bachem H-1400.0050   MW 536.59 $C_{24}H_{36}N_6O_8$ Store at -15°C

Ala-Ala-Ala-Tyr-Gly-Gly-Phe-Leu

American Peptide 30-0-72   MW 768.9 $C_{37}H_{52}N_8O_{10}$

Ala-Ala-Ala-Tyr-Gly-Gly-Phe-Met

Synonyms: Enkephalin

American Peptide 30-0-34   MW 786.9 $C_{36}H_{50}N_8O_{10}S$

Ala-Ala-Asn
Z-Ala-Ala-Asn-AMC

Bachem I-1865.0050   MW 565.58 $C_{28}H_{31}N_5O_8$ Store at -15°C

Ala-Ala-Asn
Z-Ala-Ala-Asn-MCA

Synonyms: Legumain Substrate

Peptides International MCA-3209-v   MW 565.57 $C_{28}H_{31}N_5O_8$ >98% (HPLC); lyophilized amorphous powder | Kembhavi, AA et al, *Arch Biochem Biophys*, 303:208, 1993; Chen, J-M et al, *JBC*, 272:8090, 1997; Manoury, B et al, *Nature*, 396:695, 1998; Choi, SJ et al, *JBC*, 274:27747, 1999; Dando, PM et al, *Biochem J*, 339:743, 1999

Ala-Ala-Asp
BOC-Ala-Ala-Asp-SBzl

Synonyms: Granzyme B Substrate

Alexis 260-050   MW 481.6 $C_{22}H_{31}N_3O_7S$ ≥95% (HPLC); white crystalline powder; soluble in DMSO | Odake, S et al, *Biochemistry*, 30:2217, 1991

Ala-Ala-Asp
BOC-Ala-Ala-Asp-pNA

Bachem L-1165.0050   MW 495.49 $C_{21}H_{29}N_5O_9$ Store at -15°C

Ala-Ala-Asp
Z-Ala-Ala-Asp-CMK

Synonyms: Granzyme B Inhibitor I

Calbiochem 368050   MW 441.9 $C_{19}H_{24}ClN_3O_7$ ≥95% (HPLC); solid; soluble in DMSO | Inhibits the human & murine serine protease granzyme B found in the cytoplasmic granules of cytotoxic T lymphocytes & natural killer cells; inhibits the apoptosis-related DNA fragmentation in lymphocytes by fragmentin 2, a rat lymphocyte granule protease homologous to granzyme B; Shi, L et al, *J Exp Med*, 176:1521, 1992; Odake, S et al, *Biochemistry*, 30:2217, 1991

Ala-Ala-Asp
Z-Ala-Ala-Asp-AFC

Synonyms: Granzyme B Substrate III

Calbiochem 368061   MW 620.5 $C_{28}H_{27}F_3N_4O_9$ ≥95% (TLC); lyophilized solid; soluble in DMSO; excitation max:~400 nm; emission max:~505 nm | Fluorogenic substrate for granzyme B; reaction monitored visually or quantitatively by a blue to green shift in fluorescence upon cleavage of the AFC fluorophore

Ala-Ala-Asp
BOC-Ala-Ala-Asp-S-Bzl

Synonyms: Granzyme B Substrate IV; Granzyme B Substrate

Calbiochem 368063   MW 481.6 $C_{22}H_{31}N_3O_7S$ >95% (HPLC); solid; soluble in DMSO | Odake, S et al, *Biochemistry*, 30:2217, 1991

ICN 193608   Used in apoptosis research

Ala-Ala-Asp
Z-Ala-Ala-Asp-AFC

Synonyms: Granzyme B Substrate

ICN 193609   Fluorescent substrate used in apoptosis research

Ala-Ala-Asp
Z-Ala-Ala-Asp-CMK

Synonyms: Granzyme B Inhibitor

ICN 196000   MW 441.9 $C_{19}H_{24}N_3O_7Cl$ ≥95% | Inhibits granzyme B present in cytoplasmic granules of cytotoxic T-lymphocytes & NK cells

Ala-Ala-Asp
Z-Ala-Ala-Asp-AFC

Synonyms: Granzyme B Substrate

Kamiya   MW 655   Fluorogenic

Ala-Ala-Asp
Z-Ala-Ala-Asp-CMK

Synonyms: Granzyme B Inhibitor

Kamiya   MW 441.9 $C_{19}H_{24}ClN_3O_7$

Ala-Ala-Asp-Ile-Ser-Gln-Trp-Ala-Gly-Pro-Leu
Ac-Ala-Ala-Asp-Ile-Ser-Gln-Trp-Ala-Gly-Pro-Leu

Synonyms: Hippocampal Cholinergic Neurostimulating Peptide

Bachem H-8555.0001   Store at -15°C

Ala-Ala-Cys-Lys-Cys-Asp-Asp-Glu-Gly-Pro-Asp-Ile-Arg-Thr-Ala-Pro-Leu-Thr-Gly-Thr-Val-Asp-Leu-Gly-Ser-Cys-Asn-Ala-Gly-Trp-Glu-Lys-Cys-Ala-Ser-Tyr-Tyr-Thr-Ile-Ile-Ala-Asp-Cys-Cys-Arg-Lys-Lys-Lys

Synonyms: Neurotoxin ShNa

Bachem H-1074.0100   Store at -15°C | Disulfide bonds: Cys[3]-Cys[43], Cys[5]-Cys[33], Cys[26]-Cys[44]

Ala-Ala-Cys-Lys-Cys-Asp-Asp-Glu-Gly-Pro-Asp-Ile-Arg-Thr-Ala-Pro-Leu-Thr-Gly-Thr-Val-Asp-Leu-Gly-Ser-Cys-Asn-Ala-Gly-Trp-Glu-Lys-Cys-Ala-Ser-Tyr-Tyr-Thr-Ile-Ile-Ala-Asp-Cys-Cys-Arg-Lys-Lys
Trifluoroacetate Salt

Synony Neurotoxin ShNa *ms:*

Calbiochem 569405   *Stichodactyla helianthus* (Caribbean sea anemone) synthetic   MW 5136.8 $C_{216}H_{341}N_{61}O_{72}S_6$ >90% (HPLC); lyophilized solid; soluble in water; LD$_{50}$≤2000 mg/kg | Disulfide bonds: Cys[3]-Cys[43], Cys[5]-Cys[33], Cys[26]-Cys[44]; sequence from a polypeptide neurotoxin; selectively toxic to crustaceans; acts by delaying the inactivation of sodium channels; several AA side chains in the region between positions 3 & 15 are important for binding to crab neuronal sodium channels; Kem, WR et al, *Biochemistry*, 28:3483, 1989; Pennington, MW et al, *Int J Pept Protein Res*, 36:335, 1990; Pennington, MW et al, *Peptide Res*, 3:228, 1990

Ala-Ala-Gln

Bachem H-2552.0250   MW 288.3 $C_{11}H_{20}N_4O_5$ Store at -15°C

Ala-Ala-Gly

Bachem H-1410.0250 MW 217.23 $C_8H_{15}N_3O_4$ Store at -15°C

Ala-Ala-Gly
BOC-Ala-Ala-Gly-pNA

Bachem L-1170.0050 MW 437.45 $C_{19}H_{27}N_5O_7$ Store at -15°C

Ala-Ala-Gly-Ile-Gly-Ile-Leu-Thr-Val

Synonyms: MART I (27-35); Melan A; Melanoma Associated Antigen Peptide

Bachem H-3956.0001 Human MW 813.99 $C_{37}H_{67}N_9O_{11}$ Store at -15°C

Neosystem SC1249 Human MW 813.99 Bioactive; cancer related peptide; corresponds to positions 27-35 of the MART-1 / Melan A Protein; recognized by most melanoma-specific, HLA-A0201-restricted, tumor-infiltrating lymphocytes; Kawakami, Y et al, *J Exp Med*, 180:347-352, 1994; Van den Eynde, B et al, *Cur Opin Immunol*, 7:674-681, 1995; Marincola, FM et al, *J Immunother*, 19:266-277, 1996; Loftus, DJ et al, *J Exp Med*, 184:647-657, 1996; Vanelsas A et al, *Mol Pharm of Immunol*, 26:1683-1689, 1996; Maeurer, MJ et al, *Mol Pharm of Immunol*, 26:2613-2623, 1996; Cormier, JN et al, *Cancer J Sci Amer*, 3:37-44, 1997

Ala-Ala-Gly-Met-Gly-Phe-Phe-Gly-Ala-Arg Amide

Synonyms: Urechistachykinin II

American Peptide 87-7-55 MW 983.1 $C_{44}H_{66}N_{14}O_{10}S$ Ikeda, T et al, *BBRC*, 192:1, 1993

Bachem H-1634.0001 MW 983.16 $C_{44}H_{66}N_{14}O_{10}S$ Store at -15°C

Ala-Ala-Gly-Pro-Glu-Met-Val-Arg-Gly-Gln-Val-Phe-Cys

Synonyms: Mitogen Activated Protein Kinase EFK2 Blocking Peptide (6-17), *N*-Terminal

Calbiochem 442681 Rat synthetic MW 1364.6 $C_{58}H_{93}N_{17}O_{17}S_2$ Lyophilized | Based on rat ERK2 (6-17) with a Cys C residue added; this peptide coupled to KLH was used as the immunogen for the production of Anti-MAP kinase ERK2; for use in immunoabsorption for immunoprecipitation, immunocytochemistry, Western blotting & dot blots

Ala-Ala-Leu
Z-Ala-Ala-Leu-pNA

Synonyms: Subtilisin A Substrate II; Subtilisin A Substrate; Serine Protease of *Bacillus subtilis* IFO3027 Substrate

Bachem L-1210.0050 MW 527.58 $C_{26}H_{33}N_5O_7$ Store at -15°C

ICN 195019 MW 527.6 >95%; lyophilized

Peptides International SAP-3127 MW 527.58 $C_{26}H_{33}N_5O_7$ >99% (HPLC); lyophilized amorphous powder | Stepanov, VM et al, *BBRC*, 77:298, 1977; Shimizu, Y et al, *Agric Biol Chem*, 47:1775, 1983

Ala-Ala-Leu
N-CBZ-Ala-Ala-Leu *p*-Nitroanilide

Synonyms: Subtilisin A Substrate

Sigma C 9165 FW 527.6 $C_{26}H_{33}N_5O_7$ Stepanov, VM et al, *Biochem Biophys Res Commun*, 77:298, 1977

Ala-Ala-Leu
CBZ-Ala-Ala-Leu-pNA

USBio C2098-19 MW 528.6 $C_{26}H_{34}N_5O_7$ ≥99%

Ala-Ala-Leu
Z-Ala-Ala-Leu-pNA

Synonyms: Serine Protease Substrate; Nagarse (Novo Subtilisin) Substrate; Proteinase yscE Substrate

Neosystem SC1288 *Bacillus subtilis* MW 527.56 Stepanov, VM et al, *BBRC*, 77:298-305, 1977; Shimizu, Y et al, *Agric Biol Chem*, 47:1775-1782, 1983; Emter, O & Wolf, DH, *FEBS Lett*, 166:321-325, 1984

Ala-Ala-Leu-Asp-Leu-Ser-His-Phe-Leu-Lys-Glu-Lys

Synonyms: NEF MN (85-96); HIV I Peptide

Neosystem SC649 MW 1371.60 NEF peptide from HIV-1 subtype MN

Ala-Ala-Leu-Val-Arg-Gln-Met-Ser-Val-Ala-Phe-Phe-Phe-Lys Trifluoroacetate Salt

Synonyms: Protein Kinase C$_\mu$ Peptide Substrate

Calbiochem 539564 MW 1615.0 $C_{77}H_{119}N_{19}O_{17}S$ ≥95% (HPLC); solid; soluble in water | Specific substrate for the protein kinase C$_\mu$ isozyme; Nishikawa, K et al, *J Biol Chem*, 272:952, 1997

Ala-Ala-Lys
Z-Ala-Ala-Lys

Bachem C-2380.0050 MW 422.48 $C_{20}H_{30}N_4O_6$ Store at -15°C

Ala-Ala-Lys
Z-Ala-Ala-Lys-4MβNA

Bachem J-1085.0050 MW 577.68 $C_{31}H_{39}N_5O_6$ Store at -15°C

Ala-Ala-Lys
N-CBZ-Ala-Ala-Lys 4-Methoxy-β-Naphthylamide Formate salt

Sigma C 2147 FW 623.7 $C_{31}H_{39}N_5O_6 \cdot HCOOH$ Clavin, SA et al, *Anal Biochem*, 80:355, 1977; Bigbee, WL et al, *Anal Biochem*, 88:114, 1978; substrate for plasmin

Ala-Ala-Lys Hydrochloride

Bachem H-1415.0250 MW 324.81 $C_{12}H_{24}N_4O_4 \cdot HCl$ Store at -15°C

Ala-Ala-Lys-Ile-Gln-Ala-Ser-Phe-Arg-Gly-His-Met-Ala-Arg-Lys-Lys

Synonyms: Neurogranin (28-43); Neurogranin Phosphorylation Domain; Protein Kinase C Substrate; Protein Kinase A; Protein Kinase Related Peptide

Alexis 163-004 MW 1800.2 $C_{78}H_{134}N_{28}O_{19}S$ ≥97%; white lyophilized powder; soluble in H_2O | Highly selective substrate for protein kinase C (PKC), but very poor substrate for Ca^{2+}/calmodulin-dependent kinase II & cAMP-dependent kinase; used at high substrate concentration without causing inhibition, should even be useful for measuring low PKC activity (e.g. in crude cell extracts); Chen, S-J et al, *Biochemistry*, 32:1032, 1993

Bachem H-1554.0001 MW 1800.17 $C_{78}H_{134}N_{28}O_{19}S$ Store at -15°C

Promega V5611 MW 1800

Sigma N 7279 FW 1800.2 ≥97% (HPLC) | Bioactive peptide; synthetic peptide corresponding to the phosphorylation domain of neurogranin; Km=150 *nM* (proteinkinase C); Chen, S-J et al, *Biochemistry*, 32:1032, 1993

Ala-Ala-Lys-Ile-Gln-Ala-Ser-Phe-Arg-Gly-His-Met-Ala-Arg-Lys-Lys Trifluoroacetate Salt

Synonyms: Protein Kinase C Selectide Substrate; Neurogranin (28-43)

Calbiochem 527151   MW 1800.2 $C_{78}H_{134}N_{28}O_{19}S$ ≥95% (HPLC); solid; soluble in water | Developed to overcome difficulties that can arise from using non-specific substrates when assaying PKC activity in crude cell extracts; structurally based on the sequence of neurogranin, a naturally occurring PKC substrate; allows the measurement of PKC activity even in crude cell extracts; Chang, DK et al, *Biophys J*, 72:554, 1997; Chen, SJ et al, *Biochemistry*, 32:1032, 1993

Ala-Ala-Lys-Tyr-Cys-Lys-Leu-Pro-Leu-Arg-Ile-Gly-Pro-Cys-Lys-Arg-Lys-Ile-Pro-Ser-Phe-Tyr-Tyr-Lys-Trp-Lys-Ala-Lys-Gln-Cys-Leu-Pro-Phe-Asp-Tyr-Ser-Gly-Cys-Gly-Gly-Asn-Ala-Asn-Arg-Phe-Lys-Thr-Ile-Glu-Glu-Cys-Arg-Arg-Thr-Cys-Val-Gly

Synonyms: Dendrotoxin K

Calbiochem 253709   *Dendroaspis polylepis polylepis*   MW 6559.7 $C_{294}H_{456}N_{84}O_{75}S_6$ ≥98% (HPLC); lyophilized; soluble in water & aqueous solutions; LD_{50}≤2000 mg/kg; biological activity:shown to block ^{86}Rb flux through Kv1.1 channels stably expressed in CHO cells | 57-residue protein isolated from snake venom; specific blocker of non-inactivating Kv1.1 voltage-gated K^+ channel; Bagetta, G et al, *Exp Neurol*, 147:204, 1997; Robertson, B et al, *FEBS Lett*, 383:26, 1996; Hall, A et al, *Br J Pharmacol*, 113:959, 1994; Smith, LA et al, *Biochemistry*, 32:5692, 1993

Ala-Ala-Lys-Tyr-Cys-Lys-Leu-Pro-Val-Arg-Tyr-Gly-Pro-Cys-Lys-Lys-Lys-Ile-Pro-Ser-Phe-Tyr-Tyr-Lys-Trp-Lys-Ala-Lys-Gln-Cys-Leu-Pro-Phe-Asp-Tyr-Ser-Gly-Cys-Gly-Gly-Asn-Ala-Asn-Arg-Phe-Lys-Thr-Ile-Glu-Glu-Cys-Arg-Arg-Thr-Cys-Val-Gly

Synonyms: Dendrotoxin, δ-

Alexis 630-016   *Dendroaspis angusticeps*   MW 7k   ≥98% (SDS-PAGE); lyophilized powder; dissolving 70 µg/mL gives a stock solution of 10 µM; Disulfide bonds: Cys^5-Cys^{55}, Cys^{14}-Cys^{38}, Cys^{30}-Cys^{51}; potent neurotoxin | Potent & selective blocker of voltage-gated K^+ channels in rat brain synaptosomes; binding site is different than that of α-dendrotoxin; Joubert, F & TalijAArd, N, *Hoppe-Seyler's Z Physiol Chem*, 361:661, 1980; Strong, PN, *Pharmacol Ther*, 46:137, 1990; Awan, KA & Dolly, JO, *Neuroscience*, 40:29, 1991; Benishin, CG et al, *Mol Pharmacol*, 34:152, 1988; Hu, PS et al, *Eur J Pharmacol*, 209:87, 1991; Muniz, ZM et al, *Biochemistry*, 31:12297, 1992

Ala-Ala-Phe
Ac-Ala-Ala-Phe-OMe

Bachem H-7995.0250   MW 363.41 $C_{18}H_{25}N_3O_5$ Store at -15°C

Ala-Ala-Phe
Ala-Ala-Phe-AMC Trifluoroacetate Salt

Bachem I-1035.0050   MW 578.55 $C_{25}H_{28}N_4O_5 \cdot C_2HF_3O_2$ Store at -15°C

Ala-Ala-Phe
Suc-Ala-Ala-Phe-AMC

Bachem I-1315.0050   MW 564.6 $C_{29}H_{32}N_4O_8$ Store at -15°C

Ala-Ala-Phe
Ala-Ala-Phe-AMC

Bachem I-1415.0050   MW 464.52 $C_{25}H_{28}N_4O_5$ Store at -15°C

Ala-Ala-Phe
Ala-Ala-Phe-4MβNA

Bachem J-1020.0050   MW 462.55 $C_{26}H_{30}N_4O_4$ Store at -15°C

Ala-Ala-Phe
Glutaryl-Ala-Ala-Phe-4MβNA

Bachem J-1355.0025   MW 576.65 $C_{31}H_{36}N_4O_7$ Store at -15°C

Ala-Ala-Phe
Ala-Ala-Phe-βNA

Bachem K-1050.0050   MW 432.52 $C_{25}H_{28}N_4O_3$ Store at -15°C

Bachem L-1095.0050   MW 427.46 $C_{21}H_{25}N_5O_5$ Store at -15°C

Ala-Ala-Phe
BOC-Ala-Ala-Phe-pNA

Bachem L-1175.0050   MW 527.58 $C_{26}H_{33}N_5O_7$ Store at -15°C

Ala-Ala-Phe
Suc-Ala-Ala-Phe-pNA

Bachem L-1580.0050   Store at -15°C

Ala-Ala-Phe
Ala-Ala-Phe-CMK Trifluoroacetate Salt

Bachem N-1005.0025   MW 453.85 $C_{16}H_{22}ClN_3O_3 \cdot C_2HF_3O_2$ Store at -15°C

Ala-Ala-Phe
Ala-Ala-Phe-AMC Hydrochloride

Synonyms: Aminopeptidase Substrate II, Fluorogenic

ICN 150357   MW 501.1 $C_{25}H_{28}N_4O_5 \cdot HCl$ ≥97%

Ala-Ala-Phe
N-Suc-L-Ala-L-Ala-Phe-MCA

ICN 152080   Fluorogenic substrate

Ala-Ala-Phe
Glt-Ala-Ala-Phe-MCA

Synonyms: Chymotrypsin Substrate

Peptides International MAF-3154-v   MW 578.62 $C_{30}H_{34}N_4O_8$ >98% (HPLC); lyophilized amorphous powder

Ala-Ala-Phe
Ala-Ala-Phe-MCA

Synonyms: Tripeptidyl Peptidase II Substrate

Peptides International MCA-3201-v   MW 464.51 $C_{25}H_{28}N_4O_5$ >99% (HPLC); lyophilized amorphous powder | Component of giant protease with some protease function; Glas, R et al, *Nature*, 392:618, 1998; Geier, E et al, *Science*, 283:978, 1999

Ala-Ala-Phe
Ala-Ala-Phe-AMC

Synonyms: Chymotrypsin Substrate; Tripeptidyl Peptidase II Substrate

Sigma A 3401   FW 464.5 $C_{25}H_{28}N_4O_5$ Zimmerman, M et al, *Anal Biochem*, 78:47, 1977; Balow, RM et al, *J Biol Chem*, 261:2409, 1986

Ala-Ala-Phe
Glutaryl-Ala-Ala-Phe 4-Methoxy-β-Naphthylamide

Synonyms: Enkephalinase Substrate; Neutral Endopeptidase 24.11 Substrate

Sigma G 3769   FW 576.6 $C_{31}H_{36}N_4O_7$ Spectrofluorometric; Also hydrolyzed by chymotrypsin & *E. coli* protease La

Ala-Ala-Phe
N-Suc-Ala-Ala-Phe-AMC

Sigma S 8758   FW 564.6 $C_{29}H_{32}N_4O_8$

Ala-Ala-Phe
Ala-Ala-Phe-CH₂Cl

Synonyms: Tripeptidyl Peptidase II Giant Protease Inhibitor

Peptides International IAA-3202-v   Synthetic   MW 339.82
$C_{16}H_{22}N_3O_3Cl$ >99% (HPLC); lyophilized amorphous powder | Inhibitor; some proteasome function; Glas, R et al, *Nature*, 392:618, 1998; Geier, E et al, *Science*, 283:978, 1999

Ala-Ala-Phe CMK Trifluoroacetate Salt

Synonyms: Chymotrypsin Substrate; Tripeptidyl Peptidase II Substrate

Sigma A 6892   FW 453.8 $C_{16}H_{22}ClN_3O_3 \cdot C_2HF_3O_2$
Zimmerman, M et al, *Anal Biochem*, 78:47, 1977; Balow, RM et al, *J Biol Chem*, 261:2409, 1986

Ala-Ala-Pro
BOC-Ala-Ala-Pro

Bachem A-1100.0250   MW 357.41 $C_{16}H_{27}N_3O_6$ Store at -15°C

Ala-Ala-Pro

Bachem H-1420.0250   MW 257.29 $C_{11}H_{19}N_3O_4$ Store at -15°C

Ala-Ala-Pro
Ala-Ala-Pro-pNA Hydrochloride

Bachem L-1100.0050   MW 413.86 $C_{17}H_{23}N_5O_5 \cdot$ HCl   Store at -15°C

Ala-Ala-Pro
Suc-Ala-Ala-Pro-L-Abu-pNA

Synonyms: Elastase Substrate IV

Calbiochem 324699   MW 562.6 $C_{25}H_{34}N_6O_9$ Lyophilized solid containing 95 mg D-mannitol & 5 mg substrate; ≥95% (HPLC); soluble in water & ethanol | Colorimetric substrate; Largman, L, *Biochemistry*, 22:3763, 1983

Ala-Ala-Pro
N-MeOSuc-Ala-Ala-Pro-Val p-Nitroanilide

Synonyms: Elastase Substrate

Sigma M 4765   FW 590.6 $C_{27}H_{38}N_6O_9$ Substrate for human leukocyte elastase; Nakajima, K et al, *J Biol Chem*, 254:4027, 1979

Ala-Ala-Pro-Abu
For-Ala-Ala-Pro-Abu-SBzl

Bachem H-2088.0050   Store at -15°C

Ala-Ala-Pro-Ala

Bachem H-1425.0050   MW 328.37 $C_{14}H_{24}N_4O_5$ Store at -15°C

Ala-Ala-Pro-Ala
Ac-Ala-Ala-Pro-Ala-AMC

Bachem I-1000.0050   MW 527.58 $C_{26}H_{33}N_5O_7$ Store at -15°C

Ala-Ala-Pro-Ala
Ac-Ala-Ala-Pro-Ala-βNA

Bachem K-1000.0050   MW 495.58 $C_{26}H_{33}N_5O_5$ Store at -15°C

Bachem L-1005.0050   MW 490.52 $C_{22}H_{30}N_6O_7$ Store at -15°C

Ala-Ala-Pro-Ala
BOC-Ala-Ala-Pro-Ala-pNA

Bachem L-1180.0050   MW 548.6 $C_{25}H_{36}N_6O_8$ Store at -15°C

Ala-Ala-Pro-Ala
Suc-Ala-Ala-Pro-Ala-pNA

Bachem L-1775.0050   Store at -15°C

Ala-Ala-Pro-Ala
MeOSuc-Ala-Ala-Pro-Ala-chloromethylketone

Bachem N-1280.0025   MW 474.94 $C_{20}H_{31}ClN_4O_7$ Store at -15°C

Ala-Ala-Pro-Ala
BOC-Ala-Ala-Pro-Ala-pNA

Synonyms: Elastase Substrate II

Calbiochem 324697   MW 548.6 $C_{25}H_{36}N_6O_8$ Lyophilized solid containing 95 mg D-mannitol & 5 mg substrate; ≥95% (HPLC); soluble in water & ethanol | Colorimetric substrate; Ashe, BM & Zimmerman, M, *Biochem Biophys Res Comm*, 75:194, 1977

Ala-Ala-Pro-Ala
N-Ac-Ala-Ala-Pro-Ala-β-Naphthylamide

Synonyms: Protease Substrate

Sigma A 2291   FW 495.6 $C_{26}H_{33}N_5O_5$ Chromogenic; Emter, O & Wolf, DH, *FEBS Lett*, 166:321, 1984

Ala-Ala-Pro-Ala-Ala

Bachem H-1430.0050   MW 399.45 $C_{17}H_{29}N_5O_6$ Store at -15°C

Ala-Ala-Pro-Arg
Suc-Ala-Ala-Pro-Arg-pNA

Bachem L-1720.0050   MW 633.66 $C_{27}H_{39}N_9O_9$ Store at -15°C

Ala-Ala-Pro-Asp
Suc-Ala-Ala-Pro-Asp-pNA

Bachem L-1835.0050   Store at -15°C

Ala-Ala-Pro-Asp
N-Suc-Ala-Ala-Pro-Asp p-Nitroanilide

Sigma S 6171   FW 592.6 $C_{25}H_{32}N_6O_{11}$ ≥97% (TLC)

Ala-Ala-Pro-Glu
Suc-Ala-Ala-Pro-Glu-pNA

Bachem L-1710.0050   MW 606.59 $C_{26}H_{34}N_6O_{11}$ Store at -15°C

Ala-Ala-Pro-Ile
Suc-Ala-Ala-Pro-Ile-pNA
Bachem L-1790.0050   MW 590.63 $C_{27}H_{38}N_6O_9$ Store at -15°C

Ala-Ala-Pro-Leu
Glutaryl-Ala-Ala-Pro-Leu-pNA
Bachem L-1270.0050   MW 604.66 $C_{28}H_{40}N_6O_9$ Store at -15°C

Ala-Ala-Pro-Leu
Suc-Ala-Ala-Pro-Leu-pNA
Bachem L-1390.0050   MW 590.63 $C_{27}H_{38}N_6O_9$ Store at -15°C

Ala-Ala-Pro-Leu
Ala-Ala-Pro-Leu-pNA Hydrochloride
Bachem L-2125.0050   MW 527.02 $C_{23}H_{34}N_6O_6 \cdot HCl$ Store at -15°C

Ala-Ala-Pro-Leu
Glt-Ala-Ala-Pro-Leu-pNA Hydrate
Synonyms: Pancreatic Elastase Substrate

Peptides International SGL-3129   MW 622.68 $C_{28}H_{40}N_6O_9 \cdot H_2O$ >99% (HPLC); amorphous powder | Del Mar, EG et al, *Biochem*, 19:468, 1980

Ala-Ala-Pro-Leu
N-Suc-Ala-Ala-Pro-Leu p-Nitroanilide
Synonyms: Elastase Substrate

Sigma S 8511   FW 590.6 $C_{27}H_{38}N_6O_9$ DelMar, EG et al, *Biochemistry*, 19:468, 1980

Ala-Ala-Pro-Lys
Suc-Ala-Ala-Pro-Lys-pNA
Bachem L-1725.0050   MW 605.65 $C_{27}H_{39}N_7O_9$ Store at -15°C

Ala-Ala-Pro-Met
MeOSuc-Ala-Ala-Pro-Met-AMC
Bachem I-1405.0050   MW 659.76 $C_{31}H_{41}N_5O_9S$ Store at -15°C

Ala-Ala-Pro-Met
MeOSuc-Ala-Ala-Pro-Met-pNA
Bachem L-1330.0050   MW 622.7 $C_{27}H_{38}N_6O_9S$ Store at -15°C

Ala-Ala-Pro-Met
Suc-Ala-Ala-Pro-Met-pNA
Bachem L-1395.0050   Store at -15°C

Ala-Ala-Pro-Met
N-Methoxy-Suc-Ala-Ala-Pro-Met-pNA
Synonyms: Cathepsin G Substrate

ICN 155441   MW 622.7 $C_{27}H_{38}N_6O_9S$ Soluble substrate; Nakajima, K etal *JBC*, 254:4027, 1979

Ala-Ala-Pro-Met
MeOSuc-Ala-Ala-Pro-Met-*p*NA
Synonyms: Cathepsin G Substrate I

ICN 195956   MW 622.7 $C_{27}H_{38}N_6O_9S$ >98% | Substrate for the quantitative determination of cathepsin G

Ala-Ala-Pro-Met
Suc-Ala-Ala-Pro-Met-pNA
Synonyms: Pancreatic Elastase II Substrate

Neosystem SC1289   MW 608.64   Chromogenic substrate for human pancreatic elastase 2; Del Mar, EG et al, *BBRC*, 88:346-350, 1979; Del Mar, EG et al, *Biochem*, 19:468-472, 1980

Ala-Ala-Pro-Met
N-MeOSuc-Ala-Ala-Pro-Met p-Nitroanilide
Synonyms: Cathepsin G Substrate

Sigma M 7771   FW 622.7 $C_{27}H_{38}N_6O_9S$ Nakajima, K et al, *J Biol Chem*, 254:4027, 1979

Ala-Ala-Pro-Nle
Suc-Ala-Ala-Pro-Nle-pNA
Bachem L-1765.0050   MW 590.63 $C_{27}H_{38}N_6O_9$ Store at -15°C

Ala-Ala-Pro-Nva
Suc-Ala-Ala-Pro-Nva-pNA
Bachem L-1780.0050   MW 576.61 $C_{26}H_{36}N_6O_9$ Store at -15°C

Ala-Ala-Pro-Orn
Suc-Ala-Ala-Pro-Orn-pNA
Bachem L-1760.0050   Store at -15°C

Ala-Ala-Pro-Phe
Suc-Ala-Ala-Pro-Phe-AMC
Bachem I-1465.0050   MW 661.71 $C_{34}H_{39}N_5O_9$ Store at -15°C

Ala-Ala-Pro-Phe
Ac-Ala-Ala-Pro-Phe-pNA
Bachem L-1010.0050   MW 566.61 $C_{28}H_{34}N_6O_7$ Store at -15°C

Ala-Ala-Pro-Phe
Suc-Ala-Ala-Pro-Phe-pNA
Bachem L-1400.0050   MW 624.65 $C_{30}H_{36}N_6O_9$ Store at -15°C

Ala-Ala-Pro-Phe
Suc-Ala-Ala-Pro-Phe-SBzl
Bachem M-1735.0050   MW 610.73 $C_{31}H_{38}N_4O_7S$ Store at -15°C

Ala-Ala-Pro-Phe
Suc-Ala-Ala-Pro-Phe-2,4-difluoroanilide
Bachem M-2305.0050   Store at -15°C

Ala-Ala-Pro-Phe
N-Suc-L-Ala-L-Ala-L-Pro-L-Phe-pNA
Synonyms: Chymotrypsin Substrate

Fluka 85977   MW 624.6 $C_{30}H_{36}N_6O_9$ ≥99.0% (HPLC); ≤3% water; ≤0.01% free 4-nitroaniline | Sensitive substrate; DelMar, EG et al, *Anal Biochem*, 99:316, 1979

Ala-Ala-Pro-Phe
N-Suc-L-Ala-L-Ala-L-Pro-L-Phe-MCA
Synonyms: Chymotrypsin Substrate

ICN 152081   Fluorogenic substrate

Ala-Ala-Pro-Phe
N-Suc-Ala-Ala-Pro-Phe-Thiobenzyl

ICN 156689

Ala-Ala-Pro-Phe
Suc-Ala-Ala-Pro-Phe-pNA

Synonyms: Cathepsin G Substrate

ICN 194971   MW 624.7   Lyophilized; >95% | Nakajima, JC etal, *Anal Biochem*, 254:4027, 1979

Ala-Ala-Pro-Phe
Suc-Ala-Ala-Pro-Phe-AMC

Synonyms: Chymotrypsin Substrate II; Chymotrypsin Substrate; Carboxypeptidase Y Substrate

ICN 195961   MW 661.7 $C_{34}H_{39}N_5O_9$ >95% | Fluorogenic substrate for the quantitative determination of chymotrypsin

Neosystem SC1323   MW 661.70   Oshima, G, *J Biochem*, 94:1615-1620, 1983; Oshima, G, *J Biochem*, 95:1131- 1136, 1984; Kunugi, S et al, *Mol Pharm Biochem*, 153:37-40, 1985

Ala-Ala-Pro-Phe
Suc-Ala-Ala-Pro-Phe-MCA

Synonyms: Chymotrypsin Substrate

Peptides International MAA-3114-v   MW 661.71 $C_{34}H_{39}N_5O_9$ >98% (HPLC); lyophilized amorphous powder | Sawada, S et al, *Experientia*, 39:377, 1983

Ala-Ala-Pro-Phe
N-Suc-Ala-Ala-Pro-Phe-Thiobenzyl Ester

Synonyms: Cathepsin G Substrate; Protease Substrate; Chymotrypsin Substrate

Sigma S 6518   FW 610.7 $C_{31}H_{38}N_4O_7S$ Very sensitive substrate for rat mast cell proteases & chymotrypsin; cleaved by human leukocyte cathepsin G & chymotrypsin-like serine proteases from skin

---

Ala-Ala-Pro-Phe
N-Suc-Ala-Ala-Pro-Phe p-Nitroanilide

Synonyms: Cathepsin G Substrate; Protease Substrate; Peptidyl Prolyl Isomerase Substrate; Chymotrypsin Substrate

Sigma S 7388   FW 624.6 $C_{30}H_{36}N_6O_9$ Substrate for chymotrypsin & human leukocyte cathepsin G & peptidyl prolyl isomerase; DelMar, EG et al, *Anal Biochem*, 99:316, 1979; Nakajima, K et al, *J Biol Chem*, 254:4027, 1979; Siekierka, JJ et al, *Nature*, 341:755, 1989; Fisher, G et al, *Biomed Biochem Acta*, 43:1110, 1984

Ala-Ala-Pro-Phe
N-Suc-Ala-Ala-Pro-Phe-AMC

Sigma S 9761   FW 661.7 $C_{34}H_{39}N_5O_9$ Sawada, S et al, *Experientia*, 39:377, 1983

Ala-Ala-Pro-Val
MeOSuc-Ala-Ala-Pro-Val

Bachem H-3748.0050   MW 470.52 $C_{21}H_{34}N_5O_8$ Store at -15°C

Ala-Ala-Pro-Val
Ala-Ala-Pro-Val-OMe Hydrochloride

Bachem H-6075.0250   MW 406.91 $C_{17}H_{30}N_4O_5 \cdot$ HCl   Store at -15°C

Ala-Ala-Pro-Val
MeOSuc-Ala-Ala-Pro-Val-AMC

Bachem I-1270.0050   MW 627.7 $C_{31}H_{41}N_5O_9$ Store at -15°C

Ala-Ala-Pro-Val
Suc-Ala-Ala-Pro-Val-AMC

Bachem I-1490.0050   MW 613.67 $C_{30}H_{39}N_5O_9$ Store at -15°C

Ala-Ala-Pro-Val
Ac-Ala-Ala-Pro-Val-pNA

Bachem L-1015.0050   MW 518.57 $C_{24}H_{34}N_6O_7$ Store at -15°C

Ala-Ala-Pro-Val
MeOSuc-Ala-Ala-Pro-Val-pNA

Bachem L-1335.0050   MW 590.63 $C_{27}H_{38}N_6O_9$ Store at -15°C

Ala-Ala-Pro-Val
Suc-Ala-Ala-Pro-Val-pNA

Bachem L-1770.0050   MW 576.61 $C_{26}H_{36}N_6O_9$ Store at -15°C

Ala-Ala-Pro-Val
Ala-Ala-Pro-Val-CMK

Bachem N-1010.0025   MW 388.89 $C_{17}H_{29}ClN_4O_4$ Store at -15°C

Ala-Ala-Pro-Val
MeOSuc-Ala-Ala-Pro-Val-chloromethylketone

Bachem N-1055.0025   Store at -15°C

Ala-Ala-Pro-Val
MeOSuc-Ala-Ala-Pro-Val-pNA

Synonyms: Elastase Substrate I

Calbiochem 324696   MW 590.6 $C_{27}H_{38}N_6O_9$ Lyophilized solid containing 90 mg D-mannitol & 10 mg substrate; ≥95% (HPLC); soluble in water & ethanol | Colorimetric substrate; Nakajima, K et al, *J Biol Chem*, 254:4027, 1979

Ala-Ala-Pro-Val
N-(Methyoxysuccinyl)-L-Ala-Ala-L-Pro-L-Val-CMK

Synonyms: Elastase Inhibitor, Human Granulocyte

Fluka 65348   MW 503 $C_{22}H_{35}ClN_4O_7$ ≥99.0% (TLC); ≤1% water; mp:159-160°C | Martynov, VF et al, *J Gen Chem*, 54:384, 1984; An-Zhi, W et al, *FEBS Lett*, 234:367, 1988

Ala-Ala-Pro-Val
N-Methoxy-Suc-Ala-Ala-Pro-Val-MCA

Synonyms: Pancreatic Elastase Substrate

ICN 155442   Fluorogenic substrate for assay of human leukocyte & porcine pancreatic elastase; Castillo, MJ etal, *Anal Biochem*, 99:53, 1979

Ala-Ala-Pro-Val
N-Methoxy-Suc-Ala-Ala-Pro-Val CMK

Synonyms: Granulocyte Elastase Inhibitor Human

ICN 155443   MW 503 $C_{22}H_{35}ClN_4O_7$ Martynov, VF etal, *J Gen Chem* (USSR), 54:384, 1984

Ala-Ala-Pro-Val
MeOSuc-Ala-Ala-Pro-Val-AMC

Synonyms: Elastase Substrate V

ICN 195979   MW 627.7 $C_{31}H_{41}N_5O_9$ ≥98% | Fluorogenic substrate for quantitation of human leukocyte & porcine pancreatic elastase

---

Ala-Ala-Pro-Val
MeOSuc-Ala-Ala-Pro-Val-AFC

Synonyms: Elastase Substrate VI

ICN 195980 MW 681.7 $C_{31}H_{38}N_5O_9F_3$ ≥95% | Fluorogenic substrate for quantitation of human leukocyte & porcine pancreatic elastase

Ala-Ala-Pro-Val
MeOSuc-Ala-Ala-Pro-Val-pNA

Synonyms: Pefachrome®ELA; Pefa-5811

Pentapharm 090-14, 090-08 MW 590.6 Highly sensitive chromogenic peptide substrate for leukocyte elastase

Ala-Ala-Pro-Val
MeOSuc-Ala-Ala-Pro-Val-AMC

Synonyms: Pefafluor ELA; Pefa-5802

Pentapharm 090-21, 090-05 MW 627.7 Excitation wavelength 360 nm, emission wavelength 460 nm | Fluorogenic peptide substrate for leucocyte elastase; v_{max}:0.01 μmol s^{-1}, K_M:1.0; *mM*

Ala-Ala-Pro-Val
Suc(OMe)-Ala-Ala-Pro-Val-MCA

Synonyms: Human Leukocyte Substrate; Porcine Pancreatic Elastase Substrate

Peptides International MAV-3153-v MW 627.69 $C_{31}H_{41}N_5O_9$ >99% (HPLC); lyophilized amorphous powder | Castrillo, MJ et al, *Anal Biochem*, 99:53, 1979

Ala-Ala-Pro-Val
N-MeOSuc-Ala-Ala-Pro-Val Chloromethyl Ketone

Synonyms: Elastase Inhibitor

Sigma M 0398 FW 503.0 $C_{22}H_{35}CIN_4O_7$ Inhibitor of human granulocyte elastase; Martynov, VF et al, *J Gen Chem* (USSR), 54:384, 1984; Stein, RL & Trainor, DA, *Biochemistry*, 25:5414, 1986

Ala-Ala-Pro-Val
N-MeOSuc-Ala-Ala-Pro-Val

Sigma M 2897 FW 470.5 $C_{21}H_{34}N_4O_8$

Ala-Ala-Pro-Val
N-MeOSuc-Ala-Ala-Pro-Val-AMC

Synonyms: Elastase Substrate

Sigma M 9771 FW 627.7 $C_{31}H_{41}N_5O_9$ Fluorogenic substrate for the assay of human leukocyte & porcine pancreatic elastase; Castillo, MJ et al, *Anal Biochem*, 99:53, 1979

Ala-Ala-Pro-Val
N-(Methyoxysuccinyl)-L-Ala-Ala-L-Pro-L-Val-4-Nitroanilide

Synonyms: Elastase Substrate

Fluka 65352 Human leukocytes, porcine pancreas MW 590.6 $C_{27}H_{38}N_6O_9$ ≥97.0% (HPLC); ≥99% peptide content; ≤3% water | Sensitive substrate; Nakajima, K et al, *JBC*, 254:4027, 1979

Ala-Ala-Ser-Thr-Thr-Thr-Asn-Tyr-Thr
(D-Ala^1)-D-Ala-Ser-Thr-Thr-Thr-Asn-Tyr-Thr Amide

Synonyms: Peptide T; HIV Inhibitor

ICN 153045 MW 856.9 $C_{35}H_{56}N_{10}O_{15}$ Pert, CB etal, *PNAS*, 83:9254, 1986

Ala-Ala-Pro-Val
N-MeOSuc-Ala-Ala-Pro-Val

Sigma M 2897 FW 470.5 $C_{21}H_{34}N_4O_8$

Ala-Ala-Pro-Val
N-MeOSuc-Ala-Ala-Pro-Val-AMC

Synonyms: Elastase Substrate

Sigma M 9771 FW 627.7 $C_{31}H_{41}N_5O_9$ Fluorogenic substrate for the assay of human leukocyte & porcine pancreatic elastase; Castillo, MJ et al, *Anal Biochem*, 99:53, 1979

Ala-Ala-Pro-Val
N-(Methyoxysuccinyl)-L-Ala-Ala-L-Pro-L-Val-4-Nitroanilide

Synonyms: Elastase Substrate

Fluka 65352 Human leukocytes, porcine pancreas MW 590.6 $C_{27}H_{38}N_6O_9$ ≥97.0% (HPLC); ≥99% peptide content; ≤3% water | Sensitive substrate; Nakajima, K et al, *JBC*, 254:4027, 1979

Ala-Ala-Ser-Thr-Thr-Thr-Asn-Tyr-Thr
(D-Ala^1)-D-Ala-Ser-Thr-Thr-Thr-Asn-Tyr-Thr Amide

Synonyms: Peptide T; HIV Inhibitor

ICN 153045 MW 856.9 $C_{35}H_{56}N_{10}O_{15}$ Pert, CB etal, *PNAS*, 83:9254, 1986

Ala-Ala-Trp-Phe-Lys
D-Ala-β-(2-Naphthyl)-D-Ala-Trp-D-Phe-Lys Amide

Synonyms: Growth Hormone Releasing Peptide

Sigma A 2960 FW 746.9 $C_{42}H_{50}N_8O_5$ ≥97% (HPLC) | Bioactive peptide; growth hormone-releasing peptide studied in rats; Sawade, H et al, *Regulatory Peptides*, 53:195, 1994

Ala-Ala-Trp-Phe-Pro-Pro-Nle
BOC-Ala-Ala-D-Trp-Phe-D-Pro-Pro-Nle Amide

Synonyms: GR-87389

American Peptide 62-1-28 MW 900.1 $C_{47}H_{65}N_9O_9$ Highly potent & selective NK$_2$ receptor antagonist; McElroy, AB et al, *J Med Chem*, 35:2582, 1992

Ala-Ala-Trp-Phe-Pro-Pro-Nle
PhCO-Ala-Ala-D-Trp-Phe-D-Pro-Pro-Nle Amide

Synonyms: GR-94800

American Peptide 62-1-29 MW 904.0 $C_{49}H_{61}N_9O_8$ Selective NK$_2$ tachykinin receptor antagonist; McElroy, AB et al, *J Med Chem*, 35:2582, 1992

Ala-Ala-Trp-Phe-Pro-Pro-Nle
BOC-Ala-Ala-D-Trp-Phe-D-Pro-Pro-Nle Amide

Bachem H-2794.0500 Store at -15°C

Ala-Ala-Trp-Phe-Pro-Pro-Nle
Bz-Ala-Ala-D-Trp-Phe-D-Pro-Pro-Nle Amide

Bachem H-2796.0500 Store at -15°C

Ala-Ala-Trp-Phe-Pro-Pro-Nle
PhCO-Ala-Ala-D-Trp-Phe-D-Pro-Pro-Nle Amide

Synonyms: GR-94800

Neosystem SC491 MW 904.00 Bioactive; tachykinin; highly potent & selective NK-2 receptor antagonist; McElroy, AB et al, *J Med Chem*, 35:2582-2591, 1992

Ala-Ala-Trp-Phe-Pro-Pro-Nle
BOC-Ala-Ala-D-Trp-Phe-D-Pro-Pro-Nle Amide

Synonyms: GR-87389

Neosystem SC492 MW 900.10 Bioactive; tachykinin; highly potent & selective NK-2 receptor antagonist; McElroy, AB et al, *J Med Chem*, 35:2582-2591, 1992

Ala-Ala-Tyr			
Bachem H-1445.0250	MW 323.35	$C_{15}H_{21}N_3O_5$	Store at -15°C

Ala-Ala-Tyr			
Ac-Ala-Ala-Tyr-AMC			
Bachem I-1005.0050	MW 522.56	$C_{27}H_{30}N_4O_7$	Store at -15°C

Ala-Ala-Tyr-Ala			
Bachem H-1450.0050	MW 394.43	$C_{18}H_{26}N_4O_6$	Store at -15°C

Ala-Ala-Tyr-Ala-Ala			
Bachem H-1455.0050	MW 465.51	$C_{21}H_{31}N_5O_7$	Store at -15°C

Ala-Ala-Val		
Suc-Ala-Ala-Val		
Bachem H-5005.0050	Store at -15°C	

Ala-Ala-Val			
Suc-Ala-Ala-Val-AMC			
Bachem I-1320.0050	MW 516.55	$C_{25}H_{32}N_4O_8$	Store at -15°C

Ala-Ala-Val			
Suc-Ala-Ala-Val-pNA			
Bachem L-1405.0050	MW 479.49	$C_{21}H_{29}N_5O_8$	Store at -15°C

Ala-Ala-Val	
N-Suc-Ala-Ala-Val	
ICN 156690	

Ala-Ala-Val	
N-Suc-Ala-Ala-Val-pNA	
ICN 156692	

Ala-Ala-Val
N-Suc-Ala-Ala-Val p-Nitroanilide

Synonyms: Elastase Substrate

Sigma S 1384 FW 479.5 $C_{21}H_{29}N_5O_8$ Substrate for human & rat leukocyte elastases; Wenzel, HR et al, *Hoppe-Seyler's Z Physiol Chem*, 361:1413, 1980; Stein, RL, *J Am Chem Soc*, 105:5111, 1983; Virca, ED et al, *Eur J Biochem*, 144:1, 1984

Ala-Ala-Val-Ala			
Suc-Ala-Ala-Val-Ala-pNA			
Bachem L-1410.0050	MW 550.57	$C_{24}H_{34}N_6O_9$	Store at -15°C

Ala-Ala-Val-Ala	
N-Suc-Ala-Ala-Val-Ala-pNA	
ICN 156691	

Ala-Ala-Val-Ala		
Ala-Ala-Val-Ala *p*-Nitroanilide		
Sigma A 9148	FW 427.5	$C_{21}H_{25}N_5O_5$

Ala-Ala-Val-Ala		
N-Suc-Ala-Ala-Val-Ala *p*-Nitroanilide		
Sigma S 7632	FW 550.6	$C_{24}H_{34}N_6O_9$

Ala-Ala-Val-Ala p-Nitroanilide

Synonyms: Chymotrypsin Substrate; Tripeptidyl Peptidase I Substrate

Sigma A 9273 FW 450.5 $C_{20}H_{30}N_6O_6$

Ala-Ala-Val-Ala-Leu-Leu-Pro-Ala-Val-Leu-Leu-Ala-Leu-Leu-Ala-Pro-Asp-Glu-Val-Asp
Ac-Ala-Ala-Val-Ala-Leu-Leu-Pro-Ala-Val-Leu-Leu-Ala-Leu-Leu-Ala-Pro-Asp-Glu-Val-Asp-CHO

Synonyms: Caspase III Inhibitor; Caspase III Inhibitor I; Caspase III Inhibitor I, Cell-Permeable; CPP-32/Apopain Inhibitor, Cell-Permeable

Alexis 260-046 MW 2000.5 $C_{94}H_{158}N_{20}O_{27}$ ≥97%; white lyophilized powder; soluble in DMSO | Inhibitor of Caspase III with increased cell permeability

American Peptide 81-6-35 MW 2000.4 $C_{94}H_{158}N_{20}O_{27}$ Readily cell permeable; highly specific, potent & reversible inhibitor of CPP-32; Lin, Y-Z et al, *J Biol Chem*, 271:5305, 1996; Nicholson, DW, *Nature Biotechnol*, 14:297, 1996; Tewari, M et al, *Cell*, 81:801, 1995

Calbiochem 235423 MW 2000.4 $C_{94}H_{158}N_{20}O_{27}$ ≥95% (HPLC); lyophilized solid; soluble in DMSO | Cell-permeable inhibitor of Caspase III, as well as Caspases VI, VII, VIII & X; the C-terminal DEVD-CHO sequence of this peptide is a highly specific potent & reversible inhibitor of Caspase III which strongly inhibits PARP cleavage in cultured human osteosarcoma cell extracts; the N-terminal sequence (1-16) corresponds to the hydrophobic region of the signal peptide of Kaposi fibroblast growth factor & confers cell-permeability to the peptide; Lin, Y-Z et al, *J Biol Chem*, 270:14255, 1995; Tewari, M et al, *Cell*, 81:801, 1995; Lazebnik, YA et al, *Nature*, 371:346, 1994; Nicholson, DW et al, *Nature Biotech*, 14:297, 1996; Lin, Y-Z et al, *J Biol Chem*, 271:5305, 1996; Nicholson, DW et al, *Nature*, 376:37, 1995

Ala-Ala-Val-Ala-Leu-Leu-Pro-Ala-Val-Leu-Leu-Ala-Leu-Leu-Ala-Pro-Ile-Glu-Thr-Asp
Ac-Ala-Ala-Val-Ala-Leu-Leu-Pro-Ala-Val-Leu-Leu-Ala-Leu-Leu-Ala-Pro-Ile-Glu-Thr-Asp-CHO

Synonyms: Caspase VIII Inhibitor I, Cell-Permeable; Granzyme B Inhibitor II, Cell-Permeable; IETD-CHO, Cell-Permeable

Calbiochem 218773 MW 2000.4 $C_{95}H_{162}N_{20}O_{26}$ ≥95% (HPLC); lyophilized solid; soluble in DMSO | Potent, cell-permeable & reversible inhibitor of Caspase VIII (FLICE, MACH, Mch5) & granzyme B

Ala-Ala-Val-Ala-Leu-Leu-Pro-Ala-Val-Leu-Leu-Ala-Leu-Leu-Ala-Pro-Leu-Glu-Val-Asp
Ac-Ala-Ala-Val-Ala-Leu-Leu-Pro-Ala-Val-Leu-Leu-Ala-Leu-Leu-Ala-Pro-Leu-Glu-Val-Asp-CHO

Synonyms: Caspase IV Inhibitor I, Cell-Permeable; LEVD-CHO, Cell-Permeable

Calbiochem 218766 MW 1998.5 $C_{96}H_{164}N_{20}O_{25}$ ≥95% (HPLC); lyophilized solid; soluble in DMSO | Potent, cell-permeable & reversible inhibitor of Caspase IV

Ala-Ala-Val-Ala-Leu-Leu-Pro-Ala-Val-Leu-Leu-Ala-Leu-Leu-Ala-Pro-Leu-Glu-His-Asp
Ac-Ala-Ala-Val-Ala-Leu-Leu-Pro-Ala-Val-Leu-Leu-Ala-Leu-Leu-Ala-Pro-Leu-Glu-His-Asp-CHO

Synonyms: Caspase IX Inhibitor II, Cell-Permeable; LEHD-CHO, Cell-Permeable

Calbiochem 218776 MW 2036.5 $C_{97}H_{162}N_{22}O_{25}$ ≥95% (HPLC); lyophilized solid; soluble in DMSO | Potent, cell-permeable & reversible inhibitor of Caspase IX (ICE-LAP6, Mch6); inhibits Caspase IV & Caspase V

Ala-Ala-Val-Ala-Leu-Leu-Pro-Ala-Val-Leu-Leu-Ala-Leu-Leu-Ala-Pro-Tyr-Val-Ala-Asp
Ac-Ala-Ala-Val-Ala-Leu-Leu-Pro-Ala-Val-Leu-Leu-Ala-Leu-Leu-Ala-Pro-Tyr-Val-Ala-Asp-CHO

Synonyms: Caspase I Inhibitor; Caspase I Inhibitor I; Caspase I Inhibitor I, Cell-Permeable; Interleukin Converting Enzyme Inhibitor I, Cell-Permeable

Alexis 260-047   MW 1990.5 $C_{97}H_{160}N_{20}O_{24}$ ≥97%; white lyophilized powder; soluble in DMSO | Caspase I inhibitor with increased cell permeability

American Peptide 81-6-36   MW 1990.4 $C_{97}H_{160}N_{20}O_{24}$ Cell permeable inhibitor of Caspase I; Lin, Y-Z et al, *J Biol Chem*, 271:5305, 1996; Reiter, LA, *Int J Pept Protein Res*, 43:87, 1994; Thornberry, NA et al, *Biochemistry*, 33:3934, 1994

Calbiochem 400011   MW 1990.5 $C_{97}H_{160}N_{20}O_{24}$ ≥95% (HPLC); lyophilized solid; soluble in DMSO | Cell-permeable inhibitor of Caspase I & Caspase IV; the C-terminal YVAD-CHO sequence of this peptide is a highly specific potent & reversible inhibitor of Caspase I; the N-terminal sequence (1-16) corresponds to the hydrophobic region of the signal peptide of Kaposi fibroblast growth factor & confers cell-permeability to the peptide; Lin, Y-Z et al, *J Biol Chem*, 270:14255, 1995; Reiter, LA et al, *Int J Pept Protein Res*, 43:87, 1994; Lazebnik, YA et al, *Nature*, 371:346, 1994; Thornberry, NA et al, *Biochemistry*, 33:3934, 1994; Thornberry, NA et al, *Methods Enzymol*, 244:615, 1994; Thornberry, NA et al, *Nature*, 356:768, 1992; Lin, Y-Z et al, *J Biol Chem*, 271:5305, 1996

Ala-Ala-Val-Ala-Leu-Leu-Pro-Ala-Val-Leu-Leu-Ala-Leu-Leu-Ala-Pro-Val-Glu-Ile-Asp
Ac-Ala-Ala-Val-Ala-Leu-Leu-Pro-Ala-Val-Leu-Leu-Ala-Leu-Leu-Ala-Pro-Val-Glu-Ile-Asp-CHO

Synonyms: Caspase VI Inhibitor II, Cell-Permeable

Calbiochem 218767   MW 1998.5 $C_{96}H_{164}N_{20}O_{25}$ ≥95% (HPLC); lyophilized solid; soluble in DMSO | Potent, cell-permeable & reversible inhibitor of Caspase VI (Mch-2)

Ala-Ala-Val-Ala-Leu-Leu-Pro-Ala-Val-Leu-Leu-Ala-Leu-Leu-Ala-Pro-Val-Gln-Arg-Lys-Arg-Gln-Lys-Leu-Met-Pro Trifluoroacetate Salt

Synonyms: NF-κB SN50 Cell-Permeable Inactive Control Peptide

Calbiochem 481480   MW 2781.5 $C_{129}H_{230}N_{36}O_{29}S$ ≥97% (HPLC); lyophilized solid; soluble in water | Contains the nuclear localization sequence of the transcription factor NF-κB p50 linked to the hydrophobic region of the signal peptide of Kaposi fibroblast growth factor; the peptide N-terminal K-FGF h-region confers cell-permeability while the NLS (360-369) inhibits translocation of the NF-κB active complex into the nucleus; in murine endothelial LE-II cells induced by LPS, NF-κB nuclear translocation is maximally inhibited at 18 μM; Lin, Y-Z et al, *J Biol Chem*, 270:14255, 1995

Ala-Ala-Val-Ala-Leu-Leu-Pro-Ala-Val-Leu-Leu-Ala-Leu-Leu-Ala-Pro-Val-Gln-Arg-Asn-Gly-Gln-Lys-Leu-Met-Pro Trifluoroacetate Salt

Synonyms: NF-κB SN50M Cell-Permeable Inactive Control Peptide

Calbiochem 481486   MW 2668.3 $C_{123}H_{215}N_{33}O_{30}S$ ≥97% (HPLC); lyophilized solid; soluble in water | Corresponds to the SN50 peptide sequence with substitutions of Lys[363] for Asn & Arg[364] for Gly in the NLS region; in murine endothelial LE-II cells induced by LPS, had no measurable effect on NF-κB translocation at 18 μM; Lin, Y-Z et al, *J Biol Chem*, 270:14255, 1995

Ala-Ala-Val-Asp-Leu-Ser-His-Phe-Leu-Lys-Glu-Lys

Synonyms: NEF (83-94); HTLV I Peptide

Neosystem SC572   MW 1357.57   HTLV I peptide from subtype NEF

Ala-Arg
Z-Ala-Arg

Bachem C-1060.0250   Store at -15°C

Ala-Arg
Z-Ala-Arg-OMe Hydrochloride

Bachem C-3845.0250   Store at -15°C

Ala-Arg

Bachem G-1175.0250   MW 245.28 $C_9H_{19}N_5O_3$ Store at -15°C

Ala-Arg
Ala-Arg(NO₂)-OMe Hydrochloride

Bachem G-4005.0250   MW 340.77 $C_{10}H_{20}N_6O_5 \cdot HCl$ Store at -15°C

Ala-Arg
Bz-Ala-Arg Hydrochloride

Bachem G-4145.0250   Store at -15°C

Ala-Arg
Ala-Arg-AMC Hydrochloride

Bachem I-1610.0050   MW 438.91 $C_{19}H_{26}N_6O_4 \cdot HCl$ Store at -15°C

Ala-Arg
Ala-Arg-βNA Dihydrochloride

Bachem K-1055.0050   MW 443.38 $C_{19}H_{26}N_6O_2 \cdot 2HCl$ Store at -15°C

Ala-Arg
FA-Ala-Arg

Bachem M-1810.0050   MW 365.39 $C_{16}H_{23}N_5O_5$ Store at -15°C

Ala-Arg
D-CHG-Ala-Arg-pNA·2AcOH

Synonyms: Pefachrome®TH; Pefa-5114

Pentapharm 081-20, 081-03   MW 624.7   Chromogenic peptide substrate for determination of thrombin and antithrombin III; v_{max}:4.78 μmol/min; K_M:15.9 μM

Ala-Arg-Arg
Z-Ala-Arg-Arg-AMC

Bachem I-1125.0050   MW 692.78 $C_{33}H_{44}N_{10}O_{17}$ Store at -15°C

Ala-Arg-Arg
Z-Ala-Arg-Arg-4MβNA

Bachem J-1090.0050   MW 690.8 $C_{34}H_{46}N_{10}O_6$ Store at -15°C

Ala-Arg-Arg
Z-Ala-Arg-Arg-4MβNA Dihydrochloride

Synonyms: Cathepsin B Substrate II

ICN 195953   MW 763.7 $C_{34}H_{46}N_{10}O_6 \cdot 2HCl$ Substrate for the quantitative determination of cathepsin B

Ala-Arg-Arg
N-CBZ-Ala-Arg-Arg 4-Methoxy-β-Naphthylamide Acetate Salt Free Base

Synonyms: Cathepsin B Substrate

Sigma C 8536   FW 690.8 $C_{34}H_{46}N_{10}O_6 \cdot HCOOH$ Singh, H & Kalnitsky, G, *J Biol Chem*, 253:4319, 1978

Ala-Arg-Arg
CBZ-Ala-Arg-Arg-AMC Diacetate

USBio C2098-20   MW 785.8 $C_{32}H_{45}N_{10}O_6 \cdot 2C_2H_4O_2$ ≥99%

Ala-Arg-Arg
CBZ-Ala-Arg-Arg-4MβNA Diacetate

USBio C2098-21   MW 810.9 $C_{34}H_{46}N_{10}O_6 \cdot 2C_2H_4O_2$ ≥99%

Ala-Arg-Arg-Ala

Bachem H-1208.0050   MW 472.55 $C_{18}H_{36}N_{10}O_5$ Store at -15°C

Ala-Arg-Arg-Asn-Arg-Arg-Arg-Arg-Trp-Arg-Glu-Arg-Gln-Arg

Synonyms: REV (37-50); HTLV I Peptide

Neosystem SC192   MW 2052.34   HTLV I peptide from subtype REV

Ala-Arg-Arg-Pro-Glu-Gly-Arg-Thr-Trp-Ala-Gln-Pro-Gly-Tyr

American Peptide 86-5-17   MW 1644.8 $C_{72}H_{109}N_{25}O_{20}$

Ala-Arg-Gly-Ile-Lys-Gly-Ile-Arg-Gly-Phe-Ser-Gly 3AcOH 5H₂O

Synonyms: Lysine Hydroxylase Substrate

Peptides International SAG-4166   MW 1488.67
$C_{53}H_{91}N_{19}O_{14} \cdot 3CH_3COOH \cdot 5H_2O$ >99% (HPLC); amorphous powder; bulk | Kivirikko, KI et al, *Biochem*, 11:122, 1972

Ala-Arg-Gly-Thr-Asn-Val-Gly-Arg-Glu-Cys-Cys-Leu-Glu-Tyr-Phe-Lys-Gly-Ala-Ile-Pro-Leu-Arg-Lys-Leu-Lys-Thr-Trp-Tyr-Gln-Thr-Ser-Glu-Asp-Cys-Ser-Arg-Asp-Ala-Ile-Val-Phe-Val-Thr-Val-Gln-Gly-Arg-Ala-Ile-Cys-Ser-Asp-Pro-Asn-Asn-Lys-Arg-Val-Lys-Asn-Ala-Val-Lys-Tyr-Leu-Gln-Ser-Leu-Glu-Arg-Ser

Synonyms: Thymus and Activation-Regulated Chemokine

Bachem H-4616.0010   Human   Store at -15°C

Ala-Arg-Leu-Asp-Thr-Ser-Ser-Gln-Phe-Arg-Lys-Lys-Trp-Asn-Lys-Trp-Ala-Leu-Ser-Arg Amide

Synonyms: Pro-Adrenomedullin, (N-20)-; Prodepin; Pro-Adrenomedullin N-20; Prodepin; Pro-Adrenomedullin N-Terminal 20 Peptide; Hypotensive Peptide

American Peptide 22-8-20   Rat   MW 2477.9
$C_{111}H_{177}N_{37}O_{28}$ Sakata, J et al, *BBRC*, 195:921, 1993

ICN 195614   Rat   MW ~2478   Sakata, J etal, *BBRC*, 195:921, 1993

Peptides International PAM-4292-v   Rat synthetic   MW 2477.9 $C_{111}H_{177}N_{37}O_{28}$ >99% (HPLC); lyophilized amorphous powder | Bioactive; Sakata, J et al, *BBRC*, 195:921, 1993

Ala-Arg-Leu-Asp-Val-Ala-Ala-Glu-Phe-Arg-Lys-Lys-Trp-Asn-Lys-Trp-Ala-Leu-Ser-Arg Amide

Synonyms: Pro-Adrenomedullin, (N-20)-; Prodepin

American Peptide 22-3-11   Porcine   MW 2444.9
$C_{112}H_{178}N_{36}O_{26}$

Ala-Arg-Leu-Asp-Val-Ala-Ser-Glu-Phe-Arg-Lys-Lys-Trp-Asn-Lys-Trp-Ala-Leu-Ser-Arg Amide

Synonyms: Pro-Adrenomedullin (1-20)

Bachem H-4916.0500   Human   MW 2460.87
$C_{112}H_{178}N_{36}O_{27}$ Store at -15°C

Ala-Arg-Leu-Asp-Val-Ala-Ser-Glu-Phe-Arg-Lys-Lys-Trp-Asn-Lys-Trp-Ala-Leu-Ser-Arg Amide Trifluoroacetate Salt

S Pro-Adrenomedullin N-Terminal 20 Peptide *ynonyms*:

Calbiochem 529640   Human   MW 2460.9 $C_{112}H_{178}N_{36}O_{27}$
≥98% (HPLC); lyophilized solid; soluble in water | Processed from the N-terminus of proadrenomedullin & postulated to have important physiological function because of its sequence conservation among different species; Kitamura, K et al, *Biochem Biophys Res Comm*, 194:720, 1993

Ala-Arg-Leu-Asp-Val-Ala-Ser-Glu-Phe-Arg-Lys-Lys-Trp-Asn-Lys-Trp-Ala-Leu-Ser-Arg Amide

Synonyms: Vasoactive Intestinal Peptide N-20; Prodepin; Pro-Adrenomedullin N-Terminal 20 Peptide; Hypotensive Peptide

ICN 195611   Human   MW ~2461   Kitamura, K etal, *BBRC*, 194:720, 1994

Peptides International PAM-4291-v   Human synthetic   MW 2460.9 $C_{112}H_{178}N_{36}O_{27}$ >99% (HPLC); lyophilized amorphous powder | Bioactive; Kitamura, K et al, *BBRC*, 194:720, 1993; Washimine, H et al, *BBRC*, 202:1081, 1994; Kitamura, K et al, *FEBS Lett*, 351:35, 1994; Katoh, F et al, *Neurochemistry*, 64:459, 1995

Ala-Arg-Lys-pSer-Thr-Gly-Gly-Lys-Ala-Pro-Arg-Lys-Gln-Leu-Cys

Synonyms: Histone H3, P

Upstate 12-383   MW 1680   >90% pure; lyophilized | Immunizing Peptide

ALA-ARG-LYS-SER-THR-GLY-GLY-AC-LYS-ALA-PRO-ARG-GLY-CYS

Synonyms: Histone H3 Peptide, (Ac-Lys[14])-

Upstate 12-359   MW 1329.18   >90% pure; frozen solution | Substrate for histone modifying enzymes

Ala-Arg-Pro-Ala-Lys

Synonyms: Peptide 6A

Bachem H-6905.0005   MW 541.65 $C_{23}H_{43}N_9O_6$ Store at -15°C

Ala-Arg-Pro-Ala-Lys
Ala-Arg-Pro-Ala-D-Lys Amide Trifluoroacetate Salt Free Base

Sigma A 0205   FW 540.7 $C_{23}H_{44}N_{10}O_{15}$ ≥97% (HPLC) | Bioactive peptide; potent coronary vasodilator; *J Cardiovascular Pharmacol*, 23:103, 1994

Ala-Arg-Pro-Gly-Tyr-Leu-Ala-Phe-Pro-Arg-Met Amide

Synonyms: Cardioactive Peptide A, Small

Bachem H-6925.0001   Store at -15°C

ICN 153040   MW 1277.5   Lloyd, PE etal, *J Comp Physiol*, A156:659, 1985

Neosystem SC322   MW 1053.20   Bioactive neuropeptide; Cropper, EC et al, *PNAS USA*, 74:1267-1271, 1987

Sigma A 7052   FW 1277.5   ≥95% (HPLC) | Bioactive peptide; Lloyd, PE et al, *J Comp Physiol*, A156:659, 1985

American Peptide 60-0-60   *Aplysia*   MW 1277.5
$C_{59}H_{92}N_{18}O_{12}S$ Cardioactive peptide; Lloyd, PE et al, *J Comp Physiol*, A156:659, 1985

Ala-Arg-Ser-Ala-Pro-Thr-Pro-Met-Ser-Pro-Tyr
β-Ala-Arg-Ser-Ala-Pro-Thr-Pro-Met-Ser-Pro-Tyr

ICN 153041   MW 1177.3

Ala-Arg-Thr-Lys-Gln-Thr-Ala-Arg-Ac-Lys-Ser-Thr-Gly-Cys

Synonyms: Histone H3 Peptide, (Ac-Lys[9])-

Upstate 12-358 MW 1448.6 >90% pure; frozen solution |
Substrate for histone modifying enzymes

Ala-Arg-Thr-Lys-Gln-Thr-Ala-Arg-Ac-Lys-Ser-Thr-Gly-Gly-Ac-Lys-Ala-Pro-Arg-Lys-Gln-Leu-Cys

Synonyms: Histone H3 Peptide, (Ac-Lys[9,14])-

Upstate 12-360 MW 2370.8 >90% pure; frozen solution |
Substrate for histone modifying enzymes

Ala-Arg-Thr-Lys-Gln-Thr-Ala-Arg-Ac-Lys-Ser-Thr-Gly-Gly-Ac-Lys-Ala-Pro-Arg-Lys-Gln-Leu-Ala-Gly-Gly-Lys

Ala-Arg-Thr-Lys-Gln-Thr-Ala-Arg-Ac-Lys-Ser-Thr-Gly-Gly-Ac-Lys-Ala-Pro-Arg-Lys-Gln-Leu-Ala-Gly-Gly-Lys-Biotin

S Histone H3, (Lys[9,14])-Ac-*ynonyms*:

Upstate 12-402 MW 2807 Frozen solution | Histone-Assays

Ala-Arg-Thr-Lys-Gln-Thr-Ala-Arg-Lys-Ser-Thr-Gly-Gly-Lys-Ala-Pro-Arg-Lys-Gln-Leu-Cys

Synonyms: Histone H3 Peptide

Upstate 12-357 MW 2287 >90% pure; frozen solution |
Substrate for histone modifying enzymes

Ala-Arg-Tyr-Tyr-Ser-Ala-Leu-Arg-His-Tyr-Ile-Asn-Leu-Ile-Thr-Arg-Gln-Arg-Tyr Amide

Synonyms: Neuropeptide Y (18-36); Gastrointestinal Peptide Y

Bachem H-3296.0500 MW 2456.84 $C_{112}H_{174}N_{36}O_{27}$ Store at -15°C

Neosystem SC321 MW 2456.83 Bioactive neuropeptide; hypotensive action in rat; Boublik, J et al, *Int J Pept Prot Res*, 33:11-15, 1989

Sigma N 1272 Human, porcine FW 2456.8 ≥97% (HPLC); peptide content:~70% | Bioactive peptide; competitive NPY antagonist in cardiac membranes; Balasubramanian, A, *J Biol Chem*, 265:14724, 1990

American Peptide 60-4-25 Porcine MW 2456.9 $C_{112}H_{174}N_{36}O_{27}$ Competitive NPY antagonist in cardiac membranes; Balasubramaniam, A et al, *JBC*, 1995 (in press)

Ala-Arg-Val-Leu-Ala-Glu-Ala

Ac-Ala-Arg-Val-Leu-Ala-Glu-Ala Amide

Synonyms: HIV Protease Substrate

Neosystem SC698 MW 769.89 Virus-related peptide; AIDS-related peptide; peptide substrate for a novel colorimetric assay is cleaved between Leu & Ala; the primary amino group of Leu exposed upon cleavage can be detected by the well-known reaction with 2,4,6-trinitrobenzene-sulfonic acid (TNBS), yielding yellow-colored products (ε = 10-13 000 cm-1M1); Billich, A & Winkler, G, *Peptide Res*, 3:274-276, 1990

Ala-Arg-Val-Nle-Phe-Glu-Ala-Nle

Ala-Arg-Val-Nle-*p*-Nitro-Phe-Glu-Ala-Nle Amide

Synonyms: HIV Protease Substrate V

Bachem H-1146.0001 Store at -15°C

Ala-Arg-Val-Nle-Phe-Glu-Ala-Nle

Anthranilyl-Ala-Arg-Val-Nle-(pNO₂-Phe)-Glu-Ala-Nle Amide

Synonyms: HIV Anthranilyl Substrate V

ICN 158724

Ala-Arg-Val-Nle-Phe-Glu-Ala-Nle

Ala-Arg-Val-Nle-(pNO₂-Phe)-Glu-Ala-Nle Amide

Synonyms: HIV Substrate V

ICN 158723 Synthetic

Ala-Arg-Val-Nle-Tyr-Glu-Ala-Nle

Abz-Ala-Arg-Val-Nle-Tyr(NO₂)-Glu-Ala-Nle Amide

Synonyms: HIV Substrate

American Peptide 81-0-45 MW 1097.3 $C_{50}H_{76}N_{14}O_{14}$

Ala-Asn

Z-Ala-Asn

Bachem C-1065.0250 MW 337.33 $C_{15}H_{19}N_3O_6$ Store at -15°C

Ala-Asn

Bachem G-1180.0250 MW 203.2 $C_7H_{13}N_3O_4$ Store at -15°C

Ala-Asn-Glu-Arg-Ala-Asp-Leu-Ile-Ala-Tyr-Leu-Gln-Gln-Ala-Thr-Lys

Synonyms: Cytochrome C Fragment (88-103), (Gln[99])-Moth

Bachem H-2252.0001 MW 1805.02 $C_{78}H_{129}N_{23}O_{26}$ Store at -15°C

Ala-Asn-Phe-Leu-Val-Trp-Glu-Ile-Val-Arg-Lys-Lys-Pro

Synonyms: Platelet Derived Growth Factor Antagonist

Bachem H-8650.0001 MW 1599.94 $C_{77}H_{122}N_{20}O_{17}$ Store at -15°C

Ala-Asn-Pro-Asp-Cys-Lys-Thr-Ile-Leu-Lys-Ala-Leu-Gly-Pro-Ala-Ala-Thr

Synonyms: HIV Protein p24 (326-342)

Bachem H-2986.0500 MW 1683.99 $C_{73}H_{126}N_{20}O_{23}S$ Store at -15°C

Ala-Asp

Z-Ala-Asp

Bachem C-1070.0250 MW 338.32 $C_{15}H_{18}N_2O_7$ Store at -15°C

Ala-Asp

Bachem G-1195.0250 MW 204.18 $C_7H_{12}N_2O_5$ Store at -15°C

Sigma A 0253 FW 204.2 $C_7H_{12}N_2O_5$

Ala-Asp-Ala-Gln-His-Ala-Thr-Pro-Pro-Ly-Lys-Lys-Arg-Lys-Val-Glu-Asp-Pro-Lys-Asp-Phe

Synonyms: Protein Kinase p34cdc2 Substrate; Protein Kinase Related Peptide; CSH 10

Sigma A 5688 FW 2406.7 ≥97% (HPLC); peptide content:~65% | Bioactive peptide; Marshak, DR et al, *J Cell Biochem*, 45:391, 1991

Ala-Asp-Ala-Gln-His-Ala-Thr-Pro-Pro-Lys-Lys-Lys-Arg-Lys-Val-Glu-Asp-Pro-Lys-Asp-Phe

Synonyms: Protein Kinase p34cdc2 Substrate; p34cdc2 Kinase Substrate Peptide; p34cdc2 Kinase Fragment

Alexis 165-018 MW 2406.7 $C_{106}H_{172}N_{32}O_{32}$ ≥97%; white lyophilized powder; soluble in H₂O | Highly specific; enzyme involved in mitosis & onset of the S phase; Marshak, DR et al, *J Cell Biochem*, 45:391, 1991

American Peptide 86-0-12   MW 2406.8 $C_{106}H_{172}N_{32}O_{32}$ Eliminates multiple phosphorylation sites, leaving only one specific substrate for p34cdc2 protein kinase threonine residue available for phosphorylation, thereby reducing the probability of phosphorylation by other protein kinases; Marshak, DR et al, *J Cell Biochem*, 45:391, 1991; McVey, D et al, *Nature*, 341:503, 1989

American Peptide 86-5-25   MW 978.2 $C_{39}H_{69}N_{12}O_{13}S_2$ Lee, M et al, *Nature (Lond)*, 327:31, 1987

Bachem H-3288.0500   MW 2406.73 $C_{106}H_{172}N_{32}O_{32}$ Store at -15°C

Ala-Asp-Ala-Gln-His-Ala-Thr-Pro-Pro-Lys-Lys-Lys-Arg-Lys-Val-Glu-Asp-Pro-Lys-Asp-Phe Trifluoroacetate Salt

Synonyms: p34cdc2 Substrate

Calbiochem 219425   MW 2406.7 $C_{106}H_{172}N_{32}O_{32}$ ≥98% (HPLC); solid; soluble in water | Specific substrate for p34cdc2 that is based on the simian virus 40 large antigen sequence; not phosphorylated by other kinases such as CaM kinase II, casein kinase I & II, glycogen synthase kinase 3, phosphorylase b kinase, protein kinase A & protein kinase C; Marshak, DR et al, *J Cell Biochem*, 45:391, 1991

Ala-Asp-Ala-Gln-His-Ala-Thr-Pro-Pro-Lys-Lys-Lys-Arg-Lys-Val-Glu-Asp-Pro-Lys-Asp-Phe

Synonyms: Protein Kinase p34cdc2 Substrate

ICN 159654   Synthetic   MW 2407   No activity as a substrate for proteinkinase C; Marshak, DR etal, *J Cell Biochem*, 45:391, 1991

Ala-Asp-Cys-Lys-Tyr-Lys-Phe-Glu-Asn-Trp-Gly-Ala-Cys-Asp-Gly-Gly-Thr-Gly-Thr-Lys-Val-Arg-Gln-Gly-Thr-Leu-Lys-Lys-Ala-Arg-Tyr-Asn-Ala-Gln-Gln-Glu-Thr-Ile-Arg-Val-Thr-Lys-Pro-Cys-Thr-Pro-Lys-Thr-Lys-Ala-Lys-Ala-Lys-Ala-Lys-Lys-Gly-Lys-Gly-Lys-Asp

Synonyms: Midkine (60-121); Heparin Binding Growth/Differentiation Factor; Neurotrophic Factor; Neurite Outgrowth-Promoting Factor; Plasminogen Activator Activity Enhancer

Peptides International PMK-4299-s   Human synthetic   MW 6788.9 $C_{292}H_{483}N_{91}O_{87}S_4$ >99% (HPLC); lyophilized amorphous powder | Bioactive; Disulfide bonds: Cys[62]-Cys[94], Cys[72]-Cys[104]; Muramatsu, H et al, *BBRC*, 203:1131, 1994

Ala-Asp-Cys-Ser-Ala-Thr-Gly-Asp-Thr-Cys-Asp-His-Thr-Lys-Lys-Cys-Cys-Asp-Asp-Cys-Tyr-Thr-Cys-Arg-Cys-Gly-Thr-Pro-Trp-Gly-Ala-Asn-Cys-Arg-Cys-Asp-Tyr-Tyr-Lys-Ala-Arg-Cys-Asp-Thr
Ala-Asp-Cys-Ser-Ala-Thr-Gly-Asp-Thr-Cys-Asp-His-Thr-Lys-Lys-Cys-Cys-Asp-Asp-Cys-Tyr-Thr-Cys-Arg-Cys-Gly-Thr-Pro-Trp-Gly-Ala-Asn-Cys-Arg-Cys-Asp-Tyr-Tyr-Lys-Ala-Arg-Cys-Asp-Thr(Palmitoyl) Amide

Synonyms: PLTx-II; Presynaptic Ca[2+] Channel Blocker

Peptides International PPL-4300-s   *Plectreurys tristes* (spider) synthetic   MW 5108.8 $C_{208}H_{313}N_{61}O_{70}S_{10}$ >99% (HPLC); lyophilized amorphous powder | Bioactive; Branton, WD et al, *J Neurosci*, 7:4195, 1987; Leung, H-T et al, *Neuron*, 3:767, 1989; Branton, WD et al, *Nature*, 365:496, 1993

Alexis 630-068   Synthetic   MW 5108.8   Lyophilized powder; soluble in H_2O; Disulfide bonds: are undetermined | Originally isolated from Plectreurys *tristes*; prosynaptic Ca[2+] channel blocker; potent neurotoxin; Branton, WD et al, *J Neurosci*, 7:4195, 1987; Feigenbaum, P et al, *BBRC*, 154:298, 1988; Leung, HT et al, *Neuron*, 3:767, 1989; Branton, WD et al, *Nature*, 365:496, 1993

Ala-Asp-Leu-Ile-Ala-Tyr-Leu
D-Ala-Asp-Leu-Ile-Ala-Tyr-Leu Amide

Synonyms: Neuroprotectin, α-

Bachem N-1340.0001   Store at -15°C

Ala-Asp-Lys-Ala-Asp-Val-Asn-Val-Leu-Thr-Lys-Ala-Lys-Ser-Gln

Synonyms: Parathyroid Hormone (70-84); Parathormone

American Peptide 22-1-57   Human   MW 1587.8 $C_{67}H_{118}N_{20}O_{24}$

Bachem H-6665.0001   Human   Store at -15°C

ICN 154482   Human

Ala-Asp-Lys-Pro-Asp-Met-Gly-Glu-Ile-Ala-Ser-Phe-Asp-Lys-Ala-Lys-Leu-Lys-Lys-Thr-Glu-Thr-Gln-Glu-Lys-Asn-Thr-Leu-Pro-Thr-Lys-Glu-Thr-Ile-Glu-Gln-Glu-Lys-Arg-Ser-Glu-Ile-Ser
Ac-Ala-Asp-Lys-Pro-Asp-Met-Gly-Glu-Ile-Ala-Ser-Phe-Asp-Lys-Ala-Lys-Leu-Lys-Lys-Thr-Glu-Thr-Gln-Glu-Lys-Asn-Thr-Leu-Pro-Thr-Lys-Glu-Thr-Ile-Glu-Gln-Glu-Lys-Arg-Ser-Glu-Ile-Ser

Synonyms: Thymosin β10

Bachem H-2928.0500   Human, rat   Store at -15°C

Ala-Asp-Ser-Asp-Gly-Lys

Synonyms: Hamburger Pentapeptide Analog

Bachem H-1182.0005   MW 591.58 $C_{22}H_{37}N_7O_{12}$ Store at -15°C

Ala-Asp-Ser-Gly-Glu-Gly-Asp-Phe-Leu-Ala-Glu-Gly-Gly-Gly-Val-Arg

Synonyms: Fibrinopeptide A

American Peptide 42-1-12   Human   MW 1536.6 $C_{63}H_{97}N_{19}O_{26}$ Released from fibrinogen on thrombin action; useful as a tumor marker for circumscribed & disseminated gastric cancers; higher levels found in nephrotic patients; used to detect hypercoagulable state in individuals with atrial fibrillation & in underlying activation of blood coagulation

Bachem H-1465.0001   Human   MW 1536.58 $C_{63}H_{97}N_{19}O_{26}$ Store at -15°C

Calbiochem 341662   Human   MW 1536.6 $C_{63}H_{93}N_{19}O_{26}$ Lyophilized solid; ≥97% (HPLC); soluble in 5% acetic acid | First peptide released from fibrinogen on thrombin action; useful as a tumor marker for circumscribed & disseminated gastric cancers; higher levels are reported in nephrotic patients; also useful in detection of hypercoagulable state in individuals with atrial fibrillation & in underlying activation of blood coagulation; Sagripanti, A et al, *Int J Clin Lab Res*, 24:113, 1994; Abbasciano, V et al, *Oncology*, 45:159, 1988; Uno, M et al, *Jpn Circ J*, 52:9, 1988; Blombaeck, B et al, *Biochim Biophys Acta*, 115:371, 1966

ICN 151124   Human   Johnson, BJ & WP May, *J Pharm Sci*, 58:1568, 1969

Sigma F 3254   Human   FW 1536.6   ≥97% (HPLC) | Bioactive peptide; Johnson, BJ & May, WP, *J Pharm Sci*, 58:1568, 1969

Biogenesis 4450-6002   Human synthetic   MW 1536.8   >97% (HPLC); no preservatives; lyophilized

---

Ala-Asp-Val-Asn-Val-Leu-Thr-Lys-Ala-Lys-Ser-Gln

Synonyms: Parathyroid Hormone (73-84)

Bachem H-5520.0001   Human   Store at -15°C

Ala-Asp-Val-Asp-Val-Leu-Thr-Lys-Ala-Lys-Ser-Gln

Synonyms: Parathyroid Hormone (73-84), (Asp[76])-

Sigma P 1905   Human   FW 1274.4   ≥97% (HPLC) | Bioactive peptide

---

Ala-Cys-Ala-Trp-Leu-Glu-Ala-Gln-Glu-Glu-Glu-Glu-Val-Gly

Synonyms: NEF (54-67); HTLV I Peptide

Neosystem SC219   MW 1563.65   HTLV I peptide from subtype NEF; due to the presence of a Cys this peptide can contain the dimeric form

---

Ala-Cys-Asn-Thr-Ala-Thr-Cys-Val-Thr-His-Arg-Arg-Leu-Ala-Gly-Leu-Leu-Ser-Arg-Ser-Gly-Gly-Met-Val-Lys-Ser-Asn-Phe-Val-Pro-Thr-Asn-Val-Gly-Ser-Lys-Ala-Phe Amide

Synonyms: Calcitonin Gene Related Peptide II

American Peptide 22-1-14   Human   MW 3793.4
$C_{162}H_{267}N_{51}O_{48}S_3$ Disulfide bonds: Cys[2]-Cys7; dilates blood vessel, relaxes smooth muscle & causes cardiac positive inotropic effects; Steenbergh, PH et al, *FEBS Lett*, 183:403, 1985

Ala-Cys-Asn-Thr-Ala-Thr-Cys-Val-Thr-His-Arg-Leu-Ala-Asp-Phe-Leu-Ser-Arg-Ser-Gly-Gly-Val-Gly-Lys-Asn-Asn-Phe-Val-Pro-Thr-Asn-Val-Gly-Ser-Lys-Ala-Phe Amide

Synonyms: Calcitonin Gene Regulated Peptide

American Peptide 22-7-10   Chicken   MW 3838.4
$C_{165}H_{262}N_{52}O_{50}S_2$ Disulfide bonds: Cys[2]-Cys7; Minvielle, S et al, *FEBS Lett*, 203:7, 1986

Bachem H-3352.0500   Chicken   MW 3838.35
$C_{165}H_{262}N_{52}O_{50}S_2$ Store at -15°C

Ala-Cys-Asn-Thr-Ala-Thr-Cys-Val-Thr-His-Arg-Leu-Ala-Gly-Leu-Leu-Ser-Arg-Ser-Gly-Gly-Met-Val-Lys-Ser-Asn-Phe-Val-Pro-Thr-Asn-Val-Gly-Ser-Lys-Ala-Phe Amide

Synonyms: Calcitonin Gene Regulated Peptide, α-; Calcitonin Gene Regulated Peptide II; Calcitonin Gene Related Peptide, β-; Calcitonin Gene Related Peptide II

Bachem H-6730.0500   Human   MW 3793.41
$C_{162}H_{267}N_{51}O_{48}S_3$ Store at -15°C

Calbiochem 05-23-2405   Human   MW 3793.4
$C_{162}H_{267}N_{51}O_{48}S_3$ ≥97% (HPLC); lyophilized solid; soluble in 5% acetic acid; LD$_{50}$≤2000 mg/kg | Disulfide bonds: Cys[2]-Cys[7]; potent hypotensive agent & vasodilator; Finberg, RW et al, *FEBS Lett*, 209:97, 1986; Henke, H et al, *Brain Res*, 410:404, 1987

ICN 153084   Human   Disulfide bonds: Cys[2]-Cys[7]; Steenburgh, PH et al, *FEBS Lett*, 183:403, 1985; Steenburgh, PH et al, *FEBS Lett*, 209:97, 1986

Ala-Cys-Asn-Thr-Ala-Thr-Cys-Val-Thr-His-Arg-Leu-Ala-Gly-Leu-Leu-Ser-Arg-Ser-Gly-Gly-Met-Val-Lys-Ser-Asn-Phe-Val-Pro-Thr-Asn-Val-Gly-Ser-Lys-Ala-Phe

Synonyms: Calcitonin Gene Regulated Peptide, β-

Neosystem SC984   Human   MW 3793.38   Disulfide bonds: Cys[2]-Cys[7]; bioactive; calcium metabolism peptide; Steenbergh, PH et al, *FEBS Lett*, 183:403-407, 1985; Steenbergh, PH et al, *FEBS Lett*, 209:97-103, 1986

Ala-Cys-Asn-Thr-Ala-Thr-Cys-Val-Thr-His-Arg-Leu-Ala-Gly-Leu-Leu-Ser-Arg-Ser-Gly-Gly-Met-Val-Lys-Ser-Asn-Phe-Val-Pro-Thr-Asn-Val-Gly-Ser-Lys-Ala-Phe Amide

Synonyms: Calcitonin Gene Related Peptide, β-; Calcitonin Gene Regulated Peptide, β-; Calcitonin Gene Regulated Peptide I

Sigma C 1044   Human   FW 3793.4   ≥97% (HPLC); Disulfide bonds: 2-7; peptide content:~70% | Bioactive peptide; potent hypotensive agent & vasodilator; Steenburgh, PH et al, *FEBS Lett*, 183:403, 1985; ibid, 209:97, 1986

Ala-Cys-Asp-Thr-Ala-Thr-Cys-Val-Thr-His-Arg-Leu-Ala-Gly-Leu-Leu-Ser-Arg-Ser-Gly-Gly-Val-Val-Lys-Asn-Asn-Phe-Val-Pro-Thr-Asn-Val-Gly-Ser-Lys-Ala-Phe Amide

Synonyms: Calcitonin Gene Related Peptide

American Peptide 22-1-12   Human   MW 3789.4
$C_{163}H_{267}N_{51}O_{49}S_2$ Disulfide bonds: :Cys[2]-Cys[7] | Involved in sensory, motor & autonomous functions; stimulates formation of cAMP; Morris, HR et al, *Nature*, 308:746, 1984

Ala-Cys-Asp-Thr-Ala-Thr-Cys-Val-Thr-His-Arg-Leu-Ala-Gly-Leu-Leu-Ser-Arg-Ser

Synonyms: Calcitonin Gene Related Peptide (1-19)

American Peptide 22-5-16   Human   MW 1973.3
$C_{80}H_{137}N_{27}O_{27}S_2$ Disulfide bonds: Cys[2]-Cys[7]

Ala-Cys-Asp-Thr-Ala-Thr-Cys-Val-Thr-His-Arg-Leu-Ala-Gly-Leu-Leu-Ser-Arg-Ser-Gly-Gly-Val-Val-Lys-Asn-Asn-Phe-Val-Pro-Thr-Asn-Val-Gly-Ser-Lys-Ala-Phe Amide

Synonyms: Calcitonin Gene Regulated Peptide, α-; Calcitonin Gene Regulated Peptide I; Calcitonin Gene Related Peptide I; Calcitonin Gene Related Peptide, α-

Bachem H-1470.0500   Human   MW 3789.36
$C_{163}H_{267}N_{51}O_{49}S_2$ Store at -15°C

Calbiochem 05-23-2404   Human   MW 3789.3
$C_{163}H_{267}N_{51}O_{49}S_2$ ≥97% (HPLC); lyophilized solid; soluble in 5% acetic acid; LD$_{50}$≤2000 mg/kg | Disulfide bonds: Cys[2]-Cys7; potent & long-acting vasodilator; CGRP receptors are present in pancreatic β-cells; increases plasma levels of pancreatic polypeptide; has chemotactic activity towards eosinophils; Edwards, AV & Bloom, SR, *Am J Physiol*, 267:E847, 1994; Henke, H et al, *Brain Res*, 410:404, 1987; Chiba, T et al, *Nature*, 308:746, 1984

Ala-Cys-Asp-Thr-Ala-thr-Cys-Val-Thr-His-Arg-Leu-Ala-Gly-Leu-Leu-Ser-Arg-Ser-Gly-Gly-Val-Val-Lys-Lys-Asn-Asn-Phe-Val-Pro-Thr-Asn-Val-Gly-Ser-Lys-Ala-Phe Amide

Synonyms: Calcitonin Gene Related Peptide, α-; Calcitonin Gene Related Peptide I

ICN 153082   Human   Disulfide bonds: Cys[2]-Cys7; Morris, HR et al, *Nature*, 308:746, 1984; Le Greves, P et al, *Eur J Pharmac*, 115:309, 1985

Ala-Cys-Asp-Thr-Ala-Thr-Cys-Val-Thr-His-Arg-Leu-Ala-Gly-Leu-Leu-Ser-Arg-Ser-Gly-Gly-Val-Val-Lys-Asn-Asn-Phe-Val-Pro-Thr-Asn-Val-Gly-Ser-Lys-Ala-Phe

Synonyms: Calcitonin Gene Regulated Peptide, α-

Neosystem SC113   Human   MW 3789.33   Disulfide bonds: Cys[2]-Cys[7]; bioactive; calcium metabolism peptide; Morris, HR et al, *Nature*, 308:746-748, 1984

Ala-Cys-Asp-Thr-Ala-Thr-Cys-Val-Thr-His-Arg-Leu-Ala-Gly-Leu-Leu-Ser-Arg-Ser-Gly-Gly-Val-Val-Lys-Asn-Asn-Phe-Val-Pro-Thr-Asn-Val-Gly-Ser-Lys-Ala-Phe
Ala-Cys(Et)-Asp-Thr-Ala-Thr-Cys(Et)-Val-Thr-His-Arg-Leu-Ala-Gly-Leu-Leu-Ser-Arg-Ser-Gly-Gly-Val-Val-Lys-Asn-Asn-Phe-Val-Pro-Thr-Asn-Val-Gly-Ser-Lys-Ala-Phe Amide

Synonyms: Calcitonin Gene Regulated Peptide, (Cys(Et)^{2,7})-α-

Neosystem SC1257   Human   MW 3847.46   Bioactive; calcium metabolism peptide; potent & selective CGRP 2 agonist; Dumont, Y et al, *Can J Physiol Pharmacol*, 75:671-676, 1997

Ala-Cys-Asp-Thr-Ala-Thr-Cys-Val-Thr-His-Arg-Leu-Ala-Gly-Leu-Leu-Ser-Arg-Ser-Gly-Gly-Val-Val-Lys-Asn-Asn-Phe-Val-Pro-Thr-Asn-Val-Gly-Ser-Lys-Ala-Phe Amide

Synonyms: Calcitonin Gene Related Peptide; Calcitonin Gene Related Peptide, α-; Calcitonin Gene Related Peptide I

Sigma C 0167   Human   FW 3789.3   ≥95% (HPLC); Disulfide bonds: 2-7 | Bioactive peptide; potent, long-lasting vasodilator; activation of CGRP receptors on pancreatic β-cells increases plasma levels of pancreatic enzymes

Ala-Cys-Asp-Thr-Ala-Thr-Cys-Val-Thr-His-Arg-Leu-Ala-Gly-Leu-Leu-Ser-Arg-Ser-Gly-Gly-Val-Val-Lys-Asn-Asn-Phe-Val-Pro-Thr-Asn-Val-Gly-Ser-Lys-Ala-Phe Acetate Salt

Synonyms: Calcitonin Gene Related Peptide

Biogenesis 1720-9289   Human synthetic   MW 3789.2
98.1% (HPLC); lyophilized | Disulfide bonds: Cys[2]-Cys[7]

Ala-Cys-Asp-Thr-Ala-Thr-Cys-Val-Thr-His-Arg-Leu-Ala-Gly-Leu-Leu-Ser-Arg-Ser-Gly-Gly-Val-Val-Lys-Asn-Asn-Phe-Val-Pro-Thr-Asn-Val-Gly-Ser-Lys-Ala-Phe Amide

Synonyms: Calcitonin Gene Related Peptide; Calcitonin Gene Related Peptide, α-

Peptides International PCG-4160-s, 4160-v   Human synthetic   MW 3789.4 $C_{163}H_{267}N_{51}O_{49}S_2$ >99% (HPLC); lyophilized amorphous powder | Bioactive; Disulfide bonds: Cys[2]-Cys[7]; Morris, HR et al, *Nature*, 308:746, 1984

Ala-Cys-Ser-Gly-Arg-Gly-Ser-Arg-Cys-Hyp-Hyp-Gln-Cys-Cys-Met-Gly-Leu-Arg-Cys-Gly-Arg-Gly-Asn-Pro-Gln-Lys-Cys-Ile-Gly-Ala-His-Gla-Asp-Val

Synonyms: Conotoxin GS, μ-; Na[+] Channel Blocker

American Peptide 41-0-76   MW 3615.1 $C_{139}H_{223}N_{52}O_{48}S_7$ Disulfide bonds: Cys[2]-Cys[14], Cys[9]-Cys[19], Cys[13]-Cys[27]; sodium channel blocker; Yanagawa, Y et al, *Biochemistry*, 27:6256, 1988

Peptides International PCN-4263-v   *Conus geographus* (marine snail) synthetic   MW 3618.1 $C_{139}H_{226}N_{52}O_{48}S_7$ >95% (HPLC); lyophilized amorphous powder | Bioactive; Disulfide bonds: Cys[2]-Cys[14], Cys[9]-Cys[19], Cys[13]-Cys[27]; Yanagawa, Y et al, *Biochem*, 27:6256, 1988; Nakao, M et al, *Lett Pept Sci*, 2:17, 1995

Ala-Cys-Ser-Gly-Arg-Gly-Ser-Arg-Cys-Hyp-Hyp-Gln-Cys-Cys-Met-Gly-Leu-Arg-Cys-Gly-Arg-Gly-Asn-Pro-Gln-Lys-Cys-Ile-Gly-Ala-His-Gla-Asp-Val Amide

Synonyms: Conotoxin GS, μ-

Alexis 630-047   Synthetic   MW 3612.1 $C_{139}H_{220}N_{52}O_{48}S_7$ White lyophilized powder; Disulfide bonds: Cys[2]-Cys[14], Cys[9]-Cys[19], Cys[13]-Cys[27] | Originally isolated from *Conus geographus*; Na[+] channel blocker; potent neurotoxin; Yanagawa, Y et al, *Biochemistry*, 27:6256, 1988

Ala-Cys-Tyr-Cys-Arg-Ile-Pro-Ala-Cys-Ile-Ala-Gly-Glu-Arg-Arg-Tyr-Gly-Thr-Cys-Ile-Tyr-Gln-Gly-Arg-Leu-Trp-Ala-Phe-Cys-Cys

Synonyms: Defensin I; Defensin I, α-; Human Neutrophil Peptide I; Endogenous Antibiotic Peptide; Monocyte Chemotactic Peptide

Bachem H-9855.0100   Human   Store at -15°C | Disulfide bonds: Cys[2]-Cys[30], Cys[4]-Cys[19], Cys[9]-Cys[29]

Sigma D 2043   Human   FW 3442.0   ≥80% (HPLC); peptide content:~70%; Disulfide bonds: 2-30, 4-19, 9-29 | Bioactive peptide

Peptides International PDF-4271-s   Human synthetic   MW 3442.1 $C_{150}H_{222}N_{44}O_{38}S_6$ >99% (HPLC); lyophilized amorphous powder | Bioactive; Disulfide bonds: Cys[2]-Cys[30], Cys[4]-Cys[19], Cys[9]-Cys[29]; Lehrer, RI & T Ganz, *Ann NY Acad Sci*, 797:228, 1996; Ganz, T et al, *J Clin Invest*, 76:1427, 1985; Selsted, ME et al, *J Clin Invest*, 76:1436, 1985

Ala-Cys-Tyr-Trp-Lys-Val-Cys-Thr
β-(2-Naphthyl)-D-Ala-Cys-Tyr-D-Trp-Lys-Val-Cys-Thr Amide

Synonyms: Somatostatin

Sigma N 9642   FW 1096.3   ≥97% (HPLC); Disulfide bonds: 2-7 | Bioactive peptide; cyclic somatostatin analog; high binding affinity for anterior pituitary receptors, but does not bind brain receptors; potent inhibitor of growth hormone secretion; has cytostatic effect on small lung cancer cells; Taylor, TE et al, *Biochem Biophys Res Commun*, 153:81, 1988

Ala-Gln

Bachem G-1210.0250   MW 217.23 $C_8H_{15}N_3O_4$ Store at -15°C

Ala-Gln
Ala-D-Gln

Bachem G-3685.0250   MW 217.23 $C_8H_{15}N_3O_4$ Store at -15°C

Ala-Gln
D-Ala-Gln

Bachem G-3690.0250   MW 217.23 $C_8H_{15}N_3O_4$ Store at -15°C

Ala-Gln
D-Ala-Gln-Octadecyl Ester Hydrochloride

Bachem G-4465.0100   MW 506.17 $C_{26}H_{51}N_3O_4 \cdot HCl$ Store at -15°C

Ala-Gln
N-Acetylmuramyl-L-Ala-D-Gln

Synonyms: Adjuvant Peptide

ICN 195011   MW 492.5 $C_{19}H_{32}N_4O_{11}$ Used in immunization to replace more complex proteins; Ellouz, F etal, *BBRC*, 58:1317, 1974; Lefancier, P etal, *Int J Peptide Protein Res*, 9:249, 1977; Chedid, L & E Lederer, *Biochem Pharmac*, 27:2183, 1978

Ala-Gln
N-Ac-L-Ala-L-Gln

Rexim

Ala-Gln
L-Ala-L-Gln

Synonyms: Dipeptiven

Rexim   MW 217.2 $C_8H_{15}N_3O_4$ White crystals or crystalline powder

Ala-Gln

Sigma A 0550   FW 217.2 $C_8H_{15}N_3O_4$ A substitute for glutamine in mammalian cell culture media; stable to heat-sterilization; Minamoto, Y et al, *Cytotechnology*, 5:35, 1991; Roth, E et al, *in Vitro Cell Dev Biol*, 24:696, 1988

Sigma A 8185   FW 217.2 $C_8H_{15}N_3O_4$ Cell culture tested; insect cell culture tested

Ala-Gln-Ala
Ac-Ala-Gln-Ala-pNA

Bachem L-1850.0050   MW 450.45 $C_{19}H_{26}N_6O_7$ Store at -15°C

Ala-Gln-Asp-Phe-Val-Gln-Trp-Leu-Met-Asn-Thr

Synonyms: Glucagon (19-29); Ca[2+]-Activated ATPase Inhibitor; Mg[2+]-Dependent ATPase Inhibitor

American Peptide 46-1-22   Human   MW 1352.5 $C_{61}H_{89}N_{15}O_{18}S$ Inhibits the Ca[2+] pump in liver plasma membranes with an efficiency 1000-fold higher than that of glucagon; likely to be the active peptide involved in the inhibition of liver Ca[2+] pump; Mallat, A et al, *Nature*, 325:620, 1987

ICN 154554   Human   MW 1352.7   Mallat, A etal, *Nature*, 325:620, 1987

Neosystem SC098   Human   MW 1352.52   Bioactive; brain/gut peptide; Mallat, A et al, *Nature*, 325:620, 1987

Bachem H-2758.0001   Human, bovine, porcine   MW 1352.53 $C_{61}H_{89}N_{15}O_{18}S$ Store at -15°C

Ala-Gln-Asp-Phe-Val-Gln-Trp-Leu-Nle-Asn-Thr

Synonyms: Glucagon (19-29), (Nle[27])-

Neosystem SC138   Human   MW 1334.48   Bioactive; brain/gut peptide; Le Nguyen, D et al, Proceedings of the 10th Am Peptide Symp, 1988

Ala-Gln-Gln-Phe-Phe-Gly-Leu-Met
D-Ala-Gln-Gln-Phe-Phe-Gly-Leu-Met Amide

Synonyms: Substance P (4-11), (D-Ala[4])-

Bachem H-2368.0001   Store at -15°C

ICN 153022

Ala-Gln-Glu-Pro-Val-Lys-Gly-Pro-Val-Ser-Thr-Lys-Pro-Gly-Ser-Cys-Pro-Ile-Ile-Leu-Ile-Arg-Cys-Ala-Met-Leu-Asn-Pro-Pro-Asn-Arg-Cys-Leu-Lys-Asp-Thr-Asp-Cys-Pro-Gly-Ile-Lys-Lys-Cys-Cys-Glu-Gly-Ser-Cys-Gly-Met-Ala-Cys-Phe-Val-Pro-Gln

Synonyms: Elafin; Elastase-Specific Inhibitor

Peptides International PEL-4243-v   Human synthetic   MW 5999.2 $C_{254}H_{416}N_{72}O_{75}S_{10}$ >95% (HPLC); lyophilized amorphous powder | Bioactive; Disulfide bonds: Cys[16]-Cys[45], Cys[23]-Cys[49], Cys[32]-Cys[44], Cys[38]-Cys[53]; elastase-specific inhibitor from human skin; Wiedow, O et al, *JBC*, 265:14791, 1990; Wiedow, O et al, *JBC*, 266:3356, 1991; Tsunemi, M et al, *BBRC*, 185:967, 1992

Ala-Gln-Lys
N-Acetylmuramyl-Ala-D-Isoglutaminyl-*N'*-Stearoyl-Lys

Sigma A 0936   FW 887.1 $C_{43}H_{78}N_6O_{13}$ Robinson, CP, *Drugs of the Future*, 14:432, 1989

Ala-Gln-Lys-Ala-Ala
Ala-D-Isoglutaminyl-Lys-D-Ala-D-Ala Acetate Salt Free Base

Synonyms: Peptidoglycan Peptide

Sigma A 1035   FW 487.6 $C_{20}H_{37}N_7O_7$ ≥97% (HPLC) | Bioactive peptide; a peptidoglycan peptide present in the cell walls of *Staphylococcus aureus*; Ghuysen, JM et al, *Biochem*, 5:3748, 1966

Ala-Gln-Pro-Phe
Suc-Ala-Gln-Pro-Phe-pNA

Bachem L-1600.0050   Store at -15°C

Ala-Gln-Val-Gly-Thr-Asn-Lys-Glu-Leu-Cys-Cys-Leu-Val-Tyr-Thr-Ser-Trp-Gln-Ile-Pro-Gln-Lys-Phe-Ile-Val-Asp-Tyr-Ser-Glu-Thr-Ser-Pro-Gln-Cys-Pro-Lys-Pro-Gly-Val-Ile-Leu-Leu-Thr-Lys-Arg-Gly-Arg-Gln-Ile-Cys-Ala-Asp-Pro-Asn-Lys-Lys-Trp-Val-Gln-Lys-Tyr-Ile-Ser-Asp-Leu-Lys-Leu-Asn-Ala

Synonyms: Pulmonary and Activation Regulated CC Chemokine

Bachem H-4614.0010   Human   Store at -15°C

Ala-Glu
BOC-Ala-D-Glu-OBzl

Bachem A-1105.0250   MW 408.45 $C_{20}H_{28}N_2O_7$ Store at -15°C

Ala-Glu
BOC-Ala-D-Glu(OBzl) Amide

Bachem A-1110.0250   MW 407.47 $C_{20}H_{29}N_3O_6$ Store at -15°C

Ala-Glu
BOC-Ala-D-Glu Amide

Synonyms: BOC-Ala-D-Igln

Bachem A-1165.0250   MW 317.34 $C_{13}H_{23}N_3O_6$ Store at -15°C

Ala-Glu
Z-Ala-Glu

Bachem C-1075.0250   MW 352.34 $C_{16}H_{20}N_2O_7$ Store at -15°C

Ala-Glu
Ac-Muramyl-Ala-Glu Amide

Synonyms: Ac-Muramyl-Ala-Igln

Bachem G-1050.0001   MW 492.48 $C_{19}H_{32}N_4O_{11}$ Store at -15°C

Ala-Glu
Ac-Muramyl-Ala-D-Glu Amide

Synonyms: Ac-Muramyl-Ala-D-Igln; Muramyl Dipeptide

Bachem G-1055.0001   MW 492.48 $C_{19}H_{32}N_4O_{11}$ Store at -15°C

Ala-Glu
Ac-Muramyl-D-Ala-D-Glu Amide

Synonyms: Ac-Muramyl-D-Ala-D-Igln

Bachem G-1060.0001   MW 492.48 $C_{19}H_{32}N_4O_{11}$ Store at -15°C

Ala-Glu

Bachem G-1200.0250   MW 218.21 $C_8H_{14}N_2O_5$ Store at -15°C

Ala-Glu
D-Ala-D-Glu

Bachem G-1205.0250   MW 218.21 $C_8H_{14}N_2O_5$ Store at -15°C

Ala-Glu
Ala-D-Glu Amide

Synonyms: Ala-D-Isogln

Bachem G-1255.0250   MW 217.23 $C_8H_{15}N_3O_4$ Store at -15°C

Ala-Glu
Ac-Ala-Glu

Bachem G-3970.0250   MW 260.25 $C_{10}H_{16}N_2O_6$ Store at -15°C

Ala-Glu
Ala-Glu(OtBu) Amide Hydrochloride

Synonyms: Ala-Isogln-OtBu

Bachem G-4070.0250   MW 309.79 $C_{12}H_{23}N_3O_4 \cdot HCl$ Store at -15°C

Ala-Glu
Suc-Ala-Glu-MCA

Synonyms: Ingensin Substrate; Proteasome Substrate

Peptides International MAE-3160-v   MW 475.46 $C_{22}H_{25}N_3O_9$ >98% (HPLC); lyophilized amorphous powder | Ishiura, S et al, *FEBS Letts*, 257:388, 1980

Ala-Glu

Sigma A 0378   FW 218.2 $C_8H_{14}N_2O_5$ ~98%

Ala-Glu
4-(2-Acetamido-2-Deoxy-β-D-Glucopyranosyl)-N-Acetylmuramyl-L-Ala-D-Glu Amide

Synonyms: D-GlcNAc-β-(1→4)-N-Acetylmuramyl-L-Ala-D-Igln

Sigma A 4310 FW 695.7 $C_{27}H_{45}N_5O_{16}$ ≥85% | Bioactive peptide

Ala-Glu
N-CBZ-Ala-Glu

Sigma C 2376 FW 352.3 $C_{16}H_{20}N_2O_7$

Ala-Glu
N-Suc-Ala-Glu-AMC

Synonyms: Amyloid A-4 Splitting Enzyme; Ingensin Substrate

Sigma S 8782 FW 475.5 $C_{22}H_{25}N_3O_9$ Fluorogenic; Ishiura, S et al, *FEBS Lett*, 257:388, 1989

Ala-Glu
N-Ac-Muramyl-Ala-D-Glu Amide Dihydrate

Synonyms: Muramyl Dipeptide; Adjuvant Peptide

Peptides International PAD-4031 Synthetic MW 528.51 $C_{19}H_{32}N_4O_{11}$ · $2H_2O$ >99% (HPLC); lyophilized amorphous powder | Bioactive; Ellouz, F et al, *BBRC*, 59:1317, 1974; Kotani, S et al, *Biken J*, 18:105, 1975

Ala-Glu
N-Ac-Muramyl-Ala-D-Glu Amide

Synonyms: Muramyl Dipeptide; Adjuvant Peptide

Peptides International PAD-4031-v Synthetic MW 492.48 $C_{19}H_{32}N_4O_{11}$ >99% (HPLC); lyophilized amorphous powder | Bioactive; Ellouz, F et al, *BBRC*, 59:1317, 1974; Kotani, S et al, *Biken J*, 18:105, 1975

Ala-Glu-Ala-Glu
Ala-Glu-Ac-(6-o-Stearoyl)-Muramyl-Ala-D-Glu Amide

Synonyms: Ac-(6-o-Stearoyl)-Muramyl-Ala-D-Igln

Bachem G-1065.0001 MW 758.95 $C_{37}H_{66}N_4O_{12}$ Store at -15°C

Ala-Glu-Glu-Glu-Thr-Ala-Gly-Gly-Asp-Gly-Arg-Pro-Glu-Pro-Ser-Pro-Arg-Glu Amide

Synonyms: Joining Peptide

Peptides International PJP-4288-v Rat synthetic MW 1882.9 $C_{75}H_{119}N_{25}O_{32}$ >99% (HPLC); lyophilized amorphous powder | Bioactive; pivotal neuropeptide in cardiovascular regulation; Hamakubo, T et al, *Am J Physiol*, 265, R1184, 1993; Yoshida, M et al, *Am J Physiol*, 266, R802, 1994

Ala-Glu-Glu-Lys-Glu-Lys-Leu-Ala-Tyr-Arg-Lys-Gln-Asn-Met-Asp

Synonyms: NEF (74-88); SIVmac251 Peptide

Neosystem SC609 MW 1853.08 SIVmac251 peptide from subtype NEF

Ala-Glu-Gly-Pro-Tyr-Lys-Met-Glu-His-Phe-Arg-Trp-Gly-Ser-Pro-Pro-Lys-Asp

Synonyms: Melanocyte Stimulating Hormone, β-

Sigma M 2018 Porcine FW 2176.4 ≥97% (HPLC) | Bioactive peptide

Ala-Glu-Leu-Ala-Ala-Asp-Gly-Val-Gly-Ala-Ala-Ser-Arg-Asp

Synonyms: NEF MN (25-38); HIV I Peptide

Neosystem SC640 MW 1302.36 NEF peptide from HIV-1 subtype MN

Ala-Glu-Lys-Ala-Ala
Ala-D-Glu(Lys-D-Ala-D-Ala)

Synonyms: Ala-γ-D-Glu-Lys-D-Ala-D-Ala

Bachem H-2416.0005 MW 488.54 $C_{20}H_{36}N_6O_8$ Store at -15°C

Ala-Glu-Lys-Ala-Ala
Ala-D-γ-Glu-Lys-D-Ala-D-Ala

Synonyms: Peptidoglycan-Precursor Peptide

Sigma A 0910 FW 488.5 $C_{20}H_{36}N_6O_8$ ≥97% (HPLC) | Bioactive peptide; Nieto, M & Perkins, HR, *Biochem J*, 123:789, 1971; Zeiger, AR & Maurer, pH, *Biochem*, 12:3387, 1973

Ala-Glu-Lys-Lys-Asp-Glu-Gly-Pro-Tyr-Arg-Met-Glu-His-Phe-Arg-Trp-Gly-Ser-Pro-Pro-Lys-Asp

Synonyms: Lipotropin (37-58), β-; Melanocyte Stimulating Hormone, β-

American Peptide 56-1-32 Human MW 2661.0 $C_{118}H_{174}N_{34}O_{35}S$

Bachem H-1475.0001 Human MW 2660.95 $C_{118}H_{174}N_{34}O_{35}S$ Store at -15°C

ICN 151595 Human MW 2660.9 Lemaire, S et al, *J Med Chem*, 20:155, 1977

Neosystem SC1229 Human MW 2660.94 Bioactive; melanocyte stimulating hormone-related peptide; Dixon, HBF, *Biochem Biophys Acta*, 37:38-42, 1960; Scott, AP & Lowry, PJ, *Biochem J*, 139:593-602, 1974; Schiöth, HB et al, *Pharmacology & Toxicology*, 79:161-165, 1996

Sigma M 6513 Human FW 2660.9 ≥97% (HPLC) | Bioactive peptide; hormone that stimulates melanogenesis; facilitates learning & memory; affects inflammatory & immune responses & peripheral nerve regeneration; Lemaire, S et al, *J Med Chem*, 20:155, 1977

Ala-Glu-Pro-Gln-Lys-Ser-Pro-Trp-Cys-Glu-Ala-Arg-Ser-Leu-Glu-His

Synonyms: Estrogen Receptor β N-Terminal Peptide (55-70)

Alexis 155-027 Human synthetic MW 1868.1 $C_{80}H_{122}N_{24}O_{26}S$ ≥95%; lyophilized; reconstitute with 0.1 mL distilled H_2O | Competitively binds to Ab Alexis 210-135; blocking peptide for antiserum (purified) to Erβ

Ala-Glu-Pro-Phe
Suc-Ala-Glu-Pro-Phe-AMC

Bachem I-1750.0050 Store at -15°C

Ala-Glu-Pro-Phe
Suc-Ala-Glu-Pro-Phe-pNA

Bachem L-1635.0050 Store at -15°C

Ala-Glu-Tyr-Tyr-Asn-Lys-Gln-Tyr-Leu-Glu-Gln-Thr-Arg-Ala-Glu-Leu-Asp-Thr
Ac-Ala-Glu-Tyr-Tyr-Asn-Lys-Gln-Tyr-Leu-Glu-Gln-Thr-Arg-Ala-Glu-Leu-Asp-Thr Amide

Synonyms: MHC Class II IA^S β-Chain (58-75)

Bachem H-2494.0001 MW 2276.45 $C_{100}H_{150}N_{26}O_{35}$ Store at -15°C

Ala-Glu-Tyr-Tyr-Asn-Lys-Gln-Tyr-Leu-Glu-Gln-Thr-Arg-Ala-Glu-Leu-Asp-Thr
N-Ac-Ala-Glu-Tyr-Tyr-Asn-Lys-Gln-Tyr-Leu-Glu-Gln-Thr-Arg-Ala-Glu-Leu-Asp-Thr Amide

Synonyms: MHC Class II IA^S β-Chain (58-75)

Sigma A 0710 FW 2276.4 ≥90% (HPLC) | Bioactive peptide; prevents experimental autoimmune encephalomyelitis in mice; Topham, DJ, *Proc Natl Acad Sci USA*, 91:8005, 1994

Ala-Glu-Val-Ala-Glu-Leu-Tyr-Arg-Glu-Leu-Gly-Asp-Tyr-Lys-Leu-Val-Glu-Ile-Thr

Synonyms: GP140 (486-505); SIVmac251 Peptide

Neosystem SC756   MW 2211.49   SIVmac251 peptide from subtype GP140

Ala-Glu-Val-Asp
BOC-Ala-Glu-Val-Asp-CHO Pseudo Acid

Bachem N-1755.0005   Store at -15°C

Ala-Glu-Val-Asp
Z-Ala-Glu-Val-Asp-FMK

Synonyms: Caspase VIII/VI/IX Inhibitor

Kamiya   MW 610

Ala-Gly
BOC-Ala-Gly-OSu

Bachem A-1115.0001   MW 343.34 $C_{14}H_{21}N_3O_7$ Store at -15°C

Ala-Gly
BOC-Ala-Gly

Bachem A-2670.0005   MW 246.26 $C_{10}H_{18}N_2O_5$ Store at -15°C

Ala-Gly
FMOC-Ala-Gly

Bachem B-1895.0001   MW 368.39 $C_{20}H_{20}N_2O_5$ Store at -15°C

Ala-Gly
Z-Ala-Gly

Bachem C-1080.0005   MW 280.28 $C_{13}H_{16}N_2O_5$ Store at -15°C

Ala-Gly
Z-D-Ala-Gly

Bachem C-1085.0001   MW 280.28 $C_{13}H_{16}N_2O_5$ Store at -15°C

Ala-Gly
Z-DL-Ala-Gly

Bachem C-1090.0001   MW 280.28 $C_{13}H_{16}N_2O_5$ Store at -15°C

Ala-Gly
Z-Ala-Gly Amide

Bachem C-1100.0250   MW 279.3 $C_{13}H_{17}N_3O_4$ Store at -15°C

Ala-Gly

Bachem G-1215.0001   MW 146.15 $C_5H_{10}N_2O_3$ Store at -15°C

Ala-Gly
D-Ala-Gly

Bachem G-1220.0250   MW 146.15 $C_5H_{10}N_2O_3$ Store at -15°C

Ala-Gly
DL-Ala-Gly

Bachem G-1225.0001   MW 146.15 $C_5H_{10}N_2O_3$ Store at -15°C

Ala-Gly
α-Ala-Gly

Bachem G-1230.0001   MW 146.15 $C_5H_{10}N_2O_3$ Store at -15°C

Fluka 05270   MW 146.15 $C_5H_{10}N_2O_3$ ≥99% (titration); mp:250-255°C

Ala-Gly

ICN 100311   MW 146.1 $C_5H_{10}N_2O_3$ Crystalline, ≥99%

Ala-Gly
DL-Ala-Gly

ICN 100314   MW 146.1 $C_5H_{10}N_2O_3$ Crystalline

Ala-Gly
D-Ala-Gly

Sigma A 0628   FW 146.1 $C_5H_{10}N_2O_3$

Ala-Gly
DL-Ala-Gly

Sigma A 0753   FW 146.1 $C_5H_{10}N_2O_3$

Ala-Gly

Sigma A 0878   FW 146.1 $C_5H_{10}N_2O_3$

Ala-Gly
Dansyl-D-Ala-Gly

Sigma D 3408   FW 379.4 $C_{17}H_{21}N_3O_5S$ ~95% | Florentin, D et al, *Anal Biochem*, 141:62, 1984

Ala-Gly
CBZ-Ala-Gly

USBio C2098-22   MW 280.3 $C_{13}H_{15}N_2O_5$ ≥99%

Ala-Gly Amide Hydrochloride

Bachem G-1235.0250   MW 181.62 $C_5H_{11}N_3O_2 \cdot HCl$ Store at -15°C

Ala-Gly-Ala

Bachem H-1480.0250   MW 217.23 $C_8H_{15}N_3O_4$ Store at -15°C

Ala-Gly-Ala
α-Ala-Gly-Ala

Bachem H-1485.0250   MW 217.23 $C_8H_{15}N_3O_4$ Store at -15°C

Ala-Gly-Ala
α-Ala-Gly-α-Ala

Bachem H-6170.0001   MW 217.23 $C_8H_{15}N_3O_4$ Store at -15°C

Ala-Gly-Ala-Ala

Bachem H-1206.0250   MW 288.3 $C_{11}H_{20}N_4O_5$ Store at -15°C

Ala-Gly-Ala-Val-Val-Gly-Gly-Leu-Gly-Gly-Tyr-Met-Leu-Gly-Ser-Ala-Met-Ser

Synonyms: Prion Protein (118-135)

Bachem H-4206.0001   Human   MW 1597.88 $C_{68}H_{112}N_{18}O_{22}S_2$ Store at -15°C

Ala-Gly-Arg
β-Ala-Gly-Arg-pNA·2AcOH

Synonyms: Pefachrome®TG; Pefa-5134

Pentapharm 081-40, 081-17   MW 542.6   Chromogenic peptide substrate cleaved by thrombin at a slow rate; especially suitable for determination of thrombin generation; k_{cat}:1.91 s^{-1}, K_M:1.95 mM; Prasa D, et al, *Thromb & Haemost*, 77:498, 1997; Prasa D, et al, *Thromb & Haemost*, 78:1215-1220, 1997

Ala-Gly-Arg
Bz-β-Ala-Gly-Arg-pNA·AcOH

Synonyms: Pefachrome®uPA; Pefa-5221

Pentapharm 082-20, 082-01   MW 586.6   Highly sensitive chromogenic peptide substrate for urokinase (uPA); v_{max}:7.53 µmol/min; K_M:82.0 µM; Svendsen, L in *New Methods For The Analysis Of Coagulation Using Chromogenic Substrates,* (I Witt, Ed), De Gruyter, Berlin, 251, 1977

Ala-Gly-Arg
Bz-β-Ala-Gly-Arg-AMC·AcOH

Synonyms: Pefafluor uPA; Pefa-5243

Pentapharm 082-21, 082-03   MW 623.7   Excitation wavelength 360 nm, emission wavelength 460 nm | Highly sensitive fluorogenic peptide substrate for urokinase (uPA); v_{max}:0.048 µmol s^{-1}, K_M:0.050 mM

Ala-Gly-Arg-Asn-Phe-Tyr-Asn-Val-Asp-Ile-Ser-Tyr-Leu-Lys-Lys-Leu-Cys-Gly-Thr-Val-Leu-Gly-Gly-Pro-Lys Amide

Synonyms: Pro-Cathepsin B (26-50); Cathepsin B Inhibitor

Bachem H-3942.0500   Rat   Store at -15°C

Sigma P 0350   Rat   FW 2713.2   Bioactive peptide; inhibits both human & rat; Chages, JR et al, *FEBS Lett*, 392:233, 1996

Ala-Gly-Arg-Gly-Lys-Gln-Gly-Gly-Lys-Val-Arg-Ala-Lys-Ala-Lys-Thr-Arg-Ser-Ser-Arg-Ala-Gly-Leu-Gln-Phe-Pro-Val-Gly-Arg-Val-His-Arg-Leu-Leu-Arg-Lys-Gly-Asn-Tyr

Synonyms: Buforin I

Sigma B 6173   FW 4263.0   ≥90% (HPLC); peptide content:~60% | Bioactive peptide; antimicrobial peptide; Park, Chan Bae et al, *Biochem Biophys Res Commun*, 518:408, 1996

Ala-Gly-Arg-Pro-Arg-Gln-Glu-Gly-Pro-Pro-Gln-Lys-Ser-Ala
(Ala-Gly-Arg-Pro-Arg-Gln-Glu-Gly-Pro-Pro-Gln-Lys-Ser-Ala)₈-MAP

Bachem H-7610.0001   Store at -15°C

Ala-Gly-Asn-Lys-Val-Ile-Ser-Pro-Ser-Glu-Asp-Arg-Arg-Gln-Cys

Synonyms: Protein Kinase C α Peptide (313-326)

American Peptide 86-1-10   MW 1658.8 $C_{66}H_{113}N_{24}O_{24}S$ Makowske, M et al, *JBC*, 263:3402, 1988; McLLroy, BK et al, *Anal Biochem*, 195:148, 1991

Ala-Gly-Asp-Val

American Peptide 44-0-11   MW 360.4 $C_{14}H_{24}N_4O_7$

Ala-Gly-Cys-Lys-Asn-Phe-Phe-D-Trp-Lys-Thr-Phe-Thr-Ser-Cys

Synonyms: Somatostatin, (D-Trp⁸)-

American Peptide 68-1-22   MW 1637.9 $C_{76}H_{104}N_{18}O_{19}S_2$ Disulfide bonds: Cys^3-Cys^{14}; 6-8 times more potent than somatostatin in inhibiting the release of growth hormone, glucagon & insulin; Rivier, J et al, *BBRC*, 65:746, 1975

Sigma S 2511   FW 1637.9   ≥97% (HPLC); Disulfide bonds: 3-14 | Bioactive peptide; potent somatostatin analog; may be resistant to degradation in biological fluids; Brown, M et al, *Science*, 196:1467, 1977; Reubi, J-C et al, *Endocrinology*, 110:1049, 1982; Rivier, J et al, *Biochem Biophys Res Commun*, 65:746, 1975

Ala-Gly-Cys-Lys-Asn-Phe-Phe-Trp-Lys-Thr-Phe-Thr-Ser-Cys

Synonyms: Somatostatin 14; Somatostatin

Alexis 167-004   MW 1637.9 $C_{76}H_{104}N_{18}O_{19}S_2$ ≥98%; lyophilized powder; soluble in 5% HOAc; Disulfide bonds: Cys^3-Cys^{14}

American Peptide 68-1-10   MW 1637.9 $C_{76}H_{104}N_{18}O_{19}S_2$ Disulfide bonds: Cys^3-Cys^{14}; inhibits the release of growth hormone, thyrotropin stimulating hormone & prolactic releasing hormone from pituitary; inhibits the release of calcitonin, insulin, glucagon, VIP, pancreatic polypeptide, gastrin releasing peptide & CCK; Brazeau, P et al, *Science*, 179:77, 1973

Ala-Gly-Cys-Lys-Asn-Phe-Phe-Trp-Lys-Thr-Phe-Thr-Ser-Cys
Ala-Gly-Cys-Lys-Asn-Phe-Phe-D-Trp-Lys-Thr-Phe-Thr-Ser-D-Cys

Synonyms: Somatostatin, (D-Trp⁸,D-Cys¹⁴)-

American Peptide 68-1-33   MW 1637.9 $C_{76}H_{104}N_{18}O_{19}S_2$ Disulfide bonds: Cys^3-Cys^{14}

Ala-Gly-Cys-Lys-Asn-Phe-Phe-Trp-Lys-Thr-Phe-Thr-Ser-Cys

Synonyms: Somatostatin 14

Bachem H-1490.0001   MW 1637.9 $C_{76}H_{104}N_{18}O_{19}S_2$ Store at -15°C

Ala-Gly-Cys-Lys-Asn-Phe-Phe-Trp-Lys-Thr-Phe-Thr-Ser-Cys
Ala-Gly-Cys-Lys-Asn-Phe-Phe-D-Trp-Lys-Thr-Phe-Thr-Ser-D-Cys

Synonyms: Somatostatin 14, (D-Trp⁸,D-Cys¹⁴)-

Bachem H-1500.0001   MW 1638.91 $C_{76}H_{105}N_{18}O_{19}S_2$ Store at -15°C

Ala-Gly-Cys-Lys-Asn-Phe-Phe-Trp-Lys-Thr-Phe-Thr-Ser-Cys
Ala-Gly-Cys-Lys-Asn-Phe-Phe-D-Trp-Lys-Thr-Phe-Thr-Ser-Cys

Synonyms: Somatostatin 14, (D-Trp⁸)-

Bachem H-3198.0001   MW 1637.9 $C_{76}H_{104}N_{18}O_{19}S_2$ Store at -15°C

Ala-Gly-Cys-Lys-Asn-Phe-Phe-Trp-Lys-Thr-Phe-Thr-Ser-Cys
Ala-Gly-Cys-Lys-Asn-Phe-D-Phe-Trp-Lys-Thr-Phe-Thr-Ser-Cys

Synonyms: Somatostatin 14, (D-Phe⁷)-

Bachem H-4664.0001   Store at -15°C

Ala-Gly-Cys-Lys-Asn-Phe-Phe-Trp-Lys-Thr-Phe-Thr-Ser-Cys
Ala-Gly-Cys-Lys-Asn-Phe-Phe-Trp-Lys-Thr-Phe-Thr-D-Ser-Cys

Synonyms: Somatostatin 14, (D-Ser[13])-

Bachem H-4666.0001   Store at -15°C

Ala-Gly-Cys-Lys-Asn-Phe-Phe-Trp-Lys-Thr-Phe-Thr-Ser-Cys

Synonyms: Somatostatin; Growth Hormone Release Inhibiting Factor

Calbiochem 05-23-0850   MW 1637.9 $C_{76}H_{104}N_{18}O_{19}S_2$ ≥98% (HPLC); lyophilized solid; soluble in 5% acetic acid; LD_{50}≤2000 mg/kg; may be carcinogenic/teratogenic | Disulfide bond:Cys[3]-Cys[14]; cyclic tetradecapeptide; inhibits the release of growth hormone, insulin & glucagon; inhibits voltage-gated Ca^{2+} channels; Fuji, Y et al, *FEBS Lett*, 355:117, 1994; Mandarino, L et al, *Nature*, 291:76, 1981; *Merck Index*, 12:8863

Ala-Gly-Cys-Lys-Asn-Phe-Phe-Trp-Lys-Thr-Phe-Thr-Ser-Cys
Ala-Gly-Cys-Lys-Asn-Phe-Phe-D-Trp-Lys-Thr-Phe-Thr-Ser-Cys

Synonyms: Somatostatin, (D-Trp[8])-

ICN 152994   Disulfide bonds: Cys[3]-Cys[14]; potent Somatostatin analog; Rivier, J etal, *BBRC*, 65:746, 1975; Brown, M etal, *Science*, 196:1467, 1977; Reubl, J-C, etal, *Endocrinology*, 110:1049, 1982

Ala-Gly-Cys-Lys-Asn-Phe-Phe-Trp-Lys-Thr-Phe-Thr-Ser-Cys
Ala-Gly-Cys-Lys-Asn-Phe-Phe-D-Trp-Lys-Thr-Phe-Thr-Ser-D-Cys

Synonyms: Somatostatin, (D-Trp[8],D-Cys[14])-

ICN 152997   Disulfide bonds: Cys[3]-Cys[14]; potent Somatostatin analog; Brown, M etal, *Science*, 196:1467, 1977; Reubl, J-C, etal, *Endocrinology*, 110:1049, 1982

Ala-Gly-Cys-Lys-Asn-Phe-Phe-Trp-Lys-Thr-Phe-Thr-Ser-Cys

Synonyms: Somatostatin Release Inhibiting Factor; Growth Hormone Release Inhibiting Factor; Somatostatin 14; Somatostatin 14, (D-Trp[8])-

ICN 194580   Cell culture reagent; γ-Irradiated | Disulfide bonds: Cys[3]-Cys[14]

ICN 194581   Cell culture reagent | Disulfide bonds: Cys[3]-Cys[14]; Gomez-Pan, A & MD Rodriguez-Arneo, *Endocrinol Metab*, 12:469, 1983

ICN 195505   Disulfide bonds: Cys[3]-Cys[14]; Gomez-Pan, A & MD Rodriguez-Arneo, *Endocrinol Metab*, 12:469, 1983; Brazeau, P etal, *Science*, 179:77, 1973; Gustawson, S etal, *BBRC*, 82:1229, 1978; Reubl, J-C etal, Endocrinology, 110:1049, 1982

Neosystem SC088   MW 1637.89   Disulfide bonds: Cys[3]-Cys[14]; bioactive; hypothalamic releasing hormone; Brazeau, P et al, *Science*, 179:77, 1973

Neosystem SC379   MW 1637.89   Disulfide bonds: Cys[3]-Cys[14]; bioactive; hypothalamic releasing hormone; Rivier, J et al, *BBRC*, 65:746, 1975

Ala-Gly-Cys-Lys-Asn-Phe-Phe-Trp-Lys-Thr-Phe-Thr-Ser-Cys
Ala-Gly-Cys-Lys-Asn-Phe-Phe-D-Trp-Lys-Thr-Phe-Thr-Ser-D-Cys

Synonyms: Somatostatin, (D-Trp[8],D-Cys[14])-

Sigma S 4508   FW 1637.9   ≥97% (HPLC); Disulfide bonds: 3-14 | Bioactive peptide; potent somatostatin analog; Brown, M et al, *Science*, 196:1467, 1977; Reubi, J-C et al, *Endocrinology*, 110:1049, 1982

Ala-Gly-Cys-Lys-Asn-Phe-Phe-Trp-Lys-Thr-Phe-Thr-Ser-Cys

Synonyms: Somatostatin; Growth Hormone Release Inhibiting Factor; Somatotropin Release Inhibiting Factor

Sigma S 9129   FW 1637.9   ≥97% (HPLC); Disulfide bonds: 3-14 | Bioactive peptide; inhibits the release of growth hormone, thyroid stimulating hormone, insulin & glucagon; modulates physiological functions at various sites including pituitary, pancreas, gut & brain; inhibits the release of growth hormone, insulin & glucagon; Reubi, JC et al, *Endocrinology*, 110:1049, 1982; Gustausson, S & Lundquist, G, *Biochem Biophys Res Commun*, 82:1229, 1978

Biogenesis 8330-1004   Human synthetic   MW 1638 99.4% (HPLC); lyophilized, consisting of 87% peptide, acetate counter ions and residual H_2O

Peptides International PSI-4023-v   Synthetic   MW 1637.9 $C_{76}H_{104}N_{18}O_{19}S_2$ >99% (HPLC); lyophilized amorphous powder | Bioactive; Disulfide bonds: Cys[3]-Cys[14]; human, ovine, porcine, rat, mouse; Brazeau, P et al, *Science*, 179:77, 1973; Koerker, DJ et al, *Science*, 184:482, 1974; Arimura, A et al, *Science*, 189:1007, 1975

Ala-Gly-Cys-Lys-Asn-Phe-Phe-Trp-Lys-Thr-Phe-Thr-Ser-Cys
Ala-Gly-Cys-Lys-Asn-Phe-Phe-D-Trp-Lys-Thr-Phe-Thr-Ser-Cys

Synonyms: Somatostatin, (D-Trp[8])-

Peptides International PSI-4101-v   Synthetic   MW 1637.9 $C_{76}H_{104}N_{18}O_{19}S_2$ >99% (HPLC); lyophilized amorphous powder | Bioactive; Disulfide bonds: Cys[3]-Cys[14]; Rivier, J et al, *BBRC*, 65:746, 1975;\

Ala-Gly-Cys-Lys-Asn-Phe-Phe-Trp-Lys-Thr-Phe-Thr-Ser-Cys

Synonyms: Somatostatin; Somatotropin Release Inhibiting Factor; Growth Hormone Release Inhibiting Factor

Sigma S 0885   Synthetic   FW 1637.9   Sterilized by γ-irradiation; Cell culture tested | Gustausson, S & Lundquist, G, *Biochem Biophys Res Commun*, 82:1229, 1978

Sigma S 1763   Synthetic   FW 1637.9   ≥97% (HPLC); Cell culture tested | Gustausson, S & Lundquist, G, *Biochem Biophys Res Commun*, 82:1229, 1978

Ala-Gly-Cys-Lys-Asn-Phe-Phe-Trp-Lys-Thr-Phe-Thr-Ser-Cys 2AcOH 6H2O

Synonyms: Somatostatin; Somatotropin Release Inhibiting Factor; Growth Hormone Release Inhibiting Factor

Peptides International PSI-4023   Synthetic   MW 1866.1 $C_{76}H_{104}N_{18}O_{19}S_2 \cdot 2CH_3COOH \cdot 6H_2O$ >99% (HPLC); lyophilized amorphous powder; bulk | Bioactive; Disulfide bonds: Cys[3]-Cys[14]; human, ovine, porcine, rat, mouse; Brazeau, P et al, *Science*, 179:77, 1973; Koerker, DJ et al, *Science*, 184:482, 1974; Arimura, A et al, *Science*, 189:1007, 1975

Ala-Gly-Cys-Lys-Asn-Phe-Phe-Trp-Lys-Thr-Phe-Thr-Ser-Cys Reduced

Synonyms: Somatostatin 14

Bachem H-4662.0001   Store at -15°C

Ala-Gly-Cys-Lys-Asn-Phe-Phe-Trp-Lys-Thr-Tyr-Thr-Ser-Cys

Synonyms: Somatostatin, (Tyr[11])-; Somatostatin Release Inhibiting Factor

American Peptide 68-1-14   MW 1653.9 $C_{76}H_{104}N_{18}O_{20}S_2$ Disulfide bonds: Cys[3]-Cys[14]

Ala-Gly-Cys-Lys-Asn-Phe-Phe-Trp-Lys-Thr-Tyr-Thr-Ser-Cys
Ala-Gly-Cys-Lys-Asn-Phe-Phe-D-Trp-Lys-Thr-Tyr-Thr-Ser-Cys

Synonyms: Somatostatin, (D-Trp[8],Tyr[11])-

American Peptide 68-1-37   MW 1653.9 $C_{76}H_{104}N_{18}O_{20}S_2$
Disulfide bonds: Cys[3]-Cys[14]

Ala-Gly-Cys-Lys-Asn-Phe-Phe-Trp-Lys-Thr-Tyr-Thr-Ser-Cys

Synonyms: Somatostatin 14, (Tyr[11])-

Bachem H-1495.0001   MW 1653.9 $C_{76}H_{104}N_{18}O_{20}S_2$ Store at -15°C

ICN 152996   Disulfide bond:Cys[3]-Cys[14]

Sigma S 8508   FW 1653.9   ≥90% (HPLC); Disulfide bonds: 3-14 | Bioactive peptide

---

Ala-Gly-Cys-Lys-Phe-Phe-Asp-Trp-Lys-Thr-Phe-Thr-Asp-Ser-Cys

Synonyms: Somatostatin, (des Asn[5],D-Trp[8],D-Ser[13])-

Biogenesis 8330-4009   Synthetic   MW 1524   98% (HPLC), >99% (TLC); lyophilized, consisting of 81% peptide material, acetate counter ions and residual H_2O

Ala-Gly-Cys-Tyr-Phe-Gln-Asn-Cys-Pro-Arg-Gly Amide

Synonyms: Vasopressin, (Ala-Gly-Arg[8])-

Sigma V 5378   FW 1212.4   ≥97% (HPLC); Disulfide bonds: 1-6 | Bioactive peptide; originally isolated from bovine posterior pituitary, thought to be a storage form of Arg[8]-vasopressin; has natriuretic activity equivalent to Arg[8]-vasopressin with low pressor activity; Gitelman, HJ et al, *Science*, 207:893, 1980

Ala-Gly-Gln-Trp-Phe-Gly-Asp
Ala-Gly-Gln-D-Trp-Phe-Gly-Asp(*o*-tBu)₂

Synonyms: Substance P (4-10), (Ala[4],Gly[5],D-Trp[7],Asp(di-tBu)[10])-

ICN 153020   Potent substance P antagonist; Kitada, C etal, *Peptide Chemistry 1983*, E Munekata, ed, Protein Research Foundation, p 297, 1984

Ala-Gly-Glu-Gly-Leu-Asn-Ser-Gln-Phe-Trp-Ser-Leu-Ala-Ala-Pro-Gln-Arg-Phe Amide

Synonyms: Neuropeptide AF

Bachem H-4946.0500   Human   MW 1978.2 $C_{90}H_{132}N_{26}O_{25}$ Store at -15°C

Ala-Gly-Glu-Gly-Leu-Ser-Ser-Pro-Phe-Trp-Ser-Leu-Ala-Ala-Pro-Gln-Arg-Phe Amide

Synonyms: A-18-F; Morphine Modulating Neuropeptide; Opioid Peptide; A-18-F

American Peptide 32-3-13   MW 1920.3 $C_{89}H_{130}N_{24}O_{24}$

Bachem H-5650.0500   MW 1920.16 $C_{89}H_{130}N_{24}O_{24}$ Store at -15°C

ICN 152941   MW 1920.2

Sigma A 5296   FW 1920.2   ≥97% (HPLC) | Bioactive peptide; endogenous peptide found in periaqueductal grey & in dorsal spinal cord; attenuates analgesic effects of morphine; Yang, HYT et al, *Proc Natl Acad Sci USA*, 82:7757, 1985

Ala-Gly-Glu-Pro-Lys-Leu-Asp-Ala-Gly-Val Amide

Synonyms: Pneumadin

Bachem H-8180.0001   Human   MW 955.08 $C_{41}H_{70}N_{12}O_{14}$ Store at -15°C

ICN 158827   Human   MW 955.1 $C_{41}H_{70}N_{12}O_{14}$ 97% | Batra, VK etal, *Regul Peptides*, 30:77, 1990

Sigma P 9317   Human   FW 955.1 $C_{41}H_{70}N_{12}O_{14}$ ≥97% (HPLC); peptide content:~70% | Bioactive peptide; Batra, BK et al, *Regul Peptides*, 30:77, 1990

Ala-Gly-Glu-Ser

ICN 153055   MW 362.3 $C_{13}H_{22}N_4O_8$ An inhibitor of platelet aggregation; Gartner, TK etal, *Blood*, 66 Suppl:305a, 1985

Ala-Gly-Gly
BOC-Ala-Gly-Gly

Bachem A-1120.0001   MW 303.32 $C_{12}H_{21}N_3O_6$ Store at -15°C

Ala-Gly-Gly
Z-Ala-Gly-Gly

Bachem C-1110.0001   MW 337.33 $C_{15}H_{19}N_3O_6$ Store at -15°C

Ala-Gly-Gly
Z-β-Ala-Gly-Gly

Bachem C-1115.0001   MW 337.33 $C_{15}H_{19}N_3O_6$ Store at -15°C

Ala-Gly-Gly

Bachem H-1505.0001   MW 203.2 $C_7H_{13}N_3O_4$ Store at -15°C

Ala-Gly-Gly
α-Ala-Gly-Gly

Bachem H-1515.0250   MW 203.2 $C_7H_{13}N_3O_4$ Store at -15°C

Ala-Gly-Gly
DL-Ala-Gly-Gly

Bachem H-2240.0250   MW 203.2 $C_7H_{13}N_3O_4$ Store at -15°C

Ala-Gly-Gly
D-Ala-Gly-Gly

Bachem H-5985.0250   MW 203.2 $C_7H_{13}N_3O_4$ Store at -15°C

Ala-Gly-Gly

ICN 100309   MW 203.2 $C_7H_{13}N_3O_4$ Crystalline, 99%

Ala-Gly-Gly
DL-Ala-Gly-Gly

ICN 100323   MW 203.2 $C_7H_{13}N_3O_4$ Crystalline, 99%

Ala-Gly-Gly
D-Ala-Gly-Gly

Sigma A 1128   FW 203.2 $C_7H_{13}N_3O_4$

Ala-Gly-Gly

Sigma A 1378   FW 203.2 $C_7H_{13}N_3O_4$

Ala-Gly-Gly-Ac-Lys-Gly-Gly-Ac-Lys-Gly-Met-Gly-Ac-Lys-Val-Gly-Ala-Ac-Lys-Arg-His-Ser-Cys

Synonyms: Histone H4 Peptide, (Tetra Ac-Lys)-

Upstate 12-353   MW 1954.4   >90% pure; frozen solution | Peptide inhibition assay

---

Ala-Gly-Gly-Gly
BOC-Ala-Gly-Gly-Gly

Bachem A-1125.0250   MW 360.37 $C_{14}H_{24}N_4O_7$ Store at -15°C

Ala-Gly-Gly-Gly

Bachem H-1540.0250   MW 260.25 $C_9H_{16}N_4O_5$ Store at -15°C

Ala-Gly-Gly-Gly
α-Ala-Gly-Gly-Gly

Bachem H-7595.0250   MW 260.25 $C_9H_{16}N_4O_5$ Store at -15°C

Ala-Gly-Gly-Gly-Gly

Bachem H-1545.0250   MW 317.3 $C_{11}H_{19}N_5O_6$ Store at -15°C

Ala-Gly-Ile-Val-Gln-Gln-Gln-Gln-Gln-Leu-Leu-Asp-Val-Val-Lys-Arg-Gln-Gln-Glu-Leu

Synonyms: GP140 (561-580); SIVmac251 Peptide

Neosystem SC764   MW 2321.65   SIVmac251 peptide from subtype GP140

Ala-Gly-Phe-Gly
N-Dansyl-D-Ala-Gly-p-Nitro-Phe-Gly

Synonyms: Enkephalinase Substrate

Sigma D 2155   FW 628.7 $C_{28}H_{32}N_6O_9S$ ~95% (HPLC) | Selective; fluorogenic; Florentin, D et al, *Anal Biochem*, 141:62, 1984

Ala-Gly-Phe-Leu
D-Ala-Gly-Phe-D-Leu Acetate Salt

Synonyms: Enkephalin; (des-Tyr[1],D-Ala[2],D-Leu[5])-

ICN 195651   MW 466.5 $C_{20}H_{30}N_4O_5 \cdot C_2H_4O_2$ ≥97%

Ala-Gly-Phe-Leu
D-Ala-Gly-Phe-Leu Amide Acetate Salt

Synonyms: Enkephalinamide, (des-Tyr[1],D-Ala[2],Leu[5])-;

ICN 195654   MW 465.5 $C_{20}H_{31}N_5O_4 \cdot C_2H_4O_2$ ≥97%

Ala-Gly-Phe-Leu
D-Ala-Gly-Phe-D-Leu Acetate Salt Free Base

Synonyms: Enkephalin des-Tyr[1], (D-Ala[2],D-Leu[5])-

Sigma E 5390   FW 406.5 $C_{20}H_{30}N_4O_5$ ≥97% (HPLC) | Bioactive peptide

Ala-Gly-Phe-Met
D-Ala-Gly-Phe-Met Amide Acetate Salt Free Base

Synonyms: Enkephalin, des-Tyr[1],(D-Ala[2])-; Enkephalinamide, Met-

Sigma E 5131   FW 423.5 $C_{19}H_{29}N_5O_4S$ ≥97% (HPLC) | Bioactive peptide

Ala-Gly-Phe-Met D-Ala-Gly-Phe-Met Amide Acetate Salt

Synonyms: Enkephalinamide, (des-Tyr[1],D-Ala[2],Met[5])-

ICN 195659   MW 483.5 $C_{19}H_{29}N_5O_4S \cdot C_2H_4O_2$ ≥97%

Ala-Gly-Pro-Arg
BOC-Ala-Gly-Pro-Arg-AMC Hydrochloride

Bachem I-1555.0050   MW 693.2 $C_{31}H_{44}N_8O_8 \cdot HCl$ Store at -15°C

Ala-Gly-Pro-Arg
BOC-Ala-Gly-Pro-Arg-MCA

Synonyms: Natriuretic Peptide, Atrial Precursor Processing Enzyme Substrate

Peptides International MAR-3144-v   Rat   MW 656.74 $C_{31}H_{44}N_8O_8$ >98% (HPLC); lyophilized amorphous powder | Imada, T et al, *BBRC*, 143:587, 1987; Imada, T et al, Hoppe-Seyler's *Biol Chem* Suppl, 369:113, 1988

Ala-Gly-Pro-Arg
N-t-BOC-Ala-Gly-Pro-Arg-AMC Hydrochloride

Synonyms: Natriuretic Peptide, Atrial Precursor Processing Enzyme Substrate

Sigma B 4278   Rat   FW 693.2 $C_{31}H_{44}N_8O_8 \cdot HCl$ ~95% | Imada, T et al, *Biochem Biophys Res Commun*, 143:587, 1987

Ala-Gly-Pro-Phe
Suc-Ala-Gly-Pro-Phe-pNA

Synonyms: Peptidyl Prolyl *cis-trans* Isomerase Substrate; PPIase Substrate

Bachem L-1590.0050   Store at -15°C

Neosystem SC1291   MW 611.60   Harrison, RK & Stein, RL, *Biochem*, 29:1684-1689 & 3813-3816, 1990

Ala-Gly-Sar
BOC-Ala-Gly-Sar

Bachem A-2810.0001   Store at -15°C

Ala-Gly-Ser-Ala-Met-Gly-Ala-Ala-Ser-Leu-Thr-Leu-Thr-Ala-Gln-Ser-Arg-Thr-Leu-Leu

Synonyms: GP140 (541-560); SIVmac251 Peptide

Neosystem SC762   MW 1920.20   SIVmac251 peptide from subtype GP140

Ala-Gly-Ser-Glu

Synonyms: Eosinophilotactic Peptide; Corticotropin Releasing Factor

American Peptide 40-0-64   MW 362.4 $C_{13}H_{22}N_4O_8$ Goetzl, EJ et al, *PNAS*, 72:4123, 1975

Bachem H-1550.0050   MW 362.34 $C_{13}H_{22}N_4O_8$ Store at -15°C

Ala-Gly-Ser-Glu
For-Ala-Gly-Ser-Glu

Bachem H-3015.0050   MW 390.35 $C_{14}H_{22}N_4O_9$ Store at -15°C

Ala-Gly-Ser-Glu
N-For-Ala-Gly-Ser-Glu

ICN 152761

Ala-Gly-Ser-Glu

Synonyms: Chemotactic Peptide

ICN 152798   MW 362.3 $C_{13}H_{22}N_4O_8$ An eosinophil chemotactic factor of anaphylaxis; Goetzl, EJ & F Austen, *PNAS*, 72:4123, 1975; Turnbull, LW etal, *Immunology*, 32:57, 1977; Beswick, PH & Ab Kay, *Clin Exp Immunol*, 43:399, 1981

Sigma A 5902   FW 362.3 $C_{13}H_{22}N_4O_8$ ≥90% (HPLC) | Bioactive peptide; eosinophil chemotactic factor of anaphylaxis; Goetzl, EJ & Austen, F, *Proc Natl Acad Sci USA*, 72:4123, 1975

Ala-Gly-Ser-Glu
N-Formyl-Ala-Gly-Ser-Glu

Synonyms: Chemotactic Peptide

Sigma F 8631   FW 390.3 $C_{14}H_{22}N_4O_9$ ≥80% (HPLC) | Bioactive peptide

Ala-Gly-Thr-Ala-Asp-Cys-Phe-Trp-Lys-Tyr-Cys-Val

Synonyms: Urotensin II

American Peptide 80-7-30   MW 1361.5 $C_{62}H_{84}N_{14}O_{17}S_2$ Disulfide bonds: Cys[6]-Cys[11]

ICN 153192

Sigma U 4753   FW 1361.5   ≥97% (HPLC); Disulfide bonds: 6-11 | Bioactive peptide

Bachem H-6950.0001   Goby   Store at -15°C

Ala-Gly-Tyr Amide

Bachem H-5990.0250 MW 308.34 $C_{14}H_{20}N_4O_4$ Store at -15°C

Ala-His

Synonyms: Carnosine, β-Ala-(3-^{14}C)-

ARC ARC-1278 MW 226.24
$NH_2CH_2CH_2CONCH_2(C_4H_5N_2)COOH$ 50-60 mCi/mmol; 1.85-2.22 GBq/mmol; in EtOH | Radiochemical

Ala-His
β-Ala-L-His

Synonyms: Carnosine, L-β-Ala-(3-^{3}H)-

ARC ART-584 MW 226.24 $NH_2CH_2CH_2CONCH_2(C_4H_5N_2)COOH$
30-60 Ci/mmol; 1.11-2.22 TBq/mmol; in EtOH | Radiochemical

Ala-His
Z-Ala-His

Bachem C-1120.0001 MW 360.37 $C_{17}H_{20}N_4O_5$ Store at -15°C

Ala-His
Z-Ala-His-OMe

Bachem C-1125.0001 MW 374.4 $C_{18}H_{22}N_4O_5$ Store at -15°C

Ala-His

Bachem G-1245.0250 MW 226.24 $C_9H_{14}N_4O_3$ Store at -15°C

Ala-His
α-Ala-His

Synonyms: Carnosine, L-

Bachem G-1250.0001 MW 226.24 $C_9H_{14}N_4O_3$ Store at -15°C

Ala-His
α-Ala-His(3-Me) Nitrate

Synonyms: Anserine, L-

Bachem G-4555.0025 Store at -15°C

Ala-His
β-Ala-L-His

Synonyms: Carnosine, L-; Carnosinase Substrate

Fluka 22030 MW 226.24 $C_9H_{14}N_4O_3$ ≥99% (titration); mp:258-260°C | Substrate for the assay of carnosinase; Smith, EL, *Meth Enzymol*, 2:93, 1955

Ala-His
β-Ala-His

Synonyms: Carnosine, L-; Carnosine

ICN 100936 MW 226.2 $C_9H_{14}N_4O_3$ Crystalline, ~99%

Peptides International OAH-3085 MW 226.23 $C_9H_{14}N_4O_3$ >98% (HPLC); white amorphous powder

Ala-His
β-Ala-1-Me-His

Synonyms: Anserine, L-

Sigma A 1131 FW 303.3 $C_{10}H_{16}N_4O_3 \cdot HNO_3$ Nitrate salt

Ala-His

Synonyms: Carnosine, L-; β-Ala-His

Sigma A 1503 FW 226.2 $C_9H_{14}N_4O_3$

Sigma C 9625 FW 226.2 $C_9H_{14}N_4O_3$ ~99%; crystalline

Ala-His-Ala

Bachem H-5430.0100 MW 297.13 $C_{12}H_{19}N_5O_4$ Store at -15°C

Ala-His-Lys

Bachem H-1555.0250 MW 354.41 $C_{15}H_{26}N_6O_4$ Store at -15°C

Ala-His-Pro-Phe
Suc-Ala-His-Pro-Phe-pNA

Bachem L-1595.0050 Store at -15°C

Ala-His-Ser-Asp-Gly-Thr-Phe-Thr-Ser-Glu-Leu-Ser-Arg-Leu-Arg-Asp-Ser-Ala-Arg-Leu-Gln-Arg-Leu-Leu-Gln-Gly-Leu-Val
3-(4-Hydroxyphenyl)Propionyl-β-Ala-His-Ser-Asp-Gly-Thr-Phe-Thr-Ser-Glu-Leu-Ser-Arg-Leu-Arg-Asp-Ser-Ala-Arg-Leu-Gln-Arg-Leu-Leu-Gln-Gly-Leu-Val Amide

Synonyms: Secretin, (deamino-Tyr-β-Ala)-; Gastrointestinal Peptide

Sigma S 4515 FW 3274.7 ≥90% (HPLC); peptide content:~70% | Bioactive peptide; used as a tracer in radioimmunoassay for secretin; Wunsch, E, *Hoppe-Seyler's Z Physiol Chem*, 357:1417, 1976

Ala-Igln
N-Ac-Muramyl-L-Ala-D-Igln

Synonyms: Adjuvant Peptide

Fluka 01365 MW 492.5 $C_{19}H_{32}N_4O_{11}$ ≥99% (TLC); ≤5% water | Adam, A et al, *Mol Cell Biochem*, 41:27, 1981

Ala-Igln
N-Acetylmuramyl-D-Ala-D-Igln

ICN 153037 MW 492.5 $C_{19}H_{32}N_4O_{11}$

Ala-Igln
N-Acetylmuramyl-L-Ala-L-Igln

ICN 153038 MW 492.5 $C_{19}H_{32}N_4O_{11}$

Ala-Igln
N-Acetylmuramyl-6-O-Stearoyl-L-Ala-D-Igln

ICN 195853 MW 758.9 $C_{37}H_{66}N_4O_{12}$ ≥98% | Reported to be a safer substrate than Freund's Adjuvant in malaria vaccines

Ala-Igln
N-Ac-Muramyl-L-Ala-D-Igln

Synonyms: Adjuvant Peptide

Neosystem SC355 MW 492.45 Virus-related peptide; AIDS-related peptide; inhibits HIV replication in CD-4+ H9 lymphocytes; Masihi, KN et al, *AIDS Res Human Retrovir*, 6:393, 1990

Ala-Igln
N-Acetylmuramyl-6-O-Stearoyl-L-ala-D-Igln
Stearoylated

Synonyms: Adjuvant Peptide Isomer

Sigma A 0300 FW 758.9 $C_{37}H_{66}N_4O_{12}$ ≥97% (HPLC) | Bioactive peptide

Ala-Igln
N-Acetylmuramyl-D-Ala-D-Igln

Synonyms: Adjuvant Peptide Isomer

Sigma A 0640 FW 492.5 $C_{19}H_{32}N_4O_{11}$ ≥85% (HPLC) | Bioactive peptide

Ala-Igln
N-Acetylmuramyl-L-Ala-L-Igln
Synonyms: Adjuvant Peptide Isomer

Sigma A 4773   FW 492.5 $C_{19}H_{32}N_4O_{11}$ Bioactive peptide

Ala-Igln
N-Acetylmuramyl-L-Ala-D-Igln
Synonyms: Muramyl Dipeptide; Adjuvant Peptide

Sigma A 9519   FW 492.5 $C_{19}H_{32}N_4O_{11}$ ≥98% (TLC) | Bioactive peptide; enhances antigenicity of weak antigens, increasing antibody titer; Lefancier, P et al, *Int J Peptide Protein Res*, 9:249, 1977; Ellouz, F et al, *Biochem Biophys Res Commun*, 59:1317, 1974; Chedid, L & Lederer, E, *Biochem Pharmacol*, 27:2183, 1978

Ala-Igln
N-Ac-D-Glucosaminyl-β-(1→4)-N-Acetylmuramyl-L-Ala-D-Igln
ICN 158221   Synthetic   Lyophilized; 98% | A novel synthetic analog of bacterial cell wall glycopeptide which acts as a modulator of humoral & cellular immunity reactions; possesses immunoadjuvant & protective activity against bacterial & viral (including tumorigenic) infections; differs from well-known muramyl peptides in that it contains GlcNAC attached to muramic acid via the β-(1→4) glycosidic bond; soluble in H_2O, EtOH, MeOH, DMF & physiological saline (1 g/mL); Campbell, MJ etal, *Immunology*, 145:1029, 1990; Balitsky, KP etal, *Int J Immunopharmacol*, 11:429, 1989

Ala-Ile
Z-Ala-Ile
Bachem C-1130.0001   MW 336.39 $C_{17}H_{24}N_2O_5$ Store at -15°C

Ala-Ile
Bachem G-1260.0250   MW 202.25 $C_9H_{18}N_2O_3$ Store at -15°C

ICN 100361   MW 202.3 $C_9H_{18}N_2O_3$ Crystalline

Ala-Ile
β-Ala-Ile
ICN 100933   MW 202.3 $C_9H_{18}N_2O_3$ Crystalline

Ala-Ile-Arg-Asn-Asp-Glu-Glu-Leu-Asn-Lys-Leu-Leu-Gly-Lys-Val-Thr-Ile-Ala-Gln-Gly-Gly-Val-Leu-Pro-Asn-Ile-Gln-Ala-Val-Leu-Leu-Pro-Lys-Lys-Thr
Synonyms: Histone H2A (86-120); MB-35

Sigma H 8647   FW 3755.4   ≥95% (HPLC) | Bioactive peptide; thymic peptide with 100% sequence homology to a region of histone H2A; enhances release of growth hormone & prolactin from cultured pituitary cells; Badamchian, M et al, *Endocrinology*, 128:1580, 1991

Ala-Ile-Asp-Met-Ser-His-Phe-Ile-Lys-Glu-Lys-Gly-Gly-Leu-Glu
Synonyms: NEF (116-130); SIVmac251 Peptide

Neosystem SC614   MW 1674.93   SIVmac251 peptide from subtype NEF

Ala-Ile-Phe-Gln-Ser-Ser-Met-Thr-Lys-Ile-Leu-Glu-Pro-Phe-Arg-Lys-Gln-Asn-Pro-Asp-Ile-Val-Ile-Tyr-Gln
Synony POL (325-349); RT (158-182); HTLV I Peptide *ms*:

Neosystem SC689   MW 2971.40   HTLV I peptide from subtype POL (PR/RT)

Ala-Ile-Pro-Met
MeOSuc-Ala-Ile-Pro-Met-pNA
Bachem L-1340.0050   Store at -15°C

Ala-Ile-Pro-Phe
Suc-Ala-Ile-Pro-Phe-pNA
Bachem L-1665.0050   Store at -15°C

Ala-Ile-Val-Val-Gly-Gly-Val-Met-Leu-Gly-Ile-Ile-Ala-Gly-Lys-Asn-Ser-Gly-Val-Asp-Glu-Ala-Phe-Phe-Val-Leu-Lys-Gln-His-His-Val-Glu-Tyr-Gly-Ser-Asp-His-Arg-Phe-Glu-Ala-Asp
Synonyms: Amyloid α-Protein (42-1)

Bachem H-3976.0500   MW 4514.1 $C_{203}H_{311}N_{55}O_{60}S$ Store at -15°C

Ala-Leu
D-Ala-Leu
American Peptide 79-2-24   MW 202.3 $C_9H_{18}N_2O_3$

Ala-Leu
Z-Ala-Leu
Bachem C-3155.0001   MW 336.39 $C_{17}H_{24}N_2O_5$ Store at -15°C

Ala-Leu
Z-Ala-Leu Amide
Bachem C-4010.0250   MW 335.4 $C_{17}H_{25}N_3O_4$ Store at -15°C

Ala-Leu
Bachem G-1265.0250   MW 202.25 $C_9H_{18}N_2O_3$ Store at -15°C

Ala-Leu
D-Ala-Leu
Bachem G-1270.0250   MW 202.25 $C_9H_{18}N_2O_3$ Store at -15°C

Ala-Leu
DL-Ala-DL-Leu
Bachem G-1275.0001   MW 202.25 $C_9H_{18}N_2O_3$ Store at -15°C

Ala-Leu
α-Ala-Leu
Bachem G-1280.0001   MW 202.25 $C_9H_{18}N_2O_3$ Store at -15°C

Ala-Leu
ICN 100316   MW 202.3 $C_9H_{18}N_2O_3$ Crystalline

Ala-Leu
DL-Ala-DL-Leu
ICN 100334   MW 202.3 $C_9H_{18}N_2O_3$ Crystalline

Ala-Leu
D-Ala-Leu
ICN 100338   MW 202.3 $C_9H_{18}N_2O_3$ Crystalline

Ala-Leu
β-Ala-Leu
ICN 100941   MW 202.3 $C_9H_{18}N_2O_3$ Crystalline, ~99%

Ala-Leu

Sigma A 1878	FW 202.3	$C_9H_{18}N_2O_3$

Ala-Leu
D-Ala-Leu

Sigma A 2892	FW 202.3	$C_9H_{18}N_2O_3$

Ala-Leu
DL-Ala-DL-Leu

Sigma A 9008	FW 202.3	$C_9H_{18}N_2O_3$

Ala-Leu
N-CBZ-Ala-Leu

Sigma C 3251	FW 336.4	$C_{17}H_{24}N_2O_5$

Ala-Leu-Ala

Bachem H-5975.0250   MW 273.33 $C_{12}H_{23}N_3O_4$ Store at -15°C

Sigma A 3671   FW 273.3 $C_{12}H_{23}N_3O_4$

Ala-Leu-Ala-Leu
Trt-Ala-Leu-Ala-Leu-OBzl

Bachem H-1514.0050   Store at -15°C

Ala-Leu-Ala-Leu

ICN 153042   MW 386.5 $C_{18}H_{34}N_4O_5$ Trouet, A, *Targeting of Drugs*, G Gregoriadis, J Senior & A Trouet, eds, Plenum Press, New York, p 19, 1982

Sigma A 3546   FW 386.5 $C_{18}H_{34}N_4O_5$ ≥97% (TLC) | Trouet, A et al, *Targeting of Drugs*, Gregoriadis, G, Senior, J & Trouet, A, eds, Plenum Press, New Yourk, 1982, pp 19-30

Ala-Leu-Ala-Trp-His-Ser-Ser-Ala-Tyr-Gly-Pro-Asp-Gln-Arg-Ala-Gln

Synonyms: Ca^{2+}-Sensing Receptor Protein N-Terminal Peptide (12-27)

Alexis 167-010   Rat, synthetic   MW 1757.9 $C_{77}H_{112}N_{24}O_{24}$ ≥95%; lyophilized; reconstitute with 0.1 mL distilled H_2O | Competitively binds to Ab Alexis 210-143; antiserum blocking peptide

---

Ala-Leu-Arg-Gly-Pro-Lys-Met-Met-Arg-Asp-Ser-Gly-Cys-Phe-Gly-Arg-Arg-Leu-Asp-Arg-Ile-Gly-Ser-Leu-Ser-Gly-Leu-Gly-Cys-Asn-Val-Leu-Arg-Arg-Tyr

Synonyms: Aldosterone Secretion Inhibiting Factor (1-35)

Sigma A 7555   FW 3910.6   Bovine   ≥97% (HPLC); (Disulfide bonds: 13-29) | Bioactive peptide; occurs with bovine adrenal ANF in brain cells & competes more strongly than porcine BNP for bovine adrenal ANF receptors; Nguyen, TT et al, *Endocrinology*, 124:1591, 1989

Bachem H-3116.0500   Bovine   MW 3910.64 $C_{164}H_{278}N_{58}O_{45}S_4$ Store at -15°C

Ala-Leu-Cys-Asn-Cys-Asn-Arg-Ile-Ile-Ile-Pro-His-Met-Cys-Trp-Lys-Lys-Cys-Gly-Lys-Lys Amide

Synonyms: Tertiapin; Inward Rectifier K^+ Channel Blocker

Bachem N-1745.0500   Store at -15°C | Disulfide bonds: Cys^3-Cys^{14}, Cys^5-Cys^{18}

Peptides International PTK-4364-s   *Apis mellifera* (honey bee) synthetic   MW 2455.1 $C_{106}H_{176}N_{34}O_{23}S_5$ >95% (HPLC); lyophilized amorphous powder | Bioactive; Disulfide bonds: Cys^3-Cys^{14}, Cys^5-Cys^{18}; Lu, Z & R MacKinnon, *Biochem*, 36:6936, 1997; Jin, W & Z Lu, *Biochem*, 37:13291, 1998; Xu, X & JW Nelson, *Proteins Struct Funct Genet*, 17:124, 1993

Ala-Leu-Cys-Asp-Asp-Pro-Arg-Val-Asp-Arg-Trp-Tyr-Cys-Gln-Phe-Val-Glu-Gly
Ac-Ala-Leu-Cys-Asp-Asp-Pro-Arg-Val-Asp-Arg-Trp-Tyr-Cys-Gln-Phe-Val-Glu-Gly Amide

Synonyms: E-76

Bachem H-4902.0001   MW 2211.47 $C_{97}H_{139}N_{27}O_{29}S_2$ Store at -15°C

Ala-Leu-Gly
DL-Ala-DL-Leu-Gly

Bachem H-1570.0001   MW 259.31 $C_{11}H_{21}N_3O_4$ Store at -15°C

ICN 100339   MW 259.3 $C_{11}H_{21}N_3O_4$ Crystalline

Sigma A 0759   FW 259.3 $C_{11}H_{21}N_3O_4$

Ala-Leu-Ile-Leu-Thr-Leu-Val-Ser

Synonyms: Sex Pheromone Inhibitor iPD1; cPD1 Sex Pheromone Inhibitor

Bachem H-9985.0005   MW 829.05 $C_{39}H_{72}N_8O_{11}$ Store at -15°C

Sigma A 1061   *Streptococcus faecalis*   FW 829.0 $C_{39}H_{72}N_8O_{11}$ ≥90% (HPLC) | Bioactive peptide; Mori, M et al, *J Bacteriol*, 169:1747, 1987

Ala-Leu-Ile-Thr-Pro-Lys-Lys-Ile-Lys-Pro-Pro-Leu-Pro-Ser

Synonyms: VIF (152-165); HTLV I Peptide

Neosystem SC550   MW 1502.90   HTLV I peptide from subtype VIF

Ala-Leu-Leu-Arg-Cys-Asn-Asp-Thr-Asn-Tyr-Ser-Gly-Phe-Met-Pro-Lys-Cys-Ser-Lys-Val
Ala-Leu-Leu-Arg-Cys(Acm)-Asn-Asp-Thr-Asn-Tyr-Ser-Gly-Phe-Met-Pro-Lys-Cys(Acm)-Ser-Lys-Val

Synonyms: GP140 (241-260); SIVmac251 Peptide

Neosystem SC732   MW 2389.78   SIVmac251 peptide from subtype GP140

Ala-Leu-Leu-Glu-Thr-Tyr-Cys-Ala-Thr-Pro-Ala-Lys-Ser-Glu

Synonyms: Insulin-Like Growth Factor I (57-70); Insulin-Like Growth Factor II (54-67)

American Peptide 50-1-22   MW 1495.7 $C_{65}H_{104}N_{15}O_{23}S$

Neosystem SC680   MW 1496.69   Bioactive; synthetic growth factor-related peptide; due to the presence of a Cys this peptide can contain the dimeric form

ICN 154468   Human synthetic   MW 1496.6

Ala-Leu-Lys
D-Ala-Leu-Lys-AMC

Synonyms: Plasmin Substrate

Bachem I-1040.0050   MW 487.6 $C_{25}H_{37}N_5O_5$ Store at -15°C

Sigma A 8171   FW 487.6 $C_{25}H_{37}N_5O_5$ 95% (HPLC) | Fluorogenic; Smith, RE et al, *Thromb Res*, 17:393, 1980

Ala-Leu-Lys-Arg-Gln-Gly-Arg-Thr-Leu-Tyr-Gly-Phe-Gly-Gly

Synonyms: Osteogenic Growth Peptide

Bachem H-8640.0001   Store at -15°C

Ala-Leu-Lys-Arg-Gln-Gly-Arg-Thr-Leu-Tyr-Gly-Phe-Gly-Gly Acetate Salt

Synonyms: Osteogenic Growth Peptide

Biogenesis 7060-7059   Rat synthetic   MW 1524   ≥90% (HPLC); lyophilized | The OGP peptide has identical sequence to the C-terminus of Histone H4; Bab et al, *EMBO J*, 11:1867, 1992

Ala-Leu-Pro-Phe
Suc-Ala-Leu-Pro-Phe

Bachem H-1526.0050   Store at -15°C

Ala-Leu-Pro-Phe
Suc-Ala-Leu-Pro-Phe-AMC

Bachem I-1695.0050   Store at -15°C

Ala-Leu-Pro-Phe
Suc-Ala-Leu-Pro-Phe-pNA

Bachem L-1620.0050   Store at -15°C

Ala-Leu-Pro-Phe
Suc-Ala-Leu-Pro-Phe-AMC

Synonyms: Peptidyl Prolyl *cis-trans* Isomerase Substrate

Neosystem SC1252   MW 703.78

Ala-Leu-Pro-Phe
Suc-Ala-Leu-Pro-Phe-pNA

Synonyms: Peptidyl Prolyl *cis-trans* Isomerase Substrate

Neosystem SC1253   MW 667.70   Kofron, JL et al, *Biochem*, 30:6127-6134, 1991; Nielsen, JB et al, *PNAS* USA, 89:7471-7475, 1992; Shan, XY et al, *J Cell Biol*, 126:853-862, 1994

Peptides International SAF-3162-v   MW 666.73 $C_{33}H_{42}N_6O_9$ >98% (HPLC); lyophilized amorphous powder | Kofron, JL et al, *Biochem*, 30:6127, 1991

Ala-Leu-Trp-Lys-Thr-Met-Leu-Lys-Lys-Leu-Gly-Thr-Met-Ala-Leu-His-Ala-Gly-Lys-Ala-Ala-Leu-Gly-Ala-Ala-Ala-Asp-Thr-Ile-Ser-Gln-Gly-Thr-Gln

Synonyms: Dermaseptin; Dermaseptin I

American Peptide 79-4-15   MW 3455.1 $C_{152}H_{257}N_{43}O_{44}S_2$ Antibiotic peptide; the water soluble, thermostable & non-hemolytic peptide exhibits highly potent antimicrobial activity against pathogenic fungi at micromolar concentration; Mor, A et al, *Biochemistry*, 30:8824, 1991

Bachem H-1294.0500   Store at -15°C

Neosystem SC901   MW 3455.08   Bioactive; antibacterial/antimicrobial peptide; Mor, A et al, *Biochem*, 30:8824-8830, 1991

Ala-Leu-Trp-Lys-Thr-Met-Leu-Lys-Lys-Leu-Gly-Thr-Met-Ala-Leu-His-Ala-Gly Amide

Synonyms: Dermaseptin I (1-18)

Neosystem SC902   MW 1969.47   Bioactive; antibacterial/antimicrobial peptide; analog of Dermaseptin with a full antimicrobial activity; Mor, A et Nicolas, P, *JBC*, 269:1934-1939, 1994

Ala-Leu-Trp-Lys-Thr-Met-Leu-Lys-Lys-Leu-Gly-Thr-Met-Ala-Leu-His-Ala-Gly-Lys-Ala-Ala-Leu-Gly-Ala-Ala-Ala-Asp-Thr-Ile-Ser-Gln-Gly-Thr-Gln

Synonyms: Dermaseptin

ICN 195036   *Phyllomedusa sauvagei*   Originally obtained as skin extracts from the South American arboreal frog; the first vertebrate peptide which demonstrates potent anti-microbial activity against pathogenic fungi at micromolar concentrations; Mor, A etal, *Biochem*, 30:8824, 1991

Sigma D 4671   *Phyllomedusa sauvagii*   FW 3455.1   ≥97% (HPLC) | Bioactive peptide; highly potent antifungal activity at micromolar concentration; Mor, A et al, *Biochemistry*, 30:8824, 1991

Ala-Leu-Tyr-Leu
Ala-Leu-Tyr-Leu-βNA

Bachem K-1065.0050   MW 603.76 $C_{34}H_{45}N_5O_5$ Store at RT

Ala-Lys
Z-Ala-Lys

Bachem C-1140.0250   MW 351.4 $C_{17}H_{25}N_3O_5$ Store at -15°C

Ala-Lys
α-Ala-Lys Hydrochloride

Bachem G-1295.0250   MW 253.73 $C_9H_{19}N_3O_3 \cdot HCl$ Store at -15°C

Ala-Lys
FA-Ala-Lys

Bachem M-1350.0050   MW 337.38 $C_{16}H_{23}N_3O_5$ Store at -15°C

Ala-Lys
N-(3-(2-Furyl)Acryloyl)-L-Ala-Lys

Synonyms: Carboxypeptidase N Substrate, Human Plasma

ICN 158211   MW 337.4 $C_{16}H_{23}N_3O_5$ White to light yellow powder | Plummer, TH & MT Kimmel, *Anal Biochem*, 9:2784, 1970

Ala-Lys
D-Ala-HHT-Lys-pNA·2AcOH

Synonyms: Pefachrome®PL; Pefa-5329

Pentapharm 083-20, 083-10   MW 626.8   Chromogenic peptide substrate for plasmin; used for determination of activity of plasminogen activators (tPA, uPA) as well as α_2-antiplasmin & plasminogen-activator inhibitor (PAI) is possible; v_{max}:20.7 µmol/min; K_M:0.127 *mM*

Ala-Lys
β-Ala-Lys Hydrochloride

Sigma A 6164   FW 253.7 $C_9H_{19}N_3O_3 \cdot HCl$

Ala-Lys
N-(3-(2-Furyl)Acryloyl)-Ala-Lys

Synonyms: Carboxypeptidase N Substrate

Sigma F 5882   FW 337.4 $C_{16}H_{23}N_3O_5$ White to light yellow powder | Substrate for human plasma carboxypeptidase N; Plummer, TH & Kimmel, MT, *Anal Biochem*, 108:348, 1980

Ala-Lys Hydrochloride

Bachem G-1290.0250   MW 253.73 $C_9H_{19}N_3O_3 \cdot HCl$ Store at -15°C

Ala-Lys-Arg
Z-Ala-Lys-Arg-AMC Dihydrochloride

Synonyms: ICRM-Serine Protease I Substrate

ICN 196008   MW 737.7 $C_{33}H_{44}N_8O_7 \cdot 2HCl$ ≥97% | Fluorogenic substrate

Ala-Lys-Arg-Pro-Gln-Arg-Ala-Thr-Ser-Asn-Val-Phe-Ser

Synonyms: Myosin Light Chain Kinase Substrate

ICN 159632   Skeletal muscle   MW 1460.8

Ala-Lys-Glu-Arg-Leu-Glu-Ala-Lys-His-Arg-Glu-Arg-Met-Ser-Gln-Val-Met

Synonyms: Amyloid/A₄ Precursor Protein (319-335), β-; Amyloid α/A4 Precursor Protein 770 (394-410); Amyloid β/A4 Precursor Protein (319-335)

American Peptide 62-5-10 MW 2099.5 $C_{86}H_{151}N_{31}O_{26}S_2$ Could reduce the neurologic damage in vivo in a model of central nervous system ischemia, possibly through its effect on synaptic plasticity; Ninomiya, H et al, *J Cell Biol*, 121:879, 1993; Bowes, MP et al, *Exp Neurol*, 129:112, 1994

Bachem H-2594.0001 MW 2099.47 $C_{86}H_{151}N_{31}O_{26}S_2$ Store at -15°C

Sigma A 1335 FW 2099.5 ≥97% (HPLC); peptide content:~70% | Bioactive peptide

Ala-Lys-Ile-Gly-Asn-Ser-Ile-Gly-Leu-Met-Gly

Synonyms: Amyloid β-Protein (25-35), Scrambled

Neosystem SC942 MW 1060.27 Bioactive; used as inactive control; Carette, B et al, *Neurosci Lett*, 151:111-114, 1993

Ala-Lys-Leu-Arg-Glu-Arg-Leu-Lys-Gln-Arg-Gln-Gln-Leu-Gln-Asn-Arg-Arg-Gly-Leu-Asp-Ile-Leu-Phe-Leu-Gln-Glu-Gly-Gly-Leu

Bachem H-1262.0500 MW 3478.07 $C_{151}H_{262}N_{52}O_{42}$ Store at -15°C

Ala-Lys-Lys-Leu-Ser-Lys-Asp-Arg-Met-Lys-Lys-Tyr-Met-Ala-Arg-Arg-Lys-Trp-Gln-Lys-Thr-Gly Amide

Synonyms: Myosin Light Chain Kinase (480-501)

Bachem H-3554.0500 Store at -15°C

Ala-Lys-Lys-Leu-Ser-Lys-Asp-Arg-Met-Lys-Lys-Tyr-Met-Ala-Arg-Arg-Lys-Trp-Gln-Lys-Thr-Gly Amide Acetate Salt

Synonyms: Myosin Light Chain Kinase Inhibitor Peptide (480-501); SM-1

Calbiochem 05-23-1700 MW 2738.4 $C_{120}H_{209}N_{41}O_{28}S_2$ >97% (HPLC); lyophilized solid; soluble in 5% acetic acid | Inhibitor of calmodulin-dependent activation of the smooth muscle myosin light chain kinase; reversibly decreases M-type potassium current without affecting □-type or delayed rectifier-type potassium currents; Akasu, T et al, *Neuron*, 11:1133, 1993; Kemp, BE et al, *J Biol Chem*, 262:2542, 1987

Ala-Lys-Phe-Asp-Lys-Phe-Tyr-Gly-Leu-Met Amide

Synonyms: Scyliorhinin I

Bachem H-1006.0005 MW 1218.49 $C_{59}H_{87}N_{13}O_{13}S$ Store at -15°C

American Peptide 62-7-70 *Scyliorhinus caniculus* (common dogfish) intestine MW 1218.5 $C_{59}H_{87}N_{13}O_{13}S$ High sequence homology with physalaemin; contains tachykinin-like activity & crossreacts with antisera directed against the C-terminal region of substance P; Conlon, JM et al, *FEBS Lett*, 200:111, 1986

Ala-Lys-Pro-Glu-Ala-Pro-Gly-Glu-Asp-Ala-Ser-Pro-Glu-Glu-Leu-Ser-Arg-Tyr-Tyr-Ala-Ser-Leu-Arg-His-Tyr-Leu-Asn-Leu-Val-Thr-Arg-Gln-Arg-Tyr Amide

Synonyms: Peptide YY (3-36)

Neosystem SC892 Porcine MW 3980.4 Bioactive neuropeptide; brain/gut peptide

Ala-Lys-Pro-Phe
Suc-Ala-Lys-Pro-Phe-pNA

Bachem L-1630.0050 Store at -15°C

Ala-Lys-Pro-Ser-Tyr-Hyp-Hyp-Thr-Tyr-Lys

Sigma A 7060 FW 1183.3 ≥97% (HPLC) | Bioactive peptide; bioadhesive peptide; Swerdloff, MD, *Int J Peptide Protein Res*, 33:318, 1989

Ala-Lys-Pro-Ser-Tyr-Hyp-Hyp-Thr-Tyr-Lys Ala-Lys(TFA)-Pro-Ser-Tyr-Hyp-Hyp-Thr-Tyr-Lys

Sigma A 7185 FW 1279.3 ≥97% (HPLC) | Bioactive peptide; bioadhesive peptide; Swerdloff, MD, *Int J Peptide Protein Res*, 33:318, 1989

Ala-Met
Z-Ala-Met

Bachem C-1145.0001 MW 354.43 $C_{16}H_{22}N_2O_5S$ Store at -15°C

Ala-Met

Bachem G-1300.0250 MW 220.29 $C_8H_{16}N_2O_3S$ Store at -15°C

Ala-Met
DL-Ala-DL-Met

Bachem G-4500.0001 MW 202.29 $C_8H_{16}N_2O_3S$ Store at -15°C

Sigma A 2128 FW 220.3 $C_8H_{16}N_2O_3S$

Ala-Met

Sigma A 2253 FW 220.3 $C_8H_{16}N_2O_3S$

Ala-Met
N-CBZ-Ala-Met

Sigma C 3376 FW 354.4 $C_{16}H_{22}N_2O_5S$

Ala-Nle
DL-Ala-DL-Nle

ICN 100345 MW 202.2 $C_9H_{18}N_2O_3$ Crystalline

Ala-Nle-Pro-Phe
Suc-Ala-Nle-Pro-Phe-pNA

Bachem L-1660.0050 Store at -15°C

Ala-Nva
DL-Ala-DL-Nva

ICN 100349 MW 188.2 $C_8H_{16}N_2O_3$ Crystalline

Ala-Phe
Z-Ala-Phe

Bachem C-1155.0001 MW 370.41 $C_{20}H_{22}N_2O_5$ Store at -15°C

Ala-Phe
Z-Ala-Phe-OMe

Bachem C-1160.0001 MW 384.43 $C_{21}H_{24}N_2O_5$ Store at -15°C

Ala-Phe
Z-D-Ala-Phe

Bachem C-1165.0250 MW 370.41 $C_{20}H_{22}N_2O_5$ Store at -15°C

Ala-Phe

Bachem G-1320.0250 MW 236.27 $C_{12}H_{16}N_2O_3$ Store at -15°C

Ala-Phe
Ala-D-Phe

Bachem G-1325.0250   MW 236.27 $C_{12}H_{16}N_2O_3$ Store at -15°C

Ala-Phe
D-Ala-Phe

Bachem G-1330.0250   MW 236.27 $C_{12}H_{13}N_2O_3$ Store at -15°C

Ala-Phe
α-Ala-Phe

Bachem G-1335.0250   MW 236.27 $C_{12}H_{16}N_2O_3$ Store at -15°C

Ala-Phe
D-Ala-D-Phe Hydrate

Bachem G-1340.0250   MW 254.29 $C_{12}H_{16}N_2O_3 \cdot H_2O$ Store at -15°C

Ala-Phe
DL-Ala-DL-Phe

Bachem G-3710.0001   MW 236.27 $C_{12}H_{16}N_2O_3$ Store at -15°C

Ala-Phe
p-Hydroxy-Bz-Ala-Phe

Bachem G-3830.0025   Store at -15°C

Ala-Phe
Ala-Phe-pNA Hydrochloride

Bachem L-1610.0050   MW 392.84 $C_{18}H_{20}N_4O_4 \cdot HCl$ Store at -15°C

Ala-Phe
FA-Ala-Phe Amide

Bachem M-1355.0050   Store at -15°C

Ala-Phe

ICN 100325   MW 236.3 $C_{12}H_{18}N_2O_3$ Crystalline

Ala-Phe
DL-Ala-DL-Phe

ICN 100356   MW 236.3 $C_{12}H_{18}N_2O_3$ Crystalline

Ala-Phe
N-(3-(2-Furyl)Acryloyl)-Ala-Phe Amide

ICN 158212   MW 355.4 $C_{19}H_{21}N_3O_4$

Ala-Phe
β-Ala-Phe

Sigma A 1416   FW 236.3 $C_{12}H_{16}N_2O_3$

Ala-Phe
DL-Ala-DL-Phe

Sigma A 3003   FW 236.3 $C_{12}H_{16}N_2O_3$

Ala-Phe

Sigma A 3128   FW 236.3 $C_{12}H_{16}N_2O_3$

Ala-Phe
N-CBZ-Ala-Phe

Sigma C 3876   FW 370.4 $C_{20}H_{22}N_2O_5$

Ala-Phe
N-(3-(2-Furyl)Acryloyl)Ala-Phe Amide

Synonyms: Thermolysin Substrate

Sigma F 2380   FW 355.4 $C_{19}H_{21}N_3O_4$ Feder, J & Schuck JM, *Biochemistry*, 9:2784, 1970

Ala-Phe Amide Hydrochloride

Bachem G-1345.0250   MW 271.75 $C_{12}H_{17}N_3O_2 \cdot HCl$ Store at -15°C

Ala-Phe-Ala
Ala-D-Phe-Ala

Bachem H-2224.0250   MW 307.35 $C_{15}H_{21}N_3O_4$ Store at -15°C

Ala-Phe-Ala

Bachem H-5420.0250   MW 307.35 $C_{15}H_{21}N_3O_4$ Store at -15°C

Ala-Phe-Ala-Ala
MeOSuc-Ala-Phe-Ala-Ala-pNA

Bachem L-1345.0050   Store at -15°C

Ala-Phe-Ala-Phe-Asp-Val-Phe-Tyr-Asp
Anthranilyl-Ala-Phe-Ala-Phe-Asp-Val-Phe-Nitro-Tyr-Asp

Synonyms: Aspartate-Specific Protease Substrate

Sigma A 3442   FW 1258.3   ≥95% (HPLC) | Fluorescent; Breddam, K & Meldal, M, *Eur J Biochem*, 206:103, 1992

Ala-Phe-Ala-Phe-Glu-Val-Phe-Tyr-Asp
Abz-Ala-Phe-Ala-Phe-Glu-Val-Phe-Tyr(NO₂)-Asp

Synonyms: Glu-Specific Protease Fluorescent Substrate

American Peptide 81-0-30   MW 1272.4 $C_{63}H_{73}N_{11}O_{18}$ *Eur J Biochem*, 206:103, 1992

Ala-Phe-Ala-Phe-Glu-Val-Phe-Tyr-Asp
Anthranilyl-Ala-Phe-Ala-Phe-Glu-Val-Phe-Nitro-Tyr-Asp

Synonyms: Glutamate-Specific Protease Substrate

Sigma A 3567   FW 1272.3   ≥90% (HPLC) | Fluorescent; Breddam, K & Meldal, M, *Eur J Biochem*, 206:103, 1992

Ala-Phe-Ala-Ser-Lys-Lys-Leu-Lys-Pro-Ala
Ac-D-Nal-D-(p-Chloro)Phe-D-Pal-Ser-Lys(Nicotinoyl)-D-Lys(Nicotinoyl)-Leu-Lys(Isopropyl)-Pro-D-Ala Amide

Synonyms: Antide

American Peptide 54-1-24   MW 1596.3 $C_{82}H_{110}N_{17}O_{14}Cl$ Potent GnRH antagonist; it transiently suppressed LH & FSH release at a dose of 250 µg/kg; at 1250 µg/kg, it markedly inhibited LH & FSH secretion throughout the entire study period; Weinbauer, GF et al, *Andrologia*, 25(3):141, 1993; Ljungqvist, A et al, *BBRC*, 148:849, 1987

Ala-Phe-Ala-Ser-Lys-Lys-Leu-Lys-Pro-Ala
Ac-β-(2-Naphthyl)-D-Ala-D-p-Chloro-Phe-β-(3-Pyridyl)-D-Ala-Ser-Nε-(Nicotinoyl)-Lys-Nε-(Nicotinoyl)-D-Lys-Leu-Nε-(Isopropyl)-Lys-Pro-D-Ala Amide

Synonyms: Luteinizing Hormone Releasing Hormone Antagonist; Antide; Luteinizing Hormone Releasing Hormone Antagonist

Sigma A 8802   FW 1591.3   ≥97% (HPLC) | Bioactive peptide; high anti-ovulatory & negligible histamine release activity; Ljungqvist, A et al, *Biochem Biophys Res Commun*, 148:849, 1987

Ala-Phe-Arg-Cha-Har-Tyr
Ala-p-Fluoro-Phe-Arg-Cha-HomoArg-Tyr Amide

Synonyms: Thrombin Receptor Agonist

Neosystem SC989   MW 896.08   Bioactive; potent thrombin-receptor activating peptide; EC_{50} = 001 $m\mu M$; agonist potency has been enhanced 1000-fold with this peptide compared to the 14-mer thrombin receptor peptide sequence SFLLRNPNDKYEPF; Feng; DM et al, *J Med Chem*, 38:4125-4130, 1995

Ala-Phe-Arg-His-Cys-Asn-Pro-Asn-Gly-Thr-Trp-Asp-Phe-Met-His-Ser-Leu-Asn Acetate Salt

Synonyms: Parathyroid Hormone Receptor II (99-116), Extracellular Domain 1

Biogenesis 0100-0185   Synthetic   Semi-pure; lyophilized

Ala-Phe-Cys-Asn-Leu-Arg-Met-Cys-Gln-Leu-Ser-Cys-Arg-Ser-Leu-Gly-Leu-Leu-Gly-Lys-Cys-Ile-Gly-Asp-Lys-Cys-Glu-Cys-Val-Lys-His Amide

Synonyms: Leiurotoxin I; Scyllatoxin

Calbiochem 428905   MW 3429.2 $C_{142}H_{243}N_{45}O_{39}S_7$ ≥95% (HPLC); solid lyophilized from 20 mM NH_4HCO_3; soluble in water; LD_{50}≤2000 mg/kg | Synthetic blocker of Ca^{2+}-activated apamin-sensitive K^+ channel; although their biological activities are similar, there is no sequence homology between apamin & leiurotoxin I; Calabro, V et al, *J Pept Res*, 50:39, 1997; Auguste, P et al, *J Biol Chem*, 265:4753, 1990; Goh, JW et al, *Brain Res*, 591:165, 1992

Ala-Phe-Cys-Asn-Leu-Arg-Met-Cys-Gln-Leu-Ser-Cys-Arg-Ser-Leu-Gly-Leu-Leu-Gly-Lys-Cys-Ile-Gly-Asp-Lys-Cys-Glu-Cys-Val-Lys-His

Synonyms: Leiurotoxin I

ICN 159616   Scyllatoxin; inhibits Apamin-sensitive calcium-dependent K^+ channels

Ala-Phe-Cys-Asn-Leu-Arg-Met-Cys-Gln-Leu-Ser-Cys-Arg-Ser-Leu-Gly-Leu-Leu-Gly-Lys-Cys-Ile-Gly-Asp-Lys-Cys-Glu-Cys-Val-Lys-His Amide

Synonyms: Scyllatoxin; Leiurotoxin I; Ca^{2+}-Activated K^+ Channel Blocker, Small Conductance

Sigma S 3277   FW 3423.1   ≥90% (HPLC); Disulfide bonds: undetermined | Bioactive peptide; blocks small-conductance Ca^{2+}-activated K^+ channels; Chicchi, GG et al, *J Biol Chem*, 263:10192, 1988; Margins, JC et al, *FEBS Lett*, 260:249, 1990

American Peptide 41-0-24   *Leiurus quinquestriatus hebraeus* (scorpion)   MW 3423.1 $C_{142}H_{237}N_{45}O_{39}S_7$ Three disulfide bonds; blocker of Apamin-sensitive Ca^{2+}-activated K^+ channels; component of the venom of the Israeli scorpion; inhibits apamin binding to its receptor; Chicci, GG et al, *JBC*, 263:10192, 1988; Auguste, P et al, *JBC*, 265:4753, 1990

Peptides International PSC-4260-s   *Leiurus quinquestriatus hebraeus* (scorpion) synthetic   MW 3423.1 $C_{142}H_{237}N_{45}O_{39}S_7$ >95% (HPLC); lyophilized amorphous powder | Bioactive; Disulfide bonds: Cys^3-Cys^{21}, Cys^8-Cys^{26}, Cys^{12}-Cys^{28}; Chicchi, GG et al, *JBC*, 263, 10192, 1988; Auguste, P et al, *Biochemistry*, 31:648, 1992

Ala-Phe-Cys-Asn-Leu-Arg-Met-Cys-Gln-Leu-Ser-Cys-Arg-Ser-Leu-Gly-Leu-Leu-Gly-Lys-Cys-Lle-Gly-Asp-Lys-Cys-Glu-Gys-Val-Lys-His Amide

Synonyms: Scyllatoxin; Leiurotoxin I

Alexis 630-043   Synthetic   MW 3423.1 $C_{142}H_{237}N_{45}O_{39}S_7$ ≥98%; lyophilized powder; soluble in H_2O; Disulfide bonds: are undetermined | Originally isolated from Leiurus *quinquestriatus hebraeus*; small conductance Ca^{2+}-activated K^+ channel blocker; potent neurotoxin; Chicchi, GG et al, *J Biol Chem*, 263:10192, 1988; Strong, PN et al, *Pharmacol Ther*, 46:137, 1990; Auguste, P et al, *Biochemistry*, 31:648, 1992

Ala-Phe-Gln-Asn-Cys-Pro-Arg-Gly
Mpr-D-Pyridyl-Ala-Phe-Gln-Asn-Cys-Pro-Arg-Gly Amide

Synonyms: Vasopressin, (Deamino-Cys^1,D-3-(Pyridyl)Ala^2,Arg^8)-

American Peptide 60-0-18   MW 1055.2 $C_{45}H_{64}N_{15}O_{11}S_2$ Agonist for pituitary receptors; disulfide bonds: Mpr^1-Cys^6; Schwarty, J et al, *Endocrinology*, 129:1107, 1991

Ala-Phe-Gln-Asn-Cys-Pro-Arg-Gly
3-Mercaptopropionyl-β-3-Pyridyl-D-Ala-Phe-Gln-Asn-Cys-Pro-Arg-Gly Amide

Synonyms: Vasopressin, (Deamino-Cys^1,β-(3-Pyridyl)-D-Ala^2,Arg^8)-; Vasopressin, d(D-3Pal)-

Bachem H-3058.0001   MW 1054.22 $C_{45}H_{63}N_{15}O_{11}S_2$ Store at -15°C

Ala-Phe-Gln-Asn-Cys-Pro-Arg-Gly
3-Mercaptopropionyl-D-3-(Pyridyl)-Ala-Phe-Gln-Asn-Cys-Pro-D-Arg-Gly Amide

Synonyms: Vasopressin, (deamino-Cys^1,D-3-(Pyridyl)-Ala^2,Arg^8)-

Sigma V 2257   FW 1054.2   ≥97% (HPLC); Disulfide bonds: 1-6 | Bioactive peptide; agonist for pituitary receptors; V_{1b} agonist; Schwartz, J et al, *Endocrinol*, 129:1107, 1991

Ala-Phe-Gly

Bachem H-1575.0250   MW 293.32 $C_{14}H_{19}N_3O_4$ Store at -15°C

Ala-Phe-His-His-Val-Ala-Arg-Glu-Leu-His-Pro

Synonyms: NEF (190-200); HTLV I Peptide

Neosystem SC669   MW 1313.48   HTLV I peptide from subtype NEF

Ala-Phe-Leu-Pro-Trp-His-Arg-Leu-Phe

Synonyms: Tyrosinase (206-214)

Bachem H-3848.0001   Human   Store at -15°C

Ala-Phe-Lys
Ala-Phe-Lys-AMC

Bachem I-1045.0050   MW 521.62 $C_{28}H_{35}N_5O_5$ Store at -15°C

Ala-Phe-Lys
MeOSuc-Ala-Phe-Lys-AMC Trifluoroacetate Salt

Bachem I-1275.0050   Store at -15°C

Ala-Phe-Lys
Suc-Ala-Phe-Lys-AMC

Bachem I-1325.0050   MW 621.69 $C_{32}H_{39}N_5O_8$ Store at -15°C

Ala-Phe-Lys
Suc-Ala-Phe-Lys-AMC Trifluoroacetate Salt

Bachem I-1330.0050   MW 735.71 $C_{32}H_{39}N_5O_8 \cdot C_2HF_3O_2$ Store at -15°C

Ala-Phe-Lys
N-Suc-Ala-Phe-Lys-MCA Acetate Salt

ICN 156693

Ala-Phe-Lys
Ala-Phe-Lys-AMC

Sigma A 2038   FW 521.6 $C_{28}H_{35}N_5O_5$ Substrate for fluorometric assay of plasmin, urokinase, & thrombin; Pierzchala, PA et al, *Biochem J*, 183:555, 1979

Ala-Phe-Lys
N-Suc-Ala-Phe-Lys-AMC Acetate Salt

Synonyms: Plasmin Substrate

Sigma S 0763   FW 681.7 $C_{32}H_{39}N_5O_8 \cdot C_2H_4O_2$ Fluorogenic; Pierzchala, PA et al, *Biochem J*, 183:155, 1979

Ala-Phe-Phe-Tyr-Lys-Leu-Asp-Ile-Ile-Pro-Ile-Asp-Asn-Asp-Thr

Synonyms: GP120 (179-193); HTLV I Peptide

Neosystem SC1005   MW 1785.02   HTLV I peptide from subtype GP160

Ala-Phe-Pro

Bachem H-1580.0250   MW 333.39 $C_{17}H_{23}N_3O_4$ Store at -15°C

Ala-Phe-Pro
Ala-Phe-Pro-βNA Hydrochloride

Bachem K-1070.0250   MW 495.02 $C_{27}H_{30}N_4O_3 \cdot HCl$ Store at -15°C

Ala-Phe-Pro
Ala-Phe-Pro-pNA

Bachem L-1795.0050   MW 453.5 $C_{23}H_{27}N_5O_5$ Store at -15°C

Ala-Phe-Pro-Ala
Ala-Phe-Pro-Ala-βNA

Bachem K-1075.0250   MW 529.64 $C_{30}H_{35}N_5O_4$ Store at -15°C

Ala-Phe-Pro-Leu-Glu-Phe

Synonyms: Adrenocorticotropic Hormone (34-39)

Bachem H-1220.0005   MW 722.84 $C_{37}H_{50}N_6O_9$ Store at -15°C

Neosystem SC314   Human   MW 722.84   Bioactive

Ala-Phe-Pro-Phe
Suc-Ala-Phe-Pro-Phe-pNA

Bachem L-1605.0050   Store at -15°C

Ala-Phe-Pro-Phe
Suc-Ala-Phe-Pro-Phe-AMC

Neosystem SC1312   MW 737.80

Ala-Phe-Ser-Ser-Trp-Gly Amide

Synonyms: Locustakinin I

Bachem H-1396.0001   MW 652.71 $C_{31}H_{40}N_8O_8$ Store at -15°C

Ala-Phe-Tyr-Thr-Thr-Lys-Asn-Ile-Ile-Gly-Thr-Ile-Cys

Synonyms: GP120 MN (319-331); HIV I Peptide

Neosystem SC839   MW 1444.70   GP120 peptide from HIV-1 subtype MN; due to the presence of a Cys this peptide can contain the dimeric form

Ala-Phe-Val-Thr-Ile-Gly-Lys-Ile-Gly-Asn-Met-Arg-Gln-Ala

Synonyms: GP120 (321-334); HTLV I Peptide

Neosystem SC837   MW 1505.79   HTLV I peptide from subtype GP160

Ala-Phs-Glu-Tyr-Glu-Asp-Glu-Asp-Gly-Asp-Arg-Ile-Thr-Val-Arg-Ser-Cys

Synonyms: MEK5 Blocking Peptide (59-74)

Calbiochem 444956   Human synthetic   MW 2005.1 $C_{83}H_{125}N_{23}O_{33}S$ Lyophilized solid | Based on human & rat MEK5 (59-74) with a Cys C residue added; used to raise monospecific polyclonal antibodies to MEK5; for use in immunoabsorption for immunoprecipitation, immunocytochemistry, Western blotting & dot blots

Ala-Pro
BOC-Ala-Pro

Bachem A-2740.0001   MW 286.33 $C_{13}H_{22}N_2O_5$ Store at -15°C

Ala-Pro
Z-Ala-Pro

Bachem C-1185.0001   MW 320.35 $C_{16}H_{20}N_2O_5$ Store at -15°C

Ala-Pro

Bachem G-1350.0250   MW 186.21 $C_8H_{14}N_2O_3$ Store at -15°C

Ala-Pro
Ala-Pro-OMe Hydrochloride

Bachem G-1365.0250   MW 236.7 $C_9H_{16}N_2O_3 \cdot HCl$ Store at -15°C

Ala-Pro
D-Ala-Pro

Bachem G-3865.0250   MW 186.21 $C_8H_{14}N_2O_3$ Store at -15°C

Ala-Pro
Ala-Pro-AFC

Bachem I-1680.0050   MW 397.35 $C_{18}H_{18}F_3N_3O_4$ Store at -15°C

Ala-Pro
Ala-Pro-4MβNA Hydrochloride

Bachem J-1030.0250   MW 377.87 $C_{19}H_{23}N_3O_3 \cdot HCl$ Store at -15°C

Ala-Pro
Z-Ala-Pro-4MβNA

Bachem J-1095.0250   MW 475.54 $C_{27}H_{29}N_3O_5$ Store at -15°C

Ala-Pro
Ala-Pro-pNA Hydrochloride

Bachem L-1115.0250   MW 342.78 $C_{14}H_{18}N_4O_4 \cdot HCl$ Store at -15°C

Ala-Pro
Z-Ala-Pro-pNA

Bachem L-1215.0250   MW 440.46 $C_{22}H_{24}N_4O_6$ Store at -15°C

Ala-Pro
Suc-Ala-Pro-pNA

Bachem L-2090.0050   MW 406.4 $C_{18}H_{22}N_4O_7$ Store at -15°C

Ala-Pro
L-Ala-L-Pro

Fluka 05370   MW 186.21 $C_8H_{14}N_2O_3$ Aqueous; ~99% (titration); mp:170-175°C; hydrate

Ala-Pro

ICN 190771   MW 186.2 $C_8H_{14}N_2O_3$ Crystalline

Ala-Pro
L-Ala-L-Pro

Rexim   MW 186.2 $C_8H_{14}N_2O_3$ White to near white free flowing powder

Ala-Pro

Sigma A 3253   FW 186.2 $C_8H_{14}N_2O_3$

Ala-Pro
N-CBZ-Ala-Pro

Sigma C 4001   FW 320.3 $C_{16}H_{20}N_2O_5$

Ala-Pro
CBZ-Ala-Pro

USBio C2098-23   MW 320.3 $C_{16}H_{20}N_2O_5$ ≥99%

Ala-Pro Amide Hydrochloride

Bachem G-3955.0250   MW 221.69 $C_8H_{15}N_3O_2 \cdot HCl$ Store at -15°C

Ala-Pro-Ala

Bachem H-1595.0250   MW 257.29 $C_{11}H_{19}N_3O_4$ Store at -15°C

Ala-Pro-Ala
Suc-Ala-Pro-Ala-AMC

Bachem I-1335.0050   MW 514.54 $C_{25}H_{30}N_4O_8$ Store at -15°C

Ala-Pro-Ala
Ala-Pro-Ala-βNA

Bachem K-1635.0050   MW 382.46 $C_{21}H_{26}N_4O_3$ Store at -15°C

Ala-Pro-Ala
Ac-Ala-Pro-Ala-pNA

Bachem L-1020.0050   MW 419.44 $C_{19}H_{25}N_5O_6$ Store at -15°C

Ala-Pro-Ala
N-Suc-L-Ala-L-Pro-L-Ala-MCA

Synonyms: Elastase Substrate

ICN 152082   Specific fluorogenic substrate

Ala-Pro-Ala
Suc-Ala-Pro-Ala-MCA

Synonyms: Elastase Substrate

Peptides International MAP-3100-v   MW 514.54 $C_{25}H_{30}N_4O_8$ >98% (HPLC); lyophilized amorphous powder | Oshima, G et al, *Arch Biochem Biophys*, 233:212, 1984

Ala-Pro-Ala
Suc-Ala-Pro-Ala-pNA

Synonyms: Elastase Substrate

Peptides International SAP-3118   MW 477.49 $C_{21}H_{27}N_5O_8$ >98% (HPLC); amorphous powder

Ala-Pro-Ala
N-Suc-Ala-Pro-Ala-AMC

Synonyms: Elastase Substrate

Sigma S 8383   FW 514.5 $C_{25}H_{30}N_4O_8$ Substrate for human leukocyte elastase; Oshima, G et al, *Arch Biochem Biophys*, 233:212, 1984

Ala-Pro-Ala-Ala-Ala-Ala-Glu-Ala-Val-Ala-Gly-Leu-Ala-Pro-Val-Ala-Ala-Glu-Glu-Phe
Ac-2-MeAla-L-Pro-2-MeAla-L-Ala-2-MeAla-L-Ala-L-Glu(NH₂)-2-MeAla-L-Val-2-MeAla-Gly-L-Leu-2-MeAla-L-Pro-L-Val-2-MeAla-2-MeAla-L-Glu-L- Glu(NH₂)-Phenylalaninol

Synonyms: Alamethicin

Fluka 05125   *Trichoderma viride*   MW 1964.35 $C_{92}H_{150}N_{22}O_{25}$ ≥50% (HPLC) main component | Alamethicin-mediated fusion of lecithin vesicles; Thermos, K & Lau, ALY & Chan, SI, *PNAS*, 72:2170, 1975

Ala-Pro-Ala-Arg
Ala-Pro-Ala-Arg-βNA

Bachem K-1640.0050   MW 538.65 $C_{27}H_{38}N_8O_4$ Store at -15°C

Ala-Pro-Arg
(D)-CyclohexylAla-Pro-Arg-Chloromethylketone Ditrifluoroacetate

Synonyms: Pefa-2894

Pentapharm 390-03   MW 685.1 $C_{21}H_{37}N_6O_3Cl \cdot 2TFA$ Bulk | Irreversible inhibitor of serine proteinases, ie thrombin; $k_2/K_i = 8 \times 10^8$

Ala-Pro-Arg-Gly-Asp

Bachem H-5995.0050   MW 514.54 $C_{20}H_{34}N_8O_8$ Store at -15°C

Ala-Pro-Arg-Leu-Arg-Phe-Tyr

Synonyms: Bag Cell Peptide (1-7), α-

American Peptide 87-0-94   MW 922.2 $C_{44}H_{67}N_{13}O_9$ Rothman, BS et al, *PNAS*, 80:5753, 1983

Sigma A 5058   FW 922.1 $C_{46}H_{67}N_{13}O_9$ ≥97% (HPLC); peptide content:~70% | Bioactive peptide; Rothman, BS et al, *Proc Natl Acad Sci USA*, 80:5753, 1983

Ala-Pro-Arg-Leu-Arg-Phe-Tyr-Ser

Synonyms: Bag Cell Peptide (1-8), α-

American Peptide 87-0-93   MW 1009.2 $C_{47}H_{72}N_{14}O_{11}$ Rothman, BS et al, *PNAS*, 80:5753, 1983

Sigma A 5183   FW 1009.2   ≥97% (HPLC); peptide content:~70% | Bioactive peptide; Rothman, BS et al, *Proc Natl Acad Sci USA*, 80:5753, 1983

Ala-Pro-Arg-Leu-Arg-Phe-Tyr-Ser-Leu

Synonyms: Bag Cell Peptide (1-9), α-

American Peptide 87-0-92   MW 1122.3 $C_{53}H_{83}N_{15}O_{12}$ Rothman, BS et al, *PNAS*, 80:5753, 1983

Bachem H-6725.0005   MW 1122.34 $C_{53}H_{83}N_{15}O_{12}$ Store at -15°C

Sigma A 5308   FW 1122.3   ≥97% (HPLC); peptide content:~70% | Bioactive peptide; Rothman, BS et al, *Proc Natl Acad Sci USA*, 80:5753, 1983

Ala-Pro-Arg-Leu-Glu-Ile-Val-Pro-Thr-Met-Tyr-Ile-Tyr-Lys-Leu-Ser-Pro-Thr-Gly-Ser-Glu-Lys-Leu-Gly-Asp-Glu-Arg
Ac-Ala-Pro-Arg-Leu-Glu-Ile-Val-Pro-Thr-Met-Tyr-Ile-Tyr-Lys-Leu-Ser-Pro-Thr-Gly-Ser-Glu-Lys-Leu-Gly-Asp-Glu-Arg Amide

Synonyms: Calpastatin Peptide

Calbiochem 208904   MW 3177.7 $C_{142}H_{230}N_{36}O_{44}S$ TFA; ≥95% (HPLC); lyophilized solid; soluble in water | Useful as a negative control for calpastatin peptide; Eto, A et al, *J Bio Chem*, 270:25115, 1995

Ala-Pro-Arg-Leu-Pro-Gln-Cys-Gln-Gly-Asp-Asp-Gln-Glu-Lys-Cys-Leu-Cys-Asn-Lys-Asp-Glu-Cys-Pro-Pro-Gly-Gln-Cys-Arg-Phe-Pro-Arg-Gly-Asp-Ala-Asp-Pro-Tyr-Cys-Glu

Synonyms: Decorsin; Glycoprotein IIb/IIIa Antagonist; Platelet Aggregation Inhibitor

American Peptide 41-0-74   Leech   MW 4377.8 $C_{179}H_{271}N_{55}O_{62}S_6$ Disulfide bonds: Cys^7-Cys^{15}, Cys^{17}-Cys^{27}, Cys^{22}-Cys^{38}; Seymour, JL et al, *J Biol Chem*, 265:10143, 1990; Krezel, AM et al, *Science*, 264:1944, 1994

Peptides International PDC-4269-s   *Macrobdella decora* (leech) synthetic   MW 4377.9 $C_{179}H_{271}N_{55}O_{62}S_6$ >99% (HPLC); lyophilized amorphous powder | Bioactive; Disulfide bonds: Cys^7-Cys^{15}, Cys^{17}-Cys^{27}, Cys^{22}-Cys^{38}; Seymour, JL et al, *JBC*, 265:10143, 1990; Krezel, AM et al, *Science*, 264:1944, 1994

Ala-Pro-Arg-Leu-Pro-Gln-Cys-Gln-Gly-Asp-Gln-Glu-Lys-Cys-Leu-Cys-Asn-Lys-Asp-Glu-Cys-Pro-Pro-Gly-Gln-Cys-Arg-Phe-Pro-Arg-Gly-Asp-Ala-Asp-Pro-Tyr-Cys-Glu

Synonyms: Decorsin, (des-Asp^{10})-

American Peptide 41-0-75   Leech   MW 4262.8 $C_{175}H_{266}N_{54}O_{59}S_6$ Disulfide bonds: Cys^7-Cys^{15}, Cys^{17}-Cys^{27}, Cys^{22}-Cys^{38}; inactive control

Ala-Pro-Arg-Ser-Met-Arg-Arg-Ser-Ser-Arg-Cys-Phe-Gly-Ser-Arg-Ile-Asp-Arg-Ile-Gly-Ala-Gln-Ser-Gly-Met-Gly-Cys-Gly-Arg-Phe

Synonyms: Natriuretic Peptide (1-30), Atrial

American Peptide 14-7-25   Frog   MW 3260.7 $C_{131}H_{215}N_{49}O_{41}S_4$ Disulfide bonds: Cys^{11}-Cys^{27}

Ala-Pro-Arg-Thr-Pro-Gly-Gly-Arg-Arg

Synonyms: Myelin Basic Protein Kinase (p44*mpk*) Substrate; Mitogen Activated Protein Kinase Substrate; Myelin Basic Protein Kinase Substrate; Mitogen Activated Protein Kinase Substrate; Myelin Basic Protein (95-98); Mitogen Activated Protein Kinase Substrate Peptide

Alexis 151-024   MW 967.1 $C_{39}H_{70}N_{18}O_{11}$ ≥98%; white powder | Consensus sequence for substrate recognition by the meiosis-activated p44*mpk*; Sanghera, JS et al, *FEBS Lett*, 273:223, 1990; Clark-Lewis, I et al, *J Biol Chem*, 266:15180, 1991; Daeipour, M et al, *J Immunol*, 150:4743, 1993

Bachem H-3244.0001   Store at -15°C

Neosystem SC962   MW 967.10   Bioactive; proteinkinase-related peptide; Clarke-Lewis, I et al, *JBC*, 266:15180-15184, 1991

USBio M2363-25   Bovine synthetic   97% (HPLC); lyophilized | AA 95-98 of bovine myelin basic protein

Upstate 12-125   Bovine, synthetic   MW 967.21   Lyophilized

Ala-Pro-Arg-Thr-Pro-Gly-Gly-Arg-Arg Trifluoroacetate Salt

Synonyms: Myelin Basic Protein Peptide Substrate; ERK Kinase Substrate; Mitogen Activated Protein Kinase Substrate

Calbiochem 475920   MW 967.1 $C_{39}H_{70}N_{18}O_{11}$ ≥94% (HPLC); lyophilized solid; soluble in water | Kameshita, I et al, *J Biochem*, 122:168, 1997; Lint, JV et al, *Mol Cell Biochem*, 127:171, 1993

Ala-Pro-Gln-Val-Leu-Phe-Val-Met-His-Pro-Leu
Ala-Pro-Gln-Val-Leu-*p*-Nitro-Phe-Val-Met-His-Pro-Leu

Bachem M-2370.0001   MW 1296.56 $C_{60}H_{93}N_{15}O_{15}S$ Store at -15°C

Ala-Pro-Gln-Val-Leu-Pro-Val-Met-His
Ala-Pro-Gln-Val-Leu-Pro-Val-Met-His Amide

Synonyms: HTLV I Protease Substrate

Neosystem SC824   MW 990.23   Virus-related peptide; AIDS-related peptide

Ala-Pro-Gly
Z-Ala-Pro-Gly

Bachem C-1190.0250   MW 377.4 $C_{18}H_{23}N_3O_6$ Store at -15°C

Ala-Pro-Gly

Bachem H-1600.0250   MW 243.26 $C_{10}H_{17}N_3O_4$ Store at -15°C

Sigma A 9152   FW 243.3 $C_{10}H_{17}N_3O_4$

Ala-Pro-Gly-Asp-Arg-Ile-Tyr-Val-His-Pro-Phe

Synonyms: Angiotensin II, (Ile^3, Val^5)-

American Peptide 12-1-60   MW 1271.5 $C_{60}H_{86}N_{16}O_{15}$ Khosla, MC et al, *J Med Chem*, 24:885, 1981

Ala-Pro-Gly-Asp-Ile-Tyr-Val-His-Pro-Phe Acetate Salt Free Base

Synonyms: Angiotensin II Agonist, (Ala-Pro-Gly-(Ile^3, Val^5))-

Sigma A 0289   FW 1271.4   ≥97% (HPLC) | Bioactive peptide; high binding affinity for the AT_2 receptor subtype; Khosla, MC et al, *J Med Chem*, 24:885, 1981

Ala-Pro-Gly-Gly-Gly-Gly-Gly-Glu-Pro-Gly-Thr-Ala-Gly-Cys

Synonyms: EFK1 Blocking Peptide (7-20)

Calbiochem 442671   Rat synthetic   MW 1243.3 $C_{48}H_{78}N_{18}O_{19}S$ Lyophilized solid | Based on the rat ERK1 MAP kinase (7-20) with a Cys C residue added; this peptide coupled to KLH was used as the immunogen for the production of Anti-ERK1; for use in immunoabsorption for immunoprecipitation, immunocytochemistry, Western blotting & dot blots

Ala-Pro-Gly-Ile-Val-Ala-Pro-Gly-Asp-Arg-Ile-Tyr-Val-His-Pro-Phe
Ala-Pro-Gly-(Ile^3,Val^5)-Ala-Pro-Gly-Asp-Arg-Ile-Tyr-Val-His-Pro-Phe Acetate Salt

Synonyms: Angiotensin II

ICN 152738   Khosia, M etal, *J Med Chem*, 24:885, 1981

Ala-Pro-Gly-Pro-Arg

Synonyms: Enterostatin

Bachem H-6405.0005   Human   MW 496.57 $C_{21}H_{36}N_8O_6$ Store at -15°C

Neosystem SC930   Human   MW 496.57   Bioactive; brain/gut peptide; Erlanson-Albertsson, C & Larsson, A, *Biochimie*, 70:1245-1250, 1988; Erlanson-Albertsson, C, *Nutr Rev*, 50:307-310, 1992; Erlanson-Albertsson, C, *Scand J Nutr*, 38:11-14, 1994; Bowyer, RC et al, *Clinica Chimica Acta*, 200:137-152, 1991; Bowyer, RC et al, *Gut*, 34:1520-1525, 1993

Ala-Pro-Gly-Trp Amide Acetate Salt Free Base

Sigma A 0813   FW 428.5 $C_{21}H_{28}N_6O_4$ ≥97% (HPLC) | Bioactive peptide; neurotransmitter peptide involved with snail reproduction; Minakata, H et al, *Comp Biochem Physiol*, 100C:565, 1991

Ala-Pro-Leu
Z-Ala-Pro-Leu

Bachem C-1195.0250   MW 433.51 $C_{22}H_{31}N_3O_6$ Store at -15°C

Ala-Pro-Leu

Bachem H-8280.0250   MW 299.37 $C_{14}H_{25}N_3O_4$ Store at -15°C

Ala-Pro-Leu
N-CBZ-Ala-Pro-Leu Dicyclohexylammonium salt

Sigma C 8033   FW 614.8 $C_{22}H_{31}N_3O_6 \cdot C_{12}H_{23}N$

Ala-Pro-Leu-Ala-Pro-Arg-Asp-Ala-Gly-Ser-Gln-Arg-Pro-Arg-Lys-Lys-Glu-Asp-Asn-Val-Leu-Val-Glu-Ser-His-Glu-Lys-Ser-Leu-Gly

Synonyms: Parathyroid Hormone (39-68)

American Peptide 22-1-35   Human   MW 3285.7 $C_{139}H_{234}N_{46}O_{46}$

Ala-Pro-Leu-Ala-Pro-Arg-Asp-Ala-Gly-Ser-Gln-Arg-Pro-Arg-Lys-Lys-Glu-Asp-Asn-Val-Leu-Val-Glu-Ser-His-Glu-Lys-Ser-Leu-Gly-Glu-Ala-Asp-Lys-Ala-Asp-Val-Asn-Val-Leu-Thr-Lys-Ala-Lys-Ser-Gln

Synonyms: Parathyroid Hormone (39-84)

American Peptide 22-1-40   Human   MW 4984.6 $C_{211}H_{357}N_{67}O_{72}$

Ala-Pro-Leu-Ala-Pro-Arg-Asp-Ala-Gly-Ser-Gln-Arg-Pro-Arg-Lys-Lys-Glu-Asp-Asn-Val-Leu-Val-Glu-Ser-His-Glu-Lys-Ser-Leu-Gly-Glu-Ala-Asp-Lys-Ala-Asp-Val-Asp-Val-Leu-Thr-Lys-Ala-Lys-Ser-Gln

Synonyms: Parathyroid Hormone (39-84), (Asp[76])-

Bachem H-4926.0500   Human   MW 4983.56 $C_{211}H_{356}N_{67}O_{72}$ Store at -15°C

Ala-Pro-Leu-Ala-Pro-Arg-Asp-Ala-Gly-Ser-Gln-Arg-Pro-Arg-Lys-Lys-Glu-Asp-Asn-Val-Leu-Val-Glu-Ser-His-Glu-Lys-Ser-Leu-Gly

Synonyms: Parathyroid Hormone (39-68); Parathormone

ICN 152985   Human   MW 3285.7

Ala-Pro-Leu-Ala-Pro-Arg-Asp-Ala-Gly-Ser-Gln-Arg-Pro-Arg-Lys-Lys-Glu-Asp-Asn-Val-Leu-Val-Glu-Ser-His-Glu-Lys-Ser-Leu-Gly-Glu-Ala-Asp-Lys-Ala-Asp-Val-Asn-Val-Leu-Thr-Lys-Ala-Lys-Ser-Gln

Synonyms: Parathyroid Hormone (39-84), (Asn[76])-; Parathormone

ICN 152986   Human   MW 4984.5

Ala-Pro-Leu-Ala-Pro-Arg-Asp-Ala-Gly-Ser-Gln-Arg-Pro-Arg-Lys-Lys-Glu-Asp-Asn-Val-Leu-Val-Glu-Ser-His-Glu-Lys-Ser-Leu-Gly

Synonyms: Parathyroid Hormone (39-68)

Neosystem SC397   Human   MW 3285.66   Bioactive; calcium metabolism peptide

Sigma P 4403   Human   FW 3285.7   ≥97% (HPLC) | Bioactive peptide

Ala-Pro-Leu-Ala-Pro-Arg-Asp-Ala-Gly-Ser-Gln-Arg-Pro-Arg-Lys-Lys-Glu-Asp-Asn-Val-Leu-Val-Glu-Ser-His-Glu-Lys-Ser-Leu-Gly-Glu-Ala-Asp-Lys-Ala-Asp-Val-Asn-Val-Leu-Thr-Lys-Ala-Lys-Ser-Gln

Synonyms: Parathyroid Hormone (39-84), (Asn[76])-

Sigma P 7046   Human   FW 4984.5   ≥97% (HPLC) | Bioactive peptide

Ala-Pro-Leu-Ala-Pro-Arg-Asp-Ala-Gly-Ser-Gln-Arg-Pro-Arg-Lys-Lys-Glu-Asp-Asn-Val-Leu-Val-Glu-Ser-His-Glu-Lys-Ser-Leu-Gly

Synonyms: Parathyroid Hormone (39-68)

Peptides International PTH-4124-v   Human synthetic   MW 3285.7 $C_{139}H_{234}N_{46}O_{46}$ >97% (HPLC); lyophilized amorphous powder, hydrochloride

Ala-Pro-Leu-Ala-Pro-Arg-Asp-Ala-Gly-Ser-Gln-Arg-Pro-Arg-Lys-Lys-Glu-Asp-Asn-Val-Leu-Val-Glu-Ser-His-Glu-Lys-Ser-Leu-Gly-Glu-Ala-Asp-Lys-Ala-Asp-Val-Asn-Val-Leu-Thr-Lys-Ala-Lys-Ser-Gln

Synonyms: Parathyroid Hormone (39-84)

Peptides International PTH-4169-v   Human synthetic   MW 4984.6 $C_{211}H_{357}N_{67}O_{72}$ >99% (HPLC); lyophilized amorphous powder

USBio P3109-11   Synthetic   ≥97%; lyophilized | Suitable for antigenic applications in immunological protocols

Ala-Pro-Leu-Glu-Pro-Glu-Tyr-Pro-Gly-Asp-Asn-Ala-Thr-Pro-Glu-Gln-Met-Ala-Gln-Tyr-Ala-Ala-Glu-Leu-Arg-Arg-Tyr-Ile-Asn-Met-Leu-Thr-Arg-Pro-Arg-Tyr Amide

Synonyms: Pancreatic Polypeptide, (Asn[11])-; Pancreatic Polypeptide; Gastrointestinal Peptide

Bachem H-6610.0500   Bovine   MW 4225.78 $C_{186}H_{287}N_{53}O_{56}S_2$ Store at -15°C

ICN 154557   Bovine   MW 4227.3   Kimmel, JR etal, *JBC*, 250:9369, 1975

Sigma P 9778   Bovine   FW 4225.8   ≥95% (HPLC) | Bioactive peptide; Kimmel, JR et al, *J Biol Chem*, 250:9369, 1975; Floyd, JC et al, *Rec Prog Horm Res*, 33:519, 1977

Ala-Pro-Leu-Glu-Pro-Glu-Tyr-Pro-Gly-Asp-Asp-Ala-Thr-Pro-Glu-Gln-Met-Ala-Gln-Tyr-Ala-Ala-Glu-Leu-Arg-Arg-Tyr-Ile-Asn-Met-Leu-Thr-Arg-Pro-Arg-Tyr Amide

Synonyms: Pancreatic Polypeptide

ICN 152889   Bovine   MW 4225.8   Kimmel, JR etal, *JBC*, 250:9369, 1975; Floyd, JC etal, *Rec Prog Horm Res*, 33:519, 1977

Ala-Pro-Leu-Glu-Pro-Met-Tyr-Pro-Gly-Asp-Asn-Ala-Thr-His-Glu-Gln-Arg-Ala-Gln-Tyr-Glu-Thr-Gln-Leu-Arg-Arg-Tyr-Ile-Asn-Thr-Leu-Thr-Arg-Pro-Arg-Tyr Amide

Synonyms: Pancreatic Polypeptide; Gastrointestinal Peptide

Sigma P 6410   Rat   FW 4398.9   ≥95% (HPLC) | Bioactive peptide

Ala-Pro-Leu-Glu-Pro-Met-Tyr-Pro-Gly-Asp-Tyr-Ala-Thr-His-Glu-Gln-Arg-Ala-Gln-Tyr-Glu-Thr-Gln-Leu-Arg-Arg-Tyr-Ile-Asn-Thr-Leu-Thr-Arg-Pro-Arg-Tyr Amide

Synonyms: Pancreatic Polypeptide

American Peptide 46-5-20   Rat   MW 4399.0 $C_{195}H_{298}N_{58}O_{57}S$

Bachem H-6890.0500   Rat   MW 4398.93 $C_{195}H_{298}N_{58}O_{57}S$ Store at -15°C

ICN 152891   Rat   MW 4398.4

Neosystem SC1207   Rat   MW 4398.91   Bioactive; brain/gut peptide; Kimmel, JR et al, *Endocrinology*, 114:1725-1731, 1984

Ala-Pro-Leu-Glu-Pro-Val-Tyr-Pro-Gly-Asp-Asn-Ala-Thr-Pro-Glu-Gln-Met-Ala-Gln-Tyr-Ala-Ala-Asp-Leu-Arg-Arg-Tyr-Ile-Asn-Met-Leu-Thr-Arg-Pro-Arg-Tyr Amide

Synonyms: Pancreatic Polypeptide

American Peptide 46-1-69   Human   MW 4181.7
$C_{185}H_{287}N_{53}O_{54}S_2$ Found in the pancreas, gastrointestinal tract & CNS; influences feeding behavior when injected centrally

Bachem H-1610.0500   Human   MW 4181.77
$C_{185}H_{287}N_{53}O_{54}S_2$ Store at -15°C

Ala-Pro-Leu-Glu-Pro-Val-Tyr-Pro-Gly-Asp-Asn-Ala-Thr-Pro-Glu-Gln-Met-Ala-Arg-Tyr-Tyr-Ser-Ala-Leu-Arg-His-Tyr-Ile-Asn-Leu-Ala-Aib-Arg-Gln-Arg-Tyr Amide

Synonyms: Pancreatic Polypeptide (1-17), (Ala^{31},Aib^{32})-; Neuropeptide Y (18-36)

Bachem H-5086.0500   Human   MW 4209.76
$C_{189}H_{287}N_{55}O_{53}S$ Store at -15°C

Ala-Pro-Leu-Glu-Pro-Val-Tyr-Pro-Gly-Asp-Asn-Ala-Thr-Pro-Glu-Gln-Met-Ala-Gln-Tyr-Ala-Ala-Asp-Leu-Arg-Arg-Tyr-Ile-Asn-Met-Leu-Thr-Arg-Pro-Arg-Tyr Amide

Synonyms: Pancreatic Polypeptide; Gastrointestinal Peptide

ICN 152890   Human   MW 4181.7   Floyd, JC etal, *Rec Prog Horm Res*, 33:519, 1977

Neosystem SC104   Human   MW 4181.74   Bioactive; brain/gut peptide; Kimmel et al, *JBC*, 250:9369-9373, 1975

Sigma P 9903   Human   FW 4181.7   ≥97% (HPLC) | Bioactive peptide; Floyd, JC et al, *Rec Prog Horm Res*, 33:519, 1977

Ala-Pro-Leu-Phe
Suc-Ala-Pro-Leu-Phe-pNA

Bachem L-1975.0050   MW 666.73 $C_{33}H_{42}N_6O_9$ Store at -15°C

Ala-Pro-Leu-Ser-Gly-Phe-Tyr-Gly-Val-Arg Amide

Synonyms: Locustachykinin II

American Peptide 87-7-65   Locust   MW 1065.4
$C_{50}H_{76}N_{14}O_{12}$ Schoofs, L et al, *FEBS Lett*, 266:397, 1990

Ala-Pro-Nva
BOC-Ala-Pro-Nva-4-Chloro-SBzl

Bachem M-1170.0050   MW 526.1 $C_{25}H_{36}ClN_3O_5S$ Store at -15°C

Ala-Pro-Phe

Bachem H-1615.0050   MW 333.39 $C_{17}H_{23}N_3O_4$ Store at -15°C

Ala-Pro-Phe
Z-Ala-Pro-Phe-CMK

Bachem N-1220.0025   Store at -15°C

Ala-Pro-Phe
BOC-Ala-Pro-Phe-NHO-Bz

Synonyms: Subtilisin Inhibitor V; Elastase Inhibitor; Cysteine Protease Inhibitor; Serine Protease Inhibitor

ICN 195916   MW 552.6 $C_{29}H_{36}N_4O_7$ >95% | Irreversible inhibitor

Ala-Pro-Phe
D-Ala-Pro-Phe-pNA·AcOH

Synonyms: Pefachrome®CHY; Pefa-5980

Pentapharm 090-16, 090-15   MW 513.5   Highly sensitive chromogenic peptide substrate for chymotrypsin; v_{max}:2.5 µmol/min, K_M:0.424 mM

Ala-Pro-Phe
D-Ala-Pro-Phe

Sigma A 2166   FW 333.4 $C_{17}H_{23}N_3O_4$

Ala-Pro-Phe
N-CBZ-Ala-Pro-Phe Dicyclohexylammonium salt

Sigma C 8158   FW 467.5 $C_{25}H_{29}N_3O_6$

Ala-Pro-Phe-Arg-Ser-Ala-Leu-Glu-Ser-Ser-Pro-Ala-Asp-Pro-Ala-Thr-Leu-Ser-Glu-Asp-Glu-Ala-Arg-Leu-Leu-Leu-Ala-Ala-Leu-Val-Gln-Asp-Tyr-Val-Gln-Met-Lys-Ala-Ser-Glu-Leu-Glu-Gln-Glu-Gln-Glu-Arg-Glu-Gly-Ser-Ser-Leu-Asp-Ser-Pro-Arg-Ser

Synonyms: Procalcitonin, N-proCT Amino-Terminal; Cleavage Peptide, Human; Calcitonin N-Terminal Flanking Peptide; Procalcitonin, N-; Calcitonin (1-57), (N-Pro)-

American Peptide 22-2-30   MW 6220.8 $C_{264}H_{426}N_{74}O_{97}S$
Stimulates proliferation of normal & neoplastic human osteoblasts at nanomolar concentrations; Burns, DM et al, *PNAS*, 86:9519, 1989

Bachem H-3076.0500   Human   MW 6220.79
$C_{264}H_{426}N_{74}O_{97}S$ Store at -15°C

Sigma P 8424   Human   FW 6220.8   ≥95% (HPLC) | Bioactive peptide; potent bone-cell mitogen; Burns, DM et al, *Proc Natl Acad Sci USA*, 86:9519, 1989

Ala-Pro-Pro-Arg-Leu-Ile-Asn-Asp-Ser-Arg-Val-Leu-Glu-Arg-Tyr-Leu-Leu-Glu-Ala-Lys-Glu-Ala-Glu-Lys-Ile-Thr

Synonyms: Erythropoietin (1-26), (Asn^7,Lys^{24})-

Sigma E 8138   Human   FW 3025.5   ≥97% (HPLC) | Bioactive peptide; immunogen for production of antiserum to erythropoietin; formerly listed as Erythropoietin Fragment 1-26; Jacobs, K et al, *Nature*, 313:806, 1985; Sue, JM & Sytkowski, AJ, *Proc Natl Acad Sci USA*, 80:3651, 1983; Lin, F-K et al, *Proc Natl Acad Sci USA*, 82:7580, 1985

Ala-Pro-Pro-Arg-Leu-Ile-Cys-Asp-Ser-Arg-Val-Leu-Glu-Arg-Tyr-Leu-Leu-Glu-Ala-Lys-Glu-Ala-Glu-Asn-Ile-Thr

Synonyms: Erythropoietin (1-26)

Sigma E 8013   Human   FW 3000.5   ≥97% (HPLC) | Bioactive peptide; Jacobs, K et al, *Nature*, 313:806, 1985; Lin, F-K et al, *Proc Natl Acad Sci USA*, 82:7580, 1985; Lai, P-H et al, *J Biol Chem*, 261:3116, 1986

Ala-Pro-Ser-Asp-Pro-Arg-Leu-Arg-Gln-Phe-Leu-Gln-Lys-Ser-Leu-Ala-Ala-Ala-Ala-Gly-Lys-Gln-Glu-Leu-Ala-Lys-Tyr-Phe-Leu-Ala-Glu-Leu

Synonyms: Pro-Somatostatin (1-32); Prepro-Somatostatin, N-Terminal

American Peptide 68-1-68   Porcine   MW 3532.1
$C_{161}H_{260}N_{44}O_{45}$ Schmidt, WE et al, *FEBS Lett*, 192:141, 1985

Ala-Pro-Ser-Glu-Pro-His-His-Pro-Gly-Asp-Gln-Ala-Thr-Gln-Asp-Gln-Leu-Ala-Gln-Tyr-Tyr-Ser-Asp-Leu-Tyr-Gln-Tyr-Ile-Thr-Phe-Val-Thr-Arg-Pro-Arg-Phe Amide

Synonyms: Pancreatic Polypeptide

Bachem H-4918.0500   Rana temporaria   MW 4240.62
$C_{192}H_{276}N_{52}O_{58}$ Store at -15°C

Ala-Pro-Ser-Gly-Ala-Gln-Arg-Leu-Tyr-Gly-Phe-Gly-Leu Amide

Synonyms: Allatostatin I, Type A; Allatostatin I

American Peptide 87-7-10   MW 1335.5 $C_{61}H_{94}N_{18}O_{16}$
Woodhead, AP et al, *PNAS*, 86:5997, 1989; Pratt, GE et al, *BBRC*, 163:1243, 1989

Bachem H-8065.0001   MW 1335.53 $C_{61}H_{94}N_{18}O_{16}$ Store at -15°C

Sigma A 9929   FW 1335.5   ≥95% (HPLC) | Bioactive peptide; inhibits juvenile hormone synthesis in insects; sequences are highly conserved within the six C-terminal AA; Woodhead, AP et al, *Proc Natl Acad Sci USA*, 86:5997, 1989

Ala-Pro-Ser-Gly-His-Tyr-Lys-Gly

Bachem H-1948.0001   MW 815.88 $C_{36}H_{53}N_{11}O_{11}$ Store at -15°C

Ala-Pro-Thr-Lys-Ala-Lys-Arg-Arg-Val-Val-Gln-Arg-Glu-Lys-Arg

Synonyms: GP160 (502-516); HTLV I Peptide

Neosystem SC801   MW 1823.20   HTLV I peptide from subtype GP160

Ala-Pro-Trp-Leu-Tyr-Gly-Pro-Ala

Synonyms: Carbohydrate Structure Mimicking Peptide

Bachem H-2076.0005   MW 874.01 $C_{44}H_{59}N_9O_{10}$ Store at -15°C

Ala-Pro-Tyr
Z-Ala-Pro-Tyr

Bachem C-1210.0250   MW 483.52 $C_{25}H_{29}N_3O_7$ Store at -15°C

Ala-Pro-Tyr-Ala

Bachem H-1625.0050   MW 420.47 $C_{20}H_{28}N_4O_6$ Store at -15°C

Ala-Pro-Val
Ala-Pro-Val-EDANS

Synonyms: Ala-Pro-Val-5-((2-Aminoethyl)Amino)-Naphthalene-1-Sulfonic Acid

Bachem M-2030.0001   MW 533.65 $C_{25}H_{35}N_5O_6S$ Store at -15°C

Ala-Pro-Val
N-t-BOC-Ala-Pro-Nva-D-Chlorothiobenzyl Ester

Synonyms: Elastase Substrate

Sigma B 2017   FW 526.1 $C_{25}H_{36}ClN_3O_5S$ Leucocyte elastase substrate; Harper, JW et al, *Biochemistry*, 23:2995, 1984

Ala-Pro-Val-Ala-Asn-Glu-Leu-Arg-Cys-Gln-Cys-Leu-Gln-Thr-Val-Ala-Gly-Ile-His-Phe-Lys-Asn-Ile-Gln-Ser-Leu-Lys-Val-Met-Pro-Pro-Gly-Pro-His-Cys-Thr-Gln-Thr-Glu-Val-Ile-Ala-Thr-Leu-Lys-Asn-Gly-Arg-Glu-Ala-Cys-Leu-Asp-Pro-Glu-Ala-Pro-Met-Val-Gln-Lys-Ile-Val-Gln-Lys-Met-Leu-Lys-Gly-Val-Pro-Lys

Synonyms: Cytokine-Induced Neutrophil Chemoattractant-1/Growth Related Oncogene

Peptides International PIL-4233-v   Rat synthetic   MW 7845.4 $C_{343}H_{572}N_{98}O_{97}S_7$ >95% (HPLC); lyophilized amorphous powder | Bioactive; Disulfide bonds: Cys[9]-Cys[35], Cys[11]-Cys[51]; Watanabe, K et al, *JBC*, 264:19559, 1989; Nishiuchi, Y et al, in JA Smith & JE Rivier (Ed) *Peptides:Chemistry & Biology* (Proc 12th American Peptide Symp), Escom, Lieden, 1992, pp 911-913; Nakagawa, H et al, *Biochem J*, 301:545, 1994

Ala-Pro-Val-His-Arg-Gly-Arg-Gly-Gly-Trp-Thr-Leu-Asn-Ser-Ala-Gly-Tyr-Leu-Leu-Gly-Pro-Val-Leu-His-Pro-Pro-Ser-Arg-Ala-Glu-Gly-Gly-Gly-Lys-Gly-Lys-Thr-Ala-Leu-Gly-Ile-Leu-Asp-Leu-Trp-Lys-Ala-Ile-Asp-Gly-Leu-Pro-Tyr-Pro-Gln-Ser-Gln-Leu-Ala-Ser

Synonyms: Galanin-Like Peptide (1-60)

Bachem H-4982.0500   Porcine   MW 6204.11 $C_{281}H_{443}N_{81}O_{78}$ Store at -15°C

Ala-Pro-Val-Ser-Val-Gly-Gly-Gly-Thr-Val-Leu-Ala-Lys-Met-Tyr-Pro

Synonyms: Gastrin Releasing Peptide (1-16)

American Peptide 46-4-34   Porcine   MW 1546.9 $C_{70}H_{115}N_{17}O_{20}S$

Bachem H-5930.0001   Porcine   MW 1546.85 $C_{70}H_{115}N_{17}O_{20}S$ Store at -15°C

ICN 154551   Porcine   MW 1547.1

Neosystem SC415   Porcine   MW 1546.84   Bioactive; brain/gut peptide; CCK/gastrin peptide

Ala-Pro-Val-Ser-Val-Gly-Gly-Gly-Thr-Val-Leu-Ala-Lys-Met-Tyr-Pro-Arg-Gly-Asn-His-Trp-Ala-Val-Gly-His-Leu-Met Amide

Synonyms: Gastrin Releasing Peptide; Gastrointestinal Peptide

American Peptide 46-4-40   Porcine   MW 2805.4 $C_{126}H_{198}N_{38}O_{31}S_2$ McDonald, TJ et al, *Biochem Biophys Res Comm*, 90:227, 1979

Bachem H-1635.0500   Porcine   MW 2805.33 $C_{126}H_{198}N_{38}O_{31}S_2$ Store at -15°C

ICN 152878   Porcine   McDonald, TJ etal, *Gut*, 19:767, 1978; McDonald, TJ etal, *BBRC*, 90:227, 1979; Yajima, H etal, *Chem Pharm Bull*, 28:2276, 1980; Yajima, H etal, *Peptides, Synthesis - Structure - Function*, Proc Seventh American Peptide Symp, Rich, DH & E Gross, eds, p 19, 1981

Sigma G 1649   Porcine   FW 2805.3   ≥97% (HPLC) | Bioactive peptide; mammalian equivalent of bombesin; McDonald, TJ et al, *Biochem Biophys Res Commun*, 90:227, 1979; McDonald, TJ et al, *Gut*, 19:767, 1978; Yajima, H et al, *Peptides. Synthesis-Structure-Function*. Proceedings of the Seventh American Peptide Symposium, Rich, DH & Gross, E, eds, 19, 1981; Yajima, H et al, *Chem Pharm Bull*, 28:2276, 1980

Ala-Pro-Val-Ser-Val-Gly-Gly-Gly-Thr-Val-Leu-Ala-Lys-Met-TyrPro-Arg-Gly-Asn-His-Trp-Ala-Val-Gly-His-Leu-Met Amide

Synonyms: Gastrin Releasing Peptide

Neosystem SC420   Porcine   MW 2805.31   Bioactive; brain/gut peptide; CCK/gastrin peptide; McDonald, TJ et al, *BBRC*, 90:227-233, 1979

Ala-Ser
Z-Ala-Ser

Bachem C-1215.0001   MW 310.31 $C_{14}H_{18}N_2O_6$ Store at -15°C

Ala-Ser
Z-Ala-Ser-OMe

Bachem C-1220.0001   MW 324.33 $C_{15}H_{20}N_2O_6$ Store at -15°C

Ala-Ser

Bachem G-1380.0250   MW 176.17 $C_6H_{12}N_2O_4$ Store at -15°C

Sigma A 3503   FW 176.2 $C_6H_{12}N_2O_4$

Ala-Ser
N-CBZ-Ala-Ser-OMe

Sigma C 4126   FW 324.3 $C_{15}H_{20}N_2O_6$

Ala-Ser-Ala-Ser-Ser-Leu-Met-Asp-Lys-Glu-Ala-Val-Tyr-Phe-Ala-His-Leu-Asp-Ile-Ile-Trp

Synonyms: Endothelin B Receptor Agonist; Endothelin I, (Ala[1,3,11,15])-; Endothelin I, 4 Ala

Alexis 155-010   MW 2367.7 $C_{109}H_{163}N_{25}O_{32}S$ Highly selective ligand for the ET_B receptor; Saeki, T et al, *BBRC*, 179:286, 1991; Williams Jr, DL et al, *BBRC*, 180:475, 1991; Nakamichi, K et al, *BBRC*, 182:144, 1992; Green, J, *Int J Pept Prot Res*, 41:492, 1993

Bachem H-3066.0500   MW 2367.71 $C_{109}H_{163}N_{25}O_{32}S$ Store at -15°C

Neosystem SC327   MW 2367.69   Bioactive; ETB receptor agonist; Saeki, T et al, *BBRC*, 179:286-292, 1991; Nakamichi, K et al, *BBRC*, 182:144-150, 1992

Sigma E 6877   FW 2367.7   ≥97% (HPLC); Disulfide bonds: 1-15, 3-11 | Bioactive peptide; selective agonist for ET-B receptors; Nakamichi, K et al, *Biochem Biophys Res Commun*, 182:144, 1992; Bigand, M & Pelton, JT, *J Pharmacol*, 107:912, 1992

American Peptide 88-1-11   Human   MW 2367.7 $C_{109}H_{163}N_{25}O_{32}S$ Very selective ET_B endothelin receptor agonist; Williams, DL et al, *BBRC*, 180:475, 1991; Nakamichi, K et al, *BBRC*, 182:144, 1992; Green, J, *Int J Pept Prot Res*, 41:492, 1993

Ala-Ser-Arg-Leu-Asp-Leu-Arg-Ile-Gly-Arg-Ile-Val-Thr-Ala-Lys-Tyr

Synonyms: Endothelial-Monocyte Activating Polypeptide II Derived Peptide

Bachem H-2278.0001   MW 1832.18 $C_{81}H_{142}N_{26}O_{22}$ Store at -15°C

Ala-Ser-Asn-Glu-Asn-Met-Glu-Thr-Met

Synonyms: Influenza Virus Nucleoprotein Epitope; NP (366-374)

Neosystem SC1307   MW 1026.10   Virus-related peptide; Rötzschke, O et al, *Nature*, 348:252-254, 1990

Ala-Ser-Gln-Arg-Ser-Phe-Trp-Ala-Glu-Leu-Asn-Ile-Ala-Arg-Leu-Arg-His-Asp-Asn-Ile-Val-Arg-Val-Val-Ala-Ala-Ser-Thr-Arg-Cys

Synonyms: cMos Blocking Peptide (149-177)

Calbiochem 475933   Mouse synthetic   MW 3441 $C_{148}H_{243}N_{51}O_{42}S$ Lyophilized | Based on the murine Mos (149-177) with a Cys C residue added; this peptide coupled to KLH was used as the immunogen for the production of Anti-cMos; for use in immunoabsorption for immunoprecipitation, immunocytochemistry, Western blotting & dot blots

---

Ala-Ser-Gln-Phe-Glu-Thr-Ser

Synonyms: Synaptobrevin II (74-80)

Bachem N-1345.0005   Rat   Store at -15°C

Ala-Ser-Glu-Gln-Gly-Tyr-Glu-Glu-Met-Arg-Ala-Phe-Gln-Gly

Synonyms: c-erbB-3 (1265-1278)

Sigma E 7647   FW 1602.7   ≥97% (HPLC) | Bioactive peptide; Prigent, SA & Gullik, WJ, *Emb J*, 13:2831, 1994

Ala-Ser-Gly-Ala-Asp-Thr-Ser-Gly-Val-Leu-Asp-Pro-Asp-Ser Acetate Salt

Synonyms: Proteoglycan I (4-17); Biglycan; PG-S1

Biogenesis 7870-6160   Human synthetic   MW 1291.9   >80% (HPLC); lyophilized | The N-terminal human sequence is also Bone Proteoglycan I Precursor (41-54)

Ala-Ser-His-Leu-Gly-Leu-Ala-Arg

Synonyms: C3a (70-77)

Bachem H-1645.0005   MW 823.95 $C_{35}H_{61}N_{13}O_{10}$ Store at -15°C

Ala-Ser-His-Leu-Gly-Leu-Ala-Arg
α-Ala-Ser-His-Leu-Gly-Leu-Ala-Arg

Synonyms: C3a (70-77), (α-Ala[70])-

Bachem H-1650.0005   MW 823.95 $C_{35}H_{61}N_{13}O_{10}$ Store at -15°C

Ala-Ser-His-Leu-Gly-Leu-Ala-Arg

Synonyms: Anaphylatoxin C3a (70-77)

ICN 153043   MW 823.9 $C_{35}H_{61}N_{13}O_{10}$ Hugli, TE & BW Erickson, *PNAS*, 74:1826, 1977

Ala-Ser-His-Leu-Gly-Leu-Ala-Arg
β-Ala-Ser-His-Leu-Gly-Leu-Ala-Arg

Synonyms: Anaphylatoxin C3a (70-77), (β-Ala[1])-

ICN 153044   MW 823.9 $C_{35}H_{61}N_{13}O_{10}$

Ala-Ser-His-Leu-Gly-Leu-Ala-Arg

Synonyms: Anaphylatoxin C3a (70-77)

Sigma A 8651   FW 823.9 $C_{35}H_{61}N_{13}O_{10}$ ≥97% (HPLC) | Bioactive peptide; Hugli, TE & Erickson, BW, *Proc Natl Acad Sci USA*, 74:1826, 1977

Ala-Ser-Thr-Asp
Ac-Ala-Ser-Thr-Asp-AMC

Bachem I-1785.0005   MW 591.58 $C_{26}H_{33}N_5O_{11}$ Store at -15°C

Ala-Ser-Thr-Asp
Z-Ala-Ser-Thr-DL-Asp-FMK

Bachem N-1670.0001   MW 542.52 $C_{23}H_{31}FN_4O_{10}$ Store at -15°C

Ala-Ser-Thr-Asp
Biotinyl-Ala-Ser-Thr-DL-Asp-FMK

Bachem N-1675.0001   MW 634.68 $C_{25}H_{39}FN_6O_{10}S$ Store at -15°C

Ala-Ser-Thr-Thr-Asp-Tyr-Thr

Synonyms: Peptide T; HIV Inhibitor

ICN 191453

Ala-Ser-Thr-Thr-Thr-Asn-Tyr-Thr

Synonyms: Peptide T; HIV Inhibitor

American Peptide 72-2-16   MW 857.9 $C_{35}H_{55}N_9O_{16}$ Lyophilized | Derived from the envelope glycoprotein (gp 120) of human immunodeficiency virus (HIV); named peptide T because of its high Thr content; inhibits HIV *in vitro* & blocks binding of the viral envelope to the CD-4 receptor; Pert, CB et al, *PNAS*, 83:9254, 1986

Ala-Ser-Thr-Thr-Thr-Asn-Tyr-Thr
D-Ala-Ser-Thr-Thr-Thr-Asn-Tyr-Thr Amide

Synonyms: Peptide T, (D-Ala[1])-; HIV Inhibitor

American Peptide 72-2-18   MW 856.9 $C_{35}H_{56}N_{10}O_{15}$ This peptide T analog blocks HIV infection of human T cells; Pert, CB et al, *PNAS*, 83:9254, 1986

Ala-Ser-Thr-Thr-Thr-Asn-Tyr-Thr
Ala-Ser-Thr-Thr-Thr-Asn-(3,5-Diiodo)Tyr-Thr

Synonyms: Peptide T, (3,5-Diiodo-Tyr[7])-; HIV Inhibitor

American Peptide 72-2-30   MW 1110.7 $C_{35}H_{54}N_9O_{16}I_2$

Ala-Ser-Thr-Thr-Thr-Asn-Tyr-Thr
D-Ala-Ser-Thr-Thr-Thr-Asn Amide

Synonyms: Peptide T, Tyr-(3,5-^{3}H)-Thr

ARC ART-704   MW 856.84   30-60 Ci/mmol; 1.85-2.22 TBq/mmol; in EtOH | Radiochemical; for AIDS research; Pert, CB et al, *PNAS*, 83:9254, 1986

Ala-Ser-Thr-Thr-Thr-Asn-Tyr-Thr

Synonyms: Peptide T; HIV Inhibitor

ICN 191543   MW 857.9 $C_{35}H_{55}N_9O_{16}$ Pert, CB etal, *PNAS*, 83:9254, 1986

Neosystem SC125   MW 857.83   Virus-related peptide; AIDS-related peptide; Pert, CB et al, *PNAS* USA, 83:9254, 1986

Ala-Ser-Thr-Thr-Thr-Asn-Tyr-Thr
Ala-Ser-Thr-Thr-Thr-Asn-3,5-Diiodo-Tyr-Thr

Synonyms: Peptide T, (3,5-Diiodo-Tyr[7])-

Sigma A 0800   FW 1109.7   ≥97% (HPLC) | Bioactive peptide

Ala-Ser-Thr-Thr-Thr-Asn-Tyr-Thr

Synonyms: Peptide T

Peptides International PPT-4188-v   Synthetic   MW 857.87 $C_{35}H_{55}N_9O_{16}$ >99% (HPLC); lyophilized amorphous powder | Bioactive; Ruff, MR et al, *FEBS Letts*, 211:17, 1987; Pert, CB et al, *PNAS* USA, 83:9254, 1986

Ala-Ser-Thr-Thr-Thr-Asn-Tyr-Thr 4H$_2$O

Synonyms: Peptide T

Peptides International PPT-4188   Synthetic   MW 929.93 $C_{35}H_{55}N_9O_{16} \cdot 4H_2O$ >99% (HPLC); lyophilized amorphous powder; bulk | Bioactive; Ruff, MR et al, *FEBS Letts*, 211:17, 1987; Pert, CB et al, *PNAS* USA, 83:9254, 1986

Ala-Ser-Thr-Thr-Thr-Asn-Tyr-Thr Acetate Salt Free Base

Synonyms: Peptide T

Sigma A 2297   FW 857.9 $C_{35}H_{55}N_9O_{16}$ ≥97% (HPLC) | Bioactive peptide; pert, CB et al, *Proc Natl Acad Sci USA*, 83:9254, 1986

Ala-Ser-Thr-Thr-Thr-Asn-Tyr-Thr Amide Acetate Salt Free Base

Synonyms: Peptide T, (D-Ala[1]-NH$_2$)-

Sigma A 2422   FW 856.9 $C_{35}H_{56}N_{10}O_{15}$ ≥97% (HPLC) | Bioactive peptide; Pert, CB et al, *Proc Natl Acad Sci USA*, 83:9254, 1986

Ala-Thr

Bachem G-1390.0250   MW 190.2 $C_7H_{14}N_2O_4$ Store at -15°C

Sigma A 3628   FW 190.2 $C_7H_{14}N_2O_4$

Ala-Thr-Cys-Asp-Leu-Leu-Ser-Gly-Thr-Gly-Ile-Asn-His-Ser-Ala-Cys-Ala-Ala-His-Cys-Leu-Leu-Arg-Gly-Asn-Arg-Gly-Gly-Tyr-Cys-Asn-Gly-Lys-Ala-Val-Cys-Val-Cys-Arg-Asn

Synonyms: Sapecin

Bachem H-2246.0500   Store at -15°C | Disulfide bonds: Cys[3]-Cys[30], Cys[16]-Cys[36], Cys[20]-Cys[38]

Ala-Thr-Gln-Arg-Leu-Ala-Asn-Phe-Leu-Val-Arg-Ser-Ser-Asn-Asn-Leu-Gly-Pro-Val-Leu-Pro-Pro-Thr-Asn-Val-Gly-Ser-Asn-Thr-Tyr Amide

Synonyms: Amylin (8-37); Diabetes Associated Peptide (8-37)

Bachem H-2746.0500   Rat   MW 3200.61 $C_{140}H_{227}N_{43}O_{43}$ Store at -15°C

ICN 159865   Rat   MW 3200.7   >98%

Sigma D 6170   Rat   FW 3200.6   ≥95% (HPLC) | Bioactive peptide; Deems, RO et al, *Biochem Biophys Res Commun*, 181:116, 1991

Ala-Thr-Gln-Arg-Leu-Ala-Asn-Phe-Leu-Val-His-Ser-Ser-Asn-Asn-Phe-Gly-Ala-Ile-Leu-Ser-Ser-Thr-Asn-Val-Gly-Ser-Asn-Thr-Tyr Amide

Synonyms: Amylin (8-37)

Bachem H-2742.0500   Human   MW 3183.49 $C_{138}H_{216}N_{42}O_{45}$ Store at -15°C

Sigma A 4563   Human   FW 3183.5   ≥90% (HPLC) | Bioactive peptide

Ala-Thr-Leu-Asp-Ala-Leu-Leu-Ala-Ala-Leu-Arg-Arg-Ile-Gln
Ac-Ala-Thr-Leu-Asp-Ala-Leu-Leu-Ala-Ala-Leu-Arg-Arg-Ile-Gln Amide

Synonyms: Neurotrophin Receptor (368-381), Ac-; Neurotrophin Receptor (365-378), Ac-

Bachem H-4182.0001   Human, rat   MW 1565.88 $C_{69}H_{124}N_{22}O_{19}$ Store at -15°C

Ala-Thr-Pro-Glu-Thr-Phe-Thr-Glu-Asp-Pro-Asn-Leu-Val-Asn-Cys Trifluoroacetate Salt

Synonyms: Microfibril Associated Glycoprotein II (19-32), ~C

Biogenesis 6199-2050   Synthetic   MW 1649.6   >80% (HPLC); lyophilized | An additional Cys has been conjugated to the C terminus

Ala-Trp
Z-Ala-Trp

Bachem C-1225.0001   MW 409.44 $C_{22}H_{23}N_3O_5$ Store at -15°C

Ala-Trp

Bachem G-1395.0250   MW 275.31 $C_{14}H_{17}N_3O_3$ Store at -15°C

Ala-Trp
α-Ala-Trp

Bachem G-1400.0250   MW 275.31 $C_{14}H_{17}N_3O_3$ Store at -15°C

Ala-Trp
L-Ala-L-Trp

Fluka 05400   MW 275.31 $C_{14}H_{17}N_3O_3$ ~99% (titration); mp:293-294°C

Ala-Trp

Sigma A 3878   FW 275.3 $C_{14}H_{17}N_3O_3$

Ala-Trp-Ala

Bachem H-5425.0100   MW 346.39 $C_{17}H_{22}N_4O_4$ Store at -15°C

Ala-Trp-Met-Asp
N-t-BOC-β-Ala-Trp-Met-Asp(Benzyl)-Phe Amide

Synonyms: Pentagastrin (Blocked); Gastrointestinal Peptide

Sigma B 9758　　FW 858.0　　$C_{44}H_{55}N_7O_9S$　≥97% (HPLC) |
Bioactive peptide; β-carboxyl of the aspartyl residue is protected by a benzyl group

Ala-Trp-Met-Asp-Phe
BOC-β-Ala-Trp-Met-Asp-Phe Amide

Synonyms: Pentagastrin

Bachem A-1130.0025　　MW 767.9　　$C_{37}H_{49}N_7O_9S$　　Store at -15°C

Ala-Trp-Met-Asp-Phe
des-BOC-β-Ala-Trp-Met-Asp-Phe Amide Acetate Salt

Synonyms: Pentagastrin

ICN 195671　　MW 727.8　　$C_{32}H_{41}N_7O_7S \cdot C_2H_4O_2$　≥90%

Ala-Trp-Met-Asp-Phe
N-t-BOC-β-Ala-Trp-Met-Asp-Phe Amide

Synonyms: Pentagastrin

ICN 195673　　MW 767.9　　$C_{37}H_{49}N_7O_9S$　≥95%

Ala-Trp-Met-Asp-Phe
BOC-β-Ala-Trp-Met-Asp-Phe Amide

Synonyms: Pentagastrin

Neosystem SC155　　MW 767.90　　Bioactive; brain/gut peptide; CCK/gastrin peptide

Ala-Trp-Met-Asp-Phe
β-Ala-Trp-Met-Asp-Phe Amide Acetate Salt Free Base

Synonyms: Pentagastrin, (des-BOC)-; Gastrointestinal Peptide

Sigma A 4032　　FW 667.8　　$C_{32}H_{41}N_7O_7S$　≥90% (HPLC) |
Bioactive peptide

Ala-Trp-Met-Asp-Phe
N-t-BOC-β-Ala-Trp-Met-Asp-Phe Amide

Synonyms: Pentagastrin; Gastrointestinal Peptide; Anxiogenic Polypeptide; Cholecystokinin Antagonist

Sigma B 1636　　FW 767.9　　$C_{37}H_{49}N_7O_9S$　≥95% (HPLC) |
Bioactive peptide; stimulates gastric acid secretion

Ala-Trp-Nle
N-cis-2,6-Dimethylpiperidinocarbonyl-β-tBu-Ala-D-Trp(1-Methoxycarbonyl)-D-Nle

Synonyms: BQ-788

Bachem H-2492.0001　　MW 641.81　　$C_{34}H_{51}N_5O_7$　　Store at -15°C

Ala-Trp-Phe-Lys
D-Ala-D-2-Nal-Trp-D-Phe-Lys Amide

Bachem H-2528.0005　　Store at -15°C

Ala-Trp-Pro-Phe
Suc-Ala-Trp-Pro-Phe-pNA

Bachem L-1670.0050　　Store at -15°C

Ala-Trp-Val-Lys-Val-Val-Glu-Glu-Lys-Ala-Phe-Ser-Pro-Glu-Val-Ile-Pro-Met-Phe

Synonyms: P25 (154-172); GAG P24 CA (22-40); HTLV I Peptide

Neosystem SC293　　MW 2206.62　　HTLV I peptide from subtype P25 (GAG P24 CA)

Ala-Tyr
Z-Ala-Tyr

Bachem C-1235.0001　　MW 386.41　　$C_{20}H_{22}N_2O_6$　　Store at -15°C

Ala-Tyr
Z-Ala-Tyr-OMe

Bachem C-3850.0001　　MW 400.43　　$C_{21}H_{24}N_2O_6$　　Store at -15°C

Ala-Tyr

Bachem G-1405.0250　　MW 252.27　　$C_{12}H_{16}N_2O_4$　　Store at -15°C

Ala-Tyr
Ala-Tyr-OEt Hydrochloride

Bachem G-3670.0001　　MW 316.78　　$C_{14}H_{20}N_2O_4 \cdot$ HCl　　Store at -15°C

Ala-Tyr

ICN 100360　　MW 252.3　　$C_{12}H_{16}N_2O_4$　　Crystalline
Sigma A 4003　　FW 252.3　　$C_{12}H_{16}N_2O_4$

Ala-Tyr Amide Hydrochloride

Bachem G-1410.0250　　MW 287.75　　$C_{12}H_{17}N_3O_3 \cdot$ HCl　　Store at -15°C

Ala-Tyr-Ala

Bachem G-1415.0250　　MW 323.35　　$C_{15}H_{21}N_3O_5$　　Store at -15°C

Ala-Tyr-Cys-Arg-Asp-Gly-Lys-Ile-Gly-Pro-Pro-Lys-Leu-Asp-Ile-Arg-Lys-Glu-Glu-Lys-Gln-Ile
Ala-Tyr-Cys(Acm)-Arg-Asp-Gly-Lys-Ile-Gly-Pro-Pro-Lys-Leu-Asp-Ile-Arg-Lys-Glu-Glu-Lys-Gln-Ile

Synonyms: Interferon-γ Antagonist; Interferon-γ Receptor (120-141), (Tyr[121],Cys(Acm)[122])-

Bachem H-2722.0500　　Human　　Store at -15°C

Ala-Tyr-Leu-Val
Suc-Ala-Tyr-Leu-Val-pNA

Bachem L-1530.0050　　Store at -15°C

Ala-Tyr-Ser-Tyr-Val-Ser-Glu-Tyr-Lys-Arg-Leu-Pro-Val-Tyr-Asn-Phe-Gly-Leu Amide

Synonyms: Allatostatin, Type B; Allatostatin B_2

Bachem H-1428.0001　　MW 2168.48　　$C_{104}H_{150}N_{24}O_{27}$　　Store at -15°C

Sigma A 6575　　FW 2168.5　　Cockroach　≥97% (HPLC) |
Bioactive peptide; *Proc Natl Acad Sci USA*, 88:2412, 1991

Ala-Tyr-Trp-Lys-Abu-Phe
(N-Me)Ala-Tyr-D-Trp-Lys-Abu-Phe

Synonyms: BIM-23027

American Peptide 68-1-69　　MW 778　　$C_{43}H_{53}N_8O_6$　　The sst2 receptor-preferring analog; elevates Ca^{2+} with a pEC_{50} of 8.63 in the presence of 1 μM carbachol; Connor, M, *Br J Pharmacol*, 120:455, 1997

Ala-Tyr-Val-Ala-Asp
Ethoxycarbonyl-Ala-Tyr-Val-Ala-Asp-CHO Pseudo acid

Bachem N-1545.0005　　MW 593.63　　$C_{27}H_{39}N_5O_{10}$　　Store at -15°C

Ala-Val
Z-Ala-Val

Bachem C-1245.0001 MW 322.36 $C_{16}H_{22}N_2O_5$ Store at -15°C

Ala-Val
Z-β-Ala-Val

Bachem C-1250.0001 MW 322.36 $C_{16}H_{22}N_2O_5$ Store at -15°C

Ala-Val

Bachem G-1420.0001 MW 188.23 $C_8H_{16}N_2O_3$ Store at -15°C

Ala-Val
DL-Ala-DL-Val

Bachem G-1425.0005 MW 188.23 $C_8H_{16}N_2O_3$ Store at -15°C

Ala-Val
α-Ala-Val

Bachem G-1430.0001 MW 188.23 $C_8H_{16}N_2O_3$ Store at -15°C

Ala-Val

ICN 100328 MW 188.2 $C_8H_{16}N_2O_3$ 99%

Ala-Val
DL-Ala-DL-Val

ICN 100359 MW 188.2 $C_8H_{16}N_2O_3$ Crystalline

Sigma A 0634 FW 188.2 $C_8H_{16}N_2O_3$

Ala-Val

Sigma A 4253 FW 188.2 $C_8H_{16}N_2O_3$

Ala-Val
N-CBZ-β-Ala-Val

Sigma C 4251 FW 322.4 $C_{16}H_{22}N_2O_5$

Ala-Val Amide Hydrochloride

Bachem G-4055.0250 MW 223.7 $C_8H_{17}N_3O_2 \cdot HCl$ Store at -15°C

Ala-Val-Gln-Ser-Lys-Pro-Pro-Ser-Lys-Arg-Asp-Pro-Pro-Lys-Met-Gln-Thr-Asp

Synonyms: Systemin

American Peptide 79-3-10 MW 2010.3 $C_{85}H_{144}N_{26}O_{28}S$
Pearce, G et al, *Science*, 253:895, 1991

Bachem H-8675.0001 MW 2010.3 $C_{85}H_{144}N_{26}O_{28}S$ Store at -15°C

Ala-Val-Glu-Ile-Asp
Ac-Ala-Val-Glu-Ile-Asp-CHO

Synonyms: Caspase VI Inhibitor II

Calbiochem 218758 MW 500.5 $C_{22}H_{36}N_4O_9$ ≥97% (HPLC); solid; soluble in DMSO | Reversible inhibitor of Caspase VI (Mch-2)

Ala-Val-Gly-Ile-Gly-Ala

Synonyms: GP41; HIV Peptide

Bachem H-1186.0005 MW 486.57 $C_{21}H_{38}N_6O_7$ Store at -15°C | HIV peptide

Ala-Val-Gly-Ile-Gly-Ala-Leu-Phe-Leu-Gly-Phe-Leu-Gly-Ala-Ala-Gly-Ser-Thr-Met-Gly-Ala-Arg-Ser Amide

Synonyms: GP41 (519-541); HIV Peptide (519-541)

Bachem H-2978.0500 Store at -15°C | HIV peptide

Ala-Val-Leu

Bachem H-1655.0250 MW 301.39 $C_{14}H_{27}N_3O_4$ Store at -15°C

Ala-Val-Leu-Pro-Arg-Ser-Ala-Lys-Glu-Leu

Synonyms: Interleukin VIII (-5 to +5)

Bachem H-3564.0005 Store at -15°C

Ala-Val-Pro-Phe
Suc-Ala-Val-Pro-Phe-pNA

Bachem L-1625.0050 Store at -15°C

Ala-Val-Ser-Glu-His-Gln-Leu-Leu-His-Asp-Lys-Gly-Lys-Ser-Ile-Gln

Synonyms: Parathyroid Hormone Related Peptide (1-16); Hypercalcemia of Malignancy Factor (1-16)

Sigma H 3270 Human FW 1790.0 ≥95% (HPLC); peptide content:~70% | Bioactive peptide

Bachem H-6575.0500 Human, rat MW 1790.01 $C_{77}H_{128}N_{24}O_{25}$ Store at -15°C

Ala-Val-Ser-Glu-His-Gln-Leu-Leu-His-Asp-Lys-Gly-Lys-Ser-Ile-Gln-Asp-Leu-Arg-Arg-Arg-Phe-Phe-Leu-His-His-Leu-Ile-Ala-Glu-Ile-His-Thr-Ala-Glu-Ile-Arg-Ala-Thr-Ser

Synonyms: Hypercalcemia Malignancy of Factor (1-40); Parathyroid Hormone Related Protein (1-40); Parathyroid Hormone-Like Adenylate Cyclase Stimulating Protein

ICN 153131 MW 4675 Moseky, JM etal, *PNAS*, 84:5048, 1987; Suva, LJ etal, *Science*, 237:893, 1987

Ala-Val-Ser-Glu-His-Gln-Leu-Leu-His-Asp-Lys-Gly-Lys-Ser-Ile-Gln-Asp-Leu-Arg-Arg-Arg-Phe-Phe-Leu-His-His-Leu-Ile-Ala-Glu-Ile-His-Thr-Ala-Glu-Tyr

Synonyms: Hypercalcemia Malignancy of Factor (1-36), (Tyr[36])-; Parathyroid Hormone Related Protein (1-36)

ICN 154490 MW 4310.3

Ala-Val-Ser-Glu-His-Gln-Leu-Leu-His-Asp-Lys-Gly-Lys-Ser-Ile-Gln-Asp-Leu-Arg-Arg-Arg-Phe-Phe-Leu-His-His-Leu-Ile-Ala-Glu-Ile-His-Thr-Ala Amide

Synonyms: Hypercalcemia Malignancy of Factor (1-34); Parathyroid Hormone Related Protein (1-34)

ICN 154491 MW 4017.1

Ala-Val-Ser-Glu-His-Gln-Leu-Leu-His-Asp-Lys-Gly-Lys-Ser-Ile-Gln-Asp-Leu-Arg-Arg-Arg-Phe-Phe-Leu-His-His-Leu-Ile-Ala-Glu-Ile-His-Thr-Ala-Glu-Ile-Arg-Ala-Thr-Ser-Glu-Val-Ser-Pro-Asn-Ser-Lys-Pro-Ser-Pro-Asn-Thr-Lys-Asn-His-Pro-Val-Arg-Phe-Gly-Ser-Asp-Asp-Glu-Gly-Arg-Tyr-Leu-Thr-Gln-Glu-Thr-Asn-Lys-Val-Glu-Thr-Tyr-Lys-Glu-Gln-Pro-Leu-Lys-Thr-Pro

Synonyms: Parathyroid Hormone Related Protein (1-86); Hypercalcemia of Malignancy Factor (1-86)

Bachem H-9815.0250 Human Store at -15°C

Ala-Val-Ser-Glu-His-Gln-Leu-Leu-His-Asp-Lys-Gly-Lys-Ser-Ile-Gln-Asp-Leu-Arg-Arg-Arg-Phe-Phe-Leu-His-His-Leu-Ile-Ala-Glu-Ile-His-Thr-Ala-Glu-Ile

Synonyms: Parathyroid Hormone Related Protein (1-36); Humoral Hypercalcemia of Malignancy Factor (1-36)

Neosystem SC1237   Human   MW 4259.87   Bioactive; calcium metabolism peptide; For review, see Stewart, AF, *Bone*, 19:303-306, 1996

Ala-Val-Ser-Glu-His-Gln-Leu-Leu-His-Asp-Lys-Gly-Lys-Ser-Ile-Gln-Asp-Leu-Arg-Arg-Arg-Phe-Phe-Leu-His-His-Leu-Ile-Ala-Glu-Ile-His-Thr-Ala

Synonyms: Parathyroid Hormone Related Protein (1-34); Humoral Hypercalcemia of Malignancy Factor (1-34)

Neosystem SC185   Human   MW 4017.60   Bioactive; calcium metabolism peptide; Suva, LJ et al, *Science*, 237:893, 1987

Ala-Val-Ser-Glu-His-Gln-Leu-Leu-His-Asp-Lys-Gly-Lys-Ser-Ile-Gln-Asp-Leu-Arg-Arg-Arg-Phe-Phe-Leu-His-His-Leu-Ile-Ala-Glu-Ile-His-Thr-Ala-Glu-Ile-Arg-Ala-Thr-Ser

Synonyms: Parathyroid Hormone Related Peptide (1-40); Hypercalcemia of Malignancy Factor (1-40); Parathyroid Hormone-Like Adenylate Cyclase Stimulating Protein

Sigma H 4644   Human   FW 4675.3   ≥97% (HPLC) | Bioactive peptide; PTH plays a major role in the modulation of serum calcium concentration & thereby affects the physiology of mineral & bone metabolism; Fairwell, T et al, *Biochemistry*, 22:2691, 1983; Potts, JT et al, *Adv Protein Chem*, 35:323, 1982; Moseky, JM et al, *Proc Natl Acad Sci USA*, 84:5048, 1987; Suva, LJ et al, *Science*, 237:893, 1987

Ala-Val-Ser-Glu-His-Gln-Leu-Leu-His-Asp-Lys-Gly-Lys-Ser-Ile-Gln-Asp-Leu-Arg-Arg-Arg-Phe-Phe-Leu-His-His-Leu-Ile-Ala-Glu-Ile-His-Thr-Ala Amide

Synonyms: Parathyroid Hormone Related Peptide (1-34); Hypercalcemia of Malignancy Factor (1-34)

Sigma H 9148   Human   FW 4016.6   ≥97% (HPLC) | Bioactive peptide; Horvichi, N et al, *Science*, 238:1566, 1987; Kemp, BE et al, *Science*, 238:1568, 1967

Peptides International PTH-4205-v   Human synthetic   MW 4016.6 $C_{180}H_{288}N_{58}O_{47}$ >99% (HPLC); lyophilized amorphous powder | Bioactive; Suva, LJ Get al, *Science*, 237:893, 1987

Ala-Val-Ser-Glu-His-Gln-Leu-Leu-His-Asp-Lys-Gly-Lys-Ser-Ile-Gln-Asp-Leu-Arg-Arg-Arg-Phe-Phe-Leu-His-His-Leu-Ile-Ala-Glu-Ile-His-Thr-Ala-Glu-Tyr

Synonyms: Parathyroid Hormone Related Protein (1-36), (Tyr[36])-; Hypercalcemia of Malignancy Factor (1-36), (Tyr[36])-

Bachem H-3208.0500   Human, rat   Store at -15°C

Ala-Val-Ser-Glu-His-Gln-Leu-Leu-His-Asp-Lys-Gly-Lys-Ser-Ile-Gln-Asp-Leu-Arg-Arg-Arg-Phe-Phe-Leu-His-His-Leu-Ile-Ala-Glu-Ile-His-Thr-Ala

Synonyms: Parathyroid Hormone Related Protein (1-34); Hypercalcemia of Malignancy Factor (1-34)

Bachem H-6630.0500   Human, rat   MW 4017.61 $C_{180}H_{287}N_{57}O_{48}$ Store at -15°C

Ala-Val-Ser-Glu-His-Gln-Leu-Leu-His-Asp-Lys-Gly-Lys-Ser-Ile-Gln-Asp-Leu-Arg-Arg-Arg-Phe-Phe-Leu-His-His-Leu-Ile-Ala-Glu-Ile-His-Thr-Ala-Glu-Ile-Arg-Ala-Thr-Ser

Synonyms: Parathyroid Hormone Related Protein (1-40); Hypercalcemia of Malignancy Factor (1-40)

Bachem H-6810.0500   Human, rat   MW 4675.34 $C_{207}H_{334}N_{66}O_{58}$ Store at -15°C

Ala-Val-Ser-Glu-His-Gln-Leu-Leu-His-Asp-Lys-Gly-Lys-Ser-Ile-Gln-Asp-Leu-Arg-Arg-Arg-Phe-Phe-Leu-His-His-Leu-Ile-Ala-Glu-Ile-His-Thr-Ala Amide

Synonyms: Parathyroid Hormone Related Protein (1-34); Hypercalcemia of Malignancy Factor (1-34)

Bachem H-9095.0500   Human, rat   MW 4016.63 $C_{180}H_{288}N_{58}O_{47}$ Store at -15°C

Ala-Val-Ser-Glu-His-Gln-Leu-Leu-His-Asp-Lys-Gly-Lys-Ser-Ile-Gln-Asp-Leu-Arg-Arg-Arg-Phe-Phe-Leu-His-His-Leu-Ile-Ala-Glu-Ile-His-Thr-Ala Trifluoroacetate Salt

Synonyms: Parathyroid Hormone Related Peptide (1-34)

Biogenesis 7170-9355   Human, rat synthetic   MW 4017.6   >95% (HPLC); lyophilized

---

Ala-Val-Ser-Glu-His-Gln-Leu-Leu-His-Asp-Lys-Gly-Lys-Ser-Ile-Gln-Asp-Leu-Arg-Arg-Arg-Phe-Phe-Leu-His-His-Leu-Ile-Ala-Glu-Ile-His-Thr-Ala

Synonyms: Hypercalcemia Malignancy of Factor (1-34); Parathyroid Hormone Related Protein (1-34)

ICN 154493   MW 4018.1   Sulva, LJ etal, *Science*, 237:893, 1987; Moseley, JM etal, *PNAS*, 84:5048, 1987; McKee, RL etal, *J Bone & Mineral Research*, 3(Supp 1), Abstract #17, p S73, 1988

Ala-Val-Ser-Glu-Ile-Gln-Leu-Leu-His-Asp-Lys-Gly-Lys-Ser-Ile-Gln-Asp-Leu-Arg-Arg-Arg-Phe-Phe-Leu-His-His-Leu-Ile-Ala-Glu-Ile-His-Thr-Ala-Glu-Ile-Arg-Ala-Thr-Ser

Synonyms: Hypercalcemia Malignancy of Factor (1-40); Parathyroid Hormone Related Protein (1-40)

American Peptide 22-1-70   Human   MW 4675.4 $C_{207}H_{334}N_{66}O_{58}$

Ala-Val-Ser-Glu-Ile-Gln-Leu-Leu-His-Asp-Lys-Gly-Lys-Ser-Ile-Gln-Asp-Leu-Arg-Arg-Arg-Phe-Phe-Leu-His-His-Leu-Ile-Ala-Glu-Ile-His-Thr-Ala Amide

Synonyms: Hypercalcemia Malignancy of Factor (1-34); Parathyroid Hormone Related Protein (1-34)

American Peptide 22-1-71   Human   MW 4016.6 $C_{180}H_{288}N_{58}O_{47}$ Horvichi, N et al, *Science*, 238:1566, 1987; Kemp, BE et al, *Science*, 238:1568, 1987

Ala-Val-Ser-Glu-Ile-Gln-Leu-Leu-His-Asp-Lys-Gly-Lys-Ser-Ile-Gln-Asp-Leu-Arg-Arg-Arg-Phe-Phe-Leu-His-His-Leu-Ile-Ala-Glu-Ile-His-Thr-Ala

Synonyms: Hypercalcemia Malignancy of Factor (1-34); Parathyroid Hormone Related Protein (1-34)

American Peptide 22-1-74   Human   MW 4017.7 $C_{180}H_{287}N_{57}O_{48}$ Found to have anabolic effect on bone formation in rats; inhibits the rapid decline in bone formation due to denervation; Suva, LJ et al, *Science*, 237:893, 1987

Ala-Val-Ser-Glu-Ile-Gln-Leu-Leu-His-Asp-Lys-Gly-Lys-Ser-Ile-Gln-Asp-Leu-Arg-Arg-Arg-Phe-Trp-Leu-His-His-Leu-Ile-Ala-Glu-Ile-His-Thr-Ala-Glu-Ile

Synonyms: Parathyroid Hormone Related Protein (1-36),(Ile[5], Trp[23])-; Humoral Hypercalcemia of Malignancy Factor (1-36), (Ile[5],Trp[23])-

Neosystem SC1239   Human   MW 4274.93   Bioactive; calcium metabolism peptide; a parathyroid hormone receptor subtype that fully responds to PTH but not at all to PTHrP (contrary to the PTH-1 (PTH/PTHrP) receptor which responds fully to both ligands); 2 divergent residues in PTH & PTHrP account for PTH-2 receptor selectivity:Phe[23] in PTHrP/Trp[23] in PTH (which determines binding affinity), His[5] in PTHrP/Ile[5] in PTH (which determines signaling capability); changing these 2 residues of PTHrP to the corresponding residues of PTH converts PTHrP into a potent PTH-2 receptor agonist; Gardella, TJ et al, *JBC*, 271:19888-19893, 1996

Ala-Val-Ser-Glu-Ile-Gln-Leu-Leu-His-Asp-Lys-Gly-
Lys-Ser-Ile-Gln-Asp-Leu-Arg-Arg-Arg-Phe-Trp-Leu-
His-His-Leu-Ile-Ala-Glu-Ile-His-Thr-Ala-Glu-Tyr

Synonyms: Parathyroid Hormone Related Protein (1-36),
(Ile[5],Trp[23],Tyr[36])-; Hypercalcemia of Malignancy Factor (1-36),
(Ile[5],Trp[23],Tyr[36])-

Bachem H-3924.0500   Human, rat   Store at -15°C

Ala-Val-Ser-Glu-Ile-Gln-Leu-Met-His-Asn-Leu-Gly-
Lys-His-Leu-Ala-Ser-Val-Glu-Arg-Met-Gln-Trp-Leu-
Arg-Lys-Lys-Leu-Gln-Asp-Val-His-Asn-Phe

Synonyms: Parathyroid Hormone (1-34)

American Peptide 22-5-21   Rat   MW 4057.8
$C_{180}H_{291}N_{55}O_{48}S_2$ Keutmann, HT et al, *ASBMR Conf*, Abstract,
1984; Keutmann, HT et al, *Endocrinol*, 117:1230, 1985

Ala-Val-Ser-Glu-Ile-Gln-Leu-Met-His-Asn-Leu-Gly-
Lys-His-Leu-Ala-Ser-Val-Glu-Arg-Met-Gln-Trp-Leu-
Arg-Lys-Lys-Leu-Gln-Asp-Val-His-Asn-Phe-Val-Ser-
Leu-Gly-Val-Gln-Met-Ala-Ala-Arg-Glu-Gly-Ser-Tyr-
Gln-Arg-Pro-Thr-Lys-Lys-Glu-Asp-Asn-Val-Leu-Val-
Asp-Gly-Asn-Ser-Lys-Ser-Leu-Gly-Glu-Gly-Asp-Lys-
Ala-Asp-Val-Asp-Val-Leu-Val-Lys-Ala-Lys-Ser-Gln

Synonyms: Parathyroid Hormone (1-84)

Bachem H-3086.0500   Rat   Store at -15°C

Ala-Val-Ser-Glu-Ile-Gln-Leu-Met-His-Asn-Leu-Gly-
Lys-His-Leu-Ala-Ser-Val-Glu-Arg-Met-Gln-Trp-Leu-
Arg-Lys-Lys-Leu-Gln-Asp-Val-His-Asn-Phe

Synonyms: Parathyroid Hormone (1-34)

Bachem H-5460.0500   Rat   MW 4057.76 $C_{180}H_{291}N_{55}O_{48}S_2$
Store at -15°C

Ala-Val-Ser-Glu-Ile-Gln-Leu-Met-His-Asn-Leu-Gly-
Lys-His-Leu-Ala-Ser-Val-Glu-Arg-Met-Gly-Trp-Leu-
Arg-Lys-Lys-Leu-Gln-Asp-Val-His-Asn-Phe

Synonyms: Parathyroid Hormone (1-34); Parathyroid Hormone,
Rat; Parathormone

Calbiochem 512585   Rat   MW 4057.8 $C_{180}H_{291}N_{55}O_{48}S_2$
≥97% (HPLC); lyophilized solid; soluble in 5% acetic acid;
harmful:LD$_{50}$≤2000 mg/kg | Suitable for the studies of bone
metabolism & glandular disorders, characterization of specific
receptor sites & as an antigen; *Merck Index*, 12:7168

Ala-Val-Ser-Glu-Ile-Gln-Leu-Met-His-Asn-Leu-Gly-
Lys-His-Leu-Ala-Ser-Val-Glu-Arg-Met-Gln-Trp-Leu-
Arg-Lys-Lys-Leu-Gln-Asp-Val-His-Asn-Phe

Synonyms: Parathyroid Hormone (1-34)

Neosystem SC1224   Rat   MW 4057.74   Bioactive; calcium
metabolism peptide; Heinrich, G et al, *JBC*, 259:3320-3329, 1984

Sigma P 3921   Rat   FW 4057.7   ≥97% (HPLC) | Bioactive
peptide

Ala-Val-Ser-Glu-Ile-Gln-Leu-Met-His-Asn-Leu-Gly-
Lys-His-Leu-Ala-Ser-Val-Glu-Arg-Met-Gln-Trp-Leu-
Arg-Lys-Lys-Leu-Gln-Asp-Val-His-Asn-Phe
Trifluoroacetate Salt

Synonyms: Parathyroid Hormone (1-34)

Biogenesis 7170-9002   Rat synthetic   MW 4057.8   >98%
(HPLC); lyophilized

Ala-Val-Ser-Glu-Ile-Gln-Leu-Nle-His-Asn-Leu-Gly-
Lys-His-Leu-Ala-Ser-Val-Glu-Arg-Nle-Gln-Trp-Leu-
Arg-Lys-Lys-Leu-Gln-Asp-Val-His-Asn-Tyr Amide

Synonyms: Parathyroid Hormone (1-34), (Nle[8,21],Tyr[34])-;
Parathyroid Hormone (1-34), (Nle[8,21],Tyr[34],-NH$_2$)-

American Peptide 22-5-25   Rat   MW 4036.7
$C_{182}H_{296}N_{56}O_{48}$

Bachem H-5525.0500   Rat   MW 4036.7 $C_{182}H_{296}N_{56}O_{48}$
Store at -15°C

Sigma P 1803   Rat   FW 4037.7   ≥97% (HPLC) | Bioactive
peptide

Ala-Val-Ser-Glu-Ile-Gln-Phe-Met-His-Asn-Leu-Gly-
Lys-His-Leu-Ser-Ser-Met-Glu-Arg-Val-Glu-Trp-Leu-
Arg-Lys-Lys-Leu-Gln-Asp-Val-His-Asn-Phe

Synonyms: Parathyroid Hormone (1-34); Parathyroid Hormone,
Bovine; Parathormone

American Peptide 22-3-15   Bovine   MW 4108.8
$C_{183}H_{288}N_{54}O_{50}S_2$

Bachem H-1660.0500   Bovine   MW 4108.77
$C_{183}H_{288}N_{54}O_{50}S_2$ Store at -15°C

Calbiochem 512550   Bovine   MW 4108.8 $C_{183}H_{288}N_{54}O_{50}S_2$
>95% (HPLC); lyophilized solid; soluble in 5% acetic acid;
harmful:LD$_{50}$≤2000 mg/kg | N-terminal active peptide fragment of
native hormone that functions as a regulatory factor in the
homeostatic control of Ca^{2+} & phosphate metabolism; regulates
Ca^{2+} & phosphate flux across cellular membranes in bone & kidney
resulting in increased serum Ca^{2+} levels; increases the activity of
osteoblasts & osteoclasts; increases tubular reabsorption of Ca^{2+} &
stimulates renal hydroxylation of 25-(OH)-Vitamin D$_3$ to 1,25-
(OH)$_2$-Vitamin D$_3$; increases the production of inhibitory IGF
binding protein 4 in bone cells; *Merck Index*, 12:7168; Matsui, H et
al, *Am J Physiol*, 268:R21, 1995; Dua, K et al, *Exp Physiol*, 79:401,
1994

ICN 152976   Bovine   MW 4108.7   Tregear, GW et al,
Biochem, 16:2817, 1977

Neosystem SC313   Bovine   MW 4108.74   Bioactive; calcium
metabolism peptide; Tregear, GW et al, *Biochem*, 16:2817, 1977

Sigma P 3671   Bovine   FW 4108.7   ≥97% (HPLC); peptide
content:~70% | Bioactive peptide; similar to Sigma P 7399 but
produced by Sigma; Goltzmann, D et al, *J Biol Chem*, 250:3199,
1975

Ala-Val-Ser-Glu-Ile-Gln-Phe-Met-His-Asn-Leu-Gly-
Lys-His-Leu-Ser-Ser-Met-Glu-Arg-Val-Glu-Trp-Leu-
Arg-Lys-Lys-Leu-Gln-Asp-Val-His-Asn-Tyr Amide

Synonyms: Parathyroid Hormone (1-34), (Tyr[34])-

Peptides International PTH-4179-v   Bovine synthetic   MW
4123.8 $C_{183}H_{289}N_{55}O_{50}S_2$ >95% (HPLC); lyophilized amorphous
powder

Ala-Val-Ser-Glu-Ile-Gln-Phe-Nle-His-Asn-Leu-Gly-
Lys-His-Leu-Ser-Ser-Nle-Glu-Arg-Val-Glu-Trp-Leu-
Arg-Lys-Lys-Leu-Gln-Asp-Val-His-Asn-Tyr Amide

Synonyms: Parathyroid Hormone (1-34), (Nle[8,18],Tyr[34])-;
Parathormone; Parathyroid Hormone (1-34), (Nle[8,18],Tyr[34],-NH$_2$)-

Bachem H-1670.0500   Bovine   Store at -15°C

ICN 152978   Bovine   MW 4087.7   About twice as active as
unsubstituted bovine parathyroid hormone, fragment 1-34;
Rosenblatt, M & JT Potts Jr, *Endocrin Res Commun*, 4:115, 1977;
Coltrera, MD etal, *JBC*, 256:10555, 1981

Sigma P 2905   Bovine   FW 4087.7   ≥97% (HPLC) |
Bioactive peptide; Rosenblatt, M & Potts, JT Jr, *Endocr Res
Commun*, 4:115, 1977

Arg-Abz
Bz-Arg-4-Abz

Bachem M-1125.0050   MW 397.43 $C_{20}H_{23}N_5O_4$ Store at
-15°C

Arg-Ala

Bachem G-4170.0250   MW 245.28 $C_9H_{19}N_5O_3$ Store at -15°C

Arg-Ala Amide Dihydrochloride

Bachem G-3845.0250   MW 317.22 $C_9H_{20}N_6O_2 \cdot 2HCl$ Store at -15°C

Arg-Ala-Arg-Thr-Ser-Ser-Phe-Ala-Glu-Pro-Gly
Arg-Ala-Arg-Thr-Ser-pSer-Phe-Ala-Glu-Pro-Gly

Synonyms: Phospho-GSK3α, (Ser[21])-

Upstate 12-396   MW 1258   Frozen solution | Immunizing peptide

Arg-Ala-Gly-Phe-Ala-Pro-Phe-Arg

Synonyms: Bradykinin, (des-Pro[3], (Ala[2,6]))-; Angiotensin I Converting Enzyme Inhibitor

Sigma B 4791   FW 921.1 $C_{43}H_{64}N_{14}O_9$ ≥97% (HPLC); peptide content:~70% | Bioactive peptide; Chaturvedi, D et al, *Peptide Res*, 6:308, 1993

Arg-Ala-Trp-Phe-Pro-Pro-Nle
BOC-Arg-Ala-D-Trp-Phe-D-Pro-Pro-Nle Amide

Synonyms: GR-83074

American Peptide 62-1-27   MW 985.2 $C_{50}H_{72}N_{12}O_9$ Highly potent & selective NK$_2$ receptor antagonist; McElroy, AB et al, *J Med Chem*, 35:2582, 1992

Bachem H-2792.0500   Store at -15°C

Neosystem SC493   MW 985.20   Bioactive; tachykinin; highly potent & selective NK-2 receptor antagonist; McElroy, AB et al, *J Med Chem*, 35:2582-2591, 1992

Arg-Arg

Bachem G-1465.0250   MW 330.39 $C_{12}H_{26}N_8O_3$ Store at -15°C

Arg-Arg
Arg-Arg-AMC 3 Hydrochloride

Bachem I-1055.0050   Store at -15°C

Arg-Arg
Z-Arg-Arg-AMC

Bachem I-1135.0050   Store at -15°C

Arg-Arg
Arg-Arg-4MβNA 3 Hydrochloride

Bachem J-1040.0050   MW 594.97 $C_{23}H_{35}N_9O_3 \cdot 3HCl$ Store at -15°C

Arg-Arg
Z-Arg-Arg-4MβNA

Bachem J-1105.0050   MW 619.72 $C_{31}H_{41}N_9O_5$ Store at -15°C

Arg-Arg
Arg-Arg-βNA 3 Hydrochloride

Bachem K-1085.0050   MW 564.95 $C_{22}H_{33}N_9O_2 \cdot 3HCl$ Store at -15°C

Arg-Arg
Z-Arg-Arg-βNA

Bachem K-1170.0050   MW 589.7 $C_{30}H_{39}N_9O_4$ Store at -15°C

Arg-Arg
Z-Arg-Arg-pNA ·2 Hydrochloride

Bachem L-1225.0050   Store at -15°C

Arg-Arg
Z-Arg-Arg-pNA Dihydrochloride

Synonyms: Cathepsin B Substrate

ICN 194968   MW 658.6   Lyophilized; >95%

Arg-Arg
Z-Arg-Arg-AMC Dihydrochloride

Synonyms: Cathepsin B Substrate III

ICN 195954   MW 694.6 $C_{30}H_{39}N_9O_6 \cdot 2HCl$ ≥95% | Fluorogenic substrate for the quantitative determination of cathepsin B

Arg-Arg
Z-Arg-Arg-MCA

Synonyms: Cathepsin B Substrate

Peptides International MRR-3123-v   MW 621.70 $C_{30}H_{39}N_9O_6$ >98% (HPLC); lyophilized amorphous powder | Barrett, AJ & H Kirschke in *Proteolytic Enzymes Part C*, Methods in Enzymology, Vol 80, (L Lorand, Ed), Academic Press, New York, 1981, pp 535

Arg-Arg
Arg-Arg-β-Naphthylamide Trihydrochloride

Synonyms: Dipeptidyl Aminopeptidase III Substrate

Sigma A 3261   FW 564.9 $C_{22}H_{33}N_9O_2 \cdot 3HCl$ Possibly carcinogenic; Ellis, S & Nvenke, JM, *J Biol Chem*, 242:4623, 1967; Swanson, AA et al, *Biochem Biophys Res Commun*, 84:1151, 1978; Chan, SAT et al, *Biochem Biophys Res Commun*, 127:962, 1985

Arg-Arg
Arg-Arg 4-Methoxy-β-Naphthylamide Trihydrochloride

Synonyms: Dipeptidyl Aminopeptidase III Substrate

Sigma A 9398   FW 595.0 $C_{23}H_{35}N_9O_3 \cdot 3HCl$ Fluorogenic; Smith, RE & Van Frank, RM, *Lysosomes in Biology & Pathology*, Vol 4, *p* 193, JT Dingle & RT Dean, eds, North Holland/American Elsevier, 1975

Arg-Arg
*N*ᵅ-CBZ-Arg-Arg-AMC Free Base

Synonyms: Cathepsin B Substrate

Sigma C 5429   FW 621.7 $C_{30}H_{39}N_9O_6$ Contains ~4% isopropanol hydrochloride | Barrett, AJ & Kirschke, H, *Meth Enzymol*, 80:535, 1981

Arg-Arg
*N*ᵅ-CBZ-Arg-Arg 4-Methoxy-β-Naphthylamide Acetate Salt Free Base

Synonyms: Cathepsin B Substrate

Sigma C 5520   FW 619.7 $C_{31}H_{41}N_9O_5$ Mort, JS & Leduc, M, *Anal Biochem*, 119:148, 1982; Delaisse, JM et al, *Biochem Biophys Res Commun*, 125:441, 1984

Arg-Arg Amide 3 Hydrochloride

Bachem G-3985.0250   MW 438.79 $C_{12}H_{27}N_9O_2 \cdot 3HCl$ Store at -15°C

Arg-Arg-Ala-Asn-Ala-Leu-Leu-Ala-Asn-Gly-Val-Glu-Leu-Arg-Asp

Synonyms: Tumor Necrosis Factor α (31-45)

Bachem H-8250.0500   Human   MW 1667.89 $C_{69}H_{122}N_{26}O_{22}$ Store at -15°C

ICN 157173   Human   MW 1667.9   97%

Sigma T 1167   Human   FW 1667.9   ≥97% (HPLC); peptide content ~70% | Bioactive peptide

Arg-Arg-Ala-Ser-Val-Ala
Arg-Arg-Ala-pSer-Val-Ala

Synonyms: Serine Phosphopeptide

Upstate 12-220   Synthetic   MW 736   >98%; lyophilized | Phosphopeptide

USBio S1000-75   Synthetic   ≥95% (HPLC); lyophilized | Six residue phosphopeptide

Arg-Arg-Arg

Bachem H-1790.0050   MW 486.58 $C_{18}H_{38}N_{12}O_4$ Store at -15°C

Arg-Arg-Arg
Z-Arg-Arg-Arg-4MβNA

Bachem J-1110.0050   MW 775.91 $C_{37}H_{53}N_{13}O_6$ Store at -15°C

Arg-Arg-Arg
N^c-CBZ-Arg-Arg-Arg-4-Methoxy-β-Naphthylamide

Synonyms: Trypsin Substrate

Sigma C 5645   FW 775.9 $C_{37}H_{53}N_{13}O_6$

Arg-Arg-Arg-Ala-Asp-Asp-Ser-(Asp)₅

Synonyms: Casein Kinase II Substrate

Sigma C 2460   FW 1450.4   ≥97% (HPLC) | $K_m=19\mu M$; cannot be phosphorylated by caseinkinase-1; Marin, O et al, *Biochem Biophys Res Commun*, 198:898, 1994

Arg-Arg-Arg-Ala-Asp-Asp-Ser-Asp-Asp-Asp-Asp-Asp

Bachem H-2486.0001   Store at -15°C

Arg-Arg-Arg-Ala-Pro-Leu-Ser-Pro Amide

Synonyms: Serine Kinase Substrate Peptide, Biotinylated; Protein Kinase Substrate

Upstate 12-362   MW 1291   >98% pure; frozen solution

Arg-Arg-Arg-Arg

Bachem H-4464.0005   MW 642.77 $C_{24}H_{50}N_{16}O_5$ Store at -15°C

Arg-Arg-Arg-Arg-Arg-Arg

Bachem H-4622.0005   MW 955.14 $C_{36}H_{74}N_{24}O_7$ Store at -15°C

Arg-Arg-Arg-Asp-Asp-Asp-Ser-Asp-Asp-Asp

Synonyms: Casein Kinase II Substrate; Casein Kinase II Substrate Peptide

American Peptide 86-2-27   MW 1264.2 $C_{45}H_{73}N_{19}O_{24}$ Kissmehl, R et al, *FEBS Lett*, 402:227, 1997

Upstate 12-330   MW 1264   >90% (HPLC); frozen solution

USBio C2085-62   Synthetic   ≥95% (HPLC); frozen solution in 1 mL of assay dilution buffer:20 *mM* MOPS, pH 7.2; 25 *mM* β-glycerol phosphate, 5 *mM* EGTA, 1 *mM* sodium orthovanadate, 1 *mM* DTT | K_m:60 *mM*, V_{max}:2.2 mmole/min/mg with Casein Kinase 2

Arg-Arg-Arg-Glu-Glu-Glu-Thr-Glu-Glu-Glu

Synonyms: Casein Kinase II Substrate; Casein Kinase II

Alexis 153-003   MW 1362.4 $C_{52}H_{87}N_{19}O_{24}$ White lyophilized powder | Specific; derived from kinase II phosphorylation site found in casein; not phosphorylated by casein kinase I, cAMP-dependent protein kinase, protein kinase G, myosin light chain kinase, phosphorylase kinase & the EGF receptor kinase; Kuenzel, EA & Krebs, ED, *PNAS*, 82; 737, 1985

American Peptide 86-2-25   MW 1362.3 $C_{52}H_{87}N_{19}O_{24}$ Kuenzel, EA & Krebs, EG, *PNAS*, 82:737, 1985

Neosystem SC406   MW 1362.37   Bioactive; proteinkinase-related peptide; Kuenzel, EA & Krebs, ED, *PNAS USA*, 82:737-741, 1985

Promega V5661   MW 1362   See Promega V5601

Arg-Arg-Arg-Leu-Ser-Ser-Leu-Arg-Ala

Synonyms: S6 Kinase Substrate Peptide (231-239); S6 Kinase/Rsk Substrate Peptide I

Upstate 12-124   Human synthetic   MW 1114   Lyophilized | Peptide 231-239 of human 40S ribosomal protein S6

USBio S0050-25   Human synthetic   Lyophilized | Corresponds to AA 231-239 of human 40S ribosomal protein S6

Arg-Arg-Arg-Leu-Ser-Ser-Leu-Arg-Ala Amide Trifluoroacetate Salt

Synonyms: Ribosomal S6 Kinase Substrate (231-239); Protein Kinase A Substrate; Protein Kinase C Substrate; p90rsk Kinase Substrate

Calbiochem 559280   MW 1113.3 $C_{45}H_{88}N_{22}O_{11}$ ≥97% (HPLC); lyophilized solid; soluble in water | AAs 231-239 of human 40S ribosomal protein S6; Pelech, SL et al, *PNAS*, 83:5968, 1986; Gabrielli, B et al, *FEBS Lett*, 175:219, 1984

Arg-Arg-Arg-Val-Thr-Ser-Ala-Ala-Arg-Arg-Ser
Biotinyl-ε-Aminocaproyl-Arg-Arg-Arg-Val-Thr-Ser-Ala-Ala-Arg-Arg-Ser

Bachem H-4238.0001   Store at -15°C

Arg-Arg-Cys-Tyr-Arg-Lys-Lys-Pro-Tyr-Arg-Cit-Cys-Arg
Arg-Arg-2-Nal-Cys-Tyr-Arg-Lys-D-Lys-Pro-Tyr-Arg-Cit-Cys-Arg

Synonyms: Polyphemusin II Derived Peptide; T140

Bachem H-4626.0500   Store at -15°C

Arg-Arg-Gln-Arg-Arg-Arg-Pro-Pro-Gln

Synonyms: TAT (52-60); HTLV I Peptide

Neosystem SC523   MW 1249.44   HTLV I peptide from subtype TAT

Arg-Arg-Glu-Ala-Glu-Asn-Pro-Gln-Ala-Gly-Ala-Val-Glu-Leu-Gly-Gly-Gly-Leu-Gly-Gly-Leu-Gln-Ala-Leu-Ala-Leu-Glu-Gly-Pro-Pro-Gln-Lys-Arg

Synonyms: Pro-Insulin C-Peptide (31-63)

Bachem H-2142.0500   Porcine   Store at -15°C

Arg-Arg-Glu-Ala-Glu-Asp-Leu-Gln-Val-Gly-Gln-Val-Glu-Leu-Gly-Gly-Gly-Pro-Gly-Ala-Gly-Ser-Leu-Gln-Pro-Leu-Ala-Leu-Glu-Gly-Ser-Leu-Gln-Lys-Arg

Synonyms: Pro-Insulin C-Peptide (55-89); C-Peptide; Insulin Chain C; Insulin C Chain

American Peptide 20-1-15   Human   MW 3617.1 $C_{153}H_{259}N_{49}O_{52}$ Frank, BH et al, *Peptides:7th Am Pept Symp*, 729, 1981

Bachem H-1800.0500   Human   MW 3617.04 $C_{153}H_{259}N_{49}O_{52}$ Store at -15°C

ICN 153079   Human   Frank, BH etal, *Proc Seventh American Peptide Symp*, Rich, DH & E Gross, eds, p 729, 1981

Sigma C 1775   Human   FW 3617.0   ≥95% (HPLC) |
Bioactive peptide; Frank, BH et al, *Peptides, Synthesis-Structure-Function Proc 7th Amer Pept Symp,* Rich, DH & Gross, E, eds, 729, 1981

Sigma C 5051   Human   FW 3617.0   ≥95% (HPLC) |
Bioactive peptide; similar to Sigma C 1775 but produced by Sigma; Frank, BH et al, *Peptides, Synthesis-Structure-Function Proc 7th Amer Pept Symp,* Rich, DH & Gross, E, eds, 729, 1981

Arg-Arg-Glu-Glu-Glu-Thr-Glu-Glu-Glu

Synonyms: Casein Kinase II Substrate

American Peptide 86-2-26   MW 1206.2 $C_{46}H_{75}N_{15}O_{23}$
Kissmehl, R et al, *FEBS Lett*, 402:227, 1997

Bachem H-3248.0001   Store at -15°C

ICN 195842   MW 1206.3   ≥97%

Arg-Arg-Gly-Asp-Met-Glu

Bachem H-6805.0005   Store at -15°C

Arg-Arg-Hyp-Hyp-Gly-Phe-Ser-Phe-Phe-Arg
D-Arg-Arg-Hyp-Hyp-Gly-Phe-Ser-D-Phe-Phe-Arg

Synonyms: Bradykinin, (D-Arg⁰,Hyp²,³,D-Phe⁷)-

Bachem H-9090.0001   Store at -15°C

Arg-Arg-Ile-Arg-Pro-Lys-Leu-Lys-Trp-Asp-Asn-Gln

Synonyms: Dynorphin A (6-17)

American Peptide 26-4-30   Porcine   MW 1609.9
$C_{71}H_{120}N_{26}O_{17}$

Arg-Arg-Leu-Glu-Glu-Glu-Glu-Glu-Ala-Tyr-Gly

Synonyms: Gastrin (22-30), (Arg-Arg)-; Gastrointestinal Peptide; Tyrosine Kinase Substrate

Sigma G 6535   Human   FW 1380.4   ≥97% (HPLC) |
Bioactive peptide; Baldwin, GS et al, *Biochem Biophys Res Commun*, 109:656, 1982

Arg-Arg-Leu-Gly-Asn-Gln-Leu-Leu-Ile-Ala-Ile-Leu-Leu-Leu-Ser-Val-Tyr-Gly-Ile-Lys-Lys

Synonyms: GP140-Lys-Lys (4-20), Arg-Arg-; SIVmac251 Peptide

Neosystem SC708   MW 2381.97   SIVmac251 peptide from subtype GP140

Arg-Arg-Leu-Ile-Glu-Asp-Ala-Glu-Tyr-Ala-Ala-Arg-Gly

Synonyms: Protein Tyrosine Kinase Substrate; RR-*src* Peptide; RR-*src*; Tyrosine Kinase Substrate Peptide; Tyrosine Protein Kinase Substrate; Tyrosine Kinase Substrate RR-SRC; Insulin Receptor Kinase Substrate

Alexis 166-002   MW 1519.7 $C_{64}H_{106}N_{22}O_{21}$ ≥97%;
lyophilized powder; soluble in H_2O | Peptide sequence corresponds to the tyrosine phosphorylation site in the Rous sarcoma virus (RSV)-encoded transforming protein pp60v-*src*; *in vitro* tyrosine phosphorylation of the peptide is stimulated by epidermal growth factor; Casnellie, JE et al, *PNAS*, 79:282, 1982; Pike, LJ et al, *J Biol Chem*, 261:3782, 1986

American Peptide 86-0-53   MW 1519.6 $C_{64}H_{106}N_{22}O_{21}$
Derived from the sequence surrounding the tyrosine phosphorylation site of the Rous sarcoma virus-encoded transforming protein pp60; used as a substrate to monitor the purification of the insulin receptor, a protein tyrosine kinase; Czernilofsky, AP et al, *Nature (Lond)*, 287:198, 1980

Bachem H-5445.0001   MW 1519.68 $C_{64}H_{106}N_{22}O_{21}$ Store at -15°C

ICN 153049   MW 1519.7   Pike, LJ etal, *JBC*, 261:3782, 1986

ICN 154582   MW 1519.9   Casnellie, JE etal, *PNAS*, 79:282, 1982

ICN 154587   MW 1519.9   Pike, LJ etal, *JBC*, 261:3782, 1986

Peptides International SRG-4184-v   MW 1519.7
$C_{64}H_{106}N_{22}O_{21}$ >99% (HPLC); lyophilized amorphous powder |
Casnellie, JE et al, *PNAS* USA, 79:282, 1982

Arg-Arg-Leu-Ile-Glu-Asp-Ala-Glu-Tyr-Ala-Ala-Arg-Gly
Arg-Arg-Leu-Ile-Glu-Asp-Ala-Glu-pTyr-Ala-Ala-Arg-Gly

Synonyms: Tyrosine Phosphopeptide

Upstate 12-217   Synthetic   MW 1597   >90%; lyophilized |
Phosphopeptide

USBio T9245   Synthetic   ≥90% (HPLC); lyophilized | 13 residue synthetic phosphopeptide; tested as a substrate for Protein Tyrosine Phosphatase-1B

Arg-Arg-Leu-Ile-Glu-Asp-Ala-Glu-Tyr-Ala-Ala-Arg-Gly Free Base

Synonyms: Tyrosine Protein Kinase Substrate; Protein Kinase Related Peptide

Sigma A 7433   FW 1519.7   ≥97% (HPLC) | Bioactive peptide; Pike, LJ et al, *J Biol Chem*, 261 3782, 1986; Casnellie, JE et al, *Proc Natl Acad Sci USA*, 79:282, 1982

Arg-Arg-Leu-Ile-Glu-Asp-Ala-Glu-Tyr-Ala-Ala-Arg-Gly-Arg-Arg-Leu-Ile-Glu-Asp-Ala-Glu-Tyr-Ala-Ala-Arg-Gly 2AcOH 4H₂O

Synonyms: Tyrosine Protein-Kinase Substrate

Peptides International SRG-4184   MW 1711.87
$C_{64}H_{106}N_{22}O_{21}$ · 2AcOH · 4H₂O   >99% (HPLC); amorphous powder; bulk | Casnellie, JE et al, *PNAS* USA, 79:282, 1982

Arg-Arg-Leu-Ile-Glu-Asp-Asn-Glu-Tyr-Thr-Ala-Arg-Gly

Synonyms: pp60v-*src* (412-422); Protein Tyrosine Kinase Substrate; pp60*src* (412-422); Tyrosine Protein Kinase Substrate; pp60v-*src* Autophosphorylation Site; Tyrosine Kinase Substrate

Alexis 165-023   MW 1592.7 $C_{66}H_{109}N_{23}O_{23}$ White lyophilized powder | Residues 2-12 in this peptide correspond to the sequence of the reported site of tyrosine phosphorylation (412-422) in the Rous sarcoma virus (RSV)-encoded transforming protein pp60v-*src*; *in vitro* tyrosine phosphorylation of the peptide is stimulated by epidermal growth factor; Pike, LJ et al, *PNAS*, 79:1443, 1982; Pike, LJ et al, *J Biol Chem*, 261:3782, 1986

American Peptide 86-0-10   MW 1592.9 $C_{66}H_{109}N_{23}O_{23}$
Residues 2-12 of this peptide corresponding to the sequence of AA 412-422 in the Rous sarcoma virus-encoded transforming protein pp60*src* is the reported site of Tyr phosphorylation; Krebs, EG et al, *PNAS*, 79:1443, 1982

Bachem H-1795.0001   MW 1592.73 $C_{66}H_{109}N_{23}O_{23}$ Store at -15°C

ICN 153050   MW 1592.9   Pike, LJ etal, *JBC*, 261:3782, 1986

ICN 154581   MW 1592.9   Pike, LJ etal, *PNAS*, 79:1443, 1982

Arg-Arg-Leu-Ile-Glu-Asp-Asn-Glu-Tyr-Thr-Ala-Arg-Gly
Arg-Arg-Leu-Ile-Glu-Asp-Asn-Glu-Tyr(PO₃H₂)-Thr-Ala-Arg-Gly

Synonyms: pp60v-*src* Autophosphorylation Site; Tyrosine Protein Kinase Substrate

ICN 195844   MW 1671.9   Useful in non-radiolabeled tyrosine phosphatase assay & as a competitive tyrosine phosphatase inhibitor

Arg-Arg-Leu-Ile-Glu-Asp-Asn-Glu-Tyr-Thr-Ala-Arg-Gly

Synonyms: Protein Tyrosine Kinase Substrate; Protein Kinase Related Peptide

Sigma A 7907   FW 1592.7   ≥95% (HPLC) | Bioactive peptide; Krebs, EG et al, *Proc Natl Acad Sci USA*, 79:1443, 1982

Arg-Arg-Leu-Leu-Phe-Tyr-Lys-Tyr-Val-Tyr-Lys-Arg-Tyr-Arg-Ala-Gly-Lys-Gln-Arg-Gly

Synonyms: Exchange Inhibitory Peptide

Neosystem SC910   MW 2622.11   Virus-related peptide; bioactive; inhibitor of cardiac sarcolemmal Na⁺-Ca²⁺ exchanger; Li et al, *JBC*, 266:1014-1020, 1991; Chin et al, *Circ Res*, 72:497-503, 1993; Bouchard et al, *J Physiol*, 469:583-599, 1993

Arg-Arg-Leu-Leu-Ser-Val-Tyr-Gly-Ile-Tyr-Cys-Thr-Gln-Tyr-Val-Thr-Val-Lys-Lys

Synonyms: GP140-Lys-Lys (14-28), Arg-Arg-; SIVmac251 Peptide

Neosystem SC709   MW 2290.74   SIVmac251 peptide from subtype GP140; due to the presence of a Cys this peptide can contain the dimeric form

Arg-Arg-Leu-Ser-Ser-Leu-Arg-Ala

Synonyms: cAMP-Dependent Protein Kinase Substrate; Protein Kinase C Substrate; S6 Phosphate Acceptor Peptide; Ribosomal S6-1 Kinase Substrate Peptide; Phosphate Acceptor Peptide; Protein Kinase Substrate; S6 Peptide

Alexis 167-006   MW 958.1 $C_{39}H_{75}N_{17}O_{11}$ ≥98%; white lyophilized powder; soluble in H_2O | Gabrielli, B et al, *FEBS Lett*, 175:219, 1984; Pelech, SL et al, *PNAS*, 83:5968, 1986; Ferrari, S et al, *FEBS Lett*, 184:72, 1985

American Peptide 86-2-35   MW 958.1 $C_{39}H_{75}N_{17}O_{11}$ Derived from the sequence of a region that contains the major phosphorylation site for cAMP-dependent protein kinase in rat hepatic ribosomal protein S6

Bachem H-9380.0005   MW 958.13 $C_{39}H_{75}N_{17}O_{11}$ Store at -15°C

Calbiochem 05-23-4902   MW 958.1 $C_{39}H_{75}N_{17}O_{11}$ >97% (HPLC); lyophilized solid; soluble in 5% acetic acid | Corresponds to a specific region of the ribosomal protein S6 that contains both the insulin & cAMP-regulated phosphorylation sites; Gabrielli, B et al, *FEBS Lett*, 175:219, 1984

ICN 195843   MW 958.2   ≥97% | Substrate for p90 rsk (S6K) in cell extracts using the phosphocellulose assay

Arg-Arg-Lys-Ala-Ser-Gly-Pro

Synonyms: Histone H1-7 Substrate; cAMP-Dependent Protein Kinase Substrate; H1-7; Histone VII; Histone H1 Phosphorylation Site; Protein Kinase Related Peptide

Alexis 158-001   MW 770.9 $C_{31}H_{58}N_{14}O_9$ ≥98%; white lyophilized powder; soluble in H_2O | Derived from the phosphorylation site (serine-38) in calf thymus histone H1; Pomerantz, AH et al, *PNAS*, 74:4261, 1977

American Peptide 86-0-11   MW 771.0 $C_{31}H_{58}N_{14}O_9$ Pomerantz, AM et al, *PNAS*, 74:4261, 1977

Bachem H-1805.0005   MW 770.89 $C_{31}H_{58}N_{14}O_9$ Store at -15°C

ICN 153051   MW 770.9 $C_{31}H_{58}N_{14}O_9$ Pomerantz, AM etal, *PNAS*, 74:4261, 1977

ICN 159599   MW 770.9   Substrate for proteinkinase C; phosphorylation site corresponds to Ser[38] in calf thymus Histone H1

Sigma A 3651   FW 770.9 $C_{31}H_{58}N_{14}O_9$ ≥97% (HPLC) | Bioactive peptide; histone H1 phosphorylation site; Pomerantz, AM et al, *Proc Natl Acad Sci USA*, 74:4261, 1977

Arg-Arg-Lys-Ala-Ser-Gly-Pro-Pro-Val

Synonyms: Phosphate Acceptor Peptide

American Peptide 86-5-40   MW 967.1 $C_{41}H_{74}N_{16}O_{11}$ O'Brian, CA et al, *BBRC*, 124:296, 1984

Arg-Arg-Lys-Asp-Leu-His-Asp-Asp-Glu-Glu-Asp-Glu-Ala-Met-Ser-Ile-Thr-Ala

Synonyms: Casein Kinase I Substrate

Bachem H-2484.0001   MW 2131.27 $C_{85}H_{139}N_{27}O_{35}S$ Store at -15°C

Calbiochem 218730   MW 2131.3 $C_{85}H_{139}N_{27}O_{35}S$ ≥95% (HPLC); lyophilized solid; soluble in 0.1% TFA | A superior substrate compared to the non-phosphorylated casein kinase I peptide substrates; its V_{max} value is six-fold higher than that of casein; phosphorylation rate of this substrate by casein kinase II is negligible; Marin, O et al, *Biochem Biophys Res Comm*, 198:898, 1994

Sigma C 2335   FW 2131.3   ≥97% (HPLC); Peptide content:~70%; | K_m=172 μM; V_{max} is 6-fold higher than for casein; Phosphorylation rate by caseinkinase-2 negligible; Marin, O et al, *Biochem Biophys Res Commun*, 198:898, 1994

Arg-Arg-Lys-Tyr-Ile-Gln-Tyr-Gly-Ile-Tyr-Val-Val-Val-Gly-Val-Ile-Leu-Leu-Arg-Lys-Lys

Synonyms: GP140-Lys-Lys (691-707), Arg-Arg-; SIVmac251 Peptide

Neosystem SC777   MW 2565.18   SIVmac251 peptide from subtype GP140

Arg-Arg-Phe-Met-Trp-Met-Arg-Arg
Ac-Arg-Phe-Met-Trp-Met-Arg Amide

Bachem H-1992.0005   MW 967.23 $C_{44}H_{66}N_{14}O_7S_2$ Store at -15°C

Arg-Arg-Pro-Hyp-Gly-Ala-Ser-Phe-Ala-Arg
1-Aaa-D-Arg-Arg-Pro-Hyp-Gly-β-(2-Thienyl)-Ala-Ser-D-Phe-β-(2-Thienyl)-Ala-Arg

Synonyms: Bradykinin, (1-Aaa-D-Arg⁰,Hyp³,β-(2-Thienyl)-Ala⁵,⁸,D-Phe⁷)-

Bachem H-1114.0001   MW 1470.79 $C_{68}H_{99}N_{19}O_{14}S_2$ Store at -15°C

Arg-Arg-Pro-Hyp-Gly-Ala-Ser-Phe-Ala-Arg
1-Adamantanecarbonyl-D-Arg-Arg-Pro-Hyp-Gly-β-(2-Thienyl)-Ala-Ser-D-Phe-β-(2-Thienyl)-Ala-Arg

Synonyms Bradykinin, (1-Adamantanecarbonyl-D-Arg⁰,Hyp³,β-(2-Thienyl)-Ala⁵,⁸,D-Phe⁷)-:

Bachem H-1116.0001   Store at -15°C

Arg-Arg-Pro-Hyp-Gly-Ala-Ser-Phe-Ala-Arg
D-Arg-Arg-Pro-Hyp-Gly-α-(2-Thienyl)-Ala-Ser-D-Phe-α-(2-Thienyl)-Ala-Arg

Synonyms: Bradykinin, (D-Arg⁰,Hyp³,α-(2-Thienyl)-Ala⁵,⁸,D-Phe⁷)-

Bachem H-6560.0001   Store at -15°C

Arg-Arg-Pro-Hyp-Gly-Phe-Ser-Phe-Leu-Arg
D-Arg-Arg-Pro-Hyp-Gly-Phe-Ser-D-Phe-Leu-Arg

Synonyms: Bradykinin, (D-Arg⁰,Hyp³,D-Phe⁷,Leu⁸)-; Bradykinin, (D-Arg-(Hyp³,D-Phe,Leu⁸))-

Bachem H-1652.0001   MW 1248.45 $C_{57}H_{89}N_{19}O_{13}$ Store at -15°C

Neosystem SC915   MW 1248.45   Bioactive; potent antagonist that discriminates between B-2A & B-2B receptors; Rhaleb, NE et al, *Life Sci*, 51:PL125, 1992

Sigma B 4916   1248.4; ≥97% (HPLC) | Bioactive peptide; short-acting competitive antagonist that discriminates between B_{2A} & B_{2B} receptors; Weak B_1 antagonist; Rhaleb, N-E et al, *Life Sci*, 51:PL 125, 1992

Arg-Arg-Pro-Hyp-Gly-Phe-Ser-Phe-Phe-Arg
D-Arg-Arg-Pro-Hyp-Gly-Phe-Ser-D-Phe-Phe-Arg

Synonyms: Bradykinin, (D-Arg⁰,Hyp³,D-Phe⁷)-

Bachem H-6385.0005   MW 1282.47 $C_{60}H_{87}N_{19}O_{13}$ Store at -15°C

Arg-Arg-Pro-Hyp-Gly-Thi-Ser-Arg
D-Arg-Arg-Pro-Hyp-Gly-Thi-Ser-D-Tic-Oic-Arg

Synonyms: Hoe 140

American Peptide 18-1-75   MW 1304.5   Bradykinin B$_2$ receptor antagonist; prevented bradykinin-induced incapacitation; Tonussi, CR et al, *Eur J Pharmacol*, 326:61, 1997

Arg-Arg-Pro-Hyp-Gly-Thi-Ser-Phe-Thi-Arg
D-Arg-Arg-Pro-Hyp-Gly-Thi-Ser-D-Phe-Thi-Arg

Synonyms: Bradykinin, (D-Arg^0,Hyp^3,Thi^{5,8},D-Phe^7)-

Neosystem SC413   MW 1294.51   Bioactive; bradykinin antagonist; Stewart, JM & Vavrek, RJ, The Proceedings of the; 34th Colloquiem *Protides of the Biological Fluids*, 34:473, 1986

Arg-Arg-Pro-Hyp-Gly-Thi-Ser-Phe-Thi-Arg
D-Arg-Arg-Pro-Hyp-Gly-Thi-Ser-D-Phe-Thi-Arg 2AcOH 4H$_2$O

Synonyms: Bradykinin, D-Arg-(Hyp^3,Thi^{5,8},D-Phe^7)-; Bradykinin B2 Receptor Antagonist

Peptides International PBK-4202   Synthetic   MW 1486.6   C$_{56}$H$_{83}$N$_{19}$O$_{13}$S$_2$ · 2CH$_3$COOH · 4H$_2$O   >99% (HPLC); lyophilized amorphous powder; bulk | Bioactive; Schachter, M et al, *Br J Pharmacol*, 92:851, 1987; Stewart, JM & RJ Vavrek, *Adv Biosci*, 65:73, 1987; Perry, DC, *Pharmacol Biochem Behav*, 28:15, 1987

Arg-Arg-Pro-Hyp-Gly-Thi-Ser-Phe-Thi-Arg
D-Arg-Arg-Pro-Hyp-Gly-Thi-Ser-D-Phe-Thi-Arg

Synonyms: Bradykinin, D-Arg-(Hyp^3,Thi^{5,8},D-Phe^7)-; Bradykinin B2 Receptor Antagonist

Peptides International PBK-4202-v   Synthetic   MW 1294.5   C$_{56}$H$_{83}$N$_{19}$O$_{13}$S$_2$ >99% (HPLC); lyophilized amorphous powder | Bioactive; Schachter, M et al, *Br J Pharmacol*, 92:851, 1987; Stewart, JM & RJ Vavrek, *Adv Biosci*, 65:73, 1987; Perry, DC, *Pharmacol Biochem Behav*, 28:15, 1987

Arg-Arg-Pro-Hyp-Gly-Thi-Ser-Tic-Arg
D-Arg-Arg-Pro-Hyp-Gly-Thi-Ser-D-Tic-Oic-Arg 2AcOH 4H$_2$O

Synonyms: Bradykinin, D-Arg-(Hyp^5,Thi^5,D-Tic^7,Oic^8)-; Hoe 140

Peptides International PKB-4293   Synthetic   MW 1496.6   C$_{59}$H$_{89}$N$_{19}$O$_{13}$S · 2CH$_3$COOH · 4H$_2$O   >99% (HPLC); lyophilized amorphous powder; bulk | Bioactive; Hock, FJ et al, *Br J Pharmacol*, 102:769, 1991; Wirth, K et al, *Br J Pharmacol*, 102:774, 1991; Baydon, AR & B Woodward, *Br J Pharmacol*, 103:1829, 1991

Arg-Arg-Pro-Hyp-Gly-Thi-Ser-Tic-Arg
D-Arg-Arg-Pro-Hyp-Gly-Thi-Ser-D-Tic-Oic-Arg

Synonyms: Bradykinin, D-Arg-(Hyp^3,Thi^5,D-Tic^7,Oic^8)-; Hoe 140; Bradykinin B2 Receptor Antagonist

Peptides International PKB-4293-v   Synthetic   MW 1304.6   C$_{59}$H$_{89}$N$_{19}$O$_{13}$S   >99% (HPLC); lyophilized amorphous powder | Bioactive; Hock, FJ et al, *Br J Pharmacol*, 102:769, 1991; Wirth, K et al, *Br J Pharmacol*, 102:774, 1991; Baydon, AR & B Woodward, *Br J Pharmacol*, 103:1829, 1991

Arg-Arg-Pro-Phe-His-Sta-Ile-His-Lys
Z-Arg-Arg-Pro-Phe-His-Sta-Ile-His-Lys(BOC)-OMe

Bachem C-3195.0005   MW 1495.79   C$_{72}$H$_{110}$N$_{20}$O$_{15}$ Store at -15°C

Arg-Arg-Pro-Phe-His-Sta-Ile-His-Lys
N$^\alpha$-CBZ-Arg-Arg-Pro-Phe-His-N$^\alpha$-BOC-Sta-Ile-His-Lys-OMe

Synonyms: Renin Inhibitor

Sigma C 9415   FW 1495.8   ≥85% (HPLC) | Bioactive peptide; specific & long acting; Wood, JM et al, *Hypertension*, 7:797, 1985

Arg-Arg-Pro-Pro-Gln-Gly-Ser-Gln-Thr-His-Gln-Val-Ser-Leu-Ser

Synonyms: TAT (56-70); HTLV I Peptide

Neosystem SC210   MW 1677.84   HTLV I peptide from subtype TAT

Arg-Arg-Pro-Pro-Gly-Ala-Ser-Arg
D-Arg-L-Arg-L-Pro-*trans*-4-Hydroxy-L-Pro-Gly-3-(2-Thienyl)-L-Ala-L-Ser-D-1,2,3,4-Tetrahydro-3-Isoquinolinecarbonyl-L-(2α,3β,7aβ)-Octahydro-1H-Indole-2-Carbonyl-L-Arg Free Base

Synonyms: Bradykinin B2 Receptor Antagonist; Hoe 140

Sigma H-157   Synthetic   MW 1304.6   C$_{59}$H$_{89}$N$_{19}$O$_{13}$S   >99%; white solid; peptide content & salt form information are provided with each lot | Selective; Feletou et al, *Br J Pharmacol*, 112:683, 1994; Wirth et al, *Br J Pharmacol*, 102:774, 1991; Trifilieff et al, *J Pharmacol Exp Ther*, 263:1377, 1992

Arg-Arg-Pro-Pro-Gly-Ala-Ser-Phe-Ala-Arg
D-Arg-Arg-Pro-Hydroxy-Pro-Gly-β-(2-Thienyl)Ala-Ser-D-Phe-β-(2-Thienyl)Ala-Arg

Synonyms: Bradykinin, (D-Arg-(Hyp^3,Thi^{5,8},D-Phe^7))-; Bradykinin Antagonist

Sigma B 1650   1294.5; ≥97% (HPLC) | Bioactive peptide; Stewart, JM & Vavrek, RJ, *The Proceedings of the 34th Colloquium "Peptides of the Biological Fluids"*, 34:473, 1986

Arg-Arg-Pro-Pro-Gly-Ala-Ser-Phe-Ala-Arg
N$^\alpha$-Aaa-D-Arg-Arg-Pro-Hydroxy-Pro-Gly-β-(2-Thienyl)Ala-Ser-D-Phe-β-(2-Thienyl)Ala-Arg

Synonyms: Bradykinin, (N$^\alpha$-Aaa-D-Arg-(Hyp^3,Thi^{5,8},D-Phe^7))-; Bradykinin Antagonist

Sigma B 6029   1470.8; ≥97% (HPLC) | Bioactive peptide; highly potent antagonist; no agonist activity; Lammek, B et al, *Peptides*, 11:1041, 1990

Arg-Arg-Pro-Pro-Gly-Phe-Ser-Phe-Phe-Arg
D-Arg-Arg-Pro-Hydroxy-Pro-Gly-Phe-Ser-D-Phe-Phe-Arg

Synonyms: Bradykinin, (D-Arg-(Hyp^3,D-Phe^7))-; Bradykinin Antagonist

Sigma B 1775   1282.5; ≥97% (HPLC) | Bioactive peptide; Stewart, JM & Vavrek, RJ, *The Proceedings of the 34th Colloquium "Peptides of the Biological Fluids"*, 34:473, 1986

Arg-Arg-Pro-Pro-Gly-Thi-Ser-Phe-Thi-Arg
N-AdamantaneAc-D-Arg-Arg-Pro-Hyp-Gly-Thi-Ser-D-Phe-Thi-Arg

Synonyms: Bradykinin, (N-Admantane-Ac-D-Arg^0-Hyp^3,Thi^{5,8},D-Phe^7)-

American Peptide 18-1-70   MW 1469.8   C$_{68}$H$_{98}$N$_{19}$O$_{14}$S$_2$ Lammek, B et al, *Peptides*, 11:1041, 1990

Arg-Arg-Pro-Pro-Gly-Thi-Ser-Phe-Thi-Arg
N-Adamantanecarbonyl-D-Arg-Arg-Pro-Hyp-Gly-Thi-Ser-D-Phe-Thi-Arg

Synonyms: Bradykinin, (N-Admantanecarbonyl-D-Arg^0-Hyp^3,Thi^{5,8},D-Phe^7)-; Bradykinin Antagonist

American Peptide 18-1-71   MW 1455.8   C$_{67}$H$_{96}$N$_{19}$O$_{14}$S$_2$ Highly potent

Arg-Arg-Pro-Trp-Ile-Leu

Synonyms: Neurotensin

American Peptide 62-1-12   MW 840.1   C$_{40}$H$_{65}$N$_{13}$O$_7$

Arg-Arg-Pro-Tyr-Ile-Leu

Synonyms: Neurotensin (8-13)

American Peptide 62-1-24 MW 817.0 $C_{38}H_{64}N_{12}O_8$
Smallest active fragment of neurotensin; St. Pierre, S et al, *J Med Chem*, 24:370, 1981

Arg-Arg-Pro-Tyr-Ile-Leu
Ac-Arg-Arg-Pro-Tyr-Ile-Leu

Synonyms: Neurotensin (8-13), (*N*-Ac)-; Neurotensin (8-13), Ac-

American Peptide 62-1-26 MW 859.1 $C_{40}H_{66}N_{12}O_9$
Shortest sequence with full binding & pharmacological activities; Granier, C et al, *Eur J Biochem*, 124:117, 1982

Bachem H-1020.0005 MW 859.04 $C_{40}H_{66}N_{12}O_9$ Store at -15°C

Arg-Arg-Pro-Tyr-Ile-Leu

Synonyms: Neurotensin (8-13)

Bachem H-1810.0005 MW 817 $C_{38}H_{64}N_{12}O_8$ Store at -15°C

ICN 152938 MW 817 $C_{38}H_{64}N_{12}O_8$ Smallest active fragment of neurotensin; St-Pierre, S etal, *J Med Chem*, 24:370, 1981

Arg-Arg-Pro-Tyr-Ile-Leu
N-α-Ac-Arg-Arg-Pro-Tyr-Ile-Leu

Synonyms: Neurotensin (8-13), (*N*-Ac)-

ICN 152939 MW 859 $C_{40}H_{66}N_{12}O_9$ Active fragment of neurotensin; Granier, C etal, *Eur J Biochem*, 124:117, 1982

Arg-Arg-Pro-Tyr-Ile-Leu
Ac-Arg-Arg-Pro-Tyr-Ile-Leu

Synonyms: Neurotensin (8-13), Ac-

Neosystem SC065 MW 859.04 Bioactive; Granier, C et al, *Mol Pharm Biochem*, 124:117, 1982

Arg-Arg-Pro-Tyr-Ile-Leu

Synonyms: Neurotensin (8-13)

Neosystem SC357 MW 817.00 Bioactive

Arg-Arg-Pro-Tyr-Ile-Leu Acetate Salt Free Base

Synonyms: Neurotensin (8-13)

Sigma N 5266 FW 817.0 $C_{38}H_{64}N_{12}O_8$ ≥97% (HPLC) | Bioactive peptide; smallest active fragment of neurotensin; St-Pierre, S et al, *J Med Chem*, 24:370, 1981

Arg-Arg-Ser-Ser-Cys-Phe-Gly-Gly-Arg-Ile-Asp-Arg-Ile-Gly-Ala-Gln-Ser-Gly-Leu-Gly-Cys-Asn-Ser-Phe-Arg-Tyr

Synonyms: Natriuretic Peptide (3-28), Atrial; Natriuretic Peptide (8-33), Atrial; Natriuretic Peptide (125-150), Atrial ; Natriuretic Factor (3-28), Atrial α-

Sigma A 4167 Human FW 2880.2 ≥97% (HPLC) | Bioactive peptide

American Peptide 14-5-44 Rat MW 2862.2 $C_{119}H_{189}N_{43}O_{36}S_2$ Disulfide bonds: Cys^7-Cys^{23} | Produces a natriuretic effect & relaxes isolated vascular smooth muscle preparations

ICN 159890 Rat MW 2862.7 98%

Neosystem SC028 Rat MW 2862.19 Disulfide bonds: Cys^7-Cys^{23}; bioactive; Winquist, RJ et al, *Eur J Pharmacol*, 102:169, 1984

Sigma A 6791 Rat FW 2862.2 ≥97% (HPLC) | Bioactive peptide; natriuretic activity is comparable to that of atrial extracts in anesthetized rats & in isolated vascular smooth muscle preparations; Winquist, RJ et al, *Eur J Pharmacol*, 102:169, 1984

Peptides International PAF-4159-v Rat synthetic MW 2862.2 $C_{119}H_{189}N_{43}O_{36}S_2$ >99% (HPLC); lyophilized amorphous powder | Bioactive; Disulfide bonds: Cys^7-Cys^{23}; Seidah, NG et al, *PNAS USA*, 81:2640, 1984

Arg-Arg-Ser-Ser-Cys-Phe-Gly-Gly-Arg-Met-Asp-Arg-Ile-Gly-Ala-Gln-Ser-Gly-Leu-Gly-Cys-Asn-Ser-Phe-Arg-Tyr

Synonyms: Natriuretic Peptide (3-28), Atrial ; Natriuretic Factor (3-28), Atrial; hANF (3-28)

American Peptide 14-1-18 Human MW 2880.3 $C_{117}H_{187}N_{43}O_{36}S_3$ Disulfide bonds: Cys^7-Cys^{23}

Bachem H-1335.0500 Human MW 2880.25 $C_{118}H_{187}N_{43}O_{36}S_3$ Store at -15°C

ICN 159887 Human MW 2880.8 98%

Arg-Arg-Trp-Cys-Tyr-Arg-Lys-Cys-Tyr-Lys-Gly-Tyr-Cys-Tyr-Arg-Lys-Cys-Arg Amide

Synonyms: T22; Polyphemusin II Derived Peptide; Polyphemusin II, ($Tyr^{5,12}$,Lys^7)-

American Peptide 72-3-32 MW 2487.0 $C_{109}H_{164}N_{38}O_{22}S_4$ Disulfide bonds: Cys^4-Cys^{17}, Cys^8-Cys^{13}; strong anti-HIV activity; exerts its effect by blocking virus-cell fusion process through binding to both gp120 & CD-4; Tamamura, H et al, *Biochimica et Biophysica Acta*, 1298:37, 1996; Tamamura, H et al, *BBRC*, 205:1729, 1994

Bachem H-2694.0500 Store at -15°C | Disulfide bonds: Cys^4-Cys^{17}, Cys^8-Cys^{13}

Arg-Arg-Trp-Gln-Gln-Leu-Leu-Ala-Leu-Ala-Asp-Arg-Ile-Tyr-Ser

Synonyms: REV (46-60); SIVmac251 Peptide

Neosystem SC600 MW 1889.18 SIVmac251 peptide from subtype REV

Arg-Arg-Trp-Gln-Trp-Arg-Met-Lys-Lys-Leu-Gly

Synonyms: Lactoferricin; BLFC

American Peptide 72-1-31 Bovine MW 1544.9 $C_{70}H_{113}N_{25}O_{13}S$ Antibiotic peptide; considered to be a major protecting component from bacteria at the mucosal surfaces & in colostrum & milk; the basic AA-rich region of bovine lactoferricin is involved in the interaction with bacterial phospholipid membranes & plays an important role in antimicrobial activity; Kang, JH et al, *Int J Pept Protein Res*, 48(4):357, 1996

Arg-Arg-Trp-Trp-Cys-Arg
Ac-Arg-Arg-Trp-Trp-Cys-Arg Amide

Synonyms: Interleukin VIII Inhibitor

Bachem H-2268.0005 MW 1003.2 $C_{45}H_{66}N_{18}O_7S$ Store at -15°C

Arg-Asn Amide

Bachem G-3700.0250 MW 287.32 $C_{10}H_{21}N_7O_3$ Store at -15°C

Arg-Asn-Ala-Thr-Ala-Val
Dnp-Arg-Asn-Ala-Thr-Ala-Val Amide

Bachem M-2085.0025 MW 795.81 $C_{31}H_{49}N_{13}O_{12}$ Store at -15°C

Arg-Asn-Arg-Leu-Ile-Pro-Pro-Phe-Trp-Lys-Thr-Arg Amide

Synonyms: CTX IV (3-14), ($Arg^{3,14}$)-

Neosystem SC153 MW 1582.91 Bioactive; cardiotoxic peptide

Arg-Asn-Ile-Ala-Glu-Ile-Ile-Lys-Asp-Ile

Synonyms: Laminin B2 Chain Peptide R-10-I

Bachem H-1016.0005 MW 1184.4 $C_{52}H_{93}N_{15}O_{16}$ Store at -15°C

Neosystem SC337 MW 1184.40 Bioactive; cell attachment peptide; Liesi, P et al, *FEBS Lett*, 244:141-148, 1989

Arg-Asp

Bachem G-1470.0250   MW 289.29 $C_{10}H_{19}N_5O_5$ Store at -15°C

Arg-Asp
Arg-D-Asp

Bachem G-4455.0250   MW 289.29 $C_{10}H_{19}N_5O_5$ Store at -15°C

Arg-Asp
retro-Arg-malonyl-Asp

Bachem H-3918.0050   Store at -15°C

Arg-Asp

ICN 150390   MW 289.3 $C_{10}H_{19}N_5O_5$ Crystalline

Sigma A 2387   FW 289.3 $C_{10}H_{19}N_5O_5$

Arg-Asp-Ala-Gly-Ser-Gln-Arg-Pro-Arg-Lys-Lys-Glu-Asp-Asn-Val-Leu-Val-Glu-Ser-His-Glu-Lys-Ser-Leu-Gly

Synon Parathyroid Hormone (44-68)*yms:*

American Peptide 22-1-42   Human   MW 2836.1 $C_{117}H_{199}N_{41}O_{41}$ Rosenblatt, M et al, *JMC*, 20:1452, 1977

Bachem H-1820.0500   Human   MW 2836.12 $C_{117}H_{199}N_{41}O_{41}$ Store at -15°C

Arg-Asp-Ala-Gly-Ser-Gln-Arg-Pro-Arg-Lys-Lys-Glu-Asp-Asn-Val-Leu-Val-Glu-Ser-His-Glu-Lys-Ser-Leu-Gly
Biotinyl-Arg-Asp-Ala-Gly-Ser-Gln-Arg-Pro-Arg-Lys-Lys-Glu-Asp-Asn-Val-Leu-Val-Glu-Ser-His-Glu-Lys-Ser-Leu-Gly

S Parathyroid Hormone (44-68), Biotinyl-*ynonyms:*

Bachem H-5790.0500   Human   Store at -15°C

Arg-Asp-Ala-Gly-Ser-Gln-Arg-Pro-Arg-Lys-Lys-Glu-Asp-Asn-Val-Leu-Val-Glu-Ser-His-Glu-Lys-Ser-Leu-Gly

Synonyms: Parathyroid Hormone (44-68); Parathormone

ICN 152988   Human   MW 2836.1

Sigma P 3155   Human   FW 2836.1   ≥97% (HPLC) | Bioactive peptide

Arg-Asp-Cys-Cys-Thr-Hyp-Arg-Lys-Cys-Lys-Asp-Arg-Arg-Cys-Lys-Hyp-Met-Lys-Cys-Cys-Ala Amide

Synonyms: Conotoxin GIIIB, μ-; Geographutoxin II

Calbiochem 234620   *Conus geographus*   MW 2640.2 $C_{101}H_{175}N_{39}O_{30}S_7$ >99% (HPLC); lyophilized solid; soluble in water & 0.9% NaCl; LD_{50}≤50 mg/kg | Disulfide bonds: Cys[3]-Cys[15], Cys[4]-Cys[20], Cys[10]-Cys[21]; selective blocker of skeletal muscle voltage-dependent Na[+] channels; composed of 22 amino acid residues including six Cys & three *trans*-4-hydroxy-L-Pro residues; Kubo, S et al, *Peptide Res*, 6:66, 1993; Gray, WR et al, *Ann Rev Biochem*, 57:665, 1988; Robitaille, R & Charlton, MP, *J Neurosci*, 12:297, 1992; Hong, SJ & Chang, CC, *Br J Pharmacol*, 97:934, 1989

Arg-Asp-Cys-Cys-Thr-Hyp-Hyp-Arg-Lys-Cys-Lys-Asp-Arg-Arg-Cys-Lys-Hyp-Met-Lys-Cys-Cys-Ala Amide

Synonyms: Geographutoxin II ; Conotoxin GIIIB, μ-; Na[+] Channel Blocker

Bachem H-9015.0500   MW 2640.21 $C_{101}H_{175}N_{39}O_{30}S$ Store at -15°C | Disulfide bonds: Cys[3]-Cys[15], Cys[4]-Cys[20], Cys[10]-Cys[21]

Sigma C 1676   FW 2640.2   ≥95% (HPLC); Disulfide bonds: undefined; peptide content:~60% | Bioactive peptide; vertebrate skeletal muscle Na[+]-channel blocker; Kubo, S et al, *Peptide Chem*, 1989:257, 1990; Hong, SJ & Chang, CC, *Brit J Pharmacol*, 97:934, 1989; Cruz, LJ et al, *J Biol Chem*, 260:9280, 1985

American Peptide 41-0-51   *Conus geographus* (marine snail)   MW 2637.2 $C_{101}H_{172}N_{39}O_{30}S_7$ LD_{50}:500 mg/kg | Disulfide bonds: Cys[3]-Cys[15], Cys[4]-Cys[20], Cys[10]-Cys[21]; isolated from venom; selective blocker of skeletal muscle voltage-dependent Na[+] channels; Kubo, S et al, *Peptide Chem*, 1989:257, 1990

Peptides International PCN-4217-v   *Conus geographus* (marine snail) synthetic   MW 2640.2 $C_{101}H_{175}N_{39}O_{30}S_7$ >95% (HPLC); lyophilized amorphous powder | Bioactive; Disulfide bonds: Cys[3]-Cys[15], Cys[4]-Cys[20], Cys[10]-Cys[21]; specific for skeletal muscle; Sato, S et al, *FEBS Lett*, 155:277, 1983; Cruz, LJ et al, *JBC*, 260:9280, 1985; Ohizumi, Y et al, *JBC*, 261:6149, 1986; Kubo, S et al, *Pept Res*, 6:66, 1993

Alexis 630-054   Synthetic   MW 2640.2 $C_{101}H_{175}N_{39}O_{30}S_7$ White lyophilized powder; Disulfide bonds: Cys[3]-Cys[15], Cys[4]-Cys[20], Cys[10]-Cys[21]; Potent neurotoxin | Originally isolated from *Conus geographus*; see Alexis 630-023

Arg-Asp-Cys-Cys-Thr-Hyp-Hyp-Lys-Cys-Lys-Asp-Arg-Gln-Cys-Lys-Hyp-Gln-Arg-Cys-Cys-Ala Amide Trifluoroacetate Salt

Synonyms: Conotoxin GIIIA, μ-

Calbiochem 234622   *Conus geographus*   MW 2609 $C_{100}H_{170}N_{38}O_{32}S_6$ ≥95% (HPLC); lyophilized solid; soluble in water; LD_{50}≤50 mg/kg | Disulfide bonds: Cys[3]-Cys[15], Cys[4]-Cys[20], Cys[10]-Cys[21]; potent neurotoxin; selectively blocks vertebrate skeletal muscle voltage-dependent Na[+] channels; Hidaka, Y et al, *FEBS Lett*, 264:29, 1990; Olivera, BN et al, *Science*, 249:257, 1990; Cruz, LJ et al, *Biochemistry*, 28:3437, 1989; Cruz, LJ et al, *J Biol Chem*, 260:9280, 1985

Arg-Asp-Cys-Cys-Thr-Hyp-Hyp-Lys-Lys-Cys-Lys-Asp-Arg-Gln-Cys-Lys-Hyp-Gln-Arg-Cys-Cys-Ala Amide

Synony Conotoxin GIIIA, μ-*ms:*

American Peptide 41-0-50   MW 2606.1 $C_{100}H_{167}N_{38}O_{32}S_6$ Disulfide bonds: Cys[3]-Cys[15], Cys[4]-Cys[20], Cys[10]-Cys[21]; isolated from venom; skeletal muscle Na[+] channel blocker; selective blocker of voltage-dependent Na[+] channels; exhibits channel binding kinetics similar to those of tetrodotoxin & saxitoxin, but possesses a 1000-fold selectivity for muscle versus neuronal Na[+] channels; Cruz, LJ et al, *Biochemistry*, 28:3437, 1989

Bachem H-2738.0500   MW 2609.08 $C_{100}H_{170}N_{38}O_{32}S_6$ Store at -15°C | Disulfide bonds: Cys[3]-Cys[15], Cys[4]-Cys[20], Cys[10]-Cys[21]

Sigma C 7420   FW 2609.0   ≥90% (HPLC); Disulfide bonds: Undefined; peptide content:~60% | Bioactive peptide; skeletal muscle Na[+]-channel blocker; Cruz, LJ et al, *Biochem*, 28:3437, 1989

Arg-Asp-Hyp-Cys-Cys-Tyr-His-Pro-Thr-Cys-Asn-Met-Ser-Asn-Pro-Gln-Ile-Cys Amide

Synonyms: Conotoxin EI, α-

Bachem H-3394.0500   Store at -15°C | Disulfide bonds: Cys[4]-Cys[10], Cys[5]-Cys[18]

American Peptide 41-0-80   *Conus ermineus*   MW 2092.4 $C_{83}H_{124}N_{27}O_{27}S_5$ Disulfide bonds: Cys[4]-Cys[10], Cys[5]-Cys[18]; novel nicotinic acetylcholine receptor ligand; Martinez, JS et al, *Biochemistry*, 34:14519, 1995

Calbiochem 234618   *Conus ermineus*   MW 2609 $C_{83}H_{125}N_{27}O_{27}S_5$ ≥95% (HPLC); lyophilized solid; soluble in 0.1% TFA; LD_{50}≤50 mg/kg | Disulfide bonds: Cys[4]-Cys[10]; Cys[5]-Cys[18]; novel selective nicotinic acetylcholine receptor (nAChR) antagonist; selectively binds the agonist site near the α/δ interface in Torpedo nAChRs & shows high affinity for both the α/δ - & α/γ-subunit interfaces in mammalian nAChRs; Martinez, JS et al, *Biochemistry*, 34:14519, 1995

Arg-Asp-Leu-Glu-Lys-His-Gly-Ala-Ile-Thr-Ser-Ser

Synonyms: NEF (35-46); HTLV I Peptide

Neosystem SC501   MW 1313.43   HTLV I peptide from subtype NEF

Arg-Asp-Leu-Glu-Lys-His-Gly-Ala-Leu-Thr-Ser-Ser

Synonyms: NEF MN (37-48); HIV I Peptide

Neosystem SC641   MW 1313.43   NEF peptide from HIV-1 subtype MN

Arg-Asp-Leu-Pro-Phe-Phe-Pro-Val-Pro-Ile-Asp

Synonyms: Amyloid α-Protein (15-25), (Arg[15],Asp[16,25],Pro[18,21,23],Val[22],Ile[24])-

Bachem H-3904.0001   MW 1315.54 $C_{64}H_{94}N_{14}O_{16}$ Store at -15°C

Arg-Asp-Tyr-Ser-Glu-Leu-Ala-Leu-Asn-Val-Thr-Glu-Ser-Phe-Asp-Lys

Synonyms: GP140-Lys (63-76), Arg-; SIVmac251 Peptide

Neosystem SC714   MW 1887.03   SIVmac251 peptide from subtype GP140

Arg-Asp-Tyr-Thr-Gly-Trp-Nle-Asp-Phe
Arg-Asp-Tyr(SO₃H)-Thr-Gly-Trp-Nle-Asp-Phe Amide Sulfated

Synonyms: Cholecystokinin (25-33), (Thr[28],Nle[31])-

Bachem H-1825.0001   Store at -15°C

Arg-Asp-Tyr-Thr-Gly-Trp-Nle-Asp-Phe
Arg-Asp-Tyr(SO³H)-Thr-Gly-Trp-Nle-Asp-Phe Amide

Synonyms: Cholecystokinin (25-33), (Tyr(SO₃H)[27],Thr[28],Nle[31])-NH₂); Gastrointestinal Peptide

ICN 195679   MW 1252.3   ≥95%

Sigma C 2421   FW 1252.3   ≥97% (HPLC) | Bioactive peptide

Arg-Cha-Arg-Phe
Ac-Arg-D-Cha-BTD-D-Arg-*p*-Chloro-D-Phe Amide

Synonyms: Nociceptin Receptor Antagonist Peptide III-BTP

Neosystem SC1351   MW 890.50   Neuropeptide; binds with high affinity to ORL1 receptor; has antagonist acid activity to ORL1 & partial agonist activity at m, d &k-opioid receptors; Becker, JAJ et al, *JBC*, 274:27513-27522, 1999

Arg-Cys-Ala-Gly-Gly-Arg-Ile-Asp-Arg-Ile-Tic-Arg-Cys
Arg-Cys-α-Cyclohexyl-Ala-Gly-Gly-Arg-Ile-Asp-Arg-Ile-D-Tic-Arg-Cys Amide

Synonyms: Natriuretic Factor (6-18), Atrial (Arg[6],α-Cyclohexyl-Ala[8],D-Tic[16],Arg[17],Cys[18])-

Bachem H-3048.0500   Rat   Store at -15°C

Arg-Cys-Arg-Val-His-Cys-Pro
D-Arg-Cys-Arg-Val-His-Cys-Pro

Synonyms: Antistasin (32-38), (D-Arg[32])-; Antistasin Related Peptide

Bachem H-2674.0001   Store at -15°C

Arg-Cys-Gly-Val-Pro-Asp
Ac-Arg-Cys-Gly-Val-Pro-Asp Amide

Synonyms: Matrix Metalloproteinase III Inhibitor; Stromelysin I Inhibitor

Bachem H-2504.0005   MW 551.87 $C_{27}H_{46}N_{10}O_9S$ Store at -15°C

Calbiochem 444218   MW 686.8 $C_{27}H_{46}N_{10}O_9S$ Solid; ≥90% (HPLC); soluble in water | Fotouhi, N et al, *J Biol Chem*, 269:30227, 1994; Hanglow, AC et al, *Agents Actions*, 39:C148, 1993

ICN 196022   MW 686.8 $C_{27}H_{47}N_{10}O_9S$ ≥90% | Inhibits the matrix metalloproteinase stromelysin 1

Arg-Cys-Ser-Ser-Asn-Ile-Thr-Gly-Leu-Leu-Leu-Thr
Arg-Cys(Acm)-Ser-Ser-Asn-Ile-Thr-Gly-Leu-Leu-Leu-Thr

Synonyms: GP120 (449-460); HTLV I Peptide

Neosystem SC567   MW 1348.58   HTLV I peptide from subtype GP160

Arg-Cys-Ser-Ser-Asn-Ile-Thr-Gly-Leu-Leu-Leu-Thr-Arg-Asp-Gly-Gly

Synonyms: GP120 (449-464); HTLV I Peptide

Neosystem SC249   MW 1662.88   HTLV I peptide from subtype GP160; due to the presence of a Cys this peptide can contain the dimeric form

---

Arg-Cys-Val-Arg-Leu-His-Glu-Ser-Cys-Leu-Gly-Gln-Gln-Val-Pro-Cys-Cys-Asp-Pro-Cys-Ala-Thr-Cys-Tyr-Cys-Arg-Phe-Phe-Asn-Ala-Phe-Cys-Tyr-Cys-Arg-Lys-Leu-Gly-Thr-Ala-Met-Asn-Pro-Cys-Ser-Arg-Thr

Synonyms: Agouti Related Protein (86-132); Melanocortin Receptor-3/4 Antagonist; Appetite Boosting Peptide

Peptides International PAR-4366-s   Human synthetic   MW 5347.2 $C_{223}H_{339}N_{69}O_{63}S_{11}$ >95% (HPLC); lyophilized amorphous powder | Bioactive; Disulfide bonds: Cys[87]-Cys[102], Cys[94]-Cys[108], Cys[101]-Cys[119], Cys[105]-Cys[129], & Cys[110]-Cys[117]; Rosenfeld, RD et al, *Biochem*, 37:16041, 1998; Bures, EJ et al, *Biochem*, 37:12172, 1998; Shutter, JR et al, *Genes Dev*, 11:593, 1997

Arg-Dab-Pro-Tyr-Ile-Leu

Synonyms: Neurotensin (8-13), (Dab[9])-

Bachem H-3404.0005   Store at -15°C

Arg-Gln

Bachem G-4230.0250   MW 302.33 $C_{11}H_{22}N_6O_4$ Store at -15°C

Arg-Gln-Ala-Asn-Phe-Leu-Gly-Lys-Ile-Trp-Pro-Ser-Tyr-Lys-Gly-Arg

Synonyms: P15 (429-444); GAG P7 NC (52-55)/P1 (1-12); HTLV I Peptide

Neosystem SC274   MW 1921.23   HTLV I peptide from subtype P15 (GAG P7 NC/GAG P1/GAG P6)

Arg-Gln-Arg-Arg
Arg-Gln-Arg-Arg-AMC

Bachem I-1600.0025   Store at -15°C

Arg-Gln-Arg-Gln-Arg-Ser-Thr-Ser-Thr-Pro-Asn-Val-His-Met-Val-Ser-Thr-Thr-(Gly-Cys)

Synonyms: Raf1 Blocking Peptide (25-42)

Calbiochem 543503   Human synthetic   MW 2246.5 $C_{88}H_{152}N_{34}O_{31}S_2$ Lyophilized | Based on human, rat & frog Raf1 sequence (in human, 25-42); N-terminus had an Arg residue substituted for the native Ser & at the C-terminus, a Gly & a Cys residue were added to facilitate coupling to KLH; used as the immunogen for the production of Anti-Raf1; for use in immunoabsorption for immunoprecipitation, immunocytochemistry, Western blotting & dot blots

---

Arg-Gln-Asp-Arg-Val-Phe-His-Ser-Arg-Asn-Ser-Ile

Synonyms: Fibronectin (1371-1382), (Phe[1376])-; Fibronectin (1371-1382)

Bachem H-2568.0001   MW 1514.67 $C_{63}H_{103}N_{25}O_{19}$ Store at -15°C

ICN 195618   MW 1514.7   97%

Sigma F 0668   FW 1514.7   ≥97% (HPLC); peptide content:~70% | Bioactive peptide; overlapping sequences from the fibronectin cell-binding domain; inhibit platelet aggregation by interfering with fibronectin binding to platelet membrane glycoprotein IIb/IIIa; Mohri, H et al, *Peptides*, 16:263, 1995

Arg-Gln-Asp-Arg-Val-Pro-His-Ser-Arg-Asn-Ser-Ile

Synonyms: Fibronectin Fragment (1371-1382)

Bachem H-3608.0001   Store at -15°C

Arg-Gln-Asp-Ile-Leu-Asp-Leu-Trp-Ile-Tyr-His-Thr-Gln-Gly

Synonyms: NEF (106-119); HTLV I Peptide

Neosystem SC508   MW 1757.96   HTLV I peptide from subtype NEF

Arg-Gln-Asp-Ile-Leu-Asp-Leu-Trp-Val-Tyr-His-Thr-Gln-Gly

Synonyms: NEF MN (108-121); HIV I Peptide

Neosystem SC653   MW 1745.90   NEF peptide from HIV-1 subtype MN

Arg-Gln-Gln-Phe-Phe-Gly-Leu-Met
D-Arg-Gln-Gln-Phe-Phe-Gly-Leu-Met Amide

Synonyms: Substance P (4-11), (D-Ala[4])-

American Peptide 70-1-56   MW 940.1 $C_{44}H_{65}N_{11}O_{10}S$

Arg-Gln-Pro-Lys-Lys-Ser-Asn-Glu-Leu-Pro-Gln-Cys
Acetate Salt

Synonyms: Bone Morphogenetic Protein 8 (8-18), ~C

Biogenesis 1406-1390   Synthetic   >80% (HPLC); lyophilized

Arg-Glu

Bachem G-1475.0250   MW 303.32 $C_{11}H_{21}N_5O_5$ Store at -15°C

Arg-Glu Acetate Salt

ICN 150391   MW 447.5 $C_{11}H_{21}N_5O_5 \cdot C_2H_4O_2$

Arg-Glu-Arg-Met-Ser

Synonyms: Amyloid β/A4 Precursor Protein (328-332); Amyloid/A4 Precursor Protein (328-332), β-; Amyloid α/A4 Precursor Protein 770 (403-407)

Alexis 151-011   MW 677.8 $C_{25}H_{47}N_{11}O_9S$ ≥97%; white lyophilized powder; soluble in TFA/H2O | Active domain of amyloid β/A4 protein precursor (APP) that promotes fibroblast growth; Ninomiya, H et al, *J Cell Biol*, 121:879, 1993

American Peptide 62-5-15   MW 677.8 $C_{25}H_{47}N_{11}O_9S$
Growth-promoting activity on fibroblasts & may represent the only site of the secreted form of amyloid β/A4 protein precursor (sAPP-695) involved in the growth stimulation; also involved in the pathogenesis of Alzheimer's disease; Ninomiya, H et al, *J Cell Biol*, 121:879, 1993

Bachem H-1608.0005   MW 677.78 $C_{25}H_{47}N_{11}O_9S$ Store at -15°C

Arg-Glu-Asp-Val

American Peptide 44-0-12   MW 517.5 $C_{20}H_{35}N_7O_9$

Arg-Glu-Glu-Val-Asn-Leu-Asp-Ala-Glu-Phe-Lys-Arg
Arg-Glu(EDANS)-Glu-Val-Asn-Leu-Asp-Ala-Glu-Phe-Lys(DABCYL)-Arg

Synonyms: Amyloid α/A4 Precursor Protein 770 (668-675) - Lys(DABCYL)-Arg, Arg-Glu(EDANS)-(Asn[670],Leu[671])-

Bachem M-2470.0001   MW 2005.24 $C_{91}H_{129}N_{25}O_{25}S$ Store at -15°C

Arg-Glu-Glu-Val-Asn-Leu-Asp-Ala-Glu-Phe-Lys-Asp-Arg
Arg-Glu(EDANS)Glu-Val-Asn-Leu-Asp-Ala-Glu-Phe-Lys-(Asp-acyl)-Arg

Synonyms: B-Secretase Inhibitor

Kamiya   MW 1650 $C_{73}H_{118}N_{16}O_{27}$ 99±1% (HPLC)

Arg-Glu-Lys-Arg
Z-Arg-Glu-Lys-Arg-AMC

Bachem I-1590.0025   Store at -15°C

Arg-Glu-Thr-Gln-Ile-Ala-Lys-Gly-Asn-Glu-Gln-Ser-Phe-Arg-Val-Asp-Leu-Arg-Thr-Leu-Leu-Arg-Tyr-Tyr

Synonyms: MHC Class I Derived Peptide

Bachem H-1942.0500   MW 2957.34 $C_{130}H_{210}N_{40}O_{39}$ Store at -15°C

Arg-Gly
N-Ac-DL-PropArg-Gly

Sigma A 6263   FW 155.2 $C_7H_9NO_3$

Arg-Gly
L-Prop-Arg-Gly

Synonyms: Pentynoic Acid, 2-Amino-4-

USBio P9053   MW 113.1 $C_5H_7NO_2$ ≥98%

Arg-Gly Amide Sulfate

Bachem G-1485.0250   MW 328.35 $C_8H_{18}N_6O_2 \cdot H_2SO_4$ Store at -15°C

Arg-Gly Hydrochloride

Bachem G-3940.0250   MW 267.72 $C_8H_{17}N_5O_3 \cdot HCl$ Store at -15°C

Sigma A 6301   FW 267.7 $C_8H_{17}N_5O_3 \cdot HCl$

Arg-Gly-Arg
BOC-D-Arg-Gly-Arg-CHO

American Peptide 81-5-90   MW 471.6 $C_{19}H_{37}N_9O_5$

Arg-Gly-Arg
Z-D-Arg-Gly-Arg-pNA Dihydrochloride

Bachem L-2115.0025   MW 714.61 $C_{28}H_{39}N_{11}O_7 \cdot 2HCl$ Store at -15°C

Arg-Gly-Arg-Arg-Gln-Pro-Ile-Pro-Lys-Ala

Synonyms: HCV Core Protein (59-68)

American Peptide 72-3-33   MW 1178.4 $C_{50}H_{91}N_{21}O_{12}$ Very important tool for the serodiagnosis of HCV in infected patients; Sallberg, M et al, *J Med Virol*, 43:62, 1994

Bachem H-2542.0001   Store at -15°C

Arg-Gly-Asp

Synonyms: Fibronectin Fragment; Fibronectin

Bachem H-1830.0005   MW 346.34 $C_{12}H_{22}N_6O_6$ Store at -15°C

Calbiochem 03-34-0029   MW 346.4 $C_{12}H_{22}N_6O_6$
Lyophilized; ≥99% (HPLC); soluble in 5% HOAc | AA sequence within fibronectin that mediates cell attachment; also identified in other proteins; D'Souza, SE et al, *Trends Biochem Sci*, 16:246 1991; Iida, J et al, *J Cell Biol*, 118:431, 1992; Eggleston, DS et al, *Int J Pep Protein*, 36:161, 1990

ICN 153052   MW 346.3 $C_{12}H_{22}N_6O_6$

Neosystem SC341   MW 346.34   Bioactive; *cell attachment* peptide; fibronectin fragment; inactive in cell adhesion studies; Plow, EF et al, *PNAS*, 82:8057, 1985

Sigma A 8052 FW 346.3 $C_{12}H_{22}N_6O_6$ ≥97% (TLC) |
Bioactive peptide; tripeptide from the cell-attachment domain of fibronectin

Arg-Gly-Asp-Cys

American Peptide 44-0-13 MW 448.5 $C_{15}H_{26}N_7O_7S$

Bachem H-3156.0005 MW 449.49 $C_{15}H_{27}N_7O_7S$ Store at -15°C

Arg-Gly-Asp-Ser

Synonyms: Fibronectin Inhibitor; Fibronectin; Fibronectin Active Fragment

American Peptide 44-0-14 MW 433.4 $C_{15}H_{27}N_7O_8$
Supports fibroblast attachment & inhibits fibronectin binding to platelets; target sequence for syphilis spirochete cytadherence; Gartner, TK et al, *JBC*, 260:11891, 1985

Bachem H-1155.0005 MW 433.42 $C_{15}H_{27}N_7O_8$ Store at -15°C

Calbiochem 03-34-0002 MW 433.4 $C_{15}H_{27}N_7O_8$
Lyophilized; ≥97% (HPLC); soluble in 5% HOAc | Inhibits fibronectin function for binding to platelet-binding sites; Ginsberg, M et al, *J Biol Chem*, 260:3931, 1985

ICN 153053 MW 433.4 $C_{15}H_{27}N_7O_8$ Supports fibroblast attachment & inhibits fibronectin binding to platelets; a target sequence for syphilis spirochete cytadherence; Pierschbacher, MD & E Ruoslahti, *Nature*, 309:30, 1984

Neosystem SC122 MW 433.42 Bioactive; cell attachment peptide; fibronectin fragment; exhibits cell attachment-promoting activity; Gartner, TK & Bennett, JS, *JBC*, 260:11891, 1985

Sigma A 9041 FW 433.4 $C_{15}H_{27}N_7O_8$ ≥95% (HPLC); peptide content:~70% | Bioactive peptide; tripeptide which supports fibroblast attachment & inhibits fibronectin binding to platelets; target sequence for syphilis spirochete cytadherence; Pierschbacher, MD & Ruoslahti, E, *Nature*, 309:30, 1984

Peptides International PFA-4171-v Synthetic MW 433.42 $C_{15}H_{27}N_7O_8$ >99% (HPLC); lyophilized amorphous powder | Bioactive; Piershbacher, MD & E Ruoslahti, *Nature*, 309:30, 1984; Haverstick, DM et al, *Blood*, 66:946, 1985

Arg-Gly-Asp-Ser ½AcOH Dihydrate

Synonyms: Fibronectin Active Fragment

Peptides International PFA-4171 Synthetic MW 499.48 $C_{15}H_{27}N_7O_8$ ½CH_3COOH · $2H_2O$ >99% (HPLC); lyophilized amorphous powder; bulk | Bioactive; Piershbacher, MD & E Ruoslahti, *Nature*, 309:30, 1984; Haverstick, DM et al, *Blood*, 66:946, 1985

Arg-Gly-Asp-Ser-Pro-Ala-Ser-Ser-Lys-Pro

Synonyms: Fibronectin Inhibitor; Fibronectin

American Peptide 44-0-15 MW 1001.2 $C_{40}H_{68}N_{14}O_{16}$
Inhibits fibronectin adhesion to fibroblasts

Bachem H-6720.0001 MW 1001.07 $C_{40}H_{68}N_{14}O_{16}$ Store at -15°C

Calbiochem 03-34-0003 MW 1001.1 $C_{40}H_{68}N_{14}O_{16}$
Lyophilized; ≥97% (HPLC); soluble in 5% HOAc | Inhibits fibronectin binding to cells; Levesque, JP et al, *PNAS*, 83:6494, 1986

ICN 153054 MW 1001.1 Inhibits fibronectin binding to fibroblasts; Yamada, KM & DW Kennedy, *J Cell Biol*, 99:29, 1984; Levesque, JP etal, *PNAS*, 83:6494, 1986

Sigma A 6677 FW 1001.1 ≥97% (HPLC) | Bioactive peptide; inhibits fibronectin binding to fibroblasts; Yamada, KM & Kennedy, DW, *J Cell Biol*, 99:29, 1984; Levesque, JP et al, *Proc Natl Acad Sci USA*, 83:6494, 1986

Arg-Gly-Asp-Thr

American Peptide 44-0-16 MW 447.5 $C_{16}H_{29}N_7O_8$

Arg-Gly-Asp-Trp
D-Arg-Gly-Asp-Trp

Bachem H-3332.0005 Store at -15°C

Neosystem SC978 MW 532.56 Bioactive; cell attachment peptide; fibronectin fragment; causes substantial inhibition of platelet aggregation & adhesion to fibrinogen & fibrin; Hantgan, RR et al, *Thromb Haemost*, 68:694-700, 1992; Kieffer, B et al, *Int J Pep Prot Res*, 44:70-79, 1994; Braaten, JV et al, *Blood*, 83:982-993, 1994; Carr, ME et al, *Thromb Haemost*, 73:499-505, 1995; Hantgan, RR et al, *Blood*, 86:1001-1009, 1995

Arg-Gly-Asp-Val

American Peptide 44-0-17 MW 445.5 $C_{17}H_{31}N_7O_7$

Bachem H-3158.0005 MW 445.48 $C_{17}H_{31}N_7O_7$ Store at -15°C

Neosystem SC1270 MW 445.47 Bioactive; cell attachment peptide; fibronectin fragment; platelet aggregation inhibitor; Syversen, PV et al, *Thromb Res*, 76:299-305, 1994

Arg-Gly-Gln-Trp-Phe-Gly-Asp
Arg-Gly-Gln-D-Trp-Phe-Gly-Asp(OtBu)-OtBu 2TFA

Synonyms: Substance P (4-10), (Arg[4],Gly[5],D-Trp[7],Asp(OtBu)-OtBu[10])-

Bachem H-1405.0001 Store at -15°C

Arg-Gly-Glu-Arg-Thr-Ala-Phe-Ile-Lys-Asp-Gln-Ser-Ala-Leu

Synonyms: Furin Convertase C-Terminal Peptide (780-793)

Alexis 156-001 Synthetic MW 1591.8 $C_{66}H_{114}N_{22}O_{22}$ ≥95%; lyophilized; reconstitute with 0.1 mL distilled H_2O | Competitively binds to Ab Alexis 210-134; antiserum blocking peptide

Arg-Gly-Glu-Ser

Synonyms: Fibronectin

American Peptide 44-0-18 MW 447.5 $C_{16}H_{29}N_7O_8$ Gartner, TK et al, *JBC*, 260:11891, 1985

Bachem H-7745.0005 MW 447.45 $C_{16}H_{29}N_7O_8$ Store at -15°C

Sigma A 1550 FW 447.4 $C_{16}H_{29}N_7O_8$ ≥97% (HPLC) | Bioactive peptide

Arg-Gly-Glu-Ser Acetate Salt

ICN 195617 MW 507.4 $C_{16}H_{29}N_7O_8$ · $C_2H_4O_2$ ≥97% | Inhibits platelet aggregation

Arg-Gly-Glu-Ser Acetate Salt Free Base

Synonyms: Fibronectin

Sigma A 5686 FW 447.4 $C_{16}H_{29}N_7O_8$ ≥95% (HPLC) | Bioactive peptide; inhibitor of platelet aggregation; similar to Sigma A 1550 but produced by Sigma; Gartner, TK et al, *Blood*, 66 Suppl (Abstr 1104):305a, 1985

Arg-Gly-Ile-Pro-Ser-Lys-Lys-Pro-Val-Ala-Asp-Tyr-Phe-Leu

Synonyms: Protein Phosphatase X/C Blocking Peptide (294-307)

Calbiochem 539540 Rabbit polyclonal MW 1590.9 $C_{75}H_{119}N_{19}O_9$ Liquid; affinity purified | Recognizes mammalian PPX catalytic subunit (34 kDa); exhibits minor cross-reactivity with PP2A/C which can be eliminated by including the PP2A/C blocking peptide

Arg-Gly-Leu
Bz-Arg-Gly-Leu-4MβNA Hydrochloride

Bachem J-1055.0050 MW 640.18 $C_{32}H_{41}N_7O_5$ · HCl Store at -15°C

Arg-Gly-Phe-Ala-Phe-Val-Thr-Phe
(Arg-Gly-Phe-Ala-Phe-Val-Thr-Phe)₈-MAP

Bachem H-7660.0001   Store at -15°C

Arg-Gly-Phe-Phe

Synonyms: Insulin B Chain (22-25)

Bachem H-6005.0025   MW 525.61 $C_{26}H_{35}N_7O_5$ Store at -15°C

Biogenesis 5329-4102   Human synthetic   MW 585   >95% (HPLC); lyophilized

Arg-Gly-Phe-Phe-Leu
Bz-Arg-Gly-Phe-Phe-Leu-4MβNA

Bachem J-1060.0050   MW 898.08 $C_{50}H_{59}N_9O_7$ Store at -15°C

Arg-Gly-Phe-Phe-Pro
Bz-Arg-Gly-Phe-Phe-Pro

Bachem H-2140.0050   MW 726.83 $C_{38}H_{41}N_8O_7$ Store at -15°C

Arg-Gly-Phe-Phe-Pro
Bz-Arg-Gly-Phe-Phe-Pro-4MβNA Hydrochloride

Synonyms: Cathepsin D Substrate I

Bachem J-1065.0050   MW 918.49 $C_{49}H_{55}N_9O_7 \cdot HCl$ Store at -15°C

ICN 195955   MW 918.5 $C_{49}H_{55}N_9O_7 \cdot HCl$ >98% | Substrate for the quantitative determination of cathepsin D

Arg-Gly-Phe-Phe-Tyr-Thr-Pro-Lys-Ala

Synonyms: Insulin B Chain (22-30)

ICN 155047

Sigma I 5508   FW 1086.3   ≥97% (HPLC) | Bioactive peptide

Arg-Gly-Phe-Pro
Bz-Arg-Gly-Phe-Pro-4MβNA Hydrochloride

Bachem J-1330.0050   MW 771.32 $C_{40}H_{46}N_8O_6$ Store at -15°C

Arg-Gly-Pro-Cys-Arg-Ala-Phe-Ile

Synonyms: Urinary Trypsin Inhibitor Fragment

Bachem H-2692.0001   MW 919.12 $C_{40}H_{66}N_{14}O_9S$ Store at -15°C

Sigma U 4751   FW 919.1 $C_{40}H_{66}N_{14}O_9S$ ≥97% (HPLC); peptide content: ~70% | Bioactive peptide; fragment of domain II of the urinary trypsin inhibitor (UTI); corresponds to active sequence of UTI for plasmin inhibition; Kobayashi, H et al, *Cancer Res*, 55:1847, 1995

Arg-Gly-Pro-Gly-Arg-Ala-Phe-Val

Synonyms: GP120 (316-323); HTLV I Peptide

Neosystem SC815   MW 859.00   HTLV I peptide from subtype GP160

Arg-Gly-Pro-Gly-Arg-Ala-Phe-Val-Thr-Ile

Synonyms: V3 Decapeptide P18-I10; GP120 (318-327); HIV Peptide

American Peptide 72-3-34   MW 1073.3 $C_{48}H_{80}N_{16}O_{12}$ The minimal peptide presented by all four MHC class I molecules to CTL; Shirai, M et al, *J Immunology*, 158:3181, 1997

Bachem H-2696.0001   Store at -15°C

Arg-Gly-Pro-Phe-Pro-Ile

Synonyms: Substance IB, α-; Sexual Agglutination Peptide

Bachem H-1840.0005   MW 685.82 $C_{33}H_{51}N_9O_7$ Store at -15°C

ICN 152835   MW 685.8 $C_{33}H_{51}N_9O_7$ Aoyagi, H, *Experientia*, 33:870, 1977

Sigma A 2152   FW 685.8 $C_{33}H_{51}N_9O_7$ ≥90% (HPLC) | Bioactive peptide; Aoyagi, H, *Experientia*, 3:870, 1977

Arg-Gly-Tyr-Ala-Leu-Gly

Synonyms: cAMP-Dependent Protein Kinase Competitive Inhibitor

Alexis 151-022   MW 635.7 $C_{28}H_{45}N_9O_8$ Kemp, BE et al, *PNAS*, 73:1038, 1976

American Peptide 86-0-52   MW 635.8 $C_{28}H_{45}N_9O_8$ Kemp, BE et al, *PNAS*, 73:1038, 1976

Bachem M-1105.0005   MW 635.72 $C_{28}H_{45}N_9O_8$ Store at -15°C

Arg-Gly-Tyr-Ser-Leu-Gly

Synonyms: cAMP-Dependent Protein Kinase Substrate

Alexis 151-021   MW 651.7 $C_{28}H_{45}N_9O_9$ Kemp, BE et al, *PNAS*, 73:1038, 1976; Kemp, BE et al, *Fed Proc*, 35:1384, 1976

American Peptide 86-0-51   MW 651.8 $C_{28}H_{45}N_9O_9$ Kemp, BE et al, *PNAS*, 73:1038, 1976

Arg-Gly-Tyr-Ser-Leu-Gly
L-Arg-Gly-L-Tyr-L-Ser-L-Leu-Gly

Synonyms: Phosphate Acceptor Peptide

ICN 150392   MW 651.8 $C_{28}H_{45}N_9O_9$ Kemp, etal, *PNAS*, 73:1038, 1976

Arg-Gly-Tyr-Val-Tyr-Gln-Gly-Leu

Synonyms: Vesicular Stomatitis Virus Peptide; NP (52-59)

Neosystem SC1346   MW 955.08   Virus-related peptide; VSV H-2K(b)-restricted dominant CTL epitope; Van Bleek, GM & Nathenson, SG, *Nature*, 348:213-216, 1990

Arg-Gly-Val-Phe-Arg

Synonyms: Serum Albumin Propeptide Sequence

Bachem H-1845.0005   MW 633.75 $C_{28}H_{47}N_{11}O_6$ Store at -15°C

Arg-Gly-Val-Val-Asn-Ala
DABCYL-Arg-Gly-Val-Val-Asn-Ala

Bachem H-2442.0001   Store at -15°C

Arg-Gly-Val-Val-Asn-Ala-Ser-Ser-Arg-Leu-Ala
4-(4-Dimethylaminophenylazo)Benzoyl-Arg-Gly-Val-Val-Asn-Ala-Ser-Ser-Arg-Leu-Ala-5-((2-Aminoethyl)Amino)-Naphthalene-1-Sulfonic Acid

Synonyms: CMV Protease Fluorogenic Substrate, Human

American Peptide 72-3-20   MW 1631.9 $C_{72}H_{108}N_{23}O_{19}S$ The first fluorescence-based assay of the herpes virus proteases; Holskin, BP et al, *Anal Biochem*, 227:148, 1995

Arg-Gly-Val-Val-Asn-Ala-Ser-Ser-Arg-Leu-Ala
Arg-Gly-Val-Val-Asn-Ala-(®)-Ser-Ser-Arg-Leu-Ala

Synonyms: Human CMV Assemblin Protease Inhibitor

Bachem N-1420.0001   Store at -15°C

Arg-Gly-Val-Val-Asn-Ala-Ser-Ser-Arg-Leu-Ala
DABCYL-Arg-Gly-Val-Val-Asn-Ala-Ser-Ser-Arg-Leu-Ala-EDANS

Synonyms: CMV Protease Substrate

Bachem M-2060.0001   Human   Store at -15°C | Fluorogenic

Arg-Gly-Val-Val-Asn-Ala-Ser-Ser-Arg-Leu-Ala-Lys

Synonyms: Human CMV Assemblin Protease Substrate (M-site)

Bachem M-2065.0001   Store at -15°C

Arg-His Amide

Bachem G-3850.0250   Store at -15°C

Arg-His-His-Tyr-Glu-Ser-Pro-His-Pro-Arg-Ile-Ser

Synonyms: VIF (41-52); HTLV I Peptide

Neosystem SC544   MW 1515.65   HTLV I peptide from subtype VIF

Arg-His-Ile-Pro-Arg-Arg-Ile-Arg-Gln-Gly-Leu-Glu-Arg-Ile-Leu

Synonyms: GP41 (846-860); HTLV I Peptide

Neosystem SC242   MW 1913.30   HTLV I peptide from subtype GP41

Arg-His-Lys-Thr-Asp-Ser-Phe-Val-Gly-Leu-Met Amide

Synonyms: Neurokinin A, (Arg⁰)-

Neosystem SC015   MW 1289.51   Bioactive; tachykinin

Arg-His-Phe Diacetate Salt

ICN 153056   MW 578.6 $C_{21}H_{30}N_8O_4 \cdot 2C_2H_4O_2$
Sigma A 2541   FW 578.6 $C_{21}H_{30}N_8O_4 \cdot 2C_2H_4O_2$ ≥97% (HPLC) | Bioactive peptide

Arg-His-Phe-Trp-Gln-Gln

Bachem H-5360.0005   MW 901 $C_{42}H_{56}N_{14}O_9$ Store at -15°C

Arg-His-Trp-Ser-Tyr-Gly-Trp-Leu-Pro
D-Arg-His-Trp-Ser-Tyr-Gly-Trp-Leu-Pro-NHEt

Synonyms: Luteinizing Hormone Releasing Hormone, (D-Arg¹,Trp⁷,Leu⁸,des-Gly¹⁰)-

ICN 152920

Arg-Ile

Bachem G-1490.0250   MW 287.36 $C_{12}H_{25}N_5O_3$ Store at -15°C

Arg-Ile Acetate Salt

Sigma A 7036   FW 347.4 $C_{12}H_{25}N_5O_3 \cdot C_2H_4O_2$

Arg-Ile-Arg-Pro-Lys-Leu-Lys-Trp-Asp-Asn-Gln

Synonyms: Dynorphin A (7-17)

American Peptide 26-4-31   Porcine   MW 1453.7 $C_{65}H_{108}N_{22}O_{16}$

Arg-Ile-Arg-Thr-Trp-Lys-Ser-Leu-Val-Lys

Synonyms: VIF (17-26); HTLV I Peptide

Neosystem SC541   MW 1286.58   HTLV I peptide from subtype VIF

Arg-Ile-Cys-Tyr-Ile-His-Lys-Ala-Ser-Leu-Pro-Arg-Ala-Thr-Lys-Thr-Cys-Val-Glu-Asn-Thr-Cys-Tyr-Lys-Met-Phe-Ile-Arg-Thr-Gln-Arg-Glu-Tyr-Ile-Ser-Glu-Arg-Gly-Cys-Gly-Cys-Pro-Thr-Ala-Met-Trp-Pro-Tyr-Gln-Thr-Glu-Cys-Cys-Lys-Gly-Asp-Arg-Cys-Asn-Lys

Synonyms: Calciseptine

Alexis 630-006   *Dendroaspis polylepis polylepis*   MW 7036.2 $C_{299}H_{468}N_{90}O_{87}S_{10}$ ≥95%; lyophilized; soluble in H_2O | Selective blocker of L-type, voltage-dependent Ca^{2+} channels in various cell types; potent neurotoxin; in contrast to classical L-type Ca^{2+} channel drugs, does not affect L-subtype Ca^{2+} channels of skeletal muscle; Weille, JR et al, *PNAS*, 88:2437, 1991; Schweitz, H et al, *Toxicon*, 28:847, 1990; Yasuda, O et al, *BBRC*, 194:587, 1993

Calbiochem 208274   *Dendroaspis polylepis polylepis*   MW 7036.2 $C_{299}H_{468}N_{90}O_{87}S_{10}$ ≥98% (SDS-PAGE); lyophilized; soluble in H_2O; S-S linkage not determined; harmful:LD_{50} ≤2000 mg/kg | A peptidyl toxin that acts as a potent & selective blocker of L-type Ca^{2+} channels in various cell types; also blocks K^+-induced contractions in cardiac cells; Schweitz, H et al, *Toxicon*, 28:847, 1990; Yasuda, O et al, *Biochem Biophys Res Comm*, 194:587, 1993; DeWeille, JR et al, *PNAS*, 88:2437, 1991

Arg-Ile-Cys-Tyr-Ile-His-Lys-Ala-Ser-Leu-Pro-Arg-Ala-Thr-Lys-Thr-Cys-Glu-Asn-Thr-Cys-Tyr-Lys-Met-Phe-Ile-Arg-Thr-Gln-Arg-Glu-Tyr-Ile-Ser-Glu-Arg-Gly-Cys-Gly-Cys-Pro-Thr-Ala-Met-Trp-Pro-Tyr-Gln-Thr-Glu-Cys-Cys-Lys-Gly-Asp-Arg-Cys-Asn-Lys

Synonyms: Calciseptine

Sigma C 1836   *Dendroaspis polylepis polylepis* (black mamba)   FW 7036.1   ≥97% (HPLC); vial contains 100 µg; peptide content:~65%; Disulfide bonds: 3-22, 17-39, 41-52, 53-58 | Bioactive peptide; L-type Ca^{2+} channel blocker; Weille, JR De et al, *Proc Natl Acad Sci USA*, 88:2437, 1991; Kuroda, H et al, *Pept Res*, 5:265, 1992

Arg-Ile-Cys-Tyr-Ile-His-Lys-Ala-Ser-Leu-Pro-Arg-Ala-Thr-Lys-Thr-Cys-Val-Glu-Asn-Thr-Cys-Tyr-Lys-Met-Phe-Ile-Arg-Thr-Gln-Arg-Glu-Tyr-Ile-Ser-Glu-Arg-Gly-Cys-Gly-Cys-Pro-Thr-Ala-Met-Trp-Pro-Tyr-Gln-Thr-Glu-Cys-Cys-Lys-Gly-Asp-Arg-Cys-Asn-Lys

Synonyms: Calciseptine; L-Type Ca^{2+} Channel Blocker

Peptides International PCL-4255-s   *Dendroaspis polylepis polylepis* (black mamba) synthetic   MW 7036.2 $C_{299}H_{468}N_{90}O_{87}S_{10}$ >95% (HPLC); lyophilized amorphous powder | Bioactive; Disulfide bonds: Cys³-Cys²², Cys¹⁷-Cys³⁹, Cys⁴¹-Cys⁵², Cys⁵³-Cys⁵⁸; De Weille, JR et al, *PNAS* USA, 88:2437, 1991; Kuroda, H et al, *Pept Res*, 5:265, 1992; Watanabe, TX et al, *Jpn J Pharmacol*, 68:305, 1995; Teramoto, N et al, *Pflügers Arch*, 432, 1996

Arg-Ile-Cys-Tyr-Ser-His-Lys-Ala-Ser-Leu-Pro-Arg-Ala-Thr-Lys-Thr-Cys-Val-Glu-Asn-Thr-Cys-Tyr-Lys-Met-Phe-Ile-Arg-Thr-His-Arg-Glu-Tyr-Ile-Ser-Glu-Arg-Gly-Cys-Gly-Cys-Pro-Thr-Ala-Met-Trp-Pro-Tyr-Gln-Thr-Glu-Cys-Cys-Lys-Gly-Asp-Arg-Cys-Asn-Lys

Synonyms: FS2

Calbiochem 344158   *Dendroaspis polylepis polylepis*   MW 7019.1 $C_{297}H_{461}N_{91}O_{87}S_{10}$ ≥98% (HPLC); lyophilized; soluble in H_2O; biological activity:fully active as tested by its ability to block spontaneous or K^+-induced contraction of cardiac cells; harmful:LD_{50} ≤2000 mg/kg | Polypeptide toxin that selectively blocks non-skeletal muscle L-type Ca^{2+} channels; close homolog of Calciseptine (Calbiochem Cat. No. 208274) with similar *in vitro* & *in vivo* biological activities; competitively inhibits the binding of ³H-nitrendipine to rat brain synaptosomal membranes (Kd=210 nM); Schweitz, H et al, *Toxicon*, 28:847, 1990; Yasuda, O et al, *Artery*, 21:287, 1994; Albrand, J-P et al, *Biochemistry*, 34:5923, 1995

Arg-Ile-Gln-Arg-Gly-Pro-Gly-Arg

Synonyms: GP120 (313-320); HTLV I Peptide

Neosystem SC812   MW 939.08   HTLV I peptide from subtype GP160

Arg-Ile-Gln-Arg-Gly-Pro-Gly-Arg-Ala-Phe-Val-Thr-Ile-Gly-Lys

Synonyms: HIV I Inhibitory Peptide, R14K; p18 Peptide; GP120 (315-329); HIV Peptide; GP120 (313-327); HTLV I Peptide

American Peptide 72-2-39   MW 1656.0 $C_{73}H_{126}N_{26}O_{18}$
Derived from the V3 loop of HIV-1 IIIB gp120, efficiently blocks infection of several T-cell lines & normal human T cells by HIV-1 IIIB; inhibits syncytium formation in human cells; induces HIV-1 specific cytoxic T lymphocyte responses that effectively kill virus-infected cells & could fin several therapeutic applications

American Peptide 72-3-36   MW 1656 $C_{73}H_{126}N_{26}O_{18}$
Derived from HIV-1 env gp160; immunogenic & contains both T-cell & B-cell epitopes; due to its broad reactivity, P18 Peptide is one of the candidates for inclusion as a subunit vaccine against HIV; Moukrim, Z et al, *Biomed & Pharmacother*, 50:494, 1996

Bachem H-1402.0001   MW 1655.97 $C_{73}H_{126}N_{26}O_{18}$ Store at -15°C

Neosystem SC1007   MW 1655.96   HTLV I peptide from subtype GP160

Arg-Ile-His-Ile-Gly-Pro-Gly-Arg-Ala-Phe-Tyr-Thr-Thr-Lys-Asn

Synonyms: GP120 V3 Loop Peptide MN (311-325); HIV I Peptide

Neosystem SC694   MW 1730.98   HIV-1 GP120 V3 Loop peptide

Arg-Ile-Leu-Ala-Val-Glu-Arg-Tyr-Leu-Lys-Asp-Gln-Gln-Leu-Leu-Gly-Ile-Trp-Gly-Cys-Ser

Synonyms: GP41 (584-604); HTLV I Peptide

Neosystem SC802   MW 2461.90   HTLV I peptide from subtype GP41; due to the presence of a Cys this peptide can contain the dimeric form

Arg-Ile-Leu-Ala-Val-Glu-Arg-Tyr-Leu-Lys-Asp-Gln-Gln-Leu-Leu-Gly-Ile-Trp-Gly-Cys-Ser-Gly-Lys

Synonyms: GP41; HIV Envelope Protein (579-601)

Sigma H 4521   FW 2647.1   ≥97% (HPLC) | Bioactive peptide; peptide corresponding to the region of the HIV envelope protein (gp41) identified as a major epitope recognized by Ab of AIDS patients; Kemp, BE et al, *Science*, 241:1352, 1988; Wang, JJG et al, *Proc Natl Acad Sci USA*, 83:6159, 1986

Arg-Ile-Leu-Ser-Thr-Tyr-Leu-Gly-Arg-Ser-Ala-Glu

Synonyms: REV (58-69); HTLV I Peptide

Neosystem SC553   MW 1365.55   HTLV I peptide from subtype REV

Arg-Leu

Bachem G-1495.0250   MW 287.36 $C_{12}H_{25}N_5O_3$ Store at -15°C

Arg-Leu
FA-Arg-Leu

Bachem M-1995.0050   MW 407.47 $C_{19}H_{29}N_5O_5$ Store at -15°C

Arg-Leu Acetate Salt

Sigma A 8282   FW 347.4 $C_{12}H_{25}N_5O_3 \cdot C_2H_4O_2$

Arg-Leu-Ala-Phe-His-His-Val-Ala-Arg-Glu-Leu-His-Pro-Glu

Synonyms: NEF (188-201); HTLV I Peptide

Neosystem SC518   MW 1711.94   HTLV I peptide from subtype NEF

Arg-Leu-Arg-Gly-Gly
Z-Arg-Leu-Arg-Gly-Gly-AMC

Synonyms: Isopeptidase T Substrate

Bachem I-1690.0025   MW 848.96 $C_{40}H_{56}N_{12}O_9$ Store at -15°C

Neosystem SC1310   MW 848.96   Fluorogenic substrate; Stein, RL et al, *Biochem*, 34:12616-12623, 1995

Arg-Leu-Arg-Phe-Asp

Synonyms: Bag Cell Peptide Factor, γ-

American Peptide 88-0-96   MW 705.8 $C_{31}H_{51}N_{11}O_8$
Scheller, RH et al, *Cell*, 32:7, 1983

Arg-Leu-Arg-Phe-His

Synonyms: Bag Cell Peptide Factor, β-

American Peptide 88-0-95   MW 727.8 $C_{33}H_{53}N_{13}O_6$
Scheller, RH et al, *Cell*, 32:7, 1983

Arg-Leu-Cys-Arg-Ile-Val-Val-Ile-Arg-Val-Cys-Arg

Synonyms: Bactenecin

Bachem H-9585.0500   Store at -15°C

Neosystem SC183   MW 1483.90   Disulfide bonds: Cys^3-Cys^{11}; bioactive; antibacterial/antimicrobial peptide; Romeo, D et al, *JBC*, 263:9573-9575, 1988

American Peptide 73-3-10   Bovine neutrophil granules   MW 1483.9   Bovine; $C_{63}H_{118}N_{24}O_{13}S_2$ Disulfide bonds: Cys^3-Cys^{11} | Antibiotic peptide; cyclic cationic dodecapeptide; exhibits a significant antibacterial activity *in vitro* that is comparable to that of the defensins; Domenico, R et al, *J Biol Chem*, 263:No. 20, 9573-9575, 1988

Arg-Leu-His-Gln-Asn-Gly-Met-Pro-Phe-Ser-Pro-Arg-Leu Amide

Synonyms: Locustamyotropin IV; Lom-MT IV

Bachem H-1574.0001   MW 1551.84 $C_{68}H_{110}N_{24}O_{16}S$ Store at -15°C

Arg-Lys

Bachem G-1505.0250   MW 302.38 $C_{12}H_{26}N_6O_3$ Store at -15°C

Arg-Lys Acetate Salt Free Base

Sigma A 4783   FW 302.4 $C_{12}H_{26}N_6O_3$

Arg-Lys-Arg-Ala-Arg-Lys-Glu

Synonyms: Protein Kinase G Inhibitor; Histone H2B (29-35), (Ala^{32})-; cGMP-Dependent Protein Kinase Inhibitor; Protein Kinase Related Peptide

Alexis 158-002   MW 943.1 $C_{38}H_{74}N_{18}O_{10}$ ≥98%; white lyophilized powder; soluble in H_2O | Derived by a serine-for-alanine substitution at position 32 of histone H2B (29-35); for the corresponding unsubstituted histone fragment see Alexis 158-003; Glass, DB & Miller, MD, *Fed Proc*, 39:867, 1980; Glass, DB, *Biochem J*, 213:159, 1983

ICN 195838   MW 943.1   ≥97% | Histone H2B (29-35) non-phosphorylating analog of substrate peptide of PKG

Sigma A 8186   FW 943.1 $C_{38}H_{74}N_{18}O_{10}$ ≥97% (HPLC) | Bioactive peptide; Glass, D et al, *Biochem J*, 213:159, 1983

Arg-Lys-Arg-Ala-Arg-Lys-Glu Trifluoroacetate Salt

Synonyms: Protein Kinase G Inhibitor

Calbiochem 370654   MW 943.1 $C_{38}H_{74}N_{18}O_{10}$ ≥97% (HPLC); lyophilized solid; soluble in water | A more specific inhibitor of PKG relative to PKA; sequence corresponds to a non-phosphorylatable analog $(Ser^{32}$ to $Ala^{32})$ of histone H2B (residues 29-35); Glass, DB et al, *Biochem J*, 213:159, 1983

Arg-Lys-Arg-Ala-Arg-Lys-Gly

Synonyms: cGMP-Dependent Protein Kinase Inhibitor

American Peptide 86-0-56   MW 943.2 $C_{38}H_{74}N_{18}O_{10}$

Arg-Lys-Arg-Ser-Arg-Ala-Glu

Synonyms: Histone H2B (29-35), (Ala^{34})-; cGMP-Dependent Protein Kinase Substrate; cGMP Protein Kinase Substrate

Alexis 158-004   MW 902.0 $C_{35}H_{67}N_{17}O_{11}$ ≥98%; white lyophilized powder; soluble in H_2O | For the corresponding unsubstituted histone fragment see Alexis 158-003; Glass, DB & Krebs, EG, *J Biol Chem*, 257:1196, 1982

American Peptide 86-0-59   MW 902.1 $C_{35}H_{67}N_{17}O_{11}$ Glass, DB & Krebs, EG, *JBC*, 257:1196, 1982

Bachem H-3214.0001   MW 902.03 $C_{35}H_{67}N_{17}O_{11}$ Store at -15°C

Neosystem SC870   MW 902.02   Bioactive; proteinkinase-related peptide

Arg-Lys-Arg-Ser-Arg-Lys-Glu

Synonyms: Histone H2B (29-35); cGMP-Dependent Protein Kinase Substrate; cGMP Protein Kinase Substrate

Alexis 158-003   MW 959.1 $C_{38}H_{74}N_{18}O_{11}$ ≥98%; white lyophilized powder; soluble in H_2O | For the corresponding Ala-for-Ser substituted peptide see Alexis 158-002; Glass, DB & Krebs, EG, *JBC*, 254:9728, 1979

American Peptide 86-0-55   MW 959.2 $C_{38}H_{74}N_{18}O_{11}$ Glass, DB & Krebs, EG, *JBC*, 254:9728, 1979

Neosystem SC416   MW 959.12   Bioactive; proteinkinase-related peptide; Glass, DB & Krebs, EG, *JBC*, 254:9728, 1979

Arg-Lys-Arg-Thr-Leu-Arg-Arg-Leu

Synonyms: Epidermal Growth Factor Receptor (651-658); Protein Kinase C Substrate

Alexis 155-021   MW 1098.4 $C_{46}H_{91}N_{21}O_{10}$ ≥97%; white lyophilized powder; soluble in H_2O | Potent substrate; O'Brian, CA et al, *Invest New Drugs*, 9:169, 1991

Arg-Lys-Arg-Thr-Leu-Arg-Arg-Leu
Myr-Arg-Lys-Arg-Thr-Leu-Arg-Arg-Leu

Synonyms: Epidermal Growth Factor Receptor (651-658); Protein Kinase C Inhibitor

Alexis 155-022   MW 1308.7 $C_{60}H_{117}N_{21}O_{11}$ ≥97%; white lyophilized powder; soluble in H_2O; myristoylated | Potent, cell permeable & possibly selective inhibitor of protein kinase C; the *N*-terminal myristoylated form of Alexis 155-021; O'Brian, CA et al, *Invest New Drugs*, 9:169, 1991; Carmichael, WW, *J Appl Bacteriol*, 72:445, 1992; Ward, NG & O'Brian, CA, *Biochemistry*, 32:11903, 1993

Arg-Lys-Arg-Thr-Leu-Arg-Arg-Leu

Bachem M-1950.0005   MW 1098.36 $C_{46}H_{91}N_{21}O_{10}$ Store at -15°C

Arg-Lys-Arg-Thr-Leu-Arg-Arg-Leu
Myr-Arg-Lys-Arg-Thr-Leu-Arg-Arg-Leu

Bachem N-1310.0001   MW 1308.72 $C_{60}H_{117}N_{21}O_{11}$ Store at -15°C

Arg-Lys-Arg-Thr-Leu-Arg-Arg-Leu
Myr-*N*-Arg-Lys-Arg-Thr-Leu-Arg-Arg-Leu
Trifluoroacetate Salt Myristoylated

Synonyms: Protein Kinase C Inhibitor; Epidermal Growth-Factor Receptor (651-658)

Calbiochem 476475   MW 1308.7 $C_{60}H_{117}N_{21}O_{11}$ ≥97% (HPLC); lyophilized solid; soluble in water | Conserved sequence which is identical to v-*erb*B (95-102); an *N*-terminal myristoylated membrane-permeable inhibitor of protein kinase C; Ward, NE & O'Brian, CA, *Biochemistry*, 32:11903, 1993; O'Brian, CA et al, *Invest New Drugs*, 9:169, 1991

Arg-Lys-Arg-Thr-Leu-Arg-Arg-Leu

Synonyms: Epidermal Growth Factor (651-658); Protein Kinase Substrate

ICN 195834   MW 1098.5   ≥97% | Conserved sequence of EGF-R that is identical to v-*erb*-B (95-102); contains EGF-R Thr-654, a PKC phosphorylation site

Arg-Lys-Arg-Thr-Leu-Arg-Arg-Leu
N-Myr-Arg-Lys-Arg-Thr-Leu-Arg-Arg-Leu

Synonyms: Epidermal Growth Factor (651-658)

ICN 195836   MW 1308.1   ≥97%; myristoylated | Conserved sequence of EGF-R that is identical to v-*erb*-B (95-102); *N*-terminal myristoylation allows membrane permeability & inhibits proteinkinase C in intact cells

Arg-Lys-Asp

Synonyms: Thymopoietin II (32-34)

Bachem H-8760.0050   MW 417.47 $C_{16}H_{31}N_7O_6$ Store at -15°C

Arg-Lys-Asp(Asp-Val-Tyr)-Val-Tyr

Bachem H-8690.0005   Store at -15°C

Arg-Lys-Asp-Val

Synonyms: Thymopoietin II (32-35)

Bachem H-5915.0005   MW 516.6 $C_{21}H_{40}N_8O_7$ Store at -15°C

Arg-Lys-Asp-Val-Tyr

Synonyms: Thymopentin; Thymopoietin V

American Peptide 87-0-22   MW 679.8 $C_{30}H_{49}N_9O_9$

Arg-Lys-Asp-Val-Tyr
Arg-Lys(Et)-Asp-Val-Tyr

Synonyms: Thymopoietin II (32-26), (Lys(Et)[33])-

Bachem H-1032.0005   Store at -15°C

Arg-Lys-Asp-Val-Tyr
Arg-Lys-Asp-Val-Tyr-OEt

Synonyms: Thymopoietin II Ethyl Ester (32-36); Thymopoietin V Ethyl Ester

Bachem H-1034.0005   Store at -15°C

Arg-Lys-Asp-Val-Tyr

Synonyms: Thymopoietin (32-36); Thymopoietin II (32-36); Thymopentin; Thymopoietin Pentapeptide

ICN 152142   Golstein, G etal, *Science*, 204:1309, 1979

Neosystem SC944   MW 679.77   Bioactive; reproduces biological activity of thymopoietin; Goldstein, G et al, *Science*, 204:1309-1310, 1979; Audhya, T et al, *PNAS* USA, 81:2847-2849, 1984; Audhya, T et al, *PNAS* USA, 84:3545-3549, 1987

Sigma T 8806   FW 679.8 $C_{30}H_{49}N_9O_9$ ≥97% (HPLC) | Bioactive peptide

Arg-Lys-Asp-Val-Tyr Acetate Salt Free Base

Synonyms: Thymopoietin II (32-36); Thymopentin; Thymopoietin V

Sigma A 4777   FW 679.8 $C_{30}H_{49}N_9O_9$ ≥97% (HPLC) | Bioactive peptide; active fragment of thymopoietin II; Goldstein, G et al, *Science*, 204:1309, 1979

Arg-Lys-Glu-Lys-Met-Arg-Gly-Lys-Met-Glu-Asn-Glu-Lys-Arg-Arg-Glu-Lys-Tyr-Cys

Synonyms: HPK1 Blocking Peptide (473-490)

Calbiochem 403785   Mouse synthetic   MW 2500.0 $C_{102}H_{179}N_{37}O_{30}S_3$ Lyophilized | Based on mouse HPK1 (residues 473-490) with a cysteine C residue added; this peptide coupled to KLH, was used as the immunogen for the production of Anti-HPK1; for use in immunoabsorption for immunoprecipitation, immunocytochemistry, Western blotting & dot blots

Arg-Lys-Glu-Val-Tyr

Synonyms: SP-5; Splenin (32-36); Splenopentin

Bachem H-6930.0005   Store at -15°C

Neosystem SC943   MW 693.80   Bioactive; reproduces biological activity of splenin; Goldstein, G et al, *Science*, 204:1309-1310, 1979; Audhya, T et al, *PNAS USA*, 81:2847-2849, 1984; Audhya, T et al, *PNAS USA*, 84:3545-3549, 1987

Arg-Lys-Glu-Val-Tyr Acetate Salt

Synonyms: Splenopentin; Splenin (32-36), SP-5

ICN 153057   Bovine   Active fragment of bovine splenin; Goldstein, G etal, *PNAS*, 81:2847, 1984

Arg-Lys-Glu-Val-Tyr Acetate Salt Free Base

Synonyms: Splenin (32-36); Splenopentin; SP-5

Sigma A 2042   Bovine   FW 693.8 $C_{31}H_{51}N_9O_9$ ≥97% (HPLC) | Bioactive peptide; active fragment of bovine splenin; Goldstein, G et al, *Proc Natl Acad Sci USA*, 81:2847, 1984

Arg-Lys-Gly-Asp-Ile-Lys-Ser-Tyr
Arg-Lys-Gly-Asp-Ile-Lys-Ser-Tyr-pNA

Bachem L-2110.0001   MW 1086.22 $C_{48}H_{75}N_{15}O_{14}$ Store at -15°C

Arg-Lys-Ile-Asp-Arg-Leu-Ile-Asp-Arg-Leu-Ile-Glu-Arg-Ala-Glu

Synonyms: VPU (36-50); HTLV I Peptide

Neosystem SC530   MW 1896.22   HTLV I peptide from subtype VPU

Arg-Lys-Ile-Ser-Ala-Ser-Glu-Phe

Synonyms: cGMP-Dependent Protein Kinase

Promega V7451   MW 937   References are the same as for Promega V5601

Arg-Lys-Ile-Ser-Ala-Ser-Glu-Phe-Asp-Arg-Pro-Leu-Arg

Synonyms: cGMP Specific Phosphodiesterase; cGMP-Dependent Protein Kinase Substrate; BPDEtide

Alexis 152-022   MW 1574.8 $C_{68}H_{115}N_{23}O_{20}$ White lyophilized powder | Corresponds to a peptide sequence surrounding the phosphorylation site of cGMP binding cGMP specific phosphodiesterase; Corbin, JD & Doskeland, SO, *J Biol Chem*, 258:11391, 1983; Colbran, JL et al, *J Biol Chem*, 267:9589, 1992

Bachem H-3216.0001   Store at -15°C

Arg-Lys-Ile-Ser-Ala-Ser-Glu-Phe-Asp-Arg-Pro-Leu-Arg
Biotinyl-Arg-Lys-Ile-Ser-Ala-Ser-Glu-Phe-Asp-Arg-Pro-Leu-Arg

Bachem H-4308.0001   Store at -15°C

Arg-Lys-Ile-Ser-Ala-Ser-Glu-Phe-Asp-Arg-Pro-Leu-Arg

Synonyms: Protein Kinase G Substrate; BPDEtide

ICN 195837   Bovine lung   MW 1575   ≥97% | From bovine lung cGMP specific phosphodiesterase; specific cGMP dependent PKG substrate

Arg-Lys-Ile-Ser-Ala-Ser-Glu-Phe-Asp-Arg-Pro-Leu-Arg Trifluoroacetate Salt

Synonyms: Protein Kinase G Substrate, BPDEtide

Calbiochem 203678   Bovine lung   MW 1574.8 $C_{68}H_{115}N_{23}O_{20}$ ≥97% (HPLC); lyophilized solid; soluble in water | Sequence derived from bovine lung cGMP specific phosphodiesterase (cG-BPDE); specific & sensitive substrate for PKG relative to PKA; not phosphorylated by PKC or by CaM kinase II; Colbran, JL et al, *J Biol Chem*, 267:9589, 1992; Patel, AI & Diamond, J, *J Pharmacol Exp Ther*, 283:885, 1997; Corbin, JD & Doskeland, SO, *J Biol Chem*, 258:11391, 1983

Arg-Lys-Leu-Pro-Pro-Arg-Pro-Arg-Arg Amide Trifluoroacetate Salt

Synonyms: Phosphatidyl Inositol 3-Kinase SH3 Domain Binding Protein

Calbiochem 526550   MW 1174.5 $C_{51}H_{95}N_{23}O_9$ >98% (HPLC); lyophilized solid; soluble in water | Sequence is derived from a directed random library using PI3K SH3 domain as the ligand; selectively binds to PI3K SH3; Chen, JK et al, *J Am Chem Soc*, 115:12591, 1993

Arg-Lys-Lys-Gly-Cys-Trp-Lys-Cys-Gly-Lys-Glu-Gly-His-Gln

Synonyms: P15 (409-422); GAG P7 NC (32-45) HTLV I Peptide

Neosystem SC272   MW 1644.92   HTLV I peptide from subtype P15 (GAG P7 NC/GAG P1/GAG P6); due to the presence of a Cys this peptide can contain the dimeric form

Arg-Lys-Pro-Trp-Tle-Leu
N^α-Me-Arg-Lys-Pro-Trp-Tle-Leu

Synonyms: Neurotensin (8-13), (N^α-Me-Arg[8],Lys[9],Trp[11],Tle[12])-

Neosystem SC979   MW 826.16   Bioactive; this stable analog reported to have high affinity for the NT receptor & appears to possess central activity after systemic administration; agonist properties have been recently investigated; Tokumura, T et al, *Chem Pharm Bull* (Tokyo), 38:3094-3098, 1990; Yamakawa, I et al, *J Pharm Sci*, 81:808-811, 1992; Yamakawa, I et al, *Chem Pharm Bull* (Tokyo), 40:2870-2872, 1992; Tokumura, T et al, *J Pharm Sci*, 82:725-728, 1993; Akunne, HC et al, *Biochem Pharmacol*, 49:1147-1154, 1995

Arg-Met

Bachem G-2115.0250   MW 305.4 $C_{11}H_{23}N_5O_3S$ Store at -15°C

Arg-Met Amide

Bachem G-1510.0250   MW 304.42 $C_{11}H_{24}N_6O_2S$ Store at -15°C

Arg-Met-Pro-Pro-Arg-Arg-Asp-Ala-Met-Pro-Ser-Asp-Ala

Synonyms: Protein Phosphatase 2Bα Blocking Peptide (268-280); Calcineurin Blocking Peptide

Calbiochem 539531   Synthetic   MW 1499.7 $C_{60}H_{102}N_{22}O_{19}S_2$ Liquid | From PP2B catalytic subunit; ideal blocking peptide for Anti-PP2Bα

Arg-Phe

Bachem G-1515.0250   MW 321.38 $C_{15}H_{23}N_5O_3$ Store at -15°C

Arg-Phe
Arg-Phe-OBzl Sulfate

Bachem G-3640.0250   Store at -15°C

Arg-Phe
Arg(NO₂)-Phe-OMe Hydrochloride

Bachem G-4010.0250   Store at -15°C

Arg-Phe
D-Arg-Phe

Bachem G-4025.0050   MW 321.38 $C_{15}H_{23}N_5O_3$ Store at -15°C

Arg-Phe Acetate Salt

ICN 150393   MW 381.4 $C_{15}H_{23}N_5O_3 \cdot C_2H_4O_2$

Arg-Phe Amide

Bachem G-1520.0001   MW 320.4 $C_{15}H_{24}N_6O_2$ Store at -15°C

Arg-Phe-Ala

Bachem H-1865.0250   MW 452.51 $C_{18}H_{28}N_6O_4$ Store at -15°C

Arg-Phe-Ala-Arg-Lys-Gly-Ala-Leu-Arg-Gln-Lys-Asn-Val

Synonyms: Protein Kinase C Substrate Peptide (19-31), (Ser[25])-; Protein Kinase C Inhibitor (19-31); Protein Kinase C (19-31); Protein Kinase C Pseudosubstrate

American Peptide 86-0-33   MW 1559.9 $C_{67}H_{118}N_{26}O_{17}$ This peptide is high affinity protein kinase C pseudosubstrate; House, C et al, *Science*, 238:1726, 1987

Bachem H-3232.0001   MW 1543.84 $C_{67}H_{118}N_{26}O_{16}$ Store at -15°C

Calbiochem 05-23-4904   MW 1543.8 $C_{67}H_{118}N_{26}O_{16}$ >97% (HPLC); solid; soluble in 5% acetic acid | More potent inhibitor of PKC than PKC Inhibitor 19-36; House, C & Kemp, BE, *Science*, 238:1726, 1987

ICN 159655   MW 1542.8

Arg-Phe-Ala-Arg-Lys-Gly-Ala-Leu-Arg-Gln-Lys-Asn-Val
Myr-Arg-Phe-Ala-Arg-Lys-Gly-Ala-Leu-Arg-Gln-Lys-Asn-Val Myristoylated

Synonyms: Protein Kinase C Peptide Inhibitor; Protein Kinase C, Myr-ψ-

Promega; V5691   MW 1754   ≥70% peptide (FAB/MS) | Inhibits PKC-g, since the sequence of this isozyme differs from the pseudosubstrate only in that the Phe residue is replaced with Cys; the addition of myristic acid enhances its permeability to the cell plasma membrane, allowing it to be used as an inhibitor for intracellular PKC in insect cells in culture; inhibits the phosphorylation of the myristoylated alanine-rich Ckinase substrate protein (MARCKS) when the PKC system is activated; does not affect the tyrosinekinase activity of the EGF receptor; Eicholtz, T et al, *JBC*, 268:1982, 1993

Arg-Phe-Ala-Arg-Lys-Gly-Ala-Leu-Arg-Gln-Lys-Asn-Val

Synonyms: Protein Kinase C Inhibitor Peptide (19-31)

Upstate 12-121   Synthetic   MW 1544   Lyophilized | Pseudosubstrate region of proteinkinase C,[1] amino acids 19-31

USBio P9103-50   Synthetic   ≥97% (HPLC); lyophilized | Corresponds to the pseudosubstrate region of proteinkinase C, AA 19-31; a specific PKC inhibitor for *in vitro* assays; inhibits phosphorylation of PKC substrate peptide by proteinkinase C (PKC) isoforms

Arg-Phe-Ala-Arg-Lys-Gly-Ala-Leu-Arg-Gln-Lys-Asn-Val Amide

Synonyms: Protein Kinase C (19-31)

Neosystem SC448   MW 1542.85   Bioactive; proteinkinase-related peptide

Arg-Phe-Ala-Arg-Lys-Gly-Ala-Leu-Arg-Gln-Lys-Asn-Val-His-Glu-Val-Lys

Synonyms: Protein Kinase C Peptide (19-35)

American Peptide 86-1-20   MW 2037.4 $C_{89}H_{153}N_{33}O_{22}$

Arg-Phe-Ala-Arg-Lys-Gly-Ala-Leu-Arg-Gln-Lys-Asn-Val-His-Glu-Val-Lys-Asn

Synonyms: Protein Kinase C Inhibitor Peptide Peptide (19-36); Protein Kinase C (19-36)

American Peptide 86-1-25   MW 2151.6 $C_{93}H_{159}N_{35}O_{24}$ Acts as pseudosubstrate by binding to the active sites of protein kinases; inhibits both the phosphorylation of protein kinase C substrates & the autophosphorylation of protein kinase C; House, C et al, *Science*, 238:1726, 1987

Bachem H-9370.0001   MW 2151.51 $C_{93}H_{159}N_{35}O_{24}$ Store at -15°C

Arg-Phe-Ala-Arg-Lys-Gly-Ala-Leu-Arg-Gln-Lys-Asn-Val-His-Glu-Val-Lys-Asn Trifluoroacetate Salt

Synonyms: Protein Kinase C Inhibitor (19-36)

Calbiochem 539560   MW 2151.5 $C_{93}H_{159}N_{35}O_{24}$ ≥95% (HPLC); lyophilized solid; soluble in water | Acts as pseudo-substrate by binding to the active sites of protein kinases; potent inhibitor of protein kinase C but not of protein kinase A; House, C & Kemp, BE, *Science*, 238:1726, 1987

Arg-Phe-Ala-Arg-Lys-Gly-Ala-Leu-Arg-Gln-Lys-Asn-Val-His-Glu-Val-Lys-Asn

Synonyms: Protein Kinase C (19-36); Protein Kinase C Pseudosubstrate

ICN 154585   MW 2151.8   House, C & BE Kemp, *Science*, 238:1726, 1987

Arg-Phe-Ala-Arg-Lys-Gly-Ala-Leu-Arg-Gln-Lys-Asn-Val-His-Glu-Val-Lys-Asn Amide

Synonyms: Protein Kinase C Inhibitor (19-36)

ICN 191458

Arg-Phe-Ala-Arg-Lys-Gly-Ala-Leu-Arg-Gln-Lys-Asn-Val-His-Glu-Val-Lys-Asn

Synonyms: Protein Kinase C (19-36)

Neosystem SC377   MW 2151.50   Bioactive; proteinkinase-related peptide; specific inhibitor of proteinkinase C; House, C & Kemp, BE, *Science*, 238:1726, 1987

Arg-Phe-Ala-Arg-Lys-Gly-Ala-Leu-Glu-Gln-Lys-Asn-Val-His-Glu-Val-Lys-Asn

Synonyms: Protein Kinase C Non-Inhibiting Analog Peptide (19-36), (Glu[27])-

American Peptide 86-1-22   MW 2124.5 $C_{92}H_{154}N_{32}O_{26}$ House, C et al, *Science*, 238:1726, 1987

Arg-Phe-Ala-Arg-Lys-Gly-Ser-Leu-Arg-Gln-Lys-Asn-Val

Synonyms: Protein Kinase C Substrate (19-31), (Ser[25])-; Protein Kinase C (19-31), (Ser[25])-; Protein Kinase C Substrate

Bachem H-3286.0001   MW 1559.84 $C_{67}H_{118}N_{26}O_{17}$ Store at -15°C

ICN 154584   MW 1560   House, C & BE Kemp, *Science*, 238:1726, 1987

Neosystem SC378   MW 1559.83   Bioactive; proteinkinase-related peptide; potent substrate for proteinkinase C; House, C & Kemp, BE, *Science*, 238:1726, 1987

Sigma P 1835   FW 1559.8   ≥97% (HPLC) | Bioactive peptide; Km=0.2 μM; House, C & Kemp, BE, *Science*, 238:1726, 1987

Arg-Phe-Ala-Arg-Lys-Gly-Ser-Leu-Arg-Gln-Lys-Asn-Val-His-Glu-Val-Lys-Asn

Synonyms: Protein Kinase C (19-36); Protein Kinase C Inhibitor

Sigma P 8462 FW 2151.5 ≥97% (HPLC) | Bioactive peptide; specific inhibitor of both autophosphorylation & protein substrate phosphorylation; House, C & Kemp, BE, *Science*, 238:1726, 1987

Arg-Phe-Ala-Val-Arg-Asp-Met-Arg-Gln-Thr-Val-Ala-Val-Gly-Val-Ile-Lys-Ala-Val-Asp-Lys-Lys Trifluoroacetate Salt

S Protein Kinase C$_\delta$ Peptide Substrate *ynonyms*:

Calbiochem 539563 MW 2488.0 $C_{109}H_{191}N_{35}O_{29}S$ ≥95% (HPLC); lyophilized solid; soluble in water | Highly specific substrate for the protein kinase C$_\delta$ isozymes; the sequence corresponds to residues 422-443 of murine eEF-1α-containing Thr431 that is phosphorylated by PKC$_\delta$; readily phosphorylated by PKC; Kielbassa, K et al, *J Biol Chem*, 270:6156, 1995

Arg-Phe-Arg-Tyr-Cys-Ala-Pro-Pro-Gly-Tyr-Ala-Leu-Leu-Arg-Cys-Asn-Asp-Thr-Asn-Tyr Arg-Phe-Arg-Tyr-Cys(Acm)-Ala-Pro-Pro-Gly-Tyr-Ala-Leu-Leu-Arg-Cys(Acm)-Asn-Asp-Thr-Asn-Tyr

Synonyms: GP140 (231-250); SIVmac251 Peptide

Neosystem SC731 MW 2535.87 SIVmac251 peptide from subtype GP140

Arg-Phe-Asp-Ser

Synonyms: Fibronectin

Bachem H-7720.0025 MW 523.55 $C_{22}H_{33}N_7O_8$ Store at -15°C

ICN 153058 MW 523.5 $C_{22}H_{33}N_7O_8$ Auffray, C & M Novotny, *Human Immunol*, 15:381, 1986

Neosystem SC146 MW 523.5 Bioactive; cell attachment peptide; fibronectin fragment; Auffray, C & Novotny, J, *Human Immunology*, 15:381, 1986

Sigma A 1675 FW 523.5 $C_{22}H_{33}N_7O_8$ ≥97% (HPLC) | Bioactive peptide; Auffray, C & Novotny, J, *Human Immunology*, 15:381, 1986

Arg-Phe-Asp-Ser-Arg-Leu-Ala-Phe-His-His-Val-Ala

Synonyms: NEF (184-195); HTLV I Peptide

Neosystem SC517 MW 1455.64 HTLV I peptide from subtype NEF

Arg-Phe-Met-Trp-Met-Lys Ac-Arg-Phe-Met-Trp-Met-Lys Amide

Synonyms: Opioid Receptor Antagonist

Bachem H-1994.0005 MW 939.22 $C_{44}H_{66}N_{12}O_7S_2$ Store at -15°C

Neosystem SC926 MW 939.20 Bioactive; opioid peptide; potent mu receptor antagonist in the guinea pig ileum assay; Dooley, CT, Chung, NN, Schiller, PW & Houghten, R, *PNAS* USA, 90:10811-10815, 1993

Arg-Phe-Met-Trp-Met-Thr Ac-Arg-Phe-Met-Trp-Met-Thr Amide

Bachem H-1996.0005 MW 912.15 $C_{42}H_{61}N_{11}O_8S_2$ Store at -15°C

Arg-Phe-Phe-Pro-Leu-Met Ac-Arg-Phe-Phe-Pro-Leu-Met Amide

Synonyms: Septide, (Ac-Arg^1)

Neosystem SC391 MW 851.07 Bioactive; tachykinin

Arg-Phe-Phe-Sar-Leu-Met Ac-Arg-Phe-Phe-Sar-Leu-Met(O$_2$) Amide

Synonyms: Substance P (6-11), Ac-(Arg^6,Sar^9,Met(O$_2$)^{11})-; NK-1 Receptor Agonist

Sigma S 2275 FW 857.0 $C_{40}H_{60}N_{10}O_9S$ ~98% (HPLC) | Bioactive peptide; selective; Regoli, D et al, *Regulatory Peptides*, 22:153, 1988

Arg-Phe-Tyr-Val-Val-Met

Bachem H-1418.0005 MW 814.02 $C_{39}H_{59}N_9O_8S$ Store at -15°C

Arg-Phe-Tyr-Val-Val-Met-Trp-Lys

Bachem H-1414.0005 Store at -15°C

Arg-Pro BOC-Arg-Pro-Lauryl Ester Laurate

Synonyms: Micelle-Forming Thrombin Inhibitor

Bachem A-3820.0250 MW 740.08 $C_{28}H_{53}N_5O_5 \cdot C_{12}H_{23}O_2$ Store at -15°C

Arg-Pro

Bachem G-3675.0050 MW 271.32 $C_{11}H_{21}N_5O_3$ Store at -15°C

Arg-Pro Arg-Pro-OMe Hydrochloride

Bachem G-4150.0250 Store at -15°C

Arg-Pro Arg-Pro-pNA

Bachem L-1125.0250 Store at -15°C

Arg-Pro Arg-Pro-p-Nitroanilide Acetate Salt Free Base

Synonyms: Dipeptidyl Aminopeptidase IV Substrate

Sigma A 1204 FW 391.4 $C_{17}H_{25}N_7O_4$ Nagatsu, T et al, *Anal Biochem*, 74:466, 1976; Hino, M et al, *Clin Chem*, 22:1256, 1976; Hama, T et al, *Mol Cell Biochem*, 43:35, 1982

Arg-Pro Amide Dihydrochloride

Bachem G-3660.0250 Store at -15°C

Arg-Pro-Arg-Ala-Ala-Thr-Phe

Synonyms: Akt/SGK Substrate Peptide

Upstate 12-340 MW 818 >95% pure; frozen solution

USBio A1125-40 ≥95% after Sephadex G-10 chromatography in 10% HOAc; frozen solution in sterile deionized H$_2$O; stable for 2 years at -20°C from date of shipment | Used to determine Akt/PKB activity in immunoprecipitates

Arg-Pro-Arg-Pro-Gln-Gln-Phe-Phe-Gly-Leu-Met Amide

Synonyms: Substance P, (Arg^3)-

American Peptide 70-1-32 MW 1375.7 $C_{63}H_{98}N_{20}O_{13}S$

Arg-Pro-Asp-Phe-Cys-Leu-Glu-Pro-Pro-Tyr-Thr-Gly-
Pro-Cys-Lys-Ala-Arg-Ile-Ile-Arg-Tyr-Phe-Tyr-Asn-
Ala-Lys-Ala-Gly-Leu-Cys-Gln-Thr-Phe-Val-Ile-Ile-
Arg-Tyr-Phe-Tyr-Asn-Ala-Lys-Ala-Gly-Leu-Cys-Gln-
Thr-Phe-Val-Tyr-Gly-Gly-Cys-Arg-Ala-Lys-Arg-Asn-
Asn-Phe-Lys-Ser-Ala-Glu-Asp-Cys-Met-Arg-Thr-Cys-
Gly-Gly-Ala

Synonyms: Aprotinin

Peptides International IAT-3830-PI   Synthetic   MW
6511.57 $C_{284}H_{432}N_{84}O_{79}S_7$ Lyophilized white solid; 6.7 TIU/mg =
6.048 KIU/mg; bulk | Inhibitor; Disulfide bonds: Cys^5-Cys^{55},
Cys^{14}-Cys^{38}, Cys^{30}-Cys^{51}; Trautschold, I et al, *Biochem Pharm*,
16:59, 1967

Arg-Pro-Gly
(Z-Arg-Pro-Gly)₂ MR

Synonyms: Serine Protease Substrate

Kamiya   MW 1222   >99% (HPLC)

Arg-Pro-Gly-Gly-Gly-Asp-Met-Arg-Asp-Asn-Trp-Arg-Ser-Glu-Leu-Tyr

Synonyms: GP120 (474-489); HTLV I Peptide

Neosystem SC250   MW 1909.06   HTLV I peptide from
subtype GP160

Arg-Pro-Gly-Leu-Leu-Asp-Leu-Lys

Synonyms: Octaneuropeptide

Neosystem SC163   MW 911.11   Bioactive neuropeptide;

Arg-Pro-Gly-Phe-Ser-Pro-Phe-Arg

Synonyms: Angiotensin I Converting Enzyme Inhibitor; Bradykinin,
(des-Pro²)-; Peptidyl-Dipeptidase A Inhibitor; Kininase II Inhibitor;
Peptidyl-Dipeptidase A, Kininase II Inhibitor

American Peptide 12-1-87   MW 963.1 $C_{45}H_{66}N_{14}O_{10}$
Naruse, M et al, *Chem Pharm Bull*, 29:3369, 1981

Sigma B 2026   FW 963.1 $C_{45}H_{66}N_{14}O_{10}$ ≥97% (HPLC) |
Bioactive peptide; potentiates activity on bradykinin-induced
contraction of guinea pig ileum & hypertension in rats; Naruse, M et
al, *Chem Pharm Bull*, 29:3369, 3734, 1981

Peptides International IBK-4097-v   Synthetic   MW 963.11
$C_{45}H_{66}N_{14}O_{10}$ >99% (HPLC); lyophilized amorphous powder |
Inhibitor; Naruse, M et al, *Chem Pharm Bull*, 29, 3369, 1981

Peptides InternationalIBK-4097-v   Synthetic   MW 963.11
$C_{45}H_{66}N_{14}O_{10}$ >99% (HPLC); lyophilized amorphous powder |
Bioactive; Naruse, M et al, Chem *Pharm Bull*, 29:3369, 1981

Arg-Pro-Gly-Phe-Ser-Pro-Phe-Arg 2AcOH 3H₂O

Synonyms: Bradykinin, (des-Pro²)-; Peptidyl-Dipeptidase A
Inhibitor; Kininase II Inhibitor; Angiotensin I Converting Enzyme
Inhibitor; Peptidyl-Dipeptidase A, Kininase II Inhibitor

Peptides International IBK-4097   Synthetic   MW 1137.27
$C_{45}H_{66}N_{14}O_{10}$ · 2CH₃COOH · 3H₂O   >99% (HPLC); amorphous
powder; bulk | Inhibitor; Naruse, M et al, *Chem Pharm Bull*, 29,
3369, 1981

Peptides InternationalIBK-4097   Synthetic   MW 1137.2
$C_{45}H_{66}N_{14}O_{10}$ · 2CH₃COOH · 3H₂O   >99% (HPLC); lyophilized
amorphous powder; bulk | Bioactive; Naruse, M et al, Chem
Pharm Bull, 29:3369, 1981

Arg-Pro-Gly-Pro-Pro-Gly-Leu-Gln-Gly-Arg-Leu-Gln-Arg-Leu-Leu-Gln-Ala-Asn-Gly-Asn-His-Ala-Ala-Gly-Ile-Leu-Thr-Met Amide

Synonyms: Orexin B; Appetite Boosting Peptide

Alexis 164-003   Rat, mouse   MW 2908.5 $C_{125}H_{215}N_{45}O_{33}S$
≥97%; lyophilized powder | See Alexis 164-002

Bachem H-4176.0500   Rat, mouse   MW 2936.44
$C_{126}H_{215}N_{45}O_{34}S$ Store at -15°C

Neosystem SC1339   Rat, mouse   MW 2936/43   Bioactive
neuropeptide; one of two new hypothalamic peptides regulating
feeding behavior (the other is orexin-A); both come from the same
precursor by proteolytic processing; stimulate food consumption in
a dose-dependent manner when administered centrally to rats;
Sakurai, T et al, *Cell*, 92:573-585, 1998; De Lecea, L et al, *PNAS*,
USA, 95:322-327, 1998

Peptides International POR-4347-s   Rat, mouse synthetic
MW 2936.4 $C_{126}H_{215}N_{45}O_{34}S$ >95% (HPLC); lyophilized
amorphous powder | Bioactive; Sakurai, T et al, *Cell*, 92:573,
1998; de Lecea, L et al, *PNAS* USA, 95:322, 1998

Arg-Pro-Hyp-Gly-Phe-Ser-Pro-Phe-Arg

Synonyms: Bradykinin, (Hyp³)-

Bachem H-5465.0001   Store at -15°C

Sigma B 7775   1076.2; ≥97% (HPLC) | Bioactive peptide;
Kato, H et al, *FEBS Lett*, 232:252, 1988

Peptides International PBK-4193-v   Human synthetic   MW
1076.2 $C_{50}H_{73}N_{15}O_{12}$ >99% (HPLC); lyophilized amorphous
powder | Bioactive; Kato, H et al, *FEBS Lett*, 232:252, 1988

Arg-Pro-Hyp-Gly-Phe-Ser-Pro-Phe-Arg 2AcOH 3H₂O

Synonyms: Bradykinin, (Hyp³)-

Peptides International PBK-4193   Human synthetic   MW
1250.3 $C_{50}H_{73}N_{15}O_{12}$ · 2CH₃COOH · 3H₂O   >99% (HPLC);
lyophilized amorphous powder; bulk | Bioactive; Kato, H et al,
FEBS Lett, 232:252, 1988

Arg-Pro-Hyp-Gly-Phe-Ser-Pro-Tyr-Arg
Arg-Pro-Hyp-Gly-Phe-Ser-Pro-Tyr(Me)-Arg

Synonyms: Bradykinin, (Hyp³,Tyr(Me)⁸)-; B₂ Receptor Agonist

Calbiochem 05-23-0514   MW 1106.2 $C_{51}H_{75}N_{15}O_{13}$ ≥98%
(HPLC); lyophilized; soluble in 5% HOAc | Selective; Galizzi, JP et
al, *Br J Pharmacol*, 113:389, 1994; Rhaleb, NE & Carretero, OA,
Life Sci, 55:1351, 1994

Arg-Pro-Ile-Abu
2,5-DiHydroxy-Bz-Arg-Pro-Ile-Abu

Synonyms: Nazumamide A

Bachem N-1335.0001   MW 605.69 $C_{28}H_{43}N_7O_8$ Store at
-15°C

Arg-Pro-Ile-Abu
2,5-Dihydroxybenzoyl-Arg-Pro-Ile-Abu

Synonyms: Nazumamide A; Thrombin Inhibitor

Sigma N 1898   FW 605.7 $C_{28}H_{43}N_7O_8$ ≥97% (HPLC) |
Bioactive peptide; Hayashi, K et al, *THL*, 33:5075, 1992

Arg-Pro-Leu-Ala-Leu-Trp-Arg
Mca-Arg-Pro-Leu-Ala-Leu-Trp-Arg-Dap(Dnp) Amide

Bachem M-2390.0001   MW 1368.43 $C_{64}H_{77}N_{19}O_{16}$ Store at
-15°C

Arg-Pro-Leu-Ala-Leu-Trp-Arg-Ser
Dnp-Arg-Pro-Leu-Ala-Leu-Trp-Arg-Ser

Synonyms: Matrix Metalloproteinase VII Substrate; Matrilysin
Substrate; PUMP I Substrate

Bachem M-2205.0005   Store at -15°C

Calbiochem 444228   MW 1164.3 $C_{52}H_{77}N_{17}O_{14}$ Crystalline
solid; ≥98% (hplc); excitation max:280 nm; emission max:360 nm;
soluble in 10% acetic acid | Excellent fluorogenic substrate for
MMP-7; Welch, AR et al, *Arch Biochem Biophys*, 324:59 1995

Arg-Pro-Lys-Leu-Lys-Trp-Asp-Asn-Gln

Synonyms: Dynorphin A (9-17)

American Peptide 26-4-46   Porcine   MW 1184.4
$C_{53}H_{85}N_{17}O_{14}$

Arg-Pro-Lys-Pro

Synonyms: Substance P (4-11); Substance P (1-4)

American Peptide 70-1-85   MW 496.6 $C_{22}H_{40}N_8O_5$
Hydrolyzed from substance P by a prolyl endopeptidase; derived from substance P & resists proteolytic degradation; modulates the activity of substance P on various tissues & exerts a stimulatory effect on phagocytes

Bachem H-1875.0005   MW 496.61 $C_{22}H_{40}N_8O_5$ Store at -15°C

ICN 153017   Acts on phagocytosis; Blumberg, S & VI Teichberg, *BBRC*, 90:347, 1979; Bar-Shavit, Z etal, *BBRC*, 94:1445, 1980; Blumberg, S etal, *Brain Res*, 192:477, 1980; Kato, T etal, *J Neurochem*, 35:527, 1980; Teichberg, VI etal, *Regul Peptides*, 1:327, 1981; Mazvrek, M etal, *Neuropharmacology*, 20:1025, 1981

Arg-Pro-Lys-Pro Acetate Salt

Synonyms: Substance P (1-4)

Sigma S 3011   FW 496.6 $C_{22}H_{40}N_8O_5$ ≥97% (HPLC) | Stabilizes Substance P against proteolytic degradation; stimulates release of histamine from mast cells; Blumberg, S & Teichberg, VI, *Biochem Biophys Res Commun*, 90:347, 1979; Bar-Shavit, Z et at, ibid, 94:1445, 1980

Arg-Pro-Lys-Pro-Gln
Mca-Arg-Pro-Lys-Pro-Gln

Bachem M-2250.0005   Store at -15°C

Arg-Pro-Lys-Pro-Gln-Gln-Phe

Synonyms: Substance P (1-7)

American Peptide 70-1-83   MW 900.1 $C_{41}H_{65}N_{13}O_{10}$

Bachem H-1582.0005   MW 900.05 $C_{41}H_{65}N_{13}O_{10}$ Store at -15°C

Neosystem SC008   MW 900.05   Bioactive; tachykinin

Sigma S 6272   FW 900.0 $C_{41}H_{65}N_{13}O_{10}$ ≥97% (HPLC); peptide content:~70% | Bioactive peptide; gives depressor & bradycardic effects when applied to the nucleus tractus solitarius; Hall, MER et al, *Regulatory Peptides*, 46:102, 1993

Arg-Pro-Lys-Pro-Gln-Gln-Phe-Phe-Ala-Leu-Met Amide

Synonyms: Substance P, (Ala[9])

Neosystem SC049   MW 1361.67   Bioactive; tachykinin

Arg-Pro-Lys-Pro-Gln-Gln-Phe-Phe-Gly

Synonyms: Substance P (1-9)

American Peptide 70-1-80   MW 1104.3 $C_{52}H_{77}N_{15}O_{12}$

Bachem H-1880.0001   MW 1104.28 $C_{52}H_{77}N_{15}O_{12}$ Store at -15°C

ICN 153018   Liang, T etal, *J Neurosci*, 10:1133, 1981

Sigma S 1761   FW 1104.3   ≥97% (HPLC) | Bioactive peptide; Liang, T et al, *J Neurosci*, 10:1133, 1981

Arg-Pro-Lys-Pro-Gln-Gln-Phe-Phe-Gly-Leu-Met

Synonyms: Substance P

American Peptide 70-1-13   MW 1348.7 (free acid)
$C_{63}H_{97}N_{17}O_{14}S$

Arg-Pro-Lys-Pro-Gln-Gln-Phe-Phe-Gly-Leu-Met
Arg-Pro-Lys-Pro-Gln-Gln-Phe-Phe-Gly-Leu-Met-OMe

Synonyms: Substance P, OMe-

American Peptide 70-1-27   MW 1362.7 $C_{64}H_{99}N_{17}O_{14}S$
Watson, SP et al, *Eur J Pharmacol*, 87:77, 1983

Bachem H-1895.0001   MW 1362.66 $C_{64}H_{99}N_{17}O_{14}S$ Store at -15°C

Arg-Pro-Lys-Pro-Gln-Gln-Phe-Phe-Gly-Leu-Met
Arg-Pro-Lys-Pro-Gln-Gln-*p*-Chloro-Phe-*p*-Chloro-Phe-Gly-Leu-Met Amide

Synonyms: Substance P, (*p*-Chloro-Phe[7,8])-

Bachem H-1900.0001   MW 1416.54 $C_{63}H_{96}Cl_2N_{18}O_{13}S$
Store at -15°C

Arg-Pro-Lys-Pro-Gln-Gln-Phe-Phe-Gly-Leu-Met
Arg-3,4-Dehydro-Pro-Lys-3,4-Dehydro-Pro-Gln-Gln-Phe-Phe-Gly-Leu-Met Amide

Synonyms: Substance P, (3,4-Dehydro-Pro[2,4])-

Bachem H-2824.0001   Store at -15°C

Arg-Pro-Lys-Pro-Gln-Gln-Phe-Phe-Gly-Leu-Met
Arg-Pro-Lys-Pro-Gln-Gln-Phe-*p*-Bz-Phe-Gly-Leu-Met Amide

Synonyms: Substance P, (*p*-Bz-Phe[8])-; Substance P, (Bpa[8])-

Bachem H-3334.0001   MW 1451.76 $C_{70}H_{102}N_{18}O_{14}S$ Store at -15°C

Arg-Pro-Lys-Pro-Gln-Gln-Phe-Phe-Gly-Leu-Met

Synonyms: Substance P, (Sar[9],Met(O[2])[11])-

Calbiochem 05-23-0601   MW 1348.6 (free acid)
$C_{63}H_{97}N_{17}O_{14}S$ ≥98% (HPLC); lyophilized solid; soluble in 5% acetic acid | Regulatory peptide that plays a role in immune regulation; Reubi, JC et al, *Blood*, 92:191, 1998; *Merck Index*, 12:9032

Calbiochem 05-23-0657   MW 1393.6 $C_{64}H_{100}N_{18}O_{15}S$
≥98% (HPLC); lyophilized solid; soluble in 5% acetic acid | Selective NK-1 receptor agonist; stimulates the formation of inositol phosphate & enhances taurine release from human astrocytoma cell line; Oury-Donat, F et al, *J Neurochem*, 62:1399, 1994; Rouissi, N et al, *Life Sci*, 52:L103, 1993; Giuliani, S et al, *Eur J Pharmacol*, 150:377, 1988

Arg-Pro-Lys-Pro-Gln-Gln-Phe-Phe-Gly-Leu-Met
Arg-Pro-Lys-Pro-Gln-Gln-Phe-Phe-Gly-Leu-Met-OMe

Synonyms: Substance P, (Met-OMe[11])-

ICN 153005   Cascieri, MA etal, *Molec Pharmac*, 20:457, 1981; Fournier, A, *J Med Chem*, 25:64, 1982

Arg-Pro-Lys-Pro-Gln-Gln-Phe-Phe-Gly-Leu-Met
Arg-Pro-Lys-Pro-Gln-Gln-*p*-Chloro-Phe-*p*-Chloro-Phe-Gly-Leu-Met Amide

Synonyms: Substance P, (*p*-Chloro-Phe[7,8])-

ICN 153008   Morgat, JL etal, *FEBS Lett*, 111:19, 1980

Arg-Pro-Lys-Pro-Gln-Gln-Phe-Phe-Gly-Leu-Met
Arg-ΔPro-Lys-ΔPro-Gln-Gln-Phe-Phe-Gly-Leu-Met Amide

Synonyms: Substance P, (Dehydro-Pro[2,4])-

ICN 153009

Arg-Pro-Lys-Pro-Gln-Gln-Phe-Phe-Gly-Leu-Met
Biotinyl-Arg-Pro-Lys-Pro-Gln-Gln-Phe-Phe-Gly-Leu-Met Amide

Synonyms: Substance P, Biotinyl-

Neosystem SC1210   MW 1573.93   >80% (HPLC) | Bioactive; tachykinin

Arg-Pro-Lys-Pro-Gln-Gln-Phe-Phe-Gly-Leu-Met

Synonyms: Substance P, (Met-OH[11])-; NK-1 Agonist

Sigma S 2136   FW 1348.6   ≥97% (HPLC); free acid | Bioactive peptide; Cascieri, MA et al, *Mol Pharmacol*, 20:457, 1981; Fournier, A et al, *J Med Chem*, 25:64, 1982

Arg-Pro-Lys-Pro-Gln-Gln-Phe-Phe-Gly-Leu-Met
Arg-ΔPro-Lys-ΔPro-Gln-Gln-Phe-Phe-Gly-Leu-Met
Amide

Synonyms: Substance P, (Dehydro-Pro[2,4])-; Substance P Antagonist

Sigma S 3258   FW 1343.6   ≥97% (HPLC) | Bioactive peptide

Arg-Pro-Lys-Pro-Gln-Gln-Phe-Phe-Gly-Leu-Met
Arg-Pro-Lys-Pro-Gln-Gln-*p*-Chloro-Phe-*p*-Chloro-
Phe-Gly-Leu-Met Amide

Synonyms: Substance P, (*p*-Chloro-Phe[7,8])-

Sigma S 5382   FW 1416.5   ≥95% (HPLC) | Bioactive peptide; substance P analog that can be tritiated; Morgat, JL et al, *FEBS Lett*, 111:19, 1980

Arg-Pro-Lys-Pro-Gln-Gln-Phe-Phe-Gly-Leu-Met

Synonyms: Substance P

Biogenesis 8450-1004   Synthetic   MW 1347.5   99.6% (HPLC), >99% (TLC); lyophilized, consisting of 76% peptide material, acetate counter ions and residual H_2O | To avoid Met oxidation, it is best to dissolve the peptide under a nitrogen stream

---

Arg-Pro-Lys-Pro-Gln-Gln-Phe-Phe-Gly-Leu-Met
Amide

Synonyms: Substance P; G_1-Protein Activator

Alexis 167-003   MW 1347.6 $C_{63}H_{98}N_{18}O_{13}$ ≥97%; lyophilized powder; soluble in H_2O | Neuropeptide inducing Ca^{2+} mobilization; activates G_1-proteins directly; induces superoxide production in neurophils

American Peptide 70-1-10   MW 1347.7 $C_{63}H_{98}N_{18}O_{13}S$ Found in the gut as well as in the brain; responsible for a number of excitatory effects on both central & peripheral neurons; contracts smooth muscle, constricts bronchioles & increases capillary permeability; when released from afferent nerves, causes neurogenic inflammatory response, including mast cell degranulation; Roberts, AI et al, *Cell Immunol*, 141:457, 1992

Bachem H-1890.0001   MW 1347.65 $C_{63}H_{98}N_{18}O_{13}S$ Store at -15°C

Calbiochem 05-23-0600   MW 1347.6 $C_{63}H_{98}N_{18}O_{13}S$ >97% (HPLC); lyophilized; soluble in 5% HOAc | Neuropeptide found in high concentrations in the gut; when released from afferent nerves, causes a neurogenic inflammatory response, including mast cells degranulation; induces Ca^{2+} mobilization; activates G_α, $G_{q/11}\alpha$ & $G_s\alpha$ subunits in CHO cells via activation of substance P receptors; induces superoxide production in neutrophils; Chiwakata, CB et al, *Infect Immun*, 64:5106, 1996; Tanabe, T et al, *Eur J Pharmacol*, 299:187, 1996; Zhu, J et al, *Eur J Pharmacol*, 268:279, 1994; Heath, MJ et al, *J Neurophysiol*, 72:1192, 1994; *Merck Index*, 12:9032; Roush, ED & Kwatra, MM, *FEBS Lett*, 428:291, 1998; Garland, AM et al, *Soc Neurosci*, 19:Abstract 104.6, 1993; Wei, JY et al, *Soc Neurosci*, 19:Abstract 136.14, 1993

ICN 152077   Chang, M etal, *Nature New Biol*, 232:86, 1971; Marx, JL etal, *Science*, 205:886, 1979; Pernow, B, *Pharmac Rev*, 35:85, 1983

Neosystem SC000   MW 1347.64   Bioactive; tachykinin

Peptides International PSP-4014-v   Synthetic   MW 1347.7 $C_{63}H_{98}N_{18}O_{13}S$ >95% (HPLC); lyophilized amorphous powder | Bioactive; human, bovine, rat, mouse; Regoli, D et al, *Pharmacol Rev*, 46:551, 1994; von Euler, US & JH Gaddum, *J Physiol*, 72:74, 1931; Chang, MM et al, *Nature New Biol*, 232:86, 1971; Marx, JL, *Science*, 205, 886, 1979

Arg-Pro-Lys-Pro-Gln-Gln-Phe-Phe-Gly-Leu-Met
Amide 3AcOH 5H₂O

Synonyms: Substance P

Peptides International PSP-4014   Synthetic   MW 1617.74 $C_{63}H_{98}N_{18}O_{13}S \cdot 3CH_3COOH \cdot 5H_2O$ >95% (HPLC); lyophilized amorphous powder; bulk | Bioactive; human, bovine, rat, mouse; Regoli, D et al, *Pharmacol Rev*, 46:551, 1994; von Euler, US & JH Gaddum, *J Physiol*, 72:74, 1931; Chang, MM et al, *Nature New Biol*, 232:86, 1971; Marx, JL, *Science*, 205, 886, 1979

Arg-Pro-Lys-Pro-Gln-Gln-Phe-Phe-Gly-Leu-Met
Amide Acetate Salt Free Base

Synonyms: Neurokinin I Agonist; Substance P

Sigma S 6883   FW 1347.6   ≥98% (HPLC) | Bioactive peptide; potent vasodilator & hypotensive agent; induces salivation; increases capillary permeability; induces mast cell degranulation; putative neurotransmitter in sensory (pain) afferents; Substance P has been proposed as a neuromodulator involved in the transmission of pain; also affects contraction of smooth muscle, reduction of blood pressure & stimulation of secretory tissue; Nicoll, RA et al, *Ann Rev Neurosci*, 3:227, 1980; Marx, JL, *Science*, 205:886, 1979; Chang, M et al, *Nature New Biol*, 232:86, 1971

Arg-Pro-Lys-Pro-Gln-Gln-Phe-Phe-Gly-Leu-Met Free
Acid

Synonyms: Substance P, (Met[11])-

Bachem H-1885.0001   MW 1348.64 $C_{63}H_{97}N_{17}O_{14}S$ Store at -15°C

ICN 153004   Cascieri, MA etal, *Molec Pharmac*, 20:457, 1981; Fournier, A, *J Med Chem*, 25:64, 1982

Arg-Pro-Lys-Pro-Gln-Gln-Phe-Phe-Gly-Leu-Met-Gly-
Lys-Arg

Synonyms: Substance P-Gly-Lys-Arg; Prepro-Tachykinin (58-71), β-

American Peptide 70-1-38   MW 1690.0 $C_{77}H_{124}N_{24}O_{17}S$

Arg-Pro-Lys-Pro-Gln-Gln-Phe-Phe-Gly-Leu-Met-OMe

Synonyms: Substance P, (Met-OMe[11])-; NK-1 Agonist

Sigma S 2011   FW 1362.7   ≥97% (HPLC) | Bioactive peptide; Cascieri, MA et al, *Mol Pharmacol*, 20:457, 1981; Fournier, A et al, *J Med Chem*, 25:64, 1982

Arg-Pro-Lys-Pro-Gln-Gln-Phe-Phe-Gly-Leu-Nle
Arg-Pro-Lys-Pro-Gln-Gln-Phe(NO₂)-Phe-Gly-Leu-Nle
Amide

Synonyms: Substance P, ((pNO₂) Phe[7],Nle[11])-

American Peptide 70-9-11   MW 1375.6 $C_{64}H_{100}N_{19}O_{15}$

Arg-Pro-Lys-Pro-Gln-Gln-Phe-Phe-Gly-Leu-Nle Amide

Synonyms: Substance P, (Nle[11])-

American Peptide 70-1-21   MW 1329.6 $C_{64}H_{100}N_{18}O_{13}$ Substance P analog that avoids methionine oxidation problems; Chipkin RE et al, *Arch Int Pharmacodyn*, 240:193, 1979

Bachem H-1905.0001   MW 1329.61 $C_{64}H_{100}N_{18}O_{13}$ Store at -15°C

ICN 153006   Substance P analog with high biological activity; avoids methionine oxidation problems; Chipkin, RE etal, *Arch Int Pharmacodyn*, 240:193, 1979; Escher E etal, *J Med Chem*, 25:470, 1982

Neosystem SC052   MW 1329.60   Bioactive; tachykinin

Sigma S 1136   FW 1329.6   ≥97% (HPLC) | Bioactive peptide; avoids Met oxidation problems; Chipkin, RE et al, *Arch Int Pharmacodyn*, 240:193, 1979; Escher, E et al, *J Med Chem*, 25:470, 1982

Arg-Pro-Lys-Pro-Gln-Gln-Phe-Phe-His-Leu-Met
D-Arg-D-Pro-Lys-Pro-Gln-Gln-D-Phe-Phe-D-His-Leu-
Met Amide

Synonyms: Substance P, (D-Arg[1],D-Pro[2],D-Phe[7],D-His[9])-

American Peptide 70-9-21   MW 1427.7 $C_{67}H_{102}N_{20}O_{13}S$ Substance P antagonist; Post, C et al, *Eur J Pharmacol*, 113:335, 1985

Bachem H-1385.0001   MW 1427.74 $C_{67}H_{102}N_{20}O_{13}S$ Store at -15°C

ICN 153013   Potent substance P antagonist; Post, C & K Folkers, *Eur J Pharmac*, 113:335, 1985

Arg-Pro-Lys-Pro-Gln-Gln-Phe-Phe-His-Leu-Met
D-Arg-D-Pro-Lys-Pro-Gln-Gln-D-Phe-Phe-D-His-Leu-Met Amide Free Base

Synonyms: Substance P, (D-Arg[1],D-Pro[2],D-Phe[7],D-His[9])-; Substance P Antagonist

Sigma S 3639   FW 1427.7   ≥95% (HPLC); peptide content:~70% | Bioactive peptide; Post, C & Folkers, K, *Eur J Pharmacol*, 113:335, 1985

Arg-Pro-Lys-Pro-Gln-Gln-Phe-Phe-Pro-Leu-Met
Arg-(Dehydro)Pro-Lys-(Dehydro)Pro-Gln-Gln-Phe-Phe-Pro-Leu-Met Amide

Synonyms: Substance P, ((Dehydro) Pro[2,4],Pro[9])-

American Peptide 70-9-30   MW 1383.5 $C_{66}H_{98}N_{18}O_{13}S$

Arg-Pro-Lys-Pro-Gln-Gln-Phe-Phe-Pro-Leu-Met
Arg-Pro-Lys-Pro-Gln-Gln-Phe-Phe-Pro-Leu-Met Amide

Synonyms: Substance P, (Pro[9])-

Sigma S 7029   FW 1387.7   ≥97% (HPLC) | Bioactive peptide

Arg-Pro-Lys-Pro-Gln-Gln-Phe-Phe-Pro-Leu-Met Amide

Synonyms: Substance P (6-11), (Pro[9])-

American Peptide 70-1-31   MW 1387.7 $C_{66}H_{102}N_{18}O_{13}S$

Neosystem SC430   MW 1387.70   Bioactive; tachykinin; potent & selective ligand of NK-1 tachykinin receptors; Petitet, F et al, *J Neurochem*, 56:879-889, 1991

Arg-Pro-Lys-Pro-Gln-Gln-Phe-Phe-Pro-Leu-Trp
Arg-Pro-Lys-Pro-Gln-Gln-Phe-Phe-D-Pro-MeLeu-Trp Amide

Synonyms: Substance P, (D-Pro[9],MeLeu[10], Trp[11])-

Neosystem SC497   MW 1456.70   Bioactive; tachykininTachykinin antagonist; Chassaing, G, *Neuropeptides*, 23:73-79, 1992

Arg-Pro-Lys-Pro-Gln-Gln-Phe-Phe-Pro-Leu-Trp

Synonyms: Substance P (1-11), (((S,S)Pro-Leu(Spiro-γ-Lactam))[9,10],Trp[11]); GR-71251

Neosystem SC925   MW 1468.88   Bioactive; tachykinin; potent & highly selective NK-1 receptor antagonist; Hagan, RM et al, *Br J Pharmacol*, 99(Suppl):62P, 1990; Ward, P et al, *J Med Chem*, 33:1848-1851, 1990

Arg-Pro-Lys-Pro-Gln-Gln-Phe-Phe-Pro-Leu-Trp
Arg-Pro-Lys-Pro-Gln-Gln-Phe-Phe-D-Pro-Leu-Trp Amide

Synonyms: Substance P, (D-Pro[9]-(Spiro-γ-Lactam)[9,10]-Trp[11])-; GR-71251; Substance P NK-1 Antagonist

Sigma S 2421   FW 1468.8   ≥97% (HPLC); peptide content:~60% | Bioactive peptide; potent & highly selective; the unusual linkage is formed by an ethylene bond from the α-position of Pro[9] to the N of Leu[10]; Hagan, RM et al, *Br J Pharmacol* 99:Suppl, 62P, 1990; Ward, P et al, *J Med Chem*, 33:1848, 1990

Arg-Pro-Lys-Pro-Gln-Gln-Phe-Phe-Pro-Pro-Trp
Arg-Pro-Lys-Pro-Gln-Gln-Phe-Phe-D-Pro-Pro-Trp Amide

Synonyms: Substance P, (D-Pro[9], Pro[10],Trp[11])-

Neosystem SC496   MW 1426.68   Bioactive; tachykinin antagonist; Chassaing, G, *Neuropeptides*, 23:73-79, 1992

Arg-Pro-Lys-Pro-Gln-Gln-Phe-Phe-Sar-Leu
Arg-Pro-Lys-Pro-Gln-Gln-Phe-Phe-Sar-Leu-Me(O₂) Amide

Synonyms: Substance P, (Sar[9],Met(O)₂[11])-; Substance P NK-1 Receptor Agonist

Sigma S 3672   FW 1393.7   ≥97% (HPLC) | Bioactive peptide; highly selective; evokes concentration-dependent mucus secretion; more potent than Substance P; Drapeau et al, *Neuropeptides*, 10:43, 1987; Quirion, R & Dam, TV, *Regulatory Peptides*, 22:18, 1988

Arg-Pro-Lys-Pro-Gln-Gln-Phe-Phe-Sar-Leu-Met
Arg-Pro-Lys-Pro-Gln-Gln-Phe-(N-Me)Phe-Sar-Leu-Met Amide

Synonyms: Substance P, (Me-Phe[8],Sar[9])-

American Peptide 70-1-30   MW 1376.7 $C_{65}H_{103}N_{18}O_{13}S$

Arg-Pro-Lys-Pro-Gln-Gln-Phe-Phe-Sar-Leu-Met
Arg-Pro-Lys-Pro-Gln-Gln-Phe-Phe-Sar-Leu-Met(O₂) Amide

Synonyms: Substance P, (Sar[9],Met(O₂)[11])-

Bachem H-9410.0001   MW 1393.68 $C_{64}H_{100}N_{18}O_{15}S$ Store at -15°C

ICN 158739   MW 1393.7   >97% | Highly specific neurokinin NK_1 receptor agonist; Quirion, R & TV Dam, *Regulatory Peptides*, 22:18, 1988

Neosystem SC439   MW 1393.66   Bioactive; tachykinin; highly selective NK-1 receptor agonist; Quirion, R & Dam, TV, *Regul Peptides*, 22:18, 1988

Arg-Pro-Lys-Pro-Gln-Gln-Phe-Phe-Sar-Leu-Met Amide

Synonyms: Substance P, (Sar[9])-; Substance P NK-1 Agonist

American Peptide 70-1-33   MW 1361.7 $C_{64}H_{100}N_{18}O_{13}S$
Sandberg, BE et al, *Eur J Biochem*, 114:329, 1981

ICN 153007   Sandberg, BEB etal, *Eur J Biochem*, 114:329, 1981

Sigma S 6636   FW 1361.7   ≥97% (HPLC) | Bioactive peptide; Sandberg, BEB et al, *Eur J Biochem*, 114:329, 1981

Arg-Pro-Lys-Pro-Gln-Gln-Phe-Phe-Trp-Leu-Met
Arg-D-Pro-Lys-Pro-Gln-Gln-D-Phe-Phe-D-Trp-Leu-Met Amide

Synonyms: Substance P, (D-Pro[2],D-Phe[7],D-Trp[9])-; Substance P Antagonist

American Peptide 70-1-92   MW 1476.8 $C_{72}H_{105}N_{19}O_{13}S$
One of the first effective antagonists of substance P shows agonist activity in several *in vitro* preparations; Folkers, K et al, *Acta Physiol Scan*, 111:505, 1981

Bachem H-1910.0001   MW 1476.81 $C_{72}H_{105}N_{19}O_{13}S$ Store at -15°C

ICN 153011   Effective substance P antagonist; Folkers, K etal, *Acta Physiol Scand*, 111:505, 1981; Lembeck, F etal, *BBRC*, 103:1318, 1981; Piercey, MF etal, *Science*, 214:1361, 1981; Posell, S etal, *Acta Physiol Scand*, 111:381, 1981; Fuxe, K etal, *Eur J Pharmac*, 77:171, 1982; Hawcook, Ab etal, *Eur J Pharmac*, 80:135, 1982; Jiang, Z-G, *Science*, 217:741, 1982

Sigma S 5635   FW 1476.8   ≥95% (HPLC) | Bioactive peptide; Jiang, Z-G et al, *Science*, 217:739, 1982

Arg-Pro-Lys-Pro-Gln-Gln-Phe-Tyr-Gly-Leu-Met

Synonyms: Substance P, (Tyr[8])-

Biogenesis 8450-1204   Synthetic   MW 1363.6   99% (HPLC)/TLC; lyophilized, consisting of 79.2% peptide material, acetate counter ions and residual H_2O | To avoid Met oxidation, it is best to dissolve the peptide under a nitrogen stream; my be iodinated for use in RIA; Fisher et al, *J Med Chem*, 19:325, 1976

Arg-Pro-Lys-Pro-Gln-Gln-Phe-Tyr-Gly-Leu-Met Amide

Synonyms: Substance P, (Tyr[8])-; T-Kinin

American Peptide 70-1-18 MW 1363.7 $C_{63}H_{98}N_{18}O_{14}S$
Fischer, G et al, *JMC*, 325:457, 1976

Bachem H-1915.0001 MW 1363.65 $C_{63}H_{98}N_{18}O_{14}S$ Store at -15°C

ICN 152078 Fisher, GH etal, *J Med Chem*, 19:325, 1976

Neosystem SC051 MW 1363.64 Bioactive; tachykinin;
Fisher, GH & Folken, K, *J Med Chem*, 19:325, 1976

Sigma S 6008 FW 1363.6 ≥97% (HPLC) | Bioactive peptide;
can be iodinated; Fisher, GH et al, *J Med Chem*, 19:325, 1976

Peptides International PSP-4059-v Synthetic MW 1363.7
$C_{63}H_{98}N_{18}O_{14}S$ >97% (HPLC); lyophilized amorphous powder |
Bioactive; for radioimmunoassay; see Peptides International PBK-
4130 (Isoleucyl-Seryl-Bradykinin)

Arg-Pro-Lys-Pro-Gln-Gln-Phe-Tyr-Gly-Leu-Nle Amide

Synonyms: Substance P, (Tyr[8], Nle[11])-

American Peptide 70-1-24 MW 1345.6 $C_{64}H_{100}N_{18}O_{14}$

Neosystem SC054 MW 1345.60 Bioactive; tachykinin

Arg-Pro-Lys-Pro-Gln-Gln-Trp-Phe-Trp-Leu-Leu
D-Arg-Pro-Lys-Pro-Gln-Gln-D-Trp-Phe-D-Trp-Leu-Leu Amide

Synonyms: Substance P, (D-Arg[1],D-Trp[7,9],Leu[11])-; Spantide

American Peptide 70-1-88 MW 1497.8 $C_{75}H_{108}N_{20}O_{13}$ Non-
selective tachykinin antagonist; widely used as a pharmacological
tool for elucidating the roles of endogenous NKs in physiological
processes; Rossel, S et al, *Int Substance P Symp*, Dublin, 1983

Arg-Pro-Lys-Pro-Gln-Gln-Trp-Phe-Trp-Leu-Leu
D-Arg-D-Pro-Lys-Pro-Gln-Gln-D-Trp-Phe-D-Trp-Leu-Leu Amide

Synonyms: Substance P, (D-Arg[1],D-Pro[2],D-Trp[7,9],Leu[11])-

American Peptide 70-1-90 MW 1497.8 $C_{75}H_{108}N_{20}O_{13}$
Potent antagonist of substance P & bombesin; Jensen, RT et al,
Nature, 209:61, 1984

Arg-Pro-Lys-Pro-Gln-Gln-Trp-Phe-Trp-Leu-Leu
D-Arg-Pro-Lys-Pro-Gln-Gln-D-Trp-Phe-D-Trp-Leu-Leu Amide

Synonyms: Substance P, (D-Arg[1],D-Trp[7,9],Leu[11])-; Spantide I

Bachem H-1925.0001 MW 1497.81 $C_{75}H_{108}N_{20}O_{13}$ Store at -15°C

Arg-Pro-Lys-Pro-Gln-Gln-Trp-Phe-Trp-Leu-Leu
D-Arg-D-Pro-Lys-Pro-Gln-Gln-D-Trp-Phe-D-Trp-Leu-Leu Amide

Synonyms: Substance P, (D-Arg[1],D-Pro[2],D-Trp[7,9],Leu[11])-

Bachem H-1930.0001 MW 1497.81 $C_{75}H_{108}N_{20}O_{13}$ Store at -15°C

Arg-Pro-Lys-Pro-Gln-Gln-Trp-Phe-Trp-Leu-Leu
D-Arg-Pro-Lys-Pro-Gln-Gln-D-Trp-Phe-D-Trp-Leu-Leu Amide

Synonyms: Substance P, (D-Arg[1],D-Trp[7,9],Leu[11])-; Spantide

ICN 153014 Potent substance P antagonist; Folkers, K etal,
International Substance P Symposium, Dublin, 1983

Arg-Pro-Lys-Pro-Gln-Gln-Trp-Phe-Trp-Leu-Leu
D-Arg-D-Pro-Lys-Pro-Gln-Gln-D-Trp-Phe-D-Trp-Leu-Leu Amide

Synonyms: Substance P, (D-Arg[1],D-Pro[2],D-Trp[7,9],Leu[11])-

ICN 153015 Substance P antagonist; inhibits reflex response in
the spinal cord of the newborn rat; bombesin receptor antagonist;
Yanagisawa, M etal, *Acta Physiol Scand*, 116:109, 1982; Rosell, S
etal, *Acta Physiol Scand*, 117:445, 1983; Jensen, RT etal, *Nature*,
209:61, 1984

Neosystem SC394 MW 1497.80 Bioactive; tachykinin;
substance P antagonist & bombesin receptor antagonist; Lundberg,
JM et al, *PNAS* USA, 80:1120-1124, 1983; Jensen, RT et al, *Nature*,
209:61, 1984

Arg-Pro-Lys-Pro-Gln-Gln-Trp-Phe-Trp-Leu-Leu
D-Arg-Pro-Lys-Pro-Gln-Gln-D-Trp-Phe-D-Trp-Leu-Leu Amide

Synonyms: Substance P, (D-Arg[1],D-Trp[7,9],Leu[11])-; Spantide I

Neosystem SC396 MW 1497.80 Bioactive; tachykinin;
substance P antagonist; Rosell, S et al, International Substance P
Symposium, Dublin, 1983; Folkers, K et al, *Br J Pharmacol*,
83:449-456, 1984

Arg-Pro-Lys-Pro-Gln-Gln-Trp-Phe-Trp-Leu-Leu
D-Arg-Pro-Lys-Pro-Gln-Gln-D-Trp-Phe-D-Trp-Leu-Leu Amide Acetate Salt Free Base

Synonyms: Substance P, (D-Arg[1],D-Trp[7,9],Leu[11])-; Substance P
Antagonist; Spantide I

Sigma S 0274 FW 1497.8 ≥97% (HPLC) | Bioactive peptide;
potent with weak histamine-releasing action; similar to Sigma S
3641 but produced by Sigma; Folkers, K et al, *Brit J Pharmacol*,
83:449, 1984; Hakanson, R et al, *Regulatory Peptides*, 31:75, 1990

Sigma S 3641 FW 1497.8 ≥97% (HPLC) | Bioactive peptide;
potent with weak histamine-releasing action; Folkers, K et al, *Brit J
Pharmacol*, 83:449, 1984; Hakanson, R et al, *Regulatory Peptides*,
31:75, 1990

Arg-Pro-Lys-Pro-Gln-Gln-Trp-Phe-Trp-Leu-Leu
D-Arg-D-Pro-Lys-Pro-Gln-Gln-D-Trp-Phe-D-Trp-Leu-Leu Amide Acetate Salt Free Base

Synonyms: Substance P, (D-Arg[1],D-Pro[2],D-Trp[7,9],Leu[11])-;
Substance P Antagonist

Sigma S 4152 FW 1497.8 ≥97% (HPLC) | Bioactive peptide;
Jensen, RT et al, *Nature*, 209:61, 1984

Arg-Pro-Lys-Pro-Gln-Gln-Trp-Phe-Trp-Leu-Leu
D-Arg-D-Pro-Lys-Pro-Gln-Gln-D-Trp-Phe-D-Trp-Leu-Leu Amide Trihydrochloride Octahydroxy

Synonyms: Substance P, (D-Arg[1],D-Pro[2],D-Trp[7,9],Leu[11])-; Bombesin
Receptor Antagonist

Peptides International PSP-4172 Synthetic MW 1751.3
$C_{75}H_{108}N_{20}O_{13} \cdot 3HCl \cdot 8H_2O$ 99% (HPLC); lyophilized amorphous
powder; bulk | Bioactive; Jensen, RT et al, *Nature*, 309, 61, 1984

Arg-Pro-Lys-Pro-Gln-Gln-Trp-Phe-Trp-Leu-Leu
D-Arg-D-Pro-Lys-Pro-Gln-Gln-D-Trp-Phe-D-Trp-Leu-Leu Amide Hydrochloride

Synonyms: Substance P, (D-Arg[1],D-Pro[2],D-Trp[7,9],Leu[11])-; Bombesin
Receptor Antagonist

Peptides International PSP-4172-v Synthetic MW 1497.8
$C_{75}H_{108}N_{20}O_{13}$ >99% (HPLC); lyophilized amorphous powder |
Bioactive; Jensen, RT et al, *Nature*, 309, 61, 1984

Arg-Pro-Lys-Pro-Gln-Gln-Trp-Phe-Trp-Leu-Leu
D-Arg-Pro-Lys-Pro-Gln-Gln-D-Trp-Phe-D-Trp-Leu-Leu
Amide Trihydrochloride Octahydroxy

Synonyms: Substance P, (D-Arg[1],D-Trp[7,9],Leu[11])-; Spantide; Substance P Antagonist

Peptides International PSP-4173 Synthetic MW 1751.3 $C_{75}H_{108}N_{20}O_{13} \cdot 3HCl \cdot 8H_2O$ 99% (HPLC); lyophilized amorphous powder; bulk | Bioactive; Folkers, K et al, *Br J Pharmacol*, 83:449, 1984

Arg-Pro-Lys-Pro-Gln-Gln-Trp-Phe-Trp-Leu-Leu
D-Arg-Pro-Lys-Pro-Gln-Gln-D-Trp-Phe-D-Trp-Leu-Leu
Amide Hydrochloride

Synonyms: Substance P, (D-Arg[1],D-Trp[7,9],Leu[11])-; Spantide; Substance P Antagonist

Peptides International PSP-4173-v Synthetic MW 1497.8 $C_{75}H_{108}N_{20}O_{13}$ >99% (HPLC); lyophilized amorphous powder | Bioactive; Folkers, K et al, *Br J Pharmacol*, 83:449, 1984

Arg-Pro-Lys-Pro-Gln-Gln-Trp-Phe-Trp-Leu-Met
Arg-D-Pro-Lys-Pro-Gln-Gln-D-Trp-Phe-D-Trp-Leu-Met
Amide

Synonyms: Substance P, (D-Pro[2],D-Trp[7,9])-

American Peptide 70-9-23 MW 1515.9 $C_{74}H_{106}N_{20}O_{13}S$ Long-term administration of this CNS substance P antagonist to the eye inhibits the ocular response to infrared irradiations of the iris; Endberg, G et al, *Nature*, 293:222, 1981

Bachem H-1920.0001 MW 1515.85 $C_{74}H_{106}N_{20}O_{13}S$ Store at -15°C

ICN 153010 CNS antagonist of substance P; long-term eye administration inhibits ocular response to iris infrared irradiation; Engberg, G etal, *Nature*, 293:222, 1981; Bynke, G etal, *Experentia*, 40:368, 1984

Arg-Pro-Lys-Pro-Gln-Gln-Trp-Phe-Trp-Leu-Met
Arg-D-Pro-Lys-Pro-Gln-Gln-D-Trp-Phe-D-Trp-Leu-Met
Amide Acetate Salt Free Base

Synonyms: Substance P, (D-Pro[2],D-Trp[7,9])-; Substance P Antagonist

Sigma S 0145 FW 1515.8 ≥97% (HPLC) | Bioactive peptide; similar to Sigma S 4509 but produced by Sigma; Engberg, G et, *Nature*, 293:222, 1981; Lembeck, F et al, *Biochem Biophys Res Commun*, 103:1318, 1981

Sigma S 4509 FW 1515.8 ≥97% (HPLC) | Bioactive peptide; Engberg, G et al, *Nature*, 293:222, 1981; Lembeck, F et al, *Biochem Biophys Res Commun*, 103:1318, 1981

Arg-Pro-Lys-Pro-Gln-Gln-Trp-Phe-Trp-Leu-Met
Arg-D-Pro-Lys-Pro-Gln-Gln-D-Trp-Phe-D-Trp-Leu-Met
Amide Trihydrochloride 6H2O

Synonyms: Substance P, (D-Pro[2],D-Trp[7,9])-; Substance P Antagonist

Peptides International PSP-4113 Synthetic MW 1733.3 $C_{74}H_{106}N_{20}O_{13}S \cdot 3HCl \cdot 6H_2O$ 95% (HPLC); lyophilized amorphous powder; bulk | Bioactive; Engberg, G et al, *Nature*, 293:222, 1981

Arg-Pro-Lys-Pro-Gln-Gln-Trp-Phe-Trp-Leu-Met
Arg-D-Pro-Lys-Pro-Gln-Gln-D-Trp-Phe-D-Trp-Leu-Met
Amide Hydrochloride

Synonyms: Substance P, (D-Pro[2],D-Trp[7,9])-; Substance P Antagonist

Peptides International PSP-4113-v Synthetic MW 1515.8 $C_{74}H_{106}N_{20}O_{13}S$ >95% (HPLC); lyophilized amorphous powder | Bioactive; Engberg, G et al, *Nature*, 293:222, 1981

Arg-Pro-Lys-Pro-Gln-Gln-Trp-Phe-Trp-Trp-Met
Arg-Pro-Lys-Pro-Gln-Gln-D-Trp-Phe-D-Trp-D-Trp-Met
Amide

Synonyms: Substance P, (D-Trp[7,9,10])-

Alexis 165-002 MW 1588.9 $C_{79}H_{105}N_{21}O_{13}S$ ≥97%; lyophilized powder; soluble in H_2O | Potent, selective inhibitor of M1 mAChR-promoted GTP hydrolysis by $G_{q/11}$, whereas not effective in inhibiting M2 mAChR-promoted GTP hydrolysis by G_i; Folkers, K et al, *Br J Pharmacol*, 83:449, 1984; Mukai, H et al, *J Biol Chem*, 267:16237, 1992

Arg-Pro-Lys-Pro-Gln-Gln-Trp-Phe-Trp-Trp-Met
Arg-Pro-Lys-Pro-Gln-Gln-D-Trp-Phe-D-Trp-D-Trp-Met
Amide Trifluoroacetate Salt

Synonyms: G Protein Antagonist; GP Antagonist-2A

Calbiochem 371780 MW 1588.9 $C_{79}H_{105}N_{21}O_{13}S$ >98% (TLC); lyophilized; soluble in H_2O | Selectively inhibits the activation of G_q by M_1-muscarinic cholinergic receptors; Mukai, H et al, *J Biol Chem*, 267:16237, 1992

Arg-Pro-Lys-Pro-Gln-Gln-Trp-Phe-Trp-Trp-Met
Arg-Pro-Lys-Pro-Gln-Gln-D-Trp-Phe-D-Trp-D-Trp-Met
Amide

Synonyms: Protein G Antagonist IIA; Substance P, (D-Trp[7,9,10])-; Substance P Antagonist

ICN 158348 MW 1589.3 99% | Reversible & competitive inhibitor of G proteins; Mukai, H etal, *JBC*, 267:16237, 1992

Sigma S 2546 FW 1588.9 ≥97% (HPLC) | Bioactive peptide

Arg-Pro-Lys-Pro-Leu-Ala
Arg-Pro-Lys-Pro-Leu-Ala-SBzl

Bachem H-1154.0005 Store at -15°C

Arg-Pro-Lys-Pro-Leu-Ala-Nva-Trp-Lys
6-(7-Nitrobenzo(1,2,5)Oxadiazol-4-yl-Amino)-
Hexanoyl-Arg-Pro-Lys-Pro-Leu-Ala-Nva-Trp-Lys-AMC

Bachem M-2300.0001 MW 1598.87 $C_{78}H_{111}N_{21}O_{16}$ Store at 2-8°C

Arg-Pro-Lys-Pro-Phe-Gln-Trp-Phe-Gly-Leu-Trp
D-Arg-Pro-Lys-Pro-D-Phe-Gln-D-Trp-Phe-Gly-Leu-D-Trp Amide

Synonyms: Substance P, (D-Arg[1],D-Phe[5],D-Trp[7,11])-

Bachem H-2822.0001 MW 1460.75 $C_{75}H_{101}N_{19}O_{12}$ Store at -15°C

Arg-Pro-Lys-Pro-Phe-Gln-Trp-Phe-Trp-Leu-Leu
D-Arg-Pro-Lys-Pro-D-Phe-Gln-D-Trp-Phe-D-Trp-Leu-Leu Amide

Synonyms: Antagonist D; Substance P, (D-Arg[1],D-Phe[5],D-Trp[7,9],Leu[11])-

American Peptide 70-9-18 MW 1516.9 $C_{79}H_{109}N_{19}O_{12}$ Inhibits the binding of GRP, vasopressin, bradykinin, cholecystokinin, galanin & neurotensin to their receptors *in vitro*; Fabregat, I et al, *J Cell Physiol*, 145:88, 1990

Bachem H-6935.0001 MW 1516.86 $C_{79}H_{109}N_{19}O_{12}$ Store at -15°C

ICN 153016 Substance P antagonist; Tsou, K etal, *Eur J Pharmac*, 110:155, 1985; Heinz-Erian, P etal, *Gastroenterology*, 90:1455, 1986

Neosystem SC395 MW 1516.85 Bioactive; tachykinin; Woll, PJ & Rozengurt, E, *PNAS USA*, 85:1859, 1988

Sigma S 3144   FW 1516.9   ≥97% (HPLC) | Bioactive peptide; inhibits binding of bradykinin, cholecystokinin, galanin, growth hormone releasing factor, neurotensin, & Substance P to their cell surface receptors; Tsou, K et al, *Eur J Pharmacol*, 110:155, 1985; Heinz-Erian, P et al, *Gastroenterology*, 90:1455, 1986; Woll, PJ & Rozengurt, E, *Proc Natl Acad Sci USA*, 85:1859, 1988; Houben, H & Denef, C, *Peptides*, 14:109, 1993; Reeve, RG & Bleehen, NM, *Biochem Biophys Res Commun*, 199:1313, 1994; Jones, DA et al, *Peptides*, 16:777, 1995

Arg-Pro-Lys-Pro-Trp-Gln-Trp-Phe-Trp-Leu-Leu
D-Arg-Pro-Lys-Pro-D-Trp-Gln-D-Trp-Phe-D-Trp-Leu-Leu Amide

Synonyms: Substance P, (D-Arg[1],D-Trp[5,7,9],Leu[11])-

Bachem H-3992.0001   Store at -15°C

Arg-Pro-Lys-Pro-Tyr-Ala-Nva-Trp-Met-Lys
Mca-Arg-Pro-Lys-Pro-Tyr-Ala-Nva-Trp-Met-Lys(Dnp) Amide

Bachem M-2105.0001   MW 1656.89 $C_{79}H_{105}N_{19}O_{19}S$ Store at -15°C

Arg-Pro-Lys-Pro-Tyr-Ala-Nva-Trp-Met-Lys
MOCAc-Arg-Pro-Lys-Pro-Tyr-Ala-Nva-Trp-Met-Lys(Dnp) Amide

Synonyms: Matrix Metalloproteinase Fluorescence-Quenching Substrate

Peptides International SMO-3167-v   MW 1656.9
$C_{79}H_{105}N_{19}O_{19}S$ >96% (HPLC); lyophilized amorphous powder | Nagase, H et al, *JBC*, 269:20952, 1994

Arg-Pro-Lys-Pro-Tyr-Ala-Nva-Trp-Met-Lys
AMC-Arg-Pro-Lys-Pro-Tyr-Ala-Nva-Trp-Met-(2,4-Dinitrophenyl)-Lys Amide

Sigma M 9420   FW 1656.9   ≥97% (HPLC) | Nagase, H et al, *J Biol Chem*, 269:20952, 1994

Arg-Pro-Lys-Pro-Val-Glu-Nva-Trp-Arg-Lys
Mca-Arg-Pro-Lys-Pro-Val-Glu-Nva-Trp-Arg-Lys(Dnp) Amide

Bachem M-2110.0001   MW 1675.87 $C_{78}H_{110}N_{22}O_{20}$ Store at -15°C

Arg-Pro-Lys-Pro-Val-Glu-Nva-Trp-Arg-Lys
1MOCAc-Arg-Pro-Lys-Pro-Val-Glu-Nva-Trp-Arg-Lys(Dnp) Amide

Synonyms: Matrix Metalloproteinase III Fluorescence-Quenching Substrate; Stromelysin Fluorescence-Quenching Substrate

Peptides International SMO-3168-v   MW 1675.9
$C_{78}H_{110}N_{22}O_{20}$ >96% (HPLC); lyophilized amorphous powder | Nagase, H et al, *JBC*, 269:20952, 1994

Arg-Pro-Lys-Pro-Val-Glu-Nva-Trp-Arg-Lys
AMC-Arg-Pro-Lys-Pro-Tyr-Ala-Nva-Trp-Met-(2,4-Dinitrophenyl)-Lys Amide

Sigma M 9545   FW 1675.9   ≥97% (HPLC)

Arg-Pro-Phe-His-Leu-Leu-Val-Tyr
Suc-Arg-Pro-Phe-His-Leu-Leu-Val-Tyr-AMC

Bachem I-1340.0005   MW 1301.51 $C_{66}H_{88}N_{14}O_{14}$ Store at -15°C

Arg-Pro-Phe-His-Leu-Leu-Val-Tyr
N-Suc-L-Arg-L-Pro-L-Phe-L-His-L-Leu-L-Leu-L-Val-L-Tyr-MCA

Synonyms: Renin Substrate

ICN 152083   Fluorogenic substrate; Murakami, K et al, *Seikagaku*, 50:762, 1978

Arg-Pro-Phe-His-Leu-Leu-Val-Tyr
Suc-Arg-Pro-Phe-His-Leu-Leu-Val-Tyr-MCA

Synonyms: Renin Substrate; Proteinase A Substrate

Peptides International MRP-3110-v   MW 1301.5
$C_{66}H_{88}N_{14}O_{14}$ >98% (HPLC); lyophilized amorphous powder | Murakami, K et al, *Anal Biochem*, 110:232, 1981; Yokosawa, H et al, *Anal Biochem*, 134:210, 1983

Arg-Pro-Phe-His-Leu-Leu-Val-Tyr
N-Suc-Arg-Pro-Phe-His-Leu-Leu-Val-Tyr-AMC

Sigma S 8883   FW 1301.5   Sakakibara, S et al, *Anal Biochem*, 110:232, 1981; substrate for renin

---

Arg-Pro-Pro

Synonyms: Bradykinin (1-3)

American Peptide 12-1-90   MW 368.4 $C_{16}H_{28}N_6O_4$

Bachem H-1935.0025   MW 368.44 $C_{16}H_{28}N_6O_4$ Store at -15°C

Arg-Pro-Pro-Gly-Ala-Ser-Phe-Ala-Arg
Arg-Pro-Pro-Gly-α-(2-Thienyl)-Ala-Ser-D-Phe-α-(2-Thienyl)-Ala-Arg

Synonyms: Bradykinin, (α-(2-Thienyl)-Ala[5,8],D-Phe[7])-

Bachem H-9080.0001   Store at -15°C

Arg-Pro-Pro-Gly-Ala-Ser-Pro-Arg
Arg-Pro-Pro-Gly-α-(2-Thienyl)-Ala-Ser-Pro-Arg

Bachem H-9620.0005   Store at -15°C

Arg-Pro-Pro-Gly-Phe

Synonyms: Bradykinin (1-5)

American Peptide 18-1-55   MW 572.7 $C_{27}H_{40}N_8O_6$

Bachem H-1945.0005   MW 572.67 $C_{27}H_{40}N_8O_6$ Store at -15°C

ICN 152786   MW 572.2 $C_{27}H_{40}N_8O_6$

Sigma B 1401   FW 572.7 $C_{27}H_{40}N_8O_6$ ≥97% (HPLC) | Bioactive peptide; Barabe, J et al, *Can J Phys Pharm*, 55:855 & 1270, 1977; *Can J Biochem*, 57:1084, 1979; Redman, LW et al, *Can J Biochem*, 57:529, 1979

Arg-Pro-Pro-Gly-Phe-Ser

Synonyms: Bradykinin (1-6)

American Peptide 18-1-56   MW 659.8 $C_{30}H_{45}N_9O_8$

Bachem H-1950.0005   MW 659.74 $C_{30}H_{45}N_9O_8$ Store at -15°C

ICN 152787   MW 659.7 $C_{30}H_{45}N_9O_8$

Sigma B 1526   FW 659.7 $C_{30}H_{45}N_9O_8$ ≥97% (HPLC) | Bioactive peptide; see Sigma B 1401

Arg-Pro-Pro-Gly-Phe-Ser-Ala-Phe-Lys
Mca-Arg-Pro-Pro-Gly-Phe-Ser-Ala-Phe-Lys(Dnp)

Synonyms: Bradykinin, Mca-(Ala[7],Lys(Dnp)[9])-

Bachem M-2405.0001   MW 1388.46 $C_{66}H_{81}N_{15}O_{19}$ Store at -15°C

Arg-Pro-Pro-Gly-Phe-Ser-Phe-Phe-Arg
Arg-Pro-Pro-Gly-Phe-Ser-D-Phe-Phe-Arg

Synonyms: Bradykinin, (D-Phe[7])-; Bradykinin Antagonist; B₂ Receptor Antagonist

American Peptide 18-1-66   MW 1110.3 $C_{54}H_{75}N_{15}O_{11}$
Although it is a weak agonist in rat uterus (D-Phe[7]), acts as a competitive B₂ receptor antagonist in guinea pig ileum by inhibiting bradykinin-induced contractions; Vavrek, RJ et al, *Peptides*, 6:161, 1985

Arg-Pro-Pro-Gly-Phe-Ser-Phe-Phe-Arg
Arg-Pro-Pro-Gly-Phe-Ser-N-Me-D-Phe-Phe-Arg

Synonyms: Bradykinin, (N-Me-D-Phe[7])-

Bachem H-5094.0001   Store at -15°C

Arg-Pro-Pro-Gly-Phe-Ser-Phe-Phe-Arg
Arg-Pro-Pro-Gly-Phe-Ser-D-Phe-Phe-Arg

Synonyms: Bradykinin, (D-Phe[7])-

Bachem H-9085.0001   MW 1110.28 $C_{54}H_{75}N_{15}O_{11}$ Store at -15°C

Arg-Pro-Pro-Gly-Phe-Ser-Phe-Phe-Arg
Arg-Pro-Pro-Gly-Phe-Ser-D-Phe-Phe-Arg Acetate Salt

Synonyms: Bradykinin, (D-Phe[7])-; Bradykinin Antagonist

ICN 152759   MW 1319.6   Vavrek, RJ & JM Stewart, *Peptides*, 6:161, 1985

Arg-Pro-Pro-Gly-Phe-Ser-Phe-Phe-Arg
Arg-Pro-Pro-Gly-Phe-Ser-D-Phe-Phe-Arg

Synonyms: Bradykinin, (D-Phe[7])-

Neosystem SC171   MW 1110.28   Bioactive; bradykinin B2 receptor antagonist

Arg-Pro-Pro-Gly-Phe-Ser-Phe-Phe-Arg
Arg-Pro-Pro-Gly-Phe-Ser-D-Phe-Phe-Arg Acetate Salt Free Base

Synonyms: Bradykinin, (D-Phe[7])-; B₂ Antagonist

Sigma B 7894   FW 1110.3   ≥97% (HPLC) | Bioactive peptide; competitive in guinea pig ileum; weak agonist in rat uterus; Vavrek, RJ & Stewart, JM, *Peptides*, 6:161, 1985

Arg-Pro-Pro-Gly-Phe-Ser-Pro

Synonyms: Bradykinin (1-7)

American Peptide 18-1-57   MW 756.9 $C_{35}H_{52}N_{10}O_{9}$

Bachem H-1955.0005   MW 756.86 $C_{35}H_{52}N_{10}O_{9}$ Store at -15°C

ICN 152788   MW 756.9 $C_{35}H_{52}N_{10}O_{9}$

Sigma B 1651   FW 756.9 $C_{35}H_{52}N_{10}O_{9}$ ≥97% (HPLC) | Bioactive peptide; see Sigma B 1401

Arg-Pro-Pro-Gly-Phe-Ser-Pro-Leu

Synonyms: Bradykinin, (des-Arg[9],Leu[8])-; Bradykinin, des-Arg[9]-(Leu[8])-; Bradykinin B1 Receptor Agonist

Bachem H-1960.0005   MW 870.02 $C_{41}H_{63}N_{11}O_{10}$ Store at -15°C

Neosystem SC921   MW 870.02   Bioactive; bradykinin B1 receptor antagonist; Regoli, D et al, *Can J Phys Pharm*, 55:855, 1977

Peptides International PBK-4065-v   Synthetic   MW 870.02 $C_{41}H_{63}N_{11}O_{10}$ >99% (HPLC); lyophilized amorphous powder | Bioactive; Regoli, D et al, *Can J Physiol Pharmacol*, 55:855, 1977

Arg-Pro-Pro-Gly-Phe-Ser-Pro-Leu Acetate Salt Free Base

Synonyms: Bradykinin, (des-Arg[9],Leu[8])-; Bradykinin Antagonist

Sigma B 6769   FW 870.0 $C_{41}H_{63}N_{11}O_{10}$ ≥97% (HPLC) | Bioactive peptide

Arg-Pro-Pro-Gly-Phe-Ser-Pro-Leu AcOH 3H₂O

Synonyms: Bradykinin, des-Arg[9]-(Leu[8])-; Bradykinin B1 Receptor Agonist

Peptides International PBK-4065   Synthetic   MW 989.12 $C_{41}H_{63}N_{11}O_{10} \cdot CH_{3}COOH \cdot 3H_{2}O$ >99% (HPLC); lyophilized amorphous powder; bulk | Bioactive; Regoli, D et al, *Can J Physiol Pharmacol*, 55:855, 1977

Arg-Pro-Pro-Gly-Phe-Ser-Pro-Phe

Synonyms: Bradykinin, (des-Arg[9])-

Bachem H-1965.0005   MW 904.04 $C_{44}H_{61}N_{11}O_{10}$ Store at -15°C

Arg-Pro-Pro-Gly-Phe-Ser-Pro-Phe
Arg-3,4-Dehydro-Pro-3,4-Dehydro-Pro-Gly-Phe-Ser-Pro-Phe

Synonyms: Bradykinin, (3,4-Dehydro-Pro[2,3],des-Arg[9])-

Bachem H-3124.0001   Store at -15°C

Arg-Pro-Pro-Gly-Phe-Ser-Pro-Phe

Synonyms: Bradykinin, (des-Arg[9])-; Bradykinin, (des-Arg[9])-; Bradykinin B1 Receptor Agonist

Neosystem SC911   MW 904.03   Bioactive; bradykinin B1 receptor agonist; Regoli, D et al, *Can J Phys Pharm*, 55:855, 1977

Peptides International PBK-4067-v   Synthetic   MW 904.04 $C_{44}H_{61}N_{11}O_{10}$ >99% (HPLC); lyophilized amorphous powder | Bioactive; Regoli, D et al, *Can J Physiol Pharmacol*, 55:855, 1977

Arg-Pro-Pro-Gly-Phe-Ser-Pro-Phe Acetate Salt

Synonyms: Bradykinin, (des-Arg[9])-

Biogenesis 1500-0254   Synthetic   MW 904.1 $C_{44}H_{61}N_{11}O_{10}$ 99% (HPLC); lyophilized | Regoli et al, *Can J Phys Pharm*, 55:855, 1977

ICN 152792   MW 964.1 $C_{44}H_{61}N_{11}O_{10} \cdot C_{2}H_{4}O_{2}$ Regoli, D etal, *Can J Physiol Pharmac*, 55:855, 1977; Marceau, F etal, *Can J Physiol Pharmac*, 59:131, 1981

Arg-Pro-Pro-Gly-Phe-Ser-Pro-Phe Acetate Salt Free Base

Synonyms: Bradykinin, (des-Arg[9])-; Selective B; Bradykinin Agonist

Sigma B 4397   FW 904.0 $C_{44}H_{61}N_{11}O_{10}$ ≥97% (HPLC) | Bioactive peptide; Regoli, D et al, *Can J Physiol Pharmacol*, 55:855, 1977; Marceau, F et al, *ibid*, 59:131, 1981

Arg-Pro-Pro-Gly-Phe-Ser-Pro-Phe AcOH 3H₂O

Synonyms: Bradykinin, (des-Arg[9])-; Bradykinin B1 Receptor Agonist

Peptides International PBK-4067   Synthetic   MW 1023.1 $C_{44}H_{61}N_{11}O_{10} \cdot CH_{3}COOH \cdot 3H_{2}O$ >99% (HPLC); lyophilized amorphous powder; bulk | Bioactive; Regoli, D et al, *Can J Physiol Pharmacol*, 55:855, 1977

Arg-Pro-Pro-Gly-Phe-Ser-Pro-Phe-Arg

Synonyms: Bradykinin; Nitric Oxide Synthase Activator

Alexis 152-006   MW 1060.3 $C_{50}H_{73}N_{15}O_{11}$ White lyophilized powder | Activator of Ca²⁺-dependent reconstitutive nitric oxide synthases; Richard, V et al, *Am J Physiol*, 259:H1433, 1990; Bogle, RG et al, *BBRC*, 180:926, 1991

American Peptide 18-1-10   MW 1060.2 $C_{50}H_{73}N_{15}O_{11}$ This peptide contracts smooth muscles, relaxes blood vessels & increases vascular permeability; other physiological functions include stimulation of pain receptors, inhibition of cAMP accumulation & induction of the release of nitric oxide; involved in edema resulting from trauma or injury; Regoli, D et al, *J Pharmacol Rev*, 32:1, 1980

Arg-Pro-Pro-Gly-Phe-Ser-Pro-Phe-Arg
Arg-Pro-Pro-Gly-(p-Chloro)Phe-Ser-Pro-(p-Chloro)Phe-Arg

Synonyms: Bradykinin, ((p-Chloro)Phe[5,8])-

American Peptide 18-1-64   MW 1129.1 $C_{50}H_{71}N_{15}O_{11}Cl_2$

Arg-Pro-Pro-Gly-Phe-Ser-Pro-Phe-Arg
Arg-([3H])Pro-([3H])Pro-Gly-Phe-Ser-Pro-Phe-Arg

Synonyms: Bradykinin, (Prolyl[2,4]-3,4(n)-[3H])-

Amersham TRK943   0.2 M triethylammonium acetate solution, pH 4.0, containing 18% acetonitrile; 1.48-4.07 TBq/mmol, 40-110 Ci/mmol; 9.25 MBq/mL, 250 µCi/mL

Arg-Pro-Pro-Gly-Phe-Ser-Pro-Phe-Arg
Arg-Pro-Pro-Gly-p-Chloro-Phe-Ser-Pro-p-Chloro-Phe-Arg

Synonyms: Bradykinin, (p-Chloro-Phe[5,8])-

Bachem H-1940.0001   MW 1129.11 $C_{50}H_{71}Cl_2N_{15}O_{11}$ Store at -15°C

Arg-Pro-Pro-Gly-Phe-Ser-Pro-Phe-Arg

Synonyms: Bradykinin

Bachem H-1970.0025   MW 1060.22 $C_{50}H_{73}N_{15}O_{11}$ Store at -15°C

Arg-Pro-Pro-Gly-Phe-Ser-Pro-Phe-Arg
Arg-3,4-Dehydro-Pro-3,4-Dehydro-Pro-Gly-Phe-Ser-Pro-Phe-Arg

Synonyms: Bradykinin, (3,4-Dehydro-Pro[2,3])-

Bachem H-3132.0001   Store at -15°C

Arg-Pro-Pro-Gly-Phe-Ser-Pro-Phe-Arg

Synonyms: Bradykinin

Calbiochem 05-23-0500   MW 1060.2 $C_{50}H_{73}N_{15}O_{11}$ ≥97% (HPLC); lyophilized; soluble in 5% HOAc; LD_{50}≤2000 mg/kg; may be carcinogenic/teratogenic | Induces the release of nitric oxide; other physiological functions include stimulation of pain receptors, inhibition of cAMP accumulation, induction of smooth muscle contraction & vasodilation; also involved in edema resulting from trauma or injury; improves post-ischemic recovery of heart via a nitric oxide-dependent mechanism; Mombouli, JV & Vanhoutte, PM, *Am J Hypertens*, 8:195, 1995; Zhu, P et al, *Cardiovasc Res*, 29:658, 1995; Bedarida, GV et al, *Horm Metab Res*, 26:109, 1994; Altiok, N & Fredholm, BB, *Cell Signal*, 5:279, 1993; *Merck Index*, 12:1386

Neosystem SC090   MW 1060.22   Bioactive; Boissonnas, RA et al, *Helv Chim Acta*, 43:1349-1350, 1960

Peptides International PBK-4002-v   Synthetic   MW 1060.2 $C_{50}H_{73}N_{15}O_{11}$ >99% (HPLC); lyophilized amorphous powder | Bioactive; human, bovine, rat, mouse; Erdös, EG (Ed), *Bradykinin, Kallidin & Kallikrein*, Handbook of Experimental Pharmacology, Vol 25 & 25(Suppl), Springer-Verlag, Berlin, 1979; Elliott, DF et al, *Biochem J*, 74:15, 1960; Nicolaides, ED & HA DeWald, *J Org Chem*, 26:3872, 1961; Pierce, JV & ME Webster, *BBRC*, 5:353, 1961

Arg-Pro-Pro-Gly-Phe-Ser-Pro-Phe-Arg 2AcOH 3H₂O

Synonyms: Bradykinin

Peptides International PBK-4002   Synthetic   MW 1234.3 $C_{50}H_{73}N_{15}O_{11} \cdot 2CH_3COOH \cdot 3H_2O$ >99% (HPLC); lyophilized amorphous powder; bulk | Bioactive; human, bovine, rat, mouse; Erdös, EG (Ed), *Bradykinin, Kallidin & Kallikrein*, Handbook of Experimental Pharmacology, Vol 25 & 25(Suppl), Springer-Verlag, Berlin, 1979; Elliott, DF et al, *Biochem J*, 74:15, 1960; Nicolaides, ED & HA DeWald, *J Org Chem*, 26:3872, 1961; Pierce, JV & ME Webster, *BBRC*, 5:353, 1961

Arg-Pro-Pro-Gly-Phe-Ser-Pro-Phe-Arg Acetate Salt

Synonyms: Bradykinin

Biogenesis 1500-0059   Human synthetic   MW 1060   >99% (HPLC); lyophilized

Arg-Pro-Pro-Gly-Phe-Ser-Pro-Phe-Arg Acetate Salt Free Base

Synonyms: Bradykinin; Nitric Oxide Synthase Activator

Sigma B 3259   FW 1060.2   ≥98% (HPLC) | Bioactive peptide; activates pain receptor; induces smooth muscle contraction; activates Ca^{2+}-dependent nitric oxide synthase; increases capillary permeability; believed to play important role in regulation of fluid & electrolyte balance, smooth muscle contraction, vasodilation & capillary permeability. Bradykinin potentiators possess no bradykinin activity themselves but enhance the activity of bradykinin; Kato, H & Suzuki, T, *Biochemistry*, 10:972, 1971

Arg-Pro-Pro-Gly-Phe-Ser-Pro-Phe-Arg Triacetate Salt

Synonyms: Bradykinin

Fluka 15859   MW 1240.39 $C_{50}H_{73}N_{15}O_{11} \cdot 3C_2H_4O_2$ ~98% (HPLC) | Regoli, D & Barabe, J, *Pharmacol Rev*, 32:1, 1980

ICN 101112   MW 1240.4 $C_{50}H_{73}N_{15}O_{11} \cdot 3C_2H_4O_2$ Lyophilized | Fox, RH et al, *J Physiol*, 157:589, 1961

Arg-Pro-Pro-Gly-Phe-Ser-Pro-Tyr-Arg

Synonyms: Bradykinin, (Tyr[8])-

American Peptide 18-1-28   MW 1076.2 $C_{50}H_{73}N_{15}O_{12}$ Useful for determination of human urinary kinin levels; Mashford, ML et al, *Biochem Pharmacol*, 20:969, 1971

Bachem H-1975.0001   Store at -15°C

ICN 152781   MW 1076.2   Used in the determination of urinarykinin; Mashford, ML etal, *Biochem Pharmac*, 20:969, 1971

Sigma B 7885   FW 1076.2   ≥97% (HPLC) | Bioactive peptide; useful for determination of human urinarykinin levels; Mashford, ML et al, *Biochem Pharmacol*, 20:969, 1971

Peptides International PBK-4075-v   Synthetic   MW 1076.2 $C_{50}H_{73}N_{15}O_{12}$ >99% (HPLC); lyophilized amorphous powder | Bioactive; for radioimmunoassay; Nielsen, MD et al, *Clinica Chimica Acta*, 125:145, 1982; Fredrick, MJ et al, *Life Sci*, 37:331, 1985

Arg-Pro-Pro-Gly-Phe-Thr-Pro-Phe-Arg

Synonyms: Bradykinin, (Thr[6])-

Bachem H-6325.0005   MW 1074.25 $C_{51}H_{75}N_{15}O_{11}$ Store at -15°C

ICN 152780   MW 1074.2   Pick, T etal, *Comp Biochem Physiol*, C87:287, 1987; Yashura, T etal, *Toxicon*, 25:527, 1987

Sigma B 1400   FW 1074.2   ≥97% (HPLC) | Bioactive peptide; Yasuhara, T et al, *Toxicon*, 25:527, 1987; Pick, T et al, *Comp Biochem Physiol*, C87:287, 1987

Arg-Pro-Pro-Gly-Thi-Ser-Phe-Thi-Arg
Arg-Pro-Pro-Gly-Thi-Ser-D-Phe-Thi-Arg

Synonyms: Bradykinin, (Thi[5,8],D-Phe[7])-; Bradykinin Antagonist

American Peptide 18-1-68   MW 1122.4 $C_{50}H_{71}N_{15}O_{11}S_2$ Vavrek, RJ et al, *Peptides*, 6:161, 1985

Arg-Pro-Pro-Gly-Thi-Ser-Phe-Thi-Arg
Arg-Pro-Pro-Gly-Thi-Ser-D-Phe-Thi-Arg Acetate Salt

Synonyms: Bradykinin, (Thi[5,8],D-Phe[7])-; Bradykinin Antagonist

ICN 152784   Potent; Vavrek, RJ & JM Stewart, *Peptides*, 6:161, 1985

Arg-Pro-Pro-Gly-Thi-Ser-Phe-Thi-Arg
Arg-Pro-Pro-Gly-Thi-Ser-D-Phe-Thi-Arg 2AcOH 4H$_2$O

Synonyms: Bradykinin, (Thi^{5,8},D-Phe^7)-; Bradykinin B2 Receptor Antagonist

Peptides International PBK-4175   Synthetic   MW 1314.4   C$_{50}$H$_{71}$N$_{15}$O$_{11}$S$_2$ · 2CH$_3$COOH · 4H$_2$O   >99% (HPLC); lyophilized amorphous powder; bulk | Bioactive; Vavrek, RJ & JM Stewart, *Peptides*, 6:61, 1985

Arg-Pro-Pro-Gly-Thi-Ser-Phe-Thi-Arg
Arg-Pro-Pro-Gly-Thi-Ser-D-Phe-Thi-Arg

Synonyms: Bradykinin, (Thi^{5,8},D-Phe^7)-; Bradykinin B2 Receptor Antagonist

Peptides International PBK-4175-v   Synthetic   MW 1122.3   C$_{50}$H$_{71}$N$_{15}$O$_{11}$S$_2$ >99% (HPLC); lyophilized amorphous powder | Bioactive; Vavrek, RJ & JM Stewart, *Peptides*, 6:61, 1985

Arg-Pro-Pro-Gly-Tyr-Ser-Pro-Phe-Arg

Synonyms: Bradykinin, (Tyr^5)-

American Peptide 18-1-30   MW 1076.2   C$_{50}$H$_{73}$N$_{15}$O$_{12}$

ICN 159900   MW 1076.3   98%

Arg-Pro-Tyr-Ile-Leu

Synonyms: Neurotensin (9-13)

Bachem H-3830.0005   Store at -15°C

ICN 154509   MW 660.9

Neosystem SC356   MW 660.81   Bioactive

Arg-Pro-Val-Lys-Val-Tyr-Pro-Asn-Gly-Ala-Glu-Asp-Glu-Ser-Ala-Glu-Ala-Phe-Pro-Leu-Glu-Phe

Synonyms: Adrenocorticotropic Hormone (18-39); Corticotropin-Like Intermediate Peptide

American Peptide 10-1-32   Human   MW 2465.7   C$_{112}$H$_{165}$N$_{27}$O$_{36}$ Larsson, LI, *Histochem*, 55:225, 1978

Bachem H-1215.0001   Human   MW 2465.7   C$_{112}$H$_{165}$N$_{27}$O$_{36}$ Store at -15°C

Neosystem SC063   Human   MW 2465.69   Bioactive; Scott, AP et al, *Nature*, 244:65-67, 1973

Sigma A 0673   Human   FW 2465.7   ≥97% (HPLC) | Bioactive Peptide; Larsson, LI, *Histochemistry*, 55:225, 1978

Arg-Pro-Val-Lys-Val-Tyr-Pro-Asn-Gly-Ala-Glu-Asp-Glu-Ser-Ala-Glu-Ala-Phe-Pro-Leu-Glu-Phe Acetate Salt

Synon Adrenocorticotropic Hormone (18-39)*yms*:

Biogenesis 0178-0539   Human synthetic   MW 2465   98.2% (HPLC); lyophilized

Arg-Pro-Val-Lys-Val-Tyr-Pro-Asn-Val-Ala-Glu-Asn-Glu-Ser-Ala-Glu-Ala-Phe-Pro-Leu-Glu-Phe

Synonyms: Adrenocorticotropic Hormone (18-39); Corticotropin A

ICN 152723   Human   Larsson, LI, *Histochemistry*, 55:225, 1978

Arg-Ser

Bachem G-1525.0250   MW 261.28   C$_9$H$_{19}$N$_5$O$_4$ Store at -15°C

Arg-Ser-Ala-Thr-Glu-Thr-Leu-Ala-Gly-Ala-Trp-Arg-Asp-Leu-Trp-Glu-Thr-Leu-Arg-Arg

Synonyms: GP140 (841-860); SIVmac251 Peptide

Neosystem SC792   MW 2388.66   SIVmac251 peptide from subtype GP140

Arg-Ser-Arg

American Peptide 87-1-76   MW 417.5   C$_{15}$H$_{31}$N$_9$O$_5$

Bachem H-1980.0005   MW 417.47   C$_{15}$H$_{31}$N$_9$O$_5$ Store at -15°C

Sigma A 2416   FW 417.5   C$_{15}$H$_{31}$N$_9$O$_5$

Arg-Ser-Arg-Gly-Pro-Ser-Pro-Arg-Arg

Bachem H-1656.0001   Store at -15°C

Arg-Ser-Arg-His-Phe

Bachem H-1985.0005   MW 701.79   C$_{30}$H$_{47}$N$_{13}$O$_7$ Store at -15°C

ICN 153059   MW 701.8   C$_{30}$H$_{47}$N$_{13}$O$_7$

Arg-Ser-Cys-Ile-Asp-Thr-Ile-Pro-Gln-Ser-Arg-Cys-Thr-Ala-Phe-Gln-Cys-Lys-His-Ser-Met-Lys-Tyr-Arg-Leu-Ser-Phe-Cys-Arg-Lys-Thr-Cys-Gly-Thr-Cys

Synonyms: Neurotoxin ShK, (Gln^9)-

Bachem H-3518.0100   *Stichodactyla helianthus*   Store at -15°C | Disulfide bonds: Cys^3-Cys^{35}, Cys^{12}-Cys^{28}, Cys^{17}-Cys^{32}

Arg-Ser-Cys-Ile-Asp-Thr-Ile-Pro-Lys-Ser-Arg-Cys-Thr-Ala-Phe-Gln-Cys-Lys-His-Ser-Met-Lys-Tyr-Arg-Leu-Ser-Phe-Cys-Arg-Lys-Thr-Cys-Gly-Thr-Cys

Synonyms: Neurotoxin ShK

Bachem H-2358.0100   Store at -15°C | Disulfide bonds: Cys^3-Cys^{35}, Cys^{12}-Cys^{28}, Cys^{17}-Cys^{32}

Arg-Ser-Cys-Ile-Asp-Thr-Ile-Pro-Lys-Ser-Arg-Cys-Thr-Ala-Phe-Gln-Cys-Lys-His-Ser-Met-Ala-Tyr-Arg-Leu-Ser-Phe-Cys-Arg-Lys-Thr-Cys-Gly-Thr-Cys

Synonyms: Neurotoxin ShK, (Ala^{22})-

Bachem H-3512.0100   *Stichodactyla helianthus*   Store at -15°C | Disulfide bonds: Cys^3-Cys^{35}, Cys^{12}-Cys^{28}, Cys^{17}-Cys^{32}

Arg-Ser-Cys-Ile-Asp-Thr-Ile-Pro-Lys-Ser-Arg-Cys-Thr-Ala-Phe-Gln-Cys-Lys-His-Ser-Met-Dap-Tyr-Arg-Leu-Ser-Phe-Cys-Arg-Lys-Thr-Cys-Gly-Thr-Cys

Synonyms: Neurotoxin ShK, (Dap^{22})-

Bachem H-4218.0100   *Stichodactyla helianthus*   Store at -15°C | Disulfide bonds: Cys^3-Cys^{35}, Cys^{12}-Cys^{28}, Cys^{17}-Cys^{32}

Arg-Ser-Cys-Ile-Asp-Thr-Ile-Pro-Lys-Ser-Arg-Cys-Thr-Ala-Phe-Gln-Cys-Lys-His-Ser-Met-Lys-Tyr-Arg-Leu-Ser-Phe-Cys-Arg-Lys-Thr-Cys-Gly-Thr-Cys

Synonyms: Voltage-Dependent K$^+$ Channel (A Channel) Blocker; Neurotoxin ShK

Calbiochem 569400   *Stichodactyla helianthus*   MW 4054.8   C$_{169}$H$_{274}$N$_{54}$O$_{48}$S$_7$ >99% (HPLC); lyophilized powder; soluble in water & 0.9% NaCl; LD$_{50}$≤2000 mg/kg | Novel voltage-gated K$^+$ channel blocker that inhibits the specific binding of Dendrotoxin I to rat brain membranes; forms oligomeric pores in plasma membrane; Tejuca, M et al, *Biochemistry*, 35:14947, 1996; Karlsson, E et al, *Toxicon*, 31:504, 1993

Peptides International PSK-4287-s   *Stichodactyla helianthus* (sea anemone) synthetic   MW 4054.8   C$_{169}$H$_{274}$N$_{54}$O$_{48}$S$_7$ >95% (HPLC); lyophilized amorphous powder | Bioactive; Disulfide bonds: Cys^3-Cys^{35}, Cys^{12}-Cys^{28}, Cys^{17}-Cys^{32}; Karlsson, E Aet al, *Toxicon*, 31:504, 1993

Alexis 630-044   Synthetic   MW 4054.8   C$_{169}$H$_{274}$N$_{54}$O$_{48}$S$_7$ Lyophilized powder; Disulfide bonds: Cys^3-Cys^{35}, Cys^{12}-Cys^{28} & Cys^{17}-Cys^{32} | Originally isolated from sea anemone Stichodactyla *helianthus*; voltage dependent K$^+$ channel (A channel) blocker; potent neurotoxin; Karlsson, E et al, *Toxicon*, 31:504, 1993

Arg-Ser-Cys-Ile-Asp-Thr-Ile-Pro-Lys-Ser-Gln-Cys-Thr-Ala-Phe-Gln-Cys-Lys-His-Ser-Met-Lys-Tyr-Arg-Leu-Ser-Phe-Cys-Arg-Lys-Thr-Cys-Gly-Thr-Cys

Synonyms: Neurotoxin ShK, (Gln[11])-

Bachem H-3516.0100 *Stichodactyla helianthus* Store at -15°C | Disulfide bonds: Cys^3-Cys^{35}, Cys^{12}-Cys^{28}, Cys^{17}-Cys^{32}

Arg-Ser-Gly-Pro-Pro-Gly-Leu-Gln-Gly-Arg-Leu-Gln-Arg-Leu-Leu-Gln-Ala-Ser-Gly-Asn-His-Ala-Ala-Gly-Ile-Leu-Thr-Met Amide

Synonyms: Orexin B; Appetite Boosting Peptide

Alexis 164-002 Human MW 2871.4 $C_{122}H_{212}N_{44}O_{34}S$ ≥97%; lyophilized powder | New hypothalamic neuropeptide that stimulates food intake in rats in a similar manner to orexin A, but the effect does not last as long as that of orexin A; affinity of orexin B for the OX_1 receptor is significantly lower than that of orexin A; binding affinities for the OX_2 receptor are similar for orexin A & B; Sakurai, T et al, *Cell*, 92:573, 1998; Schwartz, MW, *Nature Med*, 4:385, 1998

Bachem H-4174.0500 Human MW 2899.38 $C_{123}H_{212}N_{44}O_{35}S$ Store at -15°C

Neosystem SC1338 Human MW 2899.36 Bioactive neuropeptide; one of two new hypothalamic peptides regulating feeding behavior (the other is orexin-A); both come from the same precursor by proteolytic processing; stimulate food consumption in a dose-dependent manner when administered centrally to rats; Sakurai, T et al, *Cell*, 92:573-585, 1998; De Lecea, L et al, *PNAS, USA*, 95:322-327, 1998

Peptides International POR-4348-s Human synthetic MW 2899.4 $C_{123}H_{212}N_{44}O_{35}S$ >95% (HPLC); lyophilized amorphous powder | Bioactive; rat, mouse, bovine; Sakurai, T et al, *Cell*, 92:573, 1998; de Lecea, L et al, *PNAS* USA, 95:322, 1998

Arg-Ser-Gly-Pro-Pro-Gly-Leu-Gln-Gly-Cys Trifluoroacetate Salt

Synonyms: Orexin B (1-9), ~C

Biogenesis 7049-5150 Synthetic >80% (HPLC); lyophilized | Represents AA 1-9 of orexin B; an additional Cys has been conjugated to the C-terminus

Arg-Ser-Leu-Arg-Arg-Ser-Ser-Cys-Phe-Gly-Gly-Arg-Ile-Asp-Arg-Ile-Gly-Ala-Gln-Ser-Gly-Leu-Gly-Cys-Asn-Ser-Phe-Arg-Tyr

Synonyms: Natriuretic Factor (1-28), Atrial (Arg[0])-α-

Neosystem SC427 Rat MW 3218.61 Bioactive

Arg-Ser-Ser-Cys-Phe-Gly-Gly-Arg-Ile-Asp-Arg-Ile-Gly-Ala-Cys Amide

Synonyms: Natriuretic Factor (4-18), Atrial (Cys[18])-; Natriuretic Factor (4-23), Atrial C-; Natriuretic Peptide (4-23), Atrial (des-Gln[18],des-Ser[19],des-Gly[20,22],des-Leu[21])-; Natriuretic Peptide (4-23), Atrial des-(Gln[18],Ser[19],Gly[20],Leu[21],Gly[22])-

American Peptide 14-1-45 Rat MW 1594.9 $C_{64}H_{107}N_{25}O_{19}S_2$ Disulfide bonds: Cys^7-Cys^{18} | Specifically binds with ~99% of ANF receptors in the isolated perfused rat kidney; competes effectively with biologically active atrial natriuretic peptides for binding sites; MAAck, T et al, *Science*, 238:675, 1987

Bachem H-3134.0500 Rat Store at -15°C

Sigma A 1802 Rat FW 1594.8 ≥95% (HPLC); Disulfide bonds: :7-23 | Bioactive peptide

Arg-Ser-Ser-Cys-Phe-Gly-Gly-Arg-Ile-Asp-Arg-Ile-Gly-Ala-Gln-Ser-Gly-Leu-Gly-Cys-Asn-Ser-Phe-Arg-Tyr

Synon Natriuretic Peptide (4-28), Atrial *yms:*

Sigma A 6916 Human FW 2724.0 ≥97% (HPLC) | Bioactive peptide; useful as an immunogen for producing Ab to hANF

Arg-Ser-Ser-Cys-Phe-Gly-Gly-Arg-Ile-Asp-Arg-Ile-Gly-Ala-Gln-Ser-Gly-Leu-Gly-Cys-Asn-Ser-Phe-Arg

Synonyms: Natriuretic Peptide (4-27), Atrial; Auriculin A

American Peptide 14-5-45 Rat MW 2542.9 $C_{104}H_{168}N_{38}O_{33}S_2$ Disulfide bonds: :Cys^7-Cys^{23} | Atlas, SA et al, *Nature*, 309:717, 1984

Arg-Ser-Ser-Cys-Phe-Gly-Gly-Arg-Ile-Asp-Arg-Ile-Gly-Ala-Gln-Ser-Gly-Leu-Gly-Cys-Asn-Ser-Phe-Arg-Tyr

Synon Natriuretic Peptide (4-28), Atrial; Auriculin B *yms:*

American Peptide 14-5-47 Rat MW 2706.0 $C_{113}H_{177}N_{39}O_{35}S_2$ Disulfide bonds: :Cys^7-Cys^{23} | Atlas, SA et al, *Nature*, 309:717, 1984; Misono, KS et al, *BBRC*, 119:524, 1984

Arg-Ser-Ser-Cys-Phe-Gly-Gly-Arg-Ile-Asp-Arg-Ile-Gly-Ala-Gln-Ser-Gly-Leu-Gly-Cys-Asn-Ser-Phe-Arg

Synonyms: Natriuretic Peptide (126-149), Atrial

Bachem H-7750.0500 Rat MW 2542.85 $C_{104}H_{168}N_{38}O_{33}S_2$ Store at -15°C

Arg-Ser-Ser-Cys-Phe-Gly-Gly-Arg-Ile-Asp-Arg-Ile-Gly-Ala-Gln-Ser-Gly-Leu-Gly-Cys-Asn-Ser-Phe-Arg-Tyr

Synonyms: Natriuretic Peptide (126-150), Atrial ; Natriuretic Peptide (4-28), Atrial; Auriculin B

Bachem H-7755.0500 Rat MW 2706.02 $C_{113}H_{177}N_{39}O_{35}S_2$ Store at -15°C

ICN 153063 Rat MW 2706 Atlas, SA, *Nature*, 309:717, 1984; Maki, M et al, *Nature*, 309:722, 1984, Misono, KS etal, *BBRC*, 119:524, 1984

Arg-Ser-Ser-Cys-Phe-Gly-Gly-Arg-Ile-Asp-Arg-Ile-Gly-Ala-Gln-Ser-Gly-Leu-Gly-Cys Amide

Synonyms: Natriuretic Peptide (4-23), Atrial des-

ICN 159889 Rat MW 1595.3 98%

Arg-Ser-Ser-Cys-Phe-Gly-Gly-Arg-Ile-Asp-Arg-Ile-Gly-Ala-Gln-Ser-Gly-Leu-Gly-Cys-Asn-Ser-Phe-Arg-Tyr

Synon Natriuretic Peptide (4-28), Atrial *yms:*

Sigma A 2677 Rat FW 2706.0 ≥97% (HPLC) | Bioactive peptide; Misono, KS et al, *Biochem Biophys Res Commun*, 199:524, 1984; Atlas, SA et al, *Nature*, 309:717, 1984; Maki, M et al, *Nature*, 309:722, 1984

Arg-Ser-Ser-Cys-Phe-Gly-Gly-Arg-Ile-Asp-Arg-Ile-Gly-Ala-Gln-Ser-Gly-Leu-Gly-Cys-Asn-Ser-Phe-Arg

Synonyms: Natriuretic Peptide (4-27), Atrial

Sigma A 2802 Rat FW 2542.8 ≥97% (HPLC) | Bioactive peptide; Atlas, SA et al, *Nature*, 309:717, 1984

Arg-Ser-Ser-Cys-Phe-Gly-Gly-Arg-Met-Asp-Arg-Ile-Gly-Ala-Gln-Ser-Gly-Leu-Gly-Cys-Asn-Ser-Phe-Arg-Tyr

Synonyms: Natriuretic Peptide (4-28), Atrial

American Peptide 14-1-38 Human MW 2724.1 $C_{112}H_{175}N_{39}O_{35}S_3$ Disulfide bonds: Cys^7-Cys^{23} | Tanaka, I et al, *BBRC*, 124:663, 1984

Bachem H-1990.0500 Human MW 2724.06 $C_{112}H_{175}N_{39}O_{35}S_3$ Store at -15°C

ICN 159888 Human MW 2725.9 98% | Tanaka, I etal, *BBRC*, 124:663, 1984

Arg-Ser-Ser-Cys-Tyr-Gly-Gly-Arg-Ile-Asp-Arg-Ile-Gly-Ala-Gln-Ser-Gly-Leu-Gly-Cys-Asn-Ser-Phe-Arg

Synonyms: Natriuretic Peptide (4-27), Atrial; Auriculin A

ICN 153062   Rat   MW 2542.8   Atlas, SA, *Nature*, 309:717, 1984

Arg-Thr Amide Dihydrochloride

Bachem G-3895.0250   Store at -15°C

Arg-Thr-Lys-Arg-Ser-Gly-Ser-Val-Tyr-Glu-Pro-Leu-Lys-Ile

Synonyms: Malantide; cAMP-Dependent Protein Kinase Substrate; Malantide; cAMP-Dependent Protein Kinase Substrate; cAMP-Dependent Protein Kinase C Substrate; Protein Kinase Substrate

Alexis 162-003   MW 1633.9 $C_{72}H_{124}N_{22}O_{21}$ Very potent; useful in measuring activity of this enzyme via phosphorylation-induced decrease in fluorescence; peptide sequence corresponds to the phosphorylation site on the β-subunit phosphorylase kinase; Malencik, DA & Anderson, SR, *Anal Biochem*, 132:32, 1983; Murray, KJ, *Biochem J*, 267:703, 1990

American Peptide 86-0-42   MW 1633.9 $C_{72}H_{124}N_{22}O_{21}$ cAMP-dependent protein kinase & protein kinase C substrate in various tissues; Murray, KJ et al, *JBC*, 261:25, 1990

Bachem H-3262.0001   MW 1633.91 $C_{72}H_{124}N_{22}O_{21}$ Store at -15°C

ICN 159801   MW 1634.1   Very potent proteinkinase substrate

Sigma A 3317   FW 1633.9   ≥97% (HPLC); peptide content:~65% | Bioactive peptide; high-affinity substrate for cAMP-dependent proteinkinase; Murray, KJ et al, *Biochem J*, 267:703, 1990

Arg-Trp

Bachem G-1530.0250   MW 360.42 $C_{17}H_{24}N_6O_3$ Store at -15°C

Arg-Trp Amide Dihydrochloride

Bachem G-3650.0100   MW 432.35 $C_{17}H_{25}N_7O_2 \cdot 2HCl$ Store at -15°C

Arg-Trp-Lys-Pro-Gln-Gln-Trp-Phe-Trp-Leu-Met
Arg-D-Trp-Lys-Pro-Gln-Gln-D-Trp-Phe-D-Trp-Leu-Met Amide

Synonyms: Substance P, (D-Trp²,⁷,⁹)-

American Peptide 70-1-95   MW 1605.0 $C_{80}H_{109}N_{21}O_{13}S$ Substance P antagonist

Bachem H-1995.0001   Store at -15°C

ICN 153012

Arg-Trp-Phe-Trp-Leu-Met
Arg-D-Trp-(nMe)Phe-D-Trp-Leu-Met Amide

Synonyms: Antagonist G

American Peptide 70-9-32   MW 952.2 $C_{49}H_{67}N_{12}O_6S$ Anticancer agent, developed for the treatment of small cell lung cancer

---

Arg-Trp-Phe-Trp-Leu-Met
Arg-D-Trp-*N*-Me-Phe-D-Trp-Leu-Met Amide

Synonyms: Substance P (6-11), (Arg⁶,D-Trp⁷,⁹,*N*-Me-Phe⁸)-

Bachem H-1510.0001   Store at -15°C

ICN 153032   Substance P receptor antagonist; Laufer, R etal, *PNAS*, 82:7444, 1985

Arg-Trp-Phe-Trp-Leu-Met
Arg-D-Trp-*N*-Methyl-Phe-D-Trp-Leu-Met Amide

Synonyms: Substance P (6-11), (Arg⁶,D-Trp⁷,⁹,*N*-Me-Phe⁸)-; G Antagonist

Sigma S 6392   FW 951.2 $C_{49}H_{66}N_{12}O_6S$ ≥97% (HPLC) | Bioactive peptide; anticancer agent; broad-spectrum growth factor antagonist; resistant to degradation by peptidases; developed for treatment of small cell lung cancer; Laufer, R et al, *Proc Natl Acad Sci USA*, 82:7444, 1985

Arg-Tyr

Bachem G-1535.0250   MW 337.38 $C_{15}H_{23}N_5O_4$ Store at -15°C

Arg-Tyr Amide Dihydrochloride

Bachem G-4095.0250   Store at -15°C

Arg-Tyr-Gly-Gly-Phe-Met

Synonyms: Enkephalin, ((Arg⁰)-Met)-; Lipotropin (60-65), β-

American Peptide 30-0-26   MW 729.9 $C_{33}H_{47}N_9O_8S$

Arg-Tyr-Gly-Gly-Phe-Met-Thr-Ser-Glu-Lys-Ser-Gln-Thr-Pro-Leu-Val-Thr-Leu-Phe-Lys-Asn-Ala-Ile-Ile-Lys-Asn-Ala-Tyr-Lys-Lys-Gly-Glu

Synonyms: Endorphin, (Arg⁰)-β-; Lipotropin (60-91), β-

ICN 154531   Human   MW 3621.8

Arg-Tyr-Gly-Gly-Phe-Met-Thr-Ser-Glu-Lys-Ser-Gln-Thr-Pro-Leu-Val-Thr-Leu-Phe-Lys-Asn-Ala-Ile-Ile-Lys-Asn-Ala-Tyr-Lys-Lys-Gly-Glu
D-Arg-Tyr-Gly-Gly-Phe-Met-Thr-Ser-Glu-Lys-Ser-Gln-Thr-Pro-Leu-Val-Thr-Leu-Phe-Lys-Asn-Ala-Ile-Ile-Lys-Asn-Ala-Tyr-Lys-Lys-Gly-Glu

Synonyms: Endorphin, (D-Arg⁰)-β-

ICN 154532   Human   MW 3621.8

Arg-Tyr-Gly-Gly-Phe-Met-Thr-Ser-Glu-Lys-Ser-Gln-Thr-Pro-Leu-Val-Thr-Leu-Phe-Lys-Asn-Ala-Ile-Ile-Lys-Asn-Ala-Tyr-Lys-Lys-Gly-Glu

Synonyms: Endorphin, Arg-β-

Sigma E 7638   Human   FW 3621.2   ≥97% (HPLC) | Bioactive peptide

---

Arg-Tyr-Leu-Gly-Tyr-Leu

Synonyms: Casein, α- (90-95); Opioid Peptide

American Peptide 32-3-25   MW 783.9 $C_{38}H_{57}N_9O_9$ The most potent opioid based on examining the bioactivity of various opioid peptides in different bioassays

Bachem H-2000.0005   MW 783.93 $C_{38}H_{57}N_9O_9$ Store at -15°C

ICN 152944   MW 783.9 $C_{38}H_{57}N_9O_9$ Loukas, S etal, *Biochem*, 22:4567, 1983

Sigma C 1658   FW 783.9 $C_{38}H_{57}N_9O_9$ ≥97% (HPLC) | Bioactive peptide; Loukas, S et al, *Biochem*, 22:4567, 1983

Arg-Tyr-Leu-Gly-Tyr-Leu-Glu

Synonyms: Casein, α- (90-96); Opioid Peptide

Bachem H-2005.0005   MW 913.04 $C_{43}H_{64}N_{10}O_{12}$ Store at -15°C

ICN 152945   MW 912 $C_{43}H_{64}N_{10}O_{12}$

Neosystem SC495   MW 913.20   Virus-related peptide; bioactive

Sigma C 1783   FW 913.0 $C_{43}H_{64}N_{10}O_{12}$ ≥97% (HPLC) | Bioactive peptide; Loukas, S et al, *Biochem*, 22:4567, 1983

Arg-Tyr-Leu-Pro-Thr

Synonyms: Proctolin

American Peptide 60-0-52   MW 648.8 $C_{30}H_{48}N_8O_8$ The first structurally characterized myotropic insect neuromodulator; detected in different parts of the insect body such as the central nervous system & digestive tract, & in the brain of other invertebrates; Konopinska, D et al, *J Pep Res*, 49:457, 1997

Bachem N-1015.0005   Store at -15°C

ICN 151949   Insect neurotransmitter; Starratt, AN & BE Brown, *Life Sci*, 17:1253, 1976; O'Shea, J etal, *Science*, 23:567, 1981

Arg-Tyr-Leu-Pro-Thr Acetate Salt Free Base

Synonyms: Proctolin; Aminoenkephalinase Inhibitor

Sigma P 4280   FW 648.8 $C_{30}H_{48}N_8O_8$ ≥97% (HPLC) | Bioactive peptide; potent, selective inhibitor; may be an insect neurotransmitter; O'Shea, M et al, *Science*, 213:567, 1981; Starratt, AN & Brown, BE, *Life Sci*, 17:1253, 1976

Arg-Tyr-Pro-Leu-Thr-Phe-Gly-Trp-Cys-Tyr-Lys

Synonyms: NEF (134-144); HTLV I Peptide

Neosystem SC634   MW 1433.69   HTLV I peptide from subtype NEF; due to the presence of a Cys this peptide can contain the dimeric form

Arg-Tyr-Ser-Thr-Gln-Val-Asp-Pro-Glu-Leu-Ala-Asp-Gln-Leu

Synonyms: VIF (93-106); HTLV I Peptide

Neosystem SC548   MW 1634.76   HTLV I peptide from subtype VIF

Arg-Tyr-Tyr-Arg-Ile-Lys
Ac-Arg-Tyr-Tyr-Arg-Ile-Lys Amide

Bachem H-5096.0005   MW 939.13 $C_{44}H_{70}N_{14}O_9$ Store at -15°C

Arg-Tyr-Val-Val-Leu-Pro-Arg-Pro-Val-Cys-Phe-Glu-Lys-Gly-Met-Asn-Tyr-Thr-Val-Arg

Synonyms: Laminin B1 Chain (641-660)

Sigma A 8430   FW 2427.9   ≥97% (HPLC) | Bioactive peptide; heparin-binding & cell adhesion-promoting activities; Charonis, AS et al, *J Cell Biol*, 107:1253, 1988

Arg-Val

Bachem G-1540.0250   MW 273.34 $C_{11}H_{23}N_5O_3$ Store at -15°C

Arg-Val Acetate Salt

Sigma A 7161   FW 333.4 $C_{11}H_{23}N_5O_3 \cdot C_2H_4O_2$

Arg-Val-Arg
((S)-1-Carboxy-2-Phenylethyl)-Carbamoyl-Arg-Val-Arg-CHO

Synonyms: Antipain

Bachem H-1765.0050   MW 604.71 $C_{27}H_{44}N_{10}O_6$ Store at -15°C

Arg-Val-Arg
((S)-1-Carboxy-2-Phenylethyl)Carbamoyl-L-Arg-L-Val-Arg-al

Synonyms: Antipain; Protease Inhibitor

Fluka 10791   MW 604.71 $C_{27}H_{44}N_{10}O_6$ ~1000 U/mg; 1 U is the amount of inhibitor which reduces the trypsin activity by 1 BAEE-U; 1 BAEE-U is the amount of enzyme which increases the absorbance at 253 nm by 0.001/min at pH 7.6, 25°C | Suda, H et al, *J Antibiotics*, 25:263, 1972

Arg-Val-Arg
((S)-1-Carboxy-2-Phenylethyl)-Carbamoyl-Arg-Val-Arg Dihydrate Hydrochloride

Synonyms: Antipain; Protease Inhibitor

Alexis 260-004   Microbial   MW 677.2 $C_{27}H_{44}N_{10}O_6 \cdot HCl \cdot 2H_2O$ Pale yellow powder; soluble in H_2O | Suda, H et al, *J Antibiot*, 25:263, 1972; Umezawa, H, *Meth Enzymol*, 55:678, 1976

Arg-Val-Arg
((S)-1-Carboxy-2-Phenylethyl)-Carbamoyl-Arg-Val-Arg-al

Synonyms: Antipain; Trypsin Inhibitor; Papain Inhibitor; Cathepsin Inhibitor

ICN 152843   Microbial   MW 604.7 $C_{27}H_{44}N_{10}O_6$ Inhibitor for trypsin, papain & cathepsins A&B; Suda, H etal, *J Antibiot*, 25:263, 1972; Umezawa, H etal, *J Antibiot*, 25:267, 1972; Ikezawa, H etal, *J Antibiot*, 25:738, 1972

Arg-Val-Arg
((S)-1-Carboxy-2-Phenylethyl)-Carbamoyl-Arg-Val-Arg-al Hydrochloride

Synonyms: Antipain; Trypsin Inhibitor; Papain Inhibitor; Cathepsin Inhibitor

ICN 198585   Microbial   MW 641.2 $C_{27}H_{44}N_{10}O_6 \cdot HCl$ Inhibitor for trypsin, papain & cathepsins A&B; Suda, H etal, *J Antibiot*, 25:263, 1972; Umezawa, H etal, *J Antibiot*, 25:267, 1972; Ikezawa, H etal, *J Antibiot*, 25:738, 1972

Arg-Val-Arg
((S)-1-Carboxy-2-Phenylethyl)-Carbamoyl-L-Arg-L-Val-Arg-al Hydrochloride Dihydrate

Synonyms: Antipain; Trypsin Inhibitor; Papain Inhibitor; Cathepsin A/B Inhibitor

Peptides International IAP-4062   Microbial   MW 677.20 $C_{27}H_{44}N_{10}O_6 \cdot HCl \cdot 2H_2O$ Amorphous powder; integrity assessed by activity; bulk | Inhibitor; Suda, H et al, *J Antibiotics*, 25:263, 1972; Umezawa, S et al, *J Antibiotics*, 25:267, 1972

Arg-Val-Arg
((S)-1-Carboxy-2-Phenylethyl)-Carbamoyl-L-Arg-L-Val-Arg-al

Synonyms: Antipain; Trypsin Inhibitor; Papain Inhibitor; Cathepsin A/B Inhibitor

Peptides International IAP-4062-v   Microbial   MW 604.71 $C_{27}H_{44}N_{10}O_6$ Lyophilized amorphous powder; integrity assessed by activity

Arg-Val-Arg-Arg
BOC-Arg-Val-Arg-Arg-AMC

Synonyms: Furin Convertase Substrate; Furin

Alexis 260-040   MW 843.0 $C_{38}H_{62}N_{14}O_8$ ≥97%; white solid; soluble in H_2O | Fluorogenic; Hatsuzawa, K et al, *J Biol Chem*, 267:16094, 1992; Molloy, SS et al, *J Biol Chem*, 267:16396, 1992

American Peptide 81-7-40   MW 843.0 $C_{38}H_{62}N_{14}O_8$ Calcium-dependent serine protease associated with Golgi membranes; responsible for secretory processing of precursor proteins at paired basic residues; Hatsuzawa, K et al, *JBC*, 267:16094, 1992

Bachem I-1645.0025   MW 843.01 $C_{38}H_{62}N_{14}O_8$ Store at -15°C

Arg-Val-Arg-Arg
BOC-Arg-Val-Arg-Arg-MCA

Synonyms: Furin Substrate

Peptides International MRR-3155-v   MW 843.00 $C_{38}H_{62}N_{14}O_8$ >98% (HPLC); lyophilized amorphous powder | Hatsuzawa, K et al, *J Biochem*, 111:296, 1992; Hatsuzawa, K et al, *JBC*, 267:1609, 1992

Arg-Val-Arg-Arg
N-t-BOC-Arg-Val-Arg-Arg-AMC Acetate Salt Free Base

Synonyms: Furin Convertase Substrate

Sigma B 2158 FW 843.0 $C_{38}H_{62}N_{14}O_8$ Fluorogenic; Hatsuzawa, K et al, *J Biol Chem*, 267:16094, 1992; Molloy, SS et al, *J Biol Chem*, 267:16396, 1992

Arg-Val-Arg-Lys
Decanoyl-Arg-Val-Arg-Lys-CMK

Bachem N-1435.0005 MW 744.42 $C_{34}H_{66}ClN_{11}O_5$ Store at -15°C

Arg-Val-Arg-Phe
N-(-N'-Carbonyl-Arg-Val-Arg-al)-Phe

Synonyms: Antipain

Sigma A 6191 Microbial FW 604.7 $C_{27}H_{44}N_{10}O_6$ Free base, hydrochloride | Bioactive peptide; protease inhibitor; Suda, H et al, *J Antibiot*, 25:263, 1972

Arg-Val-Asp-Ser-Ala-Asp-Glu-Ser-Asn-Asp-Asp-Gly-Phe-Asp

Synonyms: Calfluxin

Bachem H-8165.0001 MW 1541.46 $C_{60}H_{88}N_{18}O_{30}$ Store at -15°C

Arg-Val-Gly-Arg-Pro-Glu

Synonyms: Bovine Adrenal Medulla 12P (7-12)

Bachem H-5365.0001 Store at -15°C

Arg-Val-Leu-Phe-Glu-Ala-Nle
Arg-Val-Leu-(®)-Phe-Glu-Ala-Nle Amide

Bachem N-1270.0001 Store at -15°C

Arg-Val-Lys-Arg
Decanoyl-Arg-Val-Lys-Arg-CMK

Synonyms: Furin Convertase Inhibitor

Alexis 260-022 MW 744.3 $C_{34}H_{66}N_{11}O_5Cl$ White lyophilized powder | When purified furin or PACE4 (another mammalian convertase) was incubated in the presence of this peptide, proendothelin-1 (proET-1) processing was completely abolished *in vitro* & *in vivo*; Garten, W et al, *Biochimie*, 76:217, 1994; Denault, J-B et al, *J Cardiovasc Pharmacol*, 26:S47, 1995; Denault, J-B et al, *FEBS Lett*, 362:276, 1995

Bachem N-1505.0005 MW 744.42 $C_{34}H_{66}ClN_{11}O_5$ Store at -15°C

Arg-Val-Lys-Arg-Gly-Leu-Ala-Tyr-Asp
Abz-Arg-Val-Lys-Arg-Gly-Leu-Ala-Tyr(NO₂)-Asp

Synonyms: Furin Substrate

American Peptide 81-6-20 MW 1241.4 $C_{54}H_{84}N_{18}O_{16}$ Internally quenched fluorogenic peptide substrate; contains anthranilic acid as fluorescent donor & 3-nitro-4-hydroxy-L-phenylalanine as acceptor (quencher); Angliker, H et al, *Anal Biochem*, 224:409, 1995

Arg-Val-Lys-Arg-Gly-Leu-Ala-Tyr-Asp
Aminobenzoyl-Arg-Val-Lys-Arg-Gly-Leu-Ala-Tyr(NO₂)-Asp

Sigma A 3210 FW 1241.4 ≥97% (HPLC); peptide content:~65% | Fluorogenic substrate; Angliker, H et al, *Anal Biochem*, 224:409, 1995

Arg-Val-Nle-Phe-Glu-Ala-Nle
Arg-Val-Nle-p-Nitro-Phe-Glu-Ala-Nle Amide

Synonyms: HIV Protease Substrate VI

Bachem H-1148.0001 Store at -15°C

Arg-Val-Thr-Ala-Ile-Glu-Lys-Tyr-Leu-Gln-Asp-Gln-Ala-Arg-Leu-Asn-Ser-Trp-Gly-Cys-Ala-Phe-Arg-Gln-Val-Cys-His-Thr-Thr-Val-Pro-Trp-Val-Asn-Asp-Ser Amide

Synonyms: GP41; HIV Antigenic Peptide V

Bachem H-2984.0500 Store at -15°C | HIV peptide

Arg-Val-Tyr-Ile-His-Pro-Ile

Synonyms: Angiotensin III, (Ile⁷)-; Angiotensin II, (des-Asp¹,Ile⁸)-; Angiotensin II, (des-Asp¹-(Ile⁸))-; Angiotensin III Selective Antagonist

American Peptide 12-1-65 MW 897.1 $C_{43}H_{68}N_{12}O_9$ Bravo, EL et al, *J Clin Endocrinol Metab*, 40:530, 1975

Bachem H-1710.0005 MW 897.09 $C_{43}H_{68}N_{12}O_9$ Store at -15°C

Peptides International PAN-4037-v Synthetic MW 897.09 $C_{43}H_{68}N_{12}O_9$ >99% (HPLC); lyophilized amorphous powder

Arg-Val-Tyr-Ile-His-Pro-Ile 2AcOH 4H₂O

Synonyms: Angiotensin II, (des-Asp¹-(Ile⁸))-; Angiotensin III Selective Antagonist

Peptides International PAN-4037 Synthetic MW 1089.2 $C_{43}H_{68}N_{12}O_9 \cdot 2CH_3COOH \cdot 4H_2O$ >99% (HPLC); lyophilized amorphous powder | Bioactive; Kono, T et al, *J Clin Endocrinol Metab*, 52:354, 1981

Arg-Val-Tyr-Ile-His-Pro-Ile Acetate Salt

Synonyms: Angiotensin III, (Ile⁷)-

ICN 152739 Inhibitor of Angiotensin III; Bravo, IL et al, *J Clin Endocrinol Metab*, 40:530, 1975

Arg-Val-Tyr-Ile-His-Pro-Ile Acetate Salt Free Base

Synonyms: Angiotensin III, (Ile⁷)-

Sigma A 0911 FW 897.1 $C_{43}H_{68}N_{12}O_9$ ≥97% (HPLC) | Bioactive peptide; bravo, EL et al, *J Clin Endocrinol Metab*, 40:530, 1975

Arg-Val-Tyr-Ile-His-Pro-Phe

Synonyms: Angiotensin II, (des-Asp¹)-; Angiotensin III

Bachem H-1755.0005 MW 931.11 $C_{46}H_{66}N_{12}O_9$ Store at -15°C

Neosystem SC059 MW 931.10 Human Bioactive; Chiv, AT et al, *PNAS USA*, 71:341, 1974

American Peptide 12-1-61 Human MW 931.2 $C_{46}H_{66}N_{12}O_9$ Stimulates release of aldosterone from adrenal glands; Chiu, AT et al, *PNAS*, 71:341, 1974

Calbiochem 05-23-0102 Human MW 931.1 $C_{46}H_{66}N_{12}O_9$ ≥95% (HPLC); lyophilized; soluble in 5% HOAc; LD₅₀≤2000 mg/kg; may be carcinogenic/teratogenic | Inhibits degradation of enkephlin by aminopeptidase & dipeptidylaminopeptidase; potentiates analgesic activity of Met-enkephlain; Sasaki, S et al, *Brain Res*, 600:335, 1993

Biogenesis 0560-1509 Human synthetic MW 930.2 99.3% (HPLC), >99% (TLC); 76.3% peptide material, acetate counter ions and residual H₂O; lyophilized

Peptides International PAN-4028-v Human synthetic MW 931.11 $C_{46}H_{66}N_{12}O_9$ >99% (HPLC); lyophilized amorphous powder

Arg-Val-Tyr-Ile-His-Pro-Phe Acetate Salt

Synonyms: Angiotensin II, (des-Asp¹)-; Angiotensin III; Angiotensin II Heptapeptide

Fluka 10385 MW 1048.2 $C_{46}H_{66}N_{12}O_{19} \cdot 5CH_3COOH \cdot 1½H_2O$ ≥98% (HPLC)

ICN 191237 Human Stimulates aldosterone release from adrenal glands; Chiu, AT & MJ Peach, *PNAS*, 71:341, 1974

Arg-Val-Tyr-Ile-His-Pro-Phe Acetate Salt Free Base

Synonyms: Angiotensin II, (des-Asp[1])-; Angiotensin III

Sigma A 0903 FW 931.1 $C_{46}H_{66}N_{12}O_9$ ≥98% (HPLC) |
Bioactive peptide; putative neurotransmitter; less potent than
angiotensin II; induces release of aldosterone; inhibits degradation
of enkephalins & potentiates analgesic activity of Met-enkephalin;
Chiu, AT & Peach, MJ, *Proc Natl Acad Sci USA*, 71:341, 1974

Arg-Val-Tyr-Ile-His-Pro-Phe-His-Leu

Synonyms: Angiotensin I, (des-Asp[1])-

Bachem H-1700.0005 MW 1181.41 $C_{58}H_{84}N_{16}O_{11}$ Store at
-15°C

Biogenesis 0560-0139 Human synthetic MW 1182 >99%
(HPLC); lyophilized, containing 84.2% peptide material, acetate
counter ions and residual H_2O

Arg-Val-Tyr-Ile-His-Pro-Phe-His-Leu Acetate Salt Free Base

Synonyms: Angiotensin I, (des-Asp[1])-; Angiotensin

Sigma A 5778 FW 1181.4 Human ≥97% (HPLC) |
Bioactive peptide; see Sigma A 9650

Arg-Val-Tyr-Ile-His-Pro-Phe-His-Leu Free Base

Synonyms: Angiotensin I, (des-Asp[1])-

ICN 159874 MW 1181.4 98%

Arg-Val-Tyr-Ile-His-Pro-Phe 2AcOH 4H₂O

Synonyms: Angiotensin III

Peptides International PAN-4028 Human synthetic MW
1123.2 $C_{46}H_{66}N_{12}O_9 \cdot 2CH_3COOH \cdot 4H_2O$ >99% (HPLC);
lyophilized amorphous powder

Arg-Val-Tyr-Val-His-Pro-Ile Acetate Salt

Synonyms: Angiotensin III, (Val[4],Ile[7])-

ICN 194134 MW 883.1 $C_{42}H_{66}N_{12}O_9$; 97% | Angiotensin
III antagonist; Tabrizchi, R etal, *Life Sciences*, 43:537, 1988

Arg-Val-Tyr-Val-His-Pro-Ile Acetate Salt Free Base

Synonyms: Angiotensin III, (Val[4],Ile[7])-

Sigma A 1036 FW 883.1 $C_{42}H_{66}N_{12}O_9$ ≥97% (HPLC) |
Bioactive peptide

Arg-Val-Tyr-Val-His-Pro-Phe

Synonyms: Angiotensin III, (Val[4])-

American Peptide 12-1-64 MW 917.1 $C_{45}H_{64}N_{12}O_9$ Mann, J
et al, *Amer J Physiol*, 241:R124, 1981

Bachem H-1760.0005 Store at -15°C

Arg-Val-Tyr-Val-His-Pro-Phe Acetate Salt

Synonyms: Angiotensin III, (Val[4])-

ICN 152741

Arg-Val-Tyr-Val-His-Pro-Phe Acetate Salt Free Base

Synonyms: Angiotensin III, (Val[4])-

Sigma A 6277 FW 917.1 $C_{45}H_{64}N_{12}O_9$ ≥97% (HPLC) |
Bioactive peptide; Mann, J et at, *Amer J Physiol*, 241:R124, 1981

Asn-Ala-Gln-Leu-Glu-Lys-Arg-Ser-Phe-Cys-Ser-Ala-Met-Val Acetate Salt

Synonyms: Glutamate Receptor VI (818-831)

Biogenesis 4670-5259 Synthetic >80% (HPLC); lyophilized

Asn-Ala-Gln-Phe-Arg-His-Asn-Ser-Gly-Tyr-Gln-Val-His-His-Gln-Lys-Leu-Val-Phe-Phe-Ala-Gln-Asn-Val-Gly-Ser-Asn-Lys-Gly-Ala-Ile-Ile-Gly-Leu-Met-Val-Gly-Gly-Val-Val

Asn(4-Aminobutyl)-Ala-Gln(4-Aminobutyl)-Phe-Arg-His-Asn(4-Aminobutyl)-Ser-Gly-Tyr-Gln(4-Aminobutyl)-Val-His-His-Gln-Lys-Leu-Val-Phe-Phe-Ala-Gln(4-Aminobutyl)-Asn(4-Aminobutyl)-Val-Gly-Ser-Asn-Lys-Gly-Ala-Ile-Ile-Gly-Leu-Met-Val-Gly-Gly-Val-Val

Synonyms: Amyloid α-Protein (1-40), (Asn(4-Aminobutyl)[1,7,23],
Gln(4-Aminobutyl)[3,11,22])-

Bachem H-4984.0500 MW 4750.69 $C_{218}H_{355}N_{65}O_{52}S$ Store
at -15°C

Asn-Ala-Gln-Thr-Ser-Val-Ser-Pro-Ser-Lys-Val-Ile-Leu-Pro-Arg-Gly-Gly-Ser-Val-Leu-Val-Thr-Cys

Synonyms: Intercellular Adhesion Molecule I (1-23)

Bachem H-2078.0500 MW 2313.72 $C_{99}H_{173}N_{29}O_{32}S$ Store
at -15°C

Asn-Ala-Ile-Gln-Glu-Ala-Arg-Arg-Leu-Leu-Asn-Leu-Ser-Arg-Asp

Synonyms: GM-CSF (17-31)

Bachem H-3436.0001 Store at -15°C

Asn-Ala-Pro-Val-Ser-Ile-Pro-Gln

Synonyms: Neuroprotective Protein (74-81), Activity-Dependent

Bachem H-5066.0001 Mouse, rat MW 824.93
$C_{36}H_{60}N_{10}O_{12}$ Store at -15°C

Asn-Arg 2,4-Dihydroxyphenyl Ac-L-Asn-*N''*-(L-Arg-Puteanyl)-Cadaverine

Synonyms: Neurotensin NSTX III

Alexis 630-087 Synthetic MW 664.80 $C_{30}H_{52}N_{10}O_7$
≥98%; soluble in H_2O; neurotoxin | Originally isolated from
Nephila maculata; specific, irreversible spider neurotoxin that
inhibits both glutamate & postsynaptic potentials in the
neuromuscular junction; Aramaki, Y et al, *Proc Jpn Acad*, 62B:359,
1986; Teshima, T et al, *THL*, 28:3509, 1987; Ino, H et al, *Neurosci
Res*, 8:29, 1990; Teshima, T et al, *Tetrahedron*, 47:3305, 1991

Asn-Arg-Asn-Phe-Leu-Arg-Phe Amide

Synonyms: FMRF Amide Related Peptide

American Peptide 60-9-30 MW 965.1 $C_{44}H_{68}N_{16}O_9$ Mercier,
AJ et al, *Peptides*, 14:137, 1993

Bachem H-1364.0005 MW 365.13 $C_{44}H_{68}N_{16}O_9$ Store at
-15°C

Asn-Arg-Asn-Phe-Leu-Arg-Phe Amide Trifluoroacetate Salt Free Base

Synonyms: FMRF Related Peptide

Sigma A 8826 FW 965.1 $C_{44}H_{68}N_{16}O_9$ ≥97% (HPLC) |
Bioactive peptide; Mercier, AJ et al, *Peptides*, 14:137, 1993

Asn-Arg-Cys-Ser-Gln-Gly-Ser-Cys-Trp-Asn

Bachem M-2210.0001 Store at -15°C

Bachem M-2215.0001 Store at -15°C | Disulfide bonds:
Cys[3]-Cys[8]

Asn-Arg-Pro-Tyr-Ile-His-Pro-Phe-Gln-Leu Trifluoroacetate Salt

Synonyms: Angiotensin I

ICN 195626 Elasmobranch fish MW 1284.5 Takei, Y etal,
Endocrin, 139:281, 1993

Asn-Arg-Pro-Tyr-Ile-His-Pro-Phe-Gln-Leu Trifluoroacetate Salt Free Base

Synonyms: Angiotensin I

Sigma A 0833   Elasmobranch fish   FW 1284.5   ≥97% (HPLC) | Bioactive peptide; Takei, Y et al, *Endocrin*, 139:281, 1993; see Sigma A 9650

Asn-Arg-Tyr-Tyr-Ala-Ser-Leu-Arg-His-Tyr-Leu-Asn-Leu-Val-Thr-Arg-Gln-Arg-Tyr Amide

Synonyms: Peptide YY (18-36)

Neosystem SC418   Human   MW 2485.83   Bioactive neuropeptide; brain/gut peptide

Asn-Arg-Val-Tyr-Val-His-Pro-Phe

Synonyms: Angiotensin II, (Asn^1,Val^5)-

American Peptide 12-1-54   MW 1031.2 $C_{49}H_{70}N_{14}O_{11}$ Galardy, RE et al, *J Med Chem*, 21:1279, 1978

Bachem H-6010.0005   MW 1031.18 $C_{49}H_{70}N_{14}O_{11}$ Store at -15°C

Peptides International PAN-4036-v   Synthetic   MW 1031.2 $C_{49}H_{70}N_{14}O_{11}$ >99% (HPLC); lyophilized amorphous powder

Asn-Arg-Val-Tyr-Val-His-Pro-Phe Acetate Salt

Synonyms: Angiotensin II, (Asn^1,Val^5)-

ICN 152731   Galardy, RE etal, *Biochemistry*, 15:2303, 1976; Galardy, RE etal, *J Med Chem*, 21:1279, 1978

Asn-Arg-Val-Tyr-Val-His-Pro-Phe Acetate Salt Free Base

Synonyms: Angiotensin II, (Asn^1,Val^5)-

Sigma A 6402   FW 1031.2   ≥97% (HPLC) | Bioactive peptide; Galardy, RE et al, *J Med Chem*, 21:1279, 1978; Galardy, RE et al, *Biochemistry*, 15:2303, 1976

Asn-Arg-Val-Tyr-Val-His-Pro-Phe AcOH 4H$_2$O

Synonyms: Angiotensin II, (Asn^1,Val^5)-

Peptides International PAN-4036   Synthetic   MW 1163.3 $C_{49}H_{70}N_{14}O_{11} \cdot CH_3COOH \cdot 4H_2O$ >99% (HPLC); lyophilized amorphous powder

Asn-Arg-Val-Tyr-Val-His-Pro-Phe-Asn-Leu

Synonyms: Angiotensin I, (Asn^1,Val^5,Asn^9)-

American Peptide 12-7-25   Salmon   MW 1258.5 $C_{59}H_{87}N_{17}O_{14}$ Takemoto, Y et al, *Gen Comp Endocrinol*, 51:219, 1983

ICN 152727   Salmon   MW 1258.4   Takemoto, Y etal, *Gen Comp Endocrinol*, 51:219, 1983

Sigma A 2928   Salmon   FW 1258.4   ≥97% (HPLC) | Bioactive peptide; Takemoto, Y et al, *Gen Comp Endocrinol*, 51:219, 1983; see Sigma A 9650

Asn-Arg-Val-Tyr-Val-His-Pro-Phe-His-Leu

Synonyms: Angiotensin I

Sigma A 3178   Goosefish   FW 1281.5   ≥97% (HPLC) | Bioactive peptide; Hayashi, T et al, *Chem Pharm Bull*, 26:215, 1978; see Sigma A 9650

Asn-Asn-Asn-Asn-Gly-Ser-Glu-Ile-Phe-Arg-Pro-Gly-Gly-Gly-Asp

Synonyms: GP120 (465-479); HTLV I Peptide

Neosystem SC252   MW 1547.56   HTLV I peptide from subtype GP160

Asn-Asn-Gln-Lys-Ile-Val-Asn-Leu-Lys-Glu-Lys-Val-Ala-Gln-Leu-Glu-Ala

Synonyms: Fibrinogen γ-Chain (117-133)

Bachem H-2506.0001   MW 1939.24 $C_{84}H_{147}N_{25}O_{27}$ Store at -15°C

Asn-Asn-Leu-Leu-Arg-Ala-Ile-Glu-Ala-Gln-Gln-His-Leu-Leu-Gln-Leu-Thr-Val-Trp-Gly-Ile-Lys-Gln-Leu-Gln-Ala-Arg-Ile-Leu-Ala-Val-Glu-Arg-Tyr-Leu-Lys-Asp-Gln
Ac-Asn-Asn-Leu-Leu-Arg-Ala-Ile-Glu-Ala-Gln-Gln-His-Leu-Leu-Gln-Leu-Thr-Val-Trp-Gly-Ile-Lys-Gln-Leu-Gln-Ala-Arg-Ile-Leu-Ala-Val-Glu-Arg-Tyr-Leu-Lys-Asp-Gln Amide

Synonyms: T21; HIV I gp41 (558-595)

American Peptide 72-3-35   Synthetic   MW 4526.4 $C_{204}H_{339}N_{61}O_{55}$ Strong inhibitor of HIV-1 viral mediated cell-to-cell fusion with EC$_{50}$ values of 1 µg/mL; Lawless, MK et al, *Biochemistry*, 35:13697, 1996

Asn-Asn-Thr-Arg-Lys-Ser-Ile-Arg-Ile-Gln-Arg-Gly-Pro-Gln-Arg-Ala-Phe-Val-Thr-Ile-Gly-Lys-Ile-Gly

Synonyms GP120; HIV Envelope Protein (307-330):

ICN 153128   Completely blocks the fusion inhibition activity of both antisera & serum from a chimpanzee infected with HTLV IIIB; Rusche, JR etal, *PNAS*, 85:3198, 1988

Asn-Asn-Thr-Arg-Lys-Ser-Ile-Arg-Ile-Gln-Arg-Gly-Pro-Gly-Arg

Synonyms: GP120 (306-320); HTLV I Peptide

Neosystem SC835   MW 1752.92   HTLV I peptide from subtype GP160

Asn-Asn-Thr-Arg-Lys-Ser-Ile-Arg-Ile-Gln-Arg-Gly-Pro-Gly-Arg-Ala-Phe-Val-Thr-Ile-Gly-Lys-Ile-Gly

Synonyms: GP120 (308-331); HIV Peptide; GP120; HIV Envelope Protein (308-331)

Bachem H-7005.0500   MW 2640.09 $C_{114}H_{199}N_{41}O_{31}$ Store at -15°C

Sigma H 7894   FW 2640.1   ≥97% (HPLC); peptide content:~70% | Bioactive peptide; blocks fusion inhibition activity of both antiserum & HTLV IIIβ-infected chimpanzee serum; Rusche, JR et al, *Proc Natl Acad Sci USA*, 85:3198, 1988

Asn-Asp-Asp-Cys-Glu-Leu-Cys-Val-Asn-Val-Ala-Cys-Thr-Gly-Cys-Leu

Synonyms: Guanylate Cyclase Stimulator; Uroguanylin; Guanylate Cyclase C Activator

American Peptide 43-0-14   Human   MW 1667.9 $C_{64}H_{102}N_{18}O_{26}S_4$ Disulfide bonds: Cys^4-Cys^{12}, Cys^7-Cys^{15}; endogenous guanylate cyclase C activator; Kita, T et al, *Am J Physiol*, 266:F342, 1994

Sigma U 5131   Human   FW 1667.9   ≥95% (HPLC); Disulfide bonds: 4-12, 7-15 | Bioactive peptide; stimulates intestinal guanylate cyclase; Hamra, FK et al, *Proc Natl Acad Sci USA*, 90:10464, 1993

Peptides International PUG-4295-s   Human synthetic   MW 1667.9 $C_{64}H_{102}N_{18}O_{26}S_4$ >99% (HPLC); lyophilized amorphous powder | Bioactive; Disulfide bonds: Cys^4-Cys^{12}, Cys^7-Cys^{15}; Kita, T Cet al, *Am J Physiol*, 266, F342 (1994)

Asn-Asp-Asp-Gly-Gly-Phe-Ser-Glu-Glu-Trp-Glu-Ala-Gln-Arg-Asp-Ser-His-Leu-Gly-Cys

Synonyms: Presenilin I-Cys, (331-349)

Bachem H-3988.0500   Human, mouse   Store at -15°C

Asn-Asp-Asp-Pro-Pro-Ile-Ser-Ile-Asp-Leu-Thr-Phe-His-Leu-Leu-Arg-Asn-Met-Ile-Glu-Met-Ala-Arg-Ile-Glu-Asn-Glu-Arg-Glu-Gln-Ala-Gly-Leu-Asn-Arg-Lys-Tyr-Leu-Asp-Glu-Val Amide

Synonyms: Urotensin I

American Peptide 80-7-15   MW 4869.1 $C_{210}H_{340}N_{62}O_{67}S_2$
Contains a homologous sequence with the hypothalamic corticotropin-releasing factor (CRF) & the frog skin peptide sauvagine; Lederis, K et al, *Science*, 218:162, 1982

Bachem H-5500.0500   Store at -15°C

Asn-Asp-Asp-Pro-Pro-Ile-Ser-Ile-Asp-Leu-Thr-Phe-His-Leu-Leu-Arg-Asn-Met-Ile-Glu-Met-Ala-Arg-Ile-Glu-Asn-Glu-Asr-Glu-Gln-Ala-Gly-Leu-Asn-Arg-Lys-Tyr-Leu-Asp-Glu-Val Amide

Synonyms: Urotensin I

ICN 153191

Asn-Asp-Asp-Pro-Pro-Ile-Ser-Ile-Asp-Leu-Thr-Phe-His-Leu-Leu-Arg-Asn-Met-Ile-Glu-Met-Ala-Arg-Ile-Glu-Asn-Glu-Arg-Glu-Gln-Ala-Gly-Leu-Asn-Arg-Lys-Tyr-Leu-Asp-Glu-Val Amide

Synonyms: Urotensin I

Sigma U 7253   Teleost fish   FW 4869.5   ≥95% (HPLC) |
Bioactive peptide

Asn-Gln-Arg-Lys-Ile-Val-Lys-Cys-Phe-Asn-Cys-Gly-Lys-Glu-Gly-His-Ile

Synonyms: P15 (385-401); GAG p7 NC (8-24); HTLV I Peptide

Neosystem SC270   MW 1972.31   HTLV I peptide from subtype P15 (GAG P7 NC/GAG P1/GAG P6); due to the presence of a Cys this peptide can contain the dimeric form

---

Asn-Gln-Asn-Arg-Asn-Lys-Ser-Ser-Ser-His-Gln-Asp-Cys Acetate Salt

Synonyms: Bone Morphogenetic Protein 5 (1-12), ~C

Biogenesis 1406-0930   Synthetic   MW 1517.3   86% (HPLC); lyophilized

---

Asn-Gln-Gly-Arg-His-Phe-Cys-Gly-Gly-Ala-Leu-Ile-His-Ala-Arg-Phe-Val-Met-Thr-Ala-Ala-Ser-Cys-Phe-Gln

Synon CAP₃₇ (20-44)yms:

American Peptide 72-2-22   MW 2721.2 $C_{118}H_{178}N_{38}O_{31}S_3$
Disulfide bond:Cys²⁶-Cys42; antibiotic peptide; multifunctional protein isolated from granules of human neutrophils; it is antibiotic & chemotactic & binds lipopolysaccharide; mimics CAP₃₇ antibiotic & lipopolysaccharide binding action; Pereira, HA et al, *PNAS*, 90:4733, 1993

Asn-Gln-Gly-Gln-Tyr-Met-Asn-Thr-Pro-Trp-Arg-Asn-Pro-Ala-Glu

Synonyms: NEF (61-75); SIVmac251 Peptide

Neosystem SC608   MW 1805.94   SIVmac251 peptide from subtype NEF

Asn-Gln-Ile-Arg-Met-Lys-Ile-Gly-Val-Met-Phe-Gly-Asn-Pro-Glu-Thr-Thr-Thr-Gly-Gly

Synonyms: FECO Peptide

Neosystem SC999   MW 2151.48   Virus-related peptide; bioactive; derived from RecA, a bacterial protein that repairs & recombines DNA; not only binds to single-stranded but also has other activities, including the pairing of homologous DNAs; Voloshin, ON et al, *Science*, 272:868-872, 1996; Stasiak, A, *Science*, 272:828-829, 1996

Asn-Gln-Ile-Arg-Met-Lys-Ile-Gly-Val-Met-Trp-Gly-Asn-Pro-Glu-Thr-Thr-Thr-Gly-Gly

Synonyms: rec A-Like Protein (193-212)

Bachem H-3912.0001   Store at -15°C

Sigma R 6772   Bioactive peptide; useful in DNA pairing studies of rec A protein reactions; Volshin, ON et al, *Science*, 272:869, 1996

Asn-Gln-Leu-Val-Val-Pro-Ser-Glu-Gly-Leu-Tyr-Leu-Ile-Tyr-Ser-Gln-Val-Leu-Phe-Lys

Synonyms: Tumor Necrosis Factor α (46-65)

Bachem H-8245.0500   Human   MW 2310.72 $C_{110}H_{172}N_{24}O_{30}$ Store at -15°C

ICN 157548   Human   MW 2310.7   97%

Sigma T 1292   Human   FW 2310.7   ≥97% (HPLC) |
Bioactive peptide

Asn-Glu

Bachem G-4075.0250   MW 261.24 $C_9H_{15}N_3O_6$ Store at -15°C

Asn-Glu-Ala-Tyr-Val-His-Asp-Ala-Pro-Val-Arg-Ser-Leu-Asn

Synonyms: Interleukin IB Convertase Substrate; Interleukin Converting Enzyme Substrate; Interleukin Iβ Converting Enzyme Substrate I; Interleukin Iβ Precursor (110-123); Interleukin Iβ Converting Enzyme Substrate

American Peptide 81-6-80   MW 1584.7 $C_{68}H_{105}N_{21}O_{23}$
Contains the native cleavage site of the IL-1B precursor protein; inhibits the cleavage of the IL-1B precursor by ICE

Bachem M-1890.0001   MW 1584.71 $C_{68}H_{105}N_{21}O_{23}$ Store at -15°C

ICN 195848   MW 1584.8   ≥97% | Competitive inhibitor & substrate

Sigma I 2775   FW 1584.7   ≥97% (HPLC) | Bioactive peptide; Thornberry, NA et al, *Nature*, 356:768, 1992

Asn-Glu-Ala-Tyr-Val-His-Asp-Ala-Pro-Val-Arg-Ser-Leu-Asn Trifluoroacetate Salt

Synonyms: Caspase I Substrate I; Interleukin Iβ Converting Enzyme Substrate I

Calbiochem 400016   MW 1584.7 $C_{68}H_{105}N_{21}O_{23}$ >97% (HPLC); solid; soluble in water | Contains the native cleavage site of the IL-1β precursor protein; is hydrolyzed by Caspase I & inhibits the cleavage of the IL-1β precursor by Caspase I; a substrate of Caspase IV; Thornberry, NA et al, *Nature*, 356:768, 1992

Asn-Glu-Ala-Tyr-Val-His-Asp-Ala-Pro-Val-Ser-Leu-Asn

Synonyms: Caspase I Substrate I

Oncogene 400016   For inhibition of cleavage of IL Iβ precursor by Caspase I

Asn-Glu-Gly-Leu-Gly-Trp-Ala-Gly-Trp

Synonyms: HCV Nucleoprotein (88-96)

Bachem H-3522.0001   Store at -15°C

Asn-Glu-Leu-Glu-Pro-Leu-Asn-Arg-Pro-Glu-Leu-Lys-Cys-Glu-Arg

Synonyms: Estrogen Receptor α N-Terminal Peptide (21-32)

Alexis 155-032   Synthetic   MW 1553.8 $C_{66}H_{112}N_{20}O_{21}S$ ≥90% (HPLC); synthetic peptide dissolved in 10 *mM* sodium acetate, pH 4.5 | Competitively binds to Ab Alexis 210-202; blocking peptide for antiserum (purified) to ERα

Asn-Gly-Ala-Lys-Ala-Leu-Met-Gly-Gly-His-Gly-Ala-Thr-Lys-Val-Met-Val-Gly-Ala-Ala-Ala

Synonyms: Prion Protein (106-126), Scrambled

Bachem H-4882.0500 Human MW 1912.27
$C_{80}H_{138}N_{26}O_{24}S_2$ Store at -15°C

Asn-Gly-Thr
Bz-Asn-Gly-Thr Amide

Bachem M-2080.0050 Store at -15°C

Asn-His-Cys-Asn-Thr-Ser-Val-Ile-Gln-Glu-Ser-Cys-Asp-Lys-His-Tyr-Trp-Asp-Thr-Ile
Asn-His-Cys(Acm)-Asn-Thr-Ser-Val-Ile-Gln-Glu-Ser-Cys(Acm)-Asp-Lys-His-Tyr-Trp-Asp-Thr-Ile

Synonyms: GP140 (211-230); SIVmac251 Peptide

Neosystem SC729 MW 2535.74 SIVmac251 peptide from subtype GP140

Asn-Ile-Asp-Trp-Thr-Asp-Gly-Asn-Gln-Thr-Ser-Ile-Thr-Met-Ser-Ala-Glu-Val-Ala-Glu

Synonyms: GP140 (471-490); SIVmac251 Peptide

Neosystem SC755 MW 2182.29 SIVmac251 peptide from subtype GP140

Asn-Leu-Gly-Leu-Asp-Cys-Asp-Glu-His-Ser-Ser-Glu-Ser-Arg

Synonyms: Bone Morphogenetic Protein 11 (1-14)

Biogenesis 1406-1690 Human synthetic MW 1561.6
>80% (HPLC); acetate buffer; lyophilized

Asn-Lys-Arg-Lys-Arg-Ile-His-Ile-Gly-Pro-Gly-Arg-Ala-

Synonyms: GP120 V3 Loop Peptide MN N-25-C

Neosystem SC686 MW 2858.39 HIV-1 GP120 V3 Loop peptide; due to the presence of a Cys this peptide can contain the dimeric form

Asn-Lys-Arg-Lys-Arg-Ile-His-Ile-Gly-Pro-Gly-Arg-Ala-Phe

Synonyms: GP120 MN (307-320); HIV I Peptide

Neosystem SC838 MW 1649.96 GP120 peptide from HIV-1 subtype MN

Asn-Lys-Leu-Ile-Val-Arg-Arg-Gly-Gln-Ser-Phe-Tyr-Val-Gln-Ile-Asp-Phe-Ser-Arg-Pro-Tyr-Asp-Pro-Arg-Arg-Asp

Synonyms: Factor XIIIa (72-97)

Sigma F 0166 FW 3226.6 ≥90% (HPLC) | Bioactive peptide; inhibitor of fibrin cross-linking catalyzed by coagulation factor XIIIa; Achyuthan, KE et al, *J Biol Chem*, 268:21284, 1993

Asn-Lys-Lys-Lys-Asp-Asp-Glu-Val-Asp-Arg-Asp-Ala-Pro-Ser-Arg-Lys-Lys-Ala-Lys-Glu

Synonyms: Inositol I/IV/V-Trisphosphate Receptor (Type-1) Cytoplasmic C-Terminal Peptide (1829-1848); IP3R-I

Alexis 159-003 Human synthetic MW 2357.6
$C_{97}H_{169}N_{33}O_{35}$ ≥95%; lyophilized; reconstitute with 0.1 mL distilled H_2O | Competitively binds to Ab Alexis 210-115; antiserum blocking peptide

Asn-Orn
Indol-3-yl-Ac-Asn-1,5-Diaminopentane-Retro-Orn-H

Synonyms: Nephilatoxin 11

Bachem H-8605.0001 MW 487.6 $C_{24}H_{37}N_7O_4$ Store at -15°C

Asn-Orn-Arg
Indol-3-yl-Ac-Asn-1,5-Diaminopentane-Retro-Orn-Arg

Synonyms: Nephilatoxin 9

Bachem H-8600.0001 MW 643.79 $C_{30}H_{49}N_{11}O_5$ Store at -15°C

Asn-Phe
Z-Asn-Phe-OMe

Bachem C-3545.0001 MW 427.46 $C_{22}H_{25}N_3O_6$ Store at -15°C

Asn-Phe-Asp-Glu-Ile-Asp-Arg-Ser-Gly-Phe-Gly-Phe-Asn

Synonyms: Gastrointestinal Peptide; Orcokinin

Bachem H-8830.0001 MW 1515.58 $C_{67}H_{92}N_{18}O_{23}$ Store at -15°C

Sigma O 2011 FW 1517.6 ≥90% (HPLC) | Bioactive peptide; potent stimulator of hindgut contractions; Stangier, J et al, *Peptides*, 13:859, 1992

Asn-Phe-Pro
BOC-Asn-Phe-Pro-CHO

Bachem A-2860.0025 Store at -15°C

Asn-Phe-Thr-Gln-Glu-Val-Ser-Arg-Leu-Asn-Ile-Asn-Leu-His-Phe-Ser

Synonyms: GP46 (140-155); HTLV I Peptide

Neosystem SC826 MW 1919.12 HTLV I peptide

Asn-Pro-Asn
(-Asn-Pro-Asn-Ala)₂

Bachem H-5480.0001 MW 810.82 $C_{32}H_{50}N_{12}O_{13}$ Store at -15°C

Asn-Pro-Asn-Ala
(-Asn-Pro-Asn-Ala)₆

Bachem H-6340.0001 Store at -15°C

Asn-Pro-Asn-Ala-Asn-Pro-Asn-Ala

Synonyms: Circumsporozoite Protein, Repetitive Sequence

American Peptide 79-2-22 MW 810.8 $C_{32}H_{50}N_{12}O_{13}$

ICN 153069 *Plasmodium falciparum* Useful as Ag for raising Ab to the CS protein—such Ab recognize native CS protein & block *P. falciparum* sporozoite invasion of human hepatoma cells *in vitro*; Dame, JB etal, *Science*, 225:593, 1984; Enea, V etal, *Science*, 225:628, 1984; Ballou, WR etal, *Science*, 228:996, 1985; Young, JF etal, *Science*, 228:958, 1985; Zavala, F etal, *Science*, 228:1436, 1985; Miller, LH etal, *Science*, 234:1349, 1985

Asn-Pro-Asn-Ala-Asn-Pro-Asn-Ala-Asn-Pro-Asn-Ala

Synonyms: Circumsporozoite Protein, Repetitive Sequence

Sigma A 6546 FW 1207.2 ≥97% (HPLC) | Bioactive peptide; repetitive sequence from the circumsporozoite (CS) protein of *Plasmodium falciparum*; antigen for raising Ab to the CS protein; such Ab recognize native CS protein & block *Plasmodium falciparum* sporozoite invasion of human hepatoma cells *in vitro*; Zavala, F et al, *Science*, 228:1436, 1985

ICN 153070 *Plasmodium falciparum* Useful as Ag for raising Ab to the CS protein—such Ab recognize native CS protein & block *P. falciparum* sporozoite invasion of human hepatoma cells *in vitro*; Dame, JB etal, *Science*, 225:593, 1984; Enea, V etal, *Science*, 225:628, 1984; Ballou, WR etal, *Science*, 228:996, 1985; Young, JF etal, *Science*, 228:958, 1985; Zavala, F etal, *Science*, 228:1436, 1985; Miller, LH etal, *Science*, 234:1349, 1985

Asn-Pro-Glu-Tyr
Asn-Pro-Glu-Tyr(PO₃H₂)

Bachem H-2706.0001   Store at -15°C

Asn-Pro-Met-Tyr-Asn-Ala-Val-Ser-Asn-Ala-Asp-Leu-Met-Asp-Phe-Lys

Synonyms: Cardiodilatin (1-16)

American Peptide 14-1-25   Human   MW 1830.1
$C_{79}H_{120}N_{20}O_{26}S_2$

Asn-Pro-Met-Tyr-Asn-Ala-Val-Ser-Asn-Ala-Asp-Leu-Met-Asp-Phe-Lys-Asn-Leu-Leu-Asp-His-Leu-Glu-Glu-Lys-Met-Pro-Leu-Glu-Asp

Synonyms: Natriuretic Peptide (26-55), Prepro-Atrial

American Peptide 14-1-32   Human   MW 3508.0
$C_{152}H_{236}N_{38}O_{51}S$ Vesely, DL et al, *BBRC*, 148:1540, 1987

Asn-Pro-Phe-His-Ser-Trp-Gly Amide

Synonyms: Culekinin Depolarizing Peptide

Bachem H-2248.0005   MW 842.91 $C_{40}H_{50}N_{12}O_9$ Store at -15°C

Asn-Pro-Thr-Asn-Leu-His

Bachem H-1972.0001   Fleshfly   MW 694.75 $C_{29}H_{46}N_{10}O_{10}$
Store at -15°C

Asn-Ser
FMOC-Asn(Mtt)-Ser(ψ(Me,Me)pro)

Synonyms: (4S)-3-(FMOC-Asn(Mtt))-2,2-Dimethyl-Oxazolidine-4-Carboxylic Acid

Bachem B-3445.0001   MW 737.85 $C_{45}H_{43}N_3O_7$ Store at -15°C

Asn-Ser-Arg-Gly-Asp-Lys-Gln-Arg-Gly-Ser-Lys-Pro-Pro-Thr-Lys-Gly-Tyr

Synonyms: VIF (185-200)-(Tyr); SIVmac251 Peptide

Neosystem SC580   MW 1876.06   SIVmac251 peptide from subtype VIF

Asn-Ser-Lys-Met-Ala-His-Ser-Ser-Ser-Cys-Phe-Gly-Gln-Lys-Ile-Asp-Arg-Ile-Gly-Ala-Val-Ser-Arg-Leu-Gly-Cys-Asp-Gly-Leu-Arg-Leu-Phe

Synonyms: Natriuretic Peptide (1-32), Brain

American Peptide 14-5-11   Rat   MW 3453.0
$C_{146}H_{239}N_{47}O_{44}S_3$ Disulfide bonds: :Cys¹⁰-Cys²⁶ | Kojima, M et al, *BBRC*, 159:1420, 1989

Asn-Ser-Lys-Met-Ala-His-Ser-Ser-Ser-Cys-Phe-Gly-Gln-Lys-Ile-Asp-Arg-Ile-Gly-Ala-Val-Ser-Arg-Leu-Gly-Cys-Asp-Gly-Leu-Arg-Leu-Phe
NH₂-Asn-Ser-Lys-Met-Ala-His-Ser-Ser-Ser-Cys-Phe-Gly-Gln-Lys-Ile-Asp-Arg-Ile-Gly-Ala-Val-Ser-Arg-Leu-Gly-Cys-Asp-Gly-Leu-Arg-Leu-Phe

Synonyms: Natriuretic Peptide, Brain

ICN 154501   Rat   MW 3453.4   Kojima, M etal, *BBRC*, 159:1420, 1989

Asn-Ser-Lys-Met-Ala-His-Ser-Ser-Ser-Cys-Phe-Gly-Gln-Lys-Ile-Asp-Arg-Ile-Gly-Ala-Val-Ser-Arg-Leu-Gly-Cys-Asp-Gly-Leu-Arg-Leu-Phe

Synonyms: Natriuretic Peptide 32, Brain; Natriuretic Peptide, Brain; Natriuretic Peptide 32, B-Type; Natriuretic Peptide 32, Brain

Sigma B 9901   Rat   FW 3453.0   ≥97% (HPLC); peptide content:~70% | Bioactive peptide, ; disulfide bonds: 10-26; Kojima, M et al, *Biochem Biophys Res Commun*, 159:1420, 1989

Biogenesis 1505-0652   Rat synthetic   MW 3453.4   100% (HPLC); lyophilized, consisting of 76% peptide material, TFA counter ions and residual H_2O | To avoid Met oxidation, it is best to dissolve the peptide under a nitrogen stream; Kojima et al, *BBRC*, 159:1420, 1989

Peptides International PBN-4213-v   Rat synthetic   MW 3453.0 $C_{146}H_{239}N_{47}O_{44}S_3$ >95% (HPLC); lyophilized amorphous powder | Bioactive; Disulfide bonds: Cys¹⁰-Cys²⁶; Kojima, M et al, *BBRC*, 159:1420, 1989

Asn-Ser-Lys-Pro-Ser-Pro-Asn-Thr-Lys-Asn-His-Pro-Val-Arg-Phe Trifluoroacetate Salt

Synonyms: Parathyroid Hormone Related Peptide (45-59)

Biogenesis 7170-9550   Synthetic   MW 1723.8   97% (HPLC); lyophilized

---

Asn-Ser-Met-Gln-Pro-Val-Lys-Glu-Thr-Pro-Gly-Asn-Ala Acetate Salt

Synonyms: Glucose Transporter III

Biogenesis 4670-1690   Mouse synthetic   >80% (HPLC); lyophilized

---

Asn-Ser-Ser-Ser-Gly-Glu-Met-Met-Met-Glu-Lys-Gly-Glu-Ile-Lys

Synonyms: GP120 (146-160); HTLV I Peptide

Neosystem SC1002   MW 1657.88   HTLV I peptide from subtype GP160

Asn-Ser-Ser-Trp-Pro-Trp-Gln-Ile-Glu-Tyr-Ile-His-Phe-Leu-Ile-Arg-Gln-Leu-Ile-Arg

Synonyms: GP140 (761-780); SIVmac251 Peptide

Neosystem SC784   MW 2600.01   SIVmac251 peptide from subtype GP140

Asn-Ser-Thr-Val-Thr-Ser-Leu-Ile-Ala-Asn-Ile-Asp-Trp-Thr-Asp-Gly-Arg-Lys

Synonyms: GP140-Arg-Lys (462-477); SIVmac251 Peptide

Neosystem SC754   MW 1991.18   SIVmac251 peptide from subtype GP140

Asn-Thr-Glu-His-Leu-Val-Asp-Ser-Phe-Gln-Glu-Met-Gly

Synonyms: Gonadotropin Releasing Hormone Associated Peptide (1-13); Gonadotropin Releasing Hormone Precursor Peptide (14-26); GAP

American Peptide 54-5-41   Rat   MW 1506.6 $C_{63}H_{95}N_{17}O_{24}S$

Asn-Thr-Trp-His-Lys-Val-Gly-Lys-Asn-Val-Tyr-Leu-Pro-Pro-Arg-Glu-Gly-Asp-Leu-Thr

Synonyms: GP140 (441-460); SIVmac251 Peptide

Neosystem SC752   MW 2324.62   SIVmac251 peptide from subtype GP140

Asn-Trp-Cys-Lys-Arg-Gly-Arg-Lys-Gln-Cys-Lys-Thr-His-Pro-His
Ac-Asn-Trp-Cys-Lys-Arg-Gly-Arg-Lys-Gln-Cys-Lys-Thr-His-Pro-His Amide

Synonyms: Amyloid/A₄ Precursor Protein (96-110), β-; Amyloid β/A4 Precursor Protein 770 (96-110), Ac- Cyclized

American Peptide 62-1-15   MW 1918.3 $C_{81}H_{128}N_{32}O_{19}S_2$
Disulfide bonds: Cys⁹⁸-Cys¹⁰⁵; cyclized peptide amide which is homologous to the heparin-binding domain of APP; strong affinity to heparin; inhibits binding of ¹²⁵I-labeled APP to heparin; blocks the heparan sulfate proteoglycan-dependent stimulatory effect of APP on neurite outgrowth; Small, DH et al, *J Neurosci*, 14:2117, 1994

Bachem H-2232.0001   MW 1918.24 $C_{81}H_{128}N_{32}O_{19}S_2$ Store at -15°C

Asn-Trp-Thr-His-Cys-Phe-Asp-Pro-Gln-Ile-Gln-Ala-Ile-Val-Ser-Ser-Pro-Cys-His-Asn-Ser
Asn-Trp-Thr-His-Cys(Acm)-Phe-Asp-Pro-Gln-Ile-Gln-Ala-Ile-Val-Ser-Ser-Pro-Cys(Acm)-His-Asn-Ser

Synonyms: GP46 (272-292); HTLV I Peptide

Neosystem SC830   MW 2527.78   HTLV I peptide

Asn-Trp-Val-Glu-Asp-Arg-Asp-Val-Thr-Thr-Gln-Arg-Pro-Lys-Glu-Arg-His-Arg-Arg-Asn

Synonyms: GP140 (411-430); SIVmac251 Peptide

Neosystem SC749   MW 2592.81   SIVmac251 peptide from subtype GP140

Asn-Tyr-Pro-Leu-Glu-Leu-Tyr-Glu-Arg-Val-Arg-Thr-Gly-Cys

Synonyms: Protein Kinase C γ Peptide

American Peptide 86-0-25   MW 1712.0 $C_{75}H_{116}N_{21}O_{23}S$

Asn-Tyr-Thr-Pro-Gly-Pro-Gly-Ile-Arg-Tyr-Pro-Leu-Thr

Synonyms: NEF MN (128-140); HIV I Peptide

Neosystem SC657   MW 1449.62   NEF peptide from HIV-1 subtype MN

Asn-Tyr-Thr-Pro-Gly-Pro-Gly-Val-Arg-Tyr-Pro-Leu-Thr

Synonyms: NEF (126-138); HTLV I Peptide

Neosystem SC511   MW 1434.61   HTLV I peptide from subtype NEF

Asn-Tyr-Tyr-Gly-Trp-Met-Asp-Phe
Asn-Tyr(SO₃H)-Tyr(SO₃H)-Gly-Trp-Met-Asp-Phe Amide

Synonyms: Cionin

Bachem H-1978.0001   Store at -15°C

Asn-Val

Bachem G-3720.0250   MW 231.25 $C_9H_{17}N_3O_4$ Store at -15°C

Asn-Val-Ile-Gln-Ile-Ser-Asn-Asp-Leu-Glu-Asn-Leu-Arg

Synonyms: Leptin (93-105); Obese Gene Peptide (93-105)

American Peptide 46-2-10   Human   MW 1527.7 $C_{64}H_{110}N_{20}O_{23}$ Zhang, Y et al, *Nature*, 372:425, 1994; Pelleymounter, MA et al, *Science*, 269:541, 1995

Bachem H-3426.0001   Human   MW 1527.7 $C_{64}H_{110}N_{20}O_{23}$ Store at -15°C

Asn-Val-Leu-Gly-Ala-Pro-Lys-Lys-Leu-Asn-Glu-Ser-Gln-Ala-Val

Synonyms: Protein A Derived Peptide

Bachem H-2228.0001   *Staphylococcus aureus*   Store at -15°C

Asn-Val-Thr-Glu-Asn-Phe-Asn-Met-Trp-Lys-Asn

Synonyms: GP120 (88-98); HTLV I Peptide

Neosystem SC261   MW 1396.53   HTLV I peptide from subtype GP160

Asn-Val-Trp-Ala-Thr-His-Ala-Cys-Val

Synonyms: GP120 (67-75); HTLV I Peptide

Neosystem SC538   MW 1000.14   HTLV I peptide from subtype GP160; due to the presence of a Cys this peptide can contain the dimeric form

---

Asp-Ala

Bachem G-1550.0250   MW 204.18 $C_7H_{12}N_2O_5$ Store at -15°C

Asp-Ala
Asp-α-Ala

Bachem G-1555.0250   MW 204.18 $C_7H_{12}N_2O_5$ Store at -15°C

Asp-Ala
α-Asp-Ala

Synonyms: Asp(Ala-OH)-OH

Bachem G-1560.0250   MW 204.18 $C_7H_{12}N_2O_5$ Store at -15°C

Asp-Ala
N-Cyanuryl-Asp-Ala-Anilide

Synonyms: N-(2,4-Dichloro-1,3,5-triazinyl)-Asp-Ala-Anilide

Bachem G-4420.0050   Store at -15°C

Asp-Ala
Asp-Ala-βNA

Bachem K-1100.0250   MW 329.36 $C_{17}H_{19}N_3O_4$ Store at RT

Asp-Ala
FA-Asp-Ala

Bachem M-2140.0050   Store at -15°C

Asp-Ala

Sigma A 1277   FW 204.2 $C_7H_{12}N_2O_5$

Asp-Ala-Ala-Arg-Glu-Gly-Phe-Leu-Ala-Thr-Leu-Val-Val-Leu-His-Arg-Ala-Gly-Ala-Arg

Synonyms: Peptide 6, (Ala⁹²)-; p16 (84-103), (Ala⁹²)-; Peptide 6 (84-103), (Ala⁹²)-; p16, (Ala⁹²)-

Bachem H-3718.0500   MW 2123.45 $C_{93}H_{155}N_{31}O_{26}$ Store at -15°C

Sigma P 7967   FW 2123.4   Bioactive peptide; inhibits phosphorylation of the retinoblastoma gene product pRb mediated by Cdk4-cyclin D1; Fahraeus, R, *Curr Biol*, 6:84, 1996

Asp-Ala-Ala-Arg-Glu-Gly-Phe-Leu-Asp-Thr-Leu-Val-Val-Leu-His-Arg-Ala-Gly-Ala-Arg

Synonyms: p16 (84-103); Peptide 6; HTLV I Peptide

Neosystem SC1250   MW 2167.45   Bioactive; cancer related peptide; this 20-residue peptide derived from p16CDKN2/INK4A can mediate 3 of the known functions of p16: it interacts with Cdk4 & Cdk6, it inhibits pRb phosphorylation *in vitro/in vivo* & it blocks entry into S phase; these results suggest a novel & exciting means by which the function of the p16 suppressor gene can be restored in human tumors & an attractive approach for future peptidomimetic drug design; Fahraeus, R et al, *Cur Biol*, 6:84-91, 1996

Asp-Ala-Ala-Leu-Arg-Gln-Leu-Arg-Ser-Pro-Arg-Arg-Thr-Gln-Ala-Pro-Ser-Ala-Gln-Glu

Synonyms: Rab4 (191-210)

Bachem H-3694.0500   Store at -15°C

Asp-Ala-Asp-Glu-Tyr-Leu
Asp-Ala-Asp-Glu-Tyr(PO₃H₂)-Leu Amide
Phosphorylated

Synonyms: Protein Tyrosine Phosphatase 1B Substrate; Epidermal Growth Factor Receptor (988-993)

Neosystem SC977   MW 803.71   Bioactive; synthetic growth factor-related peptide; a high affinity substrate for *Yersinia* protein tyrosine phosphatase; Zhang, ZY et al, *Biochem*, 33:2285-2290, 1994; Jia, Z et al, *Science*, 268:1754-1758, 1995

Bachem H-2702.0001   Human   Store at -15°C

Asp-Ala-Asp-Ser-Ser-Ile-Glu-Lys-Gln-Val-Ala-Leu-Leu-Lys-Ala-Leu-Tyr-Gly-His-Gly-Gln-Ile-Ser-His-Lys-Arg-His-Lys-Thr-Asp-Ser-Phe-Val-Gly-Leu-Met
Amide

Synonyms: Gastrointestinal Peptide; Neuropeptide K; Substance K Precursor Peptide

American Peptide 60-4-16   Porcine   MW 3980.6 $C_{175}H_{284}N_{52}O_{52}S$ Tatemoto, K et al, *BBRC*, 128:947, 1985

ICN 191467   Porcine   MW 3980.5   Brain tachykinin peptide with contactive & hypotensive activity; Tatemoto, K etal, *BBRC*, 128:947, 1985; Theodorsson-Norheim, E etal, *BBRC*, 131:77, 1985

Neosystem SC330   Porcine   MW 3980.55   Bioactive neuropeptide; Theodorsson-Norheim, E et al, *BBRC*, 131:77, 1985

Sigma N 7392   Porcine   FW 3980.5   ≥97% (HPLC); peptide content:~70% | Bioactive peptide; brain tachykinin that induces contraction of the gall bladder, vasodilation & increased capillary permeability; Tatemoto, K et al, *Biochem Biophys Res Commun*, 128:947, 1985; Theodorsson-Norheim, E et al, *Biochem Biophys Res Commun*, 131:77, 1985

Asp-Ala-Glu-Ala-Val-Gly-Pro-Glu-Ala-Phe-Ala-Asp-Gln-Asp-Leu-Asp-Glu-Arg-Glu-Val-Arg

Synonyms: Magainin Space Peptide

American Peptide 87-9-50   MW 2332.3 $C_{97}H_{150}N_{28}O_{39}$
Antibiotic peptide

Asp-Ala-Glu-Asn-Leu-Ile-Asp-Ser-Phe-Gln-Glu-Ile-Val

Synonyms: Gonadotropin Releasing Hormone Associated Peptide (1-13); Gonadotropin Releasing Hormone Precursor Peptide (14-26); GAP; GnRH Precursor Peptide (14-26); Prepro Human GnRH Precursor Peptide (14-26); Prepro-Gonadotropin Releasing Hormone (14-26); GAP (1-13); GnRH Precursor (14-26)

American Peptide 54-1-40   Human   MW 1492.5 $C_{65}H_{101}N_{15}O_{25}$ Millar, RP et al, *Science*, 232:68, 1986

Bachem H-5605.0500   Human   MW 1492.6 $C_{65}H_{101}N_{15}O_{25}$ Store at -15°C

ICN 153060   Human   MW 1492.6   Millar, RP, *Science*, 232:68, 1986

Sigma A 5667   Human   FW 1492.6   ≥97% (HPLC); peptide content:~65% | Bioactive peptide; Millar, RP et al, *Science*, 232:68, 1986

Neosystem SC354   Rat   MW 1492.60   Bioactive; hypothalamic releasing hormone; LHRH-related peptide; Adelman, JP et al, *PNAS USA*, 83:179, 1986; Millar, RP et al, *Science*, 232:68-70, 1986

Asp-Ala-Glu-Asn-Leu-Ile-Asp-Ser-Phe-Gln-Glu-Ile-Val-Lys-Glu-Val-Gly-Gln-Leu-Ala-Glu-Thr-Gln-Arg

Synonyms: Gonadotropin Releasing Hormone Associated Peptide (1-24); Gonadotropin Releasing Hormone Precursor Peptide (14-37); GAP

American Peptide 54-1-39   Human   MW 2732.9 $C_{114}H_{190}N_{32}O_{43}$

Asp-Ala-Glu-Phe-Arg-His-Asp-Ser-Gly-Tyr-Gln-Val-His-His-Gln-Lys

Synonyms: Amyloid β-Protein (1-16), (Gln¹¹)-

Alexis 151-007   MW 1955.0 $C_{84}H_{119}N_{27}O_{28}$

Asp-Ala-Glu-Phe-Arg-His-Asp-Ser-Gly-Tyr-Gln-Val-His-His-Gln-Lys
(Gln¹¹) β-Asp-Ala-Glu-Phe-Arg-His-Asp-Ser-Gly-Tyr-Gln-Val-His-His-Gln-Lys

Synonyms: Amyloid (1-16)

American Peptide 62-0-70   MW 1954.1 $C_{84}H_{120}N_{28}O_{27}$

Asp-Ala-Glu-Phe-Arg-His-Asp-Ser-Gly-Tyr-Gln-Val-His-His-Gln-Lys

Synonyms: Amyloid Peptide (1-16), (Gln¹¹)-β-; Amyloid β-Protein (1-16), (Gln¹¹)-

ICN 159866   MW 1954.1   >98%

Sigma A 4309   FW 1954.0   ≥97% (HPLC) | Bioactive peptide

Asp-Ala-Glu-Phe-Arg-His-Asp-Ser-Gly-Tyr-Gln-Val-His-His-Gln-Lys-Leu-Val-Phe-Phe-Ala-Glu-Asp-Val-Gly-Ser-Asn-Lys-Gly-Ala-Ile-Ile-Gly-Leu-Met-Val-Gly-Gly-Val-Val

Synonyms: Amyloid β-Protein (1-40), (Gln¹¹)-

Alexis 151-006   MW 4328.9 $C_{194}H_{296}N_{54}O_{57}S$

Asp-Ala-Glu-Phe-Arg-His-Asp-Ser-Gly-Tyr-Gln-Val-His-His-Gln-Lys-Leu-Val-Phe-Phe-Ala-Glu-Asp-Val-Gly-Ser-Asn-Lys
β-Asp-Ala-Glu-Phe-Arg-His-Asp-Ser-Gly-Tyr-Gln-Val-His-His-Gln-Lys-Leu-Val-Phe-Phe-Ala-Glu-Asp-Val-Gly-Ser-Asn-Lys

Synonyms: Amyloid (1-28)

American Peptide 62-0-75   MW 3261.5 $C_{145}H_{210}N_{42}O_{45}$ Castano, EM et al, *Biochem Biophys Res Comm*, 141:782, 1986; Whitson, JS et al, *Science*, 243:1488, 1989

Asp-Ala-Glu-Phe-Arg-His-Asp-Ser-Gly-Tyr-Gln-Val-His-His-Gln-Lys-Leu-Val-Phe-Phe-Ala-Glu-Asp-Val-Gly-Ser-Asn-Lys-Gly-Ala-Ile-Ile-Gly-Leu-Met-Val-Gly-Gly-Val-Val

Synonyms: Amyloid (1-40), (Gln¹¹)-β-

American Peptide 62-0-79   MW 4328.9 $C_{194}H_{296}N_{54}O_{57}S$ Klegeris, A et al, *Biochem Biophys Res Comm*, 199:984, 1994; Kowalska, MA & Badellino, K, *Biochem Biophys Res Comm*, 205:1829, 1994; Paradis, E et al, *J Neurosci*, 16:7533, 1996

Asp-Ala-Glu-Phe-Arg-His-Asp-Ser-Gly-Tyr-Gln-Val-His-His-Gln-Lys-Leu-Val-Phe-Phe-Ala-Glu-Asp-Val-Gly-Ser-Asn-Lys

Synonyms: Amyloid α-Protein (1-28), (Gln¹¹)-

Bachem H-2362.0500   MW 3261.52 $C_{145}H_{210}N_{42}O_{45}$ Store at -15°C

Asp-Ala-Glu-Phe-Arg-His-Asp-Ser-Gly-Tyr-Gln-Val-His-His-Gln-Lys-Leu-Val-Phe-Phe-Ala-Glu-Asp-Val-Gly-Ser-Asn-Lys-Gly-Ala-Ile-Ile-Gly-Leu-Met-Val-Gly-Gly-Val-Val

Synonyms: Amyloid Peptide (1-40), (Gln¹¹)-β-

ICN 159870   MW 4066.9   >98%

Asp-Ala-Glu-Phe-Arg-His-Asp-Ser-Gly-Tyr-Gln-Val-His-His-Gln-Lys-Leu-Val-Phe-Phe-Ala-Glu-Asp-Val-Gly-Ser-Asn-Lys

Synonyms: Amyloid β-Protein (1-28), (Gln[11])-

Sigma A 1434 FW 3261.5 ≥97% (HPLC) | Bioactive peptide; β-Amyloid peptide that forms fibrils *in vitro* that have the same structure & antigenicity as those found in Alzheimer's patients; Castano, EM et al, *Biochem Biophys Res Commun*, 141:782, 1986; Whitson, JS et al, *Science*, 243:1488, 1989

Asp-Ala-Glu-Phe-Arg-His-Asp-Ser-Gly-Tyr-Gln-Val-His-His-Gln-Lys-Leu-Val-Phe-Phe-Ala-Glu-Asp-Val-Gly-Ser-Asn-Lys-Gly-Ala-Ile-Ile-Gly-Leu-Met-Val-Gly-Gly-Val-Val
β-Asp-Ala-Glu-Phe-Arg-His-Asp-Ser-Gly-Tyr-Gln-Val-His-His-Gln-Lys-Leu-Val-Phe-Phe-Ala-Glu-Asp-Val-Gly-Ser-Asn-Lys-Gly-Ala-Ile-Ile-Gly-Leu-Met-Val-Gly-Gly-Val-Val Trifluoroacetate Salt

Synonyms: Amyloid (1-40), β-

Calbiochem 171590 Human MW 4328.9 $C_{194}H_{296}N_{54}O_{57}S$ ≥95% (HPLC); solid; soluble in 5% HOAc | Neurotrophic & neurotoxic effects depending upon neuronal age & concentration of β-protein; major component of senile & Alzheimer's plaques; promotes down-regulation of *bcl*-2 & enhances phosphorylation of Tau; Paradis, E et al, *J Neurosci*, 16:7533, 1996; Klegeris, A et al, *Biochem Biophys Res Comm*, 199:984, 1994; Kowalska, MA & Badellino, K, *Biochem Biophys Res Comm*, 205:1829, 1994; Yankner, BA et al, *Science*, 250:279, 1990

Asp-Ala-Glu-Phe-Arg-His-Asp-Ser-Gly-Tyr-Gln-Val-His-His-Gln-Lys-Leu-Val-Phe-Phe-Ala-Glu-Asp-Val-Gly-Ser-Asn-Lys
H2N-Asp-Ala-Glu-Phe-Arg-His-Asp-Ser-Gly-Tyr-Gln-Val-His-His-Gln-Lys-Leu-Val-Phe-Phe-Ala-Glu-Asp-Val-Gly-Ser-Asn-Lys

Synonyms: Amyloid Peptide (1-28), β-

Biogenesis 0490-1902 Human synthetic Purified; lyophilized | Castano et al, *BBRC*, 141:782, 1986

Asp-Ala-Glu-Phe-Arg-His-Asp-Ser-Gly-Tyr-Glu
β-Asp-Ala-Glu-Phe-Arg-His-Asp-Ser-Gly-Tyr-Glu

Synonyms: Amyloid (1-11)

American Peptide 62-0-10 MW 1325.3 $C_{56}H_{76}N_{16}O_{22}$

Asp-Ala-Glu-Phe-Arg-His-Asp-Ser-Gly-Tyr-Glu

Synonyms: Amyloid α-Protein (1-11)

Bachem H-2956.0500 MW 1325.31 $C_{56}H_{76}N_{16}O_{22}$ Store at -15°C

Asp-Ala-Glu-Phe-Arg-His-Asp-Ser-Gly-Tyr-Glu-Val-His-His-Gln-Lys

Synonyms: Amyloid α-Protein (1-16)

Bachem H-2958.0500 MW 1955.03 $C_{84}H_{119}N_{27}O_{28}$ Store at -15°C

Peptides International PAB-4359-v Human synthetic MW 1955.0 $C_{84}H_{119}N_{27}O_{28}$ >99% (HPLC); lyophilized amorphous powder | Bioactive; blocker for plaque-induced microgliosis; reducer for brain inflammation; Giulian, D et al, *JBC*, 273:29719, 1998

Asp-Ala-Glu-Phe-Arg-His-Asp-Ser-Gly-Tyr-Glu-Val-His-His-Gln-Lys-Leu-Val-Phe-Phe-Ala-Glu-Asp-Val-Gly-Ser-Asn-Lys-Gly-Ala-Ile-Ile-Gly-Leu-Met-Val-Gly-Gly-Val-Val-Ile-Ala-Thr

Synonyms: Amyloid β-Protein (1-43)

Alexis 151-001 MW 4615.3 $C_{207}H_{318}N_{56}O_{62}S$ ≥95%; white lyophilized powder; soluble in H2O | Kang, J et al, *Nature*, 325:733, 1987; Goldgaber, D et al, *Science*, 235:877, 1987; Pike, CJ et al, *J Neurosci*, 13:1676, 1993

Asp-Ala-Glu-Phe-Arg-His-Asp-Ser-Gly-Tyr-Glu-Val-His-His-Gln-Lys-Leu-Val-Phe-Phe-Ala-Glu-Asp-Val-Gly-Ser-Asn-Lys-Gly-Ala-Ile-Ile-Gly-Leu-Met-Val-Gly-Gly-Val-Val-Ile-Ala

Synonyms: Amyloid β-Protein (1-42)

Alexis 151-002 MW 4514.1 $C_{203}H_{311}N_{55}O_{60}S$ ≥98%; white lyophilized powder; soluble in DMSO | Kang, J et al, *Nature*, 325:733, 1987; Goldgaber, D et al, *Science*, 235:877, 1987; Murrell, J et al, *Science*, 254:97, 1991; Barrow, CJ & Zagorski, MG, *Science*, 253:179, 1991

Asp-Ala-Glu-Phe-Arg-His-Asp-Ser-Gly-Tyr-Glu-Val-His-His-Gln-Lys-Leu-Val-Phe-Phe-Ala-Glu-Asp-Val-Gly-Ser-Asn-Lys-Gly-Ala-Ile-Ile-Gly-Leu-Met-Val-Gly-Gly-Val-Val

Synonyms: Amyloid β-Protein (1-40)

Alexis 151-003 MW 4329.9 $C_{194}H_{295}N_{53}O_{58}S$ ≥97%; white lyophilized powder | Jarrett, JT et al, *Biochemistry*, 32:4693, 1993; Sticht, H et al, *Eur J Biochem*, 233:293, 1995; Yankner, BA et al, *Science*, 250:279, 1990

Asp-Ala-Glu-Phe-Arg-His-Asp-Ser-Gly-Tyr-Glu-Val-His-His-Gln-Lys-Leu-Val-Phe-Phe-Ala-Glu-Asp-Val-Gly-Ser-Asn-Lys-Gly-Ala-Ile-Ile-Gly-Leu-Met-Val-Gly-Gly

Synonyms: Amyloid β-Protein (1-38)

Alexis 151-004 MW 4328.9 $C_{184}H_{277}N_{51}O_{56}S$ Mattson, MP et al, *J Neurosci*, 12:376, 1992

Asp-Ala-Glu-Phe-Arg-His-Asp-Ser-Gly-Tyr-Glu-Val-His-His-Gln-Lys-Leu-Val-Phe-Phe-Ala-Glu-Asp-Val-Gly-Ser-Asn-Lys

Synonyms: Amyloid β-Protein (1-40)

Alexis 151-005 MW 3262.5 $C_{145}H_{209}N_{41}O_{46}$ ≥97%; white lyophilized powder; soluble in H2O | Kang, J et al, *Nature*, 325:733, 1987; Kirshner, DA et al, *PNAS*, 84:6953, 1987; Hollosi, M et al, *peptide Res*, 2:109, 1989; Whitson, JS et al, *Science*, 243:1488, 1989; Marx, JL et al, *Science*, 243:1664, 1989

Asp-Ala-Glu-Phe-Arg-His-Asp-Ser-Gly-Tyr-Glu-Val-His-His-Gln-Lys-Leu-Val-Phe-Phe-Ala-Glu-Asp-Val-Gly-Ser-Asn-Lys
β-Asp-Ala-Glu-Phe-Arg-His-Asp-Ser-Gly-Tyr-Glu-Val-His-His-Gln-Lys-Leu-Val-Phe-Phe-Ala-Glu-Asp-Val-Gly-Ser-Asn-Lys

Synonyms: Amyloid (1-28)

American Peptide 62-0-74 MW 3262.5 $C_{145}H_{209}N_{41}O_{46}$ This synthetic peptide formed fibrils that have the same antigenic & structural features as the AD amyloid filaments; Kang, J et al, *Nature*, 325:773, 1987

Asp-Ala-Glu-Phe-Arg-His-Asp-Ser-Gly-Tyr-Glu-Val-His-His-Gln-Lys-Leu-Val-Phe-Phe-Ala-Glu-Asp-Val-Gly-Ser-Asn-Lys-Gly-Ala-Ile-Ile-Gly-Leu-Met-Val-Gly-Gly
β-Asp-Ala-Glu-Phe-Arg-His-Asp-Ser-Gly-Tyr-Glu-Val-His-His-Gln-Lys-Leu-Val-Phe-Phe-Ala-Glu-Asp-Val-Gly-Ser-Asn-Lys-Gly-Ala-Ile-Ile-Gly-Leu-Met-Val-Gly-Gly

Synonyms: Amyloid (1-38)

American Peptide 62-0-76 MW 4131.6 $C_{184}H_{277}N_{51}O_{56}S$ Similar to the amyloid β-protein fragment (25-35); destabilizes calcium homeostasis & makes neurons vulnerable to environmental insults; Mattson, MP et al, *J Neurosci*, 12:376, 1992

Asp-Ala-Glu-Phe-Arg-His-Asp-Ser-Gly-Tyr-Glu-Val-His-His-Gln-Lys-Leu-Val-Phe-Phe-Ala-Glu-Asp-Val-Gly-Ser-Asn-Lys-Gly-Ala-Ile-Ile-Gly-Leu-Met-Val-Gly-Gly-Val-Val
β-Asp-Ala-Glu-Phe-Arg-His-Asp-Ser-Gly-Tyr-Glu-Val-His-His-Gln-Lys-Leu-Val-Phe-Phe-Ala-Glu-Asp-Val-Gly-Ser-Asn-Lys-Gly-Ala-Ile-Ile-Gly-Leu-Met-Val-Gly-Gly-Val-Val

Synonyms: Amyloid (1-40)

American Peptide 62-0-78   MW 4329.9 $C_{194}H_{295}N_{53}O_{58}S$
Exhibits neurotrophic & neurotoxic effects that depend on the neuronal age & concentration of the β-protein; Yankner, BA et al, *Science*, 250:279, 1990

Asp-Ala-Glu-Phe-Arg-His-Asp-Ser-Gly-Tyr-Glu-Val-His-His-Gln-Lys-Leu-Val-Phe-Phe-Ala-Glu-Asp-Val-Gly-Ser-Asn-Lys-Gly-Ala-Ile-Ile-Gly-Leu-Met-Val-Gly-Gly-Val-Val-Ile-Ala
β-Asp-Ala-Glu-Phe-Arg-His-Asp-Ser-Gly-Tyr-Glu-Val-His-His-Gln-Lys-Leu-Val-Phe-Phe-Ala-Glu-Asp-Val-Gly-Ser-Asn-Lys-Gly-Ala-Ile-Ile-Gly-Leu-Met-Val-Gly-Gly-Val-Val-Ile-Ala

Synonyms: Amyloid (1-42)

American Peptide 62-0-80   MW 4514.1 $C_{203}H_{311}N_{55}O_{60}S$
Believed to be the major subunit of vascular & plaque filaments in individuals with Alzheimer's disease, elderly people & patients with trisomy 21 (Down's Syndrome); *Neurobiology of Aging*, Vol. 13:No. 5, 1992

Asp-Ala-Glu-Phe-Arg-His-Asp-Ser-Gly-Tyr-Glu-Val-His-His-Gln-Lys-Leu-Val-Phe-Phe-Ala-Glu-Asp-Val-Gly-Ser-Asn-Lys-Gly-Ala-Ile-Ile-Gly-Leu-Met-Val-Gly-Gly-Val-Val-Ile-Ala-Thr
β-Asp-Ala-Glu-Phe-Arg-His-Asp-Ser-Gly-Tyr-Glu-Val-His-His-Gln-Lys-Leu-Val-Phe-Phe-Ala-Glu-Asp-Val-Gly-Ser-Asn-Lys-Gly-Ala-Ile-Ile-Gly-Leu-Met-Val-Gly-Gly-Val-Val-Ile-Ala-Thr

Synonyms: Amyloid (1-43)

American Peptide 62-0-83   MW 4615.2 $C_{207}H_{318}N_{56}O_{62}S$
Goldgaber, D et al, *Science*, 235:877, 1987; Yankner, BA et al, *Science*, 250:279, 1990; Pike, CJ et al, *J Neurosci*, 13:1676, 1993

Asp-Ala-Glu-Phe-Arg-His-Asp-Ser-Gly-Tyr-Glu-Val-His-His-Gln-Lys-Leu-Val-Phe-Phe-Ala-Glu-Asp-Val-Gly-Ser-Asn-Lys-Gly-Ala-Ile-Ile-Gly-Leu-Met-Val-Gly-Gly-Val-Val
(^{125}I)β-Asp-Ala-Glu-Phe-Arg-His-Asp-Ser-Gly-Tyr-Glu-Val-His-His-Gln-Lys-Leu-Val-Phe-Phe-Ala-Glu-Asp-Val-Gly-Ser-Asn-Lys-Gly-Ala-Ile-Ile-Gly-Leu-Met-Val-Gly-Gly-Val-Val

Synonyms: Amyloid (1-40), (^{125}I)-β-

Amersham IM294   Lyophilized; ~74 TBq/mmol; ~2000 Ci/mmol

Asp-Ala-Glu-Phe-Arg-His-Asp-Ser-Gly-Tyr-Glu-Val-His-His-Gln-Lys-Leu-Val-Phe-Phe-Ala-Glu-Asp-Val-Gly-Ser-Asn-Lys-Gly-Ala-Ile-Ile-Gly-Leu-Met-Val-Gly-Gly-Val-Val

Synonyms: Amyloid α-Protein (1-40)

Bachem H-1194.0500   MW 4329.86 $C_{194}H_{295}N_{53}O_{58}S$ Store at -15°C

Asp-Ala-Glu-Phe-Arg-His-Asp-Ser-Gly-Tyr-Glu-Val-His-His-Gln-Lys-Leu-Val-Phe-Phe-Ala-Glu-Asp-Val-Gly-Ser-Asn-Lys-Gly-Ala-Ile-Ile-Gly-Leu-Met-Val-Gly-Gly-Val-Val-Ile-Ala

Synonyms: Amyloid α-Protein (1-42)

Bachem H-1368.0500   MW 4514.1 $C_{203}H_{311}N_{55}O_{60}S$ Store at -15°C

Asp-Ala-Glu-Phe-Arg-His-Asp-Ser-Gly-Tyr-Glu-Val-His-His-Gln-Lys-Leu-Val-Phe-Phe-Ala-Glu-Asp-Val-Gly-Ser-Asn-Lys-Gly-Ala-Ile-Ile-Gly-Leu-Met-Val-Gly-Gly-Val-Val-Ile-Ala-Thr

Synonyms: Amyloid α-Protein (1-43)

Bachem H-1586.0500   MW 4615.21 $C_{207}H_{318}N_{56}O_{62}S$ Store at -15°C

Asp-Ala-Glu-Phe-Arg-His-Asp-Ser-Gly-Tyr-Glu-Val-His-His-Gln-Lys-Leu-Val-Phe-Phe-Ala-Glu-Asp-Val-Gly-Ser-Asn-Lys-Gly-Ala-Ile-Ile-Gly-Leu-Met-Val-Gly-Gly

Synonyms: Amyloid α-Protein (1-38)

Bachem H-2966.0500   MW 4131.6 $C_{184}H_{277}N_{51}O_{56}S$ Store at -15°C

Asp-Ala-Glu-Phe-Arg-His-Asp-Ser-Gly-Tyr-Glu-Val-His-His-Gln-Lys-Leu-Val-Phe-Phe-Ala-Glu-Asp-Val-Gly-Ser-Asn-Lys-Gly-Ala-Ile-Ile-Gly-Leu-Met-Val-Gly-Gly-Val-Val-Ile-Ala
D-Asp-Ala-Glu-Phe-Arg-His-Asp-Ser-Gly-Tyr-Glu-Val-His-His-Gln-Lys-Leu-Val-Phe-Phe-Ala-Glu-Asp-Val-Gly-Ser-Asn-Lys-Gly-Ala-Ile-Ile-Gly-Leu-Met-Val-Gly-Gly-Val-Val-Ile-Ala

Synonyms: Amyloid α-Protein (1-42), (D-Asp^1)-

Bachem H-4854.0500   MW 4514.1 $C_{203}H_{311}N_{55}O_{60}S$ Store at -15°C

Asp-Ala-Glu-Phe-Arg-His-Asp-Ser-Gly-Tyr-Glu-Val-His-His-Gln-Lys-Leu-Val-Phe-Phe-Ala-Glu-Asp-Val-Gly-Ser-Asn-Lys

Synonyms: Amyloid α-Protein (1-28); Alzheimer's Disease α-Protein (SP28); Amyloid Peptide (1-28), β-; Alzheimer's Disease Protein

Bachem H-7865.0500   MW 3262.5 $C_{145}H_{209}N_{41}O_{46}$ Store at -15°C

ICN 154474   MW 3261.5   >97% (HPLC) | Castano, EM etal, *BBRC*, 141:782, 1986; Whitson, JS etal, *Science*, 243:1488, 1989

Asp-Ala-Glu-Phe-Arg-His-Asp-Ser-Gly-Tyr-Glu-Val-His-His-Gln-Lys-Leu-Val-Phe-Phe-Ala-Glu-Asp-Val-Gly-Ser-Asn-Lys-Gly-Ala-Ile-Ile-Gly-Leu-Met-Val-Gly-Gly

Synonyms: Amyloid Peptide (1-38), β-

ICN 159868   MW 4132.2   >98% | Mattson, MP etal, *J Neurosci*, 12:376, 1992

Asp-Ala-Glu-Phe-Arg-His-Asp-Ser-Gly-Tyr-Glu-Val-His-His-Gln-Lys-Leu-Val-Phe-Phe-Ala-Glu-Asp-Val-Gly-Ser-Asn-Lys-Gly-Ala-Ile-Ile-Gly-Leu-Met-Val-Gly-Gly-Val-Val

Synonyms: Amyloid Peptide (1-40), β-

ICN 159869   MW 4330.4   >98% | Yanker, BA etal, *Science*, 250:279, 1990

Asp-Ala-Glu-Phe-Arg-His-Asp-Ser-Gly-Tyr-Glu-Val-His-His-Gln-Lys-Leu-Val-Phe-Phe-Ala-Glu-Asp-Val-Gly-Ser-Asn-Lys-Gly-Ala-Ile-Ile-Gly-Leu-Met-Val-Gly-Gly-Val-Val-Ile-Ala

Synonyms: Amyloid Peptide (1-42), β-

ICN 159871   MW 4514.1   >98%

Asp-Ala-Glu-Phe-Arg-His-Asp-Ser-Gly-Tyr-Glu-Val-His-His-Gln-Lys-Leu-Val-Phe-Phe-Ala-Glu-Asp-Val-Gly-Ser-Asn-Lys

Synonyms: Amyloid β Protein (1-28)

Neosystem SC488   MW 3262.49   Bioactive; Kang J et al, *Nature*, 325:733-736, 1987

Asp-Ala-Glu-Phe-Arg-His-Asp-Ser-Gly-Tyr-Glu-Val-His-His-Gln-Lys-Leu-Val-Phe-Phe-Ala-Glu-Asp-Val-Gly-Ser-Asn-Lys-Gly-Ala-Ile-Ile-Gly-Leu-Met-Val-Gly-Gly-Val-Val

Synonyms: Amyloid β Protein (1-40)

Neosystem SC875   MW 4329.84   Bioactive; Amyloid β protein sequence; Yankner, BA et al, *Science*, 250:279-282, 1990

Asp-Ala-Glu-Phe-Arg-His-Asp-Ser-Gly-Tyr-Glu-Val-His-His-Gln-Lys-Leu-Val-Phe-Phe-Ala-Glu-Asp-Val-Gly-Ser-Asn-Lys

Synonyms: Amyloid β-Protein (1-28), (Gln[11])-

Sigma A 0184   FW 3262.5   ≥97% (HPLC) | Bioactive peptide; β-Amyloid peptide that forms fibrils *in vitro* that have the same structure & antigenicity as those found in Alzheimer's patients

---

Asp-Ala-Glu-Phe-Arg-His-Asp-Ser-Gly-Tyr-Glu-Val-His-His-Gln-Lys-Leu-Val-Phe-Phe-Ala-Glu-Asp-Val-Gly-Ser-Asn-Lys-Gly-Ala-Ile-Ile-Gly-Leu-Met-Val-Gly-Gly

Synonyms: Amyloid β-Protein (1-38)

Sigma A 0189   FW 4131.6   ≥97% (HPLC) | Bioactive peptide; destabilizes calcium homeostasis & renders human cortical neurons vulnerable to environmental insults; Yankner, BA et al, *Science*, 250:279, 1990; Mattson, MP et al, *J Neurosci*, 12:376, 1992

Asp-Ala-Glu-Phe-Arg-His-Asp-Ser-Gly-Tyr-Glu-Val-His-His-Gln-Lys-Leu-Val-Phe-Phe-Ala-Glu-Asp-Val-Gly-Ser-Asn-Lys-Gly-Ala-Ile-Ile-Gly-Leu-Met-Val-Gly-Gly-Val-Val

Synonyms: Amyloid β-Protein (1-40)

Sigma A 1075   FW 4329.8   ≥90% (HPLC) | Bioactive peptide; neurotoxin *in vivo* & in neuronal cell culture, also has neurotrophic effects depending on age of the cells; prepared by Sigma; Yankner, BA et al, *Science*, 250:279, 1990

Sigma A 5813   FW 4329.8   ≥90% (HPLC) | Bioactive peptide; neurotoxin *in vivo* & in neuronal cell culture, also has neurotrophic effects depending on age of the cells

---

Asp-Ala-Glu-Phe-Arg-His-Asp-Ser-Gly-Tyr-Glu-Val-His-His-Gln-Lys-Leu-Val-Phe-Phe-Ala-Glu-Asp-Val-Gly-Ser-Asn-Lys-Gly-Ala-Ile-Ile-Gly-Leu-Met-Val-Gly-Gly-Val-Val-Ile-Ala-Thr

Synonyms: Amyloid β-Protein (1-43)

Sigma A 7712   FW 4615.2   Bioactive peptide; primary constituent of senile plaques & cerebrovascular deposits in Alzheimer's disease & Down's syndrome; Pike, CJ et al, *J Neurosci*, 13:1676, 1993

Asp-Ala-Glu-Phe-Arg-His-Asp-Ser-Gly-Tyr-Glu-Val-His-His-Gln-Lys-Leu-Val-Phe-Phe-Ala-Glu-Asp-Val-Gly-Ser-Asn-Lys-Gly-Ala-Ile-Ile-Gly-Leu-Met-Val-Gly-Gly-Val-Val-Ile-Ala

Synonyms: Amyloid β-Protein (1-42)

Sigma A 9810   FW 4514.1   Bioactive peptide; predominant fragment of amyloid β-protein found in the brains of patients with Alzheimer's disease & Down's syndrome; Jarrett, JT et al, *Biochem*, 32:4693, 1993

Asp-Ala-Glu-Phe-Arg-His-Asp-Ser-Gly-Tyr-Glu-Val-His-His-Gln-Lys-Leu-Val-Phe-Phe-Ala-Glu-Asp-Val-Gly-Ser-Asn-Lys-Gly-Ala-Ile-Ile-Gly-Leu-Met-Val-Gly-Gly-Val-Val

Synonyms: Amyloid (1-40), β-

Biogenesis 0490-1916   Human synthetic   MW 4330.4 >95% (HPLC); lyophilized, consisting of 72% peptide material, TFA counter ions and residual H_2O; no preservatives | To avoid Met oxidation, it is best to dissolve peptide under a nitrogen stream

---

Asp-Ala-Glu-Phe-Arg-His-Asp-Ser-Gly-Tyr-Glu-Val-His-His-Gln-Lys-Leu-Val-Phe-Phe-Ala-Glu-Asp-Val-Gly-Ser-Asn-Lys-Gly-Ala-Ile-Ile-Gly-Leu-Met-Val-Gly-Gly-Val-Val-Ile-Ala-Thr

Synonyms: Amyloid β-Protein (1-43)

Peptides International PAB-4370-v   Human synthetic   MW 4615.1 $C_{207}H_{318}N_{56}O_{62}S$ >95% (HPLC); lyophilized amorphous powder; trifluoroacetate | Bioactive; major plaque component in Alzheimer's Disease; Tamaoka, A et al, *BBRC*, 205:834, 1994; Hsiao, K et al, *Science*, 274:99 1996; Duff, K et al, *Nature*, 383:710, 1996; Kawarabayashi, T et al, *Brain Res*, 765:343, 1997

Asp-Ala-Glu-Phe-Arg-His-Asp-Ser-Gly-Tyr-Glu-Val-His-His-Gln-Lys-Leu-Val-Phe-Phe-Ala-Glu-Asp-Val-Gly-Ser-Asn-Lys-Gly-Ala-Ile-Ile-Gly-Leu-Met-Val-Gly-Gly-Val-Val Hydrochloride

Synonyms: Amyloid β-Protein (1-40)

Peptides International PAB-4379-v   Human synthetic   MW 4329.9 $C_{194}H_{295}N_{53}O_{58}S$ >95% (HPLC); lyophilized amorphous powder | Bioactive; major plaque component in Alzheimer's Disease; specific form easily transferrable to β-structure

---

Asp-Ala-Glu-Phe-Arg-His-Asp-Ser-Gly-Tyr-Glu-Val-His-His-Gln-Lys-Leu-Val-Phe-Phe-Ala-Glu-Asp-Val-Gly-Ser-Asn-Lys-Gly-Ala-Ile-Ile-Gly-Leu-Met-Val-Gly-Gly-Val-Val

Synonyms: Amyloid β-Protein (1-40)

Peptides International PAM-4307-v   Human synthetic   MW 4329.9 $C_{194}H_{295}N_{53}O_{58}S$ >95% (HPLC); lyophilized amorphous powder; trifluoroacetate | Bioactive; peptide deposited in the brain of Alzheimer's Disease patients; Yankner, BA et al, *Science*, 250:279, 1990

Asp-Ala-Glu-Phe-Arg-His-Asp-Ser-Gly-Tyr-Glu-Val-His-His-Gln-Lys-Leu-Val-Phe-Phe-Ala-Glu-Asp-Val-Gly-Ser-Asn-Lys-Gly-Ala-Ile-Ile-Gly-Leu-Met-Val-Gly-Gly-Val-Val-Ile-Ala

Synonyms: Amyloid β-Protein (1-42)

Peptides International PAM-4349-v   Human synthetic   MW 4514.1 $C_{203}H_{311}N_{55}O_{60}S$ >95% (HPLC); lyophilized amorphous powder; trifluoroacetate | Bioactive; major plaque component in Alzheimer's Disease; Goldgaber, D et al, *Science*, 235:877, 1987; Roher, AE et al, *PNAS USA*, 90:10836, 1993; Suzuki, N et al, *Science*, 264:1336, 1994; Citron, M et al, *PNAS USA*, 93:13170, 1996

Asp-Ala-Glu-Phe-Gly-His-Asp-Ser-Gly-Phe-Glu-Val-Arg-His-Gln-Lys-Leu-Val-Phe-Phe-Ala-Glu-Asp-Val-Gly-Ser-Asn-Lys-Gly-Ala-Ile-Val-Ile-Gly-Leu-Met-Val-Gly-Gly-Val-Val-Ile-Ala
β-Asp-Ala-Glu-Phe-Gly-His-Asp-Ser-Gly-Phe-Glu-Val-Arg-His-Gln-Lys-Leu-Val-Phe-Phe-Ala-Glu-Asp-Val-Gly-Ser-Asn-Lys-Gly-Ala-Ile-Ile-Gly-Leu-Met-Val-Gly-Gly-Val-Val-Ile-Ala

Synonyms: Amyloid (1-42)

American Peptide 62-0-84 Rat MW 4418.0
$C_{199}H_{307}N_{53}O_{59}S$ Paradis, E et al, *J Neurosci*, 16:7533, 1996; Murrell, J et al, *Science*, 254:97, 1991; Goldgaber, D et al, *Science*, 235:877, 1987

Asp-Ala-Glu-Phe-Gly-His-Asp-Ser-Gly-Phe-Glu-Val-Arg-His-Gln-Lys-Leu-Val-Phe-Phe-Ala-Glu-Asp-Val-Gly-Ser-Asn-Lys-Gly-Ala-Ile-Ile-Gly-Leu-Met-Val-Gly-Gly-Val-Val
β-Asp-Ala-Glu-Phe-Gly-His-Asp-Ser-Gly-Phe-Glu-Val-Arg-His-Gln-Lys-Leu-Val-Phe-Phe-Ala-Glu-Asp-Val-Gly-Ser-Asn-Lys-Gly-Ala-Ile-Ile-Gly-Leu-Met-Val-Gly-Gly-Val-Val

Synonyms: Amyloid (1-40)

American Peptide 62-0-86 Rat MW 4233.8
$C_{190}H_{291}N_{51}O_{57}S$ Klegeris, A et al, *Biochem Biophys Res Comm*, 199:984, 1994; Kowalska, MA & Badellino, K, *Biochem Biophys Res Comm*, 205:1829, 1994; Yankner, BA et al, *Science*, 250:279, 1990

Asp-Ala-Glu-Phe-Gly-His-Asp-Ser-Gly-Phe-Glu-Val-Arg-His-Gln-Lys-Leu-Val-Phe-Phe-Ala-Glu-Asp-Val-Gly-Ser-Asn-Lys-Gly-Ala-Ile-Ile-Gly-Leu-Met-Val-Gly-Gly-Val-Val
β-Asp-Ala-Glu-Phe-Gly-His-Asp-Ser-Gly-Phe-Glu-Val-Arg-His-Gln-Lys-Leu-Val-Phe-Phe-Ala-Glu-Asp-Val-Gly-Ser-Asn-Lys-Gly-Ala-Ile-Ile-Gly-Leu-Met-Val-Gly-Gly-Val-Val Trifluoroacetate salt

Synonyms: Amyloid (1-40), β-

Calbiochem 171593 Rat MW 4233.8 $C_{190}H_{291}N_{51}O_{57}S$
≥95% (HPLC); lyophilized; soluble in 5% HOAc | Neurotrophic & neurotoxic effects depending upon neuronal age & concentration of β-protein; Klegeris, A et al, *Biochem Biophys Res Comm*, 199:984, 1994; Kowalska, MA & Badellino, K, *Biochem Biophys Res Comm*, 205:1829, 1994; Yankner, BA et al, *Science*, 250:279, 1990

Asp-Ala-Glu-Phe-Gly-His-Asp-Ser-Gly-Phe-Glu-Val-Arg-His-Gln-Lys-Leu-Val-Phe-Phe-Ala-Glu-Asp-Val-Gly-Ser-Asn-Lys-Gly-Ala-Ile-Ile-Gly-Leu-Met-Val-Gly-Gly-Val-Val-Ile-Ala
β-Asp-Ala-Glu-Phe-Gly-His-Asp-Ser-Gly-Phe-Glu-Val-Arg-His-Gln-Lys-Leu-Val-Phe-Phe-Ala-Glu-Asp-Val-Gly-Ser-Asn-Lys-Gly-Ala-Ile-Ile-Gly-Leu-Met-Val-Gly-Gly-Val-Val-Ile-Ala Trifluoroacetate salt

Synonyms: Amyloid (1-42), β-

Calbiochem 171596 Rat MW 4418 $C_{199}H_{307}N_{53}O_{59}S$
≥80% (HPLC); lyophilized; soluble in 5% HOAc | Predominant peptide found in the brain of patients with Alzheimer's Disease & Down's Syndrome; promotes down-regulation of bcl-2 & increases levels of *bax* in neurons; Paradis, E et al, *J Neurosci*, 16:7533, 1996; Goldgabber, D et al, *Science*, 235:877, 1987; Kang, J et al, *Nature*, 325:733, 1987; Murrell, J et al, *Science*, 254:97, 1991

Asp-Ala-Glu-Phe-Lys
Asp-Ala-Glu-Phe-(Lys-DNP)-Amide

BioSource International 03-402

Asp-Ala-Gly-His-Gly-Gln-Ile-Ser-His-Lys-Arg-His-Lys-Thr-Asp-Ser-Phe-Val-Gly-Leu-Met Amide

Synonyms: Neuropeptide, γ-; Prepro-Tachykinin (71-92), γ-

Bachem H-7455.0500 Store at -15°C

American Peptide 60-7-20 Rabbit MW 2320.6
$C_{99}H_{158}N_{34}O_{29}S$ Neuropeptide located within AA 72-92 of γ-preprotachykinin; has been found in the rabbit intestinal extract

Asp-Ala-His-Lys

Bachem H-2020.0100 MW 469.5 $C_{19}H_{31}N_7O_7$ Store at -15°C

Asp-Ala-Phe-Val-Ala-Leu-Met
Asp-Ala-Phe-Val-β-Ala-Leu-Met Amide

Synonyms: Neurokinin A (4-10), (Ala^5,β-Ala^8)-; Neurokinin A (4-10), (Ala^5,β-Ala^8)-α-; Neurokinin (4-10), (Ala^5,β-Ala^8)-α-; Neurokinin II Receptor Agonist

American Peptide 62-1-45 MW 765.0 $C_{35}H_{56}N_8O_9S$
Potent & selective NK$_2$ tachykinin receptor agonist

ICN 159978 MW 780.9 $C_{35}H_{56}N_8O_9S$ 97% | NK$_2$ receptor agonist; Evangelista, S etal, *Peptides*, 11:293, 1990

Sigma N 6144 FW 764.9 $C_{35}H_{56}N_8O_9S$ ≥97% (HPLC) | Bioactive peptide; anti-ulcer activity; Evangelista, S et al, *Peptides*, 11:293, 1990

Asp-Ala-Pro-Ala-Ala-Pro-Ala-Gly-Pro-Ala-Val-Pro-Val

Synonyms: Sheet Breaker Peptide, α-

Bachem H-5082.0001 MW 1132.28 $C_{51}H_{81}N_{13}O_{16}$ Store at -15°C

Asp-Ala-Ser-Gly-Glu

Bachem H-2025.0100 MW 477.43 $C_{17}H_{27}N_5O_{11}$ Store at -15°C

Asp-Ala-Ser-Phe-His-Ser-Trp-Gly Amide

Synonyms: Leukokinin IV

American Peptide 33-1-21 MW 904.9 $C_{41}H_{52}N_{12}O_{12}$
Holman, GM et al, *Comp Biochem Physiol*, 84C:271, 1986

Bachem H-9245.0001 MW 904.94 $C_{41}H_{52}N_{12}O_{12}$ Store at -15°C

ICN 154478 MW 905

Asp-Ala-Val-Tyr-Ile-His-Pro-Phe-His-Leu

Synonyms: Angiotensin I

Calbiochem 05-23-0100 MW 1296.5 Human; $C_{62}H_{89}N_{17}O_{14}$
≥99% (HPLC); lyophilized; soluble in 5% HOAc; LD$_{50}$≤2000 mg/kg; may be carcinogenic/teratogenic | Precursor of angiotensin II; produced by renin cleavage of angiotensinogen in the plasma; *Merck Index*, 12:689

Asp-Als-Glu-Phe-Arg-His-Asp-Ser-Gly-Tyr-Glu-Val-His-His-Gln-Lys-Leu-Val-Phe-Phe-Ala-Glu-Asp-Val-Gly-Ser-Asn-Lys-Gly-Ala-Ile-Ile-Gly-Leu-Met-Val-Gly-Gly-Val-Val Free Base

Synonyms: Amyloid β-Protein (1-40)

Sigma A-189 Synthetic MW 4329.86 $C_{194}H_{295}N_{53}O_{58}S$
Lyophilized; peptide content & salt form information are provided with each lot | Major constituent of senile plaques & neurofibrillary tangles that occur in the hippocampus, neocortex & amygdala of patients with Alzheimer's disease; Emre et al, *Neurobiol Aging*, 13:553, 1992

Asp-Arg
Asp-Arg-βNA

Bachem K-1105.0250 MW 414.46 $C_{20}H_{26}N_6O_4$ Store at -15°C

Asp-Arg-Asn-Phe-Leu-Arg-Phe Amide

Bachem H-1362.0005 MW 966.11 $C_{44}H_{67}N_{15}O_{10}$ Store at -15°C

Asp-Arg-Asn-Phe-Leu-Arg-Phe Amide
Trifluoroacetate Salt Free Base

Synonyms: FMRF Related Peptide

Sigma A 8701 Crayfish pericardial organs FW 966.1 $C_{44}H_{67}N_{15}O_{10}$ ≥95% (HPLC) | Bioactive peptide; cardioexcitatory; augments neuromuscular synaptic transmission; Mercier, LJ et al, *Peptides*, 14:137, 1993

Asp-Arg-Gly-Asp-Ser
Ac-Asp-Arg-Gly-Asp-Ser

Bachem H-3528.0005 Store at -15°C

Asp-Arg-Gly-Asp-Ser

Bachem H-3532.0005 Store at -15°C

Asp-Arg-Gly-Asp-Ser
Trimesyl(-Asp-Arg-Gly-Asp-Ser)₃

Synonyms: Trimesyl(DRGDS)₃

Bachem H-3538.0005 Store at -15°C

Asp-Arg-Leu-Asp-Ser
Ac-Asp-Arg-Leu-Asp-Ser

Bachem H-3534.0005 Store at -15°C

Asp-Arg-Leu-Asp-Ser

Bachem H-3536.0005 Store at -15°C

Asp-Arg-Leu-Tyr-Ser-Phe-Gly-Leu Amide

Synonyms: Allatostatin IV, Type A; Allatostatin IV

American Peptide 87-7-40 MW 969.1 $C_{45}H_{68}N_{12}O_{12}$ Woodhead, AP et al, *PNAS*, 86:5997, 1989

Bachem H-8080.0001 MW 969.11 $C_{45}H_{68}N_{12}O_{12}$ Store at -15°C

Sigma A 9554 FW 969.1 $C_{45}H_{68}N_{12}O_{12}$ ≥97% (HPLC) | Bioactive peptide

Asp-Arg-Pro-Glu-Gly-Ile-Glu-Glu-Glu-Gly-Gly-Glu-Arg-Asp-Arg-Asp-Arg-Ser

Synonyms: GP41 (733-750); HTLV I Peptide

Neosystem SC241 MW 2102.10 HTLV I peptide from subtype GP41

Asp-Arg-Val-Tyr

Synonyms: Angiotensin II (1-14)

ICN 159876 MW 551.6 98% | Goetzl, EJ etal, *BBRC*, 97:1097, 1980

American Peptide 12-1-37 Human MW 551.6 $C_{24}H_{37}N_7O_8$ Goetzl, EJ et al, *BBRC*, 97:1097, 1980

Asp-Arg-Val-Tyr-Ile

Synonyms: Angiotensin I/II (1-5)

Bachem H-2878.0005 MW 664.76 $C_{30}H_{48}N_8O_9$ Store at -15°C

Asp-Arg-Val-Tyr-Ile-Amino-Phe-Pro-Phe
Asp-Arg-Val-Tyr-Ile-*p*-Amino-Phe-Pro-Phe

Synonyms: Angiotensin II, (*p*-Amino-Phe⁶)-; Angiotensin II Agonist

Sigma A 1811 FW 1071.2 ≥97% (HPLC) | Bioactive peptide; angiotensin II agonist with high binding affinity for the AT₂ receptor subtype; Speth, RC & Kim, KH, *Biochem Biophys Res Commun*, 169:997, 1990

Asp-Arg-Val-Tyr-Ile-His

Synonyms: Angiotensin I/II (1-6)

Bachem H-2882.0005 MW 801.9 $C_{36}H_{55}N_{11}O_{10}$ Store at -15°C

Asp-Arg-Val-Tyr-Ile-His-Ala
Asp-Arg-Val-Tyr-Ile-His-D-Ala

Synonyms: Angiotensin Antagonist (1-7); A-779; Angiotensin I/II (1-7), (D-Ala⁷)-

American Peptide 12-7-26 MW 873 $C_{39}H_{60}N_{12}O_{11}$ Potent & selective; antagonizes antidiuretic effect & changes in MAP produced by Ang-(1-7) microinjection into the dorsomedial or ventrolateral medulla; Santos, RAS et al, *Hypertension*, 27(4):875, 1996

Bachem H-2888.0005 MW 872.98 $C_{39}H_{60}N_{12}O_{11}$ Store at -15°C

Asp-Arg-Val-Tyr-Ile-His-Pro

Synonyms: Angiotensin I/II (1-7)

Bachem H-1715.0005 MW 899.02 $C_{41}H_{62}N_{12}O_{11}$ Store at -15°C

Asp-Arg-Val-Tyr-Ile-His-Pro
Biotinyl-Asp-Arg-Val-Tyr-Ile-His-Pro

Synonyms: Angiotensin I/II (1-7), Biotinyl-

Bachem H-4046.0005 MW 1125.32 $C_{51}H_{76}N_{14}O_{13}S$ Store at -15°C

Asp-Arg-Val-Tyr-Ile-His-Pro

Synonyms: Angiotensin II (1-7); Angiotensin I/II (1-7); Angiotensin (1-7)

ICN 194132 MW 899.0 95% | Osei, SY etal, *Eur J Pharmacol*, 234:35, 1993

Neosystem SC374 MW 899.01 Human Bioactive; Schiavone MT et al, *PNAS USA*, 85:4095, 1988

Peptides International PAN-4332-v Human synthetic MW 899.02 $C_{41}H_{62}N_{12}O_{11}$ >99% (HPLC); lyophilized amorphous powder | Bioactive; canine, rat; Schiavone, MT et al, *PNAS USA*, 85:4095, 1988; Ferrario, CM et al, *Hypertension*, 18(III):III-126, 1991; Santos, RAS et al, *Hypertension*, 19(II):II-56, 1992; DelliPizzi, A et al, *Br J Pharmacol*, 111:1, 1994

Asp-Arg-Val-Tyr-Ile-His-Pro Acetate Salt Free Base

Synonyms: Angiotensin (1-7)

Sigma A 9202 Human FW 899.0 $C_{41}H_{62}N_{12}O_{11}$ ≥97% (HPLC) | Bioactive peptide; angiotensin I (AI) metabolite, A(1-7), elicited a concentration-dependent dilator response (ED₅₀≥2 µM) in porcine coronary artery rings which was markedly attenuated by the nitric oxide (NO) synthase inhibitor; Osei, SY et al, *Eur J Pharmacol*, 234:35, 1993

Asp-Arg-Val-Tyr-Ile-His-Pro-Ala

Synonyms: Angiotensin II, (Ala⁸)-

Neosystem SC400 MW 970.09 Human Bioactive

Asp-Arg-Val-Tyr-Ile-His-Pro-Cys-His-Leu-Leu-Tyr-Tyr-Ser

Synonyms: Renin Substrate Tetradecapeptide, (Cys⁸)-

American Peptide 12-2-20 Rat MW 1778.1 $C_{83}H_{118}N_{21}O_{21}S$

Asp-Arg-Val-Tyr-Ile-His-Pro-Ile

Synonyms: Angiotensin II, (Ile[8])-

Neosystem SC401   Human   MW 1012.17   Bioactive

Asp-Arg-Val-Tyr-Ile-His-Pro-Leu-His-Leu-Leu-Tyr-Tyr-Ser

Synonyms: Renin Substrate Tetradecapeptide, (Leu[8])-

American Peptide 12-2-30   Rat   MW 1789.1 $C_{86}H_{125}N_{21}O_{21}$

Asp-Arg-Val-Tyr-Ile-His-Pro-Phe

Synonyms: Angiotensin II

Bachem H-1705.0005   MW 1046.19 $C_{50}H_{71}N_{13}O_{12}$ Store at -15°C

Asp-Arg-Val-Tyr-Ile-His-Pro-Phe
Asp-Arg-Val-3,5-diiodo-Tyr-Ile-His-Pro-Phe

Synonyms: Angiotensin II, (3,5-Diiodo-Tyr[4])-

Bachem H-2886.0001   MW 1297.99 $C_{50}H_{69}I_{2}N_{13}O_{12}$ Store at -15°C

Asp-Arg-Val-Tyr-Ile-His-Pro-Phe
β-Asp-Arg-Val-Tyr-Ile-His-Pro-Phe Acetate Salt

Synonyms: Angiotensin II, (β-Asp[1])-

ICN 194130   MW 1046.2   97%

Asp-Arg-Val-Tyr-Ile-His-Pro-Phe

Synonyms: Nitric Oxide Synthase Inhibitor; 2-Iminobiotin; Guanidinobiotin, Hexahydro-2-Imino-1H-Thienol(3,4-d)Imidazole-4-Pentanoic Acid; Angiotensin II

ICN 195057   MW 243.3 $C_{10}H_{17}N_3O_2S$ ≥98% | Heney, G et al, *Anal Biochem*, 114:92, 1981

Neosystem SC058   MW 1046.19   Human   Bioactive; Ramsey, DJ, *Neuroscience*, 4:313-321, 1979

Asp-Arg-Val-Tyr-Ile-His-Pro-Phe
β-Asp-Arg-Val-Tyr-Ile-His-Pro-Phe Acetate Salt

Synonyms: Angiotensin II, (β-Asp[1])-

Sigma A 9410   FW 1046.2 (free base)   ≥97% (HPLC) | Bioactive peptide; Rioux, F et al, *Can J Physiol Pharmacol*, 53:383, 1975

Asp-Arg-Val-Tyr-Ile-His-Pro-Phe

Synonyms: Angiotensin II

American Peptide 12-1-34   Human   MW 1046.3 $C_{50}H_{71}N_{13}O_{12}$ Inagami, T et al, *Cardiovase Drugs Ther*, 2:453, 1988

Calbiochem 05-23-0101   Human   MW 1046.2 $C_{50}H_{71}N_{13}O_{12}$ ≥98% (HPLC); lyophilized; soluble in 5% HOAc; LD$_{50}$≤2000 mg/kg; may be carcinogenic/teratogenic | Functions in blood pressure maintenance; stimulates release of aldosterone from the adrenal gland; strong vasoconstrictive effects; increases entry of Ca^{2+} in heart muscle via voltage-sensitive channels & activates myosin light chain kinase; activates JAK2 in smooth muscle cells; activates p125[FAK] & a cytosolic 115-120 kDa calcium-dependent tyrosine kinase in rat epithelial cells also activates p60c-*src* in vascular smooth muscle cells; inhibits adenylate cyclase activity in spontaneously hypertensive rats; *Merck Index*, 12:689; Mazzolai, L et al, *Hypertension*, 31:1324, 1998; Earp, HS et al, *J Biol Chem*, 270:28440, 1995; Ishida, M et al, *Circ Res*, 77:1053, 1995; Vyas, SJ & Jackson, EK, *J Pharmacol Exp Ther*, 273:768, 1995

Asp-Arg-Val-Tyr-Ile-His-Pro-Phe
Asp-Arg-Val-Tyr(PO$_3$H$_2$)-Ile-His-Pro-Phe

Synonyms: Angiotensin II, (Tyr(PO$_3$H$_2$))[4]

Calbiochem 05-23-0111   Human   MW 1126.2 $C_{50}H_{72}N_{13}O_{15}P$ ≥97% (HPLC); lyophilized; soluble in 5% HOAc; LD$_{50}$≤2000 mg/kg; may be carcinogenic/teratogenic | Inhibits renin release in a mean arterial pressure assay with similar potency to angiotensin II amide; has a longer duration of action which is attributed to either stronger binding to the angiotensin II receptor or slower degradation by recirculating peptidases; Kitas, EA et al, *Peptide Res*, 6:205, 1993; Shin, YA & Yoo, SE, *Biopolymers*, 38:183, 1996

Asp-Arg-Val-Tyr-Ile-His-Pro-Phe

Synonyms: Angiotensin II; Hypertensin II

Fluka 10383   Human   MW 1184.38 $C_{50}H_{71}N_{13}O_{12} \cdot$ 2CH$_3$COOH •H$_2$O   ≥98% (HPLC) | Inagami, T et al, *Cardiovasc Drugs Ther*, 2:453, 1988; Ganong, WF, *Horm Res*, 31:24, 1989

Peptides International PAN-4001-v   Human synthetic   MW 1046.2 $C_{50}H_{71}N_{13}O_{12}$ >99% (HPLC); lyophilized amorphous powder

Asp-Arg-Val-Tyr-Ile-His-Pro-Phe Acetate Salt

Synonyms: Angiotensin II

ICN 195051   Human synthetic   MW 1166.4 $C_{50}H_{71}N_{13}O_{12} \cdot$ 2C$_2$H$_4$O$_2$

Asp-Arg-Val-Tyr-Ile-His-Pro-Phe Acetate Salt Free Base

Synonyms: Angiotensin II

Sigma A 9525   Human   FW 1046.2   ≥97% (HPLC), | Bioactive peptide; putative neurotransmitter; hormone involved in regulation of fluid volume & which induces release of aldosterone; in nature, produced by the action of angiotensin converting enzyme on angiotensin I, the C-terminal –His-Leu is cleaved; Inagami, T et al, *Cardiovasc Drugs Ther*, 2:453, 1988 (review)

Asp-Arg-Val-Tyr-Ile-His-Pro-Phe AcOH 4H$_2$O

Synonyms: Angiotensin II

Peptides International PAN-4001   Human synthetic   MW 1178.3 $C_{50}H_{71}N_{13}O_{12} \cdot$ CH$_3$COOH $\cdot$ 4H$_2$O   >99% (HPLC); lyophilized amorphous powder

Asp-Arg-Val-Tyr-Ile-His-Pro-Phe-His

Synonyms: Angiotensin I (1-9)

Bachem H-5038.0005   MW 1183.34 $C_{56}H_{78}N_{16}O_{13}$ Store at -15°C

Asp-Arg-Val-Tyr-Ile-His-Pro-Phe-His-Leu

Synonyms: Hypertensin I; Angiotensin I, ((^{125}I)-Tyr[4])-

Bachem H-1680.0005   MW 1296.51 $C_{62}H_{89}N_{17}O_{14}$ Store at -15°C

Neosystem SC057   MW 1296.49   Human   Bioactive

American Peptide 12-1-10   Human   MW 1296.5 $C_{62}H_{89}N_{17}O_{14}$ Felix, D et al, *Hypertension*, 6:111, 1991

Fluka 10382   Human   MW 1524.78 $C_{62}H_{89}N_{17}O_{14} \cdot$ 2CH$_3$COOH •H$_2$O   ≥99% (HPLC) | Precursor of angiotensin II, produced by the action of renin on angiotensinogen; Negro-Vilar, A et al, *Ann NY Acad Sci*, 512:218, 1987

ICN 68130   Human   ~2000 Ci/mmol, ~74 TBq/mmol; lyophilized from 0.1M NaPO$_4$, 5% BSA

Biogenesis 0560-0109   Human synthetic   MW 1296.7   99.3% (HPLC), >99% (TLC); lyophilized

Peptides International PAN-4007-v   Human synthetic   MW 1296.5 $C_{62}H_{89}N_{17}O_{14}$ >99% (HPLC); lyophilized amorphous powder | Bioactive; porcine, canine, rat, rabbit, guinea pig; Page IH & FM Bumpus (Eds), *Angiotensin*, Handbook of Experimental Pharmacology, Vol 37, Springer-Verlag, Berlin, 1974; Peach, MJ, *Physiol Rev*, 57:313, 1997

Asp-Arg-Val-Tyr-Ile-His-Pro-Phe-His-Leu 2AcOH 4H₂O

Synonyms: Angiotensin I

Peptides International PAN-4007   Human synthetic   MW 1488.6 $C_{62}H_{89}N_{17}O_{14} \cdot 2CH_3COOH \cdot 4H_2O$ >99% (HPLC); lyophilized amorphous powder | Bioactive; porcine, canine, rat, rabbit, guinea pig; Arakawa, K et al, *Biochem J*, 104:900, 1967; Akagi, H et al, *Chem Pharm Bull*, 30:2498, 1982

Asp-Arg-Val-Tyr-Ile-His-Pro-Phe-His-Leu Acetate Salt

Synonyms: Angiotensin I; Hypertensin I

ICN 152724   Human   MW 1416.8 $C_{62}H_{89}N_{17}O_{14} \cdot 2C_2H_4O_2$

Asp-Arg-Val-Tyr-Ile-His-Pro-Phe-His-Leu Acetate Salt Free Base

Synonyms: Angiotensin I; Hypertensin I

Sigma A 9650   Human   FW 1296.5   ≥97% (HPLC) | Bioactive peptide; putative neurotransmitter; the precursor of angiotensin II, the hormone involved in regulation of fluid volume & which induces release of aldosterone; angiotensin analogs, related enzyme substrates & inhibitors & renin, angiotensinogen & angiotensin converting enzymes comprise the major components of the renin-angiotensin system of renal hypertension; Cushman, DW & Cheung, HS, *Biochem Pharm*, 20:1637, 1971

Asp-Arg-Val-Tyr-Ile-His-Pro-Phe-His-Leu-Leu-Tyr-Tyr-Ser

Synonyms: Renin Substrate Tetradecapeptide; Angiotensinogen (1-14); Pre-Angiotensinogen (1-14)

American Peptide 12-2-10   Rat   MW 1823.1 $C_{89}H_{123}N_{21}O_{21}$ Bouhnik, J et al, *Biochemistry*, 20:7010, 1981

Bachem M-2150.0001   Rat   MW 1823.08 $C_{89}H_{123}N_{21}O_{21}$ Store at -15°C

Asp-Arg-Val-Tyr-Ile-His-Pro-Phe-His-Leu-Leu-Val-Tyr-Ser

Synonyms: Angiotensinogen (1-14)

Neosystem SC424   MW 1759.03   Porcine   Bioactive

Asp-Arg-Val-Tyr-Ile-His-Pro-Phe-His-Leu-Leu-Val-Tyr-Ser
Ac-Asp-Arg-Val-Tyr-Ile-His-Pro-Phe-His-Leu-Leu-Val-Tyr-Ser

Synonyms: Renin Substrate Tetradecapeptide, Ac-; Angiotensinogen (1-14), Ac-; Pre-Angiotensinogen (1-14), Ac-

Bachem M-1025.0001   Horse   Store at -15°C

Asp-Arg-Val-Tyr-Ile-His-Pro-Phe-His-Leu-Leu-Val-Tyr-Ser

Synonyms: Renin Substrate Tetradecapeptide; Angiotensinogen (1-14); Pre-Angiotensinogen (1-14); Renin Substrate; Cathepsin D Substrate

Bachem M-1115.0005   Horse   Store at -15°C

American Peptide 12-1-67   Porcine   MW 1759.1 $C_{85}H_{123}N_{21}O_{20}$ Lyophilized | Split by renin to generate angiotensin I; readily cleaved by human cathepsin D; Kageyama, R et al, *Biochemistry*, 23:3603, 1984

ICN 152751   Porcine   MW 1759   Skeggs, LT etal, *J Exp Med*, 128:131, 1968; Galen, FX etal, *BBA*, 523:485, 1978

Asp-Arg-Val-Tyr-Ile-His-Pro-Phe-His-Leu-Leu-Val-Tyr-Ser
Ac-Asp-Arg-Val-Tyr-Ile-His-Pro-Phe-His-Leu-Leu-Val-Tyr-Ser

Synonyms: Angiotensinogen (1-14), (N-Ac)-; Renin Substrate Tetradecapeptide, N-Ac-

ICN 152754   Porcine   MW 1801.1

Sigma R 5380   Porcine   FW 1801.1   ≥97% (HPLC) | Bioactive peptide

Asp-Arg-Val-Tyr-Ile-His-Pro-Phe-His-Leu-Leu-Val-Tyr-Ser Acetate Salt Free Base

Synonyms: Angiotensinogen (1-14); Renin Substrate Tetradecapeptide

Sigma R 8380   Porcine   FW 1759.0   ≥97% (HPLC) | Bioactive peptide; similar to Sigma R 8129, but prepared by Sigma; Skeggs, LT et al, *J Exp Med*, 128:131, 1968; Galen, FX et al, *Biochim Biophys Acta*, 523:485, 1978

Asp-Arg-Val-Tyr-Ile-His-Pro-Phe-His-Leu-Val-Ile-His

Synonyms: Renin Substrate; Angiotensinogen (1-13); Pre-Angiotensinogen (1-13)

Peptides International SDH-4133-v   MW 1645.9 $C_{79}H_{116}N_{22}O_{17}$ >98% (HPLC); lyophilized amorphous powder | Tewksbury, DA et al, *BBRC*, 99:1311, 1981

American Peptide 12-1-71   Human   MW 1645.9 $C_{79}H_{116}N_{22}O_{17}$ Teksbury, D et al, *BBRC*, 99:1311, 1981

Bachem M-2230.0001   Human   Store at -15°C

ICN 159880   Human   MW 1645.9   98% | Tewksbury, DA etal, *BBRC*, 99:1311, 1981

Sigma A 4057   Human   FW 1645.9   ≥97% (HPLC) | Bioactive peptide

Asp-Arg-Val-Tyr-Ile-His-Pro-Phe-His-Leu-Val-Ile-His-Asn

Synonyms: Angiotensinogen (1-14); Renin Substrate

American Peptide 12-1-77   Human   MW 1760.1 $C_{83}H_{122}N_{24}O_{19}$ Kageyama, R et al, *Biochemistry*, 23:3603, 1984

Asp-Arg-Val-Tyr-Ile-His-Pro-Phe-His-Leu-Val-Ile-His-Asn
Ac-Asp-Arg-Val-Tyr-Ile-His-Pro-Phe-His-Leu-Val-Ile-His-Asn

Synonyms: Renin Substrate Tetradecapeptide, Ac-; Angiotensinogen (1-14), Ac-; Pre-Angiotensinogen (1-14), Ac-

Bachem M-1030.0001   Human   Store at -15°C

Asp-Arg-Val-Tyr-Ile-His-Pro-Phe-His-Leu-Val-Ile-His-Asn

Synonyms: Angiotensinogen (1-14); Pre-Angiotensinogen (1-14); Renin Substrate Tetradecapeptide

Bachem M-1120.0001   Human   MW 1760.03 $C_{83}H_{122}N_{24}O_{19}$ Store at -15°C

ICN 152750   Human   MW 1760   Kageyama, R etal, *Biochemistry*, 23:3603, 1984

Asp-Arg-Val-Tyr-Ile-His-Pro-Phe-His-Leu-Val-Ile-His-Asn
Ac-Asp-Arg-Val-Tyr-Ile-His-Pro-Phe-His-Leu-Val-Ile-His-Asn

Synonyms: Renin Substrate Tetradecapeptide, N-Ac-

Sigma R 5755   Human   FW 1802.1   ≥90% (HPLC) | Bioactive peptide

Asp-Arg-Val-Tyr-Ile-His-Pro-Phe-His-Leu-Val-Ile-His-Asn

Synonyms: Angiotensinogen (1-14); Renin Substrate Tetradecapeptide

Sigma R 5880   Human   FW 1760.0   ≥95% (HPLC) | Bioactive peptide

Asp-Arg-Val-Tyr-Ile-His-Pro-Thr

Synonyms: Angiotensin II, (Thr⁸)-

Neosystem SC402   MW 1000.12   Human   Bioactive

Asp-Arg-Val-Tyr-Ile-His-Pro-Val-His-Leu-Leu-Tyr-Tyr-Ser

Synonyms: Renin Substrate Tetradecapeptide, (Val[8])-

American Peptide 12-2-40   Rat   MW 1775.1 $C_{85}H_{123}N_{21}O_{21}$

Asp-Arg-Val-Tyr-Ile-*p*-Amino-Phe-Pro-Phe
Asp-Arg-Val-Tyr-Ile-*p*-Amino-Phe-Pro-Phe

Synonyms: Angiotensin II, (*p*-Amino-Phe[6])-

Neosystem SC450   MW 1071.23   Human   Bioactive; Angiotensin II agonist with a high selectivity & affinity for AII-β receptor subtype; Speth, RC & Kim, K, *BBRC*, 169:997-1006, 1990

Asp-Arg-Val-Tyr-Ile-Phe-Pro-Phe
Asp-Arg-Val-Tyr-Ile-*p*-Amino-Phe-Pro-Phe

Synonyms: Angiotensin II, (*p*-Amino-Phe[6])-

Bachem H-1022.0005   MW 1071.24 $C_{53}H_{74}N_{12}O_{12}$ Store at -15°C

ICN 194129   MW 1071.2   97% | Speth, RC & KH Kim, *BBRC*, 169:997, 1990

Asp-Arg-Val-Tyr-Val-His-Pro-Phe

Synonyms: Angiotensin II, (Val[5])-

Bachem H-1750.0005   MW 1032.17 $C_{49}H_{69}N_{13}O_{12}$ Store at -15°C

ICN 152730

Asp-Arg-Val-Tyr-Val-His-Pro-Phe
β-Asp-Arg-Val-Tyr-Val-His-Pro-Phe Acetate Salt

Synonyms: Angiotensin II, (Val[5])-

ICN 194131   MW 1032.2   97%

Asp-Arg-Val-Tyr-Val-His-Pro-Phe

Synonyms: Angiotensin II, (Val[5])-

American Peptide 12-1-41   Human   MW 1032.2 $C_{46}H_{69}N_{13}O_{12}$

Calbiochem 05-23-0106   Human   MW 1032.2 $C_{49}H_{69}N_{13}O_{12}$ ≥97% (HPLC); lyophilized; soluble in 5% HOAc; LD$_{50}$≤2000 mg/kg; may be carcinogenic/teratogenic | Modulates progesterone & prostaglandin F$_{2\alpha}$; Bramucci, M et al, *Am J Physiol*, 273:2089, 1997; Zou. LX et al, *Hypertension*, 27:658, 1996

Peptides International PAN-4034-v   Synthetic   MW 1032.2 $C_{49}H_{69}N_{13}O_{12}$ >99% (HPLC); lyophilized amorphous powder

Asp-Arg-Val-Tyr-Val-His-Pro-Phe Acetate Salt Free Base

Synonyms: Angiotensin II, (Val[5])-

Sigma A 2900   FW 1032.2   ≥97% (HPLC) | Bioactive peptide

Asp-Arg-Val-Tyr-Val-His-Pro-Phe AcOH 4H$_2$O

Synonyms: Angiotensin II, (Val[5])-

Peptides International PAN-4034   Synthetic   MW 1164.3 $C_{49}H_{69}N_{13}O_{12} \cdot CH_3COOH \cdot 4H_2O$ >99% (HPLC); lyophilized amorphous powder | Bioactive; Elliot, DF & WS Peart, *Nature*, 177:527, 1956; Akagi, H et al, *Chem Pharm Bull*, 30:2498, 1982

Asp-Arg-Val-Tyr-Val-His-Pro-Phe-Asn-Leu

Synonyms: Angiotensin I, (Val[5],Asn[9])-

Bachem H-1695.0001   Bullfrog   MW 1259.43 $C_{59}H_{86}N_{16}O_{15}$ Store at -15°C

ICN 152725   Bullfrog   MW 1259.4   Hasegawa, Y etal, *Gen Comp Endocrinol*, 50:75, 1983

Sigma A 3053   Bullfrog   FW 1259.4   ≥97% (HPLC) | Bioactive peptide; Hasegawa, Y et al, *Gen Comp Endocrinol*, 50:75, 1983; see Sigma A 9650

Asp-Arg-Val-Tyr-Val-His-Pro-Phe-His-Leu

Synonyms: Angiotensin I, (Val[5])-

ICN 152728   MW 1282.5

Peptides International PAN-4069-v   Bovine synthetic   MW 1282.5 $C_{61}H_{87}N_{17}O_{14}$ >99% (HPLC); lyophilized amorphous powder | Bioactive; Takai, M et al, *Peptide Chem*, 1979:187, 1980

ICN 152726   Goosefish   MW 1281.5   Hayashi, T etal, *Chem Pharm Bull*, 26:215, 1978

American Peptide 12-1-27   Human   MW 1282.5 $C_{61}H_{87}N_{17}O_{14}$

Sigma A 9402   Human   FW 1282.5   ≥97% (HPLC) | Bioactive peptide; see Sigma A 9650

Asp-Arg-Val-Tyr-Val-His-Pro-Phe-His-Leu AcOH 5H$_2$O

Synonyms: Angiotensin I, (Val[5])-

Peptides International PAN-4069   Bovine synthetic   MW 1432.6 $C_{61}H_{87}N_{17}O_{14} \cdot CH_3COOH \cdot 5H_2O$ >99% (HPLC); lyophilized amorphous powder | Bioactive; Takai, M et al, *Peptide Chem*, 1979:187, 1980

Asp-Arg-Val-Tyr-Val-His-Pro-Phe-Ser-Leu

Synonyms: Angiotensin I, (Val[5],Ser[9])-

ICN 159875   MW 1232.6   98%

Asp-Asn-Gln

Bachem H-1176.0050   Store at -15°C

Asp-Asn-Pro-Ser-Leu-Ser-Ile-Asp-Leu-Thr-Phe-His-Leu-Leu-Arg-Thr-Leu-Leu-Glu-Leu-Ala-Arg-Thr-Gln-Ser-Gln-Arg-Glu-Arg-Ala-Glu-Gln-Asn-Arg-Ile-Ile-Phe-Asp-Ser-Val Amide

Synonyms: Corticotropin Releasing Factor; Urocortin; Corticotropin Releasing Factor Receptor Ligand, Type-2

American Peptide 34-7-16   Human   MW 4697.3 $C_{204}H_{336}N_{62}O_{65}$ Binds with high affinity to CRF receptor types 1, 2a & 2b as well as stimulates cAMP accumulation from cells transfected with those receptors; Donaldson, CJ et al, *Endocrinology*, 137:2167, 1996

Bachem H-3722.0500   Human   MW 4696.31 $C_{204}H_{337}N_{63}O_{64}$ Store at -15°C

Neosystem SC1218   Human   MW 4696.29   Bioactive; hypothalamic releasing hormone; CRF/urocortin peptide; Donaldson, CJ et al, *Endocrinology*, 137:2167-2170, 1996

Neosystem SC1218   Human   MW 4696.29   Bioactive neuropeptide; Donaldson, CJ et al, *Endocrinology*, 137:2167-2170, 1996

Sigma U 4127   Human   FW 4696.3   ≥90% (HPLC) | Bioactive peptide; binds CRF receptor types 1, 2α & 2β; stimulates cAMP accumulation from cells transfected with these receptors; acts *in vitro* to release ACTH from rat anterior pituitary cells; Donaldson, CJ et al, *Endocrinology*, 137:2167, 1996

Peptides International PUC-4328-s   Human synthetic   MW 4696.3 $C_{204}H_{337}N_{63}O_{64}$ >99% (HPLC); lyophilized amorphous powder | Bioactive; Donaldson, CJ et al, *Endocrinology*, 137:2167, 1996; Behan, DP et al, *Brain Res*, 725:263, 1996; Murakami, Y et al, *Endocr J*, 44:627, 1997; Takahashi, K et al, *Peptides*, 19:643, 1998

American Peptide 34-7-17   Rat   MW 4707.4 $C_{206}H_{338}N_{62}O_{64}$ More potent than CRF at binding & activating type-2 CRF receptors; shares ~95% sequence with human version; Vaughan, J et al, *Nature*, 378:287, 1995

Asp-Asp
Asp-FMOC-Asp Amide

Synonyms: Ias, FMOC-; Ias, FMOC-L-

Bachem B-1985.0001   MW 354.36 $C_{19}H_{18}N_2O_5$ Store at RT

Asp-Asp
Asp-FMOC-D-Asp Amide

Synonyms: Ias, FMOC-D-

Bachem B-2420.0001   MW 354.36 $C_{19}H_{18}N_2O_5$ Store at RT

Asp-Asp

Bachem G-1565.0250   MW 248.19 $C_8H_{12}N_2O_7$ Store at -15°C

Sigma A 6416   FW 248.2 $C_8H_{12}N_2O_7$

Asp-Asp-Ala-Ser-Asp-Arg-Ala-Lys-Lys-Phe-Tyr-Gly-Leu-Met Amide

Synonyms: Ranamargarin

Bachem H-9335.0001   Store at -15°C

ICN 154460   MW 1616   Tang, YQ etal, *Regulatory Peptides*, 22:182, 1988

Asp-Asp-Ala-Val-Tyr-Leu-Asp-Asn-Glu-Lys-Glu-Arg-Glu-Glu-Tyr-Val-Leu-Asn-Asp-Ile-Gly-Val-Ile-Phe-Tyr-Gly-Glu-Val-Asn-Asp-Ile-Lys-Thr-Arg-Ser-Trp-Ser-Tyr-Gly-Gln-Phe

Synonyms: Coagulation Factor XIIIa (190-230)

Bachem H-2994.0500   Store at -15°C

Asp-Asp-Asp

Bachem H-8750.0050   MW 363.28 $C_{12}H_{17}N_3O_{10}$ Store at -15°C

Sigma A 4315   FW 363.3 $C_{12}H_{17}N_3O_{10}$ ≥97% (TLC)

Asp-Asp-Asp-Asp

Bachem H-8755.0050   MW 478.37 $C_{16}H_{22}N_4O_{13}$ Store at -15°C

Sigma A 4440   FW 478.4 $C_{16}H_{22}N_4O_{13}$ ≥97% (TLC)

Asp-Asp-Asp-Asp-Asp

American Peptide 87-0-71   MW 593.4 $C_{20}H_{27}N_5O_{16}$

Sigma A 4690   FW 708.5 $C_{24}H_{32}N_6O_{19}$ ≥97% (HPLC)

Asp-Asp-Asp-Asp-Asp-Asp

American Peptide 87-0-74   MW 708.5 $C_{24}H_{32}N_6O_{19}$

Asp-Asp-Glu-Glu-Ser-Ile-Thr-Arg-Arg

Synonyms: Casein Kinase I

Promega V7441   MW 1235   See Promega V5601

Asp-Asp-Phe-Me-Phe-Gly-Leu-Met Amide

Synonyms: Substance P, (Asp[5,6],Me-Phe[8])-; Senktide Analog (5-11)

American Peptide 70-1-97   MW 858.0 $C_{40}H_{57}N_8O_{11}S$

Asp-Asp-Phe-NMe-Phe-Gly-Leu-Met Amide

Synonyms: Substance P (5-11), (Asp[5,6], MePhe[8])-; Senktide, NH_2-

Neosystem SC1235   MW 856.99   Bioactive; tachykinin; NK-3 receptor agonist; Laufer, R et al, *JBC*, 261:10257-10263, 1986; Polidori, C et al, *Physiology & Behavior*, 56:877-882, 1994; Polidori, C et al, *Regul Peptides*, 66:101-104, 1996

Asp-Asp-Pro-Pro-Leu-Ser-Ile-Asp-Leu-Thr-Phe-His-Leu-Leu-Arg-Thr-Leu-Leu-Glu-Leu-Ala-Arg-Thr-Gln-Ser-Gln-Arg-Glu-Arg-Ala-Glu-Gln-Asn-Arg-Ile-Ile-Phe-Asp-Ser-Val Amide

Synonyms: Urocortin

Bachem H-3362.0500   Rat   MW 4707.33 $C_{206}H_{338}N_{62}O_{64}$ Store at -15°C

Asp-Asp-Pro-Pro-Leu-Ser-Ile-Asp-Leu-Thr-Phe-His-Leu-Leu-Arg-Thr-Leu-Leu-Glu-Leu-Ala-Arg-Thr-Gln-Ser-Gln-Arg-Glu-Arg-Ala-Glu-Gln-Asn-Arg-Ile-Ile-Phe-Asp-Ser-Val Free Acid

Synonyms: Urocortin

Bachem H-3602.0500   Rat   MW 4708.31 $C_{206}H_{337}N_{61}O_{65}$ Store at -15°C

Asp-Asp-Pro-Pro-Leu-Ser-Ile-Asp-Leu-Thr-Phe-His-Leu-Leu-Arg-Thr-Leu-Leu-Glu-Leu-Ala-Arg-Thr-Gln-Ser-Gln-Ser-Gln-Arg-Glu-Arg-Ala-Glu-Gln-Asn-Arg-Ile-Ile-Phe-Asp-Ser-Val Amide Trifluoroacetate Salt

Synonyms: Urocortin

Calbiochem 671400   Rat   MW 4707.4 $C_{206}H_{338}N_{62}O_{64}$ ≥98% (HPLC); solid; soluble in 5% acetic acid | Mammalian neuropeptide related to fish urotensin & to corticotropin-releasing factor; binds to CRF_1 & CRF_2 receptors with high affinity resulting in adenylate cyclase activation; more potent than CRF at binding & activating CRF_2 receptors as well as at inducing *c-fos* in regions enriched in CRF2 receptors; Kozicz, T et al, *J Comp Neurol*, 391:1, 1998; Vaughan, J et al, *Nature*, 378:287, 1995

Asp-Asp-Pro-Pro-Leu-Ser-Ile-Asp-Leu-Thr-Phe-His-Leu-Leu-Arg-Thr-Leu-Leu-Glu-Leu-Ala-Arg-Thr-Gln-Ser-Gln-Arg-Glu-Arg-Ala-Glu-Gln-Asn-Arg-Ile-Ile-Phe-Asp-Ser-Val Amide

Synonyms: Urocortin; Corticotropin Releasing Factor Receptor Ligand, Type-2

Neosystem SC986   Rat   MW 4707.31   Bioactive neuropeptide; hypothalamic releasing hormone; CRF/urocortin peptide; related to fish urotensin I & corticotropin-releasing factor; evokes secretion of ACTH both *in vitro* & *in vivo*; binds & activates transfected type-1 CRF receptors, the subtype expressed by pituitary corticotropes; could be an endogenous ligand for type-2 CRF receptors; Vaughan, J et al, *Nature*, 378:287-292, 1995

Sigma U 6631   Rat   FW 4707.3   ≥97% (HPLC) | Bioactive peptide; causes secretion of ACTH & has hypotensive effects; binds & activates transfected Type-2β CRF receptors (CRF-R2β); Vaughan, J et al, *Nature*, 378:292, 1995

Peptides International PUC-4327-s   Rat synthetic   MW 4707.3 $C_{206}H_{338}N_{62}O_{64}$ >99% (HPLC); lyophilized amorphous powder | Bioactive; Vaughan, J et al, *Nature*, 378:278, 1995; Turnbull, AV et al, *Eur J Pharmacol*, 303:213, 1996; Spina, M et al, *Science*, 273:1561, 1996; Zhao, LY et al, *Genomics*, 50:23, 1998

Asp-Asp-Ser-Leu-Tyr-Pro-Ile-Ala-Val-Leu-Ile-Asp-Glu

Synonyms: Protein Phosphatase 2A/A Blocking Peptide (7-19)

Calbiochem 539519   Synthetic   MW 1462.6 $C_{66}H_{103}N_{13}O_{24}$ Liquid | From PP2A/A regulatory subunit; ideal blocking peptide for Anti-PP2A/A

Asp-Cys-Ala-Trp-Leu-Glu-Ala-Gln-Glu-Glu-Glu-Glu-Val-Gly

Synonyms: NEF MN (56-69); HIV I Peptide

Neosystem SC645   MW 1608.65   NEF peptide from HIV-1 subtype MN; due to the presence of a Cys this peptide can contain the dimeric form

Asp-Cys-Lys-Thr-Ile-Leu-Lys-Ala-Leu-Gly-Pro-Ala-Ala-Thr-Leu-Glu

Synonyms: P25 (329-344); GAG P24 CA (197-212); HTLV I Peptide

Neosystem SC308   MW 1643.95   HTLV I peptide from subtype P25 (GAG P24 CA); due to the presence of a Cys this peptide can contain the dimeric form

Asp-Gln

Bachem G-1570.0250   MW 261.24 $C_9H_{15}N_3O_6$ Store at -15°C

Sigma A 1791 FW 261.2 $C_9H_{15}N_3O_6$

Asp-Gln-Gly-Phe-Asn-Ser-Trp-Gly Amide

Synonyms: Leukokinin III

American Peptide 33-1-19 MW 908.9 $C_{40}H_{52}N_{12}O_{13}$
Holman, GM et al, *Comp Biochem Physiol*, 84C:271, 1986

Bachem H-9240.0001 MW 908.93 $C_{40}H_{52}N_{12}O_{13}$ Store at -15°C

ICN 154477 MW 909

Asp-Gln-Ile-Leu-Ile-Glu-Ile-Cys-Gly-His-Lys-Ala-Ile-Gly-Thr-Val

Synonyms: POL (60-75); PR (60-75); HTLV I Peptide

Neosystem SC705 MW 1710.01 HTLV I peptide from subtype POL (PR/RT); due to the presence of a Cys this peptide can contain the dimeric form

---

Asp-Gln-Lys-Pro-Ile-Phe-Asn-Val-Ile-Pro-Pro-Ile-Pro-Val-Gly-Ser-Glu-Asn-Trp-Asn-Arg-Cys

Synonyms: Glucocorticoid Receptor N-Termnial Peptide (346-367)

Alexis 157-015 Synthetic MW 2523.9 $C_{114}H_{175}N_{31}O_{32}S$
≥95%; lyophilized; reconstitute with 0.1 mL distilled H_2O | Competitively binds to Ab Alexis 210-131; antiserum blocking peptide

---

Asp-Gln-Thr-Asp
Ac-Asp-Gln-Thr-Asp-AMC

Synonyms: Caspase III Substrate; Caspase VII Substrate

Neosystem SC1319 MW 676.64 Deduced from the cleavage site of focal adhesionkinase & gelsolin; Wen, LP et al, *JBC*, 272:26056-26061, 1997; Kothakota, S et al, *Science*, 278:294-298, 1997

Asp-Gln-Thr-Asp
Ac-Asp-Gln-Thr-Asp-MCA

Synonyms: Caspase VII/III Substrate

Peptides International MCA-3193-v MW 676.64
$C_{29}H_{36}N_6O_{13}$ >98% (HPLC); lyophilized amorphous powder | Deduced from the cleavage site of focal adhesion (kinase & gelsolin); Wen, L-P et al, *JBC*, 272:26056, 1997; Kothakota, S et al, *Science*, 278:294, 1997

Asp-Gln-Thr-Asp
Ac-Asp-Gln-Thr-Asp-CHO

Synonyms: Caspase VII/III Inhibitor

Peptides International ICA-3194-v Synthetic MW 503.47
$C_{19}H_{29}N_5O_{11}$ >99% (HPLC); lyophilized amorphous powder | Inhibitor; deduced from the cleavage site of focal adhesionkinase & gelsolin; Wen, L-P et al, *JBC*, 272:26056, 1997; Kothakota, S et al, *Science*, 278:294, 1997

Asp-Glu
N-Ac-Asp-Glu

Synonyms: Spaglumic Acid

Alexis 151-020 MW 304.3 $C_{11}H_{16}N_2O_8$ ≥98%; soluble in H_2O | Endogenous neurotransmitter with high affinity for the brain glutamate receptor; *N*-Acated α-linked acidic dipeptidase (NAALADase), a membrane-bound peptidase, hydrolyzes α-NAAG, a major brain peptide, to *N*-Acaspartate & glutamate; mGluR3 activator; Zaczek et al, *PNAS*, 80:1116, 1983; Blakely, RD & Coyle, JT, *Int Rev Neurobiol*, 30:39, 1988; Jackson, PF et al, *J Med Chem*, 39:619, 1996; Wroblewska, B et al, *J Neurochem*, 69:174, 1997

Asp-Glu
Ac-Asp-Glu

American Peptide 60-0-55 MW 304.3 $C_{11}H_{16}N_2O_8$
Endogenous neurotransmitter with high affinity for brain glutamate; Zaczek, R et al, *PNAS*, 80:1116, 1983

Bachem G-1015.0100 MW 304.26 $C_{11}H_{16}N_2O_8$ Store at -15°C

Asp-Glu

Bachem G-1575.0250 MW 262.22 $C_9H_{14}N_2O_7$ Store at -15°C

Asp-Glu
Ac-β-Asp-Glu

Synonyms: Ac-Asp(Glu-OH)

Bachem G-4590.0100 MW 304.26 $C_{11}H_{16}N_2O_8$ Store at -15°C

Asp-Glu
N-Ac-Asp-Glu

ICN 153036 MW 304.3 $C_{11}H_{16}N_2O_8$ An endogenous neuropeptide with high affinity for a brain "glutamate" receptor; Zaczek, R, etal, *PNAS*, 80, 1116, 1983

Asp-Glu

Sigma A 1916 FW 262.2 $C_9H_{14}N_2O_7$ 95-98% (TLC)

Asp-Glu
N-Ac-Asp-Glu

Sigma A 5930 FW 304.3 $C_{11}H_{16}N_2O_8$ ≥97% (TLC) | Bioactive peptide; endogenous neurotransmitter with high affinity for brain glutamate receptor; Zaczek, R et al, *Proc Natl Acad Sci USA*, 80:1116, 1983

Asp-Glu
N-Ac-β-Asp-Glu

Sigma A 9436 FW 304.3 $C_{11}H_{16}N_2O_8$

Asp-Glu
Ac-Asp-Glu Monohydrate

Synonyms: Endogenous Excitatory Neurotransmitter

Peptides International PDE-4167 Synthetic MW 322.28
$C_{11}H_{16}N_2O_8 \cdot H_2O$ >99% (HPLC); lyophilized amorphous powder | Bioactive; Reichert KL & F Fonnum, *J Neurochem*, 16:1409, 1969; Koller KJ & JT Coyle, *Eur J Pharm*, 98:193, 1984; Koller, KJ etal, *J Neurochem*, 43:1136, 1984

Asp-Glu-Asp-Glu-Glu-Abu-Ala-Ser-Lys
Ac-Asp-Glu-Asp(Edans)-Glu-Glu-Abu-ψ(COO)-Ala-Ser-Lys(DABCYL) Amide

Synonyms: HCV NS3 Protease Fluorogenic Substrate

Neosystem SC1248 MW 1548.49 Derived from the sequence NS4A/NS4B cleavage site; well suited for continuous monitoring of NS3 protease activity; useful in a continuous assay for evaluation of NS3 inhibitors as antivirals; Taliani, M et al, *Anal Biochem*, 240:60-67, 1996

Asp-Glu-Asp-Leu-Leu-Lys-Ala-Val-Arg-Leu-Ile-Lys-Phe-Leu-Tyr-Gln-Ser-Asn

Synonyms: REV (9-26); HTLV I Peptide

Neosystem SC189 MW 2165.51 HTLV I peptide from subtype REV

Asp-Glu-Glu-Ala-Val-Tyr-Phe-Ala-His-Leu-Asp-Ile-Ile-Trp
Suc-Asp-Glu-Glu-Ala-Val-Tyr-Phe-Ala-His-Leu-Asp-Ile-Ile-Trp

Synonyms: Endothelin I (8-21), Suc-(Glu[9],Ala[11,15])-; Endothelin I (8-21), *N*-Suc-(Glu[9],Ala[11,15])-; IRL-1620

Bachem H-1372.0500 MW 1820.97 $C_{86}H_{117}N_{17}O_{27}$ Store at -15°C

Neosystem SC873 MW 1820.96   Bioactive; ETB receptor agonist; Takai, M et al, *BBRC*, 184:953-959, 1992; Sakamoto, A et al, *JBC*, 268:8547-8553, 1993; Karaki, H et al, *Br J Pharmacol*, 109:486-490, 1993; Fujitani, Y et al, *J Pharmacol Exp Ther*, 267:683-689, 1993

Asp-Glu-Glu-Ala-Val-Tyr-Phe-Ala-His-Leu-Asp-Ile-Ile-Trp
N-Suc-Asp-Glu-Glu-Ala-Val-Tyr-Phe-Ala-His-Leu-Asp-Ile-Ile-Trp

Synonyms: Endothelin I (8-21) *N*-Suc-(Glu[9],Ala[11,15])-; Endothelin B Receptor Agonist

Sigma E 4518 FW 1821.0   ≥97% (HPLC) | Bioactive peptide; potent, highly selective; Takai, M et al, *Biochem Biophys Res Commun*, 184:953, 1992

Asp-Glu-Glu-Asp-Asp-Asp-Leu-Val-Gly-Val-Ser-Val-Arg-Pro-Lys

Synonyms: NEF (91-105); SIVmac251 Peptide

Neosystem SC611 MW 1672.76   SIVmac251 peptide from subtype NEF

Asp-Glu-Gly-Pro-Tyr-Arg-Met-Glu-His-Phe-Arg-Trp-Gly-Ser-Pro-Pro-Lys-Asp

Synonyms: Melanocyte Stimulating Hormone, α-; Melanocyte Stimulating Hormone, β-

Bachem H-2030.0001 Monkey   MW 2204.41
$C_{98}H_{138}N_{28}O_{29}S$ Store at -15°C

ICN 151596 Monkey

Asp-Glu-Gly-Pro-Tyr-Lys-Met-Glu-His-Phe-Arg-Trp-Gly-Ser-Pro-Pro-Lys-Asp

Synonyms: Melanocyte Stimulating Hormone, β-; Lipotropin (41-58), β-

American Peptide 56-4-31 Porcine   MW 2176.4
$C_{98}H_{138}N_{26}O_{29}S$ Lemaire, S, *J Med Chem*, 20:155, 1977

Bachem H-2035.0001 Porcine   MW 2176.4 $C_{98}H_{138}N_{26}O_{29}S$
Store at -15°C

ICN 151597 Porcine   MW 2176.4

Asp-Glu-Gly-Pro-Tyr-Lys-Met-Glu-Tyr-Phe-Arg-Trp-Gly-Ser-Pro-Pro-Lys-Asp

Synonyms: Melanocyte Stimulating Hormone, (Tyr[9])-α-; Lipotropin (41-58), (Tyr[49])-α-

Bachem H-2918.0001 Porcine   Store at -15°C

Asp-Glu-His-Asp
Ac-Asp-Glu-His-Asp-AMC

BioSource International 78-103   Fluorescing substrate

Asp-Glu-Leu-Pro-Gln-Leu-Val-Thr-Leu-Pro-His-Pro-Asn-Leu-His-Gly-Pro-Glu-Ile-Leu-Asp-Val-Pro-Ser-Thr

Synon Fibronectin Type III Connecting Segment (1-25)*yms*:

ICN 195621 MW 2732.1   ≥90%

Sigma F 5007 FW 2732.1   ≥90% (HPLC) | Bioactive peptide

Asp-Glu-Pro-Asn-Ser-Asp-Gln-Phe-Ile-Gly-Leu-Met Amide

Synonyms: Entero-Kassinin; Kassinin, Entero-

American Peptide 46-0-10 MW 1364.5 $C_{58}H_{89}N_{15}O_{21}S$
Negri, L et al, *Reg Peptides*, 22:13, 1988

Bachem H-9330.0005   Store at -15°C

ICN 154453 MW 1364.6   Negri, L etal, *Regulatory Peptides*, 22:13, 1988

Asp-Glu-Val-Asp
Ac-Asp-Glu-Val-Asp

Synonyms: Caspase III Control Peptide

Alexis 151-031 MW 518.5 $C_{20}H_{30}N_4O_{12}$ ≥97%; white lyophilized powder; soluble in H_2O

Asp-Glu-Val-Asp
Ac-Asp-Glu-Val-Asp-CHO

Synonyms: Caspase III Inhibitor; Poly ADP-Ribose Polymerase Inhibitor

Alexis 260-030 MW 502.5 $C_{20}H_{30}N_4O_{11}$ ≥97%; white lyophilized powder; soluble in DMSO | Potent, specific & competitive inhibitor of the poly(ADP-ribose) polymerase (PARP) cleavage by Caspase III; PARP cleavage occurs at the onset of apoptosis; attenuates apoptotic events *in vitro*; Schlegel, J et al, *J Biol Chem*, 271:1841, 1996; Enari, M et al, *Nature*, 380:723, 1996; Nicholson, DW et al, *Nature*, 376:37, 1995; Nicholson, DW et al, *Nature Biotech*, 14:297, 1996

Asp-Glu-Val-Asp
Ac-Asp-Glu-Val-Asp-AMC

Synonyms: Caspase III Fluorogenic Substrate

Alexis 260-031 MW 675.7 $C_{30}H_{37}N_5O_{13}$ ≥97%; white lyophilized powder; soluble in DMSO or H_2O | Sequence based on PARP cleavage site Asp[216] for Caspase III; AMC has an excitation maximum of 380 nm & an emission maximum of 460 nm; Nicholson, DW et al, *Nature*, 376:37, 1995

Asp-Glu-Val-Asp
Ac-Asp-Glu-Val-Asp-AFC

Synonyms: Caspase III Fluorogenic Substrate

Alexis 260-032 MW 729.6 $C_{30}H_{34}N_5O_{13}F_3$ ≥95%; Off-white powder; soluble in DMSO, DMF or MeOH | Similar to Caspase III substrate fluorogenic, but the AFC fluorophore has a greater Stokes' shift upon cleavage; Xiang, J et al, *Proc Natl Acad Sci USA*, 93:14559, 1996

Asp-Glu-Val-Asp
Biotinyl-Asp-Glu-Val-Asp-CHO Biotin

Synonyms: Caspase III Inhibitor

Alexis 260-034 MW 686.7 $C_{28}H_{42}N_6O_{12}S$ ≥93%; white lyophilized powder; soluble in 50% HOAc with agitation | Highly specific affinity reagent for Caspase III; useful for purification or extract depletion of Caspase III in conjunction with streptavidin conjugates; Nicholson, DW et al, *Nature*, 376:37, 1995

Asp-Glu-Val-Asp
Ac-Asp-Glu-Val-Asp-pNA

Synonyms: Caspase III Chromogenic Substrate

Alexis 260-048 MW 730.7 $C_{32}H_{38}N_6O_{14}$ ≥97%; white lyophilized powder; soluble in H_2O | Increased cell permeability

---

Asp-Glu-Val-Asp
Ac-Asp-Glu-Val-Asp-AMC

Synonyms: Apopain Substrate

American Peptide 81-6-30 MW 676.7 $C_{30}H_{38}N_5O_{13}$
Nicholson, DW, *Nature*, 376:1995; Xiang, J et al, *PNAS*, 93:14559, 1996

Asp-Glu-Val-Asp
Ac-Asp-Glu-Val-Asp-CHO

Synonyms: CPP-32/Caspase III Inhibitor

American Peptide 81-6-31 MW 502.5 $C_{20}H_{30}N_4O_{11}$ Specific inhibitor of TNF-induced activation of cPLA₂ & apoptosis; Wissing, D et al, *PNAS*, 94(10):5973-5077, 1997

Asp-Glu-Val-Asp
Ac-Asp-Glu-Val-Asp-CMK

American Peptide 81-6-32   MW 551 $C_{21}H_{31}N_4O_{11}Cl$

Asp-Glu-Val-Asp
Ac-Asp-Glu-Val-Asp-pNA

American Peptide 81-6-40   MW 638.6 $C_{26}H_{34}N_6O_{13}$
Nicholson, DW, *Nature*, 376:1995

Asp-Glu-Val-Asp
Z-Asp-Glu-Val-Asp-AFC

Synonyms: Caspase III Substrate IV

American Peptide 81-7-09   MW 821.6 $C_{36}H_{38}N_5O_{14}F_3$
Selective inhibitor of CPP-32-like Caspases; Inayat-Hussain, SH et al, *Hepatology*, 25(6):1516, 1997

Asp-Glu-Val-Asp
Ac-Asp-Glu-Val-Asp-CHO Pseudo Acid

Bachem H-2496.0005   MW 502.48 $C_{20}H_{30}N_4O_{11}$ Store at -15°C

Asp-Glu-Val-Asp
Ac-Asp-Glu-Val-Asp-AMC

Bachem I-1660.0005   MW 675.65 $C_{30}H_{37}N_5O_{13}$ Store at -15°C

Asp-Glu-Val-Asp
Ac-Asp-Glu-Val-Asp-AFC

Bachem I-1725.0005   MW 729.62 $C_{30}H_{34}F_3N_5O_{13}$ Store at -15°C

Asp-Glu-Val-Asp
Ac-Asp-Glu-Val-Asp-βNA

Bachem K-1670.0005   MW 643.65 $C_{30}H_{37}N_5O_{11}$ Store at -15°C

Bachem L-1945.0005   MW 638.59 $C_{26}H_{34}N_6O_{13}$ Store at -15°C

Asp-Glu-Val-Asp
Biotinyl-Asp-Glu-Val-Asp-CHO Pseudo Acid

Bachem N-1470.0005   MW 686.74 $C_{28}H_{42}N_6O_{12}S$ Store at -15°C

Asp-Glu-Val-Asp
Z-Asp-OMe-Glu-OMe-Val-DL-Asp-OMe-FMK

Bachem N-1555.0001   MW 688.67 $C_{30}H_{41}FN_4O_{12}$ Store at -15°C

Asp-Glu-Val-Asp
Z-Asp-Glu-Val-Asp-CMK

Bachem N-1580.0005   MW 643.05 $C_{27}H_{35}ClN_4O_{12}$ Store at -15°C

Asp-Glu-Val-Asp
Z-Asp-Glu-Val-Asp-pNA

BioSource International 77-870

Asp-Glu-Val-Asp
Ac-Asp-Glu-Val-Asp

BioSource International 77-897

Asp-Glu-Val-Asp
Ac-Asp-Glu-Val-Asp-pNA

BioSource International 77-900

Asp-Glu-Val-Asp
Ac-Asp-Glu-Val-Asp-AMC

BioSource International 77-910

Asp-Glu-Val-Asp
Ac-Asp-Glu-Val-Asp-CHO

BioSource International 77-920   Inhibitor

Asp-Glu-Val-Asp
Z-Asp-Glu-Val-Asp-AFC

BioSource International 77-931

Asp-Glu-Val-Asp
Ac-Asp-Glu-Val-Asp-AFC

BioSource International 77-934

Asp-Glu-Val-Asp
Asp-Glu-Val-Asp-CHO

BioSource International 77-950   Cell permeable inhibitor

Asp-Glu-Val-Asp
Biotin-X-Asp(OMe)-Glu(OMe)-Val-Asp(OMe)-CH₂F

Synonyms: Caspase III Inhibitor II, Biotin Conjugate

Calbiochem 218747   MW 873.0 $C_{38}H_{59}FN_7O_{13}S$ ≥95% (TLC); solid; soluble in DMSO; may be carcinogenic/teratogenic | Biotinylated derivative of Caspase III Inhibitor II; esterified to increase cell-permeability; inhibits Caspase VI, Caspase VII & Caspase X; used to purify Caspase III from cell lysates; esterase treatment increases inhibitory activity in cell-free systems; Nicholson, DW et al, *Nature*, 376:37, 1995

Asp-Glu-Val-Asp
Ac-Asp-Glu-Val-Asp-CMK

Synonyms: Caspase III Inhibitor III

Calbiochem 218750   MW 551.0 $C_{21}H_{31}ClN_4O_{11}$ ≥98% (HPLC); solid; soluble in DMSO | Potent, cell-permeable & irreversible inhibitor of Caspases III, VI, VII, VIII & X

Asp-Glu-Val-Asp
Ac-Asp-Glu-Val-Asp-pNA

Synonyms: Caspase III Colorimetric Substrate I; CPP-32/Apopain Colorimetric Substrate

Calbiochem 235400   MW 638.6 $C_{26}H_{34}N_6O_{13}$ ≥97% (HPLC); lyophilized solid; soluble in DMSO | Colorimetric substrate for Caspase III & related cysteine proteases; sequence is based on the P_1-P_4 tetrapeptide cleavage site of poly(ADP-ribose) polymerase (PARP) & includes Asp[216]; Caspase III is the protease responsible for the cleavage of PARP; a substrate for Caspases VI, VII, VIII & X; cleavage of pNA is monitored colorimetrically at ~405 nm; Nicholson, DW et al, *Nature*, 376:37, 1995; Lazebnik, YA et al, *Nature*, 371:346, 1994

Asp-Glu-Val-Asp
Ac-Asp-Glu-Val-Asp-CHO

Synonyms: Caspase III Inhibitor I; CPP-32/Apopain Inhibitor

Calbiochem 235420   MW 502.5 $C_{20}H_{30}N_4O_{11}$ ≥90% (HPLC); solid; soluble in DMSO & water | Very potent, specific & reversible inhibitor of Caspases III, VI, VII, VIII & X; Nicholson, DW et al, *Nature Biotech*, 14:297, 1996; Schlegel, J et al, *J Biol Chem*, 271:1841, 1996; Nicholson, DW et al, *Nature*, 376:37, 1995

Asp-Glu-Val-Asp
Biotin-Asp-Glu-Val-Asp-CHO

Synonyms: Caspase III Inhibitor I, Biotin Conjugate; CPP-32/Apopain Inhibitor, Biotin Conjugate

Calbiochem 235422 MW 686.7 $C_{28}H_{42}N_6O_{12}S$ ≥90% (HPLC); solid; soluble in DMSO & water; may be carcinogenic/teratogenic | Affinity ligand for Caspases III, VI, VII, VIII & X; inhibits Caspase III with similar potency to the parent compound; Nicholson, DW et al, *Nature*, 376:37, 1995

Asp-Glu-Val-Asp
Ac-Asp-Glu-Val-Asp-AMC

Synonyms: Caspase III Fluorogenic Substrate II; CPP-32/Apopain Fluorogenic Substrate

Calbiochem 235425 MW 675.6 $C_{30}H_{37}N_5O_{13}$ ≥95% (HPLC); solid; soluble in DMSO; excitation max:~365-380 nm; emission max:~430-460 nm | Fluorogenic substrate for Caspase III; cleavage of this substrate by Caspase III show Michaelis-Menten kinetics; a substrate for Caspases VI, VII, VIII & X; Nicholson, DW et al, *Nature*, 376:37, 1995; Lazebnik, YA et al, *Nature*, 371:346, 1994

Asp-Glu-Val-Asp
Z-Asp-Glu-Val-Asp-AFC

Synonyms: Caspase III Fluorogenic Substrate IV

Calbiochem 264150 MW 821.7 $C_{36}H_{38}F_3N_5O_{14}$ Single spot purity (TLC); solid; soluble in DMSO; excitation max:~400 nm; emission max:~505 nm | Fluorogenic substrate for Caspase III; reaction monitored visually or quantitatively by a blue to green shift in fluorescence upon cleavage of the AFC fluorophore; substrate for Caspases VI, VII, VIII & X; Nicholson, DW et al, *Nature*, 376:37, 1995; Chinnaiyan, AM et al, *Curr Biol*, 6:897, 1996

Asp-Glu-Val-Asp
Ac-Asp-Glu-Val-Asp-AFC

Synonyms: Caspase III Fluorogenic Substrate VII

Calbiochem 264151 MW 728.6 $C_{30}H_{33}F_3N_5O_{13}$ ≥95% (HPLC); solid; soluble in DMSO; excitation max:~400 nm; emission max:~505 nm | Fluorogenic Caspase III substrate; also acts as a substrate for granzyme B-processed ICE-LAP3 & Caspases VI, VII, VIII & X

Asp-Glu-Val-Asp
Z-Asp(OCH₃)-Glu(OCH₃)-Val-Asp(OCH₃)-FMK

Synonyms: Caspase III Inhibitor II

Calbiochem 264155 MW 668.7 $C_{30}H_{41}FN_4O_{12}$ Single spot purity (TLC); solid; soluble in DMSO; sold under license of US Patents 5,344,939 & 5,210,272 issued to Prototek, Inc | Potent, cell-permeable & irreversible inhibitor of Caspase III as well as Caspases VI, VII, VIII & X; Masuda, Y et al, *Biochem Biophys Res Comm*, 234:641, 1997; Nicholson, DW et al, *Nature Biotech*, 14:297, 1996; Schlegel, J et al, *J Biol Chem*, 271:1841, 1996; Nicholson, DW et al, *Nature*, 376:37, 1995

Asp-Glu-Val-Asp
Z-Asp(OMe)-Glu(OMe)-Val-Asp(OMe)-CH₂F

Synonyms: Interleukin Iβ Converting Enzyme-Like Inhibitor

ICN 193605 Useful in apoptosis research

Asp-Glu-Val-Asp
Ac-Asp-Glu-Val-Asp-CHO

Synonyms: Interleukin Iβ Converting Enzyme Inhibitor; CPP-32/Apopain Inhibitor

ICN 195865 MW 502.5 $C_{20}H_{30}N_4O_{11}$ ≥97% | Potent, specific & reversible inhibitor

Asp-Glu-Val-Asp
Ac-Asp-Glu-Val-Asp-CHO Biotinylated

Synonyms: Interleukin Iβ Converting Enzyme Inhibitor; CPP-32/Apopain Inhibitor

ICN 195866 MW 686.7 $C_{28}H_{42}N_6O_{12}S$ ≥97% | Potent, specific & reversible CPP-32/Apopain inhibitor

Asp-Glu-Val-Asp
Ac-Asp-Glu-Val-Asp-pNA

Synonyms: Interleukin Iβ Converting Enzyme Substrate; CPP-32/Apopain Substrate

ICN 195867 MW 638.6 $C_{26}H_{34}N_6O_{13}$ ≥97% | Colorimetric substrate

Asp-Glu-Val-Asp
Ac-Asp-Glu-Val-Asp-AMC

Synonyms: Interleukin Iβ Converting Enzyme Substrate; CPP-32/Apopain Substrate

ICN 195868 MW 675.6 $C_{30}H_{37}N_5O_{13}$ ≥97% | Fluorometric substrate

Asp-Glu-Val-Asp
Z-Asp-Glu-Val-Asp-FMK

Synonyms: Caspase III Inhibitor; CPP-32 Inhibitor

Kamiya MW 668

Asp-Glu-Val-Asp
Ac-Asp-Glu-Val-Asp-CHO

Synonyms: Caspase III Inhibitor; CPP-32 Inhibitor

Kamiya MW 502.5 $C_{20}H_{30}N_4O_{11}$ >95% (HPLC)

Asp-Glu-Val-Asp
Biotin-Asp-Glu-Val-Asp-FMK

Synonyms: Caspase III Inhibitor; CPP-32 Inhibitor

Kamiya MW 872

Asp-Glu-Val-Asp
Ac-Asp-Glu-Val-Asp-AFC

Synonyms: Caspase III Substrate; CPP-32 Substrate

Kamiya MW 729 Fluorogenic

Asp-Glu-Val-Asp
Ac-Asp-Glu-Val-Asp-pNA

Synonyms: Caspase III Substrate; CPP-32 Substrate

Kamiya MW 638.6 $C_{26}H_{34}N_6O_{13}$ >97% (HPLC)

Asp-Glu-Val-Asp
Ac-Asp-Glu-Val-Asp-AMC

Synonyms: Caspase III Substrate

Neosystem SC1208 MW 675.65 Nicholson, DW et al, *Nature*, 376:37-43, 1995; Thornberry, NA et al, *JBC*, 272:17907-17911, 1997

Asp-Glu-Val-Asp
Ac-Asp-Glu-Val-Asp-AFC

Synonyms: Caspase III Substrate

Neosystem SC1281 MW 729.62 Xiang, JL et al, *PNAS USA*, 93:14559-14563, 1996

Asp-Glu-Val-Asp
Ac-Asp-Glu-Val-Asp-pNA

Synonyms: Caspase III Substrate

Neosystem SC1282 MW 638.52 Datta, R et al, *Blood*:88,1936-1943, 1996

Asp-Glu-Val-Asp
Ac-Asp-Glu-Val-Asp-CHO

Synonyms: Caspase III Inhibitor

Neosystem SC973   MW 502.48   Potent inhibitor of Caspase III; inhibits the cleavage of poly(ADP-ribose) polymerase (PARP) by Caspase III that occurs at the onset of apoptosis; also prevents apoptotic events *in vitro*; Nicholson, DW et al, *Nature*, 376:37-43, 1995

Asp-Glu-Val-Asp
Ac-Asp-Glu-Val-Asp-MCA

Synonyms: Caspase III/VII Substrate

Peptides International SAP-3171-v   MW 675.65 $C_{30}H_{37}N_5O_{13}$ >98% (HPLC); lyophilized amorphous powder | Nicholson, DW et al, *Nature*, 376:37, 1995; Thornberry, NA et al, *JBC*, 272:17907, 1997

Asp-Glu-Val-Asp
N-Ac-Asp-Glu-Val-Asp-7-Amido-4-Trifluoromethylcoumarin

Synonyms: Caspase III Fluorescent Substrate; Apopain Substrate

Sigma A 0466   Xiang, J, et al, *PNAS USA*, 93:14559, 1996

Asp-Glu-Val-Asp
N-Ac-Asp-Glu-Val-Asp-al

Synonyms: Interleukin Iβ Converting Enzyme Inhibitor

Sigma A 0835   FW 502.5 $C_{20}H_{30}N_4O_{11}$ ≥95% | Reversible inhibitor; inhibits poly(ADP-ribose) polymerase cleavage by apopain (CPP-32); Nicholson, DW, et al, *Nature*, 376:37, 1995

Asp-Glu-Val-Asp
Ac-Asp-Glu-Val-Asp-7-Amido-4-Methylcoumarin

Synonyms: Caspase III Fluorogenic Substrate; Apopain Substrate

Sigma A 1086   FW 675.6 $C_{30}H_{37}N_5O_{13}$ ≥97% (HPLC); peptide content ~65% | Nicholson, DW, et al, *Nature*, 376:37, 1995

Asp-Glu-Val-Asp
N-Ac-Asp-Glu-Val-Asp p-Nitroanilide

Synonyms: Caspase III Substrate; Apopain Substrate

Sigma A 2559   FW 638.6 $C_{26}H_{34}N_6O_{13}$ ≥97% (HPLC); | Nicholson, DW, et al, *Nature*, 376:37, 1995

Asp-Glu-Val-Asp
Ac-Asp-Glu-Val-Asp-AMC

Synonyms: Caspase III Fluorometric Substrate; Apopain Substrate

Upstate 12-323   MW 676   Lyophilized

Asp-Glu-Val-Asp
Ac-Asp-Glu-Val-Asp-pNA

Synonyms: Caspase III Substrate, Chromogenic

Upstate 12-390   MW 639   Lyophilized | Caspase III assay

Asp-Glu-Val-Asp
Ac-Asp-Glu-Val-Asp-CHO

Synonyms: Caspase III/VII Inhibitor

Peptides International IAP-3172-v   Synthetic   MW 502.48 $C_{20}H_{30}N_4O_{11}$ >99% (HPLC); lyophilized amorphous powder | Inhibitor; Nicholson, DW et al, *Nature*, 376:37, 1995; Enari, M et al, *Nature*, 380:723, 1996; Thornberry, NA et al, *JBC*, 272:17907, 1997

Asp-Glu-Val-Asp
Biotinyl-Asp-Glu-Val-Asp-CHO

Synonyms: Caspase III/VII Inhibitor

Peptides International IBA-3173-v   Synthetic   MW 686.74 $C_{28}H_{42}N_6O_{12}S$ >99% (HPLC); lyophilized amorphous powder | Inhibitor; Nicholson, DW et al, *Nature*, 376:37, 1995

Asp-Glu-Val-Asp-Ala-Arg-Lys
MCA-Asp-Glu-Val-Asp-Ala-Arg-(Lys-DNP)-Amide

BioSource International 77-966

Asp-Glu-Val-Asp-Ala-Arg-Lys
MCA-Asp-Glu-Val-Asp-Ala-Arg-Lys(DNP) Amide

Synonyms: Caspase III Fluorogenic Substrate VI

Calbiochem 218752   MW 1213.2 $C_{51}H_{68}N_{14}O_{21}$ ≥97% (HPLC); lyophilized solid; soluble in DMSO; excitation max:~325 nm; emission max:~392 nm | Fluorogenic resonance energy transfer substrate for Caspase III & Caspase III-like enzymes, including Caspases VI, VII, VIII & X

Asp-Glu-Val-Asp-Ala-Pro-Lys
Mca-Asp-Glu-Val-Asp-Ala-Pro-Lys (Dnp)

Synonyms: Caspase III Fluorogenic Substrate

Alexis 260-018   MW 1155.1 $C_{50}H_{62}N_{10}O_{22}$ ≥96%; Yellow lyophilized powder; soluble in 0.1 *M* NaHCO₃ with agitation | Specific, highly fluorogenic substrate for the determination of Caspase III & Caspase III-like enzyme activities; poor substrate for Caspase I; Enari, M et al, *Nature*, 380:723, 1996

Asp-Glu-Val-Asp-Ala-Pro-Lys
Mca-Asp-Glu-Val-Asp-Ala-Pro-Lys (Dnp) Amide

Synonyms: Caspase III Fluorogenic Substrate

Alexis 260-051   MW 1154.1 $C_{50}H_{63}N_{11}O_{21}$ ≥97%; lyophilized powder; amide | Specific, highly fluorogenic substrate for the determination of Caspase III & Caspase III-like enzyme activities; poor substrate for Caspase I; Enari, M et al, *Nature*, 380:723, 1996

Asp-Glu-Val-Asp-Ala-Pro-Lys
Mca-Asp-Glu-Val-Asp-Ala-Pro-Lys(DNP)

Synonyms: Caspase III Substrate III; Caspase III Fluorogenic Substrate III; CPP-32/Apopain Fluorogenic Substrate II

American Peptide 81-7-93   MW 1155.3 $C_{50}H_{62}N_{10}O_{22}$ Fluorogenic; cleavage of this substrate at position P_1 Asp results in continuous fluorescent assay monitored at emission wavelength 393 nm; Caspase I has very little activity for this substrate; Enari, M et al, *Nature*, 380:723, 1996

Bachem M-2200.0001   MW 1155.09 $C_{50}H_{62}N_{10}O_{22}$ Store at -15°C

Calbiochem 235426   MW 1155.1 $C_{50}H_{62}N_{10}O_{22}$ ≥98% (HPLC); crystalline solid; soluble in NaHCO₃; excitation max:~325 nm; emission max:~392 nm | Fluorogenic resonance energy transfer substrate for Caspases VI, VII, VIII & X; Caspase I has very little effect on this substrate; Enari, M et al, *Nature*, 380:723, 1996

Asp-Glu-Val-Asp-Ala-Pro-Lys
MOCAc-Asp-Glu-Val-Asp-Ala-Pro-Lys(Dnp) Amide

Synonyms: Caspase III Fluorescence-Quenching Substrate

Peptides International MOC-3184-v   MW 1154.1 $C_{50}H_{63}N_{11}O_{21}$ >98% (HPLC); lyophilized amorphous powder | Enari, M et al, *Nature*, 380:72, 1996

Asp-Glu-Val-Asp-Ala-Pro-Lys
AMC-Asp-Glu-Val-Asp-Ala-Pro-(2,4-Dinitrophenyl)-Lys

Synonyms: Caspase III Fluorogenic Substrate; Apopain Substrate; CPP-32 Substrate

Sigma M 1169   FW 1155.1   Fluorogenic substrate for apopain (CPP-32) & CPP-32 like enzymes; cleavage at P1 Asp residue results in continuous fluorescence at 392 nm; Enari, M, *Nature*, 380:723, 1996

Asp-Gly

Bachem G-1580.0250   MW 190.16 $C_6H_{10}N_2O_5$ Store at -15°C

Asp-Gly
α-Asp-Gly

Synonyms: Asp(Gly-OH)-OH

Bachem G-1585.0250   MW 190.16 $C_6H_{10}N_2O_5$ Store at -15°C

Asp-Gly

Sigma A 1521   FW 190.2 $C_6H_{10}N_2O_5$
Sigma A 8634   FW 190.2 $C_6H_{10}N_2O_5$

Asp-Gly-Asp-Phe-Glu-Glu-Ile-Pro-Glu-Glu-Tyr-Leu-Gln
Ac-Asp-Gly-Asp-Phe-Glu-Glu-Ile-Pro-Glu-Glu-Tyr(SO₃H)-Leu-Gln Sulfated

Synonyms: Hirudin (53-65), Ac-

Bachem H-8190.0001   MW 1705.73 $C_{72}H_{100}N_{14}O_{32}S$ Store at -15°C

Asp-Gly-Asp-Phe-Glu-Glu-Ile-Pro-Glu-Glu-Tyr-Leu-Gln
Ac-Asp-Gly-Asp-Phe-Glu-Glu-Ile-Pro-Glu-Glu-Tyr(SO₃H)-Leu-Gln

Synonyms: Hirudin (55-65), N-Ac-

Sigma H 9019   FW 1453.5   ≥97% (HPLC); non-sulfated | Bioactive peptide

Asp-Gly-Glu-Ala

Synonyms: Integrin Recognition Sequence, α2β1-

Bachem H-1376.0025   MW 390.35 $C_{14}H_{22}N_4O_9$ Store at -15°C

Asp-His
α-Asp-His

Synonyms: Asp(His-OH)-OH

Bachem G-3980.0250   MW 270.25 $C_{10}H_{14}N_4O_5$ Store at -15°C

Sigma A 2762   FW 270.2 $C_{10}H_{14}N_4O_5$

Asp-His-Tyr-Asn-Cys-Val-Ser-Ser-Gly-Gly-Gln-Cys-Leu-Tyr-Ser-Ala-Cys-Pro-Ile-Phe-Thr-Lys-Ile-Gln-Gly-Thr-Cys-Tyr-Arg-Gly-Lys-Ala-Lys-Cys-Cys-Lys

Synonyms: Defensin I, β-; Antibacterial Peptide

Peptides International PDF-4337-s   Human synthetic   MW 3928.6 $C_{167}H_{256}N_{48}O_{50}S_6$ >99% (HPLC); lyophilized amorphous powder | Bioactive; Disulfide bonds: Cys⁵-Cys³⁴, Cys¹²-Cys²⁷, Cys¹⁷-Cys³⁵; inactivated in Cystic Fibrosis; Bensch, KW et al, *FEBS Lett*, 368:331, 1995; Goldman, MJ et al, *Cell*, 88:553, 1997

Asp-Leu

Bachem G-1590.0250   MW 246.26 $C_{10}H_{18}N_2O_5$ Store at -15°C

Asp-Leu
α-Asp-Leu

Synonyms: Asp(Leu-OH)-OH

Bachem G-1600.0250   MW 246.26 $C_{10}H_{18}N_2O_5$ Store at -15°C

Sigma A 6291   FW 246.3 $C_{10}H_{18}N_2O_5$

Asp-Leu Amide

Bachem G-1595.0250   MW 245.28 $C_{10}H_{19}N_3O_4$ Store at -15°C

Asp-Leu-Arg-Val-Asp-Thr-Lys-Ser-Arg-Ala-Ala-Trp-Ala-Arg-Leu-Leu-Gln-Glu-His-Pro-Asn-Ala-Arg-Lys-Tyr-Lys-Gly-Ala-Asn-Lys-Lys-Gly-Leu-Ser-Lys-Gly-Cys-Phe-Gly-Leu-Lys-Leu-Asp-Arg-Ile-Gly-Ser-Met-Ser-Gly-Leu-Gly-Cys

Synonyms: Natriuretic Peptide (1-53), C-Type; Natriuretic Peptide 53, C-Type

ICN 159907   MW 5801.7   Human   97% | Disulfide bonds: 37-53

American Peptide 14-1-42   Human   MW 5801.8 $C_{251}H_{417}N_{81}O_{71}S_3$ Disulfide bonds: Cys³⁷-Cys⁵³; Minamino, N et al, *BBRC*, 170:9730, 1990

Bachem H-8420.0200   Human   MW 5801.77 $C_{251}H_{417}N_{81}O_{71}S_3$ Store at -15°C

Sigma N 1021   Human   FW 5801.7   ≥95% (HPLC); peptide content:~65%; Disulfide bonds: :37-53 | Bioactive peptide; Tawaragi, Y et al, *Biochem Biophys Res Commun*, 175:645, 1991

Peptides International PCT-4241-s   Human synthetic   MW 5801.8 $C_{251}H_{417}N_{81}O_{71}S_3$ >95% (HPLC); lyophilized amorphous powder | Bioactive; Disulfide bonds: Cys³⁷-Cys⁵³; Tawaragi, Y et al, *BBRC*, 175:645, 1991

Asp-Leu-Arg-Val-Asp-Thr-Lys-Ser-Arg-Ala-Ala-Trp-Ala-Arg-Leu-Leu-His-Glu-His-Pro-Asn-Ala-Arg-Lys-Tyr-Lys-Gly-Gly-Asn-Lys-Lys-Gly-Leu-Ser-Lys-Gly-Cys-Phe-Gly-Leu-Lys-Leu-Asp-Arg-Ile-Gly-Ser-Met-Ser-Gly-Leu-Gly-Cys

Synonyms: Natriuretic Peptide (1-53), C-Type ; Natriuretic Peptide 53, C-Type

Bachem H-8425.0200   Porcine, rat   Store at -15°C

Peptides International PCT-4240-s   Porcine, rat synthetic   MW 5796.7 $C_{251}H_{414}N_{82}O_{70}S_3$ >95% (HPLC); lyophilized amorphous powder | Bioactive; Disulfide bonds: Cys³⁷-Cys⁵³; Minamino, N et al, *BBRC*, 170:973, 1990; Tawaragi, Y et al, *BBRC*, 172:627, 1990; Kojima, M et al, *FEBS Lett*, 276:209, 1990

Asp-Leu-Asp-Val-Pro-Ile-Pro-Gly-Arg-Phe-Asp-Arg-Arg-Val-Ser-Val-Ala-Ala-Glu

Synonyms: cAMP-Dependent Protein Kinase Substrate

Alexis 151-019   MW 2112.4 $C_{92}H_{150}N_{28}O_{29}$ ≥95%; white powder; soluble in H_2O | From the RII subunit; useful in a non-radioactive assay of calcineurin (protein phosphatase 2B) for screening purposes; Blumenthal, DK et al, *J Biol Chem*, 261:8140, 1986; Enz, A et al, *Anal Biochem*, 216:147, 1994

Asp-Leu-Asp-Val-Pro-Ile-Pro-Gly-Arg-Phe-Asp-Arg-Arg-Val-Ser-Val-Ala-Ala-Cys
Asp-Leu-Asp-Val-Pro-Ile-Pro-Gly-Arg-Phe-Asp-Arg-Arg-Val-Ser-Val-Ala-Ala-Cys-Acrylodan

Synonyms: cAMP-Dependent Protein Kinase Substrate; AR II

Alexis 260-068   MW 2311.6 $C_{105}H_{163}N_{29}O_{28}S$ ≥90% (HPLC); pale yellow crystals; soluble in H_2O | Cell permeable, fluorescent cAMP-dependent protein kinase (PKA) peptide probe for visualization of PKA activities in living cells; AR II is an acrylodon-labeled 19-AA peptide which contains a part of a regulatory domain II of bovine cardiac muscle PKA; Shows 524 nm emission when excited at 36 nm; When cAMP is added to the solution, the fluorescence intensity of AR II at 524 is gradually decreased by the phosphorylation of AR II with activated PKA; Higashi, H et al, *FEBS Lett*, 414:55, 1997

Asp-Leu-Asp-Val-Pro-Ile-Pro-Gly-Arg-Phe-Asp-Arg-Arg-Val-Ser-Val-Ala-Ala-Glu

Synonyms: cAMP-Dependent Protein Kinase RII Subunit; Calcineurin Substrate; Protein Kinase A Substrate RII, (Ala⁹⁷)-; Protein Kinase Related Peptide

American Peptide 81-6-10   MW 2112.4 $C_{92}H_{150}N_{28}O_{29}$ A fragment of the RII subunit of the cAMP-dependent protein kinase; used in a nonradioactive assay for detecting & characterizing novel inhibitors of calcineurin & for determining protein phosphatase 2B (calcineurin) activity; Enz, A et al, *Anal Biochem*, 216:147, 1994

Bachem H-2084.0001   MW 2112.37 $C_{92}H_{150}N_{28}O_{29}$ Store at -15°C

ICN 195809   MW 2112.4   >97% | c-AMP-dependent proteinkinase regulatory subunit type II (pK$_a$)

Neosystem SC964   MW 2112.37   Partial sequence of the regulatory subunit of cAMP-dependent proteinkinase; used in a nonradioactive enzyme assay allowing measurement of enzymekinetics and characterization of potential inhibitors; Enz, A et al, *Anal Biochem*, 216:147-153, 1994

Sigma C 5207   FW 2112.4   ≥97% (HPLC) | Bioactive peptide; from the regulatory RII subunit of cGMP dependent proteinkinase; useful in assays for protein phosphatase-2B (calcineurin) activity; Enz, A et al, *Anal Biochem*, 216:147, 1994

Asp-Leu-Asp-Val-Pro-Leu-Pro-Ala-Lys-Ala-Asp-Arg-Arg-Val-Ser-Val-Ala-Ala-Cys
Asp-Leu-Asp-Val-Pro-Leu-Pro-Ala-Lys-Ala-Asp-Arg-Arg-Val-Ser-Val-Ala-Ala-Cys-DACM

Synonyms: cAMP-Dependent Protein Kinase Substrate; DR II

Alexis 260-069   MW 2294.6 $C_{101}H_{160}N_{28}O_{31}S$ ≥90% (HPLC); Pale yellow crystals; soluble in H$_2$O | Cell permeable, fluorescent cAMP-dependent protein kinase (PKA) peptide probe for visualization of PKA in living cells; DR II is an DACM-labeled 19-AA peptide analog to AR II (Alexis 260-068) with increased lipophilicity, therefore DR II is more readily incorporated into a primary culture cell than AR II; DR II shows 470 nm emission when excited at 400 nm; Higashi, H et al, *FEBS Lett*, 414:55, 1997

Asp-Leu-Glu-Ile-Gly-Gln-His-Arg-Thr-Lys-Ile-Glu-Glu-Leu-Arg-Gln-His-Leu-Leu-Arg-Trp-Gly-Leu-Thr-Thr

Synonyms: POL (359-383); RT (192-216); HTLV I Peptide

Neosystem SC691   MW 3043.50   HTLV I peptide from subtype POL (PR/RT)

Asp-Leu-Ile-Ala-Cys
Ac-Asp-D-Gla-Leu-Ile-β-Cyclohexyl-Ala-Cys

Bachem N-1725.0001   MW 830.95 $C_{36}H_{58}N_6O_{14}S$ Store at -15°C

Asp-Leu-Ser-His-Phe-Leu-Lys-Glu-Lys-Gly-Gly-Leu-Asp-Gly-Leu

Synonyms: NEF MN (88-102); HIV I Peptide

Neosystem SC650   MW 1628.84   NEF peptide from HIV-1 subtype MN

Asp-Leu-Ser-His-Phe-Leu-Lys-Glu-Lys-Gly-Gly-Leu-Glu-Gly-Leu

Synonyms: NEF (86-100); HTLV I Peptide

Neosystem SC506   MW 1642.87   HTLV I peptide from subtype NEF

Asp-Leu-Ser-Lys-Gln-Met-Glu-Glu-Glu-Ala-Val-Arg-Leu-Phe-Ile-Glu-Trp-Leu-Lys-Asn-Gly-Gly-Pro-Ser-Ser-Gly-Ala-Pro-Pro-Pro-Ser

Synonyms: Exendin (9-39)

American Peptide 46-3-10   MW 3370.8 $C_{149}H_{223}N_{39}O_{48}S$ A potent glucagon-like peptide-1 receptor antagonist, blocks the stimulatory action of GLP-1 (7-36) & of exendin-4 on cAMP production in pancreatic acini; Eng, J et al, *J Biol Chem*, 267:7402, 1992; Raufman, JP et al, *J Biol Chem*, 267:21432, 1992; Fehmann, HC et al, *Peptides*, 15:453, 1994

Asp-Leu-Ser-Lys-Gln-Met-Glu-Glu-Glu-Ala-Val-Arg-Leu-Phe-Ile-Glu-Trp-Leu-Lys-Asn-Gly-Gly-Pro-Ser-Ser-Gly-Ala-Pro-Pro-Pro-Ser Amide

Synonyms: Exendin (9-39)

Bachem H-8740.0500   Store at -15°C

Neosystem SC980   MW 3369.79   Bioactive; brain/gut peptide; glucagon-like peptide-I; Raufman, JP et al, *JBC*, 266:2897-2902, 1991; Eng, J et al, *JBC*, 267:7402-7405, 1992; Raufman, JP et al, *JBC*, 267:21432-21437, 1992; Göke, R et al, *JBC*, 268:19650-19655, 1993

Asp-Leu-Ser-Lys-Gln-Met-Glu-Glu-Glu-Ala-Val-Arg-Leu-Phe-Ile-Glu-Trp-Leu-Lys-Asn-Gly-Gly-Pro-Ser-Ser-Gly-Ala-Pro-Pro-Ser Amide

Synonyms: Exendin (9-39)

Sigma E 7269   FW 3369.8   ≥90% (HPLC) | Bioactive peptide; competitive inhibitor of exendin-3 & exendin-4; Eng, J et al, *J Biol Chem*, 267:7402, 1992

Asp-Leu-Thr-Phe-His-Leu-Leu-Arg-Glu-Met-Leu-Glu-Met-Ala-Lys-Ala-Glu-Gln-Glu-Ala-Glu-Gln-Ala-Ala-Leu-Asn-Arg-Leu-Leu-Leu-Glu-Glu-Ala Amide

Synonyms: Corticotropin Releasing Factor (9-41), α-Helical; Corticotropin Releasing Factor Antagonist (9-41)

American Peptide 34-0-15   MW 3826.4 $C_{166}H_{274}N_{46}O_{53}S_2$ CRF antagonist; acts as a potent antagonist to CRF receptors to block the behavioral & physiological responses to a stressor when administered in the CNS; Rivier, J et al, *Science*, 224:889, 1984

Bachem H-2040.0500   MW 3826.41 $C_{166}H_{274}N_{46}O_{53}S_2$ Store at -15°C

ICN 153078   Rivier, J etal, *Science*, 224:889, 1984

Neosystem SC1205   MW 3826.39   Bioactive; hypothalamic releasing hormone; CRF/urocortin peptide; Rivier, J et al, *Science*, 224:889-891, 1984

Asp-Leu-Thr-Phe-His-Leu-Leu-Arg-Glu-Met-Leu-Glu-Met-Ala-Lys-Ala-Glu-Gln-Glu-Ala-Glu-Gln-Ala-Ala-Leu-Asn-Arg-Leu-Leu-Leu-Glu-Glu-Ala
Asp-Leu-Thr-Phe-His-Leu-Leu-Arg-Glu-Met-Leu-Glu-Met-Ala-Lys-Ala-Glu-Gln-Glu-Ala-Glu-Gln-Ala-Ala-Leu-Asn-Arg-Leu-Leu-Leu-Glu-Glu-Ala Amide

Synonyms: Corticotropin Releasing Factor (9-41); Corticotropin Releasing Factor, α-Helical

Sigma C 2917   FW 3827.4   ≥97% (HPLC) | Bioactive peptide; antagonist; Rivier, J et al, *Science*, 224:889, 1984

Asp-Leu-Thr-Phe-His-Leu-Leu-Arg-Glu-Met-Leu-Glu-Met-Ala-Lys-Ala-Glu-Gln-Glu-Ala-Glu-Gln-Ala-Ala-Leu-Asn-Arg-Leu-Leu-Leu-Glu-Glu-Ala Amide

Synonyms: Corticotropin Releasing Factor (9-41), α-Helical; Corticotropin Releasing Factor Antagonist

Calbiochem 05-23-0070   Helical   MW 3826.4 $C_{166}H_{274}N_{46}O_{53}S_2$ ≥98% (HPLC); lyophilized solid; soluble in 5% acetic acid | Stable analog of porcine motilin with full biological activity; acts as a potent antagonist of CRF receptors; Milton, NG et al, *J Physiol*, 456:415, 1993; River, J et al, *Science*, 224:889, 1984; Moody, TW et al, *Peptides*, 15:281, 1994

Asp-Leu-Trp-Gln-Lys

Synonyms: Uremic Pentapeptide; U$_5$-Peptide; U5-Peptide

American Peptide 87-0-12   MW 688.8 $C_{32}H_{48}N_8O_9$ Isolated from the ultra-filtrate of a uremic patient, exhibits immunomodulating properties; Abiko, T et al, *BBRC*, 89:813, 1979

Bachem H-9465.0001   MW 688.78 $C_{32}H_{48}N_8O_9$ Store at -15°C

Sigma A 4182   FW 688.8 $C_{32}H_{48}N_8O_9$ ≥97% (HPLC); peptide content:~70% | Bioactive peptide

---

Asp-Leu-Trp-Ile-Tyr-His-Thr-Gln-Gly-Tyr-Phe-Pro

Synonyms: NEF (111-122); HTLV I Peptide

Neosystem SC509   MW 1539.71   HTLV I peptide from subtype NEF

Asp-Leu-Trp-Val-Tyr-His-Thr-Gln-Gly-Tyr-Phe-Pro

Synonyms: NEF MN (113-124); HIV I Peptide

Neosystem SC654   MW 1526.66   NEF peptide from HIV-1 subtype MN

Asp-Leu-Tyr-Tyr-Leu-Met-Asp-Leu

Synonyms: Fibronectin Receptor Peptide (124-131); Integrin β1 Subunit (124-131)

Bachem H-3562.0001   Store at -15°C

Asp-Lys

Bachem G-1605.0250   MW 261.28 $C_{10}H_{19}N_3O_5$ Store at -15°C

Asp-Lys
α-Asp-Lys

Synonyms: Asp(Lys-OH)-OH

Bachem G-1615.0250   MW 261.28 $C_{10}H_{19}N_3O_5$ Store at -15°C

Asp-Lys Acetate Salt Free Base

Sigma A 1554   FW 261.3 $C_{10}H_{19}N_3O_5$

Asp-Lys-Glu-Ala-Val-Tyr-Phe-Ala-His-Leu-Asp-Ile-Ile-Trp
Suc-Asp-Lys-Glu-Ala-Val-Tyr-Phe-Ala-His-Leu-Asp-Ile-Ile-Trp

Synonyms: Endothelin B Receptor Agonist, Suc-(Glu[9],Ala[11,15])-; Endothelin I (8-21); IRL-1620

Alexis 155-006   MW 1821.0 $C_{86}H_{117}N_{17}O_{27}$ Most potent & selective ligand for the ET$_B$ receptor; IRL-1620 together with IRL-1038 (Alexis 155-007) is an indispensable pharmacological tool to assess the participation of the ET$_B$ receptor in some pathophysiological phenomena, by selectively stimulating ET$_B$ receptor-mediated responses in various pathophysiological models established *in vitro* & *in vivo*; Haynes, WG et al, *TIPS*, 14:225, 1993; Sakamoto, A et al, *J Biol Chem*, 268:8547, 1993; Takai, M et al, *BBRC*, 184:953, 1992

Asp-Lys-Glu-Ala-Val-Tyr-Phe-Ala-His-Leu-Asp-Ile-Ile-Trp
Ac-Asp-Lys-Glu-Ala-Val-Tyr-Phe-Ala-His-Leu-Asp-Ile-Ile-Trp

Synonyms: Endothelin (8-21), Ac-(Ala[11,15])-; IRL-1720

American Peptide 88-2-53   MW 1762.0 $C_{85}H_{115}N_{18}O_{23}$ Kuwahara, M et al, *Eur J Pharmacol*, 296:55, 1996

Asp-Lys-Phe-Val-Gly-Me-Leu-Nle
Asp-Lys-Phe-Val-Gly-N-Me-Leu-Nle Amide

Synonyms: Neurokinin A (4-10), (Lys[5],MeLeu[9],Nle[10])-

Neosystem SC429   MW 803.99   Bioactive; tachykinin; water soluble, highly potent & specific competitor at NK-2 binding sites; Chassaing, G et al, *Neuropeptides*, 19:91-95, 1991

Asp-Lys-Pro-Val-Ala-His-Val-Val-Ala-Asn-Pro-Gln-Ala-Glu-Gly-Gln-Leu-Gln-Trp-Leu-Asn-Arg-Arg-Ala-Asn-Ala-Leu

S Tumor Necrosis Factor α (10-36)ynonyms:

Bachem H-8255.0500   Human   Store at -15°C

ICN 157035   Human   MW 2996.4   97%

Sigma T 1042   Human   FW 2996.4   ≥97% (HPLC) | Bioactive peptide

Asp-Met
Z-Asp-Met

Bachem C-1375.0250   Store at -15°C

Asp-Met-Ala-Lys-Asp-Leu-Glu-Thr-Asn-His-His-Pro-Tyr-Phe-Gly-Asn

Synonyms: Calcitonin C-Terminal Adjacent Peptide

American Peptide 22-5-20   Rat   MW 1889.1 $C_{82}H_{117}N_{23}O_{27}S$ Birnbaum, R et al, *J Biol Chem*, 257:241, 1982

Bachem H-6735.0500   Rat   MW 1889.04 $C_{82}H_{117}N_{23}O_{27}S$ Store at -15°C

Asp-Met-Gln-Asp
Ac-Asp-Met-Gln-Asp-CHO Pseudo Acid

Bachem H-3358.0005   MW 533.56 $C_{20}H_{31}N_5O_{10}S$ Store at -15°C

Asp-Met-Gln-Asp
Ac-Asp-Met-Gln-Asp-AMC

Synonyms: Caspase III Substrate

Bachem I-1815.0005   MW 706.73 $C_{30}H_{38}N_6O_{12}S$ Store at -15°C

Neosystem SC1318   MW 706.72   Deduced from the cleavage site of proteinkinase Cδ; Ghayur, T et al, *J Exp Med*, 184:2399-2404, 1996

Asp-Met-Gln-Asp
Ac-Asp-Met-Gln-Asp-CHO

Synonyms: Caspase III Inhibitor

Peptides International ICA-3192-v   Synthetic   MW 533.56 $C_{20}H_{31}N_5O_{10}S$ >96% (HPLC); lyophilized amorphous powder | Inhibitor; Takahashi, A et al, *Oncogene*, 14:2741, 1997; Hirata, H et al, *J Exp Med*, 187:587, 1998

Asp-Met-His-Asp-Phe-Phe-Me-Phe-Gly-Leu-Met Amide

Synonyms: Neurokinin B, (MePhe[7])-

ICN 154456   MW 1272.6   Drapeau, G etal, *Neuropeptides*, 10:43, 1987

Asp-Met-His-Asp-Phe-Phe-Phe-Gly-Leu-Met
Asp-Met-His-Asp-Phe-Phe-N-Me-Phe-Gly-Leu-Met Amide

Synonyms: Neurokinin B, (N-Me-Phe[7])-

Bachem H-9280.0001   MW 1272.51 $C_{60}H_{81}N_{13}O_{14}S_2$ Store at -15°C

Asp-Met-His-Asp-Phe-Phe-Phe-Gly-Leu-Met
Asp-Met-His-Asp-Phe-Phe-Me-Phe-Gly-Leu-Met Amide

Synonyms: Neurokinin B, (MePhe[7])-

Calbiochem 05-23-0711   MW 1272.6 $C_{60}H_{81}N_{13}O_{14}S_2$ ≥97% (HPLC); solid; soluble in 5% acetic acid | Selective NK-3 receptor agonist that binds to NK-3 receptors with high affinity in a saturable & irreversible manner; Suman-Chauhan, N et al, *Eur J Pharmacol*, 269:65, 1994; Dion, S et al, *Neuropeptides*, 11:83, 1988; Regoli, D et al, *Trends Pharmacol Sci*, 9:290, 1988

Asp-Met-His-Asp-Phe-Phe-Phe-Gly-Leu-Met
Asp-Met-His-Asp-Phe-Phe-NMe-Phe-Gly-Leu-Met Amide

Synonyms: Neurokinin B, (MePhe[7])-

Neosystem SC981   MW 1272.50   Bioactive; tachykinin; potent & selective NK-3 agonist; Drapeau, G et al, *Neuropeptides*, 10:43-54, 1987

Asp-Met-His-Asp-Phe-Phe-Pro-Gly-Leu-Met Amide

Synonyms: Neurokinin III Receptor Agonist; Neurokinin B, (Pro[7])-

Bachem H-9285.0001   MW 1208.43 $C_{55}H_{77}N_{13}O_{14}S_2$ Store at -15°C

ICN 154457 MW 1208.5 Lavielle, S etal, *Reg Pep*, 22:108, 1988

Neosystem SC388 MW 1208.41 Bioactive; tachykinin; water soluble NK-3 receptor agonist; Lavielle, S et al, *Regul Peptides*, 22:108, 1988

Asp-Met-His-Asp-Phe-Phe-Val-Gly-Leu-Met

Synonyms: Neurokinin B

Biogenesis 6690-0204 Synthetic MW 1210.6 99.4% (HPLC); lyophilized, consisting of 76.2% peptide material, TFA counter ions and residual H_2O | To avoid Met oxidation, it is best to dissolve the peptide under a nitrogen stream

---

Asp-Met-His-Asp-Phe-Phe-Val-Gly-Leu-Met Amide

Synonyms: Neuromedin K; Neurokinin B; Neurokinin, α-; Neurokinin, β-; Neurokinin III Receptor Selective Agonist

American Peptide 62-1-42 MW 1210.6 $C_{55}H_{79}N_{13}O_{14}S_2$ Belongs to the tachykinin family; plays an important role in the olfactory, gustatory, visceral & neuroendocrine processing information; a potent bronchio-constrictor & has neuromodulatory roles in various brain functions; Kimura, S et al, *Proc Japan Acad*, 59:Ser B 101, 1983; Kangawa, K et al, *BBRC*, 114:533, 1983

Bachem H-2045.0001 MW 1210.44 $C_{55}H_{79}N_{13}O_{14}S_2$ Store at -15°C

Calbiochem 05-23-0700 MW 1210.4 $C_{55}H_{79}N_{13}O_{14}S_2$ ≥96% (HPLC); lyophilized solid; soluble in 3% NH_4OH; LD_{50}≤2000 mg/kg | Peptide belonging to the tachykinin family; potent bronchioconstrictor; potent NK-3 receptor ligand; may play an important role in the olfactory, gustatory, visceral & neuroendocrine processing of information; has a neuromodulatory role in various brain functions; Donaldson, LF et al, *Biochem J*, 320:1, 1996; Lucas, LR et al, *Neuroscience*, 51:317, 1992; Marksteiner, J et al, *J Comp Neurol*, 317:341, 1992; Kanagawa, K et al, *Biochem Biophys Res Comm*, 114:533, 1983; Kimura, S et al, *Proc Jpn Acad*, 59B:101, 1983

ICN 153152 MW 1210.4

Neosystem SC082 MW 1210.43 Bioactive; tachykinin; Kimura, S et al, *Proc Japan Academy*, 59(SerB):101, 1983

Sigma N 4143 FW 1210.4 ≥95% (HPLC) | Bioactive peptide; potent bronchoconstrictor; may have neuromodulatory role in brain; involved in processing olfactory, gustatory, visceral & neuroendocrine information

Peptides International PNP-4317-v Synthetic MW 1210.4 $C_{55}H_{79}N_{13}O_{14}S_2$ >95% (HPLC); lyophilized amorphous powder | Bioactive; human, porcine, rat, mouse; Kimura, S et al, *Proc Japan Acad*, 95B:101, 1983; Kangawa, K et al, *BBRC*, 114:533, 1983

Asp-Met-His-Asp-Phe-Phe-Val-Gly-Leu-Met Amide Hemiacetate 4H₂O

Synonyms: Neurokinin B; Neuromedin K; Neurokinin III Receptor Selective Agonist

Peptides International PNP-4317 Synthetic MW 1312.49 $C_{55}H_{79}N_{13}O_{14}S_2 \cdot \frac{1}{2}CH_3COOH \cdot 4H_2O$ >95% (HPLC); lyophilized amorphous powder; bulk | Bioactive; human, porcine, rat, mouse; Kimura, S et al, *Proc Japan Acad*, 95B:101, 1983; Kangawa, K et al, *BBRC*, 114:533, 1983

Asp-Met-Ser-Lys-Asp-Glu-Ser-Val-Asp-Tyr-Val-Pro-Met-Leu-Asp-Met-Lys

Synonyms: Platelet Derived Growth Factor (742-758)

Sigma P 5963 FW 2003.3 ≥97% (HPLC) | Bioactive peptide

Asp-Met-Ser-Ser-Asp-Leu-Glu-Arg-Asp-Arg-Pro-His-Val-Ser-Met-Pro-Gln-Asn-Ala-Asn

Synonyms: Katacalcin

ICN 153134 Human Calcitonin Precursor Peptide, PDN-21

Asp-Met-Ser-Ser-Asp-Leu-Glu-Arg-Asp-His-Arg-Pro-His-Val-Ser-Met-Pro-Gln-Asn-Ala-Asn

Synonyms: Katacalcin; Calcitonin Precursor Peptide, Human; C-Procalcitonin; PDN-21; Calcitonin C-Terminal Flanking Peptide

American Peptide 22-0-15 MW 2436.6 $C_{97}H_{154}N_{34}O_{36}S_2$ Flanking calcitonin on the C-terminal side within the human calcitonin precursor, can significantly lower plasma calcium level; MacIntyre, I et al, *Nature*, 300:460, 1982

Sigma K 2627 FW 2436.6 ≥95% (HPLC) | Bioactive peptide

Bachem H-2050.0500 Human MW 2436.63 $C_{97}H_{154}N_{34}O_{36}S_2$ Store at -15°C

Asp-Phe
Asp-Phe-OMe

Synonyms: Aspartame

Bachem G-1545.0001 MW 294.31 $C_{14}H_{19}N_2O_5$ Store at -15°C

Asp-Phe

Bachem G-1620.0250 MW 280.28 $C_{13}H_{16}N_2O_5$ Store at -15°C

Asp-Phe
α-Asp-Phe-OMe

Synonyms: Aspartame, α-; Asp(Phe-OMe)

Bachem G-3725.0250 MW 294.31 $C_{14}H_{19}N_2O_5$ Store at -15°C

Asp-Phe
α-Asp-Phe

Synonyms: Asp(Phe-OH)-OH

Bachem G-4750.0250 MW 280.28 $C_{13}H_{16}N_2O_5$ Store at -15°C

Asp-Phe
N-L-α-Asp-L-Phe Acid

Synonyms: Aspartame

Fluka 11290 MW 280.28 $C_{13}H_{16}N_2O_5$ ≥98% (TLC)

Asp-Phe
N-L-α-Asp-L-Phe-OMe

Synonyms: Aspartame

Fluka 11300 MW 294.31 $C_{14}H_{18}N_2O_5$ ≥99% (HPLC); ~2% H_2O; mp:243-248°C

Asp-Phe
L-Asp-L-Phe-OMe

Synonyms: Aspartame

ICN 190007 MW 294.3 $C_{14}H_{18}N_2O_5$ Crystalline | Potent synthetic sweetener

---

Asp-Phe
β-Asp-Phe-OMe

Synonyms: Aspartame

Sigma A 1050 FW 294.3 $C_{14}H_{18}N_2O_5$

Asp-Phe
Asp-Phe-OMe

Synonyms: Aspartame

Sigma A 5139 FW 294.3 $C_{14}H_{18}N_2O_5$

Asp-Phe

Synonyms: Aspartame

Sigma A 7660 FW 280.3 $C_{13}H_{16}N_2O_5$

Asp-Phe Amide

Bachem G-1625.0250   MW 279.3 $C_{13}H_{17}N_3O_4$ Store at -15°C

Asp-Phe-Arg-Leu-Phe-Ala-Phe-Tyr-Asp
Abz-Asp-Phe-Arg-Leu-Phe-Ala-Phe-Tyr(NO₂)-Asp

Synonyms: Subtilisin Fluorescent Substrate

American Peptide 81-0-10   MW 1357.5 $C_{66}H_{80}N_{14}O_{18}$
Medal, M et al, *Anal Biochem*, 195:141, 1991

Asp-Phe-Arg-Leu-Phe-Ala-Phe-Tyr-Asp
Anthranilyl-Asp-Phe-Arg-Leu-Phe-Ala-Phe-Nitro-Tyr-Asp

Synonyms: Subtilisin Substrate

Sigma A 0324   FW 1357.4   ≥97% (HPLC) | Fluorescent;
Meldal, M & Breddam, K, *Anal Biochem*, 195:141, 1991

Asp-Phe-Asp-Met-Leu-Arg-Cys-Met-Leu-Gly-Arg-Val-Phe-Arg-Pro-Cys-Trp-Gln-Tyr

Synonyms: Melanin Concentrating Hormone, (Phe[13],Tyr[19])-

Bachem H-2218.0500   Human, mouse, rat   MW 2434.92
$C_{109}H_{160}N_{30}O_{26}S_4$ Store at -15°C

Asp-Phe-Asp-Met-Leu-Arg-Cys-Met-Leu-Gly-Arg-Val-Phe-Arg-Pro-Cys-Trp-Gln-Tyr
Asp-Phe-Asp-Met-Leu-Arg-Cys-Met-Leu-Gly-Arg-Val-*p*-Bz-Phe-Arg-Pro-Cys-Trp-Gln-Tyr

Synonyms: Melanin Concentrating Hormone, (*p*-Bz-D-Phe[13],Tyr[19])-; Melanin Concentrating Hormone, (D-Bpa[13],Tyr[19])-

Bachem H-2222.0500   Human, mouse, rat   MW 2539.03
$C_{116}H_{164}N_{30}O_{27}S_4$ Store at -15°C

Asp-Phe-Asp-Met-Leu-Arg-Cys-Met-Leu-Gly-Arg-Val-Phe-Arg-Pro-Cys-Trp-Gln-Tyr

Synonyms: Melanin Concentrating Hormone, (Phe[13],Tyr[19])-

Neosystem SC1284   Human, mouse, rat   MW 2434.89
Disulfide bonds: Cys[7]-Cys[16]; rat, mouse; bioactive neuropeptide; essential
analog for preparation of a potent & Bioactive MCH iodinated
radioligand; Drozdz, R et al, *Peptides* (Proceedings of the 23rd
EPS), 1994; HLS Maia, ed, Escom, Leiden, 785-786, 1995

Asp-Phe-Asp-Met-Leu-Arg-Cys-Met-Leu-Gly-Arg-Val-Tyr-Arg-Pro-Cys-Trp-Gln-Val

Synonyms: Melanin Concentrating Hormone; Appetite Boosting
Peptide

Peptides International PMC-4369-v   Human synthetic   MW
2386.8 $C_{105}H_{160}N_{30}O_{26}S_4$ >95% (HPLC); lyophilized amorphous
powder | Bioactive; Disulfide bonds: Cys[7]-Cys[16]; rat, mouse;
Kawauchi, H et al, *Nature*, 305:321, 1983; Vaughan, JM et al,
Endocrinology, 125:1660, 1989; Knigge, KM et al, *Peptides*,
17:1063, 1996; Qu, D et al, *Nature*, 380:243, 1996; Shimada, M et
al, *Nature*, 396:670, 1998; Chambers, J et al, *Nature*, 400:261,
1999; Saito, Y et al, *Nature*, 400:265, 1999

Bachem H-1482.0500   Human, mouse, rat   MW 2386.88
$C_{105}H_{160}N_{30}O_{26}S_4$ Store at -15°C

ICN 159964   Human, mouse, rat   MW 2354.8   90% |
Disulfide bonds: 7-16; Vaughan, JH etal, *Endocrinology*, 125:1660,
1989

Neosystem SC320   Human, mouse, rat   MW 2386.84
Disulfide bonds: Cys[7]-Cys[16]; bioactive neuropeptide;

American Peptide 72-0-62   Rat   MW 2386.9
$C_{105}H_{106}N_{30}O_{26}S_4$ Disulfide bonds: Cys[7]-Cys[16]; Vaughan, JH et al,
Endocrinology, 125:1660, 1989

Sigma M 4542   Rat   FW 2386.8   ≥90% (HPLC); Disulfide
bonds: 7-16; peptide content:~70% | Bioactive peptide;
vaughan, JH et al, *Endocrinology*, 125:1660, 1989

Asp-Phe-Glu-Glu-Ile-Pro-Glu-Glu-Tyr-Leu-Gln
Ac-Asp-Phe-Glu-Glu-Ile-Pro-Glu-Glu-Tyr-Leu-Gln
Desulfated

Synonyms: Hirudin (55-65), Ac-

Bachem H-7430.0001   MW 1453.52 $C_{66}H_{92}N_{12}O_{25}$ Store at -15°C

Asp-Phe-Glu-Glu-Ile-Pro-Glu-Glu-Tyr-Leu-Gln
Ac-Asp-Phe-Glu-Glu-Ile-Pro-Glu-Glu-Tyr(SO₃H)-Leu-Gln

Synonyms: Hirudin (55-65), Ac-(Tyr(SO₃H)[63])-

Sigma H 9144   FW 1533.6   ≥97% (HPLC) | Bioactive peptide

Asp-Phe-Glu-Glu-Ile-Pro-Glu-Glu-Tyr-Leu-Gln
Ac-Asp-Phe-Glu-Glu-Ile-Pro-Glu-Glu-Tyr(SO₃H)-Leu-Gln Sulfated

Synonyms: Hirudin (55-65), Ac-

Bachem H-7435.0001   MW 1533.59 $C_{66}H_{92}N_{12}O_{28}S$ Store
at -15°C

Asp-Phe-Glu-Glu-Ile-Pro-Glu-Glu-Tyr-Leu-Gln
Asp-Phe-Glu-Glu-Ile-Pro-Glu-Glu-Tyr(SO₃H)-Leu-Gln Sulfated

Synonyms: Hirudin (55-65), (Tyr(SO₃H)[63])-

Bachem H-7445.0001   MW 1491.55 $C_{64}H_{90}N_{12}O_{27}S$ Store
at -15°C

ICN 153121   MW 1491.5

Asp-Phe-Glu-Glu-Ile-Pro-Glu-Glu-Tyr-Leu-Gln
Ac-Asp-Phe-Glu-Glu-Ile-Pro-Glu-Glu-Tyr-Leu-Gln
Non-Sulfated

Synonyms: Hirudin (55-65), Ac-

ICN 153124   MW 1453.5

Asp-Phe-Glu-Glu-Ile-Pro-Glu-Glu-Tyr-Leu-Gln
Ac-Asp-Phe-Glu-Glu-Ile-Pro-Glu-Glu-Tyr(SO₃H)-Leu-Gln Sulfated

Synonyms: Hirudin (55-65), Ac-(Tyr(SO₃H)[63])-

ICN 153125   MW 1533.6

Asp-Phe-Glu-Glu-Ile-Pro-Glu-Glu-Tyr-Leu-Gln

Synonyms: Hirudin (55-65)

Sigma H 7019   FW 1411.5   ≥97% (HPLC); non-sulfated |
Bioactive peptide; see Sigma H 6894

Asp-Phe-Glu-Glu-Ile-Pro-Glu-Glu-Tyr-Leu-Gln
Asp-Phe-Glu-Glu-Ile-Pro-Glu-Glu-Tyr(SO₃H)-Leu-Gln

Synonyms: Hirudin (55-65), (Tyr(SO₃H)[63])-

Sigma H 8894   FW 1491.5   ≥97% (HPLC) | Bioactive
peptide; see Sigma H 6894

Asp-Phe-Glu-Glu-Ile-Pro-Glu-Glu-Tyr-Leu-Gln Non-Sulfated

Synonyms: Hirudin (55-65)

ICN 153120   MW 1411.5

Asp-Phe-Me-Phe-Gly-Leu-Met
Suc-Asp-Phe-Me-Phe-Gly-Leu-Met Amide

Synonyms: Substance P, (Suc-Asp[6],Me-Phe[8])-; Senktide (6-11)

American Peptide 70-1-99   MW 842.1 $C_{40}H_{55}N_7O_{11}S$
Specific agonist for neurokinin-3 receptor; Laufer, R et al, *JBC*,
261:10257, 1986

Asp-Phe-Phe-Gly-Leu-Met
Suc-Asp-Phe-N-Me-Phe-Gly-Leu-Met Amide

Synonyms: Substance P (6-11), Suc-(Asp⁶,N-Me-Phe⁸)-; Senktide

Bachem H-5600.0001   MW 841.98   C₄₀H₅₅N₇O₁₁S   Store at -15°C

Asp-Phe-Phe-Gly-Leu-Met
Suc-Asp-Phe-MePhe-Gly-Leu-Met Amide

Synonyms: Senktide; Neurokinin B Receptor Peptide

Calbiochem 05-23-0613   MW 842.0   C₄₀H₅₅N₇O₁₁S   >95% (HPLC); lyophilized solid; soluble in 3% NH₄OH | Selective; specific agonist for the neurokinin-3 receptor; Bertland, PP & Galligan, JJ, *J Physiol*, 481:47, 1994; Laufer, R et al, *J Pharmacol Exp Ther*, 245:889, 1988

Asp-Phe-Phe-Gly-Leu-Met
Succinyl-Asp-Phe-(N-Me)-Phe-Gly-Leu-Met Amide

Synonyms: Substance P (6-11), (Suc-Asp⁶,N-Me-Phe⁸)-; Senktide

ICN 158729   MW 842.1   >97% | Highly specific neurokinin NK₃ receptor agonist; Laufer, R etal, *JBC*, 261:10257, 1986

Asp-Phe-Phe-Gly-Leu-Met
Suc-Asp-Phe-MePhe-Gly-Leu-Met Amide

Synonyms: Substance P (6-11), Suc-(Asp⁶,MePhe⁸)-; Senktide

Neosystem SC392   MW 841.94   Bioactive; tachykinin; selective for neurokinin B receptor; Wormser, U et al, *The EMBO Journal*, 5:2805, 1986

Asp-Phe-Phe-Gly-Leu-Met
Suc-Asp-Phe-N-Methyl-Phe-Gly-Leu-Met Amide

Synonyms: Substance P (6-11), Suc-(Asp⁶,N-Me-Phe⁸)-; Senktide; Neurokinin B Receptor Peptide; Substance P N-Receptor Agonist

Sigma S 6772   FW 842.0   C₄₀H₅₅N₇O₁₁S   ≥97% (HPLC) | Bioactive peptide; selective; specific agonist of the Substance P N-receptor; Wormser, U et al, *EMBO J*, 5:2805, 1986

Asp-Pro
Asp-Pro-pNA

Bachem L-1715.0250   MW 350.33   C₁₅H₁₈N₄O₆   Store at -15°C

Asp-Pro-Ala-Leu-Pro-Thr-Arg-Glu-Gly-Lys-Glu-Gly-Asp-Gly-Gly-Glu-Gly-Gly-Gly

Synonyms: GP140 (742-760); SIVmac251 Peptide

Neosystem SC782   MW 1798.84   SIVmac251 peptide from subtype GP140

Asp-Pro-Ala-Phe-Asn-Ser-Trp-Gly Amide

Synonyms: Leukokinin I

American Peptide 33-1-14   MW 892.0   C₄₁H₅₃N₁₁O₁₂   Neuropeptide originally isolated from head extracts of the Madiera cockroach, *Leucophaea maderae*; Holman, GM et al, *Comp Biochem Physiol*, 84C:205, 1986

Bachem H-6835.0001   Store at -15°C

ICN 153138   MW 891.9   C₄₁H₅₃N₁₁O₁₂   Holman, GM etal, *Comp Biochem Physiol C*, 84:205, 1986

Sigma L 2896   *Leucophaea maderae* (Madeira cockroach)   FW 891.9   C₄₁H₅₃N₁₁O₁₂   ≥97% (HPLC) | Bioactive peptide; neuropeptide originally isolated from head of Madeira cockroach; stimulates contraction of lower digestive tract; Holman, GM et al, *Comp Biochem Physiol C*, 84:205, 1986

Asp-Pro-Ala-Phe-Ser-Ser-Trp-Gly Amide

Synonyms: Leukokinin VII

American Peptide 33-1-28   MW 864.9   C₄₀H₅₂N₁₀O₁₂   Holman, GM et al, *Comp Biochem Physiol*, 88C:31, 1987

Asp-Pro-Arg
BOC-Asp(OBzl)-Pro-Arg-AMC Hydrochloride

Bachem I-1560.0050   MW 770.28   C₃₇H₄₇N₇O₉ · HCl   Store at -15°C

Asp-Pro-Arg
BOC-Asp(OBzl)-Pro-Arg-MCA

Synonyms: Thrombin Substrate, α-

Peptides International MDR-3139-v   MW 733.82   C₃₇H₄₇N₇O₉   >99% (HPLC); lyophilized amorphous powder | Kawabata, S et al, *Eur J Biochem*, 172:17, 1988

Asp-Pro-Arg
N-t-BOC-β-Benzyl-Asp-Pro-Arg-AMC Hydrochloride

Synonyms: Thrombin Substrate, α-

Sigma B 4028   FW 770.3   C₃₇H₄₇N₇O₉ · HCl   Kawabata, S et al, *Eur J Biochem*, 172:17, 1988

Asp-Pro-Gln-Ile-Gln-Ala-Ile-Val-Ser-Ser-Pro-Cys-His-Asn-Ser

Synonyms: GP46 (278-292); HTLV I Peptide

Neosystem SC831   MW 1595.74   HTLV I peptide; due to the presence of a Cys this peptide can contain the dimeric form

Asp-Pro-Gln-Phe-Tyr Hydrochloride

Bachem H-9880.0005   Store at -15°C

Asp-Pro-Glu-Arg-Glu-Val-Leu-Glu-Trp-Arg-Phe-Asp-Ser-Arg-Leu-Ala

Synonyms: NEF (175-190); HTLV I Peptide

Neosystem SC225   MW 2018.21   HTLV I peptide from subtype NEF

Asp-Pro-Gly-Phe-Ser-Ser-Trp-Gly Amide

Synonyms: Leukokinin II

American Peptide 33-1-17   MW 850.9   C₃₉H₅₀N₁₀O₁₂   Neuropeptide originally isolated from head extracts of the Madiera cockroach, *Leucophaea maderae*; Holman, GM et al, *Comp Biochem Physiol*, 54C:205, 1986

Bachem H-6830.0001   Store at -15°C

ICN 153139   MW 850.9   C₄₁H₅₃N₁₁O₁₂   Holman, GM etal, *Comp Biochem Physiol C*, 84:205, 1986

Asp-Pro-His-Asp-Phe-Trp-Val-Trp-Leu-Nle
Asp-D-Pro-His-Asp-Phe-D-Trp-Val-D-Trp-Leu-Nle Amide

Synonyms: Neurokinin B, (D-Pro²,D-Trp⁶,⁸,Nle¹⁰)-; Neurokinin Antagonist

American Peptide 62-0-45   MW 1326.5   C₆₇H₈₇N₁₅O₁₄   Neurokinin antagonist; Vaught, JL et al, *Substance P & Neurokinins*, abstr, 1986

Bachem H-9290.0001   Store at -15°C

ICN 154458   MW 1326.6   Vaught, JL etal, *Substance P & Neurokinins*, 1986 (abstract)

Asp-Pro-Met-Ser-Ser-Thr-Tyr-Ile-Glu-Glu-Leu-Gly-Lys-Arg-Glu-Val-Thr-Ile-Pro-Pro-Lys-Tyr-Arg-Glu-Leu-Leu-Ala

S Calpain Inhibitor Peptide ynonyms:

American Peptide 81-5-40   MW 3136.6   C₁₄₀H₂₂₇N₃₅O₄₄S   Selective inhibitor of calpain I & II which exhibits no effect on papain or trypsin activity; Maki, M et al, *JBC*, 264:18866, 1989

Asp-Pro-Met-Ser-Ser-Thr-Tyr-Ile-Glu-Glu-Leu-Gly-Lys-Arg-Glu-Val-Thr-Ile-Pro-Pro-Lys-Tyr-Arg-Glu-Leu-Leu-Ala
Ac-Asp-Pro-Met-Ser-Ser-Thr-Tyr-Ile-Glu-Glu-Leu-Gly-Lys-Arg-Glu-Val-Thr-Ile-Pro-Pro-Lys-Tyr-Arg-Glu-Leu-Leu-Ala Amide

Synonyms: Calpastatin Peptide

Calbiochem 208902 MW 3177.7 $C_{142}H_{230}N_{36}O_{44}S$ TFA; ≥95% (HPLC); lyophilized solid; soluble in water | A 27-residue peptide encoded by exon 1B of human calpastatin that acts as a cell-permeable & potent inhibitor of calpain I & II (IC_{50}=20 nM for purified rabbit calpain II); does not inhibit either papain or trypsin; blocks the down-regulation of protein kinase Cε (PKCε) in rat pituitary GH_4C_1 cells stimulated by thyrotropin-releasing hormone (TRH); Eto, A et al, *J Bio Chem*, 270:25115, 1995; Kawasaki, J et al, *J Biochem*, 106:274, 1989; Maki, M et al, *J Biol Chem*, 264:18866, 1989

Asp-Pro-Met-Ser-Ser-Thr-Tyr-Ile-Glu-Glu-Leu-Gly-Lys-Arg-Glu-Val-Thr-Ile-Pro-Pro-Lys-Tyr-Arg-Glu-Leu-Leu-Ala

S Calpain Inhibitor; Calpain I Inhibitor; Calpain II Inhibitor
ynonyms:

Sigma C 9181 FW 3136.6 ≥95% (HPLC) | Bioactive peptide; does not inhibit papain or trypsin; Maki, M et al, *J Biol Chem*, 264:18866, 1989

Asp-Pro-Met-Ser-Ser-Thr-Tyr-Ile-Glu-Glu-Leu-Gly-Lys-Arg-Glu-Val-Thr-Ile-Pro-Pro-Lys-Tyr-Arg-Glu-Leu-Leu-Ala
Ac-Asp-Pro-Met-Ser-Ser-Thr-Tyr-Ile-Glu-Glu-Leu-Gly-Lys-Arg-Glu-Val-Thr-Ile-Pro-Pro-Lys-Tyr-Arg-Glu-Leu-Leu-Ala Amide

Synonyms: Calpastatin (184-210), Ac-; Calpain Inhibitor (184-210), Ac-; Sperm BS-17 Component (184-210), Ac-

Bachem H-4076.0500 Human MW 3177.67 $C_{142}H_{230}N_{36}O_{44}S$ Store at -15°C

Sigma C 4285 Human Bioactive peptide; Yamazaki, T et al, *Biochemistry*, 36:8377, 1997

Asp-Pro-Pro-Asn-Pro-Asp-Arg-Phe-Tyr-Gly-Met-Met Amide

Synonyms: Entero-Hylambatin; Hylambatin, Entero-

Bachem H-9325.0005 Store at -15°C

ICN 154452 MW 1438.8 Negri, L etal, *Regulatory Peptides*, 22:13, 1988

Asp-Pro-Pro-Asp-Pro-Asp-Arg-Phe-Tyr-Gly-Met-Met Amide

Synonyms: Hylambatin

Bachem H-9320.0005 Store at -15°C

ICN 154451 MW 1439.7 Negri, L etal, *Regulatory Peptides*, 22:13, 1988

Asp-Pro-Val-Ile-Tyr-Phe-His-Arg

Synonyms: Angiotensin II

Biogenesis 0560-0609 Human synthetic MW 1045 99.1% (HPLC); lyophilized

Asp-Ser-Ala-Pro-Asn-Pro-Val-Leu-Asp-Ile-Asp-Gly-Glu-Lys-Leu-Arg-Thr-Gly-Thr-Asn

Synonyms: Miraculin (1-20)

American Peptide 82-8-10 MW 2112.5 $C_{88}H_{146}N_{26}O_{34}$ Theeraslip, S etal, *JBC*, 263:11536, 1988

Bachem H-9455.0001 Store at -15°C

ICN 154573 MW 2112.5 Theeraslip, S etal, *JBC*, 263:11536, 1988

Asp-Ser-Arg-Leu-Ala-Phe-His-His-Val-Ala

Synonyms: NEF (186-195); HTLV I Peptide

Neosystem SC668 MW 1152.27 HTLV I peptide from subtype NEF

Asp-Ser-Asp-Pro-Arg

Synonyms: Hamburger Pentapeptide; IgE (330-334); Human IgE Pentapeptide; IgE Peptide III; IgE Pentapeptide HEPP

Bachem H-6270.0005 MW 588.58 $C_{22}H_{36}N_8O_{11}$ Store at -15°C

ICN 153061 MW 588.6 $C_{22}H_{36}N_8O_{11}$ Crystalline | Hamburger, RN, *Science*, 189:389, 1975; Stanworth, DR etal, *Int Arch Allergy Appl Immunol*, 56:409, 1978

Sigma A 3526 FW 588.6 $C_{22}H_{36}N_8O_{11}$ ≥90% (HPLC) | Bioactive peptide; Hamburger, RN, *Science*, 189:389, 1975; Stanworth, DR et al, *Int Arch Allergy Appl Immunol*, 56:409, 1978

American Peptide 82-0-51 Human MW 588.6 $C_{22}H_{36}N_8O_{11}$ Inhibits IgE-mediated allergic reactions in humans

Asp-Ser-Gln-Arg-Lys-Leu-Gln-Phe-Tyr-Glu-Asp-Lys-His-Gln-Leu-Pro-Ala-Pro-Lys-Cys

Synonyms: JAK2 Blocking Peptide (765-783)

Calbiochem 420094 Mouse synthetic MW 2431.8 $C_{107}H_{167}N_{31}O_{32}S$ Lyophilized | Based on the mouse JAK2 (residues 765-783) with a cys C residue added; this peptide coupled to KLH, was used as the immunogen for the production of Anti-JAK2; for use in immunoabsorption for immunoprecipitation, immunocytochemistry, Western blotting & dot blots

Asp-Ser-Gln-Ile-Leu-Lys-Glu-Leu-Glu-Glu-Ser-Ser-Phe-Arg

Synonyms: Insulin Receptor (689-702)

Bachem H-4212.0001 Store at -15°C

Asp-Ser-Gly-Cys-Phe-Gly-Arg-Arg-Leu-Asp-Arg-Ile-Gly-Ser-Leu-Ser-Gly-Leu-Gly-Cys-Asn-Val-Leu-Arg-Arg-Tyr

Synonyms: Natriuretic Peptide (7-32), Brain ; Natriuretic Peptide 26, Brain

American Peptide 14-4-15 Porcine MW 2869.3 $C_{120}H_{198}N_{42}O_{36}S_2$ Disulfide bonds: :Cys^{10}-Cys^{26} | Endogenous neuropeptide with diuretic & hypotensive properties

Bachem H-2948.0500 Porcine MW 2869.29 $C_{120}H_{198}N_{42}O_{36}S_2$ Store at -15°C

Asp-Ser-Gly-Cys-Phe-Gly-Arg-Arg-Leu-Asp-Arg-Ile-Gly-Ser-Leu-Ser-Gly-Leu-Gly-Cys-Asn-Val-Leu-Arg-Arg-Tyr
H₂N-Asp-Ser-Gly-Cys-Phe-Gly-Arg-Arg-Leu-Asp-Arg-Ile-Gly-Ser-Leu-Ser-Gly-Leu-Gly-Cys-Asn-Val-Leu-Arg-Arg-Tyr

Natriuretic Peptide, Brain *Synonyms*:

ICN 154503 Porcine MW 2869.3 Sudoh, T etal, *Nature*, 332:78, 1988

Asp-Ser-Gly-Cys-Phe-Gly-Arg-Arg-Leu-Asp-Arg-Ile-Gly-Ser-Leu-Ser-Gly-Leu-Gly-Cys-Asn-Val-Leu-Arg-Arg-Tyr

Synonyms: Natriuretic Peptide, Brain

Neosystem SC186 Porcine MW 2869.26 Disulfide bonds: Cys^4-Cys^{20}; bioactive; ANF-related peptide; Sudoh, T et al, *Nature*, 332:78-81, 1988

Asp-Ser-Gly-Cys-Phe-Gly-Arg-Arg-Leu-Asp-Arg-Ile-Gly-Ser-Leu-Ser-Gly-Leu-Gly-Cys-Asn-Val-Leu-Arg-Arg-Tyr

Synonyms: Natriuretic Peptide 26, B-Type; Natriuretic Peptide 26, Brain

Peptides International PBN-4200-v Porcine synthetic MW 2869.3 $C_{120}H_{198}N_{42}O_{36}S_2$ >99% (HPLC); lyophilized amorphous powder | Bioactive; disulfide bonds: Cys^4-Cys^{20}; Rosenzweig, A & CE Seidman, *Ann Rev Biochem*, 60:229, 1991; Sudoh, T et al, *Nature*, 332:78, 1988

Asp-Ser-His-Ala-Lys-Arg-His-His-Gly-Tyr-Lys-Arg-Lys-Phe-His-Glu-Lys-His-His-Ser-His-Arg-Gly-Tyr

Synonyms: Histatin V; Clostripain Inhibitor; *Bacteroides gingivalis* Protease Inhibitor ; Parotid Histidine-Rich Protein

American Peptide 72-2-25 MW 3036.3 $C_{133}H_{195}N_{51}O_{33}SCl$ Antibiotic peptide; Nishikata, M et al, *Biochem Biophys Res Comm*, 174:625, 1991

Bachem H-3144.0500 MW 3036.34 $C_{133}H_{195}N_{51}O_{33}$ Store at -15°C

Sigma H 6027 Human saliva FW 3036.3 ≥97% (HPLC) | Bioactive peptide; Nishikata, M et al, *Biochem Biophys Res Commun*, 174:625, 1991

ICN 195667 Human salivary MW 3036.3 97% | A gingivalis & clostripain protease inhibitor

Peptides International PHS-4270-s Human synthetic MW 3036.3 $C_{133}H_{195}N_{51}O_{33}$ >99% (HPLC); lyophilized amorphous powder | Bioactive; Oppenheim, FG et al, *JBC*, 263:7472, 1988; Raj, PA et al, *JBC*, 265:3898, 1990; Murakami, Y et al, *Arch Oral Biol*, 35:775, 1990; Nishikata, M et al, *BBRC*, 174:625, 1991

Asp-Ser-Met-Ser-Cys-Ser-Thr-Ser-Leu-Ala-Pro-Val-Phe-Pro Acetate Salt

Synonyms: Glutamate Receptor VI (875-888)

Biogenesis 4670-5289 Synthetic >80% (HPLC); lyophilized

Asp-Ser-Phe-Trp-Ala-Leu-Met
Asp-Ser-Phe-Trp-α-Ala-Leu-Met Amide

Synonyms: Neurokinin A (4-10), $(Trp^7,\alpha-Ala^8)$-

Bachem H-2788.0001 MW 868.02 $C_{41}H_{57}N_9O_{10}S$ Store at -15°C

Asp-Ser-Phe-Val-Ala-Leu-Met
Asp-Ser-Phe-Val-α-Ala-Leu-Met Amide

Synonyms: Neurokinin A (4-10), $(\alpha-Ala^8)$-

Bachem H-2786.0001 Store at -15°C

Calbiochem 05-23-0713 MW 780.9 $C_{35}H_{56}N_8O_{10}S$ ≥96% (HPLC); lyophilized solid; soluble in 3% NH_4OH | Potent & selective NK-2 receptor agonist; stimulates ileal motility & increases intracellular Ca^{2+} levels; Subramanian, N et al, *Biochem Biophys Res Comm*, 200:1512, 1994; Rovero, P et al, *Neuropeptides*, 13:263, 1989; Rovero, P et al, *Peptides*, 10:593, 1989

Neosystem SC922 MW 780.94 Bioactive; tachykinin; potent & selective NK-2 receptor agonist; Evangelista, S et al, *Peptides*, 11:293-297, 1990

Asp-Ser-Phe-Val-Ala-Leu-Met-Asp-Ser-Phe-Val-Ala-Leu-Met
Asp-Ser-Phe-Val-Ala-Leu-Met-Asp-Ser-Phe-Val-β-Ala-Leu-Met Amide

Synonyms: Neurokinin A (4-10), $(\beta-Ala^8)$-; Neurokinin II Receptor Agonist

ICN 159977 MW 780.9 >97% | Highly specific

Asp-Ser-Phe-Val-Gly-Leu-Met Amide

Synonyms: Neurokinin A (4-10); Neurokinin (4-10), α-; Neurokinin α (4-10); Neurokinin II Agonist

American Peptide 62-1-50 MW 766.9 $C_{34}H_{54}N_8O_{10}S$

Bachem H-5955.0001 MW 766.92 $C_{34}H_{54}N_8O_{10}S$ Store at -15°C

ICN 159976 MW 766.9 $C_{34}H_{54}N_8O_{10}S$

Sigma N 5141 FW 766.9 $C_{34}H_{54}N_8O_{10}S$ ≥97% (HPLC) | Bioactive peptide; Osakada, F et al, *Eur J Pharmacol*, 120:210, 1986; Munekata, E et al, *Peptides*, 8:169, 1987

Asp-Ser-Phe-Val-Gly-Leu-Nle Amide

Synonyms: Neurokinin II Receptor Agonist; Neurokinin A (4-10), (Nle^{10})-

Bachem H-9275.0001 MW 748.88 $C_{35}H_{56}N_8O_{10}$ Store at -15°C

Calbiochem 05-23-0712 MW 748.9 $C_{35}H_{56}N_8O_{10}$ ≥94% (HPLC); solid; soluble in dilute NH_4OH | Highly selective NK-2 receptor agonist; increases motor activity by a mechanism independent of vasoactive peptide output; Watson, EG et al, *Can J Physiol Pharmacol*, 72:109, 1994; Dion, S et al, *Neuropeptides*, 11:83, 1988; Regoli, D et al, *Trends Pharmacol Sci*, 9:290, 1988

ICN 154455 MW 748.9 Drapeau, G etal, *Neuropeptides*, 10:43, 1987; Quirion, R etal, *Regulatory Peptides*, 22:18, 1988

Neosystem SC389 MW 748.87 Bioactive; tachykinin; selective NK-2 receptor agonist; Quirion, R & Dans, T, *Regul Peptides*, 22:18, 1988

Asp-Thr-Ala-Ser-Asp-Ala-Ala-Ala-Ala-Ala-Ala-Leu-Thr-Ala-Ala-Asn-Ala-Lys-Ala-Ala-Ala-Glu-Leu-Thr-Ala-Ala-Asn-Ala-Ala-Ala-Ala-Ala-Ala-Thr-Ala-Arg HPLC-6

Synonyms: Antifreeze Polypeptide

American Peptide 79-2-10 Winter flounder MW 3242.5 $C_{133}H_{225}N_{43}O_{51}$ Yang, D et al, *Nature*, 333:232, 1988

Asp-Thr-Asn-Leu-Ala-Ser-Ser-Thr-Ile-Ile-Lys-Glu-Gly-Ile-Asp-Lys-Thr-Val

Synonyms: Peptide M

Bachem H-1018.0001 MW 1905.13 $C_{81}H_{141}N_{21}O_{31}$ Store at -15°C

Asp-Thr-Gly-Ala-Asp-Asp-Thr-Val-Leu-Glu-Glu-Met-Ser-Leu-Pro

Synonyms: POL (25-39); PR (25-39); HTLV I Peptide

Neosystem SC702 MW 1592.69 HTLV I peptide from subtype POL (PR/RT)

Asp-Thr-Gly-His-Gly-Leu-Arg-Leu-Ile-His-Tyr-Ser-Tyr-Gly-Ala-Gly-Ser-Thr-Glu-Lys-Gly

Synonyms: T-Cell Receptor Peptide Vβ8.2 (39-59)

Bachem H-1984.0500 Murine MW 2219.4 $C_{96}H_{147}N_{29}O_{32}$ Store at -15°C

Asp-Thr-Gly-His-Ser-Asn-Gln-Val-Ser-Gln-Asn-Tyr

Synonyms: P18 (121-132); GAG P17 MA (121-132); HTLV I Peptide

Neosystem SC534 MW 1349.33 HTLV I peptide from subtype P25 (GAG P17 MA)

Asp-Thr-Gly-Ser-Ser-Ser-Lys-Val-Ser-Gln-Asn-Tyr-Pro-Ile-Val-Gln-Asn-Ala-Gln-Gly

Synonyms: HIV I gag Polyprotein (121-140)

Bachem H-4776.0500 MW 2080.2 $C_{86}H_{138}N_{26}O_{34}$ Store at -15°C

Asp-Thr-Met-Arg-Cys-Met-Val-Gly-Arg-Val-Tyr-Arg-Pro-Cys-Trp-Glu-Val

Synonyms: Melanin Concentrating Hormone

American Peptide 72-0-61   Salmon   MW 2099.5
$C_{89}H_{139}N_{270}O_{24}S_4$ Disulfide bonds: Cys[5]-Cys[14]; Kawauchi, H et al, *Nature*, 305:321, 1983

Bachem H-8560.0500   Salmon   MW 2099.51
$C_{89}H_{139}N_{27}O_{24}S_4$ Store at -15°C

ICN 159965   Salmon   MW 2099.8   Kawauchi, H etal, *Nature*, 305:321, 1983

Neosystem SC425   Salmon   MW 2097.93   Disulfide bonds: Cys[5]-Cys[14]; bioactive neuropeptide; Kawauchi, H et al, *Nature*, 305:321, 1983

Asp-Thr-Ser-His-His-Asp-Gln-Asp-His-Pro-Thr-Phe-Asp Amide

Synonyms: Gonadotropin Releasing Peptide, Follicular

American Peptide 54-1-51   Human   MW 1550.5
$C_{64}H_{87}N_{21}O_{25}$

Asp-Trp

Bachem G-3705.0250   MW 319.32 $C_{15}H_{17}N_3O_5$ Store at -15°C

Asp-Trp-Gln-Asp-Tyr-Thr-Ser-Gly-Pro-Gly-Ile-Arg-Tyr-Pro-Lys

Synonyms: NEF (155-169); SIVmac251 Peptide

Neosystem SC618   MW 1782.93   SIVmac251 peptide from subtype NEF

Asp-Trp-Leu-Lys-Ala-Phe-Tyr-Asp-Lys-Val-Ala-Glu-Lys-Leu-Lys-Glu-Ala-Phe Acetate Salt

Synonyms: Apolipoprotein A-1

Biogenesis 0650-0349   Human synthetic   MW 2201.8
>97% (HPLC), >99% (TLC); lyophilized

Asp-Tyr
BOC-Asp(OBzl)-Tyr(Bzl)-NHNH-Z

Bachem A-2660.0001   Store at -15°C

Asp-Tyr Non-Sulfated

Synonyms: Cholecystokinin Octapeptide (1-2)

Bachem G-1630.0250   MW 296.28 $C_{13}H_{16}N_2O_6$ Store at -15°C

Asp-Tyr-Lys-Asp-Asp-Asp-Asp-Lys Acetate Salt

Synonyms: FLAG Epitope (78-85)

Biogenesis 4497-1060   Synthetic   >80% (HPLC); lyophilized | Represents AA 78-85 of the 98-mer coded by the cloning vector pVP-FLAG

Asp-Tyr-Met

Synonyms: Cholecystokinin (26-28)

ICN 154542   MW 427.5

Asp-Tyr-Met Non-Sulfated

Synonyms: Cholecystokinin Octapeptide (1-3)

Bachem H-2052.0050   MW 427.48 $C_{18}H_{25}N_3O_7S$ Store at -15°C

Asp-Tyr-Met-Gly
Asp-Tyr(SO₃H)-Met-Gly Sulfated

Synonyms: Cholecystokinin Octapeptide (1-4); Cholecystokinin (26-29)

Bachem H-2060.0005   Store at -15°C

ICN 154543   MW 564.6

Asp-Tyr-Met-Gly

Synonyms: Cholecystokinin (26-29)

ICN 154544   MW 484.6

Asp-Tyr-Met-Gly
Asp-Tyr(SO₃H)-Met-Gly Amide

Synonyms: Cholecystokinin (26-29), (Tyr(SO₃H)[27])-; Gastrointestinal Peptide

Sigma C 5033   FW 563.6 $C_{20}H_{29}N_5O_{10}S_2$ ≥97% (HPLC) | Bioactive peptide

Asp-Tyr-Met-Gly Non-Sulfated

Synonyms: Cholecystokinin Octapeptide (1-4)

Bachem H-2055.0050   MW 484.53 $C_{20}H_{28}N_4O_8S$ Store at -15°C

Asp-Tyr-Met-Gly-Trp
Ac-Asp-Tyr(SO₃H)-Met-Gly-Trp Amide Sulfated

Synonyms: Cholecystokinin (26-30), N-Ac-

American Peptide 46-1-41   MW 791.9 $C_{33}H_{41}N_7O_{12}S_2$

Asp-Tyr-Met-Gly-Trp
Ac-Asp-Tyr-Met-Gly-Trp Amide

Synonyms: Cholecystokinin (26-30), Ac-; Gastrointestinal Peptide

Sigma C 4533   FW 711.8 $C_{33}H_{41}N_7O_9S$ ≥90% (HPLC); non-sulfated | Bioactive peptide

Asp-Tyr-Met-Gly-Trp Amide Sulfated

Synonyms: Cholecystokinin (26-30)

ICN 154545   MW 670.8

Asp-Tyr-Met-Gly-Trp Non-Sulfated

Synonyms: Cholecystokinin Octapeptide (1-5)

Bachem H-2065.0025   MW 670.74 $C_{31}H_{38}N_6O_9S$ Store at -15°C

Asp-Tyr-Met-Gly-Trp-Met
Ac-Asp-Tyr(SO₃H)-Met-Gly-Trp-Met

Synonyms: Cholecystokinin (26-31), N-Ac-

American Peptide 46-1-38   MW 923.1 $C_{38}H_{50}N_8O_{13}S_3$

Asp-Tyr-Met-Gly-Trp-Met
Ac-Asp-Tyr-Met-Gly-Trp-Met Amide Non-Sulfated

Synonyms: Cholecystokinin (26-31), N-Ac-

American Peptide 46-1-39   MW 843.0 $C_{38}H_{50}N_8O_{10}S_2$

Asp-Tyr-Met-Gly-Trp-Met
Ac-Asp-Tyr-Met-Gly-Trp-Met Amide

Synonyms: Cholecystokinin (26-31), Ac-; Gastrointestinal Peptide

Sigma C 2546   FW 843.0 $C_{38}H_{50}N_8O_{10}S_2$ ≥97% (HPLC); non-sulfated | Bioactive peptide

Asp-Tyr-Met-Gly-Trp-Met
Ac-Asp-Tyr(SO₃H)-Met-Gly-Trp-Met Amide

Synonyms: Cholecystokinin (26-31), Ac-(Tyr(SO₃H)[27])-; Gastrointestinal Peptide

Sigma C 2795   FW 923.0 $C_{38}H_{50}N_8O_{13}S_3$ ≥97% (HPLC) | Bioactive peptide

Asp-Tyr-Met-Gly-Trp-Met Amide Non-Sulfated

Synonyms: Cholecystokinin (26-31)

ICN 154546   MW 802.0

314

Asp-Tyr-Met-Gly-Trp-Met Non-Sulfated

Synonyms: Cholecystokinin Octapeptide (1-6)

Bachem H-2070.0005   Store at -15°C

Asp-Tyr-Met-Gly-Trp-Met-Asp-Phe

Synonyms: Cholecystokinin Octapeptide (26-33)

American Peptide 46-1-33   MW 1063.3 $C_{49}H_{62}N_{10}O_{16}S_3$

Asp-Tyr-Met-Gly-Trp-Met-Asp-Phe
BOC-Asp-Tyr-Met-Gly-Trp-Met-Asp-Phe Amide Desulfated

Synonyms: Cholecystokinin Octapeptide, BOC-

Bachem A-1295.0001   MW 1163.34 $C_{54}H_{70}N_{10}O_{15}S_2$ Store at -15°C

Asp-Tyr-Met-Gly-Trp-Met-Asp-Phe
Asp-Tyr(SO₃H)-Met-Gly-Trp-Met-Asp-Phe Amide Sulfated

Synonyms: Cholecystokinin Octapeptide (26-33); Sincalide

Bachem H-2080.0001   MW 1143.29 $C_{49}H_{62}N_{10}O_{16}S_3$ Store at -15°C

Asp-Tyr-Met-Gly-Trp-Met-Asp-Phe
Asp-Tyr(SO₃H)-Met-Gly-Trp-Met-Asp-Phe Amide

Synonyms: Cholecystokinin (26-33), (27-Tyr(SO₃H)); Cholecystokinin 8; Pancreozymin C-Terminal Octapeptide; Sincalide

Fluka 26690   MW 1143.3 $C_{49}H_{62}N_{10}O_{16}S_3$ ≥98% (HPLC); ≥85% peptide content | Rodriguez, RE & Sacristan, MP, *FEBS Lett*, 250:215, 1989; Emson, PC & Sandberg, BEB, *Ann Rep Med Chem*, 18:31, 1983

Asp-Tyr-Met-Gly-Trp-Met-Asp-Phe
Asp-Tyr(SO₃H)-Met-Gly-Trp-Met-Asp-Phe Amide Sulfated

Synonyms: Cholecystokinin (26-33); Cholecystokinin 8

ICN 190684   MW 1143.3   Ondetti, MA etal, *Digest Dis*, 15:149, 1970; Innis, RB & SH Snyder, *PNAS*, 77:6917, 1980; Saito H etal, *Science*, 208:1155:1980

Neosystem SC047   MW 1143.26   Bioactive; brain/gut peptide; CCK/gastrin peptide; Ondetti, MA et al, *J Am Chem Soc*, 92:195, 1970

Asp-Tyr-Met-Gly-Trp-Met-Asp-Phe
Asp-Tyr(SO₃H)-Met-Gly-Trp-Met-Asp-Phe Amide

Synonyms: Cholecystokinin (26-33), (Tyr(SO₃H)²⁷)-; Gastrointestinal Peptide

Sigma C 2175   FW 1143.3   ≥97% (HPLC) | Bioactive peptide; C-terminal octapeptide; neurotransmitter; predominant form of CCK in CNS & gastrointestinal tract; may play a role in satiety; Ondetti, MA et al, *Digestive Diseases*, 15:149, 1970; Saito, H et al, *Science*, 208:1155, 1980; Innis, RB & Snyder, SH, *Proc Natl Acad Sci USA*, 77:6917, 1980

Asp-Tyr-Met-Gly-Trp-Met-Asp-Phe
Asp-Tyr(SO₃H)-Met-Gly-Trp-Met-Asp-Phe Sulfated

Synonyms: Cholecystokinin 8 (26-33)

Biogenesis 2050-0404   Human synthetic   MW 1143.3   98.5% (HPLC); dilute PBS buffer, pH 7.2; lyophilized

---

Asp-Tyr-Met-Gly-Trp-Met-Asp-Phe
Asp-Tyr(SO₃H)-Met-Gly-Trp-Met-Asp-Phe Amide

Synonyms: Cholecystokinin Octapeptide (26-33)

Peptides International PCK-4100-v   Synthetic   MW 1143.3 $C_{49}H_{62}N_{10}O_{16}S_3$ >95% (HPLC); lyophilized amorphous powder; sulfated ammonium form | Bioactive; Ondetti, MA et al, *J Am Chem Soc*, 92:195, 1970

Asp-Tyr-Met-Gly-Trp-Met-Asp-Phe Amide

Synonyms: Cholecystokinin Octapeptide (26-33)

Peptides International PCK-4087-v   Synthetic   MW 1063.2 $C_{49}H_{62}N_{10}O_{13}S_2$ >95% (HPLC); lyophilized amorphous powder; non-sulfated ammonium form

---

Asp-Tyr-Met-Gly-Trp-Met-Asp-Phe Amide Non-Sulfated

Synonyms: Cholecystokinin Octapeptide; Cholecystokinin (26-33); Cholecystokinin 8

Bachem H-2085.0001   MW 1063.22 $C_{49}H_{62}N_{10}O_{13}S_2$ Store at -15°C

ICN 190328   CCK 26-33 Unsulfated

Neosystem SC126   MW 1063.21   Bioactive; brain/gut peptide; CCK/gastrin peptide

Sigma C 2901   FW 1063.2   ≥95% (HPLC); non-sulfated | Biologically inactive form of CCK-8

Asp-Tyr-Met-Gly-Trp-Met-Asp-Phe Free Acid Non-Sulfated

Synonyms: Cholecystokinin Octapeptide

Bachem H-2075.0001   MW 1064.21 $C_{49}H_{61}N_9O_{14}S_2$ Store at -15°C

Asp-Tyr-Met-Gly-Trp-Met-Asp-Phe-Tyr
Asp-Tyr(SO₃H)-Met-Gly-Trp-Met-Asp-Phe-Tyr Sulfated

Synonyms: Cholecystokinin Octapeptide, (Tyr⁹)-

Bachem H-9770.0001   Store at -15°C

Asp-Tyr-Trp-Val-Trp-Trp-Arg
Asp-Tyr-D-Trp-Val-D-Trp-D-Trp-Arg Amide

Synonyms: Neurokinin A (4-10), (Tyr⁵,D-Trp⁶,⁸,⁹,Ala¹⁰)-; Neurokinin II Receptor Antagonist; Neurokinin A (4-10), (Tyr⁵,D-Trp⁶,⁸,⁹,Arg-NH₂¹⁰)-

American Peptide 62-1-30   MW 1109.3 $C_{57}H_{68}N_{14}O_{10}$
Blocked intrathecal neurokinin A-induced facilitation of the spinal nociceptive flexor reflex in the rat; Patacchini, R et al, *NY Acad Sci Conf on Substance P*, Abstr 20-12, 1990

Bachem H-2072.0001   MW 1109.26 $C_{57}H_{68}N_{14}O_{10}$ Store at -15°C

Asp-Tyr-Trp-Val-Trp-Trp-Lys
Asp-Tyr-D-Trp-Val-D-Trp-D-Trp-Lys Amide

Synonyms: MEN-10376; Neurokinin A (4-10), (Tyr⁵,D-Trp⁶,⁸,⁹,Lys-NH₂¹⁰)-; Neurokinin A (4-10), (Tyr⁵,D-Trp⁶,⁸,⁹,Lys¹⁰)-α-; Neurokinin A (4-10), (Tyr⁵,D-Trp⁶,⁸,⁹, Lys¹⁰)-; Neurokinin II Agonist

American Peptide 62-1-39   MW 1081.2 $C_{57}H_{68}N_{12}O_{10}$
Potent & selective NK₂ tachykinin receptor antagonist; Maggi, CA et al, *J Pharm Exp Ther*, 257:1172, 1991; Monteau, R et al, *Eur J Pharmacol*, 314:41, 1996

Bachem H-8565.0001   MW 1081.24 $C_{57}H_{68}N_{12}O_{10}$ Store at -15°C

ICN 159979   MW 1081.2   97% | Maggi, CA etal, *J Pharm & Exp Ther*, 247:1172, 1991

Neosystem SC438   MW 1081.23   Bioactive; tachykinin; potent & highly selective NK-2 receptor antagonist; Maggi, CA et al, *J Pharmacol Exp Ther*, 257:1172-1178, 1991

Sigma N 2272   FW 1081.2   ≥97% (HPLC) | Bioactive peptide; potent, high-affinity; Maggi, CA et al, *J Pharm & Exp Ther*, 257:1172, 1991

Asp-Tyr-Val-Pro-Met-Leu
N-Ac-Asp-Tyr(2-Malonyl)-Val-Pro-Met-Leu Amide

Synonyms: Protein Tyrosine Kinase Inhibitor; Phosphatidyl Inositol 3-Kinase Inhibitor

Alexis 151-026   MW 881.0 $C_{39}H_{58}N_7O_{14}$ ≥96%; soluble in H_2O | Contains a phosphotyrosyl mimetic; inhibits PI-3 kinase C-terminal p85 SH2 domain; may have potential application for treatment of breast cancer & diabetes; Ye, B et al, *J Med Chem*, 38:4270, 1995

Asp-Tyr-Val-Pro-Met-Leu
N-Ac-Asp-Tyr(PO₃H₂)-Val-Pro-Met-Leu Amide

Synonyms: Protein Tyrosine Kinase Substrate

Alexis 151-027   MW 858.9 $C_{36}H_{57}N_7O_{13}$ ≥96%; soluble in H_2O | Phosphotyrosine containing peptide; competes with protein ligands for binding to SH2 domains of protein tyrosine kinase; Ye, B et al, *J Med Chem*, 38:4270, 1995

Asp-Tyr-Val-Pro-Met-Leu
Ac-Asp-Tyr(2-Malonyl)-Val-Pro-Met-Leu Amide

Synonyms: Tyrosine Kinase Inhibitor

American Peptide 86-2-31   MW 879.9 $C_{35}H_{57}N_7O_{14}S$ Very effective; inhibits PI-3 kinase C-terminal p85 SH2 domain binding with IC_{50} of 14.2 & might have potential application for the treatment of breast cancer & diabetes; Ye, B et al, *J Med Chem*, 38:4270, 1995

Asp-Tyr-Val-Pro-Met-Leu
Ac-Asp-Tyr(PO₃H₂)-Val-Pro-Met-Leu Amide

Synonyms: Tyrosine Kinase Inhibitor

American Peptide 86-2-32   MW 857.9 $C_{36}H_{56}N_7O_{13}SP$
Good reagent for SH2 domain inhibitor design with the ability to compete with protein ligands for binding to SH2 domains; Ye, B et al, *J Med Chem*, 38:4270, 1995

Bachem N-1480.0001   Store at -15°C

Asp-Tyr-Val-Pro-Met-Leu
Ac-Asp-Tyr(2-malonyl)-Val-Pro-Met-Leu Amide

Bachem N-1485.0001   MW 879.99 $C_{39}H_{57}N_7O_{14}S$ Store at -15°C

Asp-Tyr-Val-Pro-Met-Leu
Ac-Asp-Tyr(PO₃H₂)-Val-Pro-Met-Leu Amide

Synonyms: SH2 Domain Inhibitor Peptide

Calbiochem 566805   MW 857.9 $C_{36}H_{56}N_7O_{13}PS$ ≥95% (HPLC); solid; soluble in water | Short phosphotyrosine-containing peptide that competes with larger phosphotyrosine-containing peptides & protein ligands for binding to SH2 domains; used as a starting point for SH2 domain inhibitor design; Ye, B et al, *J Med Chem*, 38:4270, 1995

Asp-Val

Bachem G-1635.0250   MW 232.24 $C_9H_{16}N_2O_5$ Store at -15°C

Asp-Val-Ala-His-Glu-Ile-Leu-Asn-Glu-Ala-Tyr-Arg-Lys-Val-Leu-Asp-Gln-Leu-Ser-Ala-Arg-Lys-Tyr-Leu-Gln-Ser-Met-Val-Ala

Synonyms: Pituitary Adenylate Cyclase Activating Polypeptide Related Peptide (1-29)

American Peptide 34-0-52   Rat   MW 3361.9 $C_{148}H_{242}N_{42}O_{45}S$ Ogi, K et al, *BBRC*, 173:1271, 1990

Bachem H-3874.0500   Rat   Store at -15°C

Asp-Val-Ala-His-Gly-Ile-Leu-Asn-Glu-Ala-Tyr-Arg-Lys-Val-Leu-Asp-Gln-Leu-Ser-Ala-Gly-Lys-His-Leu-Gln-Ser-Leu-Val-Ala

Synonyms: Pituitary Adenylate Cyclase Activating Peptide Related Peptide

American Peptide 34-0-22   Human   MW 3146.6 $C_{139}H_{229}N_{41}O_{42}$ Ogi, K et al, *BBRC*, 173:1271, 1990

Asp-Val-Pro-Lys-Ser-Asp-Gln-Phe-Val-Gly-Leu-Met
Amide

Synonyms: Kassinin

American Peptide 62-0-55   MW 1334.6 $C_{59}H_{95}N_{15}O_{18}S$
Tachykinin peptide with substance P-like activity; Anastasi, A et al, *Experientia*, 33:857, 1977

Bachem H-2090.0001   MW 1334.56 $C_{59}H_{95}N_{15}O_{18}S$ Store at -15°C

ICN 151387   Anastasi, A et al, *Experentia*, 33:857, 1977

Neosystem SC140   MW 1334.55   Bioactive; bombesin-related peptide; tachykinin

Asp-Val-Ser-Asp-Gly-Ser-Ala-Glu-Arg-Arg-Pro-Tyr-Thr-Arg-Met-Gly-Ser-Gly-Gly-Leu-Lys-Leu-His-Cys-Val-His-Pro-Ala-Asn-Cys-Pro-Gly-Gly-Leu-Met-Val-Tyr

Synonyms: Neuron Specific Peptide

American Peptide 60-0-75   MW 3868.4 $C_{161}H_{260}N_{52}O_{51}S_4$ Disulfide bonds: Cys^{24}-Cys^{30}; Buck, LB et al, *Cell*, 51:127, 1987

Asp-Val-Ser-Thr-Pro-Pro-Thr-Val-Leu-Pro-Asp-Asn-Phe-Pro-Arg-Tyr

Synonyms: Insulin-Like Growth Factor II (69-84); Prepro-Insulin-Like Growth Factor II (69-84)

American Peptide 50-1-37   MW 1817.9 $C_{83}H_{124}N_{20}O_{26}$

Neosystem SC683   MW 1818.01   Bioactive; synthetic growth factor-related peptide

ICN 154470   Human synthetic   MW 1817.9

Asp-Val-Tyr

Synonyms: Thymopoietin II (34-36)

Bachem H-8765.0050   MW 395.41 $C_{18}H_{25}N_3O_7$ Store at -15°C

Asp-Val-Val-Asp-Ala-Asp-Glu-Tyr-Leu-Ile-Pro-Gln

Synonyms: Epidermal Growth Factor Receptor Peptide (985-996); Epidermal Growth Factor Receptor (985-996); Epidermal Growth Factor Receptor (1005-1016); Epidermal Growth Factor Receptor Peptide (1005-1016)

American Peptide 50-0-24   MW 1376.4 $C_{61}H_{93}N_{13}O_{23}$ Ullrich, A et al, *Nature*, 309:418, 1984

Neosystem SC381   MW 1376.40   Bioactive; synthetic growth factor-related peptide; Ullrich, A et al, *Nature*, 309:418, 1984

Bachem H-5620.0001   Human   MW 1376.48 $C_{61}H_{93}N_{13}O_{23}$ Store at -15°C

Sigma E 1886   Human   FW 1376.5   ≥95% (HPLC) | Bioactive peptide; fragment from the intracellular tyrosinekinase domain of EGF that has a homologous sequence to the tyrosinekinase coded by viral oncogene *erb-B*; used to raise Ab to native EGF & *v-erb-B*; Gullik, WJ et al, *Proc Royal Soc Lond B*, 226:127, 1985

ICN 151514   Human synthetic   Ullrich, A et al, *Nature*, 309:418, 1984; Gullik, WC et al, *Proc Roy Soc Lond B*, 226:127, 1985

Cit-Phe

Bachem G-3995.0050   Store at -15°C

Cpd-Gln-Ala-Leu
N-(-Nα-Carbonyl-Cpd-Gln-Ala-al)-Leu

Synonyms: Elastatinal; Elastase Inhibitor

Sigma E 0881 Microbial FW 512.6 $C_{21}H_{36}N_8O_7$ ~50% | Bioactive peptide; Umezawa, H, *Meth Enzymol*, 45:678, 1976; Barrett, AJ et al, *Meth Enzymol*, 80:581, 1981

Cyclo(-Ala-Ala)

Bachem G-1655.0250 MW 142.16 $C_6H_{10}N_2O_2$ Store at -15°C

Cyclo(-Ala-Ala)
Cyclo(-D-Ala-D-Ala)

Bachem G-4040.0250 MW 142.16 $C_6H_{10}N_2O_2$ Store at -15°C

Cyclo(-Ala-Arg-Gly-Asp)
Cyclo(-Ala-Arg-Gly-Asp-3-Aminomethylbenzoyl)

Synonyms: XJ735

Bachem H-4772.0001 MW 532.56 $C_{23}H_{32}N_8O_7$ Store at -15°C

Cyclo(-Ala-Gly)

Bachem G-1660.0250 MW 128.13 $C_5H_8N_2O_2$ Store at -15°C

Cyclo(-Ala-Gly-Ala-Gly)
Cyclo(-β-Ala-Gly-β-Ala-Gly)

Bachem H-2475.0050 MW 256.26 $C_{10}H_{16}N_4O_4$ Store at -15°C

Cyclo(-Ala-His)

Bachem G-1665.0250 MW 208.22 $C_9H_{12}N_4O_2$ Store at -15°C

Cyclo(-Ala-Pro)
Cyclo(-D-Ala-Pro)

Bachem G-4165.0250 MW 168.2 $C_8H_{12}N_2O_2$ Store at -15°C

Cyclo(-Ala-Ser)

Bachem G-1670.0250 MW 158.16 $C_6H_{10}N_2O_3$ Store at -15°C

Cyclo(Ala-Trp-Leu-Val-Asp-Cys-Pro)
Cyclo(Ala-2-Mercapto-Trp-4,5-Dihydroxy-Leu-Val-Erythro-3-Hydroxy-D-Asp-Cys-cis-4-Hydroxy-Pro)

Synonyms: Phallacidin

ICN 156125 *Amanita phalloides* MW 846.9 $C_{37}H_{48}N_8O_{11}S$ >90%

Cyclo(-Ala-Val)
Cyclo(-D-Ala-Val)

Bachem G-4710.0250 MW 170.21 $C_8H_{14}N_2O_2$ Store at -15°C

Cyclo(-Arg-Ala-Asp-Phe-Val)
Cyclo(-Arg-Ala-Asp-D-Phe-Val)

Bachem H-4088.0001 Store at -15°C

Cyclo(Arg-Gly-Asp-Phe-Val)
Cyclo(Arg-Gly-Asp-D-Phe-Val)

Synonyms: RGD Peptide, Cyclic

ICN 195967 MW 574.6 $C_{26}H_{38}N_8O_7$ ≥97% | Inhibits tumor cell adhesion to laminin & vitronectin substrates

Cyclo(-Arg-Gly-Asp-Phe-Val)
Cyclo(-Arg-Gly-Asp-D-Phe-Val)

Synonyms: Peptide, Cyclic

Bachem H-2574.0001 MW 574.64 $C_{26}H_{38}N_8O_7$ Store at -15°C

Calbiochem 182015 MW 574.6 $C_{26}H_{38}N_8O_7$ Solid; ≥95% (HPLC); soluble in 1 N HOAc | Potent inhibitor of cell adhesion; inhibits tumor cell adhesion to laminin & vitronectin substrates; *in vivo* investigations in rats showed it to be useful in the amelioration of ischemic acute renal failure in rats; inhibits tubular obstruction by preventing cell-cell adhesion; Matsuno, H et al, *Circulation*, 90:2203 1994; Noiri, E et al, *Kidney Int*, 46:1050, 1994; Gurrath, M et al, *Eur J Biochem*, 210:911, 1992

Neosystem SC1221 MW 574.64 Bioactive; *cell attachment* peptide; fibronectin fragment; a highly potent & selective inhibitor for the αvβ3 integrin; prevents retinal neovascularization in a mouse model *in vivo*; this peptide could trigger apoptosis & ICE expression in cultured glomerular mesangial cell (GMC), implicating the survival roles of both fibronectin & vitronectin for GMC's & the potential application of this compound for regulating apoptosis *in vivo*; Aumailley, M et al, *FEBS Lett*, 291:50-54, 1991; Noiri, E et al, *Kidney Int*, 46:1050-1058, 1994; Brooks, PC et al, *Cell*, 79:1157-1164, 1994; Hammes, HP et al, *Nature Medicine*, 2:529-533, 1996; Friedlander, M et al, *PNAS USA*, 93:9764-9769, 1996; Wermuth, J et al, *J Am Chem Soc*, 119:1328-1335, 1997; Chen, X et al, *BBRC*, 234:594-599, 1997

Sigma C 6581 FW 574.6 $C_{26}H_{38}N_8O_7$ ≥97% (HPLC) | Brooks, PC et al, *Cell*, 79:1157, 1994

Cyclo(-Arg-Gly-Asp-Phe-Val)
Cyclo(-Arg-Gly-Asp-D-Phe-Val) AcOH Dihydrate

Synonyms: Angiogenesis Inhibitor

Peptides International ICA-4304 Synthetic MW 670.72 $C_{26}H_{38}N_8O_7 \cdot CH_3COOH \cdot 2H_2O$ >99% (HPLC); amorphous powder; bulk | Inhibitor; Brooks, PC et al, *Cell*, 79:1157, 1994; Friedlander, M et al, *PNAS USA*, 93:9764, 1996

Cyclo(-Arg-Gly-Asp-Phe-Val)
Cyclo(-Arg-Gly-Asp-D-Phe-Val)

Synonyms: Angiogenesis Inhibitor

Peptides International ICA-4304-v Synthetic MW 574.64 $C_{26}H_{38}N_8O_7$ >99% (HPLC); lyophilized amorphous powder | Inhibitor; Brooks, PC et al, *Cell*, 79:1157, 1994; Friedlander, M et al, *PNAS USA*, 93:9764, 1996

Peptides International PCA-3618-PI Synthetic MW 588.68 $C_{27}H_{40}N_8O_7$ >98% (HPLC); white powder | Bioactive; negative control for Peptides International ICA-4304

Cyclo(-Arg-Lys-Pro-Trp-Glu)-Leu

Synonyms: Neurotensin (8-13), (Lys[9],Trp[11],Glu[12])-

Bachem H-2554.0001 MW 809.97 $C_{39}H_{59}N_{11}O_8$ Store at -15°C | Cyclic analog

Cyclo(-Asp-Asp)

Bachem G-1680.0050 MW 230.18 $C_8H_{10}N_2O_6$ Store at -15°C

Cyclo(-Asp-Asp)
Cyclo(-Asp(OMe)-Asp(OMe))

Bachem G-1685.0050 Store at -15°C

Cyclo(-Asp-Gly)

Bachem G-1690.0050 MW 172.14 $C_6H_8N_2O_4$ Store at -15°C

Cyclo(-Asp-Phe)

Bachem G-1695.0050 MW 262.27 $C_{13}H_{14}N_2O_4$ Store at -15°C

Cyclo(Asp-Pro-Ile-Leu-Trp)
Cyclo(D-Asp-Pro-D-Ile-Leu-D-Trp) Trifluoroacetate Salt

Synonyms: JKC-301; JDC-301

Calbiochem 420050   MW 624.7 $C_{32}H_{44}N_6O_7$ ≥98% (HPLC); solid; soluble in water | Highly potent ET_A receptor antagonist; Fekete, Z et al, *J Pharmacol Exp Ther*, 275:215, 1995

ICN 196010   MW 624.7 (free acid) $C_{32}H_{44}N_6O_7$ ≥98% | Highly potent endothelin ET_A receptor antagonist

Cyclo(-Asp-Pro-Ile-Leu-Trp)
Cyclo(-D-Asp-Pro-D-Ile-Leu-D-Trp)

Synonyms: Endothelin Receptor Antagonist; JKC-301

Alexis 155-012   MW 624.7 $C_{32}H_{44}N_6O_7$ Very potent; Felete, Z et al, *Pharmacol Exp Ther*, 275:215, 1995; Lippton, et al, *American Heart Assoc, 66th Scientific Sessions*, Georgia (World Congress, Nov 8-11), 1993; Widdowson, PS & Kirk, CN, *Br J Pharmacol*, 118:2126, 1996

Bachem H-3008.0001   MW 624.74 $C_{32}H_{44}N_6O_7$ Store at -15°C

Sigma J 4125   FW 624.7 $C_{32}H_{44}N_6O_7$ ≥97% (HPLC) | Bioactive peptide

Cyclo(-Asp-Pro-Val-Leu-Trp)
Cyclo(-D-Asp-Pro-D-Val-Leu-D-Trp) Sodium Salt

Synonyms: Endothelin Receptor Antagonist; BQ-123

Alexis 155-004   MW 633.7 $C_{31}H_{42}N_6O_7Na$ Weak but highly selective antagonist, originally isolated from the fermentation products of Streptomyces *misakiensis*; Ishikawa, K et al, *J Med Chem*, 35:2139, 1992; Eguchi, S et al, *FEBS Lett*, 302:243, 1992; Hiley, CR et al, *FEBS Lett*, 311:179, 1992; Ihara, M et al, *Life Sci*, 50:247, 1992; Haynes, WG et al, *TIPS*, 14:225, 1993; Atkinson, RA & Pelton, JT, *FEBS Lett*, 296:1, 1992

Calbiochem 05-23-3831   MW 632.7 $C_{31}H_{41}N_6O_5 \cdot$ Na >95% (HPLC); solid; soluble in H_2O; sold under license of Banyu Pharmaceuticals | Highly potent & selective; Ihara, M et al, *Biochem Biophys Res Comm*, 178:132, 1991; Nambi, P et al, *J Pharmacol Exp Ther*, 277:1567, 1996; Bax, AW & Saxena, PR, *Trends Pharmacol Sci*, 15:379, 1994

Cyclo(-Asp-Pro-Val-Leu-Trp)
Cyclo(-D-Asp-Pro-D-Val-Leu-D-Trp)

Synonyms: Endothelin Receptor Antagonist; BQ-123

Sigma C 1306   FW 610.7 $C_{31}H_{42}N_6O_7$ ≥97% (HPLC) | Bioactive peptide; selective antagonist; Ihara, M et al, *Life Sci*, 50:247, 1992

Cyclo(-Cys-Asn-Cyclo(-Cys-Lys-Ala-Pro-Glu-Thr-Ala-Leu-Cys)-Ala-Arg-Arg-Cys)-Gln-Gln-His Amide

Synonyms: Apamin:

Fluka 10796   Honey bee venom   MW 2027.3 $C_{79}H_{131}N_{31}O_{24}S_4$ ≥98% (HPLC); ≥85% peptide content | Neurotoxin which blocks a class of Ca^{2+}-activated K^+ channels; Blatz, AL & Magleby, KL, *Nature*, 323:718, 1986; Habermann, E, *Science*, 177:314, 1972

Cyclo(Cys-Gly-Asn-Leu-Ser-Thr-Cys)-Met-Leu-Gly-Thr-Tyr-Thr-Gln-Asp-Phe-Asn-Lys-Phe-His-Thr-Phe-Pro-Gln-Thr-Ala-Ile-Gly-Val-Gly-Ala-Pro Amide

Synonyms: Calcitonin; Calcitonin, Reduced Cyclic (1→7)-Disulfide; Calcitonin M, Human C Carcinoma

Fluka 53672   Human   MW 3417.87 $C_{151}H_{226}N_{40}O_{45}S_3$ ~95% (HPLC) | Deftos, LJ, *Hormones in Blood*, Gray, CH & James, VHT, eds, 2 (3rd ed):97, 1979, Academic Press London; MacIntyre, I et al, *Clin Orthop Res*, 217:45, 1987; Wolfe, HJ, *J Endocrinol Invest*, 5:423, 1982

Cyclo(Cys-Ser-Asn-Leu-Ser-Thr-Cys)-Val-Leu-Gly-Lys-Leu-Ser-Gln-Glu-Leu-His-Lys-Leu-Gln-Thr-Tyr-Pro-Arg-Thr-Asn-Thr-Gly-Ser-Gly-Thr-Pro Amide

Synonyms: Calcitonin

Fluka 21051   Salmon   MW 3431.9 $C_{145}H_{240}N_{44}O_{48}S_2$ ≥98% (HPLC); ≥75% peptide content

Cyclo(-Cys-Thr-Arg-Pro-Asn-Asn-Asn-Thr-Arg-Lys-Ser-Ile-His-Ile-Gly-Pro-Gly-Arg-Ala-Phe-Tyr-Thr-Thr-Gly-Glu-Ile-Ile-Gly-Asp-Ile-Arg-Gln-Ala-His-Cys)

Synonyms: GP120 V3 Loop Universal; HIV I Peptide

Neosystem SC697   MW 3896.37   HIV-1 GP120 V3 Loop peptide

Cyclo(-Gln-Trp-Phe-Gly-Leu-Met)

Bachem H-9315.0001   MW 762.93 $C_{38}H_{50}N_8O_7S$ Store at -15°C

Cyclo(Glu-Ala-Ile-Leu-Trp)
Cyclo(D-Glu-Ala-allo-D-Ile-Leu-D-Trp)

Synonyms: Endothelin A Receptor Antagonist; BE-18257B

Alexis 155-009   MW 612.3 $C_{31}H_{44}N_6O_7$ Weak but highly selective ET_A antagonist, originally isolated from the fermentation products of *Streptomyces misakiensis*; Kojiri, K et al, *J Antibiot*, 44:1342, 1991; Miyata, S et al, *J Antibiot*, 45:74 & 788, 1992; Nakajima, S et al, *J Antibiot*, 44:1348, 1991; Ihara, M et al, *BBRC*, 178:132, 1991

Cyclo(Glu-Ala-Ile-Leu-Trp)
Cyclo(D-Glu-Ala-D-allo-Ile-Leu-D-Trp)

Synonyms: Selective Endothelin A Antagonist

Sigma C 1307   FW 612.7 $C_{31}H_{44}N_6O_7$ ≥97% (HPLC); peptide content:>60% | Bioactive peptide; Ihara, M et al, *BBRC*, 178:132, 1991

Cyclo(-Glu-Ala-Ile-Leu-Trp)
Cyclo(D-Glu-Ala-allo-D-Ile-Leu-D-Trp)

Synonyms: Endothelin Receptor Antagonist; BE-18257B

American Peptide 88-2-20   MW 612.7 $C_{31}H_{44}N_6O_7$ Selective ET-1 antagonist, originally isolated from the fermentation products of *Streptomyces misakiensis*; Miyata, S et al, *J Antibot*, 46:788, 1992; Ihara, M et al, *BBRC*, 178:132, 1991

Cyclo(-Glu-Ala-Ile-Leu-Trp)
Cyclo(-D-Glu-Ala-D-allo-Ile-Leu-D-Trp)

Bachem H-8405.0001   MW 612.73 $C_{31}H_{44}N_6O_7$ Store at -15°C

Cyclo(Glu-Ala-Val-Leu-Trp)
Cyclo(D-Glu-Ala-D-Val-Leu-D-Trp)

Synonyms: Endothelin Receptor Antagonist; BE-18257A

Alexis 155-005   MW 598.7 $C_{30}H_{42}N_6O_7$ Weak & non-selective endothelin antagonist, originally isolated from the fermentation product of Streptomyces *misakiensis*; thereafter the same product was also isolated from Streptomyces sp 7338; Kojiri, K et al, *J Antibiot*, 4:1342, 1991; Nakajima, S, *J Antibiot*, 44:1348, 1991; Ihara, M et al, *BBRC*, 178:132, 1991; Miyata, S et al, *J Antibiot*, 45:74 & 788, 1992

Cyclo(-Glu-Ala-Val-Leu-Trp)
Cyclo(D-Glu-Ala-D-Val-Leu-D-Trp)

Synonyms: Endothelin Receptor Antagonist; BE-18257A/W-7338A

American Peptide 88-2-40   MW 598.7 $C_{30}H_{42}N_6O_7$ Weak & non-selective endothelin antagonist, originally isolated from the fermentation products of *Streptomyces misakiensis*; Miyata, S et al, *J Antibot*, 46:788, 1992; Miyata, S et al, *J Antibot*, 45:74, 1992

Cyclo(-Glu-Glu)

Bachem G-1700.0050   MW 258.23 $C_{10}H_{14}N_2O_6$ Store at -15°C

Cyclo(-Gly-Arg-Gly-Asp-Ser-Pro-Ala)

Bachem H-1986.0001   MW 640.65 $C_{25}H_{40}N_{10}O_{10}$ Store at -15°C

Cyclo(-Gly-Asn-Trp-His-Gly-Thr-Ala-Pro-Asp)-Trp-Phe-Phe-Asn-Tyr-Tyr-Trp

Bachem H-2508.0500   MW 2043.19 $C_{103}H_{115}N_{23}O_{23}$ Store at -15°C

Cyclo(Gly-Asn-Trp-His-Gly-Thr-Ala-Pro-Asp)-Trp-Phe-Phe-Asn-Tyr-Tyr-Trp

Synonyms: RES-701-1

Calbiochem 05-23-3840   MW 2043.2 $C_{103}H_{115}N_{23}O_{23}$ >95% (HPLC); solid; soluble in 5% acetic acid | Selectively blocks endothelin-1 binding to the ET_B receptor; Ogawa, T et al, *J Antibiot*, 48:1213, 1995; Morishita, Y, et al, *J Antibiot*, 47:269, 1994

Cyclo(-Gly-Asn-Trp-His-Gly-Thr-Ala-Pro-Asp)-Trp-Val-Tyr-Phe-Ala-His-Leu-Asp-Ile-Ile-Trp

Bachem H-4074.0500   MW 2376.62 $C_{117}H_{146}N_{28}O_{27}$ Store at -15°C

Cyclo(Gly-Asn-Trp-His-Gly-Thr-Ala-Pro-Asp-Trp-Phe-Phe-Asn-Tyr-Tyr-Trp)
Cyclo(Gly-Asn-Trp-His-Gly-Thr-Ala-Pro-β-Asp)-Trp-Phe-Phe-Asn-Tyr-Tyr-Trp)

Sigma C 2585   *Streptomyces* sp RE-701   FW 2043.2   ≥90% (HPLC) | Bioactive peptide; cyclic peptide originally isolated from the culture broth of Streptomyces sp RE-701; exhibits high selectivity for ET_B receptors, inhibits ET-1 binding & blocks ET_B receptor-mediated responses; Tanaka, T et al, *Mol Pharmacol*, 45:724, 1994

Cyclo(-Gly-Gln)

Bachem G-3945.0250   MW 185.18 $C_7H_{11}N_3O_3$ Store at -15°C

Cyclo(-Gly-Glu)

Bachem G-4290.0050   MW 186.17 $C_7H_{10}N_2O_4$ Store at -15°C

Cyclo(-Gly-Gly)

Bachem E-1965.0005   MW 114.1 $C_4H_6N_2O_2$ Store at -15°C

Cyclo(-Gly-His)

Bachem G-1705.0250   MW 194.19 $C_8H_{10}N_4O_2$ Store at -15°C

Cyclo(-Gly-Phe)

Bachem G-1715.0250   MW 204.23 $C_{11}H_{12}N_2O_2$ Store at -15°C

Cyclo(Gly-Phe-Ile-Gly-Trp-Gly-Asn-Asp)-Ile-Phe-Gly-His-Tyr-Ser-Gly-Asp-Phe
Cyclo(Gly-Phe-Ile-Gly-Trp-Gly-Asn-β-Asp)-Ile-Phe-Gly-His-Tyr-Ser-Gly-Asp-Phe

Synonyms: Anantin

Sigma A 4316   *Streptomyces coerulescens*   FW 1871.0   ≥90% (HPLC) | Bioactive peptide; cyclic peptide; competitive antagonist at atrial natriuretic factor (ANF) receptors; Weber, W et al, *J Antibiot*, 44:164, 1991

Cyclo(-Gly-Pro)

Bachem G-1720.0250   MW 154.17 $C_7H_{10}N_2O_2$ Store at -15°C

Cyclo(-Gly-Ser)

Bachem G-1725.0250   MW 144.13 $C_5H_8N_2O_3$ Store at -15°C

Cyclo(-Gly-Trp)

Bachem G-1730.0250   MW 243.27 $C_{13}H_{13}N_3O_2$ Store at -15°C

Cyclo(Gly-Tyr-Val-Pro-Met-Leu)
Cyclo(Gly-Tyr(PO₃H₂)-Val-Pro-Met-Leu) Phosphorylated

Synonyms: Platelet Derived Growth Factor

Alexis 165-019   MW 740.8 $C_{32}H_{49}N_6O_{10}PS$ Cyclic hexapeptide inhibitor of SH_2 interactions; sequence represents the autophosphorylation region around Tyr[751] of the PDGF β-receptor; Nomizu, M et al, *Tetrahedron*, 50:2691, 1994

Cyclo(-Gly-Tyr-Val-Pro-Met-Leu)
Cyclo(-Gly-Tyr(PO₃H₂)-Val-Pro-Met-Leu)

Bachem H-2062.0001   MW 740.82 $C_{32}H_{49}N_6O_{10}PS$ Store at -15°C

Cyclo(His-Phe)

ICN 153085   MW 284.3 $C_{15}H_{16}N_4O_2$

Sigma C 2651   FW 284.3 $C_{15}H_{16}N_4O_2$ Bioactive peptide

Cyclo(-His-Phe)

Bachem G-1740.0250   MW 284.32 $C_{15}H_{16}N_4O_2$ Store at -15°C

Cyclo(His-Pro)

ICN 153086   MW 234.3 $C_{11}H_{14}N_4O_2$ Degradation product of TRH metabolism; inhibits prolactin secretion *in vitro*; Bauer, K et al, *Nature*, 274:174, 1978; Matsui, T et al, *JBC*, 254:2439, 1979

Sigma C 3772   FW 234.3 $C_{11}H_{14}N_4O_2$ ≥97% (HPLC) | Bioactive peptide; degradation product of TRH metabolism; Matsui, T et al, *J Biol Chem,* 254:2439, 1979; Bauer, K et al, *Nature,* 274:174, 1978

Cyclo(-His-Pro)

Bachem G-1745.0050   MW 234.26 $C_{11}H_{14}N_4O_2$ Store at -15°C

Cyclo(-His-Pro)
Cyclo(-D-His-Pro)

Bachem G-4375.0050   Store at -15°C

Cyclo(-His-Pro)
Cyclo(-His-D-Pro)

Bachem G-4380.0050   Store at -15°C

Cyclo(-His-Pro)

Neosystem SC071   MW 234.26   Bioactive; thyrotropin releasing hormone-related peptide; Prasad, C et al, *Nature*, 268:142-144, 1977

Cyclo(Ile-Leu-Trp-Asp-Pro)
Cyclo(D-Ile-Leu-D-Trp-D-Asp-Pro)

Synonyms: JKC-301

American Peptide 88-2-30   MW 624.7 $C_{32}H_{44}N_6O_7$ Very potent ET_A antagonist, 5-10 times more potent than BQ-123; Lippton et al, *American Heart Assoc, 66th Scientific Sessions, GA World Congress*, 1993

Cyclo(Leu-Ala)

Sigma C 9038   FW 184.2 $C_9H_{16}N_2O_2$ 3-Isobutyl-6-methyl-2,5-piperazinedione

Cyclo(Leu-Gly)

Synonyms: Morphine Tolerance Peptide; 3-Isobutyl-2,5-Piperazinedione; Opioid Peptide

American Peptide 32-0-60   MW 170.3 $C_8H_{14}N_2O_2$ Waiter, R et al, *PNAS*, 76:518, 1979

ICN 152951   Blocks narcotic-induced dopamine receptor super sensitivity; Walter, R etal, *PNAS*, 75:4573, 1978

Sigma C 3526   FW 170.2 $C_8H_{14}N_2O_2$ ≥97% (HPLC); crystalline | Bioactive peptide; blocks narcotic-induced dopamine receptor super-sensitivity; Walter, R et al, *Proc Natl Acad Sci USA*, 76:518, 1979

Cyclo(-Leu-Gly)

Synonyms: Morphine Tolerance Peptide

Bachem G-1710.0050   MW 170.21 $C_8H_{14}N_2O_2$ Store at -15°C

Peptides International PMT-4070   Synthetic   MW 170.21 $C_8H_{14}N_2O_2$ >99% (HPLC); lyophilized amorphous powder; bulk | Bioactive; Walter, R et al, *PNAS USA*, 76:518, 1979

Cyclo(-Leu-Phe)

Bachem G-4385.0250   Store at -15°C

Cyclo(-Leu-Pro)

Bachem G-1750.0250   MW 210.28 $C_{11}H_{18}N_2O_2$ Store at -15°C

Cyclo(-Leu-Pro)
Cyclo(-D-Leu-D-Pro)

Bachem G-1755.0050   Store at -15°C

Cyclo(-Leu-Trp)

Bachem G-4505.0250   Store at -15°C

Cyclo(-Met-Met)

Bachem G-1760.0250   MW 262.4 $C_{10}H_{18}N_2O_2S_2$ Store at -15°C

Cyclo(-Met-Pro)

Bachem G-4725.0250   MW 228.32 $C_{10}H_{16}N_2O_2S$ Store at -15°C

Cyclo(-Phe-His-Trp-Ala-Val-Gly-His-Leu-Leu)
Cyclo(-D-Phe-His-Trp-Ala-Val-Gly-His-Leu-Leu)

Bachem H-8470.0001   Store at -15°C

Cyclo(-Phe-Phe)

Bachem G-1765.0250   MW 294.35 $C_{18}H_{18}N_2O_2$ Store at -15°C

Cyclo(Phe-Pro)
Cyclo(α-Aminobutyric Acid-L-Phe-D-Pro-L-2-Amino-8-Oxo-9,10-Epoxydecanoic Acid)

Synonyms: Chlamydocin

Calbiochem 220555   *Diheterospora chlamydospora*   MW 526.6 $C_{28}H_{38}N_4O_6$ ≥95% (HPLC); lyophilized solid; soluble in methanol; LD_{50}≤2000 mg/kg; may be carcinogenic/teratogenic | Cyclic tetrapeptide analog of HC-Toxin; potent, non-competitive & reversible inhibitor of histone deacetylase; does not inhibit histone acetyltransferases; possesses anti-mitogenic & cytostatic activities; interferes with the control of fundamental cellular processes such as chromatin structure, cell cycle progression & gene expression; Brosch, G et al, *Plant Cell*, 7:1941, 1995; Nikolskaya, AN, *Gene*, 165:207, 1995; Walton, JD et al, *Experientia*, 41:348, 1984

Cyclo(-Phe-Pro)

Bachem G-4720.0250   MW 244.29 $C_{14}H_{16}N_2O_2$ Store at -15°C

Cyclo(Phe-Ser)

Sigma C 2524   *Trichoderma viride*   FW 234.3 $C_{12}H_{14}N_2O_3$ ≥97% (TLC) | Bioactive peptide; intermediate in the biosynthesis of gliotoxin by *Trichoderma viride*; Kirby, G et al, *J Chem Soc, Perkin Trans,* 1:1336, 1978

Cyclo(-Phe-Ser)

Bachem G-1770.0250   MW 234.26 $C_{12}H_{14}N_2O_3$ Store at -15°C

Cyclo(Phe-Ser-Phe-Gly-Pro-Leu-Ala-Pro)

Synonyms: Pseudostellarin G; Tyrosinase Inhibitor

Sigma P 3334   FW 817.0 $C_{42}H_{56}N_8O_9$ ≥97% (HPLC) | Bioactive peptide; potent inhibitor of tyrosinase activity & melanin formation; Morita, H et al, *Tetrahedron Lett*, 35:3563, 1994

Cyclo(-Phe-Ser-Phe-Gly-Pro-Leu-Ala-Pro)

Synonyms: Pseudostellarin G

Neosystem SC933   *Pseudostellaria heterophylla* roots   MW 816.95   Potent tyrosinase & melanin formation inhibitory activities; Morita, H et al, *Tetrahedron Lett*, 35:3563-3564, 1994

Cyclo(-Phe-Trp)

Bachem G-1775.0250   MW 333.39 $C_{20}H_{19}N_3O_2$ Store at -15°C

Cyclo(Phe-Trp-Lys-Thr)
Cyclo((7-Aminoheptanoyl)-Phe-D-Trp-Lys-O-Benzyl-Thr)

Synonyms: Somatostatin, Cyclo-

ICN 153003   Somatostatin antagonist; Fries, JL etal, *Peptides*, 3:811, 1982

Cyclo(Phe-Trp-Lys-Thr)
Cyclo(7-Aminoheptanoyl-Phe-D-Trp-Lys-Thr(Bzl))

Synonyms: Somatostatin; Cyclo-Somatostatin

Sigma C 4801   FW 780.0 $C_{44}H_{57}N_7O_6$ ≥97% (HPLC) | Bioactive peptide; somatostatin antagonist; Fries, JL et al, *Peptides*, 3:811, 1982

Cyclo(-Phe-Trp-Lys-Thr)
Cyclo(-7-Aminoheptanoyl-Phe-D-Trp-Lys-Thr(Bzl))

Synonyms: Cyclo-Somatostatin

Bachem H-2485.0001   MW 779.98 $C_{44}H_{57}N_7O_6$ Store at -15°C

Cyclo(Pro-Ala-Ala)
Cyclo(D-Pro-L-Ala-D-Ala-L-2-Amino-8-Oxo-9,10-Epoxydecanoic Acid)

Calbiochem 373205   *Cochliobolus carbonum*   MW 436.5
$C_{21}H_{32}N_4O_6$ ≥95% (HPLC); lyophilized solid; soluble in methanol; LD_{50}≤2000 mg/kg; may be carcinogenic/teratogenic | Cyclic tetrapeptide that is a potent, non-competitive & reversible inhibitor of histone deacetylase; does not inhibit histone acetyltransferases; possesses anti-mitogenic & cytostatic activities; interferes with the control of fundamental cellular processes such as chromatin structure, cell cycle progression & gene expression; Brosch, G et al, *Plant Cell*, 41:348, 1995; Pope, MR et al, *Biochemistry*, 22:3502, 1993; Walton, JD et al, *Plant Cell*, 8:887, 1996

Cyclo(Pro-Gly)

Sigma C 7280   FW 154.2 $C_7H_{10}N_2O_2$

Sigma C 7297   FW 462.5 $C_{21}H_{30}N_6O_6$ ≥97% (HPLC) | Bioactive peptide; model peptide that binds Ca^{2+} with an affinity comparable to that of naturally-occurring Ca^{2+}-binding proteins; Sussman, F & Weinstein, H, *Proc Natl Acad Sci USA*, 86 (20):7880, 1989

Cyclo(-Pro-Gly)₃

Bachem H-9690.0005   MW 462.51 $C_{21}H_{30}N_6O_6$ Store at -15°C

Cyclo(-Pro-Pro-Phe-Phe-Val-Ile-Met-Leu-Ile)

Synonyms: Cyclo-Linopeptide B

Bachem H-4056.0001   Store at -15°C

Cyclo(Pro-Pro-Tyr-Val-Pro-Leu-Ile-Ile)

Synonyms: Hymenistatin

Alexis 158-005   MW 903.2 $C_{47}H_{82}N_8O_9$ ≥97%; white solid; soluble in methanol | Immunosuppressive effect in the humoral & cellular immune responses comparable with that of cyclosporin A; comparison of the influence of hymenistatin I & cyclosporin A on cytokine production suggests that the mechanism of the interaction with the immunological system are substantially different for the two compound tested; Cebrat, M et al, *Peptides*, 17:191, 1996

Cyclo(-Pro-Pro-Tyr-Val-Pro-Leu-Ile-Ile)

Synonyms: Hymenistatin I

Bachem H-3526.0001   Store at -15°C

Cyclo(-Pro-Tyr)

Synonyms: Maculosin

Bachem G-4715.0250   MW 260.29 $C_{14}H_{16}N_2O_3$ Store at -15°C

Cyclo(-Pro-Val)

Bachem G-4730.0250   MW 196.25 $C_{10}H_{16}N_2O_2$ Store at -15°C

Cyclo(-Ser-Asn-Leu-Ser-Thr-Asu)-Val-Leu-Gly-Lys-Leu-Ser-Gln-Glu-Leu-His-Lys-Leu-Gln-Thr-Tyr-Pro-Arg-Thr-Asp-Val-Gly-Ala-Gly-Thr-Pro Amide

Synonyms: Calcitonin, (Asu¹,⁷)-; Elcatonin

Bachem H-2214.0001   Eel   MW 3363.82 $C_{148}H_{244}N_{42}O_{47}$ Store at -15°C

Cyclo(Ser-Pro-Val-Leu-Trp)
Cyclo(D-Ser-Pro-D-Val-Leu-D-Trp)

Synonyms: Endothelin Receptor Antagonist; JKC-302

Alexis 155-013   MW 582.7 $C_{30}H_{42}N_6O_6$ Bioactive peptide; very potent ET_A antagonist; Lippton, et al, *American Heart Assoc, 66th Scientific Sessions*, Georgia (World Congress, Nov 8-11), 66:8, 1993

Calbiochem 05-23-3835   MW 582.7 $C_{30}H_{42}N_6O_6$ ≥97% (HPLC); solid; soluble in isopropanol:water (2:1) | Potent ET_A receptor antagonist that diminishes the vasoconstrictive effects of endothelin-1

ICN 196013   MW 582.7   ≥95% | Potent endothelin ET_A receptor antagonist that reduces the vasoconstrictive effects of endothelin 1

Sigma J 4250   FW 582.7 $C_{30}H_{42}N_6O_6$ ≥97% (HPLC) | Bioactive peptide

Cyclo(-Ser-Pro-Val-Leu-Trp)
Cyclo(-D-Ser-Pro-D-Val-Leu-D-Trp)

Bachem H-3064.0001   MW 582.7 $C_{30}H_{42}N_6O_6$ Store at -15°C

Cyclo(-Ser-Ser)

Bachem G-1785.0250   MW 174.16 $C_6H_{10}N_2O_4$ Store at -15°C

Cyclo(-Ser-Tyr)

Bachem G-1790.0250   MW 250.25 $C_{12}H_{14}N_2O_4$ Store at -15°C

Cyclo(-Trp-Asp-Pro-Val-Leu)
Cyclo(D-Trp-D-Asp-Pro-D-Val-Leu)

Synonyms: Endothelin Receptor Antagonist; BQ-123

American Peptide 88-2-10   MW 610.7 $C_{31}H_{42}N_6O_7$ Highly potent & selective; Ishikawa, K et al, *J Med Chem*, 35:2139, 1992; Haynes, VG et al, *TIPS*, 14:225, 1993; Hiley, CR et al, *FEBS Lett*, 311:179, 1992

Cyclo(-Trp-Asp-Pro-Val-Leu)
Cyclo(-D-Trp-D-Asp-Pro-D-Val-Leu)

Synonyms: BQ-123

Bachem H-1252.0001   MW 610.71 $C_{31}H_{42}N_6O_7$ Store at -15°C

Cyclo(-Trp-Asp-Pro-Val-Leu)
Cyclo(-D-Trp-D-Asp-Pro-D-Val-Leu) Sodium Salt

Synonyms: BQ-123

Neosystem SC454   MW 610.71   Bioactive; endothelin antagonist with a high selectivity for ETA receptor subtype; Ihara, M et al, *Life Sci*, 50:247-255, 1992; Atkinson, RA et al, *FEBS Lett*, 296:1-6, 1992; Nakamichi, K et al, *BBRC*, 182:144-150, 1992; Eguchi, S et al, *FEBS Lett*, 302:243-246, 1992; Ishikawa, K et al, *J Med Chem*, 35:2139-2142, 1992; Vigne P et al, *Eur J Pharmacol*, 245:229-232, 1993

Cyclo(-Trp-Asp-Pro-Val-Leu)
Cyclo(-D-Trp-D-Asp-Pro-D-Val-Leu)

Synonyms: BQ-123; Endothelin Antagonist

Peptides International PED-3512-PI   Synthetic   MW 610.72
$C_{31}H_{42}N_6O_7$ >98% (HPLC); white powder

Cyclo(Trp-Lys-Thr-Phe-Pro-Phe)
Cyclo(D-Trp-Lys-Thr-Phe-Pro-Phe)

Sigma C 6273   FW 807.0 $C_{44}H_{54}N_8O_7$ ≥97% (HPLC) | Bioactive peptide

Cyclo(Trp-Lys-Thr-Phe-Pro-Tyr)
Cyclo(D-Trp-Lys-Thr-Phe-Pro-Tyr)

Sigma C 6398   FW 823.0 $C_{44}H_{54}N_8O_8$ ≥97% (HPLC) | Bioactive peptide

Cyclo(-Trp-Trp)

Bachem G-1795.0250   MW 372.43 $C_{22}H_{20}N_4O_2$ Store at -15°C

Cyclo(-Trp-Tyr)

Bachem G-1800.0250   MW 349.39 $C_{20}H_{19}N_3O_3$ Store at -15°C

Cyclo(-Trp-Tyr)
Cyclo(-D-Trp-Tyr)

Bachem G-1805.0050   Store at -15°C

Cyclo(-Tyr-Arg-Gly-Asp-Cys)
Cyclo(-D-Tyr-Arg-Gly-Asp-Cys(carboxymethyl)) sulfoxide

Bachem N-1365.0001   MW 668.69 $C_{26}H_{36}N_8O_{11}S$ Store at -15°C

Cyclo(-Tyr-Ile-Gln-Asn-Asu)-Pro-Arg-Gly Amide

Synonyms: Vasotocin, (Asu[1,6],Arg[8])-; Vasotocin, (Deamino-Dicarba-Arg)-

Peptides International PVP-4027-v   Synthetic   MW 999.13
$C_{45}H_{70}N_{14}O_{12}$ >99% (HPLC); lyophilized amorphous powder |
Bioactive; cyclic form between Asu ω-carboxyl group & Tyr α-amino group; Hase, S et al, *J Amer Chem Soc*, 94:3590, 1972

Cyclo(-Tyr-Ile-Gln-Asn-Asu)-Pro-Leu-Gly Amide

Synonyms: Oxytocin, (Asu[1,6])-; Oxytocin, Deamino-Dicarba-

Peptides International POX-4025-v   Synthetic   MW 956.11
$C_{45}H_{69}N_{11}O_{12}$ >99% (HPLC); lyophilized amorphous powder |
Bioactive; cyclic form between Asu ω-carboxyl group & Tyr α-amino group; Yamanaka, T et al, *Mol Pharmacol*, 6:474, 1970

Cyclo(Tyr-Phe-Gln-Asn-Asu)-Pro-Arg-Gly Amide

Synonyms: Vasopressin, (Asu[1,6],Arg[8])-

Sigma V 6254   FW 1033.2   ≥97% (HPLC) | Bioactive peptide; analog of arginine vasopressin; Hase, S et al, *J Am Chem Soc*, 94:3590, 1972

Cyclo(-Tyr-Phe-Gln-Asn-Asu)-Pro-Arg-Gly Amide

Synonyms: Vasopressin, (Asu[1,6], Arg[8])-; Vasopressin, (Deamino-Dicarba-Arg)-

Peptides International PVP-4026-v   Synthetic   MW 1033.2
$C_{48}H_{68}N_{14}O_{12}$ >97% (HPLC); lyophilized amorphous powder |
Bioactive; cyclic form between Asu ω-carboxyl group & Tyr α-amino group; Hase, S et al, *J Amer Chem Soc*, 94:3590, 1972

Cyclo(Val-Leu-Trp-Ser-Pro)
Cyclo(D-Val-Leu-D-Trp-D-Ser-Pro)

Synonyms: JKC-302

American Peptide 88-2-31   MW 582.7 $C_{30}H_{42}N_6O_6$ Potent ET_A receptor antagonist that inhibits the vasoconstrictive effects of ET-1; very potent BQ-123 analog, 5-10 times more potent than BQ-123; Lippton et al, *American Heart Assoc, 66th Scientific Sessions, GA World Congress*, 1993

Cyclo(-Val-Val)

Bachem G-3760.0250   MW 198.27 $C_{10}H_{18}N_2O_2$ Store at -15°C

Cys-Abz-Met
Cys-4-Abz-Met

Bachem H-3548.0005   MW 371.48 $C_{15}H_{21}N_3O_4S_2$ Store at -15°C

Cys-Ala
(Cys-Ala)₂

Bachem G-1810.0250   MW 382.46 $C_{12}H_{22}N_4O_6S_2$ Store at -15°C

Cys-Ala-Asn-Asn-Leu-Arg-Gly-Cys-Gly-Leu-Tyr Acetate Salt

Synonyms: G Protein Alpha 0 (345-354), C~

Biogenesis 4737-1020   Synthetic   MW 1184.7   99% (HPLC); lyophilized

Cys-Ala-Cys-Phe-Thr-Tyr-Lys-Asp-Lys-Glu-Cys-Val-Tyr-Tyr-Cys-His-Leu-Asp-Ile-Ile-Trp

Synonyms: Endothelin, (Ala²)-

American Peptide 88-2-39   Human   MW 2613.0
$C_{120}H_{166}N_{26}O_{32}S_4$ Disulfide bonds: Cys[1]-Cys[15], Cys[3]-Cys[11]

Cys-Ala-Gly
Palmitoyl-Cys((RS)-2,3-Di(Palmitoyloxy)-Propyl)-Ala-Gly

Synonyms: Pam₃-Cys-Ala-Gly

Bachem H-8820.0005   MW 1038.61 $C_{59}H_{111}N_3O_9S$ Store at -15°C

Cys-Ala-Gly-Phe-Ala-Asn-Arg-Gly-Asp-Val-Leu-Thr-Gly-Arg Acetate Salt

Synonyms: Fibrillin II (1055-1067)

Biogenesis 4439-1081   Human synthetic   MW 1436   >80% (HPLC); lyophilized

Cys-Ala-His-Ser-Phe-Phe-Asp-Glu-Leu-Arg-Asp-Pro-Asn-Val-Lys

Synonyms: Glycogen Synthase Kinase 3β; C-Terminal Blocking Peptide (334-348)

Calbiochem 361529   Rat synthetic   MW 1778.0
$C_{78}H_{116}N_{22}O_{24}S$ Lyophilized | Based on the rat glycogen synthase kinase 3β isoform-kinase subdomain XI region (334-348); this peptide coupled to KLH, was used as the immunogen for the production of Anti-GSK3β; for use in immunoabsorption for immunoprecipitation, immunocytochemistry, Western blotting & dot blots

Cys-Ala-Ser-Asp-Glu-Glu-Asp-Ala-Pro-Ser-Thr-Asp-Ile-Tyr-Phe-Pro-Thr-Asp-Glu-Asg-Ser

Synonyms: Phosphatidyl-4-Phosphate 5-Kinase γ C-Terminal Peptide

Alexis 165-032   Synthetic   MW 2511.6 $C_{106}H_{151}N_{25}O_{44}S$
≥90% (HPLC); dissolved in 10 *mM* sodium acetate, pH 4.5 | Competitively binds to Ab Alexis 210-745; antiserum blocking peptide

Cys-Ala-Ser-Gly-Trp-Gln-Pro-Gly-Thr-Glu-Tyr-Asp-Asn-Val-Val-Leu-Lys-Lys-Gly-Pro-Lys

Synonyms: p56*dok-2* C-Teminal Peptide

Alexis 165-033   Synthetic   MW 2220.6 $C_{99}H_{154}N_{26}O_{30}S$
≥90% (HPLC); dissolved in 10 *mM* sodium acetate, pH 4.5 | Competitively binds to Ab Alexis 210-746; antiserum blocking peptide

Cys-Ala-Ser-Leu-Ser-Thr-Cys-Val-Leu-Gly-Lys-Leu-Ser-Gln-Glu-Leu-His-Lys-Leu-Gln-Thr-Tyr-Pro-Arg-Thr-Asp-Val-Gly-Ala-Gly-Thr-Pro Amide

Synonyms: Thyrocalcitonin; Calcitonin

Sigma T 9907   FW 3371.9   Chicken   ≥97% (HPLC) |
Disulfide bonds: 1-7; bioactive peptide;Kurihara, T et al, *Peptide Chemistry*, 173:1985, 1986

American Peptide 22-7-12   Chicken   MW 3371.9
$C_{145}H_{240}N_{42}O_{46}S_2$ Disulfide bonds: :Cys[1]-Cys[7] | Single chain peptide; stimulates bone formation by inhibiting osteoblasts; induces bone resorption; increases cAMP levels in osteoclasts; Kurihara, T et al, *peptide Chemistry*, 1985:173, 1986

Bachem H-3074.0001   Chicken   MW 3371.89
$C_{145}H_{240}N_{42}O_{46}S_2$ Store at -15°C

Cys-Ala-Ser-Leu-Ser-Thr-Cys-Val-Leu-Gly-Lys-Leu-Ser-Gln-Glu-Leu-His-Lys-Leu-Gln-Thr-Tyr-Pro-Arg-Thr-Asp-Val-Gly-Ala-Gly-Thr-Pro NH2-Cys-Ala-Ser-Leu-Ser-Thr-Cys-Val-Leu-Gly-Lys-Leu-Ser-Gln-Glu-Leu-His-Lys-Leu-Gln-Thr-Tyr-Pro-Arg-Thr-Asp-Val-Gly-Ala-Gly-Thr-Pro Amide

Synonyms: Calcitonin

ICN 154497 Chicken MW 3372.3 Kurihara, T etal, *peptide Chemistry*, 1985:173, 1986

Cys-Ala-Ser-Leu-Ser-Thr-Cys-Val-Leu-Gly-Lys-Leu-Ser-Gln-Glu-Leu-His-Lys-Leu-Gln-Thr-Tyr-Pro-Arg-Thr-Asp-Val-Gly-Ala-Gly-Thr-Pro

Synonyms: Calcitonin

Biogenesis 1720-7729 Chicken synthetic MW 3374.4 99.2% (HPLC); lyophilized, consisting of 88.35% peptide material, TFA counter ions and residual H_2O

Cys-Ala-Thr-Gln-Ile-Ile-Thr-Phe-Glu-Ser-Phe-Lys-Glu-Asn-Leu-Lys-Asp

Synonyms: GM-CSF (96-112)

Bachem H-3442.0001 Store at -15°C

Cys-Arg-Gly-Asp-Phe-Pro-Ala-Ser-Ser-Cys

Bachem H-2672.0001 Store at -15°C | Disulfide bonds: Cys[1]-Cys[10]

Cys-Arg-Lys-Gln-Ala-Ala-Ser-Ile-Lys-Val-Ala-Val-Ser

Synonyms: Laminin A Chain Peptide C-13-S

Neosystem SC329 MW 1360.63 Bioactive; cell attachment peptide; due to the presence of a Cys this peptide can contain the dimeric formSephel, GC et al, *BBRC*, 162:821-829, 1989

Cys-Arg-Met-His-Leu-Arg-Gln-Tyr-Glu-Leu

Synonyms: C-Terminal Blocking Peptide (385-394), $G_s\alpha$-Subunit

Calbiochem 371782 MW 1461.8 $C_{63}H_{104}N_{20}O_{16}S_2$ Lyophilized solid | Distinct C-terminal region of $G_s\alpha$-subunit; Ag used to produce antibody for Calbiochem 371732; useful as a blocking peptide; Jones, DT & Reed, RR, *J Biol Chem*, 262:14241, 1987

Cys-Arg-Phe-Gly-Thr-Cys-Thr-Val-Gln-Lys-Leu-Ala-His-Gln-Ile-Tyr Amide

Synonyms: Adrenomedullin (16-31)

Bachem H-4064.0001 Human, porcine MW 1865.21 $C_{82}H_{129}N_{25}O_{21}S_2$ Store at -15°C

Cys-Asn-Ala-Ile-Gln-Glu-Ala-Arg-Arg-Leu-Leu-Asn-Leu-Ser-Arg-Asp

Synonyms: GM-CSF (17-31), Cys-

Bachem H-3474.0001 Store at -15°C

Cys-Asn-Arg-Arg-Arg-Arg-Ile-Thr-Ser-Gly-Pro-Gly-Lys-Val-Leu-Tyr-Thr-Thr-Gly-Glu

Synonyms: GP120 V3 Loop Peptide SF33

Neosystem SC684 MW 2264.58 HIV-1 GP120 V3 Loop peptide; due to the presence of a Cys this peptide can contain the dimeric form

---

Cys-Asn-Cys-Lys-Ala-Pro-Glu-Thr-Ala-Leu-Cys-Ala-Arg-Arg-Cys-Gln-Gln-His

Synonyms: Apamin

American Peptide 41-0-23 MW 2027.3 $C_{79}H_{131}N_{31}O_{24}S_4$ Disulfide bonds: Cys[1]-Cys[11]; Cys[3]-Cys[15] | Ca^{2+}-Activated K^+ channels blocker; centrally acting neurotoxic peptide, originally isolated from the venom of *Apis melifica*, is the smallest neurotoxic polypeptide known & is the only one that passes the blood-brain barrier; blocks small-conductance Ca^{2+}-activated K^+ channels at nanomolar concentrations; induces motor hyperactivity & causes clonic convulsions & respiratory failure; Rietschoten, JV et al, *Eur J Biochem*, 56:35, 1975; Hughes, M et al, *PNAS*, 79:1308, 1982

Cys-Asn-Cys-Lys-Ala-Pro-Glu-Thr-Ala-Leu-Cys-Ala-Arg-Arg-Cys-Gln-Gln-His Amide

Synonyms: Apamin; Small Conductance Ca^{2+}-Activated K^+ Channel Blocker

Bachem H-8010.0500 MW 2027.37 $C_{79}H_{131}N_{31}O_{24}S_4$ Store at -15°C | Disulfide bonds: Cys[1]-Cys[11], Cys[3]-Cys[15]

Peptides International PAP-4257-v *Apis mellifera* (honeybee) synthetic MW 2027.3 $C_{79}H_{131}N_{31}O_{24}S_4$ >99% (HPLC); lyophilized amorphous powder | Bioactive; Disulfide bonds: Cys[1]-Cys[11], Cys[3]-Cys[15]; Haberman, E, *Pharmacol Ther*, 25:255, 1984; Blatz AL & KL Magleby, *Nature*, 323:718, 1986; Garcia, ML et al, *J Bioenerg Biomembr*, 23:615, 1991

Alexis 151-013 *Apis mellifera* synthetic MW 2027.3 $C_{131}H_{131}N_{31}O_{24}S$ ≥98% Purity; lyophilized; soluble in 0.05 M acetic acid & slightly soluble in water; Disulfide bonds: Cys[1]-Csy[11] & Cys[3]-Cys[15]; CAUTION: toxic | A small conductance Ca^{2+}-activated K^+ channel blocker; Castle, NA et al, *TINS*, 12:59, 1989; Strong, PN, *Pharmacol Ther*, 46:137, 1990

Sigma A 1289 Bee venom FW 2027.3 ≥95% (HPLC) | Disulfide bonds: 1-11, 3-15; bioactive peptide; the only polypeptide neurotoxin that isknown to pass the blood-brain barrier; blocks ATP-type Ca^{2+}-activated K^+ channels; Van Rietschoten, J et al, *Eur J Biochem*, 56:35, 1975; Hughes, M et al, *Proc Natl Acad Sci USA*, 79:1308, 1982

Sigma A 9459 Synthetic FW 2027.3 ≥97% (HPLC) | Disulfide bonds: 1-11, 3-15; bioactive peptide; the only polypeptide neurotoxin that isknown to pass the blood-brain barrier; blocks ATP-type Ca^{2+}-activated K^+ channels; Van Rietschoten, J et al, *Eur J Biochem*, 56:35, 1975; Hughes, M et al, *Proc Natl Acad Sci USA*, 79:1308, 1982

Cys-Asn-Cys-Lys-Ala-Pro-Glu-Thr-Ala-Leu-Cys-Ala-Arg-Cys-Gln-Gln-His Amide

Synonyms: Apamin

ICN 190153 Bee venom MW 2027.3 Research grade | Toxic; S-S bond between Cys[1]-Cys[11] & Cys[3]-Cys[17]; Haberman, E, *Science*, 177:314, 1972; Gauldie, J etal, *Eur J Biochem*, 61:369, 1976

Cys-Asn-Ile-Arg-Gln-Arg-Thr-Ser-Ile-Gly-Leu-Gly-Gln-Ala-Leu-Tyr-Thr-Thr-Lys-Thr-Arg-Ser

Synonyms: GP120 V3 Loop Peptide Z2

Neosystem SC632 MW 2467.82 HIV-1 GP120 V3 Loop peptide; due to the presence of a Cys this peptide can contain the dimeric form

---

Cys-Asn-Leu-Ala-Val-Ala-Ala-Ala-Ser-His-Ile-Tyr-Gln-Asn-Gln-Phe-Val-Gln

Synonyms: CD-36 Peptide P (139-155); CD-36 (139-155), Cys-

American Peptide 79-1-10 MW 1976.2 $C_{87}H_{132}N_{25}O_{26}S$ Antibiotic peptide; partially blocked collagen-induced platelet aggregation & inhibited the interaction of CD-36 to immobilized thrombospondin; Leung, L et al, *JBC*, 267:18244, 1992

Bachem H-2974.0001 Store at -15°C

Cys-Asn-Leu-Lys-Glu-Asp-Gly-Ile-Ser-Ala-Ala-Lys-Asp-Val-Lys

Synonyms: GTP Binding Protein Fragment Go Alpha

American Peptide 77-0-03   MW 1589.8 $C_{66}H_{114}N_{19}O_{24}S$
Mumby, SM et al, *PNAS*, 83:598, 1986

Cys-Asn-Lys-Arg-Lys-Arg-Ile-His-Ile-Gly-Pro-Gly-Arg-Ala-Phe-Tyr-Thr-Thr-Lys-Asn

Synonyms: GP120 V3 Loop Peptide MN

Neosystem SC625   MW 2360.76   HIV-1 GP120 V3 Loop peptide; due to the presence of a Cys this peptide can contain the dimeric form

Cys-Asn-Pro-Val-Gln-Arg-Ile-Ser-Ala-Glu-Glu-Ala-Leu-Gln-His-Pro

Synonyms: Cdk5; C-Terminal Blocking Peptide (268-283)

Calbiochem 219450   Human   MW 1794 $C_{75}H_{124}N_{24}O_{25}S$
Lyophilized | Synthetic peptide based on the human Cdk5-encoded protein kinase (268-283); this peptide coupled to KLH, was used as the immunogen for the production of Anti-Cdk5; for use in immunoabsorption for immunoprecipitation, immunocytochemistry, Western blotting & dot blots

Cys-Asn-Thr-Arg-Arg-Gly-Ile-His-Phe-Gly-Pro-Gly-Gln-Ala-Leu-Tyr-Thr-Thr-Gly-Ile

Synonyms: GP120 V3 Loop Peptide Mal

Neosystem SC848   MW 2162.45   HIV-1 GP120 V3 Loop peptide; due to the presence of a Cys this peptide can contain the dimeric form

Cys-Asn-Thr-Arg-Gln-Arg-Thr-Pro-Ile-Gly-Leu-Gly-Gln-Ser-Leu-Tyr-Thr-Thr-Arg-Ser-Arg-Ser

Synonyms: GP120 V3 Loop Peptide ELI

Neosystem SC631   MW 2498.75   HIV-1 GP120 V3 Loop peptide; due to the presence of a Cys this peptide can contain the dimeric form

Cys-Asn-Thr-Arg-Gln-Ser-Thr-Pro-Ile-Gly-Leu-Gly-Gln-Ala-Leu-Tyr-Thr-Thr-Arg-Gly-Arg-Thr-Lys

Synonyms: GP120 V3 Loop Peptide Z6

Neosystem SC633   MW 2525.81   HIV-1 GP120 V3 Loop peptide; due to the presence of a Cys this peptide can contain the dimeric form

Cys-Asn-Thr-Arg-Lys-Arg-Ile-Arg-Ile-Gln-Arg-Gly-Pro-Gly-Arg-Ala-Phe-Val-Thr-Ile-Gly-Lys

Synonyms: GP120 V3 Loop Peptide HBX2

Neosystem SC623   MW 2528.01   HIV-1 GP120 V3 Loop peptide; due to the presence of a Cys this peptide can contain the dimeric form

Cys-Asn-Thr-Arg-Lys-Ser-Ile-Arg-Ile-Gln-Arg-Gly-Pro-Gly-Arg-Ala-Phe-Val-Thr-Ile-Gly-Lys

Synonyms: GP120 V3 Loop Peptide BRU C-22-K

Neosystem SC806   MW 2458.90   HIV-1 GP120 V3 Loop peptide; due to the presence of a Cys this peptide can contain the dimeric form

Cys-Asn-Thr-Arg-Lys-Ser-Ile-Thr-Lys-Gly-Pro-Gly-Arg-Val-Ile-Tyr-Ala-Thr-Gly-Gln

Synonyms: GP120 V3 Loop Peptide RF

Neosystem SC629   MW 2150.48   HIV-1 GP120 V3 Loop peptide; due to the presence of a Cys this peptide can contain the dimeric form

Cys-Asn-Thr-Arg-Lys-Ser-Ile-Tyr-Ile-Gly-Pro-Gly-Arg-Ala-Phe-His-Thr-Thr-Gly-Arg

Synonyms: GP120 V3 Loop Peptide SF2

Neosystem SC624   MW 2235.55   HIV-1 GP120 V3 Loop peptide; due to the presence of a Cys this peptide can contain the dimeric form

Cys-Asn-Thr-Lys-Lys-Gly-Ile-Ala-Ile-Gly-Pro-Gly-Arg-Thr-Leu-Tyr-Ala-Arg-Glu-Lys

Synonyms: GP120 V3 Loop Peptide NY/5

Neosystem SC630   MW 2176.55   HIV-1 GP120 V3 Loop peptide; due to the presence of a Cys this peptide can contain the dimeric form

Cys-Asn-Thr-Ser-Val-Ile-Thr-Gln-Ala-Cys-Pro-Lys-Val-Ser-Phe-Glu

Synonyms: GP120 (201-216); HTLV I Peptide

Neosystem SC264   MW 1726.97   HTLV I peptide from subtype GP160; due to the presence of a Cys this peptide can contain the dimeric form

Cys-Asn-Thr-Thr-Arg-Ser-Ile-His-Ile-Gly-Pro-Gly-Arg-Ala-Phe-Tyr-Ala-Thr-Gly-Asp

Synonyms: GP120 V3 Loop SC

Neosystem SC626   MW 2137.35   HIV-1 GP120 V3 Loop peptide; due to the presence of a Cys this peptide can contain the dimeric form

Cys-Asn-Trp-Ala-Val-Ala-His-Leu-Cys D-Cys-Asn-Trp-Ala-Val-D-Ala-His-Leu-Cys Amide

Synonyms: Bombesin (6-14), (D-Cys[6],Asn[7],D-Ala[11],Cys[14])-

Bachem H-8465.0001   Store at -15°C

Cys-Asn-Tyr-Glu-Phe-Lys-Lys-Ile-Thr-Glu-Asp-Thr-Val-Glu-Phe-Gly-Ser

Synonyms: Peroxisomal Membrane Protein 70 C-Terminal Peptide

Alexis 165-036   Synthetic   MW 2010.2 $C_{89}H_{132}N_{20}O_{31}S$
≥90% (HPLC); dissolved in 10 *mM* sodium acetate, pH 4.5 | Competitively binds to Ab Alexis 210-205; antiserum blocking peptide

Cys-Asn-Tyr-Tyr-Ser-Asn-Ser-Tyr-Ser-Phe-Trp-Leu-Ala-Ser-Leu-Asn-Pro-Glu-Arg

Synonyms: Collagen Type IV α3 Chain (185-203)

Bachem H-4208.0500   Store at -15°C

Cys-Asn-Val-Arg-Arg-Ser-Leu-Ser-Ile-GlyPro-Gly-Arg-Ala-Phe-Arg-Thr-Arg-Glu

Synonyms: GP120 V3 Loop Peptide WMJ2

Neosystem SC628   MW 2176.48   HIV-1 GP120 V3 Loop peptide; due to the presence of a Cys this peptide can contain the dimeric form

Cys-Asn-Val-Val-Pro-Leu-Tyr-Asp-Leu-Leu-Leu-Glu Cys-Asn-Val-Val-Pro-Leu-Tyr(PO₃H₂)-Asp-Leu-Leu-Leu-Glu

Synonyms: Antiestrogen; Yp 537 Fragment

Bachem H-2762.0001   MW 1470.64 $C_{64}H_{104}N_{13}O_{22}PS$
Store at -15°C

Sigma C 3436   FW 1470.6   Phosphotyrosine-containing peptide that blocks the dimerization of the human estrogen receptor; bioactive peptide; Arnold, SF & Notides, AC, *Proc Natl Acad Sci USA*, 92:7475, 1995

Cys-Asp-Arg-Leu-Thr-Ala-Glu-Glu-Ala-Leu-Ser-His-Pro-Tyr-Met-Ser-Ile-Tyr-Ser-Phe-Pro-Thr-Asp-Glu

Synonyms: Mitogen Activated Protein Kinase EFK3 Blocking Peptide (303-325)

Calbiochem 442684 Rat synthetic MW 2776.1 $C_{121}H_{179}N_{29}O_{42}S_2$ Lyophilized | Based on rat ERK3 MAP kinase (303-325) with a Cys C added; this peptide coupled to KLH was used as the immunogen for the production of Anti-MAP kinase ERK3; for use in immunoabsorption for immunoprecipitation, immunocytochemistry, Western blotting & dot blots

Cys-Asp-Leu-Ile-Tyr-Tyr-Asp-Tyr-Glu-Glu-Asp-Tyr-Tyr-Phe-Asp-Cys

Synonyms: CDR-H3, C2; HIV I Replication Inhibitor

Bachem H-1588.0500 Store at -15°C

Sigma C 1582 FW 2130.2 ≥97% (HPLC) | Disulfide bonds: 1-16; bioactive peptide; Levi, M et al, *Proc Natl Acad Sci USA*, 90:4374, 1993

Cys-Asp-Pro-Gly-Tyr-Ile-Gly-Ser-Arg

Synonyms: Laminin (925-933)

Bachem H-2798.0001 MW 967.07 $C_{40}H_{62}N_{12}O_{14}S$ Store at -15°C

Neosystem SC080 MW 967.06 Bioactive; cell attachment peptide; laminin fragment; due to the presence of a Cys this peptide can contain the dimeric form; Iwamoto, Y et al, *Science*, 238:1132, 1987

Sigma C 0668 FW 967.1 $C_{40}H_{62}N_{12}O_{14}S$ ≥90% (HPLC) | Bioactive peptide; inhibits experimental metastasis formation; Iwamoto, Y et al, *Science*, 238:1132, 1987

Cys-Asp-Pro-Gly-Tyr-Ile-Gly-Ser-Arg Amide

Synonyms: Laminin βI Chain (925-933); Laminin (925-933)

Bachem H-1224.0001 MW 966.09 $C_{40}H_{63}N_{13}O_{13}S$ Store at -15°C

ICN 153087 Iwamoto, Y etal, *Science*, 238:1132, 1987

Neosystem SC187 MW 966.08 Bioactive; cell attachment peptide; laminin fragment; due to the presence of a Cys this peptide can contain the dimeric form; Iwamoto, Y et al, *Science*, 238:1132, 1987

Sigma C 1668 FW 966.1 $C_{40}H_{63}N_{13}O_{13}S$ ≥95% (HPLC); BME as stabilizer | Bioactive peptide; Iwamoto, Y et al, *Science*, 238:1132, 1987

Cys-Cys-Gly-Gly Cys(Cys-Gly)-Gly Acetate Salt

Sigma C 9187 FW 354.4 (free base) $C_{10}H_{18}N_4O_6S_2$ ≥97% (HPLC)

Cys-Cys-Ser-Thr-Glu-Asp-Ser-Lys-Asn-Lys-Glu-Gly-Ser-Gln-Asn-Leu-Gln-Ser-Gln

Synonyms: Estrogen Receptor β C-Terminal Peptide (467-485)

Alexis 155-030 Mouse synthetic MW 2086.2 $C_{79}H_{132}N_{26}O_{36}S_2$ ≥95%; lyophilized; reconstitute with 0.1 mL distilled H_2O | Competitively binds to Ab Alexis 210-178; blocking peptide for antiserum (purified) to Erβ

Cys-Gln-Asp-Ser-Glu-Thr-Arg-Thr-Phe-Tyr

Synonyms: Collagen Binding Fragment; Fibronectin Related Peptide

American Peptide 44-0-20 MW 1248.3 $C_{52}H_{75}N_{14}O_{20}S$ Boucaut, J et al, *J Embryology & Exp Morph*, 89:213, 1985

Bachem H-5685.0001 MW 1249.32 $C_{52}H_{76}N_{14}O_{20}S$ Store at -15°C

Sigma C 8287 FW 1249.3 ≥90% (HPLC); BME as stabilizer | Bioactive peptide

Cys-Gln-Asp-Ser-Glu-Thr-Arg-Tyr-Phe-Tyr

Synonyms: Fibronectin Related Peptide

ICN 152332 Enhances the attachment & growth of anchorage dependent cells in vitro; may be useful in structure-activity studies; Boucaut, JC etal, *J Embryol Exp Morphol*, 89:213, 1985

Cys-Gln-Cys-Ala-Ser-Gln-Lys-Asp-Lys-Lys-Cys-Trp-Ser-Tyr-Cys-Gln-Ala-Gly-Lys-Glu-Ile Amide

Synonyms: Prepro-Endothelin (110-130), (Tyr[123])-

American Peptide 88-1-13 Human MW 2402.8 $C_{100}H_{156}N_{30}O_{31}S_4$ Disulfide bonds: Cys[110]-Cys[124], Cys[112]-Cys[120]

Cys-Gln-Glu-Asn-Leu-Lys-Asp-Ile-Met-Leu-Gln Acetate Salt

Synonyms: G Protein Alpha 12 (370-379), C~

Biogenesis 4737-2220 Synthetic >80% (HPLC); lyophilized

Cys-Gln-Leu-Asn-Leu-Lys-Glu-Tyr-Asn-Leu-Val Acetate Salt

Synonyms: G Protein Alpha q/11 (344-353), C~

Biogenesis 4737-5220 Synthetic MW 1338.4 97% (HPLC); lyophilized

Cys-Gln-Phe-Val-His-Pro-Ile-Leu-Gln-Ser-Ser-Val

Synonyms: Protein Kinase C$_\alpha$ Blocking Peptide

Calbiochem 539633 MW 1357.6 $C_{61}H_{96}N_{16}O_{17}S$ Liquid; use 0.25-2.5 µg/mL of peptide to block the antibody under the specified assay conditions | Distinct C-terminal region of protein kinase C$_\alpha$; used to confirm the specificity of anti-PKC$_\alpha$ in Western blotting

Cys-Glu-Arg-Ala-Arg-Arg-Leu-Gln-Ala-Lys-Met-Met-Thr-Asn-Leu Trifluoroacetate Salt

Synonyms: Spastin (171-185)

Biogenesis 8350-1150 Human synthetic MW 1820 >80% (HPLC); lyophilized

Cys-Glu-Asp-Lys-Glu-Arg-Trp-Glu-Asp-Val-Lys-Glu-Glu-Met-Thr-Ser-Ala-Leu

Synonyms: Mitogen Activated Protein Kinase Activated Protein Kinase II Blocking Peptide (344-360), C-Terminal

Calbiochem 442691 Human synthetic MW 2198.4 $C_{91}H_{144}N_{24}O_{35}S_2$ Lyophilized | Based on human MAPKAPK-2 (344-360) with a Cys C residue added; this peptide coupled to KLH was used as the immunogen for the production of Anti-MAPKAPK-2, CT; for use in immunoabsorption for immunoprecipitation, immunocytochemistry, Western blotting & dot blots

Cys-Glu-Asp-Ser-Asp-Glu-Pro-Leu-Glu-Arg-Arg-Leu-Ser-Leu-Val Trifluoroacetate Salt

Synonyms: Cystic Fibrosis Transmembrane Conductance Regulator Protein (726-739), C~

Biogenesis 2410-5050 Synthetic MW 1761.3 90% (HPLC); lyophilized

Cys-Glu-Gln-Glu-Arg-Glu-Gly-Ser-Ser-Leu-Asp-Ser-Pro-Arg-Ser

Synonyms: Calcitonin Precursor (69-82), C~

Biogenesis 1720-9700 Human synthetic 80% (HPLC); lyophilized | Represents AA 69-82 of the human calcitonin precursor; an additional Cys has been conjugated to the N-terminus

Cys-Glu-Gln-Lys-Leu-Ile-Ser-Glu-Glu-Asp-Leu

Synonyms: c-Myc Peptide

Alexis 162-021 Synthetic MW 1575.8 $C_{54}H_{91}N_{13}O_{22}S$ ≥90% (HPLC); dissolved in 10 m*M* sodium acetate, pH 4.5 | Competitively binds to Ab Alexis 210-208; blocking peptide for antiserum (purified) to c-Myc epitope tag

Cys-Glu-Glu-Gln-Lys-Glu-Arg-Ala-Lys-Met-Gln-Lys-Gly-Tyr-Asn Trifluoroacetate Salt

Synonyms: Neuronal Apoptosis Inhibitory Protein (41-54), C~

Biogenesis 6705-5050 Human synthetic MW 1842.3 94% (HPLC); lyophilized

Cys-Glu-Gly-Asn-Val-Arg-Val-Ser-Arg-Glu-Leu-Ala-Gly-His-Thr-Gly-Tyr

Synonyms: GTP Binding Protein Fragment G Beta

American Peptide 77-0-04 MW 1847.0 $C_{76}H_{121}N_{26}O_{26}S$ Mumby, SM et al, *PNAS*, 83:598, 1986

Cys-Glu-His-D-Phe-Arg-Trp-Cys-Lys-Pro-Val
Ac-Cys-Glu-His-D-Phe-Arg-Trp-Cys-Lys-Pro-Val Amide

Synonyms: Melanocyte Stimulating Hormone (4-13), (Ac-Cys[4],D-Phe[7],Cys[10])-α-

American Peptide 56-0-12 MW 1343.6 $C_{61}H_{86}N_{18}O_{13}S_2$ Disulfide bonds: Cys[1]-Cys[7]; very potent MSH agonist with prolonged duration of action; Cody, WL et al, *J Med Chem*, 28:683, 1985

Cys-Glu-His-Nal$_2$-Arg-Trp-Gly-Cys-Pro-Pro-Lys-Asp
Ac-Cys-Glu-His-D-Nal$_2$-Arg-Trp-Gly-Cys-Pro-Pro-Lys-Asp

Synonyms: HS014

Neosystem SC1278 MW 1563.83 Disulfide bonds: Cys[1]-Cys[8]; bioactive; melanocyte stimulating hormone-related peptide; MC4 receptor antagonist that increases food intake in free-feeding rats; Schiöth, HB et al, *Br J Pharmacol*, 124:75-82, 1998; Kask, A et al, *BBRC*, 245:90-93, 1998

Cys-Glu-His-Nal-Arg-Trp-Gly-Cys-Pro-Pro-Lys-Asp
Ac-Cys-Glu-His-D-2-Nal-Arg-Trp-Gly-Cys-Pro-Pro-Lys-Asp Amide

Synonyms: Melanocyte Stimulating Hormone (11-22), β-(Ac-Cys[11],D-2-Nal[14],Cys[18])-; HS014

Bachem H-4352.0001 MW 1563.78 $C_{71}H_{94}N_{20}O_{17}S_2$ Store at -15°C

Cys-Glu-His-Phe-Arg-Trp-Cys-Lys-Pro-Val
Ac-Cys-Glu-His-D-Phe-Arg-Trp-Cys-Lys-Pro-Val Amide

Synonyms: Melanocyte Stimulating Hormone (4-13), α-(Ac-Cys[4],D-Phe[7],Cys[10])-; Melanocyte Stimulating Hormone Amide (4-13), *N*-Ac-(Cys[4,10],D-Phe[7])-α-

Bachem H-9220.0001 MW 1343.6 $C_{61}H_{86}N_{18}O_{13}S_2$ Store at -15°C

Sigma M 7907 FW 1343.6 ≥95% (HPLC); peptide content:~70% | Disulfide bonds: 4-10; bioactive peptide; cyclic MSH sequence that is resistant to tryptic degradation; potent agonist with prolonged melanotropic activity in the frog & lizard skin bioassays; Cody, WL et al, *J Med Chem*, 28:583, 1985

Cys-Glu-His-Phe-Arg-Trp-Gly-Cys-Pro-Pro-Lys-Asp
Ac-Cys-Glu-His-diCl-D-Phe-Arg-Trp-Gly-Cys-Pro-Pro-Lys-Asp

Synonyms: HS028

Neosystem SC1332 MW 1582.60 Disulfide bonds: Cys[1]-Cys[8]; bioactive; melanocyte stimulating hormone-related peptide; newly discovered high selective MC4 receptor antagonist; chronic intracerebroventricularly injection of HS028 in free feeding rats increases both food intake & body weight without inducing any signs of tachyphylaxis; Skuladottir, GV et al, *Br J Pharmacol*, 126:27-34, 1999

Cys-Gly
Cys(Bzl)-Gly

Bachem G-1042.0250 MW 268.34 $C_{12}H_{16}N_2O_3S$ Store at -15°C

Cys-Gly
(Cys-Gly)$_2$

Bachem G-1815.0250 MW 354.41 $C_{10}H_{18}N_4O_6S_2$ Store at -15°C

Cys-Gly
(TFA-Cys-Gly-OMe)$_2$

Bachem G-1850.0250 MW 574.48 $C_{16}H_{20}F_6N_4O_8S_2$ Store at -15°C

Cys-Gly

Synonyms: Glutathione Fragment

Bachem G-3755.0250 MW 178.21 $C_5H_{10}N_2O_3S$ Store at -15°C

Sigma C 0166 FW 178.2 $C_5H_{10}N_2O_3S$ ≥85% (TLC)

Cys-Gly-Ala-Gly-Glu-Ser-Gly-Lys-Ser-Thr-Ile-Val-Lys-Gln-Met-Lys

Synonyms: GTP Binding Protein Fragment G Alpha

American Peptide 77-0-00 MW 1622.9 $C_{66}H_{117}N_{20}O_{23}S_2$ Mumby, SM et al, *PNAS*, 83:598, 1986

Cys-Gly-Arg-Gly-Asp-Ser-Pro-Cys
1-Adamantane-Ac-Cys-Gly-Arg-Gly-Asp-Ser-Pro-Cys

ICN 195616 MW 968.1 $C_{40}H_{61}N_{17}O_{13}S_2$ Crystalline, ≥97% | Inhibits integrin α4β1 binding to CS-1 & VCAM-1

Cys-Gly-Arg-Gly-Asp-Ser-Pro-Cys
1-Adamantaneacetyl-Cys-Gly-Arg-Gly-Asp-Ser-Pro-Cys

Sigma A 1460 FW 968.1 $C_{40}H_{61}N_{11}O_{13}S_2$ ≥97% (HPLC) | Disulfide bonds: 1-8; bioactive peptide; inhibits the binding of the integrin α4β1 to the fibronectin connecting segment 1 (CS-1) & to the vascular cell adhesion molecule 1 (VCAM 1); Cardarelli, PM et al, *J Biol Chem*, 269:18668, 1994

Cys-Gly-Asn-Leu-Ser-Thr-Cys

Synonyms: Calcitonin (1-7)

American Peptide 22-2-20 Human MW 694.8 $C_{25}H_{42}N_8O_{11}S_2$ Disulfide bonds: :Cys[1]-Cys[7]

Cys-Gly-Asn-Leu-Ser-Thr-Cys-Met-Leu-Gly-Gly-Gly-Tyr Trifluoroacetate Salt

Synonyms: Calcitonin (1-10), ~GGY

Biogenesis 1720-8650 Human synthetic MW 1276.6 99% (HPLC); lyophilized

Cys-Gly-Asn-Leu-Ser-Thr-Cys-Met-Leu-Gly-Thr-Tyr-Gln-Asp-Phe-Asn-Lys-Phe-His-Thr-Phe-Pro-Gln-Thr-Ala-Ile-Gly-Val-Gly-Ala-Pro Amide

Synonyms: Calcitonin

American Peptide 22-1-10   Human   MW 3417.9
$C_{151}H_{226}N_{40}O_{45}S_3$ Disulfide bonds: :Cys^1-Cys^7 | Leads to a lowering of plasma calcium & phosphate levels via the inhibition of bone resorption & excretion through the kidneys; Nakagawa, Y et al, *peptide Chemistry*, 1976:189, 1977

Cys-Gly-Asn-Leu-Ser-Thr-Cys-Met-Leu-Gly-Thr-Tyr-Thr-Gln-Asp-Leu-Asn-Lys-Phe-His-Thr-Phe-Pro-Gln-Thr-Ser-Ile-Gly-Val-Gly-Ala-Pro Amide

Synonyms: Calcitonin

Bachem H-3072.0001   Rat   MW 3399.88 $C_{148}H_{228}N_{40}O_{46}S_3$
Store at -15°C

Cys-Gly-Asn-Leu-Ser-Thr-Cys-Met-Leu-Gly-Thr-Tyr-Thr-Gln-Asp-Leu-Asn-Lys-Phe-His-Thr-Phe-Pro-Gln-Thr-Ser-Ile-Gly-Val-Gly-Ala-Pro NH$_2$-Cys-Gly-Asn-Leu-Ser-Thr-Cys-Met-Leu-Gly-Thr-Tyr-Thr-Gln-Asp-Leu-Asn-Lys-Phe-His-Thr-Phe-Pro-Gln-Thr-Ser-Ile-Gly-Val-Gly-Ala-Pro Amide

Synonyms: Calcitonin

ICN 154494   Rat   MW 3399.8   Byfield, PGH etal, *FEBS Lett*, 65:242, 1976

Cys-Gly-Asn-Leu-Ser-Thr-Cys-Met-Leu-Gly-Thr-Tyr-Thr-Gln-Asp-Leu-Asn-Lys-Phe-His-Thr-Phe-Pro-Gln-Thr-Ser-Ile-Gly-Val-Gly-Ala-Pro Amide

Synonyms: Thyrocalcitonin; Calcitonin

Sigma T 0283   Rat   FW 3399.8   ≥95% (HPLC) | Disulfide bonds: 1-7; bioactive peptide; Byfield, PGH et al, *FEBS Lett*, 65:242, 1976

Cys-Gly-Asn-Leu-Ser-Thr-Cys-Met-Leu-Gly-Thr-Tyr-Thr-Gln-Asp-Phe-Asn-Lys-Phe-His-Thr-Phe-Pro-Gln-Thr-Ala-Ile-Gly-Val-Gly-Ala-Pro Amide

Synonyms: Calcitonin

Alexis 153-021   Human   MW 3417.9 $C_{151}H_{226}N_{40}O_{45}S_3$
≥96%; white lyophilized powder; soluble in 5% acetic acid |
Disulfide bondsa:Cys^1-Cys^7

Bachem H-2250.0001   Human   MW 3417.9
$C_{151}H_{226}N_{40}O_{45}S_3$ Store at -15°C

Cys-Gly-Asn-Leu-Ser-Thr-Cys-Met-Leu-Gly-Thr-Tyr-Thr-Gln-Asp-Phe-Asn-Lys-Phe-His-Thr-Phe-Pro-Gln-Thr-Ala-Ile-Gly-Val-Gly-Ala-Pro Biotinyl-Cys-Gly-Asn-Leu-Ser-Thr-Cys-Met-Leu-Gly-Thr-Tyr-Thr-Gln-Asp-Phe-Asn-Lys(biotinyl)-Phe-His-Thr-Phe-Pro-Gln-Thr-Ala-Ile-Gly-Val-Gly-Ala-Pro Amide

Synonyms: Calcitonin, (Biotinyl-Cys^1,Lys(Biotinyl)^{18})-

Bachem H-6670.0100   Human   MW 3870.5
$C_{171}H_{254}N_{44}O_{49}S_5$ Store at -15°C

Cys-Gly-Asn-Leu-Ser-Thr-Cys-Met-Leu-Gly-Thr-Tyr-Thr-Gln-Asp-Phe-Asn-Lys-Phe-His-Thr-Phe-Pro-Gln-Thr-Ala-Ile-Gly-Val-Gly-Ala-Pro

Synonyms: Calcitonin

Neosystem SC112   Human   MW 3417.87   Disulfide bonds: Cys^1-Cys^7; bioactive; Calcium metabolism peptide; Nakagawa, Y et al, *Peptide Chemistry*, 1976:189, 1977

Cys-Gly-Asn-Leu-Ser-Thr-Cys-Met-Leu-Gly-Thr-Tyr-Thr-Gln-Asp-Phe-Asn-Lys-Phe-His-Thr-Phe-Pro-Gln-Thr-Ala-Ile-Gly-Val-Gly-Ala-Pro Amide

Synonyms: Thyrocalcitonin; Calcitonin

Sigma T 3535   Human   FW 3417.9   ≥97% (HPLC) |
Disulfide bonds: 1-7; bioactive peptide; Sieber, P et al, *Helv Chim Acta*, 51:2057, 1968

Cys-Gly-Asn-Leu-Ser-Thr-Cys-Met-Leu-Gly-Thr-Tyr-Thr-Gln-Asp-Phe-Asn-Lys-Phe-His-Thr-Phe-Pro-Gln-Thr-Ala-Ile-Gly-Val-Gly-Ala-Pro Acetate Salt

Synonyms: Calcitonin

Biogenesis 1720-8019   Human synthetic   MW 3418.4
96.6% (HPLC); lyophilized | Disulfide bonds: Cys^1-Cys^7; to avoid Met oxidation, it is best to dissolve the peptide under a nitrogen stream; sterile and pyrogen free

---

Cys-Gly-Asn-Leu-Ser-Thr-Cys-Met-Leu-Gly-Thr-Tyr-Thr-Gln-Asp-Phe-Asn-Lys-Phe-His-Thr-Phe-Pro-Gln-Thr-Ala-Ile-Gly-Val-Gly-Ala-Pro Amide

Synonyms: Calcitonin

ICN 152143   Human synthetic   Disulfide bonds: Cys^1-Cys^7 | A synthetic hormone that possesses hypocalcemic & immunological activity; Nakagawa, Y etal, *peptide Chemistry*, p 189, 1977

Peptides International PCL-4051-s, 4051-v   Human synthetic   MW 3417.9 $C_{151}H_{226}N_{40}O_{45}S_3$ >95% (HPLC); lyophilized amorphous powder | Bioactive; Disulfide bonds: Cys^1-Cys^7; Neher, R et al, *Helv Chim Acta*, 51:1900, 1968; Nakagawa, Y et al, *Peptide Chem*, 1977:189 1978

USBio C0115-11   Human synthetic   ≥96%; no contaminants detected; lyophilized, white powder | Suitable for antigenic applications in immunological protocols; leads to a lowering of plasma calcium & phosphate levels via the inhibition of bone resorption & excretion through thekidneys

---

Cys-Gly-Asn-Leu-Ser-Thr-Cys-Met-Leu-Gly-Thr-Tyr-Tyr-Thr-Gln-Asp-Leu-Asn-Lys-Phe-His-Thr-Phe-Pro-Gln-Thr-Ser-Ile-Gly-Val-Gly-Ala-Pro Amide

Synonyms: Calcitonin

American Peptide 22-5-30   Rat   MW 3399.9
$C_{148}H_{228}N_{40}O_{46}S_3$ Disulfide bonds: :Cys^1-Cys^7 | Byfield, PGH et al, *FEBS Lett*, 65:242, 1976

Cys-Gly-Gly-Gln-Lys-Gly-Arg-Gly-Ser-Arg-Gly-His-Gln Acetate Salt

Synonyms: Myelin Proteolipid Protein (117-129)

Biogenesis 6420-5150   Synthetic   >80% (HPLC); lyophilized

Cys-Gly-Gly-Pro-Phe-Thr-Phe-Asp-Met-Glu-Leu-Asp-Asp-Leu-Pro-Lys-Glu-Arg-Leu-Lys-Glu-Leu-Ile-Phe-Gln-Glu-Thr-Ala-Arg-Phe-Gln-Pro-Gly-Ala-Pro-Glu-Ala-Pro

Synonyms: Mitogen Activated Protein Kinase ERK1/ERK2 Blocking Peptide (333-367)

Calbiochem 442676   Rat synthetic   MW 4294.9
$C_{193}H_{294}N_{48}O_{59}S_2$ Lyophilized | Based on rat ERK1 MAP kinase (333-367) with the CGG spacer group added & the peptide coupled to KLH; by Western blotting recognizes the 43 kDa MAP kinase encoded by the ERK1 gene as well as the doubly phosphorylated ERK1 (44 kDa) & the 42 kDa MAP kinase encoded by the ERK2 gene; recognizes ERK1 & ERK2 in rat adipose, brain, heart, intestine, kidney, liver, lung, muscle, spleen, testis & thymus tissue extracts; Also recognizes various MAP kinases in tissue extracts from *Drosophila*, sea star, clam, frog, chicken, mouse, rat, sheep & human; Hardie, F & Hanks, S, *The Protein Kinase Facts Book, Protein-Serine Kinases*, Academic Press, San Diego, CA, p. 418, 1995; Charest, DL et al, *Mol Cell Biol*, 13:4679, 1993; Boulton, TG et al, *Science*, 249:64, 1990

Cys-Gly-Ile-Lys-Tyr-Ile-Lys-Asp-Asp-Val-Ile-Leu-Asn-Glu-Pro-Ser-Ala-Asp

Synonyms: Amyloid Bri Protein Precursor 277 (89-106)

Bachem H-5048.0001   MW 1993.27 $C_{87}H_{141}N_{21}O_{30}S$ Store at -15°C

Cys-Gly-Leu-Lys-Arg-His-His-Thr-Gly-Tyr-Glu-Gln-Phe

Synonyms: Lysosome Associated Membrane Protein II; Cytoplasmic Domain Lysosome Associated Membrane Protein II Peptide

Alexis 161-004   Synthetic   MW 1575.8 $C_{69}H_{102}N_{22}O_{19}S$ ≥90% (HPLC); dissolved in 10 *mM* sodium acetate, pH 4.5 | Competitively binds to Ab Alexis 210-206; antiserum blocking peptide

---

Cys-Gly-Leu-Phe-His-Pro-Leu-Gly-Ala-Asp-Ser-Gln-Val

Synonyms: Glucose Transporter I

Biogenesis 4670-1617   Rat synthetic   Semi-pure; lyophilized

---

Cys-Gly-Lys-Ile-Glu-Pro-Leu-Gly-Val-Ala-Pro-Thr-Lys-Ala-Lys-Arg-Arg-Val-Val-Gln-Arg-Glu-Lys-Arg

Synonyms: GP120; HIV Antigenic Peptide

Bachem H-2982.0500   Store at -15°C

Cys-Gly-Tyr-Gly-Pro-Lys-Lys-Lys-Arg-Lys-Val-Gly-Gly

Synonyms: SV40 Tumor Antigen Homolog; SV40 Nuclear Transport Signal Peptide Analog

American Peptide 79-2-11   MW 1376.7 $C_{60}H_{103}N_{20}O_{15}S$ Capable of inducing nuclear transport; Lanford, RE et al, *Cell*, 46:575, 1986

Bachem H-8120.0001   MW 1377.68 $C_{60}H_{104}N_{20}O_{15}S$ Store at -15°C

Sigma C 4547   FW 1377.7   ≥90% (HPLC); ~65% peptide content | Capable of inducing nuclear transport; Lanford, RE et al, *Cell*, 46:575, 1986

Cys-Gly-Tyr-His-Phe-Gly-Gly-Ser-Asp-Gly-Gln-Gly-Ser-Asp-Gly-Gly-Val-Ser-Trp-Gly-Leu-Gly-Gly-Asp-Gly-Ala-Ala-His-Cys

Synonyms: Pepzyme

Bachem H-2604.0500   Store at -15°C | Disulfide bonds: Cys[1]-Cys[29]

Cys-Gly-Val-Asn-Val-Asn-Asp-Ser-Ser-Asn-Glu-Lys-Arg-His-Ser-Tyr-Leu-Leu-Lys-Leu-Lys Acetate Salt

Synony Secretin Receptor (123-143)*ms:*

Biogenesis 0100-0176   Synthetic   Semi-pure; lyophilized

Biogenesis 0100-0177   Synthetic   KLH conjugated; lyophilized

---

Cys-His-His-Phe-Phe-Trp-Lys-Thr-Phe-Thr-Ser-Cys
Cys-His-His-Phe-Phe-D-Trp-Lys-Thr-Phe-Thr-Ser-Cys

Synonyms: Somatostatin, (des-Ala[1],des-Gly[2],His[4,5],D-Trp[8])-; Somatostatin 14, (des-Ala[1],des-Gly[2],His[4,5],D-Trp[8])-; Somatostatin, des-Ala[1],Gly[2],(His[4,5],D-Trp[8])-

American Peptide 68-1-41   MW 1541.8 $C_{73}H_{92}N_{18}O_{16}S_2$ Disulfide bonds: Cys[1]-Cys[12]

Bachem H-2495.0001   Store at -15°C

Sigma S 2261   FW 1541.8   ≥95% (HPLC) | Disulfide bonds: 3-14; bioactive peptide

---

Cys-His-His-Phe-Phe-Trp-Lys-Thr-Tyr-Thr-Ser-Cys
Cys-His-His-Phe-Phe-D-Trp-Lys-Thr-Tyr-Thr-Ser-Cys

Synonyms: Somatostatin, (des-Ala[1],Gly[2],(His[4,5],D-Trp[8]))-

ICN 152998   Disulfide bonds: Cys[3]-Cys[14]

Cys-His-Pro-Phe-Phe-Gln-Asp-Val-Thr-Lys-Pro-Val-Pro-His-Leu-Arg-Leu

Synonyms: Cdk2; C-Terminal Blocking Peptide (283-298)

Calbiochem 219444   Human   MW 2034.4 $C_{95}H_{144}N_{26}O_{22}S$ Lyophilized | Synthetic peptide based on the human Cdk2 (283-298) with a Cys C residue added; this peptide coupled to KLH, was used as the immunogen for the production of Anti-Cdk2; for use in immunoabsorption for immunoprecipitation, immunocytochemistry, Western blotting & dot blots

---

Cys-His-Ser-Gly-Tyr-Val-Gly-Ala-Arg-Cys
Ac-Cys-His-Ser-Gly-Tyr-Val-Gly-Ala-Arg-Cys-NHEt

Synonyms: Transforming Growth Factor α-Ethylamide (34-43), *N*-Ac-

Sigma A 2553   FW 1119.3   Human   ≥97% (HPLC) | Disulfide bonds: 34-43; bioactive peptide; Nester, JJ et al, *Biochem Biophys Res Commun*, 129:226, 1985

Cys-His-Ser-Gly-Tyr-Val-Gly-Arg-Gly-Ala-Arg-Cys
N-Ac-Cys-His-Ser-Gly-Tyr-Val-Gly-Arg-Gly-Ala-Arg-Cys-NHEt

Synonyms: Transforming Growth Factor α (34-43), Ac-

ICN 152157   Human synthetic   Disulfide bonds: Cys[34]-Cys[43]; Nestor, JJ etal, *BBRC*, 129:226, 1985

Cys-His-Ser-Gly-Tyr-Val-Gly-Arg-Gly-Ala-Arg-Cys
Ac-Cys-His-Ser-Gly-Tyr-Val-Gly-Arg-Gly-Ala-Arg-Cys-OMe

Synonyms: Transforming Growth Factor α (34-43), Ac-

ICN 152158   Human synthetic   Disulfide bonds: Cys[34]-Cys[43]; Nestor, JJ etal, *BBRC*, 129:226, 1985

Cys-His-Ser-Gly-Tyr-Val-Gly-Val-Arg-Cys

Synonyms: Transforming Growth Factor α

American Peptide 50-5-22   MW 1078.2 $C_{44}H_{67}N_{15}O_{13}S_2$ Disulfide bonds: Cys[34]-Cys[43]; Nestor, et al, *BBRC*, 129:226, 1985

Cys-His-Ser-Gly-Tyr-Val-Gly-Val-Arg-Cys
Ac-Cys-His-Ser-Gly-Tyr-Val-Gly-Val-Arg-Cys-OMe

Synonyms: Transforming Growth Factor α (34-43), (*N*-Ac)-

American Peptide 50-5-25   MW 1134.3 $C_{47}H_{71}N_{15}O_{14}S_2$ Disulfide bonds: Cys[34]-Cys[43]; Nestor, et al, *BBRC*, 129:226, 1985

Cys-His-Ser-Gly-Tyr-Val-Gly-Val-Arg-Cys

Synonyms: Transforming Growth Factor α (34-43)

Sigma T 8661   FW 1078.2   Rat   ≥97% (HPLC); peptide content ~65% | Disulfide bonds: 34-43; bioactive peptide; Nester, JJ et al, *Biochem Biophys Res Commun*, 129:226, 1985

Bachem H-9195.0001   Rat   MW 1078.24 $C_{44}H_{67}N_{15}O_{13}S_2$ Store at -15°C

ICN 158758   Rat   MW 1078.2   97% | Disulfide bonds: Cys[34]-Cys[43]; 3rd disulfide loop fragment of TGF-α; competes with EGF for receptor binding; acts as an antagonist of EGF-induced mitogenesis; Ab against this fragment cross react with native TGF-α & not EGF; Nestor, JJ etal, *BBRC*, 129:226, 1985

Cys-His-Thr-Arg-Lys-Arg-Val-Thr-Leu-Gly-Pro-Gly-Arg-Val-Trp-Tyr-Thr-Thr-Gly-Glu

Synonyms: GP120 V3 Loop Peptide CDC4

Neosystem SC627   MW 2317.65   HIV-1 GP120 V3 Loop peptide; due to the presence of a Cys this peptide can contain the dimeric form

---

Cys-Ile-Ala-Gln-Asn-Asn-Cys-Thr-Gly-Leu-Glu-Gln-Glu-Gln-Met-Ile-Lys-Lys
Cys(Acm)-Ile-Ala-Gln-Asn-Asn-Cys(Acm)-Thr-Gly-Leu-Glu-Gln-Glu-Gln-Met-Ile-Lys-Lys

Synonyms: GP140 (153-168)-Lys-Lys; SIVmac251 Peptide

Neosystem SC723   MW 2193.53   SIVmac251 peptide from subtype GP140

Cys-Ile-Ile-Arg-Asn-Cys-Pro-Arg-Gly

Synonyms: Conopressin S

Calbiochem 234617   *Conus striatus* venom   MW 1028.3 $C_{41}H_{73}N_{17}O_{10}S_2$ Amide; ≥95% (HPLC); lyophilized solid; soluble in water; LD_{50}≤200 mg/kg but >50 mg/kg | Disulfide bonds: Cys^1-Cys^6; member of the vasopressin/oxytocin family; induces intense scratching behavior when injected into mice; enhances the activity & distribution of paralytic toxins within the host due to its effects on smooth muscle; Cruz, LJ et al, *J Biol Chem*, 262:15821, 1987

Cys-Ile-Ile-Arg-Asn-Cys-Pro-Arg-Gly Amide

Synonyms: Conopressin S

Bachem H-9955.0001   Store at -15°C

Cys-Ile-Ile-Arg-Asn-Cys-Pro-Lys-Gly Amide

Synonyms: Conopressin S, (Lys^8)-

Bachem H-9930.0001   Store at -15°C

Sigma C 1807   *Conus striatus* venom   FW 1000.2   ≥90% (HPLC); peptide content:~65% | Disulfide bonds: 1-6; bioactive peptide; Lys or Arg at position 8 correlated with pressor activity; Cruz, LJ et al, *J Biol Chem*, 262:15821, 1987

Cys-Ile-Ser-Lys-Ile-Ser-Glu-His-Val-Val-Leu-Thr-Ser Trifluoroacetate Salt

Synonyms: Orexin Receptor I (408-419), ~C

Biogenesis 7049-5450   Synthetic   MW 1417.8   91% (HPLC); lyophilized | Homologous sequence in orexin receptor-1 and 2

Cys-Leu
(Cys-Leu)₂

Bachem G-1820.0250   MW 466.62 $C_{18}H_{34}N_4O_6S_2$ Store at -15°C

Cys-Leu-Arg-Arg-Ala-Ser-Leu-Gly

Synonyms: Kemptide, (Cys)-

American Peptide 86-0-49   MW 874.1 $C_{35}H_{65}N_{14}O_{10}S$ Kemp, BE et al, *Fed Proc*, 35:1384, 1976

Cys-Leu-Arg-Asp-Leu-Gln-Trp-Ala-Leu-Gln-Glu-Lys-Ile-Glu-Glu

Synonyms: Protein Kinase G *N*-Terminal Blocking Peptide (4-17)

Calbiochem 539726   Bovine, human, synthetic   MW 1874.2 $C_{82}H_{132}N_{22}O_{26}S$ Lyophilized | Based on human & bovine PKG 1α & β (residues 4-17) with a one AA substitution of Trp residue for the Tyr residue at position 10; a Cys C residue added & the peptide coupled to KLH, was used as the immunogen for the production of Anti-PKG; for use in immunoabsorption for immunoprecipitation, immunocytochemistry, Western blotting & dot blots

Cys-Leu-Asn-Arg-Gln-Leu-Ser-Ser-Gly-Val-Ser-Glu-Ile-Arg
Cys-Leu-Asn-Arg-Gln-Leu-Ser-(PO₃H₂)-Ser-Gly-Val-Ser-Glu-Ile-Arg

Synonyms: Heat Shock Protein 25kD (81-93), (Cys)-

Sigma H 2654   FW 1642.8   Mouse   ≥85% (HPLC) | Bioactive peptide; Gaestel, M et al, *J Biol Chem*, 266:14721, 1991

Cys-Leu-Asn-Arg-Gln-Leu-Ser-Ser-Gly-Val-Ser-Glu-Ile-Arg
Cys-Leu-Asn-Arg-Gln-Leu-Ser(PO₃H₂)-Ser-Gly-Val-Ser-Glu-Ile-Arg

Synonyms: Heat Shock Protein 27 (81-93), Cys-; Heat Shock Protein 27 (85-97), Cys-; Heat Shock Protein 27 (77-89), Cys-(Asn^{78})-; Heat Shock Protein 25 (81-93), Cys-

Bachem H-8400.0001   Human, hamster, mouse   Store at -15°C

Cys-Leu-Asp-Lys-Gly-Val-Val-Thr-Tyr-Lys-Phe-Lys Trifluoroacetate Salt

Synonyms: Bone Morphogenetic Protein 10 (86-96)

Biogenesis 1406-1590   Synthetic   83% (HPLC); lyophilized | Represents AA 86-96 of human BMP-10; an additional Cys has been conjugated to the N-terminus

Cys-Leu-Gln-Gly-Ser-Glu-Lys-Gly-Phe-Gln-Ser-Arg-His-Leu-Ala

Synonyms: Tissue Inhibitor of Metalloproteinase I

Biogenesis 9013-1661   Human synthetic   >80% (HPLC); lyophilized

Cys-Leu-His-Asp-Asn-Leu-Lys-Gln-Leu-Met-Leu-Gln

Synonyms: C-Terminal Blocking Peptide (367-377), G₁₃α-Subunit

Calbiochem 371786   MW 1455.8 $C_{62}H_{106}N_{18}O_{18}S_2$ Liquid | Antigen used to produce Anti-G₁₃α (Calbiochem 371784); useful as a blocking peptide

Cys-Leu-Lys-Asp-Arg-His-Asp

Synonyms: Interferon-α Receptor Recognition Peptide I

Bachem H-2724.0001   Store at -15°C

Cys-Leu-Ser-Lys-Lys-His-Leu-Asn-Trp-Ile-Ile-Met-Thr-Ile Acetate Salt

Synonyms: Granzyme A (246-258), C~

Biogenesis 4741-5050   Human synthetic   80% (HPLC); lyophilized | Represents AA 246-258 of human granzyme A; an additional Cys has been conjugated to the N-terminus

Cys-Leu-Ser-Ser-Arg-Leu-Asp-Ala-Cys

Synonyms: Brain Binding Peptide

Bachem H-3558.0001   Store at -15°C

Cys-Leu-Thr-Lys-Asp-Glu-Ser-Lys-Arg-Pro-Lys-Tyr-Lys-Glu-Leu-Leu-Lys

Synonyms: C-Terminal Blocking Peptide (347-363)

Calbiochem 409313   Human synthetic   MW 2079.5 $C_{92}H_{159}N_{25}O_{27}S$ Lyophilized | Based on the human JKK1 (residues 347-363); this peptide coupled to KLH, was used as the immunogen for the production of Anti-JKK1; for use in immunoabsorption for immunoprecipitation, immunocytochemistry, Western blotting & dot blots

Cys-Lys-Arg-Gln-His-Pro-Gly-Lys-Arg-Cys

Synonyms: Pro-Thyrotropin Releasing Hormone, -SH

American Peptide 58-0-90   MW 1210.5 $C_{48}H_{83}N_{21}O_{12}S_2$ Disulfide bonds: Cys^1-Cys^{10}

Cys-Lys-Arg-Lys-Arg-Thr-Glu-Ala-Leu-Glu-Gln-Gly-Gly-Leu-Pro-Lys-Leu-Ile-Phe

Synonyms: Cdk7/MO15; C-Terminal Blocking Peptide (328-346)

Calbiochem 219456 Human MW 2315.8 $C_{103}H_{179}N_{31}O_{27}S$ Lyophilized | Synthetic peptide based on the human Cdk7/MO15 (328-346) with a Cys C residue added; this peptide coupled to KLH, was used as the immunogen for the production of Anti-Cdk7/MO15; for use in immunoabsorption for immunoprecipitation, immunocytochemistry, Western blotting & dot blots

Cys-Lys-Asn-Asn-Leu-Lys-Glu-Cys-Gly-Leu-Tyr

Synonyms: C-Terminal Blocking Peptide (345-354), G$_I\alpha$-3-Subunit

Calbiochem 371779 MW 1284.5 $C_{54}H_{89}N_{15}O_{17}S_2$ Lyophilized | Distinct C-terminal region of G$_I\alpha$-3-subunit; A used to produce antibody for Calbiochem 371729; useful as a blocking peptide; Jones, DT & Reed, RR, *J Biol Chem*, 262:14241, 1987

Cys-Lys-Asn-Phe-Phe-Trp-Lys-Thr

Synonyms: Somatostatin 14 (3-10)

Bachem H-4702.0005 MW 1073.28 $C_{52}H_{72}N_{12}O_{11}S$ Store at -15°C

Cys-Lys-Asn-Phe-Phe-Trp-Lys-Thr-Phe-Thr-Ser-Cys

Synonyms: Somatostatin 14 (3-14)

Bachem H-4774.0001 MW 1509.77 $C_{71}H_{96}N_{16}O_{17}S_2$ Store at -15°C

Cys-Lys-Asp-Asp-Met-Gly-Val-Pro-Thr-Leu-Lys-Tyr-His Acetate Salt

Synonyms: Bone Morphogenetic Protein 9 (87-98), C~

Biogenesis 1406-1490 Human synthetic MW 1506.7 93% (HPLC); lyophilized

Cys-Lys-Gln-Leu-Gln-Arg-Asp-Lys-Gln-Val-Tyr-Arg-Ala-Thr-His-Arg

Synonyms: GTP Binding Protein Fragment Gs Alpha, (Arg⁶)-

American Peptide 77-0-02 MW 2057.4 $C_{85}H_{143}N_{34}O_{24}S$ Mumby, SM et al, *PNAS*, 83:598, 1986

Cys-Lys-Gln-Leu-Gln-Lys-Asp-Lys-Gln-Val-Tyr-Arg-Ala-Thr-His-Arg

Synonyms: GTP Binding Protein Fragment Gs Alpha

American Peptide 77-0-01 MW 2001.3 $C_{85}H_{143}N_{30}O_{24}S$ Mumby, SM et al, *PNAS*, 83:598, 1986

Cys-Lys-Gln-Lys-Leu-Met-Pro-Arg-Val-Leu-Thr-Met-Ile-Gln Acetate Salt

Synonyms: Glutathione-S-Transferase Theta-1 (227-239), C~

Biogenesis 4690-6030 Synthetic >80% (HPLC); lyophilized

Cys-Lys-Glu-Asn-Leu-Asp-Ser-His-Leu-Pro-Pro-Ser-Gln-Asn-Thr-Ser-Glu-Leu-Asn-Thr-Ala

Synonyms: Cdk6; C-Terminal Blocking Peptide (306-326)

Calbiochem 219453 Human synthetic MW 2298.5 $C_{94}H_{152}N_{28}O_{37}S$ Lyophilized | This peptide coupled to KLH, was used as the immunogen for the production of Anti-Cdk6; for use in immunoabsorption for immunoprecipitation, immunocytochemistry, Western blotting & dot blots

Cys-Lys-Gly-Lys-Gly-Ala-Lys-Cys-Ser-Arg-Leu-Met-Tyr-Asp-Cys-Cys-Thr-Gly-Ser-Cys-Arg-Ser-Gly-Lys-Cys Amide

Synonyms: Conotoxin MVIIA, σ-; Conotoxin MVIIA, ω-; Ca²⁺ Channel Blocker, Reversible N-Type

Bachem H-8210.0500 MW 2639.17 $C_{102}H_{172}N_{36}O_{32}S_7$ Store at -15°C | Disulfide bonds: Cys¹-Cys¹⁶, Cys⁸-Cys²⁰, Cys¹⁵-Cys²⁵

Sigma C 1182 FW 2639.1 ≥95% (HPLC); peptide content:~70% | Disulfide bonds: 1-16, 8-20, 15-25; bioactive peptide; neuronal *N*-type Ca²⁺ channel blocker in mammalian & amphibian brain; blocks release of GABA & glutamate at neuronal synapses; probe of calcium channel receptors; selective for different receptor subtypes; Olivera, BM et al, *Biochemistry*, 26:2086, 1987

American Peptide 41-0-60 *Conus magus* MW 2639.2 $C_{102}H_{172}N_{36}O_{32}S_7$ Disulfide bonds: Cys¹-Cys¹⁶, Cys⁸-Cys²⁰, Cys¹⁵-Cys²⁵; N-Type neuronal Ca²⁺ channel blocker that competes for the same sites as ω-CgTx GVIA in mammalian brain; ω-CgTx has a higher binding affinity & narrower specificity in amphibian brain; also blocks the release of glutamate & GABA at neuronal synapses; Olivera, BM et al, *Biochemistry*, 26:2086, 1987

Peptides International PCN-4289-v *Conus magus* (marine snail) synthetic MW 2639.2 $C_{102}H_{172}N_{36}O_{32}S_7$ >95% (HPLC); lyophilized amorphous powder | Bioactive; Disulfide bonds: Cys¹-Cys¹⁶, Cys⁸-Cys²⁰, Cys¹⁵-Cys²⁵; Olivera, BM et al, *Biochem*, 26:2086, 1987; Valentino, K et al, *PNAS USA*, 90:7894, 1993; Fox, JA, *Pflügers Arch*, 429:873, 1995

Alexis 630-056 *Conus magus* synthetic MW 2636.2 $C_{102}H_{169}N_{36}O_{32}S_7$ White lyophilized powder; soluble in water | Disulfide bonds: Cys¹-Cys¹⁶, Cys⁸-Cys²⁰, Cys¹⁵-Cys²⁵; potent neurotoxin

Cys-Lys-Gly-Lys-Gly-Ala-Pro-Cys-Arg-Lys-Thr-Met-Tyr-Asp-Cys-Cys-Ser-Gly-Ser-Cys-Gly-Arg-Arg-Gly-Lys-Cys Amide

Synonyms: Conotoxin MVIIC, σ-; Conotoxin MVIIC, ω-; Ca²⁺ Channel Blocker, P/Q-Type

Bachem H-8835.0500 MW 2749.29 $C_{106}H_{178}N_{40}O_{32}S_7$ Store at -15°C | Disulfide bonds: Cys¹-Cys¹⁶, Cys⁸-Cys²⁰, Cys¹⁵-Cys²⁶

Sigma C 4188 FW 2749.2 ≥95% (HPLC); peptide content:~65%; bioactive peptide | Disulfide bonds: 1-16, 8-20, 15-26; potent, selective blocker of mammalian Q-type Ca²⁺ channels; blocker of glutamate & GABA release; Hillyard, DR et al, *Neuron*, 9:69, 1992

American Peptide 41-0-62 *Conus geographus* MW 2749.3 $C_{106}H_{178}N_{40}O_{32}S_7$ Disulfide bonds: Cys¹-Cys¹⁶, Cys⁸-Cys²⁰, Cys¹⁵-Cys²⁶; Ca²⁺ channels resistant to ω-CgTx GVIA; blocks hippocampal excitatory postsynaptic potential (EPSPs); may interact with Q-type voltage-sensitive Ca²⁺ channels (VSCC); Hillyard, R et al, *Neuron*, 9:69, 1992

Calbiochem 234630 *Conus magus* MW 2749.3 $C_{106}H_{178}N_{40}O_{32}S_7$ ≥99% (HPLC); lyophilized solid; soluble in water; highly toxic:LD₅₀ ≤50 mg/kg | Disulfide bonds Cys¹-Cys¹⁶, Cys⁸-Cys²⁰, Cys¹⁵-Cys²⁶; potent selective blocker of mammalian Q-type Ca²⁺ channels; potent blocker of glutamate (IC₅₀=35 nM) & GABA release (IC₅₀=200 nM); Turner, TJ & Dunlop, K, *Neuropharmacology*, 34:1469, 1995; Hyllyard, DR et al, *Neuron*, 9:69, 1992; Newcomb, R & Palma, A, *Brain Res*, 638:95, 1994

Peptides International PCN-4283-s, 4283-v *Conus magus* (marine snail) synthetic MW 2749.3 $C_{106}H_{178}N_{40}O_{32}S_7$ >95% (HPLC); lyophilized amorphous powder | Bioactive; Disulfide bonds: Cys¹-Cys¹⁶, Cys⁸-Cys²⁰, Cys¹⁵-Cys²⁶; Hillyard, DR et al, *Neuron*, 9:69, 1992; Adams, ME et al, *Biochem*, 32:12566, 1993; Sather, WA et al, *Neuron*, 11:291, 1993; Wheeler, DB et al, *Science*, 264:107, 1994

Alexis 630-057 *Conus magus* synthetic MW 2749.3 $C_{106}H_{178}N_{40}O_{32}S_7$ White lyophilized powder; soluble in water | Disulfide bonds: Cys¹-Cys¹⁶, Cys⁸-Cys²⁰, Cys¹⁵-Cys²⁵; potent neurotoxin

Cys-Lys-Gly-Thr-Asp-Val-Ala-Ser-Phe-Val-Lys-Leu-Ile-Leu-Gly-Asp

Synonyms: MEK6 Blocking Peptide (320-334)

Calbiochem 444960 Human synthetic MW 1666.0 $C_{74}H_{124}N_{18}O_{23}S$ Lyophilized solid | Based on human MEK6 (320-334) with a Cys C residue added; this peptide coupled to KLH was used as the immunogen for the production of Anti-MEK6; for use in immunoabsorption for immunoprecipitation, immunocytochemistry, Western blotting & dot blots

Cys-Lys-Ile-Leu-Asp-Gln-Met-Pro-Ala-Thr-Pro-Ser-Ser-Pro-Met-Tyr-Val-Asp

Synonyms: SEK1 C-Terminal Blocking Peptide (378-395)

Calbiochem 563107 Mouse synthetic MW 1996.4
$C_{83}H_{138}N_{20}O_{28}S_3$ Lyophilized | Identical to the human SEK1 sequence over these residues; this peptide coupled to KLH, was used as the immunogen for the production of Anti-SEK1; for use in immunoabsorption for immunoprecipitation, immunocytochemistry, Western blotting & dot blots

Cys-Lys-Leu-Lys-Gly-Gln-Ser-Cys-Arg-Lys-Thr-Ser-Tyr-Asp-Cys-Cys-Ser-Gly-Ser-Cys-Gly-Arg-Ser-Gly-Lys-Cys

Synonyms: Conotoxin SVIB, ω-

American Peptide 41-0-77 MW 2739.2 $C_{105}H_{175}N_{37}O_{37}S_6$
Disulfide bonds: Cys^1-Cys^{16}, Cys^8-Cys^{20}, Cys^{15}-Cys^{26}; novel calcium channel blocker; exhibits strong binding to N-type channels & receptors in neuronal Na^+ channels; Ramilo, CA et al, *Biochemistry*, 31:9919, 1992

Cys-Lys-Leu-Lys-Gly-Gln-Ser-Cys-Arg-Lys-Thr-Ser-Tyr-Asp-Cys-Cys-Ser-Gly-Ser-Cys-Gly-Arg-Ser-Gly-Lys-Cys Amide

Synonyms: Ca^{2+} Channel Blocker, N-Type; Conotoxin SVIB, ω-

Sigma C 2586 FW 2739.1 ≥97% (HPLC) | Disulfide bonds: 1-16, 8-20, 15-26; bioactive peptide presynaptic Ca^{2+} channel blocker; binds N-type channels & receptors in neuronal membranes

Calbiochem 243635 *Conus striatus* MW 2739.2
$C_{105}H_{176}N_{38}O_{36}S_6$ >99% (HPLC); lyophilized solid; soluble in 0.9% NaCl/water; highly toxic:$LD_{50} ≤50$ mg/kg | Disulfide bonds: Cys^1-Cys^{16}, Cys^8-Cys^{20}, Cys^{15}-Cys^{26}; inhibits K^+-induced ^3H-GABA release from primary rat hippocampal neurons; novel presynaptic Ca^{2+} channel blocker; exhibits strong binding to N-type channels & receptors in neuronal membranes; differs from ω-conotoxins MVIIA & GVIA in its binding site & toxicity in mammals; Ramilo, CA et al, *Biochemistry*, 31:9919, 1992; Woppmann, A et al, *Mol Cell Neurosci*, 5:350, 1994; Fox, JA et al, *Neurosci Lett*, 165:157, 1994

Peptides International PCN-4284-v *Conus striatus* (marine snail) synthetic MW 2739.2 $C_{105}H_{176}N_{38}O_{36}S_6$ >99% (HPLC); lyophilized amorphous powder | Bioactive; Disulfide bonds: Cys^1-Cys^{16}, Cys^8-Cys^{20}, Cys^{15}-Cys^{26}; Ramilo, CA et al, *Biochem*, 31:9919, 1992)

Alexis 630-053 *Conus striatus* synthetic MW 2739.2
$C_{105}H_{176}N_{38}O_{36}S_6$ White lyophilized powder; soluble in water | Disulfide bonds: Cys^1-Cys^{16}, Cys^8-Cys^{20}, Cys^{15}-Cys^{26}; potent neurotoxin

Cys-Lys-Lys-Cys-Cys-Tyr-His-Cys-Gln-Phe-Cys-Phe-Leu-Lys-Lys
Cys(Acm)-Lys-Lys-Cys(Acm)-Cys(Acm)-Tyr-His-Cys(Acm)-Gln-Phe-Cys(Acm)-Phe-Leu-Lys-Lys

Synonyms: TAT (56-70); SIVmac251 Peptide

Neosystem SC588 MW 2237.75 SIVmac251 peptide from subtype TAT

Cys-Lys-Lys-Gly-Ser-Ala-Leu-Glu-Glu-Pro-Lys-Ala-Thr-Arg-Leu
Ac-(Cys)-Lys-Lys-Gly-Ser-Ala-Leu-Glu-Glu-Pro-Lys-Ala-Thr-Arg-Leu Amide Trifluoroacetate Salt

Synonyms: Nitric Oxide Synthase Blocking Peptide (1131-1144)

Calbiochem 482729 Mouse macrophage MW 1672
$C_{71}H_{126}N_{22}O_{22}S$ ≥95% (HPLC); lyophilized solid; soluble in water | Immunization & blocking peptide for Anti-iNOS (1131-1144), mouse (rabbit); Lowenstein, CJ et al, *PNAS*, 89:6711, 1992; Lyons, CR et al, *J Biol Chem*, 267:6370, 1992

Cys-Lys-Lys-Gly-Ser-Ala-Leu-Glu-Glu-Pro-Lys-Ala-Thr-Arg-Leu

Synonyms: Nitric Oxide Synthase Macrophage C-Terminal Peptide (1131-1144); Nitric Oxide Synthase II

Alexis 163-005 Mouse synthetic MW 1630.9
$C_{69}H_{123}N_{21}O_{22}S$ ≥95%; lyophilized; reconstitute with 0.1 mL distilled H_2O | Competitively binds to Ab Alexis 210-504; antiserum blocking peptidep iNOS

Cys-Lys-Phe-Phe-Trp-Thr-Phe-Thr-Ser-Cys
Cys-Lys-Phe-Phe-D-Trp-IAmp-Thr-Phe-Thr-Ser-Cys

Synonyms: Somatostatin 14, (des-Ala[1], des-Gly[2], des-Asn[5], D-Trp[8], IAmp[9])-

Neosystem SC1308 MW 1485.77 Disulfide bonds: Cys^3-Cys^{14}; bioactive; hypothalamic releasing hormone; selective agonist at the somatostatin receptor subtype SSTR1; Liapakis, G et al, *J Pharmacol Exp Ther*, 276:1089-1094, 1996

Cys-Lys-Pro-Leu-Ile-Leu-Pro-Asp-Thr-Lys-Pro-Lys-Ile-Lys-Asp

Synonyms: Glucocorticoid Receptor Protein N-Terminal Peptide (245-259)

Alexis 157-016 Human synthetic MW 1709.1
$C_{78}H_{137}N_{19}O_{21}S$ ≥95%; lyophilized; reconstitute with 0.1 mL distilled H_2O | Competitively binds to Ab Alexis 210-153; antiserum blocking peptide

Cys-Lys-Ser-Hyp-Gly-Ser-Ser-Cys-Ser-Hyp-Thr-Ser-Tyr-Asn-Cys-Cys-Arg-Ser-Cys-Asn-Hyp-Tyr-Thr-Lys-Arg-Cys-Tyr Amide

Synonyms: Conotoxin GVIA, σ-; Conotoxin GVIA, ω-; Ca^{2+} Channel Blocker, N-Type

Bachem H-6615.0500 MW 3037.39 $C_{120}H_{182}N_{38}O_{43}S_6$ Store at -15°C | Disulfide bonds: Cys^1-Cys^{16}, Cys^8-Cys^{19}, Cys^{15}-Cys^{26}

Bachem H-9490.0500 Store at -15°C | Disulfide bonds: Cys^1-Cys^8, Cys^{15}-Cys^{19}, Cys^{16}-Cys^{26}

Sigma C 9915 FW 3037.3 ≥97% (HPLC); peptide content:~70% | Disulfide bonds: 1-16, 8-19, 15-26; bioactive peptide; powerful probe for exploring the vertebrate pre-synaptic terminal; blocks voltage-activated N-type Ca^{2+} channels; Olivera, BM et al, *Biochemistry*, 23:5087, 1984; Rivier, J et al, *J Biol Chem*, 262:1194, 1987

American Peptide 41-0-30 *Conus geographus* MW 3034.4
$C_{120}H_{179}N_{38}O_{43}S_6$ Disulfide bonds: Cys^1-Cys^{16}, Cys^8-Cys^{19}, Cys^{15}-Cys^{26}; blocks specific voltage-dependent Ca^{2+} channels in neurons, but not in muscles; does not bind to either the dihydropyridine or verapamil binding site; its antagonistic activity is attributed to a tyrosine residue at position 13; Rivier, J et al, *JBC*, 262:1194, 1987

Calbiochem 343781 *Conus geographus* MW 3037.4
$C_{120}H_{182}N_{38}O_{43}S_6$ ≥98% (HPLC); lyophilized solid; soluble in aqueous buffers & water; highly toxic:$LD_{50} ≤50$ mg/kg | Disulfide bonds: Cys^1-Cys^{16}, Cys^8-Cys^{19}, Cys^{15}-Cys^{26}; neurotoxin that acts as an antagonist of voltage-activated N-type Ca^{2+} channels & of neurotransmitter release at neuronal synapses; antagonistic activity is attributed to a Tyr residue at position 13; does not bind to dihydropyridine- or verapamil-binding sites; Kim, JI et al, *Biochem Biophys Res Comm*, 206:449, 1995; Abbott, JR et al, *Int J Dev Neurosci*, 12:43, 1994; Olivera, BM et al, *J Biol Chem*, 266:22067, 1991; Protti, DA et al, *Brain Res*, 557:336, 1991; Werth, JL et al, *Mol Pharmacol*, 40:742, 1991; Koyano, K et al, *Eur J Pharmacol*, 135:337, 1987; Abe, T et al, *Neurosci Lett*, 71:203, 1986

Peptides International PCN-4161-v *Conus geographus* (marine snail) synthetic MW 3037.4 $C_{120}H_{182}N_{38}O_{43}S_6$ >99% (HPLC); lyophilized amorphous powder | Bioactive; Olivera, BM et al, *Biochem*, 23:5087, 1984; Nishiuchi, Y et al, *Biopolymers*, 25:S61, 1986

Alexis 630-055 *Conus geographus* synthetic MW 3036.3
$C_{120}H_{181}N_{38}O_{43}S_6$ White lyophilized powder; soluble in water | Disulfide bonds: Cys^1-Cys^{16}, Cys^8-Cys^{20}, Cys^{15}-Cys^{25}; potent neurotoxin

Cys-Met
Cys-4-Abz-Met

Synonyms: Farnesyltransferase Inhibitor II

Alexis 260-021 MW 371.5 C$_3$H$_{21}$N$_3$O$_4$S$_2$ ≥92%; white lyophilized powder; soluble in water | Potent inhibitor of Ras farnesyltransferase; Hamilton, AD & Sebti, SM, *Drug News Perspect*, 8:138, 1995

Cys-Met
Cys-4-Abz-Met Trifluoroacetate Salt

Synonyms: Farnesyltransferase Inhibitor II

Calbiochem 344512 MW 371.5 C$_{15}$H$_{21}$N$_3$O$_4$S$_2$ ≥90% (HPLC); white solid; soluble in water | IC$_{50}$=50 *nM*; prevents the farnesylation of Ras thus resulting in its inability to associate with other cell signaling components in the cell; Hamilton, AD & Sebti, SM, *Drug News Perspect*, 8:138, 1995

Cys-Met-His-Ile-Glu-Ser-Asp-Ser-Tyr-Thr-Cys
Cys(Acm)-Met-His-Ile-Glu-Ser-Asp-Ser-Tyr-Thr-Cys(Acm)

Synonyms: Epidermal Growth Factor (20-31), (des-Leu[26],Cys(Acm)[20,31])-

Bachem H-6175.0001 Store at -15°C

Cys-Met-His-Ile-Glu-Ser-Leu-Asp-Ser-Tyr-Thr-Cys
Cys(Acm)-Met-His-Ile-Glu-Ser-Leu-Asp-Ser-Tyr-Thr-Cys(Acm)

Synonyms: Epidermal Growth Factor (20-31), (Cys(Acm)[20,31])-

American Peptide 50-0-35 MW 1544.0 C$_{63}$H$_{98}$N$_{16}$O$_{23}$S$_3$ Komoriya, A et al, *PNAS*, 81:1351, 1984

Bachem H-1000.0001 MW 1543.76 C$_{63}$H$_{98}$N$_{16}$O$_{23}$S$_3$ Store at -15°C

Neosystem SC382 MW 1543.74 Bioactive; synthetic growth factor-related peptide; Komoriya, A et al, *PNAS USA*, 81:1351, 1984

Cys-Met-His-Ile-Glu-Ser-Leu-Asp-Ser-Tyr-Thr-Cys
Cys(S-Acm)-Met-His-Ile-Glu-Ser-Leu-Asp-Ser-Tyr-Thr-Cys(Acm)

Synonyms: Epidermal Growth Factor (20-31), (Cys(Acm)[20,31])-

Sigma E 9384 FW 1543.7 ≥97% (HPLC) | Bioactive peptide; Komoriya, A et al, *Proc Natl Acad Sci USA*, 81:1351, 1984

ICN 151513 Human synthetic Komoriya, A etal, PNAS, 81:1351, 1984

Cys-Met-Ser-Ser-Asp-Leu-Glu-Arg-Asp-His-Arg-Pro-His Acetate Salt

Synonyms: Calcitonin Precursor (131-143), C~

Biogenesis 1720-9730 Human synthetic 90% (HPLC); lyophilized

Cys-Nle-Arg-His-Nal-Arg-Trp-Gly-Cys
Ac-Cys-Nle-Arg-His-D-2-Nal-Arg-Trp-Gly-Cys Amide

Synonyms: Melanocyte Stimulating Hormone (3-11), α-(Ac-Cys[3],Nle[4],Arg[5],D-2-Nal[7],Cys[11])-

Bachem H-4598.0001 MW 1266.52 C$_{58}$H$_{79}$N$_{19}$O$_{10}$S$_2$ Store at -15°C

Cys-Nle-Arg-His-Nal-Nal-Arg-Trp-Gly-Cys
Ac-Cys-Nle-Arg-His-D-Nal$_2$-Arg-Trp-Gly-Cys

Synonyms: HS024

Neosystem SC1331 MW 1266.56 Disulfide bonds: Cys[1]-Cys[9]; bioactive; melanocyte stimulating hormone-related peptide; high affinity & selective MC4 receptor antagonist; similar potencies for increasing food intake as SHU9119 but with fewer side-effects; Kask, A et al, *Endocrinology*, 139:5006-5014, 1998

Cys-Phe
(Cys-Phe)$_2$

Bachem G-1825.0250 MW 534.66 C$_{24}$H$_{30}$N$_4$O$_6$S$_2$ Store at -15°C

Cys-Phe-Glu-Gly-Phe-Glu-Thr-Tyr-Ile-Ala-Asn-Pro-Leu-Leu-Leu-Ser-Thr-Glu-Glu-Ser-Val

Synonyms: Protein Kinase C$_ξ$ Blocking Peptide

Calbiochem 539647 MW 2362.7 C$_{107}$H$_{160}$N$_{22}$O$_{36}$S Liquid; use 0.25-2.5 µg/mL of peptide to block the antibody under the specified assay conditions | Distinct C-terminal region of protein kinase C$_ξ$; used to confirm the specificity of anti-PKC$_ξ$ in Western blotting

Cys-Phe-Gly-Gly-Arg-Met-Asp-Arg-Ile-Gly-Ala-Gln-Ser-Gly-Leu-Gly-Cys-Asn-Ser-Phe-Arg-Tyr

Synonyms: Natriuretic Peptide (7-28), Atrial

Peptides International PAF-4139-v Human synthetic MW 2393.7 C$_{100}$H$_{153}$N$_{33}$O$_{30}$S$_3$ >95% (HPLC); lyophilized amorphous powder | Bioactive; Disulfide bonds: Cys[7]-Cys[23]; Watanabe, TX et al, *Eur J Pharmacol*, 147:49, 1988

American Peptide 14-1-23 Human, canine MW 2393.7 C$_{100}$H$_{153}$N$_{33}$O$_{30}$S$_3$ Disulfide bonds: Cys[7]-Cys[23] | Chino, N et al, *Peptide Chemistry*, 1984:241, 1985

Biogenesis 0780-0259 Synthetic MW 2393.7 >96.7% (HPLC); lyophilized, consisting of 81% peptide material, acetate counter ions and residual H$_2$O | To avoid Met oxidation, it is best to dissolve the peptide under a nitrogen stream

Cys-Phe-Gly-Ser-Arg-Ile-Asp-Arg-Ile-Gly-Ala-Gln-Ser-Gly-Met-Gly-Cys-Gly-Arg-Phe

Synonyms: Natriuretic Peptide (11-30), Atrial

American Peptide 14-7-21 Frog MW 2116.5 C$_{87}$H$_{138}$N$_{30}$O$_{26}$S$_3$ Disulfide bonds: Cys[11]-Cys[27]

Cys-Phe-Gly-Ser-Arg-Ile-Asp-Arg-Ile-Gly-Ala-Gln-Ser-Gly-Met-Gly-Cys-Gly-Arg-Arg-Phe
H$_2$N-Cys-Phe-Gly-Ser-Arg-Ile-Asp-Arg-Ile-Gly-Ala-Gln-Ser-Gly-Met-Gly-Cys-Gly-Arg-Arg-Phe

Synonyms: Natriuretic Peptide, Atrial

ICN 154506 Frog MW 2272.6 Sakata, M etal, *BBRC*, 144:1338, 1988

Cys-Phe-Gly-Ser-Arg-Ile-Asp-Arg-Ile-Gly-Ala-Gln-Ser-Gly-Met-Gly-Cys-Gly-Arg-Arg-Phe

Synonyms: Natriuretic Peptide (4-24), Atrial

Sigma A 0804 Frog FW 2272.6 ≥97% (HPLC) | Disulfide bonds: 4-20; bioactive peptide; structurally similar to rat ANF; diuretic-natriuretic activity; vascular muscle relaxant; Sakata, J et al, *Biochem Biophys Res Commun*, 155:1338, 1988

Cys-Phe-Ile-Arg-Asn-Cys-Pro-Arg-Gly Amide

Synonyms: Conopressin G, (Arg[8])-

Bachem H-9925.0001 Store at -15°C

Cys-Phe-Ile-Arg-Asn-Cys-Pro-Lys-Gly Amide

Synonyms: Conopressin G

Calbiochem 234612 *Conus geographus* MW 1034.3 C$_{44}$H$_{71}$N$_{15}$O$_{10}$S$_2$ ≥95% (HPLC); lyophilized solid; soluble in water; LD$_{50}$≤200 mg/kg but >50 mg/kg | Disulfide bonds: Cys[1]-Cys[6]; member of the vasopressin/oxytocin family; activates low- & high-voltage activated currents in the anterior lobe of molluscan cerebral ganglion; induces intense scratching behavior when injected into mice; enhances the activity & distribution of paralytic toxins within the host due to its effects on smooth muscle; Cruz, LJ et al, *J Biol Chem*, 262:15821, 1987; van Soest, PF & Kits, KS, *J Neurophysiol*, 79:1619, 1998

Cys-Phe-Ile-Gln-Asn-Cys-Pro-Orn-Gly Amide

Synonyms: Vasopressin, (Phe[2],Ile[3],Orn[8])-; Vasotocin, (Phe[2],Orn[8])-; Oxytocin, (Phe[2],Orn[8])-

American Peptide 66-0-11   MW 992.2 $C_{42}H_{65}N_{13}O_{11}S_2$
Disulfide bonds: Cys[1]-Cys[6]

Bachem H-3178.0001   Store at -15°C

Cys-Phe-Leu-Lys-Lys-Gly-Leu-Gly-Ile-Cys-Tyr-Glu-Gln-Ser-Arg

Cys(Acm)-Phe-Leu-Lys-Lys-Gly-Leu-Gly-Ile-Cys(Acm)-Tyr-Glu-Gln-Ser-Arg

Synonyms: TAT (66-80); SIVmac251 Peptide

Neosystem SC590   MW 1887.24   SIVmac251 peptide from subtype TAT

Cys-Phe-Leu-Ser-Lys-Met-Leu-Val-Tyr-Asp-Pro-Ala-Lys-Arg-Ile-Ser-Gly-Lys-Met-Ala-Leu-Lys-His-Pro-Tyr-Phe-Asp-Asp-Leu-Asp-Asn-Gln-Ile-Lys-Lys-Met

Synonyms: Cdc2 Blocking Peptide (263-297)

Calbiochem 217701   Mouse synthetic   MW 4246.1
$C_{192}H_{307}N_{49}O_{51}S_4$ Lyophilized | Synthetic peptide based on mouse 34 kDa cdc2-encoded protein kinase (263-297) with a Cys C residue added; this peptide coupled to KLH, was used as the immunogen for the production of Anti-Cdc2; for use in immunoabsorption for immunoprecipitation, immunocytochemistry, Western blotting & dot blots

Cys-Phe-Phe-Asp-Trp-Lys-Thr-Phe-Cys

Synonyms: Somatostatin, (des Ala[1],Gly[2],Lys[4],Asn[5],Thr[12],Ser[13],D-Trp[8])-

Biogenesis 8330-3509   Synthetic   MW 1079   97% (HPLC), >99% (TLC); lyophilized, consisting of 83% peptide material, acetate counter ions and residual H_2O

Cys-Phe-Ser-Arg-Ala-Tyr-Pro-Thr-Pro-Leu-Arg-Ser-Lys-Lys-Thr Amide

Synonyms: Glycoprotein Hormone α (32-46)

Bachem H-3386.0001   Store at -15°C

Cys-Phe-Val-Ala-Asn-Ser-Glu-Phe-Leu-Lys-Pro-Glu-Val-Lys-Ser

Synonyms: Protein Kinase C$_{\beta II}$ Blocking Peptide

Calbiochem 539637   MW 1698 $C_{77}H_{120}N_{18}O_{23}S$ Liquid; use 0.25-2.5 µg/mL of peptide to block the antibody under the specified assay conditions | Distinct C-terminal region of protein kinase C$_{\beta II}$; used to confirm the specificity of anti-PKC$_{\beta II}$ in Western blotting

Cys-Pro-Asp-Asn-Thr-Arg-Lys-Pro-Val-Asp-Lys-Phe-Lys-Asp-Cys

Synonyms: Lactoferrin *N*-Lobe (231-245)

Bachem H-3378.0500   Human   Store at -15°C

Cys-Pro-Ser-Val-Ser-Pro-Ser-Ser-Val-Glu-Asn-Ser-Gly-Val-Ser-Gln-Ser-Pro-Leu-Leu-Gln

Synonyms: Retinoic Acid Receptor β C-Terminal Peptide (429-448)

Alexis 166-004   Mouse synthetic   MW 2102.3
$C_{87}H_{144}N_{24}O_{34}S$ ≥95%; lyophilized; reconstitute with 0.1 mL distilled H_2O | Competitively binds to Ab Alexis 210-157 & 804-103; antiserum & monoclonal Ab blocking peptide

Cys-Ser
BOC-Cys(Bzl)-Ser-OMe

Bachem A-2840.0001   MW 412.51 $C_{19}H_{28}N_2O_6S$ Store at -15°C

Cys-Ser-Ala-Phe-Ala-Gly-Phe-Ser-Phe-Val-Ala-Asn-Pro

Synonyms: Protein Kinase C$_\delta$ Blocking Peptide

Calbiochem 539639   MW 1903.3 $C_{61}H_{84}N_{14}O_{17}S$ Liquid; use 0.25-2.5 µg/mL of peptide to block the antibody under the specified assay conditions | Distinct C-terminal region of protein kinase C$_\delta$; used to confirm the specificity of anti-PKC$_\delta$ in Western blotting

Cys-Ser-Ala-Ser-Ser-Leu-Nle-Asp-Lys-Glu-Ala-Val-Tyr-Phe-Cys-His-Leu-Ala-Ile-Ile-Trp

Synonyms: Endothelin I, (Ala[3,11,18],Nle[7])-

American Peptide 88-1-12   Human   MW 2367.7
$C_{109}H_{163}N_{25}O_{30}S_2$ Disulfide bonds: Cys[1]-Cys[15]; high affinity for the Endothelin I binding sites without demonstrating agonist activity; Hunt, JT et al, *Bioorg & Medicinal Chem Lett*, 1:33, 1991

Cys-Ser-Arg-Ala-Arg-Lys-Gln-Ala-Ala-Ser-Ile-Lys-Val-Ala-Val-Ser-Ala-Asp-Arg

Synonyms: Laminin A Chain (2091-2108), Cys-

American Peptide 87-0-54   MW 2016.3 $C_{82}H_{148}N_{31}O_{26}S$
Supports neuron cell outgrowth & stimulates neuron-like process formation; Kanemoto, T et al, *PNAS*, 87:2279, 1990

Bachem H-8030.0001   MW 2017.34 $C_{82}H_{149}N_{31}O_{26}S$ Store at -15°C

Sigma C 6171   FW 2017.3   ≥90% (HPLC) | Supports neurite outgrowth & stimulates neuronal-like process formation; bioactive peptide; Sephel, GC et al, *Biochem Biophys Res Commun*, 162:821, 1989

Cys-Ser-Asn-Leu-Ser-Thr-Cys-Val-Leu-Gly-Lys-Leu-Ser-Gln-Glu-Leu-His-Lys-Leu-Gln-Thr-Tyr-Pro-Arg-Thr-Asn-Thr-Gly-Ser-Gly-Thr-Pro Amide

Synonyms: Calcitonin I

Bachem H-2260.0001   MW 3431.9 $C_{145}H_{240}N_{44}O_{48}S_2$ Store at -15°C | Salmon

Cys-Ser-Asn-Leu-Ser-Thr-Cys-Val-Leu-Gly-Lys-Leu-Ser-Gln-Glu-Leu-His-Lys-Leu-Gln-Thr-Tyr-Pro-Arg-Thr-Asp-Val-Gly-Ala-Gly-Thr-Pro Amide

Synonyms: Thyrocalcitonin; Calcitonin

Bachem H-2255.0001   Eel   MW 3414.91 $C_{146}H_{241}N_{43}O_{47}S_2$ Store at -15°C

ICN 159901   Eel   MW 3414.9   97%

Sigma T 1284   Eel   FW 3414.9   ≥97% (HPLC) | Disulfide bonds: 1-7; bioactive peptide 8135

Cys-Ser-Asn-Leu-Ser-Thr-Cys-Val-Leu-Gly-Lys-Leu-Ser-Gln-Glu-Leu-His-Lys-Leu-Gln-Thr-Tyr-Pro-Arg-Thr-Asp-Val-Gly-Ala-Gly-Thr-Pro

Synonyms: Calcitonin

Biogenesis 1720-7829   Eel synthetic   MW 3414.9   97.9% (HPLC), 99% (TLC); lyophilized, consisting of 90.1% peptide material, acetate counter ions and residual H_2O | Disulfide bonds: Cys[1]-Cys[7]

Cys-Ser-Asn-Leu-Ser-Thr-Cys-Val-Leu-Gly-Lys-Leu-Ser-Gln-Glu-Leu-His-Lys-Leu-Gln-Thr-Tyr-Pro-Arg-Thr-Asn-Thr-Gly-Ser-Gly-Thr-Pro Amide

Synonyms: Calcitonin

Alexis 153-001   Salmon   MW 3431.9 $C_{145}H_{240}N_{44}O_{48}S_2$
≥98%; synthetic; white to off-white powder; soluble in 0.05 *M* acetic acid; biological activity: ≥5000 IU/mg | Disulfide bonds: Cys[1]-Cys[7]

American Peptide 22-7-11   Salmon   MW 3431.9
$C_{145}H_{240}N_{44}O_{48}S_2$ Disulfide bonds: :Cys[1]-Cys[7] | Has the ability to cross mucus membranes & has no significant effect on inositol phosphate accumulation; Hamilton, CR, *Am J Med*, 56:858, 1974

Cys-Ser-Asn-Leu-Ser-Thr-Cys-Val-Leu-Gly-Lys-Leu-Ser-Gln-Glu-Leu-His-Lys-Leu-Gln-Thr-Tyr-Pro-Arg-Thr-Asn-Thr-Gly-Ser-Gly-Thr-Pro NH₂-Cys-Ser-Asn-Leu-Ser-Thr-Cys-Val-Leu-Gly-Lys-Leu-Ser-Gln-Glu-Leu-His-Lys-Leu-Gln-Thr-Tyr-Pro-Arg-Thr-Asn-Thr-Gly-Ser-Gly-Thr-Pro Amide

Synonyms: Calcitonin

ICN 154495   Salmon   MW 3431.9   Hamilton, CR, *Am J Med*, 56:858, 1974

Cys-Ser-Asn-Leu-Ser-Thr-Cys-Val-Leu-Gly-Lys-Leu-Ser-Gln-Glu-Leu-His-Lys-Leu-Gln-Thr-Tyr-Pro-Arg-Thr-Asn-Thr-Gly-Ser-Gly-Thr-Pro

Synonyms: Calcitonin

Neosystem SC484   Salmon   MW 3431.88   Disulfide bonds: Cys^1-Cys^7; bioactive; calcium metabolism peptide

---

Cys-Ser-Asn-Leu-Ser-Thr-Cys-Val-Leu-Gly-Lys-Leu-Ser-Gln-Glu-Leu-His-Lys-Leu-Gln-Thr-Tyr-Pro-Arg-Thr-Asn-Thr-Gly-Ser-Gly-Thr-Pro Amide

Synonyms: Thyrocalcitonin; Calcitonin

Sigma T 3660   Salmon   FW 3431.9   ≥97% (HPLC); | Disulfide bonds: 1-7; bioactive peptide

---

Cys-Ser-Asn-Leu-Ser-Thr-Cys-Val-Leu-Gly-Lys-Leu-Ser-Gln-Glu-Leu-His-Lys-Leu-Gln-Thr-Tyr-Pro-Arg-Thr-Asn-Thr-Gly-Ser-Gly-Thr-Pro Acetate Salt

Synonyms: Calcitonin

Biogenesis 1720-8502   Salmon synthetic   MW 3431.9   98.7% (HPLC), >99% (TLC); lyophilized | Disulfide bonds: Cys^1-Cys^7

Cys-Ser-Asn-Leu-Ser-Thr-Cys-Val-Leu-Ser-Ala-Tyr-Trp-Arg-Asn-Leu-Asn-Asn-Phe-His-Arg-Phe-Ser-Gly-Met-Gly-Phe-Gly-Pro-Glu-Thr-Pro Amide

Synonyms: Calcitonin

Bachem H-3068.0001   Porcine   MW 3604.07   C₁₅₉H₂₃₂N₄₆O₄₅S₃   Store at -15°C

Cys-Ser-Asn-Leu-Ser-Thr-Cys-Val-Leu-Ser-Ala-Tyr-Trp-Arg-Asn-Leu-Asn-Asn-Phe-His-Arg-Phe-Ser-Gly-Met-Gly-Phe-Gly-Pro-Glu-Thr-Pro NH₂-Cys-Ser-Asn-Leu-Ser-Thr-Cys-Val-Leu-Ser-Ala-Tyr-Trp-Arg-Asn-Leu-Asn-Asn-Phe-His-Arg-Phe-Ser-Gly-Met-Gly-Phe-Gly-Pro-Glu-Thr-Pro Amide

Synonyms: Calcitonin

ICN 154496   Porcine   MW 3604

Cys-Ser-Asn-Leu-Ser-Thr-Cys-Val-Leu-Ser-Ala-Tyr-Trp-Arg-Asn-Leu-Asn-Asn-Phe-His-Arg-Phe-Ser-Gly-Met-Gly-Phe-Gly-Pro-Glu-Thr-Pro Amide

Synonyms: Thyrocalcitonin; Calcitonin

Sigma T 0158   Porcine   FW 3604.0   ≥97% (HPLC) | Disulfide bonds: 1-7; bioactive peptide

---

Cys-Ser-Asn-Leu-Ser-Thr-Cys-Val-Leu-Ser-Ala-Tyr-Trp-Arg-Asn-Leu-Asn-Asn-Phe-His-Arg-Phe-Ser-Gly-Met-Gly-Phe-Gly-Pro-Glu-Thr-Pro Acetate Salt

Synonyms: Calcitonin

Biogenesis 1720-8139   Porcine synthetic   MW 3606.6   >98% (HPLC); lyophilized, consisting of 74.4% peptide material, acetate counter ions and residual H₂O | Disulfide bonds: Cys^1-Cys^7

Cys-Ser-Asp-Ser-Lys-Gln-Asp-Lys-Ser-Arg-Leu-Asn-Glu Trifluoroacetate Salt

Synonyms: Somatostatin Receptor II (341-352), C~

Biogenesis 8330-7100   Human synthetic   MW 1509.8   95% (HPLC); lyophilized | Represents AA 341-352 of somatostatin receptor 2; an additional Cys has been conjugated to the N-terminus

---

Cys-Ser-Cys-Asn-Ser-Trp-Leu-Asp-Lys-Glu-Cys-Val-Tyr-Phe-Cys-His-Leu-Asp-Ile-Ile-Trp

Synonyms: Vasoactive Intestinal Contractor Peptide; Vasoactive Intestinal Contractor; Endothelin, β-; Endothelial Isopeptide; Gastrointestinal Peptide; Endothelial Peptide

Neosystem SC328   MW 2573.94   Disulfide bonds: Cys^1-Cys^{15}, Cys^3-Cys^{11}; bioactive; endothelin-related peptide; Ishida, N et al, *FEBS Lett*, 247:337-340, 1989

Alexis 170-001   Mouse   MW 2574.0   C₁₁₆H₁₆₁N₂₇O₃₂S₄   ≥97%; lyophilized powder | Disulfide bonds: Cys^1-Cys^{15}, Cys^3-Cys^{11}; Ishida, N et al, *FEBS Lett*, 247:337, 1989; Saida, K et al, *J Biol Chem*, 264:14613, 1989; Simonson, MS & Dunn, MJ, *FASEB J*, 4:2989, 1990

Bachem H-9390.0500   Mouse   Store at -15°C | Disulfide bonds: Cys^1-Cys^{11}, Cys^3-Cys^{15}

Calbiochem 05-23-3806   Mouse   MW 2574   C₁₁₆H₁₆₁N₂₇O₃₂S₄   >95% (HPLC); lyophilized solid; soluble in 5% acetic acid; LD₅₀≤2000 mg/kg | Disulfide bonds: Cys^1-Cys^{15}, Cys^3-Cys^{11}; direct effect on the pituitary; increases pulmonary arterial pressure; Taniyama, K et al, *Eur J Pharmacol*, 235:149, 1993

ICN 152864   Mouse   Disulfide bonds: Cys^1-Cys^{15}, Cys^3-Cys^{11}; Itoh, Y et al, *FEBS Lett*, 247:337, 1989

Sigma E 9512   Mouse   FW 2573.9   ≥97% (HPLC); peptide content:~65% | Disulfide bonds: 1-15, 3-11; bioactive peptide; increases pulmonary arterial pressure

---

Cys-Ser-Cys-Asn-Ser-Trp-Leu-Asp-Lys-Glu-Cys-Val-Tyr-Phe-Cys-His-Leu-Asp-Ile-Ile-Trp

Synonyms: Vasoactive Intestinal Contractor Peptide

Peptides International PED-4211-s   Mouse synthetic   MW 2574.0   C₁₁₆H₁₆₁N₂₇O₃₂S₄   >99% (HPLC); lyophilized amorphous powder | Bioactive; Disulfide bonds: Cys^1-Cys^{15}, Cys^3-Cys^{11}; Ishida, N et al, *FEBS Letts*, 247:337, 1989; Saida, K et al, *JBC*, 264:14613, 1989

Cys-Ser-Cys-Asn-Ser-Trp-Leu-Asp-Lys-Glu-Cys-Val-Tyr-Phe-Cys-His-Leu-Asp-Ile-Ile-Trp

Synonyms: Vasoactive Intestinal Contractor; Endothelin, β-; Endothelial Isopeptide

American Peptide 88-1-15   Murine   MW 2574.0   C₁₁₆H₁₆₁N₂₇O₃₂S₄   Lyophilized solid | Disulfide bonds: Cys^1-Cys^{15}, Cys^3-Cys^{11}; direct effect on the pituitary & increases pulmonary arterial pressure; Saida, K et al, *J Biol Chem*, 264:14613, 1989; Ishida, N et al, *FEBS Lett*, 247:337, 1989; Simonson, MS et al, *FASEB J*, 4:2989, 1990

Cys-Ser-Cys-Lys-Asp-Met-Thr-Asp-Lys-Glu-Cys-Leu-Asn-Phe-Cys-His-Gln-Asp-Val-Ile-Trp

Synonyms: Sarafotoxin A; Sarafotoxin S6a1; Sarafotoxin S6a

Bachem H-1046.0500   Store at -15°C | Disulfide bonds: Cys^1-Cys^{15}, Cys^3-Cys^{11}

ICN 158726   MW 2514.9   97%

Sigma S 1522   FW 2514.9   ≥97% (HPLC) | Disulfide bonds: 1-15, 3-11; bioactive peptide

American Peptide 88-9-20   *Atractaspis engaddensis*   MW 2514.9   C₁₀₅H₁₅₆N₂₈O₃₄S₅   Disulfide bonds: Cys^1-Cys^{15}, Cys^3-Cys^{11}; cardiotoxic isotoxin; Nayler, WG et al, *BBRC*, 161:89, 1989

Cys-Ser-Cys-Lys-Asp-Met-Thr-Asp-Lys-Glu-Cys-Leu-Tyr-Phe-Cys-His-Gln-Asp-Val-Ile-Trp

Synonyms: Sarafotoxin S6b; Sarafotoxin B; Endothelin Related Peptide

Bachem H-7980.0500   Store at -15°C | Disulfide bonds: Cys[1]-Cys[15], Cys[3]-Cys[11]

ICN 154527   MW 2563.9   Kloog, Y etal, *Science*, 242:268, 1988

Neosystem SC456   MW 2563.92   Disulfide bonds: Cys[1]-Cys[15], Cys[3]-Cys[11]; bioactive; endothelin-related peptide; Kloog, Y, *Science*, 242:268-270, 1988

Alexis 167-001   *Atractaspis engaddensis*   MW 2564.0   C[110]H[159]N[27]O[34]S[5]   ≥97%; lyophilized powder; soluble in 5% acetic acid; toxic | Disulfide bonds: Cys[1]-Cys[15], Cys[3]-Cys[11]; Takasaki, C et al, *Toxicon*, 26:543, 1988; Kloog, Y et al, *Science*, 242:268, 1988; Takasaki, C et al, *Nature*, 335:303, 1988; Nakajima, N et al, *J Cardiovasc Pharmacol*, 13:S8, 1989; Ambar, I et al, *BBRC*, 158:195, 1989; Nayler, WG et al, *BBRC*, 161:89, 1989; Wollberg, Z et al, *BBRC*, 162:371, 1989; Hirata, Y et al, *BBRC*, 162:441, 1989; Bousso-Mittler, D et al, *BBRC*, 162:952, 1989; Fleminger, G et al, *BBRC*, 162:1317, 1989; Galron, R et al, *BBRC*, 163:936, 1989

American Peptide 88-9-10   *Atractaspis engaddensis*   MW 2564.0   C[110]H[159]N[27]O[34]S[5]   Disulfide bonds: Cys[1]-Cys[15], Cys[3]-Cys[11]; vasoconstrictor activity, potent coronary constrictor activity; Fleminger, G et al, *BBRC*, 162:1317, 1989; Galron, R et al, *BBRC*, 163:936, 1989; Takasaki, C et al, *Toxicon*, 26:543, 1988

Sigma S 4146   *Atractaspis engaddensis*   FW 2563.9   ≥90% (HPLC) | Disulfide bonds: 1-15, 3-11; bioactive peptide; ET[A] agonist; increases intracellular Ca[2+]; Takasaki, C et al, *Toxicon*, 26:543, 1988; Kloog, Y et al, *Science*, 242:268, 1988

Peptides International PSF-4206-s   *Atractaspis engaddensis* (snake) synthetic   MW 2564.0   C[110]H[159]N[27]O[34]S[5]   >95% (HPLC) | Bioactive; Takasaki, C et al, *Toxicon*, 26:543, 1988; Kloog, Y et al, *Science*, 242:268, 1988; Nakajima, K et al, *J Cardiovasc Pharmacol*, 13:58, 1989; Watanabe, TX et al, *J Cardiovasc Pharmacol*, 17(7), S8

Calbiochem 05-23-3802   *Atractaspsis engaddensis*   MW 2564   C[110]H[159]N[27]O[34]S[5]   ≥97% (HPLC); lyophilized solid; soluble in 5% acetic acid; LD[50]≤2000 mg/kg | Disulfide bonds: Cys[1]-Cys[15], Cys[3]-Cys[11]; endothelin-related peptide that binds to the endothelin receptor in various tissues; increases intracellular Ca[2+] levels by activating ET[A] receptors; has vasoconstrictive activity & stimulates protein synthesis; Bacon, CR & Davenport, AP, *Br J Pharmacol*, 117:986, 1996; Bax, WA et al, *Br J Pharmacol*, 113:1471, 1994; Marsault, R et al, *Am J Physiol*, 261:C987, 1991; Sugden, PH et al, *Biochim Biophys Acta*, 1175:327, 1993; Koyama, Y et al, *Brain Res*, 600:81, 1993; Watanabe, C et al, *Br J Pharmacol*, 108:30, 1993

Cys-Ser-Cys-Ser-Ser-Leu-Met-Asp-Lys-Glu-Cys-Val-Tyr-Phe-Cys

Synonyms: Endothelin I (1-15)

American Peptide 88-2-35   Human   MW 1714.0   C[70]H[104]N[16]O[24]S[5]   Disulfide bonds: Cys[1]-Cys[15], Cys[3]-Cys[11]

Cys-Ser-Cys-Ser-Ser-Leu-Met-Asp-Lys-Glu-Cys-Val-Tyr-Phe-Cys Amide

Synonyms: Endothelin (1-15)

American Peptide 88-2-38   Human   MW 1713.4   C[70]H[105]N[17]O[23]S[5]   Disulfide bonds: Cys[1]-Cys[15], Cys[3]-Cys[11]

Cys-Ser-Cys-Ser-Ser-Leu-Met-Asp-Lys-Glu-Cys-Val-Tyr-Phe-Cys-His-Leu-Asp-Ile-Ile-Trp

Synonyms: Endothelin I

ICN 153092   MW 2491.9   Disulfide bonds: Cys[1]-Cys[15], Cys[3]-Cys[11]

Cys-Ser-Cys-Ser-Ser-Leu-Met-Asp-Lys-Glu-Cys-Val-Tyr-Phe-Cys-His-Leu-Asp-Ile-Ile-Trp-Val-Asn-Thr-Pro-Glu-His-Val-Val-Pro-Tyr-Gly-Leu-Gly-Ser-Pro-Ser-Arg-Ser

Synonyms: Endothelin I (1-39), Big; Endothelin 39, Big

Alexis 152-007   Bovine   MW 4370   C[192]H[287]N[49]O[58]S[5]   Disulfide bonds: Cys[1]-Cys[15], Cys[3]-Cys[11]; Price, GJ & Malone, L, *Nucl Acid Res*, 18:3658, 1990

American Peptide 88-3-10   Bovine   MW 4370.7   C[192]H[287]N[49]O[58]S[5]   Disulfide bonds: Cys[1]-Cys[15], Cys[3]-Cys[11]; Price, GJ et al, *Nucleic Acid Res*, 18:3658, 1990

Sigma E 6389   Bovine   FW 4370.0   ≥97% (HPLC) | Disulfide bonds: 1-15, 3-11; bioactive peptide; Price, GJ & Malone, L, *Nucleic Acid Research*, 18:3658, 1990

Cys-Ser-Cys-Ser-Ser-Leu-Met-Asp-Lys-Glu-Cys-Val-Tyr-Phe-Cys-His-Leu-Asp-Ile-Ile-Trp-Val-Asn-Thr-Pro-Glu-His-Val-Val-Pro-Tyr-Gly-Leu-Gly-Ser-Pro-Arg-Ser

Synonyms: Endothelin I (1-38), Big

Alexis 152-001   Human   MW 4283   C[189]H[282]N[49]O[56]S[5]   Disulfide bonds: Cys[1]-Cys[15], Cys[3]-Cys[11]; Itoh, T et al, *FEBS Lett*, 231:440, 1988; Kashiwabara, T et al, *FEBS Lett*, 247:73, 1989

Cys-Ser-Cys-Ser-Ser-Leu-Met-Asp-Lys-Glu-Cys-Val-Tyr-Phe-Cys-His-Leu-Asp-Ile-Ile-Trp

Synonyms: Endothelin I

American Peptide 88-1-10   Human   MW 2491.9   C[109]H[159]N[25]O[32]S[5]   Disulfide bonds: Cys[1]-Cys[15], Cys[3]-Cys[11]; potent vasoconstrictive 21-AA peptide produced by endothelia cells; acts on endothelia cells & on the underlying smooth muscle cells as a modulator of vascular tone; displays selectivity for ET[A] receptor; Inoue, A et al, *PNAS*, 86:2863, 1989; Yanagisawa, M et al, *TIPS*, 10:374, 1989; Price, GJ et al, *Nucleic Acids Res*, 18:3658, 1991

Cys-Ser-Cys-Ser-Ser-Leu-Met-Asp-Lys-Glu-Cys-Val-Tyr-Phe-Cys-His-Leu-Asp-Ile-Ile-Trp-Val-Asn-Thr-Pro-Glu-His-Val-Val-Pro-Tyr-Gly-Leu-Gly-Ser-Pro-Arg-Ser

Synonyms: Endothelin I (1-38), Big

American Peptide 88-1-30   Human   MW 4282.9   C[189]H[282]N[48]O[56]S[5]   Disulfide bonds: Cys[1]-Cys[15], Cys[3]-Cys[11]; Itoh, Y et al, *FEBS Lett*, 231:440, 1988; Kashiwabara, T et al, *FEBS Lett*, 247:73, 1989

Cys-Ser-Cys-Ser-Ser-Leu-Met-Asp-Lys-Glu-Cys-Val-Tyr-Phe-Cys-His-Leu-Ala-Ile-Ile-Trp

Synonyms: Endothelin I, (Ala[18])-

American Peptide 88-2-36   Human   MW 2447.9   C[108]H[159]N[25]O[30]S[5]   Disulfide bonds: Cys[1]-Cys[15], Cys[3]-Cys[11]

American Peptide 88-2-37   Human   MW 2490.8   C[109]H[160]N[26]O[31]S[5]   Disulfide bonds: Cys[1]-Cys[15], Cys[3]-Cys[11]

Cys-Ser-Cys-Ser-Ser-Leu-Met-Asp-Lys-Glu-Cys-Val-Tyr-Phe-Cys-His-Leu-Asp-Ile-Ile-Trp-Val-Asn-Thr-Pro-Glu-His-Val-Val-Pro-Tyr-Gly-Leu-Gly-Ser-Pro-Arg-Ser

Synonyms: Endothelin, Big 38; Endothelin I (1-38), Big; Endothelin 38, Big

Bachem H-9030.0500   Human   MW 4282.94   C[189]H[282]N[48]O[56]S[5]   Store at -15°C | Disulfide bonds: Cys[1]-Cys[15], Cys[3]-Cys[11]

Calbiochem 05-23-3803   Human   MW 4283   C[189]H[282]N[48]O[56]S[5]   ≥97% (HPLC); lyophilized solid; soluble in 5% acetic acid; LD[50]≤2000 mg/kg | Disulfide bonds: Cys[1]-Cys[15], Cys[3]-Cys[11]; Kashiwabara, T et al, *FEBS Lett*, 247:73, 1989; Itoh, Y et al, *FEBS Lett*, 231:440, 1988

ICN 153090   Human   MW 4282.9   Disulfide bonds: Cys[1]-Cys[15], Cys[3]-Cys11; Itoh, Y etal, *FEBS Lett*, 231:440, 1988; Yanagiswa, M etal, *Nature*, 332:411, 1988

Neosystem SC918   Human   MW 4282.89   Disulfide bonds: Cys^1-Cys^{15}, Cys^3-Cys^{11}; bioactive; Kashiwabara, T et al, *FEBS Lett*, 247:73-76, 1989

Sigma E 9387   Human   FW 4282.9   ≥97% (HPLC) | Disulfide bonds: 1-15, 3-11; bioactive peptide; Compared to ET-1, big endothelins show less vasoconstrictor activity *in vitro* but similar pressor effects *in vivo*; Itoh, Y et al, *FEBS Lett*, 231:440, 1988; Yanagisawa, M et al, *Nature*, 332:411, 1988

Cys-Ser-Cys-Ser-Ser-Leu-Met-Asp-Lys-Glu-Cys-Val-Tyr-Phe-Cys-His-Leu-Asp-Ile-Ile-Trp

Synonyms: Endothelin I

Peptides International PED-4198-s, 4198-v   Human synthetic   MW 2491.9 $C_{109}H_{159}N_{25}O_{32}S_5$ >95% (HPLC); lyophilized amorphous powder | Bioactive; Disulfide bonds: Cys^1-Cys^{15}, Cys^3-Cys^{11}; porcine, canine, rat, mouse, bovine; Yanagisawa, M & T Masaki, *Trends Pharmacol Sci*, 10:374, 1989; Sakurai, T et al, *Trends Pharmacol Sci*, 13, 1992; James, AF et al, *Cardiovasc Drug Rev*, 11:253, 1993; Yanagisawa, M et al, *Nature*, 332:411, 1988; Itoh, Y et al, *FEBS Lett*, 231:440, 1988; Kumagaye, S et al, *Int J Peptide Protein Res*, 32:519, 1988; Inoue, A et al, *PNAS USA*, 86:2863, 1989; Wantanabe, TX et al, *J Cardiovasc Pharmacol*, 17(Suppl 7):S5, 1991

Cys-Ser-Cys-Ser-Ser-Leu-Met-Asp-Lys-Glu-Cys-Val-Tyr-Phe-Cys-His-Leu-Asp-Ile-Ile-Trp-Val-Asn-Thr-Pro-Glu-His-Val-Val-Pro-Tyr-Gly-Leu-Gly-Ser-Pro-Arg-Ser

Synonyms: Endothelin I (1-38), Big

Peptides International PED-4208-s, 4208-v   Human synthetic   MW 4282.9 $C_{189}H_{282}N_{48}O_{56}S_5$ >95% (HPLC); lyophilized amorphous powder | Bioactive; Disulfide bonds: Cys^1-Cys^{15}, Cys^3-Cys^{11}; Itoh, Y et al, *FEBS Letts*, 231:440, 1988; Kashiwabara, T et al, *FEBS Lett*, 247:73, 1989

Cys-Ser-Cys-Ser-Ser-Leu-Met-Asp-Lys-Glu-Cys-Val-Tyr-Phe-Cys-His-Leu-Asp-Ile-Ile-Trp-Val-Asn-Thr-Pro-Glu-His-Val-Val-Pro-Tyr

Synonyms: Endothelin I (1-31)

Peptides International PED-4360-s   Human synthetic   MW 3628.2 $C_{162}H_{236}N_{38}O_{47}S_5$ >95% (HPLC); lyophilized amorphous powder | Bioactive; Disulfide bonds: Cys^1-Cys^{15}, Cys^3-Cys^{11}; new endogenous form of Endothelin I; Nakao, A et al, *J Immunol*, 159:1987, 1997; Kishi, F et al, *BBRC*, 248:387, 1998; Yoshizumi, M et al, *Eur J Pharmacol*, 348:305, 1998; Yoshizumi, M et al, *Br J Pharmacol*, 125:1019, 1998

Cys-Ser-Cys-Ser-Ser-Leu-Met-Asp-Lys-Glu-Cys-Val-Tyr-Phe-Cys-His-Leu-Asp-Ile-Ile-Trp

Synonyms: Endothelin I

Calbiochem 05-23-3800   Human, porcine   MW 2492 $C_{109}H_{159}N_{25}O_{32}S_5$ ≥96% (HPLC); lyophilized solid; soluble in 5% acetic acid; LD_{50}≤2000 mg/kg | Disulfide bonds: Cys^1-Cys^{15}, Cys^3-Cys^{11}; 21 AA polypeptide with potent vasoconstrictive action; produces convulsions, arterial hypertension & metabolic activation of periventricular white & gray matter; effects are mediated via L-type Ca^{2+} channels; activates phospholipase C in fibroblasts expressing ET_A receptors; Itoh, Y et al, *FEBS Lett*, 231:440, 1988; Murata, S et al, *J Pharmacol Exp Ther*, 274:1524, 1995; Cesari, M et al, *Am Heart J*, 132:1236, 1996; Gresser, O et al, *Biochem Biophys Res Comm*, 224:169, 1996; Hexum, TD et al, *Biochem Biophys Res Comm*, 167:294, 1990; Yanagisawa, M et al, *Nature*, 332:411, 1988

Neosystem SC324   Human, porcine   MW 2491.9   Disulfide bonds: Cys^1-Cys^{15}, Cys^3-Cys^{11}; bioactive; Yanagisawa, M et al, *Nature*, 332:411-415, 1988; Itoh, Y et al, *FEBS Lett*, 231:440-444, 1988

Sigma E 7764   Human, porcine   FW 2491.9   ≥97% (HPLC) | Disulfide bonds: 1-15, 3-11; bioactive peptide; potent vasoconstrictor from vascular endothelial cells; has greatest affinity toward ET_A receptors

---

Biogenesis 4113-2005   Human, porcine synthetic   MW 2491.8   >95% (HPLC); also contains some TFA counter ions and residual H_2O; lyophilized | Ile-Ile bond partially hydrolyzed during standard HCL hydrolysis; To avoid Met oxidation, it is best to dissolve the peptide under a nitrogen stream

Alexis 155-001   Human, porcine, canine, rat, mouse, bovine   MW 2492.0 $C_{109}H_{159}N_{25}O_{32}S_5$ Disulfide bonds: Cys^5-Cys^{15}, Cys^3-Cys^{11}; Yanagisawa, M & Masaki, T, *TIPS*, 10:374, 1989; Yanagisawa, M et al, *Nature*, 332:411, 1988; Price, GJ & Malone, L, *Nucl Acids Res*, 18:3658, 1991; Itoh, Y et al, *FEBS Lett*, 231:440, 1988; Kumagaye, S et al, *Int J Pept Prot Res*, 32:519, 1988; Inoue, A et al, *PNAS*, 86:2863, 1989

Bachem H-6995.0500   Human, porcine, dog, rat   MW 2491.94 $C_{109}H_{159}N_{25}O_{32}S_5$ Store at -15°C | Disulfide bonds: Cys^1-Cys^{15}, Cys^3-Cys^{11}

Cys-Ser-Cys-Ser-Ser-Leu-Met-Asp-Lys-Glu-Cys-Val-Tyr-Phe-Cys-His-Leu-Asp-Ile-Ile-Trp-Val-Asn-Thr-Pro-Glu-His-Ile-Val-Pro-Tyr-Gly-Leu-Gly-Ser-Pro-Ser-Arg-Ser

Synonyms: Endothelin I (1-39), Big

Alexis 152-002   Porcine   MW 4384.1 $C_{193}H_{289}N_{49}O_{58}S_5$ Disulfide bonds: Cys^1-Cys^{15}, Cys^3-Cys^{11}; Yanagisawa, M et al, *Nature*, 332:411, 1988; Itoh, T et al, *FEBS Lett*, 231:440, 1988; Kashiwabara, T et al, *FEBS Lett*, 247:73, 1989

American Peptide 88-4-15   Porcine   MW 4384.0 $C_{193}H_{289}N_{49}O_{58}S_5$ Disulfide bonds: Cys^1-Cys^{15}, Cys^3-Cys^{11}; Itoh, Y et al, *FEBS Lett*, 231:440, 1988; Kashiwabara, T et al, *FEBS Lett*, 247:73, 1989

Cys-Ser-Cys-Ser-Ser-Leu-Met-Asp-Lys-Glu-Cys-Val-Tyr-Phe-Cys-His-Leu-Asp-Ile-Ile-Trp-Val-Asn-Thr-Pro-Glu-His-Ile-Val-Pro-Tyr-Gly-Leu-Gly-Ser-Pro-Ser-Arg-Ser Amide

Synonyms: Endothelin I (1-39), Big

American Peptide 88-4-17   Porcine   MW 4383.0 $C_{193}H_{290}N_{50}O_{57}S_5$ Disulfide bonds: Cys^1-Cys^{15}, Cys^3-Cys^{11}

Cys-Ser-Cys-Ser-Ser-Leu-Met-Asp-Lys-Glu-Cys-Val-Tyr-Phe-Cys-His-Leu-Asp-Ile-Ile-Trp-Val-Asn-Thr-Pro-Glu-His-Ile-Val-Pro-Tyr-Gly-Leu-Gly-Ser-Pro-Ser-Arg-Ser

Synonyms: Endothelin I (1-39), Big

Bachem H-9135.0500   Porcine   MW 4384.04 $C_{193}H_{289}N_{49}O_{58}S_5$ Store at -15°C | Disulfide bonds: Cys^1-Cys^{15}, Cys^3-Cys^{11}

Calbiochem 05-23-3804   Porcine   MW 4384.1 $C_{193}H_{289}N_{49}O_{58}S_5$ ≥97% (HPLC); lyophilized solid; soluble in 5% acetic acid; LD_{50}≤2000 mg/kg | Disulfide bonds: Cys^1-Cys^{15}, Cys^3-Cys^{11}; Kashiwabara, T et al, *FEBS Lett*, 247:73, 1989; Allcock, GH et al, *J Pharmacol Exp Ther*, 275:120, 1995

Cys-Ser-Cys-Ser-Ser-Leu-Met-Asp-Lys-Glu-Cys-Val-Tyr-Phe-Cys-His-Leu-Asp-Ile-Trp-Val-Asn-Thr-Pro-Glu-His-Ile-Val-Pro-Tyr-Gly-Leu-Gly-Ser-Pro-Ser-Arg-Ser

Synonyms: Endothelin, Big 39

ICN 153091   Porcine   MW 4384   Disulfide bonds: Cys^1-Cys^{15}, Cys^3-Cys^{11}; Itoh, Y etal, *FEBS Lett*, 231:440, 1988; Yanagiswa, M etal, *Nature*, 332:411, 1988

Cys-Ser-Cys-Ser-Ser-Leu-Met-Asp-Lys-Glu-Cys-Val-Tyr-Phe-Cys-His-Leu-Asp-Ile-Ile-Trp-Val-Asn-Thr-Pro-Glu-His-Ile-Val-Pro-Tyr-Gly-Leu-Gly-Ser-Pro-Ser-Arg-Ser

Synonyms: Endothelin 39, Big; Endothelin I (1-39), Big

Sigma E 8887   Porcine   FW 4384.0   ≥97% (HPLC) | Disulfide bonds: 1-15, 3-11; bioactive peptide; compared to ET-1, big endothelins show less vasoconstrictor activity *in vitro* but similar pressor effects *in vivo*; Itoh, Y et al, *FEBS Lett*, 231:440, 1988; Yanagisawa, M et al, *Nature*, 332:411, 1988

Peptides International PED-4207-s, 4207-v   Porcine synthetic   MW 4384.0 $C_{193}H_{289}N_{49}O_{58}S_5$ >95% (HPLC); lyophilized amorphous powder | Bioactive; Yanagisawa, M et al, *Nature*, 332:411, 1988; Kashiwabara, T et al, *FEBS Letts*, 247:73, 1989

Cys-Ser-Cys-Ser-Ser-Leu-Met-Asp-Lys-Glu-Cys-Val-Tyr-Phe-Cys-His-Leu-Asp-Ile-Ile-Trp-Val-Asn-Thr-Pro-Glu-Arg-Val-Val-Pro-Tyr-Gly-Leu-Gly-Ser-Pro-Ser-Arg-Ser

Synonyms: Endothelin 39, Big; Endothelin I (1-39), Big

Alexis 152-003   Rat   MW 4389.1 $C_{192}H_{292}N_{50}O_{58}S_5$ Disulfide bonds: Cys[1]-Cys[15], Cys[3]-Cys[11]; Sakurai, T et al, *BBRC*, 175:44, 1991

American Peptide 88-5-50   Rat   MW 4389.1 $C_{192}H_{292}N_{50}O_{58}S_5$ Disulfide bonds: Cys[1]-Cys[15], Cys[3]-Cys[11]; Sakurai, T et al, *BBRC*, 175:44, 1991

Calbiochem 05-23-3811   Rat   MW 4389.1 $C_{192}H_{292}N_{50}O_{58}S_5$ >97% (HPLC); lyophilized solid; soluble in 5% acetic acid; LD$_{50}\leq$2000 mg/kg | Disulfide bonds: Cys[1]-Cys[15], Cys[3]-Cys[11]; Sakurai, T et al, *Biochem Biophys Res Comm*, 175:44, 1991

Sigma E 7265   Rat   FW 4389.0   ≥97% (HPLC) | Disulfide bonds: 1-15, 3-11; bioactive peptide; compared to ET-1, big endothelins show less vasoconstrictor activity *in vitro* but similar pressor effects *in vivo*; Sakurai, T, *Biochem Biophys Res Commun*, 175:44, 1991

Peptides International PED-4266-s   Rat synthetic   MW 4389.1 $C_{192}H_{292}N_{50}O_{58}S_5$ >95% (HPLC); lyophilized amorphous powder | Bioactive; Disulfide bonds: Cys[1]-Cys[15], Cys[3]-Cys[11]; Sakurai, T et al, *BBRC*, 175:44, 1991

Cys-Ser-Cys-Ser-Ser-Trp-Leu-Asp-Lys-Glu-Cys-Val-Tyr-Phe-Cys-His-Leu-Asp-Ile-Ile-Trp-Val-Asn-Thr-Pro-Glu-Gln-Thr-Ala-Pro-Tyr-Gly-Leu-Gly-Asn-Pro-Pro

Synonyms: Endothelin II (1-37), Big

Alexis 152-004   Human   MW 4183.8 $C_{188}H_{269}N_{45}O_{56}S_4$ Disulfide bonds: Cys[1]-Cys[15], Cys[3]-Cys[11]

Cys-Ser-Cys-Ser-Ser-Trp-Leu-Asp-Lys-Glu-Cys-Val-Tyr-Phe-Cys-His-Leu-Asp-Ile-Ile-Trp

Synonyms: Endothelin II

American Peptide 88-1-20   Human   MW 2547.0 $C_{115}H_{160}N_{26}O_{32}S_4$ Disulfide bonds: Cys[1]-Cys[15], Cys[3]-Cys[11]; produced predominantly within the kidney & intestine; has no unique physiologic functions, as compared to ET-1; displays selectivity for ET$_A$ & ET$_B$ receptor; Inoue, A et al, *PNAS*, 86:2863, 1989

Cys-Ser-Cys-Ser-Ser-Trp-Leu-Asp-Lys-Glu-Cys-Val-Tyr-Phe-Cys-His-Leu-Asp-Ile-Ile-Trp-Val-Asn-Thr-Pro-Glu-Gln-Thr-Ala-Pro-Tyr-Gly-Leu-Gly-Asn-Pro-Pro-Arg

Synonyms: Endothelin II (1-38), Big

American Peptide 88-1-39   Human   MW 4339.9 $C_{194}H_{281}N_{49}O_{57}S_4$ Disulfide bonds: Cys[1]-Cys[15], Cys[3]-Cys[11]; Suzuki, N et al, *The 3rd Endothelin Symp, Tsukaba, Japan*, 1991

Cys-Ser-Cys-Ser-Ser-Trp-Leu-Asp-Lys-Glu-Cys-Val-Tyr-Phe-Cys-His-Leu-Asp-Ile-Ile-Trp-Val-Asn-Thr-Pro-Glu-Gln-Thr-Ala-Pro-Tyr-Gly-Leu-Gly-Asn-Pro-Pro

Synonyms: Endothelin II (1-37), Big

American Peptide 88-1-40   Human   MW 4183.7 $C_{188}H_{269}N_{45}O_{56}S_4$ Disulfide bonds: Cys[1]-Cys[15], Cys[3]-Cys[11]; Yanagisawa, M et al, *The 2nd Endothelin Symp, Tsukaba, Japan*, 1990

Cys-Ser-Cys-Ser-Ser-Trp-Leu-Asp-Lys-Glu-Cys-Val-Tyr-Phe-Cys-His-Leu-Asp-Ile-Ile-Trp

Synonyms: Endothelin II

Bachem H-9020.0500   Human   MW 2546.96 $C_{115}H_{160}N_{26}O_{32}S_4$ Store at -15°C | Disulfide bonds: Cys[1]-Cys[15], Cys[3]-Cys[11]

Calbiochem 05-23-3805   Human   MW 2547 $C_{115}H_{160}N_{26}O_{32}S_4$ ≥97% (HPLC); lyophilized solid; soluble in 5% acetic acid; LD$_{50}\leq$2000 mg/kg | Disulfide bonds: Cys[1]-Cys[15], Cys[3]-Cys[11]; vasoactive peptide & potent constrictor of intestinal smooth muscle; de la Monte, SM et al, *J Histochem Cytochem*, 43:203, 1995; Inoue, A, M et al, *PNAS*, 86:2863, 1989

ICN 153093   Human   MW 2546.9   Disulfide bonds: Cys[1]-Cys[15], Cys[3]-Cys[11]

Neosystem SC325   Human   MW 2546.92   Disulfide bonds: Cys[1]-Cys[15], Cys[3]-Cys[11]; bioactive; Inoue, A et al, *PNAS USA*, 86:2863, 1989

Cys-Ser-Cys-Ser-Ser-Trp-Leu-Asp-Lys-Glu-Cys-Val-Tyr-Phe-Cys-His-Leu-Asp-Ile-Ile-Trp-Val-Asn-Thr-Pro-Glu-Gln-Thr-Ala-Pro-Tyr-Gly-Leu-Gly-Asn-Pro-Pro

Synonyms: Endothelin II, Big

Sigma E 6139   Human   FW 4183.7   ≥95% (HPLC) | Disulfide bonds: 1-15, 3-11; bioactive peptide

---

Cys-Ser-Cys-Ser-Ser-Trp-Leu-Asp-Lys-Glu-Cys-Val-Tyr-Phe-Cys-His-Leu-Asp-Ile-Ile-Trp

Synonyms: Endothelin II

Sigma E 9012   Human   FW 2546.9   ≥97% (HPLC) | Disulfide bonds: 1-15, 3-11; bioactive peptide; potent vasoconstrictor from vascular endothelial cells; displays similar selectivity for ET$_A$ & ET$_B$ receptors; Inoue, A et al, *Proc Natl Acad Sci USA*, 86:2863, 1989

Biogenesis 4113-2055   Human synthetic   MW 2547.5 99.4% (HPLC); lyophilized, consisting of 82% peptide material, TFA counter ions and residual H_2O | Ile-Ile bond partially hydrolyzed during standard HCl hydrolysis

Peptides International PED-4209-s   Human synthetic   MW 2547.0 $C_{115}H_{160}N_{26}O_{32}S_4$ >99% (HPLC); lyophilized amorphous powder | Bioactive; Disulfide bonds: Cys[1]-Cys[15], Cys[3]-Cys[11]; canine; Inoue, A et al, *PNAS USA*, 86:2863, 1989; Itoh, Y et al, *Nucleic Acids Res*, 17:5386, 1989

Cys-Ser-Cys-Ser-Ser-Trp-Leu-Asp-Lys-Glu-Cys-Val-Tyr-Phe-Cys-His-Leu-Asp-Ile-Ile-Trp-Val-Asn-Thr-Pro-Glu-Gln-Thr-Ala-Pro-Tyr-Gly-Leu-Gly-Asn-Pro-Pro

Synonyms: Endothelin II (1-37), Big

Peptides International PED-4222-s   Human synthetic   MW 4183.7 $C_{188}H_{269}N_{45}O_{56}S_4$ >99% (HPLC); lyophilized amorphous powder | Bioactive; Disulfide bonds: Cys[1]-Cys[15], Cys[3]-Cys[11]; Ohkubo, S et al, *FEBS Letts*, 274:136, 1990

Cys-Ser-Cys-Ser-Ser-Trp-Leu-Asp-Lys-Glu-Cys-Val-Tyr-Phe-Cys-His-Leu-Asp-Ile-Ile-Trp-Val-Asn-Thr-Pro-Glu-Gln-Thr-Ala-Pro-Tyr-Gly-Leu-Gly-Asn-Pro-Pro-Arg

Synonyms: Endothelin II (1-38), Big

Peptides International PED-4253-s   Human synthetic   MW 4339.9 $C_{194}H_{281}N_{49}O_{57}S_4$ >99% (HPLC); lyophilized amorphous powder | Bioactive; Disulfide bonds: Cys[1]-Cys[15], Cys[3]-Cys[11]; Kosaka, T et al, *J Biochem*, 116:443, 1994

Cys-Ser-Cys-Ser-Ser-Trp-Leu-Asp-Lys-Glu-Cys-Val-Tyr-Phe-Cys-His-Leu-Asp-Ile-Ile-Trp

Synonyms: Endothelin II

Alexis 155-002   Human, porcine, canine   MW 2547.0 $C_{115}H_{160}N_{26}O_{32}S_4$ Disulfide bonds: Cys[5]-Cys[15], Cys[3]-Cys[11]; Inoue, A et al, *PNAS*, 86:2863, 1989

Cys-Ser-Gly-Ala-Glu-Asn-Gln-Gln-Ser-Gly-Asp-Ala-Ala-Val-Thr-Glu-Ala-Glu-Asn-Gln-Gln

Synonyms: CREB

Upstate 12-377 MW 2136 >90% pure; lyophilized | Immunizing peptide

Cys-Ser-Gly-Leu-Thr-Leu-Glu-Pro-Pro-Gly-Asp-Pro-Pro-Pro

Synonyms: Kinesin Protein KIF2 N-Terminal Peptide

Alexis 160-010 Synthetic MW 1379.6 $C_{60}H_{94}N_{14}O_{21}S$ ≥90% (HPLC); dissolved in 10 *mM* sodium acetate, pH 4.5 | Competitively binds to Ab Alexis 210-204; antiserum blocking peptide

Cys-Ser-Ile-Val-Arg-Ala-Val-Gly-Val-Val-Pro-Gly Trifluoroacetate Salt

Synonyms: Bone Morphogenetic Protein 3 (59-69)

Biogenesis 1406-0730 Synthetic 92% (HPLC); lyophilized | Represents AA 59-69 of human BMP-3; an additional Cys has been conjugated to the N-terminus

Cys-Ser-Lys-Gln-Glu-Gly-Ser-Glu-Val-Val-Lys-Arg-Pro-Arg Acetate Salt

Synonyms: Osteocalcin Propeptide (38-50), Precursor

Biogenesis 7060-4030 Human synthetic MW 1603.4 >80% (HPLC); lyophilized

Cys-Ser-Phe-Leu-Ser-Pro-Glu-His-Gln-Arg-Val-Gln-Gln Acetate Salt

Synonyms: Ghrelin (3-14), C~

Biogenesis 0100-0091 Human synthetic Semi-pure; lyophilized

Cys-Ser-Pro-Ala-Glu-Asp-Ser-Lys-Ser-Lys-Glu-Gly-Ser-Gln-Asn-Leu-Gln-Ser-Gln

Synonyms: Estrogen Receptor β C-Terminal Peptide (467-485)

Alexis 155-029 Human synthetic MW 2023.1 $C_{79}H_{131}N_{25}O_{35}S$ ≥95%; lyophilized; reconstitute with 0.1 mL distilled H_2O | Competitively binds to Ab Alexis 210-180; blocking peptide for antiserum (purified) to Erβ

Cys-Ser-Pro-Ser-Leu-Ser-Pro-Ser-Ser-His-Arg-Ser-Ser-Pro-Ala-Thr-Gln-Ser-Pro

Synonyms: Retinoic Acid Receptor α C-Terminal Peptide (444-462)

Alexis 166-003 Mouse synthetic MW 1913.1 $C_{77}H_{125}N_{25}O_{30}S$ ≥95%; lyophilized; reconstitute with 0.1 mL distilled H_2O | Competitively binds to Ab Alexis 210-156 & 804-102; antiserum & monoclonal Ab blocking peptide

Cys-Ser-Ser-Asn-Ala
Palmitoyl-Cys((*RS*)-2,3-di(palmitoyloxy)-propyl)-Ser-Ser-Asn-Ala

Synonyms: Mitogenic Pentapeptide; Pam³-Cys-Ser-Ser-Asn-Ala

Bachem H-9460.0005 MW 1269.82 $C_{67}H_{124}N_6O_{14}S$ Store at -15°C

Cys-Ser-Ser-Thr-Glu-Asp-Ser-Lys-Asn-Lys-Glu-Ser-Ser-Gln-Asn-Leu-Gln-Ser-Gln

Synonyms: Estrogen Receptor β C-Terminal Peptide (467-485)

Alexis 155-025 Rat synthetic MW 2100.2 $C_{80}H_{134}N_{26}O_{38}S$ ≥95%; lyophilized; reconstitute with 0.1 mL distilled H_2O | Competitively binds to Ab Alexis 210-132; blocking peptide for antiserum (purified) to Erβ

Cys-Ser-Thr-Tyr-Thr-Ala-Asn-Pro-Glu-Phe-Val-Ile-Ala-Asn-Val

Synonyms: Protein Kinase C_βI Blocking Peptide

Calbiochem 539635 MW 1628.8 $C_{72}H_{109}N_{17}O_{24}S$ Liquid; use 0.25-2.5 μg/mL of peptide to block the antibody under the specified assay conditions | Distinct C-terminal region of protein kinase C_βI; used to confirm the specificity of anti-PKC_βI in Western blotting

Cys-Ser-Trp-Lys-Thr-Lys-Val-Asn-Trp-Leu-Ala-His-Asn-Val-Ser-Lys-Asp-Asn-Arg-Gln

Synonyms: Phosphoinositide 3-Kinase p110δ C-Terminal Peptide

Alexis 165-034 Synthetic MW 2414.8 $C_{105}H_{164}N_{34}O_{30}S$ ≥90% (HPLC); dissolved in 10 *mM* sodium acetate, pH 4.5 | Competitively binds to Ab Alexis 210-747; antiserum blocking peptide

Cysteine (Ac-Cys)₂

Bachem E-2965.0001 MW 324.38 $C_{10}H_{16}N_2O_6S_2$ Store at -15°C

Cys-Thr-Arg-Pro-Asn-Asn-Asn-Thr-Arg-Lys-Ser-Ile-Arg-Ile-Gln-Arg-Gly-Pro-Gly-Arg-Ala-Phe-Val-Thr-Ile-Gly-Lys-Ile-Gly-Asn-Met-Arg-Gln-Ala

Synonyms: GP120 V3 Loop Peptide BRU C-34-A

Neosystem SC685 MW 3812.42 HIV-1 GP120 V3 Loop peptide; due to the presence of a Cys this peptide can contain the dimeric form

Cys-Thr-Cys-Asn-Asp-Met-Thr-Asp-Glu-Glu-Cys-Leu-Asn-Phe-Cys-His-Gln-Asp-Val-Ile-Trp

Synonyms: Sarafotoxin S6c; Sarafotoxin C; ETB Receptor Agonist

Bachem H-7985.0500 MW 2515.79 $C_{103}H_{147}N_{27}O_{37}S_5$ Store at -15°C | Disulfide bonds: Cys¹-Cys¹⁵, Cys³-Cys¹¹

ICN 154528 MW 2515.8 Takasaki, C etal, *Toxicol*, 26:543, 1988

Neosystem SC457 MW 2515.75 Disulfide bonds: Cys¹-Cys¹⁵, Cys³-Cys¹¹; bioactive; endothelin-related peptide; Takasaki, C et al, *BBRC*, 189:1527-1533, 1992

Sigma S 6545 FW 2515.8 ≥97% (HPLC) | Disulfide bonds: 1-15, 3-11; bioactive peptide; ET_B agonist; toxin with strong vasoconstrictor activity; Takasaki, C et al, *Toxicon*, 26:543, 1988; Nayles, WG et al, *Biochem Biophys Res Commun*, 161:89, 1989

Alexis 167-002 *Atractaspis engaddensis* MW 2515.8 $C_{103}H_{147}N_{27}O_{37}S_5$ ≥97%; lyophilized powder; soluble in 5% acetic acid; toxic | Disulfide bonds: Cys¹-Cys¹⁵, Cys³-Cys¹¹; Takasaki, C et al, *Toxicon*, 26:543, 1988

American Peptide 88-9-35 *Atractaspis engaddensis* MW 2515.8 $C_{103}H_{147}N_{27}O_{37}S_5$ Disulfide bonds: Cys¹-Cys¹⁵, Cys³-Cys¹¹; most acidic; causes strong vasoconstriction of the coronary vessels as well as various changes in the ECG that lead to A-V block & cardiac arrest; Takasaki, C et al, *Toxicon*, 26:543, 1988

Peptides International PSF-4246-s *Atractaspis engaddensis* (snake) synthetic MW 2515.8 $C_{103}H_{147}N_{27}O_{37}S_5$ >95% (HPLC); lyophilized amorphous powder | Bioactive; Disulfide bonds: Cys¹-Cys¹⁵, Cys³-Cys¹¹; selective ETB receptor agonist; Takasaki, C et al, *Toxicon*, 26:543, 1988; Nayler, WG et al, *BBRC*, 161:89, 1989; Williams, DL et al, *BBRC*, 175:556, 1991

Calbiochem 05-23-3813 *Atractaspsis engaddensis* MW 2515.8 $C_{103}H_{147}N_{27}O_{37}S_5$ ≥97% (HPLC); lyophilized solid; soluble in 5% acetic acid; LD₅₀≤2000 mg/kg | Disulfide bonds: Cys¹-Cys¹⁵, Cys³-Cys¹¹; highly selective ET_B receptor agonist; Rockey, DC et al, *Biochem Biophys Res Comm*, 207:725, 1995; Williams, DL et al, *Biochem Biophys Res Comm*, 175:556, 1991

Cys-Thr-Cys-Asn-Asp-Nle-Thr-Asp-Glu-Glu-Cys-Leu-Asn-Phe-Cys-His-Gln-Asp-Val-Ile-Trp

Synonyms: Sarafotoxin C, (Nle⁶)-

Bachem H-9850.0500 Store at -15°C | Disulfide bonds: Cys¹-Cys¹⁵, Cys³-Cys¹¹

Cys-Thr-Cys-Lys-Asp-Met-Thr-Asp-Glu-Glu-Cys-Leu-Asn-Phe-Cys-His-Gln-Asp-Val-Ile-Trp

Synonyms: Sarafotoxin S6c, (Lys[4])-

ICN 158727 MW 2529.8 95%

Sigma S 1397 FW 2529.8 ≥95% (HPLC) | Disulfide bonds: 1-15, 3-11; bioactive peptide; ET$_B$ agonist

American Peptide 88-9-30 *Atractaspis engaddensis* MW 2529.9 $C_{105}H_{153}N_{27}O_{36}S_5$ Disulfide bonds: Cys[1]-Cys[15], Cys[3]-Cys[11]; selective ETB endothelin receptor agonist; Takasaki, C et al, *Nature*, 335:303, 1988

Cys-Thr-Cys-Lys-Asp-Met-Thr-Asp-Lys-Glu-Cys-Leu-Tyr-Phe-Cys-His-Gln-Asp-Ile-Ile-Trp

Synonyms: Sarafotoxin S6d

American Peptide 88-9-40 *Atractaspis engaddensis* MW 2592.0 $C_{112}H_{163}N_{27}O_{34}S_5$ Disulfide bonds: Cys[1]-Cys[15], Cys[3]-Cys[11]

Cys-Thr-Cys-Phe-Thr-Tyr-Lys-Asp-Lys-Glu-Cys-Val-Tyr-Tyr-Cys-His-Leu-Asp-Ile-Ile-Trp-Ile-Asn-Thr-Pro-Glu-Gln-Thr-Val-Pro-Tyr-Gly-Leu-Ser-Asn-Tyr-Arg-Gly-Ser-Phe-Arg Amide

Synonyms: Endothelin III (1-41), Big

Alexis 152-005 Human MW 4923.6 $C_{223}H_{322}N_{56}O_{63}S_4$ Disulfide bonds: Cys[1]-Cys[15], Cys[3]-Cys[11]; Bloch, KD et al, *J Biol Chem*, 264:18156, 1989; Shiba, R et al, *BBRC*, 186:588, 1992

Cys-Thr-Cys-Phe-Thr-Tyr-Lys-Asp-Lys-Glu-Cys-Val-Tyr-Tyr-Cys-His-Leu-Asp-Ile-Ile-Trp

Synonyms: Endothelin III

American Peptide 88-5-10 Human MW 2643.1 $C_{121}H_{168}N_{26}O_{33}S_4$ Disulfide bonds: Cys[1]-Cys[15], Cys[3]-Cys[11]; potent constrictor of rat blood vessels *in vitro* & an effective pressor agent *in vivo*; found in high concentrations in brain & may regulated important functions in neurons & astrocytes, such as proliferation & development; displays selectivity for ET$_C$ receptor; Inoue, A et al, *PNAS*, 86:2863, 1989; Nakajima, K et al, *J Cardiovasc Pharmacol*, 13:58, 1989; Yanagisawa M et al, *PNAS*, 85:6964, 1988

Cys-Thr-Cys-Phe-Thr-Tyr-Lys-Asp-Lys-Glu-Cys-Val-Tyr-Tyr-Cys-His-Leu-Asp-Ile-Ile-Trp-Ile-Asn-Thr-Pro-Glu-Gln-Thr-Val-Pro-Tyr-Gly-Leu-Ser-Asn-Tyr-Arg-Gly-Ser-Phe-Arg Amide

Synonyms: Endothelin III (31-41), Big

American Peptide 88-5-17 Human MW 4923.7 $C_{223}H_{322}N_{56}O_{63}S_4$ Disulfide bonds: Cys[1]-Cys[15], Cys[3]-Cys[11]; Shiba, R et al, *BBRC*, 186:588, 1992

Cys-Thr-Cys-Phe-Thr-Tyr-Lys-Asp-Lys-Glu-Cys-Val-Tyr-Tyr-Cys-His-Leu-Asp-Ile-Ile-Trp

Synonyms: Endothelin III

Peptides International PED-4199-s, 4199-v Human synthetic MW 2643.1 $C_{121}H_{168}N_{26}O_{33}S_4$ >99% (HPLC); lyophilized amorphous powder | Bioactive; Disulfide bonds: Cys[1]-Cys[15], Cys[3]-Cys[11]; porcine, rat, rabbit; Yanagisawa, M et al, *PNAS USA*, 85:6964, 1988; Inoue, A et al, *PNAS USA*, 86:2863, 1989; Nakajima, K et al, *J Cardiovasc Pharmacol*, 13(Suppl 5):S8, 1989

Cys-Thr-Cys-Phe-Thr-Tyr-Lys-Asp-Lys-Glu-Cys-Val-Tyr-Tyr-Cys-His-Leu-Asp-Ile-Ile-Trp-Ile-Asn-Thr-Pro-Glu-Gln-Thr-Val-Pro-Tyr-Gly-Leu-Ser-Asn-Tyr-Arg-Gly-Ser-Phe-Arg Amide

Synonyms: Endothelin III (1-41), Big

Peptides International PED-4223-s Human synthetic MW 4923.6 $C_{223}H_{322}N_{56}O_{63}S_4$ >99% (HPLC); lyophilized amorphous powder; amide | Bioactive; Bloch, KD et al, *JBC*, 264:18156, 1989; Kosaka, T et al, *J Biochem*, 116:443 1994

Cys-Thr-Cys-Phe-Thr-Tyr-Lys-Asp-Lys-Glu-Cys-Val-Tyr-Tyr-Cys-His-Leu-Asp-Ile-Ile-Trp

Synonyms: Endothelin III

ICN 153094 Human, porcine MW 2643.1 Disulfide bonds: Cys[1]-Cys[15], Cys[3]-Cys[11]

Alexis 155-003 Human, porcine, rat MW 2643.1 $C_{121}H_{168}N_{26}O_{33}S_4$ Disulfide bonds: Cys[5]-Cys[15], Cys[3]-Cys[11]; Yanagisawa, M et al, *PNAS*, 85:6964, 1988; Nakajima, K et al, *J Cardiovasc Pharmacol*, 13:S8, 1989; Inoue, A et al, *PNAS*, 86:2863, 1989

Bachem H-9025.0500 Human, rat MW 2643.08 $C_{121}H_{168}N_{26}O_{33}S_4$ Store at -15°C | Disulfide bonds: Cys[1]-Cys[15], Cys[3]-Cys[11]

Calbiochem 05-23-3801 Human, rat MW 2643.1 $C_{121}H_{168}N_{26}O_{33}S_4$ ≥97% (HPLC); lyophilized solid; soluble in 5% acetic acid; LD$_{50}$≤2000 mg/kg | Disulfide bonds: Cys[1]-Cys[15], Cys[3]-Cys[11]; attenuates thrombin-evoked aggregation of human platelets; potent stimulator of interleukin-6 production by endothelial cell lines; activates phospholipase C in fibroblasts expressing ET$_A$ receptors; Astarie-Dequeker et al, *Br J Pharmacol*, 114:524, 1995; Brunner, F et al, *J Mol Cell Cardiol*, 24:1294, 1995; Xin, X et al, *Endocrinology*, 136:132, 1995; Gresser, O et al, *Biochem Biophys Res Comm*, 224:169, 1996; Lamers, JM et al, *Mol Cell Biochem*, 116:59, 1992

Neosystem SC326 Human, rat MW 2643.05 Disulfide bonds: Cys[1]-Cys[15], Cys[3]-Cys[11]; bioactive; Yanagisawa, M et al, *PNAS USA*, 85:6964, 1988

Sigma E 9137 Human, rat FW 2643.1 ≥97% (HPLC); peptide content:~60%; bioactive peptide | Potent vasoconstrictor from vascular endothelial cells; preferred agonist for ET$_C$ receptors; Yanagisawa, M et al, *Proc Natl Acad Sci USA*, 85:6964, 1988

Biogenesis 4113-2105 Human, rat synthetic MW 2643.6 Lyophilized, consisting of 83% peptide material, TFA counter ions and residual H_2O; purified | Ile-Ile bond partially hydrolyzed during standard HCL hydrolysis

Cys-Thr-Cys-Phe-Thr-Tyr-Lys-Asp-Lys-Glu-Cys-Val-Tyr-Tyr-Cys-His-Leu-Asp-Ile-Ile-Trp-Ile-Asn-Thr-Pro-Glu-Gln-Thr-Val-Pro-Tyr-Gly-Leu-Ser-Asn-His-Arg-Gly-Ser-Leu-Arg Amide

Synonyms: Endothelin III (1-41), Big

American Peptide 88-5-15 Rat MW 4863.5 $C_{217}H_{322}N_{58}O_{62}S_4$ Disulfide bonds: Cys[1]-Cys[15], Cys[3]-Cys[11]; Shiba, R et al, *BBRC*, 186:588, 1992; Bloch, KD et al, *J Biol Chem*, 264:18156, 1989

Peptides International PED-4267-s Rat synthetic MW 4863.6 $C_{217}H_{322}N_{58}O_{62}S_4$ >99% (HPLC); lyophilized amorphous powder; amide | Bioactive; Shiba, R et al, *BBRC*, 186:588, 1992

Cys-Thr-Glu-Glu-Glu-Val-Gln-Asp-Thr-Arg-Leu Trifluoroacetate Salt

Synonyms: Cystic Fibrosis Transmembrane Conductance Regulator Protein (1471-1480), C~

Biogenesis 2410-5100 Synthetic MW 1324.6 90% (HPLC); lyophilized

Cys-Thr-Glu-Leu-Glu-Tyr-Leu-Gly-Pro-Glu-Asn-Asp Acetate Salt

Synonyms: Glucose Transporter IV; IRGT

Biogenesis 4670-1727 Human synthetic Purified; lyophilized | James, et al, *PNAS*, 86:8368, 1989

Cys-Thr-His-Gly-Ile-Arg-Pro-Val-Val-Ser-Thr-Gln-Leu-Leu-Leu-Asn-Gly-Ser-Leu-Ala-Glu

Synonyms: GP120 (254-274); HIV Peptide; HIV Envelope Protein (254-274); HTLV I Peptide

American Peptide 72-3-38 MW 2208.6 $C_{95}H_{161}N_{28}O_{30}S$ Ho, DD et al, *Science*, 239:1021, 1988

Bachem H-7015.0500 MW 2208.57 $C_{95}H_{162}N_{28}O_{30}S$ Store at -15°C

ICN 153127   From the 2nd conserved domain of gp120; important for HIV infectivity & Ab neutralization; Ho, DD etal, *Science*, 239:1021, 1988

Neosystem SC672   MW 2208.55   HTLV I peptide from subtype GP160; due to the presence of a Cys this peptide can contain the dimeric form

---

Cys-Thr-His-Gly-Ile-Arg-Pro-Val-Val-Ser-Thr-Gln-Leu-Leu-Leu-Asn-Gly-Ser-Leu-Ala-Glu Trifluoroacetate Salt

Synonyms: GP120 (254-274); HIV Envelope Protein (254-274)

Sigma H 7769   FW 2208.6   ≥95% (HPLC); packaged under argon | Bioactive peptide; fragment from the second conserved domain of gp120; important for HIV infectivity & antibody neutralization; Ho, DD et al, *Science*, 239:1021, 1988

Cys-Thr-Leu-Leu-Ser-Asn-Leu-Glu-Glu-Ala-Lys-Lys-Lys-Lys-Glu-Asp Trifluoroacetate Salt

Synonyms: Apolipoprotein J (46-60), C~ Beta Chain; Clusterin; SP-40; 40

Biogenesis 0650-7131   Human synthetic   MW 1849.11
Purified; lyophilized

Cys-Thr-Phe-Asn-Pro-His-Lys-Arg-Ile-Ser-Ala-Phe-Arg-Ala-Leu-Gln-His-Ser-Tyr-Leu-His-Lys

Synonyms: Cdk4; C-Terminal Blocking Peptide (277-297)

Calbiochem 219447   Human synthetic   MW 2655.1
$C_{120}H_{184}N_{38}O_{29}S$ Lyophilized | Synthetic peptide based on the human Cdk4 (277-297) with a Cys C residue added; this peptide coupled to KLH, was used as the immunogen for the production of Anti-Cdk4; for use in immunoabsorption for immunoprecipitation, immunocytochemistry, Western blotting & dot blots

---

Cys-Thr-Phe-Pro-Gln-Thr-Ala-Ile-Gly-Val-Gly-Ala-Pro

Synonyms: Calcitonin (21-32), C~

Biogenesis 1720-8750   Human synthetic   Purified; lyophilized | Represents AA 21-32 of human calcitonin; an additional Cys has been conjugated to the N-terminus

---

Cys-Thr-Thr-Ala-Val-Pro-Trp-Asn-Ala-Ser-Trp-Ser-Asn-Lys-Ser

Synonyms: GP41 (609-623); HTLV I Peptide

Neosystem SC256   MW 1651.81   HTLV I peptide from subtype GP41; due to the presence of a Cys this peptide can contain the dimeric form

---

Cys-Thr-Thr-His-Trp-Gly-Phe-Thr-Leu-Cys

Bachem H-4736.0001   MW 1166.35 $C_{52}H_{71}N_{13}O_{14}$ Store at -15°C

Cys-Thr-Tyr-Asp-Glu-Val-Ile-Ser-Phe-Val-Pro-Pro-Pro-Leu-Asp-Gln-Glu-Glu-Met-Glu-Ser

Synonyms: p38/HOG1; C-Terminal Blocking Peptide (341-360)

Calbiochem 506117   Human synthetic   MW 2428.7
$C_{106}H_{158}N_{22}O_{39}S_2$ Lyophilized | Based on the human HOG1 (341-360) with a Cys C residue added; this peptide coupled to KLH was used as the immunogen for the production of Anti-p38/HOG1; for use in immunoabsorption for immunoprecipitation, immunocytochemistry, Western blotting & dot blots; Han, J et al, *Nature*, 386:296, 1997; Derijard, B et al, *Science*, 267:682, 1995; Freshely, NW et al, *Cell*, 78:1039, 1995

Cys-Thr-Tyr-Thr-Gln-Asp-Phe-Asn-Lys-Phe-His

Synonyms: Calcitonin (11-20), C~

Biogenesis 1720-8700   Human synthetic   Purified; lyophilized | AA 11-20 of human calcitonin; an additional cysteine added to the N-terminus

---

Cys-Trp-Arg-Nva-Arg-Tyr-Cys-Trp-Arg-Nva-Arg-Tyr Cys-Trp-Arg-Nva-Arg-Tyr AmideCys-Trp-Arg-Nva-Arg-Tyr Amide

Synonyms: Neuropeptide Y (31-36), (Cys³¹, Trp³², Nva³⁴)-

Neosystem SC993   MW 1760.14   Dimerized through Cys³¹-Cys³¹; bioactive neuropeptide; specific neuropeptide Y1 receptor antagonist; binds almost exclusively to Y1 receptors; antagonizes NPY effects on Ca^{2+} in HEL cells; the inhibitory effect of NPY on isoproterenol-stimulated cAMP synthesis by SK-N-MC cells is dose-dependently abolished by this compound with an IC_{50} value of 266 ± 84*nM*; Balasubramaniam, A et al, *J Med Chem*, 39:811-813, 1996

Cys-Tyr (Cys-Tyr)₂

Bachem G-1835.0250   MW 566.66 $C_{24}H_{30}N_4O_8S_2$ Store at -15°C

Cys-Tyr-Ala-Ala-Pro-Leu-Lys-Pro-Ala-Lys-Ser-Cys

Synonyms: Insulin-Like Growth Factor I Analog

Bachem H-1356.0001   MW 1249.52 $C_{55}H_{88}N_{14}O_{15}S_2$ Store at -15°C

Cys-Tyr-Cys-Arg-Ile-Pro-Ala-Cys-Ile-Ala-Gly-Glu-Arg-Arg-Tyr-Gly-Thr-Cys-Ile-Tyr-Gln-Gly-Arg-Leu-Trp-Ala-Phe-Cys-Cys

Synonyms: Defensin I HNP-1; Defensin HNP-2; Defensin II; Human Neutrophil Peptide II

American Peptide 72-2-10   Human   MW 3442.1
$C_{150}H_{222}N_{44}O_{38}S_6$ Disulfide bonds: Cys²-Cys³⁰; Cys⁴-Cys¹⁹; Cys⁹-Cys²⁹; endogenous antibiotic peptide; displays a range of prominent anti-microbial activities against bacteria, fungi & certain enveloped viruses; Ganz, T et al, *J Clin Invest*, 76:1427, 1985; Selsted, ME et al, *J Clin Invest*, 76:1436, 1985; Ganz, T et al, *Eur J Haematol*, 44:1, 1990

American Peptide 72-2-11   Human   MW 3371.0
$C_{147}H_{217}N_{43}O_{37}S_6$ Disulfide bonds: Cys¹-Cys²⁹, Cys³-Cys¹⁸, Cys⁸-Cys²⁸; antibiotic peptide

Bachem H-9005.0100   Human   MW 3371 $C_{147}H_{217}N_{43}O_{37}S_6$
Store at -15°C | Disulfide bonds: Cys¹-Cys²⁹, Cys³-Cys¹⁸, Cys⁸-Cys²⁸

Sigma D 6790   Human   FW 3371.0   ≥95% (HPLC) |
Disulfide bonds: 1-29, 3-18, 8-28; bioactive peptide; Selsted, ME & Harwig, SSL, *Infect Immun*, 55:2281, 1987

Cys-Tyr-Gly-Glu-Glu-Asn-Val (Lauroyl-Cys-Tyr-Gly(-Glu-Glu-Asn-Val)₆)₂

Bachem H-6215.0005   MW 6702.98 $C_{280}H_{428}N_{66}O_{120}S_2$
Store at -15°C

Cys-Tyr-Ile-Gln-Asn-Cys

Synonyms: Tocinoic Acid; Pressinoic Acid, (Ile³)-

American Peptide 66-0-53   MW 740.9 $C_{30}H_{44}N_8O_{10}S_2$
Disulfide bonds: Cys¹-Cys⁶

Bachem H-2500.0001   MW 740.86 $C_{30}H_{44}N_8O_{10}S_2$ Store at -15°C

ICN 152961   Disulfide bonds: Cys¹-Cys⁶

Sigma T 3149   FW 740.8 $C_{30}H_{44}N_8O_{10}S_2$ ≥97% (HPLC) |
Disulfide bonds: 1-6; bioactive peptide

Cys-Tyr-Ile-Gln-Asn-Cys-Pro-Arg-Gly

Synonyms: Vasotocin, (Arg⁸)-

Neosystem SC384   MW 1050.21   Disulfide bonds: Cys¹-Cys⁶; bioactive; pituitary peptide

---

Cys-Tyr-Ile-Gln-Asn-Cys-Pro-Arg-Gly Amide

Synonyms: Vasotocin, (Arg⁸)-

Bachem H-1785.0001   MW 1050.23 $C_{43}H_{67}N_{15}O_{12}S_2$ Store at -15°C

Cys-Tyr-Ile-Gln-Asn-Cys-Pro-Arg-Gly Amide Acetate Salt

Synonyms: Vasotocin, (Arg[8])-

ICN 152189 Disulfide bonds: Cys[1]—Cys[6]; pituitary hormone that stimulates prolactin secretion; Hanew, K etal, *Proc Soc Exp Biol Med*, 164:257, 1980

Sigma V 0130 FW 1050.2 (free base) ≥97% (HPLC) | Disulfide bonds: 1-6; bioactive peptide; stimulator of prolactin secretion; Hanew, K et al, *Proc Soc Exp Biol Med*, 164:257, 1980

Cys-Tyr-Ile-Gln-Asn-Cys-Pro-Arg-Gly-Gly

Synonyms: Vasotocin, (Arg[8],Gly[10])-; Hydrin II; Amphibian Osmoregulatory Peptide

Bachem H-8845.0001 MW 1108.27 $C_{45}H_{69}N_{15}O_{14}S_2$ Store at -15°C

Sigma V 1133 FW 1108.3 ≥97% (HPLC) | Disulfide bonds: 1-6; bioactive peptide

ICN 154580 *Rana* MW 1108.4

Cys-Tyr-Ile-Gln-Asn-Cys-Pro-Arg-Gly-Gly-Lys

Synonyms: Hydrin I'; Vasotocin, (Arg[8],Gly[10],Lys[11])-; Vasotocin-Gly-Lys, (Arg[8])-

Bachem H-1558.0001 MW 1366.44 $C_{51}H_{81}N_{17}O_{15}S_2$ Store at -15°C

Sigma V 7009 FW 1236.4 ≥97% (HPLC) | Disulfide bonds: 1-6; bioactive peptide; Iwamuro, S et al, *Biochim Biophys Acta*, 1176:143, 1993

Cys-Tyr-Ile-Gln-Asn-Cys-Pro-Arg-Gly-Gly-Lys-Arg

Synonyms: Vasotocin, (Arg[8],Gly[10],Lys[11],Arg[12])-; Vasotocin-Gly-Lys-Arg, (Arg[8])-; Hydrin I; Amphibian Osmoregulatory Peptide

Bachem H-8840.0001 MW 1392.63 $C_{57}H_{93}N_{21}O_{16}S_2$ Store at -15°C

Sigma V 1258 FW 1392.6 ≥97% (HPLC) | Disulfide bonds: 1-6; bioactive peptide

ICN 154579 *Xenopus* MW 1392.8

Cys-Tyr-Ile-Gln-Asn-Cys-Pro-Ile-Gly Amide

Synonyms: Oxytocin, (Ile[8])-; Mesotocin

Bachem H-2505.0001 Store at -15°C

Cys-Tyr-Ile-Gln-Asn-Cys-Pro-Leu-Gly
Cys-Tyr-Ile-Gln-Asn-Cys-Pro-Leu-Gly-NH-CH₂CH₂F

Synonyms: Oxytocin-2-Fluoroethylamide

Bachem H-4236.0001 Store at -15°C

Cys-Tyr-Ile-Gln-Asn-Cys-Pro-Leu-Gly

Synonyms: Oxytocin

Neosystem SC061 MW 1007.19 Disulfide bonds: Cys[1]-Cys[6]; bioactive; pituitary peptide

Cys-Tyr-Ile-Gln-Asn-Cys-Pro-Leu-Gly Amide

Synonyms: Hypophamine, α-; Oxytocin

American Peptide 66-0-52 MW 1007.2 $C_{43}H_{66}N_{12}O_{12}S_2$ Disulfide bonds: Cys[1]-Cys[6]; synthesized in the hypothalamus & involved in the control of lactation & parturition

Bachem H-2510.0005 MW 1007.2 $C_{43}H_{66}N_{12}O_{12}S_2$ Store at -15°C

Calbiochem; 05-23-0151 MW 1007.2 $C_{43}H_{66}N_{12}O_{12}S_2$ ≥95% (HPLC); lyophilized solid; soluble in 5% acetic acid; harmful:LD₅₀≥2000 mg/kg; may be carcinogenic/teratogenic | Disulfide bond:Cys[1]-Cys[6]; principal uterus-contracting & lactation-stimulating hormone of posterior pituitary gland that causes a marked but transient relaxation of vascular smooth muscles; stimulates myometrial GTPase & phospholipase C vial coupling to $G_{q/11\alpha}$; known to increase Na[+] excretion; *Merck Index*, 12:7114; Ku, CY et al, *Endocrinology*, 136:1509, 1995

ICN 100896 MW 1007.2 $C_{43}H_{66}N_{12}O_{12}S_2$ 0.25% acetic acid, 0.5% chlorobutanol (preservative); ~20 U/mL | Disulfide bond:Cys[1]-Cys[6]

ICN 191057 MW 1007.2 $C_{43}H_{66}N_{12}O_{12}S_2$ Lyophilized; ~500 IU/mg | Disulfide bond:Cys[1]-Cys[6]

Sigma O 2882 FW 1007.2 (free base) Aqueous solution containing 0.9% NaCl & 0.5% chlorobutanol; 170-240 IU/mL | Disulfide bonds: 1-6; bioactive peptide; stimulates uterine contraction & lactation; increases Na[+] excretion; stimulates myometrial GTPase & phospholipase C

Biogenesis 7090-1004 Synthetic MW 1007.3 Lyophilized, consisting of 95% peptide material, acetate counter ions and residual H₂O; purified | Disulfide bonds: Cys[1]-Cys[6]

Peptides International POX-4084-v Synthetic MW 1007.2 $C_{43}H_{66}N_{12}O_{12}S_2$ >99% (HPLC); lyophilized amorphous powder | Bioactive; disufide bonds Cys[1]-Cys[6]; human, porcine, bovine, rat, ovine; Berde, B (Ed), *Neurohypophysial Hormones & Similar Polypeptides*, Handbook of Experimental Pharmacology, Vol 23, Springer-Verlag, Berlin, 1968; du Vigneaud, V et al, *JBC*, 205:949, 1953; Boissonnas, RA et al, *Helv Chim Acta*, 38:1491, 1955; Zaoral, M & J Rudinger, *Collection Czech Chem Commun*, 20:1183, 1955

Sigma O 3251 Synthetic oxytocin FW 1007.2 (free base) Lyophilized powder; ~50 IU/mg solid; 10% oxytocin, balance mannitol & buffer salts | Disulfide bonds: 1-6; bioactive peptide; stimulates uterine contraction & lactation; increases Na[+] excretion; stimulates myometrial GTPase & phospholipase C

Sigma O 4375 Synthetic oxytocin FW 1007.2 (free base) Lyophilized powder; ~15 IU/mg solid; 3% oxytocin, balance mannitol & buffer salts | Disulfide bonds: 1-6; bioactive peptide; stimulates uterine contraction & lactation; increases Na[+] excretion; stimulates myometrial GTPase & phospholipase C

Cys-Tyr-Ile-Gln-Asn-Cys-Pro-Leu-Gly Amide Acetate Salt

Synonyms: Oxytocin

Sigma O 6379 FW 1007.2 (free base) ≥97% (HPLC) | Disulfide bonds: 1-6; bioactive peptide; stimulates uterine contraction & lactation; increases Na[+] excretion; stimulates myometrial GTPase & phospholipase C

Cys-Tyr-Ile-Gln-Asn-Cys-Pro-Leu-Gly Free Acid

Synonyms: Oxytocin

Bachem H-6885.0005 Store at -15°C

Cys-Tyr-Ile-Gln-Asn-Cys-Pro-Lys-Gly Amide Free Acid

Synonyms: Vasotocin, (Lys[8])-

ICN 152974

Cys-Tyr-Ile-Gln-Asn-Cys-Pro-Lys-Gly Free Acid

Synonyms: Vasotocin, (Lys[8])-

Bachem H-2515.0001 Store at -15°C

Cys-Tyr-Ile-Ser-Asn-Cys-Pro-Ile-Gly Amide

Synonyms: Oxytocin, (Ser[4],Ile[8])-; Isotocin

Bachem H-2520.0001 Store at -15°C

ICN 151373 MW 966.1 $C_{41}H_{63}N_{11}O_{12}S_2$ Disulfide bond:Cys[1]-Cys[6]; Guttman, S etal, *Experentia*, 18:445, 1962

Sigma I 6131 FW 966.1 $C_{41}H_{63}N_{11}O_{12}S_2$ ≥97% (HPLC) | Disulfide bonds: 1-6; bioactive peptide; Guttmann, S et al, *Experientia*, 18:445, 1962

Biogenesis 5455-0002 Synthetic MW 96 Lyophilized, consisting of 78% peptide material, acetate counter ions and residual H₂O; purified

Cys-Tyr-Ile-Thr-Asn-Cys-Gly-Leu-Gly

Synonyms: Oxytocin, (Thr[4],Gly[7])-

Neosystem SC385   MW 940.10   Disulfide bonds: Cys[1]-Cys[6]; bioactive; pituitary peptide; selective oxytocin agonist; Lowbridge, J et al, *J Med Chem*, 20:120, 1977

Cys-Tyr-Ile-Thr-Asn-Cys-Gly-Leu-Gly Amide

Synonyms: Oxytocin, (Thr[4],Gly[7])-

American Peptide 66-0-59   MW 940.1 $C_{39}H_{61}N_{11}O_{12}S_2$
Disulfide bonds: Cys[1]-Cys[6]; shows a 640-fold increase in oxytocic/antidiuretic selectivity compared with oxytocin; Lowbridge, J et al, *J Med Chem*, 20:120, 1977

Bachem H-7710.0001   Store at -15°C

ICN 152959   MW 940.1 $C_{39}H_{61}N_{11}O_{12}S_2$ Disulfide bond:Cys[1]-Cys[6]; selective oxytocin agonist; Lowbridge, M etal, *J Med Chem*, 20:120, 1977

Sigma O 6380   FW 940.1 $C_{39}H_{61}N_{11}O_{12}S_2$ ≥97% (HPLC) |
Disulfide bonds: 1-6; bioactive peptide; selective ligand for central & peripheral oxytocin receptors; Lowbridge, J et al, *J Med Chem*, 20:120, 1977

Cys-Tyr-Ile-Gln-Asn-Cys-Pro-Arg-Gly Amide

Synonyms: Vasotocin, (Arg[8])-

Peptides International PVP-4192-v   Synthetic   MW 1050.2
$C_{43}H_{67}N_{15}O_{12}S_2$ >99% (HPLC); lyophilized amorphous powder |
Bioactive; Disulfide bonds: Cys1-Cys6; frog, chicken

Cys-Tyr-Phe-Gln-Asn-Cys

Synonyms: Pressinoic Acid

American Peptide 66-0-69   MW 775.0 $C_{33}H_{42}N_8O_{10}S_2$
Disulfide bonds: Cys[1]-Cys[6]

Cys-Tyr-Phe-Gln-Asn-Cys
BOC-Cys-Tyr-Phe-Gln-Asn-Cys

Synonyms: Pressinoic Acid, BOC-

Bachem A-1545.0050   Store at -15°C

Cys-Tyr-Phe-Gln-Asn-Cys

Synonyms: Pressinoic Acid

Bachem H-2525.0005   MW 774.88 $C_{33}H_{42}N_8O_{10}S_2$ Store at -15°C

ICN 151947   A pituitary hormone; the cyclic portion of vasopressin

Cys-Tyr-Phe-Gln-Asn-Cys-Pro-Arg

Synonyms: Vasopressin Desglycinamide, (Arg[8])-; Vasopressin, (Arg[8],des-Gly-NH$_2$[9])-; Vasopressin, (des-Gly-NH$_2$[9],(Arg[8]))-; Vasopressin, (Arg[8],des-Gly)-

American Peptide 66-0-06   MW 1028.2 $C_{44}H_{61}N_{13}O_{12}S_2$
Disulfide bonds: Cys[1]-Cys[6]

Bachem H-3184.0001   MW 1028.18 $C_{44}H_{61}N_{13}O_{12}S_2$ Store at -15°C

Sigma V 0380   FW 1028.2   ≥97% (HPLC) | Disulfide bonds: 1-6; bioactive peptide

Biogenesis 9536-0709   Synthetic   MW 1013   Lyophilized, consisting of 87% peptide material, acetate counter ions and residual H_2O; purified | Disulfide bonds: Cys[1]-Cys[6]

Cys-Tyr-Phe-Gln-Asn-Cys-Pro-Arg Amide

Synonyms: Vasopressin, (des-Glycinamide[9],Arg[8])-

ICN 152970   Disulfide bonds: Cys[1]—Cys[6]

Cys-Tyr-Phe-Gln-Asn-Cys-Pro-Arg-Gly
Cys-3,5-diiodo-Tyr-Phe-Gln-Asn-Cys-Pro-Arg-Gly Amide

Synonyms: Vasopressin, (3,5-Diiodo-Tyr[2],Arg[8])-

Bachem H-3638.0001   Store at -15°C

Cys-Tyr-Phe-Gln-Asn-Cys-Pro-Arg-Gly

Synonyms: Vasopressin, (Arg[8])-

Neosystem SC062   MW 1084.23   Disulfide bonds: Cys[1]-Cys[6]; bioactive; pituitary peptide

Cys-Tyr-Phe-Gln-Asn-Cys-Pro-Arg-Gly
Biotinyl-Cys-Tyr-Phe-Gln-Asn-Cys-Pro-Arg-Gly

Synonyms: Vasopressin, Biotinyl-(Arg[8])-

Neosystem SC1236   MW 1310.52   >80% (HPLC) | Disulfide bonds: Cys[1]-Cys[6]; bioactive; pituitary peptide

Cys-Tyr-Phe-Gln-Asn-Cys-Pro-Arg-Gly Amide

Synonyms: Antidiuretic Hormone; Hypophamine, β-; Vasopressin, (Arg[8])-

American Peptide 66-0-03   MW 1084.3 $C_{46}H_{65}N_{15}O_{12}S_2$
Disulfide bonds: Cys[1]-Cys[6]; antidiuretic peptide hormone released from the hypothalamus during stress; implicated in a variety of physiological processes including diuresis, vasoregulation, thermoregulation & memory; affects the immune system; acts via the activation of the adenylate cyclase system resulting in increased cAMP levels; Bell, J et al, *Int J Immunopharmacol*, 14:92, 1992

Bachem H-1780.0001   MW 1084.25 $C_{46}H_{65}N_{15}O_{12}S_2$ Store at -15°C

Calbiochem 05-23-0150   MW 1084.2 $C_{46}H_{65}N_{15}O_{12}S_2$ >97% (HPLC); lyophilized solid; soluble in 5% acetic acid; harmful:LD$_{50}$≤2000 mg/kg | Disulfide bonds: Cys[1]-Cys[6]; antidiuretic peptide hormone; released from the hypothalamus during stress; although noted primarily for its hemodynamic & hemostatic properties, it appears to affect the immune system; acts via the activation of the adenylate cyclase system resulting in increased cAMP levels (EC$_{50}$=34 *nM*); *Merck Index*, 12:10073; Kobayashi, H et al, *Brain Res*, 647:145, 1994; Monaghan, ML et al, *Gen Pharmacol*, 24:1013, 1993

ICN 152962   Antidiuretic Hormone; disulfide bonds: Cys[1]-Cys[6]

Sigma V 0377   FW 1084.2 (free base)   Grade VI; aqueous solution; ~100 IU/mL; containing 0.9% NaCl, 0.5% chlorobutanol, pH ~3.5; prepared from vasopressin with an activity of 367 IU/mg | Disulfide bonds: 1-6; bioactive peptide; regulates water balance by antidiuretic action; contracts arterioles (vasopressor action)

Peptides International PVP-4085-v   Synthetic   MW 1084.2
$C_{46}H_{65}N_{15}O_{12}S_2$ >99% (HPLC); lyophilized amorphous powder |
Bioactive; Disulfide bonds: Cys[1]-Cys[6]; human, bovine, ovine, rat, mouse; Berde, B (Ed), *Neurohypophysial Hormones & Similar Polypeptides*, Handbook of Experimental Pharmacology, Vol 23, Springer-Verlag, Berlin, 1968

Cys-Tyr-Phe-Gln-Asn-Cys-Pro-Arg-Gly Amide Acetate Salt

Synonyms: Vasopressin, (Arg[8])-; Antidiuretic Hormone; Hypophamine, β-

Sigma V 9879   FW 1084.2 (free base)   ≥98% (HPLC) |
Disulfide bonds: 1-6; bioactive peptide; regulates water balance by antidiuretic action; contracts arterioles (vasopressor action)

Cys-Tyr-Phe-Gln-Asn-Cys-Pro-Arg-Gly Free Acid

Synonyms: Vasopressin, (Arg[8])-; Vasopressin, (Arg[8],Gly-OH[9])-

Bachem H-1775.0001   MW 1085.23 $C_{46}H_{64}N_{14}O_{13}S_2$ Store at -15°C

Sigma V 2380   FW 1085.2   ≥97% (HPLC) | Disulfide bonds: 1-6; bioactive peptide; reduced antidiuretic activity & almost no vasopressor activity compared to AVP; Manning, M et al, *Nature*, 308:652, 1984

Cys-Tyr-Phe-Gln-Asn-Cys-Pro-Lys

Synonyms: Vasopressin Desglycinamide, (Lys[8])-

American Peptide 66-0-05   MW 1000.2 $C_{44}H_{61}N_{11}O_{12}S_2$
Disulfide bonds: Cys[1]-Cys[6]

Cys-Tyr-Phe-Gln-Asn-Cys-Pro-Lys-Gly

Synonyms: Vasopressin, (Lys[8])

Neosystem SC948   MW 1056.22   Disulfide bonds: Cys[1]-Cys[6]; bioactive; pituitary peptide

Cys-Tyr-Phe-Gln-Asn-Cys-Pro-Lys-Gly Amide

Synonyms: Lyspressin; Vasopressin, (Lys[8])-

Bachem H-2530.0001   Store at -15°C

Calbiochem 05-23-0153   MW 1056.2 $C_{46}H_{65}N_{13}O_{12}S_2$ >95% (HPLC); lyophilized solid; soluble in 5% acetic acid; harmful:$LD_{50} \leq 2000$ mg/kg | Disulfide bonds: Cys[1]-Cys[6]; synthetic analog of vasopressin; Less potent than (Arg[8])-Vasopressin; *Merck Index*, 12:10073; Gorbulev, V et al, *Eur J Biochem*, 215:1, 1993

ICN 152963   Disulfide bonds: Cys[1]—Cys[6]

Sigma V 6879   FW 1056.2   ≥98% (HPLC) | Disulfide bonds: 1-6; bioactive peptide; lower vasopressor & antidiuretic activity than AVP

American Peptide 66-0-02   Synthetic   MW 1056.3 $C_{46}H_{65}N_{13}O_{12}S_2$ Lyophilized | Less potent analog than (Arg[8])-Vasopressin; disulfide bonds: Cys[1]-Cys[6]

Cys-Tyr-Trp-Lys-Val-Cys
D-2-Nal-Cys-Tyr-D-Trp-Lys-Val-Cys-2-Nal Amide

Bachem H-2126.0001   MW 1192.47 $C_{63}H_{73}N_{11}O_9S_2$ Store at -15°C | Disulfide bonds: Cys[2]-Cys[7]

Cys-Tyr-Trp-Orn-Thr-Pen-Thr
Cys-Tyr-D-Trp-Orn-Thr-Pen-Thr Amide

Bachem H-2384.0001   Store at -15°C | Disulfide bonds: Cys[1]-Pen[6]

Cys-Tyr-Trp-Orn-Val-Cys
D-2-Nal-Cys-Tyr-D-Trp-Orn-Val-Cys-2-Nal Amide

Synonyms: BIM-23127

Bachem H-4906.0001   MW 1178.45 $C_{62}H_{71}N_{11}O_9S_2$ Store at -15°C

Cys-Val
(Cys-Val)₂

Bachem G-1840.0250   MW 438.57 $C_{16}H_{30}N_4O_6S_2$ Store at -15°C

Cys-Val
δ-(L-α-Aminoadipyl)-Cys-D-Val

Synonyms: ACV

Bachem H-4204.0005   MW 363.44 $C_{14}H_{25}N_3O_6S$ Store at -15°C

Cys-Val
Bis-δ-(L-α-Aminoadipyl)-L-Cys-bis-D-Val

Synonyms: ACV, Bis-

Bachem H-6015.0025   MW 724.85 $C_{28}H_{48}N_6O_{12}S_2$ Store at -15°C

Cys-Val
δ-(L-α-Aminoadipyl)-L-Cys-Bis-D-Val

Synonyms: ACV, Bis-

Peptides International PAC-3860-PI   Synthetic   MW 724.27   >98% (HPLC); white powder

Cys-Val-Ala-Met
Cys-Val-2-Naphthyl-3-Alanyl-Met

Synonyms: Farnesyltransferase Inhibitor, p21[ras]

Sigma C 4433   FW 548.7 $C_{26}H_{36}N_4O_5S_2$ ≥90% (HPLC) | Bioactive peptide; Hamilton, D & Sebti, SM, *Drug News Perspect*, 8:138, 1995

Cys-Val-Asp-Ile-Asp-Tyr-Phe-Met-Lys-His-Ser-Lys-Asp-His
(Cys)-Val-Asp-Ile-Asp-Tyr-Phe-Met-Lys-His-Ser-Lys-Asp-His

Synonyms: Ankyrin (1862-1874)

Biogenesis 0580-0150   Human erythroid 2.1 RBCs synthetic >80% (HPLC); lyophilized

Cys-Val-Gln-Met-Glu-Phe-Leu-Gly-Ser-Ser-Glu-Thr-Val Acetate Salt

Synonyms: Glucose Transporter II

Biogenesis 4670-1667   Rat synthetic   Semi-pure; lyophilized | Thorens, et al, *Cell*, 55:281, 1988

Cys-Val-His-Pro-Asp-Ala-Arg-Ser-Pro-Ile-Ser-Pro-Thr-Pro-Val-Pro-Val-Met

Synonyms: Protein Kinase C, Blocking Peptide

Calbiochem 539643   MW 1903.3 $C_{83}H_{135}N_{23}O_{24}S_2$ Liquid; use 0.25-2.5 µg/mL of peptide to block the antibody under the specified assay conditions | Distinct C-terminal region of protein kinase C_γ; used to confirm the specificity of anti-PKC$_\gamma$ in Western blotting

Cys-Val-Ile-Gly-Tyr-Ser-Gly-Asp-Arg-Cys
Ac-Cys(Acm)-Val-Ile-Gly-Tyr-Ser-Gly-Asp-Arg-Cys(Acm) Amide

Synonyms: Epidermal Growth Factor (33-42), Ac-(Cys(Acm)[33,42])-

Bachem H-3906.0001   Mouse   Store at -15°C

Cys-Val-Ile-Met
Ac-Cys(Farnesyl)-Val-Ile-Met

Synonyms: Cysteine Tetrapeptide, S-Farnesylated

Alexis 290-002   MW 711.0 $C_{36}H_{62}N_4O_6S_2$ ≥95%; yellow solid; soluble in chloroform | The isoprenylation pathway requires an endoprotease that cleaves the modified protein at the isoprenylated Cys residue; the endoprotease was readily assayed with this tetrapeptide substrate; Ma, Y-T et al, *Biochemistry*, 32:2386, 1993

Bachem M-2040.0005   Store at -15°C

Cys-Val-Ile-Pro-Ala-Pro-Thr-Pro-Glu-Pro-Gly-Asn-Ala-Glu-Leu Trifluoroacetate Salt

Synonyms: Microfibril Associated Glycoprotein I (57-70), C~

Biogenesis 6199-1050   Synthetic   MW 1506.7   Purified; lyophilized | An additional Cys has been conjugated to the N terminus

Cys-Val-Ile-Ser
Cys(Farnesyl)-Val-Ile-Ser

Bachem H-1808.0005   Store at -15°C

Cys-Val-Ile-Ser
FA-Cys(Farnesyl)-Val-Ile-Ser

Bachem M-1930.0005   Store at -15°C

Cys-Val-Lys-Leu-Ser-Pro-Leu-Cys-Ile-Thr-Met-Arg-Cys-Asn-Lys-Ser-Glu-Thr-Asp-Arg
Cys(Acm)-Val-Lys-Leu-Ser-Pro-Leu-Cys(Acm)-Ile-Thr-Met-Arg-Cys(Acm)-Asn-Lys-Ser-Glu-Thr-Asp-Arg

Synonyms: GP140 (101-120); SIVmac251 Peptide

Neosystem SC718   MW 2510.98   SIVmac251 peptide from subtype GP140

Cys-Val-Met
Cys-Val-2-Nal-Met

Bachem H-3552.0005 MW 548.73 $C_{26}H_{36}N_4O_5S_2$ Store at -15°C

Cys-Val-Nal-Met-Cys-Val-Nal-Met

Synonyms: Farnesyltransferase Inhibitor III, p21[ras]

Calbiochem 344514 MW 548.7 $C_{26}H_{36}N_4O_5S_2$ ≥95% (HPLC); white solid; soluble in 50% acetic acid | Most potent inhibitor of p21[ras] farnesyltransferase (FTase); IC_{50}=12 nM); Hamilton, AD & Sebti, SM, *Drug News Perspect*, 8:138, 1995; Leftheris, K et al, *Bioorg Med Chem Lett*, 4:887, 1994

Cys-Val-Phe-Met
Cys-(®)-Val-(®)-Phe-Met

Synonyms: N-((S)-2-((R)-2-Amino-3-Mercaptopropylamino)-3-Methylbutyl)-Phe-Met; B581

Bachem N-1390.0005 Store at -15°C

Cys-Val-Pro-Ala-Asp-Ile-Asn-Lys-Glu-Glu-Glu-Phe Trifluoroacetate Salt

Synonyms: X Linked Inhibitor of Apoptosis Protein (12-23)

Biogenesis 9701-1030 Human synthetic MW 697.5 (M+2H) Purified; lyophilized

Cys-Val-Tyr-Phe-Cys-His-Leu-Asp-Ile-Ile-Trp

Synonyms: Endothelin B Receptor Antagonist; Endothelin I (11-21), (Cys[11,15])-; IRL-1038, (Cys[11],Cys[15])-

Alexis 155-007 MW 1409.7 $C_{68}H_{92}N_{14}O_{15}S_2$ Disulfide bonds: Cys[1]-Cys[5]; antagonist for the ET_B receptor with a controversial pattern in activity; Haynes, WG et al, *TIPS*, 14:225, 1993; Urade, Y et al, *FEBS Lett*, 311:12, 1992; Takai, M et al, *BBRC*, 184:953, 1992; Urade, Y et al, *FEBS Lett*, 342:103, 1994

Bachem H-1658.0001 MW 1409.7 $C_{68}H_{92}N_{14}O_{15}S_2$ Store at -15°C

Calbiochem; 05-23-3883 MW 1409.7 $C_{68}H_{92}N_{14}O_{15}S_2$ >95% (HPLC); solid; soluble in 5% acetic acid & DMF | Disulfide bonds: Cys[11]-Cys[15]; highly potent ET_B-receptor antagonist; Urade, N et al, *FEBS Lett*, 342:103, 1994; Urade, Y et al, *FEBS Lett*, 311:12, 1992

Neosystem SC861 MW 1409.67 Disulfide bonds: Cys[11]-Cys[15]; bioactive; ETB antagonist; Urade, Y et al, *FEBS Lett*, 311:12-16, 1992; Urade, Y et al, *FEBS Lett*, 342:103, 1994

Sigma E 7519 FW 1409.7 ≥97% (HPLC) | Disulfide bonds: 11-15; bioactive peptide; inhibits ET-B receptor-mediated contraction of guinea pig ileal & tracheal smooth muscle without any significant agonist activity, but does not affect the ET-A receptor-mediated contraction of rat aortic smooth muscle; Urade, Y et al, *FEBS Lett*, 311:12, 1992

American Peptide 88-2-41 Human MW 1410.0 $C_{68}H_{92}N_{14}O_{15}S_2$ Disulfide bonds: Cys[11]-Cys[15]; inhibits ET_B receptor mediated contraction of guinea pig ideal & tracheal smooth muscle without any significant agonist activity, but does not affect the ET_A receptor-mediated contraction of rat aortic smooth muscle; Haynes, WG et al, *TIPS*, 14:225, 1993; Urade, N et al, *FEBS Lett*, 342:103, 1994

Cys-Val-Tyr-Val-Arg-Ser-Ala-Ile-Gln-Leu-Gly-Asn-Tyr-Lys

Synonyms: Proteoglycan II (316-329); Decorin; PG-32; PG-40

Biogenesis 7870-6200 Human synthetic Purified; lyophilized | This C-terminal human sequence is also bone proteoglycan II precursor (346-359); rat, mouse and bovine proteoglycan II and human proteoglycan I exhibit high homology in this region

Cys-Val-Val-Ile-Val-Gly-Arg-Ile-Val-Leu-Ser-Gly

Synonyms: HCV NS4A Protein (22-33)

Bachem H-4006.0001 Store at -15°C | FDA strain

Cys-Val-Val-Ile-Val-Gly-Arg-Val-Val-Leu-Ser-Gly-Lys

Synonyms: HCV NS4A Protein (22-34)

Bachem H-2764.0001 Store at -15°C | H strain

Cys-Val-Val-Leu-Lys-Ser-Asp-Thr-Glu-Gln-Ser-Glu-Asp
(Cys)-Val-Val-Leu-Lys-Ser-Asp-Thr-Glu-Gln-Ser-Glu-Asp Acetate Salt

Synonyms: Ankyrin (3910-3921)

Biogenesis 0580-0250 Human brain variant synthetic >80% (HPLC); lyophilized

D-CHA-Ala-Arg-AMC·2AcOH

Synonyms: Pefafluor TH; Pefa-15865

Pentapharm 081-23, 081-19 MW 675.8 Excitation wavelength 360 nm, emission wavelength 460 nm | Sensitive fluorogenic peptide substrate for thrombin; used for determination of thrombin activity for research, in-process and quality control; K_M:1.93 μM,k_{cat}:53.9 s^{-1}

Dimer (Ser-Leu-Arg-Arg-Ser-Ser-Cys7-Phe-Gly-Gly-Arg-Met-Asp-Arg-Ile-Gly-Ala-Gln-Ser-Gly-Leu-Gly-Cys²³-Asn-Ser-Phe-Arg-Tyr)→(Tyr-Arg-Phe-Ser-Asn-Cys23'-Gly-Leu-Gly-Ser-Gln-Ala-Gly-Ile-Arg-Asp-Met-Arg-Gly-Gly-Phe-Cys7'-Ser-Ser-Arg-Arg-Leu-Ser)←

Synonyms: Natriuretic Peptide, Atrial β-

Peptides International PAF-4168-s Human synthetic MW 6161.0 $C_{254}H_{406}N_{90}O_{78}S_6$ >95% (HPLC); lyophilized amorphous powder | Bioactive; Disulfide bonds: Cys⁷-Cys²³, Cys⁷'-Cys²³; antiparallel dimer of human ANP, Fragment 1-28; Kangawa, K et al, *Nature*, 313:397, 1985; Chino, N et al, *BBRC*, 141:665, 1986

Dip-Leu-Asp-Ile-Ile-Trp
Ac-D-Dip-Leu-Asp-Ile-Ile-Trp

Synonyms: PD 142893; Endothelin I (16-21), (Ac-D-Dip[16])-

American Peptide 88-2-43 MW 924.1 $C_{50}H_{65}N_7O_{10}$ Non-selective ET_A & ET_B receptor antagonist; Cody, WL et al, *J Med Chem*, 35:3301, 1992; Haynes, VG et al, *TIPS*, 14:225, 1993; Hingorani et al, *J Med Chem*, 36:2585, 1993

Calbiochem 513015 MW 924.1 $C_{50}H_{65}N_7O_{10}$ ≥98% (HPLC); solid; soluble in phosphate buffer, pH 7.5 | Highly potent but non-selective endothelin receptor antagonist; displays higher potency towards the ET_A receptor compared to the ET_B receptor; Nishiyama, M et al, *Jpn J Pharmacol*, 69:391, 1995; Doherty, AM et al, *J Med Chem*, 36:2585, 1993; Warner, TD et al, *Br J Pharmacol*, 110:777, 1993; Cody, WL et al, *J Med Chem*, 35:3301, 1992

Neosystem SC959 MW 923.9 Bioactive; endothelin-related peptide; ETA/ETB receptor antagonist; Cody, WL et al, *J Med Chem*, 35:3301-3303, 1992; Doherty, AM et al, *J Cardiovasc Pharmacol*, 22 Suppl 8:98-102, 1993; Warner, TD et al, *J Cardiovasc Pharmacol* 22 Suppl 8:117-120, 1993; Warner, TD et al, *Br J Pharmacol* 110:777-782, 1993; Doherty, AM et al, *J Med Chem*, 36:2585-2594, 1993; Wu-Wong, JR et al, *Life Sci*, 54:1727-1734, 1994; Cody, WL et al, *J Med Chem*, 38:2809-2819, 1995

Gln-Ala-Arg
BOC-Gln-Ala-Arg-AMC Hydrochloride

Bachem I-1550.0050 MW 667.16 $C_{29}H_{42}N_8O_8$ · HCl Store at -15°C

Gln-Ala-Arg
BOC-Gln-Ala-Arg-MCA

Synonyms: Trypsin Substrate

Peptides International MQR-3135-v MW 630.70 $C_{29}H_{42}N_8O_8$ >98% (HPLC); lyophilized amorphous powder | Kawabata, S et al, *Eur J Biochem*, 172:17, 1988

Gln-Ala-Arg
N-t-BOC-Gln-Ala-Arg MCA Hydrochloride

Synonyms: Trypsin Substrate

Sigma B 4153   FW 667.2 $C_{29}H_{42}N_8O_8 \cdot HCl$ ~95% |
Kawabata, S et al, *Eur J Biochem*, 172:17, 1988

Gln-Ala-Gly-His-Asn-Lys-Val-Gly-Ser-Leu-Gln-Tyr-Leu

Synonyms: VIF (136-148); HTLV I Peptide

Neosystem SC549   MW 1414.58   HTLV I peptide from
subtype VIF; due to the presence of a Gln in N-terminal position
this peptide can contain the pyroglutamic form

Gln-Ala-Lys-His-Lys-Gln-Arg-Lys-Arg-Leu-Lys-Ser-Ser-Cys Trifluoroacetate Salt

Synonyms: Bone Morphogenetic Protein 2 (1-14)

Biogenesis 1406-0630   Human synthetic   MW 1697.8
Purified; lyophilized | Represents AA 1-14 of human BMP-2

Gln-Ala-Lys-Ser-Gln-Gly-Gly-Ser-Asn

Synonyms: Thymus Factor

American Peptide 87-0-03   MW 875.9 $C_{33}H_{57}N_{13}O_{15}$

Gln-Ala-Thr-Val-Gly-Asp-Ile-Asn-Thr-Glu-Arg-Pro-Gly-Met-Leu-Asp-Phe-Thr-Gly-Lys

Synonyms: Diazepam Binding Inhibitor Fragment; Diazepam
Binding Inhibitor

American Peptide 60-1-70   Human   MW 2150.5
$C_{91}H_{148}N_{26}O_{32}S$ Ferrero, P et al, *GABAergic Trans & Anxiety*, E.
Costa, Raven Press, 177, 1986

Bachem H-6760.0001   Human   MW 2150.4 $C_{91}H_{148}N_{26}O_{32}S$
Store at -15°C

Calbiochem 05-23-1500   Human   MW 2150.4
$C_{91}H_{148}N_{26}O_{32}S$ ≥90% (HPLC); lyophilized; soluble in 5% acetic
acid | A peptide fragment from the diazepam binding inhibitory
protein; displaces ligands bound to the β-carboline/benzodiazepine
recognition site, an allosteric modulatory site of the GABA$_A$ receptor
complex; Gray, PW et al, *PNAS*, 83:7547, 1986

ICN 193417   Human   MW 2150.4   Gray, PW etal, *PNAS*,
83:7547, 1986

Gln-Ala-Thr-Val-Gly-Asp-Ile-Asn-Thr-Glu-Arg-Pro-Gly-Met-Leu-Asp-Phe-Thr

Synonyms: Octadecane Neuropeptide-Diazepam Binding Inhibitor

Neosystem SC366   Human   MW 1965.16   Bioactive
neuropeptide; due to the presence of a Gln in N-terminal position
this peptide can contain the pyroglutamic form; Anxiety peptide e

Gln-Ala-Thr-Val-Gly-Asp-Ile-Asn-Thr-Glu-Arg-Pro-Gly-Met-Leu-Asp-Phe-Thr-Gly-Lys

Synonyms: Diazepam Binding Inhibitor (51-70); Anxiety Peptide
(1-20)

Sigma G 9898   Human   FW 2150.4   ≥95% (HPLC) |
Bioactive peptide; Gray, PW et al, *Proc Natl Acad Sci USA*, 83:7547,
1986

Biogenesis 0644-5105   Human synthetic   MW 2150.6
Purified; lyophilized

Gln-Ala-Thr-Val-Gly-Asp-Val-Asn-Thr-Asp-Arg-Pro-Gly-Leu-Leu-Asp-Leu-Lys

Synonyms: Diazepam Binding Inhibitor; Anxiety Peptide

American Peptide 60-0-62   MW 1912.2 $C_{81}H_{138}N_{24}O_{29}$
Tryptic peptide of the diazepam binding inhibitor; binds to the
benzodiazepine receptor & increases anxiety; Ferrero, P et al,
Neuropharmacology, 23:1359, 1984; Marx, JL, *Science*, 227:934,
1985

Gln-Ala-Thr-Val-Gly-Asp-Val-Asn-Thr-Asp-Arg-Pro-Gly-Leu-Leu-Asp-Leu-Lys-Tyr

Synonyms: Diazepam Binding Inhibitor, (Tyr[19])-

American Peptide 60-0-64   MW 2075.2 $C_{90}H_{147}N_{25}O_{31}$

Gln-Ala-Thr-Val-Gly-Asp-Val-Asn-Thr-Asp-Arg-Pro-Gly-Leu-Leu-Asp-Leu-Lys

Synonyms: Anxiety Peptide; Octadecane Neuropeptide; Diazepam
Binding Inhibitor; Octadecane Neuropeptide-Diazepam Binding
Inhibitor

Bachem H-1770.0001   MW 1912.13 $C_{81}H_{138}N_{24}O_{29}$ Store at
-15°C

ICN 151475

ICN 151777   Tryptic peptide of the diazepam binding inhibitor;
binds to the benzodiazepine receptor & increases anxiety; Ferrero,
P etal, *Neuropharmacology*, 23:1359, 1984; Marx, JL etal, *Science*,
227:934, 1985

Sigma G 3642   FW 1912.1   ≥90% (HPLC) | Bioactive
peptide; tryptic peptide of the rat diazepam binding inhibitor (DBI);
binds to the benzodiazepine receptor & appears to increase
anxiety; Marx, JL, *Science*, 227:934, 1985; Ferrero, P et al,
Neuropharmacology, 23:1359, 1984

Neosystem SC128   Rat   MW 1912.12   Bioactive
neuropeptide; due to the presence of a Gln in N-terminal position
this peptide can contain the pyroglutamic form; Anxiety peptide;
Marx, JL, *Science*, 227:934, 1985

Gln-Ala-Thr-Val-Gly-Asp-Val-Asn-Thr-Asp-Arg-Pro-Gly-Leu-Leu-Asp-Leu-Lys-Tyr

Synonyms: Octadecane Neuropeptide-Diazepam Binding Inhibitor-
(Tyr[19])

Neosystem SC129   Rat   MW 2075.30   Bioactive
neuropeptide; due to the presence of a Gln in N-terminal position
this peptide can contain the pyroglutamic form e

Gln-Ala-Thr-Val-Gly-Asp-Val-Asn-Thr-Asp-Arg-Pro-Gly-Leu-Leu-Asp-Leu-Lys

Synonyms: Anxiety Peptide (1-18); Octadecane Neuropeptide

Biogenesis 0644-5125   Rat synthetic   MW 1912.1
Purified; lyophilized

Gln-Arg-Arg
BOC-Gln-Arg-Arg-AMC

Bachem I-1655.0025   MW 715.81 $C_{32}H_{49}N_{11}O_8$ Store at
-15°C

Gln-Arg-Arg
BOC-Gln-Arg-Arg-MCA

Synonyms: Carboxyl Side of Paired Basic Residue Cleaving Enzyme
Substrate

Peptides International MQR-3122-v   MW 715.81
$C_{32}H_{49}N_{11}O_8$ >98% (HPLC); lyophilized amorphous powder |
Mizuno, K et al, *BBRC*, 144:807, 1987

Gln-Arg-Arg
N-t-BOC-Gln-Arg-Arg MCA Acetate Salt

Synonyms: Paired Basic Residue Cleaving Enzyme Substrate

Sigma B 4653   FW 715.8 (free base) $C_{32}H_{49}N_{11}O_8$ ~95% |
Substrate for carboxyl side; Mizuno, K et al, *Biochem Biophys Res
Commun*, 144:807, 1987

Gln-Arg-Arg-Gln-Arg-Lys-Ser-Arg-Arg-The-Ile

ICN 153100   The C-terminal sequence of the human
Interleukin-2 receptor; may be useful in the assay of proteinkinase
C; Gallis, B, *JBC*, 261:5075, 1986

Gln-Arg-Arg-Gln-Arg-Lys-Ser-Arg-Arg-Thr-Ile

Synonyms: Protein Kinase C Substrate; Interleukin II Receptor

Bachem H-9685.0001 MW 1484.73 $C_{59}H_{113}N_{29}O_{16}$ Store at -15°C

ICN 159657 MW 1484.8 From the human IL-2 receptor; Gallis, B etal, *JBC*, 26:5075, 1986

Alexis 159-001 Human MW 1484.7 Derived from the cytoplasmic C-terminus of the human interleukin-2 receptor; Gallis, B et al, *J Biol Chem*, 261:5075, 1986

Sigma G 9016 Human FW 1484.7 ≥95% (HPLC); peptide content:~65% | Bioactive peptide; C-terminal sequence; may be useful in the assay of proteinkinase; C Gallis, B et al, *J Biol Chem*, 261:5075, 1986

Gln-Arg-Asn-Gln-Leu-Leu-Gln-Lys-Glu-Pro-Asp-Leu-Arg-Leu Acetate Salt

Synonyms: Chromogranin C (1-14); Secretogranin II

Biogenesis 2095-0420 Human synthetic Semi-pure; lyophilized | The C-terminal human sequence is unique to chromogranin C; it is also chromogranin C precursor (31-44)

Gln-Arg-Gly-Pro-Gly-Arg-Ala-Phe

Synonyms: GP120 (315-322); HTLV I Peptide

Neosystem SC814 MW 887.99 HTLV I peptide from subtype GP160; due to the presence of a Gln in N-terminal position this peptide can contain the pyroglutamic form

Gln-Arg-Pro-Arg-Leu-Ser-His-Lys-Gly-Pro-Met-Pro-Phe

Synonyms: Apelin-13

Bachem H-4566.0001 Human, bovine MW 1550.85 $C_{69}H_{111}N_{23}O_{16}S$ Store at -15°C

Gln-Arg-Pro-Lys-Glu-Arg-His-Arg-Arg-Asn-Tyr-Val-Pro-Cys-His-Ile-Arg-Gln-Ile-Ile

Synonyms: GP140 (421-440); SIVmac251 Peptide

Neosystem SC750 MW 2600.04 SIVmac251 peptide from subtype GP140; due to the presence of a Cys this peptide can contain the dimeric form; due to the presence of a Gln in N-terminal position this peptide can contain the pyroglutamic form

Gln-Asn-Cys-Pro-Arg-Gly Amide

Synonyms: Vasopressin (4-9), (Arg[8])-

Bachem H-4092.0005 Store at -15°C

Gln-Asp-Leu-Thr-Met-Lys-Tyr-Gln-Ile-Phe

Synonyms: Peptide 810

Bachem H-2586.0001 Store at -15°C

Gln-Asp-Val-His

Synonyms: Parathyroid Hormone (29-32)

Bachem H-2272.0005 Human Store at -15°C

Gln-Gln
BOC-Gln-Gln

Bachem A-3210.0001 MW 374.39 $C_{15}H_{26}N_4O_7$ Store at -15°C

Gln-Gln

Bachem G-1885.0250 MW 274.28 $C_{10}H_{18}N_4O_5$ Store at -15°C

Gln-Gln-Phe-Phe-Gly-Leu-Met Amide

Synonyms: Substance P, Hepta-; Substance P (5-11)

American Peptide 70-1-51 MW 869.1 $C_{41}H_{60}N_{10}O_9S$

ICN 153026 Yajima, H etal, *Chem Pharm Bull*, 21:2500, 1973

Sigma S 1261 FW 869.0 $C_{41}H_{60}N_{10}O_9S$ ≥95% (HPLC) | Bioactive peptide

Gln-Glu

Bachem G-4080.0250 MW 275.26 $C_{10}H_{17}N_3O_6$ Store at -15°C

Gln-Glu-Glu-Glu-Glu-Glu-Thr-Ala-Gly-Ala-Pro-Gln-Gly-Leu-Phe-Arg-Gly Amide

Synonyms: Chromogranin A (272-288); Pancreastatin (33-49)

American Peptide 46-4-42 Porcine MW 1846.9 $C_{77}H_{119}N_{23}O_{30}$ The C-terminal fragment of pancreastatin is a potent inhibitor of insulin release; Tatemoto, K et al, *Nature*, 324:476, 1986

Bachem H-5905.0001 Porcine Store at -15°C

ICN 152887 Porcine MW 1846.9 Tatemoto, K etal, *Nature*, 324:476, 1986

Neosystem SC056 Porcine MW 1846.92 Bioactive; brain/gut peptide; due to the presence of a Gln in N-terminal position this peptide can contain the pyroglutamic form; Tatemoto, K et al, *Nature*, 324:476-478, 1986

Sigma P 2919 Porcine FW 1846.9 ≥95% (HPLC) | Gastrointestinal peptide; Bioactive peptide; Tatemoto, K et al, *Nature*, 324:476, 1986

Gln-Glu-Lys-Gln-Asn-Thr-Val-Ala-Thr-Ala-His-Ala-Gly-Phe-Phe-Leu-Arg-Glu-Asn-Glu-Gly

Bachem H-3736.0500 Store at -15°C

Gln-Gly
Z-Gln-Gly

Bachem C-1635.0250 MW 337.33 $C_{15}H_{19}N_3O_6$ Store at -15°C

Gln-Gly

Bachem G-1890.0250 MW 203.2 $C_7H_{13}N_3O_4$ Store at -15°C

Gln-Gly
Z-Gln-Gly

Synonyms: Transglutaminase Substrate

Peptides International SZQ-3190 MW 337.33 $C_{15}H_{19}N_3O_6$ >99% (HPLC); amorphous powder | Folk, JE & PW Cole, *JBC*, 241:5518, 1966; Ando, H et al, *Agric Biol Chem*, 53:2613, 1989

Gln-Gly
N^ε-CBZ-Gln-Gly

Sigma C 6154 FW 337.3 $C_{15}H_{19}N_3O_6$

Gln-Gly
CBZ-Gln-Gly

USBio C2098-69 MW 337.3 $C_{15}H_{19}N_3O_6$ ≥99%

Gln-Gly-Arg
BOC-Gln-Gly-Arg-AMC Hydrochloride

Bachem I-1595.0050 MW 653.14 $C_{28}H_{40}N_8O_8 \cdot HCl$ Store at -15°C

Gln-Gly-Arg
BOC-Gln-Gly-Arg-MCA

Synonyms: Factor XIIa Substrate

Peptides International MQR-3136-v MW 616.67 $C_{28}H_{40}N_8O_8$ >98% (HPLC); lyophilized amorphous powder | Kawabata, S et al, *Eur J Biochem*, 172:17, 1988

Gln-Gly-Arg
N-t-BOC-Gln-Gly-Arg MCA Hydrochloride

Synonyms: Factor XIIa Substrate

Sigma B 4778   FW 653.1 $C_{28}H_{40}N_8O_8 \cdot HCl$ ~95% |
Kawabata, S et al, *Eur J Biochem*, 172:17, 1988

Gln-Gly-Ile-Ala-Gly-Gln-Arg
Dnp-Gln-Gly-Ile-Ala-Gly-Gln-D-Arg

Synonyms: Collagenase Substrate

Peptides International SDQ-3088-v   MW 894.90
$C_{35}H_{54}N_{14}O_{14}$ >96% (HPLC); lyophilized amorphous powder |
Reference substrate for Peptides International assay SDP-3087;
Masui, Y et al, *Biochem Med*, 17:215, 1977

Gln-Gly-Pro Trifluoroacetate Salt

Bachem H-3135.0050   Store at -15°C

Gln-His-Trp-Ser-Tyr-Gly-Leu-Arg-Pro-Gly-Gln-His-Trp-Ser-Tyr-Gly-Leu-Arg-Pro-Gly
Gln-His-Trp-Ser-Tyr-Gly-Leu-Arg-Pro-Gly-Biotinyl-Gln-His-Trp-Ser-Tyr-Gly-Leu-Arg-Pro-Gly Amide

Synonyms Luteinizing Hormone Releasing Hormone, (Biotinyl-Gln[1])-:

Bachem H-4792.0001   MW 1425.64 $C_{65}H_{92}N_{20}O_{15}S$ Store
at -15°C

Gln-Leu-Gln-Leu-Gln-Ala-Ala-Ser-Asn-Phe-Lys-Ser-Pro-Val-Lys-Thr-Ile-Arg

Synonyms: Casein Kinase II (198-215), β-; Cell Cycle-Dependent
Kinase Substrate

ICN 195841   MW 2028.1   ≥97% | *cdc2* substrate &
competitive inhibitor of apoptosis induced by perforin & fragmentin-
2 in YAC-1 lymphoma cells (peptide A)

Gln-Leu-Lys-Thr-Ala-Asp-Leu-Pro-Ala-Gly-Arg-Asp-Glu-Thr-Thr-Ser-Phe-Val-Leu-Val
Ac-Gln-Leu-Lys-Thr-Ala-Asp-Leu-Pro-Ala-Gly-Arg-Asp-Glu-Thr-Thr-Ser-Phe-Val-Leu-Val Amide

Synonyms: Adhesin (1025-1044), Ac-; P1025

Bachem H-4538.0500   MW 2202.49 $C_{97}H_{160}N_{26}O_{32}$ Store at
-15°C

Gln-Lys-Arg-Pro-Ser-Gln-Arg-Ser-Lys-Tyr-Leu

Synonyms: Bovine Myelin Basic Protein (4-14); Protein Kinase C
Substrate

American Peptide 86-0-21   MW 1390.5 $C_{60}H_{103}N_{21}O_{17}$
Described as the most specific substrate for the selective assay of
protein kinase C; Yasuda, I et al, *BBRC*, 166:1220, 1990

Gln-Lys-Arg-Pro-Ser-Gln-Arg-Ser-Lys-Tyr-Leu
Ac-Gln-Lys-Arg-Pro-Ser-Gln-Arg-Ser-Lys-Tyr-Leu

Synonyms: Myelin Basic Protein (4-14), Ac-; Protein Kinase C
Substrate Peptide

American Peptide 86-2-45   MW 1432.7 $C_{62}H_{105}N_{21}O_{18}$
Acetylated analog; Yasuda, I et al, *BBRC*, 166:1220, 1990

Gln-Lys-Arg-Pro-Ser-Gln-Arg-Ser-Lys-Tyr-Leu

Synonyms: Myelin Basic Protein (4-14)

Bachem H-1072.0001   MW 1390.61 $C_{60}H_{103}N_{21}O_{17}$ Store
at -15°C

Gln-Lys-Arg-Pro-Ser-Gln-Arg-Ser-Lys-Tyr-Leu
Ac-Gln-Lys-Arg-Pro-Ser-Gln-Arg-Ser-Lys-Tyr-Leu

Synonyms: Myelin Basic Protein (4-14), (Ac-Gln[4])-

Bachem H-3238.0001   Store at -15°C

Gln-Lys-Arg-Pro-Ser-Gln-Arg-Ser-Lys-Tyr-Leu

Synonyms: Myelin Basic Protein (4-14); Protein Kinase C Substrate

ICN 159627   MW 1390.6   Reportedly the most specific
substrate for PKC; Yasuda, I etal, *BBRC*, 166:1220, 1990

Gln-Lys-Arg-Pro-Ser-Gln-Arg-Ser-Lys-Tyr-Leu
N-Ac-Gln-Lys-Arg-Pro-Ser-Gln-Arg-Ser-Lys-Tyr-Leu

Synonyms: Myelin Basic Protein (4-14), *N*-Ac-; Protein Kinase C α-
Substrate; Protein Kinase C β-Substrate; Protein Kinase C γ-
Substrate

ICN 195833   MW 1432.8   ≥97% | Specific substrate for
isoforms of PKC; resistant to phosphatases in brain hippocampal
extracts

Gln-Lys-Arg-Pro-Ser-Gln-Arg-Ser-Lys-Tyr-Leu

Synonyms: Myelin Basic Protein (4-14); Protein Kinase Related
Peptide; Bovine Myelin Basic Protein (4-14); Protein Kinase C
Substrate; Protein Kinase C Substrate Inhibitor

Neosystem SC376   MW 1390.60   Bioactive; proteinkinase-
related peptide; due to the presence of a Gln in N-terminal position
this peptide can contain the pyroglutamic form; very selective
substrate for proteinkinase C; Yasuda, T et al, *BBRC*, 166:1220-
1227, 1990

Sigma G 5777   FW 1390.6   ≥97% (HPLC); peptide
content:~70% | Bioactive peptide; Yasuda, I et al, *Biochem
Biophys Res Commun*, 166:1220, 1990

Sigma M 6913   FW 1390.6   ≥95% (HPLC); peptide
content:~65% | Bioactive peptide; Yasuda, I et al, *Biochem
Biophys Res Commun*, 166:1220, 1990

Upstate 12-123   Synthetic   MW 1390   Lyophilized

Gln-Lys-Leu-Val-Phe-Phe
Ac-Gln-Lys-Leu-Val-Phe-Phe Amide

Synonyms: Amyloid β-Protein (15-20), Ac-

Bachem H-3684.0005   MW 822.02 $C_{42}H_{63}N_9O_8$ Store at
-15°C

Sigma A 6933   FW 822.0 $C_{42}H_{63}N_9O_8$ ≥97% (HPLC) |
Bioactive peptide; inhibits polymerization of amyloid β-protein (1-
40) into amyloid fibrils; Tjernberg, LO et al, *J Biol Chem*, 271:8545,
1996

Gln-Phe-Glu-Thr

Synonyms: Synaptobrevin II (76-79)

Bachem N-1350.0005   Rat   Store at -15°C

Gln-Phe-Ile-Asn-Met-Trp-Gln-Glu-Val-Gly

Synonyms: GP120 (427-436); HTLV I Peptide

Neosystem SC569   MW 1251.42   HTLV I peptide from
subtype GP160; due to the presence of a Gln in N-terminal position
this peptide can contain the pyroglutamic form

Gln-Phe-Phe-Gly-Leu-Met Amide

Synonyms: Substance P (6-11); Substance P, Hexa-

American Peptide 70-1-67   MW 741.0 $C_{36}H_{52}N_8O_7S$

ICN 153029

Gln-Pro
BOC-Gln-Pro

Synonyms: Peptidoglutaminase Substrate

Bachem A-3170.0001   MW 343.38 $C_{15}H_{25}N_3O_6$ Store at
-15°C

Peptides International SQP-3151   MW 343.38 $C_{15}H_{25}N_3O_6$
>99% (HPLC); amorphous powder | Hamada, JS & WE Marshall, *J
Food Sci*, 53:1132, 1988

Gln-Pro
N-t-BOC-Gln-Pro

Synonyms: Peptidoglutaminase Substrate

Sigma B 4403   FW 343.4 $C_{15}H_{25}N_3O_6$ Hamada, JS & Marshall, WE, *J Food Sci*, 53:1132, 1988

Gln-Pro-Glu-Lys-Ala-Lys-Lys-Glu-Thr-Val-Glu-Lys-Ala-Val-Ala-Thr

Synonyms: TAT (110-125); SIVmac251 Peptide

Neosystem SC594   MW 1757.01   SIVmac251 peptide from subtype TAT; due to the presence of a Gln in N-terminal position this peptide can contain the pyroglutamic form

Gln-Ser-Arg-Asn-Arg-Ser-Thr-Gln-Ser-Gln-Asp-Val-Ala-Arg-Gly-Cys

Synonyms: Bone Morphogenetic Protein 6

Biogenesis 1406-1059   Rat synthetic   MW 1793   Semi-pure; no preservatives; lyophilized

Gln-Ser-Arg-Lys-Arg-Arg-Arg-Thr-Pro-Lys-Lys-Ala-Lys-Ala-Asn

Synonyms: TAT (78-92); SIVmac251 Peptide

Neosystem SC589   MW 1825.15   SIVmac251 peptide from subtype TAT; due to the presence of a Gln in N-terminal position this peptide can contain the pyroglutamic form

Gln-Ser-Leu-Gly-Gly-Leu-Gly-Lys-Gly-Leu-Ser-Ser-Arg-Ser

Synonyms: NEF (41-54); SIVmac251 Peptide

Neosystem SC607   MW 1346.50   SIVmac251 peptide from subtype NEF; due to the presence of a Gln in N-terminal position this peptide can contain the pyroglutamic form

Gln-Trp-Leu
Ac-Gln-Trp-Leu Amide

Bachem H-1025.0050   MW 486.57 $C_{24}H_{34}N_6O_5$ Store at -15°C

Gln-Trp-Phe
BOC-Gln-D-Trp(For)-Phe-OBzl

Bachem H-5092.0025   Store at -15°C

Gln-Val-Glu-Leu-Leu-Ile-Lys-Gln-Ala-Thr-Ser-His-Glu-Asn-Leu-Cys

Synonyms: FRAP Blocking Peptide

Calbiochem 344287   Rat synthetic   MW 1825.1 $C_{78}H_{131}N_{22}O_{26}S$ Lyophilized | Based on the rat FRAP with a Cys (C) residue added & the peptide coupled to KLH; this sequence is 100% conserved in the human FRAP sequence; coupled to KLH, is used as the immunogen for the production of Anti-FRAP; suitable for use in immunoabsorption for immunoprecipitation, immunocytochemistry, Western blotting & dot blots

Gln-Val-Pro-Leu-Arg-Pro-Met-Thr-Tyr-Lys

Synonyms: NEF (73-82); HTLV I Peptide

Neosystem SC505   MW 1232.50   HTLV I peptide from subtype NEF; due to the presence of a Gln in N-terminal position this peptide can contain the pyroglutamic form

Glp-Ala

Synonyms: Pyroglutamyl Peptidase Substrate; L-Pyrrolidone-Carbonyl-L-Ala

Bachem G-3090.0250   MW 200.19 $C_8H_{12}N_2O_4$ Store at -15°C

ICN 156122   MW 200.2 $C_8H_{12}N_2O_4$

Peptides International OVA-3079   MW 200.19 $C_8H_{12}N_2O_4$ >99% (HPLC); amorphous powder | Doolittle, RF in *Proteolytic Enzymes*, Methods in Enzymology, Vol 19, (GE Perlmann & L Lorand, Eds), Academic Press, New York, 1970, pp 555-569

Sigma P 2788   FW 200.2 $C_8H_{12}N_2O_4$

Glp-Ala-Asp-Pro-Asn-Lys-Phe-Tyr-Gly-Leu-Met Amide

Synonyms: Physalaemin Tachykinin; Physalaemin

American Peptide 62-0-57   MW 1265.5 $C_{58}H_{84}N_{14}O_{16}S$ Potent vasodilator & hypotensive agent; Erspamer, V et al, *Experientia*, 20:489, 1964

Bachem H-4585.0001   MW 1265.46 $C_{58}H_{84}N_{14}O_{16}S$ Store at -15°C

ICN 151883   MW 1265.4   Powerful vasodilator & hypotensive agent; Bertaccini, G etal, *Br J Pharmac*, 25:380, 1965

Neosystem SC077   MW 1265.45   Bioactive; tachykinin

Sigma P 2149   *Physalaemus fuscumaculatus*   FW 1265.4   ≥97% (HPLC) | Bioactive peptide; structural homology to Substance P; potent hypotensive; induces salivation, intestinal contraction & vasodilation; Bertaccini, G et al, *Br J Pharmacol*, 25:380, 1965

Peptides International PPY-4030-v   *Physalaemus fuscumaculatus* (frog) Synthetic   MW 1265.5 $C_{58}H_{84}N_{14}O_{16}S$ >95% (HPLC); lyophilized amorphous powder | Bioactive; Ersapmer, V et al, *Experientia*, 20:489, 1964; Bernardi, L et al, *Experientia*, 20:490, 1964

Glp-Ala-Asp-Pro-Asn-Lys-Phe-Tyr-Gly-Leu-Met Amide AcOH 3H$_2$O

Synonyms: Physalaemin

Peptides International PPY-4030   *Physalaemus fuscumaculatus* (frog) Synthetic   MW 1379.61 $C_{58}H_{84}N_{14}O_{16}S$ •CH$_3$COOH · 3H$_2$O   >95% (HPLC); lyophilized amorphous powder; bulk | Bioactive; Ersapmer, V et al, *Experientia*, 20:489, 1964; Bernardi, L et al, *Experientia*, 20:490, 1964

Glp-Ala-Glu

ICN 153170   MW 329.3 $C_{13}H_{19}N_3O_7$

Sigma P 3037   FW 329.3 $C_{13}H_{19}N_3O_7$ Bioactive peptide

Glp-Ala-Lys-Ser-Gln-Gly-Gly-Ser-Asn

Synonyms: Serum Thymic Factor

American Peptide 87-0-04   MW 858.9 $C_{33}H_{54}N_{12}O_{15}$ Pleau, JM et al, *JBC*, 252:8045, 1977

Bachem H-4865.0005   Store at -15°C

ICN 152056   Bach, JF etal, *Nature*, 266:55, 1977; Pleau, JM etal, *J Biochem*, 252:8045, 1977

Sigma S 8256   FW 858.9 $C_{33}H_{54}N_{12}O_{15}$ ≥95% (HPLC) | Bioactive peptide; Bach, JF et al, *Nature*, 266:55, 1977; Pleau, JM et al, *J Biol Chem*, 252:8045, 1977; Strachan, RG et al, *J Med Chem*, 22:586, 1979

Biogenesis 8278-0002   Synthetic   Lyophilized, consisting of 84.34% peptide material, TFA counter ions and residual H$_2$O; purified

Peptides International PST-4058-v   Synthetic   MW 858.86 $C_{33}H_{54}N_{12}O_{15}$ >99% (HPLC); lyophilized amorphous powder | Bioactive; Bach, JF et al, *Nature*, 266:55, 1977

Glp-Ala-Lys-Ser-Gln-Gly-Gly-Ser-Asn AcOH Dihydrate

Synonyms: Serum Thymic Factor

Peptides International PST-4058   Synthetic   MW 954.94 $C_{33}H_{54}N_{12}O_{15}$ •CH$_3$COOH · 2H$_2$O   >99% (HPLC); lyophilized amorphous powder; bulk | Bioactive; Bach, JF et al, *Nature*, 266:55, 1977

Glp-Ala-Ser-Asn-Cys-Phe-Ala-Ile-Arg-His-Phe-Glu-Asn-Lys-Phe-Ala-Val-Glu-Thr-Leu-Ile-Cys-Ser-Arg-Thr-Val-Lys-Lys-Asn-Ile-Ile-Glu-Glu-Asn

Synonyms: Amyloid Bri Protein (1-34); Amyloid Bri Protein Precursor 277 (244-277)

Bachem H-4728.0500 MW 3937.52 $C_{173}H_{275}N_{49}O_{52}S_2$ Store at -15°C

Glp-Arg
Glp-CHG-Arg-pNA·AcOH

Synonyms: Pefachrome®PCa3297; Pefa-3297

Pentapharm 089-21, 089-08 MW 604.7 Sensitive chromogenic peptide substrate with improved selectivity for activated protein C; no interference with Protac®; v_{max}:0.014 µmol/min; K_M:0.899 mM; Stocker, K et al, *Folia Haematol*, 115:260, 1988; Takahashi, H et al, *Clin Chim Acta*, 175:217, 1988; Wikstroem, P et al, Poster GTH, Frankfurt, Germany, 1998

Glp-Arg
Glp-CHG-Arg-AMC·AcOH

Synonyms: Pefafluor PCa3342; Pefa-3342

Pentapharm 089-26, 089-10 MW 641.7 Excitation wavelength 360 nm, emission wavelength 460 nm | Sensitive chromogenic peptide substrate with significantly improved selectivity for activated protein C; k_{cat}:8.5 s^{-1}, K_M:0.26 mM

Glp-Arg-Pro-Arg-Leu-Ser-His-Lys-Gly-Pro-Met-Pro-Phe

Synonyms: Apelin (24-36), (Glp^1)-; Apelin-13, (Glp^1)-; APJ Receptor Ligand

Bachem H-4568.0001 Human, bovine MW 1533.82 $C_{69}H_{108}N_{22}O_{16}S$ Store at -15°C

Neosystem SC1341 Human, bovine MW 1533.81 Virus-related peptide; bioactive; corresponds to the carboxy-terminal 13 residues of apelin with a pyroglutamyl at the N-terminus; 66-fold more active than apelin, the newly discovered endogenous ligand for the APJ receptor; Tatemoto, K et al, *BBRC*, 251:471-476, 1998

Peptides International PAP-4361-v Human, bovine synthetic MW 1533.8 $C_{69}H_{108}N_{22}O_{16}S$ >95% (HPLC); lyophilized amorphous powder | Bioactive; human, bovine; Tatemoto, K et al, *BBRC*, 251:471, 1998

Glp-Arg-Pro-Pro-Met-Glu-Glu-Glu-Glu-Glu-Ala-Tyr-Gly-Trp-Met-Asp-Phe Amide

Synonyms: Gastrin I

Bachem H-9165.0001 Rat MW 2126.31 $C_{94}H_{128}N_{22}O_{31}S_2$ Store at -15°C

ICN 154549 Rat MW 2126.2 Reeves, JR et al, *Peptides*, 2:45, 1981

Glp-Arg-Pro-Pro-Met-Glu-Glu-Glu-Glu-Glu-Ala-Tyr-Gly-Trp-Met-Asp-Phe Trifluoroacetate Salt

Synonyms: Gastrin (1-17)

Biogenesis 4620-1809 Rat synthetic MW 126.3 Purified; lyophilized

Glp-Arg-Pro-Pro-Met-Glu-Glu-Glu-Glu-Glu-Ala-Tyr-Gly-Trp-Met-Asp-Phe
Glp-Arg-Pro-Pro-Met-Glu-Glu-Glu-Glu-Glu-Ala-Tyr-(SO₃H)-Gly-Trp-Met-Asp-Phe Sulfated

Synonyms: Gastrin (1-17)

Biogenesis 4620-1819 Rat synthetic MW 2207 Lyophilized, consisting of 74.78% peptide material, acetate counter ions and residual H_2O; purified | To avoid Met oxidation, it is best to dissolve the peptide under a nitrogen stream

Glp-Arg-Thr-Lys-Arg MCA

Synonyms: Furin Substrate

Sigma P 4454 FW 827.9 $C_{37}H_{57}N_{13}O_9$ ≥97% (HPLC) | Fluorogenic substrate for furin

Glp-Arg-Thr-Lys-Arg-AMC

Bachem I-1650.0005 Store at -15°C

Glp-Arg-Thr-Lys-Arg-MCA

Synonyms: Furin Substrate

Peptides International MPR-3159-v MW 827.94 $C_{37}H_{57}N_{13}O_9$ >98% (HPLC); lyophilized amorphous powder | Hatsuzawa, K et al, *JBC*, 267:16094, 1992

Glp-Asn-Cys-Cys-Pro-Arg
Glp-Asn-Cys(H-Cys)-Pro-Arg

Synonyms: Vasopressin (4-8), (Glp^4,Cys(Cys)^6,Arg^8)-

Bachem H-2456.0001 MW 718.81 $C_{26}H_{42}N_{10}O_{10}S_2$ Store at -15°C

Glp-Asn-Cys-Cys-Pro-Arg
Glp-Asn-Cys-(S-S-Cys)-Pro-Arg

Synonyms: Vasopressin (4-8), (Glp^4,(Cys$_2$)^6,Arg^8)-

Sigma V 9009 FW 718.8 $C_{26}H_{42}N_{10}O_{10}S_2$ ≥97% (HPLC) | Bioactive peptide; found in brain as a metabolite of AVP; induces long-term potentiation of synaptic transmission in hippocampal slices; may be a memory-enhancing peptide; Range, XW et al, *Neuroreport*, 4:1135, 1993; Zhou, AW et al, *Peptides*, 16:581, 1995

Glp-Asn-Cys-Cys-Pro-Arg-Gly Amide Acetate Salt

Synonyms: Vasopressin (4-8), (Glp^4,(Cys$_2$)^6,Arg^8)-

Sigma V 0255 FW 774.9 (free base) $C_{28}H_{46}N_{12}O_{10}S_2$ ≥95% (HPLC) | Bioactive peptide; more potent neuropeptide than arginine vasopressin with selective effects on memory & related processes; DeWeid, D et al, *Science*, 221:1310, 1983

Glp-Asn-Glu

Sigma P 5148 FW 300.3 $C_{11}H_{16}N_4O_6$ ≥97% (HPLC) | Bioactive peptide

Glp-Asn-Gly

Synonyms: Aggression Inducing Peptide

Bachem H-4870.0050 MW 300.27 $C_{11}H_{16}N_4O_6$ Store at -15°C

ICN 153171 MW 300.3 $C_{11}H_{16}N_4O_6$ From patients with congenital generalized lipodystrophy; Reichelt, KL et al, *Psychopharmacology of Aggression*, Sandler, M ed, Raven Press, New York, 1979

Glp-Asn-Phe-His-Leu-Arg-Pro Amide Hydrochloride

Synonyms: Antho-RP Amide II

Bachem H-9765.0005 Store at -15°C

Glp-Asn-Pro-Asn-Arg-Phe-Ile-Gly-Leu-Met Amide

Synonyms: Phyllomedusin

American Peptide 62-0-68 MW 1171.4 $C_{52}H_{82}N_{16}O_{13}S$ Erspamer, V et al, *In Substance P*, ed, Von Euler & Pernow, Raven Press, NY, 67, 1979

Glp-Asp-Asp-Ser-Asp-Glu-Glu-Asn
Glp-Asp-Asp-Ser(PO₃H₂)-Asp-Glu-Glu-Asn

Synonyms: DNA Transcription Inhibitory Peptide

Bachem H-1938.0001 Store at -15°C

Glp-Asp-Pro-Phe-Leu-Arg-Phe Amide

Synonyms: FMRF-Like Peptide

Bachem H-9260.0005 MW 904.04 $C_{44}H_{61}N_{11}O_{10}$ Store at -15°C

ICN 158422 MW 904 $C_{44}H_{61}N_{11}O_{10}$ 97% | Originally isolated from *Helix aspersa*; Price, DA etal, *Biol Bull*, 169:256, 1985

American Peptide 60-9-40 *Helix aspersa* (snail) MW 904.0 $C_{44}H_{61}N_{11}O_{10}$ Price, DA et al, *Biol Bull*, 169:256, 1985

Neosystem SC118 *Helix aspersa* (snail) MW 904.03 Bioactive neuropeptide; cardioregulatory hormone; Price, DA et al, *Biol Bull*, 169:256, 1985

Sigma P 3807 *Helix aspersa* (snail) FW 904.0 $C_{44}H_{61}N_{11}O_{10}$ ≥97% (HPLC) | Bioactive peptide; Price, DA et al, *Biol Bull*, 169:256, 1985

Glp-Asp-Ser-Gly-Asp-Glu-Trp-Pro-Gln-Gln-Pro-Phe-Val-Pro-Arg-Leu Amide

Synonyms: Locustapyrokinin

Sigma L 1658 *Locusta migratoria* FW 1881.0 ≥97% (HPLC) | Bioactive peptide; Schoofs, L et al, *Gen Comp Endocrinol*, 81:97, 1991

Glp-Asp-Ser-Gly-Asp-Gly-Trp-Pro-Gln-Gln-Pro-Phe-Val-Pro-Arg-Leu Amide

Synonyms: Locustapyrokinin, (Gly[6])-

Sigma L 7279 FW 1809.0 ≥97% (HPLC) | Bioactive peptide

Glp-Asp-Val-Asp-His-Val-Phe-Leu-Arg-Phe Amide

Synonyms: Leucomyosuppressin

Bachem H-6840.0001 Store at -15°C

Glp-Gln

Bachem G-4195.0250 MW 257.25 $C_{10}H_{15}N_3O_5$ Store at -15°C

Glp-Gln-Ala

Synonyms: Eisenin

Bachem H-1962.0025 MW 328.33 $C_{13}H_{20}N_4O_6$ Store at -15°C

Glp-Gln-Arg-Leu-Gly-Asn-Gln-Trp-Ala-Val-Gly-His-Leu-Leu
Glp-Gln-Arg-Leu-Gly-Asn-Gln-Trp-Ala-Val-Gly-His-Leu-(®)-Leu Amide

Synonyms: Bombesin, (Leu[13]-(®)-Leu[14])-

Bachem H-7075.0500 Store at -15°C

Glp-Gln-Arg-Leu-Gly-Asn-Gln-Trp-Ala-Val-Gly-His-Leu-Leu
Glp-Gln-Arg-Leu-Gly-Asn-Gln-Trp-Ala-Val-Gly-His-Leu-ψ(CH₂NH)-Leu Amide

Synonyms: Bombesin, (Leu[13]-ψ(CH₂NH)-Leu[14])-; Bombesin Receptor Antagonist

ICN 152855 MW 1588.8 Specific & competitive; Coy, DH etal, *JBC*, 263:5056, 1988; Cowan, A, *Trends Pharmac Sci*, 9:1, 1988

Neosystem SC451 MW 1587.84 Bioactive bombesin receptor antagonist; Coy, DH et al, *JBC* 263, 5056-5060, 1988; Mahmoud, S et al:*Cancer Res* 51:1798, 1991

Glp-Gln-Arg-Leu-Gly-Asn-Gln-Trp-Ala-Val-Gly-His-Leu-Leu
Glp-Gln-Arg-Leu-Gly-Asn-Gln-Trp-Ala-Val-Gly-His-Leu-ψ-(CH₂NH)-Leu

Synonyms: Gastrointestinal Peptide; Bombesin, (Leu[13]-ψ(CH₂NH)-Leu[14])-

Sigma B 1025 FW 1588.8 ≥97% (HPLC) | Bioactive peptide; specific & competitive bombesin receptor antagonist; Cowan, A, *Trends Pharmacol Sci*, 9:1, 1988

Glp-Gln-Arg-Leu-Gly-Asn-Gln-Trp-Ala-Val-Gly-His-Leu-Met Amide

Synonyms: Bombesin

Alexis 152-018 MW 1619.9 $C_{71}H_{110}N_{24}O_{18}S$ White lyophilized powder

Bachem H-2155.0001 MW 1619.87 $C_{71}H_{110}N_{24}O_{18}S$ Store at -15°C

Fluka 15560 MW 1619.86 $C_{71}H_{110}N_{24}O_{18}S$ ≥99% (HPLC); ≥85% peptide content; ≤9% H₂O; ≤6% acetate | Mitogen which acts through the inositol lipid pathway; Millar, JBA & Rozengurt, E, *PNAS*, 86:3204, 1989; Lloyd, AC et al, *Biochem J*, 260:813, 1989

ICN 190307 MW 1619.9 $C_{71}H_{110}N_{24}O_{18}S$ Tetradecapeptide with biological activities in the central nervous system & gastrointestinal tract; thermoregulatory effects; Bertaccini, G etal, *Br J Pharmac*, 52:219, 1984; Erspamer, V etal, *Br J Pharmac*, 42:227, 1984; Rivier, JE & MR Brown, *Biochemistry*, 17:1776, 1978

Neosystem SC079 MW 1619.86 Bioactive; Anastasi, A et al, *Experientia*, 27:166, 1971

American Peptide 16-7-10 *Bombina bombina* & *Boombina variegata variegata* MW 1619.9 $C_{71}H_{110}N_{24}O_{18}S$ Originally isolated from the skins of the amphibians *Boombina bombina* & *Boombina variegata variegata*; a potent stimulant of gastric acid secretion & has shown strong biological activity in the central nervous system; Anastasi, A et al, *Experientia*, 27:166, 1971

Peptides International PBM-4086-v *Bombina bombina* (frog) synthetic MW 1619.9 $C_{71}H_{110}N_{24}O_{18}S$ >95% (HPLC); lyophilized amorphous powder | Bioactive; Anastasi, A et al, *Experientia*, 27:166, 1971

Glp-Gln-Arg-Leu-Gly-Asn-Gln-Trp-Ala-Val-Gly-His-Leu-Met Amide Acetate Salt

Synonyms: Gastrointestinal Peptide; Bombesin

Sigma B 4272 FW 1619.9 (free base) ≥97% (HPLC) | Bioactive peptide; tetradecapeptide that has biological activities in the central nervous system & gastrointestinal tract, as well as thermoregulatory effects; Rivier, JE & Brown, MR, *Biochemistry*, 17:1766, 1978; Bertaccini, G et al, *Br J Pharmacol*, 52:219, 1974; Erspamer, V et al, *Br J Pharmacol*, 52:227, 1974

Biogenesis 1400-0604 Pan-species synthetic MW 1619.8 Purified; lyophilized | Brown et al, *Science*, 196:998, 1977

Glp-Gln-Arg-Leu-Gly-Asn-Gln-Trp-Ala-Val-Gly-His-Leu-Met Amide AcOH 7H₂O

Synonyms: Bombesin

Peptides International PBM-4086 *Bombina bombina* (frog) synthetic MW 1806.0 $C_{71}H_{110}N_{24}O_{18}S \cdot CH_3COOH \cdot 7H_2O$ >95% (HPLC); lyophilized amorphous powder; bulk | Bioactive; Anastasi, A et al, *Experientia*, 27:166, 1971

Glp-Gln-Arg-Leu-Gly-Asn-Gln-Trp-Ala-Val-Gly-His-Leu-Met Amide Free Base

Synonyms: Bombesin

Calbiochem 203675   MW 1619.9 (free base) $C_{71}H_{110}N_{24}O_{18}S$ ≥97% (HPLC); white lyophilized solid; soluble in 5% acetic acid; harmful:LD_{50} ≤2000 mg/kg; may be carcinogenic/teratogenic | Tetradecapeptide that activates a cascade of biochemical reactions including Ins(1,4,5)P3-induced mobilization of intracellular Ca^{2+}, Na^+ & K^+ fluxes, protein kinase C activation & cAMP production; potent mitogen of Swiss 3T3 cells; stimulates phospholipase D in Swiss 3T3 fibroblasts; *Merck Index*, 12:1359; Wang, JL et al, *Biochem J*, 320:87, 1996; Briscoe, CP et al, *Biochem J*, 306:115, 1995

Glp-Gln-Arg-Leu-Gly-Asn-Gln-Trp-Ala-Val-Gly-Phe-Leu-Leu
Glp-Gln-Arg-Leu-Gly-Asn-Gln-Trp-Ala-Val-Gly-D-Phe-Leu-Leu Amide

Synonyms: Bombesin, (D-Phe[12],Leu[14])-; Bombesin Receptor Antagonist

American Peptide 16-7-25   MW 1611.9 $C_{75}H_{114}N_{22}O_{18}$ Specific antagonist; Heinz-Erian, P et al, *Am J Physiol*, 252:G439, 1987

Bachem H-7070.0001   MW 1611.87 $C_{75}H_{114}N_{22}O_{18}$ Store at -15°C

Neosystem SC169   MW 1611.86   Bioactive; bombesin receptor antagonist; Heinz-Erian et al, *Am J Physiol*, 252:G439-G442, 1987

Glp-Gln-Arg-Leu-Gly-Asn-Gln-Trp-Ala-Val-Gly-Phe-Leu-Met
Glp-Gln-Arg-Leu-Gly-Asn-Gln-Trp-Ala-Val-Gly-D-Phe-Leu-Met Amide

Synonyms: Bombesin Receptor Antagonist; Bombesin, (D-Phe[12])-; Gastrointestinal Peptide

American Peptide 16-7-24   MW 1629.9 $C_{74}H_{112}N_{22}O_{18}S$ New; because of its specificity, useful for defining the role of bombesin in various physiological or pathological processes; Heinz-Erian, P et al, *Am J Physiol*, 252:G439, 1987

Bachem H-3038.0001   MW 1629.91 $C_{74}H_{112}N_{22}O_{18}S$ Store at -15°C

Neosystem SC168   MW 1629.90   Bioactive; bombesin receptor antagonist; Heinz-Erian et al, *Am J Physiol*, 252:G439-G442, 1987

Sigma B 0775   FW 1629.9   ≥97% (HPLC) | Bioactive peptide; antagonist that interacts selectively with the bombesin receptor; Heinz-Erian, P et al, *Amer J Physiol*, 252:G439, 1987

Glp-Gln-Arg-Tyr-Gly-Asn-Gln-Trp-Ala-Val-Gly-His-Leu-Met Amide

Synonyms: Gastrointestinal Peptide; Bombesin, (Tyr[4])-

American Peptide 16-7-22   MW 1669.9 $C_{74}H_{108}N_{24}O_{19}S$ Rivier, JE et al, *Biochemistry*, 17:1766, 1978

Bachem H-2165.0001   Store at -15°C

Sigma B 5397   FW 1669.9   ≥97% (HPLC) | Bioactive peptide; Rivier, JE & Brown, MT, *Biochemistry*, 17:1766, 1978

Glp-Gln-Arg-Tyr-Gly-Asn-Gln-Trp-Ala-Val-Gly-Phe-Leu-Met
Glp-Gln-Arg-Tyr-Gly-Asn-Gln-Trp-Ala-Val-Gly-D-Phe-Leu-Met Amide

Synonyms: Bombesin Receptor Antagonist; Bombesin, (Tyr[4],D-Phe[12])-; Gastrointestinal Peptide

American Peptide 16-7-26   MW 1679.9 $C_{77}H_{110}N_{22}O_{19}S$ Specific

Bachem H-9065.0001   MW 1679.92 $C_{77}H_{110}N_{22}O_{19}S$ Store at -15°C

Sigma B 0650   FW 1679.9   ≥95% (HPLC) | Bioactive peptide; bombesin receptor antagonist; Heinz-Erian, P et al, *Amer J Physiol*, 252:G439, 1987

Glp-Gln-Arg-Tyr-Thr-Gly-Trp-Met-Asp-Phe
Glp-Gln-Arg-Tyr(SO₃H)-Thr-Gly-Trp-Met-Asp-Phe Amide Sulfated

Synonyms: Gastrointestinal Peptide; Caerulein

Sigma C 9026   FW 1352.4   ≥97% (HPLC) | Bioactive peptide; Bernardi, L et al, *Experientia*, 23:702, 1967

Glp-Gln-Asp-Tyr(SO₃H)-Thr-Gly-Trp-Met-Asp-Phe Amide

Synonyms: Caerulein

ICN 152860   MW 1352.4 $C_{58}H_{73}N_{13}O_{21}S_2$ Sulfated | Anastasi, A et al, *Experientia*, 23:699, 1967; Bernardi, L et al, *Experentia*, 23:702, 1967

Glp-Gln-Asp-Tyr-Thr-Gly-Trp-Met-Asp-Phe
Glp-Gln-Asp-Tyr(SO₃H)-Thr-Gly-Trp-Met-Asp-Phe Amide

Synonyms: Cholecystokinin; Caerulein

American Peptide 46-1-50   MW 1352.4 $C_{58}H_{73}N_{13}O_{21}S_2$ Anastasi, A et al, *Experientia*, 23:699, 1967

Bachem H-3220.0001   MW 1352.42 $C_{58}H_{73}N_{13}O_{21}S_2$ Store at -15°C

Glp-Gln-Asp-Tyr-Thr-Gly-Trp-Met-Asp-Phe Amide

Synonyms: Caerulein

ICN 154541   MW 1272.5   Non-sulfated

Glp-Gln-Asp-Tyr-Thr-Gly-Trp-Met-Asp-Phe Amide Desulfated

Synonyms: Caerulein

Bachem H-2245.0001   MW 1272.36 $C_{58}H_{73}N_{13}O_{18}S$ Store at -15°C

Glp-Gln-Lys-Leu-Gly-Asn-Gln-Trp-Ala-Val-Gly-His-Leu-Met Amide

Synonyms: Gastrointestinal Peptide; Bombesin, (Lys[3])-

American Peptide 16-7-12   MW 1591.9 $C_{71}H_{110}N_{22}O_{18}S$

Bachem H-2160.0001   MW 1591.86 $C_{71}H_{110}N_{22}O_{18}S$ Store at -15°C

Sigma B 1647   FW 1591.8   ≥97% (HPLC) | Bioactive peptide

Glp-Gln-Phe-Phe-Gly-Leu-Met Amide

Synonyms: Substance P (5-11), (Glp[5])-

American Peptide 70-1-54   MW 852.0 $C_{41}H_{57}N_9O_9S$

Bachem H-4875.0005   MW 852.03 $C_{41}H_{57}N_9O_9S$ Store at -15°C

ICN 153027   Yajima, H et al, *Chem Pharm Bull*, 21:2500, 1973; Bergmann, J et al, *Experientia*, 30:401, 1974; Bury, RW & ML Mashford, *J Med Chem*, 19:854, 1976; Blumberg, S et al, *Brain Res*, 192:477, 1980; Kato, T et al, *J Neurochem*, 35:527, 1980; Kato, T etal, *Proc Jpn Acad B*, 56:388, 1980; Brown, CL & MR Hanley, *Br J Pharmac*, 73:517, 1981; Nakata, N etal, *J Neurochem*, 37:1529, 1981; Teichberg, VI etal, *Reg Peptides*, 1:327, 1981

Sigma S 2886   FW 852.0 $C_{41}H_{57}N_9O_9S$ ≥95% (HPLC) | Bioactive peptide; Bury, RW & Mashford, ML, *J Med Chem*, 19:854, 1976

Glp-Gln-Phe-Phe-Sar-Leu-Met
Glp-Gln-Phe-(N-Me)Phe-Sar-Leu-Met Amide

Synonyms: Substance P (5-11), (Glp[5],Me-Phe[8],Sar[9])-

American Peptide 70-1-62   MW 881.1 $C_{43}H_{62}N_9O_9S$ Resistant to the digestion of substance P-degrading enzyme; Sandberg, BE et al, *Eur J Biochem*, 114:329, 1981

Glp-Gln-Phe-Phe-Sar-Leu-Met
Glp-Gln-Phe-*N*-Me-Phe-Sar-Leu-Met Amide

Synonyms: Substance P (5-11), (Glp[5],*N*-Me-Phe[8],Sar[9])-

Bachem H-4880.0001   MW 895.09 $C_{43}H_{61}N_9O_9S$ Store at -15°C

ICN 153028   Substance P analog resistant to attack by a substance P degrading enzyme; Sandberg, BEB etal, *Eur J Biochem*, 114:329, 1981; Eison, AS, *Brain Res*, 238:137, 1982; Eison, AS etal, *Science*, 215:188, 1982

Sigma S 4391   FW 880.1 $C_{43}H_{61}N_9O_9S$ ≥97% (HPLC) | Bioactive peptide; analog which is resistant to attack by a Substance P degrading enzyme; similar to Sigma S 8385 but produced by Sigma Sandberg, BEB et al, *Eur J Biochem*, 114:329, 1981; Eison, AS et al, *Science*, 215:188, 1982; Eison, AS et al, *Brain Res*, 238:137, 1982

Sigma S 8385   FW 880.1 $C_{43}H_{61}N_9O_9S$ ≥97% (HPLC) | Bioactive peptide

Glp-Gln-Phe-Phe-Sar-Leu-Met Amide

Synonyms: Substance P (5-11), (Glp[5],Sar[9])-

American Peptide 70-1-64   MW 866.1 $C_{42}H_{59}N_9O_9S$

Glp-Gln-Pro-Gly-Leu-Trp Amide

Synonyms: Metamorphosin A

Bachem H-3054.0005   MW 709.8 $C_{34}H_{47}N_9O_8$ Store at -15°C

Sigma M 1045   FW 709.8 $C_{34}H_{47}N_9O_8$ ≥97% (HPLC) | Bioactive peptide; novel peptide controlling development of the lower metazoan *hydractinia echinata*; Leitz, T et al, *Developmental Biology*, 163:440, 1994

Glp-Gln-Trp-Ala-Val-Gly-His-Phe-Met Amide

Synonyms: Litorin

American Peptide 62-0-62   MW 1085.3 $C_{51}H_{68}N_{14}O_{11}S$
Anastasi, A et al, *Experientia*, 31:510, 1975

Bachem H-4000.0001   Store at -15°C

ICN 151563

Biogenesis 5688-9009   Synthetic   >95%; no preservatives; lyophilized | Anastasi et al, *Experimentia*, 31:510, 1975

Glp-Gln-Trp-Phe-Trp-Trp-Met
Glp-Gln-D-Trp-Phe-D-Trp-D-Trp-Met Amide

Synonyms: G Protein Antagonist; Protein G Antagonist II; Substance P (5-11), (Glp[5],D-Trp[7,9,10])-

Bachem H-8595.0001   MW 1093.28 $C_{57}H_{64}N_{12}O_9S$ Store at -15°C

ICN 158346   MW 1093.6   99% | Reversible & competitive inhibitor of G proteins; Mukai, H etal, *JBC*, 267:16237, 1992

Sigma S 8031   FW 1093.3   ≥90% (HPLC) | Bioactive peptide; inhibits activation of the G proteins, G_i & G_o by muscarinic receptors & activation of G_s by β-adrenergic receptors; Mukai, H et al, *J Biol Chem*, 267:16237, 1992

Glp-Glu-Arg-Pro-Pro-Leu-Gln-Gln-Pro-Pro-His-Arg-Asp-Lys-Lys-Pro-Cys-Lys-Asn-Phe-Phe-Trp-Lys-Thr-Phe-Ser-Ser-Cys-Lys

Synonyms: Cortistatin 29

American Peptide 34-7-30   MW 3556.1 $C_{161}H_{242}N_{47}O_{41}S_2$
Disulfide bonds: Cys[17]-Cys[28]; de Leccea, L et al, *Nature*, 381:242, 1996

Glp-Glu-Asp-Ser-Gly

Synonyms: Epidermal Mitosis Inhibiting Pentapeptide

Bachem H-6770.0005   MW 517.45 $C_{19}H_{27}N_5O_{12}$ Store at -15°C

Sigma E 1638   FW 517.4 $C_{19}H_{27}N_5O_{12}$ ≥97% (HPLC) | Bioactive peptide; Reichelt, KL, *Biochem Biophys Res Commun*, 146:1493, 1987

Glp-Glu-Glu-Glu-Glu-Glu-Thr-Ala-Gly-Ala-Pro-Gln-Gly-Leu-Phe-Arg-Gly Amide

Synonyms: Pancreastatin (33-49), (Glp[33])-

Peptides International PPP-4186-v   Porcine synthetic   MW 1829.9 $C_{77}H_{116}N_{22}O_{30}$ >99% (HPLC); lyophilized amorphous powder | Bioactive; Tatemoto, K et al, *Nature*, 324:476, 1986

Glp-Glu-Pro Amide

Synonyms: Thyrotropin Releasing Hormone, (Glu[2])-; FPP; Thyrotropin Releasing Hormone Related Peptide; Thyrotropin Releasing Hormone Analog; Fertilization Promoting Peptide

American Peptide 79-2-25   MW 355.4 $C_{15}H_{23}N_4O_6$ TRH-related peptide originally isolated from rabbit; Cockle, SM et al, *JBC*, 264:7788, 1989

Bachem H-2464.0050   MW 354.36 $C_{15}H_{22}N_4O_6$ Store at -15°C

ICN 158452   MW 354.4 $C_{15}H_{22}N_4O_6$ 97% | Cockle, SM etal, *JBC*, 264:7788, 1989

Sigma P 2053   FW 354.4 $C_{15}H_{22}N_4O_6$ ≥97% (HPLC) | Bioactive peptide; originally isolated from rabbit prostate & human semen; Cockle, SM et al, *J Biol Chem*, 264:7788, 1989

Glp-Gly

Bachem G-3695.0250   MW 186.17 $C_7H_{10}N_2O_4$ Store at -15°C

Glp-Gly-Arg
Glp-Gly-Arg-pNA Acetate Salt

ICN 156123

Glp-Gly-Arg
Z-Glp-Gly-Arg-MCA Hydrochloride

Synonyms: Factor Xa Substrate

Peptides International MPR-3138-v   MW 633.66 $C_{31}H_{35}N_7O_8$ >98% (HPLC); lyophilized amorphous powder | Kawabata, S et al, *Eur J Biochem*, 172:17, 1988

Glp-Gly-Arg
Glp-Gly-Arg-MCA

Synonyms: TPA Substrate; Urokinase Substrate

Peptides International MPR-3145-v   MW 499.53 $C_{23}H_{29}N_7O_6$ >98% (HPLC); lyophilized amorphous powder

Glp-Gly-Arg
Glp-Gly-Arg *p*-NA Hydrochloride

Sigma P 4080   FW 498.9 $C_{19}H_{26}N_8O_6 \cdot HCl$

Glp-Gly-Arg-Leu-Gly-Thr-Gln-Trp-Ala-Val-Gly-His-Leu-Met Amide

Synonyms: Gastrointestinal Peptide; Alytesin; Bombesin, (Gly[2],Thr[6])-

American Peptide 62-0-66   MW 1535.8 $C_{68}H_{106}N_{22}O_{17}S$
Biological activity similar to bombesin; Anastasi, A et al, *Experientia*, 27:166, 1971; Melchiorri, P, *Gut Hormone*, Churchill-Livingston, Edinburgh, 534:1978

Bachem H-6715.0001   MW 1535.79 $C_{68}H_{106}N_{22}O_{17}S$ Store at -15°C

Sigma A 2803   FW 1535.8   ≥95% (HPLC) | Bioactive peptide; bombesin agonist originally isolated from the skin of the European amphibian, Alytes obstetricans; biological activity very similar to bombesin; Anastasi, A et al, *Experientia*, 27:166, 1971

Glp-Gly-Arg-Phe Amide

Synonyms: Antho-RF Amide

Bachem H-6035.0025   MW 488.55 $C_{22}H_{32}N_8O_5$ Store at -15°C

ICN 153102   A neuropeptide originally isolated from the pennatulid *Renillakoellikeri* & the sea anemone *Anthopleura elegantissima*; Grimmelikhuijzen, CJP & D Graff, *PNAS*, 83:9817, 1986; Grimmelikhuijzen, CJP & Gregor, A, *FEBS Lett*, 211:105, 1987

Glp-Gly-Arg-Phe Amide Acetate Salt

Synonyms: Antho-RF Amide Neuropeptide

Sigma P 9799   FW 488.5 (free base) $C_{22}H_{32}N_8O_5$ ≥97% (HPLC) | Bioactive peptide; originally isolated from the pennatulid *Renillakoellikeri* & the sea anemone *Anthopleura elegantissima*; may be a neurotransmitter at neuromuscular junctions; Grimmelikhuijzen, CJP & Graff, D, *Proc Natl Acad Sci USA*, 83:9817, 1986; Grimmelikhuijzen, CJP & Gregor, A, *FEBS Lett*, 211:105, 1987

Glp-Gly-Gly-Leu-Arg-Trp Amide

Bachem H-6490.0025   Store at -15°C

Glp-Gly-Leu-Arg-Trp Amide

Synonyms: Antho-RW Amide II

Bachem H-6485.0025   MW 640.74 $C_{30}H_{44}N_{10}O_6$ Store at -15°C

Glp-Gly-Leu-Pro-Pro-Arg-Pro-Lys-Ile-Pro-Pro

Synonyms: Bradykinin Potentiator B; Bradykinin Destroying Plasma Kinase Inhibitor; Bradykinin; Peptidyl-Dipeptidase A Inhibitor; Kininase II Inhibitor; Angiotensin I Converting Enzyme Inhibitor; Peptidyl-Dipeptidase A, Kininase II Inhibitor

American Peptide 12-1-81   MW 1182.4 $C_{56}H_{91}N_{15}O_{13}$ Kato, H et al, *Biochemistry*, 10:972, 1971

Bachem H-2205.0001   MW 1182.43 $C_{56}H_{91}N_{15}O_{13}$ Store at -15°C

ICN 152796   MW 1182.7   Kato, H & T Suzuki, *Biochemistry*, 10:972, 1971

Sigma B 0507   FW 1182.4   ≥97% (HPLC) | Bioactive peptide; inhibitor of angiotensin-converting enzyme & bradykinin-destroying plasmakinases; Kato, H & Suzuki, T, *Biochemistry*, 10:972, 1971

Peptides International IAB-4009-v   *Agkistrodon halys blomhoffii* (Mamushi) synthetic   MW 1182.4 $C_{56}H_{91}N_{15}O_{13}$ >99% (HPLC); lyophilized amorphous powder | Inhibitor; Kato, H & T Sazuki, *Biochem*, 10:972, 1971

Peptides InternationalIAB-4009-v   *Agkistrodon halys blomhoffii* (Mamushi) synthetic   MW 1182.4 $C_{56}H_{91}N_{15}O_{13}$ >99% (HPLC); lyophilized amorphous powder | Bioactive; Kato, H & T Suzuki, *Biochem*, 10:972, 1971

Glp-Gly-Leu-Pro-Pro-Arg-Pro-Lys-Ile-Pro-Pro AcOH 4H₂O

Synonyms: Bradykinin Potentiator B; Peptidyl-Dipeptidase A, Kininase II Inhibitor; Angiotensin I Converting Enzyme Inhibitor

Peptides InternationalIAB-4009   *Agkistrodon halys blomhoffii* (Mamushi) synthetic   MW 1314.5 $C_{56}H_{91}N_{15}O_{13} \cdot CH_3COOH \cdot 4H_2O$ >99% (HPLC); lyophilized amorphous powder; bulk | Bioactive; Kato, H & T Suzuki, *Biochem*, 10:972, 1971

Glp-Gly-Leu-Pro-Pro-Gly-Pro-Pro-Ile-Pro-Pro

Synonyms: Angiotensin I Converting Enzyme Inhibitor; Bradykinin Potentiator C; Bradykinin Destroying Plasma Kinase Inhibitor; Bradykinin; Peptidyl-Dipeptidase A Inhibitor; Kininase II Inhibitor; Peptidyl-Dipeptidase A, Kininase II Inhibitor

American Peptide 12-1-84   MW 1052.3 $C_{51}H_{77}N_{11}O_{13}$ Kato, H et al, *Biochemistry*, 10:972, 1971

Bachem H-2210.0001   MW 1052.24 $C_{51}H_{77}N_{11}O_{13}$ Store at -15°C

ICN 152797   MW 1052.4   Kato, H & T Suzuki, *Biochemistry*, 10:972, 1971

Sigma B 0632   FW 1052.2   ≥97% (HPLC) | Bioactive peptide; inhibitor of angiotensin-converting enzyme & bradykinin-destroying plasmakinases; Kato, H & Suzuki, T, *Biochemistry*, 10:972, 1971

Peptides International IAC-4010-v   *Agkistrodon halys blomhoffii* (Mamushi) synthetic   MW 1052.2 $C_{51}H_{77}N_{11}O_{13}$ >99% (HPLC); lyophilized amorphous powder | Inhibitor; Kato, H & T Sazuki, *Biochem*, 10:972, 1971

Peptides InternationalIAC-4010-v   *Agkistrodon halys blomhoffii* (Mamushi) synthetic   MW 1106.2 $C_{51}H_{77}N_{11}O_{13} \cdot 3H_2O$ >99% (HPLC); lyophilized amorphous powder | Bioactive; Kato, H & T Suzuki, *Biochem*, 10:972, 1971

Glp-Gly-Leu-Pro-Pro-Gly-Pro-Pro-Ile-Pro-Pro 3H₂O

Synonyms: Bradykinin Potentiator C; Peptidyl-Dipeptidase A, Kininase II Inhibitor; Angiotensin I Converting Enzyme Inhibitor

Peptides InternationalIAC-4010   *Agkistrodon halys blomhoffii* (Mamushi) synthetic   MW 1106.2 $C_{51}H_{77}N_{11}O_{13} \cdot 3H_2O$ >99% (HPLC); lyophilized amorphous powder; bulk | Bioactive; Kato, H & T Suzuki, *Biochem*, 10:972, 1971

Glp-Gly-Lys-Arg-Pro-Trp-Ile-Leu

Synonyms: Xenopsin (XP); Xenopsin 25; Xenopsin

Bachem H-4885.0001   MW 980.18 $C_{47}H_{73}N_{13}O_{10}$ Store at -15°C

ICN 154461   MW 980.3   Araki, K etal, *Chem Pharm Bull*, 21:2801, 1973

Neosystem SC073   MW 980.18   Bioactive; neurotensin-related peptide; natural analog of neurotensin (8-13)

Sigma X 6375   FW 980.2 $C_{47}H_{73}N_{13}O_{10}$ ≥97% (HPLC) | Biologically active octapeptide with potent contractile activity on rat stomach strips *in vitro*; Araki, K et al, *Chem Pharm Bull*, 21:2801, 1973; ibid, 23:3132, 1975

American Peptide 62-0-64   *Xenopus laevis* (South African frog) skin extracts   MW 980.2 $C_{47}H_{73}N_{13}O_{10}$ Sequence homology to mammalian neurotensin & shares a number of its biological properties; Araki, K et al, *Chem Pharm Bull*, 21:2801, 1973; Araki, K et al, *Chem Pharm Bull*, 23:3132, 1975

Glp-Gly-Met-Ile-Gly-Thr-Leu-Thr-Ser-Lys-Arg-Ile-Lys-Gln Amide

Synonyms: Levitide

Bachem H-9250.0001   MW 1542.87 $C_{66}H_{119}N_{21}O_{19}S$ Store at -15°C

ICN 154445   MW 1543.1   Poulter, L etal, *JBC*, 263:3279, 1988

American Peptide 62-7-15   *Xenopus laevis*   MW 1543.0 $C_{66}H_{119}N_{21}O_{19}S$ Neurohormone-like peptide found in the skin of *Xenopus laevis*; Poulter et al, *JBC*, 263:3279, 1988

Sigma L 5522   *Xenopus laevis*   FW 1542.9   ≥97% (HPLC) | Bioactive peptide; neurohormone-like peptide found in the skin of *Xenopus laevis*; Poulter, L et al, *J Bio Chem*, 263:3279, 1988

Glp-Gly-Pro-Pro-Ile-Ser-Ile-Asp-Leu-Ser-Leu-Glu-Leu-Leu-Arg-Lys-Met-Ile-Glu-Ile-Glu-Lys-Gln-Glu-Lys-Glu-Lys-Gln-Gln-Ala-Ala-Asn-Asn-Arg-Leu-Leu-Leu-Asp-Thr-Ile Amide

Synonyms: Sauvagine; Corticotropin Releasing Factor

Bachem H-4890.0500   Store at -15°C

ICN 153176   Causes release of ACTH & endorphins; Montecucchi, PC, Hoppe Seyler's *Z Physiol Chem*, 360:1178, 1979; Erspamer, V & P Melchiorri, *Trends Pharmac Sci*, 20:391, 1980

Sigma S 3884   *Phyllomedusa sauvagei*   FW 4599.4   ≥97% (HPLC) | Bioactive peptide; originally isolated from the skin of the frog, Phyllomedusa *sauvagei*; affects diuresis & the cardiovascular system; reported to cause release of ACTH & endorphins; Montecucchi, PC et al, *Hoppe-Seyler's Z Physiol Chem*, 360:1178, 1979; Erspamer, V & Melchiorri, P, *Trends Pharmacol Sci*, 20:391, 1980

American Peptide 34-7-11 *Phyllomedusa sauvagei* (frog) skin MW 4599.4 $C_{202}H_{346}N_{56}O_{63}S$ Presents a number of pharmacological actions on diuresis, the cardiovascular system & the endocrine glands; Montecucchi, PC et al, *Int J Pept Protein Res*, 18:113, 1981

Glp-Gly-Pro-Pro-Ile-Ser-Ile-Asp-Leu-Ser-Leu-Glu-Leu-Leu-Arg-Lys-Met-Ile-Glu-Ile-Glu-Lys-Gln-Glu-Lys-Glu-Lys-Gln-Gln-Ala-Ala-Asn-Arg-Leu-Leu-Leu-Asp-Thr-Ile Amide

Synonyms: Sauvagine

Alexis 167-009 *Phyllomedusa sauvagei* (neotropical frog) skin, synthetic MW 4599.4 $C_{202}H_{346}N_{56}O_{63}S$ ≥98% (HPLC); lyophilized | Montecucchi, PC & Henschen, A, *Int J Pept Prot Res*, 18:113, 1981; Vale, W et al, *Science*, 213:1394, 1981; Melchiorri, P & Negri, L, *Regul Pept*, 2:1, 1981

Glp-Gly-Pro-Pro-Ile-Ser-Ile-Asp-Leu-Ser-Leu-Glu-Leu-Leu-Arg-Lys-Met-Ile-Glu-Ile-Glu-Lys-Gln-Glu-Lys-Glu-Lys-Gln-Gln-Ala-Ala-Asn-Asn-Arg-Leu-Leu-Leu-Asp-Thr-Ile Amide

Synonyms: Sauvagine

Neosystem SC1214 *Phyllomedusa sauvagei* skin (Central/South American frog) MW 4599.35 Bioactive; hypothalamic releasing hormone; CRF/urocortin peptide; binds to & activates CRF1 receptors with higher affinity & equal potency than CRF; although CRF is significantly weaker at the CRF2 receptor, sauvagine retains its high affinity interactions with this receptor subtype; Montecucchi, PC & Henschen, A, *Int J Pept Prot Res*, 18:113-120, 1981; Grigoriadis, DE et al, *Mol Pharmacology*, 50:679-686, 1996

Glp-Gly-Pro-Trp-Leu-Glu

Synonyms: Gastrin I (1-6)

Biogenesis 4620-0129 Human synthetic MW 711 Lyophilized, consisting of 77% peptide material, acetate counter ions and residual H_2O; purified

---

Glp-Gly-Pro-Trp-Leu-Glu-Glu-Glu-Glu-Glu-Ala-Tyr-Gly

Synonyms: Gastrointestinal Peptide; Gastrin I (1-13)

ICN 152874 Human MW 1519.5

Sigma G 6261 Human FW 1519.5 ≥75% (HPLC) | Bioactive peptide

Glp-Gly-Pro-Trp-Leu-Glu-Glu-Glu-Glu-Glu-Ala-Tyr-Gly-Trp

Synonyms: Gastrointestinal Peptide; Gastrin I (1-14)

Bachem H-3095.0001 Human Store at -15°C

ICN 152875 Human

Sigma G 6386 Human FW 1705.8 ≥90% (HPLC) | Bioactive peptide

Glp-Gly-Pro-Trp-Leu-Glu-Glu-Glu-Glu-Glu-Ala-Tyr-Gly-Trp-Leu-Asp-Phe Amide

Synonyms: Gastrointestinal Peptide; Gastrin I, (Leu[15])-

Bachem H-3090.0001 Human MW 2080.19 $C_{98}H_{126}N_{20}O_{31}$ Store at -15°C

ICN 152872 Human Useful for biological & immunological studies; potent stimulator of gastric secretion; Walsh, JH & MI Grossman, *New Engl J Med*, 292:1324, 1975

Sigma G 9145 Human FW 2080.2 ≥97% (HPLC) | Bioactive peptide

Glp-Gly-Pro-Trp-Leu-Glu-Glu-Glu-Glu-Glu-Ala-Tyr-Gly-Trp-Met-Asp-Phe Amide

Synonyms: Gastrin I

American Peptide 46-1-58 Human MW 2098.3 $C_{97}H_{124}N_{20}O_{31}S$ This endogenous peptide mainly found in the gut can activate gastric acid secretion, delay gastric emptying & increase blood flow in the gastric mucosa; Gregory, RA et al, *Nature*, 209:583, 1966

Bachem H-3085.0001 Human MW 2098.23 $C_{97}H_{124}N_{20}O_{31}S$ Store at -15°C

Glp-Gly-Pro-Trp-Leu-Glu-Glu-Glu-Glu-Glu-Ala-Tyr-Gly-Trp-Met-Asp-Phe Glp-Gly-Pro-Trp-Leu-Glu-Glu-Glu-Glu-Glu-Ala-Tyr(SO₃H)-Gly-Trp-Met-Asp-Phe Amide Sulfated

Synonyms: Gastrin I; Gastrin II

Bachem H-9170.0500 Human Store at -15°C

Glp-Gly-Pro-Trp-Leu-Glu-Glu-Glu-Glu-Glu-Ala-Tyr-Gly-Trp-Met-Asp-Phe Amide

Synonyms: Gastrin I; HG-17; Gastrin I, Heptadecapeptide; Gastrin I, Little; Gastrin I (18-34), (Glp[18])-Big; Gastrin I

Calbiochem 05-23-2301 Human MW 2098.2 $C_{97}H_{124}N_{20}O_{31}S$ ≥90% (HPLC); white solid; soluble in 5% acetic acid | Peptide hormone produced in the stomach; responsible for stimulation of gastric acid secretion; delays gastric emptying, increases blood flow in the gastric mucosa; *Merck Index*, 12:4383; Walsh, JH, *Fed Proc*, 36:1948, 1977

Fluka 53673 Human MW 2098.23 $C_{97}H_{124}N_{20}O_{31}S$ ~95% (HPLC) | Bertaccini, G, *Handb Exp Pharmacol*, 59:11, 1982; Dockray, GJ, *Hormones in Blood* (CH Gray, VHT James, eds), Vol 2 3rd Ed:357, 1979, Academic Press London; Walsh, JH, *Federation Proc*, 36:1948, 1977

ICN 152871 Human MW 2098.2 Gregory, RA etal, *Nature*, 209:583, 1966; Walsh, JH, *Fed Proc*, 36:1948, 1977

Glp-Gly-Pro-Trp-Leu-Glu-Glu-Glu-Glu-Glu-Ala-Tyr-Gly-Trp-Met-Asp-Phe Glp-Gly-Pro-Trp-Leu-Glu-Glu-Glu-Glu-Glu-Ala-Tyr(SO₃H)-Gly-Trp-Met-Asp-Phe Sulfated Amide

Synonyms: Gastrin II; Gastrin I

ICN 152877 Human

Glp-Gly-Pro-Trp-Leu-Glu-Glu-Glu-Glu-Glu-Ala-Tyr-Gly-Trp-Met-Asp-Phe Amide

Synonyms: Gastrin I

Neosystem SC102 Human MW 2098.22 Bioactive; brain/gut peptide; CCK/gastrin peptide; Gregory, H et al, *Nature*, 204:931, 1964

Glp-Gly-Pro-Trp-Leu-Glu-Glu-Glu-Glu-Glu-Ala-Tyr-Gly-Trp-Met-Asp-Phe-Gly

Synonyms: Gastrin-Gly (1-17); Gastrin, Glycine-Extended

Neosystem SC953 Human MW 2156.26 Bioactive; brain/gut peptide; CCK/gastrin peptide; Gly-extended gastrin is the immediate precursor of the bioactive carboxyamidated gastrin; only amidated gastrin exerts a physiological effect that results in gastric acid secretion; both amidated gastrin & Gly-extended gastrin exert growth promoting effects on AR4-2J cells (derived from rat pancreas) via interaction with distinct receptors; Singh, P et al, *Am J Physiol*, 266:G459-G468, 1994; Seva, C et al, *Science*, 265:410-412, 1994; Singh, P et al, *JBC*, 270:8429-8438, 1995; Varro, A et al, *J Clin Invest*, 95:1642-1649, 1995; Kaise, M et al, *JBC*, 270:11155-11160, 1995; Todisco, A et al, *JBC*, 270:28337-28341, 1995; Hansen, CP et al, *Digestion*, 57:22-29, 1996

Glp-Gly-Pro-Trp-Leu-Glu-Glu-Glu-Glu-Glu-Ala-Tyr-Gly-Trp-Met-Asp-Phe Amide

Synonyms: Gastrointestinal Peptide; Gastrin I

Sigma G 9020 Human FW 2098.2 ≥97% (HPLC) | Bioactive peptide; produced in the stomach; stimulates gastric acid secretion; Gregory, RA et al, *Nature*, 209:583, 1966; Walsh, JH, *Fed Proc*, 36:1948, 1977

Glp-Gly-Pro-Trp-Leu-Glu-Glu-Glu-Glu-Glu-Ala-Tyr-Gly-Trp-Met-Asp-Phe Ammonium Salt

Synonyms: Gastrin I (1-17)

Biogenesis 4620-0504 Human synthetic MW 2098.5 Purified; lyophilized | Gregory et al, *Nature*, 209:583, 1966

Glp-Gly-Pro-Trp-Leu-Glu-Glu-Glu-Glu-Glu-Ala-Tyr-Gly-Trp-Met-Asp-Phe
Glp-Gly-Pro-Trp-Leu-Glu-Glu-Glu-Glu-Glu-Ala-Tyr(SO₃H)-Gly-Trp-Met-Asp-Phe

Synonyms: Gastrin I; G17

Biogenesis 4620-0604 Human synthetic MW 2178.5 Lyophilized, consisting of 88.78% peptide material, TFA counter ions and residual H_2O; purified | To avoid Met oxidation, it is best to dissolve the peptide under a nitrogen stream; used to coat microtitre plate wells in ELISA & for blocking studies in IHC

Glp-Gly-Pro-Trp-Leu-Glu-Glu-Glu-Glu-Glu-Ala-Tyr-Gly-Trp-Met-Asp-Phe Amide

Synonyms: Gastrin I

Peptides International PGR-4143-v Human synthetic MW 2098.2 $C_{97}H_{124}N_{20}O_{31}S$ >95% (HPLC); lyophilized amorphous powder; ammonium form | Bioactive; Gregory, H et al, *Nature*, 204:931, 1964; Anderson, JC et al, *Nature*, 204:933, 1964

Glp-Gly-Pro-Trp-Leu-Glu-Glu-Glu-Glu-Glu-Ala-Tyr-Gly-Trp-Mox-Asp-Phe Amide

Synonyms: Gastrin I, (Mox¹⁵)-

Fluka 64683 Human MW 2082.2 (free peptide) $C_{97}H_{124}N_{20}O_{32}$ Bertaccini, G, *Handb Exp Pharmacol*, 59:11, 1982; Dockray, GJ, *Hormones in Blood* (CH Gray, VHT James, eds), Vol 2 3rd Ed:357, 1979, Academic Press London; Walsh, JH, *Federation Proc*, 36:1948, 1977

Glp-Gly-Val-Asn-Asp-Asn-Glu-Glu-Gly-Phe-Phe-Ser-Ala-Arg

Synonyms: Fibrinopeptide B

Bachem H-2947.0001 MW 1552.58 $C_{66}H_{93}N_{19}O_{25}$ Store at -15°C

Bachem H-2947.0005 MW 1552.58 $C_{66}H_{93}N_{19}O_{25}$ Store at -15°C

American Peptide 42-1-21 Human MW 1552.6 $C_{66}H_{93}N_{19}O_{25}$ Released from fibrinogen on thrombin action; release of fibrinopeptide B from fibrinogen is necessary for Ca^{2+} uptake & subsequently bind to fibrin

ICN 151125 Human MW 1552.6

Sigma F 3379 Human FW 1552.6 ≥97% (HPLC) | Bioactive peptide

Biogenesis 4450-6302 Human synthetic Purified; n/a; lyophilized

Glp-Gly-Val-Asn-Asp-Asn-Glu-Glu-Gly-Phe-Phe-Ser-Ala-Arg-Tyr

Synonyms: Fibrinopeptide B, (Tyr¹⁵)-

Bachem H-2955.0001 MW 1715.75 $C_{75}H_{102}N_{20}O_{27}$ Store at -15°C

Sigma F 7767 FW 1715.7 ≥97% (HPLC) | Bioactive peptide

American Peptide 42-1-27 Human MW 1715.8 $C_{75}H_{102}N_{20}O_{27}$

ICN 153099 Human MW 1715.7

Glp-His

Synonyms: Thyrotropin Releasing Hormone (1-2)

Biogenesis 8940-1039 Synthetic MW 266.11 Lyophilized, consisting of 94.9% peptide material, acetate counter ions and residual H_2O

Glp-His Trifluoroacetate Salt Free Acid

Synonyms: Luteinizing Hormone Releasing Hormone (1-2)

Bachem G-4700.0250 MW 380.28 $C_{11}H_{14}N_4O_4 \cdot C_2HF_3O_2$ Store at -15°C

Glp-His-Gly

Synonyms: Anorexigenic Peptide

American Peptide 87-0-10 MW 323.4 $C_{13}H_{17}N_5O_5$ Trygstad, O et al, *Acta Endocrinol*, 89:196, 1978

Bachem H-4905.0025 MW 323.31 $C_{13}H_{17}N_5O_5$ Store at -15°C

ICN 153047 Suppresses the cephalic phase rise in serum insulin & gastrin levels; Trygstad, O etal, *Acta Endocrin*, 89:176, 1978; Coy, D etal, *J Physiol*, 314:225, 1981; Harding, DRK etal, *Int J Pept Prot Res*, 18:214, 1981; Schally, AV etal, *Proc Soc Exp Biol Med*, 170:264, 1982

Glp-His-Gly Acetate Salt

Sigma P 4147 FW 323.3 (free base $C_{13}H_{17}N_5O_5$ ≥97% (HPLC) | Bioactive peptide; reported to suppress the cephalic phase rise in serum insulin & gastrin levels; Coy, D et al, *J Physiol*, 314:225, 1981; Schally, AV et al, *Proc Soc Exp Biol Med*, 170:264, 1982

Glp-His-Gly Amide Trifluoroacetate Salt

Sigma P 4022 FW 322.3 (free base) $C_{13}H_{18}N_6O_4$ ≥97% (HPLC) | Bioactive peptide

Glp-His-Gly-Thr-Ala-Pro-Glu-Cys-Phe-Trp-Lys-Tyr-Cys-Ile

Synonyms: Prepro-Urotensin II (110-123), (Glp¹¹⁰)-

Bachem H-4892.0500 Rat MW 1663.9 $C_{77}H_{102}N_{18}O_{20}S_2$ Store at -15°C

Peptides International PUT-4371-v Rat synthetic MW 1663.9 $C_{77}H_{102}N_{18}O_{20}S_2$ >99% (HPLC); lyophilized amorphous powder | Bioactive; Disulfide bonds: Cys⁸-Cys¹³; potent vasoconstrictor; Coulouarn, Y et al, *FEBS Letts*, 457:28, 1999

Glp-His-Pro

Synonyms: Thyrotropin Releasing Hormone

American Peptide 52-0-81 MW 363.4 (free acid) $C_{16}H_{21}N_5O_5$

Glp-His-Pro
Glp-3-Me-His-Pro Amide

Synonyms: Thyrotropin Releasing Hormone, (3-Me-His²)-

American Peptide 58-0-85 MW 376.4 $C_{17}H_{24}N_6O_4$

Glp-His-Pro
Glp-His-3,4-Dehydro-Pro Amide

Synonyms: Thyrotropin Releasing Hormone, (3,4-Dehydro-Pro-NH23)-

Bachem H-4900.0005 Store at -15°C

Glp-His-Pro
Glp-His(1-Me)-Pro Amide

Synonyms: Thyrotropin Releasing Hormone, (His(1-Me)²)-; Thyrotropin Releasing Hormone, (His(τ-Me)²)-

Bachem H-7665.0005 Store at -15°C

Glp-His-Pro
Glp-His-Pro-AMC

Synonyms: Thyrotropin Releasing Hormone-AMC

Bachem I-1440.0050   MW 520.6   Store at -15°C

Glp-His-Pro
Glp-His-Pro-4MβNA

Synonyms: Thyrotropin Releasing Hormone-4MβNA

Bachem J-1380.0050   Store at -15°C

Glp-His-Pro
Glp-His-Pro-βNA

Synonyms: Thyrotropin Releasing Hormone-βNA

Bachem K-1490.0050   MW 488.55 $C_{26}H_{28}N_6O_4$ Store at -15°C

Glp-His-Pro
L-Glp-L-His-L-Pro Amide

Synonyms: Thyroliberin; Thyrotropin Releasing Hormone

Fluka 83177   MW 362.4 $C_{16}H_{22}N_6O_4$ ≥98.0% (HPLC); ≥85% peptide content; ≤3% water | Schally, AV et al, *Ann Rev Biochem*, 47:89, 1978; Kolesnick, RN, *JBC*, 264:11688, 1989

Glp-His-Pro

Synonyms: Thyrotropin Releasing Hormone

ICN 153172   MW 363.4 (free acid) $C_{16}H_{21}N_5O_5$

Glp-His-Pro
Glp-3-Me-His-Pro Amide

Synonyms: Thyrotropin Releasing Hormone, (3-Me-His[2])-

ICN 153174   MW 376.4 $C_{17}H_{24}N_6O_4$

Glp-His-Pro

Synonyms: Thyrotropin Releasing Hormone

Neosystem SC044   MW 363.37   Bioactive; thyrotropin releasing hormone-related peptide

Glp-His-Pro
Glp-3-Me-His-Pro Amide

Synonyms: Thyrotropin Releasing Hormone, (3-Me-His[2])-

Sigma P 5173   FW 376.4 $C_{17}H_{24}N_6O_4$ ≥97% (HPLC) | Bioactive peptide; analog that is ten-times more potent than TRH in releasing TSH from the anterior pituitary

Glp-His-Pro

Synonyms: Thyrotropin Releasing Hormone

Biogenesis 8940-1019   Synthetic   MW 363.4   Lyophilized, consisting of 65.0% peptide material, acetate counter ions and residual H_2O

Glp-His-Pro Amide

Synonyms: Thyrotropin Releasing Hormone; Thyrotropin Releasing Factor

American Peptide 52-0-80   MW 362.4 $C_{16}H_{22}N_6O_4$ Involved in the regulation of the hypothalamic pituitary-thyroid axis; stimulates the synthesis & release of thyrotropin & prolactin; Bowers, CY et al, *JMC*, 14:477, 1971

Bachem H-4915.0050   MW 362.39 $C_{16}H_{22}N_6O_4$ Store at -15°C

Calbiochem 609401   MW 362.4 $C_{16}H_{22}N_6O_4$ ≥98% (HPLC); white solid; soluble in DMSO & methanol; may be carcinogenic/teratogenic | Hypothalamic neurohormone widely distributed throughout the CNS where it functions as a transmitter; Stimulates the release of thyrotropin & prolactin from the anterior pituitary via the hypophyseal portal system; Actions of TRF are mediated by specific receptors in thyrotrophs & in various regions of the CNS; Shown to mobilize intracellular Ca^{2+} by increasing the production of inositol trisphosphate; *Merck Index*, 12:9720; Cui, ZJ et al, *Endocrinology*, 134:2245, 1994; Berwaer, M et al, *Mol Cell Endocrinol*, 92:1, 1993

ICN 153173   MW 362.4 $C_{16}H_{22}N_6O_4$ Bowers, CY etal, *J Med Chem*, 14:477, 1971

Neosystem SC042   MW 362.39   Bioactive; thyrotropin releasing hormone-related peptide; Guillemin, R, *Science*, 202:390-402, 1978; Schally, AV, *Science*, 202:18-28, 1978

Peptides International PTR-4011-v   Synthetic   MW 362.39 $C_{16}H_{22}N_6O_4$ >99% (HPLC); lyophilized amorphous powder | Bioactive; human, ovine, porcine, rat; Burgus, R et al, *CR Acad Sci Paris*, 269:1870, 1969; Boler, J et al, *BBRC*, 37:705, 1969

Glp-His-Pro Amide Acetate Salt

Synonyms: Thyrotropin Releasing Hormone

Biogenesis 8940-1004   Human synthetic   MW 362   purified; lyophilized

Glp-His-Pro Amide Hydrate

Synonyms: Thyrotropin Releasing Hormone

Peptides International PTR-4011   Synthetic   MW 380.41 $C_{16}H_{22}N_6O_4 \cdot H_2O$ >99% (HPLC); lyophilized amorphous powder | Bioactive; human, ovine, porcine, rat; Burgus, R et al, *CR Acad Sci Paris*, 269:1870, 1969; Boler, J et al, *BBRC*, 37:705, 1969

Glp-His-Pro Free Acid

Synonyms: Thyrotropin Releasing Hormone

Bachem H-4910.0025   MW 363.37 $C_{16}H_{21}N_5O_5$ Store at -15°C

Sigma P 3905   FW 363.4 $C_{16}H_{21}N_5O_5$ ≥97% (HPLC) | Bioactive peptide

Glp-His-Pro Free Base Amide

Synonyms: Thyrotropin Releasing Hormone

Sigma P 1319   FW 362.4 $C_{16}H_{22}N_6O_4$ ≥98% (HPLC) | Bioactive peptide; Boler, J, *Biochem Biophys Res Comm*, 37:705, 1969; Bowers, CY et al, *J Med Chem*, 14:477, 1971

Glp-His-Pro Free Base Amide Acetate Salt

Synonyms: Thyrotropin Releasing Hormone

Sigma P 2161   FW 362.4 (free base) $C_{16}H_{22}N_6O_4$ ≥97% (HPLC) | Bioactive peptide; Boler, J, *Biochem Biophys Res Comm*, 37:705, 1969; Bowers, CY et al, *J Med Chem*, 14:477, 1971

Glp-His-Pro Free Base Amide Trifluoroacetate Salt

Synonyms: Thyrotropin Releasing Hormone

Sigma P 3588   FW 362.4 (free base) $C_{16}H_{22}N_6O_4$ ≥98% (HPLC) | Bioactive peptide; Boler, J, *Biochem Biophys Res Comm*, 37:705, 1969; Bowers, CY et al, *J Med Chem*, 14:477, 1971

Glp-His-Pro-Gly

Synonyms: Thyrotropin Releasing Hormone-Gly

American Peptide 58-0-98   MW 420.4 $C_{18}H_{24}N_6O_6$

Bachem H-1036.0025   MW 420.43 $C_{18}H_{24}N_6O_6$ Store at -15°C

ICN 158453   MW 420.4 $C_{18}H_{24}N_6O_6$ 97% | Mori, M etal, *Neuropeptides*, 16:57, 1990

Sigma P 2687   FW 420.4 $C_{18}H_{24}N_6O_6$ ≥97% (HPLC); peptide content:~70% | Bioactive peptide; direct precursor of TRH; found in significant concentrations in hypothalamus, pituitary, thyroid & blood of rats & in human CSF; Mori, M et al, *Neuropeptides*, 16:57, 1990

Glp-His-Pro-Gly Amide

American Peptide 58-0-84 MW 419.5 $C_{18}H_{25}N_7O_5$

Neosystem SC043 MW 419.44 Bioactive; thyrotropin releasing hormone-related peptide

Glp-His-Pro-Gly-Lys

American Peptide 58-0-97 MW 548.6 $C_{24}H_{36}N_8O_7$

Glp-His-Trp-Ser Free Acid

Synonyms: Luteinizing Hormone Releasing Hormone (1-4)

Bachem H-4814.0250 MW 539.55 $C_{25}H_{29}N_7O_7$ Store at -15°C

Glp-His-Trp-Ser-His-Ala-Trp-Tyr-Pro
Glp-His-Trp-Ser-His-D-Ala-Trp-Tyr-Pro-NHEt

Synonyms: Luteinizing Hormone Releasing Hormone, (des-Gly[10],D-Ala[6],Pro-NHEt[9])-

Bachem H-4286.0001 Chicken Store at -15°C

Glp-His-Trp-Ser-His-Arg-Trp-Tyr-Pro
Glp-His-Trp-Ser-His-D-Arg-Trp-Tyr-Pro-NHEt

Synonyms: Luteinizing Hormone Releasing Hormone, (des-Gly[10],D-Arg[6],Pro-NHEt[9])-

Bachem H-4288.0001 Chicken Store at -15°C

Glp-His-Trp-Ser-His-Asp-Trp-Lys-Pro-Gly Amide

Synonyms: Luteinizing Hormone Releasing Hormone III

Bachem H-4258.0001 Lamprey Store at -15°C

Neosystem SC1350 Sea lamprey MW 1259.35 Bioactive; hypothalamic releasing hormone; a potent & specific follicle-stimulating hormone (FSH)-releasing peptide; antitumor activity superior to other LHRH analogs; Yu, WH et al, *PNAS USA*, 94:9499-9503, 1997; Mezö, I et al, *J Med Chem*, 40:3353-3358, 1997

Glp-His-Trp-Ser-His-Gly-Trp-Tyr-Pro-Gly Amide

Synonyms: Luteinizing Hormone Releasing Hormone II

American Peptide 54-8-24 Chicken MW 1236.5 $C_{60}H_{69}N_{17}O_{13}$ Miyamoto, K et al, *Japan Peptide Symp Proc*, 99, 1983

Bachem H-4278.0005 Chicken Store at -15°C

Glp-His-Trp-Ser-Tyr
Glp-His-Trp-Ser-Tyr-NHNH₂

Synonyms: Luteinizing Hormone Releasing Hormone (1-5) Hydrazide

Bachem H-4080.0250 Store at -15°C

Glp-His-Trp-Ser-Tyr Free Acid

Synonyms: Luteinizing Hormone Releasing Hormone (1-5)

Neosystem SC351 MW 702.72 Bioactive; hypothalamic releasing hormone; Mc Dermott, JR et al, *Peptides*, 4:25-30, 1983

Glp-His-Trp-Ser-Tyr-Ala-Leu-Arg-Pro
Glp-His-Trp-Ser-Tyr-D-Ala-Leu-Arg-Pro-NHEt

Synonyms: Luteinizing Hormone Releasing Hormone, (D-Ala[6],des-Gly[10])-; Luteinizing Hormone Releasing Hormone, (des-Gly[10],D-Ala[6],Pro-NHEt[9])-; Luteinizing Hormone Releasing Hormone Antagonist

American Peptide 54-0-33 MW 1167.4 $C_{56}H_{78}N_{16}O_{12}$

Bachem H-4070.0005 Store at -15°C

ICN 152917 Marked stimulatory effect on spawning in fish; Barnabe, G & R Barnabe-Quet, *Aquaculture*, 49:125, 1985

Glp-His-Trp-Ser-Tyr-Ala-Leu-Arg-Pro
Glp-His-Trp-Ser-Tyr-D-Ala-Leu-Arg-Pro-NHEt Acetate Salt

Synonyms: Luteinizing Hormone Releasing Hormone Ethylamide, (des-Gly[10],(D-Ala[6]))-

Sigma L 4513 FW 1167.3 (free base) ≥98% (HPLC) | Bioactive peptide; potent LRH agonist in rats & mice; reversibly delays sexual maturation in rats, stimulates spawning activity in fish; Barnabe, G & Barnabe-Quet, R, *Aquaculture*, 49:125, 1985; CA, 89:84988b, 1978

Glp-His-Trp-Ser-Tyr-Ala-Leu-Arg-Pro-Gly
Glp-His-Trp-Ser-Tyr-D-Ala-Leu-Arg-Pro-Gly Amide

Synonyms: Luteinizing Hormone Releasing Hormone, (D-Ala[6])-

American Peptide 54-0-29 MW 1196.3 $C_{56}H_{77}N_{17}O_{13}$ Monahan, MW et al, *Biochemistry*, 12:4616, 1973

Bachem H-4020.0001 MW 1196.33 $C_{56}H_{77}N_{17}O_{13}$ Store at -15°C

ICN 152907 Potent LH-RH analog; Monahan, MW etal, *Biochem*, 12:4616, 1973

Glp-His-Trp-Ser-Tyr-Ala-Leu-Arg-Pro-Gly
Glp-His-Trp-Ser-Tyr-D-Ala-N-Me-Leu-Arg-Pro-Gly Amide

Synonyms: Luteinizing Hormone Releasing Hormone, (D-Ala[6],N-Me-Leu[7])-

ICN 154603

Glp-His-Trp-Ser-Tyr-Ala-Leu-Arg-Pro-Gly
Glp-His-Trp-Ser-Tyr-D-Ala-Leu-Arg-Pro-Gly Amide Acetate Salt

Synonyms: Luteinizing Hormone Releasing Hormone, (D-Ala[6])-

Sigma L 0512 FW 1196.3 (free base) ≥97% (HPLC) | Bioactive peptide; potent LRH agonist

Sigma L 1898 FW 1196.3 (free base) ≥97% (HPLC) | Bioactive peptide; potent LRH agonist; Monahan, MW et al, *Biochem*, 12:4616, 1973

Glp-His-Trp-Ser-Tyr-Ala-Leu-Arg-Pro-Gly
Glp-His-Trp-Ser-Tyr-D-Ala-N-Me-Leu-Arg-Pro-Gly Amide

Synonyms: Luteinizing Hormone Releasing Hormone, (D-Ala[6],N-Me-Leu[7])-

Sigma L 6884 FW 1210.4 ≥97% (HPLC) | Bioactive peptide; LRH analog with <5X the potency of LRH; Ling, N & Vale, W, *Biochem Biophys Res Commun*, 63:801, 1975

Glp-His-Trp-Ser-Tyr-Ala-Trp-Leu-Pro
Glp-His-Trp-Ser-Tyr-D-Ala-Trp-Leu-Pro-NHEt

Synonyms: Luteinizing Hormone Releasing Hormone, (des-Gly[10],D-Ala[6],Pro-NHEt[9])-

Bachem H-7525.0001 Salmon Store at -15°C

Glp-His-Trp-Ser-Tyr-Arg-Trp-Leu-Pro
Glp-His-Trp-Ser-Tyr-D-Arg-Trp-Leu-Pro-NHEt

Synonyms: Luteinizing Hormone Releasing Hormone, (des-Gly[10],D-Arg[6],Pro-NHEt[9])-

Bachem H-9205.0005 Salmon Store at -15°C

Glp-His-Trp-Ser-Tyr-Gly Amide

Synonyms: Luteinizing Hormone Releasing Hormone (1-6)

Bachem H-4494.0005 MW 758.79 $C_{36}H_{42}N_{10}O_9$ Store at -15°C

Glp-His-Trp-Ser-Tyr-Gly-Leu-Arg-Hyp-Gly Amide

Synonyms: Luteinizing Hormone Releasing Hormone, (Hyp[9])-

Neosystem SC471　MW 1198.30　Bioactive; hypothalamic releasing hormone

Sigma L 7154　FW 1198.3　≥97% (HPLC) | Bioactive peptide

Glp-His-Trp-Ser-Tyr-Gly-Leu-Arg-Hyp-Gly Free Acid

Synonyms: Luteinizing Hormone Releasing Hormone, (Hyp[9])-

Neosystem SC472　MW 1199.29　Bioactive; hypothalamic releasing hormone

Glp-His-Trp-Ser-Tyr-Gly-Leu-Arg-Hyp-Gly-Gly Free Acid

Synonyms: Luteinizing Hormone Releasing Hormone, (Hyp[9],Gly[11])-

Neosystem SC478　MW 1256.34　Bioactive; hypothalamic releasing hormone

Glp-His-Trp-Ser-Tyr-Gly-Leu-Arg-Hyp-Gly-Gly-Lys-Arg Free Acid

Synonyms: Luteinizing Hormone Releasing Hormone, (Hyp[9],Gly[11],Lys[12],Arg[13])-

Neosystem SC487　MW 1540.70　Bioactive; hypothalamic releasing hormone

Glp-His-Trp-Ser-Tyr-Gly-Leu-Arg-Pro-NHEt

Synonyms: Luteinizing Hormone Releasing Hormone, (des-Gly[10],Pro-NHEt[9])-; Fertirelin; Luteinizing Hormone Releasing Hormone, (des-Gly[10])-

American Peptide 54-0-32　MW 1153.3　$C_{55}H_{76}N_{16}O_{12}$

Bachem H-4055.0001　MW 1153.31　$C_{55}H_{76}N_{16}O_{12}$　Store at -15°C

ICN 152916

Glp-His-Trp-Ser-Tyr-Gly-Leu-Arg-Pro-Gly
Glp-His-Trp-Ser-3,5-diiodo-Tyr-Gly-Leu-Arg-Pro-Gly Amide

Synonyms: Luteinizing Hormone Releasing Hormone, (3,5-Diiodo-Tyr[5])-

Bachem H-1375.0001　MW 1434.1　$C_{55}H_{73}I_2N_{17}O_{13}$　Store at -15°C

Glp-His-Trp-Ser-Tyr-Gly-Leu-Arg-Pro-Gly
Glp-His(3-Me)-Trp-Ser-Tyr-Gly-Leu-Arg-Pro-Gly Amide

Synonyms: Luteinizing Hormone Releasing Hormone, (His(3-Me)[2])-; Luteinizing Hormone Releasing Hormone, (His(π-Me)[2])-

Bachem H-4492.0001　MW 1196.33　$C_{56}H_{77}N_{17}O_{13}$　Store at -15°C

Glp-His-Trp-Ser-Tyr-Gly-Leu-Arg-Pro-Gly
Glp-His-Trp-D-Ser-Tyr-Gly-Leu-Arg-Pro-Gly Amide

Synonyms: Luteinizing Hormone Releasing Hormone, (D-Ser[4])-; Gonadorelin, (D-Ser[4])-

Bachem H-4706.0001　MW 1182.31　$C_{55}H_{75}N_{17}O_{13}$　Store at -15°C

Glp-His-Trp-Ser-Tyr-Gly-Leu-Arg-Pro-Gly
Glp-His(1-Me)-Trp-Ser-Tyr-Gly-Leu-Arg-Pro-Gly Amide ·TFA

Synonyms: Luteinizing Hormone Releasing Hormone, (His(1-Me)[2])-; Luteinizing Hormone Releasing Hormone, (His(τ-Me)[2])-

Bachem H-5405.0001　MW 1196.33　$C_{56}H_{77}N_{17}O_{13}$　Store at -15°C

Glp-His-Trp-Ser-Tyr-Gly-Leu-Arg-Pro-Gly
Glp-His-Trp-Ser-3,5-Diiodo-Tyr-Gly-Leu-Arg-Pro-Gly Amide

Synonyms: Luteinizing Hormone Releasing Hormone, (3,4-Diiodo-Tyr[5])-

ICN 152905

Glp-His-Trp-Ser-Tyr-Gly-Leu-Arg-Pro-Gly Amide

Synonyms: Luteinizing Hormone Releasing Hormone; Gonadorelin; Luteinizing Hormone Releasing Factor; Gonadotropin Releasing Hormone

Bachem H-4005.0005　MW 1182.31　$C_{55}H_{75}N_{17}O_{13}$　Store at -15°C

Calbiochem 05-23-1050　MW 1182.3　$C_{55}H_{75}N_{17}O_{13}$　>90% (HPLC); lyophilized solid; soluble in 5% acetic acid; may be carcinogenic/teratogenic | Neurohumoral hormone present in hypothalamus; decapeptide with blocked N- & C-termini; stimulates secretion of pituitary hormones LH & FSH, which in turn leads to induction of ovulation; *Merck Index*, 12:5500

ICN 151567　Hypothalamic peptide which stimulates pituitary release of LH & FSH

Neosystem SC087　MW 1182.30　Bioactive; hypothalamic releasing hormone

American Peptide 54-1-11　Human　MW 1182.4　$C_{55}H_{75}N_{17}O_{13}$　Involved in the regulation of reproductive functions by controlling the synthesis & release of Luteinizing Hormone (LH) & Follicle Stimulating Hormone (FSH) from the anterior pituitary gland; Rivier, JE et al, *Life Sci*, 23:869, 1978; Tilemans, D et al, *Endocrinology*, 130:887, 1992

ICN 152904　Human

Biogenesis 5730-1009　Human synthetic　MW 1182　Lyophilized, consisting of 91.7% peptide material, acetate counter ions and residual H_2O; purified

Peptides International PLR-4013-v　Human synthetic　MW 1182.3　$C_{55}H_{75}N_{17}O_{13}$　>99% (HPLC); lyophilized amorphous powder | Bioactive; porcine, rat; Matsuo, H et al, *BBRC*, 43:1334, 1971

Glp-His-Trp-Ser-Tyr-Gly-Leu-Arg-Pro-Gly Amide 2CH₃COOH 4H₂O

Synonyms: Luteinizing Hormone Releasing Hormone; Gonadotropin Releasing Hormone

Peptides International PLR-4013　Synthetic　MW 1374.47　$C_{55}H_{75}N_{17}O_{13} \cdot 2CH_3COOH \cdot 4H_2O$　>99% (HPLC); lyophilized amorphous powder; bulk | Bioactive; porcine, rat; Matsuo, H et al, *BBRC*, 43:1334, 1971

Glp-His-Trp-Ser-Tyr-Gly-Leu-Arg-Pro-Gly Amide Acetate

Synonyms: Luteinizing Hormone Releasing Hormone; Gonadoliberin; Gonadotropin Releasing Hormone

Fluka 62690　MW 1182.33 (free peptide)　$C_{55}H_{75}N_{17}O_{13}$　≥98.0% (HPLC); ≥80% peptide content; ≤12% acetate; ≤10% water | Principal regulator of the Ca^{2+}-dependent release of gonadotropins, LH & follicle-stimulating hormone; Conn, PM et al, *Recent Prog Horm Res*, 43:29, 1987

Glp-His-Trp-Ser-Tyr-Gly-Leu-Arg-Pro-Gly Amide Acetate Salt

Synonyms: Luteinizing Hormone Releasing Hormone

Sigma L 7134　Human　FW 1182.3 (free base)　≥98% (HPLC) | Bioactive peptide; hypothalamic peptide that stimulates release of luteinizing hormone & follicle stimulating hormone from anterior pituitary, regulating reproductive functions; Vale, WW et al, *Peptides, Structure & Biological Function, Proc Sixth American Peptide Symposium*, Gross, E & Meienhofer, M, eds, 781, 1979

Glp-His-Trp-Ser-Tyr-Gly-Leu-Arg-Pro-Gly Free Acid

Synonyms: Luteinizing Hormone Releasing Hormone, (Gly[10])-

American Peptide 54-1-12   MW 1183.4 $C_{55}H_{74}N_{16}O_{14}$
Rivier, JE et al, *Life Sci*, 23:869, 1978

Bachem H-4010.0005   MW 1183.29 $C_{55}H_{74}N_{16}O_{14}$ Store at -15°C

Sigma L 8008   FW 1183.3   ≥97% (HPLC) | Bioactive peptide;
Schally, AV, *Science*, 202:18, 1978; Rivier, JE & Vales, WW, *Life Sci*, 23:869, 1978

Glp-His-Trp-Ser-Tyr-Gly-Leu-Arg-Pro-Gly-Gly Free Acid

Synonyms: Luteinizing Hormone Releasing Hormone, (Gly[11])-

Neosystem SC479   MW 1240.34   Bioactive; hypothalamic releasing hormone

Glp-His-Trp-Ser-Tyr-Gly-Leu-Arg-Pro-Gly-Gly-Lys-Arg Free Acid

Synonyms: Luteinizing Hormone Releasing Hormone, (Gly[11],Lys[12],Arg[13])-

Neosystem SC486   MW 1524.70   Bioactive; hypothalamic releasing hormone

Glp-His-Trp-Ser-Tyr-Gly-Leu-Arg-Pro-NHEt

Synonyms: Luteinizing Hormone Releasing Hormone Ethylamide, (des-Gly[10])-

Sigma L 0262   FW 1153.3   ≥97% (HPLC) | Bioactive peptide; LRH agonist

Glp-His-Trp-Ser-Tyr-Gly-Leu-Gln-Pro-Gly
Glp-His-Trp-Ser-3,4-Diiodo-Tyr-Gly-Leu-Gln-Pro-Gly Amide

Synonyms: Luteinizing Hormone Releasing Hormone, (Gln[8])-

ICN 152906   Chicken   King, JA & RP Millar, *JBC*, 257:10729, 1982

Glp-His-Trp-Ser-Tyr-Gly-Leu-Gln-Pro-Gly Amide

Synonyms: Luteinizing Hormone Releasing Hormone, (Gln[8])-

Sigma L 0637   Avian   FW 1154.2; ≥97% (HPLC)   Bioactive peptide; hypothalamic peptide that stimulates release of gonadotropins from anterior pituitary, regulating reproductive functions

American Peptide 54-8-23   Chicken   MW 1154.3
$C_{54}H_{71}N_{15}O_{14}$ King, JA et al, *JBC*, 257:10729, 1982

Bachem H-3106.0001   Chicken   MW 1154.25 $C_{54}H_{71}N_{15}O_{14}$
Store at -15°C

Glp-His-Trp-Ser-Tyr-Gly-Leu-Lys-Pro-Gly Amide

Synonyms: Luteinizing Hormone Releasing Hormone, (Lys[8])-

American Peptide 54-0-28   MW 1154.3 $C_{55}H_{75}N_{15}O_{13}$
Chang, JK et al, *J Med Chem*, 15:623, 1972

Glp-His-Trp-Ser-Tyr-Gly-Leu-Ser-Pro-Gly Amide

Synonyms: Luteinizing Hormone Releasing Hormone

Bachem H-4262.0005   Sea bream   Store at -15°C

Glp-His-Trp-Ser-Tyr-Gly-Trp-Leu-Pro-Gly

Synonyms: Luteinizing Hormone Releasing Hormone, (Trp[7], Leu[8])-

American Peptide 54-0-27   MW 1213.3 (free acid)
$C_{60}H_{72}N_{14}O_{14}$

Glp-His-Trp-Ser-Tyr-Gly-Trp-Leu-Pro-Gly Amide

Synonyms: Luteinizing Hormone Releasing Hormone

American Peptide 57-7-25   Salmon   MW 1212.5
$C_{60}H_{73}N_{15}O_{13}$

Bachem H-6845.0001   Salmon   MW 1212.33 $C_{60}H_{73}N_{15}O_{13}$
Store at -15°C

Sigma L 4897   Salmon   FW 1212.3; ≥98% (HPLC)   Bioactive peptide; hypothalamic peptide that stimulates release of gonadotropins from anterior pituitary, regulating reproductive functions

Biogenesis 5730-3009   Salmon synthetic   MW 1213.5
Lyophilized, consisting of 89.05% peptide material, acetate counter ions and residual H_2O; purified

Glp-His-Trp-Ser-Tyr-His-Leu
Glp-His-Trp-Ser-Tyr-D-His(Bzl)-Leu Free Acid

Synonyms: Luteinizing Hormone Releasing Hormone (1-7), (D-His(Bzl)[6])-

Bachem H-4804.0001   MW 1043.15 $C_{53}H_{62}N_{12}O_{11}$ Store at -15°C

Glp-His-Trp-Ser-Tyr-His-Leu-Arg-Pro
Glp-D-His-Trp-Ser-Tyr-D-His(Bzl)-Leu-Arg-Pro-NHEt

Synonyms: Luteinizing Hormone Releasing Hormone, (des-Gly[10],D-His[2],D-His(Bzl)[6],Pro-NHEt[9])-; Histrelin, (D-His[2])-

Bachem H-4652.0001   MW 1323.52 $C_{66}H_{86}N_{18}O_{12}$ Store at -15°C

Glp-His-Trp-Ser-Tyr-His-Leu-Arg-Pro
Glp-His-Trp-Ser-D-Tyr-D-His(Bzl)-Leu-Arg-Pro-NHEt

Synonyms: Luteinizing Hormone Releasing Hormone, (des-Gly[10],D-Tyr[5],D-His(Bzl)[6],Pro-NHEt[9])-; Histrelin, (D-Tyr[5])-

Bachem H-4654.0001   MW 1323.52 $C_{66}H_{86}N_{18}O_{12}$ Store at -15°C

Glp-His-Trp-Ser-Tyr-His-Leu-Arg-Pro
Glp-His-Trp-Ser-Tyr-His(Bzl)-Leu-Arg-Pro-NHEt

Synonyms: Luteinizing Hormone Releasing Hormone, (des-Gly[10],His(Bzl)[6],Pro-NHEt[9])-; Histrelin, (His(Bzl)[6])-

Bachem H-4656.0001   MW 1323.52 $C_{66}H_{86}N_{18}O_{12}$ Store at -15°C

Glp-His-Trp-Ser-Tyr-His-Leu-Arg-Pro
Glp-His-Trp-Ser-Tyr-D-His(Bzl)-D-Leu-Arg-Pro-NHEt

Synonyms: Luteinizing Hormone Releasing Hormone, (des-Gly[10],D-His(Bzl)[6],D-Leu[7],Pro-NHEt[9])-; Histrelin, (D-Leu[7])-

Bachem H-4658.0001   MW 1323.52 $C_{66}H_{86}N_{18}O_{12}$ Store at -15°C

Glp-His-Trp-Ser-Tyr-His-Leu-Arg-Pro
Glp-His-Trp-D-Ser-Tyr-D-His(Bzl)-Leu-Arg-Pro-NHEt

Synonyms: Luteinizing Hormone Releasing Hormone, (des-Gly[10],D-Ser[4],D-His(Bzl)[6],Pro-NHEt[9])-; Histrelin, (D-Ser[4])-

Bachem H-4704.0001   MW 1323.52 $C_{66}H_{86}N_{18}O_{12}$ Store at -15°C

Glp-His-Trp-Ser-Tyr-His-Leu-Arg-Pro
Glp-His-Trp-Ser-Tyr-D-His(Bzl)-Leu-Arg-Pro-NHEt

Synonyms: Luteinizing Hormone Releasing Hormone, (des-Gly[10],D-His(Bzl)[6],Pro-NHEt[9])-; Histrelin

Bachem H-9210.0005   MW 1323.52 $C_{66}H_{86}N_{18}O_{12}$ Store at -15°C

Glp-His-Trp-Ser-Tyr-His-Leu-Arg-Pro
Glp-His-Trp-Ser-Tyr-D-His(Bzl)-Leu-Arg-Pro-NHEt

Synonyms: Luteinizing Hormone Releasing Hormone Ethylamide, (des-Gly[10],(D-His(Bzl)[6]))-; Luteinizing Hormone Releasing Hormone, (D-His(Bzl)[6],des-Gly[10])-; Luteinizing Hormone Releasing Hormone Antagonist

ICN 152919   Sundaram, K et al, *Life Sci*, 28:83, 1981; Lecomte, P et al, *Endocrinology*, 110:1, 1982; McCreary, BR et al, *Gen Comp Endocrinol*, 46:511, 1962

Sigma L 2761   FW 1323.5   ≥97% (HPLC); peptide content:~70% | Bioactive peptide; potent LRH agonist; inhibits actions of sex steroids on the male & female reproductive tracts; has greater lipophilicity than other LRH agonists, but retains high water solubility; Sundaram, K et al, *Life Sci*, 28:83, 1981; Lecomte, P et al, *Endocrinology*, 110:1, 1982; McCreary, BR et al, *Gen Comp Endocrinol*, 46:511, 1982

Glp-His-Trp-Ser-Tyr-Leu-Leu-Arg-Pro
Glp-His-Trp-Ser-Tyr-D-Leu-Leu-Arg-Pro-NHEt

Synonyms: Luteinizing Hormone Releasing Hormone, (des-Gly[10],D-Leu[6],Pro[9])-; Leuprolide; Luteinizing Hormone Releasing Hormone, (des-Gly[10],D-Leu[6],Pro-NHEt[9])-; Leuprolide; Leuprorelin

American Peptide 54-1-22   MW 1209.4 $C_{59}H_{84}N_{16}O_{12}$

Bachem H-4060.0005   MW 1209.42 $C_{59}H_{84}N_{16}O_{12}$ Store at -15°C

Glp-His-Trp-Ser-Tyr-Leu-Leu-Arg-Pro
Glp-D-His-Trp-Ser-Tyr-D-Leu-Leu-Arg-Pro-NHEt

Synonyms: Luteinizing Hormone Releasing Hormone, (des-Gly[10],D-His[2],D-Leu[6],Pro-NHEt[9])-; Leuprolide, (D-His[2])-

Bachem H-4316.0001   Store at -15°C

Glp-His-Trp-Ser-Tyr-Leu-Leu-Arg-Pro
Et-Glp-His-Trp-Ser-Tyr-D-Leu-D-Leu-Arg-Pro-NHEt

Synonyms: Luteinizing Hormone Releasing Hormone, (des-Gly[10],D-Leu[6],D-Leu[7],Pro-NHEt[9])-; Leuprolide, (D-Leu[7])-

Bachem H-4636.0001   MW 1209.42 $C_{59}H_{84}N_{16}O_{12}$ Store at -15°C

Glp-His-Trp-Ser-Tyr-Leu-Leu-Arg-Pro
Glp-His-Trp-Ser-D-Tyr-D-Leu-Leu-Arg-Pro-NHEt

Synonyms: Luteinizing Hormone Releasing Hormone, (des-Gly[10],D-Tyr[5],D-Leu[6],Pro-NHEt[9])-; Leuprolide, (D-Tyr[5])-

Bachem H-4638.0001   Store at -15°C

Glp-His-Trp-Ser-Tyr-Leu-Leu-Arg-Pro
Glp-His-Trp-Ser-Tyr-D-Leu-Leu-Arg-Pro-NHEt

Synonyms: Luteinizing Hormone Releasing Hormone Ethylamide, (des-Gly[10],(D-Leu[6]))-; Leuprolide

Sigma L 0399   FW 1209.4   ≥98% (HPLC) | Bioactive peptide

Glp-His-Trp-Ser-Tyr-Leu-Val-Arg-Pro-Gly
Glp-His-Trp-Ser-Tyr-D-Leu-Val-Arg-Pro-Gly-NHEt

Synonyms: Luteinizing Hormone Releasing Hormone (1-9), (D-Leu[6],Val[7])-

American Peptide 54-1-25   MW 1195.4 $C_{58}H_{82}N_{16}O_{12}$
Rivier, JE et al, *Life Sci*, 23:869, 1978

Glp-His-Trp-Ser-Tyr-Lys-Leu-Arg-Pro-Gly
Glp-His-Trp-Ser-Tyr-D-Lys-Leu-Arg-Pro-Gly Amide

Synonyms: Luteinizing Hormone Releasing Hormone, (D-Lys[6])-

Bachem H-4025.0001   Store at -15°C

Sigma L 5022   FW 1253.4   ≥95% (HPLC) Bioactive peptide;

American Peptide 54-0-31   Human   MW 1253.5
$C_{59}H_{84}N_{18}O_{13}$

Glp-His-Trp-Ser-Tyr-Phe-Leu-Arg-Pro
Glp-His-Trp-Ser-Tyr-D-Phe-Leu-Arg-Pro-NHEt

Synonyms: Luteinizing Hormone Releasing Hormone Ethylamide, (des-Gly[10],(D-Phe[6]))-

Sigma L 8886   FW 1243.4   ≥97% (HPLC) | Bioactive peptide

Glp-His-Trp-Ser-Tyr-Ser-Leu-Arg-Pro
Glp-His-Trp-Ser-Tyr-D-Ser(tBu)-Leu-Arg-Pro-NHEt

Synonyms: Luteinizing Hormone Releasing Hormone, (D-Ser(tBu)[6],des-Gly[10])-; Busereline

American Peptide 54-1-23   MW 1239.5 $C_{60}H_{86}N_{16}O_{13}$

Glp-His-Trp-Ser-Tyr-Ser-Leu-Arg-Pro-Gly
Glp-His-Trp-Ser-Tyr-D-Ser(tBu)-Leu-Arg-Pro-(Aza)Gly Amide

Synonyms: Luteinizing Hormone Releasing Hormone, ((tBu)D-Ser[6],(Aza)Gly[10])-; Goserelin

American Peptide 54-1-20   MW 1283.5 $C_{60}H_{86}N_{18}O_{14}$

Glp-His-Trp-Ser-Tyr-Trp
Glp-His-Trp-Ser-Tyr-D-Trp Amide

Synonyms: Luteinizing Hormone Releasing Hormone (1-6), (D-Trp[6])-; Triptorelin (1-6)

Bachem H-4574.0005   MW 887.95 $C_{45}H_{49}N_{11}O_9$ Store at -15°C

Glp-His-Trp-Ser-Tyr-Trp-Leu-Arg-Pro
Glp-His-Trp-Ser-Tyr-D-Trp-Leu-Arg-Pro-NHEt

Synonyms: Luteinizing Hormone Releasing Hormone, (des-Gly[10],D-Trp[6],Pro-NHEt[9])-; Deslorelin

American Peptide 54-0-34   MW 1282.5 $C_{64}H_{83}N_{17}O_{12}$
Klibanski, A et al, *J Clin Endocrinol Metab*, 68:81, 1989; Coy, DV & Schally, AV, *Ann Clin Res*, 10:139, 1978; Foster, CM et al, *J Clin Endocrinol*, 1984; *Metab*, 59:801,

Bachem H-4065.0005   MW 1282.47 $C_{64}H_{83}N_{17}O_{12}$ Store at -15°C

Glp-His-Trp-Ser-Tyr-Trp-Leu-Arg-Pro
Glp-D-His-Trp-Ser-Tyr-D-Trp-Leu-Arg-Pro-NHEt

Synonyms: Luteinizing Hormone Releasing Hormone, (des-Gly[10],D-His[2],D-Trp[6],Pro-NHEt[9])-

Bachem H-4986.0001   MW 1282.47 $C_{64}H_{83}N_{17}O_{12}$ Store at -15°C

Glp-His-Trp-Ser-Tyr-Trp-Leu-Arg-Pro
Glp-His-Trp-Ser-D-Tyr-D-Trp-Leu-Arg-Pro-NHEt

Synonyms: Luteinizing Hormone Releasing Hormone, (des-Gly[10],D-Ser[4],D-Trp[6],Pro-NHEt[9])-

Bachem H-4988.0001   MW 1282.47 $C_{64}H_{83}N_{17}O_{12}$ Store at -15°C

Glp-His-Trp-Ser-Tyr-Trp-Leu-Arg-Pro
Glp-His-Trp-Ser-D-Tyr-D-Trp-Leu-Arg-Pro-NHEt

Synonyms: Luteinizing Hormone Releasing Hormone, (des-Gly[10],D-Tyr[5],D-Trp[6],Pro-NHEt[9])-

Bachem H-4992.0001   MW 1282.47 $C_{64}H_{83}N_{17}O_{12}$ Store at -15°C

Glp-His-Trp-Ser-Tyr-Trp-Leu-Arg-Pro
Glp-His-Trp-Ser-Tyr-D-Trp-D-Leu-Arg-Pro-NHEt

Synonyms: Luteinizing Hormone Releasing Hormone, (des-Gly[10],D-Trp[6],D-Leu[7],Pro-NHEt[9])-

Bachem H-4994.0001   MW 1282.47 $C_{64}H_{83}N_{17}O_{12}$ Store at -15°C

Glp-His-Trp-Ser-Tyr-Trp-Leu-Arg-Pro
Glp-His-Trp-Ser-Tyr-D-Trp-Leu-Arg-Pro-NHEt

Synonyms: Luteinizing Hormone Releasing Hormone, (D-Trp[6],des-Gly[10])-; Luteinizing Hormone Releasing Hormone Ethylamide, (des-Gly[10],(D-Trp[6]))-

ICN 152918

Sigma L 5386   FW 1282.5   ≥97% (HPLC) | Bioactive peptide; long-acting LRH used in the treatment of precocious puberty

Glp-His-Trp-Ser-Tyr-Trp-Leu-Arg-Pro-Gly
Glp-His-Trp-Ser-Tyr-D-Trp-Leu-Arg-Pro-Gly-NHEt

Synonyms: Luteinizing Hormone Releasing Hormone, (D-Trp[6])-; Triptorelin

American Peptide 54-1-21   MW 1311.5 $C_{64}H_{82}N_{18}O_{13}$

Bachem H-4075.0005   MW 1311.47 $C_{64}H_{82}N_{18}O_{13}$ Store at -15°C

Glp-His-Trp-Ser-Tyr-Trp-Leu-Arg-Pro-Gly
Glp-D-His-Trp-Ser-Tyr-D-Trp-Leu-Arg-Pro-Gly Amide

Synonyms: Luteinizing Hormone Releasing Hormone, (D-His[2],D-Trp[6])-; Triptorelin, (D-His[2])-

Bachem H-4642.0001   Store at -15°C

Glp-His-Trp-Ser-Tyr-Trp-Leu-Arg-Pro-Gly
Glp-His-Trp-D-Ser-Tyr-D-Trp-Leu-Arg-Pro-Gly Amide

Synonyms: Luteinizing Hormone Releasing Hormone, (D-Ser[4],D-Trp[6])-; Triptorelin, (D-Ser[4])-

Bachem H-4644.0001   MW 1311.47 $C_{64}H_{82}N_{18}O_{13}$ Store at -15°C

Glp-His-Trp-Ser-Tyr-Trp-Leu-Arg-Pro-Gly
Glp-His-Trp-Ser-D-Tyr-D-Trp-Leu-Arg-Pro-Gly Amide

Synonyms: Luteinizing Hormone Releasing Hormone, (D-Tyr[5],D-Trp[6])-; Triptorelin, (D-Tyr[5])-

Bachem H-4646.0001   Store at -15°C

Glp-His-Trp-Ser-Tyr-Trp-Leu-Arg-Pro-Gly
Glp-His-Trp-Ser-Tyr-D-Trp-D-Leu-Arg-Pro-Gly Amide

Synonyms: Luteinizing Hormone Releasing Hormone, (D-Trp[6],D-Leu[7])-; Triptorelin, (D-Leu[7])-

Bachem H-4648.0001   Store at -15°C

Glp-His-Trp-Ser-Tyr-Trp-Leu-Arg-Pro-Gly
Glp-His-Trp-Ser-Tyr-D-Trp-Leu-Arg-Pro-Gly Amide

Synonyms: Luteinizing Hormone Releasing Hormone, (D-Trp[6])-

ICN 152908

Neosystem SC161   MW 1311.47   Bioactive; hypothalamic releasing hormone; Schally, AV et al, *PNAS USA*, 82:2498-2502, 1985

Sigma L 9761   FW 1311.5   ≥97% (HPLC) | Bioactive peptide; potent LRH agonist; used in treatment of prostate cancer

Glp-His-Trp-Ser-Tyr-Trp-Leu-Arg-Pro-Gly Amide

Synonyms: Luteinizing Hormone Releasing Hormone, (Trp[6])-; Triptorelin, (Trp[6])-

Bachem H-4578.0001   MW 1311.47 $C_{64}H_{82}N_{18}O_{13}$ Store at -15°C

Glp-His-Trp-Ser-Tyr-Trp-Leu-Arg-Pro-Gly-Leu-Arg-Pro-Gly
Glp-His-Trp-Ser-Tyr-D-Trp-Leu-Arg-Pro-Gly-Leu-Arg-Pro-Gly Amide

Synonyms: Luteinizing Hormone Releasing Hormone-Leu-Arg-Pro-Gly, (D-Trp[6])-; Triptorelin-Leu-Arg-Pro-Gly

Bachem H-4582.0001   MW 1734.98 $C_{83}H_{115}N_{25}O_{17}$ Store at -15°C

Glp-His-Trp-Ser-Tyr-Trp-Leu-Ser-Pro
Glp-His-Trp-Ser-Tyr-D-Trp-Leu-Ser-Pro-NHEt

Synonyms: Luteinizing Hormone Releasing Hormone, (des-Gly[10],D-Trp[6],Pro-NHEt[9])-

Bachem H-4284.0001   Sea bream   Store at -15°C

Glp-His-Tyr-Ser-Leu-Glu-Trp-Lys-Pro-Gly Amide

Synonyms: Luteinizing Hormone Releasing Hormone

American Peptide 54-7-26   Lamprey   MW 1226.3 $C_{58}H_{79}N_{15}O_{15}$ Sherwood, NM et al, *JBC*, 261:4812, 1986

Glp-Leu-Asn-Phe-Ser-Ala-Gly-Trp Amide

Synonyms: Adipokinetic Hormone II-L; Adipokinetic Hormone II

American Peptide 60-9-23   *Locusta migratoria*   MW 904.0 $C_{43}H_{57}N_{11}O_{11}$ Siegert, K et al, *Hoppe Seyler's Zeitschrift fur Physiologische Cheme*, 1986; Gade, G et al, *BBRC*, 134:723, 1986

Bachem H-5610.0001   Locusta migratoria   MW 903.99 $C_{43}H_{57}N_{11}O_{11}$ Store at -15°C

ICN 151427   *Locusta migratoria*   MW 904 $C_{43}H_{57}N_{11}O_{11}$ Regulator of energy metabolism during long flight; Siegert, K, *Biol Chem* (Hoppe-Seyler), 366:723, 1985

Neosystem SC120   Locusta migratoria   MW 903.99   Bioactive neuropeptide; regulator of energy metabolism during flight; Gäde, G et al, *BBRC*, 134:723, 1986

Sigma A 8421   *Locusta migratoria*   FW 904.0 $C_{43}H_{57}N_{11}O_{11}$ ≥97% (HPLC) | Bioactive peptide; regulator of energy metabolism during long flight; Siegert, K et al, *Biol Chem Hoppe-Seyler*, 366:723, 1985

Glp-Leu-Asn-Phe-Ser-Pro-Gly-Trp Amide

Synonyms: Crustacean Erythrophore Concentrating Hormone

Bachem H-6750.0005   Store at -15°C

Glp-Leu-Asn-Phe-Ser-Thr-Gly-Trp Amide

Synonyms: Adipokinetic Hormone II-S; Adipokinetic Hormone II

Neosystem SC119   Schistocera gregaria   MW 934.02   Bioactive; regulator of energy metabolism during flight; Gäde, G et al, *BBRC*, 134:723, 1986

American Peptide 60-9-21   *Schistocera gregaris*   MW 934.0 $C_{44}H_{59}N_{11}O_{12}$ Gade, G et al, *BBRC*, 134:723, 1986

Bachem H-5615.0001   Schistocerca gregaria   MW 934.02 $C_{44}H_{59}N_{11}O_{12}$ Store at -15°C

ICN 151428   *Schistocerca gregaria*   MW 934 $C_{44}H_{59}N_{11}O_{12}$ Regulator of energy metabolism during long flight; Siegert, K, *Biol Chem* (Hoppe-Seyler), 366:723, 1985

Sigma A 3797   *Schistocerca gregaria*   FW 934.0 $C_{44}H_{59}N_{11}O_{12}$ ≥90% (HPLC) | Bioactive peptide; regulator of energy metabolism during long flight; Siegert, K et al, *Biol Chem Hoppe-Seyler*, 366:723, 1985

Glp-Leu-Asn-Phe-Thr-Pro-Asn-Trp-Gly-Thr Amide

Synonyms: Adipokinetic Hormone

American Peptide 60-9-18   Locust   MW 1159.4 $C_{54}H_{74}N_{14}O_{15}$ Stone, JV et al, *Nature*, 263:207, 176

Glp-Leu-Gly-Pro-Gln-Gly-Pro-Pro-His-Leu-Val-Ala-Asp-Pro-Ser-Lys-Lys-Gln-Gly-Pro-Trp-Leu-Glu-Glu-Glu-Glu-Glu-Ala-Tyr-Gly-Trp-Met-Asp-Phe Amide

Synonyms: Gastrin I, Big

American Peptide 46-1-55   Human   MW 3849.3 $C_{176}H_{251}N_{43}O_{53}S$ Choudhury, AM et al, *Hoppe-Seyler's Z Physiol Chem*, 361:1719, 1980

Bachem H-7320.0500   Human   Store at -15°C

Glp-Leu-Gly-Pro-Gln-Gly-Pro-Pro-His-Leu-Val-Ala-Asp-Pro-Ser-Lys-Lys-Gln-Gly-Pro-Trp-Leu-Glu-Glu-Glu-Glu-Glu-Ala-Tyr-Gly-Trp-Met-Asp-Phe Amide

Synonyms: Gastrin I, Big; HG-34; Gregory Structure

ICN 152870   Human   MW 3849.2   Choudhury, AM etal, *Hoppe-Seyler's Z Physiol Chem*, 361:1719, 1980

Glp-Leu-Gly-Pro-Gln-Gly-Pro-Pro-His-Leu-Val-Ala-Asp-Pro-Ser-Lys-Lys-Gln-Gly-Pro-Trp-Leu-Glu-Glu-Glu-Glu-Glu-Ala-Tyr-Gly-Trp-Met-Asp-Phe Amide

Synonyms: Gastrointestinal Peptide; Gastrin I, Big; HG-34; Gregory Structure

Sigma G 5024   Human   FW 3849.2   ≥97% (HPLC) | Bioactive peptide; storage form of gastrin that also has biological activity; Choudhury, AM et al, *Hoppe-Seyler's Z Physiol Chem*, 361:1719, 1980

Sigma G 7264   Human   FW 3849.2   ≥97% (HPLC) | Bioactive peptide; storage form of gastrin that also has biological activity; Choudhury, AM et al, *Hoppe-Seyler's Z Physiol Chem*, 361:1719, 1980

Glp-Leu-Gly-Pro-Gln-Gly-Pro-Pro-His-Leu-Val-Ala-Asp-Pro-Ser-Lys-Lys

Synonyms: Gastrin (1-17), Big; G34

Biogenesis 4620-1309   Human synthetic   MW 1752
Lyophilized, consisting of 88% peptide material, acetate counter ions and residual H_2O; purified

Glp-Leu-Gly-Pro-Gln-Gly-Pro-Pro-His-Leu-Val-Ala-Asp-Pro-Ser-Lys-Lys-Gln-Gly-Pro-Trp-Leu-Glu-Glu-Glu-Glu-Glu-Ala-Tyr-Gly-Trp-Met-Asp-Phe Amide

Synonyms: Gastrin, Big; G34

Biogenesis 4620-1504   Human synthetic   MW 3849.3
Lyophilized, consisting of 85% peptide material, TFA counter ions and residual H_2O; purified | To avoid Met oxidation, it is best to dissolve the peptide under a nitrogen stream; Bonato et al, *Life Sci*, 39:959, 1986

Glp-Leu-Gly-Pro-Gln-Gly-Pro-Pro-His-Leu-Val-Ala-Asp-Pro-Ser-Lys-Lys-Gln-Gly-Pro-Trp-Leu-Glu-Glu-Glu-Glu-Glu-Ala-Tyr-Gly-Trp-Met-Asp-Phe Amide

Synonyms: Gastrin, Big Ammonium Form

Peptides International PGR-4183-s   Human synthetic   MW 3849.3 $C_{176}H_{251}N_{43}O_{53}S$ >95% (HPLC); lyophilized amorphous powder | Bioactive; Jorpes, JE & V Mutt (Eds), *Secretin, Cholecystokinin, Pancreozymin & Gastrin*, Handbook of Experimental Pharmacology, Vol 34, Springer-Verlag, Berlin, 1973; Choudhury, AM et al, Hoppe-Seyler's *Z Physiol Chem*, 361:1719, 1980

Glp-Leu-Leu-Gly-Gly-Arg-Phe Amide

Synonyms: Pol-RF Amide

Bachem H-5585.0025   Store at -15°C

Glp-Leu-Leu-Gly-Gly-Arg-Phe-Gly-Tyr

Bachem H-5970.0005   Store at -15°C

Glp-Leu-Thr-Glu-Asn-Lys-Pro-Arg

Synonyms: Neurotensin (1-8), (Thr[3])-

ICN 152936   MW 968.1 $C_{41}H_{69}N_{13}O_{14}$

Glp-Leu-Thr-Phe-Ser-Ser-Gly-Trp-Gly-Asn Amide

Synonyms: Hypertrehalosaemic Neuropeptide

Bachem H-9255.0001   Heliothis zea   MW 1078.15 $C_{49}H_{67}N_{13}O_{15}$ Store at -15°C

ICN 154448   *Heliothis zea*   MW 1078.3   Jaffe, H et al, *BBRC*, 155:344, 1988

Glp-Leu-Thr-Phe-Thr-Pro-Asn-Trp Amide

Synonyms: Myoactive Peptide II; Periplanetin CC-2

American Peptide 60-0-50   Cockroach   MW 988.1 $C_{48}H_{65}N_{11}O_{12}$

Glp-Leu-Thr-Phe-Thr-Pro-Asn-Trp-Gly-Thr Amide

Synonyms: Stick Insect Hypertrehalosaemic Factor II

American Peptide 60-9-47   MW 1146.2 $C_{54}H_{75}N_{13}O_{15}$

Glp-Leu-Thr-Phe-Thr-Ser-Trp-Gly Amide

Synonyms: Adipokinetic Hormone

American Peptide 60-9-20   *Heliothis zea, Manduca sexta*   MW 921.0 $C_{44}H_{60}N_{10}O_{12}$ Ziegler, R et al, *BBRC*, 133:337, 1985; Jaffe, H et al, *BBRC*, 135:622, 1986

Glp-Leu-Tyr-Gln-Asn-Lys-Pro-Arg-Arg-Pro-Tyr-Ile-Leu

Synonyms: Neurotensin, (Gln[4])-

American Peptide 62-1-20   MW 1672.0 $C_{78}H_{122}N_{22}O_{19}$ Folkers, K et al, *PNAS*, 73:3833, 1976

Bachem H-4460.0005   MW 1671.96 $C_{78}H_{122}N_{22}O_{19}$ Store at -15°C

ICN 152929   MW 1672   Folkers, K etal, *PNAS*, 73:3833, 1976; Garcia-Sevilla, JA etal, Naunyn-Schmiedeberg's *Arch Pharmac*, 305:213, 1978; Yajima, H etal, *Chem Pharm Bull*, 27:2238, 1979; Andersson, S etal, Scand *J Gastroenterol*, 15:253, 1980

Neosystem SC359   MW 1671.96   Bioactive; Folkers, K et al, *PNAS USA*, 73:3833, 1976

Sigma N 9885   FW 1672.0   ≥97% (HPLC) | Bioactive peptide; potent agonist in CNS; Folkers, K et al, *Proc Natl Acad Sci USA*, 73:3833, 1976; Yajima, H et al, *Chem Pharm Bull*, 27:2238, 1979; Andersson, S et al, *Scan J Gastroent*, 15:253, 1980

Glp-Leu-Tyr-Glu-Asn-Lys

Synonyms: Neurotensin (1-6)

American Peptide 62-1-21   MW 776.8 $C_{35}H_{52}N_8O_{12}$

Bachem H-4440.0005   MW 776.85 $C_{35}H_{52}N_8O_{12}$ Store at -15°C

ICN 152934   MW 776.8 $C_{35}H_{52}N_8O_{12}$

Sigma N 9760   FW 776.8 $C_{35}H_{52}N_8O_{12}$ ≥97% (HPLC) | Bioactive peptide; used to identify antibody specificities in cross-reactivity studies; Carraway, R, *Ann NY Acad Sci*, 400:17, 1982

Glp-Leu-Tyr-Glu-Asn-Lys-Pro-Arg

Synonyms: Neurotensin (1-8)

Bachem H-4445.0005   Store at -15°C

ICN 152935   MW 1030.1

Sigma N 0509   FW 1030.1   ≥97% (HPLC) | Bioactive peptide; used to identify antibody specificities in cross-reactivity studies; Carraway, R, *Ann NY Acad Sci*, 400:17, 1982

Glp-Leu-Tyr-Glu-Asn-Lys-Pro-Arg-Arg-Pro

Synonyms: Neurotensin (1-10)

Neosystem SC364   MW 1283.45   Bioactive

Glp-Leu-Tyr-Glu-Asn-Lys-Pro-Arg-Arg-Pro-Phe-Ile-Leu
Glp-Leu-Tyr-Glu-Asn-Lys-Pro-Arg-Arg-Pro-D-Phe-Ile-Leu

Synonyms: Neurotensin, (D-Phe[11])-

ICN 152930   MW 1656.9

Glp-Leu-Tyr-Glu-Asn-Lys-Pro-Arg-Arg-Pro-Phe-Ile-Leu

Synonyms: Neurotensin, (Phe[11])-

ICN 152931   MW 1656.9

Glp-Leu-Tyr-Glu-Asn-Lys-Pro-Arg-Arg-Pro-Phe-Ile-Leu

Glp-Leu-Tyr-Glu-Asn-Lys-Pro-Arg-Arg-Pro-D-Phe-Ile-Leu

Synonyms: Neurotensin, (D-Phe[11])-

Sigma N 0136   FW 1656.9   ≥97% (HPLC) | Bioactive peptide

Glp-Leu-Tyr-Glu-Asn-Lys-Pro-Arg-Arg-Pro-Phe-Ile-Leu

Synonyms: Neurotensin, (Phe[11])-

Sigma N 0386   FW 1656.9   ≥95% (HPLC) | Bioactive peptide

Glp-Leu-Tyr-Glu-Asn-Lys-Pro-Arg-Arg-Pro-Trp-Ile-Leu

Glp-Leu-Tyr-Glu-Asn-Lys-Pro-Arg-Arg-Pro-D-Trp-Ile-Leu

Synonyms: Neurotensin, (D-Trp[11])-

American Peptide 62-1-23   MW 1696.0 $C_{80}H_{122}N_{22}O_{19}$
Neurotensin antagonist; Quirion, R et al, *Eur J Pharm*, 61:309, 1980

Bachem H-4475.0005   Store at -15°C

Glp-Leu-Tyr-Glu-Asn-Lys-Pro-Arg-Arg-Pro-Trp-Ile-Leu

Synonyms: Neurotensin, (Trp[11])-

Bachem H-7130.0005   Store at -15°C

Glp-Leu-Tyr-Glu-Asn-Lys-Pro-Arg-Arg-Pro-Trp-Ile-Leu

Glp-Leu-Tyr-Glu-Asn-Lys-Pro-Arg-Arg-Pro-D-Trp-Ile-Leu

Synonyms: Neurotensin, (D-Trp[11])-

ICN 152932   Neurotensin antagonist; Quirion, R etal, *Eur J Pharmac*, 61:309, 1980; Quirion, R etal, *Neuropeptides*, 1:253, 1981

Sigma N 2760   FW 1696.0   ≥97% (HPLC) | Bioactive peptide; neurotensin antagonist; Quirion, R et al, *Eur J Pharmacol*, 61:309, 1980; Quirion, R et al, *Neuropeptides*, 1:253, 1981

Glp-Leu-Tyr-Glu-Asn-Lys-Pro-Arg-Arg-Pro-Tyr

Synonyms: Neurotensin (1-11)

American Peptide 62-1-22   MW 1446.7 $C_{66}H_{99}N_{19}O_{18}$ St Pierre, S et al, *J Med Chem*, 24:370, 1981

Bachem H-4455.0005   Store at -15°C

ICN 152937   MW 1446.6   St-Pierre, S etal, *J Med Chem*, 24:370, 1981

Neosystem SC399   MW 1446.63   Bioactive

Sigma N 0634   FW 1446.6   ≥97% (HPLC) | Bioactive peptide; used to identify antibody specificities in cross-reactivity studies; Carraway, R, *Ann NY Acad Sci*, 400:17, 1982; St-Pierre, S et al, *J Med Chem*, 24:370, 1981

Glp-Leu-Tyr-Glu-Asn-Lys-Pro-Arg-Arg-Pro-Tyr-Ile-Leu

Synonyms: Neurotensin

Alexis 163-002   MW 1673.0 $C_{78}H_{121}N_{21}O_{20}$ ≥98%; lyophilized powder; soluble in water | Bissette, G et al, *Life Sci*, 23:2173, 1978; Carraway, R & Leeman, SE, *J Biol Chem*, 248:6854, 1973

American Peptide 60-1-12   MW 1673.0 $C_{78}H_{121}N_{21}O_{20}$
Isolated primarily from the nervous system of the hypothalamus, median eminence, substantia nigra & brain stem, has many neurotransmitter-like characteristics; lowers blood pressure, increases vascular permeability & increases plasma levels of glucagon, glucose, growth hormone & prolactin in rats; Carraway, RE et al, *J Biol Chem*, 248:6854, 1973; Rivier, JE et al, *J Med Chem*, 20:1409, 1977

Glp-Leu-Tyr-Glu-Asn-Lys-Pro-Arg-Arg-Pro-Tyr-Ile-Leu

Glp-Leu-Tyr-Glu-Asn-Lys-Pro-Arg-Arg-Pro-D-Tyr-Ile-Leu

Synonyms: Neurotensin, (D-Tyr[11])-

American Peptide 62-1-10   MW 1673.0 $C_{78}H_{121}N_{21}O_{20}$ A neurotensin analog which induces significant hypothermic & motor activities; Loosen, PT et al, *Neuropharmacology*, 17:109, 1978

Glp-Leu-Tyr-Glu-Asn-Lys-Pro-Arg-Arg-Pro-Tyr-Ile-Leu

Synonyms: Neurotensin

Bachem H-4435.0001   Store at -15°C

Glp-Leu-Tyr-Glu-Asn-Lys-Pro-Arg-Arg-Pro-Tyr-Ile-Leu

Glp-Leu-Tyr-Glu-Asn-Lys-Pro-Arg-Arg-Pro-D-Tyr-Ile-Leu

Synonyms: Neurotensin, (D-Tyr[11])-

Bachem H-4480.0005   MW 1672.95 $C_{78}H_{121}N_{21}O_{20}$ Store at -15°C

Glp-Leu-Tyr-Glu-Asn-Lys-Pro-Arg-Arg-Pro-Tyr-Ile-Leu

Synonyms: Neurotensin

Calbiochem 05-23-1200   MW 1673 $C_{78}H_{121}N_{21}O_{20}$ >99% (HPLC); lyophilized solid; soluble in 5% acetic acid | Potent hypothalamic peptide neurotransmitter; lowers blood pressure, increases vascular permeability & increases plasma levels of glucagon, glucose, growth hormone & prolactin in rats; a modulator of noradrenergic activity; heterogeneously distributed in the mammalian central nervous system & has many neurotransmitter-like characteristics; Tyler, BM et al, *Brain Res*, 792:246, 1998; Tsuda, K & Masuyuma, Y, *Am J Hypertens*, 6:473, 1993; *Merck Index*, 12:6569

ICN 151740   Carraway, R & S Leeman, *JBC*, 250:1912, 1975; Rivier, JE etal, *J Med Chem*, 20:1409, 1977

Glp-Leu-Tyr-Glu-Asn-Lys-Pro-Arg-Arg-Pro-Tyr-Ile-Leu

Glp-Leu-Tyr-Glu-Asn-Lys-Pro-Arg-Arg-Pro-3,5-Dibromo-Tyr-Ile-Leu

Synonyms: Neurotensin, (3,5-Dibromo-Tyr[11])-

ICN 152928   MW 1830.7

Glp-Leu-Tyr-Glu-Asn-Lys-Pro-Arg-Arg-Pro-Tyr-Ile-Leu

Glp-Leu-Tyr-Glu-Asn-Lys-Pro-Arg-Arg-Pro-D-Tyr-Ile-Leu

Synonyms: Neurotensin, (D-Tyr[11])-

ICN 152933   MW 1672.9   Neurotensin analog which induces significant hypothermic & motor activity; Loosen, PT etal, *Neuropharmacology*, 17:109, 1978; Jolicoeur, FB etal, *Peptides*, 2:171, 1981

Glp-Leu-Tyr-Glu-Asn-Lys-Pro-Arg-Arg-Pro-Tyr-Ile-Leu

Synonyms: Neurotensin

Neosystem SC066   MW 1672.94   Bioactive; Carraway, R & Leeman, SE, *JBC*, 248:6854, 1973

Glp-Leu-Tyr-Glu-Asn-Lys-Pro-Arg-Arg-Pro-Tyr-Ile-Leu

Glp-Leu-Tyr(I)-Glu-Asn-Lys-Pro-Arg-Arg-Pro-Tyr-Ile-Leu

Synonyms: Neurotensin, (Monoiodo-Tyr[3])-

Neosystem SC480   MW 1798.84   Bioactive; for specific Tyr[3] iodination

---

Glp-Leu-Tyr-Glu-Asn-Lys-Pro-Arg-Arg-Pro-Tyr-Ile-Leu
Glp-Leu-Tyr-Glu-Asn-Lys-Pro-Arg-Arg-Pro-D-Tyr-Ile-Leu

Synonyms: Neurotensin, (D-Tyr[11])-

Sigma N 0261   FW 1672.9   ≥95% (HPLC) | Bioactive peptide; neurotensin analog which induces significant hypothermic & motor activities; Loosen, PT et al, *Neuropharmacology*, 17:109, 1978; Nolicoeur, FB et al, *Peptides*, 2:171, 1981

Glp-Leu-Tyr-Glu-Asn-Lys-Pro-Arg-Arg-Pro-Tyr-Ile-Leu
Glp-Leu-Tyr-Glu-Asn-Lys-Pro-Arg-Arg-Pro-3,5-Dibromo-Tyr-Ile-Leu

Synonyms: Neurotensin, (3,5-DiBromo-Tyr[11])-

Sigma N 3010   FW 1830.7   ≥97% (HPLC) | Bioactive peptide

Glp-Leu-Tyr-Glu-Asn-Lys-Pro-Arg-Arg-Pro-Tyr-Ile-Leu

Synonyms: Neurotensin (1-13)

Biogenesis 6750-1004   Synthetic   MW 1672.9   Lyophilized, consisting of 88% peptide material, acetate counter ions and residual H_2O; purified | Carraway et al, *JBC*, 248:6854, 1973

Peptides International PNT-4029-v   Synthetic   MW 1672.9 $C_{78}H_{121}N_{21}O_{20}$ >99% (HPLC); lyophilized amorphous powder | Bioactive; human, bovine, canine; Carraway, R & SE Leeman, *JBC*, 248:6854, 1973

Glp-Leu-Tyr-Glu-Asn-Lys-Pro-Arg-Arg-Pro-Tyr-Ile-Leu 2AcOH 6H₂O

Synonyms: Neurotensin

Peptides International PNT-4029   Synthetic   MW 1901.1 $C_{78}H_{121}N_{21}O_{20} \cdot 2CH_3COOH \cdot 6H_2O$ >99% (HPLC); lyophilized amorphous powder; bulk | Bioactive; human, bovine, canine; Carraway, R & SE Leeman, *JBC*, 248:6854, 1973

Glp-Leu-Tyr-Glu-Asn-Lys-Pro-Arg-Arg-Pro-Tyr-Ile-Leu Acetate Salt

Synonyms: Neurotensin

Sigma N 6383   FW 1672.9 (free base)   ≥95% (HPLC) | Bioactive peptide; neurotransmitter-like hypothalamic peptide that lowers blood pressure, increases vascular permeability & increases blood levels of glucagon, glucose, growth hormone & prolactin in rats; Carraway, R & Leeman, S, *J Biol Chem*, 250:1912, 1975; Rivier, JE et al, *J Med Chem*, 20:1409, 1977

Glp-Leu-Tyr-Glu-Asn-Lys-Ser-Arg-Arg-Pro-Tyr-Ile-Leu

Synonyms: Neurotensin, (Ser[7])-

American Peptide 62-6-30   Guinea pig   MW 1662.9 $C_{76}H_{119}N_{21}O_{21}$

Neosystem SC358   Guinea pig   MW 1662.91   Bioactive; Shaw, C et al, *FEBS Lett*, 202:187, 1986

Glp-Lys-Arg-Pro-Ser-Gln-Arg-Ser-Lys-Tyr-Leu

Synonyms: Myelin Basic Protein (4-14), (Glp[4])-; Protein Kinase C Substrate; Protein Kinase Related Peptide

American Peptide 86-2-40   MW 1373.6 $C_{60}H_{100}N_{20}O_{17}$ Derived from Myelin Basic Protein (MBP), this peptide is selective substrate for protein kinase C; Yasuda, I et al, *BBRC*, 166:1220, 1990

Peptides International SKL-4237-v   MW 1373.6 $C_{60}H_{100}N_{20}O_{17}$ >98% (HPLC); lyophilized amorphous powder | Yasuda, I et al, *BBRC*, 166:1220, 1990

Sigma P 2186   FW 1373.6   ≥97% (HPLC) | Bioactive peptide; specific substrate for assay of proteinkinase C; Yasuda, I et al, *Biochem Biophys Res Commun*, 166:1220, 1990

Glp-Lys-Arg-Pro-Ser-Gln-Arg-Ser-Lys-Tyr-Leu Acetate Salt

Synonyms: Protein Kinase C Substrate-Myelin Basic Protein; Myelin Basic Protein (4-14)

Calbiochem 539492   Synthetic   MW 1373.6 $C_{60}H_{100}N_{20}O_{17}$ ≥95% (HPLC); lyophilized solid; soluble in 5% acetic acid | Derived from MBP; acts as a specific substrate for protein kinase C; Farrar, YJK et al, *Biochem Biophys Res Comm*, 180:694, 1991; Yasuda, I et al, *Biochem Biophys Res Comm*, 166:1220, 1990

Glp-Lys-Trp-Ala-Pro

Synonyms: Bradykinin Potentiating Pentapeptide; Bradykinin Potentiating Pentapeptide 5a

Bachem H-2225.0005   MW 611.7 $C_{30}H_{41}N_7O_7$ Store at -15°C

ICN 152795   MW 611.7 $C_{30}H_{41}N_7O_7$ Ferreira, SH etal, *Biochemistry*, 9:2583, 1970; Ufkes, JGN etal, *Eur J Pharmac*, 40:137, 1976

Sigma P 8772   FW 611.7 $C_{30}H_{41}N_7O_7$ ≥95% (HPLC) | Bioactive peptide; competitive inhibitor of angiotensin-I converting enzyme; potentiates bradykinin; Ferreira, SH et al, *Biochemistry*, 9:2583, 1970

Glp-Met-Ala-Val-Lys-Lys-Tyr-Leu-Asn-Ser-Ile-Leu-Asn Amide

Synonyms: Vasoactive Intestinal Peptide (16-28), (Glp[16])-

Bachem H-5635.0001   Human, bovine, porcine, rat   MW 1503.83 $C_{68}H_{114}N_{18}O_{18}S$ Store at -15°C

American Peptide 48-6-17   Porcine   MW 1503.8 $C_{68}H_{114}N_{18}O_{18}S$

Glp-Met-Ala-Val-Lys-Lys-Tyr-Leu-Asn-Ser-Val-Leu-Thr Amide

Synonyms: Vasoactive Intestinal Peptide (16-28), (Glp[16])-; Gastrointestinal Peptide; Vasoactive Intestinal Peptide (16-28), (Glp[16])-

ICN 152899

Bachem H-2818.0001   Chicken   Store at -15°C

Sigma V 0879   Chicken   FW 1476.8   ≥97% (HPLC) | Bioactive peptide; Bodanszky, M et al, *Bioorg Chem*, 8:399, 1979

Glp-Phe

Bachem G-3910.0250   MW 276.29 $C_{14}H_{16}N_2O_4$ Store at -15°C

Glp-Phe-Arg-His-Asp-Ser-Gly-Tyr-Glu-Val-His-His-Gln-Lys-Leu-Val-Phe-Phe-Ala-Glu-Asp-Val-Gly-Ser-Asn-Lys-Gly-Ala-Ile-Ile-Gly-Leu-Met-Val-Gly-Gly-Val-Val-Ile-Ala

Synonyms: Amyloid β-Protein (3-42), (Glp[3])-

Bachem H-4796.0500   MW 4309.92 $C_{196}H_{299}N_{53}O_{55}S$ Store at -15°C

Peptides International PAB-4367-v   Human synthetic   MW 4309.9 $C_{196}H_{299}N_{53}O_{55}S$ >95% (HPLC); lyophilized amorphous powder; trifluoroacetate | Bioactive; major neuritic plaque component in Alzheimer's Disease; dominant deposition in senile plaques; Saido, TC et al, *Neuron*, 14:457, 1995; Iwatsubo, T et al, *Am J Pathol*, 149:1823, 1996; Kuo, Y-M et al, *BBRC*, 237:188, 1997

Glp-Phe-Gly Amide

Bachem H-3856.0025   Store at -15°C

Glp-Phe-Ile-Asn-Asp-Ala-Glu-Thr-Glu-Leu-Met-Met-
Ser-Lys-Leu-Pro-Leu-Glu-Asn-Pro-Val-Val-Leu-Asn-
Ser-Phe-His-Phe-Ala-Ala-Asp-Cys-Cys-Thr-Ser-Tyr-
Ile-Ser-Gln-Ser-Ile-Pro-Cys-Ser-Leu-Met-Lys-Ser-
Tyr-Phe-Glu-Thr-Ser-Ser-Glu-Cys-Ser-Lys-Pro-Gly-
Val-Ile-Phe-Leu-Thr-Lys-Lys-Gly-Arg-Gln-Val-Cys-
Ala-Lys-Pro-Ser-Gly-Pro-Gly-Val-Gln-Asp-Cys-Met-
Lys-Lys-Leu-Lys-Pro-Tyr-Ser-Ile

Synonyms: Macrophage Inflammatory Protein-5

Bachem H-4612.0010   Human   Store at -15°C

Glp-Phe-Leu
Glp-Phe-Leu-pNA

Synonyms: Thiol Protease Substrate; Papain Substrate; Ficin Substrate; Bromelain Substrate

Bachem L-1460.0050   Store at -15°C

ICN 156124   MW 509.6 $C_{26}H_{31}N_5O_6$ Chromogenic substrate; Filippova, IY etal, *Anal Biochem*, 143:298, 1984

Peptides International SPL-3130   MW 509.56 $C_{26}H_{31}N_5O_6$ >99% (HPLC); lyophilized amorphous powder | Filippova, LY et al, *Anal Biochem*, 143:293, 1984

Sigma P 3169   FW 509.6 $C_{26}H_{31}N_5O_6$ Chromogenic substrate for thiol proteases; Filippova, IY et al, *Anal Biochem*, 143:293, 1984

Glp-Phe-Phe-Gly-Leu-Met Amide

Synonyms: Substance P (6-11), (Glp[6])-

American Peptide 70-1-70   MW 723.9 $C_{36}H_{49}N_7O_7S$ This fragment increased the blood pressure after injection into the cerebral ventricles

Bachem H-4920.0005   MW 723.89 $C_{36}H_{49}N_7O_7S$ Store at -15°C

ICN 153030   Increases blood pressure when injected into rat cerebral ventricles; Bury, RW & ML Mashford, *J Med Chem*, 19:854, 1976; Traczyk, WZ & J Kubicki, *Neuropharmacology*, 19:607, 1980

Sigma S 3136   FW 723.9 $C_{36}H_{49}N_7O_7S$ ≥95% (HPLC) | Bioactive peptide; Substance P agonist; increased blood pressure when injected into cerebral ventricles of rat; Traczyk, WZ & Kubicki, J, *Neuropharmacology*, 19:607, 1980

Glp-Phe-Phe-Pro-Leu-Met Amide

Synonyms: Substance P (6-11), (Glp[6],Pro[9])-; Septide

American Peptide 70-9-15   MW 764.0 $C_{39}H_{53}N_7O_7S$ Specific agonist for substance P receptor; Wormser, U et al, *JMC*, 29:1284, 1986

Bachem H-4925.0001   MW 763.96 $C_{39}H_{53}N_7O_7S$ Store at -15°C

Calbiochem 05-23-0612   MW 763.9 $C_{39}H_{53}N_7O_7S$ >95% (HPLC); lyophilized solid; soluble in 10% acetic acid | Agonist for neurokinin-1 receptors; binds at a site that is distinct from substance P binding site; Pradier, L et al, *Mol Pharmacol*, 45:287, 1994; Laufer, R et al, *J Pharmacol Exp Ther*, 245:889, 1988

ICN 153031   Potent & selective agonist for the substance P *P*-receptor subtype; Laufer, R etal, *J Med Chem*, 29:1284, 1986; Wormser, U etal, *EMBO J*, 5:2805, 1986

Neosystem SC390   MW 763.95   Bioactive; tachykinin; selective substance P receptor peptide; Wormser, U et al, *The EMBO Journal*, 5:2805, 1986

Sigma S 6764   FW 764.0 $C_{39}H_{53}N_7O_7S$ ≥95% (HPLC) | Bioactive peptide; potent & selective agonist for Substance P *P*-receptor subtype; Wormser, U et al, *EMBO J*, 5:2805, 1986; Laufer, R et al, *J Med Chem*, 29:1284, 1986

Glp-Phe-Pro Amide

Synonyms: Thyrotropin Releasing Hormone, (Phe[2])-

American Peptide 52-0-83   MW 372.4 $C_{19}H_{24}N_4O_4$

Bachem H-2462.0050   MW 372.42 $C_{19}H_{24}N_4O_4$ Store at -15°C

Glp-Phe-Pro-Ser-Tyr-Phe-Leu-Arg-Pro-Gly
Glp-D-Phe-Pro-Ser-Tyr-D-Phe-Leu-Arg-Pro-Gly Amide

Synonyms: Luteinizing Hormone Releasing Hormone, (D-Phe[2,6],Pro[3])-; Luteinizing Hormone Releasing Hormone Antagonist

American Peptide 54-0-36   MW 1193.5 $C_{59}H_{80}N_{14}O_{13}$ Potent antagonist; Humphries, J et al, *J Med Chem*, 21:120, 1978

Bachem H-4045.0001   MW 1193.37 $C_{59}H_{80}N_{14}O_{13}$ Store at -15°C

ICN 152912   Potent LH-RH antagonist; Humphries, J etal, *J Med Chem*, 21:120, 1978

Sigma L 4261   FW 1193.4   ≥97% (HPLC) | Bioactive peptide; potent LRH antagonist; Humphries, J et al, *J Med Chem*, 21:120, 1978

Glp-Phe-Thr-Asn-Val-Ser-Cys-Thr-Thr-Ser-Lys-Gln-Cys-Trp-Ser-Val-Cys-Gln-Arg-Leu-His-Asn-Thr-Ser-Arg-Gly-Lys-Cys-Met-Asn-Lys-Lys-Cys-Arg-Cys-Tyr-Ser

Synonyms: Charybdotoxin, (Gln[12])-

ICN 159567   MW 4294.2

Sigma C 8089   FW 4294.9   ≥85% (HPLC) | Bioactive peptide; disulfide bonds: 7-28, 13-33, 17-35

Glp-Phe-Thr-Asn-Val-Ser-Cys-Thr-Thr-Ser-Lys-Glu-Cys-Trp-Ser-Val-Cys-Gln-Arg-Leu-His-Asn-Thr-Ser-Arg-Gly-Lys-Cys-Met-Asn-Lys-Glu-Cys-Arg-Cys-Tyr-Ser

Synonyms: Charybdotoxin, (Glu[32])-

Bachem H-4734.0100   MW 4296.89 $C_{175}H_{272}N_{56}O_{57}S_7$ Store at -15°C | Disulfide bonds: Cys[7]-Cys[28], Cys[13]-Cys[33], Cys[17]-Cys[35]

Glp-Phe-Thr-Asn-Val-Ser-Cys-Thr-Thr-Ser-Lys-Glu-Cys-Trp-Ser-Val-Cys-Gln-Arg-Leu-His-Asn-Thr-Ser-Arg-Gly-Lys-Cys-Met-Asn-Lys-Lys-Cys-Arg-Cys-Tyr-Ser

Synonyms: Charybdotoxin

Bachem H-9595.0100   Store at -15°C | Disulfide bonds: Cys[7]-Cys[28], Cys[13]-Cys[33], Cys[17]-Cys[35]

Glp-Phe-Thr-Asn-Val-Ser-Cys-Thr-Thr-Ser-Lys-Glu-Cys-Trp-Ser-Val-Cys-Gln-His-Leu-His-Asn-Thr-Ser-Arg-Gly-Lys-Cys-Met-Asn-Lys-Lys-Cys-Arg-Cys-Tyr-Ser

Synonyms: Charybdotoxin, (His[19])-

ICN 159568   MW 4276.2

Glp-Phe-Thr-Asn-Val-Ser-Cys-Thr-Thr-Ser-Lys-Glu-Cys-Trp-Ser-Val-Cys-Gln-Arg-Leu-His-Asn-Thr-Ser-Arg-Gly-Lys-Cys-Met-Asn-Lys-Lys-Cys-Arg-Cys-Tyr-Ser

Synonyms: Charybdotoxin, (Lys(Ac)[11])-

ICN 159569   MW 4337.2

Glp-Phe-Thr-Asn-Val-Ser-Cys-Thr-Thr-Ser-Lys-Glu-Cys-Trp-Ser-Val-Cys-Gln-His-Leu-His-Asn-Thr-Ser-Arg-Gly-Lys-Cys-Met-Asn-Lys-Lys-Cys-Arg-Cys-Tyr-Ser

Synonyms: Charybdotoxin, (His[19])-

Sigma C 7839   FW 4276.8;   ≥85% (HPLC) | Disulfide bonds: 7-28, 13-33, 17-35; bioactive peptide

Glp-Phe-Thr-Asn-Val-Ser-Cys-Thr-Thr-Ser-Lys-Glu-Cys-Trp-Ser-Val-Cys-Gln-Arg-Leu-His-Asn-Thr-Ser-Arg-Gly-Lys-Cys-Met-Asn-Lys-Lys-Cys-Arg-Cys-Tyr-Ser

Glp-Phe-Thr-Asn-Val-Ser-Cys-Thr-Thr-Ser-Lys(Ac)-Glu-Cys-Trp-Ser-Val-Cys-Gln-Arg-Leu-His-Asn-Thr-Ser-Arg-Gly-Lys-Cys-Met-Asn-Lys-Lys-Cys-Arg-Cys-Tyr-Ser

Synonyms: Charybdotoxin, (Lys(Ac)[11])-

Sigma C 8214   FW 4337.9   ≥85% (HPLC) | Disulfide bonds: 7-28, 13-33, 17-35; bioactive peptide

---

Glp-Phe-Thr-Asn-Val-Ser-Cys-Thr-Thr-Ser-Lys-Glu-Cys-Trp-Ser-Val-Cys-Gln-Arg-Leu-His-Asn-Thr-Ser-Arg-Gly-Lys-Cys-Met-Asn-Lys-Lys-Cys-Arg-Cys-Tyr-Ser

Synonyms: Ca^{2+}-Activated K^+ Channel Blocker; Charybdotoxin

ICN 158887   *Leiurus quinquestriatus* (scorpion)   MW 4295.9 >98% | Potent & specific inhibitor of Ca^{2+} activated K^+ channels in cells of the anterior pituitary & bovine aortic smooth muscle; demonstrates inhibition of K^+ conductance & human T lymphocyte mitogen-stimulated proliferation; Gimenez-Gallego, G et al, *PNAS*, 85:3329, 1988; Mackinnon, R & C Miller, *Science*, 245:1382, 1989; Deutsch, C et al, *JBC*, 266:3668, 1991; Price, M et al, *PNAS*, 86:10171, 1989

Sigma C 7802   *Leiurus quinquestriatus* (scorpion) venom   FW 4295.9   ≥95% (HPLC) | Disulfide bonds: 7-28, 13-33, 17-35; bioactive peptide; peptide found in scorpion venom; a potent & selective inhibitor of the Ca^{2+} activated K^+ channel present in GH_3 anterior pituitary cells & primary bovine aortic smooth muscle cells; Gimenez-Gallego, G et al, *PNAS USA*, 85:3329, 1988; MacKinnon, R & Miller, C, Science, 245:1382, 1989

Calbiochem 220332   *Leiurus quinquestriatus hebraeus*   MW 4295.9 $C_{176}H_{277}N_{57}O_{55}S_7$ >98% (HPLC); lyophilized solid; LD_{50}≤2000 mg/kg | Disulfide bonds: Cys^7-Cys^{28}; Cys^{13}-Cys^{33}; Cys^{17}-Cys^{35}; high-affinity ligand & potent blocker of the tetraethylammonium & voltage-sensitive Ca^{2+}-activated K^+ channel; Russell, SN et al, *Am J Physiol*, 267:C1729, 1994; Asano, M et al, *Br J Pharmacol*, 108:214, 1993; Price, M et al, *PNAS*, 86:19171, 1989; Gimenez-Gallego, G et al, *PNAS*, 85:3329, 1988; Moczydlowsky, E et al, *J Membr Biol*, 109:95, 1988; *Merck Index*, 12:2087

Peptides International PCB-4227-s   *Leiurus quinquestriatus hebraeus* (scorpion) synthetic   MW 4295.9 $C_{176}H_{277}N_{57}O_{55}S_7$ >95% (HPLC); lyophilized amorphous powder | Bioactive; Disulfide bonds: Cys^7-Cys^{28}, Cys^{13}-Cys^{33}, Cys^{17}-Cys^{35}; Garcia, ML et al, *Am J Physiol*, 269:C1, 1995; Gimenez-Gallego, G et al, *PNAS USA*, 85:3329, 1988; Lambert, P et al, *BBRC*, 170:684, 1990

American Peptide 41-0-10   *Leiurus quinquestriatus hebraeus* (scorpion) toxin   MW 4296.0 $C_{176}H_{277}N_{57}O_{55}S_7$ Disulfide bonds: Cys^7-Cys^{28}, Cys^{13}-Cys^{33}, Cys^{17}-Cys^{35}; potent & selective inhibitor of Ca^{2+}-activated K^+ channels; 37 amino acid peptide can block variety of Ca^{2+}-activated K^+ channels of varying conductance & transient increase blood pressure after intravenous injection; Gimenez-Gallego, G et al, *PNAS*, 82:3329, 1988; Miller et al, *Nature*, 313:316, 1985

Alexis 630-059   Synthetic   MW 4295.9 $C_{176}H_{277}N_{57}O_{55}S_7$ Lyophilized powder; potent neurotoxin | Disulfide bonds: Cys^7-Cys^{28}, Cys^{13}-Cys^{33}, Cys^{17}-Cys^{35}

Glp-Phe-Thr-Asp-Val-Asp-Cys-Ser-Val-Ser-Lys-Glu-Cys-Trp-Ser-Val-Cys-Lys-Asp-Leu-Phe-Gly-Val-Asp-Arg-Gly-Lys-Cys-Met-Gly-Lys-Lys-Cys-Arg-Cys-Tyr-Gln

Synonyms: Iberiotoxin, (Asp[19])-; Ca^{2+}-Activated K^+ Channel Blocker; Maxi-K^+ Channel Blocker

Bachem H-9940.0100   MW 4230.91 $C_{179}H_{274}N_{50}O_{55}S_7$ Store at -15°C | Disulfide bonds: Cys^7-Cys^{28}, Cys^{13}-Cys^{33}, Cys^{17}-Cys^{35}

ICN 159600   MW 4230.2

Peptides International PIB-4235-s   *Buthus tamulus* (scorpion) synthetic   MW 4230.9 $C_{179}H_{274}N_{50}O_{55}S_7$ >95% (HPLC); lyophilized amorphous powder | Bioactive; disulfide bonds Cys^7-Cys^{28}, Cys^{13}-Cys^{33}, Cys^{17}-Cys^{35}; Galvez, A et al, *JBC*, 265:11083, 1990

American Peptide 41-0-20   *Buthus tamulus* (scorpion) venom   MW 4230.9 $C_{179}H_{274}N_{50}O_{55}S_7$ Disulfide bonds: Cys^7-Cys^{28}, Cys^{13}-Cys^{33}, Cys^{17}-Cys35; selective inhibitor of Ca^{2+} -activated K^+ channels; shows 68% sequence homology with charybdotoxin & another scorpion-derived peptide toxin; selectively & reversibly blocks the high-conductance Ca^{2+}-activated K^+ channel; transiently increases blood pressure after intraventricular injection; Galvez, A et al, *JBC*, 265:11083, 1990; Candia et al, *Biophys J*, 63:583, 1992

Calbiochem 401001   *Buthus tamulus* (scorpion) venom   MW 4230.9 $C_{179}H_{274}N_{50}O_{55}S_7$ >99% (HPLC); lyophilized powder; LD_{50}≤2000 mg/kg | Disulfide bonds: Cys^7-Cys^{28}, Cys^{13}-Cys^{35}, Cys^{17}-Cys^{33}; highly selective & potent blocker of the high conductance Ca^{2+}-activated K^+ channel; does not affect other types of voltage-dependent K^+ channels that are inhibited by Charybdotoxin; modulates ChTx binding by an allosteric mechanism; Vanelli, G et al, *J Appl Physiol*, 76:1098, 1994; Galvez, A et al, *J Biol Chem*, 265:11083, 1990

ICN 158935   *Buthus tamulus* (scorpion) venom   MW 4231.8 98% | More specific than Charybdotoxin at inhibiting high conductance Ca^{2+}-activated K^+ channel via allosteric inhibition; Galvez, A et al, *JBC*, 265:11083, 1990

Sigma I 2141   *Buthus tamulus* (scorpion) venom   MW 4231.8 $C_{179}H_{274}N_{50}O_{55}S_7$ ≥85% (HPLC); white powder; peptide content & salt form information are provided with each lot; potent neurotoxin | Disulfide bonds: Cys^7-Cys^{28}, Cys^{13}-Cys^{35}, Cys^{17}-Cys^{33}; highly selective & potent blocker of the high conductance Ca^{2+}-activated K^+ channel; does not affect other types of voltage-dependent K^+ channels that are inhibited by charybdotoxin; modulates ChTx binding by an allosteric mechanism; Vanelli, G et al, *J Appl Physiol*, 76:1098, 1994; Galvez, A et al, *J Biol Chem*, 265:11083, 1990

Glp-Phe-Trp-Lys-Tyr Hydrochloride

Bachem H-1136.0005   Store at -15°C

Glp-Phe-Trp-Ser-Tyr-Ala-Leu-Arg-Pro-Gly
Glp-D-Phe-Trp-Ser-Tyr-D-Ala-Leu-Arg-Pro-Gly Amide

Synonyms: Luteinizing Hormone Releasing Hormone, (D-Phe[2],D-Ala[6])-

American Peptide 54-0-30   MW 1206.4 $C_{59}H_{79}N_{15}O_{13}$

ICN 152911   Beattie, CW et al, *J Med Chem*, 18:244, 1975

Sigma L 0387   FW 1206.4   ≥97% (HPLC) | Bioactive peptide; LRH antagonist; Yardley, JP et al, *J Med Chem*, 18:1244, 1975

Glp-Phe-Trp-Ser-Tyr-Trp-Leu-Arg-Pro-Gly
D-Glp-D-Phe-D-Trp-Ser-Tyr-D-Trp-Leu-Arg-Pro-Gly Amide

Synonyms: Luteinizing Hormone Releasing Hormone Antagonist; Luteinizing Hormone Releasing Hormone, (D-Glp[1],D-Phe[2],D-Trp[3,6])-

American Peptide 54-0-35   MW 1321.5 $C_{67}H_{84}N_{16}O_{13}$ Rivier, JE et al, *Life Sci*, 23:869, 1978

Bachem H-4040.0005   MW 1321.5 $C_{67}H_{84}N_{16}O_{13}$ Store at -15°C

ICN 152913   Potent LH-RH antagonist; Rivier, JE & WW Vale, *Life Sci*, 23:869, 1978

Glp-Phe-Trp-Ser-Tyr-Trp-Leu-Arg-Pro-Gly
D-Glp-D-Phe-D-Trp-Ser-Tyr-D-Trp-Leu-Arg-Pro-Gly Amide Acetate Salt

Synonyms: Luteinizing Hormone Releasing Hormone, (D-Glp[1],D-Phe[2],D-Trp[3,6])-

Sigma L 2636   FW 1321.5 (free base)   ≥97% (HPLC) | Bioactive peptide; Substance P antagonist; inhibits spinal cord vasomotor responses

Sigma L 6524   FW 1321.5 (free base)   ≥97% (HPLC) | Bioactive peptide; Substance P antagonist; inhibits spinal cord vasomotor responses; JE & Vale, WW, *Life Sci*, 23:869, 1978

Glp-Phg-Arg
Glp-Phg-Arg-pNA AcOH

Synonyms: Pefachrome®FXIa; Pefa-3371

Pentapharm 090-45, 090-40   MW 598.62   Highly sensitive chromogenic peptide substrate for factor XIa

Glp-Pro-Arg
Glp-Pro-Arg-pNA Hydrochloride

Synonyms: Tryptase Substrate

Neosystem SC1300   MW 502.51   Chromogenic substrate; McEuen, AR et al, *Biochem Pharmacol*, 52:331-340, 1996

Glp-Pro-Arg
Glp-Pro-Arg-AMC·AcOH

Synonyms: Pefafluor PCa; Pefa-5791

Pentapharm 089-25, 089-05   MW 599.6   Excitation wavelength 360 nm, emission wavelength 460 nm | Sensitive fluorogenic peptide substrate for activated protein C;k_{cat}:62.0 s^{-1}, K_M:0.56 mM

Glp-Pro-Arg-Arg-Lys-Leu-Cys-Ile-Leu-His-Arg-Asn-Pro-Gly-Arg-Cys-Tyr-Asp-Lys-Ile-Pro-Ala-Phe-Tyr-Tyr-Asn-Gln-Lys-Lys-Lys-Gln-Cys-Glu-Arg-Phe-Asp-Trp-Ser-Gly-Cys-Gly-Gly-Asn-Ser-Asn-Arg-Phe-Lys-Thr-Ile-Glu-Glu-Cys-Arg-Arg-Thr-Cys-Ile-Gly

Synonyms: Dendrotoxin, α-

Bachem H-1088.0100   Store at -15°C | Disulfide bonds: Cys^7-Cys^{57}, Cys^{16}-Cys^{40}, Cys^{32}-Cys^{53}

Alexis 630-013   *Dendroaspis angusticeps*   MW 7048.1 ≥97% (SDS-PAGE); lyophilized; dissolving 140 µg/mL gives a stock solution of 20 µM | Disulfide bonds: Cys^7-Cys^{57}, Cys^{16}-Cys^{40}, Cys^{32}-Cys^{53}; potent neurotoxin; potent & selective blocker of voltage-gated K^+ channels; Joubert, F & Talijaard, N, *Hoppe-Seyler's Z Physiol Chem*, 361:661, 1980; Strong, PN, *Pharmacol Ther*, 46:137, 1990; Stansfield, CE et al, *Neurosci Lett*, 64:299, 1986; Schulenburg, P et al, *Eur J Biochem*, 210:257, 1992; Harvey, AL & Karlsson, E, *Br J Pharmacol*, 77:153, 1982; Halliwell, JV et al, *PNAS*, 83:493, 1986; Dreyer, F & Penner, R, *J Physiol*, 386:455, 1987; Schauf, CL, *J Pharmacol Exp Ther*, 241:793, 1987; Stansfield, CE et al, *Neuroscience*, 23:893, 1987; Benishin, CG et al, *Mol Pharmacol*, 34:152, 1988; Castle, NA et al, *TINS*, 12:59, 1989; Lambert, P et al, *BBRC*, 170:684, 1990; Brau, J et al, *J Physiol*, 420:365, 1990; Hu, PS et al, *Eur J Pharmacol*, 209:87, 1991; Muniz, ZM et al, *Biochemistry*, 31:12297, 1992; Baker, M et al, *J Physiol*, 464:321, 1993; McGivern, J et al, *Br J Pharmacol*, 109:535, 1993

Calbiochem 253702   *Dendroaspis angusticeps*   MW 7048.1 $C_{305}H_{472}N_{98}O_{84}S_6$ >90% (SDS-PAGE); lyophilized; soluble in water; Disulfide bonds: Cys^7-Cys^{57}; Cys^{16}-Cys^{40}; Cys^{32}-Cys^{53}; LD_{50}≤2000 mg/kg | Presynaptic neurotoxin that selectively blocks fast activating voltage-gated K^+ channels & enhances neuromuscular transmission due to acetylcholine release at the neuromuscular junction; Smith, LA et al, *Biochem*, 32:5692, 1993; Tibbs, GR et al, *FEBS Lett*, 255:159, 1989; Benishin, CG et al, *Mol Pharmacol*, 34:152, 1988

Calbiochem 253704   *Dendroaspis angusticeps*   MW 7048.1 $C_{305}H_{472}N_{98}O_{84}S_6$ >97% (SDS-PAGE); lyophilized; soluble in water; Cys^7-Cys^{57}; Cys^{16}-Cys^{40}; Cys^{32}-Cys^{53}; LD_{50}≤2000 mg/kg; bioassay:confirmed via release of ^{3}H-GABA in primary culture of rat hippocampus in dose-dependent manner | Blocks inactivating voltage-gated K^+ channels in rat brain synaptosomes; exhibits higher selectivity for slowly inactivating outward K^+ currents; Muniz, ZM et al, *Biochemistry*, 31:12297, 1992; Hall, A et al, *Br J Pharmacol*, 113:959, 1994; Benishin, CG et al, *Mol Pharmacol*, 34:152, 1988; Brau, J et al, *J Physiol*, 420:365, 1990

Glp-Pro-Asn-Pro-Asp-Glu-Phe-Val-Gly-Leu-Met Amide

Synonyms: PG-KII; Neurokinin

American Peptide 62-0-43   MW 1229.4 $C_{55}H_{82}N_{13}O_{17}S$ PG-KII is neither an NK_1 nor an NK_2 receptor agonist but has a spectrum of biological actions close to that of the NK_3 receptor agonists; elicits strong contractile activity in guinea pig ileum & regulates inhibition of gastric acid secretion after administration centrally; Improta, G et al, *Peptides*, 17(6):1003, 1996

Glp-Pro-Asp-Pro-Asn-Ala-Phe-Tyr-Gly-Leu-Met Amide

Synonyms: Uperolein

American Peptide 62-0-67   MW 1234.4 $C_{57}H_{79}N_{13}O_{16}S$ Anastasi, A et al, *Experientia*, 31:394, 1975

Glp-Pro-Leu-Arg-Lys-Leu-Cys-Ile-Leu-His-Arg-Asn-Pro-Gly-Arg-Cys-Tyr-Gln-Lys-Ile-Pro-Ala-Phe-Tyr-Tyr-Asn-Gln-Lys-Lys-Lys-Gln-Cys-Glu-Gly-Phe-Thr-Trp-Ser-Gly-Cys-Gly-Gly-Asn-Ser-Asn-Arg-Phe-Lys-Thr-Ile-Glu-Glu-Cys-Arg-Arg-Thr-Cys-Ile-Arg-Lys

Synonyms: Dendrotoxin II

Alexis 630-017   *Dendroaspis p. polylepis*   MW 7149.4 ≥98% (SDS-PAGE); lyophilized powder; soluble in water | Disulfide bonds: Cys^7-Cys^{57}, Cys^{16}-Cys^{40}, Cys^{32}-Cys^{53}; potent neurotoxin; most potent member of the dendrotoxin group; Benoit, E & Dubois, JM, *Brain Res*, 377:374, 1986; Strong, PN, *Pharmacol Ther*, 46:137, 1990; Kondo, T et al, *Neurosci Res*, 13:207, 1992; Hopkins, WF et al, *Eur J Physiol*, 428:382, 1994

Glp-Pro-Leu-Arg-Lys-Leu-Cys-Ile-Leu-His-Arg-Asp-Pro-Gly-Arg-Cys-Tyr-Gln-Lys-Ile-Pro-Ala-Phe-Tyr-Tyr-Asn-Gln-Lys-Lys-Lys-Gln-Cys-Glu-Gly-Phe-Thr-Trp-Ser-Gly-Cys-Gly-Gly-Asn-Ser-Asn-Arg-Phe-Lys-Thr-Ile-Glu-Glu-Cys-Arg-Arg-Thr-Cys-Ile-Arg-Lys

Synonyms: Dendrotoxin I; Voltage-Dependant K^+ Channel Blocker

Peptides International PDN-4330-s   *Dendroaspis polylepis polylepis* (black mamba) synthetic   MW 7133.3 $C_{312}H_{487}N_{97}O_{84}S_6$ >99% (HPLC); lyophilized amorphous powder | Bioactive; Disulfide bonds: Cys^7-Cys^{57}, Cys^{16}-Cys^{40}, Cys^{32}-Cys^{53}; Strydom, DJ, *Nature New Biol*, 243:88, 1973; Bidard, J-N et al, *BBRC*, 143:383, 1987; Harvey, AL et al, *BBRC*, 163:394, 1989; Nishio, H et al, *J Pept Res*, 51:355, 1998

Glp-Pro-Leu-Pro-Asp-Cys-Cys-Arg-Gln-Lys-Thr-Cys-Ser-Cys-Arg-Leu-Tyr-Glu-Leu-Leu-His-Gly-Ala-Gly-Asn-His-Ala-Ala-Gly-Ile-Leu-Thr-Leu Amide

Synonyms: Orexin A; Appetite Boosting Peptide

Peptides International POR-4346-s   Human synthetic   MW 3561.2 $C_{152}H_{243}N_{47}O_{44}S_4$ >99% (HPLC); lyophilized amorphous powder | Bioactive; Disulfide bonds: Cys^6-Cys^{12}, Cys^7-Cys^{14}; rat, mouse, bovine; Sakurai, T et al, *Cell*, 92:573, 1998; de Lecea, L et al, *PNAS USA*, 95:322, 1998

Alexis 164-001   Human, bovine, rat, mouse   MW 3537.2 $C_{151}H_{247}N_{47}O_{43}S_4$ ≥96%; lyophilized | Disulfide bonds: Cys^6-Cys^{12}, Cys^7-Cys^{14}; new hypothalamic neuropeptide that stimulates food intake in rats with impressive potency; specifically binds to the new OX_1 receptor; Sakurai, T et al, *Cell*, 92:573, 1998; Schwartz, MW, *Nature Med*, 4:385, 1998

Bachem H-4172.0500   Human, bovine, rat, mouse   MW 3561.16 $C_{152}H_{243}N_{47}O_{44}S_4$ Store at -15°C | Disulfide bonds: Cys^6-Cys^{12}, Cys^7-Cys^{14}

Glp-Pro-Leu-Pro-Asp-Cys-Cys-Arg-Gln-Lys-Thr-Cys-Ser-Cys-Arg-Leu-Tyr-Glu-Leu-Leu-His-Gly-Ala-Gly-Asn-His-Ala-Ala-Gly-Ile-Leu-Thr-Leu

Synonyms: Orexin A

Neosystem SC1337 Human, bovine, rat, mouse MW 3561.12 Disulfide bonds: Cys[6]-Cys[12], Cys[7]-Cys[14]; bioactive neuropeptide; one of two new hypothalamic peptides regulating feeding behavior (the other is orexin-B); both come from the same precursor by proteolytic processing; stimulate food consumption in a dose-dependent manner when administered centrally to rats; Sakurai, T et al, *Cell*, 92:573-585, 1998; De Lecea, L et al, *PNAS*, USA, 95:322-327, 1998

Glp-Pro-Pro-Gly-Gly-Ser-Lys-Val-Ile-Leu-Phe

Synonyms: Neuropeptide, Head Activator; Hydra Peptide; Head Activator

American Peptide 60-0-48 MW 1125.5 $C_{54}H_{84}N_{12}O_{14}$ Bodenmuller, H et al, *Nature*, 293:579, 1981

Bachem H-3790.0001 Store at -15°C

ICN 153129 Hydra MW 1125.3 Morphogenic peptide from hydra; Birr, C etal, *FEBS Lett*, 131:317, 1981; Schaller, HC & H Bodenmueller, *Nature*, 293:579, 1981; Schaller, HC & H Bodenmueller, *Naturwissenschaften*, 68:252, 1981; Schaller, HC & H Bodenmueller, *PNAS*, 78:7000, 1981; Bodenmuller, H, *Nachr Chem Tech Lab*, 30:263, 1982

Glp-Pro-Pro-Gly-Gly-Ser-Lys-Val-Ile-Leu-Phe Acetate Salt

Synonyms: Hydra Peptide; Hydra Morphogenic Peptide

Sigma H 2136 FW 1125.3 (free base) ≥97% (HPLC) | Bioactive peptide; Schaller, HC & Bodenmuller, H, *Proc Natl Acad Sci USA*, 78:7000, 1981

Sigma H 4770 FW 1125.3 (free base) ≥97% (HPLC) | Bioactive peptide; Schaller, HC & Bodenmuller, H, *Proc Natl Acad Sci USA*, 78:7000, 1981

Glp-Pro-Pro-Gly-Gly-Ser-Lys-Val-Ile-Leu-Tyr

Synonyms: Head Activator Neuropeptide, (Tyr[11])-

ICN 154447 MW 1141.5 Bodenmuller, H etal, *Int J Peptide Prot Res*, 29:140, 1987

Glp-Pro-Ser-Lys-Asp-Ala-Phe-Ile-Gly-Leu-Met Amide

Synonyms: Eledoisin

American Peptide 62-0-56 MW 1188.4 $C_{54}H_{86}N_{13}O_{15}S$

Bachem H-2685.0001 MW 1188.42 $C_{54}H_{85}N_{13}O_{15}S$ Store at -15°C

Neosystem SC078 MW 1188.40 Bioactive; tachykinin

Sigma E 7880 FW 1188.4 ≥97% (HPLC) | Bioactive peptide; tachykinin originally isolated from the salivary gland of octopus with structural homology to substance P; potent vasodilator & hypotensive agent; induces salivation & increases capillary permeability

ICN 151028 Octopus salivary glands MW 1188.4 $C_{54}H_{85}N_{13}O_{15}S$

Glp-Pro-Val
Glp-Pro-Val-pNA

Synonyms: Elastase Substrate III

Calbiochem 324698 MW 445.5 $C_{21}H_{27}N_5O_6$ Lyophilized solid containing 45 mg D-mannitol & 5 mg substrate; ≥95% (HPLC); soluble in water & ethanol | Colorimetric substrate; Kramps, JA et al, *J Clin Lab Invest*, 43:427, 1983

Glp-Ser-Leu-Arg-Trp Amide

Synonyms: Antho-RW Amide I

Bachem H-6230.0025 MW 670.77 $C_{31}H_{46}N_{10}O_7$ Store at -15°C

ICN 153103 A neuropeptide originally isolated from sea anemones; Graff, D & Grimmelikhuijzen, CJP, *Brain Res*, 422:354, 1988

Sigma P 9924 Sea anemone FW 670.8 $C_{31}H_{46}N_{10}O_7$ ≥97% (HPLC) | Bioactive peptide; neuropeptide originally isolated from sea anemones; inhibits contraction of ectodermal tentacle muscles but induces contraction of endodermal muscles Graff, D & Grimmelikhuijzen, CJP, *Brain Res*, 422:354, 1988

Glp-Thr-Phe-Gln-Tyr-Ser-Arg-Gly-Trp-Thr-Asn Amide

Synonyms: Corazonin

Bachem H-8050.0001 MW 1369.46 $C_{62}H_{84}N_{18}O_{18}$ Store at -15°C

Sigma C 2921 American cockroach FW 1369.5 ≥97% (HPLC) | Bioactive peptide; cardioaccelerating peptide originally isolated from the corpus cardiacum of the American cockroach; the most potent insect cardioactive neuropeptide; Veenstra, JA, *FEBS Lett*, 250:231, 1989

Glp-Thr-Ser-Phe-Thr-Pro-Arg-Leu Amide

Synonyms: Leukopyrokinin

American Peptide 33-1-11 MW 931.1 $C_{42}H_{66}N_{12}O_{12}$ Myotropic substance was originally isolated from head extracts of the Madiera cockroach, *Leucophaea maderae*

Bachem H-6245.0001 Store at -15°C

ICN 153140 MW 931.1 $C_{42}H_{66}N_{12}O_{12}$ Nachman, RJ etal *BBRC*, 137:936, 1986

Sigma L 6268 FW 931.1 $C_{42}H_{66}N_{12}O_{12}$ ≥97% (HPLC) | Bioactive peptide

Glp-Trp

Bachem G-4545.0250 Store at -15°C

Glp-Trp
Glp-Trp-OEt

Bachem G-4550.0250 Store at -15°C

Glp-Trp-Gly Amide

Bachem H-3858.0025 Store at -15°C

Glp-Trp-Leu-Arg-Gly-Arg-Phe Amide Hydrochloride

Bachem H-1056.0025 Store at -15°C

Glp-Trp-Leu-Gly-Gly-Arg-Phe Amide Hydrochloride

Bachem H-1058.0025 Store at -15°C

Glp-Trp-Lys-Leu-Gly-Arg-Phe Amide

Bachem H-7310.0025 MW 916.09 $C_{45}H_{65}N_{13}O_8$ Store at -15°C

Glp-Trp-Lys-Leu-Gly-Arg-Phe-Gly-Tyr Hydrochloride

Bachem H-7315.0025 Store at -15°C

Glp-Trp-Pro-Arg-Pro-Gln-Ile-Pro-Pro

Synonyms: Bradykinin Potentiator; Bradykinin Potentiating Pentapeptide 9a; Angiotensin Converting Enzyme Inhibitor, (Sar[1], Ile[7])-; Bradykinin Potentiating Factor; SQ 20881; Angiotensin Converting Enzyme Inactivator

American Peptide 12-1-80 MW 1101.3 $C_{53}H_{76}N_{14}O_{12}$ Ondetti, MA et al, *Biochemistry*, 10:4033, 1971

Bachem H-2215.0010 MW 1101.27 $C_{53}H_{76}N_{14}O_{12}$ Store at -15°C

ICN 152743 MW 1101.3 Greene, LJ etal, *Adv Exp Med Biol*, 8:81, 1970; Ondetti, MA etal, *Biochemistry*, 10:4033, 1971; Cheung, DW & DW Cushman, *Biochim Biophys Acta*, 293:451, 1973; Martin, LC etal, *Biochem J*, 184:713, 1979; Malborough, DI etal, *Arch Biochem Biophys*, 210:43, 1981

Sigma A 0773 FW 1101.3 ≥95% (HPLC) | Bioactive peptide; Ondetti, MA et al, *Biochemistry*, 10:4033, 1971

Glp-Tyr-Thr-Asn-Val-Ser-Cys-Thr-Thr-Ser-Lys-Glu-Cys-Trp-Ser-Val-Cys-Gln-Arg-Leu-His-Asn-Thr-Ser-Arg-Gly-Lys-Cys-Met-Asn-Lys-Lys-Cys-Arg-Cys-Tyr-Ser

Synonyms: Charybdotoxin, (Tyr[2])-

ICN 159570 MW 4311.2

Sigma C 7964 FW 4311.9 ≥85% (HPLC) | Disulfide bonds: 7-28, 13-33, 17-35; bioactive peptide

---

Glp-Val

Bachem G-1310.0250 Store at -15°C

Glp-Val-Asn-Phe-Ser-Pro-Gly-Trp-Gly-Thr Amide

Synonyms: Hypertrehalosemic Neuropeptide

Bachem H-6815.0001 *Nauphoeta cinerea* MW 1074.16 $C_{50}H_{67}N_{13}O_{14}$ Store at -15°C

ICN 153104 *Nauphoeta cinerea* (cockroach) MW 1074.2 A neuropeptide with an AA sequence similar to that of adipokinetic hormone I; Gaede, G & KL Rinehart, *BBRC*, 141:774, 1986

Sigma P 0175 *Nauphoeta cinerea* (cockroach) FW 1074.2 ≥95% (HPLC) | Bioactive peptide; AA sequence similar to that of Adipokinetic Hormone I; involved in carbohydrate metabolism & increases the trehalose content of hemolymph; Gaede, G & Rinehart, KL, *Biochem Biophys Res Commun*, 141:774, 1986

Glp-Val-Asn-Phe-Ser-Thr-Gly-Trp Amide

Synonyms: Adipokinetic Hormone G

Bachem H-9230.0005 *Gryllus bimaculatus* MW 919.99 $C_{43}H_{57}N_{11}O_{12}$ Store at -15°C

ICN 154473 *Gryllus bimaculatus* MW 920.1 Grade, G & KL Rinehart, *BBRC*, 149:908, 1987

Glp-Val-Asp-Pro-Asn-Ile-Gln

Bachem H-1012.0005 MW 795.85 $C_{34}H_{53}N_9O_{13}$ Store at -15°C

Glp-Val-Asp-Pro-Asn-Ile-Gln-Ala

Bachem H-1008.0005 MW 866.93 $C_{37}H_{58}N_{10}O_{14}$ Store at -15°C

Glp-Val-Lys-Leu-Tyr-Arg-Pro Amide Hydrochloride

Bachem H-9760.0005 Store at -15°C

Glp-Val-Pro-Gln-Trp-Ala-Val-Gly-His-Phe-Met Amide

Synonyms: Ranatensin

American Peptide 62-0-60 MW 1281.5 $C_{61}H_{84}N_{16}O_{13}S$ Nakajima, T et al, *Fed Proc*, 29:282, 1970

Bachem H-4930.0001 Store at -15°C

ICN 153175 Nakajama, T etal, *Fed Proc*, 29:282, 1970

Neosystem SC164 MW 1281.50 Bioactive; bombesin-related peptide; Nakajima, T et al, *Fed Proc*, 29:282, 1970

Glu-Abu
γ-Glu-Abu

Synonyms: Glu(Abu-OH)-OH

Bachem G-3900.0250 MW 232.25 $C_9H_{16}N_2O_5$ Store at -15°C

Glu-Abu-Gly
γ-Glu-Abu-Gly

Synonyms: Ophthalmic Acid; Glu(Abu-Gly-OH)-OH

Bachem H-3145.0050 MW 289.29 $C_{11}H_{19}N_3O_6$ Store at -15°C

Glu-Abz
γ-Glu-4-Abz

Synonyms: Glu(4-Abz-OH)-OH

Bachem M-1435.0050 MW 266.25 $C_{12}H_{14}N_2O_5$ Store at RT

Glu-Ala

Bachem G-1900.0250 MW 218.21 $C_8H_{14}N_2O_5$ Store at -15°C

Glu-Ala
γ-Glu-Ala

Synonyms: Glu(Ala-OH)-OH

Bachem G-1905.0250 MW 218.21 $C_8H_{14}N_2O_5$ Store at -15°C

Glu-Ala
Glu-Ala-pNA

Bachem L-1260.0025 MW 338.32 $C_{13}H_{19}N_3O_6$ Store at -15°C

Glu-Ala

Sigma G 3376 FW 218.2 $C_8H_{14}N_2O_5$

Glu-Ala-Ala-Leu-Lys-Leu-Ala-Arg
FMOC-Glu-Ala-Ala-Leu-Lys-Leu-Ala-Arg

Synonyms: C3a (70-77), (FMOC-Glu[70],Ala[71,72],Lys[74])-

Bachem B-2280.0005 MW 1093.29 $C_{53}H_{80}N_{12}O_{13}$ Store at -15°C

Glu-Ala-Arg
BOC-Glu(OBzl)-Ala-Arg-AMC Hydrochloride

Bachem I-1575.0050 MW 758.27 $C_{36}H_{47}N_7O_9 \cdot$ HCl Store at -15°C

Glu-Ala-Arg
BOC-Glu(OBzl)-Ala-Arg-MCA

Synonyms: Factor XIa Substrate

Peptides International MER-3134-v MW 721.81 $C_{36}H_{47}N_7O_9$ >99% (HPLC); lyophilized amorphous powder | Kawabata, S et al, *Eur J Biochem*, 172:17, 1988

Glu-Ala-Arg
N-t-BOC-γ-Bzl-Glu-Ala-Arg MCA Hydrochloride

Synonyms: Factor XIa Substrate

Sigma B 4903 FW 758.3 $C_{36}H_{47}N_7O_9 \cdot$ HCl ~95% | Kawabata, S et al, *Eur J Biochem*, 172:17, 1988

Glu-Ala-Asn-Lys-Gly-Glu-Asn-Asn-Cys-Leu-Leu-His-Pro-Met

Synonyms: NEF MN (157-170); HIV I Peptide

Neosystem SC661 MW 1572.72 NEF peptide from HIV-1 subtype MN; due to the presence of a Cys this peptide can contain the dimeric form

---

Glu-Ala-Asn-Lys-Gly-Glu-Asn-Thr-Ser-Leu-Leu-His-Pro-Val

Synonyms: NEF (155-168); HTLV I Peptide

Neosystem SC513 MW 1508.65 HTLV I peptide from subtype NEF

Glu-Ala-Asp-Lys-Ala-Asp-Val-Asn-Val-Leu-Thr-Lys-Ala-Lys-Ser-Gln

Synonyms: Parathyroid Hormone (69-84), (Asp[76])-; Parathormone

American Peptide 22-1-46 Human MW 1716.9 $C_{72}H_{125}N_{21}O_{27}$

ICN 152992 Human

Peptides International PTH-4170-v Human synthetic MW 1716.9 $C_{72}H_{125}N_{21}O_{27}$ >97% (HPLC); lyophilized amorphous powder, hydrochloride

Glu-Ala-Asp-Pro-Asn-Lys-Phe-Tyr-Pro-Leu-Trp

Synonyms: Physalaemin (1-11), (((S,S)Pro-Leu(Spiro-γ-Lactam))[9,10],Trp[11]); GR-82334

Neosystem SC434 MW 1386.57 Bioactive; tachykinin; water soluble specific reversible antagonist at NK-1 receptors; Ward, P et al, *J Med Chem*, 33:1848-1851, 1990

Glu-Ala-Asp-Pro-Thr-Gly-His-Ser-Tyr

Synonyms: Melanoma Associated Antigen 1 (161-169)

Bachem H-3636.0001 Human Store at -15°C

Glu-Ala-Glu

Synonyms: Thymosin α (25-27)

Sigma G 2637 FW 347.3 $C_{13}H_{21}N_3O_8$ ≥97% (HPLC) | Bioactive peptide; Birr, C & Stollenwerk, U, *Angew Chem Int Ed Engl*, 18:394, 1979; Birr, C et al, *Peptides, Synthesis-Structure-Function, Proc 7th Amer Pept Symp*, Rich, D & Gross, E, eds, 541, 1981

Glu-Ala-Glu-Asn Ammonium Salt

Synonyms: Thymosin α (25-28)

Sigma G 5775 FW 461.4 (free acid) $C_{17}H_{27}N_5O_{10}$ ≥95% (HPLC) | Bioactive peptide; Birr, C & Stollenwerk, U, *Angew Chem Int Ed Engl*, 18:394, 1979; Birr, C et al, *Peptides, Synthesis-Structur-Function, Proc 7th Amer Pept Symp*, Rich, D & Gross, E, eds, 541, 1981

Glu-Ala-Glu-Asp-Leu-Gln-Gly-Gln-Glu-Leu-Gly-Gly-Gly-Pro-Gly-Ala-Gly-Ser-Leu-Gln-Pro-Leu-Ala-Leu-Glu-Gly-Ser-Leu-Gln
Glu-Ala-Glu-Asp-Leu-Gln-(D8)Val-Gly-Gln-(D8)Val-Glu-Leu-Gly-Gly-Gly-Pro-Gly-Ala-Gly-Ser-Leu-Gln-Pro-Leu-Ala-Leu-Glu-Gly-Ser-Leu-Gln

Synonyms: C-Peptide, ((D[8])Val[7,10])-

Bachem H-4242.0500 Human Store at -15°C

Glu-Ala-Glu-Asp-Leu-Gln-Val-Gly-Gln-Val-Glu-Leu-Gly-Gly-Gly-Pro-Gly-Ala-Gly-Ser-Leu-Gln-Pro-Leu-Ala-Leu-Glu-Gly-Ser-Leu-Gln

Synonyms: C-Peptide (33-63), Pro-Insulin; C-Peptide (33-63); Pro-Insulin C Peptide (33-63); C-Peptide (3-33)

American Peptide 20-1-11 Human MW 3020.3 $C_{129}H_{211}N_{35}O_{48}$

Bachem H-2470.0500 Human Store at -15°C

ICN 153081 Human Igano, K etal, *Bull Chem Soc Jpn*, 54:3088, 1981

Sigma C 8662 Human FW 3020.3; ≥85% (HPLC) Bioactive peptide; Igano, K, *Bull Chem Soc Japan*, 54:3088, 1981

Glu-Ala-Gly
γ-Glu-Ala-Gly-pNA

Synonyms: Glu(Ala-Gly-pNA)

Bachem L-2070.0050 MW 395.37 $C_{16}H_{21}N_5O_7$ Store at -15°C

Glu-Ala-Gly-Asp-Asp-Ile-Val-Pro-Cys-Ser-Met-Ser-Tyr-Thr-Trp-Thr-Gly-Ala

Synonyms: S5A; S5B

Bachem M-2185.0001 Store at -15°C

Glu-Ala-Leu
Z-Glu(OtBu)-Ala-Leu-pNA

Bachem L-2040.0050 MW 641.72 $C_{32}H_{43}N_5O_9$ Store at -15°C

Glu-Ala-Leu-Cys-Asp-Pro-Thr-Lys-Gly-Ser-Arg-Ser-Pro-Gln-Asp

Synonyms: REV (94-108); SIVmac251 Peptide

Neosystem SC603 MW 1603.72 SIVmac251 peptide from subtype REV; due to the presence of a Cys this peptide can contain the dimeric form

Glu-Ala-Leu-Glu-Leu-Ala-Arg-Gly-Ala-Ile-Phe-Gln-Ala

Synonyms: Brain Injury Derived Neurotrophic Peptide

Neosystem SC987 MW 1388.58 Bioactive neuropeptide; due to the presence of a Glu in N-terminal position this peptide can contain the pyroglutamic form; this fragment of a brain injury-derived protein promotes neuronal survival & rescues neurons from injury caused by glutamate; this finding might open the possibility of therapeutic application of neurotrophic peptide to the injured brain; Hama, T et al, *JBC*, 270:29067-29070, 1995

Glu-Ala-Leu-Glu-Leu-Ala-Arg-Gly-Ala-Ile-Phe-Gln-Ala Amide

Synonyms: Brain Injury Derived Neurotrophic Peptide

Bachem H-3914.0001 Store at -15°C

Glu-Ala-Leu-Phe-Gln-pNA

Bachem L-2050.0005 MW 726.79 $C_{34}H_{46}N_8O_{10}$ Store at -15°C

Glu-Ala-Ser-Asn-Cys-Phe-Ala-Ile-Arg-His-Phe-Glu-Asn-Lys-Phe-Ala-Val-Glu-Thr-Leu-Ile-Cys-Ser

Synonyms: Amyloid Bri Protein (1-23); Amyloid Bri Protein Precursor 277 (244-266)

Bachem H-5052.0001 MW 2627.99 $C_{116}H_{175}N_{31}O_{35}S_2$ Store at -15°C

Glu-Ala-Tyr-Gly-Trp-Met-Asp-Phe
D-Glu-Ala-Tyr(SO₃H)-Gly-Trp-Met-Asp-Phe Amide Sulfated

Synonyms: Gastrin I (10-17), (D-Glu[10])-

Bachem H-5054.0001 Human MW 1097.19 $C_{48}H_{60}N_{10}O_{16}S_2$ Store at -15°C

Glu-Ala-Val-Ser-Leu-Lys-Pro-Thr Trifluoroacetate Salt

Synonyms: Protein Kinase C_ε Translocation Inhibitor Peptide

Calbiochem 539522 MW 844.0 $C_{37}H_{65}N_9O_{13}$ ≥97% (HPLC); lyophilized solid; soluble in water | Selectively inhibits the translocation of PKC_ε to subcellular sites; inhibition of PKC_ε translocation is known to specifically block phorbol ester- or norepinephrine-mediated regulation of contraction in cardiomyocytes; Johnson, JA et al, *J Biol Chem*, 271:24962, 1996

Glu-Arg-Ala-Glu-Asp-Ser-Gly-Asn-Glu-Ser-Glu-Gly-Glu-Ile

Synonyms: VPU (47-60); HTLV I Peptide

Neosystem SC531 MW 1521.47 HTLV I peptide from subtype VPU

Glu-Arg-Arg
N-α-t-BOC-L-Glu-L-Arg-L-Arg-MCA

ICN 150494 MW 715.8 $C_{32}H_{49}N_{11}O_8$ Fluorogenic substrate

Glu-Arg-Met-Arg-Pro-Arg-Lys-Arg-Gln-Gly-Ser-Val-Arg-Arg-Arg-Val

Synonyms: Pepstatin ε;; Peptide ε;; Protein Kinase C$_\epsilon$ Substrate

Alexis 165-006   MW 2067.5 $C_{83}H_{155}N_{39}O_{21}S$ ≥98%; white lyophilized powder; soluble in water | Specific substrate for PKCε, derived from the pseudosubstrate site of PKCε (149-164), with Ala[159] replaced by Ser; also phosphorylated by PKCε; Schaap, D et al, *FEBS Lett*, 243:351, 1989; Schaap, D & Parker, PJ, *J Biol Chem*, 265:7301, 1990

Bachem H-3236.0500   MW 2067.46 $C_{83}H_{155}N_{39}O_{21}S$ Store at -15°C

ICN 195832   MW 2066.2   ≥97% | Phosphorylation substrate

Glu-Arg-Met-Arg-Pro-Arg-Lys-Arg-Gln-Gly-Ser-Val-Arg-Arg-Arg-Val Trifluoroacetate Salt

Synonyms: Protein Kinase C$_\epsilon$ Peptide Substrate; Peptide ε

Calbiochem 539562   MW 2067.5 $C_{83}H_{155}N_{39}O_{21}S$ ≥99% (HPLC); solid; soluble in water | Pseudosubstrate region of the e-isotype of protein kinase C (PKCε) with an Ala to Ser substitution; exhibits an apparent specificity for this peptide; Schaap, D & Parker, PJ, *J Biol Chem*, 265:7301, 1990; Schaap, D et al, *FEBS Let*, 243:351, 1989

Glu-Arg-Nle-Phe-Leu-Ser-Phe-Pro DABCYL-Glu-Arg-Nle-Phe-Leu-Ser-Phe-Pro-EDANS

Bachem M-2120.0001   Store at -15°C

Glu-Arg-Pro-Gly-Met-Leu-Asp-Phe-Thr

Synonyms: Diazepam Binding Inhibitor (42-50)

Neosystem SC365   Human   MW 1065.21   Bioactive neuropeptide;

Glu-Arg-Pro-Pro-Leu-Gln-Gln-Pro-Pro-His-Arg-Asp Glp-Glu-Arg-Pro-Pro-Leu-Gln-Gln-Pro-Pro-His-Arg-Asp

Synonyms: Cortistatin 29 (1-13)

American Peptide 34-7-32   MW 1596.8 $C_{68}H_{107}N_{24}O_{21}$ de Leccea, L et al, *Nature*, 381:242, 1996

Glu-Asn-Asn-Cys-Leu-Leu-His-Pro-Met-Ser-Gln-His-Gly-Trp-Met

Synonyms: NEF MN (162-176); HIV I Peptide

Neosystem SC662   MW 1800.00   NEF peptide from HIV-1 subtype MN; due to the presence of a Cys this peptide can contain the dimeric form

Glu-Asn-Asp-Tyr-Ile-Asn-Ala-Ser-Leu Glu-Asn-Asp-Tyr(PO$_3$H$_2$)-Ile-Asn-Ala-Ser-Leu

Synonyms: Protein Tyrosine Phosphatase Substrate

Bachem M-2010.0001   MW 1118.06 $C_{44}H_{68}N_{11}O_{21}P$ Store at -15°C

Calbiochem 539750   MW 1118.1 $C_{44}H_{68}N_{11}O_{21}P$ ≥90% (HPLC); solid; soluble in 10% aqueous TFA | An excellent substrate for the detection & characterization of a wide variety of intracellular & receptor-linked protein tyrosine phosphatases, particularly when limiting amounts of tissue extracts or immunoprecipitates are available; Daum, G et al, *Anal Biochem*, 211:50, 1993

Glu-Asn-Gly-Leu-Pro-Val-His-Leu-Asp-Gln-Ser-Ile-Phe-Arg-Arg

Synonyms: Angiogenin C-Terminal Peptide; Angiogenin (108-122)

American Peptide 12-0-15   MW 1781.1 $C_{78}H_{125}N_{25}O_{23}$ inhibits enzymatic & biological activities of angiogenin

Bachem H-8095.0001   MW 1781.01 $C_{78}H_{125}N_{25}O_{23}$ Store at -15°C

Sigma A 9429   FW 1781.0   ≥97% (HPLC) | Bioactive peptide; inhibits the enzymatic & biological activity of Angiogenin; Rybak, SM et al, *Biochem Biophys Res Commun*, 162:535, 1989

Glu-Asn-Gly-Leu-Pro-Val-His-Leu-Asp-Gln-Ser-Ile-Phe-Arg-Arg-Pro

Synonyms: Angiogenin (108-123)

American Peptide 12-0-25   MW 1878.1 $C_{83}H_{132}N_{26}O_{24}$ Inhibits enzymatic & biological activity of angiogenin, including neovascularization

Bachem H-8020.0001   MW 1878.12 $C_{83}H_{132}N_{26}O_{24}$ Store at -15°C

Sigma A 9304   FW 1878.1; ≥97% (HPLC)   Bioactive peptide; inhibits the enzymatic & biological activity of angiogenin, including neovascularization; Rybak, SM et al, *Biochem Biophys Res Commun*, 162:535, 1989

Glu-Asn-Ile-Thr-Ala-Leu-Leu-Glu-Glu-Ala-Gln-Ile-Gln-Gln-Glu-Lys-Asn-Met-Tyr-Glu

Synonyms: GP140 (651-670); SIVmac251 Peptide

Neosystem SC773   MW 2394.63   SIVmac251 peptide from subtype GP140

Glu-Asn-Pro-Val-Val-His-Phe-Phe-Lys-Asn-Ile-Val-Thr-Pro-Arg-Thr-Pro

Synonyms: Myelin Basic Protein (83-99)

Bachem H-4306.0001   Bovine   Store at -15°C

Glu-Asn-Thr-Ser-Leu-Leu-His-Pro-Val-Ser-Leu-His-Gly-Met-Asp

Synonyms: NEF (160-174); HTLV I Peptide

Neosystem SC228   MW 1649.83   HTLV I peptide from subtype NEF

Glu-Asp

Bachem G-1910.0250   MW 262.22 $C_9H_{14}N_2O_7$ Store at -15°C

Glu-Asp-Ala-Ser-Thr-Pro-Cys Mca-Glu-Asp-Ala-Ser-Thr-Pro-Cys

Bachem H-2776.0001   Store at -15°C

Glu-Asp-Ala-Ser-Thr-Pro-Cys-Ser-Gly-Ser-Phe-Leu Mca-Glu-Asp-Ala-Ser-Thr-Pro-Cys-Ser-Gly-Ser-*p*-Nitro-Phe-Leu

Bachem M-2135.0001   Store at -15°C

Glu-Asp-Asn-Cys-Ile-Ala-Glu-Asp-Tyr-Gly-Lys-Cys-Thr-Trp-Gly-Gly-Thr-Lys-Cys-Cys-Arg-Gly-Arg-Pro-Cys-Arg-Cys-Ser-Met-Ile-Gly-Thr-Asn-Cys-Glu-Cys-Thr-Pro-Arg-Leu-Ile-Met-Glu-Gly-Leu-Ser-Phe-Ala

Synonyms: Specific P-Type Calcium Channel Blocker; Agatoxin TK, ω-

American Peptide 41-0-69   MW 5273.1 $C_{215}H_{337}N_{65}O_{70}S_{10}$ Disulfide bonds: Cys⁴-Cys²⁰; Cys¹²-Cys²⁵; Cys¹⁹-Cys³⁶; Cys²⁷-Cys³⁴ | Teramoto, T et al, *BBRC*, 196:134, 1993

Calbiochem 122302   *Agelenopsis aperta*   MW 5273 $C_{215}H_{337}N_{65}O_{70}S_{10}$ >99% (HPLC); lyophilized solid; soluble in water; harmful:LD$_{50}$ ≤2000 mg/kg | Disulfide bonds: Cys⁴-Cys²⁰, Cys¹²-Cys²⁵, Cys¹⁹-Cys³⁶ & Cys²⁷-Cys³⁴; inhibits K⁺ -induced ³H-GABA release from primary rat hippocampal neurons; specific P-type calcium channel blocker with no apparent effect on L-type, N-type & T-type channels; useful tool for investigating the role of P-type calcium channels in the peripheral & central nervous systems; also inhibits the release of AA transmitters & monoamines; Kimura, M et al, *Neuroscience*, 66:609, 1995; Teramoto, T et al, *Biochem Biophys Res Comm*, 196:134, 1993

Glu-Asp-Asn-Cys-Ile-Ala-Glu-Asp-Tyr-Gly-Lys-Cys-
Thr-Trp-Gly-Gly-Thr-Lys-Cys-Cys-Arg-Gly-Arg-Pro-
Cys-Arg-Cys-Ser-Met-Ile-Gly-Thr-Asn-Cys-Glu-Cys-
Thr-Pro-Arg-Leu-Ile-Met-Glu-Gly-Leu-Ser-Phe-Ala
Glu-Asp-Asn-Cys-Ile-Ala-Glu-Asp-Tyr-Gly-Lys-Cys-
Thr-Trp-Gly-Gly-Thr-Lys-Cys-Cys-Arg-Gly-Arg-Pro-
Cys-Arg-Cys-Ser-Met-Ile-Gly-Thr-Asn-Cys-Glu-Cys-
Thr-Pro-Arg-Leu-Ile-Met-Glu-Gly-Leu-D-Ser-Phe-Ala

Synonyms: Agatoxin TK, ω-

ICN 195883 *Agelenopsis aperta* MW 5273
$C_{215}H_{337}N_{65}O_{70}S_{10}$ ≥95% | Potent P-type calcium channel blocker

Glu-Asp-Asn-Cys-Ile-Ala-Glu-Asp-Tyr-Gly-Lys-Cys-
Thr-Trp-Gly-Gly-Thr-Lys-Cys-Cys-Arg-Gly-Arg-Pro-
Cys-Arg-Cys-Ser-Met-Ile-Gly-Thr-Asn-Cys-Glu-Cys-
Thr-Pro-Arg-Leu-Ile-Met-Glu-Gly-Leu-Ser-Phe-Ala

Synonyms: Agatoxin TK, ω-; Agatoxin IVB, ω-

Peptides International PAG-4294-s *Agelenopsis aperta*
(funnel web spider) synthetic MW 5273.1 $C_{215}H_{337}N_{65}O_{70}S_{10}$
>95% (HPLC); lyophilized amorphous powder | Bioactive;
Disulfide bonds: Cys[4]-Cys[20], Cys[12]-Cys[25], Cys[19]-Cys[36], Cys[27]-Cys[34];
P-type Ca[2+] channel selective blocker; Kuwada, M et al, *Mol Pharmacol*, 46:587, 1994; Shikata, Y et al, *JBC*, 270:16719, 1995;
Kozaki, T et al, *Peptide Chem*, 1995:245, 1996; Teramoto, T et al,
Brain Res, 756:225, 1997

Glu-Asp-Asn-Cys-Ile-Ala-Glu-Asp-Tyr-Gly-Lys-Cys-
Thr-Trp-Gly-Gly-Thr-Lys-Cys-Cys-Arg-Gly-Arg-Pro-
Cys-Arg-Cys-Ser-Met-Ile-Gly-Thr-Asn-Cys-Glu-Cys-
Thr-Pro-Arg-Leu-Ile-Met-Glu-Gly-Leu-Ser-Phe-Ala
Glu-Asp-Asn-Cys-Ile-Ala-Glu-Asp-Tyr-Gly-Lys-Cys-
Thr-Trp-Gly-Gly-Thr-Lys-Cys-Cys-Arg-Gly-Arg-Pro-
Cys-Arg-Cys-Ser-Met-Ile-Gly-Thr-Asn-Cys-Glu-Cys-
Thr-Pro-Arg-Leu-Ile-Met-Glu-Gly-Leu-D-Ser-Phe-Ala

Synonyms: Agatoxin TK, ω-

Sigma A 2202 *Agelenopsis aptera* (funnel web spider) venom
FW 5273.0; Disulfide bonds: 4-20, 12-25, 19-36, 27-34;
bioactive peptide; found in the venom of the funnel web spider;
potent P-type calcium-channel blocker; Kuwada, M et al, *Mol Pharmacol*, 46:587, 1994

Alexis 630-067 *Agelenopsis aptera* (funnel web spider) venom,
synthetic MW 5273.1 $C_{215}H_{337}N_{65}O_{20}S_{10}$ ≥98%; lyophilized;
soluble in water; toxic | Disulfide bonds: Cys[4]-Cys[20], Cys[12]-Cys[25],
Cys[19]-Cys[36], Cys[27]-Cys[34]; selective P-type Ca[2+] channel blocker with
no apparent effect on L-type, N-type & T-type channels; useful tool
for investigating the role of P-type channels in the peripheral &
central nervous systems; Kuwada, M et al, *Mol Pharmacol*, 46:587,
1994; Shikata, Y et al, *J Bio Chem*, 270:16719, 1995; Kozaki, T et
al, *Peptide Chem*, 33:245, 1996

Glu-Asp-Cys-Gly-Thr-Ser-Gly-Thr-Gln-Gly-Val-Gly-
Ser-Pro

Synonyms: REV (87-100); HTLV I Peptide

Neosystem SC195 MW 1294.31 HTLV I peptide from
subtype REV; due to the presence of a Cys this peptide can contain
the dimeric form

Glu-Asp-Leu-Leu-Lys-Ala-Val-Arg-Leu-Ile-Lys

Synonyms: REV (10-20); HTLV I Peptide

Neosystem SC552 MW 1297.60 HTLV I peptide from
subtype REV

Glu-Asp-Lys-Pro-Ile-Leu-Phe-Phe-Arg-Leu-Gly-Lys-
Glu
Ac-Glu-Asp(EDANS)-Lys-Pro-Ile-Leu-Phe-Phe-Arg-
Leu-Gly-Lys(DABCYL)-Glu Amide

Bachem M-2295.0001 MW 2132.52 $C_{104}H_{146}N_{24}O_{23}S$ Store
at -15°C

Glu-Asp-Val-Gly-Ser-Asn-Lys-Gly-Ala-Ile-Ile-Gly-
Leu-Met

Synonyms: Amyloid Protein (22-35), β-

Alexis 151-010 MW 1403.6 $C_{59}H_{102}N_{16}O_{21}S$ Takadera, T et
al, *Neurosci Lett*, 161:41, 1993

Glu-Asp-Val-Gly-Ser-Asn-Lys-Gly-Ala-Ile-Ile-Gly-
Leu-Met
β-Glu-Asp-Val-Gly-Ser-Asn-Lys-Gly-Ala-Ile-Ile-Gly-
Leu-Met

Synonyms: Amyloid (22-35)

American Peptide 62-0-35 MW 1403.6 $C_{59}H_{102}N_{16}O_{21}S$
Produces both neurotrophic & neurotoxic effects of β-amyloid
protein on cultured neurons; forms aggregates & typical amyloid
fibrils resembling those of the β-amyloid protein in senile plaque
cores; Takadera, T et al, *Neurosci Lett*, 161:41, 1993

Glu-Asp-Val-Gly-Ser-Asn-Lys-Gly-Ala-Ile-Ile-Gly-
Leu-Met

Synonyms: Amyloid α-Protein (22-35); Amyloid Protein (22-35), β-

Bachem H-1976.0001 MW 1403.62 $C_{59}H_{102}N_{16}O_{21}S$ Store
at -15°C

Sigma A 5080 FW 1403.6 ≥95% (HPLC) | Bioactive
peptide; cytotoxic effect on cultured neurons from rat hippocampus
in serum free medium; Takadera, T et al, *Neurosci Lett*, 161:41,
1993

Glu-Cys
Glu(Cys)

Synonyms: Glutathione Reduced Form, (des-Gly)-

Bachem G-4305.0250 Store at -15°C

Glu-Cys
Glu(Cys-OEt)

Synonyms: Glutathione Reduced Form Ethyl Ester, (des-Gly)-;
GCE, γ-

Bachem G-4430.0250 Store at -15°C

Glu-Cys
Glu(Cys) Trifluoroacetate Salt

Synonyms: Glutathione Reduced Form, (des-Gly)-

Bachem G-4565.0250 Store at -15°C

Glu-Cys
(Glu(Cys-βNA))₂

Synonyms: (γ-Glu-Cys-βNA)₂

Bachem K-1650.0050 MW 748.88 $C_{36}H_{40}N_6O_8S_2$ Store at
-15°C

Glu-Cys
γ-Glu-Cys Trifluoroacetate Salt

Synonyms: Glutathione Reduced Form, (des-Gly)-

Sigma G 0903 FW 250.3 (free base) $C_8H_{14}N_2O_5S$ ≥80%
(HPLC); peptide content: ~70%

Glu-Cys
(γ-Glu-Cys-βNA)₂

Synonyms: Glutathione-βNA, bis(des-Gly)-

Sigma G 3410 FW 748.9 $C_{36}H_{40}N_6O_8S_2$ Oxidized; ≥97%
(TLC) | Disulfide bonds: 2[A]-2[B]

Glu-Cys-Ala
Glu(Cys-α-Ala)

Synonyms: Homoglutathione

Bachem H-3944.0050 MW 321.35 $C_{11}H_{19}N_3O_6S$ Store at
-15°C

Glu-Cys-Cys-Asn-Pro-Ala-Cys-Gly-Arg-His-Tyr-Ser-Cys

Glu-Cys-Cys-Asn-Pro-Ala-Cys-Gly-Arg-His-Tyr-Ser-Cys Amide

Synonyms: Conotoxin GI Hydrochloride, α-; Nicotinic Acetylcholine Receptor Blocker

Peptides International PCN-4126-v   *Conus geographus* (marine snail) synthetic   MW 1437.6 $C_{55}H_{80}N_{20}O_{18}S_4$ >97% (HPLC); lyophilized amorphous powder | Bioactive; Disulfide bonds: Cys^2-Cys^7, Cys^3-Cys^{13}; Olivera, BM et al, *Science*, 230:1338, 1985; Gray, WR & BM Olivera, *Ann Rev Biochem*, 57:665, 1998; Myers, RA et al, *Chem Rev*, 93:1923, 1993; Olivera, BM et al, *Ann Rev Biochem*, 63:823, 1994; Gray, WR et al, *JBC*, 256:4734, 1981; Nishiuchi, Y & S Sakakibara, *FEBS Lett*, 148:260 1982

Glu-Cys-Cys-Asn-Pro-Ala-Cys-Gly-Arg-His-Tyr-Ser-Cys Amide

Synonyms: Conotoxin GI; Conotoxin GI, α-

Bachem H-7890.0001   MW 1437.63 $C_{55}H_{80}N_{20}O_{18}S_4$ Store at -15°C | Disulfide bonds: Cys^2-Cys^7, Cys^3-Cys^{13}

Sigma C 8653   FW 1437.6   ≥97% (HPLC) | Disulfide bonds: 2-7, 3-13; bioactive peptide; postsynaptic inhibitor at the neuromuscular junction; Gray, WR et al, *J Biol Chem*, 256:4734, 1981

American Peptide 41-0-26   *Conus geographus* (marine snail) MW 1437.6 $C_{55}H_{80}N_{20}O_{18}S_4$ Disulfide bonds: Cys^2-Cys^7, Cys^3-Cys^{13}; postsynaptic inhibitor at the neuromuscular junction; isolated from the venom of the marine snail & is a potent antagonist of the postsynaptic nicotinic acetylcholine receptor; Gray, WR et al, *PNAS*, 85:3329, 1988; Lambret, P et al, *BBRC*, 170:684, 1990

Glu-Cys-Cys-Asn-Pro-Ala-Cys-Gly-Arg-His-Tyr-Ser-Cys Amide Hydrochloride

Synonyms: Conotoxin GI, α-

Alexis 630-046   *Conus geographus* synthetic   MW 1436.6 $C_{55}H_{79}N_{20}O_{18}S_4$ ≥97%; white lyophilized powder; soluble in water; potent neurotoxin | Disulfide bonds: Cys^2-Cys^7, Cys^3-Cys^{13}; postsynaptic inhibitor at the neuromuscular junction; Gray, WR et al, *J Biol Chem*, 256:4734, 1981; Hann, RM et al, *Biochemistry*, 33:14058, 1994; Nishiuchi, Y & Sakakibara, S, *FEBS Lett*, 148:260, 1992

Glu-Cys-Glu
(γ-Glu-Cys-Glu)₂ Trifluoroacetate Salt

Synonyms: Glutathione Analog

Sigma G 6909   FW 756.8 (free base) $C_{26}H_{40}N_6O_{16}S_2$ Oxidized; ≥99% (HPLC) | Disulfide bonds: 2_A-2_B; found in maize seedlings exposed to cadmium; a phytochelatin believed to be involved in heavy metal detoxification; Meuwly, P, Thibault, P & Rauser, WE, *FEBS Lett*, 336:472, 1993

Glu-Cys-Glu-Ser-Gly-Pro-Cys-Cys-Arg-Asn-Cys-Lys-Phe-Leu-Lys-Glu-Gly-Thr-Ile-Cys-Lys-Arg-Ala-Arg-Gly-Asp-Asp-Met-Asp-Asp-Tyr-Cys-Asn-Gly-Lys-Thr-Cys-Asp-Cys-Pro-Arg-Asn-Pro-His-Lys-Gly-Pro-Ala-Thr

Synonyms: Echistatin

Bachem H-9010.0100   Store at -15°C | Disulfide bonds: Cys^2-Cys^{11}, Cys^7-Cys^{32}, Cys^8-Cys^{37}, Cys^{20}-Cys^{39}

Glu-Cys-Glu-Ser-Gly-Pro-Cys-Cys-Arg-Asn-Cys-Lys-Phe-Leu-Lys-Glu-Gly-Thr-Ile-Cys-Lys-Arg-Ala-Arg-Gly-Asp-Asp-Met-Asp-Asp-Tyr-Cys-Asn-Gly-Lys-Thr-Cys-Asp-Cys-Pro-Arg-Asn-Pro-His-Lys-Gly-Pro-Ala-Thr (Cyclic)

Synony Echistatin *ms*:

ICN 154539   MW 5418.2   Gan, ZR etal, *JBC*, 263:19827, 1988

Glu-Cys-Glu-Ser-Gly-Pro-Cys-Cys-Arg-Asn-Cys-Lys-Phe-Leu-Lys-Glu-Gly-Thr-Ile-Cys-Lys-Arg-Ala-Arg-Gly-Asp-Asp-Met-Asp-Asp-Tyr-Cys-Asn-Gly-Lys-Thr-Cys-Asp-Cys-Pro-Arg-Asn-Pro-His-Lys-Gly-Pro-Ala-Thr

Synonyms: Fibronectin; Echistatin

Sigma E 2138   FW 1249.3; ≥95% (HPLC)   Bioactive peptide; potent inhibitor of platelet aggregation; Gan, Z-R et al, *J Biol Chem*, 263:19827, 1988; Garsky, VM et al, *PNAS USA*, 86:4022, 1989

Glu-Cys-Gly

Synonyms: Glutathione L-Oxidized Form; Glutathione L-Reduced Form

Alexis 157-001   MW 612.6 (free acid) $C_{20}H_{32}N_6O_{12}S_2$ ~99%; white solid

Alexis 157-002   MW 307.3 (free acid) $C_{10}H_{17}N_3O_6S$ 99%; white solid

Glu-Cys-Gly
FMOC-Glu-(Cys-FMOC-Gly)

Synonyms: Glutathione, N,S-Bis-FMOC-

Bachem B-2855.0250   Store at -15°C

Glu-Cys-Gly
(Glu(Cys-Gly-(3-(4-Aminobutylamino)-Propyl)-Amide))₂

Synonyms: Spermidine, N^1-Glutathionyl-

Bachem H-1126.0005   MW 867.1 $C_{34}H_{66}N_{12}O_{10}S_2$ Store at -15°C

Glu-Cys-Gly
Glu(Cys-Gly-OEt) Reduced

Synonyms: Glutathione

Bachem H-1298.0250   MW 335.38 $C_{12}H_{21}N_3O_6S$ Store at -15°C

Glu-Cys-Gly
Glu(Cys(1,2-Dicarboxyethyl)-Gly)

Synonyms: Glutathione, S-(1,2-Dicarboxyethyl)-

Bachem H-1556.0250   Store at -15°C

Glu-Cys-Gly
Glu(Cys-Gly-Isopropyl Ester) Reduced

Synonyms: Glutathione

Bachem H-2236.0250   MW 349.41 $C_{13}H_{23}N_3O_6S$ Store at -15°C

Glu-Cys-Gly
Glu(Cys-Gly-OEt)-OEt Reduced

Synonyms: Glutathione

Bachem H-2238.0250   MW 363.44 $C_{14}H_{25}N_3O_6S$ Store at -15°C

Glu-Cys-Gly
Glu(Cys-(^{15}N)Gly)

Synonyms: Glutathione Reduced Form, ((^{15}N)Gly)-

Bachem H-4584.0025   MW 308.32 $C_{10}H_{17}^{15}NN_2O_6S$ Store at -15°C

Glu-Cys-Gly
(Glu(Cys-(^{15}N)Gly))₂

Synonyms: Glutathione, ((^{15}N)Gly)-

Bachem H-4586.0025   MW 614.63 $C_{20}H_{32}^{15}N_2N_4O_{12}S_2$ Store at -15°C

Glu-Cys-Gly
L-γ-Glu-L-Cys-Gly Monoethyl Ester

Synonyms: Glutathione Monoethyl Ester

Calbiochem 353905   MW 335.4 $C_{12}H_{21}N_3O_6S$ >90% (HPLC); solid; soluble in water | Cell-permeable derivative of glutathione that undergoes hydrolysis by intracellular esterases thereby increasing intracellular GSH concentration in many tissues & cell types; effective transport used to protect cells against radiation damage, oxidants & various toxic compounds including heavy metals; a protective agent against cellular damage such as in cataracts & mitochondrial degeneration; Grattagliano, I et al, *J Pharmacol Exp Ther*, 272:484, 1995; Martensson, J et al, *PNAS*, 90:317, 1993; Anderson, ME & Meister, A, *Anal Biochem*, 183:16, 1989; Anderson, ME et al, *Arch Biochem Biophys*, 239:538, 1985; Wellner, VP et al, *PNAS*, 81:4732, 1984; Puri, RN & Meister, A, *PNAS*, 80:5258, 1983

Glu-Cys-Gly
γ-Glu-Cys-Gly Free Acid

Synonyms: Glutathione Reduced Form

Calbiochem 3541   MW 307.3 $C_{10}H_{17}N_3O_6S$ >98% (titration); solid; soluble in DMF, ethanol & water; may be carcinogenic/teratogenic | A tripeptide that serves as a component of the γ-glutamyl AA transport system; endogenous antioxidant that provides protection against autooxidation; Shivakumar, BR et al, *J Pharmacol Exp Ther*, 247:1167, 1995; *Merck Index*, 12:4483

Glu-Cys-Gly
γ-L-Glu-L-Cys-Gly

Synonyms: Glutathione Reduced Form

Fluka 49750   MW 307.33 $C_{10}H_{17}N_3O_6S$ ≥97.0% (HPLC); ≤0.05% Na; ≤0.0005% Cd, Co, Cr, Cu, Fe, K, Mg, Mn, Ni, Pb, Zn; ≤0.001% Ca

Glu-Cys-Gly
Glu(Cys-Gly)

Synonyms: Glutathione; γ-Glu-Cys-Gly

Peptides International OEG-3050   MW 307.33 $C_{10}H_{17}N_3O_6S$ >98% (HPLC); white powder

Glu-Cys-Gly

Synonyms: Glutathione; γ-Glu-Cys-Gly; Glu(Cys-Gly); Glutathione, *S*-(*p*-Azidophenylacyl)-; Glutathione, *S*-Butyl-; Glutathione, *S*-(*p*-Chlorophenacyl)-

Peptides International OEG-3050   MW 307.33 $C_{10}H_{17}N_3O_6S$ >96% (HPLC); amorphous powder

Sigma A 1782   FW 466.5 $C_{18}H_{22}N_6O_7S$ ≥95.0% | Photoaffinity derivative of glutathione; reported to inhibit glyoxalase & glutathione-*S*-transferase; Seddon, AP & Douglas, KT, *FEBS Lett*, 110:262, 1980; Seddon, AP et al, *Biochem Biophys Res Commun*, 95:446, 1980 also see Sigma references & comments for Glutathione

Sigma B 7886   FW 363.4 $C_{14}H_{25}N_3O_6S$

Sigma C 9898   FW 459.9 $C_{18}H_{22}ClN_3O_7S$

Glu-Cys-Gly
γ-Glu-Cys(Suc)-Gly

Synonyms: Glutathione, *S*-(1,2-Dicarboxyethyl)-

Sigma D 2804   FW 423.4 $C_{14}H_{21}N_3O_{10}S$ ≥97% (HPLC) | Natural anti-inflammatory peptide that inhibits histamine release *in vitro*; Sakaue, T eat al, *Arzneim-Forsch*, 42:1482, 1992

Glu-Cys-Gly

Synonyms: Glutathione, *S*-Decyl-; Glutathione, *S*-Ethyl-; Glutathione Oxidized, Agarose

Sigma D 3275   FW 447.6 $C_{20}H_{37}N_3O_6S$

Sigma E 0508   FW 335.4 $C_{12}H_{21}N_3O_6S$

Sigma G 0387   Attached through the amino group to epoxy activated 4% cross-linked beaded agarose; spacer:12 atoms (10 carbon); 3-7 μmoles/mL gel; lyophilized powder stabilized with lactose; 1 g powder swells to ~6 mL gel | Useful in affinity chromatography of glutathione-*S*-transferases; Simons, PC & Vander Jagt, DL, *Anal Biochem*, 82:334, 1977

Glu-Cys-Gly
γ-Glu-Cys-Gly-OEt

Synonyms: Glutathione Reduced Form Ethyl Ester

Sigma G 1404   FW 335.4 $C_{12}H_{21}N_3O_6S$ ≥95% (TLC)

Glu-Cys-Gly

Synonyms: Glutathione, Agarose

Sigma G 3907   3 Pre-packed columns, 2.5 mL each; Sigma G 4510 glutathione suspended in 0.5 M NaCl & 20% ethanol | Useful in affinity chromatography of glutathione-*S*-transferases; Simons, PC & Vander Jagt, DL, *Anal Biochem*, 82:334, 1977

Glu-Cys-Gly
γ-Glu-Cys-Gly Free Acid

Synonyms: Glutathione Reduced Form

Sigma G 4251   FW 307.3 $C_{10}H_{17}N_3O_6S$ ≥98%

Glu-Cys-Gly
(γ-Glu-Cys-Gly) (γ-Glu-Cys-Gly) Free Acid

Synonyms: Glutathione Oxidized Form

Sigma G 4376   FW 612.6 $C_{20}H_{32}N_6O_{12}S_2$ ≤6% EtOH | Disulfide bonds: 2_A-2_B

Sigma G 4501   FW 612.6 $C_{20}H_{32}N_6O_{12}S_2$ ~98%; lyophilized powder; EtOH-free | Disulfide bonds: 2_A-2_B

Glu-Cys-Gly

Synonyms: Glutathione, Agarose

Sigma G 4510   Attached through the sulfur to epoxy activated 4% cross-linked beaded agarose; spacer:12 atoms (10 carbon); 10-20 μmoles/mL gel; lyophilized powder stabilized with lactose; 1 g powder swells to ~14 mL gel | Useful in affinity chromatography of glutathione-*S*-transferases; Smith, DB & Johnson, SK *Gene*, 67:31, 1988; Guan, K & Dixon, JE, *Anal Biochem*, 192:262, 1991; Simons, PC & Vander Jagt, DL, *Anal Biochem*, 82:334, 1977

Glu-Cys-Gly
(γ-Glu-Cys-Gly) (γ-Glu-Cys-Gly) Free Acid

Synonyms: Glutathione Oxidized Form

Sigma G 4626   FW 656.6 $C_{20}H_{30}N_6O_{12}S_2Na_2$ 99+%; disodium salt; white powder; substantially EtOH-free | Disulfide bonds: 2_A-2_B

Glu-Cys-Gly
γ-Glu-Cys-Gly

Synonyms: Glutathione Reduced Form

Sigma G 6013   FW 307.3 (free acid) $C_{10}H_{17}N_3O_6S$ ≥99%; crystalline; cell culture tested

Glu-Cys-Gly
γ-Glu-Cys-Gly Free Acid

Synonyms: Glutathione Reduced Form

Sigma G 6529   FW 307.3 $C_{10}H_{17}N_3O_6S$ SigmaUltra; 98-100%; residue on ignition <0.1%; solubility (0.1 M in water, 20°C):complete, colorless; insoluble matter <0.1%; NH_4^+,SO_4:<0.05%; K:<0.005%; Mg, Al, Ca, P, Zn, Cu, Fe:<0.0005%; Pb:<0.001%

Glu-Cys-Gly
(γ-Glu-Cys-Gly) (γ-Glu-Cys-Gly) Free Acid

Synonyms: Glutathione Oxidized Form

Sigma G 6654   FW 612.6 $C_{20}H_{32}N_6O_{12}S_2$ SigmaUltra; ~98%; residue on ignition <0.5%; solubility (0.1 M in water, 20°C):complete, colorless; insoluble matter <0.1%; SO_4, Cl:<0.05%; K:<0.005%; Al, Ca, Cu, Fe:<0.0005%; ≤6% EtOH | Disulfide bonds: 2_A-2_B

Sigma G 9027   FW 612.6 $C_{20}H_{32}N_6O_{12}S_2$ ~98%; crystallized & lyophilized; EtOH-free | Disulfide bonds: 2_A-2_B

Glu-Cys-Gly

Synonyms: Glutathione, *S*-(Hexyl)-; Glutathione, *S*-(Methyl)-; Glyoxalase Inhibitor; Glutathione, *S*-(*p*-Nitrobenzyl)-; Glutathione, *S*-(Octyl)-; Glutathione, *S*-(Propyl)-

Sigma H 6886   FW 391.5 $C_{16}H_{29}N_3O_6S$ Ligand useful for affinity chromatography of glutathione-*S*-transferase & glutathione peroxidase

Sigma M 4139   FW 321.3 $C_{11}H_{19}N_3O_6S$

Sigma N 1509   FW 442.4 $C_{17}H_{22}N_4O_8S$

Sigma O 5502   FW 419.5 $C_{18}H_{33}N_3O_6S$

Sigma P 5406   FW 349.4 $C_{13}H_{23}N_3O_6S$

Glu-Cys-Gly Disulfide

Synonyms: Spermidine, N^1,N^8-Bis(Glutathionyl)-; Trypanothione

Bachem H-7510.0025   Store at -15°C

Glu-Cys-Gly Free Acid

Synonyms: Glutathione, *S*-(Lactoyl)-; Glyoxalase II Substrate

Sigma L 7140   FW 379.4 $C_{13}H_{21}N_3O_8S$ Oxidized form; ~95%

Glu-Cys-Gly Sodium Salt

Synonyms: Glutathione Disulfide, Coenzyme A

Sigma C 5018   FW 1072.8 (free acid)   ~90% | Enzymatically assayed using glutathione reductase

Glu-Cys-Gly Sulfonic Acid

Synonyms: Glutathione Sulfonic Acid

Sigma G 9267   FW 355.3 $C_{10}H_{17}N_3O_9S$

Glu-Cys-Lys Disodium Salt

Synonyms: Glutathione Oxidized Form

Sigma G 2140   FW 656.6 $C_{20}H_{30}N_6O_{12}S_2Na_2$ ≥99%; substantially ethanol-free; cell culture tested

Glu-Gln
γ-Glu-Gln

Synonyms: Glu(Gln-OH)-OH

Bachem G-1930.0250   MW 275.26 $C_{10}H_{17}N_3O_6$ Store at -15°C

ICN 157203   MW 275.3 $C_{10}H_{17}N_3O_6$ Crystalline

Sigma G 2634   FW 275.3 $C_{10}H_{17}N_3O_6$ Crystalline

Glu-Gln-Ile-Gly-Trp-Met-Thr-Asn-Asn-Pro-Pro-Ile-Pro-Val-Gly-Glu-Ile

Synonyms: P25 (245-261); GAG P24 CA (113-129); HTLV I Peptide

Neosystem SC300   MW 1895.15   HTLV I peptide from subtype P25 (GAG P24 CA)

Glu-Gln-Phe-Asp-Asp-Tyr-Gly-His-Met-Arg-Phe
Glu-Gln-Phe-Asp-Asp-Tyr(SO₃H)-Gly-His-Met-Arg-Phe Amide

Synonyms: Perisulfakinin

Bachem H-1038.0001   Store at -15°C

Glu-Gln-Trp-Phe-Trp-Trp-Met
D-Glu-Gln-D-Trp-Phe-D-Trp-D-Trp-Met Amide

Synonyms: Substance P (5-11), (D-Glu[5],D-Trp[7,9,10])-

Bachem H-2184.0001   Store at -15°C

Glu-Gln-Val-Thr-Asn-Val-Gly-Gly-Ala-Val-Val-Thr-Gly-Val-Thr-Ala-Val-Ala-Gln-Lys-Thr-Val-Glu-Gly-Ala-Gly-Ser-Ile-Ala-Ala-Ala-Thr-Gly-Phe-Val

Synonyms: Non-β-Amyloid Component of Alzheimer's Disease; Amyloid Protein, Non-Aβ Component of Alzheimer's Disease; Amyloid Protein, Non-Aβ Component (61-95); Synuclein

American Peptide 62-5-25   MW 3260.6 $C_{141}H_{235}N_{39}O_{49}$ Isolated from the insoluble core of Alzheimer's disease (AD) amyloid plaque; represents the fragment (61-95) of a 140 amino acid precursor known as NACP or α-synuclein found to form amyloid fibrils via a nucleation-dependent polymerization mechanism; Han, H et al, *Chemistry & Biology*, 2:163, 1995

Bachem H-2598.0500   MW 3260.65 $C_{141}H_{235}N_{39}O_{49}$ Store at -15°C

Sigma A 2835   FW 3260.6   ≥80% (HPLC) | Bioactive peptide; found in the insoluble core of amyloid plaque; forms fibrils via a nucleation-dependent polymerization mechanism; Han, H et al, *Chemistry & Biology*, 2:163, 1995

Glu-Glp
BOC-Glu-(NHO-Bz)-Glp

Synonyms: Prolyl Endopeptidase Inhibitor

ICN 195932   MW 419.5 $C_{21}H_{29}N_3O_6$ ≥95% | Novel acylating Pro-specific peptidase

Glu-Glu

Bachem G-1915.0250   MW 276.25 $C_{10}H_{16}N_2O_7$ Store at -15°C

Glu-Glu
γ-Glu-Glu

Synonyms: Glu(Glu-OH)-OH

Bachem G-1920.0250   MW 276.25 $C_{10}H_{16}N_2O_7$ Store at -15°C

Glu-Glu
D-Glu-D-Glu

Bachem G-3750.0250   MW 276.25 $C_{10}H_{16}N_2O_7$ Store at -15°C

Glu-Glu
Glu-Glu-βNA

Bachem K-1250.0050   MW 401.42 $C_{20}H_{23}N_3O_6$ Store at -15°C

Glu-Glu
FA-Glu-Glu

Bachem M-2385.0050   MW 412.4 $C_{18}H_{24}N_2O_9$ Store at -15°C

Glu-Glu

ICN 157204   MW 276.2 $C_{10}H_{16}N_2O_7$ Crystalline

Glu-Glu
γ-Glu-Glu

ICN 157205   MW 276.2 $C_{10}H_{16}N_2O_7$ Crystalline

Glu-Glu

Peptides International OEE-3080   MW 276.25 $C_{10}H_{16}N_2O_7$
>98% (HPLC); white amorphous powder

Sigma G 3640   FW 276.2 $C_{10}H_{16}N_2O_7$

Glu-Glu
γ-Glu-Glu

Sigma G 5267   FW 276.2 $C_{10}H_{16}N_2O_7$

Glu-Glu-Ala-Pro-Ser-Leu-Arg-Pro-Ala-Pro-Pro-Pro-Ile-Ser-Gly-Gly-Gly-Tyr-Arg

Synonyms: Fibrinogen β-Chain (24-42); Fibrinogen α-Chain (10-28)

American Peptide 42-1-33   MW 1951.2 $C_{50}H_{80}N_{18}O_{16}$

Bachem H-3034.0001   Store at -15°C

Glu-Glu-Asn
H(-Glu-Glu-Asn-Val)6

Bachem H-6140.0005   Store at -15°C

Glu-Glu-Asn-Val-Glu-His-Asp-Glu-Glu-Asn-Val-Glu-Glu-Asn-Val
(-Glu-Glu-Asn-Val-Glu-His-Asp-Ala)₂-Glu-Glu-Asn-Val-Glu-Glu-Asn-Val

Bachem H-6155.0005   Store at -15°C

Glu-Glu-Asp

Bachem H-3155.0050   MW 391.34 $C_{14}H_{21}N_3O_{10}$ Store at -15°C

Glu-Glu-Asp-Ser-Gly-Glp-Glu-Asp-Ser-Gly

Synonyms: Epidermal Mitosis Inhibiting Pentapeptide; Chalone

ICN 153095   Reichelt, KL etal, BBRC, 146:1493, 1987

Glu-Glu-Gln
Glu(Glu(Gln))

Synonyms: γ-Glu-γ-Glu-Gln

Bachem H-3170.0050   MW 404.38 $C_{15}H_{24}N_4O_9$ Store at -15°C

Glu-Glu-Glu
4-Nitro-Bz-Glu(Glu(Glu))

Synonyms: 4-Nitro-Bz-γ-Glu-γ-Glu-Glu

Bachem H-1166.0025   MW 554.47 $C_{22}H_{26}N_4O_{13}$ Store at -15°C

Glu-Glu-Glu

Bachem H-3160.0050   Store at -15°C

Glu-Glu-Glu
Glu(Glu(Glu))

Synonyms: γ-Glu-γ-Glu-Glu

Bachem H-3165.0050   MW 405.36 $C_{15}H_{23}N_3O_{10}$ Store at -15°C

Glu-Glu-Glu-Glu-Glu-Glu-Ala-Tyr-Gly-Trp-Met-Asp-Phe
D-Glu-Glu-Glu-Glu-Glu-Glu-Ala-Tyr(SO₃H)-Gly-D-Trp-Met-Asp-Phe Amide

Bachem H-5046.0001   MW 1742.77 $C_{73}H_{95}N_{15}O_{31}S_2$ Store at -15°C

Glu-Glu-Glu-Glu-Glu-Met-Ala-Val-Val-Pro-Gln-Gly-Leu-Phe-Arg-Gly Amide

Synonyms: Gastrointestinal Peptide; Pancreastatin (37-52); Chromograinin A (286-301)

Sigma P 9809   Human   FW 1819.0   ≥97% (HPLC) |
Bioactive peptide; Konecki, DS et al, *J Biol Chem*, 262:17026, 1987

Peptides International PCR-4214-v   Human synthetic   MW 1819.0 $C_{78}H_{123}N_{21}O_{27}S$ >97% (HPLC); lyophilized amorphous powder, hydrochloride | Bioactive; Konecki, DS et al, *JBC*, 262:17026, 1987

Glu-Glu-Glu-Glu-Lys-Asp-Ile-Glu-Ala-Glu-Glu-Arg-Gly-Asp-Leu-Gly-Glu-Gly-Gly-Ala-Trp-Arg-Leu-His

Synonyms: Prepro-Thyrotropin Releasing Hormone (83-106)

American Peptide 58-0-91   MW 2754.9 $C_{115}H_{176}N_{34}O_{45}$

Glu-Glu-Glu-Glu-Val-Gly-Phe-Pro-Val-Lys-Pro-Gln

Synonyms: NEF MN (64-75); HIV I Peptide

Neosystem SC646   MW 1388.49   NEF peptide from HIV-1 subtype MN

Glu-Glu-Glu-Glu-Val-Gly-Phe-Pro-Val-Thr

Synonyms: NEF (62-71); HTLV I Peptide

Neosystem SC503   MW 1135.19   HTLV I peptide from subtype NEF

Glu-Glu-His-Ser-Lys-Gln-Tyr-Arg-Cys-Leu-Ser-Phe-Gln-Pro-Gln-Cys-Ser-Met-Lys Trifluoroacetate Salt

Synonyms: Vitamin D Receptor (395-413), C-Terminal

Biogenesis 9580-3110   Human synthetic   MW 2330.24   Purified; lyophilized

Glu-Glu-Leu
BOC-Glu-Glu-Leu-OMe

Bachem A-1700.0025   MW 503.55 $C_{22}H_{37}N_3O_{10}$ Store at -15°C

Glu-Glu-Leu

Bachem H-1924.0050   Store at -15°C

Glu-Glu-Leu
N-t-BOC-Glu-Glu-Leu Methyl Ester

Sigma B 1011   FW 503.5 $C_{22}H_{37}N_3O_{10}$ ≥97% (HPLC) |
Bioactive peptide; possible substrate for Vitamin K-dependent carboxylation studies

Glu-Glu-Leu-Arg-Ser-Leu-Tyr-Asn-Thr-Val-Ala-Thr-Leu-Tyr

Synonyms: P18 (73-86); GAG P17 MA (73-86); HTLV I Peptide

Neosystem SC284   MW 1671.86   HTLV I peptide from subtype P25 (GAG P17 MA)

Glu-Glu-Leu-Leu-Lys-Gln-Ala-Leu-Gln-Gln-Ala-Gln-Gln-Leu-Leu-Gln-Gln-Ala-Gln-Glu-Leu-Ala-Lys-Lys
Ac-Glu-Glu-Leu-Leu-Lys-Gln-Ala-Leu-Gln-Gln-Ala-Gln-Gln-Leu-Leu-Gln-Gln-Ala-Gln-Glu-Leu-Ala-Lys-Lys Amide

Synonyms: Peptitergent PD1

Bachem H-2998.0500   Store at -15°C

Neosystem SC917 MW 2819.25 Virus-related peptide; bioactive; maintains solubility of membrane proteins; when mixed with this peptide, 85% of bacteriorhodopsin & 60% of rhodopsin remained in solution over a period of 2 days in their native forms; Schafmeister, CE et al, *Science*, 262:734-738, 1993

Glu-Glu-Lys
Glu(Glu(Lys))

Synonyms: γ-Glu-γ-Glu-Lys

Bachem H-3175.0050 MW 404.42 $C_{16}H_{28}N_4O_8$ Store at -15°C

Glu-Glu-Lys-Leu-Ile-Val-Val-Ala-Phe

Bachem H-3688.0001 Store at -15°C

Glu-Glu-Met-Leu-Phe-Ile-Tyr-Gly-His-Tyr-Lys-Gln-Ala-Thr-Val-Gly-Asp-Ile-Asn-Thr-Glu-Arg-Pro-Gly-Met-Leu-Asp-Phe-Thr

Synonyms: Diazepam Binding Inhibitor (22-50)

Neosystem SC367 Human MW 3376.79 Bioactive neuropeptide;

Glu-Glu-Val
N-t-BOC-Glu-Glu-Val Methyl Ester

Sigma B 9886 FW 489.5 $C_{21}H_{35}N_3O_{10}$ ≥97% (HPLC) | Bioactive peptide

---

Glu-Glu-Val-Val-Ala-Cys
Ac-Glu-Glu-Val-Val-Ala-Cys-pNA

Synonyms: HCV NS3 Protease Substrate

American Peptide 81-0-14 MW 809.9 $C_{34}H_{49}N_8O_{13}S$ Substrate for continuous spectrophotometric assay

---

Glu-Glu-Val-Val-Ala-Cys
Ac-Glu-Glu-Val-Val-Ala-Cys-AMC

Bachem I-1675.0001 MW 847.94 $C_{38}H_{53}N_7O_{13}S$ Store at -15°C

Glu-Glu-Val-Val-Ala-Cys
Ac-Glu-Glu-Val-Val-Ala-Cys-pNA

Bachem L-1910.0001 Store at -15°C

Glu-Gly
Z-Glu-Gly

Bachem C-1640.0250 MW 338.32 $C_{15}H_{18}N_2O_7$ Store at -15°C

Glu-Gly

Bachem G-1935.0250 MW 204.18 $C_7H_{12}N_2O_5$ Store at -15°C

Glu-Gly
γ-Glu-Gly

Synonyms: Glu(Gly-OH)-OH

Bachem G-1940.0250 MW 204.18 $C_7H_{12}N_2O_5$ Store at -15°C

Glu-Gly
γ-D-Glu-Gly

Synonyms: D-Glu(Gly-OH)-OH

Bachem G-1945.0250 Store at -15°C

ICN 153101 MW 204.2 $C_7H_{12}N_2O_5$ An excitatory AA antagonist; Collingridge, DL etal, *J Physiol*, 334:19, 1983

Glu-Gly
γ-Glu-Gly

ICN 157206 MW 204.2 $C_7H_{12}N_2O_5$

Glu-Gly
γ-D-Glu-Gly

Synonyms: Excitatory AA Antagonist

Sigma G 3765 FW 204.2 $C_7H_{12}N_2O_5$ ≥97% (HPLC) | Bioactive peptide; Collingridge, GL et al, *J Physiol*, 334:19, 1983

Glu-Gly
γ-Glu-Gly

Sigma G 8390 FW 204.2 $C_7H_{12}N_2O_5$

Glu-Gly-Arg
BOC-Glu(OBzl)-Gly-Arg-AMC Hydrochloride

Bachem I-1545.0050 MW 744.25 $C_{35}H_{45}N_7O_9$ · HCl Store at -15°C

Glu-Gly-Arg
Glu-Gly-Arg-pNA

Bachem L-1455.0050 MW 480.48 $C_{19}H_{28}N_8O_7$ Store at -15°C

Glu-Gly-Arg
Glu-Gly-Arg-CMK

Bachem N-1325.0005 Store at -15°C

Glu-Gly-Arg
N-α-t-BOC-γ-Bzl-L-Glu-Gly-L-Arg-MCA Hydrochloride

ICN 150485 MW 744.2 $C_{35}H_{45}N_7O_9$ · HCl Fluorogenic substrate

---

Glu-Gly-Arg
Nᵃ-Glu-Gly-Arg-Chloromethylketone

Synonyms: Pefa-3591

Pentapharm 385-02 MW 620.9 $C_{14}H_{25}N_6O_5Cl$·2TFA Bulk | Irreversible inhibitor of serine proteinases, ie urokinase & factor Xa; $k_2/K_I = 994$

Glu-Gly-Arg
BOC-Glu(OBzl)-Gly-Arg-MCA

Peptides International MEG-3115-v MW 707.78 $C_{35}H_{45}N_7O_9$ >98% (HPLC); lyophilized amorphous powder | Iwanaga, S et al, *Kinins-II:Biochemistry, Pathophysiology, & Clinical Aspects*, (S Fujii, H Moriya, & T Suzuki, Eds), Plenum Publishing Co, 1979, pp 147-163

Glu-Gly-Arg
N-t-BOC-γ-Benzyl-Glu-Gly-Arg MCA Hydrochloride

Sigma B 5028 FW 744.2 $C_{35}H_{45}N_7O_9$ · HCl ~95% | Iwanaga, S et al, *KININS-II:Biochem, Pathophysiol, & Clin Aspects*, Fujii, S et al, ed, Plenum Publ Co, p 147, 1979

Glu-Gly-Gly
γ-Glu-Gly-Gly

Synonyms: Glu(Gly-Gly-OH)-OH

Bachem H-3180.0250 MW 261.24 $C_9H_{15}N_3O_6$ Store at -15°C

Glu-Gly-His-Ile-Ala-Arg-Asn-Cys-Arg-Ala-Pro-Arg-Lys-Lys-Gly

Synonyms: P15 (398-412); GAG P7 NC (21-35) HTLV I Peptide

Neosystem SC271   MW 1692.96   HTLV I peptide from subtype P15 (GAG P7 NC/GAG P1/GAG P6); due to the presence of a Cys this peptide can contain the dimeric form

Glu-Gly-Phe

Bachem H-3185.0250   MW 351.36 $C_{16}H_{21}N_3O_6$ Store at -15°C

ICN 157207   MW 351.4 $C_6H_{21}N_3O_6$ Crystalline

Sigma G 3501   FW 351.4 $C_{16}H_{21}N_3O_6$

Glu-Gly-Pro-Trp-Leu-Glu-Glu-Glu-Glu-Glu-Ala-Tyr-Gly-Trp-Met-Asp-Phe
Glu-Gly-Pro-Trp-Leu-Glu-Glu-Glu-Glu-Glu-Ala-Tyr(SO₃H)-Gly-Trp-Met-Asp-Phe Amide Sulfated

Synonyms: Gastrointestinal Peptide; Gastrin I

Sigma G 1260   FW 2178.3   ≥95% (HPLC) | Bioactive peptide

Glu-Gly-Ser-Asp-Thr-Ile-Thr-Leu-Pro-Cys-Arg-Ile-Lys-Gln-Phe-Ile-Asn-Met-Trp-Gln-Glu

Synonyms: GP120 (414-434); HTLV I Peptide

Neosystem SC266   MW 2509.87   HTLV I peptide from subtype GP160; due to the presence of a Cys this peptide can contain the dimeric form

Glu-Gly-Thr-Asp-Arg-Val-Ile-Glu

Synonyms: GP41 (829-836); HTLV I Peptide

Neosystem SC259   MW 917.97   HTLV I peptide from subtype GP41

Glu-Gly-Thr-Phe-Thr-Ser-Asp-Leu-Ser-Lys-Gln-Met-Glu-Glu-Glu-Ala-Val-Arg-Leu-Phe-Ile-Glu-Trp-Leu-Lys-Asn-Gly-Gly-Pro-Ser-Ser-Gly-Ala-Pro-Pro-Pro-Ser Amide

Synonyms: Exendin IV (3-39)

Bachem H-3864.0500   Store at -15°C

Glu-Gly-Val-Asn-Asp-Asn-Glu-Glu-Gly-Phe-Phe-Ser-Ala-Arg

Synonyms: Fibrinopeptide B, (Glu¹)-

Bachem H-2950.0001   MW 1570.59 $C_{66}H_{95}N_{19}O_{26}$ Store at -15°C

American Peptide 42-1-24   Human   MW 1570.6 $C_{66}H_{95}N_{19}O_{26}$

ICN 153098   Human

Sigma F 3261   Human   FW 1570.6   ≥95% (HPLC) | Bioactive peptide

Glu-Gly-Val-Pro-Ser-Thr-Ala-Ile-Arg-Glu-Ile-Ser-Leu-Leu-Lys-Glu

Synonyms: p34*cdc*2 Peptide

American Peptide 86-5-30   MW 1742.0 $C_{76}H_{132}N_{20}O_{26}$
Norbury, CJ et al, *Biochim Biophys Acta*, 989:85, 1989

ICN 195847   MW 1740.9   ≥97% | Cyclic-dependent proteinkinase which causes meiotic maturation & Ca rise; Ab blocking peptide

Glu-Gly-Val-Pro-Ser-Thr-Ala-Ile-Arg-Glu-Ile-Ser-Leu-Leu-Lys-Glu-(Gly-Gly-Cys)

Synonyms: p34*cdc*2 Peptide Blocking Peptide (42-57)

Calbiochem 541331   Human synthetic   MW 1959.3 $C_{83}H_{143}N_{23}O_{29}S$ Lyophilized | Based on bovine, human, mouse & yeast 34 kDa Cdc2-encoded protein kinase (1-37) with a three AA GGC linker arm attached; this peptide coupled to KLH, was used as the immunogen for the production of Anti-PSTAIRE; for use in immunoabsorption for immunoprecipitation, immunocytochemistry, Western blotting & dot blots

Glu-Gly-Val-Tyr-Val-His-Pro-Val

Synonyms: Angiotensin Receptor Antagonist; Angiotensin II Antipeptide

American Peptide 12-1-40   MW 899.0 $C_{42}H_{62}N_{10}O_{12}$ "Anti-angiotensin II" peptide deduced from the antisense mRNA complementary to the human angiotensin II mRNA; the sequence of this octapeptide is 50% homologous to angiotensin II; it acts as an antagonist of angiotensin II causing uterine contractions or hypertensive responses in the rat; Moore, GJ et al, *Biochem Biophys Res Comm*, 160:1387, 1989

Bachem H-8160.0001   MW 899.01 $C_{42}H_{62}N_{10}O_{12}$ Store at -15°C

ICN 158848   Moore, GJ etal, *BBRC*, 160:1387, 1989

Neosystem SC373   MW 899.01   Bioactive; Angiotensin II antagonist; Moore, GJ et al, *BBRC*, 160:1387, 1989

Sigma A 4184   FW 899.0 $C_{42}H_{62}N_{10}O_{12}$ ≥97% (HPLC) | Bioactive peptide; an angiotensin receptor antagonist; inhibits uterine contractions & hypertensive responses in rat; Moore, GJ et al, *Biochem Biophys Res Commun*, 160:1387, 1989

Glu-His
γ-Glu-His

Synonyms: Glu(His-OH)-OH

Bachem G-3950.0250   MW 284.27 $C_{11}H_{16}N_4O_5$ Store at -15°C

Glu-His
Glu-His-βNA

Bachem K-1260.0050   MW 409.45 $C_{21}H_{23}N_5O_4$ Store at -15°C

Glu-His
γ-Glu-His

ICN 157208   MW 284.3 $C_{11}H_{16}N_4O_5$ ≥95%; crystalline

Glu-His
γ-Glu-His

Sigma G 6882   FW 284.3 $C_{11}H_{16}N_4O_5$ 95-97%

Glu-His-Gly

Bachem H-9615.0050   Store at -15°C

Glu-His-Ile-Pro-Ala

Synonyms: Fibrinogen Binding Peptide

Bachem H-8635.0005   Store at -15°C

Neosystem SC912   MW 565.63   Bioactive; cell attachment peptide; fibrinogen-related peptide; binds fibrinogen & inhibits platelet aggregation & adhesion to fibrinogen & vitronectin; Gartner, TK & DB Taylor, *Proc Soc Exp Biol Med*, 198:649-655, 1991

Glu-His-Nal-Arg-Trp-Gly-Cys-Pro-Pro-Lys-Asp
3-Mercaptopropionyl-Glu-His-D-2-Nal-Arg-Trp-Gly-Cys-Pro-Pro-Lys-Asp Amide

Synonyms: Melanocyte Stimulating Hormone (11-22), (Deamino-Cys[11],D-2-Nal[14],Cys[18])-β-; JKC-363

Bachem H-4942.0001 MW 1506.73 $C_{69}H_{91}N_{19}O_{16}S_2$ Store at -15°C

Glu-His-Phe-Arg-Trp-Gly

Synonyms: Adrenocorticotropic Hormone (5-10)

Bachem H-1195.0005 MW 830.9 $C_{39}H_{50}N_{12}O_9$ Store at -15°C

Glu-His-Phe-Lys-Phe
Met(O)-Glu-His-Phe-D-Lys-Phe

Synonyms: Adrenocorticotropic Hormone (4-9), (Met(O)[4],D-Lys[8],Phe[9])-; Corticotropin A

ICN 159853 MW 854.1 >98%

Glu-His-Phe-Lys-Phe
Met(O₂)-Glu-His-Phe-D-Lys-Phe

Synonyms: Adrenocorticotropic Hormone (4-9), (Met(O₂)[4],D-Lys[8],Phe[9])-; Corticotropin A

ICN 159854 MW 870.1 >97%

American Peptide 10-1-42 Human MW 870.1 $C_{40}H_{55}N_9O_{11}S$ Sandman, C et al, *Peptides*, 1:104, 1980

Glu-His-Phe-Lys-Phe
Met(O)-Glu-His-Phe-D-Lys-Phe

Synonyms: Adrenocorticotropic Hormone (4-9), (Met(O)[4],D-Lys[8],Phe[9])-

American Peptide 10-1-45 Human MW 854.1 $C_{40}H_{55}N_9O_{10}S$ Van Riesen, VH et al, *Arzneim-Forsch*, 28:1294, 1978

Glu-His-Phe-Lys-Phe-Gly
Met(O₂)-Glu-His-Phe-D-Lys-Phe-Gly

Synonyms: Adrenocorticotropic Hormone (4-10), (Met(O₂)[4],D-Lys[8],Phe[9])-; Corticotropin A

ICN 159855 MW 927.2 >98%

Glu-His-Phe-Lys-Phe-Gly
Met(O)-Glu-His-Phe-D-Lys-Phe-Gly

Synonyms: Adrenocorticotropic Hormone (4-10), (Met(O)[4],D-Lys[8],Phe[9])-; Corticotropin A

ICN 159856 MW 911.2 >98%

Glu-His-Pro Amide

Synonyms: Thyrotropin Releasing Hormone, (Glu[1])-

American Peptide 52-0-82 MW 379.4 $C_{16}H_{23}N_6O_5$

Glu-His-Trp-Ser-Tyr-Gly-Leu-Arg-Pro
Glu-His-Trp-Ser-Tyr-Gly-Leu-Arg-Pro-NHEt

Synonyms: Luteinizing Hormone Releasing Hormone, (des-Gly[10],Pro-NHEt[9])-

Neosystem SC470 MW 1153.30 Bioactive; hypothalamic releasing hormone; potent LHRH agonist; Okada, J & Kondo, S, *Chem Pharm Bull*, 33:4464, 1985

Glu-Ile-Cys-Thr-Glu-Met-Glu-Lys-Glu-Gly-Lys-Ile-Ser-Lys-Ile-Gly-Pro

Synonyms: POL (203-219); RT (36-52); HTLV I Peptide

Neosystem SC688 MW 1892.21 HTLV I peptide from subtype POL (PR/RT); due to the presence of a Cys this peptide can contain the dimeric form

Glu-Ile-Gln-Lys-Gln-Gly-Gln-Gly-Gln-Trp-Thr-Tyr-Gln-Ile-Tyr-Gln-Glu-Pro-Phe-Lys-Asn-Leu-Lys-Thr-Gly

Synony POL (495-519); RT (328-352); HTLV I Peptide *ms*:

Neosystem SC693 MW 3013.35 HTLV I peptide from subtype POL (PR/RT)

Glu-Ile-Gly-Asp-Glu-Glu-Asn-Ser-Ala-Lys-Phe-Pro-Ile Amide

Synonyms: Neuropeptide EI

Bachem H-2396.0001 Rat MW 1447.57 $C_{63}H_{98}N_{16}O_{23}$ Store at -15°C

Glu-Ile-Ile-Val-Thr-His-Phe-Pro-Phe-Asp-Glu-Gln-Asn-Cys-Ser-Met-Lys

Synonyms: Acetylcholine Receptor α1 (129-145)

Bachem H-4186.0500 Human, bovine, rat, mouse MW 2038.33 $C_{90}H_{136}N_{22}O_{28}S_2$ Store at -15°C

Glu-Ile-Leu-Asp-Val

Synonyms: Fibronectin CS-1 (1978-1982)

Bachem H-2592.0005 MW 587.67 $C_{26}H_{45}N_5O_{10}$ Store at -15°C

American Peptide 44-0-41 Human, bovine, rat MW 587.7 $C_{26}H_{45}N_5O_{10}$ Derived from the type III connecting segment domain (IIICS) of fibronectin; promotes tumor cell attachment after surface immobilization; to inhibits experimental tumor metastasis of B16-BL6 melanoma

Glu-Ile-Leu-Asp-Val-Pro-Ser-Thr

Synonyms: Fibronectin CS-1 Peptide; Fibronectin CS-1 (1978-1985); Fibronectin CS-1

American Peptide 44-0-42 MW 873.0 $C_{38}H_{64}N_8O_{15}$ The connecting segment 1 (CS-1) is a cell attachment domain located in the type III homology connecting segment (IIICS) of fibronectin; Wayner, EA et al, *JCB*, 109:1321, 1989

Bachem H-2094.0001 MW 872.97 $C_{38}H_{64}N_8O_{15}$ Store at -15°C

Neosystem SC913 MW 872.97 Bioactive; cell attachment peptide; Wayner, EA et al, *J Cell Biol*, 109:1321-1330, 1989

Glu-Ile-Leu-Glu-Val-Pro-Ser-Thr

American Peptide 44-0-44 MW 887.0 $C_{39}H_{66}N_8O_{15}$ Derived from the type III connecting segment domain (IIICS) of fibronectin; promotes tumor cell attachment after surface immobilization; inhibits experimental tumor metastasis of B16-BL6 melanoma

Glu-Ile-Leu-Ser-Gln-Leu-Tyr-Arg-Pro-Leu-Glu-Ala-Cys-Tyr

Synonyms: TAT (38-51); SIVmac251 Peptide

Neosystem SC587 MW 1697.96 SIVmac251 peptide from subtype TAT; due to the presence of a Cys this peptide can contain the dimeric form

Glu-Ile-Lys-Asp-Thr-Lys-Glu-Ala-Leu-Asp-Lys-Ile-Glu-Glu-Glu

Synonyms: P18 (93-107); GAG P17 MA (93-107); HTLV I Peptide

Neosystem SC286 MW 1789.95 HTLV I peptide from subtype P25 (GAG P17 MA)

Glu-Leu
Z-Glu-Leu ·DCHA

Bachem C-1645.0250 MW 575.75 $C_{19}H_{26}N_2O_7 \cdot C_{12}H_{23}N$ Store at -15°C

Glu-Leu
γ-Glu-Leu

Synonyms: Glu(Leu-OH)-OH

Bachem G-1950.0250 MW 260.29 $C_{11}H_{20}N_2O_5$ Store at -15°C

ICN 157209 MW 260.3 $C_{11}H_{20}N_2O_5$ Crystalline

Glu-Leu
γ-Glu-Leu

Sigma G 7007 FW 260.3 $C_{11}H_{20}N_2O_5$

Glu-Leu-Ala-Gly-Ala-Pro-Pro-Glu-Pro-Ala

Synonyms: Lipotropin (88-10), β-; Lipotropin (1-10), α-

American Peptide 30-0-83 Porcine MW 951.1 $C_{42}H_{66}N_{10}O_{15}$

Bachem H-6315.0005 Porcine MW 951.04 $C_{42}H_{66}N_{10}O_{15}$ Store at -15°C

ICN 152811 Porcine MW 951 $C_{42}H_{66}N_{10}O_{15}$

Glu-Leu-Arg-Met-Ser-Ser-Ser-Tyr-Pro-Thr-Gly-Leu-Ala-Asp-Val-Lys-Ala-Gly-Pro-Ala-Gln-Thr-Leu-Ile-Arg-Pro-Gln-Asp-Met-Lys-Gly-Ala-Ser-Arg-Ser-Pro-Glu-Asp-Ser-Ser-Pro-Asp-Ala-Ala-Arg-Ile-Arg-Val

Synonyms: Pro-Adrenomedullin (45-92)

American Peptide 22-2-12 Human MW 5114.8 $C_{215}H_{358}N_{67}O_{73}S_2$

Glu-Leu-Asp-Lys-Trp-Ala

American Peptide 72-3-40 MW 760.9 $C_{35}H_{52}N_8O_{11}$ Located within the immunosuppressive domain of HIV-1gp41; recognized by human monoclonal antibody 2F5, which has the broad-spectrum neutralizing capacity against HIV-1 laboratory strains & primary isolates; Chen, YH et al, *Immuno Lett*, 52:153, 1996

Glu-Leu-Glu-Pro-Glu-Asp-Glu-Ala-Arg-Pro-Gly-Gly-Phe-Asp-Arg-Leu-Gln-Ser-Glu-Asp-Lys-Ala-Ile-Arg-Thr-Ile-Met-Glu-Phe-Leu-Ala-Phe-Leu-His-Leu-Lys-Glu-Ala-Gly-Ala-Leu Amide

Synonyms: Gastrointestinal Peptide; Galanin Message Associated Peptide (1-41); Prepro-Galanin (65-105)

American Peptide 41-1-20 MW 4643.3 $C_{206}H_{326}N_{56}O_{64}S$ Rokaeus, A et al, *PNAS*, 83:6287, 1986

Bachem H-6780.0500 MW 4643.26 $C_{206}H_{326}N_{56}O_{64}S$ Store at -15°C

ICN 152866 Porcine Rokaeus, A & J Brownstein, *PNAS*, 83:6287, 1986

Sigma G 4646 Porcine FW 4643.2 ≥95% (HPLC) | Bioactive peptide; galanin precursor protein composed of a leader sequence, a single copy of galanin, & a 59 AA galanin message associated protein; Rokaeus, A & Brownstein, MJ, *Proc Natl Acad Sci USA*, 83:6287, 1986

Glu-Leu-His-Pro-Glu-Tyr-Phe-Lys-Asn-Cys

Synonyms: NEF (197-206); HTLV I Peptide

Neosystem SC671 MW 1271.43 HTLV I peptide from subtype NEF; due to the presence of a Cys this peptide can contain the dimeric form

Glu-Leu-Lys-Glu-Leu-Pro-Pro-Val-Thr-Ser-Glu-Gln Acetate Salt

Synonyms: Glucose Transporter V

Biogenesis 4670-1764 Human synthetic Semi-pure; lyophilized

Glu-Leu-Thr-Tyr-Leu-Gln-Tyr-Gly-Trp-Ser-Tyr-Phe-His-Glu-Ala-Val-Gln-Ala-Gly-Trp

Synonyms: GP140 (821-840); SIVmac251 Peptide

Neosystem SC790 MW 2448.67 SIVmac251 peptide from subtype GP140

Glu-Leu-Tyr-Lys-Tyr-Lys-Val-Val-Lys-Ile-Glu-Pro-Leu-Gly-Val-Ala

Synonyms: GP120 (487-502); HTLV I Peptide

Neosystem SC251 MW 1849.24 HTLV I peptide from subtype GP160

Glu-Leu-Tyr-Lys-Tyr-Lys-Val-Val-Lys-Ile-Glu-Pro-Leu-Gly-Val-Ala-Pro-Thr-Lys-Ala-Lys-Arg-Arg

Synonyms: GP120 (487-509); HTLV I Peptide

Neosystem SC695 MW 2687.26 HTLV I peptide from subtype GP160

Glu-Lys
BOC-γ-Glu-Lys

Synonyms: BOC-Glu(Lys-OH)

Bachem A-3155.0250 Store at -15°C

Glu-Lys
Glu(BOC-Lys-OMe)-OMe Hydrochloride

Synonyms: BOC-Lys(retro-MeO-Glu-H)-OMe

Bachem A-4475.0250 MW 439.94 $C_{18}H_{33}N_3O_7 \cdot HCl$ Store at -15°C

Glu-Lys
Glu(OMe)-(BOC-Lys-OMe) Hydrochloride

Synonyms: BOC-Lys(retro-Glu(OMe)-H)-OMe

Bachem A-4480.0250 MW 439.94 $C_{18}H_{33}N_3O_7 \cdot HCl$ Store at -15°C

Glu-Lys

Bachem G-1955.0250 MW 275.31 $C_{11}H_{21}N_3O_5$ Store at -15°C

Glu-Lys
Glu(H-Lys)

Synonyms: Lys(retro-HO-Glu-H); γ-Glu-ε-Lys

Bachem G-1970.0250 MW 275.31 $C_{11}H_{21}N_3O_5$ Store at -15°C

Glu-Lys

ICN 157210 MW 275.3 $C_{11}H_{21}N_3O_5$

Glu-Lys
γ-Glu-ε-Lys

ICN 157211 MW 275.3 $C_{11}H_{21}N_3O_5$

Glu-Lys

Sigma G 3390 FW 275.3 $C_{11}H_{21}N_3O_5$

Glu-Lys
γ-Glu-ε-Lys

Sigma G 5136 FW 275.3 $C_{11}H_{21}N_3O_5$

Glu-Lys-Ala-His-Asp-Gly-Gly-Arg

Synonyms: C-Telopeptide

American Peptide 89-0-30   MW 868.9 $C_{34}H_{56}N_{14}O_{13}$ An RIA based on monoclonal antibody raised against this 8-AA sequence was developed; represents a valuable tool for assessing bone resorption; Bonde, M et al, *Clin Chem*, 42(10):1639, 1996

Glu-Lys-Gly-Gly-Leu-Asp-Gly-Leu-Ile-Tyr-Ser-Gln-Lys-Arg

Synonyms: NEF MN (95-108); HIV I Peptide

Neosystem SC651   MW 1563.77   NEF peptide from HIV-1 subtype MN

Glu-Lys-Gly-Gly-Leu-Glu-Gly-Ile-Tyr-Tyr-Ser-Ala-Arg-Arg

Synonyms: NEF (125-138); SIVmac251 Peptide

Neosystem SC615   MW 1598.77   SIVmac251 peptide from subtype NEF

Glu-Lys-Gly-Gly-Leu-Glu-Gly-Leu-Ile-His-Ser-Gln-Arg-Arg

Synonyms: NEF (93-106); HTLV I Peptide

Neosystem SC222   MW 1579.77   HTLV I peptide from subtype NEF

Glu-Lys-Lys
BOC-Glu-Lys-Lys-AMC

Bachem I-1095.0025   MW 660.77 $C_{32}H_{48}N_6O_9$ Store at -15°C

Glu-Lys-Lys
N-α-*t*-BOC-L-Glu-L-Lys-L-Lys-MCA

Synonyms: Plasmin Substrate

ICN 150495   MW 660.8 $C_{32}H_{48}N_6O_9$ Fluorogenic substrate

Glu-Lys-Lys
BOC-Glu-Lys-Lys-MCA

Synonyms: Plasmin Substrate

Peptides International MEK-3105-v   MW 660.77 $C_{32}H_{48}N_6O_9$ >98% (HPLC); lyophilized amorphous powder | Kato, H et al, *J Biochem*, 88:183, 1980

Glu-Lys-Lys
N-*t*-BOC-Glu-Lys-Lys MCA

Synonyms: Plasmin Substrate

Sigma B 6513   FW 660.8 $C_{32}H_{48}N_6O_9$ 90-95% (HPLC) | Kato, H et al, *J Biochem*, 88:183, 1988

Glu-Lys-Pro-Leu-Gln-Asn-Phe-Thr-Leu-Cys-Phe-Arg Amide

Synonyms: Amyloid P Component (27-38)

Bachem H-2942.0001   MW 1494.78 $C_{68}H_{107}N_{19}O_{17}S$ Store at -15°C

Glu-Lys-Ser-Leu-Gly-Glu-Ala-Asp-Lys-Ala-Asp-Val-Asn-Val-Leu-Thr-Lys-Ala-Lys-Ser-Gln

Synonyms: Parathyroid Hormone (64-84), (Asn^{76})-

American Peptide 22-1-41   Human   MW 2231.5 $C_{94}H_{163}N_{27}O_{35}$

Bachem H-3190.0500   Human   MW 2231.49 $C_{94}H_{163}N_{27}O_{35}$ Store at -15°C

Glu-Lys-Ser-Leu-Gly-Glu-Ala-Asp-Lys-Ala-Asp-Val-Asn-Val-Leu-Thr-Lys-Ala-Lys-Ser-Gln
Biotinyl-Glu-Lys-Ser-Leu-Gly-Glu-Ala-Asp-Lys-Ala-Asp-Val-Asn-Val-Leu-Thr-Lys-Ala-Lys-Ser-Gln

Synonyms: Parathyroid Hormone (64-84), Biotinyl-

Bachem H-6605.0500   Human   Store at -15°C

Glu-Lys-Ser-Leu-Gly-Glu-Ala-Asp-Lys-Ala-Asp-Val-Asn-Val-Leu-Thr-Lys-Ala-Lys-Ser-Gln

Synonyms: Parathyroid Hormone (64-84); Parathormone

ICN 154484   Human   MW 2231.8

Glu-Lys-Ser-Leu-Gly-Glu-Ala-Asp-Lys-Ala-Asp-Val-Asp-Val-Leu-Thr-Lys-Ala-Lys-Ser-Gln

Synonyms: Parathyroid Hormone (64-84), (Asp^{76})-

Sigma P 6157   Human   FW 2232.5   ≥95% (HPLC) | Bioactive peptide

Glu-Met
Z-Glu-Met

Bachem C-1650.0250   MW 412.46 $C_{18}H_{24}N_2O_7S$ Store at -15°C

Glu-Met
γ-Glu-Met

Synonyms: Glu(Met-OH)-OH

Bachem H-3195.0250   MW 278.33 $C_{10}H_{18}N_2O_5S$ Store at -15°C

Glu-Phe
γ-Glu-Phe

Synonyms: Glu(Phe-OH)-OH

Bachem G-1975.0250   MW 294.31 $C_{14}H_{18}N_2O_5$ Store at -15°C

Bachem M-1240.0001   MW 428.44 $C_{22}H_{24}N_2O_7$ Store at -15°C

Glu-Phe
γ-Glu-Phe

ICN 157212   MW 294.3 $C_{14}H_{18}N_2O_5$

Glu-Phe-Thr-Asn-Val-Ser-Cys-Thr-Thr-Ser-Lys-Glu-Cys-Trp-Ser-Val-Cys-Gln-Arg-Leu-His-Asn-Thr-Ser-Arg-Gly-Lys-Cys-Met-Asn-Lys-Lys-Cys-Arg-Cys-Tyr-Ser

Synonyms: Charybdotoxin

Sigma C-133   *Leiurus quinquestriatus hebraeus* synthetic   MW 4.35k (peptide free base)   >98%; white solid; peptide content & salt form information are provided with each lot; packaged under argon; should be reconstituted in saline solution (~150 *mM* salt); bioassayed product; potent neurotoxin | Specific inhibitor of the high conductance Ca^{2+}-activated K^+ channel; 37 AA peptide present in the venom of the scorpion; Miller et al, *Nature*, 313:316, 1985; Gimenez-Gallego et al, *Proc Natl Acad Sci USA*, 85:3329, 1988; Strong, *Pharmac Ther*, 46:137, 1990

Glu-Phe-Tyr

Bachem H-3200.0250   MW 457.48 $C_{23}H_{27}N_3O_7$ Store at -15°C

Glu-*p*NA Hydrate

Peptides International SEN-3067   MW 285.26 $C_{11}H_{13}N_3O_5 \cdot$ H_2O >98% (HPLC); amorphous powder

Glu-Pro-Gln-Tyr-Glu-Glu-Ile-Pro-Ile-Tyr-Leu
Biotinyl-ε-Aminocaproyl-Glu-Pro-Gln-Tyr(PO₃H₂)-Glu-Glu-Ile-Pro-Ile-Tyr-Leu

Bachem H-1546.0001   Store at -15°C

Glu-Pro-Gln-Tyr-Glu-Glu-Ile-Pro-Ile-Tyr-Leu
Glu-Pro-Gln-Tyr(PO₃H₂)-Glu-Glu-Ile-Pro-Ile-Tyr-Leu

Bachem H-8905.0001   Store at -15°C

Glu-Pro-Pro-Leu-Ser-Gln-Glu-Ala-Phe-Ala-Asp-Leu-Trp-Lys-Lys

Synonyms: DNA-Dependent Protein Kinase

Promega V5671   MW 1759

Glu-Pro-Trp-Lys-His-Pro-Gly-Ser-Gln-Pro-Lys-Thr

Synonyms: TAT (9-20); HTLV I Peptide

Neosystem SC520   MW 1391.55   HTLV I peptide from subtype TAT

Glu-Ser

Bachem G-1980.0250   MW 234.21 $C_8H_{14}N_2O_6$ Store at -15°C

Glu-Ser-Ala-Asn-Leu-Gly-Glu-Glu-Ile-Leu-Ser-Gln-Leu-Tyr-Arg

Synonyms: TAT (31-45); SIVmac251 Peptide

Neosystem SC586   MW 1721.88   SIVmac251 peptide from subtype TAT

Glu-Ser-Leu-Phe

Bachem H-1212.0050   Store at -15°C

Glu-Ser-Met-Asp
Ac-Glu-Ser-Met-Asp-CHO Aldehyde

Synonyms: Caspase III Processing Enzyme Inhibitor II

Alexis 260-056   MW 506.5 $C_{19}H_{30}N_4O_{10}S$ Corresponds to one of the cleavage sites of the inactive 32 kDa Caspase III precursor (AA 25-28); blocks the formation of the p17 subunit & concomitantly induces the accumulation of the p20 peptide; Han, Z et al, *J Biol Chem*, 272:13432, 1997

Glu-Ser-Met-Asp
Ac-Glu-Ser-Met-Asp-CHO Pseudo Acid

Bachem N-1625.0005   MW 506.53 $C_{19}H_{30}N_4O_{10}S$ Store at -15°C

Glu-Ser-Met-Asp
N-Ac-Glu-Ser-Met-Asp-CHO

Sigma A 1341   Bioactive peptide; Han, Z et al, *J Biol Chem*, 272:13432, 1997

Glu-Ser-Pro-Leu-Ile-Ala-Lys-Val-Leu-Thr-Thr-Glu-Pro-Pro-Ile-Ile-Thr-Pro-Val-Arg-Arg

Synonyms: Phospholipase A₂ Activating Peptide

Alexis 165-007   MW 2330.8; ≥97%; White lyophilized powder; Soluble in water $C_{106}H_{184}N_{28}O_{30}$ Peptide fragment from phospholipase A₂ activating protein (PLAP) that activates phospholipase A₂ in a dose-dependent manner; AA sequence bears considerable homology to melittin; Clark, MA et al, *PNAS*, 88:5418, 1991

Bachem H-8440.0500   MW 2330.8 $C_{106}H_{184}N_{28}O_{30}$ Store at -15°C

ICN 159653   MW 2330.8   Activates phospholipase A2; AA sequence shows similarity to melittin; Clark, MA etal, *PNAS*, 88:5418, 1991

ICN 195682   >98% | A PLA₂ activating protein fragment which spans the PLAP-melittin homology sequence; activates phospholipase A₂ (PLA₂) by a factor of 10 at 1 µg/ml in BC3H cell sonicates; a member of the β-transducin (G$_β$) superfamily

Sigma G 1153   FW 2330.8   ≥95% (HPLC); peptide content:~70% | Bioactive peptide; induced in smooth muscle & endothelial cells treated with leukotriene D4; activates phospholipase A2; Clark, MA et al, *Proc Natl Acad Sci USA*, 88:5418, 1991

Glu-Ser-Val-Trp-Gly-Asp-Glu-Lys-Ser-Ser-Phe-Ile-Cys

Synonyms: Connexin 32 (41-52)

Biogenesis 2260-1509   Human synthetic   Semi-pure; lyophilized

---

Glu-Thr

Bachem G-1985.0100   MW 248.24 $C_9H_{16}N_2O_6$ Store at -15°C

Glu-Thr-Gly-Gln-Glu-Thr-Ala-Tyr-Phe-Leu-Leu-Lys-Leu-Ala-Gly-Arg-Trp-Pro-Val-Lys

Synonyms: Endonuclease Antigenic Site

Neosystem SC821   MW 2307.67   Virus-related peptide; AIDS-related peptide

---

Glu-Thr-Leu-Asp-Ser-Leu-Gly-Gly-Val-Leu-Glu-Ala-Ser-Gly-Tyr

Synonyms: Leptin (126-140); Obese Gene Peptide (126-140)

Bachem H-3492.0001   Human   Store at -15°C

Glu-Thr-Leu-Leu-Val-Gln-Asn-Ala-Asn-Pro-Asp-Cys-Lys-Thr-Ile-Leu-Lys-Ala-Leu

Synonyms: P25 (319-337); GAG P24 CA (187-205); HTLV I Peptide

Neosystem SC307   MW 2084.45   HTLV I peptide from subtype P25 (GAG P24 CA); due to the presence of a Cys this peptide can contain the dimeric form

---

Glu-Thr-Pro-Asp-Cys-Phe-Trp-Lys-Tyr-Cys-Val

Synonyms: Urotensin II

Bachem H-4768.0001   Human   MW 1388.59 $C_{64}H_{85}N_{13}O_{18}S_2$ Store at -15°C

Glu-Thr-Pro-Asp-Cys-Phe-Trp-Lys-Tyr-Cys-Val
Hydrochloride

Synonyms: Urotensin II

Peptides International PUT-4365-v   Human synthetic   MW 1388.6 $C_{64}H_{85}N_{13}O_{18}S_2$ >99% (HPLC); lyophilized amorphous powder | Bioactive; Disulfide bonds: Cys[5]-Cys[10]; potent vasoconstrictor; Donaldson, CJ et al, *Endocrinology*, 137:2167, 1996; Coulouarn, Y et al, *PNAS USA*, 95:15803, 1998; Ames, RS et al, *Nature*, 401:282, 1999

Glu-Thr-Tyr

Bachem H-3210.0250   MW 411.41 $C_{18}H_{25}N_3O_8$ Store at -15°C

Glu-Thr-Tyr-Ser-Lys 2TFA

Bachem H-3215.0050   Store at -15°C

Glu-Trp

Bachem G-1990.0250   MW 333.35 $C_{16}H_{19}N_3O_5$ Store at -15°C

Glu-Trp
γ-Glu-Trp

Synonyms: Glu(Trp-OH)-OH

Bachem G-1995.0250   MW 333.34 $C_{16}H_{19}N_3O_5$ Store at -15°C

Glu-Trp
γ-D-Glu-Trp

Synonyms: D-Glu(Trp-OH)-OH

Bachem G-4745.0250   MW 333.34 $C_{16}H_{19}N_3O_5$ Store at -15°C

Glu-Trp

ICN 157225   MW 333.3 $C_{16}H_{19}N_3O_5$

Glu-Trp
γ-Glu-Trp

ICN 157226   MW 333.3 $C_{16}H_{19}N_3O_5$

Sigma G 0517   FW 333.3 $C_{16}H_{19}N_3O_5$

Glu-Trp-Arg-Phe-Asp-Ser-Arg-Leu-Ala-Phe-His-His-Val-Ala-Arg-Glu-Leu

Synonyms: NEF (182-198); HTLV I Peptide

Neosystem SC226   MW 2169.43   HTLV I peptide from subtype NEF

Glu-Trp-Glu-Arg-Lys-Val-Asp-Phe-Leu-Glu-Glu-Asn-Ile-Thr-Ala-Leu-Leu-Glu-Glu-Ala

Synonyms: GP140 (641-660); SIVmac251 Peptide

Neosystem SC772   MW 2434.68   SIVmac251 peptide from subtype GP140

Glu-Tyr
Z-Glu-Tyr-OEt

Bachem C-1652.0001   MW 472.5 $C_{24}H_{28}N_2O_8$ Store at -15°C

Glu-Tyr
Z-D-Glu-Tyr

Bachem C-3050.0250   MW 444.44 $C_{22}H_{24}N_2O_8$ Store at -15°C

Glu-Tyr

Bachem G-2000.0250   MW 310.31 $C_{14}H_{18}N_2O_6$ Store at -15°C

Glu-Tyr
γ-Glu-Tyr

Synonyms: Glu(Tyr-OH)-OH

Bachem G-2005.0250   MW 310.31 $C_{14}H_{18}N_2O_6$ Store at -15°C

Bachem M-1245.0001   MW 444.44 $C_{22}H_{24}N_2O_8$ Store at -15°C

Glu-Tyr
Z-L-Glu-L-Tyr

Fluka 96150   MW 444.44 $C_{22}H_{24}N_2O_8$ ≥99.0% (titration); mp:185-187°C

Glu-Tyr
γ-Glu-Tyr

ICN 157227   MW 310.3 $C_{14}H_{18}N_2O_6$

Glu-Tyr

ICN 193418   MW 310.3 $C_{14}H_{18}N_2O_6$

Glu-Tyr
N-CBZ-Glu-Tyr

Sigma C 0257   FW 444.4 $C_{22}H_{24}N_2O_8$

Glu-Tyr
γ-Glu-Tyr

Sigma G 5010   FW 310.3 $C_{14}H_{18}N_2O_6$

Glu-Tyr
CBZ-Glu-Tyr

USBio C2098-65   MW 444.4 $C_{22}H_{24}N_2O_8$ ≥99%

Glu-Tyr-Arg-Lys-Ile-Leu-Arg-Gln-Arg-Lys-Ile-Asp-Arg

Synonyms: VPU (28-40); HTLV I Peptide

Neosystem SC528   MW 1774.10   HTLV I peptide from subtype VPU

Glu-Tyr-Asp-Pro-Thr-Ile-Asp-Ser-Tyr-Arg-Lys-Glu-(Cys)

Synonyms: Ras Blocking Peptide (31-43)

Calbiochem 553573   MW 1746.9 $C_{74}H_{111}N_{19}O_{28}S$
Lyophilized; soluble in H_2O | From the effector binding Loop (L2) of H-, K-, & N-Ras with a Cys residue added; this peptide, coupled to KLH, was used as the immunogen for the production of Anti-*Ras*; Suitable for use in immunoabsorption for immunoprecipitation, immunocytochemistry, Western blotting & dot blots

Glu-Tyr-Glu

Bachem H-3225.0250   MW 439.42 $C_{19}H_{25}N_3O_9$ Store at -15°C

Glu-Val

Bachem G-2010.0250   MW 246.26 $C_{10}H_{18}N_2O_5$ Store at -15°C

Glu-Val
γ-Glu-Val

Synonyms: Glu(Val-OH)-OH

Bachem G-2015.0250   MW 246.26 $C_{10}H_{18}N_2O_5$ Store at -15°C

Glu-Val

ICN 157228   MW 246.3 $C_{10}H_{18}N_2O_5$ Crystalline

Glu-Val
γ-Glu-Val

ICN 157229   MW 246.3 $C_{10}H_{18}N_2O_5$

Sigma G 0392   FW 246.3 $C_{10}H_{18}N_2O_5$

Glu-Val

Sigma G 3005   FW 246.3 $C_{10}H_{18}N_2O_5$

Glu-Val-Asp-Pro-Ile-Gly-His-Leu-Tyr

Bachem H-3634.0001   Store at -15°C

Glu-Val-Glu-Asp-Leu-Gln-Val-Arg-Asp-Val-Glu-Leu-Ala-Gly-Ala-Pro-Gly-Glu-Gly-Gly-Leu-Gln-Pro-Leu-Ala-Leu-Glu-Gly-Ala-Leu-Gln

Synonyms: C-Peptide

Bachem H-2912.0500   Dog   Store at -15°C

ICN 154500   Dog   MW 3175   Kwok, SCM etal, *JBC*, 258:2361, 1983

Glu-Val-Gly-Met-Met-Lys-Gly-Gly-Ile-Arg-Lys-Asp-Arg-Arg-Gly

Synonyms: Estrogen Receptor (247-261); Estrogen Receptor α Peptide (247-261)

Biogenesis 7025-1209   Human synthetic   Purified; lyophilized | The synthetic sequence is derived from the DNA binding domain of this receptor

Alexis 155-028   Synthetic   MW 1690.0 $C_{68}H_{124}N_{26}O_{20}S_2$ ≥95%; lyophilized; reconstitute with 0.1 mL distilled H_2O | From ER-DNA binding domain; competitively binds to Ab Alexis 803-004; antiserum blocking peptide

---

Glu-Val-Lys-Met

MeOSuc-Glu-Val-Lys-Met-pNA

Bachem L-1740.0025   Store at -15°C

Glu-Val-Lys-Met-Asp-Ala-Glu-Phe-Lys

MCA-Glu-Val-Lys-Met-Asp-Ala-Glu-Phe-(Lys-DNP)

BioSource International 03-408

Glu-Val-Lys-Tyr-Asp-Pro-Cys-Phe-Gly-His-Lys-Ile-Asp-Arg-Ile-Asn-His-Val-Ser-Asn-Leu-Gly-Cys-Pro-Ser-Leu-Arg-Asp-Pro-Arg-Pro-Asn-Ala-Pro-Ser-Thr-Ser-Ala

Synonyms: Natriuretic Peptide

Bachem H-4904.0500   MW 4190.69 $C_{180}H_{282}N_{56}O_{56}S_2$ Store at -15°C | *Dendroaspis*

Glu-Val-Lys-Tyr-Asp-Pro-Cys-Phe-Gly-His-Lys-Ile-Asp-Arg-Ile-Asn-His-Val-Ser-Asn-Leu-Gly-Cys-Pro-Ser-Leu-Arg-Asp-Pro-Asn-Ala-Pro-Ser-Thr-Ser-Ala

Synonyms: Natriuretic Peptide, (des-Arg[30],des-Pro[31])-

Bachem H-4888.0500   Dendroaspis   MW 3937.39 $C_{169}H_{263}N_{51}O_{54}S_2$ Store at -15°C

Glu-Val-Phe

Bachem H-3230.0250   MW 393.44 $C_{19}H_{27}N_3O_6$ Store at -15°C

ICN 157230   MW 393.4 $C_{19}H_{27}N_3O_6$ Crystalline

Sigma G 3751   FW 393.4 $C_{19}H_{27}N_3O_6$

Glu-Val-Pro-Gln-Trp-Ala-Val-Gly-His-Phe-Met Amide

Synonyms: Ranatensin

Sigma R 9002   FW 1281.5   ≥95% (HPLC) | Bioactive peptide; Nakajama, T et al, *Fed Proc*, 29:282, 1970

Glu-Val-Tyr-Leu-Lys-Ala-Ser-Gln-Phe-Pro-Ala-Gly-Ile-Lys-Gly

Glu(EDANS)-Val-Tyr-Leu-Lys-Ala-Ser-Gln-Phe-Pro-Ala-Gly-Ile-Lys(DABCYL)-Gly

Bachem M-2355.0001   Store at -15°C

Glu-Val-Val-Pro-Pro-Gln-Val-Leu-Ser-Glu-Pro-Asn-Glu-Glu-Ala-Gly-Ala-Ala-Leu-Ser-Pro-Leu-Pro-Glu-Val-Pro-Pro-Trp-Thr-Gly-Glu-Val-Ser-Pro-Ala-Gln-Arg

Synonyms: Natriuretic Peptide (56-92), Prepro-Atrial

American Peptide 14-1-33   Human   MW 3878.3 $C_{173}H_{270}N_{44}O_{57}$ Vesely, DL et al, *BBRC*, 148:1540, 1987

Glx-Pro-Leu-Arg-Lys-Leu-Cys-Ile-Leu-His-Arg-Asn-Pro-Gly-Arg-Cys-Tyr-Gln-Lys-Ile-Pro-Ala-Phe-Tyr-Tyr-Asn-Gln-Lys-Lys-Lys-Gln-Cys-Glu-Gly-Phe-Thr-Trp-Ser-Gly-Cys-Gly-Gly-Asn-Ser-Asn-Arg-Phe-Lys-Thr-Ile-Glu-Glu-Cys-Arg-Arg-Thr-Cys-Ile-Arg-Lys

Synonyms: Dendrotoxin I

Calbiochem 253703   *Dendroaspis polylepis polylepis*   MW 7149.4 $C_{312}H_{491}N_{99}O_{83}S_6$ ≥98% (SDS-PAGE); lyophilized; soluble in water; Cys^7-Cys^{57}; Cys^{16}-Cys^{40}; Cys^{32}-Cys^{53}; LD_{50}≤2000 mg/kg | Evokes transmitter release by blocking fast activating and slow inactivating K^+ channels; Mourre, C et al, *Brain Res*, 762:223, 1997; Hopkins, WF et al, *Soc Neurosci*, 19:Abstract 707, 1993; Kondo, T et al, *Neurosci Res*, 13:207, 1992

Gly-Abu

Gly-γ-Abu

Bachem G-2045.0001   MW 160.17 $C_6H_{12}N_2O_3$ Store at -15°C

Gly-Abu

Gly-DL-Abu

Bachem G-3740.0005   MW 160.17 $C_6H_{12}N_2O_3$ Store at -15°C

Gly-Abu

FA-Gly-Abu Amide

Bachem M-1835.0050   Store at -15°C

Gly-Ala

Z-Gly-Ala

Bachem C-1670.0001   MW 280.28 $C_{13}H_{16}N_2O_5$ Store at -15°C

Gly-Ala

Z-Gly-DL-Ala

Bachem C-1680.0001   MW 280.28 $C_{13}H_{16}N_2O_5$ Store at -15°C

Gly-Ala

Z-Gly-Ala Amide

Bachem C-1685.0001   MW 279.3 $C_{13}H_{17}N_3O_4$ Store at -15°C

Gly-Ala

Bachem G-2020.0005   MW 146.15 $C_5H_{10}N_2O_3$ Store at -15°C

Gly-Ala

Gly-α-Ala

Bachem G-2030.0005   MW 146.15 $C_5H_{10}N_2O_3$ Store at -15°C

Gly-Ala

Gly-DL-Ala

Bachem G-3745.0005   MW 146.15 $C_5H_{10}N_2O_3$ Store at -15°C

Gly-Ala

Gly-Ala-AMC Hydrochloride

Bachem I-1065.0050   MW 339.78 $C_{15}H_{17}N_3O_4 \cdot HCl$ Store at -15°C

Gly-Ala

Gly-Ala-βNA

Bachem K-1300.0250   MW 271.32 $C_{15}H_{17}N_3O_2$ Store at -15°C

Gly-Ala
FA-Gly-Ala Amide

Bachem M-1825.0050 MW 265.27 $C_{12}H_{15}N_3O_4$ Store at -15°C

Gly-Ala
Gly-DL-Ala

ICN 101842 MW 146.1 $C_5H_{10}N_2O_3$

Gly-Ala

ICN 101844 MW 146.1 $C_5H_{10}N_2O_3$

Gly-Ala
Gly-β-Ala

ICN 101852 MW 146.1 $C_5H_{10}N_2O_3$ Crystalline
ICN 105570 MW 146.1 $C_5H_{10}N_2O_3$ Crystalline

Gly-Ala
Borotrimethyl-Gly-Ala-OMe

ICN 154636 MW 216.09 $BC_9H_{21}N_2O_3$ Low melting solid; >97%

Gly-Ala
N-CBZ-Gly-Ala

Sigma C 7876 FW 280.3 $C_{13}H_{16}N_2O_5$

Gly-Ala

Sigma G 0502 FW 146.1 $C_5H_{10}N_2O_3$

Gly-Ala
Gly-DL-Ala

Sigma G 1504 FW 146.1 $C_5H_{10}N_2O_3$

Gly-Ala
Gly-D-Ala

Sigma G 3128 FW 146.1 $C_5H_{10}N_2O_3$

Gly-Ala
CBZ-Gly-Ala

USBio C2098-71 MW 280.3 $C_{13}H_{16}N_2O_5$ ≥99%

Gly-Ala
N-(R)-(2-(Hydroxyaminocarbonyl)Methyl)-4-
Methylpentanoyl-L-t-Butyl-Gly-L-Ala 2-Aminoethyl

Synonyms: TAPI-2; Matalloprotease Inhibitor

Peptides International INH-3852-PI Synthetic MW 415.54 $C_{19}H_{37}N_5O_5$ >98% (HPLC); white powder; amide | Inhibitor; Hooper, NM et al, *Biochem J*, 321:265-279, 1997

Gly-Ala Amide Hydrochloride

Bachem G-2035.0250 MW 181.62 $C_5H_{11}N_3O_2 \cdot$ HCl Store at -15°C

Gly-Ala-Ala
Gly-α-Ala-α-Ala

Bachem H-6225.0250 Store at -15°C

Gly-Ala-Ala

ICN 157231 MW 217.2 $C_8H_{15}N_3O_4$
Sigma G 4504 FW 217.2 $C_8H_{15}N_3O_4$

Gly-Ala-Ala-Ala-Ala
Gly-Ala-Ala-D-Ala-D-Ala

Synonyms: Immunogenic Pentapeptide

Bachem H-6820.0050 Store at -15°C
ICN 157232 MW 359.4 $C_{14}H_{25}N_5O_6$

Gly-Ala-Ala-Pro-Phe-Nitro-Tyr-Asp
Anthranilyl-Gly-Ala-Ala-Pro-Phe-Nitro-Tyr-Asp

Synonyms: Proline-Specific Protease Substrate

Sigma A 0449 FW 903.9 $C_{42}H_{49}N_9O_{14}$ ≥95% (HPLC) | Fluorescent substrate; Szwajcer-Dey, E et al, *J Bact*, 174:2454, 1992

Gly-Ala-Ala-Pro-Phe-Tyr-Asp
Abz-Gly-Ala-Ala-Pro-Phe-Tyr(NO₂)-Asp

Synonyms: Pro-Specific Protease Fluorescent Substrate

American Peptide 81-0-20 MW 903.9 $C_{42}H_{49}N_9O_{14}$ *J Bact*, 174:2454, 1992

Gly-Ala-Arg-Ala-Ser-Val-Leu-Ser-Gly-Gly-Glu-Leu-
Asp-Arg-Trp-Glu-Lys-Ile-Arg-Leu

Synonyms: P18 (2-21); GAG P17 MA (2-21); HTLV I Peptide

Neosystem SC664 MW 2213.51 HTLV I peptide from subtype P25 (GAG P17 MA)

Gly-Ala-Asp

Bachem H-3245.0250 MW 261.24 $C_9H_{15}N_3O_6$ Store at -15°C

Gly-Ala-Gln
Z-Gly-Ala-Gln-AMC

Bachem I-1520.0050 Store at -15°C

Gly-Ala-Gln
Z-Gly-Ala-Gln-βNA

Bachem K-1590.0050 Store at -15°C

Gly-Ala-Gly

Bachem H-3250.0250 MW 203.2 $C_7H_{13}N_3O_4$ Store at -15°C

Gly-Ala-Gly
Gly-α-Ala-Gly

Bachem H-3255.0250 MW 203.2 $C_7H_{13}N_3O_4$ Store at -15°C

Gly-Ala-Gly-Gly
Gly-α-Ala-Gly-Gly

Bachem H-7570.0250 Store at -15°C

Gly-Ala-His
Z-Gly-Ala-His-AMC

Bachem I-1515.0050 Store at -15°C

Gly-Ala-His
Z-Gly-Ala-His-βNA

Bachem K-1580.0050 Store at -15°C

Gly-Ala-His-Tyr-Met-Arg-Ala-Leu-Ser-Asn-Val-Glu

Synonyms: Glycoprotein IIb (656-667)

Bachem H-1982.0001 MW 1347.52 $C_{57}H_{90}N_{18}O_{18}S$ Store at -15°C

Gly-Ala-Hyp

Bachem H-3260.0250 MW 259.26 $C_{10}H_{17}N_3O_5$ Store at -15°C

Gly-Ala-Ile-Ile-Gly-Leu-Met-Val-Gly-Gly-Val-Val

Synonyms: Amyloid α-Protein (29-40)

Bachem H-3984.0001 MW 929.39 $C_{49}H_{88}N_{12}O_{13}S$ Store at -15°C

Gly-Ala-Leu

Bachem H-3265.0250 MW 259.31 $C_{11}H_{21}N_3O_4$ Store at -15°C

Gly-Ala-Met
Z-Gly-Ala-Met-AMC

Bachem I-1525.0050 Store at -15°C

Gly-Ala-Met
Z-Gly-Ala-Met-βNA

Bachem K-1575.0050 Store at -15°C

Gly-Ala-Phe

Bachem H-3270.0250 MW 293.32 $C_{14}H_{19}N_3O_4$ Store at -15°C

Gly-Ala-Phe
Z-Gly-Ala-Phe-βNA

Bachem K-1585.0050 Store at -15°C

Gly-Ala-Pro
Gly-Ala-Pro-AFC

Bachem I-1700.0025 MW 454.41 $C_{20}H_{21}F_3N_4O_5$ Store at -15°C

Gly-Ala-Pro
Z-Gly-Ala-Pro-βNA

Bachem K-1570.0050 Store at -15°C

Gly-Ala-Pro
Bz-Gly-Ala-Pro

Synonyms: Angiotensin I Converting Enzyme Substrate; Hippuryl-Ala-Pro

Peptides International SGP-3126 MW 347.37 $C_{17}H_{21}N_3O_5$ >99% (HPLC); amorphous powder | Cheung, et al, *JBC*, 255:401, 1980

Gly-Ala-Pro-Gln-Gly-Arg-Val-Pro-Glu-Ala-Arg-Pro-Asn-Ser Acetate Salt

Synonyms: Gelsolin (11-24)

Biogenesis 4628-6120 Human plasma synthetic Semi-pure; lyophilized

Gly-Ala-Pro-Val-Pro-Tyr-Pro-Asp-Pro-Leu-Glu-Pro-Arg

Synonyms: Osteocalcin (7-19); Calcium Binding Protein Fragment

ICN 152421 MW 1407.6 Calcium binding protein indigenous to the organic matrix of bone, dentin & other mineralized tissues; a specific osteoblasts marker produced during bone formation; Hauschka, PV, *Homeostasis*, 16:258, 1986; Jueppner, H etal, *Calcif Tissue Int*, 39:310, 1986

Bachem H-6195.0001 Human MW 1407.59 $C_{65}H_{98}N_{16}O_{19}$ Store at -15°C

Sigma O 3632 Human FW 1407.6 (free base) ≥97% (HPLC) | Bioactive peptide; Hauschka, PV, *Haemostasis*, 16:258, 1986; Jueppner, H et al, *Calcif Tissue Int*, 39:310, 1986; Pastoureau, P & Delmas, PD, *Clin Chem*, 36:1620, 1990

Gly-Ala-Ser-Phe-Tyr-Ser-Trp-Gly Amide

Synonyms: Leukokinin VIII

American Peptide 33-1-31 MW 872.9 $C_{42}H_{52}N_{10}O_{11}$ Holman, GM et al, *Comp Biochem Physiol*, 88C:31, 1987

Gly-Ala-Tyr

American Peptide 79-3-15 MW 309.3 $C_{14}H_{19}N_3O_5$

Bachem H-3275.0250 MW 309.32 $C_{14}H_{19}N_3O_5$ Store at -15°C

ICN 157233 MW 309.3 $C_{14}H_{19}N_3O_5$

Sigma G 1017 FW 309.3 $C_{14}H_{19}N_3O_5$

Gly-Ala-Val-Ser-Thr-Ala

Synonyms: Fibrin, Necro; Nectofibrin Hexapeptide

American Peptide 42-1-10 Human MW 504.6 $C_{20}H_{36}N_6O_9$ Brentani, RR et al, *PNAS*, 85:364, 1988

Bachem H-9150.0005 Human Store at -15°C

Gly-Arg
BOC-Gly-Arg

Bachem A-1745.0001 MW 331.37 $C_{13}H_{25}N_5O_5$ Store at -15°C

Gly-Arg

Bachem G-2050.0250 MW 231.26 $C_8H_{17}N_5O_3$ Store at -15°C

Gly-Arg
Bz-Gly-Arg

Synonyms: Hippuryl-Arg; Carboxypeptidase B Substrate; Carboxypeptidase N Substrate

Bachem G-2265.0250 MW 335.36 $C_{15}H_{21}N_5O_4$ Store at -15°C

Gly-Arg
Glutaryl-Gly-Arg-AMC Hydrochloride

Bachem I-1195.0050 MW 538.99 $C_{23}H_{30}N_6O_7 \cdot HCl$ Store at -15°C

Gly-Arg
Gly-Arg-AMC

Bachem I-1215.0050 MW 388.43 $C_{18}H_{24}N_6O_4$ Store at -15°C

Gly-Arg
Gly-Arg-4MβNA Dihydrochloride

Bachem J-1200.0250 MW 459.38 $C_{19}H_{26}N_6O_3 \cdot 2HCl$ Store at -15°C

Gly-Arg
Gly-Arg-βNA Hydrochloride

Bachem K-1305.0250 MW 392.89 $C_{18}H_{24}N_6O_2 \cdot HCl$ Store at -15°C

Gly-Arg
Gly-Arg-pNA

Bachem L-1285.0050 MW 351.37 $C_{14}H_{21}N_7O_4$ Store at -15°C

Gly-Arg
CH₃OCO-D-CHA-Gly-Arg-pNA·AcOH

Synonyms: Pefachrome®FXa; Pefa-5523

Pentapharm 085-20, 085-06 MW 622.7 Chromogenic peptide substrate for factor Xa; k_{cat}:140 s^{-1}, K_M:0.106 *mM*

Gly-Arg
CH₃SO₂-D-CHA-Gly-Arg-AMC·AcOH

Synonyms: Pefafluor FXa; Pefa-5534

Pentapharm 085-21, 085-12   MW 679.8   Excitation wavelength 360 nm, emission wavelength 460 nm | Highly sensitive fluorogenic peptide substrate for factor Xa; k_{cat}:162.0 s⁻¹, K_M:0.22 mM

Gly-Arg
CH₃SO₂-D-CHA-Gly-Arg-pNA·AcOH

Synonyms: Pefachrome®LAL; Pefa-5517

Pentapharm 086-20, 091-02   MW 642.7   Highly sensitive chromogenic peptide substrate for determination of bacterial endotoxins

Gly-Arg
CH₃SO₂-D-HHT-Gly-Arg-AMC·AcOH

Synonyms: Pefafluor LAL; Pefa-5589

Pentapharm 086-21, 086-05   MW 695.8   Excitation wavelength 360 nm, emission wavelength 460 nm | Highly sensitive fluorogenic peptide substrate for determination of bacterial endotoxins

Gly-Arg
CH₃SO₂-D-HHT-Gly-Arg-pNA·AcOH

Synonyms: Pefachrome®tPA; Pefa-5937

Pentapharm 091-20, 091-01   MW 658.9   Highly sensitive chromogenic peptide substrate for tissue-type plasminogen activator (tPA); Kluft, C (Ed), Tissue Type Plasminogen Activator (tPA):Physiological & Clinical Aspects, Vol I & II, CRC Press, Boca Raton, 1988

Gly-Arg
D-HHT-Gly-Arg-pNA·2AcOH

Synonyms: Pefachrome®FXIIa; Pefa-5963

Pentapharm 092-21, 092-02, 092-32   MW 640.7   Highly sensitive chromogenic peptide substrate for factor XIIa; K_M:0.8 mM; Stürzebecher, J et al, *Thromb Res*, 55:709, 1989

Gly-Arg
CH₃SO₂-D-CHG-Gly-Arg-AMC· AcOH

Synonyms: Pefafluor FIXa; Pefa-10148

Pentapharm 095-03   MW 665.7   Excitation wavelength 360 nm, emission wavelength 460 nm | Fluorogenic peptide substrate for factor IXa with improved sensitivity; used for determination of factor IXa activity for in-process and quality control of factor IX preparations; v_{max}:28.1 μmol/min, K_M:0.23mM (determined in the presence of 33% ethylene glycol), sensitivity significantly increased in the presence of 33% ethylene glycol

Gly-Arg
CH₃SO₂-D-CHG-Gly-Arg-pNA·AcOH

Synonyms: Pefachrome®FIXa; Pefa-3107

Pentapharm 095-20, 095-02   MW 628.7   Chromogenic substrate for factor IXa with improved sensitivity; used for determination of factor IXa activity for in-process and quality control of factor IX preparations; k_{cat}:4.4 s⁻¹; K_M:1.3 mM (determined in the presence of 33% ethylene glycol); sensitivity significantly increased in the presence of 33% ethylene glycol; Stürzebecher J, et al, *FEBS Letts*, 412:295, 1997; Prasa D, & J Stürzebecher, *Throm Res*, 92:99, 1998

Gly-Arg
Glt-Gly-Arg-MCA

Synonyms: Urokinase Substrate

Peptides International MGG-3097-v   MW 502.53   C₂₃H₃₀N₆O₇   >98% (HPLC); lyophilized amorphous powder | Morita, T et al, *J Biochem*, 82:1495, 1977

Gly-Arg
Bz-Gly-Arg

Synonyms: Hippuryl-Arg

Peptides International SGR-3059   MW 335.36   C₁₅H₂₁N₅O₄   >99% (HPLC); amorphous powder

Gly-Arg
Gly-Arg 4-Methoxy-β–Naphthylamide Dihydrochloride

Synonyms: Dipeptidyl Aminopeptidase I Substrate

Sigma G 1013   FW 459.4   C₁₉H₂₆N₆O₃ · 2HCl   McGuire, MJ et al, *Arch Biochem Biophys*, 295:280, 1992

Gly-Arg
N-Glutaryl-Gly-Arg MCA Hydrochloride

Synonyms: Urokinase Substrate

Sigma G 2386   FW 539.0;   C₂₃H₃₀N₆O₇ · HCl   Fluorogenic; Morita, T et al, *J Biochem*, 82:1495, 1977; Lottenberg, R et al, *Meth Enzymol*, 80:341, 1981

Gly-Arg
Gly-Arg 4-p-NA Dihydrochloride

Synonyms: Plasmin Substrate; Urokinase Substrate

Sigma G 8148   FW 424.3   C₁₄H₂₁N₇O₄ · 2HCl   Nieuwenhuizen, W et al, *Anal Biochem*, 83:143, 1977

Gly-Arg
Bz-Gly-Arg

Synonyms: Hippuryl-Arg; Carboxypeptidase B Substrate; Carboxypeptidase N Substrate

Sigma H 2508   C₁₅H₂₁N₅O₄

Gly-Arg-Ala-Asp-Ser-Pro

Synonyms: Fibronectin Inhibitor

American Peptide 44-0-21   MW 601.6   C₂₃H₃₉N₉O₁₀   Inactive control for inhibition of cell attachment to various extracellular matrix components

Calbiochem 03-34-0052   MW 601.6   C₂₃H₃₉N₉O₁₀   Solid; ≥98% (HPLC); soluble in 5% HOAc | Inactive control peptide for other fibronectin inhibitors; Torimoto, Y et al, *J Exp Med*, 172:1315, 1990

Gly-Arg-Ala-Asp-Ser-Pro-Lys

Synonyms: Fibronectin Analog; Fibronectin

American Peptide 44-0-22   MW 729.8   C₂₉H₅₁N₁₁O₁₁   Pytela, R et al, *Science*, 231:1559, 1986

Bachem H-5690.0001   MW 729.79   C₂₉H₅₁N₁₁O₁₁   Store at -15°C

ICN 153105   MW 729.8   C₂₉H₅₁N₁₁O₁₁

Neosystem SC148   MW 729.79   Bioactive; cell attachment peptide; fibronectin fragment

Sigma G 4144   FW 729.8   C₂₉H₅₁N₁₁O₁₁   ≥97% (HPLC); peptide content:~70% | Bioactive peptide

Gly-Arg-Ala-Phe-Val-Thr-Ile-Gly-Lys

Synonyms: GP120 (319-327); HTLV I Peptide

Neosystem SC540   MW 948.13   HTLV I peptide from subtype GP160

Gly-Arg-Arg
BOC-Gly-Arg-Arg-AMC

Bachem I-1565.0025   MW 644.73   C₂₉H₄₄N₁₀O₇   Store at -15°C

Gly-Arg-Arg
BOC-Gly-Arg-Arg-MCA

Synonyms: Carboxyl Side of Paired Basic Residue Cleaving Enzyme Substrate

Peptides International MGR-3142-v   MW 644.73
$C_{29}H_{44}N_{10}O_7$ >98% (HPLC); lyophilized amorphous powder | Mizuno, K et al, *BBRC*, 144:807, 1987

Gly-Arg-Arg
N-t-BOC-Gly-Arg-Arg MCA Dihydrochloride

Synonyms: Protease Substrate

Sigma B 7778   FW 717.7 $C_{29}H_{44}N_{10}O_7 \cdot 2HCl$ ~95% | Substrate for proteases which cleave on the carboxyl side of paired basic residues; Mizuno, K et al, *Biochem Biophys Res Commun*, 144:807, 1987

Gly-Arg-Asp-Gly-Ser

Bachem H-3172.0001   MW 490.47 $C_{17}H_{30}N_8O_9$ Store at -15°C

Gly-Arg-Cys-Cys-His-Pro-Ala-Cys-Gly-Lys-Asn-Tyr-Ser-Cys Amide

Synonyms: Conotoxin MI; Conotoxin MI, α-; Nicotinic Acetylcholine Receptor Blocker

Bachem H-7765.0500   Store at -15°C | Disulfide bonds: Cys[3]-Cys[8], Cys[4]-Cys[14]

Sigma C 9790   *Conus magus* (marine snail)   FW 1493.7 ≥97% (HPLC); peptide content:~70% | Disulfide bonds: 3-8, 4-14; bioactive peptide; peptide toxin; blocks the acetylcholine receptor at the neuromuscular junction; McIntosh, M et al, Arch *Biochem Biophys*, 218:329, 1982

Peptides International PCN-4140-v   *Conus magus* (marine snail) synthetic   MW 1493.7 $C_{58}H_{88}N_{22}O_{17}S_4$ >99% (HPLC); lyophilized amorphous powder | Bioactive; Disulfide bonds: Cys[3]-Cys[8], Cys[4]-Cys[14]); McIntosh, M et al, *Arch Biochem Biophys*, 218:329, 1982; Nishiuchi, Y & S Sakakibara, *Peptide Chem*, 1983:191, 1984

Alexis 630-048   *Conus magus* synthetic   MW 1491.7; $C_{58}H_{86}N_{22}O_{17}S_4$ White lyophilized powder; soluble in water | Disulfide bonds: Cys[3]-Cys[8], Cys[4]-Cys[14]; potent neurotoxin; postsynaptic inhibitor at the neuromuscular junction; McIntosh, M et al, *Arch Biochem Biophys*, 218:329, 1982; Hann, RM et al, *Biochemistry*, 33:14058, 1994; Nishiuchi, Y & Sakakibara, S, *FEBS Lett*, 148:260, 1992

Gly-Arg-Gly

Bachem H-1306.0250   MW 288.31 $C_{10}H_{20}N_6O_4$ Store at -15°C

Gly-Arg-Gly
Ac-Gly-Arg-Gly Amide

Bachem H-7845.0025   MW 329.36 $C_{12}H_{23}N_7O_4$ Store at -15°C

Gly-Arg-Gly
Hippuryl–Arg-Gly

Sigma H 4146   FW 392.4 $C_{17}H_{24}N_6O_5$ ≥97% (HPLC); peptide content:~70%

Gly-Arg-Gly-Asp

Bachem H-2960.0005   MW 403.4 $C_{14}H_{25}N_7O_7$ Store at -15°C

Gly-Arg-Gly-Asp Acetate Salt

Synonyms: Fibronectin

Sigma G 3892   FW 403.4 (free base) $C_{14}H_{25}N_7O_7$ ≥97% (HPLC) | Bioactive peptide

Sigma G 9022   FW 403.4 (free base) $C_{14}H_{25}N_7O_7$ ≥97% (HPLC) | Bioactive peptide; tetrapeptide overlapping the cell-attachment domain of fibronectin; Haverstick, DM et al, *Blood*, 66:946, 1985; Yamada, KM et al, *J Cell Biochem*, 28:99, 1985

Gly-Arg-Gly-Asp-Asn-Pro

Synonyms: Fibronectin Analog; Fibronectin

American Peptide 44-0-65   MW 614.6 $C_{23}H_{38}N_{10}O_{10}$ Superior inhibitor of cell attachment to fibronectin & a weak inhibitor of cell attachment to vitronectin; also inhibits binding of fibrinogen, fibronectin, vitronectin & von Willebrand factor to platelets & resists to carboxypeptidases

Bachem H-3174.0001   MW 614.62 $C_{23}H_{38}N_{10}O_{10}$ Store at -15°C

ICN 195669   MW 614.6 $C_{23}H_{38}N_{10}O_{10}$ ≥97%

Sigma G 2529   FW 614.6 $C_{23}H_{38}N_{10}O_{10}$ ≥97% (HPLC); peptide content:~70% | Bioactive peptide

Gly-Arg-Gly-Asp-Ser

Synonyms: Fibronectin Analog; Fibronectin Inhibitor; Fibronectin; Fibronectin Active Fragment

American Peptide 44-0-23   MW 490.5 $C_{17}H_{30}N_8O_9$ Derived from the cell binding region of fibronectin competitively inhibits direct binding of fibroblasts to fibronectin; also an inhibitor of thrombin-induced platelet aggregation; Humphries, MJ et al, *Science*, 233:467, 1986

Bachem H-1345.0005   MW 490.47 $C_{17}H_{30}N_8O_9$ Store at -15°C

Calbiochem 03-34-0027   MW 490.5 $C_{17}H_{30}N_8O_9$ Lyophilized; ≥95% (HPLC); soluble in 5% HOAc | Competitively inhibits direct binding of fibroblasts to fibronectin; Pierschbacher, MD & Ruoslahti, E, *Nature*, 309:30, 1984

ICN 153107   MW 490.5 $C_{17}H_{30}N_8O_9$ Cell-attachment domain of fibronectin; Akiyama, SK & KM Yamada, *JBC*, 260:10402, 1985

Neosystem SC347   MW 490.47   Bioactive; cell attachment peptide; fibronectin fragment; Humphries, MJ et al, *Science*, 233:467, 1986

Sigma G 4391   FW 490.5 $C_{17}H_{30}N_8O_9$ ≥97% (HPLC); peptide content:~70% | Bioactive peptide

Peptides International PFA-4189-v   Synthetic   MW 490.47 $C_{17}H_{30}N_8O_9$ >99% (HPLC); lyophilized amorphous powder | Bioactive; Akiyama SK & KM Yamada, *JBC*, 260:10402, 1985; Olden, K et al, *Ann NY Acad Sci*, 551:421, 1988

Gly-Arg-Gly-Asp-Ser ½AcOH Dihydrate

Synonyms: Fibronectin Active Fragment

Peptides International PFA-4189   Synthetic   MW 556.53 $C_{17}H_{30}N_8O_9 \cdot ½CH_3COOH \cdot 2H_2O$ >99% (HPLC); lyophilized amorphous powder; bulk | Bioactive; Akiyama SK & KM Yamada, *JBC*, 260:10402, 1985; Olden, K et al, *Ann NY Acad Sci*, 551:421, 1988

Gly-Arg-Gly-Asp-Ser-Cys

Bachem H-7255.0005   MW 593.62 $C_{20}H_{35}N_9O_{10}S$ Store at -15°C

Gly-Arg-Gly-Asp-Ser-Pro

Synonyms: Fibronectin Analog

American Peptide 44-0-24   MW 587.6 $C_{22}H_{37}N_9O_{10}$ Used for the affinity purification of fibronectin receptors, as it contains the RGD integrin recognition site of the fibronectin cell binding domain; also inhibits fibronectin binding to platelet sites; Pytela, R et al, *Science*, 231:1559, 1986

Gly-Arg-Gly-Asp-Ser-Pro
Gly-Arg-Gly-Asp-ᴅ-Ser-Pro

Synonyms: Fibronectin Analog

American Peptide 44-0-38   MW 587.6 $C_{22}H_{37}N_9O_{10}$ Inhibits cell attach with fibronectin but not vitronectin

Bachem H-3164.0001 MW 587.59 $C_{22}H_{37}N_9O_{10}$ Store at -15°C

Gly-Arg-Gly-Asp-Ser-Pro

Synonyms: Fibronectin Inhibitor

Bachem H-7630.0005 MW 587.59 $C_{22}H_{37}N_9O_{10}$ Store at -15°C

Calbiochem 03-34-0035 MW 587.6 $C_{22}H_{37}N_9O_{10}$ Lyophilized; ≥98% (HPLC); soluble in 5% HOAc | Inhibits fibronectin binding to platelet-binding sites; Pierschbacher, MD & Ruoslahti, E, *Nature*, 309:30, 1984

Neosystem SC349 MW 587.59 Bioactive; cell attachment peptide; fibronectin fragment; Pytela, R et al, *PNAS*, 82:5766, 1985

Gly-Arg-Gly-Asp-Ser-Pro-Ala-Ser-Ser-Lys
Gly-D-Arg-Gly-Asp-Ser-Pro-Ala-Ser-Ser-Lys

American Peptide 44-0-48 MW 961.0 $C_{37}H_{64}N_{14}O_{16}$

Gly-Arg-Gly-Asp-Ser-Pro-Cys

Synonyms: Fibronectin Analog

American Peptide 44-0-25 MW 689.7 $C_{25}H_{41}N_{10}O_{11}S$ Pieschbacher, MD et al, *PNAS*, 81:5985, 1984

Bachem H-7245.0005 MW 690.74 $C_{25}H_{42}N_{10}O_{11}S$ Store at -15°C

Gly-Arg-Gly-Asp-Ser-Pro-Lys

Synonyms: Fibronectin Analog

American Peptide 44-0-39 MW 715.8 $C_{28}H_{49}N_{11}O_{11}$ Pytela, R et al, *Science*, 231:1559, 1986

Bachem H-6190.0001 MW 715.77 $C_{28}H_{49}N_{11}O_{11}$ Store at -15°C

Neosystem SC149 MW 715.76 Bioactive; cell attachment peptide; fibronectin fragment; Pytela, R et al, *Science*, 231:1559, 1986

Sigma G 1269 FW 715.8 $C_{28}H_{49}N_{11}O_{11}$ ≥97% (HPLC) | Bioactive peptide; Pytela, R et al, *Science*, 231:1559, 1986

ICN 151526 Synthetic MW 715.8 $C_{28}H_{49}N_{11}O_{11}$ Pytela, R et al, *Science*, 231:1559, 1986

Gly-Arg-Gly-Asp-Thr-Pro

Synonyms: Fibronectin Analog; Fibronectin Inhibitor; Fibronectin

American Peptide 44-0-46 MW 601.6 $C_{23}H_{39}N_9O_{10}$ Inhibits binding of fibrinogen, fibronectin, vitronectin & von Willebrand factor to platelets; also inhibits cell attachment to collagen, fibronectin & vitronectin & resists to carboxypeptidases; Dedhar, S et al, *J Cell Biol*, 104:585, 1987; Gehlsen, KR et al, *J Cell Biol*, 106:925, 1988

Bachem H-7725.0005 MW 601.62 $C_{23}H_{39}N_9O_{10}$ Store at -15°C

Calbiochem 03-34-0055 MW 601.6 $C_{23}H_{39}N_9O_{10}$ Lyophilized; ≥97% (HPLC); soluble in 5% HOAc | Inhibits binding of fibronectin, fibronogen, vitronectin & von Willebrand factor to platelets; inhibits cell attachment to collagen, fibronectin & vitronectin; resistant to carboxypeptidases; Pierschbacher, MD & Ruoslahti, E, *PNAS*, 81:5985, 1984; Gehlsen, KR et al, *J Cell Biol*, 106:925, 1988

ICN 153108 MW 601.6 $C_{23}H_{39}N_9O_{10}$ Inhibits the cell attachment of fibronectin, vitronectin & collagen Type I; Dedhar, S et al, *J Cell Biol*, 104:585, 1987; Gehisen, KR et al, *J Cell Biol*, 106:925, 1988

Sigma G 5646 FW 601.6 $C_{23}H_{39}N_9O_{10}$ ≥97% (HPLC) | Bioactive peptide; inhibits cell attachment of fibronectin, vitronectin & Type I collagen; Dedhar, S et al, *J Cell Biol*, 104:585, 1987; Gehlsen, KR et al, *J Cell Biol*, 106:925, 1988

Gly-Arg-Gly-Glu-Ser

Synonyms: Fibronectin Analog

American Peptide 44-0-51 MW 504.5 $C_{18}H_{32}N_8O_9$

Bachem H-3166.0001 MW 504.5 $C_{18}H_{32}N_8O_9$ Store at -15°C

Neosystem SC348 MW 504.50 Bioactive; cell attachment peptide; fibronectin fragment; inactive form for control; Humphries, HJ et al, *Science*, 233:467, 1986

Gly-Arg-Gly-Glu-Ser-Pro

Synonyms: Fibronectin Analog

American Peptide 44-0-26 MW 601.6 $C_{23}H_{39}N_9O_{10}$ Inactive fibronectin analog, used as control for tests with GRGDSP; Pytela, R et al, *Science*, 231:1559, 1986

Bachem H-3136.0001 MW 601.62 $C_{23}H_{39}N_9O_{10}$ Store at -15°C

Gly-Arg-Gly-Glu-Thr-Pro

Bachem H-3752.0001 Store at -15°C

Gly-Arg-Gly-Leu-Ser-Leu-Ser-Arg

Synonyms: cAMP-Dependent Protein Kinase Substrate

Alexis 157-014 MW 845.0; $C_{34}H_{64}N_{14}O_{11}$ ≥98%; white lyophilized powder; soluble in water | Corresponds to AA 106-113 of the human myelin basic protein; Daile, P et al, *Nature*, 257:416, 1975

Bachem H-7405.0005 MW 844.97 $C_{34}H_{64}N_{14}O_{11}$ Store at -15°C

ICN 153109 MW 845 $C_{34}H_{64}N_{14}O_{11}$ Daile, P etal, *Nature*, 257:416, 1975

Gly-Arg-Gly-Phe-Ser-Pro-Lys

Bachem H-6185.0001 Store at -15°C

Gly-Arg-Pro-Arg-Thr-Ser-Ser-Phe-Ala-Glu-Gly

Synonyms: Crosstide; Akt Substrate

Upstate 12-331 MW 1163.6 >95% pure; frozen solution

Gly-Arg-Thr-Gly-Arg-Arg-Asn-Ala-Ile-His-Asp Amide

Synonyms: cAMP-Dependent Protein Kinase Inhibitor (14-24)

Alexis 151-017 MW 1251.4; ≥95%; White powder; Soluble in water $C_{49}H_{86}N_{24}O_{15}$ Adenosine 3',5'-Cyclic Monophosphate-dependent Protein Kinase Inhibitor (14-24) Amide

American Peptide 86-1-55 MW 1251.3 $C_{49}H_{86}N_{24}O_{15}$ Cheng, HC et al, *JBC*, 261:989, 1986

Gly-Arg-Thr-Gly-Arg-Arg-Asn-Ser-Ile Amide

Synonyms: Protein Kinase A Substrate Peptide

Upstate 12-152 Synthetic MW 1016 >95%; lyophilized

USBio P9102-91D Synthetic ≥95% (HPLC); lyophilized | 9 residue synthetic peptide, which is a specific substrate of PKA (K_m ~0.11 μM); compared to Kemptide, the Km of PKA Substrate Peptide is 40 fold lower & V_{max} is 2x higher

Gly-Arg-Tyr-Asp-Ser Acetate Salt

ICN 195670 MW 656.6 $C_{24}H_{36}N_8O_{10} \cdot C_2H_4O_2$

Gly-Asn

Bachem G-2055.0001 MW 189.17 $C_6H_{11}N_3O_4$ Store at -15°C

Gly-Asn
Gly-DL-Asn

ICN 101848 MW 189.2 $C_6H_{11}N_3O_4$ Crystalline

Gly-Asn

ICN 101849 MW 189.2 $C_6H_{11}N_3O_4$ Crystalline

Gly-Asn
Gly-D-Asn

ICN 157234   MW 189.2 $C_6H_{11}N_3O_4$ Crystalline

Gly-Asn

Sigma G 0627   FW 189.2 $C_6H_{11}N_3O_4$

Gly-Asn-Ala-Thr-Asn-Thr-Asn-Ser-Ser-Asn-Thr-Asn-Ser-Ser-Ser

Synonyms: GP120 (135-149); HTLV I Peptide

Neosystem SC1001   MW 1455.37   HTLV I peptide from subtype GP160

Gly-Asn-Asn-Arg-Pro-Ile-Tyr-Ile-Pro-Gln-Pro-Arg-Pro-Pro-His-Pro-Arg-Leu

Synonyms: Apidecin II

Biogenesis 0647-7059   Honeybee synthetic   MW 2122.5   Lyophilized, containing 90.9% peptide material, acetate counter ions and residual H_2O; purified

Gly-Asn-Asn-Arg-Pro-Val-Tyr-Ile-Pro-Gln-Pro-Arg-Pro-Pro-His-Pro-Arg-Ile

Synonyms: Apidecin Ia

Biogenesis 0647-7009   Honeybee synthetic   MW 2108.4   Lyophilized, containing 83.2% peptide material, acetate counter ions and residual H_2O; purified

Gly-Asn-His-Trp-Ala-Val-Gly-His-Leu-Met Amide

Synonyms: Gastrin Releasing Peptide (18-27); Neuromedin C

ICN 151179   MW 1120.3   Roth etal, *BBRC*, 112:528, 1983

Bachem H-3120.0001   Human, porcine, canine   Store at -15°C

Neosystem SC135   Porcine   MW 1120.29   Bioactive; brain/gut peptide; CCK/gastrin peptide; bombesin-related peptide; Minamino, N et al, *BBRC*, 119:14-20, 1984

American Peptide 62-0-49   Porcine spinal cord   MW 1120.3 $C_{50}H_{73}N_{17}O_{11}S$ Bombesin-like peptide that exhibits a potent stimulant effect on smooth muscle of rat uterus; Roth, KA et al, *BBRC*, 112:528, 1983; Minamino, N et al, *BBRC*, 119:14, 1984

Sigma N 6388   Porcine spinal cord   FW 1121.3;   ≥97% (HPLC); peptide content:~70% | Bioactive peptide; bombesin-like peptide which stimulates rat uterine smooth muscle; Minamino, N et al, *Biochem Biophys Res Commun*, 119:14, 1984

Biogenesis 6700-0204   Synthetic   MW 1120.3   Purified; lyophilized | Roth et al, *BBRC*, 112:528, 1983; Minamino et al, *BBRC*, 119:14, 1984

Peptides International PNM-4153-v   Synthetic   MW 1120.3 $C_{50}H_{73}N_{17}O_{11}S$ >95% (HPLC); lyophilized amorphous powder | Bioactive; human, porcine, canine; Roth, KA et al, *BBRC*, 112:528, 1983; Minamino, N et al, *BBRC*, 119:14, 1984

Gly-Asn-His-Trp-Ala-Val-Gly-His-Leu-Met Amide 2AcOH 5H₂O

Synonyms: Neuromedin C

Peptides International PNM-4153   Synthetic   MW 1330.49 $C_{50}H_{73}N_{17}O_{11}S \cdot 2CH_3COOH \cdot 5H_2O$ >95% (HPLC); lyophilized amorphous powder; bulk | Bioactive; human, porcine, canine; Roth, KA et al, *BBRC*, 112:528, 1983; Minamino, N et al, *BBRC*, 119:14, 1984

Gly-Asn-Ile-Phe-Ala-Asn-Leu-Phe-Lys-Gly-Leu-Phe-Gly-Lys-Lys-Glu

Synonyms: ADP-Ribosylation Factor Inhibitory Peptide P-13

ICN 159885   MW 1782   Inhibits cholera toxin ARF-dependent ADP-ribosylation, the transport from the endoplasmic reticulum to Golgi, intra-Golgi transport & *in vitro* endosome fusion at IC_{50} = 10-40*mM*; Randazzo, PA etal, *JBC*, 268:9555, 1993; Kahn, RA etal, *JBC*, 267:13039, 1992

Gly-Asn-Leu-Trp-Ala-Thr-Gly-His-Phe-Met Amide

Synonyms: Neuromedin B; Neuromedin B Agonist

Bachem H-3280.0001   MW 1132.31 $C_{52}H_{73}N_{15}O_{12}S$ Store at -15°C

ICN 153151   MW 1132.3

American Peptide 62-0-47   Mammalian   MW 1132.3 $C_{52}H_{73}N_{15}O_{12}S$ Bombesin-like peptide that binds with high affinity to the neuromedin B bombesin receptor; Kangawa, K et al, *BBRC*, 114:541, 1983

Neosystem SC134   Porcine   MW 1132.30   Bioactive; bombesin-related peptide; Minamino, N et al, *BBRC*, 114:541-548, 1983

Sigma N 3762   Porcine   FW 1132.3   ≥90% (HPLC) | Bioactive peptide; increases intracellular Ca^{2+} & phosphatidylinositol in glioblastoma cells; per Want et al, *Biochem J*, 286:641, 1992

Peptides International PNM-4152-v   Synthetic   MW 1132.3 $C_{52}H_{73}N_{15}O_{12}S$ >95% (HPLC); lyophilized amorphous powder | Bioactive; human, porcine, rat; Minamino, N et al, *BBRC*, 114:541, 1983

Gly-Asn-Leu-Trp-Ala-Thr-Gly-His-Phe-Met Amide AcOH 5H₂O

Synonyms: Neuromedin B

Peptides International PNM-4152   Synthetic   MW 1282.43 $C_{52}H_{73}N_{15}O_{12}S \cdot CH_3COOH \cdot 5H_2O$ >95% (HPLC); lyophilized amorphous powder; bulk | Bioactive; Minamino, N et al, *BBRC*, 114:541, 1983

Gly-Asn-Phe-Leu-Gln-Ser-Arg-Pro-Glu-Pro-Thr-Ala-Pro-Pro-Phe

Synonyms: P15 (446-460); GAG P1 (13-15)/P6 (1-12); HTLV I Peptide

Neosystem SC275   MW 1657.84   HTLV I peptide from subtype P15 (GAG P7 NC/GAG P1/GAG P6)

Gly-Asn-Trp-His-Gly-Thr-Ala-Pro-Asp-Trp-Phe-Phe-Asn-Tyr-Tyr-Trp

Synonyms: Endothelin B Receptor Antagonist; RES-701-1

Alexis 155-018   MW 2043.2 $C_{103}H_{115}N_{23}O_{23}$ Beta peptide bond:Gly[1] & Asp9; inhibits selectively ET-1 binding to the ET_B receptor; Morishita, Y et al, *J Antibiot*, 47:269, 1994; He, JX et al, *Bioorg Med Chem Lett*, 5:621, 1995; Sudjarwo, SA et al, *BBRC*, 200:627, 1994

American Peptide 88-2-51   MW 2643.2 $C_{103}H_{115}N_{23}O_{23}$ β Peptide bond:Gly[1]-Asp[9]; selectively inhibits the ET-1 binding to the ET_B-receptor; inhibited ET-1 binding to ET_B-receptor with an IC_{50}=10 nm; Morishita, Y et al, *J Antibiot*, 47:269, 1994; Ogawa, T et al, *J Antibiot*, 48(11):1213, 1995; *Trend in Pharmaceutical Sci*, 16:217, 1995

Gly-Asn-Trp-His-Gly-Thr-Ala-Pro-Asp-Trp-Val-Tyr-Phe-Ala-His-Leu-Asp-Ile-Ile-Trp

Synonyms: KT7421

Bachem H-3596.0500   Store at -15°C

Gly-Asn-Trp-His-Gly-Thr-Ser-Pro-Asp-Trp-Phe-Phe-Asn-Tyr-Tyr-Trp

Synonyms: Endothelin B Receptor Antagonist; RES-701-3

Alexis 155-019   MW 2059.2 $C_{103}H_{115}N_{23}O_{24}$ Beta peptide bond:Gly[1] & Asp[9]

American Peptide 88-2-52   MW 2059.2 $C_{103}H_{115}N_{23}O_{24}$ β Peptide bond:Gly[1]-Asp[9]; inhibits selectively the ET-1 binding to the ET_B-receptor

Gly-Asn-Trp-Phe-Asp-Leu-Ala-Ser-Trp-IleLys-Tyr-Ile-Gln-Tyr-Gly-Ile-Tyr-Val-Val

Synonyms: GP140 (681-700); SIVmac251 Peptide

Neosystem SC776   MW 2435.80   SIVmac251 peptide from subtype GP140

Gly-Asp
Z-Gly-Asp

Bachem C-1690.0001   MW 324.29 $C_{14}H_{16}N_2O_7$ Store at -15°C

Gly-Asp

Bachem G-2065.0001   MW 190.16 $C_6H_{10}N_2O_5$ Store at -15°C

Gly-Asp
Gly-D-Asp

Bachem G-2070.0001   MW 190.16 $C_6H_{10}N_2O_5$ Store at -15°C

Gly-Asp
Gly-DL-Asp

Bachem G-2075.0001   MW 190.16 $C_6H_{10}N_2O_5$ Store at -15°C

Gly-Asp

ICN 100336   MW 190.2 $C_6H_{10}N_2O_5$ Crystalline

Gly-Asp
Gly-D-Asp

ICN 101850   MW 190.2 $C_6H_{10}N_2O_5$ Crystalline

Gly-Asp
Gly-DL-Asp

ICN 101851   MW 190.2 $C_6H_{10}N_2O_5$ Crystalline

Gly-Asp

Sigma G 0752   FW 190.2 $C_6H_{10}N_2O_5$

Gly-Asp
Gly-D-Asp

Sigma G 2629   FW 190.2 $C_6H_{10}N_2O_5$

Gly-Asp
Gly-DL-Asp

Sigma G 2754   FW 190.2 $C_6H_{10}N_2O_5$

Gly-Asp-Arg-Ala-Asp-Gly-Gln-Pro-Ala-Gly-Asp-Arg-Ala-Asp-Gly-Gln-Pro-Ala

Bachem H-6260.0005   MW 1753.76 $C_{68}H_{108}N_{26}O_{29}$ Store at -15°C

Gly-Asp-Asp-Asp-Asp-Lys
Gly-Asp-Asp-Asp-Asp-Lys-βNA

Bachem K-1310.0025   MW 788.77 $C_{34}H_{44}N_8O_{14}$ Store at -15°C

Gly-Asp-Asp-Asp-Asp-Lys
Gly-Asp-Asp-Asp-Asp-Lys β-Naphthylamide

Synonyms: Enterokinase Substrate

ICN 157235   MW 789.8 $C_{34}H_{45}N_8O_{14}$ Sensitive & specific fluorogenic substrate; Antonowicz, I etal, *Clin Chim Acta*, 101:69, 1980

Sigma G 5261   FW 788.8 $C_{34}H_{44}N_8O_{14}$ ≥95% (HPLC) | Sensitive & specific fluorogenic substrate; Antonowicz, I et al, *Clin Chim Acta*, 101:69, 1980

Gly-Asp-Cys-Leu-Pro-His-Leu-Lys-Arg-Cys-Lys-Ala-Asp-Asn-Asp-Cys-Cys-Gly-Lys-Lys-Cys-Lys-Arg-Arg-Gly-Thr-Asn-Ala-Glu-Lys-Arg-Cys-Arg

Synonyms: Imperatoxin A

Bachem H-4098.0100   MW 3758.4 $C_{148}H_{254}N_{58}O_{45}S_6$ Store at -15°C

Peptides International PIM-4343-s   *Pandinus imperator* (scorpion) synthetic   MW 3758.4 $C_{148}H_{254}N_{58}O_{45}S_6$ >99% (HPLC); lyophilized amorphous powder | Bioactive; activator of Ca^{2+} release channels/ryanodine receptors; Valdivia, HH et al, *PNAS USA*, 89:12185, 1992; El-Hayek, R et al, *JBC*, 270:28696, 1995; Zamudio, FZ et al, *FEBS Letts*, 405:385, 1997

Gly-Asp-Gly

Bachem H-1308.0250   MW 247.21 $C_8H_{13}N_3O_6$ Store at -15°C

Gly-Asp-Gly-Arg-His-Asp-Leu-Leu-Val-Gly-Ala-Pro-Leu

Synonyms: Glycoprotein IIb (300-312); Peptide G13

Bachem H-1366.0001   MW 1319.48 $C_{57}H_{94}N_{18}O_{18}$ Store at -15°C

Gly-Asp-Gly-Arg-Leu-Tyr-Ala-Phe-Gly-Leu Amide

Synonyms: Allostatin II; Allatostatin II, Type A

American Peptide 87-7-20   MW 1067.2 $C_{49}H_{74}N_{14}O_{13}$ Woodhead, AP et al, *PNAS*, 86:5997, 1989

Bachem H-8070.0001   MW 1067.21 $C_{49}H_{74}N_{14}O_{13}$ Store at -15°C

Sigma A 9804   FW 1067.2   ≥97% (HPLC) | Bioactive peptide

Gly-Asp-Lys-Thr-Pro-Gly-Gly-Gly-Gly-Ala-Asn-Leu-Lys-Gly-Asp-Arg-Ser-Arg-Leu-Leu-Arg

Synonyms: Natriuretic Peptide (30-50), Prepro-C-Type

American Peptide 14-6-20   Porcine, rat   MW 2125.4 $C_{87}H_{153}N_{33}O_{29}$ C-Type Natriuretic Peptide

Gly-Asp-Phe-Glu-Glu-Ile-Pro-Glu-Glu-Tyr-Leu-Gln Ac-Gly-Asp-Phe-Glu-Glu-Ile-Pro-Glu-Glu-Tyr(SO₃H)-Leu-Gln Sulfated

Synonyms: Hirudin (54-65), Ac-

Bachem H-7415.0001   MW 1590.64 $C_{68}H_{95}N_{13}O_{29}S$ Store at -15°C

Gly-Asp-Phe-Glu-Glu-Ile-Pro-Glu-Glu-Tyr-Leu-Gln Gly-Asp-Phe-Glu-Glu-Ile-Pro-Glu-Glu-Tyr(SO₃H)-Leu-Gln Sulfated

Synonyms: Hirudin (54-65), (Tyr(SO₃H)[63])-

Bachem H-7425.0001   MW 1548.6 $C_{66}H_{93}N_{13}O_{28}S$ Store at -15°C

ICN 153119   MW 1548.6

Gly-Asp-Phe-Glu-Glu-Ile-Pro-Glu-Glu-Tyr-Leu-Gln Ac-Gly-Asp-Phe-Glu-Glu-Ile-Pro-Glu-Glu-Tyr-Leu-Gln Non-Sulfated

Synonyms: Hirudin (54-65), Ac-

ICN 153122   MW 1510.6

Gly-Asp-Phe-Glu-Glu-Ile-Pro-Glu-Glu-Tyr-Leu-Gln
Ac-Gly-Asp-Phe-Glu-Glu-Ile-Pro-Glu-Glu-Tyr(SO₃H)-Leu-Gln Sulfated

Synonyms: Hirudin (54-65), Ac-(Tyr(SO₃H)⁶³)-

ICN 153123   MW 1590.6

Gly-Asp-Phe-Glu-Glu-Ile-Pro-Glu-Glu-Tyr-Leu-Gln
Gly-Asp-Phe-Glu-Glu-Ile-Pro-Glu-Glu-Tyr(SO₃H)-Leu-Gln

Synonyms: Hirudin (54-65), (Tyr(SO₃H)⁶³)-

Sigma H 6894   FW 1548.6   ≥95% (HPLC) | Bioactive peptide; 65-residue polypeptide; acts as an anticoagulant by inhibiting thrombin; the seven C-terminal residues (59-65) are essential for the hirudin-thrombin interaction; Krstenansky, JL & Mao, SJT, *FEBS Lett*, 211:10, 1987

Gly-Asp-Phe-Glu-Glu-Ile-Pro-Glu-Glu-Tyr-Leu-Gln
Ac-Gly-Asp-Phe-Glu-Glu-Ile-Pro-Glu-Glu-Tyr(SO₃H)-Leu-Gln

Synonyms: Hirudin (54-65), Ac-(Tyr(SO₃H)⁶³)-

Sigma H 7144   FW 1590.6   ≥95% (HPLC) | Bioactive peptide

Gly-Asp-Phe-Glu-Glu-Ile-Pro-Glu-Glu-Tyr-Leu-Gln Non-Sulfated

Synonyms: Hirudin (54-65)

Bachem H-7420.0001   MW 1468.54 $C_{66}H_{93}N_{13}O_{25}$ Store at -15°C

ICN 153118   MW 1468.5

Sigma H 6769   FW 1468.5   ≥97% (HPLC) | Bioactive peptide

Gly-Cys
(Gly-Cys)₂

Bachem G-1845.0250   MW 354.41 $C_{10}H_{18}N_4O_6S_2$ Store at -15°C

Gly-Cys

Bachem G-4440.0250   Store at -15°C

Gly-Cys
Bz-Gly-Cys-2-Aminoethyl

Synonyms: Hippuryl-Cys-2-Aminoethyl; Bz-Gly-4-Thia-Lys

Bachem G-4490.0250   Store at -15°C

Gly-Cys-Arg-His-Ser-Arg-Ile-Gly-Val-Thr-Gln-Gln-Arg-Arg-Ala

Synonyms: VPR (75-89); HTLV I Peptide

Neosystem SC560   MW 1724.96   HTLV I peptide from subtype VPR; due to the presence of a Cys this peptide can contain the dimeric form

Gly-Cys-Cys-Ser-Asp-Pro-Arg-Cys-Ala-Trp-Arg-Cys Amide

Synonyms: Conotoxin MI, α-; Conotoxin Im I, α-; Nicotinic Acetylcholine Receptor Blocker

American Peptide 41-0-41   MW 1351.6 $C_{52}H_{78}N_{20}O_{15}S_4$ Disulfide bonds: Cys²-Cys⁸; Cys³-Cys¹²

Bachem H-2448.0500   MW 1351.58 $C_{52}H_{78}N_{20}O_{15}S_4$ Store at -15°C | Disulfide bonds: Cys²-Cys⁸, Cys³-Cys¹²

Sigma C 2461   FW 1351.6   ≥97% (HPLC); vial contains 500 µg; peptide content:~70% | Disulfide bonds: 2-8, 3-12; bioactive peptide; blocker for nicotinic acetylcholine receptor in CNS

American Peptide 41-0-40   *Conus imperialis*   MW 1493.7 $C_{58}H_{88}N_{22}O_{17}S_4$ Disulfide bonds: Cys²-Cys⁸; Cys³-Cys¹²; isolated from the venom of the snail; a ligand for nicotinic acetylcholine receptors; highly active against the neuromuscular receptor in frogs but not in mice; McIntosh, JM et al, *JBC*, 269:16733, 1994

Calbiochem 234629   *Conus imperialis*   MW 1351.6 $C_{52}H_{78}N_{20}O_{15}S_4$ TFA; >95% (HPLC); lyophilized solid; soluble in water; LD₅₀≤50 mg/kg⁸ | Disulfide bonds: Cys²-Cys¹²; competitive antagonist of the neuronal α-bungarotoxin-sensitive nicotinic receptors; does not affect the activation of currents gated by GABA, glycine or NMDA; Codignola, A et al, *Neurosci Lett*, 206:53, 1996; Pereira, EF et al, *J Pharmacol Exp Ther*, 278:1472, 1996; McIntosh, JM et al, *J Biol Chem*, 269:16733, 1994

Peptides International PCN-4311-v   *Conus imperialis* (marine snail) synthetic   MW 1351.6 $C_{52}H_{78}N_{20}O_{15}S_4$ >99% (HPLC); lyophilized amorphous powder | Bioactive; Disulfide bonds: Cys²-Cys⁸, Cys³-Cys¹²; McIntosh, JM et al, *JBC*, 269:16733, 1994; Johnson, DS et al, *Mol Pharmacol*, 48:194, 1995; Pereira, EFR et al, *J Pharmacol Exp Ther*, 278:1472, 1996

Gly-Cys-Gly

Bachem H-1552.0250   MW 235.26 $C_7H_{13}N_3O_4S$ Store at -15°C

Gly-Cys-Gly
4-Nitro-Z-Gly-Cys(4-Nitrobenzo(1,2,5)Oxadiazol-4-yl)-Gly

Bachem M-1590.0025   MW 577.49 $C_{21}H_{19}N_7O_{11}S$ Store at -15°C

Gly-Cys-Hyp-Trp-Glu-Pro-Trp-Cys
Gly-Cys-Hyp-D-Trp-Glu-Pro-Trp-Cys Amide

Synonyms: Contryphan

Bachem H-3998.0500   Store at -15°C

Gly-Cys-Lys-Asn-Phe-Phe-Trp-Lys

Synonyms: Somatostatin 14 (2-9)

Bachem H-4696.0005   MW 1029.23 $C_{50}H_{68}N_{12}O_{10}S$ Store at -15°C

Gly-Gln
Z-Gly-Gln

Bachem C-3180.0001   MW 337.33 $C_{15}H_{19}N_3O_6$ Store at -15°C

Gly-Gln
Gly-D-Gln

Bachem G-3730.0250   MW 203.2 $C_7H_{13}N_3O_4$ Store at -15°C

Gly-Gln
Gly-L-Gln Hydrate

Rexim   MW 221.2 $C_7H_{15}N_3O_5$ White crystals or crystalline powder

Gly-Gln
N-Ac-L-Gly-L-Gln

Rexim

Gly-Gln

Synonyms: Endorphin (30-31), α-; Lipotropin (90-91), α-

Bachem N-1070.0001   Camel   MW 203.2 $C_7H_{13}N_3O_4$ Store at -15°C

Gly-Gln Hydrate

Sigma G 5149   FW 203.2 $C_7H_{13}N_3O_4$ ≥97% (TLC) | Bioactive peptide; inhibitory neuropeptide; heat-stable substitute for glutamine in mammalian cell culture media; Parish, DC & Smyth, DG, *Biochem Soc Trans*, 10:221, 1982; Parish, DC et al, *Nature*, 306:267, 1983; Koelle, GB et al, *Proc Natl Acad Sci USA*, 82:5213, 1985; Roth, E et al, *In Vitro Cell Dev Biol*, 24:696, 1988

Gly-Gln-Gln-His-His-Leu-Gly-Gly-Ala-Lys-Gln-Ala-Gly-Asp-Val

Synonyms: Fibrinogen γ-Chain (397-411); Fibrinogen Related Peptide

American Peptide 42-1-31 MW 1502.6 $C_{62}H_{99}N_{23}O_{21}$ This carboxy-terminal fragment of human fibrinogen γ-chain contains a recognition site for the human platelet receptor; not involved in the fibrin polymerization reaction; Kloczewiak, M et al, *Thrombosis Res*, 29:249, 1983

Bachem H-3044.0001 MW 1502.61 $C_{62}H_{99}N_{23}O_{21}$ Store at -15°C

ICN 153096 Strong, DD etal, *Biochem*, 21:1414, 1982; Kloczewiak, M etal, *Thromb Res*, 29:249, 1983

Sigma F 3643 FW 1502.6 ≥97% (HPLC); peptide content:~70% | Bioactive peptide; Strong, DD et al, *Biochem*, 21:1414, 1982

Gly-Gln-Gly

Bachem H-1304.0250 MW 260.25 $C_9H_{16}N_4O_5$ Store at -15°C

Gly-Gln-Gly-Gly-Gly-Thr-His-Asn-Gln-Trp-Asn-Lys-Pro-Gly-Gly-Cys

Synonyms: Prion Protein; Progesterone Receptor Peptide (533-547); PR-AT 4.14

Alexis 165-035 Human synthetic MW 1597.7 $C_{65}H_{96}N_{24}O_{22}S$ ≥95%; lyophilized; reconstitute with 0.1 mL distilled H_2O | Competitively binds to Ab Alexis 210-200; antiserum blocking peptide

Alexis 165-027 Synthetic MW 1790.1 $C_{87}H_{128}N_{20}O_{21}$ ≥95%; lyophilized; reconstitute with 0.1 mL distilled H_2O | From PR DNA binding domain; competitively binds to Ab Alexis 803-007; monoclonal Ab blocking peptide

Gly-Gln-Leu-Gln-Pro-Ser-Leu-Gln-Thr-Gly-Ser-Glu-Glu-Leu-Arg-Ser-Leu

Synonyms: P18 (62-78); GAG P17 MA (62-78); HTLV I Peptide

Neosystem SC283 MW 1843.02 HTLV I peptide from subtype P25 (GAG P17 MA)

Gly-Gln-Lys-Leu-Val-Phe-Phe-Ala-Glu-Asp-Val-Gly-Gly-Lys-Lys-Lys-Lys-Lys-Lys
Gly-Gln-Lys-Leu-Val-Phe-Phe-Ala-Glu-Asp-Val-Gly-Gly-ε-Aminocaproyl-Lys-Lys-Lys-Lys-Lys-Lys

Synonyms: Amyloid α-Protein (15-25)-Gly-ε-Aminocaproyl(-Lys)₆,Gly-

Bachem H-3978.0500 MW 2248.74 $C_{105}H_{178}N_{28}O_{26}$ Store at -15°C

Gly-Gln-Met-Arg-Glu-Pro-Arg-Gly-Ser-Asp-Ile-Ala

Synonyms: P25 (226-237); GAG P24 CA (94-105); HTLV I Peptide

Neosystem SC849 MW 1316.45 HTLV I peptide from subtype P25 (GAG P24 CA)

Gly-Gln-Pro-Arg

Synonyms: Rigin

American Peptide 87-0-66 MW 456.6 $C_{18}H_{32}N_8O_6$

Bachem H-6920.0005 Store at -15°C

ICN 191515 Phagocytosis stimulating tetrapeptide originally isolated from human IgG

Gly-Glu
Z-Gly-Glu

Bachem C-1695.0001 MW 338.32 $C_{15}H_{18}N_2O_7$ Store at -15°C

Gly-Glu
Gly-Glu-pNA

Bachem L-1730.0050 MW 324.29 $C_{13}H_{16}N_4O_6$ Store at -15°C

Gly-Glu

ICN 101855 MW 204.2 $C_7H_{12}N_2O_5$

Gly-Glu
N-Ac-Gly-D-Glu

Synonyms: Excitatory Peptide

ICN 159846 MW 246.2 98% | More potent excitatory peptide than L-Glu; Garyaev, AP etal, *Eur J Pharmacol*, 187:157, 1991

Gly-Glu

Synonyms: Endorphin (30-31), α-; Lipotropin (90-91), α-

Sigma G 0877 FW 204.2 $C_7H_{12}N_2O_5$

Bachem G-2080.0001 Human MW 204.18 $C_7H_{12}N_2O_5$ Store at -15°C

Gly-Glu-Asn-Pro-Thr-Trp-Lys-Gln-Trp-Arg-Arg-Asp-Asn-Arg-Arg-Gly-Leu-Arg-Met-Tyr

Synonyms: VIF (163-181)-(Tyr); SIVmac251 Peptide

Neosystem SC579 MW 2619.94 SIVmac251 peptide from subtype VIF

Gly-Glu-Gla-Gla-Leu-Gln-Gla-Asn-Gln-Gla-Leu-Ile-Arg-Gla-Lys-Ser-Asn Free Acid

Synonyms: Conantokin G

Bachem H-2156.0500 Store at -15°C

Gly-Glu-Gla-Gla-Leu-Gln-Gla-Asn-Gln-Gla-Leu-Ile-Arg-Gla-Lys-Ser-Asn Amide

Synonyms: Conantokin G

Bachem H-9960.0100 Store at -15°C

Neosystem SC976 MW 2264.13 Bioactive; toxin; McIntosh, JM et al, *JBC*, 259:14343-14346, 1984; Chandler, P et al, *JBC*, 268:17173-17178, 1993; Zhou, LM et al, *J Neurochem*, 66:620-628, 1996

Gly-Glu-Gla-Gla-Leu-Gln-Gla-Asn-Gln-Gla-Leu-Ile-Arg-Gla-Lys-Ser-Asn Amide Acetate Salt

Synonyms: Conantokin G; Sleeper Peptide

Calbiochem 234550 *Conus geographus* MW 2264.2 $C_{88}H_{138}N_{26}O_{44}$ >90% (HPLC); lyophilized solid; soluble in water; LD₅₀≤200 mg/kg but >50 mg/kg | Non-competitive NMDA receptor antagonist that interacts with the glutamate binding site; this activity has been attributed to a potent non-competitive inhibition of polyamine responses; causes either sleep or hyperactivity in mice, depending on their age; Chandler, P et al, *J Biol Chem*, 268:17173, 1993; Nishiuchi, Y et al, *Int J Pept Protein Res*, 42:533, 1993; Hammerland, LJ et al, *Eur J Pharmacol*, 226:239, 1992; Gray, WR et al, *Ann Rev Biochem*, 57:665, 1988; Rivier, J et al, *Biochemistry*, 26:8508, 1987; Cruz, LJ et al, *J Toxin Rev*, 249:257, 1985; Olivera, BM et al, *Science*, 230:1338, 1985; McIntosh, J et al, *J Biol Chem*, 259:14343, 1984

Gly-Glu-Gla-Gla-Leu-Gln-Gla-Asn-Gln-Gla-Leu-Ile-Arg-Gla-Lys-Ser-Asn Amide

Synonyms: Conantokin G; Sleeper Peptide; Aspartate Receptor Antagonist, *N*-Methyl-D-

Peptides International PCO-4265-v   *Conus geographus* (marine snail) synthetic   MW 2264.2 $C_{88}H_{138}N_{26}O_{44}$ >99% (HPLC); lyophilized amorphous powder | Bioactive; Gray, WR & BM Olivera, *Ann Rev Biochem*, 57:665, 1998; McIntosh, JM et al, *JBC*, 259:14343, 1984; Hammerland, LG et al, *Eur J Pharmacol*, 226:239, 1992; Nishiuchi, Y et al, *Int J Pept Protein Res*, 42:533, 1993

American Peptide 41-0-73   *Conus geographus*, (marine snail) MW 2264.1 $C_{88}H_{138}N_{26}O_{44}$ NMDA antagonist; novel toxin isolated from the venom of the fish-hunting cone snail; antagonist of brain NMDA receptors, attributed to a potent non-competitive inhibition of polyamine responses at the NMDA receptor complex; Hammerland, LJ et al, *Eur J Pharmacol*, 226:239, 1992

Gly-Glu-Gla-Gla-Tyr-Gln-Lys-Asn-Met-Leu-Gla-Asn-Leu-Arg-Gla-Ala-Glu-Val-Lys-Lys-Asn-Ala Amide Acetate Salt

S Conantokin T; Sleeper Peptide *ynonyms*:

Calbiochem 234555   *Conus tulipa*   MW 2683.8 $C_{110}H_{175}N_{31}O_{45}S$ >90% (HPLC); lyophilized solid; LD$_{50}$≤200 mg/kg but >50 mg/kg | NMDA receptor antagonist; this activity has been attributed to a potent non-competitive inhibition of polyamine responses; causes either sleep or hyperactivity in mice, depending on their age; Haack, J et al, *J Biol Chem*, 265:6025, 1990; Nishiuchi, Y et al, *Int J Pept Protein Res*, 42:533, 1993; Skolnick, P et al, *J Neurochem*, 59:1516, 1992

Gly-Glu-Gla-Gla-Tyr-Gln-Lys-Met-Leu-Gla-Asn-Leu-Arg-Gla-Ala-Glu-Val-Lys-Lys-Asn-Ala Amide

Synonyms: Conantokin T; Sleeper Peptide; Aspartate Receptor Antagonist, *N*-Methyl-D-

Peptides International PCO-4264-v   *Conus tulipa* (marine snail) synthetic   MW 2683.8 $C_{110}H_{175}N_{31}O_{45}S$ >95% (HPLC); lyophilized amorphous powder | Bioactive; Haack, JA et al, *JBC*, 265:6025, 1990; Nishiuchi, Y et al, *Int J Pept Protein Res*, 42:533, 1993

American Peptide 41-0-72   *Conus tullpa* (marine snail)   MW 2683.9 $C_{110}H_{175}N_{31}O_{45}S$ NMDA Antagonist; Haack, JA et al, *JBC*, 265:6025, 1990

Gly-Glu-Gln-Arg-Lys-Asp-Val-Tyr-Val-Gln-Leu-Tyr-Leu

Synonyms: Thymopoietin II (29-41)

American Peptide 74-1-12   MW 1610.8 $C_{73}H_{115}N_{19}O_{22}$ Domain containing the active site of thymopoietin II fragment 32-36

Sigma G 3774   FW 1610.8; ≥97% (HPLC)   Bioactive peptide; domain containing the active site of thymopoietin II (32-36); Schlesinger, DH et al, *Cell*, 5:367, 1975; Goldstein, G et al, *Science*, 204:1309, 1979

Gly-Glu-Gln-Arg-Lys-Asp-Val-Tyr-Val-Glu-Leu-Tyr-Leu

Synonyms: Thymopoietin II (29-41)

Bachem H-3285.0001   MW 1611.82 $C_{73}H_{114}N_{18}O_{23}$ Store at -15°C

Gly-Glu-Glu-Glu-Leu-Gln-Glu-Asn-Gln-Glu-Leu-Ile-Arg-Glu-Lys-Ser-Asn Amide

Synonyms: Conantokin G, (Glu^{3,4,7,10,14})-

Bachem H-1236.0500   Store at -15°C

Gly-Glu-Glu-Glu-Leu-Gln-Glu-Asn-Gln-Glu-Leu-Ile-Arg-Glu-Lys-Ser-Asn
Gly-Glu-γ-CarboxyGlu-γ-CarboxyGlu-Leu-Gln-γ-CarboxyGlu-Asn-Gln-γ-CarboxyGlu-Leu-Ile-Arg-γ-CarboxyGlu-Lys-Ser-Asn Free Acid

Synonyms: Conantokin G, (Asn^{17})-; Sleep Inducing Peptide

Sigma C 2051   FW 2264.2   ≥97% (HPLC) | Bioactive peptide; Olivera, BM et al, *Science*, 249:257, 1990; Haack, JA et al, *J Biol Chem*, 265:6025, 1990

Gly-Glu-Glu-Glu-Leu-Gln-Glu-Asn-Gln-Glu-Leu-Ile-Arg-Glu-Lys-Ser-Asn
Gly-Glu-γ-CarboxyGlu-γ-CarboxyGlu-Leu-Gln-γ-CarboxyGlu-Asn-Gln-γ-CarboxyGlu-Leu-Ile-Arg-γ-CarboxyGlu-Lys-Ser-Asn Amide

Synonyms: Conantokin G A; Sleep Inducing Peptide

Sigma C 4311   Cone snail   FW 2264.2   ≥75% (HPLC) | Bioactive peptide; McIntosh, JM et al, *J Biol Chem*, 259:14343, 1984; Rivier, J et al, *Biochemistry*, 26:8508, 1987

Gly-Glu-Glu-Ile-Gln-Ile-Gly-His-Ile-Pro-Arg-Glu-Asp-Val-Asp-Tyr-His-Leu-Tyr-Pro

Synonyms: Fibronectin Type III Connecting Segment (90-109)

ICN 195622   MW 2380.6   ≥97%

Sigma F 6398   FW 2380.6   ≥97% (HPLC) | Bioactive peptide

Gly–Glu–Gly

Bachem H-8770.0250   MW 261.24 $C_9H_{15}N_3O_6$ Store at -15°C

Gly-Glu-Gly-Phe-Leu-Gly-Phe-Leu
Gly-Glu-Gly-Phe-Leu-Gly-D-Phe-Leu

Bachem N-1075.0005   MW 838.96 $C_{41}H_{58}N_8O_{11}$ Store at -15°C

Gly-Glu-Ile-Lys-Asn-Cys-Ser-Phe-Asn-Ile-Ser-Thr-Ser-Ile-Arg

Synonyms: GP120 (157-171); HTLV I Peptide

Neosystem SC1003   MW 1668.88   HTLV I peptide from subtype GP160; due to the presence of a Cys this peptide can contain the dimeric form

Gly-Glu-Leu-Asp-Arg-Trp-Glu-Lys-Ile-Arg-Leu-Arg-Pro-Gly-Gly

Synonyms: P18 (11-25); GAG P17 MA (11-25); HTLV I Peptide

Neosystem SC278   MW 1782.03   HTLV I peptide from subtype P25 (GAG P17 MA)

Gly-Glu-Pro-Pro-Pro-Gly-Lys-Pro-Ala-Asp-Asp-Ala-Gly-Leu-Val

Synonyms: Gastric Juice Peptide Fragment

Bachem H-2074.0500   Store at -15°C

Gly-Gly
(Phenyl-4(n)-³H)Hippuryl-Gly-Gly

Amersham TRK806   EtOH:H$_2$O (1:1); 307-18.5 GBq/mmol, 100-500 mCi/mmol; 18.5 MBq/mL, 500 µCi/mL

Gly-Gly

Synonyms: Good's Buffer

Amersham US16507   ≥99.0%; ≤0.02% free Gly; pK$_a$ 8.25 (20 °C)

Gly-Gly
BOC-Gly-Gly

Bachem A-1750.0005   MW 232.24 $C_9H_{16}N_2O_5$ Store at -15°C

Gly-Gly
FMOC-Gly-Gly

Bachem B-1485.0001   MW 354.36 $C_{19}H_{18}N_2O_5$ Store at -15°C

Gly-Gly
Z-Gly-Gly

Bachem C-1700.0005   MW 266.25 $C_{12}H_{14}N_2O_5$ Store at -15°C

Gly-Gly
Z-Gly-Gly Amide

Bachem C-1705.0001   MW 265.27 $C_{12}H_{15}N_3O_4$ Store at -15°C

Gly-Gly
Z-Gly-Gly-OSu

Bachem C-1710.0001   MW 363.33 $C_{16}H_{17}N_3O_7$ Store at -15°C

Gly-Gly
Z-Gly-Gly-ONp

Bachem C-1720.0001   MW 387.35 $C_{18}H_{17}N_3O_7$ Store at -15°C

Gly-Gly
Ac-Gly-Gly

Bachem G-1025.0001   MW 174.16 $C_6H_{10}N_2O_4$ Store at -15°C

Gly-Gly

Bachem G-2090.0005   MW 132.12 $C_4H_8N_2O_3$ Store at -15°C

Gly-Gly
Gly-Gly-OBzl *p*-Tosylate

Bachem G-2105.0005   MW 394.45 $C_{11}H_{14}N_2O_3 \cdot C_7H_8O_3S$ Store at -15°C

Gly-Gly
Bz-Gly-Gly

Synonyms: Hippuryl-Gly

Bachem G-2270.0005   MW 236.23 $C_{11}H_{12}N_2O_4$ Store at -15°C

Gly-Gly
Gly-Gly-OEt Hydrochloride

Bachem G-4220.0025   MW 196.63 $C_6H_{12}N_2O_3 \cdot HCl$ Store at -15°C

Gly-Gly
Biotinyl-Gly-Gly

Bachem G-4485.0250   MW 358.42 $C_{14}H_{22}N_4O_5S$ Store at -15°C

Gly-Gly
Gly-Gly-AMC Hydrochloride

Bachem I-1470.0050   Store at -15°C

Gly-Gly
Gly-Gly-βNA Hydrobromide

Bachem K-1560.0250   MW 338.2 $C_{14}H_{15}N_3O_2 \cdot HBr$ Store at -15°C

Gly-Gly
Gly-Gly-pNA Hydrochloride

Bachem L-1935.0250   MW 288.69 $C_{10}H_{12}N_4O_4 \cdot HCl$ Store at -15°C

Gly-Gly

Synonyms: Diglycine

Fluka 50199   MW 132.12 $C_4H_8N_2O_3$ ≥99.5% (titration); ≤0.05% residue on ignition; ≤0.1% loss on drying; ≤0.00001% As; ≤0.0005% Al, Ba, Bi, Cd, Co, Cr, Cu, Fe, Li, Mg, Mn, Mo, Ni, Pb, Sr, Zn; ≤0.001% Ca; ≤0.02% NH_4^+; ≤0.005% K, Na, Cl, SO_4 ₁ Superior buffer for the growth of *A. halophytica* (Chroococcales) in hypersaline media; Tindall, DR et al, *Phycologia*, 17:179, 1978

Fluka 50200   MW 132.12 $C_4H_8N_2O_3$ ≥99.0% (titration); ≤0.1% residue on ignition; ≤1% loss on drying; ≤0.0005% Cd, Co, Cr, Cu, Fe, Mg, Mn, Ni, Pb, Zn; ≤0.002% Ca; ≤0.005% K, Na, Cl, SO_4; mp:220-240°C

Gly-Gly
N-CBZ-Gly-Gly

ICN 100217   MW 266.3 $C_{12}H_{14}N_2O_6$ Crystalline

Gly-Gly
N-CBZ-Gly-Gly-*p*-Nitrophenyl Ester

ICN 101258   MW 387.3 $C_{18}H_{17}N_3O_7$ Crystalline

Gly-Gly
N-Chloro-Ac-Gly-Gly

ICN 101336   MW 208.6 $C_6H_9ClN_2O_4$ Crystalline

Gly-Gly
Dinitropyridyl-Gly-Gly

ICN 101831   MW 299.2 $C_9H_9N_5O_7$

Gly-Gly
Gly-Gly-β-Naphthylamide Hydrobromide

ICN 101859   MW 338.2 $C_{14}H_{15}N_3O_2 \cdot HBr$ Crystalline

Gly-Gly
Gly-Gly-OEt hydrochloride

ICN 101860   MW 196.6 $C_6H_{12}N_2O_3 \cdot HCl$ Crystalline

Gly-Gly

ICN 101862   MW 168.6 $C_4H_8N_2O_3 \cdot HCl$ Crystalline; hydrochloride

Gly-Gly
Gly-Gly-OMe Hydrochloride

ICN 101863   MW 182.6 $C_5H_{10}N_2O_3 \cdot HCl$

Gly-Gly
Hippuryl-Gly-Gly

Synonyms: Angiotensin Converting Enzyme Substrate; Benzoyl-Gly-Gly

ICN 152744   Crystalline | Yang, HYT etal, *J Pharm Exp Ther*, 177:291, 1971; Nakajima, T etal, *BBA*, 315:430, 1973; Oshima, G etal, *J Biochem*, 80:477, 1976; Oshima, G etal, *BBA*, 566:128, 1979

Gly-Gly
Borotrimethyl-Gly-GlyOMe

ICN 154635   MW 188.04 $BC_7H_{17}N_2O_3$ >97% | Potent hypocholesterolemic activity

Gly-Gly
Boro-Gly-Gly Amide

ICN 154645   MW 130.94 $BC_3H_{10}N_3O_2$ >97%

Gly-Gly
Gly-Gly-OBz p-Toluenesulfonate Salt

ICN 157251   MW 376.4 $C_{11}H_{12}N_2O_2 \cdot C_7H_8O_3S$

Gly-Gly

Peptides International OGG-3028   MW 132.12 $C_4H_8N_2O_3$ >98% (HPLC); colorless crystalline powder

Gly-Gly
N-CBZ-Gly-Gly

Sigma C 8001   FW 266.3 $C_{12}H_{14}N_2O_5$

Gly-Gly
N-CBZ-Gly-Gly p-Nitrophenyl Ester

Sigma C 8387   FW 387.3 $C_{18}H_{17}N_3O_7$

Gly-Gly
Gly-Gly Benzyl Ester p-Toluenesulfonate salt

Sigma G 1141   FW 394.4 $C_{11}H_{14}N_2O_3 \cdot C_7H_8O_3S$

Gly-Gly
Gly-Gly β-Naphthylamide Hydrobromide

Sigma G 1772   FW 338.2 $C_{14}H_{15}N_3O_2 \cdot HBr$

Gly-Gly
Gly-Gly 7-Amido-4-Methylcoumarin

Sigma G 1788   FW 289.3 (free base) $C_{14}H_{15}N_3O_4$ ≥97% (TLC); peptide content: ~65%

Gly-Gly
Gly-Gly Ethyl Ester Hydrochloride

Sigma G 3129   FW 196.6 $C_6H_{12}N_2O_3 \cdot HCl$

Gly-Gly

Sigma G 7278   FW 132.1; Free base $C_4H_8N_2O_3$ SigmaUltra; >99.5% (titration); pH (1 M in water, 20°C):4.5-6.0; loss on drying (110°C):<0.05%; residue on ignition (900°C):<0.05%; solubility (1 M in water, 20°C):complete, colorless; insoluble matter:passes filter test; SO_4, Cl, K, Na:<0.005%; Ca:<0.001%; Al, Ba, Bi, Cd, Co, Cr, Cu, Fe, Li, Mg, Mn, Mo, Ni, Pb, Sr, Zn:<0.0005%; As:<0.00001%; NH_4^+:<0.02%; A_{260}<0.075; A_{280}<0.072 (1 M in water) | pKa=8.2 at 25°C; useful pH range 7.5-8.9

Gly-Gly
Hippuryl-Gly-Gly

Synonyms: Angiotensin I Converting Enzyme Substrate

Sigma H 1763   FW 293.3 $C_{13}H_{15}N_3O_5$ Yang, HYT et al, *J Pharmacol Exp Ther*, 177:291, 1971; Filipovic, N et al, *Clin Chim Acta*, 88:173, 1978; Stewart, TA et al, *Peptides*, 2:145, 1981

Gly-Gly
CBZ-Gly-Gly

USBio C2098-72   MW 266.3 $C_{12}H_{14}N_2O_5$ ≥99%

Gly-Gly
CBZ-Gly-Gly-ONp

USBio C2098-73   MW 387.4 $C_{18}H_{17}N_3O_7$ ≥99%

Gly-Gly Amide

Bachem G-2095.0001   MW 131.13 $C_4H_9N_3O_2$ Store at -15°C

Gly-Gly Amide Hydrochloride

Bachem G-2100.0001   MW 167.6 $C_4H_9N_3O_2 \cdot HCl$ Store at -15°C

ICN 157250   MW 167.6 $C_4H_9N_3O_2 \cdot HCl$ Crystalline

Sigma G 6879   FW 167.6 $C_4H_9N_3O_2 \cdot HCl$

Gly-Gly Free Base

ICN 101856   MW 132.1 $C_4H_8N_2O_3$ Crystalline; pK$_a$ = 8.2 (25°C) | Useful pH range 7.5-8.9

ICN 194548   MW 132.1 $C_4H_8N_2O_3$ Cell culture reagent; crystalline; pK$_a$ = 8.2 (25°C) | Useful pH range 7.5-8.9

Sigma G 1002   FW 132.1 $C_4H_8N_2O_3$ ≥99% (titration) | pKa=8.2 at 25°C; useful pH range 7.5-8.9

Gly-Gly Hydrochloride

Sigma G 1127   FW 168.6 $C_4H_8N_2O_3 \cdot HCl$ pKa=8.2 at 25°C; useful pH range 7.5-8.9

Gly-Gly-Abu
Gly-Gly-γ-Abu

Bachem H-5580.0250   Store at -15°C

Gly-Gly-Ala
Z-Gly-Gly-Ala

Bachem C-1725.0001   MW 337.33 $C_{15}H_{19}N_3O_6$ Store at -15°C

Gly-Gly-Ala

Bachem H-3290.0250   MW 203.2 $C_7H_{13}N_3O_4$ Store at -15°C

Gly-Gly-Ala
Gly-Gly-α-Ala

Bachem H-3295.0250   MW 203.2 $C_7H_{13}N_3O_4$ Store at -15°C

Gly-Gly-Ala
Gly-L-Gly-Ala

Fluka 50150   MW 146.15 $C_5H_{10}N_2O_3$ ~99% (titration); ≤0.05% residue on ignition; mp:230°C

Gly-Gly-Ala
Gly-Gly-β-Ala

ICN 101843   MW 203.2 $C_7H_{13}N_3O_4$ Crystalline

Gly-Gly-Ala

ICN 101845   MW 203.2 $C_7H_{13}N_3O_4$ Crystalline

Gly-Gly-Ala
N-CBZ-Gly-Gly-Ala

Sigma C 8126   FW 337.3 $C_{15}H_{19}N_3O_6$

Gly-Gly-Ala

Sigma G 0254   FW 203.2 $C_7H_{13}N_3O_4$

Gly-Gly-Ala-Gly

Bachem H-3300.0250 MW 260.25 $C_9H_{16}N_4O_5$ Store at -15°C

Gly-Gly-Ala-Gly
Gly-Gly-α-Ala-Gly

Bachem H-7590.0250 Store at -15°C

Gly-Gly-Ala-Ile-Ser-Met-Arg-Arg-Ser-Lys-Pro-Ala-Gly-Asp-Leu-Arg-Gln-Lys-Leu-Leu-Arg-Ala-Tyr

Synonyms: NEF (2-23)-(Tyr); SIVmac251 Peptide

Neosystem SC578 MW 2545.00 SIVmac251 peptide from subtype NEF

Gly-Gly-Arg

Bachem H-3305.0250 MW 288.31 $C_{10}H_{20}N_6O_4$ Store at -15°C

Gly-Gly-Arg
Gly-Gly-Arg-Alat

Bachem H-3310.0250 MW 359.39 $C_{13}H_{25}N_7O_5$ Store at -15°C

Gly-Gly-Arg
Z-Gly-Gly-Arg-AMC Hydrochloride

Bachem I-1140.0050 MW 616.07 $C_{28}H_{33}N_7O_7 \cdot HCl$ Store at -15°C

Gly-Gly-Arg
Z-Gly-Gly-Arg-4MβNA Hydrochloride

Bachem J-1115.0050 MW 614.1 $C_{29}H_{35}N_7O_6 \cdot HCl$ Store at -15°C

Gly-Gly-Arg
Z-Gly-Gly-Arg-βNA Hydrochloride

Bachem K-1175.0050 MW 584.08 $C_{28}H_{33}N_7O_5 \cdot HCl$ Store at RT

Gly-Gly-Arg
Gly-Gly-Arg-Anilide

Bachem M-1450.0050 MW 363.42 $C_{16}H_{25}N_7O_3$ Store at -15°C

Gly-Gly-Arg
N-CBZ-Gly-Gly-L-Arg-MCA

Synonyms: Urokinase Substrate

ICN 150572 MW 579.6 $C_{28}H_{33}N_7O_7$ Fluorogenic substrate

Gly-Gly-Arg
Z-Gly-Gly-Arg-AMC Hydrochloride

Synonyms: Urokinase Substrate II

ICN 195901 MW 616.1 $C_{28}H_{33}N_7O_7 \cdot HCl$ ≥95% | Fluorescent substrate; suitable for plasminogen activators

Gly-Gly-Arg
N-CBZ-Gly-Gly-Arg 4-Methoxy-β-Naphthylamide Hydrochloride

Synonyms: Urokinase Substrate

Sigma C 5770 FW 614.1 $C_{29}H_{35}N_7O_6 \cdot HCl$ Fluorogenic; Huseby, RM et al, *Thromb Res*, 10:679, 1977; Bigbee, WL et al, *Anal Biochem*, 88:114, 1978

Gly-Gly-Arg
N-CBZ-Gly-Gly-Arg β-Naphthylamide Hydrochloride

Synonyms: Trypsin Substrate

Sigma C 7784 FW 584.1 $C_{28}H_{33}N_7O_5 \cdot HCl$ ~90% | Sensitive & specific fluorogenic substrate for the assay of trypsin in human serum; Kramer, S et al, *J Surg Res*, 8:253, 1968

Gly-Gly-Arg
N-CBZ-Gly-Gly-Arg 7-Amido-4-Methylcoumarin Hydrochloride

Synonyms: Urokinase Substrate

Sigma C 9396 FW 616.1 $C_{28}H_{33}N_7O_7 \cdot HCl$ Fluorogenic; Zimmerman, M et al, *Proc Natl Acad Sci USA*, 75:750, 1978

Gly-Gly-Arg
CBZ-Gly-Gly-Arg-4MβNA Hydrochloride

USBio C2098-74 MW 614.1 $C_{29}H_{35}N_7O_6$ ≥99%

Gly-Gly-Arg
CBZ-Gly-Gly-Arg-AMC

USBio C2098-78 MW 579.6 $C_{28}H_{33}N_7O_7$ ≥99%

Gly-Gly-Arg Acetate Salt

ICN 153110

Sigma G 6887 FW 288.3 (free base) $C_{10}H_{20}N_6O_4$ ≥97% (HPLC); peptide content:~70% | Bioactive peptide

Gly-Gly-Arg-Trp-Ile-Leu-Ala-Ile-Pro-Arg-Arg-Ile-Arg-Gln-Gly-Leu-Glu-Leu-Thr-Leu-Leu

Synonyms: GP140 (861-881); SIVmac251 Peptide

Neosystem SC794 MW 2431.95 SIVmac251 peptide from subtype GP140

Gly-Gly-Asn-Ala
Ac-tBu-Gly-tBu-Gly-Asn(Me)₂-Ala-AMC

Bachem I-1870.0001 MW 656.78 $C_{33}H_{48}N_6O_8$ Store at -15°C

Gly-Gly-Asp
Gly-L-Gly-Asp

Fluka 50170 MW 226.19 $C_6H_{10}N_2O_5 \cdot 2H_2O$ ≥99% (titration); mp:205°C; dihydrate

Gly-Gly-Asp-Ala

Bachem H-3320.0250 MW 318.29 $C_{11}H_{18}N_4O_7$ Store at -15°C

Gly-Gly-Cys

Bachem H-3325.0250 MW 235.26 $C_7H_{13}N_3O_4S$ Store at -15°C

Gly-Gly-Cys-Leu-Pro-His-Asn-Arg-Phe-Cys-Asn-Ala-Leu-Ser-Gly-Pro-Arg-Cys-Cys-Ser-Gly-Leu-Lys-Cys-Lys-Glu-Leu-Ser-Ile-Trp-Asp-Ser-Arg-Cys-Leu Amide

Synonyms: Presynaptic Calcium Channel Antagonist; Agelenin

American Peptide 41-0-25 *Agelena opulenta* (spider) MW 3818.5 $C_{160}H_{254}N_{52}O_{45}S_6$ Disulfide bonds: Cys[3]-Cys[19]; Cys[10]-Cys[24]; Cys[18]-Cys[34] | Hagiwara, K et al, *Biomed Res*, 11:181, 1990

Peptides International PAG-4247-s *Agelena opulenta* (spider) synthetic MW 3818.5 $C_{160}H_{254}N_{52}O_{45}S_6$ >99% (HPLC); lyophilized amorphous powder | Bioactive; Disulfide bonds: Cys[3]-Cys[19], Cys[10]-Cys[24] & Cys[18]-Cys[34]; presynaptic Ca[2+] channel antagonist; Hagiwara, K et al, Biomedical *Res*, 11:181, 1990; Inui, T et al, *Pept Res*, 5:140, 1992

Gly-Gly-Gln

Bachem H-2382.0250   Store at -15°C

Gly-Gly-Glu

Bachem H-3330.0250   MW 261.24 $C_9H_{15}N_3O_6$ Store at -15°C

Gly-Gly-Glu-Ala

Bachem H-3340.0250   MW 332.31 $C_{12}H_{20}N_4O_7$ Store at -15°C

Gly-Gly-Glu-Ala
Gly-Gly-Glu-Ala-OMe

Bachem H-3345.0250   MW 346.34 $C_{13}H_{22}N_4O_7$ Store at -15°C

Gly-Gly-Glu-Val-Leu-Gly-Lys-Arg-Tyr-Gly-Gly-Phe-Met

Synonyms: Enkephalin; Prepro-Enkephalin (128-140)

Bachem H-2855.0001   Bovine, ovine   MW 1370.6 $C_{61}H_{95}N_{17}O_{17}S$ Store at -15°C

ICN 152833   Ovine, bovine   Carboxy terminal portion of peptide F; Micanovic, R et al, BBRC, 118:299, 1984

Sigma P 7162   Ovine, bovine   FW 1370.6   ≥97% (HPLC) | Bioactive peptide; originally isolated from adrenal chromaffin granules; carboxyl-terminal portion of peptide *M*; Micanovic, R et al, *Biochem Biophys Res Commun*, 118:299, 1984

Gly-Gly-Gly
Benzoyl-Gly-Gly-Gly

Synonyms: Angiotensin I Converting Enzyme Substrate

American Peptide 12-1-95   MW 293.3 $C_{13}H_{15}N_3O_5$ Oshima, G et al, *J Biochem*, 80:477, 1976

Gly-Gly-Gly
BOC-Gly-Gly-Gly

Bachem A-4380.0005   MW 289.29 $C_{11}H_{19}N_3O_6$ Store at -15°C

Gly-Gly-Gly
FMOC-Gly-Gly-Gly

Bachem B-1490.0001   Store at -15°C

Gly-Gly-Gly
Z-Gly-Gly-Gly

Bachem C-1730.0001   MW 323.31 $C_{14}H_{17}N_3O_6$ Store at -15°C

Gly-Gly-Gly
Z-Gly-Gly-Gly-OSu

Bachem C-1735.0001   MW 420.38 $C_{18}H_{20}N_4O_8$ Store at -15°C

Gly-Gly-Gly
Z-Gly-Gly-Gly-ONp

Bachem C-1740.0001   MW 444.4 $C_{20}H_{20}N_4O_8$ Store at -15°C

Gly-Gly-Gly

Bachem H-3350.0005   MW 189.17 $C_6H_{11}N_3O_4$ Store at -15°C

Gly-Gly-Gly
Gly-Gly-Gly-OEt Hydrochloride

Bachem H-3365.0250   MW 253.69 $C_8H_{15}N_3O_4 \cdot HCl$ Store at -15°C

Gly-Gly-Gly
Bz-Gly-Gly-Gly

Synonyms: Hippuryl-Gly-Gly

Bachem M-1480.0250   MW 293.28 $C_{13}H_{15}N_3O_5$ Store at -15°C

Gly-Gly-Gly

Synonyms: Triglycine

Fluka 50239   MW 189.17 $C_6H_{11}N_3O_4$ ≥99.5% (titration); ≤0.05% residue on ignition; ≤0.1% loss on drying; ≤0.00001% As; ≤0.0005% Al, Ba, Bi, Cd, Co, Cr, Cu, Fe, Li, Mg, Mn, Mo, Ni, Pb, Sr, Zn; ≤0.001% Ca; ≤0.005% K, Na, Cl, SO_4

Fluka 50240   MW 189.17 $C_6H_{11}N_3O_4$ ≥99.5% (titration); ≤0.1% residue on ignition; ≤1% loss on drying; ≤0.0005% Cd, Co, Cr, Cu, Fe, Mg, Mn, Ni, Pb, Zn; ≤0.002% Ca; ≤0.005% K, Na, Cl, SO_4; mp:240-250°C

Gly-Gly-Gly
Borotrimethyl-Gly-Gly-Gly-OEt

ICN 154651   MW 259.11 $BC_{10}H_{22}N_3O_4$ Oil; >97%

Gly-Gly-Gly
Boro-Gly-Gly-Gly Amide

ICN 154652   MW 188.00 $BC_5H_{13}N_4O_3$ >97%

Gly-Gly-Gly

Peptides International OGG-3061   MW 189.17 $C_6H_{11}N_3O_4$ >98% (HPLC); colorless crystalline powder

Gly-Gly-Gly
Bz-Gly-Gly-Gly

Synonyms: Angiotensin I Converting Enzyme Substrate; Hippuryl-Gly-Gly

Peptides International SGG-3128   MW 293.28 $C_{13}H_{15}N_3O_5$ >99% (HPLC); amorphous powder | Yang, HYT et al, *J Pharmacol Exp Ther*, 177:291, 1971; Nakajima, T et al, *BBA*, 315:430, 1973; Oshima, G et al, *J Biochem (Tokyo)*, 80:477, 1976

Gly-Gly-Gly
N-CBZ-Gly-Gly-Gly

Sigma C 8251   FW 323.3 $C_{14}H_{17}N_3O_6$

Gly-Gly-Gly

Synonyms: Triglycine

Sigma G 1377   FW 189.2 $C_6H_{11}N_3O_4$

Gly-Gly-Gly
N-Succinyl-Gly-Gly-Gly p-NA

Sigma S 5546   FW 409.4 $C_{16}H_{19}N_5O_8$

Gly-Gly-Gly
CBZ-Gly-Gly-Gly

USBio C2098-79   MW 323.3 $C_{14}H_{17}N_3O_6$ ≥99%

Gly-Gly-Gly Amide Hydrochloride

Bachem H-3355.0250   MW 224.65 $C_6H_{12}N_4O_3 \cdot HCl$ Store at -15°C

Gly-Gly-Gly Chloride Free

Synonyms: Triglycine

ICN 101864   MW 189.2 $C_6H_{11}N_3O_4$ Crystalline

Gly-Gly-Gly-Ala

Bachem H-3370.0250   MW 260.25 $C_9H_{16}N_4O_5$ Store at -15°C

Gly-Gly-Gly-Ala
Gly-Gly-Gly-α-Ala

Bachem H-7575.0250   MW 260.25 $C_9H_{16}N_4O_5$ Store at -15°C

Gly-Gly-Gly-Gly
Z-Gly-Gly-Gly-Gly

Bachem C-1745.0001   MW 380.36 $C_{16}H_{20}N_4O_7$ Store at -15°C

Gly-Gly-Gly-Gly

Synonyms: Tetraglycine; Triglycyl-Glycine

Bachem H-3380.0250   MW 246.22 $C_8H_{14}N_4O_5$ Store at -15°C

Fluka 87300   MW 246.22 $C_8H_{14}N_4O_5$ ~99.0%

ICN 157252   MW 246.2 $C_8H_{14}N_4O_5$ Crystalline

Sigma G 3882   FW 246.2 $C_8H_{14}N_4O_5$

Gly-Gly-Gly-Gly-Gly
Z-Gly-Gly-Gly-Gly-Gly

Bachem C-1755.0250   MW 437.41 $C_{18}H_{23}N_5O_8$ Store at -15°C

Gly-Gly-Gly-Gly-Gly

Synonyms: Pentaglycine

Bachem H-3395.0250   MW 303.28 $C_{10}H_{17}N_5O_6$ Store at -15°C

Fluka 76790   MW 303.28 $C_{10}H_{17}N_5O_6$ ≥99.0% (TLC)

ICN 157253   MW 303.3 $C_{10}H_{17}N_5O_6$ Crystalline

Sigma G 5755   FW 303.3 $C_{10}H_{17}N_5O_6$

Gly-Gly-Gly-Gly-Gly
CBZ-Gly-Gly-Gly-Gly-Gly

USBio C2098-80   MW 437.4 $C_{18}H_{23}N_5O_8$ ≥99%

Gly-Gly-Gly-Gly-Gly Amide Hydrobromide

Bachem H-3400.0050   Store at -15°C

Gly-Gly-Gly-Gly-Gly-Gly
Z-Gly-Gly-Gly-Gly-Gly-Gly

Bachem C-1760.0250   MW 494.46 $C_{20}H_{26}N_6O_9$ Store at -15°C

Gly-Gly-Gly-Gly-Gly-Gly

Synonyms: Hexaglycine

Bachem H-3405.0250   MW 360.33 $C_{12}H_{20}N_6O_7$ Store at -15°C

ICN 157254   MW 360.4 $C_{12}H_{20}N_6O_7$ Crystalline

Sigma G 5630   FW 360.3 $C_{12}H_{20}N_6O_7$

Gly-Gly-Gly-Lys
BOC-Gly-Gly-Gly-Lys

Bachem A-3275.0250   Store at -15°C

Gly-Gly-His
Z-Gly-Gly-His

Bachem C-1765.0250   MW 403.4 $C_{18}H_{21}N_5O_6$ Store at -15°C

Gly-Gly-His

Synonyms: Copper Binding Peptide

Bachem H-3417.0250   MW 269.26 $C_{10}H_{15}N_5O_4$ Store at -15°C

ICN 153111   MW 269.3 $C_{10}H_{15}N_5O_4$ Lau, SJ etal, *JBC*, 249:5878, 1974

Peptides International OGH-3076   MW 269.26 $C_{10}H_{15}N_5O_4$ >98% (HPLC); colorless crystalline powder | Lau, Show-Jy et al, *JBC*, 249:5878, 1974

Gly-Gly-His Acetate Salt

Synonyms: Copper Binding Peptide

Sigma G 5772   FW 269.3 (free base) $C_{10}H_{15}N_5O_4$ ≥97% (TLC) | Bioactive peptide; Lau, SJ et al, *J Biol Chem*, 249:5878, 1974; Iyer, S et al, *Biochem J,* 169:61, 1978

Gly-Gly-His-Ala

Bachem H-3420.0250   MW 340.34 $C_{13}H_{20}N_6O_5$ Store at -15°C

Gly-Gly-His-Gly

Bachem H-3425.0250   MW 326.31 $C_{12}H_{18}N_6O_5$ Store at -15°C

Gly-Gly-Ile

Bachem H-3430.0250   MW 245.28 $C_{10}H_{19}N_3O_4$ Store at -15°C

ICN 104889   MW 245.3 $C_{10}H_{19}N_3O_4$ Crystalline

Sigma G 4384   FW 245.3 $C_{10}H_{19}N_3O_4$

Gly-Gly-Leu
Z-Gly-Gly-Leu

Bachem C-1775.0250   MW 379.41 $C_{18}H_{25}N_3O_6$ Store at -15°C

Gly-Gly-Leu

Bachem H-3435.0250   MW 245.28 $C_{10}H_{19}N_3O_4$ Store at -15°C

Gly-Gly-Leu
Z-Gly-Gly-Leu-AMC

Bachem I-1425.0050   MW 536.59 $C_{28}H_{32}N_4O_7$ Store at -15°C

Gly-Gly-Leu
Z-Gly-Gly-Leu-βNA

Bachem K-1180.0050   MW 504.59 $C_{28}H_{32}N_4O_5$ Store at RT

Gly-Gly-Leu
BOC-Gly-Gly-Leu-pNA

Bachem L-1190.0050   MW 465.51 $C_{21}H_{31}N_5O_7$ Store at -15°C

Gly-Gly-Leu
Z-Gly-Gly-Leu-pNA

Bachem L-1230.0050   MW 499.52 $C_{24}H_{29}N_5O_7$ Store at -15°C

Gly-Gly-Leu
Z-Gly-Gly-Leu Amide

Bachem M-1255.0250   MW 378.43 $C_{18}H_{26}N_4O_5$ Store at -15°C

Gly-Gly-Leu

ICN 101867   MW 245.3 $C_{10}H_{19}N_3O_4$ Crystalline; 99%

Gly-Gly-Leu
Gly-Gly-D-Leu

ICN 105569   MW 245.3 $C_{10}H_{19}N_3O_4$ Crystalline

Gly-Gly-Leu
BOC-Gly-Gly-Leu-pNA

Synonyms: Subtilisin A Substrate I

ICN 195018   MW 465.5   >95%; lyophilized

Gly-Gly-Leu
Z-Gly-Gly-Leu-AMC

Synonyms: Subtilisin BPN' Substrate; Subtilisin Carlsberg Substrate

Neosystem SC1287   MW 536.58   Kanaoka, Y et al, *Chem Pharm Bull*, 33:1721-1724, 1985

Gly-Gly-Leu
Z-Gly-Gly-Leu-pNA

Synonyms: Neutral Endopeptidase 245 Substrate; Proteinase yscE Substrate; Subtilisin Substrate; Neutral Endopeptidase Substrate

Neosystem SC1294   MW 499.50   Emter, O & Wolf, DH, *FEBS Lett*, 166:321-325, 1984; Kanaoka, Y et al, *Chem Pharm Bull*, 33:1721, 1985; Zolfaghari, R et al, *Biochem J*, 241:129-135, 1987; Heinemeyer, W et al, *EMBO J*, 10:555-562, 1991

Peptides International SGL-3111   MW 499.52 $C_{24}H_{29}N_5O_7$ >99% (HPLC); amorphous powder | Orlowski, M & S Wilk in *Peptides, Structure & Biological Functions*, (E Gross & J Meienhofer, Eds), Pierce Chemical Co, 1980, p 925

Gly-Gly-Leu
N-CBZ-Gly-Gly-Leu β-Naphthylamide

Sigma C 1908   FW 504.6 $C_{28}H_{32}N_4O_5$

Gly-Gly-Leu
N-CBZ-Gly-Gly-Leu *p*-NA

Synonyms: Subtilisin Substrate; Neutral Endopeptidase Substrate

Sigma C 3022   FW 499.5 $C_{24}H_{29}N_5O_7$ Sensitive chromogenic substrate; Lyublinskaya, L et al, *Anal Biochem*, 62:371, 1974; Wilk, S & Orlowski, M, *Biochem Biophys Res Commun*, 90:1, 1979

Gly-Gly-Leu
N-CBZ-Gly-Gly-Leu

Sigma C 8501   FW 379.4 $C_{18}H_{25}N_3O_6$

Gly-Gly-Leu

Sigma G 9503   FW 245.3 $C_{10}H_{19}N_3O_4$

Gly-Gly-Leu
CBZ-Gly-Gly-Leu-pNA

USBio C2098-75   MW 499.5 $C_{24}H_{29}N_5O_7$ ≥99%

Gly-Gly-Lys

Bachem H-3445.0250   MW 260.29 $C_{10}H_{20}N_4O_4$ Store at -15°C

Gly-Gly-Lys-Ala-Ala

Bachem H-3450.0050   MW 402.45 $C_{16}H_{30}N_6O_6$ Store at -15°C

Gly-Gly-Lys-Arg

Bachem H-3455.0250   MW 416.48 $C_{16}H_{32}N_8O_5$ Store at -15°C

Gly-Gly-Lys-Pro-Asp-Leu-Arg-Pro-Cys-His-Pro-Pro-Cys-His-Tyr-Ile-Pro-Arg-Pro-Lys-Pro-Arg

Synonyms: Peptide I of *T. wagleri* Venom; Waglerin I

Sigma P 2312   *T. wagleri* venom   FW 2520.0   ≥80% (HPLC); peptide content:~60% | Disulfide bonds: 9-13; bioactive peptide

Calbiochem 681655   *Trimeresurus wagleri*   MW 2520 $C_{112}H_{175}N_{37}O_{26}S_2$ >95% (HPLC); lyophilized solid; soluble in water; LD_{50}≤2000 mg/kg | Disulfide bonds: Cys^9-Cys^{13}; potent neurotoxin that blocks neuromuscular transmission in mice while having limited effect in rats; acts on both presynaptic & postsynaptic sites of the mouse motor endplate; acts on the benzodiazepine site of $GABA_A$ receptor/channel complex to increase its affinity for GABA agonists; Lin, WW et al, *Toxicon*, 33:111, 1995; Tsai, MC et al, *Toxicon*, 33:363, 1995; McArdle, JJ et al, *Soc Neurosci Abstr*, 18:969, 1992; Ye, JH & McArdle, JJ, *J Pharmacol Exp Ther*, 282:74, 1997; Schmidt, JJ & Weinstein, SA, *Toxicon*, 33:1043, 1995

Gly-Gly-Lys-Pro-Asp-Leu-Arg-Pro-Cys-Tyr-Pro-Pro-Cys-His-Tyr-Ile-Pro-Arg-Pro-Lys-Pro-Arg

Synonyms: Peptide II of *T. wagleri* Venom

Sigma P 2437   *T. wagleri* venom   FW 2546.0   ≥97% (HPLC); peptide content:~65% | Disulfide bonds: 9-13; bioactive peptide; homologous to Peptide I except at position 10 (Tyr for His)

Gly-Gly-Met

Bachem H-3460.0250   MW 263.32 $C_9H_{17}N_3O_4S$ Store at -15°C

Gly-Gly-Nva
Z-Gly-Gly-Nva

Bachem C-1790.0250   MW 365.39 $C_{17}H_{23}N_3O_6$ Store at -15°C

Gly-Gly-Phe
Z-Gly-Gly-Phe

Bachem C-1795.0001   MW 413.43 $C_{21}H_{23}N_3O_6$ Store at -15°C

Gly-Gly-Phe

Bachem H-3465.0001   MW 279.3 $C_{13}H_{17}N_3O_4$ Store at -15°C

Gly-Gly-Phe
Glutaryl-Gly-Gly-Phe-AMC

Bachem I-1200.0050   MW 550.57 $C_{28}H_{30}N_4O_8$ Store at -15°C

Gly-Gly-Phe
Z-Gly-Gly-Phe-βNA

Bachem K-1185.0100   Store at RT

Gly-Gly-Phe
Glutaryl-Gly-Gly-Phe-βNA

Bachem K-1280.0050   MW 518.57 $C_{28}H_{30}N_4O_6$ Store at -15°C

Gly-Gly-Phe
Suc-Gly-Gly-Phe-pNA

Bachem L-1415.0050   Store at -15°C

Gly-Gly-Phe
Z-Gly-Gly-Phe-CMK

Bachem N-1030.0025   MW 445.9 $C_{22}H_{24}ClN_3O_5$ Store at -15°C

Gly-Gly-Phe

ICN 157255   MW 279.3 $C_{13}H_{17}N_3O_4$ Crystalline

Gly-Gly-Phe
N-CBZ-Gly-Gly-Phe CMK

Synonyms: Chymotrypsin Inhibitor

Sigma C 7533   FW 445.9 $C_{22}H_{24}N_3O_5Cl$ Site-specific inhibitor of chymotrypsin; Segal, DM et al, *Biochemistry*, 10:3728, 1971

Gly-Gly-Phe
N-CBZ-Gly-Gly-Phe

Sigma C 9001   FW 413.4 $C_{21}H_{23}N_3O_6$

Gly-Gly-Phe

Sigma G 1379   FW 279.3 $C_{13}H_{17}N_3O_4$

Gly-Gly-Phe
N-Glutaryl-Gly-Gly-Phe β-Naphthylamide

Synonyms: Chymotrypsin Substrate

Sigma G 4636   FW 518.6 $C_{28}H_{30}N_4O_6$ Rinderknecht, H et al, *Clin Chim Acta*, 59:139, 1975; Horsthemke, B & Bauer, K, *Biochemistry*, 19:2867, 1980

Gly-Gly-Phe
N-Succinyl-Gly-Gly-Phe p-NA

Synonyms: Chymotrypsin Substrate; Protease B Substrate, *S. griseus*

Sigma S 1899   FW 499.5 $C_{23}H_{25}N_5O_8$ Achstetter, T et al, *Arch Biochem Biophys*, 207:445, 1981

Gly-Gly-Phe-Gly
BOC-Gly-Gly-Phe-Gly

Bachem A-4450.0250   Store at -15°C

Gly-Gly-Phe-Leu

Synonyms: Enkephalin, ((des-Tyr[1])-Leu)-; Enkephalin, (des-Tyr[1])-Leu-; Enkephalin, (des-Tyr[1],Leu[5])-; Enkephalinase Inhibitor

American Peptide 30-0-73   MW 392.6 $C_{19}H_{28}N_4O_5$ A tetrapeptide important for the study of enkephalinase inhibition

BACHEM N-1175.0025   MW 392.46 $C_{19}H_{28}N_4O_5$ Store at -15°C

ICN 190425   Malfroy, B etal, *Nature*, 276:523, 1978; Fournie-Zaluski, MC, etal, *BBRC*, 91:130, 1979; Malfroy, B etal, *Eur J Pharmac*, 52:209, 1979; Malfroy, B etal, *Eur J Pharmac*, 57:79, 1979; Schwartz, C etal, *Life Sci*, 29:1715, 1981

Gly-Gly-Phe-Leu Acetate Salt

Synonyms: Enkephalin, (des-Tyr[1]-Leu)-; Enkephalinase Inhibitor

Sigma E 7255   FW 392.5 (free base) $C_{19}H_{28}N_4O_5$ ≥97% (HPLC) | Bioactive peptide; Malfroy, B et al, *Nature*, 276:523, 1978; Fournie-Zaluski, MC et al, *Biochem Biophys Res Commun*, 91:130, 1979

Gly-Gly-Phe-Leu Amide Acetate Salt

Synonyms: Enkephalinamide, (des-Tyr[1],Leu[5])-; Enkephalin

ICN 195653   MW 451.5 $C_{19}H_{29}N_5O_4 \cdot C_2H_4O_2$ ≥97%

Sigma E 3506   FW 391.5 (free base); $C_{19}H_{29}N_5O_4$ ≥95% (HPLC) | Bioactive peptide

Gly-Gly-Phe-Leu-Arg-Arg-Ile

Synonyms: Dynorphin A (2-8), (des-Tyr[1])-

Bachem H-3788.0001   Store at -15°C

American Peptide 26-4-69   Porcine   MW 818.0 $C_{37}H_{63}N_{13}O_8$

ICN 154523   Porcine   MW 818.1

Gly-Gly-Phe-Leu-Arg-Arg-Ile-Arg-Pro-Lys-Leu

Synonyms: Dynorphin A (2-12)

American Peptide 26-4-55   Porcine   MW 1312.6 $C_{60}H_{105}N_{21}O_{12}$

Gly-Gly-Phe-Leu-Arg-Arg-Ile-Arg-Pro-Lys-Leu-Lys

Synonyms: Dynorphin A (2-13), (des-Tyr[1])-; Dynorphin κ-Agonist (2-13)

ICN 154525   Porcine   MW 1441

ICN 195610   Porcine   MW 1440.8   97%; ~65% peptide

Sigma D 4399   Porcine   FW 1440.8   ≥97% (HPLC); peptide content:~65% | Bioactive peptide

Gly-Gly-Phe-Leu-Arg-Arg-Ile-Arg-Pro-Lys-Leu-Lys-Trp-Asp-Asn-Gln

Synonyms: Dynorphin A (2-17)

Bachem H-2988.0001   MW 1984.34 $C_{90}H_{146}N_{30}O_{21}$ Store at -15°C

American Peptide 26-4-26   Porcine   MW 1983.4 $C_{90}H_{147}N_{31}O_{20}$

ICN 154520   Porcine   MW 1984.6

Gly-Gly-Phe-Leu-Arg-Arg-Ile-Arg-Pro-Lys-Leu-Lys-Trp-Asp-Asn-Gln Amide

Synonyms: Dynorphin A (2-17)

American Peptide 26-4-22   Porcine   MW 1983.4 $C_{90}H_{147}N_{31}O_{20}$

Gly-Gly-Phe-Met

Synonyms: Enkephalin, ((des-Tyr[1])-Met)-; Lipotropin (62-65), β-; Enkephalin, (des-Tyr[1])-Met-; Enkephalin, (des-Tyr[1],Met[5])-; Enkephalinase Inhibitor

American Peptide 30-0-36   MW 410.5 $C_{18}H_{26}N_4O_5S$

Bachem N-1180.0025   MW 410.49 $C_{18}H_{26}N_4O_5S$ Store at -15°C

ICN 190433   Malfroy, B etal, *Nature*, 276:523, 1978; Fournie-Zaluski, MC etal, *BBRC*, 91:130, 1979; Swerts, JP etal, *Eur J Pharmac*, 53:209, 1979; Swerts, JP etal, *Eur J Pharmac*, 57:279, 1979; Schwartz, JC etal, *Life Sci*, 29:1715, 1981

Gly-Gly-Phe-Met Acetate Salt

Synonyms: Enkephalin, (des-Tyr[1]-Met)-

Sigma E 2256   FW 410.5 (free base); $C_{18}H_{26}N_4O_5S$ ≥97% (HPLC) | Bioactive peptide; important for the study of enkephalinase inhibition; Malfroy, B et al, *Nature*, 276:523, 1978; Fournie-Zaluski, MC et al, *Biochem Biophys Res Commun*, 91:130, 1979

Gly-Gly-Phe-Met Amide Acetate Salt

Synonyms: Enkephalinamide, (des-Tyr[1],Met[5])-; Enkephalin

ICN 195658   MW 469.5 $C_{18}H_{27}N_5O_4S \cdot C_2H_4O_2$ ≥97%

Sigma E 3631   FW 409.5 (free base); $C_{18}H_{27}N_5O_4S$ ≥97% (HPLC) | Bioactive peptide

Gly-Gly-Phe-Met-Thr-Ser-Glu-Lys-Ser-Gln-Thr-Pro-Leu-Val-Thr-Leu

Synonyms: Endorphin, (des-Tyr[1])-γ-; Endorphin, (des-Tyr[1]-γ-)-; Lipotropin (62-77), β-

ICN 190419   MW 1695.9   Reported to have neuroleptic-like activity in rats & in schizophrenic patients; de Weid, D etal, *Eur J Pharmac*, 49:427, 1978; de Weid, D etal, *Lancet*, May 13:1046, 1978; de Weid, D etal, *Brain Res*, 179:85, 1979; Pedigo, NW, etal, *Life Sci*, 24:1645, 1979; Versteed, DHG etal, *Brain Res*, 179:85, 1979; Verhoef, J etal, *Life Sci*, 26:851, 1980

Sigma E 3005   FW 1695.9   ≥97% (HPLC) | Bioactive peptide; Verhoef, J et al, *Life Sci*, 26:851, 1980; DeWied, D et al, *Eur J Pharmacol*, 49:427, 1978; Versteed, DHG, *Brain Res*, 179:85, 1979

Gly-Gly-Phe-Met-Thr-Ser-Glu-Lys-Ser-Gln-Thr-Pro-Leu-Val-Thr-Leu-Phe-Lys-Asn-Ala-Ile-Ile-Lys-Asn-Ala-Tyr-Lys-Lys-Gly-Glu

Synonyms: Endorphin, (des-Tyr[1])-β-; Lipotropin (62-91), β-

American Peptide 28-1-30   Human   MW 3302.0
$C_{149}H_{242}N_{38}O_{44}S$

Gly-Gly-Phe-Phe
Gly-Gly-Phe-Phe-OEt

Bachem M-1455.0250   MW 454.53 $C_{24}H_{30}N_4O_5$ Store at -15°C

Gly-Gly-Pro

Bachem H-3470.0250   MW 229.24 $C_9H_{15}N_3O_4$ Store at -15°C

Gly-Gly-Pro
N-CBZ-Gly-Gly-Pro

Sigma C 9126   FW 363.4 $C_{17}H_{21}N_3O_6$

Gly-Gly-Pro-Ala

Bachem H-3475.0250   MW 300.32 $C_{12}H_{20}N_4O_5$ Store at -15°C

Gly-Gly-Sar

Bachem H-3480.0250   MW 203.2 $C_7H_{13}N_3O_4$ Store at -15°C

ICN 157256   MW 203.2 $C_{17}H_{13}N_3O_4$ Crystalline

Gly-Gly-Ser
N-CBZ-Gly-Gly-Ser

Sigma C 9251   FW 353.3 $C_{15}H_{19}N_3O_7$

Gly-Gly-Ser-Ala

Bachem H-1536.0250   Store at -15°C

Gly-Gly-Ser-Leu-Tyr-Ser-Phe-Gly-Leu
Gly-Gly-Ser-Leu-Tyr-Ser-Phe-Gly-Leu Amide

Synonyms: Allostatin II

American Peptide 87-7-30   MW 899.0 $C_{42}H_{62}N_{10}O_{12}$
Woodhead, AP et al, *PNAS*, 86:5997, 1989

Gly-Gly-Ser-Leu-Tyr-Ser-Phe-Gly-Leu Amide

Synonyms: Allatostatin III, Type A

Bachem H-8075.0001   MW 899.01 $C_{42}H_{62}N_{10}O_{12}$ Store at -15°C

Gly-Gly-Trp
Z-Gly-Gly-Trp

Bachem C-1800.0250   MW 452.47 $C_{23}H_{24}N_4O_6$ Store at -15°C

Gly-Gly-Trp

Bachem H-3485.0250   MW 318.33 $C_{15}H_{18}N_4O_4$ Store at -15°C

Gly-Gly-Trp-Ser-His-Trp

Synonyms: Thrombospondin Analog

Bachem H-1578.0005   MW 728.77 $C_{35}H_{40}N_{10}O_8$ Store at -15°C

Gly-Gly-Tyr

Bachem H-3495.0250   MW 295.3 $C_{13}H_{17}N_3O_5$ Store at -15°C

ICN 101868   MW 295.3 $C_{13}H_{17}N_3O_5$

Gly-Gly-Tyr-Ala

Bachem H-1538.0250   MW 366.37 $C_{16}H_{22}N_4O_6$ Store at -15°C

Gly-Gly-Tyr-Arg

Synonyms: Papain Inhibitor

American Peptide 81-5-50   MW 451.5 $C_{19}H_{29}N_7O_6$ Used for affinity chromatography; Funk, M et al, *Int J Peptide & Protein Res*, 13:296, 1979

Bachem N-1080.0050   MW 451.48 $C_{19}H_{29}N_7O_6$ Store at -15°C

ICN 153112   MW 451.5 $C_{19}H_{29}N_7O_6$ Useful ligand for affinity chromatographic purification of papain; Funk, MO etal, *Int J Pept Prot Res*, 13:296, 1979

Sigma G 5386   FW 451.5 $C_{19}H_{29}N_7O_6$ ≥97% (HPLC) | Bioactive peptide; useful for affinity chromatography; Funk, MO et al, *Int J Pept Prot Res*, 13:296, 1979

Gly-Gly-Tyr-Arg AcOH Hydrate

Synonyms: Papain Affinity Ligand

Peptides International OGG-3119   MW 529.55 $C_{19}H_{29}N_7O_6 \cdot CH_3COOH \cdot H_2O$ >98% (HPLC); colorless crystalline powder | Funk, MO et al, *Int J Peptide Protein Res*, 13:29, 1979

Gly-Gly-Val
Z-Gly-Gly-Val

Bachem C-1810.0250   MW 365.39 $C_{17}H_{23}N_3O_6$ Store at -15°C

Gly-Gly-Val

Bachem H-3500.0250   MW 231.25 $C_9H_{17}N_3O_4$ Store at -15°C

ICN 101869   MW 231.3 $C_9H_{17}N_3O_4$ Crystalline

Gly-Gly-Val
N-CBZ-Gly-Gly-Val

Sigma C 9376   FW 365.4 $C_{17}H_{23}N_3O_6$

Gly-Gly-Val-Val-Lys-Asn-Asn-Phe-Val-Pro-Thr-Asn-Val-Gly-Ser-Lys-Ala-Phe Amide

Synonyms: Calcitonin Gene Related Peptide (20-37)

American Peptide 22-5-15   Human   MW 1834.1
$C_{83}H_{132}N_{24}O_{23}$

Gly-His
Z-Gly-His

Bachem C-1815.0250   MW 346.34 $C_{16}H_{18}N_4O_5$ Store at -15°C

Gly-His
Z-Gly-His-NHNH₂

Bachem C-1820.0001   MW 360.37 $C_{16}H_{20}N_6O_4$ Store at -15°C

Gly-His

Bachem G-2120.0250   MW 212.21 $C_8H_{12}N_4O_3$ Store at -15°C

Gly-His
Gly-L-His Hydrochloride Hydrate

Fluka 50280   MW 248.67 $C_8H_{12}N_4O_3 \cdot HCl$ ≥99% (titration); contains 1 mole water

Gly-His

Sigma G 1627   FW 212.2 (free base) $C_8H_{12}N_4O_3$

Gly-His Free Base

ICN 157257   MW 212.2 $C_8H_{12}N_4O_3$ Crystalline

Gly-His Hydrochloride

Bachem G-2125.0001   MW 248.67 $C_8H_{12}N_4O_3 \cdot HCl$ Store at -15°C

ICN 157258   MW 248.7 $C_8H_{12}N_4O_3 \cdot HCl$ Crystalline

Sigma G 1502   FW 248.7 $C_8H_{12}N_4O_3 \cdot HCl$

Gly-His-Arg-AMC

Bachem I-1855.0050   MW 515.49 $C_{24}H_{21}N_9O_5$ Store at -15°C

Gly-His-Arg-Pro

Bachem H-2940.0025   MW 465.51 $C_{19}H_{31}N_9O_5$ Store at -15°C

Gly-His-Arg-Pro Acetate Salt

Synonyms: Fibrin β-Chain (1-4)

Sigma G 8636   Human   FW 465.5 (free base) $C_{19}H_{31}N_9O_5$ ≥97% (HPLC) | Bioactive peptide; Laudano, AP & Doolittle, RF, *Proc Natl Acad Sci USA*, 75:3085, 1978; Laudano, AP & Doolittle, RF, *Biochemistry*, 19:1013, 1980

Gly-His-Gln-Ala-Ala-Met-Gln-Met-Leu-Lys-Glu-Thr-Ile-Asn-Glu-Glu

Synonyms: P25 (193-208); GAG P24 CA (61-76); HTLV I Peptide

Neosystem SC296   MW 1830.06   HTLV I peptide from subtype P25 (GAG P24 CA)

Gly-His-Gly

Bachem H-3505.0250   MW 269.26 $C_{10}H_{15}N_5O_4$ Store at -15°C

ICN 157259   MW 285.3 $C_{10}H_{15}N_5O_5$ Crystalline | For NMR studies of DNA-oligopeptide binding

Sigma G 5504   FW 269.3 $C_{10}H_{15}N_5O_4$

Gly-His-Leu
Benzoyl-Gly-L-His-L-Leu

Synonyms: Angiotensin Converting Enzyme Substrate; Hippuryl-L-His-L-Leu

ICN 151262   MW 429.5 $C_{21}H_{27}N_5O_5$ Cushman, DW & HS Cheung, *Biochem Pharmacol*, 20:1637, 1971

Gly-His-Leu
Bz-Gly-His-Leu Hydrate

Synonyms: Angiotensin I Converting Enzyme Substrate; Hippuryl-His-Leu

Peptides International SGL-3064   MW 447.5 $C_{21}H_{27}N_5O_5 \cdot H_2O$ >99% (HPLC); amorphous powder | Cushman, DW & HS Cheung, *Biochem Pharmacol*, 20:1637, 1971

Gly-His-Lys

Synonyms: Liver Cell Growth Factor; Growth Factor, Liver Cell

American Peptide 83-0-02   MW 340.4 $C_{14}H_{24}N_6O_4$ Growth-modulating plasma tripeptide; produces a variety of responses ranging from the stimulation of growth & differentiation of outright toxicity when added at nanomolar concentrations to the cultured systems

Bachem H-3510.0250   MW 340.38 $C_{14}H_{24}N_6O_4$ Store at -15°C

Calbiochem 05-22-1401   MW 340.4 $C_{14}H_{24}N_6O_4$ ≥95% (HPLC); lyophilized solid; soluble at pH 2.0 | Serum factor used in mammalian tissue culture; active at nanogram levels; Pickart, L, *In Vitro*, 17:459, 1981

ICN 153114   Pickart, L etal, *BBRC*, 54:562, 1973

Gly-His-Lys Acetate Salt

Synonyms: Growth Factor, Liver Cell

Sigma G 1887   FW 340.4 (free base) $C_{14}H_{24}N_6O_4$ ≥97% (TLC); peptide content:~70% | Bioactive peptide; Pickart, L et al, *Biochem Biophys Res Commun*, 54:562, 1973; Pickart, L et al, *J Chromat*, 175:65, 1979

Sigma G 7387   Synthetic $C_{14}H_{24}N_6O_4C_2H_4O_2$ Sterilized by γ-irradiation; cell culture tested

Gly-His-Lys AcOH Hydrate

Synonyms: Liver Cell Growth Factor

Peptides International PLC-4022   Synthetic   MW 418.45 $C_{14}H_{24}N_6O_4 \cdot CH_3COOH \cdot H_2O$ >99% (HPLC); lyophilized amorphous powder; bulk | Bioactive; Pickart, L et al, *BBRC*, 54:562, 1973;

Gly-His-Lys-Ala-Arg-Val-Leu-Ala-Glu-Ala-Met-Ser-Gln-Val-Thr-Asn

Synonyms: P25 (357-372); GAG P24 CA (224-231)/P2 (1-9); HTLV I Peptide

Neosystem SC311   MW 1711.95   HTLV I peptide from subtype P25 (GAG P24 CA); due to the presence of a Cys this peptide can contain the dimeric form

Gly-Hyp

Synonyms: Prolidase Substrate

Bachem G-2130.0250   MW 188.18 $C_7H_{12}N_2O_4$ Store at -15°C

ICN 157260   MW 188.2 $C_7H_{12}N_2O_4$ Crystalline

Gly-Hyp-Glu

Bachem H-3520.0250   Store at -15°C

Gly-Ile
Z-Gly-Ile

Bachem C-1830.0001   MW 322.36 $C_{16}H_{22}N_2O_5$ Store at -15°C

Gly-Ile

Bachem G-2135.0001   MW 188.23 $C_8H_{16}N_2O_3$ Store at -15°C

Gly-Ile
Gly-DL-Ile

ICN 101875   MW 188.2 $C_8H_{16}N_2O_3$ 99%

Gly-Ile

ICN 101876   MW 188.2 $C_8H_{16}N_2O_3$ Crystalline

Gly-Ile
Borotrimethyl-Gly-Ile-OMe

ICN 154639   MW 244.14 $BC_{11}H_{25}N_2O_3$ Liquid; >97%

Gly-Ile
Boro-Gly-Ile Amide

ICN 154648   MW 202.06 $BC_8H_{19}N_2O_3$ >97%

Gly-Ile
N-CBZ-Gly-Ile

Sigma C 9501   FW 322.4 $C_{16}H_{22}N_2O_5$

Gly-Ile

Sigma G 1752   FW 188.2 $C_8H_{16}N_2O_3$

Gly-Ile-Ala
Z-Gly-Ile-Ala

Bachem C-3245.0250   Store at -15°C

Gly-Ile-Ala-Gly-His-Thr-Tyr-Leu-Gln-Ala-Ser-Glu-Lys-Phe-Lys-Nle-Trp-Gly-Ala-Glu

Synonyms: HSV I UL 26 Open Reading Frame (238-257), (Nle^{25}3)-; HSV I Proteinase Substrate II

Bachem M-1965.0500   Store at -15°C

Sigma H 2029   FW 2206.5   ≥90% (HPLC); peptide content:~60% | Bioactive peptide; contains the putative cleavage site essential for HSV-I proteinase maturational release; Liu, F & Roizman, B, *J Virol*, 65:206, 1991; Welch, AR et al, *J Virol*, 65:4091, 1991; Welch, AR et al, *Proc Natl Acad Sci USA*, 88:10792, 1991

Gly-Ile-Arg-Tyr-Pro-Leu-Thr-Phe-Gly-Trp-Cys-Phe-Lys-Leu-Val-Pro

Synonyms: NEF MN (134-149); HIV I Peptide

Neosystem SC658   MW 1897.00   NEF peptide from HIV-1 subtype MN; due to the presence of a Cys this peptide can contain the dimeric form

Gly-Ile-Arg-Tyr-Pro-Lys-Thr-Phe-Gly-Trp-Leu-Trp-Lys-Leu-Val

Synonyms: NEF (164-178); SIVmac251 Peptide

Neosystem SC619   MW 1864.26   SIVmac251 peptide from subtype NEF

Gly-Ile-Cys-Ala-Cys-Arg-Arg-Arg-Phe-Cys-Pro-Asn-Ser-Glu-Arg-Phe-Ser-Gly-Tyr-Cys-Arg-Val-Asn-Gly-Ala-Arg-Tyr-Val-Arg-Cys-Cys-Ser-Arg-Arg

Synonyms: CSI; Corticostatin

Bachem H-9045.0500   MW 4003.69 $C_{163}H_{265}N_{63}O_{44}S_6$ Store at -15°C

American Peptide 34-7-25   Rabbit   MW 3997.7 $C_{163}H_{259}N_{63}O_{44}S_6$ The cationic Arg- & Cys-rich peptide was isolated from rabbit fetal & adult lung; inhibitor of ACTH-stimulated corticosterone production in rat

Gly-Ile-Gly-Ala-Ser-Ile-Leu-Ser-Ala-Gly-Lys-Ser-Ala-Leu-Lys-Gly-Leu-Ala-Lys-Gly-Leu-Ala-Glu-His-Phe-Ala-Asn Amide

Synonyms: Bombinin-Like Peptide I

Bachem H-8625.0500   Store at -15°C

Gly-Ile-Gly-Ala-Val-Leu-Lys-Val-Leu-Thr-Thr-Gly-Leu-Pro-Ala-Leu-Ile-Ser-Trp-Ile-Lys-Arg-Lys-Arg-Gln-Gln

Syno Melittin *nyms*:

American Peptide 86-4-11   MW 2847.5 (free acid) $C_{131}H_{228}N_{38}O_{32}$

Gly-Ile-Gly-Ala-Val-Leu-Lys-Val-Leu-Thr-Thr-Gly-Leu-Pro-Ala-Leu-Ile-Ser-Trp-Ile-Lys-Arg-Lys-Arg-Gln-Gln Amide

Synonyms: Melittin; Protein Kinase C Inhibitor

Bachem H-4310.0001   MW 2846.5 $C_{131}H_{229}N_{39}O_{31}$ Store at -15°C

Neosystem SC1271   MW 2846.49   Bioactive; antibacterial/antimicrobial peptide; the principal toxic component of bee venom; a cationic, amphipathic linear peptide of high antimicrobial activity;kinase inhibitor, selective for proteinkinase C & AMP-dependent proteinkinases; Habermann, VE & Jentsch, J, Hoppe-Seyler's *Z Physiol Chem*, 348:37-50, 1967; Raynor, RL et al, *JBC*, 266:2753-2758, 1991; Beven, L & Wroblewski, H, *Res Microbiol*, 148:163-175, 1997

Alexis 162-006   Bee venom   MW 2846.5 $C_{131}H_{229}N_{39}O_{31}$ ≥85% (HPLC); phospholipase A$_2$ <5 U/mg solid; faint yellow powder; soluble in water; toxic | Causes disruption of normal cellular activity & cell lysis; binds calmodulin in a calcium-dependent manner, activates phospholipase A2 & inhibits protein kinase C by binding to the catalytic domain in a Mg-ATP sensitive manner; used for affinity purification of several Ca^{2+}-binding proteins; Kincaid, RL & Coulson, CC, *BBRC*, 133:256, 1985; Baudier, J et al, *Biochemistry*, 26:2886, 1987

American Peptide 86-4-10   Bee venom   MW 2846.5 $C_{131}H_{229}N_{39}O_{31}$ Binds calmodulin in a Ca^{2+}-dependent manner, activates phospholipase A$_2$ & inhibits protein kinase C by binding to the catalytic domain in a Mg-ATP sensitive manner; Raynor et al, *JBC*, 266:2758, 1991; Schweitz, *Toxicon*, 22:308, 1984

Sigma M 2272   Bee venom   ~85% (HPLC); phospholipase A$_2$ impurity:<5 U/mg solid | The principle hemolytic component of honey bee venom; binds calmodulin in a Ca^{++} dependent manner; inhibits Na$^+$ -K$^+$-ATPase; Jarrett, HW & Madhavan, R, *J Biol Chem*, 266:362, 1991; Cuppoletti, J & Abbott, AJ, *Arch Biochem Biophys*, 283:249, 1990; Baudier, J et al, *Biochem*, 26:2886, 1987

Sigma M 7391   Bee venom   ~70% (HPLC); phospholipase A$_2$ impurity:<20 U/mg solid

Fluka 63650   Honey bee venom   MW 2846.53 $C_{131}H_{229}N_{39}O_{31}$ ~70.0% (HPLC) | Main component of the honey bee venom; basic hydrophobic peptide which alters membrane permeability; Dawson, CR et al, *Biochim Biophys Acta*, 510:75, 1978; Vesely, DL, *Science*, 213:359, 1981; Maulet, Y & Cox, JA, *Biochemistry*, 22:61, 1983; Kincaid, RL, *Meth Enzymol*, 139:3, 1987

Sigma M 4171   Honey bee venom, synthetic   ≥97% (HPLC)

Alexis 162-007   Synthetic   MW 2846.5 $C_{131}H_{229}N_{39}O_{31}$ ≥98%; white lyophilized powder; toxic | Causes disruption of normal cellular activity & cell lysis; binds calmodulin in a calcium-dependent manner, activates phospholipase A2 & inhibits protein kinase C by binding to the catalytic domain in a Mg-ATP sensitive manner; used for affinity purification of several Ca^{2+}-binding proteins; Kincaid, RL & Coulson, CC, *BBRC*, 133:256, 1985; Baudier, J et al, *Biochemistry*, 26:2886, 1987

Gly-Ile-Gly-Ala-Val-Leu-Thr-Thr-Gly-Leu-Pro-Ala-Leu-Ile-Ser-Trp-Ile-Lys-Arg-Lys-Arg-Gln-Gln Amide

Synonyms: Melittin

Calbiochem 444605   Bee venom   MW 2847 $C_{131}H_{229}N_{39}O_{31}$ >97% (HPLC); lyophilized solid; soluble in 5% acetic acid; harmful:LD$_{50}$ ≤2000 mg/kg | 26-residue polypeptide that binds calmodulin in a Ca^{2+}-dependent manner, activates phospholipase A$_2$ & inhibits protein kinase C (IC$_{50}$=5-7 μM) by binding to the catalytic domain in a Mg^{2+} -ATP sensitive manner; used for affinity purification of several Ca^{2+}-binding proteins; activates G$_i$α-1 & G$_{11}$α activities & inhibits G$_s$ activity; Gravitt, KR et al, *Biochem Pharmacol*, 47:425, 1994; Fukushima, N et al, *Peptides*, 19:811, 1998; Zhu, J et al, *Eur J Pharmacol*, 268:279, 1994; Baudier, J et al, *Biochemistry*, 26:2886, 1987; *Merck Index*, 12:5867; Kincaid, RL & Coulson, CC, *Biochem Biophys Res Commun*, 133:256, 1985

Gly-Ile-Gly-Asp-Pro-Val-Thr-Cys-Leu-Lys-Ser-Gly-Ala-Ile-Cys-His-Pro-Val-Phe-Cys-Pro-Arg-Arg-Tyr-Lys-Gln-Ile-Gly-Thr-Cys-Gly-Leu-Pro-Gly-Thr-Lys-Cys-Cys-Lys-Lys-Pro

Synonyms: Defensin II, β-; Antibacterial Peptide

Peptides International PDF-4338-s   Human synthetic   MW 4328.2 $C_{188}H_{305}N_{55}O_{50}S_6$ >99% (HPLC); lyophilized amorphous powder | Bioactive; specific for gram-negative bacteria; effective for *Candida albicans*; Harder, J et al, *Nature*, 387:861, 1997; Hiratsuka, T et al, *BBRC*, 249:943, 1998

Gly-Ile-Gly-Gly-Phe-Ile-Lys-Val-Arg-Gln-Tyr-Asp-Gln-Ile-Leu

Synonyms: POL (49-63); PR (49-63); HTLV I Peptide

Neosystem SC704   MW 1707.00   HTLV I peptide from subtype POL (PR/RT)

Gly-Ile-Gly-Lys-Phe-Leu-His-Ala-Ala-Lys-Lys-Phe-Ala-Lys-Ala-Phe-Val-Ala-Glu-Ile-Met-Asn-Ser Amide

Synony Magainin II, (Ala[8,13,18])-*ms:*

Sigma M 8155   FW 2478.0   ≥97% (HPLC) | Bioactive peptide; magainin analog with greater anti-microbial activity; Hao-Chia Chen et al, *FEBS Lett*, 236:462, 1988

Gly-Ile-Gly-Lys-Phe-Leu-His-Ser-Ala-Gly-Lys-Phe-Gly-Lys-Ala-Phe-Val-Gly-Glu-Ile-Met-Lys-Ser

Synonyms: Magainin I

Bachem H-6565.0500   Store at -15°C

ICN 191465   Giovannini, MG etal, *Biochem J*, 243:113, 1987; Zasloff, M, *PNAS*, 84:5449, 1987

Gly-Ile-Gly-Lys-Phe-Leu-His-Ser-Ala-Gly-Lys-Phe-Gly-Lys-Ala-Phe-Val-Gly-Glu-Ile-Met-Asn-Ser

Synonyms: Magainin II

ICN 191466   Cannon, M, *Nature*, 328:478, 1987; Zasloff, M, *PNAS*, 84:5449, 1987

Gly-Ile-Gly-Lys-Phe-Leu-His-Ser-Ala-Gly-Lys-Phe-Gly-Lys-Ala-Phe-Val-Gly-Glu-Ile-Met-Lys-Ser

Synonyms: Magainin I

Sigma M 7152   FW 2409.9   ≥97% (HPLC) | Bioactive peptide; antibiotic peptide; Zasloff, M, *Proc Natl Acad Sci USA*, 84:5449, 1987; Giovannini, MG et al, *Biochem J*, 243:113, 1987

Peptides International PMG-4196-v   *Xenopus laevis* (frog) synthetic   MW 2409.9 $C_{112}H_{177}N_{29}O_{28}S$ >95% (HPLC); lyophilized amorphous powder | Bioactive; potent antimicrobial peptide; Zasloff, *M, PNAS* USA, 84:5449, 1987

Gly-Ile-Gly-Lys-Phe-Leu-His-Ser-Ala-Lys-Lys-Phe-Gly-Lys-Ala-Phe-Val-Gly-Glu-Ile-Met-Asn-Ser

Synonyms: Magainin II

Bachem H-6570.0500   Store at -15°C

Sigma M 7402   FW 2466.9   ≥97% (HPLC) | Bioactive peptide; antibiotic peptide; Zasloff, M, *Proc Natl Acad Sci USA*, 84:5449, 1987; Giovannini, MG et al, *Biochem J*, 243:113, 1987

American Peptide 72-2-26   Frog skin   MW 2466.9 $C_{114}H_{180}N_{30}O_{29}S$ Antibiotic peptide; a hemolytic & antimicrobial peptide; Mor, A et al, *Biochemistry*, 30:8824, 1991

Gly-Ile-Leu-Gly-Phe-Val-Phe-Thr-Leu

Synonyms: Influenza Virus Matrix Protein (58-66)

Neosystem SC899   MW 966.19   Virus-related peptide; Bednarek, MA et al, *J Immunol*, 147:4047-4053, 1991

Gly-Ile-Pro-Cys-Leu-Cys-Asp-Ser-Asp-Gly-Pro-Ser-Val-Arg-Gly-Asn-Thr-Leu-Ser-Gly-Ile-Ile-Trp-Leu-Ala-Gly-Cys-Pro-Ser-Gly-Trp-His-Asn-Cys-Lys-Lys-His-Gly-Pro-Thr-Ile-Gly-Trp-Cys-Cys-Lys-Gln

Synonyms: Toxin II, Isoleucine Isotoxin; ATX II-Ile

Calbiochem 616393   *Anemonia sulcata*   MW 4954.8 $C_{214}H_{331}N_{63}O_{61}S_6$ ≥97% (HPLC); lyophilized solid; soluble in water; LD$_{50}$≤2000 mg/kg | Purified isoleucine isomer of Toxin II

Gly-Ile-Trp-His-His-Tyr
Gly-Ile-2-Nal-Trp-His-His-Tyr

Bachem H-4084.0001   Store at -15°C

Gly-Ile-Val-Cys-Leu-Cys-Asp-Ser-Asp-Gly-Pro-Ser-Val-Arg-Gly-Asn-Thr-Leu-Ser-Gly-Ile-Ile-Trp-Leu-Ala-Gly-Cys-Pro-Ser-Gly-Trp-His-Asn-Cys-Lys-Lys-His-Gly-Pro-Thr-Ile-Gly-Trp-Cys-Cys-Lys-Gln

Synonyms: Toxin II, Valine Isotoxin; ATX II-Val

Calbiochem 616391   *Anemonia sulcata*   MW 4940.8 $C_{213}H_{329}N_{63}O_{61}S_6$ >92% (HPLC); lyophilized solid; soluble in water; LD$_{50}$≤2000 mg/kg

Gly-Ile-Val-Glu-Gln-Cys-Cys-Ala-Ser-Val-Cys-Ser-Leu-Tyr-Gln-Leu-Glu-Asn-Tyr-Cys-Asn
Gly-Ile-Val-Glu-Gln-Cys(SO₃H)-Cys(SO₃H)-Ala-Ser-Val-Cys(SO₃H)-Ser-Leu-Tyr-Gln-Leu-Glu-Asn-Tyr-Cys(SO₃H)-Asn Ammonium Salt

Synonyms: Insulin Chain A, Oxidized

Sigma I 1633   Bovine insulin   FW 2531.6 (free acid)   ≥80% (HPLC); <1% chain B | Bioactive peptide; prepared by a modification of Sanger, F et al, *Biochem J*, 44:126, 1949

Gly-Ile-Val-Glu-Gln-Cys-Cys-Thr-Ser-Ile-Cys-Ser-Leu-Tyr-Gln-Leu-Glu-Asn-Tyr-Cys-Asn-Phe-Val-Asn-Gln-His-Leu-Cys-Gly-Ser-His-Leu-Val-Glu-Ala-Leu-Tyr-Leu-Val-Cys-Gly-Glu-Arg-Gly-Phe-Phe-Tyr-Thr-Pro-Lys-Thr
(A Chain:Gly-Ile-Val-Glu-Gln-Cys-Cys-Thr-Ser-Ile-Cys-Ser-Leu-Tyr-Gln-Leu-Glu-Asn-Tyr-Cys-Asn)-(B Chain:Phe-Val-Asn-Gln-His-Leu-Cys-Gly-Ser-His-Leu-Val-Glu-Ala-Leu-Tyr-Leu-Val-Cys-Gly-Glu-Arg-Gly-Phe-Phe-Tyr-Thr-Pro-Lys-Thr)

Synonyms: Insulin

American Peptide 20-1-10   Human   MW 5807.6 $C_{257}H_{383}N_{65}O_{77}S_6$ Disulfide bonds: Cys[6]-Cys[11], Cys[7]-Cys[28], Cys[20]-Cys[40]

Gly-Ile-Val-Glu-Gln-Cys-Cys-Thr-Ser-Ile-Cys-Ser-Leu-Tyr-Gln-Leu-Glu-Asn-Tyr-Cys-Asn-Phe-Val-Asn-Gln-His-Leu-Cys-Gly-Ser-His-Leu-Val-Glu-Ala-Leu-Tyr-Leu-Val-Cys-Gly-Glu-Arg-Gly-Phe-Phe-Tyr-Thr-Pro-Lys-Thr

Synonyms: Insulin A Chain

Peptides International PIN-4088-s, 4088-v   Human synthetic   MW 5807.7 $C_{257}H_{383}N_{65}O_{77}S_6$ >97% (HPLC); lyophilized amorphous powder | Bioactive; Disulfide bonds: Cys[A6]-Cys[A11], Cys[A7]-Cys[B7], Cys[A20]-Cys[B19]; Dorzbach, E (Ed), *Insulin I*, Handbook of Experimental Pharmacology, Vol 32(1), Springer-Verlag, Berlin, 1971; Hasselblatt, A & FV Bruchhausen (Eds), *Insulin II*, Handbook of Experimental Pharmacology, Vol 32(2), Springer-Verlag, Berlin, 1975; Morihara, K et al, *Nature*, 280:412, 1979; Morihara, K et al, *BBRC*, 92:396, 1980

Gly-L-Arg
Glutaryl-Gly-L-Arg-MCA

Synonyms: Urokinase Substrate

ICN 151192   MW 502.5 $C_{23}H_{30}N_6O_7$ Morita, T etal, *J Biochem*, 82:1495, 1977

Gly-Leu
Z-Gly-Leu

Bachem C-1835.0001   MW 322.36 $C_{16}H_{22}N_2O_5$ Store at -15°C

Gly-Leu
Z-Gly-Leu Amide

Bachem C-1845.0001   MW 321.38 $C_{16}H_{23}N_3O_4$ Store at -15°C

Gly-Leu
Ac-Gly-Leu

Bachem G-1030.0250   MW 230.26 $C_{10}H_{18}N_2O_4$ Store at -15°C

Gly-Leu
Ac-Gly-Leu Amide

Bachem G-1035.0250   MW 229.3 $C_{10}H_{19}N_3O_3$ Store at -15°C

Gly-Leu
Gly-DL-Leu

Bachem G-2145.0005   MW 188.23 $C_8H_{16}N_2O_3$ Store at -15°C

Gly-Leu
FA-Gly-Leu

Bachem M-1370.0050   MW 308.33 $C_{15}H_{20}N_2O_5$ Store at -15°C

Gly-Leu
FA-Gly-Leu Amide

Synonyms: FAGLA

Bachem M-1375.0050   MW 307.35 $C_{15}H_{21}N_3O_4$ Store at -15°C

Gly-Leu

Bachem M-1460.0005   MW 188.23 $C_8H_{16}N_2O_3$ Store at -15°C

Gly-Leu
Gly-L-Leu

Fluka 50300   MW 188.23 $C_8H_{16}N_2O_3$ ≥99% (titration); ≤0.1% residue on ignition; mp:250°C | Substrate for the assay of glycyl-L-leucine dipeptidase; Smith, EL, *Meth Enzymol*, 2:93, 1955

Gly-Leu
Gly-D-Leu

ICN 101877   MW 188.2 $C_8H_{16}N_2O_3$ Crystalline

Gly-Leu
Gly-DL-Leu

ICN 101879   MW 188.2 $C_8H_{16}N_2O_3$ Crystalline

Gly-Leu
Borotrimethyl-Gly-Leu-OMe

ICN 154638   MW 244.14 $BC_{11}H_{25}N_2O_3$ Liquid; >97%

Gly-Leu
Boro-Gly-Leu Amide

ICN 154647   MW 187.05 $BC_7H_{18}N_3O_2$ >97%

Gly-Leu

ICN 157261   MW 188.2 $C_8H_{16}N_2O_3$ Crystalline

Peptides International OGL-3022   MW 188.23 $C_8H_{16}N_2O_3$ >98% (HPLC); colorless crystalline powder

Gly-Leu
Z-Gly-Leu

Peptides International SGL-3019   MW 322.36 $C_{16}H_{22}N_2O_5$ >99% (HPLC); amorphous powder

Gly-Leu
Z-Gly-Leu Amide

Peptides International SGL-3037   MW 321.38 $C_{16}H_{23}N_3O_4$ >99% (HPLC); amorphous powder

Gly-Leu
N-CBZ-Gly-Leu

Sigma C 9626   FW 322.4 $C_{16}H_{22}N_2O_5$

Gly-Leu
N-FA-Gly-Leu Amide

Synonyms: Thermolysin Substrate

Sigma F 7383   FW 307.3 $C_{15}H_{21}N_3O_4$ Feder, J & Schuck JM, *Biochemistry*, 9:2784, 1970

Gly-Leu

Sigma G 2002   FW 188.2 $C_8H_{16}N_2O_3$

Gly-Leu
Gly-DL-Leu

Sigma G 8253   FW 188.2 $C_8H_{16}N_2O_3$

Gly-Leu
CBZ-Gly-Leu

USBio C2098-76   MW 322.4 $C_{16}H_{22}N_2O_5$ ≥99%

Gly-Leu Amide Hydrochloride

Bachem G-2150.0001   MW 223.7 $C_8H_{17}N_3O_2 \cdot HCl$ Store at -15°C

ICN 157262   Crystalline

Sigma G 4629   FW 223.7 $C_8H_{17}N_3O_2 \cdot HCl$

Gly-Leu-Ala
Z-Gly-Leu-Ala

Bachem C-3260.0250   MW 393.44 $C_{19}H_{27}N_3O_6$ Store at -15°C

Gly-Leu-Ala
FA-Gly-Leu-Ala

Bachem M-1800.0050   MW 379.42 $C_{18}H_{25}N_3O_6$ Store at -15°C

Gly-Leu-ala
Gly-DL-Leu-DL-Ala

Sigma G 3254   FW 259.3 $C_{11}H_{21}N_3O_4$

Gly-Leu-Arg-Hyp-Gly Amide

Synonyms: Luteinizing Hormone Releasing Hormone (6-10), (Hyp[9])-

Neosystem SC475   MW 513.59   Bioactive; hypothalamic releasing hormone

Gly-Leu-Arg-Pro-Gly
Gly-Leu-Arg-Dehydro-Pro-Gly Amide

Synonyms: Luteinizing Hormone Releasing Hormone (6-10), (Dehydro-Pro[9])-

Neosystem SC476   MW 496.59   Bioactive; hypothalamic releasing hormone

Gly-Leu-Arg-Pro-Gly Amide

Synonyms: Luteinizing Hormone Releasing Hormone (6-10)

Neosystem SC353   MW 497.60   Bioactive; hypothalamic releasing hormone; Mc Dermott, JR et al, *Peptides*, 4:25-30, 1983

Gly-Leu-Arg-Thr-Gln-Ser-Phe-Ser
DABCYL-Gly-Leu-Arg-Thr-Gln-Ser-Phe-Ser-EDANS

Synonyms: HAV 3C Protease Substrate; 3C Protease

Bachem M-1900.0001   MW 1394.58 $C_{65}H_{87}N_{17}O_{16}S$ Store at -15°C

Sigma D 2921   Hepatitis A virus   FW 1394.6; ≥90% (HPLC) Fluorogenic substrate; Pennington, MW et al, *Peptides* 1992, 936, 1993

Gly-Leu-Asn-Lys-Ile-Val-Arg-Met-Tyr-Ser-Pro-Thr-Ser-Ile-Leu-Asp-Ile-Arg

Synonyms: HIV Protein p24 (269-286)

Bachem H-3006.0500   Store at -15°C

Gly-Leu-Asp-Ile-Ile-Trp
Ac-D-Bhg-Leu-Asp-Ile-Ile-Trp

Synonyms: Endothelin I (16-21), (Ac-D-Bhg[16])-; PD 145065

American Peptide 88-2-33   MW 950.1 $C_{52}H_{67}N_7O_{10}$ Non-selective ET_A & ET_B receptor antagonist; Cody, WL et al, *Peptides*, 1992; Schneider, CH et al, *22nd Eur Pept Symp*, ESCOM, p. 687, 1993

Calbiochem 513020   MW 994.1 $C_{52}H_{65}N_7O_{10}$ · 2Na   ≥98% (HPLC); solid; soluble in phosphate buffer, pH 7.5 | Highly potent but non-selective endothelin receptor antagonist; Doherty, AM et al, *J Cardiovasc Pharmacol*, 22:S98 (suppl), 1993; Allcock, GH et al, *Br J Pharmacol*, 116:2482, 1995; Cody, WL et al, *Med Chem Res*, 3:154, 1993

Gly-Leu-Asp-Ile-Ile-Trp
Ac-D-Bhg-Leu-Asp-Ile-(Nme)Ile-Trp

Synonyms: PD 156252

Calbiochem 513025   MW 1006.2 $C_{53}H_{65}N_7O_{10}$ · 2Na   ≥98% (HPLC); solid; soluble in phosphate buffer, pH 7.5 | Highly potent but non-selective endothelin receptor antagonist; exhibits increased proteolytic stability & cell-permeability; Doherty, AM et al, *J Cardiovasc Pharmacol*, 22:S98 (suppl), 1993; Allcock, GH et al, *Br J Pharmacol*, 116:2482, 1995; Cody, WL et al, *J Med Chem*, 40:2228, 1997; Cody, WL et al, *Med Chem Res*, 3:154, 1993

Gly-Leu-Asp-Ile-Ile-Trp
N-Ac-α-(10,11-Dihydro-5 Dibenzoyl(a,d)Cycloheptadien-5-yl)-D-Gly-Leu-Asp-Ile-Ile-Trp

Synonyms: Endothelin Receptor Antagonist; PD 145065

Sigma P 3084   FW 948.1 $C_{52}H_{65}N_7O_{10}$ ≥97% (HPLC) | Bioactive peptide

Gly-Leu-Asp-Ile-Ile-Trp
N-Ac-α-(10,11-Dihydro-5 Dibenzoyl(a,d)Cycloheptadien-5-yl)-D-Gly-Leu-Asp-Ile-N-Me-Ile-Trp

Synonyms: Endothelin Receptor Antagonist; PD 156252

Sigma P 3209   FW 962.2 $C_{53}H_{67}N_7O_{10}$ ≥97% (HPLC); peptide content:~70% | Bioactive peptide

Gly-Leu-Glu-Gln-Glu-Gln-Met-Ile-Ser-Cys-Lys-Phe-Thr-Met-Thr-Gly-Leu-Lys-Arg-Asp

Synonyms: GP140 (161-180); SIVmac251 Peptide

Neosystem SC724   MW 2315.70   SIVmac251 peptide from subtype GP140; due to the presence of a Cys this peptide can contain the dimeric form

Gly-Leu-Gly
Z-Gly-Leu-Gly

Bachem C-1850.0250   MW 379.41 $C_{18}H_{25}N_3O_6$ Store at -15°C

Gly-Leu-Gly

Bachem H-3540.0250   MW 245.28 $C_{10}H_{19}N_3O_4$ Store at -15°C

Gly-Leu-Gly-Gln-His-Ile-Tyr-Glu-Thr-Tyr-Gly-Asp-Thr-Trp-Ala

Synonyms: VPR (41-55); HTLV I Peptide

Neosystem SC200   MW 1710.82   HTLV I peptide from subtype VPR

Gly-Leu-Gly-Gly

Bachem M-1465.0250   MW 302.33 $C_{12}H_{22}N_4O_5$ Store at -15°C

Gly-Leu-Gly-Leu

Bachem H-3545.0250   MW 358.44 $C_{16}H_{30}N_4O_5$ Store at -15°C

Gly-Leu-His-Thr-Gly-Glu-Arg-Asp-Trp-His-Leu-Gly-Gln-Gly-Val

Synonyms: VIF (71-85); HTLV I Peptide

Neosystem SC232   MW 1661.79   HTLV I peptide from subtype VIF

Gly-Leu-Leu-Gly

Bachem H-3550.0250   MW 358.44 $C_{16}H_{30}N_4O_5$ Store at -15°C

Gly-Leu-Met Amide

Synonyms: Substance P (9-11)

American Peptide 70-1-78   MW 318.4 $C_{13}H_{26}N_4O_3S$

ICN 153035

Sigma S 8261   FW 318.4 $C_{13}H_{26}N_4O_3S$ ≥95% (HPLC) | Bioactive peptide

Gly-Leu-Met Amide Acetate Salt

Synonyms: Substance P (9-11)

Sigma S 1022   FW 318.4 (free base); $C_{13}H_{26}N_4O_3S$ ≥97% (HPLC) | Bioactive peptide

Gly-Leu-Met Amide Hydrochloride

Synonyms: Substance P (9-11)

Bachem H-5880.0250   MW 354.9 $C_{13}H_{26}N_4O_3S$ · HCl   Store at -15°C

Gly-Leu-Phe

Bachem H-3555.0250   MW 335.4 $C_{17}H_{25}N_3O_4$ Store at -15°C

Gly-Leu-Phe
MeOSuc-Gly-Leu-Phe-AMC

Bachem I-1430.0050   Store at -15°C

Gly-Leu-Phe

ICN 157263   MW 335.4 $C_{17}H_{25}N_3O_4$

Sigma G 1142   FW 335.4 $C_{17}H_{25}N_3O_4$

Gly-Leu-Sar

Bachem H-7115.0250   Store at -15°C

Gly-Leu-Ser-Arg-Ser-Cys-Phe-Gly-Val-Lys-Leu-Asp-Arg-Ile-Gly-Ser-Met-Ser-Gly-Leu-Gly-Cys

Synonyms: Natriuretic Peptide, C-Type

American Peptide 14-6-10 Chicken MW 2241.7 $C_{93}H_{157}N_{29}O_{29}S_3$ Disulfide bonds: Cys^6-Cys^{22}; Arimura, JJ et al, *BBRC*, 174:142, 1991

Gly-Leu-Ser-Lys-Gly-Cys-Phe-Gly-Leu-Lys-Leu-Asp-Arg-Ile-Gly-Ser-Met-Ser-Gly-Leu-Gly-Cys-Asn-Ser-Phe-Arg-Tyr

Synonyms: Vasonatrin Peptide (1-27)

American Peptide 14-5-10 MW 2865.4 $C_{124}H_{198}N_{36}O_{36}S_3$ Disulfide bonds: Cys^6-Cys^{22};chimera of atrial natriuretic peptide (ANP) & a C-type natriuretic peptide (CNP); contains the 22 AA of CNP plus the 5 C-terminal AA of ANP; possesses the venodilating actions of CNP, the natriuretic actions of ANP & unique arterial vasodilating actions associated neither with ANP nor with CNP; Wei, CM et al, *J Clin Invest*, 92:2048, 1993

Bachem H-2502.0500 Store at -15°C

Gly-Leu-Ser-Lys-Gly-Cys-Phe-Gly-Leu-Lys-Leu-Asp-Arg-Ile-Gly-Ser-Met-Ser-Gly-Leu-Gly-Cys

Synonyms: Natriuretic Peptide (1-22), C-Type; Natriuretic Peptide 22, C-Type; Natriuretic Peptide, C-Type; Natriuretic Peptide (32-53), C-Type

ICN 159908 MW 2197.6 $C_4H_6N_2O_2$ 97% | Disulfide bonds: 6-22; Sudoh, T etal, *BBRC*, 168:863, 1990

Sigma N 8768 FW 2197.6 ≥97% (HPLC); peptide content:~65% | Disulfide bonds: 6-22; bioactive peptide; potent, selective agonist at BNP receptors; activates guanylate cyclase; may regulate fluid homeostasis in CNS; Sudoh, T et al, *Biochem Biophys Res Commun*, 168:863, 1990

American Peptide 14-1-48 Human MW 2197.6 $C_{93}H_{157}N_{27}O_{28}S_3$ Disulfide bonds: Cys^6-Cys^{22}; Sudoh, T et al, *BBRC*, 168:863, 1990

Peptides International PCT-4229-v Human synthetic MW 2197.6 $C_{93}H_{157}N_{27}O_{28}S_3$ >95% (HPLC); lyophilized amorphous powder | Bioactive; Disulfide bonds: Cys^6-Cys^{22}; porcine, rat, mouse; Sudoh, T et al, *BBRC*, 168:863, 1990

Calbiochem 05-23-0310 Human, porcine MW 2197.7 $C_{93}H_{157}N_{27}O_{28}S_3$ >98% (HPLC); lyophilized solid; soluble in 5% acetic acid | Disulfide bond:Cys^6-Cys^{22}; member of the natriuretic peptide family that shares structural homology with ANP & plays a role in the regulation of body fluid homeostasis, vascular tone & vascular growth; inhibits adenylate cyclase activity in a concentration-dependent manner; stimulates particulate guanylate cyclase activity; exerts neuromodulatory effects independently of guanylate cyclase activation or adenylate cyclase inhibition; Kelley, T et al, *Am J Respir Cell Mol Biol*, 16:464, 1997; Savoie, P et al, *FEBS Lett*, 370:6, 1995; Inoue, A et al, *Biochem Biophys Res Comm*, 221:703, 1996; Trachte, GJ et al, *Am J Physiol*, 268:C987, 1995; Koller, KJ et al, *Science*, 252:120, 1991; Sudoh, T et al, *Biochem Biophys Res Comm*, 168:863, 1990

Bachem H-1296.0500 Human, porcine, rat Store at -15°C

Neosystem SC1200 Human, porcine, rat MW 2197.61 Disulfide bonds: Cys^6-Cys^{22}; bioactive; ANF-related peptide; produced in vascular endothelial cells; acts as an endothelium-derived relaxing peptide; plays a role in human renal function; Sudoh, T et al, *BBRC*, 168:863-870, 1990; Kojima, M et al, *FEBS Lett*, 276:209-213, 1990; Tawaragi, Y et al, *BBRC*, 175:645-651, 1991; Igaki, T et al, *Kidney Int*, Suppl 55:S144-S147, 1996

Gly-Leu-Tyr

Bachem H-3560.0250 MW 351.4 $C_{17}H_{25}N_3O_5$ Store at -15°C

Gly-Leu-Tyr
Gly-L-Leu-L-Tyr

Fluka 50340 MW 351.41 $C_{17}H_{25}N_3O_5$ ≥98% (TLC); mp:235-240°C

Gly-Leu-Tyr

ICN 157264 MW 351.4 $C_{17}H_{25}N_3O_5$ Crystalline

Sigma G 5629 FW 351.4 $C_{17}H_{25}N_3O_5$

Gly-L-Pro
Gly-L-Pro-MCA

Synonyms: X-Prolyl Dipeptidyl Aminopeptidase Substrate

ICN 151203 TFA Salt | Imai, K etal, *J Biochem* (Tokyo), 93:431, 1983

Gly-Lys
Bz-Gly-D-Lys

Synonyms: Hippuryl-D-Lys

Bachem G-1240.0050 MW 307.35 $C_{15}H_{21}N_3O_4$ Store at -15°C

Gly-Lys
Bz-Gly-Lys

Synonyms: Hippuryl-Lys

Bachem G-2275.0250 MW 307.35 $C_{15}H_{21}N_3O_4$ Store at -15°C

Gly-Lys
Phenylac-Gly-Lys

Bachem G-4100.0250 Store at -15°C

Gly-Lys
Ac-Gly-Lys-βNA

Bachem K-1005.0050 MW 370.45 $C_{20}H_{26}N_4O_3$ Store at -15°C

Gly-Lys
Ac-Gly-Lys-OMe

Synonyms: AGLME

Bachem M-1035.0001 MW 259.31 $C_{11}H_{21}N_3O_4$ Store at -15°C

Gly-Lys
N-Ac-Gly-Lys-OMe Acetate Salt

Synonyms: Urokinase Substrate

ICN 100182 MW 319.4 $C_{11}H_{21}N_3O_4$-$C_2H_4O_2$ Crystalline | Useful for substrate for measurement of urokinase activity; Walton, PL, *Biochem Biophys Acta*, 132:104, 1967

Gly-Lys
Ac-Gly-Lys-OMe Hydrochloride

Synonyms: Urokinase Substrate I

ICN 195902 MW 295.8 $C_{11}H_{21}N_3O_4 \cdot$ HCl ≥95% | Substrate for urokinase quantitation

Gly-Lys
Bz-Gly-Lys

Synonyms: Hippuryl-Lys

Peptides International SGK-3047 MW 307.35 $C_{15}H_{21}N_3O_4$ >99% (HPLC); amorphous powder

Gly-Lys
Ac-Gly-Lys-OMe AcOH

Synonyms: Urokinase Substrate; C1s Substrate

Peptides International SGK-3058 MW 319.36 $C_{11}H_{21}N_3O_4 \cdot$ CH_3COOH >98% (HPLC); lyophilized amorphous powder | Walton, PL, *BBA*, 132:104, 1967; Miwa, N et al, *BBRC*, 112:754, 1983; Sim, RB in, *Proteolytic Enzymes Part C*, Methods in Enzymology, Vol 80, L Lorand, Ed, Academic Press, New York, 1981, pp 26-42

Gly-Lys
N-Ac-Gly-Lys β-Naphthyl Ester Hydrochloride
Sigma A 1913   FW 407.9 $C_{20}H_{25}N_3O_4 \cdot HCl$

Gly-Lys
N-Ac-Gly-Lys Methyl Ester Acetate Salt
Sigma A 8261   FW 319.4 $C_{11}H_{21}N_3O_4 \cdot C_2H_4O_2$

Gly-Lys
Bz-Gly-D-Lys
Synonyms: Hippuryl-D-Lys; Carboxypeptidase B Substrate; Carboxypeptidase N Substrate
Sigma H 4265   FW 307.3 $C_{15}H_{21}N_3O_4$
Sigma H 6750   FW 307.3 $C_{15}H_{21}N_3O_4$

Gly-Lys
N-Phenylacetyl-Gly-Lys
Sigma P 7053   FW 321.4 $C_{16}H_{23}N_3O_4$ Peptide that is 50X sweeter than sucrose; Nosho, Y & Ohfuji, T, *Chem & Eng News*, 68(2):25, 1990

Gly-Lys Hydrochloride
Bachem G-3630.0250   MW 239.7 $C_8H_{17}N_3O_3 \cdot HCl$ Store at -15°C
ICN 157265   MW 239.7 $C_8H_{17}N_3O_3 \cdot HCl$ Crystalline
Sigma G 2127   FW 239.7 $C_8H_{17}N_3O_3 \cdot HCl$

Gly-Lys-Arg
BOC-Gly-Lys-Arg-AMC Hydrochloride
Bachem I-1580.0050   MW 653.18 $C_{29}H_{44}N_8O_7 \cdot HCl$ Store at -15°C

Gly-Lys-Arg
BOC-Gly-Lys-Arg-MCA
Synonyms: Carboxyl Side of Paired Basic Residue Cleaving Enzyme Substrate
Peptides International MGK-3143-v   MW 616.72 $C_{29}H_{44}N_8O_7$ >98% (HPLC); lyophilized amorphous powder | Mizuno, K et al, *BBRC*, 144:807, 1987

Gly-Lys-Arg
N-t-BOC-Gly-Lys-Arg MCA Hydrochloride
Synonyms: Protease Substrate
Sigma B 5153   FW 653.2 $C_{29}H_{44}N_8O_7 \cdot HCl$ ~95% | Substrate for proteases which cleave on the carboxyl side of paired basic residues; Mizuno, K et al, *Biochem Biophys Res Commun*, 144:807, 1987

Gly-Lys-Arg-Trp-Gly
Bachem H-3575.0025   MW 602.69 $C_{27}H_{42}N_{10}O_6$ Store at -15°C

Gly-Lys-Gly
Gly-Lys(retro-Gly-H)
Bachem H-6380.0250   MW 260.29 $C_{10}H_{20}N_4O_4$ Store at -15°C

Gly-Lys-Gly
Bachem H-8510.0250   MW 260.29 $C_{10}H_{20}N_4O_4$ Store at -15°C

Gly-Lys-Gly-Ala-Gly-Leu-Ser-Leu-Ser-Arg-Phe-Ser-Trp-Gly-Ala
Synonyms: Myelin Basic Protein (104-118), (Ala[107])-
ICN 159629   MW 1493.7

Neosystem SC176
Neosystem SC176   MW 1493.68   Bioactive; proteinkinase-related peptide proteinkinase C inhibitor; De Su, H et al, *BBRC*, 134:78-84, 1986

Gly-Lys-Gly-Arg-Gly-Leu-Ser-Leu-Ser-Ala-Phe-Ser-Trp-Gly-Ala
Synonyms: Myelin Basic Protein (104-118), (Ala[113])-
ICN 159630   MW 1493.7
Neosystem SC177   MW 1493.68   Bioactive; proteinkinase-related peptide proteinkinase C inhibitor; De Su, H et al, *BBRC*, 134:78-84, 1986

Gly-Lys-Gly-Arg-Gly-Leu-Ser-Leu-Ser-Arg-Phe-Ser-Trp-Gly-Ala
Synonyms: Myelin Basic Protein (104-118)
ICN 159628   MW 1578.8
Neosystem SC175   MW 1578.79   Bioactive; proteinkinase-related peptide proteinkinase C substrate; Turner, R et al, *JBC*, 260:11503-11507, 1985

Gly-Lys-His
Bachem H-3580.0050   MW 336.35 $C_{14}H_{20}N_6O_4$ Store at -15°C

Gly-Lys-Lys-Glu-Lys-Pro-Glu-Lys-Lys-Val-Lys-Lys-Ser-Asp-Cys-Gly-Glu-Trp-Gln-Trp-Ser-Val-Cys-Val-Pro-Thr-Ser-Gly-Asp-Cys-Gly-Leu-Gly-Thr-Arg-Glu-Gly-Thr-Arg-Thr-Gly-Ala-Glu-Cys-Lys-Gln-Thr-Met-Lys-Thr-Gln-Arg-Cys-Lys-Ile-Pro-Cys-Asn-Trp-Lys-Lys-Gln-Phe-Gly-Ala-Glu-Cys-Lys-Tyr-Gln-Phe-Gln-Ala-Trp-Gly-Glu-Cys-Asp-Leu-Asn-Thr-Ala-Leu-Lys-Thr-Arg-Thr-Gly-Ser-Leu-Lys-Arg-Ala-Leu-His-Asn-Ala-Glu-Cys-Gln-Lys-Thr-Val-Thr-Ile-Ser-Lys-Pro-Cys-Gly-Lys-Leu-Thr-Lys-Pro-Lys-Pro-Gln-Ala-Glu-Ser-Lys-Lys-Lys-Lys-Lys-Glu-Gly-Lys-Lys-Gln-Glu-Lys-Met-Leu-Asp
Synonyms: Pleiotrophin; Heparin Binding Growth Factor; Neurite Outgrowth-Promoting Factor
Peptides International PTN-4335-v   Human synthetic   MW 15302. $C_{658}H_{1079}N_{197}O_{198}S_{12}$ >95% (HPLC); lyophilized amorphous powder | Bioactive; Disulfide bonds: Cys[15]-Cys[44], Cys23-Cys[53], Cys[30]-Cys57, Cys[67]-Cys[99], Cys[77]-Cys[109]; Li, Y-S et al, *Science*, 250:1690, 1990; Milner, PG et al, *Biochemistry*, 31:12023, 1992; Czubayko, F et al, *PNAS USA*, 93:14753, 1996; Inui, T et al, in preparation

Gly-Lys-Pro-Ile-Leu-Phe-Phe-Arg-Leu-Lys-Arg
Mca-Gly-Lys-Pro-Ile-Leu-Phe-Phe-Arg-Leu-Lys(Dnp)-D-Arg Amide
Bachem M-2455.0001   MW 1756.04 $C_{85}H_{122}N_{22}O_{19}$ Store at -15°C

Gly-Lys-Pro-Ile-Leu-Phe-Phe-Arg-Leu-Lys-Arg
MOCAc-Gly-Lys-Pro-Ile-Leu-Phe-Phe-Arg-Leu-Lys(Dnp)-D-Arg Amide
Synonyms: Cathepsin D/E Fluorescence-Quenching Substrate
Peptides International SMO-3200-v   MW 1756.0 $C_{85}H_{122}N_{22}O_{19}$ >98% (HPLC); lyophilized amorphous powder | Yasuda, Y et al, *J Biochem*, 125:1137, 1999

Gly-Lys-Pro-Ile-Pro-Asn-Pro-Leu-Leu-Gly-Leu-Asp-Ser-Thr Trifluoroacetate Salt
Synonyms: RNA Polymerase-α (95-108); SV5; PK-Tag
Biogenesis 8120-5060   Synthetic   Semi-pure; lyophilized

Gly-Met
Z-Gly-Met

Bachem C-1855.0001   MW 340.4 $C_{15}H_{20}N_2O_5S$ Store at -15°C

Gly-Met

Bachem G-2155.0001   MW 206.27 $C_7H_{14}N_2O_3S$ Store at -15°C

ICN 101882   MW 206.3 $C_7H_{14}N_2O_3S$ Crystalline

Gly-Met
Borotrimethyl-Gly-Met-OMe

ICN 154644   MW 262.18 $BC_{10}H_{23}N_2O_3S$ Low melting solid; >97% | Potent hypocholesterolemic activity

Gly-Met
Boro-Gly-Met Amide

ICN 154650   MW 205.08 $BC_6H_{16}N_3O_2S$ >97%

Gly-Met

Sigma G 2252   FW 206.3 $C_7H_{14}N_2O_3S$

Gly-Met
Gly-DL-Met

Sigma G 6754   FW 206.3 $C_7H_{14}N_2O_3S$

Gly-Met-Ala-Ser-Lys-Ala-Gly-Ala-Ile-Ala-Gly-Lys-Ile-Ala-Lys-Val-Ala-Leu-Lys-Ala-Leu Amide

Synonyms: PGLa

Neosystem SC184   MW 1968.47   Bioactive; antibacterial/antimicrobial peptide; Soravia, E et al, *FEBS Lett*, 228:337-340, 1988

Gly-Met-Asp-Ser-Leu-Ala-Phe-Ser-Gly-Gly-Leu
Gly-Met-Asp-Ser-Leu-Ala-Phe-Ser-Gly-Gly-Leu Amide

Synonyms: Buccalin; Modulating Neuropeptide

American Peptide 87-0-20   MW 1053.5 $C_{45}H_{72}N_{12}O_{15}S$
Unlike the small cardioactive peptides, this peptide acts presynaptically on the nerve terminals to inhibit Accholine release; Cropper, EC et al, *PNAS*, 85:6177, 1988

Gly-Met-Asp-Ser-Leu-Ala-Phe-Ser-Gly-Gly-Leu

Synonyms: Buccalin; Modulating Neuropeptide

ICN 154475   MW 1053.3   Cropper, EC etal, *PNAS*, 85:6177, 1988

Gly-Met-Asp-Ser-Leu-Ala-Phe-Ser-Gly-Gly-Leu
Gly-Met-Asp-Ser-Leu-Ala-Phe-Ser-Gly-Gly-Leu Amide

Synonyms: Buccalin

Neosystem SC332   MW 1053.20   Bioactive neuropeptide

Gly-Met-Asp-Ser-Leu-Ala-Phe-Ser-Gly-Gly-Leu Amide

Synonyms: Buccalin

Bachem H-9235.0005   MW 1053.21 $C_{45}H_{72}N_{12}O_{15}S$ Store at -15°C

Sigma A 4528   FW 1053.2   ≥97% (HPLC) | Bioactive peptide; Cooper, EC et al, *Proc Natl Acad Sci USA*, 85:6177, 1988

Gly-Met-Gly

Bachem H-3585.0250   MW 263.32 $C_9H_{17}N_3O_4S$ Store at -15°C

Gly-Nle
Gly-DL-Nle-OMe Hydrochloride

Bachem G-2165.0001   Store at -15°C

Gly-Nle
Gly-DL-Nle

ICN 101883   MW 188.2 $C_8H_{16}N_2O_3$ Crystalline

Gly-Nva
Z-Gly-Nva

Bachem C-1860.0001   MW 308.33 $C_{15}H_{20}N_2O_5$ Store at -15°C

Gly-Nva
FA-Gly-Nva Amide

Bachem M-1830.0050   MW 293.32 $C_{14}H_{19}N_3O_4$ Store at -15°C

Gly-Nva
Gly-DL-Nva

ICN 101884   MW 174.2 $C_7H_{14}N_2O_3$ Crystalline

Gly-Pen-Gly-Arg-Gly-Asp-Ser-Pro-Cys-Ala

American Peptide 44-0-36   MW 948.1 $C_{35}H_{57}N_{13}O_{14}S_2$
Disulfide bonds: Pen[2]-Cys[9]

Bachem H-3964.0001   Store at -15°C | Disulfide bonds: Pen[2]-Cys[9]

Sigma G 5275   FW 948.0 $C_{35}H_{57}N_{13}O_{14}S_2$ ≥97% (HPLC) | Disulfide bonds: 2-9; bioactive peptide

Gly-Phe
BOC-Gly-Phe-OBzl

Bachem A-1755.0001   MW 412.49 $C_{23}H_{28}N_2O_5$ Store at -15°C

Gly-Phe
FMOC-Gly-Phe

Bachem B-1495.0001   MW 444.49 $C_{26}H_{24}N_2O_5$ Store at -15°C

Gly-Phe
Z-Gly-Phe

Bachem C-1865.0001   MW 356.38 $C_{19}H_{20}N_2O_5$ Store at -15°C

Gly-Phe
Z-Gly-D-Phe

Bachem C-1870.0001   MW 356.38 $C_{19}H_{20}N_2O_5$ Store at -15°C

Gly-Phe
Z-Gly-Phe Amide

Bachem C-1875.0001   MW 355.39 $C_{19}H_{21}N_3O_4$ Store at -15°C

Gly-Phe

Bachem G-2175.0005   MW 222.24 $C_{11}H_{14}N_2O_3$ Store at -15°C

Gly-Phe
Gly-D-Phe

Bachem G-2180.0001   MW 222.24 $C_{11}H_{14}N_2O_3$ Store at -15°C

Gly-Phe
Gly-DL-Phe

Bachem G-3735.0005   MW 222.24 $C_{11}H_{14}N_2O_3$ Store at -15°C

Gly-Phe
Gly-Phe-AMC
Bachem I-1220.0050   MW 379.42 $C_{21}H_{21}N_3O_4$ Store at -15°C

Gly-Phe
Glutaryl-Gly-Phe-4MβNA
Bachem J-1192.0050   MW 491.54 $C_{27}H_{29}N_3O_6$ Store at -15°C

Gly-Phe
Gly-Phe-4MβNA
Bachem J-1205.0050   MW 377.44 $C_{22}H_{23}N_3O_3$ Store at -15°C

Gly-Phe
Gly-Phe-βNA
Bachem K-1325.0050   MW 347.42 $C_{21}H_{21}N_3O_2$ Store at -15°C

Bachem L-1290.0050   MW 342.35 $C_{17}H_{18}N_4O_4$ Store at -15°C

Gly-Phe
Bz-Gly-Phe
Synonyms: Hippuryl-Phe; Carboxypeptidase A Substrate
Bachem M-1490.0250   MW 326.35 $C_{18}H_{18}N_2O_4$ Store at RT

Gly-Phe
FA-Gly-Phe Amide
Synonyms: FAGPA
Bachem M-1770.0250   MW 341.37 $C_{18}H_{19}N_3O_4$ Store at -15°C

Gly-Phe
Gly-L-Phe
Fluka 50370   MW 222.25 $C_{11}H_{14}N_2O_3$ ≥99% (titration); mp:264°C

Gly-Phe
Bz-Gly-Phe
Synonyms: Hippuryl-Phe; Carboxypeptidase A Substrate
Fluka 53284   MW 326.35 $C_{18}H_{18}N_2O_4$ ≥99% (TLC); mp:140-144°C | Everitt, MT & Neurath, H, *FEBS Lett*, 110:292, 1980

Gly-Phe
Z-L-Gly-L-Phe
Fluka 96260   MW 356.38 $C_{19}H_{20}N_2O_5$ ≥99.0% (titration); mp:125-126°C

Gly-Phe
Gly-L-Phe Amide Acetate
Synonyms: Cathepsin C Substrate
ICN 101873   MW 281.3 $C_{11}H_{15}N_3O_2 \cdot C_2H_4O_2$ Crystalline

Gly-Phe
Gly-DL-Phe
ICN 101885   MW 222.2 $C_{11}H_{14}N_2O_3$ Crystalline

Gly-Phe
ICN 101886   MW 222.2 $C_{11}H_{14}N_2O_3$ Crystalline

Gly-Phe
Benzoyl-Gly-L-Phe
Synonyms: Hippuryl-L-Phe
ICN 101947   MW 326.4 $C_{18}H_{18}N_2O_4$ Crystalline

Gly-Phe
Gly-D-Phe
ICN 105574   MW 222.2 $C_{11}H_{14}N_2O_3$ Crystalline

Gly-Phe
Borotrimethyl-Gly-Phe-OMe
ICN 154641   MW 278.16 $BC_{14}H_{23}N_2O_3$ >97%

Gly-Phe
Nᵃ-Tosylglycyl-3-Amidinophenylalanine-OMe Acetate
Synonyms: Antithrombin III; Factor Xa Inhibitor; TAPA
ICN 194105   2.5 μmoles/vial; a factor Xa concentration of 60 U (ICTH)/L causes an inhibition of 52% by a Xa inhibitor concentration of 2.5 μmoles/L

Gly-Phe
Gly-Phe-pNA Hydrochloride
Synonyms: Cathepsin C Substrate
ICN 194969   MW 378.9   Lyophilized; >95% | Planta, RJ & MA Gruber, *Anal Biochem*, 5:360, 1963

Gly-Phe
Z-Gly-Phe-NHO-Bz-p-Me
Synonyms: Subtilisin Inhibitor III; Serine Protease Inhibitor
ICN 195918   MW 505.5 $C_{27}H_{27}N_3O_7$ >95%

Gly-Phe
Z-Gly-Phe-NHO-Bz
Synonyms: Subtilisin Inhibitor II; Serine Protease Inhibitor
ICN 195919   MW 475.5 $C_{26}H_{25}N_3O_6$ >95%

Gly-Phe
Gly-Phe-pNA Hydrochloride
Synonyms: Dipeptidyl Aminopeptidase I Substrate; Cathepsin C Substrate
Neosystem SC1293   MW 342.34   Chromogenic substrate

Gly-Phe
Nα-TosylGly-3-Amidino-(D,L)-Phe Methylester Hydrochloride
Synonyms: Pefabloc®Xa; Pefa-3000
Pentapharm 385-12, 385-03   MW 469.0 $C_{20}H_{24}O_5N_4S \cdot HCl$
Readily soluble in water (200 mg/mL, ~400 *mM*) | Used to exclude undesired activity of factor Xa in the presence of other serine proteases (eg other clotting factors); Hauptmann J et al, *Thromb Haemost*, 63:220, 1990; Vieweg H et al, *Pharmazie*, 38:818, 1983

Gly-Phe
Gly-Phe Amide AcOH
Peptides International OGF-3023   MW 281.3 $C_{11}H_{15}N_3O_2 \cdot CH_3COOH$ >98% (HPLC); colorless crystalline powder

Gly-Phe
Peptides International OGF-3053   MW 222.24 $C_{11}H_{14}N_2O_3$ >98% (HPLC); colorless crystalline powder

Gly-Phe
Z-Gly-Phe
Peptides International SGF-3020   MW 356.38 $C_{19}H_{20}N_2O_5$ >99% (HPLC); amorphous powder

Gly-Phe
Z-Gly-Phe Amide

Peptides International SGF-3021   MW 355.39 $C_{19}H_{21}N_3O_4$
>99% (HPLC); amorphous powder

Gly-Phe
N-CBZ-Gly-Phe

Sigma C 0252   FW 356.4 $C_{19}H_{20}N_2O_5$

Gly-Phe
N-CBZ-Gly-Phe Amide

Sigma C 0377   FW 355.4 $C_{19}H_{21}N_3O_4$

Gly-Phe
Gly-Phe p-NA

Synonyms: Cathepsin C Substrate

Sigma G 0142   FW 342.4 $C_{17}H_{18}N_4O_4$ Chromogenic; Planta, RJ & Gurber, M, *Anal Biochem*, 5:360, 1963

Gly-Phe
Gly-DL-Phe

Sigma G 2627   FW 222.2 $C_{11}H_{14}N_2O_3$

Gly-Phe

Sigma G 2752   FW 222.2 $C_{11}H_{14}N_2O_3$

Gly-Phe
Gly-Phe β-Naphthylamide

Synonyms: Cathepsin C Substrate

Sigma G 4255   FW 347.4 $C_{21}H_{21}N_3O_2$ Crystalline | Jadot, M et al, *Biochem J*, 219:965, 1984; Doughty, MJ & Gruenstein, EI, *Biochem & Cell Biol*, 64:772, 1986

Gly-Phe
Gly-D-Phe

Sigma G 4879   FW 222.2 $C_{11}H_{14}N_2O_3$

Gly-Phe
N-Glutaryl-Gly-Phe 4-Methoxy-β-Naphthylamide

Sigma G 7135   FW 491.5 $C_{27}H_{29}N_3O_6$

Gly-Phe
Gly-Phe β-Naphthylamide

Synonyms: Cathepsin C Substrate

Sigma G 9512   FW 347.4 $C_{21}H_{21}N_3O_2$ Crystalline | Jadot, M et al, *Biochem J*, 219:965, 1984; Doughty, MJ & Gruenstein, EI, *Biochem & Cell Biol*, 64:772, 1986

Gly-Phe
N-Benzoyl-Gly-Phe

Synonyms: Carboxypeptidase A Hippuryl-Phe Substrate

Sigma H 6875   FW 326.4 $C_{18}H_{18}N_2O_4$ Sebastian, JF & Lo, WY, *Can J Biochem*, 56:329, 1978; Woodbury, RG et al, *Meth Enzymol*, 80:588, 1981

Gly-Phe
CBZ-Gly-Phe

USBio C2098-81   MW 356.4 $C_{19}H_{20}N_2O_5$ ≥99%

Gly-Phe
((S)-1-Carboxyl-2-phenylethyl)-carbamoyl-α-(2-iminohexahydro-4(S)-pyrimidyl)-(S)-Gly-X-Phe-al

Synonyms: Chymostatin, Mixture

ICN 152845   Microbial   A mixture of type A, B & C; X = Leu (Type A), Ile (Type B), Val (Type C); Umezawa, H etal, *J Antibiot*, 23:425, 1970; Tatsuta, K etal, *J Antibiot*, 26:625, 1973

Gly-Phe
Nα-(2-Naphthylsulfonyl-Gly)-4-Amidino-(D,L)-Phe-Piperidide Acetate

Synonyms: NAPAP; Pefabloc®TH; Pefa-3204

Pentapharm 381-12, 381-01   Synthetic   MW 581.7 $C_{27}H_{31}O_4N_5S$-AcOH   Solubility:1.5 mg/mL (2.6 *mM*) in H_2O | One of the most potent & selective competitive inhibitors of thrombin; used to exclude undesired thrombin activity; potent anticoagulant in *in vitro* test systems; Stürzebecher J et al, *Thromb Res*, 36:457, 1984

Gly-Phe Amide

Bachem M-1470.0001   MW 221.26 $C_{11}H_{15}N_3O_2$ Store at -15°C

Gly-Phe Amide Acetate Salt

Sigma G 2877   FW 281.3 $C_{11}H_{15}N_3O_2 \cdot C_2H_4O_2$

Gly-Phe Amide Hydrochloride

Bachem G-2185.0001   MW 257.72 $C_{11}H_{15}N_3O_2 \cdot$ HCl   Store at -15°C

Gly-Phe-Ala
Z-Gly-Phe-Ala

Bachem C-3250.0250   MW 427.46 $C_{22}H_{25}N_3O_6$ Store at -15°C

Gly-Phe-Ala

Bachem H-3590.0250   MW 293.32 $C_{14}H_{19}N_3O_4$ Store at -15°C

ICN 157266   MW 293.3 $C_{14}H_{19}N_3O_4$ Crystalline

Sigma G 4754   FW 293.3 $C_{14}H_{19}N_3O_4$

Gly-Phe-Ala-Asp
Gly-D-Phe-Ala-Asp Ammonium Salt

Synonyms: Achatin I; Achatin II Stereoisomer

Sigma A 2705   FW 408.4 (free acid) $C_{18}H_{24}N_4O_7$ ≥95% (HPLC) | Bioactive peptide; stereoisomer of Gly-Phe-Ala-Asp, Achatin II; excites the achatina neuron by increasing membrane conductance of sodium ions; Akamatsu, M et al, *Biosci Biotech Biochem*, 58:1123, 1994

Gly-Phe-Ala-Asp

Synonyms: Achatin

American Peptide 86-5-15   *Achatina fulica Ferussac*   MW 407.4 $C_{18}H_{23}N_4O_7$ Excitatory neurotransmitter of Achatina neurones; Emaduddin, M et al, *Eur J Pharmacol*, 302:129, 1996

Gly-Phe-Arg

American Peptide 87-0-73   MW 378.4 $C_{17}H_{26}N_6O_4$

Gly-Phe-Arg Dihydrochloride

ICN 157267

Sigma G 7388   FW 451.4 $C_{17}H_{26}N_6O_4 \cdot$ 2HCl

Gly-Phe-Arg-Gly-Asp-Gly-Gln

Bachem H-1606.0001   MW 735.76 $C_{30}H_{45}N_{11}O_{11}$ Store at -15°C

Gly-Phe-Asn-Ser-Ala-Leu-Met-Phe Amide

Synonyms: SALMF 1; S_1

American Peptide 60-0-10 *Asterias rubens* & *Asterias forbesi* (starfish)   MW 885.1 $C_{41}H_{60}N_{10}O_{10}S$ Novel neuropeptide identified in species belongs to the phylum *Echnodermata; Elphick, MR et al, Proc R Soc Lon,* B243:121, 1991

Gly-Phe-Asp-Leu-Asn-Gly-Gly-Gly-Val-Gly

Synonyms: Amyloid P I; Speract

American Peptide 87-0-90   MW 891.9 $C_{38}H_{57}N_{11}O_{14}$ Originally isolated from sea urchin eggs, stimulates the respiration, motility & cyclic nucleotide metabolism of sea urchin spermatozoa; Garbers, DL et al, *JBC,* 257:2734, 1982

Bachem H-6940.0001   Store at -15°C

Sigma G 9773   FW 891.9 $C_{38}H_{57}N_{11}O_{14}$ ≥97% (HPLC) | Bioactive peptide; stimulates the respiration, motility & cyclic nucleotide metabolism of sea urchin spermatozoa Garbers, DL et al, *J Biol Chem,* 257:2734, 1982

Gly-Phe-Gln-Glu-Ala-Tyr-Arg-Arg-Phe-Tyr-Gly-Pro-Val

Synonyms: Bone Gla Protein; Osteocalcin (37-49)

ICN 159983   MW 1589.8   97%

American Peptide 22-1-63   Human   MW 1589.9 $C_{75}H_{104}N_{20}O_{19}$ Poser, JW et al, *PNAS,* 255:8685, 1980

Bachem H-5875.0001   Human   MW 1589.77 $C_{75}H_{104}N_{20}O_{19}$ Store at -15°C

ICN 154481   Human   MW 1590.0   Poser, JW etal, *PNAS,* 255:8685, 1980

Gly-Phe-Gln-Glu-Ala-Tyr-Arg-Arg-Phe-Tyr-Gly-Pro-Val Acetate Salt

Synonyms: Bone Gla Protein (37-49)

Biogenesis 1403-1004   Human synthetic   MW 1590 Purified; lyophilized | Poser et al, *PNAS,* 255:8685, 1980

Gly-Phe-Gln-Glu-Ala-Tyr-Arg-Arg-Phe-Tyr-Gly-Pro-Val Trifluoroacetate Salt

Synonyms: Osteocalcin (37-49)

Biogenesis 7060-1579   Human synthetic   MW 1589.77 Purified; lyophilized | Poser et al, *PNAS,* 255:8685, 1980

Gly-Phe-Gly
Z-Gly-Phe-Gly-CHO Semicarbazone

Bachem C-3085.0050   MW 454.49 $C_{22}H_{26}N_6O_5$ Store at -15°C

Gly-Phe-Gly

Bachem H-3595.0250   MW 279.3 $C_{13}H_{17}N_3O_4$ Store at -15°C

Gly-Phe-Gly
Gly-Phe-Gly-CHO Semicarbazone

Bachem H-7650.0025   MW 320.35 $C_{14}H_{20}N_6O_3$ Store at -15°C

Gly-Phe-Gly
Suc-Gly-Phe-Gly-pNA

Bachem L-1685.0050   Store at -15°C

Gly-Phe-Gly
N-Succinyl-Gly-Phe-Gly p-NA

Sigma S 6296   FW 499.5 $C_{23}H_{25}N_5O_8$ ≥97% (TLC)

Gly-Phe-Gly-Lys
BOC-Gly-Phe-Gly-Lys

Bachem A-3280.0250   Store at -15°C

Gly-Phe-Gly-Lys
Gly-Phe-Gly-(BOC-Lys)

Synonyms: BOC-Lys(retro-Gly-Phe-Gly-H)

Bachem A-3285.0250   MW 507.59 $C_{24}H_{37}N_5O_7$ Store at -15°C

Gly-Phe-Ile-Gly-Trp-Gly-Asn-Asp-Ile-Phe-Gly-His-Tyr-Ser-Gly-Asp-Phe

Synonyms: Anantin

American Peptide 14-1-41   MW 1871.0 $C_{90}H_{111}N_{21}O_{24}$ β-peptide bond between: Gly^1-Asp^8 | A cyclic peptide from *Streptomyces coerulescens;* competitive antagonist at atrial natriuretic factor receptors; Weber, W et al, *J Antibiot,* 44:164, 1991

Bachem H-8140.0001   MW 1889.01 $C_{90}H_{113}N_{21}O_{25}$ Store at -15°C | Linear sequence

Gly-Phe-Leu

Bachem H-3600.0250   MW 335.4 $C_{17}H_{25}N_3O_4$ Store at -15°C

Gly-Phe-Leu
FA-Gly-Phe-Leu

Bachem M-1845.0050   Store at -15°C

Gly-Phe-Leu

ICN 157268   MW 335.4 $C_{17}H_{25}N_3O_4$

Sigma G 0892   FW 335.4 $C_{17}H_{25}N_3O_4$

Gly-Phe-Leu-Arg-Arg-Ile

Synonyms: Dynorphin A (3-8)

American Peptide 26-4-71   Porcine   MW 760.9 $C_{35}H_{60}N_{12}O_7$

Gly-Phe-Leu-Arg-Arg-Ile-Arg-Pro-Lys-Leu-Lys

Synonyms: Dynorphin A (3-13)

American Peptide 26-4-56   Porcine   MW 1383.8 $C_{64}H_{114}N_{22}O_{12}$

Gly-Phe-Leu-Arg-Arg-Ile-Arg-Pro-Lys-Leu-Lys-Trp-Asp-Asn-Gln

Synonyms: Dynorphin A (3-17)

American Peptide 26-4-27   Porcine   MW 1927.3 $C_{88}H_{143}N_{29}O_{20}$

Gly-Phe-Leu-Asn-Gly-Ser-Cys-Ser-Gly-Leu-Asp-Glu-Glu-Ala-Ser-Gly-Pro-Glu-Arg Acetate Salt

Synonyms: Parathyroid Hormone Receptor I (562-580), Cytoplasmic Domain 4

Biogenesis 0100-0183   Synthetic   Semi-pure; lyophilized

Gly-Phe-Lys-Asn-Val-Glu-Met-Met-Thr-Ala-Arg-Gly-Phe Amide

Synonyms: Allatotropin; Neurohormone

Bachem H-1014.0001   MW 1486.78 $C_{65}H_{103}N_{19}O_{17}S_2$ Store at -15°C

Sigma A 1686   *Manduca sexta*   FW 1486.8   ≥97% (HPLC); peptide content:~70% | Bioactive peptide; novel neurohormone which stimulates the release of juvenile hormone from the corpora allata *in vitro;* Kataoka, H et al, *Science,* 243:1481, 1989

Gly-Phe-Phe

Bachem H-3605.0250	MW 369.42	$C_{20}H_{23}N_3O_4$	Store at -15°C
ICN 157269	MW 369.4	$C_{20}H_{23}N_3O_4$	Crystalline
Sigma G 5379	FW 369.4	$C_{20}H_{23}N_3O_4$	

Gly-Phe-Phe-Ala-Leu-Ile-Pro-Lys-Ile-Ile-Ser-Ser-Pro-Leu-Phe-Lys-Thr-Leu-Leu-Ser-Ala-Val-Gly-Ser-Ala-Leu-Ser-Ser-Ser-Gly-Gly-Gln-Glu

Synonyms: Pardaxin P1, (Gly[31])-

ICN 159987 MW 3323.9 97%

Sigma O 0435 *Pardachirus marmoratus* (moses sole) FW 3323.9 ≥97% (HPLC) | Bioactive peptide; induces neurotransmitter release; Shai, Y et al, *FEBS Lett*, 242:161, 1988; Thompson, SA et al, *Science*, 233:341, 1986

Gly-Phe-Pro
D-Aminobenzoyl)-Gly-*p*-Nitro-Phe-Pro

Synonyms: Angiotensin Converting Enzyme

Sigma A 4408 FW 483.5 $C_{23}H_{25}N_5O_7$ Fluorogenic; Carmel, A & Yaron, A, *Eur J Biochem*, 87:265, 1978

Gly-Phe-Ser

Bachem H-3610.0250	MW 309.32	$C_{14}H_{19}N_3O_5$	Store at -15°C
ICN 157270	MW 309.3	$C_{14}H_{19}N_3O_5$	
Sigma G 0767	FW 309.3	$C_{14}H_{19}N_3O_5$	

Gly-Phe-Trp
Gly-*p*-Iodo-Phe-Trp

Bachem H-7475.0050 Store at -15°C

Gly-Phe-Tyr

Bachem N-1085.0250 MW 385.42 $C_{20}H_{23}N_3O_5$ Store at -15°C

Gly-Pro
BOC-Gly-Pro

Bachem A-1760.0005 MW 272.3 $C_{12}H_{20}N_2O_5$ Store at -15°C

Gly-Pro
FMOC-Gly-Pro

Bachem B-3315.0001 Store at -15°C

Gly-Pro
Z-Gly-Pro

Bachem C-1880.0001 MW 306.32 $C_{15}H_{18}N_2O_5$ Store at -15°C

Gly-Pro
Z-Gly-Pro-OSu

Bachem C-1885.0001 MW 403.39 $C_{19}H_{21}N_3O_7$ Store at -15°C

Gly-Pro

Bachem G-2190.0001 MW 172.18 $C_7H_{12}N_2O_3$ Store at -15°C

Gly-Pro
Tos-Gly-Pro

Bachem G-3310.0005 MW 326.37 $C_{14}H_{18}N_2O_5S$ Store at -15°C

Gly-Pro
Gly-3,4-Dehydro-Pro

Bachem G-3890.0050 Store at -15°C

Gly-Pro
Z-Gly-Pro-AMC

Bachem I-1145.0050 MW 463.49 $C_{25}H_{25}N_3O_6$ Store at -15°C

Gly-Pro
Gly-Pro-AMC Hydrobromide

Bachem I-1225.0050 MW 410.27 $C_{17}H_{19}N_3O_4 \cdot HBr$ Store at -15°C

Gly-Pro
Suc-Gly-Pro-AMC

Bachem I-1345.0050 MW 429.43 $C_{21}H_{23}N_3O_7$ Store at -15°C

Gly-Pro
Gly-Pro-4MβNA

Bachem J-1210.0250 MW 327.38 $C_{18}H_{21}N_3O_3$ Store at -15°C

Gly-Pro
Z-Gly-Pro-4MβNA

Bachem J-1365.0250 Store at -15°C

Gly-Pro
Z-Gly-Pro-βNA

Bachem K-1190.0250 MW 431.49 $C_{25}H_{25}N_3O_4$ Store at RT

Gly-Pro
Gly-Pro-βNA

Bachem K-1335.0250 MW 297.36 $C_{17}H_{19}N_3O_2$ Store at -15°C

Gly-Pro
Z-Gly-Pro-pNA

Bachem L-1235.0250 MW 426.43 $C_{21}H_{22}N_4O_6$ Store at -15°C

Gly-Pro
Gly-Pro-pNA *p*-Tosylate

Bachem L-1295.0050 MW 464.5 $C_{13}H_{16}N_4O_4 \cdot C_7H_8O_3S$ Store at -15°C

Gly-Pro
Suc-Gly-Pro-pNA

Bachem L-1585.0050 Store at -15°C

Gly-Pro
Gly-Pro-pNA Hydrochloride

Bachem L-1880.0050 MW 328.76 $C_{13}H_{16}N_4O_4 \cdot HCl$ Store at -15°C

Gly-Pro
Z-L-Gly-L-Pro

Fluka 96270 MW 306.32 $C_{15}H_{18}N_2O_5$ ~99% (TLC); mp:155-156°C

Gly-Pro
Z-L-Gly-L-Pro-MCA

Synonyms: Post-Proline Cleaving Enzyme Substrate

Fluka 96280   MW 463.49 $C_{25}H_{25}N_3O_6$ ≥97.0%; mp:96-102°C | Fluorogenic substrate; Momand, J & Clarke, S, *Biochemistry*, 26:7798, 1987

Gly-Pro
Z-L-Gly-L-Pro-2-Naphthylamide

Synonyms: Endopeptidase Substrate, Prolyl

Fluka 96283   MW 431.49 $C_{25}H_{25}N_3O_4$ ≥97.0% (TLC); mp:138-141°C | Specific fluorogenic substrate; Hauzer, K et al, *Coll Czech Chem Comm*, 47:1139, 1982

Gly-Pro
Z-L-Gly-L-Pro-pNA

Synonyms: Endopeptidase Substrate, Prolyl

Fluka 96286   MW 426.43 $C_{21}H_{22}N_4O_6$ ≥99.0% (TLC) | Fluorescence substrate; Hauzer, K et al, *Coll Czech Chem Comm*, 47:1139, 1982

Gly-Pro

ICN 101887   MW 172.2 $C_7H_{12}N_2O_3$ Crystalline

Gly-Pro
N-Suc-Gly-L-Pro-MCA

Synonyms: Post-Proline Cleaving Enzyme Substrate

ICN 152084   Fluorogenic substrate; Nakano, K etal, *Seikagaku*, 51:642, 1979

Gly-Pro
Borotrimethyl-Gly-Pro-OMe

ICN 154640   MW 228.1 $BC_{10}H_{21}N_2O_3$ >97% | Potent triglyceridemic activity

Gly-Pro
Z-Gly-Pro-pNA

Synonyms: Prolyl Oligopeptidase Substrate; Post-Proline Cleaving Enzyme Substrate

Neosystem SC1290   MW 426.36   Blumberg, S et al, *Brain Res*, 192:477-486, 1980; Mäkinen, PL et al, *Infection & Immunity*, 62:4938-4947, 1994

Gly-Pro
Gly-Pro-MCA

Synonyms: X-Prolyl Dipeptidyl Aminopeptidase Substrate

Peptides International MGP-3090-v   MW 329.35 $C_{17}H_{19}N_3O_4$ >99% (HPLC); lyophilized amorphous powder | Kato, T et al, *Biochem Med*, 19:351, 1978

Gly-Pro
Suc-Gly-Pro-MCA

Synonyms: Post-Proline Cleaving Enzyme Substrate

Peptides International MGP-3109-v   MW 429.43 $C_{21}H_{23}N_3O_7$ >99% (HPLC); lyophilized amorphous powder | Kato, T et al, *J Neurochem*, 35:527, 1980

Gly-Pro

Peptides International OGP-3052   MW 172.18 $C_7H_{12}N_2O_3$ >98% (HPLC); colorless crystalline powder

Gly-Pro
Z-Gly-Pro

Peptides International SGP-3055   MW 306.32 $C_{15}H_{18}N_2O_5$ >99% (HPLC); amorphous powder

Gly-Pro
Gly-Pro-pNA Tos

Synonyms: X-Prolyl Dipeptidyl Aminopeptidase Substrate; GPNT

Peptides International SGP-3074-v   MW 464.49 $C_{13}H_{16}N_4O_4$ · $C_7H_8O_3S$ >99% (HPLC); lyophilized amorphous powder | Nagatsu, T et al, *Anal Biochem*, 74:466, 1976; Fujita, K et al, *Clin Chim Acta*, 88:15, 1978

Gly-Pro
N-t-BOC-Gly-Pro

Sigma B 9260   FW 272.3 $C_{12}H_{20}N_2O_5$

Gly-Pro
N-CBZ-Gly-Pro

Sigma C 0502   FW 306.3 $C_{15}H_{18}N_2O_5$

Gly-Pro
N-CBZ-Gly-Pro-Citrulline p-NA

Sigma C 8151   FW 583.6 $C_{27}H_{33}N_7O_8$

Gly-Pro
Gly-Pro p-NA Hydrochloride

Synonyms: Dipeptidyl Peptidase IV Substrate

Sigma G 0513   FW 328.8 $C_{13}H_{16}N_4O_4$ · HCl   Kojima, K et al, *J Chromatog*, 189:233, 1980; Nagatsu, T et al, *Anal Biochem*, 74:466, 1976

Gly-Pro
Gly-Pro MCA Hydrobromide

Synonyms: Dipeptidyl Aminopeptidase IV Substrate

Sigma G 2761   FW 410.3 $C_{17}H_{19}N_3O_4$ · HBr   Kato, T et al, *Biochem Med*, 19:351, 1978

Gly-Pro
Gly-Pro p-NA p-Toluenesulfonate salt

Sigma G 2901   FW 464.5 $C_{13}H_{16}N_4O_4$ · $C_7H_8O_3S$

Gly-Pro

Sigma G 3002   FW 172.2 $C_7H_{12}N_2O_3$

Gly-Pro
Gly-Hydroxy-Pro

Sigma G 7129   FW 188.2 $C_{17}H_{12}N_2O_4$

Gly-Pro
Gly-Pro 4-Methoxy-β-Naphthylamide

Sigma G 9137   FW 327.4 (free base) $C_{18}H_{21}N_3O_3$

Gly-Pro
Gly-Pro 4-Methoxy-β-Naphthylamide Hydrochloride

Synonyms: Dipeptidyl Aminopeptidase IV Substrate

Sigma G 9262   FW 363.8 $C_{18}H_{21}N_3O_3$ · HCl   Off-white powder | Lojda, Z, *Histochem*, 59:153, 1979; Gossrau, R et al, *Histochem*, 80:183, 1984

Gly-Pro
N-Succinyl-Gly-Pro MCA

Synonyms: Prolyl Endopeptidase Substrate

Sigma S 4383   FW 429.4 $C_{21}H_{23}N_3O_7$ Fluorogenic; Kato, T et al, *Neurohem*, 35:527, 1980; Yokosawa, H et al, *J Biochem (Tokyo)*, 94:1067, 1983; Soeda, S et al, *Chem Pharm Bull*, 33:2445, 1985

Gly-Pro
CBZ-Gly-Pro

USBio C2098-77   MW 306.3 $C_{15}H_{18}NNO_5$ ≥99%

Gly-Pro
CBZ-Gly-Pro-AMC

USBio C2098-83   MW 463.5 $C_{25}H_{25}NO$ ≥99%

Gly-Pro-Ala
Z-Gly-Pro-Ala

Bachem C-1890.0250   MW 377.4 $C_{18}H_{23}N_3O_6$ Store at -15°C

Gly-Pro-Ala

Bachem H-3615.0250   MW 243.26 $C_{10}H_{17}N_3O_4$ Store at -15°C

Gly-Pro-Ala
Isoamylphosphonyl-Gly-L-Pro-L-Ala

Synonyms: Collagenase Inhibitor

ICN 150706   95% (TLC, NMR); a 1 m*M* concentration completely inhibits 20 μmole of *Clostridium histolyticum* collagenase (K_I = 20 *μM*)

Gly-Pro-Ala

ICN 157271   MW 243.3 $C_{10}H_{17}N_3O_4$ Crystalline

Sigma G 7004   FW 243.3 $C_{10}H_{17}N_3O_4$

Gly-Pro-Ala-Ser-Val-Pro-Thr-Thr-Cys-Cys-Phe-Asn-Leu-Ala-Asn-Arg-Lys-Ile-Pro-Leu-Gln-Arg-Leu-Glu-Ser-Tyr-Arg-Arg-Ile-Thr-Ser-Gly-Lys-Cys-Pro-Gln-Lys-Ala-Val-Ile-Phe-Lys-Thr-Lys-Leu-Ala-Lys-Asp-Ile-Cys-Ala-Asp-Pro-Lys-Lys-Lys-Trp-Val-Gln-Asp-Ser-Met-Lys-Tyr-Leu-Asp-Gln-Lys-Ser-Pro-Thr-Pro-Lys-Pro

Synonyms: Eotaxin

Bachem H-4602.0010   Human   MW 8360.9 $C_{372}H_{609}N_{105}O_{103}S_5$ Store at -15°C

Gly-Pro-Arg

American Peptide 87-0-80   MW 328.4 $C_{13}H_{24}N_6O_4$

Bachem H-2930.0025   MW 328.37 $C_{13}H_{24}N_6O_4$ Store at -15°C

Gly-Pro-Arg
Z-Gly-Pro-Arg-AMC Hydrochloride

Bachem I-1150.0025   Store at -15°C

Gly-Pro-Arg
Tos-Gly-Pro-Arg-AMC Hydrochloride

Bachem I-1365.0010   MW 676.19 $C_{30}H_{37}N_7O_7S \cdot HCl$ Store at -15°C

Gly-Pro-Arg
Z-Gly-Pro-Arg-4MβNA

Bachem J-1120.0050   MW 617.71 $C_{32}H_{39}N_7O_6$ Store at -15°C

Gly-Pro-Arg
N-Tosyl-Gly-L-Pro-L-Arg pNA Acetate

Synonyms: Thrombin Substrate

Fluka 90175   MW 662.7 $C_{26}H_{34}N_8O_7S \cdot C_2H_4O_2$ ~99.0% (TLC)

Gly-Pro-Arg
N-p-Tosyl-Gly-Pro-Arg-MCA

ICN 156947

Gly-Pro-Arg
N-p-Tosyl-Gly-Pro-Arg-pNA Acetate Salt

Synonyms: Thrombin Substrate

ICN 156948   Crystalline | Chromogenic substrate; Abilgaard, U etal, *Thromb Res*, 11:549, 1977; Lottenberg, R etal, *Methods Enzymol*, 80:341, 1981

Gly-Pro-Arg
Tos-Gly-Pro-Arg-pNA

Synonyms: Thrombin Substrate

Neosystem SC1299   MW 602.65   Chromogenic substrate; Lottenberg, R et al, *Meth Enz*, 80:341-361, 1981

Gly-Pro-Arg
Z-Gly-Pro-Arg-MCA

Synonyms: Cathepsin K Substrate

Peptides International MCA-3208-v   MW 619.67 $C_{31}H_{37}N_7O_7$ >99% (HPLC); lyophilized amorphous powder | Aibe, K et al, *Biol Pharm Bull*, 19:1026, 1996; Bühling, F et al, *Am J Respir Cell Mol Biol*, 20:612, 1999

Gly-Pro-Arg
N-CBZ-Gly-Pro-Arg p-NA Acetate Salt

Sigma C 2276   FW 582.6 (free acid) $C_{27}H_{34}N_8O_7$

Gly-Pro-Arg
N-p-Tosyl-Gly-Pro-Arg MCA Hydrochloride

Sigma T 0273   FW 676.2 $C_{30}H_{37}N_7O_7S \cdot HCl$

Gly-Pro-Arg
N-p-Tosyl-Gly-Pro-Arg p-NA Acetate Salt

Synonyms: Thrombin Substrate

Sigma T 1637   FW 662.7 $C_{26}H_{34}N_8O_7S \cdot C_2H_4O_2$ Chromogenic; Abildgaard, U et al, *Thromb Res*, 11:549, 1977; Lottenberg, R et al, *Meth Enz*, 80:341, 1981

Gly-Pro-Arg
CBZ-Gly-Pro-Arg-AMC Hydrochloride

USBio C2098-84   MW 656.1 $C_{31}H_{37}N_7O_7 \cdot HCl$ ≥99%

Gly-Pro-Arg-Pro

Synonyms: Fibrinolysis Inhibiting Factor

American Peptide 44-0-28   MW 425.5 $C_{18}H_{31}N_7O_5$ Plow, E et al, *PNAS*, 29:3711, 1982

Bachem H-2935.0025   MW 425.49 $C_{18}H_{31}N_7O_5$ Store at -15°C

Neosystem SC133   MW 425.49   Bioactive; cell attachment peptide; fibronectin fragment; Plow, EF & Marguerie, G, *PNAS USA*, 79:3711, 1982

Gly-Pro-Arg-Pro
Gly-Pro-Arg-Pro-AcOH

Synonyms: Pefabloc®FG; Pefa-6003

Pentapharm 099-11, 099-01   MW 485.5   Readily soluble in H_2O | Added to fibrinogen-containing reaction mixtures to inhibit disturbing fibrin-related turbidity, gel formation & fibrin deposition in diagnostic & preparative procedures (eg, measurement of thrombin generation after activation in plasma); inhibits fibrin formation during purification & processing of clotting factors & other plasma proteins; Furlan M, et al, *Thromb Haemost*, 47:118, 1982; Prasa D, et al, *Thromb Haemost*, 77:498, 1997

Gly-Pro-Arg-Pro

Synonyms: Fibrin Polymerization Inhibitor

Biogenesis 4440-1509 Synthetic MW 425.5 Lyophilized, consisting of 73.75% peptide material, acetate counter ions and residual H_2O; purified

Gly-Pro-Arg-Pro Acetate Salt

Synonyms: Fibrin Polymerization Inhibitor

ICN 153115 $C_{18}H_{31}N_7O_5 \cdot C_2H_3O_2$ Inhibitor of fibrin polymerization; Plow, EF etal, *Biochem Pharmac*, 36:4035, 1987

Sigma G 1895 FW 425.5 (free base) $C_{18}H_{31}N_7O_5$ ≥97% (HPLC) | Bioactive peptide; Laudano, AP & Doolittle, RF, *Biochemistry*, 19:1012, 1980

Gly-Pro-Arg-Pro Amide

Bachem H-1998.0025 MW 424.5 $C_{18}H_{32}N_8O_4$ Store at -15°C

Gly-Pro-Arg-Pro Amide Acetate Salt

Synonyms: Fibrin Polymerization Inhibitor

Sigma G 5779 FW 424.5 (free base) $C_{18}H_{32}N_8O_4$ ≥97% (HPLC) | Bioactive peptide; potent inhibitor; Kawasaki, K et al, *Chem Pharm Bull*, 40:3253, 1992

Gly-Pro-Arg-Pro-Pro Amide

Bachem H-2612.0025 Store at -15°C

Gly-Pro-Arg-Pro-Pro-Glu-Arg-His-Gln-Ser Amide

Bachem H-2002.0001 MW 1159.27 $C_{48}H_{78}N_{20}O_{14}$ Store at -15°C

Gly-Pro-Gln-Arg-Glu-Pro-His-Asn-Glu-Trp-Thr-Leu

Synonyms: VPR (9-20); HTLV I Peptide

Neosystem SC556 MW 1463.57 HTLV I peptide from subtype VPR

Gly-Pro-Glu

Synonyms: Insulin-Like Growth Factor I (1-3)

Bachem H-2468.0250 Store at -15°C

Gly-Pro-Gly

Bachem H-9745.0250 Store at -15°C

Gly-Pro-Gly Amide Hydrochloride

Bachem H-9865.0250 MW 264.71 $C_9H_{16}N_4O_3 \cdot HCl$ Store at -15°C

Gly-Pro-Gly-Arg-Ala-Phe-Val-Thr

Synonyms: GP120 (317-324); HTLV I Peptide

Neosystem SC816 MW 803.91 HTLV I peptide from subtype GP160

Gly-Pro-Gly-Gly

Synonyms: Dipeptidyl Peptidase IV Inhibitor

Bachem H-3625.0250 Store at -15°C

ICN 153116 $C_{11}H_{18}N_4O_5$

Sigma G 6011 FW 286.3; ≥97% (HPLC) $C_{11}H_{18}N_4O_5$ Bioactive peptide; inhibits entry of HIV-1 or HIV-2 into T lymphoblastoid & monocytoid cell lines; Callebaut, C et al, *Science*, 262:2045, 1993

Gly-Pro-Gly-Gly-Pro-Ala
Z-Gly-Pro-Gly-Gly-Pro-Ala

Bachem M-1260.0100 MW 588.62 $C_{27}H_{36}N_6O_9$ Store at -15°C

Gly-Pro-Gly-Gly-Pro-Ala
N-CBZ-Gly-Pro-Gly-Gly-Pro-Ala

Synonyms: Collagenase Substrate

Sigma C 8008 FW 588.6 $C_{27}H_{36}N_6O_9$ Grassmann, W & Nordwig, A, *Hoppe-Seyler's Z Physiol Chem*, 322:267, 1960

Gly-Pro-Gly-Gly-Pro-Ala
CBZ-Gly-Pro-Gly-Gly-Pro-Ala

USBio C2098-85 $C_{27}H_{36}N_6O_9$ ≥99%

Gly-Pro-Hyp
FMOC-Gly-Pro-Hyp

Bachem B-2080.0001 MW 507.54 $C_{27}H_{29}N_3O_7$ Store at -15°C

Gly-Pro-Hyp

Bachem H-3630.0250 MW 285.3 $C_{12}H_{19}N_3O_5$ Store at -15°C

Sigma G 6027 FW 285.3 $C_{12}H_{19}N_3O_5$ ~6% isopropanol

Gly-Pro-Hyp Acetate Salt

ICN 157272 Crystalline

Gly-Pro-Hyp-Gly-Ala-Gly

Synonyms: Antiarrhythmic Peptide

American Peptide 14-1-80 MW 469.5 $C_{19}H_{29}N_6O_8$ Aonuma, S et al, *Chem Pharm Bull*, 32:219, 1984

Gly-Pro-Ile-Ser

Bachem H-1028.0025 MW 372.42 $C_{16}H_{28}N_4O_6$ Store at -15°C

Gly-Pro-Leu
Gly-Pro-Leu-βNA Hydrochloride

Bachem K-1340.0050 MW 446.98 $C_{23}H_{30}N_4O_3 \cdot HCl$ Store at -15°C

Gly-Pro-Leu
Z-Gly-Pro-Leu

Peptides International SGL-3039 MW 419.48 $C_{21}H_{29}N_3O_6$ >99% (HPLC); amorphous powder

Gly-Pro-Leu β-Naphthylamide Hydrochloride

Synonyms: Triaminopeptidase Substrate

Sigma G 1138 FW 447.0 $C_{23}H_{30}N_4O_3 \cdot HCl$ Aoyagi, T et al, *Biochem Biophys Res Commun*, 80:435, 1978

Gly-Pro-Leu-Ala
p-Aminobenzoyl-Gly-Pro-D-Leu-D-Ala-Hydroxamic Acid

Synonyms: Matrix Metalloproteinase Inhibitor, Human; Collagenase Inhibitor; Gelatinase Inhibitor; Stromelysin Inhibitor

ICN 195664 MW 490.6 $C_{23}H_{34}N_6O_6$ 97% | Broad spectrum protease inhibitor

Gly-Pro-Leu-Ala
p-Aminobenzoyl-Gly-Pro-D-Leu-D-Ala Hydroxamic Acid

Synonyms: Protease Inhibitor

Sigma A 4336 FW 490.6 $C_{23}H_{34}N_6O_6$ ≥97% (HPLC) | Bioactive peptide; broad-spectrum protease inhibitor, active against human matrix metalloproteinases, collagenase, gelatinase & stromelysin; Odake, S et al, *Biochem Biophys Res Commun*, 199:1442, 1994

Gly-Pro-Leu-Asp
Ac-Gly-Pro-Leu-Asp-AMC

Bachem I-1845.0005   MW 599.64 $C_{29}H_{37}N_5O_9$ Store at -15°C

Gly-Pro-Leu-Gly
Z-Gly-Pro-Leu-Gly

Peptides International SGG-3040   MW 476.53 $C_{23}H_{32}N_4O_7$ >99% (HPLC); amorphous powder

Gly-Pro-Leu-Gly-Pro
Suc-Gly-Pro-Leu-Gly-Pro-AMC

Bachem I-1350.0025   MW 696.76 $C_{34}H_{44}N_6O_{10}$ Store at -15°C

Gly-Pro-Leu-Gly-Pro
Z-Gly-Pro-Leu-Gly-Pro

Bachem M-1265.0100   MW 573.65 $C_{28}H_{39}N_5O_8$ Store at -15°C

Gly-Pro-Leu-Gly-Pro
N-Suc-Gly-L-Pro-L-Leu-Gly-L-Pro-MCA

Synonyms: Collagenase-Like Peptidase Substrate

ICN 152085   Fluorogenic substrate; Kojima, K etal, *Anal Biochem*, 99, 1979

Gly-Pro-Leu-Gly-Pro
Cbo-Gly-Pro-Leu-Gly-Pro-pNA

Synonyms: Pefachrome®Col; Pefa-15821

Pentapharm 090-30   MW 693.7   Chromogenic peptide substrate for determination of *Clostridium histolyticum* collagenase activity

Gly-Pro-Leu-Gly-Pro
Suc-Gly-Pro-Leu-Gly-Pro-MCA

Synonyms: Collagenase-Like Peptidase Substrate

Peptides International MGP-3108-v   MW 696.76 $C_{34}H_{44}N_6O_{10}$ >98% (HPLC); lyophilized amorphous powder | Kojima, K et al, *Anal Biochem*, 100:43, 1979

Gly-Pro-Leu-Gly-Pro
Z-Gly-Pro-Leu-Gly-Pro Hydrate AcOEt

Synonyms: Bacterial Collagenase Crystalline Substrate

Peptides International SGP-3029   MW 679.78 $C_{28}H_{39}N_5O_8$ · H_2O · $CH_3COOC_2H_5$ >99% (HPLC); amorphous powder | Nagai, Y et al, *BBA*, 37:567, 1960

Gly-Pro-Leu-Gly-Pro
N-CBZ-Gly-Pro-Leu-Gly-Pro

Synonyms: Collagenase Substrate

Sigma C 0627   FW 573.6 $C_{28}H_{39}N_5O_8$ Nagai, Y et al, *J Biochem*, 82:1495, 1977

Gly-Pro-Leu-Gly-Pro
N-Suc-Gly-Pro-Leu-Gly-Pro MCA

Synonyms: Collagenase-Like Peptidase Substrate

Sigma S 4258   FW 696.8 $C_{34}H_{44}N_6O_{10}$ Kojima, K et al, *Anal Biochem*, 100:43, 1979

Gly-Pro-Lys
Tos-Gly-Pro-Lys-AMC

Bachem I-1370.0050   MW 611.72 $C_{30}H_{37}N_5O_7S$ Store at -15°C

Gly-Pro-Lys
N-Tosyl-Gly-L-Pro-L-Lys-pNA Acetate

Fluka 90178   MW 634.8 $C_{26}H_{34}N_6O_7S$ · $C_2H_4O_2$ ≥97.0% (TLC); ≤4% water

Gly-Pro-Lys
N-p-Tosyl-Gly-Pro-Lys-pNA Acetate Salt

Synonyms: Plasmin Substrate

ICN 156949   Chromogenic substrate; Lottenberg, R etal, *Methods Enzymol*, 80:341, 1981

Gly-Pro-Lys
Tos-Gly-Pro-Lys-pNA

Synonyms: Plasmin Substrate

Neosystem SC1295   MW 574.64   Chromogenic substrate; Lottenberg, R et al, *Meth Enz*, 80:341-361, 1981

Gly-Pro-Lys
Tos-Gly-Pro-Lys-pNA AcOH

Synonyms: Pefachrome®Tryp; Pefa-3675

Pentapharm 090-36, 090-35   MW 634.7   Chromogenic peptide substrate for tryptase & other uses; v_{max}:35.6 µmol/min, K_M:0.014 *mM*; alpha lytic protease from *Achromobacter lyticus*; Eriksen, J & KA Holm, *Analyst*, 122:169, 1997; Ren, S et al, *J Immunol*, 3540-3548, 1997

Gly-Pro-Lys
N-p-Tosyl-Gly-Pro-Lys p-NA Acetate Salt

Synonyms: Plasmin Substrate

Sigma T 6140   FW 574.7 (free base) $C_{26}H_{34}N_6O_7S$ Chromogenic; Lottenberg, R et al, *Meth Enz*, 80:341, 1981

Gly-Pro-Lys-Thr-Pro-Glu-Glu-Lys-Thr-Ala-Asn-Thr-Ile-Ser-Lys-Phe-Asp-Cys

Synonyms: Protein Kinase C β Peptide

American Peptide 86-1-15   MW 1966.2 $C_{84}H_{135}N_{22}O_{30}S$

Gly-Pro-Phe-Pro-Leu
Z-Gly-Pro-Phe-Pro-Leu

Bachem C-1910.0050   MW 663.77 $C_{35}H_{45}N_5O_8$ Store at -15°C

Gly-Pro-Pro

Bachem H-6350.0250   MW 269.3 $C_{12}H_{19}N_3O_4$ Store at -15°C

Gly-Pro-Pro-Gly-Pro-Pro-Ser-Ala-Gly-Phe-Asp-Phe-Ser-Phe-Leu-Pro-Gln-Pro-Pro-Gln-Glu-Lys-Ala-His-Asp-Gly-Gly-Arg-Tyr-Tyr-Arg-Ala

Synonyms: Pro-Collagen α1 (1187-1218)

Bachem H-4156.0500   Human   Store at -15°C

Gly-Pro-Ser-Gln-Pro-Thr-Tyr-Pro-Gly-Asp-Asn-Ala-Thr-Pro-Glu-Gln-Met-Ala-Arg-Tyr-Tyr-Ser-Ala-Leu-Arg-Arg-Tyr-Ile-Asn-Met-Ala-Aib-Arg-Gln-Arg-Tyr Amide

Synonyms: Pancreatic Polypeptide; Neuropeptide Y (19-23), (Gly[1],Ser[3],Gln[4],Thr[6],Ala[31],Aib[32],Gln[34])-

Bachem H-5088.0500   Human   MW 4207.73 $C_{183}H_{281}N_{57}O_{54}S_2$ Store at -15°C

Gly-Pro-Ser-Gln-Pro-Thr-Tyr-Pro-Gly-Asp-Asp-Ala-Pro-Val-Glu-Asp-Leu-Ile-Arg-Phe-Tyr-Asp-Asn-Leu-Gln-Gln-Tyr-Leu-Asn-Val-Val-Thr-Arg-His-Arg-Tyr Amide

Synonyms: Gastrointestinal Peptide; Pancreatic Polypeptide

American Peptide 46-8-25 Avian MW 4237.7 $C_{190}H_{283}N_{53}O_{58}$ Kimmel, JR et al, *JBC*, 250:9369, 1975

ICN 152888 Avian MW 4237.6 Hazelwood, RC etal, *Gen Comp Endocrin*, 21:485, 1973; Kimmel, JR etal, *JBC*, 250:9369, 1975; Floyd, JC etal, *Rec Prog Horm Res*, 33:519, 1977

Sigma P 9653 Avian FW 4237.6 ≥97% (HPLC) | Bioactive peptide; Kimmel, JR et al, *J Biol Chem*, 250:9369, 1975; Floyd, JC et al, *Rec Prog Horm Res*, 33:519, 1977; Hazelwood, RC et al, *Gen Comp Endocrinol*, 21:485, 1973

Gly-Pro-Ser-Gly-Phe-Tyr-Gly-Val-Arg Amide

Synonyms: Locustachykinin I

American Peptide 87-7-60 Locust MW 938.0 $C_{43}H_{63}N_{13}O_{11}$ Schoofs, L et al, *FEBS Lett*, 266:397, 1990

Gly-Pro-Val-Glu-Asp-Ala-Ile-Thr-Ala-Ala-Ile-Gly-Arg-Val-Ala-Cys

Synonyms: Coxsackie B3 Virus Epitope

Bachem H-2096.0500 MW 1542.78 $C_{65}H_{111}N_{19}O_{22}S$ Store at -15°C

Gly-Sar
Gly-Sar (Gly 1-^{14}C)

ARC ARC-1158 MW 146.1 $H_2NCH_2CON(CH_3)CH_2COOH$ 50-60 mCi/mmol; 1.85-2.22 GBq/mmol; in EtOH | Radiochemical

Gly-Sar
Gly-Sar (Gly 2-^{3}H)

ARC ART-585 MW 146.15 $H_2NCH_2CON(CH_3)CH_2COOH$ 30-60 Ci/mmol; 1.85-2.22 TBq/mmol; in EtOH:H$_2$O (1:1) | Radiochemical

Gly-Sar
Z-Gly-Sar

Bachem C-1915.0001 MW 280.28 $C_{13}H_{16}N_2O_5$ Store at -15°C

Gly-Sar

Bachem G-2195.0001 MW 146.15 $C_5H_{10}N_2O_3$ Store at -15°C

Gly-Sar
Gly-L-Sar

Fluka 50430 MW 146.15 $C_5H_{10}N_2O_3$ ≥99% (titration); mp:198-202°C

Gly-Sar

ICN 105243 MW 146.2 $C_5H_{10}N_2O_3$

Sigma G 3127 FW 146.1 $C_5H_{10}N_2O_3$

Gly-Sar-Sar

Bachem H-6650.0250 Store at -15°C

Gly-Ser
Z-Gly-Ser

Bachem C-1920.0001 MW 296.28 $C_{13}H_{16}N_2O_6$ Store at -15°C

Gly-Ser

Bachem G-2200.0001 MW 162.15 $C_5H_{10}N_2O_4$ Store at -15°C

Gly-Ser
Gly-DL-Ser

Bachem G-4280.0001 MW 162.15 $C_5H_{10}N_2O_4$ Store at -15°C

Gly-Ser

ICN 101890 MW 162.1 $C_5H_{10}N_2O_4$ Crystalline

Gly-Ser
Borotrimethyl-Gly-Ser-OMe

ICN 154642 MW 218.06 $BC_8H_{19}N_2O_4$ Low melting solid; >97%

Gly-Ser
Boro-Gly-Ser Amide

ICN 154649 MW 160.97 $BC_4H_{22}N_3O_3$ >97%

Gly-Ser
Gly-DL-Ser

Sigma G 0379 FW 162.1 $C_5H_{10}N_2O_4$

Gly-Ser

Sigma G 3252 FW 162.1 $C_5H_{10}N_2O_4$

Gly-Ser
Gly-D-Ser

Sigma G 3754 FW 162.1 $C_5H_{10}N_2O_4$

Gly-Ser-Ala

Bachem H-3640.0250 MW 233.22 $C_8H_{15}N_3O_5$ Store at -15°C

Gly-Ser-Ala-Lys-Val-Ala-Phe-Ser-Ala-Ile-Arg-Ser-Thr-Asn-His

Synonyms: Cerebellin, (des-Ser[1])-

American Peptide 88-0-46 MW 1545.7 $C_{66}H_{108}N_{22}O_{21}$ Siemmon, JR et al, *PNAS*, 81:6866, 1984

Bachem H-5535.0500 MW 1545.72 $C_{66}H_{108}N_{22}O_{21}$ Store at -15°C

ICN 154476 MW 1545.6 Slemmon, JR etal, *PNAS*, 81:6866, 1984

Gly-Ser-Arg-Ala-His-Ser-Ser-His-Leu-Lys-Ser-Lys-Lys-Gly-Gln-Ser-Thr-Ser-Arg-His-Lys-Lys

Synonyms: Peptide 46; Cellular Tumor Antigen p53 (361-382); p53 (361-382)

Bachem H-4054.0500 Human Store at -15°C

Gly-Ser-Asn-Lys-Gly-Ala-Ile-Ile-Gly-Leu-Met

Synonyms: Amyloid β-Protein (25-35)

Alexis 151-008 MW 1060.3 $C_{45}H_{81}N_{13}O_{14}S$ ≥98%; white lyophilized powder; soluble in water | Yankner, BA et al, *Science*, 250:279, 1990

Gly-Ser-Asn-Lys-Gly-Ala-Ile-Ile-Gly-Leu-Met
β-Gly-Ser-Asn-Lys-Gly-Ala-Ile-Ile-Gly-Leu-Met

Synonyms: Amyloid (25-35)

American Peptide 62-0-72 MW 1060.3 $C_{45}H_{81}N_{13}O_{14}S$
The biologically active domain of the amyloid β-protein for neurotrophic & neurotoxic effects is located within this region, which is homologous to the sequence of the peptides in the tachykinin family; Yankner, BA et al, *Science*, 250:279, 1990

Gly-Ser-Asn-Lys-Gly-Ala-Ile-Ile-Gly-Leu-Met

Synonyms: Amyloid α-Protein (25-35)

Bachem H-1192.0001 MW 1060.28 $C_{45}H_{81}N_{13}O_{14}S$ Store at -15°C

Gly-Ser-Asn-Lys-Gly-Ala-Ile-Ile-Gly-Leu-Met
Gly-Ser-Asn-Lys-Gly-Ala-Ile-Ile-Gly-Leu-Met(O)

Synonyms: Amyloid α-Protein (25-35), (Met(O)[35])-

Bachem H-2962.0001 MW 1076.28 $C_{45}H_{81}N_{13}O_{15}S$ Store at -15°C

Gly-Ser-Asn-Lys-Gly-Ala-Ile-Ile-Gly-Leu-Met

Synonyms: Amyloid Peptide (25-35), β-; Amyloid β-Protein (25-35); Neurotrophic/Neurodegenerative Peptide

ICN 159867 MW 1060.6 >98% | Yanker, BA etal, *Science*, 250:279, 1990

Neosystem SC489 MW 1060.27 Bioactive; Yankner, BA et al, *Science*, 250:279-282, 1990

Sigma A 4559 FW 1060.3 ≥97% (HPLC); peptide content:~70% | Bioactive peptide; functional domain of β-amyloid required for both neurotrophic & neurotoxic effects; Yankner, BA et al, *Science*, 250:279, 1990

Peptides International PAM-4309-v Human synthetic MW 1060.3 $C_{45}H_{81}N_{13}O_{14}S$ >95% (HPLC); lyophilized amorphous powder; trifluoroacetate | Bioactive; Yankner, BA et al, *Science*, 250:279, 1990; Meda, L et al, *Nature*, 374:647 1995

Gly-Ser-Asn-Lys-Gly-Ala-Ile-Ile-Gly-Leu-Met Amide

Synonyms: Amyloid α-Protein (25-35)

Bachem H-4222.0001 MW 1059.3 $C_{45}H_{82}N_{14}O_{13}S$ Store at -15°C

Gly-Ser-Asp-Ile-Ala-Gly-Thr-Thr-Ser-Thr-Leu

Synonyms: P25 (233-243); GAG P24 CA (101-111); HTLV I Peptide

Neosystem SC666 MW 1022.07 HTLV I peptide from subtype P25 (GAG P24 CA)

Gly-Ser-Gly-Met-Met-Ser-Lys-Ser-Asp-Asn-Phe-Gly-Glu-Lys-Cys Trifluoroacetate Salt

Synonyms: Calretinin (54-67), ~C

Biogenesis 1741-1060 Human, rat synthetic MW 1578.1
Purified; lyophilized

Gly-Ser-Gly-Phe-Ser-Ser-Trp-Gly Amide

Synonyms: Leukokinin V

American Peptide 33-1-23 MW 782.8 $C_{35}H_{46}N_{10}O_{11}$

Gly-Ser-His-Trp-Ala-Val-Gly-His-Leu-Met Amide

Synonyms: Neuromedin C, (Ser[2])-

American Peptide 62-3-10 MW 1093.3 $C_{49}H_{72}N_{16}O_{11}S$
BBRC, 178:529, 1991

ICN 159167 MW 1093.3 97%

Sigma N 1773 Porcine spinal cord FW 1093.3 ≥97% (HPLC); peptide content:~70% | Bioactive peptide; bombesin-like peptide originally isolated from the brain of the European green frog, Rana *ridibunda*; *BBRC*, 178:529, 1991

Gly-Ser-Leu-Lys-Gln-Gln-Leu-Arg-Glu-Tyr-Ile-Arg

Bachem H-9780.0001 Store at -15°C

Gly-Ser-Phe

ICN 157273 MW 309.3 $C_{14}H_{19}N_3O_5$

Sigma G 0642 FW 309.3 $C_{14}H_{19}N_3O_5$

Gly-Ser-Phe-Leu-Val-Arg-Glu-Ser

Synonyms: Tyrosine Phosphorylation Site Inhibitor; Tyrosine Phosphorylation Site Inhibitor Peptide

Alexis 168-003 MW 894.0 $C_{39}H_{63}N_{11}O_{13}$ ≥98%; lyophilized powder; soluble in water | Effective inhibitor of EGF receptor binding to SH₂ of PLCγ1; corresponds to a highly conserved AA sequence common in a variety of SH₂-containing proteins; the minimum essential sequence for recognition of the phosphotyrosine site; Hidaka, M et al, *BBRC*, 180:1490, 1991

ICN 193727 MW 894 Blocks binding at phosphorylated tyrosine sites

Gly-Ser-Ser-Phe-Leu-Ser-Pro-Glu-His-Gln-Arg-Val-Gln-Gln-Arg-Lys-Glu-Ser-Lys-Lys-Pro-Pro-Ala-Lys-Leu-Gln-Pro-Arg
Gly-Ser-Ser(Octanoyl)-Phe-Leu-Ser-Pro-Glu-His-Gln-Arg-Val-Gln-Gln-Arg-Lys-Glu-Ser-Lys-Lys-Pro-Pro-Ala-Lys-Leu-Gln-Pro-Arg

Synonyms: Ghrelin

Bachem H-4864.0500 Human MW 3370.91 $C_{149}H_{249}N_{47}O_{42}$ Store at -15°C

Gly-Ser-Ser-Phe-Leu-Ser-Pro-Glu-His-Gln-Arg-Val-Gln-Gln-Arg-Lys-Glu-Ser-Lys-Lys-Pro-Pro-Ala-Lys-Leu-Gln-Pro-Arg
Gly-Ser-Ser(n-Octanoyl)-Phe-Leu-Ser-Pro-Glu-His-Gln-Arg-Val-Gln-Gln-Arg-Lys-Glu-Ser-Lys-Lys-Pro-Pro-Ala-Lys-Leu-Gln-Pro-Arg

Synonyms: Ghrelin; Growth Hormone Releasing Peptide

Peptides International PGH-4372-s Human synthetic MW 3370.9 $C_{149}H_{249}N_{47}O_{42}$ >99% (HPLC); lyophilized amorphous powder | Bioactive; endogenous growth-hormone releasing peptide with novel regulatory mechanism; Kojima, M et al, *Nature*, 402:656, 1999

Gly-Ser-Ser-Phe-Leu-Ser-Pro-Glu-His-Gln-Lys-Ala-Gln-Gln-Arg-Lys-Glu-Ser-Lys-Lys-Pro-Pro-Ala-Lys-Leu-Gln-Pro-Arg
Gly-Ser-Ser(Octanoyl)-Phe-Leu-Ser-Pro-Glu-His-Gln-Lys-Ala-Gln-Gln-Arg-Lys-Glu-Ser-Lys-Lys-Pro-Pro-Ala-Lys-Leu-Gln-Pro-Arg

Synonyms: Ghrelin

Bachem H-4862.0500 Rat MW 3314.84 $C_{147}H_{245}N_{45}O_{42}$ Store at -15°C

Gly-Ser-Ser-Phe-Leu-Ser-Pro-Glu-His-Gln-Lys-Ala-Gln-Gln-Arg-Lys-Glu-Ser-Lys-Lys-Pro-Pro-Ala-Lys-Leu-Gln-Pro-Arg
Gly-Ser-Ser(n-Octanoyl)-Phe-Leu-Ser-Pro-Glu-His-Gln-Lys-Ala-Gln-Gln-Arg-Lys-Glu-Ser-Lys-Lys-Pro-Pro-Ala-Lys-Leu-Gln-Pro-Arg

Synonyms: Ghrelin; Growth Hormone Releasing Peptide

Peptides International PGH-4373-s Rat synthetic MW 3314.8 $C_{147}H_{245}N_{45}O_{42}$ >99% (HPLC); lyophilized amorphous powder | Bioactive; endogenous growth-hormone releasing peptide with novel regulatory mechanism; Kojima, M et al, *Nature*, 402:656, 1999

Gly-Ser-Val-Ala-Phe-Pro-Ala-Glu-Asn-Gly-Val-Gln-Asn-Thr-Glu-Ser-Thr-Gln-Glu

Synonyms: Neuropeptide GE

Bachem H-4572.0500 MW 1965.02 $C_{81}H_{125}N_{23}O_{34}$ Store at -15°C

Gly-Ser-Val-Val-Ile-Val-Gly-Arg-Ile-Ile-Leu-Ser-Gly-Arg

Synonyms: HCV NS4A Protein (21-34), JT Strain; HCV NS4A Protein (21-34)

American Peptide 72-3-39 MW 1424.7 $C_{63}H_{115}N_{20}O_{17}$
Shimizu, Y et al, *J Virol*, 70:127, 1996; Steinkuhler, C et al, *JBC*, 271:6367, 1996

Bachem H-3486.0001 Store at -15°C | JT strain

Gly-Thr

Bachem G-2205.0001 MW 176.17 $C_6H_{12}N_2O_4$ Store at -15°C

Gly-Thr
Gly-DL-Thr

Bachem G-2215.0001 MW 176.17 $C_6H_{12}N_2O_4$ Store at -15°C

ICN 101892 MW 176.2 $C_6H_{12}N_2O_4$ Crystalline

Gly-Thr

ICN 101893 MW 176.2 $C_6H_{12}N_2O_4$ Crystalline

Gly-Thr
Gly-D-Tyr

Sigma G 2379 FW 176.2 $C_6H_{12}N_2O_4$

Gly-Thr-Arg-Ala-Glu-Asn-Arg-Thr-Tyr-Ile-Tyr-Trp-His-Gly-Arg-Asp-Asn-Arg-Thr-Ile

Synonyms: GP140 (281-300); SIVmac251 Peptide

Neosystem SC736 MW 2479.69 SIVmac251 peptide from subtype GP140

Gly-Thr-Asn-Asn-Thr-Asp-Lys-Ile-Asn-Leu-Thr-Ala-Pro-Gly-Gly-Gly-Asp-Pro-Glu-Val

Synonyms: GP140 (371-390); SIVmac251 Peptide

Neosystem SC745 MW 1970.07 SIVmac251 peptide from subtype GP140

Gly-Thr-Lys-Gly-Gly-Gln-Asp

Bachem H-6985.0005 Store at -15°C

Gly-Thr-Ser-Arg-Asn-Lys-Arg-Gly-Val-Phe-Val-Leu-Gly-Phe-Leu-Gly-Phe-Leu-Ala-Thr

Synonyms: GP140 (521-540); SIVmac251 Peptide

Neosystem SC760 MW 2140.51 SIVmac251 peptide from subtype GP140

Gly-Thr-Val-Leu-Val-Gly-Pro-Thr-Pro-Val-Asn-Ile-Ile-Gly-Arg

Synonyms: POL (73-87); PR (73-87); HTLV I Peptide

Neosystem SC706 MW 1492.78 HTLV I peptide from subtype POL (PR/RT)

Gly-Trp
Z-Gly-Trp

Bachem C-1925.0001 MW 395.42 $C_{21}H_{21}N_3O_5$ Store at -15°C

Gly-Trp
Z-Gly-D-Trp

Bachem C-1930.0250 MW 395.42 $C_{21}H_{21}N_3O_5$ Store at -15°C

Gly-Trp

Bachem G-2220.0001 MW 261.28 $C_{13}H_{15}N_3O_3$ Store at -15°C

Gly-Trp
Gly-D-Trp

Bachem G-2225.0001 MW 261.28 $C_{13}H_{15}N_3O_3$ Store at -15°C

Gly-Trp
Gly-Trp-βNA Hydrochloride

Bachem K-1345.0250 MW 422.91 $C_{23}H_{22}N_4O_2 \cdot HCl$ Store at -15°C

Gly-Trp

ICN 101894 MW 261.3 $C_{13}H_{15}N_3O_3$ Crystalline

Gly-Trp
DANSYL-Gly-Trp

Sigma D 0511 FW 494.6 $C_{25}H_{26}N_4O_5S$

Gly-Trp Amide Hydrochloride

Bachem G-2230.0250 MW 296.76 $C_{13}H_{16}N_4O_2 \cdot HCl$ Store at -15°C

Gly-Trp-Gly

Bachem H-3655.0250 MW 318.33 $C_{15}H_{18}N_4O_4$ Store at -15°C

Gly-Trp-Gly
4-Nitro-Z-Gly-Trp-Gly

Bachem M-1595.0050 MW 497.46 $C_{23}H_{23}N_5O_8$ Store at -15°C

Gly-Trp-Gly-Gly

Bachem H-3660.0100 MW 375.38 $C_{17}H_{21}N_5O_5$ Store at -15°C

Gly-Trp-Met-Asp-Phe
N-t-BOC-β-Gly-Trp-Met-Asp-Phe Amide

Synonyms: Cholecystokinin (29-33), N-t-BOC

ICN 195674 MW 753.9 $C_{36}H_{47}N_7O_9S$ ≥95%

Gly-Trp-Met-Asp-Phe
N-t-BOC-Gly-Trp-Met-Asp-Phe Amide

Synonyms: Gastrointestinal Peptide (29-33)

Sigma B 1386 FW 753.9 $C_{36}H_{47}N_7O_9S$ ≥97% (HPLC) | Bioactive peptide

Gly-Trp-Met-Asp-Phe
BOC-Gly-Trp-Met-Asp(OBzl)-Phe Amide

Synonyms: Gastrin I (13-17), BOC-(Asp(OBzl)[16])-; Cholecystokinin Octapeptide (4-8), BOC-(Asp(OBzl)[7])-; Caerulein (6-10), BOC-(Asp(OBzl)[9])-

Bachem A-4310.0025 Human Store at -15°C

Gly-Trp-Pro-Gln-Ala-Pro-Ala-Met-Asp-Gly-Ala-Gly-Lys-Thr-Gly-Ala-Glu-Glu-Ala-Gln-Pro-Pro-Glu-Gly-Lys-Gly-Ala-Arg-Glu-His-Ser-Arg-Gln-Glu-Glu-Glu-Glu-Glu-Thr-Ala-Gly-Ala-Pro-Gln-Gly-Leu-Phe-Arg-Gly Amide

Synonyms: Pancreastatin (1-49); Chromogranin A (240-288)

American Peptide 46-4-60   Porcine   MW 5103.5
$C_{214}H_{330}N_{68}O_{76}S$ Potent inhibitor of glucose-induced & gastric inhibitory peptide-induced insulin secretion from rat pancreas; shows a high sequence homology to the central part of bovine, human & rat chromogranin A; Tatemoto, K et al, *Nature*, 324:476, 1986

Bachem H-6165.0500   Porcine   MW 5103.45
$C_{214}H_{330}N_{68}O_{76}S$ Store at -15°C

ICN 152886   Porcine   MW 5103.4   Pancreatic peptide that inhibits insulin secretion; Tatemoto, K etal, *Nature*, 324:476, 1986

Neosystem SC055   Porcine   MW 5103.43   Bioactive; brain/gut peptide; Tatemoto, K et al, *Nature*, 324:476-478, 1986

Gly-Trp-Pro-Gln-Ala-Pro-Ala-Nle-Asp-Gly-Ala-Gly-Lys-Thr-Gly-Ala-Glu-Glu-Ala-Gln-Pro-Pro-Glu-Gly-Lys-Gly-Ala-Arg-Glu-His-Ser-Arg-Gln-Glu-Glu-Glu-Glu-Glu-Thr-Ala-Gly-Ala-Pro-Gln-Gly-Leu-Phe-Arg-Gly Amide

Synonyms Pancreastatin, (Nle^8)-:

American Peptide 46-4-61   Porcine   MW 5085.2
$C_{215}H_{332}N_{68}O_{76}$

Gly-Trp-Thr-Leu-Asn-Ala-Ala-Trp-Tyr-Leu-Leu-Gly-Pro-His
Gly-Trp-Thr-Leu-Asn-Ala-Ala-D-Trp-Tyr-Leu-Leu-Gly-Pro-His-L-Alaninol

Synonyms: Galanin (1-15), $(Ala^6, D-Trp^8, L-Alaninol^{15})$-

Bachem H-4066.0500   MW 1655.92 $C_{81}H_{114}N_{20}O_{18}$ Store at -15°C

Gly-Trp-Thr-Leu-Asn-Ser-Ala-Gly-Tyr-Leu-Leu-Gly-Pro-Arg-Pro-Lys-Pro-Gln-Gln-Trp-Phe-Trp-Leu-Leu
Gly-Trp-Thr-Leu-Asn-Ser-Ala-Gly-Tyr-Leu-Leu-Gly-Pro-D-Arg-Pro-Lys-Pro-Gln-Gln-D-Trp-Phe-D-Trp-Leu-Leu Amide

Galanin (1-13); Spantide; C7*Synonyms:*

Neosystem SC885   MW 2828.31   Bioactive; brain/gut peptide; highly potent galanin antagonist that blocks galanin-induced feeding; Langel, U Land,T & Bartfai, T, *Int J Pep Prot Res*, 39:516-522, 1992; Bartfai, T et al, *TIPS*, 13:312-317, 1992; Crawley, JN et al, *Brain Res*, 600:268-272, 1993; Corwin, RL et al, *Eur J Neurosci*, 5:1528-1533, 1993

Gly-Trp-Thr-Leu-Asn-Ser-Ala-Gly-Tyr-Leu-Leu-Gly-Pro-Arg-His-Tyr-Ile-Asn-Leu-Ile-Thr-Arg-Gln-Arg-Tyr Amide

Synonyms: Galanin (1-13)-Neuropeptide Y (25-36); M-32

Bachem H-3374.0500   MW 2962.41 $C_{136}H_{209}N_{41}O_{34}$ Store at -15°C

Gly-Trp-Thr-Leu-Asn-Ser-Ala-Gly-Tyr-Leu-Leu-Gly-Pro-Arg-Pro-Lys-Pro-Gln-Gln-Trp-Phe-Trp-Leu-Leu
Gly-Trp-Thr-Leu-Asn-Ser-Ala-Gly-Tyr-Leu-Leu-Gly-Pro-D-Arg-Pro-Lys-Pro-Gln-Gln-D-Trp-Phe-D-Trp-Leu-Leu Amide

Synonyms: Galanin (1-13); Spantide I ; Galanin (1-13)-Spantide I; C7

American Peptide 46-1-06   MW 2828.3 $C_{138}H_{199}N_{35}O_{30}$ A galanin receptor antagonist which inhibits galanin-induced feeding in rats; Crawley, JN et al, *Brain Res*, 600:268, 1993

Bachem H-2578.0500   MW 2828.32 $C_{138}H_{199}N_{35}O_{30}$ Store at -15°C

Sigma G 7160   FW 2828.3; ≥97% (HPLC)   Bioactive peptide; tandem-linked peptide which acts as a potent galanin receptor antagonist; Crawley, JN et al, *Brain Res*, 600:268, 1993

Gly-Trp-Thr-Leu-Asn-Ser-Ala-Gly-Tyr-Leu-Leu-Gly-Pro-Gln-Gln-Phe-Phe-Gly-Leu-Met Amide

Synonyms: Galanin (1-13)-Substance P (5-11); Galantide; M-15; Galanin (1-13); Substance P (5-11); Gastrointestinal Peptide

American Peptide 46-1-08   MW 2199.6 $C_{104}H_{151}N_{25}O_{26}S$ Galanin antagonist; causes widespread antagonism to the action of galanin & reversibly blocks the neuronal actions of galanin in a dose-dependent manner; inhibits release of acetylcholine, reduces the response to galanin in the hypothalamus & blocks the effect of galanin on glucose-induced insulin secretion

Bachem H-1312.0500   MW 2199.56 $C_{104}H_{151}N_{25}O_{16}S$ Store at -15°C

Neosystem SC883   MW 2199.55   Bioactive; brain/gut peptide; high affinity galanin antagonist; Bartfai, T et al, *PNAS USA*, 88:10961-10965, 1991; Lindskog, S et al, *Eur J Pharmacol*, 210:183-188, 1992; Langel, U et al, *Int J Pept Prot Res*, 39:516-522, 1992

Sigma G 1278   FW 2199.5   ≥90% (HPLC) | Bioactive peptide; galanin receptor antagonist; Bartfai, T et al, *Proc Natl Acad Sci USA*, 88:10961, 1991; Lindskog, S et al, *Eur J Pharmacol*, 210:183, 1992

Gly-Trp-Thr-Leu-Asn-Ser-Ala-Gly-Tyr-Leu-Leu-Gly-Pro-His-Ala-Ile

Synonyms: Galanin (1-16)

Bachem H-3138.0500   Porcine, rat   MW 1669.9
$C_{78}H_{116}N_{20}O_{21}$ Store at -15°C

Neosystem SC884   Porcine, rat   MW 1669.9   Bioactive; brain/gut peptide; high affinity full agonist at the hippocampal galanin receptor; Fisone, G et al, *PNAS USA*, 86:9588-9591, 1989

Gly-Trp-Thr-Leu-Asn-Ser-Ala-Gly-Tyr-Leu-Leu-Gly-Pro-His-Ala-Ile-Asp-Asn-His-Arg-Ser-Phe-His-Asp-Lys-Tyr-Gly-Leu-Ala Amide

Synonyms: Gastrointestinal Peptide; Galanin

Alexis 157-004   Porcine   MW 3210.6 $C_{146}H_{213}N_{43}O_{40}$ Tatemoto, K et al, *FEBS Lett*, 164:124, 1983

American Peptide 46-1-10   Porcine   MW 3210.6
$C_{146}H_{213}N_{43}O_{40}$ Originally isolated from porcine upper small intestine inhibits the release of acetylcholine, dopamine, norepinephrine & other neurotransmitters; inhibits secretion of somatostatin, insulin & pancreatic polypeptide; Tatemoto, K et al, *FEBS Lett*, 164:124, 1983

Bachem H-1365.0500   Porcine   MW 3210.56
$C_{146}H_{213}N_{43}O_{40}$ Store at -15°C

Calbiochem 05-23-2350   Porcine   MW 3210.6
$C_{146}H_{312}N_{43}O_{40}$ ≥97% (HPLC); lyophilized solid; soluble in 5% acetic acid; harmful:LD_{50} ≤2000 mg/kg | Biologically active neuropeptide consisting of 20 AAs; inhibits secretion of somatostatin, insulin & pancreatic polypeptide; induces repolarization & reduces free Ca^{2+} due to interference with voltage-activated Ca^{2+} channels; inhibits ω-conotoxin GVIA-sensitive Ca^{2+} channels in parasympathetic neurons; Mirriam, LA & Parsons, RL, *Neurophysiol*, 73:1374, 1995; Homaidan, FR, *PNAS*, 88:8744, 1991; Ahren, B et al, *FEBS Lett*, 299:233, 1988; Lindskog, S & Ahren, B, *Eur J Pharmacol*, 205:21, 1991; Tatemoto, K et al, *FEBS Lett*, 164:124, 1983

ICN 191462   Porcine   MW 3210.6   Tatemoto, K etal, *FEBS Lett*, 164:124

Neosystem SC092   Porcine   MW 3210.55   Bioactive; brain/gut peptide; Tatemoto, K et al, *FEBS Lett*, 164:124-128, 1983; Rökaeus, A, TINS, 10:158-164, 1987; Ahren, B et al, *FEBS Lett*, 229:233-237, 1988

Sigma G 5773   Porcine   FW 3210.6   ≥97% (HPLC) | Bioactive peptide; inhibits neurotransmitter release; inhibits secretion of somatostatin, insulin & pancreatic polypeptide; blocks voltage-activated Ca^{2+} channels; Haynes, LW, *TIPS*, 6:214, 1986; Tatemoto, K et al, *FEBS Lett*, 164:124, 1983

Gly-Trp-Thr-Leu-Asn-Ser-Ala-Gly-Tyr-Leu-Leu-Gly-Pro-His-Ala-Ile-Asp-Asn-His-Arg-Ser-Phe-His-Asp-Lys-Tyr-Gly-Leu-Ala

Synonyms: Galanin

Biogenesis 4600-5204   Porcine synthetic   MW 3211.2
Lyophilized, consisting of 90.7% peptide material, acetate counter ions and residual H_2O; purified

Gly-Trp-Thr-Leu-Asn-Ser-Ala-Gly-Tyr-Leu-Leu-Gly-Pro-His-Ala-Ile-Asp-Asn-His-Arg-Ser-Phe-Ser-Asp-Lys-His-Gly-Leu-Thr Amide

Synonyms: Gastrointestinal Peptide; Galanin

American Peptide 46-5-10   Rat   MW 3164.5
$C_{141}H_{211}N_{43}O_{41}$ Synthesized initially as part of a 124-AA precursor that includes a signal peptide, galanin & a 60-AA galanin mRNA-associated peptide; Vrontakis, ME et al, *JBC*, 262:16755, 1987; Kaplan, LM et al, *PNAS*, 85:1065, 1988

Bachem H-7450.0500   Rat   MW 3164.49 $C_{141}H_{211}N_{43}O_{41}$
Store at -15°C

ICN 152865   Rat   MW 3164.5 $C_{141}H_{211}N_{43}O_{41}$ Vrontakis, ME etal, *JBC*, 262:16755, 1987; Kaplan, LM etal, *PNAS*, 1065, 1988

Neosystem SC936   Rat   MW 3164.48   Bioactive; brain/gut peptide; Vrontakis, ME et al, *JBC*, 262:16755-16758, 1987; Kaplan, LM et al, *PNAS USA*, 85:1065-1069, 1988

Sigma G 8272   Rat   FW 3164.5   ≥97% (HPLC); peptide content:~65% | Bioactive peptide; inhibits neurotransmitter release; blocks voltage-activated Ca^{2+} channels; Haynes, LW, *TIPS*, 6:214, 1986; Kaplan, LM et al, *Proc Natl Acad Sci USA*, 85:1065, 1988; Vrontakis, ME et al, *J Biol Chem*, 262:16755, 1987

Peptides International PGA-4244-v   Rat synthetic   MW 3164.5 $C_{141}H_{211}N_{43}O_{41}$ >99% (HPLC); lyophilized amorphous powder | Bioactive; Vrontakis, ME et al, *JBC*, 262:16755, 1987; Kaplan, LM et al, *PNAS USA*, 85:1065, 1988

Gly-Trp-Thr-Leu-Asn-Ser-Ala-Gly-Tyr-Leu-Leu-Gly-Pro-His-Ala-Val-Gly-Asn-His-Arg-Ser-Phe-Ser-Asp-Lys-Asn-Gly-Leu-Thr-Ser

Synonyms: Galanin

Alexis 157-003   Human   MW 3157.4 $C_{139}H_{210}N_{42}O_{43}$
Bersani, M et al, *FEBS Lett*, 283:189, 1991; Schmidt, WE et al, *PNAS*, 88:11435, 1991; Rokaeus, A, *TINS*, 10:158, 1987; Haynes, LW, *TIPS*, 214:1986; Ahren, B et al, *FEBS Lett*, 229:233, 1988; Dupre, J, *Pancreas*, 3:119, 1988; Crawley, JN & Wenk, GL, *TINS*, 12:278, 1989; Rossowski, WJ & Coy, DH, *Life Sci*, 44:1807, 1989; Messell, T et al, *Regul Pept*, 28:161, 1990

American Peptide 46-1-01   Human   MW 3157.5
$C_{139}H_{210}N_{42}O_{43}$ Shown to inhibit acetylcholine release, glutamate release & carbachol-stimulated phosphatidylinositol hydrolysis in the hippocampus; also linked to behavioral & cognitive deficits in Alzheimer's disease; Schmidt, WE et al, *PNAS*, 88:11435, 1991

Gly-Trp-Thr-Leu-Asn-Ser-Ala-Gly-Tyr-Leu-Leu-Gly-Pro-His-Ala-Val-Gly-Asn-His-Arg-Ser-Phe-Ser-Asp-Lys-Asn-Gly-Leu-Thr

Gly-D-Trp-Thr-Leu-Asn-Ser-Ala-Gly-Tyr-Leu-Leu-Gly-Pro-His-Ala-Val-Gly-Asn-His-Arg-Ser-Phe-Ser-Asp-Lys-Asn-Gly-Leu-Thr

Synonyms: Galanin (1-29), (D-Trp[2])-

Bachem H-4122.0500   Human   Store at -15°C

Gly-Trp-Thr-Leu-Asn-Ser-Ala-Gly-Tyr-Leu-Leu-Gly-Pro-His-Ala-Val-Gly-Asn-His-Arg-Ser-Phe-Ser-Asp-Lys-Asn-Gly-Leu-Thr-Ser

Synonyms: Gastrointestinal Peptide; Galanin

Bachem H-8230.0500   Human   MW 3157.45
$C_{139}H_{210}N_{42}O_{43}$ Store at -15°C

Neosystem SC882   Human   MW 3157.44   Bioactive; brain/gut peptide; Bersani, M et al, *FEBS Lett*, 283:189-194, 1991; Schmidt, WE et al, *PNAS USA*, 88:11435-11439, 1991

Sigma G 0278   Human   FW 3157.4   ≥97% (HPLC); peptide content:~70% | Bioactive peptide; co-located with acetylcholine & inhibits its release; inhibits the release of Glu, decreases excitability of spinal neurons; blocks voltage-activated Ca^{2+} channels; linked to behavioral & cognitive deficits in Alzheimer's disease; Bersani, M et al, *FEBS Lett*, 283:189, 1991

Peptides International PGA-4245-v   Human synthetic   MW 3157.5 $C_{139}H_{210}N_{42}O_{43}$ >99% (HPLC); lyophilized amorphous powder | Bioactive; Crawley, JN & GL Wenk, *Trends Neurosci*, 12:278, 1989; Bartfai, T et al, *FEBS Letts*, 283:189, 1991

Gly-Trp-Thr-Leu-Asn-Ser-Ala-Gly-Tyr-Leu-Leu-Gly-Pro-Ile-Asn-Leu-Lys-Ala-Leu-Ala-Ala-Leu-Ala-Lys-Lys-Ile-Leu Amide

Synonyms: Galanin (1-13)-Mastoparan

Bachem H-4188.0500   MW 2809.44 $C_{133}H_{222}N_{34}O_{32}$ Store at -15°C

Gly-Trp-Thr-Leu-Asn-Ser-Ala-Gly-Tyr-Leu-Leu-Gly-Pro-Pro-Pro-Ala-Leu-Ala-Leu-Ala Amide

Synonyms: Galanin (1-13); Galanin Inhibitor; Galanin Antagonist; M-40; Galanin (1-13)-Pro-Pro-(Ala-Leu-)₂Ala; Galanin (1-13)-Pro-Pro-Ala-Leu-Ala-Leu-Ala

American Peptide 46-1-19   MW 1981.3 $C_{94}H_{145}N_{23}O_{24}$
Significantly antagonizes galanin-induced feeding behavior at doses approximately equimolar to the active doses of rat galanin; inhibits the binding of (125I-Tyr[26])-galanin (porcine) in rat hypothalamic membranes with an IC_{50} value of 15 *nM*; Leibowitz, SF et al, *Brain Res*, 599:148, 1992; Crawley, JN et al, *Brain Res*, 600:268, 1993; Xu, XJ et al, *Br J Pharmacol*, 116(3):2076, 1995

Bachem H-2576.0500   MW 1981.33 $C_{94}H_{145}N_{23}O_{24}$ Store at -15°C

Sigma G 7285   FW 1981.3   ≥97% (HPLC) | Bioactive peptide; galanin receptor antagonist; Crawley, JN et al, *Brain Res*, 600:268, 1993

Gly-Trp-Thr-Leu-Asn-Ser-Ala-Gly-Tyr-Leu-Leu-Gly-Pro-Pro-Pro-Gly-Phe-Ser-Pro-Phe-Arg Amide

Synonyms: Galanin Receptor Agonist; Galanin Antagonist; M-35; Galanin (1-13)-Bradykinin (2-9); Galanin (1-13); Bradykinin (2-9)

American Peptide 46-5-14   MW 2233.6 $C_{107}H_{153}N_{27}O_{26}$ At low concentrations, this chimeric peptide antagonizes the effect of galanin, whereas at concentrations above 10 *nM*, acts as a galanin receptor agonist; Kask, K et al, *Reg Peptides*, 59:341, 1995

Bachem H-1346.0500   MW 2233.56 $C_{107}H_{153}N_{27}O_{26}$ Store at -15°C

Neosystem SC886   MW 2233.55   Bioactive; brain/gut peptide; high affinity galanin antagonist; Oegren, SO et al, *Neuroscience*, 51:1-5, 1992; Bartfai, T et al, TIPS, 13:312-317, 1992

Sigma G 6409   FW 2233.6   ≥97% (HPLC) | Bioactive peptide; tandem-linked peptide fragments produce high affinity competitive galanin antagonist; Oegren, SO et al, *Neuroscience*, 51:1, 1992; Bartfai, T et al, *TIPS*, 13:312, 1992

Gly-Trp-Thr-Leu-Asn-Thr-Ala-Trp-Trp-Leu-Leu-Gly-Pro-His

Gly-Trp-Thr-Leu-Asn-D-Thr-Ala-D-Trp-D-Trp-Leu-Leu-Gly-Pro-His-L-Alaninol

Synonyms: Galanin (1-15), (D-Thr[6],D-Trp[8,9],L-Alaninol[15])-

Bachem H-1576.0500   MW 1708.98 $C_{84}H_{117}N_{21}O_{18}$ Store at -15°C

Gly-Trp-Thr-Leu-Asn-Thr-Ala-Trp-Trp-Leu-Leu-Gly-Pro-His-Ala

Gly-Trp-Thr-Leu-Asn-D-Thr-Ala-D-Trp-D-Trp-Leu-Leu-Gly-Pro-His-Ala-ol

Synonyms: Gastrointestinal Peptide; Galanin (1-15), (D-Thr[6],D-Trp[8,9],15-ol)-

Sigma G 3535   FW 1709.0   ≥97% (HPLC) | Bioactive peptide; potent galanin antagonist down to sub-micromolar concentrations; Yanaihara, N et al, *Regul Peptides*, 46:93, 1993

Gly-Tyr
Z-Gly-Tyr

Bachem C-1935.0001   MW 372.38 $C_{19}H_{20}N_2O_6$ Store at -15°C

Gly-Tyr
Z-Gly-Tyr Amide

Bachem C-1940.0001   MW 371.39 $C_{19}H_{21}N_3O_5$ Store at -15°C

Gly-Tyr

Bachem G-2235.0005   MW 238.24 $C_{11}H_{14}N_2O_4$ Store at -15°C

Gly-Tyr
Gly-D-Tyr

Bachem G-4580.0001   Store at -15°C

Gly-Tyr
Gly-L-Tyr

Fluka 50471   MW 238.25 $C_{11}H_{14}N_2O_4$ ≥98% (titration); contains 1-2 mol water; mp:282-285°C

Gly-Tyr

ICN 101895   MW 238.2 $C_{11}H_{14}N_2O_4$ Crystalline

Gly-Tyr
Borotrimethyl-Gly-Tyr-OMe

ICN 154643   MW 294.16 $BC_{14}H_{23}N_2O_4$ >97%

Gly-Tyr
Gly-L-Tyr Dihydrate

Rexim   MW 274.3 $C_{11}H_{18}N_2O_6$ White crystals or crystalline powder

Gly-Tyr
N-CBZ-Gly-Tyr

Sigma C 4262   FW 372.4 $C_{19}H_{20}N_2O_6$ ~95%

Gly-Tyr

Sigma G 3502   FW 238.2 $C_{11}H_{14}N_2O_4$

Gly-Tyr Amide

Bachem G-2240.0250   MW 297.31 $C_{13}H_{19}N_3O_5$ Store at -15°C

Gly-Tyr Amide Acetate Salt

Sigma G 3627   FW 297.3 $C_{11}H_{15}N_3O_3 \cdot C_2H_4O_2$

Gly-Tyr Amide Hydrochloride

Bachem G-2245.0250   MW 273.72 $C_{11}H_{15}N_3O_3 \cdot HCl$ Store at -15°C

Sigma G 3752   FW 273.7 $C_{11}H_{15}N_3O_3 \cdot HCl$

Gly-Tyr-Ala

Bachem H-3665.0250   MW 309.32 $C_{14}H_{19}N_3O_5$ Store at -15°C

ICN 157274   MW 309.3 $C_{14}H_{19}N_3O_5$

Sigma G 1267   FW 309.3 $C_{14}H_{19}N_3O_5$

Gly-Tyr-Arg-Pro-Val-Phe-Ser-Ser-Pro-Pro-Ser-Tyr-Phe-Gln-Gln-Thr-His-Thr-Gln

Synonyms: GP140 (722-740); SIVmac251 Peptide

Neosystem SC780   MW 2227.41   SIVmac251 peptide from subtype GP140

Gly-Tyr-Gly

Bachem H-3670.0250   MW 295.3 $C_{13}H_{17}N_3O_5$ Store at -15°C

Gly-Tyr-Gly-Gly-Phe-Met

Synonyms: Enkephalin, (Gly[0])-Met-

Bachem H-2850.0025   MW 750.83 $C_{39}H_{38}N_6O_8S$ Store at -15°C

Gly-Tyr-Gly-Ser-Ser-Ser-Arg-Arg-Ala-Pro-Gln-Thr

Synonyms: Insulin-Like Growth Factor I C-Peptide; Insulin-Like Growth Factor I (30-41)

American Peptide 50-1-24   MW 1266.5 $C_{51}H_{83}N_{19}O_{19}$ Used to generate an antibody to the C-peptide region of IGF-I; Hintz, RL et al, *J Clin Endo Metab*, 50:405, 1980

Bachem H-7460.0001   MW 1266.34 $C_{51}H_{83}N_{19}O_{19}$ Store at -15°C

Neosystem SC679   MW 1266.33   Bioactive; synthetic growth factor-related peptide; Hintz, RL et al, *J Clin Endo Metab*, 50:405, 1980

ICN 191469   Human synthetic   Corresponds to the C-peptide region of IGF-1 & may be used to generate Ab reactive against that region; Hintz, RL etal, *J Clin Endocrinol Metab*, 50:405, 1980

Gly-Tyr-Gly-Ser-Ser-Ser-Arg-Arg-Ala-Pro-Gln-Thr Trifluoroacetate Salt

Synonyms: Insulin-Like Growth Factor I (30-41)

Biogenesis 5345-0809   Human synthetic   MW 1266.5   Purified; lyophilized

Gly-Tyr-Pro-Gly-Gln-Val

Synonyms: Proteinase Activated Receptor IV Agonist Peptide

Bachem H-4348.0005   MW 616.68 $C_{28}H_{41}N_7O_9$ Store at -15°C

Neosystem SC1306   Human   MW 619.67   Bioactive; thrombin receptor-derived peptide; corresponds to the tethered ligand sequence exposed from the amino-terminus of PAR-4 after proteolysis by thrombin; Xu, WF et al, *PNAS* USA, 95:6642-6646, 1998

Gly-Tyr-Pro-Gly-Lys-Phe

Synonyms: Proteinase Activated Receptor IV Agonist Peptide

Bachem H-4404.0005   MW 667.77 $C_{33}H_{45}N_7O_8$ Store at -15°C

Neosystem SC1343   Mouse   MW 667.76   Bioactive; thrombin receptor-derived peptide; corresponds to the tethered ligand of the mouse PAR-4 recently identified; Kahn, ML et al, *Nature*, 394:690-694, 1998

Gly-Tyr-Tyr-Pro-Thr

Synonyms: Gluten Exorphin A5

Bachem H-1668.0005   MW 599.64 $C_{29}H_{37}N_5O_9$ Store at -15°C

Gly-Val
FMOC-Gly-Val

Bachem B-1500.0001   MW 396.44 $C_{22}H_{24}N_2O_5$ Store at -15°C

Gly-Val
Z-Gly-Val
Bachem C-3140.0001   MW 308.33 $C_{15}H_{20}N_2O_5$ Store at -15°C

Gly-Val
Bachem G-2250.0001   MW 174.2 $C_7H_{14}N_2O_3$ Store at -15°C

Gly-Val
Gly-D-Val
Bachem G-2255.0001   MW 174.2 $C_7H_{14}N_2O_3$ Store at -15°C

Gly-Val
Gly-DL-Val
Bachem G-2260.0001   MW 174.2 $C_7H_{14}N_2O_3$ Store at -15°C

Gly-Val
FA-Gly-Val Amide
Bachem M-1820.0050   MW 293.32 $C_{14}H_{19}N_3O_4$ Store at -15°C

Gly-Val
Gly-DL-Val
ICN 101898   MW 174.2 $C_7H_{14}N_2O_3$ Crystalline

Gly-Val
ICN 101899   MW 174.2 $C_7H_{14}N_2O_3$ Crystalline

Gly-Val
Borotrimethyl-Gly-Val-OMe
ICN 154637   MW 230.11 $BC_{10}H_{23}N_2O_3$ Liquid; >97%

Gly-Val
Boro-Gly-Val-OMe
ICN 154646   MW 188.04 $BC_7H_{17}N_2O_3$ >97%

Gly-Val
Gly-D-Val
Sigma G 3877   FW 174.2 $C_7H_{14}N_2O_3$

Gly-Val
Sigma G 4127   FW 174.2 $C_7H_{14}N_2O_3$

Gly-Val-Arg-Tyr-Pro-Leu-Thr-Phe-Gly-Trp-Cys
Synonyms: NEF (132-142); HTLV I Peptide

Neosystem SC636   MW 1298.52   HTLV I peptide from subtype NEF; due to the presence of a Cys this peptide can contain the dimeric form

Gly-Val-Arg-Tyr-Pro-Leu-Thr-Phe-Gly-Trp-Cys-Tyr-Lys
Synonyms: NEF (132-144); HTLV I Peptide

Neosystem SC635   MW 1589.87   HTLV I peptide from subtype NEF; due to the presence of a Cys this peptide can contain the dimeric form

Gly-Val-Asp-Lys-Ala-Gly-Cys-Arg-Tyr-Met-Phe-Gly-Gly-Cys-Ser-Val-Asn-Asp-Asp-Cys-Cys-Pro-Arg-Leu-Gly-Cys-His-Ser-Leu-Phe-Ser-Tyr-Cys-Ala-Trp-Asp-Leu-Thr-Phe-Ser-Asp
Synonyms: SNX-482; Class E (R-Type) Ca^{2+} Channel Blocker

Peptides International PCB-4363-s   *Hysterocrates gigas* (tarantula) synthetic   MW 449 $C_{192}H_{274}N_{52}O_{60}S_7$ >95% (HPLC); lyophilized amorphous powder | Bioactive; Disulfide bonds: Cys^7-Cys^{21}, Cys^{14}-Cys^{26}, Cys^{20}-Cys^{33}; Newcomb, R et al, Biochemistry, 37:15353, 1998; Ürge, L et al, *Peptides* 1998:748-749, 1998; Tottene, A et al, *J Neurosci*, 20:171, 2000

Gly-Val-Gln-Val-Glu-Thr-Ile-Ser-Pro-Gly-Asp-Gly-Arg-Cys
Synonyms: FKBP12 Protein N-Terminal Peptide (1-13); Blocking Peptide for Antiserum FKBP12

Alexis 156-002   Human synthetic   MW 1417.6 $C_{57}H_{96}N_{18}O_{22}S$ ≥95%; lyophilized; reconstitute with 0.1 mL distilled H_2O | Competitively binds to Ab Alexis 210-142; antiserum blocking peptide

Gly-Val-Glu-Ile-Asn-Val-Lys-Cys-Ser-Gly-Ser-Pro-Gln-Cys-Leu-Lys-Pro-Cys-Lys-Asp-Ala-Gly-Met-Arg-Phe-Gly-Lys-Cys-Met-Asn-Arg-Lys-Cys-His-Cys-Thr-Pro-Lys
Synonyms: Kaliotoxin

American Peptide 41-0-79   MW 4150.0 $C_{171}H_{283}N_{55}O_{49}S_8$ Disulfide bonds: Cys^8-Cys^{28}, Cys^{14}-Cys^{33}, Cys^{18}-Cys^{35}; high conductance Ca^{2+}-activated K^+ channel blocker; Crest, M et al, *JBC*, 267:1640, 1992; Fernandez, I et al, *Biochemistry*, 33:14256, 1994

Bachem H-1004.0500   MW 4150.01 $C_{171}H_{283}N_{55}O_{49}S_8$ Store at -15°C | Disulfide bonds: Cys^8-Cys^{28}, Cys^{14}-Cys^{33}, Cys^{18}-Cys^{35}

Gly-Val-Glu-Ile-Asn-Val-Lys-Cys-Ser-Gly-Ser-Pro-Gln-Cys-Leu-Lys-Pro-Cys-Lys-Asp-Ala-Gly-Met-Arg-Phe-Gly-Lys-Cys-Met-Asn-Arg-Lys-Cys-His-Cys-Thr-Pro
Synonyms: Kaliotoxin (1-37)

American Peptide 41-0-78   *Androctonus mauretanicus*   MW 4020.8 $C_{165}H_{272}N_{54}O_{47}S_8$ Disulfide bonds: Cys^8-Cys^{28}, Cys^{14}-Cys^{33}, Cys^{18}-Cys^{35}; high conductance Ca^{2+}-activated K^+ channel blocker; Crest, M et al, *JBC*, 267:1640, 1992

Gly-Val-Glu-Ile-Asn-Val-Lys-Cys-Ser-Gly-Ser-Pro-Gln-Cys-Leu-Lys-Pro-Cys-Lys-Asp-Ala-Gly-Met-Arg-Phe-Gly-Lys-Cys-Met-Asn-Arg-Lys-Cys-His-Cys-Thr-Pro-Lys
Synonyms: Kaliotoxin

Calbiochem 420312   *Androctonus mauretanicus*   MW 4021.8 $C_{165}H_{271}N_{53}O_{48}S_8$ ≥98% (HPLC); crystalline solid; soluble in saline & water; LD_{50}≤2000 mg/kg | Disulfide bonds: Cys^8-Cys^{28}, Cys^{14}-Cys^{33}; Cys^{18}-Cys^{35}; blocker of Ca^{2+}-activated high conductance K^+ channels; selectively inhibits voltage-dependent K^+ channels in human & murine β-lymphocytes; blocks the invertebrate & mammalian sympathetic BK-type channels in neurons; inhibits the Shaker K^+ current resulting from mRNA expression in *Xenopus* oocyte; binds to rat brain synaptosomal membrane & is displaced by dendrotoxin; Fernandez, I et al, *Biochemistry*, 33:14256, 1994; Romi, R et al, *J Biol Chem*, 268:26302, 1993; Crest, M et al, *J Biol Chem*, 267:1640, 1992

Gly-Val-Glu-Ile-Asn-Val-Lys-Cys-Ser-Gly-Ser-Pro-Gln-Cys-Leu-Lys-Pro-Cys-Lys-Asp-Ala-Gly-Met-Arg-Phe-Gly-Lys-Cys-Met-Asn-Arg-Lys-Cys-His-Cys-Thr-Pro

Synonyms: Kaliotoxin (1-37); Ca^{2+}-Activated K$^+$ Channel Blocker, High Conductance

Sigma K 3630 *Androctonus mauretanicus* (scorpion) FW 4021.8 ≥90% (HPLC) | Disulfide bonds: 8-28, 14-33, 18-35; bioactive peptide; scorpion toxin; inhibits calcium-activated potassium channels in mollusk & rabbit nerve cells; Crest, M et al, *J Biol Chem*, 267:1640, 1992

Peptides International PKL-4259-s *Androctonus mauretanicus mauretanicus* (scorpion) synthetic MW 4021.8 C$_{165}$H$_{271}$N$_{53}$O$_{48}$S$_8$ >95% (HPLC); lyophilized amorphous powder | Bioactive; Disulfide bonds: Cys^8-Cys^{28}, Cys^{14}-Cys^{33}, Cys^{18}-Cys^{35}; Crest, M et al, *JBC*, 267:1640, 1992; Romi, R et al, *JBC*, 268:26302, 1993

Gly-Val-Gly-Ser-Pro-Gln-Ile-Leu-Val-Glu-Ser-Pro-Thr-Val-Leu

Synonyms: REV (96-110); HTLV I Peptide

Neosystem SC196 MW 1495.73 HTLV I peptide from subtype REV

Gly-Val-Leu-Ser-Asn-Val-Ile-Gly-Tyr-Leu-Lys-Lys-Leu-Gly-Thr-Gly-Ala-Leu-Asn-Ala-Val-Leu-Lys-Gln

American Peptide 72-2-27 MW 2456.9 C$_{112}$H$_{194}$N$_{30}$O$_{31}$ Antimicrobial peptide originally isolated from *Xenopus laevis*

Gly-Val-Leu-Ser-Asn-Val-Ile-Gly-Tyr-Leu-Lys-Lys-Leu-Gly-Thr-Gly-Ala-Leu-Asn-Ala-Val-Leu-Lys-Gln Trifluoroacetate Salt

Synonyms: PDQ

Sigma G 9031 FW 2456.9 ≥95% (HPLC) | Bioactive peptide; antimicrobial peptide originally isolated from *Xenopus laevis*

Gly-Val-Phe

Bachem H-3675.0050 MW 321.38 C$_{16}$H$_{23}$N$_3$O$_4$ Store at -15°C

Gly-Val-Pro-Ala-Trp-Arg-Asn-Ala-Thr-Ile-Pro-Leu-Phe-Cys-Ala-Thr-Lys-Asn-Arg-Asp

Synonyms: GP140 (31-50); SIVmac251 Peptide

Neosystem SC711 MW 2230.57 SIVmac251 peptide from subtype GP140; due to the presence of a Cys this peptide can contain the dimeric form

Gly-Val-Pro-Ile-Asn-Val-Ser-Cys-Thr-Gly-Ser-Pro-Gln-Cys-Ile-Lys-Pro-Cys-Lys-Asp-Ala-Gly-Met-Arg-Phe-Gly-Lys-Cys-Met-Asn-Arg-Lys-Cys-His-Cys-Thr-Pro-Lys

Synonyms: Agitoxin II

Calbiochem 123000 *Leiurus quinquestriatus hebraeus* MW 4090.9 C$_{169}$H$_{278}$N$_{54}$O$_{48}$S$_8$ >98% (SDS-PAGE); lyophilized; bioassay:inhibition of Kv 1.3 K$^+$ channels expressed in *Xenopus* oocytes:<1 n*M*; LD$_{50}$≤2000 mg/kg | Potent inhibitor of the Shaker K$^+$ channel as well as the mammalian homolog of Shaker; Garcia, ML et al, *Biochemistry*, 33:6834, 1994

Alexis 630-011 *Leiurus quinquestriatus hebraeus* (scorpion venom) MW 4090.9; C$_{169}$H$_{278}$N$_{54}$O$_{48}$S$_8$ ≥98% (SDS-PAGE); lyophilized; soluble in water; stock solution should be made in 0.1% BSA, 100 m*M* NaCl, 10 m*M* Tris, pH 7.5, & 1 m*M* EDTA; addition of 0.01% BSA to experimental solutions before applying the toxin is essential; potent neurotoxin | Super potent inhibitor of the *Shaker* K$^+$ channel & mammalian homologs of *Shaker*; Garcia, ML et al, *Biochemistry*, 33:6834, 1994; Gross, A et al, *Neuron*, 13:961, 1994

Gly-Val-Ser-Cys-Leu-Cys-Asp-Ser-Asp-Gly-Pro-Ser-Val-Arg-Gly-Asn-Thr-Leu-Ser-Gly-Thr-Leu-Trp-Leu-Tyr-Pro-Ser-Gly-Cys-Pro-Ser-Gly-Trp-His-Asn-Cys-Lys-Ala-His-Gly-Pro-Thr-Ile-Gly-Trp-Cys-Cys-Lys-Gln

Synonyms: Anthopleurin A; Anthopleurin; Anthopleura Toxin A; Toxin A

Bachem H-9590.0100 MW 5131.8 C$_{220}$H$_{326}$N$_{64}$O$_{67}$S$_6$ Store at -15°C | Disulfide bonds: Cys^4-Cys^{46}, Cys^6-Cys^{36}, Cys^{29}-Cys^{47}

Calbiochem 178005 Synthetic MW 5131.8 C$_{220}$H$_{326}$N$_{64}$O$_{67}$S$_6$ ≥95% (HPLC); lyophilized; soluble in H$_2$O; Disulfide bonds: Cys^4-Cys^{46}, Cys^6-Cys^{36}; Cys^{29}-Cys^{47}; LD$_{50}$≤200 mg/kg but >50 mg/kg | Sequence from toxin isolated from the sea anemone *Anthopleura xanthogrammica*; inhibits the inactivation of voltage-gated Na$^+$ channels; has inotropic but no chronotropic effects on mammalian heart preparations; Kelso, GJ et al, *Biochemistry*, 35:14157, 1996; Pennington, MW et al, *Int J Pept Protein Res*, 43:463, 1994; Scriabine, A et al, *J Cardiovasc Pharmacol*, 1:571, 1979; Tanaka, M et al, *Biochemistry*, 16:204, 1977

Gly-Val-Tyr-Val-His-Pro-Val

Synonyms: Angiotensin III Antipeptide

American Peptide 12-1-63 MW 769.9 C$_{37}$H$_{55}$N$_9$O$_9$ An angiotensin receptor antagonist; Moore, GJ et al, *Biochem Biophys Res Comm*, 160:1387, 1989

ICN 158849 Angiotensin receptor antagonist; Moore, GJ etal, *BBRC*, 160:1387, 1989

Gly-Val-Tyr-Val-His-Pro-Val Acetate Salt

Synonyms: Angiotensin III Antipeptide

Sigma A 4809 FW 769.9 (free base); C$_{37}$H$_{55}$N$_9$O$_9$ ≥97% (HPLC) | Bioactive peptide; angiotensin receptor antagonist; Moore, GJ et al, *Biochem Biophys Res Commun*, 160:1387, 1989

Gly-X-Phe
((S)-1-Carboxy-2-Phenylethyl)-Carbamoyl-α-(2-Iminohexahydro-4(S)-Pyrimidyl)-(S)-Gly-X-Phe-al

Synonyms: Chymostatin; Chymotrypsin Inhibitor; Papain Inhibitor; Cathepsin A/B/D Inhibitor

Peptides International ICY-4063 Microbial ~50%; amorphous powder; bulk | Inhibitor; x:L-Leu (Type A), L-Val (Type B), L-Ile (Type C); Umezawa, H et al, *J Antibiotics*, 23:425, 1970; Tatsuta, K et al, *J Antibiotics*, 26:625, 1973

Peptides International ICY-4063-v Microbial ~50%; lyophilized amorphous powder | Inhibitor; x:L-Leu (Type A), L-Val (Type B), L-Ile (Type C); Umezawa, H et al, *J Antibiotics*, 23:425, 1970; Tatsuta, K et al, *J Antibiotics*, 26:625, 1973

Gpl-Arg-Pro-Pro-Met-Glu-Glu-Glu-Glu-Glu-Ala-Tyr-Gly-Trp-Met-Asp-Phe Amide

Synonyms: Gastrointestinal Peptide; Gastrin; Gastrin I

Sigma G 1276 Rat FW 2126.3 ≥97% (HPLC) | Bioactive peptide; produced in the stomach; stimulates gastric acid secretion; Schaffer, MH et al, *Peptides*, 3:693, 1982

His
N,N-Dimethyl-His

Bachem F-3625.0001 MW 183.21 C$_8$H$_{13}$N$_3$O$_2$ Store at 2-8°C

His
N,N-Dimethyl-His-OMe

Bachem F-3630.0001 MW 197.24 C$_9$H$_{15}$N$_3$O$_2$ Store at 2-8°C

His-Ala-Asp-Gly-Ser-Phe-Ser-Asp-Glu-Met-Asn-Thr-Ile-Leu-Asp-Asn-Leu-Ala-Ala-Arg-Asp-Phe-Ile-Asn-Trp-Leu-Ile-Gln-Thr-Lys-Ile-Thr-Asp

Synonyms: Glucagon-Like Peptide II

Peptides International PGL-4376-v   Human synthetic   MW 3766.1 $C_{165}H_{254}N_{44}O_{55}S$ >95% (HPLC); lyophilized amorphous powder | Bioactive; food intake regulator; Drucker, DJ, *Trends Endocrinol Metabl*, 10:153, 1999; Hartmann, B et al, *Peptides*, 21:73, 2000; Tang-Christensen, M et al, *Nat Med*, 6:802, 2000

His-Ala
Z-His-Ala

Bachem C-1965.0250   MW 360.37 $C_{17}H_{20}N_4O_5$ Store at -15°C

His-Ala

Bachem G-2285.0250   MW 226.24 $C_9H_{14}N_4O_3$ Store at -15°C

ICN 157374   MW 226.2 $C_9H_{14}N_4O_3$ Crystalline

Sigma H 2754   FW 226.2 $C_9H_{14}N_4O_3$

His-Ala-Asp-Ala-Ile-Phe-Thr-Asn-Ser-Tyr-Arg-Lys-Val-Leu-Gly-Gln-Leu-Ser-Ala-Arg-Lys-Leu-Leu-Gln-Asp-Ile-Nle-Ser-Arg-Gln-Gln-Gly Amide

Synonyms: Growth Hormone Releasing Factor (1-32), (His[1],Nle[27])-

American Peptide 52-1-59   Human   MW 3627.2 $C_{159}H_{265}N_{51}O_{46}$

His-Ala-Asp-Ala-Ile-Phe-Thr-Ser-Ser-Tyr-Arg-Arg-Ile-Leu-Gly-Gln-Leu-Tyr-Ala-Arg-Lys-Leu-Leu-His-Glu-Ile-Met-Asn-Arg-Gln-Gln-Gly-Glu-Arg-Asn-Gln-Glu-Gln-Arg-Ser-Arg-Phe-Asn

Synonyms: Growth Hormone Releasing Factor

American Peptide 52-5-13   Rat   MW 5232.9 $C_{225}H_{361}N_{77}O_{66}S$ Spiess, J et al, *Nature*, 303:532, 1983

His-Ala-Asp-Ala-Ile-Phe-Thr-Ser-Ser-Tyr-Arg-Arg-Ile-Leu-Gly-Gln-Leu-Tyr-Ala-Arg-Lys-Leu-Leu-His-Glu-Ile-Met-Asn-Arg Amide

Synonyms: Growth Hormone Releasing Factor (1-29)

American Peptide 52-5-14   Rat   MW 3473.1 $C_{155}H_{251}N_{49}O_{40}S$

Bachem H-1902.0500   Rat   MW 3473.07 $C_{155}H_{251}N_{49}O_{40}S$ Store at -15°C

His-Ala-Asp-Ala-Ile-Phe-Thr-Ser-Ser-Tyr-Arg-Arg-Ile-Leu-Gly-Gln-Leu-Tyr-Ala-Arg-Lys-Leu-Leu-His-Glu-Ile-Met-Asn-Arg-Gln-Gln-Gly-Glu-Arg-Asn-Gln-Glu-Gln-Arg-Ser-Arg-Phe-Asn

Synonyms: Growth Hormone Releasing Factor (1-43)

Bachem H-5440.0500   Rat   MW 5232.89 $C_{225}H_{361}N_{77}O_{66}S$ Store at -15°C

His-Ala-Asp-Ala-Ile-Phe-Thr-Ser-Ser-Tyr-Arg-Arg-Ile-Leu-Gly-Gln-Leu-Tyr-Ala-Arg-Lys-Leu-Leu-His-Glu-Ile-Met-Asn-Arg Amide

Synonyms: Growth Hormone Releasing Factor (1-29)

Sigma G 0519   Rat   FW 3473.1   ≥90% (HPLC) | Bioactive peptide; equipotent to GRF in releasing somatotropin from anterior pituitary

His-Ala-Asp-Ala-Ile-Phe-Thr-Ser-Ser-Tyr-Arg-Arg-Ile-Leu-Gly-Gln-Leu-Tyr-Ala-Arg-Lys-Leu-Leu-His-Glu-Ile-Met-Asn-Arg-Gln-Gln-Gly-Glu-Arg-Asn-Gln-Glu-Gln-Arg-Ser-Arg-Phe-Asn

Synonyms: Growth Hormone Releasing Factor

Sigma G 6646   Rat   FW 5232.9   ≥95% (HPLC); peptide content:~70% | Bioactive peptide; hypothalamic peptide that stimulates release of somatotropin from anterior pituitary; Spiess, J et al, *Nature*, 303:532, 1983

His-Ala-Asp-Gly-Ser-Phe-Ser-Asp-Glu-Met-Asn-Thr-Ile-Leu-Asp-Asn-Leu-Ala-Ala-Arg-Asp-Phe-Ile-Asn-Trp-Leu-Ile-Gln-Thr-Lys-Ile-Thr-Asp-Arg

Synonyms: Glucagon-Like Peptide II; Prepro-Glucagon (126-159)

American Peptide 46-1-14   Human   MW 3922.9 $C_{171}H_{266}N_{48}O_{56}S$

Bachem H-4766.0500   Human   MW 3922.35 $C_{171}H_{266}N_{48}O_{56}S$ Store at -15°C

His-Ala-Asp-Gly-Ser-Phe-Ser-Asp-Glu-Met-Asn-Thr-Ile-Leu-Asp-Asn-Leu-Ala-Thr-Arg-Asp-Phe-Ile-Asn-Trp-Leu-Ile-Gln-Thr-Lys-Ile-Thr-Asp

Synonyms: Glucagon-Like Peptide II; Prepro-Glucagon (126-158)

American Peptide 46-4-18   Rat   MW 3796.2 $C_{166}H_{256}N_{44}O_{56}S$

Bachem H-5002.0500   Rat   MW 3796.19 $C_{166}H_{256}N_{44}O_{56}S$ Store at -15°C

His-Ala-Asp-Gly-Val-Phe-Thr-Ser-Asp-Phe-Ser-Arg-Leu-Leu-Gly-Gln-Leu-Ser-Ala-Lys-Lys-Tyr-Leu-Glu-Ser-Leu-Ile Amide

Synonyms: Peptide Histidine Isoleucine 27; Peptide Histidine Isoleucine; Gastrointestinal Peptide

Bachem H-3730.0500   Porcine   MW 2995.43 $C_{136}H_{216}N_{36}O_{40}$ Store at -15°C

Neosystem SC334   Porcine   MW 2995.42   Bioactive; brain/gut peptide; VIP/PHI/secretin/helodermin peptide; Tatemoto, K & Mutt, V, *PNAS USA*, 78:6603, 1981; Tatemoto, K, *Peptides*, 5:151, 1984

Sigma P 5048   Porcine   FW 2995.4   ≥97% (HPLC); peptide content:~70% | Bioactive; biological activities similar to vasoactive intestinal peptide & secretin; Moroder, L et al, *Peptides, Synthesis-Structure-Function, Proceedings of the Seventh American Peptide Symposium*, Rich, DH & Gross, E, eds, 49, 1981; Tatemoto, K & Mutt, V, *Proc Natl Acad Sci USA*, 78:6603, 1981

Sigma P 8028   Porcine   FW 2995.4   ≥90% (HPLC) | Bioactive peptide

ICN 152892   Porcine intestine   MW 2995.4   Biological activities similar to vasoactive intestinal peptide & secretin; Moroder, L etal, *Peptides, Synthesis – Structure – Function, Proceedings of the 7th American Peptide Symposium*, DH Rich & E Gross, eds, p 49, 1981; Tatemoto, K & V Mutt, *PNAS*, 78:6603, 1981; Ahren, B & J Lundquist, *Neuropeptides*, 11:159, 1988

His-Ala-Asp-Gly-Val-Phe-Thr-Ser-Asp-Phe-Ser-Lys-Leu-Leu-Gly-Gln-Leu-Ser-Ala-Lys-Lys-Tyr-Leu-Glu-Ser-Leu-Met Amide

Synonyms: Gastrointestinal Peptide; Peptide Histidine Methionine 27; Peptide Histidine Methionine, Human

Sigma P 4295   FW 2985.4   ≥97% (HPLC) | Bioactive peptide; Itoh, N et al, *Nature*, 304:547, 1983

His-Ala-Asp-Gly-Val-Phe-Thr-Ser-Asp-Phe-Ser-Lys-Leu-Leu-Gly-Gln-Leu-Ser-Ala-Lys-Lys-Tyr-Leu-Glu-Ser-Leu-Met-Gly-Lys-Arg-Val-Ser-Ser-Asn-Ile-Ser-Glu-Asp-Pro-Val-Pro-Val

Synonyms: Prepro-Vasoactive Intestinal Peptide (81-122)

American Peptide 48-1-15   Human   MW 4552.2 $C_{202}H_{325}N_{53}O_{64}S$

His-Ala-Asp-Gly-Val-Phe-Thr-Ser-Asp-Phe-Ser-Lys-Leu-Leu-Gly-Gln-Leu-Ser-Ala-Lys-Lys-Tyr-Leu-Glu-Ser-Leu-Met Amide

Synonyms: Peptide Histidine Methionine 27

Bachem H-6355.0500   Human   MW 2985.46
$C_{135}H_{214}N_{34}O_{40}S$ Store at -15°C

His-Ala-Asp-Gly-Val-Phe-Thr-Ser-Asp-Phe-Ser-Lys-Leu-Leu-Gly-Gln-Leu-Ser-Ala-Lys-Lys-Tyr-Leu-Glu-Ser-Leu-Met-Gly-Lys-Arg-Val-Ser-Ser-Asn-Ile-Ser-Glu-Asp-Pro-Val-Pro-Val

Synonyms: Prepro-Vasoactive Intestinal Peptide (81-122)

Bachem H-6910.0500   Human   Store at -15°C

His-Ala-Asp-Gly-Val-Phe-Thr-Ser-Asp-Phe-Ser-Lys-Leu-Leu-Gly-Gln-Leu-Ser-Ala-Lys-Lys-Tyr-Leu-Glu-Ser-Leu-Met Amide

Synonyms: Peptide Histidine Methionine 27; Peptide Histidine Isoleucine; Peptide Histidine Methionine

ICN 152894   Human   MW 2985.4   Differs by only 2 AA from PHI-27; closely related to vasoactive intestinal peptide; Itoh, N et al, *Nature*, 304:547, 1983

Neosystem SC336   Human   MW 2985.44   Bioactive; brain/gut peptide; VIP/PHI/secretin/helodermin peptide; Tatemoto, K et al, *FEBS Lett*, 174:258, 1984; Itoh, N et al, *Nature*, 304:547-549, 1983

Peptides International PPM-4177-v   Human synthetic   MW 2985.5 $C_{135}H_{214}N_{34}O_{40}S$ >95% (HPLC); lyophilized amorphous powder | Bioactive; Itoh, N et al, *Nature*, 304:547, 1983

His-Ala-Asp-Gly-Val-Phe-Thr-Ser-Asp-Ser-Arg-Leu-Leu-Gly-Gln-Leu-Ser-Ala-Lys-Lys-Tyr-Leu-Glu-Ser-Leu-Ile Amide

Peptide Histidine Isoleucine 27*Synonyms*:

American Peptide 48-4-10   Porcine   MW 2995.5 $C_{136}H_{216}N_{36}O_{40}$ Porcine intestinal peptide which exhibits biological activities similar to vasoactive intestinal peptide & secretin; Tatemoto, K et al, *PNAS*, 78:6603, 1981

His-Ala-Asp-Gly-Val-Phe-Thr-Ser-Asp-Ser-Lys-Leu-Leu-Gly-Gln-Leu-Ser-Ala-Lys-Lys-Tyr-Leu-Glu-Ser-Leu-Met Amide

Synonyms: Peptide Histidine Methionine 27; Peptide Histidine Isoleucine

American Peptide 48-1-11   Human   MW 2985.5 $C_{135}H_{214}N_{34}O_{40}S$ Itoh, N et al, *Nature*, 304:547, 1983

His-Ala-Asp-Gly-Val-Phe-Thr-Ser-Asp-Tyr-Ser-Arg-Leu-Leu-Gly-Gln-Ile-Ser-Ala-Lys-Lys-Tyr-Leu-Glu-Ser-Leu-Ile Amide

Synonyms: Peptide Histidine Isoleucine; Peptide Histidine Isoleucine 27; Gastrointestinal Peptide

American Peptide 48-5-10   Rat   MW 3011.5 $C_{136}H_{216}N_{36}O_{41}$ Nishizawa, M et al, *FEBS Lett*, 183:55, 1985

Bachem H-7760.0500   Rat   MW 3011.43 $C_{136}H_{216}N_{36}O_{41}$ Store at -15°C

ICN 152893   Rat   MW 3011.4   Nishizawa, M et al, *FEBS Lett*, 183:55, 1985

Sigma P 4420   Rat   FW 3011.4   ≥95% (HPLC) | Bioactive peptide; Nishizawa, M et al, *FEBS Lett*, 183:55, 1985

His-Ala-Asp-Phe-Val-Phe-Thr-Ser-Asp-Phe-Ser-Arg-Leu-Leu-Gly-Gln-Leu-Ser-Ala-Lys-Lys-Tyr-Leu-Glu-Ser-Leu-Ile
His-Ala-Asp-D-Phe-Val-Phe-Thr-Ser-Asp-Phe-Ser-Arg-Leu-Leu-Gly-Gln-Leu-Ser-Ala-Lys-Lys-Tyr-Leu-Glu-Ser-Leu-Ile Amide

Synonyms: Peptide Histidine Isoleucine, (D-Phe⁴)-

Neosystem SC335   Porcine   MW 3085.54   Bioactive; brain/gut peptide; VIP/PHI/secretin/helodermin peptideA highly selective VIP agonist; Robberecht, P et al, *Mol Pharm Biochem*, 165:243-249, 1987

His-Ala-Glu-Gly-Thr-Phe-Thr-Ser-Asp-Val-Ser-Ser-Tyr-Leu-Glu-Gly-Gln-Ala-Ala-Lys-Glu-Phe-Ile-Ala-Trp-Leu-Val-Lys-Gly-Arg-Gly

Synonyms: Glucagon-Like Peptide I (7-37); Prepro-Glucagon (78-108)

Bachem H-5102.0500   Human   MW 3355.71 $C_{151}H_{228}N_{40}O_{47}$ Store at -15°C

His-Ala-Glu-Gly-Thr-Phe-Thr-Ser-Asp-Val-Ser-Ser-Tyr-Leu-Glu-Gly-Gln-Ala-Ala-Lys-Glu-Phe-Ile-Ala-Trp-Leu-Val-Lys-Gly-Arg Amide

Synonyms: Glucagon-Like Peptide I (7-36); Prepro-Glucagon (78-107)

Bachem H-6795.0500   Human   MW 3297.68 $C_{149}H_{226}N_{40}O_{45}$ Store at -15°C

ICN 152883   Human   Effector in the hormonal control of insulin secretion; Bell, GI et al, *Nature*, 304:368, 1983

His-Ala-Glu-Gly-Thr-Phe-Thr-Ser-Asp-Val-Ser-Ser-Tyr-Leu-Glu-Gly-Gln-Ala-Ala-Lys-Glu-Phe-Ile-Ala-Trp-Leu-Val-Lys Amide

Synonyms: Glucagon-Like Peptide I (7-34)

Neosystem SC1258   Human   MW 3084.43   Bioactive; brain/gut peptide; circulating form isolated recently from human blood filtrate; Richter, R et al, 3rd International Symposium on VIP, PACAP & Related Peptides, Freiburg (Germany), September 1997

His-Ala-Glu-Gly-Thr-Phe-Thr-Ser-Asp-Val-Ser-Ser-Tyr-Leu-Glu-Gly-Gln-Ala-Ala-Lys-Glu-Phe-Ile-Ala-Trp-Leu-Val-Lys-Gly-Arg Amide

Synonyms: Glucagon-Like Peptide I (7-36)

Neosystem SC905   Human   MW 3297.67   Bioactive; brain/gut peptide; Bell, GI et al, *Nature*, 304:368-371, 1983; Orskov, C & Nielsen, JH, *FEBS Lett*, 229:175-178, 1988

His-Ala-Glu-Gly-Thr-Phe-Thr-Ser-Asp-Val-Ser-Ser-Tyr-Leu-Glu-Gly-Gln-Ala-Ala-Lys-Glu-Phe-Ile-Ala-Trp-Leu-Val-Lys-Gly-Arg-Gly

Synonyms: Glucagon-Like Peptide I (7-37)

Neosystem SC908   Human   MW 3355.7   Bioactive; brain/gut peptide; Gefel, D et al, *Endocrinol*, 126:2164, 1990

His-Ala-Glu-Gly-Thr-Phe-Thr-Ser-Asp-Val-Ser-Ser-Tyr-Leu-Glu-Gly-Gln-Ala-Ala-Lys-Glu-Phe-Ile-Ala-Trp-Leu-Val-Lys-Gly-Arg Amide

Synonyms: Gastrointestinal Peptide; Glucagon-Like Peptide I (7-36); Prepro-Glucagon (78-107)

Sigma G 8147   Human   FW 3297.7   ≥97% (HPLC) | Bioactive peptide; effector in the hormonal control of insulin secretion; Bell, GI et al, *Nature*, 304:368, 1983

Peptides International PGL-4344-v   Synthetic   MW 3297.7 $C_{149}H_{226}N_{40}O_{45}$ >99% (HPLC); lyophilized amorphous powder | Bioactive; Turton, MD et al, *Nature*, 379:69, 1996; Tang-Christensen, M et al, *Am J Physiol*, 271, R848 (1996); van Dijk, G et al, *Nature*, 385:214, 1997

His-Ala-Glu-Gly-Thr-Phe-Thr-Ser-Asp-Val-Ser-Ser-Tyr-Leu-Glu-Gly-Gln-Ala-Ala-Lys-Glu-Phe-Ile-Ala-Trp-Leu-Val-Lys-Gly-Arg-Gly

Synonyms: Glucagon-Like Peptide I (7-37)

Peptides International PGP-4280-v   Synthetic   MW 3355.8
$C_{151}H_{228}N_{40}O_{47}$ >99% (HPLC); lyophilized amorphous powder |
Bioactive; Holz IV, GG et al, *Nature*, 361:362, 1993

His-Ala-Glu-Lys-His-Trp-Phe-Val-Gly-Leu

Synonyms: FGF Acidic I (102-111)

Bachem H-6710.0001   Bovine brain   Store at -15°C

His-Ala-Gly-Pro-Ile-Ala-Pro-Gly-Gln-Met-Arg-Glu-Pro-Arg-Gly

Synonyms: P25 (219-233); GAG P24 CA (87-101); HTLV I Peptide

Neosystem SC298   MW 1573.79   HTLV I peptide from
subtype P25 (GAG P24 CA)

His-Ala-Phe-Ile-Lys-Arg-Ser-Asp-Ala-Glu-Glu-Val-Asp-Phe-Ala-Gly-Trp-Leu-Cys

Synonyms: MEK1 C-Terminal Blocking Peptide (359-377)

Calbiochem 444948   Human synthetic   MW 2194.5
$C_{99}H_{144}N_{26}O_{29}S$ Lyophilized solid | Based on human MEK1 (359-
377); this peptide coupled to KLH was used as the immunogen for
the production of Anti-MEK1; for use in immunoabsorption for
immunoprecipitation, immunocytochemistry, Western blotting & dot
blots

His-Arg

Bachem G-2290.0050   MW 311.34 $C_{12}H_{21}N_7O_3$ Store at
-15°C

His-Arg-Asp-Ala-Ile-Phe-Thr-Asn-Ser-Tyr-Arg-Lys-Val-Leu-Gln-Leu-Ser-Ala-Arg-Lys-Leu-Leu-Gln-Asp-Ile-Nle-Ser-Arg
Naphthyl Ac-His-D-Arg-Asp-Ala-Ile-Phe(p-Chloro)-Thr-Asn-Ser-Tyr-Arg-Lys-Val-Leu-(2-Aminobutyric Acid)-Gln-Leu-Ser-Ala-Arg-Lys-Leu-Leu-Gln-Asp-Ile-Nle-Ser-Arg Amide Acetate Salt

Synonyms: Growth Hormone Releasing Hormone Inhibitor; MZ-4-
181

Calbiochem 476506   MW 3629.7 $C_{164}H_{264}ClN_{49}O_{42}$ ≥98%
(HPLC); white solid; soluble in 5% acetic acid; may be
carcinogenic/teratogenic | Highly potent long-acting inhibitor *in
vitro* & *in vivo*; Zarandi, M et al, *PNAS*, 91:12298, 1994

His-Arg-Leu-Arg-Tyr Amide

Synonyms: Neuropeptide Y (32-36), (His[32],Leu[34])-

Bachem H-3544.0005   MW 742.88 $C_{33}H_{54}N_{14}O_6$ Store at
-15°C

His-Asn-Lys-Gln-Glu-Gly-Arg-Asp-His-Asp-Lys-Ser-Lys-Gly-His-Phe-His-Arg-Val-Val-Ile-His-His-Lys-Gly-Gly-Lys-Ala-His-Arg-Gly

Synonyms: Inhibin-Like Peptide

Sigma I 9638   Human   FW 3591.0   ≥97% (HPLC); peptide
content:~65% | Bioactive peptide; selectively inhibits the
secretion of FSH; Sairan, MRet al, *Proc Natl Acad Sci USA*,
84:2043, 1987; Yamashiro, D, ibid, 81:5399, 1984

His-Asn-Thr-Asn-Gly-Val-Thr-Ala-Ala-Cys-Ser-His-Glu

Synonyms: Haemagglutinin Peptide H (130-142)

Neosystem SC805   MW 1340.39   Virus-related peptide; due
to the presence of a Cys this peptide can contain the dimeric form

His-Asp

Bachem G-2295.0250   MW 270.25 $C_{10}H_{14}N_4O_5$ Store at
-15°C

His-Asp-Glu-Gly-Thr-Phe-Thr-Ser-Asp-Val-Ser-Ser-Tyr-Leu-Glu-Gly-Gln-Ala-Ala-Lys-Glu-Phe-Ile-Ala-Trp-Leu-Val-Lys-Gly-Arg Amide

Synonyms: Glucagon-Like Peptide I; Prepro-Glucagon (78-107)

American Peptide 46-1-13   Human   MW 3297.5
$C_{149}H_{226}N_{40}O_{45}$ Bell, GI et al, *Nature*, 304:368, 1983

His-Asp-Glu-Phe-Glu-Arg-His-Ala-Glu-Gly-Thr-Phe-Thr-Ser-Asp-Val-Ser-Ser-Tyr-Leu-Glu-Gly-Gln-Ala-Ala-Lys-Glu-Phe-Ile-Ala-Trp-Leu-Val-Lys-Gly-Arg-Gly

Synonyms: Glucagon-Like Peptide I; Prepro-Glucagon (72-108)

American Peptide 46-1-11   Human   MW 4169.6
$C_{186}H_{275}N_{51}O_{59}$ Bell, GI et al, *Nature*, 304:368, 1983

His-Asp-Glu-Phe-Glu-Arg-His-Ala-Glu-Gly-Thr-Phe-Thr-Ser-Asp-Val-Ser-Ser-Tyr-Leu-Glu-Gly-Gln-Ala-Ala-Lys-Glu-Phe-Ile-Ala-Trp-Leu-Val-Lys-Gly-Arg Amide

Synonyms: Prepro-Glucagon (72-107); Glucagon-Like Peptide I
(72-107)

American Peptide 46-1-12   Human   MW 4111.5
$C_{184}H_{273}N_{51}O_{57}$ Bell, GI et al, *Nature*, 304:368, 1983
Bachem H-6025.0500   Human   MW 4111.5 $C_{184}H_{273}N_{51}O_{57}$
Store at -15°C

His-Asp-Glu-Phe-Glu-Arg-His-Ala-Glu-Gly-Thr-Phe-Thr-Ser-Asp-Val-Ser-Ser-Tyr-Leu-Glu-Gly-Gln-Ala-Ala-Lys-Glu-Phe-Ile-Ala-Trp-Leu-Val-Lys-Gly-Arg-Gly

Synonyms: Glucagon-Like Peptide I; Prepro-Glucagon (72-108)

ICN 152881   Human   Bell, GI etal, *Nature*, 304:368, 1983

His-Asp-Glu-Phe-Glu-Arg-His-Ala-Glu-Gly-Thr-Phe-Thr-Ser-Asp-Val-Ser-Ser-Tyr-Leu-Glu-Gly-Gln-Ala-Ala-Lys-Glu-Phe-Ile-Ala-Trp-Leu-Val-Lys-Gly-Arg Amide

Synonyms: Glucagon-Like Peptide I (1-36); Prepro-Glucagon (72-
107)

ICN 152882   Human   Bell, GI etal, *Nature*, 304:368, 1983
Neosystem SC904   Human   MW 4111.49   Bioactive;
brain/gut peptide; Holst, JJ et al, *FEBS Lett*, 211:169-174, 1987;
Drucker, DJ et al, *PNAS USA*, 84:3434-3438, 1987

His-Asp-Glu-Phe-Glu-Arg-His-Ala-Glu-Gly-Thr-Phe-Thr-Ser-Asp-Val-Ser-Ser-Tyr-Leu-Glu-Gly-Gln-Ala-Ala-Lys-Glu-Phe-Ile-Ala-Trp-Leu-Val-Lys-Gly-Arg-Gly

Synonyms: Gastrointestinal Peptide; Glucagon-Like Peptide I;
Prepro-Glucagon (72-108)

Sigma G 3265   Human   FW 4169.5   ≥97% (HPLC) |
Bioactive peptide; Bell, GI et al, *Nature*, 304:368, 1983

His-Asp-Glu-Phe-Glu-Arg-His-Ala-Glu-Gly-Thr-Phe-Thr-Ser-Asp-Val-Ser-Ser-Tyr-Leu-Glu-Gly-Gln-Ala-Ala-Lys-Glu-Phe-Ile-Ala-Trp-Leu-Val-Lys-Gly-Arg Amide

Synonyms: Gastrointestinal Peptide; Glucagon-Like Peptide I (1-
36); Prepro-Glucagon (72-107)

Sigma G 4397   Human   FW 4111.5   ≥97% (HPLC) |
Bioactive peptide; Bell, GI et al, *Nature*, 304:368, 1983

His-Asp-Met-Asn-Lys-Val-Leu-Asp-Leu

Synonyms: Anti-Inflammatory Peptide II; Peptide II; Antiflammin II

Bachem H-9440.0005 MW 1084.26 $C_{46}H_{77}N_{13}O_{15}S$ Store at -15°C

Neosystem SC180 MW 1084.25 Bioactive; anti-inflammatory peptide

Sigma H 3021 FW 1084.3 ≥97% (HPLC); peptide content:~65% | Bioactive peptide; inhibitor of phospholipase A_2; inhibits synthesis of platelet activating factor; Miele, L et al, *Nature*, 335:726, 1988

His-Asp-Ser-Gly-Tyr-Glu-Val-His-His-Gln-Lys-Leu-Val-Phe-Phe-Ala-Gln-Asp-Val-Gly-Ser-Asn-Lys-Gly-Ala-Ile-Ile-Gly-Leu-Met-Val-Gly-Gly-Val-Val
(Gln^{22}) β-His-Asp-Ser-Gly-Tyr-Glu-Val-His-His-Gln-Lys-Leu-Val-Phe-Phe-Ala-Gln-Asp-Val-Gly-Ser-Asn-Lys-Gly-Ala-Ile-Ile-Gly-Leu-Met-Val-Gly-Gly-Val-Val

Synonyms: Amyloid (6-40)

American Peptide 62-0-90 MW 3710.3 $C_{167}H_{258}N_{46}O_{48}S$

His-Cys-Lys-Phe-Trp-Trp

Synonyms: HIV Integrase Protein Inhibitor; HIV Integrase Inhibitor

Bachem H-3524.0005 MW 906.07 $C_{46}H_{55}N_{11}O_7S$ Store at -15°C

Neosystem SC1246 MW 906.07 >80% (HPLC) | Virus-related peptide; AIDS-related peptide; due to the presence of a Cys this peptide can contain the dimeric form; identified using a combinatorial chemical library; inhibits virus-encoded integrase (IN) protein-mediated 3'-processing & integration with an IC_{50} of 2mμM; active on IN proteins from other retroviruses such as HIV-2, feline immunodeficiency virus & Moloney murine leukemia virus, suggesting a conserved region of IN is targeted; inhibits the phosphoryl-transfer disintegration reaction, suggesting that it acts at or near the catalytic site of IN; may be useful for structure-function analysis of IN; Lutzke, RAP et al, *PNAS USA*, 92:11456-11460, 1995

Sigma H 6387 FW 906.1 $C_{46}H_{55}N_{11}O_7S$ ≥95% (HPLC); peptide content:~65% | Bioactive peptide; inhibits IN-mediated 3'-processing & integration of HIV DNA; also active on IN proteins from HIV-2, feline immunodeficiency virus & Moloney murine leukemia virus; Lutzke, RAP et al, *Proc Natl Acad Sci USA*, 92:11456, 1995

His-D-Phe-Asp-Ala-Val-Phe-Thr-Asp-Asn-Tyr-Thr-Arg-Leu-Arg-Lys-Gln-Met-Ala-Val-Lys-Lys-Tyr-Leu-Asn-Ser-Ile-Leu-Asn Amide

Synonyms: Vasoactive Intestinal Peptide, (D-Phe^2)-

ICN 154563 MW 3385.9

His-Gln-Lys-Leu-Val-Phe-Phe-Ala-Lys
MCA-His-Gln-Lys-Leu-Val-Phe-Phe-Ala-(Lys-DNP)

BioSource International 03-410

His-Gln-Val-Ser-Leu-Ser-Lys-Gln-Pro-Thr-Ser-Gln-Pro-Arg-Gly-Asp

Synonyms: TAT (65-80); HTLV I Peptide

Neosystem SC211 MW 1764.91 HTLV I peptide from subtype TAT

His-Gln-Val-Val-Ser-Ser-Asp-Phe-Asn-Ser-Asp-Thr Acetate Salt

Synonyms: Glyceraldehyde-3-Phosphate Dehydrogenase

Biogenesis 4700-0059 Human, rat synthetic Semi-pure; lyophilized | A 12 AA sequence from a highly conserved region of human and rat glyceraldehyde-3-phosphate dehydrogenase

His-Glu

Bachem G-2300.0250 MW 284.27 $C_{11}H_{16}N_4O_5$ Store at -15°C

His-Glu-Arg-Glu-Glu-Glu-Leu-Arg-Lys-Arg-Leu-Arg-Leu-Ile

Synonyms: REV (4-17); SIVmac251 Peptide

Neosystem SC596 MW 1877.17 SIVmac251 peptide from subtype REV

His-Glu-Asp-Ile-Ile-Ser-Leu-Trp-Asp-Gln-Ser-Leu-Lys

Synonyms: GP120 (105-117); HTLV I Peptide

Neosystem SC243 MW 1583.76 HTLV I peptide from subtype GP160

His-Gly
Z-His-Gly

Bachem C-1970.0250 MW 346.34 $C_{16}H_{18}N_4O_5$ Store at -15°C

His-Gly

Bachem G-2305.0250 MW 212.21 $C_8H_{12}N_4O_3$ Store at -15°C

ICN 157375 MW 212.2 $C_8H_{12}N_4O_3$ Crystalline

Sigma H 9000 FW 212.2 $C_8H_{12}N_4O_3$

His-Gly-Glu-Gly-Thr-Phe-Thr-Ser-Asp-Leu-Ser-Lys-Gln-Met-Glu-Glu-Glu-Ala-Val-Arg-Leu-Phe-Ile-Glu-Trp-Leu-Lys-Asn-Gly-Gly-Pro-Ser-Ser-Gly-Ala-Pro-Pro-Pro-Ser Amide

Synonyms: Exendin IV

American Peptide 46-3-12 MW 4186.7 $C_{184}H_{282}N_{50}O_{60}S$ Stimulates a monophasic increase in cAMP beginning at 100 pM that plateaus at 10 nM; exendin-4-induced increase in cAMP is inhibited progressively by increasing concentrations of the exendin receptor antagonist, exendin-(9-39) amide; Eng, J et al, *J Biol Chem*, 267:7402, 1992

Bachem H-8730.0500 MW 4186.6 $C_{184}H_{282}N_{50}O_{60}S$ Store at -15°C

Sigma E 7144 FW 4186.6 ≥97% (HPLC) | Bioactive peptide; originally isolated from the venom of the lizard, Heloderma suspectum; this polypeptide stimulates a monophasic cAMP increase in pancreatic acini by interacting exclusively with exendin receptors; Eng, J et al, *J Biol Chem*, 267:7402, 1992

His-Gly-Glu-Phe-Ala-Pro-Gly-Asn-Tyr-Pro-Ala-Leu-Trp-Ser-Tyr-Ala

Synonyms: Sendai Virus Nucleoprotein (321-336)

Bachem H-1394.0001 MW 1179.93 $C_{85}H_{110}N_{20}O_{23}$ Store at -15°C

Sigma S 5670 FW 1780.9 ≥90% (HPLC) | Bioactive peptide; recognized by cytotoxic T lymphocytes in vivo; mice immunized with fragment were protected from lethal virus infection; Kast, WM et al, *Proc Natl Acad Sci USA*, 88:2283, 1991

His-Gly-Gly

Bachem H-3735.0250 Store at -15°C

His-Gly-His
Ac-His-Gly-His-NHMe

Bachem H-1030.0050 MW 404.43 $C_{17}H_{24}N_8O_4$ Store at -15°C

His-Gly-His
Ac-His-Gly-His

Bachem H-1560.0050 MW 391.39 $C_{16}H_{21}N_7O_5$ Store at -15°C

His-His

Bachem G-4595.0250   Store at -15°C

His-His-Gly-His
Ac-His-His-Gly-His-NHMe

Bachem H-1035.0050   MW 541.57 $C_{23}H_{31}N_{11}O_5$ Store at -15°C

His-His-Gly-His
Ac-His-His-Gly-His

Bachem H-1630.0050   MW 528.53 $C_{22}H_{28}N_{10}O_6$ Store at -15°C

His-His-Gly-Val-Val-Glu-Val-Asp-Ala-Ala-Val-Thr-Pro-Glu-Glu-Arg-His-Leu-Ser-Lys

Synonyms: Amyloid Protein Precursor (657-676), β-; Peptide 20; Amyloid α/A4 Precursor Protein 770 (732-751)

American Peptide 62-5-20   MW 2210.5 $C_{95}H_{152}N_{30}O_{31}$ Cytoplasmic fragment of the amyloid protein precursor; believed to be a specific GTP-binding protein Go activator & provides the necessary sequence for the formation of the protein Go/APP complex; Nishimoto, I et al, *Nature*, 362:75, 1993

Bachem H-2968.0500   MW 2210.44 $C_{95}H_{152}N_{30}O_{31}$ Store at -15°C

His-His-Leu-Gly-Gly-Ala-Lys-Gln-Ala-Gly-Asp-Val

Synonyms: Fibrinogen γ-Chain Dodecapeptide; Fibrinogen Binding Inhibitor Peptide; Fibrinogen γ-Chain (400-411), Human

American Peptide 42-1-40   MW 1189.3 $C_{50}H_{80}N_{18}O_{16}$ This synthetic dodecapeptide represents the specific platelet receptor recognition site of the human fibrinogen g chain (residues 400-411); also a potent inhibitor of the binding of fibrinogen, fibronectin & von Willebrand factor to thrombin- or ADP-stimulated platelets; used in studying phosphorylation processes associated with platelet activation; Ferrell, JE et al, *PNAS*, 86:2234, 1989

Bachem H-9140.0001   MW 1189.3 $C_{50}H_{80}N_{18}O_{16}$ Store at -15°C

ICN 154538   MW 1189.5   Ferrell, JE etal, *PNAS*, 86:2234, 1989

Neosystem SC350   MW 1189.29   Bioactive; cell attachment peptide; fibrinogen-related peptide; fibrinogen binding inhibitor; Ferrell, JE et al, *PNAS USA*, 86:2234, 1989

Sigma F 9145   FW 1189.3   ≥97% (HPLC); peptide content:~70% | Bioactive peptide

His-His-Val-Ala-Arg-Glu-Leu-His-Pro-Glu-Tyr-Phe-Lys-Asn-Cys

Synonyms: NEF (192-206); HTLV I Peptide

Neosystem SC227   MW 1880.11   HTLV I peptide from subtype NEF; due to the presence of a Cys this peptide can contain the dimeric form

His-Ile

Synonyms: Peptide Histidine Isoleucine

Biogenesis 7260-0504   Porcine synthetic   >95%; lyophilized

His-Ile-Pro-Leu-Gly-Asp-Ala-Arg-Leu-Val-Ile-Thr

Synonyms: VIF (56-67); HTLV I Peptide

Neosystem SC545   MW 1304.55   HTLV I peptide from subtype VIF

His-Leu
Z-His-Leu

Bachem C-1975.0250   MW 402.45 $C_{20}H_{26}N_4O_5$ Store at -15°C

His-Leu

Bachem G-2310.0250   MW 268.32 $C_{12}H_{20}N_4O_3$ Store at -15°C

His-Leu
Bz-Gly-His-Leu

Synonyms: Hippuryl-His-Leu

Bachem M-1485.0250   MW 429.48 $C_{21}H_{27}N_5O_5$ Store at -15°C

His-Leu
p-Hydroxyhippuryl-His-Leu

Bachem M-1505.0050   MW 461.48 $C_{21}H_{27}N_5O_7$ Store at -15°C

His-Leu
N-Hippuryl-L-His-L-Leu Tetrahydrate

Fluka 53285   MW 501.58 $C_{21}H_{27}N_5O_5 \cdot 4H_2O$ ≥99% (TLC); ≥95% peptide content | Saharov, I Yu, et al, *Anal Biochem*, 166:14, 1987

His-Leu
Hippuryl-His-Leu Acetate Salt

Synonyms: Angiotensin Converting Enzyme Substrate; Benzoyl-Gly-His-Leu

ICN 152745   Cushman, DW & HS Cheung, *Biochem Pharmac*, 20:1637, 1971

His-Leu
Hydroxyhippuryl-His-Leu

Synonyms: Angiotensin Converting Enzyme Substrate; p-Hydroxybenzoyl-Gly-His-Leu

ICN 152746

His-Leu

ICN 157376   MW 268.3 $C_{12}H_{20}N_4O_3$ Crystalline

Peptides International OHL-3065   MW 268.32 $C_{12}H_{20}N_4O_3$ >98% (HPLC); colorless crystalline powder

Sigma H 2504   FW 268.3 $C_{12}H_{20}N_4O_3$

His-Leu
p-Hydroxyhippuryl-His-Leu

Synonyms: Angiotensin I Converting Enzyme Substrate

Sigma H 3135   FW 445.5 $C_{21}H_{27}N_5O_6$ Kasahara, Y & Ashimara, Y, *J Clin Chem & Clin Biochem*, 19:726, 1981

His-Leu
Hippuryl-His-Leu

Synonyms: Angiotensin Converting Enzyme Substrate

Sigma H 1635   Synthetic   FW 429.5 (free base) $C_{21}H_{27}N_5O_5$ Cushman, DW & Cheung, HS, *Biochem Pharmacol*, 20:1637, 1971

His-Leu
Hippuryl-His-Leu Acetate Salt

Synonyms: Angiotensin Converting Enzyme Substrate

Sigma H 4884   Synthetic   FW 429.5 (free base) $C_{21}H_{27}N_5O_5$ Cushman, DW & Cheung, HS, *Biochem Pharmacol*, 20:1637, 1971

His-Leu-Asp-Ile-Ile-Trp
Ac-His-Leu-Asp-Ile-Ile-Trp

Synonyms: Endothelin I (16-21)

American Peptide 88-2-34   Human   MW 838.0 $C_{41}H_{59}N_9O_{10}$

His-Leu-Asp-Ile-Ile-Trp-D-Val-Asn-Thr-Pro-Glu-His-Val-Val-Pro-Tyr-Gly-Phe-Gly-Ser-Pro-Arg-Ser

Synonyms: Endothelin I (16-38), (D-Val[22],Phe[33])-Big; Endothelin Converting Enzyme Inhibitor

American Peptide 88-1-34   Human   MW 2621.0
$C_{122}H_{178}N_{32}O_{33}$

His-Leu-Asp-Ile-Ile-Trp-D-Val-Asn-Thr-Pro-Glu-His-Val-Val-Pro-Tyr-Gly-Leu-Gly-Ser-Pro-Arg-Ser

Synonyms: Endothelin Converting Enzyme Inhibitor, (D-Val[22])-Big; Endothelin I (16-38), (D-Val[22])-Big

American Peptide 88-1-35   Human   MW 2586.9
$C_{119}H_{180}N_{32}O_{33}$ Strongly inhibits the ECE activity; Morita, A et al, *FEBS Lett*, 353, 84, 1994

Sigma E 9895   Human   FW 2586.9   ≥95% (HPLC); peptide content:~70% | Bioactive peptide; antagonist of big endothelin 1-mediated dopamine release; Morita, A et al, *FEBS Lett*, 353:84, 1994

His-Leu-Asp-Ile-Ile-Trp-Val-Asn-Thr-Pro-Glu-His-Val-Val-Pro-Tyr-Gly-Leu-Gly-Ser-Pro-Arg-Ser
His-Leu-Asp-Ile-Ile-Trp-D-Val-Asn-Thr-Pro-Glu-His-Val-Val-Pro-Tyr-Gly-Leu-Gly-Ser-Pro-Arg-Ser

Synonyms: Endothelin I (16-38), (D-Val[22])-Big; Endothelin, Big

Bachem H-2526.0500   Human   MW 2586.93
$C_{119}H_{180}N_{32}O_{33}$ Store at -15°C

His-Leu-Gly-Leu-Ala-Arg

Synonyms: Anaphylatoxin C3a (70-77); C3a (72-77); Anaphylatoxic Peptide C3a (72-77)

ICN 153126   MW 665.8   Human; $C_{29}H_{51}N_{11}O_7$ Hartung, HP et al, *Agents & Actions*, 15, 1984

Bachem H-2235.0005   Human   MW 665.79 $C_{29}H_{51}N_{11}O_7$ Store at -15°C

Sigma H 0765   Human   FW 665.8 $C_{29}H_{51}N_{11}O_7$ ≥97% (HPLC) | Bioactive peptide; stimulates thromboxane B_2 production; Hartung, HP et al, *Agents & Actions*, 15:14, 1984

His-Leu-His
Z-His-Leu-His-βNA

Bachem K-1195.0050   Store at RT

His-Leu-His
His-Leu-His-βNA

Bachem K-1355.0050   MW 530.6   Store at -15°C

His-Leu-Leu-Val-Phe
His-Leu-Leu-Val-Phe-OMe

Bachem H-3740.0050   MW 841.81 $C_{33}H_{51}N_7O_6$ Store at -15°C

His-Leu-Pro-Pro-Pro-Val

Bachem H-2254.0005   MW 658.8 $C_{32}H_{50}N_8O_7$ Store at -15°C

His-Leu-Pro-Pro-Pro-Val-His-Leu-Pro-Pro-Pro-Val

Bachem H-2256.0001   MW 1299.59 $C_{64}H_{98}N_{16}O_{13}$ Store at -15°C

His-Lys
Z-His-Lys-BOC

Bachem C-3665.0001   Store at -15°C

His-Lys
Z-His-Lys-BOC Hydrazide

Bachem C-3670.0001   MW 531.61 $C_{25}H_{37}N_7O_6$ Store at -15°C

His-Lys
His-Lys-OMe 3 Hydrochloride

Bachem G-4110.0250   Store at -15°C

His-Lys Hydrobromide

Bachem G-2320.0250   MW 364.24 $C_{12}H_{21}N_5O_3$ · HBr   Store at -15°C

ICN 157377   MW 364.3 $C_{12}H_{21}N_5O_3$ · HBr

Sigma H 0775   FW 364.2 $C_{12}H_{21}N_5O_3$ · HBr

His-Lys-Ala-Arg-Val-Leu-Ala-Glu-Ala-Met-Ser Amide

Synonyms: HIV Protease Substrate III-B (Native Sequence); HIV Protease Substrate III-B

Bachem H-9650.0001   Store at -15°C

Neosystem SC674   MW 1211.44   Virus-related peptide; AIDS-related peptide

His-Lys-Ala-Arg-Val-Leu-Phe-Glu-Ala-Nle-Ser
His-Lys-Ala-Arg-Val-Leu-p-Nitro-Phe-Glu-Ala-Nle-Ser Amide

Synonyms: HIV Protease Substrate III

Bachem H-9035.0001   MW 1314.51 $C_{58}H_{95}N_{19}O_{16}$ Store at -15°C

His-Lys-Ala-Arg-Val-Leu-Phe-Glu-Ala-Nle-Ser
Anthranilyl-His-Lys-Ala-Arg-Val-Leu-(pNO₂-Phe)-Glu-Ala-Nle-Ser Amide

Synonyms: HIV Anthranilyl Substrate III

ICN 158720

His-Lys-Ala-Arg-Val-Leu-Phe-Glu-Ala-Nle-Ser
His-Lys-Ala-Arg-Val-Leu-p-Nitro-Phe-Glu-Ala-Nle-Ser Amide

Synonyms: HIV Protease Substrate III; HIV Protease Substrate

Neosystem SC673   MW 1314.70   Virus-related peptide; AIDS-related peptide; Pennington, MW et al, *Peptides*, Proceedings of the 22nd EPS, Interlaken, p 936, CH Schneider & AN Eberle, Eds Escom Leiden, 1993

Sigma H 5397   FW 1314.5   ≥97% (HPLC) | Bioactive peptide; Pennington, MW et al, *Peptides 1990*, Proc 21st Eur Pept Symp, Platja d'Aro, Spain, 787:E Giralt & D Andreu, eds, Escom, Leiden, 1991

His-Lys-Ala-Arg-Val-Leu-Phe-Glu-Ala-Nle-Ser
His-Lys-Ala-Arg-Val-Leu-(pNO₂-Phe)-Glu-Ala-Nle-Ser Amide

Synonyms: HIV Substrate III

ICN 158719   Synthetic   Synthetic peptide analog useful in measuring activity of the HIV protease; HIV protease K_m = 25 μM for HIV Substrate III; enzymatic activity can be monitored chromatographically by reverse phase HPLC or spectrophotometrically at 300 nm due to the p-nitro-Phe residue

His-Lys-Ala-Arg-Val-Leu-p-Nitro-Phe-Glu-Ala-Nle-Ser
Anthranilyl-His-Lys-Ala-Arg-Val-Leu-p-Nitro-Phe-Glu-Ala-Nle-Ser Amide

Synonyms: HIV Protease Substrate

Sigma A 0811   FW 1433.6   Peptide content ~70% | Dunn, BM et al, Abstract – 12th American Peptide Symposium, Boston, 1991

His-Lys-Ala-Arg-Val-Leu-Tyr-Glu-Ala-Nle-Ser Abz-His-Lys-Ala-Arg-Val-Leu-Tyr(NO₂)-Glu-Ala-Nle-Ser Amide

Synonyms: HIV Fluorescent Substrate

American Peptide 81-0-40   MW 1449.6 $C_{65}H_{100}N_{20}O_{18}$

His-Lys-Cys-Asn-Thr-Ala-Thr-Cys-Ala-Thr-Gln-Arg-Leu-Ser-Thr-Asn-Val-Gly-Ser-Asn-Thr-Tyr Amide

Synonyms: Amylin

Biogenesis 0486-5106   Human synthetic   MW 3898
Purified; lyophilized

His-Lys-Phe-Ser-Val-Ser-Gly-Glu-Gly-Glu-Gly-Asp-Ala-Thr-Cys

Synonyms: Green Fluorescent Protein (26-39)

Alexis 157-018   *Aequorea Victoria* (jellyfish) synthetic   MW 1523.6 $C_{62}H_{94}N_{18}O_{25}S$ ≥95%; lyophilized; reconstitute with 0.1 mL distilled H_2O | Competitively binds to Ab Alexis 210-199; antiserum blocking peptide

His-Lys-Thr-Asp-Ser-Phe-Val-Gly-Leu-Met

Synonyms: Neurokinin A

Biogenesis 6690-0004   Synthetic   MW 1133.25   90.8% peptide material, acetate counter ions and residual H_2O; lyophilized | To avoid Met oxidation, it is best to dissolve the peptide under a nitrogen stream

His-Lys-Thr-Asp-Ser-Phe-Val-Gly-Leu-Met Amide

Synonyms: Neurokinin A/Substance K; Neuromedin L; Neurokinin A; Substance K; Neurokinin, α-; Neurokinin II Agonist; Neurokinin II Receptor Selective Agonist

American Peptide 62-1-40   MW 1133.3 $C_{50}H_{80}N_{14}O_{14}S$
Belongs to the tachykinin family; a more potent bronchio-constrictor than substance P & may regulate neutrophil recruitment in the lower respiratory tract; Kimura, S et al, *Proc Japan Acad*, 59:Ser B 101, 1983; Nawa, H et al, *Nature*, 306:32, 1983

Bachem H-3745.0001   MW 1133.34 $C_{50}H_{80}N_{14}O_{14}S$ Store at -15°C

Calbiochem 05-23-0801   MW 1133.3 $C_{50}H_{80}N_{14}O_{14}S$ ≥99% (HPLC); lyophilized solid; soluble in 5% acetic acid; LD_{50}≤2000 mg/kg | Peptide belonging to the tachykinin family; more potent bronchioconstrictor than substance P; released in association with the activation of polymodal C-nociceptors; regulates neutrophil recruitment into lower respiratory tract; exhibits higher affinity for NK-2 receptors compared to NK-1 or NK-3 & plays a role in hematopoietic regulation; Nagahisa, A et al, *Eur J Pharmacol*, 217:191, 1992; Rameshwar, P & Gascom, P, *Blood*, 88:98, 1996; Chan, CC et al, *Can J Physiol Pharmacol*, 72:11, 1993; von Essen, SG et al, *Am J Physiol*, 263:L226, 1992; Xu, XJ & Wiesenfeld-Hallin, A, *Acta Physiol Scand*, 144:63, 1992

ICN 153150   MW 1133.3

Neosystem SC012   MW 1133.33   Bioactive; tachykinin; Nawa, H et al, *Nature*, 306:32, 1983

Sigma N 4267   FW 1133.3   ≥95% (HPLC); peptide content:~70% | Bioactive peptide; more potent bronchoconstrictor than Substance P; Kimura, S et al, *Proc Japan Academy*, 59:101 (Ser B), 1983; Nawa, H et al, *Nature*, 306:32, 1983

Peptides International PNK-4154-v   Synthetic   MW 1133.3 $C_{50}H_{80}N_{14}O_{14}S$ >95% (HPLC); lyophilized amorphous powder | Bioactive; human, porcine, rat, mouse; Kimura, S et al, *Proc Japan Acad*, 59B:101, 1983; Kangawa, K et al, *Peptide Chem,* 1983:309, 1984; Nawa, H et al, *Nature*, 306:32, 1983

His-Lys-Thr-Asp-Ser-Phe-Val-Gly-Leu-Met Amide 2AcOH 5H₂O

Synonyms: Neurokinin A; Neuromedin L Substance K; Neurokinin II Receptor Selective Agonist

Peptides International PNK-4154   Synthetic   MW 1343.49 $C_{50}H_{80}N_{14}O_{14}S \cdot {}_2CH_3COOH$ >95% (HPLC); lyophilized amorphous powder; bulk | Bioactive; human, porcine, rat, mouse; Kimura, S et al, *Proc Japan Acad*, 59B:101, 1983; Kangawa, K et al, *Peptide Chem*, 1983:309, 1984; Nawa, H et al, *Nature*, 306:32, 1983

His-Met
BOC-His-Met

Bachem A-1785.0001   MW 386.47 $C_{16}H_{26}N_4O_5S$ Store at -15°C

His-Met
Z-His-Met

Bachem C-1980.0001   MW 420.49 $C_{19}H_{24}N_4O_5S$ Store at -15°C

His-Met

Synonyms: Peptide Histidine Methionine

Bachem G-2325.0250   MW 286.36 $C_{11}H_{18}N_4O_3S$ Store at -15°C

Biogenesis 7260-2504   Synthetic   >95%; lyophilized

His-Met-Arg-Ser-Ala-Met-Ser-Gly-Leu-His-Leu-Val-Lys-Arg-Arg

Synonyms: SAMS Peptide; AMP-Activated Protein Kinase Substrate

Upstate 12-355   MW 1777.9   >80% pure; frozen solution

His-Phe
Z-His-Phe

Bachem C-1985.0250   MW 436.47 $C_{23}H_{24}N_4O_5$ Store at -15°C

His-Phe
His-Phe-βNA Dihydrochloride

Bachem K-1360.0250   MW 500.43 $C_{25}H_{25}N_5O_2 \cdot 2HCl$ Store at -15°C

His-Phe

Bachem N-1095.0250   MW 302.33 $C_{15}H_{18}N_4O_3$ Store at -15°C

ICN 157378   MW 302.3 $C_{15}H_{18}N_4O_3$ Crystalline

Sigma H 2629   FW 302.3 $C_{15}H_{18}N_4O_3$

His-Phe Amide Hydrochloride

Bachem G-2330.0250   MW 337.81 $C_{15}H_{19}N_5O_2 \cdot HCl$ Store at -15°C

His-Phe-Arg-Trp

Synonyms: Characteristic Melanocyte Stimulating Hormone Tetrapeptide

Bachem H-3750.0025   MW 644.73 $C_{32}H_{40}N_{10}O_5$ Store at -15°C

His-Phe-Arg-Trp-Gly-Lys-Pro-Val-Gly-Lys-Lys-Arg-Arg-Pro-Val-Lys-Val-Tyr-Pro

Synonyms: Corticotropin A; Adrenocorticotropic Hormone (6-24)

ICN 159857

American Peptide 10-1-55   Human   MW 2335.9 $C_{111}H_{175}N_{35}O_{21}$

Neosystem SC016   Human   MW 2335.83   Bioactive

His-Phe-Asp-Ala-Val-Phe-Thr-Asp-Asn-Tyr-Thr-Arg-Leu-Arg-Lys-Gln-Met-Ala-Val-Lys-Lys-Tyr-Leu-Asn-Ser-Ile-Leu-Asn
His-D-Phe-Asp-Ala-Val-Phe-Thr-Asp-Asn-Tyr-Thr-Arg-Leu-Arg-Lys-Gln-Met-Ala-Val-Lys-Lys-Tyr-Leu-Asn-Ser-Ile-Leu-Asn Amide

Synonyms: Vasoactive Intestinal Peptide, (D-Phe²)-

Bachem H-5640.0001   Human, bovine, porcine, rat   MW 3385.94 $C_{153}H_{242}N_{44}O_{41}S$ Store at -15°C

American Peptide 48-0-13   Porcine   MW 3386.2 $C_{153}H_{242}N_{44}O_{41}S$ Partial VIP agonist with low intrinsic activity

His-Phe-Met-Pro-Thr

Synonyms: Murine CMV pp 89 (170-174)

Bachem H-9420.0005   Store at -15°C

His-Phe-Phe
Z-His-*p*-Nitro-Phe-Phe-OMe

Bachem C-1990.0250   MW 642.67 $C_{33}H_{34}N_6O_8$ Store at -15°C

His-Phe-Phe
Z-His-Phe-Phe-OEt

Bachem M-1270.0250   MW 611.7 $C_{34}H_{37}N_5O_6$ Store at -15°C

His-Phe-Trp
Z-His-Phe-Trp-OEt

Bachem M-1275.0250   MW 650.74 $C_{36}H_{38}N_6O_6$ Store at -15°C

His-Phe-Tyr
Z-His-Phe-Tyr-OEt

Bachem M-1280.0250   Store at -15°C

His-Pro

Synonyms: Thyrotropin Releasing Hormone (2-3), Diketopiperazine

Bachem G-2335.0250   MW 252.27 $C_{11}H_{16}N_4O_3$ Store at -15°C

Biogenesis 8940-1059   Synthetic   MW 234.23   Lyophilized, consisting of 90.63% peptide material, acetate counter ions and residual H_2O

His-Pro Amide Dihydrobromide

Bachem G-4185.0250   MW 413.11 $C_{11}H_{17}N_5O_2 \cdot$ 2HBr   Store at -15°C

His-Pro-Gln-Tyr-Asn-Gln-Arg

Synonyms: Cathepsin G (77-83)

Bachem H-1266.0005   MW 942 $C_{40}H_{59}N_{15}O_{12}$ Store at -15°C

His-Pro-Gln-Tyr-Asn-Gln-Arg Amide

Synonyms: Cathepsin G (77-83)

Bachem H-8240.0005   Store at -15°C

His-Pro-Leu-Gln-Lys-Thr-Tyr

Synonyms: Band 3 Protein (547-553)

Bachem H-1596.0005   Human   Store at -15°C

His-Pro-Lys-Arg-Pro-Trp-Ile-Leu

Synonyms: Xenopsin Related Peptide II; Xenopsin Related Peptide I

American Peptide 63-0-65   MW 1046.3 $C_{51}H_{79}N_{15}O_9$ Carraway, RE et al, *Reg Peptides*, 22:303, 1988

Bachem H-9345.0005   MW 1046.28 $C_{51}H_{79}N_{15}O_9$ Store at -15°C

ICN 154463   MW 1046.4   Carraway, RE etal, *Regulatory Peptides*, 22:303, 1988

His-Pro-Phe
Ac-His-Pro-Phe

Bachem H-7275.0025   MW 441.49 $C_{22}H_{27}N_5O_5$ Store at -15°C

His-Pro-Phe-His-Leu-Leu-Val-Tyr
His-Pro-Phe-His-Leu-D-Leu-Val-Tyr

Synonyms: Renin Inhibitor

Bachem N-1100.0005   MW 1025.22 $C_{52}H_{72}N_{12}O_{10}$ Store at -15°C

Sigma H 5396   FW 1025.2   ≥95% (HPLC); peptide content:~65% | Bioactive peptide; Poulsen, K et al, *Biochemistry*, 12:3877, 1973

His-Pro-Phe-His-Leu-Leu-Val-Tyr
D-His-Pro-Phe-His-Leu-ψ-(CH₂NH)-Leu-Val-Tyr

Synonyms: Renin Inhibitor

Sigma H 6137   FW 1011.2   ≥90% (HPLC) | Bioactive peptide; potent inhibitor of human renin; Szelke, M et al, *Nature*, 299:555, 1982

His-Ser

Bachem G-2340.0250   MW 242.24 $C_9H_{14}N_4O_4$ Store at -15°C

His-Ser
His-Ser-βNA

Bachem K-1365.0050   MW 367.41 $C_{19}H_{21}N_5O_3$ Store at -15°C

His-Ser

ICN 157379   MW 242.2 $C_9H_{14}N_4O_4$ Crystalline

Sigma H 3129   FW 242.2 $C_9H_{14}N_4O_4$

His-Ser
His-Ser 4-Methyoxy-β-Naphthylamide Acetate Salt

Synonyms: Cathepsin C Substrate; Dipeptidyl Aminopeptidase I Substrate

Sigma H 9758   FW 397.4 (free acid) $C_{20}H_{23}N_5O_4$

His-Ser-4MβNA

Bachem J-1225.0050   MW 397.43 $C_{20}H_{23}N_5O_4$ Store at -15°C

His-Ser-Arg-Asn-Ser-Ile-Thr-Leu-Thr-Asn-Leu-Thr

Synonyms: Fibronectin (1377-1388)

Bachem H-2572.0001   MW 1356.5 $C_{56}H_{97}N_{19}O_{20}$ Store at -15°C

ICN 195619   MW 1356.5   97%

Sigma F 0793   FW 1356.5   ≥97% (HPLC); peptide content:~70% | Bioactive peptide; overlapping sequences from the fibronectin cell-binding domain; they both inhibit platelet aggregation by interfering with fibronectin binding to platelet membrane glycoprotein IIb/IIIa; Mohri, H et al, *Peptides*, 16:263, 1995

His-Ser-Asp-Ala-Ile-Phe-Thr-Gln-Gln-Tyr-Ser-Lys-Leu-Leu-Ala-Lys-Ala-Leu-Gln-Lys-Tyr-Leu-Als-Ser-Ile-Leu-Gly-Ser-Arg-Thr-Ser-Pro-Pro-Pro Amide

Synonyms: Helodermin

American Peptide 73-0-12   MW 3843.5   C₁₇₆H₂₈₅N₄₇O₄₉
Hoshino, M et al, *FEBS Lett*, 178:233, 1984

His-Ser-Asp-Ala-Ile-Phe-Thr-Gln-Gln-Tyr-Ser-Lys-Leu-Leu-Ala-Lys-Leu-Ala-Leu-Gln-Lys-Tyr-Leu-Ala-Ser-Ile-Leu-Gly-Ser-Arg-Thr-Ser-Pro-Pro-Pro Amide

Synonyms: Helodermin

Neosystem SC903   MW 3843.47   Bioactive; brain/gut peptide; VIP/PHI/secretin/helodermin peptide; VIP-secretin-like peptide isolated from Gila monster venom; Robberecht, P et al, *FEBS Lett*, 166:277-282, 1984; Hoshino, M et al, *FEBS Lett*, 178:233-239, 1984

His-Ser-Asp-Ala-Ile-Phe-Thr-Glu-Glu-Tyr-Ser-Lys-Leu-Leu-Ala-Lys-Ala-Leu-Gln-Lys-Tyr-Leu-Ala-Ser-Ile-Leu-Gly-Ser-Arg-Thr-Ser-Pro-Pro-Pro Amide

Synonyms: Helodermin

Bachem H-5062.0001   MW 3732.3   C₁₇₀H₂₇₂N₄₄O₅₀   Store at -15°C

His-Ser-Asp-Ala-Ile-Phe-Thr-Glu-Glu-Tyr-Ser-Lys-Leu-Leu-Ala-Lys-Leu-Ala-Leu-Ser-Ile-Leu-Gly-Ser-Arg-Thr-Ser-Pro-Pro-Pro Amide Sodium Salt

Synonyms: Hainanmycin

ICN 196002   MW 907.1   C₄₇H₇₉O₁₅Na   Polyether antibiotic with broad antiseptic properties

---

His-Ser-Asp-Ala-Ile-Phe-Thr-Glu-Glu-Tyr-Ser-Lys-Leu-Leu-Ala-Lys-Leu-Ala-Leu-Gln-Lys-Tyr-Leu-Ala-Ser-Ile-Leu-Gly-Ser-Arg-Thr-Ser-Pro-Pro-Pro Amide

Synonyms: Helodermin

ICN 196004   MW 3845.5   C₁₇₆H₂₈₃N₄₅O₅₁   ≥97% | VIP family peptide shown to stimulate adenylate cyclase in rat pancreas

---

His-Ser-Asp-Ala-Ile-Phe-Thr-Gly-Glu-Tyr-Ser-Lys-Leu-Leu-Ala-Lys-Leu-Ala-Leu-Gln-Lys-Tyr-Leu-Ala-Ser-Ile-Leu-Gly-Ser-Arg-Thr-Ser-Pro-Pro-Pro Amide

Synonyms: Helodermin

Calbiochem 05-23-2800   MW 3845.5   C₁₇₆H₂₈₃N₄₅O₅₁   ≥97% (HPLC); lyophilized solid; soluble in 5% acetic acid | 35-residue peptide of the vasoactive intestinal polypeptide (VIP) family that exhibits vasodilatory properties; stimulates adenylate cyclase in rat pancreas; inhibitor of phospholipase A₂; Tanaka, Y et al, *Res Commun Mol Pathol Pharmacol*, 98:141, 1997; Kashimura, J et al, *Pancreas*, 10:161, 1995; Trotz, ME et al, *Regul Pept*, 48:301, 1993; Robberecht, P et al, *Regul Pept*, 26:117, 1989

His-Ser-Asp-Ala-Leu-Phe-Thr-Asp-Thr-Tyr-Thr-Arg-Leu-Arg-Lys-Gln-Met-Ala-Met-Lys-Lys-Tyr-Leu-Asn-Ser-Val-Leu-Asn Amide

Synonyms: Vasoactive Intestinal Peptide

American Peptide 48-6-15   Guinea pig   MW 3344.9   C₁₄₇H₂₃₉N₄₃O₄₂S₂   Du, BH et al, *BBRC*, 128:1093, 1985

His-Ser-Asp-Ala-Thr-Phe-Thr-Ala-Glu-Tyr-Ser-Lys-Leu-Leu-Ala-Lys-Leu-Ala-Leu-Gln-Lys-Tyr-Leu-Glu-Ser-Ile-Leu-Gly-Ser-Ser-Thr-Ser-Pro-Arg-Pro-Pro-Ser-Ser

Synonyms: Helospectin I

American Peptide 71-0-48   MW 4095.7   C₁₈₃H₂₉₃N₄₇O₅₉
Parker, DS et al, *JBC*, 259:11751, 1984

His-Ser-Asp-Ala-Thr-Phe-Thr-Ala-Glu-Tyr-Ser-Lys-Leu-Leu-Ala-Lys-Leu-Ala-Leu-Gln-Lys-Tyr-Leu-Glu-Ser-Ile-Leu-Gly-Ser-Ser-Thr-Ser-Pro-Arg-Pro-Pro-Ser

Synonyms: Helospectin II

American Peptide 71-0-49   MW 4008.6   C₁₈₀H₂₈₈N₄₆O₅₇
Parker, DS et al, *JBC*, 159:11751, 1984

His-Ser-Asp-Ala-Val-pCl-Phe-Thr-Asp-Asn-Tyr-Thr-Arg-Leu-Arg-Lys-Gln-Leu-Ala-Val-Lys-Lys-Tyr-Leu-Asn-Ser-Ile-Leu-Asn Amide

Synonyms: Vasoactive Intestinal Peptide, (D-p-Chloro-Phe⁶,Leu¹⁷)-

ICN 152896   Porcine   VIP antagonist; Pandol, SJ etal, *Am J Physiol*, 250:G553, 1986

His-Ser-Asp-Ala-Val-Phe-Thr-Asp-Asn-Tyr-Thr-Arg

Synonyms: Gastrointestinal Peptide; Vasoactive Intestinal Peptide (1-12); CD-4 Receptor Ligand

ICN 152897   Human   Sacerdote, P etal, *J Neurosci Res*, 18:102, 1987

Sigma V 0131   Human, porcine, rat   FW 3467.1   ≥97% (HPLC) | Bioactive peptide; Sacerdote, P et al, *J Neurosci Res*, 18:102, 1987

American Peptide 48-1-12   Human, porcine, rat, ovine   MW 1425.6   C₆₁H₈₈N₁₈O₂₂   Sacerdote, P et al, *Neuroimmunomodulation*, Perezpolo, JR (ed), 102, 1987

His-Ser-Asp-Ala-Val-Phe-Thr-Asp-Asn-Tyr-Thr-Arg-Leu-Arg-Arg-Gln-Leu-Ala-Val-Arg-Arg-Tyr-Leu-Asn-Ser-Ile-Leu-Asn-Gly-Lys-Arg Amide

Synonyms: Vasoactive Intestinal Peptide-Gly-Lys-Arg, (Arg¹⁵,²⁰,²¹,Leu¹⁷)-

American Peptide 48-1-75   MW 3733.3   C₁₆₂H₂₆₇N₅₇O₄₅
Kashimoto, K et al, *Peptide Chemistry*, 361, 1995

His-Ser-Asp-Ala-Val-Phe-Thr-Asp-Asn-Tyr-Thr-Arg-Leu-Arg-Arg-Gln-Leu-Ala-Val-Arg-Arg-Tyr-Leu-Asn-Ser-Ile-Leu-Asn Amide

Synonyms: Vasoactive Intestinal Peptide, (Arg¹⁵,²⁰,²¹,Leu¹⁷)-

American Peptide 48-1-70   Human, porcine, rat, ovine   MW 3392.1   C₁₄₈H₂₄₀N₅₀O₄₂   M-VIP; Kashimoto, K et al, *Peptide Chem*, 361, 1995

His-Ser-Asp-Ala-Val-Phe-Thr-Asp-Asn-Tyr-Thr-Arg-Leu-Arg-Lys-Gln-Leu-Ala-Val-Lys-Lys-Tyr-Leu-Asn-Ser-Ile-Leu-Asn
His-Ser-Asp-Ala-Val-D-(p-Chloro)Phe-Thr-Asp-Asn-Tyr-Thr-Arg-Leu-Arg-Lys-Gln-Leu-Ala-Val-Lys-Lys-Tyr-Leu-Asn-Ser-Ile-Leu-Asn Amide

Synonyms: Vasoactive Intestinal Peptide, (p-Chloro-D-Phe⁶,Leu¹⁷)-

American Peptide 48-1-28   MW 3342.3   C₁₄₈H₂₄₀N₄₄O₄₂Cl   VIP receptor antagonist; Pandol, SJ et al, *Am J Physiol*, 250:G553, 1986

His-Ser-Asp-Ala-Val-Phe-Thr-Asp-Asn-Tyr-Thr-Arg-Leu-Arg-Lys-Gln-Leu-Ala-Val-Lys-Lys-Tyr-Leu-Asn-Ser-Ile-Leu-Asn
His-Ser-Asp-Ala-Val-p-Chloro-Phe-Thr-Asp-Asn-Tyr-Thr-Arg-Leu-Arg-Lys-Gln-Leu-Ala-Val-Lys-Lys-Tyr-Leu-Asn-Ser-Ile-Leu-Asn Amide

Synonyms: Vasoactive Intestinal Peptide, (D-p-Chloro-Phe⁶,Leu¹⁷)-

ICN 198748   Human   VIP antagonist; Pandol, SJ etal, *Am J Physiol*, 250:G553, 1986

His-Ser-Asp-Ala-Val-Phe-Thr-Asp-Asn-Tyr-Thr-Arg-
Leu-Arg-Lys-Gln-Leu-Ala-Val-Lys-Lys-Tyr-Leu-Asn-
Ser-Ile-Leu-Asn

His-Ser-Asp-Ala-Val-*p*-Chloro-D-Phe-Thr-Asp-Asn-
Tyr-Thr-Arg-Leu-Arg-Lys-Gln-Leu-Ala-Val-Lys-Lys-
Tyr-Leu-Asn-Ser-Ile-Leu-Asn Amide

Synonyms: Vasoactive Intestinal Peptide, (*p*-Chloro-D-Phe[6],Leu[17])-

Bachem H-5515.0500   Human, bovine, porcine, rat   MW
3342.25 $C_{148}H_{239}ClN_{44}O_{42}$ Store at -15°C

His-Ser-Asp-Ala-Val-Phe-Thr-Asp-Asn-Tyr-Thr-Arg-
Leu-Arg-Lys-Gln-Leu-Ala-Val-Lys-Lys-Tyr-Leu-Asn-
Ser-Ile-Leu-Asn-Gly-Lys

Synonyms: Vasoactive Intestinal Peptide, (Leu[17],Gly[29],Lys[30])-

Neosystem SC141   Human, porcine, rat   MW 3494.00
Bioactive; brain/gut peptide; VIP/PHI/secretin/helodermin peptide;
Tachibana, S & Itoh, O, *Proceedings of the 10th Amer Pept Symp*,
1988

His-Ser-Asp-Ala-Val-Phe-Thr-Asp-Asn-Tyr-Thr-Arg-
Leu-Arg-Lys-Gln-Leu-Ala-Val-Lys-Lys-Tyr-Leu-Asn-
Ser-Ile-Leu-Asn

His-Ser-Asp-Ala-Val-*p*-Chloro-D-Phe-Thr-Asp-Asn-
Tyr-Thr-Arg-Leu-Arg-Lys-Gln-Leu-Ala-Val-Lys-Lys-
Tyr-Leu-Asn-Ser-Ile-Leu-Asn Amide

Synonyms: Vasoactive Intestinal Peptide, (*p*-Chloro-D-Phe[6],Leu[17])-

Neosystem SC923   Human, porcine, rat   MW 3342.19
Bioactive; brain/gut peptide; VIP/PHI/secretin/helodermin peptide;
Pandol, SJ et al, *Am J Physiol*, 250:G553-G557, 1986

His-Ser-Asp-Ala-Val-Phe-Thr-Asp-Asn-Tyr-Thr-Arg-
Leu-Arg-Lys-Gln-Leu-Ala-Val-Lys-Lys-Tyr-Leu-Asn-
Ser-Ile-Leu-Asn

His-Ser-Asp-Ala-Val-*p*-Chloro-Phe-Thr-Asp-Asn-
Tyr-Thr-Arg-Leu-Arg-Lys-Gln-Leu-Ala-Val-Lys-Lys-
Tyr-Leu-Asn-Ser-Ile-Leu-Asn Amide

Synonyms: Gastrointestinal Peptide; Vasoactive Intestinal Peptide,
(D-*p*-Chloro-Phe[6],Leu[17])-; Vasoactive Intestinal Peptide Receptor
Antagonist

Sigma V 4380   Human, porcine, rat   FW 3342.2   ≥97%
(HPLC) | Bioactive peptide; Pandol, SJ et al, *Amer J Physiol*,
250:G553, 1986

His-Ser-Asp-Ala-Val-Phe-Thr-Asp-Asn-Tyr-Thr-Arg-
Leu-Arg-Lys-Gln-Met-Ala-Val-Lys-Lys-Tyr-Leu-Asn-
Ser-Ile-Leu-Asn Amide

Synonyms: Vasoactive Intestinal Peptide, Human; Vasoactive
Intestinal Contractor; Vasoactive Intestinal Peptide;
Gastrointestinal Peptide

Bachem H-3775.0500   Human, bovine, porcine, rat   MW
3325.84 $C_{147}H_{238}N_{44}O_{42}S$ Store at -15°C

Peptides International PVA-4110-s, 4110-v   Human, porcine
synthetic   MW 3325.8 $C_{147}H_{238}N_{44}O_{42}S$ >95% (HPLC);
lyophilized amorphous powder | Bovine, rat, canine; Mutt, V & SI
Said, *Eur J Biochem*, 42:581, 1974; Bodner, M et al, *PNAS* USA,
82:3548, 1985

Calbiochem 05-23-2101   Human, porcine, rat   MW 3325.9
$C_{147}H_{238}N_{44}O_{42}S$ ≥95% (HPLC); lyophilized solid; soluble in 5%
acetic acid; LD_{50}≤2000 mg/kg | Mitogenic factor for embryonic
neurons in the sympathetic nervous system; has powerful
hypotensive & vasodilatory effects; behaves either as a circulating
hormone or as a neurotransmitter; stimulates cAMP production in
rat peritoneal macrophages; nitric oxide stimulates the release of
this peptide; Bakker, R et al, *Am J Physiol*, 264:R362, 1993;
Grider, JR et al, *Am J Physiol*, 264:G334, 1993

Neosystem SC085   Human, porcine, rat   MW 3325.83
Bioactive; brain/gut peptide; VIP/PHI/secretin/helodermin peptide;
Said, SI & Mutt, V, *Science*, 169:1217-1218, 1970; Mutt, V & Said,
SI, *Mol Pharm Biochem*, 42:581-589, 1974

Sigma V 3628   Human, porcine, rat   FW 3325.8   ≥95%
(HPLC) | Bioactive peptide

Sigma V 6130   Human, porcine, rat   FW 3325.8   ≥95%
(HPLC) | Bioactive peptide; widely distributed in brain &
peripheral nervous system; modulates cholinergic & serotonergic
neurotransmission; vasodilator, bronchodilator, smooth muscle
relaxant; mitogen for embryonic sympathetic neurons; released by
nitric oxide; Bodanszky, M et al, *Proc Natl Acad Sci USA*, 70:382,
1973

American Peptide 48-1-10   Human, porcine, rat, ovine   MW
3325.7 $C_{147}H_{238}N_{44}O_{42}S$ Synthesized in the central nervous
system & gastrointestinal tract; behaves as a circulating hormone
to have powerful hypotensive & vasodilatory effects or acts as a
neurotransmitter to modulate both cholinergic & serotonergic
systems; stimulates cAMP production in rat peritoneal
macrophages; nitric oxide enhances release of this peptide;
Segura, JJ et al, *Reg Peptides*, 37:145, 1992

ICN 195541   Porcine   Potent vasodilator; Bodanszky, M et al,
PNAS, 70:382, 1973

His-Ser-Asp-Ala-Val-Phe-Thr-Asp-Asn-Tyr-Thr-Arg-
Leu-Arg-Lys-Gln-Met-Ala-Val-Lys-Lys-Tyr-Leu-Asn-
Ser-Ile-Leu-Asn

Synonyms: Vasoactive Intestinal Peptide

Biogenesis 9535-0702   Porcine synthetic   MW 3325.7
Contains additional acetate counter ions and residual H_2O;
lyophilized | To avoid Met oxidation, it is best to dissolve the
peptide under a nitrogen stream

His-Ser-Asp-Gly-Ile-Phe-Thr-Asp-Ser-Tyr-Ser-Arg-
Tyr-Arg-Arg-Gln-Leu-Ala-Val-Arg-Arg-Tyr-Leu-Ala-
Ala-Val-Leu Amide

Synonyms: Pituitary Adenylate Cyclase Activating Peptide (1-27),
(Arg[14,20,21],Leu[16])-; Pituitary Adenylate Cyclase Activating Peptide,
M-

American Peptide 34-0-41   Human, ovine, rat   MW 3113.7
$C_{143}H_{226}N_{46}O_{39}$

His-Ser-Asp-Gly-Ile-Phe-Thr-Asp-Ser-Tyr-Ser-Arg-
Tyr-Arg-Arg-Gln-Leu-Ala-Val-Arg-Arg-Tyr-Leu-Ala-
Ala-Val-Leu-Gly-Lys-Arg Amide

Synonyms: Pituitary Adenylate Cyclase Activating Peptide-Gly-Lys-
Arg (1-27), (Arg[14,20,21],Leu[16])-; Pituitary Adenylate Cyclase
Activating Peptide, BM-

American Peptide 34-0-42   Human, ovine, rat   MW 3555.1
$C_{157}H_{253}N_{53}O_{42}$

His-Ser-Asp-Gly-Ile-Phe-Thr-Asp-Ser-Tyr-Ser-Arg-
Tyr-Arg-Leu-Lys-Gln-Met-Ala-Val-Lys-Lys-Tyr-Leu-
Ala-Ala-Val-Leu-Gly-Lys-Arg-Tyr-Lys-Gln-Arg-Val-
Lys-Asn-Lys

Synonyms: Pituitary Adenylate Cyclase Activating Polypeptide (1-
38)

Biogenesis 7394-9156   Human, ovine synthetic   MW 4534.7
Lyophilized, consisting of 98% peptide material, TFA ions and
residual H_2O; lyophilized | To avoid Met oxidation, it is best to
dissolve the peptide under a nitrogen stream

His-Ser-Asp-Gly-Ile-Phe-Thr-Asp-Ser-Tyr-Ser-Arg-
Tyr-Arg-Lys-Gln-Met-Ala-Val-Lys-Lys-Tyr-Leu-Ala-
Ala-Val-Leu-Gly-Lys-Arg-Tyr-Lys-Gln-Arg-Val-Lys-
Asn-Lys Amide

Synonyms: Adenylate Cyclase Activating Peptide (1-38)

Fluka 02115   MW 4534.3 $C_{203}H_{331}N_{63}O_{53}S$ ≥98% (HPLC);
≥65% peptide content | Miyata, A et al, *BBRC*, 164:567, 1989

His-Ser-Asp-Gly-Ile-Phe-Thr-Asp-Ser-Tyr-Ser-Arg-Tyr-Arg-Lys-Gln-Met-Ala-Val-Lys-Lys-Tyr-Leu-Ala-Ala-Val-Leu Amide

Synonyms: Adenylate Cyclase Activating Peptide; Pituitary Adenylate Cyclase Activating Polypeptide (1-27)

ICN 195879   MW 3147.6   ≥97%

His-Ser-Asp-Gly-Ile-Phe-Thr-Asp-Ser-Tyr-Ser-Arg-Tyr-Arg-Lys-Gln-Met-Ala-Val-Lys-Lys-Tyr-Leu-Ala-Ala-Val-Leu-Gly-Lys-Arg-Tyr-Lys-Gln-Arg-Val-Lys-Asn-Lys Amide

Synonyms: Adenylate Cyclase Activating Peptide (1-38); Pituitary Adenylate Cyclase Activating Polypeptide 38

ICN 195881   MW 4534.3   More potent than VIP at stimulating adenylate cyclase

Sigma A 1439   FW 4534.3   Peptide content:~70% | Bioactive peptide; hypothalamic polypeptide with substantial sequence homology to vasoactive intestinal peptide (VIP); more potent than VIP in stimulating adenylate cyclase in rat pituitary cells; Miyata, A et al, *Biochem Biophys Res Commun*, 164:567, 1989

His-Ser-Asp-Gly-Ile-Phe-Thr-Asp-Ser-Tyr-Ser-Arg-Tyr-Arg-Lys-Gln-Met-Ala-Val-Lys-Lys-Tyr-Leu-Ala-Ala-Val-Leu-Gly-Lys-Arg-Tyr-Lys-Gln-Arg-Ile-Lys-Asn-Lys Amide

Synonyms: Pituitary Adenylate Cyclase Activating Peptide; Pituitary Adenylate Cyclase Activating Peptide 38

American Peptide 34-0-50   Frog   MW 4548.4 $C_{204}H_{333}N_{63}O_{53}S$

His-Ser-Asp-Gly-Ile-Phe-Thr-Asp-Ser-Tyr-Ser-Arg-Tyr-Arg-Lys-Gln-Met-Ala-Val-Lys-Lys-Tyr-Leu-Ala-Ala-Val-Leu Amide

Synonyms: Pituitary Adenylate Cyclase Activating Polypeptide (1-27)

Peptides International PPA-4231-v   Human synthetic   MW 3147.6 $C_{142}H_{224}N_{40}O_{39}S$ >95% (HPLC); lyophilized amorphous powder | Bioactive; ovine, rat; Miyata, A et al, *BBRC*, 170:643, 1990; Kimura, C et al, *BBRC*, 166:81, 1990; Ogi, K et al, *BBRC*, 173:1271, 1990

His-Ser-Asp-Gly-Ile-Phe-Thr-Asp-Ser-Tyr-Ser-Arg-Tyr-Arg-Lys-Gln-Met-Ala-Val-Lys-Lys-Tyr-Leu-Ala-Ala-Val-Leu-Gly-Lys-Arg-Tyr-Lys-Gln-Arg-Val-Lys-Asn-Lys Amide

Synonyms: Pituitary Adenylate Cyclase Activating Peptide (1-38)

American Peptide 34-0-20   Human, ovine, rat   MW 4534.4 $C_{203}H_{331}N_{63}O_{53}S$ Lyophilized | More active than vasoactive intestinal peptide (VIP) in stimulating adenylate cyclase; Miyata, A et al, *BBRC*, 164:567, 1989; Ogi, K et al, *BBRC*, 173:1271, 1990

His-Ser-Asp-Gly-Ile-Phe-Thr-Asp-Ser-Tyr-Ser-Arg-Tyr-Arg-Lys-Gln-Met-Ala-Val-Lys-Lys-Tyr-Leu-Ala-Ala-Val-Leu Amide

Synonyms: Pituitary Adenylate Cyclase Activating Peptide (1-27); Pituitary Adenylate Cyclase Activating Polypeptide (1-27)

American Peptide 34-0-40   Human, ovine, rat   MW 3147.7 $C_{142}H_{224}N_{40}O_{39}S$ PACAP-38 & its N-terminal fragment PACAP-27 are novel neuropeptides originally isolated from bovine hypothalamus; also found in humans & rats; both showing considerable homology with vasoactive intestinal polypeptide (VIP), can stimulate adenylate cyclase much more potently than VIP; Miyata, A et al, *BBRC*, 164:567, 1989; Mirata, A et al, *Reg Peptides*, 26:170, 1989

Bachem H-1172.0500   Human, ovine, rat   MW 3147.65 $C_{142}H_{224}N_{40}O_{39}S$ Store at -15°C

His-Ser-Asp-Gly-Ile-Phe-Thr-Asp-Ser-Tyr-Ser-Arg-Tyr-Arg-Lys-Gln-Met-Ala-Val-Lys-Lys-Tyr-Leu-Ala-Ala-Val-Leu-Gly-Lys-Arg-Tyr-Lys-Gln-Arg-Val-Lys-Asn-Lys Amide

Synonyms: Pituitary Adenylate Cyclase Activating Polypeptide (1-38); Pituitary Adenylate Cyclase Activating Polypeptide 38

Bachem H-8430.0200   Human, ovine, rat   MW 4534.32 $C_{203}H_{331}N_{63}O_{53}S$ Store at -15°C

Neosystem SC877   Human, ovine, rat   MW 4534.3   Bioactive; pituitary peptide; Miyata, A et al, *BBRC*, 164:567-574, 1989; Ogi, K et al, *BBRC*, 173:1271-1279, 1990

His-Ser-Asp-Gly-Ile-Phe-Thr-Asp-Ser-Tyr-Ser-Arg-Tyr-Arg-Lys-Gln-Met-Ala-Val-Lys-Lys-Tyr-Leu-Ala-Ala-Val-Leu Amide

Synonyms: Pituitary Adenylate Cyclase Activating Polypeptide 27

Neosystem SC887   Human, ovine, rat   MW 3147.64   Bioactive; pituitary peptide; Tatsuno, I et al, *BBRC*, 168:1027-1033, 1993; Ogi, K et al, *BBRC*, 173:1271-1279, 1990

His-Ser-Asp-Gly-Ile-Phe-Thr-Asp-Ser-Tyr-Ser-Arg-Tyr-Arg-Lys-Gln-Met-Ala-Val-Lys-Lys-Tyr-Leu-Ala-Ala-Val-Leu-Gly-Lys-Arg-Tyr-Lys-Gln-Arg-Val-Lys-Asn-Lys Amide

Synonyms: Pituitary Adenylate Cyclase Activating Peptide; Pituitary Adenylate Cyclase Activating Peptide 38

Calbiochem; 05-23-2150   Ovine   MW 4534.3 $C_{203}H_{331}N_{63}O_{53}S$ ≥95% (HPLC); lyophilized solid; soluble in 5% acetic acid | More active than vasoactive intestinal peptide (VIP) in stimulating adenylate cyclase; Kobayashi, H et al, *Brain Res*, 647:145, 1994; Kimura, C et al, *Biochem Biophys Res Commun*, 166:81, 1990; Chastre, E et al, *J Biol Chem*, 266:21239, 1991; Lemeuth, V et al, *Am J Physiol*, 260:G625, 1991

His-Ser-Asp-Gly-Ile-Phe-Thr-Asp-Ser-Tyr-Ser-Arg-Tyr-Arg-Lys-Gln-Met-Ala-Val-Lys-Lys-Tyr-Leu-Ala-Ala-Val-Leu Amide

Synonyms: Pituitary Adenylate Cyclase Activating Peptide; Pituitary Adenylate Cyclase Activating Peptide 27

Calbiochem; 05-23-2151   Ovine   MW 3147.7 $C_{142}H_{224}N_{40}O_{39}S$ ≥98% (HPLC); lyophilized solid; soluble in 5% acetic acid | Increases cAMP levels in a dose-dependent manner; increases tyrosine hydroxylase expression in chromaffin cells; Kobayashi, H et al, *Brain Res*, 647:145, 1994; Tatsuno, I et al, *Biochem Biophys Res Commun*, 168:1027, 1990

Sigma A 9808   Ovine   FW 3147.6   ≥97% (HPLC); peptide content:~70% | Bioactive peptide; increases cAMP levels; increases tyrosine hydroxylase expression in chromaffin cells; Miyata, A et al, *Biochem Biophys Res Commun*, 164:567, 1989; Miyata, A et al, *Regulatory Peptides*, 26:170, 1989

His-Ser-Asp-Gly-Ile-Phe-Thr-Asp-Ser-Tyr-Ser-Arg-Tyr-Arg-Lys-Gln-Met-Ala-Val-Lys-Lys-Tyr-Leu-Ala-Ala-Val-Leu-Gly-Lys-Arg-Tyr-Lys-Gln-Arg-Val-Lys-Asn-Lys Amide

Synonyms: Pituitary Adenylate Cyclase Activating Polypeptide (1-38)

Peptides International PPA-4221-v   Human synthetic   MW 4534.3 $C_{203}H_{331}N_{63}O_{53}S$ >95% (HPLC); lyophilized amorphous powder | Bioactive; ovine, rat; Miyata, A et al, *BBRC*, 164:567, 1989

His-Ser-Asp-Gly-Thr-Phe-Thr-Ser-Asp-Leu-Ser-Lys-
Gln-Met-Glu-Glu-Glu-Ala-Val-Arg-Leu-Phe-Ile-Glu-
Trp-Leu-Lys-Asn-Gly-Gly-Pro-Ser-Ser-Gly-Ala-Pro-
Pro-Pro-Ser

Synonyms: Exendin III

American Peptide 46-3-14   MW 4203.7 $C_{184}H_{281}N_{49}O_{62}S$ A new member of the glucagon superfamily originally isolated from the venom of the lizard, *Heloderma horridum*; at low concentration it interacts with putative exendin receptors causing an increase in pancreatic acinar cAMP, while at higher concentration it interacts with VIP receptor to stimulate an increase in cellular cAMP & amylase releasing; Eng, J et al, *J Biol Chem*, 267:7402, 1992

His-Ser-Asp-Gly-Thr-Phe-Thr-Ser-Asp-Leu-Ser-Lys-
Gln-Met-Glu-Glu-Glu-Ala-Val-Arg-Leu-Phe-Ile-Glu-
Trp-Leu-Lys-Asn-Gly-Gly-Pro-Ser-Ser-Gly-Ala-Pro-
Pro-Pro-Ser Amide

Synonyms: Exendin III

Bachem H-8735.0500   MW 4202.63 $C_{184}H_{282}N_{50}O_{61}S$ Store at -15°C

Sigma E 7019   FW 4202.6   ≥97% (HPLC) | Bioactive peptide; originally isolated from the venom of the lizard, *Heloderma horridum*; at low concentrations, the peptide interacts with exendin receptors, whereas at higher concentrations it interacts with VIP receptors; Eng, J et al, *J Biol Chem*, 267:7402, 1992

His-Ser-Asp-Gly-Thr-Phe-Thr-Ser-Glu-Leu-Ser-Arg-
Leu-Arg-Asp-Ser-Ala-Arg-Leu-Gln-Arg-Leu-Leu-Gln-
Gly-Leu-Val Amide

Synonyms: Gastrointestinal Peptide; Secretin

Sigma S 0137   FW 3039.4   ≥97% (HPLC) | Bioactive peptide; highly purified synthetic peptide believed to be the active principle in secretin preparations

Sigma S 0648   FW 3039.4   Partially purified; soluble in water or saline; activity:~25 units/mg solid; vial contains ~100 Crick-Harper-Raper units (determined by RIA using pure synthetic porcine secretin as standard) | Bioactive peptide

Sigma S 5014   FW 3039.4   ≥97% (HPLC) | Bioactive peptide

Bachem H-3780.0001   Porcine   MW 3055.45 $C_{130}H_{220}N_{44}O_{41}$ Store at -15°C

His-Ser-Asp-Gly-Thr-Phe-Thr-Ser-Glu-Leu-Ser-Arg-
Leu-Arg-Asp-Ser-Ala-Arg-Leu-Gln-Arg-Leu-Leu-Gln-
Gly-Leu-Val-Asp-Asp-Asp-Asp-Asp-Asp
His-Ser-Asp-Gly-Thr-Phe-Thr-Ser-Glu-Leu-Ser-Arg-
Leu-Arg-Asp-Ser-Ala-Arg-Leu-Gln-Arg-Leu-Leu-Gln-
Gly-Leu-Val Asp₆ Amide

Synonyms: Secretin Hexaaspartate

Fluka 84873   Porcine   MW 3055.44 $C_{130}H_{22}N_{44}O_{41}$ ~98.0% (TLC) | Mutt, V, *Brain Peptides* (DT Krieger, et al, eds), 871, 1983, John Wiley & Sons, New York; Bertaccini, G, *Handbook Exp Pharmacol (Mediators Drugs Gastro Intest Motil, Part 2)*, 59:11, 1982; Dockray, GJ, *Hormones in Blood* (CH Gray, VHT James, eds), vol 2, 3rd ed, 357, 1979, Academic Press, London

His-Ser-Asp-Gly-Thr-Phe-Thr-Ser-Glu-Leu-Ser-Arg-
Leu-Arg-Asp-Ser-Ala-Arg-Leu-Gln-Arg-Leu-Leu-Gln-
Gly-Leu-Val Amide

Synonyms: Secretin

ICN 190272   Porcine   May contain 25-30 U pancreozymin/100 U secretin | Crick, J et al, *J Physiol* (London), 110:367, 1949; Jorpes, JE et al, *BBRC*, 9:275, 1962

Neosystem SC099   Porcine   MW 3055.44   Bioactive; brain/gut peptide; VIP/PHI/secretin/helodermin peptide; Jorpes, JE & Mutt, V, in *Handbook of Experimental Pharmacology*, 34:1-178, 1973

His-Ser-Asp-Gly-Thr-Phe-Thr-Ser-Glu-Leu-Ser-Arg-
Leu-Arg-Asp-Ser-Ala-Arg-Leu-Gln-Arg-Leu-Leu-Gln-
Gly-Leu-Val Acetate Salt

Synonyms: Secretin

Biogenesis 8240-1002   Porcine synthetic   MW 3055.5   Purified; lyophilized

His-Ser-Asp-Gly-Thr-Phe-Thr-Ser-Glu-Leu-Ser-Arg-
Leu-Arg-Asp-Ser-Ala-Arg-Leu-Gln-Arg-Leu-Leu-Gln-
Gly-Leu-Val Amide

Synonyms: Secretin

Peptides International PSE-4112-v   Porcine synthetic   MW 3055.4 $C_{130}H_{220}N_{44}O_{41}$ >99% (HPLC); lyophilized amorphous powder

His-Ser-Asp-Gly-Thr-Phe-Thr-Ser-Glu-Leu-Ser-Arg-
Leu-Arg-Glu-Gly-Ala-Arg-Leu-Gln-Arg-Leu-Leu-Gln-
Gly-Leu-Val Amide

Synonyms: Gastrointestinal Peptide; Secretin

American Peptide 46-1-75   Human   MW 3039.5 $C_{130}H_{220}N_{44}O_{40}$ Lyophilized | Stimulates the release of a pancreatic, NaHCO₃, containing aqueous secretion, which, in turn, neutralizes the hydrochloric acid that has found its way into the small intestine; the gene of secretin is also expressed in brain, heart, lung, kidney & testis; Carlquist, M et al, *IRCS Med Sci*, 13:217, 1985

Bachem H-3022.0001   Human   MW 3039.45 $C_{130}H_{220}N_{44}O_{40}$ Store at -15°C

Calbiochem 05-23-2102   Human   MW 3039.5 $C_{130}H_{220}N_{44}O_{40}$ >95% (HPLC); lyophilized solid; soluble in 5% acetic acid | A strongly basic gastrointestinal peptide hormone that stimulates the secretion of bicarbonate-rich pancreatic fluid; *Merck Index*, 12:8564; Villanger, O et al, *Gastroenterology*, 108:850, 1995

ICN 154561   Human   MW 3039.8   Carlquist, M et al, *IRCS Med Science*, 13:217, 1985

Neosystem SC1203   Human   MW 3039.44   Bioactive; brain/gut peptide; VIP/PHI/secretin/helodermin peptide; Carlquist, M et al, *IRCS Med Sci*, 13:217-218, 1985

Sigma S 7147   Human   FW 3039.4   ≥97% (HPLC) | Bioactive peptide; stimulates secretion of carbonate-rich pancreatic fluid; relaxes smooth muscle; Carlquist, M et al, *IRCS Med Science*, 13:217, 1985; Other general references on Secretin:Crick, J et al, *J Physiol* (London), 110:367, 1949; Bodanszky, M et al, *J Am Chem Soc*, 89:6753, 1967; Gutierrez, LV & Baron, JH, *Gut*, 13:721, 1972; Jorpes, JE et al, *Biochem Biophys Res Commun*, 9:275, 1962; Kofod et al, *Acta Endocrinol*, 126:1, 1992; Mutt et al, *Eur J Biochem*, 15:513, 1970

His-Ser-Asp-Gly-Thr-Phe-Thr-Ser-Glu-Leu-Ser-Arg-
Leu-Arg-Glu-Gly-Ala-Arg-Leu-Gln-Arg-Leu-Leu-Gln-
Gly-Leu-Val Acetate Salt

Synonyms: Secretin

Biogenesis 0100-0175   Human synthetic   Semi-pure; lyophilized

His-Ser-Asp-Gly-Thr-Phe-Thr-Ser-Glu-Leu-Ser-Arg-
Leu-Arg-Glu-Gly-Ala-Arg-Leu-Gln-Arg-Leu-Leu-Gln-
Gly-Leu-Val Amide

Synonyms: Secretin

Peptides International PSE-4165-v   Human synthetic   MW 3039.4 $C_{130}H_{220}N_{44}O_{40}$ >99% (HPLC); lyophilized amorphous powder | Bioactive; Jorpes, JE & V Mutt (Eds), *Secretin, Cholecystokinin, Pancreozymin & Gastrin*, Handbook of Experimental Pharmacology, Vol 34, Springer-Verlag, Berlin, 1973; Carlquist, M et al, *IRCS Med Sci*, 13:217, 1985; Iguchi, K et al, *Peptide Chem*, 1985:191, 1986;

His-Ser-Gln-Gly-Thr-Phe-Thr-Ser-Asp-Tyr-Ser-Lys-Tyr-Leu-Asp-Ser-Arg-Arg-Ala-Gln-Asp-Phe-Val-Gln-Trp-Leu-Met-Asn-Thr-Lys-Arg-Asn-Lys-Asn-Asn-Ile-Ala

Synonyms: Glucagon 37; Oxyntomodulin

American Peptide 46-1-18   MW 4424.9 $C_{192}H_{292}N_{59}O_{60}S$

ICN 152880   Bataille, D etal, *Peptides*, 2:41, 1981

His-Ser-Gln-Gly-Thr-Phe-Thr-Ser-Asp-Tyr-Ser-Lys-Tyr-Leu-Asp-Ser-Arg-Arg-Ala-Gln-Asp-Phe-Val-Gln-Trp-Leu-Met-Asn-Thr

Synonyms: Glucagon (1-29); Gastrointestinal Peptide; Hyperglycemic-Glycogenolytic Factor

American Peptide 46-1-15   Human   MW 3482.6 $C_{153}H_{225}N_{43}O_{49}S$

Calbiochem 05-23-2700   Human   MW 3482.8 $C_{153}H_{225}N_{43}O_{49}S$ ≥97% (HPLC); lyophilized solid; soluble in 5% acetic acid; harmful:LD_{50}≤2000 mg/kg; may be carcinogenic/teratogenic | Peptide hormone secreted by pancreatic α-cells; important hyperglycemic agent that is released when blood glucose levels are low; *Merck Index*, 12:4455

Neosystem SC097   Human   MW 3482.78   Bioactive; brain/gut peptide

ICN 152879   Human pancreas   MW 3982.8   Antihypoglycemic agent

Biogenesis 4660-0404   Human synthetic   MW 3482.6   Purified; HCl salt; lyophilized

Peptides International PGL-4098-s   Human synthetic   MW 3482.8 $C_{153}H_{225}N_{43}O_{49}S$ >95% (HPLC); lyophilized amorphous powder | Bioactive; Lefebvre, PJ & RH Unger (Eds), *Glucagon*, Pergamon Press, Oxford, 1972

Bachem H-6790.0500   Human, bovine, porcine   MW 3482.8 $C_{153}H_{225}N_{43}O_{49}S$ Store at -15°C

Sigma G 1774   Human, bovine, porcine   FW 3482.8   ≥90% (HPLC) | Bioactive peptide; highly-conserved 29-AA polypeptide hormone produced by the pancreatic cells; released in response to low blood glucose; activation of liver glucagon receptors induces cyclic-AMP-mediated phosphorylation of enzymes involved in gluconeogenesis & glycogenolysis; Unger, RH, *Diabetologia*, 28:574, 1985; Foster, DW, *Diabetes*, 33:1188, 1984

His-Ser-Gln-Gly-Thr-Phe-Thr-Ser-Asp-Tyr-Ser-Lys-Tyr-Leu-Asp-Ser-Arg-Arg-Ala-Gln-Asp-Phe-Val-Gln-Trp-Leu-Met-Asn-Thr-Lys-Arg-Asn-Lys-Asn-Asn-Ile-Ala

Synonyms: Glucagon (1-37); Oxyntomodulin

Bachem H-6880.0500   Porcine   MW 4421.88 $C_{192}H_{295}N_{59}O_{60}S$ Store at -15°C

His-Ser-Gln-Gly-Thr-Phe-Thr-Ser-Asp-Tyr-Ser-Lys-Tyr-Leu-Asp-Ser-Arg-Arg-Ala-Gln-Asp-Phe-Val-Gln-Trp-Leu-Met-Asn-Thr

Synonyms: Gastrointestinal Peptide; Glucagon

Sigma G 9410   Porcine pancreas   FW 3482.8   ≥90% (HPLC); peptide content:~70% | Bioactive peptide

---

His-Ser-Glu-Gly-Thr-Phe-Thr-Ser-Asp-Tyr-Ser-Lys-Tyr-Leu-Asp-Ser-Arg-Arg-Ala-Gln-Asp-Phe-Val-Gln-Trp-Leu-Met-Asn-Thr

Synonyms: Glucagon

Fluka 49090   Porcine pancreas   MW 3482.78 $C_{153}H_{225}N_{43}O_{49}S$ ~90% (HPLC) | Foa, PP, *Biomed Res*, 6:3, 1985

His-Ser-Glu-Gly-Thr-Phe-Thr-Ser-Asp-Val-Ser-Ser-Tyr-Leu-Glu-Gly-Gln-Ala-Ala-Lys-Glu-Phe-Ile-Ala-Trp-Leu-Val-Lys-Gly-Arg Amide

Synonyms: Glucagon-Like Peptide I (7-36), (Ser[8])-; Prepro-Glucagon (78-107), (Ser[79])-

Bachem H-4592.0500   Human   MW 3313.68 $C_{149}H_{225}N_{40}O_{46}$ Store at -15°C

His-Ser-Gly-Ile-Asn-Ser-Ser-Asn-Ala-Glu-Val-Leu-Ala-Leu-Phe-Asn-Val-Thr-Glu-Met-Asp-Ala-Gly-Glu-Tyr

Synon Keratinocyte Growth Factor Receptor Peptide *yms*:

Bachem H-1562.0500   MW 2668.88 $C_{114}H_{174}N_{30}O_{42}S$ Store at -15°C

His-Ser-Leu-Gly-Lys-Leu-Leu-Gly-Arg-Pro-Asp-Lys-Phe

Synonyms: Myelin Proteolipid Protein (139-151), (Leu[144],Arg[147])-

Bachem H-2482.0500   Store at -15°C

His-Ser-Leu-Gly-Lys-Trp-Leu-Gly-His-Pro-Asp-Lys-Phe

Synonyms: Myelin Proteolipid Protein (139-151); Proteolipid Protein (139-151); Experimental Allergic Encephalomyelitis Inducing Peptide; Encephalitogenic Determinant

Bachem H-2478.0001   Store at -15°C

Neosystem SC974   MW 1521.74   Bioactive; this myelin proteolipid protein fragment is encephalitogenic in SJL mice; Kuchroo, VK et al, *Pathobiology*, 59:305-312, 1991; Kuchroo, VK et al, *J Immunol*, 148:3776-3782, 1992; Kuchroo, VK et al, *J Immunol*, 153:3326-3336, 1994

Peptides International PLP-3602-PI   Synthetic   MW 1521.76 $C_{72}H_{104}N_{20}O_{17}$ >98% (HPLC); white powder | Bioactive; Kuchroo, VK et al, *J Immunol*, 153:3326, 1994

His-Ser-Phe
Ac-His-Ser-Phe

Bachem H-7280.0025   MW 431.45 $C_{20}H_{25}N_5O_6$ Store at -15°C

His-Thr-Ile-Ser-Arg-Ile-Ala-Val-Ser-Tyr-Gln-Thr-Lys-Val-Asn-Leu-Leu-Ser-Ala

Synonyms: Tumor Necrosis Factor α (78-96)

Bachem H-2704.0500   Human   Store at -15°C

His-Thr-Tyr-Leu-Gln-Ala-Ser-Glu-Lys-Phe-Lys-Met-Trp-Gly Amide

Synonyms: HSV Amide UL 26 Open Reading Frame (242-255)

Bachem M-2160.0500   Store at -15°C

His-Trp
Z-His-Trp

Bachem C-2000.0250   MW 475.5 $C_{25}H_{25}N_5O_5$ Store at -15°C

His-Trp

Bachem G-2350.0250   MW 341.37 $C_{17}H_{19}N_5O_3$ Store at -15°C

His-Trp-Ala-Trp-Phe-Lys
His-D-Trp-Ala-Trp-D-Phe-Lys Amide

Synonyms: Melanocyte Stimulating Hormone (6-11), (D-Trp[7],Ala[8],D-Phe[10])-α-; GHRP-6; GHRP, (His[1],Lys[6])-

American Peptide 52-1-80   MW 873.0 $C_{46}H_{56}N_{12}O_6$ Stimulates GH release *in vitro* & *in vivo*; Villas-Boas Weffort, RF et al, *Metabolism*, 46(6):706, 1997

Bachem H-9990.0005   MW 873.03 $C_{46}H_{56}N_{12}O_6$ Store at -15°C

Neosystem SC949   MW 873.03   Bioactive; hypothalamic releasing hormone; specifically stimulates growth hormone secretion *in vitro* & *in vivo*; Bowers, CY et al, *Endocrinology*, 114:1537, 1984; Momany, FA et al, *Endocrinology*, 114:1531, 1984; Smith, RG et al, *Science*, 260:1640-1643, 1993; Renner, U et al, *J Clin Endocrinol Metab*, 78:1090-1096, 1994

Sigma M 2910   FW 873.0 $C_{46}H_{56}N_{12}O_6$ ≥97% (HPLC) | Bioactive peptide; selective inhibitor of α-MSH activity in frog skin bioassay; identical to (His[1], Lys[8])-GHRH which selectively releases growth hormone *in vitro* & *in vivo*; somatostatin antagonist

His-Trp-Ala-Val-Gly-His-Leu
Ac-His-Trp-Ala-Val-Gly-His-Leu Amide

Synonyms: Gastrin Releasing Peptide (20-26), Ac-

Bachem H-6705.0001   Human, porcine, canine   MW 857.97 $C_{41}H_{55}N_{13}O_8$ Store at -15°C

His-Trp-Ala-Val-Gly-His-Leu-Met
Ac-His-Trp-Ala-Val-Gly-His-Leu-Met Amide

Synonyms: Gastrin Releasing Peptide (20-27), Ac-

Bachem H-1040.0001   Human, porcine, canine   MW 991.19 $C_{46}H_{66}N_{14}O_9S$ Store at -15°C

American Peptide 46-4-37   Porcine   MW 991.2 $C_{46}H_{66}N_{14}O_9S$

His-Trp-Lys

Bachem H-3785.0050   Store at -15°C

His-Trp-Lys-Trp-Phe-Lys
His-D-Trp-D-Lys-Trp-D-Phe-Lys Amide

Synonyms: Melanocyte Stimulating Hormone (6-11), (D-Trp[7],D-Lys[8],D-Phe[10])-α-; GHRP-6, (D-Lys[3])-

Bachem H-3108.0005   MW 930.12 $C_{49}H_{63}N_{13}O_6$ Store at -15°C

Neosystem SC961   MW 930.12   Bioactive; hypothalamic releasing hormone; GHRP-6 antagonist; Veeraragavan, K et al, *Life Sci*, 50:1149-1155, 1992; Smith, RG et al, *Science*, 260:1640-1643, 1993

His-Trp-Ser-Tyr-Gly-Leu-Arg-Pro-Gly
For-His-Trp-Ser-Tyr-Gly-Leu-Arg-Pro-Gly Amide

Synonyms: Luteinizing Hormone Releasing Hormone (2-10), Formyl-

Bachem H-1380.0001   Store at -15°C

His-Trp-Ser-Tyr-Gly-Leu-Arg-Pro-Gly Amide

Synonyms: Luteinizing Hormone Releasing Hormone, (des-Glp[1])-

Bachem H-9200.0005   Store at -15°C

ICN 152909

Sigma L 8762   FW 1071.2   ≥97% (HPLC) | Bioactive peptide

His-Trp-Ser-Tyr-His-Leu-Arg-Pro
His-Trp-Ser-Tyr-D-His(Bzl)-Leu-Arg-Pro-NHEt

Synonyms: Luteinizing Hormone Releasing Hormone (2-9), (D-His(Bzl)[6],Pro-NHEt[9])-

Bachem H-4806.0001   MW 1212.42 $C_{61}H_{81}N_{17}O_{10}$ Store at -15°C

His-Trp-Ser-Tyr-Trp-Leu-Arg-Pro-Gly
For-His-Trp-Ser-Tyr-D-Trp-Leu-Arg-Pro-Gly Amide

Synonyms: Luteinizing Hormone Releasing Hormone (2-10), Formyl-(D-Trp[6])-; Triptorelin (2-10), Formyl-

Bachem H-4576.0001   MW 1228.38 $C_{60}H_{77}N_{17}O_{12}$ Store at -15°C

His-Tyr

Bachem G-2355.0250   MW 318.33 $C_{15}H_{18}N_4O_4$ Store at -15°C

ICN 157381   MW 318.3 $C_{15}H_{18}N_4O_4$ Crystalline

His-Tyr-Ile-Asn-Leu-Ile-Thr-Arg-Gln-Arg-Tyr Amide

Synonyms: Neuropeptide Y (26-36)

Neosystem SC369   MW 1475.71   Bioactive neuropeptide;

His-Tyr-Tyr
Z-His-Tyr-Tyr-OEt

Bachem M-1285.0250   MW 643.7 $C_{34}H_{37}N_5O_8$ Store at -15°C

His-Val
Z-His-Val

Bachem C-2010.0250   MW 388.42 $C_{19}H_{24}N_4O_5$ Store at -15°C

His-Val

Bachem G-2360.0250   MW 254.29 $C_{11}H_{18}N_4O_3$ Store at -15°C

His-Val-Asp-Ala-Ile-Phe-Thr-Thr-Asn-Tyr-Arg-Lys-Leu-Leu-Ser-Gln-Leu-Tyr-Ala-Arg-Lys-Leu-Ile-Gln-Asp-Ile-Met-Asn-Lys-Gln-Gly-Glu-Arg-Ile-Gln-Glu-Arg-Ala-Arg-Leu-Ser

Synonyms: Growth Hormone Releasing Factor

American Peptide 52-1-10   Murine   MW 5032.9 $C_{220}H_{365}N_{69}O_{64}S$ Frohman, MA et al, *Mol Endo*, 3:1529, 1989

Hyp-Gly

Bachem G-2365.0250   MW 188.18 $C_7H_{12}N_2O_4$ Store at -15°C

Hyp-Gly-Phe-Lys-Gly-Ile-Arg-Gly-His

Bachem H-8415.0001   MW 984.13 $C_{44}H_{69}N_{15}O_{11}$ Store at -15°C

Ile-Ala

Bachem G-2370.0250   MW 202.25 $C_9H_{18}N_2O_3$ Store at -15°C

ICN 102081

Ile-Ala
Suc-Ile-Ala-MCA

Synonyms: Amyloid A4-Generating Enzyme Substrate

Peptides International MIA-3158-v   MW 459.50 $C_{23}H_{29}N_3O_7$ >98% (HPLC); lyophilized amorphous powder | Ishiura, S et al, *FEBS Letts*, 260:131, 1990; Ishiura, S et al, *Neurosci Lett*, 115:329, 1990

Ile-Ala
N-Succinyl-Ile-Ala MCA

Sigma S 6407   FW 459.5 $C_{23}H_{29}N_3O_7$

Ile-Ala-Ala-Gly-Arg-Thr-Gly-Arg-Arg-Gln-Ala-Ile-His-Asp-Ile-Leu-Val-Ala-Ala

Synonyms: cAMP-Dependent Protein Kinase Inhibitor; Protein Kinase Inhibitor Peptide

Alexis 165-024   MW 1989.3 $C_{85}H_{149}N_{31}O_{24}$ ≥98%; white lyophilized powder; soluble in water | Specific inhibitor; Hashimoto, Y & Soderling, TR, *Arch Biochem Biophys*, 252:418, 1987

Bachem H-3234.0001   MW 1989.31 $C_{85}H_{149}N_{31}O_{24}$ Store at -15°C

ICN 154517   MW 1989.6

Ile-Ala-Arg-Arg-His-Pro-Tyr-Phe
Synonyms: Kinetensin, (des-Leu⁹)-
Bachem H-1358.0005   MW 1059.24 $C_{50}H_{74}N_{16}O_{10}$ Store at -15°C

Ile-Ala-Arg-Arg-His-Pro-Tyr-Phe-Leu
Synonyms: Neurotensin Related Peptide; Neurotensin; Kinetensin
American Peptide 62-0-32   MW 1172.4 $C_{56}H_{85}N_{17}O_{11}$
Structure similar to neurotensin & angiotensin I; obtained by pepsin treatment of human, bovine & dog plasma albumins; increases vascular permeability when injected intradermally into rats & releases histamine from rat mast cells *in vitro*; Mogard, MH et al, *BBRC*, 136:983, 1986; Carraway, RE et al, *JBC*, 262:5968, 1987
Bachem H-9350.0005   Store at -15°C
Fluka; 60776   MW 1172.4 $C_{56}H_{85}N_{17}O_{11}$ ≥98% (HPLC); ≥65% peptide content | Increases vascular permeability & releases histamine from rat mast cells; Sydbom, A et al, *Agents Actions*, 27:68, 1989
Sigma K 1879   FW 1172.4   ≥97% (HPLC); peptide content:~70% | Bioactive peptide; neurotensin-related peptide which appears in human plasma treated with pepsin; increases vascular permeability & releases histamine from rat mast cells; Mogard, MH et al, *Biochem Biophys Res Commun*, 13:983, 1986; Carraway, RE et al, *J Biol Chem*, 262:5968, 1987
ICN 154464   Human recombinant, expressed in *E. coli*   MW 1172.5   Neurotensin-related peptide which increases vascular permeability & releases histamine from rat mast cells; Carraway, RE etal, *JBC*, 262:5968, 1987

Ile-Ala-Gly-Gln-Arg
Dnp-Ile-Ala-Gly-Gln-D-Arg
Bachem M-1850.0025   Store at -15°C

Ile-Arg
Bachem G-2375.0250   MW 287.36 $C_{12}H_{25}N_5O_3$ Store at -15°C

Ile-Arg-Cys-Phe-Ile-Thr-Pro-Asp-Ile-Thr-Ser-Lys-Asp-Cys-Pro-Asn-Gly-His-Val-Cys-Tyr-Thr-Lys-Thr-Trp-Cys-Asp-Ala-Phe-Cys-Ser-Ile-Arg-Gly-Lys-Arg-Val-Asp-Leu-Gly-Cys-Ala-Ala-Thr-Cys-Pro-Thr-Val-Lys-Thr-Gly-Cal-Asp-Ile-Gln-Cys-Cys-Ser-Thr-Asp-Asn-Cys-Asn-Pro-Phe-Pro-Thr-Arg-Lys-Arg-Pro
Synonyms: Cobrotoxin, α-; Neurotoxin III
Calbiochem 233605   *Naja naja kaouthia*   MW 7821 $C_{332}H_{520}N_{98}O_{101}S_{10}$ Single band purity (IEF); lyophilized solid; soluble in water; no detectable contaminants (IEF); LD₅₀≤2000 mg/kg; not available for sale outside of the United States | Disulfide bonds: Cys³-Cys²⁰, Cys¹⁴-Cys⁴¹, Cys²⁶-Cys³⁰, Cys⁴⁵-Cys⁵⁶, Cys⁵⁷-Cys⁶²; "long" neurotoxin that blocks nerve transmission postsynaptically in skeletal muscle; binds to the nicotinic acetylcholine receptor on the postsynaptic membrane of the neuromuscular junction preventing the binding of acetylcholine; unlike "short" neurotoxins, its binding is irreversible; activates endogenous phospholipase C; Fletcher, JE et al, *Biochem Cell Biol*, 69:274, 1991; Vernon, LP & Rogers, A, *Toxin*, 30:701, 1992; *Merck Index*, 12:2514

Ile-Arg-Gln-Gly-Pro-Lys-Glu-Pro-Phe-Arg-Asp-Tyr-Val-Asp-Arg-Phe-Tyr-Lys-Thr-Leu
Synonyms: P25 (285-304); GAG P24 CA (153-172); HTLV I Peptide
Neosystem SC304   MW 2528.89   HTLV I peptide from subtype P25 (GAG P24 CA)

Ile-Arg-Ile-Gln-Arg-Gly-Pro-Gly
Synonyms: GP120 (312-319); HTLV I Peptide
Neosystem SC811   MW 896.06   HTLV I peptide from subtype GP160

Ile-Arg-Ile-Gln-Arg-Gly-Pro-Gly-Arg-Ala-Phe-Val-Thr-Ile-Gly-Lys
Synonyms: GP120 (312-327); HTLV I Peptide
Neosystem SC244   MW 1769.12   HTLV I peptide from subtype GP160

Ile-Arg-Ile-Leu-Gln-Gln-Leu-Leu-Phe-Ile-His-Phe-Arg-Ile
Synonyms: VPR (61-74); HTLV I Peptide
Neosystem SC840   MW 1812.2   HTLV I peptide from subtype VPR

Ile-Arg-Pro
Bachem H-1178.0050   MW 384.48 $C_{17}H_{32}N_6O_4$ Store at -15°C

Ile-Arg-Pro-Lys-Leu-Lys-Trp-Asp-Asn-Gln
Synonyms: Dynorphin A (8-17)
American Peptide 26-4-32   Porcine   MW 1297.5 $C_{59}H_{96}N_{18}O_{15}$

Ile-Arg-Val-Val-Met
Bachem H-1416.0005   MW 616.83 $C_{27}H_{52}N_8O_6S$ Store at -15°C

Ile-Asn
Bachem G-2385.0250   MW 245.28 $C_{10}H_{19}N_3O_4$ Store at -15°C
ICN 155006   MW 245.3 $C_{10}H_{19}N_3O_4$
Sigma I 3635   FW 245.3 $C_{10}H_{19}N_3O_4$

Ile-Asn-Leu-His-Phe-Ser-Lys-Cys-Gly-Phe-Pro-Phe-Ser-Leu
Synonyms: GP46 (150-163); HTLV I Peptide
Neosystem SC828   MW 1609.90   HTLV I peptide; due to the presence of a Cys this peptide can contain the dimeric form

Ile-Asn-Leu-Lys-Ala-Ile-Ala-Ala-Leu-Val-Lys-Lys-Val-Leu Amide
Synonyms: Mast Cell Degranulating Peptide HR-2; Mast Cell Degranulating Peptide HR-1
American Peptide 41-9-10   MW 1493.0 $C_{71}H_{133}N_{19}O_{15}$
Linear low-molecular-weight peptides HR-1 & HR-2 isolated from the venom of the giant hornet *Vespa orientalis*, differ structurally from the MCD peptide from honeybee venom but possess similar biological effects; these are capable of degranulating rat mast cells, initiating histamine release; Tuichibaev, MU et al, *Biochem USSR*, 53:183, 1988
Bachem H-8375.0001   MW 1492.96 $C_{71}H_{133}N_{19}O_{15}$ Store at -15°C
ICN 159961   MW 1493   97% | Tuichibaev, MU, *Biochem USSR*, 53:219, 1988
Sigma M 8655   *Vespa orientalis* (giant hornet)   FW 1493.0   ≥97% (HPLC); peptide content:~70% | Bioactive peptide; similar biological effects as MCD peptide from bee venom; Tuichibaev, MU, *Biochem USSR*, 53:219, 1988; CA, 108:16299e, 1988

Ile-Asn-Leu-Lys-Ala-Leu-Ala-Ala-Leu-Ala-Lys-Ala-Leu-Leu Amide

Synonyms: Mastoparan 7

American Peptide 87-0-62   MW 1421.9 $C_{67}H_{124}N_{18}O_{15}$ This mastoparan analog has five-fold greater potency than mastoparan; directly activates G-proteins; Higashijirna, T et al, *JBC*, 265:14176, 1990

Bachem H-3002.0001   MW 1421.83 $C_{67}H_{124}N_{18}O_{15}$ Store at -15°C

ICN 159023   MW 1635.9   98% | Mastoparan analog with greater potency; Higashijima, T etal, *JBC*, 265:14176, 1990

Ile-Asn-Leu-Lys-Ala-Leu-Ala-Ala-Leu-Ala-Lys-Ala-Leu-Leu Amide Trifluoroacetate Salt

Synonyms: Mastoparan 7

Calbiochem 444896   MW 1421.8 $C_{67}H_{124}N_{18}O_{15}$ >98% (HPLC); white solid; soluble in water; harmful:LD$_{50}$ ≤2000 mg/kg | Cell-permeable mastoparan analog with five-fold greater potency than mastoparan; directly activates G-proteins; evokes rapid upregulation of FAK tyrosine phosphorylation via activation of G-proteins; Zhang, C et al, *FEBS Lett*, 386:185, 1996; Leyte, A et al, *EMBO J*, 11:4795, 1992; Higashijima, T et al, *J Biol Chem*, 265:14176, 1990

Ile-Asn-Leu-Lys-Ala-Leu-Ala-Ala-Leu-Ala-Lys-Lys-Ile-Leu Amide

Synonyms: Mastoparan

American Peptide 87-0-61   MW 1478.8 $C_{70}H_{131}N_{19}O_{15}$ Originally isolated from the wasp venom of *Vespula lewisii*, stimulates vesicles secretion by stimulating Go; Hirai, Y et al, *Chem Pharm Bull*, 27:1942, 1979

Bachem H-3810.0001   MW 1478.93 $C_{70}H_{131}N_{19}O_{15}$ Store at -15°C

Neosystem SC182   MW 1478.92   Bioactive; toxin; Hirai, Y et al, *Chem Pharm Bull*, 27:1942, 1979

Alexis 162-001   Synthetic   MW 1478.9 $C_{70}H_{131}N_{19}O_{15}$ ≥98%; white lyophilized powder; potent neurotoxin | Originally isolated from Wasp venom; Hirai, Y et al, *Chem Pharm Bull*, 27:1942, 1979; Higashijima, T, *J Biol Chem*, 268:6491, 1998

Sigma M 5280   *Vespula lewisii*   FW 1478.9   ≥97% (HPLC) | Bioactive peptide; mast cell degranulating peptide; activates phospholipase A2; stimulates exocytosis & phosphoinositide breakdown; inhibits calmodulin; cell permeable; Hirai, Y et al, *Chem Pharm Bull*, 27:1942, 1979

ICN 151587   *Vespula lewisii*   MW 1478.9   A mast cell degranulating peptide & γ-protein activator; Hirai, Y etal, *Chem Pharm Bull*, 27:1942, 1979

Peptides International PMS-4107-v   *Vespula lewisii* (wasp) synthetic   MW 1478.9 $C_{70}H_{131}N_{19}O_{15}$ >99% (HPLC); lyophilized amorphous powder | Bioactive; Hirai, Y et al, *Chem Pharm Bull*, 27:1942, 1979

Ile-Asn-Leu-Lys-Ala-Leu-Ala-Ala-Leu-Ala-Lys-Lys-Ile-Leu Amide 4AcOH 6H$_2$O

Synonyms: Mastoparan

Peptides International PMS-4107   *Vespula lewisii* (wasp) synthetic   MW 1827.2 $C_{70}H_{131}N_{19}O_{15} \cdot 4CH_3COOH \cdot 6H_2O$ >99% (HPLC); lyophilized amorphous powder; bulk | Bioactive; Hirai, Y et al, *Chem Pharm Bull*, 27:1942, 1979

Ile-Asn-Leu-Lys-Ala-Leu-Ala-Ala-Leu-Ala-Lys-Lys-Ile-Leu Amide Trifluoroacetate Salt

Synonyms: Mastoparan

Calbiochem 444898   *Vespula lewisii* (wasp) venom, synthetic MW 1478.9 $C_{70}H_{131}N_{19}O_{15}$ >98% (HPLC); white solid; soluble in water; harmful:LD$_{50}$ ≤2000 mg/kg | Cell-permeable synthetic peptide with sequence identical to *Vespula lewisii*; amphiphilic wasp venom tetradecapeptide capable of directly activating pertussis toxin-sensitive G-proteins by a mechanism analogous to that of G-proteins-coupled receptors; acts preferentially on G$_i$ & G$_o$ rather than G$_s$; potent facilitator of the mitochondrial permeability transition pore; acts as a calmodulin antagonist & activates phospholipase A$_1$; can increase levels of intracellular free Ca^{2+}; also reported to inhibit Na$^+$-K$^+$ ATPase activity (IC$_{50}$=7.5 μM); Klinker, JF et al, *Biochem Pharmacol*, 51:217, 1996; Langel, U et al, *Regul Peptides*, 62:47, 1996; Pfeiffer, DR et al, *J Biol Chem*, 270:4923, 1995; Igarashi, M et al, *Science*, 259:77, 1993; Komatsu, M et al, *Endocrinology*, 130:221, 1992; Adolfo Garcia-Sainz, J et al, *Biochem Biophys Res Commun*, 179:852, 1991

Ile-Asn-Leu-Lys-Ala-Lys-Ala-Ala-Leu-Ala-Lys-Lys-Leu-Leu

Synonyms: Mastoparan 17

Bachem H-3004.0001   Store at -15°C

Ile-Asn-Leu-Lys-Ala-Lys-Ala-Ala-Leu-Ala-Lys-Lys-Leu-Leu Amide

Synonyms: Mastoparan 17

American Peptide 87-0-64   MW 1493.0 $C_{70}H_{132}N_{20}O_{15}$ Inactive analog of mastoparan & used as a negative control; Higashijirna, T et al, *JBC*, 265:14176, 1990

ICN 159960   MW 1493   98% | Inactive mastoparan analog for use as a negative control; Higashijima, T etal, *JBC*, 265:14176, 1990

Ile-Asn-Pro-Ile-Tyr-Arg-Leu-Arg-Tyr Amide

Synonyms: Neuropeptide Y (28-36), (Pro^{30},Tyr^{32},Leu^{34})-

Bachem H-3546.0001   MW 1206.46 $C_{57}H_{91}N_{17}O_{12}$ Store at -15°C

Neosystem SC972   MW 1206.45   Bioactive neuropeptide; Y1 receptor antagonist; an antagonist of NPY in HEL cells (IC$_{50}$ = 8 ± 12nM); effectively displaced radiolabeled NPY from Y1-type receptor in SK-N-MC cells (IC$_{50}$ = 3 ± 07nM); may provide further insight into the physiological role(s) for NPY in the mammalian & peripheral nervous system; Leban, JJ et al, *J Med Chem*, 38:1150-1157, 1995

Ile-Asn-Pro-Ser-Ser-Ser-Gln-Asn-Ser-Gln-Asn-Phe Acetate Salt

Synonyms: Glutamate Receptor II (834-845)

Biogenesis 4670-5144   Synthetic   Semi-pure; lyophilized

Ile-Asn-Thr-Pro-Glu-Gln-Thr-Val-Pro-Tyr-Gly-Leu-Ser-Asn-Tyr-Arg-Gly-Ser-Phe-Arg Amide

Synonyms: Endothelin III (22-41), Big

American Peptide 88-5-16   Human   MW 2298.6 $C_{102}H_{156}N_{30}O_{31}$ Onda, H et al, *FEBS Lett*, 261:327, 1990; Bloch, KD et al, *J Biol Chem*, 264:18156, 1989

Ile-Asn-Trp-Lys-Gly-Ile-Ala-Ala-Met-Ala-Lys-Lys-Leu Amide

Synonyms: Mastoparan X

ICN 159024   MW 1555.9   98% | Similar action to mastoparan; Higashijima, T etal, *JBC*, 265:14176, 1990

Ile-Asn-Trp-Lys-Gly-Ile-Ala-Ala-Met-Ala-Lys-Lys-Leu-Leu Amide

Synonyms: Mastoparan X

Bachem H-9445.0001   Store at -15°C

American Peptide 87-0-67 *Vespa xanthoptera* MW 1556.0
$C_{73}H_{126}N_{20}O_{15}S$ Enhances the transport of Ca^{2+}, Na^+ & K^+ ions through black membrane; Hirai, Y et al, *Chem Pharm Bull*, 27:1745, 1979

Ile-Asp
D-Ile-Asp
Bachem G-3840.0250 MW 246.26 $C_{10}H_{18}N_2O_5$ Store at -15°C

Ile-Asp-Asn-Asp-Thr-Thr-Ser-Tyr-Thr-Leu-Thr-Ser-Cys
Synonyms: GP120 (189-201); HTLV I Peptide

Neosystem SC1006 MW 1433.5 HTLV I peptide from subtype GP160; due to the presence of a Cys this peptide can contain the dimeric form

Ile-Asp-Ile-Ser-Ile-Glu-Leu-Asn-Lys-Ala-Lys-Ser-Asp-Leu-Glu-Glu-Ser-Lys-Glu-Trp-Ile-Arg-Arg-Ser-Asn-Gln-Lys-Leu-Asp-Ser-Ile-Gly-Asn-Trp-His
Ac-Ile-Asp-Ile-Ser-Ile-Glu-Leu-Asn-Lys-Ala-Lys-Ser-Asp-Leu-Glu-Glu-Ser-Lys-Glu-Trp-Ile-Arg-Arg-Ser-Asn-Gln-Lys-Leu-Asp-Ser-Ile-Gly-Asn-Trp-His Amide
Synonyms: Human Parainfluenza Virus Type 3 Fusion Protein (454-488); HPIV-3 Fusion Protein (454-488), Ac-

Bachem H-3588.0500 Store at -15°C

Ile-Cys-Cys-Asn-Pro-Ala-Cys-Gly-Pro-Lys-Tyr-Ser-Cys Amide
Synonyms: Conotoxin SI, α-; Nicotinic Acetylcholine Receptor Blocker

Bachem H-1112.0500 Store at -15°C | Disulfide bonds: Cys^2-Cys^7, Cys^3-Cys^{13}

American Peptide 41-0-27 *Conus striatus* MW 1353.6 $C_{55}H_{84}N_{16}O_{16}S_4$ Disulfide bonds: Cys^2-Cys^7; Cys^3-Cys^{13}; isolated from the piscivorous venom of *Conus striatus*; blocks nicotinic acetylcholine receptors in fish

Calbiochem 234624 *Conus striatus* MW 1353.6 $C_{55}H_{84}N_{16}O_{16}S_4$ TFA; ≥95% (HPLC); lyophilized solid; soluble in water; highly toxic:LD_{50}≤50 mg/kg | Disulfide bonds: Cys^2-Cys^7, Cys^3-Cys^{13}; potent neurotoxin; blocks nicotinic acetylcholine receptors in fish; used to differentiate the nicotinic cholinergic receptors in different vertebrate systems; Lambert, P et al, *Biochem Biophys Res Comm*, 170:684, 1990; Gray, WR et al, *PNAS*, 85:3329, 1988; Hann, RM et al, *Biochemistry*, 33:14058, 1994; Munson, MC & Barany, G, *J Am Chem Soc*, 115:10203, 1993; Zafaralla, GC et al, *Biochemistry*, 27:7102, 1988

Sigma C 7677 Conus striatus (cone snail) FW 1353.6 ≥97% (HPLC) | Disulfide bonds: 2-7, 3-13; bioactive peptide; toxin which blocks nicotinic acetylcholine receptors in fish; Zafarella, GC et al, *Biochemistry*, 27:7102 1988

Peptides International PCN-4228-v *Conus striatus* (marine snail) synthetic MW 1353.6 $C_{55}H_{84}N_{16}O_{16}S_4$ >99% (HPLC); lyophilized amorphous powder | Bioactive; Disulfide bonds: Cys^2-Cys^7, Cys^3-Cys^{13}; Zafaralla, GC et al, *Biochem*, 27:7102, 1988

Alexis 630-049 Synthetic MW 1352.6 $C_{55}H_{83}N_{16}O_{16}S_4$ White lyophilized powder; soluble in water; potent neurotoxin | Disulfide bonds between Cys^2-Cys^7, Cys^3-Cys^{13}; postsynaptic inhibitor at the neuromuscular junction; Zafaralla, GC et al, *Biochemistry*, 27:7102, 1988; Hann, RM et al, *Biochemistry*, 33:14058, 1994; Munson, MC & Barany, G, *JACS*, 115:10203, 1993

Ile-Gln
Bachem G-2390.0250 MW 259.31 $C_{11}H_{21}N_3O_4$ Store at -15°C

Ile-Gln-Arg-Gly-Pro-Gly-Arg-Ala
Synonyms: GP120 (314-321); HTLV I Peptide
Neosystem SC813 MW 853.98 HTLV I peptide from subtype GP160

Ile-Gln-Arg-Gly-Pro-Gly-Arg-Ala-Phe-Val-Thr-Ile-Gly-Lys-Ile-Gly-Asn
Synonyms: GP120 (314-330); HTLV I Peptide
Neosystem SC836 MW 1784.09 HTLV I peptide from subtype GP160

Ile-Gln-Gln-Glu-Lys-Asn-Met-Tyr-Glu-Leu-Gln-Lys-Leu-Asn-Ser-Trp-Asp-Val-Phe
Synonyms: GP140 (662-680); SIVmac251 Peptide
Neosystem SC774 MW 2413.72 SIVmac251 peptide from subtype GP140

Ile-Glu
Bachem G-4045.0250 MW 260.29 $C_{11}H_{20}N_2O_5$ Store at -15°C

Ile-Glu-Ala-Arg
N-Ac-Ile-Glu-Ala-Arg pNA Acetate Salt
Sigma A 0180 FW 649.7 (free base $C_{28}H_{43}N_9O_9$

Ile-Glu-Ala-Leu
Z-Ile-Glu(OtBu)-Ala-Leu-CHO
Synonyms: Proteasome Inhibitor I
Bachem C-3900.0005 Store at -15°C

Calbiochem 539160 MW 618.8 $C_{32}H_{50}N_4O_8$ ≥95% (TLC); solid; soluble in methanol | Cell-permeable inhibitor of the chymotrypsin-like activity of the multicatalytic proteinase complex (MCP; 20S proteasome) in HT4 cells; causes the accumulation of ubiquintinated proteins in neuronal cells; prevents the activation of NF-κB in response to TNF-α & okadaic acid through inhibition of IκB-α degradation, thereby interfering with induction of iNOS in macrophages; Griscavage, JM et al, *PNAS*, 93:3308, 1996; Figueiredo-Pereira, ME et al, *J Neurochem*, 63:1578, 1994; Mori, S et al, *J Biol Chem*, 270:29447, 1995; Traenckner, EBM et al, *EMBO J*, 13:5433, 1994

ICN 195931 MW 618.8 $C_{32}H_{50}N_4O_8$ ≥97% | Cell permeable inhibitor of chymotrypsin-like activity of 20S proteasome

Ile-Glu-Ala-Leu
N-CBZ-Ile-Glu(O-t-Butyl)-Ala-Leu-CHO
Sigma C 6831 FW 618.8 $C_{32}H_{50}N_4O_8$ ≥80% (HPLC) | Figueiredo-Pereira, ME et al, *J Neurochem*, 63:5433, 1994

Ile-Glu-Ala-Leu
Carbobenzoxy-Ile-Glu(OtBu)-Ala-Leu-CHO
Synonyms: Proteasome Inhibitor
Peptides International IAT-3169-v Synthetic MW 618.77 $C_{32}H_{50}N_4O_8$ >99% (HPLC); lyophilized amorphous powder | Inhibitor; Figueiredo-Pereira, ME, *J Neurochem*, 63:1578, 1994; Traenckner, EB-M et al, EMBO J, 13:5433, 1994

Ile-Glu-Glu-Glu-Gln-Asn-Lys-Ser-Lys-Lys-Lys-Ala
Synonyms: P18 (104-115); GAG P17 MA (104-115); HTLV I Peptide
Neosystem SC287 MW 1431.60 HTLV I peptide from subtype P25 (GAG P17 MA)

Ile-Glu-Gly
BOC-Ile-Glu-Gly-L-argininol-p-methoxybenzoyl ester
Synonyms: BOC-Ile-Glu-Gly-L-Argininol-p-Anisic Acid Ester
Bachem N-1615.0005 Store at -15°C

Ile-Glu-Gly-Arg
BOC-Ile-Glu-Gly-Arg-AMC
Bachem I-1100.0025 MW 730.82 $C_{34}H_{50}N_8O_{10}$ Store at -15°C

Ile-Glu-Gly-Arg
N-α-t-BOC-L-Ile-L-Glu-Gly-L-Arg-MCA Hydrochloride

Synonyms: Factor Xa Substrate

ICN 150498   MW 767.3 $C_{34}H_{50}N_8O_{10} \cdot HCl$ Fluorogenic substrate

Ile-Glu-Gly-Arg
BOC-Ile-Glu-Gly-Arg-MCA

Synonyms: Factor Xa Substrate

Peptides International MIE-3094-v   MW 730.82
$C_{34}H_{50}N_8O_{10}$ >98% (HPLC); lyophilized amorphous powder | Morita, T et al, *J Biochem*, 82:1495, 1977

Ile-Glu-Gly-Arg
N-Benzoyl-L-Ile-Glu-Gly-Arg pNA Hydrochloride

Sigma B 2291   FW 734.2 $C_{32}H_{43}N_9O_9 \cdot HCl$

Ile-Glu-Gly-Arg
N-t-BOC-Ile-Glu-Gly-Arg MCA Hydrochloride

Synonyms: Factor Xa Substrate

Sigma B 9760   FW 767.3 $C_{34}H_{50}N_8O_{10} \cdot HCl$ Sensitive & specific fluorogenic substrate for factor Xa; Morita, T et al, J Biochem (Tokyo), 82:1495, 1977; Lottenberg, R et al, *Meth Enz*, 80:341, 1981

Ile-Glu-Pro-Asp
Ac-Ile-Glu-Pro-Asp-AMC

Bachem I-1835.0005   MW 671.71 $C_{32}H_{41}N_5O_{11}$ Store at -15°C

Ile-Glu-Pro-Asp
Ac-Ile-Glu-Pro-Asp-pNA

Bachem L-2105.0005   MW 634.64 $C_{28}H_{38}N_6O_{11}$ Store at -15°C

Ile-Glu-Pro-Asp
Ac-Ile-Glu-Pro-Asp-CHO Pseudo Acid

Bachem N-1740.0005   MW 498.53 $C_{22}H_{34}N_4O_9$ Store at -15°C

Ile-Glu-Pro-Asp
Ac-Ile-Glu-Pro-Asp-pNA

Synonyms: Granzyme B Substrate II

Kamiya   MW 634   >97% (HPLC)

Ile-Glu-Pro-Dpr-Tyr-Arg-Leu-Arg-Tyr Amide

Synonyms: Neuropeptide YY$_1$ Receptor Antagonist, (Cyclic (2,4'), (2',4)-Diamide)-

American Peptide 60-1-23   MW 2352.8 $C_{110}H_{170}N_{34}O_{34}$ Potent & selective; Kanatani, A et al, *Endocrinology*, 137(8):3177, 1996

Ile-Glu-Pro-Tyr-Arg-Leu-Arg-Tyr
Ile-Glu-Pro-Dapa-Tyr-Arg-Leu-Arg-Tyr Amide, Cyclic (2,4') (2',4) Diamide

Synonyms: Neuropeptide Y Antagonist; 1229U91

Neosystem SC1226   MW 2352.82   Bioactive neuropeptide; high affinity & selective neuropeptide Y-Y1 receptor antagonist; Hegde, SS et al, *J Pharmacol Exp Ther*, 275:1261-1266, 1995; Tadepalli, AS et al, *J Cardiovasc Pharmacol*, 27:712-718, 1996; Lew, MJ et al, *Br J Pharmacol*, 117:1768-1772, 1996; Gehlert, DR et al, *Mol Pharmacol*, 50:112-118, 1996; Kanatani, A et al, *Endocrinology*, 137:3177-3182, 1996; Kennedy, B et al, *J Pharmacol Exp Ther*, 281:291-296, 1997; Angus, JA, *Clinical & Exp Pharmacol & Physiol*, 24:297-304, 1997

Ile-Glu-Ser-Ile-Pro-Asp-Pro-Pro-Thr-Asn-Thr-Pro-Glu-Ala-Leu

Synonyms: REV (82-96); SIVmac251 Peptide

Neosystem SC602   MW 1593.75   SIVmac251 peptide from subtype REV

Ile-Glu-Thr-Asp
Ac-Ile-Glu-Thr-Asp AMC

Synonyms: Caspase III Processing Enzyme Substrate; Granzyme B Substrate II

Alexis 260-042   MW 675.7 $C_{31}H_{41}N_5O_{12}$ ≥98%; white lyophilized powder; soluble in DMSO | Fluorogenic substrate for Caspase III processing enzyme & granzyme B; AMC has an excitation maximum of 380 nm & an emission maximum of 460 nm

Ile-Glu-Thr-Asp
Ac-Ile-Glu-Thr-Asp-CHO

Synonyms: Caspase III Processing Enzyme Inhibitor I; Granzyme B Inhibitor I

Alexis 260-043   MW 502.5 $C_{21}H_{34}N_4O_{10}$ ≥95%; white lyophilized powder; soluble in water

Ile-Glu-Thr-Asp
Z-Ile-Glu-Thr-Asp-pNA

Synonyms: Caspase III Processing Enzyme Substrate I; Granzyme B Substrate I

Alexis 260-044   MW 730.7 $C_{33}H_{42}N_6O_{13}$ White lyophilized powder; soluble in water | Chromogenic substrate with increased cell permeability; sequence corresponds to one of the cleavage sites of the inactive 32 kDa Caspase III precursor (AA 172-175)

Ile-Glu-Thr-Asp
Ac-Ile-Glu-Thr-Asp-pNA

Synonyms: Caspase III Processing Enzyme Substrate II; Granzyme B Substrate II

Alexis 260-045   MW 638.6 $C_{27}H_{38}N_6O_{12}$ White lyophilized powder; soluble in water | Chromogenic substrate; corresponds to one of the cleavage sites of the inactive 32 kDa Caspase III precursor (AA 172-175); Han, Z et al, *J Biol Chem*, 272:13432, 1997

Ile-Glu-Thr-Asp
Z-Ile-Glu-Thr-Asp-AFC

Bachem I-1760.0005   MW 821.76 $C_{37}H_{42}F_3N_5O_{13}$ Store at -15°C

Ile-Glu-Thr-Asp
Ac-Ile-Glu-Thr-Asp-AMC

Bachem I-1810.0005   MW 675.69 $C_{31}H_{41}N_5O_{12}$ Store at -15°C

Ile-Glu-Thr-Asp
Ac-Ile-Glu-Thr-Asp-pNA

Bachem L-1985.0005   MW 638.63 $C_{27}H_{38}N_6O_{12}$ Store at -15°C

Ile-Glu-Thr-Asp
Ac-Ile-Glu-Thr-Asp-CHO Pseudo Acid

Bachem N-1620.0005   MW 502.52 $C_{21}H_{34}N_4O_{10}$ Store at -15°C

Ile-Glu-Thr-Asp
Ac-Ile-Glu-Thr-Asp-CHO

BioSource International 77-849   Inhibitor

Ile-Glu-Thr-Asp
Z-Ile-Glu-Thr-Asp-pNA

BioSource International 77-852

Ile-Glu-Thr-Asp
Ac-Ile-Glu-Thr-Asp-pNA

BioSource International 77-855

BioSource International 77-856

Ile-Glu-Thr-Asp
Ac-Ile-Glu-Thr-Asp-AFC

BioSource International 77-980   Fluorescing substrate

Ile-Glu-Thr-Asp
Z-Ile-Glu-Thr-Asp-AFC

BioSource International 77-983

Ile-Glu-Thr-Asp
Ac-Ile-Glu-Thr-Asp-AMC

BioSource International 78-100   Fluorescing substrate

Ile-Glu-Thr-Asp
Ile-Glu-Thr-Asp-CHO

BioSource International 78-112   Cell permeable inhibitor

Ile-Glu-Thr-Asp
Z-Ile-Glu(OMe)-Thr-Asp(OMe)-CH$_2$F

Synonyms: Caspase VIII Inhibitor II; Granzyme B Inhibitor III

Calbiochem 218759   MW 654.7 $C_{30}H_{43}FN_4O_{11}$ Single spot purity (TLC); solid; soluble in DMSO | Potent, cell-permeable & irreversible inhibitor of Caspase VIII (MACH, FLICE, Mch5); also inhibits granzyme B; Martin, DA et al, *J Biol Chem*, 273:4345, 1998; Sweeney, EA et al, *FEBS Lett*, 425:61, 1998

Ile-Glu-Thr-Asp
Ac-Ile-Glu-Thr-Asp-CHO

Synonyms: Granzyme B Inhibitor II; Caspase VIII Inhibitor I

Calbiochem 368055   MW 502.5 $C_{21}H_{34}N_4O_{10}$ ≥95% (HPLC); lyophilized solid; soluble in DMSO | Potent, reversible granzyme B & Caspase VIII inhibitor; Han, Z et al, *J Biol Chem*, 272:13422, 1997

Ile-Glu-Thr-Asp
Ac-Ile-Glu-Thr-Asp-*p*NA

Synonyms: Granzyme B Substrate I; Caspase VIII Substrate I

Calbiochem 368057   MW 638.6 $C_{27}H_{38}N_6O_{12}$ ≥97% (HPLC); lyophilized solid; soluble in water | Colorimetric substrate for detection of granzyme B & Caspase VIII activity; cleavage is monitored colorimetrically at ~405 nm

Ile-Glu-Thr-Asp
Z-Ile-Glu-Thr-Asp-AFC

Synonyms: Granzyme B Substrate II; Caspase VIII Substrate II,

Calbiochem 368059   MW 821.8 $C_{37}H_{42}F_3N_5O_{13}$ ≥97% (HPLC); lyophilized solid; soluble in DMSO; excitation max:~400 nm; emission max:~505 nm | Fluorogenic substrate for granzyme B & Caspase VIII activity; reaction monitored visually or quantitatively by a blue to green shift in fluorescence upon cleavage of the AFC fluorophore

Ile-Glu-Thr-Asp
Z-Ile-Glu-Thr-Asp-FMK

Synonyms: Caspase VIII Inhibitor; FLICE Inhibitor

Kamiya   MW 654.5

Ile-Glu-Thr-Asp
Ac-Ile-Glu-Thr-Asp-CHO

Synonyms: Caspase VIII Inhibitor; FLICE Inhibitor

Kamiya   MW 502.5 $C_{21}H_{34}N_4O_{10}$ >95%

Ile-Glu-Thr-Asp
Ac-Ile-Glu-Thr-Asp-pNA

Synonyms: Caspase VIII Substrate

Kamiya   MW 638 $C_{27}H_{38}N_6O_{12}$ >97% (HPLC)

Ile-Glu-Thr-Asp
Ac-Ile-Glu-Thr-Asp-AFC

Synonyms: Caspase VIII Substrate; FLICE Substrate

Kamiya   MW 654   Fluorogenic

Ile-Glu-Thr-Asp
Ac-Ile-Glu-Thr-Asp-AMC

Synonyms: Pro-Caspase III Cleaving Enzyme Substrate

Neosystem SC1313   MW 675.69   Fluorogenic substrate; Caspase III is expressed in cells as an inactive 32k precursor (pro-Caspase III) from which 17kDa (p17) & 12kDa (p12) subunits of the mature Caspase III are proteotically generated during apoptosis; pro-Caspase III is first cleaved yielding the mature p12 subunit & a p20 intermediate product which is subsequently cleaved generating the mature p17 subunit; corresponds to the cleavage site of pro-Caspase III producing p12 & p20; granzyme B, Caspase VI & Caspase VIII also hydrolyze this peptide; Han, Z et al, *JBC*, 272:13432-13436, 1997; Thornberry, NA et al, *JBC*, 272:17907-17911, 1997

Ile-Glu-Thr-Asp
Ac-Ile-Glu-Thr-Asp-pNA

Synonyms: Pro-Caspase III Cleaving Enzyme Substrate

Neosystem SC1314   MW 638.61   Chromogenic substrate; deduced from the same cleavage site described for Neosystem SC1313; Han, Z et al, *JBC*, 272:13432-13436, 1997

Ile-Glu-Thr-Asp
Ac-Ile-Glu-Thr-Asp-MCA

Synonyms: Pro-Caspase III Cleaving Enzyme Substrate

Peptides International MCA-3195-v   MW 675.69 $C_{31}H_{41}N_5O_{12}$ >98% (HPLC); lyophilized amorphous powder | Deduced from the cleavage site of ProCaspase III; Thornberry, NA et al, *JBC*, 272:17907, 1997

Ile-Glu-Thr-Asp
N-Ac-Ile-Glu-Thr-Asp-CHO

Sigma A 1216   Bioactive peptide; Han, Z et al, *J Biol Chem*, 272:13432, 1997

Ile-Glu-Thr-Asp
Ac-Ile-Glu-Thr-Asp-pNA

Synonyms: Caspase VIII Substrate, Chromogenic

Upstate 12-391   MW 639.6   Lyophilized | Caspase VIII assay

Ile-Glu-Thr-Asp
Ac-Ile-Glu-Thr-Asp-CHO

Synonyms: Caspase VIII/VI Inhibitor; Pro-Caspase III Cleaving Enzyme Inhibitor; Granzyme B Inhibitor

Peptides International ICA-3196-v   Synthetic   MW 502.52 $C_{21}H_{34}N_4O_{10}$ >99% (HPLC); lyophilized amorphous powder | Inhibitor; Han, Z et al, *JBC*, 272:13432, 1997

Ile-Glu-Thr-Asp
Ac-Ile-Glu-Thr-Asp-AMC

Synonyms: Caspase VIII Fluorometric Substrate

USBio C2087-43   Synthetic   ≥95% (HPLC); frozen solution | Contains the recognition sequence for Caspase VIII (FLICE); used to measure Caspase activity *in vitro*

Ile-Glu-Thr-Asp-Ala
Ac-Ile-Glu-Thr-Asp-Ala-MC

Synonyms: Caspase VIII Fluorometric Substrate

Upstate 12-351   MW 676   >95% pure; frozen solution | Used to assay for Caspase activity

Ile-Glu-Thr-Val-Pro-Val-Lys-Leu-Lys-Pro-Gly-Met-Asp-Gly-Pro-Lys-Val-Lys-Gln-Trp-Pro-Leu-Thr-Glu-Glu

Synony POL (172-196); RT (5-29); HTLV I Peptide *ms*:

Neosystem SC687   MW 2820.33   HTLV I peptide from subtype POL (PR/RT)

Ile-Gly
BOC-Ile-Gly

Bachem A-1845.0001   MW 288.34 $C_{13}H_{24}N_2O_5$ Store at -15°C

Ile-Gly
FMOC-Ile-Gly

Bachem B-3495.0001   MW 410.47 $C_{23}H_{26}N_2O_5$ Store at -15°C

Ile-Gly
Z-Ile-Gly

Bachem C-3855.0001   MW 322.36 $C_{16}H_{22}N_2O_5$ Store at -15°C

Ile-Gly

Bachem G-2395.0250   MW 188.23 $C_8H_{16}N_2O_3$ Store at -15°C

ICN 102087

Ile-Gly
N-t-BOC-Ile-Gly

Sigma B 7768   FW 288.3 $C_{13}H_{24}N_2O_5$

Ile-Gly-Arg-Asn-Leu-Leu-Thr-Gln-Ile-Gly-Cys-Thr-Leu-Asn-Phe

Synonyms: POL (85-99); PR (85-99); HTLV I Peptide

Neosystem SC707   MW 1662.96   HTLV I peptide from subtype POL (PR/RT); due to the presence of a Cys this peptide can contain the dimeric form

Ile-Gly-Gly
Z-Ile-Gly-Gly

Bachem C-2050.0001   MW 379.41 $C_{18}H_{25}N_3O_6$ Store at -15°C

Ile-Gly-Gly

Bachem H-3815.0250   MW 245.28 $C_{10}H_{19}N_3O_4$ Store at -15°C

ICN 102086

Ile-Gly-Leu-Met
β-Ile-Gly-Leu-Met

Synonyms: Amyloid (32-35)

American Peptide 62-0-12   MW 432.6

Ile-Gly-Leu-Met Acetate Salt

Synonyms: Amyloid β-Protein (32-35)

Sigma A 4950   FW 432.6 (free base) $C_{19}H_{36}N_4O_5S$ ≥95% (HPLC) | Bioactive peptide

Ile-Gly-Pro-Glu-Val-Pro-Asp-Asp-Arg-Asp-Phe-Glu-Pro-Ser Acetate Salt

Synonyms: Proteoglycan II (6-19); Decorin; PG-S2; PG-40

Biogenesis 7870-6180   Human synthetic   MW 1573.1   Purified; lyophilized | This N-terminal human sequence is also Bone Proteoglycan II Precursor (36-49)

Ile-Gly-Val-Thr-Gln-Gln-Arg-Arg-Ala-Arg-Asn-Gly-Ala-Ser-Arg-Ser

Synonyms: VPR (81-96); HTLV I Peptide

Neosystem SC201   MW 1756.94   HTLV I peptide from subtype VPR

Ile-His
Z-Ile-His

Bachem C-2055.0001   MW 402.45 $C_{20}H_{26}N_4O_5$ Store at -15°C

Ile-His

Bachem G-2400.0250   Store at -15°C

Ile-His-Phe-Leu-Ile-Arg-Gln-Leu-Ile-Arg-Leu-Leu-Thr-Trp-Leu-Phe-Ser-Asn-Cys-Arg

Synonyms: GP140 (771-790); SIVmac251 Peptide

Neosystem SC785   MW 2543.11   SIVmac251 peptide from subtype GP140; due to the presence of a Cys this peptide can contain the dimeric form

Ile-His-Pro-Phe

Synonyms: Angiotensin I/II (5-8); Angiotensin II (5-8)

Bachem H-3846.0005   MW 512.61 $C_{26}H_{36}N_6O_5$ Store at -15°C

ICN 159878   MW 512.6   98% | Goetzl, EJ etal, *BBRC*, 97:1097, 1980

American Peptide 12-1-39   Human   MW 512.6 $C_{26}H_{36}N_6O_5$ Goetzl, EJ et al, *BBRC*, 97:1097, 1980

Ile-His-Ser-Ile-Ser-Glu-Arg-Ile-Leu-Ser-Thr-Tyr-Leu

Synonyms: REV (52-64); HTLV I Peptide

Neosystem SC193   MW 1531.77   HTLV I peptide from subtype REV

Ile-Ile
Z-Ile-Ile

Bachem C-2060.0001   MW 378.47 $C_{20}H_{30}N_2O_5$ Store at -15°C

Ile-Ile

Bachem G-2405.0250   MW 244.33 $C_{12}H_{24}N_2O_3$ Store at -15°C

Ile-Ile
N-CBZ-Ile-Ile

Sigma C 1752   FW 378.5 $C_{20}H_{30}N_2O_5$

Ile-Ile-Gly-Leu
Propionyl-Ile-Ile-Gly-Leu Amide

Synonyms: Amyloid β-Protein (31-34), Propionyl-

Bachem H-4124.0005   MW 469.63 $C_{23}H_{43}N_5O_5$ Store at -15°C

Ile-Ile-Gly-Leu-Met
β-Ile-Ile-Gly-Leu-Met

Synonyms: Amyloid (31-35)

American Peptide 62-0-31   MW 545.7   Neurotoxic activity in brain tissue culture similar to that of fragment 1-40; Penke, B et al, *13th Am Pept Symp, Abstract P413*, 1993

Ile-Ile-Gly-Leu-Met Acetate Salt

Synonyms: Amyloid β-Protein (31-35)

Sigma A 5075   FW 545.7 (free base) $C_{25}H_{47}N_5O_6S$ ≥95% (HPLC) | Bioactive peptide; neurotoxic activity in brain tissue culture similar to that of (1-40); Penke, B et al, Abstract P413, 13th American Peptide Symposium, Edmonton, 1993

Ile-Ile-Ile

Bachem H-3820.0250   MW 357.49 $C_{18}H_{35}N_3O_4$ Store at -15°C

Ile-Ile-Ser-Ala-Val-Val-Gly-Ile-Leu

Synonyms: HER2/neu (654-662)

Neosystem SC1274   MW 884.12   Bioactive; cancer related peptide; cytotoxic T lymphocytes (CTL) derived from breast, ovarian & non-small cell lung cancer recognized this HER[2]/neu derived peptide; recently identified as a tumor-associated antigen in human pancreatic cancer recognized by CTL; Yoshino, I et al, *Cancer Res*, 54:3387-3390, 1994; Peoples, GE et al, *PNAS USA*, 92:432-436, 1995; Linehan, DC et al, *J Immunol*, 155:4486-4491, 1995; Peiper, M et al, *Mol Pharm Immunol*, 27:1115-1123, 1997; Peiper, M et al, *Surgery*, 122:235-241, 1997

Ile-Ile-Thr-Gly-Leu-Leu-Glu-Phe-Glu-Val-Tyr-Leu-Glu-Tyr-Leu-Gln-Asn-Arg-Phe-Glu-Ser-Ser-Glu-Glu-Gln-Ala-Arg-Ala-Val-Gln-Met-Ser-Thr-Lys

Synonyms: Interleukin VI (8-121)

Bachem H-1398.0500   Human   MW 4023.53 $C_{179}H_{281}N_{45}O_{58}S$ Store at -15°C

Sigma I 3271   Human   FW 4023.5   ≥80% (HPLC); peptide content:~70% | Bioactive peptide; competitively inhibits binding of IL-6 to its receptor; 20% of Met residues may be present as the sulfoxide; Ekida, T et al, *Biochem Biophys Res Commun*, 189:211, 1992

Ile-Ile-Thr-Leu
(2R)-2-(Ac-(N-Me-L-Ile)-L-Ile-L-Thr-L-Leu)-2-Methyloxirane

Synonyms: Epoxomicin; Proteasome Inhibitor

Peptides International IEP-4381-v   Synthetic   MW 554.72 $C_{28}H_{50}N_4O_7$ >99% (HPLC); lyophilized amorphous powder | Inhibitor; Meng, L et al, *PNAS USA*, 96:10403, 1999; Sin, N et al, *Bioorg Med Chem Lett*, 9:2283, 1999; Kim, KB et al, *Bioorg Med Chem Lett*, 9:3335, 1999; Groll, M et al, *J Am Chem Soc*, 122:1237, 2000

Ile-Ile-Trp
N-Suc-Ile-Ile-Trp-AMC

American Peptide 80-0-10   MW 686.8 $C_{37}H_{44}N_5O_8$

Ile-Ile-Trp
Suc-Ile-Ile-Trp-MCA

Peptides International MIW-3150-v   MW 687.79 $C_{37}H_{45}N_5O_8$ >98% (HPLC); lyophilized amorphous powder

Ile-Ile-Trp
N-Succinyl-Ile-Ile-Trp MCA

Sigma S 8524   FW 687.8 $C_{37}H_{45}N_5O_8$

Ile-Ile-Trp-Phe-Asn-Thr-Pro-Glu-His-Val-Val-Pro-Tyr-Gly-Leu-Gly-Ser-Pro-Arg

Synonyms: Endothelin Converting Enzyme Inhibitor; Endothelin I (19-37), (Phe[22])-Big; Endothelin I, Big

American Peptide 88-1-36   Human   MW 2182.5 $C_{104}H_{152}N_{26}O_{26}$ This N- & C-terminally truncated big endothelin-1 analog is a potent & selective ECE inhibitor & is at least 30-fold more potent than phosphoramidon in blocking the big ET-1 induced vasoconstriction in the kidney; Claing, A et al, *4th Int Conf on Endothelin, London*, 1995

Bachem H-3614.0500   Human   MW 2182.51 $C_{104}H_{152}N_{26}O_{26}$ Store at -15°C

Ile-Ile-Trp-Val-Asn-Thr-Pro-Glu-His-Val-Val-Pro-Tyr-Gly-Leu-Gly-Ser-Pro-Arg

Synonyms: Endothelin I (19-37), (Phe[22])-Big; Endothelin Converting Enzyme Inhibitor

Alexis 152-014   Human   MW 2134.5 $C_{100}H_{152}N_{26}O_{26}$ Potent & selective inhibitor; useful tool in the development of more selective & enzyme-resistant inhibitors of membrane-bound ECE

Ile-Ile-Trp-Val-Asn-Thr-Pro-Glu-His-Val-Val-Pro-Tyr-Gly-Leu-Gly-Ser-Pro-Arg-Ser

Synonyms: Endothelin I (19-38), Big

American Peptide 88-1-33   Human   MW 2221.7 $C_{103}H_{157}N_{27}O_{28}$

Ile-Ile-Trp-Val-Asn-Thr-Pro-Glu-His-Val-Val-Pro-Tyr-Gly-Leu-Gly-Ser-Pro-Arg

Synonyms: Endothelin I (19-37), (Phe[22])-Big; Endothelin Converting Enzyme Inhibitor

Sigma E 9397   Human   FW 2182.5   ≥95% (HPLC) | Bioactive peptide; Claing, A et al, *J Cardiovasc Pharmacol*, 26:S72, 1995

Ile-Leu
Z-Ile-Leu

Bachem C-2065.0001   MW 378.47 $C_{20}H_{30}N_2O_5$ Store at -15°C

Ile-Leu

Bachem G-2410.0250   MW 244.33 $C_{12}H_{24}N_2O_3$ Store at -15°C

Ile-Leu
Ile-Leu-OMe Hydrochloride

Bachem G-3665.0001   MW 294.82 $C_{13}H_{26}N_2O_3 \cdot HCl$ Store at -15°C

Ile-Leu
N-(3-((Hydroxyamino)Carbonyl)-1-Oxo 2(R)-Benzylpropyl)-Ile-Leu

Synonyms: JMV-390-1

Neosystem SC894   MW 449.55   Bioactive; neurotensin-related peptide; potent inhibitor of multiple neurotensin/neuromedin N degrading enzymes; Kitabgi, P et al, *J Neurosci Letters*, 142:200-204, 1992; Doulut, S et al, *J Med Chem*, 36:1369-1379, 1993

Ile-Leu
N-CBZ-Ile-Leu

Sigma C 1877   FW 378.5 $C_{20}H_{30}N_2O_5$

Ile-Leu-Asn-Gly-Ile-Asn-Asn-Tyr-Lys-Asn-Pro-Lys-Leu

Synonyms: Interleukin IIβ (44-56)

American Peptide 85-1-20   MW 1501.0 $C_{68}H_{113}N_{19}O_{19}$

Ile-Leu-Lys-Glu-Pro-Val-His-Gly-Val

Synonyms: HIV I POL (476-484); HTLV I Peptide

Neosystem SC1344   MW 991.19   >97 % (HPLC) | Virus-related peptide; AIDS-related peptide; HIV-1 HLA-A*0201-restricted dominant CTL epitope; Tsomides, TJ et al, *J Exp Med*, 180:1283-1293, 1994

Ile-Leu-Met-Glu-Lys-Pro-Ser-Arg-Pro-Met-Glu-Ser-Asn-Pro-Asp-Thr-Glu-Gly-Cys Acetate Salt

Synonyms: Parathyroid Hormone Receptor II (524-542), Cytoplasmic Domain 4

Biogenesis 0100-0187   Synthetic   Semi-pure; lyophilized

Ile-Leu-Pro-Trp-Lys-Trp-Pro-Trp-Trp-Pro-Trp-Arg-Arg Amide

Synonyms: Indolicidin

American Peptide 72-2-28   MW 1906.3 $C_{100}H_{132}N_{26}O_{13}$
Antibiotic peptide; 13-residue peptide amide was isolated & characterized from the cytoplasmic granules of bovine neutrophils, having potent antibacterial activity; the primary structure is characterized by 5 tryptophan residues, which is the highest mole percentage of any known protein sequence & is unique among known endogenous antibacterial peptides; Subbalakshmi, C et al, *FEBS Lett*, 395:48, 1996

Bachem H-1234.0500   MW 1906.32 $C_{100}H_{132}N_{26}O_{13}$ Store at -15°C

Sigma I 0144   FW 1906.3   ≥97% (HPLC) | Bioactive peptide; potent antimicrobial activity *in vitro* against bacteria & fungi; Selsted, ME et al, *J Biol Chem*, 267:4292, 1992

Ile-Leu-Val-Glu-Ser-Pro-Thr-Val-Leu-Glu-Ser-Gly-Thr-Lys-Glu

Synonyms: REV (102-116); HTLV I Peptide

Neosystem SC197   MW 1601.81   HTLV I peptide from subtype REV

Ile-L-Glu-Gly-L-Arg-Ala
N$^\alpha$-Benzoyl-L-Ile-L-Glu-Gly-L-Arg-Ala-4-Nitroanilide Hydrochloride

Synonyms: Trypsin Substrate

Fluka 13000   MW 734.22 $C_{32}H_{43}N_9O_9 \cdot HCl$ ≥97% (HPLC) | Chromogenic; liberated 4-nitroanilide is measured spectrophotometrically at 405 nm; Geiger, R & Fritz, H, *Methods of Enzymatic Analysis*, Bergmeyer, HU, ed, Verlag Chemie Weinheim, vol 5:3rd Edition, 119, 1984

Ile-Lys-Asn-Leu-Gln-Ser-Leu-Asp-Pro-Ser-His

Synonyms: Cholecystokinin (10-20); Gastrointestinal Peptide

Bachem H-2415.0001   MW 1251.41 $C_{54}H_{90}N_{16}O_{18}$ Store at -15°C

ICN 152861   MW 1252.4

Sigma C 9269   FW 1251.4   ≥97% (HPLC) | Bioactive peptide

Biogenesis 2050-1209   Synthetic   MW 1251.4   Lyophilized, consisting of 75% peptide material, acetate counter ions and residual H_2O; purified

Ile-Lys-Cys-Asn-Cys-Lys-Arg-His-Val-Ile-Lys-Pro-His-Ile-Cys-Arg-Lys-Ile-Cys-Gly-Lys-Asn Amide

Synonyms: Voltage-Dependent K$^+$ Channel Blocker; Mast Cell Degranulating Peptide

Bachem H-7885.0500   Store at -15°C | Disulfide bonds: Cys^3-Cys^{15}, Cys^5-Cys^{19}

ICN 154446   MW 2587.6   Taylor, JW etal, *JBC*, 259:13957, 1984; Bidard, JN etal, *Brain Res*, 418:235, 1987

American Peptide 41-9-30   *Apis mellifera*   MW 2587.3 $C_{110}H_{192}N_{40}O_{24}S_4$ Disulfide bonds: Cys^3-Cys^{15}, Cys^5-Cys^{19}; blocks voltage-gated K$^+$ channels & causes convulsions & hyperactivity; causes mast cell degranulation & subsequent histamine release; Billingham, MEJ et al, *Nature*, 245:163, 1973

Calbiochem 516490   *Apis mellifera*   MW 2587.3 $C_{110}H_{192}N_{40}O_{24}S_4$ >98% (SDS-PAGE); lyophilized powder; soluble in water; LD$_{50}$≤2000 mg/kg | Disulfide bonds: Cys^3-Cys^{15}, Cys^5-Cys^{19}; toxic component of bee venom that selectively blocks some neuronal voltage-gated K$^+$ channels at submicromolar concentrations; induces long-term potentiation of the excitatory postsynaptic potentials in CA1 hippocampal neurons; Buku, A et al, *Int J Pep Protein Res*, 44:410, 1994; Neuman, R et al, *Eur J Neurosci*, 3:253, 1991

Peptides International PMC-4258-v   *Apis mellifera* (honeybee) synthetic   MW 2587.3 $C_{110}H_{192}N_{40}O_{24}S_4$ >99% (HPLC); lyophilized amorphous powder | Bioactive; disulfide bonds Cys^3-Cys^{15}, Cys^5-Cys^{19}; Haberman, E, *Science*, 177:314, 1972; Ziai, MR et al, *J Pharmacol*, 42:457, 1990

Alexis 162-002   *Apis mellifera* synthetic   MW 2587.2 $C_{110}H_{192}N_{40}O_{24}S_4$ ≥98%; white lyophilized powder; potent neurotoxin | Disulfide bonds: Cys^3-Cys^{15}, Cys^5-Cys^{19}; voltage-dependent K$^+$ channel blocker; Haberman, E et al, *Science*, 177:314, 1972; Ziai, MR et al, *J Pharm Pharmacol*, 42:457, 1990; Strong, PN, *Pharmacol Ther*, 46:137, 1990

Sigma M 8036   Bee venom   FW 2587.2   Disulfide bonds: 3-15, 5-19; bioactive peptide; toxic component of venom that selectively blocks some neuronal voltage-gated K$^+$ channels; degranulates mast cells & releases histamine; Taylor, JW et al, *J Biol Chem*, 259:13957, 1984; Bidard, JN et al, *Brain Res*, 418:235, 1987

Ile-Lys-Gln-Leu-Gln-Ala-Arg-Ile-Leu-Ala-Val-Glu-Arg-Tyr-Leu-Lys-Asp

Synonyms: GP41 (578-594); HTLV I Peptide

Neosystem SC255   MW 2057.46   HTLV I peptide from subtype GP41

Ile-Lys-Gln-Phe-Ile-Asn-Met-Trp-Gln-Glu-Val-Gly

Synonyms: GP120 (425-436); HTLV I Peptide

Neosystem SC566   MW 1492.75   HTLV I peptide from subtype GP160

Ile-Lys-Ile-Gly-Gly-Gln-Leu-Lys-Glu-Ala-Leu-Leu-Asp-Thr-Gly

Synonyms: POL (13-27); PR (13-27); HTLV I Peptide

Neosystem SC701   MW 1555.83   HTLV I peptide from subtype POL (PR/RT)

Ile-Lys-Pro-Glu-Ala-Pro-Gly-Glu-Asp-Ala-Ser-Pro-Glu-Glu-Leu-Asn-Arg-Tyr-Tyr-Ala-Ser-Leu-Arg-His-Tyr-Leu-Asn-Leu-Val-Thr-Arg-Gln-Arg-Tyr Amide

Synonyms: Peptide YY (3-36)

American Peptide 48-0-33   Human   MW 4049.6 $C_{180}H_{279}N_{53}O_{54}$ Selectively binds to Y2 receptors; found in human intestine & circulating blood; Grandt, D et al, *Reg Peptides*, 40:161, 1992

Bachem H-8585.0500   Human   Store at -15°C

Neosystem SC896   Human   MW 4049.51   Bioactive neuropeptide; brain/gut peptide; Eberlein, GA et al, *Peptides*, 10:797-803, 1989; Grandt, D et al, *BBRC*, 186:1299-1306, 1992; Grandt, D et al, *Regul Peptides*, 40:161, 1992

Ile-Lys-Thr-Glu-Glu-Ile-Ser-Glu-Val-Asn-Leu-Asp-Ala-Glu-Phe
DABCYL-Ile-Lys-Thr-Glu-Glu-Ile-Ser-Glu-Val-Asn-Leu-Asp-Ala-Glu-Phe-EDANS

Synonyms: Amyloid β/A4 Precursor Protein 770 (661-675)-EDANS, DABCYL-(Asn^{670},Leu^{671})-

Bachem M-2445.0001   MW 2236.49 $C_{103}H_{146}N_{22}O_{32}S$ Store at -15°C

Ile-Lys-Val-Ala-Val

Synonyms: Laminin A Chain Peptide I-5-V

Neosystem SC338   MW 528.69   Bioactive; cell attachment peptide; Sephel, GC et al, *BBRC*, 162:821-829, 1989

Ile-Met
Z-Ile-Met

Bachem C-2070.0001   MW 396.51 $C_{19}H_{28}N_2O_5S$ Store at -15°C

Ile-Met

Bachem G-2415.0250   MW 262.37 $C_{11}H_{22}N_2O_3S$ Store at -15°C

Ile-Met-Asp-Gln-Val-Pro-Phe-Ser-Val

Synonyms: Melanocyte Protein PMEL 17 (209-217), (Met[21]0)-; Melanocyte Protein PMEL 17 (52-60), (Met[53])-

Bachem H-4938.0001   Human, bovine, mouse   MW 1035.23 $C_{47}H_{74}N_{10}O_{14}S$ Store at -15°C

Ile-Met-Gln-Val-Pro-Phe-Ser-Val

Synonyms: Melanocyte Protein PMEL 17 (209-217), (des-Asp[211],Met[210])-; Melanocyte Protein PMEL 17 (52-60), (des-Asp[54],Met[53])-

Bachem H-4362.0001   Human, bovine, mouse   MW 920.14 $C_{43}H_{69}N_9O_{11}S$ Store at -15°C

Ile-Met-Glu-Phe-Leu-Ala-Phe-Leu-His-Leu-Lys-Glu-Ala-Gly-Ala-Leu Amide

Synonyms: Gastrointestinal Peptide; Galanin Message Associated Peptide (25-41); Prepro-Galanin (89-105)

Sigma G 4396   FW 1903.3   ≥97% (HPLC) | Bioactive peptide

Ile-Phe
Z-Ile-Phe

Bachem C-2075.0001   MW 412.49 $C_{23}H_{28}N_2O_5$ Store at -15°C

Ile-Phe
Z-Ile-Phe-OMe

Bachem C-3540.0001   MW 426.51 $C_{24}H_{30}N_2O_5$ Store at -15°C

Ile-Phe

Bachem G-2420.0250   MW 278.35 $C_{15}H_{22}N_2O_3$ Store at -15°C

ICN 102088

Ile-Phe
N-CBZ-Ile-Phe

Sigma C 2127   FW 412.5 $C_{23}H_{28}N_2O_5$

Ile-Phe-Ile-Asn-Cys-Pro-Arg-Ala
β-Mercapto-β,β-Cyclopentamethylenepropionyl-D-Ile-Phe-Ile-Asn-Cys-Pro-Arg-Ala Amide

Synonyms: Vasopressin, (d(CH₂)⁵¹,D-Ile²,Ile⁴,Arg⁸,Ala-NH₂⁹)-

Bachem H-3056.0001   Store at -15°C

Ile-Phe-Ile-Asn-Cys-Pro-Arg-Gly
β-Mercapto-β,β-Cyclopentamethylenepropionyl-D-Ile-Phe-Ile-Asn-Cys-Pro-Arg-Gly Amide

Synonyms: Vasopressin, (d(CH₂)⁵¹,D-Ile²,Ile⁴,Arg⁸)-

Bachem H-2404.0001   Store at -15°C

Ile-Phe-Lys
D-Ile-Phe-Lys-pNA

Bachem L-1300.0050   MW 526.64 $C_{27}H_{38}N_6O_5$ Store at -15°C

Ile-Phe-Lys
D-Ile-Phe-Lys pNA

Synonyms: Plasmin Substrate

Sigma I 6886   FW 526.6 $C_{27}H_{38}N_6O_5$ ≥95% (HPLC); peptide content:~70% | Specific substrate for human plasmin; Szabo, GC et al, *Thromb Res*, 20:199, 1980

Ile-Phe-Lys-Lys-Glu-Asp-Cys-Lys-Tyr-Ile-Val-Val-Glu-Lys-Lys-Asp-Pro-Lys-Lys-Thr-Cys-Ser-Val-Ser-Glu-Trp-Gly-Ile

Synonyms: Inhibin β-Subunit (67-94); Inhibin (67-94), α-

Sigma I 1398   Human   FW 3299.9   ≥95% (HPLC) | Disulfide bonds: 73-87; bioactive peptide; suppressed the circulating levels of FSH in adult male rats; Mahale, SD et al, *Int J Peptide Protein Res*, 42:132, 1993

Bachem H-1602.0500   Human seminal plasma   MW 3299.9 $C_{150}H_{240}N_{36}O_{43}S_2$ Store at -15°C

Ile-Pro
FMOC-Ile-Pro

Bachem B-2135.0001   MW 450.54 $C_{26}H_{30}N_2O_5$ Store at -15°C

Ile-Pro
Z-Ile-Pro

Bachem C-3145.0250   MW 362.43 $C_{19}H_{26}N_2O_5$ Store at -15°C

Ile-Pro

Bachem G-2425.0250   MW 228.29 $C_{11}H_{20}N_2O_3$ Store at -15°C

Ile-Pro
L-trans-Epoxysuccinyl-Ile-Pro Propylamide

Synonyms: CA-074; L-trans-Epoxysuccinyl(Propylamide)-Ile-Pro; N-(L-3-trans-Propylcarbamoyl-Oxirane-2-Carbonyl)-Ile-Pro

Bachem N-1475.0001   MW 383.45 $C_{18}H_{29}N_3O_6$ Store at -15°C

Ile-Pro
L-trans-Epoxysuccinyl-Ile-Pro-OMe Propylamide

Synonyms: CA-074-OMe; L-trans-Epoxysuccinyl(Propylamide)-Ile-Pro-OMe; N-(L-3-trans-Propylcarbamoyl-Oxirane-2-Carbonyl)-Ile-Pro-OMe

Bachem N-1660.0001   Store at -15°C

Ile-Pro
N-CBZ-Ile-Pro

Sigma C 2252   FW 362.4 $C_{19}H_{26}N_2O_5$

Ile-Pro
(L-3-trans-(Propylcarbamoyl)Oxirane-2-Carbonyl)-L-Ile-L-Pro

Synonyms: CA-074; Cathepsin B Inhibitor

Peptides International IEC-4322-v   Synthetic   MW 383.44 $C_{18}H_{29}N_3O_6$ >99% (HPLC); lyophilized amorphous powder | Inhibitor; Murata, M et al, *FEBS Letts*, 280:307, 1991; Towatari, T et al, *FEBS Letts*, 280:311, 1991; Inubushi, T et al, *J Biochem*, 116:282, 1994

Ile-Pro
(L-3-*trans*-(Propylcarbamoyl)Oxirane-2-Carbonyl)-L-Ile-L-Pro Me

Synonyms: CA-074; Cathepsin B Proinhibitor

Peptides International IEC-4323-v Synthetic MW 397.47 $C_{19}H_{31}N_3O_6$ >99% (HPLC); lyophilized amorphous powder; methyl ester | Inhibitor; membrane permeable analog of CA-074; Buttle, DJ et al, *Arch Biochem Biophys*, 299:377, 1992

Ile-Pro-Arg
D-Ile-Pro-Arg *p*NA Dihydrochloride

Synonyms: Tissue Plasminogen Activator Substrate

Sigma I 0898 FW 577.5 $C_{23}H_{36}N_8O_5 \cdot 2HCl$ ≥97% (HPLC) | Chromogenic substrate for tissue plasminogen activator (t-PA); Verheijen, JH et al, *Methods of Enzymatic Analysis*, 3rd ed (Bergmeyer, J & Grassi, M, eds), Vol 5:425, 1984

Ile-Pro-Asn-Arg-Thr-Arg-His-Cys-Gln-Pro-Glu-Lys-Ala-Lys-Lys

Synonyms: TAT (102-116); SIVmac251 Peptide

Neosystem SC593 MW 1806.11 SIVmac251 peptide from subtype TAT; due to the presence of a Cys this peptide can contain the dimeric form

Ile-Pro-Glu-Pro-Tyr-Val-Trp-Asp

Synonyms: Leech Osmoregulatory Factor

Bachem H-4596.0001 Store at -15°C

Ile-Pro-Ile

Synonyms: Dipeptidyl Aminopeptidase IV Inhibitor; Diprotin A; Dipeptidyl Peptidase IV Inhibitor

Bachem H-3825.0050 MW 341.45 $C_{17}H_{31}N_3O_4$ Store at -15°C

ICN 153132 MW 341.4 $C_{17}H_{31}N_3O_4$ Umezawa, H etal, *J Antibiot*, 37:422, 1985

Sigma I 9759 FW 341.4 $C_{17}H_{31}N_3O_4$ ≥97% (HPLC) | Bioactive peptide; inhibits entry of HIV-1 or HIV-2 into T lymphoblastoid & monocytoid cell lines; Umezawa, H et al, *J Antibiot*, 37:422, 1984; Callebaut, C et al, *Science*, 262:2045, 1993

Peptides International IDP-4132-v Synthetic MW 341.45 $C_{17}H_{31}N_3O_4$ >99% (HPLC); lyophilized amorphous powder

Ile-Pro-Ile Hydrate

Synonyms: Diprotin A; Dipeptidyl Aminopeptidase IV Inhibitor

Peptides International IDP-4132 Synthetic MW 359.47 $C_{17}H_{31}N_3O_4 \cdot H_2O$ >99% (HPLC); amorphous powder; bulk | Inhibitor; Umezawa, H et al, *J Antibiotics*, 37:422, 1984

Ile-Pro-Ile Monohydrate

Synonyms: Diprotin A; Dipeptidyl Peptidase IV Inhibitor

Alexis 260-036 Synthetic MW 359.5 $C_{17}H_{31}N_3O_4 \cdot H_2O$ ≥98%; white lyophilized powder; soluble in water | Umezawa, H et al, *J Antibiot*, 37:422, 1984; Kuda, M et al, *J Biochem*, 97:1211, 1985; Bauvois, B, *Eur J Immunol*, 20:459, 1990

Ile-Pro-Ile-Tyr-Glu-Lys-Lys-Tyr-Gly-Gln-Val-Pro-Met-Cys-Asp-Ala-Gly-Glu-Gln-Cys-Ala-Val-Arg-Lys-Gly-Ala-Arg-Ile-Gly-Lys-Leu-Cys-Asp-Cys-Pro-Arg-Gly-Thr-Ser-Cys-Asn-Ser-Phe-Leu-Leu-Lys-Cys-Leu

Synonyms: Cocaine- & Amphetamine-Regulated Transcript (55-102); Food Intake Inhibitor; Anorectic Peptide, New

Peptides International PCA-4350-s, 4351-s Human synthetic MW 5245.2 $C_{225}H_{365}N_{65}O_{65}S_7$ >95% (HPLC); lyophilized amorphous powder | Bioactive; Disulfide bonds: Cys[68]-Cys[86], Cys[74]-Cys[94], Cys[88]-Cys[101]); Kristensen, P et al, *Nature*, 393:72, 1998; Douglass, J & S Daoud, *Gene*, 169:241, 1996

Bachem H-4446.0100 Rat MW 5259.26 $C_{226}H_{367}N_{65}O_{65}S_7$ Store at -15°C | Disulfide bonds: Cys[74]-Cys[94], Cys[68]-Cys[86], Cys[88]-Cys[101]

Ile-Pro-Met-Phe-Ser-Ala-Leu-Ser-Glu-Gly-Ala-Thr-Pro-Gln-Asp-Leu

Synonyms: P25 (169-184); GAG P24 CA (37-52); HTLV I Peptide

Neosystem SC294 MW 1676.90 HTLV I peptide from subtype P25 (GAG P24 CA)

Ile-Pro-Val-Gly-Glu-Ile-Tyr-Lys-Arg-Trp-Ile-Ile-Leu-Gly-Leu

Synonyms: P25 (256-270); GAG P24 CA (124-138); HTLV I Peptide

Neosystem SC301 MW 1770.19 HTLV I peptide from subtype P25 (GAG P24 CA)

Ile-Pro-Val-Lys-Gln-Ala-Asp-Ser-Gly-Ser-Ser-Cys Acetate Salt

Synonyms: Osteopontin (1-11), ~C

Biogenesis 7060-9031 Human synthetic Semi-pure; lyophilized

Ile-Ser
Z-Ile-Ser

Bachem C-2080.0001 MW 352.39 $C_{17}H_{24}N_2O_6$ Store at -15°C

Ile-Ser

Bachem G-2430.0250 MW 218.25 $C_9H_{18}N_2O_4$ Store at -15°C

Ile-Ser-Arg-Pro-Pro-Gly-Phe-Ser-Pro-Phe-Arg

Synonyms: Bradykinin, (Ile-Ser[0])-; T-Kinin

American Peptide 18-1-34 MW 1260.5 $C_{59}H_{89}N_{17}O_{14}$ Okamoto, H et al, *BBRC*, 112:701, 1983

Peptides International PBK-4130-v Rat synthetic MW 1260.4 $C_{59}H_{89}N_{17}O_{14}$ >99% (HPLC); lyophilized amorphous powder | Bioactive; Okamoto, H & LM Greenbaum, *BBRC*, 112:701, 1983

Ile-Ser-Arg-Pro-Pro-Gly-Phe-Ser-Pro-Phe-Arg 2AcOH 5H₂O

Synonyms: Bradykinin, (Ile-Ser)-; T-Kinin

Peptides International PBK-4130 Rat synthetic MW 1470.5 $C_{59}H_{89}N_{17}O_{14} \cdot 2CH_3COOH \cdot 5H_2O$ >99% (HPLC); lyophilized amorphous powder; bulk | Bioactive; Okamoto, H & LM Greenbaum, *BBRC*, 112:701, 1983

Ile-Ser-Arg-Pro-Pro-Gly-Phe-Ser-Pro-Phe-Arg Acetate Salt

Synonyms: Bradykinin, (Ile-Ser)-; T-Kinin

ICN 152755 Okamoto, H & M Greenbaum, *BBRC*, 112:701, 1983

Sigma B 1643 1260.5 (free base) ≥95% (HPLC) | Bioactive peptide; agonist; Okamoto, H et al, *Biochem Biophys Res Commun*, 112:701, 1983

Ile-Ser-Gln-Ala-Val-His-Ala-Ala-His-Ala-Glu-Ile-Asn-Glu-Ala-Gly-Arg

Synonyms: Ovalbumin (323-339)

Neosystem SC1303 MW 1773.92 Virus-related peptide; bioactive

Ile-Ser-Ile-Asn-Gln-Asp-Leu-Lys-Ala-Ile-Thr-Asp-Met-Leu-Leu-Thr-Glu-Gln-Ile-Arg-Glu-Arg-Gln-Arg-Tyr-Leu-Ala-Asp-Leu-Arg-Gln-Arg-Leu-Leu-Glu-Lys Amide

Synonyms: Egg Laying Hormone

Bachem H-3282.0500   Aplysia california   MW 4384.13
$C_{190}H_{329}N_{59}O_{57}S$ Store at -15°C

American Peptide 79-1-19   *Aplysia californica*   MW 4384.2
$C_{190}H_{329}N_{59}O_{57}S$ Neurotransmitter on cells of the abdominal ganglion; Scheller, RH et al, *Cell*, 32:7, 1993

Ile-Ser-Leu-Asn-Lys-Tyr-Tyr-Asn-Leu-Thr-Met-Lys-Cys-Arg-Arg-Pro-Gly-Asn-Lys-Thr

Synonyms: GP140 (301-320); SIVmac251 Peptide

Neosystem SC738   MW 2400.83   SIVmac251 peptide from subtype GP140; due to the presence of a Cys this peptide can contain the dimeric form

Ile-Ser-Leu-Asp-Leu-Thr-Phe-His-Leu-Leu-Arg-Glu-Val-Leu-Glu-Met-Ala-Arg-Ala-Glu-Gln-Leu-Ala-Gln-Gln-Ala-His-Ser

Synonyms: Corticotropin Releasing Factor (6-33)

Bachem H-3456.0500   Human, rat   Store at -15°C

Sigma C 0961   Human, rat   FW 3220.7   ≥97% (HPLC) | Bioactive peptide; CRF fragment with high affinity to CRF-binding protein (CRF-BP) with very low affinity to CRF-R; displaces CRF from CRF-BP; has cognition-enhancing effects in learning & memory models in animals

Ile-Thr
FMOC-Ile-Thr(ψ(Me,Me)pro)

Synonyms: (4S,5R)-3-(FMOC-Ile)-2,2,5-Trimethyl-Oxazolidine-4-Carboxylic Acid

Bachem B-3440.0001   MW 494.59 $C_{28}H_{34}N_2O_6$ Store at -15°C

Ile-Thr-Asp-Gln-Val-Pro-Phe-Ser-Val

Synonyms: Melanocyte Protein PMEL 17 (209-217); Melanocyte Protein PMEL 17 (52-60)

Bachem H-4106.0001   Human, bovine, mouse   Store at -15°C

Ile-Thr-Gln-Val-Pro-Phe-Ser-Val

Synonyms: Melanocyte Protein PMEL 17 (209-217), (des-Asp[211])-; Melanocyte Protein PMEL 17 (52-60), (des-Asp[54])-

Bachem H-4364.0001   Human, bovine, mouse   MW 890.05
$C_{42}H_{67}N_9O_{12}$ Store at -15°C

Ile-Thr-Leu-Pro-Cys-Arg-Ile-Lys-Gln-Phe-Ile-Asn
Ile-Thr-Leu-Pro-Cys(Acm); HTLV I Peptide-Arg-Ile-Lys-Gln-Phe-Ile-Asn

Synonyms: GP120 (419-430); HTLV I Peptide

Neosystem SC563   MW 1516.86   HTLV I peptide from subtype GP160

Ile-Thr-Ser-Phe-Glu-Glu-Ala-Lys-Gly-Leu-Asp-Arg-Ile-Asn-Glu-Arg-Met-Pro-Pro-Arg-Arg-Asp-Als-Met-Pro

Synon Calcineurin Autoinhibitory Fragment *yms*:

Alexis 153-002   MW 2930.4 $C_{124}H_{205}N_{39}O_{39}S_2$ Potent inhibitor of calcineurin (protein phosphatase 2B) corresponding to the C-terminal region of calcineurin; peptide does not interfere with calmodulin binding, does not inhibit Ca^{2+}/calmodulin-dependent protein kinase II but inhibits the catalytic subunits of protein phosphatases 1 & 2A; inhibits Mn^{2+}-stimulated calcineurin activity, but has no effect on Ni^{2+}-stimulated activity; Hashimoto, Y et al, *J Biol Chem*, 265:1924, 1990; Hendley, B et al, *Science*, 258:296, 1992; Yokohama, Y & Wang, JH, *FEBS Lett*, 337:128, 1994; Berrino, BA et al, *J Biol Chem*, 270:340, 1995

Ile-Thr-Ser-Phe-Glu-Glu-Ala-Lys-Gly-Leu-Asp-Arg-Ile-Asn-Glu-Arg-Met-Pro-Pro-Arg-Arg-Asp-Ala-Met-Pro

Synonyms: Calcineurin Autoinhibitory Peptide ; Calcineurin Inhibitor ; Calcineurin Autoinhibitory Fragment

American Peptide 22-2-15   MW 2930.4 $C_{124}H_{205}N_{39}O_{39}S_2$
Inhibits Mn^{2+}-stimulated calcineurin activity using (^{32}P)-Myosin light chain as a substrate but has no effect on Ni^{2+}-stimulated activity; AA sequence is identical with the C-terminal region of the calmodulin-binding domain; Hashimoto, Y et al, *JBC*, 265:1924, 1990

Bachem H-8910.0500   MW 2930.36 $C_{124}H_{205}N_{39}O_{39}S_2$ Store at -15°C

Calbiochem 207000   MW 2930.4 $C_{124}H_{205}N_{39}O_{39}S_2$ ≥97% (HPLC); lyophilized; soluble in H_2O | Corresponds to the C-terminal domain (457-482) of the calmodulin-binding domain of calcineurin; specific calcineurin inhibitor; does not inhibit CaM kinase II, protein phosphatase 1 or 2A; inhibits Mn^{2+}-stimulated calcineurin activity but has no effect on Ni^{2+}-stimulated activity; Perrino, BA et al, *J Biol Chem*, 270:340, 1995; Yokoyama, Y & Wang, JH, *FEBS Lett*, 337:128, 1994; Hashimoto, Y et al, *J Biol Chem*, 265:1924, 1990

ICN 195820   MW 2930.7   ≥97% | Inhibits calcineurin & Mn^{2+} stimulated calcineurin activity but has no effect on Ni^{2+} stimulated activity

Neosystem SC937   MW 2930.34   Gives complete inhibition of the protein phosphatase activity of calcineurin; this sequence constitutes the autoinhibitory domain which interacts with the active site & is responsible for the low basal phosphatase activity in the absence of Ca^{2+}/calmodulin; represents a minimal sequence; Hashimoto, Y et al, *JBC*, 265:1924-1927, 1990; Perrino, BA et al, *JBC*, 270:340-346, 1995; Chaudhuri, B et al, *FEBS Lett*, 357:221-226, 1995

Sigma C 3937   FW 2930.3   ≥90% (HPLC) | Bioactive peptide; inhibits the phosphatase activity of calcineurin; Hashimoto, Y et al, *J Biol Chem*, 265:1924, 1990

Ile-Trp
Z-Ile-Trp

Bachem C-2085.0001   MW 451.52 $C_{25}H_{29}N_3O_5$ Store at -15°C

Ile-Trp

Bachem G-2435.0250   MW 317.39 $C_{17}H_{23}N_3O_3$ Store at -15°C

Ile-Trp
1-Naphthalenylsulfonyl-Ile-Trp-CHO

Bachem N-1760.0005   Store at -15°C

Ile-Trp-Val-Asn

Bachem H-7170.0005   MW 530.63 $C_{26}H_{38}N_6O_6$ Store at -15°C

Ile-Tyr
Z-Ile-Tyr

Bachem C-2090.0001   MW 428.49 $C_{23}H_{28}N_2O_6$ Store at -15°C

Ile-Tyr

Bachem G-2440.0250   MW 294.35 $C_{15}H_{22}N_2O_4$ Store at -15°C

Ile-Tyr-Gly-Glu-Phe
Ac-Ile-Tyr-Gly-Glu-Phe Amide

Synonyms: Protein Tyrosine Kinase pp60c-*src* Substrate; p60c-*src* Substrate II

Alexis 151-028   MW 668.7 $C_{33}H_{43}N_5O_{10}$ ≥98%; white lyophilized powder | Cantley, LC et al, *Cell*, 64:281, 1991; Nair, SA et al, *J Med Chem*, 38:4276, 1995

American Peptide 86-2-33   MW 668.8 $C_{33}H_{44}N_6O_9$
Excellent small peptide substrate for the protein tyrosine kinase pp60c-*src*; Nair, SA et al, *J Med Chem*, 38:4276, 1995

Ile-Tyr-Gly-Glu-Phe
Ac-Ile-Tyr(PO$_3$H$_2$)-Gly-Glu-Phe Amide

Synonyms: p60c-*src* Substrate II, Phosphorylated

American Peptide 86-2-34   MW 748.8 $C_{33}H_{45}N_6O_{12}P$
Product of the phosphorylation of P60c-*src* Substrate II; Nair, SA et al, *J Med Chem*, 38:4276, 1995

Ile-Tyr-Gly-Glu-Phe
Ac-Ile-Tyr-Gly-Glu-Phe Amide

Bachem M-2165.0005   Store at -15°C

Ile-Tyr-Gly-Glu-Phe
Ac-Ile-Tyr(PO$_3$H$_2$)-Gly-Glu-Phe Amide

Bachem M-2170.0001   Store at -15°C

Ile-Tyr-Gly-Glu-Phe
Ac-Ile-Tyr-Gly-Glu-Phe Amide

Synonyms: p60c-*src* Substrate II

Calbiochem 567812   MW 668.7 $C_{33}H_{44}N_6O_9$ ≥97% (HPLC); solid; soluble in 0.1% TFA | Excellent pentapeptide substrate for p60c-*src*; Nair, SA et al, *J Med Chem*, 38:4276, 1995

Ile-Tyr-Gly-Glu-Phe
Ac-Ile-Tyr(PO$_3$H$_2$)-Gly-Glu-Phe Amide

Synonyms: p60c-*src* Substrate II, Phosphorylated

Calbiochem 567814   MW 748.7 $C_{33}H_{45}N_6O_{12}P$ ≥95% (HPLC); solid; soluble in 0.1% TFA | Product of the phosphorylation of p60c-*src* substrate II by the protein tyrosine kinase p60c-*src*; Nair, SA et al, *J Med Chem*, 38:4276, 1995

Ile-Tyr-Leu-Gly-Gly-Pro-Phe-Ser-Pro-Asn-Val-Leu

Synonyms: C-Reactive Protein (174-185)

Bachem H-1344.0001   MW 1276.5 $C_{62}H_{93}N_{13}O_{16}$ Store at -15°C

Sigma C 3705   FW 1276.5   ≥97% (HPLC) | Bioactive peptide; enhances tumoricidal activity of human monocytes & alveolar macrophages *in vitro*; Thomassen, MJ et al, *J Immunotherapy*, 13:1, 1993

Ile-Tyr-Pro-Arg-Tyr

Synonyms: Angiotensin Converting Enzyme Inhibiting Peptide

Sigma I 3771   FW 710.8 $C_{35}H_{50}N_8O_8$ ≥97% (HPLC) | Bioactive peptide; reduces blood pressure in hypertensive rats; Saito, Y et al, *Biosci Biotech Biochem*, 58:1767, 1994

Ile-Val
Z-Ile-Val

Bachem C-2095.0001   MW 364.44 $C_{19}H_{28}N_2O_5$ Store at -15°C

Ile-Val

Bachem G-2445.0250   MW 230.31 $C_{11}H_{22}N_2O_3$ Store at -15°C

ICN 102091

Ile-Val-Cys-His-Thr-Thr-Ala-Thr-Ser-Pro-Ile-Ser-Ala-Val-Thr-Cys-Pro-Pro-Gly-Glu-Asn-Leu-Cys-Tyr-Arg-Lys-Met-Trp-Cys-Asp-Ala-Phe-Cys-Ser-Ser-Arg-Gly-Lys-Val-Val-Glu-Leu-Gly-Cys-Ala-Ala-Thr-Cys-Pro-Ser-Lys-Lys-Pro-Tyr-Glu-Glu-Val-Thr-Cys-Cys-Ser-Thr-Asp-Lys-Cys-Asn-Pro-His-Pro-Lys-Gln-Arg-Pro-Gly

Synonyms: Bungarotoxin, α-

Calbiochem 203980   *Bungarus multicinctus*   MW 7984.3 $C_{338}H_{529}N_{97}O_{105}S_{11}$ Lyophilized; soluble in H_2O; no detectable contaminants (IEF); Disulfide bonds: Cys[3]-Cys[23], Cys[16]-Cys[44]; Cys[29]-Cys[33]; Cys[48]-Cys[59]; Cys[60]-Cys[65]; LD$_{50}$≤2000 mg/kg | Blocks neuromuscular transmission by irreversible binding to motor end-plate accholine receptor but does not depress accholine release from motor nerve endings; blocks nicotine-induced increase of intracellular Ca^{2+} in PC12 cells; prevents opening of nicotinic receptor-associated ion channels; Zhang, ZW et al, *Neuron*, 12:167, 1994; *Merck Index*, 12:1513

Ile-Val-Gln-Gln-Phe-Gly-Phe-Gln-Arg-Arg-Ala-Ser-Asp-Asp-Gly-Lys-Leu-Thr-Asp

Synonyms: c-*raf* Kinase

Promega V2201   MW 2181   References are the same as for Promega V5601

Ile-Val-Gln-Pro-Ile-Ile-Ser-Lys-Leu-Tyr-Gly-Ser-Gly-Gly-Pro-Pro-Pro-Thr-Gly-Glu-Glu-Asp-Thr-Asp-Glu-Lys-Lys-Asp-Glu-Leu

Synonyms: Steroidogenesis-Activator Polypeptide

Bachem H-8855.0500   Rat   MW 3213.54 $C_{141}H_{226}N_{34}O_{51}$ Store at -15°C

Ile-Val-Pro-Phe-Leu-Gly-Pro-Leu-Leu-Gly-Leu-Leu-Thr Amide

Synonyms: Icaria Chemotactic Peptide; Chemotactic Peptide; Chemotactic Peptide, Icaria

ICN 195642   MW 1351.7   97%

Sigma I 3516   FW 1351.7   ≥97% (HPLC) | Bioactive peptide

Leu-Ala
Z-Leu-Ala

Bachem C-2130.0001   MW 336.39 $C_{17}H_{24}N_2O_5$ Store at -15°C

Leu-Ala

Bachem G-2460.0001   MW 202.25 $C_9H_{18}N_2O_3$ Store at -15°C

Leu-Ala
DL-Leu-DL-Ala

Bachem G-2470.0001   MW 202.25 $C_9H_{18}N_2O_3$ Store at -15°C

Leu-Ala
Leu-Ala-βNA

Bachem K-1395.0250   MW 327.43 $C_{19}H_{25}N_3O_2$ Store at -15°C

Leu-Ala
Z-L-Leu-L-Ala
Fluka 96750   MW 336.40 $C_{17}H_{24}N_2O_5$ ≥99.0% (titration); mp:150-152°C

Leu-Ala
DL-Leu-DL-Ala
ICN 102167   MW 202.3 $C_9H_{18}N_2O_3$ 99%; crystalline
ICN 102168   MW 202.3 $C_9H_{18}N_2O_3$ Crystalline

Leu-Ala
Leu-β-Ala
ICN 155185   MW 202.3 $C_9H_{18}N_2O_3$ Crystalline

Leu-Ala
N-CBZ-Leu-Ala
Sigma C 0170   FW 336.4 $C_{17}H_{24}N_2O_5$

Leu-Ala
Leu-β-Ala
Sigma L 3002   FW 202.3 $C_9H_{18}N_2O_3$

Leu-Ala
DL-Leu-DL-Ala
Sigma L 4627   FW 202.3 $C_9H_{18}N_2O_3$

Leu-Ala
Sigma L 9250   FW 202.3 $C_9H_{18}N_2O_3$

Leu-Ala
CBZ-Leu-Ala
USBio C2099-01   MW 336.4 $C_{17}H_{24}N_2O_5$ ≥99%

Leu-Ala Amide Hydrochloride
Bachem G-2475.0250   MW 237.73 $C_9H_{19}N_3O_2$ · HCl   Store at -15°C

Leu-Ala-Gln-Ala-Val-Arg-Ser-Ser-Ser-Arg
DABCYL-Leu-Ala-Gln-Ala-Val-Arg-Ser-Ser-Ser-Arg-EDANS
Synonyms: Tumor Necrosis Factor α-EDANS (-4 to +6), DABCYL-
Bachem M-2155.0001   Human   Store at -15°C

Leu-Ala-Glu-Glu-Glu-Val-Val-Ile-Arg-Ser-Ala-Asn-Phe-Thr-Asp-Asn
Synonyms: GP120 (270-285); HTLV I Peptide
Neosystem SC265   MW 1806.94   HTLV I peptide from subtype GP160

Leu-Ala-Gly
Cl-Ac-(OH)Leu-Ala-Gly Amide
Synonyms: Elastase Inhibitor
Peptides International ICL-4146-v   *P. aeruginosa* synthetic   MW 350.80 $C_{13}H_{23}N_4O_5Cl$ >99% (HPLC); lyophilized amorphous powder | Inhibitor; Nishino, N & JC Powers, *JBC*, 255:3482, 1980

Leu-Ala-Gly
N-Chloro-Ac-N-Hydroxy-L-Leu-L-Ala-Gly Amide
Alexis 260-006   Synthetic   MW 350.8 $C_{13}H_{23}N_4O_5$ White powder; soluble in methanol | Inhibitor of P. *aeruginosa* elastase; Nishino, N & Powers, JC, *J Biol Chem*, 255:3482, 1980

Leu-Ala-Gly
Cl-Ac-(OH)Leu-Ala-Gly Amide
Synonyms: Elastase Inhibitor
Peptides International ICL-4146   Synthetic   MW 350.80 $C_{13}H_{23}N_4O_5Cl$ >99% (HPLC); amorphous powder; bulk | Inhibitor; Nishino, N & JC Powers, *JBC*, 255:3482, 1980

Leu-Ala-His-Gln-Ile-Tyr-Gln-Phe-Thr-Asp-Lys-Asp-Lys-Asp-Asn-Val-Ala-Pro-Arg-Ser-Lys-Ile-Ser-Pro-Gln-Gly-Tyr Amide
Synonyms: Adrenomedullin (26-52)
Bachem H-4138.0500   Human   MW 3119.49 $C_{139}H_{216}N_{40}O_{42}$ Store at -15°C

Leu-Ala-Leu
Trt-Leu-Ala-Leu
Bachem H-5090.0250   Store at -15°C

Leu-Ala-Leu-Ala
Leu-Ala-Leu-Ala-OEt Hydrochloride
Bachem H-3855.0050   MW 451.01 $C_{20}H_{38}N_4O_5$ · HCl   Store at -15°C

Leu-Ala-Leu-Ala-Asp-Arg-Ile-Tyr-Ser-Phe-Pro-Asp-Pro-Pro-Thr
Synonyms: REV (52-66); SIVmac251 Peptide
Neosystem SC601   MW 1675.90   SIVmac251 peptide from subtype REV

Leu-Ala-Pro
Bachem H-3860.0250   MW 299.37 $C_{14}H_{25}N_3O_4$ Store at -15°C

Leu-Ala-Trp-Lys-Phe-Asp-Pro-Thr-Leu-Ala-Tyr-Thr-Tyr-Glu-Ala
Synonyms: NEF (211-225); SIVmac251 Peptide
Neosystem SC621   MW 1789.01   SIVmac251 peptide from subtype NEF

Leu-Arg
Bachem G-2480.0250   MW 287.36 $C_{12}H_{25}N_5O_3$ Store at -15°C

Leu-Arg
Z-Leu-Arg-4MβNA
Bachem J-1390.0050   Store at -15°C

Leu-Arg
Z-Leu-Arg-βNA Hydrochloride
Bachem K-1205.0050   Store at -15°C

Leu-Arg
Z-Leu-Arg-MCA Hydrochloride
Synonyms: Cathepsin K/S/V Substrate
Peptides International MCA-3210-v   MW 578.66 $C_{30}H_{38}N_6O_6$ >99% (HPLC); lyophilized amorphous powder | Brömme, D et al, *JBC*, 271:2126, 1996; Brömme, D et al, *Biochem*, 38:2377, 1999

Leu-Arg Acetate Salt
ICN 155186   MW 287.4 $C_{12}H_{25}N_5O_3$
Sigma L 2634   FW 287.4 (free base) $C_{12}H_{25}N_5O_3$

Leu-Arg-Ala-Arg-Gly-Glu-Thr-Tyr-Gly-Arg-Leu-Leu-Gly-Glu-Val

Synonyms: NEF (21-35); SIVmac251 Peptide

Neosystem SC605 MW 1689.93 SIVmac251 peptide from subtype NEF

Leu-Arg-Ala-His-Ala-Val-Asp-Val-Asn-Gly Amide

Synonyms: Cadherin Peptide

American Peptide 79-1-20 Avian MW 1050.2
$C_{44}H_{75}N_{17}O_{13}$ Blaschuk, OW et al, *Dev Biol*, 139:227, 1990

Leu-Arg-Ala-Met-Thr-Tyr-Lys-Leu-Ala-Ile-Asp-Met-Ser-His-Phe-Ile

Synonyms: NEF (108-123); SIVmac251 Peptide

Neosystem SC613 MW 1909.33 SIVmac251 peptide from subtype NEF

Leu-Arg-Arg
BOC-Leu-Arg-Arg-AMC Hydrochloride

Bachem I-1585.0050 MW 737.3 $C_{33}H_{52}N_{10}O_7 \cdot$ HCl Store at -15°C

Leu-Arg-Arg
BOC-Leu-Arg-Arg-MCA

Synonyms: Carboxyl Side of Paired Basic Residue Cleaving Enzyme Substrate; Proteasome Substrate

Peptides International MLR-3140-v MW 700.84
$C_{33}H_{52}N_{10}O_7$ >98% (HPLC); lyophilized amorphous powder |
Mizuno, K et al, *BBRC*, 144:807, 1987; Aki, M et al, *J Biochem*, 115:257, 1994

Leu-Arg-Arg
N-t-BOC-Leu-Arg-Arg MCA Dihydrochloride

Synonyms: Protease Substrate

Sigma B 5403 FW 773.8 $C_{33}H_{52}N_{10}O_7 \cdot$ 2HCl ~95% |
Substrate for proteases which cleave on the carboxyl side of paired basic residues; Mizuno, K et al, *Biochem Biophys Res Commun*, 144:807, 1987

Leu-Arg-Arg-Ala-Ser-Leu-Gly

Synonyms: Kemptide; Phosphate Acceptor Peptide; cAMP-Dependent Protein Kinase Substrate; Protein Kinase A; S6 Kinase Substrate

Alexis 160-006 MW 769.9 $C_{32}H_{61}N_{14}O_8$ ≥98%; white lyophilized powder; soluble in water | AA sequence defines the phosphorylation site of porcine liver pyruvate kinase; Kemp, BE et al, *Fed Proc*, 35:1384, 1976; Kemp, BE et al, *J Biol Chem*, 252:4888, 1977; Pomerantz, AH et al, *PNAS*, 74:4261, 1977; Kemp, BE & Clark, MG, *J Biol Chem*, 253:5147, 1978; Maller, JL et al, *PNAS*, 75:248, 1978; Feramisco, JR et al, *J Biol Chem*, 255:4240, 1980

American Peptide 86-0-50 MW 771.9 $C_{32}H_{61}N_{13}O_9$ Protein Kinase Substrate (cAMP); a synthetic peptide substrate corresponding to part of the phosphorylation site sequence in porcine liver pyruvate kinase; Kemp, BE et al, *Fed Proc*, 35:1384, 1976

Bachem M-1510.0005 MW 771.92 $C_{32}H_{61}N_{13}O_9$ Store at -15°C

Calbiochem; 05-23-4900 MW 771.9 $C_{32}H_{61}N_{13}O_9$ >97% (HPLC); lyophilized solid; soluble in 5% acetic acid | Phosphate acceptor peptide; synthetic substrate for PKA; for steady state kinetic analysis of PKA in quench flow techniques; Zelada, A et al, *Eur J Biochem*, 252:245, 1998; Grant, BD & Adams, JA, *Biochemistry*, 35:2022, 1996; Kemp, BE et al, *J Biol Chem*, 252:4888, 1977; Kemp, BE et al, *Fed Proc*, 35:1384, 1976

Fluka; 60645 MW 771.9 $C_{32}H_{61}N_{13}O_9$ ≥98% (HPLC); ≤5% water; ≤15% acetate; ≥80% peptide content | Substrate for cAMP-dependent protein kinase; Kubler, D et al, *JBC*, 264:14549, 1989; Slice, LW & Taylor, SS, *JBC*, 264:20940, 1989

ICN 151389 Kemp, BE etal, *Fed Proc*, 35:1384, 1976

Neosystem SC408 MW 771.92 Bioactive; proteinkinase-related peptide; Kemp, B, *Fed Proc*, 35:1384, 1976; Kemp, B et al, *JBC*, 252:4888, 1977

Promega; V5601 MW 772 Kemp, BE et al, *JBC*, 252:4888, 1977; Casnellie, JE, *Meth* Enzymol, 200:115, 1991; Chen, S-J et al, *Biochemistry*, 32:1032, 1993; Lees-Miller, S et al, *Mol Cell Biol*, 12:5041, 1992; Beaudette, K et al, *JBC*, 268:20825, 1993; App, H et al, *Mol Cell Biol*, 11:913, 1991; Colbran, JL et al, *JBC*, 267:9589, 1992; Agostinis, P et al, *FEBS Lett*, 259:75, 1989; Kuenzel, EA & Krebs, EG, *PNAS*, 82:737, 1985

Upstate 12-257 MW 771 Lyophilized

USBio K0150 HPLC purified; lyophilized | Phosphorylated as efficiently as natural protein substrates

Leu-Arg-Arg-Ala-Ser-Leu-Gly Acetate Salt

Synonyms: Protein Kinase Related Peptide; Kemptide; Phosphate Acceptor Peptide; Protein Kinase A Substrate

Sigma K 1127 Synthetic FW 771.9 (free base) $C_{32}H_{61}N_{13}O_9$ ≥97% (HPLC) | Bioactive peptide; Kemp, BE et al, *Fed Proc*, 35:1384, 1976; Maller, JE et al, *Proc Natl Acad Sci USA*, 75:248, 1978

Leu-Arg-Arg-Ala-Ser-Leu-Gly Amide

Synonyms: Kemptamide; Kemptide

Alexis 160-002 MW 770.9 $C_{32}H_{62}N_{14}O_8$; | Carboxy-terminal amide analog of Kemptide

American Peptide 86-0-58 MW 771.0 $C_{32}H_{62}N_{14}O_8$

Leu-Arg-Arg-Ala-Ser-Val-Ala

Synonyms: cAMP-Dependent Protein Kinase Substrate; Kemptide, (Val[6],Ala[7])-; Protein Kinase Related Peptide

Alexis 160-007 MW 771.9 $C_{32}H_{61}N_{13}O_9$ ≥98%; white lyophilized powder; soluble in water | Peptide sequence defines the phosphorylation site of rat liver pyruvate kinase; Zetterqvist, O et al, *BBRC*, 70:696, 1976

Bachem M-1515.0005 MW 771.92 $C_{32}H_{61}N_{13}O_9$ Store at -15°C

ICN 153136 Active substrate for PKA

Sigma K 2877 FW 771.9 $C_{32}H_{61}N_{13}O_9$ ≥97% (HPLC) | Bioactive peptide; phosphorylation site of liver pyruvatekinase; substrate for proteinkinase A; Edlund, B et al, *Biochem Biophys Res Commun*, 79:139, 1977; Berglund, L et al, *J Biol Chem*, 252:613, 1977

Leu-Arg-Arg-Arg-Arg-Phe-Ala-Phe-Cys
Leu-Arg-Arg-Arg-Arg-Phe-D-Ala-Phe-Cys(NPys) Amide

Bachem H-3696.0001 Store at -15°C

Leu-Arg-Arg-Asp-Leu-Asp-Ala-Ser-Arg-Glu-Ala-Lys-Lys-Gln-Val-Glu-Lys-Ala-Leu-Glu

Synonyms: M Protein Epitope of Group A Streptococci; Group A Streptococcal M Protein Peptide

Bachem H-2488.0500 Store at -15°C

Sigma L 7783 FW 2355.7 ≥97% (HPLC); peptide content:~65% | Bioactive peptide from the conserved region of Group A streptococcal M protein; useful as an antigen to produce a rheumatic fever vaccine; Pruksakorn, S et al, *Lancet*, 344:639, 1994

Leu-Arg-Arg-Trp-Ser-Leu-Gly

Synonyms: Kemptide, (Trp[4])-; cAMP-Dependent Protein Kinase Substrate; Protein Kinase Related Peptide; Phosphoprotein Phosphatase Substrate

Alexis 160-008 MW 887.1 $C_{40}H_{66}N_{14}O_9$ ≥98%; white lyophilized powder; soluble in water | Phosphorylated product is an active substrate for phosphoprotein phosphatase; Wright, DE et al, *PNAS*, 78:6048, 1981

Bachem M-1525.0005 MW 887.05 $C_{40}H_{66}N_{14}O_9$ Store at -15°C

ICN 159614 Substrate for cAMP-dependent proteinkinase; Wright, DE etal, *PNAS*, 78:6048, 1981

Sigma L 3523 FW 887.0 $C_{40}H_{66}N_{14}O_9$ ≥97% (HPLC) | Bioactive peptide; peptide fluorescence intensity changes as a function of its phosphorylation state; Wright, DE et al, *Proc Natl Acad Sci USA*, 78:6048, 1981

Leu-Arg-Arg-Val-Arg-Glu-Val-Leu-Arg-Thr-Glu-Leu-Thr-Tyr-Leu-Gln-Tyr-Gly-Trp-Ser

Synonyms: GP140 (811-830); SIVmac251 Peptide

Neosystem SC789 MW 2538.93 SIVmac251 peptide from subtype GP140

Leu-Arg-Asn
L-β-Phenyllactoyl-Leu-Arg-Asn Amide Hydrochloride

Synonyms: Antho-RN Amide

Bachem H-7920.0005 Store at -15°C

Leu-Arg-Gln-Ser-Gln-Phe-Val-Gly-Ser-Arg Amide

Synonyms: Urechistachykinin I

American Peptide 87-7-50 MW 1176.3 $C_{50}H_{85}N_{19}O_{14}$ Ikeda, T et al, *BBRC*, 192:1, 1993

Bachem H-1632.0001 MW 1176.35 $C_{50}H_{85}N_{19}O_{14}$ Store at -15°C

Leu-Arg-Gly-Gly
Z-Leu-Arg-Gly-Gly-AMC

Bachem I-1685.0025 Store at -15°C

Leu-Arg-Gly-Gly
Z-Leu-Arg-Gly-Gly-MCA Hydrochloride

Synonyms: Proteasome Substrate

Peptides International MLG-3176-v MW 692.77 $C_{34}H_{44}N_8O_9$ >98% (HPLC); lyophilized amorphous powder | Stein, RL et al, *Biochem*, 34:12616, 1995

Leu-Arg-His-Tyr-Leu-Asn-Leu-Leu-Thr-Arg-Gln-Arg-Tyr
Ac-Leu-Arg-His-Tyr-Leu-Asn-Leu-Leu-Thr-Arg-Gln-Arg-Tyr Amide

Synonyms: Neuropeptide Y (24-36), *N*-Ac-(Leu[28], Leu[31])-; Neuropeptide Y (24-36), Ac-(Leu[28,31])-

Neosystem SC1269 MW 1787.10 Bioactive neuropeptide; Potter, EK et al, *Eur J Pharmacol*, 267:253-262, 1994; Lacroix, JS et al, *Br J Pharmacol*, 113:479-484, 1994; Tracey, DJ et al, *Brain Res*, 669:245-254, 1995; Mccloskey, MJD et al, *Neuropeptides*, 31:193-197, 1997

American Peptide 60-1-19 Human MW 1787.1 $C_{81}H_{131}N_{27}O_{19}$ Selective Y_2 agonist

---

Leu-Arg-Hyp-Gly Amide

Synonyms: Luteinizing Hormone Releasing Hormone (7-10), (Hyp[9])-

Neosystem SC477 MW 456.54 Bioactive; hypothalamic releasing hormone

---

Leu-Arg-Leu

Bachem H-3865.0025 MW 400.52 $C_{18}H_{36}N_6O_4$ Store at -15°C

Leu-Arg-Leu-Thr-Val-Trp-Gly-Thr-Lys-Asn-Leu-Gln-Thr-Arg-Val-Thr-Ala-Ile-Glu-Lys

Synonyms: GP140 (581-600); SIVmac251 Peptide

Neosystem SC766 MW 2327.75 SIVmac251 peptide from subtype GP140

Leu-Arg-Lys-Arg-Leu-Arg-Leu-Ile-His-Leu-Leu-His-Gln-Thr

Synonyms: REV (10-23); SIVmac251 Peptide

Neosystem SC597 MW 1797.22 SIVmac251 peptide from subtype REV

Leu-Arg-Pro

Bachem H-1274.0100 MW 384.48 $C_{17}H_{32}N_6O_4$ Store at -15°C

Leu-Arg-Pro-Gly Amide

Synonyms: Luteinizing Hormone Releasing Hormone (7-10)

American Peptide 54-0-26 MW 440.6 $C_{19}H_{36}N_8O_4$

Leu-Arg-Pro-Gly Amide Dihydrochloride

Synonyms: Luteinizing Hormone Releasing Hormone (7-10)

Bachem H-3870.0250 MW 513.47 $C_{19}H_{36}N_8O_4$ · 2HCl Store at -15°C

Sigma L 3398 FW 440.5 (free base) $C_{19}H_{36}N_8O_4$ ≥97% (HPLC) | Bioactive peptide

---

Leu-Arg-Pro-Gly-Gly-Lys-Lys-Lys-Tyr-Lys-Leu-Lys-His-Ile-Val

Synonyms: P18 (21-35); GAG P17 MA (21-35); HTLV I Peptide

Neosystem SC279 MW 1765.21 HTLV I peptide from subtype P25 (GAG P17 MA)

Leu-Asn

Bachem G-2485.0250 MW 245.28 $C_{10}H_{19}N_3O_4$ Store at -15°C

ICN 155187 MW 245.3 $C_{10}H_{19}N_3O_4$

Sigma L 0641 FW 245.3 $C_{10}H_{19}N_3O_4$

Leu-Asn-Lys-Ile-Val-Arg-Met-Tyr-Ser

Synonyms: P25 (270-278); GAG P24 CA (138-146); HTLV I Peptide

Neosystem SC537 MW 1123.37 HTLV I peptide from subtype P25 (GAG P24 CA)

Leu-Asn-Lys-Ile-Val-Arg-Met-Tyr-Ser-Pro-Thr-Ser-Ile-Leu-Asp-Ile-Arg-Gln

Synonyms: P25 (270-287); GAG P24 CA (138-155); HTLV I Peptide

Neosystem SC303 MW 2147.56 HTLV I peptide from subtype P25 (GAG P24 CA)

Leu-Asn-Phe-Ser-Pro-Gly-Trp
Glp-Leu-Asn-Phe-Ser-Pro-Gly-Trp Amide

Synonyms: Erythrophore Concentrating Hormone, Crustacean

American Peptide 60-9-22 MW 930.0 $C_{45}H_{59}N_{11}O_{11}$ Fernlund, P et al, *Biochem Biophys Acta*, 371:304, 1974

Leu-Asp

Bachem G-3835.0250 MW 246.26 $C_{10}H_{18}N_2O_5$ Store at -15°C

ICN 155188 MW 245.3 $C_{10}H_{19}N_3O_4$

Sigma L 3020 FW 246.3 $C_{10}H_{18}N_2O_5$

Leu-Asp-Gln-Trp-Phe-Gly
Ac-Leu-Asp-Gln-Trp-Phe-Gly Amide

Synonyms: Neurokinin II Receptor Antagonist; R396

Bachem H-1258.0001 MW 805.89 $C_{39}H_{51}N_9O_{10}$ Store at -15°C

Neosystem SC481   MW 805.90   Bioactive; tachykinin; specific & selective NK-2 receptor antagonist; Dion, S et al, *Pharmacology*, 41:184, 1990; Rovero, P et al, *Neuropeptides*, 23:143-145, 1992; Quartara, L et al, *Medicinal Res Rev*, 15:139-155, 1995

Leu-Asp-Ile-Ile-Trp
Ac-D-Bhg-Leu-Asp-Ile-Ile-Trp

Synonyms: Endothelin Receptor Receptor Antagonist; PD 145065

Alexis 155-016   MW 950.1 $C_{52}H_{67}N_7O_{10}$ Potent ligand for both the ET$_A$ & ET$_B$ receptor subtypes; Cody, WL, Doherty, AM, He, JX, Topliss, JG, Haleen, SJ, LaDouceur, D, Flynn, MA, Hill, KE & Reynolds, EE, *Peptides 1992:Proceedings of the 22nd European Peptide Symposium* (CH Schneider & AN Eberle, eds), ESCOM, p. 687, 1993

Leu-Asp-Leu-Leu-Phe-Leu

Synonyms: CKS-17 (7-12)

Bachem H-1442.0005   Store at -15°C

Leu-Asp-Phe-Pro
Phenylac-Leu-Asp-Phe-D-Pro Amide

Synonyms: VLA-4 Inhibitor

Bachem H-3376.0005   Store at -15°C

Leu-Asp-Val-Pro-
(4-((2-Methylphenyl)Aminocarbonyl)-
Aminophenyl)acetyl-Leu-Asp-Val-Pro

Synonyms: Fibronectin CS-1 (1980-1983), (4-((2-Methylphenyl)Aminocarbonyl)-Aminophenyl)Acetyl-

Bachem N-1765.0005   Store at -15°C

Leu-Asp-Val-Val-Lys-Arg-Gln-Gln-Glu-Leu-Leu-Arg-Leu-Thr-Val-Trp-Gly-Thr-Lys-Asn

Synonyms: GP140 (571-590); SIVmac251 Peptide

Neosystem SC765   MW 2396.81   SIVmac251 peptide from subtype GP140

Leu-Cys-Gly-Arg-Thr-Gly-Arg-Arg-Asn-Ser-Ile
Biotin-Leu-Cys-Gly-Arg-Thr-Gly-Arg-Arg-Asn-Ser-Ile Amide

Synonyms: Protein Kinase A Substrate Peptide, Biotinylated

Upstate 12-394   MW 1355   >95% pure; lyophilized | Kinase assay

Leu-Gln-Ala-Ala-Pro-Ala-Leu-Asp-Lys-Leu

Bachem H-9995.0005   Store at -15°C

Leu-Gln-Ala-Arg-Ile-Leu-Ala-Val-Glu-Arg-Tyr-Leu-Lys-Asp-Gln-Gln-Leu

Synonyms: GP41 (581-597); HTLV I Peptide

Neosystem SC240   MW 2057.42   HTLV I peptide from subtype GP41

Leu-Gln-Asn-Arg-Arg-Gly-Leu-Asp-Leu-Leu-Phe-Leu-Lys-Glu-Gly-Gly-Leu

Synonyms: CKS-17

American Peptide 86-5-18   MW 1942.3 $C_{87}H_{148}N_{26}O_{24}$ The major immunosuppressive site of retroviral TM protein; can suppress T effector cell function *in vitro*; Oostendorp, RAJ et al, *Eur J Immunol*, 22:1505, 1992

Bachem H-7600.0500   MW 1942.29 $C_{87}H_{148}N_{26}O_{24}$ Store at -15°C

Leu-Gln-Asp-Val-His-Asn-Phe-Val-Ala-Leu-Gly-Ala-Pro-Leu-Ala-Pro-Arg-Asp-Ala-Gly-Ser

Synonyms: Parathormone; Parathyroid Hormone (28-48)

American Peptide 22-1-29   Human   MW 2148.4 $C_{95}H_{150}N_{28}O_{29}$ Rosenblatt, M et al, *Biochemistry*, 17:2811, 1977

Bachem H-3875.0500   Human   MW 2148.41 $C_{95}H_{150}N_{28}O_{29}$ Store at -15°C

ICN 152984   Human   MW 2148.4

Sigma P 5519   Human   FW 2148.4   ≥97% (HPLC) | Bioactive peptide

Leu-Gln-Gly-Ser-Leu-Gln-Asp-Met-Leu-Trp-Gln-Leu-Asp-Leu-Ser-Pro-Gly-Cys

Synonyms: Leptin (150-167); Obese Gene Peptide (150-167)

Bachem H-3432.0001   Human   MW 2004.32 $C_{87}H_{138}N_{22}O_{28}S_2$ Store at -15°C

Leu-Gln-Lys-Leu-Asn-Ser-Trp-Asp-Val-Phe-Gly-Asn-Trp-Phe-Asp-Leu-Ala-Ser-Trp-Ile

Synonyms: GP140 (671-690); SIVmac251 Peptide

Neosystem SC775   MW 2439.75   SIVmac251 peptide from subtype GP140

Leu-Gln-Ser-Glu-Asp-Lys-Ala-Ile-Arg-Thr-Ile-Met-Glu-Phe-Leu-Ala-Phe-Leu-His-Leu-Lys-Glu-Ala-Gly-Ala-Leu Amide

Synonyms: Gastrointestinal Peptide; Galanin Message Associated Peptide (16-41); Prepro-Galanin (80-105)

American Peptide 46-1-25   MW 2944.5 $C_{134}H_{219}N_{35}O_{37}S$

Bachem H-9725.0001   MW 2944.49 $C_{134}H_{219}N_{35}O_{37}S$ Store at -15°C

Sigma G 4521   FW 2944.5   ≥90% (HPLC) | Bioactive peptide

Leu-Gln-Thr-Arg-Val-Thr-Ala-Ile-Glu-Lys-Tyr-Leu-Lys-Asp-Gln-Ala-Gln-Leu-Asn-Ala

Synonyms: GP140 (591-610); SIVmac251 Peptide

Neosystem SC767   MW 2303.64   SIVmac251 peptide from subtype GP140

Leu-Gln-Val-Gln-Leu-Ser-Ile-Arg

Bachem H-4588.0001   MW 956.15 $C_{42}H_{77}N_{13}O_{12}$ Store at -15°C

Leu-Gln-Val-Ser-Tyr-Glu-Glu-Tyr-Leu-Cys-Met-Lys-Thr-Leu Acetate Salt

Synonyms: Glucocorticoid Receptor (656-669), Steroid Binding Domain

Biogenesis 4663-5159   Synthetic   Semi-pure; lyophilized

Leu-Glu

Bachem G-4065.0250   MW 260.29 $C_{11}H_{20}N_2O_5$ Store at -15°C

Leu-Glu-Arg-Phe-Ala-Val-Asn-Pro-Gly-Leu-Leu-Glu-Thr-Ser-Glu

Synonyms: P18 (41-55); GAG P17 MA (41-55); HTLV I Peptide

Neosystem SC281   MW 1674.87   HTLV I peptide from subtype P25 (GAG P17 MA)

Leu-Glu-Asp-Gly-Pro-Lys-Phe-Leu

Synonyms: Thymic Humoral γ-2 Factor

American Peptide 87-0-30   MW 918.0 $C_{43}H_{67}N_9O_{13}$ Brustein, Y et al, *Biochemistry*, 27:4066, 1988

Leu-Glu-Glu-Asp
Z-Leu-Glu-Glu-Asp-FMK

Synonyms: Caspase XIII Inhibitor

Kamiya   MW 696.5

Leu-Glu-Glu-Glu-Glu-Glu-Ala-Tyr-Gly-Trp-Met-Asp-Phe Amide

Synonyms: Minigastrin I; Gastrin I, Mini-; Gastrointestinal Peptide; HG-13; Gastrin I (5-17)

Bachem H-3105.0001   Human   MW 1646.75 $C_{74}H_{99}N_{15}O_{26}S$ Store at -15°C

ICN 152873   Human   Isolated from tumor tissue; corresponds to human gastrin I, fragment 5-17; Gregory, RA & HJ Tracy, *Gut*, 15:683, 1974

Sigma G 0267   Human tumor tissue   FW 1646.7   ≥97% (HPLC) | Bioactive peptide; Gregory, RA & Tracy, HJ, *Gut*, 15:683, 1974

Leu-Glu-Glu-Leu-Lys-Asn-Glu-Ala-Val-Arg-His-Phe-Pro-Arg-Ile

Synonyms: VPR (23-37); HTLV I Peptide

Neosystem SC557   MW 1851.13   HTLV I peptide from subtype VPR

Leu-Glu-Glu-Leu-Lys-Glu-Glu-Ala-Leu-Lys-His-Phe-Asp-Pro-Tyr

Synonyms: VPR (24-37)-(Tyr); SIVmac251 Peptide

Neosystem SC577   MW 1861.08   SIVmac251 peptide from subtype VPR

Leu-Glu-His-Asp
Ac-Leu-Glu-His-Asp-AFC

Bachem I-1820.0005   MW 765.7 $C_{33}H_{38}F_3N_7O_{11}$ Store at -15°C

Leu-Glu-His-Asp
Ac-Leu-Glu-His-Asp-AMC

Bachem I-1825.0005   MW 711.73 $C_{33}H_{41}N_7O_{11}$ Store at -15°C

Leu-Glu-His-Asp
Ac-Leu-Glu-His-Asp-CHO Pseudo Acid

Bachem N-1720.0005   MW 538.56 $C_{23}H_{34}N_6O_9$ Store at -15°C

Leu-Glu-His-Asp
Ac-Leu-Glu-His-Asp-CMK

Bachem N-1750.0005   MW 587.03 $C_{24}H_{35}ClN_6O_9$ Store at -15°C

Leu-Glu-His-Asp
Ac-Leu-Glu-His-Asp-AMC

BioSource International 78-109   Fluorescing substrate

Leu-Glu-His-Asp
Leu-Glu-His-Asp-CHO

BioSource International 78-121   Cell permeable inhibitor

Leu-Glu-His-Asp
Ac-Leu-Glu-His-Asp-pNA

BioSource International 78-136

BioSource International 78-137

Leu-Glu-His-Asp
Z-Leu-Glu(OMe)-His-Asp(OMe)-CH₂F Trifluoroacetate Salt

Synonyms: Caspase IX Inhibitor I

Calbiochem 218761   MW 690.7 $C_{32}H_{43}FN_6O_{10}$ single spot purity (TLC); solid; soluble in DMSO | Potent, cell-permeable & irreversible inhibitor of Caspase IX (ICE-LAP6, Mch6); also inhibits Caspases IV & V

Leu-Glu-His-Asp
Ac-Leu-Glu-His-Asp-AFC

Synonyms: Caspase IX Fluorogenic Substrate I

Calbiochem 218765   MW 765.7 $C_{33}H_{38}F_3N_7O_{11}$ ≥95% (HPLC); solid; soluble in DMSO; excitation max:~400 nm; emission max:~505 nm | Fluorogenic substrate for Caspase IX; reaction monitored visually or quantitatively by a blue to green shift in fluorescence upon cleavage of the AFC fluorophore; Thornberry, NA et al, *J Biol Chem*, 272:17907, 1997

Leu-Glu-His-Asp
Z-Leu-Glu-His-Asp-FMK

Synonyms: Caspase IV/V/IX Inhibitor

Kamiya   MW 804

Leu-Glu-His-Asp
Ac-Leu-Glu-His-Asp-AFC

Synonyms: Caspase IV/V/IX Substrate

Kamiya   MW 879   Fluorogenic

Leu-Glu-His-Asp
Ac-Leu-Glu-His-Asp-pNA

Synonyms: Caspase IV/V/IX Substrate

Kamiya   MW 788   >97% (HPLC).

Leu-Glu-His-Asp
Ac-Leu-Glu-His-Asp-AMC

Synonyms: Caspase IX Substrate

Neosystem SC1309   MW 711.73   Thornberry, NA et al, *JBC*, 272:17907-17911, 1997

Leu-Glu-His-Asp
Ac-Leu-Glu-His-Asp-MCA

Synonyms: Caspase IX Substrate

Peptides International MCA-3198-v   MW 711.73 $C_{33}H_{41}N_7O_{11}$ >98% (HPLC); lyophilized amorphous powder | Thornberry, NA et al, *JBC*, 272:17907, 1997

Leu-Glu-His-Asp
Ac-Leu-Glu-His-Asp-AMC

Synonyms: Caspase IX Fluorogenic Substrate; Caspase Substrate

Upstate 12-364   MW 712   >98% pure; lyophilized

Leu-Glu-His-Asp
Ac-Leu-Glu-His-Asp-pNA

Synonyms: Caspase IX Chromogenic Substrate

Upstate 12-393   MW 675.5   Lyophilized | Caspase IX assay

Leu-Glu-His-Asp
Ac-Leu-Glu-His-Asp-AMC

Synonyms: Caspase IX Fluorogenic Substrate

USBio C2088-12   ≥98%; lyophilized | Contains the recognition sequence for Caspase IX (ICE-LAP6, Mch6); used to measure Caspase activity *in vitro*; emits with a I_{max} of 380nm, whereas free AMC generated by Caspase IX activity emits with a I_{max} of 460nm

Leu-Glu-His-Asp
Ac-Leu-Glu-His-Asp-CHO

Synonyms: Caspase IX Inhibitor

Peptides International ICA-3199-v   Synthetic   MW 538.56
$C_{23}H_{34}N_6O_9$ >99% (HPLC); lyophilized amorphous powder |
Inhibitor; Thornberry, NA et al, *JBC*, 272:17907, 1997

Leu-Glu-Lys-Glu-Glu-Gly-Ile-Ile-Pro-Asp-Trp-Gln-Asp-Tyr-Thr

Synonyms: NEF (146-160); SIVmac251 Peptide

Neosystem SC617   MW 1835.00   SIVmac251 peptide from
subtype NEF

Leu-Glu-Lys-His-Gly-Ala-Ile-Thr-Ser-Ser-Asn-Thr-Ala-Ala

Synonyms: NEF (37-50); HTLV I Peptide

Neosystem SC218   MW 1399.52   HTLV I peptide from
subtype NEF

Leu-Glu-Lys-His-Gly-Ala-Leu-Thr-Ser-Ser-Asn-Thr-Ala-Ala

Synonyms: NEF MN (39-52); HIV I Peptide

Neosystem SC642   MW 1400.50   NEF peptide from HIV-1
subtype MN

Leu-Glu-Ser-Ser-Asn-Glu-Arg-Ser-Ser-Cys-Ile-Leu-Glu-Ala-Asp

Synonyms: TAT (12-26); SIVmac251 Peptide

Neosystem SC584   MW 1652.75   SIVmac251 peptide from
subtype TAT; due to the presence of a Cys this peptide can contain
the dimeric form

Leu-Glu-Thr-Ser-Glu-Gly-Cys-Arg-Gln-Ile-Leu-Gly-Gln-Leu-Gln

Synonyms: P18 (51-65); GAG P17 MA (51-65); HTLV I Peptide

Neosystem SC282   MW 1674.88   HTLV I peptide from
subtype P25 (GAG P17 MA); due to the presence of a Cys this
peptide can contain the dimeric form

Leu-Glu-Trp-Arg-Phe-Asp-Ser-Arg-Leu-Ala-Phe

Synonyms: NEF (181-191); HTLV I Peptide

Neosystem SC516   MW 1439.63   HTLV I peptide from
subtype NEF

Leu-Glu-Val-Asp
Ac-Leu-Glu-Val-Asp-pNA

Synonyms: Caspase IV Substrate; Caspase VI Substrate; Caspase
I Substrate

Alexis 260-061   MW 636.7 $C_{28}H_{40}N_6O_{11}$ ≥98%; white
powder; soluble in basic buffers | Chromogenic substrate;
Talanian, RV et al, *J Biol Chem*, 272:9677, 1997

Leu-Glu-Val-Asp
Ac-Leu-Glu-Val-Asp-CHO

Alexis 260-065   MW 500.6 $C_{22}H_{36}N_4O_9$ ≥95%; white
powder; soluble in water | Caspase IV Inhibitor (Aldehyde); Ac-
LEVD-CHO; Inhibitor of Caspase I, -4 & -6 respectively; Most
potent for Caspase IV; Talanian, RV et al, *J Biol Chem*, 272:9677,
1997

Leu-Glu-Val-Asp
Ac-Leu-Glu-Val-Asp-CHO Pseudo Acid

Bachem N-1700.0005   Store at -15°C

Leu-Glu-Val-Asp
Ac-Leu-Glu-Val-Asp-CHO

BioSource International 77-846   Inhibitor

Leu-Glu-Val-Asp
Leu-Glu-Val-Asp-CHO

BioSource International 78-115   Cell permeable inhibitor

Leu-Glu-Val-Asp
Ac-Leu-Glu-Val-Asp-AFC

Synonyms: Caspase IV Fluorogenic Substrate II

Calbiochem 218748   MW 727.7 $C_{32}H_{40}F_3N_5O_{11}$ ≥95%
(HPLC); solid; soluble in DMSO; excitation max:~400 nm; emission
max:~505 nm | fluorogenic substrate for Caspase IV; reaction
monitored visually or quantitatively by a blue to green shift in
fluorescence upon cleavage of the AFC fluorophore; Thornberry, NA
et al, *J Biol Chem*, 272:17907, 1997

Leu-Glu-Val-Asp
Ac-Leu-Glu-Val-Asp-CHO

Synonyms: Caspase IV Inhibitor I

Calbiochem 218755   MW 500.6 $C_{22}H_{36}N_4O_9$ ≥97% (HPLC);
lyophilized solid; soluble in DMSO | Caspase IV inhibitor; Talanian,
RV et al, *J Biol Chem*, 272:9677, 1997

Leu-Glu-Val-Asp-Gly-Trp-Lys
Mca-Leu-Glu-Val-Asp-Gly-Trp-Lys(Dnp) Amide

Bachem M-2315.0001   Store at -15°C

Leu-Glu-Val-Asp-Gly-Trp-Lys
MCA-Leu-Glu-Val-Asp-Gly-Trp-(Lys-DNP)-Amide

BioSource International 77-882

Leu-Glu-Val-Asp-Gly-Trp-Lys
MCA-Leu-Glu-Val-Asp-Gly-Trp-Lys-(DNP) Amide

Synonyms: Caspase IV Fluorogenic Substrate I

Calbiochem 218756   MW 1227.3 $C_{57}H_{70}N_{12}O_{19}$ ≥97%
(HPLC); lyophilized solid; soluble in DMSO; excitation max:~325
nm; emission max:~392 nm | Fluorogenic resonance energy
transfer substrate for Caspase IV activity; Talanian, RV et al, *J Biol
Chem*, 272:9677, 1997

Leu-Gly
BOC-Leu-Gly

Bachem A-1875.0001   MW 288.34 $C_{13}H_{24}N_2O_5$ Store at
-15°C

Leu-Gly
FMOC-Leu-Gly

Bachem B-3150.0001   Store at -15°C

Leu-Gly
Z-Leu-Gly

Bachem C-2140.0001   MW 322.36 $C_{16}H_{22}N_2O_5$ Store at
-15°C

Leu-Gly
Z-Leu-Gly-OMe

Bachem C-3115.0001   MW 336.39 $C_{17}H_{24}N_2O_5$ Store at
-15°C

Leu-Gly
Ac-Leu-Gly

Bachem G-1040.0500   MW 230.26 $C_{10}H_{18}N_2O_4$ Store at
-15°C

Leu-Gly			
Leu-Gly			
Bachem G-2495.0001	MW 188.23	$C_8H_{16}N_2O_3$	Store at -15°C

Leu-Gly			
D-Leu-Gly			
Bachem G-2500.0001	MW 188.23	$C_8H_{16}N_2O_3$	Store at -15°C

Leu-Gly			
DL-Leu-Gly			
Bachem G-2505.0001	MW 188.23	$C_8H_{16}N_2O_3$	Store at -15°C

Leu-Gly			
Leu-Gly Amide Hydrobromide			
Bachem G-2510.0250	MW 268.15	$C_8H_{17}N_3O_2 \cdot$ HBr	Store at -15°C

Leu-Gly	
Leu-Gly-OtBu Hydrochloride	
Bachem G-2515.0250	Store at -15°C

Leu-Gly			
Leu-Gly-βNA			
Bachem K-1400.0250	MW 313.4	$C_{18}H_{23}N_3O_2$	Store at -15°C

Leu-Gly			
L-Leu-Gly Hydrate			
Fluka 61970	MW 188.23	$C_8H_{16}N_2O_3$	≥99.0% (TLC); mp:245°C

| **Leu-Gly** | | |
| ICN 102174 | MW 188.2 | $C_8H_{16}N_2O_3$ | 99%; crystalline |

Leu-Gly		
D-Leu-Gly		
ICN 155198	MW 188.2	$C_8H_{16}N_2O_3$

Leu-Gly		
DL-Leu-Gly		
ICN 155199	MW 188.2	$C_8H_{16}N_2O_3$

Leu-Gly		
Peptides International OLG-3024	MW 188.23	$C_8H_{16}N_2O_3$
>98% (HPLC); colorless crystalline powder		

Leu-Gly		
N-CBZ-Leu-Gly		
Sigma C 2752	FW 322.4	$C_{16}H_{22}N_2O_5$

Leu-Gly		
D-Leu-Gly		
Sigma L 3502	FW 188.2	$C_8H_{16}N_2O_3$

Leu-Gly		
L-Leu-Gly β-Naphthylamide		
Sigma L 6377	FW 313.4	$C_{18}H_{23}N_3O_2$

Leu-Gly		
DL-Leu-Gly		
Sigma L 9500	FW 188.2	$C_8H_{16}N_2O_3$

| **Leu-Gly** | | |
| Sigma L 9625 | FW 188.2 | $C_8H_{16}N_2O_3$ |

Leu-Gly		
CBZ-Leu-Gly		
USBio C2099-02	MW 322.4	$C_{16}H_{22}N_2O_5$	≥99%

Leu-Gly-Ala-Ser-Trp-His-Arg-Pro-Asp-Lys-Cys-Cys-Leu-Gly-Tyr-Gln-Lys-Arg-Pro-Leu-Pro-Gln-Val-Leu-Leu-Ser-Ser-Trp-Tyr-Pro-Thr-Ser-Gln-Leu-Cys-Ser-Lys-Pro-Gly-Val-Ile-Phe-Leu-Thr-Lys-Arg-Gly-Arg-Gln-Val-Cys-Ala-Asp-Lys-Ser-Lys-Asp-Trp-Val-Lys-Lys-Leu-Met-Gln-Gln-Leu-Pro-Val-Thr-Ala

Synonyms: Macrophage Inflammatory Protein-2

| Bachem H-4608.0010 | Virus | Store at -15°C |

Leu-Gly-Arg			
BOC-Leu-Gly-Arg-AMC			
Bachem I-1105.0050	MW 601.7	$C_{29}H_{43}N_7O_7$	Store at -15°C

| **Leu-Gly-Arg** | |
| **BOC-Leu-Gly-Arg-pNA** | |

Synonyms: Endotoxin Substrate

| Bachem L-1195.0050 | MW 564.64 | $C_{25}H_{40}N_8O_7$ | Store at -15°C |

| **Leu-Gly-Arg** | |
| **N-α-t-BOC-L-Leu-Gly-L-Arg-MCA** | |

Synonyms: Clotting Enzyme, Horseshoe Crab

| ICN 150499 | MW 601.7 | $C_{29}H_{43}N_7O_7$ | Fluorogenic substrate |

| **Leu-Gly-Arg** | |
| **N-t-BOC-Leu-Gly-Arg pNA** | |

Synonyms: Horseshoe Crab Clotting Enzyme Substrate

| Sigma B 2516 | FW 564.6 | $C_{25}H_{40}N_8O_7$ | Chromogenic substrate |

used in quantitative assays of endotoxin

| **Leu-Gly-Arg** | |
| **N-t-BOC-Leu-Gly-Arg MCA Acetate Salt** | |

Synonyms: Convertases Substrate

| Sigma B 4511 | FW 601.7 (free base); | $C_{29}H_{43}N_7O_7$ |

Fluorogenic substrate; Caporak, LH et al, *J Immunol*, 126:1963, 1981

| **Leu-Gly-Arg** | |
| **BOC-Leu-Gly-Arg-MCA** | |

Synonyms: Clotting Enzyme Substrate, Horseshoe Crab

Peptides International MLG-3102-v   Horseshoe Crab   MW 601.70 $C_{29}H_{43}N_7O_7$ >99% (HPLC); lyophilized amorphous powder | Iwanaga, S et al, *Haemostasis*, 7:183, 1978

| **Leu-Gly-Arg** | |
| **BOC-Leu-Gly-Arg-pNA Acetate Salt** | |

Synonyms: Endotoxin Substrate

| ICN 195981 | Mouse | MW 624.7 | $C_{25}H_{40}N_8O_7 \cdot C_2H_4O_2$ | ≥98% |

| Fluorogenic substrate for quantification of endotoxins

Leu-Gly-Arg-Arg-Gly-Trp-Glu-Ala-Leu-Lys-Tyr-Trp-Trp-Asn-Leu

Synonyms: GP41 (790-804); HTLV I Peptide

Neosystem SC258   MW 1948.25   HTLV I peptide from subtype GP41

Leu-Gly-Arg-Ser-Gly-Gly-Asp-Ile-Ile-Lys-Lys-Met-Gln-Thr-Leu

Bachem H-8655.0001   MW 1616.95 $C_{69}H_{125}N_{21}O_{21}S$ Store at -15°C

Leu-Gly-Gly
D-Leu-Gly-Gly

Bachem H-3880.0250   MW 245.28 $C_{10}H_{19}N_3O_4$ Store at -15°C

Leu-Gly-Gly
DL-Leu-Gly-Gly

Bachem H-3885.0001   MW 245.28 $C_{10}H_{19}N_3O_4$ Store at -15°C

Leu-Gly-Gly
Leu-Gly-Gly-4MβNA Hydrochloride

Bachem J-1240.0050   MW 436.94 $C_{21}H_{28}N_4O_4 \cdot HCl$ Store at -15°C

Leu-Gly-Gly

Bachem M-1530.0001   MW 245.28 $C_{10}H_{19}N_3O_4$ Store at -15°C

Leu-Gly-Gly
L-Leu-Gly-Gly

Synonyms: Aminopeptidase Substrate

Fluka 61990   MW 245.28 $C_{10}H_{19}N_3O_4$ ≥99.0% (TLC); mp:220°C; ≤1% Leu+Gly | Hayashi, M & Oshima, K, *J Biochem (Tokyo)*, 87:1403, 1980; Smith, EL, *Meth Enzymol*, 2:83, 1955; Marks, N & Lajtha, A, *Meth Enzymol*, 19:534, 1970

Leu-Gly-Gly

ICN 102175   MW 245.3 $C_{10}H_{19}N_3O_4$

Leu-Gly-Gly
D-Leu-Gly-Gly

ICN 155200   MW 245.3 $C_{10}H_{19}N_3O_4$ Crystalline

Leu-Gly-Gly
DL-Leu-Gly-Gly

ICN 155201   MW 245.3 $C_{10}H_{19}N_3O_4$ Crystalline

Leu-Gly-Gly

Peptides International OLG-3025   MW 245.28 $C_{10}H_{19}N_3O_4$ >98% (HPLC); colorless crystalline powder

---

Leu-Gly-Gly
DL-Leu-Gly-Gly

Sigma L 4127   FW 245.3 $C_{10}H_{19}N_3O_4$

Leu-Gly-Gly

Sigma L 9750   FW 245.3 $C_{10}H_{19}N_3O_4$

Leu-Gly-His-Ile-Val-Ser-Pro-Arg-Cys-Glu-Tyr-Gln-Ala-Gly

Synonyms: VIF (125-138); HTLV I Peptide

Neosystem SC571   MW 1529.73   HTLV I peptide from subtype VIF; due to the presence of a Cys this peptide can contain the dimeric form

---

Leu-Gly-Ile-Ala-Gly-Arg
Dnp-Leu-Gly-Ile-Ala-Gly-Arg Amide

Synonyms: Serum Peptidase Substrate

Peptides International SDL-3083-v   MW 750.81 $C_{31}H_{50}N_{12}O_{10}$ >98% (HPLC); lyophilized amorphous powder | Masui, Y et al, *Biochem Med*, 17:215, 1977

Leu-Gly-Ile-Trp-Gly-Cys-Ser-Gly-Lys-Leu-Ile-Cys

Synonyms: HIV I (598-609), (Ile[600])-

ICN 191464

Leu-Gly-Leu

Bachem H-5380.0250   MW 301.39 $C_{14}H_{27}N_3O_4$ Store at -15°C

Leu-Gly-Leu-Trp-Gly-Cys-Ser-Gly-Lys-Leu-Ile-Cys

Synonyms: HIV I (598-609)

ICN 191463

Leu-Gly-Phe
DL-Leu-Gly-DL-Phe

Bachem H-3890.0250   MW 335.4 $C_{17}H_{25}N_3O_4$ Store at -15°C

ICN 155202   MW 335.4 $C_{17}H_{25}N_3O_4$ Crystalline

Sigma L 4502   FW 335.4 $C_{17}H_{25}N_3O_4$

Leu-Gly-Pro

Bachem H-3895.0250   MW 285.34 $C_{13}H_{23}N_3O_4$ Store at -15°C

Leu-Gly-Pro-Ala
FA-Leu-Gly-Pro-Ala

Synonyms: FALGPA

Bachem M-1385.0050   MW 476.53 $C_{23}H_{32}N_4O_7$ Store at -15°C

Leu-Gly-Pro-Ala
N-(3-(2-Furyl)Acryloyl)-L-Leu-Gly-L-Pro-L-Ala

Synonyms: Collagenase Substrate

Fluka 48173   MW 476.5 $C_{23}H_{32}N_4O_7$ ≥99% (HPLC); ≥90% peptide content; ≤1% water | Substrate for the continuous spectrophotometric assay of *Cl histolyticum* collagenase; Van Wart, HE & Steinbrink, DR, *Anal Biochem*, 113:356, 1981

Leu-Gly-Pro-Ala
N-(3-(2-Furyl)Acryloyl)-Leu-Gly-Pro-Ala

Synonyms: Collagenase Substrate

ICN 158213   MW 476.5 $C_{23}H_{32}N_4O_7$ Van Wart, HE & DR Steinbrink, *Anal Biochem*, 113:356, 1981

Sigma F 5135   FW 476.5 $C_{23}H_{32}N_4O_7$ Van Wart, HE & Steinbrink, DR, *Anal Biochem*, 113:356, 1981

Leu-Gly-Thr-Ile-Pro-Gly

Synonyms: Laminin Binding Inhibitor

Neosystem SC990   MW 556.66   Bioactive; cell attachment peptideThis hexapeptide from domain V inhibits binding of a cellular receptor to laminin; Hunter, DD et al, *Nature*, 338:229-234, 1989

Leu-Gly-Tyr

Bachem H-6495.0250   MW 351.4 $C_{17}H_{25}N_3O_5$ Store at -15°C

Leu-His

Bachem G-2520.0250   MW 268.32 $C_{12}H_{20}N_4O_3$ Store at -15°C

Leu-His-Leu

Bachem H-5385.0100   MW 381.48 $C_{18}H_{31}N_5O_4$ Store at -15°C

Leu-Ile
Z-Leu-Ile

Bachem C-2145.0001   MW 378.47 $C_{20}H_{30}N_2O_5$ Store at -15°C

Leu-Ile

Bachem G-2525.0001   MW 244.33 $C_{12}H_{24}N_2O_3$ Store at -15°C

Leu-Ile-Gly-Arg-Lys-Lys

Synonyms: Fibronectin (1954-1959)

Bachem H-2068.0005   Store at -15°C

Leu-Ile-His-Leu-Leu-His-Gln-Thr-Ile-Asp-Ser-Tyr-Pro-Thr-Gly

Synonyms: REV (16-30); SIVmac251 Peptide

Neosystem SC598   MW 1707.94   SIVmac251 peptide from subtype REV

Leu-Ile-His-Ser-Gln-Arg-Arg-Gln-Asp-Ile-Leu-Asp-Leu-Trp-Ile

Synonyms: NEF (100-114); HTLV I Peptide

Neosystem SC507   MW 1906.21   HTLV I peptide from subtype NEF

Leu-Ile-Lys-Phe-Leu-Tyr-Gln-Ser-Asn-Pro-Pro-Pro-Asn

Synonyms: REV (18-30); HTLV I Peptide

Neosystem SC190   MW 1530.78   HTLV I peptide from subtype REV

Leu-Ile-Pro-Pro-Phe-Trp-Lys
Leu-Ile-Pro-Pro-Phe-Trp-Lys Amide

Synonyms: CTX IV (6-12)

Neosystem SC152   MW 899.14   Bioactive; cardiotoxic peptide

Leu-Ile-Tyr-Ser-Gln-Lys-Arg-Gln-Asp-Ile-Leu-Asp-Leu-Trp-Val

Synonyms: NEF MN (102-116); HIV I Peptide

Neosystem SC652   MW 1892.18   NEF peptide from HIV-1 subtype MN

Leu-Leu
L-Leu-L-Leu (4,5-^{3}H)

ARC ART-550   MW 244.3   30-60 Ci/mmol; 1.11-2.22 TBq/mmol; in EtOH | Radiochemical

Leu-Leu
BOC-Leu-(®)-Leu

Bachem A-1880.0250   MW 330.47 $C_{17}H_{34}N_2O_4$ Store at -15°C

Leu-Leu
BOC-Leu-(®)-Leu-OBzl

Bachem A-1885.0250   Store at -15°C

Leu-Leu
BOC-Leu-(®-(2-Chloro-Z))-Leu

Bachem A-1890.0250   Store at -15°C

Leu-Leu
BOC-Leu-Leu

Bachem A-3230.0001   MW 344.45 $C_{17}H_{32}N_2O_5$ Store at -15°C

Leu-Leu
Z-Leu-Leu

Bachem C-2150.0001   MW 378.47 $C_{20}H_{30}N_2O_5$ Store at -15°C

Leu-Leu
Leu-D-Leu

Bachem G-2530.0250   MW 244.33 $C_{12}H_{24}N_2O_3$ Store at -15°C

Leu-Leu
D-Leu-Leu

Bachem G-2535.0250   MW 244.33 $C_{12}H_{24}N_2O_3$ Store at -15°C

Leu-Leu
D-Leu-D-Leu

Bachem G-2540.0250   MW 244.33 $C_{12}H_{24}N_2O_3$ Store at -15°C

Leu-Leu
Leu-Leu-OMe Hydrobromide

Bachem G-2550.0001   MW 339.27 $C_{13}H_{26}N_2O_3 \cdot$ HBr   Store at -15°C

Leu-Leu

Bachem M-1535.0001   MW 244.33 $C_{12}H_{24}N_2O_3$ Store at -15°C

ICN 102164   MW 244.3 (free base) $C_{12}H_{24}N_2O_3$

Leu-Leu
Leu-Leu-OMe Hydrobromide

ICN 153142   MW 339.2 $C_{13}H_{26}N_2O_3 \cdot$ HBr   Toxic effects on naturalkiller cells; Thiele, DL & PE Lipsky, *Fed Proc*, 44:590, 1985

Leu-Leu
Leu-D-Leu

ICN 155209   MW 244.3 $C_{12}H_{24}N_2O_3$ Crystalline

Leu-Leu
D-Leu-Leu

ICN 155210   MW 244.3 $C_{12}H_{24}N_2O_3$ Crystalline

Leu-Leu
D-Leu-D-Leu

ICN 155211   MW 244.3 $C_{12}H_{24}N_2O_3$ Crystalline

Leu-Leu
DL-Leu-DL-Leu

ICN 155212   MW 244.3 $C_{12}H_{24}N_2O_3$ Crystalline

Leu-Leu
N-CBZ-Leu-Leu Methyl Ester

Sigma C 7131   FW 392.5 $C_{21}H_{32}N_2O_5$

Leu-Leu

Sigma L 2752   FW 244.3 (free base) $C_{12}H_{24}N_2O_3$

Leu-Leu
D-Leu-D-Leu

Sigma L 3252   FW 244.3 $C_{12}H_{24}N_2O_3$

Leu-Leu
Leu-D-Leu

Sigma L 3377   FW 244.3 $C_{12}H_{24}N_2O_3$

Leu-Leu
DL-Leu-DL-Leu

Sigma L 3627   FW 244.3 $C_{12}H_{24}N_2O_3$

Leu-Leu
Leu-Leu-OMe Hydrobromide

Sigma L 7393   FW 339.3 $C_{13}H_{26}N_2O_3 \cdot HBr$ ≥97% (HPLC); | Bioactive peptide; peptide derivative toxic to naturalkiller cells; Thiele, DL & Lipsky, PE, *Fed Proc*, 44:590 (Abstr 1049), 1985

Leu-Leu
Carbobenzoxy-Leu-Leu-CHO

Synonyms: Calpain Inhibitor

Peptides International IZL-3178-v   Synthetic   MW 362.47 $C_{20}H_{30}N_2O_4$ >99% (HPLC); lyophilized amorphous powder | Inhibitor; does not inhibit Proteasome at the level of 10-6 *M* concentration; Saito, Y et al, *Neurosci Letts*, 120:1, 1990; Tsubuki, S et al, *J Biochem*, 119:572, 1996

Leu-Leu Acetate

ICN 155208   MW 304.4 $C_{12}H_{24}N_2O_3 \cdot C_2H_4O_2$

Leu-Leu Acetate Salt

Sigma L 4634   FW 304.4 $C_{12}H_{24}N_2O_3$

Leu-Leu Amide Hydrochloride

Bachem G-2545.0250   MW 279.81 $C_{12}H_{25}N_3O_2 \cdot HCl$ Store at -15°C

ICN 155213   MW 279.8 $C_{12}H_{24}N_3O_2 \cdot HCl$

Sigma L 8015   FW 279.8 $C_{12}H_{25}N_3O_3 \cdot HCl$

Leu-Leu-Ala

Bachem H-3905.0250   MW 315.41 $C_{15}H_{29}N_3O_4$ Store at -15°C

Leu-Leu-Ala
L-Leu-L-Leu-Ala Hydrate

Fluka 61960   MW 202.25 $C_9H_{18}N_2O_2$ ≥95.0% (TLC) | Used as fluorescence intensifier in fluorimetric methods for the determination of phenylalanine; McCaman, MW & Robins, E, *J Lab Clin Med*, 59:885, 1962; Hill, JB et al, *Clin Chem*, 11:541, 1965

Leu-Leu-Arg
BOC-Leu-Leu-Arg-CHO

Synonyms: Leupeptin

American Peptide 81-5-80   MW 484.6 $C_{23}H_{44}N_6O_5$ Protease Inhibitor; Borin, G et al, *Hoppe-Seyler's Z Physiol Chem*, 362:1435, 1981

Leu-Leu-Arg
Z-Leu-Leu-Arg-AMC

Bachem I-1615.0050   MW 691.83 $C_{36}H_{49}N_7O_7$ Store at -15°C

Leu-Leu-Arg
Ac-Leu-Leu-Arg-CHO hemisulfate

Synonyms: Leupeptin

Bachem N-1000.0005   MW 426.56 $C_{20}H_{38}N_6O_4$ Store at -15°C

Calbiochem 108975   MW 475.6 $C_{20}H_{38}N_6O_4 \cdot ½ H_2SO_4$ ≥95% (HPLC); lyophilized solid; soluble in water; LD_{50}≤2000 mg/kg | Reversible inhibitor of trypsin-like protease & cysteine proteases; inhibits activation-induced programmed cell death & restores defective immune responses of HIV+ donors; Sarin, A et al, *J Immunol*, 153:862, 1994; Montenez, JP et al, *Toxicol Lett*, 73:201, 1994; *Merck Index*, 12:5483

Leu-Leu-Arg
N-Ac-L-Leu-L-Leu-L-Arg-CHO Hemisulfate

Synonyms: Leupeptin; Trypsin Inhibitor; Plasmin Inhibitor; Papain Inhibitor; Cathepsin B Inhibitor

Fluka 62070   MW 493.62 $C_{20}H_{38}N_6O_4 \cdot 5 H_2SO_4 \cdot H_2O$ 4000 U/mg; 1 U corresponds to the amount of inhibitor which reduces the trypsin activity by 1 BAEE-U; 1 BAEE-U is the amount of enzyme which increases the absorbance at 253 nm by 0.001/min at pH 7.6, 25°C | Aoyagi, T et al, *J Antibiotics*, 22:283, 1969; *ibid*, 22:558, 1969; Umezawa, H, *Meth Enzymol*, 45:678, 1976

Leu-Leu-Arg
Ac-Leu-Leu-Arg-CHO Hemisulfate

Synonyms: Leupeptin; Trypsin Inhibitor; Plasmin Inhibitor; Papain Inhibitor; Cathepsin B Inhibitor

ICN 151553   MW 475.6 $C_{20}H_{38}N_6O_4 \cdot ½H_2SO_4$ Kondo, S et al, *Chem Pharm Bull*, 17:1869, 1969

Leu-Leu-Arg
Ac-Leu-Leu-Arg-CHO Trifluroacetate Salt

Synonyms: Leupeptin

ICN 195623   MW 540.6 $C_{20}H_{38}N_6O_4 \cdot C_2HF_3O_2$ ≥90% | Inhibitor for trypsin, plasmin, papain & cathepsin B

Leu-Leu-Arg
Ac-Leu-Leu-Arg-CHO Hydrochloride

Synonyms: Leupeptin

ICN 195624   MW 463.0 $C_{20}H_{38}N_6O_4 \cdot HCl$ ≥60% | Protease inhibitor

Leu-Leu-Arg
Propionyl-Leu-Leu-Arg-CHO Hemisulfate

Synonyms: Leupeptin

ICN 195625   MW 489.6 $C_{21}H_{40}N_6O_4 \cdot ½ H_2SO_4$ ≥90% | Protease inhibitor

Leu-Leu-Arg
Nα-t-BOC-Leu-Leu-Arg-CHO

Synonyms: Leupeptin, Nα-t-BOC-Deacetyl-

Sigma B 7530   FW 484.6 $C_{23}H_{44}N_6O_5$ ≥97% (TLC) | Bioactive peptide; non-selective Ser & Cys protease inhibitor; Borin, G et al, *Hoppe-Seyler's Z Physiol Chem*, 362:1435, 1981

Leu-Leu-Arg
Propionyl-Leu-Leu-Arg-CHO Hemisulfate

Synonyms: Leupeptin, Propionyl-

Sigma L 3402   FW 489.6 $C_{21}H_{40}N_6O_4 \cdot ½ H_2SO_4$ ≥90% (HPLC) | Bioactive peptide; inhibitor of Ser & Cys proteases; Aoyagi, T et al, *J Antibiot*, 22:558, 1969

Leu-Leu-Arg
Ac-Leu-Leu-Arg-CHO

Synonyms: Leupeptin; Trypsin Inhibitor; Plasmin Inhibitor; Papain Inhibitor; Cathepsin B Inhibitor; Kallikreins Inhibitor

American Peptide 81-5-70 Microbial MW 426.6 $C_{20}H_{38}N_6O_4$ Inhibits plasma & tissue kallikreins; Umezawa, HJ, *Antibiotics*, 22:283, 1969

Leu-Leu-Arg
Ac-L-Leu-L-Leu-L-Arg-al ½H₂SO₄ Hydrate

Synonyms: Leupeptin; Trypsin Inhibitor; Plasmin Inhibitor; Papain Inhibitor; Cathepsin B Inhibitor

Peptides International ILP-4041 Microbial MW 493.62 $C_{20}H_{38}N_6O_4 \cdot$ ½$H_2SO_4 \cdot H_2O$ Amorphous powder; integrity assessed by activity; bulk | Inhibitor; Kondo, S et al, *Chem Pharm Bull*, 17:1896, 1969

Leu-Leu-Arg
Ac-L-Leu-L-Leu-L-Arg-al

Synonyms: Leupeptin; Trypsin Inhibitor; Plasmin Inhibitor; Papain Inhibitor; Cathepsin B Inhibitor

Peptides International ILP-4041-v Microbial MW 426.56 $C_{20}H_{38}N_6O_4$ Lyophilized amorphous powder; integrity assessed by activity; sulfate | Inhibitor; Kondo, S et al, *Chem Pharm Bull*, 17:1896, 1969

Leu-Leu-Arg
Ac-Leu-Leu-Arg-CHO Hydrochloride

Synonyms: Leupeptin; Serine Protease Inhibitor; Cysteine Protease Inhibitor

Sigma L 0649 Microbial FW 463.0 $C_{20}H_{38}N_6O_4 \cdot$ HCl ≥70% (HPLC); gives multiple peaks on HPLC due to equilibria among 3 forms in solution; purity determined using 3 main peaks; majority of contaminating peptide is racemized leupeptin | Aoyagi, T, *J Antibiot*, 22:283, 1969; Saino, T et al, *Chem Pharm Bull*, 30:2319, 1982

Leu-Leu-Arg
Ac-Leu-Leu-Arg-CHO Trifluoroacetate Salt Free Base

Synonyms: Leupeptin; Serine Protease Inhibitor; Cysteine Protease Inhibitor

Sigma L 2023 Microbial FW 426.6 $C_{20}H_{38}N_6O_4$ ≥90% (HPLC); gives multiple peaks on HPLC due to equilibria among 3 forms in solution; purity determined using 3 main peaks; majority of contaminating peptide is racemized leupeptin | Inhibitor of Ser & Cys proteases; Aoyagi, T, *J Antibiot*, 22:283, 1969; Saino, T et al, *Chem Pharm Bull*, 30:2319, 1982

Leu-Leu-Arg
Ac-Leu-Leu-Arg-CHO Hydrochloride

Synonyms: Leupeptin

Sigma L 9783 Microbial FW 463.0 $C_{20}H_{38}N_6O_4 \cdot$ HCl ≥90% (HPLC); gives multiple peaks on HPLC due to equilibria among 3 forms in solution; purity determined using 3 main peaks; majority of contaminating peptide is racemized leupeptin | Inhibitor of Ser & Cys proteases; Aoyagi, T, *J Antibiot*, 22:283, 1969; Saino, T et al, *Chem Pharm Bull*, 30:2319, 1982

Leu-Leu-Arg
Ac-Leu-Leu-Arg-CHO Hemisulfate Salt

Synonyms: Leupeptin; Serine Protease; Cysteine Protease

Sigma L 2884 Microbial source FW 475.6 $C_{20}H_{38}N_6O_4 \cdot$ ½H_2SO_4 ≥90% (HPLC); HPLC purified; gives multiple peaks on HPLC due to equilibria among 3 forms in solution; purity determined using 3 main peaks; majority of contaminating peptide is racemized leupeptin | Bioactive peptide; Aoyagi, T et al, *J Antibiot*, 22:283, 1969; Saino, T et al, *Chem Pharm Bull*, 30:2319, 1982

Leu-Leu-Arg
Ac-L-Leu-L-Leu-L-Arg-CHO Sulfate

Synonyms: Leupeptin; Trypsin Inhibitor; Plasmin Inhibitor; Papain Inhibitor; Cathepsin B Inhibitor

Alexis 260-009 Synthetic MW 493.6 $C_{20}H_{38}N_6O_4 \cdot$ 0.5$H_2SO_4 \cdot H_2O$ ≥95%; white lyophilized powder; soluble in water | Kondo, S et al, *Chem Pharm Bull*, 17:1896, 1969

Leu-Leu-Arg
Ac-Leu-Leu-Arg-CHO Hemisulfate Salt

Synonyms: Leupeptin

Sigma L 8511 Synthetic FW 475.6 $C_{20}H_{38}N_6O_4 \cdot$ ½H_2SO_4 ~70% (HPLC); gives multiple peaks on HPLC due to equilibria among 3 forms in solution; purity determined using 3 main peaks; majority of contaminating peptide is racemized leupeptin | Inhibitor of Ser & Cys proteases; Aoyagi, T, *J Antibiot*, 22:283, 1969; Saino, T et al, *Chem Pharm Bull*, 30:2319, 1982

Leu-Leu-Arg-Gly-Pro-Ser-Trp-Asp-Pro-Phe-Arg-Cys
Leu-Leu-Arg-Gly-Pro-pSer-Trp-Asp-Pro-Phe-Arg-Cys

Synonyms: Phospho-HSP27, (Ser[15])-

Upstate 12-397 MW 1526 Frozen solution | Immunizing peptide

Leu-Leu-Glu
Z-Leu-Leu-Glu-βNA

Bachem K-1285.0050 MW 632.86 $C_{35}H_{44}N_4O_7$ Store at -15°C

Leu-Leu-Glu
Z-Leu-Leu-Glu-AMC

Synonyms: Proteasome Substrate II

Calbiochem 539141 MW 664.8 $C_{35}H_{44}N_4O_9$ ≥98% (HPLC); solid; soluble in DMSO; excitation max:~380 nm; emission max:~460 nm | Fluorogenic substrate

Leu-Leu-Glu
Z-Leu-Leu-Glu-MCA

Synonyms: Proteasome Substrate

Peptides International MLG-3179-v MW 664.76 $C_{35}H_{44}N_4O_9$ >98% (HPLC); lyophilized amorphous powder

Leu-Leu-Glu
N-CBZ-Leu-Leu-Glu β-Naphthylamide

Synonyms: Endopeptidase Substrate

Sigma C 0788 FW 632.8 $C_{35}H_{44}N_4O_7$ Substrate for a cation-sensitive endopeptidase; Wilk, S & Orlowski, M, *J Neurochem*, 40:842, 1983

Leu-Leu-Gly

Bachem H-3910.0250 MW 301.39 $C_{14}H_{27}N_3O_4$ Store at -15°C

Leu-Leu-Gly-Glu-Val-Glu-Asp-Gly-Ser-Ser-Gln-Ser-Leu-Gly-Gly

Synonyms: NEF (31-45); SIVmac251 Peptide

Neosystem SC606 MW 1447.51 SIVmac251 peptide from subtype NEF

Leu-Leu-His-Asn-Leu-Trp-Lys-Ser-Ile-Gln-Asp-Leu-
Arg-Arg-Arg-Phe-Phe-Leu-His-His-Leu-Ile-Ala-Glu-
Ile-His-Thr-Ala
Leu-Leu-His-Asn-Leu-D-Trp-Lys-Ser-Ile-Gln-Asp-Leu-
Arg-Arg-Arg-Phe-Phe-Leu-His-His-Leu-Ile-Ala-Glu-
Ile-His-Thr-Ala Amide

Synonyms: Parathyroid Hormone Related Protein (7-34), (Asn^{10},Leu^{11},D-Trp^{12})-; Hypercalcemia of Malignancy Factor (7-34), (Asn^{10},Leu^{11},D-Trp^{12})-

Bachem H-3274.0500   Human, rat   Store at -15°C

Leu-Leu-His-Asp-Lys-Gly-Lys-Ser-Ile-Gln-Asp-Leu-
Arg-Arg-Arg-Phe-Phe-Leu-His-His-Leu-Ile-Ala-Glu-
Ile-His-Thr-Ala Amide

Synonyms: Parathyroid Hormone Related Protein (7-34); Humoral Hypercalcemia of Malignancy Factor (7-34); Parathyroid Hormone; Parathyroid Hormone Related Protein Antagonist; Hypercalcemia of Malignancy Factor (7-34)

ICN 154488   MW 3366.3   Antagonist; McKee, RL et al, *J Bone & Mineral Research*, 3(Suppl 1), Abstract #17, p S73, 1988

American Peptide 22-1-73   Human   MW 3365.0 $C_{153}H_{247}N_{49}O_{37}$ Potent antagonist of pTH-related protein (1-34) *in vitro* & *in vivo*; McKee, RL et al, *J Bone & Mineral Res*, 3:(Suppl 1, Abst 17), S73, 1988

Neosystem SC897   Human   MW 3364.94   Bioactive; calcium metabolism peptide; *in vitro* & *in vivo* antagonist against PTHrp; Nagasaki, K et al, *BBRC*, 158:1036-1042, 1989

Sigma H 4022   Human   FW 3364.9   ≥97% (HPLC); peptide content:~60%; | Bioactive peptide; potent antagonist of hypercalcemia malignancy factor fragment 1-34 *in vitro* & *in vivo*; Nagasaki, K et al, *Biochem Biophys Res Commun*, 158:1036, 1989

Peptides International PTH-4215-v   Human synthetic   MW 3365.0 $C_{153}H_{247}N_{49}O_{37}$ >99% (HPLC); lyophilized amorphous powder | Bioactive; Nagasaki, K et al, *BBRC*, 158:1036, 1989; Suva, LJ et al, *Science*, 237:893, 1987

Bachem H-9100.0500   Human, rat   MW 3364.95 $C_{153}H_{247}N_{49}O_{37}$ Store at -15°C

Leu-Leu-Leu
Z-Leu-Leu-Leu-CHO

Synonyms: MG-132; Proteasome Inhibitor

American Peptide 81-5-15   MW 475.6 $C_{26}H_{41}N_3O_5$ Potent & reversible cell permeable proteasome inhibitor; Lee, DH et al, *JBC*, 271:27280, 1996; Rock, KL et al, *Cell*, 78:761, 1994

Leu-Leu-Leu

Bachem H-3915.0250   MW 357.49 $C_{18}H_{35}N_3O_4$ Store at -15°C

Leu-Leu-Leu
Z-Leu-Leu-4,5-Dehydro-Leu-CHO

Bachem N-1590.0001   MW 473.61 $C_{26}H_{39}N_3O_5$ Store at -15°C

Leu-Leu-Leu
Z-Leu-Leu-Leu-CHO

Synonyms: Calpain Inhibitor IV; MG-132; Proteasome Inhibitor

Bachem N-1635.0001   MW 475.63 $C_{26}H_{41}N_3O_5$ Store at -15°C

Calbiochem 474790   MW 475.6 $C_{26}H_{41}N_3O_5$ ≥90% (HPLC); solid; soluble in DMSO | Potent reversible & cell-permeable proteasome inhibitor; blocks protein breakdown in mammalian cells & in permeable strains of yeast; reduces the degradation of ubiquitin-conjugated proteins by the 26S complex without affecting its ATPase or isopeptidase activities; used to implicate the proteasome in the breakdown of membrane proteins, including CFTR within the endoplasmic reticulum; activates c-Jun N-terminal kinase which initiates apoptosis; inhibits NF-κB activation; Wiertz, EJHJ et al, *Cell*, 84:769, 1996; Rock, KL et al, *Cell*, 78:761, 1994; Meriin, AB et al, *J Biol Chem*, 273:6373, 1998; Lee, DH & Goldberg, AL, *J Biol Chem*, 271:27280, 1996; Jensen, TJ et al, *Cell*, 83:129, 1995; Read, MA et al, *Immunity*, 2:493, 1995; Adams, J & Stein, R, *Ann Rep Med Chem*, 31:279, 1996

Leu-Leu-Leu
4-Hydroxy-5-Iodo-3-Nitrophenylacetyl-Leu-Leu-Leu-Vinylsulfone

Synonyms: NLVS; NIP-L$_3$VS; Proteasome Inhibitor

Calbiochem 482240   MW 722.6 $C_{28}H_{43}IN_4O_8S$ ≥90% (HPLC); solid; soluble in DMSO | Irreversibly inhibits trypsin-like & chymotrypsin-like activity in isolated proteasomes & blocks their function in living cells; in contrast to Lactacystin, it also inhibits the peptidylglutamylpeptidase activity of isolated proteasomes; covalently modifies the active site threonine of the catalytic β-subunit of the proteasome; Bogyo, M et al, *PNAS*, 94:6629, 1997

Leu-Leu-Leu
4-Hydroxy-3-Nitrophenylacetyl-Leu-Leu-Leu-Vinylsulfone

Calbiochem 492025   MW 596.7 $C_{28}H_{44}N_4O_8S$ ≥97% (HPLC); solid; soluble in DMSO | Intermediate that can be used to prepare radiolabeled ^{125}I-NIP-L$_3$VS for proteasome inhibition studies; Bogyo, M et al, *PNAS*, 94:6629, 1997

Leu-Leu-Leu
Z-Leu-Leu-Leu-AMC

Synonyms: Proteasome Substrate I

Calbiochem 539140   MW 648.8 $C_{36}H_{48}N_4O_7$ ≥98% (HPLC); solid; soluble in DMSO; excitation max:~380 nm; emission max:~460 nm | Fluorogenic substrate; Tsubuki, S et al, *Biochem Biophys Res Comm*, 196:1195, 1993

Leu-Leu-Leu

ICN 155214   MW 357.5 $C_{18}H_{35}N_3O_4$ Crystalline

Leu-Leu-Leu
Z-Leu-Leu-Leu-CHO

Synonyms: Calpain Inhibitor; Proteasome Inhibitor

Neosystem SC1326   MW 475.62   Strong inhibitor; Saito, Y et al, *Neurosci Lett*, 120:1-4, 1990; Jensen, TJ et al, *Cell*, 83:129-135, 1995; Tsubuki, S et al, *J Biochem*, 119:572-576, 1996

Leu-Leu-Leu
Z-Leu-Leu-Leu-MCA

Synonyms: Proteasome Substrate

Peptides International MLL-3177-v   MW 648.80 $C_{36}H_{48}N_4O_7$ >98% (HPLC); lyophilized amorphous powder | Tsubuki, S et al, *BBRC*, 196:1195, 1993

Leu-Leu-Leu

Sigma L 0879   FW 357.5 $C_{18}H_{35}N_3O_4$

Leu-Leu-Leu
Carbobenzoxy-Leu-Leu-Leu-CHO

Synonyms: MG-132; Proteasome Inhibitor

Peptides International IZL-3175-v   Synthetic   MW 475.63 $C_{26}H_{41}N_3O_5$ >99% (HPLC); lyophilized amorphous powder | Inhibitor; Saito, Y et al, *Neurosci Letts*, 120:1, 1990; Jensen, TJ et al, *Cell*, 83:129, 1995

Leu-Leu-Leu Amide

Bachem H-3920.0050   MW 356.51 $C_{18}H_{36}N_4O_3$ Store at -15°C

ICN 155215   MW 356.5 $C_{18}H_{36}N_4O_3$

Sigma L 0391   FW 356.5 $C_{18}H_{36}N_4O_3$

Leu-Leu-Leu-Ala
N-CBZ-Leu-Leu-Leu-Ala

Sigma C 2211   FW 475.6 $C_{26}H_{41}N_3O_5$ ≥80% (HPLC) | Inhibitor of proteasome

Leu-Leu-Leu-Phe
Leu-Leu-Leu-Phe-OMe Hydrochloride

Bachem H-3925.0250   MW 555.16 $C_{28}H_{46}N_4O_5 \cdot$ HCl   Store at -15°C

Leu-Leu-Met
Ac-Leu-Leu-Met-CHO

Synonyms: Calpain I Inhibitor; Calpain II Inhibitor; Cathepsin B Inhibitor; Cathepsin L Inhibitor; Calpain Inhibitor II

Alexis 260-038   MW 401.6 $C_{19}H_{35}N_3O_4S$ ≥98%; white powder; soluble in ethanol; protect from light & moisture | Inhibits activation-induced programmed cell death & restores defective immune response in HIV+ donors

American Peptide 81-5-30   MW 401.6 $C_{19}H_{35}N_3O_4S$ A competitive inhibitor of cathepsin B, cathepsin L & calpain; inhibits activation-induced programmed cell death & restores defective immune responses in HIV+ donors; Sarin, A, *J Exp Med*, 178:1693, 1993; Sarin, A et al, *J Immunol*, 153:862, 1994

Bachem N-1315.0001   MW 401.57 $C_{19}H_{35}N_3O_4S$ Store at -15°C

Leu-Leu-Met
N-Ac-Leu-Leu-Met-CHO

Synonyms: Calpain Inhibitor II

Calbiochem 208721   MW 401.6 $C_{19}H_{35}N_3O_4S$ Single spot purity (TLC); white solid; soluble in DMSO & EtOH | Inhibitor of calpain I (K_I=120 nM), calpain II (K_I=230 nM), cathepsin B (K_I=100 nM) & cathepsin L (K_I=0.6 nM); inhibits activation-induced programmed cell death & restores defective immune responses in HIV+ donors; prevents nitric oxide production by activated macrophages by interfering with transcription of the inducible nitric oxide synthase gene; Sarin, A et al, *J Immunol*, 153:862, 1994; Griscavage, JM et al, *Biochem Biophys Res Comm*, 215:721, 1995; Sarin, A et al, *J Exp Med*, 178:1693, 1993

Leu-Leu-Met
N-Ac-L-Leu-L-Leu-L-Met-CHO

Synonyms: Calpain Inhibitor II

Fluka 21278   MW 401.56 $C_{19}H_{35}N_3O_4S$ ≥98% (TLC) | Shows marked inhibition of cathepsin B; Sasaki, T et al, *J Enz Inhib*, 3:195, 1990

Leu-Leu-Met
N-Ac-Leu-Leu-Met-CHO

Synonyms: Calpain Inhibitor II; Cysteine Endopeptidase Inhibitor; Calpain Inhibitor I; Cathepsin L Inhibitor; Cathepsin B Inhibitor

ICN 158835   MW 403.6   >98% | An inhibitor of both Calpain I & II; useful in the study of cytoskeletal & muscle protein turnover; Murachi, T, *Trends Biochem Sci*, 8:167, 1983; Yoshimura etal, *JBC*, 258:8883, 1983

Sigma A 6060   FW 401.6 $C_{19}H_{35}N_3O_4S$ ≥97% (TLC) | Bioactive peptide; potent inhibitor; does not inhibit trypsin; Sasaki, T et al, *Enzyme Inhibition*, 3:195, 1990

Leu-Leu-Nle
Ac-Leu-Leu-Nle-CHO

Synonyms: Calpain I Inhibitor; Calpain II Inhibitor; Cathepsin B Inhibitor; Cathepsin L Inhibitor; Calpain Inhibitor I

Alexis 260-037   MW 383.5 $C_{20}H_{37}N_3O_4$ ≥98%; white powder; soluble in ethanol; protect from light & moisture | Protects against neuronal damage caused by hypoxia & ischemia; Inhibits apoptosis in thymocytes & metamyelocytes

American Peptide 81-5-10   MW 383.5 $C_{20}H_{37}N_3O_4$ Sasaki et al, *J Enzyme Inhibition*, 3:195, 1990

Leu-Leu-Nle
Ac-Leu-Leu-Nle-ol

Synonyms: Calpain Inhibitor I

American Peptide 81-5-20   MW 385.5 $C_{20}H_{39}N_3O_4$

Leu-Leu-Nle

Bachem H-2086.0250   Store at -15°C

Leu-Leu-Nle
Ac-Leu-Leu-Nle-CHO

Synonyms: Calpain Inhibitor I

Bachem N-1320.0001   MW 383.53 $C_{20}H_{37}N_3O_4$ Store at -15°C

Leu-Leu-Nle
Z-Leu-Leu-Nle-CHO

Bachem N-1695.0001   MW 475.63 $C_{26}H_{41}N_3O_5$ Store at -15°C

Leu-Leu-Nle
N-Ac-Leu-Leu-Nle-CHO

Synonyms: Calpain Inhibitor I; MG-101

Calbiochem 208719   MW 383.5 $C_{20}H_{37}N_3O_4$ ≥98% (TLC); white solid; soluble in DMSO & EtOH | Inhibitor of calpain I (K_I=190 nM), calpain II (K_I=220 nM), cathepsin B (K_I=150 nM) & cathepsin L (K_I=0.5 nM); inhibits neutral cysteine proteases & the proteasome (K_I=6 μM); protects against neuronal damage caused by hypoxia & ischemia; inhibits apoptosis in thymocytes & metamyelocytes; inhibits the proteolysis of IκB-α & IκB-β by the ubiquitin-proteasome complex; inhibits cell cycle progression at G_1/S & metaphase/anaphase in CHO cells by inhibiting cyclin B degradation; prevents nitric oxide production by activated macrophages by interfering with transcription of the inducible nitric oxide synthase gene; Milligan, SA et al, *Arch Biochem Biophys*, 335:388, 1996; Griscavage, JM et al, *Biochem Biophys Res Comm*, 215:721, 1995; Squier, MK et al, *J Cell Physiol*, 159:229, 1994; Rami, J & Kreiglstein, J, *Brain Res*, 609:67, 1993; Sherwood, SW et al, *PNAS*, 90:3353, 1993; Vinitsky, A et al, *Biochemistry*, 31:9421, 1992; Sasaki, T et al, *J Enzyme Inhib*, 3:195, 1990

Leu-Leu-Nle
N-Ac-L-Leu-L-Leu-L-Nle-CHO

Synonyms: Calpain Inhibitor I

Fluka 21277   MW 383.53 $C_{20}H_{37}N_3O_4$ ≥98% (TLC) | Shows marked inhibition of cathepsin L; Sasaki, T et al, *J Enz Inhib*, 3:195, 1990

Leu-Leu-Nle
N-Ac-Leu-Leu-Nle-CHO

Synonyms: Calpain Inhibitor I

ICN 158834   MW 383.5   98% | An inhibitor of both Calpain I & II; useful in the study of cytoskeletal & muscle protein turnover; Murachi, T, *Trends Biochem Sci*, 8:167, 1983; Yoshimura etal, *JBC*, 258:8883, 1983

Leu-Leu-Nle
Ac-Leu-Leu-Nle-CHO

Synonyms: Cathepsin B Inhibitor; Cathepsin L; Calpain Inhibitor

Neosystem SC1327   MW 383.52   Competitive inhibitor of cathepsins B & L, & calpain; Sasaki, T et al, *J Enzyme Inhibition*, 3:195-201, 1990

Leu-Leu-Nva
Z-Leu-Leu-Nva-CHO

Synonyms: MG-115; Proteasome Inhibitor

Calbiochem 474780   MW 461.6 $C_{25}H_{39}N_3O_5$ >90% (HPLC); solid; soluble in DMSO | Potent reversible inhibitor; specifically inhibits the chymotrypsin-like activity of the proteasome & induces apoptosis in Rat-1 & PC12 cells; Palombella, VJ et al, *Cell*, 78:773, 1994; Rock, KL et al, *Cell*, 78:761, 1994; Lopes, UG et al, *J Biol Chem*, 272:12893, 1997; Lee, DH & Goldberg, AL, *J Biol Chem*, 271:27280, 1996; Vinitsky, A et al, *Biochemistry*, 31:9421, 1992

Leu-Leu-Nva
N-CBZ-Leu-Leu-Nva-CHO

Sigma C 6706   FW 461.6 $C_{25}H_{39}N_3O_5$ ≥90% (HPLC) | Rock, KL et al, *Cell*, 78:761, 1994; Palombella, VJ et al, *Cell*, 78:773, 1994

Leu-Leu-Nva
Carbobenzoxy-Leu-Leu-Nva-CHO

Synonyms: MG-115; Proteasome Inhibitor

Peptides International IAT-3170-v   Synthetic   MW 461.60 $C_{25}H_{39}N_3O_5$ >99% (HPLC); lyophilized amorphous powder | Inhibitor; Saito, Y et al, *Neurosci Letts*, 120, 1, 1990; Vinitsky, A et al, *Biochem*, 31:9421, 1992; Rock, KL et al, *Cell*, 78:761, 1994; Palombella, VJ et al, *Cell*, 78:773, 1994

Leu-Leu-Phe

Bachem H-6065.0250   MW 391.51 $C_{21}H_{33}N_3O_4$ Store at -15°C

Leu-Leu-Phe
Z-Leu-Leu-Phe-CHO

Synonyms: Proteasome Inhibitor II

Calbiochem 539162   MW 509.7 $C_{29}H_{39}N_3O_5$ ≥95% (HPLC); lyophilized solid; soluble in DMSO | Potent & cell-permeable proteasome inhibitor; inhibits the chymotrypsin-like activity of the pituitary multicatalytic proteinase complex (MPC); does not inhibit the peptidylglutamyl-peptide hydrolyzing (PGPH) activity of MPC even at concentrations of 200 μM; blocks the decay of IκB-α & IκB-β proteins in exponentially growing WEHI 231 cells; Schauer, SL et al, *J Immunol*, 160:4398, 1998; Orlowski, M et al, *Biochemistry*, 36:13946, 1997; Orlowski, M et al, *Biochemistry*, 32:1563, 1993

Leu-Leu-Phe Amide

Bachem H-3935.0050   MW 390.53 $C_{21}H_{34}N_4O_3$ Store at -15°C

ICN 155216   MW 390.5 $C_{21}H_{34}N_4O_3$

Leu-Leu-Ser-Lys-Arg-Gly-His-Cys-Pro-Arg-Ile-Leu-Phe-Arg-Cys-Pro-Leu-Ser-Asn-Pro-Ser-Asn-Lys-Cys-Trp-Arg-Asp-Tyr-Asp-Cys-Pro-Gly-Val-Lys-Lys-Cys-Cys-Glu-Gly-Phe-Cys-Gly-Lys-Asp-Cys-Leu-Tyr-Pro-Lys

Synonyms: Sodium Potassium ATPase Inhibitor I

Peptides International PSP-4216-s   Porcine synthetic   MW 5628.7 $C_{245}H_{378}N_{72}O_{65}S_8$ >99% (HPLC); lyophilized amorphous powder | Bioactive; Disulfide bonds: Cys[8]-Cys[37], Cys[15]-Cys[41], Cys[24]-Cys[36], Cys[30]-Cys[45]; Araki, K et al, *BBRC*, 164:496, 1989; Araki, K et al, *BBRC*, 172:42, 1990; Nishio, H et al, *Pept Res*, 5:227, 1992

Leu-Leu-Thr-Trp-Leu-Phe-Ser-Asn-Cys-Arg-Thr-Leu-Leu-Ser-Arg-Ala-Tyr-Gln-Ile-Leu

Synonyms: GP140 (781-800); SIVmac251 Peptide

Neosystem SC786   MW 2411.88   SIVmac251 peptide from subtype GP140; due to the presence of a Cys this peptide can contain the dimeric form

Leu-Leu-Tyr

Bachem H-3940.0050   MW 407.51 $C_{21}H_{33}N_3O_5$ Store at -15°C

Leu-Leu-Tyr
Z-Leu-Leu-Tyr-α-Keto Aldehyde

Bachem N-1715.0005   MW 553.66 $C_{30}H_{39}N_3O_7$ Store at -15°C

Leu-Leu-Tyr
Z-Leu-Leu-Tyr-CH₂F

Synonyms: Calpain Inhibitor IV

Calbiochem 208724   MW 557.7 $C_{30}H_{40}FN_3O_6$ Single spot purity (TLC); white solid; soluble in DMSO | A potent cell-permeable & irreversible calpain inhibitor; Angliker, H et al, *J Med Chem*, 35:216, 1992

Leu-Leu-Tyr

ICN 155217   MW 407.5 $C_{21}H_{33}N_3O_5$

Leu-Leu-Tyr
Z-Leu-Leu-Tyr-FMK

Synonyms: Calpain Inhibitor I

Kamiya   MW 557.5

Leu-Leu-Tyr

Sigma L 9890   FW 407.5 $C_{21}H_{33}N_3O_5$

Leu-Leu-Tyr-Glu-Met-Leu-Ala-Gly-Gln-Ala-Pro-Phe-Glu-Gly-Glu-Asp-Glu-Asp-Glu-Leu-Phe-Gln-Ser-Ile-Met-Glu-His-Asn-Val

Synonyms: Protein Kinase C (530-558)

American Peptide 86-0-24   MW 3354.7 $C_{148}H_{220}N_{34}O_{51}S_2$ Potent activator of PKC with a Ka of ~10 *mM*; House, C et al, *FEBS Lett*, 249:243, 1989

Leu-Leu-Tyr-Glu-Met-Leu-Ala-Gly-Gln-Ala-Pro-Phe-Glu-Gly-Glu-Asp-Glu-Asp-Glu-Leu-Phe-Gln-Ser-Ile-Met-Glu-His-Asn-Val Amide

Synonyms: Protein Kinase C (530-558)

Bachem H-8045.0500   MW 3354.72 $C_{148}H_{221}N_{35}O_{50}S_2$ Store at -15°C

Leu-Leu-Tyr-Glu-Met-Leu-Ala-Gly-Gln-Ala-Pro-Phe-Glu-Gly-Glu-Asp-Glu-Asp-Glu-Leu-Phe-Gln-Ser-Ile-Met-Glu-His-Asn-Val

Synonyms: Protein Kinase C Activating Peptide (530-558)

ICN 159656   MW 3355   Activates proteinkinase C

Leu-Leu-Tyr-Glu-Met-Leu-Ala-Gly-Gln-ala-Pro-Phe-Glu-Gly-Glu-Asp-Glu-Asp-Glu-Leu-Phe-Gln-Ser-Ile-Met-Glu-His-Asn-Val Amide

Synonyms: Protein Kinase C (530-558)

Sigma P 2303   FW 3354.7   ≥97% (HPLC) | Bioactive peptide; part of the catalytic domain of proteinkinase C & a potent activator of the enzyme; House, C et al, *FEBS Lett*, 249:243, 1989

Leu-Leu-Val-Phe

Bachem H-3945.0050   MW 490.64 $C_{26}H_{42}N_4O_5$ Store at -15°C

Leu-Leu-Val-Tyr
Suc-Leu-Leu-Val-Tyr AMC

Synonyms: Calpain Substrate; Chymotrypsin-Like Enzymes Substrate; Calpain I Substrate; Calpain II Substrate

Alexis 260-070   Fluorogenic substrate; Edelstein, CL et al, *Kidney Int*, 50:1150, 1996; Wang, KKW et al, *PNAS*, 93:6687, 1996

Leu-Leu-Val-Tyr

Bachem H-3950.0250   MW 506.64 $C_{26}H_{42}N_4O_6$ Store at -15°C

Leu-Leu-Val-Tyr
Suc-Leu-Leu-Val-Tyr-AMC

Synonyms: Proteasome Substrate III; Calpain Substrate; Chymotrypsin Substrate; Ingensin Substrate

Bachem I-1395.0025   Store at -15°C

Calbiochem 539142   MW 763.9 $C_{40}H_{53}N_5O_{10}$ ≥98% (HPLC); solid; soluble in DMSO; excitation max:~380 nm; emission max:~460 nm | Fluorogenic substrate; Tsubuki, S et al, *Biochem Biophys Res Comm*, 196:1195, 1993; Rock, KL et al, *Cell*, 78:761, 1994; Sasaki, T et al, *J Biol Chem*, 259:12489, 1984

Leu-Leu-Val-Tyr
N-Suc-L-Leu-L-Leu-L-Val-L-Tyr-MCA

Synonyms: Chymotrypsin Substrate

ICN 152086   Fluorogenic substrate

Leu-Leu-Val-Tyr
Suc-Leu-Leu-Val-Tyr-AMC

Synonyms: Chymotrypsin Substrate II; Chymotrypsin Substrate; Carboxypeptidase Y Substrate; Calpain Substrate; Proteasome Substrate; Ingensin Substrate

ICN 195962   MW 763.9 $C_{40}H_{53}N_5O_{10}$ >95% | Fluorogenic substrate for the quantitative determination of chymotrypsin

Neosystem SC1267   MW 763.90   Sensitive fluorogenic substrate for chymotrypsin, carboxypeptidase Y, calpain & proteasome; Sawada, H et al, *Experientia*, 39:377-378, 1983; Sasaki, T et al, *JBC*, 259:12489-12494, 1984; Kunugi, S et al, *Mol Pharm Biochem*, 153:37-40, 1985; Ishiura, S et al, *FEBS Lett*, 189:119-123, 1985; Heinemeyer, W et al, *EMBO J*, 10:555-562, 1991

Peptides International MLL-3120-v   MW 763.89 $C_{40}H_{53}N_5O_{10}$ >98% (HPLC); lyophilized amorphous powder | Sawada, H et al, *Experientia*, 39:377, 1983; Sasaki, T et al, *JBC*, 259:12849, 1984; Ishiura, S et al, *FEBS Letts*, 189:119, 1985; Tsukahara, T et al, *Eur J Biochem*, 177:26, 1988

Leu-Leu-Val-Tyr
N-Suc-Leu-Leu-Val-Tyr MCA

Sigma S 6510   FW 763.9 $C_{40}H_{53}N_5O_{10}$ ≥90% (HPLC) | Sawada, S et al, *Experientia*, 39:377, 1983; Substrate for chymotrypsin-like enzyme from ascidian sperm

Leu-Leu-Val-Tyr-Ser
Leu-Leu-Val-Tyr-Ser-βNA

Bachem K-1410.0050   MW 718.89 $C_{39}H_{54}N_6O_7$ Store at -15°C

Leu-Lue-Ser-Lys-Arg-Gly-His-Cys-Pro-Arg-Ile-Leu-Phe-Arg-Cys-Pro-Leu-Ser-Asn-Pro-Ser-Asn-Lys-Cys-Trp-Arg-Asp-Tyr-Asp-Cys-Pro-Gly-Val-Lys-Lys-Cys-Cys-Glu-Gly-Phe-Cys-Gly-Lys-Asp-Cys-Leu-Tyr-Pro-Lys

Synonyms: Sodium Potassium ATPase Inhibitor I

American Peptide 79-2-20   Porcine   MW 5628.7 $C_{245}H_{378}N_{72}O_{65}S_8$ Disulfide bonds: Cys[8]-Cys[37], Cys[15]-Cys[41], Cys[24]-Cys[36], Cys[30]-Cys[45]; Araki, K et al, *BBRC*, 164:496, 1989

Leu-Lys-Arg
BOC-Leu-Lys-Arg-AMC

Bachem I-1570.0050   MW 672.83 $C_{33}H_{52}N_8O_7$ Store at -15°C

Leu-Lys-Arg
BOC-Leu-Lys-Arg-MCA

Synonyms: Carboxyl Side of Paired Basic Residue Cleaving Enzyme Substrate

Peptides International MLK-3141-v   MW 672.83 $C_{33}H_{52}N_8O_7$ >98% (HPLC); lyophilized amorphous powder | Mizuno, K et al, *BBRC*, 144:807, 1987

Leu-Lys-Arg
N-t-BOC-Leu-Lys-Arg MCA Hydrochloride

Synonyms: Protease Substrate

Sigma B 5278   FW 709.3 $C_{33}H_{52}N_8O_7 \cdot HCl$ ~95% | Substrate for proteases which cleave on the carboxyl side of paired basic residues; Mizuno, K et al, *Biochem Biophys Res Commun*, 144:807, 1987

Leu-Lys-Glu-Lys-Asn-Leu-Tyr-Leu-Ser-Cys-Val-Leu-Lys-Asp-Asp-Lys-Pro-Thr-Leu-Gln-Leu-Glu-Ser-Val-Asp-Pro-Lys-Asn-Tyr-Pro

Synonyms: Interleukin Iβ (178-207)

Bachem H-8300.0500   Human   Store at -15°C

Leu-Lys-Glu-Thr-Ile-Asn-Glu-Glu-Ala-Ala-Glu-Trp-Asp-Arg-Val-His-Pro-Val

Synonyms: P25 (201-218); GAG P24 CA (69-86); HTLV I Peptide

Neosystem SC297   MW 2136.34   HTLV I peptide from subtype P25 (GAG P24 CA)

Leu-Lys-His-Ile-Val-Trp-Ala-Ser-Arg-Glu-Leu-Glu-Arg-Phe-Ala-Val

Synonyms: P18 (31-46); GAG P17 MA (31-46); HTLV I Peptide

Neosystem SC280   MW 1954.30   HTLV I peptide from subtype P25 (GAG P17 MA)

Leu-Lys-Lys-Phe-Asn-Ala-Arg-Arg-Lys-Leu-Lys-Gly-Ala-Ile-Leu-Thr-Thr-Met-Leu-Ala

Synonyms: Calmodulin-Dependent Protein Kinase II (290-309)

American Peptide 73-0-45   MW 2273.9 $C_{103}H_{185}N_{31}O_{24}S$ Potent calmodulin antagonist; Ca[2+]/calmodulin-dependent protein kinase II inhibitor; Payne, ME et al, *JBC*, 263:7190, 1989

Bachem H-9365.0001   MW 2273.86 $C_{103}H_{185}N_{31}O_{24}S$ Store at -15°C

Leu-Lys-Lys-Phe-Asn-Ala-Arg-Arg-Lys-Leu-Lys-Gly-Ala-Ile-Leu-Thr-Thr-Met-Leu-Ala Trifluoroacetate Salt

Synon Calmodulin Binding Domain; Calmodulin Kinase II (290-309)*yms*:

Calbiochem 208734 MW 2274.3 $C_{103}H_{185}N_{31}O_{24}S$ >97% (HPLC); lyophilized solid; soluble in water | Potent calmodulin antagonist; inhibits activation of calmodulin-dependent protein kinase II (IC_{50}=52 nM); Payne, MF et al, *J Biol Chem*, 263:7190, 1988; James, P et al, *Trends Biochem Sci*, 20:38, 1995; Yazawa, M et al, *Biochemistry*, 31:3171, 1992

Leu-Lys-Lys-Phe-Asn-Ala-Arg-Arg-Lys-Leu-Lys-Gly-Ala-Ile-Leu-Thr-Thr-Met-Leu-Ala

Synonyms: Protein Kinase Related Peptide; Calmodulin-Dependent Protein Kinase II (290-309)

Sigma C 4926 FW 2273.8 ≥97% (HPLC); peptide content:~70% | Bioactive peptide; potent calmodulin antagonist; Payne, ME et al, *J Biol Chem*, 263:7190, 1989

Leu-Lys-Phe-Ser-Lys-Lys-Phe
Ac-Leu-Lys-Phe-Ser-Lys-Lys-Phe

Bachem H-3224.0001 Store at -15°C

Leu-Lys-Pro-Val-Lys-Lys-Lys-Lys-Ile-Lys-Arg-Glu-Ile-Lys-Ile-Leu-Glu-Asn-Leu-Arg-Gly-Gly-Cys

Synonyms: Casein Kinase II Blocking Peptide, α-Subunit (70-89)

Calbiochem 231575 Human synthetic MW 2692.4 $C_{121}H_{223}N_{37}O_{29}S$ Lyophilized | Based on the human CK2α sequence (70-89); synthesized with a C-terminal GGC linker peptide & the peptide coupled to KLH; this sequence is identical to that found in the α-subunit of CK2 from chicken, mouse, rabbit, rat, *C. elegans, Drosophila, Xenopus* & zebra fish; the peptide coupled to KLH was used as the immunogen for the production of Anti-CK2α; for use in immunoabsorption for immunoprecipitation, immunocytochemistry, Western blotting & dot blots

Leu-Met
Z-Leu-Met

Bachem C-2155.0001 MW 396.51 $C_{19}H_{28}N_2O_5S$ Store at -15°C

Leu-Met

Bachem G-2555.0250 MW 262.37 $C_{11}H_{22}N_2O_3S$ Store at -15°C

ICN 155218 MW 262.4 $C_{11}H_{22}N_2O_3S$ Crystalline

Sigma L 9875 FW 262.4 $C_{11}H_{22}N_2O_3S$

Leu-Met Amide

Bachem G-2560.0250 MW 261.39 $C_{11}H_{23}N_3O_2S$ Store at -15°C

Leu-Met-Asp-Lys-Glu-Ala-Val-Tyr-Phe-Ala-His-Leu-Asp-Ile-Ile-Trp
Ac-Leu-Met-Asp-Lys-Glu-Ala-Val-Tyr-Phe-Ala-His-Leu-Asp-Ile-Ile-Trp

Synonyms: Endothelin B Receptor Agonist; BQ-3020; Endothelin I (6-21), *N*-Ac-(Ala[11,15])-; Endothelin Receptor Agonist (6-21), Ac-(Ala[11,15])-; Endothelin I (6-21), Ac-(Ala[11,15])-; Endothelin Receptor Agonist (6-21), *N*-Ac-(Ala[11,15])-

Alexis 155-015 MW 2006.3 $C_{96}H_{140}N_{20}O_{25}S$ Selective ligand for the ET_B receptor; Ihara, M et al, *Life Sci*, 51:47, 1992; Sakamoto, A et al, *J Biol Chem*, 268:8547, 1993; Adachi, M et al, *Eur J Biochem*, 220:37, 1994

American Peptide 88-2-42 MW 2006.6 $C_{96}H_{140}N_{20}O_{25}S$ Highly potent, selective receptor agonist; does not exhibit any significant binding to ET_A-receptors; Ihara, M et al, *Life Sci*, 61:47, 1992; Sakamoto, A et al, *J Biol Chem*, 268:8547, 1993; Adachi, M et al, *Eur J Biochem*, 220:37, 1994

Bachem H-8520.0500 MW 2006.35 $C_{96}H_{140}N_{20}O_{25}S$ Store at -15°C

Calbiochem 05-23-3839 MW 2006.4 $C_{96}H_{140}N_{20}O_{25}S$ >95% (HPLC); solid; soluble in 2.5% NH_4OH | Highly potent, selective receptor agonist; does not exhibit any significant binding to ET_A-receptors; Saeki, T et al, *Biochem Biophys Res Comm*, 179:286, 1991; Bacon, CR et al, *J Endocrinol*, 144:127, 1995; Adachi, M et al, *Eur J Biochem*, 220:37, 1994; Sakamoto, A et al, *J Biol Chem*, 268:8547, 1993

Neosystem SC872 MW 2006.34 Bioactive ETB receptor agonist; Ihara, M et al, *Life Sci*, 51:PL47-52, 1992; Takei, K et al, *Life Sci*, 53:PL111-115, 1993; Sakamoto, A et al, *JBC*, 268:8547-8553, 1993; Karet, FE et al, *Kidney Int*, 44:36-42, 1993; Tsunoda, H et al, *Br J Pharmacol*, 110:1437-1440, 1993

Leu-Met-His-Asn-Leu-Gly-Lys-His-Leu-Asn-Ser-Met-Glu-Arg-Val-Glu-Trp-Leu-Arg-Lys-Lys-Leu-Gln-Asp-Val-His-Asn-Phe-Val-Ala-Leu-Gly-Ala-Pro-Leu-Ala-Pro-Arg-Asp-Ala-Gly-Ser-Gln-Arg-Pro-Arg-Lys-Lys-Glu-Asp-Asn-Val-Leu-Val-Glu-Ser-His-Glu-Lys-Ser-Leu-Gly-Glu-Ala-Asp-Lys-Ala-Asp-Val-Asn-Val-Leu-Thr-Lys-Ala-Lys-Ser-Gln

Synonyms: Parathyroid Hormone (7-84)

Bachem H-3084.0500 Human Store at -15°C

Leu-Met-Ile-Trp-Asp-Gln-Lys-Ser-Leu-Lys-Arg-Cys Trifluoroacetate Salt

Synonyms: Granzyme B (18-29)

Biogenesis 4741-5100 Human synthetic MW 1521 Purified; lyophilized

Leu-Met-Tyr-Pro-Thr-Tyr-Leu-Lys

Synonyms: Vasoactive Intestinal Peptide Receptor Binding Inhibitor; L-8-K

American Peptide 48-1-35 MW 1028.2 $C_{50}H_{77}N_9O_{12}S$ Singh, H et al, *NY Acad Sci*, Abstract 33, 1987

ICN 153143 Singh, H et al, *NY Acad Sci*, Abst 33, 1987

Sigma L 0146 FW 1028.3 ≥97% (HPLC) | Bioactive peptide

Leu$_n$
L-Leu-(L-Leu)$_n$-L-Leu

Fluka 81361 MW 3-15k

Leu-Nle
N-Benzyloxycarbonyl-L-Leu-Nle-CHO

Synonyms: Calpeptin; Calpain Inhibitor; Cathepsin L Inhibitor

Alexis 260-014 MW 362.5 $C_{20}H_{30}N_2O_4$ ≥98%; white lyophilized powder; soluble in DMSO or DMF | Membrane-permeable inhibitor; promotes neurite elongation in differentiating PC12 cells; prevents Ca^{2+}-ionophore induced degradation of actin binding protein & P235 in platelets; inhibits myosin light chain phosphorylation in platelets stimulated by collagen, ionomycin or thrombin; Sasaki, T et al, *J Enz Inhib*, 3:195, 1990; Mehdi, S, *TIBS*, 16:150, 1991; Saito, K & Nixon, RA, *Neurochem Res*, 18:231, 1993; Pinter, M et al, *Neurosci Lett*, 170:91, 1994

Leu-Nle
Z-Leu-Nle-CHO

Synonyms: Calpeptin; Benzyloxycarbonyldipeptidyl Aldehyde

Calbiochem 03-34-0051   MW 362.5 $C_{20}H_{30}N_2O_4$ ≥95% (TLC); lyophilized solid; soluble in DMSO & DMF; sold under license of U.S. Patent 5,081,284 issued to Suntory | A cell-permeable calpain inhibitor; Inactivates calpain I (ID_{50}=52 nM), calpain II (ID_{50}=34 nM) & papain (ID_{50}=138 nM); promotes neurite elongation in differentiating PC12 cells; prevents Ca^{2+}-ionophore-induced degradation of actin binding protein & P_{235} in platelets; inhibits myosin light chain phosphorylation in platelets stimulated by collagen, ionomycin or thrombin; inhibits the growth of estrogen receptor positive breast cancer cells; Shiba, E et al, *Anticancer Res*, 16:773, 1996; Pinter, M et al, *Neurosci Lett*, 170:93, 1994; Saito, K & Nixon, RA, *Neurochem Res*, 18:231, 1993; Tsujinaka, T et al, *Biochem Biophys Res Comm*, 153:1201, 1988

Leu-Nle
N-CBZ-L-Leu-Nle-CHO

Synonyms: Calpeptin; Calpain Inhibitor; Cathepsin L Inhibitor

ICN 159563   MW 362.5 $C_{20}H_{30}N_2O_4$ Sasaki etal, *J Enz Inhib*, 3:195, 1990

Leu-Phe
BOC-Leu-(®)-Phe

Bachem A-3030.0250   Store at -15°C

Leu-Phe
Z-Leu-Phe

Bachem C-2165.0001   MW 412.49 $C_{23}H_{28}N_2O_5$ Store at -15°C

Leu-Phe
Z-Leu-Phe Amide

Bachem C-2170.0001   MW 411.5 $C_{23}H_{29}N_3O_4$ Store at -15°C

Leu-Phe

Bachem G-2565.0001   MW 278.35 $C_{15}H_{22}N_2O_3$ Store at -15°C

Leu-Phe
DL-Leu-DL-Phe

Bachem G-2570.0001   MW 278.35 $C_{15}H_{22}N_2O_3$ Store at -15°C

Leu-Phe
((2*R*,4*R*,5*S*)-2-Benzyl-5-(BOC-Amino)-4-Hydroxy-6-Phenyl-Hexanoyl)-Leu-Phe Amide

Synonyms: L-686,458

Bachem H-5106.0001   Store at -15°C

Leu-Phe

ICN 102176   MW 278.4 $C_{15}H_{22}N_2O_3$ 99%; crystalline

Leu-Phe
DL-Leu-DL-Phe

ICN 102192   MW 278.4 $C_{15}H_{22}N_2O_3$ Crystalline

Leu-Phe

Sigma L 3127   FW 278.4 $C_{15}H_{22}N_2O_3$

Leu-Phe
DL-Leu-DL-Phe

Sigma L 4752   FW 278.4 $C_{15}H_{22}N_2O_3$

Leu-Phe Amide Hydrochloride

Bachem G-2575.0001   MW 313.83 $C_{15}H_{23}N_3O_2 \cdot HCl$ Store at -15°C

Leu-Phe-Ile-His-Phe-Arg-Ile-Gly-Cys-Arg-His-Ser-Arg

Synonyms: VPR (68-80); HTLV I Peptide

Neosystem SC253   MW 1641.96   HTLV I peptide from subtype VPR; due to the presence of a Cys this peptide can contain the dimeric form

Leu-Phe-Leu

Bachem H-5390.0250   MW 391.51 $C_{21}H_{33}N_3O_4$ Store at -15°C

Leu-Phg-Arg
D-Leu-Phg-Arg-AMC 2AcOH

Synonyms: Pefafluor FIXa3688; Pefa-3688

Pentapharm 095-04   MW 696.8   Excitation wavelength 360 nm, emission wavelength 460 nm | Highly sensitive fluorogenic peptide substrate for factor IXa; v_{max}:1.12 µmol/min, K_M:0.028mM

Leu-Phg-Arg
D-Leu-Phg-Arg-pNA 2AcOH

Synonyms: Pefachrome®FIXa; Pefa-3960

Pentapharm 095-11, 095-01   MW 660.8   Highly sensitive chromogenic peptide substrate for factor IXa

Leu-Pro
BOC-Leu-Pro

Bachem A-1900.0001   MW 328.41 $C_{16}H_{28}N_2O_5$ Store at -15°C

Leu-Pro
FMOC-Leu-Pro

Bachem B-2185.0001   MW 450.54 $C_{26}H_{30}N_2O_5$ Store at -15°C

Leu-Pro Hydrochloride

Bachem G-2585.0250   MW 264.75 $C_{11}H_{20}N_2O_3 \cdot HCl$ Store at -15°C

ICN 155220   MW 264.8 $C_{11}H_{20}N_2O_3 \cdot HCl$ Crystalline

Sigma L 8753   FW 264.8 $C_{11}H_{20}N_2O_3 \cdot HCl$

Leu-Pro Trifluoroacetate Salt

Sigma L 6006   FW 342.3 $C_{11}H_{20}N_2O_3 \cdot C_2HF_3O_2$ White powder

Leu-Pro-Gly-Leu-Pro-Ser-Ala-Ala-Ser-Ser-Glu-Asp-Ala-Gly-Gln-Ser Amide

Synonyms: Galanin Message Associated Peptide (44-59); Prepro-Galanin (108-123)

Bachem H-7715.0001   MW 1485.57 $C_{61}H_{100}N_{18}O_{25}$ Store at -15°C

ICN 152867   MW 1485.6

Leu-Pro-Leu-Arg-Phe
Leu-Pro-Leu-Arg-Phe Amide 2AcOH Dihydrate

Synonyms: Chicken Brain Peptide

Peptides International PBC-4144   Synthetic   MW 799.97 $C_{32}H_{53}N_9O_5 \cdot 2CH_3COOH \cdot 2H_2O$ >99% (HPLC); lyophilized amorphous powder | Bioactive; Dockray, GJ et al, *Nature*, 305:328, 1983;

Leu-Pro-Leu-Arg-Phe Amide

American Peptide 60-0-42   MW 643.9 $C_{32}H_{53}N_9O_5$

Leu-Pro-Leu-Glu-Glu-Ser-Glu-His-Leu-Pro-Leu-Ser-Thr-Val Acetate Salt

Synonyms: Neuropeptide Y Receptor IV (345-357)

Biogenesis 6732-0470   Synthetic   >80% (HPLC); lyophilized

Leu-Pro-Phe-Phe-Asp

Synonyms: Amyloid α-Protein (17-21), (Pro^{18},Asp^{21})-; Sheet Breaker Peptide iAβ5 β-

Bachem H-4876.0005   MW 637.73 $C_{33}H_{43}N_5O_8$ Store at -15°C

Peptides International PAB-3615-PI   Synthetic   MW 637.73 $C_{33}H_{43}N_5O_8$ >98% (HPLC); white powder | Bioactive; inhibitor of amyloid deposition; Soto, C et al, *Nat Med*, 4:822, 1998

Peptides International PAB-4358-v   Synthetic   MW 637.73 $C_{33}H_{43}N_5O_8$ >99% (HPLC); lyophilized amorphous powder | Bioactive; inhibitor of amyloid deposition; Soto, C et al, *Nat Med*, 4:822, 1998

Leu-Pro-Pro-Gly-Pro-Leu-Pro-Arg-Pro Amide Dihydrochloride

Synonyms: Antho-RP Amide I

Bachem H-9730.0001   Store at -15°C

Leu-Pro-Pro-Leu-Glu-Arg-Leu-Thr-Leu-Asp-Cys-Asn-Glu-Asp

Synonyms: REV (75-88); HTLV I Peptide

Neosystem SC194   MW 1627.83   HTLV I peptide from subtype REV; due to the presence of a Cys this peptide can contain the dimeric form

Leu-Pro-Pro-Ser

Bachem H-7635.0005   MW 412.49 $C_{19}H_{32}N_4O_6$ Store at -15°C

Leu-Pro-Pro-Ser-Arg

Synonyms: IgG$_1$, Human Fc (351-355); Lymphocyte Activating Pentapeptide

Bachem H-6850.0005   MW 568.67 $C_{25}H_{44}N_8O_7$ Store at -15°C

ICN 153144   MW 568.7 $C_{25}H_{44}N_8O_7$ Hobbs, MV etal, *J Immunol*, 138:2581, 1987

Sigma L 3146   FW 568.7 $C_{25}H_{44}N_8O_7$ ≥97% (HPLC) | Bioactive peptide; fragment from the Fc region of human IgG1; Hobbs, MV et al, *J Immunol*, 138:2581, 1987

Leu-Pro-Val-Asn-Ser-Pro-Met-Asn-Lys-Gly-Asp-Thr-Glu-Val-Met Acetate Salt

Synonyms: Chromogranin A (1-15)

Biogenesis 2095-0230   Synthetic   Semi-pure; lyophilized

Leu-Ser
Z-Leu-Ser

Bachem C-2175.0001   MW 352.39 $C_{17}H_{24}N_2O_6$ Store at -15°C

Leu-Ser
Z-Leu-Ser-OMe

Bachem C-3305.0001   MW 366.41 $C_{18}H_{26}N_2O_6$ Store at -15°C

Leu-Ser

Bachem G-2595.0250   MW 218.25 $C_9H_{18}N_2O_4$ Store at -15°C

ICN 155221   MW 218.3 $C_9H_{18}N_2O_4$ White powder

Sigma L 2502   FW 218.3 $C_9H_{18}N_2O_4$ White powder

Leu-Ser Amide Hydrochloride

Bachem G-2600.0250   Store at -15°C

Leu-Ser-Ala-Leu

Bachem H-2144.0050   Store at -15°C

Leu-Ser-Arg-Leu-Phe-Asp-Asn-Ala

Synonyms: Growth Hormone (6-13

American Peptide 87-0-70   Human   MW 935.2 $C_{41}H_{66}N_{12}O_{13}$

Leu-Ser-Glu-Leu-Ile-Ile-Asn-Asn-Ala-Thr-Glu-Glu-Leu-Leu-Ile-Lys-Gly-Leu

Synonyms: Saposin C (15-32)

Bachem H-3692.0500   Rat   Store at -15°C

Leu-Ser-Glu-Thr-Lys-Pro-Ala-Val Trifluoroacetate Salt

Synonyms: Protein Kinase C$_\varepsilon$ Translocation Inhibitor Peptide

Calbiochem 539542   MW 844.0 $C_{37}H_{65}N_9O_{13}$ ≥98% (HPLC); lyophilized solid; soluble in water | A scrambled peptide with an identical AA composition to that of PKC$_\varepsilon$ Translocation Inhibitor Peptide; used as a negative control; Yedovitzky, M et al, *J Biol Chem*, 272:1417, 1997

Leu-Ser-Lys-Leu

Bachem H-4756.0250   MW 459.59 $C_{21}H_{41}N_5O_6$ Store at -15°C

Leu-Ser-Lys-Leu Amide

Bachem H-4786.0250   MW 458.6 $C_{21}H_{42}N_6O_5$ Store at -15°C

Leu-Ser-Phe

Bachem H-3960.0250   MW 365.43 $C_{18}H_{27}N_3O_5$ Store at -15°C

ICN 155222   MW 365.4 $C_{18}H_{27}N_3O_5$

Sigma L 8140   FW 365.4 $C_{18}H_{27}N_3O_5$

Leu-Ser-Phe-Nle-Ala-Ile
Leu-Ser-Phe-Nle-Ala-Ile-OMe Trifluoroacetate Salt

Bachem M-1545.0025   Store at -15°C

Leu-Ser-Phe-Nle-Ala-Leu
Leu-Ser-Phe(NO₂)-Nle-Ala-Leu-OMe

American Peptide 79-3-20   MW 722.9 $C_{34}H_{56}N_7O_{10}$

Leu-Ser-Phe-Nle-Ala-Leu
Leu-Ser-p-Nitro-Phe-Nle-Ala-Leu-OMe Trifluoroacetate Salt

Synonyms: Pepsin Substrate; Chymosin Substrate

Bachem M-1540.0025   MW 835.88 $C_{34}H_{55}N_7O_{10} \cdot C_2HF_3O_2$ Store at -15°C

ICN 153145 $C_{15}H_{22}N_2O_3$ Useful for chymosin activity determination; Martin, P etal, *Eur J Biochem*, 122:31, 1982

Leu-Ser-Phe-Nle-Ala-Leu
Leu-Ser-p-Nitro-Phe-Nle-Ala-Leu Methyl Ester

Sigma L 8138   FW 721.8 $C_{34}H_{55}N_7O_{10}$ ≥97% (HPLC) | Bioactive peptide

Leu-Ser-Pro-Glu-Met-Lys-Thr-Phe-Val-Asp-Gln-Tyr-Gly-Cys

Synonyms: Calbindin (53-65), ~C

Biogenesis 1716-7060   Human, rat synthetic   Semi-pure; lyophilized

Leu-Ser-Ser-Phe-Val-Arg-Ile Amide

Synonyms: LSSFVRI Amide

Bachem H-1644.0001   MW 820 $C_{38}H_{65}N_{11}O_9$ Store at -15°C

Leu-Ser-Ser-Phe-Val-Arg-Ile Amide Acetate Salt

Sigma L 2031   FW 820.0 (free base) $C_{38}H_{65}N_{11}O_9$ ≥97% (HPLC) | Bioactive peptide; inhibits Helix central neurons; Pedder, S et al, *Comp Biochem Physiol*, 103C:441, 1992

Leu-Ser-Thr-Arg
N-α-t-BOC-L-Leu-L-Ser-L-Thr-L-Arg-MCA

Synonyms: Protein C Activated

ICN 150500   MW 732.8 $C_{34}H_{52}N_8O_{10}$ Fluorogenic substrate

Leu-Ser-Thr-Arg
D-Leu-Ser-Thr-Arg-pNA

ICN 155223   MW 595.7 $C_{25}H_{41}N_9O_8$

Leu-Ser-Thr-Arg
BOC-Leu-Ser-Thr-Arg-MCA

Synonyms: Protein C Activated Substrate

Peptides International MLS-3112-v   MW 732.83 $C_{34}H_{52}N_8O_{10}$ >98% (HPLC); lyophilized amorphous powder | Ohno, Y et al, *J Biochem*, 90:1387, 1981

Leu-Ser-Thr-Arg
BOC-Leu-Ser-Thr-Arg-pNA AcOH Hydrate

Synonyms: Protein C Activated Substrate

Peptides International SLR-3125   MW 773.84 $C_{30}H_{49}N_9O_{10}$ · CH_3COOH · H_2O >98% (HPLC); amorphous powder

Leu-Ser-Thr-Arg
N-t-BOC-Leu-Ser-Thr-Arg MCA

Synonyms: Protein C Substrate

Sigma B 4636   FW 732.8 $C_{34}H_{52}N_8O_{10}$ Fluorogenic substrate for activated protein C; Ohno, Y et al, *J Biochem*, 90:1387, 1981

Leu-Ser-Thr-Arg
N-t-BOC-Leu-Ser-Thr-Arg pNA

Sigma B 8666   FW 695.8 $C_{30}H_{49}N_9O_{10}$ ≥97% (HPLC); peptide content:≥80%

Leu-Ser-Thr-Arg
D-Leu-Ser-Thr-Arg pNA

Sigma L 1391   FW 595.7 $C_{25}H_{41}N_9O_8$

Leu-Ser-Thr-Arg-
BOC-Leu-Ser-Thr-Arg-AMC Acetate Salt

Synonyms: Protein C Substrate

ICN 195930   MW 792.9 $C_{34}H_{52}N_8O_{10}$ · $C_2H_4O_2$ ≥99% | Fluorogenic substrate for activated Protein C quantitation

Leu-Ser-Trp-Asp-Leu-Pro-Glu-Pro-Arg-Ser-Arg-Ala-Gly-Lys-Ile-Arg-Val-His-Pro-Arg-Gly-Asn-Leu-Trp-Ala-Thr-Gly-His-Phe-Met Amide

Synonyms: Neuromedin B (1-30)

American Peptide 62-0-52   MW 3485.1 $C_{157}H_{243}N_{51}O_{38}S$ Minamino, N et al, *BBRC*, 130:685, 1985

Leu-Thr-Arg
D-Leu-Thr-Arg-pNA

Bachem L-1380.0050   MW 508.58 $C_{22}H_{36}N_8O_6$ Store at -15°C

Leu-Thr-Arg
N-α-t-BOC-L-Leu-L-Thr-L-Arg-MCA

ICN 150501   MW 645.8 $C_{31}H_{47}N_7O_8$ Fluorogenic substrate

Leu-Thr-Arg
BOC-Leu-Thr-Arg-MCA

Synonyms: Factor VIIa-Tf Substrate

Peptides International MLT-3106-v   MW 645.76 $C_{31}H_{47}N_7O_8$ >98% (HPLC); lyophilized amorphous powder

Leu-Thr-Arg-Pro-Arg-Tyr

Synonyms: Pancreatic Polypeptide (31-36)

American Peptide 46-1-73   Human   MW 805.0 (free acid) $C_{36}H_{60}N_{12}O_9$

Leu-Thr-Arg-Pro-Arg-Tyr Amide

Synonyms: Pancreatic Polypeptide (31-36)

American Peptide 46-1-71   Human   MW 804.0 $C_{36}H_{61}N_{13}O_9$

Bachem H-6895.0005   Human   MW 803.96 $C_{36}H_{61}N_{13}O_8$ Store at -15°C

ICN 154558   Human   MW 804

Leu-Thr-Arg-Pro-Arg-Tyr Free Acid

Synonyms: Pancreatic Polypeptide (31-36)

Bachem H-6900.0005   Human   Store at -15°C

ICN 154559   Human   MW 805

Leu-Thr-Cys-Val-Lys-Ser-Asn-Ser-Ile-Trp-Phe-Pro-Thr-Ser-Glu-Asp-Cys-Pro-Asp-Gly-Gln-Asn-Leu-Cys-Phe-Lys-Arg-Trp-Gln-Tyr-Ile-Ser-Pro-Arg-Met-Tyr-Asp-Phe-Thr-Arg-Gly-Cys-Ala-Ala-Thr-Cys-Pro-Lys-Ala-Glu-Tyr-Arg-Asp-Val-Ile-Asn-Cys-Cys-Gly-Thr-Asp-Lys-Cys-Asn-Lys

Synonyms: Muscarinic Toxin VII; Muscarinic Acetylcholine Receptor I Agonist

Peptides International PMT-4340-s   *Dendroaspis angusticeps* (green mamba) synthetic   MW 7472.5 $C_{322}H_{484}N_{90}O_{98}S_9$ >95% (HPLC); lyophilized amorphous powder | Bioactive; Adem, A & E Karlsson, *Life Sci*, 60:1069, 1997

Leu-Thr-Cys-Val-Thr-Ser-Lys-Ser-Ile-Phe-Gly-Ile-Thr-Thr-Glu-Asn-Cys-Pro-Asp-Gly-Gln-Asn-Leu-Cys-Phe-Lys-Lys-Trp-Tyr-Tyr-Ile-Val-Pro-Arg-Tyr-Ser-Asp-Ile-Thr-Trp-Gly-Cys-Ala-Ala-Thr-Cys-Pro-Lys-Pro-Thr-Asn-Val-Arg-Glu-Thr-Ile-Arg-Cys-Cys-Glu-Thr-Asp-Lys-Cys-Asn-Glu

Synonyms: Muscarinic Toxin I; Muscarinic Acetylcholine Receptor I Agonist

Peptides International PMT-4341-s   *Dendroaspis angusticeps* (green mamba) synthetic   MW 7509.6 $C_{326}H_{499}N_{87}O_{101}S_8$ >99% (HPLC); lyophilized amorphous powder | Bioactive; Jolkkonen, M et al, *Toxicon*, 33:399, 1995; Jerusalinsky, D & AL Harvey, *Trends Pharmacol Sci*, 15:424, 1994; Adem, A & E Karlsson, *Life Sci*, 60:1069, 1997

Leu-Thr-Phe-Lys-Phe-Tyr-Met-Pro-Lys-Lys-Ala

Synonyms: Interleukin II (60-70)

American Peptide 85-1-15   MW 1373.7 $C_{68}H_{104}N_{14}O_{14}S$

Leu-Thr-Thr-Gly-Ser-Val-Val-Ile-Val-Gly-Arg-Ile-Ile-Leu-Ser-Gly-Arg-Pro-Ala-Val-Val-Pro-Asp

Synonyms: HCV NS4A Protein (18-40), JT Strain

American Peptide 72-3-42   MW 2319.8 $C_{104}H_{183}N_{29}O_{30}$
Kca/Km coefficient is increased by ~40 times after the addition of ~10 fold molar excess of this peptide over the NS3 protease; enhances the cleavage rate of the S5A/5B peptide substrate by the NS3 protease; Shimizu, Y et al, *J Virol*, 70:127, 1996

Leu-Thr-Val-Val-Cys-Ala-Ala-Gly-Ala-Glu-Pro-Leu-Pro-Arg-Asn-Gly-Asp-Gln-Trp-His-Glu-Ile-Arg-Gln-Gly-Arg-Leu-Pro-(Gly-Gly-Cys)

Synonyms: WEE1 Protein Kinase Blocking Peptide (505-532)

Calbiochem 683703   Human synthetic   MW 3301.8
$C_{141}H_{226}N_{46}O_{42}S_2$ Lyophilized | Based on kinase subdomain X (505-532) of human WEE1 protein kinase with a GGC linker arm added; this sequence is identical to rat WEE1 sequence over these residues; this peptide coupled to KLH, was used as the immunogen for the production of Anti-WEE1; for use in immunoabsorption for immunoprecipitation, immunocytochemistry, Western blotting & dot blots

Leu-Trp
N-(α-Rhamnopyranosyloxyhydroxyphosphinyl)-L-Leu-L-Trp

Synonyms: Phosphoramidon; Endothelin Converting Enzyme Inhibitor; Thermolysin Inhibitor; Neutral Endopeptidase 24.11 Inhibitor; Natriuretic Peptide Degradation Enzyme, Atrial

American Peptide 88-1-37   MW 543.5 $C_{23}H_{32}N_3O_{10}P$ Non-selective inhibitor; Suda, H et al, *J Antibiotics*, 26:621, 1973; Stephenson, SL et al, *Biochem J*, 243:183, 1987; Roques, BP & Beaumont, A, *Trends Pharmacol Sci*, 11:245, 1990

Leu-Trp
Z-Leu-Trp

Bachem C-2180.0001   MW 451.52 $C_{25}H_{29}N_3O_5$ Store at -15°C

Leu-Trp

Bachem G-2605.0250   MW 317.39 $C_{17}H_{23}N_3O_3$ Store at -15°C

ICN 155224   MW 317.4 $C_{17}H_{23}N_3O_3$ Light tan crystals

Leu-Trp
(N,N-Hexamethylene) Carbamoyl-Leu-D-Trp(N-Me)-D-2-Pya

Synonyms: FR-139317

Neosystem SC490   MW 604.71   Bioactive; endothelin-related peptide; ETA receptor antagonist; Aramori, I et al, *Mol Pharmacol*, 43:127-131, 1993; Sogabe, K et al, *J Pharmacol Exp Ther*, 264:1040-1046, 1993; Cardell, LO et al, *Br J Pharmacol*, 108:448-452, 1993; Nirei, H et al, *Life Sci*, 52:1869-1874, 1993; Benigni, A et al, *Kidney Int*, 44:440-444, 1993; Doherty, AM et al, *J Med Chem*, 36:2585-2594, 1993; Schoeffter, P et al, *Eur J Pharmacol*, 241:165-169, 1993; Battistini, B et al, *Br J Pharmacol*, 111:1009-1016, 1994

Leu-Trp

Sigma L 0503   FW 317.4 $C_{17}H_{23}N_3O_3$ Light tan crystals

Leu-Trp
N-(α-Rhamnopyranosyloxy-Hydroxyphosphinyl)-Leu-Trp Free Acid Sodium Salt

Synonyms: Phosphoramidon; Enkephalinase Inhibitor; Metallo-Endopeptidase Inhibitor; Thermolysin Inhibitor

Sigma R 7385   FW 543.5 $C_{23}H_{34}N_3O_{10}P$ ≥97% (TLC) | Bioactive peptide; potent inhibitor of thermolysin & other bacterial metallo-endopeptidases; Umezawa, H, *Meth Enzymol*, 45:678, 1976

Leu-Trp
N-(α-Rhamnopyranosyloxyhydroxyphosphinyl)-L-Leu-L-Trp

Synonyms: Phosphoramidon

Alexis 260-011   Microbial   MW 623.5 $C_{23}H_{32}N_3O_{10}P \cdot Na \cdot 2H_2O$ ≥98%; slightly yellow powder; soluble in water | Inhibitor of thermolysin & neutral endopeptidase-24.11 (ANP degradation enzyme) endothelin converting enzyme inhibitor; Suda, H et al, *J Antibiot*, 26:621, 1973; Stephenson, SL & Kenny, AJ, *Biochem J*, 243:183, 1987; Roques, BP & Beaumont, A, *TIPS*, 11:245, 1990; Matsumura, Y et al, *Eur J Pharmacol*, 185:103, 1990

Leu-Trp
N-(α-Rhamnopyranosyloxyhydroxyphosphinyl)-Leu-Trp Sodium Salt

Synonyms: Phosphoramidon; Thermolysin Inhibitor; Collagenase Inhibitor

ICN 152851   Microbial   MW 588.5 $C_{23}H_{34}N_3O_{10}PNa_2$ Suda, H et al, *J Antibiot*, 26:621, 1963

Leu-Trp
N-(α-Rhamnopyranosyloxyhydroxyphosphinyl)-L-Leu-L-Trp Disodium Dihydrate

Synonyms: Phosphoramidon; Thermolysin Inhibitor; Neutral Endopeptidase 24.11 Inhibitor; Natriuretic Peptide Degradation Enzyme Inhibitor, Atrial; Endothelin Converting Enzyme Inhibitor

Peptides International IPO-4082   Microbial   MW 623.01
$C_{23}H_{32}N_3O_{10}P \cdot 2Na \cdot 2H_2O$ Amorphous powder; integrity assessed by activity; bulk | Inhibitor; Suda, H et al, *J Antibiotics*, 26:621, 1973; *Stephenson, SL & AJ Kenny, Biochem J, 243:183, 1987*; Roques, BP & A Beaumont, *Trends Pharmacol Sci*, 11:245, 1990; Matsumura, Y et al, *Eur J Pharmacol*, 185:103, 1990

Leu-Trp
N-(α-Rhamnopyranosyloxyhydroxyphosphinyl)-L-Leu-L-Trp Sodium Salt

Synonyms: Phosphoramidon; Thermolysin Inhibitor; Neutral Endopeptidase 24.11 Inhibitor; Natriuretic Peptide Degradation Enzyme Inhibitor, Atrial; Endothelin Converting Enzyme Inhibitor

Peptides International IPO-4082-v   Microbial   MW 543.51
$C_{23}H_{34}N_3O_{10}P$ Lyophilized amorphous powder; integrity assessed by activity | Inhibitor; Suda, H et al, *J Antibiotics*, 26:621, 1973; *Stephenson, SL & AJ Kenny, Biochem J, 243:183, 1987*; Roques, BP & A Beaumont, *Trends Pharmacol Sci*, 11:245, 1990; Matsumura, Y et al, *Eur J Pharmacol*, 185:103, 1990

Leu-Trp
Homopiperidinyl-CO-Leu-D-Nim-Formyl-D-Trp

Synonyms: BQ-610; ETA-Selective Antagonist; Endothelin A Receptor Antagonist

Peptides International PED-3610-PI   Synthetic   MW 656.79
$C_{36}H_{44}N_6O_6$ >98% (HPLC); white powder | Inhibitor; Ishikawa, K et al in CH Schneider & AN Eberle (Eds), *Peptides*, 1992, ESCOM, Leiden, 1993, p 685

Leu-Trp-Ala
(Hexahydro-1H-Azepinyl)Carbonyl-Leu-D-Trp(Me)-D-Ala(2-Pyridyl)

Synonyms: Endothelin A Receptor Antagonist; FR-139317

Alexis 155-017   MW 605 $C_{32}H_{43}N_6O_5$ Selective ET_A antagonist; Loffler, B-M et al, *FEBS Lett*, 333:108, 1993; Doherty, AM et al, *J Med Chem*, 36:2585, 1993

Leu-Trp-Ala
N,N-Hexamethylene-Carbamoyl-Leu-N-Methyl-D-Trp-D-2-Pyridyl-Ala

Synonyms: Endothelin Receptor Antagonist; FR-139317

Sigma H 8403   FW 604.7 $C_{33}H_{44}N_6O_5$ ≥97% (HPLC) | Bioactive peptide; Aramori, I et al, *Mol Pharmacol*, 43:127, 1993

Leu-Trp-Ala-Ala-Tyr-Phe
Hexamethyleniminocarbonyl-Leu-D-Trp-D-Ala-β-Ala-Tyr-D-Phe

Synonyms: TTA-386

American Peptide 88-2-48   MW 895.1 $C_{48}H_{62}N_8O_9$ Kitada, C et al, *J Cardiovasc Pharmacol*, 22 Suppl 8:S128, 1993

Leu-Trp-Ala-Ala-Tyr-Phe
Hexamethyleneiminocarbonyl-Leu-D-Trp-D-Ala-β-Ala-Tyr-D-Phe

Synonyms: Endothelin A Receptor Antagonist; TTA-386

Sigma T 8677   FW 917.0 $C_{48}H_{61}N_8O_9Na$ ≥97% (HPLC) | Bioactive peptide; endothelin, ET$_A$, receptor-selective antagonist to ET-1; Kitada, C et al, *J Cardiovasc Pharmacol*, 22:S128, 1993; Jerome, EH et al, *FASEB J*, 8:PA331, 1994

Leu-Trp-Leu
N-cis-2,6-Dimethylpiperidinocarbonyl-L-γ-Methylleucyl-D-1-Methoxycarbonyl Tryptophanyl-D-Nle

Synonyms: Endothelin Receptor Antagonist; BQ-788

American Peptide 88-2-55   MW 663.7 $C_{34}H_{50}N_5O_7Na$ Highly potent & selective; Fukuroda, T et al, *BBRC*, 199:1461, 1994; Ishikawa, K et al, *PNAS*, 91:4892, 1994

Leu-Trp-Leu

Bachem H-3965.0250   MW 430.55 $C_{23}H_{34}N_4O_4$ Store at -15°C

Leu-Trp-Leu
N-cis-2,6-Dimethylpiperidinocarbonyl-L-γ-MeLeu-D-Trp(MeOCO)-D-Nle Sodium Salt

Synonyms: Endothelin Receptor Antagonist; BQ-788

Calbiochem 05-23-3838   MW 663.7 $C_{34}H_{50}N_5O_7 \cdot Na$ >95% (HPLC); solid; soluble in H_2O | Highly selective; Fukuroda, T et al, *Biochem Biophys Res Comm*, 199:1461, 1994; Ishikawa, K et al, *PNAS*, 91:4892, 1994

Leu-Trp-Met

Bachem H-3970.0250   MW 448.59 $C_{22}H_{32}N_4O_4S$ Store at -15°C

Leu-Trp-Met-Arg

Bachem H-3975.0050   MW 604.77 $C_{28}H_{44}N_8O_5S$ Store at -15°C

Leu-Trp-Met-Arg-Phe

Bachem H-3980.0050   MW 751.95 $C_{37}H_{53}N_9O_6S$ Store at -15°C

Leu-Trp-Met-Arg-Phe-Ala

Bachem H-3985.0050   MW 823.03 $C_{40}H_{58}N_{10}O_7S$ Store at -15°C

Leu-Trp-Nle
N-cis-2,6-Dimethylpeperidinocarbonyl-L-γ-Me-Leu-D-Trp(COOMe)-D-Nle-ONa Sodium Salt

Synonyms: Endothelin B Receptor Antagonist; BQ-788

Alexis 155-020   MW 663.7 $C_{34}H_{50}N_5O_7Na$ Highly potent & selective ET$_B$ antagonist; Fukuroda, T et al, *BBRC*, 199:1461, 1994; Ishikawa, K et al, *PNAS*, 91:4892, 1994

Leu-Trp-Nle
Dmpc-γ-MeLeu-D-Trp(1-CO$_2$CH$_3$)-D-Nle Sodium Salt

Synonyms: BQ-788

Neosystem SC928   MW 663.79   Bioactive; endothelin-related peptide; potent & selective ETB receptor antagonist; a powerful tool for investigating the role of ET in physiological & pathological processes; Ishikawa, K et al, *PNAS USA*, 91:4892-4896, 1994; Fukuroda, T et al, *BBRC*, 199:1461-1465, 1994; Karaki, H et al, *BBRC*, 205:168-173, 1994; D'Orléans-Juste, P, *Br J Pharmacol*, 113:1257-1262, 1994

Leu-Trp-Nle
2,6-Dimethylpiperidinecarbonyl-γ-Me-Leu-N$_{in}$-(Methoxycarbonyl)-D-Trp-D-Nle Sodium Salt

Synonyms: Endothelin B Receptor Antagonist; BQ-788

Sigma B 6791   FW 663.8 $C_{34}H_{50}N_5O_7Na$ ≥90% (HPLC) | Bioactive peptide; Ishikawa, K et al, *Proc Natl Acad Sci USA*, 91:4892, 1994

Leu-Trp-Nle
N-cis-2,6-Dimethylpiperidinocarbonyl-L-γ-Me-Leu-D-Trp(COOMe)-D-Nle Sodium Salt

Synonyms: BQ-788; ETB-Selective Antagonist; Endothelin B Receptor Antagonist

Peptides International PED-3788-PI   Synthetic   MW 663.80 $C_{34}H_{50}N_5O_7Na$ >98% (HPLC); white powder | Inhibitor; Fukuroda, T et al, *BBRC*, 199:1461, 1994

Leu-Trp-Pal
(Hexahydro-1-Azepinyl)Carbonyl-Leu-(1-Me)-D-Trp-D-Pal

Synonyms: FR-139317

American Peptide 88-2-46   MW 606.8 $C_{33}H_{46}N_6O_5$ A modified tripeptide with potent ET$_A$ antagonist function; Sogabe, K et al, *J Pharmacol Exp Ther*, 264:1040, 1993; Endoh, M et al, *J Pharmacol Exp Ther*, 277:61, 1996; Fujita, K et al, *Japanese Journal of Pharmacology*, 70:313, 1996

Leu-Trp-Trp
(Hexahydro-1H-Azepinyl)Carbonyl-Leu-D-Trp(CHO)-D-Trp

Synonyms: Endothelin A Receptor Antagonist; BQ-610

Alexis 155-011   MW 656.8 $C_{36}H_{44}N_6O_6$ 30 times more potent ET$_A$ antagonist than BQ-123 in porcine aortic smooth muscle membrane assay, same activity as BQ-123 when tested on porcine cerebellum membrane

Leu-Trp-Trp
Homopiperidinyl-Carbonyl-Leu-D-Trp(CHO)-D-Trp

Synonyms: Endothelin Receptor Antagonist; BQ-610

American Peptide 88-2-32   MW 656.8 $C_{36}H_{44}N_6O_6$ 30 times more potent than BQ-123 in porcine aortic smooth muscle membrane assay; same activity as BQ-123 when tested on porcine cerebellum membrane; Nagase, T et al, *2nd Japan Symp on peptide Chem*, Shizuoka, 1992; Ishikawa, K et al, *Peptides*, Schneider & Eberle, eds, ESCOM, p. 685, 1992

Leu-Trp-Trp
Hexahydro-1H-Azepinylcarbonyl-Leu-D-Trp-D-Trp

Synonyms: Endothelin Receptor Antagonist; BQ-485

American Peptide 88-2-45   MW 650.8 $C_{35}H_{43}N_6O_5NA$ Selective; Itoh, S et al, *Biochem Biophys Res Comm*, 195:969, 1993

Leu-Trp-Trp
Azepane-1-Carbonyl-Leu-D-Trp(For)-D-Trp

Synonyms: BQ-610

Bachem H-4914.0500   MW 656.78 $C_{36}H_{44}N_6O_6$ Store at -15°C

Leu-Trp-Trp
(N,N-Hexamethylene) Carbamoyl-Leu-D-Trp(CHO)-D-Trp

Synonyms: BQ-610

Neosystem SC860 MW 656.80 Bioactive; endothelin ETA antagonist; Ishikawa, K et al, *Peptides* (Proceedings of the 22nd EPS) ESCOM, Leiden:685-686, 1992; Verheyden, P et al, *FEBS Lett*, 344:55-60, 1994

Leu-Trp-Trp
N,N-Hexamethylene-Carbamoyl-Leu-N-For-D-Trp-D-Trp

Synonyms: Endothelin A Receptor Antagonist; BQ-610

Sigma H 8278 FW 656.8 $C_{36}H_{44}N_6O_6$ Bioactive peptide; Verheyden, P et al, *FEBS Lett*, 344:55, 1994

Leu-Trp-Tyr
(Hexahydro-1H-Azepinyl)carbonyl-Leu-(1-Me)-D-Trp-D-Tyr)

Synonyms: PD 151242

American Peptide 88-2-47 MW 620.8 $C_{34}H_{46}N_5O_6$ Peter, MG et al, *British Journal of Pharmacology*, 114:297, 1995

Leu-Tyr
Z-Leu-Tyr

Bachem C-2185.0001 MW 428.49 $C_{23}H_{28}N_2O_6$ Store at -15°C

Leu-Tyr
Z-Leu-Tyr Amide

Bachem C-2190.0001 MW 427.5 $C_{23}H_{29}N_3O_5$ Store at -15°C

Leu-Tyr

Bachem G-2610.0001 MW 294.35 $C_{15}H_{22}N_2O_4$ Store at -15°C

Leu-Tyr
D-Leu-Tyr

Bachem G-2615.0001 MW 294.35 $C_{15}H_{22}N_2O_4$ Store at -15°C

Leu-Tyr
DL-Leu-DL-Tyr

Bachem G-4310.0001 MW 294.35 $C_{15}H_{22}N_2O_4$ Store at -15°C

Leu-Tyr
Suc-Leu-Tyr-AMC

Bachem I-1355.0050 Store at -15°C

Leu-Tyr
Z-Leu-Tyr-CMK

Bachem N-1255.0025 Store at -15°C

Leu-Tyr
Suc-Leu-Tyr-AMC

Synonyms: Calpain Fluorogenic Substrate I

Calbiochem 208723 MW 551.6 $C_{29}H_{33}FN_3O_8$ ≥98% (TLC); lyophilized solid; soluble in DMSO | Fluorogenic substrate for the quantitative determination of porcine calpain I & II; Woo, KM et al, *J Bio Chem*, 264:2088, 1989; Sasaki, T et al, *J Biol Chem*, 259:12489, 1984

Leu-Tyr

ICN 102177 MW 294.3 $C_{15}H_{22}N_2O_4$ 99%; crystalline

Leu-Tyr
D-Leu-Tyr

ICN 155225 MW 294.3 $C_{15}H_{22}N_2O_4$ Crystalline

Leu-Tyr
DL-Leu-DL-Tyr

ICN 155226 MW 294.3 $C_{15}H_{22}N_2O_4$

Leu-Tyr
Suc-Leu-Tyr-AMC

Synonyms: Calpain Fluorogenic Substrate I; Calpain Substrate; *E. coli* Protease Ti Substrate

ICN 195898 MW 551.6 $C_{29}H_{33}N_3O_8$ ≥98%; lyophilized | Fluorogenic substrate for porcine Calpain I & II

Neosystem SC1268 MW 551.59 Sasaki, T et al, *JBC*, 259:12489-12494, 1984; Woo, KM et al, *JBC*, 264:2088-2091, 1989

Leu-Tyr
N-CBZ-Leu-Tyr

Sigma C 1010 FW 428.5 $C_{23}H_{28}N_2O_6$

Leu-Tyr

Sigma L 0501 FW 294.3 $C_{15}H_{22}N_2O_4$

Leu-Tyr
D-Leu-Tyr

Sigma L 3877 FW 294.3 $C_{15}H_{22}N_2O_4$

Leu-Tyr
DL-Leu-DL-Tyr

Sigma L 8632 FW 294.3 $C_{15}H_{22}N_2O_4$ Composition of the 4 diasteromers varies from lot to lot

Leu-Tyr
N-Suc-Leu-Tyr-MCA

Synonyms: Calpain I Substrate; Calpain II Substrate

Sigma S 1153 FW 551.6 $C_{29}H_{33}N_3O_8$ Sasaki, T et al, *J Biol Chem*, 259:12489, 1984; Substrate for porcine calpain I & II

Sigma S 3389 FW 551.6 $C_{29}H_{33}N_3O_8$

ICN 156696 Porcine Sasaki, T et al, *JBC*, 259:12489, 1984

Leu-Tyr-Arg-Leu-Glu-Leu-Gly-Asp-Tyr-Lys-Leu-Val-Glu-Ile-Thr-Pro-Ile-Gly-Leu-Ala

Synonyms: GP140 (491-510); SIVmac251 Peptide

Neosystem SC757 MW 2276.69 SIVmac251 peptide from subtype GP140

Leu-Tyr-Asn-Thr-Val-Ala-Thr-Leu-Tyr-Cys-Val-His-Gln-Arg-Ile

Synonyms: P18 (78-92); GAG P17 MA (78-92); HTLV I Peptide

Neosystem SC285 MW 1794.10 HTLV I peptide from subtype P25 (GAG P17 MA); due to the presence of a Cys this peptide can contain the dimeric form

Leu-Tyr-Asp-Arg

Synonyms: Kyotorphin, D-Arg²-

Biogenesis 5596-0039 Synthetic MW 337.3 Lyophilized, consisting of 71% peptide material, acetate counter ions and residual H_2O; purified; lyophilized

Leu-Tyr-Ile-Asp-Phe-Arg-Lys-Asp-Leu-Gly

Bachem H-2730.0001   Store at -15°C

Leu-Tyr-Leu

Bachem H-3990.0250   MW 407.51 $C_{21}H_{33}N_3O_5$ Store at -15°C

Leu-Tyr-Leu-Arg

Synonyms: Kyotorphin

Biogenesis 5596-0009   Bovine synthetic   MW 409.4   90% peptide material, dihydrochloride counter ions and residual H_2O; purified; lyophilized

Leu-Val
BOC-Leu-(®)-Val

Bachem A-1905.0250   MW 316.44 $C_{16}H_{32}N_2O_4$ Store at -15°C

Leu-Val
Z-Leu-Val

Bachem C-2195.0001   MW 364.44 $C_{19}H_{28}N_2O_5$ Store at -15°C

Leu-Val

Bachem G-2620.0001   MW 230.31 $C_{11}H_{22}N_2O_3$ Store at -15°C

Leu-Val
Leu-Val-4MβNA Hydrochloride

Bachem J-1245.0050   MW 421.97 $C_{22}H_{31}N_3O_3 \cdot$ HCl   Store at -15°C

Leu-Val

ICN 102178   MW 230.3 $C_{11}H_{22}N_2O_3$ Crystalline

Leu-Val
DL-Leu-DL-Val

ICN 155227   MW 230.3 $C_{11}H_{22}N_2O_3$ Crystalline

Leu-Val

Sigma L 1377   FW 230.3 $C_{11}H_{22}N_2O_3$

Leu-Val-Arg-Cys-Gly-Lys-His-Ser-Arg
Leu-Val-Arg-D-Cys-Gly-Lys-His-Ser-Arg

Synonyms: Thrombin Receptor Agonist L9R

Neosystem SC1219   MW 1055.26   >80% (HPLC) | Bioactive; due to the presence of a Cys this peptide can contain the dimeric form; an antagonist in human platelet aggregation; inhibits activation & inhibition by TRAP-14 of choline acetyltransferase activity in septal neuron/glial cell cultures; antagonizes thrombin effects on astrocyte morphology; Debeir, T, Benavides, J & Vigé, X, *Brain Res*, 708:159-166, 1996; Debeir, T, Vigé, X & Benavides, J, *Eur J Pharmacol*, 323:111-117, 1997

Leu-Val-Cys-Asp
Ac-Leu-Val-Cys-Asp-AMC

Bachem I-1790.0005   MW 647.75 $C_{30}H_{41}N_5O_9S$ Store at -15°C

Leu-Val-Gln-Pro-Arg-Gly-Ser-Arg-Asn-Gly-Pro-Gly-Pro-Trp-Gln-Gly-Gly-Arg-Arg-Lys-Phe-Arg-Arg-Gln-Arg-Pro-Arg-Leu-Ser-His-Lys-Gly-Pro-Met-Pro-Phe

Synonyms: Apelin-36; APJ Receptor Ligand

Bachem H-4896.0500   Human   MW 4195.89 $C_{184}H_{297}N_{69}O_{43}S$ Store at -15°C

Peptides International PAP-4362-s   Human synthetic   MW 4195.9 $C_{184}H_{297}N_{69}O_{43}S$ >95% (HPLC); lyophilized amorphous powder | Bioactive; human, bovine; Tatemoto, K et al, *BBRC*, 251:471, 1998

Leu-Val-Glu-Ile-Thr-Pro-Ile-Gly-Leu-Ala-Pro-Thr-Asp-Val-Lys-Arg-Tyr-Thr-Thr-Gly

Synonyms: GP140 (501-520); SIVmac251 Peptide

Neosystem SC758   MW 2144.49   SIVmac251 peptide from subtype GP140

Leu-Val-Gly
Z-Leu-Val-Gly-Diazomethylketone

Bachem C-3455.0025   MW 445.52 $C_{22}H_{31}N_5O_5$ Store at -15°C

Leu-Val-Gly
CBZ-Leu-Val-Gly Diazomethyl Ketone

Synonyms: Cysteine Proteinase Inhibitor

Sigma C 9546   FW 445.5 $C_{22}H_{31}N_5O_5$ ≥97% (HPLC) | Bioactive peptide; strong inhibitor with antibacterial activity; Bjorck, L et al, *Nature*, 337:385, 1989

Leu-Val-Leu-Ala
Leu-Val-Leu-Ala-pNA

Bachem L-2045.0025   MW 534.66 $C_{26}H_{42}N_6O_6$ Store at -15°C

Leu-Val-Lys
Ac-Leu-Val-Lys-CHO

Synonyms: Cathepsin B Inhibitor

Bachem N-1380.0001   MW 384.52 $C_{19}H_{36}N_4O_4$ Store at -15°C

ICN 195952   MW 384.5 $C_{19}H_{36}N_4O_4$ Lysinal analog of leupeptin; a more potent cathepsin B inhibitor than leupeptin

Leu-Val-Phe
Ac-Leu-Val-Phe-CHO

Bachem N-1395.0001   MW 403.52 $C_{22}H_{33}N_3O_4$ Store at -15°C

Leu-Val-Phe
N-Ac-Leu-Val-Phe-CHO

Synonyms: HIV Protease Inhibitor

ICN 196005   MW 403.5 $C_{22}H_{33}N_3O_4$ ≥95% | Potent inhibitor

Leu-Val-Phe-Ala
N-Ac-Leu-Val-Phe-Ala-CHO

Synonyms: HIV Protease Inhibitor

Sigma A 5205   FW 403.5 $C_{22}H_{33}N_3O_4$ Bioactive peptide; potent inhibitor; Sarubbi, E et al, *FEBS Lett*, 319:253, 1993

Leu-Val-Phe-Phe-Ala-Glu-Asp-Val-Gly-Ser-Asn-Lys-Gly-Ala-Ile-Ile-Gly-Leu-Met-Val-Gly-Gly-Val-Val
β-Leu-Val-Phe-Phe-Ala-Glu-Asp-Val-Gly-Ser-Asn-Lys-Gly-Ala-Ile-Ile-Gly-Leu-Met-Val-Gly-Gly-Val-Val

Synonyms: Amyloid (17-40)

American Peptide 62-0-85   MW 2392.8 $C_{110}H_{178}N_{26}O_{31}S$

Leu-Val-Tyr
Leu-Val-Tyr-AMC

Bachem I-1250.0050   MW 550.66 $C_{30}H_{38}N_4O_6$ Store at -15°C

Leu-Val-Val-Asp-Leu-Thr-Asp-Ile-Asp-Pro-Asp-Val-Ala-Tyr-Ser-Ser-Val-Pro-Tyr-Glu-Lys

Synonyms: Leukotriene A4 Hydrolase (365-385)

Bachem H-3396.0001   Human   Store at -15°C

Leu-Val-Val-Tyr-Pro-Trp

Synonyms: Myelopeptide II; Bone Gla Protein

American Peptide 22-1-64   MW 776.0 $C_{41}H_{57}N_7O_8$ Able to restore the T-cell phenotype altered by HL-60 CM; the MP-2 recovery effect is connected with its influence on the CD-3 & CD-4 antigen expression on T lymphocytes; Strelkov, LA et al, *Immunology Lett*, 50:143, 1996

Leu-Val-Val-Tyr-Pro-Trp-Thr-Gln-Arg

Synonyms: Valorphin-Arg, Leu-; Hemorphin-6, Leu-Val-Val-

Bachem H-8880.0001   Human   MW 1161.37 $C_{56}H_{84}N_{14}O_{13}$ Store at -15°C

Lys-Abz
FMOC-Lys-(Retro-Abz-*N*-Me-BOC)

Synonyms: BOC-*N*-Me-Abz-(FMOC-Lys)

Bachem B-2515.0001   MW 601.7 $C_{34}H_{39}N_3O_7$ Store at -15°C

Lys-Abz
FMOC-Lys-(Retro-Abz-BOC)

Synonyms: BOC-Abz-(FMOC-Lys); Ablysin-BOC, FMOC-

Bachem B-3180.0250   MW 587.67 $C_{33}H_{37}N_3O_7$ Store at -15°C

Lys-Ala

Bachem G-2630.0250   MW 217.27 $C_9H_{19}N_3O_3$ Store at -15°C

Lys-Ala
Ac-Lys-D-Ala-D-lactic acid acetate

Bachem G-4410.0050   MW 391.42 $C_{14}H_{25}N_3O_6 \cdot C_2H_4O_2$ Store at -15°C

Lys-Ala
Lys-Ala-AMC Hydrochloride

Bachem I-1260.0050   MW 410.9 $C_{19}H_{26}N_4O_4 \cdot HCl$ Store at -15°C

Lys-Ala
Lys-Ala-4MβNA Dihydrochloride

Bachem J-1250.0050   MW 445.39 $C_{20}H_{28}N_4O_3 \cdot 2HCl$ Store at -15°C

Lys-Ala
Lys-Ala-βNA

Bachem K-1420.0050   MW 342.44 $C_{19}H_{26}N_4O_2$ Store at -15°C

Lys-Ala
Ac-Lys-Ala-βNA

Bachem K-1630.0050   MW 384.48 $C_{21}H_{28}N_4O_3$ Store at -15°C

Lys-Ala
Lys-Ala-pNA Dihydrochloride

Bachem L-2085.0050   MW 410.3 $C_{15}H_{23}N_5O_4 \cdot 2HCl$ Store at -15°C

Lys-Ala
Ac-Lys(Ac)-D-Ala-D-lactic acid

Bachem M-1330.0050   MW 373.41 $C_{16}H_{27}N_3O_7$ Store at -15°C

Lys-Ala
FA-Lys-Ala

Bachem M-2000.0050   MW 337.38 $C_{16}H_{23}N_3O_5$ Store at -15°C

Lys-Ala
Lys-Ala 4-Methoxy-β-Naphthylamide Dihydrochloride

ICN 155282   MW 445.4 $C_{20}H_{28}N_4O_3 \cdot 2HCl$

Lys-Ala
Lys-Ala β-Naphthylamide

ICN 155283   MW 342.4 $C_{19}H_{26}N_4O_2$

Lys-Ala
*N*ᵅ,*N*ᵋ-Diacetyl-Lys-D-Ala-D-Lactic Acid

Synonyms: Penicillin Sensitive D-Alanine Carboxypeptidase Substrate

ICN 157572   MW 373.4 $C_{16}H_{27}N_3O_7$ Yocum, RR et al, *PNAS*, 76:2730, 1979

Lys-Ala
Lys-Ala-AMC

Synonyms: Dipeptidyl Aminopeptidase II Substrate

Neosystem SC1292   MW 374.44   Nagatsu, T et al, *Anal Biochem*, 147:80-85, 1985

Lys-Ala
Lys-Ala-MCA

Synonyms: Dipeptidyl Aminopeptidase II Substrate

Peptides International MKA-3124-v   MW 374.44 $C_{19}H_{26}N_4O_4$ >98% (HPLC); lyophilized amorphous powder | Mantle, D et al, *Biochem J*, 211:567, 1983

Lys-Ala
*N*ᵅ,*N*ᵋ-Diacetyl-Lys-D-Ala-D-Lactic Acid

Synonyms: Penicillin Sensitive D-Alanine Carboxypeptidase Substrate

Sigma D 2279   FW 373.4 $C_{16}H_{27}N_3O_7$ Yocum, RR et al, *Proc Natl Acad Sci USA*, 76:2730, 1979

Lys-Ala
Lys-Ala-4-Methoxy-β-Naphthylamide Dihydrochloride

Synonyms: Dipeptidyl Aminopeptidase II Substrate

Sigma L 2270   FW 445.4 $C_{20}H_{28}N_4O_3 \cdot 2HCl$ Fluorogenic substrate; Dolbeare, FA & Smith, RE, *Clin Chem*, 23:1485, 1977; Gossrau, R & Lojda, Z, *Histochemistry*, 70:53, 1980

Lys-Ala
Lys-Ala-β-Naphthylamide

Synonyms: Dipeptidyl Aminopeptidase II Substrate

Sigma L 5138   FW 342.4 $C_{19}H_{26}N_4O_2$ Imai, K et al, *J Biochem*, 93:431, 1983; McDonald, JK et al, *J Biol Chem*, 243:4143, 1968

Lys-Ala
Lys-Ala-MCA Dihydrochloride

Synonyms: Dipeptidyl Aminopeptidase II Substrate

ICN 155281   Rat brain   MW 447.4 $C_{19}H_{26}N_4O_4 \cdot 2HCl$ Imai, K et al, *J Biochem*, 93:431, 1983

Sigma L 8139   Rat Brain   FW 447.4 $C_{19}H_{26}N_4O_4 \cdot$ 2HCl
Substrate for dipeptidyl aminopeptidase; Imai, K et al, *J Biochem*, 93:431, 1983

Lys-Ala Amide Dihydrochloride
Bachem M-1550.0050   Store at -15°C

Lys-Ala Dihydrobromide
ICN 155280   MW 379.1 $C_9H_{19}N_3O_3 \cdot$ 2HBr   Crystalline
Sigma L 5127   FW 379.1 $C_{19}H_{26}N_4O_4 \cdot$ 2HBr

Lys-Ala-Ala
N,N,Diacetyl-Lys-D-Ala-D-Ala
Synonyms: Penicillin Sensitive D-Alanine Carboxypeptidase Substrate
American Peptide 80-0-13   MW 372.4 $C_{16}H_{28}N_4O_6$ Nguyen-Disteche, M et al, *Biochem J*, 207:109, 1982

Lys-Ala-Ala
BOC-Lys-Ala-Ala
Bachem A-1930.0250   Store at -15°C

Lys-Ala-Ala
Bachem H-4085.0250   MW 288.35 $C_{12}H_{24}N_4O_4$ Store at -15°C

Lys-Ala-Ala
Ac-Lys-D-Ala-D-Ala
Bachem H-8005.0050   MW 330.38 $C_{14}H_{26}N_4O_4$ Store at -15°C

Lys-Ala-Ala
Ac-Lys(Ac)-D-Ala-D-Ala
Bachem M-1325.0050   MW 372.42 $C_{16}H_{28}N_4O_6$ Store at -15°C

Lys-Ala-Ala
N$^\epsilon$,N$^\epsilon$-Diacetyl-Lys-D-Ala-D-Ala
Synonyms: Penicillin Sensitive D-Alanine Carboxypeptidase Substrate
ICN 157571   MW 372.4 $C_{16}H_{28}N_4O_6$ Nguyes-Disteche, M etal, *Biochem J*, 207:109, 1982

Lys-Ala-Ala
N$^\alpha$,N$^\epsilon$-Diacetyl-Lys-D-Ala-D-Ala
Synonyms: Penicillin Sensitive D-Alanine Carboxypeptidase Substrate
Sigma D 9904   FW 372.4 $C_{16}H_{28}N_4O_6$ Peptide content ≥90% | Nguyen-Disteche, M et al, *Biochem J*, 207:109, 1982

Lys-Ala-Ala
Ac-Lys-D-Ala-D-Ala
Synonyms: DD-Carboxypeptidase Substrate
Sigma A 6950   *S albus*   FW 330.4 $C_{14}H_{26}N_4O_5$ ≥95% (HPLC) | Lehy-Bouille, M et al, *Biochem*, 9:2961, 1970

Lys-Ala-Arg-Gly-Trp-Phe-Tyr-Arg-His-His-Tyr-Glu-Ser-Pro
Synonyms: VIF (34-47); HTLV I Peptide
Neosystem SC230   MW 1834.02   HTLV I peptide from subtype VIF

Lys-Ala-Arg-Val-Nle-Nitro-Phe-Glu-Ala-Nle
Anthranilyl-Lys-Ala-Arg-Val-Nle-*p*-Nitro-Phe-Glu-Ala-Nle Amide
Synonyms: HIV Protease Substrate
Sigma A 0686   FW 1209.4   ≥90% (HPLC) | Dunn, BM et al, Abstract – 12th American Peptide Symposium, Boston, 1991

Lys-Ala-Arg-Val-Nle-Phe-Glu-Ala-Nle
Lys-Ala-Arg-Val-Nle-Phe(NO^2)-Glu-Ala-Nle Amide
Synonyms: HIV Substrate
American Peptide 72-2-45   MW 1091.3 $C_{49}H_{84}N_{15}O_{13}$
Pennington, M et al, *APS Symp*, Boston, 1991

Lys-Ala-Arg-Val-Nle-Phe-Glu-Ala-Nle
Lys-Ala-Arg-Val-Nle-*p*-Nitro-Phe-Glu-Ala-Nle Amide
Synonyms: HIV Protease Substrate IV
Bachem H-1048.0001   MW 1090.29 $C_{49}H_{83}N_{15}O_{13}$ Store at -15°C

Lys-Ala-Arg-Val-Nle-Phe-Glu-Ala-Nle
Anthranilyl-Lys-Ala-Arg-Val-Nle-(pNO$_2$-Phe)-Glu-Ala-Nle Amide
Synonyms: HIV Anthranilyl Substrate IV
ICN 158722

Lys-Ala-Arg-Val-Nle-Phe-Glu-Ala-Nle
Lys-Ala-Arg-Val-Nle-(pNO$_2$-Phe)-Glu-Ala-Nle Amide
Synonyms: HIV Substrate IV
ICN 158721   Synthetic

Lys-Ala-Arg-Val-Nle-Phe-Glu-Ala-Nle Amide
Synonyms: HIV Protease Substrate
Sigma L 6525   FW 1090.3   ≥97% (HPLC) | Bioactive peptide; Richards, AD et al, *J Bio Chem*, 265:7733, 1990; Phylip, LH et al, *Biochem Biophys Res Commun*, 171:439, 1990

Lys-Ala-Arg-Val-Nle-*p*-Nitro-Phe-Glu-Ala-Nle
Lys-Ala-Arg-Val-Nle-*p*-Nitro-Phe-Glu-Ala-Nle Amide
Synonyms: HIV Protease Substrate IV
Neosystem SC483   MW 1090.40   Virus-related peptide; AIDS-related peptide

Lys-Ala-Arg-Val-Tyr-Phe-Glu-Ala-Nle
Lys-Ala-Arg-Val-Tyr-*p*-Nitro-Phe-Glu-Ala-Nle Amide
Synonyms: HIV Protease Substrate VII
Bachem H-1286.0001   Store at -15°C

Lys-Ala-Leu-Gly-Ile-Ser-Tyr-Gly-Arg-Lys
Synonyms: TAT (41-50); HTLV I Peptide
Neosystem SC521   MW 1092.30   HTLV I peptide from subtype TAT

Lys-Ala-Leu-Gly-Ile-Ser-Tyr-Gly-Arg-Lys-Lys-Arg-Gln
Synonyms: TAT (41-54); HTLV I Peptide
Neosystem SC522   MW 1660.98   HTLV I peptide from subtype TAT

Lys-Ala-Leu-Gly-Pro-Ala-Ala-Thr-Leu-Glu-Glu-Met-Met-Thr-Ala-Cys-Gln
Synonyms: P25 (335-351); GAG P24 CA (203-219); HTLV I Peptide
Neosystem SC309   MW 1765.08   HTLV I peptide from subtype P25 (GAG P24 CA); due to the presence of a Cys this peptide can contain the dimeric form

Lys-Ala-Lys
Ac-Lys(Ac)-D-Ala-D-Lys

Bachem H-7260.0050   MW 429.52 $C_{19}H_{35}N_5O_6$ Store at -15°C

Lys-Ala-Met-Tyr-Ala-Pro-Pro-Ile-Ser-Gly-Gln-Ile

Synonyms: GP120 (437-448); HTLV I Peptide

Neosystem SC565   MW 1275.52   HTLV I peptide from subtype GP160

Lys-Ala-Phe-Gly

Bachem H-5375.0025   Store at -15°C

Lys-Ala-Pro

Bachem H-4090.0250   Store at -15°C

Lys-Ala-Pro-Ser-Gly-Arg-Met-Ser-Ile-Val-Lys-Asn-Leu-Gln-Asn-Leu-Asp-Pro-Ser-His-Arg-Ile-Ser-Asp-Arg-Asp-Tyr-Met-Gly-Trp-Met-Asp-Phe
Lys-Ala-Pro-Ser-Gly-Arg-Met-Ser-Ile-Val-Lys-Asn-Leu-Gln-Asn-Leu-Asp-Pro-Ser-His-Arg-Ile-Ser-Asp-Arg-Asp-Tyr(SO₃H)-Met-Gly-Trp-Met-Asp-Phe Amide

Synonyms: Cholecystokinin 33

Peptides International PCK-4201-s   Human synthetic   MW 3945.5 $C_{167}H_{263}N_{51}O_{52}S_4$ >95% (HPLC); lyophilized amorphous powder | Bioactive; Takahashi, Y et al, *PNAS USA*, 82:1931, 1987; Kurano, Y et al, in *Peptides*, Proc 10th American Peptide Symp, (GR Marshall, Ed), ESCOM *Science* Publishers BV 1988, pp 162-165

Lys-Ala-Pro-Ser-Gly-Arg-Val-Ser-Met-Ile-Lys-Asn-Leu-Gln-Ser

Synonyms: Cholecystokinin (1-15)

Biogenesis 2050-1009   Synthetic   MW 1616   Lyophilized, comprising 89% peptide material, acetate counter ions and residual H_2O; purified | To avoid Met oxidation, it is best to dissolve the peptide under a nitrogen stream

---

Lys-Ala-Pro-Ser-Gly-Arg-Val-Ser-Met-Ile-Lys-Asn-Leu-Gln-Ser-Leu-Asp-Pro-Ser-His-Arg

Synonyms: Gastrointestinal Peptide; Cholecystokinin (1-21); Cholecystokinin 12

Bachem H-2410.0500   Store at -15°C

ICN 195678   MW 2321.7   ≥95%

Sigma C 2296   FW 2321.7   ≥95% (HPLC) | Bioactive peptide; amino terminal fragment of CCK-33 cleaved to form CCK-12

Lys-Ala-Pro-Ser-Gly-Arg-Val-Ser-Met-Ile-Lys-Asn-Leu-Gln-Ser-Leu-Asp-Pro-Ser-His-Arg-Ile-Ser-Asp-Arg-Asp-Tyr-Met-Gly-Trp-Met-Asp-Phe
Lys-Ala-Pro-Ser-Gly-Arg-Val-Ser-Met-Ile-Lys-Asn-Leu-Gln-Ser-Leu-Asp-Pro-Ser-His-Arg-Ile-Ser-Asp-Arg-Asp-Tyr(SO₃H)-Met-Gly-Trp-Met-Asp-Phe Amide

Synonyms: Cholecystokinin 33

Peptides International PCK-4176-s   Porcine synthetic   MW 3918.5 $C_{166}H_{262}N_{50}O_{52}S_4$ >95% (HPLC); lyophilized amorphous powder | Bioactive; Mutt, V & JE Jorpes, *Eur J Biochem*, 6:156, 1968; Mutt, V & JE Jorpes, *Biochem J*, 125:57P, 1971; Kurano, Y et al, *J Chem Soc Chem Commun*, 5:323, 1987

Lys-Ala-Pro-Val
MeOSuc-Lys(2-Picolinoyl)-Ala-Pro-Val-pNA

Bachem L-1350.0025   Store at -15°C

Lys-Ala-Ser-Gln-Asn-Phe-Pro-Val-Val
Ac-Lys-Ala-Ser-Gln-Asn-Phe(NO₂)-Pro-Val-Val Amide

Synonyms: HIV Protease Substrate I

American Peptide 81-7-45   MW 1076.2 $C_{47}H_{75}N_{14}O_{15}$

Lys-Ala-Ser-Gln-Asn-Phe-Pro-Val-Val
Ac-Lys-Ala-Ser-Gln-Asn-p-Nitro-Phe-Pro-Val-Val Amide

Synonyms: HIV Protease Substrate

Neosystem SC667   MW 1076.19   Virus-related peptide; AIDS-related peptide

---

Lys-Ala-Val-Tyr-Asn-Phe-Ala-Thr-Met

Synonyms: Lymphocytic Choriomeningitis Virus Peptide; GP (33-41)

Neosystem SC1347   MW 1044.23   Virus-related peptide; corresponds to the LCMV dominant CTL epitope; restricted both by H-2D(b) & H-2K(b) molecules; the LCMV virus can escape CTL recognition by mutations altering this epitope; Pircher, H et al, *Nature*, 346:629-633, 1990; Hudrisier, D et al, *Virology*, 234:62-73, 1997

Lys-Arg

Bachem G-2635.0250   MW 302.38 $C_{12}H_{26}N_6O_3$ Store at -15°C

Lys-Arg
Z-Lys-Arg-pNA ·2 Hydrochloride

Bachem L-1240.0050   MW 629.54 $C_{26}H_{36}N_8O_6 \cdot 2HCl$
Store at -15°C

Lys-Arg
Lys-Arg Iron Agar

Fluka 17159   Composition: 5 g/L peptic digest of animal tissue, 3 g/L yeast extract, 10 g/L L-arginine, 10 g/L L-lysine, 1 g/L glucose, 0.5 g/L ferric ammonium citrate, 0.04 g/L sodium thiosulfate, 0.02 g/L bromocresol purple, 15 g/L agar | For the isolation & presumptive identification of *Yersinia* species from milk & milk products

---

Lys-Arg-Arg-Glu-Ile-Leu-Ser-Arg-Arg-Pro-Ser-Tyr-Arg-Lys

Synonyms: PCREB

Upstate 12-378   MW 1925   >90% pure; lyophilized | Immunizing peptide

---

Lys-Arg-Arg-Trp-Lys-Lys-Asn-Phe-Ile-Ala-Val

Synonyms: Myosin Light Chain Kinase Inhibitor

ICN 159631   MW 1444.9

Lys-Arg-Arg-Trp-Lys-Lys-Asn-Phe-Ile-Ala-Val Amide

Synonyms: Myosin Light Chain Kinase (342-352)

Alexis 162-012   MW 1444.9 $C_{68}H_{113}N_{23}O_{12}$ Corresponds to a part of the calmodulin-binding sequence in skeletal muscle myosin light chain kinase; Kemp, BE et al, *J Biol Chem*, 262:2542, 1987

Lys-Arg-Asn-Lys-Asn-Asn-Ile-Ala

Synonyms: Gastrointestinal Peptide; Glucagon 37 (30-37); Oxyntomodulin (30-37)

Bachem H-5910.0001   MW 957.1 $C_{39}H_{72}N_{16}O_{12}$ Store at -15°C

Sigma G 5899   FW 957.1 $C_{39}H_{72}N_{16}O_{12}$ ≥97% (HPLC); peptide content:~70% | Bioactive peptide; Bataille, E et al, *FEBS Lett*, 146:79, 1982

ICN 154556   Porcine   MW 957.2   Audousset-Puech, MP et al, *J Med Chem*, 28:1529, 1985

Lys-Arg-Asn-Pro-Gly-Ser-Gln-Lys-Arg-Phe-Pro-Ser-Asn-Cys-Gly-Arg-Asp

Synonyms: Phosphorylase Kinase β-Subunit (420-436)

Alexis 165-008   MW 1947.2 $C_{79}H_{131}N_{31}O_{25}S$ Increases phospholipase A_2 activity in murine smooth muscle cell; more effective than synthetic melittin; activity of the phosphorylase kinase γ-subunit is regulated by the α- & β-subunits; this β-subunit-derived peptide inhibits the activity of the γ-subunit/calmodulin complex & is also a substrate for this complex; homologous to phosphorylase β, the natural substrate for the kinase; Sanchez, VE & Carlson, GM, *J Biol Chem*, 268:17889, 1993

American Peptide 86-2-56   MW 1946.2 $C_{79}H_{130}N_{31}O_{25}S$ Inhibit activity of the γ-subunit calmodulin complex; Sanchez, VE et al, *JBC*, 268:17889, 1993

Bachem H-1968.0500   MW 1947.18 $C_{79}H_{131}N_{31}O_{25}S$ Store at -15°C

Lys-Arg-Gln-Gly-Arg-Phe-Gly-Lys Hydrochloride

Bachem H-7395.0005   Store at -15°C

Lys-Arg-Gln-His-Pro-Gly

Synonyms: Thyrotropin Releasing Hormone Precursor Peptide (1-6)

Sigma L 4647   FW 721.8 $C_{30}H_{51}N_{13}O_8$ ≥97% (HPLC); peptide content:~70% | Bioactive peptide; Jackson, IMD et al, *Science*, 229:1097, 1985

Lys-Arg-Gln-His-Pro-Gly-Lys-Arg

Synonyms: Thyrotropin Releasing Hormone Precursor Peptide (1-8)

American Peptide 58-0-96   MW 1006.2 $C_{42}H_{75}N_{19}O_{10}$ Jackson, IMD et al, *Science*, 229:1097, 1985

ICN 159954   MW 1006.2   97% | Jackson, IMD etal, *Science*, 229:1097, 1985

Sigma L 4772   FW 1006.2   ≥97% (HPLC); peptide content:~60% | Bioactive peptide; Jackson, IMD et al, *Science*, 229:1097, 1985

Lys-Arg-Glu-Leu-Val-Glu-Pro-Leu-Thr-Pro-Ser-Gly-Glu-Ala-Pro-Asn-Gln-Ala-Leu-Leu-Arg

Synonyms: ERT Protein Kinase Substrate

Alexis 155-024   MW 2318.7 $C_{101}H_{179}N_{31}O_{31}$ ≥97%; white lyophilized powder; soluble in water | Substrate for the growth factor-stimulated (MAP2-related) protein kinase (ERT protein kinase); ERT protein kinase phosphorylates the epidermal growth factor receptor at Thr[669]; Alvarez, E et al, *J Biol Chem*, 266:15277, 1991

Lys-Arg-Glu-Leu-Val-Glu-Pro-Leu-Thr-Pro-Ser-Gly-Glu-Ala-Pro-Asn-Gln-Ala-Leu-Leu-Arg Amide

Synonyms: ERT Protein Kinase Substrate

Alexis 155-026   MW 2317.7 $C_{101}H_{173}N_{31}O_{31}$ ≥97%; white lyophilized powder; soluble in water | Substrate for the growth factor-stimulated (MAP2-related) protein kinase (ERT protein kinase); ERT protein kinase phosphorylates the epidermal growth factor receptor at Thr[669]; Alvarez, E et al, *J Biol Chem*, 266:15277, 1991

Lys-Arg-Glu-Leu-Val-Glu-Pro-Leu-Thr-Pro-Ser-Gly-Glu-Ala-Pro-Asn-Gln-Ala-Leu-Leu-Arg

Synonyms: Epidermal Growth Factor Receptor Thr[669] Protein Kinase Substrate; Epidermal Growth Factor Receptor (661-681); Epidermal Growth Factor Receptor T669 Peptide (661-681); Mitogen Activated Protein Kinase Substrate

Bachem H-3242.0001   Store at -15°C

Neosystem SC963   MW 2318.65   Bioactive; proteinkinase-related peptide; Alvarez, E et al, *JBC*, 266:15277-15285, 1991

Sigma E 9520   FW 2318.7   ≥97% (HPLC); | Bioactive peptide; phosphorylated at Thr[669] by MAPkinase; Alvarez, E et al, *J Biol Chem*, 266:15277, 1991; Gonzalez, FA et al, *J Biol Chem*, 266:22159, 1991

American Peptide 86-5-35   Synthetic   MW 2315.0 $C_{101}H_{172}N_{30}O_{32}$ Substrate for ERT protein kinase, a growth factor-stimulated protein kinase that phosphorylates epidermal growth factor receptor at Thr[669]

Lys-Arg-Lys-Arg-Lys-Arg-Arg-Gly-Asp-Val (-Lys-Arg)₃-Arg-Gly-Asp-Val

Synonyms: Bifunctional Antiplatelet Agent

Bachem H-1662.0001   Store at -15°C

Lys-Arg-Phe-Ala-Arg-Lys-Gly-Ser-Leu-Arg-Gln-Lys-Asn-Val
Nᵉ-Biotinyl-Lys-Arg-Phe-Ala-Arg-Lys-Gly-Ser-Leu-Arg-Gln-Lys-Asn-Val

Synonyms: Protein Kinase C Fragment; Biocytin (19-31), (Ser[25])-

Sigma P 8963   FW 1914.3   ≥97% (HPLC) | Bioactive peptide

Lys-Arg-Pro-Ala-Gly-Phe-Ser-Pro-Phe-Arg

Synonyms: Bradykinin, (Lys⁰-Ala³)-; Bradykinin, Lys-(Ala³)-; Bradykinin, (Lys-(Ala³))-

American Peptide 18-1-24   MW 1162.4 $C_{54}H_{83}N_{17}O_{12}$ Mindroiu, T et al, *JBC*, 261:7407, 1966

Bachem H-9535.0005   Store at -15°C

ICN 159899   MW 1162.3   98% | Mindroiu, T etal, *Biol Chem*, 261:7407, 1986

Sigma B 1525   MW 1162.4   ≥97% (HPLC) | Bioactive peptide; Mindroiu, T et al, *J Biol Chem*, 261:7407, 1986

Lys-Arg-Pro-Hyp-Gly-Phe-Ser-Pro-Phe-Arg

Synonyms: Bradykinin, (Lys⁰-Hyp³)-; Bradykinin, Lys-(Hyp³)-

American Peptide 18-1-51   MW 1203.4 $C_{56}H_{84}N_{17}O_{13}$ New kinin recently isolated from human urine; also found in the mixture of human plasma protein & hog pancreatic kallikrein; Sasaguri, M et al, *BBRC*, 150:511, 1988

Bachem H-9075.0001   MW 1204.4 $C_{56}H_{85}N_{17}O_{13}$ Store at -15°C

Peptides International PBK-4191-v   Human synthetic   MW 1204.4 $C_{56}H_{85}N_{17}O_{13}$ >99% (HPLC); lyophilized amorphous powder | Bioactive; Sasaguri, M et al, *BBRC*, 150:511, 1988

Lys-Arg-Pro-Hyp-Gly-Phe-Ser-Pro-Phe-Arg 3AcOH 4H₂O

Synonyms: Bradykinin, Lys-(Hyp³)-

Peptides International PBK-4191   Human synthetic   MW 1456.6 $C_{56}H_{85}N_{17}O_{13} \cdot 3CH_3COOH \cdot 4H_2O$ >99% (HPLC); lyophilized amorphous powder; bulk | Bioactive; Sasaguri, M et al, *BBRC*, 150:511, 1988

Lys-Arg-Pro-Pro-Gly-Phe-Ser-Pro-Leu

Synonyms: Bradykinin, Lys-(des-Arg⁹,Leu⁸)-; Bradykinin B1 Receptor Antagonist; Kallidin, (des-Arg¹⁰,Leu⁹)-

American Peptide 18-1-11   MW 998.2 $C_{47}H_{75}N_{13}O_{11}$ Menke, JG et al, *J Biol Chem*, 269:21583, 1994

Bachem H-2582.0005   Store at -15°C

Neosystem SC966   MW 998.19   Bioactive, (Lys⁰, des-Arg⁹, Leu⁸)-; bradykinin B1 receptor antagonist; Menke, JG, *JBC*, 269:21583-21586, 1994

Lys-Arg-Pro-Pro-Gly-Phe-Ser-Pro-Phe

Synonyms: Kallidin, (des-Arg¹⁰)-; Bradykinin, Lys-(des-Arg⁹)-; Bradykinin B1 Receptor Selective Agonist

Bachem H-3122.0005   MW 1032.21 $C_{50}H_{73}N_{13}O_{11}$ Store at -15°C

ICN 195632   MW 1032.2

Neosystem SC965   MW 1032.21   Bioactive; selective bradykinin B1 receptor agonist; Regoli, D & Barabe, J, *Pharmacol Rev*, 32:1-46, 1980; Menke, JG et al, *JBC*, 269:21583-21586, 1994; Schneck, KA et al, *Eur J Pharmacol*, 266:277-282, 1994; Galizzi, JP et al, *Br J Pharmacol*, 113:389-394, 1994

Sigma B 1542   FW 1032.2   ≥95% (HPLC) | Bioactive peptide

Peptides International PBK-4303-v   Synthetic   MW 1032.2 $C_{50}H_{73}N_{13}O_{11}$ >99% (HPLC); lyophilized amorphous powder | Bioactive; Galizzi, JP et al, *Br J Pharmacol*, 113:389, 1994; Menke, JG et al, *JBC*, 269:21583, 1994; Zuzack, JS et al, *J Pharmacol Exp Ther*, 277:1337, 1996

Lys-Arg-Pro-Pro-Gly-Phe-Ser-Pro-Phe 2AcOH 4H2O

Synonyms: Kallidin, (des-Arg[10])-; Bradykinin, (Lys-des-Arg[9])-; Bradykinin B1 Receptor Selective Agonist

Peptides International PBK-4303   Synthetic   MW 1224.3 $C_{50}H_{73}N_{13}O_{11} \cdot 2CH_3COOH \cdot 4H_2O$ >99% (HPLC); lyophilized amorphous powder; bulk | Bioactive; Galizzi, JP et al, *Br J Pharmacol*, 113:389, 1994; Menke, JG et al, *JBC*, 269:21583, 1994; Zuzack, JS et al, *J Pharmacol Exp Ther*, 277:1337, 1996

Lys-Arg-Pro-Pro-Gly-Phe-Ser-Pro-Phe-Arg

Synonyms: Bradykinin, (Lys[0])-; Kallidin, (Lys)-

American Peptide 18-1-21   MW 1187.7 $C_{56}H_{85}N_{17}O_{12}$

Bachem H-2180.0005   MW 1188.4 $C_{56}H_{85}N_{17}O_{12}$ Store at -15°C

ICN 151574

Neosystem SC091   MW 1188.39   Bioactive; bradykinin analog

Biogenesis 1500-0309   Synthetic   MW 1188.2   Lyophilized, containing 78.15% peptide material, acetate counter ions and residual H_2O; purified

Peptides International PBK-4008-v   Synthetic   MW 1188.4 $C_{56}H_{85}N_{17}O_{12}$ >99% (HPLC); lyophilized amorphous powder | Bioactive; human, bovine; Pierce, JV & ME Webster, *BBRC*, 5:353, 1961; Elliott DF & GP Lewis, *Biochem J*, 95:437, 1965

Lys-Arg-Pro-Pro-Gly-Phe-Ser-Pro-Phe-Arg 3AcOH 4H2O

Synonyms: Bradykinin, (Lys)-; Kallidin

Peptides International PBK-4008   Synthetic   MW 1440.6 $C_{56}H_{85}N_{17}O_{12} \cdot 3CH_3COOH \cdot 4H_2O$ >99% (HPLC); lyophilized amorphous powder; bulk | Bioactive; human, bovine; Pierce, JV & ME Webster, *BBRC*, 5:353, 1961; Elliott DF & GP Lewis, *Biochem J*, 95:437, 1965

Lys-Arg-Pro-Pro-Gly-Phe-Ser-Pro-Phe-Arg Acetate Salt

Synonyms: Bradykinin, (Lys)-; Kallidin

ICN 152756

Sigma B 4889   1188.4 (free base)   ≥95% (HPLC) | Bioactive peptide; induces smooth muscle contraction; peripheral vasodilator

Lys-Arg-Pro-Pro-Gly-Phe-Ser-Pro-Phe-Arg-Ser-Val-Gln-Val-Ser

Synonyms: Bradykinin-Ser-Val-Gln-Val-Ser, (Lys)-

Bachem H-5925.0001   MW 1688.95 $C_{77}H_{121}N_{23}O_{20}$ Store at -15°C

Lys-Arg-Pro-Pro-Gly-Phe-Ser-Pro-Tyr-Arg

Synonyms: Bradykinin, Lys-(Tyr[8])-; Kallidin, (Tyr[9])-

Bachem H-4378.0005   MW 1204.4 $C_{56}H_{85}N_{17}O_{13}$ Store at -15°C

Lys-Arg-Pro-Ser-Lys-Asn-Leu-Lys-Ala-Arg-Cys Acetate Salt

Synonyms: Bone Morphogenetic Protein 14 (9-19)

Biogenesis 1406-1890   Human synthetic   Purified; lyophilized

Lys-Arg-Pro-Trp-Ile-Leu
N-Me-Lys-Arg-Pro-Trp-Ile-Leu

Bachem H-9550.0005   Store at -15°C

Lys-Arg-Ser-Arg

Synonyms: Osteoblast Adhesive Peptide

Bachem H-4594.0005   Store at -15°C

Lys-Arg-Thr-Gly-Gln-Tyr-Lys-Leu

Synonyms: FGF Basic (119-126)

Bachem H-1952.0001   Human   MW 993.18 $C_{44}H_{76}N_{14}O_{12}$ Store at -15°C

Lys-Arg-Thr-Ile-Arg-Arg
Lys-Arg-pThr-Ile-Arg-Arg

Synonyms: Threonine Phosphopeptide

Upstate 12-219   Synthetic   MW 908   >98%; lyophilized | Phosphopeptide

USBio T5025   Synthetic   Purity determined (HPLC); lyophilized | 6 residue synthetic phosphopeptide substrate for protein phosphatase-1

Lys-Arg-Thr-Leu-Arg
N-Myr-Lys-Arg-Thr-Leu-Arg

Alexis 162-017   MW 883.2 $C_{42}H_{82}N_{12}O_8$ Effective inhibitor of protein kinase C (PKC)-catalyzed histone phosphorylation; Ioannides, CG et al, *Cell Immunol*, 131:242, 1990; O'Brian, CA et al, *Biochem Pharmacol*, 39:49, 1990

Lys-Arg-Thr-Leu-Arg

Bachem M-1945.0005   MW 672.83 $C_{28}H_{56}N_{12}O_7$ Store at -15°C

Lys-Arg-Thr-Leu-Arg
Myr-Lys-Arg-Thr-Leu-Arg

Bachem N-1305.0001   MW 883.19 $C_{42}H_{82}N_{12}O_8$ Store at -15°C

Lys-Arg-Thr-Leu-Arg-Arg

Synonyms: Epidermal Growth Factor Receptor (652-657); Protein Kinase C Substrate; Protein Kinase Substrate; Protein Kinase Related Peptide

Alexis 155-023   MW 829.0 $C_{34}H_{68}N_{16}O_8$ ≥97%; white lyophilized powder; soluble in water | Specific substrate for protein kinase C in PC12 pheochromocytoma; Heasley, L & Johnson, GL, *J Biol Chem*, 264:8646, 1989

American Peptide 86-2-50   MW 829.0 $C_{34}H_{68}N_{16}O_8$ Highly specific substrate for protein kinase C in PC12 pheochromocytoma cells & represents a fragment of the EGF receptor; Heasley, L et al, *JBC*, 264:8646, 1989

ICN 159658   MW 829.1

Sigma L 7897   FW 829.0 (free base) $C_{34}H_{68}N_{16}O_8$ ≥97% (HPLC); peptide content:~60% | Bioactive peptide; derived from EGF receptor; Heasley, L & Johnson, GL, *J Biol Chem*, 264:8646, 1989

Lys-Arg-Thr-Leu-Arg-Arg Trifluoroacetate Salt

Synonyms: Protein Kinase Related Peptide; Protein Kinase C Substrate

Sigma L 9905   FW 829.0 (free base) $C_{34}H_{68}N_{16}O_8$ ≥97% (HPLC); peptide content:~60% | Bioactive peptide; derived from EGF receptor; Heasley, L & Johnson, GL, *J Biol Chem*, 264:8646, 1989

Lys-Arg-Trp-Ile-Ile-Leu-Gly-Leu-Asn-Lys

Synonyms: P25 (263-272); GAG P24 CA (131-140); HTLV I Peptide

Neosystem SC536   MW 1240.55   HTLV I peptide from subtype P25 (GAG P24 CA)

Lys-Arg-Trp-Ile-Ile-Leu-Gly-Leu-Asn-Lys-Ile-Val-Arg-Met-Tyr-Cys

Bachem H-7955.0001   MW 2006.56 $C_{93}H_{156}N_{26}O_{19}S_2$ Store at -15°C

Lys-Asn-Arg-Trp-Glu-Asp-Pro-Gly-Lys-Gln-Leu-Tyr-Asn-Val-Glu-Ala

Synonyms: C3 Peptide P16; C3d Peptide P16

American Peptide 79-4-10   MW 1947.4 $C_{86}H_{131}N_{25}O_{27}$ C_3 peptide involved in cell proliferation regulation; Frade, R et al, *BBRC*, 188:833, 1992

Bachem H-1374.0500   MW 1947.14 $C_{86}H_{131}N_{25}O_{27}$ Store at -15°C

Sigma C 1457   FW 1947.1   ≥97% (HPLC) | Bioactive peptide; induces Tyr phosphorylation of phosphoproteins 100 & 105, & *in vitro* proliferation of human B lymphocytes; Frade, R et al, *Biochem Biophys Res Commun*, 188:833, 1992

Lys-Asn-Asn-Gln-Lys-Ser-Glu-Pro-Leu-Ile-Gly-Arg-Lys-Lys-Thr

Bachem H-3142.0001   Store at -15°C

Lys-Asn-Glu-Ala-Val-Arg-His-Phe-Pro-Arg-Ile-Trp-Leu

Synonyms: VPR (27-39)

Neosystem SC199   MW 1665.96   HTLV I peptide from subtype VPR

Lys-Asn-Glu-Phe-Ile-Arg-Phe Amide

Synonyms: AF-1

Bachem H-3338.0001   MW 952.12 $C_{45}H_{69}N_{13}O_{10}$ Store at -15°C

Lys-Asn-Pro-Tyr-Ile-Leu

Synonyms: Neurotensin (8-13), (Lys⁸,Asn⁹)-; LANT-6

American Peptide 62-0-28   MW 746.9 $C_{36}H_{58}N_8O_9$ Carraway, RE & Ferris, CF, *JBC*, 258:2475, 1983

Neosystem SC387   MW 746.90   Bioactive

Lys-Asp
Lys(retro-Asp-H)

Synonyms: Asp-ε-Lys

Bachem G-1610.0250   MW 261.28 $C_{10}H_{19}N_3O_5$ Store at -15°C

Lys-Asp

Bachem G-2640.0250   MW 261.28 $C_{10}H_{19}N_3O_5$ Store at -15°C

ICN 155284   MW 261.3 $C_{10}H_{19}N_3O_5$ Crystalline

Sigma L 1627   FW 261.3 $C_{10}H_{19}N_3O_5$

Lys-Asp-Asn-Lys-Lys-Tyr-Tyr-Ser-Asp-Thr-Lys-Lys-Leu-Asn-Tyr-Arg-Cys

Synonyms: Cdc2 Related Protein Kinase KKIALRE; C-Terminal Blocking Peptide (339-354)

Calbiochem 219424   Human   MW 2152.5 $C_{95}H_{150}N_{26}O_{29}S$ Lyophilized | Synthetic peptide based on the human KKIALRE kinase (339-354) with a Cys C residue added; Ser was substituted for Cys in the peptide sequence to avoid the formation of extraneous KLH-peptide disulfide linkages during coupling; this peptide coupled to KLH, was used as the immunogen for the production of Anti-KKIALRE Kinase; for use in immunoabsorption for immunoprecipitation, immunocytochemistry, Western blotting & dot blots

Lys-Asp-Ser-Phe-Val-Gly-Leu-Met

Synonyms: Neurokinin A (3-10), (Lys³,Gly⁸-R-γ-Lactam-Leu⁹); GR-64349

Neosystem SC433   MW 921.10   Bioactive; tachykinin; water soluble specific agonist at NK-2 receptors; Sheldrick, RL et al, *Agents Actions Suppl*, 31:205-209, 1990; Hagan, RM et al, *Neuropeptides*, 19:127-135, 1991

Lys-Asp-Ser-Ser-Leu-Tyr-Pro-Ala-Leu-Thr-Phe-Asp-Lys
Suc-Lys-Asp-Ser-Ser-Leu-Tyr-Pro-Ala-Leu-Thr-Phe-Asp-Lys

Synonyms: HIV Protease Substrate II

Bachem H-7180.0005   MW 1584.74 $C_{72}H_{109}N_{15}O_{25}$ Store at -15°C

Lys-Asp-Ser-Ser-Leu-Tyr-Pro-Ala-Leu-Thr-Phe-Asp-Lys
N-Suc-Lys-Asp-Ser-Ser-Leu-Tyr-Pro-Ala-Leu-Thr-Phe-Asp-Lys

ICN 156697   MW 1584.7

Sigma S 4021   FW 1584.7   Bioactive peptide

Lys-Asp-Val-Tyr

Synonyms: Thymopoietin II (33-36)

Bachem H-2408.0050   Store at -15°C

Lys-Cys-Asn-Thr-Ala-Thr-Cys-Ala-Thr-Gln-Arg-Leu-Ala-Asn-Phe-Leu-Val-His-Ser-Ser-Asn-Asn-Phe-Gly-Ala-Ile-Leu-Ser-Ser-Thr-Asn-Val-Gly-Ser-Asn-Thr-Tyr

Synonyms: Amylin

American Peptide 74-5-15   Human   MW 3904.3   Disulfide bonds: Cys²-Cys⁷ | Secreted from the β-cells of the pancreas & structurally related to calcitonin; anoretic effects in rats; responsible for the etiology of insulin resistance of type II diabetes mellitus through its modulation of peripheral effects of insulin; blocks activation of glycogen synthase by insulin

Lys-Cys-Asn-Thr-Ala-Thr-Cys-Ala-Thr-Gln-Arg-Leu-Ala-Asn-Phe-Leu-Val-His-Ser-Ser-Asn-Asn-Phe-Gly-Ala-Ile-Leu-Ser-Ser-Thr-Asn-Val-Gly-Ser-Asn-Thr-Tyr Amide

Synonyms: Amylin; Diabetes Associated Peptide; Insulinoma or Islet Amyloid Peptide; Amyloid Protein Precursor; Insulinoma or Islet Amyloid Polypeptide; Islet Amyloid Polypeptide

Bachem H-7905.0500   Human   MW 3903.33 $C_{165}H_{261}N_{51}O_{55}S_2$ Store at -15°C

ICN 153088   Human   Clark, A etal, *Lancet*, ii:231, 1987; Cooper, JGS etal, PNAS, 84:8626, 1987; Westermark, P etal, *PNAS*, 84:3881, 1987

Sigma D 2162   Human   FW 3903.3   ≥97% (HPLC) | Disulfide bonds: 2-7; bioactive peptide; Westermark, P et al, *Proc Natl Acad Sci USA*, 84:3881, 1987; Cooper, JGS et al, *Proc Natl Acad Sci USA*, 84:8628, 1987

Peptides International PAM-4219-v   Human synthetic   MW 3903.3 $C_{165}H_{261}N_{51}O_{55}S_2$ >97% (HPLC); lyophilized amorphous powder | Bioactive; Disulfide bonds: Cys²-Cys⁷; Cooper, GJS, *Endocrinol Rev*, 15:163, 1994; Westermark, P et al, *PNAS USA*, 84:3881, 1987; Clark, A et al, *Lancet*, 2:231, 1987

Lys-Cys-Asn-Thr-Ala-Thr-Cys-Ala-Thr-Gln-Arg-Leu-Ala-Asn-Phe-Leu-Val-Arg-Ser-Ser-Asn-Asn-Leu-Gly-Pro-Val-Leu-Pro-Pro-Thr-Asn-Val-Gly-Ser-Asn-Thr-Tyr Amide

Synonyms: Amylin; Diabetes Associated Peptide; Islet Amyloid Polypeptide

American Peptide 74-5-10 Rat MW 3920.5 $C_{167}H_{272}N_{52}O_{53}S_2$ Disulfide bonds: Cys^2-Cys^7 | A diabetes associated peptide; Leffert, JD et al, *PNAS*, 86:3127, 1989; Asai, J et al, *BBRC*, 164:400, 1989

Bachem H-9475.0500 Rat MW 3920.45 $C_{167}H_{272}N_{52}O_{53}S_2$ Store at -15°C

ICN 154499 Rat MW 3921 Leffert, JD etal, *PNAS*, 86:3127, 1989

ICN 159864 Rat MW 3920.7 >98% | Leffert, JD etal, *PNAS*, 86:3127, 1989

Biogenesis 0486-5156 Rat synthetic MW 3917 Purified; TFA counter ions; lyophilized | Disulfide bonds: Cys^2-Cys^7; Leffert et al, *PNAS*, 86:3127, 1989

Peptides International PAM-4220-v Rat synthetic MW 3920.4 $C_{167}H_{272}N_{52}O_{53}S_2$ >99% (HPLC); lyophilized amorphous powder | Bioactive; Disulfide bonds: Cys^2-Cys^7; Leffert, JD et al, *PNAS* USA, 86:3127, 1989; Asai, J et al, *BBRC*, 164:400, 1989

Lys-Cys-Asp-Ile-Cys-Thr-Asp-Glu-Tyr

Synonyms: Tyrosinase (243-251)

Bachem H-3852.0001 Human Store at -15°C

Lys-Cys-Thr-Cys-Cys-Ala

Synonyms: Metallothionein I (56-61)

Bachem H-7775.0005 MW 627.81 $C_{22}H_{41}N_7O_8S_3$ Store at -15°C

ICN 159955 MW 627.8 $C_{22}H_{41}N_7O_8S_3$ 75% | C-terminal portion of mouse liver protein; high affinity binding to Ca^{2+} & Zn^{2+}; Yoshida, A etal, *PNAS*, 76:486, 1979

Lys-Cys-Thr-Cys-Cys-Ala Acetate Salt

Synonyms: Metallothionein I (56-61)

Sigma L 4512 Mouse liver FW 627.8 (free base) $C_{22}H_{41}N_7O_8S_3$ ≥75% (HPLC) | Bioactive peptide; used in metal binding & detoxification studies; Yoshida, A et al, *Proc Natl Acad Sci USA*, 76:486, 1979

Lys-Gln-Ala-Glu-Ala-Val-Thr-Ser-Pro-Arg

Synonyms: Tyrosine Hydroxylase (24-33)

Alexis 168-005 Rat MW 1086.2 $C_{45}H_{79}N_{15}O_{16}$ Represents consensus site for MAP kinase phosphorylation; includes the Ser^{31} phosphorylation site; Haycock, JW et al, *PNAS*, 89:2365, 1992

Lys-Gln-Ala-Glu-Ala-Val-Thr-Ser-Pro-Arg Trifluoroacetate Salt

Synonyms: Mitogen Activated Protein Kinase Substrate Tyrosine Hydroxylase (24-33)

Calbiochem 454860 MW 1086.2 $C_{45}H_{79}N_{15}O_{16}$ ≥97% (HPLC); lyophilized; soluble in water | Corresponds to the consensus site for MAP kinase phosphorylation & includes the Ser^{31} phosphorylation site; Haycock, JW et al, *PNAS*, 89:2365, 1992

Lys-Gln-Ala-Gly-Asp-Val

Synonyms: Platelet gpIIb/IIIa Binding Peptide; RGD Related Peptide

American Peptide 79-2-35 MW 616.7 $C_{25}H_{44}N_8O_{10}$ Binds to the platelet protein gp IIb/IIIa

Bachem H-3314.0001 MW 616.67 $C_{25}H_{44}N_8O_{10}$ Store at -15°C

ICN 159956 MW 616.7 $C_{25}H_{44}N_8O_{10}$ 97% | Related to RGD

Sigma L 2776 FW 616.7 $C_{25}H_{44}N_8O_{10}$ ≥97% (HPLC) | Bioactive peptide; binds to platelet gpIIb/IIIa; Ruoslahti, E, *J Clin Invest*, 87:1, 1991

Lys-Gln-Ala-Lys-Glu-Lys-Arg-Gln-Glu-Gln-Ile-Ala-Lys-Arg-Arg-Arg-Leu-Ser-Ser-Leu-Arg-Ala-Ser-Thr-Ser-Lys-Ser-Gly-Gly-Ser-Gln-Lys

Synonyms: Ribosomal S6 Kinase Substrate 32

American Peptide 86-0-13 MW 3630.1 $C_{149}H_{270}N_{56}O_{49}$ Used to measure the activity of kinases that phosphorylate ribosomal protein S6, though it can also be used as substrate

Lys-Gln-Ala-Ser-Pro-Val-Ala-Phe-Lys-Lys-Ile-Asn-Asn-Asn Acetate Salt

Synonyms: Neuropeptide Y Receptor I (365-378)

Biogenesis 6732-0170 Synthetic Semi-pure; lyophilized

Lys-Gln-Asn-Met-Asp-Asp-Ile-Asp-Glu-Glu-Asp-Asp-Asp-Leu-Val

Synonyms: NEF (84-98); SIVmac251 Peptide

Neosystem SC610 MW 1793.83 SIVmac251 peptide from subtype NEF

Lys-Gln-Asn-Phe-Ala-Thr-Tyr-Lys-Glu-Gly-Tyr-Asn-Val-Tyr-Gly-Ile-Glu-Ser-Val-Lys-Ile

Synonyms: GluR 2/3

Upstate 12-382 MW 2452 >90% pure; lyophilized | Immunizing Peptide

Lys-Gln-Lys

Bachem H-4095.0050 MW 402.49 $C_{17}H_{34}N_6O_5$ Store at -15°C

Lys-Gln-Phe-Ile-Asn-Met-Trp-Gln-Glu-Val-Gly-Lys-Ala-Met-Tyr-Ala-Pro-Pro

Synonyms: GP120 (421-438); HIV Peptide; HIV Envelope Protein (421-438)

Bachem H-1354.0500 MW 2138.54 $C_{99}H_{148}N_{24}O_{25}S_2$ Store at -15°C

Sigma H 6153 FW 2138.5 ≥90% (HPLC) | Bioactive peptide; derived from the CD-4 attachment region of HIV gp120; inhibits syncytial formation *in vitro* & generates high titers of anti-gp120 antibodies in mice, rabbits & goats; Morrow, WJ et al, *Immunology*, 75:557, 1992

Lys-Glu
Lys(Retro-Glu-H)

Synonyms: Glu-ε-Lys

Bachem G-1960.0250 MW 275.31 $C_{11}H_{21}N_3O_5$ Store at -15°C

Lys-Glu

Bachem G-2645.0250 MW 275.31 $C_{11}H_{21}N_3O_5$ Store at -15°C

Lys-Glu-Ala-Lys-Ser-Gln-Gly-Gly-Ser-Asn

Synonyms: Serum Thymic Factor, (Lys)-

Bachem H-4100.0001 Store at -15°C

ICN 158735 MW 1005 97%

Lys-Glu-Glu-Ala-Glu

Synonyms: Thymosin α_1 (23-27), (Lys^{23})-

ICN 159957 MW 604.6 $C_{24}H_{40}N_6O_{12}$ 97%

Sigma L 3760 FW 604.6 $C_{24}H_{40}N_6O_{12}$ ≥97% (HPLC) | Bioactive peptide

Lys-Glu-Gly

Bachem H-4105.0250 MW 332.36 $C_{13}H_{24}N_4O_6$ Store at -15°C

Lys-Glu-Gly-Asp-Gly-Gly-Glu-Gly-Gly-Gly-Asn-Ser-Ser-Trp-Pro-Trp-Gln-Ile-Glu-Tyr

Synonyms: GP140 (751-770); SIVmac251 Peptide

Neosystem SC783   MW 2153.20   SIVmac251 peptide from subtype GP140

Lys-Glu-Gly-His-Gln-Met-Lys-Asp-Cys-Thr-Glu-Arg-Gln-Ala-Asn-Phe

Synonyms: P15 (418-433); GAG P7 NC (41-55)/P1; HTLV I Peptide

Neosystem SC273   MW 1922.12   HTLV I peptide from subtype P15 (GAG P7 NC/GAG P1/GAG P6); due to the presence of a Cys this peptide can contain the dimeric form

Lys-Glu-Ile-Gln-Trp-Lys-Thr-His-Glu-Val-Phe-Asp-Ala-Lys-Ser-Lys-Ser-Ala

Synonyms: Leptin Receptor Extracellular Domain Peptide (577-594)

Alexis 161-003   Synthetic   MW 2132.4 $C_{96}H_{150}N_{26}O_{29}$ ≥95%; lyophilized; reconstitute with 0.1 mL distilled H_2O | Competitively binds to Ab Alexis 210-168; antiserum blocking peptide

Lys-Glu-Phe-Ile-Leu-Thr-Asp-Glu-Glu-Val-Gln-Arg-Lys-Arg-Glu-Met-Ile-Leu-Lys-Arg Trifluoroacetate Salt

Synon Vitamin D Receptor (91-110), Hinge *yms:*

Biogenesis 9580-3050   Human synthetic   MW 2560.7 Purified; lyophilized

Lys-Glu-Thr-Tyr-Ser-Lys

Bachem H-4110.0050   MW 754.84 $C_{33}H_{54}N_8O_{12}$ Store at -15°C

Lys-Glu-Thr-Val-Glu-Lys-Ala-Val-Ala-Thr-Ala-Pro-Gly-Leu-Gly-Arg

Synonyms: TAT (116-131); SIVmac251 Peptide

Neosystem SC595   MW 1626.87   SIVmac251 peptide from subtype TAT

Lys-Gly
BOC-Lys(Z)-Gly-OMe

Bachem A-1485.0001   MW 451.52 $C_{22}H_{33}N_3O_7$ Store at -15°C

Lys-Gly
BOC-Lys(BOC)-Gly

Bachem A-2750.0001   MW 403.48 $C_{18}H_{33}N_3O_7$ Store at -15°C

Lys-Gly
Lys(Retro-Gly-H)

Synonyms: Gly-ε-Lys

Bachem G-3860.0250   MW 203.24 $C_8H_{17}N_3O_3$ Store at -15°C

Lys-Gly Dihydrobromide

ICN 155285   MW 365.1 $C_8H_{17}N_3O_3 \cdot 2HBr$ White powder

Lys-Gly Hydrochloride

Bachem G-2650.0250   MW 239.7 $C_8H_{17}N_3O_3 \cdot HCl$ Store at -15°C

ICN 155286   MW 239.7 $C_8H_{17}N_3O_3 \cdot HCl$ Crystalline

Sigma L 5752   FW 239.7 $C_8H_{17}N_3O_3 \cdot HCl$

Lys-Gly-Arg
Methoxycarbonyl-Lys(Z)-Gly-Arg-pNA

Bachem L-1970.0050   Store at -15°C

Lys-Gly-Arg
CH_3CO-Lys(Cbo)-Gly-Arg-pNA·AcOH

Synonyms: Pefachrome®C1E; Pefa-5603

Pentapharm 087-31, 087-03   MW 715.8   Highly sensitive chromogenic substrate for C1-esterase; used for determination of C1-esterase inhibitor (C1-INH) in plasma

Lys-Gly-Arg-Gly
BOC-Lys(BOC)-Gly-Arg-Gly-OMe

Bachem A-4410.0100   Store at -15°C

Lys-Gly-Arg-Gly-Lys-Gln-Gly-Gly-Lys-Val-Arg-Ala-Lys-Ala-Lys-Thr-Arg-Ser-Ser

Synonyms: Parasin I

Bachem H-4542.0500   MW 2000.34 $C_{82}H_{154}N_{34}O_{24}$ Store at -15°C

Lys-Gly-Asp-Glu-Glu-Ser-Leu-Ala

Synonyms: Delicious Peptide

American Peptide 87-0-69   MW 848.0 $C_{34}H_{57}N_9O_{16}$ Yamasaki, Y et al, *Agri Biol Chem*, 42:1761, 1978

Lys-Gly-Asp-Ser

American Peptide 44-0-31   MW 405.4 $C_{15}H_{27}N_5O_8$

Bachem H-3168.0005   MW 405.41 $C_{15}H_{27}N_5O_8$ Store at -15°C

Lys-Gly-Glu

Bachem H-4115.0050   MW 332.36 $C_{13}H_{24}N_4O_6$ Store at -15°C

Lys-Gly-Glu-Lys-Val-Asp-Leu-Asn-Thr-Lys-Arg-Thr-Lys-Lys-Ser-Gln-His-Thr-Ser-Glu-Gly

Synonyms: Toxic Shock Syndrome Toxin I (58-78)

Bachem H-2522.0500   Store at -15°C

Lys-Gly-Gly

Bachem H-4120.0250   MW 260.29 $C_{10}H_{20}N_4O_4$ Store at -15°C

Lys-Gly-Gly-Lys

Bachem H-4125.0100   MW 388.47 $C_{16}H_{32}N_6O_5$ Store at -15°C

Lys-Gly-His-Lys

Synonyms: SPARC (119-122)

Bachem H-2472.0050   Mouse   MW 468.56 $C_{20}H_{36}N_8O_5$ Store at -15°C

Sigma S 8531   Mouse   FW 468.6 $C_{20}H_{36}N_8O_5$ ≥97% (HPLC); Peptide content:~60% | Bioactive peptide; stimulates angiogenesis; Lane, TF et al, *J Cell Biol*, 125:929, 1994

Lys-Gly-Lys

Bachem H-4130.0050   MW 331.42 $C_{14}H_{29}N_5O_4$ Store at -15°C

Lys-Gly-Ser-Gly-Ser-Gly-Arg-Pro-Arg-Thr-Ser-Ser-Phe-Ala-Glu-Gly Biotinylated

Synonyms: Crosstide

Upstate 12-385 MW 1801 >90% pure; frozen solution | Proteinkinase assay

Lys-Gly-Trp-Lys

Bachem H-4135.0050 MW 517.63 $C_{25}H_{39}N_7O_5$ Store at -15°C

Lys-Gly-Trp-Lys
Lys-Gly-Trp-Lys-OtBu

Bachem H-4140.0050 MW 573.74 $C_{29}H_{47}N_7O_5$ Store at -15°C

Lys-His-Glu-Tyr-Leu-Arg-Phe Amide

Synonyms: AF-2

Bachem H-1642.0001 Store at -15°C

Lys-His-Gly Amide

Synonyms: Bursin

American Peptide 87-0-49 MW 339.4 $C_{14}H_{25}N_7O_3$ Audhya, T et al, *Science*, 231:997, 1986

Bachem H-5920.0010 Avian MW 339.4 $C_{14}H_{25}N_7O_3$ Store at -15°C

ICN 152415 Avian MW 448.8 $C_{14}H_{25}N_7O_3$ Audhya, T et al, *Science*, 231:997, 1986

Lys-His-Gly Amide Trihydrochloride

Synonyms: Bursin

Sigma B 5644 FW 448.8 $C_{14}H_{25}N_7O_3 \cdot 3HCl$ ≥97% (HPLC); peptide content:~60%; | Bioactive peptide; Audhya et al, *Science*, 231:997, 1986

Lys-His-Leu-Asn-Ser-Met-Glu-Arg-Val-Glu-Trp-Leu-Arg-Lys-Lys-Leu-Gln-Asp-Val-His-Asn-Phe

Synonyms: Parathyroid Hormone (13-34); Parathormone

American Peptide 22-1-32 Human MW 2808.3 $C_{125}H_{199}N_{39}O_{33}S$ Nakamura, R et al, *Endocrinologia Japonica*, 28:547, 1981

Bachem H-4145.0500 Human MW 2808.26 $C_{125}H_{199}N_{39}O_{33}S$ Store at -15°C

ICN 152982 Human MW 2808.2 Hypotensive & hypertensive activity; Goltzmann, D et al, JBC, 250:3199, 1975; Nakamura, R et al, *Endocrin Japan*, 28:547, 1981

Sigma P 2780 Human FW 2808.2 ≥97% (HPLC) | Bioactive peptide; Goltzmann, D et al, *J Biol Chem*, 250:3199, 1975

Peptides International PTH-4106-v Human synthetic MW 2808.3 $C_{125}H_{199}N_{39}O_{33}S$ >95% (HPLC); lyophilized amorphous powder | Bioactive; Nakamura, R et al, *Endocrinol Jpn*, 28:547, 1981

Lys-His-Phe-Pro-Gln-Phe-pSer-Tyr-Ser-Ala-Ser

Synonyms: PAKT1/PKB, (Ser[473])-

Upstate 12-386 MW 1376 >90% pure; lyophilized | Immunizing Peptide

Lys-His-Ser-Gly-Pro-Glu-Asn-Asn-Gln-Cys Acetate Salt

Synonyms: Bone Morphogenetic Protein 15 (15-24)

Biogenesis 1406-1990 Human synthetic MW 1113.1 Purified; lyophilized

Lys-His-Thr-Phe-Asn-Asp-Arg-Arg-Leu-Pro-Gly-Lys-Glu-Thr-Met-Ala

Synonyms: GluR 6/7

Upstate 12-381 MW 1901 >90% pure; lyophilized | Immunizing peptide

Lys-Ile

Bachem G-2655.0250 MW 259.35 $C_{12}H_{25}N_3O_3$ Store at -15°C

Lys-Ile-Cys-Ile-Arg-Ile-Gln-Ile-Ser

Synonyms: Interleukin I Receptor Peptide

Bachem H-1386.0001 Human MW 1073.37 $C_{47}H_{88}N_{14}O_{12}S$ Store at -15°C

Lys-Ile-Glu-Pro-Leu-Gly-Val-Ala-Pro-Thr-Lys-Ala-Lys-Arg-Arg-Val-Val-Gln-Arg-Glu-Lys-Arg

Synonyms: GP160 (495-516); HTLV I Peptide

Neosystem SC800 MW 2560.10 HTLV I peptide from subtype GP160

Lys-Ile-Leu-Arg-Gln-Arg-Lys-Ile-Asp-Arg-Leu-Ile-Asp-Arg-Leu

Synonyms: VPU (31-45); HTLV I Peptide

Neosystem SC529 MW 1936.37 HTLV I peptide from subtype VPU

Lys-Ile-Leu-Asn-Asp-Leu-Ser-Ser-Asp-Ala-Pro-Gly-Val-Pro-Arg
(Lys-Ile-Leu-Asn-Asp-Leu-Ser-Ser-Asp-Ala-Pro-Gly-Val-Pro-Arg)₈-MAP

Bachem H-7655.0001 Store at -15°C

Lys-Ile-Leu-Gly-Asn-Gln-Gly-Ser-Phe-Leu-Thr-Lys-Gly-Pro-Ser-Lys-Leu

Synonyms: CD-4 (37-53)

ICN 153068 MW 1788.1

Sigma C 5167 FW 1788.1 ≥97% (HPLC) | Bioactive peptide

Lys-Ile-Phe-Met-Lys
Ac-Lys-Ile-Phe-Met-Lys Amide

Synonyms: Inactivation Gate Peptide

Bachem H-2476.0005 MW 706.95 $C_{34}H_{58}N_8O_6S$ Store at -15°C

Lys-Ile-Pro-Tyr-Ile-Leu

Synonyms: Neuromedin N

Bachem H-4150.0005 MW 745.96 $C_{38}H_{63}N_7O_8$ Store at -15°C

Calbiochem 05-23-1202 MW 746.0 $C_{38}H_{63}N_7O_8$ ≥97% (HPLC); lyophilized solid; soluble in 5% acetic acid | Neurotensin-like peptide that induces contraction of guinea pig ileum & induces hypotension in the rat; acts as a potent modulator of dopamine D_2 receptor agonist binding; Minamino, N et al, *Biochem Biophys Res Comm*, 122:542, 1984; Li, XM et al, *Neurosci Lett*, 155:121, 1993

ICN 153153 MW 746 $C_{38}H_{63}N_7O_8$ Exhibits contractile activity on guinea pig ileum; induces hypotension in rats; Minamino, N et al, BBRC, 122:542, 1984

Neosystem SC072 Porcine MW 745.96 Bioactive; neurotensin-related peptide; Minamino, N et al, BBRC, 122:542, 1984

American Peptide 62-4-31 Porcine spinal cord MW 746.0 $C_{38}H_{63}N_7O_8$ Endogenous neurotensin-like neuropeptide; exhibits contractile activity on guinea pig ileum & induces hypotension in the rat; a potent modulator of dopamine D_2 receptor agonist binding; Minamino, N et al, BBRC, 122:542, 1984

Sigma N 6513   Porcine spinal cord   FW 746.0 $C_{38}H_{63}N_7O_8$
≥97% (HPLC) | Bioactive peptide; neurotensin-like peptide that exhibits contractile activity on guinea pig ileum & induces hypotension in rat; modulates dopamine D_2 agonist binding; Evangilista, S et al, *Peptides*, 11:293, 1990

Biogenesis 6700-0604   Synthetic   MW 746.0   Purified; lyophilized | Minamino et al, *BBRC*, 122:542, 1984

Lys-Leu
FMOC-Lys(BOC)-Leu

Bachem B-1480.0001   Store at -15°C

Lys-Leu
Z-Lys(BOC)-Leu-OMe

Bachem C-3405.0005   MW 507.63 $C_{26}H_{41}N_3O_7$ Store at -15°C

Lys-Leu
Z-Lys-Leu

Bachem C-3505.0001   Store at -15°C

Lys-Leu

Bachem G-2660.0250   MW 259.35 $C_{12}H_{25}N_3O_3$ Store at -15°C

Lys-Leu
FA-Lys-Leu

Bachem M-1915.0050   MW 379.46 $C_{19}H_{29}N_3O_5$ Store at -15°C

Lys-Leu

Sigma L 1755   FW 340.3 $C_{12}H_{25}N_3O_3 \cdot$ HBr
Monohydrobromide

Sigma L 1879   FW 259.3 (free base) $C_{12}H_{25}N_3O_3$ Acetate salt

Lys-Leu Acetate Salt

ICN 155293   MW 319.4 $C_{12}H_{25}N_3O_3 \cdot C_2H_4O_2$ Crystalline

Lys-Leu Amide Dihydrochloride

Bachem G-2670.0250   MW 331.29 $C_{12}H_{26}N_4O_2 \cdot$ 2HCl
Store at -15°C

Lys-Leu Hydrobromide

Bachem G-2665.0250   MW 340.26 $C_{12}H_{25}N_3O_3 \cdot$ HBr   Store at -15°C

ICN 155294   MW 340.3 $C_{12}H_{25}N_3O_3 \cdot$ HBr   Crystalline

Lys-Leu Hydrochloride

ICN 155295   MW 347.3 $C_{12}H_{26}N_4O_3 \cdot$ 2HCl   Crystalline

Lys-Leu-Ala-Leu
BOC-Lys(FMOC)-Leu-Ala-Leu

Bachem A-3400.0100   Store at -15°C

Lys-Leu-Lys

Bachem H-4155.0250   Store at -15°C

Lys-Leu-Lys-Lys-Thr-Glu-Thr-Gln-Glu-Lys-Asn-Pro-Leu-Pro-Ser-Lys-Glu-Thr-Ile-Glu-Gln-Glu-Lys

Synonyms: Thymosin β₄ (16-38)

Bachem H-2926.0500   MW 2727.11 $C_{118}H_{204}N_{32}O_{41}$ Store at -15°C

Lys-Leu-Ser-Tyr-Asp-Asp-Lys-Val-Phe-Glu-Asn-Val-Glu-Phe-Thr-Pro-Arg-Leu Amide

Synonyms: Pheromonotropic Neuropeptide; Pheromonotropin; MRCH, *Pseudaletia*

Bachem H-8660.0001   *Pseudaletia separata*   MW 2199.49
$C_{101}H_{155}N_{25}O_{30}$ Store at -15°C

ICN 159997   *Pseudaletia separata*   95% | Matsumoto, S etal, *BBRC*, 182:534, 1992

Sigma P 0701   *Pseudaletia separata*   FW 2199.5   ≥95% (HPLC) | Bioactive peptide; neuropeptide involved in pheromone excretion; *Biochem Biophys Res Commun*, 182:534, 1992

Lys-Leu-Thr-Pro-Leu-Cys-Val-Ser-Leu

Synonyms: GP120 (121-129); HTLV I Peptide

Neosystem SC1009   MW 973.24   HTLV I peptide from subtype GP160; due to the presence of a Cys this peptide can contain the dimeric form

Lys-Leu-Val-Phe-Phe

Synonyms: Amyloid β-Protein (16-20)

Bachem H-3682.0005   MW 652.84 $C_{35}H_{52}N_6O_6$ Store at -15°C

Neosystem SC994   MW 652.83   Bioactive; serves as a binding sequence during Aβ polymerization & fibril formation; peptide may serve as a lead compound for development of peptide & non-peptide agents aimed at inhibiting Aβ amyloidogenesis *in vivo*; Tjernberg, LO et al, *JBC*, 271:8545-8548, 1996

Lys-Lys
Lys-Lys-βNA

Bachem K-1425.0250   MW 399.54 $C_{22}H_{33}N_5O_2$ Store at -15°C

Lys-Lys
Lys-ε-N-(2-Furoyl-Me)-L-Lys Dihydrochloride

Synonyms: Furosine

Neosystem SC494   MW 254.28 · 72.92   Quantification of Furosine with HPLC methods proves to be a powerful analytical tool with which to determine the thermal history of a food, resulting in interesting applications & quality control perspectives; Heynes, K et al, *Angew Chem*, 80:627, 1968; Finot, PA et al, *Experientia*, 24:1097, 1968; Finot, PA et al, *Separatum Experientia*, 25:134, 1969; Resmini, P & Pellegrino, L, *Int Chrom Lab*, 6:7-11, 1991

Lys-Lys Dihydrochloride

Synonyms: Dilysine

Bachem G-2675.0250   MW 347.29 $C_{12}H_{26}N_4O_3 \cdot$ 2HCl
Store at -15°C

Sigma L 5502   FW 347.3 $C_{12}H_{26}N_4O_3 \cdot$ 2HCl

Lys-Lys-Ala
Fmoc4-Lys2-Lys-β-Ala-® (200-400 mesh)

Bachem D-1750.0250   Store at 2-8°C

Lys-Lys-Ala
Boc⁴-Lys²-Lys-β-Ala-MBHA Resin

Neosystem RA00002   100-200 mesh; 1% DVB; substitution:0.15-0.5 mmol BOC/g

Lys-Lys-Ala
Fmoc⁴-Lys²-Lys-β-Ala-WANG Resin

Neosystem RA00006   100-200 mesh; 1% DVB; substitution:0.15-0.5 mmol BOC/g

Lys-Lys-Ala-Leu-Arg-Arg-Gln-Glu-Ala-Val-Asp-Ala-Leu

Synonyms: Autocamtide II Related Inhibitory Peptide; Ca^{2+}/Calmodulin-Dependent Protein Kinase II Inhibitor

Alexis 151-029 MW 1497.8 $C_{64}H_{118}N_{22}O_{19}$ ≥98%; white lyophilized powder; soluble in water | Highly specific & potent inhibitor of Ca^{2+}/calmodulin-dependent protein kinase II without affecting cAMP-dependent protein kinase, protein kinase C & Ca^{2+}/calmodulin-dependent protein kinase IV; Ishida, A et al, *BRRC*, 212:806, 1995; Ishida, A & Fujisawa, H, *J Biol Chem*, 270:2163, 1995

Lys-Lys-Ala-Leu-Arg-Arg-Gln-Glu-Ala-Val-Asp-Ala-Leu
Myr-Lys-Lys-Ala-Leu-Arg-Arg-Gln-Glu-Ala-Val-Asp-Ala-Leu Myristoylated

Synonyms: Autocamtide II Related Inhibitory Peptide

Alexis 151-030 MW 1708.1 $C_{78}H_{142}N_{22}O_{20}$ ≥98%; white lyophilized powder; soluble in water | *N*-terminal myristoylated to increase cell-permeability

Lys-Lys-Ala-Leu-Arg-Arg-Gln-Glu-Ala-Val-Asp-Ala-Leu

Synonyms: Calmodulin-Dependent Protein Kinase II Inhibitor; Autocamtide II Related Inhibitory Peptide; Autocamtide II, (Ala⁹)-

American Peptide 86-5-10 MW 1497.7 $C_{64}H_{116}N_{22}O_{19}$ Highly specific & potent inhibitor; does not affect cycle AMP-dependent protein kinase, protein kinase C & calmodulin-dependent protein kinase IV

Bachem H-3384.0001 MW 1497.76 $C_{64}H_{116}N_{22}O_{19}$ Store at -15°C

Calbiochem 189480 MW 1497.8 $C_{64}H_{116}N_{22}O_{19}$ ≥97% (HPLC); solid; soluble in H_2O | Non-phosphorylatable analog of Autocamtide II that is a highly specific & potent inhibitor of calmodulin-dependent protein kinase II; 50 times more potent that (Ala²⁸⁶)-CaM kinase II inhibitor, 281-301; has no effect on PKA, PKC & CaM kinase IV; Ishida, A et al, *Biochem Biophys Res Comm*, 212:806, 1995

Lys-Lys-Ala-Leu-Arg-Arg-Gln-Glu-Ala-Val-Asp-Ala-Leu
Myr-*N*-Lys-Lys-Ala-Leu-Arg-Arg-Gln-Glu-Ala-Val-Asp-Ala-Leu

Synonyms: Autocamtide II Related Inhibitory Peptide, Myristoylated

Calbiochem 189482 MW 1708.1 $C_{78}H_{142}N_{22}O_{20}$ ≥98% (HPLC); lyophilized; soluble in H_2O | AIP which has been myristoylated at the *N*-terminus, enhancing its cell-permeability

Lys-Lys-Ala-Leu-Arg-Arg-Gln-Glu-Ala-Val-Asp-Ala-Leu

Synonyms: Autocamtide II Related Inhibitory Peptide, (Ala⁹)-

Sigma A 4308 FW 1497.8 ≥97% (HPLC); peptide content:~70% | Bioactive peptide; potent inhibitor of calmodulin-dependent proteinkinase II; Ishida, A et al, *Biochem Biophys Res Commun*, 212:806, 1995

Lys-Lys-Ala-Leu-Arg-Arg-Gln-Glu-Thr-Val-Asp-Ala-Leu

Synonyms: Autocamtide II Related Inhibitory Peptide; Ca^{2+}/Calmodulin-Dependent Protein Kinase II Substrate; Autocamtide II; Calmodulin Kinase Substrate; Ca^{2+}/Calmodulin-Dependent Protein Kinase Substrate; Calmodulin-Dependent Protein Kinase II Substrate

Alexis 151-023 MW 1527.8 $C_{65}H_{118}N_{22}O_{20}$ ≥98%; white lyophilized powder; soluble in water | Very selective substrate for Ca^{2+}/calmodulin-dependent protein kinase II; Hanson, PI et al, *Neuron*, 3:59, 1989

American Peptide 86-0-40 MW 1527.8 $C_{65}H_{118}N_{22}O_{20}$ Highly selective & potent substrate; Hanson, PI et al, *Neuron*, 3:59, 1989

Bachem H-3218.0001 MW 1527.79 $C_{65}H_{118}N_{22}O_{20}$ Store at -15°C

ICN 193614 MW 1526.9 >97% | Very selective substrate which exhibits no activity towards PKA; Hanson, PI etal, *Neuron*, 3:59, 1989

Sigma A 0591 Bioactive peptide; highly selective Ca^{2+}/Calmodulin-dependent proteinkinase substrate; Hanson, PI et al, *Neuron*, 3:59, 1989

Lys-Lys-Ala-Leu-Arg-Arg-Gln-Glu-Thr-Val-Asp-Ala-Leu Trifluoroacetate Salt

Synonyms: Autocamtide II; Ca^{2+}/Calmodulin-Dependent Protein Kinase Substrate

Calbiochem 189475 MW 1527.8 $C_{65}H_{118}N_{22}O_{20}$ ≥95% (HPLC); solid; soluble in H_2O | Highly selective substrate; Jones, PM & Persaud, SJ, *Am J Physiol*, 274:E708, 1998; Hanson, PI et al, *Neuron*, 3:59, 1989

Lys-Lys-Ala-Leu-His-Arg-Gln-Glu-Thr-Val-Asp-Ala-Leu

Synonyms: Autocamtide II, (His⁵)-

American Peptide 86-5-11 MW 1508.8 $C_{65}H_{113}N_{21}O_{20}$

Lys-Lys-Arg-Ala-Ala-Arg-Ala-Thr-Ser Amide

Synonyms: Myosin Light Chain Kinase (11-19); Myosin Light Chain Kinase Inhibitor, Smooth Muscle; Myosin Kinase Inhibiting Peptide; Myosin Kinase Inhibitor

Alexis 162-011 MW 987.2 $C_{40}H_{78}N_{18}O_{11}$ Effective inhibitor of the smooth muscle myosin light chain kinase; Pearson, RB et al, *J Biol Chem*, 261:25, 1986

American Peptide 79-2-40 MW 987.2 $C_{40}H_{78}N_{18}O_{11}$ Pearson, RB et al, *JBC*, 261:25, 1986

ICN 154586 MW 988.2 Pearson, RB etal, *JBC*, 261:25, 1986

Sigma L 2275 FW 987.2 $C_{40}H_{78}N_{18}O_{11}$ ≥97% (HPLC) | Bioactive peptide; Pearson, RB, *J Biol Chem*, 261:25, 1985

Lys-Lys-Arg-Ala-Ala-Arg-Ala-Thr-Ser-Asn-Val-Phe-Ala

Synonyms: Myosin Light Chain Kinase Substrate

ICN 159633 Smooth muscle MW 1418.8

Lys-Lys-Arg-Ala-Ala-Arg-Ala-Thr-Ser-Asn-Val-Phe-Ala Amide

Synonyms: Myosin Light Chain Kinase Substrate; Myosin Light Chain Kinase Substrate, Smooth Muscle

Alexis 162-013 MW 1418.8 $C_{61}H_{107}N_{23}O_{16}$ Kemp, BE & Pearson, RB, *J Biol Chem*, 260:3355, 1985

Bachem H-3252.0001 Store at -15°C

Lys-Lys-Arg-Asn-Arg-Thr-Leu-Thr-Lys

Synonyms: p70 Ribosomal S6 Kinase Substrate Peptide; Ribosomal S6 Kinase II Substrate; p90 Ribosomal S6 Kinase II Substrate

Upstate 12-243 MW 1141 >95%; lyophilized | Substrate for p90 Ribosomal S6 Kinase-2 (Rsk-2) and p70 Ribosomal S6 Kinase

Lys-Lys-Arg-Glu-Leu-Val-Glu-Pro-Leu-Thr-Pro-Ser-Gly-Glu-Ala-Pro-Asn-Gln-Ala-Leu-Leu-Arg
N'-Biotinyl-Lys-Lys-Arg-Glu-Leu-Val-Glu-Pro-Leu-Thr-Pro-Ser-Gly-Glu-Ala-Pro-Asn-Gln-Ala-Leu-Leu-Arg Biotinylated

Synonyms: Biocytin-Epidermal Growth Factor Receptor (661-681); Mitogen Activated Protein Kinase Substrate

Sigma E 2646 FW 2673.1 ≥97% (HPLC) | Bioactive peptide

Lys-Lys-Arg-Phe-Ser-Phe-Lys-Lys-Ser-Phe-Lys-Leu-Ser-Gly-Phe-Ser-Phe-Lys-Lys-Asn-Lys-Lys Amide Trifluoroacetate Salt

Synonyms: MARCKS PSD Derived Peptide PKC Substrate

Calbiochem 454880 MW 1542.9 $C_{75}H_{123}N_{21}O_{14}$ ≥95% (HPLC); lyophilized solid; soluble in water | Peptide sequence (154-165) from myristoylated Ala-rich C-kinase substrate (MARCKS) protein phosphorylation site domain (PSD); phosphorylated by protein kinase C (including PKC_ϵ) at Ser5 & Ser9; poor substrate for CaM kinase I, II & III, protein kinase A & G; has a strong affinity for rat brain PKC; Blackshear, PJ et al, *J Biol Chem*, 268:1501, 1993; Amess, B et al, *FEBS Lett*, 297:285, 1992; Graff, JM et al, *J Biol Chem*, 266:14390, 1991

Lys-Lys-Arg-Pro-Gln-Arg-Ala-Thr-Ser-Asn-Val-Phe-Ser Amide

Synonyms: Kemptamide

American Peptide 86-0-54 MW 1517.9 $C_{65}H_{112}N_{24}O_{18}$

ICN 154516 MW 1517.9

Lys-Lys-Arg-Pro-Hyp-Gly-Ala-Ser-Phe-Ala-Arg
Lys-Lys-Arg-Pro-Hyp-Gly-β-(2-Thienyl)-Ala-Ser-D-Phe-β-(2-Thienyl)-Ala-Arg

Synonyms: Bradykinin, Lys-Lys-(Hyp³,β-(2-Thi)-Ala⁵,⁸,D-Phe⁷)-

Bachem H-9070.0001 Store at -15°C

Lys-Lys-Arg-Pro-Hyp-Gly-Thi-Ser-Phe-Thi-Arg
Lys-Lys-Arg-Pro-Hyp-Gly-Thi-Ser-D-Phe-Thi-Arg

Synonyms: Bradykinin, Lys-Lys-(Hyp³,Thi⁵,⁸,D-Phe⁷)-

Neosystem SC414 MW 1394.67 Bioactive; bradykinin antagonist; Griesbacher, T & Lembeck, F, *Br J Pharmacol*, 92:333, 1987

Lys-Lys-Asp-Ser-Gly-Pro-Tyr

Synonyms: Lipotropin (39-45), β-; Endorphin

ICN 151557 MW 739.9 $C_{35}H_{55}N_9O_{12}$

Sigma L 2759 Human FW 793.9 $C_{35}H_{55}N_9O_{12}$ ≥97% (HPLC) | Bioactive peptide

---

Lys-Lys-Asp-Val-Val-Ile-Gln-Asp-Asp-Asp-Val-Glu-Ser-Thr-Met-Val-Glu-Lys-Cys

Synonyms: Protein Kinase $C_{\alpha,\beta}$ Blocking Peptide (371-388)

Calbiochem 539534 Synthetic MW 2181.5 $C_{90}H_{153}N_{23}O_{35}S_2$ Lyophilized | Based on the rat, mouse, rabbit, human, *Xenopus* & bovine kinase C α- & β-isoforms (371-388) with a cys C residue added; this peptide coupled to KLH was used as the immunogen for the production of Anti-cPKC; for use in immunoabsorption for immunoprecipitation, immunocytochemistry, Western blotting & dot blots

---

Lys-Lys-Glu-Asp-Asn-Val-Leu-Val-Glu-Ser-His-Glu-Lys-Ser-Leu-Gly-Glu-Ala-Asp-Lys-Ala-Asp-Val-Asn-Val-Leu-Thr-Lys-Ala-Lys-Ser-Gln

Synonyms: Parathyroid Hormone (53-84), (Asn⁷⁶)-; Parathormone

American Peptide 22-1-45 Human MW 3510.9 $C_{149}H_{253}N_{43}O_{54}$ Rosenblatt, M et al, *Endo*, 103:978, 1978

Bachem H-4165.0500 Human MW 3510.91 $C_{149}H_{253}N_{43}O_{54}$ Store at -15°C

ICN 152991 Human MW 3511.9 Rosenblatt, M etal, *Endocrinology*, 103:978, 1978

ICN 154486 Human MW 3512.4 Rosenblatt, M etal, *Endocrinology*, 103:978, 1978

Lys-Lys-Glu-Asp-Asn-Val-Leu-Val-Glu-Ser-His-Glu-Lys-Ser-Leu-Gly-Glu-Ala-Asp-Lys-Ala-Asp-Val-Asp-Val-Leu-Thr-Lys-Ala-Lys-Ser-Gln

Synonyms: Parathyroid Hormone (53-84), (Asp⁷⁶)-

Neosystem SC398 Human MW 3511.88 Bioactive; calcium metabolism peptide; Rosenblatt, M et al, *Endocrinology*, 103:978, 1978

Lys-Lys-Glu-Asp-Asn-Val-Leu-Val-Glu-Ser-His-Glu-Lys-Ser-Leu-Gly-Glu-Ala-Asp-Lys-Ala-Asp-Val-Asn-Val-Leu-Thr-Lys-Ala-Lys-Ser-Gln

Synonyms: Parathyroid Hormone (53-84), (Asn⁷⁶)-

Sigma P 1921 Human FW 3510.9 ≥95% (HPLC) | Bioactive peptide

Lys-Lys-Glu-Asp-Val-Val-Abu-Cys

Bachem H-2766.0001 Store at -15°C

Lys-Lys-Glu-Asp-Val-Val-Abu-Cys
Mca-Lys-Lys-Glu-Asp-Val-Val-Abu-Cys

Bachem H-2772.0001 Store at -15°C

Lys-Lys-Glu-Asp-Val-Val-Abu-Cys-Ser-Abu-Ser-Phe-Lys-Lys
Mca-Lys-Lys-Glu-Asp-Val-Val-Abu-Cys-Ser-Abu-Ser-p-Nitro-Phe-Lys-Lys Amide

Bachem M-2125.0001 Store at -15°C

Lys-Lys-Glu-Asp-Val-Val-Abu-Cys-Ser-Abu-Ser-Tyr-Lys-Lys
Lys(biotinyl)-Lys-Glu-Asp-Val-Val-Abu-Cys-Ser-Abu-Ser-Tyr-Lys-Lys Amide

Bachem M-2130.0001 Store at -15°C

Lys-Lys-Glu-Asp-Val-Val-Abu-Cys-Ser-Abu-Ser-Tyr-Lys-Lys Amide

Bachem M-2095.0001 Store at -15°C

Lys-Lys-Gly-Glu

Synonyms: Lipotropin (88-91), β-; Melanotropin Potentiating Factor; Endorphin (28-31), β-

American Peptide 28-0-29 MW 460.5 $C_{19}H_{36}N_6O_7$ The C-terminal tetrapeptide of β-endorphin potentiates the interaction of β-endorphin with its brain opiate receptors; also potentiates the melanotropic activity of melanocyte-stimulating hormones (MSH) & seems to be involved in other biological systems

ICN 151559 $C_{19}H_{36}N_6O_7$ Unique affinity for brain opiate receptors; Smyth, DG, *Nature*, 269:167, 1977; Carte, RJ etal, *Nature*, 279:74, 1979

Neosystem SC143 MW 460.53 Bioactive; opioid peptide

Bachem H-4170.0050 Human MW 460.53 $C_{19}H_{36}N_6O_7$ Store at -15°C

Sigma L 0134 Human FW 460.5 $C_{19}H_{36}N_6O_7$ ≥95% (HPLC) | Bioactive peptide; gives a 5-fold enhancement of α-MSH activity at 0.5 nM; Carter, RJ et al, *Nature*, 279:74, 1979

Lys-Lys-Gly-Ser-Glu-Gln-Glu-Ser-Val-Lys-Glu-Phe-Leu-Ala-Lys-Cys

Synonyms: Protein Kinase A *N*-Terminal Blocking Peptide (7-21)

Calbiochem 539230 Human synthetic MW 1181.1 $C_{78}H_{131}N_{21}O_{26}S$ Lyophilized | Based on the human 41 kDa isoform of the catalytic subunit of cAMP-dependent protein kinase (PKA) (7-21) with a Cys C residue added; this peptide coupled to KLH was used as the immunogen for the production of Anti-PKA; for use in immunoabsorption for immunoprecipitation, immunocytochemistry, Western blotting & dot blots; Hardie, F & Hanks, S, *The Protein Kinase Facts Book, Protein-Serine Kinases*, Academic Press, San Diego, CA, p. 418, 1995; Knighton, DR et al, *Science*, 253:414, 1991

Lys-Lys-Leu-Arg-Arg-Thr-Leu-Ser-Val-Ala

Synonyms: PRAK Substrate Peptide; Protein Kinase Substrate

Upstate 12-363 MW 1171 >98% pure; frozen solution

USBio P5600-06 Synthetic ≥98% (HPLC, mass spectroscopy); 0.1 mg/mL; frozen solution

Lys-Lys-Leu-Asn-Arg-Thr-Leu-Ser-Val-Ala

Synonyms: MAPKAP Kinase II Substrate Peptide

USBio M2355-09 ≥95%; lyophilized | Suitable as a substrate for active MAPKAP Kinase-2 & active GST-MAPKAP Kinase-2; also serves as a substrate for inactive GST-MAPKAP Kinase-2 upon enzyme activation; a substrate for p90 S6 Kinase & CaM Kinase II; peptide related to human glycogen synthase residues 1–9

Upstate 12-240 Synthetic MW 1129 >95%; lyophilized | Peptide related to human glycogen synthase

Lys-Lys-Leu-Val-Phe-Phe-Ala

Synonyms: Amyloid β-Protein (15-21), (Lys[15])-

Bachem H-4062.0005 MW 852.09 $C_{44}H_{69}N_9O_8$ Store at -15°C

Lys-Lys-Lys

Synonyms: Trilysine

Bachem H-4175.0050 MW 402.54 $C_{18}H_{38}N_6O_4$ Store at -15°C

Sigma L 8901 FW 402.5 $C_{18}H_{38}N_6O_4$ ≥97% (TLC); peptide content:~60%

Lys-Lys-Lys-Ala
Boc[8]-Lys[4]-Lys[2]-Lys-β-Ala-PAM-® (200-400 mesh)

Synonyms: Mitogen Activated Protein Matrix

Bachem D-1655.0250 Store at 2-8°C

Lys-Lys-Lys-Ala
BOC-Lys(BOC)-Lys(BOC-Lys(BOC))-β-Ala-O-PAM Resin

Fluka 09851 ~1.44 mmol total N/g resin; 200-400 mesh particle size; crosslinked with 1% DVB

Lys-Lys-Lys-Ala
Boc[8]-Lys[4]-Lys[2]-Lys-β-Ala-MBHA Resin

Neosystem RA00001 100-200 mesh; 1% DVB; substitution:0.15-0.5 mmol BOC/g

Lys-Lys-Lys-Ala
Fmoc[8]-Lys[4]-Lys[2]-Lys-β-Ala-MBHA Resin

Neosystem RA00003 100-200 mesh; 1% DVB; substitution:0.15-0.5 mmol BOC/g

Lys-Lys-Lys-Ala
BOC-Lys(FMOC)[4]-Lys[2]-Lys-β-Ala-MBHA Resin

Neosystem RA00004 100-200 mesh; 1% DVB; substitution:0.15-0.5 mmol BOC/g; Posnett, DN et al, *JBC*, 263:1719, 1988; Tam, JP, *PNAS*, 85:5404, 1988

Lys-Lys-Lys-Ala
Fmoc[8]-Lys[4]-Lys[2]-Lys-β-Ala-Rink Resin

Neosystem RW00005 100-200 mesh; 1% DVB; substitution:0.15-0.5 mmol BOC/g

Lys-Lys-Lys-Asp-Lys-Val-Lys-Lys-Gly-Gly-Pro-Gly-Ser-Glu-Cys-Ala-Glu-Trp-Ala-Trp-Gly-Pro-Cys-Thr-Pro-Ser-Ser-Lys-Asp-Cys-Gly-Val-Gly-Phe-Arg-Glu-Lys-Lys-Glu-Phe-Gly-Ala-Asp-Cys-Lys-Tyr-Lys-Phe-Glu-Asn-Trp-Gly-Ala-Cys-Asn-Ala-Gln-Cys-Gln-Glu-Thr-Ile-Arg-Val-Thr-Lys-Pro-Cys-Thr-Pro-Lys-Thr-Lys-Ala-Lys-Ala-Lys-Ala-Lys-Lys-Gly-Lys-Gly-Lys-Asp

Synonyms: Midkine; Heparin Binding Growth/Differentiation Factor; Neurotrophic Factor; Neurite Outgrowth-Promoting Factor; Plasminogen Activator Activity Enhancer

Peptides International PMK-4298-v Human synthetic MW 13240.3 $C_{570}H_{915}N_{177}O_{167}S_{10}$ >99% (HPLC); lyophilized amorphous powder | Bioactive; Disulfide bonds: Cys[15]-Cys[39], Cys23-Cys[48], Cys[30]-Cys[52], Cys[62]-Cys[94], Cys[72]-Cys[104]; Tsutsui, J-I et al, *BBRC*, 176:792, 1991;

Lys-Lys-Lys-Cys
FMOC-Lys(FMOC)-Lys(FMOC-Lys(FMOC))-β-Cys-o-Wang Resin

Fluka 09850 ~0.2 mmol/g resin; crosslinked with 1% DVB; 100-200 mesh particle size

Lys-Lys-Lys-Cys-Ala
BOC-Lys(BOC)-Lys(BOC-Lys(BOC))-Cys(Acm)-β-Ala-O-PAM Resin

Fluka 09852 ~1.6 mmol total N/g resin; 200-400 mesh particle size; crosslinked with 1% DVB

Lys-Lys-Lys-Cys-Ile-Ala-Lys-Asp-Tyr-Gly-Arg-Cys-Lys-Trp-Gly-Gly-Thr-Pro-Cys-Cys-Arg-Gly-Arg-Gly-Cys-Ile-Cys-Ser-Ile-Met-Gly-Thr-Asn-Cys-Glu-Cys-Lys-Pro-Arg-Leu-Ile-Met-Glu-Gly-Leu-Gly-Leu-Ala

Synonyms: Agatoxin IVA, σ-; Agatoxin IVA, ω-

Bachem H-1544.0100 MW 5202.33 $C_{217}H_{360}N_{68}O_{60}S_{10}$ Store at -1F5°C | Disulfide bonds: Cys[4]-Cys[20], Cys[12]-Cys[25], Cys[19]-Cys[36], Cys[27]-Cys[34]

Calbiochem 121975 *Agelenopsis aperta* MW 5202.3 $C_{217}H_{360}N_{68}O_{60}S_{10}$ >99% (HPLC); lyophilized solid; soluble in water; harmful:LD$_{50}$≤2000 mg/kg | Disulfide bonds: Cys[4]-Cys[20], Cys[12]-Cys[25], Cys[19]-Cys[36], Cys[27]-Cys[34]; potent selective high affinity blocker of P-type Ca^{2+} channels (Kd=2 nM); fully inhibits Ca^{2+} current in a whole-cell patch-clamp assay at 50 *nM* using Purkinje neurons; Nooney, JM & Lodge, D, *Eur J Pharmacol*, 306:138, 1996; Mintz, IM et al, *Nature*, 355:827, 1992

Peptides International PAG-4256-s *Agelenopsis aperta* (funnel web spider) synthetic MW 5202.3 $C_{217}H_{360}N_{68}O_{60}S_{10}$ >95% (HPLC); lyophilized amorphous powder | Bioactive; Olivera, BM et al, *Ann Rev Biochem,* 63:823, 1994; Uchitel, OD, *Toxicon*, 35:1161, 1997; Mintz, IM et al, *Nature*, 355:827, 1992; Turner, TJ et al, *Science*, 258:310, 1992; Nishio, H et al, *BBRC*, 196:1447, 1993

Alexis 630-001 *Agelenopsis aperta* synthetic MW 5202.3 $C_{217}H_{360}N_{68}O_{60}S_{10}$ ≥98%; lyophilized; soluble in water; toxic | Disulfide bonds Cys[4]-Cys[20], Cys[12]-Cys[25], Cys[19]-Cys[36], Cys[27]-Cys[34]; selective P-type Ca^{2+} channel blocker; Mintz, IM et al, *Nature*, 355:827, 1992; Turner, TJ et al, *Science*, 258:310, 1992; Nishio, H et al, *BBRC*, 196:1447, 1993; Takahasi, T & Momiyama, A, *Nature*, 366:156, 1993; Yu, H et al, *Biochemistry*, 32:13123, 1993; Scott, RH et al, *TINS*, 16:153, 1993; Olivera, BM et al, *Ann Rev Biochem*, 63:823, 1994; Dunlap, K et al, *TINS*, 18:89, 1995

Lys-Lys-Lys-Leu-Arg-Arg-Gln-Glu-Ala-Phe-Asp-Ala-Tyr

Synonyms: Autocamtide II Related Inhibitory Peptide, (Lys[3], Phe[10],Tyr[13])-; Calmodulin-Dependent Protein Kinase II Inhibitor; Autocamtide II Related Inhibitory Peptide

Peptides International IKK-4374-v   Synthetic   MW 1652.9 $C_{74}H_{121}N_{23}O_{20}$ >99% (HPLC); lyophilized amorphous powder | Inhibitor; Ishida, A & H Fujisawa, *JBC*, 270:2163, 1995; Ishida, A et al, *FEBS Letts*, 427:115, 1998

Lys-Lys-Lys-Lys
BOC-Lys(Z)-Lys(Z)-Lys(Z)-Lys(Z)-OBzl

Bachem A-3185.0001   MW 1257.49 $C_{68}H_{88}N_{80}O_{15}$ Store at -15°C

Lys-Lys-Lys-Lys

Synonyms: Tetralysine

Bachem H-4180.0050   MW 530.71 $C_{24}H_{50}N_8O_5$ Store at -15°C

Sigma L 9026   FW 530.7 $C_{24}H_{50}N_8O_5$ ≥95% (TLC); peptide content:~60%

Lys-Lys-Lys-Lys-Ala
Boc[16]-Lys[8]-Lys[4]-Lys[2]-Lys-β-Ala-MBHA Resin

Neosystem RA00007   100-200 mesh; 1% DVB; substitution:0.15-0.5 mmol BOC/g

Lys-Lys-Lys-Lys-Lys
Z-Lys(BOC)-Lys(BOC)-Lys(BOC)-Lys(BOC)-Lys(BOC)

Bachem C-1445.0050   Store at -15°C

Lys-Lys-Lys-Lys-Lys

Synonyms: Pentalysine

Bachem H-4185.0050   MW 658.89 $C_{30}H_{62}N_{10}O_6$ Store at -15°C

Sigma L 9151   FW 658.9 $C_{30}H_{62}N_{10}O_6$ ≥95% (TLC); peptide content:~60%

Lys-Lys-Lys-Lys-Lys-Ala
(BOC-Lys(BOC))₄-Lys²-β-Ala-CM Resin

American Peptide RMAP100   1% DVB cross-linked:200-400 mesh | Precursor resin for solid phase peptide synthesis

Lys-Lys-Lys-Lys-Lys-Arg-Phe-Ser-Phe-Lys-Lys-Ser-Phe-Lys-Leu-Ser-Gly-Phe-Ser-Phe-Lys-Lys-Asn-Lys-Lys

Synonyms: Protein Kinase C Substrate; Myristoylated Alanine Rich C-Kinase Substrate PSD Peptide ; Myristoylated Alanine Rich C-Kinase Substrate Protein (151-175); PSD Peptide

ICN 195811   MW 3078.8   >97% | From myristoylated Ala-rich Ckinase substrate (MARCKS) protein; poor PKA, PKG, CaMKI, II, III substrate; useful in phosphocellulose, SDS-PAGE or anion exchange assays

Alexis 162-004   Bovine brain   MW 3081.8 $C_{147}H_{243}N_{41}O_{31}$ ≥97%; white lyophilized powder; soluble in water | AA residues 151-175 of the myristoylated ala-rich C-kinase substrate (MARCKS) protein phosphorylation site domain (psd); protein kinase C isotypes (including PKCε) phosphorylate serines 8, 12 & 19 of this peptide; very weak substrate for cAMP- & cGMP dependent protein kinase & CaM kinases I, II & III; Graff, JM et al, *J Biol Chem*, 266:14390, 1991; Blackshear, PJ, *J Biol Chem*, 268:1501, 1993

Lys-Lys-Lys-Lys-Lys-Lys-Ala
(BOC-Lys(FMOC))₄-Lys²-Lys-β-Ala-CM Resin

American Peptide RMAP120   1% DVB cross-linked:200-400 mesh | Precursor resin for solid phase peptide synthesis

Lys-Lys-Lys-Lys-Lys-Lys-Ala
(BOC-Lys(BOC))₄-Lys²-Lys-β-Ala-PAM Resin

American Peptide RMAP130   1% DVB cross-linked:200-400 mesh | Precursor resin for solid phase peptide synthesis

Lys-Lys-Lys-Lys-Lys-Lys-Ala
(BOC-Lys(FMOC))₄-Lys²-Lys-β-Ala-PAM Resin

American Peptide RMAP150   1% DVB cross-linked:200-400 mesh | Precursor resin for solid phase peptide synthesis

Lys-Lys-Lys-Lys-Lys-Lys-Ala
(FMOC-Lys(FMOC))₄-Lys²-Lys-(-Ala-Wang Resin

American Peptide RMAP160   1% DVB cross-linked:200-400 mesh | Precursor resin for solid phase peptide synthesis

Lys-Lys-Lys-Lys-Lys-Lys-Cys-Ala
(BOC-Lys(BOC))₄-Lys²-Lys-Cys(Acm)-β-Ala-CM Resin

American Peptide RMAP110   1% DVB cross-linked:200-400 mesh | Precursor resin for solid phase peptide synthesis

Lys-Lys-Lys-Lys-Lys-Lys-Cys-Ala
(BOC-Lys(BOC))₄-Lys²-Lys-Cys(Acm)-(-Ala-PAM Resin

American Peptide RMAP140   1% DVB cross-linked:200-400 mesh | Precursor resin for solid phase peptide synthesis

Lys-Lys-Lys-Lys-Lys-Lys-Cys-Ala
(FMOC-Lys(FMOC))₄-Lys²-Lys-Cys(Acm)-(-Ala-Wang Resin

American Peptide RMAP170   1% DVB cross-linked:200-400 mesh | Precursor resin for solid phase peptide synthesis

Lys-Lys-Lys-Lys-Lys-Lys-Ala
BOC-Lys(BOC)-Lys(BOC-Lys(BOC))-Lys(BOC-Lys(BOC)-Lys(BOC-Lys(BOC)))-β-Ala-O-PAM Resin

Fluka 09854   ~1.8 mmol total N/g resin; 200-400 mesh particle size; crosslinked with 1% DVB

Lys-Lys-Lys-Lys-Lys-Lys-Ala
(Lys)₄-(Lys)₂-Lys-β-Ala

Synonyms: Multiple Antigenic Peptide

Sigma M 3672   Synthetic   FW 986.3 $C_{45}H_{91}N_{15}O_9$ ≥90% (HPLC) | Water-soluble peptide carrier; peptide antigens can be conjugated to MAP using a variety of cross-linking reagents

Lys-Lys-Lys-Lys-Lys-Lys-Ala
BOC₈-Lys₄-Lys₂-Lys-β-Ala-PAM

Synonyms: Multiple Antigenic Peptide Core, PAM Resin

Sigma M 8038   Synthetic   1% Divinylbenzene cross-linked polystyrene:200-400 mesh; substitution:0.4-0.8 mmol BOC/g resin | Water-soluble peptide carrier; peptide antigens can be conjugated to MAP using a variety of cross-linking reagents

Lys-Lys-Lys-Lys-Lys-Lys-Lys-Cys
FMOC-Lys(FMOC)-Lys(FMOC-Lys(FMOC))-Lys(FMOC-Lys(FMOC)-Lys(FMOC-Lys(FMOC)))-β-Cys-o-Wang Resin

Fluka 09865   ~0.8 mmol total N/g resin; crosslinked with 1% DVB; 100-200 mesh particle size

Lys-Lys-Lys-Lys-Lys-Lys-Cys-Ala
BOC-Lys(BOC)-Lys(BOC-Lys(BOC))-Lys(BOC-Lys(BOC)-Lys(BOC-Lys(BOC)))-Cys(Acm)-β-Ala-O-PAM Resin

Fluka 09864   ~2.0 mmol total N/g resin; 200-400 mesh particle size; crosslinked with 1% DVB

Lys-Lys-Lys-Met-Glu-Lys-Arg-Phe-Val-Phe-Asn-Lys-Ile-Glu-Ile-Asn-Asn-Lys-Leu-Glu-Phe-Glu-Ser-Ala-Gln-Phe-Pro-Asn-Trp-Tyr-Ile-Ser-Thr

Synonyms: Interleukin Iβ (208-240)

Bachem H-8285.0500   Human   Store at -15°C

Lys-Lys-Lys-Trp-Lys-Lys-Lys

Bachem H-4190.0005   MW 1029.3 $C_{47}H_{84}N_{18}O_8$ Store at -15°C

Lys-Lys-Lys-Val-Ser-Arg-Ser-Gly-Leu-Tyr-Arg-Ser-Pro-Ser-Met-Pro-Glu-Asn-Leu-Asn-Arg-Pro-Arg

Synonyms: CHKtide

Upstate 12-373   MW 2701   Frozen solution | Proteinkinase assay

Lys-Lys-Pro-Tyr-Ile-Leu
Lys-(®)-Lys-Pro-Tyr-Ile-Leu

Synonyms: Neurotensin (8-13), (Lys[8]-(®)-Lys[9])-

Bachem H-8370.0005   Store at -15°C

Lys-Lys-Pro-Tyr-Ile-Leu

Synonyms: Neurotensin (8-13), (Lys[8],Lys[9])-

Bachem H-8380.0005   MW 760.98 $C_{38}H_{64}N_8O_8$ Store at -15°C

Lys-Lys-Pro-Tyr-Ile-Leu
BOC-Lys-ψ(CH₂NH)-Lys-Pro-Tyr-Ile-Leu

Synonyms: Neurotensin (8-13), BOC(Lys[8]-ψ (CH₂NH)-Lys[9])-

Neosystem SC458   MW 847.11   Bioactive; full agonist with high receptor binding properties & high resistance to enzyme hydrolysis; very interesting tool for pharmacological properties studies of neurotensin receptor; Lugrin, D et al, *Eur J Pharmacol*, 205:191-198, 1991; Doulut, S et al, *Pept Res*, 5:30-38, 1992

Lys-Lys-Pro-Tyr-Ile-Leu
Lys-ψ(CH₂NH)-Lys-Pro-Tyr-Ile-Leu

Synonyms: Neurotensin (8-13), (Lys[8]-ψ(CH₂NH)-Lys[9])-; JMV-449

Neosystem SC459   MW 747.00   Bioactive; full agonist with high receptor binding properties & high resistance to enzyme hydrolysis; very interesting tool for pharmacological properties studies of neurotensin receptor; Lugrin, D et al, *Eur J Pharmacol*, 205:191-198, 1991; Doulut, S et al, *Pept Res*, 5:30-38, 1992; Dubuc, I et al, *Eur J Pharmacol*, 219:327-329, 1992

Lys-Lys-Pro-Tyr-Ile-Leu

Synonyms: Neurotensin (8-13), (Lys[8,9])-

Neosystem SC469   MW 760.97   Bioactive; neurotensin analog equipotent to neurotensin (8-13); full agonist with high receptor binding properties & high resistance to enzyme hydrolysis; Very interesting tool for pharmacological properties studies of neurotensin receptor; Lugrin, D et al, *Eur J Pharmacol*, 205:191-198, 1991; Doulut, S et al, *Pept Res*, 5:30-38, 1992

Lys-Lys-Pro-Tyr-Ile-Leu
Lys-ψ-(CH₂NH)-Lys-Pro-Tyr-Ile-Leu

Synonyms: Neurotensin (8-13), (Lys[8]-ψ-(CH₂NH)-Lys[9])-

Sigma N 0522   FW 747.0 $C_{38}H_{66}N_8O_7$ ≥97% (HPLC); peptide content:~60% | Bioactive peptide; degradation-resistant analog exhibits higher binding & biological potencies than neurotensin; Lugrin, D et al, *Eur J Pharmacol*, 205:191, 1991; Doulut, S et al, *Peptide Res*, 5:30, 1992

Lys-Lys-Sar-Glu
Ac-Lys-ᴅ-Lys-Sar-Glu

Synonyms: Melanotropin Potentiating Factor, Ac-(ᴅ-Lys[2],Sar[3])-; Lipotropin (88-91), Ac-(ᴅ-Lys[89],Sar[90])-β-; Endorphin (28-31), Ac-(ᴅ-Lys[29],Sar[30])-β-

Bachem H-2512.0050   Human   Store at -15°C

Lys-Lys-Tyr-Arg-Tyr-Tyr-Leu-Lys-Pro-Leu-Cys-Lys-Lys

Synonyms: Myotoxin II (105-117)

Bachem H-4346.0001   MW 1731.18 $C_{83}H_{135}N_{21}O_{17}S$ Store at -15°C

Lys-Met

Bachem G-2680.0250   MW 277.39 $C_{11}H_{23}N_3O_3S$ Store at -15°C

Lys-Met Formate Salt

ICN 155296

Sigma L 8265   FW 277.4 (free base) $C_{11}H_{23}N_3O_3S$

Lys-Phe Dihydrobromide

Sigma L 8885   FW 455.2 $C_{15}H_{23}N_3O_3 \cdot 2HBr$ 1 mole of methanol as solvent of crystallization

Lys-Phe Hydrochloride

Bachem G-2685.0250   MW 329.83 $C_{15}H_{23}N_3O_3 \cdot HCl$ Store at -15°C

ICN 155298   MW 329.8 $C_{15}H_{23}N_3O_3 \cdot HCl$ Crystalline

Sigma L 6002   FW 329.8 $C_{15}H_{23}N_3O_3 \cdot HCl$

Lys-Phe-Arg-Arg-Gln-Arg-Pro-Arg-Leu-Ser-His-Lys-Gly-Pro-Met-Pro-Phe

Synonyms: Apelin (20-36); Apelin-17

Neosystem SC1340   Human, bovine   MW 2138.56   Virus-related peptide; bioactive; recently identified as the endogenous ligand for the APJ receptor; this G protein-coupled receptor has been reported to support the efficient entry of HIV as a coreceptor with CD-4, suggesting an important role of the APJ receptor & its endogenous ligand in HIV infection; C-terminal fragments of apelin exhibit much higher activity than apelin; corresponds to the carboxy-terminal 17 residues of apelin; 8-fold more active than apelin; Tatemoto, K et al, *BBRC*, 251:471-476, 1998

Lys-Phe-Gly-Lys

Bachem H-4205.0050   MW 478.59 $C_{23}H_{38}N_6O_5$ Store at -15°C

Lys-Phe-His-Glu-Lys-His-His-Ser-His-Arg-Gly-Tyr

Synonyms: Histatin VIII; Hemagglutination Inhibiting Peptide

American Peptide 79-2-41   MW 1562.7 $C_{70}H_{99}N_{25}O_{17}$ Originally isolated from human parotid saliva

Bachem H-1422.0001   MW 1562.71 $C_{70}H_{99}N_{25}O_{17}$ Store at -15°C

ICN 159958   MW 1562.7   95% | Hemagglutination inhibiting peptide; Oppenheim, FG etal, *JBC*, 263:7472, 1988

Sigma L 7404   FW 1562.7   ≥95% (HPLC) | Bioactive peptide; originally isolated from human parotid saliva

Lys-Phe-Ile-Gly-Leu-Met
Lys-Phe-Ile-Gly-Leu-Met Amide

Synonyms: Eledoisin Related Peptide

Peptides International PEL-4003-v   Synthetic   MW 706.95 $C_{34}H_{58}N_8O_6S$ >95% (HPLC); lyophilized amorphous powder | Bioactive; Sakakibara, S & M Fujino, *Bull Chem Soc Japan*, 39:947 1966

Lys-Phe-Ile-Gly-Leu-Met Amide

Synonyms: Eledoisin Related Peptide; Eledoisin (6-11), (Lys[6])-

American Peptide 62-0-58   MW 707.0 $C_{34}H_{58}N_8O_6S$

Bachem H-2690.0005   MW 706.95 $C_{34}H_{58}N_8O_6S$ Store at -15°C

ICN 154526   MW 707

ICN 190454   MW 779.9 $C_{34}H_{58}N_8O_6S \cdot 2HCl$ Vasodilator, hypotensive agent & stimulator of extravascular smooth muscle; Erspamer, V & A Arastosi, *Experientia*, 18:58, 1962

Lys-Phe-Ile-Gly-Leu-Met Amide 2AcOH 3H₂O

Synonyms: Eledoisin Related Peptide

Peptides International PEL-4003   Synthetic   MW 881.11 $C_{34}H_{58}N_8O_6S \cdot 2CH_3COOH \cdot 3H_2O$ >95% (HPLC); lyophilized amorphous powder; bulk | Bioactive; Sakakibara, S & M Fujino, *Bull Chem Soc Japan*, 39:947 1966

Lys-Phe-Ile-Gly-Leu-Met Amide Dihydrochloride

Synonyms: Eledoisin Related Peptide

Sigma E 3253   FW 779.9 $C_{34}H_{58}N_8O_6S \cdot 2HCl$ ≥90% (HPLC) | Bioactive peptide; vasodilator, hypotensive agent & stimulator of extravascular smooth muscle; Erspamer, V & Arastosi, A, *Experientia*, 18:58, 1962

Lys-Phe-Lys

Bachem H-4210.0250   MW 421.54 $C_{21}H_{35}N_5O_4$ Store at -15°C

Lys-Phe-Thr-Met-Thr-Gly-Leu-Lys-Arg-Asp-Lys-Thr-Lys-Glu-Tyr-Asn-Glu-Thr-Trp-Tyr-Cys

Synonyms: GP140 (171-190); SIVmac251-Cys Peptide

Bachem H-8385.0500   Store at -15°C

Lys-Phe-Thr-Met-Thr-Gly-Leu-Lys-Arg-Asp-Lys-Thr-Lys-Glu-Tyr-Asn-Glu-Thr-Trp-Tyr

Synonyms: GP140 (171-190); SIVmac251 Peptide

Neosystem SC725   MW 2539.88   SIVmac251 peptide from subtype GP140

Lys-Phe-Thr-Met-Thr-Gly-Leu-Lys-Arg-Asp-Lys-Thr-Lys-Glu-Tyr-Asn-Glu-Thr-Trp-Tyr-Cys

Synonyms: GP140 (171-190)-(Cys); SIVmac251 Peptide

Neosystem SC847   MW 2643.02   SIVmac251 peptide from subtype GP140

Lys-Phe-Tyr

Bachem H-4215.0250   MW 456.54 $C_{24}H_{32}N_4O_5$ Store at -15°C

Lys-Pro
BOC-Lys(BOC)-Pro

Bachem A-1035.0001   Store at -15°C

Lys-Pro

Bachem G-4190.0250   MW 243.31 $C_{11}H_{21}N_3O_3$ Store at -15°C

Lys-Pro
Z-Lys-Pro-4MβNA

Bachem J-1135.0050   MW 532.64 $C_{30}H_{36}N_4O_5$ Store at -15°C

Lys-Pro
Lys-Pro-4MβNA Dihydrochloride

Bachem J-1255.0050   MW 471.43 $C_{22}H_{30}N_4O_3 \cdot 2HCl$ Store at -15°C

Lys-Pro
Lys-Pro-4-Methoxy-β-Naphthylamide Dihydrochloride

Synonyms: Dipeptidyl-Peptidase II Substrate

ICN 155299   MW 471.4 $C_{22}H_{30}N_4O_3 \cdot 2HCl$ McDonald, JK etal, *BBA*, 616:68, 1980

Sigma L 2386   FW 471.4 $C_{22}H_{30}N_4O_3 \cdot 2HCl$ McDonald, JK et al, *Biochim Biophys Acta*, 616:68, 1980

Lys-Pro
Nε-TFA-L-Lys-Pro

Synonyms: Angiotensin Converting Enzyme Inhibitor

Sigma T 7025   FW 339.3 $C_{13}H_{20}F_3N_3O_4$ Blacklack, TJ et al, *9th Amer Peptide Symp*, Toronto, 1985

Lys-Pro Amide Dihydrochloride

Bachem H-4220.0050   Store at -15°C

Lys-Pro Hydrochloride

Sigma L 9276   FW 243.3 (free base) $C_{11}H_{21}N_3O_3$

Lys-Pro-Ala-Gly-Asp-Leu-Arg-Gln-Lys-Leu-Leu-Arg-Ala-Arg-Gly

Synonyms: NEF (11-25); SIVmac251 Peptide

Neosystem SC604   MW 1679.00   SIVmac251 peptide from subtype NEF

Lys-Pro-Ala-Pro-Phe-Asn-Trp-Phe-Ala-Leu-Nle
D-Lys(Nicotinoyl)-Pro-α-(3-Pyridyl)-Ala-Pro-3,4-dichloro-D-Phe-Asn-D-Trp-Phe-α-(3-Pyridyl)-D-Ala-Leu-Nle Amide

Synonyms: Substance P, (D-Lys(Nicotinoyl)[1],α-(3-Pyridyl)-Ala[3],3,4-Dichloro-D- Phe[5],Asn[6],D-Trp[7],α-(3-Pyridyl)-D-Ala[9],Nle[11])-; Spantide III

Bachem H-2954.0001   Store at -15°C

Lys-Pro-Ala-Pro-Phe-Asn-Trp-Phe-Trp-Leu-Nle
Nε-Nicotinoyl-D-Lys-Pro-(3-Pyridyl)Ala-Pro-(3,4-Dichloro)D-Phe-Asn-D-Trp-Phe-D-Trp-Leu-Nle Amide

Synonyms: Spantide II

American Peptide 70-1-87   MW 1670.8 $C_{86}H_{106}N_{18}O_{13}Cl_2$ Tachykinin antagonist; even more potent substance P antagonist than spantide I & less effective in releasing histamine from mast cells; shows negligible neurotoxicity compared with spantide I; Folkers et al, *PNAS*, 87:4833, 1990; Kakanson, R et al, *BBRC*, 178:297, 1991

Lys-Pro-Ala-Pro-Phe-Asn-Trp-Phe-Trp-Leu-Nle
D-Lys(Nicotinoyl)-Pro-α-(3-Pyridyl)-Ala-Pro-3,4-dichloro-D-Phe-Asn-D-Trp-Phe-D-Trp-Leu-Nle Amide

Synonyms: Substance P, (D-Lys(Nicotinoyl)[1],α-(3-Pyridyl)-Ala[3],3,4-Dichloro-D-Phe[5],Asn[6],D-Trp[7,9],Nle[11])-; Spantide II

Bachem H-8310.0001   Store at -15°C

Lys-Pro-Ala-Pro-Phe-Asn-Trp-Phe-Trp-Leu-Nle
Nε-Nicotinoyl-D-Lys-Pro-(3-Pyridyl)Ala-Pro-(3,4-Dichloro)D-Phe-Asn-D-Trp-Phe-D-Trp-Leu-Nle Amide

Synonyms: Spantide II

ICN 158736   MW 1670.2   >97% | Potent substance P antagonist; Folkers, K etal, *AAs*, 5:233, 1993

Lys-Pro-Ala-Pro-Phe-Asn-Trp-Phe-Trp-Leu-Nle
D-Lys(Nicotinoyl)-Pro-3-(3-Pyridyl)-Ala-Pro-3,4-
Dichloro-D-Phe-Asn-D-Trp-Phe-D-Trp-Leu-Nle Amide

Synonyms: Spantide II

Neosystem SC920 MW 1668.8 Bioactive; tachykinin; highly potent tachykinin antagonist; more potent than Spantide I; negligible neurotoxicity; Folkers, K et al, *PNAS USA*, 87:4833-4835, 1990; Hakanson, R et al, *BBRC*, 178:297-301, 1991

Lys-Pro-AMC

Bachem I-1745.0050 MW 400.49 $C_{21}H_{28}N_4O_4$ Store at -15°C

Lys-Pro-Arg

Bachem H-6055.0050 MW 399.49 $C_{17}H_{33}N_7O_4$ Store at -15°C

Lys-Pro-Arg
D-Lys(Cbo)-Pro-Arg-pNA·2AcOH

Synonyms: Pefachrome®PCa; Pefa-5773

Pentapharm 089-20, 089-02 MW 773.9 Highly sensitive chromogenic peptide substrate for activated protein C; K_M:0.303 mM

Lys-Pro-Arg-Arg-Pro-Tyr-Thr-Asp-Asn-Tyr-Thr-Arg-
Leu-Arg-Lys-Gln-Met-Ala-Val-Lys-Lys-Tyr-Leu-Asn-
Ser-Ile-Leu-Asn Amide

Synonyms: Vasoactive Intestinal Peptide Antagonist; Gastrointestinal Peptide; Vasoactive Intestinal Peptide, (Lys[1],Pro[2,5],Arg[3,4],Tyr[6])-

Bachem H-9935.0500 MW 3467.11 $C_{154}H_{257}N_{49}O_{40}S$ Store at -15°C

Sigma V 2132 Human, porcine, rat FW 3467.1 ≥97% (HPLC); Peptide content:~70% | Bioactive peptide; Gozes, I et al, *Endocrinology*, 125:2945, 1989

Lys-Pro-Asn-Pro-Glu-Arg-Phe-Tyr-Gly-Leu-Met
Amide

Synonyms: Ranakinin; Neurokinin I Tachykinin Receptor Antagonist

Bachem H-1424.0001 MW 1350.61 $C_{62}H_{95}N_{17}O_{15}S$ Store at -15°C

ICN 158558 MW 1350.6 97% | Antagonist of NK[1] tachykinin receptor; *J Neurochem*, 56:2086, 1991

Sigma R 4892 FW 1350.6 ≥97% (HPLC) | Bioactive peptide; originally isolated from the brain of frog Rana *ridibunda*; O'Harte, F et al, *J Neurochem*, 57:2086, 1991

Lys-Pro-Asn-Val-Thr-Tyr-Ala-Ser-Val-Ile-Leu-Arg-
Asp-Tyr-Lys-Gln-Ser-Ser-Ser-Thr

Synonyms: mGluR1

Upstate 12-380 MW 2371 >90% pure; lyophilized | Immunizing peptide

---

Lys-Pro-Asp-Arg-Pro-Tyr-Thr-Asp-Asn-Tyr-Thr-Arg-
Leu-Arg-Lys-Gln-Met-Ala-Val-Lys-Lys-Tyr-Leu-Asn-
Ser-Ile-Leu-Asn Amide

Synonyms: Vasoactive Intestinal Peptide, (Lys[1],Pro[2,5],Arg[3,4],Tyr[6])-

American Peptide 48-6-31 Human, porcine, rat, ovine MW 3467.1 $C_{154}H_{257}N_{49}O_{40}S$ VIP antagonist; hybrid of neurotensin (6-11) & VIP (7-28) is a potent & competitive antagonist of VIP-binding to glial cell; markedly inhibits VIP-stimulated sexual behavior; inhibits tumor growth *in vitro* & *in vivo* by antagonizing VIP receptors on non-small cell lung cancer cells; Gozes, I et al, *Endocrinology*, 125:2945, 1989

Lys-Pro-Asp-Asn-Pro-Gly-Glu-Asp-Ala-Pro-Ala-Glu-
Asp-Leu-Ala-Arg-Tyr-Tyr-Ser-Ala-Leu-Arg-His-Tyr-
Ile-Asn-Leu-Leu-Thr-Arg-Pro-Arg-Tyr Amide

Synonyms: Gastrointestinal Peptide; Neuropeptide Y, (Leu[31],Pro[34])-; Neuropeptide Y[1] Receptor Agonist

Sigma N 7768 Porcine FW 4222.7 ≥97% (HPLC) | Bioactive peptide; Fuhlendorff, J et al, *Proc Natl Acad Sci USA*, 87:182, 1990

Lys-Pro-Asp-Asn-Pro-Gly-Glu-Asp-Ala-Pro-Ala-Glu-
Asp-Met-Ala-Arg-Tyr-Tyr-Ser-Ala-Leu-Arg-His-Tyr-
Ile-Asn-Leu-Leu-Thr-Arg-Pro-Arg-Tyr Amide

Synonyms: Gastrointestinal Peptide; Neuropeptide Y, (Leu[31],Pro[34])-; Neuropeptide Y[1] Receptor Agonist

Sigma N 6146 Human FW 4240.7 ≥97% (HPLC) | Bioactive peptide; Fuhlendorff, J et al, *Proc Natl Acad Sci USA*, 87:182, 1990

Lys-Pro-Gln-Gln-Phe-Phe-Gly-Leu-Met Amide

Synonyms: Substance P (3-11); Substance P, Nona-

American Peptide 70-1-43 MW 1094.4 $C_{52}H_{79}N_{13}O_{11}S$

Lys-Pro-Gln-Gln-Phe-Phe-Gly-Leu-Met Amide Acetate
Salt

Synonyms: Substance P (3-11); Substance P Antagonist

Sigma S 7511 FW 1094.3 (free base) ≥95% (HPLC) | Bioactive peptide; Piotrowski et al, *Agents Actions*, 20:178, 1987

Lys-Pro-Gln-Leu-Trp-Pro

Synonyms: C-Reactive Protein (201-206)

Bachem H-1438.0005 MW 767.93 $C_{38}H_{57}N_9O_8$ Store at -15°C

Sigma C 3830 FW 767.9 $C_{38}H_{57}N_9O_8$ ≥97% (HPLC) | Bioactive peptide; interferes with the membrane-associated oxidative metabolism in human neutrophils; Shephard, EG et al, *Immunology*, 76:79, 1992

Lys-Pro-Gln-Lys-Thr-Lys-Gly-His-Arg-Gly-Ser-His-
Thr-Met-Asn-Gly-His

Synonyms: VIF (176-192); HTLV I Peptide

Neosystem SC236 MW 1901.13 HTLV I peptide from subtype VIF

Lys-Pro-Gly-Lys-Arg-Lys-Glu-Gln-Glu-Lys-Lys-Lys-
Arg-Arg-Thr Trifluoroacetate Salt

Synonyms: Parathyroid Hormone Related Peptide (93-107)

Biogenesis 7170-9650 Synthetic Purified; lyophilized

Lys-Pro-Gly-Thr-Pro-Pro-Lys-Val-Pro-Arg-Thr-Pro-
Pro-Gly-Glu-Glu-Leu-Ala-Glu-Pro-Gln-Ala-Ala-Gly-
Gly-Asn-Gln

*S*ynonyms: Natriuretic Peptide (1-27), Prepro-C-Type

American Peptide 14-6-30 Rat MW 2724.1 $C_{118}H_{191}N_{35}O_{39}$ C-Type Natriuretic Peptide

---

Lys-Pro-Ile Amide Hydrochloride

Bachem H-7970.0025 Store at -15°C

Lys-Pro-Pro
Lys(Abz)-Pro-Pro-pNA

Bachem L-1980.0050 Store at -15°C

Lys-Pro-Pro-Gly-Leu-Trp Amide

Synonyms: He-LW Amide II; LW Amide II

Bachem H-3678.0005 Store at -15°C

Lys-Pro-Pro-Gly-Phe-Ser-Pro-Phe-Arg Acetate Salt

Synonyms: Bradykinin, (Lys[1])-; Fibrinopeptide B, (des-Arg[14])-; Follicle Stimulating Hormone Receptor (48-61); Glucocorticoid Receptor (151-164)

ICN 195629 MW 1092.2 Nicolaides, ED etal, *J Med Chem*, 6:739, 1963

Biogenesis 4450-6410 Human synthetic >80% (HPLC); lyophilized

Biogenesis 4561-8509 Human, rat synthetic >80% (HPLC); lyophilized

Biogenesis 4663-5059 Synthetic >80% (HPLC); lyophilized

Lys-Pro-Pro-Thr-Pro-Pro-Pro-Glu-Pro-Glu-Thr

Synonyms: SV40 Tumor Antigen C-Terminus

American Peptide 79-2-42 MW 1189.3 $C_{54}H_{84}N_{12}O_{18}$ Carboxy-terminal sequence of simian virus 40 large tumor antigen; synthesized to raise specific antibodies; Walter, G et al, *PNAS*, 77:5197, 1980

Bachem H-4225.0005 MW 1189.33 $C_{54}H_{84}N_{12}O_{18}$ Store at -15°C

ICN 153146 C-terminal sequence of SV40 large tumor Ag; used to prepare specific Ab; Walter, G etal, *PNAS*, 77:5197, 1980

Sigma L 0765 FW 1189.3 ≥95% (HPLC) | Bioactive peptide; carboxy-terminal sequence of SV40 tumor antigen; Walter, G et al, *Proc Natl Acad Sci USA*, 77:5197, 1980

Lys-Pro-Ser-Pro-Asp-Arg-Phe-Tyr-Gly-Leu-Met Amide

Synonyms: Ranatachykinin A

Bachem H-1572.0001 MW 1309.56 $C_{60}H_{92}N_{16}O_{15}S$ Store at -15°C

Sigma R 1645 FW 1309.5 ≥97% (HPLC) | Bioactive peptide; Kangawa, K et al, *Regul Peptides*, 46:81, 1993

Lys-Pro-Thr
Lys-D-Pro-Thr

Synonyms: Interleukin Iβ (193-195), (D-Pro[194])-

Bachem H-7230.0025 Human MW 344.41 $C_{15}H_{28}N_4O_5$ Store at -15°C

Lys-Pro-Tyr

Bachem H-2558.0050 Store at -15°C

Lys-Pro-Tyr-Ile-Leu
BOC-Lys-Pro-Tyr-Ile-Leu-OMe Hydrochloride

Synonyms: Neurotensin (9-13)-OMe, (BOC-Lys[9])-

Bachem A-2590.0050 Store at -15°C

Lys-Pro-Val
Lys-D-Pro-Val 2TFA Free Acid

Synonyms: Melanocyte Stimulating Hormone (11-13), (D-Pro[12])-α-

Bachem H-6590.0025 MW 342.4 $C_{16}H_{30}N_4O_4$ Store at -15°C

Lys-Pro-Val
Ac-Lys-Pro-D-Val Amide

Synonyms: Melanocyte Stimulating Hormone (11-13), Ac-(D-Val[13])-α-

Bachem H-8610.0050 MW 383.49 $C_{18}H_{33}N_5O_4$ Store at -15°C

Lys-Pro-Val
Ac-D-Lys-Pro-D-Val Amide

Synonyms: Melanocyte Stimulating Hormone (11-13), Ac-(D-Lys[11],D-Val[13])-α-

Bachem H-8615.0050 MW 383.49 $C_{18}H_{33}N_5O_4$ Store at -15°C

Lys-Pro-Val
Ac-Lys-Pro-Val Amide

Synonyms: Melanocyte Stimulating Hormone (11-13), Ac-α-

Bachem H-8685.0050 MW 383.49 $C_{18}H_{33}N_5O_4$ Store at -15°C

Lys-Pro-Val Amide Dihydrochloride

Synonyms: Melanocyte Stimulating Hormone (11-13), α-

Bachem H-7265.0025 Store at -15°C

Lys-Pro-Val Free Acid

Synonyms: Melanocyte Stimulating Hormone (11-13), α-

Bachem H-4230.0050 MW 374.44 $C_{16}H_{30}N_4O_6$ Store at -15°C

Lys-Pro-Val-Gly-Lys-Lys-Arg-Arg-Pro-Val-Lys-Val-Tyr-Pro

Synonyms: Corticotropin A; Adrenocorticotropic Hormone (11-24)

Bachem H-1210.0001 MW 1652.06 $C_{77}H_{134}N_{24}O_{16}$ Store at -15°C

ICN 152722 ~60% peptide | Kastin, AJ etal, *Nature*, 207:978, 1965; Jolles, J etal, *Brain Res*, 224:315, 1981; Veldhuis, HD & ER DeKloet, *Neuroendocrinology*, 34:374, 1982

Neosystem SC403 Human MW 1652.06 Bioactive; Jolles, J et al, *Brain Res*, 224:315-326, 1981

Sigma A 2532 Human FW 1652.1 ≥97% (HPLC); peptide content:~60% | Bioactive peptide; ACTH fragment that lacks adrenocorticotropic activity; Kastin, AJ et al, *Nature*, 207:978, 1965; Jolles, J et al, *Brain Res*, 224:315, 1981; Veldhuis, HD & DeKloet, ER, *Neuroendocrinology*, 34:374, 1982

Biogenesis 0178-0459 Human synthetic MW 1651.07 Purified; lyophilized

Lys-Ser
Z-Lys(Z)-Ser

Bachem C-2955.0001 MW 501.54 $C_{25}H_{31}N_3O_8$ Store at -15°C

Lys-Ser
Lys-Ser-4MβNA

Bachem J-1260.0050 MW 388.47 $C_{20}H_{28}N_4O_4$ Store at -15°C

Lys-Ser Hydrochloride

Bachem G-2690.0250 MW 269.73 $C_9H_{19}N_3O_4 \cdot HCl$ Store at -15°C

Lys-Ser-Asp-Gly-Gly-Val-Lys-Lys-Arg-Lys-Ser-Ser-Ser-Ser

Synonyms: Ca^{2+}/Calmodulin-Dependent Protein Kinase IV Substrate; Ca^{2+}/Calmodulin-Dependent Protein Kinase II-δ (345-358); Calmodulin-Dependent Protein Kinase IV Substrate; Peptide γ; Protein Kinase Related Peptide; Ca^{2+}/Calmodulin-Dependent Protein Kinase IV Substrate, Brain-Specific

Alexis 153-008 MW 1450.6 $C_{58}H_{107}N_{21}O_{22}$ ≥98%; white lyophilized powder; soluble in water | Very selective & potent substrate for a brain-specific Ca^{2+}/calmodulin-dependent protein kinase IV over Ca^{2+}/calmodulin-dependent protein kinase II; Miyano, O et al, *J Biol Chem*, 267:1198, 1992

ICN 195839 MW 1450.8 ≥97% | From Calmodulin-dependent proteinkinase II-γ (345-358; specific for CamKIV & useful for purified enzyme & crude extracts using phosphocellulose assay

Sigma C 0843 FW 1450.6 Bioactive peptide; selective & potent substrate for brain-specific Ca^{2+}/calmodulin-dependent proteinkinase IV; Miyano, O et al, *J Biol Chem*, 267:1198, 1992

Lys-Ser-Leu-Val-Lys-His-His-Met-Tyr-Val-Ser

Synonyms: VIF (22-32); HTLV I Peptide

Neosystem SC542   MW 1328.59   HTLV I peptide from subtype VIF

Lys-Ser-Met-Gln-Val-Pro-Phe-Ser-Arg-Cys-Cys-Phe-Ser-Phe-Ala-Glu-Gln-Glu-Ile-Pro-Leu-Arg-Ala-Ile-Leu-Cys-Tyr-Arg-Asn-Thr-Ser-Ser-Ile-Cys-Ser-Asn-Glu-Gly-Leu-Ile-Phe-Lys-Leu-Lys-Arg-Gly-Lys-Glu-Ala-Cys-Ala-Leu-Asp-Thr-Val-Gly-Trp-Val-Gln-Arg-His-Arg-Lys-Met-Leu-Arg-His-Cys-Pro-Ser-Lys-Arg-Lys

Synonyms: I-309

Bachem H-4604.0010   Human   MW 8484.06
$C_{370}H_{598}N_{114}O_{99}S_8$ Store at -15°C

Lys-Ser-Met-Leu-Thr-Val-Ser-Asn-Ser-Cys-Cys-Leu-Asn-Thr-Leu-Lys-Lys-Glu-Leu-Pro-Leu-Lys-Phe-Ile-Gln-Cys-Tyr-Arg-Lys-Met-Gly-Ser-Ser-Cys-Pro-Asp-Pro-Pro-Ala-Val-Val-Phe-Arg-Leu-Asn-Lys-Gly-Arg-Glu-Ser-Cys-Ala-Ser-Thr-Asn-Lys-Thr-Trp-Val-Gln-Asn-His-Leu-Lys-Lys-Val-Asn-Pro-Cys

Synonyms: T-Cell Activation Gene III

Bachem H-4618.0010   Mouse   Store at -15°C

Lys-Thr

Bachem G-4255.0250   MW 247.3 $C_{10}H_{21}N_3O_4$ Store at -15°C

Lys-Thr-Asn-Met-Lys-His-Met-Ala-Gly-Ala-Ala-Ala-Ala-Gly-Ala-Cys Acetate Salt

Synonyms: Prion Protein (106-120), ~C

Biogenesis 7672-5509   Human synthetic   Semi-pure; lyophilized | Forloni et al, *Nature,* 362:543, 1993

Lys-Thr-Asn-Met-Lys-His-Met-Ala-Gly-Ala-Ala-Ala-Ala-Gly-Ala-Val-Val-Gly-Gly-Leu-Gly

Synonyms: Prion Protein (106-126)

Bachem H-1566.0500   Human   MW 1912.27
$C_{80}H_{138}N_{26}O_{24}S_2$ Store at -15°C

Lys-Thr-Glu-Glu-Ile-Ser-Glu-Val-Asn-Sta-Val-Ala-Glu-Phe

Synonyms: Amyloid α/A4 Precursor Protein 770 (662-675), (Asn[670],Sta[671],Val[672])-; Secretase Inhibitor, β-

Bachem H-4848.0001   MW 1651.83 $C_{73}H_{118}N_{16}O_{27}$ Store at -15°C

Peptides International IBS-4378-v   Synthetic   MW 1651.80
$C_{73}H_{118}N_{16}O_{27}$ >99% (HPLC); lyophilized amorphous powder | Inhibitor; Sinha, S et al, *Nature*, 402:537, 1999

Lys-Thr-Glu-Glu-Ile-Ser-Glu-Val-Lys-Met
Lys-Thr-Glu-Glu-Ile-Ser-Glu-Val-Lys-Met-pNA

Bachem L-1905.0001   Store at -15°C

Lys-Thr-Leu-Arg-Ala-Glu-Gln-Ala-Ser-Gln-Glu-Val-Lys-Asn-Trp-Met-Thr-Glu-Thr

Synonyms: P25 (302-320); GAG P24 CA (170-188); HTLV I Peptide

Neosystem SC306   MW 2250.51   HTLV I peptide from subtype P25 (GAG P24 CA)

Lys-Thr-Lys-Cys-Lys-Phe-Leu-Lys-Lys-Cys

Synonyms: Endotoxin Inhibitor

American Peptide 74-1-26   MW 1224.6 $C_{55}H_{97}N_{15}O_{12}S_2$
Disulfide bonds: Cys[4]-Cys[10]; antibiotic peptide; endotoxin inhibitor; Rustici, A et al, *Science*, 259:361, 1993

Bachem H-1382.0001   MW 1224.6 $C_{55}H_{97}N_{15}O_{12}S_2$ Store at -15°C

Lys-Thr-Lys-Gln-Leu-Arg
Ac-Lys-Thr-Lys-Gln-Leu-Arg-MCA

Synonyms: Hepatocyte Growth Factor Activator Substrate

Peptides International MCA-3185-v   MW 972.16
$C_{45}H_{73}N_{13}O_{11}$ >99% (HPLC); lyophilized amorphous powder | Mizuno, K et al, *BBRC*, 198:1161, 1994

Lys-Thr-Lys-Glu-Tyr-Asn-Glu-Thr-Trp-Tyr-Ser-Thr-Asp-Leu-Val-Cys-Glu-Gln-Gly-Asn

Synonyms: GP140 (181-200); SIVmac251 Peptide

Neosystem SC726   MW 2408.57   SIVmac251 peptide from subtype GP140; due to the presence of a Cys this peptide can contain the dimeric form

Lys-Thr-Lys-Gly-Ser-Gly-Phe-Phe-Val-Phe Amide

Bachem H-9980.0005   Store at -15°C

Lys-Thr-Phe-Cys-Gly-Thr-Pro-Glu-Tyr-Leu-Ala-Pro-Glu-Val-Arg-Arg-Glu-Pro-Arg-Ile-Leu-Ser-Glu-Glu-Glu-Gln-Glu-Met-Phe-Arg-Asp-Phe-Tyr-Ile-Ala-Asp-Trp-Cys

Synonyms: Protein Kinase D Peptide

Upstate 12-401   MW 4770   Frozen solution | PKD substrate

Lys-Thr-Thr-Lys-Ser

Synonyms: Pro-Collagen Type I (212-216); Pro-Collagen (212-216), Type I

Bachem H-1592.0005   Store at -15°C

Neosystem SC995   MW 563.65   Bioactive; cell attachment peptide; vitronectin fragment; the minimum sequence necessary for potent stimulation of collagen & fibronectin production in a variety of mesenchymal cells; Katayama, K et al, *JBC*, 268:9941-9944, 1993

Lys-Thr-Trp-Gly-Gln-Tyr-Trp-Gln-Val

Synonyms: Melanocyte Protein PMEL 17 (154-162)

Bachem H-3958.0001   Human   MW 1195.34 $C_{58}H_{78}N_{14}O_{14}$
Store at -15°C

Lys-Thr-Tyr

Bachem H-4245.0250   MW 410.47 $C_{19}H_{30}N_4O_6$ Store at -15°C

Lys-Trp

Bachem G-2695.0250   MW 332.4 $C_{17}H_{24}N_4O_3$ Store at -15°C

Lys-Trp-Asn-Lys-Trp-Ala-Leu-Ser-Arg Amide

Synonyms: Pro-Adrenomedullin (12-20)

Bachem H-3994.0001   Human   MW 1187.41 $C_{56}H_{86}N_{18}O_{11}$
Store at -15°C

Lys-Trp-Asp-Asn-Gln

Synonyms: Dynorphin A (13-17)

American Peptide 26-4-48   Porcine   MW 689.7
$C_{30}H_{43}N_9O_{10}$

Lys-Trp-Cys-Phe-Arg-Val-Cys-Tyr-Arg-Gly-Ile-Cys-Tyr-Arg-Arg-Cys-Arg Amide

Synonyms: Tachyplesin I

Bachem H-1202.0500   MW 2263.78 $C_{99}H_{151}N_{35}O_{19}S_4$ Store at -15°C | Disulfide bonds: Cys^3-Cys^{16}, Cys^7-Cys^{12}

Lys-Trp-Gly-Lys

Bachem H-4255.0050   MW 517.63 $C_{25}H_{39}N_7O_5$ Store at -15°C

Lys-Trp-Gly-Lys
Lys-Trp-Gly-Lys-OtBu

Bachem H-4260.0050   MW 573.74 $C_{29}H_{47}N_7O_5$ Store at -15°C

Lys-Trp-Lys

Bachem H-4265.0050   MW 460.58 $C_{23}H_{36}N_6O_4$ Store at -15°C

Lys-Trp-Lys Acetate Salt

ICN 153147   Recognizes & cleaves DNA at apurinic sites; binds to DNA & various ss copolymers of adenine & cytosine; Maurizot, JC etal, *Biochem*, 17:2096, 1978

ICN 153148   Nicks super coiled or relaxed DNA at apurinic/apyrimidinic sites; Pierre, J & J Laval, *JBC*, 256:10217, 1981

Sigma L 5384   FW 460.6 (free base) $C_{23}H_{36}N_6O_4$ ≥97% (TPLC) | Bioactive peptide; reported to bind DNA; Maurizot, JC, *Biochemistry*, 17:2096, 1978

Lys-Trp-Lys-Asp-Ala-Ile-Lys-Glu-Val-Lys-Gln-Thr-Ile-Val-Lys-His-Pro-Arg-Tyr-Thr

Synonyms: GP140 (351-370); SIVmac251 Peptide

Neosystem SC743   MW 2468.92   SIVmac251 peptide from subtype GP140

Lys-Trp-Lys-Leu-Phe-Lys-Lys-Ile-Glu-Lys-Val-Gly-Gln-Asn-Ile-Arg-Asp-Gly-Ile-Ile-Lys-Ala-Gly-Pro-Ala-Val-Ala-Val-Val-Gly-Gln-Ala-Thr-Gln-Ile-Ala-Lys Amide

Synonyms: Cecropin A

Bachem H-3094.0500   MW 4003.84 $C_{184}H_{313}N_{53}O_{46}$ Store at -15°C

Sigma C 6830   FW 4003.8   ≥97% (HPLC) | Bioactive peptide; antibacterial peptide originally identified in moths (*Hyalophora cecropia*) & later in pig intestine; Boman, HG & Hultmark, D, *Ann Rev Microbiol*, 41:103, 1987; Lee, J-Y et al, *Proc Natl Acad Sci USA*, 86:9159, 1989

American Peptide 87-9-59   Porcine   MW 4003.8 $C_{184}H_{313}N_{53}O_{46}$ Antibiotic peptide; 37-residue peptide with antibacterial activity against *E. coli*, *Pseudomonas aeruginosa*, *Bacillus megaterium* & *Micrococcus luteus* at micromolar concentrations; Andreau, D et al, *Proc 20th Euro Pept Symp*, p. 361, 1988

Lys-Trp-Lys-Leu-Phe-Lys-Lys-Ile-Gly-Ala-Val-Leu-Lys-Val-Leu Amide

Synonyms: Cecropin A-Melittin Hybrid Peptide

American Peptide 87-9-63   MW 1770.3 $C_{89}H_{152}N_{22}O_{15}$ Antibiotic peptide; CA(1-7)M(2-9)NH₂; 15-AA residue hybrid peptide incorporating partial sequences of cecropin A & melittin; causes the release of carboxyfluoresceine encapsulated in phosphatidylcholine liposomes; Fernandez, I et al, *Biopolymers*, 34(9):1251, 1994; Juvvadi, P et al, *J Peptide Sci*, 2(4):223, 1996

Lys-Trp-Lys-Leu-Phe-Lys-Lys-Ile-Gly-Ile-Gly-Ala-Val-Leu-Lys-Val-Leu-Thr-Thr-Gly-Leu-Pro-Ala-Leu-Ile-Ser Amide

Cecropin A (1-8)-Melittin (1-18)*Synonyms*:

Bachem H-4314.0001   Store at -15°C

Lys-Trp-Lys-Val-Phe-Lys-Lys-Ile-Glu-Lys-Met-Gly-Arg-Asn-Ile-Arg-Asn-Gly-Ile-Val-Lys-Ala-Gly-Pro-Ala-Ile-Ala-Val-Leu-Gly-Glu-Ala-Lys-Ala-Leu Amide

Synonyms: Cecropin B

American Peptide 87-9-58   MW 3834.7 $C_{176}H_{302}N_{52}O_{41}S$ Antibiotic peptide; 35-residue peptide with activity against different tumor cell lines; Andreau, D et al, *PNAS*, 80:6475, 1983; Andreau, D et al, *Biochem*, 24:1683, 1985

Lys-Trp-Lys-Val-Phe-Lys-Lys-Ile-Glu-Lys-Met-Gly-Arg-Asn-Ile-Arg-Asn-Gly-Ile-Val-Lys-Ala-Gly-Pro-Ala-Ile-Ala-Val-Leu-Gly-Glu-Ala-Lys-Ala-Leu Free Acid

Synonyms: Cecropin B

American Peptide 87-9-60   MW 3835.7 $C_{176}H_{301}N_{51}O_{42}S$

Lys-Trp-Lys-Val-Phe-Lys-Lys-Ile-Glu-Lys-Met-Gly-Arg-Asn-Ile-Arg-Asn-Gly-Ile-Val-Lys-Ala-Gly-Pro-Ala-Ile-Ala-Val-Leu-Gly-Glu-Ala-Lys-Ala-Leu Amide

Synonyms: Cecropin B

Bachem H-3096.0500   MW 3834.73 $C_{176}H_{302}N_{52}O_{41}S$ Store at -15°C

Sigma C 1796   FW 3835.7   ≥97% (HPLC) | Bioactive peptide

Lys-Trp-Ser-Lys-Arg-Val-Thr-Gly-Trp-Pro-Thr-Val-Arg-Glu

Synonyms: NEF MN (4-17); HIV I Peptide

Neosystem SC637   MW 1730.00   NEF peptide from HIV-1 subtype MN

Lys-Trp-Ser-Lys-Ser-Ser-Val-Val-Gly-Trp-Pro-Thr-Val-Arg-Glu

Synonyms: NEF (4-18); HTLV I Peptide

Neosystem SC215   MW 1745.99   HTLV I peptide from subtype NEF

Lys-Tyr

Bachem G-2700.0250   MW 309.37 $C_{15}H_{23}N_3O_4$ Store at -15°C

Lys-Tyr-Arg-Pro-Gly-Arg-Lys Amide

Bachem H-9565.0005   Store at -15°C

Lys-Tyr-Gly-Gln-Val-Pro-Met-Cys-Asp-Ala-Gly-Glu-Gln-Cys-Ala-Val-Arg-Lys-Gly-Ala-Arg-Ile-Gly-Lys-Leu-Cys-Asp-Cys-Pro-Arg-Gly-Thr-Ser-Cys-Asn-Ser-Phe-Leu-Leu-Lys-Cys-Leu

Synonyms: CART (61-102)

Bachem H-4448.0100   Human, rat   MW 4515.36 $C_{189}H_{310}N_{58}O_{56}S_7$ Store at -15°C | Disulfide bonds: $Cys^{74}-Cys^{94}$, $Cys^{68}-Cys^{86}$, $Cys^{88}-Cys^{101}$

Lys-Tyr-Gly-Gly-Phe-Met-Thr-Ser-Glu-Lys-Ser-Gln-Thr-Pro-Leu-Val-Thr-Leu-Phe-Lys-Asn-Ala-Ile-Ile-Lys-Asn-Ala-Tyr-Lys-Lys-Gly-Glu
Nε-Biotinyl-Lys-Tyr-Gly-Gly-Phe-Met-Thr-Ser-Glu-Lys-Ser-Gln-Thr-Pro-Leu-Val-Thr-Leu-Phe-Lys-Asn-Ala-Ile-Ile-Lys-Asn-Ala-Tyr-Lys-Lys-Gly-Glu

Synonyms: Endorphin, Biocytin-β-

American Peptide 28-1-17   Human   MW 3593.2 $C_{162}H_{255}N_{41}O_{47}S_2$

ICN 195647   Human   MW 3818.5   97%

Sigma E 8139   Human   FW 3818.5   ≥97% (HPLC) | Bioactive peptide; biotinylated probe detected with streptavidin-linked enzymes; Fischer, PM et al, *J Immunoassay*, 11 (3):311, 1990

Lys-Tyr-Gly-Lys

Bachem H-4275.0050   MW 494.59 $C_{23}H_{38}N_6O_6$ Store at -15°C

Lys-Tyr-Lys

Bachem H-4280.0050   MW 437.54 $C_{21}H_{35}N_5O_5$ Store at -15°C

Lys-Tyr-Lys Acetate Salt

Sigma L 3271   FW 437.5 (free base) $C_{21}H_{35}N_5O_5$ ≥97% (HPLC) | Bioactive peptide; induces nicks at apurinic/apyrimidinic sites in circular DNA; Pierre, J & Laval, J, *J Biol Chem*, 256:10217, 1981

Lys-Tyr-Pro-Ser-Lys-Pro-Asp-Asn-Pro-Gly-Glu-Asp-Ala-Pro-Ala-Glu-Asp-Leu-Ala-Arg-Tyr-Tyr-Ser-Ala-Leu-Arg-His-Tyr-Ile-Asn-Leu-Ile-Thr-Arg-Gln-Arg-Tyr
ε-Biotinyl-Lys-Tyr-Pro-Ser-Lys-Pro-Asp-Asn-Pro-Gly-Glu-Asp-Ala-Pro-Ala-Glu-Asp-Leu-Ala-Arg-Tyr-Tyr-Ser-Ala-Leu-Arg-His-Tyr-Ile-Asn-Leu-Ile-Thr-Arg-Gln-Arg-Tyr Amide

Synonyms: Gastrointestinal Peptide; Neuropeptide Y, Biocytin-; Neuropeptide Y, Porcine

Sigma B 8530   FW 4608.2   ≥95% (HPLC) | Bioactive peptide; labeled brain peptide which possesses both gastrointestinal & vasoconstrictor properties; Fischer, PM et al, *J Immunoassay*, 11(3):311, 1990; Balasubramanian, A et al, *Peptides*, 11:1151, 1990

Lys-Tyr-Pro-Ser-Lys-Pro-Asp-Asn-Pro-Gly-Glu-Asp-Ala-Pro-Ala-Glu-Asp-Leu-Ala-Arg-Tyr-Try-Ser-Ala-Leu-Arg-His-Tyr-Ile-Asn-Leu-Ile-Thr-Arg-Gln-Arg-Tyr
ε-Biotinyl-Lys-Tyr-Pro-Ser-Lys-Pro-Asp-Asn-Pro-Gly-Glu-Asp-Ala-Pro-Ala-Glu-Asp-Leu-Ala-Arg-Tyr-Try-Ser-Ala-Leu-Arg-His-Tyr-Ile-Asn-Leu-Ile-Thr-Arg-Gln-Arg-Tyr Amide

Synonyms: Neuropeptide Y; Neuropeptide Y, Biocytin Porcine

ICN 195672   Porcine   MW 4608.2   ≥95% | Biotin-labeled brain peptide demonstrating gastrointestinal & vasoconstrictor effects

Lys-Tyr-Ser

Bachem H-4285.0250   MW 396.44 $C_{18}H_{28}N_4O_6$ Store at -15°C

Lys-Tyr-Thr

Bachem H-4290.0250   MW 410.47 $C_{19}H_{30}N_4O_6$ Store at -15°C

Lys-Tyr-Val-Met-Gly-His-Phe-Arg-Trp-Asp-Arg-Phe Amide

Synonyms: Melanocyte Stimulating Hormone, (Lys⁰)-γ-1-

American Peptide 56-0-36   MW 1641.1 $C_{78}H_{109}N_{23}O_{15}S$

Lys-Tyr-Val-Met-Gly-His-Phe-Arg-Trp-Asp-Arg-Phe-Gly-Arg-Arg-Asn-Gly-Ser-Ser-Ser-Ser-Gly-Val-Gly-Gly-Ala-Ala-Gln

Synonyms: Melanocyte Stimulating Hormone, Lys-γ-3-

Neosystem SC074   Bovine   MW 3071.38   Bioactive; melanocyte stimulating hormone-related peptide

Lys-Tyr-Val-Met-Gly-His-Phe-Arg-Trp-Asp-Arg-Phe-Gly-Pro-Arg-Asn-Ser-Ser-Ser-Ala-Gly-Gly-Ser-Ala-Gln

Synon Melanocyte Stimulating Hormone, Lys-γ-3-*yms*:

Neosystem SC070   Rat   MW 2799.07   Bioactive; melanocyte stimulating hormone-related peptide; Ling, N et al, *Life Sci*, 25:1773-1780, 1979

Lys-Val

Bachem G-2705.0250   MW 245.32 $C_{11}H_{23}N_3O_3$ Store at -15°C

Lys-Val
Bz-Gly-Lys-Val

Synonyms: Hippuryl-Lys-Val

Bachem H-3720.0100   MW 406.48 $C_{20}H_{30}N_4O_5$ Store at -15°C

Lys-Val Dihydrobromide

ICN 155300   MW 407.1 $C_{11}H_{23}N_3O_3 \cdot 2HBr$ Crystalline

Lys-Val Hydrochloride

ICN 155301   MW 281.8 $C_{11}H_{23}N_3O_3 \cdot HCl$ Crystalline

Sigma L 6252   FW 281.8 $C_{11}H_{23}N_3O_3 \cdot HCl$ Crystalline

Lys-Val-Glu-Lys-Ile-Gly-Glu-Gly-Thr-Tyr-Gly-Val-Val-Tyr-Lys

Synonyms: *src* Substrate Peptide (6-20)

Upstate 12-140   Synthetic   MW 1670   >97%; powder | p34*cdc2* substrate for *in vitro src*-familykinase assays

Lys-Val-Ile-Leu-Phe

Synonyms: Head Activator (7-11); Hydra Peptide (7-11)

Bachem H-3805.0005   MW 618.82 $C_{32}H_{54}N_6O_6$ Store at -15°C

ICN 153130   MW 618.8 $C_{32}H_{54}N_6O_6$ Useful for the development of an ELISA; Schaller, HC etal, *Eur J Biochem*, 138:365, 1987

Sigma H 6263   FW 618.8 $C_{32}H_{54}N_6O_6$ ≥97% (HPLC) | Bioactive peptide; Schaller, HC et al, *Eur J Biochem*, 138:365, 1984

Lys-Val-Pro-Gln-Thr-Pro-Leu-His-Thr-Ser-Arg-Val-Leu-Lys

Synonyms: Protein Kinase II (326-339), Mitogen Activated; P38RK Substrate

Sigma M 4046   FW 1603.9   ≥97% (HPLC) | Bioactive peptide; Stokoe, D et al, *EMBO J*, 11:3985, 1992

Melanocyte Stimulating Hormone (1-10), β- ~C

Biogenesis 6045-1520   Human synthetic melanocytes   Semi-pure; acetate salt; lyophilized

Met-Ala
Z-Met-Ala

Bachem C-2220.0001   MW 354.43 $C_{16}H_{22}N_2O_5S$ Store at -15°C

Met-Ala
For-Met-Ala

Bachem G-1855.0250   MW 248.3 $C_9H_{16}N_2O_4S$ Store at -15°C

Met-Ala

Bachem G-2710.0250   MW 220.29 $C_8H_{16}N_2O_3S$ Store at -15°C

Met-Ala
N-For-Met-Ala

ICN 152762

Met-Ala
N-Ac-Met-Ala

Synonyms: Acylamino Acid Releasing Enzyme Substrate

Sigma A 1450   FW 262.3 $C_{10}H_{18}N_2O_4S$ ≥97% (HPLC) | Gagnon, J et al, *JBC*, 253:7464, 1978

Met-Ala
N-For-Met-Ala

Synonyms: Chemotactic Peptide

Sigma F 8502   FW 248.3 $C_9H_{16}N_2O_4S$ ≥95% (HPLC) | Bioactive peptide

Met-Ala

Sigma M 1251   FW 220.3 (free base) $C_8H_{16}N_2O_3S$

Met-Ala
Met-β-Ala

Sigma M 2273   FW 220.3 $C_8H_{16}N_2O_3S$

Met-Ala-Glu-Lys-Leu-Lys-Glu-Glu-Asp-Gly-Glu-Asp-Gly-Ser-Cys

Synonyms: EFK5 N-Terminal Blocking Peptide (1-14)

Calbiochem 442687   Human synthetic   MW 1640.8 $C_{64}H_{105}N_{17}O_{29}S_2$ Lyophilized solid | Based on the human ERK5 (1-14) with a Cys C residue added; at residue 4 a Lys residue was substituted for the Pro found in the original sequence; this peptide coupled to KLH was used as the immunogen for the production of Anti-ERK5; for use in immunoabsorption for immunoprecipitation, immunocytochemistry, Western blotting & dot blots

Met-Ala-Gly-Arg-Ser-Gly-Asp-Ser-Asp-Glu-Asp-Leu-Leu-Lys-Ala-Val

Synonyms: REV (1-16); HTLV I Peptide

Neosystem SC188   MW 1663.82   HTLV I peptide from subtype REV

Met-Ala-Gly-Pro-His-Pro-Val-Ile-Val-Ile-Thr-Gly-Pro-His-Glu-Glu

Synonyms: NFAT Inhibitor

Bachem H-4874.0500   MW 1683.95 $C_{75}H_{118}N_{20}O_{22}S$ Store at -15°C

Met-Ala-Pro-Asp-Ala-Ala-Pro-Ala-Ser-Cys Acetate Salt

Synonyms: Bone Morphogenetic Protein 12 (59-68)

Biogenesis 1406-1780   Human synthetic   MW 933.6   Purified; lyophilized

Met-Ala-Ser

Bachem H-4315.0250   MW 307.37 $C_{11}H_{21}N_3O_5S$ Store at -15°C

Met-Ala-Ser
Ac-Met-Ala-Ser

Bachem H-6135.0050   MW 349.41 $C_{13}H_{23}N_3O_6S$ Store at -15°C

Met-Ala-Ser
For-Met-Ala-Ser

Bachem H-6210.0050   MW 335.38 $C_{12}H_{21}N_3O_6S$ Store at -15°C

Met-Ala-Ser

Sigma M 1004   FW 307.4 $C_{11}H_{21}N_3O_5S$

Met-Arg

Bachem G-2720.0250   MW 305.4 $C_{11}H_{23}N_5O_3S$ Store at -15°C

Met-Arg-Arg-Ala-Glu-Pro-Ala-Ala-Asp-Gly-Val-Gly

Synonyms: NEF (20-31); HTLV I Peptide

Neosystem SC500   MW 1229.37   HTLV I peptide from subtype NEF

Met-Arg-Arg-Ala-Glu-Pro-Ala-Glu-Leu-Ala-Ala-Asp-Gly-Val-Gly

Synonyms: NEF MN (19-33); HIV I Peptide

Neosystem SC639   MW 1542.73   NEF peptide from HIV-1 subtype MN

Met-Arg-Cys-Asn-Lys-Ser-Glu-Thr-Asp-Arg-Trp-Gly-Leu-Thr-Lys-Ser-Ser-Thr-Thr-Ile

Synonyms: GP140 (111-130); SIVmac251 Peptide

Neosystem SC719   MW 2314.61   SIVmac251 peptide from subtype GP140; due to the presence of a Cys this peptide can contain the dimeric form

Met-Arg-Leu-Phe-Val

Synonyms: Erythromycin Resistance Peptide

Neosystem SC1349   MW 664.86   Bioactive; antibiotic resistance peptide; Tenson, T et al, *JBC*, 272:17425-17430, 1997

Met-Arg-Phe

Bachem H-2965.0050   MW 452.58 $C_{20}H_{32}N_6O_4S$ Store at -15°C

Met-Arg-Phe-Ala

Bachem H-4320.0050   MW 523.66 $C_{23}H_{37}N_7O_5S$ Store at -15°C

Met-Arg-Phe-Ala Acetate Salt

Sigma M 1170   FW 523.6 (free base) $C_{23}H_{37}N_7O_5S$ ≥90% (HPLC) | Marker in an unambiguous method for sequencing tetrapeptides using FAB, MI & collisional activation spectra in combination; Kulik W & Heerma, W, *Biol Mass Spectrom*, 20(9):553, 1991

Met-Arg-Phe-Phe-Val

Synonyms: Ketolide Resistance Peptide

Neosystem SC1348   MW 698.88   Bioactive; antibiotic resistance peptide; Tripathi, S et al, *JBC*, 273:20073-20077, 1998

Met-Asn

Bachem G-2725.0050   MW 263.32 $C_9H_{17}N_3O_4S$ Store at -15°C

ICN 155374   MW 263.3 $C_9H_{17}N_3O_4S$ Crystalline

Sigma M 6754   FW 263.3 $C_9H_{17}N_3O_4S$

Met-Asn-Tyr-Ala-Leu-Lys-Gly-Gln-Gly-Arg-Thr-Leu-Tyr-Gly-Phe

Synonyms: Histogranin

Bachem H-2266.0500   MW 1719 $C_{78}H_{119}N_{21}O_{21}S$ Store at -15°C

Met-Asn-Tyr-Leu-Ala-Phe-Pro-Arg-Met Amide

Synonyms: Cardioactive Peptide B, Small

Bachem H-3005.0001 MW 1141.43 $C_{52}H_{80}N_{14}O_{11}S_2$ Store at -15°C

ICN 153149 MW 1141.4 Morris, HR, *Nature*, 300:643, 1982

Neosystem SC323 MW 1141.41 Bioactive neuropeptide; Cropper, EC et al, *PNAS USA*, 74:1267-1271, 1987

Sigma M 6779 FW 1141.4 ≥97% (HPLC) | Bioactive peptide; Morris, HR et al, *Nature*, 300:643, 1982

American Peptide 60-0-58 *Aplysia* MW 1141.4 $C_{52}H_{80}N_{14}O_{11}S_2$ Insect neuropeptide; cardioactive peptide; Morris, HR et al, *Nature*, 300:643, 1982

Met-Asp

Bachem G-2730.0250 MW 264.3 $C_9H_{16}N_2O_5S$ Store at -15°C

Met-Asp-Arg-Val-Leu-Ser-Arg-Tyr
Ac-Met-Asp-Arg-Val-Leu-Ser-Arg-Tyr

Bachem H-1045.0005 MW 1081.26 $C_{46}H_{76}N_{14}O_{14}S$ Store at -15°C

Met-Asp-Arg-Val-Leu-Ser-Arg-Tyr
N-Ac-Met-Asp-Arg-Val-Leu-Ser-Arg-Tyr

Sigma A 9924 FW 1081.3 ≥97% (HPLC) | Bioactive peptide

Met-Asp-Lys-Val-Leu-Asn-Arg-Glu
Ac-Met-Asp-Lys-Val-Leu-Asn-Arg-Glu

Bachem H-1440.0005 MW 1046.21 $C_{43}H_{75}N_{13}O_{15}S$ Store at -15°C

Met-Asp-Lys-Val-Leu-Asn-Arg-Tyr
N-Ac-Met-Asp-Lys-Val-Leu-Asn-Arg-Tyr

Synonyms: SV40 Tumor Antigen Antigenic Peptide

ICN 154576 MW 1080.3

Met-Asp-Phe
N-t-BOC-β-Met-Asp-Phe Amide

ICN 195675 MW 510.6 $C_{23}H_{34}N_4O_7S$ ≥95%

Met-Asp-Phe
N-t-BOC-Met-Asp-Phe Amide

Synonyms: Gastrointestinal Peptide

Sigma B 3884 FW 510.6 $C_{23}H_{34}N_4O_7S$ ≥97% (HPLC); crystalline | Bioactive peptide

Met-Asp-Pro-Val-Asp-Pro-Asn-Ile-Glu

Synonyms: HIV I tat Protein (1-9)

Bachem H-3974.0001 MW 1029.14 $C_{43}H_{68}N_{10}O_{17}S$ Store at -15°C

Met-Asp-Tyr-Asp-Phe-Lys-Val-Lys-Leu-Ser-Ser-Glu-Arg-Glu-Arg-Cys

Synonyms: Cdk8; N-Terminal Blocking Peptide (1-15)

Calbiochem 217706 Human MW 2006.3 $C_{85}H_{136}N_{24}O_{28}S_2$ Lyophilized | Synthetic peptide based on the human Cdk8 (1-15) with a Cys C residue added; this peptide coupled to KLH, was used as the immunogen for the production of Anti-Cdk8; for use in immunoabsorption for immunoprecipitation, immunocytochemistry, Western blotting & dot blots

Met-Cys-Glu-Lys

Bachem H-4325.0005 Store at -15°C

Met-Cys-His-Phe-Gly-Gly-Arg-Met-Asp-Arg-Ile-Ser-Cys-Tyr-Arg Amide

Synonyms: Natriuretic Peptide (5-19), Atrial mini-(Met[5],Cys[6,17],His[7],Ser[16],Tyr[18],Arg[19])-

Bachem H-3372.0500 Store at -15°C

Met-Cys-Met-Pro-Cys-Phe-Thr-Thr-Asp-His-Gln-Met-Ala-Arg-Lys-Cys-Asp-Asp-Cys-Cys-Gly-Gly-Lys-Gly-Arg-Gly-Lys-Cys-Tyr-Gly-Pro-Gln-Cys-Leu-Cys-Arg

Synonyms: Chlorotoxin

Calbiochem 230750 *Leiurus quinquestriatus* MW 3996.7 $C_{158}H_{249}N_{53}O_{47}S_{11}$ >95% (HPLC); solid; soluble in normal saline; LD_{50}≤2000 mg/kg | One of the most effective intracellular blocker of small-conductance chloride ion channels derived from epithelial cells; De Bin, JA et al, *Am J Physiol*, 264:C361, 1993; Lippens, G et al, *Biochemistry*, 34:13, 1995; Ullrich, N & Sontheimer, H, *Am J Physiol*, 270:C1511, 1996

Met-Cys-Met-Pro-Cys-Phe-Thr-Thr-Asp-His-Gln-Met-Ala-Arg-Lys-Cys-Asp-Asp-Cys-Cys-Gly-Gly-Lys-Gly-Arg-Gly-Lys-Cys-Tyr-Gly-Pro-Gln-Cys-Leu-Cys-Arg Amide

Synonyms: Chlorotoxin; Small Conductance Cl⁻ Channel Blocker

Peptides International PCN-4282-v *Leiurus quinquestriatus* (scorpion) synthetic MW 3995.8 $C_{158}H_{249}N_{53}O_{47}S_{11}$ >95% (HPLC); lyophilized amorphous powder | Bioactive; Disulfide bonds: Cys[2]-Cys[19],Cys[5]-Cys[28],Cys[16]-Cys[33], Cys[20]-Cys[35]; DeBin, JA et al, *Am J Physiol*, 264:C361, 1993; Najib, J et al, in *Innovation & Perspective in Solid Phase Synthesis*, (R Epton, Ed), Mayflower Worldwide, Birmingham, 1994, pp 615-618; Lippens, G et al, *Biochem*, 34:13, 1995

Alexis 630-069 Synthetic MW 3995.8 $C_{158}H_{249}N_{52}O_{48}S_{11}$ White lyophilized powder; toxic | Disulfide bonds: Cys[2]-Cys[19], Cys[5]-Cys[28], Cys[16]-Cys[33] and Cys[20]-Cys[35]; small conductance Cl⁻ channel blocker; DeBin, JA et al, *Am J Physiol*, 264:C361, 1993; Lippens, G et al, *Biochemistry*, 34:13, 1995

Met-Gln

Bachem G-2740.0250 MW 277.35 $C_{10}H_{19}N_3O_4S$ Store at -15°C

Met-Gln-Arg-Gly-Asn-Phe-Arg-Asn-Gln-Arg-Lys-Ile-Val-Lys

Synonyms: P15 (378-391); GAG p7 NC (1-14); HTLV I Peptide

Neosystem SC269 MW 1775.10 HTLV I peptide from subtype P15 (GAG P7 NC/GAG P1/GAG P6)

Met-Gln-Ile-Phe-Val-Lys-Thr-Leu-Thr-Gly-Lys-Thr-Ile-Thr-Leu-Glu-Val-Glu-Pro-Ser-Asp-Thr-Ile-Glu-Asn-Val-Lys-Ala-Lys-Ile-Gln-Asp-Lys-Glu-Gly-Ile-Pro-Pro-Asp-Gln-Gln-Arg-Leu-Ile-Phe-Ala-Gly-Lys-Gln-Leu-Glu-Asp-Gly-Arg-Thr-Leu-Ser-Asp-Tyr-Asn-Ile-Gln-Lys-Glu-Ser-Thr-Leu-His-Leu-Val-Leu-Arg-Leu-Arg-Gly-Gly

Met-Gln-Ile-Phe-Val-Lys-Thr-Leu-Thr-Gly-Lys-Thr-Ile-Thr-Leu-Glu-Val-Glu-Pro-Ser-Asp-Thr-Ile-Glu-Asn-Val-Lys-Ala-Lys-Ile-Gln-Asp-Lys-Glu-Gly-Ile-Pro-Pro-Asp-Gln-Gln-Arg-Leu-Ile-Phe-Ala-Gly-Lys-Gln-Leu-Glu-Asp-Gly-Arg-Thr-Leu-Ser-Asp-Tyr-Asn-Ile-Gln-Lys-Glu-Ser-Thr-Leu-His-Leu-Val-Leu-Arg-Leu-Arg-Gly-Gly-CHO

Synonyms: Ubiquitin-H Aldehyde Semisynthetic; Deubiquitinating Enzyme Inhibitor

Peptides International IUB-3207-v Synthetic MW 8546.7 $C_{378}H_{629}N_{105}O_{118}S$ >99% (HPLC); lyophilized amorphous powder | Inhibitor; Met at position 1 is oxidized to Met(O); Schaeffer, JR & RE Cohen, *Biochem*, 35:10886, 1996; Melandri, F et al, *Biochem*, 35:12893, 1996; Baek, SH et al, *JBC*, 272:25560, 1997; Dang, LC et al, *Biochem*, 37:1868, 1998

Met-Gln-Met-Asn-Lys-Val-Leu-Asp-Ser

Synonyms: Anti-Inflammatory Peptide III; Peptide III

Bachem H-2806.0001   MW 1065.28 $C_{43}H_{76}N_{12}O_{15}S_2$ Store at -15°C

Neosystem SC181   MW 1065.26   Bioactive; anti-inflammatory peptide

Met-Gln-Met-Lys-Lys-Val-Leu-Asp-Ser

Synonyms: Peptide I; Anti-Inflammatory Peptide I; Antiflammin I; Phospholipase A_2 Inhibitor

Bachem H-9435.0005   MW 1079.35 $C_{45}H_{82}N_{12}O_{14}S_2$ Store at -15°C

ICN 159966   MW 1079.3   97%

Neosystem SC179   MW 1079.33   Bioactive; anti-inflammatory peptide

Sigma M 7782   FW 1079.3   ≥97% (HPLC); peptide content:~70% | Bioactive peptide; inhibits synthesis of platelet activating factor

Met-Gln-Pro-Ile-Gln-Ile-Ala-Ile-Ala-Ala

Synonyms: VPU (1-10); HTLV I Peptide

Neosystem SC526   MW 1055.30   HTLV I peptide from subtype VPU

Met-Gln-Trp-Asn-Ser-Thr-Ala-Phe-His-Gln-Thr

Synonyms: HBV Pre-S Region (120-131), (Ala[127])-

American Peptide 72-2-19   MW 1350.5 $C_{59}H_{83}N_{17}O_{18}S$

Met-Gln-Trp-Asn-Ser-Thr-Ala-Phe-His-Gln-Thr-Leu-Gln-Asp-Pro-Arg-Val-Arg-Gly-Leu-Tyr-Leu-Pro-Ala-Gly-Gly-Ser-Ser-Ser-Gly-Thr-Val

Synonyms: Pre-S2 (1-32); Pre-S (120-151)

Bachem H-1352.0500   Store at -15°C

Met-Gln-Trp-Asn-Ser-Thr-Ala-Phe-His-Gln-Thr-Leu-Gln-Asp-Pro-Arg-Val-Arg-Gly-Leu-Tyr-Leu-Pro-Ala-Gly-Gly

Synonyms: Pre-S2 (1-26); Pre-S (120-145)

Bachem H-9000.0500   MW 2944.33 $C_{131}H_{199}N_{39}O_{37}S$ Store at -15°C

Met-Gln-Trp-Asn-Ser-Thr-Thr-Phe-His-Gln-Thr-Leu-Gln-Asp-Pro-Arg-Val-Arg-Gly-Leu-Tyr-Phe-Pro-Ala-Gly-Gly

Synonyms: HBV Pre-S Region (120-145)

American Peptide 72-2-17   MW 3008.4 $C_{135}H_{199}N_{39}O_{38}S$ Neurath, AR et al, *Science*, 244:392, 1984

Sigma H 7395   FW 3008.4   ≥95% (HPLC) | Bioactive peptide; Neurath, AR et al, *Vaccine*, 4:35, 1986

Met-Glu
Ac-Met-Glu

Bachem G-1045.0100   MW 320.37 $C_{12}H_{20}N_2O_6S$ Store at -15°C

Met-Glu

Bachem G-2735.0250   MW 278.33 $C_{10}H_{18}N_2O_5S$ Store at -15°C

ICN 155375   MW 278.3 $C_{10}H_{18}N_2O_5S$ Crystalline

Sigma M 6504   FW 278.3 $C_{10}H_{18}N_2O_5S$

Met-Glu-Arg-Val-Glu-Trp-Leu-Arg-Lys-Lys-Leu-Gln-Asp-Val-His-Asn-Phe-Val-Ala-Leu-Gly-Ala-Pro-Leu-Ala-Pro-Arg-Asp-Ala-Gly-Ser

Synonyms: Parathyroid Hormone (18-48)

Bachem H-3412.0500   Human   Store at -15°C

Met-Glu-Gln-Pro-Gln-Glu-Glu-Thr-Pro-Glu-Ala-Arg-Glu-Glu-Cys

Synonyms: PPARδ C-Terminal Peptide

Alexis 165-030   Mouse synthetic   MW 1805.9 $C_{71}H_{112}N_{20}O_{31}S_2$ ≥95%; lyophilized; reconstitute with 0.1 mL distilled H_2O | Competitively binds to Ab Alexis 210-191; antiserum blocking peptide

Met-Glu-Glu-Arg-Pro-Pro-Glu-Asn-Glu-Gly-Pro-Gln-Arg-Glu-Pro-Trp-Asp-Glu-Tyr

Synonyms: VPR (1-18)-(Tyr); SIVmac251 Peptide

Neosystem SC576   MW 2388.50   SIVmac251 peptide from subtype VPR

Met-Glu-His-Phe-Arg-Trp

Synonyms: Adrenocorticotropic Hormone (4-9)

Bachem H-1165.0005   MW 905.05 $C_{42}H_{56}N_{12}O_9S$ Store at -15°C

Met-Glu-His-Phe-Arg-Trp-Gly

Synonyms: Corticotropin A; Adrenocorticotropic Hormone (4-10)

Bachem H-1180.0005   MW 962.1 $C_{44}H_{59}N_{13}O_{10}S$ Store at -15°C

ICN 152720   Greven, HM & D DeWied, *Eur J Pharmacol*, 2:14, 1967; Wolthius, OL & D DeWied, *Pharmacol Biochem Behavior*, 4:273, 1978

American Peptide 10-1-37   Human   MW 962.2 $C_{44}H_{59}N_{13}O_{10}S$ Schwyzer, R et al, *FEBS Lett*, 19:229, 1971

Neosystem SC018   Human   MW 962.09   Bioactive

Met-Glu-His-Phe-Arg-Trp-Gly
Met-Glu-His-D-Phe-Arg-Trp-Gly

Synonyms: Adrenocorticotropic Hormone (4-10), (D-Phe[7])-

Neosystem SC426   Human   MW 962.09   Bioactive

Met-Glu-His-Phe-Arg-Trp-Gly

Synonyms: Adrenocorticotropic Hormone (1-4)

Sigma A 0401   Human   FW 962.1 $C_{44}H_{59}N_{13}O_{10}S$ ≥97% (HPLC) | Bioactive peptide; core fragment of ACTH, α- & β-MSH, β-lipotropin; facilitates active & passive avoidance acquisition, improves short term memory consolidation & promotes memory retrieval; lacks adrenocorticotropic activity; Wolthuis, OL & DeWied, D, *Pharmacol Biochem Behav*, 4:273, 1978; Greven, HM & DeWied, D, *Eur J Pharmacol*, 2:14, 1967

Met-Glu-His-Phe-Arg-Trp-Gly-Lys

Synonyms: Corticotropin A; Adrenocorticotropic Hormone (4-11)

Bachem H-1190.0001   MW 1090.27 $C_{50}H_{71}N_{15}O_{11}S$ Store at -15°C

ICN 159039

American Peptide 10-1-40   Human   MW 1090.4 $C_{50}H_{71}N_{15}O_{11}S$

Met-Glu-His-Phe-Lys-Phe
Met(O)-Glu-His-Phe-D-Lys-Phe

Synonyms: Adrenocorticotropic Hormone (4-9), (Met(O)[4],D-Lys[8],Phe[9])-

Bachem H-1175.0005   MW 854 $C_{40}H_{55}N_9O_{10}S$ Store at -15°C

Met-Glu-His-*p*-Iodo-Phe-Arg-Trp-Gly

Synonyms: Adrenocorticotropic Hormone (4-10), (*p*-Iodo-Phe[7])-

Bachem H-2784.0005   MW 1088 $C_{44}H_{58}IN_{13}O_{10}S$ Store at -15°C

Met-Glu-Leu-Arg-Asp-Val-Ser-Leu-Gln-Asp-Pro-Arg-Asp-Arg-Cys

Synonyms: Blocking Peptide (1-14), *N*-Terminal; GCK (1-14)

Calbiochem 371891   Human synthetic   MW 1833.1
$C_{72}H_{125}N_{25}O_{26}S_2$ Lyophilized | Based on the human GCK (1-14) with a Cys C residue added; this peptide coupled to KLH, was used as the immunogen for the production of Anti-GCK; for use in immunoabsorption for immunoprecipitation, immunocytochemistry, Western blotting & dot blots

Met-Glu-Pro-Ser-Ala-Thr-Pro-Gly-Ala-Gln-Met-Gly-Val-Pro-Cys Trifluoroacetate Salt

Synonyms: Orexin Receptor I (1-14), ~C

Biogenesis 7049-5400   Human synthetic   MW 1477.7
Purified; lyophilized

Met-Glu-Ser-Asp-Lys-Gly-Trp-Thr-Ser-Ala-Ser-Thr-Ser-Gly-Lys-Pro-Arg-Lys-Asp-Lys Acetate Salt

Synonyms: Parathyroid Hormone Receptor I (63-82), Extracellular Domain 1

Biogenesis 0100-0181   Synthetic   Semi-pure; lyophilized

Met-Glu-Ser-Glu-His-Glu-Val-Ser-Thr-Pro-Val-Ser-Ala-Glu-Glu-Leu-Lys-Ala-Glu-Ile-Ala-Val-Leu-Gln-Glu-Lys-Leu-Ala-Ala-Gly-Glu-Asp-Val-Ser-His-Glu-Leu-Glu-Glu-Lys-Glu-Lys-Ala-Leu-Ala-Asn-His-Ser-Glu

Bachem H-4852.0001   MW 5372.86 $C_{228}H_{369}N_{61}O_{86}S$ Store at -15°C

Met-Glu-Thr-Pro-Leu-Arg-Glu-Gln-Glu-Asn-Ser-Leu-Glu-Ser-Ser

Synonyms: TAT (1-15); SIVmac251 Peptide

Neosystem SC583   MW 1749.86   SIVmac251 peptide from subtype TAT

Met-Glu-Tyr-Met-Ser-Thr-Gly-Ser-Asp-Asn-Lys-Glu-Glu-Ile-Asp-Cys

Synonyms: COT; N-Terminal Blocking Peptide (1-15)

Calbiochem 235261   Human synthetic   MW 1852.0
$C_{73}H_{114}N_{18}O_{32}S_3$ Lyophilized | Based on the human cancer Osaka thyroid (COT) protein kinase sequence (1-15) with a Cys C residue added; this peptide coupled to KLH, was used as the immunogen for the production of Anti-COT; for use in immunoabsorption for immunoprecipitation, immunocytochemistry, Western blotting & dot blots

Met-Glu-Val-Asp-Pro-Ile-Gly-His-Leu-Tyr

Synonyms: Melanoma Associated Antigen 3 (167-176)

Bachem H-3686.0001   Human   Store at -15°C

Met-Glu-Val-Gln-Leu-Gly-Leu-Gly-Arg-Val-Tyr-Pro-Arg-Pro-Pro-Ser-Lys-Thr-Tyr-Arg-Gly-Cys

Synonyms: Androgen Receptor Protein N-Terminal Peptide

Alexis 151-033   Mouse, rat, human synthetic   MW 2508.0
$C_{110}H_{179}N_{33}O_{30}S_2$ ≥95%; lyophilized; reconstitute with 0.1 mL distilled H_2O | Competitively binds to Ab Alexis 210-152; antiserum blocking peptide

Met-Glu-Val-Gly-Trp-Tyr-Arg-Ser-Pro-Phe-Ser-Arg-Val-Val-His-Leu-Tyr-Arg-Asn-Gly-Lys

Synonyms: Myelin Oligodendrocyte Glycoprotein (35-55); Experimental Allergic Encephalomyelitis Inducing Peptide; Myelin Oligodendrocyte Glycoprotein Peptide (35-55)

Neosystem SC1272   Rat   MW 2581.97   Bioactive; a single injection of MOG (35-55) produces a relapsing-remitting neurologic disease with extensive plaque-like demyelination; this peptide is highly encephalitogenic & can induce strong T & B cell responses; Gardinier, MV et al, *J Neurosci Res*, 33:177-187, 1992; Kerlero de Rosbo, N et al, *Mol Pharm Immunol*, 25:985-993, 1995; Mendel, I et al, *Mol Pharm Immunol*, 25:1951-1959, 1995; Bennun, A et al, *J Neurology*, 243:S14-S22, 1996; Ichikawa, M et al, *J Immunol*, 157:919-926, 1996; Mendel, I et al, *Mol Pharm Immunol*, 26:2470-2479, 1996; Willenborg, DO et al, *J Immunol*, 157:3223-3227, 1996; Ichikawa, M et al, *Int Immunol*, 8:1667-1674, 1996; Suen, WE et al, *J Exp Med*, 186:1233-1240, 1997; Korner, H et al, *J Exp Med*, 186:1585-1590, 1997

Bachem H-4184.0500   Rat, mouse   Store at -15°C

Met-Gly
BOC-Met-Gly

Bachem A-1970.0001   MW 306.38 $C_{12}H_{22}N_2O_5S$ Store at -15°C

Met-Gly
BOC-D-Met-Gly

Bachem A-1975.0001   MW 306.38 $C_{12}H_{22}N_2O_5S$ Store at -15°C

Met-Gly
BOC-Met-Gly-OSu

Bachem A-1980.0001   MW 403.46 $C_{16}H_{25}N_3O_7S$ Store at -15°C

Met-Gly
Z-Met-Gly

Bachem C-2225.0001   MW 340.4 $C_{15}H_{20}N_2O_5S$ Store at -15°C

Met-Gly
Z-Met-Gly-OEt

Bachem C-3910.0001   MW 368.45 $C_{17}H_{24}N_2O_5S$ Store at -15°C

Met-Gly

Bachem G-2745.0250   MW 206.27 $C_7H_{14}N_2O_3S$ Store at -15°C
ICN 155376   MW 206.3 $C_7H_{14}N_2O_3S$ Crystalline

Met-Gly
N-CBZ-Met-Gly-OEt

Sigma C 3877   FW 368.4 $C_{17}H_{24}N_2O_5S$

Met-Gly

Sigma M 3129   FW 206.3 $C_7H_{14}N_2O_3S$

Met-Gly-Ala-Arg-Ala-Ser-Val-Leu-Ser-Gly-Gly-Glu-Leu-Asp-Arg

Synonyms: P18 (1-15); GAG P17 MA (1-15); HTLV I Peptide

Neosystem SC277   MW 1518.70   HTLV I peptide from subtype P25 (GAG P17 MA)

Met-Gly-Glu-Thr-Leu-Gly-Asp-Ser-Pro-Ile-Asp-Pro-Glu-Ser-Asp-Ser-Cys

Synonyms: PPARγ2 C-Terminal Peptide (1-16)

Alexis 165-031   Mouse synthetic   MW 1752.9 $C_{69}H_{109}N_{17}O_{32}S_2$ ≥95%; lyophilized; reconstitute with 0.1 mL distilled H_2O | Blocking peptide for Antiserum to PPARγ2; sequence is absent in PPARγ1; competitively binds to Ab Alexis 210-192; antiserum blocking peptide

Met-Gly-Gly

Bachem H-4330.0250   MW 263.32 $C_9H_{17}N_3O_4S$ Store at -15°C

Met-Gly-His-Phe-Arg-Trp

Synonyms: Melanocyte Stimulating Hormone (3-8), γ-

Bachem H-4335.0001   MW 832.98 $C_{39}H_{52}N_{12}O_7S$ Store at -15°C

Met-Gly-Met

Bachem H-4340.0250   MW 337.46 $C_{12}H_{23}N_3O_4S_2$ Store at -15°C

Met-Gly-Met-Met

Bachem H-4345.0100   MW 468.66 $C_{17}H_{32}N_4O_5S_3$ Store at -15°C

Sigma M 4786   FW 468.6 $C_{17}H_{32}N_4O_5S_3$ ≥97%

Met-Gly-Thr-Asn-Leu-Ser-Val-Pro-Asn-Pro-Leu-Gly-Phe-Phe-Pro-Asp-His-Gln-Leu-Asp-Pro

Synonyms: Pre-S1 (12-32); Pre-S (12-32)

Bachem H-1348.0500   MW 2296.59 $C_{104}H_{154}N_{26}O_{31}S$ Store at -15°C

Met-Gly-Trp-Met-Asp-Phe
BOC-Met-Gly-Trp-Met-Asp-Phe Amide

Synonyms: Cholecystokinin Octapeptide (3-8), BOC-

Bachem A-2650.0010   MW 885.08 $C_{41}H_{56}N_8O_{10}S_2$ Store at -15°C

Met-Gly-Trp-Met-Asp-Phe Amide

Synonyms: Cholecystokinin Octapeptide (3-8)

Bachem H-2425.0010   MW 784.96 $C_{36}H_{48}N_8O_8S_2$ Store at -15°C

Met-His

Bachem G-2750.0250   MW 286.36 $C_{11}H_{18}N_4O_3S$ Store at -15°C

Met-His-Arg-Gln-Glu-Ala-Val-Asp-Cys-Leu-Lys-Lys-Phe-Asn-Ala-Arg-Arg-Lys-Leu-Lys-Gly-Ala

Synonyms: Calmodulin-Dependent Protein Kinase II (281-302), (Ala[286])-

Bachem H-3246.0500   Store at -15°C

Met-His-Arg-Gln-Glu-Ala-Val-Asp-Cys-Leu-Lys-Lys-Phe-Asn-Ala-Arg-Arg-Lys-Leu-Lys-Gly Amide Trifluoroacetate Salt

Synonyms: Ca²⁺/Calmodulin Kinase II Inhibitor (281-301), (Ala[286])-

Calbiochem 208710   MW 2528.1 $C_{108}H_{187}N_{39}O_{27}S_2$ ≥97% (HPLC); lyophilized; soluble in H_2O | Corresponds to residues 281-301 of the CaM kinase II α-subunit; potent inhibitor of CaM kinase II catalytic fragment; the inhibition is competitive with respect to ATP; Smith, MK et al, *J Biol Chem*, 267:1761, 1992

Met-His-Arg-Gln-Glu-Thr-Val-Asp-Cys-Leu-Lys Amide Trifluoroacetate Salt

Synonyms: Ca²⁺/Calmodulin Kinase II Substrate (281-291)

Calbiochem 208708   Synthetic   MW 1358.6 $C_{55}H_{95}N_{19}O_{17}S_2$ ≥95% (HPLC); lyophilized; soluble in H_2O | Synthetic peptide from the autophosphorylation region of CaM kinase II that serves as an excellent substrate for native CaM kinase II or for the active fragment; corresponds to residues 281-291 of the α-subunit of CaM kinase II; Yamagata, Y et al, *J Biol Chem*, 266:15391, 1991

Met-His-Arg-Gln-Glu-Thr-Val-Asp-Cys-Leu-Lys-Lys-Phe-Asn-Ala-Arg-Arg-Lys-Leu-Lys-Gly-Ala-Ile-Leu-Thr-Thr-Met-Leu-Ala

Synonyms: Calmodulin-Dependent Protein Kinase II (281-309);

Bachem H-3254.0500   Store at -15°C

Met-His-Arg-Gln-Glu-Thr-Val-Asp-Cys-Leu-Lys-Lys-Phe-Asn-Ala-Arg-Arg-Lys-Leu-Lys-Gly-Ala-Ile-Leu-Thr-Thr-Met-Leu-Ala Trifluoroacetate Salt

Synonyms: Ca²⁺/Calmodulin Kinase II Inhibitor (281-309)

Calbiochem 208711   Synthetic   MW 3374.1 $C_{146}H_{254}N_{46}O_{39}S_3$ ≥97% (HPLC); lyophilized; soluble in H_2O | Synthetic peptide containing the calmodulin binding site (290-309) & the autophosphorylation site (Thr[286]) of CaM kinase II; can be the phosphorylated at Thr[286] by PKC; useful as a calmodulin binding peptide; inhibits CaM kinase II by blocking Ca²⁺/calmodulin activation; Waxham, MN et al, *Biochemistry*, 32:2923, 1993; Waxham, MN et al, *Brain Res*, 609:1, 1993; Colbran, RJ et al, *J Biol Chem*, 264:4800, 1989; Colbran, RJ et al, *J Biol Chem*, 263:18145, 1988

Met-Ile

Bachem G-2755.0250   MW 262.37 $C_{11}H_{22}N_2O_3S$ Store at -15°C

Met-Leu

Bachem G-2760.0250   Store at -15°C

Met-Leu
Met-Leu-AMC Trifluoroacetate Salt

Bachem I-1885.0050   MW 533.57 $C_{21}H_{29}N_3O_4S \cdot C_2HF_3O_2$ Store at -15°C

Met-Leu
For-Met-Leu-AMC

Bachem I-1895.0050   MW 447.56 $C_{22}H_{29}N_3O_5S$ Store at -15°C

Met-Leu
For-Met-Leu-pNA

Bachem L-2030.0050   MW 410.49 $C_{18}H_{26}N_4O_5S$ Store at -15°C

Met-Leu
Met-Leu-pNA

Bachem L-2055.0050   MW 382.48 $C_{17}H_{26}N_4O_4S$ Store at -15°C

Met-Leu

ICN 155705   MW 262.4 $C_{11}H_{22}N_2O_3S$ Crystalline

Met-Leu
For-Met-Leu-pNA

Peptides International SFM-3624-PI   MW 410.50 $C_{18}H_{26}N_4O_5S$ >96% (HPLC); white crystalline powder | Wei, Y & D Pei, *Anal Biochem*, 250:29, 1997

Met-Leu
Sigma M 9630   FW 262.4 $C_{11}H_{22}N_2O_3S$

Met-Leu-Glu
For-Met-Leu-Glu
Bachem H-3020.0050   MW 419.5 $C_{17}H_{29}N_3O_7S$ Store at -15°C

Met-Leu-Gly
Bachem H-4350.0250   MW 319.43 $C_{13}H_{25}N_3O_4S$ Store at -15°C

Met-Leu-Gly-Ile-Ile-Ala-Gly-Lys-Asn-Ser-Gly
β-Met-Leu-Gly-Ile-Ile-Ala-Gly-Lys-Asn-Ser-Gly
Synonyms: Amyloid (35-25)

American Peptide 62-0-33   MW 1060.3 $C_{45}H_{81}N_{13}O_{14}S$
Inactive control; Yankner, BA et al, *Science*, 250:279, 1990

Met-Leu-Gly-Ile-Ile-Ala-Gly-Lys-Asn-Ser-Gly
Synonyms: Amyloid β-Protein (35-25)

Bachem H-2964.0001   MW 1060.28 $C_{45}H_{81}N_{13}O_{14}S$ Store at -15°C

Sigma A 2201   FW 1060.3   ≥97% (HPLC) | Bioactive peptide; inactive control; the N-C inverted sequence of (25-35); Yankner, BA et al, *Science*, 250:279, 1990

Met-Leu-Phe
For-Met-Leu-Phe
Synonyms: Chemotactic Peptide

American Peptide 15-1-10   MW 437.6 $C_{21}H_{31}N_3O_5S$

Met-Leu-Phe
BOC-Met-Leu-Phe
Bachem A-2690.0050   MW 509.67 $C_{25}H_{39}N_3O_6S$ Store at -15°C

Met-Leu-Phe
For-Met-Leu-p-Iodo-Phe
Bachem H-3025.0050   MW 563.46 $C_{21}H_{30}IN_3O_5S$ Store at -15°C

Met-Leu-Phe
For-Met-Leu-Phe
Bachem H-3030.0050   MW 437.56 $C_{21}H_{31}N_3O_5S$ Store at -15°C

Met-Leu-Phe
For-Met-Leu-Phe-OMe
Bachem H-3035.0050   MW 451.59 $C_{22}H_{33}N_3O_5S$ Store at -15°C

Met-Leu-Phe
Ethoxycarbonyl-Met-Leu-Phe-OMe
Bachem H-3542.0025   Store at -15°C

Met-Leu-Phe
Benzylureido-Met-Leu-Phe
Bachem H-3708.0050   Store at -15°C

Met-Leu-Phe
4-MethoxyPhenylureido-Met-Leu-Phe
Bachem H-3712.0050   Store at -15°C

Met-Leu-Phe
N-For-L-Met-L-Leu-L-Phe
Synonyms: Chemotactic Peptide

Fluka 47729   MW 437.56 $C_{21}H_{31}N_3O_5S$ ~99% (TLC); ≥95% peptide content; ≤1% water | Potent chemotactic peptide; induces a metabolic burst in macrophages accompanied by increase in respiratory rate, the secretion of lysosomal enzymes & the production of superoxide anion; Showell, HJ et al, *J Exp Med*, 143:1154, 1976; Homma, Y et al, *Biochem J*, 229:643, 1985

Fluka 47732   MW 437.56 $C_{21}H_{31}N_3O_5S$ ≥97% (HPLC); ≥95% peptide content; ≤1% water

Met-Leu-Phe
N-For-Met-Leu-Phe
ICN 151170   MW 437.6 $C_{21}H_{31}N_3O_5S$ Aswanikumar, S etal, *J Exp Med*, 143:1154, 1976; Aswanikumar, S etal, *BBRC*, 74:810, 1977; O'Dea, RF, *Nature*, 272:462 1978; Niedel, J etal, *JBC*, 254:10700, 1979; Naccache, P etal, *Science*, 203:461, 1979

Met-Leu-Phe
N-For-Met-Leu-p-Iodo-Phe
ICN 152763   MW 563.4 $C_{21}H_{30}IN_3O_5S$

Met-Leu-Phe
N-For-Met-Leu-Phe-OBz
ICN 152764   MW 437.6 $C_{21}H_{31}N_3O_5S$ Freer, RJ etal, *Biochem*, 21:257, 1982

Met-Leu-Phe
N-For-Met-Leu-Phe-OMe
Synonyms: Chemotactic Peptide

ICN 152765   Potent chemotactic peptide for human blood monocytes; Ho, PP etal, *Arthritis Rheum*, 21:133, 1978

Met-Leu-Phe
N-t-BOC-Met-Leu-Phe
Synonyms: Chemotactic Peptide Antagonist

ICN 152799   MW 509.7 $C_{25}H_{39}N_3O_6S$ ≥95% | Freer, RJ etal, *Peptides, Structure & Biological Function*, <u>Proceedings of the Sixth American Peptide Symposium</u>, E Gross & M Meienhofer eds, 749, 1979

Met-Leu-Phe
N-Ac-Met-Leu-Phe
ICN 195634   MW 451.6 $C_{22}H_{33}N_3O_5S$ For chemoattractant structural activity & requirement research of synthetic peptides; Freer, RJ etal, *Biochem*, 19:2404, 1980

Met-Leu-Phe
N-For-Met-Leu-Phe-Benzylamide
Synonyms: Chemotactic Peptide

ICN 195636   MW 526.7 $C_{28}H_{38}N_4O_4S$ ≥97% | Potent chemotactic peptide

Met-Leu-Phe
N-For-Met-Leu-Phe-o-Fluorobenzylamide
ICN 195637   MW 528.7 $C_{28}H_{37}FN_4O_3S$

Met-Leu-Phe
N-For-Met-Leu-Phe-p-Fluorobenzylamide
ICN 195638   MW 528.7 $C_{28}H_{37}FN_4O_3S$

Met-Leu-Phe
N-Formyl-Met-Leu-Phe

Neosystem SC121   MW 437.55   Bioactive; cell attachment peptide; chemotactic peptide; Schowell, HJ et al, *J Exp Med*, 143:1154-1169, 1976

Met-Leu-Phe
N-Ac-Met-Leu-Phe

Synonyms: Chemotactic Peptide

Sigma A 4536   FW 451.6 $C_{22}H_{33}N_3O_5S$ ≥97% (HPLC) | Bioactive peptide; used in studies on the structural requirements & specificity of synthetic peptide chemoattractants; Freer, RJ et al, *Biochem*, 19:2404, 1980

Met-Leu-Phe
N-t-BOC-Met-Leu-Phe

Synonyms: Chemotactic Peptide Antagonist

Sigma B 0511   FW 509.7 $C_{25}H_{39}N_3O_6S$ ≥97% (HPLC) | Bioactive peptide; Freer, RJ et al, *Peptides, Structure & Biological Function, Proceedings of the Sixth American Peptide Symposium*, Gross, E & Meienhofer, M, eds, 749, 1979

Met-Leu-Phe
N-For-Met-Leu-Phe o-Fluorobenzylamide

Synonyms: Chemotactic Peptide

Sigma F 0762   FW 528.7 $C_{28}H_{37}FN_4O_3S$ Bioactive peptide

Met-Leu-Phe
N-For-Met-Leu-Phe p-Fluorobenzylamide

Synonyms: Chemotactic Peptide

Sigma F 1012   FW 544.7 $C_{28}H_{37}FN_4O_4S$ Bioactive peptide

Met-Leu-Phe
N-For-Met-Leu-Phe

Synonyms: Chemotactic Peptide

Sigma F 3506   FW 437.6 $C_{21}H_{31}N_3O_5S$ ≥97% (HPLC) | Bioactive peptide; potent inducer of chemotaxis in leukocytes; Aswanikumar, S et al, *Biochem Biophys Res Commun*, 74:810, 1977; Naccache, P et al, *Science*, 203:461, 1979; Obrist, R et al, *Cancer Immunol Immunother*, 32:406, 1991

Met-Leu-Phe
N-For-Met-Leu-Phe-OMe

Synonyms: Chemotactic Peptide

Sigma F 6632   FW 451.6 $C_{22}H_{33}N_3O_5S$ ≥97% (HPLC) | Bioactive peptide; potent chemotactic peptide for human blood monocytes; Ho, PP et al, *Arthritis Rheum*, 21:133, 1978

Met-Leu-Phe
N-For-Met-Leu-Phe-OBz

Synonyms: Chemotactic Peptide

Sigma F 6758   FW 527.7 $C_{28}H_{37}N_3O_5S$ ≥97% (HPLC) | Bioactive peptide; Freer, RJ et al, *Biochem*, 21:257, 1982

Met-Leu-Phe
N-For-Met-Leu-p-Iodo-Phe

Synonyms: Chemotactic Peptide

Sigma F 6882   FW 563.4 $C_{21}H_{30}IN_3O_5S$ ≥85% (HPLC) | Bioactive peptide

Met-Leu-Phe
N-For-Met-Leu-Phe Benzylamide

Synonyms: Chemotactic Peptide

Sigma F 9758   FW 526.7 $C_{28}H_{38}N_4O_4S$ ≥97% (HPLC) | Bioactive peptide; potent; Freer, RJ et al, *Biochem*, 21:257, 1982

Met-Leu-Phe
For-Met-Leu-Phe

Synonyms: Chemotactic Peptide

Peptides International PCT-4066   Synthetic   MW 437.56 $C_{21}H_{31}N_3O_5S$ >95% (HPLC); lyophilized amorphous powder; bulk | Bioactive; Williams, LT et al, *PNAS USA*, 74:1204, 1977

Peptides International PCT-4066-v   Synthetic   MW 437.56 $C_{21}H_{31}N_3O_5S$ >95% (HPLC); lyophilized amorphous powder | Bioactive; Williams, LT et al, *PNAS USA*, 74:1204, 1977

Met-Leu-Phe Acetate Salt

Synonyms: Chemotactic Peptide

ICN 195643   MW 469.5 $C_{20}H_{31}N_3O_4S \cdot C_2H_4O_2$ ≥97% | Weak chemotactic properties

Sigma M 6014   FW 409.5 (free base) $C_{20}H_{31}N_3O_4S$ ≥97% (HPLC) | Bioactive peptide; exhibits very weak chemotactic properties; Freer, RJ et al, *Biochem*, 19:2404, 1980

Met-Leu-Phe-Lys
For-Met-Leu-Phe-Lys

Bachem H-3040.0050   MW 565.73 $C_{27}H_{43}N_5O_6S$ Store at -15°C

Met-Leu-Phe-Lys
N-For-Met-Leu-Phe-Lys

ICN 152766   MW 565.7 $C_{27}H_{43}N_5O_6S$ Freer, RJ et al, *Biochem*, 19:2404, 1980

Met-Leu-Phe-Lys
N-For-Met-Leu-Phe-Lys Acetate Salt

Synonyms: Chemotactic Peptide

Sigma F 2385   FW 565.7 (free base) $C_{27}H_{43}N_5O_6S$ ≥90% (HPLC) | Bioactive peptide; Freer, RJ et al, *Biochem*, 19:2404, 1980

Met-Leu-Phe-Phe
For-Met-Leu-Phe-Phe

Bachem H-4294.0050   MW 584.74 $C_{30}H_{40}N_4O_6S$ Store at -15°C

Met-Leu-Phe-Phe
N-For-Met-Leu-Phe-Phe

Synonyms: Chemotactic Peptide

ICN 195639   MW 584.7 $C_{30}H_{40}N_4O_6S$ Potent chemotactic peptide

Sigma F 2009   FW 584.7 $C_{30}H_{40}N_4O_6S$ ≥90% (HPLC) | Bioactive peptide; potent; Freer, RJ et al, *Biochem*, 21:257, 1982

Met-Leu-pNA Hydrochloride

Peptides International SML-3622-PI   MW 418.95 $C_{17}H_{26}N_4O_4S \cdot HCl$ >96% (HPLC); amorphous powder | Y Wei & D Pei, *Anal Biochem*, 250:29, 1997

Met-Leu-Thr-Lys-Phe-Glu-Thr-Lys-Ser-Ala-Arg-Val-Lys-Gly-Leu-Ser-Phe-His-Pro-Lys-Arg-Pro-Trp-Ile-Leu

Synonyms: Xenopsin 25; Xenin 25; Xenopsin Related Peptide

Bachem H-8860.0500   MW 2971.61 $C_{139}H_{224}N_{38}O_{32}S$ Store at -15°C

ICN 156754   Human   MW 2971.6   97%

Sigma X 2127   Human   FW 2971.6   ≥97% (HPLC) | Bioactive peptide

American Peptide 46-4-73   Human mucosa   MW 2971.6 $C_{139}H_{224}N_{38}O_{32}S$ Released into circulation after a meal; may stimulate exocrine pancreatic secretion; Feurle, GE et al, *J Biol Chem*, 267:22305, 1992

Peptides International PXN-4279-v   Human synthetic   MW 2971.6 $C_{139}H_{224}N_{38}O_{32}S$ >95% (HPLC); lyophilized amorphous powder | Bioactive; Feurle, GE et al, *JBC*, 267:22305, 1992

Met-Leu-Tyr
For-Met-Leu-Tyr DCHA

Bachem H-3045.0050   MW 453.56 $C_{21}H_{31}N_3O_6S$ Store at -15°C

Met-Leu-Tyr
N-For-Met-Leu-Tyr

ICN 152767   MW 453.6 $C_{21}H_{31}N_3O_6S$

Met-Leu-Tyr
N-For-Met-Leu-Tyr Dicyclohexylammonium Salt

Synonyms: Chemotactic Peptide

Sigma F 5256   FW 453.6 (free acid) $C_{21}H_{31}N_3O_6S$ ≥95% (HPLC) | Bioactive peptide

Met-Lys
For-Met-Lys

Bachem G-1860.0050   MW 305.4 $C_{12}H_{23}N_3O_4S$ Store at -15°C

Met-Lys

Bachem G-2765.0250   MW 277.39 $C_{11}H_{23}N_3O_3S$ Store at -15°C

Met-Lys Formate Salt

ICN 155706   MW 323.4 $C_{11}H_{23}N_3O_3S \cdot CH_2O_2$

Sigma M 7632   FW 323.4 $C_{11}H_{23}N_3O_3S \cdot CH_2O_2$

Met-Lys-Arg-Pro-Pro-Gly-Phe-Ser-Pro-Phe-Arg

Synonyms: Bradykinin, (Met-Lys⁰)-; Bradykinin, (Met-Lys)-

American Peptide 18-1-23   MW 1319.6 $C_{61}H_{94}N_{18}O_{13}S$ Elliot, DF et al, *J Biochem*, 95:437, 1965

Bachem H-2190.0005   MW 1319.6 $C_{61}H_{94}N_{18}O_{13}S$ Store at -15°C

ICN 151620   MW 1319.6   Naturally occurringkinin that is converted to bradykinin in plasma; Elliot, DF & GP Lewis, *J Biochem*, 95:437, 1965; Araujo-Viel, MS etal, *Hoppe-Seyler's Z Physiol Chem*, 362:337, 1981

Peptides International PBK-4012-v   Synthetic   MW 1319.6 $C_{61}H_{94}N_{18}O_{13}S$ >95% (HPLC); lyophilized amorphous powder | Bioactive; human, bovine; Elliott, DF & GP Lewis, *Biochem J*, 95:437, 1965); Ohkubo, I et al, *Biochem*, 23:5691, 1984

Met-Lys-Arg-Pro-Pro-Gly-Phe-Ser-Pro-Phe-Arg
3AcOH Dihydrate

Synonyms: Bradykinin, (Met-Lys)-

Peptides International PBK-4012   Synthetic   MW 1535.7 $C_{61}H_{94}N_{18}O_{13}S \cdot 3CH_3COOH \cdot 2H_2O$ >95% (HPLC); lyophilized amorphous powder; bulk | Bioactive; human, bovine; Elliott, DF & GP Lewis, *Biochem J*, 95:437, 1965); Ohkubo, I et al, *Biochem*, 23:5691, 1984

Met-Lys-Arg-Pro-Pro-Gly-Phe-Ser-Pro-Phe-Arg
Acetate Salt

Synonyms: Bradykinin, (Met-Lys)-

Sigma B 5014   MW 1319.6 (free base)   ≥97% (HPLC) | Bioactive peptide; Elliot, DF & Lewis, GP, *J Biochem*, 95:437, 1965; Araujo-Viel, MS et al, *Hoppe-Seyler's Z Physiol Chem*, 362:337, 1981

Met-Lys-Arg-Ser-Arg-Gly-Pro-Ser-Pro-Arg-Arg

Synonyms: Bradykinin, (Met-Lys-(Ser², Arg³, Pro⁵, Arg⁸))-; Bradykinin, Met-Lys-(Ser², Arg³, Pro⁵, Arg⁸)-

Bachem H-1654.0001   MW 1327.58 $C_{53}H_{98}N_{24}O_{14}S$ Store at -15°C

Sigma M 0421   MW 1327.6   ≥97% (HPLC); peptide content:~65% | Bioactive peptide; neuropeptide found in *Aplysia californica* with high sequence homology to bradykinin &kallidin; Wickham, L & Desgroseillers, L, *DNA & Cell Biol*, 10:249, 1991

ICN 195633   *Aplysia californica* (mollusk)   MW 1327.6   A bradykinin-like neuropeptide with high sequence homology to mammalian Bradykinin, (kallidin) & Met-Lys-bradykinin

Met-Lys-Cys-Arg-Arg-Pro-Gly-Asn-Lys-Thr-Val-Leu-Pro-Val-Thr-Ile-Met-Ser-Gly-Leu

Synonyms: GP140 (311-330); SIVmac251 Peptide

Neosystem SC739   MW 2201.73   SIVmac251 peptide from subtype GP140; due to the presence of a Cys this peptide can contain the dimeric form

Met-Met
BOC-Met-Met

Bachem A-3235.0001   MW 380.53 $C_{15}H_{28}N_2O_5S_2$ Store at -15°C

Met-Met

Bachem G-2770.0250   MW 280.41 $C_{10}H_{20}N_2O_3S_2$ Store at -15°C

Met-Met
Met-D-Met

Bachem G-2775.0250   MW 280.41 $C_{10}H_{20}N_2O_3S_2$ Store at -15°C

Met-Met
D-Met-Met

Bachem G-2780.0250   MW 280.41 $C_{10}H_{20}N_2O_3S_2$ Store at -15°C

Met-Met
D-Met-D-Met

Bachem G-2785.0250   MW 280.41 $C_{10}H_{20}N_2O_3S_2$ Store at -15°C

Met-Met

ICN 155707   MW 280.4 $C_{10}H_{20}N_3O_3S_2$ Crystalline

Peptides International OMM-3152   MW 280.41 $C_{10}H_{20}N_2O_3S_2$ >96% (HPLC); colorless powder

Sigma M 3004   FW 280.4 $C_{10}H_{20}N_2O_3S$ Crystalline

Met-Met-Ala

Bachem H-4355.0250   MW 351.49 $C_{13}H_{25}N_3O_4S_2$ Store at -15°C

Met-Met-Arg-Asp-Ser-Gly-Cys-Phe-Gly-Arg-Arg-Ile-Asp-Arg-Ile-Gly-Ser-Leu-Ser-Gly-Met-Gly-Cys-Asn-Gly-Ser-Arg-Lys-Asn

Synonyms: Natriuretic Factor (1-29), Atrial

Bachem H-3062.0500   Chicken   Store at -15°C

Met-Met-Arg-Asp-Ser-Gly-Cys-Phe-Gly-Arg-Arg-Ile-Asp-Arg-Ile-Gly-Ser-Leu-Ser-Gly-Met-Gly-Cys-Asn-Gly-Ser-Arg-Lys-Asn
H₂N-Met-Met-Arg-Asp-Ser-Gly-Cys-Phe-Gly-Arg-Arg-Ile-Asp-Arg-Ile-Gly-Ser-Leu-Ser-Gly-Met-Gly-Cys-Asn-Gly-Ser-Arg-Lys-Asn

Synonyms: Natriuretic Peptide, Atrial

ICN 154507 Chicken MW 3160 Miyata, A etal, *BBRC*, 155:1220, 1988

Met-Met-Arg-Asp-Ser-Gly-Cys-Phe-Gly-Arg-Arg-Ile-Asp-Arg-Ile-Gly-Ser-Leu-Ser-Gly-Met-Gly-Cys-Asn-Gly-Ser-Arg-Lys-Asn

Synonyms: Natriuretic Peptide (1-29), Atrial

Sigma A 9052 Chicken FW 3160.6 ≥97% (HPLC) | Bioactive peptide; disulfide bonds: 7-23; originally isolated from chicken atrium; has diuretic-natriuretic & hypotensive activity similar to mammalian ANF; Myata, A et al, *Biochem Biophys Res Commun*, 155:1330, 1988

Met-Met-Met
For-Met-Met-Met

Bachem H-3050.0050 MW 439.62 $C_{16}H_{29}N_3O_5S_3$ Store at -15°C

Met-Met-Met

Bachem H-4360.0250 MW 411.61 $C_{15}H_{29}N_3O_4S_3$ Store at -15°C

Met-Met-Met
D-Met-Met-Met

Bachem H-4365.0250 MW 411.61 $C_{15}H_{29}N_3O_4S_3$ Store at -15°C

Met-Met-Met
N-For-Met-Met-Met

Synonyms: Chemotactic Peptide

ICN 152768 MW 439.6 $C_{16}H_{29}N_3O_5S$

Sigma F 2635 FW 439.6 $C_{16}H_{29}N_3O_5S_3$ ≥97% (TLC) | Bioactive peptide

---

Met-Met-Phe
For-Met-Met-Phe

Bachem H-7195.0050 MW 455.6 $C_{20}H_{29}N_3O_5S_2$ Store at -15°C

Met-Met-Phe
N-For-L-Met-L-Met-L-Phe

Fluka 47730 MW 455.59 $C_{20}H_{29}N_3O_5S_2$ ≥99% (TLC)

Met-Met-Thr-Ala-Cys-Gln-Gly-Val-Gly-Gly-Pro-Gly-His-Lys-Ala

Synonyms: P25 (346-360); GAG P24 CA (214-228); HTLV I Peptide

Neosystem SC310 MW 1444.70 HTLV I peptide from subtype P25 (GAG P24 CA); due to the presence of a Cys this peptide can contain the dimeric form

---

Met-Phe
For-Met-Phe

Bachem G-1865.0250 MW 324.4 $C_{15}H_{20}N_2O_4S$ Store at -15°C

Met-Phe

Bachem G-2790.0250 MW 296.39 $C_{14}H_{20}N_2O_3S$ Store at -15°C

Met-Phe
Met-Phe-OMe Hydrochloride

Bachem G-2795.0250 Store at -15°C

Met-Phe
N-For-Met-Phe

Synonyms: Neutrophil Dysfunction Test Peptide

ICN 152769 MW 324.4 $C_{15}H_{20}N_2O_4S$ Williams, LT etal, *PNAS*, 74:1204, 1977; Jadwin, DF etal, *Am J Clin Pathol*, 76:395, 1981

Met-Phe

ICN 155708 MW 296.4 $C_{14}H_{20}N_2O_3S$ ~98%; crystalline

Met-Phe
N-For-Met-Phe

Synonyms: Neutrophil Dysfunction Test Peptide

Sigma F 8506 FW 324.4 $C_{15}H_{20}N_2O_4S$ ≥97% (HPLC) | Bioactive, chemotactic peptide; Jadwin, DF et al, *Am J Clin Pathol*, 76:395, 1981; Williams, LT et al, *Proc Natl Acad Sci USA*, 74:1204, 1977

Met-Phe

Sigma M 9505 FW 296.4 $C_{14}H_{20}N_2O_3S$ ~98%

Met-Phe-Gly

Bachem H-4370.0250 MW 353.44 $C_{16}H_{23}N_3O_4S$ Store at -15°C

Met-Phe-Met
N-For-Met-Phe-Met

Synonyms: Chemotactic Peptide

ICN 152770 MW 455.6 $C_{20}H_{29}N_3O_5S_2$ Showell, HJ etal, *J Exp Med*, 14:1154, 1976

Sigma F 2510 FW 455.6 $C_{20}H_{29}N_3O_5S_2$ ≥97% (TLC) | Bioactive peptide

---

Met-Pro
BOC-Met-Pro

Bachem A-1985.0001 MW 346.45 $C_{15}H_{26}N_2O_5S$ Store at -15°C

Met-Pro

ICN 155709 $C_{10}H_{18}N_2O_3S$ Crystalline

Met-Pro Hydrochloride

Bachem G-2800.0250 MW 282.79 $C_{10}H_{18}N_2O_3S \cdot HCl$ Store at -15°C

Sigma M 0912 FW 282.8 $C_{10}H_{18}N_2O_3S \cdot HCl$

Met-Pro-Arg-Trp-Arg-Leu-Phe-Arg-Arg-Ile-Asp-Arg-Val-Gly-Lys-Gln-Ile-Lys-Gln-Gly-Ile-Leu-Arg-Ala-Gly-Pro-Ala-Ile-Ala-Leu-Val-Gly-Asp-Ala-Arg-Ala-Val-Gly

Synonyms: Lytic Peptide Shiva-1

Bachem H-2752.0500 Store at -15°C

Met-Pro-Arg-Val-Arg-Ser-Leu-Phe-Gln-Glu-Gln-Glu-Glu-Pro-Glu-Pro-Gly-Met-Glu-Glu-Ala-Gly-Glu-Met-Glu-Gln-Lys-Gln-Leu-Gln

Synonyms: Nocistatin; Allodynia/Hyperalgesia-Blocking Peptide; Nociceptin Action Blocking Peptide

Neosystem SC1342 Human MW 3561.95 Bioactive neuropeptide; 30-mer peptide that may be the human counterpart of bovine nocistatin; Minami, T et al, *Br J Pharmacol*, 124:1016-1018, 1998; Mollereau, C et al, *PNAS USA*, 93:8666-8670, 1998

Peptides International PNO-4355-v   Human synthetic   MW 3562.0 $C_{149}H_{238}N_{42}O_{53}S_3$ >95% (HPLC); lyophilized amorphous powder | Bioactive; Minami, T et al, *Br J Pharmacol*, 124:1016, 1998; Mollereau, C et al, *PNAS* USA, 93:8666, 1996

Met-Pro-Gly

Bachem H-5755.0250   Store at -15°C

Met-Pro-Lys-Lys-Lys-Pro-Thr-Pro-Ile-Gln-Leu-Asn-Cys

Synonyms: MEK1 N-Terminal Blocking Peptide (1-12)

Calbiochem 444943   Rat synthetic   MW 1497.9   Lyophilized solid | Based on rat MEK1 (1-12) with a Cys C residue added; this peptide coupled to KLH was used as the immunogen for the production of Anti-MEK1; for use in immunoabsorption for immunoprecipitation, immunocytochemistry, Western blotting & dot blots

Met-Pro-Lys-Trp-Lys-Val-Phe-Lys-Lys-Ile-Glu-Lys-Val-Gly-Arg-Asn-Ile-Arg-Asn-Gly-Ile-Val-Lys-Ala-Gly-Pro-Ala-Ile-Ala-Val-Leu-Gly-Glu-Ala-Lys-Ala-Leu-Gly

Synonyms: Lytic Peptide SB-37

Bachem H-2748.0500   Store at -15°C

Met-Pro-Phe-Arg-Trp-Phe-Lys-Pro-Val
Met-Pro-D-Phe-Arg-D-Trp-Phe-Lys-Pro-Val Amide

Synonyms: Melanocyte Stimulating Hormone (5-13), (Met[5],Pro[6],D-Phe[7],D-Trp[9],Phe[10])-α-

Bachem H-2716.0001   Store at -15°C

Met-Pro-Val-Asp-Asn-Arg-Asn-His-Asn-Glu-Gly-Met-Val-Thr Acetate Salt

Synonyms: Chromogranin B (1-14); Secretogranin I

Biogenesis 2095-0320   Human synthetic   Purified; lyophilized | The C-terminal sequence is unique to chromogranin B; it is also chromogranin B precursor (21-34)

Met-Ser

Bachem G-2805.0250   MW 236.29 $C_8H_{16}N_2O_4S$ Store at -15°C

ICN 155711   MW 236.3 $C_8H_{16}N_2O_4S$

Sigma M 9380   FW 236.3 $C_8H_{16}N_2O_4S$

Met-Ser-Arg-Pro-Ala-Cys-Pro-Asn-Asp-Lys-Tyr-Glu

Synonyms: T1; Thrombin Receptor Antagonist; Peptide T1

Bachem H-2514.0500   MW 1410.6 $C_{58}H_{91}N_{17}O_{20}S_2$ Store at -15°C

Neosystem SC969   MW 1410.58   Bioactive; due to the presence of a Cys this peptide can contain the dimeric form; antagonizes platelet aggregation triggered by the agonist peptide; inhibits serotonin release & Tyr phosphorylation triggered by either thrombin or the agonist peptide; anti-aggregatory activity was about ten-fold higher than that of previously reported peptide antagonists of the thrombin receptor; Doorbar, J & Winter, G, *J Mol Biol* 244:361-369, 1994

Met-Ser-Asn-Asn-Gly-Leu-Asp-Val-Gln-Asp-Lys-Pro-Cys

Synonyms: PAK1 N-Terminal Blocking Peptide (1-12)

Calbiochem 506276   Rat synthetic   MW 1420.6 $C_{56}H_{93}N_{17}O_{22}S_2$ Lyophilized | Based on rat PAK1 (1-12) with a Cys C residue added; this peptide coupled to KLH, was used as the immunogen for the production of Anti-PAK1; for use in immunoabsorption for immunoprecipitation, immunocytochemistry, Western blotting & dot blots

Met-Ser-Asp-Ser-Leu-Asp-Asn-Glu-Glu-Lys-Pro-Pro-Ala-Cys

Synonyms: PAK3 N-Terminal Blocking Peptide (1-13)

Calbiochem 506278   Mouse synthetic   MW 1535.7 $C_{61}H_{98}N_{16}O_{26}S_2$ Lyophilized | Based on mouse PAK3 (1-13) with a Cys C residue added; this peptide coupled to KLH, was used as the immunogen for the production of Anti-PAK3; for use in immunoabsorption for immunoprecipitation, immunocytochemistry, Western blotting & dot blots

Met-Ser-Asp-Ser-Lys-Ser-Asp-Gly-Gln-Phe-Tyr-Ser-Val-Gln-Val-Ala-(Cys)

Synonyms: SAPKα N-Terminal Blocking Peptide (1-16)

Calbiochem 559308   Rat synthetic   MW 1852.0 $C_{77}H_{118}N_{20}O_{29}S_2$ Lyophilized | Based on rat SAPKα (1-16) with a cys C residue added & the peptide coupled to KLH; this peptide coupled to KLH, was used as the immunogen for the production of Anti-SAPKα; for use in immunoabsorption for immunoprecipitation, immunocytochemistry, Western blotting & dot blots

Met-Ser-Gly

Bachem H-5760.0250   MW 293.34 $C_{10}H_{19}N_3O_5S$ Store at -15°C

Met-Ser-Gly-Thr-Lys-Leu-Glu-Asp-Ser-Pro-Pro-Cys-Arg-Asn-Cys Trifluoroacetate Salt

Synonyms: Orexin Receptor II (1-14), ~C

Biogenesis 7049-5500   Synthetic   MW 1638.6   Purified; lyophilized | An additional Cys has been conjugated to the C-terminus

Met-Thr

Bachem G-2810.0250   MW 250.32 $C_9H_{18}N_2O_4S$ Store at -15°C

Met-Thr-Met-Thr-Leu-His-Thr-Lys-Ala-Ser-Gly-Met-Ala-Leu-Leu-His-Gln-Ile-Glu-Gly-Asn-Cys

Synonyms: Estrogen Receptor α N-Terminal Peptide (1-21)

Alexis 155-031   Synthetic   MW 2386.9 $C_{99}H_{168}N_{30}O_{30}S_4$ ≥90%; synthetic peptide dissolved in 10 *mM* sodium acetate, pH 4.5 | Competitively binds to Ab Alexis 210-201; blocking peptide for antiserum (purified) to ERα

Met-Thr-Tyr-Lys-Ala-Ala-Val-Asp-Leu-Ser-His-Phe-Leu-Lys-Glu-Lys

Synonyms: NEF (79-94); HTLV I Peptide

Neosystem SC221   MW 1881.21   HTLV I peptide from subtype NEF

Met-Trp
For-Met-Trp

Bachem G-1870.0250   MW 363.44 $C_{17}H_{21}N_3O_4S$ Store at -15°C

Met-Trp

Bachem G-2815.0250   MW 335.43 $C_{16}H_{21}N_3O_3S$ Store at -15°C

Met-Trp
N-For-Met-Trp

Synonyms: Chemotactic Peptide

ICN 152771   MW 363.4 $C_{17}H_{21}N_3O_4S$

Sigma F 6128   FW 363.4 $C_{17}H_{21}N_3O_4S$ ≥90% (HPLC); off-white to tan crystals | Bioactive peptide

Met-Trp-Asp-Phe-Asp-Asp-Leu-Asn-Phe-Thr-Gly-Met-Pro-Pro-Ala-Asp-Glu-Asp-Tyr-Ser-Pro Ac-Met-Trp-Asp-Phe-Asp-Asp-Leu-Asn-Phe-Thr-Gly-Met-Pro-Pro-Ala-Asp-Glu-Asp-Tyr-Ser-Pro Amide

Synonyms: Interleukin VIII Receptor Antagonist (9-29)

Bachem N-1585.0500   Store at -15°C

Met-Trp-Gln-Glu-Val-Gly-Lys-Ala-Met-Tyr-Ala-Pro

Synonyms: GP120 (431-442); HTLV I Peptide

Neosystem SC564   MW 1410.66   HTLV I peptide from subtype GP160

Met-Trp-Tyr-Arg-Pro-Asp-Leu-Asp-Glu-Arg-Lys-Gln-Gln-Lys-Arg-Glu

Bachem H-2176.0500   MW 2178.48 $C_{94}H_{148}N_{30}O_{28}S$ Store at -15°C

Met-Tyr
Z-Met-Tyr

Bachem C-2235.0001   MW 446.52 $C_{22}H_{26}N_2O_6S$ Store at -15°C

Met-Tyr

Bachem G-2820.0250   MW 312.39 $C_{14}H_{20}N_2O_4S$ Store at -15°C

Met-Tyr-Lys

Bachem H-4375.0050   MW 440.56 $C_{20}H_{32}N_4O_5S$ Store at -15°C

Met-Tyr-Phe Amide

Bachem H-4380.0050   MW 458.58 $C_{23}H_{30}N_4O_4S$ Store at -15°C

ICN 155712   MW 458.6 $C_{23}H_{30}N_4O_4S$

Met-Tyr-Pro-Arg-Gly-Asn-His-Trp-Ala-Val-Gly-His-Leu-Met Amide

Synonyms: Gastrin Releasing Peptide (14-27)

ICN 154552   MW 1668.2   Yanaihara, N etal, *Regulatory Peptides*, Suppl. 1:123, 1980

Neosystem SC167   MW 1667.97   Bioactive; brain/gut peptide; CCK/gastrin peptide; Yanaihara, W et al, *Regulatory Peptides*, Abstract of the 3rd Int Symp of Gut Hormones, Cambridge, UK, 1980

Bachem H-3115.0001   Human, porcine, canine   MW 1667.98 $C_{75}H_{110}N_{24}O_{16}S_2$ Store at -15°C

American Peptide 46-4-32   Porcine   MW 1668.0 $C_{75}H_{110}N_{24}O_{16}S_2$ Yanaihara, N et al, *Reg Peptides, Suppl*, 1:123, 1980

Met-Val

Bachem G-2825.0250   MW 248.35 $C_{10}H_{20}N_2O_3S$ Store at -15°C

Met-Val
For-Met-Val

Bachem G-3825.0250   MW 276.36 $C_{11}H_{20}N_2O_4S$ Store at -15°C

Met-Val
N-For-Met-Val

ICN 152772   MW 276.4 $C_{11}H_{20}N_2O_4S$

Met-Val

ICN 155713   MW 248.3 $C_{10}H_{20}N_2O_3S$ Crystalline

Met-Val
N-For-Met-Val

Synonyms: Chemotactic Peptide

Sigma F 5003   FW 276.4 $C_{11}H_{20}N_2O_4S$ ≥97% (HPLC); crystalline | Bioactive peptide

Met-Val

Sigma M 1506   FW 248.3 $C_{10}H_{20}N_2O_3S$

Met-Val-Asp-Thr-Glu-Ser-Pro-Ile-Cys-Pro-LLeu-Ser-Pro-Leu-Glu-Ala-Asp-Asp-Cys

Synonyms: PPARα Peptide

Alexis 165-029   Synthetic   MW 2035.3 $C_{84}H_{135}N_{19}O_{33}S_3$ ≥95%; lyophilized; reconstitute with 0.1 mL distilled H_2O | Competitively binds to Ab Alexis 210-190; antiserum blocking peptide

Met-Val-His-Gln-Ala-Ile-Ser-Pro-Arg-Thr-Leu-Asn-Ala-Trp-Val

Synonyms: P25 (142-156); GAG P24 CA (10-24); HTLV I Peptide

Neosystem SC291   MW 1723.02   HTLV I peptide from subtype P25 (GAG P24 CA)

Met-Val-Thr-Ser-Leu-Asn-Glu-Asp-Asn-Glu-Ser-Cys

Synonyms: Kinesin Protein KIF2 N-Terminal Peptide

Alexis 160-009   Synthetic   MW 1341.5 $C_{51}H_{84}N_{14}O_{24}S_2$ ≥90% (HPLC); dissolved in 10 *mM* sodium acetate, pH 4.5 | Competitively binds to Ab Alexis 210-203; antiserum blocking peptide

Mixture

Synonyms: Caspase Substrate

Calbiochem 218780   Contains 5 mg each of Caspase I Substrate IV (Ac-Tyr-Val-Ala-Asp-*p*NA); Caspase III Substrate I (Ac-Asp-Glu-Val-Asp-*p*NA); Caspase I Substrate VII (Ac-Trp-Glu-His-Asp-*p*NA); Caspase VI Substrate II (Ac-Val-Glu-Ile-Asp-*p*NA); Granzyme B Substrate I (Ac-Ile-Glu-Thr-Asp-*p*NA) & 100 mg of *p*-Nitroaniline (as reference standard); LD_{50}≤200 mg/kg but >50 mg/kg | Colorimetric Set I

Calbiochem 218782   Contains 1 mg each of Caspase III Substrate II (Ac-Asp-Glu-Val-Asp-AMC); Caspase IV Substrate II (Ac-Leu-Glu-Val-Asp-AFC); Caspase IX Substrate I (Ac-Leu-Glu-His-Asp-AFC); Caspase I Substrate III (Ac-Tyr-Val-Ala-Asp-AMC); Caspase II Substrate I (Z-Val-Asp-Val-Ala-Asp-AFC); Caspase V Substrate II (Ac-Trp-Glu-His-Asp-AFC); Caspase VI Substrate I (Ac-Val-Glu-Ile-Asp-AMC); Granzyme B Substrate II (Z-Ile-Glu-Thr-Asp-AFC) & 10 mg of AMC reference standard & 50 mg AFC reference standard | Fluorogenic Set II

Mixtures
L-Glu-(L-Glu)n -L-Glu

Synonyms: Poly Glu

Fluka 81326   MW 2-15k

Mixtures
L-Glu-(L-Glu)n -L-Glu·Nan+3 Sodium Salt

Synonyms: Poly Glu

Fluka 81327   MW 15-50k

Mixtures
L-Glu-(L-Glu)n -L-Glu Nan+3 Sodium Salt

Synonyms: Poly Glu

Fluka 81328   MW 50-100k

Mixtures
Poly-(L-Glutamic Acid-L-Tyrosine, 4:1) Sodium Salt
Synonyms: Poly Glu-Tyr

Fluka 81357   MW 20-50k

Mixtures
Synonyms: Cysteine, Poly-S-Benzyl-L-

ICN 102680   MW 2-10k

Mixtures
Poly-γ-Benzyl-L-Glu

ICN 102683   MW 15k-30k

Mixtures
Poly(L-His,L-Glu)-Poly-DL-Ala-Poly-L-Lys
Synonyms: Multichain Polyamino Acid

Sigma M 3774   MW 50-250k

Mixtures
(S)-1-Carboxy-2-Phenylethyl-Carbamoyl-α-(2-Iminohexahydro-4(S)-Pyrimidyl)-(S)-Glycyl-X-Phenylalaninal
Synonyms: Chymostatin

Alexis 260-005   Microbial   MW 474.6 $C_{21}H_{38}N_4O_8$ Mixture of type A (X=L-Leu), type B (X=L-Val), type C (X=L-Ile); inhibitor of chymotrypsin, papain & cathepsin A, B & D; Umezawa, H et al, *J Antibiot*, 23:425, 1970

Mpr-Tyr-Phe-Gln-Asn-Cys-Pro-Arg
Synonyms: Vasopressin Desglycinamide, (Arg[8])-Deamino-

American Peptide 66-0-07   MW 1013.2 $C_{44}H_{60}N_{12}O_{12}S_2$
Disulfide bonds: Mpr[1]-Cys[6]

Mpr-Tyr-Phe-Gln-Asn-Cys-Pro-Lys
Synonyms: Vasopressin Desglycinamide, (Lys[8])-Deamino-

American Peptide 66-0-08   MW 985.2 $C_{44}H_{60}N_{10}O_{12}S_2$
Disulfide bonds: Mpr[1]-Cys[6]

Nal-Abu-Phe-Abu-Abu-Nal
Ac-1-Nal-Abu-Phe-(®)-Abu-Abu-1-Nal Amide

Bachem N-1705.0001   Store at -15°C

Nal-Ala
N-(R)-(2-(Hydroxyaminocarbonyl)Methyl)-4-Methylpentanoyl-L-Nal-Ala Amide
Synonyms: TAPI-0; Collagenase Inhibitor; Matrix Metalloproteinase Inhibitor; Tumor Necrosis Factor Inhibitor, α-

Peptides International INH-3850-PI   Synthetic   MW 456.55
$C_{24}H_{32}N_4O_5$ >98% (HPLC); white powder | Inhibitor; Darlak et al, *JBC*, 265:5199, 1990; Spatola, AF et al, *Peptides:Chemistry & Biology*, JA Smith & JE Rivier (Eds), ESCOM, Leiden, 1992, p 820; Mohler KM et al, *Nature*, 370:218, 1994

Nal-Ala
N-(R)-(2-(Hydroxyaminocarbonyl)Methyl)-4-Methylpentanoyl-L-Nal-Ala-2-Aminoethyl
Synonyms: TAPI-1; Collagenase Inhibitor; Matrix Metalloproteinase Inhibitor; Tumor Necrosis Factor Inhibitor, α-

Peptides International INH-3855-PI   Synthetic   MW 499.62
$C_{26}H_{37}N_5O_5$ >98% (HPLC); off-white powder; amide | Inhibitor; Darlak et al, *JBC*, 265:5199, 1990; Spatola, AF et al, *Peptides:Chemistry & Biology*, JA Smith & JE Rivier (Eds), ESCOM, Leiden, 1992, p 820; Mohler KM et al, *Nature*, 370:218, 1994

Nal-Ala
(2R)-S-Ac-2-Mercaptomethyl-4-Methylpentanoyl-L-β-Nal-Ala Amide
Synonyms: SIMP II, Ac-; Collagenase Inhibitor

Peptides International ISN-3831-PI   Synthetic   MW 471.62
$C_{25}H_{33}N_3O_4S$ >98% (HPLC); white powder | Inhibitor; protected precursor

---

Nal-Ala
(2R)-2-Mercaptomethyl-4-Methylpentanoyl-L-β-Nal-Ala Amide
Synonyms: SIMP II; Collagenase Inhibitor

Peptides International ISN-3835-PI   Synthetic   MW 429.59
$C_{23}H_{31}N_3O_3S$ >98% (HPLC); white powder; free thiol

Nal-Cyclo(Glu-Tyr-Trp-Lys-Val)-Thr
D-2-Nal-Cyclo(γ-Glu-Tyr-D-Trp-Lys-Val-L-α,γ-Diaminobutyryl)-Thr Amide
Synonyms: Somatostatin Analog

Calbiochem 567685   MW 1103.3 $C_{57}H_{74}N_{12}O_{11}$ ≥95% (HPLC); solid; soluble in water | Exerts a strong inhibitory effect on myointimal proliferation in response to vascular injury without changing the serum level of growth hormone; this is the first compound described to act on smooth muscle cell proliferation without interfering with growth hormone secretion; Thurieau, C et al, *Eur J Med Chem*, 30:115, 1995

Nal-Cys-Tyr-D-Trp-Lys-Val-Cys-Thr
D-Nal-Cys-Tyr-D-Trp-Lys-Val-Cys-Thr Amide
Synonyms: Somatostatin Tumor Inhibiting Analog

American Peptide 68-1-45   MW 1097.4 $C_{54}H_{70}N_{11}O_{10}S_2$
Disulfide bonds: Cys[2]-Cys[7]; Taylor, JE, *BBRC*, 153:81, 1988

Nal-Phe-Ala-Gly-Arg-Pro-Ala
Ac-D-2-Nal-4-Chloro-D-Phe-β-(3-Pyridyl)-D-Ala-Gly-Arg-Pro-D-Ala Amide
Synonyms: Luteinizing Hormone Releasing Hormone Antagonist

Bachem H-5076.0001   Store at -15°C

Nal-Phe-Ala-Ser-Lys-Lys-Leu-Lys-Pro-Ala
Ac-D-2-Nal-4-Chloro-D-Phe-β-(3-Pyridyl)-D-Ala-Ser-Lys(Nicotinoyl)-D-Lys(Nicotinoyl)-Leu-Lys(Isopropyl)-Pro-D-Ala Amide
Synonyms: Antide

Bachem H-9215.0001   MW 1587.93 $C_{82}H_{108}N_{17}O_{14}S$ Store at -15°C

Nle-Arg
D-Nle-CHA-Arg-pNA·2AcOH
Synonyms: Pefachrome®PL-Strept; Pefa-5321

Pentapharm 083-21, 083-05   MW 680.8   Highly sensitive chromogenic peptide substrate for the plasminogen-streptokinase complex; v_{max}:0.024 μmol/min; K_M:0.4 mM; Svendsen, LG et al, *Semin Thromb Haemost*, 9:250,1983

Nle-Arg-His-Nal-Arg-Trp-Gly-Cys
3-Mercaptopropionyl-Nle-Arg-His-D-2-Nal-Arg-Trp-Gly-Cys Amide
Synonyms: Melanocyte Stimulating Hormone (3-11), (Deamino-Cys[3],Nle[4],Arg[5],D-2-Nal[7],Cys[11])-α-; JKC-366

Bachem H-4944.0001   MW 1209.47 $C_{56}H_{76}N_{18}O_9S_2$ Store at -15°C

Nle-Arg-Phe Amide
Synonyms: Molluscan Cardioexcitatory Neuropeptide Analog; FMRF Amide Analog

Bachem H-2970.0025   Store at -15°C

ICN 153156   MW 433.6 $C_{21}H_{35}N_7O_3$

Sigma N 3637   FW 433.6 $C_{21}H_{35}N_7O_3$ ≥95% (HPLC) |
Bioactive peptide

Nle-Asp-His-Nal₂-Arg-Trp-Lys
Ac-Nle-Asp-His-D-Nal₂-Arg-Trp-Lys Amide

Synonyms: Melanocyte Stimulating Hormone (4-10), Ac-Nle⁴-
c(Asp⁵,D-Nal₂⁷,Lys¹⁰)-α-; SHU9119

Neosystem SC1232   MW 1074.31   Amide bond Asp⁵-Lys¹⁰;
bioactive; melanocyte stimulating hormone-related peptide; agouti-
mimetic peptide is a potent antagonist of MC4-R & a less potent
antagonist of the MC3-R; co-administration of SHU9119 & MTII (full
agonist of the MC3-R & MC4-R) blocks the inhibition of feeding
induced by MTII; administration of SHU9119 significantly enhances
nocturnal feeding or feeding stimulated by a prior fast; when
SHU9119 is co-injected, intracerebroventricularly in rats with α-
MSH, it prevents the antipyretic action of exogenous α-MSH;
Hruby, VJ et al, *J Med Chem*, 38:3454-3461, 1995; Li, SJ et al, *J
Neurosci*, 16:5182-5188, 1996; Fan, W et al, *Nature*, 385:165-168,
1997; Huang, QH et al, *J Neurosci*, 17:3343-3351, 1997

Nle-Asp-His-Phe-Arg-Trp-Lys Amide
Ac-Nle-Asp-His-D-Phe-Arg-Trp-Lys Amide

Synonyms: Melanocyte Stimulating Hormone (4-10), Ac-Nle⁴-
c(Asp⁵,D-Phe⁷,Lys¹⁰)-α-; Melanotan II

Neosystem SC1217   MW 1024.18   Amide bond Asp⁵-Lys¹⁰;
bioactive; melanocyte stimulating hormone-related peptide; a full
agonist of the MC3-R & MC4-R; intracerebroventricular
administration inhibits feeding in four models of hyperphagia:fasted
C57BL/6J, ob/ob, & AY mice, & mice injected with neuropeptide Y;
shown to have tanning activity & to stimulate erectile activity in the
human male; Al-Obeidi, F et al, *J Med Chem*, 32:2555-2561, 1989;
Lan, EL et al, *J Pharma Sci*, 83:1081-1084, 1994; Haskell-Luevano
et al, *BBRC*, 204:1137-1142, 1994; Haskell-Luevano et al, *J Med
Chem*, 39:432-435, 1996; Dorr, RT et al, *Life Sci*, 58:1777-1784,
1996; Fan W et al, *Nature*, 385:165-168, 1997; Hruby, J, P370
15th American Peptide Symposium, Nashville, 1997

Nle-Cyclo(-Asp-His-Nal-Arg-Trp-Lys)
Ac-Nle-Cyclo(-Asp-His-D-2-Nal-Arg-Trp-Lys Amide)

Synonyms: Melanocyte Stimulating Hormone (4-10), (Ac-
Nle⁴,Asp⁵,D-2-Nal⁷,Lys¹⁰)-Cyclo-α-; SHU9119

Bachem H-3952.0001   MW 1074.25 $C_{54}H_{71}N_{15}O_9$ Store at
-15°C

Nle-Cyclo(-Asp-His-Phe-Arg-Trp-Lys)
Ac-Nle-Cyclo(-Asp-His-D-Phe-Arg-Trp-Lys Amide)

Synonyms: Melanocyte Stimulating Hormone (4-10), (Ac-
Nle⁴,Asp⁵,D-Phe⁷,Lys¹⁰)-Cyclo-α-; Melanotan II

Bachem H-3902.0001   MW 1024.19 $C_{50}H_{69}N_{15}O_9$ Store at
-15°C

Nle-Gln-His-Phe-Arg-Trp-Gly
Ac-Nle-Gln-His-D-Phe-Arg-D-Trp-Gly Amide

Synonyms: Melanocyte Stimulating Hormone (4-10)-NH₂, (Ac-Nle⁴,
Gln⁵,D-Phe⁷,D-Trp⁹)-α-; Melanocyte Stimulating Hormone (4-10),
(Ac-Nle⁴,Gln⁵,D-Phe⁷,D-Trp⁸)-α-; HP-228 ; Nitric Oxide Synthase
Inhibitor

Bachem H-3594.0001   MW 984.13 $C_{47}H_{65}N_{15}O_9$ Store at
-15°C

Neosystem SC1244   MW 984.12   Bioactive; melanocyte
stimulating hormone-related peptide; a more potent protective
agent against *E. coli* lipopolysaccharide (LPS) in mice & rabbits & to
prevent the rise in TNF-α & IL-1 in animals given LPS; inhibits the
induction by LPS of nitric oxide synthase (NOS) *in vivo* but not *in
vitro*; capable of blocking type II NOS induction in rats given LPS
HP-228 does not prevent NOS induction due to LPS, IL-1β or TNF-α
in rat aortic smooth muscle cells; Girten, B et al, *FASEB J*, 9:A955,
1995; Abou-Mohamed, G et al, *J Pharmacol Exp Ther*, 275:584-
591, 1995

Sigma M 7903   FW 984.1 $C_{47}H_{65}N_{15}O_9$ Bioactive peptide;
Abou-Mohamed, G et al, *J Pharm Exp Ther*, 275:584, 1995

Nle-Glu-His-Phe-Arg-Trp-Gly
Ac-Nle-Glu-His-D-Phe-Arg-Trp-Gly Amide

Synonyms: Melanocyte Stimulating Hormone (4-10), N-Ac,(Nle⁴,D-
Phe⁷)-α-

American Peptide 56-0-29   MW 985.1 $C_{47}H_{64}N_{14}O_{10}$
Sawyer, TK et al, *PNAS*, 25:1022, 1982

Nle-Leu-Phe
For-Nle-Leu-Phe

Bachem H-3060.0050   MW 419.52 $C_{22}H_{33}N_3O_5$ Store at
-15°C

Nle-Leu-Phe
N-For-Nle-Leu-Phe

ICN 152773   MW 419.5 $C_{22}H_{33}N_3O_5$ Day, AR etal, *FEBS Lett*,
77:291, 1977; Sha'afi, RI etal, *FEBS Lett*, 91:305, 1978; Sha'afi,
RI etal, *BBA*, 541:150, 1978; Freer, RJ etal, *Peptides, Structure &
Biological Function*, Proceedings of the Sixth American Peptide
Symposium, Gross, E & M Meienhofer, eds, 749, 1979

Nle-Leu-Phe
N-t-BOC-Nle-Leu-Phe

Synonyms: Chemotactic Peptide Antagonist

ICN 195635   MW 491.6 $C_{26}H_{41}N_3O_6$ ≥97%

Nle-Leu-Phe
N-For-Nle-Leu-Phe-OMe

ICN 195640   MW 433.5 $C_{23}H_{35}N_3O_5$ ≥97% | For
chemoattractant antagonistic studies

Nle-Leu-Phe
N-t-BOC-Nle-Leu-Phe

Synonyms: Chemotactic Peptide Antagonist

Sigma B 3886   FW 491.6 $C_{26}H_{41}N_3O_6$ ≥97% (HPLC) |
Bioactive peptide; Freer, RJ et al, *Peptides, Structure & Biological
Function*, Proceedings of the Sixth American Peptide Symposium,
Gross, E & Meienhofer, M, eds, 749, 1979

Nle-Leu-Phe
N-For-Nle-Leu-Phe-OMe

Synonyms: Chemotactic Peptide

Sigma F 1760   FW 433.5 $C_{23}H_{35}N_3O_5$ ≥97% (HPLC) |
Bioactive peptide; Used in the study of chemoattractant
antagonists; Fruchtmann, R et al, *Hoppe-Seyler's Z Physiol Chem*,
362:163, 1981

Nle-Leu-Phe
N-For-Nle-Leu-Phe

Synonyms: Chemotactic Peptide

Sigma F 3631   FW 419.5 $C_{22}H_{33}N_3O_5$ ≥97% (HPLC) |
Bioactive peptide; Freer, RJ et al, *Peptides, Structure & Biological
Function*, Proceedings of the Sixth American Peptide Symposium,
Gross, E & Meienhofer, M, eds, 749, 1979; Shaafi, RI et al, *Biochim
Biophys Acta*, 541:150, 1978

Nle-Leu-Phe-Nle-Tyr-Lys
For-Nle-Leu-Phe-Nle-Tyr-Lys

Bachem H-3065.0025   MW 824.03 $C_{43}H_{65}N_7O_9$ Store at
-15°C

Nle-Leu-Phe-Nle-Tyr-Lys
N-For-Nle-Leu-Phe-Nle-Tyr-Lys

Synonyms: Chemotactic Peptide

ICN 152774   MW 824 $C_{43}H_{65}N_7O_9$ Potent chemoattractant for human neutrophils; radioiodination yields a peptide with specific radioactivity & full biological activity; Niedel, J etal, *Science*, 205a:1412, 1979; Niedel, J etal, *JBC*, 254:10700, 1979; Bonser, RW etal, *BBRC*, 102:1269, 1981; Sklar, LA etal, *JBC*, 256:9909, 1981; Niedel, J etal, *JBC*, 256:9295, 1981; Sklar, LA etal, *PNAS*, 78:7540, 1981

Sigma F 0267   FW 824.0 $C_{43}H_{65}N_7O_9$ ≥97% (HPLC) | Bioactive peptide; radioiodinated molecule has full biological activity; potent chemoattractant for human neutrophils; Niedel, J et al, *J Biol Chem*, 254:10700, 1979; Niedel JE, *J Biol Chem*, 256:9295, 1981; Sklar, LA et al, *Proc Natl Acad Sci USA*, 78:7540, 1981; Johansson, B et al, *J Cell Biol*, 121:1281, 1993; Remes JJ et al, *Exp Cell Res*, 209:26, 1993

Nle-Leu-Phe-Tyr
N-For-Nle-Leu-Phe-Tyr

Synonyms: Chemotactic Peptide

ICN 195641   MW 582.7 $C_{31}H_{42}N_4O_7$

Sigma F 2134   FW 582.7 $C_{31}H_{42}N_4O_7$ ≥90% (HPLC) | Bioactive peptide

Nle-Nle
FMOC-Nle-Nle

Bachem B-2465.0001   Store at -15°C

Nle-Nle
Z-Nle-Nle

Bachem C-3730.0001   MW 378.47 $C_{20}H_{30}N_2O_5$ Store at -15°C

Nle-Pro-Nle-Asp
Ac-Nle-Pro-Nle-Asp-AMC

Bachem I-1850.0005   MW 655.75 $C_{33}H_{45}N_5O_9$ Store at -15°C

Nle-Sta-Ala-Sta

Synonyms: Pepstatin Analog; Renin Inhibitor

Sigma N 7390   FW 516.7 $C_{25}H_{48}N_4O_7$ ≥95% (HPLC) | Bioactive peptide; Guegan, R et al, *J Med Chem*, 29:1152, 1986

Nva-Nva
Z-Nva-Nva

Bachem C-3745.0001   Store at -15°C

Nα-t-BOC-Leu-Leu-Arg-CHO

Synonyms: Leupeptin

ICN 195666   MW 484.6 $C_{23}H_{44}N_6O_5$ 97% | N^α-t-BOC-Deacetylleupeptin; protease inhibitor

Orn-Ala
Orn-β-Ala

Bachem G-4105.0250   MW 203.24 $C_8H_{17}N_3O_3$ Store at -15°C

Orn-Ala
FA-Orn-Ala Hydrochloride

Bachem M-2020.0050   Store at -15°C

Orn-Ala
Orn-β-Ala Dihydrochloride

Synonyms: Salty Peptide

Sigma O 0506   FW 276.2 $C_8H_{17}N_3O_3 \cdot 2HCl$ Okai, H & Tamuya, M, *Chem Eng News*, 68(2):26, 1990

Orn-Asp

Bachem G-2840.0250   MW 247.25 $C_9H_{17}N_3O_5$ Store at -15°C

ICN 155998   MW 247.3 $C_9H_{17}N_3O_5$

Orn-Orn

Bachem G-2845.0250   MW 246.31 $C_{10}H_{22}N_4O_3$ Store at -15°C

Orn-Orn-Orn

Bachem H-4495.0100   MW 360.46 $C_{15}H_{32}N_6O_4$ Store at -15°C

Orn-Orn-Orn Acetate Salt

Sigma O 3631   FW 360.5 (free base) $C_{15}H_{32}N_6O_4$

Orn-Ser
(3S,6S)-3-(3-(N-(N-(Nα-Ac-Nδ-For-Nδ- Hydroxyl-L-Orn)-L-Ser)-N-(Hydroxy)Amino)Propyl)-6-(3-(N-For-N-Hydroxyamino)Propyl)-2,5-Piperazinedione

Synonyms: Foroxymithine; Angiotensin I Converting Enzyme Inhibitor

Alexis 260-008   Microbial   MW 575.6 $C_{22}H_{37}N_7O_{11}$ ≥98%; off-white solid; Soluble in water | Umezawa, H et al, *J Antibiot*, 38:1813, 1985; Aoyagi, T et al, *J Appl Biochem*, 7:388, 1985

Orn-Ser
(3S,6S)-3-(3(N-(N-(Nα-Ac-N-for-N-Hydroxy-L-Orn)-L-Ser)-N-(Hydroxy)Amino)Propyl)-6-(3-(N-Formyl-N-Hydroxyamino)-Propyl)-2,5-Piperazinedione

Synonyms: Foroxymithine; Angiotensin Converting Enzyme Inhibitor

ICN 152850   Microbial   MW 575.6 $C_{22}H_{37}N_7O_{11}$ Aoyagi, T etal, *J Appl Biochem*, 7:388, 1975; Umezawa, H etal, *J Antibiot*, 38:1813, 1975

Orn-Ser
(3S,6S)-3-(3-(N-(N-(Nα-Ac-Nδ-Formyl-Nδ-Hydroxy-L-Orn)-L-Ser)-Piperazinedione

Synonyms: Foroxymithine; Angiotensin I Converting Enzyme Inhibitor

Peptides International IFR-4190   Microbial   MW 575.58 $C_{22}H_{37}N_7O_{11}$ Amorphous powder; integrity assessed by activity; bulk | Inhibitor; Umezawa, H et al, *J Antibiotics*, 38:1813, 1985; Aoyagi, T et al, *J Appl Biochem*, 7:388, 1985

Peptides International IFR-4190-v   Microbial   MW 575.58 $C_{22}H_{37}N_7O_{11}$ Lyophilized amorphous powder; integrity assessed by activity

Pen-Arg-Gly-Asp-Cys
Ac-Pen-Arg-Gly-Asp-Cys

Synonyms: Fibrinogen Receptor Antagonist

Bachem H-1614.0001   Store at -15°C

Pen-Arg-Gly-Asp-Cys
N-Ac-Pen-Arg-Gly-Asp-Cys

Synonyms: Fibrinogen Receptor Antagonist

ICN 195615   MW 620.7 $C_{22}H_{36}N_8O_9S_2$ ≥97% | Strongly inhibits platelet aggregation

Sigma A 5582   FW 620.7 $C_{22}H_{36}N_8O_9S_2$ ≥97% (HPLC) |
Disulfide bonds: 1-5; bioactive peptide; potent inhibitor of blood
platelet aggregation; Bogusky, MJ et al, *Biopolymers*, 33:1287,
1993

Pen-Gly-Phe-Pen
D-Pen-Gly-Phe-D-Pen
Synonyms: Enkephalin, (des-Tyr[1],D-Pen[2,5])-
American Peptide 30-0-78   MW 482.6 $C_{21}H_{30}N_4O_5S_2$
Disulfide bonds: D-Pen[2]-D-Pen[5]

Pen-Gly-Phe-Pen
D-Pen-Gly-Phe-Pen
Synonyms: Enkephalin, (des-Tyr[1],D-Pen[2],Pen[5])-
American Peptide 30-0-79   MW 482.6 $C_{21}H_{30}N_4O_5S_2$
Disulfide bonds: D-Pen[2]-Pen[5]

Pen-Tyr-Phe-Val-Asn-Cys-Pro-Arg-Gly
Deamino-Pen-Tyr-Phe-Val-Asn-Cys-Pro-D-Arg-Gly
Amide
Synonyms: Vasotocin, (Deamino-Pen[1],Val[4],D-Arg[8])-
American Peptide 66-0-21   MW 1068.3 $C_{48}H_{69}N_{13}O_{11}S_2$
Arginine vasopressin antagonist; disulfide bonds: Pen[1]-Cys[6]

Phe₂-Leu-Asp-Ile-Ile-Trp
Ac-D-Dip-Leu-Asp-Ile-Ile-Trp
Synonyms: Endothelin Receptor Antagonist; PD 142893
Alexis 155-008   MW 924.1 $C_{50}H_{65}N_7O_{10}$ Non-selective
endothelin-1 antagonist with high affinity for both the ET$_A$ & ET$_B$
receptor subtypes; Cody, WL et al, *J Med Chem*, 35:3301, 1992;
Haynes, WG et al, *TIPS*, 14:225, 1993; Doherty, AM et al, *J Med
Chem*, 36:2585, 1993

Phe-Ala
D-Phe-L-Ala (2,3-³H)
ARC ART-694   MW 236.0
$C_6H_5CH_2CH(NH_2)CONHCH(CH_3)COOH$ 30-60 Ci/mmol; 1.11-2.22
TBq/mmol; in EtOH | Radiochemical

Phe-Ala
Z-Phe-Ala
Bachem C-2360.0001   MW 370.41 $C_{20}H_{22}N_2O_5$ Store at
-15°C

Phe-Ala
Z-Phe-Ala Amide
Bachem C-2365.0001   MW 369.42 $C_{20}H_{23}N_3O_4$ Store at
-15°C

Phe-Ala
Bachem G-2850.0250   MW 236.27 $C_{12}H_{16}N_2O_3$ Store at
-15°C

Phe-Ala
Phe-β-Ala
Bachem G-2855.0250   MW 236.27 $C_{12}H_{16}N_2O_3$ Store at
-15°C

Phe-Ala
D-Phe-Ala
Bachem G-4390.0250   MW 236.27 $C_{12}H_{16}N_2O_3$ Store at
-15°C

Phe-Ala
Phe-Ala-βNA
Bachem K-1620.0250   Store at -15°C

Phe-Ala
FA-Phe-Ala
Bachem M-1955.0050   MW 356.38 $C_{19}H_{20}N_2O_5$ Store at
-15°C

Phe-Ala
Z-Phe-Ala-Diazomethylketone
Bachem N-1040.0050   MW 394.43 $C_{21}H_{22}N_4O_4$ Store at
-15°C

Phe-Ala
Z-Phe-DL-Ala-FMK
Bachem N-1780.0005   MW 386.42 $C_{21}H_{23}FN_2O_4$ Store at
-15°C

Phe-Ala
Z-Phe-Ala-CH₂F
Synonyms: Cathepsin B Inhibitor I
Calbiochem 342000   MW 386.4 $C_{21}H_{23}FN_2O_4$ ≥98% (HPLC);
solid; soluble in DMSO; sold under license of US Patents 5,344,939
& 5,210,272 issued to Prototek, Inc | suitable as a negative
control for Caspase I; Rosenthal, PJ et al, *J Clin Invest*, 91:1057,
1993; Rauber, P et al, *Biochem J*, 239:633, 1986; Shaw, E et al,
Biomed Biochim Acta, 45:1397, 1986; Rasmick, D, *Anal Biochem*,
149:461, 1985

Phe-Ala
Z-L-Phe-L-Ala
Fluka 97029   MW 370.40 $C_{20}H_{22}N_2O_5$ ≥99.0% (titration);
mp:151-154°C

Phe-Ala
Phe-β-Ala
ICN 102634   MW 236.3 $C_{12}H_{16}N_2O_3$

Phe-Ala
ICN 156126   MW 236.3 $C_{12}H_{16}N_2O_3$ Crystalline

Phe-Ala
Z-Phe-Ala-CH₂F
Synonyms: Interleukin Iβ Converting Enzyme-Like Inhibitor
ICN 193607   Useful in apoptosis research

Phe-Ala
Biotin-Phe-Ala-FMK
Synonyms: Caspase Inhibitor Negative Control; Cathepsin B
Inhibitor
Kamiya   MW 386

Phe-Ala
Z-Phe-Ala-FMK
Synonyms: Caspase Inhibitor Negative Control; Cathepsin B
Inhibitor
Kamiya   MW 386

Phe-Ala
N-CBZ-Phe-Ala
Sigma C 1634   FW 370.4 $C_{20}H_{22}N_2O_5$

Phe-Ala
(2*R*)-2-Mercaptomethyl-4-Methylpentanoyl-Phe-Ala
Amide
Synonyms: Collagenase Inhibitor
Sigma M 3906   FW 379.5 $C_{19}H_{29}N_3O_3S$ ≥90% (TLC) |
Bioactive peptide

Phe-Ala
Sigma P 3251   FW 236.3 $C_{12}H_{16}N_2O_3$

Phe-Ala
CBZ-Phe-Ala-Diazomethylketone
USBio C2099-40   MW 395.4 $C_{21}H_{23}N_4O_4$ ≥99%

Phe-Ala
(2R)-S-Ac-2-Mercaptomethyl-4-Methylpentanoyl-L-Phe-L-Ala Amide
Synonyms: SIMP I, Ac-; Collagenase Inhibitor
Peptides International ISN-3821-PI   Synthetic   MW 421.56 $C_{21}H_{31}N_3O_4S$ >98% (HPLC); white powder | Inhibitor; protected precursor

Phe-Ala
(2R)-2-Mercaptomethyl-4-Methypentanoyl-L-Phe-Ala Amide
Synonyms: SIMP I; Collagenase Inhibitor
Peptides International ISN-3825-PI   Synthetic   MW 379.52 $C_{19}H_{29}N_3O_3S$ >98% (HPLC); white powder; free thiol

Phe-Ala Amide Hydrochloride
Bachem G-2860.0250   MW 271.75 $C_{12}H_{17}N_3O_2 \cdot HCl$ Store at -15°C

Phe-Ala-Ala-Abz
Phe-Ala-Ala-4-Abz
Bachem H-9790.0050   Store at -15°C

Phe-Ala-Ala-Phe
Glutaryl-Phe-Ala-Ala-Phe-AMC
Bachem I-1535.0025   MW 725.8 $C_{39}H_{43}N_5O_9$ Store at -15°C

Phe-Ala-Ala-Phe
Suc-Phe-Ala-Ala-Phe-pNA
Bachem L-1675.0050   Store at -15°C

Phe-Ala-Ala-Phe-Phe-Val-Leu
BOC-Phe-Ala-Ala-p-Nitro-Phe-Phe-Val-Leu-pyridin-4-ylmethyl ester
Bachem M-1175.0025   MW 1050.22 $C_{55}H_{71}N_9O_{12}$ Store at -15°C

Phe-Ala-Ala-Phe-Phe-Val-Leu
Phe-Ala-Ala-p-Nitro-Phe-Phe-Val-Leu-pyridin-4-yl-OMe
Bachem M-1690.0025   MW 950.11 $C_{50}H_{63}N_9O_{10}$ Store at -15°C

Phe-Ala-Ala-Phe-Phe-Val-Leu
Phe-Ala-Ala-Phe(4-NO₂)-Phe-Val-Leu-(4-Pyridylmethyl) Ester
Synonyms: Cathepsin D Substrate
Fluka 77431   MW 950.1 $C_{50}H_{63}N_9O_{10}$ ≥98.0% (TLC) | Agarwal, N & Rich, DH, *Anal Biochem*, 130:158, 1983

Phe-Ala-Ala-Phe-Phe-Val-Leu
Phe-Ala-Ala-p-Nitro-Phe-Phe-Val-Leu 4-Pyridylmethyl Ester
Synonyms: Cathepsin D Substrate
ICN 156127   MW 950.1 $C_{50}H_{63}N_9O_{10}$ Agarwal, N & D Rich, *Anal Biochem*, 130:158, 1983

Phe-Ala-Ala-Phe-Phe-Val-Leu
N-t-BOC-Phe-Ala-Ala-p-Nitro-Phe-Phe-Val-Leu 4-Hydroxymethylpyridine Ester
Synonyms: Cathepsin D Substrate
Sigma B 1644   FW 1050.2   Substrate for continuous assay; Agarwal, N & Rich, DH, *Anal Biochem*, 130:158, 1983

Phe-Ala-Ala-Phe-Phe-Val-Leu
Phe-Ala-Ala-p-Nitro-Phe-Phe-Val-Leu 4-Pyridylmethyl Ester
Synonyms: Cathepsin D Substrate
Sigma P 8168   FW 950.1 $C_{50}H_{63}N_9O_{10}$ ≥97% (HPLC) | Agarwal, N & Rich, D, *Anal Biochem*, 130:158, 1983

Phe-Ala-Arg
Bz-Phe-Ala-Arg
Synonyms: Enkephalin
American Peptide 32-0-81   MW 496.6 $C_{25}H_{32}N_6O_5$

Phe-Ala-Arg
Benzoyl-Phe-Ala-Arg
Synonyms: Carboxypeptidase E Substrate
Neosystem SC144   MW 496.56   Stack, G et al, *Life Sciences*, 34:113-121, 1984

Phe-Ala-Arg-Lys-Gly-Ala-Leu-Arg-Gln
N-Myr-Phe-Ala-Arg-Lys-Gly-Ala-Leu-Arg-Gln
Synonyms: Protein Kinase C Inhibitor
Alexis 162-018   MW 1256.6 $C_{60}H_{105}N_{17}O_{12}$ ≥97%; white lyophilized powder; soluble in water | Selective & cell-permeable inhibitor; corresponds to the pseudosubstrate domain of PKCα & PKCβ subtypes; Eichholtz, T et al, *J Biol Chem*, 268:1982, 1993

Phe-Ala-Arg-Lys-Gly-Ala-Leu-Arg-Gln
Myr-Phe-Ala-Arg-Lys-Gly-Ala-Leu-Arg-Gln
Bachem N-1370.0001   MW 1256.6 $C_{60}H_{105}N_{17}O_{12}$ Store at -15°C

Phe-Ala-Arg-Lys-Gly-Ala-Leu-Arg-Gln
Bachem N-1375.0005   MW 1046.24 $C_{46}H_{79}N_{17}O_{11}$ Store at -15°C

Phe-Ala-Arg-Lys-Gly-Ala-Leu-Arg-Gln
Myr-N-Phe-Ala-Arg-Lys-Gly-Ala-Leu-Arg-Gln Amide Trifluoroacetate Salt Myristoylated
Synonyms: Protein Kinase C Inhibitor (19-27)
Calbiochem 476480   MW 1255.6 $C_{60}H_{106}N_{18}O_{11}$ ≥98% (HPLC); lyophilized solid; soluble in water | Pseudosubstrate sequence from protein kinase C$_\alpha$ (PKC$_\alpha$) & PKC$_\beta$; N-terminus is myristoylated to allow membrane permeability; highly specific inhibitor of TPA activation of MARCKS phosphorylation in fibroblast primary cultures; exhibits 98% inhibition at 100 μM; Ward, NE & O'Brian, CA, *Biochemistry*, 32:11903, 1993; Eicholtz, T et al, *J Biol Chem*, 268:1982, 1993

Phe-Ala-Asn
Bachem H-4510.0250   MW 350.38 $C_{16}H_{22}N_4O_5$ Store at -15°C

Phe-Ala-Glu-Asp-Val-Gly-Ser-Asn-Lys-Gly
Synonyms: Amyloid β-Protein (20-29)
Bachem H-3808.0001   MW 1023.07 $C_{43}H_{66}N_{12}O_{17}$ Store at -15°C

Phe-Ala-Glu-Pro-Leu-Pro-Ser-Glu-Glu-Glu-Gly-Glu-Ser-Tyr-Ser-Lys-Glu-Val-Pro-Glu-Met-Glu-Lys-Arg-Tyr-Gly-Gly-Phe-Met-Arg-Phe

Synonyms: Peptide B

ICN 154515   Bovine   MW 3657.5

Phe-Ala-Glu-Pro-Leu-Pro-Ser-Glu-Glu-Glu-Gly-Glu-Ser-Tyr-Ser-Lys-Glu-Val-Pro-Glu-Met-Glu-Lys-Arg-Tyr-Gly-Gly-Phe-Met-Arg-Phe

Synonyms: Peptide B

American Peptide 30-3-10   Bovine   MW 3657.1
$C_{163}H_{239}N_{39}O_{53}S_2$ Micanovic, R et al, *Peptides*, 5:853, 1984

Phe-Ala-Gly-Arg-Ile-Asp-Arg-Ile-Gly-Ala-Gln-Ser-Gly-Leu-Gly-Cys-Asn-Ser-Phe-Arg-Tyr
Mpr-Phe-D-Ala-Gly-Arg-Ile-Asp-Arg-Ile-Gly-Ala-Gln-Ser-Gly-Leu-Gly-Cys-Asn-Ser-Phe-Arg-Tyr Amide

Synonyms: Natriuretic Peptide (7-28), Atrial (Mpr[7],D-Ala[9])-

American Peptide 14-1-50   Rat   MW 2373.7
$C_{102}H_{157}N_{33}O_{29}S_2$ Disulfide bonds: Mpr[7]-Cys[23]

Phe-Ala-Gly-Arg-Ile-Asp-Arg-Ile-Gly-Ala-Gln-Ser-Gly-Leu-Gly-Cys-Asn-Ser-Phe-Arg-Tyr
3-Mercaptopropionyl-Phe-D-Ala-Gly-Arg-Ile-Asp-Arg-Ile-Gly-Ala-Gln-Ser-Gly-Leu-Gly-Cys-Asn-Ser-Phe-Arg-Tyr Amide

Synonyms: Natriuretic Peptide (3-24), Atrial (Deamino-Cys[3],D-Ala[5])-; Atriopeptin III (3-24), (Deamino-Cys[3],D-Ala[5])-; Atriopeptin III (3-24), (Deamino-Cys[3],D-Ala[5])-

ICN 195628   Rat   MW 2373.7

Sigma A 7171   Rat   FW 2373.7   ≥97% (HPLC) | Bioactive peptide; disulfide bonds: 3-19

Phe-Ala-Leu-Ala-Leu-Lys-Ala-Leu-Lys-Lys-Ala-Leu-Lys-Lys-Leu-Lys-Lys-Ala-Leu-Lys-Lys-Ala-Leu

Synonyms: Hecate

Bachem H-4094.0500   Store at -15°C

Phe-Ala-Nle
N-t-BOC-D-Phe-Ala-Nle-pNA

Synonyms: Human Granulocyte Substrate

Sigma B 8397   FW 569.7 $C_{29}H_{39}N_5O_7$ Marossy, K et al, *Biochem Biophys Res Commun*, 96:762, 1980

Phe-Ala-Pro
Bz-Phe-Ala-Pro

Synonyms: Angiotensin I Converting Enzyme Substrate

Bachem H-9050.0050   MW 437.5 $C_{24}H_{27}N_3O_5$ Store at -15°C

Phe-Ala-Pro
N-Bz-Phe-Ala-Pro

Synonyms: Angiotensin I Converting Enzyme Substrate

Sigma B 3903   FW 437.5 $C_{24}H_{27}N_3O_5$ Ryan, JW et al, *Adv Exp Med Biol*, 156B:805, 1983

Phe-Arg
D-Phe-Pip-Arg-pNA

American Peptide 81-6-50   MW 552.6 $C_{27}H_{36}N_8O_5$

Phe-Arg
FMOC-Phe-Arg

Bachem B-1510.0001   Store at -15°C

Phe-Arg
Z-Phe-Arg-OMe Hydrochloride

Bachem C-2375.0250   MW 506.01 $C_{24}H_{31}N_5O_5 \cdot HCl$ Store at -15°C

Phe-Arg

Bachem G-2865.0250   MW 321.38 $C_{15}H_{23}N_5O_3$ Store at -15°C

Phe-Arg
Hippuryl-Phe-Arg

Synonyms: Bz-Gly-Phe-Arg

Bachem H-3725.0100   MW 482.54 $C_{24}H_{30}N_6O_5$ Store at -15°C

Phe-Arg
Ac-Phe-Arg-AMC Hydrochloride

Bachem I-1015.0050   MW 557.05 $C_{27}H_{32}N_6O_5 \cdot HCl$ Store at -15°C

Phe-Arg
Z-Phe-Arg-AMC Hydrochloride

Bachem I-1160.0050   MW 649.15 $C_{33}H_{36}N_6O_6 \cdot HCl$ Store at -15°C

Phe-Arg
Z-Phe-Arg-4MβNA Hydrochloride

Bachem J-1140.0050   MW 647.17 $C_{34}H_{38}N_6O_5 \cdot HCl$ Store at -15°C

Phe-Arg
Suc-Phe-Arg-4MβNA

Bachem J-1385.0050   Store at -15°C

Phe-Arg
Phe-Arg-βNA Dihydrochloride

Bachem K-1450.0050   MW 519.47 $C_{25}H_{30}N_6O_2 \cdot 2HCl$ Store at -15°C

Phe-Arg
Z-Phe-Arg-pNA Hydrochloride

Bachem L-1242.0050   MW 612.09 $C_{29}H_{33}N_7O_6 \cdot HCl$ Store at -15°C

Phe-Arg
D-Phe-Homopro-Arg-pNA Diacetate

Bachem L-1490.0050   Store at -15°C

Phe-Arg
Ac-Phe-Arg-OEt

Bachem M-1060.0250   MW 391.47 $C_{19}H_{29}N_5O_4$ Store at -15°C

Phe-Arg
N-CBZ-L-Phe-L-Arg-MCA

Synonyms: Kallikrein Substrate, Plasma; Cathepsin Substrate; Cathepsin B Substrate

ICN 150579   MW 612.7 $C_{33}H_{36}N_6O_6$ Fluorogenic substrate

---

Phe-Arg
Phe-Arg-β-Naphthylamide Dihydrochloride

ICN 156128   MW 519.5 $C_{25}H_{30}N_6O_2 \cdot 2HCl$

Phe-Arg
D-Phe-L-Pipecolyl-Arg-pNA Triacetate Salt

Synonyms: Thrombin Substrate

ICN 156215   MW 732.8 $C_{27}H_{36}N_8O_5 \cdot 3C_2H_4O_2$ Chromogenic substrate; useful in the assay of Antithrombin III; Abildgaard, U et al, *Thromb Res*, 11:549, 1977; Lottenberg, R et al, *Methods Enzymol*, 80:341, 1981

Phe-Arg
Z-Phe-Arg-AMC Trifluoroacetate Salt

Synonyms: Plasma Kallikrein Substrate; Cathepsin B Substrate; Cathepsin L Substrate

Neosystem SC1275   MW 612.68   Morita, T et al, *J Biochem*, 82:1495, 1977

Phe-Arg
D-Phe-Pip-Arg-pNA

Synonyms: Thrombin Substrate

Neosystem SC1297   MW 552.58   Chromogenic substrate for thrombin & for various snake venom proteases; Lottenberg, R et al, *Meth Enz*, 80:341-361, 1981; Teng, C-M et al, *Toxicon*, 27:161-167, 1989; Rijkers, DTS et al, *Thromb Res*, 79:491-499, 1995; Serrano, SM et al, *Biochem*, 34:7186-7133, 1995

Phe-Arg
Z-Phe-Arg-MCA Hydrochloride

Synonyms: Plasma Kallikrein Substrate; Plasma Cathepsin B/L Substrate

Peptides International MFR-3095-v   MW 612.69 $C_{33}H_{36}N_6O_6$ >99% (HPLC); lyophilized amorphous powder | Morita, T et al, *J Biochem*, 82:1495, 1977; Barret, AJ. *J Biochem*, 187:909, 1980

Phe-Arg
N-CBZ-Phe-Arg-4-Methoxy-β-Naphthylamide Hydrochloride

Synonyms: Cathepsin B Substrate; Cathepsin L Substrate

Sigma C 3282   FW 647.2 $C_{34}H_{38}N_6O_5 \cdot HCl$ Inactivation of cathepsin B by 4 *M* urea led to a selective assay of cathepsin L activity in goat brain homogenates

Phe-Arg
N-CBZ-Phe-Arg-MCA Hydrochloride

Synonyms: Plasma Kallikrein Substrate

Sigma C 9521   FW 649.1 $C_{33}H_{36}N_6O_6 \cdot HCl$ Fluorogenic substrate; Morita, T et al, *J Biochem*, 82:1495, 1977; Lottenberg, R et al, *Meth Enz*, 80:341, 1981

Phe-Arg
D-Phe-L-Pipecolyl-Arg-pNA Dihydrochloride

Sigma P 3955   FW 625.6 $C_{27}H_{36}N_8O_5 \cdot 2HCl$

Phe-Arg
Phe-Arg-β-Naphthylamide Dihydrochloride

Synonyms: Dipeptidyl Aminopeptidase I Substrate; Cathepsin C Substrate

Sigma P 4157   FW 519.5 $C_{25}H_{30}N_6O_2 \cdot 2HCl$ Substrate for dipeptidyl aminopeptidase I (cathepsin C)

Phe-Arg
CBZ-Phe-Arg-AMC Hydrochloride

USBio C2099-41   MW 649.1 $C_{33}H_{36}N_6O_6 \cdot HCl$ ≥99%

Phe-Arg-Ala-Asp-His-Pro-Phe-Leu

Synonyms: Ovokinin

Bachem H-2676.0001   Store at -15°C

Phe-Arg-Arg

Bachem H-1988.0025   MW 477.57 $C_{21}H_{35}N_9O_4$ Store at -15°C

Phe-Arg-Arg-Leu-Ser-Ile-Ser-Thr

Synonyms: Serine/Threonine Kinase Substrate Peptide, Biotinylated

Upstate 12-366   MW 1317   >98% pure; frozen solution

Phe-Arg-Asp-Tyr-Val-Asp-Arg-Phe-Tyr-Lys-Thr-Leu-Arg-Ala-Glu-Gln-Ala-Ser

Synonyms: P25 (293-310); GAG P24 CA (161-178); HTLV I Peptide

Neosystem SC305   MW 2265.51   HTLV I peptide from subtype P25 (GAG P24 CA)

Phe-Arg-Lys-Lys-Trp-Asn-Lys-Trp-Ala-Leu-Ser-Arg Amide

Synonyms: Pro-Adrenomedullin N-Terminal 20 Peptide (9-20); PAMP-12; Hypotensive Peptide

Peptides International PAM-4339-v   Human synthetic   MW 1619.0 $C_{77}H_{119}N_{25}O_{14}$ >99% (HPLC); lyophilized amorphous powder | Bioactive; major endogenous form of PAMP; Kuwasato, K et al, 414:105, 1997

Phe-Arg-Ser-Val-Gln
Cyclohexylacetyl-Phe-Arg-Ser-Val-Gln Amide

Synonyms: KKI-7; Kallikrein Inhibitor

Bachem H-9885.0005   Store at -15°C

Neosystem SC158   MW 758.92   Virus-related peptide; bioactive; Benton, J, Proceedings of the 10th Am Pept Symposium, 1987

Sigma C 6922   FW 758.9 $C_{36}H_{58}N_{10}O_8$ ≥97% (HPLC) | Bioactive peptide; inhibits human, porcine, rat & caninekallikrein; Burton, J, *Proc 10th Amer Pept Symp*, Marshall, GR, ed, ESCOM Sci Publ, Leiden, Neth, 647, 1988

Phe-Arg-Trp-Gly-Lys-Pro-Val-Gly-Lys-Lys Amide

Synonyms: Adrenocorticotropic Hormone (7-16)

Neosystem SC372   Human   MW 1201.48   Bioactive; Wolterink, G & Van Ree, JM, *Life Sci*, 45:703-710, 1989

Phe-Arg-Trp-Gly-Lys-Pro-Val-Gly-Lys-Lys-Arg-Arg-Pro-Val-Lys-Val-Tyr-Pro-Asn-Val-Ala-Glu-Asn-Glu-Ser-Ala-Glu-AA-Phe-Pro-Leu-Glu

Synonyms: Adrenocorticotropic Hormone (7-38); Corticotropin A

ICN 152721   Li, CH et al, *PNAS*, 75:4306, 1978

Phe-Arg-Trp-Gly-Lys-Pro-Val-Gly-Lys-Lys-Arg-Arg-Pro-Val-Lys-Val-Tyr-Pro-Asn-Gly-Ala-Glu-Asp-Glu-Ser-Ala-Glu-Ala-Phe-Pro-Leu-Glu

Synonyms: Corticotropin Inhibiting Peptide; Adrenocorticotropic Hormone (7-38)

Bachem H-1205.0500   Human   MW 3659.17 $C_{167}H_{257}N_{47}O_{46}$ Store at -15°C

Neosystem SC037   Human   MW 3659.15   Bioactive; Li, CH et al, *PNAS* USA, 75:4306, 1978

Sigma A 1527   Human   FW 3659.2   ≥97% (HPLC) | Bioactive peptide; potent inhibitor of ACTH-stimulated adenylate cyclase; Li, CH et al, *Proc Natl Acad Sci USA*, 75:4306, 1978

Phe-Arg-Trp-Gly-Lys-Pro-Val-Gly-Tyr

Synonyms: Adrenocorticotropic Hormone (7-15), (Tyr[15])-

Bachem H-1200.0005   MW 1109.3 $C_{55}H_{76}N_{14}O_{11}$ Store at -15°C

Phe-Asn-Leu-Pro-Leu-Gly-Asn-Tyr-Lys-Lys-Pro

Synonyms: Fibroblast Growth Factor (1-11), Brain Derived Acidic; Fibroblast Growth Factor Acidic Fragment (1-11)

American Peptide 50-0-27   MW 1290.4 $C_{62}H_{95}N_{15}O_{15}$
Gimenez-Gallego, G et al, *Science*, 230:1385, 1985

Sigma F 3635   Bovine   FW 1290.5   ≥97% (HPLC) |
Bioactive peptide; Esch, F et al, *Biochem Biophys Res Commun*, 133:554, 1985

Bachem H-5665.0001   Bovine brain   Store at -15°C

Phe-Asn-Leu-Pro-Leu-Gly-Asn-Tyr-Lys-Lys-Pro Acidic

Synonyms: Fibroblast Growth Factor (1-11), α-

ICN 151524   BOVINE SYNTHETIC   Corresponds to that of αFGF in bovine brain tissue; Esch, F etal, *BBRC*, 133:554, 1985

Phe-Asn-Lys-His-Thr-Glu-Ile-Ile-Glu-Glu-Asp-Thr-Asn-Lys-Asp-Lys-Pro-Ser-Tyr-Gln-Phe-Gly-Gly-His-Asn-Ser-Val-Asp-Phe-Glu-Glu-Asp-Thr-Leu-Pro-Lys-Val

Synonyms: Fibronectin Binding Protein; Fibronectin Binding Protein Peptide D3; Peptide D₃

American Peptide 44-0-45   MW 4309.7 $C_{190}H_{283}N_{49}O_{66}$ The synthetic peptide mimics the structure of a 38-AA unit from a staphylococcal fibronectin-binding protein; this unit represents the protein domain to which the fibronectin-binding activity has been localized; inhibits binding of fibronectin to bacterial cells; Signas, C et al, *PNAS*, 86:699, 1989

Bachem H-9145.0500   MW 4309.63 $C_{190}H_{283}N_{49}O_{66}$ Store at -15°C

ICN 154540   MW 4310.2   Signas, C etal, *PNAS*, 86:699, 1989

Sigma F 4647   FW 4309.6   ≥97% (HPLC) | Bioactive peptide; inhibits binding of fibronectin to bacterial cells; Signas, C et al, *Proc Natl Acad Sci USA*, 86:699, 1989

Phe-Asp
Z-Phe-Asp

Bachem C-2385.0001   MW 414.42 $C_{21}H_{22}N_2O_7$ Store at -15°C

Phe-Asp

Bachem G-2870.0250   MW 280.28 $C_{13}H_{16}N_2O_5$ Store at -15°C

Phe-Asp-Gly-Cys-Glu-Asp-Asp-Tyr-Asn-Tyr-Tyr-Ser-Arg-Ser Salt Free

Synonyms: Acetylcholine Transporter Protein (526-539); VAT

Biogenesis 0030-5059   Rat vesicular synthetic   Semi-pure; lyophilized

Phe-Cit
Z-Phe-Cit-AMC

Bachem I-1640.0025   MW 613.67 $C_{33}H_{35}N_5O_7$ Store at -15°C

Phe-Cys-Phe-Trp-Lys-Thr-Cys-Thr
D-Phe-Cys-Phe-D-Trp-Lys-Thr-Cys-Thr-ol

Synonyms: Octreotide; SMS 201-995; Somatostatin Analog

American Peptide 68-1-47   Synthetic   MW 1019.2
$C_{48}H_{66}N_{10}O_9S_2$ Disulfide bonds: Cys²-Cys⁷; three times more potent than the native hormone in inhibiting the secretion of growth hormone; Maouyo, D et al, *Pancreas*, 14:47, 1997; Bauer, W et al, *Life Sci*, 31:1133, 1982

Phe-Cys-Tyr-Trp-Arg-Thr-Pen-Thr
D-Phe-Cys-Tyr-D-Trp-Arg-Thr-Pen-Thr Amide

Synonyms: CTAP

American Peptide 68-1-70   MW 1104.3 $C_{51}H_{69}N_{13}O_{11}S_2$
Disulfide bonds: Cys²-Pen⁷; analog of octreotide containing penicillamine instead of Cys⁷; poor ligand for SRIF receptor, but is the potent antagonist of the μ-type opiate receptor; Dawson-Basoa, M et al, *Brain Res*, 757:37, 1997

Bachem H-3698.0001   Store at -15°C | :Cys²-Pen⁷

Phe-Cys-Tyr-Trp-Lys-Val-Cys-Trp
D-Phe-Cys-Tyr-D-Trp-Lys-Val-Cys-Trp Amide

Synonyms: RC-160; Vapreotide

American Peptide 68-1-71   MW 1131.4 $C_{57}H_{70}N_{12}O_9S_2$
Disulfide bonds: Cys²-Cys⁷; somatostatin analog; Re-188-RC-160 (radioisotope Rhenium-188-labeled RC-160) has been successfully used for the local/regional treatment of experimental breast cancer & other cancers; Zamora, PO et al, *Hybridoma*, 16:85, 1997

Phe-Cys-Tyr-Trp-Lys-Val-Cys-Trp
D-Phe-Cys-Tyr-D-Trp-Lys-Val-Cys-Trp

Synonyms: RC-160

Neosystem SC950   MW 1131.38   Disulfide bonds: Cys²-Cys⁷; bioactive; hypothalamic releasing hormone; somatostatin peptide; shows a high potency & a long duration of action for inhibition of growth hormone release; Cai, RZ et al, *PNAS USA*, 83:1896-1900, 1986; Mason-Garcia, M et al, *PNAS USA*, 85:5688-5692, 1988; Hofland, LJ et al, *Endocrinology*, 134:301-306, 1994; Varnum, JM et al, *JBC*, 269:12583-12588, 1994; Pinski, J et al, *Intl J Cancer*, 57:574-580, 1994; Buscail, L et al, *PNAS USA*, 92:1580-1584, 1995

Phe-Cys-Tyr-Trp-Orn-Thr-Pen-Thr
D-Phe-Cys-Tyr-D-Trp-Orn-Thr-Pen-Thr Amide

Synonyms: NTB; Naltriben; Somatostatin Analog, (Cys²,Tyr³,Orn⁵,Pen⁷)-

American Peptide 68-1-65   MW 1062.3 $C_{50}H_{67}N_{11}O_{11}S_2$
Disulfide bonds: Cys²-Pen⁷; very potent & highly selective ligand for μ-opioid receptors; Gulya, K et al, *Life Sci*, 38:2221, 1986; Pelton, JT et al, *JMC*, 29:2370, 1986; Tseng, LF et al, *J Pharmacol Exp Ther*, 280:600, 1997

Phe-Cys-Tyr-Trp-Orn-Thr-Pen-Thr
D-Phe-Cys-Tyr-D-Trp-Orn-Thr-Pen-Thr-ol

Synonyms: CTOP

American Peptide 68-1-66   MW 1049.3 $C_{50}H_{68}N_{10}O_{11}S_2$
Disulfide bonds: Cys²-Pen⁷; analog of octreotide containing penicillamine instead of Cys⁷; poor ligand for SRIF receptor, but is the potent antagonist of the μ-type opiate receptor; Knapp Jr, FF et al, *Anticancer Res*, 17:1783, 1997; Zamorta, PO et al, *Hybridoma*, 16:85, 1997

Phe-Cys-Tyr-Trp-Orn-Thr-Pen-Thr
D-Phe-Cys-Tyr-D-Trp-Orn-Thr-Pen-Thr Amide

Synonyms: CTOP, (Cys²,Tyr³,Orn⁵,Pen⁷)-

Bachem H-2186.0001   MW 1062.28 $C_{50}H_{67}N_{11}O_{11}S_2$ Store at -15°C | Disulfide bonds: Cys²-Pen⁷

Phe-Cys-Tyr-Trp-Orn-Thr-Pen-Thr
D-Phe-Cys-Tyr-D-Trp-Orn-Thr-Pen-Thr

Synonyms: CTOP

Neosystem SC941   MW 1062.27   Disulfide bonds: Cys²-Pen⁷; bioactive; hypothalamic releasing hormone; somatostatin peptide; mu-opioid receptor specific antagonist; Gulya, K et al, *Life Sci*, 38:2221, 1986; Pelton, JT et al, *J Med Chem*, 29:2370, 1986; Kramer, TH et al, *J Pharmacol Exp Ther*, 249:544-551, 1989; Fanselow, MS et al, *J Pharmacol Exp Ther*, 250:825-830, 1989; Czlonkowski, A et al, *Eur J Pharmacol*, 242:229-235, 1993; Tanaka, E et al, *J Neurosci*, 14:1106-1113, 1994

Phe-Cys-Tyr-Trp-Orn-Thr-Pen-Thr
D-Phe-Cys-Tyr-D-Trp-Orn-Thr-Pen-Thr Amide

Synonyms: CTOP

Sigma P 5296   FW 1062.3   ≥97% (HPLC) | Disulfide bonds: 2-7; bioactive peptide; selective ligand for μ-opioid receptors; somatostatin analog; Hawkins, KN et al, *J Pharmacol Exp Ther*, 248:73, 1989

Sigma C-225   Synthetic   MW 1062 $C_{50}H_{67}N_{11}O_{11}S_2$ >97%; white powder; peptide content & salt form information are provided with each lot | Selective μ-opioid receptor antagonist; Toll, *J Pharmacol Exp Ther*, 260:316, 1992; Derrick et al, *J Neurosci*, 14:4359, 1994; Devine et al, *Eur J Pharmacol*, 243:55, 1993

Phe-Gln-Gly-Pro

Bachem H-4520.0050   Store at -15°C

Phe-Gln-Trp-Ala-Val-Gly-His-Leu
D-Phe-Gln-Trp-Ala-Val-Gly-His-Leu-NHEt

Synonyms: Bombesin (6-14), (D-Phe[6],Leu-NHEt[13],des-Met[14])-

Bachem H-3042.0001   Store at -15°C

Phe-Gln-Trp-Ala-Val-Gly-His-Leu-Phe
D-Phe-Gln-Trp-Ala-Val-Gly-His-Leu-(®)-*p*-Chloro-Phe Amide

Synonyms: Bombesin (6-14), (D-Phe[6],Leu[13]-(®)-*p*-Chloro-Phe[14])-

Bachem H-3028.0500   Store at -15°C

Phe-Gln-Val-Val-Cys-Gly
Phe-Gln-Val-Val-Cys(NPys)-Gly Amide

Bachem H-1946.0001   MW 804.95 $C_{34}H_{48}N_{10}O_9S_2$ Store at -15°C

Sigma P 3708   FW 804.9 $C_{34}H_{48}N_{10}O_9S_2$ ≥95% (HPLC) | Bioactive peptide; selectively blocks thrombin-induced platelet aggregation by binding to platelet calpain; does not inhibit platelet aggregation induced by hemostasis factors other than thrombin; Puri, RN et al, *Throm Res*, 72:183, 1993

Phe-Gln-Val-Val-Cys-Gly Amide

Bachem H-2282.0001   MW 650.8 $C_{29}H_{46}N_8O_7S$ Store at -15°C

Phe-Glu
D-Phe-L-Glu (Glu U-^{14}C)

ARC ARC-1051   MW 358.71   200-300 mCi/mmol; 7.4-11.1 GBq/mmol; in EtOH:H$_2$O (7:3) | Radiochemical

Phe-Glu
Z-Phe-Glu

Bachem C-2390.0001   MW 428.44 $C_{22}H_{24}N_2O_7$ Store at -15°C

Phe-Glu

Bachem G-2875.0250   MW 294.31 $C_{14}H_{18}N_2O_5$ Store at -15°C

Phe-Glu-Cys-Thr-Thr-His-Gln-Pro-Arg-Ser-Pro-Leu-Arg-Asp-Leu-Lys-Gly-Ala-Leu-Glu-Ser-Leu-Ile-Glu-Glu-Glu-Thr-Gly-Gln

Synonyms: Gonadotropin Releasing Hormone Associated Peptide (25-53); Gonadotropin Releasing Hormone Precursor Peptide (38-66)

American Peptide 54-1-38   Human   MW 3284.7 $C_{140}H_{225}N_{40}O_{49}S$

Phe-Glu-Gln-Asn-Thr-Ala-Gln-Pro

Synonyms: Connexin 37 (52-59), (Gln[54])-; MUT 1

Bachem H-3458.0001   Store at -15°C

Phe-Glu-Pro-Ile-Pro-Glu-Glu-Tyr-Leu-Glu
Suc-Phe-Glu-Pro-Ile-Pro-Glu-Glu-Tyr(SO$_3$H)-Leu-D-Glu Sulfated

Synonyms: Hirudin (56-65), Suc-(Pro[58],D-Glu[65])-

Bachem H-8145.0001   MW 1445.52 $C_{64}H_{88}N_{10}O_{26}S$ Store at -15°C

Phe-Glu-Trp-Thr-Pro-Gly-Trp-Tyr-Gln-Tyr-Ala-Leu-Pro-Leu
Ac-Phe-Glu-Trp-Thr-Pro-Gly-Trp-Tyr-Gln-L-Azetidine-2-Carbonyl-Tyr-Ala-Leu-Pro-Leu Amide

Synonyms: AF12198

Bachem H-4146.0500   MW 1895.15 $C_{96}H_{123}N_{19}O_{22}$ Store at -15°C

Phe-Glu-Trp-Thr-Pro-Gly-Tyr-Trp-Gln-Pro-Tyr-Ala-Leu-Pro-Leu

Synonyms: AF11377

Bachem H-3738.0001   Store at -15°C

Phe-Gly
BOC-Phe-Gly

Bachem A-2745.0001   MW 322.36 $C_{16}H_{22}N_2O_5$ Store at -15°C

Phe-Gly
FMOC-Phe-Gly

Bachem B-3060.0001   Store at -15°C

Phe-Gly
Z-Phe-Gly

Bachem C-2395.0001   MW 356.38 $C_{19}H_{20}N_2O_5$ Store at -15°C

Phe-Gly
Z-Phe-Gly Amide

Bachem C-2400.0001   MW 355.39 $C_{19}H_{21}N_3O_4$ Store at -15°C

Phe-Gly

Bachem G-2880.0001   MW 222.24 $C_{11}H_{14}N_2O_3$ Store at -15°C

Phe-Gly
Phe-Gly-βNA

Bachem K-1940.0250   Store at -15°C

Phe-Gly
Ac-Phe-Gly-pNA

Bachem L-1060.0050   MW 384.39 $C_{19}H_{20}N_4O_5$ Store at -15°C

Phe-Gly

ICN 102635   MW 222.2 $C_{11}H_{14}N_2O_3$ Crystalline

Phe-Gly
Z-Phe-Gly-NHO-Bz Hydrochloride

Synonyms: Cathepsin Inhibitor I

ICN 195948   MW 475.5 $C_{26}H_{25}N_3O_6$ >95%; lyophilized | Cysteine protease inhibitor which selectively inhibits cathepsins B, L, S & papain

Phe-Gly
Z-Phe-Gly-NHO-Bz-*p*Me

Synonyms: Cathepsin Inhibitor II

ICN 195949   MW 489.5 $C_{27}H_{27}N_3O_6$ Cysteine protease inhibitor which selectively inhibits cathepsins B, L, S & papain

Phe-Gly
Z-Phe-Gly-NHO-Bz-pOMe

Synonyms: Cathepsin Inhibitor III

ICN 195950   MW 505.5 $C_{27}H_{27}N_3O_7$ Cysteine protease inhibitor which selectively inhibits cathepsins B, L, S & papain

Phe-Gly
N-CBZ-Phe-Gly

Sigma C 3147   FW 356.4 $C_{19}H_{20}N_2O_5$

Phe-Gly

Sigma P 3376   FW 222.2 $C_{11}H_{14}N_2O_3$

Phe-Gly
CBZ-Phe-Gly

USBio C2099-42   MW 356.4 $C_{19}H_{20}N_2O_5$ ≥99%

Phe-Gly Amide Hydrochloride

Bachem G-2885.0250   MW 221.26 $C_{11}H_{15}N_3O_2$ Store at -15°C

Phe-Gly-Arg
N-Methylsulfonyl-D-Phe-Gly-Arg-pNA

Synonyms: Chromozym-tPA

American Peptide 81-0-13   MW 576.5 $C_{24}H_{32}N_8O_7S$ Determination of t-PA (tissue plasminogen activator) in purified preparation & in cell culture supernatants in the microgram scale

Phe-Gly-Arg
MeSO₂-D-Phe-Gly-Arg-pNA

Synonyms: Factor VIIa Substrate

Neosystem SC1324   MW 498.52   Chromogenic substrate; Neuenschwander, PF et al, *Thrombosis & Haemostasis*, 70:970–977, 1993

Phe-Gly-Arg
CH₃SO₂-D-Phe-Gly-Arg-AMC·AcOH

Synonyms: Pefafluor tPA; Pefa-5954

Pentapharm 091-21, 091-06   MW 673.8   Excitation wavelength 360 nm, emission wavelength 460 nm | Highly sensitive fluorogenic peptide substrate for tissue-type plasminogen activator (tPA); k_{cat}:11.0 s⁻¹, K_M:0.14 *mM*

Phe-Gly-Gly
BOC-Phe-Gly-Gly

Bachem A-2985.0250   MW 379.41 $C_{18}H_{25}N_3O_6$ Store at -15°C

Phe-Gly-Gly
Z-Phe-Gly-Gly

Bachem C-2405.0001   MW 413.43 $C_{21}H_{23}N_3O_6$ Store at -15°C

Phe-Gly-Gly

Bachem H-4525.0250   MW 279.3 $C_{13}H_{17}N_3O_4$ Store at -15°C

Phe-Gly-Gly
FA-Phe-Gly-Gly

Synonyms: FAPGG

Bachem M-1400.0050   MW 399.4 $C_{20}H_{21}N_3O_6$ Store at -15°C

Phe-Gly-Gly
N-(3-(2-Furyl)Acryloyl)-L-Phe-Gly-Gly

Synonyms: Angiotensin Converting Enzyme Substrate

Fluka 48176   MW 399.4 $C_{20}H_{21}N_3O_6$ ≥99% (HPLC); ≥90% peptide content; ≤1% water | Substrate for the continuous spectrophotometric assay of angiotensin converting enzyme; Holmquist, B et al, *Anal Biochem*, 95:540 1979

Phe-Gly-Gly

ICN 102632   MW 279.3 $C_{13}H_{17}N_3O_4$ Crystalline

Phe-Gly-Gly
N-(3-(2-Furyl)Acryloyl)-Phe-Gly-Gly

Synonyms: Angiotensin Converting Enzyme Substrate

ICN 152747   Useful in the continuous spectrophotometric assay of angiotensin-converting enzyme; Holmquist, B etal, *Anal Biochem*, 95:540, 1979

Sigma F 7131   FW 399.4 $C_{20}H_{21}N_3O_6$ Substrate for continuous spectrophotometric assay of angiotensin converting enzyme; Holmquist, B et al, *Anal Biochem*, 95:540, 1979; Maguire, GA & Price, CP, *Ann Clin Biochem*, 22:204, 1985; Harjanne, A, *Clin Chem*, 30:901, 1984

Phe-Gly-Gly

Sigma P 3501   FW 279.3 $C_{13}H_{17}N_3O_4$

Phe-Gly-Gly-Gly
BOC-Phe-Gly-Gly-Gly

Bachem A-2990.0250   Store at -15°C

Phe-Gly-Gly-Phe

Bachem H-4530.0050   MW 426.47 $C_{22}H_{26}N_4O_5$ Store at -15°C

ICN 153157   MW 426.5 $C_{22}H_{26}N_4O_5$

Sigma P 3626   FW 426.5 $C_{22}H_{26}N_4O_5$ ≥97% (HPLC) | Bioactive peptide

Phe-Gly-Gly-Phe-Thr-Gly-Ala-Arg-Lys-Ser-Ala-Arg-Lys
Phe-(®)-Gly-Gly-Phe-Thr-Gly-Ala-Arg-Lys-Ser-Ala-Arg-Lys Amide

Synonyms: Nociceptin (1-13), (Phe¹-(®)-Gly²)-; Orphanin FQ (1-13), (Phe¹-(®)-Gly²)-

Bachem H-4564.0500   MW 1367.62 $C_{61}H_{102}N_{22}O_{14}$ Store at -15°C

Phe-Gly-Gly-Phe-Thr-Gly-Ala-Arg-Lys-Ser-Ala-Arg-Lys
Phe-ψ(CH₂-NH)-Gly-Gly-Phe-Thr-Gly-Ala-Arg-Lys-Ser-Ala-Arg-Lys Amide

Synonyms: Nociceptin (1-13), (Phe¹-ψ(CH₂-NH)-Gly²)-

Neosystem SC1266   MW 1367.62   Bioactive neuropeptide; pseudopeptide analog of Nociceptin; claimed to be both antagonist & agonist of ORL-1 receptor in different *in vitro* assays; acts as an ORL-1 receptor agonist *in vivo*; Guerrini, R et al, *Br J Pharmacol*, 123:163-165, 1998; Calo, G et al, *J Med Chem*, 41:3360-3366, 1998; Butour, JL et al, *Eur J Pharmacol*, 349:R5-R6, 1998; Calo, G et al, *Br J Pharmacol*, 125:373-378, 1998; Grisel, JE et al, *Eur J Pharmacol*, 357:R1-R3, 1998

Phe-Gly-Gly-Phe-Thr-Gly-Ala-Arg-Lys-Ser-Ala-Arg-Lys Amide

Synonyms: Orphanin FQ (1-13); Nociceptin (1-13)

Bachem H-4072.0001   Store at -15°C

Neosystem SC1234   MW 1381.60   Bioactive neuropeptide; new potent nociceptin receptor agonist; the smallest peptide fragment of Nociceptin reported to date maintaining the same efficacy & potency as the natural peptide in mouse vas deferens & in guinea pig ileum bioassays; Calo, G et al, *Eur J Pharmacol*, 311:R3-R5, 1996; Guerrini, R et al, *J Med Chem*, 40:1789-1793, 1997; Calo, G et al, *Can J Physiol Pharmacol*, 75:713-718, 1997

Phe-Gly-Gly-Phe-Thr-Gly-Ala-Arg-Lys-Ser-Ala-Arg-Lys-Leu-Ala-Asn-Gln

Synonyms: Nociceptin/Orphanin FQ; Opioid Receptor-Like I Peptide

American Peptide 32-0-10   MW 1809.1 $C_{79}H_{129}N_{27}O_{22}$ Putative endogenous agonist for ORL_1 receptors; acts as a transmitter in the brain by modulating nociceptive & locomotor behavior; Rainer, K et al, *Science*, 270:792, 1995

Phe-Gly-Gly-Phe-Thr-Gly-Ala-Arg-Lys-Ser-Ala-Arg-Lys-Leu-Ala-Asn-Gln
Phe-Gly-Gly-Phe-Thr-Gly-Ala-Arg-Lys-Ser-Ala-Arg-Lys-Leu(3,4,5-³H)-Ala-Asn-Gln

Synonyms: Nociceptin, (Leu³,⁴,⁵-³H)-

ARC ART-726   MW 1808.1   80-100 Ci/mmol; 2.96-3.7 TBq/mmol; In 0.1 *N* AcOH:EtOH (8:2) | Radiochemical

Phe-Gly-Gly-Phe-Thr-Gly-Ala-Arg-Lys-Ser-Ala-Arg-Lys-Leu-Ala-Asn-Gln

Synonyms: Nociceptin; Orphanin FQ; Opioid Receptor-Like I Receptor Agonist

Bachem H-3036.0001   MW 1809.06 $C_{79}H_{129}N_{27}O_{22}$ Store at -15°C

Neosystem SC985   MW 1809.06   Bioactive neuropeptide; an endogenous agonist of the ORL1 (Opioid Receptor-Like 1) receptor & may be endowed with pro-nociceptive properties; shares structural similarity with otherknown peptides, in particular Dynorphin A; Meunier, JC et al, *Nature*, 377:532-535, 1995; Reinscheid, RK et al, *Science*, 270:792-794, 1995; Saito, Y et al, *BBRC*, 217:539-545, 1995

Sigma N-184   MW 1809 (free base) $C_{79}H_{129}N_{27}O_{22}$ White powder; peptide content & salt form information are provided with each lot | Putative endogenous agonist for opioid receptor-like (ORL) receptors; Reinscheid et al, *Science*, 270:792, 1995; Meunier et al, *Nature*, 377:532, 1995; Julius, *Nature*, 377:476, 1995

Sigma O 4011   FW 1809.1   ≥97% (HPLC); peptide content:~70% | Bioactive peptide; Reinsheid, RK et al, *Science*, 270:792, 1995; Meunier, J et al, *Nature*, 377:532, 1995

Peptides International PNO-4318-v   Human synthetic   MW 1809.1 $C_{79}H_{129}N_{27}O_{22}$ >99% (HPLC); lyophilized amorphous powder | Bioactive; rat, mouse, bovine, porcine; Meunier, J-C et al, *Nature*, 377:532, 1995; Reinscheid, RK et al, *Science*, 270:792, 1995

Phe-Gly-Gly-Phe-Thr-Gly-Ala-Arg-Lys-Ser-Ala-Arg-Lys-Tyr-Ala-Asn-Glu
Phe-Gly-Gly-Phe-Thr-Gly-Ala-Arg-Lys-Ser-Ala-Arg-Lys-Tyr(Diiodo)-Ala-Asn-Glu

Synonyms: Nociceptin, (Tyr¹⁴-¹²⁵I)-

ARC ARI-104   MW 1859.1   2200 Ci/mmol; 81.4 TBq/mmol; In 0.1 *N* AcOH:EtOH (8:2) | Radiochemical

Phe-Gly-Gly-Phe-Thr-Gly-Ala-Arg-Lys-Ser-Ala-Arg-Lys-Tyr-Ala-Asn-Glu
Phe-Gly-Gly-Phe-Thr-Gly-Ala-Arg-Lys-Ser-Ala-Arg-Lys-Tyr(3,5-³H)-Ala-Asn-Glu

Synonyms: Nociceptin, (Tyr¹⁴)(Tyr³,⁵-³H)-

ARC ART-725   MW 1808.1   25-50 Ci/mmol; 925 GBq-1.85 TBq/mmol; In 0.1 *N* AcOH:EtOH (8:2) | Radiochemical

Phe-Gly-Gly-Phe-Thr-Gly-Ala-Arg-Lys-Ser-Ala-Arg-Lys-Tyr-Ala-Asn-Gly
Phe(³H)-Gly-Gly-Phe-Thr-Gly-Ala-Arg-Lys-Ser-Ala-Arg-Lys-Tyr-Ala-Asn-Gly

Synonyms: Nociceptin, (*p*-³H-Phe¹)-

ARC ART-736   MW 1808.1   25-50 Ci/mmol; 925 GBq-1.85 TBq/mmol; In 0.1 *N* AcOH:EtOH (8:2) | Radiochemical

Phe-Gly-His-Phe-Phe-Ala
Phe-Gly-His-*p*-Nitro-Phe-Phe-Ala-OMe

Bachem H-4535.0025   Store at -15°C

Phe-Gly-His-Phe-Phe-Ala-Phe
Phe-Gly-His-*p*-Nitro-Phe-Phe-Ala-Phe-OMe

Bachem H-4540.0025   Store at -15°C

Phe-Gly-Leu
Suc-Phe-Gly-Leu-βNA

Bachem K-1510.0050   Store at -15°C

Phe-Gly-Leu
N-Suc-Phe-Gly-Leu-β-Naphthylamide

Synonyms: Luteinizing Hormone Releasing Hormone-Degrading Neutral Endopeptidase Substrate

ICN 156698

Sigma S 6768   FW 560.6 $C_{31}H_{36}N_4O_6$ Horsthemke, B & Bauer, K, *Biochemistry*, 19:2867, 1980

Phe-Gly-Leu-Met Amide

Synonyms: Substance P (8-11)

ICN 153034

Phe-Gly-Leu-Met Amide Acetate Salt

Synonyms: Substance P (8-11)

Sigma S 0897   FW 465.6 (free base) $C_{22}H_{35}N_5O_4S$ ≥97% (HPLC) | Bioactive peptide

Phe-Gly-Phe-Gly
Z-Phe-Gly-Phe-Gly

Bachem C-3020.0050   MW 560.61 $C_{30}H_{32}N_4O_7$ Store at -15°C

Phe-Gly-Phe-Gly

Bachem H-4550.0050   MW 426.47 $C_{22}H_{26}N_4O_5$ Store at -15°C

ICN 153158   MW 426.5 $C_{22}H_{26}N_4O_5$

Sigma P 3751   FW 426.5 $C_{22}H_{26}N_4O_5$ ≥97% (TLC) | Bioactive peptide

Phe-Gly-Phe-Leu-Pro-Ile-Tyr-Arg-Arg-Pro-Ala-Ser Amide

Synonyms: Granuliberin R

Alexis 157-005   MW 1422.7 $C_{69}H_{103}N_{19}O_{14}$ White lyophilized powder; soluble in water | Mast cell degranulating peptide originally isolated from the skin of the frog Rana *rugosa*; Nakajima, T & Yasuhara, T, *Chem Pharm Bull*, 25:2464, 1977

American Peptide 87-0-72   MW 1422.9 $C_{69}H_{103}N_{19}O_{14}$ Originally isolated from the skin of the frog *Rana rugosa*; a mast cell degranulating peptide; Nakajima, T et al, *Chem Phar Bull*, 25:2464, 1977

Bachem H-6800.0001   MW 1422.7 $C_{69}H_{103}N_{19}O_{14}$ Store at -15°C

ICN 153117   Mast cell degranulating peptide originally isolated from the skin of the frog *Rana rugosa*; Nakajima, T & T Yasuhara, *Chem Pharm Bull*, 25:2464, 1977

Sigma G 7897   *Rana rugosa* (frog)   FW 1422.7   ≥97% (HPLC) | Bioactive peptide; mast cell degranulating peptide found in the skin of the frog; Nakajima, T & Yasuhara, T, *Chem Pharm Bull,* 25:2464, 1977

Phe-His
Phe-His-OMe Dihydrochloride

Bachem G-2890.0250   MW 389.28 $C_{16}H_{20}N_4O_3 \cdot$ 2HCl Store at -15°C

Phe-His-Leu
Z-Phe-His-Leu

Bachem M-1305.0250   MW 549.63 $C_{29}H_{35}N_5O_6$ Store at -15°C

Phe-His-Leu
FA-Phe-His-Leu

Bachem M-1405.0050   Store at -15°C

Phe-His-Leu
CBZ-Phe-His-Leu

USBio C2099-43   MW 549.6 $C_{29}H_{35}N_5O_6$ ≥99%

Phe-His-Leu-Leu-Arg-Glu-Met-Leu-Glu-Met-Ala-Lys-Ala-Glu-Gln-Glu-Ala-Glu-Gln-Ala-Ala-Leu-Asn-Arg-Leu-Leu-Leu-Glu-Glu-Ala Amide

Synonyms: Corticotropin Releasing Factor Antagonist; Corticotropin Releasing Factor (12-41), α-Helical

American Peptide 34-0-16   MW 3497.1 $C_{153}H_{251}N_{43}O_{47}S_2$

Bachem H-3268.0500   Store at -15°C

Phe-His-Leu-Leu-Arg-Glu-Val-Leu-Glu-Nle-Ala-Arg-Ala-Glu-Gln-Leu-Ala-Gln-Glu-Ala-His-Lys-Asn-Arg-Lys-Leu-Nle-Glu-Ile-Ile
D-Phe-His-Leu-Leu-Arg-Glu-Val-Leu-Glu-Nle-Ala-Arg-Ala-Glu-Gln-Leu-Ala-Gln-Glu-Ala-His-Lys-Asn-Arg-Lys-Leu-Nle-Glu-Ile-Ile Amide

Synonyms: Corticotropin Releasing Factor Antagonist; Astressin

American Peptide 34-0-14   MW 3563.3 $C_{161}H_{251}N_{49}O_{42}$ Peptide β bond:Glu[30]-Lys[33] | New CRF antagonist (Cyclo(30/33)); injected into the CSF at low doses has an antagonistic action against CRF & stress-related alterations of gastrointestinal motor function, without an intrinsic effect in these *in vivo* systems; ~100 times more potent than α-helical CRF (9-41) at inhibiting ACTH secretion *in vitro*; Martinez, V et al, *J Pharmacol Exp Ther*, 280(2):754, 1997

Phe-His-Leu-Leu-Arg-Glu-Val-Leu-Glu-Nle-Ala-Arg-Ala-Glu-Gln-Leu-Ala-Gln-Glu-Ala-His-Lys-Asn-Arg-Lys-Leu-Met-Glu-Ile-Ile
D-Phe-His-Leu-Leu-Arg-Glu-Val-Leu-Glu-Nle-Ala-Arg-Ala-Glu-Gln-Leu-Ala-Gln-Glu-Ala-His-Lys-Asn-Arg-Lys-Leu-Met-Glu-Ile-Ile Amide

Synonyms: Corticotropin Releasing Factor (12-41), (D-Phe[12],Nle[21,38],Glu[30],Lys[33])-

American Peptide 34-0-13   Human, rat   MW 3581.3 $C_{161}H_{271}N_{49}O_{43}$ CRF receptor antagonist

Phe-His-Leu-Leu-Arg-Glu-Val-Leu-Glu-Nle-Ala-Arg-Ala-Glu-Gln-Leu-Ala-Gln-Gln-Ala-His-Ser-Asn-Arg-Lys-Leu-Nle-Glu-Ile-Ile
D-Phe-His-Leu-Leu-Arg-Glu-Val-Leu-Glu-Nle-Ala-Arg-Ala-Glu-Gln-Leu-Ala-Gln-Gln-Ala-His-Ser-Asn-Arg-Lys-α-Me-Leu-Nle-Glu-Ile-Ile Amide

Synonyms: Corticotropin Releasing Factor (12-41), (D-Phe[12],Nle[21,38],α-Me-Leu[37])-

Bachem H-3266.0500   Human, rat   Store at -15°C

Phe-His-Leu-Leu-Arg-Glu-Val-Leu-Glu-Nle-Ala-Arg-Ala-Glu-Gln-Leu-Ala-Gln-Cyclo(-Glu-Ala-His-Lys)-Asn-Arg-Lys-Leu-Nle-Glu-Ile-Ile
D-Phe-His-Leu-Leu-Arg-Glu-Val-Leu-Glu-Nle-Ala-Arg-Ala-Glu-Gln-Leu-Ala-Gln-Cyclo(-Glu-Ala-His-Lys)-Asn-Arg-Lys-Leu-Nle-Glu-Ile-Ile Amide

Synonyms: Astressin; Corticotropin Releasing Factor (12-41), (D-Phe[12],Nle[21,38],Glu[30],Lys[33])-

Bachem H-3422.0500   Human, rat   Store at -15°C

Phe-His-Leu-Leu-Arg-Glu-Val-Leu-Glu-Nle-Ala-Arg-Ala-Glu-Gln-Leu-Ala-Gln-Cyclo(Glu-Ala-His-Lys)-Asn-Arg-Lys-Leu-Nle-Glu-Ile-Ile
D-Phe-His-Leu-Leu-Arg-Glu-Val-Leu-Glu-Nle-Ala-Arg-Ala-Glu-Gln-Leu-Ala-Gln-Cyclo(-γ-Glu-Ala-His-ε-Lys)-Asn-Arg-Lys-Leu-Nle-Glu-Ile-Ile Amide

Synonyms: Astressin; Corticotropin Releasing Factor (12-41), (D-Phe[12],Nle[21,38],Glu[30],Lys[33])-

Sigma A 4933   Human, rat   FW 3563.2   ≥90% (HPLC) | Bioactive peptide; highly potent CRF antagonist; Gulyas, J et al, *Proc Natl Acad Sci USA*, 92:10575, 1995

Phe-His-Pro-Lys-Arg-Pro-Trp-Ile-Leu

Synonyms: Xenopsin Related Peptide I; Xenopsin Related Peptide II

American Peptide 62-0-63   MW 1193.6 $C_{60}H_{88}N_{16}O_{10}$ Carraway, RE et al, *Reg Peptides*, 22:303, 1988

Bachem H-9340.0005   MW 1193.46 $C_{60}H_{88}N_{16}O_{10}$ Store at -15°C

ICN 154462   MW 1193.6   Carraway, RE etal, *Regulatory Peptides*, 22:303, 1988

Phe-His-Trp-Ala-Val-Ala-His-Pro-Phe
Phe-His-Trp-Ala-Val-D-Ala-His-D-Pro-σ(CH₂NH)-Phe

Synonyms: Bombesin (6-14), (Deamino-Phe[6],His[7],D-Ala[11],D-Pro[13]-σ(CH₂NH)-Phe[14])-; Bombesin Receptor Antagonist

ICN 195677   MW 1081.3   ≥97%

Phe-His-Trp-Ala-Val-Ala-His-Pro-Phe
Deamino-Phe-His-Trp-Ala-Val-D-Ala-His-D-Pro-ψ(CH₂NH)-Phe Amide

Synonyms: Bombesin (6-14), (Deamino-Phe[6],His[7],D-Ala[11],D-Pro[13]-ψ(CH₂NH)-Phe[14])-

Sigma B 7033   FW 1081.3   ≥97% (HPLC) | Gastrointestinal peptide; bioactive peptide; antagonist that inhibits bombesin evoked release of gastrointestinal hormones *in vivo* & *in vitro*; Singh, P et al, *Regulatory Peptides*, 40:75, 1992

Phe-His-Val-Glu-Thr-Pro-Glu-Glu-Arg-Glu-Glu-Trp-Thr-Cys

Synonyms: Protein Kinase B Blocking Peptide (88-100)

Calbiochem 530312   Human synthetic   MW 1791.9 $C_{78}H_{110}N_{20}O_{27}S$ Lyophilized | Based on the α-isoform of human PKB kinase (88-100) with a Cys residue added; this peptide coupled to KLH was used as the immunogen for the production of Anti-PKB; for use in immunoabsorption for immunoprecipitation, immunocytochemistry, Western blotting & dot blots; Hardie, F & Hanks, S, *The Protein Kinase Facts Book, Protein-Serine Kinases*, Academic Press, San Diego, CA, p. 418, 1995; Knighton, DR et al, *Science*, 253:414, 1991

Phe-Ile
Z-Phe-Ile

Bachem C-2410.0001   MW 412.49 $C_{23}H_{28}N_2O_5$ Store at -15°C

Phe-Ile

Bachem G-2895.0250   MW 278.35 $C_{15}H_{22}N_2O_3$ Store at -15°C

Phe-Ile-Asn-Met-Trp-Gln-Glu-Val-Gly-Lys-Ala-Met-Tyr-Ala-Pro-Pro-Ile-Ser

Synonyms: GP120 (428-445); HTLV I Peptide

Neosystem SC247   MW 2082.46   HTLV I peptide from subtype GP160

Phe-Ile-Asp-Pro-Glu-Leu-Gln-Arg-Ser-Trp-Glu-Glu-Lys-Glu-Gly-Glu-Gly-Val-Leu-Met-Pro-Glu

Synonyms: Prepro-Thyrotropin Releasing Hormone (178-199); Corticotropin Release Inhibiting Factor

American Peptide 58-0-95   MW 2618.8 $C_{116}H_{176}N_{28}O_{39}S$ Inhibits basal & CRF-stimulated ACTH synthesis & secretion; fulfills the criteria for a physiological corticotropin release-inhibiting factor (CRIF)

Bachem H-3598.0500   MW 2618.9 $C_{116}H_{176}N_{28}O_{39}S$ Store at -15°C

Sigma P 7842   FW 2618.9   Bioactive peptide; inhibits basal & CRF-stimulated ACTH synthesis & secretion in cultured primary anterior pituitary cells; Redei, E et al, *Endocrinology*, 136:3557, 1995

Biogenesis 8940-1519   Synthetic   MW 2618.8   Lyophilized, consisting of 82.0% peptide material, TFA counter ions and residual H_2O | To avoid Met oxidation, it is best to dissolve the peptide under a nitrogen stream

Phe-Leu

Bachem G-2900.0001   MW 278.35 $C_{15}H_{22}N_2O_3$ Store at -15°C

Phe-Leu
Phe-D-Leu Amide Hydrochloride

Bachem G-3780.0250   MW 313.83 $C_{15}H_{23}N_3O_2 \cdot HCl$ Store at -15°C

Phe-Leu
Z-Phe-Leu

Bachem M-1310.0001   MW 412.49 $C_{23}H_{28}N_2O_5$ Store at -15°C

Phe-Leu
Carboxymethyl-Phe-Leu

Bachem N-1190.0025   MW 336.39 $C_{17}H_{24}N_2O_5$ Store at -15°C

Phe-Leu
N-Carboxymethyl-Phe-Leu

Synonyms: Enkephalinase Inhibitor; Angiotensin Converting Enzyme Inhibitor

ICN 152831   MW 336.4 $C_{17}H_{24}N_2O_5$ A highly potent inhibitor; Fournier-Zaluski, MC etal, *Life Sci*, 31:2947, 1982; Fournie-Zaluski, MC etal, *J Med Chem*, 26:60, 1983

Phe-Leu

ICN 156131   MW 278.4 $C_{15}H_{22}N_2O_3$ Crystalline

Phe-Leu
N-CBZ-Phe-Leu

Sigma C 1141   FW 412.5 $C_{23}H_{28}N_2O_5$

Phe-Leu
N-Carboxymethyl-Phe-Leu

Synonyms: Enkephalin; Enkephalinase Inhibitor

Sigma C 7030   FW 336.4 $C_{17}H_{24}N_2O_5$ ≥97% (HPLC) | Bioactive peptide; highly potent inhibitor of enkephalinase; Fournie-Zaluski, MC et al, *Life Sci*, 31:2947, 1982

Phe-Leu

Sigma P 3876   FW 278.4 $C_{15}H_{22}N_2O_3$ Crystalline

Phe-Leu
CBZ-Phe-Leu

USBio C2099-44   MW 412.5 $C_{23}H_{28}N_2O_5$ ≥99%

Phe-Leu Amide Hydrobromide

Bachem G-2905.0250   MW 358.28 $C_{15}H_{23}N_3O_2 \cdot HBr$ Store at -15°C

ICN 156132   MW 358.3 $C_{15}H_{23}N_3O_2 \cdot HBr$ Crystalline

ICN 156133   MW 313.8 $C_{15}H_{23}N_3O_2 \cdot HCl$ Crystalline

Sigma P 4001   FW 358.3 $C_{15}H_{23}N_3O_2 \cdot HBr$

Phe-Leu Amide Hydrochloride

Bachem G-2910.0250   MW 313.83 $C_{15}H_{23}N_3O_2 \cdot HCl$ Store at -15°C

Sigma P 1125   FW 313.8 $C_{15}H_{23}N_3O_2 \cdot HCl$

Phe-Leu-Ala
Z-Phe-Leu-Ala

Bachem C-3280.0250   MW 483.57 $C_{26}H_{33}N_3O_6$ Store at -15°C

Phe-Leu-Arg
Bz-Phe-Leu-Arg

Synonyms: Enkephalinase Substrate

American Peptide 32-0-80   MW 538.7 $C_{28}H_{38}N_6O_5$

Phe-Leu-Arg-Phe Amide

Synonyms: Molluscan Cardioexcitatory Neuropeptide Analog

Bachem H-2985.0025   MW 580.73 $C_{30}H_{44}N_8O_4$ Store at -15°C

ICN 153159   MW 580.7 $C_{30}H_{44}N_8O_4$

Sigma P 5652   FW 580.7 $C_{30}H_{44}N_8O_4$ ≥97% (HPLC) | Bioactive peptide

Phe-Leu-Asp-Ile-Ile-Trp
N-Ac-β-Phenyl-D-Phe-Leu-Asp-Ile-Ile-Trp

Synonyms: Endothelin Receptor Antagonist PD 142893

Sigma P 2959   FW 924.1 $C_{50}H_{65}N_7O_{10}$ ≥97% (HPLC) |
Bioactive peptide; non-selective endothelin receptor antagonist;
contains the His-Leu-Asp-Ile-Ile-Trp sequence found in all
endothelins with the His replaced by a bulky aromatic moiety

Phe-Leu-Glu-Glu-Ile

Synonyms: Prothrombin Precursor (5-9)

Bachem M-1695.0005   MW 649.74 $C_{31}H_{47}N_5O_{10}$ Store at
-15°C

ICN 153160   Rat   MW 649.7 $C_{31}H_{47}N_5O_{10}$ Useful substrate
for studies of the γ-carboxylation of Glu by the Vit K-dependent
carboxylation system

Phe-Leu-Glu-Glu-Ile Ammonium Salt

Synonyms: Prothrombin Precursor (5-9)

Sigma P 9396   Rat   FW 649.7 (free acid) $C_{31}H_{47}N_5O_{10}$
≥97% (HPLC) | Bioactive peptide; reported to be useful as an
artificial substrate for Vitamin K-dependent carboxylation studies;
Houser, RM et al, *FEBS Lett*, 75:226, 1977

Phe-Leu-Glu-Glu-Leu

Synonyms: Prothrombin Precursor; Vitamin K-Dependent
Carboxylase Substrate

Bachem M-1700.0005   MW 649.74 $C_{31}H_{47}N_5O_{10}$ Store at
-15°C

ICN 153161   MW 649.7 $C_{31}H_{47}N_5O_{10}$ Useful substrate for
studies of the γ-carboxylation of Glu by the Vit K-dependent
carboxylation system

Sigma P 5523   FW 649.7 (free acid) $C_{31}H_{47}N_5O_{10}$ ≥97%
(HPLC) | Bioactive peptide; Suttie, JW et al, *Arch Biochem
Biophys*, 202:515, 1980; Rich, DH et al, *Int J Pept Prot Res*, 18:41,
1981

Phe-Leu-Glu-Glu-Val

Synonyms: Vitamin K-Dependent Carboxylase Substrate;
Prothrombin Precursor; Vitamin K-Dependent Carboxylation
Substrate

American Peptide 87-0-58   MW 635.8 $C_{30}H_{45}N_5O_{10}$ This
peptide is a substrate for the microsomal vitamin K-dependent
carboxylase, which catalyzes the formation of peptide-bound γ-
carboxyglutamyl (Gla) residues

Bachem M-1705.0005   Store at -15°C

ICN 153162   MW 635.7 $C_{30}H_{45}N_5O_{10}$ Useful substrate for
studies of the γ-carboxylation of Glu by the Vit K-dependent
carboxylation system

Sigma P 5398   FW 635.7 (free acid) $C_{30}H_{45}N_5O_{10}$
Ammonium salt; ≥97% (HPLC) | Bioactive peptide; Houser, RM et
al, *FEBS Lett*, 75:226, 1977; Suttie, JW et al, *Biochem Biophys Res
Commun*, 86:500, 1979

Phe-Leu-Gly-Gly-Leu-Ile-Lys-Ile-Val-Pro-Ala-Met-Ile-Cys-Ala-Val-Thr-Lys-Lys-Cys

Synonyms: Ranalexin

Bachem H-1612.0500   MW 2103.73 $C_{97}H_{167}N_{23}O_{22}S_3$ Store
at -15°C

Sigma R 1770   FW 2103.7   ≥97% (HPLC) | Disulfide bonds:
14-20; bioactive peptide; novel amphibian antimicrobial peptide;
Clark, DP et al, *J Biol Chem*, 269:10849, 1994

Phe-Leu-Gly-Phe-Leu-Gly-Ala-Ala-Gly-Ser-Thr

Synonyms: GP41 (524-534); HTLV I Peptide

Neosystem SC239   MW 1040.18   HTLV I peptide from
subtype GP41

Phe-Leu-His-Gln-Glu-Arg-Met-Asp-Val-Cys-Glu-Thr-His-Leu-His-Trp-His-Thr-Val-Ala-Lys

Synonyms: Amyloid β/A4 Precursor Protein 770 (135-155)

Bachem H-3726.0500   MW 2618 $C_{116}H_{173}N_{35}O_{31}S_2$ Store at
-15°C

Phe-Leu-Leu-Arg-Asn
DL-Isoser-Phe-Leu-Leu-Arg-Asn

Bachem H-1944.0001   MW 748.88 $C_{34}H_{56}N_{10}O_9$ Store at
-15°C

Phe-Leu-Leu-Arg-Asn

Synonyms: Thrombin Receptor (43-47)

Bachem H-8325.0005   MW 661.8 $C_{31}H_{51}N_9O_7$ Store at
-15°C

Neosystem SC853   Human   MW 661.90   Bioactive; Vu, T et
al, *Cell*, 64:1057-1068, 1991; Vassallo, RR Jr et al, *JBC*, 267:6081-
6085, 1992

Phe-Leu-Phe
Suc-Phe-Leu-Phe-4MβNA

Bachem J-1320.0050   Store at -15°C

Phe-Leu-Phe
Phe-Leu-Phe-4MβNA Trifluoroacetate Salt

Bachem J-1350.0050   Store at -15°C

Phe-Leu-Phe
Suc-Phe-Leu-Phe-pNA

Bachem L-1425.0050   Store at -15°C

Phe-Leu-Phe
Suc-Phe-Leu-Phe-SBzl

Bachem M-1740.0050   MW 631.79 $C_{35}H_{41}N_3O_6S$ Store at
-15°C

Phe-Leu-Phe
N-Suc-Phe-Leu-Phe Thiobenzyl Ester

Synonyms: Cathepsin G Substrate

ICN 156699   Harper, JW etal, *Anal Biochem*, 118:382, 1981

Sigma S 6643   FW 631.8 $C_{35}H_{41}N_3O_6S$ Harper, JW et al,
Anal Biochem, 118:382, 1981

Phe-Leu-Phe-Gln-Pro-Gln-Arg-Phe Amide

Synonyms: Neuropeptide FF Morphine Modulating Neuropeptide;
Morphine Modulating Neuropeptide; NPFF; Neuropeptide FF; Opioid
Peptide; F-8-F-NH₂

American Peptide 32-0-64   MW 1081.3 $C_{54}H_{76}N_{14}O_{10}$
Increases blood pressure & heart rate after I.V. administration; the
effects of NPFF are caused by the activation of peripheral NPFF
receptors that are functionally coupled to catecholamine receptors;
Yang, H et al, *PNAS*, 82:7757, 1985; Allard, M et al, *J Pharmacol
Exp Ther*; 274(1):577, 1995

Bachem H-5655.0001   MW 1081.29 $C_{54}H_{76}N_{14}O_{10}$ Store at
-15°C

ICN 152958   Yang, HYT etal, *PNAS*, 82:7757, 1985

Sigma P 3293   FW 1081.3   ≥95% (HPLC) | Bioactive
peptide; endogenous peptide found in periaqueductal grey & in
dorsal spinal cord; attenuates analgesic effects of morphine; Yang,
HYT et al, *Proc Natl Acad Sci USA*, 82:7757, 1985

Neosystem SC946   Bovine central nervous system   MW 1081.28   Bioactive; brain neuropeptide; modulates the action of morphine F-8-F amide; immunoreactivity is present in rat central nervous system & at particular high concentrations in the dorsal spinal cord; F-8-F amide receptors in the human spinal cord & lower medulla oblongata are mainly concentrated within spinal areas; implicated in the analgesic action of opiates; Yang, H-YT et al, *PNAS* USA, 82:7757-7761, 1985; Majane, EA & Yang, H-YT, *Peptides*, 8:657-662, 1987; Kivipelto, L et al, *J Comp Neurol*, 286:269, 1989; Allard, M et al, *Brain Res*, 500:169, 1989; Allard, M et al, *Neuroscience*, 40:81, 1990; Lake, JR et al, *Neurosci Lett* 132:29, 1991; Allard, M et al, *Brain Res*, 633:127-132, 1994; Gicquel, S et al, *J Med Chem*, 37:3477-3481, 1994; Marco, N et al, *Neuroscience*, 64:1035-1044, 1995; Aarnisalo, AA & Panula, P, *Neuroscience*, 65:175-192, 1995

Phe-Leu-Phe-Leu-Phe
BOC-Phe-Leu-Phe-Leu-Phe

Bachem A-2200.0025   MW 785.98 $C_{44}H_{59}N_5O_8$ Store at -15°C

Phe-Leu-Phe-Leu-Phe
BOC-Phe-D-Leu-Phe-D-Leu-Phe

Bachem A-2205.0025   MW 785.98 $C_{44}H_{59}N_5O_8$ Store at -15°C

Phe-Leu-Phe-Leu-Phe
N-t-BOC-Phe-Leu-Phe-Leu-Phe

Synonyms: Chemotactic Peptide Antagonist

ICN 152760   MW 786 $C_{44}H_{59}N_5O_8$ Competitive antagonist of the binding of N-For-Met-Leu-Phe to neutrophils; O'Dea, RF et al, *Nature*, 272:462, 1978; Naccache, P et al, *Science*, 203:461, 1979; Freer, RJ et al, *Peptides*, 1:289, 1980

Phe-Leu-Phe-Leu-Phe
N-t-BOC-Phe-D-Leu-Phe-D-Leu-Phe

Synonyms: Chemotactic Peptide Inhibitor

ICN 152800   MW 786 $C_{44}H_{59}N_5O_8$ Specifically inhibits the chemotactic response of neutrophils to peptide agonists; completely blocks the specific protein carboxymethylation promoted by N-For-Met-Leu-Phe; Aswanikumar, S et al, *PNAS*, 73:2439, 1976; Aswanikumar, S et al, *Nature*, 272:462, 1978; Naccache, P et al, *Science*, 203:461, 1979

Sigma B 2386   FW 786.0 $C_{44}H_{59}N_5O_8$ ≥90% (HPLC) | Bioactive peptide; chemotactic peptide inhibitor; Naccache, P et al, *Science*, 203:461, 1979

Phe-Leu-Pro-His-Val-Phe-Ala-Glu-Leu-Ser-Asp-Arg-Lys-Gly-Phe-Val-Gln-Gly-Asn-Gly-Ala-Val-Glu-Ala-Leu-His-Asp-Phe-Tyr-Pro-Asp-Trp-Met-Asp-Phe Amide

Synonyms: Gastrin

American Peptide 46-8-10   Chicken   MW 4055.6 $C_{190}H_{265}N_{47}O_{51}S$

Phe-Leu-Pro-Leu-Ile-Leu-Gly-Lys-Leu-Val-Lys-Gly-Leu-Leu Amide

Synonyms: Mast Cell Degranulating Peptide HR-2

American Peptide 41-9-20   MW 1523.0 $C_{77}H_{135}N_{17}O_{14}$

Bachem H-8390.0001   MW 1523.03 $C_{77}H_{135}N_{17}O_{14}$ Store at -15°C

ICN 159962   MW 1523   97%

Sigma M 8780   *Vespa orientalis* (giant hornet)   FW 1523.0 ≥97% (HPLC) | Bioactive peptide; similar biological effects as MCD peptide from bee venom; Tuichibaev, MU, *Biochem USSR*, 53:219, 1988; CA, 108:16299e, 1988

Phe-Leu-Pro-Val-Leu-Ala-Gly-Ile-Ala-Ala-Lys-Val-Val-Pro-Ala-Leu-Phe-Cys-Lys-Ile-Thr-Lys-Lys-Cys

Synonyms: Brevinin-1

Bachem H-1584.0500   MW 2529.24 $C_{121}H_{202}N_{28}O_{26}S_2$ Store at -15°C

Phe-Leu-Ser-Tyr-Lys

Synonyms: Phospholipase A_2 (70-74)

Bachem H-3968.0005   Store at -15°C

Phe-Leu-Trp-Gly-Pro-Arg-Ala-Leu-Val

Synonyms: Melanoma Associated Antigen 3 (271-279)

Bachem H-3714.0001   Human   Store at -15°C

Phe-Leu-Trp-Lys-Asp-Leu-Gln-Arg-Val-Arg-Gly-Asp-Leu-Gly-Ala-Ala-Leu-Asp-Ser-Trp-Ile-Thr

Synonyms: Prepro-Thyrotropin Releasing Hormone (53-74)

American Peptide 58-0-92   MW 2560.8 $C_{118}H_{182}N_{32}O_{32}$

Phe-Leu-Tyr-Cys-Lys-Met-Asn-Trp-Phe-Leu-Asn-Trp-Val-Glu-Asp-Arg-Asp-Val-Thr-Thr

Synonyms: GP140 (401-420); SIVmac251 Peptide

Neosystem SC748   MW 2580.95   SIVmac251 peptide from subtype GP140; due to the presence of a Cys this peptide can contain the dimeric form

Phe-Lys
Ac-Phe-Lys

Bachem G-4090.0250   MW 335.4 $C_{17}H_{25}N_3O_4$ Store at -15°C

Phe-Lys
Isovaleryl-Phe-Lys-pNA Hydrochloride

Bachem L-1960.0050   Store at -15°C

Phe-Lys
FA-Phe-Lys Hydrochloride

Bachem M-2275.0050   Store at -15°C

Phe-Lys
Z-Phe-Lys-2,4,6-Trimethylbenzoyloxy-Methylketone

Bachem N-1400.0005   MW 587.72 $C_{34}H_{41}N_3O_6$ Store at -15°C

Phe-Lys
N-Ac-Phe-Lys

Sigma A 0305   FW 335.4 $C_{17}H_{25}N_3O_4$ 23 times sweeter than sucrose; Nosho, Y & Ohfuji, T, *Chem Eng News*, 68(2):25, 1990

Phe-Lys-Ala-Ala-Ala-Leu-Ala-Arg
Phe-Lys-Ala-β-Cyclohexyl-Ala-β-Cyclohexyl-Ala-Leu-D-Ala-Arg

Synonyms: C5a Inhibitory Sequence

Bachem H-8135.0005   MW 1011.32 $C_{51}H_{86}N_{12}O_9$ Store at -15°C

ICN 159991   MW 1011.3   97%

Sigma P 9067   FW 1011.3   ≥97% (HPLC); peptide content:~65% | Bioactive peptide

Phe-Lys-Leu-Val-Pro-Val-Glu-Pro-Glu-Lys-Ile-Glu-Glu

Synonyms: NEF MN (145-157); HIV I Peptide

Neosystem SC659   MW 1556.81   NEF peptide from HIV-1 subtype MN

Phe-Lys-Lys-Ser-Phe-Lys-Leu Amide

Synonyms: Protein Kinase C Substrate; Myristoylated Alanine Rich C-Kinase Protein (159-165)

American Peptide 86-0-23   MW 896.2 $C_{45}H_{73}N_{11}O_8$
Mehrani, H et al, *Neurochem Int*, 31:139, 1997

Bachem H-1638.0001   MW 896.14 $C_{45}H_{73}N_{11}O_8$ Store at -15°C

Alexis 162-005   Bovine brain   MW 896.1 $C_{45}H_{73}N_{11}O_8$
≥97%; white lyophilized powder; soluble in water | Sufficiently selective for protein kinase C to permit specific assay of this enzyme in crude cell extracts & permeabilized cells; Spencker, T et al, *Liebigs*, 1993:237; Blackshear, PJ, *J Biol Chem*, 268:1501, 1993

Phe-Lys-Pro-Ala-Pro-Ala-Thr-Asn-Thr-Gln-Asn-Tyr
Acetate Salt

Synonyms: Glutamate Receptor III (838-849)

Biogenesis 4670-5159   Synthetic   Semi-pure; lyophilized

Phe-Lys-Val-Asp-Glu-Glu-Phe-Gln-Gly-Pro-Ile-Val-Ser-Gln-Asn-Arg-Arg-Tyr-Phe-Leu-Phe-Arg-Pro-Arg-Asn Amide

S Neuromedin U-25synonyms:

ICN 153155   MW 3142.6   Minamino, N etal, *BBRC*, 130:1078, 1985

Bachem H-5510.0500   Porcine   MW 3142.57 $C_{144}H_{217}N_{43}O_{37}$ Store at -15°C

American Peptide 62-0-53   Porcine spinal cord   MW 3142.6 $C_{144}H_{217}N_{43}O_{37}$ Potent stimulator of rat uterus smooth muscle; influences blood pressure in rats & dogs; Minamino, N et al, *BBRC*, 130:1078, 1985

Sigma N 8138   Porcine spinal cord   FW 3142.6   ≥97% (HPLC) | Bioactive peptide; exhibits potent contractile hypertensive effect & uterine smooth muscle stimulation in rat; Minamino, N et al, *Biochem Biophys Res Commun*, 130:1078, 1985

Phe-Met
Z-Phe-Met

Bachem C-2415.0001   MW 430.53 $C_{22}H_{26}N_2O_5S$ Store at -15°C

Phe-Met
For-Phe-Met

Bachem G-1875.0250   MW 324.4 $C_{15}H_{20}N_2O_4S$ Store at -15°C

Phe-Met

Bachem G-2915.0250   MW 296.39 $C_{14}H_{20}N_2O_3S$ Store at -15°C

Phe-Met
N-(2(S)-(2(R)-Amino-3-Mercaptopropylamino)-3-Methylbutyl)-Phe-Met Trifluoroacetate Salt

Synonyms: Farnesyltransferase Inhibitor I; B581

Calbiochem 344510   MW 470.7 $C_{22}H_{38}N_4O_3S_2$ ≥95% (HPLC); white solid; soluble in DMSO & water | Potent, cell-permeable inhibitor of farnesyltransferase (FTase) that is 37-fold more active against FTase (IC_{50}=21 nM *in vitro*) than against geranylgeranyltransferase (GGTase; IC_{50}=790 nM); Very resistant to proteolysis; Garcia, AM et al, *J Biol Chem*, 268:18415, 1993

Phe-Met
N-CBZ-Phe-Met

Sigma C 9146   FW 430.5 $C_{22}H_{26}N_2O_5S$

Phe-Met

Sigma P 7647   FW 296.4 $C_{14}H_{20}N_2O_3S$

Phe-Met-Arg-Phe

American Peptide 60-0-32   MW 599.8 $C_{29}H_{41}N_7O_5S$

Phe-Met-Arg-Phe
Phe-D-Met-Arg-Phe Amide

Synonyms: Molluscan Neuropeptide Cardioexcitatory Analog, (D-Met[2])-

American Peptide 60-0-33   MW 598.8 $C_{29}H_{42}N_8O_4S$

Phe-Met-Arg-Phe
Phe-Met-Arg-D-Phe Amide

Synonyms: Molluscan Cardioexcitatory Neuropeptide Analog, (D-Phe[4])-

American Peptide 60-0-34   MW 598.8 $C_{29}H_{42}N_8O_4S$

Phe-Met-Arg-Phe
D-Phe-Met-Arg-Phe Amide

Synonyms: Molluscan Neuropeptide Cardioexcitatory Analog, (D-Phe[1])-

American Peptide 60-0-35   MW 598.8 $C_{29}H_{42}N_8O_4S$

Phe-Met-Arg-Phe
Phe-Met-Arg-D-Phe Amide

Synonyms: FMRF Amide, (D-Phe[4])-

Bachem H-3342.0005   MW 598.77 $C_{29}H_{42}N_8O_4S$ Store at -15°C

Phe-Met-Arg-Phe
Phe-D-Met-Arg-Phe Amide

Synonyms: FMRF Amide, (D-Met[2])-

Bachem H-3344.0005   MW 598.77 $C_{29}H_{42}N_8O_4S$ Store at -15°C

Phe-Met-Arg-Phe
D-Phe-Met-Arg-Phe Amide

Synonyms: FMRF Amide, (D-Phe[1])-

Bachem H-3346.0005   MW 598.77 $C_{29}H_{42}N_8O_4S$ Store at -15°C

Phe-Met-Arg-Phe
Phe-Met-D-Arg-Phe Amide

Synonyms: FMRF Amide, (D-Arg[3])-; Molluscan Cardioexcitatory Neuropeptide Analog, (D-Arg[3])-

Bachem H-7185.0005   MW 598.77 $C_{29}H_{42}N_8O_4S$ Store at -15°C

ICN 153164   MW 598.8 $C_{29}H_{42}N_8O_4S$ 97% | Mues, G etal, *Life Sci*, 31:2555, 1982

Phe-Met-Arg-Phe
Phe-Met-Arg-D-Phe Amide

Synonyms: Molluscan Cardioexcitatory Neuropeptide Analog, (D-Phe[4])-

ICN 159992   MW 598.8 $C_{29}H_{42}N_8O_4S$ 97% | Mues, G etal, *Life Sci*, 31:2555, 1982

Phe-Met-Arg-Phe
Phe-D-Met-Arg-Phe Amide

Synonyms: Molluscan Cardioexcitatory Neuropeptide Analog, (D-Met[2])-

ICN 159993   MW 598.8 $C_{29}H_{42}N_8O_4S$ 97% | Mues, G etal, *Life Sci*, 31:2555, 1982

Phe-Met-Arg-Phe
D-Phe-Met-Arg-Phe Amide

Synonyms: Molluscan Cardioexcitatory Neuropeptide Analog, (D-Phe[1])-

ICN 159994   MW 598.8 $C_{29}H_{42}N_8O_4S$ 97% | Mues, G etal, *Life Sci*, 31:2555, 1982

Sigma P 6535   FW 598.8 $C_{29}H_{42}N_8O_4S$ ≥97% (HPLC); peptide content:~65% | Bioactive peptide; Mues, G et al, *Life Sci*, 31:2555, 1982

Phe-Met-Arg-Phe
Phe-D-Met-Arg-Phe Amide

Synonyms: Molluscan Cardioexcitatory Neuropeptide Analog, (D-Met[2])-

Sigma P 6660   FW 598.8 $C_{29}H_{42}N_8O_4S$ ≥97% (HPLC); peptide content:~65% | Bioactive peptide; Mues, G et al, *Life Sci*, 31:2555, 1982

Phe-Met-Arg-Phe
Phe-Met-Arg-D-Phe Amide

Synonyms: Molluscan Cardioexcitatory Neuropeptide Analog, (D-Phe[4])-

Sigma P 6785   FW 598.8 $C_{29}H_{42}N_8O_4S$ ≥97% (HPLC); peptide content:~65% | Bioactive peptide; Mues, G et al, *Life Sci*, 31:2555, 1982

Phe-Met-Arg-Phe
Phe-Met-D-Arg-Phe Amide

Synonyms: Molluscan Cardioexcitatory Neuropeptide Analog, (D-Arg[3])-

Sigma P 6910   FW 598.8 $C_{29}H_{42}N_8O_4S$ ≥97% (HPLC); peptide content:~70% | Bioactive peptide; Mues, G et al, *Life Sci*, 31:2555, 1982

Phe-Met-Arg-Phe Amide

Synonyms: Molluscan Cardioexcitatory Neuropeptide Analog; FMRF Amide; Molluscan Cardioexcitatory Neuropeptide; Molluscan Cardioexcitatory Peptide

American Peptide 60-9-17   MW 598.8 $C_{29}H_{42}N_8O_4S$ Lyophilized solid | Appears to localize with neuropeptide Y in some regions of the brain; inhibits Na^+-Ca^{2+} exchange in cardiac sarcolemmal vesicles; Price, DA & Greenberg, MJ, *Science*, 197:670, 1977

Bachem H-2975.0005   MW 598.77 $C_{29}H_{42}N_8O_4S$ Store at -15°C

Calbiochem 05-22-5850   MW 598.7 $C_{29}H_{42}N_8O_4S$ >95% (HPLC); lyophilized solid; soluble in 5% acetic acid | Appears to localize with neuropeptide Y in some regions of the brain; inhibits Na^+-Ca^{2+} exchange in cardiac sarcolemmal vesicles; Kits, KS et al, *J Gen Physiol*, 110:611, 1997; Wright, DE & Demski, LS, *Brain Behav Evol*, 47:267, 1996; *Merck Index*, 12:3528

ICN 151703   Price, *Science*, 197:670, 1977

ICN 153163   MW 598.8 $C_{29}H_{42}N_8O_4S$ Price, D & MJ Greenberg, *Science*, 197:670, 1977; Greenberg, MJ etal, *Neuropeptides*, 1:309, 1981

Neosystem SC117   MW 598.76   Bioactive neuropeptide; Price, DA & Greenberg, MJ, *Science*, 197:670, 1977

Sigma P 4898   FW 598.8 $C_{29}H_{42}N_8O_4S$ ≥97% (HPLC); peptide content:~65% | Bioactive peptide; Price, D & Greenberg, MJ, *Science*, 197:670, 1977; Greenberg, MJ et al, *Neuropeptides*, 1:309, 1981

Peptides International PFM-4142-v   Synthetic   MW 598.77 $C_{29}H_{42}N_8O_4S$ >95% (HPLC); lyophilized amorphous powder | Bioactive; Price, DA & MJ Greenberg, *Science*, 197:670, 1977

Phe-Met-Arg-Phe Amide 1½AcOH Dihydrate

Synonyms: Molluscan Cardioexcitatory Neuropeptide; FMRF Amide

Peptides International PFM-4142   Synthetic   MW 724.88 $C_{29}H_{42}N_8O_4S \cdot 1\frac{1}{2}CH_3COOH \cdot 2H_2O$ >95% (HPLC); lyophilized amorphous powder; bulk | Bioactive; Price, DA & MJ Greenberg, *Science*, 197:670, 1977

Phe-Met-His-Asn-Leu-Gly-Lys-His-Leu-Ser-Ser-Met-Glu-Arg-Val-Glu-Trp-Leu-Arg-Lys-Lys-Leu-Gln-Asp-Val-His-Asn-Tyr Amide

Synonyms: Parathyroid Hormone (7-34), (Tyr[34])-; Parathyroid Hormone (7-34), (Tyr[34])-; Parathyroid Hormone Antagonist

American Peptide 22-3-25   Bovine   MW 3496.1 $C_{156}H_{244}N_{48}O_{40}S_2$ Horiuchi, N et al, *Science*, 220:1053, 1983

Bachem N-1110.0500   Bovine   MW 3496.08 $C_{156}H_{244}N_{48}O_{40}S_2$ Store at -15°C

Peptides International PTH-4185-v   Bovine synthetic   MW 3496.1 $C_{156}H_{244}N_{48}O_{40}S_2$ >95% (HPLC); lyophilized amorphous powder | Bioactive; Horiuchi, N et al, *Science*, 220:1053, 1987

Phe-Met-His-Asn-Leu-Trp-Lys-His-Leu-Ser-Ser-Met-Glu-Arg-Val-Glu-Trp-Leu-Arg-Lys-Lys-Leu-Gln-Asp-Val-His-Asn-Tyr
Phe-Met-His-Asn-Leu-D-Trp-Lys-His-Leu-Ser-Ser-Met-Glu-Arg-Val-Glu-Trp-Leu-Arg-Lys-Lys-Leu-Gln-Asp-Val-His-Asn-Tyr Amide

Synonyms: Parathyroid Hormone (7-34), (D-Trp[12],Tyr[34])-

Bachem H-9115.0500   Bovine   Store at -15°C

Phe-Nle-Arg-Phe
Ac-Phe-Nle-Arg-Phe Amide

Bachem H-1055.0005   MW 622.77 $C_{32}H_{46}N_8O_5$ Store at -15°C

Phe-Nle-Arg-Phe
N-Ac-Phe-Nle-Arg-Phe Amide

Synonyms: Molluscan Cardioexcitatory Neuropeptide Analog

ICN 153039   MW 622.8 $C_{23}H_{46}N_8O_5$ A molluskan cardio-excitatory neuropeptide analog

Sigma A 2163   FW 622.8 $C_{32}H_{46}N_8O_5$ ≥95% (HPLC) | Bioactive peptide

Phe-Nle-His-Asn-Leu-Gly-Lys-His-Leu-Ser-Ser-Nle-Glu-Arg-Val-Glu-Trp-Leu-Arg-Lys-Lys-Leu-Gln-Asp-Val-His-Asn-Tyr Amide

Synonyms: Parathyroid Hormone (7-34), (Nle[8,18],Tyr[34])-

American Peptide 22-3-21   Bovine   MW 3460.0 $C_{158}H_{248}N_{48}O_{40}$ Antagonist of PTrelated protein; stimulates epidermal proliferation & hair growth & regulates the hair cycle

Bachem H-9120.0500   Bovine   Store at -15°C

Phe-Nle-Sta-Ala-Sta
Isovaleryl-Phe-Nle-Sta-Ala-Sta

Bachem H-5435.0001   Store at -15°C

Phe-Pal-Ser-Tyr-Cit-Leu-Arg-Pro-Ala
Ac-D-Nal-D-(pCl)Phe-D-Pal-Ser-Tyr-D-Cit-Leu-Arg-Pro-D-Ala Amide

Synonyms: Cetrorelix D-20761

American Peptide 54-1-26   MW 1434.1 $C_{70}H_{95}N_{17}O_{14}Cl$ Potent GnRH antagonist; the most active inhibitor of LH & FSH secretion compared to the GnRH antagonists Nal-Glu & Antide in terms of the effective dose & duration of action in the experimental model; Weinbauer, GF et al, *Andrologia*, 25(3):141, 1993

Phe-Phe
BOC-Phe-(®)-Phe

Bachem A-2210.0250   MW 398.5 $C_{23}H_{30}N_2O_4$ Store at -15°C

Phe-Phe
BOC-Phe-(®-(2-Chloro-Z))-Phe

Bachem A-2220.0250   Store at -15°C

Phe-Phe
BOC-Phe-Phe

Bachem A-3205.0001   MW 412.49 $C_{23}H_{28}N_2O_5$ Store at -15°C

Phe-Phe
FMOC-Phe-Phe

Bachem B-2150.0001   MW 534.61 $C_{33}H_{30}N_2O_5$ Store at -15°C

Phe-Phe
Z-Phe-Phe

Bachem C-2420.0001   MW 446.5 $C_{26}H_{26}N_2O_5$ Store at -15°C

Phe-Phe
Z-Phe-Phe-CHO Semicarbazone

Bachem C-3380.0050   MW 487.56 $C_{27}H_{29}N_5O_4$ Store at -15°C

Phe-Phe
Ac-Phe-Phe

Bachem G-1070.0250   MW 354.41 $C_{20}H_{22}N_2O_4$ Store at -15°C

Phe-Phe

Bachem G-2925.0250   MW 312.37 $C_{18}H_{20}N_2O_3$ Store at -15°C

Phe-Phe
D-Phe-D-Phe

Bachem G-3805.0250   MW 312.37 $C_{18}H_{20}N_2O_3$ Store at -15°C

Phe-Phe
FA-Phe-Phe

Synonyms: FAPP

Bachem M-1760.0050   MW 432.48 $C_{25}H_{24}N_2O_5$ Store at -15°C

Phe-Phe
Z-Phe-Phe-Diazomethylketone

Bachem N-1045.0050   Store at -15°C

Phe-Phe
Cyanoac-Phe-Phe

Synonyms: Angiotensin I Converting Enzyme Inactivator

Bachem N-1360.0250   MW 379.42 $C_{21}H_{21}N_3O_4$ Store at -15°C

Phe-Phe

ICN 102639   MW 312.4 $C_{18}H_{20}N_2O_3$ Crystalline

Phe-Phe
N-(3-(2-Furyl)Acryloyl)-Phe-Phe

ICN 158214   MW 432.5 $C_{25}H_{24}N_2O_5$

Phe-Phe
Cyanoacetyl-Phe-Phe

Synonyms: Angiotensin Converting Enzyme Inactivator

ICN 194135   MW 379.4 $C_{21}H_{21}N_3O_4$ 97% | Ghosh, SS etal, *J Med Chem*, 35:4175, 1992

Phe-Phe
Mu-Phe-hPhe-FMK

Synonyms: Calpain Inhibitor II

Kamiya   MW 455

Phe-Phe
Cyanoacetyl-Phe-Phe

Synonyms: Angiotensin I Converting Enzyme Inactivator; ACE Inactivator

Sigma A 5457   FW 379.4 $C_{21}H_{21}N_3O_4$ ≥97% (HPLC) | Bioactive peptide; Ghosh, SS et al, *J Med Chem*, 35:4175, 1992

Phe-Phe
N-(3-(2-Furyl)Acryloyl)-Phe-Phe

Synonyms: Carboxypeptidase Substrate Serum

Sigma F 7133   FW 432.5 $C_{25}H_{24}N_2O_5$ Peterson, LM et al, *Anal Biochem*, 125:420, 1982

Phe-Phe
Di-L-Phe

Sigma P 4126   FW 312.4 $C_{18}H_{20}N_2O_3$

Phe-Phe
N-(-N'-Carbonyl-Cpd-X-Phe-CHO)-Phe

Synonyms: Chymostatin; Serine Protease Inhibitor; Cysteine Protease Inhibitor; Chymotrypsin Inhibitor; Papain Inhibitor; Cathepsin A Inhibitor; Cathepsin B Inhibitor; Cathepsin C Inhibitor

Sigma C 7268   Microbial   Mixture of A (major), B & C components; A:X=Leu, B:X=Val, C:X=Ile

Phe-Phe Amide Hydrochloride

Bachem G-2930.0250   MW 347.84 $C_{18}H_{21}N_3O_2 \cdot HCl$ Store at -15°C

Phe-Phe-Arg
Z-D-Phe-Phe-Arg(NO₂)

Bachem C-2425.0050   MW 647.69 $C_{32}H_{37}N_7O_8$ Store at -15°C

Phe-Phe-Arg
FA-Phe-Phe-Arg

Bachem M-1410.0050   Store at -15°C

Phe-Phe-Arg
D-Phe-Phe-Arg-CMK

Bachem N-1215.0005   MW 501.03 $C_{25}H_{33}ClN_6O_3$ Store at -15°C

Phe-Phe-Arg
(D)-Phe-Phe-Arg-Chloromethylketone Hydrochloride Trifluoroacetate

Synonyms: Pefa-2858

Pentapharm 390-02   MW 669.5 $C_{25}H_{33}N_6O_3Cl \cdot HCl \cdot TFA$ Irreversible inhibitor of serine proteinases, especially PK; k_2/K_i = 1981

Phe-Phe-Arg
CBZ-Phe-Phe-Arg(NO₂)

USBio C2099-45   MW 647.7 $C_{32}H_{37}N_7O_8$ ≥99%

Phe-Phe-Gly
BOC-Phe-Phe-Gly

Bachem A-2195.0250   MW 469.54 $C_{25}H_{31}N_3O_6$ Store at -15°C

Phe-Phe-Gly
Z-D-Phe-Phe-Gly

Bachem H-9430.0050   MW 503.56 $C_{28}H_{29}N_3O_6$ Store at -15°C

Phe-Phe-Gly
N-CBZ-D-Phe-Phe-Gly

Synonyms: Virus Replication Inhibiting Peptide

ICN 154575   MW 503.6   Richardson, C etal, *Virology*, 105:205, 1980

Phe-Phe-Gly
N-t-BOC-Phe-Phe-Gly

Sigma B 8143   FW 469.5 $C_{25}H_{31}N_3O_6$

Phe-Phe-Gly
N-CBZ-D-Phe-Phe-Gly

Synonyms: Virus Replication Inhibitor

Sigma C 9405   FW 503.6 $C_{28}H_{29}N_3O_6$ ≥97% (HPLC) | Bioactive peptide; Richardson, CD et al, *Virology*, 105:205, 1980

Phe-Phe-Gly
Z-D-Phe-Phe-Gly

Synonyms: Virus Replication Inhibiting Peptide

Peptides International PVI-4092   Synthetic   MW 503.55 $C_{28}H_{29}N_3O_6$ >99% (HPLC); lyophilized amorphous powder; bulk | Bioactive; Richardson, CD et al, *Virology*, 105:205, 1980

Phe-Phe-Gly-Gly
Trt-Phe-Phe-Gly-Gly-OBzl

Bachem H-1518.0050   Store at -15°C

Phe-Phe-Gly-Leu-Met Amide

Synonyms: Substance P (7-11); Substance P, Penta-

American Peptide 70-1-73   MW 612.8 $C_{31}H_{44}N_6O_5S$

Bachem H-4555.0005   MW 612.79 $C_{31}H_{44}N_6O_5S$ Store at -15°C

ICN 153033   Bury, RW & ML Mashford, *J Med Chem*, 19:854, 1976

Phe-Phe-Gly-Leu-Met Amide Acetate Salt

Synonyms: Substance P (7-11)

Sigma S 2394   FW 612.8 (free base) $C_{31}H_{44}N_6O_5S$ ≥97% (HPLC) | Bioactive peptide; Bury, RW & Mashford, ML, *J Med Chem*, 19:854, 1976

Phe-Phe-His
Ac-Phe-Phe-His-OMe

Bachem H-1060.0100   MW 505.57 $C_{27}H_{31}N_5O_5$ Store at -15°C

Phe-Phe-His-Trp-Ala-Val-Ala-His-Pro-Phe
Phe-Deamino-Phe-His-Trp-Ala-Val-D-Ala-His-D-Pro-(®)-Phe Amide

Synonyms: Gastrin Releasing Peptide (19-27), (Deamino-Phe[19],D-Ala[24],D-Pro[26]-(®)-Phe[27])-; BW-10

Bachem H-2756.0500   Human, porcine, canine   Store at -15°C

Phe-Phe-Lys
Phe-Phe-Lys(BOC)

Bachem A-3175.0050   MW 540.66 $C_{29}H_{40}N_4O_6$ Store at -15°C

Phe-Phe-Phe
Z-Phe-Phe-Phe

Bachem C-3025.0050   MW 593.68 $C_{35}H_{35}N_3O_6$ Store at -15°C

Phe-Phe-Phe

Synonyms: Tri-L-Phenylalanine

Bachem H-4560.0250   MW 459.55 $C_{27}H_{29}N_3O_4$ Store at -15°C

ICN 156212   MW 459.5 $C_{27}H_{29}N_3O_4$

Sigma P 7918   FW 459.5 $C_{27}H_{29}N_3O_4$

Phe-Phe-Phe-Phe

Bachem H-4565.0250   MW 606.72 $C_{36}H_{38}N_4O_5$ Store at -15°C

Phe-Phe-Phe-Phe Acetate Salt

Synonyms: Tetra-L-Phenylalanine

ICN 156213   MW 666.8 $C_{36}H_{38}N_4O_5 \cdot C_2H_4O_2$

Sigma P 2544   FW 666.8 $C_{36}H_{38}N_4O_5 \cdot C_2H_4O_2$

Phe-Phe-Phe-Phe-Phe

Synonyms: Penta-L-Phenylalanine

Bachem H-4570.0100   MW 753.9 $C_{45}H_{47}N_5O_6$ Store at -15°C

ICN 156214   MW 753.9 $C_{45}H_{47}N_5O_6$ Crystalline

Phe-Phe-Phe-Phe-Phe Acetate Salt

Synonyms: Penta-L-Phenylalanine

Sigma P 3121   FW 753.9 (free base) $C_{45}H_{47}N_5O_6$ ≥97% (HPLC)

Phe-Phe-Pro-Leu-Met
δ-Aminovaleryl-Phe-Phe-Pro-N-Me-Leu-Met Amide

Synonyms: Substance P (7-11), δ-Aminovaleryl-(Pro[9],N-Me-Leu[10])-

Bachem H-3336.0001   Store at -15°C

Phe-Phe-Pro-Leu-Met
δ-Aminovaleryl-Phe-Phe-Pro-MeLeu-Met Amide

Synonyms: Substance P (7-11), δ-Aminovaleryl-(Pro[9],MeLeu[10])-; GR-73632

Neosystem SC432   MW 766.0   Bioactive; tachykinin; water soluble specific agonist at NK-1 receptors; Sheldrick, RL et al, *Agents Actions Suppl*, 31:205-209, 1990; Hagan, RM et al, *Neuropeptides*, 19:127-135, 1991

Phe-Phe-Trp-Lys-Thr-Phe

Synonyms: Somatostatin Sequence (6-11)

Biogenesis 8330-5609   Synthetic   MW 875.1   Lyophilized, consisting of 80.6% peptide material, acetate counter ions and residual H_2O

Phe-Phe-Trp-Ser-Tyr-Arg-Leu-Arg-Pro-Ala
Ac-D-(*p*-Chloro)Phe-D-(*p*-Chloro)Phe-D-Trp-Ser-Tyr-D-Arg-Leu-Arg-Pro-D-Ala Amide

Synonyms: Luteinizing Hormone Releasing Hormone, *N*-Ac-((*p*-Chloro)D-Phe[1,2],D-Trp[3],D-Arg[6],D-Ala[10])-; Luteinizing Hormone Releasing Hormone Antagonist

American Peptide 54-0-38   MW 1454.6 $C_{69}H_{94}N_{18}O_{13}Cl_2$
Coy, D et al, *Science*, 218:160, 1982

Phe-Phe-Trp-Ser-Tyr-Arg-Leu-Arg-Pro-Ala
Ac-D-*p*-Chloro-Phe-D-*p*-Chloro-Phe-D-Trp-Ser-Tyr-D-Arg-Leu-Arg-Pro-D-Ala Amide

Synonyms: Luteinizing Hormone Releasing Hormone, (Ac-D-*p*-Chloro-Phe[1,2],D-Trp[3],D-Arg[6],D-Ala[10])-

Sigma L 9011   FW 1452.5   ≥95% (HPLC) | Bioactive peptide; LH-RH antagonist; Coy, D et al, *Science,* 218:160, 1982

Phe-Phe-Tyr-Glu-Thr-His-Gly-Thr-Lys-Asn-Tyr-Phe-Thr-Ser-Val-Ala-His-Pro-Asn-Leu-Phe-Ile-Ala-Thr-Lys-Gln-Asp-Tyr

Synonyms: Interleukin Iα (223-250)

Bachem H-8290.0500   Human   MW 3340.7 $C_{158}H_{219}N_{37}O_{44}$
Store at -15°C

Phe-Pro
BOC-Phe-Pro

Bachem A-2730.0005   MW 362.43 $C_{19}H_{26}N_2O_5$ Store at -15°C

Phe-Pro
BOC-D-Phe-Pro-OSu

Bachem A-3095.0001   Store at -15°C

Phe-Pro
BOC-D-Phe-Pro

Bachem A-3100.0001   MW 362.43 $C_{19}H_{26}N_2O_5$ Store at -15°C

Phe-Pro
FMOC-Phe-Pro

Bachem B-2610.0001   Store at -15°C

Phe-Pro
Z-Phe-Pro

Bachem C-2430.0001   MW 396.44 $C_{22}H_{24}N_2O_5$ Store at -15°C

Phe-Pro
Z-D-Phe-Pro

Bachem C-3485.0001   MW 396.44 $C_{22}H_{24}N_2O_5$ Store at -15°C

Phe-Pro
Z-D-Phe-Pro-Boro-Mpg-Pinanediol Ester

Synonyms: Z-D-Phe-Pro-(3-Methoxypropyl)Boroglycine-Pinanediol Ester; Z-D-Phe-Pro-*N*-((1*S*)-4-Methoxy-1-((1*S*,2*S*,6*R*,8*S*)-2,9,9-Trimethyl-3,5-Dioxa-4-Bora-Tricyclo(6.1.1.02,6)Dec-4-yl)-Butyl)Amide

Bachem C-3720.0005   MW 659.63 $C_{37}H_{50}BN_3O_7$ Store at -15°C

Phe-Pro

Bachem G-2935.0250   MW 262.31 $C_{14}H_{18}N_2O_3$ Store at -15°C

Phe-Pro
Phe-D-Pro Hydrochloride

Bachem G-3355.0250   MW 298.77 $C_{14}H_{18}N_2O_3 \cdot$ HCl   Store at -15°C

Phe-Pro
D-Phe-Pro

Bachem G-4115.0250   MW 262.31 $C_{14}H_{18}N_2O_3$ Store at -15°C

Phe-Pro
Phe-Pro-βNA

Bachem K-1455.0050   MW 387.48 $C_{24}H_{25}N_3O_2$ Store at -15°C

Phe-Pro
α-Aminobenzoyl-Gly-*p*-Nitro-Phe-Pro

Synonyms: Angiotensin Converting Enzyme Substrate (Fluorogenic)

ICN 152748   Carmel, A & A Yaron, *Eur J Biochem*, 87:265, 1978

Phe-Pro

ICN 156216   MW 262.3 $C_{14}H_{18}N_2O_3$ Crystalline
Sigma P 6258   FW 262.3 $C_{14}H_{18}N_2O_3$

Phe-Pro
CBZ-Phe-Pro

USBio C2099-46   MW 396.5 $C_{22}H_{24}N_2O_5$ ≥99%

Phe-Pro-Ala

Bachem H-7485.0250   MW 333.39 $C_{17}H_{23}N_3O_4$ Store at -15°C

Phe-Pro-Ala
Phe-Pro-Ala-βNA

Bachem K-1460.0050   MW 458.56 $C_{27}H_{30}N_4O_3$ Store at -15°C
Bachem L-1800.0050   Store at -15°C

Phe-Pro-Ala-Met
Phe-Pro-Ala-Met-4-Methoxy-β-Naphthylamide

ICN 156217
Sigma P 7146   FW 619.8 $C_{33}H_{41}N_5O_5S$ Substrate for cathepsin B

Phe-Pro-Arg
D-Phe-Pro-Arg-CMK Dihydrochloride

Synonyms: Thrombin Inhibitor

Alexis 260-001   MW 524.0 $C_{21}H_{31}ClN_6O_3 \cdot$ 2HCl   ≥90%; white powder; soluble in dilute acetic acid | Extremely potent & selective irreversible inhibitor of thrombin; Kettner, C & Shaw, E, *Thromb Res*, 14:969, 1979

Phe-Pro-Arg

Bachem H-3916.0050   MW 418.5 $C_{20}H_{30}N_6O_4$ Store at -15°C

Phe-Pro-Arg
D-Phe-Pro-Arg-5-amido-isophthalic acid-dimethyl ester

Bachem M-1710.0050   MW 609.68 $C_{30}H_{39}N_7O_7$ Store at -15°C

Phe-Pro-Arg
BOC-D-Phe-Pro-Arg

Bachem N-1020.0050 MW 518.61 $C_{25}H_{38}N_6O_6$ Store at -15°C

Phe-Pro-Arg
D-Phe-Pro-Arg-CMK

Bachem N-1065.0005 MW 450.97 $C_{21}H_{31}ClN_6O_3$ Store at -15°C

Phe-Pro-Arg
Biotinyl-ε-Aminocaproyl-D-Phe-Pro-Arg-CMK

Bachem N-1565.0005 MW 790.43 $C_{37}H_{56}ClN_9O_6S$ Store at -15°C

Phe-Pro-Arg
D-Phe-Pro-Arg-5-Amidoisophthalic Acid Dimethyl Ester

Synonyms: Thrombin Substrate

ICN 156218

Sigma P 3398 FW 609.7 (free base) $C_{30}H_{39}N_7O_7$ A highly sensitive thrombin substrate that can be used for convenient, indirect assays of antithrombin III, heparin, & coagulation factors VIII & IX

Phe-Pro-Asp-Trp-Gln-Asn-Tyr-Thr-Pro-Gly-Pro-Gly-Ile-Arg-Tyr

Synonyms: NEF MN (123-137); HIV I Peptide

Neosystem SC656 MW 1810.00 NEF peptide from HIV-1 subtype MN

Phe-Pro-Met-Phe-Lys-Arg-Gly-Arg-Cys-Leu-Cys-Ile-Gly-Pro-Gly-Val-Lys-Ala-Val-Lys-Val-Ala-Asp-Ile-Glu-Lys-Ala-Ser-Ile-Met-Tyr-Pro-Ser-Asn-Asn-Cys-Asp-Lys-Ile-Glu-Val-Ile-Ile-Thr-Leu-Lys-Glu-Asn-Lys-Gly-Gln-Arg-Cys-Leu-Asn-Pro-Lys-Ser-Lys-Gln-Ala-Arg-Leu-Ile-Ile-Lys-Lys-Val-Glu-Arg-Lys-Asn-Phe

Synonyms: Interferon-Inducible T-Cell α-Chemoattractant

Bachem H-4606.0010 Human Store at -15°C

Phe-Pro-Phe
Suc-Phe-Pro-Phe-pNA

Bachem L-1430.0050 MW 629.67 $C_{33}H_{35}N_5O_8$ Store at -15°C

Phe-Pro-Thr-Ile-Pro-Leu-Ser-Arg-Leu-Phe-Asp-Asn-Ala-Met-Leu-Arg-Ala-His-Arg-Leu-His-Gln-Leu-Ala-Phe-Asp-Thr-Tyr-Gln-Glu-Phe-Glu-Glu-Ala-Tyr-Ile-Pro-Lys-Glu-Gln-Lys-Tyr-Ser

Synonyms: Human Growth Hormone (1-43); Growth Hormone (1-43)

Bachem H-3146.0500 MW 5215.93 $C_{240}H_{358}N_{62}O_{67}S$ Store at -15°C

American Peptide 87-1-21 Human MW 5216.2 $C_{240}H_{358}N_{62}O_{67}S$ Using hGH (1-43) to treat obese yellow Avy/A mice with hGh (1-43) enhanced the *in vitro* sensitivity of their adipose tissue to insulin

Phe-Ser

Bachem G-2940.0250 MW 252.27 $C_{12}H_{16}N_2O_4$ Store at -15°C

Phe-Ser Amide Hydrochloride

Bachem G-2945.0250 Store at -15°C

Phe-Ser-Arg
BOC-Phe-Ser-Arg-AMC

Bachem I-1400.0005 MW 665.75 $C_{33}H_{43}N_7O_8$ Store at -15°C

Phe-Ser-Arg
N-α-t-BOC-L-Phe-L-Ser-L-Arg-MCA

Synonyms: Trypsin Substrate

ICN 150505 MW 665.8 $C_{33}H_{43}N_7O_8$ Fluorogenic substrate

Phe-Ser-Arg
BOC-Phe-Ser-Arg-AMC Trifluoroacetate Salt

Synonyms: Coagulation Factor XIa Substrate; Tryptase Substrate

Neosystem SC1277 MW 665.74 Iwanaga, S et al, *Kinins-II:Biochem, Pathophysiology & Clinical Aspects*, Fujii, S et al eds, Plenum Press, New York, p 147, 1979; Muramatsu, M et al, *Biol Chem* Hoppe-Seyler, 369:617-625, 1988

Phe-Ser-Arg
BOC-Phe-Ser-Arg-MCA

Synonyms: Trypsin Substrate; Tryptase Substrate; 73K Protease Substrate

Peptides International MFS-3107-v MW 665.75 $C_{33}H_{43}N_7O_8$ >99% (HPLC); lyophilized amorphous powder | Iwanaga, S et al, *Kinins-II:Biochemistry, Pathophysiology, & Clinical Aspects*, (S Fujii, H Moriya & T Suzuki, Eds), Plenum Publishing Co, 1979, pp 147-163; Muramatsu, M et al, Hoppe-Seyler's *Biol Chem*, 369:61, 1988; Molla, A et al, *J Biochem*, 104:616, 1988

Phe-Ser-Arg
N-t-BOC-Phe-Ser-Arg-MCA

Sigma B 6388 FW 665.7 $C_{33}H_{43}N_7O_8$

Phe-Ser-Gln-Val-Lys-Gly-Ala-Val-Asp-Asp-Asp-Val-Ala-Glu

Synonyms: Protein Phosphatase 2A/Bα Blocking Peptide (14-27)

Calbiochem 539524 Synthetic MW 1479.6 $C_{63}H_{98}N_{16}O_{25}$ Liquid | From PP2A/Bα regulatory subunit; ideal blocking peptide for Anti-PP2A/Bα

Phe-Ser-Met-Lys-Asn-Leu-His-Arg-Arg-Val-Lys-Ile-Cys Acetate Salt

Synonyms: Bone Sialoprotein II (1-12), ~C

Biogenesis 1407-0131 Human synthetic Semi-pure; lyophilized

Phe-Ser-Phe
D-Phe-Ser(Bzl)-Phe

Bachem H-9810.0250 Store at -15°C

Phe-Ser-Phe-Phe-Ala-Ala-Abz
D-Phe-Ser(Bzl)-Phe-Phe-Ala-Ala-4-Abz

Bachem H-1226.0005 Store at -15°C

Phe-Ser-Trp-Gly-Ala-Glu-Gly-Gln-Arg

Synonyms: Experimental Allergic Encephalitogenic Peptide; Experimental Allergic Encephalogenic Peptide; Myelin Basic Protein, Active Fragment

American Peptide 87-0-01 MW 1037.2 $C_{46}H_{64}N_{14}O_{14}$

ICN 153165 MW 1037.1 Active fragment of myelin basic protein; causes experimental allergic encephalomyelitis (inflammatory demyelinating disease of the CNS) via a cell-mediated immune response; Shapira, R etal, *Science*, 78:736, 1971; Westall, FC etal, *Nature*, 229:22, 1971

Sigma P 6272   FW 1037.1   ≥97% (HPLC) | Bioactive peptide; induces experimental allergic encephalomyelitis by a mechanism involving cell-mediated immunity; Shapira, R et al, *Science*, 173:736, 1971; Suzuki, K et al, *Chem Pharm Bull*, 21:2627, 1973; ibid, 21:2634, 1973

Bachem H-2680.0001   Human   MW 1037.1 $C_{46}H_{64}N_{14}O_{14}$ Store at -15°C

Phe-Ser-Val

Bachem H-4575.0250   MW 351.4 $C_{17}H_{25}N_3O_5$ Store at -15°C

Phe-Thiaphe
Ac-Phe-Thiaphe

Synonyms: Carboxypeptidase A Dipeptide Mimetic Substrate

Peptides International STP-3621-PI   MW 372.45 $C_{19}H_{20}N_2O_4S$ >98% (HPLC); white crystalline powder | Shamamian, P et al, Annual Meeting of the Society for Surgery of the Alimentary Tract, May 1998; Brown, KS et al, *Anal Biochem*, 161:219, 1987; Hwang, SY et al, *Anal Biochem*, 154:552, 1986

Phe-Thr-Asp-Asn-Tyr-Thr-Arg-Leu-Arg-Lys-Gln-Met-Ala-Val-Lys-Lys-Tyr-Leu-Asn-Ser-Ile-Leu-Asn Amide

Synon Vasoactive Intestinal Peptide (6-28)*yms*:

Bachem H-2066.0500   Human, bovine, porcine, rat   MW 2816.32 $C_{126}H_{207}N_{37}O_{34}S$ Store at -15°C

Sigma V 4508   Porcine   FW 2816.3   ≥97% (HPLC) | Bioactive peptide; gastrointestinal peptide; potent VIP receptor antagonist; Fishbein, VA et al, *Peptide*, 15:95, 1994

Phe-Thr-Asp-Ser-Tyr-Ser-Arg-Tyr-Arg-Lys-Gln-Met-Ala-Val-Lys-Lys-Tyr-Leu-Ala-Ala-Val-Leu Amide

Synonyms: Adenylate Cyclase Activating Peptide (6-27); Pituitary Adenylate Cyclase Activating Polypeptide (6-27); Pituitary Adenylate Cyclase Activating Polypeptide 27 Inhibitor

ICN 195880   MW 2638.1   ≥97% | Inhibits PACAP-27

Sigma A 6938   FW 2638.1   Peptide content:~65% | Bioactive peptide; Robberecht, P et al, *FEBS Lett*, 286:133, 1991

Phe-Thr-Asp-Ser-Tyr-Ser-Arg-Tyr-Arg-Lys-Gln-Met-Ala-Val-Lys-Lys-Tyr-Leu-Ala-Ala-Val-Leu-Gly-Lys-Arg-Tyr-Lys-Gln-Arg-Val-Lys-Asn-Lys Amide

Synonyms: Pituitary Adenylate Cyclase Activating Polypeptide Selective Antagonist; Pituitary Adenylate Cyclase Activating Peptide (6-38); Pituitary Adenylate Cyclase Activating Peptide 38 Antagonist

Peptides International PPA-4286-v   Human synthetic   MW 4024.8 $C_{182}H_{300}N_{56}O_{45}S$ >95% (HPLC); lyophilized amorphous powder | Bioactive; ovine, rat; Robberecht, P et al, *Eur J Biochem*, 207:239, 1992; Vandermeers, A et al, *Eur J Biochem*, 208:815, 1992

Alexis 165-026   Human, ovine, rat   MW 4024.8 $C_{182}H_{300}N_{56}O_{45}S$ Lyophilized | Potent & selective inhibitor of PACAP (1-27) amide stimulated pituitary adenylate cyclase; Robberecht, P et al, *Eur J Biochem*, 207:239, 1992; Robberecht, P et al, *Mol Pharmacol*, 42:347, 1992

American Peptide 34-0-19   Human, ovine, rat   MW 4024.8 $C_{182}H_{300}N_{56}O_{45}S$ Potent antagonist of PACAP 38; much more potent & selective than PACAP (6-27) in inhibiting PACAP-27-stimulated pituitary adenylate cyclase; Robberecht, P et al, *Eur J Biochem*, 207:239, 1992; Robberecht, P et al, *Mol Pharmacol*, 42:347, 1992

Phe-Thr-Asp-Ser-Tyr-Ser-Arg-Tyr-Arg-Lys-Gln-Met-Ala-Val-Lys-Lys-Tyr-Leu-Ala-Ala-Val-Leu Amide

Synonyms: Pituitary Adenylate Cyclase Activating Peptide (6-27); Pituitary Adenylate Cyclase Activating Peptide 27 Inhibitor

American Peptide 34-0-45   Human, ovine, rat   MW 2638.3 $C_{121}H_{193}N_{33}O_{31}S$ Specific inhibitor of PACAP-27 stimulated adenylate cyclase; allows a clear functional discrimination between PACAP-27 receptors & PACAP-38 receptors; Robberecht, P et al, *FEBS Lett*, 286:133, 1991

Phe-Thr-Asp-Ser-Tyr-Ser-Arg-Tyr-Arg-Lys-Gln-Met-Ala-Val-Lys-Lys-Tyr-Leu-Ala-Ala-Val-Leu-Gly-Lys-Arg-Tyr-Lys-Gln-Arg-Val-Lys-Asn-Lys Amide

Synonyms: Pituitary Adenylate Cyclase Activating Polypeptide (6-38)

Bachem H-2734.0500   Human, ovine, rat   MW 4024.8 $C_{182}H_{300}N_{56}O_{45}S$ Store at -15°C

Phe-Thr-Asp-Ser-Tyr-Ser-Arg-Tyr-Arg-Lys-Gln-Met-Ala-Val-Lys-Lys-Tyr-Leu-Ala-Ala-Val-Leu Amide

Synonyms: Pituitary Adenylate Cyclase Activating Polypeptide (6-27)

Bachem H-8435.0200   Human, ovine, rat   Store at -15°C

Phe-Thr-Asp-Ser-Tyr-Ser-Arg-Tyr-Arg-Lys-Gln-Met-Ala-Val-Lys-Lys-Tyr-Leu-Ala-Ala-Val-Leu-Gly-Lys-Arg-Tyr-Lys-Gln-Arg-Val-Lys-Asn-Lys Amide

Synonyms: Pituitary Adenylate Cyclase Activating Polypeptide (6-38)

Neosystem SC1255   Human, ovine, rat   MW 4024.78 Bioactive; pituitary peptide; Robberecht, P et al, *Mol Pharmacology*, 42:347-355, 1992; Robberecht, P et al, *Mol Pharm Biochem*, 207:239-246, 1992

Phe-Thr-Asp-Ser-Tyr-Ser-Arg-Tyr-Arg-Lys-Met-Ala-Val-Lys-Lys-Tyr-Leu-Ala-Ala-Val-Leu Amide

Synonyms: Pituitary Adenylate Cyclase Activating Peptide (6-27), (des-Gln[16])-; Pituitary Adenylate Cyclase Activating Peptide 27 Inhibitor

American Peptide 34-0-46   Human, ovine, rat   MW 2510.0 $C_{116}H_{185}N_{31}O_{29}S$

Phe-Thr-Gly-Pro-Val

Synonyms: Osteocalcin (45-49), (Thr[46])-

Bachem H-9720.0005   Human   Store at -15°C

Phe-Thr-Leu-Cys-Phe-Arg Amide

Synonyms: Amyloid P Component (33-38)

Bachem H-2946.0001   MW 784.98 $C_{37}H_{56}N_{10}O_7S$ Store at -15°C

Phe-Thr-Pro-Arg-Leu Amide

Synonyms: Leucopyrokinin (4-8)

Bachem H-6240.0001   Store at -15°C

ICN 153141   MW 632.8 $C_{30}H_{48}N_8O_7$ C-terminal pentapeptide with 30% the activity of leucopyrokinin; Nachman, RJ etal *BBRC*, 137:936, 1986

Sigma L 5018   FW 631.8 $C_{30}H_{49}N_9O_6$ ≥97% (HPLC) | Bioactive peptide; C-terminal pentapeptide of leucopyrokinin, retaining 30% of the activity of the parent; Nachman, RJ et al, *Biochem Biophys Res Commun*, 137:936, 1986

Phe-Trp
Z-Phe-Trp

Bachem C-2435.0001   MW 485.54 $C_{28}H_{27}N_3O_5$ Store at -15°C

Phe-Trp
Ac-Phe-Trp
Bachem G-1080.0250   MW 393.44 $C_{22}H_{23}N_3O_4$ Store at -15°C

Phe-Trp
Bachem G-2950.0250   MW 351.41 $C_{20}H_{21}N_3O_3$ Store at -15°C

Phe-Trp Amide Hydrochloride
Bachem G-4000.0250   MW 386.88 $C_{20}H_{22}N_4O_2 \cdot HCl$ Store at -15°C

Phe-Trp-Lys-Thr
Cyclo(Aminoheptanoyl-Phe-D-Trp-Lys-Thr(Bzl))
Synonyms: Somatostatin Antagonist
American Peptide 68-1-48   MW 780.0 $C_{44}H_{57}N_7O_6$ Fries, JL et al, *Peptides*, 3:811, 1982

Phe-Trp-Lys-Thr-Phe-Thr-Ser-Cys
Synonyms: Somatostatin 14 (7-14)
Bachem H-4698.0005   MW 1019.19 $C_{49}H_{66}N_{10}O_{12}S$ Store at -15°C

Phe-Trp-Phe-Trp-Leu-Leu
N-Me-D-Phe-D-Trp-Phe-D-Trp-Leu-(®)-Leu Amide
Synonyms: Substance P (6-11), (N-Me-D-Phe[6],D-Trp[7,9],Leu[10]-(®)-Leu[11])-
Bachem H-2584.0005   Store at -15°C

Phe-Trp-Phe-Trp-Leu-Leu
Me-D-Phe-D-Trp-Phe-D-Trp-Leu-ψ(CH₂NH)-Leu Amide
Synonyms: Substance P (6-11), (Me-D-Phe[6],D-Trp[7,9],Leu[10]-ψ(CH₂NH)Leu[11])-
Sigma S 3922   FW 910.2 $C_{53}H_{67}N_9O_5$ ≥97% (HPLC); peptide content:~70% | Bioactive peptide; this Substance P analog inhibited small-cell lung cancer proliferation *in vitro* & *in vivo*; Orosz, A et al, *Int J Cancer*, 60:82, 1995

Phe-Trp-Ser-Phe-Gly-Ser-Glu-Asp-Gly-Ser-Gly-Asp-Ser-(Cys) Acetate Salt
Synonyms: Glutamic Acid Decarboxylase (8-20)
Biogenesis 4670-6621   Human 65 k isoform synthetic   MW 1480.5   Semi-pure; liquid

Phe-Tyr
BOC-Phe-Tyr
Bachem A-2580.0001   MW 428.49 $C_{23}H_{28}N_2O_6$ Store at -15°C

Phe-Tyr
Z-Phe-Tyr(tBu)-Diazomethylketone
Bachem C-3895.0050   Store at -15°C

Phe-Tyr
Ac-Phe-Tyr
Bachem G-1085.0250   MW 370.41 $C_{20}H_{22}N_2O_5$ Store at -15°C

Phe-Tyr
Ac-Phe-Tyr Amide
Bachem G-1090.0250   MW 369.42 $C_{20}H_{23}N_3O_4$ Store at -15°C

Phe-Tyr
Bachem G-2955.0250   MW 328.37 $C_{18}H_{20}N_2O_4$ Store at -15°C

Phe-Tyr
Ac-D-Phe-Tyr
Bachem G-3855.0250   MW 370.41 $C_{20}H_{22}N_2O_5$ Store at -15°C

Phe-Tyr
Ac-Phe-3,5-diiodo-Tyr
Bachem M-1070.0250   MW 622.2 $C_{20}H_{20}I_2N_2O_5$ Store at RT

Phe-Tyr
Z-Phe-Tyr-CHO
Bachem N-1540.0025   Store at -15°C

Phe-Tyr
N-Ac-L-Phe-3,5-Diiodo-L-Tyr
Fluka 01408   MW 622.2 $C_{20}H_{20}I_2N_2O_5$ ≥97% (elemental analysis)
ICN 100122   MW 622.2 $C_{20}H_{20}I_2N_2O_5$ White powder

Phe-Tyr
ICN 156219   MW 328.4 $C_{18}H_{20}N_2O_4$ Crystalline

Phe-Tyr
Z-Phe-Tyr
Peptides International SFY-3044   MW 462.50 $C_{26}H_{26}N_2O_6$ >99% (HPLC); amorphous powder

Phe-Tyr
N-Ac-D-Phe-Tyr
Sigma A 3387   FW 370.4 $C_{20}H_{22}N_2O_5$ Crystalline

Phe-Tyr
N-t-BOC-Phe-Tyr Dicyclohexylammonium Salt
Sigma B 7897   FW 609.8 $C_{23}H_{28}N_2O_6 \cdot C_{12}H_{23}N$

Phe-Tyr
Sigma P 4876   FW 328.4 $C_{18}H_{20}N_2O_4$

Phe-Tyr Amide Hydrochloride
Bachem G-2960.0250   MW 363.84 $C_{18}H_{21}N_3O_3 \cdot HCl$ Store at -15°C

Phe-Tyr-Gly-Pro-Val
Synonyms: Bone Gla Protein (45-19); Osteocalcin (45-49)
American Peptide 22-1-62   MW 581.7 $C_{30}H_{39}N_5O_7$
ICN 159984   MW 581.7 $C_{30}H_{39}N_5O_7$ 97%
Bachem H-9125.0005   Human   MW 581.67 $C_{30}H_{39}N_5O_7$ Store at -15°C
ICN 154479   Human   MW 581.7
Sigma O 4382   Human   FW 581.7 $C_{30}H_{39}N_5O_7$ ≥97% (HPLC) | Bioactive peptide

Phe-Tyr-Leu
Z-Phe-Tyr-Leu
Synonyms: Metalloproteinase Substrate
Peptides International SFL-3131   MW 575.66 $C_{32}H_{37}N_3O_7$ >99% (HPLC); amorphous powder

Phe-Val
BOC-Phe-(®)-Val

Bachem A-3010.0250   MW 350.46 $C_{19}H_{30}N_2O_4$ Store at -15°C

Phe-Val
BOC-Phe-(®-(2-Chloro-Z))-Val

Bachem A-3040.0250   Store at -15°C

Phe-Val
Z-Phe-Val

Bachem C-2440.0001   MW 398.46 $C_{22}H_{26}N_2O_5$ Store at -15°C

Phe-Val
Z-D-Phe-Val

Bachem C-2445.0001   MW 398.46 $C_{22}H_{26}N_2O_5$ Store at -15°C

Phe-Val

Bachem G-2965.0250   MW 264.32 $C_{14}H_{20}N_2O_3$ Store at -15°C

Phe-Val
FA-Phe-Val Amide

Bachem M-2005.0050   MW 383.45 $C_{21}H_{25}N_3O_4$ Store at -15°C

Phe-Val

ICN 102629   MW 264.3 $C_{14}H_{20}N_2O_3$ 98.5%; crystalline

Phe-Val
D-Phe-Val-pNA

ICN 156220   MW 384.4 $C_{20}H_{24}N_4O_4$

Sigma P 3144   FW 384.4 $C_{20}H_{24}N_4O_4$

Phe-Val

Sigma P 5001   FW 264.3 $C_{14}H_{20}N_2O_3$

Phe-Val-Arg
Bz-Phe-Val-Arg-AMC

Bachem I-1080.0050   MW 681.79 $C_{37}H_{43}N_7O_6$ Store at -15°C

Phe-Val-Arg
Bz-Phe-Val-Arg-4MβNA Hydrochloride

Bachem J-1070.0050   MW 716.28 $C_{38}H_{45}N_7O_5$ Store at -15°C

Phe-Val-Arg
Bz-Phe-Val-Arg-pNA Hydrochloride

Bachem L-1150.0050   MW 681.19 $C_{33}H_{40}N_8O_6 \cdot HCl$ Store at -15°C

Phe-Val-Arg
Z-Phe-Val-Arg-pNA Hydrochloride

Bachem L-1245.0050   MW 711.22 $C_{34}H_{42}N_8O_7 \cdot HCl$ Store at -15°C

Phe-Val-Arg
Bz-Phe-Val-Arg-pNA Hydrochloride

Synonyms: Thrombin Substrate I

Calbiochem 605210   MW 681.2 $C_{33}H_{40}N_8O_6$ Lyophilized solid containing 45 mg D-mannitol & 5 mg substrate; ≥95% (HPLC); soluble in water | Colorimetric; Lottenberg, R et al, *Methods Enzymol*, 80:341 1981

Phe-Val-Arg
Bz-Phe-Val-Arg-AMC Hydrochloride

Synonyms: Thrombin Substrate III

Calbiochem 605211   MW 718.3 $C_{37}H_{43}N_7O_6 \cdot HCl$ Lyophilized solid; ≥98% (TLC); soluble in water & ethanol | Sensitive fluorogenic substrate for the quantitative determination of thrombin

Phe-Val-Arg
N-Benzoyl-L-Phe-L-Val-L-Arg-4-Nitroanilide Hydrochloride

Synonyms: Thrombin Substrate; Trypsin Substrate; Reptilase Substrate

Fluka 13042   MW 681.20 $C_{33}H_{40}N_8O_6 \cdot HCl$ ≥99% (TLC); ≤5% H_2O | Lottenberg, R et al, *Meth Enzymol*, 80:341, 1981

Phe-Val-Arg
Bz-Phe-Val-Arg-AMC Hydrochloride

Synonyms: Thrombin Substrate III

ICN 195915   MW 718.3 $C_{37}H_{43}N_7O_6 \cdot HCl$ ≥98% | Fluorogenic substrate for thrombin quantitation

Phe-Val-Arg
N-Benzoyl-Phe-Val-Arg-4-Methoxy-β-Naphthylamide Hydrochloride

Synonyms: Thrombin Substrate

Sigma B 1260   FW 716.3 $C_{38}H_{45}N_7O_5 \cdot HCl$

Phe-Val-Arg
N-Benzoyl-Phe-Val-Arg-pNA Hydrochloride

Synonyms: Trypsin Substrate; Thrombin Substrate; Reptilase Substrate

Sigma B 7632   FW 681.2 $C_{33}H_{40}N_8O_6 \cdot HCl$ Svendsen, L et al, *Folia Haematol*, 98:446, 1972; Lottenberg, R et al, *Meth Enz*, 80:341, 1981

Phe-Val-Arg
N-Benzoyl-Phe-Val-Arg-MCA

Sigma B 9635   FW 681.8 $C_{37}H_{43}N_7O_6$ ~95% (TLC)

Phe-Val-Asn-Gln-His-Leu-Cys-Gly-Ser-His-Leu-Val-Glu-Ala-Leu-Tyr-Leu-Val-Cys-Gly-Glu-Arg-Gly-Phe-Phe-Tyr-Thr-Pro-Lys-Ala
Phe-Val-Asn-Gln-His-Leu-Cys(SO₃H)-Gly-Ser-His-Leu-Val-Glu-Ala-Leu-Tyr-Leu-Val-Cys(SO₃H)-Gly-Glu-Arg-Gly-Phe-Phe-Tyr-Thr-Pro-Lys-Ala

Synonyms: Insulin Chain B, Oxidized; Carboxypeptidase Y Substrate

Sigma I 6383   Bovine   FW 3495.9 (free acid)   ≥80% (HPLC); contains <1% chain A; soluble at ~20 mg/mL 0.1 *M* NH_4OH | Bioactive peptide; prepared by a modification of Sanger, F et al, *Biochem J*, 44:126, 1949; Hayashi, H et al, *J Biol Chem*, 248:2296, 1973; Johansen, JT et al, *Carlsberg Res Commun*, 41:1, 1976

Phe-Val-Gln-Trp-Leu-Met-Asn-Thr

Synonyms: Glucagon (22-29)

American Peptide 46-1-20   Human   MW 1038.2 $C_{49}H_{71}N_{11}O_{12}S$ Abiko, T et al, *Chem Pharm Bull*, 27:2827, 1979

ICN 154555   Human   MW 1038.3   Abiko, T etal, *Chem Pharm Bull*, 27:2827, 1979

Phe-Val-Gly-Trp-Phe-Leu-Gly-Trp-Asp-Asp-Asp-Tyr-Trp-Ser Acetate Salt

Synonyms: Gelsolin (729-742)

Biogenesis 4628-6220   Human plasma + cytoplasmic synthetic   Semi-pure; lyophilized

---

Phe-Val-Pro-Ile-Phe-Thr-His-Ser-Glu-Leu-Gln-Lys-Ile-Arg-Glu-Lys-Glu-Arg-Asn-Lys-Gly-Gln

Synonyms: Motilin

American Peptide 48-6-10   Canine   MW 2685.0 $C_{120}H_{194}N_{36}O_{34}$

Bachem H-7680.0500   Canine   Store at -15°C

Phe-Val-Pro-Ile-Phe-Thr-His-Ser-Glu-Leu-Gln-Lys-Ile-Arg-Glu-Lys-Glu-Arg-Asn-Lys-Ile-Arg-Asn-Lys-Gly-Gln

Synonyms: Motilin

ICN 152884   Canine   MW 2685.1   Reeve, JR, *J Chromatogr*, 321:421, 1985

Phe-Val-Pro-Ile-Phe-Thr-His-Ser-Glu-Leu-Gln-Lys-Ile-Arg-Glu-Lys-Glu-Arg-Asn-Lys-Gly-Gln

Synonyms: Motilin; Gastric Motor Stimulatory Peptide

Sigma M 3278   Canine   FW 2685   ≥97% (HPLC); peptide content: ~70% | Bioactive peptide; gastrointestinal peptide; Reeve, JR et al, *J Chromatography*, 321:421, 1985

Phe-Val-Pro-Ile-Phe-Thr-His-Ser-Glu-Leu-Gln-Lys-Ile-Arg-Glu-Lys-Glu-Arg-Asn-Lys-Gly-Gln Acetate Salt

Synon Motilin *yms*:

Biogenesis 6370-1004   Canine synthetic   MW 2685   Purified; lyophilized

---

Phe-Val-Pro-Ile-Phe-Thr-Tyr-Gly-Glu-Leu-Gln-Arg-Leu-Gln-Glu-Lys-Glu-Arg-Asn-Lys-Gly-Gln

Synonyms: Motilin, (Leu[13])-

American Peptide 48-4-21   Porcine   MW 2681.0 $C_{121}H_{190}N_{34}O_{35}$ Exhibits gastric motor stimulating activity; Miyashita, E et al, *Biotech Lett*, 10:763, 1988

Sigma M 7530   Porcine   FW 2681.0   ≥97% (HPLC) | Bioactive peptide; gastrointestinal peptide; exhibits gastric motor stimulating activity; Miyashita, E et al, *Biotechnology Letters*, 10:763, 1988

Phe-Val-Pro-Ile-Phe-Thr-Tyr-Gly-Glu-Leu-Gln-Arg-Leu-Glu-Glu-Lys-Glu-Arg-Asn-Lys-Gly-Gln

Synonyms: Motilin, (Leu[13],Glu[14])-

Fluka 61895   Porcine   MW 2682.04 $C_{121}H_{189}N_{33}O_{36}$ Fox, JET, Life Sci, 35:695, 1984; Mutt, V, *Brain Peptides*, (DT Krieger et al, eds), 871, 1983, John Wiley & Sons, New York; Bertaccini, G, *Handbook Exp Pharmacol*, 59:85, 1982; Bloom, SR, *Hormones in Blood*, (CH Gray & James, VHT, eds), 321, 1979, Academic Press, London

---

Phe-Val-Pro-Ile-Phe-Thr-Tyr-Gly-Glu-Leu-Gln-Arg-Met-Gln-Glu-Lys-Glu-Arg-Asn-Lys-Gly-Gln

Synonyms: Motilin; Gastric Motor Stimulatory Peptide

Bachem H-4385.0500   Human, porcine   MW 2699.09 $C_{120}H_{188}N_{34}O_{35}S$ Store at -15°C

Peptides International PML-4147-v   Human, porcine synthetic   MW 2699.1 $C_{120}H_{188}N_{34}O_{35}S$ >95% (HPLC); lyophilized amorphous powder | Bioactive; Brown, JC et al, *Can J Biochem*, 51:533, 1973; Seino, Y et al, *FEBS Letts*, 223:74, 1987

American Peptide 48-4-20   Porcine   MW 2699.1 $C_{120}H_{188}N_{34}O_{35}S$ Schubert, H et al, *Can J Biochem*, 52:7, 1954

ICN 152885   Porcine   MW 2699.1   Exhibits gastric motor-stimulating ability; Brown, JC etal, *Can J Biochem*, 51:533, 1973; Schubert, H & JC Brown, *Can J Biochem*, 52:7, 1974; Yamada, S etal, *J Am Chem Soc*, 97:7174, 1975; Ikota, N etal *Chem Pharm Bull*, 28:3347, 1980

Neosystem SC100   Porcine   MW 2699.07   Bioactive; brain/gut peptide; motilin peptide; Brown, JC et al, *Can J Biochem*, 51:533-537, 1973; Fox, JET, *Life Sciences*, 35:695-706, 1984

Sigma M 4505   Porcine   FW 2699.1   ≥97% (HPLC) | Bioactive peptide; gastrointestinal peptide; Schubert, H et al, *Can J Biochem*, 52:7, 1954; Yamada, S et al, *J Am Chem Soc*, 97:7174, 1975; Ikota, N et al, *Chem Pharm Bull*, 28:3347, 1980

Phe-Val-Pro-Ile-Phe-Thr-Tyr-Gly-Glu-Leu-Gln-Arg-Nle-Glu-Glu-Lys-Glu-Arg-Asn-Lys-Gly-Gln

Synonyms: Motilin, (Nle[13],Glu[14])-

Bachem H-4376.0500   Human, porcine   Store at -15°C

Fluka 74590   Porcine   MW 2682.04 $C_{121}H_{189}N_{33}O_{36}$ Fox, JET, *Life Sci*, 35:695, 1984; Mutt, V, *Brain Peptides*, (DT Krieger et al, eds), 871, 1983, John Wiley & Sons, New York; Bertaccini, G, *Handbook Exp Pharmacol*, 59:85, 1982; Bloom, SR, *Hormones in Blood*, (CH Gray & James, VHT, eds), 321, 1979, Academic Press, London

---

Phg-His
4-Carboxy-3-Hydroxy-L-Phg-His

Synonyms: Glutamate Receptor II Agonist; Glutamate Receptor II Antagonist; Metabotropic Glutamate Receptor Agonist

Sigma C 3586   FW 211.2 $C_9H_9NO_5$ Metabotropic; Birse, EF et al, *Neuroscience*, 52:481, 1993; Orlando, LR et al, *Neurosci Lett*, 202:109, 1995

Phg-Leu
D-Phg-Leu

Bachem G-4335.0001   MW 264.32 $C_{14}H_{20}N_2O_3$ Store at -15°C

Pmp-Tyr-Ile-Gln-Asn-Cys-Pro-Orn-Gly
Pmp-Tyr(Me)-Ile-Gln-Asn-Cys-Pro-Orn-Gly Amide

Synonyms: Vasotocin, (Pmp[1],Tyr(OMe)[2],Orn[8])-

American Peptide 66-0-16   MW 1075.3 $C_{48}H_{74}N_{12}O_{12}S_2$ Disulfide bonds: Pmp[1]-Cys[6]

Pmp-Tyr-Phe-Gln-Asn-Cys-Pro-Arg-Gly
Pmp-Tyr(Me)-Phe-Gln-Asn-Cys-Pro-Arg-Gly Amide

Synonyms: Vasopressin, (Pmp[1],Tyr(OMe)[2],Arg[8])-

American Peptide 66-0-09   MW 1151.4 $C_{52}H_{74}N_{14}O_{12}S_2$ Selective V₁ vasopressin antagonist; disulfide bonds: Pmp[1]-Cys[6]; Kruszynski, M et al, *JMC*, 23:364, 1980

Pmp-Tyr-Phe-Val-Asn-Cys-Pro-Cit-Gly
Pmp-D-Tyr(OEt)-Phe-Val-Asn-Cys-Pro-Cit-Gly Amide

Synonyms: Vasopressin, (Pmp[1],D-Tyr(OEt)[2],Val[4],Cit[8])-

American Peptide 66-0-12   MW 1153.4 $C_{53}H_{76}N_{12}O_{13}S_2$ Disulfide bonds: Pmp[1]-Cys[6]; Manning, M et al, *Proc 18th Eur Peptide Symp*, 401, 1984

Poly(Ala)
Poly-DL-Ala

ICN 102677   MW 1-5k   MW by nonaqueous end group titration

ICN 151907   MW 1-5k

Poly(Ala)
Poly-L-Ala
Sigma P 5517   MW 10-25k based on viscosity   Based on poly-γ-benzyl L-glutamate plot:Blout, ER & Karlson, RH, *J Am Chem Soc*, 78:941, 1956; *Progress in Polymer Science, Japan*, 6:51, 1973 (Onogi, S & Uno, K, eds, John Wiley, Kodansha Ltd, Tokyo)

Poly(Ala)
Poly-DL-Ala
Sigma P 9003   MW 1-5k   MW based on viscosity; polyamino acids are used as simple models in structure-property correlation studies of natural polypeptides & proteins & are subjects in the following applications:as enzyme inhibitors, as substrates in the isolation of plasma membranes, in chromosomal preparations, in microencapsulation, in sustained release devices, as drug delivery devices; *polyamino Acids, Polypeptides & Proteins*, Proc Intern Symp, Madison, WI, 1961, Stahmann MA, ed, The Univ of Wisconsin Press, Madison (1962); Silman, IH & Sela, M, *polyamino Acids*, Fasman GD, ed, Marcel Dekker, Inc, NY, 605, 1967; Jacobson, BS & Branton, D, *Science*, 195:302, 1977; Rajendra, BR et al, *Human Genetics*, 55:363, 1980; Lim, F & Sun, AM, *Science*, 210:908, 1980; Sidman, KR et al, *J Membr Sci*, 7:277, 1980; Ryser, HJ-P & Shen, W-C, *Proc Natl Acad Sci USA*, 75:3867, 1978

Poly(Ala)
Poly-β-Ala
Synonyms: Poly(3-Aminopropionic Acid); Poly(Imino(1-Oxo-1,3-Propanediyl))
Sigma P 9313   MW 5-15k   Prepared via phosphorylation | MW based on viscosity Ogata, N et al, *Polymer J*, 8:129, 1976

Poly(Ala:Glu:Lys:Tyr) 6:2:5:1 Hydrobromide
Synonyms: Poly-(Ala,Glu,Lys,Tyr)
Sigma P 1152   MW 20-30k   MW by LALLS

Poly(Arg)
Poly-DL-Arg
ICN 102678   MW 15-50k   Sulfate

Poly(Arg)
Poly-DL-Arg Hydrochloride
ICN 151908   MW 10-20k
ICN 71103L   MW 10-20k

Poly(Arg)
Poly-L-Arg Hydrochloride
Sigma P 3892   MW 70-150k   <3% ornithine residues | MW based on precursor poly-L-ornithine
Sigma P 4663   MW 5-15k   MW based on precursor poly-L-ornithine; contains <3% ornithine residues | See Sigma P 9003 for polyamino acids (homopolymers)

Poly(Arg)
Poly-L-Arg Sulfate salt
Sigma P 7637   MW 15-50k   <3% ornithine residues | MW based on precursor poly-L-ornithine

Poly(Arg)
Poly-L-Arg Hydrochloride
Sigma P 7762   MW 15-70k   <3% ornithine residues | MW based on precursor poly-L-ornithine

Poly(Arg:Pro:Thr) 1:1:1
Sigma P 9306   MW 5-20k   <5% ornithine residues; polymer composition may deviate from the molar feed ratios | Random copolymer of L-Arg; MW based on precursor poly-L-ornithine

Poly(Arg:Pro:Thr) 6:3:1 Hydrochloride
Sigma P 9431   MW 10-30k   <5% ornithine residues; polymer composition may deviate from the molar feed ratios | Random copolymer of L-Arg; MW based on precursor poly-L-ornithine

Poly(Arg:Ser) 3:1 Hydrochloride
Sigma P 0286   MW 20-50k   <5% ornithine residues | MW based on precursor poly-L-ornithine; random copolymer of L-arginine; molar feed ratios employed in the random copolymerizations are given in the listing for each product; actual composition of product determined by AA analysis printed on product label with MW; deviations from feed ratios are the result of differences in monomer reactivities; incorporation will normally be within ±10% of feed ratios but large differences in monomer reactivities can lead to larger deviations; monomers are of the L-configuration unless otherwise noted; Amino acid copolymers have been studied in the applications:as synthetic antigens, as immunosuppressants, as substrates in the purification of proteinkinases, in biodegradable sustained release systems; Sela, M, Fuchs, S & Arnon, R, *Biochem J*, 85:223, 1962; Schneider, CJ et al, *Eur J Immunol*, 8:406, 1978; Teitelbaum, D et al, *Eur J Immunol*, 1:242, 1971; Braun, S, Raymond, WE & Racker, E, *J Biol Chem*, 259:2051, 1984; Sidman, KR et al, *J Membr Sci*, 7:277, 1980

Poly(Arg:Trp) 4:1
Sigma P 0411   MW 20-50k   <5% ornithine residues | Random copolymer of L-Arg; MW based on precursor poly-L-ornithine

Poly(Arg:Tyr) 4:1
Sigma P 7411   MW 20-50k   <5% ornithine residues | Random copolymer of L-Arg; MW based on precursor poly-L-ornithine

Poly(Arg-Gln-Arg-Tyr)
T4(Arg-Gln-Arg-Tyr)4
Synonyms: Neuropeptide Y (33-36); TASP Molecule T4-(NPY(33-36))4
American Peptide 60-1-24   Binds to the Y2 receptor with high affinity & has poor binding to the Y1 receptor; the first potent & selective Y2 antagonist

Poly(Asn)
Poly-(α,β-(N-(3-Hyppyl)-DL-Asn))
Synonyms: Aspartamide
Sigma P 0937   MW 5-20k   ≥95%; prepared via polysuccinimide | MW based on viscosity; also assayed by LALLS; Lupu-Lotan et al, *Biopolymers*, 3:625, 1965; Neri, P et al, *J Med Chem*, 16:893, 1973

Poly(Asn)
Poly-L-Asn
Sigma P 8137   MW 5-15k   May contain ≤30% aspartic acid residues in the polymer | MW based on precursor poly-β-benzyl L-aspartate

Poly(Asn)
Poly-(α,β-(N-(2-Hydroxyethyl)-DL-Asn))
Synonyms: Aspartamide
Sigma P 9061   MW 5-20k   ≥95%; prepared via polysuccinimide | MW based on viscosity; also assayed by LALLS; Lupu-Lotan et al, *Biopolymers*, 3:625, 1965; Neri, P et al, *J Med Chem*, 16:893, 1973

Poly(Asp)
Poly-Asp:L-(^{14}C(U)) Sodium Salt
ARC ARC-1117   100-300 mCi/mmol; 3.7-11.1 GBq/mmol; in sterile H_2O | Radiochemical

Poly(Asp)
Poly-DL-Asp Sodium Salt

ICN 151909   MW 5-15k

Poly(Asp)
Poly-β-Benzyl-L-Asp

Sigma P 2266   MW 50-100k   MW based on viscosity; based on poly-γ-benzyl L-glutamate plot:Blout, ER & Karlson, RH, *J Am Chem Soc*, 78:941, 1956

Poly(Asp)
Poly-(α,β)-DL-Asp Sodium Salt

Sigma P 3056   MW 10-30k   Also assayed by LALLS; prepared by thermal polycondensation | MW based on viscosity; based on poly-L-glutamic acid plot:Idelson, M & Blout, ER, *J Am Chem Soc*, 80:4631, 1958

Sigma P 3418   MW 2-10k   Also assayed by LALLS; prepared by thermal polycondensation | MW based on viscosity; based on poly-L-glutamic acid plot:Idelson, M & Blout, ER, *J Am Chem Soc*, 80:4631, 1958; Kovacs, J et al, *J Org Chem*, 26:1084, 1961

Poly(Asp)
Poly-β-Benzyl-L-Asp

Sigma P 3887   MW 5-15k   MW based on viscosity; based on poly-γ-benzyl L-glutamate plot:Blout, ER & Karlson, RH, *J Am Chem Soc*, 78:941, 1956

Poly(Asp)
Poly-L-Asp Sodium Salt

Sigma P 5387   MW 5-15k   Also assayed by LALLS | MW based on viscosity; based on poly-L-glutamic acid plot:Idelson, M & Blout, ER, *J Am Chem Soc*, 80:4631, 1958

Sigma P 6762   MW 15-50k   Also assayed by LALLS | MW based on viscosity; based on poly-L-glutamic acid plot:Idelson, M & Blout, ER, *J Am Chem Soc*, 80:4631, 1958

Poly(Asp)
Poly-β-Benzyl-L-Asp

Sigma P 7137   MW 15-50k   MW based on viscosity; based on poly-γ-benzyl L-glutamate plot:Blout, ER & Karlson, RH, *J Am Chem Soc*, 78:941, 1956

Poly(Cys)
Poly-S-CBZ-L-Cys

Sigma P 0263   MW 5-15k   MW based on viscosity

Poly(Cys)
Poly-S-Benzyl-L-Cys

Sigma P 7639   MW 2-10k   MW determined by non-aqueous end group titration

Poly(Cys:Lys)
Synonyms: Multichain Polyamino Acid; A-L

Sigma M 3524   MW 50-200k (calculated); average MW ~20k From poly-L-lysine · HBr

Poly(Cys:Lys) Hydrochloride
Synonyms: Multichain Polyamino Acid; A-L

Sigma M 9405   MW 5-15k   ~6:1 | MW by LALLS; a substrate in affinity chromatography; Parikh, I & Cuatrecasas, P, *Chem & Eng News*, Aug 26, 1985; mutichain polyamino acids are well defined synthetic antigens which have contributed to the field of immunogenetics & the definition of the immune response genes; they are branched chain polymers with linear polymeric amino acids attached to a poly-L-lysine backbone; multichain polymers are prepared by polypeptidations by the method of Sela, M et al, *Biochem J*, 93:566, 1964; composition is determined by AA analysis; MW is calculated based on the MW of the poly-L-lysine backbone & the AA composition; McDevitt, HO et al, *J Exp Med*, 135:1259, 1972; Schwartz, AH & Paul, WE, *J Exp Med*, 143:529, 1976; Isaac, R & Moses, E, *J Immunology*, 118:566, 1977; Singer, S et al, *J Exp Med*, 147:1611, 1978; Sela, M et al, *Biochem J*, 85:223, 1962

Poly(-D-Lys) Hydrobromide

Fluka 81358   MW 30-70k   Useful for promoting cell adhesion to solid substances; Jacobson, BS & Branton, D, *Science*, 195:302, 1977

Poly(Glu)
Poly(L-Glu)

ICN 102682   MW 2-15k

ICN 151917   MW 15-50k

ICN 151918   MW >50k

Poly(Glu)
Poly-γ-Methyl-L-Glu

Sigma P 1388   MW 10-50k   MW based on viscosity; also assayed by LALLS

Poly(Glu)
Poly-L-Glu Sodium Salt

Sigma P 1818   MW 1500-3k   May contain ≤15% NaBr | MW determined by capillary electrophoresis; also assayed by LALLS

Poly(Glu)
Poly-N⁵-(3-Hydroxypropyl)-L-Glu

Sigma P 1899   MW 40-60k   MW based on viscosity; also assayed by LALLS

Poly(Glu)
Poly-L-Glu Sodium Salt

Sigma P 1943   MW 750-1500   May contain ≤15% NaBr | MW determined by capillary electrophoresis; also assayed by LALLS

Poly(Glu)
Poly-γ-Methyl-L-Glu

Sigma P 2264   MW >100k   MW based on viscosity; also assayed by LALLS

Poly(Glu)
Poly-γ-Benzyl-D-Glu

Sigma P 3388   MW 150-350k   MW based on viscosity; also assayed by LALLS

Poly(Glu)
Poly-D-Glu Sodium Salt

Sigma P 4033   MW 15-50k   MW based on viscosity; also assayed by LALLS

Poly(Glu)
Poly-L-Glu Sodium Salt

Sigma P 4636   MW 3-15k   MW based on viscosity; also assayed by LALLS

Poly(Glu)
Poly-D-Glu Sodium Salt

Sigma P 4637   MW 50-100k   MW based on viscosity; also assayed by LALLS

Poly(Glu)
Poly-L-Glu Sodium Salt

Sigma P 4761   MW 15-50k   MW based on viscosity; also assayed by LALLS

Poly(Glu)
Poly-N^5-(3-Hydroxypropyl)-L-Glu

Sigma P 4774   MW 15-30k   MW based on viscosity; also assayed by LALLS

Poly(Glu)
Poly(Glu:Glu-OEt) 1:1

Sigma P 4785   MW 70-150k   MW based on precursor

Poly(Glu)
Poly-L-Glu Sodium Salt

Sigma P 4886   MW 50-100k   MW based on viscosity; also assayed by LALLS

Poly(Glu)
Poly(Glu:Glu-OEt) 4:1

Sigma P 4910   MW 70-150k   MW based on precursor

Poly(Glu)
Poly-γ-Benzyl-L-Glu

Sigma P 5011   MW 30-70k   MW based on viscosity; also assayed by LALLS

Sigma P 5136   MW 150-350k   MW based on viscosity; also assayed by LALLS

Sigma P 5261   MW 15-30k   MW based on viscosity; also assayed by LALLS

Sigma P 5386   MW 70-150k   MW based on viscosity; also assayed by LALLS

Poly(Glu)
Poly-γ-Ethyl-L-Glu

Sigma P 8035   MW >100k   MW based on viscosity; also assayed by LALLS

Poly(Glu)
Poly(Glu:Glu-OMe) 4:1

Sigma P 8160   MW 70-150k   MW based on precursor

Poly(Glu)
Poly-L-Glu

Sigma P 8202   MW 2-15k   MW based on viscosity; viscosity is derived from the viscosity of the polymer in H$_2$O & the polyhydroxypropyl-L-glutamine formula for viscosity-MW correlation

Poly(Glu)
Poly-γ-n-Hexyl-L-Glu

Sigma P 8938   MW 30-70k   Assayed by LALLS

Poly(Glu)
Poly-D-Glu Sodium Salt

Sigma P 9917   MW 2-15k   MW based on viscosity; also assayed by LALLS

Poly(Glu) Sodium Salt

Peptides International OEE-3063   MW >8k cut off by dialysis
Colorless amorphous powder | NCA polymerized product

Poly(Glu:Ala)
Poly(L-Glu:L-Ala), 6:4

ICN 152696

Poly(Glu:Ala) 6:4 Sodium Salt

Sigma P 1650   MW 20-50k   MW based on viscosity; also assayed by LALLS

Poly(Glu:Ala:Tyr)
Poly(L-Glu:L-Ala:L-Tyr), 6:3:1

ICN 152697

Poly(Glu:Ala:Tyr) 1:1:1 Sodium Salt

Sigma P 4149   MW 20-50k   MW based on viscosity; also assayed by LALLS

Poly(Glu:Ala:Tyr) 6:3:1 Sodium Salt

Sigma P 3899   MW 20-50k   MW based on viscosity; also assayed by LALLS

Poly(Glu:Leu) 4:1 Sodium Salt

Sigma P 0812   MW 30-70k   MW based on viscosity; also assayed by LALLS

Poly(Glu:Lys)
Poly(D-Glu:D-Lys), 6:4

ICN 152695

Poly(Glu:Lys)
Poly(L-Glu:L-Lys), 1:4 Hydrobromide

ICN 152698

Poly(Glu:Lys)
Poly(D-Glu:D-Lys) 6:4 Hydrobromide

Sigma P 7658   MW 20-50k   MW based on viscosity; also assayed by LALLS

Poly(Glu:Lys) 1:4 Hydrobromide

Sigma P 0650   MW 150-300k   MW based on viscosity

Poly(Glu:Lys:Tyr)
Poly(L-Glu:L-Lys:L-Tyr), 1:1:1

ICN 152699

Poly(Glu:Lys:Tyr) 6:3:1 Sodium Salt

Sigma P 4409   MW 20-50k   MW based on viscosity; also assayed by LALLS

Poly(Glu:Phe) 4:1 Sodium Salt

Sigma P 0687   MW 30-70k   MW based on viscosity; also assayed by LALLS

Poly(Glu:Tyr)
Poly(L-Glu:L-Tyr), 1:1
ICN 152700

Poly(Glu:Tyr)
Poly(L-Glu:L-Tyr), 4:1 Sodium Salt
ICN 193568   MW ~50-60k

Poly(Glu:Tyr) 1:1 Sodium Salt
Sigma P 0151   MW 20-50k   MW based on viscosity; also assayed by LALLS

Poly(Glu:Tyr) 4:1 Sodium Salt
Sigma P 0275   MW 20-50k   MW based on viscosity; also assayed by LALLS

Poly(Glu:Tyr:Ala)
Poly(L-Glu:L-Tyr:L-Ala), 1:1:1
ICN 152702

Poly(Gly)
Sigma P 3548   MW 5-10k   Prepared via phosphorylation | MW based on viscosity
Sigma P 8791   MW 2-5k   Prepared via phosphorylation | MW based on viscosity

Poly(His)
Poly(L-His)
ICN 102686   MW 5-15k

Poly(His)
Poly-L-His Hydrochloride
Sigma P 2534   MW 15-50k   MW assayed by LALLS

Poly(His)
Poly-L-His
Sigma P 9386   MW 5-15k   MW assayed by LALLS

Poly(His)
Poly-im-Benzyl-L-His
Sigma P 9637   MW 5-50k   MW based on viscosity; also assayed by LALLS

Poly(His:Glu:Ala:Lys)
Poly(L-His,L-Glu)-Poly-DL-Ala-Poly(L-Lys)(H,G(AL))
ICN 152709

Poly(Hyp)
Poly(L-Hyp)
ICN 102688   MW 10-30k

Poly(Hyp)
Poly(D-Hyp)
ICN 152688   MW 20-60k

Poly(Hyp)
Poly-L-Hyp
Sigma P 0388   MW 5-20k   MW based on viscosity; also assayed by LALLS; based on poly-L-Pro plot:Mattice, WL & Mandelkern, L, *J Am Chem Soc*, 93:1769, 1971; see Sigma P 9003 for polyamino acids (homopolymers)

Poly(Ile)
Poly-L-Ile
Sigma P 3329   MW 5-15k   MW based on viscosity; based on poly-γ-benzyl L-glutamate plot:Blout, ER & Karlson, RH, *J Am Chem Soc*, 78:941, 1956; see Sigma P 9003 for polyamino acids (homopolymers)

Poly(Leu)
Poly(L-Leu)
ICN 102689   MW 3-15k

Poly(Leu)
Poly-L-Leu
Sigma P 2020   MW 100-150k   MW based on viscosity; see Sigma P 9003 for polyamino acids (homopolymers);Based on poly-γ-benzyl L-glutamate plot:Blout, ER & Karlson, RH, *J Am Chem Soc*, 78:941, 1956
Sigma P 5637   MW 15-50k   MW based on viscosity; see Sigma P 9003 for polyamino acids (homopolymers);Based on poly-γ-benzyl L-glutamate plot:Blout, ER & Karlson, RH, *J Am Chem Soc*, 78:941, 1956
Sigma P 5762   MW 3-15k   MW based on viscosity; see Sigma P 9003 for polyamino acids (homopolymers);Based on poly-γ-benzyl L-glutamate plot:Blout, ER & Karlson, RH, *J Am Chem Soc*, 78:941, 1956

Poly(-L-Lys) Hydrobromide
Fluka 81331   MW 5-10k   Conjugation of poly-L-Lys to albumin & horseradish peroxidase enhances cellular uptake; Shen, W-C & Ryser, HJ-P, *PNAS*, 75:1872, 1978
Fluka 81332   MW 10-20k   Immobilization of living cells in biocompatible semipermeable microcapsules; Goosen, MFA et al, *Polym Sci Tech*, 34:235, 1986
Fluka 81333   MW 20-30k
Fluka 81338   MW 30-70k   A polycation which binds to DNA, red cell membrane & any negatively charged protein; useful for promoting cell adhesion; King, GA et al, *Biotech Prog*, 3:231, 1987; Bottenstein, JE, *Adv Biosciences*, 61:3, 1977; Payares, G & Evans, WH, Mol Biochem Parisitol, 23:129, 1987; Vankova, R & Borman, CH, *Physiol Plant*, 70:1, 1987; Ryser, HJ-P & Shen, C, *NATO ASI Ser*, Ser A, 113:103, 1986
Fluka 81339   MW 70-150k   Useful in promoting cell adhesion to solid substrates; Weill, BJ et al, *J Immuno Meth*, 57:327, 1983
Fluka 81355   MW 150-300k   Preparation of polycationic beads; Helmly, RB & Brown, KM, *Roux's Arch Dev Biol*, 196:262, 1987
Fluka 81356   MW ≥300k   Useful in promoting cell adhesion to solid substrates; Jacobson, BS & Branton, D, *Science*, 195:302, 1977

Poly(Lys)
Poly(L-Lys) Hydrobromide
ICN 102691   MW 4-15k

Poly(Lys)
Poly(D-Lys) Hydrobromide
ICN 102694   MW 4-15k
ICN 150175   MW 75-150k   Attachment factor in cell & tissue culture applications

Poly(Lys)
Poly(L-Lys) Hydrobromide
ICN 150176   MW 30-70k   Attachment factor for tissue culture
ICN 150177   MW >70k   Attachment factor for tissue culture
ICN 152689   MW 1.5-8k

Poly(Lys)
Poly(L-Lys) Hydrochloride
ICN 152690 MW >20k

Poly(Lys)
Poly(L-Lys)
ICN 152691 2.0 *M* NaCl, 0.02% thimerosal, agarose

Poly(Lys)
Poly(L-Lys) Hydrobromide
ICN 194543 MW 30-70k Cell culture reagent; γ-irradiated | Attachment factor for tissue culture
ICN 194544 MW >70k Cell culture reagent; γ-irradiated | Attachment factor for tissue culture
ICN 71120V MW 70k

Poly(Lys)
Poly-ε-CBZ-D-Lys
Sigma P 0168 MW 250-500k MW based on viscosity; also assayed by LALLS; Yaron, A & Berger, A, *Biochim Biophys Acta*, 69:397, 1963; see Sigma P 9003 for polyamino acids (homopolymers)

Poly(Lys)
Poly-DL-Lys Hydrobromide
Sigma P 0171 MW 1-4k MW based on viscosity; also assayed by LALLS; Yaron, A & Berger, A, *Biochim Biophys Acta*, 69:397, 1963; see Sigma P 9003 for polyamino acids (homopolymers)

Poly(Lys)
Poly-D-Lys
Sigma P 0296 MW 1-4k MW based on viscosity; also assayed by LALLS; Yaron, A & Berger, A, *Biochim Biophys Acta*, 69:397, 1963; see Sigma P 9003 for polyamino acids (homopolymers)

Poly(Lys)
Poly-L-Lys Hydrobromide
Sigma P 0879 MW 1-4k MW based on viscosity; also assayed by LALLS; Yaron, A & Berger, A, *Biochim Biophys Acta*, 69:397, 1963; see Sigma P 9003 for polyamino acids (homopolymers)

Poly(Lys)
Poly-D-Lys Hydrobromide
Sigma P 0899 MW 70-150k MW >70,000 useful in promoting cell adhesion to solid substrates; MW based on viscosity; also assayed by SEC-LALLS; Jacobson, BS & Branton, D, *Science*, 195:302, 1977; Yaron, A & Berger, A, *Biochim Biophys Acta*, 69:397, 1963; see Sigma P 9003 for polyamino acids (homopolymers)
Sigma P 1024 MW >300k MW >70,000 useful in promoting cell adhesion to solid substrates; MW based on viscosity; also assayed by SEC-LALLS; Jacobson, BS & Branton, D, *Science*, 195:302, 1977; Yaron, A & Berger, A, *Biochim Biophys Acta*, 69:397, 1963; see Sigma P 9003 for polyamino acids (homopolymers)
Sigma P 1149 MW 150-300k MW >70,000 useful in promoting cell adhesion to solid substrates; MW based on viscosity; also assayed by SEC-LALLS; Jacobson, BS & Branton, D, *Science*, 195:302, 1977; Yaron, A & Berger, A, *Biochim Biophys Acta*, 69:397, 1963; see Sigma P 9003 for polyamino acids (homopolymers)

Poly(Lys)
Poly-L-Lys Hydrobromide
Sigma P 1274 MW 70-150k MW >70,000 useful in promoting cell adhesion to solid substrates; MW based on viscosity; also assayed by SEC-LALLS; Jacobson, BS & Branton, D, *Science*, 195:302, 1977; Yaron, A & Berger, A, *Biochim Biophys Acta*, 69:397, 1963; see Sigma P 9003 for polyamino acids (homopolymers)
Sigma P 1399 MW 150-300k MW >70,000 useful in promoting cell adhesion to solid substrates; MW based on viscosity; also assayed by SEC-LALLS; Jacobson, BS & Branton, D, *Science*, 195:302, 1977; Yaron, A & Berger, A, *Biochim Biophys Acta*, 69:397, 1963; see Sigma P 9003 for polyamino acids (homopolymers)
Sigma P 1524 MW >300k MW >70,000 useful in promoting cell adhesion to solid substrates; MW based on viscosity; also assayed by SEC-LALLS; Jacobson, BS & Branton, D, *Science*, 195:302, 1977; Yaron, A & Berger, A, *Biochim Biophys Acta*, 69:397, 1963; see Sigma P 9003 for polyamino acids (homopolymers)

Poly(Lys)
Poly-L-Lys
Sigma P 2636 MW 30-70k MW based on viscosity; also assayed by SEC-LALLS; Jacobson, BS & Branton, D, *Science*, 195:302, 1977; Yaron, A & Berger, A, *Biochim Biophys Acta*, 69:397, 1963; see Sigma P 9003 for polyamino acids (homopolymers)

Poly(Lys)
Poly-L-Lys Hydrochloride
Sigma P 2658 MW 15-30k MW based on viscosity; also assayed by SEC-LALLS; Jacobson, BS & Branton, D, *Science*, 195:302, 1977; Yaron, A & Berger, A, *Biochim Biophys Acta*, 69:397, 1963; see Sigma P 9003 for polyamino acids (homopolymers)

Poly(Lys)
Poly-ε-CBZ-DL-Lys
Sigma P 2883 MW 5-20k MW based on viscosity; also assayed by LALLS; Yaron, A & Berger, A, *Biochim Biophys Acta*, 69:397, 1963; see Sigma P 9003 for polyamino acids (homopolymers)

Poly(Lys)
Poly-L-Lys FITC Labeled Hydrobromide
Sigma P 3069 MW 30-70k From poly-L-lys; degree of substitution:0.003-0.01 mole FITC/mole Lys monomer | See Sigma P 9003 for polyamino acids (homopolymers)

Poly(Lys)
Poly-L-Lys Succinylated
Sigma P 3513 MW >50k MW based on precursor poly-L-lysine; also assayed by LALLS & SEC-LALLS; intermediate for coupling proteins to poly-L-lysine; Stason, W et al, *Biochim Biophys Acta*, 133:582, 1967; Haber, E et al, *J Clin Endocr*, 29:1349, 1969; see Sigma P 9003 for polyamino acids (homopolymers)

Poly(Lys)
Poly-L-Lys Hydrobromide
Sigma P 3543 MW 15-30k FITC Labeled From poly-L-lysine; degree of substitution:0.003-0.01 mole FITC/mole lysine monomer | See Sigma P 9003 for polyamino acids (homopolymers)

Poly(Lys)
Poly-DL-Lys Hydrobromide
Sigma P 4158 MW >70k MW based on viscosity; also assayed by SEC-LALLS; Yaron, A & Berger, A, *Biochim Biophys Acta*, 69:397, 1963; see Sigma P 9003 for polyamino acids (homopolymers)

Poly(Lys)
Poly-L-Lys Succinylated

Sigma P 4283 MW 15-30k MW based on precursor poly-L-lysine; also assayed by LALLS & SEC-LALLS; intermediate for coupling proteins to poly-L-lysine; Stason, W et al, *Biochim Biophys Acta*, 133:582, 1967; Haber, E et al, *J Clin Endocr*, 29:1349, 1969; see Sigma P 9003 for polyamino acids (homopolymers)

Poly(Lys)
Poly-D-Lys Hydrobromide

Sigma P 4408 MW 15-30k MW based on viscosity; also assayed by SEC-LALLS; Jacobson, BS & Branton, D, *Science*, 195:302, 1977; Yaron, A & Berger, A, *Biochim Biophys Acta*, 69:397, 1963; see Sigma P 9003 for polyamino acids (homopolymers)

Poly(Lys)
Poly-ε-CBZ-L-Lys

Sigma P 4510 MW 1-4k See Sigma P 9003 for polyamino acids (homopolymers)

Poly(Lys)
Poly-L-Lys

Sigma P 4707 MW 70-150k 0.01% solution; prepared in tissue culture grade H$_2$O; sterilize-filtered | Endotoxin tested; cell culture tested

Sigma P 4832 MW 150-300k 0.01% solution; prepared in tissue culture grade H$_2$O; sterilize-filtered | Endotoxin tested; cell culture tested

Poly(Lys)
Poly-L-Lys Hydrobromide

Sigma P 5899 MW >300k Lyophilized; sterilized by γ-irradiation | Cell culture tested

Sigma P 6282 MW 70-150k Lyophilized; sterilized by γ-irradiation | Cell culture tested

Poly(Lys)
Poly-D-Lys Hydrobromide

Sigma P 6403 MW 4-15k MW based on viscosity; also assayed by SEC-LALLS; Jacobson, BS & Branton, D, *Science*, 195:302, 1977; Yaron, A & Berger, A, *Biochim Biophys Acta*, 69:397, 1963; see Sigma P 9003 for polyamino acids (homopolymers)

Poly(Lys)
Poly-D-Lys

Sigma P 6407 MW 70-150k Lyophilized; Sterilized by γ-irradiation | Cell culture tested

Poly(Lys)
Poly-L-Lys Hydrobromide

Sigma P 6516 MW 4-15k MW based on viscosity; also assayed by SEC-LALLS; Yaron, A & Berger, A, *Biochim Biophys Acta*, 69:397, 1963; see Sigma P 9003 for polyamino acids (homopolymers)

Poly(Lys)
Poly-D-Lys Hydrobromide

Sigma P 7280 MW 30-70k Lyophilized; sterilized by γ-irradiation | Cell culture tested

Sigma P 7405 MW >300k Lyophilized; sterilized by γ-irradiation; cell culture tested

Sigma P 7886 MW 30-70k MW based on viscosity; also assayed by SEC-LALLS; Jacobson, BS & Branton, D, *Science*, 195:302, 1977; Yaron, A & Berger, A, *Biochim Biophys Acta*, 69:397, 1963; see Sigma P 9003 for polyamino acids (homopolymers)

Poly(Lys)
Poly-L-Lys Hydrobromide

Sigma P 7890 MW 15-30k MW based on viscosity; also assayed by SEC-LALLS; Jacobson, BS & Branton, D, *Science*, 195:302, 1977; Yaron, A & Berger, A, *Biochim Biophys Acta*, 69:397, 1963; see Sigma P 9003 for polyamino acids (homopolymers)

Poly(Lys)
Poly-L-Lys

Sigma P 8920 0.1% w/v aqueous solution with preservative added | Promotes adhesion of tissue sections to glass slides; particularly useful with immunohistochemical techniques in various microwave procedures; instructions included; see Sigma P 9003 for polyamino acids (homopolymers)

Poly(Lys)
Poly-L-Lys Hydrobromide

Sigma P 8954 MW ~1k MW based on viscosity; also assayed by SEC-LALLS; Yaron, A & Berger, A, *Biochim Biophys Acta*, 69:397, 1963; see Sigma P 9003 for polyamino acids (homopolymers)

Poly(Lys)
Poly-DL-Lys Hydrobromide

Sigma P 9011 MW 30-70k MW based on viscosity; also assayed by SEC-LALLS; Yaron, A & Berger, A, *Biochim Biophys Acta*, 69:397, 1963; see Sigma P 9003 for polyamino acids (homopolymers)

Poly(Lys)
Poly-L-Lys Hydrobromide

Sigma P 9155 MW 30-70k Lyophilized; sterilized by γ-irradiation | Cell culture tested

Poly(Lys)
Poly-L-Lys Hydrochloride

Sigma P 9404 MW 30-70k MW based on viscosity; also assayed by SEC-LALLS; Jacobson, BS & Branton, D, *Science*, 195:302, 1977; Yaron, A & Berger, A, *Biochim Biophys Acta*, 69:397, 1963; see Sigma P 9003 for polyamino acids (homopolymers)

Poly(Lys)
Poly-ε-CBZ-L-Lys

Sigma P 9503 MW 200-500k MW based on viscosity; also assayed by LALLS; Yaron, A & Berger, A, *Biochim Biophys Acta*, 69:397, 1963; see Sigma P 9003 for polyamino acids (homopolymers)

Poly(Lys) Hydrobromide

Peptides International OKK-3056 MW >8k cut off by dialysis
Colorless amorphous powder | NCA polymerized product

Poly(Lys) Hydrochloride

Peptides International OKK-3075 MW >8k cut off by dialysis
Colorless amorphous powder | NCA polymerized product

Poly(Lys:Ala) 1:1 Hydrobromide

Sigma P 4024 MW 20-50k MW based on viscosity; random copolymer of L-Lysine; based on poly-L-lysine hydrobromide:Yaron, A & Berger, A, *Biochim Biophys Acta*, 69:397, 1963; see Sigma P 0286

Poly(Lys:Ala) 2:1 Hydrobromide

Sigma P 1276   MW 20-50k   MW based on viscosity; random copolymer of L-Lysine; based on poly-L-lysine hydrobromide:Yaron, A & Berger, A, *Biochim Biophys Acta*, 69:397, 1963; see Sigma P 0286

Poly(Lys:Ala) 3:1 Hydrobromide

Sigma P 1151   MW 20-50k   MW based on viscosity; random copolymer of L-Lysine; based on poly-L-lysine hydrobromide:Yaron, A & Berger, A, *Biochim Biophys Acta*, 69:397, 1963; see Sigma P 0286

Poly(Lys:Ala:Glu:Tyr) 5:6:2:1 Hydrobromide

Sigma P 1278   MW 30-70k   MW based on LALLS; random copolymer of L-Lysine; see Sigma P 0286

Poly(Lys:Phe)
Poly(L-Lys:L-Phe), 1:1 Hydrobromide

ICN 152703

Poly(Lys:Phe) 1:1 Hydrobromide

Sigma P 3150   MW 20-50k   MW based on viscosity; random copolymer of L-Lysine; based on poly-L-lysine hydrobromide:Yaron, A & Berger, A, *Biochim Biophys Acta*, 69:397, 1963; see Sigma P 0286

Poly(Lys:Ser) 3:1 Hydrobromide

Sigma P 9160   MW 20-50k   MW based on viscosity; random copolymer of L-Lysine; based on poly-L-lysine hydrobromide:Yaron, A & Berger, A, *Biochim Biophys Acta*, 69:397, 1963; see Sigma P 0286

Poly(Lys:Trp)
Poly(L-Lys:L-Trp), 1:1 Hydrobromide

ICN 152704

Poly(Lys:Trp) 4:1 Hydrobromide

Sigma P 9285   MW 20-50k   MW based on viscosity; random copolymer of L-Lysine; based on poly-L-lysine hydrobromide:Yaron, A & Berger, A, *Biochim Biophys Acta*, 69:397, 1963; see Sigma P 0286

Poly(Lys:Tyr)
Poly(L-Lys:L-Tyr), 1:1 Hydrobromide

ICN 152705

Poly(Lys:Tyr)
Poly(L-Lys:L-Tyr), 1:9 Hydrobromide

ICN 152706

Poly(Lys:Tyr) 1:1 Hydrobromide

Sigma P 4274   MW 50-150k   MW based on viscosity; random copolymer of L-Lysine; based on poly-L-lysine hydrobromide:Yaron, A & Berger, A, *Biochim Biophys Acta*, 69:397, 1963; see Sigma P 0286

Poly(Lys:Tyr) 1:9 Hydrobromide

Sigma P 2025   MW 50-150k   MW based on viscosity; random copolymer of L-Lysine; based on poly-L-glutamic acid sodium salt plot:Idelson, M & Blout, ER, *J Am Chem Soc*, 80:4631, 1958; see Sigma P 0286

Poly(Lys:Tyr) 4:1 Hydrobromide

Sigma P 4659   MW 20-50k   MW based on viscosity; random copolymer of L-Lysine; based on poly-L-lysine hydrobromide:Yaron, A & Berger, A, *Biochim Biophys Acta*, 69:397, 1963; see Sigma P 0286

Poly(Lys-Ala) Hydrochloride

Sigma P 5209   MW 5-15k (viscosity)   Sequential copolymer; based on poly-L-lysine hydrobromide:Yaron, A & Berger, A, *Biochim Biophys Acta*, 69:397, 1963; General comments & references for random copolymers are the same as for Sigma P 0286

Poly(Met)
Poly(L-Met)

ICN 102695   MW 30-50k   Prepared by polymerization of L-Met

Poly(Nle:Lys)-Nle
(Nle-(Suc-Lys)₄)₈-Nle

Synonyms: Protein Sequencing Standard; Internal Sequencing Standard Peptide

Sigma I 2274   Lyophilized; vial contains ~2 nmoles | Useful as a qualitative standard in protein sequencing to assure sequencer performance; Elliott, JI et al, *Anal Biochem*, 211:94, 1992

Poly(Orn)
Poly(L-Orn) Hydrobromide

ICN 152692   MW 5-20k

Poly(Orn)
Poly-DL-Orn Hydrobromide

Sigma P 0421   MW 15-30k   MW based on viscosity; also assayed by LALLS & SEC-LALLS; based on poly-L-lysine plot:Yaron, A & Berger, A, *BBA*, 69:397, 1963; see Sigma P 9003 for polyamino acids (homopolymers)

Sigma P 0546   MW 30-50k   MW based on viscosity; also assayed by LALLS & SEC-LALLS; based on poly-L-lysine plot:Yaron, A & Berger, A, *BBA*, 69:397, 1963; see Sigma P 9003 for polyamino acids (homopolymers)

Sigma P 0671   MW >50k   MW based on viscosity; also assayed by LALLS & SEC-LALLS; based on poly-L-lysine plot:Yaron, A & Berger, A, *BBA*, 69:397, 1963; see Sigma P 9003 for polyamino acids (homopolymers)

Poly(Orn)
Poly-L-Orn Hydrobromide

Sigma P 2533   MW 15-30k   MW based on viscosity; also assayed by LALLS & SEC-LALLS; based on poly-L-lysine plot:Yaron, A & Berger, A, *BBA*, 69:397, 1963; see Sigma P 9003 for polyamino acids (homopolymers)

Sigma P 3530   MW 15-30k   MW based on viscosity; also assayed by LALLS & SEC-LALLS; based on poly-L-lysine plot:Yaron, A & Berger, A, *BBA*, 69:397, 1963; see Sigma P 9003 for polyamino acids (homopolymers)

Sigma P 3655   MW 30-70k   MW based on viscosity; also assayed by LALLS & SEC-LALLS; based on poly-L-lysine plot:Yaron, A & Berger, A, *BBA*, 69:397, 1963; see Sigma P 9003 for polyamino acids (homopolymers)

Sigma P 4538   MW 5-15k   MW based on capillary electrophoresis; also assayed by LALLS & SEC-LALLS; based on poly-L-lysine plot:Yaron, A & Berger, A, *BBA*, 69:397, 1963; see Sigma P 9003 for polyamino acids (homopolymers)

Sigma P 4638   MW 100-200k   MW based on viscosity; also assayed by LALLS & SEC-LALLS; based on poly-L-lysine plot:Yaron, A & Berger, A, *BBA*, 69:397, 1963; see Sigma P 9003 for polyamino acids (homopolymers)

Poly(Orn)
Poly-L-Orn

Sigma P 4957   MW 30-70k   0.01% solution; prepared in tissue culture grade H₂O; sterilize-filtered | Endotoxin tested; cell culture tested

Poly(Orn)
Poly-δ-CBZ-DL-Orn

Sigma P 5143   MW 5-20k   MW based on viscosity; also assayed by LALLS; based on poly-e-CBZ-L-lysine plot:Yaron, A & Berger, A, *BBA*, 69:397, 1963; see Sigma P 9003 for polyamino acids (homopolymers)

Poly(Orn)
Poly-L-Orn Hydrobromide

Sigma P 5666   MW ~1k   MW based on viscosity; also assayed by LALLS & SEC-LALLS; based on poly-L-lysine plot:Yaron, A & Berger, A, *BBA*, 69:397, 1963; see Sigma P 9003 for polyamino acids (homopolymers)

Poly(Orn)
Poly-DL-Orn Hydrobromide

Sigma P 8638   MW 3-15k   MW based on viscosity; also assayed by LALLS & SEC-LALLS; based on poly-L-lysine plot:Yaron, A & Berger, A, *BBA*, 69:397, 1963; see Sigma P 9003 for polyamino acids (homopolymers)

Poly(Orn:Leu) 1:1 Hydrobromide

Sigma P 7414   MW 20-50k   MW based on viscosity; random copolymer of L-ornithine; Based on poly-L-lysine hydrobromide:Yaron, A & Berger, A, *BBA*, 69:397, 1963; see Sigma P 0286

Poly(Orn:Ser) 3:1 Hydrobromide

Sigma P 9035   MW 20-50k   MW based on viscosity; random copolymer of L-ornithine; Based on poly-L-lysine hydrobromide:Yaron, A & Berger, A, *BBA*, 69:397, 1963; see Sigma P 0286

Poly(Orn:Trp) 4:1 Hydrobromide

Sigma P 0536   MW 20-50k   MW based on viscosity; random copolymer of L-ornithine; Based on poly-L-lysine hydrobromide:Yaron, A & Berger, A, *BBA*, 69:397, 1963; see Sigma P 0286

Poly(Orn:Tyr) 4:1 Hydrobromide

Sigma P 4534   MW 20-50k   MW based on viscosity; random copolymer of L-ornithine; Based on poly-L-lysine hydrobromide:Yaron, A & Berger, A, *BBA*, 69:397, 1963; see Sigma P 0286

Poly(Phe)
Poly(L-Phe)

ICN 104795   MW 2-5k

Poly(Phe)
Poly-L-Phe

Sigma P 6886   MW 2-5k   MW based on viscosity; based on poly-γ-benzyl L-glutamate plot:Blout, ER & Karlson, RH, *J Am Chem Soc*, 78:941, 1956; Sela, M & Berger, A, *J Am Chem Soc*, 77:1893, 1955; Sela, M & Berger, A, *J Am Chem Soc*, 75:6350, 1953; see Sigma P 9003 for polyamino acids (homopolymers)

Sigma P 7011   MW 5-15k   MW based on non-aqueous end group titration; Sela, M & Berger, A, *J Am Chem Soc*, 77:1893, 1955; Sela, M & Berger, A, *J Am Chem Soc*, 75:6350, 1953; see Sigma P 9003 for polyamino acids (homopolymers)

Sigma P 8264   MW 15-50k   MW based on non-aqueous end group titration; Sela, M & Berger, A, *J Am Chem Soc*, 77:1893, 1955; Sela, M & Berger, A, *J Am Chem Soc*, 75:6350, 1953; see Sigma P 9003 for polyamino acids (homopolymers)

Poly(Phe:Glu:Ala:Lys:Phe)
Poly(L-Phe:L-Glu)-Poly-DL-Ala-Poly(L-Lys)(Phe,G(AL))

ICN 152710

Poly(Pro)
Poly(L-Pro)

ICN 104981   MW 1-10k

Poly(Pro)
Poly-L-Pro

Sigma P 2129   MW 10-30k   MW based on viscosity; also assayed by LALLS; based on poly-L-Pro plot:Mattice, WL & Mandelkern, L, *J Am Chem Soc*, 93:1769, 1971; see Sigma P 9003 for polyamino acids (homopolymers)

Sigma P 2254   MW 1-10k   MW based on viscosity; also assayed by LALLS; based on poly-L-Pro plot:Mattice, WL & Mandelkern, L, *J Am Chem Soc*, 93:1769, 1971; see Sigma P 9003 for polyamino acids (homopolymers)

Sigma P 3886   MW >30k   MW based on viscosity; also assayed by LALLS; based on poly-L-Pro plot:Mattice, WL & Mandelkern, L, *J Am Chem Soc*, 93:1769, 1971; see Sigma P 9003 for polyamino acids (homopolymers)

Poly(Pro-Gly-Pro)
Poly(L-Pro-Gly-L-Pro)

ICN 152707

Poly(Pro-Gly-Pro)

Sigma P 6665   MW 2-10k (LALLS)   May contain ≤5% bromide | Sequential copolymer

Poly(Pro-Hyp-Gly)
(Pro-Hyp-Gly)₅ 10H₂O

Peptides International OPG-4032   MW 1534.55
$C_{60}H_{87}N_{15}O_{21} \cdot 10H_2O$ >96% (HPLC); white powder

Poly(Pro-Hyp-Gly)
(Pro-Hyp-Gly)₁₀ 20H₂O

Peptides International OPG-4033   MW 3051.2
$C_{120}H_{172}N_{30}O_{41} \cdot 20H_2O$ >90% (HPLC); white powder

Poly(Pro-Pro-Gly)
(Pro-Pro-Gly)₅ 4H₂O

Peptides International OPG-4005   MW 1436.54
$C_{60}H_{87}N_{15}O_{16} \cdot 4H_2O$ >97% (HPLC); white powder

Poly(Pro-Pro-Gly)
(Pro-Pro-Gly)₁₀ 9H₂O

Peptides International OPG-4006   MW 2693.04
$C_{120}H_{172}N_{30}O_{31} \cdot 9H_2O$ >95% (HPLC); white powder

Poly(Sar)
Poly(L-Sar)

ICN 105198   MW 1-5k

Poly(Ser)
Poly(L-Ser)

ICN 104898   MW 5-10k
ICN 152693   MW 3-10k

Poly(Ser)
Poly-DL-Ser

Sigma P 0833   MW 5-15k   MW based on viscosity; based on poly-γ-benzyl L-glutamate plot:Blout, ER & Karlson, RH, *J Am Chem Soc*, 78:941, 1956; Based on the viscosity of its precursor, Poly-*o*-Ac-L-serine; see Sigma P 9003 for polyamino acids (homopolymers)

Sigma P 5887   MW 5-10k   MW based on viscosity; Based on poly-γ-benzyl L-glutamate plot:Blout, ER & Karlson, RH, *J Am Chem Soc*, 78:941, 1956; Based on the viscosity of its precursor, Poly-*o*-Ac-L-serine; see Sigma P 9003 for polyamino acids (homopolymers)

Poly(Ser)
Poly-o-CBZ-L-Ser
Sigma P 7512   MW 5-20k   MW based on LALLS; see Sigma P 9003 for polyamino acids (homopolymers)

Poly(Thr)
Poly-L-Thr
Sigma P 8077   MW 5-15k   MW based on viscosity; based on poly-L-glutamic acid plot:Idelson, M & Blout, ER, *J Am Chem Soc*, 80:4631, 1958; see Sigma P 9003 for polyamino acids (homopolymers)

Poly(Trp)
Poly(L-Trp)
ICN 104896   MW 1-5k

Poly(Trp)
Poly(DL-Trp)
ICN 104897   MW 1-5k

Poly(Trp)
Poly-L-Trp
Sigma P 0644   MW 5-15k   MW based on viscosity; also assayed by LALLS based on poly-γ-benzyl L-glutamate plot:Blout, ER & Karlson, RH, *J Am Chem Soc*, 78:941, 1956; see Sigma P 9003 for polyamino acids (homopolymers)
Sigma P 4647   MW 1-5k   MW based on viscosity; also assayed by LALLS based on poly-γ-benzyl L-glutamate plot:Blout, ER & Karlson, RH, *J Am Chem Soc*, 78:941, 1956; see Sigma P 9003 for polyamino acids (homopolymers)
Sigma P 4772   MW 15-50k   MW based on viscosity; also assayed by LALLS based on poly-γ-benzyl L-glutamate plot:Blout, ER & Karlson, RH, *J Am Chem Soc*, 78:941, 1956; see Sigma P 9003 for polyamino acids (homopolymers)

Poly(Trp)
Poly-DL-Trp
Sigma P 8389   MW 1-5k   MW based on viscosity; also assayed by LALLS; based on poly-γ-benzyl L-glutamate plot:Blout, ER & Karlson, RH, *J Am Chem Soc*, 78:941, 1956; see Sigma P 9003 for polyamino acids (homopolymers)
Sigma P 8514   MW 5-15k   MW based on viscosity; also assayed by LALLS based on poly-γ-benzyl L-glutamate plot:Blout, ER & Karlson, RH, *J Am Chem Soc*, 78:941, 1956; see Sigma P 9003 for polyamino acids (homopolymers)

Poly(Tyr)
Poly(L-Tyr)
ICN 104796   MW 40-100k
ICN 152694   MW 12-35k
Sigma P 1800   MW 10-40k   MW assayed by LALLS; see Sigma P 9003 for polyamino acids (homopolymers)

Poly(Tyr)
Poly-o-CBZ-D-Tyr
Sigma P 3263   MW 150-250k   MW based on viscosity; also assayed by LALLS based on poly-γ-benzyl L-glutamate plot:Blout, ER & Karlson, RH, *J Am Chem Soc*, 78:941, 1956; see Sigma P 9003 for polyamino acids (homopolymers)

Poly(Tyr)
Poly-D-Tyr
Sigma P 3638   MW 40-100k   MW assayed by LALLS; see Sigma P 9003 for polyamino acids (homopolymers)

Poly(Tyr)
Poly-L-Tyr
Sigma P 7887   MW 40-100k   MW assayed by LALLS; see Sigma P 9003 for polyamino acids (homopolymers)

Poly(Tyr:Glu:Ala:Lys)
Poly(L-Tyr:L-Glu)-Poly-DL-Ala-Poly(L-Lys)(T,G(AL))
ICN 152711

Poly(Tyr:Glu:Ala:Lys)
Poly(L-Tyr:L-Glu)-Poly(DL-Ala)-Poly(L-Lys)
Synonyms: Multichain Polyamino Acid
Sigma M 3649   MW 100-250k

Poly(Val)
Poly(L-Val)
ICN 102676   MW 12-35k

Poly(-ε-CBZ-L-Lys)
ICN 102692   MW 1-4k

Pro-Ala
Z-Pro-Ala
Bachem C-2485.0001   MW 320.35 $C_{16}H_{20}N_2O_5$ Store at -15°C

Pro-Ala
Bachem G-2985.0250   MW 186.21 $C_8H_{14}N_2O_3$ Store at -15°C
ICN 102733   MW 186.2 $C_8H_{14}N_2O_3$

Pro-Ala
N-CBZ-Pro-Ala
Sigma C 7647   FW 320.3 $C_{16}H_{20}N_2O_5$

Pro-Ala
Sigma P 0755   FW 186.2 $C_8H_{14}N_2O_3$

Pro-Ala
CBZ-Pro-Ala
USBio C2099-51   MW 320.4 $C_{16}H_{20}N_2O_5$ ≥99%

Pro-Ala-Abu-Cys-His-Ala-Lys-Abz
Dnp-Pro-β-Cyclohexyl-Ala-Abu-Cys(Me)-His-Ala-Lys(N-Me-Abz) Amide
Bachem M-1910.0001   MW 1105.29 $C_{51}H_{72}N_{14}O_{12}S$ Store at -15°C

Pro-Ala-Ala-Asp-Gly-Val-Gly-Ala-Ala-Ser-Arg-Asp-Leu-Glu-Lys
Synonyms: NEF (25-39); HTLV I Peptide
Neosystem SC217   MW 1456.57   HTLV I peptide from subtype NEF

Pro-Ala-Glu-Asp-Leu-Ala-Arg-Tyr-Tyr-Ser-Ala-Leu-Arg-His-Tyr-Ile-Asn-Leu-Ile-Thr-Arg-Gln-Arg-Tyr Amide
Synonyms: Neuropeptide Y (13-36); Neuropeptide Y Y2-Receptor Selective Agonist
Bachem H-9300.0001   Porcine   MW 2982.4 $C_{135}H_{209}N_{41}O_{36}$ Store at -15°C
ICN 154449   Porcine   MW 2982.7   Walker, MW & RJ Miller, *Mol Pharm*, 34:779, 1988

Neosystem SC371 Porcine MW 2982.39 Bioactive neuropeptide; mimics the effects of NPY & PYY at presynaptic receptors in vas deferens; Krstenansky, JL et al, *PNAS* USA, 86:4377, 1989

Sigma N 6521 Porcine FW 2982.4 ≥97% (HPLC); peptide content:~70% | Bioactive peptide; gastrointestinal peptide; Y_2 agonist; competes for NPY binding sites in rat brain; mimics the inhibitory effects of NPY at presynaptic receptors in rat vas defercus; Walker, MW & Miller, RJ, *Mol Pharm*, 34:779, 1988

Peptides International PNP-4315-s Porcine synthetic MW 2982.4 $C_{135}H_{209}N_{41}O_{36}$ >99% (HPLC); lyophilized amorphous powder | Bioactive; bovine; Walker, MW & RJ Miller, *Mol Pharmacol*, 34:779, 1988; Sheikh, SP, *Am J Physiol*, 261:G701, 1991; Fuhlendorff, J et al, *PNAS* USA, 87:182, 1990

Pro-Ala-Glu-Asp-Met-Ala-Arg-Tyr-Tyr-Ser-Ala-Leu-Arg-His-Tyr-Ile-Asn-Leu-Ile-Thr-Arg-Gln-Arg-Tyr Amide

Syno Neuropeptide Y (13-36)*nyms*:

American Peptide 60-1-16 Human MW 3000.5 $C_{134}H_{207}N_{41}O_{36}S$ Selective Y_2 agonist

---

Pro-Ala-Glu-Asp-Met-Ala-Arg-Tyr-Tyr-Ser-Ala-Leu-Arg-His-Tyr-Ile-Asn-Leu-Leu-Thr-Arg-Pro-Arg-Tyr Amide

Syno Neuropeptide Y (13-36), (Leu³¹,Pro³⁴)-*nyms*:

Bachem H-3318.0500 Human, rat Store at -15°C

Pro-Ala-Glu-Asp-Met-Ala-Arg-Tyr-Tyr-Ser-Ala-Leu-Arg-His-Tyr-Ile-Asn-Leu-Ile-Thr-Arg-Gln-Arg-Tyr Amide

Syno Neuropeptide Y (13-36)*nyms*:

Bachem H-3324.0500 Human, rat MW 3000.44 $C_{134}H_{207}N_{41}O_{36}S$ Store at -15°C

Neosystem SC370 Human, rat MW 3000.42 Bioactive neuropeptide

Pro-Ala-Gly-Cys-His-Ala-Lys Dnp-Pro-Cha-Gly-Cys(Me)-His-Ala-Lys (NMa) Amide

Peptides International SDP-3815-PI MW 1077.24 $C_{49}H_{68}N_{14}O_{12}S$ >98% (HPLC); white crystalline powder | Berman, J et al, *JBC*, 267:1434, 1992

Pro-Ala-Gly-Cys-His-Ala-Lys-Abz Dnp-Pro-β-Cyclohexyl-Ala-Gly-Cys(Me)-His-Ala-Lys(*N*-Me-Abz) Amide

Bachem M-2055.0001 MW 1077.23 $C_{49}H_{68}N_{14}O_{12}S$ Store at -15°C

Pro-Ala-Gly-Pro Z-Pro-Ala-Gly-Pro-4MβNA

Bachem J-1145.0050 MW 629.71 $C_{34}H_{39}N_5O_7$ Store at -15°C

Pro-Ala-Leu-Pro-Glu-Asp-Gly-Gly-Ser-Gly-Ala-Phe-Pro-Pro-Gly-His-Phe-Lys-Asp-Pro-Lys-Arg-Leu-Tyr

Synonyms: Fibroblast Growth Factor (1-24), Brain Derived Basic ; Fibroblast Growth Factor Basic Fragment (1-24)

American Peptide 50-0-30 MW 2553.7 $C_{118}H_{173}N_{31}O_{33}$ Gimenez-Gallego, G et al, *Science*, 230:1385, 1985

Sigma F 5895 Bovine FW 2553.9 HPLC shows two peaks which are believed to be different conformations of the same peptide | Bioactive peptide; Esch, F et al, *Proc Natl Acad Sci USA*, 82:6507, 1985

Bachem H-5660.0001 Bovine brain Store at -15°C

Pro-Ala-Leu-Pro-Glu-Asp-Gly-Gly-Ser-Gly-Ala-Phe-Pro-Pro-Gly-His-Phe-Lys-Asp-Pro-Lys-Arg-Leu-Tyr Basic

Syno Fibroblast Growth Factor (1-24), β-*nyms*:

ICN 151525 BOVINE SYNTHETIC Corresponds to AA 1-24 of bovine brain bFGF; Esch, F etal, *PNAS*, 82:6507, 1985

Pro-Arg

Bachem G-2990.0250 MW 271.32 $C_{11}H_{21}N_5O_3$ Store at -15°C

Pro-Arg Z-Pro-Arg-AMC Hydrochloride

Bachem I-1165.0050 MW 599.09 $C_{29}H_{34}N_6O_6 \cdot HCl$ Store at -15°C

Pro-Arg Pro-Arg-AMC Hydrochloride

Bachem I-1605.0050 Store at -15°C

Pro-Arg Pro-Arg-4MβNA

Bachem J-1285.0050 Store at -15°C

Pro-Arg Pro-Arg-βNA Hydrochloride

Bachem K-1470.0250 Store at -15°C

Pro-Arg Pro-Arg-4-Methoxy-β-Naphthylamide Acetate Salt

ICN 156367 Light orange powder

Pro-Arg Z-Aad-Pro-Arg-pNA AcOH

Synonyms: Pefachrome®FXIa; Pefa-3090

Pentapharm 090-46, 090-41 MW 728.77 Highly selective chromogenic peptide substrate for factor XIa

---

Pro-Arg Pro-Arg-4-Methoxy-β-Naphthylamide Acetate Salt

Synonyms: Cathepsin C Substrate

Sigma P 7521 FW 426.5 (free base) $C_{22}H_{30}N_6O_3$ Light orange powder | Fluorogenic substrate; Dolbeare, FA & Smith, RE, *Clin Chem*, 23:1485, 1977

Pro-Arg-Cys-Glu-Tyr-Gln-Ala-Gly-His-Asn-Lys-Val-Gly-Ser-Leu

Synonyms: VIF (131-145); HTLV I Peptide

Neosystem SC234 MW 1658.85 HTLV I peptide from subtype VIF; due to the presence of a Cys this peptide can contain the dimeric form

---

Pro-Arg-Gly Amide

Bachem H-4670.0025 MW 327.39 $C_{13}H_{25}N_7O_3$ Store at -15°C

Pro-Arg-Ile-Trp-Leu-His-Gly-Leu-Gly-Gln-His-Ile-Tyr-Glu

Synonyms: VPR (35-48); HTLV I Peptide

Neosystem SC558 MW 1718.97 HTLV I peptide from subtype VPR

Pro-Arg-Leu-Glu-Pro-Trp-Lys-His-Pro

Synonyms: TAT (6-14); HTLV I Peptide

Neosystem SC519   MW 1159.35   HTLV I peptide from subtype TAT

Pro-Asn

Bachem G-2995.0250   MW 229.24 $C_9H_{15}N_3O_4$ Store at -15°C

ICN 156368   MW 229.2 $C_9H_{15}N_3O_4$

Pro-Asn-Ser-Lys-Leu-Asp-Asp-Gly-Asn-Met-Ser-Val-His-Met-Gly Acetate Salt

Synonyms: Ecdysone Receptor (78-92)

Biogenesis 4000-0270   Insect synthetic   MW 1602.8   Semi-pure; lyophilized

Pro-Asn-Thr-Cys-Glu-Ile-Cys-Ala-Tyr-Ala-Ala-Cys-Thr-Gly-Cys

Synonyms: Guanylin; Guanylate Cyclase C Activator

Bachem H-1342.0500   Rat   Store at -15°C | Disulfide bonds: Cys^4-Cys^{12}, Cys^7-Cys^{15}

Alexis 157-011   Rat, mouse   MW 1515.8 $C_{60}H_{90}N_{16}O_{22}S_4$ ≥98%; white lyophilized powder; soluble in water | Disulfide bonds: Cys^4-Cys^{12}, Cys^7-Cys^{15}; endogenous intestinal guanylyl cyclase activator; de Sauvage, FJ et al, PNAS, 89:9089, 1992; Wiegand, RC et al, BBRC, 185:812, 1992; Currie, MG et al, PNAS, 89:947, 1992; Schulz, S et al, J Biol Chem, 267:16019, 1992

American Peptide 43-0-12   Rat, mouse   MW 1515.7 $C_{60}H_{90}N_{16}O_{22}S_4$ Disulfide bonds: Cys^4-Cys^{12}, Cys^7-Cys^{15}; an activate intestinal guanylate cyclase in the intestine & stimulate the production of cGMP which causes a secretory diarrhea; may play a role in regulating fluid & electrolyte absorption in intestines; Wiegand, RC et al, BBRC, 185:812, 1992

Peptides International PGN-4275-s   Rat, mouse synthetic   MW 1515.7 $C_{60}H_{90}N_{16}O_{22}S_4$ >99% (HPLC); lyophilized amorphous powder | Bioactive; Disulfide bonds: Cys^4-Cys^{12}, Cys^7-Cys^{15}; Currie, MG et al, PNAS USA, 89:947, 1992; Wiegand, RC et al, BBRC, 185:812, 1992; Schults, S et al, JBC 267:16019, 1992; de Sauvage, FJ et al, PNAS USA, 89:9089, 1992

Pro-Asp

Bachem G-3000.0250   MW 230.22 $C_9H_{14}N_2O_5$ Store at -15°C

ICN 156369   MW 230.2 $C_9H_{14}N_2O_5$

Sigma P 7037   FW 230.2 $C_9H_{14}N_2O_5$

Pro-Asp-Gly-Phe
Ac-Pro-Asp-Gly-Phe-Beta (572-589)

BioSource International 77-400

Pro-Asp-Lys-Asp-Phe-Ile-Val-Asn-Pro-Ser-Asp-Leu-Val-Leu-Asp-Asn-Lys-Ala-Ala-Leu-Arg-Asp-Tyr-Leu-Arg-Gln-Ile-Asn-Glu-Tyr-Phe-Ala-Ile-Ile-Gly-Arg-Pro-Arg-Phe Amide

Synonyms: Neuropeptide F

Bachem H-3278.0500   Store at -15°C

Pro-Asp-Val-Asp-His-Val-Phe-Leu-Arg-Phe Amide

Synonyms: Schisto FLRF Amide; FMRF Amide-Like Neuropeptide; FMRF-Like Neuropeptide

Bachem H-8040.0001   MW 1243.44 $C_{59}H_{86}N_{16}O_{14}$ Store at -15°C

Sigma P 2178   FW 1243.4   ≥97% (HPLC) | Bioactive peptide; originally isolated from locust (Schistocera gregaria); cardioinhibitory; inhibits contraction of the oviduct visceral muscle, but potentiates contraction of the extensortibiae muscle; Robb, S et al, Biochem Biophys Res Commun, 160:850, 1989

American Peptide 60-9-45   Schistocera gregaria (locust)
MW 1243.5 $C_{59}H_{86}N_{16}O_{14}$ Originally isolated from the sexually mature thoracic nervous system of the locust; functions as a potent cardioinhibitory agent; inhibits the contraction of the locust oviduct visceral muscle, but has potentiating effects on the extensortibiae muscle; Robb, S et al, BBRC, 160:850, 1989

ICN 157923   Schistocerca gregaria (locust)   MW 1243.4   97% | FMRF-amide like neuropeptide

Pro-Cha-Gly-Cys-His-Ala-Lys
DNP-Pro-Cha-Gly-Cys(Me)-His-Ala-Lys(Nma) Amide

Synonyms: Collagenase Substrate

American Peptide 81-7-90   MW 1077.2 $C_{49}H_{68}N_{14}O_{12}S$ Berman, J et al, JBC, 267:1434, 1992

Pro-Cys-Lys-Asn-Phe-Phe-Trp-Lys-Thr-Phe-Ser-Ser-Cys-Lys

Synonyms: Cortistatin 14; Cortistatin (16-29); Sleep Modulating Peptide

American Peptide 34-7-31   MW 1721.1 $C_{81}H_{113}N_{19}O_{19}S_2$ Disulfide bonds: Cys^2-Cys^{13}; de Leccea, L et al, Nature, 381:242, 1996

Neosystem SC996   MW 1721.02   Disulfide bonds: Cys^2-Cys^{13}; bioactive; a neuropeptide that exhibits strong structural similarity to somatostatin, although it is the product of a different gene; depresses neuronal electrical activity; unlike somatostatin, induces low-frequency waves in the cerebral cortex & antagonizes the effects of acetylcholine on hippocampal & cortical measures of excitability; suggests a mechanism for cortical synchronization related to sleep; De Lecea, L et al, Nature, 381:242-245, 1996

Sigma C 5808   FW 1721.0   ≥97% (HPLC); 0.5 mg/vial | Bioactive peptide; disulfide bonds: 2-13; somatostatin-like neuropeptide with neuronal depressant & sleep modulating properties; de Leccea, L et al, Nature, 381:242, 1996

Peptides International PCN-4329-v   Rat synthetic   MW 1721.0 $C_{81}H_{113}N_{19}O_{19}S_2$ >97% (HPLC); lyophilized amorphous powder | Bioactive; Disulfide bonds: Cys^2-Cys^{13}; neuronal depressant; De Lecea, L et al, Nature, 381:242, 1996; De Lecea, L et al, J Neurosci, 17:5868, 1997; Connor, M et al, Br J Pharmacol, 122:1567, 1997; Flood, JF et al, Brain Res, 775:250 1997

Pro-Gln

Bachem G-3010.0250   MW 243.26 $C_{10}H_{17}N_3O_4$ Store at -15°C

Pro-Gln-Arg-Phe Amide

Synonyms: Morphine Modulating Neuropeptide C-Terminal Fragment

American Peptide 32-0-67   MW 545.6 $C_{25}H_{39}N_9O_5$ Yang, H et al, PNAS, 82:7757, 1985

Bachem H-6865.0005   Store at -15°C

ICN 154514   MW 545.7

Pro-Gln-Arg-Phe Amide Trifluoroacetate Salt

Synonyms: Morphine Modulating Peptide

Sigma P 0208   FW 545.6 (free base) $C_{25}H_{39}N_9O_5$ ≥97% (HPLC) | Opioid peptide; bioactive peptide; Raffa, RB et al, Peptides, 15:404, 1994

Pro-Gln-Asp-Val-Lys-Phe-Pro

Synonyms: HCV Core Protein (19-25)

Bachem H-2536.0001   Store at -15°C

Pro-Gln-Gln-Phe-Phe-Gly-Leu-Met
Dehydro-Pro-Gln-Gln-Phe-Phe-Gly-Leu-Met Amide

Synonyms: Substance P, (Dehydro-Pro^4)-4-1

American Peptide 70-1-48   MW 964.3 $C_{46}H_{65}N_{11}O_{10}S$

Pro-Gln-Gln-Phe-Phe-Gly-Leu-Met
Pro-Gln-Gln-Phe-*p*-Iodo-Phe-Gly-Leu-Met Amide

Synonyms: Substance P (4-11), (*p*-Iodo-Phe[8])-

ICN 153023

Pro-Gln-Gln-Phe-Phe-Gly-Leu-Met Amide

Synonyms: Substance P, Octa-; Substance P (4-11)

American Peptide 70-1-46   MW 966.2 $C_{46}H_{67}N_{11}O_{10}S$
More active than the natural peptide in various pharmacological tests *in vitro* & *in vivo*; Bury, RW et al, *JMC*, 19:854, 1976

Bachem H-4680.0001   MW 966.17 $C_{46}H_{67}N_{11}O_{10}S$ Store at -15°C

ICN 153021   Bergmann, J et al, *Experentia*, 30:401, 1974; Bury, RW & ML Mashford, *J Med Chem*, 19:854, 1976; Bury, RW & ML Mashford, *Clin Exp Pharmac Physiol*, 4:453, 1977; Couture, R et al, *Can J Physiol Pharmac*, 57:1437, 1979

Pro-Gln-Gln-Phe-Phe-Gly-Leu-Met Amide Acetate Salt

Synonyms: Substance P (4-11)

Sigma S 0397   FW 966.2 (free base) $C_{46}H_{67}N_{11}O_{10}S$ ≥97% (HPLC) | Bioactive peptide; potent substance P antagonist *in vivo* & *in vitro*; Couture, R et al, *Can J Physiol Pharmacol*, 57:1437, 1979; Bury, RW & Mashford, ML, *Clin Exp Pharmacol Physiol*, 4:453, 1977

Pro-Gln-Gln-Trp-Phe-Trp-Leu-Met
D-Pro-Gln-Gln-D-Trp-Phe-D-Trp-Leu-Met Amide

Synonyms: Substance P (4-11), (D-Pro[4],D-Trp[7,9])-

American Peptide 70-9-26   MW 1134.4 $C_{57}H_{75}N_{13}O_{10}S$
Potent competitive antagonist of substance P; Caranikas, S et al, *Eur J Pharmacol*, 77:205, 1982

ICN 153024   One of the most potent substance P antagonists; Caranikas, S et al, *Eur J Pharmac*, 77:205, 1982; Caranikas, S et al, *J Med Chem*, 25:1313, 1982

Pro-Gln-Gln-Trp-Phe-Trp-Leu-Met
D-Pro-Gln-Gln-D-Trp-Phe-D-Trp-Leu-Met Amide AcOH 4H2O

Synonyms: Substance P (4-11), (D-Pro[4],D-Trp[7,9])-; Substance P Antagonist

Peptides International PSP-4114   Synthetic   MW 1266.51
$C_{57}H_{75}N_{13}O_{10}S \cdot CH_3COOH \cdot 4H_2O$ >95% (HPLC); lyophilized amorphous powder; bulk | Bioactive; Caranikas, S et al, *Eur J Pharmacol*, 77:205, 1982

Pro-Gln-Gln-Trp-Phe-Trp-Leu-Met
D-Pro-Gln-Gln-D-Trp-Phe-D-Trp-Leu-Met Amide

Synonyms: Substance P, (D-Pro[4],D-Trp[7,9])-; Substance P Antagonist

Peptides International PSP-4114-v   Synthetic   MW 1134.4
$C_{57}H_{75}N_{13}O_{10}S$ >95% (HPLC); lyophilized amorphous powder | Bioactive; Caranikas, S et al, *Eur J Pharmacol*, 77:205, 1982

Pro-Gln-Gln-Trp-Phe-Trp-Leu-Nle
D-Pro-Gln-Gln-D-Trp-Phe-D-Trp-Leu-Nle Amide

Synonyms: Substance P (4-11), (D-Pro[4],D-Trp[7,9],Nle[11])-

American Peptide 70-9-27   MW 1116.3 $C_{58}H_{77}N_{13}O_{10}$
Potent competitive antagonist of substance P; Bredi-Dobreva, G et al, *Acta Physiol Pharmacol*, 15:10, 1989

Bachem H-4690.0001   MW 1116.33 $C_{58}H_{77}N_{13}O_{10}$ Store at -15°C

Sigma S 6647   FW 1116.3   ≥95% (HPLC) | Bioactive peptide; potent Substance P, antagonist; Bredi-Dobreva, G et al, *Acta Physiol Pharmacol Bulg*, 15:10, 1989

Pro-Gln-Gln-Trp-Phe-Trp-Trp-Met
D-Pro-Gln-Gln-D-Trp-Phe-D-Trp-D-Trp-Met Amide

Synonyms: Substance P (4-11), (D-Pro[4],D-Trp[7,9,10])-

American Peptide 70-9-29   MW 1207.4 $C_{62}H_{74}N_{14}O_{10}S$
Competitive antagonist for substance P, bombesin & cholecystokinin; Mirzrahi, J et al, *Eur J Pharmacol*, 91:139, 1983

Bachem H-4695.0001   MW 1207.42 $C_{62}H_{74}N_{14}O_{10}S$ Store at -15°C

Sigma S 6397   FW 1207.4   ≥95% (HPLC) | Bioactive peptide; competitive antagonist of Substance P, bombesin, & cholecystokinin; Regoli, D et al, *Eur J Pharmacol*, 97:179, 1984; Mizrahi, J et al, *Dur J Pharmacol*, 99:193, 1984; Zhang, L et al, *Biochim Biophys Acta*, 972:37, 1988

Pro-Gln-Gln-Trp-Phe-Trp-Trp-Phe
D-Pro-Gln-Gln-D-Trp-Phe-D-Trp-D-Trp-Phe Amide

Synonyms: Substance P (4-11), (D-Pro[4],D-Trp[7,9,10],Phe[11])-

Sigma S 6522   FW 1223.4   ≥97% (HPLC) | Bioactive peptide; potent competitive Substance P antagonist; Regoli, D et al, *Eur J Pharmacol*, 97:179, 1984; Mizrahi, J et al, *Dur J Pharmacol*, 99:193, 1984

Pro-Gln-Gln-Trp-Val-Trp-Trp-Met
D-Pro-Gln-Gln-D-Trp-Val-D-Trp-D-Trp-Met Amide

Synonyms: Substance P (4-11), (D-Pro[4],D-Trp[7,9,10],Val[8])-

Bachem H-4705.0001   Store at -15°C

ICN 153025   Antagonist of substance P in the peripheral & central nervous systems; Stoppini, L et al, *Neurosci Lett*, 37:279, 1983

Pro-Gln-Gly
Dnp-Pro-Gln-Gly

Bachem M-1335.0025   MW 466.41 $C_{18}H_{22}N_6O_9$ Store at -15°C

Pro-Gln-Gly
DNP-L-Pro-L-Gln-Gly

ICN 151010   MW 466.4 $C_{18}H_{22}N_6O_9$ Reference compound for measurement of collagenase activity; Masui, Y et al, *Biochem Med*, 17:215, 1977

Pro-Gln-Gly
Dnp-Pro-Gln-Gly

Synonyms: Collagenase Substrate

Peptides International SDP-3089   MW 466.41 $C_{18}H_{22}N_6O_9$
>98% (HPLC); amorphous powder

Pro-Gln-Gly-Ile-Ala-Gly-Gln-Arg
Dnp-Pro-Gln-Gly-Ile-Ala-Gly-Gln-D-Arg

Synonyms: Collagenase Substrate, Animal

Bachem M-1340.0025   MW 992.02 $C_{40}H_{61}N_{15}O_{15}$ Store at -15°C

Peptides International SDP-3087-v   MW 992.02
$C_{40}H_{61}N_{15}O_{15}$ >96% (HPLC); lyophilized amorphous powder | Masui, Y et al, *Biochem Med*, 17:215, 1977

Pro-Gln-Gly-Ile-Ala-Gly-Gln-Arg
N-DNP-Pro-Gln-Gly-Ile-Ala-Gly-Gln-D-Arg

Synonyms: Collagenase Substrate

Sigma D 2393   FW 992.0 $C_{40}H_{61}N_{15}O_{15}$ ~95% (HPLC) | Fluorogenic substrate; Gray, RD & Saneii, HH, *Anal Biochem*, 120:339, 1982

Pro-Gln-Ile-Thr-Leu-Trp-Gln-Arg-Pro-Leu-Val-Thr-Ile-Lys-Ile

Synonyms: POL (1-15); PR (1-15); HTLV I Peptide

Neosystem SC700   MW 1806.22   HTLV I peptide from subtype POL (PR/RT)

Pro-Gln-Phe-Tyr Hydrochloride

Bachem H-9900.0005   Store at -15°C

Pro-Glu

Bachem G-3005.0250   MW 244.25 $C_{10}H_{16}N_2O_5$ Store at -15°C

Pro-Glu-Ala-Asn
Dnp-Pro-Glu-Ala-Asn Amide

Bachem M-1860.0025   MW 594.54 $C_{23}H_{30}N_8O_{11}$ Store at -15°C

Pro-Glu-Ala-His-Trp-Thr-Lys-Leu-Gln-His-Ser-Leu-Asp-Thr-Ala-Leu-Arg

Synonyms: Prepro-Nerve Growth Factor (99-115)

Bachem H-4425.0001   Mouse   Store at -15°C

Pro-Glu-Arg-Glu-Val-Leu-Glu-Trp-Arg-Phe

Synonyms: NEF (176-185); HTLV I Peptide

Neosystem SC515   MW 1360.53   HTLV I peptide from subtype NEF

Pro-Glu-Gly-Thr-Arg-Gln-Ala-Arg-Arg-Asn-Arg-Arg-Arg-Arg

Synonyms: REV (31-44); HTLV I Peptide

Neosystem SC191   MW 1809.03   HTLV I peptide from subtype REV

Pro-Glu-Lys-Ala-Lys-Lys-Glu-Thr-Val-Glu-Lys-Ala-Tyr

Synonyms: TAT-(Tyr) (111-122); SIVmac251 Peptide

Neosystem SC581   MW 1520.74   SIVmac251 peptide from subtype TAT

Pro-Glu-Thr

Bachem H-4715.0025   MW 345.35 $C_{14}H_{23}N_3O_7$ Store at -15°C

Pro-Gly
BOC-Pro-Gly

Bachem A-2725.0005   MW 272.3 $C_{12}H_{20}N_2O_5$ Store at -15°C

Pro-Gly
BOC-D-Pro-Gly

Bachem A-3260.0001   MW 272.3 $C_{12}H_{20}N_2O_5$ Store at -15°C

Pro-Gly
FMOC-Pro-Gly

Bachem B-1515.0001   Store at -15°C

Pro-Gly
Z-Pro-Gly

Bachem C-2490.0001   MW 306.32 $C_{15}H_{18}N_2O_5$ Store at -15°C

Pro-Gly
Z-Pro-Gly Amide

Bachem C-2495.0001   MW 305.33 $C_{15}H_{19}N_3O_4$ Store at -15°C

Pro-Gly

Bachem G-3015.0001   MW 172.18 $C_7H_{12}N_2O_3$ Store at -15°C

Pro-Gly
BOC-Pro-ψ(CH₂N-2-Chloro-Z)-Gly

Neosystem BB01502   MW 426.9

Pro-Gly

Sigma P 0880   FW 172.2 $C_7H_{12}N_2O_3$

Pro-Gly
CBZ-Pro-Gly

USBio C2099-52   MW 306.4 $C_{15}H_{18}N_2O_5$ ≥99%

Pro-Gly Amide Hydrochloride

Bachem G-3020.0250   MW 207.66 $C_7H_{13}N_3O_2$ · HCl   Store at -15°C

Pro-Gly Monohydrate

ICN 102740   MW 172.2 $C_7H_{12}N_2O_3$ · H_2O Crystalline

Pro-Gly-Ala-Ile-Pro-Gly

Synonyms: Tropoelastin Fragment

Bachem H-1936.0005   Store at -15°C

Pro-Gly-Arg-Tyr Amide Hydrochloride

Bachem H-9510.0025   Store at -15°C

Pro-Gly-Gly

Bachem M-1730.0250   MW 229.24 $C_9H_{15}N_3O_4$ Store at -15°C

ICN 102734   MW 229.2 $C_9H_{15}N_3O_4$ Crystalline

Sigma P 9382   FW 229.2 $C_9H_{15}N_3O_4$ Contains solvent of crystallization; actual content given on label

Pro-Gly-Gly-Gly-Thr-Leu-Pro-Pro-Ser-Gly

Synonyms: Sperm Peptide P10G

Bachem H-2424.0001   Store at -15°C

Pro-Gly-Leu-Pro-Ser-Ala-Ala-Ser-Ser-Glu-Asp-Ala-Gly-Gln-Ser Amide

Synonyms: Galanin Message Associated Peptide (44-59); Prepro-Galanin (108-123)

Sigma G 4271   FW 1485.6   ≥97% (HPLC) | Bioactive peptide; gastrointestinal peptide

Pro-Gly-Lys-Ala-Arg

Bachem H-4720.0050   MW 527.63 $C_{22}H_{41}N_9O_6$ Store at -15°C

Pro-Gly-Thr-Cys-Glu-Ile-Cys-Ala-Tyr-Ala-Ala-Cys-Thr-Gly-Cys

Synonyms: Myoactive Peptide I; Neurohormone D; Guanylin; Intestinal Guanylate Cyclase Activator; Guanylate Cyclase C Activator

American Peptide 60-9-24   Cockroach   MW 973.0 $C_{46}H_{60}N_{12}O_{12}$

Alexis 157-010   Human   MW 1458.7 $C_{58}H_{87}N_{15}O_{21}S_4$ ≥99%; white lyophilized powder; soluble in water | Disulfide bonds: Cys⁴-Cys¹², Cys⁷-Cys¹⁵; endogenous intestinal guanylyl cyclase activator; de Sauvage, FJ et al, *PNAS*, 89:9089, 1992; Wiegand, RC et al, *FEBS Lett*, 311:150, 1992

American Peptide 43-0-10   Human   MW 1458.7 $C_{58}H_{87}N_{15}O_{21}S_4$ Disulfide bonds: Cys⁴-Cys¹², Cys⁷-Cys¹⁵; isolated from rat intestine can activate intestinal guanylate cyclase in a human colon carcinoma-derived cell line, T8₄; resembles heat-stable endotoxins produced by *E. coli*; Wiegand, RC et al, *FEBS Lett*, 311:150, 1992

Bachem H-2996.0500   Human   Store at -15°C | Disulfide bonds: Cys⁴-Cys¹², Cys⁷-Cys¹⁵

Peptides International PGN-4274-s   Human synthetic   MW 1458.7 $C_{58}H_{87}N_{15}O_{21}S_4$ >99% (HPLC); lyophilized amorphous powder | Bioactive; Disulfide bonds: Cys⁴-Cys¹², Cys⁷-Cys¹⁵; Forte, LR & MG Currie, *FASEB J*, 9:643, 1995; Forte, LR et al, *Am J Kidney Dis*, 28:296, 1996; Wiegand, RC et al, *FEBS Letts*, 311:150, 1992; de Sauvage, FJ et al, *PNAS USA*, 89:9089, 1992

Pro-His-Ala

Bachem H-4725.0100   MW 323.35 $C_{14}H_{21}N_5O_4$ Store at -15°C

Pro-His-Cys-Lys-Arg-Met

Bachem H-2458.0005   Store at -15°C

Sigma P 0463   FW 771.0 $C_{31}H_{54}N_{12}O_7S_2$ ≥97% (HPLC); peptide content:~70% | Bioactive peptide; acts as antioxidant by reacting with active oxygen species such as superoxide & hydroxyl radicals; Ueda, J et al, *Biochem Mol Biol Int*, 33:1041, 1994

Pro-His-Gly

Bachem H-4740.0100   MW 309.33 $C_{13}H_{19}N_5O_4$ Store at -15°C

Pro-His-Leu

Bachem H-4745.0100   MW 365.43 $C_{17}H_{27}N_5O_4$ Store at -15°C

Pro-His-Phe

Bachem H-4750.0100   MW 399.45 $C_{20}H_{25}N_5O_4$ Store at -15°C

Pro-His-Pro-Phe-His-Leu-Phe-Val-Tyr

Bachem N-1115.0005   Store at -15°C

Pro-His-pro-Phe-His-Phe-Phe-Val-Tyr-Lys

Synonyms: Renin Inhibitor Octapeptidyl-Lysine, (Phe⁵,⁶)-; Renin Inhibitor

American Peptide 12-1-70   MW 1318.6 $C_{69}H_{87}N_{15}O_{12}$ Specific inhibitor of human renin *in vitro*; Burton, J et al, *PNAS*, 77:5476, 1980

Bachem N-1120.0005   MW 1318.54 $C_{69}H_{87}N_{15}O_{12}$ Store at -15°C

ICN 153166   MW 1318.5   Specific inhibitor of human renin *in vivo*; Burton, J etal, *PNAS*, 77:4476, 1980; Cody, RJ etal, *BBRC*, 97:230, 1980

Sigma P 2402   FW 1318.5   ≥97% (HPLC) | Bioactive peptide; specific inhibitor of human renin *in vivo*; Burton, J et al, *Proc Natl Acad Sci USA*, 77:5476, 1980; Cody, RJ et al, *Biochem Biophys Res Commun*, 97:230, 1980

Pro-His-Ser-Arg-Asn

Synonyms: Fibronectin (1376-1380)

Bachem H-2596.0005   MW 609.64 $C_{24}H_{39}N_{11}O_8$ Store at -15°C

Pro-His-Ser-Arg-Asn Acetylated Amidated

Synonyms: Invasion Inducing Fibronectin Peptide

Upstate 12-399   MW 650   >95% pure; frozen solution | *In vitro* cell invasion peptide

Pro-His-Ser-Cys-Asn Acetylated Amidated

Synonyms: Invasion Inhibiting Fibronectin Peptide

Upstate 12-400   MW 597   >95% pure; frozen solution | *In vitro* cell invasion peptide

Pro-His-Tyr

Bachem H-4755.0100   MW 415.45 $C_{20}H_{25}N_5O_5$ Store at -15°C

Pro-His-Val

Bachem H-4760.0100   MW 351.41 $C_{16}H_{25}N_5O_4$ Store at -15°C

Pro-His-Val-Thr-Arg-Arg-Thr-Pro-Asp-Tyr-Phe-Leu

Synonyms: Protein Phosphatase 2A/C Blocking Peptide (298-309)

Calbiochem 539528   Synthetic   MW 1501.7 $C_{69}H_{104}N_{20}O_{18}$ Liquid | From PP2A/C catalytic subunit; ideal blocking peptide for Anti-PP2A/C

Pro-Hyp
Z-Pro-Hyp

Bachem C-2505.0001   MW 362.38 $C_{18}H_{22}N_2O_6$ Store at -15°C

Pro-Hyp

Bachem G-3025.0250   MW 228.25 $C_{10}H_{16}N_2O_4$ Store at -15°C

ICN 156373   MW 228.2 $C_{10}H_{16}N_2O_4$ Crystalline

Sigma P 3760   FW 228.2 $C_{10}H_{16}N_2O_4$

Pro-Ile

Bachem G-3030.0250   MW 228.29 $C_{11}H_{20}N_2O_3$ Store at -15°C

ICN 102735   MW 228.3 $C_{11}H_{20}N_2O_3$ Crystalline

Sigma P 9507   FW 228.3 $C_{11}H_{20}N_2O_3$

Pro-Ile-Leu-Gln-Arg-Leu-Ser-Ala-Thr-Leu-Arg-Arg-Val-Arg-Glu-Val-Leu-Arg-Thr

Synonyms: GP140 (802-820); SIVmac251 Peptide

Neosystem SC788   MW 2277.74   SIVmac251 peptide from subtype GP140

Pro-Ile-Val-Gln-Asn-Ile-Gln-Gly-Gln-Met-Val-His-Gln-Ala-Ile-Ser

Synonyms: P25 (133-148); GAG P24 CA (1-16); HTLV I Peptide

Neosystem SC290   MW 1763.04   HTLV I peptide from subtype P25 (GAG P24 CA)

Pro-Leu
Z-Pro-Leu

Bachem C-3105.0001   MW 362.43 $C_{19}H_{26}N_2O_5$ Store at -15°C

Pro-Leu
Pz-Pro-Leu

Bachem G-2970.0250   Store at -15°C

Pro-Leu

Bachem G-3035.0001 MW 228.29 $C_{11}H_{20}N_2O_3$ Store at -15°C

Pro-Leu
Mca-Pro-Leu

Synonyms: Matrix Metalloproteinase II/VII Control

Bachem M-1975.0050 Store at -15°C

Calbiochem 03-32-5033 MW 444.5 $C_{23}H_{28}N_2O_7$ Solid; ≥98% (HPLC); soluble in MeOH | Reference compound for MMP-2/MMP-7 substrate; Knight, CG et al, *FEBS Lett*, 296:263 1992

Pro-Leu
N-CBZ-Pro-D-Leu

ICN 152950 MW 362.4 $C_{19}H_{26}N_2O_5$ Inhibits the development of physical dependence on morphine in mice; Walter, R etal, *PNAS*, 75:4573, 1978

Pro-Leu
FMOC-(S,S)-(Pro-Leu)-Spirolactam

Neosystem FB01501 MW 476.57 Ward, P et al, *J Med Chem*, 33:1848, 1990

Pro-Leu
MOC-Ac-Pro-Leu

Sigma M 0546 FW 444.5 $C_{23}H_{28}N_2O_7$ ≥97% (TLC) | Knight, CG et al, *FEBS Lett*, 296:263, 1992

Pro-Leu

Sigma P 1130 FW 228.3 $C_{11}H_{20}N_2O_3$

Pro-Leu
CBZ-Pro-D-Leu

USBio C2099-53 MW 362.4 $C_{19}H_{26}N_2O_5$ ≥99%

Pro-Leu-Ala
Z-Pro-Leu-Ala-NHOH

Bachem N-1285.0250 MW 448.52 $C_{22}H_{32}N_4O_6$ Store at -15°C

Pro-Leu-Ala
Z-Pro-D-Leu-D-Ala-NHOH

Bachem N-1290.0250 MW 448.52 $C_{22}H_{32}N_4O_6$ Store at -15°C

Pro-Leu-Ala-Are-Thr-Leu-Ser-Val-Ala-Gly-Leu-Pro-Gly-Lys-Lys

Synonyms: Protein Kinase Related Peptide; Syntide II; Calmodulin-Dependent Protein Kinase Substrate

Sigma S 2525 FW 1507.8 ≥97% (HPLC) | Bioactive peptide; Hashimoto, Y & Soderling, TR, *Arch Biochem Biophys*, 252:418, 1987

Pro-Leu-Ala-Arg-Thr-Leu-Ser-Val-Ala-Gly-Leu-Pro-Gly-Lys-Lys

Synonyms: Ca^{2+}/Calmodulin-Dependent Protein Kinase II Substrate; Protein Kinase C Substrate; Syntide II; Calmodulin Kinase II Substrate; Calmodulin-Dependent Protein Kinase Substrate; Ca^{2+}/Calmodulin Kinase II Substrate; Calmodulin-Dependent Protein Kinase II Substrate

Alexis 167-007 MW 1507.8 $C_{68}H_{122}N_{20}O_{18}$ Sequence is homologous to the phosphorylation site 2 in glycogen synthase; Hashimoto, Y & Soderling, TR, *Arch Biochem Biophys*, 252:418, 1987

American Peptide 86-0-30 MW 1508.1 $C_{68}H_{122}N_{20}O_{18}$ Hashimoto, Y et al, *Arch Biochem Biophys*, 252:418, 1987

Bachem H-9385.0001 MW 1507.84 $C_{68}H_{122}N_{20}O_{18}$ Store at -15°C

Calbiochem 05-23-4910 MW 1507.9 $C_{68}H_{122}N_{20}O_{18}$ ≥98% (HPLC); solid; soluble in 5% acetic acid | Phosphate acceptor peptide that serves as a substrate for CaM kinase II & protein kinase C; Inoue, S et al, *Biochem Biophys Res Comm*, 215:861, 1995; Hashimoto, Y et al, *Arch Biochem Biophys*, 252:418, 1987

ICN 159666 MW 1507.8 Selective substrate

Pro-Leu-Ala-Gln-Ala-Val-Arg-Ser-Ser-Ser-Arg
Mca-Pro-Leu-Ala-Gln-Ala-Val-Dap(Dnp)-Arg-Ser-Ser-Ser-Arg Amide

Synonyms: Tumor Necrosis Factor α (-5 to +6), Mca-(Endo-1a-Dap(Dnp))-

Bachem M-2255.0001 Human Store at -15°C

Pro-Leu-Ala-Leu-Trp-Ala-Arg
DNP-Pro-Leu-Ala-Leu-Trp-Ala-Arg

Synonyms: Matrix Metalloproteinase I Substrate; Fibroblast Collagenase Substrate

Calbiochem 444211 MW 992.1 $C_{46}H_{65}N_{13}O_{12}$ Solid; ≥95% (HPLC); excitation max:280 nm; emission max:360 nm; soluble in 0.1% aqueous TFA | Fluorogenic substrate for human fibroblast collagenase (HFC, MMP-1; K_m=130 μM); Netzel-Arnett, S et al, *Anal Biochem*, 195:86 1991

ICN 196019 MW 992.1 $C_{46}H_{65}N_{13}O_{12}$ ≥95%; soluble in 0.1% aq TFA; K_m = 130 μM; excitation max = 280 nm; emission max = 360 nm | Fluorogenic substrate for human fibroblast collagenase

Pro-Leu-Ala-Trp
Ac-Pro-Leu-Ala-((S)-2-Mercapto-Pentanoyl)-Trp Amide

Bachem H-1326.0005 MW 642.82 $C_{32}H_{46}N_6O_6S$ Store at -15°C

Pro-Leu-Ala-Tyr-Trp-Ala-Arg
DNP-Pro-Leu-Ala-Tyr-Trp-Ala-Arg

Synonyms: Collagenase Substrate; Matrix Metalloproteinases VIII Substrate; Neutrophil Collagenase Substrate

Calbiochem 444230 MW 1042.1 $C_{49}H_{63}N_{13}O_{13}$ Solid; ≥95% (HPLC); excitation max:280 nm; emission max:360 nm; soluble in 0.1% TFA | Fluorogenic substrate for human neutrophil collagenase (HNC; MMP-8); Netzel-Arnett, S et al, *Anal Biochem*, 195:86 1991

ICN 196024 MW 1042.1 $C_{49}H_{63}N_{13}O_{13}$ ≥95% | Fluorogenic substrate

Pro-Leu-Arg-Arg-Thr-Leu-Ser-Val-Ala-Ala Amide

Synonyms: Calmodulin-Dependent Protein Kinase Substrate Analog

American Peptide 86-2-60 MW 1082.3 $C_{47}H_{87}N_{17}O_{12}$ Pearson, RB et al, *JBC*, 260:14471, 1985

Pro-Leu-Gly
Z-Pro-Leu-Gly

Bachem C-2515.0250 MW 419.48 $C_{21}H_{29}N_3O_6$ Store at -15°C

Pro-Leu-Gly
Z-Pro-Leu-Gly Amide

Bachem C-2520.0001 MW 418.49 $C_{21}H_{30}N_4O_5$ Store at -15°C

Pro-Leu-Gly
Z-Pro-Leu-Gly-OEt

Bachem C-2525.0001 MW 447.53 $C_{23}H_{33}N_3O_6$ Store at -15°C

Pro-Leu-Gly
Z-Pro-Leu-Gly-NHOH

Bachem C-3205.0250   MW 434.49 $C_{21}H_{30}N_4O_6$ Store at -15°C

Pro-Leu-Gly
Ac-Pro-Leu-Gly

Bachem H-6480.0250   MW 327.38 $C_{15}H_{25}N_3O_5$ Store at -15°C

Pro-Leu-Gly
Pro-D-Leu-Gly Amide

Synonyms: Melanocyte Stimulating Hormone Release Inhibiting Factor, (D-Leu²)-; Melanocyte Stimulating Hormone Release Inhibiting Factor I

Bachem H-9225.0025   MW 284.36 $C_{13}H_{24}N_4O_3$ Store at -15°C

Pro-Leu-Gly
Pro-Leu-Gly-pNA

Bachem L-1940.0050   Store at -15°C

Pro-Leu-Gly
Pro-Leu-Gly-NHOH Hydrochloride

Bachem M-1775.0250   MW 336.82 $C_{13}H_{24}N_4O_4 \cdot HCl$ Store at -15°C

Pro-Leu-Gly
DNP-L-Pro-L-Leu-Gly

ICN 151011   MW 451.4 $C_{19}H_{25}N_5O_8$ Reference compound for measurement of collagenase activity; Masui, Y etal, *Biochem Med*, 17:215, 1977

Pro-Leu-Gly
Pro-Leu-Gly-Hydroxamate Hydrochloride

ICN 156375   MW 336.8 $C_{13}H_{24}N_4O_4 \cdot HCl$ Affinity ligand for purifying human collagenases; Moore, WM & CA Spelburg, *Biochem*, 25:5189, 1986

Pro-Leu-Gly
MOCAc-Pro-Leu-Gly

Synonyms: MOC-Ac-type Fluorescence-Quenching Substrate

Peptides International MOC-3164-s   MW 501.54 $C_{25}H_{31}N_3O_8$ >99% (HPLC); lyophilized amorphous powder | Reference compound; Knight, CG et al, *FEBS Lettss*, 296:263, 1992

Pro-Leu-Gly
Dnp-Pro-Leu-Gly

Peptides International SDP-3082   MW 451.44 $C_{19}H_{25}N_5O_8$ >99% (HPLC); amorphous powder | Reference compound for Peptides International SDP-3073-v

Pro-Leu-Gly
N-CBZ-Pro-Leu-Gly Amide

Sigma C 5002   FW 418.5 $C_{21}H_{30}N_4O_5$

Pro-Leu-Gly
N-CBZ-Pro-Leu-Gly-OEt

Sigma C 8282   FW 447.5 $C_{23}H_{33}N_3O_6$

Pro-Leu-Gly
N-CBZ-Pro-Leu-Gly-Hydroxamate

Synonyms: Collagenase Inhibitor, Human

Sigma C 8537   FW 434.5 $C_{21}H_{30}N_4O_6$ ≥97% (TLC) | Bioactive peptide; Moore, WM & Spilburg, CA, *Biochemistry*, 25:5189, 1986

Pro-Leu-Gly
MOC-Ac-Pro-Leu-Gly

Sigma M 4293   FW 501.5 $C_{25}H_{31}N_3O_8$ Fluorescent peptide for use as a reference; The product of the action of matrix metallo-proteinases on the internally quenched substrate Sigma M 6412; Knight, CG et al, *FEBS Lett*, 296:263, 1992;

Pro-Leu-Gly
Pro-Leu-Gly-Hydroxamate Hydrochloride

Sigma P 5298   FW 336.8 $C_{13}H_{24}N_4O_4 \cdot HCl$ Affinity ligand for the purification of human collagenases; Moore, WM & Spelburg, CA, *Biochemistry*, 25:5189, 1986

Pro-Leu-Gly Amide

Synonyms: Melanocyte Stimulating Hormone Release Inhibiting Factor; Melanocyte Stimulating Hormone Release Inhibiting Factor I; Melanostatin; Oxytocin (7-9)

American Peptide 56-0-70   MW 284.4 $C_{13}H_{24}N_4O_3$ Walter, R et al, *PNAS*, 76:518, 1979

Bachem H-4305.0250   MW 284.36 $C_{13}H_{24}N_4O_3$ Store at -15°C

Sigma P 9887   FW 284.4 $C_{13}H_{24}N_4O_3$ ≥97% (HPLC) | Bioactive peptide; inhibits the release of MSH; Celis, ME et al, *Proc Natl Acad Sci USA*, 68:1428, 1971; Walter, R et al, ibid, 76:518, 1979

Pro-Leu-Gly Amide ½H₂O

Synonyms: Melanocyte Stimulating Hormone Release Inhibiting Factor

Peptides International PMI-4024   Synthetic   MW 293.37 $C_{13}H_{24}N_4O_3 \cdot \frac{1}{2}H_2O$ >99% (HPLC); lyophilized amorphous powder; bulk

Pro-Leu-Gly-Cys-His-Ala-Arg
DNP-Pro-Leu-Gly-Cys(Me)-His-Ala-D-Arg Amide

Synonyms: Collagenase Substrate

American Peptide 81-7-73   MW 932.1 $C_{38}H_{57}N_{15}O_{11}S$ Berman, J et al, *JBC*, 267:1434, 1992

Pro-Leu-Gly-Cys-His-Ala-Arg
N-DNP-Pro-Leu-Gly-Cys(Me)-His-Ala-D-Arg Amide

Sigma D 0296   FW 932.0 $C_{38}H_{57}N_{15}O_{11}S$ ≥95% (HPLC) | Berman, J et al, *J Biol Chem*, 267:1434, 1992

Pro-Leu-Gly-Gly

Bachem H-4765.0050   MW 342.4 $C_{15}H_{26}N_4O_5$ Store at -15°C

Pro-Leu-Gly-His-Ala-Arg
Dnp-Pro-Leu-Gly-Cys(Me)-His-Ala-D-Arg Amide

Bachem M-1905.0005   MW 932.03 $C_{38}H_{57}N_{15}O_{11}S$ Store at -15°C

Pro-Leu-Gly-Ile-Ala-Gly-Arg
Dnp-Pro-Leu-Gly-Ile-Ala-Gly-Arg Amide

Synonyms: Collagenase Substrate, Animal

Peptides International SDP-3073-v   MW 847.93 $C_{36}H_{57}N_{13}O_{11}$ >98% (HPLC); lyophilized amorphous powder | Masui, Y et al, *Biochem Med*, 17:215, 1977

Pro-Leu-Gly-Leu-Ala-Ala-Arg
MOC-Ac-Pro-Leu-Gly-Leu-β-(2,4-Dinitrophenylamino)Ala-Ala-Arg Amide

Synonyms: Matrix Metalloproteinase Substrate

Sigma M 6412   FW 1093.2   ≥97% (HPLC) | Internally quenched substrate for assay of matrix metallo-proteinases; yields fluorescent product MOCAc-Pro-Leu-Gly; Knight, CG et al, *FEBS Lett*, 296:263, 1992

Pro-Leu-Gly-Leu-Ala-Arg
7-Methoxycourmarin-4-Ac-Pro-Leu-Gly-Leu-Dnp-(2,3-Diaminopropionyl)-Ala-Arg

Synonyms: Collagenase Substrate

American Peptide 81-0-11 MW 1093.1 $C_{49}H_{68}N_{14}O_{15}$ 50-100 times more sensitive when using PUMP as well as other matrix metalloproteases; increase in sensitivity of the assay as well as reducing background fluorescence; Knight, CG et al, *FEBS 10610*, 296:263, 1992

Pro-Leu-Gly-Leu-Ala-Arg
Mca-Pro-Leu-Gly-Leu-Dap(Dnp)-Ala-Arg Amide

Bachem M-1895.0001 MW 1093.16 $C_{49}H_{68}N_{14}O_{15}$ Store at -15°C

Pro-Leu-Gly-Leu-Ala-Arg
MCA-Pro-Leu-Gly-Leu-N-3-(2,4-Dinitrophenyl)-L-2,3-Diaminopropionyl-Ala-Arg AcOH Amide

Synonyms: Matrix Metalloproteinase II/VII Substrate

Calbiochem 03-32-5032 MW 1153.2 $C_{49}H_{68}N_{14}O_{15} \cdot CH_3COOH$ solid; ≥98% (HPLC); excitation max:325 nm; emission max:393 nm; soluble in MeOH | Extremely sensitive fluorogenic substrate for matrix metalloproteinases (MMPs); for assessment of total activity in crude preparations by continuous assays; Knight, CG et al, *FEBS Lett*, 296:263 1992

Pro-Leu-Gly-Leu-Ala-Arg
MCA-Pro-Leu-Gly-Leu-DPA-Ala-Arg Amide

Synonyms: Matrix Metalloproteinases II/VII Substrate

ICN 196020 MW 1153.2 $C_{49}H_{68}N_{14}O_{15} \cdot C_2H_4O_2$ ≥98% | Fluorogenic substrate for MMP

Pro-Leu-Gly-Leu-Ala-Arg
MOCAc-Pro-Leu-Gly-Leu-A2pr(Dnp)-Ala-Arg Amide

Synonyms: Matrix Metalloproteinase Fluorescence-Quenching Substrate

Peptides International MOC-3163-v MW 1093.2 $C_{49}H_{68}N_{14}O_{15}$ >98% (HPLC); lyophilized amorphous powder | Knight, CG et al, *FEBS Lettss*, 296:263, 1992

Pro-Leu-Gly-Leu-Gly
Ac-Pro-Leu-Gly-((S)-2-mercapto-4-Methylpentanoyl)-Leu-Gly-OEt

Bachem H-7145.0025 MW 655.86 $C_{31}H_{53}N_5O_8S$ Store at -15°C

Pro-Leu-Gly-Leu-Trp-Ala-Arg
DNP-Pro-Leu-Gly-Leu-Trp-Ala-D-Arg Amide

Synonyms: Collagenase Substrate; Gelatinase Substrate

American Peptide 81-7-70 MW 977.1 $C_{45}H_{64}N_{14}O_{11}$ Fluorogenic matrix metalloprotease substrate; Stack, MS & Gray, RD, *JBC*, 264:4277, 1989

Bachem M-1855.0005 MW 977.09 $C_{45}H_{64}N_{14}O_{11}$ Store at -15°C

Peptides International SDP-3820-PI MW 977.10 $C_{45}H_{64}N_{14}O_{11}$ >98% (HPLC); white crystalline powder | Stack, MS & RD Gray, *JBC*, 264:4277, 1989; Darlak, K et al, *JBC*, 265:5199, 1990

Pro-Leu-Gly-Met-Trp-Ser-Arg
DNP-Pro-Leu-Gly-Met-Trp-Ser-Arg

Synonyms: Matrix Metalloproteinases II/IX Substrate; Gelatinase Substrate, Fibroblast & Neutrophil

ICN 196021 MW 1012.1 $C_{44}H_{61}N_{13}O_{15}S$ ≥95%

Pro-Leu-Gly-Phe-Phe-Pro-Asp-His-Gln-Leu-Asp-Pro-Ala-Phe-Gly-Ala-Asn-Ser-Asn-Asn-Pro-Asp-Trp-Asp-Phe-Asn-Pro

Synonyms: HBV Receptor Binding Fragment

American Peptide 72-1-15 MW 3030.2 $C_{140}H_{185}N_{35}O_{42}$ Neurath, AR et al, *Cell*, 46:429, 1986

Pro-Leu-Gly-Pro-Arg
Pz-Pro-Leu-Gly-Pro-D-Arg

Synonyms: Pz-Peptide

Bachem M-1715.0050 Store at -15°C

Pro-Leu-Gly-Pro-Arg
4-Phenylazobenzyloxycarbonyl- Pro-Leu-Gly-Pro-D-Arg

Synonyms: Collagenase Substrate

ICN 156159 MW 776.9 $C_{38}H_{52}N_{10}O_8$ Wuensch, E & HG Heidrich, *Z Physiol Chem*, 333:149, 1963

Pro-Leu-Gly-Pro-D-Lys
7-Methoxycoumarin-3-Carboxylyl-Pro-Leu-Gly-Pro-D-Lys(DNP)OH

Synonyms: Clostridial Collagenase Substrate; MCC-Pro-Leu-Gly-Pro-D-Lys-(DNP); PZ-Peptidase Substrate

Calbiochem 03-32-5024 MW 878.9 $C_{41}H_{50}N_8O_{14}$ Solid; ≥99% (HPLC); soluble in MeOH & 0.1% aqueous TFA | Novel quenched-fluorescence substrate; eliminates background fluorescence normally found in tryptophan-containing substrates; Tisljar, U et al, *Anal Biochem*, 186:112 1990; Tisljar, U et al, *Biochem Biophys Res Comm*, 162:1460, 1989

Pro-Leu-Gly-Pro-Lys
Mca-Pro-Leu-Gly-Pro-D-Lys(Dnp)

Bachem M-2270.0001 Store at -15°C

Pro-Leu-Gly-Trp-Ala-Arg
N-DNP-Pro-Leu-Gly-Trp-Ala-D-Arg Amide

Synonyms: Collagenase Substrate, Vertebrate; Gelatinase Substrate, Vertebrate

Sigma D 2293 FW 977.1 $C_{45}H_{64}N_{14}O_{11}$ Fluorogenic substrate; Stack, MS & Gray, RD, *J Biol Chem*, 264:4277, 1989

Pro-Leu-Gly-Trp-Ser-Arg
DNP-Pro-Leu-Gly-Trp-Ser-Arg

Synonyms: Matrix Metalloproteinase II/IX Substrate; Gelatinase A/B Substrate

Calbiochem 444215 MW 1012.1 $C_{44}H_{61}N_{13}O_{13}S$ Solid; ≥95% (HPLC); excitation max:280 nm; emission max:360 nm; soluble in 0.1% aqueous TFA | Fluorogenic substrate for both the 72 kDa human fibroblast gelatinase (HFG; MMP-2) & the 92 kDa human neutrophil gelatinase (HNG; MMP-9); Netzel-Arnett, S et al, *Anal Biochem*, 195:86 1991

Pro-Leu-Ile-Tyr-Pro

Synonyms: Hypotensive Peptide ACE-Inhibitor

Neosystem SC137 MW 601.74 Virus-related peptide; bioactive

Pro-Leu-Phe-Cys-Ala-Thr-Lys-Asn-Arg-Asp-Thr-Trp-Gly-Thr-Thr-Gln-Cys-Leu-Pro-Asp
Pro-Leu-Phe-Cys(Acm)-Ala-Thr-Lys-Asn-Arg-Asp-Thr-Trp-Gly-Thr-Thr-Gln-Cys(Acm)-Leu-Pro-Asp

Synonyms: GP140 (41-60); SIVmac251 Peptide

Neosystem SC712 MW 2409.71 SIVmac251 peptide from subtype GP140

Pro-Leu-Pro-Asp-Cys-Cys-Arg-Gln-Lys-Thr-Cys-Ser-Cys Acetate Salt

Synonyms: Orexin A (2-14)

Biogenesis 7049-5050   Synthetic   Purified; lyophilized

Pro-Leu-Ser-Arg-Thr-Leu-Ser-Val-Ala-Ala-Lys-Lys

Synonyms: Glycogen Synthase (1-12), (Ala^{9,10},Lys^{11,12})-; Protein Kinase C Substrate Peptide; Protein Kinase C Substrate; Protein Kinase Related Peptide

Alexis 157-013   MW 1270.5 $C_{56}H_{103}N_{17}O_{16}$ $\geq$98%; white lyophilized powder; soluble in water | Comprises the sequence of the glycogen synthase phosphorylation site; House, C et al, *J Biol Chem*, 262:772, 1987; Egan, JJ et al, *Anal Biochem*, 175:552, 1988; Alexander, DR et al, *Biochem J*, 260:893, 1989

American Peptide 86-0-41   MW 1270.7 $C_{56}H_{103}N_{17}O_{16}$ Excellent substrate for assaying protein kinase C, e.g. in permeabilized T cells

Bachem H-9375.0005   MW 1270.54 $C_{56}H_{103}N_{17}O_{16}$ Store at -15°C

ICN 154583   MW 1270.7   Egan, J etal, *Analyt Biochem*, 175:552, 1988

Sigma P 5307   FW 1270.5 $\geq$97% (HPLC); peptide content:~70% | Bioactive peptide; Egan, J et al, *Anal Biochem*, 175:552, 1988; Alexander, DR et al, *Biochem J*, 268:303, 1990

Pro-Leu-Ser-Arg-Thr-Leu-Ser-Val-Ala-Ala-Lys-Lys Amide

Synonyms: Protein Kinase C Substrate; Glycogen Synthetase (1-12)

ICN 191459

Pro-Leu-Ser-Arg-Thr-Leu-Ser-Val-Ala-Ala-Lys-Lys Amide Trifluoroacetate Salt

Synonyms: Protein Kinase C Substrate; Glycogen Synthase (1-8); Calmodulin-Dependent Protein Kinase II Substrate

Calbiochem 361600   Synthetic   MW 1269.6 $C_{56}H_{104}N_{18}O_{15}$ $\geq$95% (HPLC); lyophilized solid; soluble in water | Partially derived from a sequence in glycogen synthase; readily phosphorylated by PKC; Ahmad, Z et al, *J Biol Chem*, 259:8743, 1984

Pro-Leu-Ser-Arg-Thr-Leu-Ser-Val-Ser-Ser Amide

Synonyms: Calmodulin-Dependent Protein Kinase Substrate; Ca^{2+}/Calmodulin Kinase II Substrate; Peptide I; Glycogen Synthase (1-10)

American Peptide 86-2-55   MW 1045.1 $C_{44}H_{80}N_{18}O_{15}$

Calbiochem 208699   Synthetic   MW 1045.2 $C_{44}H_{80}N_{14}O_{15}$ >95% (HPLC); lyophilized; soluble in 5% HOAc | Synthetic decapeptide that serves as a phosphorylation substrate for Ca^{2+}/calmodulin-dependent protein kinase II; Homma, T et al, *J Biol Chem*, 265:17613, 1990; Pearson, RB et al, *J Biol Chem*, 260:14471, 1985

Pro-Leu-Thr-Glu-Glu-Ala-Glu-Leu-Glu-Leu-Ala-Glu-Asn-Arg-Glu-Ile-Leu-Lys-Glu-Pro-Val-His-Gly-Val-Tyr

Synony POL (461-485); RT (294-318); HTLV I Peptide *ms:*

Neosystem SC692   MW 2879.21   HTLV I peptide from subtype POL (PR/RT)

Pro-Leu-Tyr-Lys-Lys-Ile-Ile-Lys-Lys-Leu-Leu-Glu-Ser

Synonyms: Platelet Factor IV (58-70); Angiogenesis Inhibitor

Bachem H-1174.0001   Human   MW 1573.01 $C_{76}H_{133}N_{17}O_{18}$ Store at -15°C

Sigma P 3459   Human   FW 1573.0 $\geq$97% (HPLC) | Bioactive peptide; Maione, TE et al, *Science*, 247:77, 1990

Peptides International IPF-4305-v   Human synthetic   MW 1573.0 $C_{76}H_{133}N_{17}O_{18}$ >99% (HPLC); lyophilized amorphous powder | Inhibitor; Maione, TE et al, *Science*, 247:77, 1990

Peptides InternationalIPF-4305-v   Human synthetic   MW 1573.0 $C_{76}H_{133}N_{17}O_{18}$ >99% (HPLC); lyophilized amorphous powder | Bioactive; Maione, TE et al, *Science*, 247:77, 1990

Pro-Leu-Val
BOC-Pro-Leu-Val-OMe

Bachem A-2260.0250   MW 441.57 $C_{22}H_{39}N_3O_6$ Store at -15°C

Pro-Lys

Bachem G-3040.0250   MW 243.31 $C_{11}H_{21}N_3O_3$ Store at -15°C

Pro-Lys-Gln-Ala-Trp-Gly-Trp-Phe-Gly-Gly-Lys-Trp-Lys-Asp-Ala-Ile-Lys-Glu-Tyr-Lys

Synonyms: GP140 (341-360); SIVmac251 Peptide

Neosystem SC742   MW 2423.79   SIVmac251 peptide from subtype GP140

Pro-Lys-Leu-Leu-Lys-Thr-Phe-Leu-Ser-Lys-Trp-Ile-Gly

Synonyms: Seminalplasmin Fragment Analog; SPFK

Bachem H-1636.0001   MW 1530.92 $C_{76}H_{123}N_{17}O_{16}$ Store at -15°C

Pro-Lys-Lys-Lys-Arg-Lys-Val-Glu-Asp-Pro-Tyr-Cys

Synonyms: Large Tumor Antigen

ICN 158168   97% | Useful as a nuclear localization signal peptide; Goldfarb, DS etal, *Nature*, 322:641, 1986

Sigma P 6296   Synthetic   FW 1490.8 $\geq$97% (HPLC); peptide content:~70% | Bioactive peptide; 10 residues of large T-antigen sequence; used as a nuclear localization signal peptide; Goldfarb, DS et al, *Nature*, 322:641, 1986

Pro-Lys-Pro-Gln-Gln-Phe-Phe-Gly-Leu-Met Amide

Synonyms: Substance P (2-11); Substance P, Deca-

American Peptide 70-1-40   MW 1191.5 $C_{57}H_{86}N_{14}O_{12}S$ Analog of the neuropeptide substance P; responsible for a number of excitatory effects on both central & peripheral neurons; Bury, RW et al, *JMC*, 19:854, 1976

Bachem H-4775.0001   MW 1191.46 $C_{57}H_{86}N_{14}O_{12}S$ Store at -15°C

ICN 153019   Bury, RW & ML Mashford, *J Med Chem*, 19:854, 1976

Pro-Lys-Pro-Gln-Gln-Phe-Phe-Gly-Leu-Met Amide Acetate Salt

Synonyms: Substance P (2-11)

Sigma S 0272   FW 1191.5 (free base) $\geq$97% (HPLC) | Bioactive peptide; substance P antagonist; Bury, RW & Mashford, ML, *J Med Chem*, 19:854, 1976; Piotrowski et al, *Agents Actions*, 20:178, 1987

Pro-Lys-Pro-Leu-Ala-Leu-Ala-Arg
Mca-Pro-Lys-Pro-Leu-Ala-Leu-Dap(Dnp)-Ala-Arg Amide

Bachem M-2225.0001   MW 1304.43 $C_{59}H_{85}N_{17}O_{17}$ Store at -15°C

Pro-Lys-Thr-Pro-Lys-Lys-Ala-Lys-Lys-Leu

Synonyms: Cdc2 Protein Kinase

Promega V2211   MW 1137   See Promega V5601

Pro-Met
BOC-Pro-Met

Bachem A-2600.0001   Store at -15°C

Pro-Met
Z-Pro-Met

Bachem C-2530.0001   MW 380.47 $C_{18}H_{24}N_2O_5S$ Store at -15°C

Pro-Met

Bachem G-3045.0250   MW 246.33 $C_{10}H_{18}N_2O_3S$ Store at -15°C

ICN 156384   MW 246.3 $C_{10}H_{18}N_2O_3S$ Crystalline

Sigma P 1255   FW 246.3 $C_{10}H_{18}N_2O_3S$

Pro-Met-Ser-Met-Leu-Arg-Leu Amide

Synonyms: Myomodulin

American Peptide 91-1-10   MW 846.2 $C_{36}H_{67}N_{11}O_8S_2$
Cropper, EC et al, *PNAS*, 84:5483, 1987

Pro-Phe
Carbomethoxycarbonyl-D-Pro-D-Phe-OBzl

American Peptide 72-3-10   MW 438.5 $C_{24}H_{26}N_2O_6$
Antibiotic peptide; inhibits binding of gp 120 of HIV-1 virus to CD-4 (cell surface glycoprotein); Finberg, RW et al, *Science*, 249:287, 1990

Pro-Phe
BOC-Pro-Phe

Bachem A-3195.0005   MW 362.43 $C_{19}H_{26}N_2O_5$ Store at -15°C

Pro-Phe
BOC-Pro-D-Phe

Bachem A-3245.0001   MW 362.43 $C_{19}H_{26}N_2O_5$ Store at -15°C

Pro-Phe
BOC-D-Pro-Phe

Bachem A-3250.0001   MW 362.43 $C_{19}H_{26}N_2O_5$ Store at -15°C

Pro-Phe
BOC-D-Pro-D-Phe

Bachem A-3255.0001   MW 362.43 $C_{19}H_{26}N_2O_5$ Store at -15°C

Pro-Phe
FMOC-Pro-Phe

Bachem B-1520.0001   MW 484.55 $C_{29}H_{28}N_2O_5$ Store at -15°C

Pro-Phe
Z-Pro-Phe

Bachem C-2535.0001   MW 396.44 $C_{22}H_{24}N_2O_5$ Store at -15°C

Pro-Phe

Bachem G-3050.0250   MW 262.31 $C_{14}H_{18}N_2O_3$ Store at -15°C

Pro-Phe
Carbomethoxycarbonyl-Pro-Phe-OBzl

Bachem G-4120.0050   MW 438.48 $C_{24}H_{26}N_2O_6$ Store at -15°C

Pro-Phe
4-Methoxy-Bz-Pro-Phe-pNA

Bachem L-1700.0050   Store at -15°C

Pro-Phe
Bz-Pro-Phe-pNA

Bachem L-1705.0050   MW 486.53 $C_{27}H_{26}N_4O_5$ Store at -15°C

Pro-Phe

ICN 102749   MW 262.3 $C_{14}H_{18}N_2O_3$ Crystalline

Pro-Phe
N-Carbomethoxycarbonyl-D-Pro-D-Phe Benzyl Ester

Synonyms: Harvard Peptide

ICN 159735   MW 438.5 $C_{24}H_{26}N_2O_6$ Blocks gp120 binding to CD-4 glycoprotein; inhibits infection by HIV; Finberg, RW etal, *Science*, 249:287, 1990

Pro-Phe
N-Carbomethoxycarbonyl-Pro-Phe Benzyl Ester

Synonyms: Harvard Peptide

ICN 159736   MW 438.5 $C_{24}H_{26}N_2O_6$ Blocks gp120 binding to CD-4 glycoprotein & inhibits HIV infection; less active than the D-isomer peptide; Finberg, RW etal, *Science*, 249:287, 1990

Pro-Phe
BOC-Pro-Phe-NHO-Bz-*p*-Chloro

Synonyms: Subtilisin Inhibitor IV; Serine Protease Inhibitor

ICN 195917   MW 516 $C_{26}H_{30}N_3O_6Cl$ >95%

Pro-Phe
Carbomethoxycarbonyl-D-Pro-D-Phe Benzyl Ester

Sigma C 4924   FW 438.5 $C_{24}H_{26}N_6O_6$ Bioactive peptide; inhibits binding of the envelope protein (gp 120) of the HIV-1 virus to the cell surface glycoprotein CD-4; Finberg, RW et al, *Science*, 249:287, 1990

Pro-Phe
N-CBZ-Pro-Phe

Sigma C 7772   FW 396.4 $C_{22}H_{24}N_2O_5$

Pro-Phe

Sigma P 1505   FW 262.3 $C_{14}H_{18}N_2O_3$

Pro-Phe
N-Carbomethoxycarbonyl-L-Pro-L-Phe-OBzl

Synonyms: Carbomethoxycarbonyl Dipeptide

Peptides International PCD-3801-PI   Synthetic   MW 438.48 $C_{24}H_{26}N_2O_6$ >98% (HPLC); white powder | Bioactive; prevention of HIV-1 infection; Finberg, RW et al, *Science*, 249:287, 1990

Pro-Phe
N-Carbomethoxycarbonyl-L-Pro-D-Phe-OBzl

Synonyms: Carbomethoxycarbonyl Dipeptide

Peptides International PCD-3802-PI   Synthetic   MW 438.48 $C_{24}H_{26}N_2O_6$ >98% (HPLC); white powder | Bioactive; prevention of HIV-1 infection; Finberg, RW et al, *Science*, 249:287, 1990

Pro-Phe
N-Carbomethoxycarbonyl-D-Pro-L-Phe-OBzl

Synonyms: Carbomethoxycarbonyl Dipeptide

Peptides International PCD-3803-PI Synthetic MW 438.48
$C_{24}H_{26}N_2O_6$ >98% (HPLC); white powder | Bioactive; prevention of HIV-1 infection; Finberg, RW et al, *Science*, 249:287, 1990

Pro-Phe
N-Carbomethoxycarbonyl-D-Pro-D-Phe-OBzl

Synonyms: Carbomethoxycarbonyl Dipeptide

Peptides International PCD-3804-PI Synthetic MW 438.48
$C_{24}H_{26}N_2O_6$ >98% (HPLC); white powder | Bioactive; prevention of HIV-1 infection; Finberg, RW et al, *Science*, 249:287, 1990

Pro-Phe Amide Hydrochloride

Bachem G-3055.0250 MW 297.78 $C_{14}H_{19}N_3O_2 \cdot HCl$ Store at -15°C

ICN 156394 MW 297.8 $C_{14}H_{19}N_3O_2 \cdot HCl$

Sigma P 1630 FW 297.8 $C_{14}H_{19}N_3O_2 \cdot HCl$

Pro-Phe-Arg
D-Pro-Phe-Arg-pNA Dihydrochloride

Bachem L-2120.0025 MW 611.53 $C_{26}H_{34}N_8O_5 \cdot 2HCl$ Store at -15°C

Pro-Phe-Arg
D-Pro-Phe-Arg-CMK

Bachem N-1210.0025 MW 450.97 $C_{21}H_{31}ClN_6O_3$ Store at -15°C

Pro-Phe-Arg
L-Pro-Phe-L-Arg-MCA Hydrochloride

Synonyms: Kallikrein Substrate

ICN 151953 Substrate for pancreatic & urinarykallikrein; Morita, T etal, *J Biochem*, 82:1495, 1977

Pro-Phe-Arg
3-PhPr-Pro-Phe-Arg-pNA·AcOH

Synonyms: Pefachrome®PAP; Pefa-5920

Pentapharm 090-13, 090-12 MW 730.8 Highly sensitive chromogenic peptide substrate for papain; v_{max}:35.6 µmol/min, K_M:0.20 *mM*

Pro-Phe-Arg
Pro-Phe-Arg-MCA Hydrochloride

Synonyms: Pancreatic/Urinary Kallikrein Substrate; Proteasome Substrate

Peptides International MPF-3096-v MW 575.67
$C_{30}H_{37}N_7O_5$ >98% (HPLC); lyophilized amorphous powder form | Morita, T et al, *J Biochem*, 82:1495, 1977; Ustrell, V et al, *PNAS USA*, 92:584, 1995

Pro-Phe-Arg
L-Pro-Phe-Arg-pNA Dihydrochloride

Sigma P 7959 FW 611.5 $C_{26}H_{34}N_8O_5 \cdot 2HCl$

Pro-Phe-Arg
Pro-Phe-Arg-MCA

Synonyms: Kallikrein Substrate, Pancreatic; Kallikrein Substrate, Urinary

Sigma P 9273 FW 575.7 $C_{30}H_{37}N_7O_5$ ~95%; peptide content:~70% | Fluorogenic substrate; Sakakibara, S et al, *J Biochem*, 82:1495, 1977; Lottenberg, R et al, *Meth Enz*, 80:341, 1981

Pro-Phe-Arg
N-Bz-Pro-Phe-Arg-pNA Hydrochloride

Synonyms: Thrombin-Like Enzyme Substrate, *Agkistrodon contortrix*; Kallikrein Substrate, Plasma

Sigma B 2133 *Agkistrodon contortrix* FW 679.2
$C_{33}H_{38}N_8O_6 \cdot HCl$ Chromogenic substrate; Claeson, G et al, *Haemostasis*, 7:62, 1978; Lottenberg, R et al, *Meth Enz*, 80:341, 1981

Pro-Phe-Arg-AMC

Bachem I-1295.0050 MW 575.67 $C_{30}H_{37}N_7O_5$ Store at -15°C

Pro-Phe-Arg-Ser-Val-Gln
Ac-Pro-Phe-Arg-Ser-Val-Gln Amide

Synonyms: Kallikrein Inhibitor; Kallikrein Inhibitor 5

American Peptide 87-0-24 MW 773.9 $C_{35}H_{55}N_{11}O_9$

Neosystem SC157 MW 773.89 Virus-related peptide; bioactive

Pro-Phe-Cys-Asn-Ala-Phe-Thr-Gly-Cys

Synonyms: Cardioactive Peptide, Crustacean

American Peptide 14-1-84 MW 957.1 $C_{42}H_{56}N_{10}O_{12}S_2$
Disulfide bonds: Cys^3-Cys^9; native & synthetic CCAP exhibits high ino- & chrono-tropic effects on a semi-isolated crab heart preparation; Strangier, J et al, *PNAS*, 84:575, 1986

Pro-Phe-Cys-Asn-Ala-Phe-Thr-Gly-Cys Amide

Synonyms: Crustacean Cardioactive Peptide

Bachem H-6745.0001 MW 956.11 $C_{42}H_{57}N_{11}O_{11}S_2$ Store at -15°C

Pro-Phe-Gly-Lys

Bachem H-4785.0050 MW 447.54 $C_{22}H_{33}N_5O_5$ Store at -15°C

Pro-Phe-Gly-Lys Acetate Salt

ICN 153168 Hochschwendener, SM & RA Laursen, *JBC*, 256:11166, 1981

Sigma P 6691 FW 447.5 (free base) $C_{22}H_{33}N_5O_5$ ≥97% (HPLC) | Bioactive peptide

Pro-Phe-His-Leu-Leu-Val-Tyr-Ser
Z-Pro-Phe-His-Leu-Leu-Val-Tyr-Ser-4MβNA

Bachem J-1150.0010 MW 1264.49 $C_{68}H_{85}N_{11}O_{13}$ Store at -15°C

Pro-Phe-His-Leu-Leu-Val-Tyr-Ser
Z-Pro-Phe-His-Leu-Leu-Val-Tyr-Ser-βNA

Bachem K-1220.0010 MW 1234.46 $C_{67}H_{83}N_{11}O_{12}$ Store at -15°C

Pro-Phe-Lys

Bachem H-4790.0250 MW 390.48 $C_{20}H_{30}N_4O_4$ Store at -15°C

Pro-Phe-Pro-Gly-Pro-Ile

Synonyms: Casomorphin, (des-Tyr¹)-β-

ICN 152949 MW 626.8 $C_{32}H_{46}N_6O_7$

American Peptide 32-3-22 Bovine MW 626.8 $C_{32}H_{46}N_6O_7$

Pro-Phe-Thr-Arg-Asn-Tyr-Tyr-Val-Arg-Ala-Val-Leu-His-Leu

Bachem H-2518.0001 Store at -15°C

Pro-Phe-Trp-Ser-Tyr-Trp-Leu-Arg-Pro-Gly
Ac-3,4-Dehydro-Pro-p-fluoro-D-Phe-D-Trp-Ser-Tyr-D-Trp-Leu-Arg-Pro-Gly Amide

Synonyms: Luteinizing Hormone Releasing Hormone, (Ac-3,4-Dehydro-Pro[1],p-Fluoro-D-Phe[2],D-Trp[3,6])-

Bachem H-4050.0005   Store at -15°C

Pro-Phe-Trp-Ser-Tyr-Trp-Leu-Arg-Pro-Gly
N-Ac-Δ[3]Pro[1]-D-p-Fluoro-Phe-D-Trp-Ser-Tyr-D-Trp-Leu-Arg-Pro-Gly Amide

Synonyms: Luteinizing Hormone Releasing Hormone, (N-Ac-3,4-Dehydro-Pro[1],D-p-F-Phe[2],D-Trp[3,6])-; Luteinizing Hormone Releasing Hormone Antagonist

ICN 152914   Suppresses LH & FSH secretion; Cetel, NS etal, *J Clin Endocrin Metab*, 57:62, 1983; Pineda, JL etal, *J Endocrin Metab*, 56:420, 1983; Rivier, C etal, *Biol Reprod*, 29:374, 1984; Grady, RR etal, *Neuroendocrinology*, 40:246, 1985

Pro-Pro
BOC-Pro-Pro

Bachem A-3365.0005   MW 312.37 $C_{15}H_{24}N_2O_5$ Store at -15°C

Pro-Pro
FMOC-Pro-Pro

Bachem B-2145.0001   MW 434.49 $C_{25}H_{26}N_2O_5$ Store at -15°C

Pro-Pro
Z-Pro-Pro

Bachem C-2545.0001   MW 346.38 $C_{18}H_{22}N_2O_5$ Store at -15°C

Pro-Pro
Z-Pro-Pro-CHO-Dimethyl Acetal

Bachem N-1490.0025   MW 376.45 $C_{20}H_{28}N_2O_5$ Store at -15°C

Pro-Pro
N-t-BOC-Pro-Pro

Sigma B 0754   FW 312.4 $C_{15}H_{24}N_2O_5$

Pro-Pro
CBZ-Pro-Pro

USBio C2099-55   MW 346.4 $C_{18}H_{25}N_2O_5$ ≥99%

Pro-Pro Hydrochloride

Bachem G-3060.0250   MW 248.71 $C_{10}H_{16}N_2O_3 \cdot HCl$ Store at -15°C

Pro-Pro-Ala
N-((2S,3R)-3-Amino-2-Hydroxy-4-Phenylbutanoyl)-Pro-Pro-Ala Amide Trifluoroacetate Salt

Synonyms: Apstatin; Membrane-Bound Aminopeptidase P Inhibitor

Sigma A 1395   FW 459.5 (free base) $C_{23}H_{33}N_5O_5$ ≥95% (HPLC) | Bioactive peptide; potent, specific inhibitor; potentiates the effects of bradykinin; Prechel, MM et al, *J Pharmacol Exp Therap*, 275:1136, 1995; U.S. Patent No. 8,455,281

Pro-Pro-Asp-Ser-Asp-Pro-Gln-Ile-Pro-Pro-Pro-Tyr-Val-Glu-Pro-Thr-Ala-Pro-Gln-Val-Leu

Synonyms: P19 (110-130); HTLV I Peptide

Neosystem SC832   MW 2257.51   HTLV I peptide

Pro-Pro-Gly-Phe-Ser-Pro

Synonyms: Bradykinin (2-7)

American Peptide 18-1-58   MW 600.7 $C_{29}H_{40}N_6O_8$

Bachem H-2230.0005   MW 600.67 $C_{29}H_{40}N_6O_8$ Store at -15°C

ICN 152789   MW 600.7 $C_{29}H_{40}N_6O_8$

Sigma B 2151   FW 600.7 $C_{29}H_{40}N_6O_8$ ≥97% (HPLC) | Bioactive peptide

---

Pro-Pro-Gly-Phe-Ser-Pro-Phe-Arg

Synonyms: Bradykinin (2-9); Bradykinin, (des-Arg[1])-

American Peptide 18-1-59   MW 904.0 $C_{44}H_{61}N_{11}O_{10}$

Bachem H-2200.0005   MW 904.04 $C_{44}H_{61}N_{11}O_{10}$ Store at -15°C

ICN 159896   MW 904.1   98%

Sigma B 1901   FW 904.0 $C_{44}H_{61}N_{11}O_{10}$ ≥97% (HPLC) | Bioactive peptide; Nicolaides, ED et al, *J Med Chem*, 6:739, 1963

Pro-Pro-Pro

Bachem H-4795.0050   MW 309.37 $C_{15}H_{23}N_3O_4$ Store at -15°C

Pro-Pro-Pro-Gly-Ile-Arg-Gly-Pro

Bachem H-3466.0001   Store at -15°C

Pro-Pro-Pro-Gly-Met-Arg-Pro-Pro

Bachem H-3464.0001   Store at -15°C

Pro-Pro-Pro-Pro
BOC-Pro-Pro-Pro-Pro

Bachem A-2615.0100   MW 506.6 $C_{25}H_{38}N_4O_7$ Store at -15°C

Pro-Pro-Pro-Pro

Bachem H-4800.0050   MW 406.48 $C_{20}H_{30}N_4O_5$ Store at -15°C

Pro-Pro-Pro-Pro
N-t-BOC-Pro-Pro-Pro-Pro

Sigma B 0629   FW 506.6 $C_{25}H_{38}N_4O_7$

Pro-Pro-Pro-Tyr-Val-Glu-Pro-Thr-Ala-Pro-Gln-Val-Leu

Synonyms: P19 (118-130); HTLV I Peptide

Neosystem SC833   MW 1407.62   HTLV I peptide

Pro-Ser

Bachem G-3065.0250   MW 202.21 $C_8H_{14}N_2O_4$ Store at -15°C

Pro-Ser-Hyp-Gly-Asp-Trp

Bachem H-4086.0001   Store at -15°C

Pro-Ser-Leu-His-Arg-Thr-Gln-Ala-Asp-Glu-Leu-Pro-Ala-Cys

Synonyms: A-Raf Protein Kinase C-Terminal Blocking Peptide (584-597)

Calbiochem 553002   Human synthetic   MW 1537.7 $C_{64}H_{104}N_{20}O_{22}S$ Lyophilized | Based on the human, mouse & rat A-Raf sequences (human residues 584-597); this peptide coupled to KLH was used as the immunogen for the production of Anti-A-Raf Protein Kinase; for use in immunoabsorption for immunoprecipitation, immunocytochemistry, Western blotting & dot blots

---

Pro-Ser-Lys-Pro-Asp-Asn-Pro-Gly-Glu-Asp-Ala-Pro-Ala-Glu-Asp-Leu-Ala-Arg-Tyr-Tyr-Ser-Ala-Leu-Arg-His-Tyr-Ile-Asn-Leu-Ile-Thr-Arg-Gln-Arg-Tyr Amide

Synonyms: Neuropeptide Y (2-36)

American Peptide 60-4-20   Porcine   MW 4090.6
$C_{181}H_{278}N_{54}O_{55}$

Bachem H-2216.0500   Porcine   MW 4090.53
$C_{181}H_{278}N_{54}O_{55}$ Store at -15°C

Neosystem SC891   Porcine   MW 4090.52   Bioactive neuropeptide

Pro-Ser-Lys-Pro-Asp-Asn-Pro-Gly-Glu-Asp-Ala-Pro-Ala-Glu-Asp-Met-Ala-Arg-Tyr-Tyr-Ser-Ala-Leu-Arg-His-Tyr-Ile-Asn-Leu-Ile-Thr-Arg-Gln-Arg-Tyr Amide

Synonyms: Neuropeptide Y (2-36)

Bachem H-3316.0500   Human, rat   MW 4108.57
$C_{180}H_{276}N_{54}O_{55}S$ Store at -15°C

Neosystem SC859   Human, rat   MW 4108.55   Bioactive neuropeptide

Pro-Ser-Thr-His-Val-Leu-Ile-Thr-His-Thr-Ile

Synonyms: Tumor Necrosis Factor α (70-80), (Ile[76])-

Bachem H-2698.0001   Human   Store at -15°C

Pro-Ser-Tyr-Phe-Gln-Gln-Thr-His-Thr-Gln-Gln-Asp-Pro-Ala-Leu-Pro-Thr-Arg-Glu-Gly

Synonyms: GP140 (731-750); SIVmac251 Peptide

Neosystem SC781   MW 2301.45   SIVmac251 peptide from subtype GP140

Pro-Thr-Asp-Val-Lys-Arg-Tyr-Thr-Thr-Gly-Gly-Thr-Ser-Arg-Asn-Lys-Arg-Gly-Val-Phe

Synonyms: GP140 (511-530); SIVmac251 Peptide

Neosystem SC759   MW 2240.50   SIVmac251 peptide from subtype GP140

Pro-Thr-Glu-Phe-Phe-Arg-Leu
Pro-Thr-Glu-Phe-*p*-Nitro-Phe-Arg-Leu

Bachem H-1002.0005   MW 954.05 $C_{44}H_{63}N_{11}O_{13}$ Store at -15°C

Pro-Thr-Gly-Pro-Leu-Gly-Pro-Lys-Gly-Gln-Thr-Gly-Glu-Leu-Gly-Ala-Hyp-Gly-Asn-Lys-Gly-Glu-Gln-Gly-Pro-Lys

Synonyms: Collagen Type II (245-270), (Ala[260],Hyp[261],Asn[263])-

Bachem H-1604.0500   Store at -15°C

Pro-Thr-His-Ile-Lys-Trp-Gly-Asp

Bachem N-1450.0005   MW 953.07 $C_{44}H_{64}N_{12}O_{12}$ Store at -15°C

Pro-Thr-Pro-Ser Amide

Synonyms: IgA$_1$ Proteinase Inhibitor; IgA$_1$ Protease Inhibitor

American Peptide 81-6-90   MW 399.4 $C_{17}H_{29}N_5O_6$ Analog of the hinge region of human IgA2; inhibits proteolysis of IgA by *Neisseria gonorrheae* protease type 1; Wood, SG et al, *JMC*, 32:2407, 1989

ICN 158355   MW 399.4 $C_{17}H_{29}N_5O_6$ 97% | IgA$_2$ hinge region analog

Sigma P 8179   FW 399.4 $C_{17}H_{29}N_5O_6$ ≥97% (HPLC) | Bioactive peptide; analog of the hinge region of human IgA$_2$; inhibits proteolysis of IgA by Neisseria *gonorrheae* protease type I; Wood, SG et al, *J Med Chem*, 32:2407, 1989

Pro-Trp

Bachem G-3070.0250   MW 301.35 $C_{16}H_{19}N_3O_3$ Store at -15°C

Pro-Trp Acetate Salt

ICN 156428   MW 361.4 $C_{16}H_{19}N_3O_3 \cdot C_2H_4O_2$ Light yellow to tan crystals

Pro-Tyr

Bachem G-3075.0250   MW 278.31 $C_{14}H_{18}N_2O_4$ Store at -15°C

ICN 102736   MW 278.3 $C_{14}H_{18}N_2O_4$ Crystalline

Pro-Tyr Amide Hydrochloride

Bachem G-3080.0250   Store at -15°C

Pro-Tyr-Ala-Tyr-Trp-Met-Arg
DNP-Pro-Tyr-Ala-Tyr-Trp-Met-Arg

Synonyms: Matrix Metalloproteinases III Substrate; Stromelysin Substrate

Calbiochem 444220   MW 1052.3 $C_{54}H_{65}N_{13}O_{14}S$ Solid; ≥95% (HPLC); excitation max:280 nm; emission max:360 nm; soluble in 0.1% aqueous TFA | Fluorogenic substrate for human stromelysin (MMP-3); Netzel-Arnett, S et al, *Anal Biochem*, 195:86 1991

ICN 196023   MW 1152.3 $C_{54}H_{65}N_{13}O_{14}S$ ≥95% | Fluorogenic substrate for human stromelysin

Pro-Tyr-Asn-Ser-Ser-Pro-Arg-Pro-Glu-Gln-His-Lys-Ser-Tyr-Lys-(Cys)
Ac-Pro-Tyr-Asn-Ser-Ser-Pro-Arg-Pro-Glu-Gln-His-Lys-Ser-Tyr-Lys-(Cys) Amide Trifluoroacetate Salt

Synonyms: Nitric Oxide Synthase Blocking Peptide (599-613); eNOS Blocking Peptide

Calbiochem 482727   Bovine endothelial cells   MW 1962.2 $C_{85}H_{128}N_{26}O_{26}S$ ≥98% (HPLC); lyophilized solid; soluble in water | Immunization & blocking peptide for Anti-eNOS (599-613), bovine (rabbit); Michel, T & Lamas, S, *J Cardiovasc Pharmacol*, 20:S45, 1992

Pro-Tyr-Asn-Ser-Ser-Pro-Arg-Pro-Glu-Gln-His-Lys-Ser-Tyr-Lys-Cys
Ac-Pro-Tyr-Asn-Ser-Ser-Pro-Arg-Pro-Glu-Gln-His-Lys-Ser-Tyr-Lys-Cys

Synonyms: Nitric Oxide Synthase Blocking Peptide (599-613)

American Peptide 91-1-25   Bovine endothelial cell   MW 1962.2 $C_{85}H_{126}N_{25}O_{27}S$ Michel, T et al, *J Cardiovasc Pharmacol*, 20:S45, 1992

Pro-Tyr-Asp-Arg-Ile-Ser-Asn-Ser-Ala-Phe-Ser-Asp-Phe Amide

Synonyms: LymnaDF Amide-1

Bachem H-1648.0001   Store at -15°C

Pro-Tyr-Cys-Trp-His-Tyr-Pro-Pro-Lys-Pro-Cys-Gly-Ile-Val-Pro-Ala

Synonyms: HCV-1 e2 Protein (484-499)

Bachem H-2678.0500   Store at -15°C

Pro-Tyr-Lys-Glu-Phe-Gly-Ala-Thr

Bachem H-4810.0005   Store at -15°C

Pro-Tyr-Ser-Val-Gly-Phe-Arg-Glu-Ala-Asp-Ala-Cys
Pro-Tyr-Ser-Val-Gly-Phe-Arg-Glu-Ala-Asp-Ala(-Cys)
Acetate Salt

Synonyms: Ankyrin (1-11)

Biogenesis 0580-0050   Human erythroid 2.1/2.2 synthetic
Purified; lyophilized

Pro-Tyr-Tyr-Gly-Asp-Glu-Pro-Nle
DABCYL-Pro-Tyr-Tyr-Gly-Asp-Glu-Pro-Nle-EDANS

Synonyms: Collagen Type III α1 Chain (1070-1077)-EDANS, DABCYL-(Nle[1077])-; Collagen Type III α1 Chain (1062-1069)-EDANS, DABCYL-(Asp[1067],Nle[1069])-

Bachem M-2380.0001   Human, mouse   MW 1452.61
$C_{72}H_{85}N_{13}O_{18}S$ Store at -15°C

Pro-Val

Bachem G-3085.0250   MW 214.27 $C_{10}H_{18}N_2O_3$ Store at
-15°C

ICN 102738   MW 214.3 $C_{10}H_{18}N_2O_3$ Crystalline

Pro-Val-Asp

Bachem H-4815.0250   MW 329.35 $C_{14}H_{23}N_3O_6$ Store at
-15°C

Pro-Val-Gly

Bachem H-4820.0250   MW 271.32 $C_{12}H_{21}N_3O_4$ Store at
-15°C

Pro-Val-Gly-Lys-Lys-Arg-Arg-Pro-Val-Lys-Val-Tyr-Pro-Asn-Val-Ala-Glu-Asn-Glu-Ser-Ala-Glu-Ala-Phe-Pro-Leu-Glu-Phe

Synonyms: Adrenocorticotropic Hormone (12-39)

American Peptide 10-1-16   Rat   MW 3173.9
$C_{145}H_{227}N_{39}O_{41}$

Pro-Val-Pro-Leu-Gln-Leu-Pro-Pro-Leu-Glu-Arg-Leu-Thr-Leu-Asp

Synonyms: REV (70-84); HTLV I Peptide

Neosystem SC554   MW 1701.03   HTLV I peptide from
subtype REV

Pro-Val-Thr-Gln-Glu-Phe-Trp-Asp-Asn-Leu-Glu-Lys-Glu-Thr-Glu-Gly-Leu-Arg-Glu-Glu-Met

Synonyms: Apolipoprotein A-1 (66-86)

Biogenesis 0650-0379   Human synthetic   MW 2580.8
Lyophilized, containing 68.5% peptide material, TFA counter ions
and residual H_2O | To avoid Met oxidation, it is best to dissolve
the peptide under a nitrogen stream

Pro-Val-Thr-Leu
Pro-Val-Thr-Leu-OMe Hydrochloride

Synonyms: Eglin c (42-45)-OMe

Bachem H-1184.0025   MW 442.56 $C_{21}H_{38}N_4O_6$ Store at
-15°C

Pro-Val-Thr-Lys-Pro-Gln-Ala Amide

Synonyms: Sorbin C-Terminal Peptide C-7; C7 Sorbin

Bachem H-2242.0005   MW 738.89 $C_{33}H_{58}N_{10}O_9$ Store at
-15°C

Neosystem SC997   Porcine   MW 738.88   Bioactive;
brain/gut peptide; Vagne-Descroix, M et al, *Mol Pharm Biochem*,
201:53-59, 1991; Nicol, P et al, *Peptides*, 15:1013-1019, 1994;
Marquet, F et al, *Gastroenterol Clin Biol*, 18:702-707, 1994;
Grishina, O et al, *Gastroenterol Clin Biol*, 19:487-493, 1995

Sar-Ala

Bachem G-3095.0250   MW 160.17 $C_6H_{12}N_2O_3$ Store at
-15°C

Sar-Arg-Gly-Asp-Ser-Pro

Synonyms: Protease Resistant Peptide

American Peptide 44-0-35   MW 601.6 $C_{23}H_{39}N_9O_{10}$ Highly
protease resistant analog

Bachem H-3162.0001   MW 601.62 $C_{23}H_{39}N_9O_{10}$ Store at
-15°C

Sar-Arg-Val-Tyr-Ile-His-Pro Amide

Synonyms: (Sar[1])-Angiotensin I/II (1-7)

Bachem H-2892.0005   MW 854.02 $C_{40}H_{63}N_{13}O_8$ Store at
-15°C

Sar-Arg-Val-Tyr-Ile-His-Pro-Ala

Synonyms: Angiotensin II, (Sar[1],Ala[8])-; Angiotensin II Antagonist;
Angiotensin II Selective Antagonist

American Peptide 12-1-43   MW 926.1 $C_{43}H_{67}N_{13}O_{10}$ Pals, D
et al, *Circ Res*, 29:664, 1971

Bachem H-1720.0005   MW 926.09 $C_{43}H_{67}N_{13}O_{10}$ Store at
-15°C

Calbiochem 05-23-0125   Human   MW 926.1 $C_{43}H_{67}N_{13}O_{10}$
>99% (HPLC); lyophilized; soluble in 5% HOAc; LD_{50}≤2000 mg/kg;
may be carcinogenic/teratogenic | Angiotensin antagonist with low
activity; Reed, G et al, *Neurosci Lett*, 210:209, 1996

Peptides International PAN-4035-v   Synthetic   MW 926.09
$C_{43}H_{67}N_{13}O_{10}$ >99% (HPLC); lyophilized amorphous powder

Sar-Arg-Val-Tyr-Ile-His-Pro-Ala Acetate Salt

Synonyms: Angiotensin II Inhibitor; Angiotensin II, (Sar[1],Ala[8])-;
Angiotensin II Antagonist

ICN 152732   Pals, D etal, *Circ Res*, 29:664, 1971; Geiger, R
etal, *Hoppe-Seyler's Z Physiol Chem*, 355:1083, 1974; Geiger, R
etal, *Hoppe-Seyler's Z Physiol Chem*, 357:825, 1976

Sigma A 8026   FW 926.1 (free base) $C_{43}H_{67}N_{13}O_{10}$ ≥98%
(HPLC) | Bioactive peptide; Geiger, R et al, *Hoppe-Seyler's Z
Physiol Chem*, 355:1083, 1974; ibid, 357:825, 1976

Sar-Arg-Val-Tyr-Ile-His-Pro-Ala AcOH 4H₂O

Synonyms: Angiotensin II, (Sar[1],Ala[8])-; Angiotensin II Selective
Antagonist

Peptides International PAN-4035   Synthetic   MW 1058.2
$C_{43}H_{67}N_{13}O_{10} \cdot CH_3COOH \cdot 4H_2O$ >99% (HPLC); lyophilized
amorphous powder | Bioactive; Pals, DT et al, *Circ Res*, 29:673,
1971; Bumpus, FM et al, *Circ Res*, 32 & 33(I):1, 1973

Sar-Arg-Val-Tyr-Ile-His-Pro-Gly

Synonyms: Angiotensin II, (Sar[1],Gly[8])-

Bachem H-1725.0005   MW 912.06 $C_{42}H_{65}N_{13}O_{10}$ Store at
-15°C

Sar-Arg-Val-Tyr-Ile-His-Pro-Gly Acetate Salt

Synonyms: Angiotensin II Inhibitor; Angiotensin II, (Sar[1],Gly[8])-;
Angiotensin II Antagonist

ICN 152733   Park, WK, *Can J Biochem*, 52:113, 1974

Sigma A 7401   FW 912.1 (free base) $C_{42}H_{65}N_{13}O_{10}$ ≥97%
(HPLC) | Bioactive peptide; Park, WK, *Can J Biochem*, 52:113,
1974

Sar-Arg-Val-Tyr-Ile-His-Pro-Ile

Synonyms: Angiotensin II, (Sar[1],Ile[8])-; Angiotensin II Selective
Antagonist

Bachem H-1730.0005   MW 968.17 $C_{46}H_{73}N_{13}O_{10}$ Store at
-15°C

Neosystem SC938 Human MW 968.16 Bioactive; specific angiotensin II antagonist; Jones, C et al, *Peptides*, 10:459-463, 1989; Mallorga, P et al, *Curr Eye Res*, 8:841-849, 1989; Baker, KM et al, *Am J Physiol*, 259:610-618, 1990; Rowe, BP et al, *Brain Res*, 534:129-134, 1990; Summers, C & Myers, LM, *Am J Physiol*, 260:79-87, 1991

Peptides International PAN-4016-v Synthetic MW 968.16 $C_{46}H_{73}N_{13}O_{10}$ >99% (HPLC); lyophilized amorphous powder | Bioactive; Türker, RK et al, *Science*, 177:1203, 1972; Bumpus, FM et al, *Circ Res*, 32 & 33(I):1, 1973

Sar-Arg-Val-Tyr-Ile-His-Pro-Ile Acetate Salt
Synonyms: Angiotensin II Inhibitor; Angiotensin II, (Sar[1],Ile[8])-; Angiotensin II Antagonist

ICN 152734 Khosia, MC et al, *J Med Chem*, 15:792, 1972; Kono, T et al, *J Clin Endocrinol Metab*, 52:354, 1981

Sigma A 8776 FW 968.2 (free base) $C_{46}H_{73}N_{13}O_{10}$ ≥97% (HPLC) | Bioactive peptide; Khosla, MC et al, *J Med Chem*, 15:792, 1972; Kono, T et al, *J Clin Endocrinol Metab*, 52:354, 1981

Sar-Arg-Val-Tyr-Ile-His-Pro-Ile AcOH 4H2O
Synonyms: Angiotensin II, (Sar[1],Ile[8])-; Angiotensin II Selective Antagonist

Peptides International PAN-4016 Synthetic MW 1100.2 $C_{46}H_{73}N_{13}O_{10} \cdot CH_3COOH \cdot 4H_2O$ >99% (HPLC); lyophilized amorphous powder

Sar-Arg-Val-Tyr-Ile-His-Pro-Leu Acetate Salt
Synonyms: Angiotensin II Inhibitor; Angiotensin II, (Sar[1],Leu[8])-; Angiotensin II Antagonist

ICN 152735 Sen, I et al, *Eur J Biochem*, 136:41, 1983

Sigma A 8276 FW 968.2 (free base) $C_{46}H_{73}N_{13}O_{10}$ ≥97% (HPLC) | Bioactive peptide; Sen, I et al, *Eur J Biochem*, 136:41, 1983

Sar-Arg-Val-Tyr-Ile-His-Pro-Phe
Synonyms: Angiotensin II, (Sar[1])-

American Peptide 12-1-50 MW 1002.2 $C_{49}H_{71}N_{13}O_{10}$ Mendelsohn, FAO et al, *PNAS*, 81:1575, 1984

Bachem H-1740.0005 MW 1002.18 $C_{49}H_{71}N_{13}O_{10}$ Store at -15°C

Sar-Arg-Val-Tyr-Ile-His-Pro-Phe
Sar-Arg-Val-Tyr(Me)-Ile-His-Pro-Phe
Synonyms: Angiotensin II, (Sar[1],Tyr(Me)[4])-; Sarmesin

Bachem H-4178.0005 MW 1016.21 $C_{50}H_{73}N_{13}O_{10}$ Store at -15°C

Sar-Arg-Val-Tyr-Ile-His-Pro-Phe
Synonyms: Angiotensin II, (Sar[1])-

Neosystem SC437 Human MW 1002.15 Bioactive; Israel, A et al, *Brain Res*, 322:341, 1984

Sar-Arg-Val-Tyr-Ile-His-Pro-Phe Acetate Salt
Synonyms: Angiotensin II, (Sar[1])-

ICN 152729 Israel, A et al, *Brain Res*, 322:341, 1984; Mendelsohn, FAO et al, *PNAS*, 81:1575, 1984

Sigma A 4410 FW 1002.2 (free base) ≥97% (HPLC) | Bioactive peptide; Mendelsohn, FAO et al, *Proc Natl Acad Sci USA*, 81:1575, 1984; Israel, A et al, *Brain Res*, 322:341, 1984

Sar-Arg-Val-Tyr-Ile-His-Pro-Thr
Synonyms: Angiotensin II, (Sar[1],Thr[8])-; Angiotensin II Selective Antagonist

Bachem H-1745.0005 MW 956.11 $C_{44}H_{69}N_{13}O_{11}$ Store at -15°C

Peptides International PAN-4102-v Synthetic MW 956.11 $C_{44}H_{69}N_{13}O_{11}$ >99% (HPLC); lyophilized amorphous powder | Bioactive; Khosla, MC et al, *J Med Chem*, 17:1156, 1974

Sar-Arg-Val-Tyr-Ile-His-Pro-Thr 2AcOH 4H2O
Synonyms: Angiotensin II, (Sar[1],Thr[8])-; Angiotensin II Selective Antagonist

Peptides International PAN-4102 Synthetic MW 1148.2 $C_{44}H_{69}N_{13}O_{11} \cdot 2CH_3COOH \cdot 4H_2O$ >99% (HPLC); lyophilized amorphous powder | Bioactive; Khosla, MC et al, *J Med Chem*, 17:1156, 1974

Sar-Arg-Val-Tyr-Ile-His-Pro-Thr Acetate Salt
Synonyms: Angiotensin II, (Sar[1],Thr[8])-; Angiotensin II Antagonist

ICN 152736 Potent, long-lasting; Munoz-Ramirez, H et al, *Res Commun Chem Pathol Pharmacol*, 13:649, 1976

Sigma A 9900 FW 956.1 (free base) $C_{44}H_{69}N_{13}O_{10}$ ≥97% (HPLC) | Bioactive peptide; potent, long-acting antagonist of Angiotensin II; Munoz-Ramirez, H et al, *Res Comm Chem Path Pharmacol*, 13:649, 1976

Sar-Arg-Val-Tyr-Val-His-Pro-Ala
Synonyms: Saralasin; Angiotensin II, (Sar[1],Val[5],Ala[8])-; Angiotensin II Selective Antagonist

American Peptide 12-1-45 MW 912.1 $C_{42}H_{65}N_{13}O_{10}$ As a competitive angiotensin II antagonist, saralasin could inhibit the pressor effects of angiotensin II in rats & show blood pressure lowering activity in humans; used in the diagnosis of renin-dependent hypertension; Pals, D et al, *Circ Res*, 29:664, 1971

Bachem H-1232.0005 MW 912.06 $C_{42}H_{65}N_{13}O_{10}$ Store at -15°C

Neosystem SC409 MW 912.05 Human Bioactive; specific Angiotensin II antagonist; Pellicer, A et al, *Science*, 240:1660, 1988

Peptides International PAN-4071-v Synthetic MW 912.06 $C_{42}H_{65}N_{13}O_{10}$ >99% (HPLC); lyophilized amorphous powder | Bioactive; Pals, DT et al, *Circ Res*, 29:673, 1971

Sar-Arg-Val-Tyr-Val-His-Pro-Ala Acetate Salt
Synonyms: Angiotensin II, (Sar[1],Val[5],Ala[8])-; Saralasin; Angiotensin II Inhibitor; Angiotensin II Antagonist

ICN 152737 Pals, DT et al, *Circ Res*, 29:673, 1987; Pellicer, A et al, *Science*, 240:1660, 1988

Sigma A 2275 FW 912.1 (free base) $C_{42}H_{65}N_{13}O_{10}$ ≥97% (HPLC) | Bioactive peptide; competitive angiotensin II antagonist; inhibits pressor effects of angiotensin II in rats & lowers blood pressure in humans; Pals, DT et al, *Cir Res*, 29:673, 1971

Sar-Arg-Val-Tyr-Val-His-Pro-Ala AcOH 4H2O
Synonyms: Angiotensin II, (Sar[1],Val[5],Ala[8])-; Angiotensin II Selective Antagonist

Peptides International PAN-4071 Synthetic MW 1044.1 $C_{42}H_{65}N_{13}O_{10} \cdot CH_3COOH \cdot 4H_2O$ >99% (HPLC); lyophilized amorphous powder | Bioactive; Pals, DT et al, *Circ Res*, 29:673, 1971

Sar-Gln-Gln-Phe-Phe-Gly-Leu-Met Amide
Synonyms: Substance P (4-11), (Sar[4])-

American Peptide 70-1-59 MW 940.1 $C_{44}H_{65}N_{11}O_{10}S$

Sar-Gly
Bachem G-3105.0250 MW 146.15 $C_5H_{10}N_2O_3$ Store at -15°C

Sar-Gly-Gly
Bachem H-4935.0250 MW 203.2 $C_7H_{13}N_3O_4$ Store at -15°C

Sar-Met
Z-Sar-Met
Bachem C-2580.0001 MW 354.43 $C_{16}H_{22}N_2O_5S$ Store at -15°C

Sar-Phe			
Bachem G-3125.0250	MW 236.27	$C_{12}H_{16}N_2O_3$	Store at -15°C

Sar-Pro			
Bachem G-3130.0250	MW 186.21	$C_8H_{14}N_2O_3$	Store at -15°C

Sar-Pro-Arg
Z-Sar-Pro-Arg

Bachem C-2585.0250   MW 476.53 $C_{22}H_{32}N_6O_6$ Store at -15°C

Sar-Pro-Arg
Sar-Pro-Arg-pNA

Bachem L-1990.0050   Store at -15°C

Sar-Pro-Arg
Sar-Pro-Arg-pNA Dihydrochloride

Synonyms: Thrombin Substrate

ICN 156586   Chromogenic substrate; Duncan, A etal, *Clin Chem*, 31:853, 1985

Sar-Pro-Arg
Sar-Pro-Arg-pNA

Synonyms: Thrombin Substrate, α-

Neosystem SC1320   MW 462.49   Chromogenic substrate used for human α-thrombin assay; Malikayil, JA et al, *Biochem*, 36:1034-1040, 1997

Sar-Pro-Arg
Sar-Pro-Arg-pNA Dihydrochloride

Synonyms: Thrombin Substrate

Sigma S 9009   FW 535.4 $C_{20}H_{30}N_8O_5 \cdot 2HCl$ Chromogenic substrate; Duncan, A et al, *Clin Chem*, 31:853, 1985

Sar-Sar

Bachem G-3135.0250   MW 160.17 $C_6H_{12}N_2O_3$ Store at -15°C

Sar-Val
Sar-N-Me-Val-OMe Hydrochloride

Bachem G-3122.0050   Store at -15°C

Sar-Val-Tyr-Ile-His-Pro-Ile Acetate Salt

Synonyms: Angiotensin III, (Sar[1],Ile[7])-; Angiotensin III Antagonist

ICN 194133   MW 812   97% | Angiotensin III antagonist; Tabrizchi, R etal, *Life Sciences*, 43:537, 1988

Sigma A 4455   FW 812.0 (free base) $C_{40}H_{61}N_9O_9$ 7% (HPLC) | Bioactive peptide; selective antagonist; Tabrizchi, R et al, *Life Sciences*, 43:537, 1988

Ser-Abu-Ser-Tyr-Lys-Lys Amide

Bachem H-2768.0001   Store at -15°C

Ser-Abz-His-Lys-Ala-Arg-Val-Leu-Phe-Glu-Ala-Nle-Ser
Ser-Abz-His-Lys-Ala-Arg-Val-Leu-p-Nitro-Phe-Glu-Ala-Nle-Ser Amide

Synonyms: Anthranilyl-HIV Protease Substrate III

Bachem H-1044.0001   MW 1433.63 $C_{65}H_{100}N_{20}O_{17}$ Store at -15°C

Ser-Ala
Z-Ser-Ala

Bachem C-2615.0250   MW 310.31 $C_{14}H_{18}N_2O_6$ Store at -15°C

Ser-Ala

Bachem G-3160.0250   MW 176.17 $C_6H_{12}N_2O_4$ Store at -15°C

Ser-Ala
Ser-β-Ala

Bachem G-3165.0250   MW 176.17 $C_6H_{12}N_2O_4$ Store at -15°C

Ser-Ala

ICN 156602   MW 176.2 $C_6H_{12}N_2O_4$ White powder

Ser-Ala
Ser-β-Ala

ICN 156603   MW 176.2 $C_6H_{12}N_2O_4$

Ser-Ala

Sigma S 3003   FW 176.2 $C_6H_{12}N_2O_4$ White powder

Ser-Ala
Ser-β-Ala

Sigma S 4639   FW 176.2 $C_6H_{12}N_2O_4$

Ser-Ala-Asn-Ser-Asn-Pro-Ala-Leu-Ala-Pro-Arg-Glu-Arg-Lys-Ala-Gly-Cys-Lys-Asn-Phe-Phe-D-Trp-Lys-Thr-Tyr-Thr-Ser-Cys

Synonyms: Somatostatin (1-28), (Leu[8],D-Trp[22],Tyr[25])-

American Peptide 68-1-57   MW 3146.6 $C_{138}H_{209}N_{41}O_{40}S_2$ Disulfide bonds: Cys[17]-Cys[28]; labeled by radioactive iodine; used to characterize somatostatin binding sites in rat brain & hamster insulinoma membranes; Reubi, JC et al, *Endo*, 110:1049, 1982

Ser-Ala-Asn-Ser-Asn-Pro-Ala-Leu-Ala-Pro-Arg-Glu-Arg-Lys-Ala-Gly-Cys-Lys-Asn-Phe-Phe-Trp-Lys-Thr-Tyr-Thr-Ser-Cys
Ser-Ala-Asn-Ser-Asn-Pro-Ala-Leu-Ala-Pro-Arg-Glu-Arg-Lys-Ala-Gly-Cys-Lys-Asn-Phe-Phe-D-Trp-Lys-Thr-Tyr-Thr-Ser-Cys

Synonyms: Somatostatin 28, (Leu[8],D-Trp[22],Tyr[25])-

Bachem H-3202.0500   Store at -15°C

Ser-Ala-Asn-Ser-Asn-Pro-Ala-Leu-Ala-Pro-Arg-Glu-Arg-Lys-Ala-Gly-Cys-Lys-Asn-Phe-Phe-Trp-Lys-Thr-Tyr-Thr-Ser-Cys
Biotinyl-Ser-Ala-Asn-Ser-Asn-Pro-Ala-Leu-Ala-Pro-Arg-Glu-Arg-Lys-Ala-Gly-Cys-Lys-Asn-Phe-Phe-D-Trp-Lys-Thr-Tyr-Thr-Ser-Cys

Synonyms: Somatostatin 28, Biotinyl-(Leu[8],D-Trp[22],Tyr[25])-

Bachem H-3204.0500   Store at -15°C

Ser-Ala-Asn-Ser-Asn-Pro-Ala-Leu-Ala-Pro-Arg-Glu-Arg-Lys-Ala-Gly-Cys-Lys-Asn-Phe-Phe-Trp-Lys-Thr-Tyr-Thr-Ser-Cys
Ser-Ala-Asn-Ser-Asn-Pro-Ala-Leu-Ala-Pro-Arg-Glu-Arg-Lys-Ala-Gly-Cys-Lys-Asn-Phe-Phe-D-Trp-Lys-Thr-Tyr-Thr-Ser-Cys

Synonyms: Somatostatin 28, (Leu[8],D-Trp[22],Tyr[25])-

ICN 153001   Disulfide bonds: Cys[17]-Cys[28]; Reubl, J-C etal, *Endocrinology*, 110:1049, 1982

Sigma S 2636 FW 3146.5 ≥97% (HPLC) | Disulfide bonds: 17-28; bioactive peptide; somatostatin analog that can be iodinated & used to characterize binding sites in rat brain & hamster insulinoma membranes; Reubi, J-C et al, *Endocrinology*, 110:1049, 1982

Ser-Ala-Asn-Ser-Asn-Pro-Ala-Met-Ala-Pro-Arg-Glu

Synonyms: Somatostatin 28 (1-12)

American Peptide 68-1-62 MW 1244.4 $C_{49}H_{81}N_{17}O_{19}S$
Derived from the cleavage product of prosomatostatin; Benoit, R et al, *PNAS*, 79:917, 1982; Morrison, JH et al, *Brain Res*, 62:344, 1983

Bachem H-4945.0001 Store at -15°C

Ser-Ala-Asn-Ser-Asn-Pro-Ala-Met-Ala-Pro-Arg-Glu-Arg-Lys

Synonyms: Somatostatin 28 (1-14)

American Peptide 68-1-60 MW 1528.7 $C_{61}H_{105}N_{23}O_{21}S$
Wuensch, E et al, *Biopolymers*, 20:1741, 1981

Bachem H-4950.0001 MW 1528.71 $C_{61}H_{105}N_{23}O_{21}S$ Store at -15°C

ICN 153002 Wuensch, E etal, *Biopolymers*, 20:1741, 1981

Sigma S 2386 FW 1528.7 ≥97% (HPLC) | Bioactive peptide; tetradecapeptide cleavage product of prosomatostatin; Wuensch, E et al, *Biopolymers*, 20:1741, 1981

Ser-Ala-Asn-Ser-Asn-Pro-Ala-Met-Ala-Pro-Arg-Glu-Arg-Lys-Ala-Gly-Cys-Lys-Asn-Phe-Phe-Trp-Lys-Thr-Phe-Thr-Ser-Cys

Synonyms: Somatostatin 28; Pro-Somatostatin

American Peptide 68-1-50 MW 3148.6 $C_{137}H_{207}N_{41}O_{39}S_3$
Disulfide bonds: Cys[17]-Cys[28]; from the posttranslational cleavage of prosomatostatin, which in turn is derived from a large precursor, preprosomatostatin; found in the central & peripheral nervous system, the gastrointestinal tract where it acts as a somatostatin receptor agonist; Bohlen, P et al, *BBRC*, 96:725, 1980; Pradayrol, L et al, *FEBS Lett*, 109:55, 1980

Bachem H-4955.0500 MW 3148.6 $C_{137}H_{207}N_{41}O_{39}S_3$ Store at -15°C

ICN 152999 Disulfide bonds: Cys[17]-Cys[28]; Pradyrol, L etal, *FEBS Lett*, 109:55, 1980; Schally, AV etal, *PNAS*, 77:4489, 1980

Neosystem SC095 MW 3148.57 Disulfide bonds: Cys[17]-Cys[28]; bioactive; hypothalamic releasing hormone; Pradayrol, L et al, *FEBS Lett*, 109:55, 1980

Sigma S 6135 FW 3148.6 ≥97% (HPLC) | Disulfide bonds: 17-28; bioactive peptide; inhibits the release of growth hormone, thyroid stimulating hormone, insulin & glucagon; Pradayrol, L et al, *FEBS Lett*, 109:55, 1980; Schally, AV et al, *Proc Natl Acad Sci USA*, 77:4489, 1980

Ser-Ala-Asn-Ser-Asn-Pro-Ala-Met-Ala-Pro-Arg-Glu-Arg-Lys-Ala-Gly-Cys-Lys-Asn-Phe-Phe-Trp-Lys-Thr-Phe-Thr-Ser-Cys
H₂N-Ser-Ala-Asn-Ser-Asn-Pro-Ala-Met-Ala-Pro-Arg-Glu-Arg-Lys-Ala-Gly-Cys-Lys-Asn-Phe-Phe-Trp-Lys-Thr-Phe-Thr-Ser-Cys

Synonyms: Somatostatin

Biogenesis 8330-1104 Human synthetic MW 3148.8
99% (TLC) and HPLC; lyophilized | To avoid Met oxidation, it is best to dissolve the peptide under a nitrogen stream

---

Ser-Ala-Asn-Ser-Asn-Pro-Ala-Met-Ala-Pro-Arg-Tyr-Arg-Lys

Synonyms: Somatostatin 28, (1-14), (Tyr[12])-

American Peptide 68-1-44 MW 1562.8 $C_{65}H_{107}N_{23}O_{20}S$

Bachem H-4960.0001 Store at -15°C

Ser-Ala-Asn-Ser-Asn-Pro-Ala-Nle-Ala-Pro-Arg-Glu-Arg-Lys-Ala-Gly-Cys-Lys-Asn-Phe-Phe-Trp-Lys-Thr-Phe-Thr-Ser-Cys

Synonyms: Somatostatin (1-28), (Nle[8])-

American Peptide 68-1-54 MW 3130.6 $C_{138}H_{209}N_{41}O_{39}S_2$
Disulfide bonds: Cys[17]-Cys[28]

Ser-Ala-Glu-Glu-Tyr-Glu-Tyr-Pro-Ser Non-sulfated

Synonyms: Cholecystokinin Flanking Peptide; Cholecystokinin Precursor (107-115)

American Peptide 46-1-42 MW 1074.1 $C_{47}H_{63}N_9O_{20}$

Bachem H-4908.0001 Human MW 1074.07 $C_{47}H_{63}N_9O_{20}$
Store at -15°C

Ser-Ala-Gly-Thr

Synonyms: Vitronectin (307-310)

Neosystem SC970 Human MW 334.33 Bioactive; cell attachment peptide; procollagen fragment; a ligand for integrin αvβ3; this novel binding sequence, identical to a tetrapeptide found in vitronectin, is a candidate for a synergistic site in this adhesive protein that may act in concert with RGD to promote molecular recognition; Healy, JM et al, *Biochem*, 34:3948-3955, 1995

Ser-Ala-Leu-Arg-His-Tyr-Ile-Asn-Leu-Ile-Thr-Arg-Gln-Arg-Tyr Amide

Synonyms: Neuropeptide Y (22-36)

Bachem H-9305.0001 MW 1903.22 $C_{85}H_{139}N_{29}O_{21}$ Store at -15°C

Neosystem SC345 MW 1903.21 Bioactive neuropeptide; calmodulin inhibitory peptide; Ishiguro, T et al, *Chem Pharm Bull*, 36:2720, 1988

ICN 154450 Porcine MW 1903.4 Ishiguro, T etal, *Chem Pharm Bull*, 36:2720, 1988

Ser-Ala-Leu-Pro-Leu-Glu-Ser-Gly-Pro-Thr-Gly-Gln-Asp-Ser-Val-Gln-Asp-Ala-Thr-Gly Amide

Synonyms: Pro-Cortistatin (28-47)

American Peptide 34-7-33 MW 1929.0 $C_{79}H_{129}N_{23}O_{33}$ de Leccea, L et al, *Nature*, 381:242, 1996

Ser-Ala-Lys-Leu-Cys-Pro-Gly-Gly-Asn-Cys-Val

Synonyms: Sperm Activating Peptide IIB, Ser-Ala-; Alloresact, Ser-Ala-; Sperm Activating Peptide

Bachem H-1426.0001 MW 1046.24 $C_{42}H_{71}N_{13}O_{14}S_2$ Store at -15°C

ICN 158730 MW 1046.2 97%

American Peptide 87-0-07 *Glyptocidaris crenularis* (sea urchin) egg jelly MW 1046.2 $C_{42}H_{71}N_{13}O_{14}S_2$ Causes significant increases of sperm respiration rates & sperm cyclic nucleotide concentrations; disulfide bonds: Cys[5]-Cys[10]

Sigma S 8665 Sea urchin FW 1046.2 ≥97% (HPLC) | Disulfide bonds: 5-11; bioactive peptide

---

Ser-Ala-Ser-Asn-Asn-Arg-Leu-Ile-Pro-Asn-Arg-Thr-Arg

Synonyms: TAT (95-107); SIVmac251 Peptide

Neosystem SC592 MW 1498.66 SIVmac251 peptide from subtype TAT

Ser-Arg-Ala-His-Gln-His-Ser-Met-Glu-Ile-Arg-Thr-Pro-Asp-Ile-Asn-Pro-Ala-Trp-Tyr-Ala-Gly-Arg-Gly-Ile-Arg-Pro-Val-Gly-Arg-Phe Amide

Synonyms: Prolactin Releasing Peptide (1-31); Prolactin Releasing Peptide 31; Pre-Prolactin (23-53)

Bachem H-4386.0500 Bovine MW 3576.07 $C_{157}H_{244}N_{54}O_{41}S$ Store at -15°C

Ser-Arg-Ala-His-Gln-His-Ser-Met-Glu-Thr-Arg-Thr-Pro-Asp-Ile-Asn-Pro-Ala-Trp-Tyr-Thr-Gly-Arg-Gly-Ile-Arg-Pro-Val-Gly-Arg-Phe Amide

Synonyms: Prolactin Releasing Peptide 31; Pre-Prolactin (23-53); Prolactin Releasing Peptide (1-31); Prolactin Releasing Peptide; Prolactin Releasing Peptide, Specific

Bachem H-4384.0500   Rat   MW 3594.04 $C_{156}H_{242}N_{54}O_{43}S$ Store at -15°C

Neosystem SC1333   Rat   MW 3594.02   Bioactive; hypothalamic releasing hormone

Peptides International PPR-4353-v   Rat synthetic   MW 3594.0 $C_{156}H_{242}N_{54}O_{43}S$ >95% (HPLC); lyophilized amorphous powder | Bioactive; Hinuma, S et al, *Nature*, 393:272, 1998

Ser-Arg-Asn-Arg-Cys-Asn-Asp-Gln Amide

Synonyms: Fibronectin (196-203)

Bachem H-2064.0005   MW 991.06 $C_{35}H_{62}N_{18}O_{14}S$ Store at -15°C

Ser-Arg-Thr-His-Arg-His-Ser-Met-Glu-Ile-Arg-Cys Acetate Salt

Synonyms: Prolactin Releasing Peptide 31 (1-11), ~C; Prolactin Releasing Peptide 31

Biogenesis 0100-0169   Human synthetic   Semi-pure; lyophilized

Ser-Arg-Thr-His-Arg-His-Ser-Met-Glu-Ile-Arg-Thr-Pro-Asp-Ile-Asn-Pro-Ala-Trp-Tyr-Ala-Ser-Arg-Gly-Ile-Arg-Pro-Val-Gly-Arg-Phe Amide

Synonyms: Prolactin Releasing Peptide (1-31); Prolactin Releasing Peptide 31; Pre-Prolactin (23-53); Prolactin Releasing Peptide; Prolactin Releasing Peptide, Specific

Bachem H-4382.0500   Human   MW 3664.18 $C_{160}H_{252}N_{56}O_{42}S$ Store at -15°C

Neosystem SC1334   Human   MW 3664.16   Bioactive; hypothalamic releasing hormone

Peptides International PPR-4352-v   Human synthetic   MW 3664.2 $C_{160}H_{252}N_{56}O_{42}S$ >95% (HPLC); lyophilized amorphous powder | Bioactive; Hinuma, S et al, *Nature*, 393:272, 1998

Ser-Arg-Val-Ser-Arg-Arg-Ser-Arg

Synonyms: Insulin-Like Growth Factor II (33-40); Insulin-Like Growth Factor II C-Peptide

American Peptide 50-1-31   MW 1003.2 $C_{38}H_{74}N_{20}O_{12}$

Bachem H-7250.0001   MW 1003.13 $C_{38}H_{74}N_{20}O_{12}$ Store at -15°C

Neosystem SC681   MW 1003.13   Bioactive; synthetic growth factor-related peptide

Sigma I 3263   FW 1003.1   ≥97% (HPLC); peptide content:~65% | Bioactive peptide; Hintz, RL & Lin, F, *J Clin Endocrinol Metab*, 54:442, 1982

ICN 153133   Human synthetic   Corresponds to the 33-40 peptide sequence of human IGF-II; Hintz, RL & FJ Lin, *Clin Endocrinol Metab*, 54:442, 1982

Ser-Arg-Val-Tyr-Ile-His-Pro-Phe
D-Ser-Arg-Val-Tyr-Ile-His-Pro-Phe

Synonyms: Angiotensin II, (D-Ser[1])-

Neosystem SC375   Human   MW 1018.18   Bioactive; Hui, KY et al, *Int J Pept Prot Res*, 34:177-183, 1989

Ser-Asn

Bachem G-3170.0250   MW 219.2 $C_7H_{13}N_3O_5$ Store at -15°C

ICN 156604   MW 219.2 $C_7H_{13}N_3O_5$

Ser-Asn-Asn-Phe-Gly-Ala-Ile-Leu-Ser-Ser

Synonyms: Amylin (20-29)

Bachem H-3746.0001   Human   MW 1009.08 $C_{43}H_{68}N_{12}O_{16}$ Store at -15°C

Ser-Asn-Glu-Ala-Ile-Ser-Pro-Phe-Asp-Gln-Gly-Met-Met-Gly-Tyr-Val-Ile-Lys-Thr-Asn-Lys-Asn-Ile-Pro-Arg-Met Amide

Ecdysis-Triggering Hormone, Mas-*Synonyms*:

Bachem H-3388.0500   Manduca sexta   MW 2941.45 $C_{127}H_{206}N_{36}O_{38}S_3$ Store at -15°C

Ser-Asn-Leu-Ser-Thr-Asu-Val-Leu-Gly-Lys-Leu-Ser-Gln-Glu-Leu-His-Lys-Leu-Gln-Thr-Tyr-Pro-Arg-Thr-Asp-Val-Gly-Ala-Gly-Thr-Pro Amide

Synonyms: Elcatonin; Calcitonin, (Asu[1,7])-; CarBOCalcitonin

American Peptide 22-7-13   Eel   MW 3363.9 $C_{147}H_{242}N_{42}O_{47}$ Peptide bond:Asu[1]-Asu[7] | Aminosuberic acid analog of eel calcitonin; has all the biological properties of the corresponding natural calcitonin; the substitution of the disulfide bond of natural calcitonins with an ethylene bridge in 1-7 N-terminal position gives greater stability & excellent tolerability when used *in vivo*

Ser-Asn-Lys-Ser-Leu-Glu-Gln-Ile-Trp-Asn-Asn-Met-Thr

Synonyms: GP41 (620-632); HTLV I Peptide

Neosystem SC257   MW 1564.73   HTLV I peptide from subtype GP41

Ser-Asn-Pro-Ala-Met-Ala-Pro-Arg-Glu-Arg-Lys-Ala-Gly-Cys-Lys-Asn-Phe-Phe-Trp-Lys-Thr-Phe-Thr-Ser-Cys

Syno Somatostatin 25 *nyms*:

Bachem H-9580.0500   MW 2876.34 $C_{127}H_{191}N_{37}O_{34}S_3$ Store at -15°C

Sigma S 1007   FW 2876.3   ≥97% (HPLC) | Disulfide bonds: 14-25; bioactive peptide; originally isolated from ovine hypothalamus; based on *in vitro* studies, may be more potent than somatostatin; Reubi, J-C et al, *Endocrinology*, 110:1049, 1982

American Peptide 68-1-46   Bovine hypothalamus   MW 2876.4 $C_{127}H_{191}N_{37}O_{34}S_3$ Disulfide bonds: Cys[14]-Cys[25]; somatostatin-like immunoactivity & bioactivity *in vitro*; more potent than somatostatin based on an *in vitro* bioassay; Bohlen, P et al, *BBRC*, 96:725, 1980

Ser-Asn-Thr-Ala-Ala-Thr-Asn-Ala-Asp-Cys-Ala-Trp-Leu-Glu-Ala

Synonyms: NEF MN (48-62); HIV I Peptide

Neosystem SC643   MW 1539.59   NEF peptide from HIV-1 subtype MN; due to the presence of a Cys this peptide can contain the dimeric form

Ser-Asn-Thr-Ala-Leu-Arg-Arg-Tyr-Asn-Gln-Trp-Ala-Thr-Gly-His-Phe-Met Amide

Synonyms: Ranatensin R

Neosystem SC139   MW 2052.30   Bioactive; bombesin-related peptide

Ser-Asp

Bachem G-3175.0250   MW 220.18 $C_7H_{12}N_2O_6$ Store at -15°C

ICN 156605   MW 220.2 $C_7H_{12}N_2O_6$

Sigma S 4889   FW 220.2 $C_7H_{12}N_2O_6$

Ser-Asp-Ala-Ala-Val-Asp-Thr-Ser-Ser-Glu-Ile-Thr-Thr-Lys-Asp-Leu-Lys-Glu-Lys-Lys-Glu-Val-Val-Glu-Glu-Ala-Glu-Asn

Ac-Ser-Asp-Ala-Ala-Val-Asp-Thr-Ser-Ser-Glu-Ile-Thr-Thr-Lys-Asp-Leu-Lys-Glu-Lys-Lys-Glu-Val-Val-Glu-Glu-Ala-Glu-Asn

Synonyms: Thymosin α_1, Bovine

American Peptide 74-1-13   MW 3108.3 $C_{129}H_{215}N_{33}O_{55}$
Sztein, MB et al, *Springer Semin Immunopathol*, 9:1, 1986

Bachem H-6945.0500   MW 3108.32 $C_{129}H_{215}N_{33}O_{55}$ Store at -15°C

ICN 153181   Bovine   Goldstein, AL etal, *PNAS*, 74:725, 1977

Sigma T 3410   Bovine   FW 3108.3   ≥90% (HPLC) |
Bioactive peptide; Goldstein, AL et al, *Proc Natl Acad Sci USA*, 74:725, 1977

Sigma T 3641   Bovine   FW 3108.3   ≥90% (HPLC) |
Bioactive peptide; Goldstein, AL et al, *Proc Natl Acad Sci USA*, 74:725, 1977

Ser-Asp-Arg-Asn-Phe-Leu-Arg-Phe Amide

Synonyms: FMRF Amide-Like Peptide I

American Peptide 60-9-28   Lobster   MW 1053.3 $C_{47}H_{72}N_{16}O_{12}$

Ser-Asp-Glu-Asp-Ser-Asp-Gly-Asp-Arg-Pro-Gln-Ala-Ser-Pro-Gly-Leu-Gly-Pro-Gly-Pro

Synonyms: Chromostatin 20; Chromostatin; Chromogranin A (124-143)

Sigma C 1680   FW 1953.9   ≥97% (HPLC) | Bioactive peptide; Galindo, E et al, *Natl Acad Sci USA*, 88:1426, 1991

American Peptide 46-4-70   Bovine   MW 1953.9 $C_{78}H_{120}N_{24}O_{35}$ Derived from chromogranin A, which inhibits chromaffin cell secretion; chromostatin corresponding to the first 20 N-terminal AA of bovine chromogranin A produces a dose-dependent inhibition of catecholamine secretion from cultured chromaffin cells in the range of 10-9 to 10-6 M; Galindo, E et al, *PNAS*, 88:1426, 1991

Bachem H-8475.0500   Bovine   MW 1953.95 $C_{78}H_{120}N_{24}O_{35}$ Store at -15°C

Ser-Asp-Gly-Arg

American Peptide 44-0-32   MW 433.4 $C_{15}H_{27}N_7O_8$

Neosystem SC150   MW 433.42   Bioactive; cell attachment peptide; fibronectin fragment; Yamada, KM & Kennedy, DW, *J Cell Biol*, 28:99, 1985

Ser-Asp-Gly-Arg-Gly

Synonyms: Fibronectin

American Peptide 44-0-33   MW 490.5 $C_{17}H_{30}N_8O_9$

Bachem H-7740.0005   MW 490.47 $C_{17}H_{30}N_8O_9$ Store at -15°C

Sigma S 3771   FW 490.5 $C_{17}H_{30}N_8O_9$ ≥90% (HPLC) | Bioactive peptide

Ser-Asp-Gly-Asp-Gly

ICN 153177   MW 490.5 $C_{17}H_{30}N_8O_9$

Ser-Asp-Lys-Pro
Ac-Ser-Asp-Lys-Pro

American Peptide 87-0-25   MW 487.5 $C_{20}H_{33}N_5O_9$

Bachem H-1156.0005   MW 487.51 $C_{20}H_{33}N_5O_9$ Store at -15°C

Sigma A 6433   FW 487.5 $C_{20}H_{33}N_5O_9$ ≥97% (HPLC) | Bioactive peptide

Ser-Asp-Phe-Glu-Glu-Phe-Ser-Leu-Asp-Asp-Ile-Glu-Gln

Synonyms: Hirullin

Bachem H-1966.0001   MW 1573.59 $C_{68}H_{96}N_{14}O_{29}$ Store at -15°C

Ser-Asp-Pro-Arg-Glu-Arg-Ile-Pro-Pro-Gly-Asn-Ser-Gly-Glu-Glu-Tyr

Synonyms: VPX (2-16)-(Tyr); SIVmac251 Peptide

Neosystem SC574   MW 1802.87   SIVmac251 peptide from subtype VPX

Ser-Asp-Thr-Cys-Trp-Ser-Thr-Thr-Ser-Phe-Gln-Lys-Lys-Thr-Ile-His-Cys-Lys-Trp-Arg-Glu-Lys-Pro-Leu-Met-Leu-Met

Melanin Concentrating Hormone Gene Overprinted Polypeptide 27Synonyms:

Bachem H-4748.0500   Rat   MW 3284.9 $C_{145}H_{227}N_{39}O_{40}S_4$ Store at -15°C

Ser-Cys-Asn-Thr-Ala-Thr-Cys-Val-Thr-His-Arg-Leu-Ala-Gly-Leu-Leu-Ser-Arg-Ser-Gly-Gly-Val-Val-Lys-Asp-Asn-Phe-Val-Pro-Thr-Asn-Val-Gly-Ser-Glu-Ala-Phe Amide

Synonyms: Calcitonin Gene Related Peptide

American Peptide 22-5-10   Rat   MW 3806.3 $C_{162}H_{262}N_{50}O_{52}S_2$ Disulfide bonds: Cys²-Cys⁷; Rosenfeld, MG et al, *Nature*, 304:129, 1983

Ser-Cys-Asn-Thr-Ala-Thr-Cys-Val-Thr-His-Arg-Leu-Ala-Gly-Leu-Leu-Ser-Arg-Ser-Gly-Gly-Val-Val-Lys-Asp-Asn-Phe-Val-Pro-Thr-Asn-Val-Gly-Ser-Lys-Ala-Phe Amide

Synonyms: Calcitonin Gene Related Peptide II; Calcitonin Gene Related Peptide, β-

American Peptide 22-5-35   Rat   MW 3805.4 $C_{163}H_{267}N_{51}O_{50}S_2$ Disulfide bonds: Cys²-Cys7; Steenbergh, PH et al, *FEBS Lett*, 183:403, 1985

Ser-Cys-Asn-Thr-Ala-Thr-Cys-Val-Thr-His-Arg-Leu-Ala-Gly-Leu-Leu-Ser-Arg-Ser-Gly-Gly-Val-Val-Lys-Asp-Asn-Phe-Val-Pro-Thr-Asn-Val-Gly-Ser-Glu-Ala-Phe Amide

Synonyms: Calcitonin Gene Regulated Peptide, α-; Calcitonin Gene Related Peptide, α-

Bachem H-2265.0500   Rat   MW 3806.3 $C_{162}H_{262}N_{50}O_{52}S_2$ Store at -15°C

ICN 153083   Rat   Disulfide bonds: Cys²-Cys⁷; Amara, SG etal, *Nature*, 298:240, 1982; Rosenfeld, MG etal, *Nature*, 304:129, 1983; Le Greves, P etal, *Eur J Pharmac*, 115:309, 1985

Ser-Cys-Asn-Thr-Ala-Thr-Cys-Val-Thr-His-Arg-Leu-Ala-Gly-Leu-Leu-Ser-Arg-Ser-Gly-Gly-Val-Val-Lys-Asp-Asn-Phe-Val-Pro-Thr-Asn-Val-Gly-Ser-Lys-Ala-Phe

Synonyms: Calcitonin Gene Regulated Peptide, β-

Neosystem SC1261   Rat   MW 3805.33   Disulfide bonds: Cys²-Cys⁷; bioactive; calcium metabolism peptide; Amara, SG et al, *Science*, 229:1094-1097, 1985

Ser-Cys-Asn-Thr-Ala-Thr-Cys-Val-Thr-His-Arg-Leu-Ala-Gly-Leu-Leu-Ser-Arg-Ser-Gly-Gly-Val-Val-Lys-Asp-Asn-Phe-Val-Pro-Thr-Asn-Val-Gly-Ser-Glu-Ala-Phe

Synonyms: Calcitonin Gene Regulated Peptide, α-

Neosystem SC983   Rat   MW 3806.28   Disulfide bonds: Cys²-Cys⁷; bioactive; calcium metabolism peptide; Rosenfeld, MG et al, *Nature*, 304:129-135, 1983; Brubaker, PL et Greenberg, GR, *Endocrinology*, 133:2833-2837, 1993; Tan, KKC et al, *Br J Pharmacol*, 111:703-710, 1994; Nitsos, I et al, *Neuroscience*, 62:257-264, 1994; Lopezbelmonte, J et Whittle, BJR, *Eur J Pharmacol*, 271:R15-R17, 1994; Cox, HM, *Can J* Physiol & *Pharmacol*, 73:974-980, 1995

Ser-Cys-Asn-Thr-Ala-Thr-Cys-Val-Thr-His-Arg-Leu-Ala-Gly-Leu-Leu-Ser-Arg-Ser-Gly-Gly-Val-Val-Lys-Asp-Asn-Phe-Val-Pro-Thr-Asn-Val-Gly-Ser-Glu-Ala-Phe Amide

Synonyms: Calcitonin Gene Related Peptide, α-; Calcitonin Gene Related Peptide I

Sigma C 0292   Rat   FW 3806.3   ≥97% (HPLC) | Disulfide bonds: 2-7; bioactive peptide; potent, long-lasting vasodilator; activation of CGRP receptors on pancreatic β-cells increases plasma levels of pancreatic enzymes

Ser-Cys-Asn-Thr-Ala-Thr-Cys-Val-Thr-His-Arg-Leu-Ala-Gly-Leu-Leu-Ser-Arg-Ser-Gly-Gly-Val-Val-Lys-Asp-Asn-Phe-Val-Pro-Thr-Asn-Val-Gly-Ser-Glu-Ala-Phe
H_2N-Ser-Cys-Asn-Thr-Ala-Thr-Cys-Val-Thr-His-Arg-Leu-Ala-Gly-Leu-Leu-Ser-Arg-Ser-Gly-Gly-Val-Val-Lys-Asp-Asn-Phe-Val-Pro-Thr-Asn-Val-Gly-Ser-Glu-Ala-Phe Amide

Synonyms: Calcitonin Gene Related Peptide

Biogenesis 1720-9304   Rat synthetic   MW 3806.9
Purified; no preservatives; lyophilized | Rosenfeld et al, *Nature*, 304:129, 1983

Ser-Cys-Asn-Thr-Ala-Thr-Cys-Val-Thr-His-Arg-Leu-Ala-Gly-Leu-Leu-Ser-Arg-Ser-Gly-Gly-Val-Val-Lys-Asp-Asn-Phe-Val-Pro-Thr-Asn-Val-Gly-Ser-Glu-Ala-Phe Amide

Synonyms: Calcitonin Gene Related Peptide; Calcitonin Gene Related Peptide, α-

Peptides International PCG-4163-s, 4163-v   Rat synthetic
MW 3806.3 $C_{162}H_{262}N_{50}O_{52}S_2$ >99% (HPLC); lyophilized amorphous powder | Bioactive; Disulfide bonds: Cys²-Cys⁷; Amara, SG et al, *Nature*, 298:240, 1982; Rosenfeld, MG et al, *Nature*, 304:129, 1983

Ser-Cys-Asp-Lys-His-Tyr-Trp-Asp-Thr-Ile-Arg-Phe-Arg-Tyr-Cys-Ala-Pro-Pro-Gly-Tyr
Ser-Cys(Acm)-Asp-Lys-His-Tyr-Trp-Asp-Thr-Ile-Arg-Phe-Arg-Tyr-Cys(Acm)-Ala-Pro-Pro-Gly-Tyr

Synonyms: GP140 (221-240); SIVmac251 Peptide

Neosystem SC730   MW 2620.93   SIVmac251 peptide from subtype GP140

Ser-Cys-Ile-Leu-Glu-Ala-Asp-Ala-Thr-Thr-Pro-Glu-Ser-Ala-Asn

Synonyms: TAT (20-34); SIVmac251 Peptide

Neosystem SC585   MW 1521.61   SIVmac251 peptide from subtype TAT; due to the presence of a Cys this peptide can contain the dimeric form

Ser-Cys-Ser-Leu-Pro-Gln-Thr-Ser-Gly-Leu-Gln-Lys-Pro-Glu-Ser Amide

Synonyms: Leptin (116-130); Obese Gene Peptide (116-130)

Bachem H-3966.0001   Mouse   MW 1560.75 $C_{64}H_{109}N_{19}O_{24}S$
Store at -15°C

American Peptide 46-2-12   Murine   MW 1560.7
$C_{64}H_{107}N_{18}O_{25}S$ Grasso, P et al, *Endocrinology*, 138:1413, 1997

Ser-Gln

Bachem G-3180.0250   Store at -15°C

ICN 156606   MW 233.2 $C_8H_{15}N_3O_5$

Sigma S 4764   FW 233.2 $C_8H_{15}N_3O_5$

Ser-Gln-Ala-Phe-Leu-Phe-Gln-Pro-Gln-Arg-Phe Amide

Synonyms: Neuropeptide SF

Bachem H-4948.0500   Human   MW 1367.57 $C_{65}H_{94}N_{18}O_{15}$
Store at -15°C

Ser-Gln-Asn-Phe-Pro-Ile-Val
Ser-Gln-Asn-*p*-Nitro-Phe-Pro-Ile-Val

Bachem H-7225.0005   Store at -15°C

Ser-Gln-Asn-Phe-Pro-Ile-Val-Gln
Ser-Gln-Asn-Phe-(®)-Pro-Ile-Val-Gln

Bachem N-1460.0500   MW 918.06 $C_{42}H_{67}N_{11}O_{12}$ Store at -15°C

Ser-Gln-Asn-Phe-Pro-Ile-Val-Gln
Ser-Gln-Asn-Phe-π(CH₂N)-Pro-Ile-Val-Gln

Synonyms: HIV I Protease Inhibitor

ICN 158731   MW 918.1   97% | Affinity ligand for purification; Heimbach, JC etal, *BBRC*, 164:955, 1989

Ser-Gln-Asn-Tyr
Ac-Ser-Gln-Asn-Tyr

Synonyms: Antibiotic Peptide

American Peptide 72-2-40   MW 552.6 $C_{23}H_{32}N_6O_{10}$

Bachem H-9890.0025   MW 552.54 $C_{23}H_{32}N_6O_{10}$ Store at -15°C

Ser-Gln-Asn-Tyr

Synonyms: HIV Protease Substrate Tetrapeptide

Neosystem SC423   MW 510.50   Virus-related peptide; AIDS-related peptide

Ser-Gln-Asn-Tyr
Ac-Ser-Gln-Asn-Tyr

Synonyms: HIV I Protease Substrate Cleavage Product

Sigma A 0931   FW 552.5 $C_{23}H_{32}N_6O_{10}$ ≥97% (HPLC) |
Bioactive peptide; Moore, M et al, *Biochem Biophys Res Commun*, 159:420, 1989

Ser-Gln-Asn-Tyr-Pro-Ile-Val

Synonyms: HIV I Protease Substrate (129-135)

Bachem H-7235.0005   MW 819.91 $C_{37}H_{57}N_9O_{12}$ Store at -15°C

ICN 158732   MW 819.9 $C_{37}H_{57}N_9O_{12}$ 97% | HIV-I GAG (129-135); smallest fragment with activity; Darke, PL etal, *JBC*, 264:2307, 1989

Ser-Gln-Asn-Tyr-Pro-Ile-Val
DNS-Ser-Gln-Asn-Tyr-Pro-Ile-Val

Synonyms: HIV Protease Substrate

ICN 196006 MW 1053.2 $C_{49}H_{68}N_{10}O_{14}S$ ≥99% |
Fluorogenic substrate

Ser-Gln-Asn-Tyr-Pro-Ile-Val

Synonyms: HIV gag (129-135); HIV I Protease Substrate

Peptides International SSV-4236-v MW 819.91
$C_{37}H_{57}N_9O_{12}$ >99% (HPLC); lyophilized amorphous powder |
Billich, S et al, *JBC*, 263:17905, 1988; Darke, PL et al, *BBRC*,
156:297, 1988; Darke, PL et al, *JBC*, 264:2307, 1989

Sigma S 5151 FW 819.9 $C_{37}H_{57}N_9O_{12}$ ≥97% (HPLC) |
Bioactive peptide; minimum sequence to show such activity; Darke,
PL et al, *J Biol Chem*, 264:2307, 1989

Ser-Gln-Asn-Tyr-Pro-Ile-Val Amide

Synonyms: HIV Protease Substrate Heptapeptide

Neosystem SC422 MW 818.93 Virus-related peptide; AIDS-
related peptide

Ser-Gln-Asn-Tyr-Pro-Val-Val
Ac-Ser-Gln-Asn-Tyr-Pro-Val-Val Amide

Synonyms: HIV I Protease Substrate

American Peptide 81-7-20 MW 846.7 $C_{38}H_{58}N_{10}O_{12}$ Moore,
M et al, *BBRC*, 159:420, 1989

Sigma A 0806 FW 846.9 $C_{38}H_{58}N_{10}O_{12}$ ≥97% (HPLC) |
Bioactive peptide; Moore, M et al, *Biochem Biophys Res Commun*,
159:420, 1989

Ser-Gln-Asp-Ser-Ala-Glu-Arg-Ile-Gln-Glu-Arg-Leu-Arg-Asn-Ser-Lys-Met-Ala-His-Ser-Ser-Ser-Cys-Phe-Gly-Gln-Lys-Ile-Asp-Arg-Ile-Gly-Ala-Val-Ser-Arg-Leu-Gly-Cys-Asp-Gly-Leu-Arg-Gln-Phe

Synonyms: Natriuretic Peptide, Iso-Atrial

ICN 195627 Rat MW 5037.6 Similar physiological &
pharmacological properties to rat ANP; Flynn, TG etal, *BBRC*,
161:830, 1989

Sigma A 9179 Rat FW 5037.6 ≥97% (HPLC) | Bioactive
peptide; disulfide bonds: 23-39; exhibits physiological &
pharmacological properties similar to rANP; Flynn, TG et al,
Biochem Biophys Res Commun, 161:830, 1989

Ser-Gln-Asp-Ser-Ala-Phe-Arg-Ile-Gln-Glu-Arg-Leu-Arg-Asn-Ser-Lys-Met-Ala-His-Ser-Ser-Ser-Cys-Phe-Gly-Gln-Lys-Ile-Asp-Arg-Ile-Gly-Ala-Val-Ser-Arg-Leu-Gly-Cys-Asp-Gly-Leu-Arg-Leu-Phe

Synonyms: Natriuretic Peptide 45, B-Type; Natriuretic Peptide 45,
Brain

Bachem H-8035.0500 Rat MW 5040.75 $C_{213}H_{349}N_{71}O_{65}S_3$
Store at -15°C

Sigma B 6154 Rat FW 5040.7 ≥97% (HPLC) | Disulfide
bonds: 23-39; bioactive peptide; storage form of BNP-32; Aburaya,
M et al, *Biochem Biophys Res Commun*, 163:226, 1989;
Kambayashi, Y et al, *Biochem Biophys Res Commun*, 163:233,
1989

Peptides International PBN-4218-s Rat synthetic MW
5040.8 $C_{213}H_{349}N_{71}O_{65}S_3$ >95% (HPLC); lyophilized amorphous
powder | Bioactive; Disulfide bonds: Cys[23]-Cys[39]; Aburaya, M et
al, *BBRC*, 163:226, 1989; Kambayashi, Y et al, *BBRC*, 163:233,
1989

Ser-Gln-Asp-Ser-Ala-Phe-Arg-Ile-Gln-Gly-Arg-Leu-Arg-Asn-Ser-Lys-Met-Ala-His-Ser-Ser-Ser-Cys-Phe-Gly-Gln-Lys-Ile-Asp-Arg-Ile-Gly-Ala-Val-Ser-Arg-Leu-Gly-Cys-Asp-Gly-Leu-Arg-Leu-Phe

Synonyms: Natriuretic Peptide (1-32), Brain

American Peptide 14-5-20 Rat MW 5040.8
$C_{213}H_{349}N_{71}O_{65}S_3$ Disulfide bonds: :Cys[23]-Cys[39] | Aburaya, M et al,
BBRC, 163:226, 1989

Ser-Gln-Glu-Pro-Pro-Ile-Ser-Leu-Asp-Leu-Thr-Phe-His-Leu-Leu-Arg-Glu-Val-Leu-Glu-Met-Thr-Lys-Ala-Asp-Gln-Leu-Ala-Gln-Gln-Ala-His-Asn-Asn-Arg-Lys-Leu-Leu-Asp-Ile-Ala Amide

Synonyms: Corticotropin Releasing Factor

American Peptide 34-3-11 Bovine MW 4697.4
$C_{206}H_{340}N_{60}O_{63}S$ Causes the release of ACTH & endorphins; Esch,
F et al, *BBRC*, 122:899, 1984

Bachem H-3264.0500 Bovine MW 4697.4 $C_{206}H_{340}N_{60}O_{63}S$
Store at -15°C

Sigma C 2671 Bovine FW 4697.4 ≥90% (HPLC) |
Bioactive peptide; simulates ACTH & endorphin release from
anterior pituitary; may act as a neuromodulator in CNS involved
with autonomic & endocrine responses to stress; Esch, F et al,
Biochem Biophys Res Commun, 122:899, 1984

Ser-Gln-Glu-Pro-Pro-Ile-Ser-Leu-Asp-Leu-Thr-Phe-His-Leu-Leu-Arg-Glu-Val-Leu-Glu-Met-Thr-Lys-Ala-Asp-Gln-Leu-Ala-Gln-Gln-Ala-His-Ser-Asn-Arg-Lys-Leu-Leu-Asp-Ile-Ala Amide

Synonyms: Corticotropin Releasing Factor

American Peptide 34-2-11 Ovine MW 4670.4
$C_{205}H_{339}N_{59}O_{63}S$ Vale, W et al, *Science*, 213:1394, 1981

Ser-Gln-Glu-Pro-Pro-Ile-Ser-Leu-Asp-Leu-Thr-Phe-His-Leu-Leu-Arg-Glu-Val-Leu-Glu-Met-Thr-Lys-Ala-Asp-Gln-Leu-Ala-Gln-Gln-Ala-His-Ser-Asn-Arg-Lys-Leu-Leu-Asp-Ile-Ala
Ser-Gln-Glu-Pro-Pro-Ile-Ser-Leu-Asp-Leu-Thr-Phe-His-Leu-Leu-Arg-Glu-Val-Leu-Glu-Met(O)-Thr-Lys-Ala-Asp-Gln-Leu-Ala-Gln-Gln-Ala-His-Ser-Asn-Arg-Lys-Leu-Leu-Asp-Ile-Ala Amide

Synonyms: Corticotropin Releasing Factor, (Met(O)[21])-

American Peptide 34-2-24 Ovine MW 4686.4
$C_{205}H_{339}N_{59}O_{64}S$

Ser-Gln-Glu-Pro-Pro-Ile-Ser-Leu-Asp-Leu-Thr-Phe-His-Leu-Leu-Arg-Glu-Val-Leu-Glu-Nle-Thr-Lys-Ala-Asp-Gln-Leu-Ala-Gln-Gln-Ala-Tyr-Ser-Asn-Arg-Lys-Leu-Leu-Asp-Ile-Ala Amide

Synonyms: Corticotropin Releasing Factor, (Nle[21],Tyr[32])-

American Peptide 34-2-30 Ovine MW 4678.4
$C_{209}H_{343}N_{57}O_{64}$

Ser-Gln-Glu-Pro-Pro-Ile-Ser-Leu-Asp-Leu-Thr-Phe-His-Leu-Leu-Arg-Glu-Val-Leu-Glu-Met-Thr-Lys-Ala-Asp-Gln-Leu-Ala-Gln-Gln-Ala-His-Ser-Asn-Arg-Lys-Leu-Leu-Asp-Ile-Ala Amide

Synonyms: Corticotropin Releasing Factor

Bachem H-2445.0500 Ovine Store at -15°C

Calbiochem 05-23-0051 Ovine MW 4670.4
$C_{205}H_{339}N_{59}O_{63}S$ >95% (HPLC); solid; soluble in 5% acetic acid;
sold under license of US Patent 4,415,558 issued to The Salk
Institute | Neuropeptide with immunomodulatory properties;
Okita, M et al, *J Endocrinol*, 156:359, 1998; Vale, W et al, *Science*,
213:1394, 1981

Ser-Gln-Glu-Pro-Pro-Ile-Ser-Leu-Asp-Leu-Thr-Phe-His-Leu-Leu-Arg-Glu-Val-Leu-Glu-Met-Thr-Lys-Ala-Asp-Gln-Leu-Ala-Gln-Gln-Ala-His-Ser-Asn-Arg-Lys-Leu-Leu-Asp-Ile-Ala

Synonyms: Corticotropin Releasing Factor

Biogenesis 2330-0189   Ovine synthetic   MW 4671   89.6% peptide material, TFA counter ions and residual H_2O; sterile and pyrogen free; purified; lyophilized | To avoid Met oxidation, it is best to dissolve the peptide under a nitrogen stream

Ser-Gln-Glu-Pro-Pro-Ile-Ser-Leu-Asp-Leu-Thr-Phe-His-Leu-Leu-Arg-Glu-Val-Leu-Glu-Met-Thr-Lys-Ala-Asp-Gln-Leu-Ala-Gln-Gln-Ala-His-Ser-Asn-Arg-Lys-Leu-Leu-Asp-Ile-Ala Trifluoroacetate Salt

Synonyms: Corticotropin Releasing Factor

Biogenesis 2330-0190   Ovine synthetic   MW 4671 Purified; lyophilized | To avoid Met oxidation, it is best to dissolve the peptide under a nitrogen stream

Ser-Gln-Glu-Pro-Pro-Ile-Ser-Leu-Asp-Leu-Thr-Phe-His-Leu-Leu-Arg-Glu-Val-Leu-Glu-Met-Thr-Lys-Ala-Asp-Gln-Leu-Ala-Gln-Gln-Ala-His-Ser-Asn-Arg-Lys-Leu-Leu-Asp-Ile-Ala
Ser-Gln-Glu-Pro-Pro-Ile-Ser-Leu-Asp-Leu-Thr-Phe-His-Leu-Leu-Arg-Glu-Val-Leu-Glu-Met-Thr-Lys-Ala-Asp-Gln-Leu-Ala-Gln-Gln-Ala-His-Ser-Asn-Arg-Lys-Leu-Leu-Asp-Ile-Ala Amide

Synonyms: Corticotropin Releasing Factor

Peptides International PCR-4111-s, 4111-v   Ovine synthetic MW 4670.4 $C_{205}H_{339}N_{59}O_{63}S$ >95% (HPLC); lyophilized amorphous powder | Bioactive; Vale, W et al, *Science*, 213:1394, 1981

Ser-Gln-Glu-Pro-Pro-Ile-Ser-Leu-Asp-Leu-Thr-Phe-His-Leu-Leu-Arg-Glu-Val-Leu-Glu-Met-Thr-Lys-Ala-Asp-Gln-Leu-Ala-Gln-Gln-Ala-His–Ser-Asn-Arg-Lys-Leu-Leu-Asp-Ile-Ala Amide

Synonyms: Corticotropin Releasing Factor

ICN 153075   Sheep   Vale, W etal, *Science*, 213:1394, 1981

Ser-Gln-Glu-Pro-Pro-Ile-Ser-Leu-Asp-Leu-Thr-Phe-His-Leu-Leu-Arg-Glu-Val-Leu-Glu-Met-Thr-Lys-Ala-Asp-Gln-Leu-Ala-Gln-Gln-Ala-His-Ser-Asn-Arg-Lys-Leu-Leu-Asp-Ile-Ala Amide

Synonyms: Corticotropin Releasing Factor

Sigma C 3167   Sheep   FW 4670.4   ≥95% (HPLC) | Bioactive peptide; simulates ACTH & endorphin release from anterior pituitary; may act as a neuromodulator in CNS involved with autonomic & endocrine responses to stress; Vale, W et al, *Science*, 213:1394, 1981

Ser-Gln-Gly-Ser-Thr-Leu-Arg-Val-Gln-Gln-Arg-Pro-Gln-Asn-Ser-Lys-Val-Thr-His-Ile-Ser-Ser-Cys-Phe-Gly-His-Lys-Ile-Asp-Arg-Ile-Gly-Ser-Val-Ser-Arg-Leu-Gly-Cys-Asn-Ala-Leu-Lys-Leu-Leu

Synonyms: Natriuretic Peptide (1-45), Brain

American Peptide 14-5-30   Murine   MW 4919.7 $C_{209}H_{354}N_{70}O_{63}S_2$ Disulfide bonds: :Cys[23]-Cys[39]

Ser-Gln-Gly-Thr-Phe-Thr-Ser-Glu-Tyr-Ser-Lys-Tyr-Leu-Asp-Ser-Arg-Arg-Ala-Gln-Asp-Phe-Val-Gln-Trp-Leu-Met-Asn-Thr

Synonyms: Glucagon, (des-His[1],Glu[9])-

American Peptide 46-1-16   MW 3359.6 $C_{148}H_{220}N_{40}O_{48}S$

Ser-Gln-Gly-Thr-Phe-Thr-Ser-Glu-Tyr-Ser-Lys-Tyr-Leu-Asp-Ser-Arg-Arg-Ala-Gln-Asp-Phe-Val-Gln-Trp-Leu-Met-Asn-Thr Amide

Synonyms: Glucagon (1-29), (des-His[1], Glu[9])-

American Peptide 46-1-17   MW 3358.6 $C_{148}H_{221}N_{41}O_{47}S$ Potent glucagon antagonist; inhibitor of glucagon-stimulated glycogenolysis *in vivo*; Unson, CG et al, *Peptides*, 10:1171, 1989

Sigma G 1651   FW 3358.7   ≥97% (HPLC) | Bioactive peptide; gastrointestinal peptide; inhibitor of glucagon-stimulated glycogenolysis *in vivo*; Unson, CG et al, *Peptides*, 10:1171, 1989

Bachem H-2754.0500   Human, bovine, porcine   MW 3358.7 $C_{148}H_{221}N_{41}O_{47}S$ Store at -15°C

Ser-Gln-His-Gly-Trp-Met-Thr-Arg-Arg-Glu-Lys-Cys

Synonyms: NEF MN (171-182); HIV I Peptide

Neosystem SC663   MW 1519.71   NEF peptide from HIV-1 subtype MN; due to the presence of a Cys this peptide can contain the dimeric form

Ser-Gln-Pro-Glu-Thr-Arg-Thr-Gly-Asp-Asp-Asp-Pro-His-Arg-Leu-Leu-Gln-Gln-Leu-Val-Leu-Ser-Gly-Asn-Leu-Ile-Lys-Glu-Ala-Val-Arg-Arg-Leu-His-Ser-Arg-Arg-Leu-Gln

Synonyms: FRATIDE; GSK3 INHIBITORY PEPTIDE

Upstate 12-387   MW 4534.12   >90% pure; frozen solution

Ser-Glu

Bachem G-3185.0250   MW 234.21 $C_8H_{14}N_2O_6$ Store at -15°C

ICN 156607   MW 234.2 $C_8H_{14}N_2O_6$

Sigma S 3638   FW 234.2 $C_8H_{14}N_2O_6$

Ser-Glu-Asn-Tyr-Pro-Ile-Val

Synonyms: HIV gag Fragment (129-135)

American Peptide 72-2-50   MW 820.9 $C_{37}H_{56}N_8O_{13}$

Ser-Glu-Glu-Pro-Pro-Ile-Ser-Leu-Asp-Leu-Thr-Phe-His-Leu-Leu-Arg-Glu-Val-Leu-Glu-Met-Ala-Arg-Ala-Glu-Gln-Leu-Ala-Gln-Gln-Ala-His-Ser-Asn-Arg-Lys-Leu-Met-Glu-Ile-Ile Amide

Synonyms: Corticotropin Releasing Factor

American Peptide 34-1-10   Human, rat   MW 4757.5 $C_{208}H_{344}N_{60}O_{63}S_2$ Hypothalamic hormone that stimulates the synthesis & release of ACTH from the anterior pituitary; Shibahara, S et al, *EMBO J*, 2:775, 1983

Ser-Glu-Glu-Pro-Pro-Ile-Ser-Leu-Asp-Leu-Thr-Phe-His-Leu-Leu-Arg-Glu-Val-Leu-Glu-Met-Ala-Arg-Ala-Glu-Gln-Leu-Ala-Gln-Gln-Ala-His-Ser-Asn-Arg-Lys-Leu-Met-Glu-Ile-Ile
Ser-Glu-Glu-Pro-D-Pro-Ile-Ser-Leu-Asp-Leu-Thr-Phe-His-Leu-Leu-Arg-Glu-Val-Leu-Glu-Met-Ala-Arg-Ala-Glu-Gln-Leu-Ala-Gln-Gln-Ala-His-Ser-Asn-Arg-Lys-Leu-Met-Glu-Ile-Ile Amide

Synonyms: Corticotropin Releasing Factor, (D-Pro[5])-; Corticotropin Releasing Factor Receptor Agonist

American Peptide 34-1-11   Human, rat   MW 4757.5 $C_{208}H_{344}N_{60}O_{63}S_2$

Ser-Glu-Glu-Pro-Pro-Ile-Ser-Leu-Asp-Leu-Thr-Phe-His-Leu-Leu-Arg-Glu-Val-Leu-Glu-Cys

Synonyms: Corticotropin Releasing Factor, (Cys[21])-

American Peptide 34-1-12   Human, rat   MW 2439.8 $C_{109}H_{173}N_{26}O_{35}S$

Ser-Glu-Glu-Pro-Pro-Ile-Ser-Leu-Asp-Leu-Thr-Phe-
His-Leu-Leu-Arg-Glu-Val-Leu-Glu-Met-Ala-Arg-Ala-
Glu-Gln-Leu-Ala-Gln-Gln-Ala-His-Ser-Asn-Arg-Lys-
Leu-Met-Glu-Ile-Ile Amide

Synonyms: Corticotropin Releasing Factor

Bachem H-2435.0500   Human, rat   Store at -15°C

ICN 153074   Human, rat   Causes the release of ACTH &
endorphins; Vale, W etal, *Science*, 213:1394, 1981; Shibahara, S
etal, *EMBO J*, 2:775, 1983; Vale W etal, 65th Endocrine Society
Meeting, San Antonio, 1983

Ser-Glu-Glu-Pro-Pro-Ile-Ser-Leu-Asp-Leu-Thr-Phe-
His-Leu-Leu-Arg-Glu-Val-Leu-Glu-Met-Ala-Arg-Ala-
Glu-Gln-Leu-Ala-Gln-Gln-Ala-His-Ser-Asn-Arg-Lys-
Leu-Met-Glu-Ile-Ile
Ser-Glu-Glu-Pro-Pro-Ile-Ser-Leu-Asp-Leu-Thr-Phe-
His-Leu-Leu-Arg-Glu-Val-Leu-Glu-Met-Ala-Arg-Ala-
Glu-Gln-Leu-Ala-Gln-Gln-Ala-His-Ser-Asn-Arg-Lys-
Leu-Met-Glu-Ile-Ile Amide

Synonyms: Corticotropin Releasing Factor

Neosystem SC060   Human, rat   MW 4757.49   Bioactive;
hypothalamic releasing hormone; CRF/urocortin peptide; Vale, W et
al, *Science*, 213:1394-1397, 1981; Rivier, CL, *Ann Rev Physiol*,
48:475-494, 1986; Suda, T et al, *J Clin Endocrinol & Metab*,
59:861-866, 1984

Ser-Glu-Glu-Pro-Pro-Ile-Ser-Leu-Asp-Leu-Thr-Phe-
His-Leu-Leu-Arg-Glu-Val-Leu-Glu-Met-Ala-Arg-Ala-
Glu-Gln-Leu-Ala-Gln-Gln-Ala-His-Ser-Asn-Arg-Lys-
Leu-Met-Glu-Ile-Ile Amide

Synonyms: Corticotropin Releasing Factor

Sigma C 3042   Human, rat   FW 4757.5   ≥95% (HPLC) |
Bioactive peptide; simulates ACTH & endorphin release from
anterior pituitary; may act as a neuromodulator in CNS involved
with autonomic & endocrine responses to stress

Ser-Glu-Glu-Pro-Pro-Ile-Ser-Leu-Asp-Leu-Thr-Phe-
His-Leu-Leu-Arg-Glu-Val-Leu-Glu-Met-Ala-Arg-Ala-
Glu-Gln-Leu-Ala-Gln-Gln-Ala-His-Ser-Asn-Arg-Lys-
Leu-Met-Glu-Ile-Ile

Synonyms: Corticotropin Releasing Factor

Biogenesis 2330-0089   Human, rat synthetic   MW 4575.8
Lyophilized, consisting of 90.6% peptide material, acetate counter
ions and residual H_2O; sterile and pyrogen free; purified | To
avoid Met oxidation, it is best to dissolve the peptide under a
nitrogen stream

Ser-Glu-Glu-Pro-Pro-Ile-Ser-Leu-Asp-Leu-Thr-Phe-
His-Leu-Leu-Arg-Glu-Val-Leu-Glu-Met-Ala-Arg-Ala-
Glu-Gln-Leu-Ala-Gln-Gln-Ala-His-Ser-Asn-Arg-Lys-
Leu-Met-Glu-Ile-Ile
Ser-Glu-Glu-Pro-Pro-Ile-Ser-Leu-Asp-Leu-Thr-Phe-
His-Leu-Leu-Arg-Glu-Val-Leu-Glu-Met-Ala-Arg-Ala-
Glu-Gln-Leu-Ala-Gln-Gln-Ala-His-Ser-Asn-Arg-Lys-
Leu-Met-Glu-Ile-Ile Amide

Synonyms: Corticotropin Releasing Factor

Peptides International PCR-4136-s, 4136-v   Human, rat
synthetic   MW 4757.5 $C_{208}H_{344}N_{60}O_{63}S_2$ >95% (HPLC);
lyophilized amorphous powder | Bioactive; Rivier, CL & PM
Plotsky, *Ann Rev Physiol*, 48:475, 1986; Antoni, FA, *Endocrinol
Rev*, 7:351, 1986; Owens, MJ & CB Nemeroff, *Pharmacol Rev*,
43:425, 1991; Schaefer, M et al, *Eur J Pharmacol*, 323:1, 1997;
Spiess, J et al, *Biochem*, 22:4341, 1983; Shibahara, S et al, *The
EMBO Journal*, 2:775, 1983

Ser-Glu-Glu-Pro-Pro-Ile-Ser-Leu-Asp-Leu-Thr-Phe-
His-Leu-Leu-Arg-Glu-Val-Leu-Glu-Met-Ala-Arg-Ala-
Glu-Gln-Leu-Ala-Gln-Gln-Ala-His-Ser-Asn-Arg-Lys-
Leu-Met-Glu-Asn-Phe Amide

Synonyms: Corticotropin Releasing Factor

American Peptide 34-4-11   Porcine   MW 4792.6
$C_{209}H_{337}N_{61}O_{64}S_2$

Ser-Glu-Gly

Bachem H-4965.0050   MW 291.26 $C_{10}H_{17}N_3O_7$ Store at
-15°C

Ser-Glu-Gly-Ala-Thr-Pro-Gln-Asp-Leu-Asn-Thr-Met-
Leu-Asn-Thr-Val-Gly

Synonyms: P25 (176-192); GAG P24 CA (44-60); HTLV I Peptide

Neosystem SC295   MW 1747.89   HTLV I peptide from
subtype P25 (GAG P24 CA)

Ser-Glu-His-Gln-Leu-Leu-His-Asp-Lys-Gly-Lys-Ser-
Ile-Gln-Asp-Leu-Arg-Arg-Arg-Phe-Phe-Leu-His-His-
Leu-Ile-Ala-Glu-Ile-His-Thr-Ala-Glu-Ile

Synonyms: Parathyroid Hormone Related Protein (3-36); Humoral
Hypercalcemia of Malignancy Factor (3-36)

Neosystem SC1304   Human   MW 4089.66   Bioactive;
calcium metabolism peptide

Ser-Glu-Ile-Gln-Leu-Leu-His-Asp-Lys-Gly-Lys-Ser-
Ile-Gln-Asp-Leu-ArgArg-Arg-Phe-Trp-Leu-His-His-
Leu-Ile-Ala-Glu-Ile-His-Thr-Ala-Glu-Ile

Synonyms: Parathyroid Hormone Related Protein (3-36),
(Ile[5],Trp[23])-; Humoral Hypercalcemia of Malignancy Factor (3-36),
(Ile[5],Trp[23])-

Neosystem SC1305   Human   MW 4104.72   Bioactive;
calcium metabolism peptide

Ser-Glu-Ile-Gln-Leu-Met-His-Asn-Leu-Gly-Lys-His-
Leu-Ala-Ser-Val-Glu-Arg-Met-Gln-Trp-Leu-Arg-Lys-
Lys-Leu-Gln-Asp-Val-His-Asn-Phe

Synonyms: Parathyroid Hormone (3-34)

Neosystem SC1225   Rat   MW 3887.53   Bioactive; calcium
metabolism peptide

Ser-Glu-Ile-Gln-Leu-Met-His-Asn-Leu-Gly-Lys-His-
Leu-Asn-Ser-Met-Glu-Arg-Val-Glu-Trp-Leu-Arg-Lys-
Lys-Leu-Gln-Asp-Val-His-Asn-Phe

Synonyms: Parathyroid Hormone (3-34)

Neosystem SC1265   Human   MW 3931.54   Bioactive;
calcium metabolism peptide; Lee, CW, *Mol Endocrinology*, 9:1269-
1278, 1995

Ser-Glu-Ile-Gln-Leu-Nle-His-Asn-Leu-Gly-Lys-His-
Leu-Asn-Ser-Nle-Glu-Arg-Val-Glu-Trp-Leu-Arg-Lys-
Lys-Leu-Gln-Asp-Val-His-Asn-Tyr Amide

Synonyms: Parathyroid Hormone (3-34), (Nle[8,18],Tyr[34])-

American Peptide 22-1-33   Human   MW 3910.5
$C_{175}H_{282}N_{54}O_{48}$

Peptides International PTH-4181-v   Human synthetic   MW
3910.5 $C_{175}H_{282}N_{54}O_{48}$ >99% (HPLC); lyophilized amorphous
powder

Ser-Glu-Ile-Gln-Phe-Met-His-Asn-Leu-Gly-Lys-His-
Leu-Ser-Ser-Met-Glu-Arg-Val-Glu-Trp-Leu-Arg-Lys-
Lys-Leu-Gln-Asp-Val-His-Asn-Phe

Synonyms: Parathyroid Hormone (3-34)

American Peptide 22-3-22   Bovine   MW 3938.6
$C_{175}H_{274}N_{52}O_{48}S_2$ Goltzman, D et al, *JBC*, 250:3199, 1975

Bachem H-3088.0500   Bovine   Store at -15°C

Ser-Glu-Ile-Gln-Phe-Nle-His-Asn-Leu-Gly-Lys-His-Leu-Ser-Ser-Nle-Glu-Arg-Val-Glu-Trp-Leu-Arg-Lys-Lys-Leu-Gln-Asp-Val-His-Asn-Tyr Amide

Synonyms: Parathyroid Hormone (3-34), (Nle[8,18],Tyr[34])-; Parathormone

American Peptide 22-3-20   Bovine   MW 3917.5
$C_{177}H_{279}N_{53}O_{48}$ Rosenblat et al, *JBC*, 252:5847, 1977

Bachem N-1130.0500   Bovine   Store at -15°C

ICN 152981   Bovine   MW 3917.5   Hormonal inhibitor lacking agonist activity; Rosenblatt, M & JT Potts Jr, *Endocrin Res Commun*, 4:115, 1977

Neosystem SC888   Bovine   MW 3917.46   Bioactive; calcium metabolism peptide; Rosenblatt, M et al, *JBC*, 252:5847-5851, 1977; Segre, GV et al, *JBC*, 254:6980-6986, 1979

Sigma P 3030   Bovine   FW 3917.5   ≥97% (HPLC); peptide content:~70% | Bioactive peptide; Rosenblatt, M et al, *J Biol Chem*, 252:5847, 1977

Ser-Glu-Ile-Trp-Arg-Asp-Ile-Asp-Phe

Synonyms: Tyrosinase (192-200)

Bachem H-3812.0001   Human, mouse   Store at -15°C

Ser-Glu-Lys-Ile-Asp-Met-Val-Asn-Glu-Thr-Ser-Ser-Cys-Ile-Ala-Lys-Lys

Synonyms: GP140 (141-155)-Lys-Lys; SIVmac251 Peptide

Neosystem SC722   MW 1883.16   SIVmac251 peptide from subtype GP140; due to the presence of a Cys this peptide can contain the dimeric form

Ser-Glu-Val-Asn-Leu-Asp-Ala-Glu
Mca-Ser-Glu-Val-Asn-Leu-Asp-Ala-Glu-Dap(Dnp)

Synonyms: Amyloid β/A4 Precursor Protein 770-Dap(Dnp) (667-674), Mca-(Asn[670],Leu[671])-

Bachem M-2425.0001   MW 1344.27 $C_{56}H_{73}N_{13}O_{26}$ Store at -15°C

Ser-Glu-Val-Asn-Leu-Asp-Ala-Glu-Phe

Synonyms: Amyloid β/A4 Precursor Protein 770 (667-675), (Asn[670],Leu[671])-; APP770 (667-675), (Asn[670],Leu[671])-

Bachem H-4836.0001   MW 1023.07 $C_{44}H_{66}N_{10}O_{18}$ Store at -15°C

Ser-Glu-Val-Asn-Leu-Asp-Ala-Glu-Phe
DABCYL-Ser-Glu-Val-Asn-Leu-Asp-Ala-Glu-Phe-EDANS

Synonyms: Amyloid β/A4 Precursor Protein 770 (667-675)-EDANS, DABCYL-(Asn[670],Leu[671])-

Bachem M-2435.0001   MW 1522.66 $C_{71}H_{91}N_{15}O_{21}S$ Store at -15°C

Ser-Glu-Val-Asn-Leu-Asp-Ala-Glu-Phe-Arg

Synonyms: Amyloid β/A4 Precursor Protein 770 (667-676), (Asn[670],Leu[671])-; APP770 (667-676), (Asn[670],Leu[671])-

Bachem H-4834.0001   MW 1179.25 $C_{50}H_{78}N_{14}O_{19}$ Store at -15°C

Ser-Glu-Val-Asn-Leu-Asp-Ala-Glu-Phe-Arg-Lys-Arg-Arg
Mca-Ser-Glu-Val-Asn-Leu-Asp-Ala-Glu-Phe-Arg-Lys(Dnp)-Arg-Arg Amide

Synonyms: Amyloid β/A4 Precursor Protein 770-Lys(Dnp)-Arg-Arg (667-676), Mca-(Asn[670],Leu[671])-

Bachem M-2465.0001   MW 2001.1 $C_{86}H_{125}N_{27}O_{29}$ Store at -15°C

Ser-Glu-Val-Asn-Leu-Asp-Ala-Glu-Phe-Arg-Lys-Arg-Arg
MOCAc-Ser-Glu-Val-Asn-Leu-Asp-Ala-Glu-Phe-Arg-Lys(Dnp)-Arg-Arg Amide

Synonyms: Fluorescence Quenching Substrate

Peptides International SMO-3212-v   MW 2001.1
$C_{86}H_{125}N_{27}O_{29}$ >96% (HPLC); lyophilized amorphous powder | Flanks β-cleavage site of Swedish-type amyloid precursor protein; Koike, H et al, *J Biochem*, 126:235, 1999

Ser-Glu-Val-Asn-Leu-Asp-Ala-Glu-Phe-Lys
Mca-Ser-Glu-Val-Asn-Leu-Asp-Ala-Glu-Phe-Lys(Dnp)

Synonyms: Amyloid β/A4 Precursor Protein 770-Lys(Dnp) (667-675), Mca-(Asn[670],Leu[671])-

Bachem M-2420.0001   MW 2677.42 $C_{68}H_{88}N_{14}O_{27}$ Store at -15°C

Ser-Glu-Val-Lys-Met-Asp-Ala-Glu-Phe-Arg

Synonyms: Amyloid β/A4 Precursor Protein 770 (667-676)

Bachem H-4842.0001   MW 1211.36 $C_{51}H_{82}N_{14}O_{18}S$ Store at -15°C

Ser-Glu-Val-Lys-Met-Asp-Ala-Glu-Phe-Arg-Lys-Arg-Arg
Mca-Ser-Glu-Val-Lys-Met-Asp-Ala-Glu-Phe-Arg-Lys(Dnp)-Arg-Arg Amide

Synonyms: Amyloid β/A4 Precursor Protein 770-Lys(Dnp)-Arg-Arg (667-676), Mca-

Bachem M-2460.0001   MW 2033.21 $C_{87}H_{129}N_{27}O_{28}S$ Store at -15°C

Ser-Glu-Val-Lys-Val-Asp-Ala-Glu-Phe-Arg

Synonyms: Amyloid β/A4 Precursor Protein 770 (667-676), (Val[671])-

Bachem H-4838.0001   MW 1179.3 $C_{51}H_{82}N_{14}O_{18}$ Store at -15°C

Ser-Gly
Z-Ser-Gly

Bachem C-2620.0250   MW 296.28 $C_{13}H_{16}N_2O_6$ Store at -15°C

Ser-Gly
Z-Ser-Gly-OEt

Bachem C-3310.0001   MW 324.33 $C_{15}H_{20}N_2O_6$ Store at -15°C

Ser-Gly
Ac-Ser-Gly

Bachem G-1100.0050   MW 204.18 $C_7H_{12}N_2O_5$ Store at -15°C

Ser-Gly

Bachem G-3190.0250   MW 162.15 $C_5H_{10}N_2O_4$ Store at -15°C

ICN 156608   MW 162.1 $C_5H_{10}N_2O_4$ 99%; crystalline

Sigma S 5250   FW 162.1 $C_5H_{10}N_2O_4$ ~99%

Ser-Gly-Arg
N-t-BOC-o-Bzl-Ser-Gly-Arg-pNA

Sigma B 8266   FW 628.7 $C_{29}H_{40}N_8O_8$

Ser-Gly-Arg-Gly-Ac-Lys-Gly-Gly-Lys-Gly-Leu-Gly-Lys-Gly-Gly-Ala-Lys-Arg-His-Arg-Cys

Synonyms: Histone H4 Peptide, (Ac Lys[5])-; HAT Substrate

Upstate 12-343   MW 2,009.5   >90% pure; frozen solution

Ser-Gly-Arg-Gly-Lys-Gly-Gly-Ac-Lys-Gly-Leu-Gly-Lys-Gly-Gly-Ala-Lys-Arg-His-Arg-Cys

Synonyms: Histone H4 Peptide, (Ac-Lys[8])-; HAT Substrate

Upstate 12-344   MW 2,009.6   >90% pure; frozen solution

Ser-Gly-Arg-Gly-Lys-Gly-Gly-Lys-Gly-Leu-Gly-Ac-Lys-Gly-Gly-Ala-Lys-Arg-His-Arg-Cys

Synonyms: Histone H4 Peptide, (Ac-Lys[12])-; HAT Substrate

Upstate 12-345   MW 2,009.6   >90% pure; frozen solution

Ser-Gly-Arg-Gly-Lys-Gly-Gly-Lys-Gly-Leu-Gly-Lys-Gly-Gly-Ala-Ac-Lys-Arg-His-Arg-Cys

Synonyms: Histone H4 Peptide, (Ac-Lys[16])-; HAT Substrate

Upstate 12-346   MW 2,009.5   >90% pure; frozen solution

Ser-Gly-Arg-Gly-Lys-Gly-Gly-Lys-Gly-Leu-Gly-Lys-Gly-Gly-Ala-Lys-Arg-His-Arg-Cys

Synonyms: Histone H4; HAT Substrate

Upstate 12-347   MW 1,967.6   >90% pure; frozen solution

Ser-Gly-Arg-Gly-Lys-Gly-Gly-Lys-Gly-Leu-Gly-Lys-Gly-Gly-Ala-Lys-Arg-His-Arg-Lys-Val-Leu-Arg-Gly-Ser-Gly-Ser-Lys
Ser-Gly-Arg-Gly-Lys-Gly-Gly-Lys-Gly-Leu-Gly-Lys-Gly-Gly-Ala-Lys-Arg-His-Arg-Lys-Val-Leu-Arg-Gly-Ser-Gly-Ser-Lys-Biotin

Synonyms: Histone H4 Peptide, Biotinylated

Upstate 12-372   MW 3005.6   >90% pure; lyophilized | Immunizing peptide; HAT assay

---

Ser-Gly-Asn-Glu-Ser-Glu-Gly-Glu-Ile-Ser-Ala-Leu-Val-Glu

Synonyms: VPU (52-65); HTLV I Peptide

Neosystem SC532   MW 1420.45   HTLV I peptide from subtype VPU

Ser-Gly-Gln-Ser-Trp-Arg-Pro-Gln-Gly-Arg-Phe Amide

Synonyms: ACEP-1; Cardioexcitatory Peptide

American Peptide 87-0-06   MW 1304.4 $C_{57}H_{85}N_{21}O_{15}$ The peptide originally isolated from the atria of Achatina has the potent excitatory effects on the heart ventricle & Helix neurones; its N-terminal tetrapeptide is similar to the coelenterate peptide, Antho-Rfamide; *BBRC,* 167:777, 1990

Bachem H-1646.0001   MW 1304.43 $C_{57}H_{85}N_{21}O_{15}$ Store at -15°C

ICN 158733   African giant snail   MW 1304.4   97%

Sigma S 8540   African giant snail   FW 1304.4   ≥97% (HPLC) | Bioactive peptide; *Biochem Biophys Res Commun,* 167:777, 1990

Ser-Gly-Gly

Bachem H-4970.0250   MW 219.2 $C_7H_{13}N_3O_5$ Store at -15°C

Ser-Gly-Gly-Val-Val-Lys-Asn-Asn-Phe-Val-Pro-Thr-Asn-Val-Gly-Ser-Lys-Ala-Phe Amide

Synonyms: Calcitonin Gene Regulated Peptide (19-37), α-

Bachem H-8885.0001   Human   MW 1921.19 $C_{86}H_{137}N_{25}O_{25}$ Store at -15°C

Ser-Gly-Gly-Val-Val-Lys-Asn-Asn-Phe-Val-Pro-Thr-Asn-Val-Gly-Ser-Lys-Ala-Phe
Ac-Ser-Gly-Gly-Val-Val-Lys-Asn-Asn-Phe-Val-Pro-Thr-Asn-Val-Gly-Ser-Lys-Ala-Phe Amide

Synonyms: Calcitonin Gene Regulated Peptide (19-37), Ac-α-

Bachem H-8890.0001   Human   MW 1963.22 $C_{88}H_{139}N_{25}O_{26}$ Store at -15°C

Ser-Gly-Lys-Ala-Arg-Gly-Trp-Phe-Tyr

Synonyms: VIF (32-40); HTLV I Peptide

Neosystem SC543   MW 1071.20   HTLV I peptide from subtype VIF

Ser-Gly-Phe-Met-Pro-Lys-Cys-Ser-Lys-Val-Val-Val-Ser-Ser-Cys-Thr-Arg-Met-Met-Glu
Ser-Gly-Phe-Met-Pro-Lys-Cys(Acm)-Ser-Lys-Val-Val-Val-Ser-Ser-Cys(Acm)-Thr-Arg-Met-Met-Glu

Synonyms: GP140 (251-270); SIVmac251 Peptide

Neosystem SC733   MW 2349.83   SIVmac251 peptide from subtype GP140

Ser-Gly-Pro-Tyr-Ser-Phe-Asn-Ser-Gly-Leu-Thr-Phe Amide

Synonyms: SALMF 2; S_2

American Peptide 60-0-11   *Asterias rubens* & *Asterias forbesi* (starfish)   MW 1275.4 $C_{59}H_{82}N_{14}O_{18}$ Novel neuropeptide identified in species belongs to the phylum *Echnodermata; Elphick,* MR et al, *Proc R Soc Lon,* B243:121, 1991

Ser-Gly-Ser-Ala-Lys-Val-Ala-Phe-Ser-Ala-Ile-Arg-Ser-Thr-Asn-His

Synonyms: Cerebellin

American Peptide 60-1-53   MW 1632.8 $C_{69}H_{113}N_{23}O_{23}$ Siemmon, JR et al, *PNAS,* 81:6866, 1984

Bachem H-5530.0500   MW 1632.8 $C_{69}H_{113}N_{23}O_{23}$ Store at -15°C

ICN 151451   MW 1632.8   Slemmon, JR etal, *PNAS,* 81:6866, 1984

Ser-Gly-Ser-Ala-Lys-Val-Ala-Phe-Ser-Ala-Ile-Arg-Ser-Thr-Asn-His Trifluoroacetate Salt

Synonyms: Cerebellin

Biogenesis 1937-5055   Non-species synthetic   MW 1632.8   Purified; lyophilized | Slemmon et al, *PNAS,* 81:6866, 1984

Ser-Gly-Ser-Phe-Leu
Ser-Gly-Ser-p-Nitro-Phe-Leu

Bachem H-2778.0001   Store at -15°C

Ser-Gly-Tyr-Ser-Thr-Glu-Val-Val-Ala-Leu-Ser-Arg-Leu-Gln-Gly-Ser-Leu-Gln-Asp-Met-Leu-Trp-Gln-Leu-Asp-Leu-Ser-Pro-Gly-Cys

Synonyms: Leptin (138-167); Obese Gene Peptide (138-167)

Bachem H-3428.0500   Human   MW 3254.69 $C_{141}H_{225}N_{37}O_{47}S_2$ Store at -15°C

Ser-His

Bachem G-3195.0250   MW 242.24 $C_9H_{14}N_4O_4$ Store at -15°C

Ser-His Acetate Salt

ICN 156609

Sigma S 4757   FW 242.2 (free base) $C_9H_{14}N_4O_4$

Ser-Ile-Gln-Phe-His-Trp-Lys-Asn-Ser-Asn-Gln-Ile-Lys-Ile-Leu-Gly-Asn-Gln-Gly-Ser-Phe-Leu-Thr-Lys-Gly-Pro-Ser-Lys-Leu-Asn-Asp-Arg-Ala-Asp

Synonyms: CD-4 (25-58)

ICN 153067 MW 3843.3 Putative receptor for the AIDS virus (HIV); Madden, PJ etal, *Cell*, 40:93, 1985; Bradford, A etal, *Science*, 240:1335, 1988; Sattenau, QJ & RA Weiss, *Cell*, 52:631, 1988

Sigma C 5292 FW 3843.3 ≥97% (HPLC) | Bioactive peptide; CD-4 is the cell surface glycoprotein receptor for HIV (AIDS virus); Sattenau, QJ & Weiss, RA, *Cell*, 52:631, 1988; Jameson, BA et al, *Science*, 240:1335, 1988; Maddon, PJ et al, *Cell*, 40:93, 1985

Ser-Ile-Gly-Ser-Ala-Leu-Lys-Lys-Ala-Leu-Pro-Val-Ala-Lys-Lys-Ile-Gly-Lys-Ile-Ala-Leu-Pro-Ile-Ala-Lys-Ala-Ala-Leu-Pro

Synonyms: Ceratotoxin A

Bachem H-1616.0500 Store at -15°C

Ser-Ile-Gly-Ser-Ala-Phe-Lys-Lys-Ala-Leu-Pro-Val-Ala-Lys-Lys-Ile-Gly-Lys-Ala-Ala-Leu-Pro-Ile-Ala-Lys-Ala-Ala-Leu-Pro

Synonyms: Ceratotoxin B

Bachem H-1618.0500 Store at -15°C

Ser-Ile-Gly-Ser-Leu-Ala-Lys

Synonyms: Penicillin Binding Protein 1b; Penicillin Binding Protein 1b Active Site Peptide

American Peptide 72-2-41 *E. coli* MW 674.8 $C_{29}H_{54}N_8O_{10}$ Bioactive peptide; antibiotic peptide; this tryptic peptide originally isolated from *E. coli* contains an active site of penicillin-binding protein 1b; Nicholas, RA et al, *Biochemistry*, 26:3448, 1985

ICN 153178 *E. coli* Active site peptide; Nicholas, RA etal, *Biochem*, 24:3448, 1985

Sigma S 3019 *E. coli* FW 674.8 ≥95% (HPLC) | Bioactive peptide; antibiotic peptide; Nicholas, RA et al, *Biochemistry*, 26:3448, 1985

Ser-Ile-Ile-Asn-Phe-Glu-Lys-Leu

Synonyms: Ovalbumin (257-264)

American Peptide 74-1-16 MW 963.2 $C_{45}H_{74}N_{10}O_{13}$ Subcutaneous immunization with this peptide induces OVA-specific, CD8+CTLs that lyse the OVA-expressing target; Celluzzi, CM et al, *J Invest Dermatol*, 108(5):716, 1997

Neosystem SC1302 MW 963.14 Virus-related peptide; bioactive

Bachem H-4866.0001 Chicken MW 963.14 $C_{45}H_{74}N_{10}O_{13}$ Store at -15°C

Ser-Ile-Lys-Val-Ala-Val

Synonyms: Laminin A1 Chain (2099-2104)

Bachem H-2684.0001 Store at -15°C

Calbiochem 428030 Synthetic MW 615.8 $C_{28}H_{53}N_7O_8$ ≥95% (HPLC); solid; soluble in 0.1% TFA | Modulates tumor invasion, metastasis & angiogenesis; induces the synthesis of prostaglandin E_2, interstitial collagenase & gelatinase B in monocyte cell cultures; laminin fragments containing the SIKVAV sequence may contribute to the excessive induction of metalloproteinases in chronic inflammatory lesions; Corcoran, ML et al, *J Biol Chem*, 270:10365, 1995

Ser-Ile-Pro-Ser-Lys-Asp-Ala-Leu-Leu-Lys

Synonyms: Sodefrin

Bachem H-2566.0001 MW 1071.28 $C_{48}H_{86}N_{12}O_{15}$ Store at -15°C

Ser-Ile-Pro-Ser-Lys-Asp-Ala-Leu-Leu-Lys Acetate Salt

Synonyms: Sodefrin

ICN 196007 MW 1131.3 ≥95% | Female newt attracting pheromone

Sigma S 6046 Newt cloacal glands FW 1071.3 (free base) ≥95% (HPLC) | Bioactive peptide; female-attracting peptide pheromone in newt cloacal glands; Kikuyama, S et al, *Science*, 267:1643, 1995

Ser-Ile-Val-Ile-Ile-Glu-Tyr-Arg-Lys-Ile-Leu-Arg

Synonyms: VPU (23-34); HTLV I Peptide

Neosystem SC573 MW 1502.86 HTLV I peptide from subtype VPU

Ser-Leu
Z-Ser-Leu DCHA

Bachem C-2625.0250 MW 533.71 $C_{17}H_{24}N_2O_6 \cdot C_{12}H_{23}N$ Store at -15°C

Ser-Leu

Bachem G-3200.0250 MW 218.25 $C_9H_{18}N_2O_4$ Store at -15°C

ICN 156614 MW 218.3 $C_9H_{18}N_2O_4$ Crystalline

Ser-Leu Amide Hydrochloride

Bachem G-3205.0250 MW 253.73 $C_9H_{19}N_3O_3 \cdot HCl$ Store at -15°C

Ser-Leu-Ala-Leu-Ala-Asp-Asp-Ala-Ala-Phe-Arg-Glu-Arg-Ala-Arg-Leu-Leu-Ala-Ala-Leu-Glu-Arg-Arg-His-Trp-Leu-Asn-Ser-Tyr-Met-His-Lys-Leu-Leu-Val-Leu-Asp-Ala-Pro

Synonyms: TIP-39

Bachem H-4878.0500 MW 4504.25 $C_{202}H_{325}N_{61}O_{54}S$ Store at -15°C

Ser-Leu-Arg-Arg-Ser-Ser-Cyclo(-Cys-Phe-Gly-Gly-Arg-Ile-Asp-Arg-Ile-Gly-Ala-Gln-Ser-Gly-Leu-Gly-Cys)-Asn-Ser-Phe-Arg-Tyr

Synonyms: Natriuretic Peptide (123-150), Atrial; Natriuretic Peptide (99-126), Atrial; Natriuretic Peptide, Atrial α-; Atriopeptin, α-

Fluka 11306 Rat MW 3062.4 $C_{128}H_{205}N_{45}O_{29}$ ≥98% (HPLC) | de Bold, AJ, *Science*, 230:767, 1985

Ser-Leu-Arg-Arg-Ser-Ser-Cys-Phe-Gly-Gly-Arg

Synonyms: Natriuretic Peptide (1-11), Atrial

American Peptide 14-1-70 Rat MW 1224.4 $C_{49}H_{83}N_{20}O_{18}S$

Ser-Leu-Arg-Arg-Ser-Ser-Cys-Phe-Gly-Gly-Arg-Ile-Asp-Arg-Ile-Gly-Ala-Gln-Ser-Gly-Leu-Gly-Cys-Asn-Ser-Phe-Arg-Tyr

Synonyms: Natriuretic Factor (1-28), Atrial; Guanylate Cyclase (Neuronal) Activator; Cardionatrin; Natriuretic Factor (1-28), Atrial α-

American Peptide 14-5-41 Rat MW 3062.5 $C_{128}H_{205}N_{45}O_{39}S_2$ Disulfide bonds: Cys[7]-Cys[23] | Activates neuronal guanylate cyclase & reduces the activity of Na^+-K^+ ATPase in rat kidney; Flynn, TG et al, *BBRC*, 117:859, 1983

Bachem H-2100.0500 Rat MW 3062.45 $C_{128}H_{205}N_{45}O_{39}S_2$ Store at -15°C

Calbiochem 05-23-0301 Rat MW 3062.5 $C_{128}H_{205}N_{45}O_{39}S_2$ ≥97% (HPLC); lyophilized; soluble in 5% HOAc; Disulfide bonds: :Cys^7-Cys^{23} | Reduces the activity of Na$^+$ -K$^+$ -ATPase in rat kidney; activator of neuronal guanylate cyclase; suppresses the transcription of its own guanylyl cyclase-linked receptor; Cao, L et al, *J Biol Chem*, 270:24891, 1995; Gorny, D et al, *J Physiol Pharmacol*, 45:173, 1994; Rambotti, MG et al, *Brain Res*, 644:52, 1994; Spreca, A et al, *Histochem J*, 26:778, 1994; *Merck Index*, 12:903

ICN 151438 Rat MW 3062.4 Disulfide bonds: Cys^7-Cys^{23}

Neosystem SC011 Rat MW 3062.43 Disulfide bonds: Cys^7-Cys23; bioactive; Flynn, TG et al, *BBRC*, 117:859-865, 1983

Sigma A 8208 Rat FW 3062.4 ≥97% (HPLC); peptide content:~70% | Bioactive peptide; disulfide bonds: 7-23; stimulates guanylate cyclase activity; Von Geldern, TW et al, *J Med Chem*, 35:808, 1992

Biogenesis 0780-0404 Rat synthetic MW 3062.5 Purified; lyophilized

Peptides International PAF-4151-s Rat synthetic MW 3062.4 $C_{128}H_{205}N_{45}O_{39}S_2$ >99% (HPLC); lyophilized amorphous powder | Bioactive; Disulfide bonds: Cys^7-Cys^{23}; rabbit, mouse; Flynn, TG et al, *BBRC*, 117:859, 1983; Watanabe, TX et al, *Eur J Pharmacol*, 147:49, 1988

Peptides International PAF-4151-v Rat synthetic MW 3062.4 $C_{128}H_{205}N_{45}O_{39}S_2$ >99% (HPLC); lyophilized amorphous powder | Bioactive; Disulfide bonds: Cys^7-Cys^{23}; rabbit, mouse; Flynn, TG et al, *BBRC*, 117:859, 1983; Watanabe, TX et al, *Eur J Pharmacol*, 147:49, 1988

Ser-Leu-Arg-Arg-Ser-Ser-Cys-Phe-Gly-Gly-Arg-Met-Asp-Arg-Ile-Gly-Ala-Gln-Ser-Gly-Leu-Gly-Cys-Asn-Ser-Phe-Arg-Tyr

Synonyms: Natriuretic Factor (1-28), Atrial α-

Neosystem SC025 MW 3080.46 Human Disulfide bonds: Cys^7-Cys^{23}; bioactive; Kangawa, K & Matsuo, H, *BBRC*, 118:131-138, 1984

Ser-Leu-Arg-Arg-Ser-Ser-Cys-Phe-Gly-Gly-Arg-Met-Asp-Arg-Ile-Gly-Ala-Gln-Ser-Gly-Leu-Gly-Cys-Asn-Ser-Phe-Arg-Tyr
Ser-Leu-Arg-Arg-Ser-Ser-Cys-Phe-Gly-Gly-Arg-Met(O)-Asp-Arg-Ile-Gly-Ala-Gln-Ser-Gly-Leu-Gly-Cys-Asn-Ser-Phe-Arg-Tyr

Synonyms: Natriuretic Peptide (1-28), Atrial (Met(O)$_{12}$)-

American Peptide 14-1-35 Human MW 3096.5 $C_{127}H_{203}N_{45}O_{40}S_3$ Disulfide bonds: Cys^7-Cys^{23} | Chino, N et al, *Peptide Chemistry*, 1984:241, 1985

Ser-Leu-Arg-Arg-Ser-Ser-Cys-Phe-Gly-Gly-Arg-Met-Asp-Arg-Ile-Gly-Ala-Gln-Ser-Gly-Leu-Gly-Cys-Asn-Ser-Phe-Arg-Tyr

Synonyms: Natriuretic Peptide (1-28), Atrial; Mitogen Activated Protein Kinase Inhibitor; Urodilatin

Calbiochem 05-23-0300 Human MW 3080.5 $C_{127}H_{203}N_{45}O_{39}S_3$ ≥97% (HPLC); lyophilized; soluble in 5% HOAc; Disulfide bonds: :Cys^7-Cys^{23} | Increases the rate of renal excretion; higher levels are reported in congestive heart failure; inhibits MAP kinase through the clearance receptor; Prins, BA et al, *J Biol Chem*, 271:14156, 1996; Yechieli, H et al, *Am J Physiol*, 265:F119, 1993; *Merck Index*, 12:903

ICN 151435 Human MW 3180.5 Disulfide bonds: Cys^7-Cys^{23}; potent diuretic, natriuretic & vasodilatory activity

Sigma A 1663 Human FW 3080.5 ≥97% (HPLC) | Bioactive peptide; disulfide bonds: 7-23

Biogenesis 0780-0204 Human synthetic MW 3080.5 Lyophilized, containing 88.39% peptide material, acetate counter ions and residual H_2O; sterile and pyrogen free; purified | To avoid Met oxidation, it is best to dissolve the peptide under a nitrogen stream

Peptides International PAF-4135-s Human synthetic MW 3080.5 $C_{127}H_{203}N_{45}O_{39}S_3$ >95% (HPLC); lyophilized amorphous powder | Bioactive; Disulfide bonds: Cys^7-Cys^{23}; porcine, bovine, canine; Needleman, P et al, *Ann Rev Pharmacol Toxicol*, 29:23, 1989; Rosenzweig, A & CE Seidman, *Ann Rev Biochem*, 60:229, 1991; Kangawa, K & H Matsuo, *BBRC*, 118:131, 1984; Watanabe, TX et al, *Eur J Pharmacol*, 147:49, 1988

Peptides International PAF-4135-v Human synthetic MW 3080.5 $C_{127}H_{203}N_{45}O_{39}S_3$ >95% (HPLC); lyophilized amorphous powder | Bioactive; Disulfide bonds: Cys^7-Cys^{23}; porcine, bovine, canine; Needleman, P et al, *Ann Rev Pharmacol Toxicol*, 29:23, 1989; Rosenzweig, A & CE Seidman, *Ann Rev Biochem*, 60:229, 1991; Kangawa, K & H Matsuo, *BBRC*, 118:131, 1984; Watanabe, TX et al, *Eur J Pharmacol*, 147:49, 1988

Ser-Leu-Arg-Arg-Ser-Ser-Cys-Phe-Gly-Gly-Arg-Met-Asp-Arg-Ile-Gly-Ala-Gln-Ser-Gly-Leu-Gly-Cys-Asn-Ser-Phe-Arg-Tyr
Ser-Leu-Arg-Arg-Ser-Ser-Cys-Phe-Gly-Gly-Arg-Met(O)-Asp-Arg-Ile-Gly-Ala-Gln-Ser-Gly-Leu-Gly-Cys-Asn-Ser-Phe-Arg-Tyr

Synonyms: Natriuretic Peptide (1-28), Atrial Met(O)^{12})-

Peptides International PAF-4145-v Human synthetic MW 3096.5 $C_{127}H_{203}N_{45}O_{40}S_3$ >99% (HPLC); lyophilized amorphous powder | Bioactive; Disulfide bonds: Cys^7-Cys^{23}; Watanabe, TX et al, *Eur J Pharmacol*, 147:49, 1988

Ser-Leu-Asp-Lser-Pro-Ala-Ala-Leu-Ala-Glu-Arg-Gly-Ala-Arg-Asn-Ala-Leu-Gly-Gly-His-Gln-Glu-Ala-Pro-Glu-Arg-Glu

Synonyms: S Prepro-Corticotropin Releasing Factor (125-151)

American Peptide 34-1-27 Human MW 2803.2 $C_{115}H_{188}N_{40}O_{42}$

Ser-Leu-Gln-Tyr-Leu-Ala-Leu-Ala-Ala-Leu-Ile-Thr-Pro-Lys-Lys-Ile

Synonyms: VIF (144-159); HTLV I Peptide

Neosystem SC235 MW 1743.16 HTLV I peptide from subtype VIF

Ser-Leu-His-Gly-Met-Asp-Asp-Pro-Glu-Arg-Glu-Val-Leu-Glu

Synonyms: NEF (169-182); HTLV I Peptide

Neosystem SC514 MW 1626.75 HTLV I peptide from subtype NEF

Ser-Leu-Ile-Gly-Arg-Leu

Synonyms: Proteinase Activated Receptor II Agonist Peptide

Bachem H-3586.0005 MW 657.81 $C_{29}H_{55}N_9O_8$ Store at -15°C

Neosystem SC951 Mouse MW 657.81 Bioactive; thrombin receptor-derived peptide; corresponds to the tethered ligand sequence of mouse PAR-2; functions as a PAR-2 agonist; Nystedt, S et al, *PNAS USA*, 91:9208-9212, 1994

Ser-Leu-Ile-Gly-Arg-Leu Amide

Bachem H-5078.0005 MW 656.83 $C_{29}H_{56}N_{10}O_7$ Store at -15°C

Ser-Leu-Ile-Gly-Lys-Val

Synonyms: Proteinase Activated Receptor II Agonist Peptide

Bachem H-5042.0005 MW 615.77 $C_{28}H_{53}N_7O_8$ Store at -15°C

Neosystem SC1230 Human MW 615.77 Bioactive; thrombin receptor-derived peptide; Molino, M et al, *JBC*, 272:4043-4049, 1997

Ser-Leu-Ile-Gly-Lys-Val Amide

Bachem H-4624.0005   MW 614.79 $C_{28}H_{54}N_8O_7$ Store at -15°C

Ser-Leu-Leu

Bachem H-4975.0100   MW 331.41 $C_{15}H_{29}N_3O_5$ Store at -15°C

Ser-Leu-Leu-Gln-Val-Leu-Asn-Val-Lys-Glu-Gly-Thr-Pro-Ser

Synonyms: Gastrin Releasing Peptide Gene Associated Peptide Fragment

Bachem H-9175.0001   Store at -15°C

Ser-Leu-Leu-Lys

Bachem H-4754.0250   MW 459.59 $C_{21}H_{41}N_5O_6$ Store at -15°C

Ser-Leu-Leu-Lys Amide

Bachem H-4788.0250   MW 458.6 $C_{21}H_{42}N_6O_5$ Store at -15°C

Ser-Leu-Pro-Glu-Ala-Gly-Pro-Gly-Arg-Thr-Leu-Val-Ser-Ser-Lys-Pro-Gln-Ala-His-Gly-Ala-Pro-Ala-Pro-Pro-Ser-Gly-Ser-Ala-Pro-His-Phe-Leu

Synonyms: Pro-Adrenomedullin (153-185)

American Peptide 22-2-13   Human   MW 3219.6 $C_{143}H_{224}N_{42}O_{43}$

Ser-Leu-Pro-Gly-Arg-Trp-Lys-Pro-Lys-Met-Ile-Gly-Gly-Ile-Gly

Synonyms: POL (37-51); PR (37-51); HTLV I Peptide

Neosystem SC703   MW 1596.95   HTLV I peptide from subtype POL (PR/RT)

Ser-Leu-Trp-Asp-Gln-Ser-Leu-Lys-Pro-Cys-Val-Lys-Leu-Thr-Pro-Leu

Synonyms: GP120 (110-125); HTLV I Peptide

Neosystem SC262   MW 1828.19   HTLV I peptide from subtype GP160; due to the presence of a Cys this peptide can contain the dimeric form

Ser-Leu-Tyr-Asn-Thr-Val-Ala-Thr-Leu

Synonyms: HIV I p17 gag (77-85)

Neosystem SC1345   MW 981.11   Virus-related peptide; AIDS-related peptide; corresponds to the HIV-1 optimal CTL epitope restricted by HLA-A0201; sequence variations in this immunodominant epitope have been shown to profoundly alter the pattern of CTL response; Tsomides, TJ et al, *J Exp Med*, 180:1283-1293, 1994; Goulder, PJR et al, *J Exp Med*, 185:1423-1433, 1997; Brander, C et al, *J Clin Invest*, 101:2259-2566, 1998

Ser-Leu-Val
Ac-Ser-Leu-Val

Synonyms: Fas C-Terminal Tripeptide

Bachem H-4068.0050   Store at -15°C

Ser-Lys-Lys-Lys-Ala-Gln-Gln-Ala-Ala-Ala-Asp-Thr-Gly

Synonyms: P18 (111-123); GAG P17 MA (111-123); HTLV I Peptide

Neosystem SC288   MW 1303.43   HTLV I peptide from subtype P25 (GAG P17 MA)

Ser-Lys-Pro-Asp-Asn-Pro-Gly-Glu-Asp-Ala-Pro-Ala-Glu-Asp-Leu-Ala-Arg-Tyr-Tyr-Ser-Ala-Leu-Arg-His-Tyr-Ile-Asn-Leu-Ile-Thr-Arg-Gln-Arg-Tyr Amide

Synonyms: Neuropeptide Y (3-36)

Bachem H-8570.0500   Porcine   MW 3993.41 $C_{176}H_{271}N_{53}O_{54}$ Store at -15°C

Neosystem SC932   Porcine   MW 3993.4   Bioactive neuropeptide; discovered in porcine brain where it is abundant; selectively binds to Y2 receptors; Grandt, D et al, *Regul Peptides*, 40:161, 1992

Sigma N 4279   Porcine brain   FW 3993.4   ≥95% (HPLC); peptide content:~60% | Gastrointestinal peptide; bioactive peptide; selectively binds to Y_2 receptors; Grandt, D et al, *Regul Peptides*, 40:161, 1992

Ser-Lys-Pro-Asp-Asn-Pro-Gly-Glu-Asp-Ala-Pro-Ala-Glu-Asp-Met-Ala-Arg-Tyr-Tyr-Ser-Ala-Leu-Arg-His-Tyr-Ile-Asn-Leu-Ile-Thr-Arg-Gln-Arg-Tyr Amide

Synonyms: Neuropeptide Y (3-36)

American Peptide 60-1-17   Human   MW 4011.5 $C_{175}H_{269}N_{53}O_{54}S$ Selective Y_2 agonist

Bachem H-3326.0500   Human, rat   MW 4011.45 $C_{175}H_{269}N_{53}O_{54}S$ Store at -15°C

Sigma N 9407   Human, rat   FW 4011.4   Gastrointestinal peptide; bioactive peptide; selectively binds to Y_2 receptors; Grandt, D et al, *Regul Peptides*, 40:161, 1992

Ser-Lys-Trp-Asp-Asp-Pro-Trp-Glu-Glu-Val-Leu-Ala-Trp-Lys-Phe

Synonyms: NEF (201-215); SIVmac251 Peptide

Neosystem SC620   MW 1864.08   SIVmac251 peptide from subtype NEF

Ser-Met

Bachem G-3210.0250   MW 236.29 $C_8H_{16}N_2O_4S$ Store at -15°C

ICN 156615   MW 236.3 $C_9H_{16}N_2O_4S$ Crystalline

Sigma S 4129   FW 236.3 $C_8H_{16}N_2O_4S$

Ser-Met-Glu-His-Phe-Arg-Trp-Gly-Lys-Pro-Val-Gly-Lys-Lys-Arg-Arg-Pro-Val-Lys-Val-Tyr-Pro

Synonyms: Adrenocorticotropic Hormone (3-24)

Bachem H-4716.0001   Human   MW 2683.23 $C_{124}H_{196}N_{38}O_{27}S$ Store at -15°C

Ser-Met-Glu-Val-Arg-Gly-Trp

Synonyms: Melanocyte Stimulating Hormone, δ-

Bachem H-4405.0001   MW 863.99 $C_{37}H_{57}N_{11}O_{11}S$ Store at -15°C

ICN 152927   MW 864 $C_{37}H_{57}N_{11}O_{11}S$

Ser-Phe

Bachem G-3220.0250   MW 352.27 $C_{12}H_{16}N_2O_4$ Store at -15°C

ICN 156616   MW 252.3 $C_{12}H_{16}N_2O_4$ Crystalline

Ser-Phe Amide Hydrochloride

Bachem G-3225.0250   MW 287.75 $C_{12}H_{17}N_3O_3 \cdot HCl$ Store at -15°C

Ser-Phe-Gly-Cys-Arg-Phe-Gly-Thr-Cys-Thr-Val-Gln-Lys-Leu-Ala-His-Gln-Ile-Tyr-Gln-Phe-Thr-Asp-Lys-Asp-Lys-Asp-Asn-Val-Ala-Pro-Arg-Ser-Lys-Ile-Ser-Pro-Gln-Gly-Tyr Amide

Synonyms: Adrenomedullin (13-52)

American Peptide 22-1-11   Human   MW 4533.2
$C_{200}H_{308}N_{58}O_{59}S_2$ Disulfide bonds: Cys^{16}-Cys^{21} | Perret, M et al, *Life Sciences*, 53:377, 1994

Bachem H-4936.0500   Human   MW 4533.13
$C_{200}H_{308}N_{58}O_{59}S_2$ Store at -15°C

ICN 195608   Human   MW ~4533   Perret, M etal, *Life Sciences*, 53:377, 1994

Ser-Phe-Leu-Leu-Arg

Bachem H-1408.0005   MW 634.78 $C_{30}H_{50}N_8O_7$ Store at -15°C

Ser-Phe-Leu-Leu-Arg Amide

Bachem H-2938.0005   MW 633.79 $C_{30}H_{51}N_9O_6$ Store at -15°C

Ser-Phe-Leu-Leu-Arg-Asn

Synonyms: Thrombin Receptor Activator Peptide VI; Thrombin Receptor (42-47)

Bachem H-8365.0005   MW 748.9 $C_{34}H_{56}N_{10}O_9$ Store at -15°C

Neosystem SC852   Human   MW 749.00   Bioactive

Ser-Phe-Leu-Leu-Arg-Asn Amide

Bachem H-2936.0005   MW 747.9 $C_{34}H_{57}N_{11}O_8$ Store at -15°C

Ser-Phe-Leu-Leu-Arg-Asn-Pro

Bachem H-2234.0001   MW 846.01 $C_{39}H_{63}N_{11}O_{10}$ Store at -15°C

Ser-Phe-Leu-Leu-Arg-Asn-Pro-Asn-Asp-Lys-Tyr-Glu-Pro-Phe

Synonyms: Thrombin Receptor Activator; Thrombin Receptor (42-55)

Bachem H-8105.0001   MW 1739.95 $C_{81}H_{118}N_{20}O_{23}$ Store at -15°C

ICN 158734   MW 1739.9   97%

Sigma S 7152   FW 1739.9   ≥97% (HPLC); peptide content:~70% | Bioactive peptide; Vu, TKH et al, *Cell*, 64:1057, 1991; Vassallo, RR et al, *J Biol Chem*, 267:6081, 1992

Neosystem SC449   Human   MW 1739.94   Bioactive; thrombin receptor agonist peptide

Ser-Phe-Leu-Leu-Cit

Bachem H-1406.0001   Store at -15°C

Ser-Phe-Phe-Leu-Arg-Asn

Synonyms: Thrombin Receptor (42-47)

Neosystem SC855   Hamster   MW 783.00   Bioactive; Rasmussen, UB et al, *FEBS Lett*, 288:123-128, 1991; Vouret-Craviari, V et al, *Molec Biol of the Cell*, 3:95-102, 1992

Ser-Phe-Phe-Leu-Arg-Asn-Pro-Gly-Glu-Asn-Thr-Phe-Glu-Leu

Synonyms: Thrombin Receptor (42-55)

Neosystem SC854   Hamster   MW 1671.00   Bioactive

Ser-Phe-Pro-Trp-Met-Glu-Ser-Asp-Val-Thr

Synonyms: Thyrotropin Releasing Hormone Potentiating Peptide; Prepro-Thyrotropin Releasing Hormone (160-169)

American Peptide 58-0-94   MW 1198.3 $C_{54}H_{75}N_{11}O_{18}S$
Originally isolated from bovine hypothalamus; gives the first direct evidence for non-TRH peptides originating from the TRH precursor *in vivo*

Bachem H-1434.0001   MW 1198.32 $C_{54}H_{75}N_{11}O_{18}S$ Store at -15°C

Neosystem SC441   MW 1198.30   Bioactive; thyrotropin releasing hormone-related peptide

Sigma S 6421   FW 1198.3   ≥97% (HPLC) | Bioactive peptide; originally isolated from bovine hypothalamus; potentiates the TRH-induced release of TSH from the anterior pituitary

Ser-Phe-Ser-Nle-Glu-His-Phe-Arg-Trp-Gly-Lys-Pro-Val-Gly-Lys-Lys-Arg-Arg-Pro-Val-Lys-Val-Tyr-Pro

Synonyms: Adrenocorticotropic Hormone (1-24), (Phe^2,Nle^4)-; Corticotropin A

Bachem H-6080.0500   MW 2899.44 $C_{137}H_{212}N_{40}O_{30}$ Store at -15°C

ICN 159852   Human   MW 2899.8   >98%

Neosystem SC017   Human   MW 2899.43   Bioactive

Sigma A 6552   Human   FW 2899.4   ≥97% (HPLC); peptide content:~70% | Bioactive peptide

Ser-Phe-Val-Asn-Ser-Glu-Phe-Leu-Lys-Pro-Glu-Val-Lys-Ser

Synonyms: Protein Kinase C β-1 Peptide; AB/Protein Kinase C 660-673, α-

American Peptide 86-1-35   MW 1610.8 $C_{74}H_{115}N_{17}O_{23}$

Ser-Pro

Bachem G-3230.0250   MW 202.21 $C_8H_{14}N_2O_4$ Store at -15°C

Ser-Pro-Ala-Asn-Ala-Gln-Ile-Thr-Arg-Lys-Arg-His-Lys-Ile-Asn-Ser-Phe-Val-Gly-Leu-Met Amide

Synonyms: Carassin

Sigma C 2055   FW 2367.8   ≥97% (HPLC); peptide content:~65% | Bioactive peptide

Ser-Pro-Ala-Val-Asp-Lys-Ala-Gln-Ala-Glu-Leu

Synonyms: SMCX (963-973)

Bachem H-2714.0001   Human   Store at -15°C

Ser-Pro-Arg-Thr-Leu-Asn-Ala-Trp-Val-Lys-Val-Val-Glu-Glu-Lys

Synonyms: P25 (148-162); GAG P24 CA (16-30); HTLV I Peptide

Neosystem SC292   MW 1756.03   HTLV I peptide from subtype P25 (GAG P24 CA)

Ser-Pro-Glu-Glu-Leu-Ser-Arg-Tyr-Tyr-Ala-Ser-Leu-Arg-His-Tyr-Leu-Asn-Leu-Val-Thr-Arg-Gln-Arg-TyrAmide

Syno Peptide YY (13-36)*nyms*:

ICN 191454   Porcine

Ser-Pro-Glu-Glu-Leu-Ser-Arg-Tyr-Tyr-Ala-Ser-Leu-Arg-His-Tyr-Leu-Asn-Leu-Val-Thr-Arg-Gln-Arg-Tyr Amide

Synonyms: Peptide YY (13-36)

Neosystem SC419   Porcine   MW 3014.39   Bioactive neuropeptide; brain/gut peptide; Walker, MW & Miller, RJ, *Mol Pharm*, 34:779, 1988

Bachem H-9185.0500   Porcine, rat   Store at -15°C

Ser-Pro-His-Pro-Arg-Ile-Ser-Ser-Glu-Val-His-Ile-Pro-Leu-Gly

Synonyms: VIF (46-60); HTLV I Peptide

Neosystem SC231   MW 1625.84   HTLV I peptide from subtype VIF

Ser-Pro-Leu-Ala-Gln-Ala-Val-Arg-Ser-Ser-Ser-Arg Dnp-Ser-Pro-Leu-Ala-Gln-Ala-Val-Arg-Ser-Ser-Ser-Arg Amide

Synonyms: Tumor Necrosis Factor α (71-82), Dnp-Pro-

Bachem M-2290.0001   Human   Store at -15°C

Ser-Pro-Lys-His-His-Ser-Gln-Arg-Ala-Arg-Lys-Lys-Asn-Lys-Asn-Cys Trifluoroacetate Salt

Synonyms: Bone Morphogenetic Protein 4 (1-16)

Biogenesis 1406-0830   Human synthetic   MW 1919.2
Purified; lyophilized | Represents AA 1-16 of human BMP-4

Ser-Pro-Lys-Met-Val-Gln-Gly-Ser-Gly-Cys-Phe-Gly-Arg-Lys-Met-Asp-Arg-Ile-Ser-Ser-Ser-Ser-Gly-Leu-Gly-Cys-Lys-Val-Leu-Arg-Arg-His

Synonyms: Natriuretic Peptide (1-32), Brain; Natriuretic Peptide 32, Brain

American Peptide 14-1-90   Human   MW 3464.1
$C_{143}H_{244}N_{50}O_{42}S_4$ Disulfide bonds: :Cys^{10}-Cys^{26}

Bachem H-9060.0500   Human   Store at -15°C

Calbiochem 05-23-0305   Human   MW 3464.1
$C_{143}H_{244}N_{50}O_{42}S_4$ ≥95% (HPLC); lyophilized; soluble in 5% HOAc; Disulfide bonds: :Cys^{10}-Cys^{26} | Plasma levels of BNP are used as a marker for left ventricular hypertrophy; Kohno, M et al, *Am J Med*, 98:257, 1995; *Merck Index*, 12:903

ICN 153065   Human   MW 3464.1   Disulfide bonds: Cys^{10}-Cys^{26} | Sudoh, T et al, *BBRC*, 159:1427, 1989

Sigma B 5900   Human   FW 3464.1   ≥97% (HPLC); peptide content:~70% | Disulfide bonds: 10-26; bioactive peptide; diuretic, natriuretic, hypotensive & vasorelaxant properties similar to ANF; originally purified from brain, but precursor proteins have also been found in cardiac tissue; Sudoh, T et al, *Biochem Biophys Res Commun*, 159:1427, 1989

Ser-Pro-Lys-Met-Val-Gln-Gly-Ser-Gly-Cys-Phe-Gly-Arg-Lys-Met-Asp-Arg-Ile-Ser-Ser-Ser-Ser-Gly-Leu-Gly-Cys-Lys-Val-Leu-Arg-Arg-His Trifluoroacetate Salt

Synonyms: Natriuretic Peptide (1-32), Brain

Biogenesis 1505-0502   Human synthetic   MW 3464.7
Purified; lyophilized

Ser-Pro-Lys-Met-Val-Gln-Gly-Ser-Gly-Cys-Phe-Gly-Arg-Lys-Met-Asp-Arg-Ile-Ser-Ser-Ser-Ser-Gly-Leu-Gly-Cys-Lys-Val-Leu-Arg-Arg-His

Synonyms: Natriuretic Peptide 32, B-Type; Natriuretic Peptide 32, Brain

Peptides International PBN-4212-v   Human synthetic   MW 3464.1 $C_{143}H_{244}N_{50}O_{42}S_4$ >95% (HPLC); lyophilized amorphous powder | Bioactive; Disulfide bonds: Cys^{10}-Cys^{26}; Sudoh, T et al, *BBRC*, 159:1427, 1989; Kambayashi, Y et al, *FEBS Lett*, 259:341, 1990

Ser-Pro-Lys-Thr-Met-Arg-Asp-Ser-Gly-Cys-Phe-Gly-Arg-Arg-Leu-Asp-Arg-Ile-Gly-Ser-Leu-Ser-Gly-Leu-Gly-Cys-Asn-Val-Leu-Arg-Arg-Tyr

Synonyms: Natriuretic Peptide (1-32), Brain; Natriuretic Peptide 32, Brain; Natriuretic Peptide, Brain

American Peptide 14-4-11   Porcine   MW 3570.2
$C_{149}H_{250}N_{52}O_{44}S_3$ Disulfide bonds: :Cys^{10}-Cys^{26} | Sudoh, T et al, *BBRC*, 155:726, 1988

Bachem H-2952.0500   Porcine   MW 3570.15
$C_{149}H_{250}N_{52}O_{44}S_3$ Store at -15°C

ICN 153066   Porcine   MW 3570.1   Disulfide bonds: Cys^{10}-Cys^{26} | Sudoh, T etal, *BBRC*, 155:726, 1988

Sigma B 6651   Porcine   FW 3570.1   ≥97% (HPLC); peptide content:~70% | Disulfide bonds: 10-26; bioactive peptide; Sudoh, T et al, *Biochem Biophys Res Commun*, 155:726, 1988

Ser-Pro-Ser-Asn-Ser-Lys-Cys-Pro-Asp-Gly-Pro-Asp-Cys-Phe-Val-Gly-Leu-Met Amide

Synonyms: Scyliorhinin II

Bachem H-2902.0500   MW 1851.12 $C_{77}H_{119}N_{21}O_{26}S_3$ Store at -15°C

American Peptide 62-7-72   *Scyliorhinus caniculus* (common dogfish) intestine   MW 1851.1 $C_{77}H_{119}N_{21}O_{26}S_3$ Disulfide bonds: Cys^7-Cys^{13}; contains tachykinin-like activity; Conlon, JM et al, *FEBS Lett*, 200:111, 1986

Ser-Pro-Ser-Val-Asp-Lys-Ala-Arg-Ala-Glu-Leu

Synonyms: SMCY (950-960)

Bachem H-2712.0001   Human   Store at -15°C

Ser-Pro-Val-Thr-Leu-Asp-Leu-Arg-Tyr

Synonyms: Eglin c (41-49)

American Peptide 81-7-50   MW 1063.2 $C_{48}H_{78}N_{12}O_{15}$
Nonapeptide corresponding to the active center of eglin c; inhibits leukocyte cathepsin G & α-chymotrypsin; Fujii, A et al, *Chem Pharm Bull*, 42:1518, 1994

Bachem H-2474.0001   MW 1063.22 $C_{48}H_{78}N_{12}O_{15}$ Store at -15°C

Ser-Ser

Bachem G-3235.0250   MW 192.17 $C_6H_{12}N_2O_5$ Store at -15°C

Ser-Ser-Ala-Ser-Asp-Tyr-Asn-Ser-Ser-Glu-Leu-Lys-Thr-Ala-(Cys) Acetate Salt

Synonyms: Bone Morphogenetic Protein 6

Biogenesis 1406-1159   Rat synthetic   Purified; lyophilized

Ser-Ser-Arg-Cys-Phe-Gly-Ser-Arg-Ile-Asp-Arg-Ile-Gly-Ala-Gln-Ser-Gly-Met-Gly-Cys-Gly-Arg-Phe

Synonyms: Natriuretic Peptide (8-30), Atrial

American Peptide 14-7-23   Frog   MW 2405.7
$C_{97}H_{153}N_{33}O_{33}S_3$ Disulfide bonds: Cys^{11}-Cys^{27}

Ser-Ser-Arg-Leu-Ala
Ser-Ser-Arg-Leu-Ala-EDANS

Bachem H-2444.0001   Store at -15°C

Ser-Ser-Asn-Ala
Palmitoyl-Ser-Ser-Asn-Ala

Synonyms: Mitogenic Tetrapeptide

ICN 154572   MW 615.8

Ser-Ser-Asp-Arg-Ser-Ala-Leu-Leu-Lys-Ser-Lys-Leu-Arg

Synonyms: Natriuretic Peptide (104-116), Prepro-Atrial

American Peptide 14-1-28   Human   MW 1460.7
$C_{61}H_{113}N_{21}O_{20}$

Ser-Ser-Asp-Arg-Ser-Ala-Leu-Leu-Lys-Ser-Lys-Leu-Arg-Ala-Leu-Leu-Thr-Ala-Pro-Arg

Synonyms: Natriuretic Peptide (104-123), Prepro-Atrial; Kaliuretic Peptide; Pro-Natriuretic Factor (79-98), Atrial Human

American Peptide 14-1-30   Human   MW 2183.6
$C_{94}H_{171}N_{31}O_{28}$

Bachem H-3402.0001   Human   MW 2183.58 $C_{94}H_{171}N_{31}O_{28}$
Store at -15°C

Ser-Ser-Asp-Cys-Phe-Gly-Ser-Arg-Ile-Asp-Arg-Ile-Gly-Ala-Gln-Ser-Gly-Met-Gly-Cys-Gly-Arg-Arg-Phe

Synonyms: Natriuretic Factor (1-24), Atrial

Bachem H-3052.0500   MW 2561.87 $C_{103}H_{165}N_{37}O_{34}S_3$ Store at -15°C | frog

---

Ser-Ser-Asp-Cys-Phe-Gly-Ser-Arg-Ile-Asp-Arg-Ile-Gly-Ala-Gln-Ser-Gly-Met-Gly-Cys-Gly-Arg-Arg-Phe
H₂N-Ser-Ser-Asp-Cys-Phe-Gly-Ser-Arg-Ile-Asp-Arg-Ile-Gly-Ala-Gln-Ser-Gly-Met-Gly-Cys-Gly-Arg-Arg-Phe

Synon Natriuretic Peptide, Atrial *yms:*

ICN 154505   Frog   MW 2561.8   Sakata, M etal, *BBRC*, 144:1338, 1988

Ser-Ser-Asp-Cys-Phe-Gly-Ser-Arg-Ile-Asp-Arg-Ile-Gly-Ala-Gln-Ser-Gly-Met-Gly-Cys-Gly-Arg-Arg-Phe

Synonyms: Natriuretic Peptide (1-24), Atrial

Sigma A 0929   Frog   FW 2561.8   ≥95% (HPLC) | Bioactive peptide; disulfide bonds: 4-20; structurally similar to rat ANF; Has diuretic-natriuretic activity; Vascular muscle relaxant; Sakata, J et al, *Biochem Biophys Res Commun*, 155:1338, 1988

Ser-Ser-Cys-Phe-Gly-Gly-Arg-Ile-Asp-Arg-Ile-Gly-Ala-Gln-Ser-Gly-Leu-Gly-Cys-Asn-Ser-Phe-Arg-Tyr

Synonyms: Natriuretic Peptide (5-28), Atrial; Atriopeptin III; Guanylate Cyclase Agonist

American Peptide 14-5-48   Rat   MW 2549.9 $C_{107}H_{165}N_{35}O_{34}S_2$ Disulfide bonds: :Cys^7-Cys^{23} | Delays outward K$^+$ current in cardiac cells & increases the permeability of endothelial cell monolayers along with an elevation in cGMP levels; Geller, DM et al, *BBRC*, 120:333, 1984

Ser-Ser-Cys-Phe-Gly-Gly-Arg-Ile-Asp-Arg-Ile-Gly-Ala-Gln-Ser-Gly-Leu-Gly-Cys-Asn-Ser-Phe-Arg

Synonyms: Natriuretic Peptide (5-27), Atrial; Atriopeptin II; Guanylate Cyclase (Particulate) Activator

American Peptide 14-5-51   Rat   MW 2386.7 $C_{98}H_{156}N_{34}O_{32}S_2$ Disulfide bonds: :Cys^7-Cys^{23} | Cardiac atrial peptide which relaxes both intestinal & vascular smooth muscle; Currie, MG, *Science*, 223:67, 1984

Ser-Ser-Cys-Phe-Gly-Gly-Arg-Ile-Asp-Arg-Ile-Gly-Ala-Gln-Ser-Gly-Leu-Gly-Cys-Asn-Ser

Synonyms: Natriuretic Peptide (5-25), Atrial; Atriopeptin I; Atriopeptin I

American Peptide 14-5-53   Rat   MW 2083.3 $C_{83}H_{135}N_{29}O_{30}S_2$ Disulfide bonds: :Cys^7-Cys^{23} | Weak agonist of cGMP coupled atrial natriuretic peptide receptor; Currie, MG, *Science*, 223:67, 1984

Bachem H-2105.0500   Rat   MW 2083.29 $C_{83}H_{135}N_{29}O_{30}S_2$
Store at -15°C

Ser-Ser-Cys-Phe-Gly-Gly-Arg-Ile-Asp-Arg-Ile-Gly-Ala-Gln-Ser-Gly-Leu-Gly-Cys-Asn-Ser-Phe-Arg

Synonyms: Atriopeptin II

Bachem H-2110.0500   Rat   MW 2386.66 $C_{98}H_{156}N_{34}O_{32}S_2$
Store at -15°C

Ser-Ser-Cys-Phe-Gly-Gly-Arg-Ile-Asp-Arg-Ile-Gly-Ala-Gln-Ser-Gly-Leu-Gly-Cys-Asn-Ser-Phe-Arg-Tyr

Synonyms: Atriopeptin III

Bachem H-2115.0500   Rat   MW 2549.83 $C_{107}H_{165}N_{35}O_{34}S_2$
Store at -15°C

Ser-Ser-Cys-Phe-Gly-Gly-Arg-Ile-Asp-Arg-Ile-Gly-Ala-Gln-Ser-Gly-Leu-Gly-Cys-Asn-Ser

Synonyms: Natriuretic Peptide (5-25), Atrial; Atriopeptin I

Calbiochem 05-23-0315   Rat   MW 2083.3 $C_{83}H_{135}N_{29}O_{30}S_2$ >95% (HPLC); lyophilized; soluble in 5% HOAc; Disulfide bonds: :Cys^3-Cys^{19} | Weak agonist of cGMP coupled atrial natriuretic peptide receptor; Geller, DM et al, *Biochem Biophys Res Comm*, 120:333, 1984

Ser-Ser-Cys-Phe-Gly-Gly-Arg-Ile-Asp-Arg-Ile-Gly-Ala-Gln-Ser-Gly-Leu-Gly-Cys-Asn-Ser-Phe-Arg

Synonyms: Natriuretic Peptide (5-27), Atrial; Atriopeptin II; Guanylate Cyclase (Particulate) Activator

Calbiochem 05-23-0316   Rat   MW 2386.7 $C_{98}H_{156}N_{34}O_{32}S_2$ >95% (HPLC); lyophilized; soluble in 5% HOAc; Disulfide bonds: :Cys^3-Cys^{19} | Papapetropoulos, A et al, *J Cell Physiol*, 167:213, 1996

Ser-Ser-Cys-Phe-Gly-Gly-Arg-Ile-Asp-Arg-Ile-Gly-Ala-Gln-Ser-Gly-Leu-Gly-Cys-Asn-Ser-Phe-Arg-Tyr

Synonyms: Natriuretic Peptide (5-28), Atrial; Atriopeptin III

Calbiochem 05-23-0317   Rat   MW 2549.9 $C_{107}H_{165}N_{35}O_{34}S_2$ ≥98% (HPLC); lyophilized; soluble in 5% HOAc; Disulfide bonds: :Cys^3-Cys^{19} | Delays outward K$^+$ current in cardiac cells; increases cGMP levels & the permeability of endothelial cell monolayers; Bkaily, G et al, *J Mol Cell Cardiol*, 25:1305, 1993; Mene, P et al, *Exp Neurol*, 1:245, 1993; Geller, DM et al, *Biochem Biophys Res Comm*, 120:333, 1984

Ser-Ser-Cys-Phe-Gly-Gly-Arg-Ile-Asp-Arg-Ile-Gly-Ala-Gln-Ser-Gly-Leu-Gly-Cys-Asn-Ser

Synonyms: Natriuretic Peptide (5-25), Atrial; Atriopeptin I

ICN 151439   Rat   MW 2083.3   Cardiac atrial peptide which selectively relaxes intestinal, but not vascular, smooth muscle; Disulfide bonds: Cys^7-Cys^{23}; Currie, MG etal, *Science*, 223:67, 1984

Ser-Ser-Cys-Phe-Gly-Gly-Arg-Ile-Asp-Arg-Ile-Gly-Ala-Gln-Ser-Gly-Leu-Gly-Cys-Asn-Ser-Phe-Arg

Synonyms: Natriuretic Peptide (5-27), Atrial; Atriopeptin II

ICN 151440   Rat   Cardiac atrial peptide which selectively relaxes intestinal, but not vascular, smooth muscle; Disulfide bonds: Cys^7-Cys^{23}; Currie, MG etal, *Science*, 223:67, 1984

Ser-Ser-Cys-Phe-Gly-Gly-Arg-Ile-Asp-Arg-Ile-Gly-Ala-Gln-Ser-Gly-Leu-Gly-Cys-Asn-Ser-Phe-Arg-Tyr

Synonyms: Natriuretic Peptide (5-28), Atrial; Atriopeptin III ; Natriuretic Factor (5-28), Atrial; Atriopeptin III, α-

ICN 151441   Rat   MW 2549.8   Disulfide bonds: Cys^7-Cys^{23}

Neosystem SC010   Rat   MW 2549.81   Disulfide bonds: Cys^7-Cys^{23}; bioactive; Geller, DM et al, *BBRC*, 120:333-338, 1984

Sigma A 2413   Rat   FW 2549.8   ≥90% (HPLC) | Bioactive peptide; increases guanylate cyclase activity & cGMP levels in endothelial cells in culture; delays outward K$^+$ current in cardiac cells

---

Ser-Ser-Cys-Phe-Gly-Gly-Arg-Ile-Asp-Arg-Ile-Gly-Ala-Gln-Ser-Gly-Leu-Gly-Cys-Asn-Ser-Phe-Arg

Synonyms: Natriuretic Peptide (5-27), Atrial; Atriopeptin II; Guanylate Cyclase (Particulate) Activator

Sigma A 9035   Rat   FW 2386.6   ≥97% (HPLC) | Bioactive peptide; activates particulate guanylate cyclase; cardiac atrial peptide which relaxes both intestinal & vascular smooth muscle; Currie, MG et al, *Science*, 223:67, 1984

Ser-Ser-Cys-Phe-Gly-Gly-Arg-Ile-Asp-Arg-Ile-Gly-Ala-Gln-Ser-Gly-Leu-Gly-Cys-Asn-Ser

Synonyms: Natriuretic Peptide (5-25), Atrial; Atriopeptin I

Sigma A 9160 Rat FW 2083.3 ≥95% (HPLC) | Bioactive peptide; weak agonist at the cGMP coupled ANF receptor; cardiac atrial peptide which selectively relaxes intestinal but not vascular smooth muscle; Currie, MG et al, *Science*, 223:67, 1984

Neosystem SC410 Rat, rabbit, mouse MW 2083.28
Disulfide bonds: Cys^3-Cys^{19}; bioactive; ANF-related peptide; Currie, MG et al, *Science*, 223:67, 1984

Ser-Ser-Cys-Phe-Gly-Gly-Arg-Met-Asp-Arg-Ile-Gly-Ala-Gln-Ser-Gly-Leu-Gly-Cys-Asn-Ser-Phe-Arg-Tyr

Synonyms: Natriuretic Peptide (5-28), Atrial

American Peptide 14-1-21 Human MW 2567.9
$C_{106}H_{163}N_{35}O_{34}S_3$ Disulfide bonds: Cys^7-Cys^{23} | Truncated form of ANF found to exist in the brain & thymic macrophages; Chino, N et al, *Peptide Chemistry*, 1984:241, 1985

Ser-Ser-Cys-Phe-Gly-Gly-Arg-Met-Asp-Arg-Ile-Gly-Ala-Gln-Ser-Gly-Leu-Gly-Cys-Asn-Ser-Phe-Arg

Synonyms: Natriuretic Peptide (5-27), Atrial

American Peptide 14-1-40 Human MW 2404.7
$C_{97}H_{154}N_{354}O_{32}S_3$ Disulfide bonds: Cys^7-Cys^{23}

Ser-Ser-Cys-Phe-Gly-Gly-Arg-Met-Asp-Arg-Ile-Gly-Ala-Gln-Ser-Gly-Leu-Gly-Cys-Asn-Ser-Phe-Arg-Tyr

Synonyms: Natriuretic Peptide (5-28), Atrial; hANF (5-28)

Calbiochem 05-23-0302 Human MW 2567.9
$C_{106}H_{163}N_{35}O_{34}S_3$ ≥98% (HPLC); lyophilized; soluble in 5% HOAc; Disulfide bonds: :Cys^7-Cys^{23} | Truncated form of ANF that exists in the brain; also found to occur in thymic macrophages; Throsby, M et al, *Endocrinology*, 132:2184, 1993; *Merck Index*, 12:903

Ser-Ser-Cys-Phe-Gly-Gly-Arg-Met-Asp-Arg-Ile-Gly-Ala-Gln-Ser-Gly-Leu-Gly-Cys-Asn-Ser-Phe-Arg

Synonyms: Natriuretic Peptide (5-27), Atrial

ICN 151436 Human Disulfide bonds: Cys^7-Cys^{23}

Ser-Ser-Cys-Phe-Gly-Gly-Arg-Met-Asp-Arg-Ile-Gly-Ala-Gln-Ser-Gly-Leu-Gly-Cys-Asn-Ser-Phe-Arg-Tyr

Synonyms: Natriuretic Peptide (5-28), Atrial

ICN 151437 Human Disulfide bonds: Cys^7-Cys^{23}

Peptides International PAF-4137-v Human synthetic MW 2567.9 $C_{106}H_{163}N_{35}O_{34}S_3$ >95% (HPLC); lyophilized amorphous powder | Bioactive; Disulfide bonds: Cys^7-Cys^{23}; Ueda, S et al, *BBRC*, 149:1055, 1987; Watanabe, TX et al, *Eur J Pharmacol*, 147:49, 1988

Ser-Ser-Cys-Phe-Gly-Gly-Arg-Met-Asp-Arg-Ile-Gly-Ala-Gln-Ser-Gly-Leu-Gly-Cys-Asn-Ser-Phe-Arg

Synonyms: Natriuretic Peptide (5-27), Atrial

Peptides International PAF-4138-v Human synthetic MW 2404.7 $C_{97}H_{154}N_{34}O_{32}S_3$ >95% (HPLC); lyophilized amorphous powder | Bioactive; Disulfide bonds: Cys^7-Cys^{23}; Watanabe, TX et al, *Eur J Pharmacol*, 147:49, 1988

Ser-Ser-Cys-Tyr-Gly-Gly-Arg-Ile-Asp-Arg-Ile-Gly-Ala-Gln-Ser-Gly-Leu-Gly-Cys-Asn-Ser-Phe-Arg
H_2N-Ser-Ser-Cys-Tyr-Gly-Gly-Arg-Ile-Asp-Arg-Ile-Gly-Ala-Gln-Ser-Gly-Leu-Gly-Cys-Asn-Ser-Phe-Arg

Synonyms: Natriuretic Peptide (5-27), Atrial (Tyr^8)-

ICN 154504 Rat MW 2403.0 Budzik, GP et al, *BBRC*, 144:422, 1987

Ser-Ser-Glu-Gly-Glu-Ser-Pro-Asp-Phe-Pro-Glu-Glu-Leu-Glu-Lys

Synonyms: Peptide Histidine Methionine (156-170); Prepro-Vasoactive Intestinal Peptide (156-170)

American Peptide 48-1-31 MW 1679.9 $C_{71}H_{106}N_{16}O_{31}$

Bachem H-9190.0001 Human MW 1679.71 $C_{71}H_{106}N_{16}O_{31}$ Store at -15°C

Ser-Ser-Glu-Val-Ala-Gly-Glu-Gly-Asp-Gly-Asp-Ser-Met-Gly-His-Glu-Asp-Leu-Tyr

Synonyms: Prepro-Enkephalin B (186-204)

American Peptide 26-1-12 Human MW 1955.0
$C_{78}H_{115}N_{21}O_{36}S$

Ser-Ser-Ser

Synonyms: Tri-L-Serine

Bachem H-4980.0050 MW 279.25 $C_9H_{17}N_3O_7$ Store at -15°C

ICN 156617 MW 279.2 $C_9H_{17}N_3O_7$

Sigma S 4517 FW 279.2 $C_9H_{17}N_3O_7$

Ser-Thr
Z-Ser-Thr-OMe

Bachem C-3410.0001 MW 354.36 $C_{16}H_{22}N_2O_7$ Store at -15°C

Ser-Thr-Ala-Pro-Leu-Pro-Trp-Pro-Trp-Ser-Pro-Ala-Ala-Leu-Arg-Leu-Leu-Gln-Arg-Pro-Pro-Glu-Glu-Pro-Ala-Val-His-Ala-Asp-Cys-His-Arg

Synonyms: Inhibin α-Subunit (1-32)

Sigma I 8641 Porcine FW 3600.1 ≥90% (HPLC); peptide content:~60% | Bioactive peptide; inhibin fragment used as an immunogen for raising antibodies against inhibin; Mason, AJ et al, *Nature*, 318:659, 1985

Ser-Thr-Asp-Asn-Glu-Ser-Arg-Cys-Tyr-Met-Asn-His-Cys-Asn-Thr-Ser-Val-Ile-Gln-Glu
Ser-Thr-Asp-Asn-Glu-Ser-Arg-Cys(Acm)-Tyr-Met-Asn-His-Cys(Acm)-Asn-Thr-Ser-Val-Ile-Gln-Glu

Synonyms: GP140 (201-220); SIVmac251 Peptide

Neosystem SC728 MW 2473.64 SIVmac251 peptide from subtype GP140

Ser-Thr-Asp-Leu-Val-Cys-Glu-Gln-Gly-Asn-Ser-Thr-Asp-Asn-Glu-Ser-Arg-Cys-Tyr-Met
Ser-Thr-Asp-Leu-Val-Cys(Acm)-Glu-Gln-Gly-Asn-Ser-Thr-Asp-Asn-Glu-Ser-Arg-Cys(Acm)-Tyr-Met

Synonyms: GP140 (191-210); SIVmac251 Peptide

Neosystem SC727 MW 2394.53 SIVmac251 peptide from subtype GP140

Ser-Thr-Gly-Cys-Arg-Phe-Gly-Thr-Cys-Thr-Met-Gln-Lys-Leu-Ala-His-Gln-Ile-Tyr-Gln-Phe-Thr-Asp-Lys-Asp-Lys-Asp-Gly-Met-Ala-Pro-Arg-Asn-Lys-Ile-Ser-Pro-Gln-Gly-Tyr Amide

Synonyms: Adrenomedullin (11-50)

American Peptide 22-8-15 Rat MW 4521.2
$C_{194}H_{304}N_{58}O_{59}S_4$ Disulfide bonds: Cys^4-Cys^9

Ser-Thr-Gly-Cys-Arg-Phe-Gly-Thr-Cys-Thr-Met-Gln-Lys-Leu-Ala-His-Gln-Ile-Tyr-Gln-Phe-Thr-Asp-Lys-Asp-Lys-Asp-Gly-Met-Ala-Pro-Arg-Asn-Lys-Ile-Ser-Pro-Gln-Gly-Tyr Amide Trifluoroacetate salt

Synonyms: Adrenomedullin (11-50)

Calbiochem 121706   Rat   MW 4532 $C_{194}H_{306}N_{58}O_{59}S_4$
White solid; soluble in H_2O | Novel hypotensive peptide; the C-terminal fragment of rat adrenomedullin that exhibits hypotensive properties similar to adrenomedullin; Berthiaume, N et al, *Can J Physiol Pharmacol*, 73:1080, 1995; Sakata, J et al, *Biochem Biophys Res Comm*, 195:921, 1993; Lin, B et al, *Eur J Pharmacol*, 260:1, 1994

Ser-Thr-Gly-Cys-Arg-Phe-Gly-Thr-Cys-Thr-Met-Gln-Lys-Leu-Ala-His-Gln-Ile-Tyr-Gln-Phe-Thr-Asp-Lys-Asp-Lys-Asp-Gly-Met-Ala-Pro-Arg-Asn-Lys-Ile-Ser-Pro-Gln-Gly-Tyr Amide

Synonyms: Adrenomedullin (11-50)

ICN 195613   Rat   MW ~4521

Ser-Thr-Gly-Ser-Lys-Gln-Arg-Ser-Gln-Asn-Arg-Ser-Lys-Thr-Cys

Synonyms: Bone Morphogenetic Protein 7 (1-14), ~C

Biogenesis 1406-1290   Human synthetic   Semi-pure; lyophilized | Represents AA 1-14 of human BMP-7; an additional Cys has been conjugated to the C-terminus

Ser-Thr-Phe-Trp-Ala-Tyr-Gln-Pro-Asp-Gly-Asp-Asn-Asp-Pro-Thr-Asp-Tyr-Gln-Lys-Tyr-Glu-His-Thr-Ser-Ser-Pro-Ser-Gln-Leu-Leu-Ala-Pro-Gly-Asp-Tyr

Synonyms: Luminal Cholecystokinin Releasing Factor (1-35)

Neosystem SC1259   MW 3995.14   Bioactive; hypothalamic releasing hormone; CRF/urocortin peptide; this amino-terminal region displays the biological activity of native LCRF; may be an important tool for investigating the regulation of intestinal CCK release & the actions of endogenous intestinal CCK; Spannagel, AW et al, *PNAS USA*, 93:4415-4420, 1996; Spannagel, AW et al, *Am J Physiol* (Gastrointest Liver Physiol), 36:G754-G758, 1997; Miyasaka, K et al, *Pancreas*, 15:310-313, 1997

Ser-Thr-Pro-Leu-Met-Ser-Trp-Pro-Trp-Ser-Pro-Ser-Ala-Leu-Arg-Leu-Leu-Gln-Arg-Pro-Pro-Glu-Glu-Pro-Ala-Ala-His-Ala-Asn-Cys-His-Arg

Synonyms: Inhibin α-Subunit (1-32)

Sigma I 8516   Human   FW 3637.1   ≥90% (HPLC); peptide content:~70% | Bioactive peptide; inhibin fragment used as an immunogen for raising antibodies against inhibin; Mayo, KE et al, *Proc Natl Acad Sci USA*, 83:5849, 1986

Ser-Thr-Pro-Leu-Met-Ser-Trp-Pro-Trp-Ser-Pro-Ser-Ala-Leu-Arg-Leu-Leu-Gln-Arg-Pro-Pro-Glu-Glu-Pro-Gly-Tyr

Synonyms: Inhibin-Gly-Tyr (1-24), α-

Biogenesis 5323-1309   Human synthetic   MW 2996.5
Lyophilized, consisting of 99% peptide material, acetate counter ions and residual H_2O; purified | To avoid Met oxidation, it is best to dissolve the peptide under a nitrogen stream

Ser-Trp-Leu-Ser-Lys-Thr-Ala-Lys-Lys-Leu-Glu-Asn-Ser-Ala-Lys-Lys-Arg-Ile-Ser-Glu-Gly-Ile-Ala-Ile-Ala-Ile-Gln-Gly-Gly-Pro-Arg

Synonyms: Cecropin P_1

American Peptide 87-9-61   Porcine   MW 3339.4
$C_{147}H_{253}N_{45}O_{43}$ Antibacterial peptide; Lee, JY et al, *PNAS*, 86:9159, 1989

Sigma C 7927   Porcine   FW 3338.9   ≥97% (HPLC) |
Bioactive peptide

Ser-Tyr

Bachem G-3240.0250   MW 268.27 $C_{12}H_{16}N_2O_5$ Store at -15°C

Ser-Tyr
Ser-Tyr-AMC

Bachem I-1475.0050   Store at -15°C

Ser-Tyr
Ser-Tyr-βNA

Bachem K-1505.0050   MW 393.44 $C_{22}H_{23}N_3O_4$ Store at -15°C

Ser-Tyr

ICN 156618   MW 268.3 $C_{12}H_{16}N_2O_5$

Ser-Tyr
Ser-Tyr-4-Methoxy-β-Naphthylamide

ICN 156619   MW 423.5 $C_{23}H_{25}N_3O_5$

Ser-Tyr
Ser-Tyr β-Naphthylamide

ICN 156620   MW 393.4 $C_{22}H_{23}N_3O_4$ Off-white powder

Ser-Tyr
Ser-Tyr-β-Naphthylamide

Synonyms: Cathepsin C Substrate

Sigma S 5628   FW 393.4 $C_{22}H_{23}N_3O_4$ Off-white powder | McDonald, JK et al, *J Biol Chem*, 241:1494, 1966; McGuire, MJ et al, *Arch Biochem Biophys*, 295:280, 1992

Ser-Tyr

Sigma S 8258   FW 268.3 $C_{12}H_{16}N_2O_5$

Ser-Tyr-Ala-Gly-Ala-Val-Val-Asn-Asp-Leu

Synonyms: HSV H2 (6-15)

Neosystem SC159   MW 1008.09   Virus-related peptide

Ser-Tyr-Arg-Gly-Gly-Thr-Lys-Ile-Glu-Pro-Asn-Lys-Lys-Ala
(Ser-Tyr-Arg-Gly-Gly-Thr-Lys-Ile-Glu-Pro-Asn-Lys-Lys-Ala)₈-MAP

Bachem H-7605.0001   Store at -15°C

Ser-Tyr-Gly

Synonyms: Luteinizing Hormone Releasing Hormone

Biogenesis 5730-1409   Synthetic   MW 325.3   Lyophilized, consisting of 99% peptide material, acetate ions and residual H_2O; purified

Ser-Tyr-Gly-Arg-Lys-Lys-Arg-Arg-Gln-Arg-Arg-Arg-Pro-Pro-Gln

Synonyms: TAT (46-60); HTLV I Peptide

Neosystem SC209   MW 1969.28   HTLV I peptide from subtype TAT

Ser-Tyr-Gly-Leu-Arg-Hyp-Gly Amide

Synonyms: Luteinizing Hormone Releasing Hormone (4-10), (Hyp⁹)-

Neosystem SC473   MW 763.84   Bioactive; hypothalamic releasing hormone

Ser-Tyr-Gly-Leu-Arg-Pro-Gly Amide

Synonyms: Luteinizing Hormone Releasing Hormone (4-10)

American Peptide 54-1-13   MW 748.0 $C_{33}H_{53}N_{11}O_9$

Bachem H-3728.0005   MW 747.85 $C_{33}H_{53}N_{11}O_9$ Store at -15°C

Sigma L 5387   FW 747.9 $C_{33}H_{53}N_{11}O_9$ ≥97% (HPLC) |
Bioactive peptide

Ser-Tyr-His-Leu-Arg-Pro
Ser-Tyr-D-His(Bzl)-Leu-Arg-Pro-NHEt

Synonyms: Luteinizing Hormone Releasing Hormone (4-9), (D-His(Bzl)[6],Pro-NHEt[9])-

Bachem H-4802.0250   MW 889.07 $C_{44}H_{64}N_{12}O_8$ Store at -15°C

Ser-Tyr-Leu-Gln-Asp-Ser-Val-Pro-Asp-Ser-Phe-Gln-Asp

Bachem H-4416.0005   MW 1500.54 $C_{65}H_{93}N_{15}O_{26}$ Store at -15°C

Ser-Tyr-Leu-Leu-Arg-Pro
Ser-Tyr-D-Leu-Leu-Arg-Pro-NHEt

Synonyms: Luteinizing Hormone Releasing Hormone (4-9), (D-Leu[6],Pro-NHEt[9])-

Bachem H-4008.0005   Store at -15°C

Ser-Tyr-Leu-Lys-Lys-Leu-Cys-Gly-Thr-Val-Leu-Gly-Gly-Pro-Lys Amide

Synonyms: Pro-Cathepsin B (36-50); Cathepsin B Inhibitor

Bachem H-3938.0001   Rat   Store at -15°C

Sigma P 0475   Rat   FW 1562.9   Bioactive peptide; inhibitor of both human & rat cathepsin B; Chages, JR et al, *FEBS Lett*, 392:233, 1996

Ser-Tyr-Lys

Bachem H-4985.0250   Store at -15°C

Ser-Tyr-Pro-Thr-Gly-Pro-Gly-Thr-Ala-Asn-Gln-Arg-Arg-Gln-Arg

Synonyms: REV (26-40); SIVmac251 Peptide

Neosystem SC599   MW 1688.82   SIVmac251 peptide from subtype REV

Ser-Tyr-Ser-Met

Synonyms: Corticotropin A; Adrenocorticotropic Hormone (1-4)

Bachem H-1125.0005   MW 486.55 $C_{20}H_{30}N_4O_8S$ Store at -15°C

ICN 152715

Sigma A 2782   Human   FW 486.5 $C_{20}H_{30}N_4O_8S$ ≥97% (HPLC) | Bioactive peptide; lacks adrenocorticotropic & melanocyte-stimulating activities & that does not enhance avoidance learning

---

Ser-Tyr-Ser-Met-Glu-His-Phe-Arg-Trp-Gly

Synonyms: Corticotropin A; Adrenocorticotropic Hormone (1-10)

Bachem H-1130.0001   MW 1299.43 $C_{59}H_{78}N_{16}O_{16}S$ Store at -15°C

ICN 152716

American Peptide 10-1-34   Human   MW 1299.6 $C_{59}H_{78}N_{16}O_{16}S$

Neosystem SC019   Human   MW 1299.42   Bioactive; Li, ZG et al, *Endocrinology*, 125:592-596, 1989

Sigma A 1709   Human   FW 1299.4   ≥97% (HPLC) | Bioactive peptide; facilitates avoidance acquisition; Lacks adrenocorticotropic activity

---

Ser-Tyr-Ser-Met-Glu-His-Phe-Arg-Trp-Gly-Lys-Pro-Val
Ac-Ser-Tyr-Ser-Met-Glu-His-Phe-Arg-Trp-Gly-Lys-Pro-Val

Synonyms: Melanocyte Stimulating Hormone, α-

American Peptide 56-0-25   MW 1665.9 (free acid) $C_{77}H_{108}N_{20}O_{20}S$

Ser-Tyr-Ser-Met-Glu-His-Phe-Arg-Trp-Gly-Lys-Pro-Val
Ac-Ser-Tyr-Ser-Met-Glu-His-D-Phe-Arg-Trp-Gly-Lys-Pro-Val Amide

Synonyms: Melanocyte Stimulating Hormone, (D-Phe[7])-α-

American Peptide 56-0-27   MW 1664.8 $C_{77}H_{109}N_{21}O_{19}S$

Ser-Tyr-Ser-Met-Glu-His-Phe-Arg-Trp-Gly-Lys-Pro-Val
Ac-Ser-Tyr-Ser-Met-Glu-His-Phe-Arg-Trp-Gly-Lys-Pro-Val Amide

Synonyms: Melanocyte Stimulating Hormone, α-

American Peptide 56-1-11   MW 1664.9 $C_{77}H_{109}N_{21}O_{19}S$
Stimulates skin darkening in amphibians; also involved in control of learning, feeding & motivation in mammals; Harris, JI et al, *Nature*, 179:1346, 1957

Ser-Tyr-Ser-Met-Glu-His-Phe-Arg-Trp-Gly-Lys-Pro-Val
Ac-Ser-Tyr-Ser-Met-Glu-His-Phe-Arg-Trp-Gly-Lys-Pro-Val Free Acid

Synonyms: Melanocyte Stimulating Hormone, α-; Adrenocorticotropic Hormone (1-13), Ac-

Bachem H-1070.0001   MW 1665.89 $C_{77}H_{108}N_{20}O_{20}S$ Store at -15°C

Ser-Tyr-Ser-Met-Glu-His-Phe-Arg-Trp-Gly-Lys-Pro-Val
Ac-Ser-Tyr-Ser-Met-Glu-His-Phe-Arg-Trp-Gly-Lys-Pro-Val Amide

Synonyms: Melanocyte Stimulating Hormone, α-; Adrenocorticotropic Hormone (1-13), Ac-

Bachem H-1075.0001   MW 1664.91 $C_{77}H_{109}N_{21}O_{19}S$ Store at -15°C

Ser-Tyr-Ser-Met-Glu-His-Phe-Arg-Trp-Gly-Lys-Pro-Val

Synonyms: Adrenocorticotropic Hormone (1-13)

Bachem H-1135.0001   MW 1623.86 $C_{75}H_{106}N_{20}O_{19}S$ Store at -15°C

Ser-Tyr-Ser-Met-Glu-His-Phe-Arg-Trp-Gly-Lys-Pro-Val
Ac-Ser(Ac)-Tyr-Ser-Met-Glu-His-Phe-Arg-Trp-Gly-Lys-Pro-Val Amide

Synonyms: Melanocyte Stimulating Hormone, (Diacetyl)-α-

Bachem H-7080.0001   Store at -15°C

Ser-Tyr-Ser-Met-Glu-His-Phe-Arg-Trp-Gly-Lys-Pro-Val
Ac-Ser-Tyr-Ser-Met-Glu-His-Phe-Arg-Trp-Gly-Lys-Pro-Val Amide

Synonyms: Melanocyte Stimulating Hormone, α-; Melanotropin

Calbiochem 05-23-0751   MW 1664.9 $C_{77}H_{109}N_{21}O_{19}S$ | >95% (HPLC); lyophilized solid; soluble in 5% acetic acid | Peptide hormone secreted in the anterior pituitary gland; controls the intensity of pigmentation; potent inhibitor of ICAM-1 expression in malignant melanocytes; *Merck Index*, 12:6377; Morandini, R et al, *J Cell Physiol*, 175:276, 1998

Fluka 63605   MW 1664.9 $C_{77}H_{109}N_{21}O_{19}S$ ≥98.0% (HPLC); ≥80% peptide content | Sawyer, TK et al, *Am Zool*, 23:529, 1983

Ser-Tyr-Ser-Met-Glu-His-Phe-Arg-Trp-Gly-Lys-Pro-Val
N-Ac-Ser-Tyr-Ser-Met-Glu-His-Phe-Arg-Trp-Gly-Lys-Pro-Val Amide

Synonyms: Melanocyte Stimulating Hormone, α-

ICN 151594   MW 1664.9   Harris, JJ etal, *Nature*, 179:1346, 1957; Blake, J & CH Li, *Int J Protein Res*, 111:185, 1971

Ser-Tyr-Ser-Met-Glu-His-Phe-Arg-Trp-Gly-Lys-Pro-Val
(Ac)₂-Ser-Tyr-Ser-Met-Glu-His-Phe-Arg-Trp-Gly-Lys-Pro-Val Amide

Synonyms: Melanocyte Stimulating Hormone, (Diacetyl)-α-

ICN 152924   MW 1706.9

Ser-Tyr-Ser-Met-Glu-His-Phe-Arg-Trp-Gly-Lys-Pro-Val

Synonyms: Adrenocorticotropic Hormone (1-13); Corticotropin A

ICN 159036

Ser-Tyr-Ser-Met-Glu-His-Phe-Arg-Trp-Gly-Lys-Pro-Val
Ac-Ser-Tyr-Ser-Met-Glu-His-Phe-Arg-Trp-Gly-Lys-Pro-Val Amide

Synonyms: Melanocyte Stimulating Hormone, α-

Neosystem SC004   MW 1664.90   Bioactive; melanocyte stimulating hormone-related peptide

---

Ser-Tyr-Ser-Met-Glu-His-Phe-Arg-Trp-Gly-Lys-Pro-Val
Ac-Ser-Tyr-Ser-Met-Glu-His-Phe-Arg-Trp-Gly-Lys-Pro-Val Free Acid

Synonyms: Melanocyte Stimulating Hormone, α-

Neosystem SC005   MW 1665.88   Bioactive; melanocyte stimulating hormone-related peptide

---

Ser-Tyr-Ser-Met-Glu-His-Phe-Arg-Trp-Gly-Lys-Pro-Val
N-Ac-Ser-Tyr-Ser-Met-Glu-His-Phe-Arg-Trp-Gly-Lys-Pro-Val Amide

Synonyms: Melanocyte Stimulating Hormone, α-

Sigma M 4135   FW 1664.9   ≥97% (HPLC) | Bioactive peptide; stimulates melanogenesis; facilitates learning & memory; affects inflammatory & immune responses & peripheral nerve regeneration; derived from ACTH (1-13); may have important physiological roles in the control of vertebrate pigment cell melanogenesis, neural functioning related to learning & behavior & fetal development; Sawyer, TK et al, *Proc Natl Acad Sci USA*, 79:1751, 1982

Ser-Tyr-Ser-Met-Glu-His-Phe-Arg-Trp-Gly-Lys-Pro-Val
N-Ac-Ser-Tyr-Ser-Met-Glu-His-Phe-Arg-Trp-Gly-Lys-Pro-Val

Synonyms: Melanocyte Stimulating Hormone, (Val¹³)-α-

Sigma M 7892   FW 1665.9 (free acid)   ≥97% (HPLC) | Bioactive peptide

---

Ser-Tyr-Ser-Met-Glu-His-Phe-Arg-Trp-Gly-Lys-Pro-Val
Ac-Ser(Ac)-Tyr-Ser-Met-Glu-His-Phe-Arg-Trp-Gly-Lys-Pro-Val Amide

Synonyms: Melanocyte Stimulating Hormone, (Diacetyl)-α-

Sigma M 7896   FW 1706.9   ≥97% (HPLC) | Bioactive peptide

Ser-Tyr-Ser-Met-Glu-His-Phe-Arg-Trp-Gly-Lys-Pro-Val

Synonyms: Adrenocorticotropic Hormone (1-13)

American Peptide 10-1-28   Human   MW 1623.9 $C_{75}H_{106}N_{20}O_{19}S$

Neosystem SC053   Human   MW 1623.85   Bioactive; melanocyte stimulating hormone-related peptide

Sigma A 5555   Human   FW 1623.8   ≥97% (HPLC) | Bioactive peptide; minimum sequence required for weak adrenocorticotropic activity

---

Ser-Tyr-Ser-Met-Glu-His-Phe-Arg-Trp-Gly-Lys-Pro-Val
Ac-Ser-Tyr-Ser-Met-Glu-His-Phe-Arg-Trp-Gly-Lys-Pro-Val Amide

Synonyms: Melanocyte Stimulating Hormone, α-

Biogenesis 6045-0504   Synthetic   MW 1665   98.5% (HPLC), >99% (TLC); lyophilized, consisting of 87.7% peptide material, acetate counter ions and residual H_2O | To avoid Met oxidation, it is best to dissolve the peptide under a nitrogen stream

Peptides International PMR-4057-v   Synthetic   MW 1664.9 $C_{77}H_{109}N_{21}O_{19}S$ >95% (HPLC); lyophilized amorphous powder | Bioactive; human, porcine, bovine, rat, mouse; Lee, TH & AB Lerner, *JBC*, 22:943, 1956; Biossonas, RA et al, *Helv Chim Acta*, 41:1867, 1958; Schwyzer, R et al, *Helv Chim Acta*, 46:870, 1963

Ser-Tyr-Ser-Met-Glu-His-Phe-Arg-Trp-Gly-Lys-Pro-Val Acetate Salt

Synonyms: Melanocyte Stimulating Hormone, (α-des-Ac)-

Biogenesis 6045-0709   Synthetic   MW 1623   Lyophilized, consisting of 83.5% peptide material; purified | To avoid Met oxidation, it is best to dissolve the peptide under a nitrogen stream

---

Ser-Tyr-Ser-Met-Glu-His-Phe-Arg-Trp-Gly-Lys-Pro-Val Amide

Synonyms: Adrenocorticotropic Hormone (1-13); Melanocyte Stimulating Hormone, (des-Ac)-α-

American Peptide 56-0-24   MW 1622.9 $C_{75}H_{107}N_{21}O_{18}S$

Bachem H-4390.0001   MW 1622.87 $C_{75}H_{107}N_{21}O_{18}S$ Store at -15°C

ICN 152923   MW 1622.9

Neosystem SC006   MW 1622.86   Bioactive; melanocyte stimulating hormone-related peptide

Sigma M 8267   FW 1622.9   ≥97% (HPLC) | Bioactive peptide; retains ~10% of the activity of α-MSH

Ser-Tyr-Ser-Met-Glu-His-Phe-Arg-Trp-Gly-Lys-Pro-Val-Gly
Ac-Ser-Tyr-Ser-Met-Glu-His-Phe-Arg-Trp-Gly-Lys-Pro-Val-Gly

Synonyms: Adrenocorticotropic Hormone (1-14), Ac-

Bachem H-1085.0001   MW 1722.95 $C_{79}H_{111}N_{21}O_{21}S$ Store at -15°C

Ser-Tyr-Ser-Met-Glu-His-Phe-Arg-Trp-Gly-Lys-Pro-Val-Gly

Synonyms: Corticotropin A; Adrenocorticotropic Hormone (1-14)

Bachem H-1140.0001   MW 1680.91 $C_{77}H_{109}N_{21}O_{20}S$ Store at -15°C

ICN 152717

Sigma A 8804   Human   FW 1680.9   ≥95% (HPLC) | Bioactive peptide; adrenocorticotropic activity & weak melanocyte stimulating activity

---

Ser-Tyr-Ser-Met-Glu-His-Phe-Arg-Trp-Gly-Lys-Pro-Val-Gly-Lys-D-Lys-Arg-Arg-Pro-Val-Lys-Val-Tyr-Pro

*Synony Adrenocorticotropic Hormone (1-24), (D-Lys¹⁶)-*ms*:

Bachem H-4996.0500   Human   MW 2933.48 $C_{136}H_{210}N_{40}O_{31}S$ Store at -15°C

Ser-Tyr-Ser-Met-Glu-His-Phe-Arg-Trp-Gly-Lys-Pro-Val-Gly-Lys-Lys

Synonyms: Corticotropin A; Adrenocorticotropic Hormone (1-16)

Bachem H-6050.0001   MW 1937.26 $C_{89}H_{133}N_{25}O_{22}S$ Store at -15°C

ICN 159037

American Peptide 10-1-26   Human   MW 1937.3 $C_{89}H_{133}N_{25}O_{22}S$

Sigma A 8929   Human   FW 1937.2   ≥97% (HPLC) | Bioactive peptide; weak adrenocorticotropic activity

---

Ser-Tyr-Ser-Met-Glu-His-Phe-Arg-Trp-Gly-Lys-Pro-Val-Gly-Lys-Lys-Arg-Arg-Pro-Val-Lys-Val-Tyr-Pro-Asn-Gly-Ala-Glu-Asp-Glu-Ser-Ala-Glu-Ala-Phe-Pro-Leu-Glu-Phe

Synonyms: Adrenocorticotropic Hormone, (3-(^{125}I)-Tyr^2)-

Amersham IM183   Freeze-dried; ~74 TBq/mmol; ~2000 Ci/mmol | ACTH is a 39 AA fragment peptide hormone

Amersham IM216   Freeze-dried; ~74 TBq/mmol; ~2000 Ci/mmol

Ser-Tyr-Ser-Met-Glu-His-Phe-Arg-Trp-Gly-Lys-Pro-Val-Gly-Lys-Lys-Arg-Arg-Pro-Val-Lys-Val-Tyr-Pro

Synonyms: Adrenocorticotropic Hormone (1-24), (3-(^{125}I)-Phe^2,Nle^4)-

Amersham IM312   Freeze-dried; ~74 TBq/mmol; ~2000 Ci/mmol

Ser-Tyr-Ser-Met-Glu-His-Phe-Arg-Trp-Gly-Lys-Pro-Val-Gly-Lys-Lys-Arg-Arg-Pro-Val-Lys-Val-Tyr-Pro-Asn-Gly-Ala-Glu-Asp-Glu-Ser-Ala-Glu-Ala-Phe-Pro-Leu-Glu-Phe

Synonyms: Adrenocorticotropic Hormone, Fluorescein

Amersham VACT011   Fluorescein-labeled with high activity

Ser-Tyr-Ser-Met-Glu-His-Phe-Arg-Trp-Gly-Lys-Pro-Val-Gly-Lys-Lys-Arg
Ac-Ser-Tyr-Ser-Met-Glu-His-Phe-Arg-Trp-Gly-Lys-Pro-Val-Gly-Lys-Lys-Arg

Synonyms: Adrenocorticotropic Hormone (1-17), Ac-

Bachem H-1090.0001   MW 2135.48 $C_{97}H_{147}N_{29}O_{24}S$ Store at -15°C

Ser-Tyr-Ser-Met-Glu-His-Phe-Arg-Trp-Gly-Lys-Pro-Val-Gly-Lys-Lys-Arg

Synonyms: Adrenocorticotropic Hormone (1-17)

Bachem H-1145.0001   MW 2093.44 $C_{95}H_{145}N_{29}O_{23}S$ Store at -15°C

Ser-Tyr-Ser-Met-Glu-His-Phe-Arg-Trp-Gly-Lys-Pro-Val-Gly-Lys-Lys-Arg-Arg-Pro-Val-Lys-Val-Tyr-Pro

Synonyms: Adrenocorticotropic Hormone (1-24); Tetracosactide

Bachem H-1150.0001   MW 2933.48 $C_{136}H_{210}N_{40}O_{31}S$ Store at -15°C

Ser-Tyr-Ser-Met-Glu-His-Phe-Arg-Trp-Gly-Lys-Pro-Val-Gly-Lys-Lys-Arg

Synonyms: Adrenocorticotropic Hormone (1-17); Corticotropin A

ICN 159038

Ser-Tyr-Ser-Met-Glu-His-Phe-Arg-Trp-Gly-Lys-Pro-Val-Gly-Lys-Lys-Arg
Ac-Ser-Tyr-Ser-Met-Glu-His-Phe-Arg-Trp-Gly-Lys-Pro-Val-Gly-Lys-Lys-Arg

Synonyms: Adrenocorticotropic Hormone (1-17), *N*-Ac-

ICN 159851   MW 2135.7   >98%

Ser-Tyr-Ser-Met-Glu-His-Phe-Arg-Trp-Gly-Lys-Pro-Val-Gly-Lys-Lys-Arg-Arg-Pro-Val-Lys-Val-Tyr-Ala-Asn-Gly-Ala-Glu-Glu-Glu-Ser-Ala-Glu-Ala-Phe-Pro-Leu-Glu-Phe

Synonyms: Adrenocorticotropic Hormone (1-39)

Bachem H-2894.0500   Guinea pig   MW 4529.12 $C_{206}H_{308}N_{56}O_{58}S$ Store at -15°C

Ser-Tyr-Ser-Met-Glu-His-Phe-Arg-Trp-Gly-Lys-Pro-Val-Gly-Lys-Lys-Arg

Synonyms: Adrenocorticotropic Hormone (1-17)

American Peptide 10-1-58   Human   MW 2093.5 $C_{95}H_{145}N_{29}O_{23}S$

Ser-Tyr-Ser-Met-Glu-His-Phe-Arg-Trp-Gly-Lys-Pro-Val-Gly-Lys-Lys-Arg
Ac-Ser-Tyr-Ser-Met-Glu-His-Phe-Arg-Trp-Gly-Lys-Pro-Val-Gly-Lys-Lys-Arg

Synonyms: Adrenocorticotropic Hormone (1-17), *N*-Ac-

American Peptide 10-1-59   Human   MW 2135.5 $C_{97}H_{147}N_{29}O_{24}S$

Ser-Tyr-Ser-Met-Glu-His-Phe-Arg-Trp-Gly-Lys-Pro-Val-Gly-Lys-Lys-Arg-Arg-Pro-Val-Lys-Val-Tyr-Pro-Asn-Gly-Ala-Glu-Asp-Glu-Ser-Ala-Glu-Ala-Phe-Pro-Leu-Glu-Phe

Synonyms: Adrenocorticotropic Hormone (1-39); Corticotropin

Bachem H-1160.0500   Human   MW 4541.13 $C_{207}H_{308}N_{56}O_{58}S$ Store at -15°C

Ser-Tyr-Ser-Met-Glu-His-Phe-Arg-Trp-Gly-Lys-Pro-Val-Gly-Lys-Lys-Arg-Arg-Pro-Val-Lys-Val-Tyr-Pro
D-Ser-Tyr-Ser-Met-Glu-His-Phe-Arg-Trp-Gly-Lys-Pro-Val-Gly-Lys-Lys-Arg-Arg-Pro-Val-Lys-Val-Tyr-Pro

Synony Adrenocorticotropic Hormone (1-24), (D-Ser^1)-ms:

Bachem H-4718.0500   Human   MW 2933.48 $C_{136}H_{210}N_{40}O_{31}S$ Store at -15°C

Ser-Tyr-Ser-Met-Glu-His-Phe-Arg-Trp-Gly-Lys-Pro-Val-Gly-Lys-Lys-Arg-Arg-Pro-Val-Lys-Val-Tyr-Pro-Asn-Gly-Ala-Ser-Ala-Glu-Ala-Phe-Pro-Leu-Glu-Phe

Synonyms: Adrenocorticotropic Hormone (1-39); Corticotropin

Calbiochem 05-23-0574   Human   MW 4541.1 $C_{207}H_{308}N_{56}O_{58}S$ ≥97% (HPLC); lyophilized; soluble in 5% HOAc; harmful:LD$_{50}$≤2000 mg/kg; may be carcinogenic/teratogenic | *N*-terminal synthetic peptide fragment of pituitary hormone that stimulates the secretion of adrenal corticosteroids; *Merck Index*, 12:136

Ser-Tyr-Ser-Met-Glu-His-Phe-Arg-Trp-Gly-Lys-Pro-Val-Gly-Lys-Lys-Arg-Arg-Pro-Val-Lys-Val-Tyr-Pro-Asn-Gly-Ala-Glu-Asp-Glu-Ser-Ala-Glu-Ala-Phe-Pro-Leu-Glu-Phe

Synonyms: Adrenocorticotropic Hormone; Corticotropin

Fluka 02275   Human   MW 4541.1 $C_{207}H_{308}N_{56}O_{58}S$ ≥98% (HPLC) | Pituitary hormone that stimulates glucocorticoid production; Thermos, K & Reisine, T, *Ann NY Acad Sci*, 512:187, 1987; Labrie, F et al, *ibid*, 512:97, 1987; Jones, MT & Gillham, B, *Physiol Rev*, 68:743, 1988

Ser-Tyr-Ser-Met-Glu-His-Phe-Arg-Trp-Gly-Lys-Pro-Val-Gly-Lys-Lys-Arg-Arg-Pro-Val-Lys-Val-Tyr-Pro

Synonyms: Corticotropin A ; Adrenocorticotropic Hormone (1-24)

ICN 152718   Human   90-95%

Neosystem SC009   Human   MW 2933.46   Bioactive; Jacquet, YF, *Science*, 201, 1032, 1978; Florijn, WJ & Versteeg, DHG, *Brain Res*, 494:247-254, 1989

Ser-Tyr-Ser-Met-Glu-His-Phe-Arg-Trp-Gly-Lys-Pro-Val-Gly-Lys-Lys-Arg-Arg-Pro-Val-Lys-Val-Tyr-Pro-Asn-Gly-Ala-Glu-Asp-Glu-Ser-Ala-Glu-Ala-Phe-Pro-Leu-Glu-Phe

Synonyms: Adrenocorticotropic Hormone (1-39)

Neosystem SC038   Human   MW 4541.11   Bioactive; Schally, AV, *Science*, 202:18, 1978; Tranchand-Bunel, D et al, *Medicine/Sciences*, 3:128, 1987

Ser-Tyr-Ser-Met-Glu-His-Phe-Arg-Trp-Gly-Lys-Pro-Val-Gly-Lys-Lys-Arg

Synonyms: Adrenocorticotropic Hormone (1-17)

Neosystem SC317   Human   MW 2093.43   Bioactive

Ser-Tyr-Ser-Met-Glu-His-Phe-Arg-Trp-Gly-Lys-Pro-Val-Gly-Lys-Lys-Arg-Arg-Pro-Val-Lys-Val-Tyr-Pro

Synonyms: Adrenocorticotropic Hormone (1-24); Tetracosactide

Sigma A 0298   Human   FW 2933.5   ≥97% (HPLC); peptide content:~70% | Bioactive peptide; sequence that is conserved across species; ~75% of the potency of ACTH on a molar basis; activates G proteins; Jacquet, YF, *Science*, 201:1032, 1978; Kappeler, H & Schwyzer, R, *Helv Chim Acta*, 44:1136, 1961

Ser-Tyr-Ser-Met-Glu-His-Phe-Arg-Trp-Gly-Lys-Pro-Val-Gly-Lys-Lys-Arg-Arg-Pro-Val-Lys-Val-Tyr-Pro-Asn-Gly-Ala-Glu-Asp-Glu-Ser-Ala-Glu-Ala-Phe-Pro-Leu-Glu-Phe

Synonyms: Adrenocorticotropic Hormone

Sigma A 0423   Human   FW 4541.1   ≥97% (HPLC) | Bioactive peptide; Yamashiro, D & Li, C, *J Am Chem Soc*, 95:1310, 1973; Comments are the same as for Sigma A 6303

Ser-Tyr-Ser-Met-Glu-His-Phe-Arg-Trp-Gly-Lys-Pro-Val-Gly-Lys-Lys-Arg

Synonyms: Adrenocorticotropic Hormone (1-17)

Sigma A 2407   Human   FW 2093.4   ≥97% (HPLC) | Bioactive peptide; moderate adrenocorticotropic & melanocyte stimulating activities

---

Ser-Tyr-Ser-Met-Glu-His-Phe-Arg-Trp-Gly-Lys-Pro-Val-Gly-Lys-Lys-Arg-Arg-Pro-Val-Lys-Val-Tyr-Pro-Asn-Gly-Ala-Glu-Asp-Glu-Ser-Ala-Glu-Ala-Phe-Pro-Leu-Glu-Phe

Synonyms: Adrenocorticotropic Hormone (1-39)

American Peptide 10-1-10   Human synthetic   MW 4541.7 $C_{207}H_{308}N_{56}O_{58}$ May be carcinogenic/teratogenic | N-terminal synthetic fragment of pituitary hormone that stimulates the synthesis & release of glucocorticoids & mineralocorticoids in the adrenal cortex; *NY Ann Acad Sci*, 297:1-641, 1977; Yamashiro, D et al, *JACS*, 95:1310, 1973

Ser-Tyr-Ser-Met-Glu-His-Phe-Arg-Trp-Gly-Lys-Pro-Val-Gly-Lys-Lys-Arg-Arg-Pro-Val-Lys-Val-Tyr-Pro

Synonyms: Adrenocorticotropic Hormone (1-24)

American Peptide 10-1-21   Human synthetic   MW 2933.5 $C_{136}H_{210}N_{40}O_{31}S$ May be carcinogenic/teratogenic | N-terminal synthetic fragment of pituitary hormone that stimulates the synthesis & secretion of adrenal corticosteroids; more active than ACTH *in vitro* with regard to the corticosteroid releasing activity; Jacquet, YF et al, *Science*, 201:1032, 1978

Biogenesis 0178-0359   Human synthetic   MW 2933.9   Lyophilized, containing 75.1% peptide material, TFA counter ions and residual H_2O; purified | To avoid Met oxidation, it is best to dissolve the peptide under a nitrogen stream

---

Ser-Tyr-Ser-Met-Glu-His-Phe-Arg-Trp-Gly-Lys-Pro-Val-Gly-Lys-Lys-Arg-Arg-Pro-Val-Lys-Val-Tyr-Pro-Asn-Gly-Ala-Glu-Asp-Glu-Ser-Ala-Glu-Ala-Phe-Pro-Leu-Glu-Phe

Synonyms: Adrenocorticotropic Hormone (1-39)

Biogenesis 0178-0409   Human synthetic   MW 4540.7   Lyophilized, containing 89.8% peptide material, acetate counter ions and residual H_2O; purified

---

Ser-Tyr-Ser-Met-Glu-His-Phe-Arg-Trp-Gly-Lys-Pro-Val-Gly-Lys-Lys-Arg-Arg-Pro-Val-Lys-Val-Tyr-Pro

Synonyms: Adrenocorticotropic Hormone (1-24)

Calbiochem 05-23-0753   Human synthetic   MW 2933.5 $C_{136}H_{210}N_{40}O_{31}S$ ≥97% (HPLC); lyophilized solid; soluble in 5% acetic acid; harmful:LD_{50} ≤2000 mg/kg; may be carcinogenic/teratogenic | N-terminal fragment of pituitary hormone that stimulates secretion of adrenal corticosteroids & induces growth of adrenal cortex; Directly activates G-proteins; Zhu, J et al, *Eur J Pharmacol*, 268:279, 1994; *Merck Index*, 12:136

Peptides International PAC-4109-v   Human synthetic   MW 2933.5 $C_{136}H_{210}N_{40}O_{31}S$ >95% (HPLC); lyophilized amorphous powder | Bioactive; Riniker, B et al, *Nature (New Biol)*, 235:114, 1972

Ser-Tyr-Ser-Met-Glu-His-Phe-Arg-Trp-Gly-Lys-Pro-Val-Gly-Lys-Lys-Arg-Arg-Pro-Val-Lys-Val-Tyr-Pro-Asn-Val-Ala-Glu-Asn-Glu-Ser-Ala-Glu-Ala-Phe-Pro-Leu-Glu-Phe

Synonyms: Adrenocorticotropic Hormone (1-39)

American Peptide 10-1-15   Rat   MW 4582.5 $C_{210}H_{315}N_{57}O_{57}S$ Drouin, J, *Nature*, 288:610, 1980

Bachem H-4998.0500   Rat   MW 4582.23 $C_{210}H_{315}N_{57}O_{57}S$ Store at -15°C

Ser-Tyr-Ser-Met-Glu-His-Phe-Arg-Trp-Gly-Lys-Pro-Val-Gly-Lys-Lys-Arg-Arg-Pro-Val-Lys-Val-Tyr-Pro-Asn-Val-Ala-Glu-Asn-Glu-Ser-Ala-Glu-AA-Phe-Pro-Leu-Glu-Phe

Synonyms: Adrenocorticotropic Hormone (1-39); Corticotropin A

ICN 159850   Rat   MW 4582.5   >98%

Ser-Tyr-Ser-Met-Glu-His-Phe-Arg-Trp-Gly-Lys-Pro-Val-Gly-Lys-Lys-Arg-Arg-Pro-Val-Lys-Val-Tyr-Pro-Asn-Val-Ala-Glu-Asn-Glu-Ser-Ala-Glu-Ala-Phe-Pro-Leu-Glu-Phe

Synonyms: Adrenocorticotropic Hormone

Sigma A 7075   Rat   FW 4582.2   ≥95% (HPLC) | Bioactive peptide

---

Ser-Tyr-Ser-Nle-Glu-His-D-Phe-Arg-Trp-Gly-Lys-Pro-Val

Ac-Ser-Tyr-Ser-Nle-Glu-His-D-Phe-Arg-Trp-Gly-Lys-Pro-Val Amide

Synonyms: Melanocyte Stimulating Hormone, (Nle[4],D-Phe[7])-α-; Melanotropin

Calbiochem 05-23-0756   MW 1646.9 $C_{78}H_{111}N_{21}O_{19}$ >98% (HPLC); lyophilized solid; soluble in 5% acetic acid | Potent melanotropin with prolonged biological activity; Reiter, RJ, *BioEssays*, 14:169, 1992; Hadley, ME et al, *Science*, 213:1025, 1981; Sawyer, TK et al, *PNAS*, 77:5754, 1981

Ser-Tyr-Ser-Nle-Glu-His-Phe-Arg-Trp-Gly-Lys-Pro-Val

Ac-Ser-Tyr-Ser-Nle-Glu-His-Phe-Arg-Trp-Gly-Lys-Pro-Val Amide

Synonyms: Melanocyte Stimulating Hormone, (Nle[4])-α-

American Peptide 56-0-26   MW 1647.0 $C_{78}H_{111}N_{21}O_{19}$

Ser-Tyr-Ser-Nle-Glu-His-Phe-Arg-Trp-Gly-Lys-Pro-Val

Ac-Ser-Tyr-Ser-Nle-Glu-His-D-Phe-Arg-Trp-Gly-Lys-Pro-Val Amide

Synonyms: Melanocyte Stimulating Hormone, (Nle[4],D-Phe[7])-α-

American Peptide 56-0-28   MW 1647.0 $C_{78}H_{111}N_{21}O_{19}$
Analog is 26 times as potent as α-MSH in the adenylate cyclase assay; Sawyer, TK et al, *PNAS*, 77:5754, 1980

Ser-Tyr-Ser-Nle-Glu-His-Phe-Arg-Trp-Gly-Lys-Pro-Val

Ac-Ser-Tyr-Ser-Nle-Glu-His-Phe-Arg-Trp-Gly-Lys-Pro-Val Amide

Synonyms: Melanocyte Stimulating Hormone, (Nle[4])-α-

Bachem H-1095.0001   MW 1646.87 $C_{78}H_{111}N_{21}O_{19}$ Store at -15°C

Ser-Tyr-Ser-Nle-Glu-His-Phe-Arg-Trp-Gly-Lys-Pro-Val

Ac-Ser-Tyr-Ser-Nle-Glu-His-D-Phe-Arg-Trp-Gly-Lys-Pro-Val Amide

Synonyms: Melanocyte Stimulating Hormone, (Nle[4],D-Phe[7])-α-

Bachem H-1100.0001   MW 1646.87 $C_{78}H_{111}N_{21}O_{19}$ Store at -15°C

Ser-Tyr-Ser-Nle-Glu-His-Phe-Arg-Trp-Gly-Lys-Pro-Val

N-Ac-Ser-Tyr-Ser-Nle-Glu-His-D-Phe-Arg-Trp-Gly-Lys-Pro-Val Amide

Synonyms: Melanocyte Stimulating Hormone, (Nle[4],D-Phe[7])-α-

ICN 152925   26-fold more potent than α-MSH in the adenylate cyclase assay; Sawyer, T etal, *PNAS*, 77:5754, 1980

Ser-Tyr-Ser-Nle-Glu-His-Phe-Arg-Trp-Gly-Lys-Pro-Val

NDP-MSHAc-Ser-Tyr-Ser-Nle-Glu-His-D-Phe-Arg-Trp-Gly-Lys-Pro-Val Amide

Synonyms: Melanocyte Stimulating Hormone, (Nle[4],D-Phe[7])-α-

Neosystem SC919   MW 1646.86   Bioactive; melanocyte stimulating hormone-related peptide; highly potent α-melanotropin with ultralong biological activity & resistant to enzymatic degradation by serum enzymes; important molecular probe for studying melanotropin receptors; Sawyer, TK et al, *PNAS* USA, 77:5754-5758, 1980

Ser-Tyr-Ser-Nle-Glu-His-Phe-Arg-Trp-Gly-Lys-Pro-Val

N-Ac-Ser-Tyr-Ser-Nle-Glu-His-D-Phe-Arg-Trp-Gly-Lys-Pro-Val Amide Trifluoroacetate Salt

Synonyms: Melanocyte Stimulating Hormone, (Nle[4],D-Phe[7])-α-

Sigma M 8764   FW 1646.9   ≥95% (HPLC) | Bioactive peptide; 26 times as potent as α-MSH in the adenylate cyclase assay; Sawyer, TK et al, *Proc Natl Acad Sci USA*, 77(10):5754, 1980

Ser-Val

Bachem G-3250.0250   MW 204.23 $C_8H_{16}N_2O_4$ Store at -15°C

Ser-Val-Arg-Pro-Lys-Val-Pro-Leu-Arg-Ala-Met-Thr-Tyr-Lys-Leu

Synonyms: NEF (101-115); SIVmac251 Peptide

Neosystem SC612   MW 1759.18   SIVmac251 peptide from subtype NEF

Ser-Val-Pro-Ser-Ser-Ser-Ser-Thr-Pro-Leu-Leu-Tyr-Pro-Ser-Leu-Ala-Leu-Pro-Ala-Pro

Synonyms: GP46 (246-265); HTLV I Peptide

Neosystem SC829   MW 1984.27   HTLV I peptide

Ser-Val-Ser-Glu-Cys-Gln-Leu-Met-His-Asn-Leu-Gly-Lys-His-Leu-Asn-Ser-Met-Glu-Arg-Val-Glu-Trp-Leu-Arg-Lys-Lys-Cys-Gln-Asp-Val-His-Asn-Phe

Synonyms: Parathyroid Hormone (1-34), (Cys[5,28])-

American Peptide 22-1-28   Human   MW 4095.8
$C_{175}H_{277}N_{55}O_{51}S_4$ Disulfide bonds: Cys[5]-Cys[28]

Ser-Val-Ser-Glu-Ile-Gln-Leu-Asn-His-Asn-Leu-Gly-Lys-His-Leu-Asn-Ser-Leu-Glu-Arg-Val-Glu-Trp-Leu-Arg-Lys-Lys-Leu-Gln-Asp-Val-His-Asn-Phe

Synonyms: Parathyroid Hormone (1-34), (Asn[8],Leu[18])-

American Peptide 22-3-30   Human   MW 4082.7
$C_{181}H_{290}N_{56}O_{52}$

Ser-Val-Ser-Glu-Ile-Gln-Leu-Cys

Synonyms: Parathyroid Hormone (1-8), (Cys[8])-

Bachem H-4764.0001   Human   MW 878.01 $C_{36}H_{63}N_9O_{14}S$
Store at -15°C

Ser-Val-Ser-Glu-Ile-Gln-Leu-Met-His-Asn-Leu-Gly-Lys-His-Leu-Asn-Ser-Met-Glu-Arg-Val-Glu-Trp-Leu-Arg-Lys-Lys-Leu-Gln-Asp-Val-His-Asn-Phe-Val-Ala-Leu-Gly-Ala-Pro-Leu-Ala-Pro-Arg

Synonyms: Parathyroid Hormone (1-44)

American Peptide 22-1-22   Human   MW 5064.0
$C_{225}H_{366}N_{68}O_{61}S_2$ Kimura, T et al, *Biopolymers*, 20:1823, 1981

Ser-Val-Ser-Glu-Ile-Gln-Leu-Met-His-Asn-Leu-Gly-Lys-His-Leu-Asn-Ser-Met-Glu-Arg-Val-Glu-Trp-Leu-Arg-Lys-Lys-Leu-Gln-Asp-Val-His-Asn-Phe

Synonyms: Parathyroid Hormone (1-34)

American Peptide 22-1-25   Human   MW 4117.8
$C_{181}H_{291}N_{55}O_{51}S_2$ Takai, T et al, *Peptide Chem*, 1978:187, 1979; Tregear, GW et al, *Hoppe-Seyler's Z Physiol Chem*, 355:415, 1974

Ser-Val-Ser-Glu-Ile-Gln-Leu-Met-His-Asn-Leu-Gly-Lys-His-Leu-Asn-Ser-Met-Glu-Arg-Val-Glu-Trp-Leu-Arg-Lys-Lys-Leu-Gln-Asp-Val-His-Asn-Phe-Val-Ala-Leu-Gly

Synonyms: Parathyroid Hormone (1-38)

American Peptide 22-1-30   Human   MW 4458.2
$C_{197}H_{319}N_{59}O_{55}S_2$ Hesch, RD et al, *Horm Metab Res*, 16:559, 1984

Ser-Val-Ser-Glu-Ile-Gln-Leu-Met-His-Asn-Leu-Gly-Lys-His-Leu-Asn-Ser-Leu-Glu-Arg-Val-Glu-Trp-Leu-Arg-Lys-Lys-Leu-Gln-Asp-Val-His-Asn-Phe

Synonyms: Parathyroid Hormone (1-34), (Leu[18])-

American Peptide 22-3-31   Human   MW 4099.8
$C_{182}H_{293}N_{55}O_{51}S$

Ser-Val-Ser-Glu-Ile-Gln-Leu-Met-His-Asn-Leu-Gly-Lys-His-Leu-Asn-Ser-Met-Glu-Arg-Val-Glu-Trp-Leu-Arg-Lys-Lys-Leu-Gln-Asp-Val-His-Asn-Phe Amide

Synonyms: Parathyroid Hormone (1-34), (^{125}I)-

Amersham IM291   Human   Lyophilized; ~74 TBq/mmol, ~2000 Ci/mmol

---

Ser-Val-Ser-Glu-Ile-Gln-Leu-Met-His-Asn-Leu-Gly-
Lys-His-Leu-Asn-Ser-Met-Glu-Arg-Val-Glu-Trp-Leu-
Arg-Lys-Lys-Leu-Gln-Asp-Val-His-Asn-Phe-Val-Ala-
Leu-Gly-Ala-Pro-Leu-Ala-Pro-Arg-Asp-Ala-Gly-Ser-
Gln-Arg-Pro-Arg-Lys-Lys-Glu-Asp-Asn-Val-Leu-Val-
Glu-Ser-His-Glu-Lys-Ser-Leu-Gly-Glu-Ala-Asp-Lys-
Ala-Asp-Val-Asn-Val-Leu-Thr-Lys-Ala-Lys-Ser-Gln

Synonyms: Parathyroid Hormone (1-84)

Bachem H-1370.0100   Human   Store at -15°C

Ser-Val-Ser-Glu-Ile-Gln-Leu-Met-His-Asn-Leu-Gly-
Lys-His-Leu-Asn-Ser-Met-Glu-Arg-Val-Glu-Trp-Leu-
Arg-Lys-Lys-Leu-Gln-Asp-Val-His-Asn-Phe
Biotinyl-Ser-Val-Ser-Glu-Ile-Gln-Leu-Met-His-Asn-
Leu-Gly-Lys-His-Leu-Asn-Ser-Met-Glu-Arg-Val-Glu-
Trp-Leu-Arg-Lys-Lys-Leu-Gln-Asp-Val-His-Asn-Phe

Synonyms: Parathyroid Hormone (1-34), Biotinyl-

Bachem H-2150.0500   Human   Store at -15°C

Ser-Val-Ser-Glu-Ile-Gln-Leu-Met-His-Asn-Leu-Gly-
Lys-His-Leu-Asn-Ser-Met-Glu-Arg-Val-Glu-Trp-Leu-
Arg-Lys-Lys-Leu-Gln-Asp-Val

Synonyms: Parathyroid Hormone (1-31)

Bachem H-2274.0500   Human   Store at -15°C

Ser-Val-Ser-Glu-Ile-Gln-Leu-Met-His-Asn-Leu-Gly-
Lys-His-Leu-Asn-Ser-Met-Glu-Arg-Val-Glu-Trp-Leu-
Arg-Lys-Lys-Leu-Gln-Asp-Val Amide

Synonyms: Parathyroid Hormone (1-31)

Bachem H-3408.0500   Human   Store at -15°C

Ser-Val-Ser-Glu-Ile-Gln-Leu-Met-His-Asn-Leu-Gly-
Lys-His-Leu-Asn-Ser-Met-Glu-Arg-Val-Cyclo(-Glu-
Trp-Leu-Arg-Lys)-Leu-Leu-Gln-Asp-Val
Ser-Val-Ser-Glu-Ile-Gln-Leu-Met-His-Asn-Leu-Gly-
Lys-His-Leu-Asn-Ser-Met-Glu-Arg-Val-Cyclo(-γ-Glu-
Trp-Leu-Arg-ε-Lys)-Leu-Leu-Gln-Asp-Val Amide,
Cyclized γ-Glu^{22}-ε-Lys^{26}

Synonyms: Parathyroid Hormone (1-31), (Leu^{27})-

Bachem H-4058.0500   Human   Store at -15°C

Ser-Val-Ser-Glu-Ile-Gln-Leu-Met-His-Asn-Leu-Gly-
Lys-His-Leu-Asn-Ser-Met-Glu-Arg-Val-Glu-Trp-Leu-
Arg-Lys-Lys-Leu-Gln-Asp-Val-His-Asn-Phe-Val-Ala-
Leu-Gly-Ala-Pro-Leu-Ala-Pro-Arg

Synonyms: Parathyroid Hormone (1-44)

Bachem H-4825.0500   Human   MW 5063.93
$C_{225}H_{366}N_{68}O_{61}S_2$ Store at -15°C

Ser-Val-Ser-Glu-Ile-Gln-Leu-Met-His-Asn-Leu-Gly-
Lys-His-Leu-Asn-Ser-Met-Glu-Arg-Val-Glu-Trp-Leu-
Arg-Lys-Lys-Leu-Gln-Asp-Val-His-Asn-Phe-Val-Ala-
Leu-Gly

Synonyms: Parathyroid Hormone (1-38)

Bachem H-4830.0500   Human   Store at -15°C

Ser-Val-Ser-Glu-Ile-Gln-Leu-Met-His-Asn-Leu-Gly-
Lys-His-Leu-Asn-Ser-Met-Glu-Arg-Val-Glu-Trp-Leu-
Arg-Lys-Lys-Leu-Gln-Asp-Val-His-Asn-Phe

Synonyms: Parathyroid Hormone (1-34); Parathyroid Hormone,
Human; Parathormone

Bachem H-4835.0500   Human   MW 4117.77
$C_{181}H_{291}N_{55}O_{51}S_2$ Store at -15°C

Calbiochem; 05-23-5501   Human   MW 4117.8
$C_{181}H_{291}N_{55}O_{51}S_2$ ≥97% (HPLC); lyophilized solid; soluble in 5%
acetic acid | *Merck Index*, 12:7168

ICN 152977   Human   MW 4117.7   Tregear, GW etal, Hoppe-
Seyler's *Z Physiol Chem*, 355:415, 1974; Takai, T etal, *Peptide
Chemistry*, p 187, 1979

Ser-Val-Ser-Glu-Ile-Gln-Leu-Met-His-Asn-Leu-Gly-
Lys-His-Leu-Asn-Ser-Met-Glu-Arg-Val-Glu-Trp-Leu-
Arg-Lys-Lys-Leu-Gln-Asp-Val-His-Asn-Phe-Val-Ala-
Leu-Gly

Synonyms: Parathyroid Hormone (1-38); Parathormone

ICN 152979   Human   MW 4458.2   Hesch, RD etal, *Horm
Metab Res*, 16:559, 1984

Ser-Val-Ser-Glu-Ile-Gln-Leu-Met-His-Asn-Leu-Gly-
Lys-His-Leu-Asn-Ser-Met-Glu-Arg-Val-Glu-Trp-Leu-
Arg-Lys-Lys-Leu-Gln-Asp-Val-His-Asn-Phe-Val-Ala-
Leu-Gly-Ala-Pro-Leu-Ala-Pro-Arg

Synonyms: Parathyroid Hormone (1-44); Parathormone

ICN 152980   Human   MW 5063.9   Kimura, T etal,
Biopolymers, 20:1823, 1981

Ser-Val-Ser-Glu-Ile-Gln-Leu-Met-His-Asn-Leu-Gly-
Lys-His-Leu-Asn-Ser-Met-Glu-Arg-Val-Glu-Trp-Leu-
Arg-Lys-Lys-Leu-Gln-Asp-Val-His-Asn-Phe

Synonyms: Parathyroid Hormone (1-34)

Neosystem SC123   Human   MW 4117.75   Bioactive;
calcium metabolism peptide; Takai, T et al, *Pept Chem*, 1978:187,
1979

Ser-Val-Ser-Glu-Ile-Gln-Leu-Met-His-Asn-Leu-Gly-
Lys-His-Leu-Asn-Ser-Met-Glu-Arg-Val-Glu-Trp-Leu-
Arg-Lys-Lys-Leu-Gln-Asp-Val-His-Asn-Phe-Val-Ala-
Leu-Gly

Synonyms: Parathyroid Hormone (1-38)

Neosystem SC447   Human   MW 4458.16   Bioactive;
calcium metabolism peptide

Ser-Val-Ser-Glu-Ile-Gln-Leu-Met-His-Asn-Leu-Gly-
Lys-His-Leu-Asn-Ser-Met-Glu-Arg-Val-Glu-Trp-Leu-
Arg-Lys-Lys-Leu-Gln-Asp-Val-His-Asn-Phe-Val-Ala-
Leu-Gly-Ala-Pro-Leu-Ala-Pro-Arg

Synonyms: Parathyroid Hormone (1-44)

Sigma P 0279   Human   FW 5063.9   ≥97% (HPLC) |
Bioactive peptide

Ser-Val-Ser-Glu-Ile-Gln-Leu-Met-His-Asn-Leu-Gly-
Lys-His-Leu-Asn-Ser-Met-Glu-Arg-Val-Glu-Trp-Leu-
Arg-Lys-Lys-Leu-Gln-Asp-Val-His-Asn-Phe

Synonyms: Parathyroid Hormone (1-34)

Sigma P 3796   Human   FW 4117.7   ≥95% (HPLC) |
Bioactive peptide; Tregear, F et al, *Biochem*, 16:2817, 1977

Sigma P 7149   Human   FW 4117.7   ≥95% (HPLC) |
Bioactive peptide; Tregear, F et al, *Biochem*, 16:2817, 1977

Ser-Val-Ser-Glu-Ile-Gln-Leu-Met-His-Asn-Leu-Gly-
Lys-His-Leu-Asn-Ser-Met-Glu-Arg-Val-Glu-Trp-Leu-
Arg-Lys-Lys-Leu-Gln-Asp-Val-His-Asn-Phe Acetate
Salt

Synonyms: Parathyroid Hormone (1-34)

Biogenesis 7170-6409   Human synthetic   MW 4117.8
Purified; lyophilized | To avoid Met oxidation, it is best to dissolve
the peptide under a nitrogen stream

Ser-Val-Ser-Glu-Ile-Gln-Leu-Met-His-Asn-Leu-Gly-Lys-His-Leu-Asn-Ser-Met-Glu-Arg-Val-Glu-Trp-Leu-Arg-Lys-Lys-Leu-Gln-Asp-Val-His-Asn-Phe-Val-Ala-Leu-Gly

Synonyms: Parathyroid Hormone (1-34)

Biogenesis 7170-6479 Human synthetic   MW 4458.3 Lyophilized, consisting of 88% of peptide material, acetate counter ions and residual H_2O; sterile and pyrogen free; purified | To avoid Met oxidation, it is best to dissolve the peptide under a nitrogen stream

Ser-Val-Ser-Glu-Ile-Gln-Leu-Met-His-Asn-Leu-Gly-Lys-His-Leu-Asn-Ser-Met-Glu-Arg-Val-Glu-Trp-Leu-Arg-Lys-Lys-Leu-Gln-Asp-Val-His-Asn-Phe

Synonyms: Parathyroid Hormone (1-34)

Peptides International PTH-4068-s, 4068-v   Human synthetic   MW 4117.8 $C_{181}H_{291}N_{55}O_{51}S_2$ >95% (HPLC); lyophilized amorphous powder | Bioactive; Takai, M et al, *Peptide Chem* 1979:187, 1980

Ser-Val-Ser-Glu-Ile-Gln-Leu-Met-His-Asn-Leu-Gly-Lys-His-Leu-Asn-Ser-Met-Glu-Arg-Val-Glu-Trp-Leu-Arg-Lys-Lys-Leu-Gln-Asp-Val-His-Asn-Phe-Val-Ala-Leu-Gly-Ala-Pro-Leu-Ala-Pro-Arg-Asp-Ala-Gly-Ser-Gln-Arg-Pro-Arg-Lys-Lys-Glu-Asp-Asn-Val-Leu-Val-Glu-Ser-His-Glu-Lys-Ser-Leu-Gly-Glu-Ala-Asp-Lys-Ala-Asp-Val-Asn-Val-Leu-Thr-Lys-Ala-Lys-Ser-Gln

Synonyms: Parathyroid Hormone (1-84)

Peptides International PTH-4134-v   Human synthetic   MW 9424.7 $C_{408}H_{674}N_{126}O_{126}S_2$ >95% (HPLC); lyophilized amorphous powder | Bioactive; Hendy, GN et al, *PNAS* USA, 78:7365, 1981; Kimura, T et al, *BBRC*, 114:493, 1983

Ser-Val-Ser-Glu-Ile-Gln-Leu-Met-His-Asn-Leu-Gly-Lys-His-Leu-Asn-Ser-Met-Glu-Arg-Val-Glu-Trp-Leu-Arg-Lys-Lys-Leu-Gln-Asp-Val Amide

Synonyms: Parathyroid Hormone (1-31); Adenylate Cyclase/Bone Growth Stimulating Peptide

Peptides International PTH-4324-v   Human synthetic   MW 3718.4 $C_{162}H_{270}N_{50}O_{46}S_2$ >95% (HPLC); lyophilized amorphous powder | Bioactive; Rixon, RH et al, *J Bone Miner Res*, 9:1179, 1994; Neugebauer, W et al, *Biochemistry*, 34:8835, 1995; Whitfield, JF & P Morley, *Trends Pharmacol Sci*, 16:382, 1995; Whitfield, JF et al, *Calcif Tissue Int*, 58:81, 1996

Ser-Val-Ser-Glu-Ile-Gln-Leu-Met-His-Asn-Leu-Gly-Lys-His-Leu-Asn-Ser-Met-Glu-Arg-Val-Glu-Trp-Leu-Arg-Lys-Lys-Leu-Gln-Asp-Val-His-Asn-Phe

Synonyms: Parathyroid Hormone (1-34)

USBio P3109-06   Synthetic   ≥97%; lyophilized | Suitable for antigenic applications in immunological protocols

Ser-Val-Ser-Glu-Ile-Gln-Leu-Met-His-Asn-Leu-Gly-Lys-His-Leu-Ser-Ser-Leu-Glu-Arg-Val-Glu-Trp-Leu-Arg-Lys-Lys-Leu-Gln-Asp-Val-His-Asn-Phe

Synonyms: Parathyroid Hormone (1-34)

Bachem H-4032.0500   Porcine   MW 4072.71 $C_{181}H_{292}N_{54}O_{51}S$ Store at -15°C

Ser-Val-Ser-Glu-Ile-Gln-Leu-Nle-His-Asn-Leu-Gly-Lys-His-Leu-Asn-Ser-Nle-Glu-Arg-Val-Glu-Trp-Leu-Arg-Lys-Lys-Leu-Gln-Asp-Val-His-Asn-Tyr

Synonyms: Parathyroid Hormone (1-34), (Nle[8,18],Tyr[34])-

American Peptide 22-1-27   Human   MW 4097.7 $C_{181}H_{291}N_{55}O_{51}S_2$ Noda, T et al, *41st Chem Soc Japan Meeting*, Osaka, Japan, 4S12, 1980

Bachem H-9110.0500   Human   Store at -15°C

Peptides International PTH-4129-v   Human synthetic   MW 4097.7 $C_{183}H_{295}N_{55}O_{52}$ >99% (HPLC); lyophilized amorphous powder | Bioactive; Noda, T et al, *The 41st Ann Meeting of Chem Soc of Japan*, Osaka, April 1980, Abstr 4S12

Ser-Val-Ser-Glu-Ile-Gln-Leu-Nle-His-Asn-Leu-Gly-Lys-His-Leu-Asn-Ser-Nle-Glu-Arg-Val-Glu-Trp-Leu-Arg-Lys-Lys-Leu-Gln-Asp-Val-His-Asn-Tyr Amide

Synonyms: Parathyroid Hormone (1-34), (Nle[8,18], Tyr[34])-

Peptides International PTH-4180-v   Human synthetic   MW 4096.7 $C_{183}H_{296}N_{56}O_{51}$ >99% (HPLC); lyophilized amorphous powder

Ser-Val-Ser-Glu-Ile-Glu-Leu-Met-His-Asn-Leu-Gly-Lys-His-Leu-Asn-Ser-Met-Glu-Arg-Val-Glu-Trp-Leu-Arg-Lys-Lys-Leu-Gln-Asp-Val-His-Asn-Phe-Val-Ala-Leu-Gly-Ala-Pro-Leu-Ala-Pro-Arg-Asp-Ala-Gly-Ser-Glu-Arg-Pro-Arg-Lys-Lys-Glu-Asp-Asn-Val-Leu-Val-Glu-Ser-His-Glu-Lys-Ser-Leu-Gly-Glu-Ala-Asp-Lys-Ala-Asp-Val-Asn-Val-Leu-Thr-Lys-Ala-Lys-Ser-Glu

Synonyms: Parathyroid Hormone (1-84); Parathormone

Fluka 76263   Human   MW 9424.7 $C_{408}H_{674}N_{126}O_{126}S_2$ ≥90% (GE); ≥70% peptide content | Keutmann, HT et al, *Biochemistry*, 17:5723, 1978

Ser-Val-Ser-Glu-Ile-Gln-Leu-Met-His-Asn-Leu-Gly-Lys-His-Leu-Asn-Ser-Met-Glu-Arg-Val-Glu-Trp-Leu-Arg-Lys-Lys-Leu-Gln-Asp-Val-His-Asn-Phe-Val-Ala-Leu-Gly-Ala-Pro-Leu-Ala-Pro-Arg

Synonyms: Parathyroid Hormone (1-44)

Peptides International PTH-4094-v   Human synthetic   MW 5063.9 $C_{225}H_{366}N_{68}O_{61}S_2$ >97% (HPLC); lyophilized amorphous powder | Bioactive; Kimura, T et al, *Biopolymers*, 20:1823, 1981

Sodium Glucose Transporter I (554-637)
Biogenesis 8327-1170   Rabbit r-DNA

Suc-Asp-Glu-Glu-Ala-Val-Tyr-Phe-Ala-His-Leu-Asp-Ile-Ile-Trp

Synonyms: Endothelin I (8-21), (Suc,Glu[9],Ala[11,15])-; IRL-1038; Endothelin I (8-21), Suc-(Glu[9], Ala[11,15])-; IRL-1620; ETB Receptor Selective Agonist

Calbiochem; 05-23-3832   MW 1821 $C_{86}H_{117}N_{17}O_{27}$ ≥97% (HPLC); soluble in 2.5% NH_4OH | Potent & selective ET_B receptor agonist that causes a strong, transient inhibition of ADP-induced platelet aggregation; induces transient increases in cytosolic Ca^{2+}; increases cGMP levels via a nitric oxide dependent phenomenon; Fujitani, Y et al, *J Pharmacol Exp Ther*, 267:683, 1993; McMurdo, L et al, *Eur J Pharmacol*, 259:51, 1994; Takai, M et al, *Biochem Biophys Res Comm*, 184:953, 1992

American Peptide 88-2-50   Human   MW 1820.0 $C_{86}H_{116}N_{17}O_{27}$ Potent & selective ET_B receptor agonist which causes a strong, transient inhibition of ADP-induced platelet aggregation; Takai, M et al, *BBRC*, 184:953, 1992; Sakamoto, A et al, *J Biol Chem*, 268:8547, 1993; Haynes, WG et al, *TIPS*, 14:225, 1993

Peptides International PED-4285-v   Synthetic   MW 1821.0 $C_{86}H_{117}N_{17}O_{27}$ >99% (HPLC); lyophilized amorphous powder | Bioactive; Takai, M et al, *BBRC*, 184:953, 1992; Shetty, SS et al, *BBRC*, 191:459, 1993

Suc-Lys-Asp-Ser-Ser-Leu-Tyr-Pro-Ala-Leu-Thr-Phe-Asp-Lys
N-Suc-Lys-Asp-Ser-Ser-Leu-Tyr-Pro-Ala-Leu-Thr-Phe-Asp-Lys

Synonyms: HIV Substrate II

ICN 158718   Synthetic   Synthetic peptide analog useful in measuring activity of the HIV protease

T7 Major Capsid Protein (1-11)

Biogenesis 8492-5060   Synthetic   Semi-pure; lyophilized

Thr-Ala

Bachem G-3255.0250   MW 190.2 $C_7H_{14}N_2O_4$ Store at -15°C

Thr-Ala
Thr-β-Ala

ICN 156900   MW 190.2 $C_7H_{14}N_2O_4$

Thr-Ala-Leu-Leu-Trp-Gly-Leu-Lys-Lys-Lys-Lys-Glu-Asn-Asn-Arg-Arg-Thr-His-His-Met-Gln-Leu-Met-Ile-Ser-Leu-Phe-Lys-Ser-Pro-Leu-Leu-Leu-Leu

Synonyms: Parathyroid Hormone Related Protein (140-173); Hypercalcemia of Malignancy Factor (140-173)

ICN 154487   MW 4060.5   Yasuda, T et al, JBC, 264:7722, 1989

Bachem H-9105.0500   Human   Store at -15°C

Thr-Ala-Phe-Ala-Ser-Arg-His-Gly-Lys-Arg-His-Gly-Lys-Lys-Cys Acetate Salt

Synonyms: Bone Morphogenetic Protein 13 (1-14), ~C

Biogenesis 1406-1830   Synthetic   MW 1683.8   Purified; lyophilized

Thr-Ala-Pro-Arg-Ser-Leu-Arg-Arg-Ser-Ser-Cys-Phe-Gly-Gly-Arg-Met-Asp-Arg-Ile-Gly-Ala-Gln-Ser-Gly-Leu-Gly-Cys-Asn-Ser-Phe-Arg-Tyr

Synonyms: Cardiodilatin (95-126); Urodilatin (95-126); CDD/ANP (95-126); Natriuretic Factor (1-28), Atrial (Thr-Ala-Pro-Arg)-; Natriuretic Peptide, Atrial (Thr-Ala-Pro-Arg)-

American Peptide 14-1-15   MW 3506.0 $C_{145}H_{234}N_{52}O_{44}S_3$
Disulfide bonds: Cys[105]-Cys[121]; originally isolated from human urine; belongs to the family of natriuretic-vasorelaxant peptides earlier found in heart atria; a recent study indicated that infusion of this peptide may represent a new concept for the treatment of therapy-resistant acute renal failure after liver transplantation; Schultz-Knappe, P et al, Klin Wochenschr, 66:752, 1988

Neosystem SC945   MW 3505.95   Disulfide bonds: Cys[11]-Cys[27]; bioactive; ANF-related peptide; clinically important; exhibits potent vasorelaxant activity; Gagelmann, M et al, FEBS Lett, 233:249-254, 1988; Schulz-Knappe, P et al, Klin Wochenschr, 66:752, 1988; Valentin, JP et al, Semin Nephrol, 13:61-70, 1993; Hummel, M et al, J Heart Lung Transplant, 12:209-217, 1993; Goetz, KL, J Cardiovasc Pharmacol, 22 Suppl 2:84-85, 1993; Lazurova, I et al, Vnitr Lek, 40:194-198, 1994; Weber, J et al, BBA, 1207:231-235, 1994; Forssmann, WG, Nephron, 69:211-222, 1995

Bachem H-3046.0500   Human   MW 3505.98 $C_{145}H_{234}N_{52}O_{44}S_3$ Store at -15°C

ICN 153190   Human   Disulfide bonds: Cys[10]-Cys[26]; acts on ANP receptors in renal collecting ducts

Sigma U 5754   Human   FW 3505.9   ≥97% (HPLC); peptide content:~70% | Bioactive peptide; disulfide bonds: 11-27; acts on ANP receptors in renal collecting ducts; Gagelmann, M et al, FEBS Lett, 233:249, 1983

Thr-Ala-Pro-Gly-Gly-Gly-Asp-Pro-Glu-Val-Thr-Phe-Met-Trp-Thr-Asn-Cys-Arg-Gly-Glu

Synonyms: GP140 (381-400); SIVmac251 Peptide

Neosystem SC746   MW 2125.31   SIVmac251 peptide from subtype GP140; due to the presence of a Cys this peptide can contain the dimeric form

Thr-Arg

Bachem G-3265.0250   Store at -15°C

Thr-Arg
Thr-Arg Hemisulfate Salt

ICN 156901   MW 324.3 $C_{10}H_{21}N_5O_4 \cdot \frac{1}{2}H_2SO_4$

Thr-Arg

Sigma T 2900   FW 275.3 $C_{10}H_{21}N_5O_4$ Peptide content:~70%

Thr-Arg-Asn-Tyr-Tyr-Val-Arg-Ala-Val-Leu

Bachem H-2516.0001   Store at -15°C

Thr-Arg-Asp-Gly-Gly-Asn-Asn-Asn-Asn-Gly-Ser-Glu-Ile-Phe-Arg

Synonyms: GP120 (460-474); HTLV I Peptide

Neosystem SC268   MW 1650.68   HTLV I peptide from subtype GP160

Thr-Arg-Asp-Ile-Tyr-Glu-Thr-Asp-Tyr-Arg-Lys

Synonyms: Protein Tyrosine Phosphatase Substrate; Insulin Receptor (1142-1153), Human

Sigma P 6337   FW 1622.8   ≥97% (HPLC) | Bannwarth, W & Kitas, EA, Helv Chim Acta, 75:707, 1992; O'Hare, T & Pilch, PF, J Biol Chem, 264:602, 1989; King, MJ & Sale, GJ, FEBS Lett, 237:137, 1988; Stadtmauer, L & Rosen, OM, J Biol Chem, 261:10000, 1986

Thr-Arg-Asp-Ile-Tyr-Glu-Thr-Asp-Tyr-Tyr-Arg-Lys
Thr-Arg-Asp-Ile-Tyr(PO₃H₂)-Glu-Thr-Asp-Tyr-Tyr-Arg-Lys

Synonyms: Protein Tyrosine Phosphatase Substrate Monophosphate; Protein Tyrosine Phosphatase Substrate Phosphate

Sigma P 6462   FW 1720.7   ≥97% (HPLC) | Bannwarth, W & Kitas, EA, Helv Chim Acta, 75:707, 1992; O'Hare, T & Pilch, PF, J Biol Chem, 264:602, 1989; King, MJ & Sale, GJ, FEBS Lett, 237:137, 1988; Stadtmauer, L & Rosen, OM, J Biol Chem, 261:10000, 1986

Thr-Arg-Gln-Ala-Arg-Arg-Asn-Arg-Arg-Arg-Arg-Trp-Arg-Glu-Arg-Gln-Arg

Synonyms: HIV I rev Protein (34-50)

Bachem H-3026.0500   Store at -15°C

Thr-Arg-Leu-Arg-Lys-Gln-Met-Ala-Val-Lys-Lys-Tyr-Leu-Asn-Ser-Ile-Leu-Asn Amide

Synonyms: Vasoactive Intestinal Peptide (11-28)

American Peptide 48-1-25   Human, porcine, rat, ovine   MW 2175.7 $C_{96}H_{171}N_{31}O_{24}$

Thr-Arg-Leu-Thr-Arg-Lys-Arg-Gly-Leu-Lys-Leu-Ala-Thr-Ala-Leu Amide

Synonyms: Apolipoprotein B (3358-3372); Anti-Proliferative Peptide

Sigma A 4183   FW 1697.1   ≥97% (HPLC) | Bioactive peptide; Cardin, AD et al, BBRC, 154:741, 1988

Thr-Arg-Lys-Arg

Synonyms: Anti-Kentsin

Bachem H-5020.0005   MW 559.67 $C_{22}H_{45}N_{11}O_6$ Store at -15°C

Thr-Arg-Lys-Ser-Ile-Arg-Ile-Gln-Arg-Gly-Pro

Synonyms: GP120 (308-318); HTLV I Peptide

Neosystem SC539   MW 1311.55   HTLV I peptide from subtype GP160

Thr-Arg-Pro-Asn-Asn-Asn-Thr-Arg-Lys-Ser-Ile-Arg-Ile-Gln-Arg-Gly-Pro-Gly-Arg-Ala-Phe-Val-Thr

Synonyms: GP120 (302-324); HTLV I Peptide

Neosystem SC245   MW 2640.00   HTLV I peptide from subtype GP160

Thr-Arg-Pro-Ile-Ile-Thr-Thr-Tyr-Gly-Pro-Ser-Asp-Lys-Tyr

Thr-Arg-Pro-Ile-Ile-Thr-Thr-*m*-Nitro-Tyr-Gly-Pro-Ser-Asp-Lys(Abz)-Tyr

Bachem M-2360.0001   Store at -15°C

Thr-Arg-Ser-Ala-Trp

Synonyms: Parathyroid Hormone Related Peptide (107-111); Parathyroid Hormone Related Protein (107-111); Parathyroid Hormone; Hypercalcemia of Malignancy Factor (107-111); Osteostatin

American Peptide 22-1-60   MW 619.7 $C_{27}H_{41}N_9O_8$ Contains a highly conserved sequence within the pTH-related protein; a potent inhibitor of osteoclastic bone resorption *in vitro*

Neosystem SC499   MW 619.68   Bioactive; calcium metabolism peptide; Fenton, AJ et al, *Endocrinology*, 129:3424, 1991

Sigma H 2649   FW 619.7 $C_{27}H_{41}N_9O_8$ ≥97% (HPLC) | Bioactive peptide

Bachem H-8645.0005   Human, rat   MW 619.68 $C_{27}H_{41}N_9O_8$ Store at -15°C

Thr-Arg-Ser-Ala-Trp Amide

Synonyms: Parathyroid Hormone Related Protein (107-111)

Bachem H-2244.0005   Human, rat   MW 618.69 $C_{27}H_{42}N_{10}O_7$ Store at -15°C

Thr-Arg-Ser-Ala-Trp-Leu-Asp-Ser-Gly-Val-Thr-Gly-Ser-Gly-Leu-Glu-Gly-Asp-His-Leu-Ser-Asp-Thr-Ser-Thr-Thr-Ser-Leu-Glu-Leu-Asp-Ser-Arg

Synonyms: Parathyroid Hormone; Hypercalcemia of Malignancy Factor (107-139); Parathyroid Hormone Related Protein (107-139)

Sigma H 2774   FW 3451.6   ≥90% (HPLC) | Bioactive peptide

Bachem H-3212.0500   Human   MW 3451.62 $C_{142}H_{228}N_{42}O_{58}$ Store at -15°C

Thr-Arg-Ser-Ala-Trp-Leu-Asp-Ser-Gly-Val-Thr-Gly-Ser-Gly-Leu-Glu-Gly-Asp-His-Leu-Ser-Asp-Thr-Ser-Thr-Thr-Ser-Leu-Glu-Leu-Asp-Ser-Arg Amide

Synonyms: Parathyroid Hormone Related Protein (107-139)

Bachem H-3272.0500   Human   Store at -15°C

Thr-Arg-Ser-Ser-Arg-Ala-Gly-Leu-Gln-Phe-Pro-Val-Gly-Arg-Val-His-Arg-Leu-Leu-Arg-Lys

Synonyms: Buforin II

Sigma B 6298   FW 2434.9   ≥97% (HPLC); peptide content:~60% | Bioactive peptide; antimicrobial peptide; Park, Chan Bae et al, *Biochem Biophys Res Commun*, 518:408, 1996

Thr-Asn-Ala-Ala-Cys-Ala-Trp-Leu-Glu-Ala-Gln-Glu

Synonyms: NEF (51-62); HTLV I Peptide

Neosystem SC502   MW 1306.41   HTLV I peptide from subtype NEF; due to the presence of a Cys this peptide can contain the dimeric form

---

Thr-Asn-Ala-Asp-Cys-Ala-Trp-Leu-Glu-Ala-Gln-Glu

Synonyms: NEF MN (53-64); HIV I Peptide

Neosystem SC644   MW 1352.39   NEF peptide from HIV-1 subtype MN; due to the presence of a Cys this peptide can contain the dimeric form

---

Thr-Asn-Arg-Asn-Phe-Leu-Arg-Phe Amide

Synonyms: FMRF-Like Peptide F1; FMRF Amide-Like Peptide II

Bachem H-9265.0005   MW 1066.23 $C_{48}H_{75}N_{17}O_{11}$ Store at -15°C

American Peptide 60-9-29   Lobster   MW 1066.3 $C_{48}H_{75}N_{17}O_{11}$

Thr-Asn-Glu-Ile-Val-Glu-Glu-Gln-Tyr-Thr-Pro-Gln-Asn-Leu-Ala-Thr-Leu-Glu-Ser-Val-Phe-Gln-Glu-Leu-Gly-Lys-Leu-Thr-Gly-Pro-Asn-Ser-Gln

Synonyms: Secretoneurin

Neosystem SC879   bovine   MW 3679.00   Bioactive neuropeptide; Kirchmair, R et al, *Neuroscience*, 53:359-365, 1993; Saria, A et al, *Neuroscience*, 54:1-4, 1993

Thr-Asn-Glu-Ile-Val-Glu-Glu-Gln-Tyr-Thr-Pro-Gln-Ser-Leu-Ala-Thr-Leu-Glu-Ser-Val-Phe-Gln-Glu-Leu-Gly-Lys-Leu-Thr-Gly-Pro-Asn-Ser-Gln

Synonyms: Secretoneurin

Neosystem SC881   Human   MW 3651.97   Bioactive neuropeptide; Kirchmair, R et al, *Neuroscience*, 53:359-365, 1993; Saria, A et al, *Neuroscience*, 54:1-4, 1993; Kähler, CM et al, *Eur J Pharmacol*, 304:135-139, 1996

Thr-Asn-Glu-Ile-Val-Glu-Glu-Gln-Tyr-Thr-Pro-Gln-Ser-Leu-Ala-Thr-Leu-Glu-Ser-Val-Phe-Gln-Glu-Leu-Gly-Lys-Leu-Thr-Gly-Pro-Ser-Asn-Gln

Synonyms: Secretoneurin

Neosystem SC880   rat   MW 3651.97   Bioactive neuropeptide; Kirchmair, R et al, *Neuroscience*, 53:359-365, 1993; Saria, A et al, *Neuroscience*, 54:1-4, 1993

Thr-Asn-Val-Val

Thr-Asn-Val-Val-OMe

Synonyms: Eglin c (60-63)-OMe; Leukocyte Elastase Inhibitor, Human

American Peptide 81-7-60   MW 445.5 $C_{19}H_{35}N_5O_7$ Specific inhibitor of human leukocyte elastase, without affecting any activities of porcine pancreatic elastase, cathepsin G or chymotrypsin; Tsuboi, S et al, *Chem Pharm Bull*, 39:184, 1991

Bachem H-8150.0025   MW 445.52 $C_{19}H_{35}N_5O_7$ Store at -15°C

Sigma E 2017   FW 445.5 $C_{19}H_{35}N_5O_7$ ≥95% (HPLC) | Bioactive peptide; inhibits human leukocyte elastase, but not porcine pancreatic elastase, cathepsin G or chymotrypsin; Tsuboi, S et al, *Chem Pharm Bull*, 39:184, 1991

Thr-Asp

Bachem G-3275.0250   MW 234.21 $C_8H_{14}N_2O_6$ Store at -15°C

ICN 156902   MW 234.2 $C_8H_{14}N_2O_6$

Sigma T 3025   FW 234.2 $C_8H_{14}N_2O_6$

Thr-Asp-Glu-Cys-Glu-Leu-Cys-Ile-Asn-Val-Ala-Cys-Thr-Gly-Cys

Synonyms: Uroguanylin; Guanylate Cyclase C Activator

Peptides International PUG-4354-s   Rat synthetic   MW 1569.8 $C_{60}H_{96}N_{16}O_{25}S_4$ >99% (HPLC); lyophilized amorphous powder | Bioactive; Disulfide bonds: Cys[4]-Cys[12], Cys[7]-Cys[15]; Nakazato, M et al, *Endocrinology*, in press

---

Thr-Asp-Thr-Ser-His-His-Asp-Gln-Asp-His-Pro-Thr-Phe-Asn

Synonyms: Human Follicular Gonadotropin Releasing Peptide; Follicular Gonadotropin Releasing Peptide

Neosystem SC165   MW 1651.62   Bioactive; hypothalamic releasing hormone; Li, CH et al, *PNAS USA*, 84:959-962, 1987

Bachem H-6775.0500   Human   MW 1651.63 $C_{68}H_{94}N_{22}O_{27}$
Store at -15°C

Thr-Asp-Val-Asn-Gly-Asp-Gly-Arg-His-Asp-Leu

Synonyms: B$_{12}$ gpIIb (296-306); Peptide B12; Platelet Membrane Glycoprotein IIB Peptide (296-306)

American Peptide 44-0-40   MW 1198.2 $C_{47}H_{75}N_{17}O_{20}$ This GPIIb fragment interacts directly with fibrinogen & inhibits ADP-induced aggregation of human platelets; D'Souza, SE et al, *Nature*, 350:66, 1991

Bachem H-3032.0001   MW 1198.22 $C_{47}H_{75}N_{17}O_{20}$ Store at -15°C

ICN 158805   MW 1198.2   97% | Inhibits fibrinogen binding to platelets & platelet aggregation; *Nature*, 350:66, 1991

Neosystem SC435   MW 1198.21   Bioactive; cell attachment peptide; fibrinogen-related peptide; inhibits platelet aggregation & binding of fibrinogen to platelets; D'Souza, SE et al, *Nature*, 350:66-68, 1991

Sigma P 9684   FW 1198.2   ≥97% (HPLC); peptide content:~70% | Bioactive peptide; inhibits platelet aggregation & binding of fibrinogen to platelets; *Nature*, 350:66, 1991

Thr-Cys-Asp-Asp-Pro-Arg-Phe-Gln-Asp-Ser-Ser Amide

Synonyms: Chorionic Gonadotropin (109-119), β-; Chorionic Gonadotropin β-Subunit (109-119)

Bachem H-1378.0001   Human   MW 1269.32 $C_{50}H_{76}N_{16}O_{21}S$ Store at -15°C

Sigma C 0957   Human   FW 1269.3   ≥90% (HPLC) | Bioactive peptide; highly immunogenic; stimulates production of hCG-specific antibodies; Iyer, KSN et al, *Int J Peptide Prot Res*, 39:137, 1992

Thr-Cys-Asp-Asp-Pro-Arg-Phe-Gln-Asp-Ser-Ser-Ser-Ser-Lys-Ala-Pro-Pro-Pro-Ser-Leu-Pro-Ser-Pro-Ser-Arg-Leu-Pro-Gly-Pro-Ser-Asp-Thr-Pro-Ile-Leu-Pro-Gln

Synonyms: Chorionic Gonadotropin β-Chain (109-145)

Sigma C 9288   Human   FW 3876.3   ≥97% (HPLC); BME as stabilizer | Bioactive peptide; Stevens, VC, *CIBA Foundation Symposium #19*, 200-225, 1986

Thr-Cys-His-Gln-Arg-Arg-Thr-Gln-Arg-Lys-Glu-Thr-Val-Ala Acetate Salt

Synonyms: Glutamate Receptor V (862-875)

Biogenesis 4670-5229   Synthetic   Semi-pure; lyophilized

Thr-Cys-Val-Glu-Trp-Leu-Arg-Arg-Thr-Leu-Lys-Asn

Synonyms: MHC Antigen H-2K^b (163-174)

ICN 158756   MW 1195.3   95% | Allorecognition inhibitor; Schneck, J et al, *PNAS*, 85:4185, 1983

Thr-Cys-Val-Glu-Trp-Leu-Arg-Arg-Tyr-Leu-Lys-Asn

Synonyms: MHC Antigen 2K^b (163-174)

Sigma T 8033   FW 1580.9   ≥95% (HPLC) | Bioactive peptide; inhibitor of allorecognition; Schneck, J et al, *Proc Natl Acad Sci USA*, 86:8516, 1989

Thr-Gln

Bachem G-3277.0250   MW 247.25 $C_9H_{17}N_3O_5$ Store at -15°C

ICN 156908   MW 247.3 $C_9H_{17}N_3O_5$

Sigma T 3275   FW 247.3 $C_9H_{17}N_3O_5$

Thr-Gln-Arg-Leu-Ala-Asn-Phe-Leu-Val-Arg-Ser-Ser-Asn-Asn-Leu-Gly-Pro-Val-Leu-Pro-Pro-Thr-Asn-Val-Gly-Ser-Asn-Thr-Tyr Amide

Synonyms: Amylin (8-37)

American Peptide 74-5-13   Rat   MW 3200.6 $C_{140}H_{227}N_{43}O_{43}$ Deems, RO et al, *BBRC*, 181:116, 1991

Thr-Gln-Arg-Leu-Ala-Asn-Phe-Leu-Val-His-Ser-Ser-Asn-Asn-Phe-Gly-Ala-Ile-Leu-Ser-Ser-Thr-Asn-Val-Gly-Ser-Asn-Thr-Tyr

Synonyms: Amylin (8-37)

American Peptide 74-5-16   Human   MW 3113.4 $C_{135}H_{210}N_{40}O_{45}$

Thr-Gln-Glu-Val-Ser-Arg-Leu-Asn-Ile-Asn-Leu-His-Phe-Ser-Lys-Cys-Gly-Phe-Pro-Phe-Ser

Synonyms: GP46 (142-162); HTLV I Peptide

Neosystem SC825   MW 2424.75   HTLV I peptide; due to the presence of a Cys this peptide can contain the dimeric form

---

Thr-Gln-Gly-Tyr-Phe-Pro-Asp-Trp-Gln-Asn-Tyr-Thr-Pro-Gly-Pro-Gly

Synonyms: NEF (117-132); HTLV I Peptide

Neosystem SC510   MW 1827.92   HTLV I peptide from subtype NEF

Thr-Gln-Thr-Ser-Thr-Trp-Phe-Gly-Phe-Asn-Gly-Thr-Arg-Ala-Glu-Asn-Arg-Thr-Tyr-Ile

Synonyms: GP140 (271-290); SIVmac251 Peptide

Neosystem SC735   MW 2350.53   SIVmac251 peptide from subtype GP140

Thr-Glu

Bachem G-4050.0250   MW 248.24 $C_9H_{16}N_2O_6$ Store at -15°C

Thr-Glu-Arg-Gln-Ala-Asn-Phe-Leu-Gly-Lys

Synonyms: P15 (427-436); GAG P7 NC (50-55)/P1 (1-4); HTLV I Peptide

Neosystem SC535   MW 1163.29   HTLV I peptide from subtype P15 (GAG P7 NC/GAG P1/GAG P6)

Thr-Glu-Asn-Leu-Glu-Pro-Asn-Gly-Glu-Gly Amide

Synonyms: Follicle Stimulating Hormone Receptor Binding Inhibitor (BI-10)

Bachem H-2688.0001   Store at -15°C

Thr-Glu-Asn-Phe-Asn-Met-Trp-Lys-Asn-Asp-Met-Val-Glu-Gln-Met-His-Glu-Asp-Ile-Ile-Ser

Synonyms: GP120 (90-110); HTLV I Peptide

Neosystem SC1000   MW 2611.89   HTLV I peptide from subtype GP160

Thr-Glu-Asp-Arg-Trp-Asn-Lys-Pro-Gln-Lys-Thr-Lys-Gly-His-Arg

Synonyms: VIF (170-184); HTLV I Peptide

Neosystem SC551   MW 1881.08   HTLV I peptide from subtype VIF

Thr-Glu-Gly-Ser-Asp-Thr-Ile-Thr-Leu-Pro-Cys-Arg Thr-Glu-Gly-Ser-Asp-Thr-Ile-Thr-Leu-Pro-Cys(Acm); HTLV I Peptide-Arg

Synonyms: GP120 (413-424); HTLV I Peptide

Neosystem SC562   MW 1363.50   HTLV I peptide from subtype GP160

Thr-Glu-Phe-Gly-Ser-Glu-Leu-Lys-Ser-Phe-Pro-Glu-
Val-Val-Gly-Lys-Thr-Val-Asp-Gln-Ala-Arg-Glu-Tyr-
Phe-Thr-Leu-His-Tyr-Pro-Gln-Tyr-Asn-Val-Tyr-Phe-
Leu-Pro-Glu-Gly-Ser-Pro-Val-Thr-Leu-Asp-Leu-Arg-
Tyr-Asp-Arg-Val-Arg-Val-Phe-Tyr-Asn-Pro-Gly-Thr-
Asn-Val-Val-Asn-His-Val-Pro-His-Val-Gly
Ac-Thr-Glu-Phe-Gly-Ser-Glu-Leu-Lys-Ser-Phe-Pro-
Glu-Val-Val-Gly-Lys-Thr-Val-Asp-Gln-Ala-Arg-Glu-
Tyr-Phe-Thr-Leu-His-Tyr-Pro-Gln-Tyr-Asn-Val-Tyr-
Phe-Leu-Pro-Glu-Gly-Ser-Pro-Val-Thr-Leu-Asp-Leu-
Arg-Tyr-Asp-Arg-Val-Arg-Val-Phe-Tyr-Asn-Pro-Gly-
Thr-Asn-Val-Val-Asn-His-Val-Pro-His-Val-Gly

Synonyms: Eglin c

Alexis 201-006 *Hirudo medicinalis* (leech) recombinant, expressed in E. *coli* MW 8133.2 $C_{375}H_{552}N_{96}O_{108}$ ≥97%; lyophilized powder | 70 AA peptide with identical biological activity to native eglin C; effective inhibitor of chymotrypsin & subtilisin as well as of leukocyte elastase & cathepsin G; Seemuller, U et al, *Hoppe-Seyler's Z Physiol Chem*, 361:1841, 1980; Rink, H et al, *Nucl Acids Res*, 12:6369, 1984

Thr-Glu-Pro-Glu-Tyr-Gln-Pro-Gly-Glu
Thr-Glu-Pro-Glu-Tyr(PO₃H₂)-Gln-Pro-Gly-Glu Amide

Synonyms: Protein Tyrosine Phosphatase Substrate; Tyrosine Phosphatase Substrate

Alexis 165-020 MW 1128.1 $C_{45}H_{66}N_{11}O_{21}P$ ≥97%; white lyophilized powder; soluble in water | Suitable for continuous fluorimetric assays, based on the effect of phosphate substitution on the intrinsic fluorescence of the dephosphorylated peptide; Garcia-Esheverria, C & Rich, DH, *Bioorg Med Chem Lett*, 3:1601, 1993

Thr-Glu-Pro-Glu-Tyr-Gln-Pro-Gly-Glu
Thr-Glu-Pro-Glu(PO₃H₂)-Gln-Pro-Gly-Glu Amide

Bachem M-2035.0001 MW 1128.05 $C_{45}H_{66}N_{11}O_{21}P$ Store at -15°C

Thr-Glu-Pro-Gly-Leu-Glu-Glu-Val-Gly-Glu-Ile-Glu-
Gln-Lys-Gln-Leu-Gln

Synonyms: Nocistatin; Allodynia/Hyperalgesia-Blocking Peptide, Endogenous; Nociceptin Action Blocking Peptide

Bachem H-4312.0001 Bovine Store at -15°C

Neosystem SC1280 Bovine brain MW 1927.10 Bioactive neuropeptide; 17-mer peptide included in the precursor of nociceptin; blocker of nociceptin-induced allodynia & hyperalgesia; does not bind to the nociceptin receptor; binds the membrane of mouse brain & spinal cord with high affinity (K_D=5nM); the two peptides may play opposite roles in pain transmission; Okuda-Ashitaka, E et al, *Nature*, 392:286-289, 1998; Meunier, JC et al, *Nature*, 377:532-535, 1995; Reinscheid, RK et al, *Science*, 270:792-794, 1995

Peptides International PNO-4336-v Bovine synthetic MW 1927.1 $C_{82}H_{135}N_{21}O_{32}$ >99% (HPLC); lyophilized amorphous powder; ammonium form | Bioactive; Okuda-Ashitaka, E et al, *Nature*, 392:286, 1998; Nicol, B et al, *Eur J Pharmacol*, 356:R1, 1998

Thr-Glu-Pro-Gly-Leu-Glu-Glu-Val-Gly-Glu-Ile-Glu-
Glu-Lys-Gln-Leu-Gln

Synonyms: Nocistatin, (Pro³,4-³H)(³H PNP-3); Prepro-Orphanin F2 (111-127)

ARC ART-756 Bovine 30-60 Ci/mmol; 1.11-2.22 TBq/mmol; in EtOH | Radiochemical; opiate peptide

---

Thr-Gly
Z-Thr-Gly-OEt

Bachem C-3300.0001 MW 338.36 $C_{16}H_{22}N_2O_6$ Store at -15°C

Thr-Gly

Bachem G-3280.0250 MW 176.17 $C_6H_{12}N_2O_4$ Store at -15°C

Thr-Gly-Gln-Phe-Arg-Val-Tyr-Pro-Glu-Leu-Pro-Lys-
Pro-Ser-Ile

Synonyms: Carcinoembryonic Antigen (101-115)

Bachem H-1404.0001 MW 1732.01 $C_{81}H_{126}N_{20}O_{22}$ Store at -15°C

Sigma C 3330 FW 1732.0 ≥95% (HPLC) | Bioactive peptide; Thomas, P et al, *Biochem Biophys Res Commun*, 188:671, 1992

Thr-Gly-Gly

Bachem H-5030.0250 MW 233.22 $C_8H_{15}N_3O_5$ Store at -15°C

Thr-Gly-Leu-Leu-Leu-Thr-Arg-Asp-Gly-Gly-Asn-Asn

Synonyms: GP120 (455-466); HTLV I Peptide

Neosystem SC568 MW 1230.34 HTLV I peptide from subtype GP160

Thr-Gly-Leu-Leu-Thr-Phe-Leu-Ala-Trp-Trp-His-Glu-
Trp-Ala-Ser-Gln-Asp-Ser-Ser-Ser-Thr-Ala-Phe-Glu-
Gly-Gly-Thr-Pro-Glu-Leu-Ser

Synonyms: Pro-Cortistatin (51-81)

American Peptide 34-7-34 MW 3412.7 $C_{156}H_{219}N_{37}O_{50}$ de Leccea, L et al, *Nature*, 381:242, 1996

Thr-Gly-Lys-Ile-Cys-Asn-Asn-Pro-His-Arg-Ile-Leu-
Asp-Gly-Ile-Asp-Cys-Thr-Leu-Ile-Asp

Synonyms: Hemagglutinin (48-68)

American Peptide 72-8-10 Influenza virus MW 2309.7 $C_{97}H_{161}N_{29}O_{32}S_2$ Disulfide bonds: Cys⁵²-Cys⁶⁴

Thr-Gly-Tyr-Ile-Lys-Thr-Glu-Leu-Ile-Ser-Val

Synonyms: Signal Transducer and Activator of Transcription I-α/β (699-709)

Bachem H-4744.0001 Human, mouse MW 1223.43 $C_{56}H_{94}N_{12}O_{18}$ Store at -15°C

Thr-Ile-His-Cys-Lys-Trp-Arg-Glu-Lys-Pro-Leu-Met-
Leu-Met

Synonyms: Melanin Concentrating Hormone Gene Overprinted Polypeptide 14

Bachem H-4746.0500 Rat MW 1786.26 $C_{80}H_{132}N_{22}O_{18}S_3$ Store at -15°C

Thr-Ile-Ile-Asn-Val-Cys-Thr-Ser-Pro-Lys-Gln-Cys-
Ser-Lys-Pro-Cys-Lys-Glu-Leu-Tyr-Gly-Ser-Ser-Ala-
Gly-Ala-Lys-Cys-Met-Asn-Gly-Lys-Cys-Lys-Cys-Tyr-
Asn-Asn Amide

Synonyms: Noxiustoxin

Calbiochem 492010 *Centruroides noxius* MW 4194.9 $C_{174}H_{286}N_{52}O_{54}S_7$ >95% (HPLC); lyophilized solid; soluble in water; LD₅₀≤2000 mg/kg; biological activity:inhibits Kv1.2 & Kv1.3 K⁺ channels | Disulfide bonds: Cys⁷-Cys²⁹, Cys¹³-Cys³⁴, Cys¹⁷-Cys³⁶; specifically blocks Ca²⁺ activated K⁺ channels of small conductance obtained from cultured bovine aortic endothelial cells; Eder, C et al, *J Membr Biol*, 147:137, 1995; Drakopoulou, E et al, *Biochem Biophys Res Comm*, 213:901, 1995; Grissmer, S et al, *Mol Pharmacol*, 45:1227, 1994

Thr-Ile-Ile-Asn-Val-Lys-Cys-Thr-Ser-Pro-Lys-Gln-Cys-Leu-Pro-Pro-Cys-Lys-Ala-Gln-Phe-Gly-Gln-Ser-Ala-Gly-Ala-Lys-Cys-Met-Asn-Gly-Lys-Cys-Lys-Cys-Tyr-Pro-His

Synonyms: Margatoxin; Voltage-Dependent K$^+$ Channel Blocker

Bachem H-2192.0100 MW 4179.01 $C_{178}H_{286}N_{52}O_{50}S_7$ Store at -15°C | Disulfide bonds: Cys^7-Cys^{29}, Cys^{13}-Cys^{34}, Cys^{17}-Cys^{36}

Calbiochem 444322 *Centruroides margaritatus* MW 4179 $C_{178}H_{286}N_{52}O_{50}S_7$ >99% (HPLC); lyophilized solid; biological activity:inhibits Kv1.3 K$^+$ channels at <1 *nM*; LD$_{50}$≤2000 mg/kg | Disulfide bonds: Cys^7-Cys^{29}, Cys^{13}-Cys^{34}, Cys^{17}-Cys^{36}; Voltage-dependent K$^+$ channel blocker that is specific for the Kv1.3 channel; blocks N-type current of human T lymphocytes; has no effect on Ca^{2+} activated channels; Knaus, HG et al, *Biochemistry*, 34:13627, 1995; Calbo, M et al, *J Biol Chem*, 268:18866, 1993; Brugnara, C et al, *J Membr Biol*, 147:71, 1993; Leonard, RJ et al, *PNAS*, 89:10094, 1992

Sigma M 8278 *Centruroides margaritatus* (scorpion) FW 4179.0 ≥97% (HPLC) | Disulfide bonds: 7-29, 13-34, 17-36; bioactive peptide; potassium channel blocker; Galvo, MG et al, *J Biol Chem*, 268:18866 1993

Peptides International PAR-4290-s *Centruroides margaritatus* (scorpion) synthetic MW 4179.0 $C_{178}H_{286}N_{52}O_{50}S_7$ >95% (HPLC); lyophilized amorphous powder | Bioactive; Disulfide bonds: Cys^7-Cys^{29}, Cys^{13}-Cys^{34}, Cys^{17}-Cys^{36}; specific for Kv1.3 channel; Leonard, RJ et al, *PNAS USA*, 89:10094, 1992; Garcia-Calvo, M et al, *JBC*, 268:18866, 1993

Thr-Ile-Ile-Asn-Val-Lys-Cys-Thr-Ser-Pro-Lys-Gln-Cys-Ser-Lys-Pro-Cys-Lys-Glu-Leu-Tyr-Gly-Ser-Ser-Ala-Gly-Ala-Lys-Cys-Met-Asn-Gly-Lys-Cys-Lys-Cys-Tyr-Asn-Asn Amide

Synonyms: Noxiustoxin

Bachem H-2194.0100 Store at -15°C | Disulfide bonds: Cys^7-Cys^{29}, Cys^{13}-Cys^{34}, Cys^{17}-Cys^{36}

Thr-Ile-Met-Glu-Phe-Leu-Ala-Phe-Leu-His-Leu-Lys-Glu-Ala-Gly-Ala-Leu Amide

Synonyms: Galanin Message Associated Peptide (25-41); Prepro-Galanin (108-123); Prepro-Galanin (89-105)

American Peptide 46-1-30 MW 1903.3 $C_{90}H_{143}N_{21}O_{22}S$

Bachem H-9520.0001 MW 1903.32 $C_{90}H_{143}N_{21}O_{22}S$ Store at -15°C

Thr-Ile-Nle-Nle-Gln-Arg
Ac-Thr-Ile-Nle-(®)-Nle-Gln-Arg Amide

Bachem N-1465.0500 MW 769.99 $C_{35}H_{67}N_{11}O_8$ Store at -15°C

Thr-Ile-Nle-Nle-Gln-Arg
N-Ac-Thr-Ile-Nle-πσ-(CH$_2$NH)-Nle-Gln-Arg Amide

Synonyms: HIV I Protease Inhibitor

ICN 158550 MW 770 $C_{35}H_{67}N_{11}O_8$ White powder | Miller, M etal, *Science*, 246:1149, 1989

Thr-Ile-Nle-Nle-Gln-Arg
N-Ac-Thr-Ile-Nle-ψ-(CH$_2$NH)-Nle-Gln-Arg Amide

Synonyms: HIV I Protease Inhibitor

Sigma A 8305 FW 770.0 $C_{35}H_{67}N_{11}O_8$ ≥97% (HPLC); peptide content:~70% | Bioactive peptide; Miller, M et al, *Science*, 246:1149, 1989

Thr-Ile-Val-Lys-His-Pro-Arg-Tyr-Thr-Gly-Thr-Asn-Asn-Thr-Asp-Lys-Ile-Asn-Leu

Synonyms: GP140 (362-380); SIVmac251 Peptide

Neosystem SC744 MW 2185.46 SIVmac251 peptide from subtype GP140

Thr-Leu

Bachem G-3285.0250 MW 232.28 $C_{10}H_{20}N_2O_4$ Store at -15°C

ICN 156909 MW 232.2 $C_{10}H_{20}N_2O_4$

Sigma T 8400 FW 232.3 $C_{10}H_{20}N_2O_4$

Thr-Leu-Asn-Phe
Ac-Thr-Leu-Asn-Phe

Bachem H-8540.0025 MW 535.6 $C_{25}H_{37}N_5O_8$ Store at -15°C

Thr-Leu-Glu-Leu-Leu-Glu-Glu-Leu-Lys-Asn-Glu-Ala-Val-Arg-His-Phe-Pro

Synonyms: VPR (19-35); HTLV I Peptide

Neosystem SC198 MW 2038.33 HTLV I peptide from subtype VPR

Thr-Leu-Leu-Ser-Arg-Ala-Tyr-Gln-Ile-Leu-Gln-Pro-Ile-Leu-Gln-Arg-Leu-Ser-Ala-Thr

Synonyms: GP140 (791-810); SIVmac251 Peptide

Neosystem SC787 MW 2285.71 SIVmac251 peptide from subtype GP140

Thr-Leu-Lys-Gln-Ile-Ala-Ser-Lys-Leu-Arg-Glu-Gln-Phe-Gly

Synonyms: GP120 (346-359); HTLV I Peptide

Neosystem SC246 MW 1618.89 HTLV I peptide from subtype GP160

Thr-Leu-Thr-Ala-Gln-Ser-Arg-Thr-Leu-Leu-Ala-Gly-Ile-Val-Gln-Gln-Gln-Gln-Gln-Leu

Synonyms: GP140 (551-570); SIVmac251 Peptide

Neosystem SC763 MW 2197.51 SIVmac251 peptide from subtype GP140

Thr-Lys-Arg-Arg-Ala-Ile-Gly-Phe-Lys-Lys-Leu-Ala-Glu-Ala-Val-Lys-(Cys
Ac-Thr-Lys-Arg-Arg-Ala-Ile-Gly-Phe-Lys-Lys-Leu-Ala-Glu-Ala-Val-Lys-(Cys) Amide Trifluoroacetate Salt

*Synony*Nitric Oxide Synthase Blocking Peptide (724-739); bNOS Blocking Peptide*ms*:

Calbiochem 482725 Rat brain MW 1960.4 $C_{87}H_{154}N_{28}O_{21}S$ ≥98% (HPLC); lyophilized solid; soluble in water | Immunization & blocking peptide for Anti-bNOS (724-739), rat (rabbit); Bredt, DS et al, *Nature*, 351:714, 1991

Thr-Lys-Arg-Arg-Ala-Ile-Gly-Phe-Lys-Lys-Leu-Ala-Glu-Ala-Val-Lys-Cys
Ac-Thr-Lys-Arg-Arg-Ala-Ile-Gly-Phe-Lys-Lys-Leu-Ala-Glu-Ala-Val-Lys-Cys

Synonyms: Nitric Oxide Synthase Blocking Peptide (724-739)

American Peptide 91-1-26 Rat brain MW 1960.4 $C_{87}H_{152}N_{27}O_{22}S$ Bredt, DS et al, *Nature*, 351:714, 1991

Thr-Lys-Glu-Lys-Arg-Gly-Trp-Thr-Leu-Asn-Ser-Ala-Gly-Tyr-Leu-Leu-Gly-Pro-His-Ala-Ile-Asp-Asn-His-Arg-Ser-Phe-Ser-Asp-Lys-His-Gly-Leu-Thr-Gly-Lys-Arg-Glu-Leu-Pro

Synonyms: Prepro-Galanin (28-67)

American Peptide 46-1-03 Rat MW 4489.1 $C_{198}H_{312}N_{62}O_{58}$

Thr-Lys-Pro

Synonyms: Macrophage Inhibitory Peptide

Bachem H-4300.0050 MW 344.41 $C_{15}H_{28}N_4O_5$ Store at -15°C

Thr-Lys-Pro Acetate Salt

Synonyms: IgG (286-292), Human; Tuftsin (1-3); Macrophage Inhibitory Peptide

ICN 191175 Tripeptide of the 2nd constant domain of human IgG (peptide 286-292); an inhibitor of the macrophage function; Auriault, C etal, *FEBS Lett*, 153:11, 1983

Sigma T 4157 FW 344.4 (free base) $C_{15}H_{28}N_4O_5$ ≥97% (HPLC) | Bioactive peptide; Auriault, C et al, *FEBS Lett*, 153:11, 1983

Thr-Lys-Pro-Arg

Synonyms: Tuftsin

American Peptide 80-0-25 MW 500.6 $C_{21}H_{40}N_8O_6$ Natural immunomodulating peptide originally found as a phagocytosis-stimulating factor for poly-morpho nuclear leukocytes; induces nitric oxide synthase in macrophages & exhibits antitumor & antibacterial activities; Rishioka, K et al, *Biochem Biophys Acta*, 310:230, 1973

Thr-Lys-Pro-Arg
p-Aminophenylacetyl-Thr-Lys-Pro-Arg

Synonyms: Tuftsin, Aminophenylacetyl

American Peptide 80-0-26 MW 633.7 $C_{29}H_{47}N_9O_7$ Rishioka, K et al, *Biochim Biophys Acta*, 310:230, 1973

Thr-Lys-Pro-Arg
Thr-Lys(Z)-Pro-Arg

Synonyms: Tuftsin, (Lys(Z)[2])-

Bachem H-5025.0250 MW 634.73 $C_{29}H_{46}N_8O_8$ Store at -15°C

Thr-Lys-Pro-Arg

Synonyms: Tuftsin

Bachem H-5035.0050 Store at -15°C

Thr-Lys-Pro-Arg
Thr-Lys-3,4-Dehydro-Pro-Arg

Synonyms: Tuftsin, (3,4-Dehydro-Pro[3])-

Bachem H-8515.0005 Store at -15°C

Thr-Lys-Pro-Arg
p-Aminophenylacetyl-Thr-Lys-Pro-Arg

Synonyms: Tuftsin

ICN 158759 MW 633.7 $C_{29}H_{47}N_9O_7$ 97%

Thr-Lys-Pro-Arg

Synonyms: Tuftsin

ICN 190584 Phagocytosis stimulating peptide present in γ-globulin; reduces the number of tumor colonies appearing in lungs of mice following intravenous administration of tumor cells; Noyes, RD etal, *Cancer Treat Rep*, 65:673, 1981

Thr-Lys-Pro-Arg
p-AminophenylacetyL-Thr-Lys-Pro-Arg

Synonyms: Tuftsin, (p-Aminophenylacetyl)-

Sigma T 6153 FW 633.7 $C_{29}H_{47}N_9O_7$ ≥97% (HPLC); peptide content ~60% | Bioactive peptide

Thr-Lys-Pro-Arg

Synonyms: Tuftsin; Phagocytosis Stimulating Peptide

Biogenesis 9293-2009 Synthetic MW 500.58 Lyophilized, consisting of 81% peptide material, acetate counter ions, and residual H_2O; purified

Peptides International PTF-4020-v Synthetic MW 500.60 $C_{21}H_{40}N_8O_6$ >99% (HPLC); lyophilized amorphous powder | Bioactive; Nishioka, K et al, *BBRC*, 310:230, 1973

Thr-Lys-Pro-Arg 2AcOH 4H₂O

Synonyms: Tuftsin; Phagocytosis Stimulating Peptide

Peptides International PTF-4020 Synthetic MW 692.77 $C_{21}H_{40}N_8O_6 \cdot 2CH_3COOH \cdot 4H_2O$ >99% (HPLC); lyophilized amorphous powder | Bioactive; Nishioka, K et al, *BBRC*, 310:230, 1973

Thr-Lys-Pro-Arg Acetate Salt

Synonyms: Tuftsin

Sigma T 5897 FW 500.6 (free base) $C_{21}H_{40}N_8O_6$ ≥97% (HPLC) | Bioactive peptide; Rishioka, K et al, *Biochim Biophys Acta*, 310:230, 1973

Thr-Lys-Pro-Pro-Arg

Bachem H-5045.0025 MW 597.72 $C_{26}H_{47}N_9O_7$ Store at -15°C

Thr-Lys-Pro-Ser-Asp-Glu-Glu-Met-Leu-Phe-Ile-Tyr-Gly-His-Tyr-Lys-Gln-Ala-Thr-Val-Gly-Asp-Ile-Asn-Thr-Glu-Arg-Pro-Gly-Met-Leu-Asp-Phe-Thr

Synonyms: Diazepam Binding Inhibitor (17-50)

Neosystem SC368 Human MW 3905.36 >90 % (HPLC) | Bioactive neuropeptide

Thr-Lys-Tyr

Bachem H-5050.0100 MW 410.47 $C_{19}H_{30}N_4O_6$ Store at -15°C

Thr-Met

Bachem G-3290.0250 MW 250.32 $C_9H_{18}N_2O_4S$ Store at -15°C

Thr-Met-Arg-Lys-Pro-Arg-Cys-Gly-Asn-Pro-Asp-Val-Ala-Asn

Synonyms: Peptide 74

Bachem H-8545.0500 MW 1558.81 $C_{62}H_{107}N_{23}O_{20}S_2$ Store at -15°C

Thr-Met-Cys-Tyr-Ser-His-Thr-Thr-Thr-Ser-Arg-Ala-Ile-Leu-Thr-Asn-Cys-Gly-Glu-Asn-Ser-Cys-Tyr-Arg-Lys-Ser-Arg-Arg-His-Pro-Pro-Lys-Met-Val-Leu-Gly-Arg-Gly-Cys-Gly-Cys-Pro-Pro-Gly-Asp-Asp-Asn-Leu-Glu-Val-Lys-Cys-Cys-Thr-Ser-Pro-Asp-Lys-Cys-Asn-Tyr

Synonyms: Fasciculin II; F7; Acetylcholinesterase Inhibitor

Calbiochem 341290 *Dendroaspis angusticeps* MW 6749.7 $C_{276}H_{438}N_{88}O_{90}S_{10}$ ≥98% purity (HPLC); lyophilized solid; soluble in water; disulfide bonds: Cys[3]-Cys[22]; Cys[17]-Cys[39]; Cys[41]-Cys[52]; Cys[53]-Cys[59]; LD_{50}≤50 mg/kg | High-affinity irreversible, non-competitive inhibitor of acetylcholinesterase; binds to the peripheral anionic site on the enzyme molecule; inhibits both cholinolytic & proteolytic functions of acetylcholinesterase; Duran, R et al, *Biochim Biophys Acta*, 1201: 381, 1994; Quillfeldt, J et al, *Pharmacol Biochem Behav*, 37: 439, 1990; Karson, E et al, *J Physiol*, 79: 232, 1984; Rodriquez-Ithurralde, D et al, *Neurochem Int*, 5: 267, 1983

Thr-Phe
Bachem G-3295.0250   MW 266.3 $C_{13}H_{18}N_2O_4$ Store at -15°C

Thr-Phe-Arg-Gly-Ala-Pro
Bachem H-4452.0005   MW 647.73 $C_{29}H_{45}N_9O_8$ Store at -15°C

Thr-Phe-Gln-Ala-Tyr-Pro-Leu-Arg-Glu-Ala
Synonyms: Retroviral Protease Substrate

American Peptide 81-7-10   MW 1195.3 $C_{55}H_{82}N_{14}O_{16}$
Kotler, M et al, *PNAS*, 85:4185, 1983

ICN 158757   MW 1195.3   97% | Kotler, M etal, *PNAS*, 85:4185, 1983

Sigma T 8786   FW 1195.3   ≥97% (HPLC) | Bioactive peptide; Kotler, M et al, *Proc Natl Acad Sci USA*, 85:4185, 1983

Thr-Phe-Gln-Tyr-Ser-Arg-Gly-Trp-Thr-Asn
Glp-Thr-Phe-Gln-Tyr-Ser-Arg-Gly-Trp-Thr-Asn Amide
Synonyms: Corazonin

American Peptide 60-9-50   *Periplaneta americana*   MW 1369.5 $C_{62}H_{84}N_{18}O_{18}$ New cardioaccelerating peptide isolated from the corpus cardiacum of the American cockroach; the most potent insect cardioactive neuropeptide; Veenstra, JA et al, *FEBS Lett*, 250:231, 1989

Thr-Phe-Gln-Tyr-Ser-His-Gly-Trp-Thr-Asn
Glp-Thr-Phe-Gln-Tyr-Ser-His-Gly-Trp-Thr-Asn Amide
Synonyms: Corazonin, (His[7])-

American Peptide 60-9-54   MW 1350.4 $C_{62}H_{79}N_{17}O_{18}$

Thr-Phe-Gly-Leu-Gln-Leu-Glu-Leu-Thr
Synonyms: Heat Shock Protein 65kD *Mycobacterium bovis* BCG (180-188)

Bachem H-8315.0001   MW 1021.18 $C_{47}H_{76}N_{10}O_{15}$ Store at -15°C

Sigma H 1899   FW 1021.2   ≥95% (HPLC) | Bioactive peptide; van Eden, W et al, *Nature*, 331:171, 1988; Golden, HW et al, *Agents & Actions*, 34:148, 1991

Thr-Phe-Lys-Ala-Lys-Asn-Leu-Glu-Val-Arg-Lys-Asn-Ser-Gly Acetate Salt
Synonyms: Neuropeptide Y Receptor II (357-370)

Biogenesis 6732-0270   Synthetic   Semi-pure; lyophilized

Thr-Phe-Thr
Ac-Thr-Phe-Thr Amide
Synonyms: Somatostatin Sequence (10-12), Ac-

Biogenesis 8330-6009   Human synthetic   MW 408   >99% (TLC/HPLC); lyophilized

Thr-Phe-Thr-Ser-Asp-Leu-Ser-Lys-Gln-Met-Glu-Glu-Glu-Ala-Val-Arg-Leu-Phe-Ile-Glu-Trp-Leu-Lys-Asn-Gly-Gly-Pro-Ser-Ser-Gly-Ala-Pro-Pro-Pro-Ser Amide
Synonyms: Exendin (5-39); Glucagon-Like Peptide I Receptor Antagonist

Peptides International PEX-4345-v   *Heloderma horridum* (lizard) synthetic   MW 3806.3 $C_{169}H_{262}N_{44}O_{54}S$ >95% (HPLC); lyophilized amorphous powder | Bioactive; Montrose-Rafizadeh, C et al, *JBC*, 272:21201, 1997

Thr-Phe-Thr-Ser-Glu-Leu-Ser-Arg-Leu-Arg-Asp-Ser-Ala-Arg-Leu-Gln-Arg-Leu-Leu-Gln-Gly-Leu-Val Amide
Synon Secretin (5-27)yms:

Bachem H-4940.0001   Porcine   MW 2659.09 $C_{115}H_{200}N_{38}O_{34}$ Store at -15°C

ICN 154562   Porcine   MW 2659.4   Fink, mL etal, *JACS*, 98:974, 1976

Thr-Pro Hydrochloride
Bachem G-3960.0250   MW 252.7 $C_9H_{16}N_2O_4 \cdot HCl$ Store at -15°C

Thr-Pro-Arg-Lys
Synonyms: Kentsin; Contraceptive Tetrapeptide

American Peptide 87-0-51   MW 500.6 $C_{21}H_{40}N_8O_6$

Bachem H-3840.0005   MW 500.6 $C_{21}H_{40}N_8O_6$ Store at -15°C

ICN 153137   MW 500.6 $C_{21}H_{40}N_8O_6$ Kent, HA, *J Bio Reprod*, 12:504, 1975

Thr-Pro-Asp-Ile-Asn-Pro-Ala-Trp-Tyr-Ala-Gly-Arg-Gly-Ile-Arg-Pro-Val-Gly-Arg-Phe Amide
Synonyms: Prolactin Releasing Peptide (12-31); Prolactin Releasing Peptide 20; Pre-Prolactin (34-53)

Bachem H-4388.0500   Bovine   MW 2242.57 $C_{103}H_{156}N_{32}O_{25}$ Store at -15°C

Thr-Pro-Asp-Ile-Asn-Pro-Ala-Trp-Tyr-Ala-Ser-Arg-Gly-Ile-Arg-Pro-Val-Gly-Arg-Phe Amide
Synonyms: Pre-Prolactin (34-53); Prolactin Releasing Peptide (12-31); Prolactin Releasing Peptide 20

Bachem H-4392.0500   Human   MW 2272.6 $C_{104}H_{158}N_{32}O_{26}$ Store at -15°C

Neosystem SC1335   Human   MW 2272.59   Bioactive; hypothalamic releasing hormone; prolactin-releasing peptide

Thr-Pro-Asp-Ile-Asn-Pro-Ala-Trp-Tyr-Ala-Ser-Arg-Gly-Ile-Arg-Pro-Val-Gly-Arg-Phe Amide Acetate Salt
Synony Prolactin Releasing Peptide 31 (12-51); Prolactin Releasing Peptide 31/20ms:

Biogenesis 0100-0168   Human synthetic   Semi-pure; lyophilized

Thr-Pro-Asp-Ile-Asn-Pro-Ala-Trp-Tyr-Thr-Gly-Arg-Gly-Ile-Arg-Pro-Val-Gly-Arg-Phe Amide
Synonyms: Prolactin Releasing Peptide (12-31); Prolactin Releasing Peptide 20; Pre-Prolactin (34-53)

Bachem H-4394.0500   Rat   MW 2272.6 $C_{104}H_{158}N_{32}O_{26}$ Store at -15°C

Neosystem SC1336   Rat   MW 2272.59   Bioactive; hypothalamic releasing hormone

Thr-Pro-Asp-Trp-Asn-Asn-Asp-Thr-Trp-Gln-Glu-Trp-Glu-Arg-Lys-Val-Asp-Phe-Leu-Glu
Synonyms: GP140 (631-650); SIVmac251 Peptide

Neosystem SC771   MW 2608.76   SIVmac251 peptide from subtype GP140

Thr-Pro-Lys
Bachem H-6635.0050   MW 344.41 $C_{15}H_{28}N_4O_5$ Store at -15°C

Thr-Pro-Lys-Lys-Ala-Lys-Ala-Asn-Thr-Ser-Ser-Ala-Ser-Asn-Asn
Synonyms: TAT (85-99); SIVmac251 Peptide

Neosystem SC591   MW 1518.64   SIVmac251 peptide from subtype TAT

Thr-Pro-Phe-Ser-Ala-Leu-Gln
Mca-Thr-Pro-Phe-Ser-Ala-Leu-Gln-Dap(Dnp) Amide

Synonyms: Succinate Semialdehyde Dehydrogenase-Dap(Dnp) (186-192), Mca-(Gln^{192})-

Bachem M-2395.0001 Human, *E. coli* MW 1230.26 $C_{56}H_{71}N_{13}O_{19}$ Store at -15°C

Thr-Pro-Pro-Ala-Tyr-Arg-Pro-Pro-Asn-Ala-Pro-Ile-Leu

Synonyms: HBV Core Protein (128-140)

Neosystem SC1301 MW 1406.64 Virus-related peptide; Romieu, R et al, *Int Immunol*, 10:1273-1279, 1998

Thr-Ser

Bachem G-3300.0250 MW 206.2 $C_7H_{14}N_2O_5$ Store at -15°C

ICN 156911 MW 206.2 $C_7H_{14}N_2O_5$

Sigma T 3150 FW 206.2 $C_7H_{14}N_2O_5$

Thr-Ser-Glu-Lys-Ser-Gln-Thr-Pro-Leu-Val-Thr-Leu-Phe-Lys-Asn-Ala-Ile-Ile-Lys-Asn-Ala-Tyr-Lys-Lys-Gly-Glu

Synonyms: Endorphin (6-31), β-

American Peptide 28-1-32 Human MW 2909.4 $C_{131}H_{218}N_{34}O_{40}$

Bachem H-4024.0001 Human Store at -15°C

ICN 154534 Human MW 2899.9

Thr-Ser-Gly-Thr-Gln-Gly-Val-Gly-Ser-Pro-Gln-Ile-Leu-Val-Glu

Synonyms: REV (91-105); HTLV I Peptide

Neosystem SC555 MW 1472.61 HTLV I peptide from subtype REV

Thr-Ser-Ile-Arg-Gly-Lys-Val-Gln-Lys-Glu-Tyr-Ala-Phe-Phe-Tyr

Synonyms: GP120 (168-182); HTLV I Peptide

Neosystem SC1004 MW 1837.10 HTLV I peptide from subtype GP160

Thr-Ser-Lys

Synonyms: Bovine Pineal Antireproductive Tripeptide; Bovine Pineal Antireproductive Peptide

Bachem H-5055.0050 MW 334.37 $C_{13}H_{26}N_4O_6$ Store at -15°C

American Peptide 87-0-57 Bovine MW 334.4 $C_{13}H_{26}N_4O_6$

Thr-Ser-Lys Acetate Salt

Synonyms: Antireproductive Tripeptide

ICN 153048 Orts, RJ etal, *The Physiologist*, 21:87, 1978

Sigma T 8765 FW 334.4 (free base) $C_{13}H_{26}N_4O_6$ ≥97% (HPLC) | Bioactive peptide; Orts, RJ et al, *The Physiologist*, 21:87, 1978

Thr-Ser-Lys-Met-Asp-Gln-Leu-Ala-Lys-Glu-Leu-Thr-Ala-Glu

Synonyms: Chromogranin A (324-337)

Sigma C 6446 Human FW 1649.9 ≥95% (HPLC) | Bioactive peptide

Thr-Ser-Lys-Tyr-Arg

Synonyms: Kyotorphin, Neo-; Neokyotorphin

Bachem H-3845.0005 MW 653.74 $C_{28}H_{47}N_9O_9$ Store at -15°C

ICN 154513 MW 653.8 $C_{28}H_{47}N_9O_9$ Zhu, YX etal, *FEBS Letts*, 108:253, 1986

Thr-Ser-Met-His-Thr-Asp-Val-Ser-Lys-Thr-Ser-Leu-Lys-Gln Acetate Salt

Synonyms: Neuropeptide Y Receptor (353-366), Pan-; NPYr1

Biogenesis 6732-1070 Synthetic Semi-pure; lyophilized

Thr-Ser-Ser-Ile-Glu-Phe-Ala-Arg-Leu-Gln-Phe

Synonyms: HSV I Glycoprotein (gB) (497-507); HSV I Glycoprotein B (497-507)

Bachem H-1628.0001 MW 1298.46 $C_{59}H_{91}N_{15}O_{18}$ Store at -15°C

Sigma H 1904 FW 1298.5 ≥90% (HPLC) | Bioactive peptide; 2kb T-cell epitope of HSV-I gB; activates CD8$^+$ cytotoxic T lumphocytes; Vasilakos, JP & Michael, JG, *J Immunol*, 150:2346, 1993

Thr-Ser-Thr-Glu-Pro-Gln-Tyr-Gln-Pro-Gly-Glu-Asn-Leu
Thr-Ser-Thr-Glu-Pro-Gln-Tyr(PO$_3$H$_2$)-Gln-Pro-Gly-Glu-Asn-Leu

Synonyms: pp60c-*src* (521-533)

Alexis 165-022 MW 1543.7 $C_{62}H_{95}N_{16}O_{28}P$ Phosphorylated; white lyophilized powder | Derived from the C-terminal region of the c-*src* protooncogene pp60c-*src* & binds intramolecularly or intermolecularly to the SH2 domain of the same protein; Cantley, LE et al, *Cell*, 64:281, 1991; Roussel, RR et al, *PNAS*, 88:10696, 1991

Thr-Ser-Thr-Glu-Pro-Gln-Tyr-Gln-Pro-Gly-Glu-Asn-Leu

Synonyms: pp60c-*src* (521-533)

Bachem H-3256.0001 Store at -15°C

Thr-Ser-Thr-Glu-Pro-Gln-Tyr-Gln-Pro-Gly-Glu-Asn-Leu
Thr-Ser-Thr-Glu-Pro-Gln-Tyr(PO$_3$H$_2$)-Gln-Pro-Gly-Glu-Asn-Leu Phosphorylated

Synonyms: pp60c-*src* (521-533)

Bachem H-3258.0001 Store at -15°C

Thr-Ser-Thr-Glu-Pro-Gln-Tyr-Gln-Pro-Gly-Glu-Asn-Leu
Thr-Ser-Thr-Glu-Pro-Gln-Tyr(PO$_3$H$_2$)-Gln-Pro-Gly-Glu-Asn-Leu Trifluoroacetate Salt

Synonyms: p60c-*src* Peptide (521-533), Phosphorylated

Calbiochem 567801 MW 1543.5 $C_{62}H_{95}N_{16}O_{28}P$ ≥95% (HPLC); solid; soluble in water | C-terminal phosphoregulatory peptide of p60c-*src* phosphorylated on Tyr^{527}; involved in the regulation of kinase activity by binding intramolecularly or intermolecularly to the SH2 domain of p60c-*src*; Roussel, RR et al, *PNAS*, 88:10696, 1991

Thr-Ser-Thr-Glu-Pro-Gln-Tyr-Gln-Pro-Gly-Glu-Asn-Leu
Thr-Ser-Thr-Glu-Pro-Gln-Tyr(PO$_3$H$_2$)-Gln-Pro-Gly-Glu-Asn-Leu

Synonyms: p60c-*src* Peptide (521-533), Phosphorylated; pp60c-*src* (521-533)

ICN 195922 MW 1543.5 $C_{62}H_{95}N_{16}O_{28}P$ ≥98% | C-terminal phosphoregulatory peptide involved in regulation ofkinase activity

Neosystem SC871 MW 1543.50 Bioactive; proteinkinase-related peptide; corresponds to the C-terminal sequence of pp60c-*src*; regulateskinase activity by binding intra- or intermolecularly to the SH2 domain of the pp60c-*src* protein; Roussel, RR et al, *PNAS* USA, 88:10696-10700, 1991

Thr-Ser-Thr-Glu-Pro-Gln-Tyr-Gln-Pro-Gly-Glu-Asn-Leu

Thr-Ser-Thr-Glu-Pro-Gln-pTyr-Gln-Pro-Gly-Glu-Asn-Leu

Synonyms: Tyrosine Phosphopeptide

Upstate 12-218   Synthetic   MW 1542   Phosphopeptide

USBio T9245-05   Synthetic   MW 1542   ~95% (HPLC); lyophilized | 13 residue synthetic phosphopeptide; tested as a substrate for Protein Tyrosine Phosphatase-1B

Thr-Ser-Thr-Glu-Pro-Gln-Tyr-Gln-Pro-Gly-Glu-Asn-Leu Trifluoroacetate Salt

Synonyms: p60c-*src* Peptide

Calbiochem 567800   MW 1463.5 $C_{62}H_{94}N_{16}O_{25}$ ≥97% (HPLC); solid; soluble in water | Non-phosphorylated C-terminal phosphoregulatory peptide of p60c-*src*; involved in the regulation of tyrosine kinase activity; Roussel, RR et al, *PNAS*, 88:10696, 1991

Thr-Thr
Thr-Thr-pNA

Bachem L-2035.0050   Store at -15°C

Thr-Thr-Ala-Ala-Pro-Thr-Ser-Ala-Pro-Val-Ser-Glu-Lys-Ile-Asp-Met-Val-Asn-Glu-Thr

Synonyms: GP140 (131-150); SIVmac251 Peptide

Neosystem SC721   MW 2062.27   SIVmac251 peptide from subtype GP140

Thr-Thr-Ser-Gln-Val-Arg-Pro-Arg

Synonyms: Platelet Factor IV (15-22), (Gln[18])-; PF4; CT-112

Bachem H-1542.0005   Human   MW 944.06 $C_{38}H_{69}N_{15}O_{13}$ Store at -15°C

Thr-Thr-Thr

Bachem H-5065.0050   MW 321.33 $C_{12}H_{23}N_3O_7$ Store at -15°C

Thr-Thr-Tyr-Ala-Asp-Phe-Ile-Ala-Ser-Gly-Arg-Thr-Gly-Arg-Arg-Asn-Ala-Ile-His-Asp

Synonyms: cAMP-Dependent Protein Kinase Inhibitor (5-24)

Alexis 151-014   MW 2222.4 $C_{94}H_{148}N_{32}O_{31}$ ≥97%; White powder; soluble in water | Potent inhibitor; corresponds to the active site of the skeletal muscle inhibitor protein; Cheng, C et al, *J Biol Chem*, 261:989, 1986; Wong, YS et al, *J Biol Chem*, 261:12089, 1986

Thr-Thr-Tyr-Ala-Asp-Phe-Ile-Ala-Ser-Gly-Arg-Thr-Gly-Arg-Arg-Asn-Ala-Ile Amide

Synonyms: cAMP-Dependent Protein Kinase Inhibitor (5-22)

Alexis 151-016   MW 1969.2 $C_{84}H_{137}N_{29}O_{26}$ Cheng, C et al, *J Biol Chem*, 261:989, 1986; Glass, DB et al, *J Biol Chem*, 264:8802, 1989

Thr-Thr-Tyr-Ala-Asp-Phe-Ile-Ala-Ser-Gly-Arg-Thr-Gly-Arg-Arg-Asn-Ala-Ile-His-Asp Amide

Synonyms: cAMP-Dependent Protein Kinase Inhibitor (5-24)

Alexis 151-018   MW 2221.4 $C_{94}H_{149}N_{33}O_{30}$ ≥97%; white powder; soluble in water | Cheng, C et al, *J Biol Chem*, 261:989, 1986

Thr-Thr-Tyr-Ala-Asp-Phe-Ile-Ala-Ser-Gly-Arg-Thr-Gly-Arg-Arg-Asn-Ala-Ile Amide

Synonyms: cAMP-Dependent Protein Kinase Inhibitor (5-22)

American Peptide 86-1-40   MW 1969.2 $C_{84}H_{137}N_{29}O_{26}$ Cheng, HC et al, *JBC*, 261:984, 1986

Thr-Thr-Tyr-Ala-Asp-Phe-Ile-Ala-Ser-Gly-Arg-Thr-Gly-Arg-Arg-Asn-Ala-Ile-His-Asp

Synonyms: cAMP-Dependent Protein Kinase Inhibitor (5-24)

American Peptide 86-1-45   MW 2222.4 $C_{94}H_{108}N_{32}O_{31}$ 20-residue peptide that binds to the cAMP-dependent protein kinase with high affinity & competitively inhibits the enzyme activity; Cheng, HC et al, *Biochem J*, 231:655, 1985; Scott, JD et al, *PNAS*, 82:5732, 1985

Thr-Thr-Tyr-Ala-Asp-Phe-Ile-Ala-Ser-Gly-Arg-Thr-Gly-Arg-Arg-Asn-Ala-Ile-His-Asp Amide

Synonyms: cAMP-Dependent Protein Kinase Inhibitor (5-24)

American Peptide 86-1-50   MW 2221.4 $C_{94}H_{149}N_{33}O_{30}$ Inhibits cAMP-Dependent protein kinase; Swarup, G et al, *JBC*, 258:10341, 1983

Thr-Thr-Tyr-Ala-Asp-Phe-Ile-Ala-Ser-Gly-Arg-Thr-Gly-Arg-Arg-Asn-Ala-Ile Amide

Bachem H-3222.0001   MW 1969.19 $C_{84}H_{137}N_{29}O_{26}$ Store at -15°C

Thr-Thr-Tyr-Ala-Asp-Phe-Ile-Ala-Ser-Gly-Arg-Thr-Gly-Arg-Arg-Asn-Ala-Ile-His-Asp

Bachem H-5950.0500   MW 2222.41 $C_{94}H_{148}N_{32}O_{31}$ Store at -15°C

Thr-Thr-Tyr-Ala-Asp-Phe-Ile-Ala-Ser-Gly-Arg-Thr-Gly-Arg-Arg-Asn-Ala-Ile-His-Asp Amide

Synonyms: cAMP-Dependent Protein Kinase Peptide Inhibitor

Promega; V5681   MW 2221   ≥70% peptide (FAB/MS) | Inhibits phosphorylation of target proteins by binding to the protein substrate site of the catalytic subunit of PKA; corresponds to the region 5-24 of the naturally occurring proteinkinase A inhibitor; is the most potent & specific pseudosubstrate inhibitorknown for any proteinkinase & has a Ki of 3-5 *nM* with minimal effects on other proteinkinases; Kemp, BE et al, *Meth Enzymol*, 201:287, 1991; Walsh, DA & Glass, DB, *Meth Enzymol*, 201:304, 1991

Thr-Thr-Tyr-Ala-Asp-Phe-Ile-Ala-Ser-Gly-Arg-Thr-Gly-Arg-Arg-Asn-Ala-Ile-His-Asp

Synonyms: Protein Kinase Inhibitor, Rabbit Sequence; Protein Kinase G Inhibitor

Sigma P 0300   FW 2222.4 (free base)   ≥95% (HPLC); peptide content:~70%; activity:1.0 µg inhibits 2,000-6,000 phosphorylating units of cAMP-dependent proteinkinase | Bioactive peptide; Cheng, HC et al, *J Biol Chem*, 261:989, 1986

ICN 153169   Rabbit   1.0 µg inhibits 2000-6000 phosphorylating U of cAMP dependent proteinkinase | Cheng, HC etal, *JBC*, 261:989, 1986

Thr-Thr-Tyr-Ala-Asp-Phe-Ile-Ala-Ser-Gly-Arg-Thr-Gly-Arg-Arg-Asn-Ala-Ile-His-Asp Trifluoroacetate Salt

Synony Protein Kinase A Inhibitor (5-24)ms:

Calbiochem 116805   Synthetic   MW 2222.4 $C_{94}H_{148}N_{32}O_{31}$ ≥95% (HPLC); lyophilized solid; soluble in water | Potent peptide inhibitor; its sequence is derived from heat-stable skeletal muscle inhibitor protein PKA; binds to the catalytic subunit of PKA & displaces the regulatory subunit; mimics protein substrate by binding to the catalytic site via the Arg-cluster basic substrate; Knighton, DR et al, *Science*, 253:525, 1991; Cheng, HC et al, *J Biol Chem*, 261:989, 1986

Thr-Thr-Val-Pro-Trp-Pro-Asn-Ala-Ser-Leu-Thr-Pro-Asp-Trp-Asn-Asn-Asp-Thr-Trp-Gln

Synonyms: GP140 (621-640); SIVmac251 Peptide

Neosystem SC770   MW 2343.49   SIVmac251 peptide from subtype GP140

Thr-Trp-Gly-Thr-Thr-Gln-Cys-Leu-Pro-Asp-Asn-Gly-Asp-Tyr-Ser-Glu-Leu-Ala-Leu-Asn

Synonyms: GP140 (51-70); SIVmac251 Peptide

Neosystem SC713　　MW 2198.34　　SIVmac251 peptide from subtype GP140; due to the presence of a Cys this peptide can contain the dimeric form

---

Thr-Trp-Thr-Ala-Asn-Val-Gly-Lys-Gly-Gln-Pro-Ser

Synonyms: Thrombin B-Chain (147-158)

Bachem H-8550.0001　　Human　　MW 1245.36　　$C_{54}H_{84}N_{16}O_{18}$ Store at -15°C

Neosystem SC954　　Human　　MW 1245.35　　Bioactive; inhibits thrombin binding to thrombomodulin with an apparent Ki = 94$m\mu M$; directly blocks thrombin procoagulant activities, fibrinogen clotting (K$_I$ = 385$m\mu M$) Factor V activation (K$_I$ = 33$m\mu M$) & platelet activation (K$_I$ = 645$m\mu M$); Suzuki, K et al, *JBC*, 265:13263-13267, 1990; Suzuki, K & Nishioka, J, *JBC*, 266:18498-18501, 1991

Thr-Tyr-Ala-Asp-Phe-Ile-Ala-Ser-Gly-Arg-Thr-Gly-Arg-Arg-Asn-Ala-Ile Amide

Synonyms: Protein Kinase A Inhibitor (6-22); cAMP-Dependent Protein Kinase Inhibitor

American Peptide 86-0-20　　MW 1868.1　　$C_{80}H_{130}N_{28}O_{24}$ Glass, DB et al, *JBC*, 264:1457, 1989

Thr-Tyr-Ala-Asp-Phe-Ile-Ala-Ser-Gly-Arg-Thr-Gly-Arg-Arg-Asn-Ala-Ile Amide Trifluoroacetate Salt

Synonyms Protein Kinase A Inhibitor (6-22):

Calbiochem 539684　　MW 1868.1　　$C_{80}H_{130}N_{28}O_{24}$　　≥97% (HPLC); lyophilized solid; soluble in water | Potent & competitive inhibitor of PKA; Kemp BE et al, *Methods Enzymol*, 159:173, 1988; Glass, DB et al, *J Biol Chem*, 264:8802, 1989

Thr-Tyr-Ala-Asp-Phe-Ile-Ala-Ser-Gly-Arg-Thr-Gly-Arg-Arg-Asn-Ala-Ile Amide

Synonyms: Protein Kinase Inhibitor (6-22); cAMP-Dependent Protein Kinase Inhibitor; cAMP-Dependent Protein Kinase Inhibitor (6-22); Protein Kinase A Inhibitor Peptide

ICN 195810　　MW 1869.1　　>97%

Sigma P 6062　　FW 1868.1　　≥97% (HPLC) | Bioactive peptide; heat-stable inhibitor; Glass, DB et al, *J Biol Chem*, 264:14579, 1989

USBio P9102-91　　Human synthetic　　≥98% (HPLC); lyophilized | 17 residue synthetic peptide corresponding to residues 7-23 of both human & mouse PKA inhibitor, & 6-22 of rabbit PKA inhibitor; 200 *mM* inhibits >95% of the activity of 250 ng of PKA catalytic subunit

---

Thr-Tyr-Gly-Asp-Thr-Trp-Ala-Gly-Val-Glu-Ala-Ile

Synonyms: VPR (49-60); HTLV I Peptide

Neosystem SC559　　MW 1282.37　　HTLV I peptide from subtype VPR

Thr-Tyr-Ile-Cys-Glu-Val-Glu-Asp-Gln-Lys-Glu-Glu Thr-Tyr-Ile-Cys(bzl)-Glu-Val-Glu-Asp-Gln-Lys-Glu-Glu

Synonyms: CD-4 (81-92), (Cys(Bzl)[84])-

American Peptide 72-1-30　　MW 1575.7　　$C_{69}H_{102}N_{14}O_{26}S$ Inhibits infection of CD-4 cells by HIV & HIV-induced cell fusion; Nara, PL et al, *PNAS*, 86:9159, 1989

Thr-Tyr-Ile-Cys-Glu-Val-Glu-Asp-Gln-Lys-Glu-Glu Thr-Tyr-Ile(Bzl)-Glu-Val-Glu-Asp-Gln-Lys-Glu-Glu

Synonyms: CD-4 (81-92), (Cys(Bzl)[84])-

Bachem H-8085.0001　　MW 1575.71　　$C_{69}H_{102}N_{14}O_{26}S$　　Store at -15°C

Thr-Tyr-Ile-Cys-Glu-Val-Glu-Asp-Gln-Lys-Glu-Glu Thr-Tyr-Ile-Cys(Bzl)-Glu(OBzl)-Val-Glu-Asp-Gln-Lys-Glu-Glu

Synonyms: CD-4 (81-92), (Cys(Bzl)[84],Glu(OBzl)[85])-

Bachem H-9655.0001　　Store at -15°C

Thr-Tyr-Ile-Cys-Glu-Val-Glu-Asp-Gln-Lys-Glu-Glu Thr-Tyr-Ile-Cys(Bzl)-Glu-Val-Glu-Asp-Gln-Lys-Glu-Glu

Synonyms: CD-4 (81-92), (Cys(Bzl)[84])-

Neosystem SC676　　MW 1575.70　　Virus-related peptide; AIDS-related peptide; inhibits infection of CD-4+ cells by HIV & inhibits HIV-induced cell fusion; Nara, PL et al, *PNAS USA*, 86:7139, 1989

Thr-Tyr-Ile-Cys-Glu-Val-Glu-Asp-Gln-Lys-Glu-Glu Thr-Tyr-Ile-Cys(Bzl)-Glu(OBzl)-Val-Glu-Asp-Gln-Lys-Glu-Glu

Synonyms: CD-4 (81-92), (Cys(Bzl)[84],Glu(OBzl)[85])-

Neosystem SC677　　MW 1700.83　　Virus-related peptide; AIDS-related peptide; inhibits infection of CD-4+ cells by HIV & inhibits HIV-induced cell fusion; Nara, PL et al, *PNAS USA*, 86:7139, 1989

Thr-Tyr-Ile-Cys-Glu-Val-Glu-Asp-Gln-Lys-Glu-Glu Thr-Tyr-Ile-Cys(Bzl)-Glu-Val-Glu-Asp-Gln-Lys-Glu-Glu

Synonyms: CD-4 (81-92), (Cys(Bzl)[84])-

Sigma C 2796　　FW 1575.7　　≥97% (HPLC) | Bioactive peptide; inhibits HIV-1 induced cell fusion & infection *in vitro*; Nara, PL et al, *Proc Natl Acad Sci USA*, 86:7139, 1989

Thr-Tyr-Lys

Bachem H-5070.0050　　MW 410.47　　$C_{19}H_{30}N_4O_6$　　Store at -15°C

Thr-Tyr-Ser

Bachem H-5075.0250　　MW 369.38　　$C_{16}H_{23}N_3O_7$　　Store at -15°C

ICN 153180　　MW 369.4　　$C_{16}H_{23}N_3O_7$

Sigma T 0148　　FW 369.4　　$C_{16}H_{23}N_3O_7$　　≥95% (HPLC) | Bioactive peptide

---

Thr-Tyr-Ser-Lys

Bachem H-5080.0050　　MW 497.55　　$C_{22}H_{35}N_5O_8$　　Store at -15°C

Thr-Val

Bachem G-3305.0250　　MW 218.25　　$C_9H_{18}N_2O_4$　　Store at -15°C

Thr-Val-Gln-Ala-Arg-Gln-Leu-Leu-Ser-Gly-Ile-Val-Gln-Gln-Gln-Asn

Synonyms: GP41 (543-558); HTLV I Peptide

Neosystem SC254　　MW 1783.01　　HTLV I peptide from subtype GP41

Thr-Val-Gln-Lys-Leu-Ala-His-Gln-Ile-Tyr-Gln-Phe-Thr-Asp-Lys-Asp-Lys-Asp-Asn-Val-Ala-Pro-Arg-Ser-Lys-Ile-Ser-Pro-Gln-Gly-Tyr Amide

Synonyms: Adrenomedullin (22-52); Adrenomedullin Antagonist

American Peptide 22-1-16　　Human　　MW 3576.1 $C_{159}H_{252}N_{46}O_{48}$

Bachem H-4144.0500　　Human　　MW 3576.03 $C_{159}H_{252}N_{46}O_{48}$　　Store at -15°C

ICN 195609　　Human　　MW ~3576

Peptides International PAD-4302-v   Human synthetic   MW 3576.0 $C_{159}H_{252}N_{46}O_{48}$ >99% (HPLC); lyophilized amorphous powder | Bioactive; Eguchi, S et al, *Endocrinology*, 135:2454, 1994

Thr-Val-Leu

Synonyms: Schizophrenia Related Peptide

American Peptide 60-0-44   MW 331.4 $C_{15}H_{29}N_3O_5$

Peptides International PSC-4061   Synthetic   MW 331.41 $C_{15}H_{29}N_3O_5$ >99% (HPLC); lyophilized amorphous powder; bulk | Bioactive; Frohman, CE, *Chem Eng News*, 1977:35

Thr-Val-Leu Hydrochloride

Synonyms: Schizophrenia Related Peptide

Sigma T 3391   FW 331.4 (free base) $C_{15}H_{29}N_3O_5$ ≥97% (HPLC) | Bioactive peptide; *Chem Abstr*, 88:170493x, 1978

Thr-Val-Ser-Phe-Asn-Phe
Ac-Thr-Val-Ser-Phe-Asn-Phe

Bachem H-1956.0005   Store at -15°C

Thr-Val-Thr-Glu-Gln-Ala-Ile-Glu-Asp-Val-Trp-Gln-Leu-Phe-Glu-Thr-Ser-Ile-Lys-Pro

Synonyms: GP140 (81-100); SIVmac251 Peptide

Neosystem SC716   MW 2334.60   SIVmac251 peptide from subtype GP140

Thz-Thz
Z-Thz-Thz

American Peptide 95-0-11   MW 382.4 $C_{18}H_{22}N_2O_4S$

Trp-Ala
Z-Trp-Ala

Bachem C-2695.0001   MW 409.44 $C_{22}H_{23}N_3O_5$ Store at -15°C

Trp-Ala

Bachem G-3315.0250   MW 275.31 $C_{14}H_{17}N_3O_3$ Store at -15°C

Sigma T 8152   FW 275.3 $C_{14}H_{17}N_3O_3$

Trp-Ala-Gly-Gly-Asn-Ala-Ser-Gly-Glu

Bachem H-2555.0005   MW 847.84 $C_{35}H_{49}N_{11}O_{14}$ Store at -15°C

Trp-Ala-Gly-Gly-Asp(Ala-Ser-Gly-Glu)

Bachem H-2545.0005   MW 848.82 $C_{35}H_{48}N_{10}O_{15}$ Store at -15°C

Trp-Ala-Gly-Gly-Asp-Ala-Ser-Gly-Glu

Synonyms: Sleep Inducing Peptide, δ-

American Peptide 87-0-56   MW 848.8 $C_{35}H_{48}N_{10}O_{15}$ Schoenenberger, GA et al, *PNAS*, 74:1282, 1977

Trp-Ala-Gly-Gly-Asp-Ala-Ser-Gly-Glu
Trp-Ala-Gly-Gly-Asp-Ala-Ser(PO₃H₂)-Gly-Glu

Synonyms: Sleep Inducing Peptide, δ-Phospho

American Peptide 87-0-59   MW 928.8 $C_{35}H_{49}N_{10}O_{18}P$ A more potent analog of DSIP; Friedman, TC et al, *J Clin Endocrinol Metab*, 78:1085, 1994

Trp-Ala-Gly-Gly-Asp-Ala-Ser-Gly-Glu

Bachem H-2540.0005   MW 848.82 $C_{35}H_{48}N_{10}O_{15}$ Store at -15°C

Trp-Ala-Gly-Gly-Asp-Ala-Ser-Gly-Glu
Trp-Ala-Gly-Gly-Asp-Ala-Ser(PO₃H₂)-Gly-Glu

Synonyms: Sleep Inducing Peptide, δ-Phospho

Calbiochem; 05-23-1316   MW 928.8 $C_{35}H_{49}N_{10}O_{18}P$ ≥95% (HPLC); solid; soluble in 5% acetic acid; sold under license of Pharma Bissendorf Peptide GmbH | A more potent analog of DSIP; Friedman, TC et al, *J Clin Endocrinol Metab*, 78:1085, 1994

Trp-Ala-Gly-Gly-Asp-Ala-Ser-Gly-Glu

Synonyms: Sleep Inducing Peptide, δ-

ICN 150793   MW 848.8 $C_{35}H_{48}N_{10}O_{15}$ Schoenenberger, GA & M Monnier, *PNAS*, 74:1282, 1977

ICN 153182   Schoenenberger, GA & M Monnier, *PNAS*, 74:1282, 1977; Graf, MV & AJ Kastin, *Neurosci Behav Rev*, 8:83, 1984; Graf, MV & AJ Kastin, *Peptides*, 7:1165, 1986

Neosystem SC383   MW 848.82   Bioactive; opioid peptide; Schoenenberger, GA & Monnier, M, *PNAS USA*, 74:1282, 1977

Sigma T 1762   FW 848.8 $C_{35}H_{48}N_{10}O_{15}$ ≥97% (HPLC) | Bioactive peptide; Schoenenberger, GA & Monnier, M, *Proc Natl Acad Sci USA*, 74:1282, 1977; Kafi, S et al, *Neurosci Lett*, 13:169, 1979; Monnier, M et al, *Experientia*, 33:548, 1977

Biogenesis 2630-0004   Synthetic   MW 849   Lyophilized, consisting of 81% peptide material, acetate counter ions and residual H_2O; purified

Peptides International PDS-4054-v   Synthetic   MW 848.82 $C_{35}H_{48}N_{10}O_{15}$ >99% (HPLC); lyophilized amorphous powder | Bioactive; Schoenenberger, GA & M Monnier, *PNAS USA*, 74:1282, 1977; Monnier, M et al, *Experientia*, 33:548, 1977

Trp-Ala-Gly-Gly-Asp-Ala-Ser-Gly-Glu 4H₂O

Synonyms: Sleep Inducing Peptide, δ-

Peptides International PDS-4054   Synthetic   MW 920.88 $C_{35}H_{48}N_{10}O_{15} \cdot 4H_2O$ >99% (HPLC); lyophilized amorphous powder; bulk | Bioactive; Schoenenberger, GA & M Monnier, *PNAS USA*, 74:1282, 1977; Monnier, M et al, *Experientia*, 33:548, 1977

Trp-Ala-Trp-Phe
D-Trp-Ala-Trp-D-Phe Amide

Synonyms: Growth Hormone Releasing Factor

Sigma T 9768   FW 607.7 $C_{34}H_{37}N_7O_4$ ≥97% (HPLC) | Bioactive peptide

Trp-Ala-Val-Gly-His-Leu-Met Amide

Synonyms: Bombesin (8-14)

American Peptide 16-7-23   MW 812.0 $C_{38}H_{57}N_{11}O_7S$

Bachem H-2170.0001   Store at -15°C

ICN 152859   MW 812 $C_{38}H_{57}N_{11}O_7S$

Sigma B 1150   FW 812.0 $C_{38}H_{57}N_{11}O_7S$ ≥97% (HPLC) | Bioactive peptide; gastrointestinal peptide; agonist

Trp-Arg Dihydrochloride

Bachem G-3320.0250   Store at -15°C

Trp-Arg-Asp-Leu-Trp-Glu-Thr-Leu-Arg-Arg-Gly-Gly-Arg-Trp-Ile-Leu-Ala-Ile-Pro-Arg

Synonyms: GP140 (851-870); SIVmac251 Peptide

Neosystem SC793   MW 2550.99   SIVmac251 peptide from subtype GP140

Trp-Arg-Glu-Met-Ser-Val-Trp Amide

Synonyms: WW Amide-2

Bachem H-1624.0005   Store at -15°C

Trp-Asn

Bachem G-3325.0250   MW 318.33 $C_{15}H_{18}N_4O_4$ Store at -15°C

Trp-Asn-Arg-Gln-Leu-Tyr-Pro-Glu-Trp-Thr-Glu-Ala-Gln-Arg-Leu-Asp

Bachem H-4406.0500   MW 2105.3 $C_{95}H_{137}N_{27}O_{28}$ Store at -15°C

Trp-Asp

Bachem G-3330.0250   MW 319.32 $C_{15}H_{17}N_3O_5$ Store at -15°C

Trp-Gln-Glu-Val-Gly-Lys-Ala-Met-Tyr-Ala-Pro-Pro-Ile-Ser-Gly-Gln-Ile

Synonyms: GP120 (432-448); HTLV I Peptide

Neosystem SC248   MW 1875.17   HTLV I peptide from subtype GP160

Trp-Gln-Leu-Phe-Glu-Thr-Ser-Ile-Lys-Pro-Cys-Val-Lys-Leu-Ser-Pro-Leu-Cys-Ile-Thr
Trp-Gln-Leu-Phe-Glu-Thr-Ser-Ile-Lys-Pro-Cys(Acm)-Val-Lys-Leu-Ser-Pro-Leu-Cys(Acm)-Ile-Thr

Synonyms: GP140 (91-110); SIVmac251 Peptide

Neosystem SC717   MW 2448.96   SIVmac251 peptide from subtype GP140

Trp-Gln-Pro-Pro-Arg-Ala-Arg-Ile

Synonyms: Fibronectin Adhesion Promoting Peptide

Bachem H-1594.0005   MW 1023.21 $C_{47}H_{74}N_{16}O_{10}$ Store at -15°C

ICN 195620   MW 1023.2   97%

Sigma F 3667   FW 1023.2   ≥95% (HPLC); peptide content:~70% | Bioactive peptide; sequence found in the carboxy-terminal heparin-binding domain of fibronectin; Woods, A et al, *Mol Biol Cell*, 4:605, 1993

Trp-Gln-Pro-Pro-Trp-Tyr-Cys-Lys-Glu-Pro-Val-Arg-Ile-Gly-Ser-Cys-Lys-Lys-Gln-Phe-Ser-Ser-Phe-Tyr-Phe-Lys-Trp-Thr-Ala-Lys-Lys-Cys-Leu-Pro-Phe-Leu-Phe-Ser-Gly-Cys-Gly-Gly-Asn-Ala-Asn-Arg-Phe-Gln-Thr-Ile-Gly-Glu-Cys-Arg-Lys-Lys-Cys-Leu-Gly-Lys

Synonyms: Calcicludine

Calbiochem 207555   *Dendroaspis angusticeps*   MW 6980.2 $C_{321}H_{476}N_{86}O_{78}S_6$ ≥98% (SDS-PAGE); lyophilized solid; soluble in aqueous buffer (pH 7.5); S-S linkage not determined; bioassay:shown to block spontaneous or K^+-induced contraction of cardiac cells; harmful:LD_{50} ≤2000 mg/kg | Neurotoxin that acts as a highly potent neuronal Ca^{2+} channel blocker (L-, N- & P-types); preferentially blocks L-type Ca^{2+} channels (EC_{50}=200 pM); Schweitz, H et al, *PNAS*, 91:878, 1994

Trp-Gln-Pro-Pro-Trp-Tyr-Cys-Lys-Glu-Pro-Val-Arg-Ile-Gly-Ser-Cys-Lys-Lys-Gln-Phe-Ser-Ser-Phe-Tyr-Phe-Lys-Trp-Thr-Ala-Lys-Lys-Cys-Leu-Pro-Phe-Leu-Lys-Lys-Cys-Leu-Gly-Lys

Synonyms: Calcicludine; Neuronal L-Type Ca^{2+} Channel Blocker

Peptides International PCC-4310-s   *Dendroaspis angusticeps* (green mamba) synthetic   MW 6980.2 $C_{321}H_{476}N_{86}O_{78}S_6$ >99% (HPLC); lyophilized amorphous powder | Bioactive; Disulfide bonds: Cys^7-Cys^{57}, Cys^{16}-Cys^{40}, Cys^{32}-Cys^{53}; Schweitz, H, *PNAS* USA, 91:878, 1994; Nishio, H et al, *Peptide Chem*, 1995:113, 1996; Uchitel, OD, *Toxicon*, 35:1161, 1997

Trp-Gln-Pro-Pro-Trp-Tyr-Cys-Lys-Glu-Pro-Val-Arg-Ile-Gly-Ser-Cys-Lys-Lys-Gln-Phe-Ser-Ser-Phe-Tyr-Phe-Lys-Trp-Thr-Ala-Lys-Lys-Cys-Leu-Pro-Phe-Leu-Phe-Ser-Gly-Cys-Gly-Gly-Asn-Ala-Asn-Arg-Phe-Gln-Thr-Ile-Gly-Glu-Cys-Arg-Lys-Lys-Cys-Leu-Gly-Lys

Synonyms: Calcicludine

Alexis 630-060   Synthetic   MW 6979.0 $C_{321}H_{476}N_{86}O_{78}S_6$ Lyophilized | Disulfide bonds: Cys^7-Cys^{57}, Cys^{16}-Cys^{40}, Cys^{32}-Cys^{53}; potent neurotoxin

---

Trp-Gln-Val-Asp-Arg-Met-Arg-Ile-Arg-Thr-Trp-Lys-Ser-Leu-Val

Synonyms: VIF (11-25); HTLV I Peptide

Neosystem SC229   MW 1974.35   HTLV I peptide from subtype VIF

Trp-Glu

Bachem G-3335.0250   MW 333.34 $C_{16}H_{19}N_3O_5$ Store at -15°C

ICN 157147   MW 333.3 $C_{16}H_{19}N_3O_5$ Crystalline

Trp-Glu-Ala-Lys-Leu-Ala-Lys-Ala-Leu-Ala-Lys-Ala-Leu-Ala-Lys-His-Leu-Ala-Lys-Ala-Leu-Ala-Lys-Ala-Leu-Lys-Ala-Cys-Glu-Ala

Synonyms: KALA Amphipathic Peptide

Bachem H-4096.0500   Store at -15°C

Trp-Glu-His-Asp
Ac-Trp-Glu-His-Asp-CHO

Synonyms: Caspase I Inhibitor

Alexis 260-055   MW 611.6 $C_{28}H_{33}N_7O_9$ Solid | Very potent reversible inhibitor; optimal tetrapeptide recognition motif for this enzyme; Rano, TA et al, *Chem Biol*, 4; 149, 1997; Thornberry, NA et al, *J Biol Chem*, 272:17907, 1997

Trp-Glu-His-Asp
Ac-Trp-Glu-His-Asp-AMC

Synonyms: Caspase I Substrate

Alexis 260-057   MW 784.8 $C_{38}H_{40}N_8O_{11}$ ≥98%; white powder; soluble in DMSO | Fluorogenic substrate with an optimal tetrapeptide sequence; cleaved 50-fold more efficiently than Ac-Tyr-Val-Ala-Asp-AMC; AMC has an excitation maximum of 380 nm & an emission maximum of 460 nm; Thornberry, NA et al, *J Biol Chem*, 272:17907, 1997; Rano, TA et al, *Chem Biol*, 4:149, 1997

Bachem I-1715.0005   MW 784.78 $C_{38}H_{40}N_8O_{11}$ Store at -15°C

Trp-Glu-His-Asp
Ac-Trp-Glu-His-Asp-CHO Pseudo Acid

Bachem N-1630.0005   MW 611.61 $C_{28}H_{33}N_7O_9$ Store at -15°C

Trp-Glu-His-Asp
Ac-Trp-Glu-His-Asp-pNA

BioSource International 77-974   Colored substrate

Trp-Glu-His-Asp
Ac-Trp-Glu-His-Asp-AMC

BioSource International 77-977   Fluorescing substrate

Trp-Glu-His-Asp
Ac-Trp-Glu-His-Asp-pNA

Synonyms: Caspase I Colorimetric Substrate VII; Caspase V Colorimetric Substrate I

Calbiochem 218736   MW 747.7 $C_{34}H_{37}N_9O_{11}$ ≥97% (HPLC); lyophilized solid; soluble in DMSO | Colorimetric Caspase I substrate; cleavage of *pNA* is monitored colorimetrically at ~405 nm; a substrate for Caspase IV Caspase V; Thornberry, NA et al, *J Biol Chem*, 272:17907, 1997

Trp-Glu-His-Asp
Ac-Trp-Glu-His-Asp-AMC

Synonyms: Caspase I Fluorogenic Substrate X; Caspase V Fluorogenic Substrate III

Calbiochem 218739   MW 784.8 $C_{38}H_{40}N_8O_{11}$ ≥97% (HPLC); lyophilized solid; soluble in DMSO | excitation max: ~365-380 nm; emission max: ~430-460 nm | Fluorogenic substrate for detecting Caspase I, Caspase IV & Caspase V activity; Thornberry, NA et al, *J Biol Chem*, 272:17907, 1997

Trp-Glu-His-Asp
Z-Trp-Glu(OMe)-His-Asp(OMe)-CH₂F Trifluoroacetate Salt

Synonyms: Caspase V Inhibitor I

Calbiochem 218753   MW 763.8 $C_{37}H_{42}FN_7O_{10}$ Single spot purity (TLC); solid; soluble in DMSO | Potent, cell-permeable & irreversible inhibitor of Caspase V as well as Caspases I & IV

Trp-Glu-His-Asp
Ac-Trp-Glu-His-Asp-AFC Trifluoroacetate Salt

Synonyms: Caspase V Fluorogenic Substrate II

Calbiochem 218754   MW 838.8 $C_{38}H_{37}F_3N_8O_{11}$ ≥95% (HPLC); solid; soluble in DMSO; excitation max: ~400 nm; emission max: ~505 nm | Fluorogenic substrate for Caspase V; reaction monitored visually or quantitatively by a blue to green shift in fluorescence upon cleavage of the AFC fluorophore; Thornberry, NA et al, *J Biol Chem*, 272:17907, 1997

Trp-Glu-His-Asp
Z-Trp-Glu-His-Asp-FMK

Synonyms: Caspase I/IV/V Inhibitor

Kamiya   MW 877

Trp-Glu-His-Asp
Ac-Trp-Glu-His-Asp-pNA

Synonyms: Caspase I/IV/V Substrate

Kamiya   MW 862   >97% (HPLC)

Trp-Glu-His-Asp
Ac-Trp-Glu-His-Asp-AMC

Synonyms: Caspase I Substrate

Neosystem SC1254   MW 784.78   The optimal tetrapeptide recognition sequence for Caspase I; cleaved 50-fold more efficiently than Ac-Tyr-Val-Ala-Asp-AMC (see Neosystem SC1209); Rano, TA et al, *Chemistry & Biology*, 4:149-155, 1997; Thornberry, NA et al, *JBC*, 272:17907-17911, 1997

Trp-Glu-His-Asp
Ac-Trp-Glu-His-Asp-MCA

Synonyms: Caspase I/IV/V Substrate

Peptides International MCA-3186-v   MW 784.78 $C_{38}H_{40}N_8O_{11}$ >98% (HPLC); lyophilized amorphous powder | Rano, TA et al, *Chem Biol*, 4:149, 1997; Thornberry, NA et al, *JBC*, 272:17907, 1997

Trp-Glu-His-Asp
N-Ac-Trp-Glu-His-Asp-MCA

Synonyms: Caspase I Substrate

Sigma A 0216   Fluorogenic substrate; Rano, TA et al, *Chem Biol*, 4:149, 1997

Trp-Glu-His-Asp
N-Ac-Trp-Glu-His-Asp-CHO

Synonyms: Caspase I Inhibitor

Sigma A 1466   Bioactive peptide; extremely potent inhibitor; K_I=56 pM; Rano, TM et al, *Chem Biol*, 4:149, 1997

Trp-Glu-His-Asp
Ac-Trp-Glu-His-Asp-CHO

Synonyms: Caspase I Inhibitor

Peptides International ICA-3187-v   Synthetic   MW 611.61 $C_{28}H_{33}N_7O_9$ >99% (HPLC); lyophilized amorphous powder | Inhibitor; Rano, TA et al, *Chem Biol*, 4:149, 1997

Trp-Glu-Tyr

Bachem H-6235.0250   Store at -15°C

Trp-Gly

American Peptide 32-0-12   MW 261.3 $C_{13}H_{15}N_3O_3$ Crystalline | Found in human pituitary gland; opioid-like effects on mice; Partanen, S et al, *Acta Physiol Scand*, 107:213, 1979

Trp-Gly
Z-Trp-Gly

Bachem C-2700.0001   MW 395.42 $C_{21}H_{21}N_3O_5$ Store at -15°C

Trp-Gly

Bachem G-3340.0250   MW 261.28 $C_{13}H_{15}N_3O_3$ Store at -15°C

ICN 103159   MW 261.3 $C_{13}H_{15}N_3O_3$ Crystalline

Sigma T 1754   FW 261.3 $C_{13}H_{15}N_3O_3$ ≥97% (HPLC) | Bioactive peptide; dipeptide found in human pituitary gland; Partanen, S et al, *Acta Physiol Scand*, 107:213, 1979

Sigma T 7306   Human pituitary gland   FW 297.7 $C_{13}H_{15}N_3O_3 \cdot$ HCl   Bioactive peptide; Partanen, S et al, *Acta Physiol Scand*, 107:213, 1979

Trp-Gly-Cys-Ala-Phe-Arg-Gln-Val-Cys-His-Thr-Thr-Val-Pro-Trp-Pro-Asn-Ala-Ser-Leu
Trp-Gly-Cys(Acm)-Ala-Phe-Arg-Gln-Val-Cys(Acm)-His-Thr-Thr-Val-Pro-Trp-Pro-Asn-Ala-Ser-Leu

Synonyms: GP140 (611-630); SIVmac251 Peptide

Neosystem SC769   MW 2415.77   SIVmac251 peptide from subtype GP140

Trp-Gly-Gly

Bachem H-5095.0250   MW 318.33 $C_{15}H_{18}N_4O_4$ Store at -15°C

ICN 103160   MW 318.3 $C_{15}H_{18}N_4O_4$ Crystalline

Trp-Gly-Gly-Gly-Tyr

Bachem H-5100.0025   MW 538.56 $C_{26}H_{30}N_6O_7$ Store at -15°C

Trp-Gly-Gly-Tyr

Bachem H-5105.0025   MW 481.51 $C_{24}H_{27}N_5O_6$ Store at -15°C

Trp-Gly-Leu-Thr-Lys-Ser-Ser-Thr-Thr-Ile-Thr-Thr-Ala-Ala-Pro-Thr-Ser-Ala-Pro-Val

Synonyms: GP140 (121-140); SIVmac251 Peptide

Neosystem SC720   MW 1990.23   SIVmac251 peptide from subtype GP140

Trp-Gly-Pro-Asn-Asp-Pro-Arg-Arg

Synonyms: HCV Core Protein (107-114)

Bachem H-2544.0001   Store at -15°C

Trp-Gly-Tyr

Bachem H-5110.0050   MW 424.46 $C_{22}H_{24}N_4O_5$ Store at -15°C

Trp-His-Leu-Gly-Gln-Gly-Val-Ser-Ile-Glu-Trp-Arg-Lys

Synonyms: VIF (79-91); HTLV I Peptide

Neosystem SC547   MW 1595.82   HTLV I peptide from subtype VIF

Trp-His-Trp-Leu-Gln-Leu

Synonyms: Mating Factor (1-6), α_1-

American Peptide 79-2-47   MW 882.0 $C_{45}H_{59}N_{11}O_8$

Bachem H-6205.0005   MW 882.03 $C_{45}H_{59}N_{11}O_8$ Store at -15°C

ICN 153183

Sigma T 2903   FW 882.0 $C_{45}H_{59}N_{11}O_8$ ≥97% (HPLC) | Bioactive peptide; Partanen, S et al, *Acta Physiol Scand*, 107:213, 1979

Trp-His-Trp-Leu-Gln-Leu-Lys-Pro-Gly-Gln-Pro-Met-Tyr
FMOC-Trp-His-Trp-Leu-Gln-Leu-Lys-Pro-Gly-Gln-Pro-Met-Tyr

Synonyms: Mating Factor α, FMOC-

Bachem B-1620.0001   Store at -15°C

Trp-His-Trp-Leu-Gln-Leu-Lys-Pro-Gly-Gln-Pro-Met-Tyr

Synonyms: Mating Factor, α-; Pheromone α-Factor; Factor, α-; Mating Factor, α_1-; Mating Factor, *Saccharomyces cerevisiae*

Bachem H-2925.0001   MW 1684.01 $C_{82}H_{114}N_{20}O_{17}S$ Store at -15°C

ICN 151589   Matsui, Y etal, *BBRC*, 78:534, 1977

Sigma T 6901   FW 1684.0 (free base)   Masui, Y et al, *Biochem Biophys Res Commun*, 78:534, 1977

American Peptide 83-0-60   *Saccharomyces cerevisae* (yeast) MW 1684.2 $C_{82}H_{114}N_{20}O_{17}S$ Pheromone secreted by haploid α-cells of *Saccharomyces cerevisia*, binds to Ste2p, a 7-transmembrane, G-protein-coupled receptor presenting on haploid α cells, & activates a signal transduction pathway required for conjugation & mating; Masui, Y et al, *BBRC*, 78:534, 1977; Herskowitz, I, *Cell*, 50:995, 1987

Fluka 63591   *Saccharomyces cerevisae* (yeast)   MW 1684 $C_{82}H_{114}N_{20}O_{17}S$ ≥97.0% (HPLC); ~85% peptide content; ≤10% acetate | Pheromone secreted by α-type cells of S. cerevisiae; Gooday, GW, *Ann Rev Biochem*, 43:35, 1974; Masui, Y et al, *BBRC*, 78:534, 1977

Peptides International PMF-4076-v   *Saccharomyces cerevisiae* (yeast) synthetic   MW 1684.0 $C_{82}H_{114}N_{20}O_{17}S$ >95% (HPLC); lyophilized amorphous powder | Bioactive; Stöetzler, D et al, *Eur J Biochem*, 69:397, 1976; Tanaka, T et al, *J Biochem*, 82:1681, 1977; Masui, Y et al, *BBRC*, 78:534, 1977

Neosystem SC136   Yeast   MW 1683.99   Virus-related peptide; bioactive; Stötzler, D et al, *Mol Pharm Biochem*, 69:397-400, 1976

Trp-His-Trp-Leu-Ser-Phe-Ser-Lys-Gly-Glu-Pro-Met-Tyr

Synonyms: Factor, α-SK2-; Pheromone α–SK2-Factor; Mating Factor, α-SK2-

Bachem H-4580.0001   MW 1667.91 $C_{81}H_{106}N_{18}O_{19}S$ Store at -15°C

ICN 153184   Sakurai, A etal, *FEBS Lett*, 166:339, 1984

Sigma T 0779   ≥95% (HPLC) | Bioactive peptide; Sakurai, A et al, *FEBS Lett*, 166:339, 1984

Trp-Ile

ICN 103161   Crystalline

Trp-Leu
Z-Trp-Leu

Bachem C-2705.0001   MW 451.52 $C_{25}H_{29}N_3O_5$ Store at -15°C

Trp-Leu
Z-D-Trp-Leu

Bachem C-4005.0001   MW 451.52 $C_{25}H_{29}N_3O_5$ Store at -15°C

Trp-Leu

Bachem G-3345.0250   MW 317.39 $C_{17}H_{23}N_3O_3$ Store at -15°C

ICN 103162   MW 317.4 $C_{17}H_{23}N_3O_3$ 99%; crystalline

Sigma T 1879   FW 317.4 $C_{17}H_{23}N_3O_3$

Trp-Leu-Asp-Ile-Ile-Trp
Ac-D-Trp-Leu-Asp-Ile-Ile-Trp

Synonyms: Endothelin I (16-21), (Ac-D-Trp[16])-; Endothelin I (16-21), N-Ac-(D-Trp[16])-; Endothelin I (16-21), Ac-(D-Trp[16])-

Bachem H-8850.0001   MW 887.05 $C_{46}H_{62}N_8O_{10}$ Store at -15°C

Sigma E 7394   FW 887.0 $C_{46}H_{62}N_8O_{10}$ ≥95% (HPLC) | Bioactive peptide; ET_A antagonist; inhibits ET-1 stimulated arachidonic acid release in rabbit vascular smooth muscle cells; Cody, WL, *J Med Chem*, 35:3301, 1992

American Peptide 88-2-44   Human   MW 887.0 $C_{46}H_{62}N_8O_{10}$ Inhibits ET-1 stimulated arachidonic acid release in rabbit vascular smooth muscle cells; Cody, WL et al, *J Med Chem*, 35:3301, 1992

Trp-Lys

Bachem G-3800.0250   MW 332.4 $C_{17}H_{24}N_4O_3$ Store at -15°C

Trp-Lys-Gln-Met-Ser-Val-Trp Amide

Synonyms: WW Amide-3

Bachem H-1626.0005   Store at -15°C

Trp-Lys-Glu-Met-Ser-Val-Trp Amide

Synonyms: WW Amide-1

Bachem H-1622.0005   Store at -15°C

Trp-Lys-His-Pro-Gly-Ser-Gln-Pro-Lys-Thr-Ala-Cys-Thr-Thr

Synonyms: TAT (11-24); HTLV I Peptide

Neosystem SC207   MW 1541.74   HTLV I peptide from subtype TAT; due to the presence of a Cys this peptide can contain the dimeric form

Trp-Met
Z-Trp-Met

Bachem C-2710.0001   MW 469.56 $C_{24}H_{27}N_3O_5S$ Store at -15°C

Trp-Met

Bachem G-3350.0250   MW 335.43 $C_{16}H_{21}N_3O_3S$ Store at -15°C

Trp-Met-Asn-Phe Amide Hydrochloride

Bachem H-5115.0005   MW 632.18 $C_{29}H_{37}N_7O_5S \cdot HCl$
Store at -15°C

Trp-Met-Asn-Ser-Thr-Gly-Phe-Thr-Lys-Val-Cys-Gly-Ala-Pro-Pro-Cys

Synonyms: HCV-1 e2 Protein (554-569)

Bachem H-2682.0500   Store at -15°C

Trp-Met-Asp-Phe
N-t-Amyloxycarbonyl-Trp-Met-Asp-Phe Amide

Synonyms: Gastrin Related Peptide

ICN 152853   MW 710.8 $C_{35}H_{46}N_6O_8S$ Ishii, Y & H Shiozaki, *Japan J Pharmac*, 18:93, 1968

Trp-Met-Asp-Phe
N-t-BOC-Trp-Met-Asp-Phe Amide

Synonyms: Gastrin Related Tetrapeptide

ICN 152854   MW 696.8 $C_{34}H_{44}N_6O_8S$

Sigma B 4009   FW 696.8 $C_{34}H_{44}N_6O_8S$ ≥97% (HPLC) | Bioactive peptide; gastrointestinal peptide

Trp-Met-Asp-Phe
Aoc-Trp-Met-Asp-Phe Amide

Synonyms: Gastrin Related Peptide

Peptides International PGR-4004   Synthetic   MW 710.85 $C_{35}H_{46}N_6O_8S$ >95% (HPLC); lyophilized amorphous powder; bulk | Bioactive; Ishii, Y & H Shinozaki, *Japan J Pharmacol*, 18:93, 1968

Trp-Met-Asp-Phe Amide

Synonyms: Cholecystokinin Tetrapeptide (30-33); Gastrin Related Tetrapeptide; Cholecystokinin (30-33); Gastrin I (14-17); Gastrin Tetrapeptide; Gastrin (14-17); Cholecystokinin IV; Cholecystokinin Octapeptide (5-8)

American Peptide 46-1-40   MW 596.7 $C_{29}H_{36}N_6O_6S$ Rehfeld, JF et al, *Nature*, 284:33, 1980

ICN 152862   MW 596.7   Gastrin-related tetrapeptide; potent as a releaser of insulin & other islet hormones; Rehfeld, R etal, *Nature*, 284:33, 1980

ICN 152876   MW 1705.8   Potent releaser of insulin & other islet hormones; Larsson, LI & JF Rehfeld, *Nature*, 277:575 1979; Rehfeld, JF etal, *Nature*, 284:33, 1980

Bachem H-3110.0025   Human   MW 596.71 $C_{29}H_{36}N_6O_6S$ Store at -15°C

Trp-Met-Asp-Phe Amide Hydrochloride

Synonyms: Cholecystokinin (30-33); Cholecystokinin IV; Cholecystokinin Tetrapeptide (30-33)

Sigma T 6515   FW 633.2 $C_{29}H_{36}N_6O_6S \cdot HCl$ ≥95% (HPLC) | Bioactive peptide; gastrointestinal peptide; anxiogenic (induces panic); Rex, A et al, *Neurosci Lett*, 172:139, 1994

Peptides International PCK-4083-v   Synthetic   MW 596.71 $C_{29}H_{36}N_6O_6S$ HCl   >95% (HPLC); lyophilized amorphous powder | Bioactive; Jorpes, JE & V Mutt, Eds, *Secretin, Cholecystokinin, Pancreozymin & Gastrin*, Handbook of Experimental Pharmacology, Vol 34, Springer-Verlag, Berlin, 1973; Rehfeld, JF et al, *Nature*, 284:33, 1980

Trp-Met-Asp-Phe Amide Hydrochloride Hydrate

Synonyms: Cholecystokinin Tetrapeptide (30-33)

Peptides International PCK-4083   Synthetic   MW 651.19 $C_{29}H_{36}N_6O_6S \cdot HCl \cdot H_2O$ >95% (HPLC); lyophilized amorphous powder; bulk | Bioactive; Jorpes, JE & V Mutt, Eds, *Secretin, Cholecystokinin, Pancreozymin & Gastrin*, Handbook of Experimental Pharmacology, Vol 34, Springer-Verlag, Berlin, 1973; Rehfeld, JF et al, *Nature*, 284:33, 1980

Trp-Met-Phe
N-t-BOC-β-Trp-Met-Phe Amide

Synonyms: Gastrin Related Peptide, (des-Asp[3])-

ICN 195676   MW 581.7 $C_{30}H_{39}N_5O_5S$ ≥95%

Trp-Met-Phe
N-t-BOC-Trp-Met-Phe Amide

Synonyms: Gastrin Related Tetrapeptide, (des-Asp[3])-

Sigma B 3144   FW 581.7 $C_{30}H_{39}N_5O_5S$ ≥97% (TLC) | Bioactive peptide; gastrointestinal peptide

Trp-Met-Phe-Gla
D-Trp-D-Met-p-Chloro-D-Phe-Gla Amide

Bachem N-1665.0005   MW 689.19 $C_{31}H_{37}ClN_6O_8S$ Store at -15°C

Trp-Nle-Arg-Phe Amide

Synonyms: Molluscan Cardioexcitatory Neuropeptide Analog

American Peptide 50-5-30   MW 619.8 $C_{32}H_{45}N_9O_4$

Bachem H-3000.0005   MW 619.77 $C_{32}H_{45}N_9O_4$ Store at -15°C

ICN 153185   A molluskan cardioexcitatory neuropeptide analog

Trp-Nle-Arg-Phe Amide Acetate Salt

Synonyms: Molluscan Cardioexcitatory Neuropeptide Analog

Sigma T 9897   FW 619.8 $C_{32}H_{45}N_9O_4$ ≥97% (HPLC) | Bioactive peptide

Trp-Phe
Z-Trp-Phe

Bachem C-2715.0001   MW 485.54 $C_{28}H_{27}N_3O_5$ Store at -15°C

Trp-Phe

Bachem G-3360.0250   MW 351.41 $C_{20}H_{21}N_3O_3$ Store at -15°C

ICN 157148   MW 351.4 $C_{20}H_{21}N_3O_3$ Crystalline

Sigma T 2004   FW 351.4 $C_{20}H_{21}N_3O_3$ ≥98% (TLC)

Trp-Phe-Trp-Ser-Tyr-Arg-Leu-Arg-Pro-Ala
Ac-D-Trp-p-Chloro-D-Phe-D-Trp-Ser-Tyr-D-Arg-Leu-Arg-Pro-D-Ala Amide

Synonyms: Luteinizing Hormone Releasing Hormone, (Ac-D-Trp[1],p-Chloro-D-Phe[2],D-Trp[3],D-Arg[6],D-Ala[10])-

Bachem H-5575.0001   MW 1457.1 $C_{71}H_{94}ClN_{19}O_{13}$ Store at -15°C

Trp-Phe-Trp-Ser-Tyr-Arg-Leu-Arg-Pro-Ala
N-Ac-D-Trp-D-p-Chloro-Phe-D-Trp-Ser-Tyr-D-Arg-Leu-Arg-Pro-D-Ala Amide

Synonyms: Luteinizing Hormone Releasing Hormone, (N-Ac-D-Trp[1],D-p-Chloro-Phe[2],D-Trp[3],D-Arg[6],D-Ala[10])-

ICN 152915

Trp-Phe-Trp-Ser-Tyr-Arg-Leu-Arg-Pro-Ala
Ac-D-Trp-D-*p*-Chloro-Phe-D-Trp-Ser-Tyr-D-Arg-Leu-Arg-Pro-D-Ala Amide

Synonyms: Luteinizing Hormone Releasing Hormone, (Ac-D-Trp[1],D-*p*-Chloro-Phe[2],D-Trp[3],D-Arg[6],D-Ala[10])

Sigma L 0137　FW 1457.1　≥97% (HPLC) | Bioactive peptide; LH-RH antagonist

Trp-Phe-Tyr-Ser-Pro-Arg
Trp-Phe-Tyr-Ser(PO₃H₂)-Pro-Arg-pNA

Bachem L-2075.0001　MW 1055.05　$C_{49}H_{59}N_{12}O_{13}P$　Store at -15°C

Trp-Pro
FMOC-Trp-Pro

Bachem B-2705.0001　Store at -15°C

Trp-Pro

Bachem G-3790.0250　MW 301.35　$C_{16}H_{19}N_3O_3$　Store at -15°C

Trp-Pro-Pro-Pro-Tyr

Bachem H-5895.0050　MW 658.76　$C_{35}H_{42}N_6O_7$　Store at -15°C

Trp-Pro-Pro-Tyr

Bachem H-5890.0050　MW 561.64　$C_{30}H_{35}N_5O_6$　Store at -15°C

Trp-Pro-Tyr

Bachem H-7390.0050　MW 464.52　$C_{25}H_{28}N_4O_5$　Store at -15°C

Trp-Ser

Bachem G-3365.0250　MW 291.31　$C_{14}H_{17}N_3O_4$　Store at -15°C

Trp-Ser-Lys-Met-Asp-Gln-Leu-Ala-Lys-Glu-Leu-Thr-Ala-Glu

Synonyms: Peptide WE-14; Chromogranin A (324-337); WE-14

Bachem H-8680.0001　MW 1649.89　$C_{72}H_{116}N_{18}O_{24}S$　Store at -15°C

American Peptide 76-4-71　Human　MW 1649.9　$C_{72}H_{116}N_{18}O_{24}S$　Novel chromogranin A-derived peptide, isolated from a human ileal carcinoid tumor; shares its highly conserved sequence among many mammalian species

Trp-Ser-Tyr-Gly-Leu-Arg-Pro-Gly Amide

Synonyms: Luteinizing Hormone Releasing Hormone (3-10)

Bachem H-5735.0001　MW 934.07　$C_{44}H_{63}N_{13}O_{10}$　Store at -15°C

ICN 152921

Sigma L 5512　FW 934.1　$C_{44}H_{63}N_{13}O_{10}$　≥97% (HPLC) | Bioactive peptide

Trp-Ser-Tyr-His-Leu-Arg-Pro
Trp-Ser-Tyr-D-His(Bzl)-Leu-Arg-Pro-NHEt

Synonyms: Luteinizing Hormone Releasing Hormone (3-9), (D-His(Bzl)[6],Pro-NHEt[9])-

Bachem H-4808.0005　MW 1075.28　$C_{55}H_{74}N_{14}O_9$　Store at -15°C

Trp-Thr-Val-Pro-Thr-Ala

Synonyms: Fibrin, Necto; Nectofibrin Hexapeptide; Antifibronectin Peptide

American Peptide 42-5-10　Rat　MW 673.8　$C_{32}H_{47}N_7O_9$　Brentani, RR et al, *PNAS*, 85:364, 1988

Bachem H-9155.0005　Rat　Store at -15°C

Neosystem SC342　Rat　MW 673.76　Bioactive; cell attachment peptide; fibronectin fragment; Brentani, RR et al, *PNAS*, 85:364, 1988

Trp-Trp
Z-Trp-Trp

Bachem C-2720.0250　MW 524.58　$C_{30}H_{28}N_4O_5$　Store at -15°C

Trp-Trp

Bachem G-3370.0250　MW 390.44　$C_{22}H_{22}N_4O_3$　Store at -15°C

ICN 157149　MW 390.4　$C_{22}H_{22}N_4O_3$　Crystalline

Sigma T 2129　FW 390.4　$C_{22}H_{22}N_4O_3$

Trp-Trp-Gly-Lys-Lys-Tyr-Arg-Ala-Ser-Lys-Leu-Gly-Leu-Ala-Arg

Synonyms: C3a (63-77), (Trp[63],Trp[64])-

Bachem H-1264.0001　MW 1820.17　$C_{86}H_{134}N_{26}O_{18}$　Store at -15°C

Trp-Trp-Trp

Bachem H-6970.0050　MW 576.66　$C_{33}H_{32}N_6O_4$　Store at -15°C

Trp-Tyr

Bachem G-3375.0250　MW 367.41　$C_{20}H_{21}N_3O_4$　Store at -15°C

ICN 103164　MW 367.4　$C_{20}H_{21}N_3O_4$　Crystalline

Sigma T 2254　FW 367.4　$C_{20}H_{21}N_3O_4$

Trp-Tyr-Glu-Pro-Ile-Tyr-Leu-Gly-Gly-Val-Phe-Gln-Leu-Glu-Lys-Gly-Asp

Synonyms: Tumor Necrosis Factor (114-130); Cachectin (114-130)

ICN 152336　Synthetic　MW ~2k　Biological activity comparable with an active site of TNF; corresponds to human, mouse & rabbit TNF; Beutler, G etal, *Nature*, 316:552, 1985; Beutler, B & A Cerami, *Nature*, 320:584, 1986

Trp-Val
Z-Trp-Val

Bachem C-2725.0001　MW 437.5　$C_{24}H_{27}N_3O_5$　Store at -15°C

Trp-Val

Bachem G-3380.0250　MW 303.36　$C_{16}H_{21}N_3O_3$　Store at -15°C

Trp-Val
Trp-Val-OMe Hydrochloride

Bachem G-3385.0250　MW 353.85　$C_{17}H_{23}N_3O_3 \cdot$ HCl　Store at -15°C

Trp-Val

ICN 103331　Crystalline

Trp-Val-Tyr-His-Thr-Gln-Gly-Tyr-Phe-Pro-Asp-Trp-Gln

Synonyms: NEF MN (115-127); HIV I Peptide

Neosystem SC655   MW 1728.83   NEF peptide from HIV-1 subtype MN

Try-Glu

ICN 157164   MW 310.3 $C_{14}H_{18}N_2O_6$ Crystalline

Try-Gly

ICN 103200   MW 238.2 $C_{11}H_{14}N_2O_4$ Crystalline

Try-Gly-Gly

ICN 103334   MW 295.3 $C_{13}H_{17}N_3O_5$ 99%; crystalline | Metabolite of leucine enkephalin in rat brain

Try-Ile

ICN 103201

Try-Leu

ICN 103194   MW 294.3 $C_{15}H_{22}N_2O_4$ 99%; crystalline

Tyr(SO$_3$H)-Gly-Gly-Phe-Leu Sulfated

Synonyms: Enkephalin O, (Leu^5)-

ICN 190424   Unsworth, CD etal, *Nature*, 295:879, 1982

Tyr-Abz
Bz-Tyr-4-Abz

Bachem M-1155.0001   MW 404.42 $C_{23}H_{20}N_2O_5$ Store at RT

Tyr-Abz
Bz-Tyr-4-Abz Sodium Salt

Bachem M-1160.0001   MW 426.4 $C_{23}H_{19}N_2NaO_5$ Store at RT

Tyr-Ala
Tyr-D-Ala Amide

Synonyms: Endorphin (1-2), (D-Ala^2)-α-Neo-

American Peptide 30-0-42   MW 251.3 $C_{12}H_{17}N_3O_3$

Tyr-Ala
FMOC-Tyr-Ala

Bachem B-2095.0001   Store at -15°C

Tyr-Ala
Z-Tyr-Ala

Bachem C-2750.0250   MW 386.41 $C_{20}H_{22}N_2O_6$ Store at -15°C

Tyr-Ala

Bachem G-3390.0250   MW 252.27 $C_{12}H_{16}N_2O_4$ Store at -15°C

ICN 103182   MW 252.3 $C_{12}H_{16}N_2O_4$ Crystalline

Sigma T 5129   FW 252.3 $C_{12}H_{16}N_2O_4$

Tyr-Ala-Ala-Ala-Leu-Lys-Leu-Ala-Arg

Synonyms: C3a (69-77), (Tyr^{69},Ala^{71,72},Lys^{74})-

Bachem H-1432.0005   MW 976.19 $C_{45}H_{77}N_{13}O_{11}$ Store at -15°C

Tyr-Ala-Ala-Phe-Met
Tyr-D-Ala-D-Ala-Phe-Met Amide

Synonyms: Enkephalinamide, (D-Ala^{2,3},Met^5)-

ICN 152829   Jhamandas, K & M Sutak, *Br J Pharmac*, 71:201, 1980

Tyr-Ala-Asp-Ala-Ile-Phe-Thr-Asn-Ser-Tyr-Arg-Lys-Ile-Leu-Gly-Gln-Leu-Ser-Ala-Arg-Lys-Leu-Leu-Gln-Asp-Ile-Met-Asn-Arg-Gln-Gln-Gly-Glu-Arg-Asn-Gln-Glu-Gln-Gly-Ala-Lys-Val-Arg-Leu Amide

Synonyms: Growth Hormone Releasing Factor (1-44)

American Peptide 52-2-16   Ovine   MW 5121.9 $C_{221}H_{368}N_{72}O_{66}S$ Brazeau, P et al, *BBRC*, 125:606, 1984

ICN 151208   Ovine

Bachem H-3114.0500   Ovine, caprine   MW 5121.86 $C_{221}H_{368}N_{72}O_{66}S$ Store at -15°C

Tyr-Ala-Asp-Ala-Ile-Phe-Thr-Asn-Ser-Tyr-Arg-Lys-Val-Leu-Gly-Gln-Leu-Ser-Ala-Arg-Lys-Leu-Leu-Gln-Asp-Ile-Met-Asn-Arg-Gln-Gln-Gly-Glu-Arg-Asn-Gln-Glu-Gln-Gly-Ala-Lys-Val-Arg-Leu Amide

Synonyms: Growth Hormone Releasing Factor (1-44)

American Peptide 52-3-15   Bovine   MW 5107.9 $C_{220}H_{366}N_{72}O_{66}S$ Esch, F et al, *BBRC*, 117:772, 1983

Bachem H-3112.0500   Bovine   MW 5107.84 $C_{220}H_{366}N_{72}O_{66}S$ Store at -15°C

ICN 151207   Bovine   MW 5107.8

Tyr-Ala-Asp-Ala-Ile-Phe-Thr-Asn-Ser-Tyr-Arg-Lys-Val-Leu-Gly-Gln-Leu-Ser-Ala-Arg-Lys-Leu-Leu-Gln-Asp-Ile-Met-Asn-Arg Amide

Synonyms: Growth Hormone Releasing Factor (1-29)

Sigma G 0394   Bovine   FW 3384.9   ≥95% (HPLC) | Bioactive peptide; GRF fragment that is equipotent to GRF in releasing somatotropin from anterior pituitary

Tyr-Ala-Asp-Ala-Ile-Phe-Thr-Asn-Ser-Tyr-Arg-Lys-Val-Leu-Gly-Gln-Leu-Ser-Ala-Arg-Lys-Leu-Leu-Gln-Asp-Ile-Met-Asn-Arg-Gln-Gln-Gly-Glu-Arg-Asn-Gln-Glu-Gln-Gly-Ala-Lys-Val-Arg-Leu Amide

Synonyms: Growth Hormone Releasing Factor

Sigma G 0644   Bovine   FW 5107.8   ≥97% (HPLC) | Hypothalamic peptide that stimulates release of somatotropin from anterior pituitary; stimulates release of pituitary growth hormone (GH); some structurally unrelated short peptides also elicit growth hormone secretion by a different mechanism; Felix, AM et al, *Annual Reports in Medicinal Chem*, 20:185, 1985

Tyr-Ala-Asp-Ala-Ile-Phe-Thr-Asn-Ser-Tyr-Arg-Lys-Val-Leu-Gly-Gln-Leu-Ser-Ala-Arg-Lys-Leu-Leu-Gln-Asp-Ile-Met-Ser-Arg-Gln-Gln-Gly-Glu-Ser-Asn-Gln-Glu-Arg-Gly-Ala-Arg-Ala-Arg-Leu Amide

Synonyms: Growth Hormone Releasing Factor (1-44)

American Peptide 52-1-50   Human   MW 5039.8 $C_{215}H_{358}N_{72}O_{66}S$ Hypothalamic peptide that stimulates adenohypophyseal Growth Hormone (GH) secretion; Guillermin, R et al, *Science*, 218:585, 1982

Tyr-Ala-Asp-Ala-Ile-Phe-Thr-Asn-Ser-Tyr-Arg-Lys-Val-Leu-Gly-Gln-Leu-Ser-Ala-Arg-Lys-Leu-Leu-Gln-Asp-Ile-Met-Ser-Arg-Gln-Gln-Gly-Glu-Ser-Asn-Gln-Glu-Arg-Gly-Ala Amide

Synonyms: Growth Hormone Releasing Factor (1-40)

American Peptide 52-1-52   Human   MW 4543.1 $C_{194}H_{318}N_{62}O_{62}S$

Tyr-Ala-Asp-Ala-Ile-Phe-Thr-Asn-Ser-Tyr-Arg-Lys-
Val-Leu-Gly-Gln-Leu-Ser-Ala-Arg-Lys-Leu-Leu-Gln-
Asp-Ile-Met-Ser-Arg-Gln-Gln-Gly-Glu-Ser-Asn-Gln-
Glu-Arg-Gly-Ala

Synonyms: Growth Hormone Releasing Factor (1-40)

American Peptide 52-1-53   Human   MW 4544.1
$C_{194}H_{315}N_{61}O_{63}S$ GRF (1-44) & (1-40) originally isolated from human pancreatic tumor cells are equipotent in stimulating growth hormone release; Rivier, J et al, *Nature*, 300:276, 1982

Tyr-Ala-Asp-Ala-Ile-Phe-Thr-Asn-Ser-Tyr-Arg-Lys-
Val-Leu-Gly-Gln-Leu-Ser-Ala-Arg-Lys-Leu-Leu-Gln-
Asp-Ile-Met-Ser-Arg Amide

Synonyms: Growth Hormone Releasing Factor (1-29)

American Peptide 52-1-60   Human   MW 3358.0
$C_{149}H_{246}N_{44}O_{42}S$ The shortest GRF fragment with full biological activity; Lance, VA et al, *BBRC*, 119:265, 1984

Tyr-Ala-Asp-Ala-Ile-Phe-Thr-Asn-Ser-Tyr-Arg-Lys-
Val-Leu-Gly-Gln-Leu-Ser-Ala-Arg-Lys-Leu-Leu-Gln-
Asp-Ile-Met-Ser-Arg
Tyr-D-Ala-Asp-Ala-Ile-Phe-Thr-Asn-Ser-Tyr-Arg-Lys-
Val-Leu-Gly-Gln-Leu-Ser-Ala-Arg-Lys-Leu-Leu-Gln-
Asp-Ile-Met-Ser-Arg Amide

Synonyms: Growth Hormone Releasing Factor (1-29), (D-Ala²)-

American Peptide 52-1-61   Human   MW 3358.0
$C_{149}H_{246}N_{44}O_{42}S$ ~50 times more potent than GRF (1-29) in eliciting GH secretion in the rat; Lance, VA et al, *BBRC*, 119:265, 1984

Tyr-Ala-Asp-Ala-Ile-Phe-Thr-Asn-Ser-Tyr-Arg-Lys-
Val-Leu-Gly-Gln-Leu-Ser-Ala-Arg-Lys-Leu-Leu-Gln-
Asp-Ile-Met-Ser-Arg-Gln-Gln-Gly-Glu-Ser-Asn-Gln-
Glu-Arg-Gly-Ala-Arg-Ala-Arg-Leu Amide

Synonyms: Growth Hormone Pro-Release Factor; C-Terminal Peptide

American Peptide 52-1-67   Human   MW 3568.1
$C_{153}H_{248}N_{44}O_{50}S_2$

Tyr-Ala-Asp-Ala-Ile-Phe-Thr-Asn-Ser-Tyr-Arg-Lys-
Val-Leu-Gly-Gln-Leu-Ser-Ala-Arg-Lys-Leu-Leu-Gln-
Asp-Ile-Met-Ser-Arg
Ac-Tyr-D-Ala-Asp-Ala-Ile-Phe-Thr-Asn-Ser-Tyr-Arg-
Lys-Val-Leu-Gly-Gln-Leu-Ser-Ala-Arg-Lys-Leu-Leu-
Gln-Asp-Ile-Met-Ser-Arg Amide

Synonyms: Growth Hormone Releasing Factor (1-29), Ac-(D-Ala²)-

American Peptide 52-1-69   Human   MW 3485.1
$C_{154}H_{255}N_{47}O_{43}S$ Competitive antagonist of GHRH; Robberecht, P et al, *The Endocrine Soc*, 117:1759, 1985

Tyr-Ala-Asp-Ala-Ile-Phe-Thr-Asn-Ser-Tyr-Arg-Lys-
Val-Leu-Gly-Gln-Leu-Ser-Ala-Arg-Lys-Leu-Leu-Gln-
Asp-Ile-Met-Ser-Arg-Gln-Gln-Gly-Glu-Ser-Asn-Gln-
Glu-Arg-Gly-Ala-Arg-Ala-Arg-Leu Amide

Synonyms: Growth Hormone Releasing Factor (1-44)

Bachem H-3695.0500   Human   Store at -15°C

Tyr-Ala-Asp-Ala-Ile-Phe-Thr-Asn-Ser-Tyr-Arg-Lys-
Val-Leu-Gly-Gln-Leu-Ser-Ala-Arg-Lys-Leu-Leu-Gln-
Asp-Ile-Met-Ser-Arg Amide

Synonyms: Growth Hormone Releasing Factor (1-29)

Bachem H-3705.0500   Human   MW 3357.93
$C_{149}H_{246}N_{44}O_{42}S$ Store at -15°C

Tyr-Ala-Asp-Ala-Ile-Phe-Thr-Asn-Ser-Tyr-Arg-Lys-
Val-Leu-Gly-Gln-Leu-Ser-Ala-Arg-Lys-Leu-Leu-Gln-
Asp-Ile-Met-Ser-Arg
Tyr-D-Ala-Asp-Ala-Ile-Phe-Thr-Asn-Ser-Tyr-Arg-Lys-
Val-Leu-Gly-Gln-Leu-Ser-Ala-Arg-Lys-Leu-Leu-Gln-
Asp-Ile-Met-Ser-Arg Amide

Synonyms: Growth Hormone Releasing Factor (1-29), (D-Ala²)-

Bachem H-3715.0500   Human   MW 3357.93
$C_{149}H_{246}N_{44}O_{42}S$ Store at -15°C

Tyr-Ala-Asp-Ala-Ile-Phe-Thr-Asn-Ser-Tyr-Arg-Lys-
Val-Leu-Gly-Gln-Leu-Ser-Ala-Arg-Lys-Leu-Leu-Gln-
Asp-Ile-Met-Ser-Arg-Gln-Gln-Gly-Glu-Ser-Asn-Gln-
Glu-Arg-Gly-Ala-Arg-Ala-Arg-Leu Free Acid

Synonyms: Growth Hormone Releasing Factor (1-44)

Bachem H-4686.0500   Human   MW 5040.7
$C_{215}H_{357}N_{71}O_{67}S$ Store at -15°C

Tyr-Ala-Asp-Ala-Ile-Phe-Thr-Asn-Ser-Tyr-Arg-Lys-
Val-Leu-Gly-Gln-Leu-Ser-Ala-Arg-Lys-Leu-Leu-Gln-
Asp-Ile-Met-Ser-Arg-Gln-Gln-Gly-Glu-Ser-Asn-Gln-
Glu-Arg-Gly-Ala-Arg-Ala-Arg-Leu
Tyr-Ala-β-Asp-Ala-Ile-Phe-Thr-Asn-Ser-Tyr-Arg-Lys-
Val-Leu-Gly-Gln-Leu-Ser-Ala-Arg-Lys-Leu-Leu-Gln-
Asp-Ile-Met-Ser-Arg-Gln-Gln-Gly-Glu-Ser-Asn-Gln-
Glu-Arg-Gly-Ala-Arg-Ala-Arg-Leu Amide

Synonyms: Growth Hormone Releasing Factor (1-44), (β-Asp³)-

Bachem H-4688.0500   Human   MW 5040.7
$C_{215}H_{357}N_{71}O_{67}S$ Store at -15°C

Tyr-Ala-Asp-Ala-Ile-Phe-Thr-Asn-Ser-Tyr-Arg-Lys-
Val-Leu-Gly-Gln-Leu-Ser-Ala-Arg-Lys-Leu-Leu-Gln-
Asp-Ile-Met-Ser-Arg-Gln-Gln-Gly-Glu-Ser-Asn-Gln-
Glu-Arg-Gly-Ala-Arg-Ala-Arg-Leu
Tyr-Ala-Asp-Ala-Ile-Phe-Thr-Asn-Ser-Tyr-Arg-Lys-
Val-Leu-Gly-Gln-Leu-Ser-Ala-Arg-Lys-Leu-Leu-Gln-
Asp-Ile-Met(O)-Ser-Arg-Gln-Gln-Gly-Glu-Ser-Asn-
Gln-Glu-Arg-Gly-Ala-Arg-Ala-Arg-Leu Amide

Synonyms: Growth Hormone Releasing Factor (1-44), (Met(O)²⁷)-

Bachem H-4692.0500   Human   MW 5055.72
$C_{215}H_{358}N_{72}O_{67}S$ Store at -15°C

Tyr-Ala-Asp-Ala-Ile-Phe-Thr-Asn-Ser-Tyr-Arg-Lys-
Val-Leu-Gly-Gln-Leu-Ser-Ala-Arg-Lys-Leu-Leu-Gln-
Asp-Ile-Nle-Ser-Arg Amide

Synonyms: Growth Hormone Releasing Factor (1-29), (Nle²⁷)-

Bachem H-6030.0500   Human   MW 3339.89
$C_{150}H_{248}N_{44}O_{42}$ Store at -15°C

Tyr-Ala-Asp-Ala-Ile-Phe-Thr-Asn-Ser-Tyr-Arg-Lys-
Val-Leu-Gly-Gln-Leu-Ser-Ala-Arg-Lys-Leu-Leu-Gln-
Asp-Ile-Met-Ser-Arg-Gln-Gln-Gly-Glu-Ser-Asn-Gln-
Glu-Arg-Gly-Ala-Arg-Ala-Arg-Leu Amide

Synonyms: Growth Hormone Releasing Factor (1-44)

ICN 150219   Human   MW 5039.7   Guillemin, R, *Science*, 218:585, 1982

Tyr-Ala-Asp-Ala-Ile-Phe-Thr-Asn-Ser-Tyr-Arg-Lys-
Val-Leu-Gly-Gln-Leu-Ser-Ala-Arg-Lys-Leu-Leu-Gln-
Asp-Ile-Met-Ser-Arg-Gln-Gln-Gly-Glu-Ser-Asn-Gln-
Glu-Arg-Gly-Ala Amide

Synonyms: Growth Hormone Releasing Factor (1-40)

ICN 151206   Human

Tyr-Ala-Asp-Ala-Ile-Phe-Thr-Asn-Ser-Tyr-Arg-Lys-Val-Leu-Gly-Gln-Leu-Ser-Ala-Arg-Lys-Leu-Leu-Gln-Asp-Ile-Met-Ser-Arg Amide

Synonyms: Growth Hormone Releasing Factor (1-29)

ICN 152900   Human   MW 3357.9   Lance, VA etal, *BBRC*, 119:265, 1984

Tyr-Ala-Asp-Ala-Ile-Phe-Thr-Asn-Ser-Tyr-Arg-Lys-Val-Leu-Gly-Gln-Leu-Ser-Ala-Arg-Lys-Leu-Leu-Gln-Asp-Ile-Nle-Ser-Arg Amide

Synonyms: Growth Hormone Releasing Factor (1-29), (Nle[27])-

ICN 152901   Human   MW 3339.9

Tyr-Ala-Asp-Ala-Ile-Phe-Thr-Asn-Ser-Tyr-Arg-Lys-Val-Leu-Gly-Gln-Leu-Ser-Ala-Arg-Lys-Leu-Leu-Gln-Asp-Ile-Met-Ser-Arg Amide

Synonyms: Growth Hormone Releasing Factor (1-29)

Neosystem SC075   Human   MW 3357.91   Bioactive; hypothalamic releasing hormone; releasing factor peptide

Tyr-Ala-Asp-Ala-Ile-Phe-Thr-Asn-Ser-Tyr-Arg-Lys-Val-Leu-Gly-Gln-Leu-Ser-Ala-Arg-Lys-Leu-Leu-Gln-Asp-Ile-Met-Ser-Arg-Gln-Gln-Gly-Glu-Ser-Asn-Gln-Glu-Arg-Gly-Ala-Arg-Ala-Arg-Leu Amide

Synonyms: Growth Hormone Releasing Factor

Neosystem SC086   Human   MW 5039.70   Bioactive; hypothalamic releasing hormone; releasing factor peptide; Guillemin, R et al, *Science*, 218:585, 1982

Tyr-Ala-Asp-Ala-Ile-Phe-Thr-Asn-Ser-Tyr-Arg-Lys-Val-Leu-Gly-Gln-Leu-Ser-Ala-Arg-Lys-Leu-Leu-Gln-Asp-Ile-Met-Ser-Arg
Tyr-D-Ala-Asp-Ala-Ile-Phe-Thr-Asn-Ser-Tyr-Arg-Lys-Val-Leu-Gly-Gln-Leu-Ser-Ala-Arg-Lys-Leu-Leu-Gln-Asp-Ile-Met-Ser-Arg Amide

Synonyms: Growth Hormone Releasing Factor (1-29), (D-Ala[2])-

Neosystem SC124   Human   MW 3357.91   Bioactive; hypothalamic releasing hormone; releasing factor peptide

Tyr-Ala-Asp-Ala-Ile-Phe-Thr-Asn-Ser-Tyr-Arg-Lys-Val-Leu-Gly-Gln-Leu-Ser-Ala-Arg-Lys-Leu-Leu-Gln-Asp-Ile-Met-Ser-Arg-Gln-Gln-Gly-Glu-Ser-Asn-Gln-Glu-Arg-Gly-Ala

Synonyms: Growth Hormone Releasing Factor (1-40)

Sigma G 0263   Human   FW 4544.1   ≥90% (HPLC) | Bioactive peptide; equipotent to GRF (1-44) in stimulating release of somatotropin from anterior pituitary; endogenous peptide originally found in human pancreatic tumor cells, but also seen in hypothalamus; Guillemin, R et al, *Science*, 218:585, 1982

Tyr-Ala-Asp-Ala-Ile-Phe-Thr-Asn-Ser-Tyr-Arg-Lys-Val-Leu-Gly-Gln-Leu-Ser-Ala-Arg-Lys-Leu-Leu-Gln-Asp-Ile-Nle-Ser-Arg Amide

Synonyms: Growth Hormone Releasing Factor (1-29), (Nle[27])-

Sigma G 6521   Human   FW 3339.9   ≥97% (HPLC) | Bioactive peptide; substitution of Nle for Met avoids oxidation & gives a three-fold increase in the potency of GRF (1-29)

Tyr-Ala-Asp-Ala-Ile-Phe-Thr-Asn-Ser-Tyr-Arg-Lys-Val-Leu-Gly-Gln-Leu-Ser-Ala-Arg-Lys-Leu-Leu-Gln-Asp-Ile-Met-Ser-Arg Amide

Synonyms: Growth Hormone Releasing Factor (1-29)

Sigma G 6771   Human   FW 3357.9   ≥97% (HPLC) | Bioactive peptide; GRF fragment that is equipotent to GRF in releasing somatotropin from anterior pituitary; Lance, VA et al, *Biochem Biophys Res Commun*, 119:265, 1984

Tyr-Ala-Asp-Ala-Ile-Phe-Thr-Asn-Ser-Tyr-Arg-Lys-Val-Leu-Gly-Gln-Leu-Ser-Ala-Arg-Lys-Leu-Leu-Gln-Asp-Ile-Met-Ser-Arg-Gln-Gln-Gly-Glu-Ser-Asn-Gln-Glu-Arg-Gly-Ala

Synonyms: Growth Hormone Releasing Factor (1-40)

Sigma G 8770   Human   FW 4544.1   ≥90% (HPLC) | Bioactive peptide; equipotent to GRF (1-44) in stimulating release of somatotropin from anterior pituitary; endogenous peptide originally found in human pancreatic tumor cells, but also seen in hypothalamus; Guillemin, R et al, *Science*, 218:585, 1982

Tyr-Ala-Asp-Ala-Ile-Phe-Thr-Asn-Ser-Tyr-Arg-Lys-Val-Leu-Gly-Gln-Leu-Ser-Ala-Arg-Lys-Leu-Leu-Gln-Asp-Ile-Met-Ser-Arg-Gln-Gln-Gly-Glu-Ser-Asn-Gln-Glu-Arg-Gly-Ala-Arg-Ala-Arg-Leu Amide

Synonyms: Growth Hormone Releasing Factor

Sigma G 8895   Human   FW 5039.7   ≥95% (HPLC) | Bioactive peptide; hypothalamic peptide that stimulates release of somatotropin from anterior pituitary; Guillemin, R et al, *Science*, 218:585, 1982

Tyr-Ala-Asp-Ala-Ile-Phe-Thr-Asn-Ser-Tyr-Arg-Lys-Val-Leu-Gly-Gln-Leu-Ser-Ala-Arg-Lys-Leu-Leu-Gln-Asp-Ile-Met-Ser-Arg Amide

Synonyms: Growth Hormone Releasing Factor (1-29)

Biogenesis 4751-5759   Human synthetic   MW 3359.2
Lyophilized, consisting of 86% peptide material, acetate counter ions and residual H_2O; sterile and pyrogen free; purified | To avoid Met oxidation, it is best to dissolve the peptide under a nitrogen stream

Tyr-Ala-Asp-Ala-Ile-Phe-Thr-Asn-Ser-Tyr-Arg-Lys-Val-Leu-Gly-Gln-Leu-Ser-Ala-Arg-Lys-Leu-Leu-Gln-Asp-Ile-Met-Ser-Arg-Gln-Gln-Gly-Glu-Ser-Asn-Gln-Glu-Arg-Gly-Ala-Arg-Ala-Arg-Leu Amide

Synonyms: Growth Hormone Releasing Factor (1-44)

Biogenesis 4751-5804   Human synthetic   Purified; lyophilized

Biogenesis 4751-5809   Human synthetic   MW 5040
Lyophilized, consisting of 79.13% peptide material, TFA counter ions and residual H_2O; purified | To avoid Met oxidation, it is best to dissolve the peptide under a nitrogen stream

Tyr-Ala-Asp-Ala-Ile-Phe-Thr-Asn-Ser-Tyr-Arg-Lys-Val-Leu-Gly-Gln-Leu-Ser-Ala-Arg-Lys-Leu-Leu-Gln-Asp-Ile-Met-Ser-Arg-Gln-Gln-Gly-Glu-Ser-Asn-Gln-Glu-Arg-Gly-Ala-Arg-Ala-Arg-Leu Amide

Synonyms: Growth Hormone Releasing Factor

Peptides International PGR-4127-s, 4127-v   Human synthetic   MW 5039.7 $C_{215}H_{358}N_{72}O_{66}S$ >95% (HPLC); lyophilized amorphous powder | Bioactive; Guillemin, R et al, *Science*, 218:585, 1982; Ling, N et al, *PNAS USA*, 81:4802, 1984

Tyr-Ala-Asp-Ala-Ile-Phe-Thr-Asn-Ser-Tyr-Arg-Lys-Val-Leu-Gly-Gln-Leu-Ser-Ala-Arg-Lys-Leu-Leu-Gln-Asp-Ile-Met-Ser-Arg-Gln-Gln-Gly-Glu-Arg-Asn-Gln-Glu-Gln-Gly-Ala-Arg-Val-Arg-Leu Amide

Synonyms: Growth Hormone Releasing Factor (1-44)

American Peptide 52-4-14   Porcine   MW 5108.9 $C_{219}H_{365}N_{73}O_{66}S$ Bohlen, P et al, *BBRC*, 116:726, 1983

Bachem H-1926.0500   Porcine   MW 5108.83 $C_{219}H_{365}N_{73}O_{66}S$ Store at -15°C

ICN 151209   Porcine   MW 5108.8

Sigma G 0769   Porcine   FW 5108.8   ≥97% (HPLC) | Bioactive peptide; hypothalamic peptide that stimulates release of somatotropin from anterior pituitary

Tyr-Ala-Asp-Lys-Ala-Asp-Val-Asn-Val-Leu-Thr-Lys-
Ala-Lys-Ser-Gln

Synonyms: Parathyroid Hormone (69-84), (Tyr[69])-; Parathormone

ICN 154483   Human   MW 1752

Tyr-Ala-Asp-Phe-Ile-Ala-Ser-Gly-Arg-Thr-Gly-Arg-
Arg-Asn-Ala-Ile Amide

Synonyms: Protein Kinase A Inhibitor Peptide

Upstate 12-151   Synthetic   MW 1869   >90%; lyophilized |
Residues 7-23 of both human and mouse PKA Inhibitor and 6-22 of
rabbit PKA Inhibitor

Tyr-Ala-Asp-Ser-Gly-Glu-Gly-Asp-Phe-Leu-Ala-Glu-
Gly-Gly-Gly-Val-Arg

Synonyms: Fibrinopeptide A, (Tyr[0])-

Bachem H-2945.0001   Store at -15°C

American Peptide 42-1-16   Human   MW 1699.8
$C_{72}H_{106}N_{20}O_{28}$

ICN 153097   Human   MW 1699.7

Sigma F 5260   Human   FW 1699.7   ≥95% (HPLC) |
Bioactive peptide;

Tyr-Ala-Cys-Asn-Thr-Ala-Thr-Cys-Val-Thr-His-Arg-
Arg-Leu-Ala-Gly-Leu-Leu-Ser-Arg-Ser-Gly-Gly-Met-
Val-Lys-Ser-Asn-Phe-Val-Pro-Thr-Asn-Val-Gly-Ser-
Lys-Ala-Phe Amide

Synonyms: Calcitonin Gene Related Peptide II, (Tyr[0])-

American Peptide 22-1-18   Human   MW 3956.6
$C_{171}H_{276}N_{52}O_{50}S_{3}$ Disulfide bonds: Cys[2]-Cys[7]

Tyr-Ala-Cys-Asp-Thr-Ala-Thr-Cys-Val-Thr-His-Arg-
Leu-Ala-Gly-Leu-Leu-Ser-Arg-Ser-Gly-Gly-Val-Val-
Lys-Asn-Asn-Phe-Val-Pro-Thr-Asn-Val-Gly-Ser-Lys-
Ala-Phe Amide

Synonyms: Calcitonin Gene Related Peptide, (Tyr[0])-; Calcitonin
Gene Regulated Peptide, Tyr-α-; Calcitonin Gene Regulated Peptide
I, Tyr-

American Peptide 22-1-17   Human   MW 3952.6
$C_{172}H_{276}N_{52}O_{51}S_{2}$ Disulfide bonds: :Cys[2]-Cys[7]

Bachem H-3354.0500   Human   MW 3952.54
$C_{172}H_{276}N_{52}O_{51}S_{2}$ Store at -15°C

Tyr-Ala-Cys-Asp-Thr-Ala-Thr-Cys-Val-Thr-His-Arg-
Leu-Ala-Gly-Leu-Leu-Ser-Arg-Ser-Gly-Gly-Val-Val-
Lys-Asn-Asn-Phe-Val-Pro-Thr-Asn-Val-Gly-Ser-Lys-
Ala-Phe

Synonyms: Calcitonin Gene Regulated Peptide, (Tyr[0])-α-

Neosystem SC114   Human   MW 3952.51   Disulfide bonds:
Cys[2]-Cys[7]; bioactive; calcium metabolism peptide

Tyr-Ala-Cys-Asp-Thr-Ala-Thr-Cys-Val-Thr-His-Arg-
Leu-Ala-Gly-Leu-Leu-Ser-Arg-Ser-Gly-Gly-Val-Val-
Lys-Asn-Asn-Phe-Val-Pro-Thr-Asn-Val-Gly-Ser-Lys-
Ala-Phe
Tyr-Ala-Cys(Et)-Asp-Thr-Ala-Thr-Cys(Et)-Val-Thr-
His-Arg-Leu-Ala-Gly-Leu-Leu-Ser-Arg-Ser-Gly-Gly-
Val-Val-Lys-Asn-Asn-Phe-Val-Pro-Thr-Asn-Val-Gly-
Ser-Lys-Ala-Phe Amide

Synonyms: Calcitonin Gene Regulated Peptide, (Tyr[0],Cys(Et)[2,7])-α-

Neosystem SC1256   Human   MW 4010.63   Bioactive;
calcium metabolism peptide; when iodinated, this analog could
become a useful probe to study the CGRP 2 receptor subtype in
radioreceptor assays; Dumont, Y et al, *Can J Physiol Pharmacol*,
75:671-676, 1997

Tyr-Ala-Glu-Gly-Asp-Val-His-Ala-Thr-Ser-Lys-Pro-
Ala-Arg-Arg

Synonyms: Peptide Sequencing Standard

Sigma P 2046   Lyophilized; Vial contains ~20 nmoles | Useful
in evaluating the efficiency of protein sequencers

Tyr-Ala-Glu-Gly-Thr-Phe-Ile-Ser-Asp-Tyr-Ser-Ile-Ala-
Met-Asp-Lys-Ile-His-Gln-Gln-Asp-Phe-Val-Asn-Trp-
Leu-Leu-Ala-Gln-Lys-Gly-Lys-Lys-Asn-Asp-Trp-Lys-
His-Asn-Ile-Thr-Gln

Synonyms: Gastric Inhibitory Peptide; Gastric Inhibitory
Polypeptide; Glucose-Dependent Insulinotropic Polypeptide

American Peptide 46-1-60   Human   MW 4983.6
$C_{225}H_{338}N_{60}O_{66}S$ Moody, A et al, *FEBS Lett*, 123:205, 1981

Bachem H-5645.0500   Human   MW 4983.6
$C_{226}H_{338}N_{60}O_{66}S$ Store at -15°C

Tyr-Ala-Glu-Gly-Thr-Phe-Ile-Ser-Asp-Tyr-Ser-Ile-Ala-
Met-Asp-Lys-Ile-His-Gln-Gln-Asp-Phe-Val-sn-Trp-
Leu-Leu-Ala-Gln-Lys-Gly-Lys-Lys-Asn-Asp-Trp-Lys-
His-Asn-Ile-Thr-Gln

Synonyms: Gastric Inhibitory Peptide

ICN 152868   Human   Moody, AJ etal, *FEBS Lett*, 172:142,
1984; Yajima, H etal, *Chem Pharm Bull*, 33:3578, 1985

Tyr-Ala-Glu-Gly-Thr-Phe-Ile-Ser-Asp-Tyr-Ser-Ile-Ala-
Met-Asp-Lys-Ile-His-Gln-Gln-Asp-Phe-Val-Asn-Trp-
Leu-Leu-Ala-Gln-Lys-Gly-Lys-Lys-Asn-Asp-Trp-Lys-
His-Asn-Ile-Thr-Gln

Synonyms: Gastric Inhibitory Polypeptide

Sigma G 2269   Human   FW 4983.6   ≥95% (HPLC) |
Bioactive peptide; Moody, AJ et al, *FEBS Lett*, 172:142, 1984;
Yajima, H et al, *Chem Pharm Bull*, 33:3578, 1985

Tyr-Ala-Glu-Gly-Thr-Phe-Ile-Ser-Asp-Tyr-Ser-Ile-Ala-
Met-Asp-Lys-Ile-Arg-Gln-Gln-Asp-Phe-Val-Asn-Trp-
Leu-Leu-Ala-Gln-Lys

Synonyms: Gastric Inhibitory Peptide (1-30)

American Peptide 46-1-27   Porcine   MW 3552.1
$C_{162}H_{244}N_{40}O_{48}S$ Morrow, GW et al, *Can J Physiol Pharmacol*,
74:65, 1996

Tyr-Ala-Glu-Gly-Thr-Phe-Ile-Ser-Asp-Tyr-Ser-Ile-Ala-
Met-Asp-Lys-Ile-Arg-Gln-Gln-Asp-Phe-Val-Asn-Trp-
Leu-Leu-Ala-Gln-Lys Amide

Synonyms: Gastric Inhibitory Peptide (1-30)

American Peptide 46-4-10   Porcine   MW 3551.1
$C_{162}H_{245}N_{41}O_{47}S$ Rossowski, WJ et al, *Reg Peptides*, 39:9, 1992

Tyr-Ala-Glu-Gly-Thr-Phe-Ile-Ser-Asp-Tyr-Ser-Ile-Ala-
Met-Asp-Lys-Ile-Arg-Gln-Gln-Asp-Phe-Val-Asn-Trp-
Leu-Leu-Ala-Gln-Lys-Gly-Lys-Lys-Ser-Asp-Trp-Lys-
His-Asn-Ile-Thr-Gln

Synonyms: Gastric Inhibitory Peptide

American Peptide 46-4-30   Porcine   MW 4975.7
$C_{225}H_{342}N_{60}O_{66}S$ Joernvall, H et al, *FEBS Lett*, 123:205, 1981

Tyr-Ala-Glu-Gly-Thr-Phe-Ile-Ser-Asp-Tyr-Ser-Ile-Ala-
Met-Asp-Lys-Ile-Arg-Gln-Gln-Asp-Phe-Val-Asn-Trp-
Leu-Leu-Ala-Gln-Lys Amide

Synonyms: Gastric Inhibitory Polypeptide (1-30); Glucose-
Dependent Insulinotropic Polypeptide (1-30)

Bachem H-3824.0500   Porcine   Store at -15°C

Tyr-Ala-Glu-Gly-Thr-Phe-Ile-Ser-Asp-Tyr-Ser-Ile-Ala-Met-Asp-Lys-Ile-Arg-Gln-Gln-Asp-Phe-Val-Asn-Trp-Leu-Leu-Ala-Gln-Lys-Gly-Lys-Lys-Ser-Asp-Trp-Lys-His-Asn-Ile-Thr-Gln

Synonyms: Gastric Inhibitory Polypeptide; Glucose-Dependent Insulinotropic Polypeptide

Bachem H-6220.0500   Porcine   Store at -15°C

Tyr-Ala-Glu-Gly-Thr-Phe-Ile-Ser-Asp-Tyr-Ser-Ile-Ala-Met-Asp-Lys-Ile-Arg-Gln-Gln-Asp-Phe-Val-Asn-Trp-Leu-Leu-Ala-Gln-Lys-Gly-Lys-Lys-Ser-Asp-Trp-Lys-His-Asn-Ile-Thr-Gln

Synonyms: Gastric Inhibitory Peptide

ICN 152869   Porcine   Joernvall, H etal, *FEBS Lett*, 123:205, 1981

Tyr-Ala-Glu-Gly-Thr-Phe-Ile-Ser-Asp-Tyr-Ser-Ile-Ala-Met-Asp-Lys-Ile-Arg-Gln-Gln-Asp-Phe-Val-Asn-Trp-Leu-Leu-Ala-Gln-Lys Amide

Synonyms: Gastric Inhibitory Polypeptide (1-30)

Sigma G 5404   Porcine   FW 3551.0   Bioactive peptide; Rossowski, WJ et al, *Regulatory Peptides*, 39:9, 1992

Tyr-Ala-Glu-Gly-Thr-Phe-Ile-Ser-Asp-Tyr-Ser-Ile-Ala-Met-Asp-Lys-Ile-Arg-Gln-Gln-Asp-Phe-Val-Asn-Trp-Leu-Leu-Ala-Gln-Lys-Gly-Lys-Lys-Ser-Asp-Trp-Lys-His-Asn-Ile-Thr-Gln

Synonyms: Gastric Inhibitory Polypeptide

Sigma G 5512   Porcine   FW 4975.6   ≥97% (HPLC) | Bioactive peptide

Tyr-Ala-Glu-Gly-Thr-Phe-Ile-Ser-Asp-Tyr-Ser-Ile-Ala-Met-Asp-Lys-Ile-His-Gln-Gln-Asp-Phe-Val-Asn-Trp-Leu-Leu-Ala-Gln-Lys-Gly-Lys-Lys-Asn-Asp-Trp-Lys-His-Asn-Ile-Thr-Gln

Synonyms: Gastric Inhibitory Polypeptide

Peptides International PGR-4178-s, 4178-v   Human synthetic   MW 4983.6 $C_{226}H_{338}N_{60}O_{66}S$ >95% (HPLC); lyophilized amorphous powder | Bioactive; Moody, AJ et al, *FEBS Letts*, 172:142, 1984

Tyr-Ala-Glu-Pro-Gln-Lys-Ser-Pro-Trp-Cys-Glu-Ala-Arg-Ser-Leu-Glu-His-Thr

Synonyms: Estrogen Receptor β; Estrogen Receptor β Immunizing Peptide

Upstate 12-368   MW 2132.57   >90% pure; lyophilized | Immunizing peptide

USBio E3451   Synthetic   ≥90%; lyophilized | AA 54–71 of rat & mouse estrogen receptor β; AA 46–63 of human estrogen receptor β

Tyr-Ala-Gly
BOC-Tyr-D-Ala-Gly

Bachem A-2430.0250   MW 409.44 $C_{19}H_{27}N_3O_7$ Store at -15°C

Tyr-Ala-Gly
Tyr-D-Ala-Gly

Synonyms: Enkephalin, (D-Ala[2])-

Bachem H-5125.0050   MW 309.32 $C_{14}H_{19}N_3O_5$ Store at -15°C

ICN 195662   MW 309.3 $C_{14}H_{19}N_3O_5$ ≥97% | Amino terminus tripeptide of (D-Ala[2])-Enkephalin

Tyr-Ala-Gly
Tyr-D-Ala-Gly-Isoamylamine

Synonyms: Trimu V

Neosystem SC874   MW 378.4   Bioactive; potent mu2 opioid agonist; mu1 opioid antagonist with analgesic actions; Tive, LA et al, *Eur J Pharmacol*, 216:249-255, 1992; Pick, CG et al, *Eur J Pharmacol*, 220:275-277, 1992

Tyr-Ala-Gly
N-t-BOC-Tyr-D-Ala-Gly

Sigma B 2144   FW 409.4 $C_{19}H_{27}N_3O_7$

Tyr-Ala-Gly
Tyr-D-Ala-Gly

Synonyms: Enkephalin Amino Terminus, (D-Ala[2])-

Sigma T 2774   FW 309.3 $C_{14}H_{19}N_3O_5$ ≥97% (TLC) | Bioactive peptide

Tyr-Ala-Gly-Ala-Val-Val-Asn-Asp-Leu

Synonyms: Herpes Virus Inhibitor I; Virus Related Peptide; Herpes Virus Ribonucleotide Reductase Inhibitor; HSV H2 (7-15); HSV Ribonucleotide Reductase Inhibitor

American Peptide 72-2-21   MW 921.0 $C_{41}H_{64}N_{10}O_{14}$ Cohen, EA et al, *Nature*, 321:441, 1986

ICN 153186   Cohen, EA, *Nature*, 321:441, 1986; Dutia, BM, *Nature*, 321:439, 1986

Neosystem SC156   MW 921.02   Virus-related peptide; Cohen, EA et al, *Nature*, 321:441-443, 1986; Dutia, BM et al, *Nature*, 321:439-441, 1986

Sigma T 5028   FW 921.0 $C_{41}H_{64}N_{10}O_{14}$ ≥97% (HPLC) | Bioactive peptide; Dutia, BM et al, *Nature*, 321:439, 1986; Cohen, EA et al, *Nature*, 321:441, 1986

Tyr-Ala-Gly-Cys-Lys-Asn-Phe-Phe-Trp-Lys-Thr-Phe-Thr-Ser-Cys

Synonyms: Somatostatin, (Tyr[0])-

American Peptide 68-1-20   MW 1801.1 $C_{85}H_{113}N_{19}O_{21}S_2$ Disulfide bonds: Cys[4]-Cys[15]

Tyr-Ala-Gly-Cys-Lys-Asn-Phe-Phe-Trp-Lys-Thr-Phe-Thr-Ser-Cys
Tyr-Ala-Gly-Cys-Lys-Asn-Phe-Phe-D-Trp-Lys-Thr-Phe-Thr-Ser-Cys

Synonyms: Somatostatin, (Tyr[0],D-Trp[8])-

American Peptide 68-1-25   MW 1801.1 $C_{85}H_{113}N_{19}O_{21}S_2$ Disulfide bonds: Cys[4]-Cys[15]

Tyr-Ala-Gly-Cys-Lys-Asn-Phe-Phe-Trp-Lys-Thr-Phe-Thr-Ser-Cys

Synonyms: Somatostatin 14, (Tyr[0],D-Trp[8])-; Somatostatin, (Tyr[0])-

Bachem H-4995.0001   MW 1801.08 $C_{85}H_{113}N_{19}O_{21}S_2$ Store at -15°C

ICN 152993   Disulfide bond: Cys[3]-Cys[14]

Neosystem SC089   MW 1801.06   Disulfide bonds: Cys[3]-Cys[14]; bioactive; hypothalamic releasing hormone

Neosystem SC380   MW 1801.06   Disulfide bonds: Cys[3]-Cys[14]; bioactive; hypothalamic releasing hormone; Epelbaum, J, *Peptides*, 11:21-27, 1990

Biogenesis 8330-1189   Synthetic   MW 1801.2   Lyophilized, consisting of 81% peptide material, acetate counter ions and residual H_2O; purified

Tyr-Ala-Gly-Gly-Asp-Ala-Ser-Gly-Glu

Bachem H-2560.0005   MW 825.79 $C_{33}H_{47}N_9O_{16}$ Store at -15°C

Tyr-Ala-Gly-Me-Phe-Gly
Tyr-D-Ala-Gly-*N*-Me-Phe-Gly-ol

Synonyms: DAGO

Neosystem SC178 MW 513.59 Bioactive; opioid peptide; selective peptide for the mu-binding site; Kosterlitz, HW & Paterson, SJ, *Br J Pharmacol*, 73:299, 1981

Tyr-Ala-Gly-Phe
Tyr-D-Ala-Gly-*N*-Me-Phe-glycinol

Synonyms: Enkephalin, (D-Ala[2],*N*-Me-Phe[4],Glycinol[5])-; DAGO; DAMGO

Bachem H-2535.0005 MW 513.59 $C_{26}H_{35}N_5O_6$ Store at -15°C

Tyr-Ala-Gly-Phe
3,5-Diiodo-Tyr-D-Ala-Gly-*N*-Me-Phe-glycinol

Synonyms: Enkephalin, (3,5-Diiodo-Tyr[1],D-Ala[2],*N*-Me-Phe[4],Glycinol[5])-; DAGO, (3,5-Diiodo-Tyr[1])-

Bachem H-2595.0005 MW 765.39 $C_{26}H_{33}I_2N_5O_6$ Store at -15°C

Tyr-Ala-Gly-Phe
Tyr-D-Ala-Gly-Phe

Bachem H-2885.0025 MW 456.5 $C_{23}H_{28}N_4O_6$ Store at -15°C

Tyr-Ala-Gly-Phe
Guanyl-Tyr-D-Ala-Gly-*N*-Me-Phe-Methionin(O)-ol

Synonyms: Enkephalin, (Guanyl-Tyr[1],D-Ala[2],*N*-Me-Phe[4],Methionin(O)-ol[5])-; DAMME, Guanyl-; *N*-Amidino-Tyr-D-Ala-Gly-*N*-Me-Phe-Methionin(O)-ol

Bachem H-8275.0001 Store at -15°C

Tyr-Ala-Gly-Phe-
-D-Ala-Gly-Phe-D-Leu Amide

Synonyms: Enkephalinamide, (D-Ala[2],D-Leu[5])-

ICN 152821 MW 568.7

Tyr-Ala-Gly-Phe-Gly
Tyr-D-Ala-Gly-*N*-Me-Phe-Gly-ol

Synonyms: Enkephalin, (D-Ala[2],*N*-Me-Phe[4],Gly-ol[5])-; DAGO

American Peptide 32-0-75 MW 514.6 $C_{26}H_{36}N_5O_6$ Enkephalin analog which selectively binds to opioid receptors; Kosterlitz, HW et al, *J Pharmacol*, 73:299P, 1981

ICN 150776 Analog which selectively binds to μ opioid receptors; Handa, BK et al, *Eur J Pharmac*, 70:531, 1981, Kosterlitz, HW et al, *Br J Pharmac*, 73;229, 1981

Tyr-Ala-Gly-Phe-Gly
3,5-Diiodo–Tyr-D-Ala-Gly-*N*-Methyl-Phe-Gly-ol

Synonyms: Enkephalin, (3,5-Diiodo-Tyr[1],D-Ala[2],*N*-Me-Phe[4],Gly-ol[5])-; DAGO, (3,5-Diiodo-Tyr[1])-

ICN 152836 μ-receptor-specific enkephalin analog

Tyr-Ala-Gly-Phe-Gly
Tyr-D-Ala-Gly-*N*-Me-Phe-Gly Acetate Salt

Synonyms: Enkephalin, (D-Ala[2],*N*-Me-Phe[4],Gly[5])-; DAGO; DAMGO

Sigma E 7384 FW 513.6(free base) $C_{26}H_{35}N_5O_6$ ≥97% (HPLC) | Bioactive peptide; analog which selectively binds to μ-opioid receptors; Kosterlitz, HW et al, *Br J Pharmacol*, 73:299, 1981

Tyr-Ala-Gly-Phe-Gly
3,5-Diiodo-Tyr-D-Ala-Gly-*N*-Me-Phe-Gly

Synonyms: Enkephalin, (3,5-DiI-Tyr[1],D-Ala[2],*N*-Me-Phe[4],Gly[5])-; DAGO

Sigma E 9134 FW 765.4 $C_{26}H_{33}I_2N_5O_6$ ≥97% (HPLC) | Bioactive peptide; selective μ-agonist that can be tritiated for receptor binding studies

Tyr-Ala-Gly-Phe-Gly
Tyr-D-Ala-Gly-*N*-Me-Phe-Gly

Synonyms: Enkephalin, (D-Ala[2],*N*-Me-Phe[4],Gly[5])-; DAMGO; DAGO

Sigma D-139 Synthetic MW 513.7 (free base); $C_{26}H_{35}N_5O_6$ >98%; white solid; peptide content and salt form information are provided with each lot; bioassayed product | Opioid receptor agonist; Suh et al, *Naunyn-Schmiedeberg's Arch Pharmacol*, 342:67, 1990; Hirning et al, *Neuropeptides*, 5:383, 1985; Calcagnetti et al, *Eur J Pharmacol*, 153:117, 1988

Tyr-Ala-Gly-Phe-Leu
Tyr-D-Ala-Gly-Phe-Leu Amide

Synonyms: Enkephalin, ((D-Ala[2])-Leu)-

American Peptide 30-0-68 MW 568.7 $C_{29}H_{40}N_6O_6$

Tyr-Ala-Gly-Phe-Leu
Tyr-D-Ala-Gly-Phe-Leu

Synonyms: Enkephalin, ((D-Ala[2])-Leu)-

American Peptide 30-0-69 MW 569.7 $C_{29}H_{39}N_5O_7$

Tyr-Ala-Gly-Phe-Leu
Tyr-D-Ala-Gly-Phe-D-Leu

Synonyms: Enkephalin, (D-Ala[2],D-Leu[5])-

American Peptide 30-0-71 MW 569.7 $C_{29}H_{39}N_5O_7$ An opioid receptor agonist; Wei, ET, *Life Sci*, 21:321, 1977

Tyr-Ala-Gly-Phe-Leu
Tyr-D-Ala-Gly-Phe-Leu

Synonyms: Enkephalin, (Ala[2])-Leu-

Bachem H-1276.0005 MW 569.66 $C_{29}H_{39}N_5O_7$ Store at -15°C

Tyr-Ala-Gly-Phe-Leu
Tyr-D-Ala-Gly-Phe-Leu

Synonyms: Enkephalin, (D-Ala[2])-Leu-

Bachem H-2750.0005 MW 569.66 $C_{29}H_{39}N_5O_7$ Store at -15°C

Tyr-Ala-Gly-Phe-Leu
Tyr-D-Ala-Gly-Phe-Leu Amide

Synonyms: Enkephalin, (D-Ala[2])-Leu-

Bachem H-2755.0005 Store at -15°C

Tyr-Ala-Gly-Phe-Leu
Tyr-D-Ala-Gly-Phe-D-Leu

Synonyms: Enkephalin, (D-Ala[2],D-Leu[5])-; DADLE

Bachem H-2860.0005 MW 569.66 $C_{29}H_{39}N_5O_7$ Store at -15°C

Tyr-Ala-Gly-Phe-Leu
Tyr-D-Ala-Gly-Phe-D-Leu Amide

Synonyms: Enkephalin, (D-Ala[2],D-Leu[5])-

Bachem H-2865.0005 MW 568.67 $C_{29}H_{40}N_6O_6$ Store at -15°C

Tyr-Ala-Gly-Phe-Leu
Tyr-D-Ala-Gly-Phe-Leu

Synonyms: Enkephalin, (D-Ala[2],Leu[5])-

ICN 152814   MW 569.7 $C_{29}H_{35}N_5O_7$ Pert, C, *Science*, 194:330, 1976

Tyr-Ala-Gly-Phe-Leu
Tyr-D-Ala-Gly-Phe-D-Leu

Synonyms: Enkephalin, (D-Ala[2],Leu[5])-

ICN 152815   A potent opioid pentapeptide; Baxter, MG etal, *Proc Brit Pharmac Soc*, 455P, 1977; Baxter, MG etal, *Proc Brit Pharmac Soc*, 523P, 1977; Wei, ET, *Life Sciences*, 21:321, 1977; Hill, RG & CM Pepper, *Eur J Pharmac*, 47:223, 1978

Tyr-Ala-Gly-Phe-Leu
Tyr-D-Ala-Gly-Phe-Leu Amide

Synonyms: Enkephalinamide, (D-Ala[2], Leu[5])-

ICN 152820   MW 568.7

Tyr-Ala-Gly-Phe-Leu

Synonyms: Enkephalin, (Ala[2],Leu[5])-

ICN 195649   MW 569.7 $C_{29}H_{39}N_5O_7$ ≥97%

Tyr-Ala-Gly-Phe-Leu
Tyr-D-Ala-Gly-Phe-Leu Amide Acetate Salt

Synonyms: Enkephalinamide, (D-Ala[2],Leu[5])-

ICN 195652   MW 628.7 $C_{29}H_{40}N_6O_6 \cdot C_2H_4O_2$ ≥97%

Tyr-Ala-Gly-Phe-Leu
Tyr-D-Ala-Gly-Phe-Leu

Synonyms: Enkephalin, (D-Ala[2])-Leu-

Neosystem SC034   MW 569.66   Bioactive; opioid peptide

Tyr-Ala-Gly-Phe-Leu

Synonyms: Enkephalin, ((Ala[2])-Leu)-

Sigma E 1892   FW 596.7 $C_{29}H_{39}N_5O_7$ ≥97% (HPLC); peptide content:~70% | Bioactive peptide

Tyr-Ala-Gly-Phe-Leu
Tyr-D-Ala-Gly-Phe-Leu Amide Acetate Salt

Synonyms: Enkephalinamide, ((D-Ala[2])-Leu)-

Sigma E 3381   FW 568.7 (free base); $C_{29}H_{40}N_6O_6$ ≥97% (HPLC) | Bioactive peptide

Tyr-Ala-Gly-Phe-Leu
Tyr-D-Ala-Gly-Phe-Leu

Synonyms: Enkephalin, ((D-Ala[2])-Leu)-

Sigma E 5008   FW 596.7 $C_{29}H_{39}N_5O_7$ ≥97% (HPLC) | Bioactive peptide; potent, long-acting Leu-enkephalin analog that is resistant to enzymatic degradation

Tyr-Ala-Gly-Phe-Leu
Tyr-D-Ala-Gly-Phe-D-Leu Acetate Salt

Synonyms: Enkephalin, (D-Ala[2],D-Leu[5])-; DADLE

Sigma E 7131   FW 596.7 (free base); $C_{29}H_{39}N_5O_7$ ≥97% (HPLC) | Bioactive peptide; prototypical δ-agonist; more potent & selective than Leu- & Met-enkephalin; antinociceptive potency equivalent to that of β-endorphin; Baxter, MG et al, *Brit J Pharmacol*, 59(3):455 P & 523P, 1977; Hill, RG & Pepper, CM, *Eur J Pharmacol*, 47:223, 1978

Tyr-Ala-Gly-Phe-Leu
Tyr-D-Ala-Gly-Phe-D-Leu AcOH Hydrate

Synonyms: Enkephalin, (D-Ala[2],D-Leu[5])-

Peptides International PEK-4115   Synthetic   MW 647.73 $C_{29}H_{39}N_5O_7 \cdot CH_3COOH \cdot H_2O$ >99% (HPLC); lyophilized amorphous powder; bulk | Bioactive; Wei, ET et al, *Life Sci*, 21:321, 1977

Tyr-Ala-Gly-Phe-Leu
Tyr-D-Ala-Gly-Phe-D-Leu

Synonyms: Enkephalin, (D-Ala[2],D-Leu[5])-

Peptides International PEK-4115-v   Synthetic   MW 569.66 $C_{29}H_{39}N_5O_7$ >99% (HPLC); lyophilized amorphous powder | Bioactive; Wei, ET et al, *Life Sci*, 21:321, 1977

Tyr-Ala-Gly-Phe-Leu-Arg
Tyr-D-Ala-Gly-Phe-Leu-Arg

Synonyms: Enkephalin, (D-Ala[2],Leu[5],Arg[6])-; Dalargin; Enkephalin-Arg, (D-Ala[2],Leu[5])-

American Peptide 30-0-70   MW 725.9 $C_{35}H_{51}N_9O_8$ Originally proposed as an antiulcer agent with cytoprotective properties; also shows antioxidant, anti-ischemic & wound-healing properties; Liynsky, OB et al, *Ann NY Acad Sci*, 594:461, 1990

Bachem H-3276.0005   MW 725.85 $C_{35}H_{51}N_9O_8$ Store at -15°C

ICN 195650   MW 725.8 $C_{35}H_{51}N_9O_8$ ≥97%

Sigma E 1766   FW 725.8 $C_{35}H_{51}N_9O_8$ ≥97% (HPLC) | Bioactive peptide; Leu-enkephalin analog with antioxidant, antischemic & wound-healing properties

Tyr-Ala-Gly-Phe-Leu-Arg-Arg-Ile-Arg
Tyr-D-Ala-Gly-Phe-Leu-Arg-Arg-Ile-Arg

Synonyms: Dynorphin A (1-9), (D-Ala[2])-

Bachem H-2650.0001   MW 1151.38 $C_{53}H_{86}N_{18}O_{11}$ Store at -15°C

American Peptide 26-4-81   Porcine   MW 1151.4 $C_{53}H_{86}N_{18}O_{11}$

Tyr-Ala-Gly-Phe-Leu-Arg-Arg-Ile-Arg-Pro-Lys-Leu-Lys
Tyr-D-Ala-Gly-Phe-Leu-Arg-Arg-Ile-Arg-Pro-Lys-Leu-Lys Amide

Synonyms: Dynorphin A (1-13), (D-Ala[2])-

American Peptide 26-4-52   Porcine   MW 1617.0 $C_{76}H_{129}N_{25}O_{14}$

Tyr-Ala-Gly-Phe-Leu-Arg-Arg-Ile-Arg-Pro-Lys-Leu-Lys
Tyr-D-Ala-Gly-Phe-Leu-D-Arg-Arg-Ile-Arg-Pro-Lys-Leu-Lys

Synonyms: Dynorphin A (1-13), (D-Ala[2],D-Arg[6])-

American Peptide 26-4-91   Porcine   MW 1618.0 $C_{76}H_{128}N_{24}O_{15}$

Tyr-Ala-Gly-Phe-Leu-Lys
BOC-Tyr-D-Ala-Gly-Phe-Leu-Lys

Synonyms: Enkephalin-Lys, (BOC-Tyr[1], D-Ala[2])-Leu-

Bachem A-2435.0005   Store at -15°C

Tyr-Ala-Gly-Phe-Met
Tyr-D-Ala-Gly-Phe-Met

Synonyms: Enkephalin, ((D-Ala[2])-Met)-

American Peptide 30-0-21   MW 587.7 $C_{28}H_{37}N_5O_7S$ Potent methionine enkephalin analog; Coy, D et al, *BBRC*, 73:632, 1976

Tyr-Ala-Gly-Phe-Met
Tyr-D-Ala-Gly-Phe-Met Amide

Synonyms: Enkephalin, ((D-Ala[2])-Met)-

American Peptide 30-0-24   MW 586.7 $C_{28}H_{38}N_6O_6S$
Potent methionine enkephalin analog; Pert, CB et al, *Science*, 194:330, 1976

Tyr-Ala-Gly-Phe-Met
Tyr-D-Ala-Gly-Phe-D-Met

Synonyms: Enkephalin, (D-Ala[2],D-Met[5])-

American Peptide 30-0-25   MW 587.7 $C_{28}H_{37}N_5O_7S$

Tyr-Ala-Gly-Phe-Met
3,5-Diiodo-Tyr-D-Ala-Gly-Phe-Met Amide

Synonyms: Enkephalin, (3,5-Diiodo-Tyr[1],D-Ala[2])-Met-

Bachem H-2600.0005   MW 838.51 $C_{28}H_{36}I_2N_6O_6S$ Store at -15°C

Tyr-Ala-Gly-Phe-Met
Tyr-D-Ala-Gly-Phe-Met

Synonyms: Enkephalin, (D-Ala[2])-Met-

Bachem H-2790.0005   MW 587.7 $C_{28}H_{37}N_5O_7S$ Store at -15°C

Tyr-Ala-Gly-Phe-Met
Tyr-D-Ala-Gly-Phe-Met Amide

Bachem H-2795.0005   MW 586.71 $C_{28}H_{38}N_6O_6S$ Store at -15°C

Tyr-Ala-Gly-Phe-Met
m-Iodo-Tyr-D-Ala-Gly-N-Me-Phe-Methionin(O)-ol

Synonyms: Enkephalin, (m-Iodo-Tyr[1],D-Ala[2],N-Me-Phe[4],Methionin(O)-ol[5])-

Bachem H-3656.0001   Store at -15°C

Tyr-Ala-Gly-Phe-Met
Tyr-D-Ala-Gly-N-Me-Phe-Methionin(O)-ol

Synonyms: Enkephalin, (D-Ala[2],N-Me-Phe[4],Methionin(O)-ol[5])-; DAMME; FK 33-824

Bachem H-8270.0001   Store at -15°C

Tyr-Ala-Gly-Phe-Met
Tyr-D-Ala-Gly-Phe-Met Amide

Synonyms: Enkephalinamide, (D-Ala[2],Met[5])-; DAMA

Calbiochem 05-23-0908   MW 586.7 $C_{28}H_{38}N_6O_6S$ ≥98% (HPLC); lyophilized solid; soluble in 5% acetic acid | A long-acting analog of Met enkephalin; potent inhibitor of respiratory bursts in polymorphonuclear neutrophils; exhibits anti-inflammatory activity; Rickenger, D, JE et al, *Agents Actions*, 41:18, 1994; Pert, CB et al, *Nature*, 258:330, 1976

Tyr-Ala-Gly-Phe-Met
3,5-Diiodo-Tyr-D-Ala-Gly-Phe-Met Amide

Synonyms: Enkephalinamide, (3,5-Diiodo-Tyr[1],D-Ala[2],Met[5])-

ICN 152830   MW 838.5 $C_{28}H_{36}I_2N_6O_6S$ A long-lasting enkephalin analog, resistant to enzymatic degradation

Tyr-Ala-Gly-Phe-Met
Tyr-D-Ala-Gly-Phe-Met

Synonyms: Enkephalin, (D-Ala[2],Met[5])-

ICN 190442   A potent (Met[5])-enkephalin analog; Coy, DH et al, *BBRC*, 73:632, 1976; Pert, C, Opiates & Endogenous Opioid Peptides, Kosterlitz, HW, ed, Elsevier/North Holland, Amsterdam, p79, 1976

Tyr-Ala-Gly-Phe-Met
Tyr-D-Ala-Gly-Phe-Met Amide

Synonyms: Enkephalinamide, (D-Ala[2],Met[5])-

ICN 190443   A long-lasting analog with the potency of Met[5]-enkephalin in a number of assay systems; Pert, CB etal, *Science*, 194:330, 1976

Tyr-Ala-Gly-Phe-Met
Tyr-D-Ala-Gly-Phe-D-Met Amide Acetate Salt

Synonyms: Enkephalinamide, (D-Ala[2],D-Met[5])-

ICN 195657   MW 646.7 $C_{28}H_{38}N_6O_6S \cdot C_2H_4O_2$ ≥97%

Tyr-Ala-Gly-Phe-Met
Tyr-D-Ala-Gly-Phe-Met

Synonyms: Enkephalin, (D-Ala[2])-Met-

Neosystem SC093   MW 587.69   Bioactive; opioid peptide; Coy, DH et al, *BBRC*, 73:632, 1976

Tyr-Ala-Gly-Phe-Met
Tyr-D-Ala-Gly-Phe-Met Amide

Synonyms: Enkephalin, (D-Ala[2])-Met-

Neosystem SC094   MW 586.70   Bioactive; opioid peptide; Coy, DH et al, *BBRC*, 73:632, 1976

Tyr-Ala-Gly-Phe-Met
Tyr-D-Ala-Gly-(N-Me-Phe)-Met(O)

Synonyms: Enkephalin, (D-Ala[2],N-Me-Phe[4]-Met(O[5]))-; DAMME; FK 33-824

Sigma E 0506   FW 603.7 $C_{29}H_{41}N_5O_7S$ ≥95% (HPLC) | Bioactive peptide; potent analgesic; slight preference for μ-receptors over δ-receptors with no affinity for κ-receptors; Roemer, D et al, *Nature*, 268:547, 1977

Tyr-Ala-Gly-Phe-Met
Tyr-D-Ala-Gly-Phe-Met Amide Acetate Salt Free Base

Synonyms: Enkephalinamide, ((D-Ala[2])-Met)-

Sigma E 2006   FW 586.7 $C_{28}H_{38}N_6O_6S$ ≥97% (HPLC) | Bioactive peptide; potent methionine enkephalin analog resistant to enzymatic degradation; Pert, CB et al, *Science*, 194:330, 1976

Tyr-Ala-Gly-Phe-Met
Tyr-D-Ala-Gly-Phe-Met Acetate Salt

Synonyms: Enkephalin, ((D-Ala[2])-Met)-

Sigma E 2757   FW 587.7 (free base) $C_{28}H_{37}N_5O_7S$ ≥97% (HPLC) | Bioactive peptide; potent, long-acting Met-enkephalin analog resistant to enzymatic degradation; Coy, DH et al, *Biochem Biophys Res Comm*, 73:632, 1976

Tyr-Ala-Gly-Phe-Met
Tyr-D-Ala-Gly-Phe-D-Met Acetate Salt

Synonyms: Enkephalin, (D-Ala[2],D-Met[5])-

Sigma E 3507   FW 587.7 (free base) $C_{28}H_{37}N_5O_7S$ ≥97% (HPLC) | Bioactive peptide; Ling, N et al, *Peptides, Proceedings of the Fifth American Peptides Symposium*, Goodman, M & Meienhofer, J eds, 96, 1977; Meltzer, HY et al, *Life Sci*, 22:1931, 1978

Tyr-Ala-Gly-Phe-Met
Tyr-D-Ala-Gly-Phe-D-Met Amide Acetate Salt

Synonyms: Enkephalinamide, (D-Ala[2],D-Met[5])-

Sigma E 3882   FW 586.7 (free base) $C_{28}H_{38}N_6O_6S$ ≥97% (HPLC) | Bioactive peptide; Ling, N et al, *Peptides, Proceedings of the Fifth American Peptides Symposium*, Goodman, M & Meienhofer, J eds, 96, 1977; Belluzzi, JD et al, *Life Sci*, 23:99, 1978

Tyr-Ala-Gly-Phe-Met
3,5-Diiodo-Tyr-D-Ala-Gly-Phe-Met Amide

Synonyms: Enkephalinamide, (3,5-Diiodo-Tyr[1],D-Ala[2])-

Sigma E 9009   FW 838.5 $C_{28}H_{36}I_2N_6O_6S$ ≥97% (HPLC) |
Bioactive peptide; analgesic met-enkephalin analog resistant to
enzymatic degradation; can be tritiated

Tyr-Ala-Gly-Phe-Met
Tyr-D-Ala-Gly-Phe-Met AcOH Hydrate

Synonyms: Enkephalin, (D-Ala[2],D-Met[5])-

Peptides International PEK-4116   Synthetic   MW 665.77
$C_{28}H_{37}N_5O_7S \cdot CH_3COOH \cdot H_2O$ >95% (HPLC); lyophilized
amorphous powder; bulk | Bioactive; Coy, DH et al, *BBRC*,
73:632, 1976

Tyr-Ala-Gly-Phe-Met
Tyr-D-Ala-Gly-Phe-Met

Synonyms: Enkephalin, (D-Ala[2],D-Met[5])-

Peptides International PEK-4116-v   Synthetic   MW 587.70
$C_{28}H_{37}N_5O_7S$ >95% (HPLC); lyophilized amorphous powder |
Bioactive; Coy, DH et al, *BBRC*, 73:632, 1976

Tyr-Ala-Gly-Phe-Met
Tyr-D-Ala-Gly-Phe-Met Amide AcOH Hydrate

Synonyms: Enkephalinamide, (D-Ala[2],D-Met[5])-

Peptides International PEK-4117   Synthetic   MW 664.78
$C_{28}H_{38}N_6O_6S \cdot CH_3COOH \cdot H_2O$ >95% (HPLC); lyophilized
amorphous powder; bulk | Bioactive; Coy, DH et al, *BBRC*,
73:632, 1976

Tyr-Ala-Gly-Phe-Met
Tyr-D-Ala-Gly-Phe-Met Amide

Synonyms: Enkephalinamide, (D-Ala[2],D-Met[5])-

Peptides International PEK-4117-v   Synthetic   MW 586.71
$C_{28}H_{38}N_6O_6S$ >95% (HPLC); lyophilized amorphous powder |
Bioactive; Coy, DH et al, *BBRC*, 73:632, 1976

Tyr-Ala-Gly-Phe-Met Acetate Salt

Synonyms: Enkephalin, ((Ala[2])-Met)-

Sigma E 3757   FW 587.7 (free base) $C_{28}H_{37}N_5O_7S$ ≥97%
(HPLC) | Bioactive peptide; Ling, N et al, *Peptides, Proceedings of
the Fifth American Peptides Symposium*, Goodman, M &
Meienhofer, J eds, 96, 1977

Tyr-Ala-Gly-Phe-Met Amide

Synonyms: Enkephalin, ((Ala[2])-Met)-

American Peptide 30-0-27   MW 586.7 $C_{28}H_{38}N_6O_6S$

Tyr-Ala-Gly-Phe-Met Amide Acetate Salt

Synonyms: Enkephalinamide, (D-Ala[2],Met[5])-

ICN 152827   Pert, CB et al, *Science*, 194:330, 1976
ICN 195655   MW 647.7 $C_{28}H_{37}N_5O_7S \cdot C_2H_4O_2$ ≥97%

Tyr-Ala-Gly-Phe-Met-Thr-Ser-Glu-Lys
Tyr-D-Ala-Gly-Phe-Met-Thr-Ser-Glu-Lys

Synonyms: Lipotropin (61-69), (D-Ala[2])-β-

American Peptide 28-0-26   MW 1033.2 $C_{46}H_{68}N_{10}O_{15}S$

Tyr-Ala-Gly-Phe-Met-Thr-Ser-Glu-Lys-Ser-Gln-Thr-Pro-Leu-Val-Thr-Leu
Tyr-D-Ala-Gly-Phe-Met-Thr-Ser-Glu-Lys-Ser-Gln-Thr-Pro-Leu-Val-Thr-Leu

Synonyms: Endorphin, (D-Ala[2])-γ-

American Peptide 28-4-45   MW 1873.2 $C_{84}H_{133}N_{19}O_{27}S$

Tyr-Ala-Gly-Phe-Met-Thr-Ser-Glu-Lys-Ser-Gln-Thr-Pro-Leu-Val-Thr-Leu-Phe-Lys-Asn-Ala-Ile-Ile-Lys-Asn-Ala-Tyr-Lys-Lys-Gly-Glu
Tyr-(2-Me)-Ala-Gly-Phe-Met-Thr-Ser-Glu-Lys-Ser-Gln-Thr-Pro-Leu-Val-Thr-Leu-Phe-Lys-Asn-Ala-Ile-Ile-Lys-Asn-Ala-Tyr-Lys-Lys-Gly-Glu

Synonyms: Endorphin, (2-Me-Ala[2])-β-

American Peptide 28-1-23   Human   MW 3494.1
$C_{160}H_{252}N_{39}O_{46}S$

Tyr-Ala-Gly-Phe-Phe-Gly-Leu-Met
Tyr-D-Ala-Gly-Phe-Phe-Gly-Leu-Met Amide

Synonyms: Enkephalin (1-4), (D-Ala[2])-; Substance P (7-11)

Neosystem SC482   MW 904.09   Bioactive; tachykinin; Lei, SZ
et al, *Eur J Pharmacol*, 193:209, 1991

Tyr-Ala-Lys-Arg

Bachem H-1216.0005   MW 536.63 $C_{24}H_{40}N_8O_6$ Store at
-15°C

Tyr-Ala-Phe
Tyr-D-Ala-Phe Amide

Synonyms: Casomorphin (1-3), (D-Ala[2])-β-

Bachem H-2385.0005   MW 398.46 $C_{21}H_{26}N_4O_4$ Store at
-15°C

American Peptide 24-1-27   Bovine   MW 398.5 $C_{21}H_{26}N_4O_4$

Tyr-Ala-Phe-Ala-Ty
Tyr-D-Ala-Phe-D-Ala-Tyr Amide

Synonyms: Casomorphin (1-5), (D-Ala[2,4],Tyr[5])-β-

ICN 152947   MW 632.7 $C_{33}H_{40}N_6O_7$ Decreases the short
circuit current in rabbit ileum; this effect is rapidly reversed by
naloxone; Hautefeuille, M et al, *Reg Pept* (Suppl), 4:219, 1985

Tyr-Ala-Phe-Ala-Tyr
Tyr-D-Ala-Phe-D-Ala-Tyr Amide

Synonyms: Casomorphin (1-5), (D-Ala[2,4],Tyr[5])-β-

American Peptide 24-1-52   Bovine   MW 632.7 $C_{33}H_{40}N_6O_7$

Tyr-Ala-Phe-Asp-Val-Val-Gly
Tyr-D-Ala-Phe-Asp-Val-Val-Gly Amide

Synonyms: Deltorphin C; Deltorphin I, (D-Ala[2])-

Bachem H-8055.0005   MW 768.87 $C_{37}H_{52}N_8O_{10}$ Store at
-15°C

Neosystem SC315   MW 768.87   Bioactive; opioid peptide;
high affinity for δ-opioid receptors; Erspamer, V et al, *PNAS* USA,
86:5188-5192, 1989

Sigma T 0533   FW 768.9 $C_{37}H_{52}N_8O_{10}$ ≥97% (HPLC) |
Bioactive peptide; opioid peptide with high affinity for δ-receptors;
Erspamer, V et al, *Proc Natl Acad Sci USA*, 86:5188, 1989

American Peptide 28-9-10   *Phyllomedusa bicolor*   MW 768.9
$C_{37}H_{52}N_8O_{10}$ Selective μ-opioid receptor agonist

Tyr-Ala-Phe-Asp-Val-Val-Gly
Tyr-D-Ala-Phe-Asp-Val-Val-Gly

Synonyms: Deltorphin I; Deltorphin C

ICN 195033   *Phyllomedusa bicolor*   Selective δ-opiod receptor
agonist; Erspamer, V et al, *PNAS*, 86:5188, 1989

Tyr-Ala-Phe-Glu-Ile-Ile-Gly
([3H])Tyr-D-Ala-([3H])Phe-Glu-Ile-Ile-Gly Amide

Synonyms: Deltorphin II, ([3H])(Ile[5,6])-

Amersham TRK1049   Acetonitrile:water:trifluoroacetic acid
(50:50:0.1); 1.85-3.18 TBq/mmol, 50-86 Ci/mmol; 7.4 MBq/mL,
200 μCi/mL

Tyr-Ala-Phe-Glu-Val-Val-Gly
Tyr-D-Ala-Phe-Glu-Val-Val-Gly Amide

Synonyms: Deltorphin B; Deltorphin II, (D-Ala²)-

Bachem H-8060.0005   MW 782.9 $C_{38}H_{54}N_8O_{10}$ Store at -15°C

Neosystem SC386   MW 782.89   Bioactive; opioid peptide; high affinity for δ-opioid receptors; Erspamer, V et al, *PNAS* USA, 86:5188-5192, 1989

Sigma T 0658   FW 782.9 (free base) $C_{38}H_{54}N_8O_{10}$ ≥97% (HPLC) | Bioactive peptide; opioid peptide; selective δ₂-agonist; Erspamer, V et al, *Proc Natl Acad Sci USA*, 86:5188, 1989

Sigma T 0675   FW 782.9 (free base) $C_{38}H_{54}N_8O_{10}$ ≥97% (HPLC) | Bioactive peptide; opioid peptide; selective δ₂-agonist; Erspamer, V et al, *Proc Natl Acad Sci USA*, 86:5188, 1989

American Peptide 28-9-20   *Phyllomedusa bicolor*   MW 782.9 $C_{38}H_{54}N_8O_{10}$ Selective μ-opioid receptor agonist

Tyr-Ala-Phe-Glu-Val-Val-Gly
Tyr-D-Ala-Phe-Glu-Val-Val-Gly

Synonyms: Deltorphin II; Deltorphin B

ICN 195034   *Phyllomedusa bicolor*   Selective δ-opiod receptor agonist; Erspamer, V etal, *PNAS*, 86:5188, 1989

Tyr-Ala-Phe-Gly-Tyr-Pro-Lys
Tyr-D-Ala-Phe-Gly-Tyr-Pro-Lys Amide

Bachem H-9965.0001   Store at -15°C

Tyr-Ala-Phe-Gly-Tyr-Pro-Ser
Tyr-D-Ala-Phe-Gly-Tyr-Pro-Ser Amide

Synonyms: Dermorphin

Bachem H-2565.0001   Store at -15°C

Tyr-Ala-Phe-Gly-Tyr-Pro-Ser
Tyr-D-Ala-Phe-Gly-Tyr-Pro-Ser(Ac) Amide

Synonyms: Dermorphin, (Ser(Ac)⁷)-

Bachem H-6595.0005   Store at -15°C

Tyr-Ala-Phe-Gly-Tyr-Pro-Ser
Tyr-D-Ala-Phe-Gly-Tyr-Pro-Ser Amide Acetate Salt

Synonyms: Dermorphin; μ-Receptor Binding Peptide

ICN 152952   Montecucchi, PC etal, *Int J Pept Prot Res*, 17:275, 1981

Tyr-Ala-Phe-Gly-Tyr-Pro-Ser
Tyr-D-Ala-Phe-Gly-Tyr-Pro-Ser Amide

Synonyms: Dermorphin

Neosystem SC127   MW 802.88   Bioactive; opioid peptide; Montecucchi, PC et al, *Int J Pept Prot Res*, 17:275, 1981

Tyr-Ala-Phe-Gly-Tyr-Pro-Ser
Tyr-D-Ala-Phe-Gly-Tyr-Pro-Ser Amide Acetate Salt

Synonyms: Dermorphin

Sigma D 6160   FW 802.9 (free base) $C_{40}H_{50}N_8O_{10}$ ≥97% (HPLC) | Bioactive peptide; opioid peptide; most selective μ-agonist; Montecucchi, PC et al, *Int J Pept Prot Res*, 17:275, 1981

Tyr-Ala-Phe-Gly-Tyr-Pro-Ser
Tyr-D-Ala-Phe-Gly-Tyr-Pro-Ser Amide

Synonyms: Dermorphin

American Peptide 32-0-40   *Phyllomedusa sauvagei*   MW 802.9 $C_{40}H_{50}N_8O_{10}$ Most selective μ-opioid agonist isolated from the skin of the South American frog; Montecucchi, PC et al, *Int J Pep Res*, 17:275, 1981

Tyr-Ala-Phe-Hyp-Tyr
Tyr-D-Ala-Phe-Hyp-Tyr Amide

Synonyms: Casomorphin (1-5), (D-Ala²,Hyp⁴,Tyr⁵)-β-

American Peptide 24-1-46   MW 673.8 $C_{35}H_{41}N_6O_8$

Bachem H-2310.0005   MW 674.75 $C_{35}H_{42}N_6O_8$ Store at -15°C

ICN 152946   MW 674.8 $C_{35}H_{42}N_6O_8$

Tyr-Ala-Phe-Phe
Tyr-D-Ala-Phe-Phe Amide

Synonyms: Dermorphin (1-4), (Phe⁴)-

Bachem H-8870.0005   Store at -15°C

Tyr-Ala-Phe-Pro
Tyr-D-Ala-p-Chloro-Phe-Pro Amide

Synonyms: Casomorphin (1-4), (D-Ala²,p-Chloro-Phe³)-β-

Sigma C 2408   FW 530.0 $C_{26}H_{32}ClN_5O_5$ ≥97% (HPLC) | Bioactive peptide; opioid peptide

Tyr-Ala-Phe-Pro
Tyr-D-Ala-Phe-Pro Amide

Synonyms: Casomorphin (1-4), (D-Ala²)-β-

American Peptide 24-3-35   Bovine   MW 495.6 $C_{26}H_{33}N_5O_5$

Bachem H-2370.0005   Bovine   MW 495.58 $C_{26}H_{33}N_5O_5$ Store at -15°C

Tyr-Ala-Phe-Pro-Gly
Tyr-D-Ala-Phe-Pro-Gly

Synonyms: Casomorphin (1-5), (D-Ala²)-β-

American Peptide 24-3-41   Bovine   MW 553.6 $C_{28}H_{35}N_5O_7$

Tyr-Ala-Phe-Pro-Gly
Tyr-D-Ala-Phe-Pro-Gly Amide

Synonyms: Casomorphin (1-5), (D-Ala²)-β-

American Peptide 24-3-44   Bovine   MW 552.6 $C_{28}H_{36}N_6O_6$

Bachem H-2320.0005   Bovine   MW 552.63 $C_{28}H_{36}N_6O_6$ Store at -15°C

Tyr-Ala-Phe-Pro-Gly-Pro
Tyr-D-Ala-Phe-Pro-Gly-Pro

Synonyms: Casomorphin (1-6), (D-Ala²)-β-

American Peptide 24-3-60   Bovine   MW 650.7 $C_{33}H_{42}N_6O_8$

Bachem H-2290.0005   Bovine   Store at -15°C

ICN 154519   Bovine   MW 650.8

Tyr-Ala-Phe-Pro-Met
Tyr-D-Ala-Phe-Pro-Met

Synonyms: Casomorphin (1-5), (D-Ala²,Met⁵)-β-

American Peptide 24-3-47   Bovine   MW 627.8 $C_{31}H_{41}N_5O_7S$

Tyr-Ala-Phe-Pro-Met
Tyr-D-Ala-Phe-Pro-Met Amide

Synonyms: Casomorphin (1-5), (D-Ala²,Met⁵)-β-

Bachem H-2335.0005   Bovine   Store at -15°C

Tyr-Ala-Phe-Pro-Tyr
Tyr-D-Ala-Phe-D-Pro-Tyr Amide

Synonyms: Casomorphin (1-5), (D-Ala²,D-Pro⁴,Tyr⁵)-β-

American Peptide 24-1-49   MW 658.8 $C_{35}H_{42}N_6O_7$

ICN 152948   MW 658.8 $C_{35}H_{42}N_6O_7$

Tyr-Ala-Phe-Pro-Tyr
Tyr-D-Ala-Phe-Pro-Tyr Amide
Synonyms: Casomorphin (1-5), (D-Ala², Tyr⁵)-β-

American Peptide 24-1-50　Bovine　MW 658.8　$C_{35}H_{42}N_6O_7$

Tyr-Ala-Phe-Trp-Asn
Tyr-D-Ala-Phe-Trp-Asn Amide Trifluoroacetate Salt
Bachem H-1024.0005　Store at -15°C

Tyr-Ala-Phe-Trp-Asn
Tyr-D-Ala-Phe-Trp-Asn
Bachem H-9915.0005　Store at -15°C

Tyr-Ala-Val-Ser-Glu-His-Gln-Leu-Leu-His-Asp-Lys-Gly-Lys-Ser-Ile-Gln-Asp-Leu-Arg-Arg-Arg-Phe-Phe-Leu-His-His-Leu-Ile-Ala-Glu-Ile-His-Thr-Ala-Glu-Ile-Arg-Ala-Thr-Ser
Synonyms: Hypercalcemia Malignancy of Factor (1-40), (Tyr⁰)-; Parathyroid Hormone Related Protein (1-40)

ICN 154489　MW 4839.1

Tyr-Ala-Val-Ser-Glu-His-Gln-Leu-Leu-His-Asp-Lys-Gly-Lys-Ser-Ile-Gln-Asp-Leu-Arg-Arg-Arg-Phe-Phe-Leu-His-His-Leu-Ile-Ala-Glu-Ile-His-Thr-Ala
Synonyms: Hypercalcemia Malignancy of Factor (1-34), (Tyr⁰)-; Parathyroid Hormone Related Protein (1-34), (Tyr)-

ICN 154492　MW 4181.3

Bachem H-3206.0500　Human, rat　Store at -15°C

Tyr-Ala-Val-Ser-Glu-Ile-Gln-Leu-Leu-His-Asp-Lys-Gly-Lys-Ser-Ile-Gln-Asp-Leu-Arg-Arg-Arg-Phe-Phe-Leu-His-His-Leu-Ile-Ala-Glu-Ile-His-Thr-Ala-Glu-Ile-Arg-Ala-Thr-Ser
Synonyms: Hypercalcemia Malignancy of Factor (1-40), (Tyr⁰)-; Parathyroid Hormone Related Protein (1-40)

American Peptide 22-1-72　Human　MW 4838.6
$C_{216}H_{343}N_{67}O_{60}$

Tyr-Arg
Synonyms: Kyotorphin

American Peptide 32-0-50　MW 337.4　$C_{15}H_{23}N_5O_4$
Analgesic, opioid dipeptide was first isolated from bovine brain; has an analgesic effect that is ~4.2 times more potent than that of Met-enkephalin after intracisternally administration into mice; Takagi, H et al, *Eur J Pharmacol*, 55:109, 1979

Tyr-Arg
Tyr-D-Arg
Synonyms: Kyotorphin, (D-Arg²)-

American Peptide 32-0-51　MW 337.4　$C_{15}H_{23}N_5O_4$　Yajima, H et al, *Chem Phar Bull*, 28:1935, 1980

Tyr-Arg
Synonyms: Kyotorphin

Bachem G-2450.0250　MW 337.38　$C_{15}H_{23}N_5O_4$　Store at -15°C

Tyr-Arg
Tyr-D-Arg
Synonyms: Kyotorphin, (D-Arg²)-

Bachem G-2455.0050　MW 337.38　$C_{15}H_{23}N_5O_4$　Store at -15°C

ICN 152953　Yajima, H et al, *Chem Pharm Bull*, 28:1935, 1980

Tyr-Arg
Synonyms: Kyotorphin

ICN 195268　When administered intracisternally to mice, this dipeptide has ~4.2-fold more analgesic potency than metenkephalin; promotes the release of metenkephalin; Fournie-Zaluski, MC etal, *BBRC*, 91:130, 1979; Takagi, H etal, *Nature*, 282:410, 1979

Tyr-Arg
Tyr-D-Arg Acetate Salt
Synonyms: Kyotorphin, (D-Arg²)-

Sigma K 0252　FW 397.4　$C_{15}H_{23}N_5O_4 \cdot C_2H_4O_2$　≥97% (HPLC) | Bioactive peptide; opioid peptide; analgesic effect six times greater thankyotorphin; Yajima, H et al, *Chem Pharm Bull*, 28:1935, 1980

Tyr-Arg Acetate Salt
Synonyms: Kyotorphin

Sigma K 2001　FW 337.4 (free base)　$C_{15}H_{23}N_5O_4$　≥97% (HPLC) | Bioactive peptide; opioid peptide; analgesic dipeptide; promotes the release of methionine-enkephalin; Fournie-Zaluski, MC et al, *Biochem Biophys Res Commun*, 91:130, 1979; Takagi, H et al, *Nature*, 282:410, 1979

Tyr-Arg-Arg-Ala-Ala-Val-Pro-Pro-Ser-Pro-Ser-Leu-Ser-Arg-His-Ser-Ser-Pro-His-Gln-pSer-Glu-Asp-Glu-Glu-Glu
Synonyms: Phospho-Glycogen Synthase Peptide II

Upstate 12-241　MW 3028　>97%; lyophilized | Similar to skeletal muscle glycogen synthase

Tyr-Arg-Arg-Ala-Ala-Val-Pro-Pro-Ser-Pro-Ser-Leu-Ser-Arg-His-Ser-Ser-Pro-His-Gln-Ala-Glu-Asp-Glu-Glu-Glu
Synonyms: Glycogen Synthase Peptide II

Upstate 12-242　MW 2933　>97%; lyophilized | Negative control for GSK; similar to skeletal muscle glycogen synthase

Tyr-Arg-Arg-Ala-Ala-Val-Pro-Pro-Ser-Pro-Ser-Leu-Ser-Arg-His-Ser-Ser-Pro-His-Gln-pSer-Glu-Asp-Glu-Glu-Glu
Synonyms: Phospho-Glycogen Synthase Peptide, Biotinylated

Upstate 12-395　MW 3368　Frozen solution | GSK3 substrate

Tyr-Arg-Arg-Glu-Ala-Glu-Asp-Leu-Gln-Val-Gly-Gln-Val-Glu-Leu-Gly-Gly-Gly-Pro-Gly-Ala-Gly-Ser-Leu-Gln-Pro-Leu-Ala-Leu-Glu-Gly-Ser-Leu-Gln-Lys-Arg
Synonyms: Pro-Insulin C-Peptide (55-89), (Tyr)-; C-Peptide, (Tyr)-; Insulin Chain C, (Tyr)-; Pro-Insulin (30-65), (Tyr)-

Bachem H-2465.0500　Human　MW 3780.21　$C_{162}H_{268}N_{50}O_{54}$　Store at -15°C

ICN 153080　Human

Sigma C 9781　Human　FW 3780.2　≥97% (HPLC); peptide content:~65% | Bioactive peptide

Biogenesis 7750-0609　Human synthetic　Purified; lyophilized

Tyr-Arg-Asn Amide
Bachem H-7150.0025　MW 450.5　$C_{19}H_{30}N_8O_5$　Store at -15°C

Tyr-Arg-Asp-Ala-Gly-Ser-Gln-Arg-Pro-Arg-Lys-Lys-Glu-Asp-Asn-Val-Leu-Val-Glu-Ser-His-Glu-Lys-Ser-Leu-Gly
Synonyms: Parathyroid Hormone (43-68), (Tyr⁴³)-; Parathormone

American Peptide 22-1-43　Human　MW 2999.3
$C_{126}H_{208}N_{42}O_{43}$

Bachem H-4845.0500   Human   MW 2999.29
$C_{126}H_{208}N_{42}O_{43}$ Store at -15°C

ICN 152987   Human   MW 2999.3   Rosenblatt, M et al, *J Med Chem*, 20:1452, 1977

Sigma P 2530   Human   FW 2999.3   ≥97% (HPLC) | Bioactive peptide

---

Tyr-Arg-Asp-Ala-Ile-Phe-Thr-Asn-Arg-Tyr-Arg-Lys-Val-Leu-Abu-Gln-Leu-Ser-Ala-Arg-Lys-Leu-Leu-Gln-Asp-Ile-Nle-Arg

Phenylac-Tyr-D-Arg-Asp-Ala-Ile-*p*-Chloro-Phe-Thr-Asn-Arg-Tyr-Arg-Lys-Val-Leu-Abu-Gln-Leu-Ser-Ala-Arg-Lys-Leu-Leu-Gln-Asp-Ile-Nle-D-Arg-Har Amide

Synonyms: Growth Hormone Releasing Factor (1-29), (Phenylac-Tyr[1],D-Arg[2],*p*-Chloro-Phe[6],Arg[9],Abu[15],Nle[27],D-Arg[28], Har[29])-; JV-1-36

Bachem H-4884.0500   Human   MW 3757.88
$C_{170}H_{280}ClN_{53}O_{41}$ Store at -15°C

Tyr-Arg-Asp-Ala-Ile-Phe-Thr-Asn-Har-Tyr-Arg-Lys-Val-Leu-Abu-Gln-Leu-Ser-Ala-Arg-Lys-Leu-Leu-Gln-Asp-Ile-Nle-Arg

Phenylac-Tyr-D-Arg-Asp-Ala-Ile-*p*-Chloro-Phe-Thr-Asn-Har-Tyr(Me)-Arg-Lys-Val-Leu-Abu-Gln-Leu-Ser-Ala-Arg-Lys-Leu-Leu-Gln-Asp-Ile-Nle-D-Arg-Har Amide

Synonyms: Growth Hormone Releasing Factor (1-29), (Phenylac-Tyr[1],D-Arg[2],*p*-Chloro-Phe[6],Har[9],Tyr(Me)[10],Abu[15],Nle[27],D-Arg[28],Har[29])-; JV-1-38

Bachem H-4886.0500   Human   MW 3785.93
$C_{172}H_{284}ClN_{53}O_{41}$ Store at -15°C

Tyr-Arg-Asp-Ala-Ile-Phe-Thr-Asn-Ser-Tyr-Arg-Lys-Val-Leu-Gln-Leu-Ser-Ala-Arg-Lys-Leu-Leu-Gln-Asp-Ile-Nle-Ser

2-Methylpropionyl-Tyr-D-Arg-Asp-Ala-Ile-Phe(*p*-Chloro)-Thr-Asn-Ser-Tyr-Arg-Lys-Val-Leu-(2-Aminobutyric Acid)-Gln-Leu-Ser-Ala-Arg-Lys-Leu-Leu-Gln-Asp-Ile-Nle-Ser-Agmatine Trifluoroacetate Salt

Synonyms: Growth Hormone Releasing Hormone Inhibitor; MZ-4-71

Calbiochem 476500   MW 3514.1 $C_{158}H_{263}ClN_{46}O_{42}$ ≥98% (HPLC); white solid; soluble in 5% acetic acid; may be carcinogenic/teratogenic | Highly potent inhibitor of growth hormone releasing hormone (GHRH) that also inhibits the growth of human osteosarcoma cell lines; Horvath, JE et al, *Endocrinology*, 136:3849, 1995; Zarandi, M et al, *PNAS*, 91:12298, 1994

Tyr-Arg-Asp-Ala-Ile-Phe-Thr-Asn-Ser-Tyr-Arg-Lys-Val-Leu-Gln-Leu-Ser-Ala-Arg-Lys-Leu-Leu-Gln-Asp-Ile-Nle-Ser-Arg

Naphthyl-Ac-Tyr-D-Arg-Asp-Ala-Ile-Phe(*p*-Chloro)-Thr-Asn-Ser-Tyr-Arg-Lys-Val-Leu-(2-Aminobutyric Acid)-Gln-Leu-Ser-Ala-Arg-Lys-Leu-Leu-Gln-Asp-Ile-Nle-Ser-Arg Amide Trifluoroacetate Salt

Synonyms: Growth Hormone Releasing Hormone Inhibitor; MZ-4-169

Calbiochem 476503   MW 3655.7 $C_{167}H_{266}ClN_{47}O_{43}$ ≥98% (HPLC); white solid; soluble in 5% acetic acid; may be carcinogenic/teratogenic | Highly potent long-acting inhibitor *in vitro* & *in vivo*; Zarandi, M et al, *PNAS*, 91:12298, 1994

Tyr-Arg-Asp-Ala-Ile-Phe-Thr-Asn-Ser-Tyr-Arg-Lys-Val-Leu-Gly-Glen-Leu-Ser-Ala-Arg-Lys-Leu-Leu-Gln-Asp-Ile-Ile-Ser-Arg

N-Ac-Tyr-D-Arg-Asp-Ala-Ile-Phe-Thr-Asn-Ser-Tyr-Arg-Lys-Val-Leu-Gly-Glen-Leu-Ser-Ala-Arg-Lys-Leu-Leu-Gln-Asp-Ile-Ile-Ser-Arg Amide

Synonyms: Growth Hormone Releasing Factor (1-29), (*N*-Ac-Tyr[1],D-Arg[2])-

ICN 152903   Human   MW 3485.1   A growth hormone releasing factor antagonist on membranes; Robberecht, P et al, *The Endocrine Society*, 117:1759, 1985

Tyr-Arg-Asp-Ala-Ile-Phe-Thr-Asn-Ser-Tyr-Arg-Lys-Val-Leu-Gly-Gln-Leu-Ser-Ala-Arg-Lys-Leu-Leu-Gln-Asp-Ile-Met-Ser-Arg

Ac-Tyr-D-Arg-Asp-Ala-Ile-Phe-Thr-Asn-Ser-Tyr-Arg-Lys-Val-Leu-Gly-Gln-Leu-Ser-Ala-Arg-Lys-Leu-Leu-Gln-Asp-Ile-Met-Ser-Arg Amide

Synonyms: Growth Hormone Releasing Factor (1-29), (Ac-Tyr[1],D-Arg[2])-

Bachem H-5560.0500   Human   MW 3485.08
$C_{154}H_{255}N_{47}O_{43}S$ Store at -15°C

Tyr-Arg-Gln-Ser-Met-Asn-Asn-Gln-Gly-Ser-Arg-Ser-Thr-Gly-Cys-Arg-Phe-Gly-Thr-Cys-Thr-Met-Gln-Lys-Leu-Ala-His-Gln-Ile-Tyr-Gln-Phe-Thr-Asp-Lys-Asp-Lys-Asp-Gly-Met-Ala-Pro-Arg-Asn-Lys-Ile-Ser-Pro-Gln-Gly-Tyr Amide Trifluoroacetate Salt

Synonyms: Adrenomedullin (1-50)

Calbiochem 121703   Rat   MW 5729.5 $C_{242}H_{381}N_{77}O_{75}S_5$ ≥98% (HPLC); lyophilized; soluble in H_2O | Induces dose-dependent hypotensive effects that are mediated via binding to specific Adrenomedullin binding sites (K_d=0.32 nM) on blood vessels; potent platelet cAMP-elevating activity & vasodepressor effects on rat; Nandha, KA et al, *Regul Peptides*, 62:145, 1996; Sakata, J et al, *Biochem Biophys Res Comm*, 195:921, 1993

Tyr-Arg-Gln-Ser-Met-Asn-Asn-Phe-Gln-Gly-Leu-Arg

Synonyms: Adrenomedullin (1-12)

American Peptide 22-1-08   Human   MW 1513.7
$C_{64}H_{100}N_{22}O_{19}S$

ICN 195668   Human   MW ~1514

Tyr-Arg-Gln-Ser-Met-Asn-Asn-Phe-Gln-Gly-Leu-Arg-Ser-Phe-Gly-Cys-Arg-Phe-Gly-Thr-Cys-Thr-Val-Gln-Lys-Leu-Ala-His-Gln-Ile-Tyr-Gln-Phe-Thr-Asp-Lys-Asp-Lys-Asp-Asn-Val-Ala-Pro-Arg-Ser-Lys-Ile-Ser-Pro-Gln-Gly-Tyr Amide

Synonyms: Adrenomedullin (1-52); Hypotensive Peptide

American Peptide 22-2-10   Human   MW 6028.9
$C_{264}H_{406}N_{80}O_{77}S_3$ Disulfide bonds: Cys[16]-Cys[21] | Isolated from human pheochromocytoma; has one intramolecular Disulfide bonds: & shows slight homology with calcitonin gene related peptide; carries several mutual sequences comparing with CGRP, CGRP II & Amylin; the hypotensive activity is comparable to that of CGRP which has been established as one of the strongest vasorelaxants; produced in peripheral tissue, adrenal medulla, lung & kidney, functions as a circulating hormone participating in the blood pressure control; for cardiovascular research; Elicits a potent & long lasting hypotensive effect; Kitamura, K et al, *BBRC*, 192:533, 1993; Nuki, C et al, *BBRC*, 196:245, 1993

Bachem H-2932.0500   Human   MW 6028.82
$C_{264}H_{406}N_{80}O_{77}S_3$ Store at -15°C

Tyr-Arg-Gln-Ser-Met-Asn-Asn-Phe-Gln-Gly-Leu-Arg-Ser-Phe-Gly-Cys-Arg-Phe-Gly-Thr-Cys-Thr-Val-Gln-Lys-Leu-Ala-His-Gln-Ile-Tyr-Gln-Phe-Thr-Asp-Lys-Asp-Lys-Asp-Asn-Val-Ala-Pro-Arg-Ser-Lys-Ile-Ser-Pro-Gln-Gly-Tyr Amide Trifluoroacetate Salt

Synonyms: Adrenomedullin (1-52)

Calbiochem 121700 Human MW 6028.8 $C_{264}H_{406}N_{80}O_{77}S_3$ ≥95% (HPLC); white solid | A potent hypotensive peptide hormone with structural similarity to calcitonin gene-related peptide; Has vasodilatory properties; suppresses CRF-induced ACTH-release from cultured primary cells; augments NOS expression in IL-1β-stimulated cardiac myocytes; Zimmerman, U et al, *Brain Res*, 724:238, 1996; Ikeda, U et al, *Circulation*, 94:2560, 1995; Parks, DG & May, CN, *J Neuroendocrinol*, 7:923, 1995; Kitamura, K et al, *Biochem Biophys Res Comm*, 192:553, 1993

Tyr-Arg-Gln-Ser-Met-Asn-Asn-Phe-Gln-Gly-Leu-Arg-Ser-Phe-Gly-Cys-Arg-Phe-Gly-Thr-Cys-Thr-Val-Gln-Lys-Leu-Ala-His-Gln-Ile-Tyr-Gln-Phe-Thr-Asp-Lys-Asp-Lys-Asp-Asn-Val-Ala-Pro-Arg-Ser-Lys-Ile-Ser-Pro-Gln-Gly-Tyr Amide

Synonyms: Adrenomedullin 52

ICN 195607 Human MW ~6031 Promotes a potent, long-lasting hypotensive effect; Kitamura, K etal, *BBRC*, 192:533, 1993; Nuki, C etal, *BBRC* 196:245, 1993

Sigma A 2327 Human FW 6028.8 ≥97% (HPLC) | Bioactive peptide; elicits potent hypotensive effect, comparable to calcitonin gene-related peptide; Kitamura, K et al, *Biochem Biophys Res Commun*, 192:553, 1993

Tyr-Arg-Gln-Ser-Met-Asn-Asn-Phe-Gln-Gly-Leu-Arg-Ser-Phe-Gly-Cys-Arg-Phe-Gly-Thr-Cys-Thr-Val-Gln-Lys-Leu-Ala-His-Gln-Ile-Tyr-Gln-Phe-Thr-Asp-Lys-Asp-Lys-Asp-Asn-Val-Ala-Pro-Arg-Ser-Lys-Ile-Ser-Pro-Gln-Gly-Tyr Amide

Adrenomedullin; Hypotensive Peptide *Synonyms*:

Peptides International PAD-4278-s Human synthetic MW 6028.8 $C_{264}H_{406}N_{80}O_{77}S_3$ >95% (HPLC); lyophilized amorphous powder | Bioactive; Disulfide bonds: Cys[16]-Cys[21]; *Drugs*, 49:485, 1995; Schell, DA et al, *Trends Endocrinol Metab*, 7:7, 1996; K Kitamura, K et al, *BBRC*, 192:553, 1993

Tyr-Arg-Gln-Ser-Met-Asn-Asn-Phe-Gln-Gly-Leu-Arg-Ser-Phe-Gly-Cys-Arg-Phe-Gly-Thr-Cys-Thr-Val-Gln-Lys

Syno Adrenomedullin (1-25)nyms:

Peptides International PAD-4325-v Human synthetic MW 2927.3 $C_{125}H_{192}N_{40}O_{36}S_3$ >95% (HPLC); lyophilized amorphous powder | Bioactive; Disulfide bonds: Cys[16]-Cys[21]; vasopressor fragment of human adrenomedullin; Watanabe, TX et al, *BBRC*, 219:59, 1996

Tyr-Arg-Gln-Ser-Met-Asn-Asn-Phe-Gln-Gly-Leu-Arg-Ser-Phe-Gly-Cys-Arg-Phe-Gly-Thr-Cys-Thr-Val-Gln-Lys-Leu-Ala-His-Gln-Ile-Tyr-Gln-Phe-Thr-Asp-Lys-Asp-Lys-Asp-Gly-Val-Ala-Pro-Arg-Ser-Lys-Ile-Ser-Pro-Gln-Gly-Tyr Amide

Adrenomedullin (1-52)*Synonyms*:

American Peptide 22-3-10 Porcine MW 5971.8 $C_{262}H_{403}N_{79}O_{76}S_3$ Disulfide bonds: Cys[16]-Cys[21]

Tyr-Arg-Gln-Ser-Met-Asn-Gln-Gly-Ser-Arg-Ser-Thr-Gly-Cys-Arg-Phe-Gly-Thr-Cys-Thr-Met-Gln-Lys-Leu-Ala-His-Gln-Ile-Tyr-Gln-Phe-Thr-Asp-Lys-Asp-Lys-Asp-Gly-Met-Ala-Pro-Arg-Asn-Lys-Ile-Ser-Pro-Gln-Gly-Tyr Amide

Synonyms: Adrenomedullin (1-50); Hypotensive Peptide

American Peptide 22-8-10 Rat MW 5729.5 $C_{248}H_{381}N_{77}O_{75}S_5$ Disulfide bonds: Cys[14]-Cys[19] | Potent platelet cAMP elevating activity & vasodepressor effect on rat; Sakata, J et al, *BBRC*, 195:921, 1993

Bachem H-2934.0500 Rat MW 5729.49 $C_{242}H_{381}N_{77}O_{75}S_5$ Store at -15°C

ICN 195612 Rat MW ~5730 Potent elevator of platelet cAMP activity & vasodepressor effects on rats; Sakata, J etal, *BBRC*, 195:221, 1993

Peptides International PAD-4281-s Rat synthetic MW 5729.4 $C_{242}H_{381}N_{77}O_{75}S_5$ >95% (HPLC); lyophilized amorphous powder | Bioactive; Disulfide bonds: Cys[14]-Cys[19]; Sakata, J et al, *BBRC*, 195:921, 1993

Tyr-Arg-Gly-Asp-Ser

American Peptide 44-0-34 MW 596.6 $C_{24}H_{36}N_8O_{10}$

Bachem H-3154.0001 MW 596.6 $C_{24}H_{36}N_8O_{10}$ Store at -15°C

Neosystem SC147 MW 596.60 Bioactive; cell attachment peptide; fibronectin fragment

Tyr-Arg-His Amide Hydrochloride

Bachem H-7820.0025 Store at -15°C

Tyr-Arg-Ile Amide Hydrochloride

Bachem H-7855.0050 Store at -15°C

Tyr-Arg-Lys-His-Pro-Ile
(N^α-(Nicotinoyl-L-Tyr)-N^α-(N^ε-Z-L-Arg)-L-Lys)-L-His-L-Pro-L-Ile

Synonyms: Angiotensin II Antagonist; CGP-42112A

Fluka 10379 MW 1052.2 $C_{52}H_{69}N_{13}O_{11}$ ≥98% (HPLC); ≥80% peptide content | H_2O-soluble, specific antagonist at AT-2 receptors; Whitebread, S et al, *BBRC*, 163:284, 1989; Bumpus, FM et al, *Hypertension*, 17:720, 1991

Tyr-Arg-Lys-His-Pro-Ile
N-Nicotinoyl-L-Tyr-N^ε-(N^ε-CBZ-Arg)-Lys-His-Pro-Ile

Synonyms: Angiotensin II Inhibitor

ICN 194136 MW 1052.2 95% | Whilebread, S etal, *BBRC*, 163:284, 1989

Tyr-Arg-Lys-His-Pro-Ile
N^ε-Nicotinoyl-Tyr-(N^ε-CBZ-Arg)Lys-His-Pro-Ile

Synonyms: CGP-42112A

Neosystem SC431 MW 1052.20 Bioactive; angiotensin-related peptide; selective AT2 receptor ligand; Whitebread, S et al, *BBRC*, 163:284-291, 1989; Bumpus, FM et al, *Hypertension*, 17:720-721, 1991

Tyr-Arg-Lys-His-Pro-Ile
Nic-Tyr-N^ε-(N^ε-CBZ-Arg)-Lys-His-Pro-Ile

Synonyms: Angiotensin II Inhibitor

Sigma N 5014 FW 1052.2 ≥95% (HPLC); peptide content:~60% | Bioactive peptide; potent AT₂ receptor antagonist; whitebread, S et al, *Biochem Biophys Res Commun*, 163:284, 1989

Tyr-Arg-Phe-Gly
Tyr-D-Arg-Phe-Gly Amide

Synonyms: Dermorphin (1-4), (D-Arg[2])-

American Peptide 32-0-44   MW 540.6 $C_{26}H_{36}N_8O_5$ Most potent of the (D-Arg[2])-dermorphin fragments, produces significant antinociception in morphine-tolerant mice; Suzuki, K et al, *Pep Chemistry*, 1985:203, 1986

Bachem H-6755.0005   Store at -15°C

ICN 154511   MW 540.7

Tyr-Arg-Phe-Lys
Tyr-D-Arg-Phe-Lys Amide

Synonyms: Dermorphin (1-4), (D-Arg[2],Lys[4])-; DALDA

American Peptide 32-0-11   MW 611.8 $C_{30}H_{45}N_9O_5$ Highly selective μ-opioid receptor agonist

Bachem H-8865.0005   MW 611.75 $C_{30}H_{45}N_9O_5$ Store at -15°C

Sigma D 9424   FW 611.7 $C_{30}H_{45}N_9O_5$ ≥97% (HPLC); peptide content:~55% | Bioactive peptide; opioid peptide; most selective μ-agonist; may not cross blood-brain barrier; General Info:Schiller, PW, *Progress in Medicinal Chemistry*, GP Ellis & GB West (eds) Elsevier Science Publishers, 28:301, 1991

Tyr-Arg-Phe-Phe
Tyr-D-Arg-Phe-Phe Amide

Synonyms: Dermorphin (1-4), (Phe[4])-; TAPP

Sigma D 9549   FW 545.6 $C_{30}H_{35}N_5O_5$ ≥97% (HPLC) | Bioactive peptide; opioid peptide; selective μ-agonist; lipophilic molecule that appears to cross blood-brain barrier

Tyr-Arg-Phe-Sar
Tyr-D-Arg-Phe-Sar

Synonyms: Dermorphin (1-4), (D-Arg[2],Sar[4])-; Dermorphin Analog (1-4)

American Peptide 32-0-47   MW 555.6 $C_{27}H_{37}N_7O_6$ Sasaki, Y et al, *BBRC*, 120:214, 1984

ICN 154510   MW 555.6   Sasaki, Y etal, *BBRC*, 120:214, 1984

Neosystem SC111   MW 555.63   Bioactive; opioid peptide

Tyr-Arg-Phe-Sar-Tyr-Pro-Ser
Tyr-D-Arg-Phe-Sar-Tyr-Pro-Ser Amide

Synonyms: Dermorphin Analog

ICN 151470

Neosystem SC110   MW 902.01   Bioactive; opioid peptide; Sasaki, Y et al, *BBRC*, 120:214, 1984

Tyr-Arg-Pro-Pro-Gly-Phe-Ser-Pro-Phe-Arg

Synonyms: Bradykinin, (Tyr[0])-

American Peptide 18-1-27   MW 1223.4 $C_{59}H_{82}N_{16}O_{13}$ Shimamoto, K et al, *J Lab Clin Med*, 91:721, 1978

Bachem H-2195.0001   MW 1223.4 $C_{59}H_{82}N_{16}O_{13}$ Store at -15°C

ICN 152757   Shimamoto, D etal, *J Lab Clin Med*, 91:721, 1978; Yanaihara, C etal, *Adv Exp Med Biol*, 120A:185, 1979

Neosystem SC170   MW 1223.40   Bioactive; Shimamoto, K et al, *J Lab Clin Med*, 91:721, 1978

Peptides International PBK-4056-v   Synthetic   MW 1223.4 $C_{59}H_{82}N_{16}O_{13}$ >99% (HPLC); lyophilized amorphous powder | Bioactive; for radioimmunoassay; Lewis, RE et al, *Brain Res*, 346:263, 1985; Fredrick, MJ et al, *Life Sci*, 37:331, 1985

Tyr-Arg-Pro-Pro-Gly-Phe-Ser-Pro-Phe-Arg Acetate Salt Free Base

Synonyms: Bradykinin, (Tyr)-

Sigma B 4764   MW 1223.4   ≥97% (HPLC) | Bioactive peptide; Yanaihara, C et al, *Adv Exp Med Biol*, 120A:185, 1979

Tyr-Arg-Ser-Arg-Lys-Tyr-Ser-Ser-Trp-Tyr

Synonyms: Fibroblast Growth Factor Antagonist (106-115)

Neosystem SC407   MW 1395.54   Bioactive; synthetic growth factor-related peptide

Tyr-Arg-Ser-Arg-Lys-Tyr-Ser-Ser-Trp-Tyr-Val-Ala-Leu-Lys-Arg Acetate Salt Free Base

Synonyms: Fibroblast Growth Factor Basic Fragment (106-120)

Sigma F 3768   FW 1963.3   ≥95% (HPLC) | Bioactive peptide; heparin-binding FGF fragment that inhibits FGF binding to its receptor on BHK cells, stimulates thymidine incorporation into quiescent 3T3 fibroblasts, & inhibits the FGF-induced stimulation of thymidine incorporation into 3T3 fibroblasts; Baird, A et al, *Proc Natl Acad Sci USA*, 85:2324, 1988

Tyr-Arg-Val-Arg-Phe-Leu-Ala-Lys-Glu-Asn-Val-Thr-Gln-Asp-Ala-Glu-Asp-Asn-Cys

Synonyms: CD-36 Peptide P (93-110); CD-36-Cys (93-110)

American Peptide 79-1-15   MW 2270.5 $C_{96}H_{150}N_{29}O_{33}S$ Antibiotic; represents part of the epitope that binds the monoclonal anti-CD-36 antibody OKM₅; enhanced ADP-induced & collagen-induced aggregation & augmented the interaction of CD-36 to thrombospondin in platelet-rich plasma; Leung, L et al, *JBC*, 267:18244, 1992

Bachem H-2976.0001   Store at -15°C

Tyr-Asn-Val-Pro-His-Arg-Thr-Val-Leu-Pro-Gly-Met Amide 2TFA

Bachem H-6250.0001   Store at -15°C

Tyr-Asp-Ala-Phe-Gly-Tyr-Pro-Ser
Tyr-(Asp)-Ala-Phe-Gly-Tyr-Pro-Ser Amide

Synonyms: Dermorphin

Biogenesis 2697-0009   Synthetic   MW 803   Purified; lyophilized

Tyr-Asp-Met-His-Asp-Phe-Phe-Val-Gly-Leu-Met Amide

Synonyms: Neurokinin B, (Tyr[0])-

American Peptide 62-1-44   MW 1373.7 $C_{64}H_{88}N_{14}O_{16}S_2$

Neosystem SC083   MW 1373.60   Bioactive; tachykinin

Tyr-Asp-Met-Ser-Ser-Asp-Leu-Glu-Arg-Asp-His-Arg-Pro-His-Val-Ser-Met-Pro-Gln-Asn-Ala-Asn

Synonyms: Katacalcin, (Tyr)-; Calcitonin Precursor Peptide, Human

ICN 153135

Tyr-Asp-Pro-Thr-Lys-Gly-Ser-Arg-Ser-Pro-Gln-Asp

Synonyms: REV (98-108), (Tyr)-; SIVmac251 Peptide

Neosystem SC582   MW 1350.40   SIVmac251 peptide from subtype REV

Tyr-Cys
(Tyr-Cys Amide)₂ 2TFA

Bachem G-3620.0050   MW 792.73 $C_{24}H_{32}N_6O_6S_2 \cdot C_2HF_3O_2$ Store at -15°C

Tyr-Cys-Cys-His-Pro-Ala-Cys-Gly-Lys-Asn-Phe-Asp-Cys Amide

Synonyms: Conotoxin SIA, α-

Bachem H-8590.0500   Store at -15°C | Disulfide bonds: Cys[2]-Cys[7], Cys[3]-Cys[13]

American Peptide 41-0-28   *Conus striatus*   MW 1455.7 $C_{60}H_{82}N_{18}O_{17}S_4$ Disulfide bonds: Cys[2]-Cys[7], Cys[3]-Cys[13]; isolated from *Conus striatus* venom that blocks the nicotinic acetylcholine receptor

Calbiochem 234628 *Conus striatus* MW 1455.7
$C_{60}H_{82}N_{18}O_{17}S_4$ TFA; ≥95% (HPLC); lyophilized solid; soluble in water; highly toxic:LD_{50}≤50 mg/kg | Disulfide bonds: Cys^2-Cys^7, Cys^3-Cys^{13}; potent neurotoxin; blocks nicotinic acetylcholine receptors; in contrast to α-conotoxin SI, exhibits significant biological activity also in higher vertebrates but is less potent than the α-conotoxins GI & MI; Meyers, RA et al, *Biochemistry*, 30:9370, 1991

Tyr-Cys-Gly-Phe-Leu
Tyr-D-Cys(tBu)-Gly-Phe-Leu-Thr(tBu)
Synonyms: Enkephalin-Thr, (D-Cys(tBu)^2,Thr(tBu)^6)-Leu-; BUBUC
Bachem H-8170.0005 Store at -15°C

Tyr-Cys-Gly-Phe-Leu-Thr
Tyr-D-Cys(StBu)-Gly-Phe-Leu-Thr(OtBu)
Synonyms: Enkephalin-Thr(OtBu), (D-Cys(StBu)^2)-Leu-
Neosystem SC363 MW 702.82 Bioactive; opioid peptide; high affinity for δ-opioid receptors; Gacel, G et al, *Peptides*, 11:983-988, 1990

Tyr-Cys-Ser-Lys-Arg-His-Cys-Ile-Asn-Leu-Ile-Thr-Arg-Gln-Arg-Tyr
Tyr-Cys-Ser-Lys-8-Aminooctanoyl-Arg-His-D-Cys-Ile-Asn-Leu-Ile-Thr-Arg-Gln-Arg-Tyr Amide
Synonyms: Neuropeptide Y, (Cys^2)-; Neuropeptide Y (25-32), (1-4)-8-Aminooctanoyl-(D-Cys^{27})-; Neuropeptide Y, (Cys^{2,8}-Aminooctanoic Acid^{5,24},D-Cys^{27})-; Neuropeptide Y, C2-
Bachem H-3298.0500 MW 2192.64 $C_{96}H_{158}N_{32}O_{23}S_2$ Store at -15°C

Tyr-Cys-Ser-Lys-Arg-His-Cys-Ile-Asn-Leu-Ile-Thr-Arg-Gln-Arg-Tyr
Tyr-Cys-Ser-Lys-8-Aminooctanoyl-Arg-His-D-Cys-Ile-Asn-Leu-Ile-Thr-Arg-Gln-Arg-Tyr
Synonyms: Neuropeptide Y, C2-; Aminooctanoic Acid (5-24), Cys^{2}8-; Neuropeptide Y, (D-Cys^{27})-
Neosystem SC1206 MW 2192.62 Disulfide bonds: Cys^2-Cys^{27}; bioactive neuropeptide; potent & selective Y2 receptor agonist; binds with high affinity to pig spleen membranes; McLean, LR et al, *Biochem*, 29:2016-2022, 1990; Gerald, C et al, *Nature*, 382:168-171, 1996

Tyr-Cys-Thr-Gln-Tyr-Val-Thr-Val-Phe-Tyr-Gly-Val-Pro-Ala-Trp-Arg-Asn-Ala-Thr-Ile
Synonyms: GP140 (21-40); SIVmac251 Peptide
Neosystem SC710 MW 2352.69 SIVmac251 peptide from subtype GP140; due to the presence of a Cys this peptide can contain the dimeric form

---

Tyr-Cys-Trp-Ser-Gln-Tyr-Leu-Cys-Tyr
Synonyms: WP9QY
Bachem N-1685.0001 Store at -15°C | Disulfide bonds: Cys^2-Cys^8

Tyr-D-Ala-Asp-Ala-Ile-Phe-Thr-Asn-Ser-Tyr-Arg-Lys-Val-Leu-Gly-Gln-Leu-Ser-Arg-Lys-Leu-Leu-Gln-Asp-Ile-Nle-Ser-Arg Amide
Synonyms: Growth Hormone Releasing Factor (1-29), (D-Ala^2)-
ICN 152902 Human MW 3357.9 Lance, VA etal, *BBRC*, 119:265, 1984

Tyr-Gln
Bachem G-3400.0250 MW 309.32 $C_{14}H_{19}N_3O_5$ Store at -15°C

Tyr-Gln-Ala-Lys-Ser-Gln-Gly-Gly-Ser-Asn
Synonyms: Thymus Factor, (Tyr^0)-
American Peptide 87-0-05 MW 1039.2 $C_{24}H_{66}N_{14}O_{17}$
Pleau, JM et al, *JBC*, 252:8045, 1977

Tyr-Gln-Ala-Thr-Val-Gly-Asp-Ile-Asn-Thr-Glu-Arg-Pro-Gly-Met-Leu-Asp-Phe-Thr
Synonyms: Octadecane Neuropeptide-Diazepam Binding Inhibitor, (Tyr^0)-
Neosystem SC443 Human MW 2128.34 Bioactive neuropeptide; for iodination

---

Tyr-Gln-Arg-Pro-Arg-Leu-Ser-His-Lys-Gly-Pro-Met-Pro-Phe
Synonyms: Apelin-13, (Tyr^0)-
Bachem H-4894.0001 Human, bovine MW 1714.03
$C_{78}H_{120}N_{24}O_{18}S$ Store at -15°C

Tyr-Gln-Glu-Ala-Phe-Arg-Arg-Phe-Phe-Gly-Pro-Val
Synonyms: Osteocalcin (38-49), (Tyr^{38},Phe^{42,46})-; Bone Gla Protein (38-49), (Tyr^{38},Phe^{42,46})-
ICN 154480 MW 1516.9
American Peptide 22-1-61 Human MW 1516.7
$C_{73}H_{101}N_{19}O_{17}$ Osteocalcin analog; Podenphant, J et al, *Calcif Tissue Int*, 36:536, 1984
Bachem H-9130.0005 Human MW 1516.72 $C_{73}H_{101}N_{19}O_{17}$ Store at -15°C
ICN 159985 Human MW 1516.7 97%
Sigma O 4632 Human FW 1516.7 ≥97% (HPLC) | Bioactive peptide

Tyr-Gln-Leu-Leu-Gly-Gly-Arg-Phe Amide
Bachem H-5705.0005 MW 952.12 $C_{45}H_{69}N_{13}O_{10}$ Store at -15°C

Tyr-Gln-Ser-Leu-Arg-Trp Amide
Bachem H-6255.0005 Store at -15°C

Tyr-Glu
Z-Tyr-Glu
Bachem C-2755.0001 MW 444.44 $C_{22}H_{24}N_2O_8$ Store at -15°C

Tyr-Glu
Bachem G-3395.0250 MW 310.31 $C_{14}H_{18}N_2O_6$ Store at -15°C

Tyr-Glu
Ac-Tyr(PO$_3$H$_2$)-Glu-N(C$_5$H$_{11}$)$_2$
Neosystem SC1285 MW 571.58 Bioactive; proteinkinase-related peptide; potent dipeptide inhibitor of the pp60c-*src* SH2 domain; Pacofsky, GJ et al, *J Med Chem*, 41:1894-1908, 1998

Tyr-Glu
Z-Tyr-Glu
Peptides International SYE-3069 MW 444.44 $C_{22}H_{24}N_2O_8$ >99% (HPLC); amorphous powder

---

Tyr-Glu
Sigma T 2382 FW 310.3 $C_{14}H_{18}N_2O_6$

Tyr-Glu-Ala-Glu-Asp-Leu-Gln-Val-Gly-Gln-Val-Glu-Leu-Gly-Gly-Gly-Pro-Gly-Ala-Gly-Ser-Leu-Gln-Pro-Leu-Ala-Leu-Glu-Gly-Ser-Leu-Gln

Synonyms: Pro-Insulin C-Peptide (33-63), (Tyr)-; C-Peptide, (Tyr⁰)-; Insulin Chain C

American Peptide 20-1-12   Human   MW 3183.5 $C_{138}H_{220}N_{36}O_{50}$

Bachem H-4934.0500   Human   MW 3183.48 $C_{138}H_{220}N_{36}O_{50}$ Store at -15°C

ICN 153909   Synthetic   MW 3021   Lyophilized; >98%; <2% moisture, <1% pro-insulin | Frank, BH et al, *Proc Seventh American Peptide Symp*, Rich, DH & E Gross, eds, p 729, 1981

Tyr-Glu-Gln-Leu-Arg-Asn-Ser-Abu-Ala

Bachem H-5140.0001   MW 1065.15 $C_{45}H_{72}N_{14}O_{16}$ Store at -15°C

Tyr-Glu-Gln-Leu-Arg-Asn-Ser-Arg-Ala

Synonyms: MycC Peptide

Bachem H-5145.0001   MW 1136.23 $C_{47}H_{77}N_{17}O_{16}$ Store at -15°C

Tyr-Glu-Glu-Ile
Biotinyl-ε-Aminocaproyl-Tyr(PO₃H₂)-Glu-Glu-Ile

Bachem H-8900.0001   Store at -15°C

Tyr-Glu-Glu-Ile
Tyr-Glu-Glu-Ile Biotinylated

BioSource International 77-122

Tyr-Glu-Glu-Ile

Synonyms: YEEI Peptide

BioSource International 77-124

Tyr-Glu-Glu-Ile-Glu
Ac-Tyr(PO₃H₂)-Glu-Glu-Ile-Glu

Bachem H-3724.0001   MW 803.71 $C_{32}H_{46}N_5O_{17}P$ Store at -15°C

Neosystem SC924   MW 803.70   Bioactive; proteinkinase-related peptide; inhibits EGFR binding to *src*-SH2; Luttrell, DK et al, *PNAS USA*, 91:83-87, 1994

Tyr-Glu-Glu-Ser-His-Leu-Leu-Ala

Synonyms: Synenkephalin (63-70), (Tyr⁶³)-

Neosystem SC405   MW 961.04   Bioactive; opioid peptide; Stell, WK et al, *J Neurochem*, 54:434, 1990

Tyr-Glu-Glu-Trp

Bachem H-5770.0050   MW 625.64 $C_{30}H_{35}N_5O_{10}$ Store at -15°C

Tyr-Glu-Gly-Pro-Pro-Ile-Ser-Ile-Asp-Leu-Ser-Leu-Glu-Leu-Leu-Arg-Lys-Met-Ile-Glu-Ile-Glu-Lys-Gln-Glu-Lys-Glu-Lys-Gln-Gln-Ala-Ala-Asn-Asn-Arg-Leu-Leu-Leu-Asp-Thr-Ile Amide

Synonyms: Sauvagine, Tyr-(Glu¹)-

Bachem H-5355.0500   Store at -15°C

Tyr-Glu-Lys-Pro-Gly-Ser-Pro-Pro-Arg-Glu-Val-Val-Pro-Arg-Pro-Arg-Gly-Val

Bachem H-3356.0001   Store at -15°C

Tyr-Glu-Lys-Pro-Leu-Gln-Asn-Phe-Thr-Leu-Cys-Phe-Arg Amide

Synonyms: Amyloid P Component (27-38), (Tyr)-

Bachem H-2944.0001   MW 1657.96 $C_{77}H_{116}N_{20}O_{19}S$ Store at -15°C

Tyr-Glu-Lys-Ser-Leu-Gly-Glu-Ala-Asp-Lys-Ala-Asp-Val-Asn-Val-Leu-Thr-Lys-Ala-Lys-Ser-Gln

Synonyms: Parathyroid Hormone (63-84), (Tyr⁶³)-; Parathormone

American Peptide 22-1-49   Human   MW 2394.7 $C_{103}H_{172}N_{28}O_{37}$

ICN 154485   Human   MW 2394

Tyr-Glu-Pro-Gly-Lys-Ser-Ser-Ile-Leu-Gln-His-Glu-Arg-Pro-Val-Thr-Lys-Pro-Gln-Ala Amide

Synonyms: C-20 Sorbin

Neosystem SC998   Porcine   MW 2264.56   Bioactive; brain/gut peptide; Vagne-Descroix, M et al, *Mol Pharm Biochem*, 201:53-59, 1991; Nicol, P et al, *Peptides*, 15:1013-1019, 1994; Marquet, F et al, *Gastroenterol Clin Biol*, 18:702-707, 1994; Grishina, O et al, *Gastroenterol Clin Biol*, 19:487-493, 1995

Tyr-Glu-Trp

Bachem H-5800.0050   MW 496.52 $C_{25}H_{28}N_4O_7$ Store at -15°C

Tyr-Glu-Val-Asp
Z-Tyr-Glu-Val-Asp-AFC

Synonyms: Interleukin Iβ Converting Enzyme Substrate

ICN 193601   Valuable fluorescent substrate; ideal for apoptosis research

Tyr-Glu-Val-Asp-Gly-Trp-Lys
MCA-Tyr-Glu-Val-Asp-Gly-Trp-(Lys-DNP)-Amide

BioSource International 77-879

Tyr-Glu-Val-Asp-Gly-Trp-Lys
MCA-Tyr-Glu-Val-Asp-Gly-Trp-Lys-(DNP) Amide

Synonyms: Caspase I Fluorogenic Substrate VIII

Calbiochem 218737   MW 1277.3 $C_{60}H_{68}N_{12}O_{20}$ ≥97% (HPLC); lyophilized solid; soluble in DMSO; excitation max:~325 nm; emission max:~392 nm | Fluorogenic resonance energy transfer substrate for Caspase I activity; Talanian, RV et al, *J Biol Chem*, 272:9677, 1997

Tyr-Glu-Val-Asp-Gly-Trp-Lys-Lys
MCA-Tyr-Glu-Val-Asp-Gly-Trp-Lys-(Lys-DNP)-Amide

BioSource International 77-880

Tyr-Glu-Val-Glu-Asp-Leu-Gln-Val-Arg-Asp-Val-Glu-Leu-Ala-Gly-Ala-Pro-Gly-Glu-Gly-Gly-Leu-Gln-Pro-Leu-Ala-Leu-Glu-Gly-Ala-Leu-Gln

Synonyms: C-Peptide, (Tyr⁰)-

Bachem H-2914.0500   Dog   Store at -15°C

Tyr-Glu-Val-His-His-Gln-Lys-Leu-Val-Phe-Phe

Synonyms: Amyloid β-Protein (10-20)

Alexis 151-009   MW 1446.7 $C_{71}H_{99}N_{17}O_{16}$ ≥98%; white lyophilized powder; soluble in water | Miyazaki, K et al, *Nature*, 362:839, 1993

Tyr-Glu-Val-His-His-Gln-Lys-Leu-Val-Phe-Phe
β-Tyr-Glu-Val-His-His-Gln-Lys-Leu-Val-Phe-Phe

Synonyms: Amyloid (10-20)

American Peptide 62-0-20   MW 1446.7 $C_{71}H_{99}N_{17}O_{16}$ The amyloid precursor protein (APP) is cleaved by a specific APP secretase within this amyloid β-protein region to release its extracellular portion; gelatinase A, the matrix metalloproteinase MMP-2, possessing such APP secretase-like activity, also hydrolyze the amyloid β-protein fragment (10-20) at the normal processing site, the Lys[16]-Leu[17]; Yankner, BA et al, *Science*, 250:279, 1990; Mattson, MP et al, *J Neurosci*, 12:376, 1992

Tyr-Glu-Val-His-His-Gln-Lys-Leu-Val-Phe-Phe

Synonyms: Amyloid β-Protein (10-20)

Bachem H-1388.0001   MW 1446.67 $C_{71}H_{99}N_{17}O_{16}$ Store at -15°C

Sigma A 6825   FW 1446.7   ≥97% (HPLC); peptide content:~70% | Bioactive peptide; contains the cleavage site (Lys[16]-Leu[17]) for the putative amyloid precursor protein (APP) secretase; a substrate for matrix metalloproteinase MMP-2 (gelatinase A)

Tyr-Gly

Bachem G-3405.0250   MW 238.24 $C_{11}H_{14}N_2O_4$ Store at -15°C

Sigma T 5254   FW 238.2 $C_{11}H_{14}N_2O_4$

Tyr-Gly Amide Hydrochloride

Bachem G-3410.0250   MW 273.72 $C_{11}H_{15}N_3O_3 \cdot HCl$ Store at -15°C

Tyr-Gly-Ala-Val-Val-Asn-Asp-Leu

Synonyms: Herpes Virus Inhibitor II; Herpes Virus Ribonucleotide Reductase Inhibitor

American Peptide 72-2-24   MW 850.0 $C_{38}H_{59}N_9O_{13}$ Cohen, EA et al, *Nature*, 321:441, 1986

ICN 153187   Cohen, EA, *Nature*, 321:441, 1986; Dutia, BM, *Nature*, 321:439, 1986

Sigma T 5153   FW 849.9 $C_{38}H_{59}N_9O_{13}$ ≥97% (HPLC) | Bioactive peptide; Dutia, BM et al, *Nature*, 321:439, 1986

Tyr-Gly-Arg-Phe-Ser Hydrochloride

Bachem H-1134.0005   Store at -15°C

Tyr-Gly-Arg-Pro-Arg-Glu-Ser-Gly-Lys-Lys-Arg-Lys-Arg-Lys-Arg-Leu-Lys-Pro-Thr

Synonyms: Platelet Derived Growth Factor A-Chain (194-211), (Tyr)-

Bachem H-8335.0500   MW 2341.8 $C_{101}H_{181}N_{39}O_{25}$ Store at -15°C

Tyr-Gly-Cys-Lys-Asn-Phe-Phe-Trp-Lys-Thr-Phe-Thr-Ser-Cys

Synonyms: Somatostatin, (Tyr[1])-; Somatostatin 14, (Tyr[1])-; Growth Hormone

American Peptide 68-1-17   MW 1730.0 $C_{82}H_{108}N_{18}O_{20}S_2$ Disulfide bonds: Cys[3]-Cys[14]; Arimura, A et al, *Proc Soc Exp Biol Med*, 148:784, 1975

Bachem H-5000.0001   MW 1730 $C_{82}H_{108}N_{18}O_{20}S_2$ Store at -15°C

ICN 152995   Disulfide bond:Cys[3]-Cys[14]; Arimura, A etal, *Proc Soc Exp Biol Med*, 148:784, 1975

Sigma S 4633   FW 1730.0   ≥97% (HPLC); disulfide bonds: 3-14 | Bioactive peptide; Arimura, A et al, *Proc Soc Exp Biol Med*, 148:784, 1975

Biogenesis 8330-1204   Synthetic   MW 1730   Lyophilized, consisting of 81% peptide material, acetate counter ions and residual H_2O; purified

Peptides International PSI-4038-v   Synthetic   MW 1730.0 $C_{82}H_{108}N_{18}O_{20}S_2$ >97% (HPLC); lyophilized amorphous powder | Bioactive; Disulfide bonds: Cys[3]-Cys[14]; for radioimmunoassay; Arimura, AH et al, *Proc Soc Exp Biol Med*, 148:784, 1975

Tyr-Gly-D-Ala-Phe-Leu-Arg-Arg-Ile-Arg-Pro-Lys Amide

Synonyms: Dynorphin A (1-11), (D-Ala[3])-

Bachem H-5074.0001   MW 1375.69 $C_{64}H_{106}N_{22}O_{12}$ Store at -15°C

Tyr-Gly-Gln-Val-Pro-Met-Cys-Asp-Ala-Gly-Glu-Gln-Cys-Ala-Val

Synonyms: CART (62-76)

Bachem H-5098.0001   Human, rat   MW 1570.79 $C_{64}H_{99}N_{17}O_{23}S_3$ Store at -15°C

Tyr-Gly-Glu-Gla-Gla-Leu-Gln-Gla-Asn-Gln-Gla-Leu-Ile-Arg-Gla-Lys-Ser-Asn Amide

Synonyms: Conantokin G, (Tyr[0])-

Bachem H-8130.0100   Store at -15°C

Tyr-Gly-Glu-Pro-Lys-Leu-Asp-Ala-Gly-Val Amide

Synonyms: Pneumadin

Bachem H-8175.0001   Rat   MW 1047.18 $C_{47}H_{74}N_{12}O_{15}$ Store at -15°C

Sigma P 9192   Rat   FW 1047.2   ≥97% (HPLC) | Bioactive peptide; Batra, BK et al, *Regul Peptides*, 30:77, 1990

Tyr-Gly-Gly

Synonyms: Enkephalin

Bachem H-2890.0250   MW 295.3 $C_{13}H_{17}N_3O_5$ Store at -15°C

Sigma T 9005   FW 295.3 $C_{13}H_{17}N_3O_5$ ≥97% (HPLC); crystalline | Bioactive peptide; metabolite of leucine enkephalin in rat brain; Malfroy, B et al, *Neuroscience Letters*, 11:329, 1979; Swerts, JP et al, *Eur J Pharmacol*, 57:279, 1979

Tyr-Gly-Gly-Phe

Synonyms: Lipotropin (61-64), β-; Enkephalin, (des-Met[5])-

American Peptide 28-0-22   MW 442.5 $C_{22}H_{26}N_4O_6$

Bachem H-2895.0025   MW 442.47 $C_{22}H_{26}N_4O_6$ Store at -15°C

ICN 151558   Dewey, *Opiates & Endogenous Opioid Peptides*, HW Kosterlitz, ed, Elsevier/North Holland, Amsterdam, p 103, 1976

ICN 152824   MW 442.5 $C_{22}H_{26}ClN_4O_6$ Dewey, Opiates & Opioid Peptides, Kosterlitz, HE, ed, Elsevier/North Holland, Amsterdam, p 103, 1976

Tyr-Gly-Gly-Phe-Leu

Synonyms: Enkephalin, (Leu)-

American Peptide 30-0-60   MW 555.7 $C_{28}H_{37}N_5O_7$ This opioid peptide functions as a modulator of neurotransmission; induces circular muscle contractions & reduces immunoreactive vasoactive intestinal peptide secretion; Hughes, J et al, *Nature*, 258:577, 1975

Tyr-Gly-Gly-Phe-Leu
(3,5-³H)-Tyr-Gly-Gly-Phe-Leu

Synonyms: Enkephalin, (Leu)-

ARC ART-603   30-50 Ci/mmol; 1.11-1.85 TBq/mmol; In ethanol under argon | Radiochemical

Tyr-Gly-Gly-Phe-Leu
BOC-Tyr-Gly-Gly-Phe-Leu

Synonyms: Enkephalin, BOC-Leu-

Bachem A-2440.0250   MW 655.75 $C_{33}H_{45}N_5O_9$ Store at -15°C

Tyr-Gly-Gly-Phe-Leu
3,5-Dibromo-Tyr-Gly-Gly-Phe-Leu

Synonyms: Enkephalin, (3,5-Dibromo-Tyr¹)-Leu-

Bachem H-2575.0025   MW 713.42 $C_{28}H_{35}Br_2N_5O_7$ Store at -15°C

Tyr-Gly-Gly-Phe-Leu

Synonyms: Enkephalin, Leu-; Dynorphin A (1-5); Neoendorphin (1-5), α-

Bachem H-2740.0025   MW 555.63 $C_{28}H_{37}N_5O_7$ Store at -15°C

Tyr-Gly-Gly-Phe-Leu
Tyr(SO₃H)-Gly-Gly-Phe-Leu Sulfated

Synonyms: Enkephalin, Leu-

Bachem H-2760.0001   MW 635.7 $C_{28}H_{37}N_5O_{10}S$ Store at -15°C

Tyr-Gly-Gly-Phe-Leu

Synonyms: Enkephalin, (Leu⁵)-; Enkephalin L

Calbiochem 05-23-0900   MW 555.6 $C_{28}H_{37}N_5O_7$ ≥98% (HPLC); lyophilized solid; soluble in 5% acetic acid | Opiate peptide; functions as a modulator of neurotransmission; induces circular muscle contractions & reduces immunoreactive vasoactive intestinal peptide secretion; Fox-Threlkeld, JE et al, *J Pharmacol Exp Ther*, 268:689, 1994; Hughes, J et al, *Nature*, 258:577, 1975

Fluka 61885   MW 555.64 $C_{28}H_{37}N_5O_7$ ~98% (HPLC) | Goodman, RR et al, *Brain Peptides,* DT Krieger et al, eds, 827:1983, John Wiley & Sons, New York

Tyr-Gly-Gly-Phe-Leu
3,5-Dibromo-Tyr-Gly-Gly-Phe-Leu

Synonyms: Enkephalin, (3,5-Dibromo-Tyr¹,Leu⁵)-

ICN 152813

Tyr-Gly-Gly-Phe-Leu

Synonyms: Enkephalin, (Leu⁵)-

ICN 190422   Endogenous opioid pentapeptide; Hughes, J et al, *Nature*, 258:577, 1975

Tyr-Gly-Gly-Phe-Leu
N-α-t-BOC-Tyr-Gly-Gly-Phe-Leu

Synonyms: Enkephalin, N-α-t-BOC-(Leu⁵)-

ICN 190429

Tyr-Gly-Gly-Phe-Leu

Synonyms: Enkephalin, Leu⁵-

Neosystem SC036   MW 555.63   Bioactive; opioid peptide; Hugues, J et al, *Nature*, 258:577, 1975

Biogenesis 4140-0604   Human synthetic   MW 555.6   Lyophilized, consisting of 93.39% peptide material; purified

Peptides International PEK-4043-v   Synthetic   MW 555.63 $C_{28}H_{37}N_5O_7$ >99% (HPLC); lyophilized amorphous powder | Bioactive; human, porcine, bovine, rat, mouse

Tyr-Gly-Gly-Phe-Leu
Tyr(SO₃H)-Gly-Gly-Phe-Leu

Synonyms: Enkephalin, Leu-

Peptides International PEK-4118-v   Synthetic   MW 635.70 $C_{28}H_{37}N_5O_{10}S$ >99% (HPLC); lyophilized amorphous powder; sulfated | Bioactive; Unsworth, CD & J Hughes, *Nature*, 295:519, 1982

Tyr-Gly-Gly-Phe-Leu Acetate Salt

Synonyms: Enkephalin, (Leu)-

Sigma L 9133   FW 555.6 (free base) $C_{28}H_{37}N_5O_7$ ≥97% (HPLC) | Bioactive peptide; endogenous opioid neurotransmitter/neuromodulator; localization in brain parallels the localization of δ-receptors; Hughes, J et al, *Nature*, 258:577, 1975

Tyr-Gly-Gly-Phe-Leu Amide

Synonyms: Enkephalin, (Leu)-; Enkephalinamide, (Leu⁵)-

American Peptide 30-0-80   MW 554.7 $C_{28}H_{38}N_6O_6$

Bachem H-2745.0005   Store at -15°C

ICN 190423   Chang, JK et al, *Life Sci*, 18:1473, 1976; Agarwal, NS et al, *BBRC*, 76:129, 1977

Tyr-Gly-Gly-Phe-Leu Amide Acetate Salt

Synonyms: Enkephalinamide, (Leu)-

Sigma E 3756   FW 554.6 (free base) $C_{28}H_{38}N_6O_6$ ≥97% (HPLC) | Bioactive peptide; Chang, JK et al, *Life Sci*, 18:1473, 1976; Agarwal, NS et al, *Biochem Biophys Res Commun*, 76:129, 1977

Tyr-Gly-Gly-Phe-Leu Formate Salt

Synonyms: Enkephalin, (Leu⁵)-

ICN 195648   MW 671.6 $C_{28}H_{37}N_5O_7 \cdot C_4H_4O_4$ ≥95%

Tyr-Gly-Gly-Phe-Leu Hydrate

Synonyms: Enkephalin, Leu-

Peptides International PEK-4043   Synthetic   MW 573.65 $C_{28}H_{37}N_5O_7 \cdot H_2O$ >99% (HPLC); lyophilized amorphous powder; bulk | Bioactive; human, porcine, bovine, rat, mouse

Tyr-Gly-Gly-Phe-Leu-Arg

Synonyms: Enkephalin-Arg, Leu-; Neoendorphin (1-6), α-; Dynorphin A (1-6)

Bachem H-2665.0001   MW 711.82 $C_{34}H_{49}N_9O_8$ Store at -15°C

American Peptide 26-4-75   Porcine   MW 711.8 $C_{34}H_{49}N_9O_8$ Stern, AS et al, *Arch Biochem Biophys*, 205:606, 1980

Neosystem SC048   Porcine   MW 711.82   Bioactive; opioid peptide

Tyr-Gly-Gly-Phe-Leu-Arg Acetate Salt

Synonyms: Enkephalin-Arg, (Leu)-; Dynorphin A (1-6)

Sigma E 8757   FW 711.8 (free base) $C_{34}H_{49}N_9O_8$ ≥97% (HPLC) | Bioactive peptide; non-selective δ-& μ-agonist; Corbett, AD et al, *Nature*, 299:79, 1982; Matsuo, H et al, *Peptides, Structure & Biological Function, Proceedings of the Sixth American Peptide Symposium*, Meienhofer, J & Gross, E eds, 873, 1979; Stern, AS et al, *Arch Biochem Biophys*, 205:606, 1980

ICN 151019   Porcine   Matsuo, H et al, Peptides, Structure & Biological Function, Proc Sixth Am Pept Symp, Meienhofer, J & E Gross, eds, 873, 1979; Stern, AS et al, *Arch Biochem Biophys*, 205:606, 1980

Tyr-Gly-Gly-Phe-Leu-Arg-Arg

Synonyms: Enkephalin-Arg-Arg, Leu-; Dynorphin A (1-7)

Bachem H-2660.0001   MW 868 $C_{40}H_{61}N_{13}O_9$ Store at -15°C

American Peptide 26-4-74   Porcine   MW 868.0 $C_{40}H_{61}N_{13}O_9$ Potent opioid receptor agonist

ICN 151020   Porcine

Sigma D 4524   Porcine   FW 868.0 $C_{40}H_{61}N_{13}O_9$ ≥95% (HPLC); peptide content:~70% | Bioactive peptide; potent non-selective opioid agonist; highly potent family of endogenous opioid peptides which demonstrate selective affinity for the kappa receptor; Goldstein, A, *Peptides, Structure & Function, Proceedings of the Eighth American Peptides Symposium*, Hruby, VJ & Rich DH, eds, 409:1983

Tyr-Gly-Gly-Phe-Leu-Arg-Arg-Arg-Arg-Pro-Lys-Leu-Lys

Tyr-Gly-Gly-Phe-Leu-Arg-Arg-D-Arg-Arg-Pro-Lys-Leu-Lys

Synonyms: Dynorphin A (1-13), (D-Arg⁸)-

Bachem H-5540.0001   MW 1647.01 $C_{75}H_{127}N_{27}O_{15}$ Store at -15°C

Tyr-Gly-Gly-Phe-Leu-Arg-Arg-Arg-Arg-Pro-Lys-Leu-Lys

Synonyms: Dynorphin A (1-13), (D-Arg⁸)-

American Peptide 26-4-94   Porcine   MW 1647.0 $C_{75}H_{127}N_{27}O_{15}$ Opioid receptor agonist

---

Tyr-Gly-Gly-Phe-Leu-Arg-Arg-Arg-Arg-Pro-Lys-Leu-Lys

Tyr-Gly-Gly-Phe-Leu-Arg-Arg-D-Arg-Arg-Pro-Lys-Leu-Lys

Synonyms: Dynorphin A (1-13), (D-Arg⁸)-

ICN 152804   Porcine

Sigma D 9783   Porcine   FW 1647.0   ≥97% (HPLC) | Bioactive peptide; k-agonist

---

Tyr-Gly-Gly-Phe-Leu-Arg-Arg-Arg-Gln-Phe-Lys-Val-Val-Thr

Synonyms: Dynorphin B; Rimorphin; Pro-Dynorphin (228-240)

ICN 152805   Porcine   Fischli, W et al, *PNAS*, 79:5435, 1982; Kakidani, H et al, *Nature*, 298:245, 1982; Kirkpatrick, DL et al, *PNAS*, 79:6480, 1982

Tyr-Gly-Gly-Phe-Leu-Arg-Arg-Cys-Arg-Pro-Lys-Leu-Cys Amide

Synonyms: Dynorphin A (1-13), (Cys⁶,¹³)-

Bachem H-3016.0500   MW 1565.93 $C_{69}H_{112}N_{24}O_{14}S_2$ Store at -15°C

Tyr-Gly-Gly-Phe-Leu-Arg-Arg-Gln-Phe

Synonyms: Dynorphin B (1-9)

American Peptide 26-4-38   MW 1143.3 $C_{54}H_{78}N_{16}O_{12}$

Tyr-Gly-Gly-Phe-Leu-Arg-Arg-Gln-Phe-Lys-Val-Val-Thr

Synonyms: Pro-Dynorphin (228-240); Dynorphin B; Rimorphin; Pro-Dynorphin (226-238)

American Peptide 26-4-35   Porcine   MW 1570.9 $C_{74}H_{115}N_{21}O_{17}$ Porcine posterior pituitary peptide containing leucine enkephalin & possessing high opiate activity; Kilpatrick, DL et al, *Life Sci*, 31:1849, 1982

Bachem H-2675.0001   Porcine   MW 1570.86 $C_{74}H_{115}N_{21}O_{17}$ Store at -15°C

Neosystem SC046   Porcine   MW 1570.85   Bioactive; opioid peptide; Fischli, W et al, *PNAS USA*, 79:5435, 1982; Kilpatrick, DL et al, *Life Sciences*, 31:1849, 1982

Sigma D 1398   Porcine   FW 1570.9   ≥95% (HPLC) | Bioactive peptide; k-agonist; porcine posterior pituitary peptide containing leucine enkephalin & possessing high opiate activity; Kirkpatrick, DL et al, *Proc Natl Acad Sci USA*, 79:6480, 1982; Tachibana, S et al, *Nature*, 295:339, 1982; Kakidani, H et al, *Nature*, 298:245, 1982; Fischli, W et al, *Proc Natl Acad Sci USA*, 79:5435, 1982

Biogenesis 3980-0202   Porcine synthetic   MW 1570.9 $C_{74}H_{115}N_{21}O_{17}$ Purified; contains H_2O and AcOH; lyophilized | Kilpatrick et al, *Life Sci*, 31:849, 1982

Tyr-Gly-Gly-Phe-Leu-Arg-Arg-Gln-Phe-Lys-Val-Val-Thr-Arg-Ser-Gln-Glu-Asp-Pro-Asn-Ala-Tyr-Tyr-Glu-Glu-Leu-Phe-Asp-Val

Synonyms: Pro-Dynorphin (228-256); Dynorphin B 29; Leumorphin

American Peptide 26-4-41   Porcine   MW 3527.9 $C_{161}H_{236}N_{42}O_{48}$

Bachem H-6765.0500   Porcine   MW 3527.9 $C_{161}H_{236}N_{42}O_{48}$ Store at -15°C

Tyr-Gly-Gly-Phe-Leu-Arg-Arg-Ile

Synonyms: Dynorphin A (1-8)

Bachem H-2655.0001   MW 981.17 $C_{46}H_{72}N_{14}O_{10}$ Store at -15°C

American Peptide 26-4-67   Porcine   MW 981.2 $C_{46}H_{72}N_{14}O_{10}$ Opioid receptor agonist; Weber, E et al, *Nature*, 299:77, 1982

ICN 152779   Porcine   Minamino, N et al, *BBRC*, 95:1475, 1980; Weber, E et al, *Nature*, 299:77, 1982; Corbett, A et al, *Nature*, 299:79, 1982

Neosystem SC108   Porcine   MW 981.16   Bioactive; opioid peptide; Weber, E et al, *Nature*, 299:77, 1982

Sigma D 4899   Porcine   FW 981.2 $C_{46}H_{72}N_{14}O_{10}$ ≥97% (HPLC); peptide content:~70% | Bioactive peptide; potent non-selective opioid agonist; Friederich et al, *Peptides*, 8:837, 1987; Weber, E et al, Nature, 299:77, 1982; Corbett, A et al, *ibid*, 79, 1982

Tyr-Gly-Gly-Phe-Leu-Arg-Arg-Ile-Arg

Synonyms: Dynorphin A (1-9)

Bachem H-2645.0001   MW 997.29 $C_{52}H_{84}N_8O_{11}$ Store at -15°C

American Peptide 26-4-65   Porcine   MW 1137.4 $C_{52}H_{84}N_{18}O_{11}$ Potent opioid receptor agonist; Corbett, AD et al, *Nature*, 299:79, 1982

ICN 152801   Porcine   Corbett, A et al, *Nature*, 299:79, 1982

Tyr-Gly-Gly-Phe-Leu-Arg-Arg-Ile-Arg

3,5-Diiodo-Tyr-Gly-Gly-Phe-Leu-Arg-Arg-Ile-Arg

Synonyms: Dynorphin A (1-9), (3,5-Diiodo-Tyr¹)-

ICN 152802   Porcine

Tyr-Gly-Gly-Phe-Leu-Arg-Arg-Ile-Arg

Synonyms: Dynorphin A (1-9)

Sigma D 4036   Porcine   FW 1137.3   ≥97% (HPLC) | Bioactive peptide; potent k-agonist; Corbett, AD et al, *Nature*, 299:79, 1982

Tyr-Gly-Gly-Phe-Leu-Arg-Arg-Ile-Arg-Pro

Synonyms: Dynorphin A (1-10)

Bachem H-2640.0001   Store at -15°C

American Peptide 26-4-62   Porcine   MW 1234.5 $C_{57}H_{91}N_{19}O_{12}$

ICN 152803   Porcine

Sigma D 4774   Porcine   FW 1234.5   ≥97% (HPLC) | Bioactive peptide; k-agonist; Morley et al, *Peptides*, 4:797, 1983

Tyr-Gly-Gly-Phe-Leu-Arg-Arg-Ile-Arg-Pro Amide

Synonyms: Dynorphin A (1-10)

Neosystem SC393   MW 1233.48   Bioactive; opioid peptide; Woo, S et al, *Life Sciences*, 31:1817, 1982

American Peptide 26-4-63   Porcine   MW 1233.5 $C_{57}H_{92}N_{20}O_{11}$ Woo, S et al, *Life Sci*, 31:1817, 1982

ICN 154524   Porcine   MW 1233.6   Woo, S etal, *Life Sci*, 31:1817, 1982

Tyr-Gly-Gly-Phe-Leu-Arg-Arg-Ile-Arg-Pro-Arg-Leu-Arg-Gly
Tyr-Gly-Gly-Phe-Leu-Arg-Arg-Ile-Arg-Pro-Arg-Leu-Arg-Gly-5-Aminopentylamide

Synonyms: Dynorphin A (1-13)-Gly-5-Aminopentylamide, (Arg[11,13])-; DAKLI; Dynorphin A Analog, κ-Ligand

Bachem H-3014.0500   MW 1801.22 $C_{82}H_{141}N_{31}O_{15}$ Store at -15°C

ICN 195645   MW 1801.2   Multi-purpose ligand for dynorphin κ opioid receptors

---

Tyr-Gly-Gly-Phe-Leu-Arg-Arg-Ile-Arg-Pro-Arg-Leu-Arg-Gly
N-t-BOC-Tyr-Gly-Gly-Phe-Leu-Arg-Arg-Ile-Arg-Pro-Arg-Leu-Arg-Gly-5-Aminopentylamide

Synonyms: Dynorphin A Analog κ, Nᵅ-t-BOC-BOC-DAKLI; Dynorphin A; DAKLI, BOC-

ICN 195646   MW 1901.3   97% | Precursor for producing labeled DAKLI by conjugating to FITC, biotin, [125]I (Bolton-Hunter Method), etc

Sigma B 5277   FW 1901.3   ≥97% (HPLC); peptide content:~70% | Bioactive peptide; used to labelk receptors; precursor for the production of labeled DAKLI; can be coupled to ^{125}I-Bolton-Hunter reagent, FITC, biotin, etc.; Goldstein, A et al, *Proc Natl Acad Sci USA*, 85:7375, 1988

Tyr-Gly-Gly-Phe-Leu-Arg-Arg-Ile-Arg-Pro-Gly
Tyr-Gly-Gly-Phe-Leu-Arg-Arg-Ile-Arg-Pro-Gly-CMK

Synonyms: Dynorphin A Gly-CMK (1-10)

Bachem N-1605.0001   Store at -15°C

Tyr-Gly-Gly-Phe-Leu-Arg-Arg-Ile-Arg-Pro-Lys
Tyr-Gly-Gly-Phe-Leu-Arg-Arg-Ile-Arg-D-Pro-Lys

Synonyms: Dynorphin A (1-11), (D-Pro[10])-

Bachem H-3012.0001   Store at -15°C

Tyr-Gly-Gly-Phe-Leu-Arg-Arg-Ile-Arg-Pro-Lys

Synonyms: Dynorphin A (1-11)

American Peptide 26-4-58   Porcine   MW 1362.7 $C_{63}H_{103}N_{21}O_{13}$

Tyr-Gly-Gly-Phe-Leu-Arg-Arg-Ile-Arg-Pro-Lys
Tyr-Gly-Gly-Phe-Leu-Arg-Arg-Ile-Arg-D-Pro-Lys

Synonyms: Dynorphin A (1-11), (D-Pro[10])-

American Peptide 26-4-85   Porcine   MW 1362.7 $C_{63}H_{103}N_{21}O_{13}$ κ-Opioid receptor-specific analog; Gairin, JE et al, *BBRC*, 134:1142, 1986

Tyr-Gly-Gly-Phe-Leu-Arg-Arg-Ile-Arg-Pro-Lys Amide

Synonyms: Dynorphin A (1-11)

Bachem H-5068.0001   MW 1361.66 $C_{63}H_{104}N_{22}O_{12}$ Store at -15°C

Tyr-Gly-Gly-Phe-Leu-Arg-Arg-Ile-Arg-Pro-Lys-Leu

Synonyms: Dynorphin A (1-12)

American Peptide 26-4-57   Porcine   MW 1475.8 $C_{69}H_{114}N_{22}O_{14}$

Tyr-Gly-Gly-Phe-Leu-Arg-Arg-Ile-Arg-Pro-Lys-Leu-Lys

Synonyms: Dynorphin A (1-13)

Bachem H-2625.0001   MW 1603.98 $C_{75}H_{126}N_{24}O_{15}$ Store at -15°C

Tyr-Gly-Gly-Phe-Leu-Arg-Arg-Ile-Arg-Pro-Lys-Leu-Lys
Tyr-Gly-Gly-Phe-Leu-D-Arg-Arg-Ile-Arg-Pro-Lys-Leu-Lys

Synonyms: Dynorphin A (1-13), (D-Arg[6])-

Bachem H-2630.0001   MW 1603.98 $C_{75}H_{126}N_{24}O_{15}$ Store at -15°C

Tyr-Gly-Gly-Phe-Leu-Arg-Arg-Ile-Arg-Pro-Lys-Leu-Lys
Tyr-Gly-Gly-Phe-Leu-Arg-Arg-Ile-Arg-Pro-Lys-Leu-Lys 5AcOH 6H₂O

Synonyms: Dynorphin A (1-13)

Peptides International PDY-4080   Human synthetic   MW 2012.3 $C_{75}H_{126}N_{24}O_{15}$ · $5CH_3COOH$ · $6H_2O$ >99% (HPLC); lyophilized amorphous powder; bulk | Bioactive; porcine, rat, bovine; Hughes, J, *Br Med Bull*, 39:17, 1983; Smith, AP & NM Lee, *Ann Rev Pharmacol Toxicol*, 28:123, 1988; Simonato, M & P Romualdi, *Prog Neurobiol*, 50:557, 1996; Goldstein, A et al, *PNAS USA*, 76:6666, 1979

Tyr-Gly-Gly-Phe-Leu-Arg-Arg-Ile-Arg-Pro-Lys-Leu-Lys

Synonyms: Dynorphin A (1-13)

Peptides International PDY-4080-v   Human synthetic   MW 1604.0 $C_{75}H_{126}N_{24}O_{15}$ >99% (HPLC); lyophilized amorphous powder | Bioactive; porcine, rat, bovine; Hughes, J, *Br Med Bull*, 39:17, 1983; Smith, AP & NM Lee, *Ann Rev Pharmacol Toxicol*, 28:123, 1988; Simonato, M & P Romualdi, *Prog Neurobiol*, 50:557, 1996; Goldstein, A et al, *PNAS USA*, 76:6666, 1979

American Peptide 26-4-50   Porcine   MW 1604 $C_{75}H_{126}N_{24}O_{15}$ Endogenous κ-opioid receptor agonist; attenuates galanin-induced impairment of memory processes through the mediation of κ-opioid receptors; Goldstein, A et al, *PNAS*, 76:6666, 1979

Tyr-Gly-Gly-Phe-Leu-Arg-Arg-Ile-Arg-Pro-Lys-Leu-Lys
(nMe)Tyr-Gly-Gly-Phe-Leu-Arg-Arg-Ile-Arg-Pro-Lys-Leu-Lys Amide

Synonyms: Dynorphin A (1-13), ((nMe)Tyr[1])-

American Peptide 26-4-54   Porcine   MW 1617 $C_{76}H_{129}N_{25}O_{14}$

Tyr-Gly-Gly-Phe-Leu-Arg-Arg-Ile-Arg-Pro-Lys-Leu-Lys
Tyr-Gly-Gly-Phe-Leu-D-Arg-Arg-Ile-Arg-Pro-Lys-Leu-Lys

Synonyms: Dynorphin A (1-13), (D-Arg[6])-

American Peptide 26-4-80   Porcine   MW 1604.2 $C_{75}H_{126}N_{24}O_{15}$ κ-Opioid receptor agonist; anticonvulsant

---

Tyr-Gly-Gly-Phe-Leu-Arg-Arg-Ile-Arg-Pro-Lys-Leu-Lys

Synonyms: Dynorphin A (1-13)

Calbiochem 05-23-0800   Porcine   MW 1604 $C_{75}H_{126}N_{24}O_{15}$ ≥97% (HPLC); lyophilized solid; soluble in 5% acetic acid; sold under license of US Patent 4,396,606 issued to Research Corporation | An endogenous κ-opioid agonist that improves galanin-induced impairment of memory in mice; 700 times more potent than (Leu[5])-enkephalin, 200 times more potent than normorphine & 50 times more potent than β-endorphin in the guinea pig ileum assay; Kameyama, T et al, *Neuropharmacology*, 33:1167, 1994; Goldstein, A, *PNAS*, 76:6666, 1979; *Merck Index*, 12:3528

Tyr-Gly-Gly-Phe-Leu-Arg-Arg-Ile-Arg-Pro-Lys-Leu-Lys

Tyr-Gly-Gly-Phe-Leu-D-Arg-Arg-Ile-Arg-Pro-Lys-Leu-Lys

Synonyms: Dynorphin A (1-13), (D-Arg⁶)-

ICN 151506   Porcine   Wuster, M etal, *Neurosci Lett*, 20:79, 1980

Tyr-Gly-Gly-Phe-Leu-Arg-Arg-Ile-Arg-Pro-Lys-Leu-Lys

Synonyms: Dynorphin A (1-13)

Neosystem SC109   Porcine   MW 1603.97   Bioactive; opioid peptide; Goldstein, A et al, *PNAS USA*, 76:6666, 1979

Sigma D 7017   Porcine   FW 1604.0   ≥97% (HPLC); peptide content:~60% | Bioactive peptide; potent, endogenousk-agonist that attenuates galanin-induced impairment of memory processes; resistant to enzymatic degradation; Goldstein, A et al, *Proc Natl Acad Sci USA*, 76:6666, 1979

Tyr-Gly-Gly-Phe-Leu-Arg-Arg-Ile-Arg-Pro-Lys-Leu-Lys

Tyr-Gly-Gly-Phe-Leu-D-Arg-Arg-Ile-Arg-Pro-Lys-Leu-Lys

Synonyms: Dynorphin A (1-13), (D-Arg⁶)-

Sigma D 9148   Porcine   FW 1604.0   ≥97% (HPLC); peptide content:~70% | Bioactive peptide;k-agonist; anticonvulsant; Wuster, M et al, *Neurosci Lett*, 20:79, 1980

Tyr-Gly-Gly-Phe-Leu-Arg-Arg-Ile-Arg-Pro-Lys-Leu-Lys

Synonyms: Dynorphin A (1-13)

Biogenesis 3980-0099   Porcine synthetic   MW 1603.9
Lyophilized, consisting of 84.9% peptide material, acetate counter ions and residual H_2O; purified

ICN 151021   Synthetic   A potent opioid peptide; Goldstein, A etal, *PNAS*, 76:6666, 1979

Tyr-Gly-Gly-Phe-Leu-Arg-Arg-Ile-Arg-Pro-Lys-Leu-Lys Amide

Synonyms: Dynorphin A (1-13)

American Peptide 26-4-51   Porcine   MW 1603.0
$C_{75}H_{127}N_{25}O_{14}$

Tyr-Gly-Gly-Phe-Leu-Arg-Arg-Ile-Arg-Pro-Lys-Leu-Lys-Trp-Asp-Asn-Gln

Synonyms: Dynorphin A (1-17); Pro-Dynorphin Opioid Receptor Agonist (209-225); Pro-Dynorphin (207-223); Pro-Dynorphin (209-225)

American Peptide 26-4-12   MW 2147.5 $C_{99}H_{155}N_{31}O_{23}$
Goldstein, A et al, *PNAS*, 78:7219, 1981; Tachibana, S et al, *Nature*, 295:339, 1982

Peptides International PDY-4108-v   Human synthetic   MW 2147.5 $C_{99}H_{155}N_{31}O_{23}$ >99% (HPLC); lyophilized amorphous powder | Bioactive; porcine, rat, bovine; Tachibana, S et al, *The 1981 International Narcotic Research Conference*, Kyoto, July 1981; Goldstein, A et al, *PNAS USA*, 78:7219 1981

Bachem H-2620.0001   Human, porcine   Store at -15°C

Tyr-Gly-Gly-Phe-Leu-Arg-Arg-Ile-Arg-Pro-Lys-Leu-Lys-Trp-Asp-Asn-Gln Amide

Synonyms: Dynorphin A

American Peptide 26-4-21   Porcine   MW 2146.6
$C_{99}H_{156}N_{32}O_{22}$ Goldstein, A et al, *PNAS*, 78:7219, 1981; Tachibana, S et al, *Nature*, 295:339, 1982; Faden, AI, *J Neurosci*, 12:425, 1992

Tyr-Gly-Gly-Phe-Leu-Arg-Arg-Ile-Arg-Pro-Lys-Leu-Lys-Trp-Asp-Asn-Gln

Synonyms: Pro-Dynorphin (209-225); Dynorphin A (1-17)

ICN 152778   Porcine   MW 2148 $C_{99}H_{155}N_{31}O_{23}$ Goldstein, A etal, *PNAS*, 78:7219, 1981; Tachibana, S etal, *Nature*, 295:339, 1982

Neosystem SC107   Porcine   MW 2147.51   Bioactive; opioid peptide; Tadribana, S et al, *Nature*, 295:339, 1982

Sigma D 8147   Porcine   FW 2147.5   ≥95% (HPLC) | Bioactive peptide; endogenousk-agonist that is resistant to enzymatic degradation; Goldstein, A et al, *Proc Natl Acad Sci USA*, 78:7219, 1981; Tachibana, S et al, *Nature*, 295:339, 1982; Kakidani, H et al, *Nature*, 298:245, 1982; Horikawa, S et al, *Nature*, 306:611, 1983; Corbett, AD et al, *Nature*, 299:79, 1982

Tyr-Gly-Gly-Phe-Leu-Arg-Arg-Leu

(nMe)Tyr-Gly-Gly-Phe-Leu-Arg-(nMe)Arg-D-Leu-NHEt

Synonyms: Dynorphin A (1-8); E-2078

American Peptide 26-4-49   MW 1038.3 $C_{50}H_{83}N_{15}O_9$
Metabolically stable analog of dynorphin A; not only retains opioid receptor selectivity similar to that of dynorphin A, but also produces a more potent analgesic effect than morphine even when administered subcutaneously to mice; Yoshino, H et al, *J Med Chem*, 33:206, 1990; Yoshino, H et al, *Chem Phar Bill*, 38:404, 1990

Tyr-Gly-Gly-Phe-Leu-Arg-Arg-Leu

N-Me-Tyr-Gly-Gly-Phe-Leu-Arg-N-Me-Arg-D-Leu-NHEt

Synonyms: Dynorphin A (1-8), (N-Me-Tyr¹,N-Me-Arg⁷,D-Leu-NHEt⁸)-

Bachem H-3018.0500   Store at -15°C

Tyr-Gly-Gly-Phe-Leu-Arg-Lys

Synonyms: Endorphin (1-7), α-Neo-

American Peptide 30-0-75   MW 840.0 $C_{40}H_{61}N_{11}O_9$

Tyr-Gly-Gly-Phe-Leu-Arg-Lys-Arg

Synonyms: Endorphin (1-8), (Arg⁸)-α-Neo-; Neoendorphin (1-8), (Arg⁸)-α-

American Peptide 28-4-21   MW 996.2 $C_{46}H_{73}N_{15}O_{10}$

ICN 152812   MW 996.2 $C_{46}H_{73}N_{15}O_{10}$

Tyr-Gly-Gly-Phe-Leu-Arg-Lys-Tyr

Synonyms: Neoendorphin (1-8), α-

Bachem H-4415.0001   MW 1003.17 $C_{49}H_{70}N_{12}O_{11}$ Store at -15°C

ICN 154535   MW 1003.3   Kangawa, K etal, *BBRC*, 86:153, 1979

Tyr-Gly-Gly-Phe-Leu-Arg-Lys-Tyr-Arg-Pro-Lys Amide

Synonyms: Endorphin Analog, α-Neo

American Peptide 28-4-24   MW 1383.7 $C_{66}H_{102}N_{20}O_{13}$

Tyr-Gly-Gly-Phe-Leu-Arg-Lys-Tyr-Pro

Synonyms: Endorphin, β-Neo-; Neoendorphin, β-; Neoendorphin, (des-Lys₁₀-α)-; Pro-Dynorphin (175-183)

American Peptide 28-4-26   MW 1100.3 $C_{54}H_{77}N_{13}O_{12}$
Minamino, N et al, *BBRC*, 89:864, 1981

ICN 190421   MW 1100.3   Hypothalamic opioid peptide with potent activity in the guinea pig ileum assay; Minamino, N etal, *BBRC*, 99:864, 1981

Sigma N 9758   FW 1100.3   ≥97% (HPLC) | Bioactive peptide; Leu-enkephalin precursor; 1/3 as potent as α-neoendorphin in inhibiting guinea pig ileum contraction; Minamino, N et al, *Biochem Biophys Res Commun*, 99:864, 1981

Bachem H-4420.0001   Human, porcine   Store at -15°C

Peptides International PEN-4091-v   Porcine synthetic   MW 1100.3 $C_{54}H_{77}N_{13}O_{12}$ >99% (HPLC); lyophilized amorphous powder | Bioactive; Minamino, N et al, *BBRC*, 99:864, 1981

Tyr-Gly-Gly-Phe-Leu-Arg-Lys-Tyr-Pro-Lys

Synonyms: Neoendorphin, α-; Pro-Dynorphin (175-184); Endorphin, α-Neo-

Bachem H-4410.0001   Human, porcine   Store at -15°C

American Peptide 28-4-22   Porcine   MW 1228.5 $C_{60}H_{89}N_{15}O_{13}$ Hypothalamic opioid peptide with potent activity in the guinea pig ileum assay; Kangawa, K et al, *BBRC*, 99:871, 1982

ICN 190420   Porcine   MW 1228.5   Hypothalamic opioid peptide with potent activity in the guinea pig ileum assay; Kangawa, K etal, *BBRC*, 99:871, 1981

Neosystem SC045   Porcine   MW 1228.46   Bioactive; opioid peptide; opioid activity in guinea pig ileum assay; Chino, N et al, *Pept Chem*, 1979:215, 1980

Sigma N 9633   Porcine   FW 1228.5   ≥97% (HPLC); peptide content:~70% | Bioactive peptide; endorphin; leu-enkephalin precursor; 21-fold more potent than Leu-enkephalin in inhibiting guinea pig ileum contraction; non-selective δ-, κ- & μ-agonist; Kanagawa, K et al, *Biochem Biophys Res Commun*, 99:871, 1981

Peptides International PEN-4090-v   Porcine synthetic   MW 1228.5 $C_{60}H_{89}N_{15}O_{13}$ >99% (HPLC); lyophilized amorphous powder | Bioactive; Kangawa, K Net al, *BBRC*, 99:871, 1981

Tyr-Gly-Gly-Phe-Leu-Arg-Phe

Synonyms: Enkephalin-Arg-Phe, Leu-; Dynorphin A (1-7), (Phe[7])-

Bachem H-5150.0001   MW 859 $C_{43}H_{58}N_{10}O_9$ Store at -15°C

American Peptide 26-4-88   Porcine   MW 859.0 $C_{43}H_{58}N_{10}O_9$

ICN 154521   Porcine   MW 859.1

Tyr-Gly-Gly-Phe-Leu-Arg-Phe Amide

Synonyms: Enkephalin-Arg-Phe, Leu-; Dynorphin A (1-7), (Phe[7])-

Bachem H-5155.0001   MW 858.01 $C_{43}H_{59}N_{11}O_8$ Store at -15°C

American Peptide 26-4-97   Porcine   MW 858.0 $C_{43}H_{59}N_{11}O_8$

ICN 154522   Porcine   MW 858.1

Tyr-Gly-Gly-Phe-Leu-Lys

Synonyms: Enkephalin-Lys, (Leu)-

Bachem H-2765.0001   MW 683.81 $C_{34}H_{49}N_7O_8$ Store at -15°C

Sigma E 0134   FW 683.8 $C_{34}H_{49}N_7O_8$ ≥97% (HPLC) | Bioactive peptide; non-selective δ- & μ-agonist

---

Tyr-Gly-Gly-Phe-Leu-Thr-Ser-Glu-Lys-Ser-Gln-Thr-Pro-Leu-Val-Thr-Leu-Phe-Lys-Asn-Ala-Ile-Ile-Lys-Asn-Ala-Tyr-Lys-Lys-Gly-Glu

Synonyms: Endorphin, (Leu[5])-β-

ICN 152808   Human   MW 3447   Found in renal dialysate from schizophrenic patients; Palmour, R etal, *Abst Soc for Neurosci*, Anaheim, CA, Nov 1977

Tyr-Gly-Gly-Phe-Met

Synonyms: Enkephalin, (Met)-

American Peptide 30-0-10   MW 573.8 $C_{27}H_{35}N_5O_7S$ This opioid agonist inhibits vas deferens contractions mediated through δ-receptors; Hughes, J et al, *Nature*, 258:577, 1975

Tyr-Gly-Gly-Phe-Met
BOC-Tyr-Gly-Gly-Phe-Met

Synonyms: Enkephalin, BOC-Met-

Bachem A-2445.0250   MW 673.79 $C_{32}H_{43}N_5O_9S$ Store at -15°C

Tyr-Gly-Gly-Phe-Met
BOC-Tyr-Gly-Gly-Phe-Met-OtBu

Synonyms: Enkephalin-OtBu, BOC-Met-

Bachem A-2815.0050   MW 729.9 $C_{36}H_{51}N_5O_9S$ Store at -15°C

Tyr-Gly-Gly-Phe-Met

Synonyms: Enkephalin, Met-; Endorphin (1-5)

Bachem H-2785.0025   MW 573.67 $C_{27}H_{35}N_5O_7S$ Store at -15°C

Tyr-Gly-Gly-Phe-Met
Tyr-Gly-Gly-Phe-Met(O)

Synonyms: Enkephalin, (Met(O)[5])-

Bachem H-5160.0005   MW 589.67 $C_{27}H_{35}N_5O_8S$ Store at -15°C

Tyr-Gly-Gly-Phe-Met

Synonyms: Enkephalin M; Enkephalin, (Met[5])-

Calbiochem 05-23-0901   MW 573.6 $C_{27}H_{35}N_5O_7S$ >95% (HPLC); lyophilized solid; soluble in 5% acetic acid | Opiate peptide that binds to opiate receptors; inhibits vas deferens contractions mediated through delta receptors; Tanaka, E & North, RA, *J Neurosci*, 14:1106, 1994; Martin, MI et al, *Naunyn Schon Arch Pharmacol*, 347:324, 1993; *Merck Index*, 12:3613

Fluka 64365   MW 573.67 $C_{27}H_{35}N_5O_7S$ Goodman, RR et al, *Brain Peptides*, DT Krieger et al, eds, 827:1983, John Wiley & Sons, New York

---

Tyr-Gly-Gly-Phe-Met
Tyr-Gly-Gly-Phe-Met(S=O)

Synonyms: Enkephalin, Met(S=O)[5]-

ICN 152822

Tyr-Gly-Gly-Phe-Met

Synonyms: Enkephalin, (Met[5])-

ICN 190430   Endogenous opioid pentapeptide

Tyr-Gly-Gly-Phe-Met
N-α-t-BOC-Tyr-Gly-Gly-Phe-Met

Synonyms: Enkephalin, N-α-t-BOC-(Met[5])-

ICN 190441

Tyr-Gly-Gly-Phe-Met

Synonyms: Enkephalin, Met-

Neosystem SC002   MW 573.66   Bioactive; opioid peptide; Hugues, J et al, *Nature*, 258:577, 1975

Tyr-Gly-Gly-Phe-Met
Tyr-Gly-Gly-Phe-Met(O)

Synonyms: Enkephalin, (Met(O)[5])-; Enkephalin Sulfoxide, (Met)-

Sigma E 7133   FW 589.7 $C_{27}H_{35}N_5O_8S$ ≥97% (HPLC) | Bioactive peptide; mixture of R & S sulfoxides

---

Tyr-Gly-Gly-Phe-Met

Synonyms: Enkephalin, Met-

Peptides International PEK-4042-v   Synthetic   MW 573.67 $C_{27}H_{35}N_5O_7S$ >95% (HPLC); lyophilized amorphous powder | Bioactive; human, porcine, bovine, rat, mouse

---

Tyr-Gly-Gly-Phe-Met Acetate Salt

Synonyms: Enkephalin, (Met)-

Sigma M 6638   FW 573.7 (free base) $C_{27}H_{35}N_5O_7S$ ≥97% (HPLC) | Bioactive peptide; endogenous opioid neurotransmitter/neuromodulator; localization in brain parallels the localization of μ-receptors; Akil, H et al, *Ann Rev Neurosci*, 7:223, 1984; Hughes, J et al, *Nature*, 258:577, 1975

Biogenesis 4140-5504   Human synthetic   MW 537.5 Purified; lyophilized | To avoid Met oxidation, it is best to dissolve it under a nitrogen stream; Hughes et al, *Nature*, 258:277, 1975

Tyr-Gly-Gly-Phe-Met Amide

Synonyms: Enkephalinamide, (Met[5])-

American Peptide 30-0-22   MW 572.7 $C_{27}H_{36}N_6O_6S$ Chang, JK et al, *Life Sci*, 18:1473, 1976

ICN 190431 $C_{20}H_{31}N_5O_4 \cdot C_2H_4O_2$ Chang, JK etal, *Life Sci*, 18:1473, 1976

Neosystem SC003   MW 572.68   Bioactive; opioid peptide; Chang, JK, *Life Sciences*, 18:1473, 1976

Tyr-Gly-Gly-Phe-Met Amide Acetate Salt

Synonyms: Enkephalinamide, (Met)-

Sigma E 5381   FW 572.7 (free base) $C_{27}H_{36}N_6O_6S$ ≥97% (HPLC) | Bioactive peptide; Enkephalin; Chang, JK et al, *Life Sci*, 18:473, 1976

Tyr-Gly-Gly-Phe-Met Hydrate

Synonyms: Enkephalin, Met-

Peptides International PEK-4042   Synthetic   MW 591.69 $C_{27}H_{35}N_5O_7S \cdot H_2O$ >95% (HPLC); lyophilized amorphous powder; bulk | Bioactive; human, porcine, bovine, rat, mouse

Tyr-Gly-Gly-Phe-Met-Tyr(SO₃H)-Gly-Gly-Phe-Met Sulfated

Synonyms: Enkephalin O, (Met[5])-

ICN 190432

Tyr-Gly-Gly-Phe-Met-Arg

Synonyms: Enkephalin, (Met[5],Arg[6])-; Enkephalin-Arg, Met-

American Peptide 30-1-27   MW 729.9 $C_{33}H_{47}N_9O_8S$ Robberecht, P et al, *FEBS Lett*, 286:133, 1991

Bachem H-2805.0001   MW 729.86 $C_{33}H_{47}N_9O_8S$ Store at -15°C

Tyr-Gly-Gly-Phe-Met-Arg-Arg

Synonyms: Enkephalin-Arg, (Met[5], Arg[6]); Enkephalin-Arg-Arg, Met-

American Peptide 30-0-46   MW 886.1 $C_{39}H_{59}N_{13}O_9S$

Bachem H-2815.0001   MW 886.05 $C_{39}H_{59}N_{13}O_9S$ Store at -15°C

Tyr-Gly-Gly-Phe-Met-Arg-Arg-Val

Synonyms: Metorphamide; Adrenorphin; Enkephalin-Arg-Arg-Val, (Met)-

Sigma T 8156   FW 985.2 (free acid) $C_{44}H_{68}N_{14}O_{10}S$ ≥95% (HPLC) | Bioactive peptide; opioid peptide

Tyr-Gly-Gly-Phe-Met-Arg-Arg-Val Amide

Synonyms: Metorphamide; Adrenorphin; Pro-Enkephalin Precursor (206-213)

Bachem H-6855.0001   MW 984.19 $C_{44}H_{69}N_{15}O_9S$ Store at -15°C

ICN 152954

American Peptide 30-0-50   Bovine   MW 984.2 $C_{44}H_{69}N_{15}O_9S$ Amidated octapeptide found in bovine brain & adrenal tissue; also related to the sequence of the bovine adrenal medulla peptides

Sigma T 8281   Bovine brain   FW 984.2 $C_{44}H_{69}N_{15}O_9S$ ≥97% (HPLC) | Bioactive peptide; opioid peptide; high affinity for μ-receptors; Matsuo, H et al, *Nature*, 305:721, 1983; Weber, E et al, *Proc Natl Acad Sci USA*, 80:7362, 1983

Tyr-Gly-Gly-Phe-Met-Arg-Arg-Val Free acid

Synonyms: Adrenorphin; Metorphinamide

American Peptide 30-0-51   MW 985.2 $C_{44}H_{68}N_{14}O_{10}S$ free acid

Bachem H-6860.0001   MW 985.18 $C_{44}H_{68}N_{14}O_{10}S$ Store at -15°C

ICN 152955

Tyr-Gly-Gly-Phe-Met-Arg-Arg-Val-Gly

Synonyms: Enkephalin, (Met[5],Arg[6,7],Val[8],Gly[9])-

American Peptide 30-1-31   MW 1042.2 $C_{46}H_{71}N_{15}O_{11}S$

Tyr-Gly-Gly-Phe-Met-Arg-Arg-Val-Gly-Arg-Pro-Glu

Synonyms: Bovine Adrenal Medulla 12P; Bovine Adrenal Medulla Docosapeptide

Bachem H-2125.0001   MW 1424.65 $C_{62}H_{97}N_{21}O_{16}S$ Store at -15°C

ICN 152943   MW 1424.6   Potent opiate activity; Mizuno, K etal, *BBRC*, 95:1482, 1980; Mizuno, K etal, *BBRC*, 97:1283, 1980

Sigma B 4139   Bovine adrenal medulla   FW 1424.6   ≥97% (HPLC) | Bioactive peptide; opioid peptide; Mizuno, K et al, *Biochem Biophys Res Commun*, 97:1283, 1980; Mizuno, K et al, *Biochem Biophys Res Commun*, 95:1482, 1980

American Peptide 30-3-20   Bovine adrenomedullin   MW 1424.9 $C_{62}H_{97}N_{21}O_{16}S$ Met-enkephalin precursor; Mizuno, K et al, *BBRC*, 97:1283, 1980

Peptides International PBM-4119-v   Synthetic   MW 1424.7 $C_{62}H_{97}N_{21}O_{16}S$ >95% (HPLC); lyophilized amorphous powder | Bioactive; Mizuno, K et al, *BBRC*, 95:1482, 1980

Tyr-Gly-Gly-Phe-Met-Arg-Arg-Val-Gly-Arg-Pro-Glu-Trp-Trp-Met-Asp-Tyr-Gln

Synonyms: Bovine Adrenal Medulla 18P; Bovine Adrenal Medulla Octadecapeptide

American Peptide 30-3-25   MW 2334.7 $C_{107}H_{148}N_{30}O_{26}S_2$ Mizuno, K et al, *BBRC*, 97:1283, 1980

Tyr-Gly-Gly-Phe-Met-Arg-Arg-Val-Gly-Arg-Pro-Glu-Trp-Trp-Met-Asp-Tyr-Gln-Lys-Arg-Tyr-Gly

Synonyms: Bovine Adrenal Medulla 22P

American Peptide 30-3-30   MW 2839.3 $C_{130}H_{184}N_{38}O_{31}S_2$ Mizuno, K et al, *BBRC*, 97:1283, 1980

Bachem H-2130.0500   MW 2839.26 $C_{130}H_{184}N_{38}O_{31}S_2$ Store at -15°C

Tyr-Gly-Gly-Phe-Met-Arg-Arg-Val-Gly-Arg-Pro-Glu-Trp-Trp-Met-Asp-Tyr-Gln-Lys-Arg-Tyr-Gly-Gly-Phe-Leu

Synonyms: Bovine Adrenal Medulla 3200; Peptide E; Adipokinetic Peptide E

Bachem H-4500.0500   MW 3156.65 $C_{147}H_{207}N_{41}O_{34}S_2$ Store at -15°C

ICN 152940   MW 3156.6   Kilpatrick, DL et al, *PNAS*, 78:3265, 1981

Tyr-Gly-Gly-Phe-Met-Arg-Arg-Val-Gly-Arg-Pro-Glu-Trp-Trp-Met-Asp-Tyr-Gln-Lys-Arg-Tyr-Gly

Synonyms: Bovine Adrenal Medulla 22P; Bovine Adrenal Medulla Docosapeptide

ICN 152942   MW 2839.2   Extraordinarily potent opiate activity; Mizuno, K etal, *BBRC*, 97:1283, 1980

Tyr-Gly-Gly-Phe-Met-Arg-Arg-Val-Gly-Arg-Pro-Glu-Trp-Trp-Met-Asp-Tyr-Gln-Lys-Arg-Tyr-Gly-Gly-Phe-Leu

Synonyms: Adrenal Peptide E; Bovine Adrenal Medulla 3200

American Peptide 30-3-35   Bovine   MW 3156.7
$C_{147}H_{207}N_{41}O_{34}S_2$ Kalpatrick, D et al, *PNAS*, 78:3265, 1981

Sigma A 2159   Bovine   FW 3156.6   ≥97% (HPLC) |
Bioactive peptide; opioid peptide; Kilpatrick, DL et al, *Proc Natl Acad Sci USA*, 78:3265, 1981

Tyr-Gly-Gly-Phe-Met-Arg-Gly-Leu

Synonyms: MERGL; Pro-Enkephalin; Enkephalin, (Met[5],Arg[6],Gly[7],Leu[8])-; Enkephalin-Arg-Gly-Leu, (Met[5])-

Bachem H-2820.0005   MW 900.07 $C_{41}H_{61}N_{11}O_{10}S$ Store at -15°C

Sigma E 6515   FW 900.1 $C_{41}H_{61}N_{11}O_{10}S$ ≥97% (HPLC) |
Bioactive peptide; originally extracted from bovine adrenal medulla; Gubler, U et al, *Nature*, 295:206, 1982

American Peptide 30-1-33   Bovine adrenal medulla   MW 900.1 $C_{41}H_{61}N_{11}O_{10}S$ Kilpatrick, DL et al, *BBRC*, 103:698, 1981

ICN 190440   Bovine medulla   MW 900.1 $C_{41}H_{61}N_{11}O_{10}S$
Kirkpatrick, DL et al, *BBRC*, 103:698, 1981; Gubler, U et al, *Nature*, 295:206, 1982; Jones, BN et al, *PNAS*, 79:1313, 1982; Noda, M et al, *Nature*, 295:202, 1982; Noda, M et al, *Nature*, 297:431, 1982

Tyr-Gly-Gly-Phe-Met-Arg-Lys

Synonyms: Enkephalin-Arg-Lys, Met-

Bachem H-2825.0001   MW 858.03 $C_{39}H_{59}N_{11}O_9S$ Store at -15°C

Tyr-Gly-Gly-Phe-Met-Arg-Met
Tyr-Gly-Gly-Phe-Met(O)-Arg

Synonyms: Enkephalin-Arg, (Met(O)[5])-

Bachem H-2810.0001   Store at -15°C

Tyr-Gly-Gly-Phe-Met-Arg-Phe

Synonyms: Enkephalin, (Met[5],Arg[6],Phe[7])-; MERF; Enkephalin-Arg-Phe, (Met[5])-

American Peptide 30-1-34   MW 877.0 $C_{42}H_{56}N_{10}O_9S$
Heptapeptide with opioid receptor activity

Bachem H-2830.0001   MW 877.04 $C_{42}H_{56}N_{10}O_9S$ Store at -15°C

ICN 152825   MW 877 $C_{42}H_{56}N_{10}O_9S$ Heptapeptide with opiate receptor activity; Stern, AS et al, *PNAS*, 76:6680, 1979; Greenberg, MJ et al, *Neuropeptides*, 1:309, 1981

Tyr-Gly-Gly-Phe-Met-Arg-Phe Acetate Salt

Synonyms: Enkephalin-Arg-Phe, (Met)-; MERF

Sigma E 5757   FW 877.0 (free base) $C_{42}H_{56}N_{10}O_9S$ ≥97% (HPLC) | Bioactive peptide; opioid receptor activity; Stern, AS et al, *Proc Natl Acad Sci USA*, 76:6680, 1979

Tyr-Gly-Gly-Phe-Met-Arg-Phe Amide

Synonyms: Enkephalin, (Met[5],Arg[6],Phe[7])-; Enkephalin-Arg-Phe, Met-

American Peptide 30-1-37   MW 876.1 $C_{42}H_{57}N_{11}O_8S$
Greenberg, MJ et al, *Neuropeptides*, 1:309, 1981

Bachem H-2835.0001   MW 876.05 $C_{42}H_{57}N_{11}O_8S$ Store at -15°C

Tyr-Gly-Gly-Phe-Met-Lys

Synonyms: Endorphin (1-6), (Met[5],Lys[6])-α-Neo-; Enkephalin-Lys, (Met[5])-; Enkephalin Analog, Met-

American Peptide 30-0-44   MW 701.9 $C_{33}H_{47}N_7O_8S$

Bachem H-1340.0005   MW 701.84 $C_{33}H_{47}N_7O_8S$ Store at -15°C

ICN 152826   MW 701.8 $C_{33}H_{47}N_7O_8S$ Stern, AS et al, *Arch Biochem Biophys*, 205:606, 1980

Sigma E 1760   FW 701.8 $C_{33}H_{47}N_7O_8S$ ≥95% (HPLC) |
Bioactive peptide; displaces ³H-Leu-enkephalin bound to neuroblastoma-blioma hybrid cells; Stern, AS et al, *Arch Biochem Biophys*, 205:606, 1980

Tyr-Gly-Gly-Phe-Met-Lys-Arg

Synonyms: Endorphin (1-7), (Met[5],Lys[6],Arg[7])-α-Neo-; Enkephalin-Lys-Arg, Met-

American Peptide 30-0-56   MW 858.1 $C_{39}H_{59}N_{11}O_9S$

Bachem H-2840.0001   MW 858.03 $C_{39}H_{59}N_{11}O_9S$ Store at -15°C

Tyr-Gly-Gly-Phe-Met-Lys-Lys

Synonyms: Endorphin (1-7), (Met[5],Lys[6,7])-α-Neo-; Enkephalin-Lys-Lys, Met-

American Peptide 30-0-55   MW 830.1 $C_{39}H_{59}N_9O_9S$

Bachem H-2845.0001   MW 830.02 $C_{39}H_{59}N_9O_9S$ Store at -15°C

Tyr-Gly-Gly-Phe-Met-Lys-Lys-Met-Asp-Glu-Leu-Tyr-Pro-Leu-Glu-Val-Glu-Glu-Glu-Ala-Asn-Gly-Gly-Glu-Val-Leu-Glu-Lys-Arg-Tyr-Gly-Gly-Phe-Met

Synonyms: Peptide F

American Peptide 30-3-40   Bovine   MW 3845.4
$C_{172}H_{259}N_{41}O_{53}S_3$

Tyr-Gly-Gly-Phe-Met-Thr-Leu-Phe-Lys-Asn-Ala-Ile-Ile-Lys-Asn-Ala-Tyr-Lys-Lys-Gly-Glu

Synonyms: Endorphin (1-5 + 16-31), β-

American Peptide 28-1-80   Human   MW 2393.9
$C_{112}H_{173}N_{27}O_{29}S$

Tyr-Gly-Gly-Phe-Met-Thr-Ser-Glu-Lys

Synonyms: Lipotropin (61-69), β-

American Peptide 28-0-25   MW 1019.2 $C_{45}H_{66}N_{10}O_{15}S$

Tyr-Gly-Gly-Phe-Met-Thr-Ser-Glu-Lys-Ser-Gln-Thr-Pro-Leu-Val-Thr

Synonyms: Endorphin, α-; Lipotropin (61-76), β-

American Peptide 24-4-30   MW 1746.0 $C_{77}H_{120}N_{18}O_{26}S$
Guillemin, R, *CR Acad Sci Paris*, 282:783, 1976

Tyr-Gly-Gly-Phe-Met-Thr-Ser-Glu-Lys-Ser-Gln-Thr-Pro-Leu-Val-Thr
Ac-Tyr-Gly-Gly-Phe-Met-Thr-Ser-Glu-Lys-Ser-Gln-Thr-Pro-Leu-Val-Thr

Synonyms: Endorphin, α-; Lipotropin (61-76), β-

American Peptide 28-4-31   MW 1788.0 $C_{79}H_{122}N_{18}O_{27}S$

Tyr-Gly-Gly-Phe-Met-Thr-Ser-Glu-Lys-Ser-Gln-Thr-Pro-Leu-Val-Thr

Synonyms: Endorphin (1-16), β-; Endorphin, α-; Lipotropin (61-76), β-; LPH (61-76)

Bachem H-2695.0001   MW 1745.97 $C_{77}H_{120}N_{18}O_{26}S$ Store at -15°C

Sigma E 6136   Human   FW 1746.0   ≥97% (HPLC) |
Bioactive peptide; β-lipotropin fragment found in intermediate & anterior lobes of the pituitary; opioid agonist that does not produce generalized rigidity or loss of tail pinch reflex when injected intraventricularly; Bloom, F et al, *Science*, 194:630, 1976; Guillemin, R et al, *Proc Natl Acad Sci USA*, 73:3942, 1976

Fluka 45176   Human synthetic   MW 1746 $C_{77}H_{120}N_{18}O_{26}S$
≥97% (HPLC) | Bloom, FE, *Ann Rev Pharmacol Toxicol*, 23:151, 1983

Biogenesis 4100-0004 Synthetic >99% (HPLC) and TLC; 95.81% of peptide material, acetate counter ions and residual H_2O; lyophilized | To avoid Met oxidation, it is best to dissolve the peptide under a nitrogen stream

Peptides International PEN-4055-v Synthetic MW 1746.0 $C_{77}H_{120}N_{18}O_{26}S$ >95% (HPLC); lyophilized amorphous powder | Bioactive; Goldstein, A, *Ann NY Acad Sci*, 311:49, 1978; Bloom, F et al, *Adv Biochem Psychopharm*, 22:619, 1980; Berger, PA et al, *Ann Rev Med*, 33:397, 1982; Bloom, FE, *Ann Rev Pharmacol Toxicol*, 23:151, 1983; Ling, N et al, *PNAS USA*, 73:3942, 1976

Tyr-Gly-Gly-Phe-Met-Thr-Ser-Glu-Lys-Ser-Gln-Thr-Pro-Leu-Val-Thr-Leu

Synonyms: Endorphin, γ-; Lipotropin (61-77), β-; Endorphin (1-17), β-

American Peptide 28-4-40 MW 1859.1 $C_{83}H_{131}N_{19}O_{27}S$ Ling, N et al, *PNAS*, 73:3942, 1976

Bachem H-2725.0001 MW 1859.13 $C_{83}H_{131}N_{19}O_{27}S$ Store at -15°C

Tyr-Gly-Gly-Phe-Met-Thr-Ser-Glu-Lys-Ser-Gln-Thr-Pro-Leu-Val-Thr-Leu-Phe-Lys-Asn-Ala-Ile-Ile-Lys-Asn-Ala-Tyr
Ac-Tyr-Gly-Gly-Phe-Met-Thr-Ser-Glu-Lys-Ser-Gln-Thr-Pro-Leu-Val-Thr-Leu-Phe-Lys-Asn-Ala-Ile-Ile-Lys-Asn-Ala-Tyr

Synonyms: Endorphin, β-

ICN 152810

Tyr-Gly-Gly-Phe-Met-Thr-Ser-Glu-Lys-Ser-Gln-Thr-Pro-Leu-Val-Thr-Leu

Synonyms: Endorphin, γ-; Lipotropin (61-77), β-

ICN 190418 Guillemin, R, *C & E News*, Aug 16, 1976, p18; Ling, N, *BBRC*, 74:48, 1977; Austen, BM & DG Smyth, *BBRC*, 76:477, 1977

Sigma E 6386 FW 1859.1 ≥95% (HPLC) | Bioactive peptide; β-lipotropin fragment found in intermediate & anterior lobes of the pituitary; opioid agonist that does not produce generalized rigidity or loss of tail pinch reflex when injected intraventricularly; Bloom, F et al, *Science*, 194:630, 1976; Austen, BM & Smyth, DG, *Biochem Biophys Res Commun*, 76:477, 1977; Ling, N, ibid, 74:248, 1977

Tyr-Gly-Gly-Phe-Met-Thr-Ser-Glu-Lys-Ser-Gln-Thr-Pro-Leu-Val-Thr-Leu-Phe-Lys-Asn-Ala-Ile-Ile-Lys-Asn-Ala-His-Lys-Lys-Gly-Gln

Synonyms: Endorphin, β-; Lipotropin (61-91), β-

Sigma E 0637 Bovine, camel, ovine FW 3438.0 ≥97% (HPLC) | Bioactive peptide; neurohormone secreted by the anterior pituitary gland; most potent opioid peptide; Li, CH & Chung, D, *Proc Natl Acad Sci USA*, 73:1145, 1976

American Peptide 28-6-10 Camel MW 3438.0 $C_{155}H_{250}N_{42}O_{44}S$ Camel; Li, CH et al, *PNAS*, 73:1145, 1976

Tyr-Gly-Gly-Phe-Met-Thr-Ser-Glu-Lys-Ser-Gln-Thr-Pro-Leu-Val-Thr-Leu-Phe-Lys-Asn-Ala-Ile-Ile-Lys-Asn-Ala-His
Ac-Tyr-Gly-Gly-Phe-Met-Thr-Ser-Glu-Lys-Ser-Gln-Thr-Pro-Leu-Val-Thr-Leu-Phe-Lys-Asn-Ala-Ile-Ile-Lys-Asn-Ala-His

Synonyms: Endorphin, Ac-δ-

Bachem H-1105.0001 Camel MW 3038.52 $C_{138}H_{217}N_{35}O_{40}S$ Store at -15°C

Tyr-Gly-Gly-Phe-Met-Thr-Ser-Glu-Lys-Ser-Gln-Thr-Pro-Leu-Val-Thr-Leu-Phe-Lys-Asn-Ala-Ile-Ile-Lys-Asn-Ala-His-Lys-Lys-Gly-Gln

Synonyms: Endorphin, β-; Lipotropin C Fragment; Lipotropin (61-91), β-

Bachem H-2705.0001 Camel MW 3438.01 $C_{155}H_{250}N_{42}O_{44}S$ Store at -15°C

Tyr-Gly-Gly-Phe-Met-Thr-Ser-Glu-Lys-Ser-Gln-Thr-Pro-Leu-Val-Thr-Leu-Phe-Lys-Asn-Ala-Ile-Ile-Lys-Asn-Ala-His

S Endorphin, δ-; Endorphin (1-27), β-; Lipotropin (61-87), β-; Lipotropin C' Fragment *ynonyms*:

Bachem H-2715.0001 Camel Store at -15°C

Tyr-Gly-Gly-Phe-Met-Thr-Ser-Glu-Lys-Ser-Gln-Thr-Pro-Leu-Val-Thr-Leu-Phe-Lys-Asn-Ala-Ile-Ile-Lys-Asn-Ala-His-Lys-Lys-Gly-Gln

Synonyms: Endorphin, β-; Lipotropin (61-91), β-

ICN 151033 Camel MW 3621.8 Li, CH & D Chung, *PNAS*, 73:1145, 1976; Li, C, *Central & Peripheral Endorphins:Basic & Clinical Aspects*, Mueller, E & AR Genazzani, eds, Raven Press, New York, p17, 1984

Tyr-Gly-Gly-Phe-Met-Thr-Ser-Glu-Lys-Ser-Gln-Thr-Pro-Leu-Val-Thr-Leu-Phe-Lys-Asn-Ala-Ile-Ile-Lys-Asn-Ala-His

S Endorphin (1-27), β-*ynonyms*:

ICN 154533 Camel MW 2996.9

Tyr-Gly-Gly-Phe-Met-Thr-Ser-Glu-Lys-Ser-Gln-Thr-Pro-Leu-Val-Thr-Leu-Phe-Lys-Asn-Ala-Ile-Ile-Lys-Asn-Ala-His-Lys-Lys-Gly-Gln
Ac-Tyr-Gly-Gly-Phe-Met-Thr-Ser-Glu-Lys-Ser-Gln-Thr-Pro-Leu-Val-Thr-Leu-Phe-Lys-Asn-Ala-Ile-Ile-Lys-Asn-Ala-His-Lys-Lys-Gly-Gln

Synonyms: Endorphin, β-

American Peptide 28-6-23 Camel, bovine, ovine MW 3480.1 $C_{157}H_{252}N_{42}O_{45}S$

Tyr-Gly-Gly-Phe-Met-Thr-Ser-Glu-Lys-Ser-Gln-Thr-Pro-Leu-Val-Thr-Leu-Phe-Lys-Asn-Ala-Ile-Ile-Lys-Asn-Ala-His

Synonyms: Endorphin (1-27), β-

American Peptide 28-6-28 Camel, bovine, ovine MW 2996.5 $C_{136}H_{215}N_{35}O_{39}S$

Tyr-Gly-Gly-Phe-Met-Thr-Ser-Glu-Lys-Ser-Gln-Thr-Pro-Leu-Val-Thr-Leu-Phe-Lys-Asn-Ala-Ile-Ile-Lys-Asn-Ala-His
Ac-Tyr-Gly-Gly-Phe-Met-Thr-Ser-Glu-Lys-Ser-Gln-Thr-Pro-Leu-Val-Thr-Leu-Phe-Lys-Asn-Ala-Ile-Ile-Lys-Asn-Ala-His

Synonyms: Endorphin (1-27), β-

American Peptide 28-6-29 Camel, bovine, ovine MW 3038.5 $C_{138}H_{217}N_{35}O_{40}S$

Tyr-Gly-Gly-Phe-Met-Thr-Ser-Glu-Lys-Ser-Gln-Thr-Pro-Leu-Val-Thr-Leu-Phe-Lys-Asn-Ala-Ile-Ile-Lys-Asn-Ala-Tyr-Lys-Lys-Gly-Glu

Synonyms: Endorphin, β-; Lipotropin (61-91), β-

American Peptide 28-1-10 Human MW 3465.1 $C_{158}H_{251}N_{39}O_{46}S$ Neurohormone secreted by the pituitary gland; most potent of the opiate peptides & more effective than morphine in tests of opiate receptor binding, analgesia & catatonia, high doses of β-endorphin cause rigidity & hypothermia in rats; Li, CH et al, *JMC*, 20:325, 1977

Tyr-Gly-Gly-Phe-Met-Thr-Ser-Glu-Lys-Ser-Gln-Thr-Pro-Leu-Val-Thr-Leu-Phe-Lys-Asn-Ala-Ile-Ile-Lys-Asn-Ala-Tyr-Lys-Lys-Gly-Glu
Ac-Tyr-Gly-Gly-Phe-Met-Thr-Ser-Glu-Lys-Ser-Gln-Thr-Pro-Leu-Val-Thr-Leu-Phe-Lys-Asn-Ala-Ile-Ile-Lys-Asn-Ala-Tyr-Lys-Lys-Gly-Glu

Synonyms: Endorphin, β-

American Peptide 28-1-14   Human   MW 3507.1
$C_{160}H_{253}N_{39}O_{47}S$

Tyr-Gly-Gly-Phe-Met-Thr-Ser-Glu-Lys-Ser-Gln-Thr-Pro-Leu-Val-Thr-Leu-Phe-Lys-Asn-Ala-Ile-Ile-Lys-Asn-Ala-Tyr

S Endorphin (1-27), β-*ynonyms*:

American Peptide 28-1-15   Human   MW 3022.5
$C_{139}H_{217}N_{33}O_{40}S$

Tyr-Gly-Gly-Phe-Met-Thr-Ser-Glu-Lys-Ser-Gln-Thr-Pro-Leu-Val-Thr-Leu-Phe-Lys-Asn-Ala-Ile-Ile-Lys-Asn-Ala

Synonyms: Endorphin (1-26), β-

American Peptide 28-1-16   Human   MW 2859.4
$C_{130}H_{208}N_{32}O_{38}S$

Tyr-Gly-Gly-Phe-Met-Thr-Ser-Glu-Lys-Ser-Gln-Thr-Pro-Leu-Val-Thr-Leu-Phe-Lys-Asn-Ala-Ile-Ile-Lys-Asn-Ala-Tyr
Ac-Tyr-Gly-Gly-Phe-Met-Thr-Ser-Glu-Lys-Ser-Gln-Thr-Pro-Leu-Val-Thr-Leu-Phe-Lys-Asn-Ala-Ile-Ile-Lys-Asn-Ala-Tyr

Synonyms: Endorphin (1-27), β-

American Peptide 28-1-35   Human   MW 3064.6
$C_{141}H_{219}N_{33}O_{41}S$

Tyr-Gly-Gly-Phe-Met-Thr-Ser-Glu-Lys-Ser-Gln-Thr-Pro-Leu-Val-Thr-Leu-Phe-Lys-Asn-Ala-Ile-Ile-Lys-Asn-Ala
Ac-Tyr-Gly-Gly-Phe-Met-Thr-Ser-Glu-Lys-Ser-Gln-Thr-Pro-Leu-Val-Thr-Leu-Phe-Lys-Asn-Ala-Ile-Ile-Lys-Asn-Ala

Synonyms: Endorphin (1-26), β-

American Peptide 28-1-36   Human   MW 2901.4
$C_{132}H_{210}N_{32}O_{39}S$

Tyr-Gly-Gly-Phe-Met-Thr-Ser-Glu-Lys-Ser-Gln-Thr-Pro-Leu-Val-Thr-Leu-Phe-Lys-Asn-Ala-Ile-Ile-Lys-Asn-Ala-Tyr-Lys-Lys-Gly-Glu
Ac-Tyr-Gly-Gly-Phe-Met-Thr-Ser-Glu-Lys-Ser-Gln-Thr-Pro-Leu-Val-Thr-Leu-Phe-Lys-Asn-Ala-Ile-Ile-Lys-Asn-Ala-Tyr-Lys-Lys-Gly-Glu

Synonyms: Endorphin, Ac-β-

Bachem H-1115.0001   Human   Store at -15°C

Tyr-Gly-Gly-Phe-Met-Thr-Ser-Glu-Lys-Ser-Gln-Thr-Pro-Leu-Val-Thr-Leu-Phe-Lys-Asn-Ala-Ile-Ile-Lys-Asn-Ala-Tyr-Lys-Lys-Gly-Glu

Synonyms: Endorphin, β-; Lipotropin C Fragment; Lipotropin (61-91), β-

Bachem H-2700.0001   Human   MW 3465.03
$C_{158}H_{251}N_{39}O_{46}S$ Store at -15°C

Tyr-Gly-Gly-Phe-Met-Thr-Ser-Glu-Lys-Ser-Gln-Thr-Pro-Leu-Val-Thr-Leu-Phe-Lys-Asn-Ala-Ile-Ile-Lys-Asn-Ala-Tyr

S Endorphin, δ-; Endorphin (1-27), β-; Lipotropin (61-87), β-; Lipotropin C' Fragment *ynonyms*:

Bachem H-2710.0001   Human   Store at -15°C

Tyr-Gly-Gly-Phe-Met-Thr-Ser-Glu-Lys-Ser-Gln-Thr-Pro-Leu-Val-Thr-Leu-Phe-Lys-Asn-Ala-Ile-Ile-Lys-Asn-Ala-Tyr-Lys-Lys-Gly-Glu

Synonyms: Endorphin, β-; Lipoprotein (61-91), β-

Calbiochem 05-23-0930   Human   MW 3465
$C_{158}H_{251}N_{39}O_{46}O$ ≥95% (HPLC); solid; soluble in 5% acetic acid; LD_{50}≤2000 mg/kg; may be carcinogenic/teratogenic; sold under license of US Reissue Patent Re 29842 issued to Research Corporation | Neurohormone secreted by the pituitary gland; the most potent of the opiate peptides; more effective than morphine in tests of opiate receptor binding, analgesia & catatonia; high doses cause rigidity & hypothermia in rats; Dalayeun, JF et al, *Biomed Pharmacother*, 47:311, 1993; Akil, H et al, *Ann Rev Neurosci*, 7:223, 1984; *Merck Index*, 12:3613

Tyr-Gly-Gly-Phe-Met-Thr-Ser-Glu-Lys-Ser-Gln-Thr-Pro-Leu-Val-Thr-Leu-Phe-Lys-Asn-Ala-Ile-Ile-Lys-Asn-Ala-Tyr-Lys-Lys-Gly-Glu
Ac-Tyr-Gly-Gly-Phe-Met-Thr-Ser-Glu-Lys-Ser-Gln-Thr-Pro-Leu-Val-Thr-Leu-Phe-Lys-Asn-Ala-Ile-Ile-Lys-Asn-Ala-Tyr-Lys-Lys-Gly-Glu

Synonyms: Endorphin, β-

ICN 152807   Human   MW 3621.2

Tyr-Gly-Gly-Phe-Met-Thr-Ser-Glu-Lys-Ser-Gln-Thr-Pro-Leu-Val-Thr-Leu-Phe-Lys-Asn-Ala-Ile-Ile-Lys-Asn-Ala-Tyr

S Endorphin (1-27), β-*ynonyms*:

ICN 152809   Human   MW 3022.5

Tyr-Gly-Gly-Phe-Met-Thr-Ser-Glu-Lys-Ser-Gln-Thr-Pro-Leu-Val-Thr-Leu-Phe-Lys-Asn-Ala-Ile-Ile-Lys-Asn-Ala-Tyr-Lys-Lys-Gly-Glu

Synonyms: Lipotropin (61-91), β-; Endorphin, β-

ICN 190417   Human   Li, CH etal, *J Med Chem*, 20:325, 1977; Chung, WC etal, *Int J Pept Prot Res*, 13:278, 1979; Akie, H etal, *Ann Rev Neurol*, 7:223, 1984

Neosystem SC105   Human   MW 3465.01   Bioactive; opioid peptide; Li, CH et al, *J Med Chem*, 20:325, 1977

Tyr-Gly-Gly-Phe-Met-Thr-Ser-Glu-Lys-Ser-Gln-Thr-Pro-Leu-Val-Thr-Leu-Phe-Lys-Asn-Ala-Ile-Ile-Lys-Asn-Ala-Tyr-Lys-Lys-Gly-Glu
Ac-Tyr-Gly-Gly-Phe-Met-Thr-Ser-Glu-Lys-Ser-Gln-Thr-Pro-Leu-Val-Thr-Leu-Phe-Lys-Asn-Ala-Ile-Ile-Lys-Asn-Ala-Tyr-Lys-Lys-Gly-Glu

Synonyms: Endorphin, N-Ac-β-

Sigma A 1660   Human   FW 3507.1   ≥95% (HPLC) | Bioactive peptide; predominant form of β-endorphin stored in & released from the intermediate lobe of the pituitary under conditions of foot shock or swim stress

---

Tyr-Gly-Gly-Phe-Met-Thr-Ser-Glu-Lys-Ser-Gln-Thr-Pro-Leu-Val-Thr-Leu-Phe-Lys-Asn-Ala-Ile-Ile-Lys-Asn-Ala-Tyr
Ac-Tyr-Gly-Gly-Phe-Met-Thr-Ser-Glu-Lys-Ser-Gln-Thr-Pro-Leu-Val-Thr-Leu-Phe-Lys-Asn-Ala-Ile-Ile-Lys-Asn-Ala-Tyr

Synonyms: Endorphin (1-27), N-Ac-β-

Sigma E 0762   Human   FW 3064.5   ≥97% (HPLC) | Bioactive peptide; predominant form of β-endorphin in the intermediate lobe of the pituitary under normal conditions

Tyr-Gly-Gly-Phe-Met-Thr-Ser-Glu-Lys-Ser-Gln-Thr-Pro-Leu-Val-Thr-Leu-Phe-Lys-Asn-Ala-Ile-Ile-Lys-Asn-Ala-Tyr-Lys-Lys-Gly-Glu

Synonyms: Endorphin, β-; Lipotropin (61-91), β-

Sigma E 6261 Human FW 3465.0 ≥95% (HPLC) | Bioactive peptide; neurohormone secreted by the anterior pituitary gland; most potent opioid peptide; Akil H et al, *Ann Rev Neurosci*, 7:223, 1984; Chung, WC et al, *Int J Pept Prot Res*, 13:278, 1979; Li, CH et al, *J Med Chem*, 20:325, 1977

Tyr-Gly-Gly-Phe-Met-Thr-Ser-Glu-Lys-Ser-Gln-Thr-Pro-Leu-Val-Thr-Leu-Phe-Lys-Asn-Ala-Ile-Ile-Lys-Asn-Ala-Tyr

S Endorphin (1-27), β-*ynonyms:*

Sigma E 6636 Human FW 3022.5 ≥97% (HPLC) | Bioactive peptide

Tyr-Gly-Gly-Phe-Met-Thr-Ser-Glu-Lys-Ser-Gln-Thr-Pro-Leu-Val-Thr-Leu-Phe-Lys-Asn-Ala-Ile-Ile-Lys-Asn-Ala-Tyr-Lys-Lys-Gly-Glu Acetate Salt

Synonyms: Endorphin, β-

Biogenesis 4100-0504 Human synthetic MW 3466 Purified; lyophilized | To avoid Met oxidation, it is best to dissolve the peptide under a nitrogen stream

Tyr-Gly-Gly-Phe-Met-Thr-Ser-Glu-Lys-Ser-Gln-Thr-Pro-Leu-Val-Thr-Leu-Phe-Lys-Asn-Ala-Ile-Ile-Lys-Asn-Ala-Tyr-Lys-Lys-Gly-Glu

Synonyms: Lipotropin C Fragment, β-; Lipotropin (61-91), β-; Endorphin, β-

Fluka 45177 Human synthetic MW 3465 $C_{158}H_{251}N_{39}O_{46}S$ ≥96% (HPLC) | Most potent of the opioid peptides; Akie, H et al, *Ann Rev Neurosci*, 7:223, 1984; Chang, WC et al, *Int J Pep Prot Res*, 13:278, 1979; Li, CH et al, *JMC*, 20:325, 1977

Peptides International PEN-4060-v Human synthetic MW 3465.0 $C_{158}H_{251}N_{39}O_{46}S$ >95% (HPLC); lyophilized amorphous powder

Tyr-Gly-Gly-Phe-Met-Thr-Ser-Glu-Lys-Ser-Gln-Thr-Pro-Leu-Val-Thr-Leu-Phe-Lys-Asn-Ala-Ile-Val-Lys-Asn-Ala-His-Lys-Lys-Gly-Gln

Synonyms: Endorphin, β-

American Peptide 28-4-10 Porcine MW 3424.0 $C_{154}H_{248}N_{42}O_{44}S$

ICN 154529 Porcine MW 3414.8

Tyr-Gly-Gly-Phe-Met-Thr-Ser-Glu-Lys-Ser-Gln-Thr-Pro-Leu-Val-Thr-Leu-Phe-Lys-Asn-Ala-Ile-Val-Lys-Asn-Ala-His-Lys-Lys-Gly-Gln
Ac-Tyr-Gly-Gly-Phe-Met-Thr-Ser-Glu-Lys-Ser-Gln-Thr-Pro-Leu-Val-Thr-Leu-Phe-Lys-Asn-Ala-Ile-Val-Lys-Asn-Ala-His-Lys-Lys-Gly-Gln

Synonyms: Endorphin, β-

ICN 154530 Porcine MW 3456.8

Tyr-Gly-Gly-Phe-Met-Thr-Ser-Glu-Lys-Ser-Gln-Thr-Pro-Leu-Val-Thr-Leu-Phe-Lys-Asn-Ala-Ile-Ile-Lys-Asn-Val-His-Lys-Lys-Gly-Gln

Synonyms: Lipotropin (61-91), β-; Endorphin, β-

American Peptide 28-5-10 Rat MW 3466.1 $C_{157}H_{254}N_{42}O_{44}S$

Bachem H-2814.0500 Rat MW 3466.07 $C_{157}H_{254}N_{42}O_{44}S$ Store at -15°C

ICN 152806 Rat MW 3466.1

Neosystem SC106 Rat MW 3466.05 Bioactive; opioid peptide

Sigma E 1142 Rat FW 3466.0 ≥97% (HPLC) | Bioactive peptide; neurohormone secreted by the anterior pituitary gland; most potent opioid peptide

Tyr-Gly-Gly-Phe-Met-Thr-Ser-Glu-Lys-Ser-Gln-Thr-Pro-Leu-Val-Thr-Leu

Synonyms: Endorphin, γ-; Lipotropin (61-77), β-

Biogenesis 4100-1056 Synthetic MW 1859 Lyophilized, consisting of 94.58% peptide material, acetate counter ions and residual H_2O | To avoid Met oxidation, it is best to dissolve the peptide under a nitrogen stream

Peptides International PEN-4089-v Synthetic MW 1859.1 $C_{83}H_{131}N_{19}O_{27}S$ >95% (HPLC); lyophilized amorphous powder | Bioactive; Ling, N et al, *PNAS USA*, 73:3942, 1976

Tyr-Gly-Gly-Trp-Leu

Synonyms: Gluten Exorphin B5; Enkephalin, (Trp[4])-Leu-

Bachem H-1666.0005 MW 594.67 $C_{30}H_{38}N_6O_7$ Store at -15°C

Tyr-Gly-Leu
Suc-Tyr-Gly-Leu-βNA

Bachem K-1515.0050 Store at -15°C

Tyr-Gly-Leu-Arg-Hyp-Gly
Ac-Tyr-Gly-Leu-Arg-Hyp-Gly Amide

Synonyms: Luteinizing Hormone Releasing Hormone (5-10), (Hyp[9])Ac-

Neosystem SC474 MW 718.81 Bioactive; hypothalamic releasing hormone

Tyr-Gly-Leu-Arg-Pro-Gly
Ac-Tyr-Gly-Leu-Arg-Pro-Gly Amide

Synonyms: Luteinizing Hormone Releasing Hormone (5-10), Ac-

Neosystem SC352 MW 702.81 Bioactive; hypothalamic releasing hormone; Mc Dermott, JR et al, *Peptides*, 4:25-30, 1983

Tyr-Gly-Leu-Ser-Lys-Gly-Cys-Phe-Gly-Leu-Lys-Leu-Asp-Arg-Ile-Gly-Ser-Met-Ser-Gly-Leu-Gly-Cys

Synonyms: Natriuretic Peptide 22, Tyr-C-Type

Peptides International PCT-4251-v Human synthetic MW 2360.9 $C_{102}H_{166}N_{28}O_{30}S_3$ >95% (HPLC); lyophilized amorphous powder | Bioactive; Disulfide bonds: Cys[6]-Cys[22]; for radioimmunoasay

Tyr-Gly-Lys-Val-Glu-Gln-Leu-Ser-Pro-Glu-Glu-Glu-Glu-Lys-Arg-Ile-Arg-Arg-Glu-Arg-Asn-Lys-Met-Ala-Ala-Ala

Synonyms: fos Oncogene Fragment ; M Peptide

Bachem H-3348.0500 Store at -15°C

ICN 154574 MW 3118.9 Curran, T etal, *Mol Cell Biol*, 5:167, 1985

Tyr-Gly-Pro-Phe-Leu-Arg-Arg-Ile-Arg-Pro-Lys Amide

Synonyms: Dynorphin A (1-11), (Pro[3])-

Bachem H-5072.0001 MW 1401.73 $C_{66}H_{108}N_{22}O_{12}$ Store at -15°C

Tyr-Gly-Ser-Leu-Pro-Gln-Lys-Ala-Gln-Arg-Pro-Gln-Asp-Glu-Asn

Synonyms: Myelin Basic Protein (68-84), des(Gly[77],His[78])-

American Peptide 91-0-49 Bovine MW 1730.8 $C_{73}H_{115}N_{23}O_{26}$ Mannie, M et al, *PNAS*, 82:5515, 1985

Bachem H-6870.0001 Bovine MW 1730.86 $C_{73}H_{115}N_{23}O_{26}$ Store at -15°C

ICN 159975 Bovine MW 1730.9 95% | Mannie, M etal, *PNAS*, 82:5515, 1985

Sigma M 3630 Bovine FW 1730.9 ≥95% (HPLC) | Bioactive peptide; induces experimental allergic encephalomyelitis; Mannie, M et al, *Proc Natl Acad Sci USA*, 82:5515, 1985; Hashim, GA et al, *J Immunol*, 121:665, 1978

Tyr-Gly-Ser-Leu-Pro-Gln-Lys-Ser-Gln-Arg-Ser-Gln-Asp-Glu-Asn

Synonyms: Experimental Allergic Encephalomyelitis Inducing Peptide; Myelin Basic Protein (68-82), (des-Gly[77],His[78])-Ser[75,80]-

American Peptide 91-0-48 Guinea pig MW 1736.8 $C_{71}H_{113}N_{23}O_{28}$ Mannie, M et al, *PNAS*, 82:5515, 1985

Bachem H-6875.0001 Guinea pig MW 1736.82 $C_{71}H_{113}N_{23}O_{28}$ Store at -15°C

ICN 155746 Guinea Pig MW 1736.8 Mannie, M etal, *PNAS*, 82:5515, 1985

Neosystem SC1273 Guinea Pig MW 1736.81 Bioactive; Mannie, MD et al, *PNAS USA*, 82:5515-5519, 1985

Sigma M 3755 Guinea pig FW 1736.8 ≥97% (HPLC) | Bioactive peptide; induces experimental allergic encephalomyelitis; Mannie, M et al, *Proc Natl Acad Sci USA*, 82:5515, 1985

Sigma M 5167 Guinea pig FW 1736.8 ≥97% (HPLC) | Bioactive peptide; induces experimental allergic encephalomyelitis; similar to Sigma M 3755 but produced by Sigma; Mannie, M et al, *Proc Natl Acad Sci USA*, 82:5515, 1985

Tyr-Gly-Trp

Bachem H-5475.0250 MW 424.46 $C_{22}H_{24}N_4O_5$ Store at -15°C

Tyr-Gly-Trp-Pro-Gln-Ala-Pro-Ala-Nle-Asp-Gly-Ala-Gly-Lys-Thr-Gly-Ala-Glu-Glu-Ala-Gln-Pro-Pro-Glu-Gly-Lys-Gly-Ala-Arg-Glu-His-Ser-Arg-Gln-Glu-Glu-Glu-Glu-Glu-Thr-Ala-Gly-Ala-Pro-Gln-Gly-Leu-Phe-Arg-Gly Amide

Synony Pancreastatin, (Tyr[0],Nle[8])-*ms*:

American Peptide 46-4-45 Porcine MW 5248.6 $C_{224}H_{341}N_{69}O_{78}$

Tyr-Gly-Val-Tyr-Thr-Lys-Val-Ser-Arg-Tyr-Leu-Asp-Trp-Ile-His

Synonyms: Protein C (390-404), Activated

Bachem H-8330.0500 Human MW 1900.17 $C_{91}H_{130}N_{22}O_{23}$ Store at -15°C

Tyr-His

Bachem G-3415.0250 MW 318.33 $C_{15}H_{18}N_4O_4$ Store at -15°C

Tyr-His-His-Leu-Gly-Gly-Ala-Lys-Gln-Ala-Gly-Asp-Val-Gly-Asp-Ser
Tyr-His-His-Leu-Gly-Gly-Ala-Lys-Gln-Ala-Gly-Asp-Val-(Gly-)₉Arg-Gly-Asp-Ser

Synonyms: Fibrinogen Bridged Peptide

Bachem H-2262.0500 Store at -15°C

Tyr-His-Leu-Arg-Pro
Tyr-D-His(Bzl)-Leu-Arg-Pro-NHEt

Synonyms: Luteinizing Hormone Releasing Hormone (5-9), (D-His(Bzl)[6],Pro-NHEt[9])-

Bachem H-4812.0250 MW 801.99 $C_{41}H_{59}N_{11}O_6$ Store at -15°C

Tyr-His-Lys-Thr-Asp-Ser-Phe-Val-Gly-Leu-Met Amide

Synonyms: Neurokinin A, (Tyr[0])-

American Peptide 62-1-43 MW 1296.6 $C_{59}H_{89}N_{15}O_{16}S$

Bachem H-9270.0001 Store at -15°C

ICN 154454 MW 1296.6

Neosystem SC014 MW 1296.50 Bioactive; tachykinin

Tyr-Ile

Bachem G-3425.0250 MW 394.35 $C_{15}H_{22}N_2O_4$ Store at -15°C

Tyr-Ile-Arg-Ile-Gln-Arg-Gly-Pro-Gly-Arg-Ala-Phe-Val-Thr-Ile-Gly-Lys

Synonyms: GP120 (312-327); HTLV I Peptide, (Tyr)-

Neosystem SC665 MW 1933.30 HTLV I peptide from subtype GP160

Tyr-Ile-Asn-Leu-Cys-Thr-Arg-Nva-Arg-Tyr
(Tyr-Ile-Asn-Leu-Cys-Thr-Arg-Nva-Arg-Tyr Amide)₂

Synonyms: Neuropeptide Y (27-36))₂, ((Cys[31],Nva[34])-

Bachem H-3704.0500 MW 2597.11 $C_{116}H_{186}N_{36}O_{28}S_2$ Store at -15°C

Tyr-Ile-Asn-Leu-Ile-Thr-Arg-Gln-Arg-Tyr
D-Tyr-Ile-Asn-Leu-Ile-D-Thr-Arg-Gln-Arg-D-Tyr Amide

Synonyms: Neuropeptide Y (27-36), (D-Tyr[27,36],D-Thr[32])-

Bachem H-3328.0500 MW 1338.58 $C_{61}H_{99}N_{19}O_{15}$ Store at -15°C

Neosystem SC1201 MW 1338.57 Bioactive neuropeptide; NPY analog; inhibits feeding evoked in the rat by hypothalamic NPY as well as the natural eating response to food deprivation; may act as an antagonist at one or more subtypes of the NPY receptor in the rat brain; NPY participates vitally in numerous regulatory processes in the brain; could be used to elucidate the specific functional properties of NPY; Myers, RD et al, *Brain Res* Bull, 37:237-245, 1995; Roscoe, AK et al, *Peptides*, 16:1411-1415, 1995

Tyr-Ile-Asn-Leu-Ile-Tyr-Arg-Leu-Arg-Tyr Amide

Synonyms: Neuropeptide Y (27-36), (Tyr[32],Leu[34])-

Neosystem SC971 MW 1385.67 Bioactive neuropeptide; shows a 3700-fold improvement in affinity at Y2 (rat brain, IC_{50} = 82 ± 3nM) receptors when compared to native NPY (27-36) C-terminal fragment; an agonist at Y1 human erythroleukemia (HEL) cell (ED_{50}= 88 ± 05nM) receptors with potency comparable to that of NPY (1-36) (ED_{50}= 5nM); Leban, JJ et al, *J Med Chem*, 38:1150-1157, 1995

Tyr-Ile-Gln-Asn-Asu-Pro-Arg-Gly Amide

Synonyms: Vasotocin, (Deamino-Dicarba-Arg)-; Vasotocin, (Asu[1,6],Arg[8])-

American Peptide 66-0-58 MW 999.2 $C_{45}H_{70}N_{14}O_{12}$ Peptide bond:Asu[1]-Asu[6]

ICN 152975 Disulfide bonds: Cys[1]—Cys[6]; Hase, S etal, *J Am Chem Soc*, 94:3590, 1972

Tyr-Ile-Gln-Asn-Asu-Pro-Leu-Gly Amide

Synonyms: Oxytocin, (Asu[6])-; Oxytocin, (Asu[1,6])-; Oxytocin, Deamino-Dicarba-

American Peptide 66-0-51 MW 956.1 $C_{45}H_{69}N_{11}O_{12}$ Analog of oxytocin; peptide bond: Asu[1]-Asu[6]

ICN 151795 MW 956.1 $C_{45}H_{69}N_{11}O_{12}$ Peptide bond:Tyr-Asu; biologically active analog of oxytocin; Yamanaka, T etal, *Molec Pharmac*, 6:474, 1970

Tyr-Ile-Gln-Asn-Cys-Pro-Orn-Gly
β-Mercapto-β,β-Cyclopentamethylenepropionyl-Tyr(Me)-Ile-Gln-Asn-Cys-Pro-Orn-Gly Amide

Synonyms: Oxytocin, (d(CH₂)[51],Tyr(Me)[2],Orn[8])-

Bachem H-4928.0001 MW 1075.32 $C_{48}H_{74}N_{12}O_{12}S_2$ Store at -15°C

Tyr-Ile-Gly-Ser-Arg

Synonyms: Laminin B1 Chain (929-933); Laminin Pentapeptide; Laminin (929-933)

American Peptide 44-0-37 MW 594.7 $C_{26}H_{428}N_8O_8$ Derived from the AA sequence in domain III of the laminin B_1 chain; represents minimal sequence necessary for cell attachment & laminin receptor-binding activity; inhibits lung tumor colony formation in mice injected with melanoma cells

Bachem H-6825.0005 MW 594.67 $C_{26}H_{42}N_8O_8$ Store at -15°C

Calbiochem; 05-23-3700 MW 594.7 $C_{26}H_{42}N_8O_8$ ≥97% (HPLC); solid; soluble in 5% acetic acid | Minimal sequence necessary for cell attachment & laminin receptor-binding activity; inhibits lung tumor colony formation in mice injected with melanoma cells; this inhibition is due to competitive binding of the pentapeptide to the laminin receptor on tumor cells, thus blocking bonding of cells to laminin on basement membrane; Iwamoto, Y et al, *Science*, 238:1132, 1987

ICN 153188 MW 594.7 $C_{26}H_{42}N_8O_8$

Neosystem SC142 MW 594.67 Bioactive; cell attachment peptide; laminin fragment; inhibits experimental metastasis formation; Iwamoto, Y et al, *Science*, 238:1132, 1987

Sigma T 7154 FW 594.7 $C_{26}H_{42}N_8O_8$ ≥97% (HPLC) | Bioactive peptide; major receptor binding site in laminin; Iwamoto, Y et al, *Science*, 238:1132, 1987

Tyr-Ile-Gly-Ser-Arg Amide

Synonyms: Laminin Pentapeptide

American Peptide 87-0-55 MW 593.7 $C_{26}H_{43}N_9O_7$; | Martin, GR et al, *JCB*, Abstract 504-91A, 1987

Bachem H-2802.0001 MW 593.68 $C_{26}H_{43}N_9O_7$ Store at -15°C

Neosystem SC145 MW 593.68 Bioactive; cell attachment peptide; laminin fragment; Iwamoto, Y et al, *Science*, 238:1132, 1987

Tyr-Ile-His-Pro-Phe

Synonyms: Angiotensin I/II (4-8); Angiotensin II (4-8)

Bachem H-2884.0005 MW 675.79 $C_{35}H_{45}N_7O_7$ Store at -15°C

American Peptide 12-1-51 Human MW 675.8 $C_{35}H_{45}N_7O_7$

Tyr-Ile-Nal-Gly-Lys-Trp-His-His-Phe-Lys
Tyr-Ile-2-Nal-Gly-Lys(retro-Trp-His-His-H)-Phe-Lys

Bachem H-4082.0001 Store at -15°C

Tyr-Ile-Thr-Asn-Cys-Pro-Orn
β-Mercapto-β,β-Cyclopentamethylenepropionyl-Tyr(Me)-Ile-Thr-Asn-Cys-Pro-Orn

Synonyms: Vasotocin, (d(CH$_2$)^{51},Tyr(Me)^2,Thr^4,Orn^8,des-Gly-NH$_2$^{9})-

Bachem H-2908.0001 Store at -15°C

Tyr-Ile-Thr-Asn-Cys-Pro-Orn-Tyr
β-Mercapto-β,β-Cyclopentamethylenepropionyl-Tyr(Me)-Ile-Thr-Asn-Cys-Pro-Orn-Tyr Amide

Synonyms: Vasotocin, (d(CH$_2$)^{51},Tyr(Me)^2,Thr^4,Orn^8,Tyr-NH$_2$^{9})-

Bachem H-9405.0001 Store at -15°C

Tyr-Ile-Tyr-Gly-Ser-Phe-Lys

Synonyms: src Protein Tyrosine Kinase Substrate; p60c-src Substrate I; p60c-src Substrate

Alexis 168-004 MW 877.0 $C_{44}H_{60}N_8O_{11}$ ≥97%; lyophilized powder | Selected from a random combinatorial peptide library; much more specific & efficient substrate for the Src-family protein tyrosine kinase than cdc2 (6-20); Lam, KS et al, *Int Pept Prot Res*, 45:587, 1995

Bachem H-2686.0001 Store at -15°C

Calbiochem 567810 MW 877.0 $C_{44}H_{60}N_8O_{11}$ ≥98% (HPLC); solid; soluble in 0.1% TFA | Specific & efficient substrate for src-family protein tyrosine kinases; binds to p60c-src with ~6.4-fold lower K_m value than Cdc2 (6-20) substrate, a peptide derived from p34cdc2; Lam KS et al, *Int J Pept Protein Res*, 45:587, 1995

ICN 195921 MW 877.0 $C_{44}H_{60}N_8O_{11}$ ≥98% | Specific src-family protein tyrosinekinase substrate

Tyr-Ile-Val-Gln-Met-Leu-Ala-Lys-Leu-Arg-Gln-Gly-Tyr-Arg-Pro-Val-Phe-Ser-Ser-Pro

Synonyms: GP140 (711-730); SIVmac251 Peptide

Neosystem SC779 MW 2353.80 SIVmac251 peptide from subtype GP140

Tyr-Leu
Z-Tyr-Leu

Bachem C-2770.0250 MW 428.49 $C_{23}H_{28}N_2O_6$ Store at -15°C

Tyr-Leu
Z-Tyr-Leu Amide

Bachem C-3220.0250 MW 427.5 $C_{23}H_{29}N_3O_5$ Store at -15°C

Tyr-Leu

Bachem G-3430.0250 MW 394.35 $C_{15}H_{22}N_2O_4$ Store at -15°C

Tyr-Leu
Tyr-Leu-OBzl Hydrochloride

Bachem G-3440.0250 MW 420.94 $C_{22}H_{28}N_2O_4 \cdot HCl$ Store at -15°C

Tyr-Leu

Sigma T 9878 FW 294.3 $C_{15}H_{22}N_2O_4$ Crystalline

Tyr-Leu Amide Hydrochloride

Bachem G-3435.0250 Store at -15°C

Tyr-Leu-Asn-Phe-Thr-Pro-Asn-Trp-Gly-Thr Amide

Synonyms: Adipokinetic Hormone, (Tyr^1)-

American Peptide 60-9-19 Locust MW 1211.5 $C_{58}H_{78}N_{14}O_{15}$ Schooneveld, H et al, *Cell Tissue Res*, 230:67, 1983

Tyr-Leu-Asp-Ile-Arg-Pro-Arg-Gly-Asp-Asn-Gly-Asp-Thr-Ala-Cys Acetate Salt

Synonyms: Fibrillin I (1536-1548)

Biogenesis 4439-1031 Human synthetic Semi-pure; lyophilized

Tyr-Leu-Gln-Asp-Val-His-Asn-Phe-Val-Ala-Leu-Gly-Ala-Pro-Leu-Ala-Pro-Arg-Asp-Ala-Gly-Ser

Synonyms: Parathyroid Hormone (27-48), (Tyr^{27})-; Parathormone

American Peptide 22-1-26 Human MW 2311.6 $C_{104}H_{159}N_{29}O_{31}$

Bachem H-4840.0500 Human MW 2311.58 $C_{104}H_{159}N_{29}O_{31}$ Store at -15°C

ICN 152983 Human MW 3917.5

Sigma P 2405 Human FW 2311.6 ≥97% (HPLC) | Bioactive peptide

Tyr-Leu-Glu-Pro-Gly-Pro-Val-Thr-Ala

Synonyms: Melanocyte Protein PMEL 17 (280-288)

Bachem H-2174.0001 Human MW 946.07 $C_{44}H_{67}N_9O_{14}$ Store at -15°C

Tyr-Leu-Glu-Pro-Gly-Pro-Val-Thr-Ala Acetate Salt

Synonyms: Peptide 946

Sigma T 7552 FW 946.1 (free base) $C_{44}H_{67}N_9O_{14}$ ≥97% (HPLC) | Bioactive peptide; melanoma-specific CTLs had an exceptionally high affinity for this nine-residue peptide, which reconstituted an epitope for CTL lines from each of five different melanoma patients tested; recognition by multiple CTL lines suggests that this may be a promising candidate for use in peptide-based melanoma vaccines; Cox, AL et al, *Science*, 264:716, 1994

Tyr-Leu-Leu-Pro-Ala-Gln-Val-Asn-Ile-Asp

Synonyms: Neurotrophic Factor for Retinal Cholinergic Neurons

Bachem H-2466.0001 Store at -15°C

Tyr-Leu-Lys-Asp-Gln-Ala-Gln-Leu-Asn-Ala-Trp-Gly-Cys-Ala-Phe-Arg-Gln-Val-Cys
Tyr-Leu-Lys-Asp-Gln-Ala-Gln-Leu-Asn-Ala-Trp-Gly-Cys(Acm)-Ala-Phe-Arg-Gln-Val-Cys(Acm)

Synonyms: GP140 (601-619); SIVmac251 Peptide

Neosystem SC768 MW 2493.84 SIVmac251 peptide from subtype GP140

Tyr-Leu-Phe-Gln-Pro-Gln-Arg-Phe
D-Tyr-Leu-N-Me-Phe-Gln-Pro-Gln-Arg-Phe Amide

Synonyms: Neuropeptide FF, (D-Tyr[1],N-Me-Phe[3])-; DMe

Bachem H-4752.0001 MW 1111.31 $C_{55}H_{78}N_{14}O_{11}$ Store at -15°C

Tyr-Leu-Pro-Leu-Arg-Phe Amide

American Peptide 60-0-43 MW 807.0 $C_{41}H_{62}N_{10}O_7$

Tyr-Leu-Pro-Pro-Arg-Glu-Gly-Asp-Leu-Thr-Cys-Asn-Ser-Thr-Val-Thr-Ser-Leu-Ile-Ala

Synonyms: GP140 (451-470); SIVmac251 Peptide

Neosystem SC753 MW 2150.42 SIVmac251 peptide from subtype GP140; due to the presence of a Cys this peptide can contain the dimeric form

Tyr-Leu-Tyr-Gln-Trp-Leu-Gly-Ala-Pro-Val-Pro-Tyr-Pro-Asp-Pro-Leu-Gla-Pro-Arg-Arg-Gla-Val-Cys-Gla-Leu-Asn-Pro-Asp-Cys-Asp-Glu-Leu-Ala-Asp-His-Ile-Gly-Phe-Gln-Glu-Ala-Tyr-Arg-Arg-Phe-Tyr-Gly-Pro-Val

Synonyms: Bone Gla Protein (1-49); Osteocalcin (1-49)

American Peptide 22-6-10 Human MW 5925.6 $C_{269}H_{379}N_{68}O_{81}S_2$ Disulfide bonds: :Cys[23]-Cys[29]

Bachem H-4912.0001 Human MW 5929.52 $C_{269}H_{381}N_{67}O_{82}S_2$ Store at -15°C

Tyr-Leu-Tyr-Gln-Trp-Leu-Gly-Ala-Pro-Val-Pro-Tyr-Pro-Asp-Pro-Leu-Glu-Pro-Arg-Arg-Gla-Val-Cys-Gla-Leu-Asn-Pro-Asp-Cys-Asp-Glu-Leu-Ala-Asp-His-Ile-Gly-Phe-Gln-Glu-Ala-Tyr-Arg-Arg-Phe-Tyr-Gly-Pro-Val

Synonyms: Osteocalcin, (Glu[17],Gla[21,24])-; Bone Gla Protein

Peptides International POS-4261-s Human synthetic MW 5885.5 $C_{268}H_{381}N_{67}O_{80}S_2$ >99% (HPLC); lyophilized amorphous powder | Bioactive; disulfide bonds Cys[23]-Cys[29]; Poser, JW et al, *JBC*, 255:8685, 1980; Nakao, M et al, *Pept Res*, 7:171, 1994

Tyr-Leu-Tyr-Gln-Trp-Leu-Gly-Ala-Pro-Val-Pro-Tyr-Pro-Asp-Pro-Leu-Gla-Pro-Arg-Arg-Gla-Val-Cys-Gla-Leu-Asn-Pro-Asp-Cys-Asp-Glu-Leu-Ala-Asp-His-Ile-Gly-Phe-Gln-Glu-Ala-Tyr-Arg-Arg-Phe-Tyr-Gly-Pro-Val

Synonyms: Osteocalcin, (Gla[17,21,24])-; Bone Gla Protein

Peptides International POS-4262-s Human synthetic MW 5929.5 $C_{269}H_{381}N_{67}O_{82}S_2$ >99% (HPLC); lyophilized amorphous powder | Bioactive; disulfide bonds Cys[23]-Cys[29]; Poser, JW et al, *JBC*, 255:8685, 1980; Nakao, M et al, *Pept Res*, 7:171, 1994

Tyr-Leu-Val
Suc-Tyr-Leu-Val-pNA

Bachem L-1435.0050 Store at -15°C

Tyr-Leu-Val
N-Suc-Tyr-Leu-Val-pNA

Synonyms: Leukocyte Elastase Substrate, Human; Spleen Fibrinolytic Proteinase Substrate, Human

ICN 156701

Sigma S 7763 FW 613.7 $C_{30}H_{39}N_5O_9$ Okamoto, V et al, *Biochem Biophys Res Commun*, 97:28, 1980; Okada, Y et al, *Chem Pharm Bull*, 30:4060, 1982; Tsuda, Y et al, *Chem Pharm Bull*, 35:3576, 1987; Tsuda, Y et al, *Chem Pharm Bull*, 36:3119, 1988

Tyr-Ile-Gly-Ser-Arg Amide

Synonyms: Laminin Pentapeptide

Peptides International PLP-4194-v Synthetic MW 749.82 $C_{26}H_{43}N_9O_7$ >99% (HPLC); lyophilized amorphous powder | Bioactive; Iwamoto, Y et al, *Science*, 238:1132, 1987

Tyr-Ile-Gly-Ser-Arg Amide 2AcOH Dihydrate

Synonyms: Laminin Pentapeptide

Peptides International PLP-4194 Synthetic MW 749.82 $C_{26}H_{43}N_9O_7 \cdot 2CH_3COOH \cdot 2H_2O$ >99% (HPLC); lyophilized amorphous powder; bulk | Bioactive; Iwamoto, Y et al, *Science*, 238:1132, 1987

Tyr-Lys

Bachem G-3445.0250 MW 309.37 $C_{15}H_{23}N_3O_4$ Store at -15°C

Tyr-Lys Amide Dihydrochloride

Bachem G-3450.0250 Store at -15°C

Tyr-Lys-Ala Amide Hydrochloride

Bachem H-7990.0050 MW 415.92 $C_{18}H_{29}N_5O_4 \cdot HCl$ Store at -15°C

Tyr-Lys-Arg
Z-Tyr-Lys-Arg-4MβNA

Bachem J-1325.0050 MW 754.89 $C_{40}H_{50}N_8O_7$ Store at -15°C

Tyr-Lys-Arg
Z-Tyr-Lys-Arg-pNA ·2TFA

Bachem L-1250.0050 Store at -15°C

Tyr-Lys-Arg
FA-Tyr-Lys-Arg

Bachem M-1670.0025 Store at -15°C

Tyr-Lys-Arg-His-Pro-Ile
Nicotinoyl-Tyr-Lys(Z-Arg)-His-Pro-Ile

Synonyms: Angiotensin II Receptor Ligand; CGP-42112; Angiotensin AT2:Receptor Agonist

Bachem H-9395.0001   MW 1052.2 $C_{52}H_{69}N_{13}O_{11}$ Store at -15°C

Peptides International PAN-4296-v   Synthetic   MW 1052.2 $C_{52}H_{69}N_{13}O_{11}$ >99% (HPLC); lyophilized amorphous powder | Whitebread, SE et al, *BBRC*, 181:1365, 1991; Buisson, B et al, FEBS *Lett*, 309:161, 1992; Koike, G et al, *BBRC*, 203:1842, 1994

Tyr-Lys-Gln-Arg-Val-Lys-Asn-Lys Amide

Synonyms: Pituitary Adenylate Cyclase Activating Peptide (31-38)

American Peptide 34-0-30   Human, ovine, rat   MW 1062.2 $C_{47}H_{83}N_{17}O_{11}$

Tyr-Lys-Gly-Arg-Cyclo(-Glu-Tyr-Ile-Lys)-Leu-Ile-Thr-Arg-Pro-Arg-Tyr Amide

Synonyms: Neuropeptide Y (26-36), (Tyr-Lys-Gly-Arg-(Glu[26],Lys[29],Pro[34]))-

Bachem H-3972.0500   MW 1937.32 $C_{91}H_{145}N_{27}O_{20}$ Store at -15°C

Tyr-Lys-Gly-Arg-Pro-Gly-Asn-Phe-Leu-Gln-Ser-Arg-Pro-Glu-Pro-Thr-Ala

Synonyms: P15 (441-457); GAG P1 (9-16)/P6 (1-9); HTLV I Peptide

Neosystem SC276   MW 1918.14   HTLV I peptide from subtype P15 (GAG P7 NC/GAG P1/GAG P6)

Tyr-Lys-Gly-Ile-Leu-Gly-Phe-Val-Phe-Thr-Leu-Thr-Val

Synonyms: Influenza Virus Matrix Peptide (57-68), (Tyr)-

Neosystem SC804   MW 1457.77   Virus-related peptide

Tyr-Lys-Leu-Val-Pro-Val-Glu-Pro-Asp-Lys-Val-Glu-Glu

Synonyms: NEF (143-155); HTLV I Peptide

Neosystem SC512   MW 1544.76   HTLV I peptide from subtype NEF

Tyr-Lys-Lys-Glu-Asp-Asn-Val-Leu-Val-Glu-Ser-His-Glu-Lys-Ser-Leu-Gly-Glu-Ala-Asp-Lys-Ala-Asp-Val-Asn-Val-Leu-Thr-Lys-Ala-Lys-Ser-Gln

Synonyms: Parathyroid Hormone (52-84), (Tyr[52],Asn[76])-

American Peptide 22-1-38   Human   MW 3674.1 $C_{158}H_{262}N_{44}O_{56}$

Bachem H-4850.0500   Human   MW 3674.08 $C_{158}H_{262}N_{44}O_{56}$ Store at -15°C

Tyr-Lys-Lys-Glu-Asp-Asn-Val-Leu-Val-Glu-Ser-His-Glu-Lys-Ser-Leu-Gly-Glu-Ala-Asp-Lys-Ala-Asp-Val-Asp-Val-Leu-Thr-Lys-Ala-Lys-Ser-Gln

Synonyms: Parathyroid Hormone (52-84), (Tyr[52],Asp[76])-

Bachem H-6960.0500   Human   MW 3675.07 $C_{158}H_{261}N_{43}O_{57}$ Store at -15°C

Tyr-Lys-Lys-Glu-Asp-Asn-Val-Leu-Val-Glu-Ser-His-Glu-Lys-Ser-Leu-Gly-Glu-Ala-Asp-Lys-Ala-Asp-Val-Asn-Val-Leu-Thr-Lys-Ala-Lys-Ser-Gln

Synonyms: Parathyroid Hormone (52-84), (Tyr[52],Asn[76])-; Parathormone

ICN 152989   Human   MW 3674.1

Tyr-Lys-Lys-Glu-Asp-Asn-Val-Leu-Val-Glu-Ser-His-Glu-Lys-Ser-Leu-Gly-Glu-Ala-Asp-Lys-Ala-Asp-Val-Asp-Val-Leu-Thr-Lys-Ala-Lys-Ser-Gln

Synonyms: Parathyroid Hormone (52-84), (Tyr[52],Asp[76])-; Parathormone

ICN 152990   Human   MW 3675.1

Sigma P 2655   Human   FW 3675.1   ≥97% (HPLC) | Bioactive peptide

Tyr-Lys-Lys-Glu-Asp-Asn-Val-Leu-Val-Glu-Ser-His-Glu-Lys-Ser-Leu-Gly-Glu-Ala-Asp-Lys-Ala-Asp-Val-Asn-Val-Leu-Thr-Lys-Ala-Lys-Ser-Gln

Synonyms: Parathyroid Hormone (52-84), (Tyr[52],Asn[76])- Human

Sigma P 5046   Human   FW 3674.1   ≥97% (HPLC); peptide content:~60% | Bioactive peptide

Tyr-Lys-Lys-Gly-Glu

Synonyms: Endorphin (27-31), β-; Lipotropin (87-91), β-

Bachem H-5170.0005   Human   MW 623.71 $C_{28}H_{45}N_{7}O_{9}$ Store at -15°C

Tyr-Lys-Thr

Bachem H-5175.0050   MW 410.47 $C_{19}H_{30}N_{4}O_{6}$ Store at -15°C

Tyr-Lys-Trp

Bachem H-5490.0050   MW 495.58 $C_{26}H_{33}N_{5}O_{5}$ Store at -15°C

Tyr-Lys-Val-Asn-Glu-Tyr-Gln-Gly-Pro-Val-Ala-Pro-Ser-Gly-Gly-Phe-Phe-Leu-Phe-Arg-Pro-Arg-Asn Amide

Synonyms: Neuromedin U ; Neuromedin U-23

Bachem H-9295.0001   Rat   Store at -15°C

ICN 154459   Rat   MW 2643.3   Minamino, N etal, *BBRC*, 156:355, 1988

American Peptide 62-5-50   Rat small intestine   MW 2643.0 $C_{124}H_{180}N_{34}O_{31}$ Exerts two-fold potent uterus stimulant activity compared to pig neuromedin U-25

Peptides International PNM-4377-v   Rat synthetic   MW 2643.00 $C_{124}H_{180}N_{34}O_{31}$ >99% (HPLC); lyophilized amorphous powder | Bioactive; food intake suppressor; Conlon, JM et al, *J Neurochem*, 51:988, 1988; Minamino, N et al, *BBRC*, 156:355, 1988; Howard, AD et al, *Nature*, 406:70, 2000

Tyr-Lys-Val-Gln-Asp-Asp-Thr-Lys-Thr-Leu-Ile-Lys-Thr-Ile-Val

Synonyms: Leptin (26-39), (Tyr)-; Obese Gene Peptide (26-39), (Tyr)-

American Peptide 46-2-14   Human   MW 1765.1 $C_{80}H_{137}N_{19}O_{25}$ Zhang, Y et al, *Nature*, 372:425, 1994; Pelleymounter, MA et al, *Science*, 269:541, 1995

Bachem H-3494.0001   Human   Store at -15°C

Tyr-Met-Ala-Pro-Tyr-Asp-Asn-Tyr
Tyr(PO$_3$H$_2$)-Met-Ala-Pro-Tyr-Asp-Asn-Tyr Phosphorylated

Synonyms: Platelet Derived Growth Factor β-Receptor (739-746)

Bachem H-2014.0001   Store at -15°C

Tyr-Met-Arg-Phe Amide

American Peptide 60-0-31   MW 614.8 $C_{29}H_{42}N_{8}O_{5}S$

Tyr-Met-Asp-Gly-Thr-Met-Ser-Gln-Val

Synonyms: Tyrosinase (369-377), (Asp[371])-

Bachem H-3862.0001   Human   Store at -15°C

Neosystem SC1247 Human MW 1031.16 Bioactive; cancer related peptide; this HLA-A2-restricted tyrosinase antigen on melanoma cells results from posttranslational conversion of Asn to Asp; this change is of central importance for peptide recognition by melanoma-specific T-cells, but has no impact on peptide binding to the MHC molecule; the glycosylated form of the Asn- containing peptide is capable of loading onto MHC class I molecules within the endoplasmic reticulum of two melanoma cell lines; Wölfel, T et al, *Mol Pharm Immunol*, 24:759-764, 1994; Skipper, JCA et al, *J Exp Med*, 183:527-534, 1996; Androlewicz, MJ, *Human Immunology*, 51:81-88, 1996

Tyr-Met-Glu-His-Phe-Arg-Trp
Synonyms: Corticotropin A; Adrenocorticotropic Hormone (4-9), (Tyr)-

Bachem H-1170.0005 MW 1068.22 $C_{51}H_{65}N_{13}O_{11}S$ Store at -15°C

ICN 194127 MW 1068.2 97%

Sigma A 7181 Human FW 1068.2 ≥97% (HPLC) | Bioactive peptide; analog that can be iodinated for receptor & histochemical studies

Tyr-Met-Glu-His-Phe-Arg-Trp-Gly
Synonyms: Adrenocorticotropic Hormone (4-10), (Tyr)-; Corticotropin

Bachem H-1185.0001 MW 1125.28 $C_{53}H_{68}N_{14}O_{12}S$ Store at -15°C

ICN 194128 MW 1125.3 90%

Tyr-Met-Gly
Tyr-D-Met(O)-Gly-*N*-Me-phenylalaninol
Synonyms: Syndyphalin SD-25

Bachem H-5010.0005 Store at -15°C

Tyr-Met-Gly
Tyr-D-Met(O)-Gly-*N*-Me-phenylethylamide
Synonyms: Syndyphalin SD-33

Bachem H-5015.0005 Store at -15°C

Tyr-Met-Gly-Phe
Tyr-D-Met(S=O)-Gly-Phe-ol
Synonyms: Syndyphalin 20; Syndyphalin SD-25

ICN 152840 Highly selective ligand for the m-opiate receptor; Kiso, Y etal, *Naturwissenschaften*, 68:210, 1981

Tyr-Met-Gly-Phe-Pro
Tyr-D-Met-Gly-Phe-Pro Amide
Synonyms: Enkephalin, (D-Met²,Pro⁵)-; Enkephalinamide, (D-Met²,Pro⁵)-

American Peptide 30-0-32 MW 612.8 $C_{30}H_{40}N_6O_6S$ This active analog of Met-enkephalin is resistant to proteolytic degradation; Bajusz, S et al, *FEBS Lett*, 76:91, 1977

ICN 152834 A highly active analog of Met⁵-enkephalin that is more resistant to proteolytic degradation; this product has a molar potency ratio of 5.5 relative to morphine; Bejusz, S etal, *FEBS Lett*, 76:91, 1977

Tyr-Met-Gly-Phe-Thz
Tyr-D-Met-Gly-Phe-Thz Amide
Synonyms: Enkephalinamide, (D-Met²,Thz⁵)-

ICN 152838 MW 630.8 $C_{29}H_{38}N_6O_6S_2$ Yamashiro, D etal, BBRC, 78:1124, 1977

Tyr-Met-Gly-Trp-Met-Asp-Phe
Asp-Tyr(SO₃H)-Met-Gly-Trp-Met-Asp-Phe Amide
Synonyms: Cholecystokinin Octapeptide (26-33)

American Peptide 46-1-32 MW 1143.3 $C_{49}H_{62}N_{10}O_{16}S_3$ Ondetti, MA et al, *J Am Chem Soc*, 92:195, 1970

Tyr-Met-Gly-Trp-Met-Asp-Phe
Ac-Tyr(SO₃H)-Met-Gly-Trp-Met-Asp-Phe Amide Sulfated
Synonyms: Cholecystokinin Octapeptide (2-8), Ac-

Bachem H-1120.0001 MW 1070.24 $C_{47}H_{59}N_9O_{14}S_3$ Store at -15°C

Tyr-Met-Gly-Trp-Met-Asp-Phe
Ac-Tyr(SO³H)-Met-Gly-Trp-Met-Asp-Phe Sulfated Amide
Synonyms: Cholecystokinin (27-33), *N*-Ac-

ICN 152863 Bodanszky, M etal, *Int J Pept Prot Res*, 16:402, 1980

Tyr-Met-Gly-Trp-Met-Asp-Phe
Ac-Tyr(SO₃H)-Met-Gly-Trp-Met-Asp-Phe Amide
Synonyms: Cholecystokinin (27-33), *N*-Ac-(Tyr(SO₃H)²⁷)-

Sigma C 9524 FW 1070.2 ≥95% (HPLC) | Bioactive peptide; gastrointestinal peptide; Bodanszky, M et al, *Int J Pept Prot Res*, 16:402, 1980

Tyr-Met-Gly-Trp-Met-Asp-Phe Amide
Synonyms: Cholecystokinin (27-33); Heptapeptide

American Peptide 46-1-37 MW 948.2 $C_{45}H_{57}N_9O_{10}S_2$ Schiller, PW et al, *BBRC*, 85:1332, 1978

ICN 154547 MW 948.2 Schiller, P etal, *BBRC*, 85:1332, 1978

Tyr-Met-Gly-Trp-Met-Asp-Phe Amide Non-Sulfated
Synonyms: Cholecystokinin Octapeptide (2-8)

Bachem H-2420.0005 MW 948.13 $C_{45}H_{57}N_9O_{10}S_2$ Store at -15°C

Tyr-Met-Phe-His-Leu-Met-Asp
Tyr-D-Met-Phe-His-Leu-Met-Asp Amide
Synonyms: Deltorphin A; Dermenkephalin; Dermorphin Gene Associated Peptide; Enkephalin, Derm-; Deltorphin

Bachem H-8090.0001 MW 955.17 $C_{44}H_{62}N_{10}O_{10}S_2$ Store at -15°C

ICN 195660 MW 955.2 $C_{44}H_{62}N_{10}O_{10}S$ ≥97% | Opioid peptide with specific activity for δ receptors

Sigma D 2289 FW 955.2 $C_{44}H_{62}N_{10}O_{10}S_2$ ≥97% (HPLC) | Bioactive peptide; enkephalin; opioid peptide with selective affinity for δ-receptors; Mor, A et al, *FEBS Lett*, 255:269, 1989

American Peptide 30-0-40 *Phyllomedusa sauvagei* MW 955.3 $C_{44}H_{62}N_{10}O_{10}S_2$ The peptide corresponding to the predicted prodermorphin heptapeptide was first isolated from skin extracts of the South American frog; acts as a highly potent & fully selective agonist for the δ-opioid receptor; Amiche, M et al, *Mol Pharm*, 35:774, 1989; Sagan, S et al, *BBRC*, 163:726, 1989; Mor, A et al, *FEBS Lett*, 255:259, 1989

Neosystem SC316 *Phylomedusa sauvagei* MW 955.15 Bioactive; opioid peptide; high affinity for δ-opioid receptors; Amiche, M et al, *Mol Pharm*acol, 35:774-779, 1989

Tyr-Met-Phe-His-Leu-Met-Asp Amide
Synonyms: Deltorphin, (Met²)-; Dermorphin Gene Associated Peptide; δ-Receptor Peptide

Bachem H-9355.0005 Store at -15°C

ICN 154512 MW 955.2 Lazarus, LH etal, *JBC*, 264:3047, 1989

Tyr-Nle-Gly-Trp-Nle-Asp
BOC-Tyr(SO₃H)-Nle-Gly-Trp-Nle-Asp-2-Phenylethylester

Synonyms: Cholecystokinin 2-Phenylethylester (27-32), BOC-(Nle[28,31])-; JMV-180

Neosystem SC893   MW 1049.13   Bioactive; brain/gut peptide; CCK/gastrin peptide; CCK analog that exhibits partial agonist activity; Galas, MC et al, *Am J Physiol*, 254:G176-G182, 1988; Fulcrand, P et al, *Int J Pept Prot Res*, 32:384-395, 1988

Tyr-Nle-Gly-Trp-Nle-Asp-Phe
BOC-Tyr(SO₃H)-Nle-Gly-Trp-Nle-Asp-Phe Amide

Synonyms: Cholecystokinin 7, BOC-(Nle[28,31])-; JMV-236

Neosystem SC895   MW 1092.21   Bioactive; brain/gut peptide; CCK/gastrin peptide; stable CCK analog equipotent to CCK; Rolland, M et al, *Bioorg Med Chem Lett*, 3 (5):851-854, 1993

Tyr-Pen-Gly-Phe-Pen
Tyr-D-Pen-Gly-Phe-D-Pen

Synonyms: Enkephalin; DPDPE

American Peptide 30-0-76   MW 645.8 $C_{30}H_{39}N_5O_7S_2$
Disulfide bonds: D-Pen²-D-Pen⁵; selective γ-Opioid receptor peptide; a selective ligand for the δ-opioid receptor; Mosberg, HI et al, *PNAS*, 80:5871, 1983

Tyr-Pen-Gly-Phe-Pen
Tyr-D-Pen-Gly-Phe-Pen

Synonyms: Enkephalin, (D-Pen²,Pen⁵)-

American Peptide 30-0-77   MW 645.8 $C_{30}H_{39}N_5O_7S_2$
Disulfide bonds: D-Pen²-Pen⁵; opioid receptor agonist; Mosberg, HI et al, *PNAS*, 80:5871, 1983

Tyr-Pen-Gly-Phe-Pen

Synonyms: Enkephalin, (D-Pen²,⁵,p-Chloro-Phe⁴)-

American Peptide 30-0-81   MW 680.4 $C_{30}H_{38}N_5O_7S_2C_{1L}$
Disulfide bonds: D-Pen²-D-Pen⁵; highly selective ligand with high affinity for the δ-Opioid receptor; Vaughn, LK et al, *Life Sci*, 45:1001, 1989

Tyr-Pen-Gly-Phe-Pen
Tyr-D-Pen-Gly-Phe-Pen

Synonyms: Enkephalin, (D-Pen²,Pen⁵)-; DPLPE

Bachem H-2900.0001   MW 645.8 $C_{30}H_{39}N_5O_7S_2$ Store at -15°C

Tyr-Pen-Gly-Phe-Pen
Tyr-D-Pen-Gly-Phe-D-Pen

Synonyms: Enkephalin, (D-Pen²,D-Pen⁵)-; DPDPE

Bachem H-2905.0001   MW 645.8 $C_{30}H_{39}N_5O_7S_2$ Store at -15°C

Tyr-Pen-Gly-Phe-Pen
Tyr-D-Pen-Gly-p-Chloro-Phe-D-Pen

Synonyms: Enkephalin, (D-Pen²,p-Chloro-Phe⁴,D-Pen⁵)-; DPDPE, (p-Chloro-Phe⁴)-

Bachem H-8875.0001   Store at -15°C

Tyr-Pen-Gly-Phe-Pen
Tyr-D-Pen-Gly-Phe-D-Pen

Synonyms: Enkephalin, (D-Pen²,⁵)-

ICN 151510   An analog with increased δ-receptor activity; Mosberg, HI etal, *PNAS*, 80:5871, 1983

Tyr-Pen-Gly-Phe-Pen
Tyr-D-Pen-Gly-Phe-Pen

Synonyms: Enkephalin, (D-Pen²,L-Pen⁵)-

ICN 151511   MW 645.8 $C_{30}H_{39}N_5O_7S_2$ Mosberg, HI etal, *PNAS*, 80:5871, 1983

Tyr-Pen-Gly-Phe-Pen
Tyr-D-Pen-Gly-Phe-D-Pen

Synonyms: Enkephalin, (D-Pen²,D-Pen⁵)-

Neosystem SC900   MW 645.82   Disulfide bonds: D-Pen²-D-Pen⁵; bioactive; opioid peptide; potent δ-opioid receptor enkephalin analog; Mosberg, HI et al, *PNAS USA*, 80:5871-5874, 1983; Delay-Goyet, P et al, *FEBS Lett*, 183:439-443, 1985

Tyr-Pen-Gly-Phe-Pen
Tyr-D-Pen-Gly-p-Chloro-Phe-D-Pen

Synonyms: Enkephalin, (D-Pen²,p-Chloro-Phe⁴,D-Pen⁵)-

Neosystem SC914   MW 681.21   Disulfide bonds: D-Pen²-D-Pen⁵; bioactive; opioid peptide; highly selective δ-opioid receptor enkephalin analog; Vaughn, LK et al, *Life Sciences*, 45:1001, 1989; Toth, G et al, *J Med Chem*, 33:249-253, 1990

Tyr-Pen-Gly-Phe-Pen
Tyr-D-Pen-Gly-Phe-Pen

Synonyms: Enkephalin, (D-Pen²,⁵)-; DPLPE

Sigma E 2260   FW 645.8 $C_{30}H_{39}N_5O_7S_2$ ≥97% (HPLC); peptide content:~75% | Bioactive peptide; disulfide bonds: 2-5; much greater δ-receptor selectivity than enkephalin; Mosberg, HI et al, *Proc Natl Acad Sci USA*, 80:5871, 1983

Tyr-Pen-Gly-Phe-Pen
Tyr-D-Pen-Gly-Phe-D-Pen

Synonyms: Enkephalin, (D-Pen²,⁵)-; DPDPE

Sigma E 3888   FW 645.8 $C_{30}H_{39}N_5O_7S_2$ ≥95% (HPLC) | Bioactive peptide; disulfide bonds: 2-5; antinociceptive activity mediated through the δ₁-receptor while the modulatory activity is mediated through the δ₂-receptor, tritiated DPDPE is used as a δ₁-ligand; Akiyama, K et al, *Proc Natl Acad Sci USA*, 82:2543, 1985

Tyr-Pen-Gly-Phe-Pen
Tyr-D-Pen-Gly-p-Chloro-Phe-D-Pen

Synonyms: Enkephalin, (D-Pen²,⁵,p-Chloro-Phe⁴)-; DPDPE

Sigma E 6264   FW 680.2 $C_{30}H_{38}ClN_5O_7S_2$ ≥97% (HPLC); peptide content:~70% | Bioactive peptide; disulfide bonds: 2-5; highly selective ligand with high affinity for the δ-opioid receptor; Vaughn, LK et al, *Life Sci*, 45:1001, 1989

Tyr-Pen-Gly-Phe-Pen
Tyr-D-Pen-Gly-Phe-D-Pen

Synonyms: Enkephalin, (D-Pen²,⁵)-; DPDPE

Sigma E-119   Synthetic   MW 647.8 (free base) $C_{30}H_{41}N_5O_7S_2$ >99%; white solid; peptide content and salt form information are provided with each lot; bioassayed product | δ-Opioid receptor agonist; Suh et al, *Naunyn-Schmiedeberg's Arch Pharmacol*, 342:67, 1990; Hirning et al, *Neuropeptides*, 5:383, 1985; Tortella et al, *Peptides*, 9:1177, 1988

Tyr-Pen-Gly-Phe-Phe
Tyr-D-Pen-Gly-p-Chloro-Phe-D-Phe

Synonyms: Enkephalin, (D-Pen²,⁵,p-Chloro-Phe⁴)-

ICN 195661   MW 680.2 $C_{30}H_{38}ClN_5O_7S_2$ ≥97% | High selectivity ligand for the δ-receptor

---

Tyr-Phe
Z-Tyr-Phe

Bachem C-2775.0250   MW 462.5 $C_{26}H_{26}N_2O_6$ Store at -15°C

Tyr-Phe
Ac-Tyr-Phe

Bachem G-1105.0250   MW 370.41 $C_{20}H_{22}N_2O_5$ Store at -15°C

Tyr-Phe
Ac-Tyr-Phe-OMe

Bachem G-1110.0250   MW 384.43 $C_{21}H_{24}N_2O_5$ Store at -15°C

Tyr-Phe

Bachem G-3455.0250   MW 328.37 $C_{18}H_{20}N_2O_4$ Store at -15°C

ICN 103195   MW 328.4 $C_{18}H_{20}N_2O_4$ ≥99%; crystalline

Sigma T 5379   FW 328.4 $C_{18}H_{20}N_2O_4$

Tyr-Phe Amide Hydrochloride

Bachem G-3795.0250   Store at -15°C

Tyr-Phe-Asn-Lys-Pro-Thr-Gly-Tyr-Gly-Ser-Ser-Ser-Arg-Arg-Ala-Pro-Gln-Thr

Synonyms: Insulin-Like Growth Factor I (24-41)

American Peptide 50-1-29   MW 2017.2 $C_{88}H_{133}N_{27}O_{28}$

Bachem H-3098.0001   MW 2017.19 $C_{88}H_{133}N_{27}O_{28}$ Store at -15°C

Neosystem SC678   MW 2017.18   Bioactive; synthetic growth factor-related peptide

ICN 154467   Human synthetic   MW 2017.5   Corresponds to the 24-41 fragment of human IGF-1

Tyr-Phe-Asp-Ala-Ile-Phe-Thr-Asn-Ser-Tyr-Arg-Lys-Val-Leu-Gly-Gln-Leu-Ser-Ala-Arg-Lys-Leu-Leu-Gln-Asp-Ile-Met-Ser-Arg
Ac-Tyr-D-Phe-Asp-Ala-Ile-Phe-Thr-Asn-Ser-Tyr-Arg-Lys-Val-Leu-Gly-Gln-Leu-Ser-Ala-Arg-Lys-Leu-Leu-Gln-Asp-Ile-Met-Ser-Arg Amide

Synonyms: Growth Hormone Releasing Factor (1-29), (Ac-Tyr[1],D-Phe[2])-; Vasoactive Intestinal Peptide Antagonist

American Peptide 48-1-26   MW 3476.1 $C_{17}H_{252}N_{44}O_{43}S$ Weelbroeck et al, *Endocrinology*, 116:2643, 1985

Neosystem SC081   MW 3476.05   Bioactive; brain/gut peptide; VIP/PHI/secretin/helodermin peptide; hypothalamic releasing hormone; Waelbroeck, M et al, *Endocrinology*, 116:2643, 1985

Bachem H-5565.0500   Human   Store at -15°C

Tyr-Phe-Gln-Asn-Arg-Pro-Arg-Lys
Phenylac-D-Tyr(Me)-Phe-Gln-Asn-Arg-Pro-Arg-Lys Amide

Synonyms: Vasopressin, (Phenylac[1],D-Tyr(Me)[2],Arg[6,8],Lys-NH$_2$[9])-

Bachem H-1564.0001   Store at -15°C

Tyr-Phe-Gln-Asn-Arg-Pro-Arg-Tyr
Phenylac-D-Tyr(Me)-Phe-Gln-Asn-Arg-Pro-Arg-Tyr Amide

Synonyms: Vasopressin, (Phenylac[1],D-Tyr(Me)[2],Arg[6,8],Tyr-NH$_2$[9])-

Bachem H-3194.0001   Store at -15°C

Tyr-Phe-Gln-Asn-Arg-Pro-Arg-Tyr
3-(4-Azidophenyl)propionyl-D-Tyr(Me)-Phe-Gln-Asn-Arg-Pro-Arg-Tyr Amide

Synonyms: Vasopressin, (3-(4-Azidophenyl)propionyl[1],D-Tyr(Me)[2],Arg[6], Arg[8],Tyr-NH$_2$[9])-

Bachem H-3434.0001   MW 1329.49 $C_{63}H_{84}N_{20}O_{13}$ Store at -15°C

Tyr-Phe-Gln-Asn-Arg-Pro-Arg-Tyr
4-(4-Azidophenyl)Butyryl-D-Tyr(Me)-Phe-Gln-Asn-Arg-Pro-Arg-Tyr Amide

Synonyms: Vasopressin, (4-(4-Azidophenyl)Butyryl[1],D-Tyr(Me)[2],Arg[6], Arg[8],Tyr-NH$_2$[9])-

Bachem H-3506.0001   Store at -15°C

Tyr-Phe-Gln-Asn-Asu-Pro-Arg-Gly Amide

Synonyms: Vasopressin, (Asu[1,6],Arg[8])-

American Peptide 66-0-01   MW 1033.2 $C_{48}H_{68}N_{14}O_{12}$ Analog of arginine vasopressin; peptide bond:Asu[1]-Asu[6]

ICN 152964   Analog of arginine vasopressin; peptide bond:Tyr-Asu; Hase, S etal, *J Am Chem Soc*, 94:3590, 1972

Tyr-Phe-Gln-Asn-Cys-Pro-Arg-Gly
Mpr-Tyr-Phe-Gln-Asn-Cys-Pro-Arg-Gly Amide

Synonyms: Vasopressin, (Deamino[1],Arg[8])-

American Peptide 66-0-04   MW 1069.3 $C_{46}H_{64}N_{14}O_{12}S_2$ Disulfide bonds: Mpr[1]-Cys[6]; Weigartner, H et al, *Science*, 211:601, 1981

Tyr-Phe-Gln-Asn-Cys-Pro-Arg-Gly
3-Mercapto-3-Methylbutyryl-Tyr(Me)-Phe-Gln-Asn-Cys-Pro-Arg-Gly Amide

Synonyms: Vasopressin, (Deamino-Pen[1],Tyr(Me)[2],Arg[8])-

Bachem H-5340.0001   Store at -15°C

Tyr-Phe-Gln-Asn-Cys-Pro-Arg-Gly
β-Mercapto-β,β-Cyclopentamethylenepropionyl-Tyr(Me)-Phe-Gln-Asn-Cys-Pro-Arg-Gly Amide

Synonyms: Vasopressin, (d(CH$_2$)[51],Tyr(Me)$_2$,Arg[8])-; Manning Compound

Bachem H-5350.0001   Store at -15°C

Tyr-Phe-Gln-Asn-Cys-Pro-Arg-Gly
3-Mercaptopropionyl-Tyr-Phe-Gln-Asn-Cys-Pro-D-Arg-Gly Amide

Synonyms: Desmopressin; DDAVP; Vasopressin, (Deamino-Cys[1],D-Arg[8])-

Bachem H-7675.0001   Store at -15°C

ICN 152965   Disulfide bonds: mercaptopropionyl-Cys; arginine vasopressin analog with specific antidiuretic & memory enhancement activity; Zoral, M etal, *Coll Czech Chem Commun*, 32:1250, 1967; Weingartner, H etal, *Science*, 211:601, 1981; Pliska, V, *Front Horm Res*, 13:278, 1985

Tyr-Phe-Gln-Asn-Cys-Pro-Arg-Gly
(3-Mercapto-3-Methylbutyryl-Tyr(OMe)-Phe-Gln-Asn-Cys-Pro-Arg-Gly Amide

Synonyms: Vasopressin, (Deamino-Pen[1],OMe-Tyr[2],Arg[8])-

ICN 152967   Disulfide bonds: mercapto—Cys

Tyr-Phe-Gln-Asn-Cys-Pro-Arg-Gly
(1-Mercaptocyclohexyl)Ac-Tyr(OMe)-Phe-Gln-Asn-Cys-Pro-Arg-Gly Amide

Synonyms: Vasopressin, (β-Mercapto-β,β-Cyclopentamethylenepropionyl[1],OMe-Tyr[2],Arg[8])-; Manning Compound

ICN 152969   Disulfide bonds: mercaptocyclohexyl—Cys; antagonist of arginine vasopressin; Kruszyni, M etal, *J Med Chem*, 23:364, 1980

Tyr-Phe-Gln-Asn-Cys-Pro-Arg-Gly
Mpr-Tyr-Phe-Gln-Asn-Cys-Pro-D-Arg-Gly

Synonyms: Vasopressin, Desamino-(D-Arg[8])-; DDAVP

Neosystem SC417 MW 1069.23 Disulfide bonds: Mpr[1]-Cys[6]; bioactive; pituitary peptide; enhances human learning & memory; Weingartner, H et al, *Science*, 211:601, 1981; Zaoral, M et al, *Coll Czech Chem Commun*, 32:1250-1257, 1967

Tyr-Phe-Gln-Asn-Cys-Pro-Arg-Gly
3-Mercaptopropionyl-Tyr-Phe-Gln-Asn-Cys-Pro-D-Arg-Gly Amide Acetate Salt

Synonyms: Vasopressin, (deamino-Cys[1],D-Arg[8])-; DDAVP; Desmopressin

Sigma V 1005 FW 1069.2 (free base) ≥97% (HPLC) | Bioactive peptide; disulfide bonds: 1-6; specific antidiuretic & memory enhancement activity; Zoral, M et al, *Coll Czech Chem Commun*, 32:1250, 1967; Pliska, V, *Front Horm Res*, 13:278, 1985; Weingartner, H et al, *Science*, 211:601, 1981

Tyr-Phe-Gln-Asn-Cys-Pro-Arg-Gly
3-Mercapto-3-Methylbutyryl-Tyr(OMe)-Phe-Gln-Asn-Cys-Pro-Arg-Gly Amide

Synonyms: Vasopressin, (deamino-Pen[1],o-Me-Tyr[2],Arg[8])-

Sigma V 1880 FW 1111.3 ≥97% (HPLC) | Bioactive peptide; disulfide bonds: 1-6; potent V_1 (vasopressor) antagonist with low antidiuretic activity; Bankowski, K et al, *J Med Chem*, 21:850, 1978

Tyr-Phe-Gln-Asn-Cys-Pro-Arg-Gly
(1-Mercaptocyclohexyl)-Ac-Tyr(OMe)-Phe-Gln-Asn-Cys-Pro-Arg-Gly Amide

Synonyms: Vasopressin, (β-Mercapto-β,β-Cyclopentamethylenepropionyl[1],o-Me-Tyr[2],Arg[8])-; Manning Compound

Sigma V 2255 FW 1151.4 ≥97% (HPLC) | Bioactive peptide; disulfide bonds: 1-6; very potent V_1 (vasopressor) antagonist with almost no antidiuretic activity; Kruszynski, M et al, *J Med Chem*, 23:364, 1980

Tyr-Phe-Gln-Asn-Cys-Pro-Arg-Gly
Pmp-Tyr(Me)-Phe-Gln-Asn-Cys-Pro-Arg-Gly Amide

Synonyms: Vasopressin, (Pmp[1],Tyr(Me)[2])-Arg[8]-; Vasopressin V1 Antagonist, Arg-

Peptides International PVP-4203-v Synthetic MW 1152.4 $C_{52}H_{75}N_{14}O_{12}S_2$ >99% (HPLC); lyophilized amorphous powder | Bioactive; Disulfide bonds: Cys1-Cys6; potent arginine vasopressin V1 antagonist; Kruszynski, M et al, *J Med Chem*, 23:364, 1980

Tyr-Phe-Gln-Asn-Cys-Pro-Orn-Gly
3-Mercaptopropionyl-Tyr-Phe-Gln-Asn-Cys-Pro-D-Orn-Gly Amide

Synonyms: Vasopressin, (Deamino-Cys[1],D-Orn[8])-

Bachem H-1064.0001 Store at -15°C

Tyr-Phe-Gln-Asn-Lys-Pro-Arg
Phenylac-D-Tyr(Et)-Phe-Gln-Asn-Lys-Pro-Arg Amide

Synonyms: Vasopressin, (Phenylac[1],D-Tyr(Et)[2],Lys[6],Arg[8],des-Gly[9])-

Bachem H-3186.0001 MW 1097.29 $C_{54}H_{76}N_{14}O_{11}$ Store at -15°C

Tyr-Phe-His-Glu-Ala-Val-Gln-Ala-Gly-Trp-Arg-Ser-Ala-Thr-Glu-Thr-Leu-Ala-Gly-Ala

Synonyms: GP140 (831-850); SIVmac251 Peptide

Neosystem SC791 MW 2165.34 SIVmac251 peptide from subtype GP140

Tyr-Phe-Leu-Leu-Arg-Asn-Pro

Synonyms: Thrombin Antagonist, α-

Bachem H-1674.0001 MW 922.1 $C_{45}H_{67}N_{11}O_{10}$ Store at -15°C

Neosystem SC952 MW 922.09 Bioactive; antagonist to α-thrombin & thrombin receptor agonist peptide in human platelets; induces platelet shape change without Ca^{2+} mobilization or pleckstrin phosphorylation; potentiates other agonist-induced platelet responses; induces adrenalin-treated platelets to aggregate without secretion, only in the presence of fibrinogen; may be a useful tool for differentiating between several possible activation states of the human platelet thrombin receptor; may provide a structural template for developing more efficient antagonists targeted to the thrombin receptor; Rasmussen, UB et al, *JBC* 268:14322-14328, 1993; Negrescu, EV et al, *JBC* 270:1057-1061, 1995

Tyr-Phe-Leu-Phe-Arg-Pro-Arg-Asn Amide

Synonyms: Neuromedin U-8

ICN 153154 MW 1111.3

Bachem H-5505.0001 Porcine MW 1111.32 $C_{54}H_{78}N_{16}O_{10}$ Store at -15°C

Neosystem SC331 Porcine MW 1111.31 Bioactive neuropeptide; Minamino, N et al, *BBRC*, 130:1078, 1985

American Peptide 62-0-50 Porcine spinal cord MW 1111.3 $C_{54}H_{78}N_{16}O_{10}$ Potent stimulator of rat uterus smooth muscle; influences blood pressure in rats & dogs; Minamino, N et al, *BBRC*, 130:1078, 1985

Sigma N 4263 Porcine spinal cord FW 1111.3 ≥97% (HPLC); peptide content:~70% | Bioactive peptide; peptide which exhibits potent contractile hypertensive effect & uterine smooth muscle stimulation in rat; Minamino, N et al, *Biochem Biophys Res Commun*, 130:1078, 1985

Tyr-Phe-Lys-Ala Amide Hydrochloride

Bachem H-9495.0025 Store at -15°C

Tyr-Phe-Met-Arg-Phe Amide

Synonyms: Molluscan Cardioexcitatory Neuropeptide, (Tyr)-

American Peptide 60-0-30 MW 762.0 $C_{38}H_{51}N_9O_6S$

Bachem H-2980.0005 MW 761.95 $C_{38}H_{51}N_9O_6S$ Store at -15°C

ICN 153189

Tyr-Phe-Phe
(3,5-³H)Tyr-TicΨ(CH₂NH)-Phe-Phe

Synonyms: TIPP(Ψ)

ARC ART-583 30-50 Ci/mmol; 1.11-1.85 TBq/mmol; In ethanol | Radiochemical; δ-opioid antagonist; Nevin, ST et al, *Life Sci*, 56:225, 1995

Tyr-Phe-Phe Amide Acetate Salt

ICN 157176

Sigma T 6525 FW 474.6 (free base) $C_{27}H_{30}N_4O_4$

Tyr-Phe-Phe-His-Leu-Met
Tyr-D-Phe-Phe-D-His-Leu-Met Amide

Synonyms: Substance P (6-11), (Tyr[6],D-Phe[7],D-His[9])-; Sendide

Bachem H-1568.0001 MW 856.06 $C_{44}H_{57}N_9O_7S$ Store at -15°C

ICN 158728 MW 856.2 >97% | Neurokinin NK_1 receptor antagonist; Sakurada, T et al, *Brain Res*, 593:319, 1992

Neosystem SC916 MW 856.05 Bioactive; tachykinin; selective & extremely potent NK-1 receptor antagonist; Sakurada, T et al, *Brain Res*, 593:319-322, 1992; Sakurada, T et al, *Regul Peptides*, 46:326-328, 1993

Sigma S 5421   FW 856.1 $C_{44}H_{57}N_9O_7S$ ≥97% (HPLC) |
Bioactive peptide; potent, highly selective & competitive antagonist
of spinal substance P (NK-1) receptors; Sakurada, T et al,
Regulatory Peptides, 46:326, 1993

Tyr-Phe-Val-Asn-Abu-Pro-Arg-Arg
1-Aaa-D-Tyr(Et)-Phe-Val-Asn-Abu-Pro-Arg-Arg Amide

Synonyms: Vasopressin, (1-Adamantaneacetyl[1],D-
Tyr(Et)[2],Val[4],Abu[6], Arg[8,9])-

Bachem H-7705.0001   Store at -15°C

Tyr-Phe-Val-Asn-Abu-Pro-Arg-Arg
Propionyl-D-Tyr(Et)-Phe-Val-Asn-Abu-Pro-Arg-Arg Amide

Synonyms: Vasopressin, (Propionyl[1],D-Tyr(Et)[2],Val[4],Abu[6],Arg[8,9])-

Bachem H-9400.0001   MW 1119.34 $C_{53}H_{82}N_{16}O_{11}$ Store at
-15°C

Tyr-Phe-Val-Asn-Abu-Pro-Arg-Arg
1-Adamantaneacetyl-D-Tyr(OEt)-Phe-Val-Asn-Abu-Pro-Arg-Arg Amide

Synonyms: Vasopressin, (Adamantaneacetyl[1],OEt-D-
Tyr[2],Val[4],Aminobutyryl[6],Arg[8])-

ICN 152973   Antidiuretic antagonist; Manning, M etal, *Nature*,
329:839, 1987

Tyr-Phe-Val-Asn-Abu-Pro-Arg-Arg
(1-Adamantaneacetyl-D-Tyr(OEt)-Phe-Val-Asn-Abu-Pro-Arg-Arg Amide

Synonyms: Vasopressin, (Adamantaneacetyl[1],OEt-D-
Tyr[2],Val[4],Aminobutyryl[6],Arg[8,9])-

Sigma V 2381   FW 1239.5   ≥97% (HPLC) | Bioactive
peptide; potent V_2 (antidiuretic) antagonist; also antagonist at V_1
(vasopressor) receptors; Manning, M et al, *Nature*, 329:839, 1987

Tyr-Phe-Val-Asn-Cys-Pro-Arg
β-Mercapto-β,β-Cyclopentamethylenepropionyl-D-Tyr(Et)-Phe-Val-Asn-Cys-Pro-Arg Amide

Synonyms: Vasopressin, (d(CH₂)⁵¹,D-Tyr(Et)[2],Val[4],Arg[8],des-Gly[9])-

Bachem H-3192.0001   MW 1079.36 $C_{51}H_{74}N_{12}O_{10}S_2$ Store
at -15°C

Tyr-Phe-Val-Asn-Cys-Pro-Arg
β-Mercapto-β,β-Cyclopentamethylenepropionyl-Tyr(Et)-Phe-Val-Asn-Cys-Pro-Arg

Synonyms: Vasopressin, (d(CH₂)⁵¹,Tyr(Et)[2],Val[4],Arg[8],des-Gly-
NH₂⁹)-

Bachem H-7690.0001   Store at -15°C

Tyr-Phe-Val-Asn-Cys-Pro-Arg
(1-Mercaptocyclohexyl)Ac-Tyr(OEt)-Phe-Val-Asn-Cys-Pro-Arg Amide

Synonyms: Vasopressin, des-Gly⁹-(β-Mercapto-β,β-
Cyclopentamethylenepropionyl[1],OEt-Tyr[2],Val[4],Arg[8])-

ICN 152972   Disulfide bonds: mercaptocyclohexyl—Cys; Jard, S
etal, *Molec Pharmac*, 30:171, 1986

Sigma V 4503   FW 1079.3   ≥97% (HPLC) | Bioactive
peptide; disulfide bonds: 1-6; V_1 (vasopressor) & V_2 (antidiuretic)
antagonist; Jard, S et al, *Mol Pharmacol*, 30:171, 1986

Tyr-Phe-Val-Asn-Cys-Pro-Arg-Gly
Mpr-Tyr-Phe-Val-Asn-Cys-Pro-D-Arg-Gly Amide

Synonyms: Vasopressin, (Mpr[1],Val[4],D-Arg[8])-

American Peptide 66-0-10   MW 1040.2 $C_{46}H_{65}N_{13}O_{11}S_2$
Highly potent, long-lasting antidiuretic agonist without vasopressor
activity; disulfide bonds: Mpr[1]-Cys[6]; Kruszynski, M et al, *JMC*,
23:364, 1980

Tyr-Phe-Val-Asn-Cys-Pro-Arg-Gly
Pmp-Tyr(OEt)-Phe-Val-Asn-Cys-Pro-Arg-Gly Amide

Synonyms: Vasopressin, Arginine, (Pmp[1],Tyr(OEt)[2])-

American Peptide 66-0-13   MW 1152.4 $C_{53}H_{77}N_{13}O_{12}S_2$
Disulfide bonds: :Pmp[1]-Cys[6] | Potent vasopressor antagonist with
low antidiuretic activity; Jard, S et al, *Mole Pharmacol*, 30:171,
1986

Tyr-Phe-Val-Asn-Cys-Pro-Arg-Gly
3-Mercaptopropionyl-Tyr-Phe-Val-Asn-Cys-Pro-D-Arg-Gly Amide

Synonyms: Vasopressin, (Deamino-Cys[1],Val[4],D-Arg[8])-; dVDAVP

Bachem H-3176.0001   Store at -15°C

Tyr-Phe-Val-Asn-Cys-Pro-Arg-Gly
β-Mercapto-β,β-Cyclopentamethylenepropionyl-D-Tyr(Me)-Phe-Val-Asn-Cys-Pro-Arg-Gly Amide

Synonyms: Vasopressin, (d(CH₂)⁵¹,D-Tyr(Me)₂,Val[4],Arg[8])-

Bachem H-3182.0001   Store at -15°C

Tyr-Phe-Val-Asn-Cys-Pro-Arg-Gly
3-Mercapto-3-Methylbutyryl-Tyr-Phe-Val-Asn-Cys-Pro-D-Arg-Gly Amide

Synonyms: Vasopressin, (Deamino-Pen[1],Val[4],D-Arg[8])-

Bachem H-5345.0001   Store at -15°C

Tyr-Phe-Val-Asn-Cys-Pro-Arg-Gly
β-Mercapto-β,β-Cyclopentamethylenepropionyl-Tyr(Et)-Phe-Val-Asn-Cys-Pro-Arg-Gly Amide

Synonyms: Vasopressin, (d(CH₂)⁵¹,Tyr(Et)[2],Val[4],Arg[8])-

Bachem H-7670.0001   Store at -15°C

Tyr-Phe-Val-Asn-Cys-Pro-Arg-Gly
3-Mercapto-3-Methylbutyryl-Tyr-Phe-Val-Asn-Cys-Pro-D-Arg-Gly Amide

Synonyms: Vasopressin, (Deamino-Pen[1],Val[4],D-Arg[8])-

ICN 152966   Disulfide bonds: mercapto—Cys

Tyr-Phe-Val-Asn-Cys-Pro-Arg-Gly
(1-Mercaptocyclohexyl)Ac-Tyr(OEt)-Phe-Val-Asn-Cys-Pro-Arg-Gly Amide

Synonyms: Vasopressin, (β-Mercapto-β,β-
Cyclopentamethylenepropionyl[1],OEt-Tyr[2],Val[4],Arg[8])-

ICN 152968   Disulfide bonds: mercaptocyclohexyl—Cys;
antidiuretic antagonist; Manning M etal, *J Med Chem*, 24:701,
1981; Sawyer, W-H etal, *Science*, 212:49, 1981

Tyr-Phe-Val-Asn-Cys-Pro-Arg-Gly
3-Mercapto-3-Methylbutyryl-Tyr-Phe-Val-Asn-Cys-Pro-D-Arg-Gly Amide

Synonyms: Vasopressin, (deamino-Pen[1],Val[4],D-Arg[8])-

Sigma V 2005   FW 1068.3   ≥97% (HPLC) | Bioactive
peptide; disulfide bonds: 1-6; potent V_1 (vasopressor) antagonist
with moderate but prolonged antidiuretic activity compared to AVP;
Pang, CCY et al, *Proc Soc Exp Biol Med*, 161:41, 1979; Manning, M
et al, *J Med Chem*, 20:1228, 1977

Tyr-Phe-Val-Asn-Cys-Pro-Arg-Gly
(1-Mercaptocyclohexyl)Ac-Tyr(OEt)-Phe-Val-Asn-Cys-Pro-Arg-Gly Amide

Synonyms: Vasopressin, (β-Mercapto-β,β-Cyclopentamethylenepropionyl[1],OEt-Tyr[2],Val[4],Arg[8])-

Sigma V 4253   FW 1134.4   ≥97% (HPLC) | Bioactive peptide; disulfide bonds: 1-6; potent vasopressor antagonist with low antidiuretic activity; Manning, M et al, *J Med Chem*, 24:701, 1981; Sawyer, W-H et al, *Science*, 212:49, 1981; Mah, SC & Hofbauer, KG, *Drugs of the Future*, 12:1055, 1987

TyrpPhe-Val-Asn-Cys-Pro-Arg
β-Mercapto-β,β-Cyclopentamethylenepropionyl-Tyr(Et)-Phe-Val-Asn-Cys-Pro-Arg Amide

Synonyms: Vasopressin, (d(CH₂)[51],Tyr(Et)[2],Val[4],Arg[8],des-Gly[9])-

Bachem H-3188.0001   Store at -15°C

Tyr-Pro

Synonyms: Casomorphin (1-2), β-

Bachem G-3625.0250   MW 278.31 $C_{14}H_{18}N_2O_4$ Store at -15°C

Tyr-Pro Amide Hydrochloride

Synonyms: Casomorphin (1-2), β-

Bachem G-3457.0250   Store at -15°C

Tyr-Pro-Ala-Lys-Pro-Glu-Ala-Pro-Gly-Glu-Asp-Ala-Ser-Pro-Glu-Glu-Leu-Ser-Arg-Tyr-Tyr-Ala-Ser-Leu-Arg-His-Tyr-Leu-Asn-Leu-Val-Thr-Arg-Gln-Arg-Tyr Amide

Synonyms: Peptide YY

American Peptide 48-0-40   Porcine   MW 4240.7 $C_{190}H_{288}N_{54}O_{57}$ Tatemoto, K et al, *Nature*, 285:417, 1980; Tatemoto, K et al, *PNAS*, 79:2514, 1982; Corder, R et al, *Reg Peptides*, 21:253, 1988

ICN 191455   Porcine   MW 4240.7   Gut hormone that inhibits both secretin & cholecystokinin-stimulated pancreatic secretion; Lundberg, JM, *PNAS*, 79:4471 1982; Tatemoto, K, PNAS, 79:2514, 1982

Neosystem SC333   Porcine   MW 4240.69   Bioactive neuropeptide; brain/gut peptide; inhibits secretin- & cholescystokinin-stimulated pancreatic secretion; Tatemoto, K, *PNAS USA*, 79:2514-2518, 1982

Tyr-Pro-Ala-Lys-Pro-Glu-Ala-Pro-Gly-Glu-Asp-Ala-Ser-Pro-Glu-Glu-Leu-Ser-Arg-Tyr-Tyr-Ala-Ser-Leu-Arg-His-Tyr-Leu-Asn-Leu-Leu-Thr-Arg-Pro-Arg-Tyr Amide

Synonyms: Peptide YY, (Leu[31],Pro[34])-

Neosystem SC934   Porcine   MW 4223.71   Bioactive neuropeptide; brain/gut peptide; potent Y1 agonist with little or no affinity for the Y2 & Y3 receptor subtypes; Dumont, Y et al, Posters B29 & B30 presented at the; NPY Meeting in Cambridge (9-12 August 1993)

Tyr-Pro-Ala-Lys-Pro-Glu-Ala-Pro-Gly-Glu-Asp-Ala-Ser-Pro-Glu-Glu-Leu-Ser-Arg-Tyr-Tyr-Ala-Ser-Leu-Arg-His-Tyr-Leu-Asn-Leu-Val-Thr-Arg-Gln-Arg-Tyr Amide

Synonyms: Peptide YY

Bachem H-4505.0500   Porcine, rat   Store at -15°C

Tyr-Pro-Als-Lys-Pro-Glu-Ala-Pro-Gly-Glu-Asp-Ala-Ser-Pro-Glu-Glu-Leu-Ser-Arg-Tyr-Tyr-Ala-Ser-Leu-Arg-His-Tyr-Leu-Asn-Leu-Val-Thr-Arg-Gln-Arg-Tyr Amide

Synonyms: Peptide YY

Sigma P 5801   Porcine   FW 4240.7   ≥97% (HPLC) | Bioactive peptide; gastrointestinal peptide

Tyr-Pro-Arg
D-Tyr-Pro-Arg-CMK

Bachem N-1225.0025   MW 466.97 $C_{21}H_{31}ClN_6O_4$ Store at -15°C

Tyr-Pro-Gly-Phe-Leu-Thr
BOC-Tyr-Pro-Gly-Phe-Leu-Thr

Bachem H-1664.0005   MW 796.92 $C_{40}H_{56}N_6O_{11}$ Store at -15°C

Tyr-Pro-His-Phe-Met-Pro-Thr-Asn-Leu

Synonyms: Murine CMV pp 89 (168-176)

Bachem H-9415.0001   Store at -15°C

Tyr-Pro-Ile-Lys-Pro-Glu-Ala-Pro-Gly-Glu-Asp-Ala-Ser-Pro-Glu-Glu-Leu-Asn-Arg-Tyr-Tyr-Ala-Ser-Leu-Arg-His-Tyr-Leu-Asn-Leu-Val-Thr-Arg-Gln-Arg-Tyr Amide

Synonyms: Peptide YY

American Peptide 48-0-30   Human   MW 4309.8   Found in gastrointestinal tract; involves the feeding behavior; Tatemota, K et al, *Nature*, 285:417, 1980; Tatemoto, K et al, *PNAS*, 79:2514, 1982

Tyr-Pro-Ile-Lys-Pro-Glu-Ala-Pro-Gly-Glu-Asp-Ala-Ser-Pro-Glu-Glu-Leu-Asn-Arg-Tyr-Tyr-Ala-Ser-Leu-Arg-His-Tyr-Leu-Asn-Leu-Leu-Thr-Arg-Pro-Arg-Tyr Amide

Synonyms: Peptide YY, (Leu[31],Pro[34])-

American Peptide 48-0-31   Human   MW 4292.9 $C_{195}H_{296}N_{54}O_{56}$

Tyr-Pro-Ile-Lys-Pro-Glu-Ala-Pro-Gly-Glu-Asp-Ala-Ser-Pro-Glu-Glu-Leu-Asn-Arg-Tyr-Tyr-Ala-Ser-Leu-Arg-His-Tyr-Leu-Asn-Leu-Val-Thr-Arg-Pro-Arg-Tyr Amide

Synonyms: Peptide YY, (Pro[34])-

American Peptide 48-0-32   Human   MW 4278.8 $C_{194}H_{294}N_{54}O_{56}$

Bachem H-2808.0500   Human   Store at -15°C

Tyr-Pro-Ile-Lys-Pro-Glu-Ala-Pro-Gly-Glu-Asp-Ala-Ser-Pro-Glu-Glu-Leu-Asn-Arg-Tyr-Tyr-Ala-Ser-Leu-Arg-His-Tyr-Leu-Asn-Leu-Leu-Thr-Arg-Pro-Arg-Tyr Amide

Synonyms: Peptide YY, (Leu[31],Pro[34])-

Bachem H-2812.0500   Human   Store at -15°C

Tyr-Pro-Ile-Lys-Pro-Glu-Ala-Pro-Gly-Glu-Asp-Ala-Ser-Pro-Glu-Glu-Leu-Asn-Arg-Tyr-Tyr-Ala-Ser-Leu-Arg-His-Tyr-Leu-Asn-Leu-Val-Thr-Arg-Gln-Arg-Tyr Amide

Synonyms: Peptide YY

Bachem H-9180.0500   Human   MW 4309.81 $C_{194}H_{295}N_{55}O_{57}$ Store at -15°C

ICN 154560   Human   MW 4309.8

Neosystem SC319   Human   MW 4309.80   Bioactive neuropeptide; brain/gut peptide; Tatemoto, K et al, *BBRC*, 157:713-717, 1988

Tyr-Pro-Ile-Lys-Pro-Glu-Ala-Pro-Gly-Glu-Asp-Ala-Ser-Pro-Glu-Glu-Leu-Asn-Arg-Tyr-Tyr-Ala-Ser-Leu-Arg-His-Tyr-Leu-Asn-Leu-Val-Thr-Arg-Pro-Arg-Tyr Amide

Synonyms: Peptide YY, (Pro³⁴)-

Neosystem SC988   Human   MW 4278.78   Bioactive neuropeptide; brain/gut peptide; this analog is a highly Y-1-selective full agonist of peptide YY/ neuropeptide Y receptors; useful for studying the importance of Y receptor subtypes in mediating peptide YY physiological actions; Grandt, D et al, *Eur J Pharmacol*, 269:127-132, 1994; Lloyd, KCK et al, *Am J Physiol-Gastrointest & Liver Physiol*, 33:G123-G127, 1996

Tyr-Pro-Ile-Lys-Pro-Glu-Ala-Pro-Gly-Glu-Asp-Ala-Ser-Pro-Glu-Glu-Leu-Asn-Arg-Tyr-Tyr-Ala-Ser-Leu-Arg-His-Tyr-Leu-Asn-Leu-Val-Thr-Arg-Gln-Arg-Tyr Amide

Synonyms: Peptide YY

Sigma P 1306   Human   FW 4309.8   ≥97% (HPLC) | Bioactive peptide; gastrointestinal peptide; gut hormone that inhibits both secretin & cholecystokinin-stimulated pancreatic secretion; Tatemoto, K, *Proc Natl Acad Sci USA*, 79:2514, 1982; Lundberg, JM, *Proc Natl Acad Sci USA*, 79:4471, 1982; Tatemoto, K, *Proc Natl Acad Sci USA*, 157:713, 1988

Tyr-Pro-Ile-Lys-Pro-Glu-Ala-Pro-Gly-Glu-Asp-Ala-Ser-Pro-Glu-Glu-Leu-Asn-Arg-Tyr-Tyr-Ala-Ser-Leu-Arg-His-Tyr-Leu-Asn-Leu-Val-Thr-Arg-Gln-Arg-Tyr

Synonyms: Peptide YY

Biogenesis 7260-5504   Synthetic   MW 4307.2   Lyophilized, consisting of 82% peptide material, TFA counter ions and residual H_2O; purified | Tatemoto et al, *BBRC*, 157:713, 1988

Tyr-Pro-Ile-Ser-Leu

Synonyms: Gluten Exorphin C

Bachem H-1412.0005   MW 591.71 $C_{29}H_{45}N_5O_8$ Store at -15°C

Tyr-Pro-Ile-Ser-Leu Trifluoroacetate Salt

Synonyms: Exorphin C; Opioid Receptor Peptide, δ-

Sigma E 2019   FW 591.7 (free base) $C_{29}H_{45}N_5O_8$ ≥97% (HPLC) | Bioactive peptide; opioid peptide; Fukudome, S et al, *FEBS Lett*, 316:17, 1993

Tyr-Pro-Leu-Gly Amide

Synonyms: Melanocyte Stimulating Hormone Release Inhibiting Factor, (Tyr⁰)-; Melanocyte Stimulating Hormone Release Inhibiting Factor I; Oxytocin (6-9), (Tyr⁶)-

Bachem H-5120.0050   MW 447.54 $C_{22}H_{33}N_5O_5$ Store at -15°C

ICN 152960   C-terminal tripeptide of oxytocin; has activity in numerous behavioral tests & clinical situations; has anti-opiate activity; Zadina, JE etal, *Pharmacol Biochem Behav*, 17:1193, 1982; Kasin AJ etal, *Pharmacol Biochem Behav*, 23:1045, 1985

Sigma T 2899   FW 447.5 $C_{22}H_{33}N_5O_5$ ≥97% (HPLC) | Bioactive peptide

---

Tyr-Pro-Lys

Bachem H-2556.0050   Store at -15°C

Tyr-Pro-Lys-Gly Amide Acetate Salt

Synonyms: Melanocyte Stimulating Hormone Release Inhibiting Factor I, (Tyr-K)-

Sigma T 7427   FW 462.5 (free base) $C_{22}H_{34}N_6O_5$ ≥95% (HPLC) | Bioactive peptide; has been isolated from human cortex; opiate antagonist; interacts with GABAergic & dopaminergic systems; Hackler, L et al, *Peptides*, 15:945, 1994

Tyr-Pro-Phe

Synonyms: Casomorphin (1-3), β-

American Peptide 24-1-22   MW 425.5 $C_{23}H_{27}N_3O_5$

Bachem H-2375.0005   MW 425.49 $C_{23}H_{27}N_3O_5$ Store at -15°C

ICN 150568   MW 425.5 $C_{23}H_{27}N_3O_5$

Sigma C 3146   Bovine   FW 425.5 $C_{23}H_{27}N_3O_5$ ≥97% (HPLC) | Bioactive peptide; opioid peptide

---

Tyr-Pro-Phe Amide

Synonyms: Casomorphin (1-3), β-

American Peptide 24-1-25   MW 424.5 $C_{23}H_{28}N_4O_4$

Bachem H-2380.0005   MW 424.5 $C_{23}H_{28}N_4O_4$ Store at -15°C

Tyr-Pro-Phe Free Base Hydrochloride

Synonyms: Casomorphin (1-3), β-

Sigma C 3397   Bovine   FW 425.5 $C_{23}H_{27}N_3O_5$ ≥97% (HPLC) | Bioactive peptide; opioid peptide

---

Tyr-Pro-Phe-Phe Amide

Synonyms: Endomorphin II; Endogenous μ-Opiate Receptor Selective Agonist

American Peptide 23-1-12   MW 571.7 $C_{32}H_{37}N_5O_5$ Potent & selective endogenous agonist of the γ-opiate receptor; Zadina, J et al, *Nature*, 386:499, 1997

Bachem H-4004.0005   Store at -15°C

Neosystem SC1233   MW 571.67   Bioactive; opioid peptide; Zadina, JE et al, *Nature*, 386:499-502, 1997

Sigma E 3148   FW 571.7 $C_{32}H_{37}N_5O_5$ Bioactive peptide; opioid peptide; potent, selective μ-opioid receptor agonist; Zadina, JE et al, *Nature*, 386:449, 1997

Peptides International PEM-3605-PI   Human, bovine synthetic   MW 571.68 $C_{32}H_{37}N_5O_5$ >98% (HPLC); white powder; bulk | Bioactive; Zadina, JE et al, *Nature*, 386:499, 1997; Champion, HC et al, *BBRC*, 235:567, 1997; Stone, LS et al, *Neuroreport*, 8:3131, 1997; Hackler, L et al, *Peptides*, 18:1635, 1997

Peptides International PEM-4334-v   Human, bovine synthetic   MW 571.68 $C_{32}H_{37}N_5O_5$ >99% (HPLC); lyophilized amorphous powder | Bioactive; Zadina, JE et al, *Nature*, 386:499, 1997; Champion, HC et al, *BBRC*, 235:567, 1997; Stone, LS et al, *Neuroreport*, 8:3131, 1997; Hackler, L et al, *Peptides*, 18:1635, 1997

Tyr-Pro-Phe-Phe Amide Trifluoroacetate Salt

Synonyms: Endomorphin II

Calbiochem 324731   Synthetic   MW 571.7 $C_{32}H_{37}N_5O_5$ ≥98% (HPLC); lyophilized solid; soluble in water; LD_{50}≤2000 mg/kg | Neuropeptide related to Endomorphin-1; acts as a potent & specific agonist for the μ-opiate receptor; exhibits excellent binding affinity & selectivity for the μ-receptor (>13,000-fold preference for the μ-receptor over the δ-receptor); inhibits high-threshold Ca^{2+} channel currents in NGMO-251 cells; Higashida, H et al, *J Physiol*, 507:71, 1998; Zadina, JE et al, *Nature*, 386:499, 1997

Tyr-Pro-Phe-Pro

Synonyms: Casomorphin (1-4), β-

American Peptide 24-3-28   Bovine   MW 522.6 $C_{28}H_{34}N_4O_6SCl$

Tyr-Pro-Phe-Pro
Tyr-D-Pro-Phe-Pro Amide

Synonyms: Casomorphin (1-4), (D-Pro²)-β-

American Peptide 32-0-38   Bovine   MW 521.6 $C_{28}H_{35}N_5O_5$

Tyr-Pro-Phe-Pro
Tyr-Pro-(nMe)Phe-D-Pro Amide

Synonyms: Casomorphin (1-4), ((nMe)Phe³,D-Pro⁴)-β-;
Morphiceptin

American Peptide 32-3-34   Bovine   MW 536.7 $C_{29}H_{38}N_5O_5$
A selective agonist of μ opioid receptor; Knapp, PE et al, *Brain Res*,
743(1-2):341, 1996

Tyr-Pro-Phe-Pro
Tyr-Pro-Phe-D-Pro Amide

Synonyms: Casomorphin (1-4), (D-Pro⁴)-β-; Morphiceptin, (D-Pro⁴)-

American Peptide 32-3-37   Bovine   MW 521.6 $C_{28}H_{35}N_5O_5$

Tyr-Pro-Phe-Pro

Synonyms: Casomorphin (1-4), β-

Bachem H-2350.0005   Bovine   MW 522.6 $C_{28}H_{34}N_4O_6$
Store at -15°C

Tyr-Pro-Phe-Pro
Tyr-Pro-*N*-Me-Phe-D-Pro Amide

Synonyms: Casomorphin (1-4); PL017, (*N*-Me-Phe³,D-Pro⁴)-β-

Bachem H-4932.0005   Bovine   MW 535.64 $C_{29}H_{37}N_5O_5$
Store at -15°C

Tyr-Pro-Phe-Pro Amide

Synonyms: Casomorphin (1-4), β-; Morphiceptin

ICN 152956   Highly preferential ligand for the μ-opiate
(morphine) receptor; Chang, KJ etal, *Science*, 212:75, 1981

American Peptide 24-3-30   Bovine   MW 521.6 $C_{28}H_{35}N_5O_5$
Agonist for morphine (μ) receptors; Chang, KJ et al, *Science*,
212:75, 1981

Bachem H-2355.0005   Bovine   MW 521.62 $C_{28}H_{35}N_5O_5$
Store at -15°C

Tyr-Pro-Phe-Pro Amide Free Base Hydrochloride

Synonyms: Casomorphin (1-4), β-; Morphiceptin

Sigma M 4264   FW 521.6 $C_{28}H_{35}N_5O_5$ ≥97% (HPLC) |
Bioactive peptide; opioid peptide; agonist for morphine (μ)
receptors; Chang, KJ et al, *Science*, 212:75, 1981

Tyr-Pro-Phe-Pro Amide Hydrochloride

Synonyms: Morphiceptin; Casomorphin (1-4), β-

ICN 193534   MW 558.1 $C_{28}H_{35}N_5O_5 \cdot HCl$ ≥97% | Agonist
for μ-opiate (morphine) receptors; Chang, KJ etal, *Science*, 212:75,
1981

Tyr-Pro-Phe-Pro-Gly

Synonyms: Casomorphin (1-5), β-

ICN 150569   MW 579.7 $C_{30}H_{37}N_5O_7$

American Peptide 24-3-37   Bovine   MW 579.7 $C_{30}H_{37}N_5O_7$

Tyr-Pro-Phe-Pro-Gly
Tyr-D-Pro-Phe-Pro-Gly Amide

Synonyms: Casomorphin (1-5), (D-Pro²)-β-

American Peptide 24-3-54   Bovine   MW 578.7 $C_{30}H_{38}N_6O_6$

Tyr-Pro-Phe-Pro-Gly

Synonyms: Casomorphin (1-5), β-

Bachem H-2295.0005   Bovine   MW 579.65 $C_{30}H_{37}N_5O_7$
Store at -15°C

Tyr-Pro-Phe-Pro-Gly
Tyr-D-Pro-Phe-Pro-Gly

Synonyms: Casomorphin (1-5), (D-Pro²)-β-

Bachem H-2325.0005   Bovine   MW 579.65 $C_{30}H_{37}N_5O_7$
Store at -15°C

Tyr-Pro-Phe-Pro-Gly

Synonyms: Casomorphin 5, β-

Peptides International PEK-4079-v   Bovine synthetic   MW
579.65 $C_{30}H_{37}N_5O_7$ >99% (HPLC); lyophilized amorphous
powder | Bioactive; Brantl, V et al, Hoppe-Seyler's *Z Physiol
Chem*, 360:1211, 1979; Henschen, A et al, Hoppe-Seyler's *Z
Physiol Chem*, 360:1217, 1979

Tyr-Pro-Phe-Pro-Gly Amide

Synonyms: Casomorphin (1-5), β-

American Peptide 24-3-40   Bovine   MW 578.7 $C_{30}H_{38}N_6O_6$

Bachem H-2300.0005   Bovine   MW 578.67 $C_{30}H_{38}N_6O_6$
Store at -15°C

Tyr-Pro-Phe-Pro-Gly Dihydrate

Synonyms: Casomorphin 5, β-

Peptides International PEK-4079   Bovine synthetic   MW
615.68 $C_{30}H_{37}N_5O_7 \cdot 2H_2O$ >99% (HPLC); lyophilized
amorphous powder; bulk | Bioactive; Brantl, V et al, Hoppe-
Seyler's *Z Physiol Chem*, 360:1211, 1979; Henschen, A et al,
Hoppe-Seyler's *Z Physiol Chem*, 360:1217, 1979

Tyr-Pro-Phe-Pro-Gly Free Base Hydrochloride

Synonyms: Casomorphin (1-5), β-

Sigma C 5147   FW 579.7 $C_{30}H_{37}N_5O_7$ ≥97% (HPLC) |
Bioactive peptide; opioid peptide

Tyr-Pro-Phe-Pro-Gly-Pro

Synonyms: Casomorphin (1-6), β-

American Peptide 24-3-56   Bovine   MW 676.8 $C_{35}H_{44}N_6O_8$

Bachem H-2285.0005   Bovine   MW 676.77 $C_{35}H_{44}N_6O_8$
Store at -15°C

ICN 154518   Bovine   MW 676.8

Tyr-Pro-Phe-Pro-Gly-Pro-Ile

Synonyms: Casomorphin, β-; Casomorphin 7, β-

Bachem H-2280.0005   Bovine   MW 789.93 $C_{41}H_{55}N_7O_9$
Store at -15°C

ICN 150567   Bovine   MW 790   Brantl, V etal, Hoppe-
Seyler's *Z Physiol Chem*, 360:1211, 1979; Lottspeich, F etal,
Hoppe-Seyler's *Z Physiol Chem*, 361:1835, 1980

American Peptide 24-3-10   Bovine synthetic   MW 789.9
$C_{41}H_{55}N_7O_9$ Opioid activity; first isolated from an enzymatic casein
digest; Henschen, A et al, *Z Physiol Chem*, 360:1211 & 1217, 1979

Peptides International PEK-4078-v   Bovine synthetic   MW
789.93 $C_{41}H_{55}N_7O_9$ >99% (HPLC); lyophilized amorphous
powder | Bioactive; Brantl, V et al, Hoppe-Seyler's *Z Physiol
Chem*, 360:1211, 1979; Henschen, A et al, Hoppe-Seyler's *Z
Physiol Chem*, 360:1217, 1979

Tyr-Pro-Phe-Pro-Gly-Pro-Ile 4H$_2$O

Synonyms: Casomorphin 7, β-

Peptides International PEK-4078 Bovine synthetic MW 861.99 C$_{41}$H$_{55}$N$_7$O$_9$ · 4H$_2$O >99% (HPLC); lyophilized amorphous powder; bulk | Bioactive; Brantl, V et al, Hoppe-Seyler's *Z Physiol Chem*, 360:1211, 1979; Henschen, A et al, Hoppe-Seyler's *Z Physiol Chem*, 360:1217, 1979

Tyr-Pro-Phe-Pro-Gly-Pro-Ile Acetate Salt Free Base

Sigma C 5900 Bovine synthetic FW 789.9 C$_{41}$H$_{55}$N$_7$O$_9$ ≥97% (HPLC) | Bioactive peptide; opioid peptide first isolated from an enzymatic digest of casein; Brantl, V et al, Hoppe-Seyler's *Z Physiol Chem*, 360:1211, 1979; Lottspeich, F et al, ibid, 361:1835, 1980

Tyr-Pro-Phe-Val-Glu-Pro-Ile

Synonyms: Casomorphin, β-

American Peptide 24-1-11 Human MW 864.0 C$_{44}$H$_{61}$N$_7$O$_{11}$ Greenberg, R et al, *J Biol Chem*, 259:5132, 1984

Bachem H-2275.0005 Human MW 864.01 C$_{44}$H$_{61}$N$_7$O$_{11}$ Store at -15°C

ICN 154498 Human MW 864.1 C$_{44}$H$_{61}$N$_7$O$_{11}$ Greenberg, R etal, *JBC*, 259:132, 1984

Sigma C 0783 Human FW 864.0 C$_{44}$H$_{61}$N$_7$O$_{11}$ ≥97% (HPLC) | Bioactive peptide; opioid peptide; β-Casomorphin; Greenberg, R et al, *J Biol Chem*, 259:5132, 1984

Tyr-Pro-Pro-Pro-Pro-Trp

Bachem H-7090.0005 MW 755.87 C$_{40}$H$_{49}$N$_7$O$_8$ Store at -15°C

Tyr-Pro-Pro-Trp

Bachem H-7695.0050 MW 561.64 C$_{30}$H$_{35}$N$_5$O$_6$ Store at -15°C

Tyr-Pro-Ser-Lys-Pro-Asp-Asn-Pro-Gly-Asp-Asp-Ala-Pro-Ala-Glu-Asp-Leu-Ala-Arg-Tyr-Tyr-Ser-Ala-Leu-Arg-His-Tyr-Ile-Asn-Leu-Ile-Thr-Arg-Gln-Arg-Tyr Amide

Synonyms: Neuropeptide Y

Sigma N 6269 Ovine FW 4239.7 ≥97% (HPLC) | Bioactive peptide; gastrointestinal peptide; vasoconstrictor

Tyr-Pro-Ser-Lys-Pro-Asp-Asn-Pro-Gly-Glu-Asp-Ala-Pro-Ala-Glu-Asp-Leu-Ala-Arg-Tyr-Tyr-Ser-Ala-Leu-Arg-His-Tyr-Ile-Asn-Leu-Ile-Trp-Arg-Gln-Arg-Tyr Tyr-Pro-Ser-Lys-Pro-Asp-Asn-Pro-Gly-Glu-Asp-Ala-Pro-Ala-Glu-Asp-Leu-Ala-Arg-Tyr-Tyr-Ser-Ala-Leu-Arg-His-Tyr-Ile-Asn-Leu-Ile-D-Trp-Arg-Gln-Arg-Tyr Amide

Synonyms: Neuropeptide Y, (Leu^{17},D-Trp^{32})-

American Peptide 60-1-14 Human MW 4338.8 C$_{197}$H$_{290}$N$_{56}$O$_{56}$

Tyr-Pro-Ser-Lys-Pro-Asp-Asn-Pro-Gly-Glu-Asp-Ala-Pro-Ala-Glu-Asp-Leu-Ala-Arg-Tyr-Tyr-Ser-Ala-Leu-Arg-His-Tyr-Ile-Asn-Leu-Ile-Thr-Arg-Pro-Arg-Tyr Amide

Synonyms: Neuropeptide Y, (Pro^{34})-

American Peptide 60-4-13 Porcine MW 4222.7 C$_{190}$H$_{286}$N$_{54}$O$_{56}$ A selective neuropeptide YY$_1$ receptor agonist

Tyr-Pro-Ser-Lys-Pro-Asp-Asn-Pro-Gly-Glu-Asp-Ala-Pro-Ala-Glu-Asp-Leu-Ala-Arg-Tyr-Tyr-Ser-Ala-Leu-Arg-His-Tyr-Ile-Asn-Leu-Leu-Thr-Arg-Pro-Arg-Tyr Amide

Synonyms: Neuropeptide Y, (Leu^{31},Pro^{34})-

American Peptide 60-4-14 Porcine MW 4222.7 C$_{190}$H$_{286}$N$_{54}$O$_{56}$ White solid | Specific Y1 receptor ligand; used to detect the NPY/PYY receptor subtype distribution, e.g., in rat brain; Fuhlendorff, J et al, *PNAS*, 87:182, 1990

Tyr-Pro-Ser-Lys-Pro-Asp-Asn-Pro-Gly-Glu-Asp-Ala-Pro-Ala-Glu-Asp-Leu-Ala-Arg-Tyr-Tyr-Ser-Ala-Leu-Arg-His-Tyr-Ile-Asn-Leu-Ile-Thr-Arg-Gln-Arg-Tyr

Synonyms: Neuropeptide Y

American Peptide 60-4-18 Porcine MW 4254.7 (free acid) C$_{190}$H$_{286}$N$_{54}$O$_{58}$

Tyr-Pro-Ser-Lys-Pro-Asp-Asn-Pro-Gly-Glu-Asp-Ala-Pro-Ala-Glu-Asp-Leu-Ala-Arg-Tyr-Tyr-Ser-Ala-Leu-Arg-His-Tyr-Ile-Asn-Leu-Ile-D-Trp-Arg-Gln-Arg-Tyr Amide

Synonyms: Neuropeptide Y, (D-Trp^{32})-

Bachem H-3308.0500 Porcine MW 4338.81 C$_{197}$H$_{290}$N$_{56}$O$_{56}$ Store at -15°C

Tyr-Pro-Ser-Lys-Pro-Asp-Asn-Pro-Gly-Glu-Asp-Ala-Pro-Ala-Glu-Asp-Leu-Ala-Arg-Tyr-Tyr-Ser-Ala-Leu-Arg-His-Tyr-Ile-Asn-Leu-Ile-Thr-Arg-Gln-Arg-Tyr Amide

Synonyms: Neuropeptide Y

Bachem H-4430.0500 Porcine MW 4253.7 C$_{190}$H$_{287}$N$_{55}$O$_{57}$ Store at -15°C

Tyr-Pro-Ser-Lys-Pro-Asp-Asn-Pro-Gly-Glu-Asp-Ala-Pro-Ala-Glu-Asp-Leu-Ala-Arg-Tyr-Tyr-Ser-Ala-Leu-Arg-His-Tyr-Ile-Asn-Leu-Ala-Aib-Arg-Gln-Arg-Tyr Amide

Synonyms: Neuropeptide Y, (Ala^{31},Aib^{32})-; Neuropeptide Y

Bachem H-5084.0500 Porcine MW 4195.62 C$_{187}$H$_{281}$N$_{55}$O$_{56}$ Store at -15°C

Tyr-Pro-Ser-Lys-Pro-Asp-Asn-Pro-Gly-Glu-Asp-Ala-Pro-Ala-Glu-Asp-Leu-Ala-Arg-Tyr-Tyr-Ser-Ala-Leu-Arg-His-Tyr-Ile-Asn-Leu-Leu-Thr-Arg-Pro-Arg-Tyr Amide

Synonyms: Neuropeptide Y, (Leu^{31},Pro^{34})-

Bachem H-8575.0500 Porcine MW 4222.69 C$_{190}$H$_{286}$N$_{54}$O$_{56}$ Store at -15°C

Tyr-Pro-Ser-Lys-Pro-Asp-Asn-Pro-Gly-Glu-Asp-Ala-Pro-Ala-Glu-Asp-Leu-Ala-Arg-Tyr-Tyr-Ser-Ala-Leu-Arg-His-Tyr-Ile-Asn-Leu-Ile-Thr-Arg-Pro-Arg-Tyr Amide

Synonyms: Neuropeptide Y, (Pro^{34})-

Bachem H-8580.0500 Porcine MW 4222.69 C$_{190}$H$_{286}$N$_{54}$O$_{56}$ Store at -15°C

Tyr-Pro-Ser-Lys-Pro-Asp-Asn-Pro-Gly-Glu-Asp-Ala-Pro-Ala-Glu-Asp-Leu-Ala-Arg-Tyr-Tyr-Ser-Ala-Leu-Arg-His-Tyr-Ile-Asn-Leu-Ile-Thr-Arg-Gln-Arg-Tyr Amide

Synonyms: Neuropeptide Y

Calbiochem 05-23-2000   Porcine   MW 4253.7
$C_{190}H_{287}N_{55}O_{57}$ ≥99% (HPLC); lyophilized solid; soluble in 5% acetic acid; LD_{50}≤2000 mg/kg | Inhibitor of Ca^{2+} activated K^+ channels in vascular smooth muscle cells; stimulates Na^+-K^+ ATPase activity in renal tubules; strong inhibitor of both cholecystokinin- & secretin-stimulated pancreatic secretion; Xiong, Z & Cheung, DW, *Pflugers Arch*, 429:280, 1994; Catzeflis, C et al, *Endocrinology*, 132:224, 1993; Ohtomo, Y, *Kidney Int*, 45:1606, 1994; *Merck Index*, 12:6568; Moore, RY, *Prog Brain Res*, 93:99, 1992

Tyr-Pro-Ser-Lys-Pro-Asp-Asn-Pro-Gly-Glu-Asp-Ala-Pro-Ala-Glu-Asp-Leu-Ala-Arg-Tyr-Tyr-Ser-Ala-Leu-Arg-His-Tyr-Ile-Asn-Leu-Leu-Thr-Arg-Pro-Arg-Tyr Amide

Synonyms: Neuropeptide Y (13-36), (Leu[31],Pro[34])-

Calbiochem 05-23-2006   Porcine   MW 4222.8
$C_{190}H_{286}N_{54}O_{56}$ >97% (HPLC); solid; soluble in 5% acetic acid | Specific Y_1-receptor agonist that produces orexigenic effects; binds with high affinity to Y_1 transfected HEK-293 cells; Tong, Y et al, *Brain Res Mol Brain Res*, 34:303, 1995; Fuhlendorff, J et al, *PNAS*, 87:182, 1990; Jewett, DC et al, *Brain Res*, 631:129, 1993; Larson, PJ et al, *Eur J Neurosci*, 5:1622, 1993

Tyr-Pro-Ser-Lys-Pro-Asp-Asn-Pro-Gly-Glu-Asp-Ala-Pro-Ala-Glu-Asp-Leu-Ala-Arg-Tyr-Tyr-Ser-Ala-Leu-Arg-His-Tyr-Ile-Asn-Leu-Ile-Thr-Arg-Gln-Arg-Tyr Amide

Synonyms: Neuropeptide Y (13-36)

Calbiochem 05-23-2007   Porcine   MW 2982.4
$C_{135}H_{209}N_{41}O_{36}$ ≥97% (HPLC); solid; soluble in 5% acetic acid | Inactive at certain postsynaptic sites; mimics the effects of NPY & PYY at presynaptic receptors in the vas deferens; a Y_2 receptor ligand; Wager-Page, SA et al, *Can J Physiol Pharmacol*, 71:112, 1993; Krstenansky, JL et al, *PNAS*, 86:4377, 1989

ICN 151739   Porcine   MW 4253.7   Tatemoto, K et al, *PNAS*, 79:5485, 1982

Neosystem SC116   Porcine   MW 4253.69   Bioactive neuropeptide; Tatemoto, K et al, *PNAS USA*, 79:5485, 1982

Tyr-Pro-Ser-Lys-Pro-Asp-Asn-Pro-Gly-Glu-Asp-Ala-Pro-Ala-Glu-Asp-Leu-Ala-Arg-Tyr-Tyr-Ser-Ala-Leu-Arg-His-Tyr-Ile-Asn-Leu-Leu-Thr-Arg-Pro-Arg-Tyr Amide

Synonyms: Neuropeptide Y, (Leu[31],Pro[34])-

Neosystem SC876   Porcine   MW 4222.68   Bioactive neuropeptide; specific Y1 receptor agonist; active *in vivo*; Fuhlendorff, J et al, *PNAS USA*, 87:182-186, 1990; Gehlert, DR et al, *Neurochem Int*, 21:45-67, 1992

Tyr-Pro-Ser-Lys-Pro-Asp-Asn-Pro-Gly-Glu-Asp-Ala-Pro-Ala-Glu-Asp-Leu-Ala-Arg-Tyr-Tyr-Ser-Ala-Leu-Arg-His-Tyr-Ile-Asn-Leu-Ile-Thr-Arg-Gln-Arg-Tyr Amide

Synonyms: Neuropeptide Tyrosine; Neuropeptide Y

Sigma N 3266   Porcine   FW 4253.7   ≥97% (HPLC) | Bioactive peptide; gastrointestinal peptide; brain peptide which possesses both gastrointestinal & vasoconstrictor properties; stimulates Na^+-K^+ ATPase in renal tubules; inhibits cholecystokinin- & secretin-stimulated pancreatic secretion; Similar to Sigma N 4509 but prepared by Sigma; Tatemoto, K, *Proc Natl Acad Sci USA*, 79:5485, 1982

Sigma N 4509   Porcine   FW 4253.7   ≥97% (HPLC) | Bioactive peptide; gastrointestinal peptide; brain peptide which possesses both gastrointestinal & vasoconstrictor properties; stimulates Na^+-K^+ ATPase in renal tubules; inhibits cholecystokinin- & secretin-stimulated pancreatic secretion; Tatemoto, K, *Proc Natl Acad Sci USA*, 79:5485, 1982

American Peptide 60-4-15   Porcine brain   MW 4253.7
$C_{190}H_{287}N_{55}O_{57}$ Possesses both gastrointestinal & vasoconstrictor properties; Tatemoto, K et al, *PNAS*, 79:5485, 1982

Biogenesis 6730-0504   Porcine synthetic   MW 4254.2
79% peptide material, with TFA counter ions and residual H_2O; purified; lyophilized

Tyr-Pro-Ser-Lys-Pro-Asp-Asn-Pro-Gly-Glu-Asp-Ala-Pro-Ala-Glu-Asp-Leu-Ala-Arg-Tyr-Tyr-Ser-Ala-Leu-Arg-His-Tyr-Ile-Asn-Leu-Leu-Thr-Arg-Pro-Arg-Tyr Amide

Synonyms: Neuropeptide Y; Neuropeptide Y, (Leu[31],Pro[34])-; Neuropeptide Y Y1-Receptor Selective Agonist

Peptides International PNP-4314-s   Porcine synthetic   MW 4222.7 $C_{190}H_{286}N_{54}O_{56}$ 99% (HPLC); lyophilized amorphous powder | Bioactive; Fuhlendorff, J et al, *PNAS USA*, 87:182, 1990; Sheikh, SP, *Am J Physiol*, 261:G701, 1991; Michel, MC, *Trends Pharmacol Sci*, 12:389, 1991

Peptides International PNP-4162-s, 4162-v   Porcine, bovine synthetic   MW 4253.7 $C_{190}H_{287}N_{55}O_{57}$ 99% (HPLC); lyophilized amorphous powder | Bioactive; Tatemoto, K, *PNAS USA*, 79:5485, 1982

Tyr-Pro-Ser-Lys-Pro-Asp-Asn-Pro-Gly-Glu-Asp-Ala-Pro-Ala-Glu-Asp-Met-Ala-Arg-Tyr-Tyr-Ser-Ala-Leu-Arg-His-Tyr-Ile-Asn-Leu-Ile-Thr-Arg-Gln-Arg-Tyr Biotinyl-Tyr-Pro-Ser-Lys-Pro-Asp-Asn-Pro-Gly-Glu-Asp-Ala-Pro-Ala-Glu-Asp-Met-Ala-Arg-Tyr-Tyr-Ser-Ala-Leu-Arg-His-Tyr-Ile-Asn-Leu-Ile-Thr-Arg-Gln-Arg-Tyr Amide

Synonyms: Neuropeptide Y, Biotin-

American Peptide 60-4-19   MW 4515.1 $C_{199}H_{300}N_{57}O_{60}S_2$

Tyr-Pro-Ser-Lys-Pro-Asp-Asn-Pro-Gly-Glu-Asp-Ala-Pro-Ala-Glu-Asp-Met-Ala-Lys-Tyr-Tyr-Ser-Ala-Leu-Arg-His-Tyr-Ile-Asn-Leu-Ile-Thr-Arg-Gln-Arg-Tyr Amide

Synonyms: Melanostatin

American Peptide 60-0-20   Frog   MW 4243.8
$C_{189}H_{285}N_{53}O_{57}S$ Chartrel, N et al, *PNAS*, 88:3862, 1991

Tyr-Pro-Ser-Lys-Pro-Asp-Asn-Pro-Gly-Glu-Asp-Ala-Pro-Ala-Glu-Asp-Met-Ala-Arg-Tyr-Tyr-Ser-Ala-Leu-Arg-His-Tyr-Ile-Asn-Leu-Ile-Thr-Arg-Gln-Arg-Tyr Amide

Synonyms: Neuropeptide Y

Alexis 163-003   Human   MW 4271.7 $C_{189}H_{285}N_{55}O_{57}S$
≥97%; white lyophilized powder; soluble in water | Minth, CD et al, *PNAS*, 81:4577, 1984

Tyr-Pro-Ser-Lys-Pro-Asp-Asn-Pro-Gly-Glu-Asp-Ala-Pro-Ala-Glu-Asp-Met-Ala-Arg-Tyr-Tyr-Ser-Ala-Leu Amide

Synonyms: Neuropeptide Y (1-24)

American Peptide 60-1-10   Human   MW 2657.1
$C_{116}H_{170}N_{30}O_{40}S$ Martel, JC et al, *Mol Pharmacol*, 38:494, 1990

Tyr-Pro-Ser-Lys-Pro-Asp-Asn-Pro-Gly-Glu-Asp-Ala-
Pro-Ala-Glu-Asp-Met-Ala-Arg-Tyr-Tyr-Ser-Ala-Leu-
Arg-His-Tyr-Ile-Asn-Leu-Ile-Thr-Arg-Gln-Arg-Tyr

Synonyms: Neuropeptide Y

American Peptide 60-1-13   Human   MW 4272.8 (free acid)
$C_{189}H_{284}N_{54}O_{58}S$ Deamidation of neuropeptide Y suppresses
almost all of NPY effects

Tyr-Pro-Ser-Lys-Pro-Asp-Asn-Pro-Gly-Glu-Asp-Ala-
Pro-Ala-Glu-Asp-Met-Ala-Arg-Tyr-Tyr-Ser-Ala-Leu-
Arg-His-Tyr-Ile-Asn-Leu-Ile-Thr-Arg-Gln-Arg-Tyr
Tyr-Pro-Ser-Lys-Pro-Asp-Asn-Pro-Gly-Glu-Asp-Ala-
Pro-Ala-Glu-Asp-Met-Ala-Arg-Tyr-Tyr(OMe)-Ser-Ala-
Leu-Arg-His-Tyr-Ile-Asn-Leu-Ile-Thr-Arg-Gln-Arg-
Tyr Amide

Synonyms: Neuropeptide Y, (Tyr(OMe)[21])-

American Peptide 60-1-15   Human   MW 4301.8
$C_{190}H_{287}N_{55}O_{58}S$ Martel, JC et al, *Mol Pharmacol*, 38:494, 1990

Tyr-Pro-Ser-Lys-Pro-Asp-Asn-Pro-Gly-Glu-Asp-Ala-
Pro-Ala-Glu-Asp-Met-Ala-Arg-Tyr-Tyr-Ser-Ala-Leu-
Arg-His-Tyr-Ile-Asn-Leu-Ile-Thr-Arg-Pro-Arg-Tyr
Amide

Synonyms: Neuropeptide Y (1-36), (Pro[34])-

American Peptide 60-1-18   Human   MW 4240.8
$C_{189}H_{284}N_{54}O_{56}S$ Selective Y_1 agonist

Tyr-Pro-Ser-Lys-Pro-Asp-Asn-Pro-Gly-Glu-Asp-Ala-
Pro-Ala-Glu-Asp-Met-Ala-Arg-Tyr-Tyr-Ser-Ala-Leu-
Arg-His-Tyr-Ile-Asn-Leu-Leu-Thr-Arg-Pro-Arg-Tyr
Amide

Synonyms: Neuropeptide Y, (Leu[31], Pro[34])-

American Peptide 60-1-20   Human   MW 4240.8
$C_{189}H_{284}N_{54}O_{56}S$ Selective neuropeptide YY$_1$ receptor agonist;
Fuhlendorff, J et al, *PNAS*, 87:182, 1990

Tyr-Pro-Ser-Lys-Pro-Asp-Asn-Pro-Gly-Glu-Asp-Ala-
Pro-Ala-Glu-Asp-Met-Ala-Arg-Tyr-Tyr-Ser-Ala-Leu-
Arg-His-Tyr-Ile-Asn-Leu-Ile-Trp-Arg-Gln-Arg-Tyr
Tyr-Pro-Ser-Lys-Pro-Asp-Asn-Pro-Gly-Glu-Asp-Ala-
Pro-Ala-Glu-Asp-Met-Ala-Arg-Tyr-Tyr-Ser-Ala-Leu-
Arg-His-Tyr-Ile-Asn-Leu-Ile-D-Trp-Arg-Gln-Arg-Tyr
Amide

Synonyms: Neuropeptide Y (1-36), (D-Trp[32])-

American Peptide 60-1-21   Human   MW 4356.9
$C_{196}H_{288}N_{56}O_{56}S$ Putative NPY receptor antagonist

---

Tyr-Pro-Ser-Lys-Pro-Asp-Asn-Pro-Gly-Glu-Asp-Ala-
Pro-Ala-Glu-Asp-Met-Ala-Arg-Tyr-Tyr-Ser-Ala-Leu-
Arg-His-Tyr-Ile-Asn-Leu-Ile-Thr-Arg-Gln-Arg-Tyr
Amide

Synonyms: Neuropeptide Y

American Peptide 62-1-11   Human   MW 4271.8
$C_{189}H_{285}N_{55}O_{57}S$ Plays a role in the control of blood pressure,
sexual behavior & food intake; Minth, CD et al, *PNAS*, 81:4577,
1984; Allen, J et al, *PNAS*, 84:2532, 1987

Calbiochem 05-23-2005   Human   MW 4271.7
$C_{189}H_{285}N_{55}O_{57}S$ ≥97% (HPLC); lyophilized solid; soluble in 5%
acetic acid; LD$_{50}$≤2000 mg/kg | Potent vasoconstrictor; reversibly
inhibits Ca^{2+} activated K$^+$ channels in vascular smooth muscle cells;
Xiong, Z & Cheung, DW, *Pflugers Arch*, 429:280, 1994; Wahlestedt,
C et al, *Science*, 259:528, 1993; Leibowitz, SF, *NeuroReport*,
3:1023, 1992; *Merck Index*, 12:6568

ICN 191468   Human   MW 4271.7

Tyr-Pro-Ser-Lys-Pro-Asp-Asn-Pro-Gly-Glu-Asp-Ala-
Pro-Ala-Glu-Asp-Met-Ala-Arg-Tyr-Tyr-Ser-Ala-Leu
Amide

Synonyms: Neuropeptide Y (1-24)

Sigma N 4896   Human   FW 2656.9   ≥97% (HPLC) |
Bioactive peptide; gastrointestinal peptide; weak antagonist of
NPY-induced inhibition of the vas deferens response to electrical
stimulation; Martel, JC et al, *Mol Pharmacol*, 38:494, 1990

Tyr-Pro-Ser-Lys-Pro-Asp-Asn-Pro-Gly-Glu-Asp-Ala-
Pro-Ala-Glu-Asp-Met-Ala-Arg-Tyr-Tyr-Ser-Ala-Leu-
Arg-His-Tyr-Ile-Asn-Leu-Ile-Thr-Arg-Gln-Arg-Tyr
Amide

Synonyms: Neuropeptide Y

Sigma N 5017   Human   FW 4271.7   ≥95% (HPLC) |
Bioactive peptide; gastrointestinal peptide; vasoconstrictor; inhibits
Ca^{2+}-activated K$^+$ channels in vascular smooth muscle

Tyr-Pro-Ser-Lys-Pro-Asp-Asn-Pro-Gly-Glu-Asp-Ala-
Pro-Ala-Glu-Asp-Met-Ala-Arg-Tyr-Tyr-Ser-Ala-Leu-
Arg-His-Tyr-Ile-Asn-Leu-Ile-Thr-Arg-Gln-Arg-Tyr
Tyr-Pro-Ser-Lys-Pro-Asp-Asn-Pro-Gly-Glu-Asp-Ala-
Pro-Ala-Glu-Asp-Met-Ala-Arg-Tyr-Tyr(Me)-Ser-Ala-
Leu-Arg-His-Tyr-Ile-Asn-Leu-Ile-Thr-Arg-Gln-Arg-
Tyr

Synonyms: Neuropeptide Y, (o-Me-Tyr[21])-

Sigma N 5771   Human   FW 4285.8   ≥97% (HPLC) |
Bioactive peptide; gastrointestinal peptide; analog with a high
affinity for brain receptors; Martel, J-C et al, *Mol Pharmacol*,
38:494, 1990

Tyr-Pro-Ser-Lys-Pro-Asp-Asn-Pro-Gly-Glu-Asp-Ala-
Pro-Ala-Glu-Asp-Met-Ala-Arg-Tyr-Tyr-Ser-Ala-Leu-
Arg-His-Tyr-Ile-Asn-Leu-Ile-Thr-Arg-Gln-Arg-Tyr
Tyr-Pro-Ser-Lys-Pro-Asp-Asn-Pro-Gly-Glu-Asp-Ala-
Pro-Ala-Glu-Asp-Met-Ala-Arg-Tyr-(Me)-Ser-Ala-Leu-
Arg-His-Tyr-Ile-Asn-Leu-Ile-Thr-Arg-Gln-Arg-Tyr
Amide

Synonyms: Neuropeptide Y, (Tyr(Me)[21])-

Bachem H-3302.0500   Human, rat   MW 4285.77
$C_{190}H_{287}N_{55}O_{57}S$ Store at -15°C

Tyr-Pro-Ser-Lys-Pro-Asp-Asn-Pro-Gly-Glu-Asp-Ala-
Pro-Ala-Glu-Asp-Met-Ala-Arg-Tyr-Tyr-Ser-Ala-Leu
Amide

Synonyms: Neuropeptide Y (1-24)

Bachem H-3304.0500   Human, rat   MW 2656.87
$C_{116}H_{170}N_{30}O_{40}S$ Store at -15°C

Tyr-Pro-Ser-Lys-Pro-Asp-Asn-Pro-Gly-Glu-Asp-Ala-
Pro-Ala-Glu-Asp-Met-Ala-Arg-Tyr-Tyr-Ser-Ala-Leu-
Arg-His-Tyr-Ile-Asn-Leu-Leu-Thr-Arg-Pro-Arg-Tyr
Amide

Synonyms: Neuropeptide Y, (Leu[31],Pro[34])-

Bachem H-3306.0500   Human, rat   MW 4240.73
$C_{189}H_{284}N_{54}O_{56}S$ Store at -15°C

Tyr-Pro-Ser-Lys-Pro-Asp-Asn-Pro-Gly-Glu-Asp-Ala-
Pro-Ala-Glu-Asp-Met-Ala-Arg-Tyr-Tyr-Ser-Ala-Leu-
Arg-His-Tyr-Ile-Asn-Leu-Ile-Trp-Arg-Gln-Arg-Tyr
Tyr-Pro-Ser-Lys-Pro-Asp-Asn-Pro-Gly-Glu-Asp-Ala-
Pro-Ala-Glu-Asp-Met-Ala-Arg-Tyr-Tyr-Ser-Ala-Leu-
Arg-His-Tyr-Ile-Asn-Leu-Ile-D-Trp-Arg-Gln-Arg-Tyr
Amide

Synonyms: Neuropeptide Y, (D-Trp[32])-

Bachem H-3312.0500   Human, rat   MW 4356.85
$C_{196}H_{288}N_{56}O_{56}S$ Store at -15°C

Tyr-Pro-Ser-Lys-Pro-Asp-Asn-Pro-Gly-Glu-Asp-Ala-Pro-Ala-Glu-Asp-Met-Ala-Arg-Tyr-Tyr-Ser-Ala-Leu-Arg-His-Tyr-Ile-Asn-Leu-Ile-Thr-Arg-Gln-Arg-Tyr Free Acid

Synonyms: Neuropeptide Y

Bachem H-3322.0500   Human, rat   MW 4272.73 $C_{189}H_{284}N_{54}O_{58}S$ Store at -15°C

Tyr-Pro-Ser-Lys-Pro-Asp-Asn-Pro-Gly-Glu-Asp-Ala-Pro-Ala-Glu-Asp-Met-Ala-Arg-Tyr-Tyr-Ser-Ala-Leu-Arg-His-Tyr-Ile-Asn-Leu-Ile-Thr-Arg-Gln-Arg-Tyr Amide

Synonyms: Neuropeptide Y

Bachem H-6375.0500   Human, rat   MW 4271.74 $C_{189}H_{285}N_{55}O_{57}S$ Store at -15°C

Neosystem SC115   Human, rat   MW 4271.72   Bioactive neuropeptide; Minter, CD et al, *PNAS USA*, 81:4577, 1984

Tyr-Pro-Ser-Lys-Pro-Asp-Asn-Pro-Gly-Glu-Asp-Ala-Pro-Ala-Glu-Asp-Met-Ala-Arg-Tyr-Tyr-Ser-Ala-Leu-Arg-His-Tyr-Ile-Asn-Leu-Ile-Trp-Arg-Gln-Arg-Tyr Tyr-Pro-Ser-Lys-Pro-Asp-Asn-Pro-Gly-Glu-Asp-Ala-Pro-Ala-Glu-Asp-Met-Ala-Arg-Tyr-Tyr-Ser-Ala-Leu-Arg-His-Tyr-Ile-Asn-Leu-Ile-D-Trp-Arg-Gln-Arg-Tyr Amide

Synonyms: Neuropeptide Y, (D-Trp[32])-

Neosystem SC909   Human, rat   MW 4356.83   Bioactive neuropeptide; competitive antagonist of NPY in rat hypothalamus; Balasubramaniam, et al, *J Med Chem*, 37:811-815, 1994

Tyr-Pro-Ser-Lys-Pro-Asp-Asn-Pro-Gly-Glu-Asp-Ala-Pro-Ala-Glu-Asp-Met-Ala-Arg-Tyr-Tyr-Ser-Ala-Leu-Arg-His-Tyr-Ile-Asn-Leu-Leu-Thr-Arg-Pro-Arg-Tyr Amide

Synonyms: Neuropeptide Y, (Leu[31],Pro[34])-

Neosystem SC935   Human, rat   MW 4240.71   Bioactive neuropeptide; specific Y1 receptor agonist; Fuhlendorff, J et al, *PNAS USA*, 87:182-186, 1990; Gehlert, DR et al, *Neurochem Int*, 21:45-67, 1992

Tyr-Pro-Ser-Lys-Pro-Asp-Asn-Pro-Gly-Glu-Asp-Ala-Pro-Ala-Glu-Asp-Met-Ala-Arg-Tyr-Tyr-Ser-Ala-Leu-Arg-His-Tyr-Ile-Asn-Leu-Ile-Thr-Arg-Gln-Arg-Tyr Amide

Synonyms: Neuropeptide Y

Peptides International PNP-4158-s, 4158-v   Human, rat synthetic   MW 4271.7 $C_{189}H_{285}N_{55}O_{57}S$ 95% (HPLC); lyophilized amorphous powder | Bioactive; Dumont, Y et al, *Progr Neurobiol*, 38:125, 1992; Wahlestedt, C & DJ Reis, *Ann Rev Pharmacol Toxicol*, 32:309, 1993; Minth, CD et al, *PNAS USA*, 81:4577, 1984

Tyr-Pro-Ser-Lys-Pro-Asp-Cys-Pro-Gly-Ala-Arg-Tyr-Cys-Ser-Ala-Leu-Arg-His-Tyr-Ile-Asn-Leu-Ile-Thr-Arg-Pro-Arg-Tyr

Synonyms: Neuropeptide Y, des-AA10-17-Cyclo-7/21(Cys[7,21],Pro[34])-

Neosystem SC1202   MW 3308.82   Disulfide bonds: Cys[7]-Cys[21]; bioactive neuropeptide; high-potency Y1 receptor agonist; agonistic properties with an affinity comparable to that of the native NPY molecule when tested for its ability to inhibit norepinephrine-stimulated cAMP release in SK-N-MC human neuroblastoma cells; causes an increase in blood pressure in anesthetized rats; in two central nervous system models of Y1 receptor function, stimulation of feeding & anxiolytic activity, this analog was inactive, which suggests the presence of a new subclass of receptors; Kirby, DA et al, *J Med Chem*, 38:4579-4586, 1995

Tyr-Pro-Trp

Bachem H-6045.0050   MW 464.52 $C_{25}H_{28}N_4O_5$ Store at -15°C

Tyr-Pro-Trp-Gly Amide

Synonyms: Melanocyte Stimulating Hormone Release Inhibiting Factor I, (Tyr-W)-; Melanocyte Stimulating Hormone Release Inhibiting Factor, (Tyr[0],Trp[2])-

American Peptide 60-0-21   MW 520.6 $C_{27}H_{32}N_6O_5$ Naturally occurring neuropeptide displays high selectivity for µ-opioid receptors; Gergen, KA et al, *Eur J Pharmacol*, 316:33, 1996

Bachem H-8825.0050   MW 529.59 $C_{27}H_{32}N_6O_5$ Store at -15°C

Tyr-Pro-Trp-Gly Amide Trifluoroacetate Salt

Synonyms: Melanocyte Stimulating Hormone Release Inhibiting Factor I, Tyr-(Trp[2])-

Sigma M 8009   FW 520.6 (free base) $C_{27}H_{32}N_6O_5$ ≥97% (HPLC) | Bioactive peptide; originally isolated from human frontal cortex & from bovine hypothalamus; mixed agonist/antagonist at opiate receptors; Erchegyl, J et al, *Peptides*, 13:623, 1992; Hackler, L et al, *Neuropeptides*, 24:159, 1993

Tyr-Pro-Trp-Phe Amide

Synonyms: Endomorphin I; Endogenous µ-Opiate Receptor Selective Agonist

American Peptide 23-1-10   MW 610.7 $C_{34}H_{38}N_6O_5$ Potent & selective endogenous agonist of the µ-opiate receptor; Zadina, J et al, *Nature*, 386:499, 1997

Bachem H-4002.0005   Store at -15°C

Neosystem SC1243   MW 610.71   Bioactive; opioid peptide; high affinity (K_i = 360pM) & selectivity for the mu receptor (4,000- & 15,000-fold preference over the δ- & κ-receptors); more effective than the mu-selective analogue DAMGO *in vitro*; produces potent & prolonged analgesia in mice; Zadina, JE et al, *Nature*, 386:499-502, 1997

Sigma E 3273   FW 610.7 $C_{34}H_{38}N_6O_5$ Bioactive peptide; opioid peptide; potent, selective µ-opioid receptor agonist; Zadina, JE et al, *Nature*, 386:449, 1997

Peptides International PEM-3603-PI   Human, bovine synthetic   MW 610.71 $C_{34}H_{38}N_6O_5$ >98% (HPLC); white powder; bulk | Bioactive; Zadina, JE et al, *Nature*, 386:499, 1997; Champion, HC et al, *BBRC*, 235:567, 1997; Stone, LS et al, *Neuroreport*, 8:3131, 1997; Hackler, L et al, *Peptides*, 18:1635, 1997

Peptides International PEM-4333-v   Human, bovine synthetic   MW 610.71 $C_{34}H_{38}N_6O_5$ >99% (HPLC); lyophilized amorphous powder | Bioactive; Zadina, JE et al, *Nature*, 386:499, 1997; Champion, HC et al, *BBRC*, 235:567, 1997; Stone, LS et al, *Neuroreport*, 8:3131, 1997; Hackler, L et al, *Peptides*, 18:1635, 1997

Tyr-Pro-Trp-Phe Amide Trifluoroacetate Salt

Synonyms: Endomorphin I

Calbiochem 324730   Synthetic   MW 610.7 $C_{34}H_{38}N_6O_5$ ≥98% (HPLC); lyophilized solid; soluble in water; LD$_{50}$≤2000 mg/kg | Mammalian neuropeptide; acts as a potent & specific agonist for the µ-opiate receptor; exhibits high affinity & selectivity for the µ-receptor (4000- & 15,000-fold preference for the µ-receptor over the δ- & κ-receptors, respectively); more potent than the µ-selective analog DAMGO (Tyr-D-Ala-Gly-*N*-Me-Phe-glycinol) *in vitro*; inhibits high-threshold Ca^{2+} channel currents in NGMO-251 cells; Higashida, H et al, *J Physiol*, 507:71, 1998; Zadina, JE et al, *Nature*, 386:499, 1997

Tyr-Pro-Trp-Thr-Gln-Arg-Phe

Synonyms: Hemorphin-7

Neosystem SC1231   MW 997.12   Bioactive; opioid peptide; endogenous; derived by enzymatic degradation of the blood protein hemoglobin; mu-receptor affinity reported for hemorphin-7; may be involved indirectly (via β-endorphin) or directly in the central effects such as analgesia & euphoria observed during & after exercise; may stimulate a release of both growth hormone & prolactin in the male rat; Glämsta, EL et al, *BBRC*, 184:1060-1066, 1992; Glämsta, EL et al, *Regul Peptides*, 49:9-18, 1993; Nyberg, F et al, *Biopolymers*, 43:147-156, 1997

Tyr-Pro-Tyr-Asp-Val-Pro-Asp-Tyr-Ala

Synonyms: Influenza Hemagglutinin Peptide; Hemagglutinin Tag Peptide; Hemagglutinin I (99-107)

Sigma I 2149   FW 1102.2   ≥97% (HPLC) | Bioactive peptide; sequence used in recombinant HA epitope "tagged" proteins; epitope recognized by anti-HA monoclonal antibodies; Niman, H et al, *Proc Natl Acad Sci Usa*, 80:4949, 1983

Tyr-Pro-Val-Pro Amide

Synonyms: Morphiceptin, (Val³)-; Casomorphin (1-4), (Val³)-β-

ICN 152957

American Peptide 24-3-36   Bovine   MW 473.6 $C_{24}H_{35}N_5O_5SCl$

Bachem H-2365.0005   Bovine   MW 473.57 $C_{24}H_{35}N_5O_5$ Store at -15°C

Tyr-Ser
Z-Tyr-Ser

Bachem C-2780.0250   MW 402.4 $C_{20}H_{22}N_2O_7$ Store at -15°C

Tyr-Ser
Tyr-Ser-OMe Hydrochloride

Bachem G-3680.0250   MW 318.76 $C_{13}H_{18}N_2O_5 \cdot HCl$ Store at -15°C

Tyr-Ser
N-CBZ-Tyr-Ser-OMe

Sigma C 0908   FW 416.4 $C_{21}H_{24}N_2O_7$

Tyr-Ser Amide Hydrochloride

Bachem G-3460.0250   MW 303.75 $C_{12}H_{17}N_3O_4 \cdot HCl$ Store at -15°C

Tyr-Ser-Ala-Asn-Ser-Asn-Pro-Ala-Met-Ala-Pro-Arg-Glu-Arg-Lys-Ala-Gly-Cys-Lys-Asn-Phe-Phe-Trp-Lys-Thr-Phe-Thr-Ser-Cys

Synonyms: Somatostatin (1-28), (Tyr⁰)-; Somatostatin 28, (Tyr⁰)-

American Peptide 68-1-52   MW 3311.8 $C_{146}H_{216}N_{42}O_{41}S_3$ Disulfide bonds: Cys¹⁸-Cys²⁹

Bachem H-4990.0500   MW 3311.78 $C_{146}H_{216}N_{42}O_{41}S_3$ Store at -15°C

ICN 153000   Disulfide bond:Cys¹⁷-Cys²⁸

Neosystem SC096   MW 3311.75   Disulfide bonds: Cys¹⁷-Cys²⁸; bioactive; hypothalamic releasing hormone

Sigma S 6260   FW 3311.7   ≥97% (HPLC); disulfide bonds: 17-28 | Bioactive peptide

Tyr-Ser-Arg-Ala-Leu-Ser-Arg-Gln-Leu-Ser-Ser-Cys
Tyr-Ser-Arg-Ala-Leu-pSer-Arg-Gln-Leu-Ser-Ser-Cys

Synonyms: Phospho-HSP27, (Ser⁷⁸)-

Upstate 12-398   MW 1450   Frozen solution | Immunizing peptide

Tyr-Ser-Arg-Val-Ser-Arg-Arg-Ser-Arg

Synonyms: Insulin-Like Growth Factor II (32-40), (Tyr³²)-

American Peptide 50-1-32   MW 1166.4 $C_{47}H_{83}N_{21}O_{14}$

Neosystem SC682   MW 1166.30   Bioactive; synthetic growth factor-related peptide

ICN 154469   Human synthetic   MW 1166.4

Tyr-Ser-Cys-Asn-Thr-Ala-Thr-Cys-Val-Thr-His-Arg-Leu-Ala-Gly-Leu-Leu-Ser-Arg-Ser-Gly-Gly-Val-Val-Lys-Asp-Asn-Phe-Val-Pro-Thr-Asn-Val-Gly-Ser-Glu-Ala-Phe Amide

Synonyms: Calcitonin Gene Related Peptide, (Tyr⁰)-

American Peptide 22-5-11   Rat   MW 3970.0 $C_{171}H_{271}N_{51}O_{54}S_2$ Disulfide bonds: Cys²-Cys⁷

Tyr-Ser-Cys-His-Phe-Gly-Pro-Leu-Thr-Trp-Val-Cys-Lys

Synonyms: Erythropoietin Mimetic Peptide Sequence 20

Bachem H-4344.0001   Store at -15°C

Tyr-Ser-Gln-Glu-Pro-Pro-Ile-Ser-Leu-Asp-Leu-Thr-Phe-His-Leu-Leu-Arg-Glu-Val-Leu-Glu-Met-Thr-Lys-Ala-Asp-Gln-Leu-Ala-Gln-Gln-Ala-His-Ser-Asn-Arg-Lys-Leu-Leu-Asp-Ile-Ala Amide

Synonyms: Corticotropin Releasing Factor, (Tyr⁰)-

American Peptide 34-2-22   Ovine   MW 4833.6 $C_{214}H_{348}N_{60}O_{65}S$

Bachem H-2460.0500   Ovine   Store at -15°C

Tyr-Ser-Gln-Glu-Pro-Pro-Ile-Ser-Leu-Asp-Leu-Thr-Phe-His-Leu-Leu-Arg-Glu-Val-Leu-Glu-Met-Thr-Lys-Ala-Asp-Gln-Leu-Ala-Gln-Gln-Ala-His–Ser-Asn-Arg-Lys-Leu-Leu-Asp-Ile-Ala Amide

Synonyms: Corticotropin Releasing Factor, (Tyr)-

ICN 153077   Sheep

Tyr-Ser-Gln-Glu-Pro-Pro-Ile-Ser-Leu-Asp-Leu-Thr-Phe-His-Leu-Leu-Arg-Glu-Val-Leu-Glu-Met-Thr-Lys-Ala-Asp-Gln-Leu-Ala-Gln-Gln-Ala-His-Ser-Asn-Arg-Lys-Leu-Leu-Asp-Ile-Ala Amide

Synonyms: Corticotropin Releasing Factor, (Tyr)-

Sigma C 0922   Sheep   FW 4833.5   ≥97% (HPLC) | Bioactive peptide; used to prepare iodinated CRF for receptor studies & immunoassay

Tyr-Ser-Glu-Glu-Pro-Ile-Ser-Leu-Asp-Leu-Thr-Phe-His-Leu-Leu-Arg-Glu-Val-Leu-Glu-Met-Ala-Arg-Ala-Glu-Gln-Leu-Ala-Gln-Gln-Ala-His-Ser-Asn-Arg-Lys-Leu-Met-Glu-Ile-Ile Amide

Synonyms: Corticotropin Releasing Factor, (Tyr)-

Sigma C 6426   Human, rat   FW 4920.7   ≥97% (HPLC) | Bioactive peptide; used to prepare iodinated CRF for receptor studies & immunoassay

Sigma C 7782   Human, rat   FW 4920.7   ≥97% (HPLC) | Bioactive peptide; used to prepare iodinated CRF for receptor studies & immunoassay

Tyr-Ser-Glu-Glu-Pro-Pro-Ile-Ser-Leu-Asp-Leu-Thr-Phe-His-Leu-Leu-Arg-Glu-Val-Leu-Glu-Met-Ala-Arg-Ala-Glu-Gln-Leu-Ala-Gln-Gln-Ala-His-Ser-Asn-Arg-Lys-Leu-Met-Glu-Ile-Ile
Tyr-Ser-Glu-Glu-Pro-Pro-Ile-Ser-Leu-Asp-Leu-Thr-Phe-His-Leu-Leu-Arg-Glu-Val-Leu-Glu-Met-Ala-Arg-Ala-Glu-Gln-Leu-Ala-Gln-Gln-Ala-His-Ser-Asn-Arg-Lys-Leu-Met-Glu-Ile-Ile Amide

Synonyms: Corticotropin Releasing Factor, (Tyr)-

Peptides International PCR-4141-s Human, rat synthetic MW 4920.7 $C_{217}H_{353}N_{61}O_{65}S_2$ >97% (HPLC); lyophilized amorphous powder | Bioactive; for radioimmunoassay

Tyr-Ser-Glu-Glu-Pro-Pro-Ile-Ser-Leu-Asp-Leu-Thr-Phe-His-Leu-Leu-Arg-Glu-Val-Leu-Glu-Met-Ala-Arg-Ala-Glu-Gln-Leu-Ala-Gln-Gln-Ala-His-Ser-Asn-Arg-Lys-Leu-Met-Glu-Ile-Ile Amide

Synonyms: Corticotropin Releasing Factor, (Tyr[0])-

American Peptide 34-1-22 Human, rat MW 4920.7 $C_{217}H_{353}N_{61}O_{65}S_2$
Bachem H-2455.0500 Human, rat MW 4920.69 $C_{217}H_{353}N_{61}O_{65}S_2$ Store at -15°C

Tyr-Ser-Glu-Glu-Pro-Pro-Ile-Ser-Leu-Asp-Leu-Thr-Phe-His-Leu-Leu-Arg-Glu-Val-Leu-Glu-Met-Ala-Arg-Ala-Glu-Gln-Leu-Ala-Gln-Gln-Ala-His-Ser-Asn-Arg-Lys-Leu-Met-Leu-Ile-Ile Amide

Synonyms: Corticotropin Releasing Factor, (Tyr)-

ICN 153076 Human, rat

Tyr-Ser-Glu-Glu-Pro-Pro-Ile-Ser-Leu-Asp-Leu-Thr-Phe-His-Leu-Leu-Arg-Glu-Val-Leu-Glu-Met-Ala-Arg-Ala-Glu-Gln-Leu-Ala-Gln-Gln-Ala-His-Ser-Asn-Arg-Lys-Leu-Met-Glu-Ile-Ile Amide

Synonyms: Corticotropin Releasing Factor, (Tyr[0])-

Neosystem SC318 Human, rat MW 4920.67 Bioactive; hypothalamic releasing hormone; CRF/urocortin peptide; for radioimmunoassay

Tyr-Ser-Gly-Phe-Leu-Thr
Tyr-D-Ser-Gly-Phe-Leu-Thr

Synonyms: δ-Receptor Peptide; Enkephalin-Thr, (D-Ser[2])-Leu-; DSLET; Enkephalin-Thr, (D-Ser[2],Leu[5])-; δ-Opiate Receptor Peptide

American Peptide 30-0-64 MW 686.8 $C_{33}H_{46}N_6O_{10}$ δ-opioid receptor agonist
Bachem H-2770.0005 MW 686.76 $C_{33}H_{46}N_6O_{10}$ Store at -15°C
ICN 152816 Very high specificity for δ-opiate receptors; Gacel, G et al, *FEBS Lett*, 118:245, 1980; Fournie-Zaluski, MC et al, *Mol Pharmac*, 20:484, 1980; Gacel, G et al, *J Med Chem*, 24:1119, 1981; Chesselet, MF et al, *Nature*, 291:320, 1981; David, M et al, *Eur J Pharmac*, 78:385, 1982
Neosystem SC360 MW 686.76 Bioactive; opioid peptide; δ-receptor peptide; Gacel, G et al, *FEBS Lett*, 118:245, 1980

Tyr-Ser-Gly-Phe-Leu-Thr
Tyr-D-Ser(OtBu)-Gly-Phe-Leu-Thr

Synonyms: Enkephalin-Thr, (D-Ser(OtBu)[2])-Leu-

Neosystem SC361 MW 742.87 Bioactive; opioid peptide

Tyr-Ser-Gly-Phe-Leu-Thr
Tyr-D-Ser(OtBu)-Gly-Phe-Leu-Thr(OtBu)

Synonyms: Enkephalin-Thr(OtBu), (D-Ser(OtBu)[2])-Leu-

Neosystem SC362 MW 799.40 Bioactive; opioid peptide; high affinity for δ-opioid receptors; Gacel, G et al, *J Med Chem*, 31:1891-1897, 1988

Tyr-Ser-Gly-Phe-Leu-Thr
Tyr-D-Ser-Gly-Phe-Leu-Thr Acetate Salt

Synonyms: Enkephalin-Thr, ((D-Ser[2])-Leu)-

Sigma E 7388 FW 686.8 (free base) $C_{33}H_{46}N_6O_{10}$ ≥97% (HPLC) | Bioactive peptide; highly selective ligand for δ-opioid receptors; selective δ₂-agonist; Gacel, G et al, *FEBS Lett*, 118:245, 1980; Gacel, G et al, *J Med Chem*, 24:1119, 1981; Chesselet, MF et al, *Nature*, 291:320, 1981

Tyr-Ser-Lys

Bachem H-5200.0050 MW 396.44 $C_{18}H_{28}N_4O_6$ Store at -15°C

Tyr-Ser-Phe-Glu-Asp-Leu-Tyr-Arg-Arg-Ser-Phe-Glu-Asp-Leu-Tyr-Arg-Arg
Tyr-Ser-Phe-Glu-Asp-Leu-Tyr-Arg-Arg-(Gly-)6Tyr-Ser-Phe-Glu-Asp-Leu-Tyr-Arg-Arg

Bachem H-2258.0500 Store at -15°C

Tyr-Ser-Phe-Lys-Asp-Met-Gln-Leu-Gly-Arg

Synonyms: C5a (65-74), (Tyr[65],Phe[67])-

Bachem H-3462.0001 Human MW 1244.44 $C_{55}H_{85}N_{15}O_{16}S$ Store at -15°C

Tyr-Ser-Phe-Pro-Trp-Met-Glu-Ser-Asp-Val-Thr

Synonyms: Prepro-Thyrotropin Releasing Hormone (160-169), (Tyr[0])-

Neosystem SC442 MW 1361.48 Bioactive; thyrotropin releasing hormone-related peptide

Tyr-Ser-Phe-Val-His-His-Gly-Phe-Phe-Asn-Phe-Arg-Val-Ser-Trp-Arg-Glu-Met-Leu-Ala

Synonyms: Cyclin-Dependent Kinase II Inhibitor

Alexis 153-012 MW 2530.9 $C_{121}H_{164}N_{32}O_{27}S$ White lyophilized powder | Colas, P et al, *Nature*, 380:548, 1996
American Peptide 86-1-56 MW 2530.9 $C_{121}H_{164}N_{32}O_{27}S$ Binds an epitope on the cyclin-dependent kinase 2 surface with Kd~38 nM; Colas, P et al, *Nature*, 380:548, 1996
Bachem H-3592.0001 MW 2530.89 $C_{121}H_{164}N_{32}O_{27}S$ Store at -15°C

Tyr-Ser-Pro-Lys-Met-Val-Gln-Gly-Ser-Gly-Cys-Phe-Gly-Arg-Lys-Met-Asp-Arg-Ile-Ser-Ser-Ser-Ser-Gly-Leu-Gly-Cys-Lys-Val-Leu-Arg-Arg-His

Synonyms: Natriuretic Peptide (1-32), Brain (Tyr[0])-; Natriuretic Peptide 32, B-Type; Natriuretic Peptide 32, Brain

American Peptide 14-1-91 Human MW 3627.3 $C_{152}H_{253}N_{51}O_{44}S_4$ Disulfide bonds: :Cys[11]-Cys[27]
Peptides International PBN-4230-v Human synthetic MW 3627.3 $C_{152}H_{253}N_{51}O_{44}S_4$ >97% (HPLC); lyophilized amorphous powder | Bioactive; Disulfide bonds: Cys[10]-Cys[26]; for radioimmunoassay

Tyr-Ser-Ser-Cys-Phe-Gly-Gly-Arg-Ile-Asp-Arg-Ile-Gly-Ala-Gln-Ser-Gly-Leu-Gly-Cys-Asn-Ser

Synonyms: Atriopeptin I, (Tyr[0])-

Neosystem SC411 MW 2246.45 Disulfide bonds: Cys[3]-Cys[19]; bioactive; ANF-related peptide

Tyr-Ser-Ser-Cys-Phe-Gly-Gly-Arg-Ile-Asp-Arg-Ile-Gly-Ala-Gln-Ser-Gly-Leu-Gly-Cys-Asn-Ser-Phe-Arg

Synonyms: Natriuretic Peptide (5-27), Atrial (Tyr[126])-; Atriopeptin II, (Tyro)-; Natriuretic Peptide (126-149), Atrial (Tyr[126])-; Atriopeptin II, (Tyr[0])

American Peptide 14-1-52 Rat MW 2549.9 $C_{107}H_{165}N_{35}O_{34}S_2$ Disulfide bonds: :Cys[4]-Cys[20]
Bachem H-2120.0500 Rat MW 2549.83 $C_{107}H_{165}N_{35}O_{34}S_2$ Store at -15°C

Tyr-Ser-Thr-Glu-Pro-Gln-Tyr-Gln-Pro-Gly-Glu-Asn-Leu Trifluoroacetate Salt

Synonyms: p60c-*src* Peptide (521-533)

ICN 195923   MW 1463.5 (free acid) $C_{62}H_{94}N_{16}O_{25}$ ≥98% |
C-terminal phosphoregulatory peptide involved in regulation
ofkinase activity

Tyr-Thr Amide Hydrochloride
Bachem G-3465.0250   Store at -15°C

Tyr-Thr-Arg-Leu-Arg-Lys-Gln-Met-Ala-Val-Lys-Lys-Tyr-Leu-Asn-Ser-Ile-Leu-Asn Amide
Synonyms: Vasoactive Intestinal Peptide (10-28)

Neosystem SC084   MW 2338.84   Bioactive; brain/gut
peptide; VIP/PHI/secretin/helodermin peptide; Turner, JT et al,
Peptides, 7:849-854, 1986

Bachem H-5205.0001   Human, bovine, porcine, rat   Store at
-15°C

Tyr-Thr-Arg-Leu-Arg-Lys-Gln-Met-Ala-Val-Lys-Lys-Tyr-Leu-Asn-Ser-Ile-Leu-Asn Amide Acetate Salt
Synonyms: Vasoactive Intestinal Peptide (10-28)

Sigma V 5380   Human, porcine, rat   FW 2338.8   ≥97%
(HPLC) | Bioactive peptide; gastrointestinal peptide; Westendorf,
JM et al, *Endocrinology*, 112:550, 1983

Tyr-Thr-Arg-Leu-Arg-Lys-Gln-Met-Ala-Val-Lys-Lys-Tyr-Leu-Asn-Ser-Ile-Leu-Asn Amide
Synonyms: Vasoactive Intestinal Peptide (10-28)

American Peptide 48-1-24   Human, porcine, rat, ovine   MW
2339.1 $C_{105}H_{180}N_{32}O_{26}S$ Taylor, D et al, *6th Am Peptide Symp*,
Washington DC, 1979

ICN 152898   Porcine

Tyr-Thr-Asp-Glu-Cys-Glu-Leu-Cys-Ile-Asn-Val-Ala-Cys-Thr-Gly-Cys
Synonyms: Uroguanylin, Tyr-

Bachem H-4148.0500   Rat   Store at -15°C | Disulfide
bonds: Cys⁴-Cys¹², Cys⁷-Cys¹⁵

Tyr-Thr-Gly-Leu-Phe-Thr
Tyr-D-Thr-Gly-Leu-Phe-Thr Acetate Salt
ICN 152819

Tyr-Thr-Gly-Phe-Leu-Thr
Tyr-D-Thr-Gly-Phe-Leu-Thr
Synonyms: Enkephalin-Thr, ((D-Thr²)-Leu)-

American Peptide 30-0-66   MW 700.9 $C_{34}H_{48}N_6O_{10}$

Tyr-Thr-Gly-Phe-Leu-Thr
3,5-Diiodo-Tyr-D-Thr-Gly-Phe-Leu-Thr
Synonyms: Enkephalin-Thr, (3,5-Diiodo-Tyr¹,D-Thr²)-Leu-; DTLET,
(3,5-Diiodo-Tyr¹)-

Bachem H-2615.0005   MW 952.58 $C_{34}H_{46}I_2N_6O_{10}$ Store at
-15°C

Tyr-Thr-Gly-Phe-Leu-Thr
Tyr-D-Thr-Gly-Phe-Leu-Thr
Synonyms: Enkephalin-Thr, (D-Thr²)-Leu-; DTLET; Deltakephalin

Bachem H-2775.0005   MW 700.79 $C_{34}H_{48}N_6O_{10}$ Store at
-15°C

Tyr-Thr-Gly-Phe-Leu-Thr
3,5-Diiodo-Tyr-D-Thr-Gly-Phe-Leu-Thr
Synonyms: Enkephalin-Thr, (3,5-Diiodo-Tyr¹,D-Thr²,Leu⁵)-

ICN 152818   A δ-receptor-specific enkephalin analog

Tyr-Thr-Gly-Phe-Leu-Thr
Tyr-D-Thr-Gly-Phe-Leu-Thr
Synonyms: Enkephalin-Thr, (D-Thr²)-Leu-

Neosystem SC035   MW 700.79   Bioactive; opioid peptide;
Delay-Goyet, P et al, *FEBS Lett*, 183:499, 1985; Zajac, JM et al,
BBRC, 111:390-397, 1983

Tyr-Thr-Gly-Phe-Thz
Tyr-D-Thr-Gly-Phe-Thz Amide
Synonyms: Enkephalinamide, (D-Thr²,Thz⁵)-

ICN 152839   MW 600.7 $C_{28}H_{36}N_6O_7S_2$ Yamashiro, D et al,
BBRC, 78:1124, 1977

Tyr-Thr-Lys
Bachem H-5210.0050   Store at -15°C

Tyr-Thr-Ser-Leu-Ile-His-Ser-Leu-Ile-Glu-Glu-Ser-Gln-Asn-Gln-Gln-Glu-Lys-Asn-Glu-Gln-Glu-Leu-Leu-Glu-Leu-Asp-Lys-Trp-Ala-Ser-Leu-Trp-Asn-Trp-Phe
Ac-Tyr-Thr-Ser-Leu-Ile-His-Ser-Leu-Ile-Glu-Glu-Ser-Gln-Asn-Gln-Gln-Glu-Lys-Asn-Glu-Gln-Glu-Leu-Leu-Glu-Leu-Asp-Lys-Trp-Ala-Ser-Leu-Trp-Asn-Trp-Phe Amide
Synonyms: T20; HIV I gp41 (643-678)

American Peptide 72-3-31   Synthetic   MW 4492
$C_{204}H_{301}N_{51}O_{64}$ Strong inhibitor of HIV-1 viral mediated cell-to-cell
fusion with EC_{50} values of 1 µg/mL; interacts specifically with the
tetrameric form of T21; Lawless, MK et al, *Biochemistry*, 35:13697,
1996

Tyr-Thr-Tyr-Glu-Ala-Tyr-Ala-Arg-Tyr-Pro-Glu-Glu-Leu-Glu-Ala
Synonyms: NEF (221-235); SIVmac251 Peptide

Neosystem SC622   MW 1867.98   SIVmac251 peptide from
subtype NEF

Tyr-Trp
Bachem G-3470.0250   MW 367.41 $C_{20}H_{21}N_3O_4$ Store at
-15°C

Tyr-Trp Amide
Bachem G-3475.0250   MW 366.42 $C_{20}H_{22}N_4O_3$ Store at
-15°C

Tyr-Trp-Ala-Trp-Phe
Tyr-D-Trp-Ala-Trp-D-Phe Amide
Synonyms: Growth Hormone Releasing Factor; Momany Peptide

Sigma T 5517   FW 770.9 $C_{43}H_{46}N_8O_6$ ≥95% (HPLC) |
Bioactive peptide; Momany, FA et al, *Endocrinology*, 108:31, 1981

Tyr-Trp-Gly
D-Tyr-Trp-Gly
Bachem H-7810.0050   MW 424.46 $C_{22}H_{24}N_4O_5$ Store at
-15°C

Tyr-Trp-Gly-Phe-Met
Tyr-D-Trp-Gly-Phe-Met Amide
Synonyms: Enkephalin, ((D-Trp²)-Met)-; Enkephalinamide, (D-Trp²,Met⁵)-

American Peptide 30-0-30   MW 701.9 $C_{36}H_{43}N_7O_6S$ A
growth hormone releasing peptide as well as an enkephalin-related
peptide

ICN 152828   A growth hormone-releasing & enkephalin-related
peptide; Bowers, CY et al, Molecular Endocrinology, I MacIntyre, ed,
Elsevier/North Holland, Amsterdam, p 287, 1977; Felix, AM, et al,
Ann Rep Med Chem, 20:185, 1985

Tyr-Trp-Gly-Phe-Met
Tyr-D-Trp-Gly-Phe-Met Amide Acetate Salt
Synonyms: Enkephalin, (D-Trp[2], Met[5])-

ICN 195656   MW 762.8 $C_{36}H_{42}N_6O_7S \cdot C_2H_4O_2$ ≥97%

Tyr-Trp-Gly-Phe-Met
Tyr-D-Trp-Gly-Phe-Met Amide Acetate Salt Free Base
Synonyms: Enkephalinamide, ((D-Trp[2])-Met)-

Sigma E 3007   FW 701.8 $C_{36}H_{43}N_7O_6S$ ≥95% (HPLC) | Bioactive peptide; growth hormone releasing peptide as well as an enkephalin-related peptide; Felix, AM et al, *Ann Rep Med Chem*, 20:185, 1985; Bowers, CY et al, *Molecular Endocrinology*, 287, Elsevier, North Holland, Amsterdam, 1977, MacIntyre, I, ed.

Tyr-Trp-His-Gly-Arg-Asp-Asn-Arg-Thr-Ile-Ile-Ser-Leu-Asn-Lys-Tyr-Tyr-Asn-Leu-Thr
Synonyms: GP140 (291-310); SIVmac251 Peptide

Neosystem SC737   MW 2527.82   SIVmac251 peptide from subtype GP140

Tyr-Trp-Lys Amide Dihydrochloride
Bachem H-9515.0050   Store at -15°C

Tyr-Trp-Lys-Val-Cys-Phe
3-Mercaptopropionyl-Tyr-D-Trp-Lys-Val-Cys-Phe Amide
Bachem H-8460.0001   Store at -15°C | Disulfide bonds: Cys[1]-Cys[6]

Tyr-Trp-Lys-Val-Cys-Phe
3-Mercaptopropionyl-Tyr-D-Trp-Lys-Val-Cys-*p*-Chloro-D-Phe Amide
Bachem H-9505.0001   Store at -15°C | Disulfide bonds: Cys[1]-Cys[6]

Tyr-Tyr
Z-Tyr-Tyr
Bachem C-2790.0250   MW 478.5 $C_{26}H_{26}N_2O_7$ Store at -15°C

Tyr-Tyr
Ac-Tyr-Tyr
Bachem G-1115.0250   MW 386.41 $C_{20}H_{22}N_2O_6$ Store at -15°C

Tyr-Tyr
Bachem G-3480.0250   MW 344.37 $C_{18}H_{20}N_2O_5$ Store at -15°C

ICN 103196   MW 344.4 $C_{18}H_{20}N_2O_5$ Crystalline

Tyr-Tyr
Deamino-Tyr-Tyr-Hexyl Ester
Synonyms: Pseudopolyamino Acid; DTH

Sigma D 3667   FW 413.5 $C_{24}H_{31}NO_5$ Differ from conventional polyamino acids:dipeptide 'monomers' are polymerized through their side chains with non-amide linkages; pseudopolyamino acids have the low toxicity associated with standard polyamino acids while exhibiting improved properties for biotechnical applications; Kohn, J and Langer, R, *J Am Chem Soc*, 109:817, 1987; Pulapura, S and Kohn, J, *Polymer Preprints*, 31:233, 1990

Tyr-Tyr
Deamino-Tyr-Tyr-OEt
Synonyms: Pseudoamino Acid; DTE

Sigma D 8538   FW 357.4 $C_{20}H_{23}NO_5$ Differ from conventional polyamino acids:dipeptide 'monomers' are polymerized through their side chains with non-amide linkages; pseudopolyamino acids have the low toxicity associated with standard polyamino acids while exhibiting improved properties for biotechnical applications; Kohn, J and Langer, R, *J Am Chem Soc*, 109:817, 1987; Pulapura, S and Kohn, J, *Polymer Preprints*, 31:233, 1990

Tyr-Tyr
Poly(Deamino-Tyr-Tyr-Carbonate-Hexyl Ester)
Synonyms: Pseudopolyamino Acid; Poly(DTH Carbonate)

Sigma P 7810   MW 100-200k   Differs from conventional polyamino acids:dipeptide 'monomers' are polymerized through their side chains with non-amide linkages; pseudopolyamino acids have the low toxicity associated with standard polyamino acids while exhibiting improved properties for biotechnical applications; Kohn, J and Langer, R, *J Am Chem Soc*, 109:817, 1987; Pulapura, S and Kohn, J, *Polymer Preprints*, 31:233, 1990

Tyr-Tyr
Sigma T 5504   FW 344.4 $C_{18}H_{20}N_2O_5$

Tyr-Tyr Amide Hydrochloride
Bachem G-3485.0250   MW 379.84 $C_{18}H_{21}N_3O_4 \cdot HCl$ Store at -15°C

Tyr-Tyr-Leu
Bachem H-5715.0050   MW 457.53 $C_{24}H_{31}N_3O_6$ Store at -15°C

Tyr-Tyr-Leu Acetate Salt
Sigma T 6775   FW 457.5 (free base) $C_{24}H_{31}N_3O_6$

Tyr-Tyr-Leu Amide
Bachem H-6040.0050   Store at -15°C

Tyr-Tyr-Phe
Bachem H-5220.0050   MW 491.54 $C_{27}H_{29}N_3O_6$ Store at -15°C

Tyr-Tyr-Phe Acetate Salt
Sigma T 6900   FW 491.5 (free base) $C_{27}H_{29}N_3O_6$

Tyr-Tyr-Phe Amide
Bachem H-5225.0050   MW 490.56 $C_{27}H_{30}N_4O_5$ Store at -15°C

Tyr-Tyr-Ser-Ala-Arg-Arg-His-Arg-Ile-Leu-Asp-Met-Tyr-Leu-Glu
Synonyms: NEF (133-147); SIVmac251 Peptide

Neosystem SC616   MW 1986.27   SIVmac251 peptide from subtype NEF

Tyr-Tyr-Tyr
Bachem H-5230.0250   MW 507.54 $C_{27}H_{29}N_3O_7$ Store at -15°C

Tyr-Tyr-Tyr
Tyr-Tyr-Tyr-OMe
Synonyms: Tri-Tyrosine

Bachem H-5235.0250   MW 521.57 $C_{28}H_{31}N_3O_7$ Store at -15°C

Sigma T 1507   FW 521.6 $C_{28}H_{31}N_3O_7$

Tyr-Tyr-Tyr

Synonyms: Tri-Tyrosine

Sigma T 2007   FW 507.5 $C_{27}H_{29}N_3O_7$

Tyr-Tyr-Tyr-Ile-Glu
Ac-Tyr(PO₃H₂)-Tyr(PO₃H₂)-Tyr(PO₃H₂)-Ile-Glu

Synonyms: *src* SH2 Domain Inhibitor Peptide

Bachem H-2532.0001   Store at -15°C

Calbiochem 567815   MW 1031.8 $C_{40}H_{52}N_5O_{21}P_3$ ≥97% (HPLC); solid; soluble in 0.1% TFA | Most active compound screened in a competitive ELISA assay to detect inhibitors of SH3-SH2 domain binding to the epidermal growth factor receptor (EGFR); Gilmer, T et al, *J Biol Chem*, 269:31711, 1994

Tyr-Tyr-Tyr-Tyr-Tyr-Tyr

Synonyms: Hexa-L-Tyrosine

American Peptide 87-0-60   MW 997.1 $C_{54}H_{56}N_6O_{13}$

ICN 158338   MW 997.1 $C_{54}H_{56}N_6O_{13}$ 90%

Sigma T 1780   FW 997.1 $C_{54}H_{56}N_6O_{13}$

Tyr-Val
Z-Tyr-Val

Bachem C-2800.0250   MW 414.46 $C_{22}H_{26}N_2O_6$ Store at -15°C

Tyr-Val

Bachem G-3490.0250   MW 280.32 $C_{14}H_{20}N_2O_4$ Store at -15°C

Tyr-Val
D-Tyr-Val Amide

Bachem G-3495.0050   MW 279.34 $C_{14}H_{21}N_3O_3$ Store at -15°C

Tyr-Val
Bz-Tyr-Val Amide

Bachem G-4160.0250   MW 383.45 $C_{21}H_{25}N_3O_4$ Store at -15°C

Tyr-Val

ICN 103197   MW 280.3 $C_{14}H_{20}N_2O_4$ Crystalline

Tyr-Val
D-Tyr-Val Amide Hydrochloride

Sigma T 5273   FW 279.3 (free base) $C_{14}H_{21}N_3O_3$

Tyr-Val-Ala-Asp
Ac-Tyr-Val-Ala-Asp-2,6-Dimethylbenzoyloxymethylketone

Synonyms: Caspase I Inhibitor V

Alexis 260-016   MW 654.7 $C_{33}H_{42}N_4O_{10}$ ≥95%; white to off-white powder; soluble in DMF or methanol | Highly selective, competitive & irreversible inhibitor; Thornberry, NA et al, *Biochemistry*, 33:3934, 1994

Tyr-Val-Ala-Asp
Biotinyl-Tyr-Val-Ala-Asp-CMK

Synonyms: Caspase I Inhibitor

Alexis 260-019   MW 725.3 $C_{32}H_{45}N_6O_9SCl$ ≥97%; white to off-white powder; soluble in DMSO or 50% TFA | Irreversible inhibitor; used for isolation, identification & characterization of Caspase I; Lazebnik, YA et al, *Nature*, 371:346, 1994

Tyr-Val-Ala-Asp
Ac-Tyr-Val-Ala-Asp-AMC

Synonyms: Caspase I Substrate

Alexis 260-024   MW 665.7 $C_{33}H_{39}N_5O_{10}$ ≥97%; white to off-white powder; soluble in DMSO or acetic acid | Fluorogenic substrate; AMC has an excitation maximum of 380 nm & an emission maximum of 460 nm; Thornberry, NA et al, *Nature*, 356:768, 1992; Walker, NPC et al, *Cell*, 78:343, 1994; Thornberry, NA et al, *Meth Enzymol*, 244:615, 1994

Tyr-Val-Ala-Asp
Ac-Tyr-Val-Ala-Asp-pNA

Synonyms: Caspase I Substrate

Alexis 260-026   MW 628.6 $C_{29}H_{36}N_6O_{10}$ ≥96%; white to off-white powder; soluble in water or DMSO | Chromogenic substrate; Thornberry, NA et al, *Nature*, 356:768, 1992; Reiter, LA et al, *Int J Pept Protein Res*, 43:87, 1994; Walker, NPC et al, *Cell*, 78:343, 1994; Thornberry, NA et al, *Meth Enzymol*, 244:615, 1994

Tyr-Val-Ala-Asp
Ac-Tyr-Val-Ala-Asp-CHO

Synonyms: Caspase I Inhibitor

Alexis 260-027   MW 492.5 $C_{23}H_{32}N_4O_8$ ≥97%; white to off-white powder; soluble in DMSO | Potent, specific & reversible inhibitor; strongly inhibits anti-Fas induced apoptosis in L929-Fas cells; Chapman, KT et al, *Bioorg Med Chem Lett*, 2; 613, 1992; Thornberry, NA et al, *Nature*, 356:768, 1992; Molineaux, SM et al, *PNAS*, 90:1809, 1993; Walker, NPC et al, *Cell*, 78:343, 1994; Wilson, KP et al, Nature, 370:270, 1994; Milligan, CE et al, *Neuron*, 15:385, 1995; Los, M et al, *Nature*, 375:270, 1994; Enari, M et al, *Nature*, 380:723, 1996

Tyr-Val-Ala-Asp
Ac-Tyr-Val-Ala-Asp-CMK

Synonyms: Caspase I Inhibitor

Alexis 260-028   MW 541.0 $C_{24}H_{33}N_4O_8Cl$ ≥95%; white to off-white powder; soluble in DMSO | Cell-permeable & irreversible inhibitor; inhibits Fas-mediated apoptosis; Lazebnik, YA et al, *Nature*, 371:346, 1994; Walker, NPC et al, *Cell*, 78:343, 1994; Enari, M et al, *Nature*, 380:723, 1996; Milligan, CE et al, *Neuron*, 15:385, 1995; Enari, M et al, *Nature*, 375:78, 1995; Kundo, S et al, *FASEB J*, 10:1193, 1996; Rouquet, N et al, *Curr Biol*, 6:1192, 1996

Tyr-Val-Ala-Asp
Z-Tyr-Val-Ala-Asp-AFC

Synonyms: Caspase I Substrate

Alexis 260-035   MW 811.8 $C_{39}H_{40}N_5O_{11}F_3$ ≥95%; white to off-white powder; soluble in DMSO, DMF or methanol | Fluorogenic substrate; similar to Alexis Prod. No. 260-024 but the AFC fluorophore has a greater Stokes' shift upon cleavage & the *N*-terminal is blocked by a benzyloxycarbonyl moiety which increases cell permeability; Thornberry, NA et al, *Nature*, 356:768, 1992; Reiter, LA et al, *Int J Pept Protein Res*, 43:87, 1994; Walker, NPC et al, *Cell*, 78:343, 1994; Thornberry, NA et al, *Meth Enzymol*, 244:615, 1994

Tyr-Val-Ala-Asp
Z-Tyr-Val-Ala-Asp-pNA

Synonyms: Caspase I Substrate

Alexis 260-049   MW 720.7 $C_{35}H_{40}N_6O_{11}$ ≥97%; white lyophilized powder; soluble in water | Chromogenic substrate; increased cell permeability

---

Tyr-Val-Ala-Asp
Ac-Tyr-Val-Ala-Asp-AMC

Synonyms: Interleukin Converting Enzyme Substrate

American Peptide 81-7-30   MW 665.7 $C_{33}H_{39}N_5O_{10}$ Fluorogenic; Inayat-Hussain, SH et al, *Hepatology*, 25(6):1516, 1997; Thornberry, NA et al, *Nature*, 356:768, 1992

Tyr-Val-Ala-Asp
Ac-Tyr-Val-Ala-Asp-CHO

Synonyms: Interleukin Iβ Converting Enzyme Inhibitor I

American Peptide 81-7-31   MW 492.5 $C_{23}H_{32}N_4O_8$
Thornberry, NA et al, *Nature*, 356:768, 1992; Molineaux, SM et al, *PNAS*, 90:1809, 1993; Walker, NPC et al, *Cell*, 78:343, 1994

Tyr-Val-Ala-Asp
Ac-Tyr-Val-Ala-Asp-FMK

Synonyms: Interleukin Iβ Converting Enzyme Inhibitor III; Caspase I Inhibitor

American Peptide 81-7-32   MW 762.4 $C_{33}H_{36}N_4O_{10}F_6$
Highly selective, competitive & irreversible inhibitor; Thornberry, NA et al, *Biochemistry*, 33:3934, 1994

Tyr-Val-Ala-Asp
Ac-Tyr-Val-Ala-Asp-pNA

Synonyms: Interleukin Iβ Converting Enzyme Substrate IV

American Peptide 81-7-34   MW 628.6 $C_{29}H_{36}N_6O_{10}$
Thornberry, NA et al, *Nature*, 356:768, 1992

Tyr-Val-Ala-Asp
Biotin-Tyr-Val-Ala-Asp-CHO

American Peptide 81-7-36   MW 676.8 $C_{31}H_{44}N_6O_9S$

Tyr-Val-Ala-Asp
Biotin-Tyr-Val-Ala-Asp-FMK

Synonyms: Interleukin Iβ Converting Enzyme Inhibitor III, (des-Ac,Biotin)-

American Peptide 81-7-37   MW 945.6 $C_{41}H_{48}N_6O_{11}SF_6$
Used in conjunction with ICE inhibitor III

Tyr-Val-Ala-Asp
Tyr-Val-Ala-Asp-CHO

Synonyms: Caspase Inhibitor

American Peptide 81-7-38   MW 450.5 $C_{21}H_{30}N_4O_7$ Suzuki, A, *Exp Cell Res*, 234(2):507, 1997

Tyr-Val-Ala-Asp
Ac-*N*-Me-Tyr-Val-Ala-Asp-CHO Pseudo Acid

Bachem H-2548.0005   MW 506.55 $C_{24}H_{34}N_4O_8$ Store at -15°C

Tyr-Val-Ala-Asp
Ac-Tyr-*N*-Me-Val-Ala-*N*-Me-Asp-CHO Pseudo Acid

Bachem H-2732.0001   MW 520.58 $C_{25}H_{36}N_4O_8$ Store at -15°C

Tyr-Val-Ala-Asp
Ac-Tyr-Val-Ala-Asp-CHO Pseudo Acid

Bachem H-8410.0005   MW 492.53 $C_{23}H_{32}N_4O_8$ Store at -15°C

Tyr-Val-Ala-Asp
Ac-Tyr-Val-Ala-Asp-AMC

Bachem I-1630.0005   MW 665.7 $C_{33}H_{39}N_5O_{10}$ Store at -15°C

Tyr-Val-Ala-Asp
Suc-Tyr-Val-Ala-Asp-AMC

Bachem I-1635.0025   MW 723.74 $C_{35}H_{41}N_5O_{12}$ Store at -15°C

Tyr-Val-Ala-Asp
Ac-Tyr-Val-Ala-Asp-AFC

Bachem I-1705.0005   MW 719.67 $C_{33}H_{36}F_3N_5O_{10}$ Store at -15°C

Tyr-Val-Ala-Asp
Z-Tyr-Val-Ala-Asp-AFC

Bachem I-1720.0005   MW 811.77 $C_{39}H_{40}F_3N_5O_{11}$ Store at -15°C

Tyr-Val-Ala-Asp
Suc-Tyr-Val-Ala-Asp-pNA

Bachem L-1830.0025   MW 686.68 $C_{31}H_{38}N_6O_{12}$ Store at -15°C

Tyr-Val-Ala-Asp
Ac-Tyr-Val-Ala-Asp-pNA

Bachem L-1890.0005   MW 628.64 $C_{29}H_{36}N_6O_{10}$ Store at -15°C

Tyr-Val-Ala-Asp
FITC-Tyr-Val-Ala-Asp (Contains FITC isomer I)

Bachem M-2280.0001   MW 855.88 $C_{42}H_{41}N_5O_{13}S$ Store at -15°C

Tyr-Val-Ala-Asp
Ac-Tyr-Val-Ala-Asp-CMK

Bachem N-1330.0005   MW 541.01 $C_{24}H_{33}ClN_4O_8$ Store at -15°C

Tyr-Val-Ala-Asp
Ac-Tyr-Val-Ala-Asp-2,6-dimethylbenzoyloxymethylketone

Bachem N-1520.0005   MW 654.72 $C_{33}H_{42}N_4O_{10}$ Store at -15°C

Tyr-Val-Ala-Asp
Biotinyl-Tyr-Val-Ala-Asp-CMK

Bachem N-1525.0005   MW 725.27 $C_{32}H_{45}ClN_6O_9S$ Store at -15°C

Tyr-Val-Ala-Asp
Z-Tyr-Val-Ala-Asp-CMK

Bachem N-1530.0005   MW 633.1 $C_{30}H_{37}ClN_4O_9$ Store at -15°C

Tyr-Val-Ala-Asp
Z-Tyr-Val-Ala-DL-Asp-FMK

Bachem N-1595.0001   MW 616.64 $C_{30}H_{37}FN_4O_9$ Store at -15°C

Tyr-Val-Ala-Asp
Ac-Tyr-Val-Ala-Asp(OtBu)-CHO-Dimethyl Acetal

Bachem N-1690.0005   MW 594.71 $C_{29}H_{46}N_4O_9$ Store at -15°C

Tyr-Val-Ala-Asp
Ac-Tyr-Val-Ala-Asp-pNA

BioSource International 77-328   Colored substrate

Tyr-Val-Ala-Asp
Ac-Tyr-Val-Ala-Asp-AMC

BioSource International 77-332   Fluorescing substrate

Tyr-Val-Ala-Asp
Z-Tyr-Val-Ala-Asp-pNA

BioSource International 77-873

Tyr-Val-Ala-Asp
Ac-Tyr-Val-Ala-Asp-AFC

BioSource International 77-876

Tyr-Val-Ala-Asp
Ac-Tyr-Val-Ala-Asp-CHO

BioSource International 77-925   Inhibitor

Tyr-Val-Ala-Asp
Z-Tyr-Val-Ala-Asp-AFC

BioSource International 77-928

Tyr-Val-Ala-Asp
Tyr-Val-Ala-Asp-CHO

BioSource International 77-960   Cell permeable inhibitor

Tyr-Val-Ala-Asp
Z-Tyr-Val-Ala-Asp(OMe)-CH₂F

Synonyms: Caspase I Inhibitor VI

Calbiochem 218746   MW 630.7 $C_{31}H_{39}FN_4O_9$ Single spot purity (TLC); solid; soluble in DMSO | Potent, cell-permeable & irreversible inhibitor of Caspase I & Caspase IV; Eldadah, BA et al, *J Neurosci*, 17:6105, 1997

Tyr-Val-Ala-Asp
Ac-Tyr-Val-Ala-Asp-CMK

Synonyms: Caspase I Inhibitor II; Interleukin Iβ Converting Enzyme Inhibitor II

Calbiochem 400012   MW 541.0 $C_{24}H_{33}ClN_4O_8$ ≥95% (HPLC); solid; soluble in DMSO | Cell-permeable & irreversible inhibitor of Caspase I & Caspase IV; inhibits Fas-mediated apoptosis; Enari, M et al, *Nature*, 375:78, 1995; Walker, NPC et al, *Cell*, 78:343, 1994

Tyr-Val-Ala-Asp
Ac-Tyr-Val-Ala-Asp-AMC

Synonyms: Caspase I Fluorogenic Substrate III; Interleukin Iβ Converting Enzyme Fluorogenic Substrate III

Calbiochem 400020   MW 665.7 $C_{33}H_{39}N_5O_{10}$ ≥90% (HPLC); solid; soluble in acetic acid & DMSO; excitation max:~365-380 nm; emission max:~430-460 nm | Fluorogenic Caspase I substrate; a substrate of Caspase IV; Thornberry, NA et al, *Nature*, 356:768, 1992; Walker, NPC et al, *Cell*, 78:343, 1994

Tyr-Val-Ala-Asp
Biotin-Tyr-Val-Ala-Asp-CMK

Synonyms: Caspase I Inhibitor II; Interleukin Iβ Converting Enzyme Inhibitor II, Biotin-

Calbiochem 400022   MW 725.3 $C_{32}H_{45}ClN_6O_9S$ ≥95% (HPLC); solid; soluble in DMSO; may be carcinogenic/teratogenic | Biotinylated derivative of Caspase I Inhibitor II used to label Caspase I or Caspase IV enzymes; useful for the isolation, identification & characterization of Caspase I; Lazebnik, YA et al, *Nature*, 371:346, 1994

Tyr-Val-Ala-Asp
Biotin-Tyr-Val-Ala-Asp-FOAM

Synonyms: Caspase I Inhibitor III; Interleukin Iβ Converting Enzyme Inhibitor III, Biotin-

Calbiochem 400024   MW 946.9 $C_{41}H_{48}F_6N_6O_{11}S$ ≥90% (HPLC); solid; soluble in DMF; may be carcinogenic/teratogenic | Biotinylated derivative of Caspase I Inhibitor III used to label the Caspase I or Caspase IV; useful for the isolation, identification & characterization of Caspase I or Caspase IV

Tyr-Val-Ala-Asp
Ac-Tyr-Val-Ala-Asp-pNA

Synonyms: Caspase I Colorimetric Substrate IV; Interleukin Iβ Converting Enzyme Colorimetric Substrate

Calbiochem 400025   MW 628.6 $C_{29}H_{36}N_6O_{10}$ ≥97% (HPLC); lyophilized solid; soluble in DMSO | Colorimetric Caspase I substrate similar in sequence to Caspase I substrate III, fluorogenic; cleavage of pNA is monitored colorimetrically at ~405 nm; sequence based on the Caspase I cleavage site of the IL-1β precursor & includes Asp[116]; a substrate of Caspase IV; Thornberry, NA et al, *Nature*, 356:768, 1992; Reiter, LA et al, *Int J Pept Protein Res*, 43:87, 1994; Thornberry, NA et al, *Methods Enzymol*, 244:615, 1994

Tyr-Val-Ala-Asp
Z-Tyr-Val-Ala-Asp-AFC

Synonyms: Caspase I Fluorogenic Substrate VI; Interleukin Iβ Converting Enzyme Substrate; Aminotrifluoromethylcoumarin Tetrapeptide Conjugate

Calbiochem 688225   MW 811.8 $C_{39}H_{40}F_3N_5O_{11}$ Single spot purity (TLC); solid; soluble in DMSO; excitation max:~400 nm; emission max:~505 nm | Fluorogenic Caspase I substrate; reaction monitored visually or quantitatively by a blue to green shift in fluorescence upon cleavage of the AFC fluorophore; Thornberry, NA et al, *Nature*, 356:768, 1992; Walker, NPC et al, *Cell*, 78:343, 1994

ICN 193600   Fluorescent; ideal for apoptosis research

Tyr-Val-Ala-Asp
Ac-Tyr-Val-Ala-Asp-pNA

Synonyms: Interleukin Iβ Converting Enzyme Substrate

ICN 195869   MW 628 $C_{29}H_{35}N_6O_{10}$ ≥97% | Colorimetric substrate

Tyr-Val-Ala-Asp
Ac-Tyr-Val-Ala-Asp-CHO

Synonyms: Interleukin Iβ Converting Enzyme Inhibitor I; Caspase Inhibitor I

ICN 195870   MW 520.5 $C_{24}H_{32}N_4O_9$ ≥97%

Tyr-Val-Ala-Asp
Ac-Tyr-Val-Ala-Asp-CMK

Synonyms: Interleukin Iβ Converting Enzyme Inhibitor II

ICN 195871   MW 541   ≥97% | Inhibits irreversibly

Tyr-Val-Ala-Asp
Ac-Tyr-Val-Ala-Asp-CMK Biotinylated

Synonyms: Interleukin Iβ Converting Enzyme Inhibitor II

ICN 195872   MW 724.3   ≥97% | Inhibits irreversibly

Tyr-Val-Ala-Asp
Ac-Tyr-Val-Ala-Asp-Acyloxymethylketone

Synonyms: Interleukin Iβ Converting Enzyme Inhibitor III

ICN 195873   MW 761.7   ≥90% | Selectively & strongly inhibits

Tyr-Val-Ala-Asp
Ac-Tyr-Val-Ala-Asp-Acyloxymethylketone Biotinylated

Synonyms: Interleukin Iβ Converting Enzyme Inhibitor III

ICN 195874   MW 945.9   ≥90% | Selectively & strongly inhibits

Tyr-Val-Ala-Asp
Z-Tyr-Val-Ala-Asp-FMK

Synonyms: Caspase I Inhibitor; Interleukin Converting Enzyme Inhibitor II

Kamiya   MW 630

Tyr-Val-Ala-Asp
Ac-Tyr-Val-Ala-Asp-CHO

Synonyms: Caspase I Inhibitor; Interleukin Converting Enzyme Inhibitor II

Kamiya	MW 492.5	$C_{23}H_{32}N_4O_8$	>95% (HPLC)

Tyr-Val-Ala-Asp
Ac-Tyr-Val-Ala-Asp-AFC

Synonyms: Caspase I Substrate; Interleukin Converting Enzyme II Substrate

Kamiya	MW 720	Fluorogenic

Tyr-Val-Ala-Asp
Ac-Tyr-Val-Ala-Asp-pNA

Synonyms: Caspase I Substrate; Interleukin Converting Enzyme Substrate II

Kamiya	MW 628.6	$C_{29}H_{36}N_6O_{10}$	>97% (HPLC)

Tyr-Val-Ala-Asp
Ac-Tyr-Val-Ala-Asp-AMC

Synonyms: Caspase I Substrate

Neosystem SC1209 MW 665.70 Thornberry, NA et al, *Nature*, 356:768-774, 1992; Walker, NPC et al, *Cell*, 78:343-352, 1994

Tyr-Val-Ala-Asp
Suc-Tyr-Val-Ala-Asp-pNA

Synonyms: Caspase I Substrate

Neosystem SC1317 MW 686.65 Chromogenic substrate

Tyr-Val-Ala-Asp
Ac-Tyr-Val-Ala-Asp-CHO

Synonyms: Caspase I Inhibitor

Neosystem SC929 MW 492.53 Reversible inhibitor; inhibition of Caspase I in whole human blood prevents secretion of Bioactive IL-1β; Chapman, KT, *Bioorg & Med Chem Lett*, 2:613-618, 1992; Thornberry, NA et al, *Nature*, 356:768-774, 1992; Graybill, TL et al, *Int J Pep Prot Res*, 44:173-182, 1994; Walker, NPC et al, *Cell*, 78:343-352, 1994; Fletcher, DS et al, *J Interferon & Cytokine Res*, 15:243-248, 1995

Tyr-Val-Ala-Asp
Ac-Tyr-Val-Ala-Asp-CMK

Synonyms: Caspase I Inhibitor II

Oncogene 400012 For irreversible inhibition of Caspase I protein

Tyr-Val-Ala-Asp
Ac-Tyr-Val-Ala-Asp-AMC

Synonyms: Caspase I Fluorogenic Substrate III

Oncogene 400020 Fluorogenic substrate for assay of Caspase I

Tyr-Val-Ala-Asp
Biotin-Tyr-Val-Ala-Asp-CMK Biotinylated

Synonyms: Caspase I Inhibitor II

Oncogene 400022 For labeling of Caspase I protein; isolation of Caspase I protein

Tyr-Val-Ala-Asp
Biotin-Tyr-Val-Ala-Asp-Fluoro-Acyloxymethylketone Biotinylated

Synonyms: Caspase I Inhibitor III

Oncogene 400024 For labeling of Caspase I protein; isolation of Caspase I protein

Tyr-Val-Ala-Asp
Ac-Tyr-Val-Ala-Asp-CHO

Synonyms: Caspase I Inhibitor I

Oncogene 627610 For reversible inhibition

Tyr-Val-Ala-Asp
Ac-Tyr-Val-Ala-Asp-MCA

Synonyms: Caspase I/IV Substrate

Peptides International MAY-3161-v MW 665.70 $C_{33}H_{39}N_5O_{10}$ >98% (HPLC); lyophilized amorphous powder | Thornberry, NA et al, *Nature*, 356:768, 1992

Tyr-Val-Ala-Asp
N-Ac-N-Me-Tyr-Val-Ala-Asp-CHO

Synonyms: Apopain Inhibitor; CPP-32 Inhibitor

Sigma A 0960 FW 506.6 $C_{24}H_{34}N_4O_8$ ≥95% (TLC) | Bioactive peptide; potent inhibitor of poly(ADP-ribose) polymerase cleavage by apopain (CPP-32); Nicholson, DW et al, *Nature*, 376:37, 1995

Tyr-Val-Ala-Asp
N-Ac-Tyr-Val-Ala-Asp-MCA

Synonyms: Interleukin Iβ Converting Enzyme Substrate; Caspase I Substrate

Sigma A 2452 FW 665.7 $C_{33}H_{39}N_5O_{10}$ Peptide content ~80% | Fluorogenic substrate; Thornberry, NA et al, *Nature*, 356:768, 1992

Tyr-Val-Ala-Asp
N-Ac-Tyr-Val-Ala-Asp-CHO

Synonyms: Interleukin Iβ Converting Enzyme Inhibitor; Caspase I Inhibitor

Sigma A 3707 FW 492.5 $C_{23}H_{32}N_4O_8$ Potent, reversible inhibitor; Chapman, KT, *Bioorg Med Chem Lett*, 2:613, 1992

Tyr-Val-Ala-Asp
N-Ac-Tyr-Val-Ala-Asp-pNA

Synonyms: Interleukin Iβ Converting Enzyme Substrate; Caspase I Substrate

Sigma A 3831 FW 628.6 $C_{29}H_{36}N_6O_{10}$ ≥97% (HPLC) | Substrate for interleukin-1β converting enzyme (Caspase I); *Int J Pept Prot Res*, 43:87, 1994

Tyr-Val-Ala-Asp
N-Ac-Tyr-Val-Ala-Asp

Synonyms: Interleukin Iβ Converting Enzyme Inhibitor II

Sigma A 4211 FW 541.0 $C_{24}H_{33}CIN_4O_8$ ≥90% (HPLC) | Bioactive peptide; irreversible inhibitor of IL-1β converting enzyme (ICE); Walker, NPC et al, *Cell*, 78:343, 1994; Enari, M et al, *Nature*, 375:78, 1995; Lazebnik, YA et al, *Nature*, 371:346, 1994

Tyr-Val-Ala-Asp
N-Ac-Tyr-Val-Ala-Asp-AMC

Synonyms: Interleukin Iβ Converting Enzyme Substrate; Caspase I Substrate

Sigma A 9965 Membrane-permeable fluorogenic substrate; Zhivotovsky, B et al, *Exp Cell Res*, 221:404, 1995; Xiang, J et al, *PNAS USA*, 93:14559, 1996

Tyr-Val-Ala-Asp
Ac-Tyr-Val-Ala-Asp-CHO

Synonyms: Caspase I/IV Inhibitor

Peptides International IAT-3165-v Synthetic MW 492.53 $C_{23}H_{32}N_4O_8$ >99% (HPLC); lyophilized amorphous powder | Inhibitor; Thornberry, NA et al, *Nature*, 356:768, 1992; Molineaux, SM et al, *PNAS USA*, 90:1809, 1993; Enari, M et al, *Nature*, 380:723, 1996

Tyr-Val-Ala-Asp
Ac-Tyr-Val-Ala-Asp-CH₂Cl
Synonyms: Caspase Inhibitor

Peptides International ICL-3180-v   Synthetic   MW 541.00
$C_{24}H_{33}N_4O_8Cl$ >99% (HPLC); lyophilized amorphous powder |
Inhibitor; Lazebnik, YA et al, *Nature*, 371:346, 1994; Enari, M et al,
Nature, 375:78, 1995; Milligan, CE et al, *Neuron*, 15:385, 1995;
Fujita, E et al, *BBRC*, 224:74, 1996

Tyr-Val-Ala-Asp-
N-Ac-Tyr-Val-Ala-Asp-EDANS
Synonyms: Interleukin Iβ Converting Enzyme Substrate II

ICN 195875   MW 1225.4   ≥95% | Fluorogenic substrate

Tyr-Val-Ala-AspAc-Tyr-Val-Ala-Asp-CMK
Synonyms: Interleukin Iβ Converting Enzyme Inhibitor II

American Peptide 81-7-35   MW 541 $C_{24}H_{33}N_4O_8Cl$
Inhibitor of ICE-like protease; Schulz, JB et al, *J Neurosci*,
16(15):4696, 1996; Walker, NPC et al, *Cell*, 78:343, 1994

Tyr-Val-Ala-Asp-Ala-Pro-Lys
MCA-Tyr-Val-Ala-Asp-Ala-Pro-Lys-DNP Amide
Synonyms: Caspase I Substrate

Alexis 260-023   MW 1145.2 $C_{53}H_{64}N_{11}O_{18}$ ≥98%; yellow
solid; soluble in DMSO or 80% acetic acid | Specific, highly
fluorescent substrate for the determination of Caspase I & Caspase-
like enzymes; Caspase III is only weakly active on this substrate;
Enari, M et al, *Nature*, 380:723, 1996

Alexis 260-052   MW 1144.2 $C_{53}H_{65}N_{11}O_{18}$ ≥97%; lyophilized
powder | Specific, highly fluorescent substrate for the
determination of Caspase I & Caspase-like enzymes; Caspase III is
only weakly active on this substrate; Enari, M et al, *Nature*,
380:723, 1996

Tyr-Val-Ala-Asp-Ala-Pro-Lys
Mca-Tyr-Val-Ala-Asp-Ala-Pro-Lys(DNP)
Synonyms: Caspase I Substrate V

American Peptide 81-7-94   MW 1145.2 $C_{53}H_{64}N_{10}O_{19}$
Substitution of this substrate at position P_1 Asp results in
continuous fluorescent assay monitored at emission wavelength
393 nm; Enari, M et al, *Nature*, 380:723, 1996

Bachem M-2195.0001   Store at -15°C

Tyr-Val-Ala-Asp-Ala-Pro-Lys
FITC-Tyr-Val-Ala-Asp-Ala-Pro-Lys(Dnp)
Bachem M-2285.0001   Contains FITC isomer I; store at -15°C

Tyr-Val-Ala-Asp-Ala-Pro-Lys
MCA-Tyr-Val-Ala-Asp-Ala-Pro-(Lys-DNP)-Amide
BioSource International 77-969

Tyr-Val-Ala-Asp-Ala-Pro-Lys
MCA-Tyr-Val-Ala-Asp-Ala-Pro-Lys-(DNP) Amide
Synonyms: Caspase I Fluorogenic Substrate IX

Calbiochem 218738   MW 1144.2 $C_{53}H_{65}N_{11}O_{18}$ ≥95%
(HPLC); lyophilized solid; soluble in DMSO; excitation max:~325
nm; emission max:~392 nm | Fluorogenic resonance energy
transfer substrate for Caspase I & Caspase I-like proteases

Tyr-Val-Ala-Asp-Ala-Pro-Lys
MCA-Tyr-Val-Ala-Asp-Ala-Pro-Lys(DNP)
Synonyms: Caspase I Fluorogenic Substrate V; Interleukin
Converting Enzyme Fluorogenic Substrate V

Calbiochem 400017   MW 1145.1 $C_{53}H_{64}N_{10}O_{19}$ ≥98%
(HPLC); crystalline solid; soluble in 80% acetic acid & DMSO;
excitation max:~325 nm; emission max:~393 nm | Specific
fluorogenic resonance energy transfer substrate for Caspase I &
Caspase I-like enzyme activities including Caspase IV; Caspase III
has only a weak effect on this substrate; Enari, M et al, *Nature*,
380:723, 1996

Tyr-Val-Ala-Asp-Ala-Pro-Lys
MOCAc-Tyr-Val-Ala-Asp-Ala-Pro-Lys(Dnp) Amide
Synonyms: Caspase I Fluorescence-Quenching Substrate

Peptides International MOC-3183-v   MW 1144.2
$C_{53}H_{65}N_{11}O_{18}$ >98% (HPLC); lyophilized amorphous powder |
Enari, M et al, *Nature*, 380:72, 1996

Tyr-Val-Ala-Asp-Ala-Pro-Lys
MCA-Tyr-Val-Ala-Asp-Ala-Pro-DNP-Lys
Synonyms: Interleukin Iβ Converting Enzyme Substrate

Sigma M 5295   FW 1145.1   ≥97% (HPLC); peptide
content:~55% | Fluorogenic substrate for ICE & ICE-like
enzymes; cleavage at the P1 Asp residue results in continuous
fluorescence at 392 nm; Enari, M, *Nature*, 380:723, 1996

Tyr-Val-Ala-Asp-Ala-Pro-Val
4-(4-Dimethylaminophenylazo)Benzoyl-Tyr-Val-Ala-Asp-Ala-Pro-Val-5-((2-Aminoethyl)Amino)-Naphthalene-1-Sulfonic Acid
Synonyms: Interleukin Iβ Converting Enzyme Substrate II

American Peptide 81-0-15   MW 1236.4 $C_{60}H_{75}N_{12}O_{15}S$
Pennington, MW & Thornberry, NA, *Peptide Res*, 7:72, 1994; Los, M
et al, *Nature*, 375:81, 1995

Tyr-Val-Ala-Asp-Ala-Pro-Val
DABCYL-Tyr-Val-Ala-Asp-Ala-Pro-Val-EDANS
Synonyms: Caspase I Fluorogenic Substrate II; Interleukin Iβ
Converting Enzyme Fluorogenic Substrate II

Bachem M-1940.0001   MW 1233.41 $C_{61}H_{76}N_{12}O_{14}S$ Store
at -15°C

Calbiochem 400018   MW 1233.4 $C_{61}H_{76}N_{12}O_{14}S$ >95%
(HPLC); lyophilized solid; soluble in 50% acetic acid & DMF;
excitation max:~340 nm; emission max:~490 nm | Fluorogenic
Caspase I substrate; allows a continuous assay of Caspase I; used
in screening assays of Caspase I inhibitory compounds; a substrate
of Caspase IV

Oncogene 400018   Fluorogenic substrate for continuous assay
of Caspase I activity; for screening of Caspase I inhibitory
compounds

Sigma D 7920   FW 1233.4   Fluorogenic substrate;
Pennington, WW & Thornberry, NA, *Peptide Res*, 7:72, 1994; Los,
M et al, *Nature*, 375:81, 1995

Tyr-Val-Ala-Glu
N-Ac-Tyr-Val-Ala-Glu-CHO
Sigma A 1085   FW 506.6 $C_{24}H_{34}N_4O_8$ ≥98% (TLC)

Tyr-Val-Ala-Glu-Gly-Thr-Asp-Arg-Val-Ile-Glu-Val-Val-Gln-Gly-Ala-Cys-Arg
Synonyms: GP41 (827-843), (Tyr)-; HTLV I Peptide

Neosystem SC237   MW 1965.21   HTLV I peptide from
subtype GP41; due to the presence of a Cys this peptide can
contain the dimeric form

Tyr-Val-Asn-Thr-Pro-Glu-His-Val-Val-Pro-Tyr-Gly-Leu-Gly-Ser-Pro-Arg-Ser

Synonyms: Endothelin I (22-38), (Tyr)-Big

Bachem H-7880.0500   Human   Store at -15°C

Tyr-Val-Asn-Trp-Leu-Leu-Ala-Gln-Lys-Gly-Lys-Lys-Asn-Asp-Trp-Lys-His-Asn-Ile-Thr-Gln

Synonyms: Gastric Inhibitory Peptide (23-42), (Tyr[0])-; Gastric Inhibitory Peptide (22-42), (Tyr[22])-

American Peptide 46-1-62   Human   MW 2584.9 $C_{119}H_{182}N_{34}O_{31}$

ICN 154548   Human   MW 2584.9

Tyr-Val-Asn-Val
Tyr(PO₃H₂)-Val-Asn-Val

$Tyr(PO_3H_2)$-Val-Asn-Val

Synonyms: Grb[2] SH2 Domain Ligand

Bachem H-2708.0001   Store at -15°C

Tyr-Val-Gly
D-Tyr-Val-Gly

American Peptide 60-0-61   MW 337.4 $C_{16}H_{23}N_3O_5$

Tyr-Val-Gly
Dansyl-Tyr-Val-Gly

American Peptide 81-6-60   MW 570.7 $C_{28}H_{34}N_4O_7S$

Tyr-Val-Gly
D-Tyr-Val-Gly

Bachem H-5245.0050   MW 337.38 $C_{16}H_{23}N_3O_5$ Store at -15°C

Tyr-Val-Gly
Dansyl-L-Tyr-L-Val-Gly

Fluka 30445   MW 684.68 $C_{28}H_{38}N_4O_7S \cdot C_2HF_3O_2$ ≥99% (HPLC); ≤2% H₂O; trifluoroacetate

---

Tyr-Val-Gly
Ac-Tyr-Val-Gly

Synonyms: Amidating Enzyme Substrate, α-

Peptides International SYG-3146   MW 379.41 $C_{18}H_{25}N_3O_6$ >99% (HPLC); amorphous powder

---

Tyr-Val-Gly
D-Tyr-Val-Gly

Sigma T 5148   FW 337.4 $C_{16}H_{23}N_3O_5$

Tyr-Val-Lys-Arg-Val-Lys

Synonyms: Band 3 Protein (824-829)

Bachem H-1598.0005   Human   Store at -15°C

Tyr-Val-Lys-Asp
Ac-Tyr-Val-Lys-Asp-CHO

Synonyms: Interleukin Iβ Converting Enzyme Inhibitor IV

American Peptide 81-7-01   MW 549.6 $C_{26}H_{39}N_5O_8$ Affinity ligand for interleukin-1 β-converting enzyme; Thornberry, NA et al, *Nature*, 356:768, 1992; Graybill, TL et al, *Int J Peptide Protein Res*, 44:173, 1994

Tyr-Val-Lys-Asp
Ac-Tyr-Val-Lys-Asp-CHO Pseudo Acid

Bachem N-1355.0001   MW 549.62 $C_{26}H_{39}N_5O_8$ Store at -15°C

Tyr-Val-Lys-Asp
Ac-Tyr-Val-Lys(biotinyl)-Asp-2,6-Dimethylbenzoyloxymethylketone

Bachem N-1570.0005   MW 938.11 $C_{46}H_{63}N_7O_{12}S$ Store at -15°C

Tyr-Val-Lys-Asp
N-Ac-Tyr-Val-Lys(Biotinyl)-Asp-2,6-Dimethylbenzoyloxymethyl Ketone

Synonyms: Caspase I Inhibitor

Sigma A 4339   FW 938.1 $C_{46}H_{63}N_7O_{12}S$ ≥90% (HPLC) | Bioactive peptide; useful for affinity labeling apoptosis-associated Caspase I-related proteases in cell-free extracts; Takahashi, A et al, *Proc Natl Acad Sci USA*, 93:8395, 1996

Tyr-Val-Lys-Asp
N-Ac-Tyr-Val-Lys-Asp-CHO

Sigma A 5706   FW 549.6 $C_{26}H_{39}N_5O_8$ ~75% (HPLC); peptide content ~70%

Tyr-Val-Lys-Asp
Ac-Tyr-Val-Lys-Asp-CHO

Synonyms: Caspase I Inhibitor; Caspase I Affinity Ligand

Peptides International IAT-3166-v   Synthetic   MW 549.62 $C_{26}H_{39}N_5O_8$ >99% (HPLC); lyophilized amorphous powder | Inhibitor; Thornberry, NA et al, *Nature*, 356:768, 1992; Graybill, TL et al, *Int J Pept Prot Res*, 44:173, 1994

Tyr-Val-Lys-Asp-Asn-Phe-Val-Pro-Thr-Asn-Val-Gly-Ser-Glu-Ala-Phe Amide

Synonyms: Calcitonin Gene Related Peptide (22-37), (Tyr[22])-; Calcitonin Gene Regulated Peptide (23-37), (Tyr)-α-

American Peptide 22-5-14   Rat   MW 1786.0 $C_{82}H_{120}N_{20}O_{25}$ Rosenfeld, MG et al, *Nature*, 304:129, 1983

Bachem H-2270.0001   Rat   MW 1785.97 $C_{82}H_{120}N_{20}O_{25}$ Store at -15°C

Tyr-Val-Met-Gly-His-Phe-Arg-Trp-Asp-Arg-Phe Amide

Synonyms: Melanocyte Stimulating Hormone, γ-1-

American Peptide 56-0-35   MW 1512.9 $C_{72}H_{97}N_{21}O_{14}S$ Found in the cryptic region of the ACTH/β-lipotropin precursor protein from the intermediate lobe of bovine pituitary; Nakanishi, S et al, *Nature*, 278:423, 1979

Bachem H-4395.0001   Store at -15°C

Neosystem SC1228   MW 1512.75   Bioactive; melanocyte stimulating hormone-related peptide; Nakanishi, S et al, *Nature*, 278:423-427, 1979; Ling, N et al, *Life Sci*, 25:1773-1780, 1979

Tyr-Val-Met-Gly-His-Phe-Arg-Trp-Asp-Arg-Phe-Gly

Synonyms: Melanocyte Stimulating Hormone, γ-; Melanocyte Stimulating Hormone, γ-2-

American Peptide 56-0-34   MW 1570.8 $C_{74}H_{99}N_{21}O_{16}S$ Found in the cryptic region of the ACTH/β-lipotropin precursor protein from the intermediate lobe of bovine pituitary; Nakanishi, S et al, *Nature*, 278:423, 1979

Bachem H-4400.0001   MW 1570.8 $C_{74}H_{99}N_{21}O_{16}S$ Store at -15°C

ICN 152926   MW 1570.8   Nakanishi, S etal, *Nature*, 278:423, 1978; Lis, J etal, *J Clin Endocrin Med*, 52:1053 1981

Biogenesis 6045-2702   Synthetic   MW 1571   Lyophilized, consisting of 80% peptide material, acetate counter ions and residual H₂O; purified | To avoid Met oxidation, it is best to dissolve the peptide under a nitrogen stream; Nakonishi et al, *Nature*, 278:423, 1979

Tyr-Val-Met-Gly-His-Phe-Arg-Trp-Asp-Arg-Phe-Gly-Arg-Arg-Asn-Gly-Ser-Ser-Ser-Ser-Gly-Val-Gly-Gly-Ala-Ala-Gln

S Melanocyte Stimulating Hormone, γ-3-*ynonyms*:

Bachem H-2922.0500   Store at -15°C

Neosystem SC076   Bovine   MW 2943.20   Bioactive; melanocyte stimulating hormone-related peptide; Nakanashi, S et al, *Nature*, 278:423-427, 1979

Tyr-Val-Pro-Cys-His-Ile-Arg-Gln-Ile-Ile-Asn-Thr-Trp-His-Lys-Val-Gly-Lys-Asn-Val

Synonyms: GP140 (431-450); SIVmac251 Peptide

Neosystem SC751   MW 2405.84   SIVmac251 peptide from subtype GP140; due to the presence of a Cys this peptide can contain the dimeric form

Tyr-Val-Pro-Met-Leu
Tyr(PO₃H₂)-Val-Pro-Met-Leu Phosphorylated

Synonyms: Platelet Derived Growth Factor β-Receptor (719-723)

Alexis 165-005   MW 701.8 $C_{30}H_{48}N_5O_{10}PS$ ≥97%; white lyophilized powder | Contains the PDGF β-receptor recognition sequence which specifically interacts with the SH2 domain of phosphatidylinositol 3-kinase (PI₃-kinase); in this phosphorylated form the peptide inhibits the binding of the 85 kDa subunit of PI₃-kinase to the PDGF β-receptor; similar sequences are also high-affinity binding sites for the SH2 domain of the polyoma middle-T 85 kDa protein & the insulin receptor substrate-1; Fantl, WJ et al, *Cell*, 69:413, 1992; Birge, RB & Hanafusa, H, *Science*, 262:1522, 1993

Bachem H-2012.0001   Store at -15°C

Tyr-Val-Pro-Thr-Asn-Val-Gly-Ser-Glu-Ala-Phe Amide

Synonyms: Calcitonin Gene Related Peptide (28-37), (Tyr⁰)-

American Peptide 22-5-12   Rat   MW 1182.3 $C_{54}H_{79}N_{13}O_{17}$ Morris, HR et al, *Nature*, 308:746, 1984

Tyr-Val-Ser-Glu-Ile-Gln-Leu-Met-His-Asn-Leu-Gly-Lys-His-Leu-Ala-Ser-Val-Glu-Arg-Met-Gln-Trp-Leu-Arg-Lys-Lys-Leu-Gln-Asp-Val-His-Asn-Phe

Synonyms: Parathyroid Hormone (1-34), (Tyr¹)-

Bachem H-3082.0500   Rat   Store at -15°C

Tyr-Val-Ser-Glu-Ile-Gln-Leu-Met-His-Asn-Leu-Gly-Lys-His-Leu-Asn-Ser-Met-Glu-Arg-Val-Glu-Trp-Leu-Arg-Lys-Lys-Leu-Gln-Asp-Val-His-Asn-Phe

Synonyms: Parathyroid Hormone (1-34), (Tyr¹)-

American Peptide 22-1-24   Human   MW 4193.9 $C_{187}H_{295}N_{55}O_{51}S_2$

Bachem H-3092.0500   Human   Store at -15°C

Val-Ala
Z-Val-Ala

Bachem C-2835.0001   MW 322.36 $C_{16}H_{22}N_2O_5$ Store at -15°C

Val-Ala
Z-Val-Ala-OMe

Bachem C-2840.0001   MW 336.39 $C_{17}H_{24}N_2O_5$ Store at -15°C

Val-Ala

Bachem G-3500.0250   MW 188.23 $C_8H_{16}N_2O_3$ Store at -15°C

Val-Ala
Val-β-Ala

Bachem G-3505.0250   MW 188.23 $C_8H_{16}N_2O_3$ Store at -15°C

Val-Ala
N-2-Ethoxyethyl-Val-Ala-Anilide

Bachem G-4445.0050   Store at -15°C

Val-Ala
Val-Ala-pNA

Bachem L-1845.0250   MW 308.34 $C_{14}H_{20}N_4O_4$ Store at -15°C

Val-Ala

ICN 158260   MW 188.2 $C_8H_{16}N_2O_3$ Crystalline

Val-Ala
Val-Ala-pNA Acetate Salt

ICN 158261

Val-Ala
N-CBZ-Val-Ala-OMe

Sigma C 7256   FW 336.4 $C_{17}H_{24}N_2O_5$

Val-Ala

Sigma V 1250   FW 188.2 $C_8H_{16}N_2O_3$

Val-Ala
Val-Ala-pNA Acetate Salt

Sigma V 2503   FW 308.3 (free base) $C_{14}H_{20}N_4O_4$

Val-Ala-Ala-Phe

Synonyms: Carboxypeptidase P Substrate

Bachem H-5255.0050   MW 406.48 $C_{20}H_{30}N_4O_5$ Store at -15°C

ICN 156150   MW 406.5 $C_{20}H_{30}N_4O_5$ 95% | Hedeager-Sorensen, S & AJ Kenny, *Biochem J*, 229:251, 1985

Sigma V 8251   FW 406.5 $C_{20}H_{30}N_4O_5$ ≥97% (HPLC) | Bioactive peptide

Val-Ala-Arg-Glu-Leu-His-Pro-Glu-Tyr-Phe

Synonyms: NEF (194-203); HTLV I Peptide

Neosystem SC670   MW 1260.41   HTLV I peptide from subtype NEF

Val-Ala-Asn
Z-Val-Ala-Asn-AMC

Bachem I-1840.0050   MW 593.64 $C_{30}H_{35}N_5O_8$ Store at -15°C

Val-Ala-Asp
Z-Val-Ala-DL-Asp-FMK

Synonyms: Caspase I Inhibitor VI

Alexis 260-020   MW 453.5 $C_{21}H_{28}N_3O_7F$ ≥98%; white to off-white powder; soluble in DMSO or methanol | Inhibits apoptosis induced by diverse stimuli; Zhu, H et al, *FEBS Lett*, 374:303, 1995

Val-Ala-Asp
Z-Val-Ala-Asp-OMe-FMK

Synonyms: Caspase I Inhibitor

Alexis 260-039   MW 467.5 $C_{22}H_{30}N_3O_7F$ White to off-white solid; soluble in DMSO | Highly specific, cell-permeable & irreversible inhibitor of Caspase I like proteases; inhibits Fas-mediated apoptosis in Jurkat T cells & apoptosis in THP.1 cells induced by diverse stimuli (including cycloheximide, thapsigargin, etoposide & staurosporine); Chow, SC et al, *FEBS Lett*, 364:134, 1995; Zhu, H et al, *FEBS Lett*, 374:303, 1995; Pronk, GJ et al, *Science*, 271:808, 1996; Slee, EA et al, *Biochem J*, 315:21, 1996

Val-Ala-Asp
Ac-Val-Ala-Asp-CHO

American Peptide 81-7-02   MW 329.4 $C_{14}H_{23}N_3O_6$

Val-Ala-Asp
Ac-Val-Ala-Asp-CMK

American Peptide 81-7-03   MW 377.8 $C_{15}H_{24}N_3O_6Cl$

Val-Ala-Asp
Z-Val-Ala-Asp(OMe)-FMK

Synonyms: Caspase I Inhibitor V

American Peptide 81-7-06   MW 467.4 $C_{22}H_{30}N_3O_7F$ ICE-like protease inhibitor; blocks all features of apoptosis:morphological changes, cleavage of Caspase₃ (CPP-32/Yama/Apopain) & poly(ADP-ribose)/polymerase, lamin B degradation & DNA fragmentation; Shimizu, T et al, *Leukemia*, 11(8):1238, 1997; Hara, H et al, *PNAS*, 94(5):2007, 1997

Val-Ala-Asp
Ac-Val-Ala-Asp-CHO Pseudo Acid

Bachem H-2546.0005   MW 329.35 $C_{14}H_{23}N_3O_6$ Store at -15°C

Val-Ala-Asp
Z-Val-Ala-Asp-AMC

Bachem I-1710.0005   MW 594.62 $C_{30}H_{34}N_4O_9$ Store at -15°C

Val-Ala-Asp
Z-Val-Ala-DL-Asp-FMK

Bachem N-1510.0005   MW 453.47 $C_{21}H_{28}FN_3O_7$ Store at -15°C

Val-Ala-Asp
Z-Val-Ala-DL-Asp(OMe)-FMK

Bachem N-1560.0001   MW 467.5 $C_{22}H_{30}FN_3O_7$ Store at -15°C

Val-Ala-Asp
Ac-Val-Ala-Asp-CHO

Synonyms: Pan Capase Inhibitor; Caspase Inhibitor II

BioSource International 77-972

Calbiochem 218735   MW 329.4 $C_{14}H_{23}N_3O_6$ ≥95% (HPLC); lyophilized solid; soluble in DMSO & water | Potent & reversible inhibitor of Caspases I, III, IV & VII

Val-Ala-Asp
Biotin-X-Val-Ala-Asp(OMe)-CH₂F Biotinylated

Synonyms: Caspase Inhibitor I

Calbiochem 218742   MW 672.8 $C_{30}H_{49}FN_6O_8S$ ≥80% (TLC); lyophilized solid; soluble in DMSO; may be carcinogenic/teratogenic | Biotinylated derivative of Caspase Inhibitor I

Val-Ala-Asp
Z-Val-Ala-Asp-AFC

Synonyms: Caspase Fluorogenic Substrate I

Calbiochem 218743   MW 648.6 $C_{30}H_{31}F_3N_4O_9$ ≥95% (TLC); lyophilized solid; soluble in DMSO; excitation max:~400 nm; emission max:~505 nm | Substrate for Caspases III, IV & VII; reaction monitored visually or quantitatively by a blue to green shift in fluorescence upon cleavage of the AFC fluorophore

Val-Ala-Asp
Ac-Tyr-Val-Ala-Asp-CHO

Synonyms: Caspase I Inhibitor I; Interleukin Iβ Converting Enzyme Inhibitor I

Calbiochem 400010   MW 492.5 $C_{23}H_{32}N_4O_8$ ≥95% (TLC); solid; soluble in DMSO & water | Potent, specific & reversible inhibitor of Caspase I & Caspase IV; strongly inhibits anti-APO-1 induced apoptosis in L929-APO-1 cells; Los, M et al, *Nature*, 375:81, 1995; Walker, NPC et al, *Cell*, 78:343, 1994; Wilson, KP et al, *Nature*, 370:270, 1994; Thornberry, NA et al, *Nature*, 356:768, 1992

Val-Ala-Asp
Z-Val-Ala-Asp(OMe)-CH₂F

Synonyms: Caspase Inhibitor I; Interleukin Iβ Converting Enzyme-Like Inhibitor

Calbiochem 627610   MW 467.5 $C_{22}H_{30}FN_3O_7$ Single spot purity (TLC); solid; soluble in DMSO; sold under license of US Patents 5,344,939 & 5,210,272 issued to Prototek, Inc | Highly specific, cell-permeable & irreversible inhibitor of Caspases, including Caspase I, Caspase III, Caspase IV & Caspase VII; inhibits Fas-mediated apoptosis in Jurkat T cells; McColl, KS et al, *Mol Cell Endocrinol*, 139:229, 1998; Pronk, GJ et al, *Science*, 271:808, 1996; Slee, EA et al, *Biochem J*, 315:21, 1996; Chow, SC et al, *FEBS Lett*, 364:134, 1995; Fearnhead, HO et al, *FEBS Lett*, 375:283, 1995

ICN 193604   Useful in apoptosis research

Val-Ala-Asp
Z-Val-Ala-Asp-CH₂Cl

Synonyms: Granzyme B Inhibitor

ICN 193610   Useful in apoptosis research

Val-Ala-Asp
Biotin-Val-Ala-Asp-FMK

Synonyms: Caspase I Inhibitor; Interleukin Converting Enzyme Inhibitor

Kamiya   MW 672

Val-Ala-Asp
Z-Val-Ala-Asp-FMK

Synonyms: Caspase I Inhibitor; Interleukin Converting Enzyme Inhibitor I

Kamiya   MW 468

Val-Ala-Asp
Ac-Val-Ala-Asp-CHO

Synonyms: Caspase I Inhibitor; Interleukin Converting Enzyme Inhibitor I

Kamiya   MW 329.4 $C_{14}H_{23}N_3O_6$ >95% (HPLC)

Val-Ala-Asp
Ac-Val-Ala-Asp-AFC

Synonyms: Caspase I Substrate; Interleukin Converting Enzyme Substrate I

Kamiya   MW 557   Fluorogenic

Val-Ala-Asp
Ac-Val-Ala-Asp-pNA

Synonyms: Caspase I Substrate; Interleukin Converting Enzyme Substrate I

Kamiya   MW 558   >97% (HPLC)

Val-Ala-Asp
N-Ac-Val-Ala-Asp-CHO

Sigma A 1210   FW 329.4 $C_{14}H_{23}N_3O_6$ ≥98% (TLC)

Val-Ala-Ile-Thr-Val-Leu-Val-Lys

Synonyms: Calcium-Like Peptide

American Peptide 76-4-72   MW 842.1 $C_{40}H_{75}N_9O_{10}$
Interacts with proteins of the troponin superfamily & replaces Ca^{2+} effects on calmodulin-stimulated phosphodiesterase activity & on the ability to bind with chelating agents; used to study Ca^{2+} regulated pathways & might be a valuable pharmacologic agent; Dillon, J et al, *PNAS*, 88:9726, 1991

Bachem H-8155.0001   MW 842.09 $C_{40}H_{75}N_9O_{10}$ Store at -15°C

ICN 156201   MW 842.1 $C_{40}H_{75}N_9O_{10}$ 97%

Neosystem SC856   MW 842.10   Virus-related peptide; bioactive peptide mimetic of calcium; Dillon, J et al, *PNAS* USA, 88:9726, 1991

Sigma V 2757   FW 842.1 $C_{40}H_{75}N_9O_{10}$ ≥97% (HPLC); peptide content:~70% | Bioactive peptide; Dillon, J et al, *Proc Natl Acad Sci USA*, 88:9726, 1991

Val-Ala-Pro-Gly Acetate Salt

Synonyms: Chemotactic Peptide

ICN 195644   MW 402.4 $C_{15}H_{26}N_4O_5S \cdot C_2H_4O_2$ ≥97% | A quantitative marker for human elastin

Sigma V 0883   FW 342.4 (free base) $C_{15}H_{26}N_4O_5$ ≥97% (HPLC) | Bioactive peptide; serves as a quantitative marker for human elastins; Price, LSC et al, *Matrix*, 13:307, 1993

Val-Ala-Pro-Ser-Asp-Ser-Ile-Gln-Ala-Glu-Glu-Trp-Tyr-Phe-Gly-Lys-Ile-Thr-Arg-Arg-Glu

Synonyms: pp60v-*src* (137-157); Peptide A; *src* Receptor Tyrosine Kinase Inhibitor; Epidermal Receptor Factor Receptor Tyrosine Kinase Inhibitor; Protein Kinase Inhibitor, Tyr-Specific; Tyrosine-Specific Protein Kinase Inhibitor; p60v-*src* Inhibitor Peptide (137-157); Epidermal Growth Factor Receptor Kinase Inhibitor

Alexis 165-004   MW 2482.7 $C_{111}H_{168}N_{30}O_{35}$ White lyophilized powder | Residues of the noncatalytic domain of the transforming protein of Rous sarcoma virus (pp60v-*src*); Sato, K et al, *BBRC*, 171:1152, 1990; Fukami, Y et al, *J Biol Chem*, 268:1132, 1993

American Peptide 86-0-61   MW 2482.8 $C_{111}H_{168}N_{30}O_{35}$
Derived from the non-catalytic domain of p60v-*src* (137-157); inhibits the tyrosine kinase activity of p60v-*src*; Sato, K et al, *BBRC*, 171:1156, 1990

Bachem H-8535.0500   MW 2482.74 $C_{111}H_{168}N_{30}O_{35}$ Store at -15°C

Calbiochem 657015   MW 2482.7 $C_{111}H_{168}N_{30}O_{35}$ ≥95% (HPLC); lyophilized solid; soluble in 5% acetic acid | Sato, K et al, *Biochem Biophys Res Comm*, 171:1152, 1990

ICN 195845   MW 2482.8   ≥97% | Non-competitive *src* & EGF-R inhibitor; selectively inhibits protein tyrosinekinase phosphorylation in cell extracts

Val-Arg

Bachem G-3810.0250   MW 273.34 $C_{11}H_{23}N_5O_3$ Store at -15°C

Val-Arg
Z-Val-Arg-4MβNA Hydrochloride

Bachem J-1400.0050   MW 599.13 $C_{30}H_{38}N_6O_5 \cdot HCl$ Store at -15°C

Val-Arg
D-Val-CHA-Arg-pNA·2AcOH

Synonyms: Pefachrome®GK; Pefa-5715

Pentapharm 088-20, 088-01   MW 666.8   Highly sensitive chromogenic peptide substrate for determination of glandularkallikrein (GK) activity; K_M:23.1 μM

Val-Arg-Arg-Phe-Pro-Trp-Trp-Trp-Pro-Phe-Leu-Arg-Arg

Synonyms: Tritrpticin

Bachem H-3908.0500   MW 1902.28 $C_{96}H_{132}N_{28}O_{14}$ Store at -15°C

Sigma T 0936   FW 1902.3   ≥90% (HPLC); peptide content ~65% | Bioactive peptide; broad-spectrum bactericidal peptide potent against gram-positive & gram-negative bacteria & *Aspergillus*; the name stems from the run of three Trp residues; Lawyer, C et al, *FEBS Lett*, 390:95, 1996

Val-Arg-Lys-Arg-Thr-Leu-Arg-Arg-Leu

Synonyms: Protein Kinase C Substrate; Protein Kinase Related Peptide

Alexis 165-013   MW 1197.5 $C_{51}H_{100}N_{22}O_{11}$ White lyophilized powder | Heasley, LE & Johnson, GL, *Mol Pharmacol*, 35:331, 1989

American Peptide 86-2-30   MW 1197.5 $C_{51}H_{100}N_{22}O_{11}$

Bachem H-3284.0001   MW 1197.5 $C_{51}H_{100}N_{22}O_{11}$ Store at -15°C

ICN 191457

Sigma V 2131   FW 1197.5   ≥97% (HPLC); peptide content:~65% | Bioactive peptide; House, C et al, *J Biol Chem*, 262:772, 1987

Val-Arg-Ser-Lys-Ile-Gly-Ser-Thr-Glu-Asn-Leu-Lys-His-Gln-Pro-Gly-Gly-Gly

Synonyms: Protein τ Fragment (187-204)

Bachem H-1954.0500   MW 1865.08 $C_{78}H_{133}N_{27}O_{26}$ Store at -15°C

Val-Asn

Bachem G-3815.0250   MW 231.25 $C_9H_{17}N_3O_4$ Store at -15°C

Val-Asn-Leu-Asp-Ala-Glu
DABCYL-Val-Asn-Leu-Asp-Ala-Glu-EDANS

Synonyms: Amyloid β/A4 Precursor Protein 770 (669-674)-EDANS, DABCYL-(Asn^{670},Leu^{671})-

Bachem M-2430.0001   MW 1159.29 $C_{54}H_{70}N_{12}O_{15}S$ Store at -15°C

Val-Asn-Leu-Asp-Ala-Glu-Lys
Mca-Val-Asn-Leu-Asp-Ala-Glu-Lys(Dnp)

Synonyms: Amyloid β/A4 Precursor Protein 770-Lys(Dnp) (669-674), Mca-(Asn^{670},Leu^{671})-

Bachem M-2440.0001   MW 1170.15 $C_{51}H_{67}N_{11}O_{21}$ Store at -15°C

Val-Asn-Thr-Pro-Glu-Arg-Val-Val-Pro-Tyr-Gly-Leu-Gly-Ser-Pro-Ser-Arg-Ser

Synonyms: Endothelin I (22-39), Big

American Peptide 88-5-51   Rat   MW 1915.2 $C_{83}H_{135}N_{25}O_{27}$

Val-Asn-Thr-Pro-Glu-Gln-Thr-Ala-Pro-Tyr-Gly-Leu-Gly-Asn-Pro-Pro

Synonyms: Endothelin II (22-37), Big

Alexis 152-015   Human   MW 1654.8 $C_{73}H_{111}N_{19}O_{25}$ Onda, H et al, *FEBS Lett*, 261:327, 1990

American Peptide 88-1-41   Human   MW 1654.8
$C_{73}H_{111}N_{19}O_{25}$ Onda, H et al, *FEBS Lett*, 261:327, 1990

Val-Asn-Thr-Pro-Glu-Gln-Thr-Ala-Pro-Tyr-Gly-Leu-Gly-Asn-Pro-Pro-Arg

Synonyms: Endothelin II (22-38), Big

American Peptide 88-1-38   Human   MW 1811.0
$C_{79}H_{123}N_{23}O_{26}$ Suzuki, N et al, *The 3rd Endothelin Symp, Tsukaba, Japan*, 1991

Val-Asn-Thr-Pro-Glu-His-Ile-Val-Pro-Tyr-Gly-Leu-Gly-Ser-Pro-Ser-Arg-Ser

Synonyms: Endothelin I (22-39), Big; Prepro-Endothelin I (74-91)

American Peptide 88-4-16   Porcine   MW 1910.1
$C_{84}H_{132}N_{24}O_{27}$ Itoh, Y et al, *FEBS Lett*, 231:440, 1988;
Yanagisawa, M et al, *Nature*, 332:411, 1988

Val-Asn-Thr-Pro-Glu-His-Val-Val-Pro-Tyr-Gly-Leu-Gly-Ser-Pro-Arg-Ser

Synonyms: Endothelin I (22-38), Big

Alexis 152-012   Human   MW 1809.0 $C_{80}H_{125}N_{23}O_{25}$ Itoh, T et al, *FEBS Lett*, 231:440, 1988

American Peptide 88-1-31   Human   MW 1809.1
$C_{80}H_{125}N_{23}O_{25}$ Itoh, Y et al, *FEBS Lett*, 231:440, 1988

Bachem H-7875.0500   Human   MW 1809.01 $C_{80}H_{125}N_{23}O_{25}$
Store at -15°C

Val-Asn-Thr-Pro-Glu-His-Val-Val-Pro-Tyr-Gly-Leu-Gly-Ser-Pro-Ser-Arg-Ser

Synonyms: Endothelin I (22-39), Big; Endothelin (22-39), Big

American Peptide 88-3-11   Bovine   MW 1896.27
$C_{83}H_{130}N_{24}O_{27}$ Elshourbagy, NA et al, *Nucleic Acid Res*, 18:4273, 1990

Sigma E 6014   Bovine   FW 1896.1   ≥97% (HPLC) |
Bioactive peptide; Elshourbagy, NA et al, *Nucleic Acid Research*, 18:4273, 1990

Val-Asp

Bachem G-3510.0250   MW 232.24 $C_9H_{16}N_2O_5$ Store at -15°C

ICN 158262   MW 232.2 $C_9H_{16}N_2O_5$

Val-Asp-Asp-Asp-Asp-Lys

Synonyms: Gastrin I Inhibitor

Biogenesis 4620-2609   Human synthetic   Purified; lyophilized

---

Val-Asp-Cys-Tyr-Phe-Gln-Asn-Cys-Pro-Arg-Gly Amide

Synonyms: Vasopressin, Val-Asp-(Arg[8])-

Bachem H-5265.0001   MW 1299.45 $C_{55}H_{78}N_{16}O_{17}S_2$ Store at -15°C

ICN 152971   Disulfide bonds: Cys[1]—Cys[6]; natriuretic activity similar to AVP; Gitelman, HJ etal, *Science*, 207:893, 1980

Sigma V 5253   FW 1298.4   ≥95% (HPLC); disulfide bonds: 1-6 | Bioactive peptide; originally isolated from bovine posterior pituitary, thought to be a storage form of Arg[8]-vasopressin; has natriuretic activity equivalent to Arg[8]-vasopressin with low pressor activity; Gitelman, HJ et al, *Science*, 207:893, 1980

Val-Asp-Gln-Met-Asp-Gly-Trp-Lys
MCA-Val-Asp-Gln-Met-Asp-Gly-Trp-(Lys-DNP)-Amide

BioSource International 77-885

Val-Asp-Gln-Met-Asp-Gly-Trp-Lys
MCA-Val-Asp-Gln-Met-Asp-Gly-Trp-Lys(DNP) Amide

Synonyms: Caspase III Fluorogenic Substrate V

Calbiochem 218751   MW 1359.4 $C_{60}H_{74}N_{14}O_{21}S$ ≥95% (HPLC); lyophilized solid; soluble in DMSO; excitation max:~325 nm; emission max:~392 nm | Fluorogenic resonance energy transfer substrate for Caspase III; Talanian, RV et al, *J Biol Chem*, 272:9677, 1997

Val-Asp-Gln-Val-Asp-Gly-Trp-Lys
MCA-Val-Asp-Gln-Val-Asp-Gly-Trp-(Lys-DNP)-Amide

BioSource International 77-894

Val-Asp-Gln-Val-Asp-Gly-Trp-Lys
MCA-Val-Asp-Gln-Val-Asp-Gly-Trp-Lys-(DNP) Amide

Synonyms: Caspase VII Fluorogenic Substrate I

Calbiochem 218768   MW 1327.3 $C_{60}H_{74}N_{14}O_{21}$ ≥97% (HPLC); lyophilized solid; soluble in DMSO; excitation max:~325 nm; emission max:~392 nm | Fluorogenic resonance energy transfer substrate for the detection of Caspase VII; Talanian, RV et al, *J Biol Chem*, 272:9677, 1997

Val-Asp-Pro-Glu-Leu-Ala-Asp-Gln-Leu-Ile-His-Leu-Tyr

Synonyms: VIF (98-110); HTLV I Peptide

Neosystem SC233   MW 1525.72   HTLV I peptide from subtype VIF

Val-Asp-Trp-Lys-Lys-Ile-Gly-Gln-His-Ile-Leu-Ser-Val-Leu Amide

Synonyms: Mastoparan; Mastoparan, Polistes

Bachem H-9450.0001   Store at -15°C

ICN 159963   *Polistes jadwagae*   MW 1635.9   97% | A mast cell degranulating peptide & γ-protein activator; Nakajima, T etal, *Peptides*, 6:425, 1985

Sigma M 3545   *Polistes jadwagae*   FW 1635.0   ≥97% (HPLC) | Bioactive peptide; mast cell degranulating peptide; stimulates phosphoinositide breakdown; binds calmodulin & inhibits its activity; degranulates mast cells; Nakajima, T et al, *Peptides*, 6:425, 1985

American Peptide 87-0-68   Wasp   MW 1635.0
$C_{77}H_{127}N_{21}O_{18}$ Hirai, Y et al, *Biomed Res*, 1:185, 1980

Val-Asp-Val-Ala
Z-Val-Asp-Val-Ala-Asp-FMK

Synonyms: Caspase II/III Inhibitor

Kamiya   MW 695.6

Val-Asp-Val-Ala-Asp
Ac-Val-Asp-Val-Ala-Asp-CHO

Synonyms: Caspase II Inhibitor; Caspase III Inhibitor; Caspase VII Inhibitor

Alexis 260-058   MW 543.6 $C_{23}H_{37}N_5O_{10}$ White solid; soluble in DMSO | Potent inhibitor; Talanian, RV et al, *J Biol Chem*, 272:9677, 1997

Val-Asp-Val-Ala-Asp
Ac-Val-Asp-Val-Ala-Asp-pNA

Synonyms: Caspase II Substrate

Alexis 260-059   MW 679.7 $C_{29}H_{41}N_7O_{12}$ White solid; soluble in DMSO | Preferred chromogenic substrate; Talanian, RV et al, *J Biol Chem*, 272:9677, 1997

Val-Asp-Val-Ala-Asp
Ac-Val-Asp-Val-Ala-Asp-AMC

Synonyms: Caspase II Substrate

Alexis 260-060   MW 716.8 $C_{33}H_{44}N_6O_{12}$ ≥98%; white powder; soluble in basic pH buffers | Fluorogenic substrate; AMC has an excitation maximum of 380 nm & an emission maximum of 460 nm; Talanian, RV et al, *J Biol Chem*, 272:9677, 1997

Val-Asp-Val-Ala-Asp
Z-Val-Asp-Val-Ala-Asp-AFC

Bachem I-1770.0001   MW 862.81 $C_{39}H_{45}F_3N_6O_{13}$ Store at -15°C

Val-Asp-Val-Ala-Asp
Z-Val-Asp-Val-Ala-Asp-pNA

Bachem L-2060.0001   MW 771.78 $C_{35}H_{45}N_7O_{13}$ Store at -15°C

Val-Asp-Val-Ala-Asp
Ac-Val-Asp-Val-Ala-Asp-pNA

Bachem L-2065.0001   MW 679.68 $C_{29}H_{41}N_7O_{12}$ Store at -15°C

Val-Asp-Val-Ala-Asp
Z-Val-Asp-Val-Ala-Asp-AFC

BioSource International 77-843

Val-Asp-Val-Ala-Asp
Ac-Val-Asp-Val-Ala-Asp-pNA

BioSource International 78-133

Val-Asp-Val-Ala-Asp
Z-Val-Asp-Val-Ala-Asp-AFC

Synonyms: Caspase II Fluorogenic Substrate I

Calbiochem 218740   MW 862.8 $C_{39}H_{45}F_3N_6O_{13}$ ≥97% (HPLC); lyophilized solid; soluble in DMSO; excitation max:~400 nm; emission max:~505 nm | Fluorogenic substrate for the detection of Caspase II (ICH-1); reaction monitored visually or quantitatively by a blue to green shift in fluorescence upon cleavage of the AFC fluorophore; Talanian, RV et al, *J Biol Chem*, 272:9677, 1997

Val-Asp-Val-Ala-Asp
Z-Val-Asp(OMe)-Val-Ala-Asp(OMe)-CH2F

Synonyms: Caspase II Inhibitor I

Calbiochem 218744   MW 695.7 $C_{32}H_{46}FN_5O_{11}$ ≥95% (TLC); lyophilized solid; soluble in DMSO | Irreversible inhibitor of Caspase II

Val-Asp-Val-Ala-Asp
Ac-Val-Asp-Val-Ala-Asp-AFC

Synonyms: Caspase II/III Substrate

Kamiya   MW 772   Fluorogenic

Val-Asp-Val-Ala-Asp
Ac-Val-Asp-Val-Ala-Asp-pNA

Synonyms: Caspase II/III Substrate

Kamiya   MW 679   >97% (HPLC)

Val-Asp-Val-Ala-Asp
Ac-Val-Asp-Val-Ala-Asp-MCA

Synonyms: Caspase II Substrate

Peptides International MCA-3203-v   MW 716.74 $C_{33}H_{44}N_6O_{12}$ >98% (HPLC); lyophilized amorphous powder | Talanian, RV et al, *JBC*, 272:9677, 1997

Val-Asp-Val-Ala-Asp
Ac-Val-Asp-Val-Ala-Asp-CHO

Synonyms: Caspase II Inhibitor

Peptides International IVD-3204-v   Synthetic   MW 543.57 $C_{23}H_{37}N_5O_{10}$ >99% (HPLC); lyophilized amorphous powder | Inhibitor; Talanian, RV et al, *JBC*, 272:9677, 1997

Val-Asp-Val-Ala-Asp-Gly-Trp-Lys
MCA-Val-Asp-Val-Ala-Asp-Gly-Trp-(Lys-DNP)-Amide

BioSource International 77-888

Val-Asp-Val-Ala-Asp-Gly-Trp-Lys
MAC-Val-Asp-Val-Ala-Asp-Gly-Trp-Lys-(DNP) Amide

Synonyms: Caspase II Fluorogenic Substrate II

Calbiochem 218741   MW 1270.3 $C_{58}H_{71}N_{13}O_{20}$ ≥97% (HPLC); lyophilized solid; soluble in DMSO; excitation max:~325 nm; emission max:~392 nm | Fluorogenic resonance energy transfer substrate for the detection of Caspase II; Talanian, RV et al, *J Biol Chem*, 272:9677, 1997

Val-Cys-Phe-Thr-Thr-Lys-Ala-Leu-Gly-Ile-Ser-Tyr-Gly-Arg-Lys

Synonyms: TAT (36-50); HTLV I Peptide

Neosystem SC208   MW 1643.96   HTLV I peptide from subtype TAT; due to the presence of a Cys this peptide can contain the dimeric form

Val-Cys-Phe-Thr-Thr-Thr-Ala-Leu-Gly-Ile-Ser-Tyr-Gly-Arg-Lys

Synonyms: TAT (36-50), (Thr[41])-; HTLV I Peptide

Neosystem SC850   MW 1616.89   HTLV I peptide from subtype TAT; due to the presence of a Cys this peptide can contain the dimeric form

Val-Cys-Ser-Cys-Arg-Leu-Val-Phe-Cys-Arg-Arg-Thr-Glu-Leu-Arg-Val-Gly-Asn-Cys-Leu-Ile-Gly-Gly-Val-Ser-Phe-Thr-Tyr-Cys-Cys-Thr-Arg-Val

Synonyms: Corticostatin; Neutrophil Peptide IV, Human

American Peptide 34-1-15   Human   MW 3709.5 $C_{157}H_{255}N_{49}O_{43}S_6$ Disulfide bonds: Cys[2]-Cys[30], Cys[4]-Cys[19], Cys[9]-Cys[29]

Val-Gln

Bachem G-3925.0250   MW 245.28 $C_{10}H_{19}N_3O_4$ Store at -15°C

ICN 158269   MW 245.3 $C_{10}H_{19}N_3O_4$ Crystalline

Sigma V 5252   FW 245.3 $C_{10}H_{19}N_3O_4$

Val-Gln-Ala-Ala-Ile-Asp-Tyr-Ile-Asn-Gly

Synonyms: Acyl Carrier Protein (65-74)

Sigma A 6700   FW 1063.2   ≥95% (HPLC); peptide content:~70% | Bioactive peptide

Val-Gln-Ala-Ala-Ile-Asp-Tyr-Ile-Asn-Gly Acid

Synonyms: Acyl Carrier Protein (65-74)

Bachem H-8810.0001   MW 1063.18 $C_{47}H_{74}N_{12}O_{16}$ Store at -15°C

Val-Gln-Ala-Ala-Ile-Asp-Tyr-Ile-Asn-Gly Amide

Synonyms: Acyl Carrier Protein (65-74)

Bachem H-8815.0001   MW 1062.19 $C_{47}H_{75}N_{13}O_{15}$ Store at -15°C

Val-Gln-Gln-Phe-Asp-Pro-Thr-Ala-Lys-Asp-Leu-Gln-Asp-Leu-Leu-Gln-Tyr

Synonyms: P24 (175-191); HTLV I Peptide

Neosystem SC834   MW 2022.24   HTLV I peptide

Val-Gln-Glu-Ser-Ala-Asp-Gly-Tyr-Arg-Met-Gln-His-Phe-Arg-Trp-Gly-Gln-Pro-Leu-Pro Amide

Synonyms: Melanocyte Stimulating Hormone B

Bachem H-3566.0001   Store at -15°C

Val-Gln-Gly-Glu-Glu-Ser-Asn-Asp-Lys

Synonyms: Interleukin I (163-171), β-; Interleukin Iβ (163-171)

American Peptide 85-0-52   Human   MW 1005.0
$C_{39}H_{64}N_{12}O_{19}$ Perin, F et al, *Abstr 6th Int Symp on HPLC*, 1986

Bachem H-7010.0005   Human   MW 1005.01 $C_{39}H_{64}N_{12}O_{19}$
Store at -15°C

Neosystem SC340   Human   MW 1005.00   Bioactive;
synthetic growth factor-related peptide; immuno-stimulating
peptide; Forni, G et al, *J Immunol*, 142:712, 1989

Sigma I 4638   Human   FW 1005.0   ≥97% (HPLC) |
Bioactive peptide; T cell activator without the inflammatory
properties of Interleukin I; Antoni, G et al, *J Immunol*, 137:3201,
1986; Nencioni, L et al, *J Immunol*, 139:800, 1987

Val-Gln-Tyr-Pro-Val-Glu-His-Pro-Asp-Lys-Phe-Leu-Lys-Phe-Gly-Met-Thr-Pro-Ser-Lys-Gly-Val-Leu-Phe-Tyr

Synonyms: Peptide VQY; Valosin Peptide; Valosin

American Peptide 46-4-19   Porcine   MW 2928.5
$C_{141}H_{207}N_{31}O_{35}S$ Simulates pancreatic secretion; Schmidt, WE et
al, *FEBS Lett*, 191:264, 1985

Bachem H-5495.0500   Porcine   MW 2928.45
$C_{141}H_{207}N_{31}O_{35}S$ Store at -15°C

ICN 152895   Porcine   Simulates pancreatic secretion;
Schmidt, WE et al, *FEBS Lett*, 191:264, 1985

Sigma V 6505   Porcine   FW 2928.4   ≥97% (HPLC) |
Bioactive peptide; gastrointestinal peptide; stimulates pancreatic
secretion; Schmidt, WE et al, *FEBS Lett*, 191:264, 1985

Val-Gln-Val-Asp-Gly-Trp-Lys
MCA-Val-Gln-Val-Asp-Gly-Trp-(Lys-DNP)-Amide

BioSource International 77-891

Val-Glu

Bachem G-3520.0250   MW 246.26 $C_{10}H_{18}N_2O_5$ Store at -15°C

ICN 158270   MW 246.3 $C_{10}H_{18}N_2O_5$

Sigma V 5127   FW 246.3 $C_{10}H_{18}N_2O_5$

Val-Glu-Arg-Gly
Val-D-Glu(D-Arg-Gly)

Synonyms: Ginseng Tetrapeptide; Val-γ-D-Glu-D-Arg-Gly

Bachem H-3996.0005   MW 459.5 $C_{18}H_{33}N_7O_7$ Store at -15°C

Val-Glu-Glu-Ala-Glu

Synonyms: Thymosin α_1 (23-27)

ICN 156227   MW 575.6 $C_{23}H_{37}N_5O_{12}$ 97%

Sigma V 3004   FW 575.6 $C_{23}H_{37}N_5O_{12}$ ≥97% (HPLC) |
Bioactive peptide

Val-Glu-His-Asp
Ac-Val-Glu-His-Asp-AFC

Bachem I-1890.0005   MW 751.67 $C_{32}H_{36}F_3N_7O_{11}$ Store at -15°C

Val-Glu-His-Asp
Ac-Val-Glu-His-Asp-AMC

BioSource International 78-106   Fluorescing substrate

Val-Glu-Ile-Asp
Ac-Val-Glu-Ile-Asp-CHO

Synonyms: Caspase VI Inhibitor

Alexis 260-062   MW 500.6 $C_{22}H_{36}N_4O_9$ ≥98%; white
powder; soluble in DMSO | Potent inhibitor

Val-Glu-Ile-Asp
Ac-Val-Glu-Ile-Asp-pNA

Synonyms: Caspase VI Substrate

Alexis 260-063   MW 636.7 $C_{28}H_{40}N_6O_{11}$ ≥98%; white
powder; soluble in DMSO | Chromogenic substrate; Talanian, RV
et al, *J Biol Chem*, 272:9677, 1997

Val-Glu-Ile-Asp
Ac-Val-Glu-Ile-Asp-AMC

Synonyms: Caspase VI Substrate

Alexis 260-064   MW 673.8 $C_{32}H_{43}N_5O_{11}$ ≥98%; white
powder; soluble in DMSO | Fluorogenic substrate; AMC has an
excitation maximum of 380 nm & an emission maximum of 460
nm; Martins, LM et al, *J Biol Chem*, 272:7421, 1997; Takahashi, A
et al, *PNAS*, 93:8395, 1996; Nagata, S, *Cell*, 88:355, 1997;
Talanian, RV et al, *J Biol Chem*, 272:9677, 1997; Thornberry, NA et
al, *J Biol Chem*, 272:17907, 1997

Val-Glu-Ile-Asp
Ac-Val-Glu-Ile-Asp-CHO

Synonyms: Caspase VI

American Peptide 81-7-04   MW 500.6 $C_{22}H_{36}N_4O_9$

Val-Glu-Ile-Asp
Ac-Val-Glu-Ile-Asp-AMC

Bachem I-1755.0005   MW 673.72 $C_{32}H_{43}N_5O_{11}$ Store at -15°C

Val-Glu-Ile-Asp
Z-Val-Glu-Ile-Asp-AFC

Bachem I-1765.0005   MW 819.79 $C_{38}H_{44}F_3N_5O_{12}$ Store at -15°C

Val-Glu-Ile-Asp
Ac-Val-Glu-Ile-Asp-pNA

Bachem L-1995.0005   MW 636.66 $C_{28}H_{40}N_6O_{11}$ Store at -15°C

Val-Glu-Ile-Asp
Ac-Val-Glu-Ile-Asp-CHO Pseudo Acid

Bachem N-1640.0005   Store at -15°C

Val-Glu-Ile-Asp
Z-Val-Glu-Ile-Asp-pNA

BioSource International 77-858

Val-Glu-Ile-Asp
Ac-Val-Glu-Ile-Asp-pNA

BioSource International 77-861
BioSource International 77-862

Val-Glu-Ile-Asp
Z-Val-Glu-Ile-Asp-AFC

BioSource International 77-864

Val-Glu-Ile-Asp
Ac-Val-Glu-Ile-Asp-AFC

BioSource International 77-867

Val-Glu-Ile-Asp
Val-Glu-Ile-Asp-CHO

BioSource International 78-118   Cell permeable inhibitor

Val-Glu-Ile-Asp
Z-Val-Glu(OMe)-Ile-Asp(OMe)-CH$_2$F

Synonyms: Caspase VI Inhibitor I

Calbiochem 218757   MW 652.7 $C_{31}H_{45}FN_4O_{10}$ ≥90% (TLC); solid; soluble in DMSO | Irreversible inhibitor of Caspase VI (Mch-2)

Val-Glu-Ile-Asp
Ac-Val-Glu-Ile-Asp-AMC

Synonyms: Caspase VI Fluorogenic Substrate I

Calbiochem 218760   MW 673.7 $C_{32}H_{43}N_5O_{11}$ ≥98% (HPLC); lyophilized solid; soluble in DMSO; excitation max:~360-380 nm; emission max:~430-460 nm | Fluorogenic substrate for Caspase VI; Takahashi, A et al, *Exp Cell Res*, 231:123, 1997

Val-Glu-Ile-Asp
Ac-Val-Glu-Ile-Asp-*p*NA

Synonyms: Caspase VI Colorimetric Substrate II

Calbiochem 218762   MW 636.7 $C_{28}H_{40}N_6O_{11}$ ≥97% (HPLC); lyophilized solid; soluble in DMSO | Colorimetric substrate for Caspase VI; sequence is based on the P_1-P_4 cleavage is monitored colorimetrically at ~405 nm; Thornberry, NA et al, *J Biol Chem*, 272:17907, 1997

Val-Glu-Ile-Asp
Z-Val-Glu-Ile-Asp-AFC

Synonyms: Caspase VI Fluorogenic Substrate III

Calbiochem 218763   MW 819.8 $C_{38}H_{44}F_3N_5O_{12}$ ≥97% (HPLC); lyophilized solid; soluble in DMSO; excitation max:~400 nm; emission max:~505 nm | Fluorogenic substrate for Caspase VI; reaction monitored visually or quantitatively by a blue to green shift in fluorescence upon cleavage of the AFC fluorophore; Talanian, RV et al, *J Biol Chem*, 272:9677, 1997

Val-Glu-Ile-Asp
Z-Val-Glu-Ile-Asp-FMK

Synonyms: Caspase VI Inhibitor

Kamiya   MW 652

Val-Glu-Ile-Asp
Ac-Val-Glu-Ile-Asp-CHO

Synonyms: Caspase VI Inhibitor

Kamiya   MW 500.5 $C_{22}H_{36}N_4O_9$ >95% (HPLC)

Val-Glu-Ile-Asp
Ac-Val-Glu-Ile-Asp-AFC

Synonyms: Caspase VI Substrate

Kamiya   MW 727   Fluorogenic

Val-Glu-Ile-Asp
Ac-Val-Glu-Ile-Asp-*p*NA

Synonyms: Caspase VI Substrate

Kamiya   MW 636 $C_{28}H_{40}N_6O_{11}$ >97% (HPLC)

Val-Glu-Ile-Asp
Ac-Val-Glu-Ile-Asp-AMC

Synonyms: Caspase VI Substrate

Neosystem SC1315   MW 673.72   Deduced from the cleavage site of Lamin A; Takahashi, A et al, *PNAS USA*, 93:8395-8400, 1996; Orth, K et al, *JBC*, 271:16443-16446, 1996

Val-Glu-Ile-Asp
Ac-Val-Glu-Ile-Asp-*p*NA

Synonyms: Caspase VI Substrate

Neosystem SC1316   MW 636.64   Chromogenic substrate

Val-Glu-Ile-Asp
Ac-Val-Glu-Ile-Asp-MCA

Synonyms: Caspase VI Substrate

Peptides International MVA-3181-v   MW 673.72 $C_{32}H_{43}N_5O_{11}$ >98% (HPLC); lyophilized amorphous powder | Takahashi, A, *Exp Cell Res*, 231:123, 1997; Takahashi, A et al, *Oncogene*, 14:2741, 1997

Val-Glu-Ile-Asp
N-Ac-Val-Glu-Ile-Asp-MCA

Synonyms: Caspase VI Substrate

Sigma A 0341   Fluorogenic substrate; Martins, LM et al, *JBC*, 272:7421, 1997; Nagata, S, *Cell*, 88:355, 1997

Val-Glu-Ile-Asp
N-Ac-Val-Glu-Ile-Asp-CHO

Sigma A 6339   FW 500.5 $C_{22}H_{36}N_4O_9$ Bioactive peptide

Val-Glu-Ile-Asp
Ac-Val-Glu-Ile-Asp-*p*NA

Synonyms: Caspase VI Substrate, Chromogenic

Upstate 12-388   MW 637.5   Lyophilized | Caspase VI assay

Val-Glu-Ile-Asp
Ac-Val-Glu-Ile-Asp-CHO

Synonyms: Caspase VI Inhibitor

Peptides International IVA-3182-v   Synthetic   MW 500.55 $C_{22}H_{36}N_4O_9$ >99% (HPLC); lyophilized amorphous powder | Inhibitor; Hirata, H et al, *J Exp Med*, 187:587, 1998

Val-Glu-Met-Gly-His-His-Ala-Pro-Trp-Asp-Ile-Asp-Asp-Leu

Synonyms: VPU (68-81); HTLV I Peptide

Neosystem SC533   MW 1634.78   HTLV I peptide from subtype VPU

Val-Glu-Pro-Asp-Lys-Val-Glu-Glu-Ala-Asn-Lys-Gly

Synonyms: NEF (148-159); HTLV I Peptide

Neosystem SC224   MW 1314.41   HTLV I peptide from subtype NEF

Val-Glu-Pro-Glu-Lys-Ile-Glu-Glu-Ala-Asn-Lys-Gly

Synonyms: NEF MN (150-161); HIV I Peptide

Neosystem SC660   MW 1343.45   NEF peptide from HIV-1 subtype MN

Val-Glu-Pro-Ile-Pro-Tyr

Synonyms: Caseins Immunostimulating Peptide, Human; Immunostimulating Peptide

ICN 156360   97% | Gattogno, L etal, *Immunol Lett*, 18:27, 1988

Sigma V 2256   FW 716.8 $C_{35}H_{52}N_6O_{10}$ ≥97% (HPLC) | Bioactive peptide; Gattogno, L et al, *Immunol Lett*, 18:27, 1988

Peptides—Sequences & Modifications

American Peptide 74-1-25   Human   MW 716.8
$C_{35}H_{52}N_6O_{10}$ Antibiotic peptide

Val-Glu-Ser-Ser-Lys

Synonyms: A-VI-5; Bradykinin Potentiating Factor A-VI-5

Bachem H-2220.0005   MW 548.59 $C_{22}H_{40}N_6O_{10}$ Store at -15°C

ICN 152794   MW 548.6 $C_{22}H_{40}N_6O_{10}$ Ufkes, JGN etal, *Eur J Pharmac*, 40:137, 1976

Val-Glu-Thr-Ser-Asp-His-Asp-Asn-Ser-Leu-Ser-Val-Ser-Ile-Cys Trifluoroacetate Salt

Synonyms: Aggrecan (1-14), ~C

Biogenesis 0195-8080   Human synthetic   Purified; lyophilized | An additional Cys has been conjugated to the C-terminus

Val-Gly
BOC-Val-Gly

Bachem A-2735.0001   MW 274.32 $C_{12}H_{22}N_2O_5$ Store at -15°C

Val-Gly
Z-Val-Gly

Bachem C-2850.0001   MW 308.33 $C_{15}H_{20}N_2O_5$ Store at -15°C

Val-Gly
Z-Val-Gly-OEt

Bachem C-3290.0001   MW 336.39 $C_{17}H_{24}N_2O_5$ Store at -15°C

Val-Gly

Bachem G-3525.0001   MW 174.2 $C_7H_{14}N_2O_3$ Store at -15°C
ICN 104908   MW 174.2 $C_7H_{14}N_2O_3$ Crystalline
Sigma V 1375   FW 174.2 $C_7H_{14}N_2O_3$

Val-Gly Amide Hydrochloride

Bachem G-3530.0500   Store at -15°C

Val-Gly-Ala-Leu-Ala-Val-Val-Val-Trp-Leu-Trp-Leu-Trp-Leu-Trp
HCO-Val-Gly-Ala-D-Leu-Ala-D-Val-Val-D-Val-Trp-D-Leu-Trp-D-Leu-Trp-D-Leu-Trp-NHCH2CH2OH

Synonyms: Gramicidin A

Alexis 350-233   *Bacillus brevis*   MW 1882.3 $C_{99}H_{140}N_{20}O_{17}$
High purity; ≥95%; off-white solid; contaminants:gB and gC isomers <1%; soluble in acetic acid or ethanol | Naturally-occurring ion channel-forming pentadecapeptide; causes K^+/H^+-exchange in mitochondria in a non-voltage dependent manner; Stankovic, CJ et al, *Anal Biochem*, 184:100, 1990; Koeppe-2nd, RE et al, *Biochemistry*, 34:9299, 1995

Val-Gly-Ala-Leu-Ala-Val-Val-Val-Trp-Leu-Trp-Leu-Trp-Leu-Trp
OHC-Val-Gly-Ala-D-Leu-Ala-D-Val-Val-D-Val-Trp-D-Leu-Trp-D-Leu-Trp-D-Leu-Trp-NHCH2CH2OH

Synonyms: Gramicidin A

Calbiochem 368020   *Bacillus brevis*   MW 1882.3
$C_{99}H_{140}N_{20}O_{17}$ High purity:>95% (HPLC); solid; soluble in DMSO; gB & gC isomers:<1% (HPLC); LD_{50}≤2000 mg/kg | Naturally-occurring ion channel-forming pentadecapeptide; causes K^+/H^+ exchange in mitochondria in a non-voltage dependent manner; *Merck Index*, 12:4553; Stankovic, CJ et al, *Anal Biochem*, 184:100, 1990; Koeppe, RE et al, *Biochemistry*, 34:9299, 1995

Val-Gly-Ala-Leu-Ala-Val-Val-Val-Trp-Leu-Trp-Leu-Trp-Leu-Trp
OHC-NVal-Gly-Ala-D-Leu-Ala-D-Val-Val-D-Val-Trp-D-Leu-Trp-D-Leu-Trp-D-Leu-Trp-CONHCH2CH2OH

Synonyms: Gramicidin A

ICN 194992   *Bacillus brevis*   MW 1924.9;   Highly purified gramicidin isomer; >95% | Stankovic, CJ etal, *Anal Biochem*, 184:100, 1990

Val-Gly-Arg
Bz-Val-Gly-Arg-AMC

Bachem I-1085.0050   MW 591.67 $C_{30}H_{37}N_7O_6$ Store at -15°C

Val-Gly-Arg
BOC-Val-Gly-Arg-AMC Hydrochloride

Bachem I-1110.0050   MW 624.14 $C_{28}H_{41}N_7O_7 \cdot HCl$ Store at -15°C

Val-Gly-Arg
BOC-Val-Gly-Arg-β-Naphthylamide Acetate Salt

Synonyms: Plasminogen Activator Substrate

Calbiochem 528198   MW 615.7 $C_{28}H_{41}N_7O_5 \cdot C_2H_4O_2$ Solid; ≥95% (TLC); soluble in ethanol & water | Substrate for the quantitative determination of plasminogen activator; Nieuwenhuizen, W et al, *Anal Biochem*, 83:143 1977; Dooijewaard, G & Kluft, C, *Adv Exp Med Biol*, 156A:115, 1983; Dooijewaard, G & Kluft, C, *Thromb Haemostas*, 46:63, 1981

ICN 195934   MW 615.7 $C_{28}H_{41}N_7O_5 \cdot C_2H_4O_2$ Substrate for quantitative determination of PA

Val-Gly-Arg
Bz-Val-Gly-Arg-pNA·AcOH

Synonyms: Pefachrome®Tyr; Pefa-5400

Pentapharm 084-20, 084-04   MW 614.7   Highly sensitive chromogenic peptide substrate for trypsin; v_{max}:43.3 μmol/min, K_M:0.181 *mM*

Val-Gly-Arg
N-Bz-Val-Gly-Arg-pNA Hydrochloride

Sigma B 4758   FW 591.1 $C_{26}H_{34}N_8O_6 \cdot HCl$

Val-Gly-Arg
N-CBZ-Val-Gly-Arg-pNA Acetate Salt Free Base

Sigma C 7271   FW 584.6 $C_{27}H_{36}N_8O_7$

Val-Gly-Asp-Glu

Synonyms: Eosinophilotactic Tetrapeptide; Chemotactic Peptide

Bachem H-2910.0050   MW 418.4 $C_{16}H_{26}N_4O_9$ Store at -15°C

ICN 152775   Goetzl, EJ & F Austen, *PNAS*, 72:4123, 1975; Turnbull, LW etal, *Immunology*, 32:57, 1977; Beswick, PH & AB Kay, *Clin Exp Immunol*, 43:399, 1981

Sigma V 6253   FW 418.4 $C_{16}H_{26}N_4O_9$ ≥90% (HPLC) | Bioactive peptide

Val-Gly-Gly
Z-Val-Gly-Gly-OBzl

Bachem C-2855.0001   MW 455.51 $C_{24}H_{29}N_3O_6$ Store at -15°C

Val-Gly-Gly

Bachem H-5270.0250   MW 231.25 $C_9H_{17}N_3O_4$ Store at -15°C
ICN 103237   MW 231.3 $C_9H_{17}N_3O_4$ 99%; crystalline
Sigma V 1500   FW 231.3 $C_9H_{17}N_3O_4$

Val-Gly-Gly-Ser-Glu-Ile

Synonyms: C-Reactive Protein (77-82)

Bachem H-1436.0005   MW 560.61 $C_{23}H_{40}N_6O_{10}$ Store at -15°C

Sigma C 3580   FW 560.6 $C_{23}H_{40}N_6O_{10}$ ≥97% (HPLC) | Bioactive peptide; interferes with the membrane-associated oxidative metabolism in human neutrophils; Shephard, EG et al, *Immunology*, 76:79, 1992

Val-Gly-Gly-Tyr-Gly-Tyr-Gly-Ala-Lys

ICN 156361   MW 871 $C_{40}H_{58}N_{10}O_{12}$ 97%

Sigma V 0757   FW 871.0 $C_{40}H_{58}N_{10}O_{12}$ ≥97% (HPLC); peptide content:~70% | Bioactive peptide

Val-Gly-Phe
Z-Val-Gly-Phe

Bachem C-2860.0250   MW 455.51 $C_{24}H_{29}N_3O_6$ Store at -15°C

Val-Gly-Phe-Pro-Val-Lys-Pro-Gln-Val-Pro-Leu-Arg-Pro-Met-Thr

Synonyms: NEF MN (68-82); HIV I Peptide

Neosystem SC647   MW 1667.04   NEF peptide from HIV-1 subtype MN

Val-Gly-Phe-Pro-Val-Thr-Pro-Gln-Val-Pro-Leu-Arg-Pro-Met-Thr

Synonyms: NEF (66-80); HTLV I Peptide

Neosystem SC220   MW 1638.98   HTLV I peptide from subtype NEF

Val-Gly-Ser-Glu

Synonyms: Eosinophilotactic Peptide; Corticotropin Releasing Factor; Eosinophilotactic Tetrapeptide; Eosinophil Chemotactic Factor Of Anaphylaxis

American Peptide 40-0-01   MW 390.4 $C_{15}H_{26}N_4O_8$

Bachem H-2915.0050   MW 390.39 $C_{15}H_{26}N_4O_8$ Store at -15°C

ICN 152776   Goetzl, EJ & F Austen, *PNAS*, 72:4123, 1975; Turnbull, LW etal, *Immunology*, 32:57, 1977; Beswick, PH & AB Kay, *Clin Exp Immunol*, 43:399, 1981

Sigma V 8878   FW 390.4 $C_{15}H_{26}N_4O_8$ ≥97% (HPLC) | Bioactive peptide; Turnbull, LW et al, *Immunology*, 32:57, 1977

Val-Gly-Val-Ala-Pro-Gly

Synonyms: Chemotactic Domain of Elastin; Elastin Repeating Peptide; Chemotactic Peptide

Bachem H-2390.0005   MW 498.58 $C_{22}H_{38}N_6O_7$ Store at -15°C

ICN 152777   Chemotactic toward fibroblasts & monocytes; Senior, RM etal, *J Cell Biol*, 99:870, 1984

Sigma V 1008   FW 498.6 (free base) $C_{22}H_{38}N_6O_7$ ≥97% (HPLC) | Bioactive peptide; repeating peptide in elastin, chemotactic toward fibroblasts & monocytes; Senior, RM et al, *J Cell Biol*, 99:870, 1984

Val-Gly-Val-Arg-Val-Arg

American Peptide 89-0-20   MW 684.8 $C_{29}H_{56}N_{12}O_7$

Val-Gly-Val-Ile-Leu-Leu-Arg-Ile-Val-Ile-Tyr-Ile-Val-Gln-Met-Leu-Ala-Lys-Leu-Arg

Synonyms: GP140 (701-720); SIVmac251 Peptide

Neosystem SC778   MW 2310.99   SIVmac251 peptide from subtype GP140

Val-His

Bachem G-3535.0250   MW 254.29 $C_{11}H_{18}N_4O_3$ Store at -15°C

Val-His-His-Gln-Lys-Leu-Val-Phe-Phe-Ala-Glu-Asp-Val-Gly-Ser-Asn-Lys
β-Val-His-His-Gln-Lys-Leu-Val-Phe-Phe-Ala-Glu-Asp-Val-Gly-Ser-Asn-Lys

Synonyms: Amyloid (12-28)

American Peptide 62-0-73   MW 1955.2 $C_{89}H_{135}N_{25}O_{25}$ Impairs post-training memory with remarkable efficacy & in a dose-dependent manner after injection into the limbic system in mice; with other amyloid β-protein fragments, exerts dysregulatory cognitive effects by incoordination of K^+ channel function in neurons, glia & endothelial cells; Castano, EM et al, *BBRC*, 141:782, 1986

Val-His-His-Gln-Lys-Leu-Val-Phe-Phe-Ala-Glu-Asp-Val-Gly-Ser-Asn-Lys

Synonyms: Amyloid β-Protein (12-28); Alzheimer's Disease β-Protein (12-28); Amyloid (12-28), β-; Amyloid Peptide (12-28), β-

Bachem H-7910.0001   MW 1955.2 $C_{89}H_{135}N_{25}O_{25}$ Store at -15°C

Biogenesis 0490-1932   Human synthetic   MW 1955.5 Purified; lyophilized | Castano et al, *BBRC*, 141:782, 1986

ICN 153046   MW 1955.2   Castano, EM etal, *BBRC*, 141:782, 1986

Neosystem SC947   MW 1955.2   Bioactive; Castano, EM et al, *BBRC*, 141:782-789, 1986; Flood, JF et al, *PNAS* USA, 91:380-384, 1994; Abe, E et al, *Brain Res* 636:162-164, 1994

Sigma A 3180   FW 1955.2   ≥97% (HPLC) | Bioactive peptide; impairs post-training memory in mice upon injection into limbic system structures; Castano, EM et al, *Biochem Biophys Res Commun*, 141:782, 1986

Sigma A 8677   FW 1955.2   ≥97% (HPLC) | Bioactive peptide

Val-His-Leu-Thr-Pro

Synonyms: Sickle Cell Hemoglobin β-Chain, Amino-Terminal Pentapeptide

Bachem H-5275.0005   MW 565.67 $C_{26}H_{43}N_7O_7$ Store at -15°C

ICN 156435   MW 565.7 $C_{26}H_{43}N_7O_7$ 90%

Sigma V 3505   FW 565.7 $C_{26}H_{43}N_7O_7$ ≥90% (HPLC) | Bioactive peptide

Val-His-Leu-Thr-Pro-Val-Glu-Lys Acetate

Synonyms: Sickle Cell Hemoglobin β-Chain, Amino-Terminal Region

Sigma V 5130   FW 922.1 (free base) $C_{42}H_{71}N_{11}O_{12}$ ≥97% (HPLC) | Bioactive peptide; Ingram, VM, *Nature*, 180:326, 1957; Eastlake, A et al, *J Biol Chem*, 251:6426, 1976

Val-His-Leu-Thr-Pro-Val-Glu-Lys Acetate Salt

ICN 156496   MW 922.1 (free base) $C_{42}H_{71}N_{11}O_{12}$ 97% | Hbs β-chain amino-terminal region; Ingram VM, *Nature*, 180:326, 1957; Eastlake, A etal, *JBC*, 251:6426, 1976

Val-His-Phe-Phe-Lys-Asn-Ile-Val-Thr-Ala-Arg-Thr-Pro

Synonyms: Myelin Basic Protein (87-99), (Ala[96])-; Experimental Allergic Encephalomyelitis Inducing Peptide

Bachem H-3392.0001   Human, bovine, rat   Store at -15°C

Neosystem SC1245   Human, guinea pig   MW 1529.80 Bioactive; when clone L10C1 (a specific T-cell clone for the epitope MBP (87-99)) is tolerized *in vivo* with this peptide, established paralysis is reversed, inflammatory infiltrates regress, & the heterogeneous T-cell infiltrate disappears from the brain, with only the T-cell clones that incited disease remaining in the original lesions; Brocke, S et al, *Nature*, 379:343-346, 1996

Val-His-Phe-Phe-Lys-Asn-Ile-Val-Thr-Pro-Arg-Thr-Pro

Synonyms: Myelin Basic Protein (87-99); Experimental Allergic Encephalomyelitis Inducing Peptide

Bachem H-1964.0001 Human, bovine, rat MW 1555.84 $C_{74}H_{114}N_{20}O_{17}$ Store at -15°C

Neosystem SC975 Human, guinea pig MW 1555.84 Bioactive; immunodominant epitope of myelin basic protein; a major target of T cells in lesions of multiple sclerosis & in experimental allergic encephalomyelitis (EAE); possesses encephalitogenic activity in Buffalo & Lewis rats; Jones, RE et al, *J Neuroimmunol*, 37:203-212, 1992; Sun, D et al, *Mol Pharm Immunol* 22:591-594, 1992; Offner, H et al, *J Immunol* 148:1706-1711, 1992; Gold, DP et al, *J Immunol*, 148:1712-1717, 1992; Vandenbark, AA et al, *J Immunol*, 153:852-861, 1994; Karin, N et al, *J Exp Med*, 180:2227-2237, 1994

Val-Hph
Mu-Val-Hph-CH₂F

Synonyms: Calpain Inhibitor V

Calbiochem 208726 MW 407.5 $C_{21}H_{30}FN_3O_4$ Single spot purity (TLC); white solid; soluble in DMSO | A potent cell-permeable & irreversible calpain inhibitor; Esser, RE et al, *Arthritis Rheum*, 37:236, 1994

Val-Ile
Z-Val-Ile

Bachem C-2865.0001 MW 364.44 $C_{19}H_{28}N_2O_5$ Store at -15°C

Val-Ile

Bachem G-3540.0250 MW 230.31 $C_{11}H_{22}N_2O_3$ Store at -15°C

Val-Ile-His-Asn

Synonyms: Pre-Angiotensinogen (11-14); Angiotensinogen (11-14)

Sigma A 0417 FW 481.6 $C_{21}H_{35}N_7O_6$ ≥97% (HPLC) | Bioactive peptide; product of the action of humankidney renin on human angiotensinogen; Poe, M et al, *Anal Biochem*, 140:459, 1984

Bachem H-4665.0005 Human MW 481.55 $C_{21}H_{35}N_7O_6$ Store at -15°C

ICN 152752 Human MW 481.6 $C_{21}H_{35}N_7O_6$ Produced by the action of humankidney renin on human angiotensinogen; Poe, M etal, *Anal Biochem*, 140:459, 1984

Val-Ile-His-Ser

American Peptide 88-0-98 MW 454.5 $C_{20}H_{34}N_6O_6$

Val-Ile-Thr-Thr-Tyr-Trp-Gly-Leu-His-Thr-Gly-Glu-Arg

Synonyms: VIF (65-77); HTLV I Peptide

Neosystem SC546 MW 1532.71 HTLV I peptide from subtype VIF

Val-Leu
Z-Val-Leu

Bachem C-2870.0001 MW 364.44 $C_{19}H_{28}N_2O_5$ Store at -15°C

Val-Leu
Z-D-Val-Leu

Bachem C-2875.0001 MW 364.44 $C_{19}H_{28}N_2O_5$ Store at -15°C

Val-Leu
N-Me-Val-Leu-Anilide

Bachem G-4275.0050 Store at -15°C

Val-Leu
N-2-Cyanoethyl-Val-Leu-Anilide

Bachem G-4285.0050 Store at -15°C

Val-Leu
N-2-Hydroxyethyl-Val-Leu-Anilide

Bachem G-4295.0050 MW 349.47 $C_{19}H_{31}N_3O_3$ Store at -15°C

Val-Leu
N-(2-Carbamoyl-ethyl)-Val-Leu-Anilide

Bachem G-4320.0050 Store at -15°C

Val-Leu
N-Et-Val-Leu-Anilide

Bachem G-4325.0050 Store at -15°C

Val-Leu
N-((RS)-2-Hydroxypropyl)-Val-Leu-Anilide

Bachem G-4330.0050 Store at -15°C

Val-Leu
N-((RS)-3-Chloro-2-Hydroxypropyl)-Val-Leu-Anilide

Bachem G-4340.0050 Store at -15°C

Val-Leu
Bzl-Val-Leu-Anilide

Bachem G-4360.0050 Store at -15°C

Val-Leu
N-(2-Chloro-4-Ethoxy-1,3,5-Triazinyl)-Val-Leu-Anilide

Bachem G-4365.0050 Store at -15°C

Val-Leu
N-n-Butyl-Val-Leu-Anilide

Bachem G-4395.0050 Store at -15°C

Val-Leu
N-2-Chloroethyl-Val-Leu-Anilide

Bachem G-4400.0050 Store at -15°C

Val-Leu
N-((RS)-2-Hydroxy-2-Phenylethyl)-Val-Leu-Anilide

Bachem G-4450.0050 Store at -15°C

Val-Leu
N-2-Aminoethyl-Val-Leu-Anilide

Bachem G-4460.0050 Store at -15°C

Val-Leu
N-((RS)-2-Hydroxy-1-Phenylethyl)-Val-Leu-Anilide

Bachem G-4510.0050 Store at -15°C

Val-Leu
N-CBZ-Val-Leu

Sigma C 6752 FW 364.4 $C_{19}H_{28}N_2O_5$

Val-Leu
CBZ-D-Val-Leu

USBio C2099-84 MW 364.2 $C_{19}H_{28}N_2O_5$ ≥99%

Val-Leu Hydrochloride

Bachem G-3545.0001   MW 266.77 $C_{11}H_{22}N_2O_3 \cdot HCl$ Store at -15°C

Sigma V 4008   FW 266.8 $C_{11}H_{22}N_2O_3 \cdot HCl$

Val-Leu-Arg
D-Val-Leu-Arg-pNA

Synonyms: Kallikrein Substrate

American Peptide 81-5-65   MW 506.6 $C_{23}H_{38}N_8O_5$ Used as the chromogenic substrate for a convenient, sensitive & selective assay of glandular kallikrein activity; Fujii, S et al, *Adv Exp Med Biol*, 120A:83, 1979; Svedsen, L et al, *Thromb Res*, 1:267, 1972

Val-Leu-Arg
Z-Val-Leu-Arg-4MβNA Hydrochloride

Bachem J-1160.0050   MW 712.29 $C_{36}H_{49}N_7O_6 \cdot HCl$ Store at -15°C

Val-Leu-Arg
D-Val-Leu-Arg-4MβNA Trifluoroacetate Salt

Bachem J-1335.0050   Store at -15°C

Val-Leu-Arg
DL-Val-Leu-Arg-pNA

Bachem L-1445.0050   MW 506.61 $C_{23}H_{38}N_8O_5$ Store at -15°C

Val-Leu-Arg
D-Val-Leu-Arg-pNA

Bachem L-1885.0050   MW 506.61 $C_{23}H_{38}N_8O_5$ Store at -15°C

Val-Leu-Arg
DL-Val-Leu-Arg-pNA Acetate Salt

ICN 158278

Val-Leu-Arg
D-Val-Leu-Arg-AFC Dihydrochloride

Synonyms: Kallikrein Substrate

ICN 196011   MW 670.6 $C_{27}H_{38}N_7O_5F_3 \cdot 2HCl$ ≥97% | Sensitive fluorogenickallikrein substrate

Val-Leu-Arg
D-Val-Leu-Arg-pNA

Synonyms: Kallikrein Substrate

Neosystem SC1322   MW 506.59   Chromogenic substrate; Bönner, G & Martin-Grez, M, *J Clin Chem Clin Biochem*, 19:165-168, 1981

Val-Leu-Arg
DL-Val-Leu-Arg-pNA Acetate Salt

Sigma V 2628   FW 506.6 (free base) $C_{23}H_{38}N_8O_5$

Val-Leu-Arg
D-Val-Leu-Arg-pNA

Sigma V 6258   FW 506.6 (free base) $C_{23}H_{38}N_8O_5$ ≥95% (HPLC); peptide content:~70%; salt content will vary

Val-Leu-Gly-Arg
BOC-Val-Leu-Gly-Arg

Bachem A-2477.0050   MW 543.66 $C_{24}H_{45}N_7O_7$ Store at -15°C

Val-Leu-Gly-Arg
BOC-Val-Leu-Gly-Arg-pNA

Bachem L-1205.0025   MW 663.78 $C_{30}H_{49}N_9O_8$ Store at -15°C

Val-Leu-Gly-Arg
N-t-BOC-Val-Leu-Gly-Arg-pNA Hydrobromide

Synonyms: Horseshoe Crab Clotting Enzyme Substrate

Sigma B 8391   FW 744.7 $C_{30}H_{49}N_9O_8 \cdot HBr$ Iwanaga, S et al, *Haemostasis*, 7:183, 1978

Val-Leu-Gly-Gly-Gly-Ser-Ala-Leu-Leu-Arg-Ser-Ile-Pro-Ala

Synonyms: Heat Shock Protein 65kD Fragment (437-450), (Ser[442,447])-; Neurotrophic Factor 14, Activity-Dependent

Bachem H-3716.0001   MW 1310.56 $C_{58}H_{103}N_{17}O_{17}$ Store at -15°C

Sigma H 2271   FW 1310.6   ≥97% (HPLC) | Bioactive peptide; protects against neurotoxins related to HIV infection & Alzheimer's disease; Brenneman, DE & Gozes, I, *J Clin Invest*, 97:2299, 1996

Val-Leu-Gly-Gly-Gly-Ser-Ala-Leu-Leu-Arg-Ser-Ile-Pro-Ala-Leu-Asp-Ser-Leu-Thr-Pro-Ala-Asn-Glu-Asp

Synonyms: Heat Shock Protein 65kD Fragment (437-460), (Ser[442,447])-

Sigma H 2148   Human   FW 2366.6   Bioactive peptide; Elias, D et al, *Proc Natl Acad Sci USA*, 88:3088, 1991

Val-Leu-Gly-Lys-Leu-Ser-Gln-Glu-Leu-His-Lys-Leu-Gln-Thr-Tyr-Pro-Arg-Thr-Asn-Thr-Gly-Ser-Asn-Thr-Tyr
Ac-Val-Leu-Gly-Lys-Leu-Ser-Gln-Glu-Leu-His-Lys-Leu-Gln-Thr-Tyr-Pro-Arg-Thr-Asn-Thr-Gly-Ser-Asn-Thr-Tyr Amide

Synonyms: Calcitonin I (8-32), Ac-(Asn[30],Tyr[32])-; AC187

Bachem H-4922.0500   Salmon   MW 2890.25 $C_{127}H_{205}N_{37}O_{40}$ Store at -15°C

Val-Leu-Gly-Lys-Leu-Ser-Gln-Glu-Leu-His-Lys-Leu-Gln-Thr-Tyr-Pro-Arg-Thr-Asn-Thr-Gly-Ser-Gly-Thr-Pro Amide

Synonyms: Calcitonin (8-32)

Neosystem SC1211   Salmon   MW 2725.09   Bioactive; calcium metabolism peptide; Silvestre, RA et al, *Br J Pharmacol*, 117:347-350, 1996; Wookey, PJ et al, *Am J Physiol*, 39:F289-F294, 1996

Val-Leu-Gly-Phe-Leu-Gly-Phe-Leu-Ala-Thr-Ala-Gly-Ser-Ala-Met-Gly-Ala-Ala-Ser-Leu

Synonyms: GP140 (531-550); SIVmac251 Peptide

Neosystem SC761   MW 1854.19   SIVmac251 peptide from subtype GP140

Val-Leu-Leu-Ser-Leu-Asp-Arg-Lys-Thr-Ile-Cys Acetate Salt

Synonyms: Granzyme A (15-25)

Biogenesis 4741-5000   Human synthetic   Purified; lyophilized

Val-Leu-Lys
BOC-Val-Leu-Lys-AMC

Bachem I-1115.0050   MW 615.77 $C_{32}H_{49}N_5O_7$ Store at -15°C

Val-Leu-Lys
D-Val-Leu-Lys-AMC
Bachem I-1390.0050   Store at -15°C

Val-Leu-Lys
Val-Leu-Lys-pNA Dihydrochloride
Bachem L-1450.0050   Store at -15°C

Val-Leu-Lys
D-Val-Leu-Lys-CMK Hydrochloride
Bachem N-1385.0025   MW 427.41 $C_{18}H_{35}ClN_4O_3 \cdot HCl$
Store at -15°C

Val-Leu-Lys
D-Val-Leu-Lys-pNA Dihydrochloride
Synonyms: Plasmin Substrate

Fluka 94680   MW 551.50 $C_{23}H_{38}N_6O_5 \cdot 2HCl$ ≥99.0%
(HPLC); ≤5% water | Morris, JP et al, *Biochemistry*, 20:4811,
1981; Wu, H-L et al, *PNAS*, 84:8292, 1987

Val-Leu-Lys
N-α-t-BOC-L-Val-L-Leu-L-Lys-MCA
Synonyms: Plasmin Substrate

ICN 150509   MW 615.8 $C_{32}H_{49}N_5O_7$ Fluorogenic substrate

Val-Leu-Lys
BOC-Val-Leu-Lys-MCA
Synonyms: Plasmin Substrate; Calpain Substrate

Peptides International MVL-3104-v   MW 615.77
$C_{32}H_{49}N_5O_7$ >96% (HPLC); lyophilized amorphous powder | Kato,
H et al, *J Biochem*, 88:183, 1980; Sasaki, T et al, *JBC*, 259:12489,
1984

Val-Leu-Lys
N-t-BOC-Val-Leu-Lys-MCA
Sigma B 1136   FW 615.8 $C_{32}H_{49}N_5O_7$

Val-Leu-Lys
D-Val-Leu-Lys-pNA Dihydrochloride
Sigma V 0882   FW 551.5 $C_{23}H_{38}N_6O_5 \cdot 2HCl$

Val-Leu-Pro-Val-Thr-Ile-Met-Ser-Gly-Leu-Val-Phe-His-Ser-Gln-Pro-Ile-Asn-Asp-Arg
Synonyms: GP140 (321-340); SIVmac251 Peptide

Neosystem SC740   MW 2223.61   SIVmac251 peptide from
subtype GP140

Val-Leu-Ser
Bachem H-5280.0250   MW 317.39 $C_{14}H_{27}N_3O_5$ Store at
-15°C

Val-Leu-Ser-Glu-Gly
Bachem H-5285.0050   MW 503.55 $C_{21}H_{37}N_5O_9$ Store at
-15°C

Val-Lys
Z-D-Val-Lys(Z)
Bachem C-2845.0001   MW 513.6 $C_{27}H_{35}N_3O_7$ Store at
-15°C

Val-Lys
Bachem G-4175.0250   MW 245.32 $C_{11}H_{23}N_3O_3$ Store at
-15°C

Val-Lys Hydrochloride
Bachem G-3550.0250   MW 281.78 $C_{11}H_{23}N_3O_3 \cdot HCl$ Store
at -15°C

ICN 158279   MW 281.8 $C_{11}H_{23}N_3O_3 \cdot HCl$

Sigma V 5377   FW 281.8 $C_{11}H_{23}N_3O_3 \cdot HCl$

Val-Lys-Asn-Asn-Phe-Val-Pro-Thr-Asn-Val-Gly-Ser-Lys-Ala-Phe Amide
Synonyms: Calcitonin Gene Regulated Peptide (23-37), α-

Bachem H-8895.0001   Human   MW 1620.87 $C_{74}H_{117}N_{21}O_{20}$
Store at -15°C

Val-Lys-Leu-Thr-Pro-Leu-Cys-Val-Ser-Leu-Lys-Cys-Thr-Asp-Leu-Gly
Synonyms: GP120 (120-135); HTLV I Peptide

Neosystem SC263   MW 1690.08   HTLV I peptide from
subtype GP160; due to the presence of a Cys this peptide can
contain the dimeric form

Val-Lys-Lys-Arg
Bachem H-1218.0005   MW 529.68 $C_{23}H_{47}N_9O_5$ Store at
-15°C

Val-Lys-Lys-Arg
Bz-Val-Lys-Lys-Arg-4MβNA
Bachem J-1075.0025   MW 788.99 $C_{41}H_{60}N_{10}O_6$ Store at
-15°C

Val-Lys-Lys-Arg
Z-Val-Lys-Lys-Arg-4MβNA
Bachem J-1165.0050   MW 819.02 $C_{42}H_{62}N_{10}O_7$ Store at
-15°C

Val-Lys-Lys-Arg
N-CBZ-Val-Lys-Lys-Arg-4-Methoxy-β-Naphthylamide Trihydrochloride
Synonyms: Cathepsin B Substrate

Sigma C 2772   FW 928.4 $C_{42}H_{62}N_{10}O_7 \cdot 3HCl$ Sensitive
substrate

Val-Lys-Met
Z-Val-Lys-Met-AMC
Synonyms: Proteasome Substrate IV; Amyloid β/A4 Generating
Enzyme Substrate; Ingensin Substrate

Bachem I-1625.0025   MW 667.83 $C_{34}H_{45}N_5O_7S$ Store at
-15°C

Calbiochem 539143   MW 667.8 $C_{34}H_{45}N_5O_7S$ ≥98%
(HPLC); solid; soluble in DMSO; excitation max:~380 nm; emission
max:~460 nm | Fluorogenic substrate for Alzheimer's Disease
amyloid A4-generating enzymes & for the proteasome; Ishiura, S
et al, *Neurosci Lett*, 115:329, 1990

Neosystem SC1276   MW 667.82   Ishiura, S et al, *Neuro Lett*,
115:329-334, 1990

Val-Lys-Met
Z-Val-Lys-Met-MCA
Synonyms: Amyloid A4-Generating Enzyme Substrate; Proteasome
Substrate

Peptides International MVM-3156-v   MW 667.83
$C_{34}H_{45}N_5O_7S$ >96% (HPLC); lyophilized amorphous powder |
Ishiura, S et al, *Neurosci Lett*, 115:329, 1990

Val-Lys-Met
N-CBZ-Val-Lys-Met-MCA

Synonyms: Amyloid A4-Generating Enzyme Substrate

Sigma C 2681   FW 667.8 $C_{34}H_{45}N_5O_7S$ Peptide content ~75% | Substrate for amyloid A4-generating enzymes of Alzheimer's disease; Ishiura, S et al, *Neurosci Lett*, 115:323, 1990

Val-Met
Z-Val-Met

Bachem C-2880.0001   MW 382.48 $C_{18}H_{26}N_2O_5S$ Store at -15°C

Val-Met

Bachem G-3555.0250   MW 248.35 $C_{10}H_{20}N_2O_3S$ Store at -15°C

ICN 158280   MW 248.3 $C_{10}H_{20}N_2O_3S$

Sigma V 5502   FW 248.3 $C_{10}H_{20}N_2O_3S$

Val-Phe
Z-Val-Phe-CHO

Synonyms: MDL 28170

American Peptide 81-7-07   MW 382.5 $C_{22}H_{26}N_2O_4$
Substrate analog; specifically inhibits calpains & other Cys proteinases; Song, D-K et al, *J Neurosci Res*, 39:474, 1994

Val-Phe
Z-Val-Phe

Bachem C-2885.0001   MW 398.46 $C_{22}H_{26}N_2O_5$ Store at -15°C

Val-Phe
Z-Val-Phe-OMe

Bachem C-2890.0001   MW 412.49 $C_{23}H_{28}N_2O_5$ Store at -15°C

Val-Phe
Z-Val-D-Phe-OMe

Bachem C-3535.0001   MW 412.49 $C_{23}H_{28}N_2O_5$ Store at -15°C

Val-Phe

Bachem G-3565.0250   MW 264.32 $C_{14}H_{20}N_2O_3$ Store at -15°C

Val-Phe
Z-Val-Phe-CHO

Synonyms: Calpain Inhibitor III; MDL 28170

Bachem N-1535.0025   Store at -15°C

Calbiochem 208722   MW 382.5 $C_{22}H_{26}N_2O_4$ ≥98% (TLC); white lyophilized solid; soluble in DMSO | A cell-permeable calpain inhibitor; reduces capsaicin-mediated cell death in cultured dorsal root ganglion; blocks the Ca^{2+}-ionophore A23187-induced suppression of neurite outgrowth in isolated hippocampal pyramidal neurons; exhibits neuroprotective effect in glutamate-induced toxicity; Rami, A et al, *Neurosci Res*, 27:93, 1997; Chard, PS et al, *Neuroscience*, 65:1099, 1995; Song, DK et al, *J Neurosci Res*, 39:474, 1994

Val-Phe

ICN 103236   MW 264.3 $C_{14}H_{20}N_2O_3$ Crystalline

Val-Phe
BOC-Val-Phe-NHO-Bz-pCl

Synonyms: Cathepsin L Inhibitor; Subtilisin Inhibitor; Thermitase Inhibitor

ICN 195951   MW 518 $C_{26}H_{32}N_3O_6Cl$ Inhibits these & other members of the cysteine & serine protease families

Val-Phe
Z-Val-Phe-CHO

Synonyms: Calpain Inhibitor

Neosystem SC1325   MW 382.46   Chard, PS et al, *Neurosci*, 65:1099-1108, 1995

Val-Phe
N-CBZ-Val-Phe

Sigma C 7002   FW 398.5 $C_{22}H_{26}N_2O_5$

Val-Phe
N-CBZ-Val-Phe-OMe

Sigma C 7506   FW 412.5 $C_{23}H_{28}N_2O_5$

Val-Phe

Sigma V 1875   FW 264.3 $C_{14}H_{20}N_2O_3$

Val-Phe
(2S,3R)-3-Amino-2-Hydroxy-4-Phenylbutanoyl-L-Val-L-Phe

Synonyms: Phebestin; Aminopeptidase N Inhibitor

Peptides International IPH-4342-v   Synthetic   MW 441.53 $C_{24}H_{31}N_3O_5$ >99% (HPLC); lyophilized amorphous powder | Inhibitor; Nagai, M et al, *J Antibiotics*, 50:82, 1997

Val-Phe Amide Hydrochloride

Bachem G-3570.0250   MW 299.8 $C_{14}H_{21}N_3O_2 \cdot HCl$ Store at -15°C

Val-Phe-His-Ser-Gln-Pro-Ile-Asn-Asp-Arg-Pro-Lys-Gln-Ala-Trp-Cys-Trp-Phe-Gly-Gly

Synonyms: GP140 (331-350); SIVmac251 Peptide

Neosystem SC741   MW 2373.67   SIVmac251 peptide from subtype GP140; due to the presence of a Cys this peptide can contain the dimeric form

Val-Phe-Ile-Asn-Ala-Lys-Cys-Arg-Gly-Ser-Pro-Glu-Cys-Leu-Pro-Lys-Cys-Lys-Glu-Ala-Ile-Gly-Lys-Ala-Ala-Gly-Lys-Cys-Met-Asn-Gly-Lys-Cys-Lys-Cys-Tyr-Pro

Synonyms: Tityustoxin Kα;; Tityustoxin Kα; Voltage-Dependent K$^+$ Channel (A Channel) Blocker

Bachem H-2452.0100   Store at -15°C | Disulfide bonds: Cys^7-Cys^{28}, Cys^{13}-Cys^{33}, Cys^{17}-Cys^{35}

Calbiochem 614375   *Tityus serrulatus*   MW 3941.8 $C_{168}H_{275}N_{49}O_{46}S_7$ >98% (HPLC); lyophilized solid; biological activity:inhibits Kv1.2 K$^+$ channels; LD$_{50}$≤2000 mg/kg | Disulfide bonds: Cys^7-Cys^{28}, Cys^{13}-Cys^{33}, Cys^{17}-Cys^{35}; selectively blocks voltage-gated non-inactivating K$^+$ channels in synaptosomes; interferes with the binding of α-dendrotoxin to its receptor resulting in unblocking of α-dendrotoxin-blocked inactivating channels; Casali, TA et al, *Neuropharmacology*, 34:599, 1995; Gomez, RS et al, *Neurosci Lett*, 196:131, 1995; Rogowski, RS et al, *PNAS*, 91:1475, 1994

Peptides International PTT-4313-s   *Tityus serrulatus* (scorpion) synthetic   MW 3941.8 $C_{168}H_{275}N_{49}O_{46}S_7$ >95% (HPLC); lyophilized amorphous powder | Bioactive; Werkman, TR et al, *Mol Pharmacol*, 44:430, 1993; Rogowski, RS et al, *PNAS* USA, 91:1475, 1994; Casali, TAA et al, *Neuropharmacology*, 34:599, 1995; Hopkins, WF, *J Pharmacol Exp Ther*, 285:1051, 1998

Val-Phe-Ile-Leu-Gly-Pro-Leu-Arg-Leu-Leu-Gly

Synonyms: Experimental Autoimmune Encephalomyelitis Complementary Peptide

Bachem H-3584.0001   Store at -15°C

Val-Phe-Lys
D-Val-Phe-Lys-pNA Dihydrochloride

Sigma V 6884 FW 585.5 $C_{26}H_{36}N_6O_5 \cdot 2HCl$ ≥97% (HPLC) | Substrate for discriminating between urokinase & tissue plasminogen activator (TPA); Schnyder, J et al, *Anal Biochem*, 200:156, 1992

Val-Phe-Phe-Ala-Lys
Val-Phe-Phe-Ala-(Lys-DNP)-Amide

BioSource International 03-406

Val-Phe-Ser-Val-Arg-Val-Ser-Ile-Leu-Val-Phe

Synonyms: Angiogenin Complementary Sequence (60-70)

Bachem H-4192.0001 Human MW 1265.56 $C_{62}H_{100}N_{14}O_{14}$ Store at -15°C

Val-Pro
BOC-Val-Pro

Bachem A-2480.0001 MW 314.38 $C_{15}H_{26}N_2O_5$ Store at -15°C

Val-Pro
FMOC-Val-Pro

Bachem B-2140.0001 MW 436.51 $C_{25}H_{28}N_2O_5$ Store at -15°C

Val-Pro
Z-Val-Pro

Bachem C-3340.0001 MW 348.4 $C_{18}H_{24}N_2O_5$ Store at -15°C

Val-Pro
Val-Pro-OtBu Hydrochloride

Bachem H-7120.0001 MW 306.83 $C_{14}H_{26}N_2O_3 \cdot HCl$ Store at -15°C

Val-Pro Hydrochloride

Bachem G-3575.0250 MW 250.73 $C_{10}H_{18}N_2O_3 \cdot HCl$ Store at -15°C

ICN 158281 MW 250.7 $C_{10}H_{18}N_2O_3 \cdot HCl$

Sigma V 7878 FW 250.7 $C_{10}H_{18}N_2O_3 \cdot HCl$

Val-Pro-Arg
BOC-Val-Pro-Arg-AMC Hydrochloride

Bachem I-1120.0050 MW 664.2 $C_{31}H_{45}N_7O_7 \cdot HCl$ Store at -15°C

Val-Pro-Arg
N-α-t-BOC-L-Val-L-Pro-L-Arg-MCA

Synonyms: Thrombin Substrate, α-

ICN 150510 MW 627.7 $C_{31}H_{45}N_7O_7$ Fluorogenic substrate

Val-Pro-Arg
BOC-Val-Pro-Arg-MCA

Synonyms: Thrombin Substrate, α-

Peptides International MVP-3093-v MW 627.74 $C_{31}H_{45}N_7O_7$ >99% (HPLC); lyophilized amorphous powder | Morita, T et al, *J Biochem*, 82:1495, 1977; Kawabata, S et al, *J Biochem*, 97:1073, 1985

Val-Pro-Arg
N-t-BOC-Val-Pro-Arg-MCA Hydrochloride

Synonyms: Thrombin Substrate

Sigma B 9385 FW 627.7 (free base) $C_{31}H_{45}N_7O_7$ Fluorogenic substrate; Morita, T et al, *J Biochem*, 82:1495, 1977; Lottenberg, R et al, *Meth Enz*, 80:341, 1981

Val-Pro-Asp-Pro-Arg

Synonyms: Enterostatin

American Peptide 87-0-75 MW 582.7 $C_{25}H_{42}N_8O_8$ Appetite suppressant in rats; Erlanson-Albertson, C et al, *Reg Peptides*, 22:325, 1988

ICN 153193 Decreases food intake in rats; Erianson-Albertsson, L & A Larsson, *Regul Pept*, 22:325, 1988

Bachem H-6410.0005 Pig, rat MW 582.66 $C_{25}H_{42}N_8O_8$ Store at -15°C

Sigma V 0256 Pig, rat FW 582.7 $C_{25}H_{42}N_8O_8$ ≥97% (HPLC) | Bioactive peptide; appetite suppressant in rats; Erlanson-Albertson, C & Larsson, A, *Regulatory Peptides*, 22:325, 1988

Neosystem SC931 Porcine, rat MW 582.66 Bioactive; brain/gut peptide; Erlanson-Albertsson, C & Larsson, A, *Biochimie*, 70:1245-1250, 1988; Erlanson-Albertsson, C, *Nutr Rev*, 50:307-310, 1992; Erlanson-Albertsson, C, *Scand J Nutr*, 38:11-14, 1994

Val-Pro-Ile

Synonyms: Diprotin B; Dipeptidyl Aminopeptidase IV Inhibitor

Alexis 260-013 Synthetic MW 327.4 $C_{16}H_{29}N_3O_4$ ≥96%; white lyophilized powder; soluble in water | Umezawa, H et al, *J Antibiot*, 37:422, 1984

Val-Pro-Ile-Gln-Lys-Val-Gln-Asp-Asp-Thr-Lys-Thr-Leu-Ile-Lys-Thr-Ile-Val-Thr-Arg-Ile-Asn-Asp-Ile-Ser-His-Thr-Gln-Ser-Val-Ser-Ser-Lys-Gln-Lys

Synonyms: Obese Gene Peptide (22-56); Leptin (22-56)

American Peptide 46-2-16 Human MW 3950.6 $C_{171}H_{298}N_{50}O_{56}$ Zhang, Y et al, *Nature*, 372:425, 1994; Pelleymounter, MA et al, *Science*, 269:541, 1995

Bachem H-3424.0500 Human MW 3950.55 $C_{171}H_{298}N_{50}O_{56}$ Store at -15°C

Neosystem SC1215 Human MW 3950.54 Virus-related peptide; bioactive; acutely & reversibly inhibits feeding in the rat; Samson, WK et al, *Endocrinology*, 137:5182-5185, 1996

Sigma L 4771 Human FW 3950.5 ≥95% (HPLC) | Bioactive peptide

Val-Pro-Ile-Tyr-Glu-Lys-Lys-Tyr-Gly-Gln-Val-Pro-Met-Cys-Asp-Ala-Gly-Glu-Gln-Cys-Ala-Val-Arg-Lys-Gly-Ala-Arg-Ile-Gly-Lys-Leu-Cys-Asp-Cys-Pro-Arg-Gly-Thr-Ser-Cys-Asn-Ser-Phe-Leu-Leu-Lys-Cys-Leu

Synonyms: CART (55-102)

Bachem H-4444.0100 Human MW 5245.23 $C_{225}H_{365}N_{65}O_{65}S_7$ Store at -15°C | Disulfide bonds: Cys[74]-Cys[94], Cys[68]-Cys[86], Cys[88]-Cys[101]

Val-Pro-Leu

Synonyms: Diprotin B

Bachem H-5290.0050 MW 327.42 $C_{16}H_{29}N_3O_4$ Store at -15°C

ICN 153194

Sigma V 3255 FW 327.4 $C_{16}H_{29}N_3O_4$ ≥97% (HPLC) | Bioactive peptide

Val-Pro-Leu-Arg-Pro-Met-Thr-Tyr-Lys-Ala-Ala-Leu

Synonyms: NEF MN (76-87); HIV I Peptide

Neosystem SC648 MW 1359.69 NEF peptide from HIV-1 subtype MN

Val-Pro-Leu-Arg-Pro-Met-Thr-Tyr-Lys-Ala-Ala-Val

Synonyms: NEF (74-85); HTLV I Peptide

Neosystem SC504 MW 1345.66 HTLV I peptide from subtype NEF

Val-Pro-Leu-Pro-Ala-Gly-Gly-Gly-Thr-Val-Leu-Thr-Lys-Met-Tyr-Pro

Synonyms: Gastrin Releasing Peptide (1-16)

American Peptide 46-1-72 Human MW 1600.9
$C_{74}H_{121}N_{17}O_{20}S$

ICN 154550 Human MW 1601.2

Val-Pro-Leu-Pro-Ala-Gly-Gly-Gly-Thr-Val-Leu-Thr-Lys-Met-Tyr-Pro-Arg-Gly-Asn-His-Trp-Ala-Val-Gly-His-Leu-Met Amide

Synonyms: Gastrin Releasing Peptide

American Peptide 46-1-70 Human MW 2859.3
$C_{130}H_{204}N_{38}O_{31}S_2$ Stimulates pancreatic & gastric acid secretion; Spindel, ER et al, *PNAS*, 81:5699, 1984

Bachem H-6785.0500 Human MW 2859.42
$C_{130}H_{204}N_{38}O_{31}S_2$ Store at -15°C

ICN 151178 Human Spindel, ER etal, *PNAS*, 81:5699, 1984

Neosystem SC166 Human MW 2859.40 Bioactive; brain/gut peptide; CCK/gastrin peptide; Spindel, ER et al, *PNAS USA*, 81:5699-5703, 1984

Sigma G 8022 Human FW 2859.4 ≥97% (HPLC); peptide content:~70% | Bioactive peptide; gastrointestinal peptide; mammalian equivalent of bombesin; Spindel, ER, *Proc Natl Acad Sci USA*, 81:5699, 1984

Peptides International PGR-4164-v Human synthetic MW 2859.4 $C_{130}H_{204}N_{38}O_{31}S_2$ >95% (HPLC); lyophilized amorphous powder | Bioactive; Spindel, ER et al, *PNAS USA*, 81:5699, 1984

Val-Pro-Phe
Suc-Val-Pro-Phe-4MβNA

Bachem J-1345.0050 Store at -15°C

Val-Pro-Phe
Suc-Val-Pro-Phe-pNA

Bachem L-1755.0050 MW 581.63 $C_{29}H_{35}N_5O_8$ Store at -15°C

Val-Pro-Phe
Suc-Val-Pro-Phe-SBzl

Bachem M-1885.0050 Store at -15°C

Val-Pro-Pro-Pro-Val-Pro-Pro-Arg-Arg-Arg

Synonyms: Ras Inhibitory Peptide; Guanine Nucleotide Releasing Factor (1149-1158); hSOS (1149-1158)

Alexis 166-001 MW 1170.4 $C_{53}H_{91}N_{19}O_{11}$ ≥98%; white lyophilized powder; soluble in water | Essential for the control of Ras activity; blocks the binding of hSOS1 to human Grb2, a protein that binds to activated receptor tyrosine kinases; indicates that the Grb2/hSOS1 complex links signal transduction by Ras to receptor tyrosine kinases; Li, N et al, *Nature*, 363:85, 1993

Bachem H-1392.0001 MW 1170.43 $C_{53}H_{91}N_{19}O_{11}$ Store at -15°C

Sigma V 8626 FW 1170.4 ≥97% (HPLC); peptide content:~60% | Bioactive peptide; prevents binding of hSOS1 to human Grb2; Li, N, *Nature*, 363:85, 1993

Val-Pro-Pro-Pro-Val-Pro-Pro-Arg-Arg-Arg Amide Trifluoroacetate Salt

Synonyms: hSOS N10 SH3 Binding Domain (1149-1158)

Calbiochem 385872 MW 1169.5 $C_{53}H_{92}N_{20}O_{10}$ >98% (HPLC); lyophilized solid; soluble in water | SH3 binding sequence of guanine nucleotide exchange factor hSOS which is a ligand for the adapter protein Grb2 SH3 domain; blocks Sos/Grb2 interaction, preventing Ras activation via receptor protein tyrosine kinases; has a strong affinity for the N-terminal SH3 of Grb2; Chen, JK et al, *J Am Chem Soc*, 115:12591, 1993; Li, N et al, *Nature*, 363:85, 1993

Val-Pro-Val-Glu-Ala-Val-Asp-Pro-Met

Synonyms: Prepro-Cholecystokinin V-9-M (24-32); Cholecystokinin Precursor (24-32); V-9-M

American Peptide 46-1-45 MW 956.0 $C_{42}H_{69}N_9O_{14}S$ AA 24-32 of the rat cholecystokinin prepro-sequence, used to generate an antiserum which detected different CCK precursor peptides in rat brain; also is a putative neuromodulator possessing sedative actions & preventing experimental amnesia in rats; Beinfeld, MC et al, *Brain Res*, 344:351, 1985

Bachem H-9160.0001 Rat MW 956.13 $C_{42}H_{69}N_9O_{14}S$ Store at -15°C

Val-Ser

Bachem G-3580.0250 MW 204.23 $C_8H_{16}N_2O_4$ Store at -15°C

ICN 158282 MW 204.2 $C_8H_{16}N_2O_4$

Sigma V 5627 FW 204.2 $C_8H_{16}N_2O_4$

Val-Ser-Arg-Leu-Asn-Ile-Asn-Leu-His-Phe-Ser-Lys-Cys-Gly

Synonyms: GP46 (145-158); HTLV I Peptide

Neosystem SC827 MW 1587.85 HTLV I peptide; due to the presence of a Cys this peptide can contain the dimeric form

Val-Ser-Gln-Asn-Tyr-Pro-Ile-Val

Synonyms: HIV Protease Substrate VIII

Bachem H-8215.0001 MW 919.05 $C_{42}H_{66}N_{10}O_{13}$ Store at -15°C

Val-Ser-Glu-Ile-Gln-Leu-Met-His-Asn-Leu-Gly-Lys-His-Leu-Asn-Ser-Met-Glu-Arg-Val-Glu-Trp-Leu-Arg-Lys-Lys-Leu-Gln-Asp-Val-His-Asn-Phe-Val-Ala-Leu-Gly

Synonyms: Parathyroid Hormone (2-38)

Bachem H-1316.0500 Human MW 4491.23 $C_{204}H_{314}N_{58}O_{53}S_2$ Store at -15°C

Val-Ser-Ser-Asn-Ile-Ser-Glu-Asp-Pro-Val-Pro-Val

Synonyms: Prepro-Vasoactive Intestinal Peptide (111-122); Peptide Histidine Methionine (111-122); Vasoactive Intestinal Peptide Space Peptide; Peptide Histidine Methionine/Vasoactive Intestinal Peptide; Spacer Peptide

American Peptide 48-1-30 MW 1242.5 $C_{53}H_{87}N_{13}O_{21}$ Itoh, N et al, *Nature*, 304:547, 1983

Bachem H-6915.0001 Human Store at -15°C

Val-Thr

ICN 158283 MW 218.3 $C_9H_{18}N_2O_4$

Sigma V 5752 FW 218.3 $C_9H_{18}N_2O_4$

Val-Thr-Asp-Ala-Arg-Glu-Arg-Tyr-Gly-Pro-Asn

Synonyms: ATPase N-Terminal Peptide (29-39), SERCA3

Alexis 167-011 Synthetic $C_{53}H_{84}N_{18}O_{19}$ Lyophilized; reconstitute with 0.1 mL distilled H_2O | Sequence is conserved in mouse, rat & human species; competitively binds to Ab Alexis 210-167; antiserum blocking peptide; Montecucchi, PC & Henschen, A, *Int J Pept Prot Res*, 18:113, 1981; Vale, W et al, *Science*, 213:1394, 1981; Melchiorri, P & Negri, L, *Regul Pept*, 2:1, 1981

Val-Thr-Cys-Gly

Synonyms: Hematopoietic Cell Adhesion Peptide

Bachem H-2804.0005 MW 378.45 $C_{14}H_{26}N_4O_6S$ Store at -15°C

ICN 156498 MW 378.4 $C_{14}H_{26}N_4O_6S$ 97% | Hematopoietic cell adhesion peptide from the malarial Circumsporozoite protein; Rich, KA etal, *Science*, 249:1574, 1990

American Peptide 89-0-10 Malarial *circumsporozoite* MW 377.5 $C_{14}H_{25}N_4O_6S$ From the malarial circumsporozoite protein

Sigma V 2632 Malarial *circumsporozoite* FW 3780.4 $C_{14}H_{26}N_4O_6S$ ≥97% (HPLC); peptide content:~70% | Bioactive peptide; Rich, KA *Science*, 249:1574, 1990

Val-Thr-Glu-Ser-Phe-Asp-Ala-Trp-Glu-Asn-Thr-Val-Thr-Glu-Gln-Ala-Ile-Glu-Asp-Val

Synonyms: GP140 (71-90); SIVmac251 Peptide

Neosystem SC715 MW 2283.38 SIVmac251 peptide from subtype GP140

Val-Thr-His-Arg-Leu-Ala-Gly-Leu-Leu-Ser-Arg-Ser-Gly-Gly-Met-Val-Lys-Ser-Asn-Phe-Val-Pro-Thr-Asn-Val-Gly-Ser-Lys-Ala-Phe Amide

Synonyms: Calcitonin Gene Regulated Peptide (8-37), β-

Neosystem SC1260 Human MW 3130.65 Bioactive; calcium metabolism peptide; potent antagonist of α-CGRP, human-stimulated c-AMP accumulation in SK-N-MC cells; Longmore, J et al, *Eur J Pharmacol*, 265:53-59, 1994

Val-Thr-His-Arg-Leu-Ala-Gly-Leu-Leu-Ser-Arg-Ser-Gly-Gly-Val-Val-Lys-Asn-Asn-Phe-Val-Pro-Thr-Asn-Val-Gly-Ser-Lys-Ala-Phe Amide

Synonyms: Calcitonin Gene Related Peptide I (8-37); Calcitonin Gene Regulated Peptide (8-37), α-; Calcitonin Gene Related Peptide (8-37); Calcitonin Gene Related Peptide, α-; Corticotropin Releasing Factor I (8-37)

American Peptide 22-5-17 Human MW 3125.7 $C_{139}H_{230}N_{44}O_{38}$ Competitive antagonist of CGRP receptor; Chiba, T et al, *Am J Physiol*, 256:E331, 1989

Bachem H-9895.0001 Human MW 3125.63 $C_{139}H_{230}N_{44}O_{38}$ Store at -15°C

Calbiochem 05-23-2407 Human MW 3125.6 $C_{139}H_{230}N_{44}O_{38}$ ≥98% (HPLC); lyophilized solid; soluble in 5% acetic acid | Competitive antagonist of CGRP receptor; Nuki, C et al, *Jpn J Pharmacol*, 65:99, 1994; Gardiner, AM et al, *Biochem Biophys Res Comm*, 171:938, 1990; Chiba, T et al, *Am J Physiol*, 256:E331, 1989; Han, S-P et al, *Biochem Biophys Res Comm*, 168:786, 1990

Neosystem SC890 Human MW 3125.62 Bioactive; calcium metabolism peptide; CGRP receptor antagonist; Chiba, T et al, *Am J Physiol*, 256:E331, 1989

Sigma C 2806 Human FW 3789.3 ≥97% (HPLC); peptide content:~70% | Bioactive peptide; selective competitive antagonist at CGRP receptors but not at calcitonin receptors; Chiba, T et al, *Amer J Physiol*, 256:E331, 1989

Calbiochem 05-23-0050 Human, rat MW 4757.5 $C_{208}H_{344}N_{60}O_{63}S_2$ ≥97% (HPLC); lyophilized solid; soluble in 5% acetic acid; sold under license of US Patent 4,489,163 issued to The Salk Institute | Releases ACTH from the anterior pituitary & stimulates the sympathetic nervous system & adrenal medulla; functions by reducing hyperpolarizations; acts as a functional antagonist of inflammatory mediators; Barmack, NH & Errico, P, *J Neurosci*, 13:4647, 1993; Wei, ET et al, *Ciba Found Symp*, 172:258, 1993; Smith, EM et al, *Nature*, 321:881, 1986; Udelsman, R et al, *Nature*, 319:147, 1986

Val-Thr-His-Arg-Leu-Ala-Gly-Leu-Leu-Ser-Arg-Ser-Gly-Gly-Val-Val-Lys-Asp-Asn-Phe-Val-Pro-Thr-Asn-Val-Gly-Ser-Glu-Ala-Phe Amide

Synonyms: Calcitonin Gene Related Peptide (8-37); Calcitonin Gene Regulated Peptide (8-37), α-

American Peptide 22-5-18 Rat MW 3127.6 $C_{138}H_{224}N_{42}O_{41}$

Bachem H-4924.0001 Rat MW 3127.56 $C_{138}H_{224}N_{42}O_{41}$ Store at -15°C

Val-Thr-His-Arg-Leu-Ala-Gly-Leu-Leu-Ser-Arg-Ser-Gly-Gly-Val-Val-Lys-Asn-Asn-Phe-Val-Pro-Thr-Asn-Val-Gly-Ser-Lys-Ala-Phe Amide

Synonyms: Calcitonin Gene Related Peptide (8-37); Calcitonin Gene Related Peptide (8-37), α-; Calcitonin Gene Related Peptide Antagonist

Peptides International PCG-4232-v Synthetic MW 3125.6 $C_{139}H_{230}N_{44}O_{38}$ >99% (HPLC); lyophilized amorphous powder | Bioactive; Chiba, T et al, *Am J Physiol*, 256:E331, 1989; Han, S-P et al, *BBRC*, 168:786, 1990; Dennis, T et al, *J Pharmacol Exp Ther*, 254:123, 1990; Gardiner, SM et al, *BBRC*, 171:938, 1990

Val-Thr-Lys-Gly

ICN 156499 MW 403.5 $C_{17}H_{33}N_5O_6$ 97%

Sigma V 9006 FW 403.5 $C_{17}H_{33}N_5O_6$ ≥97% (HPLC) | Bioactive peptide

---

Val-Thr-Pro
FMOC-Val-Thr(ψ(Me,Me)Pro)

Synonyms: (4S,5R)-3-(FMOC-Val)-2,2,5-Trimethyl-Oxazolidine-4-Carboxylic Acid

Bachem B-3470.0001 MW 480.56 $C_{27}H_{32}N_2O_6$ Store at -15°C

Val-Trp
Z-Val-Trp

Bachem C-2895.0001 MW 437.5 $C_{24}H_{27}N_3O_5$ Store at -15°C

Val-Trp
Z-Val-Trp-OMe

Bachem C-3530.0001 MW 451.52 $C_{25}H_{29}N_3O_5$ Store at -15°C

Val-Trp

Bachem N-1170.0250 MW 303.36 $C_{16}H_{21}N_3O_3$ Store at -15°C

ICN 158284 MW 303.4 $C_{16}H_{21}N_3O_3$ Crystalline

Sigma V 2000 FW 303.4 $C_{16}H_{21}N_3O_3$

Val-Trp-Ile

Bachem H-5295.0050 Store at -15°C

Val-Tyr
Z-Val-Tyr

Bachem C-2900.0001 MW 414.46 $C_{22}H_{26}N_2O_6$ Store at -15°C

Val-Tyr

Bachem G-3585.0250 MW 280.32 $C_{14}H_{20}N_2O_4$ Store at -15°C

ICN 103238 MW 280.3 $C_{14}H_{20}N_2O_4$ 99%; crystalline

Val-Tyr
N-CBZ-Val-Tyr-OMe

Sigma C 7127 FW 428.5 $C_{23}H_{28}N_2O_6$

Val-Tyr

Sigma V 5626 FW 280.3 $C_{14}H_{20}N_2O_4$

Val-Tyr Amide Hydrochloride

Bachem G-3590.0250 MW 315.8 $C_{14}H_{21}N_3O_3 \cdot HCl$ Store at -15°C

Val-Tyr-Ile-His-Pro

Synonyms: Angiotensin I/II (3-7)

Bachem H-6965.0005 MW 627.74 $C_{31}H_{45}N_7O_7$ Store at -15°C

Val-Tyr-Ile-His-Pro-Phe

Synonyms: Angiotensin I/II (3-8); Angiotensin II (3-8); Angiotensin IV (3-8)

Bachem H-8125.0025 MW 774.92 $C_{40}H_{54}N_8O_8$ Store at -15°C

ICN 159877 MW 774.9 98% | Hermann, K et al, *J Neurochem*, 52:863, 1989

Neosystem SC498 MW 774.91 Human Bioactive

American Peptide 12-1-32 Human MW 774.9 $C_{40}H_{54}N_8O_8$ Major metabolite of angiotensin II; binds specifically with a new angiotensin binding site that distinguishes & separates from angiotensin II receptors; Hermann, K et al, *J Neurochem*, 52:863, 1989

Peptides International PAN-4331-v Human synthetic MW 774.92 $C_{40}H_{54}N_8O_8$ >99% (HPLC); lyophilized amorphous powder | Bioactive; Haberl, RL et al, *Circ Res*, 68:1621, 1991; Harding, JW et al, *Brain Res*, 583:340, 1992; Hanesworth, JM et al, *J Pharmacol Exp Ther*, 266:1036, 1993; de Gasparo, M et al, *Hypertension*, 25:924, 1995

Val-Tyr-Ile-His-Pro-Phe Acetate Salt

Synonyms: Angiotensin II (3-8); Angiotensin IV; Angiotensin III, (des-Arg[1])-

Sigma A 3950 Human FW 774.9 (free base) $C_{40}H_{54}N_8O_8$ ≥97% (HPLC) | Bioactive peptide; major metabolite of angiotensin II with a binding site distinct & separate from Angiotensin II receptors; Herman, K et al, *J Neurochem*, 52:863, 1989; Swanson, GN et al, *Regul Peptides*, 40:409, 1992

Biogenesis 0560-1529 Human synthetic MW 775 Lyophilized, consisting of 88% peptide material; purified

Val-Tyr-Leu-Lys-Ala
Ac-Val-Tyr-Leu-Lys-Ala-thiobenzyl ester

Bachem H-4496.0001 Store at -15°C

Val-Tyr-Pro

Bachem H-5400.0005 MW 377.44 $C_{19}H_{27}N_3O_5$ Store at -15°C

Val-Tyr-Pro-Asn-Gly-Ala-Glu-Asp-Glu-Ser-Ala-Glu-Ala-Phe-Pro-Leu-Glu-Phe

Synonyms: Adrenocorticotropic Hormone (22-39); β-Cell Tropin

Bachem H-2898.0001 MW 1985.09 $C_{90}H_{125}N_{19}O_{32}$ Store at -15°C

Val-Tyr-Ser
Val-Tyr-Ser-βNA

Bachem K-1540.0050 MW 492.58 $C_{27}H_{32}N_4O_5$ Store at -15°C

Val-Tyr-Thr-Val-Gln-Ile-Cys-Thr-Lys-Ser-Gly-Asp-Trp-Lys-Ser-Lys-Cys-Phe-Tyr-Thr-Thr

Synonyms: Tissue Factor (33-53), (Cys[39])-

Bachem H-3556.0500 Store at -15°C

Val-Tyr-Val

Bachem H-5300.0250 MW 379.46 $C_{19}H_{29}N_3O_5$ Store at -15°C

Val-Tyr-Val
Tfa-Val-Tyr-Val

Bachem N-1145.0050 MW 475.47 $C_{21}H_{28}F_3N_3O_6$ Store at -15°C

Val-Tyr-Val

ICN 158285 MW 379.5 $C_{19}H_{29}N_3O_5$ Off-white to tan crystals

Sigma V 8376 FW 379.5 $C_{19}H_{29}N_3O_5$ Off-white to tan crystals

Val-Val
BOC-Val-Val

Bachem A-3200.0001 MW 316.4 $C_{15}H_{28}N_2O_5$ Store at -15°C

Val-Val
Z-Val-Val

Bachem C-2905.0001 MW 350.42 $C_{18}H_{26}N_2O_5$ Store at -15°C

Val-Val
For-Val-Val

Bachem G-1880.0250 Store at -15°C

Val-Val

Bachem G-3595.0250 MW 216.28 $C_{10}H_{20}N_2O_3$ Store at -15°C

ICN 103239 MW 216.3 $C_{10}H_{20}N_2O_3$ Crystalline

Val-Val
N-TFA-L-Val-Val Cyclohexyl Ester

Sigma T 2257 FW 394.4 $C_{18}H_{29}F_3N_2O_4$

Val-Val

Sigma V 2125 FW 216.3 $C_{10}H_{20}N_2O_3$

Val-Val-Ala
Isovaleryl-L-Val-L-Val-AHMHA-L-Ala-AHMHA

Synonyms: Acid Protease Inhibitor; Pepsin Inhibitor; Cathepsin D Inhibitor; Renin Inhibitor; Pepstatin A

Alexis 260-010 Microbial MW 685.9 $C_{34}H_{63}N_5O_9$ ≥99%; white solid | Umezawa, H et al, *J Antibiot*, 23:259, 1970

Peptides International IPA-4039 Microbial MW 685.90 $C_{34}H_{63}N_5O_9$ Amorphous powder; integrity assessed by activity; bulk | Inhibitor; Umezawa, H et al, *J Antibiotics*, 23:259, 1970

Peptides International IPA-4039-v Microbial MW 685.90 $C_{34}H_{63}N_5O_9$ Lyophilized amorphous powder; integrity assessed by activity | Inhibitor; Umezawa, H et al, *J Antibiotics*, 23:259, 1970

Val-Val-Arg
Z-Val-Val-Arg-AMC

Bachem I-1540.0050 MW 663.77 $C_{34}H_{45}N_7O_7$ Store at -15°C

Val-Val-Arg
Z-Val-Val-Arg-MCA Hydrochloride

Synonyms: Cathepsin S/L Substrate

Peptides International MCA-3211-v MW 663.76 $C_{34}H_{45}N_7O_7$ >99% (HPLC); lyophilized amorphous powder | Brömme, D et al, *Biochem J*, 264:475, 1989; Kirschke, H & B Wiederanders in *Proteolytic Enzymes: Serine and Cysteine Peptidases*, Methods in Enzymology, Vol 244, (AJ Barret, Ed), Academic Press, New York, 1994, pp 500-511

Val-Val-Asp
((2S,3R)-3-Amino-2-Hydroxy-5-Methylhexanoyl)-Val-Val-Asp Hydrochloride

Synonyms: Amastatin

Bachem N-1410.0001 MW 511.02 $C_{21}H_{38}N_4O_8 \cdot HCl$ Store at -15°C

Val-Val-Asp
((2R,3R)-3-Amino-2-Hydroxy-5-Methylhexanoyl)-Val-Val-Asp Hydrochloride

Synonyms: Epiamastatin

Bachem N-1550.0001 MW 511.02 $C_{21}H_{38}N_4O_8 \cdot HCl$ Store at -15°C

Val-Val-Asp
N-((2S,3R)-3-Amino-2-Hydroxy-5-Methylhexanoyl)-L-Val-L-Val-L-Asp Hydrochloride

Synonyms: Amastatin

Fluka 08135 MW 529.0 $C_{21}H_{38}N_4O_8 \cdot HCl \cdot H_2O$ ≥99% (TLC); passes leucine aminopeptidase-inhibition test, monohydrate | Inhibitor of aminopeptidase B & leucine aminopeptidase; Rich, DH et al, *JMC*, 27:417, 1984; Umezawa, H, *Ann Rev Microbiol*, 36:75, 1982

Val-Val-Asp
((2S,3R)-3-Amino-2-Hydroxy-5-Methylhexanoyl)-Val-Val-Asp

Synonyms: Amastatin

ICN 152842 MW 475.5 $C_{21}H_{38}N_4O_8$ Inhibitor for aminopeptidase A & leucine aminopeptidase; Tobe, H etal, *Agric Biol Chem*, 43:591, 1979

Val-Val-Asp
((2R,3R)-3-Amino-2-Hydroxy-5-Methylhexanoyl)-Val-Val-Asp

Synonyms: Epiamastatin

ICN 152848

Val-Val-Asp
((2S,3R)-3-Amino-2-Hydroxy-5-Methylhexanoyl)-Val-Val-Asp Hydrochloride

Synonyms: Amastatin

Sigma A 1276 FW 511.0 $C_{21}H_{38}N_4O_8 \cdot HCl$ ≥97% (HPLC) | Bioactive peptide; enzyme inhibitor

Val-Val-Asp
((2R,3R)-3-Amino-2-Hydroxy-5-Methylhexanoyl)-Val-Val-Asp Hydrochloride

Synonyms: Epiamastatin; Aminopeptidases Inhibitor; Metallo-Protease Inhibitor

Sigma E 3389 FW 474.6 (free base) $C_{21}H_{38}N_4O_8$ ≥97% (HPLC) | Bioactive peptide; selectivity for aminopeptidases

Val-Val-Asp
(2S,3R)-3-Amino-2-Hydroxy-5-Me-Hexanoyl)-Val-Val-Asp

Synonyms: Amastatin; Aminopeptidase A Inhibitor; Leucine Aminopeptidase Inhibitor

Alexis 260-003 Synthetic MW 474.6 $C_{21}H_{38}N_4O_8$ ≥98%; white powder; soluble in acetic acid | Aoyagi, T et al, *J Antibiot*, 31:636, 1978; Tobe, H et al, *Agric Biol Chem*, 43:591, 1979

Val-Val-Asp
((2S,3R)-3-Amino-2-Hydroxy-5-Methylhexanoyl)-L-Val-L-Val-L-Asp

Synonyms: Amastatin; Aminopeptidase A Inhibitor; Leucine Aminopeptidase Inhibitor

Peptides International IAM-4095 Synthetic MW 474.55 $C_{21}H_{38}N_4O_8$ >99% (HPLC); amorphous powder; bulk | Inhibitor; Aoyagi, T et al, *J Antibiotics*, 31:636, 1978; Tobe, H et al, *Agric Biol Chem*, 43:591, 1979

Peptides International IAM-4095-v Synthetic MW 474.55 $C_{21}H_{38}N_4O_8$ >99% (HPLC); lyophilized amorphous powder

Val-Val-Gln

Bachem H-6465.0050 Store at -15°C

Val-Val-Gln-Arg-Glu-Lys-Arg-Ala-Val-Gly-Ile-Gly

Synonyms: GP41 (510-521); HTLV I Peptide

Neosystem SC238 MW 1311.55 HTLV I peptide from subtype GP41

Val-Val-Gly

Bachem H-6470.0050 MW 273.33 $C_{12}H_{23}N_3O_4$ Store at -15°C

Val-Val-Gly-Gly-Val-Met-Leu-Gly-Ile-Ile-Ala-Gly-Lys-Asn-Ser-Gly-Val-Asp-Glu-Ala-Phe-Phe-Val-Leu-Lys-Gln-His-His-Val-Glu-Tyr-Gly-Ser-Asp-His-Arg-Phe-Glu-Ala-Asp
β-Val-Val-Gly-Gly-Val-Met-Leu-Gly-Ile-Ile-Ala-Gly-Lys-Asn-Ser-Gly-Val-Asp-Glu-Ala-Phe-Phe-Val-Leu-Lys-Gln-His-His-Val-Glu-Tyr-Gly-Ser-Asp-His-Arg-Phe-Glu-Ala-Asp

Synonyms: Amyloid (40-1)

American Peptide 62-0-77 MW 4329.9 $C_{194}H_{295}N_{53}O_{58}S$ Inactive control; Kowall, NW et al, *PNAS*, 88:7247, 1991

Val-Val-Gly-Gly-Val-Met-Leu-Gly-Ile-Ile-Ala-Gly-Lys-Asn-Ser-Gly-Val-Asp-Glu-Ala-Phe-Phe-Val-Leu-Lys-Gln-His-His-Val-Glu-Tyr-Gly-Ser-Asp-His-Arg-Phe-Glu-Ala-Asp

Synonyms: Amyloid β-Protein (40-1)

Bachem H-2972.0500 MW 4329.86 $C_{194}H_{295}N_{53}O_{58}S$ Store at -15°C

Sigma A 2326 FW 4329.8 ≥97% (HPLC) | Bioactive peptide; inactive control; the N-C inverted sequence of fragment 1-40; Kowall, NW et al, *Proc Natl Acad Sci USA*, 88:7247, 1991

Val-Val-Gly-Trp-Pro-Thr-Val-Arg-Glu-Arg-Met-Arg-Arg-Ala-Glu-Pro

Synonyms: NEF (10-25); HTLV I Peptide

Neosystem SC216 MW 1939.26 HTLV I peptide from subtype NEF

Val-Val-Ile-Ala
MeOSuc-Val-Val-Ile-Ala-pNA

Bachem L-1745.0050 Store at -15°C

Val-Val-Lys

Bachem H-6505.0050 Store at -15°C

Val-Val-Nle
Z-Val-Val-Nle-Diazomethylketone

Bachem C-3890.0050 MW 487.6 $C_{25}H_{37}N_5O_5$ Store at -15°C

Val-Val-Phe

Bachem H-6460.0050 MW 263.46 $C_{19}H_{29}N_3O_4$ Store at -15°C

Val-Val-Ser-His-Phe-Asn-Asp-Cys-Pro-Asp-Ser-His-Thr-Gln-Phe-Cys-Phe-His-Gly-Thr-Cys-Arg-Phe-Leu-Val-Gln-Glu-Asp-Lys-Pro-Ala-Cys-Val-Cys-His-Ser-Gly-Tyr-Val-Gly-Ala-Arg-Cys-Glu-His-Ala-Asp-Leu-Leu-Ala

Synonyms: Transforming Growth Factor α

Sigma T 5403 Human FW 5546.2 ≥97% (HPLC); peptide content ~80% (not determined by Sigma) | Bioactive peptide;

Val-Val-Ser-His-Phe-Asn-Lys-Cys-Pro-Asp-Ser-His-Thr-Gln-Tyr-Cys-Phe-His-Gly-Thr-Cys-Arg-Phe-Leu-Val-Gln-Glu-Glu-Lys-Pro-Ala-Cys-Val-Cys-His-Ser-Gly-Tyr-Val-Gly-Val-Arg-Cys-Glu-His-Ala-Asp-Leu-Leu-Ala

Synonyms: Transforming Growth Factor α (1-50)

Bachem H-5545.0050   Rat   MW 5617.38 $C_{244}H_{361}N_{71}O_{71}S_6$
Store at -15°C | Disulfide bonds: Cys[8]-Cys[21], Cys[16] andCys[32], Cys[34]-Cys[43]

Sigma T 5278   Rat   FW 5623.4   ≥97% (HPLC); peptide content ~80% (not determined by Sigma) | Bioactive peptide; disulfide bonds: 8-21, 16-32, 34-43; Marquardt, H et al, *Science*, 223:1079, 1984

ICN 195719   Rat synthetic   >95% | Peptic growth factor released by cancer cells; structurally related to EGF; binds EGF receptors & mediates cell growth; useful in cell culture

---

Val-Val-Ser-Ser-Cys-Thr-Arg-Met-Met-Glu-Thr-Gln-Thr-Ser-Thr-Trp-Phe-Gly-Phe-Asn

Synonyms: GP140 (261-280); SIVmac251 Peptide

Neosystem SC734   MW 2312.61   SIVmac251 peptide from subtype GP140; due to the presence of a Cys this peptide can contain the dimeric form

---

Val-Val-Sta
Isovaleryl-Val-Val-Sta-OEt

Bachem H-6680.0001   Store at -15°C

Val-Val-Sta-Ala-Sta
Isovaleryl-Val-Val-Sta-Ala-Sta

Synonyms: Pepstatin A; Aspartic Protease Inhibitor; Cathepsin D Inhibitor; Pepsin Inhibitor; Renin Inhibitor

Amersham US20037   Ethanol/methanol soluble

Val-Val-Sta-Ala-Sta
Ac-Val-Val-Sta-Ala-Sta

Synonyms: Pepstatin, Ac-

Bachem N-1250.0001   MW 643.82 $C_{31}H_{57}N_5O_9$ Store at -15°C

Val-Val-Sta-Ala-Sta
Isovaleryl-Val-Val-Sta-Ala-Sta

Synonyms: Pepstatin A; Aspartic Protease Inhibitor; Cathepsin D Inhibitor; Pepsin Inhibitor; Renin Inhibitor

Fluka 77170   MW 685.91 $C_{34}H_{63}N_5O_9$ 120,000 U/mg; 1 U corresponds to the amount of inhibitor which reduces the activity of pepsin by 1 U; 1 U corresponds to the amount of enzyme which increases the absorbance at 280 nm by 0.001/min at pH 2.0, 37°C

Val-Val-Sta-Ala-Sta
***N*-Ac-Val-Val-(3*S*,4*S*)-Sta-Ala-(3*S*,4*S*)-Sta**

Synonyms: Pepstatin, Ac-

Neosystem SC675   Virus-related peptide; AIDS-related peptide; effective inhibitor of HIV-1 protease & HIV-2 protease; Richards, AD et al, *FEBS Lett*, 253:214, 1989

Val-Val-Sta-Ala-Sta
Isovaleryl-Val-Val-Sta-Ala-Sta

Synonyms: Pepstatin A; Aspartic Protease Inhibitor; Cathepsin D Inhibitor; Pepsin Inhibitor; Renin Inhibitor

Oncogene 516482   MW 685.9 $C_{34}H_{63}N_5O_9$ Solid; ≥98% (TLC) | Irreversible inhibitor; *Merck Index*, 12:7290

Val-Val-Sta-Ala-Sta
Ac-Val-Val-Sta-Ala-Sta

Synonyms: Pepstatin; HIV I Proteinase Inhibitor; HIV II Proteinase Inhibitor

Sigma A 4815   FW 643.8 $C_{31}H_{57}N_5O_9$ ≥95% (HPLC) | Bioactive peptide; Richards, AD et al, *FEBS Lett*, 247:113, 1989; 253:214, 1989

Val-Val-Sta-Ala-Sta
Isobutyryl-Val-Val-Sta-Ala-Sta

Synonyms: Pepsinostreptin

Sigma P 7424   FW 671.9 $C_{33}H_{61}N_5O_9$ ≥97% (HPLC) | Bioactive peptide; enzyme inhibitor; forms a 1:1 complex with pepsin & inhibits its activity; Kakinuma, A et al, *J Takeda Res Lab*, 35:123, 128, 136, 1976

Val-Val-Sta-Ala-Sta
Isovaleryl-Val-Val-Sta-Ala-Sta

Synonyms: Pepstatin A; Aspartic Protease Inhibitor; Cathepsin D Inhibitor; Pepsin Inhibitor; Renin Inhibitor; Acid Protease Inhibitor

American Peptide 81-5-60   Microbial   MW 685.9 $C_{34}H_{63}N_5O_9$ Irreversible inhibitor; Umezawa, H et al, *J Antibiotics*, 23:259, 1970

Bachem N-1125.0005   Microbial   MW 685.9 $C_{34}H_{63}N_5O_9$ Store at -15°C

Sigma P 4265   Microbial   FW 685.9 $C_{34}H_{63}N_5O_9$ ≥75% (HPLC) | Bioactive peptide; enzyme inhibitor; potent inhibitor of acid proteases; Rich, DH et al, *Biochemistry*, 24:3165, 1985

Sigma P 5318   Microbial   FW 685.9 $C_{34}H_{63}N_5O_9$ ≥90% (HPLC) | Bioactive peptide; enzyme inhibitor; potent inhibitor of acid proteases; Rich, DH et al, *Biochemistry*, 24:3165, 1985

ICN 195368   Synthetic   MW 685.9 $C_{34}H_{63}N_5O_9$ Irreversible inhibitor; Umezawa, H et al, *J Antibiotics*, 23:259, 1970

Val-Val-Sta-Val-Sta
Isobutyryl-Val-Val-Sta-Val-Sta

Synonyms: Pepsinostreptin; Pepsin Inhibitor

ICN 194175   *Streptomyces ramulosus*   MW 671.9 $C_{33}H_{61}N_5O_9$ Forms a 1:1 complex with pepsin

---

Val-Val-Tyr-Pro-Trp-Thr-Gln

Synonyms: Valorphin

Bachem H-8670.0005   MW 892.02 $C_{44}H_{61}N_9O_{11}$ Store at -15°C

Val-Val-Val

Bachem H-5305.0250   MW 315.41 $C_{15}H_{29}N_3O_4$ Store at -15°C

Val-Val-Val
D-Val-D-Val-D-Val

Bachem H-5310.0250   Store at -15°C

Val-Val-Val Amide Hydrobromide

Bachem H-5315.0250   Store at -15°C

Val-Val-Val-Pro-Pro
(5*S*)-1-((2*S*)-*o*-(*N,N*-Dimethyl-Val-Val-*N*-Me-Val-Pro-Pro)-2-Hydroxyisovaleryl)-2-Oxo-4-Methoxy-5-Benzyl-3-Pyrroline

Synonyms: Dolastatin 15

Bachem H-8630.0001   MW 837.07 $C_{45}H_{68}N_6O_9$ Store at -15°C

Val-Val-Val-Val

Bachem H-5320.0050   MW 414.55 $C_{20}H_{38}N_4O_5$ Store at -15°C

670

Part 4. Peptides

Sequences Not Specified

7B2 Peptide (23-39) Acetate Salt

Synonyms: Secretogranin V; Pituitary Polypeptide; Neuroendocrine Protein 7B2

Biogenesis 0100-0167 Synthetic Semi-pure; lyophilized

Adenosine A1 Receptor

Biotrend A1R11-P Rat, canine Control peptide

Adenosine A2a Receptor

Biotrend A2aR21-P Canine Control peptide

Adenosine A2b Receptor

Biotrend A2bR23-P Human Control peptide

Adenosine A3 Receptor

Biotrend A3R32-P Human Control peptide

Biotrend A3R33-P Human Control peptide

Biotrend A3R31-P Rat Control peptide

Adrenergic Receptor β_3

Biotrend B3AR12-P Human Control peptide; 19 AA | Near C-terminal

Biotrend B3AR13-P Murine Control peptide; 20 AA | Near C-terminal

Adrenocorticotropic Hormone

Synonyms: Corticotropin A

ICN 152714 Human

ICN 152712 Porcine pituitary 70-90 U/mg

Sigma A 6303 Porcine pituitary ~90 IU/mg by RIA (pure synthetic porcine ACTH as standard); off-white to tan powder; bioassay not run by Sigma | Bioactive peptide; stimulates synthesis & secretion of glucocorticoids by adrenal cortex; endocrine functions of the adrenal cortex are regulated by an anterior pituitary hormone, ACTH & its fragments affect motivation, learning & behavior; Gispen, WH & DeWied, D, *Peptides, Structure & Function*, Proceedings of the Eighth American Peptide Symposium, Hruby, VJ & Rich, DH, eds, 399, 1983

Adrenocorticotropic Hormone (11-13)

USBio A0758-05A Suitable for antigenic applications in immunological protocols

Adrenocorticotropic Hormone (11-24)

Synonyms: Corticotropin

USBio A0758-05B Suitable for antigenic applications in immunological protocols

Biotrend 0178-0459 Synthetic >95% | Antigen

Adrenocorticotropic Hormone (1-24)

Synonyms: Corticotropin A; Tetracosactide; Corticotropin

Sigma A 8280 FW 2933.5 Human ≥90% (HPLC) | Bioactive peptide

Biotrend 0178-0359 Synthetic >95% | Antigen

Adrenocorticotropic Hormone (1-39)

Synonyms: Corticotropin

Biotrend 0178-0399 Human pituitary >95% | Antigen

Adrenocorticotropic Hormone (18-39)

Synonyms: Corticotropin

Biotrend 0178-0539 Synthetic >95% | Antigen

Adrenocorticotropic Hormone, Agarose

Synonyms: Corticotropin

ICN 191272 Porcine 10 atoms hydrophilic spacer arm; 0.5-1.5 mg ACTH/mL gel; formed by reacting diazo functional group of spacer & ligand; suspension in PBS, 0.02% NaN_3 | Useful in studies on isolated adrenal cells

Adrenomedullin Receptor Ligand, (^{125}I)-

Amersham IM282 Rat Endogenous; K_d = 1.3 *nM* rat lung membrane

Agouti Related Peptide

Biotrend AG011-P Murine Control peptide; 21 AA | Near N-terminal

Biotrend AGRP11-P Murine Control peptide; 15 AA | Near C-terminal

A-Kinase Anchoring Protein St-Ht31 Inhibitor Peptide, InCELLect™

Promega V8211 MW 2797 >80% (HPLC) | A stearated (St) form of the peptide Ht-31 derived from the human thyroid AKAP (A-kinase anchoring protein); inhibits the interaction between the RII subunits of cAMP-dependent proteinkinase & AKAP in cell extracts; allows the study of real-time physiological effects related to PKA signaling; stearated group makes peptides cell-permeable; Vijayaraghavan, S et al, *JBC*, 272:4747, 1997

AKT-1

Synonyms: PKB-α

Biotrend AKT11-P Rat Control peptide; 15 AA | Near C-terminal

Biotrend AKT12-P Rat Control peptide; 15 AA | Near C-terminal; Ser(*p*) phosphospecific

AKT-2

Synonyms: PK-β

Biotrend AKT21-P Rat Control peptide; 16 AA | Near C-terminal

Biotrend AKT31-P Rat Control peptide; 12 AA | Near C-terminal

Amastatin Hydrochloride

Biogenesis 0355-0100 $C_{21}H_{38}N_4O_8 \cdot HCl$ Non-species synthetic >99.5% (HPLC); lyophilized

Amastatin, Epi-

Biogenesis 0355-0200 $C_{21}H_{38}N_4O_8 \cdot HCl$ Non-species synthetic >99.5% (HPLC); lyophilized

Amyloid (10-20), β-

BioSource International 03-154

Amyloid (10-35), β- Amide

BioSource International 03-152

Amyloid (1-12), β-

BioSource International 03-260

Amyloid (1-20), β-

BioSource International 03-264

Amyloid (12-28), β-

BioSource International 03-141

Amyloid (1-28), (Gln[11])-β-
BioSource International 03-143

Amyloid (1-28), β-
BioSource International 03-142

Amyloid (13-40), β-
BioSource International 03-275

Amyloid (1-40), (Arg[13])-β-
BioSource International 03-201

Amyloid (1-40), (Cys)-β-
BioSource International 03-217

Amyloid (1-40), (Gln[22])-β-
BioSource International 03-213

Amyloid (1-40), (Gly[21])-β-
BioSource International 03-209

Amyloid (1-40), (Gly[5])-β-
BioSource International 03-193

Amyloid (1-40), (Nle[35])-β-
BioSource International 03-229

Amyloid (1-40), (Phe[10])-β-
BioSource International 03-197

Amyloid (1-40), β-
BioSource International 03-136
BioSource International 03-138   Pure
BioSource International 03-188

Amyloid (1-40), β-
BioSource International 03-215

Amyloid (1-40), β- Amide
BioSource International 03-205

Amyloid (1-40), β- Biotin
BioSource International 03-243

Amyloid (1-40, All), β-
BioSource International 03-134

Amyloid (1-42), β-
BioSource International 03-111

Amyloid (1-43), β-
BioSource International 03-121

Amyloid (17-28), β-
BioSource International 03-148

Amyloid (17-40), β-
BioSource International 03-114

Amyloid (22-35), β-
BioSource International 03-239

Amyloid (25-35), β-
BioSource International 03-140

Amyloid (25-35), β- Biotin
BioSource International 03-139

Amyloid (34-42), β-
BioSource International 03-291

Amyloid (35-25), β-
BioSource International 03-244

Amyloid (35-25), β- Biotin
BioSource International 03-241

Amyloid (35-42), β-
BioSource International 03-295

Amyloid (40-1), β-
BioSource International 03-245
BioSource International 03-246

Amyloid (42-1), β-
BioSource International 03-247

Amyloid (43-1), β-
BioSource International 03-249

Amyloid (5-40), β-
BioSource International 03-270

Amyloid Peptide (12-28), β-
Oncogene PP73   MW 1953   Human   >99% (HPLC); lyophilized in TFA salt; reconstitute with degassed HPLC grade dH$_2$O | Non-neurotoxic prior to a pre-incubation step; toxicity correlates to β-sheet structure

Amyloid Peptide (1-28), (Gln[11])-β-
Oncogene PP72   MW 3263   Human synthetic   >98% (HPLC); biological activity: neurotoxic activity usually observed at 30-100 μg/mL; lyophilized | For neurotoxicity studies & substrate cleavage assay; synthetic peptide corresponding to AA 1-28 of the processed human amyloid peptide with a substitution of Gln[11]; non-neurotoxic prior to a pre-incubation step

Amyloid Peptide (1-28), β-
Oncogene PP71   MW 3260   Human   >99% (HPLC); lyophilized in TFA salt; reconstitute with degassed HPLC grade dH$_2$O | Non-neurotoxic prior to a pre-incubation step; toxicity correlates to β-sheet structure

Amyloid Peptide (1-40) Biotin Conjugate, β-
Oncogene PP64B   MW 4670   Human   >98% (HPLC); lyophilized in TFA salt; reconstitute with degassed HPLC grade dH$_2$O | Non-neurotoxic prior to a pre-incubation step

Amyloid Peptide (1-40) Fluorescein Conjugate, β-
Oncogene PP64F   MW 4823   Human synthetic   >99% (HPLC); lyophilized in TFA salt; reconstitute with degassed HPLC grade dH$_2$O | AA 1-40 of the processed human amyloid peptide; non-neurotoxic prior to a pre-incubation step

Amyloid Peptide (1-40), (Gln²²)-β-

Oncogene PP68   MW 4330   Human   >99% (HPLC); biological activity: neurotoxic activity usually observed at 30-100 μg/mL; lyophilized | For neurotoxicity studies & substrate cleavage assay; synthetic peptide corresponding to AA 1-40 of the processed human amyloid peptide with a substitution of Gln²²; this fragment contains the mutation associated with Alzheimer's disease Dutch variant; non-neurotoxic prior to a pre-incubation step

Amyloid Peptide (1-40), (Gly²¹)-β-

Oncogene PP67   MW 4316   Human   >99% (HPLC); biological activity: neurotoxic activity usually observed at 30-100 μg/mL; lyophilized | For neurotoxicity studies & substrate cleavage assay; synthetic peptide corresponding to AA 1-40 of the processed human amyloid peptide with a substitution of Gly²¹; this fragment contains the mutation associated with Alzheimer's disease Flemish variant; non-neurotoxic prior to a pre-incubation step

Amyloid Peptide (1-40), (Gly⁵,Phe¹⁰,Arg¹³)-β-

Oncogene PP66   MW 4235   Rodent   >99% (HPLC); biological activity: neurotoxic activity usually observed at 30-100 μg/mL; lyophilized | For neurotoxicity studies & substrate cleavage assay; synthetic peptide corresponding to AA 1-40 of the processed human amyloid peptide with a substitution of Gly⁵, Phe¹⁰, Arg¹³; the rat homolog to human β-amyloid, 1-40; non-neurotoxic prior to a pre-incubation step

Amyloid Peptide (1-40), Fluo-β-

Amersham VAB011   Fluorescein-labeled with high activity

Amyloid Peptide (1-40), β-

Oncogene PP64   MW 4331   Human   >95% (HPLC); lyophilized in TFA salt; reconstitute with degassed HPLC grade dH2O; biological activity: neurotoxic activity usually observed at 30-100 μg/mL; lyophilized | Non-neurotoxic prior to a pre-incubation step

Oncogene PP65   MW 4331   Human   Ultra Pure, >99% (HPLC); lyophilized in TFA salt; reconstitute with degassed HPLC grade dH2O | Non-neurotoxic prior to a pre-incubation step; toxicity correlates to β-sheet structure

Amyloid Peptide (1-42), β-

Oncogene PP69   MW 4515   Human   >95% (HPLC) | For neurotoxicity studies & substrate cleavage assay; synthetic peptide corresponding to AA 1-42 of the processed human amyloid peptide; a major constituent of plaques & tangles that occur in Alzheimer's disease patients; non-neurotoxic prior to a pre-incubation step

Amyloid Peptide (1-43), β-

Oncogene PP70   MW 4616   Human   >95% (HPLC); lyophilized in TFA salt; reconstitute with degassed HPLC grade dH2O | Non-neurotoxic prior to a pre-incubation step; toxicity correlates to β-sheet structure

Amyloid Peptide (25-35), β-

Oncogene PP74   MW 1060   Human   >97% (HPLC); lyophilized in TFA salt; reconstitute with degassed HPLC grade dH2O | Non-neurotoxic prior to a pre-incubation step; toxicity correlates to β-sheet structure

Amyloid β Precursor Protein (38-48)

Synonyms: Secretase γ Substrate; A4 Precursor Protein

Oncogene PP75   MW 1290   Human   Useful for competition studies & substrate cleavage assay

Amyloid β/A4 Precursor Protein (135-155)

BioSource International 03-504

BioSource International 03-505

Oncogene PP77   MW 2603   Human   >99% (HPLC); lyophilized in TFA salt; reconstitute with degassed HPLC grade dH2O | Contains a copper (II)-binding site, which upon binding induces oxidization & dimerization of the peptide

Amyloid β/A4 Precursor Protein (319-335)

BioSource International 03-506

Oncogene PP78   MW 2099   Human synthetic   >98% (HPLC); lyophilized in TFA salt; reconstitute with degassed HPLC grade dH2O | For neurotoxicity studies & substrate cleavage assay; synthetic peptide corresponding to AA 319-335 of human β-amyloid precursor protein (βAPP); reduces neurological damage *in vivo* in a CNS ischemia model

Amyloid β/A4 Precursor Protein (328-332)

BioSource International 03-508

Oncogene PP79   MW 678   Human   >98% (HPLC); lyophilized in TFA salt; reconstitute with degassed HPLC grade dH2O | For proliferation studies, neurotoxicity studies & substrate cleavage assay; synthetic peptide corresponding to AA 328-332 of human β-amyloid precursor protein (βAPP); stimulates fibroblast cell growth & is involved in the pathogenesis of Alzheimer's disease

Amyloid β/A4 Precursor Protein (657-676)

BioSource International 03-510

Oncogene PP80   MW 2210   Human   >98% (HPLC); lyophilized in TFA salt; reconstitute with degassed HPLC grade dH2O | A specific GTP-binding protein (G₀) activator; sequence corresponds with the G₀/amyloid precursor protein complex

Amyloid β/A4 Precursor Protein (96-110)

BioSource International 03-502

Oncogene PP76   MW 1921   Human   >98% (HPLC); lyophilized in TFA salt; reconstitute with degassed HPLC grade dH2O | Contains the heparin binding domain of bAPP, which binds strongly to heparin & inhibits (¹²⁵I)β-APP binding to heparin

Angiotensin II Receptor Ligand, ((¹²⁵I)-Tyr⁴)(Sar¹,Ile⁸)-

Amersham IM248   Non-selective; antagonist; $K_d = 0.64$ nM rat ovarian membrane

Angiotensin II Receptor Ligand, (¹²⁵I)-Tyr⁴-

Amersham IM177   Non-selective; endogenous; $K_d = 2$ nM bovine adrenal fasiculata cells

Angiotensin II Receptor Ligand, (³H)-

Amersham TRK733   Non-selective; endogenous; $K_d = 2.4$ nM rat adrenal cortex microsomes

Angiotensin II, ((¹²⁵I)-Tyr⁴)-

ICN 68131   Human   ~2000 Ci/mmol, ~74 TBq/mmol; lyophilized, 0.1M NaPO₄, 5% BSA

Angiotensin II, (Fluo)-

Amersham VAT2011   Human   Fluorescein-labeled with high activity

Angiotensin II, (Sar¹, (¹²⁵I)-Tyr⁴),Ile⁸)-

ICN 68132   Human   ~2000 Ci/mmol, ~74 TBq/mmol; lyophilized, 0.1M NaPO₄, 5% BSA

Angiotensin IV, (Fluo)-

Amersham VAT4011   Fluorescein-labeled with high activity

Angiotensin Receptor Ligand, ((^{125}I)-Tyr^4)-

Amersham IM176   Non-selective; endogenous

Angiotensinogen

Synonyms: Renin Substrate

ICN 191159   Human plasma   >95% (SDS-PAGE); lyophilized | Substrate for human renin which releases Angiotensin I; Tewksbury, DA, *Fed Proc*, 42:2724, 1983

ICN 152749   Porcine pasma   1500-3000 U/g Angiotensinogen protein; 1 U yields 1.0 *n*M angiotensin I in the presence of renin, pH 6.3 & 37°C (RIA); lyophilized, containing ~65% protein, remainder is primarily Na citrate

Antichymotrypsin, Alpha I

Scipac P159-1   Pooled serum/plasma   >96%; lyophilized; available on request | Inhibitor

Scipac P159-5   Pooled serum/plasma   ~90%; frozen in TRIS buffer; binds PSA
| Inhibitor

Antipain Dihydrochloride

Biogenesis 0619-9010   Non-species synthetic   Lyophilized

Antiplasmin, α$_2$-

ICN 194186   Human plasma   Highly purified, <0.02 PEU of plasminogen & His-rich glycoprotein (HRG) | Plow, EF & D Colleen, *Blood*, 58(6):1069, 1981

Antitrypsin, Alpha I

Scipac P165-5   Pooled serum/plasma   >96%; lyophilized; available on request | Inhibitor

Scipac P165-2   Serum/plasma   40-90%; lyophilized; available on request | Inhibitor

AP Kinase Substrate II

Synonyms: MBP (90-98)

BioSource International 77-355

Apo-Serum Amyloid A

PeproTech 300-13   Human recombinant, expressed in *E. coli* >98% by SDS-PAGE & HPLC; <0.1 ng endotoxin per mg (1EU/mg); lyophilized | 104 AA polypeptide that circulates primarily in association with high-density lipoproteins (HDL); normally 1-5 mg/mL in plasma, but increases 500-1000 fold within 24 hr of an inflammatory stimulus—the most abundant HDL apo-lipoprotein under these conditions

Aprotinin

Synonyms: Kallikrein Inactivator; Esterase Inhibitor; Protease Inhibitor; Kallikrein Inactivator

Pentapharm 073-10, 073-20   MW 6512   Polyvalent reversible inhibitor of serine proteinases; used for isolation of proteins; used for biopharmaceutical downstream purification to inhibit undesired proteolytic activity of serine proteases such as trypsin, plasmin, trypsinogen, urokinase, chymotrypsin, kallikrein, elastase & others; used during iimmunodiffusion, radioimmunoassay or enzyme-linked immuno assay procedures; used in chromogenic assays for determination of antithrombin III, heparin, α$_2$-macroglobulin, factor Xa & thrombin to inhibit disturbing kallikrein or plasmin activities; activity:6,000 KIU/mg (lyophilised), 200,000 KIU/ml (solution), sterile; Fritz H, & G Wunderer, *Arzneim Forsch/Drug Res*, 33:479, 1983; Lottenberg R, et al, *Thromb Res*, 49:549, 1988

Oncogene 616398   MW 6512   C$_{284}$H$_{432}$N$_{84}$O$_{79}$S$_7$ Bovine lung Lyophilized; activity: 6000 KIU/mg dry weight; specific activity: 7000 KIU/mg protein; 1 KIU=the quantity of protease inhibitor that has the ability to inhibit 2 KU by 50% under optimal conditions; 1 KIU=0.025 antiplasmin units (APU) or 0.0011 trypsin inhibitor units; 1 KU=1000 KIU; may cause irritation! | A single chain polypeptide of 58 AA; competitive reversible inhibitor of esterase & protease activity; forms a tight complex, blocking the active site of the enzymes; inhibits a number of different proteases, including chymotrypsin, coagulation factors involved in the prephase of blood clotting, kallikrein, plasmin, tissue & leukocytic proteinases & trypsin; does not inhibit Factor Xa & thrombin; useful as a serine protease inhibitor during purification of proteins in studies of zymogen activation systems; extends the life of cells & prevents proteolytic damage of intact cells; *Merck Index*, 12:796; Azougagh Oualane, F et al, *Thromb Res*, 68:185, 1992

Oncogene 616399   MW 6512   C$_{284}$H$_{432}$N$_{84}$O$_{79}$S$_7$ Bovine lung Isotonic NaCl, aseptically filled; activity: 9000 KIU/mL; 1 KIU=the quantity of protease inhibitor that has the ability to inhibit 2 KU by 50% under optimal conditions; 1 KU=1000 KIU; may cause irritation! | *Merck Index*, 12:796

Apstatin Hydrochloride

Biogenesis 0652-5010   Non-species synthetic   Purified; lyophilized

ATF2/GST-ATF2 (19-96)

Upstate 12-367   MW 35k   >90% pure; liquid at −20 °C | Capture assay, ELISA

Autocamtide II

BioSource International 77-132

Autocamtide II Related Inhibitor

BioSource International 77-130

BioSource International 77-404

Autocamtide II, Biotinylated

BioSource International 77-402

Autocamtide III

BioSource International 77-134

Bacitracin

ICN 100165   MW 1422.7   C$_{66}$H$_{103}$N$_{17}$O$_{16}$S   ≥60 U/mg; USP Grade | Antimicrobial

Bacitracin; Peptidoglycan Synthesis Inhibitor

ICN 190301   MW 1486.1   C$_{66}$H$_{101}$N$_{17}$O$_{16}$SZn   ≥60 U/mg; USP Grade | Main component is Bacitracin A; acts as a peptide-antibiotic & an inhibitor of peptidoglycan synthesis; Scogin, D etal, *Biochemistry*, 19:3348, 1980

Bax Peptide

Amersham VPP51   Bcl-2 family; competition for VPC66

Oncogene PP51   Lyophilized with an equal amount of BSA | Suggested starting concentration of peptide for competition is 10X the concentration of primary Ab; titration recommended

Bax Peptide (150-165)

Calbiochem PP51   Human   Lyophilized solid with BSA; soluble in H$_2$O; suggested initial concentration of peptide is 10X the concentration of primary Ab | Useful in competition assays using anti-Bax (Ab-1)

Bcl-2 Peptide

Amersham VPP52   Bcl-2 family; competition for VPC68

Bcl-2 Peptide (20-34)

Oncogene PP52   Lyophilized with an equal amount of BSA |
Suggested starting concentration of peptide for competition is 10X
the concentration of primary Ab; titration recommended

Bcl-2 Protein (20-34)

Synonyms: Bcl-2 Peptide

Calbiochem PP52   Lyophilized solid with BSA; soluble in H_2O;
suggested initial concentration of peptide is 10X the concentration
of primary Ab | Useful for competition assays using anti-Bcl-2
(Ab-2)

Bcl-x Peptide

Amersham VPP53   Bcl-2 family; competition for VPC67

Bcl-x Peptide (201-216)

Oncogene PP53   Lyophilized with an equal amount of BSA |
Suggested starting concentration of peptide for competition is 10X
the concentration of primary Ab; titration recommended

Bcl-x Protein (201-216)

Synonyms: Bcl-x Peptide

Calbiochem PP53   Human   Lyophilized solid with BSA; soluble
in H_2O; suggested initial concentration of peptide is 10X the
concentration of primary Ab | Useful for competition assays using
anti-Bcl-x (Ab-1)

Bcl-x$_S$ Peptide

Amersham VPP54   Bcl-2 family; competition for VPC89

Oncogene PP54   Lyophilized with an equal amount of BSA |
Suggested starting concentration of peptide for competition is 10X
the concentration of primary Ab; titration recommended

Bcr Peptide

Amersham VPP02   Competition for VPC02; viral & oncogene
recognition

Amersham VPP18   Competition for VOP26; viral & oncogene
recognition

Oncogene PP02   Lyophilized with an equal amount of BSA |
Suggested starting concentration of peptide for competition is 10X
the concentration of primary Ab; titration recommended

Bcr Peptide II

Synonyms: p210bcr/abl (686-698)

Oncogene PP18   Lyophilized with an equal amount of BSA;
suggested starting concentration of peptide for competition is 10X
the concentration of primary Ab | Useful competition studies & dot
blot

BD-2

Synonyms: Defensin II, β-

PeproTech 300-49   MW 4.3k   Human recombinant, expressed
in *E. coli*   >98% by SDS-PAGE & HPLC; <0.1 ng endotoxin per
mg (1EU/mg); lyophilized | Antimicrobial peptide belonging to the
distinct family of beta-defensins; 41 AA; SA determined by its
ability to chemoattract immature human dendritic cells

Bestatin

Biogenesis 1102-5000   MW 344.8 $C_{16}H_{24}N_2O_4$ Non-species
synthetic   >99% (HPLC); lyophilized

Bestatin, Epi-

Biogenesis 1102-5100   Non-species synthetic   Lyophilized

Bombesin, (D-Phe^{12})-

ICN 152856   MW 1629.9   Heinz-Erian, P, *Am J Physiol*,
252:G439, 1987

Bombesin, (D-Phe^{12},Leu^{14})-

Synonyms: Bombesin Receptor Antagonist

ICN 152857   MW 1611.9   Jensen, RT etal, *Can J Physiol
Pharmac*, 64(468):175, 1986; Heinz-Erian, P, *Am J Physiol*,
252:G439, 1987

Bombesin, (Gly^2,Thr^6)-

Synonyms: Alytesin

ICN 152852   Biological activity similar to Bombesin; Anastasi, A
etal, *Experentia*, 27:16, 1971; Melchiorri, P, *Gut Hormones*,
Churchill-Livingstone, Edinburgh, *p* 534, 1973

Bombesin, (Leu^3)-

ICN 150511   MW 1591.8   For Ab studies

Bombesin, (Tyr^4)-

ICN 150512   MW 1669.9   Rivier, JE & MT Brown,
Biochemistry, 17:1766, 1978

Bombesin, (Tyr^4,D-Phe^{12})-

Synonyms: Bombesin Receptor Antagonist

ICN 152858   MW 1679.9

BPDE-Tide

BioSource International 77-196

Bradykinin Receptor Ligand, (^{3}H)-

Amersham TRK943 B_2 selectivity; endogenous; K_d = 1.8 nM
rat myometrium cells

Bradykinin, (D-Arg^0,Hyp^3,D-Phe^7)-

ICN 152783   MW 1282.5 $C_{60}H_{87}N_{19}O_{13}$ Vavrek, RJ & JM
Stewart, *Peptides*, 6:161, 1985; Stewart, JM & RJ Vavrek,
Proceedings of the 34th Colloquium "Protides of the Biological
Fluids," 34:473, 1986

Bradykinin, (D-Arg^0,Hyp^3,Thi^{5,8},D-Phe^7)-

Synonyms: Bradykinin Antagonist

ICN 152785   MW 1294.5   Stewart, JM & RJ Vavrek,
Proceedings of the 34th Colloquium "Protides of the Biological
Fluids," 34:473, 1986

Bradykinin, (des-Arg^1)-

ICN 152790   MW 904 $C_{44}H_{61}N_{11}O_{10}$ Nicolaides, ED etal, *J
Med Chem*, 6:739, 1963

Bradykinin, (des-Arg^9,Leu^8)-

Synonyms: Bradykinin Specific Inhibitor

ICN 152793   MW 870 $C_{41}H_{63}N_{11}O_{10}$ Regoli, D etal, *Can J
Physiol Pharmac*, 55:855, 1977

Bradykinin, (des-Pro^2)-

Synonyms: Angiotensin I Converting Enzyme Inhibitor

ICN 152791   MW 962 $C_{45}H_{66}N_{14}O_{10}$ Potentiates bradykinin-
induced contraction of guinea pig ileum & hypertension in rats;
Naruse, M etal, *Chem Pharm Bull*, 29:3369, 1981

Bradykinin, (Hyp^3)-

Synonyms: Bradykinin Antagonist

ICN 152758   MW 1076.2 $C_{50}H_{73}N_{15}O_{12}$ Stewart, JM & RJ
Vavrek, *Adv Biosci*, 65:73, 1987; Kato, H etal, *FEBS Lett*, 232:252,
1988

Bradykinin, (*N*-Admantane-Ac-D-Arg⁰,Hyp³,Thi⁵,⁸,D-Phe⁷)-

Synonyms: Bradykinin Antagonist

ICN 159898   MW 1471.9   98%

Bradykinin, (*N'*-Admantane-Ac-D-Arg⁰,Hyp³,Thi⁵,⁸,D-Phe⁷)-

Synonyms: Bradykinin Antagonist

ICN 159897   MW 1456.8   98% | Highly potent

Bradykinin, (*p*-Chloro-Phe⁵,⁸)-

ICN 152782   MW 1129.1

Bradykinin, (*S*-Arg⁰,Hyp³,D-Phe⁷,Leu⁸)-

ICN 195630   MW 1248.4   Potent antagonist used to discriminate β-2 receptors; Rhaleb, NE etal, *Life Sci*, 51:PL 125, 1992

Bradykinin, des-Pro²-(Ala²,⁶)-

Synonyms: Angiotensin I Converting Enzyme Inhibitor

ICN 195631   MW 921.1 $C_{43}H_{64}N_{14}O_9$ Chaturredi, D etal, *peptide Res*, 6:308, 1993

B-Raf Protein Kinase C-Terminal Blocking Peptide (734-747)

Calbiochem 553006   MW 1607.8 $C_{71}H_{102}N_{18}O_{23}S$ Synthetic   Lyophilized | Based on the human, mouse & chicken B-Raf kinase; this peptide coupled to KLH was used as the immunogen for the production of Anti-B-Raf Protein Kinase; for use in immunoabsorption immunoprecipitation, immunocytochemistry, Western blotting & dot blots

Bungarotoxin Receptor Ligand, ((¹²⁵I)-Tyr)-α-

Synonyms: Accholine Receptor Ligand

Amersham IM209   ~2000 Ci/mMol | Nicotinic selectivity; antagonist; K_d = 1.2 nm rat brain

Bungarotoxin, (3-(¹²⁵I)-Tyr)-α-

Synonyms: Accholine Receptor Ligand

Amersham IM109   Buffered aqueous solution; ≥5.55 TBq/mmol, ≥150 Ci/mmol; 18.5 MBq/mL, 500 µCi/mL | Nicotinic selectivity; antagonist; K_d = 1.2 nM rat brain

Bungarotoxin, α-*N*-Propionyl-(³H)Propionylated

Synonyms: Bungarotoxin Receptor Ligand, (³H)-α-; Accholine Receptor Ligand

Amersham TRK603   Aqueous solution, containing 0.02% BSA; 1.48-3.89 TBq/mmol, 40-105 Ci/mmol; 1.85 MBq/mL, 50 µCi/mL | Nicotinic selectivity; antagonist; K_d = 0.5 nM rat cerebral cortex membrane

C1 Esterase Inhibitor

Synonyms: C1 Inactivator

Scipac P177-7   Serum/plasma   50-90%; liquid in Sodium Phosphate buffer; 8-10 mg/mL | Inhibitor

Ca²⁺/Calmodulin Kinase II N-Terminal Blocking Peptide (7-20)

Synonyms: Calmodulin-Dependent Kinase II δ-Isoform (7-20)

Calbiochem 208704   MW 1794.0 $C_{80}H_{116}N_{18}O_{27}S$ Rat   Lyophilized | Coupled to KLH, useful as the immunogen for the production of Anti-CaMK II; for use in immunoabsorption for immunoprecipitation, immunocytochemistry, Western blotting & dot blots

Ca²⁺/Calmodulin Kinase IV C-Terminal Blocking Peptide (465-474)

Synonyms: Calmodulin-Dependent Kinase IV

Calbiochem 208714   MW 1276.4 $C_{56}H_{85}N_{13}O_{19}S$ Rat   Lyophilized | Based on rat brain calmodulin-dependent kinase IV (465-474) with a cysteine C residue added; coupled to KLH, useful as the immunogen for the production of Anti-CaMK IV; for use in immunoabsorption for immunoprecipitation, immunocytochemistry, Western blotting & dot blots

Ca²⁺/Calmodulin Kinase Kinase C-Terminal Blocking Peptide (490-503)

Synonyms: Calmodulin-Dependent Kinase Kinase

Calbiochem 208720   MW 1558.7 $C_{64}H_{103}N_{17}O_{26}S$ Rat   Lyophilized | Based on rat calmodulin-dependent kinase kinase (490-503) with a cysteine C residue added; coupled to KLH, useful as the immunogen for the production of Anti-CaMKK; for use in immunoabsorption for immunoprecipitation, immunocytochemistry, Western blotting & dot blots

C-abl Peptide

Amersham VPP01   Competition for VPC01; viral & oncogene recognition

Oncogene PP01   Lyophilized with an equal amount of BSA; suggested starting concentration of peptide for competition is 10X the concentration of primary Ab | Adjacent to kinase domain; useful for competition studies & dot blot

Calcicludine

ICN 193933   MW 6879   *Dendroaspis augusticeps*   A polypeptide toxin consisting of 60 AA; specifically blocks high threshold voltage-gated Ca²⁺; Schweitz etal, *PNAS*, 91:878, 1994

Calcineurin Autoinhibitory Peptide

BioSource International 77-396

Calcitonin

Biogenesis 1720-8239   Rat synthetic   >95% (HPLC); lyophilized

Calcitonin Gene Regulated Peptide

Synonyms: Calcitonin Gene Regulated Peptide Receptor Ligand

Amersham IM184   Human   Lyophilized; ~74 TBq/mmol, ~2000 Ci/mmol | Endogenous; K_d = 0.32 nM rat brain membrane

Calcitonin Gene Related Peptide, Cyclic Disulfide

Synonyms: Calcitonin Gene Related Peptide, α-

Fluka 21255   MW 3806.3 $C_{162}H_{262}N_{50}O_{52}S_2$ Rat   ≥97% (HPLC); ≥70% peptide content | Neuropeptide encoded by the calcitonin gene via tissue-specific RNA processing; Rosenfeld, MG et al, *Nature*, 304:129, 1983

Calcitonin Gene Related Peptide, Fluo-

Amersham VCGR011   Human   Fluorescein-labeled with high activity

Calcitonin, (3-(¹²⁵I)-Tyr¹²)-

Synonyms: Calcitonin Gene Regulated Peptide Receptor Ligand

Amersham IM175   Human   Lyophilized; ~74 TBq/mmol, ~2000 Ci/mmol | Endogenous

Calcitonin, (3-(¹²⁵I)-Tyr²³)-

Synonyms: Calcitonin Gene Regulated Peptide Receptor Ligand

Amersham IM250   Salmon   Lyophilized; ~74 TBq/mmol, ~2000 Ci/mmol | Agonist; K_d = 0.74 nM rat kidney medulla

Calmodulin Inhibitor Peptide

Biodesign A80117S   Synthetic   >97% | Apoptosis & signal transduction

Calmodulin Kinase II Inhibitor Peptide

Biodesign A80116S   Synthetic   >97% | Apoptosis & signal transduction

Calmodulin Kinase II Substrate Peptide

Biodesign A80119S   Synthetic   >97% | Apoptosis & signal transduction

Calmodulin-Dependent Protein Kinase II

BioSource International 77-156

Calmodulin-Dependent Protein Kinase II (290-309)

Synonyms: Calmodulin Antagonist

BioSource International 77-148

Fluka 21274   ≥99.0% (HPLC); ~70% peptide content | A potent antagonist; Payne, ME et al, *JBC*, 263:7190, 1988

Calmodulin-Dependent Protein Kinase IV

BioSource International 77-144

Cannabinoid Cb I Peptide

BioSource International 04-310

Casein Kinase II Peptide

BioSource International 77-208

Casein Kinase II Substrate

BioSource International 77-210

Caseinoglycopeptide

Sigma C 7278   Bovine casein   Purified powder; essentially salt-free; prepared by a modification of the method of Alais & Jolles | Alais, C & Jolles, P, *Biochim Biophys Acta*, 51:315, 1961

Caspase I Substrate

BioSource International 77-176

Caspase Iα (134-151)

Synonyms: Interleukin Converting Enzyme α

Oncogene PP55   HUMAN   Lyophilized with an equal amount of BSA | Suggested starting concentration of peptide for competition is 10X the concentration of primary Ab; titration recommended

Caspase Iβ (134-151)

Synonyms: Interleukin Converting Enzyme β

Oncogene PP56   Human   Lyophilized with an equal amount of BSA | Used in competition studies & dot blot

Caspase Iε (17-26)

Synonyms: Interleukin Converting Enzyme ε

Oncogene PP58   Human   Lyophilized with an equal amount of BSA | Used in competition studies & dot blot

Caspase Iγ (192-204)

Synonyms: Interleukin Converting Enzyme γ

Oncogene PP57   Human   Lyophilized with an equal amount of BSA | Suggested starting concentration of peptide for competition is 10X the concentration of primary Ab; titration recommended

C-C Chemokine

Synonyms: Hemofiltrate CC Chemokine I

ICN 195785   Human recombinant, expressed in *E. coli*   ≥97%; lyophilized

C-C Chemokine Receptor I Blocking Peptide (7-24)

Calbiochem 227021   MW 2044.1 $C_{86}H_{118}N_{18}O_{38}S$ Human synthetic   Liquid | This peptide coupled to KLH, was used as immunogen for the production of Anti-CCR1 (Calbiochem No. 227020); useful as a blocking peptide

CCR1 N-Terminal Peptide

Alexis 153-019   Human synthetic   Synthetic peptide dissolved in 30 μL 10 *mM* sodium acetate, pH 4.5 | Competitively binds to Ab Alexis 210-741; blocking peptide for antiserum (purified) to CCR1

CCR4 N-Terminal Peptide

Alexis 153-020   Mouse synthetic   Synthetic peptide dissolved in 80 μL 10 *mM* sodium acetate, pH 4.5 | Competitively binds to Ab Alexis 210-742; blocking peptide for antiserum (purified) to CCR4

cdc2/cdk1

Amersham VPP33   Competition for VPC25; cyclin dependent kinase; cell proliferation & cell cycle signals

cdc2/cdk1 Peptide

Oncogene PP33   Lyophilized with an equal amount of BSA | C-terminal 8 AA; used for competition studies & dot blot

c-erb B2/c-neu Peptide I (866-873)

Oncogene PP04   Lyophilized with an equal amount of BSA | Suggested starting concentration of peptide for competition is 10X the concentration of primary Ab; titration recommended; from the kinase domain of human c-*neu*; used for competition studies & dot blot

c-erb B3 Peptide (860-873)

Oncogene PP35   Lyophilized with an equal amount of BSA; suggested starting concentration of peptide for competition is 10X the concentration of primary antibody | Used for competition studies & dot blot

c-fos

Amersham VPP09   Competition for VOP17; viral & oncogene recognition

Amersham VPP10   Competition for VPC05 & VPC38; viral & oncogene recognition

c-fos Peptide I (128-152)

Oncogene PP09   Human   Lyophilized with an equal amount of BSA; suggested starting concentration of peptide for competition is 10X the concentration of primary antibody | Used for competition studies & dot blot

c-fos Peptide II (4-17)

Calbiochem PP10   Human   Lyophilized solid with BSA; suggested initial concentration of peptide is 10X the concentration of primary antibody | Suitable for competition assays using anti-c-Fos (Ab-2)

Oncogene PP10   Human   Lyophilized with an equal amount of BSA; suggested starting concentration of peptide for competition is 10X the concentration of primary antibody | Used for competition studies & dot blot

Charybdotoxin

Alexis 260-012   MW 4295.9 $C_{176}H_{277}N_{57}O_{55}S_7$ *Leiurus quinquestriatus hebraeus*   ≥98%; lyophilized powder; an addition of 0.01% BSA to experimental solutions before applying the toxin is essential; potent neurotoxin | Highly potent blocker of a variety of Ca^{2+}-activated K^+ channels in a wide range of cell types; acts in nanomolar concentration & does not affect the apamin-sensitive channels; Vazquez, J et al, *J Biol Chem*, 265:15564, 1990; Miller, C et al, *Nature*, 313:316, 1985; Smith, C et al, *J Biol Chem*, 261:14607, 1986; Gimenez-Gallego, G et al, *PNAS*, 85:3329, 1988; Castle, NA et al, *TINS*, 12:59, 1989; Vazquez, J et al, *J Biol Chem*, 265:20902, 1989; Lambert, P et al, *BBRC*, 170:684, 1990; Strong, PN, *Pharmacol Ther*, 46:137, 1990; Bontems, F et al, *Science*, 254:1521, 1991; Park, C-S et al, *PNAS*, 88:2046, 1991; Vita, C et al, *Eur J Biochem*, 217:157, 1993

CHK1 Substrate Peptide

Synonyms: CHKtide

USBio C4200   Frozen solution in 125 µL sterile deionized H_2O; 1 mg/mL

Chlorotoxin

Alexis 630-022   MW 3995.8 $C_{158}H_{249}N_{52}O_{48}S_{11}$ *Leiurus q. quinquestriatus*   ≥99%; soluble in water; toxic | Small conductance Cl^- channel blocker; DeBin, JA et al, *Am J Physiol*, 264:C361, 1993; Lippens, G et al, *Biochemistry*, 34:13, 1995

ICN 193947   MW 3996   *Leiurus q. quinquestriatus* (scorpion) The first polypeptide toxin reported to probe chloride ion channels; it inhibits submicromolar concentration small conductance chloride ion channels intracellularly; DeBin etal, *Am J Physiol*, 264:C361, 1993

Cholecystokinin (26-29), *N*-Ac-

Synonyms: Gastrointestinal Peptide

Sigma C 4658   FW 525.6 $C_{22}H_{31}N_5O_8S$ Amide; non-sulfated; ≥90% (HPLC); | Bioactive peptide

Cholecystokinin (26-33)

Biogenesis 2050-0704   Synthetic   Purified; lyophilized

Cholecystokinin VIII Receptor Ligand, (^{3}H)-

Amersham IM159   Sulfated; B & H labeled | Non-selective; endogenous; K_d = 0.22 nM guinea pig cortex membrane

Amersham TRK755   Sulfated | Non-selective; propionylated; endogenous; K_d = 0.18 nM guinea pig cortex membrane

Cholecystokinin, Agarose

ICN 191114   ~2 µM ligand/g, 0.02% NaN_3, pH 4.5, 4°C

Cholecystokinin, Fluo-

Amersham VCCK011   Fluorescein-labeled with high activity

Cholera Enterotoxin

ICN 190329   Lyophilized; when reconstituted to 1 mL with H_2O contains 1.0 or 5.0 mg protein, pH 7.5, 0.05 *M* Tris, 0.003 *M* NaN_3, 0.001 *M* Na_2 EDTA, 0.2 *M* NaCl; single major band in disc electrophoresis | Finkelstein, RA etal, *J Immunol*, 113:145, 1974; Finkelstein, RA, CRC *Crit Rev Microbiol*, 2:553, 1973; Finkelstein, RA & Lo Spalluto, *J Exp Med*, 130:185, 1969; Hollenberg, MD etal, *PNAS*, 71:4224, 1974

Cholera Toxin

ICN 150005   *V. cholerae* 569B Inaba   Lyophilized in Tris-EDTA buffer, pH 7.5; >95% (SDS-PAGE) | Activator of adenylate cyclase

Cholera Toxin A-Subunit

ICN 159905   *Vibrio cholerae*   Lyophilized | Possesses ADP-ribosyltransferase activity; as effective as native toxin, it catalyzes ADP-ribosylation in broken cell preparations; however, without the presence of the B-subunit, it is not able to penetrate cells; Gill, DM & MJ Woolkalis, *Methods Enzymol*, 195:267, 1991

Cholera Toxin B-Subunit

ICN 150006   Lyophilized | Choleragenoid; responsible for attachment of the native toxin to the surface of eukaryotic cells, it does not possess ADP-ribosyltransferase activity; novel tool for determining Gm1 involvement on the growth & differentiation of neuronal & other cells; Masco, D etal, j Neuroscience, 11:2443, 1991

Chorionic Gonadotropin

ICN 150683   Human   USP; lyophilized 14,000 IU/mg ~98% iodination grade <0.2% hLH, hTSH & hFSH

ICN 100485   Human pregnancy urine   USP; lyophilized; 5000 IU/ampoule

ICN 198591   Human pregnancy urine   USP; lyophilized 5000 IU/mg

Scipac P111-0   Urine of pregnant women   >96%; lyophilized; >14000 IU/mg | Peptide hormone

Scipac P111-2   Urine of pregnant women   20-35%; lyophilized; ~3000-5000 IU/mg total protein | Peptide hormone

Scipac P111-5   Urine of pregnant women   35-65%; lyophilized; ~5000-9000 IU/mg total protein | Peptide hormone

Scipac P111-6   Urine of pregnant women   65-90%; lyophilized; ~9000-13000 IU/mg total protein | Peptide hormone

Chorionic Gonadotropin, β-

Scipac P112-1   Urine of pregnant women   >96%; lyophilized; beta subunit;

available on request | Peptide hormone

Chorionic Gonadotropin-α

ICN 150684   Human   Lyophilized 1000 IU/mg ~98% iodination grade <0.2% hLH, hTSH & hFSH <0.1% hCG, hCG-β

ICN 150685   Human   Lyophilized 1000 IU/mg ~98% iodination grade

Chymotrypsin Inhibitor I

ICN 194978   MW 40k   Potato   lyophilized; >95%; 1 mg inhibits 2-4 mg chymotrypsin by 50% using ATEE as substrate, pH 7.5, 25°C | Plant inhibitor of chymotrypsin

Chymotrypsin Inhibitor II

ICN 194979   MW 20k   Potato   lyophilized; >95%; 1 mg inhibits 3 mg chymotrypsin by 50% using ATEE as substrate, pH 7.5, 25°C | Plant inhibitor of various proteases including chymotrypsin & trypsin; Bryant, J etal, *Biochem*, 15:3418, 1976

c-jun (1-169)-GST

USBio C5815-06   MW ~41k   Human, expressed in *E. coli* Purified by glutathione-agarose chromatography; 40–50% by SDS-PAGE & Coomassie blue staining; liquid in 110 µL TBS, pH 8.0, 25 *mM* glutathione, 10 *mM* β-MSH containing 50% glycerol | Soluble; used as a standard for c-jun using polyclonal anti-c-jun in immunoblot analysis

c-jun, Agarose Conjugate

Synonyms: GST (1-79)

Calbiochem 420109   Human, recombinant expressed in *E. coli* Purified c-jun in a 50% agarose slurry | Permits the affinity purification of SAPK from cellular extracts; the phosphotransferase activity of bound SAPK can then be measured following the addition of γ-^{32}P-ATP; Minden, A et al, Cell, 81: 1147, 1995; Sanchez, I et al, *Nature*, 372: 794, 1994; Yan, M et al, *Nature*, 372: 798, 1994

c-jun, N-Terminal Kinase Substrate

Synonyms: GST-cJun (1-79)

Alexis 201-014   Human, recombinant expressed in *E. coli* ≥95% (SDS-PAGE); ~1 mg/mL solution in 25 m*M* Tris, pH 7.4, 50 mM NaCl, 5 m*M* β-MSH & 0.5 m*M* EDTA; activity: Tested for functionality using recombinant rat c-jun *N*-terminal kinase for phosphorylation in the presence of (γ-^{32}P)ATP | Fusion protein of gluathione-*S*-transferase (GST) & a recombinant fragment of the c-jun activation domain & purified by glutathione agarose affinity & Q-Sepharose ion-exchange chromatographyBoulton, TG et al, *Cell*, 65: 663, 1991; Blenis, J, *PNAS*, 90: 5889, 1993; Hibi, M et al, *Genes Dev*, 7: 2135, 1993; Kyriakis, JM et al, *Nature*, 369: 156, 1994; Derijard, B et al, *Cell*, 76: 1025, 1994; Minden, A et al, *Science*, 266: 1719, 1994; Sanchez, I et al, *Nature*, 372: 794, 1994; Yan, M et al, *Nature*, 372: 798, 1994; Lin, A et al, *Science*, 268: 286, 1995; Berijard, B et al, *Science*, 267: 682, 1995; Lo, YYC et al, *J Biol Chem*, 271: 15703, 1996; del Arco, PG et al, *J Biol Chem*, 271: 26335, 1996

c-jun/AP-1

Amersham VPP11   Competition for VOP6; viral & oncogene recognition

Amersham VPP12   Competition for VPC07; viral & oncogene recognition

c-jun/AP-1 Peptide I (209-225)

Calbiochem PP11   Lyophilized with BSA; suggested initial concentration of peptide is 10X the concentration of primary antibody | Suitable for competition assays using Anti-c-jun/AP-1 (Ab-1)

c-jun/AP-1 Peptide I (209-225)

Oncogene PP11   Lyophilized with an equal amount of BSA; suggested starting concentration of peptide for competition is 10X the concentration of primary antibody | AA 209-225 in the DNA binding domain of v-jun; for competition studies & dot blot

c-jun/AP-1 Peptide II (73-87)

Oncogene PP12   Lyophilized with an equal amount of BSA; suggested starting concentration of peptide for competition is 10X the concentration of primary antibody | For competition studies & dot blot

c-kit Peptide I (961-976)

Oncogene PP38   Lyophilized with an equal amount of BSA; suggested starting concentration of peptide for competition is 10X the concentration of primary antibody | For competition studies & dot blot

c-myc

Amersham VPP06   Competition for VOP10; viral & oncogene recognition

c-myc Peptide I (408-439)

Calbiochem PP06   Lyophilized with BSA; suggested initial concentration of peptide is 10X the concentration of primary antibody | Suitable for competition assays using Anti-c-myc (Ab-1)

c-myc Peptide I (410-419)

Oncogene PP06   Lyophilized with an equal amount of BSA | Suggested starting concentration of peptide for competition is 10X the concentration of primary Ab; titration recommended

c-neu Peptide II (1242-1255)

Oncogene PP13   Lyophilized with an equal amount of BSA | From thekinase domain of human c-*neu*; used for competition studies & dot blot

Conotoxin GIIIB, μ-

Alexis 630-023   MW 2640.2 $C_{101}H_{175}N_{39}O_{30}S_7$ *Conus geographus*   ≥99%; white lyophilized powder; potent neurotoxin | Specific blocker of the skeletal voltage-gated Na$^+$ channels; Cruz, LJ et al, *J Biol Chem*, 260:9280, 1985; Kimura, M et al, *J Pharmacol Exp Ther*, 256:18, 1991; Hong, SJ & Chang, CC, *Br J Pharmacol*, 97:934, 1989; Olivera, BM et al, *J Biol Chem*, 266:22067, 1991; Robitaille, R & Charlton, MP, *J Neurosci*, 12:297, 1992; Kubo, S et al, *Pept Res*, 6:66, 1993

Conotoxin GVIA Receptor Ligand, (^{125}I-Tyr^{22})-ω-

Amersham IM217   Calcium channel receptor ligand; N-type selectivity; antagonist; K_d = 0.6 pM rat neocortex membrane

Conotoxin GVIA, ω-

Alexis 630-007   MW 3036.3 $C_{120}H_{181}N_{38}O_{43}S_6$ *Conus geographus*   ≥99%; white lyophilized powder; soluble in water; potent neurotoxin | Potent & selective blocker of neuronal *N*-type, voltage-dependent Ca^{2+} channels; Cruz, LJ et al, *Biochemistry*, 26:820, 1987; Olivera, BM et al, *Science*, 230:1338, 1985; Sher, E & Clementi, F, *Neuroscience*, 42:301, 1991; Olivera, BM et al, *J Biol Chem*, 266:22067, 1991; Jones, OW & So, AP, *Anal Biochem*, 214:227, 1993; Kasai, H et al, *Neurosci Res*, 4:228, 1987; McCleskey, EW et al, *PNAS*, 84:4327, 1987; Horne, WA et al, *J Biol Chem*, 266:13719, 1991; Davis, JH et al, *Biochemistry*, 32:7396, 1993; Ellinor, PT et al, *Nature*, 372:272, 1994; Olivera, BM et al, *Ann Rev Biochem*, 63:823, 1994; Dunlap, K et al, *TINS*, 18:89, 1995

Conotoxin MVIIA, ω-

Alexis 630-008   MW 2636.2 $C_{102}H_{169}N_{36}O_{32}S_7$ *Conus magus*   ≥99%; white lyophilized powder; soluble in water; potent neurotoxin | Selectively & reversibly blocks the *N*-type, voltage-dependent Ca^{2+} channels; Valentino, K et al, *PNAS*, 90:7894, 1993; Olivera, BM et al, *Science*, 230:1338, 1985; Olivera, BM et al, *J Biol Chem*, 266:22067, 1991; Kohno, T et al, *Biochemistry*, 34:10256, 1995; Olivera, BM et al, *Ann Rev Biochem*, 63:823, 1994

Conotoxin MVIIC Receptor Ligand, (^{125}I)-ω-

Amersham IM276   Calcium channel receptor ligand; P/Q-type selectivity; antagonist; IC$_{50}$ = 1 nM

Conotoxin MVIIC, ω-

Alexis 630-009   MW 2749.3 $C_{106}H_{178}N_{40}O_{32}S_7$ *Conus magus*   ≥99%; white lyophilized powder; soluble in water; potent neurotoxin | Selective & potent blocker of P/Q-type, voltage-gated Ca^{2+} channels; Hillyard, DR et al, *Neuron*, 9:69, 1992; Wheeler, DB et al, *Science*, 264:107, 1994; Olivera, BM et al, *J Biol Chem*, 266:22067, 1991; Olivera, BM et al, *Ann Rev Biochem*, 63:823, 1994

Conotoxin Receptor Ligand MVIIA, (^{125}I-Tyr^{13})-ω-

Amersham IM258   Calcium channel receptor ligand; N-type selectivity; antagonist; K_d = 0.6 pM rat neocortex membrane

Conotoxin SVIB, ω-

Alexis 630-010   MW 2739.2 $C_{105}H_{176}N_{38}O_{36}S_6$ *Conus striatus*   ≥99%; white lyophilized powder; soluble in water; potent neurotoxin | *N*-type Ca^{2+} channel blocker; Ramilo, CA et al, *Biochemistry*, 31:9919, 1992; Newcomb, R & Palma, A, *Brain Res*, 638:95, 1994; Olivera, BM et al, *J Biol Chem*, 266:22067, 1991; Olivera, BM et al, *Ann Rev Biochem*, 63:823, 1994; Fox, JA, *Neurosci Lett*, 165:157, 1994

Control Peptide Pair, Akt.PKB pS473

BioSource International 04-622Z

Control Peptide Pair, Akt.PKB pT308

BioSource International 04-602Z

Control Peptide Pair, Caspase 9(315/316)
BioSource International 04-692Z

Control Peptide Pair, c-RAF pS259
BioSource International 04-502Z

Control Peptide Pair, c-RAF pS621
BioSource International 04-504Z

Control Peptide Pair, c-RAF pYpY340/341
BioSource International 04-506Z

Control Peptide Pair, eIF-2α, pS51
BioSource International 04-728Z

Control Peptide Pair, Epidermal Growth Factor-R pT654
BioSource International 04-782Z

Control Peptide Pair, Epidermal Growth Factor-R pY1068
BioSource International 04-788Z

Control Peptide Pair, Epidermal Growth Factor-R pY1086
BioSource International 04-790Z

Control Peptide Pair, Epidermal Growth Factor-R pY1148
BioSource International 04-792Z

Control Peptide Pair, Epidermal Growth Factor-R pY1173
BioSource International 04-794Z

Control Peptide Pair, Epidermal Growth Factor-R pY845
BioSource International 04-784Z

Control Peptide Pair, Epidermal Growth Factor-R pY992
BioSource International 04-786Z

Control Peptide Pair, ERK 1/2 pTpY185/187
BioSource International 04-680Z

Control Peptide Pair, ERK 5 pTpY218/220
BioSource International 04-612Z

Control Peptide Pair, FAK pS722
BioSource International 04-588Z

Control Peptide Pair, FAK pS910
BioSource International 04-596Z

Control Peptide Pair, FAK pY397
BioSource International 04-624Z

Control Peptide Pair, FAK pY407
BioSource International 04-650Z

Control Peptide Pair, FAK pY576
BioSource International 04-652Z

Control Peptide Pair, FAK pY577
BioSource International 04-614Z

Control Peptide Pair, FAK pY861
BioSource International 04-626Z

Control Peptide Pair, GSK-3β pS9
BioSource International 04-600Z

Control Peptide Pair, GSK-3β pY216
BioSource International 04-604Z

Control Peptide Pair, Insulin Receptor pY1158
BioSource International 04-802Z

Control Peptide Pair, Insulin Receptor pYpY1162/1163
BioSource International 04-804Z

Control Peptide Pair, IRS-1 pS616
BioSource International 04-550Z

Control Peptide Pair, IκBα pSpS32/36
BioSource International 04-726Z

Control Peptide Pair, JAK1 pYpY1022/1023
BioSource International 04-422Z

Control Peptide Pair, JAK2 pYpY1007/1008
BioSource International 04-426Z

Control Peptide Pair, JNK/SAPK 1/2 pTpY183/185
BioSource International 04-682Z

Control Peptide Pair, p38 pTpY180/182
BioSource International 04-684Z

Control Peptide Pair, p53 pS392
BioSource International 04-640Z

Control Peptide Pair, Paxillin pY118
BioSource International 04-722Z

Control Peptide Pair, Paxillin pY181
BioSource International 04-724Z

Control Peptide Pair, Paxillin pY31
BioSource International 04-720Z

Control Peptide Pair, PKR pT451
BioSource International 04-668Z

Control Peptide Pair, PLC-γ1 pY783
BioSource International 04-696Z

Control Peptide Pair, Pyk2 pY402
BioSource International 04-618Z

Control Peptide Pair, Pyk2 pY579
BioSource International 04-632Z

Control Peptide Pair, Pyk2 pY580
BioSource International 04-634Z

Control Peptide Pair, Pyk2 pY881
BioSource International 04-620Z

Control Peptide Pair, src pY215
BioSource International 04-658Z

Control Peptide Pair, src pY418
BioSource International 04-660Z

Control Peptide Pair, src pY529
BioSource International 04-662Z

Control Peptide Pair, Stat1 pS727
BioSource International 04-382Z

Control Peptide Pair, Stat1 pY701
BioSource International 04-378Z   Human

Control Peptide Pair, Stat3 pS727
BioSource International 04-384Z

Control Peptide Pair, Stat3 pY705
BioSource International 04-380Z

Control Peptide Pair, Tau pS199
BioSource International 04-734Z

Control Peptide Pair, Tau pS214
BioSource International 04-742Z

Control Peptide Pair, Tau pS262
BioSource International 04-750Z

Control Peptide Pair, Tau pS396
BioSource International 04-752Z

Control Peptide Pair, Tau pS404
BioSource International 04-758Z

Control Peptide Pair, Tau pS409
BioSource International 04-760Z

Control Peptide Pair, Tau pS422
BioSource International 04-764Z

Control Peptide Pair, Tau pSpT199/202
BioSource International 04-768Z

Control Peptide Pair, Tau pT181
BioSource International 04-732Z   Human
BioSource International 04-766Z   Mouse

Control Peptide Pair, Tau pT202
BioSource International 04-736Z

Control Peptide Pair, Tau pT205
BioSource International 04-738Z

Control Peptide Pair, Tau pT212
BioSource International 04-740Z

Control Peptide Pair, Tau pT217
BioSource International 04-744Z

Control Peptide Pair, Tau pT231
BioSource International 04-746Z

Corticotropin Releasing Factor, (^{125}I)(Tyr^0)-
Synonyms: Corticotropin Releasing Factor Receptor Ligand, (^{125}I)-

Amersham IM299   Ovine   Water:acetonitrile (65:35) containing trace amounts of BSA, trifluoroacetic acid & 2-mercaptoethanol; ~74 TBq/mmol, ~2000 Ci/mmol; 3.7 MBq/mL, 100 µCi/mL | Non-selective; endogenous; K_d = 0.13 nM human CRF$_1$/LTK cells

Corticotropin Releasing Factor, (2-(^{125}I)(His^{32})-
Synonyms: Corticotropin Releasing Factor Receptor Ligand, (^{125}I)-

Amersham IM189   Human   Lyophilized; ~74 TBq/mmol, ~2000 Ci/mmol | Non-selective; endogenous; K_i = 0.95 nM CRF$_1$/CHO cells

Corticotropin Releasing Factor, Fluo-
Amersham VCRF011   Fluorescein-labeled with high activity

C-Peptide
USBio C7905-05   ≥97% (HPLC); 1 mg/mL; lyophilized | Suitable for antigenic applications in immunological protocols

C-Peptide, (Tyr^0)-
Biogenesis 1700-0404   Human synthetic   >95% (HPLC); lyophilized

Crosstide
BioSource International 77-275

c-trk Peptide (777-790)
Oncogene PP37   Lyophilized with an equal amount of BSA; suggested starting concentration of peptide for immunoabsorption is 10X the concentration of primary antibody | From the C-terminal domain of c-trk protein; used for competition studies & dot blot

Cyclin-Dependent Kinase
BioSource International 77-106

Cystatin C
Synonyms: Post Gamma-Globulin

Scipac P175-1   Urine of patients with chronic renal tubular proteinuria   >96%; lyophilized | Inhibitor

Scipac P175-2   Urine of patients with chronic renal tubular proteinuria   10-50%; lyophilized | Inhibitor

Cytokine Induced Neutrophil Chemoattractant I
ICN 195780   Rat recombinant, expressed in E. coli Lyophilized; ≥97% | C-X-C Chemokine

Deltorphin I (δ-Opioid), Fluo-(D-Ala^2)-
Amersham VDLT011   Human   Fluorescein-labeled with high activity

Deltorphin II, (D-Ala²)

BioSource International 77-630

Dendrotoxin, β-

Alexis 630-014   MW 7k   *Dendroaspis angusticeps*   ≥97% (SDS-PAGE); lyophilized powder; dissolving 35 μg/mL gives a stock solution of 5 μM; potent neurotoxin | Potent & selective blocker of non-inactivated, voltage-gated α-dendrotoxin-insensitive K⁺ channels; Bartschat, DK & Blaustein, MP, *PNAS*, 83:189, 1986; Strong, PN, *Pharmacol Ther*, 46:137, 1990; Ren, J et al, *J Pharmacol Exp Ther*, 269:209, 1994; Sorensen, RG & Blaustein, MP, *Mol Pharmacol*, 36:689, 1989; Benishin, CG et al, *Mol Pharmacol*, 34:152, 1988; Hu, PS et al, *Eur J Pharmacol*, 209:87, 1991

Calbiochem 253706   MW 7k   *Dendroaspis angusticeps* >97% (SDS-PAGE); lyophilized; soluble in water; LD₅₀≤2000 mg/kg; bioassay: confirmed via release of ³H-GABA in primary culture of rat hippocampus in dose-dependent manner | Blocks non-inactivating voltage-gated K⁺ channels in rat brain synaptosomes; blocks outward K⁺ current in primary cultured vascular smooth muscle cells; Hu, PS et al, *Eur J Pharmacol*, 209:87, 1991; Ren, J et al, *J Pharmacol Exp Ther*, 269:209, 1994

Alexis 630-015   MW 7k   *Dendroaspis angusticeps*   ≥97% (SDS-PAGE); lyophilized powder; dissolving 70 μg/mL gives a stock solution of 10 μM; potent neurotoxin | Potent & selective blocker of β-dendrotoxin-insensitive K⁺ channels in rat brain synaptosomes; See references for Alexis 630-014

Dendrotoxin, γ-

Calbiochem 253711   MW 7k   *Dendroaspis angusticeps* >97% (SDS-PAGE); lyophilized; soluble in water; LD₅₀≤2000 mg/kg; bioassay: confirmed via release of ³H-GABA in primary culture of rat hippocampus in dose-dependent manner | Blocks non-inactivating voltage-gated K⁺ channels in rat brain synaptosomes; Hu, PS et al, *Eur J Pharmacol*, 209:87, 1991

Dermorphin (μ-Opioid), Fluo-

Amersham VDRM011   Human   Fluorescein-labeled with high activity

Diabetes Associated Peptide, (¹²⁵I)-

Synonyms: Diabetes Associated Peptide Receptor Ligand, (¹²⁵I)-

Amersham IM234   Rat   Labeled with Bolton & Hunter Reagent; lyophilized; ~74 TBq/mmol, ~2000 Ci/mmol | Amylin receptor ligand; agonist; K_d = 27 pM rat nucleus accumbens membrane

Diazepam Binding Inhibitor

Synonyms: Endozepine

Calbiochem 282500   MW 9807   Porcine intestine   ≥98% (HPLC); salt-free, lyophilized solid; soluble in water | Binds medium- & long-chain acyl-CoA esters with very high affinity & may function as an intracellular carrier of acyl-CoA esters; displaces diazepam from the benzodiazepine recognition site; acts as a neuropeptide to modulate the action of GABA_A receptor; inhibits the glucose-induced insulin secretion from perfused pancreas; Chen, ZW et al, *Eur J Biochem*, 174:239, 1988; Mogensen, IB et al, *Biochem J*, 241:189, 1987; Guidotti, A et al, *PNAS*, 80:3531, 1983

Diphtheria Toxin; PW-8

ICN 150007   *Corynebacterium diphtheriae*   Lyophilized; 99% (SDS-PAGE); 10 mM NaPO₄ buffer, pH 7.4, 5% (W/W) lactose | Toxic

DNA-Protein Kinase Blocking Peptide (4085-4096)

Calbiochem 317277   MW 1537.7 $C_{70}H_{92}N_{18}O_{18}S_2$ Human synthetic   Lyophilized | Based on the human DNA-PK catalytic subunit (4085-4096) with a Cys C residue added; this peptide coupled to KLH, was used as the immunogen for the production of Anti-DNA-PK; for use in immunoabsorption for immunoprecipitation, immunocytochemistry, Western blotting & dot blots

Dynamin I Peptide

Alexis 154-001   Synthetic   Synthetic peptide dissolved in 10 mM sodium acetate, pH 4.5 | Competitively binds to Ab Alexis 210-207; blocking peptide for antiserum (purified) to Dyn 1

Dynorphin, Fluo-

Amersham VDYN011   Fluorescein-labeled with high activity

Echistatin, (¹²⁵I)-

Amersham IM304   Lyophilized; ~74 TBq/mmol, ~2000 Ci/mmol

Elastatinal

Synonyms: Elastase Inhibitor

Peptides International IEL-4064   Microbial   ~50%; amorphous powder; bulk | Inhibitor; Umezawa, H et al, *J Antibiotics*, 26:787, 1973

Peptides International IEL-4064-v   Microbial   ~50%; lyophilized amorphous powder | Inhibitor; Umezawa, H et al, *J Antibiotics*, 26:787, 1973

Elcatonin

Synonyms: Carbacalcitonin

Biogenesis 4064-1109   Non-species synthetic   >95% (HPLC); lyophilized

Endomorphin I

BioSource International 77-606

Endomorphin II

BioSource International 77-608

Endomorphin, Fluo-

Amersham VEM1011   Fluorescein-labeled with high activity

Endorphin, (3-(¹²⁵I)Tyr²⁷)-β-

Amersham IM162   Lyophilized; ~74 TBq/mmol, ~2000 Ci/mmol

Endothelial-Monocyte Activating Polypeptide II

PeproTech 100-38   MW 18.3k   Human recombinant, expressed in *E. coli*   >98% by SDS-PAGE & HPLC; <0.1 ng endotoxin per mg (1EU/mg); lyophilized | Novel cytokine; antiangiogenic factor in tumor vascular development; strongly inhibits tumor growth; 166 AA; activity determined by the apoptotic effect of MCF-7 cells

BioSource International PHC1585   Human, recombinant

Endothelin (1-38), Big-

Biogenesis 4113-2155   Synthetic   >95% (HPLC); lyophilized

Endothelin I

Sigma E-134 MW 2492 (free base) $C_{109}H_{159}N_{25}O_{32}S_5$
Human, porcine synthetic >98%; white powder; peptide content
& salt form information are provided with each lot | Endogenous
neuropeptide found in endothelial cells which activates vascular
smooth muscle cells & various signal transduction pathways to
produce a wide range of pathophysiological effects; Has greatest
affinity toward ET_A receptors; Wong-Dusting et al, *J Cardiovasc
Pharmacol*, 17:S236, 1991; Ono et al, *Nature*, 370:252, 1994;
Takei et al, *Life Sci*, 53:PL111, 1993

Endothelin I, ((^{125}I)-Tyr^{13})-

ICN 68102 Human, porcine SA>50 µCi/µg, >1.85 MBq/µg;
0.1 *M* KPB, pH 7.5, 75 *mM* NaCl, 0.3% BSA

Endothelin I, (3-(^{125}I)Tyr)-

Amersham IM223 1:1 Solution of Tris HCl (0.08M), BSA
(0.25%) aprotinin (0.01 mg/mL):n-propanol; ~74 TBq/mmol,
~2000 Ci/mmol; 3.7 MBq/mL, 100 µCi/mL

Endothelin III, (3-(^{125}I)Tyr^6)-

Amersham IM228 Human Lyophilized; ~74 TBq/mmol,
~2000 Ci/mmol

Enkephalin, (D-Ala^2,D-Leu^5)-

BioSource International 77-600

Enkephalin, (D-Ala^2,N-Me-Phe^4,Glyol^5)(Tyr^{3,5}-^{3}H)-

Amersham TRK681 0.15 M Triethylammonium phosphate
buffer, pH 4.2:acetonitrile (22:3); 1.48-2.96 TBq/mmol, 40-80
Ci/mmol; 3.7 MBq/mL, 1 mCi/mL

Enkephalin, (D-Ala^2,N-Me-Phe^4,Met(S=O)-ol)

ICN 152837 A highly reactive enkephalin analog; Roemer, D
etal, *Nature*, 268:547, 1977

Enkephalin, (D-Ala^2-Met)-

Biogenesis 4140-0029	Synthetic	95% (HPLC); lyophilized
Biogenesis 4140-0059	Synthetic	>95% (HPLC); lyophilized

Enkephalin, (Leu^5)-

BioSource International 77-602

Enkephalin, (Met)-

Biogenesis 4140-5509 Synthetic >95% (HPLC); lyophilized

Enkephalin, (Met^5)-

BioSource International 77-604

Enkephalin, Agarose

ICN 191113 ~1 µmole ligand/g, pH 4.5 buffer, 0.02 NaN$_3$

Eotaxin

ICN 193965 Human recombinant, expressed in *E. coli* ≥95%
| A new, highly selective chemokine for mouse eosinophils

ICN 193966 Murine recombinant, expressed in *E. coli* ≥95%
| A new, highly selective chemokine for mouse eosinophils

Eotaxin, (^{125}I)-

Synonyms: Eotaxin Receptor Ligand, (^{125}I)-

Amersham IM290 Human recombinant Lyophilized; ~74
TBq/mmol, ~2000 Ci/mmol | Chemokine receptor ligand; CCR3
selectivity; endogenous

Epidermal Growth Factor

Synonyms: Epidermal Growth Factor, rh-; Epidermal Growth
Factor, nm-

Chemicon EA143 ≥95%

Calbiochem 324831 MW 6k Human recombinant, expressed
in *E. coli* >95% (SDS-PAGE); lyophilized from filter-sterilized
PBS; biological activity: ED_{50}=100-400 pg/mL; may be
carcinogenic/teratogenic | Single-chain polypeptide containing 54
AA; identical to human EGF except for an additional N-terminal
Met; human EGF is 70% homologous to mouse EGF & has been
shown to have similar biological activity; mitogenic for a wide range
of ectoderm- & endoderm-derived cells from tissue such as breast,
cornea, epidermis, dermis, liver, pancreas, nerve, amnion &
adrenal medulla; acts as a survival factor in preventing apoptosis;
Merck Index, 12: 3569; Merlo, GR et al, *J Cell Biol*, 128: 1185,
1995; Yoshida, T et al, *Brain Res Dev*, 76: 147, 1993

ICN 153481 MW 6k Human recombinant, expressed in *E. coli*
≥98%; sterile-filtered lyophilized powder, carrier free; ED_{50} ~2.0
ng/mL by ^{3}H-thymidine uptake in Balb/c 3T3 cells; <0.1 ng
endotoxin/mg protein | Contains a 53 AA peptide sequence

ICN 153508 MW 6k Human recombinant, expressed in *E. coli*
>95%; sterile filtered lyophilized powder; <0.1 ng endotoxin/mg
protein | Contains a 53 AA peptide sequence

ICN 154571 Human recombinant, expressed in *E. coli* ≥98%;
lyophilized; ED_{50} = 2.0 ng/mL by ^{3}H-thymidine uptake in Balb/c
3T3 cells; ≤0.1 ng endotoxin/mg protein | Smith, K etal, *Nucl
Acids Res*, 10:4467, 1982; Lax, I etal, Molec Cell Biol, 8:1970,
1988

PeproTech 100-15 MW 6.2k Human recombinant, expressed
in *E. coli* >98% by SDS-PAGE & HPLC; <0.1 ng endotoxin per
mg (1EU/mg); lyophilized | Polypeptide growth factor; stimulates
proliferation of a wide range of cell types; 53 AA; ED_{50} ≤ 1.0
ng/mL; SA ≥ 10^6 U/mg; SA determined by dose-dependent
stimulation of the proliferation of thymidine uptake by BALB/c 3T3
cells

ICN 160035 Mouse submaxillary gland >95%; membrane
filtered lyophilized solid; tissue culture grade | Each lot is tested
for DNA replication & absolute electrophoretic homogeneity via
HPLC

ICN 160036 Mouse submaxillary gland >98%; lyophilized;
receptor iodination grade | Each lot is tested for radioreceptor
binding, DNA replication & immunological activity via HPLC;
Savage, CR & SJ Cohen, *JBC*, 247:7609, 1972

PeproTech 315-09 MW 6.0k Murine recombinant, expressed
in *E. coli* >98% by SDS-PAGE & HPLC; <0.1 ng endotoxin per
mg (1EU/mg); lyophilized | Polypeptide growth factor that
stimulates proliferation of a wide range of epidermal & epithelial
cells; 53 AA; The ED_{50} ≤ 0.15 ng/mL; SA ≥ 6.6 x 10^6 U/mg; SA
determined by the dose-dependent stimulation of thymidine uptake
by BALB/c 3T3 cells

Calbiochem 324851 MW 6.1k Murine submaxillary glands
Culture grade; ≥95% (SDS-PAGE); purified by size-exclusion
chromatography; lyophilized solid; optimal EGF concentration in
culture: 5-50 ng/mL; may be carcinogenic/teratogenic | Mitogenic
for a wide range of ectoderm- & endoderm-derived cells from tissue
such as breast, cornea, epidermis, dermis, liver, pancreas, nerve,
amnion & adrenal medulla; *Merck Index*, 12: 3569; Pascall, IC et
al, *J Mol Endocrinol*, 12: 313, 1994

Calbiochem 324856 MW 6.1k Murine submaxillary glands
Receptor grade; ≥98% (SDS-PAGE); purified by size-exclusion
chromatography followed by DEAE ion-exchange chromatography;
lyophilized solid; effective biological range: 5-50 ng/mL; may be
carcinogenic/teratogenic | Biological activity determined by
mitogenic assay with human foreskin fibroblasts; *Merck Index*, 12:
3569

ICN 198672 Porcine recombinant, expressed in *Bacillus brevis*
>95%; lyophilized; <0.1 ng/µg endotoxin | Pascall, JC etal, *J Mol
Endocrinology*, 6(1):63, 1991

ICN 160024 Rat submandibular glands Lyophilized | Each
lot undergoes EGF receptor binding & RIA analysis & is
electrophoretically homogeneous; Schaudies, RP & CR Savage,
Comp Biochem Physiol, 84B(4):497, 1986

Epidermal Growth Factor Analog, Long
Sigma E 4269   Human recombinant   >95% (*N*-terminal sequence analysis); lyophilized from 0.1 *M* acetic acid; endotoxin tested; cell culture tested | Recombinant human growth factors with altered AA chain lengths as compared to their native protein counterparts; as a result, these factors exhibit up to 10X more growth promoting activity when compared to their counterparts; they exhibit greater stability in culture & are ideal for serum-free or reduced-serum formulations

Epidermal Growth Factor Receptor Peptide
Chemicon EA145   Murine   ≥95%

Epidermal Growth Factor Receptor Peptide (1005-1016)
Oncogene PP28   Lyophilized with an equal amount of BSA; suggested starting concentration of peptide for competition is 10X the concentration of primary antibody | Used for competition studies & dot blot

Epidermal Growth Factor, (^{125}I)-
ICN 68112   Human   ~100 μCi/μg, ~3.7 MBq/μg; 50 *mM* KPB, pH 7.5, 75 *mM* NaCl, 0.1% BSA

ICN 68060   Mouse   ~100 μCi/μg, ~3.7 MBq/μg; 50 *mM* KPB, pH 7.5, 75 *mM* NaCl, 0.1% BSA

Epidermal Growth Factor, (3-(^{125}I)Tyr)-
Amersham IM196   Human recombinant   Aqueous; >27.75 TBq/mmol, >750 Ci/mmol; 3.7 MBq/mL, 100 μCi/mL

Amersham IM124   Mouse   Phosphate buffer, 1 mg/mL BSA; ~3.7 MBq/μg, ~100 μCi/μg; 1.85 MBq/mL, 50 μCi/mL

Epidermal Growth Factor, α-
ICN 160019   Mouse Submaxillary Gland   >95%; lyophilized | Superior for radioiodination, as chromatographic standard, etc; Savage, CR & SJ Cohen, *JBC*, 247:7609, 1972

Epidermal Growth Factor-R (651-658)
BioSource International 77-205

Epidermal Growth Factor-R (652-657)
BioSource International 77-203

Epidermal Growth Factor-R Fragment, Myristoylated
BioSource International 77-173

Epithelial Neutrophil Activating Peptide 78
Chemicon GF044   Human   ≥95%

Biogenesis 0100-0139   Human r-DNA   >95% (HPLC/SDS-PAGE); lyophilized

PeproTech 300-22   MW 8.0k   Human recombinant, expressed in *E. coli*   >98%; 74 AA; lyophilized with no additives; activity determined by its ability to chemoattract CXCR2 transfected HEK cells using a concentration range of 5.0-10.0 ng/mL

BioSource International PHC1334   Human, recombinant

ICN 193967   Human, recombinant expressed in *E. coli*   ≥95% purity | A chemoattractant & activator of human neutrophils

Erythropoietin (1-12)
Oncogene PP25   Lyophilized with an equal amount of BSA | Used for competition studies & dot blot

Erythropoietin, (3-(^{125}I)Tyr)-
Amersham IM178   Human recombinant   0.1M Sodium acetate, 0.1M NaCl, 0.2% BSA; 11-33 TBq/mmol, 300-900 Ci/mmol; 740kBq/mL, 20 μCi/mL

Amersham IM219   Human recombinant   High specific activity; 0.1M sodium acetate, 0.1M NaCl, 0.2% BSA; 111-148 TBq/mmol, 3000-4000 Ci/mmol; 740kBq/mL, 20 μCi/mL

Excitatory AA Receptor Ligand, ((^{3}H)-L-Glu)-
Amersham TRK445   Non-selective; endogenous

Excitatory AA Receptor Ligand, (^{3}H)(Gly)-
Amersham TRK71   NMDA-glycine site selective; endogenous

Experimental Allergenic Encephalitogenic Peptide
Biogenesis 4398-0106   Synthetic   >95% (HPLC); purified; lyophilized

Fas Ligand
Amersham VPP61   Competition for VPC78; growth/death factor interactions

Fas Ligand (261-277)
Oncogene PP61   Murine   Lyophilized with an equal amount of BSA | Used in competition studies & dot blot

Fas Peptide (321-335)
Oncogene PP62   Lyophilized with an equal amount of BSA | Suggested starting concentration of peptide for competition is 10X the concentration of primary Ab; titration recommended

Fas Protein, Extracellular Fragment
Synonyms: Fas Antigen (1-157)
ICN 198750   Human, recombinant expressed in *E. coli*   >95% | Represents the Fas sequence corresponding to base pairs 200-713 of the cDNA sequence

Fas Protein, Intracellular Fragment
Synonyms: Fas Antigen (1-157)
ICN 198751   Human, recombinant expressed in *E. coli*   >95% | Represents the Fas sequence corresponding to base pairs 765-1190 of the cDNA sequence

Fasciculin II
ICN 193957   MW 6735   *Dendroaspis augusticeps*   An anticholinesterase polypeptide toxin that is extremely potent, selective & irreversible; inhibits cholinergic signaling *in vivo* & *in vitro*; Rodriquez-Ithurralde etal, *Neurochem Int*, 5:267, 1983

Fibrin β-Chain (1-4), Human
ICN 153113   MW 526 $C_{19}H_{31}N_9O_5 \cdot C_2H_5O_2$ Acetate salt | Laudano, AP & RF Doolittle, *PNAS*, 74:3085, 1978; Laudano, AP & RF Doolittle, *Biochem*, 19:1013, 1980

Fibrinogen, (^{125}I)-
Amersham IM53   Human   Lyophilized; 5.55-9.25 MBq/mg, 150-250 μCi/mg

Fibroblast Growth Factor (1-24), Basic
Biogenesis 4460-4504   Synthetic   Purified; lyophilized

Fibroblast Growth Factor Peptide I (147-153), β-
Oncogene PP23   Lyophilized with an equal amount of BSA; suggested starting concentration of peptide for competition is 10X the concentration of primary antibody | For competition studies & dot blot

Fibroblast Growth Factor Peptide II (40-63), β-

Oncogene PP26   Lyophilized with an equal amount of BSA; suggested starting concentration of peptide for competition is 10X the concentration of primary antibody | For competition studies & dot blot

Fibroblast Growth Factor Peptide, Acidic

Chemicon FA010   ≥95%

Fibroblast Growth Factor Peptide, Basic

Chemicon FA011   ≥95%

Fibroblast Growth Factor, (^{125}I)-

Amersham IM243   Human recombinant   Lyophilized; 29.6-44 TBq/mmol, 800-1200 Ci/mmol

Fibroblast Growth Factor, Acidic

Synonyms: Fibroblast Growth Factor, nba-; Fibroblast Growth Factor, rha-

Calbiochem 341580   MW 16k   Bovine   >97% (SDS-PAGE); lyophilized aseptically in the presence of 50 µg BSA/µg FGF; biological activity: ED_{50}=100-500 pg/mL in the absence of heparin, for stimulation of ^{3}H-thymidine incorporation by NR6-3T3 fibroblasts; may be carcinogenic/teratogenic | Mitogenic for a wide variety of mesenchymal & neuroectodermal cells; Exhibits antigenic properties; increases myocardial flow in models of chronic ischemia; induces phosphofructokinase, fatty acid synthase & Ca^{2+}-ATPase mRNA expression in 3T3 cells; *Merck Index*, 12: 4117; Lopez, JJ et al, *Am J Physiol*, 274: H930, 1998; Hsu, K et al, *Biochem Biophys Res Comm*, 197: 1483, 1993

Calbiochem 341591   MW 15k   Human recombinant, expressed in *E. coli*   ≥97% (SDS-PAGE); lyophilized from filter-sterilized PBS containing 50 µg BSA/µg FGF; biological activity: ED_{50}=100-300 pg/mL in the presence of 10 µg/mL heparin as monitored in a mitogenic assay by measuring the acidic FGF-dependent ^{3}H-thymidine incorporation in quiescent NR-6-R 3T3 fibroblasts; endotoxin: ≤100 pg/µg FGF; may be carcinogenic/teratogenic | Stimulates the proliferation of primary cultures of myoblasts & osteoblastic cells; prevents myocardial apoptosis triggered by ischemia reperfusion injury; *Merck Index*, 12: 4117; Cuevas, P et al, *Eur J Med Res*, 2: 465, 1997; Junttila, T et al, *Brain Res*, 707: 81, 1996

Fibroblast Growth Factor, Basic

Synonyms: Fibroblast Growth Factor, nbb-; Fibroblast Growth Factor, rhb-

Calbiochem 341583   MW 18k   Bovine   ≥97% (SDS-PAGE); lyophilized from filter-sterilized 20 *mM* Tris, pH 7.0 containing 50 µg BSA/µg FGF; biological activity: ED_{50}=50-300 pg/mL in the absence of heparin for stimulation of ^{3}H-thymidine incorporation by NR6-3T3 fibroblasts; may be carcinogenic/teratogenic | Multifunctional factor with chemotactic activity for numerous cell types; mitogenic to neuroectoderm & mesoderm-derived cells; potent angiogenic factor both *in vivo* & *in vitro*; activates phospholipase D in endothelial cells; *Merck Index*, 12: 4117; Murai, N et al, *J Neurosurg*, 85: 1072, 1996; Sold under license of U.S. Patent 4,956,455 issued to The Salk Institute

Calbiochem 341595   MW 16.4k   Human recombinant, expressed in *E. coli*   ≥97% (SDS-PAGE); lyophilized from filter-sterilized solution in 20 *mM* Tris, 5 *mM* DTT, pH 7.0 containing 50 µg BSA/µg FGF; biological activity: ED_{50}=100-250 pg/mL as monitored in a mitogenic assay by measuring FGF basic-dependent ^{3}H-thymidine incorporation in quiescent NR-6-R 3T3 fibroblasts; endotoxin: ≤100 pg/µg FGF; may be carcinogenic/teratogenic | Potent mitogen for bone cells; inhibits DNA synthesis & alkaline phosphatase activity in rat osteosarcoma cell lines; implicated in the development of Kaposi sarcoma *in vitro*; inhibits inducible NOS in bovine retinal pigmented epithelial cells; *Merck Index*, 12: 4117; Goureau, O et al, *Eur J Biochem*, 230: 1046, 1995; Gibran, NS et al, *J Surg Res*, 56: 226, 1994; Li, JJ et al, *Cancer*, 72: 2253, 1993

Fibronectin Inhibitor

American Peptide 44-0-47   MW 587.6 $C_{22}H_{37}N_9O_{10}$ Inhibitor of cell attachment to fibronectin & vitronectin which is resistant to trypsin & carboxypeptidase

flt3-Ligand

Synonyms: flk2-Ligand

PeproTech 300-19   MW 17.6k   Human recombinant, expressed in *E. coli*   >98% by SDS-PAGE & HPLC; <0.1 ng endotoxin per mg (1EU/mg); lyophilized | Stimulates proliferation & colony formation of certain bone marrow precursor cells including $CD34^+$ cells; soluble polypeptide containing 155 AA residues which comprise the extracellular domain of the transmembrane flt3-ligand protein; $ED_{50} \leq 5.0$ ng/mL; $SA \geq 2 \times 10^5$ U/mg

Follicle Stimulating Hormone

Synonyms: WHO IRP (75/504)

ICN 151159   Human   Iodination grade; lyophilized; >30,000 IU/mg

Scipac P217-1   Pituitary glands   >96%; lyophilized | Peptide hormone

ICN 101727   Porcine pituitary

Foroxymithine

Synonyms: Angiotensin Converting Enzyme Inhibitor

Sigma F 5017   FW 575.6 $C_{22}H_{37}N_7O_{11}$ Microbial   ≥85% (HPLC); bioactive peptide | Aoyagi, T et al, *J Appl Biochem*, 7:388, 1985

FR-1

ICN 195987   MW 980.1 $C_{35}H_{57}N_{13}O_{16}S_2$ Cyclic peptide with the cell binding domain Arg-Gly-Asp & associated cell migration Pro-Ala-Ser-Ser sequences of fibronectin; potently inhibits ADP-induced platelet aggregation

Galanin (1-30), Fluo-

Amersham VGAL011   Human   Fluorescein-labeled with high activity

Galanin, (^{125}I)-

Amersham IM287   Human   Propanol:water (80:20); ~74 TBq/mmol, ~2000 Ci/mmol; 3.7 MBq/mL, 100 µCi/mL

Galanin, (3-(^{125}I)Tyr^{26})-

Amersham IM260   Porcine   Lyophilized; ~74 TBq/mmol, ~2000 Ci/mmol

Gamma Secretase Inhibitor

Kamiya   MW 706 $C_{33}H_{57}N_5O_9F_2$ Diflouroketone

Gastric Inhibitory Peptide, (^{125}I)-

Synonyms: Gastric Inhibitory Peptide Receptor Ligand, (^{125}I)-

Amersham IM303 K_d = 16 pM rat brain; lyophilized; ~74 TBq/mmol, ~2000 Ci/mmol | Non-selective polypeptide; endogenous

Gastrin I, (3-(^{125}I)Tyr^{12})

Synonyms: Gastrin I Receptor Ligand, (^{125}I-Tyr^{12})-; Cholecystokinin B; G17

Amersham IM165   Human   Lyophilized; ~74 TBq/mmol, ~2000 Ci/mmol | Endogenous; K_d = 1.26 nM guinea pig gastric fundic mucosa membrane

Gastrin Releasing Peptide (20-27), *N*-Ac-

Synonyms: Gastrin Releasing Peptide Agonist

ICN 154553 MW 991.3 Tache, Y etal, *Regulatory Peptides*, Suppl 1:112, 1980

Gastrin Releasing Peptide, (3-(^{125}I)Tyr^{15})-

Synonyms: Gastrin Releasing Peptide Receptor Ligand, (^{125}I-Tyr^{15})-; Bombesin Receptor Ligand

Amersham IM169 Lyophilized; ~74 TBq/mmol, ~2000 Ci/mmol | BB2 selectivity; endogenous; K_d = 0.5 nM Swiss 3T3 cells

Gastrin Releasing Peptide, Fluo

Amersham VGRP011 Fluorescein-labeled with high activity

Ghrelin

Biogenesis 0100-0156 Human synthetic Purified; lyophilized

Biogenesis 0100-0157 Rat synthetic Purified; lyophilized

Glucagon

ICN 151182	MW 3482.8	Crystalline
ICN 194566	MW 3482.8	Crystalline; cell culture reagent
ICN 194567	MW 3482.8	Crystalline; cell culture reagent; γ-irradiated

Biogenesis 4660-0410 Human synthetic >98% (HPLC); lyophilized

Glucagon Like Peptide I (7-36), (^{125}I)-

Amersham IM323 Water:propan-1-ol (7:3), 0.5% BSA; ~74 TBq/mmol, ~2000 Ci/mmol; 3.7 MBq/mL, 100 µCi/mL

Glucagon, (3-(^{125}I)Tyr^{10})-

Amersham IM160 Human Lyophilized; ~74 TBq/mmol, ~2000 Ci/mmol

Glucagon, Agarose

ICN 191268 Bovine pancreas 10 atoms hydrophilic spacer arm; 0.1-0.2 mg/mL gel; suspension in PBS, 0.02% NaN₃ | Formed by reaction of diazo functional group of spacer & ligand; useful for isolation of specific cell membrane receptors

Glucose Transporter III

Biogenesis 4670-1677 Human synthetic >90% (HPLC); lyophilized; acetate salt

Glutathione Oxidized Disodium Salt, L-

Fluka 49745 MW 656.6 $C_{20}H_{30}N_6Na_2O_{12}S_2$ ≥99.0% (TLC); ≤10% water

Glutathione Oxidized, L-

Synonyms: BIOSYNTH; GSSG; Glutathiol

Fluka 39084 MW 612.64 $C_{20}H_{32}N_6O_{12}S_2$ ≥99.0% (assay)

Fluka 49740 MW 612.64; $C_{20}H_{32}N_6O_{12}S_2$ ≥99.0% (HPLC); ≤0.05% Na; ≤0.0005% Cd, Co, Cr, Cu, Fe, Mn, Ni, Pb, Zn; ≤0.001% Mg; ≤0.005% K, Ca | Meister, A & Anderson, ME, *Ann Rev Biochem*, 52:711, 1983

Fluka 49741 MW 612.64 $C_{20}H_{32}N_6O_{12}S_2$ ≥95.0% (enzym); ≤5% water

Glutathione Reduced, L- Immobilized

Fluka 49739 Immobilized on Agarose CL-4B; ligand content: 10 µmol reduced glutathione/mL packed gel; lyophilized powder stabilized with 50% lactose; 70 mg powder swells to 1 mL gel | For affinity chromatography; the reduced glutathione is attached through the sulfur to epoxy-activated 4% cross-linked beaded agarose (C_{10}-spacer)

Glycogen Synthase Peptide II

USBio G8170-26 ≥97% using reverse phase HPLC; lyophilized, phosphorylated | Similar to skeletal muscle glycogen synthase; Contains sites 3b, 3c & phosphorylated site 4 from glycogen synthase; The sequence of the peptide is: YRRAAVPPSPSLSRHSSP-HQ(pS)EDEEE; Applications: Kinase Assay: Tested at 62.5*mM*/assay using glycogen synthasekinase-3b to catalyze the incorporation of phosphate into phospho-glycogen synthase peptide-2; Glycogen synthasekinase-3a was used as the enzyme for a previous lot

Glycogen Synthase Peptide II, (Ala^{21})-

USBio G8170-25 ≥97% by reverse phase HPLC; lyophilized. | Because of an alanine for serine substitution, this peptide is not a substrate for glycogen synthasekinase 3, but instead serves as a negative control; The peptide is similar to skeletal muscle glycogen synthase; it contains sites 3b & 3c; Site 4, however, is replaced with alanine; The sequence of the peptide is: YRRAAVPPSPSLSRHSSPHQAEDEEE; Applications: Protein Phosphorylation Assays: Use at 62.5–250uM final concentration; This peptide is a negative control for assay of GSK-3b activity; GSK-3b phosphorylated phospho-GS-2 peptide, but not this lot of GS-2 (Ala21) peptide

Glycoprotein Peptide, P-

Synonyms: mdr Peptide

Oncogene PP03 Lyophilized with an equal amount of BSA | Suggested starting concentration of peptide for competition is 10X the concentration of primary Ab; titration recommended; C-terminal 21 amino acids; used for competition studies & dot blot

GP121

Synonyms: HIV Peptide I

ICN 158374 Recombinant, from the IIIB isolate of HIV-1 expressed in insect cells Highly purified | Binds to CD4(+) cells, inhibits syncitia formation & reacts with HIV(+) patient sera in Western blots; similar to the native HIV peptide in its ability to inhibit syncitia formation; reacts with Ab directed against the major neutralizing envelope epitope

GP160

Synonyms: HIV Peptide I

ICN 158373 Recombinant, from the IIIB isolate of HIV-1 expressed in insect cells Highly purified | Contains full length HIV-1 envelope protein; elicits neutralizing Ab in goats & is immunoreactive with HIV(+) patient sera & monoclonal Ab directed against gp41 & gp120

GP-Antagonist-2

BioSource International 77-450

G-Protein (2A)

BioSource International 77-454

Gramicidin D

ICN 101901 *Bacillus brevis*

Growth Hormone

ICN 160074 Bovine pituitary gland Prepared according to Dellacha, JM & J Sonenberg, *JBC*, 239:1515, 1964

Scipac P220-1 Pituitary glands >96%; lyophilized | Peptide hormone

ICN 198942 Rat recombinant, expressed in *E. coli* Receptor grade; 97%; lyophilized white powder | Flint, DJ & MJ Gardner, *J Endocrinol*, 137:203, 1993

Growth Hormone Releasing Factor (1-44), (3-(^{125}I)Tyr^{10})- Amide

Amersham IM180 Lyophilized; ~74 TBq/mmol, ~2000 Ci/mmol

Growth Related Oncogene

Synonyms: Melanoma Growth Stimulating Activity; Melanoma Growth Stimulating Activity; KC; CINC

ICN 160013 Human recombinant, expressed in *E. coli* ≥98%; lyophilized; 50 ng/mL maximal chemotactic activity as determined in a modified Boyden chamber | Promotes neutrophil chemotaxis & degranulation; stimulates mitogenesis in selected human melanoma cells

ICN 160014 Rat recombinant, expressed in *E. coli* ≥98%; lyophilized; 100 ng/mL maximal chemotactic activity as determined in a modified Boyden chamber | Promotes rat neutrophil chemotaxis & degranulation

Growth Related Oncogene α

Synonyms: Growth Related Oncogene α, rh-

ICN 160016 Human recombinant, expressed in *E. coli* ≥97%; lyophilized; ED_{50} = 0.15-0.3 µg/mL as determined by myeloperoxidase release from cytochalasin B treated human neutrophils

Growth Related Oncogene α, (^{125}I)-

Synonyms: Growth Related Oncogene α Receptor Ligand, (^{125}I)-

Amersham IM305 Human recombinant B & H labeled; lyophilized; ~74 TBq/mmol, ~2000 Ci/mmol | Chemokine receptor ligand; CXCR2 selectivity; endogenous; K_d = 0.3 nM 3 ASubE P-3 human placenta cells

Growth Related Oncogene β

Synonyms: Macrophage Inhibitory Peptide II; Growth Related Oncogene γ, rh-

ICN 160017 Human recombinant, expressed in *E. coli* ≥97%; lyophilized; ED_{50} = 0.3-0.9 µg/mL as determined by myeloperoxidase release from cytochalasin B treated human neutrophils

ICN 160015 Rat recombinant, expressed in *E. coli* ≥98%; lyophilized; 50 ng/mL maximal chemotactic activity as determined in a modified Boyden chamber | Promotes rat neutrophil chemotaxis & degranulation

ICN 160018 Human recombinant, expressed in *E. coli* ≥97%; lyophilized; ED_{50} = 0.1-0.3 µg/mL as determined by myeloperoxidase release from cytochalasin B treated human neutrophils

Guanylyl Cyclase α₁ or β₁ Subunit Peptide

Alexis 157-017 Soluble; lyophilized from both α₁ & β₁ subunits in separate vials containing 200 µg each | Competitively binds to Ab Alexis 210-724; antiserum blocking peptide

Heme Oxygenase Control Peptide

Chemicon AG255 Rat brain Semi-purified | Cellular biochemistry/regulatory protein used in immunoblotting (Western)

Chemicon AG258 Rat liver Semi-purified | Cellular biochemistry/regulatory protein used in immunoblotting (Western)

Chemicon AG257 Rat spleen Semi-purified | Cellular biochemistry/regulatory protein used in immunoblotting (Western)

Chemicon AG256 Rat testes Semi-purified | Cellular biochemistry/regulatory protein used in immunoblotting (Western)

Heme Oxygenase I Control Peptide

Chemicon AG253 ≥95% | Cellular biochemistry/regulatory protein; complementary to Chemicon AB1284

Heme Oxygenase II Control Peptide

Chemicon AG254 ≥95% | Cellular biochemistry/regulatory protein; complementary to Chemicon AB1285

Heparin Binding Epidermal Growth Factor-Like Growth Factor

ICN 195730 Human recombinant, expressed in S*f*21 ≥97%; lyophilized; ED_{50} = 2-5 ng/mL

Hepatitis Virus Core (2-192)

USBio H1920-21 Recombinant ≥95% (SDS-PAGE, 280 nm, Bradford); 1 mg/mL liquid in 8 *M* urea, 20 *mM* Tris-HCl, pH 8.0, 10 *mM* BME | Suitable for antigenic applications in immunological protocolsHCV core antigen (recombinant) AA 2-192 of HCV polyprotein; Fusions: β-galactosidase (114kD) fused at the N-terminus.

Hepatitis Virus NS3 (1209-1643)

USBio H1920-24 Recombinant ≥95% (SDS-PAGE, 280 nm, Bradford); ~1 mg/mL liquid in 20 *mM* Tris-HCl, pH 8.0, 20 *mM* BME | Suitable for antigenic applications in immunological protocolsHCV NS3 antigen (recombinant) AA 1209-1643 of HCV polyprotein; Fusions: Intein fusion at the amino-terminal end.

Hepatitis Virus NS3 (1450-1643)

USBio H1920-23 Recombinant ≥95% (SDS-PAGE, 280 nm, Bradford); ~1 mg/mL; lyophilized in 8 *M* urea, 0.02 *M* Tris-Cl, pH 8.0, 0.2 *M* Imidazole | Suitable for antigenic applications in immunological protocolsHCV NS3 antigen (recombinant) AA 1450-1643 of HCV polyprotein; Fusions: Six histidine fusion at the C-terminus.

Hepatitis Virus NS4 (1658-1863)

USBio H1920-25 Recombinant ≥95% (SDS-PAGE, 280 nm, Bradford); 1 mg/mL liquid in 8 *M* urea, 20 *mM* Tris-HCl, pH 8.0, 10 *mM* BME | Suitable for antigenic applications in immunological protocolsHCV NS4a+β antigen (recombinant) AA 1658-1863 of HCV polyprotein; Fusions: β-galactosidase (114kD) fused at the N-terminus.

Hepatocyte Growth Factor

Peptides International PHE-4801-v MW 80k Human recombinant, expressed in CHO cells Lyophilized from 10mM phosphate buffer (pH 7.0), containing 0.35 M NaCl & 250 mg of BSA | Bioactive; potent mitogen for hepatocytes & epithelial/endothelial cells; Nakamura, T et al, *Nature*, 342:440, 1989; Zarnegar, R & GK Michalopoulos, *J Cell Biol*, 129:1177, 1995; Yo, Y et al, *Kidney Int*, 53:50, 1998

ICN 160245 Human recombinant, expressed in insect cell line S*f*9 ≥90%; frozen liquid; ED_{50} = 1.0-50.0 ng/mL; 1 U is the amount of HGF needed for stimulation of primary rat Hepatocyte DNA synthesis & the scattering of epithelial cells; endotoxin tested

Hepatocyte Growth Factor

Synonyms: Hepatocyte Growth Factor, rh-; Scatter Factor; Hepatopoietin A; Mammary Growth Factor

Calbiochem 375228   MW 90k   Human recombinant, expressed in *Spodoptera frugiperda*   ≥95% (SDS-PAGE); lyophilized from filter-sterilized 350 *mM* NaCl, 20 *mM* sodium phosphate buffer, pH 7.0 containing 50 μg BSA/μg HGF; biological activity: ED$_{50}$=20-40 pg/mL as measured by its ability to stimulate ^{3}H-thymidine incorporation in an HGF-responsive epithelial cell line; endotoxin: ≤100 pg/μg HGF; may be carcinogenic/teratogenic | Through binding to its high affinity met receptor, HGF has marked & varied effects on hepatocytes & other epithelial & endothelial cell types; HGF-induced proliferation of primary hepatocytes is mediated by activation of PI 3-kinase; acts as a potent mitogen for mature hepatocytes; Kaido, T et al, *Biochem Biophys Res Comm*, 218: 1, 1996; Kobayashi, Y et al, *Biochem Biophys Res Comm*, 220: 7, 1996; Skouteris, GG et al, *Biochem Biophys Res Comm*, 218: 229, 1996; Rubin, JS et al, *Biochim Biophys Acta*, 1155: 357, 1993; Strain, AJ, *J Endocrinol*, 137: 1, 1993; Nakamura, T et al, *Nature*, 342: 440, 1989

Heregulin-α Epidermal Growth Factor Domain

ICN 195747   Human recombinant, expressed in *E. coli*   ≥97%; lyophilized; ED$_{50}$ = 20-40 ng/mL

Hirudin, r-

Pentapharm 126-05, 126-07   MW 6963.5 $C_{287}H_{440}N_{80}O_{110}S_6$ Most potent & specific thrombin inhibitorknown; used in hemostaseological test procedures; used in blood & plasma fractionation to prevent the multiple enzymatic & non-enzymatic actions of thrombin; added to test mixtures to exclude undesired thrombin actions due to contaminations of reagents with prothrombin or with prothrombin activators; selectively inhibits thrombin in certain assay conditions when cross-reactivity of thrombin & the chosen enzyme should lead to cleavage of the chromogenic substrate; Dodt J, et al, *FEBS Letts*, 165:180, 1984; Meyhack B, et al, *Thromb Res*, Suppl 7:3, 1987; Svendsen L, et al, *Thromb Res*, 34:457, 1984; Stocker K, *Semin Thromb Hemost*, 17:113, 1991; Stocker K, *Haemostasis*, 21:161, 1991

Histone (29-35), (Ala^{32})-

BioSource International 77-200

Histone H1

Alexis 202-029   Calf thymus   ≥95%; lyophilized; dissolve in deionized water or buffered solution | Binds water tenaciously so the total dry solid mass may be approximately twice the protein content; Can be used as a substrate for proteinkinases & protein phosphatases

Histone H2B (29-35)

BioSource International 77-198

Histone H4 (Tetra Ac) Peptide, Biotinylated (Ac-Lys^{5,8,12,16})-

Upstate 12-379   MW 2519   >98% pure; lyophilized | Substrate for histone-modifying enzymes

Histone H4 Peptide

USBio H5110-15M   ≥90%; lyophilized | ≥90%; lyophilized

Histone H4 Peptide, (Lys^{12})-

USBio H5110-15H   ≥90%; 0.1 *mM* supplied as 100 μg in 500 μL of sterile deionized water. | Applications: HAT Assay: Use 1 mg of peptide per assay; A previous lot was tested by using 50ng of GCN5 to acetylate the peptide; Yeast GCN5 does not effectively acetylate this peptide; The acetylation of position 12 may cause steric hindrance; HATs with different specificity may acetylate this peptide; PCAF is an effective HAT enzyme; Immunoblot Analysis: Used as a standard for antibodies specific for acetylation of lysine 12 of histone H4

Histone H4 Peptide, (Lys^{16})-

USBio H5110-15Q   ≥90%; 0.1 *mM* supplied as 100 μg in 500 μL of sterile deionized water. | Applications: HAT Assay: Use 1 mg of peptide per assay; A previous lot was tested by using 50ng of GCN5 to acetylate the peptide; Yeast GCN5 does not effectively acetylate this peptide since the enzyme's preferred acetylation site is blocked; Other HATs with different specificity's, such as PCAF enzyme may acetylate this peptide; Immunoblot Analysis: This peptide can be used as a standard for antibodies specific for acetylation of lysine 16 of histone H4

Histone H4 Peptide, (Lys^5)-

USBio H5110-15K   ≥90% frozen solution in sterile deionized water. | HAT Assay: Tested by using 500ng of PCAF to acetylate the peptide; A previous lot was tested by using 50ng of GCN5 to acetylate the peptide; Use 1 mg of peptide per assay; Immunoblot Analysis: Can be used as a standard for antibodies specific for acetylation of lysine 6 of histone H4.

Histone H4 Peptide, (Lys^8)-

USBio H5110-15L   ≥90% as detemined (HPLC) analysis; frozen solution in sterile deionized water. | SGRGKGGAcKGLGKGGAKRHR-C, which corresponds to the N terminal sequence of histone H4 acetylated at position 8 plus a C terminal cysteine; May be used to measure histone acetyltransferase (HAT) activity; May be acetylated by HAT on lysines at position 5, 12 & 16; May be used to determine the preferred acetylation site by any particular HAT that acetylates histone H4, when used with peptides of identical sequence but differing acetylated lysine residues; Dot Blot: Used as a standard to assess the specificity of anti-acetyl histone antibodies for acetyl lysine 8 of Histone H4

Histone H4 Peptide, (Tetra Ac-Lys)-

USBio H5110-15P   ≥90%; frozen solution | The acetylation of lysine residues in core histones is thought to play a role in the activation of gene transcription; The neutralization of the histone's basic charge reduces it's affinity for DNA & relaxes or "opens" the chromatin, which is believed to facilitate the access of transcription regulators to the promoter region of genes; This peptide corresponds to the N-terminal sequence of Tetrahymena Histone H4; The four lysine residues are acetylated (AGGACKGGACKGMGACKVGAACKRHSC). The peptide may be used for a dot blot assay, peptide competition, or as a negative control for HAT assays; Applications: Western Blotting Assay: 1 μg/mL

Histone H4 Peptide, Biotinylated

USBio H5110-15N   ≥90%; lyophilized | ≥90%; lyophilized

HIV II Env gp36 (390-702)

USBio H6009-09   Recombinant   ≥95% (SDS-PAGE, 280 nm, Bradford); 1 mg/mL liquid in 0.01 *M* sodium carbonate, 0.01 *M* EDTA, 0.014 *M* BME, 0.05% Tween 20 | Suitable for antigenic applications in immunological protocols HIV-2 env region AA 390-702; Fusions: β-galactosidase (114kD) fused at the N-terminus; Reacts strongly with human HIV positive serum.

HIV p121 Peptide

ICN 158375   Contains 82 AA from the immunodominant region of gp41; shown to react with nearly all HIV(+) patient sera; purified from core fragments & envelope protein fragments of the HIV-1 IIIB isolate

HIV PG2 Peptide

ICN 158376   Core & envelope protein fragments of the HIV-1 IIIB isolate; contains 153 AA from the core protein p24 & 74 AA from p15; immunoreactive with HIV(+) patient sera

hSOS (1149-1158)

BioSource International 77-252

HTLV I E1 Peptide

IBT HTIPE-1070-5 Synthetic >98%; lyophilized powder | Covers Ser[162] to Lys[209] of HTLV I envelope protein (ATK-1 isolate)

HTLV II Envelope Peptide I

IBT HTIIPE-1210-5 Synthetic >98%; lyophilized powder | Covers Pro[164] to Lys[205] of HTLV II envelope protein (lambda H6.0 isolate)

Hydrolysate
Casein

Synonyms: Casein Hydrolysate

Amersham US12855 3.7-4.8% Amino N_2; 11.5-14.2% total N_2; pH 6.6-7.5 | Used in fermentation

Hydrolysate
Gelatin

Synonyms: Gelatin Hydrolysate

Amersham US16024 Used in fermentation

Hydrolysate
Casein

Synonyms: Casein Hydrolysate

Fluka 22090 0.5 Amino N/total N; 13% total nitrogen; 5% water; 6% ash; 2% solubility in water; 38% free amino acids; 86% total amino acids; free amino acids (total amino acids): Ala 1.3(2.5), Arg 3.0(3.2), Asn 1.3, Asp 0.9(6.2), Glu 2.3(18.6), Gly 0.3(1.4), His 1.2(2.5), Ile 2.3(4.4), Leu 6.0(7.6), Lys 5.1(6.8), Met 1.8(2.5), Phe 3.3(4.1), Pro 0.7(8.1), Ser 1.3(5.0), Thr 1.5(3.8), Trp 1.3(1.0), Tyr 2.0(2.9), Val 2.6(5.5) | For microbiology; pancreatic hydrolysate of casein used in the production of antibiotics, toxins, bacteriocins, enzymes

Hydrolysate
Lactalbumin

Synonyms: Lactalbumin Hydrolysate

Fluka 61302 Amino N/Total N: 0.5; 13% total nitrogen, 4% water, 6% ash, 35% free AA, 72% total AA; % free AA/total AA: Ala 1.8/2.9, Arg 2.2/2.7, Asn 1.1/0, Asp 0.8/6.4, Ile 2.4/4.0, Leu 6.6/8.0, Lys 4.5/6.5, Met 1.6/1.9, Trp 1.0/0.9, Tyr 0.3/0.9, Val 2.5/6.5, Glu 2.2/14.4, Gly 0.3/1.3, His 1.1/1.7, Phe 2.5/3.2, Pro 0.7/5.2, Ser 1.4/3.7, Thr 1.6/3.4, Gln 0.1/0

Hydrolysate
Amicase

Fluka 82514 Amino N/total N: 0.3; 13% total nitrogen; 5% water; ≤5% ash

Hydrolysate
Edamin®

Fluka 82517 Amino N/total N: 0.56; 12.3% total nitrogen; 3.2% water; 4% residue on ignition | Registered trademark of Sheffield Products Division of Quest, International

Hydrolysate
Hy-Case® Amino

Fluka 82519 Amino N/total N: 0.8; 8% total nitrogen; 5% water; 4% ash | Registered trademark of Sheffield Products Division of Quest, International

Hydrolysate
Hy-Case® M

Fluka 82520 Amino N/total N: 0.68; 8.0% total nitrogen; 2.2% water; 40.5% residue on ignition | Registered trademark of Sheffield Products Division of Quest, International

Hydrolysate
Hy-Case® SF

Fluka 82521 Amino N/total N: 0.80; 13.8% total nitrogen; 4.6% water; 0.7% residue on ignition | Registered trademark of Sheffield Products Division of Quest, International

Hydrolysate
Hy-Soy®

Fluka 82522 Amino N/total N: 0.20; 10.2% total nitrogen; 3% water; 9.9% residue on ignition | Registered trademark of Sheffield Products Division of Quest, International

Hydrolysate
NZ-Amine® AS

Fluka 82524 Amino N/total N: 0.49; 13.4% total nitrogen; 3.9% water; 5.7% residue on ignition | Registered trademark of Sheffield Products Division of Quest, International

Hydrolysate
NZ-Amine® B

Fluka 82525 Amino N/total N: 0.41; 14.0% total nitrogen; 4.4% water; 5.2% residue on ignition | Registered trademark of Sheffield Products Division of Quest, International

Hydrolysate
NZ-Amine® HD

Fluka 82527 Amino N/total N: 0.68; 12.6% total nitrogen; 4.1% water; 7.2% residue on ignition | Registered trademark of Sheffield Products Division of Quest, International

Hydrolysate
NZ-Amine® YT

Fluka 82529 Amino N/total N: 0.50; 13.5% total nitrogen; 4.9% water; 7.8% residue on ignition | Registered trademark of Sheffield Products Division of Quest, International

Hydrolysate
NZ-Case®

Fluka 82531 Amino N/total N: 0.35; 14.0% total nitrogen; 4% water; 6% residue on ignition | Registered trademark of Sheffield Products Division of Quest, International

Hydrolysate
NZ-Case® M

Fluka 82532 Amino N/total N: 0.35; 14.0% total nitrogen; 5% water; ≤10% residue on ignition | Registered trademark of Sheffield Products Division of Quest, International; for microbiology

Hydrolysate
NZ-Case® Plus

Fluka 82533 Amino N/total N: 0.57; 12.9% total nitrogen; 4.6% water; 5.3% residue on ignition | Registered trademark of Sheffield Products Division of Quest, International

Hydrolysate
NZ-Case® TT

Fluka 82534 Amino N/total N: 0.40; 13.7% total nitrogen; 5.6% water; 6.2% residue on ignition | Registered trademark of Sheffield Products Division of Quest, International

Hydrolysate
Primatone®

Fluka 82538 Amino N/total N: 0.45; 13.0% total nitrogen; 5% water; ≤10% residue on ignition | Registered trademark of Sheffield Products Division of Quest, International; for microbiology

Hydrolysate
Primatone® HS

Fluka 82543   Amino N/total N: 0.46; 12.8% total nitrogen; 3% water; 11% residue on ignition | Registered trademark of Sheffield Products Division of Quest, International

Hydrolysate
Primatone® RL

Fluka 82544   Amino N/total N: 0.5; 12% total nitrogen; 5% water; 8% residue on ignition; 200-400 mesh particle size | Registered trademark of Sheffield Products Division of Quest, International; Production of monoclonal antibodies in culture; Reuveny, S et al, *Dev Biol Stand*, 60: 185, 1985

Hydrolysate
Primatone® SG

Fluka 82546   Amino N/total N: 0.16; 16.9% total nitrogen; 3.6% water; 2.6% residue on ignition | Registered trademark of Sheffield Products Division of Quest, International

Hydrolysate
Casein

Synonyms: Casein Hydrolysate

ICN 104778   Vitamin free; salt free; hydrochloride

Hydrolysate
Amicase

Synonyms: Casein Hydrolysate

Sigma A 2427   Mixture of free amino acids with virtually no unhydrolyzed peptides; Minimal inorganic components are present; virtually free of vitamins & growth factors | Produced under severe conditions

Hydrolysate
AA Standards

Sigma A 2908   0.5 µmole/mL of each of these AA: L-cysteic acid, Nle; taurine; Trp; L-Ala, Arg, Asp, Glu, Gly, His, Ile, Leu, Lys, Met, Phe, Pro, Ser, Thr, Tyr, Val; except L-cystine at 0.25 µmole/mL; in 0.2 *N* sodium citrate, pH 2.20 | AA standard solutions for protein hydrolysates containing Nle

Hydrolysate
Collagen

Synonyms: Collagen Hydrolysate

Sigma A 9531   2.5 µmoles/mL of each of these AA: L-α-aminoadipic acid, δ-Hyl, L-Ala, Arg, Asp, Glu, Gly, His, Ile, Leu, Lys, Met, Phe, Pro, Ser, Thr, Tyr, Val; except L-cystine at 1.25 µmoles/mL, Pro & Hyp at 12.5 µmoles/mL | AA standards

Hydrolysate
Food

Synonyms: Food Hydrolysate

Sigma A 9656   10 µg/mL of each of these AA: Trp, L-Ala, Arg, Asp, Glu, Gly, His, Ile, Leu, Lys, Met, Phe, Pro, Ser, Thr, Tyr, Val; except L-cystine at 20 µg/mL | AA standards

Hydrolysate
AA Standards

Sigma A 9781   0.5 µmole/mL of each of these AA: Trp, L-Ala, Arg, Asp, Glu, Gly, His, Ile, Leu, Lys, Met, Phe, Pro, Ser, Thr, Tyr, Val; except L-cystine at 0.25 µmole/mL; in 0.2 *N* sodium citrate, pH 2.20 | AA standards

Hydrolysate
AA Standards

Sigma AA–S-18   25 µmoles/mL of each of these L-AA: Ala, Arg, Asp, Glu, Gly, His, Ile, Leu, Lys, Met, Phe, Pro, Ser, Thr, Trp, Tyr, Val; except L-cystine at 1.25 µmoles/mL; 1 mL size package in flame sealed ampules | AA standards

Hydrolysate
L-AA, (U-¹⁴C)-

Synonyms: Algal Hydrolysate

ARC ARC-474   Algae   Algal protein hydrolysate; high specific activity; in 0.01 *N* HCl | Radiochemical

Hydrolysate
Peptone Primatone® HS

Sigma P 4838   Animal tissue   Total nitrogen: ~12.2%; amino nitrogen: ~4.8%

Hydrolysate
Peptone Primatone® RL

Sigma P 4963   Animal tissue   Total nitrogen: ~12.0%; amino nitrogen: ~6.0%

Sigma P 5088   Animal tissue   Technical grade; total nitrogen: ~13.0%; amino nitrogen: ~5.6% | Useful when undissolved solids are acceptable

Hydrolysate
Peptone Primagen

Sigma P 6088   Animal tissue   Total nitrogen: ~12.0%; amino nitrogen: ~7.5%

Hydrolysate
Peptone Primagen P

Sigma P 6213   Animal tissue   Total nitrogen: ~14.8%; amino nitrogen: ~2.2%

Hydrolysate
Albumin

Synonyms: Ovalbumin Hydrolysate

Sigma A 6710   Chicken egg

Hydrolysate
Peptone Primatone® CLT

Sigma P 4588   Collagen   Technical grade hydrolysate; total nitrogen: ~16.0%; amino nitrogen: ~1.7% | Useful in media where high clarity is not required

Hydrolysate
Lactalbumin

Synonyms: Peptone; Lactalbumin Hydrolysate

Fluka 61300   Lactalbumin   Amino N/Total N: 0.40; 12.5% total nitrogen, 4% water, 5% ash, 44% free AA, 89% total AA; % free AA/total AA: Ala 1.7/4.0, Arg 3.8/3.7, Asp 0.9/9.8, Ile 2.9/5.3, Leu 9.7/11.3, Lys 7.3/8.8, Met 2.6/3.1, Trp 1.3/1.1, Tyr 0.4/1.5, Val 2.8/5.9, Glu 1.5/4.1, Gly 0.3/2.1, His 1.1/2.1, Phe 3.8/4.4, Pro 1.3/12.1, Ser 1.1/5.6, Thr 1.8/4.9 | For microbiology; nutrient for fermentations & other applications

Hydrolysate
Gluten

Synonyms: Gluten Hydrolysate

Fluka 49760   Maize   Amino N/Total N: 0.31; 8.1% total nitrogen, 8.4% water, 5.1% ash, 10% free AA, 41% total AA; % free AA/total AA: Ala 0.6/3.2, Arg 0.9/1.3, Asn 0/0, Asp 0/2.5, Ile 0.5/1.6, Leu 1.4/6.6, Lys 0.3/0.6, Met 0.3/0.8, Trp 0/0, Tyr 1.5/2.2, Val 0.4/2.0 | For microbiology; provides the benefits of corn steep liquor for fermentation processes without the associated disadvantages

Hydrolysate
Peptone Type I

Sigma P 7750 Meat Total nitrogen: ~16.0%; amino nitrogen: ~3.0%

Hydrolysate
Peptone Peptonized Milk Nutrient

Sigma P 6838 Milk solids Refined hydrolysate; total nitrogen: 5.4%; amino nitrogen: 1.7%

Hydrolysate
Peptone N-Z Soy

Sigma P 1265 Soy Total nitrogen: ~12.5%; amino nitrogen: ~5.3% | Soy protein enzymatic hydrolysate; general use peptone; substitute for casein enzymatic hydrolysates

Hydrolysate
Peptone Hy-Soy® J

Sigma P 6338 Soy Intermediate grade hydrolysate from soy; total nitrogen: 8.0%; amino nitrogen: 1.8% | Useful in media where high clarity is not required

Hydrolysate
Peptone Hy-Soy® T

Sigma P 6463 Soy Technical grade hydrolysate from soy; total nitrogen: 8.0%; amino nitrogen: 1.7% | Useful in media where high clarity is not required

Hydrolysate
Peptone N-Z Soy BL 4

Sigma P 6588 Soy Refined hydrolysate with high clarity; total nitrogen: ~13.8%; amino nitrogen: ~2.6% | High clarity

Hydrolysate
Peptone N-Z Soy BL 7

Sigma P 6713 Soy Total nitrogen: 14.0%; amino nitrogen: 2.4%

Hydrolysate
Peptone Type IV

Sigma P 0521 Soybean Total nitrogen: ~8.0%; amino nitrogen: ~2.0%

Hydrolysate, Acid
Casein

Amersham US12852 ~5.2% Amino N_2; ~8.5% total N_2; pH 6.6-7.5 | Used in fermentation

ICN 101291 ~8.2 N_2, ~6.1 amino N_2 | Acid hydrolysate

Hydrolysate, Acid
Gelatin

ICN 101773 ~225 bloom

Hydrolysate, Acid
Casein

Synonyms: Casein Hydrolysate

ICN 104769 Vitamin free; 10%, acid hydrolyzed, highly purified free of B complex vitamins | Acid hydrolysate; for microbiological procedures

Hydrolysate, Acid
Amicase

Synonyms: Casein Hydrolysate

ICN 191023 pH 5.0 (2% solution), ~13.0% N_2, ~9.2% amino N_2 | Acid hydrolysate; sulfuric acid digest of casein; lower Salt content than the HCl digest; Trp free

Hydrolysate, Acid
Hy-Case® Amino

Synonyms: Casein Hydrolysate

Sigma C 0501 Typical analysis: total nitrogen ~8.5%, amino nitrogen ~5.9%; 30-40% NaCl | Acid hydrolysate; because of the severe conditions used for hydrolysis, this product is virtually free of vitamins & growth factors; primarily free AA with a small amount of peptides

Hydrolysate, Acid
Hy-Case® SF

Synonyms: EZMix™

Sigma C 4589 Casein acid hydrolysate; same formulation as Sigma C 9386 with the added advantage of being dust-free & fast dissolving, allowing easier weighing & handling; Virtually free of vitamins & growth factors | Produced under severe conditions

Hydrolysate, Acid
Hy-Case® M

Synonyms: Casein Hydrolysate

Sigma C 7710 ~70% free AA; typical analysis: total nitrogen ~8.1%, amino nitrogen ~5.6% | Acid hydrolysate; because of the severe conditions used for hydrolysis, this product is virtually free of vitamins & growth factors

Hydrolysate, Acid
Hy-Case® P

Synonyms: Casein Hydrolysate

Sigma C 7835 typical analysis: total nitrogen ~11.0%, amino nitrogen ~8.2% | Acid hydrolysate; recommended when an intermediate Salt content between Hy-Case® M & Hy-Case® SF is needed; because of the severe conditions used for hydrolysis, this product is virtually free of vitamins & growth factors

Hydrolysate, Acid
Amisoy

Sigma S 1674 Soy Total nitrogen: ~12.6%; amino nitrogen: ~8.7%

Hydrolysate, Enzymatic
Lactalbumin

Synonyms: Lactalbumin Hydrolysate

Amersham US18055 ~12.7% total N_2; ~4.8% amino N_2; pH 7.0 (2% soln)

Hydrolysate, Enzymatic
NZ-Amine® A

Synonyms: Casein Hydrolysate

ICN 101290 Enzymatic hydrolysate of whole casein

Hydrolysate, Enzymatic
Gelatin

Synonyms: Gelatin Hydrolysate

ICN 101772

Hydrolysate, Enzymatic
Casein

Synonyms: Casein Hydrolysate

ICN 104864 5% sterile solution; vitamin free | Enzymatic hydrolysate

Hydrolysate, Enzymatic
NZ-Amine® AS

Synonyms: Casein Hydrolysate

ICN 960138 Soluble | Enzymatic hydrolysate

Hydrolysate, Enzymatic
NZ-Amine® A

Synonyms: Casein Hydrolysate

Sigma C 0626   For production of media for the growth of Charon phage & other lambda phages; produced under mild conditions; Blattner, FR et al, *Science*, 196: 161, 1977

Hydrolysate, Enzymatic
NZ-Case®

Synonyms: Casein Hydrolysate

Sigma C 1026   Moderately digested; produced under mild conditions

Hydrolysate, Enzymatic
NZ-Amine® YT

Synonyms: Casein Hydrolysate

Sigma C 1655   Typical analysis: total nitrogen ~13.2%, amino nitrogen ~6.3% | Enzymatic hydrolysate; highly refined source of AA & peptides

Hydrolysate, Enzymatic
NZ-Amine® A

Synonyms: EZMix™

Sigma C 4464   Casein enzymatic hydrolysate; same formulation as Sigma C 0626 with the added advantage of being dust-free & fast dissolving, allowing easier weighing & handling | Recommended for the production of media for the growth of Charon phage & other lambda phages; produced under mild conditions; Blattner, FR et al, *Science*, 196: 161, 1977

Hydrolysate, Enzymatic
NZ-Amine® AS

Synonyms: Casein Hydrolysate

Sigma C 4517   Typical analysis: total nitrogen ~13.0%, amino nitrogen ~6.5% | Enzymatic hydrolysate; highly refined source of AA & peptides; a form of NZ-Amine® A with much greater solubility; concentrated solutions are clear, but contain some filterable solids

Hydrolysate, Enzymatic
NZ-Case® TT

Synonyms: Casein Hydrolysate

Sigma C 4523   Moderately digested; originally formulated for use in media for the production of tetanus toxin; produced under mild conditions

Hydrolysate, Enzymatic
NZ-Case® Plus

Synonyms: Casein Hydrolysate

Sigma C 4642   Typical analysis: total nitrogen ~13.3%, amino nitrogen ~6.6% | Enzymatic hydrolysate; highly refined source of AA & peptides; hydrolysis optimized to produce rich growth of most bacteria

Hydrolysate, Enzymatic
NZ-Amine® EKC

Synonyms: Casein Hydrolysate

Sigma C 4767   Typical analysis: total nitrogen ~12.8%, amino nitrogen ~3.1% | Enzymatic hydrolysate; highly refined source of AA & peptides

Hydrolysate, Enzymatic
NZ-Amine® B

Synonyms: Casein Hydrolysate

Sigma C 6835   Typical analysis: total nitrogen ~13.2%, amino nitrogen ~5.6% | Enzymatic hydrolysate; highly refined source of AA & peptides; a refined hydrolysate with good solubility after heating

Hydrolysate, Enzymatic
NZ-Amine® BT

Synonyms: Casein Hydrolysate

Sigma C 6960   Technical grade; analog of NZ-Amine® B | Most useful in applications where clarity is not required; produced under mild conditions

Hydrolysate, Enzymatic
NZ-Amine® E

Synonyms: Casein Hydrolysate

Sigma C 7085   Produced under mild conditions

Hydrolysate, Enzymatic
NZ-Amine® YTT

Synonyms: Casein Hydrolysate

Sigma C 7460   Technical grade of NZ-Amine® YT | Useful where complete solubility is not required; produced under mild conditions

Hydrolysate, Enzymatic
NZ-Case® M

Synonyms: Casein Hydrolysate

Sigma C 7585   typical analysis: total nitrogen ~13%, amino nitrogen ~4.5% | Enzymatic hydrolysate; highly refined source of AA & peptides; highly soluble preparation provides very clear media

Hydrolysate, Enzymatic
Hy-Case® SF

Synonyms: Casein Hydrolysate

Sigma C 9386   typical analysis: total nitrogen ~12.9%, amino nitrogen ~10.3%; similar to Sigma C 0501, but ≤1.5% NaCl | Enzymatic hydrolysate; because of the severe conditions used for hydrolysis, this product is virtually free of vitamins & growth factors

Hydrolysate, Enzymatic
Edamin® S

Synonyms: Lactalbumin Hydrolysate

Sigma L 0375   Total nitrogen: ~12.4%; Amino nitrogen: ~6.9% | Produced under mild conditions

Hydrolysate, Enzymatic
Lactalbumin

Synonyms: Lactalbumin Hydrolysate

Sigma L 9010   Total nitrogen: ~12.5%; Amino nitrogen: ~6.25%; AN:TN ratio: ~0.5; cell culture tested; insect cell culture tested

Hydrolysate, Enzymatic
Edamin® K

Synonyms: Lactalbumin Hydrolysate

Sigma L 9033   Technical grade; total nitrogen: ~12.3%; Amino nitrogen: ~3.2% | Media for use when complete solubility is not a requirement

Hydrolysate, Enzymatic
NZ-Amine® AS

Synonyms: Casein Hydrolysate

Sigma N 4517   Concentrated solutions are clear but contain some filterable solids | Form of with much greater solubility; produced under mild conditions

Hydrolysate, Enzymatic
NZ-Case® Plus

Synonyms: Casein Hydrolysate

Sigma N 4642   Hydrolysis optimized to produce rich growth of most bacteria; produced under mild condition

Hydrolysate, Enzymatic
NZ-Amine® EKC

Synonyms: Casein Hydrolysate

Sigma N 4767 Produced under mild conditions; formulated for optimal growth of acid-forming bacteria

Hydrolysate, Enzymatic
Proteose Peptone

Sigma P 0431 ~14% total nitrogen | Originally adapted for use in culture media for the production of bacterial toxins; A satisfactory nutrient for general culture needs

Hydrolysate, Enzymatic
Bacteriological Peptone

Sigma P 0556 ~16% total nitrogen in a readily available form for bacteria; Has a high peptide & amino acid content; Very soluble | Produced under mild conditions; for the preparation of general culture media

Hydrolysate, Enzymatic
Pepticase

Synonyms: Casein Hydrolysate

Sigma P 1992 Typical analysis: total nitrogen~13.3%, amino nitrogen ~4.0% | Enzymatic hydrolysate; highly refined source of AA & peptides; especially good for preparation of thioglycollate media; tryptic digest with excellent solubility & clarity of solution

Hydrolysate, Enzymatic
Sheftone D

Sigma S 3646 Not highly refined | Does not form clear solution

Hydrolysate, Enzymatic
Gluten

Sigma G 4138 Corn

Hydrolysate, Enzymatic
Peptone

Sigma P 5905 Meat Total nitrogen: ~16%; amino nitrogen: ~3.0%; AN:TN ratio: ~0.2; cell culture tested

Hydrolysate, Enzymatic
Peptone Type I

Sigma P 7296 Meat Total nitrogen: ~16%; amino nitrogen: ~3.0%; AN:TN ratio: ~0.2; plant cell culture tested

Hydrolysate, Enzymatic
Peptone Primatone®

Sigma P 8388 Meat Total nitrogen: ~13.1%; amino nitrogen: ~7.6% | Meat protein enzymatic hydrolysate general medium component

Hydrolysate, Enzymatic
Edamin® S

Synonyms: Lactalbumin Hydrolysate

ICN 102129 Milk Solubility = 130 g/L (H_2O, 25°C), 12.4% total nitrogen, 6.9% amino nitrogen | Solubilized

ICN 102131 Milk Solubility = 20 g/L (H_2O, 25°C), 12.3% total nitrogen, 6.9% amino nitrogen

I-309

ICN 195786 Human recombinant, expressed in *E. coli* ≥97%; lyophilized | C-C Chemokine

Insulin

ICN 105707 Bovine pancreas 25 IU/mg dry weight; Zn stabilized

ICN 196063 Bovine pancreas 25 IU/mg dry weight; Zn stabilized; sterilized by γ-irradiation

ICN 155041 Equine pancreas ~24 IU/mg; crystalline; ~0.5% Zn

ICN 155042, ICN 193900 Human recombinant, expressed in *E. coli* ~28 IU/mg; 0.25-1.1% Zn

ICN 155043 Human synthetic from porcine insulin ~24 IU/mg protein; crystalline | Morihara, K etal, *Nature*, 280:412, 1979

ICN 155044 Porcine pancreas ~24 IU/mg; crystalline; ~0.5% Zn

ICN 155045 Sheep pancreas ~24 IU/mg; crystalline; ~0.5% Zn

Insulin (21-26)

Biogenesis 5329-4002 Human synthetic >95%; lyophilized

Insulin, (^{125}I)-

ICN 68127 Human recombinant, >50 µCi/µg, >1.85 MBq/µg; 0.1 *M* KPB, pH 7.5, 0.5% BSA

ICN 68044 Porcine 30-100 µCi/µg, 1.11-3.7 MBq/µg; 0.1 *M* KPB, pH 7.5, 0.5% BSA | Igarashi, M etal, *Diabetes Res Clin Pract*, 15:213, 1992

Insulin, (3-(^{125}I)Tyr^{A14})-

Amersham IM166 Human recombinant Lyophilized; ~74 TBq/mmol, ~2000 Ci/mmol

Insulin, (3-(^{125}I)Tyr^{B26})-

Amersham IM167 Human recombinant Lyophilized; ~74 TBq/mmol, ~2000 Ci/mmol

Insulin, Agarose

ICN 191307 Bovine pancreas 10 atoms hydrophilic spacer arm; 2-5 mg/mL gel; formed by direct epoxy-activated agarose reaction; suspension in PBS, 0.02% NaN_3 | Useful for isolation of adipocytes

Insulin-Like Growth Factor I

Synonyms: Insulin-Like Growth Factor-I, rh-; Somatomedin C; Somatomedin C; Insulin-Like Growth Factor I sR

Calbiochem 407240 MW 7.5k Human recombinant, expressed in *E. coli* ≥97% (SDS-PAGE); lyophilized solid; biological activity: ED_{50}=1.0-3.0 ng/mL as measured in a cell proliferation assay with TF-1 cells; endotoxin: ≤100 pg/µg IGF-I; may be carcinogenic/teratogenic | Polypeptide of 70 AA; ~40% homology to insulin & produces metabolic effects similar to insulin; acts as an important local regulator of bone formation; inhibits etoposide-induced apoptosis in BALB/c 3T3 cells; stimulates the activity of protein tyrosine kinases in both normal & diethylstilbestrol-treated hamster kidneys; *Merck Index*, 12: 8862; Chen, CW ad Ray, D, *Carcinogenesis*, 16: 1339, 1995; Sell, C et al, *Cancer Res*, 55: 303, 1995; Amarani, S et al, *J Bone Min Res*, 8: 157, 1993; Mohan, S, *Growth Regul*, 3: 67, 1993

IBT Receptor Grade (CU020, CU100, CM001); Media Grade (IU100, IM001, IM010) MW 7649 Human recombinant, expressed in *E. coli* >95% (receptor grade) or >70% (media grade) purity; lyophylized; free of measurable endotoxins | 70 AA peptide; stimulates growth & differentiation in most cell types; human sequence identical to porcine & bovine IGF-I

ICN 152302 Human recombinant, expressed in *E. coli* Lyophilized; purified by sequential chromatography | Biological activities include action on cartilaginous tissues, insulin-like action in muscles & fatty tissues & facilitation of differentiation of various cells

ICN 153479   MW 7649   Human recombinant, expressed in *E. coli*   Lyophilized; ≥98%; ED_{50} = 1.0 ng/mL, by a mitogenic assay stimulating ³H-thymidine uptake into Balb/c 3T3 cells & receptor binding assays using human placental membranes; <0.1 ng endotoxin/μg protein | N-terminal Met resulting from the bacterial start codon has been enzymatically removed

ICN 153505   Human recombinant, expressed in *E. coli*   Lyophilized; ≥96%; unspecified activity assayed by the stimulation of ³H-thymidine uptake by Balb/c 3T3 cells; <0.1 ng endotoxin/mg protein | 70 AA peptide sequence suitable for receptor binding assays, immunoassays & cell culture applications

ICN 154569   Human recombinant, expressed in *E. coli*   Lyophilized; ≥97%; ED_{50} = 1.0 ng/mL, by dose-dependent stimulation of ³H-thymidine uptake by Balb/c 3T3 cells; <0.1 ng endotoxin/μg protein | Contains an additional N-terminal Met as a result of the bacterial start codon; Rinderknecht, E etal, *PNAS*, 73:2365, 1976; Li, CH etal, *PNAS*, 80:2216, 1983; Peters, MA etal, *Gene*, 35:83, 1985; Ewton, DZ etal, *Endocrinol*, 120:115, 1987

PeproTech 100-11   MW 7.6k   Human recombinant, expressed in *E. coli*   >98% by SDS-PAGE & HPLC; <0.1 ng endotoxin per mg (1EU/mg); lyophilized | Polypeptide growth factor; stimulates proliferation of a wide range of cell types including muscle, bone & cartilage tissue; 70 AA; $ED_{50} \leq 1.0$ ng/mL; SA $\geq 10^6$ U/mg; SA determined by dose-dependent stimulation of thymidine uptake by BALB/c 3T3 cells

ICN 195757   Human recombinant, NSO-expressed   Soluble receptor; lyophilized; >95% | Binds IGF-I in solution but has little effect as an IGF-I antagonist

IBT MIGF1-10, MIGF1-50   MW 7600   Mouse recombinant, expressed in *E. coli*   >95 %; lyophilized; free of measurable endotoxins | 70 AA peptide; differs from human IGF-I at 4 AA & from rat at 1

PeproTech 250-19   MW 7.6k   Murine recombinant, expressed in *E. coli*   >98% by SDS-PAGE & HPLC; <0.1 ng endotoxin per mg (1EU/mg); lyophilized | Polypeptide growth factor; stimulates proliferation of a wide range of cell types including muscle, bone, & cartilage tissue; 70 AA; $ED_{50} \leq 1.0$ ng/mL; SA $\geq 1 \times 10^6$ U/mg; SA determined by the dose-dependent stimulation of thymidine uptake by BALB/c 3T3 cells

IBT WU020, WU100   MW 7719   *Oncorhynchuskisutch/mykiss* (salmon/trout) recombinant, expressed in *E. coli*   >95 %; lyophylized; free of measurable endotoxins | 70 AA peptide

IBT AFU020, AFU100   MW 7549   *Oreochromis mossambicus* (Tilapia) recombinant, expressed in *E. coli*   >95 %; lyophylized; free of measurable endotoxins | 68 AA peptide

IBT RU020, RU100, RM010   MW 7687   Rat recombinant, expressed in *E. coli*   >95 %; lyophylized; free of measurable endotoxins | Rat IGF-I is a 70 AA protein. Three of them differ from the AA sequence of human IGF-I.

ICN 198935   Rat recombinant, expressed in *E. coli*   Receptor grade; lyophilized; >95% | Stimulates growth in rat L6 myoblasts & other cell lines; Shimatzu, A & P Rotwein, *JBC*, 262:7894, 1987; Upton, Z etal, *J Endocrinol*, 149:379, 1996

ICN 198936   Salmon recombinant, expressed in *E. coli*   Receptor grade; lyophilized; >95% | 70 AA polypeptide that stimulates growth & differentiation in many cell types

IBT AEU020, AEU100   MW 7481   *Thunnus maccoyii* (Tuna) recombinant, expressed in *E. coli*   >95 %; lyophilized; free of measurable endotoxins | 68 AA peptide; stimulates growth & differentiation in many cell types; mediates many anabolic properties of growth hormone

ICN 198937   Trout recombinant, expressed in *E. coli*   Lyophilized; 95% | Shamblott, MJ & TT Chen, *PNAS*, 89:8913, 1992

Insulin-Like Growth Factor I Analog, Long R³

Synonyms: Insulin-Like Growth Factor I

Sigma I 1271   Human recombinant   >95% (N-terminal sequence analysis); lyophilized from 0.1 *M* acetic acid; endotoxin tested; cell culture tested | Recombinant human growth factors with altered AA chain lengths as compared to their native protein counterparts; as a result, these factors exhibit up to 10X more growth promoting activity when compared to their counterparts; they exhibit greater stability in culture & are ideal for serum-free or reduced-serum formulations

Insulin-Like Growth Factor I Analog, R³

Synonyms: Insulin-Like Growth Factor I

Sigma I 1146   Human recombinant   >85% (HPLC); lyophilized from 0.1 *M* acetic acid; endotoxin tested; cell culture tested | Recombinant human growth factors with altered AA chain lengths as compared to their native protein counterparts; as a result, these factors exhibit up to 10X more growth promoting activity when compared to their counterparts; they exhibit greater stability in culture & are ideal for serum-free or reduced-serum formulations

Insulin-Like Growth Factor I Antagonist

Synonyms: Insulin-Like Growth Factor I Antagonist JB1

IBT IGF1ant   Synthetic   >95 %; lyophilized | Inhibits autophosphorylation of the IGF-1 receptor in a concentration-dependent manner; inhibits cellular proliferation of a number of cell lines: human fibroblasts prostatic carcinoma cells of epithelial origin, SV40 transformed mouse 3T3 cells

Insulin-Like Growth Factor I Potentiating Peptide

IBT IGF1PP   Human synthetic   >97 %; lyophilized | Originally purified from human plasma; potentiates the activity of IGF-I in chick embryo cartilage

Insulin-Like Growth Factor I, (¹²⁵I)-

ICN 68128   Human recombinant,   >200 μCi/μg, >7.4 MBq/μg; 0.1 *M* KPB, pH 7.5, 0.5% BSA

Insulin-Like Growth Factor I, (3-(¹²⁵I)Tyr)-

Amersham IM172   Lyophilized; ~74 TBq/mmol, ~2000 Ci/mmol

Insulin-Like Growth Factor I, (Ala³¹)-

IBT AIU020, AIU100   MW 7507   Human recombinant, expressed in *E. coli*   >95 %; lyophilized; free of measurable endotoxins | 70 AA analog of human IGF-I; complete human IGF-I sequence; Ala substitution for Tyr at position 31; binding to acid-stripped human serum binding proteins similar to IGF-I; reduced binding to IGF-I & insulin receptors; binds these receptors slightly more strongly than (Leu²⁴)IGF-I or (Leu⁶⁰)IGF-I

Insulin-Like Growth Factor I, (Ala³¹)(Leu⁶⁰)-

IBT ALU020, ALU100   MW 7507   Human recombinant, expressed in *E. coli*   >95 %; lyophilized; free of measurable endotoxins | Analog of human IGF-I; Ala substitution for Tyr at position 31, Leu for Tyr at position 60; greatly reduced binding to IGF-I receptor, normal binding to acid-stripped human serum binding proteins; very low affinity for IGF-I receptor (lower than (Leu²⁴)- or (Leu⁶⁰)IGF-I)

Insulin-Like Growth Factor I, (Arg³)-

IBT Receptor Grade (EU020, EU100, EM001); Media Grade (LU100, LM001, LM010)   MW 7676   Human recombinant, expressed in *E. coli*   >95% (receptor grade) or >70% (media grade) purity; lyophilized; free of measurable endotoxins | 70 AA analog of IGF-I; Arg substitution for Glu at position 3; increased biological activity due to reduced binding to most binding proteins

Insulin-Like Growth Factor I, (Arg³)- Biotinylated

IBT BioR3IGF1-10, BioR3IGF1-20   Recombinant, expressed in *E. coli*   Lyophilized from TBS + stabilising proteins of non-mammalian origin; carrier protein free, biotinylated peptide available upon request | Useful for non-radioactive western-ligand blotting, binding studies & immunohistochemistry;. reduced binding to IGFBP

Insulin-Like Growth Factor I, (Leu^{24})-

IBT ZU020, ZU100   MW 7599   Human recombinant, expressed in *E. coli*   >95 %; lyophilized; free of measurable endotoxins | Leu substitution for Tyr in position 24; strongly reduced affinity for IGF-I receptor, reduced affinity for some IGFBP

Insulin-Like Growth Factor I, (Leu^{24})- Biotinylated

IBT BioLeuIGF1-10, BioLeuIGF1-20   Recombinant, expressed in *E. coli*   Lyophilized from TBS + stabilising proteins of non-mammalian origin; carrier protein free, biotinylated peptide available upon request | For western-ligand blottting, binding studies or immunohistochemistry

Insulin-Like Growth Factor I, (Leu^{24},Ala^{31})-

IBT ADU020, ADU100   MW 7507   Human recombinant, expressed in *E. coli*   >95 %; lyophilized; free of measurable endotoxins | Leu substitution for Tyr at position 24, Ala for Tyr at position 31; greatly reduced binding to IGF-I receptor (lower than (Leu^{24})- or (Leu^{60})IGF-I

Insulin-Like Growth Factor I, (Leu^{60})-

IBT ABU020, ABU100   MW 7599   Human recombinant, expressed in *E. coli*   >95 %; lyophilized; free of measurable endotoxins | Leu substitution for Tyr at position 60; greatly reduced binding to both IGF-I & insulin receptors; binding to acid-stripped human serum binding proteins is unaltered; lower affinity for these receptors than (Leu^{24})IGF-I

Insulin-Like Growth Factor I, Barramundi

IBT YU020, YU100   MW 7322   *Lates calcarifer* recombinant, expressed in *E. coli*   >95 %; lyophilized; free of measurable endotoxins | 66 AA peptide

Insulin-Like Growth Factor I, Biotinylated

IBT BIO-IGF1-10, BIO-IF1-20   Human recombinant, expressed in *E. coli*   Lyophilized from TBS + stabilising proteins of non-mammalian origin; carrier protein free, biotinylated peptide available upon request | For non-radioactive detection of IGFBP in western-ligand blotting; biotinylated IGF-I has low affinity to IGFBP-6, compared with IGF-II

IBT BioRATIGF1-10, BioRatIGF1-20   Recombinant, expressed in *E. coli*   Lyophilized from TBS + stabilising proteins of non-mammalian origin; carrier protein free, biotinylated peptide available upon request | For non-radioactive western-ligand blotting, binding studies or immunohistochemistry

Insulin-Like Growth Factor I, Bream

IBT AGU020, AGU100   MW 7379   *Pagrus aratus* recombinant, expressed in *E. coli*   >95 %; lyophilized; free of measurable endotoxins | 67 AA peptide; identical to (Gly^1) gilthead sea bream (*Sparus aurata*) IGF-I & (Gly1) black bream (*Acanthopagrus butcheri*) IGF-I (Gly replaces Ser at position 1)

Insulin-Like Growth Factor I, Chicken

IBT HU020, HU100, HM010   MW 7738   Human recombinant, expressed in *E. coli*   >95 %; lyophilized; free of measurable endotoxins | 70 AA peptide; differs from human IGF-I at 8 AA; stimulates growth & differentiation in most cell types

Insulin-Like Growth Factor I, Chicken Biotinylated

IBT BioCHKIGF1-10, BioCHKIGF1-20   Recombinant, expressed in *E. coli*   Lyophilized from TBS + stabilising proteins of non-mammalian origin; carrier protein free, biotinylated peptide available upon request | 70 AA peptide; differs from human IGF-I at 8 AA; stimulates growth & differentiation in most cell types; used for non-radioactive western-ligand blotting, binding studies or immunohistochemistry

Insulin-Like Growth Factor I, des(1-3)-

IBT DU020, DU100 DM001   MW 7365   Human recombinant, expressed in *E. coli*   >95 %; lyophilized; free of measurable endotoxins | Truncated form of IGF-I, lacking the terminal tripeptide; found in human brain, bovine colostrum & porcine uterus; increased biological activity due to reduced binding to most IGF- binding proteins

Insulin-Like Growth Factor I, des(1-3)- Biotinylated

IBT BioDIGF1-10, BioDIGF1-20   Recombinant, expressed in *E. coli*   Lyophilized from TBS + stabilising proteins of non-mammalian origin; carrier protein free, biotinylated peptide available upon request | Reduced binding affinity to IGFBP; useful for selective detection of IGF-I receptors on western blots, for binding studies or immunohistochemistry; used in flow cytometry; Xu et al, *Immunology*, 85:394, 1995

Insulin-Like Growth Factor I, des(2,3)(Ala^{31})-

IBT AJU020, AJU100   MW 7330   Human recombinant, expressed in *E. coli*   >95 %; lyophilized; free of measurable endotoxins | 68 AA analog of human IGF-I; lacks Pro-Glu at N-terminus positions 2-3; Ala substitution for Tyr at position 31; reduced binding to IGFI receptor; unaltered binding to acid-stripped human serum binding proteins (predominantly IGFBP-3)

Insulin-Like Growth Factor I, des(2,3)(Leu^{24})-

IBT AKU020, AKU100   MW 7373   Human recombinant, expressed in *E. coli*   >95 %; lyophilized; free of measurable endotoxins | 68 AA analog of human IGF-I; lacks Pro-Glu at N-terminus positions 2-3; Leu substitution for Tyr at position 24; similar affinity both for binding to IGF-I receptor & acid-stripped human serum binding proteins

Insulin-Like Growth Factor I, Long (Arg^3)-

IBT Receptor Grade (BU020, BU100, BM001); Media Grade (AU100, AM001, AM010)   MW 9111.6   Human recombinant, expressed in *E. coli*   >95% (receptor grade) or >70% (media grade) purity; lyophylized; free of measurable endotoxins | 83 AA analog of IGF-I; Arg substitution for Glu at position 3 & 13 AA extension at the N-terminus; significantly more potent than IGF-I *in vitro* & *in vivo*

Insulin-Like Growth Factor I, Long (Arg^3)- Biotinylated

IBT BioLR3IGF1-10, BioLR3IGF1-20   Recombinant, expressed in *E. coli*   Lyophilized from TBS + stabilising proteins of non-mammalian origin; carrier protein free, biotinylated peptide available upon request | For non-radioactive western-ligand blotting, binding studies or immunohistochemistry; IGF-I analog with the lowest affinity in western-ligand blotting

Insulin-Like Growth Factor I, Long R₃

ICN 198780   Media grade; MW 9110; >70%; lyophilized | Potent analog of insulin-like growth factor; Francis, etal, *J Mol Endocrinol*, 213, 1992; Thomas etal, *J Endocrinol*, 137:413, 1993

Insulin-Like Growth Factor I, Mouse Biotinylated

IBT BioMURIGF1-10, BioMURIGF2-20   Recombinant, expressed in *E. coli*   Lyophilized from TBS + stabilising proteins of non-mammalian origin; carrier protein free, biotinylated peptide available upon request | For non-radioactive western-ligand blotting, binding studies or immunohistochemistry

Insulin-Like Growth Factor II

Synonyms: Insulin-Like Growth Factor II, rh-; Somatomedin A

ICN 198938   Chicken recombinant, expressed in *E. coli* Receptor grade; lyophilized; >95% | Contains 68 AA sequence, twelve of which differ from the sequence in human IGF-II; Upton, Z etal, *J Mol Endocrinol*, 14:17, 1995

Calbiochem 407245 MW 7.5k Human recombinant, expressed in *E. coli* >97% (SDS-PAGE); lyophilized from sterile-filtered acetic acid; biological activity: ED$_{50}$=5-10 ng/mL as measured in a cell proliferation assay using the bovine kidney cell line MCF-7; endotoxin: ≤100 pg/µg IGF-II; may be carcinogenic/teratogenic | Its actions are similar to IGF-I; plays an important role in fetal development; stimulates muscle & bone cell proliferation & differentiation; *Merck Index*, 12: 8862; Hill, PA et al, *Endocrinology*, 136: 124, 1995; Nielson, FC et al, *Nature*, 377: 358, 1995

IBT Receptor Grade (FU020, FU100, FM1000); Media Grade (OU100, OM001) MW 7469 Human recombinant, expressed in *E. coli* >95% (receptor grade) or >70% (media grade) purity; lyophilized; free of measurable endotoxins | 67 AA peptide; stimulates growth & differentiation in many cell types; elevated levels have been associated with different types of cancer

ICN 152279 Human recombinant, expressed in *E. coli* Lyophilized; >95%; unspecified activity, but has demonstrated the ability to stimulate human epidermal carcinoma A431 cell proliferation at a concentration of 1.0 ng/mL | Potent mitogen for a variety of cell types; appears to be an important regulator of somatic cell growth; target cells include breast tumor, glioma, hepatocytes, chondrocytes & adrenal cells; Smith, MC etal, *JBC*, 264:9314, 1989; Czech, MP etal, *Cell*, 59:235, 1989

ICN 153504 Human recombinant, expressed in *E. coli* Lyophilized; ≥96%; carrier-free; unspecified activity, but measured in a receptor binding assay using human placental membrane | 67 AA peptide sequence with 60% homology with IGF-I; a potent mitogen against a wide range of cell types & is thought to be an important regulator of somatic cell growth

ICN 154568 Human recombinant, expressed in *E. coli* Lyophilized; ≥97%; ED$_{50}$ typically 8.0-15.0 ng/mL | Scott, CD & RC Baxter, *Endocrinol*, 120:1, 1987; Ewton, DZ etal, *Endocrinol*, 120:115, 1987; Baxter, RC, *Adv Clin Chem*, 25:49, 1986; Bell, GI etal, *Nature*, 310:775, 1984

PeproTech 100-12 MW 7.5k Human recombinant, expressed in *E. coli* >98% by SDS-PAGE & HPLC; <0.1 ng endotoxin per mg (1EU/mg); lyophilized | Polypeptide growth factor; stimulates proliferation of a wide range of cell types; 67 AA; ED$_{50}$ ≤ 1.0 ng/mL; SA ≥ 10^6 U/mg; SA determined by competitive binding against plasma-derived IGF-II to human placental membranes

ICN 198939 Rat Receptor grade; lyophilized; >95% | Binds to Type 2 receptors on rat L-myoblasts with similar potency as hIGF-II; Dull, TJ etal, *Nature*, 310:777, 1986

Insulin-Like Growth Factor II (33-49), Terminal

Biogenesis 5345-3504 Synthetic >95%; lyophilized

Insulin-Like Growth Factor II Analog (1-6), des-

Sigma I 1521 Human recombinant >95% (HPLC); lyophilized from 0.1 *M* acetic acid; endotoxin tested; cell culture tested | Recombinant human growth factors with altered AA chain lengths as compared to their native protein counterparts; as a result, these factors exhibit up to 10X more growth promoting activity when compared to their counterparts; they exhibit greater stability in culture & are ideal for serum-free or reduced-serum formulations

Insulin-Like Growth Factor II, (^{125}I)-

ICN 68129 Human recombinant, >200 µCi/µg, >7.4 MBq/µg; 0.1 *M* KPB, pH 7.5, 0.5% BSA

Insulin-Like Growth Factor II, (3-(^{125}I)Tyr)-

Amersham IM238 Human recombinant Lyophilized; ~74 TBq/mmol, ~2000 Ci/mmol

Insulin-Like Growth Factor II, (Arg^6)-

IBT GU020, GU100, GM001 MW 7496 Human recombinant, expressed in *E. coli* >95%; lyophilized; free of measurable endotoxins | Arg substitution for Glu at position 6; reduced binding affinity to IGFBP; greater affinity to IGF-II receptor than IGF-II or des(1-6) IGF-II

Insulin-Like Growth Factor II, (Arg^6)- Biotinylated

IBT BioR6IGF2-10, BioR6IGF2-20 Recombinant, expressed in *E. coli* Lyophilized from TBS + stabilizing proteins of non-mammalian origin; carrier protein free, biotinylated peptide available upon request | For non-radioactive western-ligand blotting, binding studies or immunohistochemistry

Insulin-Like Growth Factor II, (Gly^1)-

IBT QU100, QM001, QM005 MW 7455 Human recombinant, expressed in *E. coli* >80% (media grade); lyophilized; free of measurable endotoxins | Similar binding affinity to IGF-II receptor; similar potency in stimulation of protein synthesis in rat L6 myoblasts

Insulin-Like Growth Factor II, (Leu^{27})-

IBT TU020, TU100 MW 7420 Human recombinant, expressed in *E. coli* >95%; lyophilized; free of measurable endotoxins | Leu substitution for Tyr at position 27;2X-reduced affinity to IGF-2 receptor, 1000X- reduced affinity to IGF-I receptor

Insulin-Like Growth Factor II, (Leu^{27})- Biotinylated

IBT BioLeuIGF2-10, BioLeuIGF2-20 Recombinant, expressed in *E. coli* Lyophilized from TBS + stabilizing proteins of non-mammalian origin; carrier protein free, biotinylated peptide available upon request | Reduced binding to some IGFBP been detected in western ligand blot; for non-radioactive western-ligand blotting, binding studies or immunohistochemistry

Insulin-Like Growth Factor II, Biotinylated

IBT BioIGF2-10, BioIGF2-20 Recombinant, expressed in *E. coli* Lyophilized from TBS + stabilizing proteins of non-mammalian origin; carrier protein free, biotinylated peptide available upon request | For the detection of IGFBP in non-radioactive western-ligand blotting, immunohistochemistry or cross-linking experiments; also tested in competitive format ELISA

Insulin-Like Growth Factor II, Chicken

IBT SU020, SU100 MW 7512 Human recombinant, expressed in *E. coli* >95%; lyophilized; free of measurable endotoxins | 9 AA difference from human IGF-II

Insulin-Like Growth Factor II, Chicken Biotinylated

IBT BioCHKIGF2-10, BioCHKIGF2-20 Recombinant, expressed in *E. coli* Lyophilized from TBS + stabilizing proteins of non-mammalian origin; carrier protein free, biotinylated peptide available upon request | For non-radioactive western-ligand blotting, binding studies or immunohistochemistry

Insulin-Like Growth Factor II, des(1-6)-

IBT MU020, MU100, MU001 MW 6765 Human recombinant, expressed in *E. coli* >95%; lyophilized; free of measurable endotoxins | Truncated IGF-II; lacks first 6 N-terminal AA; reduced binding affinity to IGFBP

Insulin-Like Growth Factor II, des(1-6)- Biotinylated

IBT BioDIGF2-10, BioDIGF2-20 Recombinant, expressed in *E. coli* Lyophilized from TBS + stabilizing proteins of non-mammalian origin; carrier protein free, biotinylated peptide available upon request | Strongly reduced binding affinity to IGFBP; useful for detection of IGF-II receptors on western blots, in binding studies or immunohistochemistry

Insulin-Like Growth Factor II, Mouse

IBT MIGF2-50 MW 7400 Human recombinant, expressed in *E. coli* >95%; lyophilized; free of measurable endotoxins | 6 AA difference from human IGF-I; 2 AA difference from rat IGF-II

Insulin-Like Growth Factor II, Mouse Biotinylated

IBT BioMURIGF2-10, BioMURIGF2-20   Recombinant, expressed in *E. coli*   Lyophilized from TBS + stabilizing proteins of non-mammalian origin; carrier protein free, biotinylated peptide available upon request | For non-radioactive western-ligand blotting, binding studies or immunohistochemistry

Insulin-Like Growth Factor II, Rat

IBT AAU020, AAU100   MW 7515   Human recombinant, expressed in *E. coli*   >95%; lyophilized; free of measurable endotoxins | 4 AA difference from human IGF-II; 2 AA difference from murine IGF-II

Insulin-Like Growth Factor II, Rat Biotinylated

IBT BioRATIGF2-10, BioRATIGF2-20   Recombinant, expressed in *E. coli*   Lyophilized from TBS + stabilizing proteins of non-mammalian origin; carrier protein free, biotinylated peptide available upon request | For non-radioactive western-ligand blotting, binding studies or immunohistochemistry

Interferon

ICN 153516   Rat   Lyophilized in glycine buffer, pH 3.5; 2.7×10^4 U/mg, 2.4×10^4 U/mL

Interferon, α-

ICN 153513   Human lymphoblastoids   Lyophilized in TNE buffer, pH 7.2; 1.0×10^5 IU/mg, 1.2×10^5 IU/mL

ICN 191380   Human recombinant, expressed in *E. coli*   Ultra pure grade; ≥95%; frozen liquid; 2×10^8 U/mg, 5 µg/50 µL; <0.1 ng endotoxin/µg

ICN 153514   Mouse cells   Lyophilized in glycine buffer, pH 3.5; 4.0×10^5 IU/mg, 1.2×10^5 IU/mL

ICN 153512   Human fibroblasts   Lyophilized in PBS buffer, pH 7.4; 1.5×10^5 IU/mg, 1.2×10^5 IU/mL

ICN 195003   MW 18k   Human recombinant   Lyophilized

ICN 153515   Mouse cells   Lyophilized in glycine buffer, pH 3.5; 1.2×10^8 IU/mg, 6.8×10^4 IU/mL

ICN 198767   Murine recombinant, expressed in silkworm larvae   Lyophilized from sterile filtered solution in PBS, 1 mg/mL BSA

ICN 191381   Human recombinant, expressed in *E. coli*   Highly purified; lyophilized; >98%; 1×10^6 U/mg, by cytopathic effect via inhibition assay; <0.1 ng endotoxin/µg protein

ICN 195769   Murine recombinant, expressed in *E. coli*   Lyophilized; ≥97%; ED_{50} = 0.1-0.4 ng/mL

ICN 195787   Rat recombinant, expressed in *E. coli*   Lyophilized; ≥97%; ED_{50} = 1.0-3.0 ng/mL

Interferon-γ Inducible Protein 10

ICN 195791   Human recombinant, expressed in *E. coli*   ≥97%; lyophilized | C-X-C Chemokine

Interleukin I

ICN 150173   Human monocytes 8×10^6 U/mg protein; endotoxin, IL-2 & INF free; 1 U doubles the proliferative response of mouse thymocytes stimulated with 1 mg/mL of phytohemagglutinin alone; 100 U/mL | Purified from glass adhered monocytes stimulated with heat-killed *S. albus*;

Interleukin II

ICN 152371   Human natural   Tissue culture grade, ultra pure; ready-to-use, sterile, frozen solution in 25 *mM* HEPES buffered RPMI-1640 medium; lectin- & γ-interferon-free; 32,000 U/50 mL vial

ICN 151342   Human recombinant, expressed in *E. coli*   ≥97%; lyophilized; $≥4 \times 10^6$ BRMP U/mg; 1 U proliferates half maximal CTLL-2 cells; 1 U ≅ 1 Biological Response Modifiers Program (BRMP) U | May be used in both mouse & human systems

ICN 153896   Human recombinant, expressed in *E. coli*   Ultra pure grade; ≥98%; lyophilized; $≥5 \times 10^6$ U/mg, via ^{3}H-thymidine uptake by CTLL cells; <0.15 ng endotoxin/µg; mycoplasma free; 2 µg (= 10KU)/vial | Produced & purified under proprietary technology; functional similarities to native hIL-2; ideal for tissue culture applications

ICN 154144   Human recombinant, expressed in *E. coli*   >97%; lyophilized; 4×10^7 U/mg, via dose-dependent stimulation of ^{3}H-thymidine incorporation into CTLL-2 (a murine cytotoxic cell line); <0.1 ng endotoxin/µg protein | Conradt, HS etal, *Carbohydrate Res*, 149:443, 1986; Taniguchi, etal, *Nature*, 302:305, 1983

ICN 154566   Human recombinant, expressed in *E. coli*   ≥97%; lyophilized; $≥4 \times 10^6$ BRMP U/mg; 1 U proliferates half maximal CTLL-2 cells; 1 U ≅ 1 Biological Response Modifiers Program (BRMP) U

ICN 151336   Murine lymphocytes   Sterile filtered from lectin-stimulated cell cultures; supernatant; Gillis assay method | Gillis, etal, *J Immunol*, 120:2027, 1978

ICN 151337   Murine lymphocytes   Sterile filtered from lectin-stimulated lymphocyte supernatant; concentrate; sterile filtered & lectin free

ICN 151338   Murine lymphocytes   Highly purified, sterile; further purification of ICN 151337 | Higher titer product offers optimal stimulation characteristics for mechanistic comparison

ICN 195758   Murine recombinant, expressed in *E. coli*   ≥95%; lyophilized; ED_{50} = 0.1-0.4 ng/mL

ICN 151343   Murine recombinant, expressed in yeast   ≥95%; lyophilized; $≥1.6 \times 10^6$ BRMP U/mg; 1 U proliferates half maximal CTLL-2 cells; 1 U BRMP ≅ 1 KU | Active only in mouse systems, not in human

ICN 151339   Rat lymphocytes   Sterile filtered from lectin-stimulated cell cultures; supernatant; Gillis assay method | IL-2; Gillis, etal, *J Immunol*, 120:2027, 1978

ICN 151340   Rat lymphocytes   Sterile filtered from lectin-stimulated lymphocyte supernatant; lectin free | IL-2

ICN 151341   Rat lymphocytes   Highly purified from concentrated, lectin-free IL-2; sterile | IL-2; high titer product offers optimal stimulation characteristics for mechanistic comparison

ICN 160023   Rat recombinant, expressed in CHO cells   Frozen liquid; unspecified activity; 1 U proliferates half maximal CTLL-2 cells | Active in both rat & mouse systems—others not yet tested

ICN 195775   Rat recombinant, expressed in *E. coli*   ≥97%; lyophilized; ED_{50} = 0.1-0.4 ng/mL

ICN 1600254   Rat splenocytes   Lyophilized; unspecified activity, via ability to stimulate mouse CTLL cell proliferation | Active only in rat systems

Interleukin II Soluble Receptor α

ICN 195721   Human recombinant, NSO-expressed   >97%; lyophilized; ~0.5-1.0 µg/mL inhibits 50% of the biological response of 30 ng/mL of IL-2

ICN 195722   Human recombinant, NSO-expressed   >97%; lyophilized; ~1.0-3.0 µg/mL inhibits 50% of the biological response of 4 ng/mL of IL-15 | IL-2sRα

ICN 195756   Human recombinant, NSO-expressed   >97%; lyophilized; ED_{50} = 3-6 µg/mL

Interleukin III

ICN 154143   Human recombinant, expressed in *E. coli*   ≥98%; lyophilized; ED_{50} = 0.4 ng/mL; 1 U incorporates half maximal ^{3}H-thymidine by human M-07E cells

ICN 160227   Murine recombinant, expressed in *E. coli*   ≥95%; lyophilized; $≥10^7$ U/mg; 1 U induces half maximal FDC-P1 cell proliferation | Active in mouse systems, but is unreactive in human systems; other systems not yet tested; supports the formation of mixed colonies (granulocyte, macrophage, megakaryocyte, mast cell, erythrocyte) from mouse bone marrow cells in a soft agar colony assay

ICN 193512   Murine recombinant, expressed in yeast   ≥80%; frozen liquid; ≥10^6 U/mg; 1 U induces half maximal FDC-P2 cell proliferation | Active in mouse & rat systems; supports the formation of mixed colonies (granulocyte, macrophage, megakaryocyte, mast cell, erythrocyte) from mouse bone marrow cells in a soft agar colony assay

Interleukin III, (^{125}I)-

ICN 68114H   Human recombinant   ~100 µCi/µg, ~3.7 MBq/µg; 0.1 M KPB, pH 7.5, 0.5% BSA

ICN 68114   Murine recombinant   ~100 µCi/µg, ~3.7 MBq/µg; 0.1 M KPB, pH 7.5, 0.5% BSA

Interleukin IV

ICN 152313   Human recombinant, expressed in *E. coli*   ≥98%; lyophilized; ≥10^7 U/mg; 1 U induces half maximal CTLL cell proliferation in a hu IL-4R1.d assay | Tested only in human systems; Grabstein, K etal, *J Exp Med*, 163:1405, 1986

ICN 154142   Human recombinant, expressed in *E. coli*   ≥98%; ED_{50} = 0.2 ng/mL, measured by the stimulation of h3-thymidine incorporation by human peripheral blood T-lymphocytes

ICN 160028   Murine recombinant, expressed in *E. coli*   ≥98%; lyophilized, carrier-free; ≥10^7 U/mg; 1 U induces half maximal CT-4S cell proliferation; <0.1 ng endotoxin/mg IL-4

ICN 160029   Rat recombinant, expressed in CHO cells   Non-purified; frozen liquid; unspecified activity | No activity in mouse or human systems; bioassay show the induction of MHC Class II Ag expression on rat splenic B-cells

ICN 195776   Rat recombinant, expressed in *E. coli*   ≥97%; lyophilized; ED_{50} = 0.7-1.5 ng/mL

Interleukin IV Soluble Receptor

ICN 160030   Human recombinant, expressed in insect cell line S*f*21   ≥97%; lyophilized with BSA; ED_{50} = 5.0=10.0 ng/mL, measured by the inhibition of 50% TF-1 cell proliferation

Interleukin IV Soluble Receptor

ICN 160031   Murine recombinant, expressed in Hela cells ≥90%; frozen liquid; unspecified activity | Neutralizes the ability of IL-4 to induce CTLL-2 cell & mouse B-cell proliferation & immunoglobulins secretion; specific for mouse IL-4; not active against human IL-4

Interleukin IX

ICN 158851   Human recombinant, expressed in *E. coli*   ≥98%; lyophilized; ≥5x10^6 U/mg, 1 U induces half maximal ^{3}H-thymidine uptake in MC-9 cells co-stimulated with murine IL-4; <0.1 ng/µg protein

ICN 158850   Murine recombinant, expressed in *E. coli*   ≥98%; lyophilized; ≥10^7 U/mg, 1 U induces half maximal TS1.C3 cell proliferation | Activity in systems than mouse has not yet been tested

Interleukin IX Soluble Receptor

ICN 195746   Human recombinant, expressed in S*f*21   ≥90%; lyophilized | Binds IL-9 with low affinity & demonstrates no antagonistic activity

Interleukin I Receptor Antagonist

ICN 195744   Human recombinant, expressed in *E. coli*   ≥95%; lyophilized; 7.0-10 ng/mL in the presence of IL-1α

ICN 195768   Murine recombinant, expressed in *E. coli*   ≥97%; lyophilized; 15-30 ng/mL in the presence of 50 pg/mL IL-1α

Interleukin I Soluble Receptor I

ICN 195733   Human recombinant, NSO-expressed   ≥97%; lyophilized; 0.5-1.0 mg/mL will inhibit half the biological response of 50 pg/mL IL-1β

Interleukin I Soluble Receptor II

ICN 195736   Human recombinant, NSO-expressed   ≥97%; lyophilized; 1-5 mg/mL will inhibit half the biological response of 50 pg/mL IL-1β

Interleukin Iα

ICN 151333   Human recombinant, expressed in *E. coli*   ≥95%; frozen liquid with BSA; ≥10^8 U/mg; 1 U inhibits 50% growth of human A375 malignant melanoma indicator cells; suggested concentration = 50-400 pg/mL; dilute with PBS or culture medium in 0.1-10% carrier protein (FBS, BSA, etc) or activity may be lost

ICN 154146   Human recombinant, expressed in *E. coli* >97%; lyophilized with BSA; ED_{50} = 3.0-10 pg/mL, via a murine helper T-cell line (D10.G4.1) cell proliferation/^{3}H-thymidine uptake assay; <0.01 ng endotoxin/µg protein

ICN 160009   Murine recombinant, expressed in *E. coli*   ≥98%; sterile filtered lyophilized powder with HAS carrier; ≥10^7 U/mg protein; 1 U induces half maximal D10 cell proliferation

ICN 160021   Murine recombinant, expressed in *E. coli*   ≥95%; frozen liquid with BSA; ≥10^7 U/mg; 1 U induces half maximal D10 cell proliferation; suggested concentration = 5-500 pg/mL; dilute with PBS or culture medium in 0.1-10% carrier protein (FBS, BSA, etc) or activity may be lost; ≥50,000 U/5 µg vial

Interleukin Iα, (3-(^{125}I)Tyr)-

Amersham IM205   Human recombinant   Lyophilized; ~74 TBq/mmol, ~2000 Ci/mmol

Interleukin Iβ

ICN 160010   Murine recombinant, expressed in *E. coli*   ≥95%; frozen liquid; ≥10^7 U/mg protein; 1 U induces half maximal D10 cell proliferation | Reactive in mouse & rat systems

ICN 195774   Rat recombinant, expressed in *E. coli*   ≥97%; lyophilized; ED_{50} = 1-3 ng/mL

Interleukin Iβ Peptide (197-215)

Oncogene PP19   Human   Lyophilized with an equal amount of BSA; suggested starting concentration of peptide for competition is 10X the concentration of primary antibody | Useful for competition studies & dot blot

Interleukin Iβ, (^{125}I)-

Amersham IM222   Human recombinant   Labeled with Bolton & Hunter Reagent; lyophilized; ~56-93 TBq/mmol, 1500-2500 Ci/mmol; >22.2 TBq/mmol, >600 Ci/mmol

Interleukin V

ICN 152402   MW 25k   Human recombinant, expressed in *E. coli*   ≥98%; lyophilized carrier-free; ED_{50} ≤ 0.15 ng/mL, measured by dose-dependent TF-1 cell proliferation | 228 AA residues

ICN 158349   Murine recombinant, expressed in COS cells Frozen liquid; unspecified activity; 1 U stimulates half maximal dextran sulfate mouse splenic B-cell proliferation; unpurified in conditioned media from COS cells transfected with IL-5 cDNA | Active in mouse systems & on human eosinophils; not yet tested in other systems

ICN 195759   Murine recombinant, S*f*21 expressed   ≥97%; lyophilized; ED_{50} = 0.04-0.15 ng/mL | IL-5

Interleukin V Soluble Receptor α

ICN 195728   Human recombinant, S*f*21 expressed   ≥97%; lyophilized; 200-300 ng/mL inhibits half of the biological response of 0.5 ng/mL Il-5

Interleukin V, (^{125}I)-

ICN 68115   ~1.5 µCi/µg, ~37-180kBq/µg; 0.1 M KPB, pH 7.5

Amersham IM265   Human recombinant   Lyophilized; 18.5-52 TBq/mmol, 500-1400 Ci/mmol

Interleukin VI

ICN 152353   Activated human peripheral blood mononuclear cells >99%; PBS buffer, 0.1% HAS, pH 7.4; ~4×10^7 U/mg; 1 U induces half maximal proliferation of an indicator lymphoblastoids cell line obtained by EBV-transformation of normal peripheral blood B-cells; endotoxin- TNF- & IL-1-free | Naturally glycosylated by mammalian cells; Tosato, G et al, *Science*, 239:502, 1988; Van Damme, J et al, *J Immunol*, 140:1534, 1988

ICN 154141   Human recombinant, expressed in *E. coli* >98%; lyophilized; ED_{50} = 0.1 ng/mL, by dose-dependent stimulation of ^{3}H-thymidine uptake by B9 cells; <0.1 ng endotoxin/μg protein

ICN 154565   Human recombinant, expressed in *E. coli* ≥98%; lyophilized; ED_{50} = 0.2 ng/mL, by stimulation of ^{3}H-thymidine uptake by B9 cells; <0.1 ng endotoxin/μg protein | Van Snick et al, *PNAS*, 83:9676, 1986

ICN 152352   Human recombinant, expressed in yeast ~80%; frozen liquid; ≥10^8 U/mg; 1 U induces half maximal proliferation of B9 hybridoma cells | Active in human, canine, rat, chicken & bovine systems; not yet tested in other systems

ICN 160033   Murine recombinant, baculovirus-derived ≥95%; frozen liquid; ~5×10^7 U/mg, 1 U induces half maximal T1165 hybridoma cell proliferation | Active in mouse but not human systems; not yet tested in other systems

ICN 160032   Murine recombinant, expressed in *E. coli* ≥98%; lyophilized; 10^7 U/mg, 1 U induces half maximal ^{3}H-thymidine uptake by B9 cells; <0.1 ng endotoxin/μg protein

ICN 195802   Rat recombinant, expressed in *E. coli* ≥95%; lyophilized; ED_{50} < 0.02 ng/mL

Interleukin VI Soluble Receptor

ICN 160034   Human recombinant, expressed in S*f*21 ≥97%; lyophilized; ED_{50} = 5.0-8.0 ng/mL, 1 U inhibits half maximal murine M1 myeloid leukemic cell proliferation | Saito, T et al, *J Immunol*, 147:168, 1991

Interleukin VI, (^{125}I)-

ICN 68116   Human recombinant 300-400 μCi/μg, 11.1-14.8 MBq/μg; 0.1 *M* KPB, pH 7.5, 0.5% BSA

Interleukin VII

ICN 153473   Human recombinant, expressed in *E. coli* ≥98%; lyophilized; ≥3×10^6 U/mg, 1 U induces half maximal Con A-stimulated mouse thymocytes proliferation | Full activity on human & murine cells; not yet tested in other systems

ICN 153472   Murine recombinant, expressed in *E. coli* ≥98%; lyophilized; ≥3×10^6 U/mg, 1 U induces half maximal Con A-stimulated mouse thymocytes proliferation | Active in mouse systems; not yet tested in other systems

Interleukin VIII

ICN 153478   Human endothelial recombinant, expressed in *E. coli* ≥98%; lyophilized; maximal chemotactic activity = 25.0 ng/mL, by modified Boyden chamber; <0.1 ng/μg protein | 77 AA

ICN 160240   Human monocyte recombinant, expressed in *E. coli* ≥98%; lyophilized; maximal chemotactic activity = 25.0 ng/mL, by modified Boyden chamber; <0.1 ng/μg protein | 72 AA; higher potency in chemotaxis & degranulation than endothelial IL-8

Interleukin VIII, (^{125}I)-

Synonyms: Interleukin VIII Receptor Ligand, (^{125}I)-

ICN 68117   50-100 μCi/μg, 1.85-3.7 MBq/μg; 0.1 *M* KPB, pH 7.5, 0.5% BSA

Amersham IM249   Human recombinant Lyophilized; ~74 TBq/mmol, ~2000 Ci/mmol | Chemokine receptor ligand; CXCR1,CSCR2 selectivity; endogenous; K_d = 3.6 nM pRK5B.iL8r1.1/COS-7 cells

Interleukin X

Synonyms: Cytokine Synthesis Inhibitory Factor

ICN 158700   Human recombinant, expressed in *E. coli* ≥98%; lyophilized; ED_{50} = 2.0 ng/mL, by ^{3}H-thymidine incorporation by MC-9 cells; <0.1 endotoxin ng/μg protein | Formerly Cytokine Synthesis Inhibitory Factor (CSIF); fully active on mast cells as a growth co-stimulator along with IL-4

ICN 158852   Murine recombinant, expressed in *E. coli* ≥98%; lyophilized; ≥5.0×10^5 U/mg, by ^{3}H-thymidine incorporation by MC-9 cells; <0.1 endotoxin ng/μg protein | Not tested in species other than mouse

Interleukin X Soluble Receptor

ICN 195734   Human recombinant, expressed in S*f*21 ≥97%; lyophilized; 0.1-0.3 μg/mL inhibits half the biological response to 2 ng/mL IL-10

ICN 195766   Murine recombinant, expressed in S*f*21 ≥97%; lyophilized; 1-3 μg/mL inhibits half the biological response to 1 ng/mL IL-10

Interleukin XI

ICN 158853   Human recombinant, expressed in *E. coli* ≥98%; lyophilized; ≥2.0×10^6 U/mg, 1 U induces half maximal T1165 murine hybridoma cell proliferation | Active in mouse & human systems; other systems not yet tested

ICN 195762   Murine recombinant, expressed in *E. coli* ≥97%; lyophilized; ED_{50} = 0.05-0.15 ng/mL

Interleukin XII

Synonyms: Natural Killer Cell Stimulatory Factor; Cytotoxic Lymphocyte Maturation Factor

ICN 160042   Human recombinant, expressed in S*f*21 >95%; lyophilized; ED_{50} = 0.05-0.2 ng/mL, 1 U induces half maximal PHA-activated human lymphoblast proliferation

ICN 195763   Murine recombinant, expressed in S*f*21 ≥97%; lyophilized; ED_{50} = 0.05-0.2 ng/mL

Interleukin XII p40

ICN 195790   Human recombinant, expressed in S*f*21 ≥97%; lyophilized | Homodimer; no antagonistic activity to IL-12; typically used as an ELISA calibrator

ICN 195789   Murine recombinant, expressed in S*f*21 ≥97%; lyophilized; ED_{50} = 1-3 ng/mL in the presence of 0.3 ng/mL IL-12 | Homodimer

Interleukin XIII

ICN 158854   Human recombinant, expressed in *E. coli* ≥98%; lyophilized; ED_{50} = 1.0 ng/mL, 1 U stimulates dose-dependent TF-1 cell proliferation | Immunoregulatory protein produced by activated T-lymphocytes; 58% homologous with murine IL-13; stimulates B-cell proliferation & immunoglobulin synthesis

ICN 160243   Murine recombinant, expressed in *E. coli* ≥97%; lyophilized with BSA; ED_{50} = 3.0-6.0 ng/mL, 1 U stimulates dose-dependent TF-1 cell proliferation | Immunoregulatory protein produced by activated T-lymphocytes; stimulates B-cell proliferation & immunoglobulin synthesis

Interleukin XV

ICN 160044   Human recombinant, expressed in *E. coli* ≥95%; lyophilized; ≥10^7 U/mg, 1 U stimulates half maximal CTLL-2 cell proliferation | Activity on systems other than human has not been tested

Interleukin XVI

ICN 195005   Human recombinant, expressed in *E. coli* MW 13.5k; ≥95%; lyophilized | Chemoattractant for CD4$^+$T cells, monocytes & eosinophils

Interleukin XVII

ICN 195753   Human recombinant, expressed in *E. coli*   ≥97%; lyophilized; ED$_{50}$ = 10-30 ng/mL

Interleukin XVIII

ICN 195798   Human recombinant, expressed in *E. coli*   ≥97%; lyophilized; ED$_{50}$ = 15 ng/mL

IR (1142-1153)

BioSource International 77-254

BioSource International 77-262

IR (1142-1153), (PY^{1146})-

BioSource International 77-256

IR (1142-1153), (PY^{1150})-

BioSource International 77-258

IR (1142-1153), (PY^{1151})-

BioSource International 77-260

IRS I (Y608) Peptide

BioSource International 77-390

IRS pS616

BioSource International 04-550

JAK1 Peptide

BioSource International 04-400

JAK2 Peptide

BioSource International 04-406

JAK3 Peptide

BioSource International 04-412

JE

Synonyms: Macrophage/Monocyte Chemoattractant Protein I Homolog

ICN 195767   Murine recombinant, expressed in *E. coli*   ≥97%; lyophilized; ED$_{50}$ = 5-20 ng/mL | C-C chemokine

jun B

Amersham VPP36   Competition for VPC28; viral & oncogene recognition

jun B Peptide (45-61)

Oncogene PP36   Human   Lyophilized with an equal amount of BSA; suggested starting concentration of peptide for competition is 10X the concentration of primary antibody | Region: AA of jun B; for competition studies & dot blot

Katacalcin

Biogenesis 5545-0202   Synthetic   >95%; lyophilized

KC

ICN 195765   Murine recombinant, expressed in *E. coli*   ≥97%; lyophilized; ED$_{50}$ = 0.3-0.9 ng/mL | C-X-C Chemokine

Kemptamide PKS I

BioSource International 77-327

Kemptide PKS I

BioSource International 77-153

Kemptide PKS II

BioSource International 77-155

Keratinocyte Growth Factor

Synonyms: Keratinocyte Growth Factor, rh-; Fibroblast Growth Factor 7, rh-

Calbiochem 422425   MW 19k   Human recombinant, expressed in *E. coli*   ≥97% (SDS-PAGE); lyophilized from sterile-filtered PBS containing 50 µg BSA/µg KGF; biological activity: ED$_{50}$=15-25 ng/mL as measured by its ability to stimulate the proliferation of a monkey epithelial cell line 4MBr-5; endotoxin: ≤100 pg/µg KGF; may be carcinogenic/teratogenic | Epithelial cell specific mitogen; KGF binding to its receptor is blocked by Suramin (Calbiochem No. 574625); thought to mediate the hypothesized mesenchyme-driven proliferation of normal epithelial cells; Finch, PW et al, *Gastroenterology*, 110: 441, 1996; Marchese, C et al, *J Exp Med*, 182: 1369, 1995; Siddiqi, I et al, *Biochem Biophys Res Comm*, 215: 309, 1995; Ron, D et al, *J Biol Chem*, 268: 2984, 1993; Werner, S et al, *PNAS*, 89: 6896, 1992; Finch, PW et al, *Science*, 245: 752, 1989

ICN 158418   Human recombinant, expressed in *E. coli*   ≥98%; lyophilized; ED$_{50}$ = 10.0-100.0 ng/mL, 1 U stimulates & maintains humankeratinocytes in chemically defined medium with IGF-I | A potent mitogen ofkeratinocytes & epithelial cells

Lactoglobulin A, β-

Sigma L 4520   Bovine milk   Lyophilized; Vial contains ~25 nmoles | Use-tested; protein sequencing standard

Laminin

ICN 150027   Basement membrane of Engelbreth Holm-Swarm transplantable mouse tumor   Purified; 1-2 mg/mL in 0.05 *M* TRIS, 0.15 *M* NaCl, pH 7.4 | A natural cellular attachment factor which significantly facilitates the culture of epithelial & endothelial cells; Ledbetter, SR etal, Isolation of Laminin, in *Methods for Preparation of media, Supplements & Substrata for Serum Free Cell Culture*, AR Liss, New York, p 231, 1984

Leptin

Biogenesis 5633-1509   Synthetic   >80%; lyophilized

Leptin, (^{125}I)-

Amersham IM311   Human   Lyophilized; ~74 TBq/mmol, ~2000 Ci/mmol

Leupeptin

Biogenesis 5648-2000   Non-species synthetic   Lyophilized

Leupeptin, Propionyl-

Biogenesis 5648-2010   Non-species synthetic   Lyophilized

Luteinizing Hormone

ICN 151566   Human   Standard grade; >98%; lyophilized | WHO IRP (68/40)

Scipac P218-1   Pituitary glands   >96%; lyophilized | Peptide hormone

Luteinizing Hormone Receptor

Biogenesis 5725-2009   Rat synthetic   >80%; lyophilized

Luteinizing Hormone Releasing Factor, (3-(^{125}I)Tyr^5)-

Amersham IM181   Lyophilized; ~74 TBq/mmol, ~2000 Ci/mmol

Luteinizing Hormone Releasing Hormone

Biogenesis 5730-1104   Synthetic

Luteinizing Hormone α-Subunit

ICN 155275   Human pituitary   <2% β-subunit (RIA)

ICN 155276   Human pituitary   <2% α-subunit (RIA)

Luteinizing Hormone, Human

ICN 151565   Human   Iodination grade; >98%; lyophilized; >10,000 IU/mg; <0.1% prolactin, hFSH, hTSH | WHO IRP (68/40)

Macroglobulin, Alpha II

Scipac P144-5   Serum/plasma   >96%; lyophilized; available on request | Inhibitor

Macrophage Inflammatory Peptide IIIα, (^{125}I)-

Amersham IM322   Lyophilized; ~74 TBq/mmol, ~2000 Ci/mmol

Macrophage Inflammatory Peptide Iα, (^{125}I)-

Synonyms: Macrophage Inflammatory Peptide Iα Receptor Ligand, (^{125}I)-

Amersham IM285   Human recombinant   Lyophilized; ~74 TBq/mmol, ~2000 Ci/mmol | Chemokine receptor ligand; non-selective; endogenous; K_d = 5 nM CCR1/HEK293 cells

Amersham IM310   Human recombinant   Lyophilized; ~74 TBq/mmol, ~2000 Ci/mmol | Chemokine receptor ligand; CCR5 selectivity; endogenous; K_d = 0.47 nM peripheral blood monocytes

Macrophage/Monocyte Chemoattractant Protein I, (^{125}I)-

Synonyms: Macrophage/Monocyte Chemoattractant Protein I Receptor Ligand, (^{125}I)-

Amersham IM280   Human recombinant   Labeled with Bolton & Hunter Reagent; lyophilized; ~74 TBq/mmol, ~2000 Ci/mmol | Chemokine receptor ligand; non-selective; endogenous; K_d = 0.12 nM CCR2B/CHO membrane

Macrophage/Monocyte Chemoattractant Protein III, (^{125}I)-

Synonyms: Macrophage/Monocyte Chemoattractant Protein III Receptor Ligand, (^{125}I)-

Amersham IM289   Human recombinant   Lyophilized; ~74 TBq/mmol, ~2000 Ci/mmol | Chemokine receptor ligand; non-selective; endogenous; K_d = 0.7 nM CCR1/HEK293 cells

Macrophage/Monocyte Chemoattractant Protein IV, (^{125}I)-

Amersham IM309   Human recombinant   Lyophilized; ~74 TBq/mmol, ~2000 Ci/mmol

Magainin I

Biogenesis 5899-1009   Synthetic   >95%; lyophilized

Magainin II

Biogenesis 5899-1109   Synthetic   >95%; lyophilized

Magainin II, D-

Sigma M 8535   FW 2465.9   Amide; ≥97% (HPLC); peptide content: ~70% | Bioactive peptide; Wade, D et al, *Proc Natl Acad Sci USA*, 87:4761, 1990

Mast Cell Degranulating Peptide

ICN 193943   MW 2478   *Apis mellifera*   Selective inhibitor of a number of neuronal voltage-gated K$^+$ channels; also induces LTP & CA1 hippocampal neurons; Cherubini etal, *Nature*, 328:70, 1987

Mastoparan 17 Negative Control

BioSource International 77-323

Mastoparan 7 Substrate

BioSource International 77-319

Mastoparan 8 Substrate

BioSource International 77-321

Mastoparan Substrate

BioSource International 77-311

Mastoparan X Substrate

BioSource International 77-315

Melanocyte Stimulating Hormone, (^{125}I)(Nle^4-D-Phe^7)-α-

Amersham IM316   Direct labeled; lyophilized; ~74 TBq/mmol, ~2000 Ci/mmol

Amersham IM317   Labeled with Bolton & Hunter Reagent; lyophilized; ~74 TBq/mmol, ~2000 Ci/mmol

Melanocyte Stimulating Hormone, ND-α- Fluo-

Amersham VMSH012   Fluorescein-labeled with high activity

Melanotan I

ICN 193958   MW 7509   *Dendroaspis augusticeps*   Potently inhibits ^{3}H-pirenzepine binding to brain synaptosomes; a strong, selective & irreversible m$_1$Ach cholinoceptor agonist; Jerusalinski etal, *Neurochem Int*, 20:237, 1992

Melanotan II

ICN 193959   MW 7040   *Dendroaspis augusticeps*   Potently inhibits ^{3}H-pirenzepine binding to brain synaptosomes; a strong, selective & irreversible m$_1$Ach cholinoceptor agonist; reduces evoked Ach release from rat brain slices; causes memory facilitation which is antagonized by scopolamine; Jerusalinski etal, *Neurochem Int*, 20:237, 1992

Melanotan III

ICN 193960   MW 7379   *Dendroaspis augusticeps*   Homologous to MT-1 & -2; the most highly potent & selective m4Ach cholinoceptor ligand available; Jolkkonen etal, *FEBS Lett*, 352:91, 1994

mGluR1α (1116-1130)

Oncogene PP99   Lyophilized | Control peptide for competition studies & immunoabsorption

Microcystin-LF

Alexis 350-081   *Microcystis aeruginosa*   Analog of microcystin-LR with Phe substituted in place of Arg; hydrophobic & believed to be more cell permeable than other microcystins; may prove useful in biochemical studies in intact cells; Lawton, LA et al, *Nat Toxins*, 3:50, 1995; Lawton, LA et al, *Analyst*, 119:1525, 1994; Azevedo, SMFO et al, *J Appl Phycol*, 6:261, 1994

Microcystin-LR

ICN 158379   MW 995.2 $C_{49}H_{74}N_{10}O_{12}$ Heptapeptide ester hepatotoxin; equally potent in inhibition of protein phosphatases 1 & 2A (PP1 & PP2A); PP2B is less sensitive & PP2C is not inhibited up to 4 μM; useful for affinity purification of PP2A; not cell permeable; *JBC*, 110:8557, 1988; *FEBS Lett*, 264:187, 1990; *JBC*, 265:19401, 1990

Alexis 350-012   *Microcystis aeruginosa*   Heptapeptide ester hepatotoxin; equally potent & selective inhibitor of PP-1 & PP-2A; Rinehart, KL et al, *JACS*, 110:8557, 1988; MacKintosh, C et al, *FEBS Lett*, 264:187, 1990; Honkanen, RE et al, *J Biol Chem*, 265:19401, 1990; Ohtake, S et al, *Appl Environ Microbiol*, 57:1241, 1991; Nishiwaki, S et al, *FEBS Lett*, 279:115, 1991; Nishiwaki-Matsushima, R et al, *J Cancer Res Clin Oncol*, 118:420, 1992; Davidson, HW et al, *J Cell Biol*, 116:1343, 1992; Amick, GD et al, *Biochem J*, 287:1019, 1992; Moorhead, G et al, *FEBS Lett*, 356:46, 1994; Nishiwaki, R et al, *Cancer Lett*, 83:283, 1994

USBio M3889   *Microcystis aeruginosa*   >95% (HPLC); lyophilized | A cyclic heptapeptide toxin isolated from the fresh water cyanobacteria, *Microcystis aeruginosa*; a potent inhibitor of PP1 & PP2A:IC_{50} = 400 *pM*; useful in reducing contaminating phosphatase activities in proteinkinase assays

Microcystin-LW

Alexis 350-080   *Microcystis aeruginosa*   Analog of microcystin-LR with Trp substituted in place of Arg; hydrophobic & believed to be more cell permeable than other microcystins; may prove useful in biochemical studies in intact cells; Has a characteristically different absorption spectrum compared to other microcystins, making it a useful reference compound for HPLC analysis; Trp confers an absorption maximum at 222 nm, whereas most microcystins have a characteristic maximum at 239 nm; Lawton, LA et al, *Nat Toxins*, 3:50, 1995; Lawton, LA et al, *Analyst*, 119:1525, 1994

Microcystin-RR

Alexis 350-043   *Microcystis aeruginosa*   Arg-Arg analog of microcystin-LR; hepatotoxic, although found to be up to 10 times less toxic than microcystin-LR on i.p. injection in mice; potent inhibitor of PP-1 & PP-2A; Shirai, M et al, *Appl Environ Microbiol*, 57:1241, 1991; Lawton, LA et al, *Analyst*, 119:1525, 1994

Mitogen Activated Protein Kinase Activated Protein Kinase II, Active

USBio M2355-05   Recombinant, expressed in *E. coli*   N-terminal, GST-tagged, fusion protein encoding AA 36–371 of human MAPKAP-K2

Mitogen Activated Protein Kinase Activated Protein Kinase II, Inactive

USBio M2355-06   ≥60% by glutathione-agarose chromatography; liquid in 50 mL of 50 *mM* Tris-HCl, pH 7.5, 0.1 *mM* EGTA, 0.03% Brij-35, 0.1% (v/v) β-MSH, 50% glycerol (0.2 mg/mL) | Residues 46–400 of human MAPKAP Kinase 2 with an N-terminal GST tag & a C-terminal myc epitope; a Ser/Thrkinase activated by phosphorylation by mitogen-activated proteinkinase (MAPK); phosphorylates substrates with the sequence H-X-R-X-X-S, where H is any hydrophobic residue; substrates include MAPKAP Kinase-2 substrate peptide, heat shock proteins (murine hsp25 & human hsp27) & tyrosine hydroxylase

USBio M2355-07   Recombinant, expressed in *E. coli*   N-terminal, GST-tagged, fusion protein encoding AA 36–371 of human MAPKAP Kinase 2 (MAPKAP-K2)

Mitogen Activated Protein Kinase Substrate

Synonyms: Epidermal Growth Factor-R (661-681)

BioSource International 77-339

Mitogen Activated Protein Kinase Substrate I

BioSource International 77-338

Mitogen Activated Protein Kinase Substrate II

BioSource International 77-354

Mitogen Activated Protein Kinase Substrate III

BioSource International 77-356

BioSource International 77-357

Mitogen Activated Protein Kinase Substrate IV

BioSource International 77-358

BioSource International 77-359

Mixture

Sigma P 2693   Lyophilized solid; Contains ~25 µg of each of 9 peptides: bradykinin, bradykinin fragment 1-5, substance P, (Arg^8)-vasopressin, luteinizing hormone releasing hormone, bombesin, leucine enkephalin, methionine enkephalin & oxytocin; Must be reconstituted with Peptide Separation Buffer (Sigma P 2188) | Suitable for use as calibration standards in uncoated capillaries

Mycobacillin

ICN 151722   MW 1528.5 $C_{65}H_{85}N_{13}O_{30}$ *Bacillus subtilis* Antifungal polypeptide antibiotic isolated from culture filtrates

Myelin Basic Protein

BioSource International 77-171

Myelin Basic Protein (4-14)

BioSource International 77-169

Myosin Light Chain Kinase

BioSource International 77-108

BioSource International 77-136

Myristoylated Alanine Rich C-Kinase Substrate (159-165)

BioSource International 77-281

Myristoylated Alanine Rich C-Kinase Substrate Protein (151-175)

BioSource International 77-284

BioSource International 77-285

Natriuretic Peptide 26 (7-32), Brain

Sigma B 0777   FW 2869.3   Porcine   ≥97% (HPLC); peptide content: ~65% | Bioactive peptide; useful as an immunogen for preparing anti-BNP Ab; Sudon, T et al, *Nature*, 322:78, 1988

Natriuretic Peptide, Atrial $(3-(^{125}I)-Tyr^{28})$-; Natriuretic Peptide Receptor Ligand, Atrial $((^{125}I)-Tyr^{28})$-

Amersham IM186   Rat   Lyophilized; ~74 TBq/mmol, ~2000 Ci/mmol | Non-selective; endogenous; K_d = 0.06 pM guinea pig brain membrane

Natriuretic Peptide, Atrial $(3-(^{125}I)-Tyr^{28})$-α-

Synonyms: Natriuretic Peptide Receptor Ligand, Atrial $((^{125}I)-Tyr^{28})$-α-

Amersham IM187   Human   Lyophilized; ~74 TBq/mmol, ~2000 Ci/mmol | Non-selective; endogenous; K_d = 0.08 pM guinea pig brain membrane

Natriuretic Peptide, Atrial α-Fluo-

Amersham VANP011   Human   Fluorescein-labeled with high activity

Natriuretic Peptide, Brain

Biogenesis 1505-0552   Porcine synthetic   >95% (HPLC); lyophilized

Biogenesis 1505-0602   Porcine synthetic   >95% (HPLC); purified; lyophilized

Neurogranin (28-43)

BioSource International 77-276

Neurokinin A, (2-(^{125}I)IodoHis^1)-

Synonyms: Substance K

Amersham IM168   Lyophilized; ~74 TBq/mmol, ~2000 Ci/mmol

Neurokinin A, Fluo-

Amersham VNKA011   Human   Fluorescein-labeled with high activity

Neuromedin C, Fluo-

Amersham VNMC011   Fluorescein-labeled with high activity

Neuropeptide Tyrosine

Synonyms: Neuropeptide Y

Biogenesis 6730-0604   Synthetic   >95%; lyophilized

Neuropeptide Y

BioSource International 03-010

BioSource International 03-042   Porcine

Neuropeptide Y (13-36)

BioSource International 03-038

Neuropeptide Y (18-36)

BioSource International 03-044

Neuropeptide Y (22-36)

BioSource International 03-046

Neuropeptide Y (3-36)

BioSource International 03-040

BioSource International 03-018   Human, rat

Neuropeptide Y, (^{125}I)-

Amersham IM170   Labeled with Bolton & Hunter Reagent; lyophilized; ~74 TBq/mmol, ~2000 Ci/mmol

Neuropeptide Y, (Leu^{31},Pro^{34})-

BioSource International 03-016

Neuropeptide Y, (Pro^{34})-

BioSource International 03-013   Human, rat

Neuropeptide Y, Fluo-

Amersham VNPY011   Human   Fluorescein-labeled with high activity

Neuropeptide Y, N-(Propionyl-^{3}H)-Propionylated

Amersham TRK814   0.15 *M* triethylammonium phosphate, pH 3.2:acetonitrile (3:1); 1.85-3.7 TBq/mmol, 50-100 Ci/mmol; 7.4 MBq/mL, 200 μCi/mL

Neurotensin (2-13), Fluo-

Amersham VNT011   Human   Fluorescein-labeled with high activity

Neurotensin, (3-(^{125}I)Tyr^3)-

Amersham IM163   Lyophilized; ~74 TBq/mmol, ~2000 Ci/mmol

Neurotoxin NSTX III

Synonyms: Cadaverine, 2,4-Dihydroxyphenylacetyl-L-Asn-N1-(L-Arg-Putreanyl)-

Peptides International PNT-4195-s   MW 664.8 $C_{30}H_{52}N_{10}O_7$ *Nephila maculata* (Papua New Guinean spider) synthetic   >95% (HPLC); lyophilized amorphous powder | Bioactive; Aramaki, Y et al, *Proc Japan Acad*, 62(B):359, 1986; Teshima, T et al, *Tetrahedron Letts*, 28:3509, 1987

Neurotrophic Factor Soluble Receptor α, Ciliary

ICN 195752   Human recombinant, *Sf*21 expressed   Lyophilized; ≥97%; ED_{50} = 0.2-0.4 μg/mL in the presence of 20 ng/mL CNTF

ICN 195778   Rat recombinant, *Sf*21 expressed   Lyophilized; ≥97%; ED_{50} = 0.05-0.15 μg/mL in the presence of 10 ng/mL CNTF

Neurotrophic Factor, Ciliary

ICN 160006   Human recombinant, expressed in *E. coli* Lyophilized; >95%; ED_{50} <0.7 ng/ml; 1 U CNTF stimulates half-maximal neurite outgrowth of chicken embryo ciliary neurons | Demonstrates cross reactivity with rat, mouse & chicken systems

ICN 158699   MW 23k   Rat recombinant, expressed in *E. coli* Lyophilized; ≥2 x 10^{7}U/mg; 1 U CNTF induces half-maximal acetylcholine transferase activity in IMR32 neuroblastoma cells | Potent neural growth factor originally characterized as a chick ciliary neuron survival factor; promotes survivability of other neuronal cell types & the differentiation factor towards sympathetic neurons & type-2 astrocytes during development of the optic nerve; a useful tool for the study of neural development & regeneration; 200 AA residues

Neurotrophic Peptide, Brain

ICN 193949   MW 28k   Human recombinant, expressed in *E. coli*   This neurotrophin promotes the survival & differentiation of sensory ganglions & motor neurons; Leibrock etal, *Nature*, 341:149, 1989

ICN 193950   MW ~28k   Human recombinant, expressed in *E. coli*   5% | Neurotrophin which promotes sensory ganglions & motor neurons survival & differentiation

Neurturin

Oncogene PF031   Lyophilized | Control peptide for dot blot & immunoabsorption; suggested starting concentration of peptide for immunoabsorption of neuturin (Ab-1) is 10X the concentration of the diluted primary antibody

Neutrophin III

ICN 193951   MW 28k   Human recombinant   Promotes neuronal survival & growth; Mainsonpierre etal, *Science*, 247:1447, 1990

ICN 193952   MW ~28k   Human recombinant, expressed in *E. coli*   ≥95% | Promotes neuronal survival

Neutrophin IV

ICN 193953   MW 28k   Human recombinant, expressed in *E. coli*   Promotes neuronal survival & growth; Hohnm etal, *Nature*, 344:339, 1990

ICN 193954   MW ~28k   Human recombinant, expressed in *E. coli*   ≥95% | Promotes neuronal survival

Nitric Oxide Synthase Blocking Peptide (1131-1144)

ICN 195939 MW 1630.9 $C_{69}H_{123}N_{21}O_{22}S$ Mouse macrophage ≥98% | Immunization & blocking peptide for Anti-Inducible NOS (1131-1144)

Nitric Oxide Synthase Blocking Peptide (599-613)

ICN 195940 MW 1921.1 $C_{83}H_{125}N_{25}O_{26}S$ Bovine endothelial cells ≥98% | Immunization & blocking peptide for Anti-eNOS (599-613)

Nitric Oxide Synthase Blocking Peptide (724-739)

ICN 195938 MW 1919.4 $C_{85}H_{151}N_{27}O_{21}S$ Rat brain ≥98% | Immunization & blocking peptide for Anti-bNOS (724-739)

Nociceptin

BioSource International 77-460

Nociceptin, (^{125}I)-

Amersham IM292 Lyophilized; ~74 TBq/mmol, ~2000 Ci/mmol

Nociceptin, (Leu-^{3}H)-

Amersham TRK1047 Acetonitrile:MeOH:H$_2$O:TFA (15:15:70:0.1); 740kBq/mL, 20 μCi/mL

Nociceptin, Fluo-

Amersham VNCP011 Fluorescein-labeled with high activity

Norleucinal, *N*-Ac-Leu-Leu-

Synonyms: Calpain I Inhibitor; Calpain II Inhibitor; Cathepsin L Inhibitor; Cysteine Endopeptidase Inhibitor

Sigma A 6185 FW 383.5 $C_{20}H_{37}N_3O_4$ ≥97% (TLC) | Bioactive peptide; inhibits Cyclin B degradation, arrests the cell cycle at G1/S & at meta-/anaphase; Sasaki, T et al, *Enzyme Inhibition*, 3:195, 1990

NP-1

Synonyms: Defensin I, α-

PeproTech 300-42 MW 3.4k Human recombinant, expressed in *E. coli* >98% by SDS-PAGE & HPLC; <0.1 ng endotoxin per mg (1EU/mg); lyophilized | Antimicrobial peptide; belongs to the distinct family of alpha-defensins; 30 AA; SA determined by its ability to chemoattract immature dendritic cells

Opioid δ-Protein (3-17), Mouse

Synonyms: Opioid δ-Receptor Control Peptide

Oncogene PP91 Lyophilized | For immunoabsorption

Opioid δ-Receptor

BioSource International 04-304

Opioid κ-Receptor, Internal

BioSource International 04-302

Opioid κ-Receptor, N-Terminal

BioSource International 04-300

Opioid μ-Protein (384-398), Rat

Synonyms: Opioid μ-Receptor Control Peptide

Oncogene PP90 Lyophilized | For competition studies & immunoabsorption

Opioid μ-Receptor, Internal

BioSource International 04-308

Opioid μ-Receptor, N-Terminal

BioSource International 04-306

Orexin B, Pro^2(3,4-^{3}H)-

ARC ART-764 MW 2936.5 Rat, mouse 30-60 Ci/mmol; 1.11-2.22 TBq/mmol; in EtOH | Radiochemical

Orexin B, Pro^{4,5}(3,4-^{3}H)-

ARC ART-754 MW 2899.1 Human 30-60 Ci/mmol; 1.11-2.22 TBq/mmol; in EtOH | Radiochemical

Oxytocin

Fluka 75968 MW 1007.23 $C_{43}H_{66}N_{12}O_{12}S_2$ 500 U/mg; 1 U is the activity contained in 1.71 μg of the international standard substance | Oxytocic hormone; Edwards, CRW, *Hormones in Blood*, (CH Gray & VHT James, eds), 401, 1979, Academic Press, London

Oxytocin, Agarose

ICN 191523 ~2 μM ligand/g, pH 4.5 buffer, 0.02% NaN$_3$, 4°C

Oxytocin, Desamino-

Biogenesis 7090-1109 Synthetic >99% (TLC); 100% (HPLC); lyophilized, consisting of 85% peptide material, acetate counter ions and residual H$_2$O

p34cdc2 Peptide

Synonyms: cdc2 Protein (42-51, 54-57)

Amersham VPP34 Competition for VPC26; cell proliferation & cell cycle signals

BioSource International 77-116

Oncogene PP34 Lyophilized with an equal amount of BSA | Used for competition studies & dot blot

Pan-Caspase I (390-404)

Synonyms: Pan-Interleukin Converting Enzyme

Oncogene PP59 Human Lyophilized with an equal amount of BSA | Used in competition studies & dot blot

Pancreastatin

Biogenesis 7098-2004 Porcine synthetic >95%; lyophilized

Biogenesis 7098-2059 Rat synthetic >95%; lyophilized

Pancreastatin (33-49)

Biogenesis 7098-2102 Porcine synthetic 95%; lyophilized

Pancreatic Polypeptide

Biogenesis 7100-2056 Rat synthetic >95%; lyophilized

Biogenesis 7100-2004 Synthetic >95%; lyophilized

Pancreatic Polypeptide (31-36)

Biogenesis 7100-0759 Synthetic >95%; lyophilized

Pan-ras Arg^{12} Peptide (5-16)

Oncogene PP39 Lyophilized with an equal amount of BSA; suggested starting concentration of peptide for competition is 10X the concentration of primary antibody | For competition studies & dot blot

Pan-ras Asp^{12} Peptide

Oncogene PP15 Lyophilized with an equal amount of BSA | Suggested starting concentration of peptide for competition is 10X the concentration of primary Ab; titration recommended

Pan-ras Val^{12} Peptide (5-16)

Oncogene PP16   Lyophilized with an equal amount of BSA; suggested starting concentration of peptide for competition is 10X the concentration of primary antibody | For competition studies & dot blot

Parathyroid Hormone

Synonyms: Parathormone

Sigma P 5028   FW 9509.8   Bovine parathyroid gland Bioactive peptide; Rasmussen, H & Craig, LC, *J Biol Chem*, 236:759, 1961; Keutmann, HT et al, *Biochemistry*, 10:2779, 1971; Brewer, HB Jr & Ronan, R, *Proc Natl Acad Sci USA*, 67:1862, 1970

Sigma P 7036   FW 9424.7   Human   Peptide content: ~70% | Bioactive peptide; Rasmussen, H & Craig, LC, *J Biol Chem*, 236:759, 1961; Keutmann, HT et al, *Biochemistry*, 10:2779, 1971; Brewer, HB Jr & Ronan, R, *Proc Natl Acad Sci USA*, 67:1862, 1970

Parathyroid Hormone (1-34)

Biogenesis 7170-1402   Bovine synthetic   >95%; lyophilized

Biogenesis 7170-6404   Synthetic   >95%; lyophilized

Parathyroid Hormone (1-84), (Asn^{76})-

Biogenesis 7170-2002   Synthetic   >95%; lyophilized

Parathyroid Hormone (53-84)

Biogenesis 7170-5202   Synthetic   >95%; lyophilized

Parathyroid Hormone Related Peptide (1-34), (Tyr^0)-

Biogenesis 7170-9376   Synthetic   >95%; lyophilized

Parathyroid Hormone Related Peptide (14-28)

Biogenesis 7170-9450   MW 1952.8   Human synthetic 98% (HPLC); lyophilized; trifluoroacetate salt

Parathyroid Hormone, Fluo-; Parathormone

Amersham VPTH011   Fluorescein-labeled with high activity

PARP 214/215

BioSource International 04-698   Lyophilized

Pc-Specific PI-D1, Internal

BioSource International 04-322   Human

Pc-Specific PI-D1, N-Terminal

BioSource International 04-320   Human

Pc-Specific PI-D2, Internal

BioSource International 04-326   Mouse

Pc-Specific PI-D2, N-Terminal

BioSource International 04-324   Mouse

PD 140376 Receptor Ligand, (^{3}H)-

Amersham TRK948   CCK$_B$; antagonist; K$_d$ = 0.11 nM guinea pig cortex membrane

PD 142308 Receptor Ligand, (^{125}I)-

Amersham IM278   CCK$_B$; antagonist; K$_d$ = 0.25 nM guinea pig cortex membrane

PEC-60

Calbiochem 515050   MW 6837.0   Porcine intestine   ≥98% (HPLC); lyophilized solid; soluble in water | 60-residue intestinal hormone-like peptide; inhibits the glucose-induced insulin secretion from perfused pancreas & plays a role in the immune system; synthesized in duodenal goblet cells & in monocytes in bone marrow & blood; increases dopamine release & activates Na$^+$, K$^+$-ATPase in brain; Rimodini, R et al, *Neurosci Lett*, 177:53, 1994; Kairane, C et al, *FEBS Lett*, 345:1, 1994; Metsis, M et al, *J Biol Chem*, 267:19829, 1992; Agerbeth, B et al, *PNAS*, 86:8590, 1989

Peptide A

BioSource International 77-248

Peptide YY

BioSource International 77-730   Human

BioSource International 77-720   Porcine

Peptide YY (13-36)

BioSource International 77-734   Human

BioSource International 77-724   Porcine

Peptide YY, (^{125}I)(Tyr)-

Amersham IM259   Porcine   Lyophilized; 148 TBq/mmol, 4000 Ci/mmol

Peptide YY, (Leu^{31},Pro^{34})-

BioSource International 77-728   Human

Peptide YY, (Pro^{34})-

BioSource International 77-726   Human

Peptide YY, Fluo-

Amersham VPYY011   Porcine   Fluorescein-labeled with high activity

Phomopsin A

Sigma P 4829   FW 789.2 $C_{36}H_{45}CIN_6O_{12}$ *Phomopsis leptostromiformis*   An antimitotic cyclic peptide; Very potent inhibitor of microtubule assembly by binding to tubulin at a site different from the colchicine binding site; Li, Y et al, *Biochem Pharmacol*, 43:219, 1992; Edgar, JA, in *"Toxicology of Plant & Fungal Compounds"*, Handbook of Natural Toxins, Vol 6:371, Keeler, RF & Tu, AT eds, Marcel Dekker, Inc, 1991

Phospholipase A2

BioSource International 77-406

Phosphorylase Kinase β

BioSource International 77-298

Phosphorylated Heat- & Acid-Stable Protein I

Synonyms: Mitogen Activated Protein Kinase Substrate; p38 Kinase Substrate; Protein Kinase C Substrate; JNK Substrate

Calbiochem 516675   MW 21k   Rat recombinant   ≥95% (SDS-PAGE); liquid in 50 *mM* NaCl, 25 *mM* Tris, 5 *mM* β-mercaptoethanol, 500 *μM* EDTA, pH 7.4; biological activity: shown to undergo phosphorylation by activated MAPkinase in the presence of ^{32}P-ATP | Regulated by insulin; 177-amino acid protein substrate; phosphorylation of PHAS-I increases in response to insulin but not to cAMP stimulation of adipocytes; excellent substrate for MAPkinase both *in vivo* & *in vitro*; Hu, C et al, *PNAS*, 91:3730, 1994; Pause, A et al, *Nature*, 371:762, 1994

PI-3-Kinase SH3 Binding

BioSource International 77-350

Pituitary Adenylate Cyclase Activating Peptide, Fluo-

Synonyms: Pituitary Adenylate Cyclase Activating Peptide 27

Amersham VPAC011   Fluorescein-labeled with high activity

Pituitary Adenylate Cyclase Activating Polypeptide (1-27)

Synonyms: Adenylate Cyclase Activator

Alexis 165-003   MW 3147.7 $C_{142}H_{224}N_{40}O_{39}S$ Human, ovine, rat   Amide; lyophilized powder | More potent than vasoactive intestinal peptide (VIP)

Miyata, A et al, *BBRC*, 164:567, 1989; Kimura, C et al, *BBRC*, 168:81, 1990; Tatsuno, I et al, *BBRC*, 168:1027, 1990; Cardell, LO et al, *Regul Pept*, 36:379, 1991; Miyata, A, *Regul Pept*, 26:170, 1989

Pituitary Adenylate Cyclase Activating Polypeptide (1-38)

Synonyms: Adenylate Cyclase Activator

Alexis 165-025   MW 4534.3 $C_{203}H_{331}N_{63}O_{53}S$ Human, ovine, rat   Amide; ≥95%; lyophilized powder; soluble in water | More potent than vasoactive intestinal peptide (VIP); Miyata, A et al, *BBRC*, 164:567, 1989; Kimura, C et al, *BBRC*, 168:81, 1990; Ogi, K et al, *BBRC*, 173:1271, 1990; Cauvin, A et al, *Peptides*, 11:773, 1990; Cardell, LO et al, *Regul Pept*, 36:379, 1991; Arimura, A, *Regul Pept*, 37:287, 1992

Pituitary Adenylate Cyclase Activating Polypeptide (25-38)

Biogenesis 7394-9179   Human, ovine, rat synthetic   90%; lyophilized

PKC (19-27), Myristoylated

BioSource International 77-165

Placental Lactogen

Scipac P115-1   Human placenta   >96%; lyophilized | Peptide hormone

Scipac P115-2   Human placenta   40-90%; lyophilized | Peptide hormone

Platelet Derived Growth Factor

Amersham VPP29   Competition for VPC21; growth/ death factor interactions

Platelet Derived Growth Factor Fragment, Cyclin

BioSource International 77-300

Platelet Derived Growth Factor Peptide (101-116)

Oncogene PP29   Human   Lyophilized with an equal amount of BSA; suggested starting concentration of peptide for competition is 10X the concentration of primary antibody | For competition studies & dot blot

Platelet Derived Growth Factor Receptor Peptide (425-446)

Oncogene PP24   Lyophilized with an equal amount of BSA; suggested starting concentration of peptide for competition is 10X the concentration of primary antibody | For competition studies & dot blot

Platelet Derived Growth Factor β-Receptor

BioSource International 77-288

Pleiotrophin

Synonyms: Heparin Affin Regulatory Peptide

PeproTech 450-15   MW 15.4k   Human recombinant, expressed in *E. coli*   >98% by SDS-PAGE & HPLC; <0.1 ng endotoxin per mg (1EU/mg); lyophilized | New member of the heparin-binding neurotrophic factor family; structural homolog with Midkine—highly conserved among species; important role in developmental processes & promotion of neurite extension; 136 AA; SA determined by its ability to enhance neurite outgrowth of cerebral cortical neurons of E10 chick embryos

Polymyxin B Sulfate

Fluka 81334   MW 1385.63 $C_{55}H_{96}N_{16}O_{13} \cdot 2H_2SO_4$ Antibiotic with bactericidal action on *E. coli*; Cornu, J, *Ann Microbiol*, 131B:121, 1980; Storm, DR et al, *Ann Rev Biochem*, 46:723, 1977; Morrison, DC & Jacobs, DM, *Immunochemistry*, 13:813, 1976; Lerner, HR et al, *Physiol Plant*, 57:90, 1983

pp60v-src Autophosphorylation Site

BioSource International 77-220

BioSource International 77-224

BioSource International 77-225

pp60v-src C-Terminal Phosphorylation

BioSource International 77-236

Pro-Adrenomedullin (N-52)

American Peptide 22-1-19   MW 2460.9 $C_{112}H_{178}N_{36}O_{27}$ Human   Kitamura, K et al, *BBRC*, 194:720, 1993

Pro-Caspase I (2-19)

Synonyms: Pro-Interleukin Converting Enzyme

Oncogene PP60   Human   Lyophilized with an equal amount of BSA | Used in competition studies & dot blot

Pro-Cathepsin B

Oncogene PP41   Lyophilized with an equal amount of BSA; suggested starting concentration of peptide for competition is 10X the concentration of primary antibody | Region: sequence within pro-peptide; used for competition studies & dot blot

Pro-Cathepsin D

Oncogene PP42   Lyophilized with an equal amount of BSA; suggested starting concentration of peptide for competition is 10X the concentration of primary antibody | Region: sequence within pro-peptide; used for competition studies & dot blot

Pro-Cathepsin L

Oncogene PP43   Lyophilized with an equal amount of BSA; suggested starting concentration of peptide for competition is 10X the concentration of primary antibody | Region: sequence within pro-peptide; used for competition studies & dot blot

Pro-Endothelin I

Synonyms: Pro-Endothelin I, rh-

Alexis 201-032   Human recombinant   ~80% (SDS-PAGE); no preservatives; lyophilized; biological activity: 4 μg rhproET-1 when cleaved with furin convertase & chymosin has been shown to produce ET-1 mediated vasoconstriction of rat vas deferens & rabbit carotid artery | His-tagged rhproET-1 corresponds to preproET-1 (18-212); AA sequence is Met-Arg-Gly-Ser-(His[6]) at its *N*-terminal replacing the signal peptide; Denault, JB et al, *FEBS Lett*, 362:276, 1995

Progesterone Receptor (533-547)

Biogenesis 7721-0509   Synthetic   >80%; lyophilized

Pro-Insulin (33-63)

Biogenesis 7750-0502   Synthetic   >95%; lyophilized

Pro-Insulin C-Peptide

Synonyms: C-Peptide

Fitzgerald 30-AC95   >97% (HPLC)

Pro-Insulin C-Peptide (33-63)

Synonyms: C-Peptide

Biogenesis 1700-0304   Human synthetic   >95% (HPLC); lyophilized

PRO-INSULIN-LIKE GROWTH FACTOR II

IBT AHU020   Three major isoforms: MW 16.1k, 17.1k, 17.6k Human recombinant, expressed in *E. coli*   >95%; lyophilized; free of measurable endotoxins | 156 AA precursor of IGF-II; 87 AA C-terminal E-domain; elevated levels have been associated with non-islet cell tumor hypoglycemia & other tumor cells

Pro-Insulin-Like Growth Factor II, Biotinylated

IBT BioPROIGF2-04, BioPROIGF2-08   Human recombinant, expressed in *E. coli*   Lyophilized from TBS + stabilizing proteins of non-mammalian origin; carrier protein free, biotinylated peptide available upon request | Affinity of the precursor to IGFBP is controversial; Zapf reports no significant difference of affinities of biotinylated IGF-II or biotinylated pro-IGF-II to IGFBP-3 found in western-ligand blotting experiments; Zapf, J, *J Int Med*, 234:543, 1993; for non-radioactive western-ligand blotting, binding studies or immunohistochemistry

Prolactin

Scipac P221-1   Pituitary glands   >96%; lyophilized | Peptide hormone

Prolactin Receptor (1-14)

Biogenesis 7770-6020   Human synthetic   >80%; lyophilized

Protein G

ICN 622661   Biotin | Useful in ELISA & immunohistochemical procedures

Protein Kinase A Inhibitor Peptide

Biodesign A80118S   Synthetic   >97% | Apoptosis & signal transduction

Protein Kinase C

BioSource International 77-192

BioSource International 77-268

Protein Kinase C Eta Pseudosubstrate

BioSource International 77-740

Protein Kinase C Eta Pseudosubstrate, Myristoylated

BioSource International 77-744

Protein Kinase C Eta Substrate, Biotinylated

BioSource International 77-752

Protein Kinase C Peptide

Oncogene PP27   Lyophilized with an equal amount of BSA; suggested starting concentration of peptide for competition is 10X the concentration of primary antibody | Within C4 conserved region; for competition studies & dot blot

Protein Kinase C Pseudosubstrate (19-31)

BioSource International 77-184

Protein Kinase C Pseudosubstrate (19-36)

BioSource International 77-188

Protein Kinase C Theta Pseudosubstrate

BioSource International 77-742

Protein Kinase C Theta Pseudosubstrate, Myristoylated

BioSource International 77-746

Protein Kinase C Theta Substrate, Biotinylated

BioSource International 77-754

Protein Kinase C Zeta Pseudosubstrate

BioSource International 77-738

Protein Kinase C Zeta Pseudosubstrate, Myristoylated

BioSource International 77-748

Protein Kinase C Zeta Substrate, Biotinylated

BioSource International 77-750

Protein Kinase C ε

BioSource International 77-120

Protein Kinase Inhibitor (14-22), Myristoylated

BioSource International 77-408

Protein Kinase Inhibitor (14-24)

BioSource International 77-246

Protein Kinase Inhibitor (5-24)

BioSource International 77-240

Protein Kinase Inhibitor (5-24), Amide

BioSource International 77-238

Protein Kinase Inhibitor (6-22), Amide

BioSource International 77-244

Protein Kinase Inhibitor Peptide

BioSource International 77-214

Protein Kinase P34CDC2 Substrate

BioSource International 77-340

Protein Kinase Substrate Peptide

USBio P9103-14A   Synthetic   ≥97% (HPLC); lyophilized

PTHLP Peptide (34-53)

Oncogene PP14   Lyophilized with an equal amount of BSA; suggested starting concentration of peptide for competition is 10X the concentration of primary antibody | Useful for competition studies & dot blot

RANTES, (^{125}I)-

Synonyms: RANTES Receptor Ligand, (^{125}I)-

Amersham IM288 Human recombinant Lyophilized; ~74 TBq/mmol, ~2000 Ci/mmol | Chemokine receptor ligand; non-selective; endogenous; K_d = 0.39 nM THP-1 cells

Raytide EL Substrate

Synonyms: Protein Tyrosine Kinase p60c-src Substrate

Calbiochem PK04 Lyophilized | Offers increased labeling efficiency compared to Raytide Substrate

Raytide Kinase Substrate

Synonyms: Protein Tyrosine Kinase Substrate

Calbiochem PK02 Lyophilized | Provided with a complete directional insert for performingkinase assays

Raytide Substrate Negative Control

Synonyms: Raytide Kinase Substrate; Raytide EL Substrate

Calbiochem PK05 Lyophilized | Non-phosphorylatable negative control of protein tyrosinekinase p60c-src

Rb Peptide I

Amersham VPP21 Competition for VPC13; cell proliferation & cell cycle signals

Oncogene PP21 Lyophilized with an equal amount of BSA | Suggested starting concentration of peptide for competition is 10X the concentration of primary Ab; titration recommended

Rb Peptide II

Amersham VPP22 Competition for VPC14; cell proliferation & cell cycle signals

Rb Peptide II (248-262)

Oncogene PP22 Human Lyophilized with an equal amount of BSA | Used for competition studies & dot blot

Renin Substrate Tetradecapeptide

Sigma R 8129 FW 1759.0 Porcine ≥95% (HPLC) | Bioactive peptide

Ribosomal Protein S6 Phosphorylation

BioSource International 77-229

RII (81-89), (Ala^{97})-

BioSource International 77-809

RII (81-99), (PSer^{95},Ala^{97})-

BioSource International 77-804

Ristomycin Monosulfate

Synonyms: Ristocetin Sulfate

Fluka 83911 MW 2166.02 $C_{95}H_{110}N_8O_{44} \cdot H_2SO_4$ ≥99.0% (TLC); crystallized | Glycopeptide antibiotic which acts by binding to bacterial cell wall precursors; Perkins, HR, *Pharmacol Ther*, 16:181, 1982

SAMS Peptide

USBio S0090 Synthetic ≥80%; ~562 µM; frozen solution | AMP-activated proteinkinase (AMPK) cascades appear to be sensors for cellular energy status; members of the AMP-activated/SNF-1-related subfamily of proteinkinases are thought to be pivotal components of these highly conserved proteinkinase cascades, which are in most eukaryotic cells; the SAMS peptide is based on the 13-residue sequence around Ser^{79}, which is the unique site for the AMP-activated proteinkinase on rat acetyl-CoA carboxylase; Ser^{77}, corresponding to the cyclic-AMP-dependent proteinkinase site in the natural sequence of rat Ac-CoA Carboxylase (ACC), was replaced by Ala at residue 5 of the synthetic sequence; 2 Arg residues were added at the C-terminus to increase binding of the peptide to phosphocellulose paper

Sarafotoxin 6c

Sigma S-172 MW 2516 (free base) $C_{103}H_{147}N_{27}O_{37}S_5$ Synthetic >99%; white powder; peptide content & salt form information are provided with each lot | Selective ET$_8$ endothelin receptor agonist; Heyl et al, *Pept Res*, 6:238, 1993; Williams et al, *Biochem Biophys Res Commun*, 175:556, 1991; El-Mowafy, *J Pharmacol Exp Ther*, 268:1343, 1994

Sauvagine, (^{125}I)(Tyr^0)-

Synonyms: Sauvagine Receptor Ligand, (^{125}I)(Tyr^0)-; Corticotropin Releasing Factor Receptor Ligand

Amersham IM307 Human recombinant H$_2$O:acetonitrile (65:35); trace BSA, TFA, β-MSH; ~74 TBq/mmol, ~2000 Ci/mmol; 3.7 MBq/mL, 100 µCi/mL | Non-selective; agonist; K_d = 0.16 nM human CRF$_{2\alpha}$/CHO membrane

Sauvagine, Fluo-

Amersham VSVG011 Fluorescein-labeled with high activity

Serine Kinase Substrate Peptide, Biotinylated

USBio S1000-50 ≥98%; 1 mg/mL; frozen solution

Serotonin (5-HT) Transporter Control Peptide (579-599)

Oncogene PP87 Lyophilized | For competition studies & immunoabsorption

Serotonin 2A (5-HT$_{2A}$) Receptor Control Peptide (22-41)

Oncogene PP88 Lyophilized | For competition studies & immunoabsorption

Serotonin BSA Conjugate Control

Oncogene PP101 Lyophilized

SGK (1-60), Active

Synonyms: S422D

USBio S1010-81 MW ~48k Human recombinant, expressed in Sf9 cells ≥95%; purified using Ni-NTA agarose, activated with PDK1 & repurified using heparin Sepharose chromatography; 2 µg of enzyme in 50 µL of 50 mM Tris-HCl, pH 7.5, 0.1 mM EGTA, 0.1% β-MSH, 0.15 M NaCl, 0.02% Brij-35, 270 mM sucrose | N-truncated (missing N-terminal 60 AA), His-tagged, human SGK fusion protein, containing a S422D mutation

SGK (1-60), Inactive

Synonyms: S422D

USBio S1010-82 Human recombinant, expressed in Sf9 cells Purified using Ni-NTA agarose | N-truncated (missing N-terminal 60 AA), His-tagged, human SGK fusion protein, containing a S422D mutation; upon phosphorylation by PDK1 on T256, this protein becomes fully activated

Sodium Glucose Transporter (402-420)

Synonyms: Sodium Glucose Transporter I

Biogenesis 8327-1109   Synthetic   >80%; lyophilized

Somatostatin (SS^{14}), Fluo-

Amersham VSRF011   Human   Fluorescein-labeled with high activity

Somatostatin 14

Synonyms: Somatostatin Release Inhibiting Factor; Growth Hormone Release Inhibiting Factor; Somatostatin, Sheep; Somatotropin Release Inhibiting Factor

Fluka 85480   MW 1637.91 $C_{76}H_{104}N_{18}O_{19}S_2$ ≥97.0% (HPLC)

Somatostatin 14, (3-(^{125}I)Tyr^{11})-

Amersham IM161   Lyophilized; ~74 TBq/mmol, ~2000 Ci/mmol

Somatostatin Sequence (4-7), (Ala)-

Biogenesis 8330-5409   MW 625.8   Human synthetic   99.5 % (TLC/HPLC); lyophilized, consisting of 87% peptide material, acetate counter ions and residual H_2O; no preservatives

Somatostatin Sequence (6-11), Ac-

Biogenesis 8330-5612   MW 917.9   Human synthetic   99% (TLC), 97% (HPLC); no preservatives; lyophilized

Somatostatin Sequence (7-11), Ac-

Biogenesis 8330-5709   MW 623   Human synthetic   >99% (TLC), >99.5% (HPLC); 93.4% peptide material, acetate counter ions and residual H_2O; no preservatives; lyophilized

src C-Terminal, SH2 Domain Ligand

BioSource International 77-128

src SH2 Protein Biotin Conjugate

Calbiochem SH01B   May be carcinogenic/teratogenic | Conjugation of the SH2 domain of c-src to biotin; reacts with any phosphoprotein that binds src via the SH2 domain

src Substrate Peptide

USBio S6504   97% (HPLC); supplied in powder form

Stat1 (pY701) Peptide

BioSource International 04-376   Mouse

Stat1α Peptide

BioSource International 04-360

Stat2 Peptide

BioSource International 04-362

Stat3 Peptide

BioSource International 04-364

Stat4 Peptide

BioSource International 04-366

Stat5 Peptide

BioSource International 04-368

Stat6 Peptide

BioSource International 04-372

St-Ht31P Control Peptide, InCELLect™

Promega V8221   MW 2765   >80% (HPLC) | Used as a negative control peptide for the study of InCELLect™ AKAP St-Ht31 inhibitor peptide; presence of the hydrophobic stearated moiety enhances the cellular uptake of the peptide, allowing it to be used to study PKA signaling in intact cells in culture; Vijayaraghavan, S et al, *JBC*, 272:4747, 1997

Substance P

USBio S8000-07   ≥95% purity; lyophilized

Substance P, (^{125}I)-

Amersham IM157   Labeled with Bolton & Hunter Reagent; lyophilized; ~74 TBq/mmol, ~2000 Ci/mmol

Substance P, (^{125}I-Tyr^8)-

ICN 68107   >1000 μCi/μg, >37 MBq/μg; 0.1 *M* KPB, pH 7.5, 0.3% BSA:n-propanol (1:1)

Substance P, (2-L-Pro(3,4-^{3}H))-

ARC ART-693   MW 1347.8   25-50 Ci/mmol; 740 GBq-2.03 TBq/mmol; in EtOH | Radiochemical

Substance P, (Leu^{3,4,5}-^{3}H)

ARC ART-692   MW 1347.8   >100 Ci/mmol; 3.70 TBq/mmol; in 0.1 *N* HOAc: EtOH (8:2) | Radiochemical

Substance P, (Pro^{2,4}-3,4(n)-^{3}H

Amersham TRK786   0.15 M Triethylammonium phosphate, pH 5.5:acetonitrile (3:1), 0.2% β-MSH; 1.11-2.4 TBq/mmol, 30-65 Ci/mmol; 7.4 MBq/mL, 200 μCi/mL

Substance P, Fluo-

Amersham VSP011   Human   Fluorescein-labeled with high activity

Taurine Receptor Ligand, (^{3}H)-

Amersham TRK573   Excitatory AA receptor ligand; non-selective; endogenous

Thiostrepton

ICN 156893   MW 1664.8 $C_{72}H_{85}N_{19}O_{18}S_5$ 90% | Peptide antibiotic that inhibits binding of EF-G & GTP to the 50S ribosomal subunit

Thyrocalcitonin

Sigma T 8135   Bovine thyroid glands   TCA Powder; activity: ~0.1 MRC U/mg solid; Unit definition: 1 IU or 1 MRC unit is the activity contained in 4.74 mg of WHO Standard Preparation; | Bioactive peptide; hypocalcemic hormone produced by the parafollicular C cells of the thyroid or by the ultimobranchial bodies of nonmammalian vertebrates; a 32 AA polypeptide, 8 of which are conserved across all species; decreases blood calcium & phosphate due to inhibition of resorption by osteoblasts & osteocytes; Tenenhouse, A et al, *Proc Natl Acad Sci USA*, 53:818, 1965; Brewer, HB et al, *J biol Chem*, 2454:4232, 1970; Bioassay:Schlueter et al, *Endocrinology*, 81:854, 1967; Farley, JR et al, *Endocrinology*, 123:159, 1988

Thyroid Stimulating Hormone

Scipac P219-1   Pituitary glands   >96%; lyophilized | Peptide hormone; thyroid function protein

Thyrotropic Hormone

Synonyms: Thyroid Stimulating Hormone

ICN 160106   Bovine pituitary   10 IU/vial; ~0.5-1 IU/mg

ICN 152144   Human   Standard grade; lyophilized; >3.5 IU/mg

Thyrotropic Hormone

Synonyms: Thyroid Stimulating Hormone

ICN 198793   MW 432.4 $C_{21}H_{20}O_{10}$ Human   Iodination grade; 98% (affinity); lyophilized; ~8 IU/mg

Thyrotropic Hormone β-Subunit

ICN 156920   Human

Thyrotropin Releasing Hormone (160-169), Prepro-

Biogenesis 8940-1289   Synthetic   >95%; lyophilized

Thyrotropin Releasing Hormone, (Pro^{3,4}-^{3}H)-

ARC ART-705   50-60 Ci/mmol; 1.85-2.22 TBq/mmol; in EtOH | Radiochemical

Tissue Inhibitor of Metalloproteinase II

Biogenesis 9013-2659   MW 1656   Synthetic   >80% (HPLC); proprietary sequence; largely salt free; lyophilized

Toxin II

Synonyms: ATX II

Calbiochem 616390   *Anemonia sulcata*   Lyophilized solid; purity: ~40% ATX II-Ile, 40% ATX II-Val & 20% other isotoxins (HPLC); soluble in water; $LD_{50} \leq 2000$ mg/kg | Mixture of Ile & Val isomers; highly toxic polypeptide; acts specifically on Na^+ channel in excitable membranes, affecting neuromuscular transmission & mammalian heart muscle function; augments Na^+ dependent increase in twitch response & the duration of action potential; has a positive inotropic effect on the heart; Alsen, C, *Fed Proc*, 42: 101, 1983; Mantegazza, M et al, *J Physiol*, 507: 105, 1998

TRAIL Receptor II N-Terminal Peptide

Alexis 168-007   Human synthetic   Dissolved in 30 μL 10 m*M* sodium acetate, pH 4.5 | Blocking peptide for antiserum (purified) to TRAIL-R2; competitively binds to Ab Alexis 210-743; antiserum blocking peptide

TRAIL Receptor III N-Terminal Peptide

Alexis 168-006   Human synthetic   Dissolved in 30 μL 10 m*M* sodium acetate, pH 4.5 | Blocking peptide for antiserum (purified) to TRAIL-R3; Competitively binds to Ab Alexis 210-744; antiserum blocking peptide

Transforming Growth Factor α

PeproTech 100-16A   MW 5.5k   Human recombinant, expressed in *E. coli*   >98% by SDS-PAGE & HPLC; <0.1 ng endotoxin per mg (1EU/mg); lyophilized | Polypeptide growth factor; stimulates proliferation of a wide range of epidermal & epithelial cells; 50 AA; $ED_{50} \leq 0.7$ ng/mL; SA $\geq 1.4 \times 10^6$ U/mg; SA determined by dose-dependent stimulation of the proliferation of thymidine uptake by BALB/c 3T3 cells

Biogenesis 9130-0404	Rat synthetic
Biogenesis 9130-0604	Synthetic
Biogenesis 9130-1004	Synthetic

ICN 150206   Human platelets   >97%; lyophilized; reconstitute with 100 μl 5 m*M* HCl & transfer to 1 mL PBS with 2 mg/mL albumin | Modulates cell growth & induces changes in phenotype; enhances formation of connective tissue; stimulates growth of non-neoplastic cells (eg, NRK-1); purifies EGF (2 ng/mL) must be present for biological activity; unstable under acidic conditions—significant losses may result from neutral buffers & glass implements

ICN 152159   Porcine platelets   >96% (silver stain gels); lyophilized; $ED_{50} = 40.0$ pg/mL | Indistinguishable by many parameters from human sources; inhibits growth of certain carcinoma cells; synergistic with EGF &/or TGF-α

Transforming Growth Factor β Soluble Receptor II

ICN 195741   Human recombinant, expressed in NSO   ≥97%; lyophilized; ED_{50} = 200-500 ng/mL

Transforming Growth Factor β Soluble Receptor II/Fc Chimera

ICN 195755   Human recombinant, expressed in NSO   ≥97%; lyophilized; ED_{50} = 5-15 ng/mL

Transforming Growth Factor β Soluble Receptor III

ICN 195742   Human recombinant, expressed in NSO   ≥97%; lyophilized; 20-50 ng/mL will inhibit half the biological response of 0.4 ng/mL of TGF-β2

Transforming Growth Factor β1

ICN 154134   Human platelets   ≥95%; frozen liquid; $\geq 5 \times 10^5$ U/mg, 1 U is the amount/mL culture medium needed for half maximal culture medium acidification by Rat-1 cells | Active across a broad range of mammalian cells

ICN 160266   Human recombinant, expressed in CHO cells   >97%; lyophilized; ED_{50} = 0.02-0.06 ng/mL, measured by the dose dependent inhibition of ^{3}thymidine uptake by murine IL-4 dependent HT-2 cell line

ICN 154133   Porcine platelets   >97%; lyophilized with BSA; ED_{50} = 0.02-0.06 ng/mL, measured by the dose dependent inhibition of ^{3}H-thymidine uptake by murine IL-4 dependent HT-2 cell line

Transforming Growth Factor β1, (^{125}I)-

Amersham IM246   Lyophilized; 29.6-81.4 TBq/mmol, 800-2200 Ci/mmol | Receptor binding grade

Transforming Growth Factor β1, Latent Form

ICN 195750   Human, recombinant CHO-expressed   ≥97%; lyophilized; ED_{50} = 20-60 ng/mL before acid activation; 0.15-0.5 ng/mL after activation

Transforming Growth Factor β1.2

ICN 195795   Human recombinant, expressed in *Sf*21   ≥97%; lyophilized; ED_{50} = 0.04-0.08 ng/mL

ICN 154136   Porcine platelets   >97%; lyophilized with BSA; ED_{50} = 0.04-0.08 ng/mL, measured by the dose dependent inhibition of ^{3}H-thymidine uptake by murine IL-4 dependent HT-2 cell line

Transforming Growth Factor β2

ICN 160267   Human recombinant, expressed in *E. coli*   ≥93%; $\geq 5 \times 10^7$ U/mg, 1 U is the amount/mL culture medium needed for half maximal inhibition of Mv1Lu cell proliferation | Active across most mammalian cells & *Xenopus*; endothelial & hematopoietic progenitor cells demonstrate a weak response to TGF-β2 but are potently inhibited by TGF-β1

ICN 154135   Porcine platelets   >97%; lyophilized with BSA; ED_{50} = 0.1-0.2 ng/mL, measured by the dose dependent inhibition of ^{3}H-thymidine uptake by murine IL-4 dependent HT-2 cell line

Transforming Growth Factor β3

ICN 160068   Human recombinant, expressed in insect *Sf*21 cells   >97%; lyophilized; ED_{50} = 0.01-0.03 ng/mL, measured by the dose dependent inhibition of ^{3}H-thymidine uptake by murine IL-4 dependent HT-2 cell line

Transforming Growth Factor β5

ICN 195743   Amphibian recombinant, expressed in *Sf*21   ≥97%; lyophilized; ED_{50} = 0.01-0.03 ng/mL

trk Peptide I

Amersham VPP37   Competition for VPC31; viral & oncogene recognition

Tumor Necrosis Factor Receptor I

Synonyms: Tumor Necrosis Factor RI, s-

ICN 195737   Human recombinant, expressed in *E. coli*   ≥97%; lyophilized; ED_{50} = 0.03-0.06 μg/mL in the presence of 0.25 ng/mL TNF-α

ICN 195796   Murine recombinant, expressed in *E. coli*   ≥97%; lyophilized; ED_{50} = 0.5-1.5 μg/mL in the presence of 0.1 ng/mL TNF-α

Tumor Necrosis Factor Receptor II

ICN 195738   Human recombinant, expressed in *E. coli*   ≥97%; lyophilized; ED_{50} = 0.5-1.5 μg/mL in the presence of 0.1 ng/mL TNF-α

ICN 195797   Murine recombinant, expressed in *E. coli*   ≥97%; lyophilized; ED_{50} = 1-3 μg/mL in the presence of 0.1 ng/mL TNF-α

Tumor Necrosis Factor α

ICN 152178   Human recombinant, expressed in *E. coli*   High purity grade; ≥98%; lyophilized; ED_{50} ≤0.2 ng/mL, by dose-dependent cytolysis of murine 929 cells with actinomycin D present; <0.1 ng/μg endotoxin

ICN 153474   Human recombinant, expressed in *E. coli*   ≥97%; lyophilized; ED_{50} = 0.2 ng/mL, by dose-dependent cytolysis of murine L 929 cells with actinomycin D present | Exerts cytotoxic effects on numerous tumor & target cells

ICN 154564   Human recombinant, expressed in *E. coli*   ≥97%; lyophilized; ED_{50} = 0.2 ng/mL, by dose-dependent cytolysis of murine L 929 cells with actinomycin D present | Neta, etal, *J Immunol*, 140:108, 1988; Beutler, G etal, *Nature*, 320:584, 1986; Paul, etal, Ann Rev Immunol, 6:407, 1988

ICN 152312   Murine recombinant, expressed in *E. coli*   >97%; lyophilized; ED_{50} = 0.1-0.5 ng/mL, by dose-dependent cytolysis of murine L 929 cells with actinomycin D present

ICN 160070   Rat recombinant, expressed in *E. coli*   ≥95%; lyophilized; 10^7 U/mg; 1 U stimulates dose-dependent cytolysis of WEHI 164 cells | Active on both rat & mouse cells

ICN 195803   Rat recombinant, expressed in *E. coli*   ≥95%; lyophilized; ED_{50} <0.02 ng/mL

Tumor Necrosis Factor α, ((^{125}I)-Tyr)-

Amersham IM206   Human recombinant   Lyophilized; 18.5-37 TBq/mmol, 500-1000 Ci/mmol

Tumor Necrosis Factor α, (^{125}I)-

ICN 68121   Human recombinant   ~100 μCi/μg in 0.1 *M* KPB, pH 7.5, 0.5% BSA

Tumor Necrosis Factor α, Truncated Form

ICN 195760   Murine recombinant, expressed in *E. coli*   ≥97%; lyophilized; ED_{50} = 5-10 pg/mL | Amino-terminal truncated

Tumor Necrosis Factor β

ICN 154139   Human recombinant, expressed in *E. coli*   ≥98%; lyophilized; ED_{50} ≤0.1 ng/mL, by dose-dependent cytolysis of L 929 cells with actinomycin D; <0.1 ng/μg

TYK2 Peptide

BioSource International 04-418

Tyrosine Kinase Substrate I

BioSource International 77-384

Tyrosine Kinase Substrate II

BioSource International 77-382

Tyrosine Kinase Substrate III

BioSource International 77-386

Tyrosine Phosphorylation Peptide

BioSource International 77-430

Ubiquitin, (^{125}I)-

Amersham IM306   Lyophilized; ~74 TBq/mmol, ~2000 Ci/mmol

Urocortin, (^{125}I)(Tyr^0)-

Synonyms: Urocortin Receptor Ligand, (^{125}I)(Tyr^0)-; Corticotropin Releasing Factor Receptor Ligand

Amersham IM295   Rat   Lyophilized; ~74 TBq/mmol, ~2000 Ci/mmol | Non-selective; agonist; K_I = 0.41 nM CRF$_{2\beta}$/CHO membrane

Urocortin, Fluo-

Amersham VUCT011   Fluorescein-labeled with high activity

Vasoactive Intestinal Polypeptide, (3-(^{125}I)-Tyr^{10})-

Amersham IM158   Lyophilized; ~74 TBq/mmol, ~2000 Ci/mmol

Vasoactive Intestinal Polypeptide, Fluo-

Amersham VVIP011   Human   Fluorescein-labeled with high activity

Vasopressin-(Arg^8), (3-(^{125}I)-Tyr^2)-

Amersham IM182   Lyophilized; ~74 TBq/mmol, ~2000 Ci/mmol

Vasopressin-(Arg^8), (Tyr-3,5(n)-^{3}H)-

Amersham TRK776   0.1 *M* Ammonium acetate buffer, pH 5.5:acetonitrile (3:1); 0.37-1.1 TBq/mmol, 10-30 Ci/mmol; 7.4 MBq/mL, 200 μCi/mL

Vasopressin, (Arg^8)-

Synonyms: Argipressin Acetate; Thyrotropin Releasing Hormone

Fluka 94836   MW 1084.2 $C_{46}H_{65}N_{15}O_{12}S_2$ Acetate salt; ≥98.0% (HPLC) | Lumpkin, MD et al, *Science*, 235:1070, 1987

Biogenesis 9536-0604   MW 1086.4   Synthetic   >99% (TLC/HPLC); lyophilized, consisting of 89.5% peptide material, acetate counter ions and residual H_2O

Vasopressin, (Arg^8-,Tyr-3,5-^{3}H)-

ARC ART-703   MW 1085.2   60-85 Ci/mmol; 2.22-3.15 TBq/mmol; in EtOH, 0.05 *N* HOAc | Radiochemical

Vasopressin, Agarose

ICN 191115   Ligand concentration ~ 2 μ*M*/g in pH 4.5 buffer, 0.02% NaN$_3$, 4°C

Vasopressin, Fluo-

Amersham VVPN011   Fluorescein-labeled with high activity

Vesicular Accholine Transporter Control Peptide (511-530)

Oncogene PP89   Rat   Lyophilized | For immunoabsorption

v-H-ras Peptide (62-76)

Oncogene PP08   Lyophilized with an equal amount of BSA;
suggested starting concentration of peptide for competition is 10X
the concentration of primary antibody | For competition studies &
dot blot

Part 5. Amino Acids & Derivatives

Alanine — (±)-3-(3-Nitrobenzoyl)-Ala Hydrochloride

Synonyms: Kynurenine-3-Hydroxylase Inhibitor

Alexis 550-102   MW 274.7 $C_{10}H_{10}N_2O_5 \cdot HCl$ ≥98%; off-white solid; soluble in H_2O | First potent & selective inhibitor; Pellicciari, R et al, *J Med Chem*, 37: 647, 1994; Carpendo et al, *Neuroscience*, 61: 237, 1994

Alanine — (-)-3-(3,4-Dihydroxyphenyl)-2-Me-L-Ala

Synonyms: 2-Me-3-(3,4-Dihydroxyphenyl)-L-Ala; L-α-Me-DOPA; L-Aromatic Amino Acid Decarboxylase Inhibitor

Sigma 85,741-6 $C_{10}H_{13}NO_4 \cdot 1\frac{1}{2}H_2O$ Converted to α-methylnorepinephrine, a potent α_2 antagonist

Alanine — (α,α-Dicyclohexyl)-DL-Ala

Bachem F-3695.0250   MW 253.39 $C_{15}H_{27}NO_2$ Store at -15°C

Alanine — (α,α-Diphenyl)-DL-Ala

Bachem F-3840.0001   MW 241.29 $C_{15}H_{15}NO_2$ Store at RT

Alanine — 1-Naphthyl-Ala

Bachem F-1840.0001   MW 215.25 $C_{13}H_{13}NO_2$ Store at RT

Alanine — 2-Me-3-(3,4-Dihydroxyphenyl)-DL-Ala

Synonyms: DL-α-Me-DOPA; L-Aromatic Amino Acid Decarboxylase Inhibitor

Sigma M 7277 $C_{10}H_{13}NO_2$ Converted to α-methylnorepinephrine, a potent α_2 antagonist

Alanine — 2-Me-Ala

Synonyms: α-Aib

Fluka 08280   MW 103.12 $C_4H_9NO_2$ ≥99% (titration); ~0.1% residue on ignition; mp: >300°C

USBio A1378   MW 103.1 $C_4H_9NO_2$ ≥99%

Alanine — 2-Naphthyl-Ala

Bachem F-1855.0001   MW 215.25 $C_{13}H_{13}NO_2$ Store at RT

Alanine — 2-Naphthyl-Ala-OBzl

Bachem F-3710.0001   MW 305.38 $C_{20}H_{19}NO_2$ Store at RT

Alanine — 3-(1-Naphthyl)-D-Ala

Fluka 70732   MW 215.26 $C_{13}H_{13}NO_2$ ≥99.0% (TLC)

Alanine — 3-(1-Naphthyl)-L-Ala

Synonyms: (S)-α-Amino-1-Naphthalenepropanoic Acid

Fluka 70728   MW 215.26 $C_{13}H_{13}NO_2$ ≥99.0% (TLC); mp: 254-257°C | In a study of analogs of the LH-RH with modifications in position 3, the potency of (3-(1-Naphthyl)-L-Ala)[3]-LH-RH was 187% of that of synthetic LH-RH; Yabe, Y et al, *Chem Pharm Bull*, 24: 3149, 1976

Alanine — 3-(2-Naphthyl)-D-Ala

Fluka 70727   MW 215.26 $C_{13}H_{13}NO_2$ ≥98.0% (HPLC)

Alanine — 3-(2-Propenylsulfinyl)-Ala

Synonyms: Alliin

ICN 194125 $C_6H_{11}NO_3S$

Alanine — 3-(2-Thienyl)-DL-Ala

Synonyms: 2-Amino-3-(2-Thienyl)-Propionic Acid

Fluka 88425   MW 171.21 $C_7H_9NO_2S$ ≥98.0% (titration); mp: 270°C

Alanine — 3-(2-Thienyl)-L-Ala

Synonyms: (S)-2-Amino-3-(2-Thienyl)-Propionic Acid

Fluka 88424   MW 171.21 $C_7H_9NO_2S$ ≥98.0%; mp: 260°C

Alanine — 3-(3,4-Dihydroxyphenyl)-2-Me-L-Ala

Synonyms: Me-L-DOPA

Fluka 37862   MW 238.24 $C_{10}H_{13}NO_4 \cdot 1\frac{1}{2}H_2O$ ≥99% (TLC); mp: >300°C; ~12% H_2O; ≤0.5% Ala | Dastidar, SG et al, *Indian J Med Res*, 84: 142, 1986

Alanine — 3-(3,4-Dihydroxyphenyl)-D-Ala

Synonyms: D-DOPA

Fluka 37840   MW 197.19 $C_9H_{11}NO_4$ ≥98% (titration); mp: 276-278°C | False dopa reaction in (cytochemical & histochemical) studies of mammalian tyrosinase; White, R et al, *Stain Tech*, 58: 13, 1983

Alanine — 3-(3,4-Dihydroxyphenyl)-DL-Ala

Synonyms: DL-DOPA

Fluka 37850   MW 197.19 $C_9H_{11}NO_4$ ≥98% (titration); mp: 270-272°C

Alanine — 3-(3,4-Dihydroxyphenyl)-L-Ala

Synonyms: L-DOPA; Tyrosinase Substrate

Fluka 37830   MW 197.19 $C_9H_{11}NO_4$ ≥99% (titration); mp: 276-278°C | Substrate for the assay of tyrosinase; Vachtenheim, J et al, *Anal Biochem*, 146: 405, 1985; Pomeranntz, SH & Li, JP-C, *Meth Enzymol*, 17A: 620, 1970; Voltattorini, CB et al, *Meth Enzymol*, 142: 179, 1987; Polis, D & Shmukler, HW, *Meth Enzymol*, 2: 813, 1957

Alanine — 3-(3-Pyridyl)-Ala

USBio P9530   MW 167.1 $C_8H_{11}NO_3$ ≥99%

Alanine — 3-(3-Pyridyl)-D-Ala

Synonyms: 3'-Aza-D-Phe

Fluka 82935   MW 166.18 $C_8H_{10}N_2O_2$ ≥99.0% (HPLC); ≤0.5% H_2O; ≤0.1% residue on ignition

Alanine — 3-(4-(4-Hydroxy-3,5-Diiodophenoxy)-3,5-Diiodophenyl)-D-Ala

Synonyms: D-Thyroxine

Sigma T 2001 $C_{15}H_{11}I_4NO_4$

Alanine — 3-(4-(4-Hydroxy-3,5-Diiodophenoxy)-3,5-Diiodophenyl)-L-Ala Free Acid

Synonyms: L-Thyroxine; T_4

Sigma T 2376   FW 776.9 $C_{15}H_{11}I_4NO_4$ L-Thyroxine, T4 & tri-Iodo-L-thyronine, T3, are iodine-containing hormones produced from thyroglobulin in the thyroid follicular cells; stimulation of metabolic rate & regulation of growth & development by these hormones are due to their effects on DNA transcription & thus protein synthesis; Oppenheimer, JH et al, *Endocr Rev*, 8: 288, 1987; DeGroot, LJ & Niepomniszcze, H, *Metabolism*, 26: 665, 1977; Dussault, JH & Ruel, *J Ann Rev Physiol*, 49: 321, 1987; Evans, RM, *Science*, 240: 889, 1988

Alanine — 3-(4-(4-Hydroxy-3,5-Diiodophenoxy)-3,5-Diiodophenyl)-L-Ala Sodium Salt Pentahydrate

Sigma T 2501 $C_{15}H_{10}I_4NO_4Na \cdot 5H_2O$

Alanine — 3-(4-(p-Hydroxyphenoxy)-Phenyl)-DL-Ala

Synonyms: DL-Thyronine

ICN 105224	MW 273.3	$C_{15}H_{15}NO_4$	Off-white to gray powder
Sigma T 1501		$C_{15}H_{15}NO_4$	
Sigma T 5780		$C_{15}H_{15}NO_4$	

Alanine — 3-(4-(*p*-Hydroxyphenoxy)-Phenyl)-L-Ala

Synonyms: L-Thyronine

Sigma T 5905 $C_{15}H_{15}NO_4$

Alanine — 3,4-Dihydroxy-Ala (1-^{14}C)

Synonyms: L-DOPA

ARC ARC-1220   MW 197.2 $3,4-(OH)_2C_6H_3CH_2CH(NH_2)COOH$
50-60 mCi/mmol; 1.85-2.22 GBq/mmol; in 0.2 *N* HOAc: EtOH (9:1)
| Radiochemical

Alanine — 3-Cyclohexyl-D-Ala Hydrate

Fluka 29295   MW 171.23 $C_9H_{17}NO_2$ ≥99% (titration); 1 mole
H_2O

Alanine — 3-Cyclohexyl-L-Ala Hydrate

Synonyms: (*S*)-α-Aminocyclohexanepropionic Acid; (*S*)-2-Amino-
3-Cyclohexylpropionic Acid; Hexahydro-L-Phe

Fluka 07623   MW 171.23 $C_9H_{17}NO_2$ ≥95% (titration); mp:
235°C

Fluka 29290   MW 171.23 $C_9H_{17}NO_2$ ≥97% (titration); 3%
H_2O

Alanine — Ac-D-2-Naphthyl-Ala

Bachem F-2420.0001   MW 257.29 $C_{15}H_{15}NO_3$ Store at 2-
8°C

Alanine — Ac-β-Ala

Senn Chem 44078   MW 131.1   Substrate for N-acetyl-α-Ala
deacetylase

Alanine — Ala

Bachem E-1285.0025   MW 89.09 $C_3H_7NO_2$ Store at RT

Alanine — Ala (^{15}N)

Bachem E-3225.0100   MW 90.09 $C_3H_715NO_2$ Store at -15°C

Alanine — Ala Amide

Bachem E-1290.0001   MW 88.11 $C_3H_8N_2O$ Store at RT

Alanine — Ala Amide Hydrobromide

Bachem E-2445.0001   MW 169.02 $C_3H_8N_2O$ · HBr   Store at
2-8°C

Alanine — Ala Amide Hydrochloride

Bachem E-1295.0001   MW 124.57 $C_3H_8N_2O$ · HCl   Store at
RT

Alanine — Ala-4MβNA Free Base

Bachem J-1000.0250   MW 244.29 $C_{14}H_{16}N_2O_2$ Store at
-15°C

Alanine — Ala-4MβNA Hydrochloride

Bachem J-1005.0250   MW 280.75 $C_{14}H_{16}N_2O_2$ · HCl   Store
at -15°C

Alanine — Ala-AFC TFA

Bachem I-1455.0050   MW 414.26 $C_{13}H_{11}F_3N_2O_3$ · $C_2HF_3O_2$
Store at -15°C

Alanine — Ala-AMC

Bachem I-1410.0050   MW 246.27 $C_{13}H_{14}N_2O_3$ Store at
-15°C

Alanine — Ala-AMC TFA

Bachem I-1020.0050   MW 360.14 $C_{13}H_{14}N_2O_3$ · $C_2HF_3O_2$
Store at -15°C

Alanine — Ala-NHMe Hydrochloride

Bachem E-2840.0001   MW 138.6 $C_4H_{10}N_2O$ · HCl   Store at
RT

Alanine — Ala-OAll *p*-Tosyl

Bachem E-3505.0001   MW 302.37 $C_6H_{12}NO_2$ · $C_7H_8O_3S$
Store at 2-8°C

Alanine — Ala-OBzl Hydrochloride

Bachem E-1310.0005   MW 215.68 $C_{10}H_{13}NO_2$ · HCl   Store
at RT

Alanine — Ala-OBzl *p*-Tosyl

Bachem E-1315.0005   MW 351.42 $C_{10}H_{13}NO_2$ · $C_7H_8O_3S$
Store at RT

Alanine — Ala-OEt Hydrochloride

Bachem E-1330.0010   MW 153.61 $C_5H_{11}NO_2$ · HCl   Store
at RT

Alanine — Ala-ol

Synonyms: L-Alaninol; *S*-(+)-2-Aminopropanol; Ethanolamine
Ammonia Lyase Substrate

Senn Chem 44091   MW 75.1   Liquid | Chiral intermediate

Alanine — Ala-OMe Hydrochloride

Bachem E-1335.0010   MW 139.58 $C_4H_9NO_2$ · HCl   Store at
RT

Alanine — Ala-ONp Hydrochloride

Bachem E-1345.0005   MW 246.65 $C_9H_{10}N_2O_4$ · HCl   Store
at -15°C

Alanine — Ala-OtBu Hydrochloride

Bachem E-1325.0005   MW 181.67 $C_7H_{15}NO_2$ · HCl   Store
at 2-8°C

Alanine — Ala-pNA

Synonyms: Aminopeptidase Substrate

Bachem L-1070.0001   MW 209.21 $C_9H_{11}N_3O_3$ Store at
-15°C

Peptides International SAN-3068   MW 209.20 $C_9H_{11}N_3O_3$
>99% (HPLC); amorphous powder | Peleiderer, G in *Proteolytic
Enzymes*, Methods in Enzymology, Vol 19, GE Perlmann & L Lorand,
Eds, Academic Press, New York, 1970, pp 514-521

Alanine — Ala-pNA Hydrochloride

Bachem L-1075.0001   MW 245.7 $C_9H_{11}N_3O_3$ · HCl   Store at
-15°C

Alanine — Ala-*p*ONb Hydrobromide

Bachem E-1340.0005   MW 305.13 $C_{10}H_{12}N_2O_4$ · HBr   Store
at -15°C

Alanine — Ala-βNA

Bachem K-1012.0001   MW 214.27 $C_{13}H_{14}N_2O$ Store at RT

Alanine — Ala-βNA Hydrobromide

Bachem K-1015.0001   MW 295.18 $C_{13}H_{14}N_2O$ · HBr   Store
at -15°C

Alanine — BOC-(1,3-Diphenyl)-L-Ala
Fluka 09896   MW 341.41 $C_{20}H_{23}NO_4$ ≥98% (HPLC)

Alanine — BOC-1-Naphthyl-Ala
Bachem A-3225.0001   MW 315.37 $C_{18}H_{21}NO_4$ Store at -15°C

USBio B2254   MW 315.2 $C_{18}H_{21}NO_4$ ≥99%

Alanine — BOC-2-Naphthyl-Ala
Bachem A-2850.0001   MW 315.37 $C_{18}H_{21}NO_4$ Store at RT

USBio B2256   MW 315.2 $C_{18}H_{21}NO_4$ ≥99%

Alanine — BOC-2-Naphthyl-Ala-® (200-400 mesh)
Bachem D-1755.0001   Store at RT

Alanine — BOC-3-(1-Naphthyl)-D-Ala
Fluka 15045   MW 315.37 $C_{18}H_{21}NO_4$ ≥98% (TLC)

Alanine — BOC-3-(2-Furyl)-D-Ala Dicyclohexylamine Salt
Synonyms: (R)-2-(BOC-Amino)-3-(2-Furyl)-Propionic Acid
Fluka 09804   MW 436.59 $C_{12}H_{17}NO_5 \cdot C_{12}H_{23}N$ ≥98% (TLC)

Alanine — BOC-3-(2-Furyl)-L-Ala Dicyclohexylamine Salt
Synonyms: (S)-2-(BOC-Amino)-3-(2-Furyl)-Propionic Acid
Fluka 09803   MW 436.59 $C_{12}H_{17}NO_5 \cdot C_{12}H_{23}N$ ≥98% (TLC)

Alanine — BOC-3-(2-Pyridyl)-L-Ala
Fluka 15026   MW 266.3 $C_{13}H_{18}N_2O_4$ ≥98% (TLC)

Alanine — BOC-3-(2-Thienyl)-L-Ala
Fluka 15501   MW 271.33 $C_{12}H_{17}NO_4S$ ≥98% (TLC); mp: 71-76°C

Alanine — BOC-3-(3-Benzothienyl)-D-Ala
Fluka 15039   MW 321.39 $C_{16}H_{19}NO_4S$ ≥97% (HPLC)

Alanine — BOC-3-(3-Benzothienyl)-L-Ala
Fluka 15040   MW 321.39 $C_{16}H_{19}NO_4S$ ~98% (TLC)

Alanine — BOC-3-(3-Pyridyl)-L-Ala
Fluka 15027   MW 266.3 $C_{13}H_{18}N_2O_4$ ≥99% (TLC); mp: 135-138°C

Alanine — BOC-3-(3-Thienyl)-D-Ala Dicyclohexylamine Salt
Fluka 03754   MW 452.65 $C_{12}H_{17}NO_4S \cdot C_{12}H_{23}N$ ≥98% (TLC)

Alanine — BOC-3-(4-Pyridyl)-D-Ala
Fluka 15030   MW 266.3 $C_{13}H_{18}N_2O_4$ ≥98% (TLC); mp: 225-229°C

Alanine — BOC-3-(4-Pyridyl)-L-Ala
Fluka 15032   MW 266.3 $C_{13}H_{18}N_2O_4$ ≥98% (TLC)

Alanine — BOC-3-Cyclohexyl-D-Ala Hydrate
Fluka 15476   MW 271.36 $C_{14}H_{25}NO_4$ ~1 mol H_2O

Alanine — BOC-3-Cyclohexyl-L-Ala Hydrate
Fluka 15477   MW 271.36 $C_{14}H_{25}NO_4$ ~1 mol H_2O

Alanine — BOC-3-Iodo-D-Ala-OBzl
Fluka 15122   MW 405.23 $C_{15}H_{20}INO_4$ ≥98% (TLC); mp: 80-84°C

Alanine — BOC-3-Iodo-D-Ala-OMe
Fluka 15124   MW 329.14 $C_9H_{16}INO_4$ ≥98% (GC); mp: 57-61°C

Alanine — BOC-3-Iodo-L-Ala-OBzl
Fluka 15123   MW 405.23 $C_{15}H_{20}INO_4$ ≥98% (TLC); mp: 80-84°C

Alanine — BOC-3-Pyridyl-Ala
USBio B2258   MW 266.3 $C_{13}H_{18}N_2O_4$ ≥98%

Alanine — BOC-3-Styryl-D-Ala Dicyclohexylamine Salt
Synonyms: (R)-2-(BOC-Amino)-5-Phenyl-4-Pentenoic Acid
Fluka 09801   MW 472.67 $C_{16}H_{21}NO_4 \cdot C_{12}H_{23}N$ ≥98% (TLC)

Alanine — BOC-3-Styryl-L-Ala Dicyclohexylamine Salt
Synonyms: (S)-2-(BOC-Amino)-5-Phenyl-4-Pentenoic Acid
Fluka 03753   MW 472.67 $C_{16}H_{21}NO_4 \cdot C_{12}H_{23}N$ ≥98% (TLC)
Fluka 09799   MW 472.67 $C_{16}H_{21}NO_4 \cdot C_{12}H_{23}N$ ≥98% (TLC)

Alanine — BOC-Ala
Synonyms: BOC-L-Ala
American Peptide BLALA15   MW 189.2
Peptides International BLA-2051   MW 189.21   >98% (HPLC); white powder

Alanine — BOC-Ala-ol
Synonyms: BOC-L-Alaninol
Senn Chem 44018   MW 175.2
USBio B2251   MW 189.2 $C_8H_{15}NO_4$ ≥99%

Alanine — BOC-Ala-O-Resin
American Peptide RMRB100   1% DVB cross-linked: 100-200 mesh | t-BOC protected AA resin; preattached resins are useful for synthesis of peptides with C-terminal carboxyl group by t-BOC chemistry

Alanine — BOC-Ala-PAM-Resin
American Peptide RPAM100   1% DVB cross-linked: 100-200 mesh | t-BOC protected AA resin; preattached resins are useful for synthesis of peptides with C-terminal carboxyl group by t-BOC chemistry

Alanine — BOC-Cyclohexyl-Ala
Synonyms: BOC-Cha
USBio B2287   MW 271.7 $C_{14}H_{26}NO_4$ ≥99%

Alanine — BOC-D-1-Naphthyl-Ala
Synonyms: BOC-D-Nal(1')
Bachem A-4305.0001   MW 315.37 $C_{18}H_{21}NO_4$ Store at RT
Peptides International BDX-5381-PI   MW 315.37   >98% (HPLC); white to off-white powder
USBio B2253   MW 315.2 $C_{18}H_{21}NO_4$ ≥99%

Alanine — BOC-D-2-Naphthyl-Ala
Synonyms: BOC-D-Nal(2')
Bachem A-2575.0001   MW 315.37 $C_{18}H_{21}NO_4$ Store at -15°C
Peptides International BDX-5319-PI   MW 315.37   >98% (HPLC); white amorphous powder | Most popular Nal

USBio B2255 MW 315.2 $C_{18}H_{21}NO_4$ ≥99%

Alanine — BOC-D-2-Pyridyl-Ala

American Peptide BDALA02 MW 267.3

Alanine — BOC-D-2-Thienylalanine

Synonyms: BOC-Thi

Peptides International BDX-5389-PI MW 271.34 >98% (HPLC); white to off-white powder

Alanine — BOC-D-3-Benzothienyl-Ala

American Peptide BDALA04 MW 321.4

Alanine — BOC-D-3-Pyridyl-Ala

Synonyms: BOC-D-3-Pal

American Peptide BDALA40 MW 267.3

Peptides International BDX-5339-PI MW 266.30 >98% (HPLC); white crystalline powder

USBio B2257 MW 266.3 $C_{13}H_{18}N_2O_4$ ≥98%

Alanine — BOC-D-4-Pyridyl-Ala

Synonyms: BOC-D-4-Pal

American Peptide BDALA03 MW 267.3

Peptides International BDX-5533-PI MW 339.2 >98% (HPLC); white crystalline powder

Alanine — BOC-D-Ala

Fluka 15048 MW 189.21 $C_8H_{15}NO_4$ ≥98% (TLC)

Neosystem BA00101 MW 189.2

Peptides International BDA-2606 MW 189.21 >98% (HPLC); white crystalline powder

USBio B2250 MW 189.2 $C_8H_{15}NO_4$ ≥99%

Alanine — BOC-D-Alaninol

Synonyms: (R)-2-(BOC-Amino)-1-Propanol

Fluka 15034 MW 175.23 $C_8H_{17}NO_3$ mp: 57-61°C

Alanine — BOC-D-Ala-O-CH₂-φ-CH₂-COOH

Neosystem LP00101 MW 337.4

Alanine — BOC-D-Ala-OMe

Fluka 15049 MW 203.24 $C_9H_{17}NO_4$ ≥98% (GC)

Alanine — BOC-D-Ala-PAM Resin

Neosystem RP00101

Alanine — BOC-L-1-Naphthyl-Ala

Synonyms: BOC-Nal(1')

Peptides International BLX-5379-PI MW 315.37 >98% (HPLC); white crystalline powder

Alanine — BOC-L-2-Naphthyl-Ala

Synonyms: BOC-Nal(2')

Peptides International BLX-5327-PI MW 315.37 >98% (HPLC); white crystalline powder | Most popular Nal

Alanine — BOC-L-2-Pyridyl-Ala

American Peptide BLALA02 MW 267.3

Alanine — BOC-L-2-Thienylalanine

Peptides International BBX-5387-PI MW 271.34 >98% (HPLC); white to off-white powder

Alanine — BOC-L-3-Benzothienyl-Ala

American Peptide BLALA04 MW 321.4

Alanine — BOC-L-3-Pyridyl-Ala

Synonyms: BOC-3-Pal

American Peptide BLALA35 MW 267.3

Peptides International BLX-5331-PI MW 266.30 >98% (HPLC); white crystalline powder

Alanine — BOC-L-4-Pyridyl-Ala

Synonyms: BOC-4-Pal

American Peptide BLALA03 MW 267.3

Peptides International BLX-5532-PI MW 339.2 >98% (HPLC); white crystalline powder

Alanine — BOC-L-Ala

Fluka 15380 MW 189.21 $C_8H_{15}NO_4$ ≥99% (titration); mp: 79-81°C

Neosystem BA00102 MW 189.2

Alanine — BOC-L-Ala Hydroxysuccinimide Ester

Fluka 15385 MW 286.29 $C_{12}H_{18}N_2O_6$ mp: 161-163°C

Alanine — BOC-L-Ala-(4-ONp)

Fluka 15052 MW 310.31 $C_{14}H_{18}N_2O_6$ ~98% (HPLC)

Alanine — BOC-L-Ala-MBHA Resin

Neosystem RN00102

Alanine — BOC-L-Alaninol

Synonyms: (S)-2-(BOC-Amino)-1-Propanol

Fluka 15394 MW 175.23 $C_8H_{17}NO_3$ mp: 57-61°C | A protected form of L-alaninol utilized as intermediate for the preparation of biologically active peptides & amino aldehydes; Stanfield, CF et al, *J Org Chem*, 46: 4799, 1981; Stanfield, CF et al, *ibid*, 47: 3016, 1982; Acton, N & Komoriya, A, *Org Prep Proc Int*, 14: 381, 1982

Alanine — BOC-L-Ala-O-CH₂-φ-CH₂-COOH

Neosystem LP00102 MW 337.4

Alanine — BOC-L-Ala-OMe

Fluka 15051 MW 203.24 $C_9H_{17}NO_4$ ≥98% (TLC)

Alanine — BOC-L-Ala-PAM Resin

Neosystem RP00102

Alanine — BOC-L-β-Hal

Synonyms: (S)-3-(BOC-Amino)-Butyric Acid

Fluka 14974 MW 203.24 $C_9H_{17}NO_4$ ≥98% (TLC)

Alanine — BOC-L-β-Hal-OAll

Synonyms: (S)-3-(BOC-Amino)-Butyric Acid-OAll

Fluka 53458 MW 243.3 $C_{12}H_{21}NO_4$ ≥98% (TLC)

Alanine — BOC-L-β-Thienylalanine

American Peptide BLALA45 MW 271.2

Alanine — BOC-Me-Ala

USBio B2348 MW 203.2 $C_9H_{17}NO_4$ ≥98%

Alanine — BOC-*N*,2-Dimethyl-Ala

Synonyms: *N*-BOC-α-Methylamino Isobutyric Acid; BOC-*N*-Me-Aib

Fluka 10927   MW 217.26 $C_{10}H_{19}NO_4$ ≥98% (HPLC); mp: 154-158°C

Alanine — BOC-*N*-Me-D-Ala

Fluka 15159   MW 203.24 $C_9H_{17}NO_4$ ≥98% (TLC)

Alanine — BOC-*N*-Me-L-Ala

Fluka 15549   MW 203.24 $C_9H_{17}NO_4$ ≥99% (TLC); mp: 88-92°C

Neosystem BA00103   MW 203.2

Alanine — BOC-α-Me-Ala

Synonyms: α-(BOC-Amino)-Isobutyric Acid

Fluka 15466   MW 203.24 $C_9H_{17}NO_4$ ≥99% (titration); mp: 118-121°C | Protected unusual & seldom-occurring achiral AA; mayr, W & Jung, G, *Liebigs Ann Chem*, 715: 1980

Alanine — BOC-β2-Thienyl-L-Ala

Synonyms: BOC-Thi

USBio B2400   MW 271.3 $C_{12}H_{17}NO_4S$ ≥99%

Alanine — BOC-β-Ala

American Peptide BLALA10   MW 189.2

Fluka 15382   MW 189.21 $C_8H_{15}NO_4$ ≥99% (titration); mp: 76-77°C

Neosystem BA02101   MW 189.2

Peptides International BBA-2131   MW 189.21   >98% (HPLC); white crystalline powder

Senn Chem 44017   MW 189.2

USBio B2252   MW 189.2 $C_8H_{15}NO_4$ ≥99%

Alanine — BOC-β-Ala-(4-ONp)

Fluka 15053   MW 310.31 $C_{14}H_{18}N_2O_6$ ≥98% (HPLC)

Alanine — BOC-β-Ala-OSu

Senn Chem 44019   MW 286.3

Alanine — BOC-β-Ala-PAM Resin

American Peptide RGEN180   Substitution: 0.1-0.2 mmol/g | Precursor resin for solid phase peptide synthesis

Fluka 09846   Crosslinked with 1% DVB; 0.4-0.6 mmol/g resin; 100-200 mesh size | PAM resin esterified with BOC-β-Ala

Alanine — BOC-β-Cyclohexyl-Ala

Bachem A-3760.0001   MW 271.36 $C_{14}H_{25}NO_4$ Store at RT

Alanine — BOC-β-Cyclohexyl-Ala Dicyclohexylamine

Bachem A-2960.0001   MW 452.68 $C_{14}H_{25}NO_4 \cdot C_{12}H_{23}N$ Store at RT

Alanine — BOC-β-Cyclohexyl-D-Ala

Synonyms: BOC-D-Cha

Bachem A-3840.0001   MW 271.36 $C_{14}H_{25}NO_4$ Store at RT

Neosystem BA02301   MW 271.4

Alanine — BOC-β-Cyclohexyl-D-Ala Dicyclohexylamine

Bachem A-2920.0001   MW 452.68 $C_{14}H_{25}NO_4 \cdot C_{12}H_{23}N$ Store at RT

Alanine — BOC-β-Cyclohexyl-L-Ala

Synonyms: BOC-L-Cha

Neosystem BA02302   MW 271.4

Alanine — BOC-β-Cyclohexyl-L-Alaninol

Bachem A-4015.0001   MW 257.37 $C_{14}H_{27}NO_3$ Store at -15°C

Alanine — BOC-β-Cyclopropyl-Ala Dicyclohexylamine

Bachem A-4150.0001   MW 410.6 $C_{11}H_{19}NO_4 \cdot C_{12}H_{23}N$ Store at -15°C

Alanine — BOC-β-Cyclopropyl-D-Ala Dicyclohexylamine

Bachem A-4155.0001   MW 410.6 $C_{11}H_{19}NO_4 \cdot C_{12}H_{23}N$ Store at -15°C

Alanine — BOC-β-tBu-Ala

Synonyms: BOC-Neopentylglycine

Bachem A-3110.0001   MW 245.32 $C_{12}H_{23}NO_4$ Store at RT

Alanine — BOC-β-tBu-D-Ala

Synonyms: BOC-D-Neopentylglycine

Bachem A-4210.0001   MW 245.32 $C_{12}H_{23}NO_4$ Store at RT

Alanine — Bpoc-Ala

Bachem E-1630.0001   MW 327.38 $C_{19}H_{21}NO_4$ Store at -15°C

Alanine — Bsmoc-Ala

Synonyms: Benzo-β-Thiophenesulfone-2-Methoxycarbonyl-L-Ala

Peptides International BLA-5612-PI   MW 311.32   >98% (HPLC); white to off-white powder | LA Carpino, et al, *JACS*, 119:9915, 1997

Alanine — Bz-Ala

Bachem E-1445.0005   MW 193.2 $C_{10}H_{11}NO_3$ Store at RT

Alanine — Bz-Ala-OMe

Bachem E-1450.0001   MW 207.23 $C_{11}H_{13}NO_3$ Store at RT

Peptides International SAM-3084   MW 207.23 $C_{11}H_{13}NO_3$ >98% (HPLC); amorphous powder

Alanine — Bz-Ala-βNA

Bachem K-1115.0250   MW 318.4 $C_{20}H_{18}N_2O_2$ Store at RT

Alanine — Bz-D-Ala

Bachem F-1282.0001   MW 193.2 $C_{10}H_{11}NO_3$ Store at RT

Alanine — Carbamoyl-DL-Ala

Synonyms: DL-2-Ureidopropionic Acid

Bachem F-2345.0001   MW 132.12 $C_4H_8N_2O_3$ Store at RT

Alanine — Carbamoyl-β-Ala

Synonyms: 3-Ureidopropionic Acid

Bachem F-1375.0001   MW 132.12 $C_4H_8N_2O_3$ Store at RT

Alanine — CBZ-Ala

USBio C2098-10   MW 223.2 $C_{11}H_{13}NO_4$ ≥99%

Alanine — CBZ-Ala β-Naphthyl Ester

USBio C2098-15   MW 214.3 $C_{13}H_{14}N_2O$ ≥99%

Alanine — CBZ-Ala-Amide

USBio C2098-14 MW 222.2 $C_{11}H_{14}N_2O_3$ ≥99%

Alanine — CBZ-Ala-ONp

USBio C2098-16 MW 344.3 $C_{17}H_{16}N_2O_6$ ≥99%

Alanine — CBZ-Ala-OSu

USBio C2098-17 MW 320.3 $C_{15}H_{16}N_2O_6$ ≥99%

Alanine — CBZ-D-Ala

USBio C2098-12 MW 223.2 $C_{11}H_{13}NO_4$ ≥99%

Alanine — CBZ-DL-Ala

USBio C2098-13 MW 223.2 $C_{11}H_{13}NO_4$ ≥99%

Alanine — CBZ-Me-Ala

USBio C2099-22 MW 237.2 $C_{12}H_{15}NO_4$ ≥99%

Alanine — CBZ-β-Ala

USBio C2098-11 MW 223.2 $C_{11}H_{13}NO_4$ ≥99%

Alanine — Chloroacetyl-β-Ala

Senn Chem 44085 MW 165.6

Alanine — D-1-Naphthyl-Ala

Bachem F-1845.0001 MW 215.25 $C_{13}H_{13}NO_2$ Store at RT

Alanine — D-2-Naphthyl-Ala

Bachem F-1860.0001 MW 215.25 $C_{13}H_{13}NO_2$ Store at RT

Alanine — D-3-(1-Naphthyl)-Ala

ICN 151728 $C_{13}H_{13}NO_2$

Sigma N 1396 $C_{13}H_{13}NO_2$ Analog of D-Phe; Mierke, DF et al, *Biopolymers*, 29: 179, 1990

Sigma N 5887 $C_{13}H_{13}NO_2$

USBio N0500-10 MW 215.3 $C_{13}H_{13}NO_2$ ≥99%

Alanine — D-3-(2-Naphthyl)-Ala

ICN 151729 $C_{13}H_{13}NO_2$

Sigma N 5387 $C_{13}H_{13}NO_2$

USBio N0501-10 MW 215.3 $C_{13}H_{13}NO_2$ ≥99%

Alanine — D-Ala

Synonyms: (R)-2-Aminopropionic Acid

Bachem F-1100.0005 MW 89.09 $C_3H_7NO_2$ Store at RT

Fluka 05140 MW 89.1 $C_3H_7NO_2$ ≥99%; ≤0.05% residue on ignition

ICN 100280 $C_3H_7NO_2$

Neosystem AA00101 MW 89.1

Peptides International ADA-2801 MW 89.09 >99.9% optical purity by RP-HPLC; white crystalline powder

Rexim MW 89.1 $C_3H_7NO_2$

Sigma A 3435 Cross-linked 4% beaded agarose; activation: CNBr; attachment: amino; spacer: 1 atom; ligand immobilized: 2-8 µmoles/mL; suspension in 2.0 M NaCl containing 0.02% thimerosal | Deutsch, DG & Mertz, ET, *Proc Fed Amer Soc Exp Biol*, 29: 647, 1970; Deutsch, DG & Mertz, ET, *Science*, 170: 1095, 1970; Vuento, M & Vaheri, A, *Biochem J*, 183: 331, 1979

Sigma A 7377 $C_3H_7NO_2$ Also available as part of a kit

USBio A1205-10 MW 89.09 $C_3H_7NO_3$ ≥99%; solubility: clear, colorless & complete; specific rotation: -13 to -15°C; loss on drying: ≤0.05%; chloride: ≤0.1%; sulfate: ≤0.2%; heavy metals (Pb): ≤0.001%

Alanine — D-Ala (1-^{14}C)

ARC ARC-341 MW 89.1 $CH_3CH(NH_2)COOH$ 50-60 mCi/mmol; 1.85-2.22 GBq/mmol; EtOH:H₂O (2:98) | Radiochemical

ICN 10010E $CH_3CH(NH_2)COOH$

Alanine — D-Ala (2,3-^{3}H)

ARC ART-179 MW 89.1 $CH_3CH(NH_2)COOH$ 30-60 Ci/mmol; 1.11-2.22 TBq/mmol; in 0.01 N HCl | Radiochemical

Alanine — D-Ala Amide Hydrochloride

Bachem F-1115.0001 MW 124.57 $C_3H_8N_2O \cdot HCl$ Store at RT

Alanine — D-Ala-AMC TFA

Bachem I-1025.0050 MW 360.3 $C_{13}H_{14}N_2O_3 \cdot C_2HF_3O_2$ Store at -15°C

Alanine — D-Ala-OBzl

Sigma A 7652 $C_{10}H_{13}NO_2 \cdot C_7H_8O_3S$

Alanine — D-Ala-OBzl p-Tosyl

Bachem F-1120.0001 MW 351.42 $C_{10}H_{13}NO_2 \cdot C_7H_8O_3S$ Store at RT

Alanine — D-Ala-ol

Senn Chem 44092 MW 75.1 Liquid | Chiral intermediate

Alanine — D-Ala-OMe Hydrochloride

Bachem F-1145.0001 MW 139.58 $C_4H_9NO_2 \cdot HCl$ Store at 2-8°C

ICN 100284 $C_4H_9NO_2 \cdot HCl$

Alanine — D-Ala-OtBu Hydrochloride

Bachem F-1130.0001 MW 181.66 $C_7H_{15}NO_2 \cdot HCl$ Store at -15°C

Alanine — D-Ala-pNA Hydrochloride

Bachem L-1080.0250 MW 245.7 $C_9H_{11}N_3O_3 \cdot HCl$ Store at -15°C

Alanine — D-Ala-βNA Hydrochloride

Bachem K-1020.0250 MW 250.73 $C_{13}H_{14}N_2O \cdot HCl$ Store at RT

Alanine — Dansyl-L-Ala

ICN 100007 $C_{15}H_{18}N_2O_4S \cdot C_6H_{13}N$

Sigma D 0125 $C_{15}H_{18}N_2O_4S \cdot C_6H_{13}N$

Alanine — Dansyl-β-Ala

Sigma D 9892 $C_{15}H_{18}N_2O_4S \cdot C_6H_{13}N$

Alanine — Dinitropyridyl-DL-Ala

ICN 100596 $C_8H_8N_4O_6$

Alanine — Dinitropyridyl-β-Ala

ICN 100930 $C_8H_8N_4O_6$

Alanine — DL-1-Naphthyl-Ala

Bachem F-1850.0001 MW 215.25 $C_{13}H_{13}NO_2$ Store at RT

Alanine — DL-2-Naphthyl-Ala
Bachem F-1865.0001   MW 215.25 $C_{13}H_{13}NO_2$ Store at RT

Alanine — DL-3-(1-Naphthyl)-Ala
ICN 155791 $C_{13}H_{13}NO_2$

Sigma N 5637 $C_{13}H_{13}NO_2$

Alanine — DL-3-(2-Naphthyl)-Ala
ICN 155792 $C_{13}H_{13}NO_2$

Sigma N 5762 $C_{13}H_{13}NO_2$

Alanine — DL-Ala
Synonyms: (±)-2-Aminopropionic Acid; DL-2-Aminopropionic Acid

Bachem F-1105.0025   MW 89.09 $C_3H_7NO_2$ Store at RT

Fluka 05150   MW 89.1 $C_3H_7NO_2$ ≥99.0%; ≤0.05% residue on ignition, ≤0.2% loss on drying; mp: 295-300°C

Rexim   MW 89.1 $C_3H_7NO_2$ White crystals or crystalline powder

Sigma A 3409   FW 89.09 $C_3H_7NO_2$ Crystalline; cell culture tested; insect cell culture tested

Sigma A 7502 $C_3H_7NO_2$ Also available as part of a kit

USBio A1205-15   MW 89.09 $C_3H_7NO_2$ ≥99%

Alanine — DL-Ala 2-(4-Chlorophenyl)-1,1-Dimethylethyl Ester Hydrochloride
Synonyms: Alaproclate

ICN 193576 $C_{13}H_{18}NO_2Cl \cdot HCl$ Powerful, selective serotonin uptake inhibitor

Alanine — DL-Ala Anhydride
ICN 104829 $C_6H_{12}N_2O_3$

Alanine — DL-Ala Hydroxamate
Sigma A 8127 $C_3H_8N_2O_2$

Alanine — DL-Ala β-Naphthylamide
Sigma A 2378 $C_{13}H_{14}N_2O$

Alanine — DL-Ala β-Naphthylamide Hydrochloride
Sigma A 2503 $C_{13}H_{14}N_2O \cdot HCl$

Alanine — DL-Ala-(1-^{13}C)
ICN 530064 $^{13}C_3H_7NO_2$; $CH_3CH(NH_2)COOH$

Alanine — DL-Ala-(^{15}N)
ICN 540119 $C_3H_7{}^{15}NO_2$; $CH_3CH(NH_2)COOH$

Sigma A 29,930-8 $(CH_3CH(^{15}NH_2)CO_2H)$ Screw cap bottle

Alanine — DL-Ala-OBzl p-Tosyl
Bachem F-2400.0005   MW 351.42 $C_{10}H_{13}NO_2S \cdot C_7H_8O_3S$ Store at RT

Alanine — DL-Ala-OEt Hydrochloride
Bachem F-2095.0025   MW 153.61 $C_5H_{11}NO_2 \cdot HCl$ Store at 2-8°C

Fluka 05180   MW 153.61 $C_5H_{11}NO_2 \cdot HCl$ ≥99% (titration); mp: 85-87°C

Sigma A 3149 $C_5H_{11}NO_2 \cdot HCl$

Alanine — DL-Ala-OMe Hydrochloride
Bachem F-1150.0025   MW 139.58 $C_4H_9NO_2 \cdot HCl$ Store at 2-8°C

ICN 104828 $C_4H_9NO_2 \cdot HCl$

Sigma A 8627 $C_4H_9NO_2 \cdot HCl$

Alanine — DL-Ala-βNA Hydrochloride
Bachem K-1030.0001   MW 250.73 $C_{13}H_{14}N_2O \cdot HCl$ Store at RT

Alanine — DL-α-Ala
ICN 100423 $C_3H_7NO_2$

ICN 194616 $C_3H_7NO_2$

Alanine — Dnp-L-Ala
ICN 100298 $C_9H_9N_3O_6$

Alanine — FA-Ala-OSu
Bachem M-1345.0001   MW 306.27 $C_{14}H_{14}N_2O_6$ Store at -15°C

Alanine — FMOC-1-Naphthyl-Ala
Bachem B-1965.0001   MW 437.5 $C_{28}H_{23}NO_4$ Store at RT

USBio F5306   MW 437.5 $C_{28}H_{23}NO_4$ ≥99%

Alanine — FMOC-1-Naphthyl-Ala-® (200-400 mesh)
Bachem D-1710.0001   Store at RT

Alanine — FMOC-2-Naphthyl-Ala
Bachem B-2100.0001   MW 437.5 $C_{28}H_{23}NO_4$ Store at RT

USBio F5308   MW 437.5 $C_{28}H_{23}NO_4$ ≥99%

Alanine — FMOC-3-(1-Naphthyl)-D-Ala
Fluka 47432   MW 437.5 $C_{28}H_{23}NO_4$ ≥98% (HPLC)

Alanine — FMOC-3-(1-Naphthyl)-L-Ala
Fluka 47433   MW 437.5 $C_{28}H_{23}NO_4$ ≥98% (HPLC)

Alanine — FMOC-3-(2-Furyl)-L-Ala
Synonyms: (S)-2-(FMOC-Amino)-3-(2-Furyl)-Propionic Acid

Fluka 00351   MW 377.4 $C_{22}H_{19}NO_5$ ≥98% (HPLC)

Alanine — FMOC-3-(2-Naphthyl)-D-Ala
Fluka 47471   MW 437.5 $C_{28}H_{23}NO_4$ ≥98% (HPLC)

Alanine — FMOC-3-(2-Naphthyl)-L-Ala
Fluka 47772   MW 437.5 $C_{28}H_{23}NO_4$ ≥98% (HPLC)

Alanine — FMOC-3-(2-Pyridyl)-D-Ala
Fluka 47291   MW 388.42 $C_{23}H_{20}N_2O_4$ ≥98% (HPLC)

Alanine — FMOC-3-(2-Pyridyl)-L-Ala
Fluka 47292   MW 388.42 $C_{23}H_{20}N_2O_4$ ≥98% (HPLC)

Alanine — FMOC-3-(3-Benzothienyl)-D-Ala
Fluka 47416   MW 443.52 $C_{26}H_{21}NO_4S$ ≥98% (HPLC)

Alanine — FMOC-3-(3-Benzothienyl)-L-Ala
Fluka 47418   MW 443.52 $C_{26}H_{21}NO_4S$ ≥98% (HPLC)

Alanine — FMOC-3-(3-Pyridyl)-D-Ala
Fluka 47435   MW 388.42 $C_{23}H_{20}N_2O_4$ ≥98% (HPLC)

Alanine — FMOC-3-(3-Pyridyl)-L-Ala

Fluka 47436　MW 388.42　$C_{23}H_{20}N_2O_4$　≥98% (HPLC)

Alanine — FMOC-3-(3-Thienyl)-D-Ala

Synonyms:　(R)-2-(FMOC-Amino)-3-(3-Theinyl)-Propionic Acid

Fluka 03752　MW 393.46　$C_{22}H_{19}NO_4S$　≥98% (HPLC)

Alanine — FMOC-3-(3-Thienyl)-L-Ala

Synonyms:　(S)-2-(FMOC-Amino)-3-(3-Theinyl)-Propionic Acid

Fluka 00346　MW 393.46　$C_{22}H_{19}NO_4S$　≥98% (HPLC)

Alanine — FMOC-3-(4-Pyridyl)-D-Ala

Fluka 47293　MW 388.42　$C_{23}H_{20}N_2O_4$　≥98% (TLC)

Alanine — FMOC-3-(4-Pyridyl)-L-Ala

Fluka 47294　MW 388.42　$C_{23}H_{20}N_2O_4$　≥98% (TLC)

Alanine — FMOC-3-Cyclohexyl-D-Ala

Fluka 47313　MW 393.48　$C_{24}H_{27}NO_4$　≥98% (TLC)

Alanine — FMOC-3-Cyclohexyl-L-Ala

Fluka 47314　MW 393.48　$C_{24}H_{27}NO_4$　≥98% (TLC)

Alanine — FMOC-3-Pyridyl-Ala

USBio F5310　MW 388.4　$C_{23}H_{20}N_2O_4$　≥99%

Alanine — FMOC-3-Styryl-D-Ala

Synonyms:　(R)-2-(FMOC-Amino)-5-Phenyl-4-Pentenoic Acid

Fluka 09802　MW 413.47　$C_{26}H_{23}NO_4$　≥98% (HPLC)

Alanine — FMOC-3-Styryl-L-Ala

Synonyms:　(S)-2-(FMOC-Amino)-5-Phenyl-4-Pentenoic Acid

Fluka 00352　MW 413.47　$C_{26}H_{23}NO_4$　≥98% (HPLC)

Alanine — FMOC-Ala

American Peptide FALA105　MW 311.3

Bachem B-1015.0005　MW 311.34　$C_{18}H_{17}NO_4$　Store at RT

USBio F5304　MW 311.3　$C_{18}H_{17}NO_4$　≥99%

Alanine — FMOC-Ala (^{15}N)

Bachem B-2645.0100　MW 312.34　$C_{18}H_{17}NO_4$　Store at -15°C

Alanine — FMOC-Ala Hydrate

Synonyms:　FMOC-L-Ala Hydrate

Peptides International FLA-1706-PI　MW 329.36　>98% (HPLC); white crystalline powder

Alanine — FMOC-Ala-® (200-400 mesh)

Bachem D-1045.0001　Store at RT

Alanine — FMOC-Ala-ol

Synonyms:　FMOC-L-Alaninol

Senn Chem 44050　MW 297.3

Alanine — FMOC-Ala-OPfp

Bachem B-1100.0001　MW 477.39　$C_{24}H_{16}F_5NO_4$　Store at -15°C

USBio F5311　MW 477.4　$C_{24}H_{16}F_5NO_4$　≥99%

Alanine — FMOC-Ala-OSu

Bachem B-1030.0001　MW 408.41　$C_{22}H_{20}N_2O_6$　Store at -15°C

USBio F5313　MW 408.4　$C_{22}H_{20}N_2O_6$　≥99%

Alanine — FMOC-Ala-SASRIN™-® (200-400 mesh)

Bachem D-1300.0001　Store at 2-8°C

Alanine — FMOC-D-(3,3-Diphenyl)-Ala

Synonyms:　FMOC-D-(3,3-Dip)

Peptides International FDX-5531-PI　MW 463.5　>98% (HPLC); white crystalline powder

Alanine — FMOC-D-1-Naphthyl-Ala

Synonyms:　FMOC-D-Nal(1')

Bachem B-3020.0001　MW 437.5　$C_{28}H_{23}NO_4$　Store at -15°C

Neosystem FA02506　MW 437.47

USBio F5305　MW 437.5　$C_{28}H_{23}NO_4$　≥99%

Alanine — FMOC-D-2-Naphthyl-Ala

Synonyms:　FMOC-D-Nal(2')

Bachem B-1950.0001　MW 437.5　$C_{28}H_{23}NO_4$　Store at RT

Neosystem FA02503　MW 437.47

Peptides International FDX-1859-PI　MW 437.49　>98% (HPLC); white crystalline powder | Most popular Nal

USBio F5307　MW 437.5　$C_{28}H_{23}NO_4$　≥99%

Alanine — FMOC-D-2-Pyridyl-Ala

American Peptide FALA144　MW 389.4

Alanine — FMOC-D-3-Benzothienyl-Ala

American Peptide FALA149　MW 443.5

Alanine — FMOC-D-3-Pyridyl-Ala

Synonyms:　FMOC-D-3-Pal

American Peptide FALA150　MW 389.4

Neosystem FA08001　MW 388.42

Peptides International FDX-1772-PI　MW 388.43　>98% (TLC); white crystalline powder

USBio F5309　MW 388.4　$C_{23}H_{20}N_2O_4$　≥99%

Alanine — FMOC-D-4-Pyridyl-Ala

Synonyms:　FMOC-D-4-Pal

American Peptide FALA147　MW 389.4

Neosystem FA13101　MW 388.42

Peptides International FDX-5535-PI　MW 461.3　>98% (HPLC); white to off-white powder

Alanine — FMOC-D-Ala

Bachem B-1020.0001　MW 311.34　$C_{18}H_{17}NO_4$　Store at RT

Fluka 47508　MW 311.34　$C_{18}H_{17}NO_4$　≥98% (HPLC); mp: 151-153°C

Neosystem FA00101　MW 311.3

USBio F5303　MW 311.3　$C_{18}H_{17}NO_4$　≥99%

Alanine — FMOC-D-Ala Hydrate

Peptides International FDA-1802-PI　MW 329.36　>98% (HPLC); white crystalline powder

Alanine — FMOC-D-Ala-OSu

USBio F5312　MW 408.4　$C_{22}H_{20}N_2O_6$　≥99%

Alanine — FMOC-D-Ala-SASRIN™-® (200-400 mesh)

Bachem D-1425.0001 Store at 2-8°C

Alanine — FMOC-D-Me-Ala

USBio F5416 MW 325.3 $C_{19}H_{19}NO_4$ ≥98%

Alanine — FMOC-D-Styryl-Ala

Neosystem FA15301 MW 413.48

Alanine — FMOC-L-(3,3-Diphenyl)-Ala

Synonyms: FMOC-Dip

Peptides International FLX-5529-PI MW 463.5 >98% (HPLC); white crystalline powder

Alanine — FMOC-L-1-Naphthyl-Ala

Synonyms: FMOC-L-Nal(1); FMOC-D-Nal(1'); FMOC-Nal(1')

Neosystem FA02505 MW 437.47

Peptides International FDX-1780-PI MW 437.49 >98% (HPLC); off-white powder

Peptides International FLX-1778-PI MW 437.49 >98% (HPLC); off-white crystalline powder

Alanine — FMOC-L-2-Naphthyl-Ala

Synonyms: FMOC-L-Nal(2); FMOC-Nal(2')

Neosystem FA02504 MW 437.47

Peptides International FLX-1758-PI MW 437.49 >98% (HPLC); white crystalline powder | Most popular Nal

Alanine — FMOC-L-2-Pyridyl-Ala

American Peptide FALA143 MW 389.4

Alanine — FMOC-L-3-Benzothienyl-Ala

American Peptide FALA148 MW 443.5

Alanine — FMOC-L-3-Pyridyl-Ala

Synonyms: FMOC-L-3-Pal

American Peptide FALA145 MW 389.4

Neosystem FA08002 MW 388.42

Peptides International FLX-1766-PI MW 388.43 >98% (HPLC); white crystalline powder

Alanine — FMOC-L-4-Pyridyl-Ala

Synonyms: FMOC-L-4-Pal; FMOC-4-Pal

American Peptide FALA146 MW 389.4

Neosystem FA13102 MW 388.42

Peptides International FLX-5534-PI MW 461.3 >98% (HPLC); off-white crystalline powder

Alanine — FMOC-L-Ala

Fluka 47616 MW 311.34 $C_{18}H_{17}NO_4$ ~97% (titration); mp: 147-150°C | FMOC-protected AA are widely used in solid phase peptide synthesis; Atherton, E & Sheppard, RC, *Solid Phase Peptide Synthesis: A Practical Approach*, 1989, Oxford UP Oxford

Neosystem FA00102 MW 311.3

Alanine — FMOC-L-Ala 4-Benzyloxybenzyl Ester Polymer Bound

Fluka 47644 0.2-1.0 mmol/g resin; carrier: polystyrene, crosslinked with 1% DVB; 100-200 mesh particle size

Alanine — FMOC-L-Ala N-Carboxy Anhydride

Fluka 47604 MW 337.32 $C_{19}H_{15}NO_5$ ≥97%; mp: 130°C | Produced by Propeptide under an exclusive license from Bioresearch

Alanine — FMOC-L-Ala-OPfp

Fluka 47438 MW 477.39 $C_{24}H_{16}F_5NO_4$ ≥98% (HPLC)

Alanine — FMOC-L-Ala-Wang Resin

American Peptide RFWAN05 Preattached Wang resins are usful for synthesis of peptides with C-terminal carboxyl groups; peptides can be cleaved from the resin with 90% TFA in the presence of scavengers

Neosystem RW00102

Alanine — FMOC-L-Styryl-Ala

Neosystem FA15302 MW 413.48

Alanine — FMOC-L-β-Hal

Synonyms: (S)-3-(FMOC-Amino)-Butyric Acid

Fluka 47935 MW 325.36 $C_{19}H_{19}NO_4$ ≥98% (TLC)

Alanine — FMOC-L-β-Thienylalanine

American Peptide FALA55 MW 393.4

Alanine — FMOC-Me-Ala

USBio F5417 MW 325.3 $C_{19}H_{19}NO_4$ ≥98%

Alanine — FMOC-N-Me-Ala

Synonyms: FMOC-N-Me-L-Ala

Peptides International FMA-1791-PI MW 325.37 >95% (HPLC); off-white powder

Alanine — FMOC-N-Me-D-Ala

Synonyms: FMOC-N-Me-D-Ala

Fluka 47346 MW 325.36 $C_{19}H_{19}NO_4$ ≥98% (TLC)

Peptides International FMA-1790-PI MW 325.37 >97% (HPLC); white crystalline powder

Alanine — FMOC-N-Me-L-Ala

Fluka 47594 MW 325.36 $C_{19}H_{19}NO_4$ ≥97% (HPLC); mp: 140°C

Alanine — FMOC-α-Me-Ala

Synonyms: FMOC-α-Aib; FMOC-Aib

Fluka 47691 MW 325.36 $C_{19}H_{19}NO_4$ ≥98% (HPLC); mp: 186-188°C

USBio F5301 MW 325.5 $C_{19}H_{19}NO_4$ ≥99%

USBio F5418 MW 325.5 $C_{19}H_{19}NO_4$ ≥99%

Alanine — FMOC-β-(2,2-Dimethyl-4H-Benzo(1,3)-Dioxin-6-yl)-Ala

Synonyms: (S)-2-(FMOC-Amino)-3-(2,2-Dimethyl-4H-Benzo-(1,3)-Dioxin-6-yl)-Propionic Acid

Bachem B-3310.0001 MW 473.53 $C_{28}H_{27}NO_6$ Store at 2-8°C

Alanine — FMOC-β-2-Quinolyl-Ala

Bachem B-3165.0250 MW 438.48 $C_{27}H_{22}N_2O_4$ Store at -15°C

Alanine — FMOC-β-2-Quinolyl-D-Ala

Bachem B-3170.0250 MW 438.48 $C_{27}H_{22}N_2O_4$ Store at -15°C

Alanine — FMOC-β-2-Thienylalanine

Bachem B-1665.0001 MW 393.46 $C_{22}H_{19}NO_4S$ Store at RT

Alanine — FMOC-β-2-Thienyl-D-Ala

Synonyms: FMOC-D-Tha

Bachem B-2120.0001   MW 393.46 $C_{22}H_{19}NO_4S$ Store at RT

Neosystem FA02501   MW 393.46

Alanine — FMOC-β-2-Thienyl-L-Ala

Synonyms: FMOC-L-Tha

Neosystem FA02502   MW 393.46

Alanine — FMOC-β-3-Benzothienyl-Ala

Bachem B-2830.0001   MW 443.52 $C_{26}H_{21}NO_4S$ Store at -15°C

Alanine — FMOC-β-3-Benzothienyl-D-Ala

Bachem B-3435.0001   MW 443.52 $C_{26}H_{21}NO_4S$ Store at -15°C

Alanine — FMOC-β-3-Pyridyl-Ala

Bachem B-2005.0001   MW 388.42 $C_{23}H_{20}N_2O_4$ Store at RT

Alanine — FMOC-β-3-Pyridyl-D-Ala

Bachem B-2040.0001   MW 388.42 $C_{23}H_{20}N_2O_4$ Store at RT

Alanine — FMOC-β-Ala

Synonyms: FMOC-β-Ala

American Peptide FALA100   MW 311.3

Bachem B-1025.0001   MW 311.34 $C_{18}H_{17}NO_4$ Store at RT

Fluka 47587   MW 311.34 $C_{18}H_{17}NO_4$ ≥99% (HPLC); mp: 142-147°C

Neosystem FA02101   MW 311.3

Peptides International FLX-1744-PI   MW 311.34   >98% (HPLC); white crystalline powder

Senn Chem 44048   MW 311.3

USBio F5302   MW 311.3 $C_{18}H_{17}NO_4$ ≥99%

Alanine — FMOC-β-Ala 4-Benzyloxybenzyl Ester Polymer Bound

Fluka 09845   ~0.6 mmol/g resin; crosslinked with 1% DVB; 100-200 mesh particle size

Alanine — FMOC-β-Ala-® (200-400 mesh)

Bachem D-1625.0001   Store at RT

Alanine — FMOC-β-Ala-OPfp

Bachem B-1765.0001   MW 477.39 $C_{24}H_{16}F_5NO_4$ Store at -15°C

Alanine — FMOC-β-Ala-OSu

Senn Chem 44049   MW 408.4

Alanine — FMOC-β-Ala-Wang Resin

Neosystem RW02101

Alanine — FMOC-β-Cyclohexyl-Ala

Bachem B-1975.0001   MW 393.48 $C_{24}H_{27}NO_4$ Store at RT

Alanine — FMOC-β-Cyclohexyl-Ala-SASRIN™-® (200-400 mesh)

Bachem D-1725.0001   Store at 2-8°C

Alanine — FMOC-β-Cyclohexyl-D-Ala

Synonyms: FMOC-D-Cha

Bachem B-2345.0001   MW 393.48 $C_{24}H_{27}NO_4$ Store at RT

Neosystem FA02301   MW 393.5

Peptides International FDX-1879-PI   MW 393.49   >98% (HPLC); white to off-white powder

Alanine — FMOC-β-Cyclohexyl-L-Ala

Synonyms: FMOC-L-Cha; FMOC-Cha

Neosystem FA02302   MW 393.5

Peptides International FLX-5545-PI   MW 393.49   >98% (HPLC); white crystalline powder

USBio F5350   MW 393.5 $C_{24}H_{27}NO_4$ ≥99%

Alanine — FMOC-β-Cyclopropyl-Ala

Bachem B-2905.0001   MW 351.4 $C_{21}H_{21}NO_4$ Store at -15°C

Alanine — FMOC-β-Cyclopropyl-D-Ala

Bachem B-2915.0001   MW 351.4 $C_{21}H_{21}NO_4$ Store at -15°C

Alanine — FMOC-β-L-Ala-Wang Resin

American Peptide RFWAN07   Preattached Wang resins are usful for synthesis of peptides with C-terminal carboxyl groups; peptides can be cleaved from the resin with 90% TFA in the presence of scavengers

Alanine — For-Ala-OMe

Bachem E-1845.0001   MW 131.13 $C_5H_9NO_3$ Store at 2-8°C

Alanine — Indole-3-Ac-L-Ala

Synonyms: IAA-L-Ala

Sigma I 9262   FW 246.3 $C_{13}H_{14}N_2O_3$ Plant cell culture tested

Alanine — Indole-3-Butyryl-β-Ala

Synonyms: IBA-Ala

Sigma I 9762   FW 274.3 $C_{15}H_{18}N_2O_3$ Plant cell culture tested

Alanine — L-3-(1-Naphthyl)-Ala

ICN 151730 $C_{13}H_{13}NO_2$

Sigma N 1521 $C_{13}H_{13}NO_2$ Analog of L-Phe; Rodriguez, M et al, *Eur J Med Chem*, 26: 245, 1991

USBio N0500   MW 215.3 $C_{13}H_{13}NO_2$ ≥99%

Alanine — L-3-(2-Naphthyl)-Ala

ICN 151731 $C_{13}H_{13}NO_2$

Sigma N 4521 $C_{13}H_{13}NO_2$ Useful for synthesis of modified peptides; Rodriguez, M et al, *Eur J Med Chem*, 26: 245, 1991

Sigma N 5512 $C_{13}H_{13}NO_2$

USBio N0501   MW 215.3 $C_{13}H_{13}NO_2$ ≥99%

Alanine — L-Ala

Synonyms: (S)-2-Aminopropionic Acid; 2-Aminopropionic Acid; L-α-Aminopropionic Acid; (S)-2-2-Aminopropanoic Acid

Fluka 05129   MW 89.1 $C_3H_7NO_2$ ≥99.5%; ≤0.05% residue on ignition, loss on drying; ≤0.005% Cl, SO_4, K, Na; ≤0.0005% Al, Ba, Bi, Cd, Co, Cr, Cu, Fe, Li, Mg, Mn, Mo, Ni, Pb, Sr, Zn; ≤0.001% Ca; ≤0.01% NH_4^+; ≤0.00001% As; ≤0.3% foreign AA | Arakawa, T & Timasheff, SN, *Meth Enzymol*, 114: 49, 1985; *CRC Handbook of Microbiology*, Laskin, AI & Lachevalier, HA, eds, CRC Press, Cleveland, Ohio, 4: 1, 1974

Fluka 05130   MW 89.1 $C_3H_7NO_2$ ≥99.0%; ≤0.1% residue on ignition, loss on drying; ≤0.005% Cl, SO₄, K, Na; ≤0.0005% Cd, Co, Cr, Cu, Fe, Mg, Mn, Ni, Pb, Zn; ≤0.001% Ca; ≤0.01% NH₄⁺; ≤0.00001% As; ≤0.3% foreign AA | Koppenhoefer, B et al, *Synthesis*, 316, 1982; Watanabe, K et al, *ibid*, 225, 1987; Narita, M et al, *Tetrahedron Lett*, 23: 525, 1982; Kogen, H & Nishi, T, *Chem Comm*, 311, 1987

ICN 100287 $C_3H_7NO_2$

Neosystem AA00102   MW 89.1

Peptides International ALA-2701   MW 89.09   >99.9% optical purity by RP-HPLC; white crystalline powder

Rexim   MW 89.1 $C_3H_7NO_2$ White crystals or crystalline powder

Sigma A 0468   FW 89.09 $C_3H_7NO_2$ USP Grade

Sigma A 3519   Cross-linked 4% beaded agarose; activation: CNBr; attachment: amino; spacer: 1 atom; ligand immobilized: 2-7 µmoles/mL; suspension in 2.0 M NaCl containing 0.02% thimerosal | Deutsch, DG & Mertz, ET, *Proc Fed Amer Soc Exp Biol*, 29: 647, 1970; Deutsch, DG & Mertz, ET, *Science*, 170: 1095, 1970; Vuento, M & Vaheri, A, *Biochem J*, 183: 331, 1979

Sigma A 3534   FW 89.09 $C_3H_7NO_2$ Crystalline; cell culture tested; insect cell culture tested

Sigma A 5824 $C_3H_7NO_2$ pH (1 M in H_2O, 20°C): 5.0-7.0; residue on ignition: <0.1%; solubility (1 M in H_2O, 20°C): complete, colorless; Cl, SO₄: <0.05%; insoluble matter, K: <0.005%; Al, Ca, Cu, Fe: <0.0005%; A_{260} <0.05; A_{280} <0.05 (1 M in H_2O)

Sigma A 7627 $C_3H_7NO_2$ Also available as part of a kit

USBio A1205   MW 89.09 $C_3H_7NO_2$ ≥98%; white, crystalline powder; specific rotation: C=10, 6 N HCl +13.7° to +15.1°; pH (2.5%,H_2O): 5.7-6.7; heavy metals (Pb): ≤0.001%; endotoxin: ≤2 EU/mg; cell culture tested: murine myeloma SP2/0-Ag14

Alanine — L-Ala (1-¹⁴C)

ARC ARC-231   MW 89.1 $CH_3CH(NH_2)COOH$ 50-60 mCi/mmol; 1.85-2.22 GBq/mmol; in 0.01 N HCl | Radiochemical

ICN 10024E $CH_3CH(NH_2)COOH$

Alanine — L-Ala (¹⁵N)

ICN 540136 $C_3H_7{}^{15}NO_2$; $CH_3CH(NH_2)COOH$

Alanine — L-Ala (2,3-³H)

Amersham TRK400

ARC ART-180   MW 89.1 $CH_3CH(NH_2)COOH$ 30-60 Ci/mmol; 1.11-2.22 TBq/mmol; in 0.01 N HCl | Radiochemical

ICN 20005E $CH_3CH(NH_2)COOH$

Alanine — L-Ala (3-³H)

Sigma A 3930   FW 89.09 $CH_3CH(NH_2)CO_2H$ 70-85 Ci/mmol; radiochemical purity ≥98% (HPLC); aqueous solution containing 2% EtOH in serum bottle | Radiochemical

Alanine — L-Ala (U-¹⁴C)

Amersham CFB62

ARC ARC-302   MW 89.1 $CH_3CH(NH_2)COOH$ 50-150 mCi/mmol; 1.85-5.55 GBq/mmol; in 0.01 N HCl | Radiochemical

ICN 10021E $CH_3CH(NH_2)COOH$

ICN 10021L $CH_3CH(NH_2)COOH$

Alanine — L-Ala (U-¹⁴C) Hydrochloride

Sigma A 7428   FW 89.09 $CH_3CH(NH_2)CO_2H$ 120-180 Ci/mmol; radiochemical purity ≥98% (HPLC); aqueous solution containing 2% EtOH in Combi-vial | Radiochemical

Alanine — L-Ala 4-Methoxy-β-Naphthylamide

Synonyms: Aminopeptidase M Substrate

Sigma A 1541 $C_{14}H_{16}N_2O_2$ Fluorogenic substrate; Gossrau, R et al, *Histochemistry*, 80: 183, 1984; loc cit, 81: 167, 1984

Alanine — L-Ala 4-Methoxy-β-Naphthylamide Hydrochloride

Synonyms: Aminopeptidase M Substrate

Sigma A 5414 $C_{14}H_{16}N_2O_2 \cdot HCl$ Fluorogenic substrate; Gossrau, R et al, *Histochemistry*, 80: 183, 1984; loc cit, 81: 167, 1984

Alanine — L-Ala Amide Hydrochloride

Neosystem AA00104   MW 124.6

Alanine — L-Ala β-Naphthylamide

Sigma A 2628 $C_{13}H_{14}N_2O$

Sigma A 7771 $C_{13}H_{14}N_2O \cdot HBr$

Alanine — L-Ala-(t-OBzl)

Sigma A 0255 $C_7H_{15}NO_2 \cdot HCl$

Alanine — L-Ala-4-AMC TFA

Synonyms: Peptidase Substrate

Fluka 05198   MW 360.3 $C_{13}H_{14}N_2O_2 \cdot C_2HF_3O_2$ ≥99% (TLC) | Fluorogenic substrate; Mantle, D et al, *Biochem J*, 211: 567, 1983

Alanine — L-Ala-AMC

Synonyms: L-Ala-4-Methylumbelliferylamide; Alanine Aminopeptidase Substrate

Fluka 05197   MW 246.27 $C_{13}H_{14}N_2O_3$ Sensitive fluorogenic substrate; Manafi, M & Kneifel, W, *J Appl Bact*, 69: 822, 1990

Sigma A 4302 $C_{13}H_{14}N_2O_3 \cdot C_2HF_3O_2$

Alanine — L-Ala-L-1-Aminoethylphosphonic Acid

Synonyms: Alafosfalin; (S)-Alanyl-(R)-1-Aminoethylphosphonic Acid; Alaphosphi

Fluka 05260   MW 196.14 $C_5H_{13}N_2O_4P$ ≥99% (titration) | Allen, JG et al, *Nature*, 272: 56, 1978

Alanine — L-Alaninamide Hydrobromide

Sigma A 3521 $C_3H_8N_2O \cdot HBr$

Alanine — L-Alaninamide Hydrochloride

Sigma A 3796 $C_3H_8N_2O \cdot HCl$

Alanine — L-Ala-OAll Hydrochloride

Neosystem AA00111   MW 165.6

Alanine — L-Ala-OBzl Hydrochloride

Neosystem AA00106   MW 215.7

Sigma A 7877 $C_{10}H_{13}NO_2 \cdot HCl$

Alanine — L-Ala-OBzl p-Tosyl

ICN 100293 $C_{10}H_{13}NO_2 \cdot C_7H_8O_3S$

Alanine — L-Ala-OBzl-Tosyl

Fluka 05183   MW 351.43 $C_{10}H_{13}NO_2 \cdot C_7H_8O_3S$ ~97% (titration); mp: 114-116°C

Alanine — L-Ala-OEt

ICN 150268 $C_5H_{11}NO_2 \cdot HCl$

Alanine — L-Ala-OEt Hydrochloride

Fluka 05170   MW 153.61 $C_5H_{11}NO_2 \cdot HCl$ ~99% (titration); ~1% H_2O; mp: ~75°C

Alanine — L-Ala-OMe Hydrochloride

Fluka 05200   MW 139.58 $C_4H_9NO_2 \cdot HCl$ ≥99% (titration); ≤1% H_2O; mp: 107-110°C

ICN 100281 $C_4H_9NO_2 \cdot HCl$

Neosystem AA00110   MW 139.6

Sigma A 8752 $C_4H_9NO_2 \cdot HCl$

Alanine — L-Ala-OtBu

ICN 191226 $C_7H_{15}NO_2 \cdot HCl$

Alanine — L-Ala-OtBu Hydrochloride

Fluka 05190   MW 181.66 $C_7H_{15}NO_2 \cdot HCl$ ≥99% (titration); mp: 170-175°C

Alanine — L-Ala-pNA

Sigma A 9325 $C_9H_{11}N_3O_3 \cdot HCl$

Alanine — L-Ala-β-Naphthylamide

ICN 100290 $C_{13}H_{14}N_2O \cdot HBr$

Alanine — L-α-Ala

ICN 194617 $C_3H_7NO_2$

Alanine — L-β-Hal Hydrochloride

Synonyms: (S)-3-Aminobutyric Acid

Fluka 03766   MW 139.58 $C_4H_9NO_2 \cdot HCl$ ≥98.0% (TLC)

Alanine — N-(L-Ala)-2-Aminoacridone

Synonyms: Alanine Aminopeptidase Substrate

Fluka 05255   MW 281.31 $C_{16}H_{15}N_3O_2$ Soluble in DMSO & DMF | Fluorogenic substrate for Ala aminopeptidase for the rapid differentiation of Gram-positive form Gram-negative bacteria; Manafi, M & Kneifel, W, *J Appl Bact*, 69: 822, 1990

Alanine — N,N-Dipropyl-L-Ala

Synonyms: (S)-(+)-N,N-Dipropyl-Ala

Sigma D 4405 $C_9H_{19}NO_2$

Alanine — N-2,4-Dnp-L-Ala

Sigma D 7254 $C_9H_9N_3O_6$

Alanine — N-2,4-Dnp-β-Ala

Sigma D 7129 $C_9H_9N_3O_6$

Alanine — N-Ac-D-Ala

ICN 150226 $C_5H_9NO_3$

Sigma A 4375 $C_5H_9NO_3$

Alanine — N-Ac-DL-Ala

ICN 100048 $C_5H_9NO_3$

Sigma A 4500 $C_5H_9NO_3$

Alanine — N-Ac-L-Ala

ICN 150227 $C_5H_9NO_3$

Sigma A 4625 $C_5H_9NO_3$

Alanine — N-Ac-L-Ala-pNA

Sigma A 2019 $C_{11}H_{13}N_3O_4$

Alanine — N-Bz-DL-Ala

Sigma B 0379 $C_{10}H_{11}NO_3$

Alanine — N-Bz-L-Ala

Sigma B 3750 $C_{10}H_{11}NO_3$

Alanine — N-Bz-L-Ala-OMe

Sigma B 3380 $C_{11}H_{13}NO_3$

Alanine — N-Bz-β-Ala

Sigma B 2879 $C_{10}H_{11}NO_3$

Alanine — N-CBZ-3-Sulfamoyl-L-Ala

Sigma C 5535 $C_{11}H_{14}N_2O_6S$

Alanine — N-CBZ-D-Ala

ICN 100202 $C_{11}H_{13}NO_4$

Sigma C 3541 $C_{11}H_{13}NO_4$

Alanine — N-CBZ-DehydroAla-OMe

Sigma C 7276 $C_{12}H_{13}NO_4$

Alanine — N-CBZ-DL-Ala

ICN 104803 $C_{11}H_{13}NO_4$

Sigma C 1626 $C_{11}H_{13}NO_4$

Alanine — N-CBZ-L-Ala

ICN 100204 $C_{11}H_{13}NO_4$

Sigma C 1751 $C_{11}H_{13}NO_4$

Alanine — N-CBZ-β-Ala

Sigma C 1376 $C_{11}H_{13}NO_4$

Alanine — N-Cyclohexyl-β-Ala

Sigma C 8881 $C_9H_{17}NO_2$

Alanine — N-Cyclohexyl-β-D-Ala

USBio C8750-10   MW 171.2 $C_9H_{17}NO_2$ ≥99%

Alanine — N-Cyclohexyl-β-L-Ala

USBio C8750   MW 171.2 $C_9H_{17}NO_2$ ≥99%

Alanine — N-FMOC-3-(3-Benzothienyl)-L-Ala

Sigma F 6295 $C_{26}H_{21}NO_4S$

Alanine — N-FMOC-D-Ala (1-¹⁴C)

ARC ARC-1110   MW 311.3   50-60 mCi/mmol; 1.85-2.22 GBq/mmol; in EtOH | Radiochemical

Alanine — N-FMOC-L-Ala

Sigma F 8632 $C_{18}H_{17}NO_4$

Alanine — N-FMOC-L-Ala (1-¹⁴C)

ARC ARC-494   MW 311.3   50-60 mCi/mmol; 1.85-2.22 GBq/mmol; in EtOH | Radiochemical

Alanine — N-FMOC-L-Ala (2,3-³H)

ARC ART-401   MW 311.3   10-20 Ci/mmol; 370-740 GBq/mmol; in EtOH | Radiochemical

Alanine — N-FMOC-L-Ala (U-¹⁴C)

Sigma F 0172   40-60 mCi/mmol; purity ≥98% (HPLC); EtOH solution in Combi-vial | Radiochemical

Alanine — *N*-FMOC-L-Ala Resin Ester

Sigma F 0761   For peptide synthesis; Wang, SS, *J Am Chem Soc*, 95: 1328, 1973; Lu, G et al, *J Org Chem*, 46: 3433, 1981

Alanine — *N*-FMOC-L-Ala SASRIN Resin Ester

Sigma F 0389   For peptide synthesis; *N*-FMOC AA acyl ester of 3-methoxy-4-oxymethyl-phenoxymethylated 1% divinylbenzene cross-linked polystyrene; peptides can be cleaved from the resin with 0.5-1% CF_3CO_2H in CH_2Cl_2; Mergler, M et al, *in Peptides: Chemistry & Biology* (Proc 10th Am Peptide Symp), Marhall, GR, ed, 259, 1988

Alanine — *N*-FMOC-β-Ala

Sigma F 6635 $C_{18}H_{17}NO_4$

Alanine — *N*-For-L-Ala

ICN 151163 $C_4H_7NO_3$

Alanine — *N*-Lauroyl-L-Ala

Fluka 61726   MW 271.4 $C_{15}H_{29}NO_3$ ≥99.0% (TLC)

Alanine — *N*-Maleoyl-β-Ala

Synonyms: N-2-Carboxyethyl-Maleimide; 3-Maleimidopropionic Acid

Fluka 63285   MW 169.14 $C_7H_7NO_4$ ≥98.0% (titration); mp: 108-111°C | Reagent for the modification of peptides, protein & polymers with an SH-label; Keller, O & Rudinger, J, *Hev Chim Acta*, 58: 531, 1975; Rich, DH et al, *FEBS Lett*, 134: 261, 1981; *Biochem Soc trans*, 11: 753, 1983; Moroder, L et al, *Biopolymers*, 22: 481, 1983; Wunsch, E et al, *Biol Chem Hoppe-Seyler*, 366: 53, 1985; Moroder, L et al, *Biol Chem Hoppe-Seyler*, 368: 855, 1987

Alanine — *N*-Me-Ala

Bachem E-2125.0001   MW 103.12 $C_4H_9NO_2$ Store at RT

Alanine — *N*-Me-D-Ala

Sigma M 7524 $C_4H_9NO_2$

Alanine — *N*-Me-DL-Ala

Bachem F-1755.0001   MW 103.12 $C_4H_9NO_2$ Store at RT
ICN 155461 $C_4H_9NO_2$
Sigma M 0506 $C_4H_9NO_2$

Alanine — *N*-Me-L-Ala

Fluka 02676   MW 103.12 $C_4H_9NO_2$ ≥98.0% (TLC)
Sigma M 7649 $C_4H_9NO_2$

Alanine — *N*-O-Nps-L-Ala Dicyclohexylammonium Salt

ICN 102520 $C_9H_{10}N_2O_4S \cdot C_{12}H_{23}N$

Alanine — *N*-Phthaloyl-β-Ala

Sigma P 1503 $C_{11}H_9NO_4$

Alanine — *N*-t-BOC-3-(3-Pyridyl)-L-Ala

Sigma B 1421 $C_{13}H_{18}N_2O_4$

Alanine — *N*-t-BOC-D-3-(2-Naphthyl)-Ala

Sigma B 6028 $C_{18}H_{21}NO_4$

Alanine — *N*-t-BOC-D-Ala

Sigma B 0512 $C_8H_{15}NO_4$

Alanine — *N*-t-BOC-L-3-(1-Naphthyl)-Ala

Sigma B 9170 $C_{18}H_{21}NO_4$

Alanine — *N*-t-BOC-L-3-(2-Naphthyl)-Ala

Sigma B 6153 $C_{18}H_{21}NO_4$

Alanine — *N*-t-BOC-L-Ala

Sigma B 0393 $C_8H_{15}NO_4$

Alanine — *N*-t-BOC-L-Ala *N*-Hydroxysuccinimide Ester

Sigma B 2130 $C_{12}H_{18}N_2O_6$

Alanine — *N*-t-BOC-L-Ala Pam Resin Ester

Sigma B 4893   For peptide synthesis by the Merrifield method; Mitchell, AR et al, *J Org Chem*, 43: 2845, 1978

Alanine — *N*-t-BOC-L-Ala Resin Ester

Sigma B 8396   For peptide synthesis by the Merrifield method

Alanine — *N*-t-BOC-L-Ala-(*p*ONp)

Sigma B 5126 $C_{14}H_{18}N_2O_6$

Alanine — *N*-t-BOC-L-Ala-CHO

Sigma B 1403 $C_8H_{15}NO_3$

Alanine — *N*-t-BOC-*N*-Me-D-Ala

Sigma B 5396 $C_9H_{17}NO_4$

Alanine — *N*-t-BOC-*N*-Me-L-Ala

Sigma B 6146 $C_9H_{17}NO_4$

Alanine — *N*-t-BOC-α-Me-Ala

Synonyms: N-t-BOC-α-Aib

Sigma B 9267 $C_9H_{17}NO_4$

Alanine — *N*-t-BOC-β-Ala

ICN 150480 $C_8H_{15}NO_4$
Sigma B 9755 $C_8H_{15}NO_4$

Alanine — *N*-t-BOC-β-Cyclohexyl-D-Ala

Synonyms: N-t-BOC-Hexahydro-D-Phe

Sigma B 7780 $C_{14}H_{25}NO_4$

Alanine — *N*-t-BOC-β-Cyclohexyl-L-Ala

Synonyms: N-t-BOC-Hexahydro-L-Phe

Sigma B 7655 $C_{14}H_{25}NO_4$

Alanine — *N*-TFA-L-Ala-OMe

Sigma T 3381 $C_6H_8F_3NO_3$

Alanine — *N*ᵃ-9-FMOC-β-2-Thienyl-D-Ala

Synonyms: FMOC-D-Thi

USBio F5462   MW 393.4 $C_{22}H_{19}NO_4S$ ≥99%

Alanine — *N*ᵃ-9-FMOC-β-2-Thienyl-L-Ala

Synonyms: FMOC-Thi

USBio F5463   MW 393.4 $C_{22}H_{19}NO_4S$ ≥99%

Alanine — N^α-BOC-β-2-Thienyl-D-Ala

Synonyms: BOC-D-Thi

USBio B2399   MW 271.3 $C_{12}H_{17}NO_4S$ ≥99%

Alanine — N^α-Bz-DL-Ala

ICN 100081 $C_{10}H_{11}NO_3$

Alanine — N^α-Bz-L-Ala

ICN 150443 $C_{10}H_{11}NO_3$

Alanine — N^α-FMOC-L-Ala

ICN 151132 $C_{18}H_{17}NO_4$

Alanine — N^α-t-BOC-L-Ala

ICN 101053 $C_8H_{15}NO_4$

Alanine — N^α-t-BOC-L-Ala-(pONp)

ICN 101062 $C_{14}H_{18}N_2O_6$

Alanine — N^α-Z-3-Sulfamoyl-L-Ala

Fluka 97200   MW 302.30 $C_{11}H_{14}N_2O_6S$ ≥98.0% (TLC)

Alanine — Pht-β-Ala

Senn Chem 44123   MW 219.2

Alanine — Pth-Ala

Sigma P 8626 $C_{10}H_{10}N_2OS$ AA derivatives used as standards in protein sequencing by the Edman degradation

Alanine — Suc-Ala-pNA

Peptides International SAN-3116   MW 309.28 $C_{13}H_{15}N_3O_6$ >99% (HPLC); amorphous powder

Alanine — TentaGel S PHB-Ala-FMOC

Synonyms: FMOC-L-Ala 4-(Poly(Ethylenoxy))-OBzl Polymer Bound

Fluka 86367   0.20 mmol protected AA/g; ~90 μm particle size

Alanine — TentaGel S Trt-Ala-FMOC

Fluka 86408   0.18 mmol protected AA/g; ~90 μm particle size

Alanine — Tosyl-Ala

Bachem E-3490.0005   MW 243.28 $C_{10}H_{13}NO_4S$ Store at RT

Alanine — Uracilyl-Ala

Synonyms: (±)-Willardiine; Uracil-1-(2-Aminopropionic Acid)

Sigma W 1379 $C_7H_9N_3O_4$ Neuroactive analog of Phe; relatively inactive at pure kainate receptors; Dewar & Shaw, *J Chem Soc*, 583, 1962

Alanine — Z-(Dehydro)-Ala

Bachem C-1535.0001   MW 221.21 $C_{11}H_{11}NO_4$ Store at -15°C

Fluka 96075   MW 221.21 $C_{11}H_{11}NO_4$ ~99% (TLC)

Alanine — Z-(Dehydro)-Ala-OMe

Bachem C-1540.0001   MW 235.24 $C_{12}H_{13}NO_4$ Store at -15°C

Fluka 96077   MW 235.24 $C_{12}H_{13}NO_4$ ≥99% (TLC)

Alanine — Z-(N,2-Dimethyl)-Ala

Fluka 96079   MW 251.28 $C_{13}H_{17}NO_4$ ≥98% (TLC)

Alanine — Z-1-Naphthyl-Ala

Bachem C-3010.0001   MW 349.39 $C_{21}H_{19}NO_4$ Store at RT

Alanine — Z-2-Me-Ala

Fluka 96920   MW 237.26 $C_{12}H_{15}NO_4$ ≥98.0% (TLC)

Alanine — Z-2-Naphthyl-Ala

Bachem C-3500.0001   MW 349.39 $C_{21}H_{19}NO_4$ Store at RT

Alanine — Z-2-Naphthyl-Ala-CMK

Bachem C-3495.0250   MW 381.86 $C_{22}H_{20}ClNO_3$ Store at -15°C

Alanine — Z-3-(2-Naphthyl)-D-Ala

Fluka 96828   MW 349.39 $C_{21}H_{19}NO_4$ ≥98.0% (TLC)

Alanine — Z-3-Cyclohexyl-L-Ala

Fluka 96060   MW 305.37 $C_{17}H_{23}NO_4$ ≥98.0% (HPLC)

Alanine — Z-Ala

Synonyms: Benzyloxycarbonyl-L-Ala

Bachem C-1000.0025   MW 223.23 $C_{11}H_{13}NO_4$ Store at RT

Peptides International ZLA-2007-PI   MW 223.23   >98% (TLC); white powder

Alanine — Z-Ala Amide

Bachem C-1020.0001   MW 222.24 $C_{11}H_{14}N_2O_3$ Store at RT

Alanine — Z-Ala β-Naphthyl Ester

Bachem M-1235.0250   MW 349.39 $C_{21}H_{19}NO_4$ Store at -15°C

Alanine — Z-Ala-4MβNA

Bachem J-1080.0250   MW 378.4 $C_{22}H_{22}N_2O_4$ Store at -15°C

Alanine — Z-Ala-ol

Synonyms: Z-L-Alaninol

Senn Chem 44072   MW 209.2

Alanine — Z-Ala-ONp

Bachem C-1040.0005   MW 344.32 $C_{17}H_{16}N_2O_6$ Store at -15°C

Alanine — Z-Ala-OSu

Bachem C-1025.0005   MW 320.3 $C_{15}H_{16}N_2O_6$ Store at -15°C

Alanine — Z-D-1-Naphthyl-Ala

Bachem C-3950.0001   MW 349.39 $C_{21}H_{19}NO_4$ Store at RT

Alanine — Z-D-2-Naphthyl-Ala

Bachem C-2255.0001   MW 349.39 $C_{21}H_{19}NO_4$ Store at RT

Alanine — Z-D-Ala

Bachem C-1005.0001   MW 223.23 $C_{11}H_{13}NO_4$ Store at RT

Alanine — Z-D-Ala Amide

Bachem C-3255.0001   MW 222.24 $C_{11}H_{14}N_2O_3$ Store at RT

Alanine — Z-D-Ala-OSu
Bachem C-1030.0001 MW 320.3 $C_{15}H_{16}N_2O_6$ Store at -15°C

Alanine — Z-DL-Ala
Bachem C-1010.0025 MW 223.23 $C_{11}H_{13}NO_4$ Store at RT

Fluka 95860 MW 223.23 $C_{11}H_{13}NO_4$ ≥98.0% (TLC)

Alanine — Z-DL-Ala-OSu
Bachem C-1035.0005 MW 320.3 $C_{15}H_{16}N_2O_6$ Store at -15°C

Alanine — Z-L-Ala
Fluka 95850 MW 223.23 $C_{11}H_{13}NO_4$ ≥99.0% (titration); mp: 84-87°C

Neosystem ZA00102 MW 223.2

Alanine — Z-L-Ala Hydroxysuccinimide Ester
Fluka 95870 MW 320.30 $C_{15}H_{16}N_2O_6$ ≥99.0%; mp: 119-123°C

Alanine — Z-L-Ala-(4-ONp)
Fluka 95877 MW 344.32 $C_{17}H_{16}N_2O_6$ ≥99.0% (TLC)

Alanine — Z-N-Me-DL-Ala Dicyclohexylamine
Bachem C-3915.0001 MW 418.58 $C_{12}H_{15}NO_4 \cdot C_{12}H_{23}N$ Store at RT

Alanine — Z-N-Me-L-Ala
Fluka 00418 MW 237.26 $C_{12}H_{15}NO_4$ ≥98.0% (HPLC)

Alanine — Z-α-Me-Ala
Synonyms: Z-Aib

Bachem C-3680.0005 MW 237.26 $C_{12}H_{15}NO_4$ Store at RT

Alanine — Z-β-Ala
Bachem C-1015.0005 MW 223.23 $C_{11}H_{13}NO_4$ Store at RT

Senn Chem 44068 MW 223.2

Alanine — Z-β-Ala Amide
Senn Chem 44069 MW 222.2

Alanine — Z-β-Ala-OMe
Senn Chem 44070 MW 237.3

Alanine — Z-β-Ala-OSu
Bachem C-3355.0005 MW 320.3 $C_{15}H_{16}N_2O_6$ Store at -15°C

Senn Chem 44071 MW 320.3

Alanine — Z-β-Cyclohexyl-D-Ala Dicyclohexylamine
Bachem C-3920.0001 MW 486.7 $C_{17}H_{23}NO_4 \cdot C_{12}H_{23}N$ Store at RT

Alanine — Z-β-tBu-Ala Dicyclohexylamine
Synonyms: Z-D-Neopentyl-Gly DCHA

Bachem C-2260.0001 MW 460.66 $C_{15}H_{21}NO_4 \cdot C_{12}H_{23}N$ Store at -15°C

Alanine — Z-β-tBu-D-Ala Dicyclohexylamine
Synonyms: Z-D-Neopentyl-Gly DCHA

Bachem C-2265.0001 MW 460.66 $C_{15}H_{21}NO_4 \cdot C_{12}H_{23}N$ Store at -15°C

Alanine — α-(1,2,4-Triazol-1-yl)-DL-Ala
Bachem F-3225.0001 MW 156.14 $C_5H_8N_4O_2$ Store at -15°C

Alanine — α-(1-Cyclopentenyl)-DL-Ala
Bachem F-1470.0100 MW 155.2 $C_8H_{13}NO_2$ Store at 2-8°C

Alanine — α-2-Pyridyl-Ala
Bachem F-2820.0001 MW 166.18 $C_8H_{10}N_2O_2$ Store at 2-8°C

Alanine — α-2-Pyridyl-D-Ala
Bachem F-2790.0001 MW 166.18 $C_8H_{10}N_2O_2$ Store at -15°C

Alanine — α-2-Pyridyl-DL-Ala
Bachem F-2825.0001 MW 166.18 $C_8H_{10}N_2O_2$ Store at 2-8°C

Alanine — α-2-Quinolyl-Ala
Bachem F-3655.0250 MW 216.24 $C_{12}H_{12}N_2O_2$ Store at 2-8°C

Alanine — α-2-Quinolyl-D-Ala
Bachem F-3660.0250 MW 216.24 $C_{12}H_{12}N_2O_2$ Store at 2-8°C

Alanine — α-2-Thiazolyl-DL-Ala
Bachem F-2955.0100 MW 172.21 $C_6H_8N_2O_2S$ Store at -15°C

Alanine — α-2-Thienylalanine
Bachem F-2110.0001 MW 171.22 $C_7H_9NO_2S$ Store at RT

Alanine — α-2-Thienyl-D-Ala
Bachem F-2115.0001 MW 171.22 $C_7H_9NO_2S$ Store at RT

Alanine — α-2-Thienyl-DL-Ala
Bachem N-1150.0001 MW 171.22 $C_7H_9NO_2S$ Store at RT

Alanine — α-3-Benzothienyl-Ala
Bachem F-2490.0001 MW 221.28 $C_{11}H_{11}NO_2S$ Store at RT

Alanine — α-3-Benzothienyl-D-Ala
Bachem F-2485.0001 MW 221.28 $C_{11}H_{11}NO_2S$ Store at RT

Alanine — α-3-Pyridyl-Ala
Bachem F-3195.0001 MW 166.18 $C_8H_{10}N_2O_2$ Store at 2-8°C

Alanine — α-3-Pyridyl-D-Ala
Bachem F-2640.0001 MW 166.18 $C_8H_{10}N_2O_2$ Store at 2-8°C

Alanine — α-3-Pyridyl-DL-Ala
Bachem F-3705.0001 MW 166.18 $C_8H_{10}N_2O_2$ Store at 2-8°C

Alanine — α-Ala

Bachem F-1110.0025   MW 89.09 $C_3H_7NO_2$ Store at RT

Alanine — α-Ala-4MβNA Hydrochloride

Bachem J-1375.0250   MW 280.75 $C_{14}H_{16}N_2O_2 \cdot HCl$ Store at -15°C

Alanine — α-Ala-AMC TFA

Bachem I-1030.0050   MW 360.3 $C_{13}H_{14}N_2O_3 \cdot C_2HF_3O_2$ Store at -15°C

Alanine — α-Ala-Histamine

Synonyms: Carcinine

Bachem G-4425.0250   MW 182.23 $C_8H_{14}N_4O$ Store at -15°C

Alanine — α-Ala-OBzl *p*-Tosyl

Bachem F-1125.0005   MW 351.42 $C_{10}H_{13}NO_2 \cdot C_7H_8O_3S$ Store at RT

Alanine — α-Ala-OEt Hydrochloride

Bachem F-1140.0025   MW 153.61 $C_5H_{11}NO_2 \cdot HCl$ Store at RT

Alanine — α-Ala-OMe Hydrochloride

Bachem F-1155.0005   MW 139.58 $C_4H_9NO_2 \cdot HCl$ Store at 2-8°C

Alanine — α-Ala-OtBu Hydrochloride

Bachem F-1135.0005   MW 181.66 $C_7H_{15}NO_2 \cdot HCl$ Store at -15°C

Alanine — α-Ala-βNA Hydrobromide

Bachem K-1035.0001   MW 295.18 $C_{13}H_{14}N_2O \cdot HBr$ Store at -15°C

Alanine — α-AMC-Ala

Synonyms: (*S*)-2-Amino-3-AMC-Propionic Acid

Bachem M-2345.0025   MW 263.25 $C_{13}H_{13}NO_5$ Store at -15°C

Alanine — α-Chloro-Ala

Bachem F-1425.0001   MW 123.54 $C_3H_6ClNO_2$ Store at -15°C

Alanine — α-Chloro-Ala Hydrochloride

Bachem F-1430.0001   MW 160.01 $C_3H_6ClNO_2 \cdot HCl$ Store at -15°C

Alanine — α-Chloro-Ala-NHOH

Bachem F-3380.0250   MW 138.55 $C_3H_7ClN_2O_2$ Store at -15°C

Alanine — α-Chloro-Ala-OMe Hydrochloride

Bachem F-3465.0001   MW 174.03 $C_4H_8ClNO_2 \cdot HCl$ Store at -15°C

Alanine — α-Chloro-D-Ala Hydrochloride

Bachem F-1435.0001   MW 160.01 $C_3H_6ClNO_2 \cdot HCl$ Store at -15°C

Alanine — α-Chloro-DL-Ala

Bachem F-1440.0001   MW 123.54 $C_3H_6ClNO_2$ Store at -15°C

Alanine — α-Chloro-DL-Ala Hydrochloride

Bachem F-2325.0001   MW 160.01 $C_3H_6ClNO_2 \cdot HCl$ Store at -15°C

Alanine — α-Cyano-Ala

Bachem F-1460.0250   MW 114.1 $C_4H_6N_2O_2$ Store at -15°C

Alanine — α-Cyclohexyl-Ala Hydrochloride

Bachem F-2500.0001   MW 207.7 $C_9H_{17}NO_2 \cdot HCl$ Store at RT

Alanine — α-Cyclohexyl-D-Ala Hydrochloride

Bachem F-2505.0001   MW 207.7 $C_9H_{17}NO_2 \cdot HCl$ Store at RT

Alanine — α-Cyclopentyl-DL-Ala

Bachem F-1465.0100   MW 157.21 $C_8H_{15}NO_2$ Store at 2-8°C

Alanine — α-Cyclopropyl-Ala

Bachem F-3470.0001   MW 129.16 $C_6H_{11}NO_2$ Store at -15°C

Alanine — α-Cyclopropyl-D-Ala

Bachem F-3475.0001   MW 129.16 $C_6H_{11}NO_2$ Store at -15°C

Alanine — α-Fluoro-DL-Ala

Bachem F-2530.0250   MW 107.09 $C_3H_6FNO_2$ Store at RT

Alanine — α-Me-β-Ala

Synonyms: DL-β-Aib

Fluka 08290   MW 103.12 $C_4H_9NO_2$ ≥99% (titration); ~0.1% residue on ignition; mp: 178-180°C

Alanine — α-tBu-Ala

Synonyms: Neopentyl-Gly

Bachem F-1315.0001   MW 145.2 $C_7H_{15}NO_2$ Store at RT

Alanine — α-tBu-D-Ala

Synonyms: D-Neopentyl-Gly

Bachem F-1320.0001   MW 145.2 $C_7H_{15}NO_2$ Store at RT

Alanine — α-tBu-DL-Ala

Synonyms: DL-Neopentyl-Gly

Bachem F-1325.0001   MW 145.2 $C_7H_{15}NO_2$ Store at RT

Alanine — β-(1,2,4-Triazol-3-yl)-DL-Ala

Synonyms: (±)-2-Amino-3-(1,2,4-Triazol-3-yl)-Propionic Acid; DL-1,2,4-Triazole-3-Ala

Sigma T 9299 $C_5H_8N_4O_2$ Inhibitory His analog; Levin, DE & Ames, BN, *Environ Mutagen*, 8: 9, 1986; Strauss, A et al, *Plant Cell Tissue Organ Cult*, 3: 123, 1984

Alanine — β-(3,5-Dioxo-1,2,4-Oxadiazolidin-2-yl)-L-Ala

Synonyms: Quisqualic Acid

Sigma Q 2128 $C_5H_7N_3O_5$ Glutamate analog

Alanine — β-2-Thiazolyl-DL-Ala

Synonyms: (±)-2-Amino-3-(2-Thiazolyl)-Propionic Acid
Sigma T 2634 $C_6H_8N_2O_2S$

Alanine — β-2-Thienyl-D-Ala

Sigma T 8910 $C_7H_9NO_2S$

Alanine — β-2-Thienyl-DL-Ala

ICN 100481   MW 171.2 $C_7H_9NO_2S$ ~98%; crystalline |
Phenylalanine antagonist

Sigma T 4875 $C_7H_9NO_2S$

Alanine — β-2-Thienyl-L-Ala

ICN 156851   MW 171.2 $C_7H_9NO_2S$

Alanine — β-5-Fluorouracilyl-L-Ala

Synonyms: (S)-(-)-α-Amino-5-Fluoro-(3,4-Dihydro)-2,4-Dioxo-
1(2H)-Pyrimidinepropanoic Acid; (S)-5-Fluoro-Willardiine

Sigma F 2417   AMPA/kainate agonist $C_7H_8FN_3O_4$ Potent;
Parneu, et al, *J Neurosci*, 12: 595,1992

Alanine — β-Ala

Synonyms: 3-Aminopropionic Acid; β-Aminopropionic Acid

Alexis 101-003   MW 89.1 $C_3H_7NO_2$ White powder; soluble in
H_2O

Amersham US10665

Fluka 05159   MW 89.1 $C_3H_7NO_2$ ≥99.0%; ≤0.1% residue on
ignition, loss on drying; ≤0.005% Cl, SO_4, K, Na; ≤0.0005% Al, Ba,
Bi, Cd, Co, Cr, Cu, Fe, Li, Mg, Mn, Mo, Ni, Pb, Sr, Zn; ≤0.001% Ca;
≤0.00001% As | Yagi, T et al, *Electrophresis '83*, Hirai, H, ed, de
Gruyter, Berlin, 503, 1984; Hine, T, *ibid*, 541

Fluka 05160   MW 89.1 $C_3H_7NO_2$ ≥99.0%; ≤0.1% residue on
ignition, loss on drying; ≤0.005% Cl, SO_4, K, Na; ≤0.0005% Cd, Co,
Cr, Cu, Fe, Mg, Mn, Ni, Pb, Zn; ≤0.001% Ca; mp: ~200°C

ICN 100929 $C_3H_7NO_2$

ICN 194618 $C_3H_7NO_2$

Neosystem AA02101   MW 89.1

Senn Chem 44088   MW 89.1   Used in the synthesis of
pantothenic acid & derivatives; buffer in electroplating

Sigma A 7752 $C_3H_7NO_2$ Nonselective endogenous agonist

Sigma A 9920   FW 89.09 $C_3H_7NO_2$ Crystalline | Cell culture
tested; insect cell culture tested

USBio A1205-05   MW 89.09 $C_3H_7NO_2$ 99+%

Biogenesis 0200-0051   Pan-species   BSA conjugate;
lyophilized

Alanine — β-Ala (1-¹⁴C)

ARC ARC-183   MW 89.1 $H_2NCH_2CH_2COOH$ 40-60
mCi/mmol; 1.48-2.22 GBq/mmol; in 0.01 *N* HCl | Radiochemical

Alanine — β-Ala (1-¹⁴C)-OEt Hydrochloride

ARC ARC-809   MW 153.61 $H_2NCH_2CH_2COOC_2H_5 \cdot HCl$ 50-
60 mCi/mmol; 1.85-2.22 GBq/mmol; EtOH:H_2O (9:1) |
Radiochemical

Alanine — β-Ala (3-¹⁴C)

ARC ARC-705   MW 89.1 $H_2NCH_2CH_2COOH$ 50-60
mCi/mmol; 1.85-2.22 GBq/mmol; in 0.01 *N* HCl | Radiochemical

Alanine — β-Ala (3-³H(N))

ARC ART-205   MW 89.1 $H_2NCH_2CH_2COOH$ 30-60 Ci/mmol;
1.11-2.22 TBq/mmol; EtOH:H_2O (2:98) | Radiochemical

Alanine — β-Ala (β-¹⁴C)

Sigma A 0562   FW 89.09 $H_2NCH_2CH_2CO_2H$ 40-60 Ci/mmol;
radiochemical purity ≥98% (HPLC); aqueous solution containing 2%
EtOH in Combi-vial | Radiochemical

Alanine — β-Ala 4-Methoxy-β-Naphthylamide

Sigma A 9664 $C_{14}H_{16}N_2O_2$

Alanine — β-Ala 4-Methoxy-β-Naphthylamide Hydrochloride

Sigma A 8773 $C_{14}H_{16}N_2O_2 \cdot HCl$

Alanine — β-Ala Amide Hydrochloride

Senn Chem 44089   MW 124.6

Alanine — β-Ala Hydroxamate Hydrochloride

Sigma A 8252 $C_3H_8N_2O_2 \cdot HCl$

Alanine — β-Ala-(t-OBzl) Hydrochloride

Sigma A 3041 $C_7H_{15}NO_2 \cdot HCl$

Alanine — β-Ala-OBg TFA

Neosystem AA02104   MW 426.4

Alanine — β-Ala-OBzl

Sigma A 1139 $C_{10}H_{13}NO_2 \cdot C_7H_8O_3S$

Alanine — β-Ala-OBzl-p-Tosyl

ICN 100934 $C_{10}H_{13}NO_2 \cdot C_7H_8O_3S$

Alanine — β-Ala-OBzl-Tosyl

Senn Chem 44039   MW 351.4

Alanine — β-Ala-OEt (2,3-³H) Hydrochloride

ARC ART-730   MW 153.61 $H_2NCH_2CH_2COOC_2H_5 \cdot HCl$ 20-
40 Ci/mmol; 0.74-1.48 TBq/mmol; EtOH:H_2O (9:1) |
Radiochemical

Alanine — β-Ala-OEt Hydrochloride

Fluka 05182   MW 153.61 $C_5H_{11}NO_2 \cdot HCl$ ≥98% (titration);
≤2% H_2O; mp: 67-70°C

Senn Chem 44090   MW 153.6

Sigma A 2774 $C_5H_{11}NO_2 \cdot HCl$

Alanine — β-Ala-OMe Hydrochloride

Fluka 05210   MW 139.58 $C_4H_9NO_2 \cdot HCl$ ≥98% (titration);
mp: 103-105°C

Sigma A 9515 $C_4H_9NO_2 \cdot HCl$

Alanine — β-Ala-OtBu Hydrochloride

Senn Chem 44093   MW 181.7

Alanine — β-Anthraniloyl-D-Ala

Synonyms: D-Kynurenine; D-2-Amino-4-(2-Aminophenyl)-4-
Oxobutanoic Acid

Sigma K 2380 $C_{10}H_{12}N_2O_3$

Alanine — β-Anthraniloyl-DL-Ala

Synonyms: DL-Kynurenine; DL-2-Amino-4-(2-Aminophenyl)-4-
Oxobutanoic Acid

Sigma K 3500 $C_{10}H_{12}N_2O_3$

Sigma K 3625 $C_{10}H_{12}N_2O_3 \cdot H_2SO_4$

Alanine — β-Anthraniloyl-L-Ala

Synonyms: L-Kynurenine; L-2-Amino-4-(2-Aminophenyl)-4-Oxobutanoic Acid

Sigma K 3750	$C_{10}H_{12}N_2O_3 \cdot H_2SO_4$
Sigma K 8625	$C_{10}H_{12}N_2O_3$

Alanine — β-Chloro-D-Ala Hydrochloride

Synonyms: Alanine Racemase Inhibitor

Sigma C 4284 $C_3H_6ClNO_2 \cdot HCl$ Antibacterial agent; Manning, J et al, *Proc Natl Acad Sci USA*, 71: 417, 1974

Alanine — β-Chloro-DL-Ala

ICN 154958	$C_3H_6ClNO_2$
Sigma C 9517	$C_3H_6ClNO_2$

Alanine — β-Chloro-DL-Ala Free Base Hydrochloride

Sigma C 9262	$C_3H_6ClNO_2 \cdot HCl$

Alanine — β-Chloro-DL-Ala Hydrochloride

ICN 154959	$C_3H_6ClNO_2 \cdot HCl$

Alanine — β-Chloro-L-Ala Hydrochloride

Synonyms: Alanine Aminotransferase Inhibitor

Sigma C 9033 $C_3H_6ClNO_2 \cdot HCl$ Morino, Y et al, *J Biol Chem*, 254: 279, 1979

Alanine — β-Cyano-L-Ala

ICN 100948	
Sigma C 9151	$C_4H_6N_2O_2$
Sigma C 9650	$C_4H_6N_2O_2$

Alanine — β-*N*-Methylamino-L-Ala Hydrochloride

Synonyms: L-BMAA

ICN 155462 $C_4H_{10}N_2O_2 \cdot HCl$ Neuroexcitotoxic AA; Spencer, PS etal, *Science*, 237:517, 1987

Sigma M 4154 $C_4H_{10}N_2O_2 \cdot HCl$ Neuroexcitotoxic AA; may be involved in Guam amyotrophic lateral sclerosis; Spencer, PS et al, *Sci*, 237: 517, 1987

Alanine — β-*N*-Oxalylamino-L-Ala

Synonyms: BOAA; L-ODAP

Sigma O 5382 *Lathyrus sativus* $C_5H_8N_2O_5$ Neuroexcitotoxic AA; may be involved in the etiology of the motoneuron disease, lathyrism; Spencer, PS et at, *Sci*, 237: 517, 1987; Ross, SM et al, *Brain Res*, 425: 120, 1987

Alanine — β-Sulfo-Ala

Synonyms: Pth-Cys; Cysteic Acid

Sigma P 1894 $C_{10}H_9N_2O_4S_2Na$

Amino Acid Mixtures — (^{35}S)-L-(Cys Met)

Amersham AGQ0080 Redivue stabilized & colored, aqueous; >37 TBq/mmol, >1000 Ci/mmol; 530 MBq/mL, 14.3 mCi/mL

Amersham SJQ0079 Pro-mix L-(^{35}S) *in vitro* cell labeling mix; >37 TBq/mmol, >1000 Ci/mmol; 530 MBq/mL, 14.3 mCi/mL

ARC ARS-110 *E. coli* ^{35}S hydrolysate >1000 Ci/mmol; >37 TBq/mmol; in sterile H_2O under nitrogen; hydrolysate contains 70% L-Met ^{35}S & 25% L-Cys ^{35}S & 5% miscellaneous AA; 2% BME solution | Radiochemical reagent for metabolic labeling

Amino Acid Mixtures — (^{3}H)-(Ala Arg Asp Glu Gly His Ile Leu Lys Phe Pro Ser Thr Tyr Val)

Amersham TRK440 Aqueous EtOH (2%) solution, sterilized; 37 GBq/mL, 1 mCi/mL | Tritiated

Amino Acid Mixtures — (^{3}H)-(Leu Lys Phe Pro Tyr)

Amersham TRK550 Aqueous EtOH (2%) solution, sterilized; 37 GBq/mL, 1 mCi/mL | Tritiated, high SA

Amino Acid Mixtures — (^{3}H)-L-(Ala Glu Gly Ile Leu Phe Ser Tyr Val)

ARC ART-328 High specific activity; in EtOH:H_2O (2:98) | Radiochemical

Amino Acid Mixtures — (^{3}H)-L-(Arg Ala Asp Glu Gly His Ile Leu Lys Phe Pro Ser Thr Tyr Val)

ICN 20063 1 mCi/mL, 37 MBq/mL; 0.01 *N* HCl; 1.5% His; 4% Gly, Ser, Tyr; 5% Ile, Pro, Thr; 6% Lys; 7% Arg; 8% Ala, Asp, Phe, Val; 12.5% Glu; 14% Leu

Amino Acid Mixtures — (U-^{14}C)-L-(Ala Arg Asp Glu Gly His Ile Leu Lys Phe Pro Ser Thr Tyr Val)

Amersham CFB104 Aqueous EtOH (2%) solution, sterilized; >1.85 GBq/mg atom carbon, >50 mCi/mg atom carbon; 185 MBq/mL, 50 μCi/mL | Radiolabeled L-AA mixture excludes L-(U-^{14}C)-Met

Amino Acid Mixtures — (U-^{14}C)-L-(Ala Arg Asp Glu Gly His Ile Leu Lys Met Phe Pro Ser Thr Tyr Val)

Amersham CFB25 Aqueous EtOH (2%) solution, sterilized; >1.85 GBq/mg atom carbon, >50 mCi/mg atom carbon; 185 MBq/mL, 50 μCi/mL | Radiolabeled L-AA mixture

Amino Acid Mixtures — (U-^{14}C)-L-(Arg Ala Asp Glu Gly His Ile Leu Lys Phe Pro Ser Thr Tyr Val)

ICN 10147 100 μCi/mL, 3.7 MBq/mL; 0.01 *N* HCl; 1.5% His; 4% Gly, Ser, Tyr; 5% Ile, Pro, Thr; 6% Lys; 7% Arg; 8% Ala, Asp, Phe, Val; 12.5% Glu; 14% Leu

Amino Acid Mixtures — Ala Arg Asn Asp Cys Glu Gly His Ile Leu Lys Met Phe Pro Ser Thr Tyr Val

Fluka 09418	Mixture of 0.01 *mM* of each AA in 0.1 *M* HCl	AA standard
Fluka 09421	Mixture of 0.05 *mM* of each AA in 0.1 *M* HCl
Fluka 09423	Mixture of 0.1 *mM* of each AA in 0.1 *M* HCl
Fluka 09425	Mixture of 0.5 *mM* of each AA in 0.1 *M* HCl
Fluka 09428	Mixture of 1 *mM* of each AA in 0.1 *M* HCl

Amino Acid Mixtures — Amino Acids

ICN 1012616 For molecular biology; dehydrated media

Amino Acid Mixtures — Amino Acids Diet

ICN 960361	Synthetic, modified, powdered
ICN 960362	Synthetic, modified, powdered

Amino Acid Mixtures — Amino Acids Phenylthiohydantoin

Fluka 09420 Unstable in solution | Reference for sequential analysis of peptides

Amino Acid Mixtures — Asn Pro

Fluka 17129 2 g/L DL-Asn, 1 g/L L-Pro, 1 g/L dipotassium phosphate anhydrous, 0.5 g/L magnesium sulfate, 10 g/L potassium sulfate | Asparagine proline broth for the cultivation of *Pseudomonas aeruginosa* using membrane filter technique

Amino Acid Mixtures — BME Amino Acids

Sigma B 6766 50x solution; without Gln; sterile-filtered; endotoxin tested; cell culture tested

Amino Acid Mixtures — Dansyl L-(Ala Abu Arg Asn Asp Cysteic Acid Cys Glu Gly His Hyp+ Ile Leu Lys Met Nva Phe Pro Ser Thr Trp Tyr Val) Cyclohexylammonium Salt

Sigma DAN-L-23   100 mg each of 23 Dansyl L-AA; prepared from high purity AA; usual purity 99% (TLC) although a few items might be available only with lower purity | Derivatized AA; protein analysis reagents

Amino Acid Mixtures — Dnp-L-(Ala Arg Asn Asp Cysteine Cystine Glu Gly His Hyp+ Ile Leu Lys Met Nle Nva Orn Phe Pro Ser Trp Tyr Val) Hydrochlorides

Sigma Dnp-25   25 mg each of 25 dinitrophenyl L-AA; prepared from high purity AA; usual purity 99% (TLC) although a few items might be available only with lower purity | Derivatized AA; protein analysis reagents

Amino Acid Mixtures — Fluorescence Detection Amino Acids Standard

Sigma A 2161   25 nmoles/mL of each of these L-AA: Ala, Arg, Asp, Glu, Gly, His, Ile, Leu, Lys, Met, Phe, Pro, Ser, Thr, Trp, Tyr, Val; except L-cystine at 12.5 nmoles/mL

Amino Acid Mixtures — Grace's Amino Acids Solution

Sigma G 0273   Without L-Gln; sterile-filtered; 10X; endotoxin tested; insect cell culture tested

Amino Acid Mixtures — Grace's Amino Acids Solution-Modified for TC-100

Sigma G 0148   Without β-Ala, L-Gln & D-Ser; with Ser; sterile-filtered; 10X; endotoxin tested; insect cell culture tested

Amino Acid Mixtures — L-(Ala Arg Asn Asp Cys Cystine Glu Gln Gly His Hyp Ile Leu Lys Met Phe Pro Ser Thr Trp Tyr Val)

Fluka 09416   ≥99% (titration); ≤0.3% foreign AA

Amino Acid Mixtures — L-Arg L-Asp (1:1)

Rexim   MW 307.3 $C_{10}H_{21}N_5O_6$ White crystals or crystalline powder

Amino Acid Mixtures — L-Arg L-Glu (1:1)

Rexim   MW 321.3 $C_{11}H_{23}N_5O_6$ White crystals or crystalline powder

Amino Acid Mixtures — L-Asp L-Lys

Rexim   MW 279.3 $C_{10}H_{21}N_3O_6$ White crystals or crystalline powder

Amino Acid Mixtures — L-Asp L-Orn

Rexim   MW 265.3 $C_9H_{19}N_3O_6$

Amino Acid Mixtures — L-Glu L-Lys

Rexim   MW 329.4 $C_{11}H_{27}N_3O_8$ Dihydrate; white crystals or crystalline powder

Amino Acid Mixtures — MEM Amino Acids

Sigma M 5550   50X solution; without L-Gln; sterile-filtered | Endotoxin tested; cell culture tested

Amino Acid Mixtures — MEM Non-Essential Amino Acids

Sigma M 7145   100X solution; sterile-filtered | Endotoxin tested; cell culture tested

Amino Acid Mixtures — MEM Non-Essential Amino Acids Supplement

Sigma M 2025   Powdered mixture; each package contains sufficient material to supplement the number of liters indicated (81.4 mg/L); sterilized by γ-irradiation; cell culture tested

Amino Acid Mixtures — Physiological Acidic & Neutral Amino Acids Standard

Sigma A 6407   2.5 µmoles/mL of each of these AA except L-cystine at 1.25 µmoles/mL: β-Ala, L-Ala, L-α-Abu, DL-β-Aib, Asn, Asp, Cit, cystathionine, Glu, Gly, hydroxy-Pro, Ile, Leu, Met, Phe, o-phospho-Ser, o-phosphoethanolamine, Sar, Ser, taurine, Pro, Thr, Tyr, urea, Val

Sigma PAA-11   Includes 11 compounds: β-Ala, L-α-aminoadipic acid, L-α-Abu, L-β-Aib, Cit, cystathionine, L-Hcy, o-phosphorylethanolamine, o-phosphoserine, Sar, urea | Acidic & neutral physiological AA & related compounds

Amino Acid Mixtures — Physiological Amino Acids Standard

Sigma PAA-20   Includes 20 compounds: γ-, urea | Physiological AA & related compounds

Amino Acid Mixtures — Physiological Basic & Neutral Amino Acids Standard

Sigma A 9906   0.5 µmole/mL of each of these AA & related compounds in 0.2 N lithium citrate: β-Ala, L-Ala, L-α-Abu, γ-Abu, DL-β-Aib, Asp, L-carnosine, creatine, L-cystine, ethanolamine, Cit, cystathionine, Glu, Gly, His, Hcy, δ-Hyl, Hyp, Ile, Leu, Lys, Met, 1-Me-His, 3-Me-His, Orn, Phe, Pro, Sar, Ser, taurine, Thr, Trp, Tyr, urea, Val, L-anserine, Arg

Amino Acid Mixtures — Physiological Basic Amino Acids Standard

Sigma A 1585   2.5 µmoles/mL of each of these L-AA & related compounds except anserine & Hcy at 1.25 µmole/mL: GABA, anserine, Arg, carnosine, ethanolamine, His, δ-Hyl, Lys, 1-Me-His, 3-Me-His, Orn

Sigma A 6282   2.5 µmoles/mL of each of these AA & related compounds: GABA, L-anserine, Arg, L-carnosine, creatinine, ethanolamine, His, δ-Hyl, Lys, 1-Me-His, 3-Me-His, Orn, Trp

Sigma PAA-9   Includes 9 compounds: γ-Abu, ethanolamine (2-aminoethanol), L-anserine nitrate, L-carnosine, creatinine, δ-Hyl (mixed DL & DL-allo), 1-Me-His, 3-Me-His, Orn | Basic physiological AA & related compounds

Amino Acid Mixtures — Pth-(Ala Arg Asn Asp Gln Glu Gly His Ile Leu) N$^{\alpha}$-Pth-N$^{\epsilon}$-PTC-(Lys Met Phe Pro Ser Thr Trp Tyr Val)

Sigma P 7185   Dry stabilized film in an amber screw cap vial containing 0.1 µmole of each AA | Complete AA standard; designed for quantitative & qualitative identification of Pth-AA by HPLC; each lot is tested by HPLC to ensure the components will be within ±5% of the stated concentration

Amino Acid Mixtures — RPMI 1640 Amino Acids

Sigma R 7131   50X solution; without L-Gln; sterile-filtered; endotoxin tested; cell culture tested

Aminobenzoic Acid — BOC-2-Abz

Senn Chem 44015   MW 237.3

Aminobenzoic Acid — BOC-p-Abz

USBio B2259   MW 237.3 $C_{12}H_{15}NO_4$ ≥98%

Aminobenzoic Acid — FMOC-2-Abz

Senn Chem 44045   MW 359.4

Aminobenzoic Acid — FMOC-4-Abz

Senn Chem 44046   MW 359.4

Aminobenzoic Acid — FMOC-*p*-Abz

USBio F5314   MW 359.4 $C_{22}H_{17}NO_4$ ≥98%

Aminobenzoic Acid — FMOC-α-Abz

USBio F5315   MW 325.4 $C_{19}H_{19}NO_4$ ≥98%

Aminobutyric Acid — 4-Abu

Synonyms: GABA

Senn Chem 44094   MW 103.1   Most abundant CNS inhibitory neurotransmitter

Aminobutyric Acid — 4-Abu-OtBu Hydrochloride

Synonyms: GABA-OtBu; GABA

Senn Chem 44095   MW 195.7

Aminobutyric Acid — Abu

Synonyms: GABA

Bachem F-1190.0005   MW 103.12 $C_4H_9NO_2$ Store at RT

Aminobutyric Acid — Abu Amide Hydrochloride

Synonyms: GABA

Bachem F-2440.0250   MW 138.6 $C_4H_{10}N_2O \cdot HCl$ Store at 2-8°C

Aminobutyric Acid — Abu-OtBu Hydrochloride

Synonyms: GABA

Bachem F-3035.0001   MW 195.69 $C_8H_{17}NO_2 \cdot HCl$ Store at 2-8°C

Aminobutyric Acid — Ac-4-Abu

Synonyms: Ac-GABA; GABA

Senn Chem 44079   MW 145.2

Aminobutyric Acid — BOC-4-Abu

Synonyms: GABA

USBio B2247   MW 203.2 $C_9H_{17}NO_4$ ≥99%

Aminobutyric Acid — BOC-4-Abu CHA Salt

Synonyms: BOC-GABA CHA; GABA

Senn Chem 44032   MW 284.3

Aminobutyric Acid — BOC-Abu

Synonyms: GABA

USBio B2246   MW 203.2 $C_9H_{17}NO_4$ ≥99%

Aminobutyric Acid — BOC-D-2-Abu-Dicyclohexylamine

Synonyms: GABA

USBio B2245   MW 384.5 $C_9H_{17}NO_4 \cdot C_{12}H_{23}N$ ≥99%

Aminobutyric Acid — CBZ-Abu

Synonyms: GABA

USBio C2098-24   MW 237.3 $C_{12}H_{15}NO_4$ ≥99%

Aminobutyric Acid — D-α-Amino-*N*-Butyric Acid

Synonyms: GABA

USBio A1376-10   MW 103.1 $C_4H_9NO_2$ ≥99%

Aminobutyric Acid — FMOC-4-Abu

Synonyms: FMOC-GABA; GABA

Senn Chem 44059   MW 325.4

USBio F5316   MW 325.4 $C_{19}H_{19}NO_4$ ≥98%

Aminobutyric Acid — L-α-Amino-*N*-Butyric Acid

Synonyms: GABA

USBio A1376-11   MW 103.1 $C_4H_9NO_2$ ≥99%

Aminobutyric Acid — Z-4-Abu

Synonyms: Z-GABA; GABA

Senn Chem 44075   MW 237.3

Aminobutyric Acid — Z-Abu

Synonyms: GABA

Bachem C-1260.0001   MW 237.26 $C_{12}H_{15}NO_4$ Store at RT

Aminobutyric Acid — Z-γ-Abu

Synonyms: GABA

Bachem C-3160.0005   MW 237.26 $C_{12}H_{15}NO_4$ Store at RT

Aminobutyric Acid — γ-Abu-OtBu Hydrochloride

Synonyms: GABA

Bachem F-3755.0001   MW 195.69 $C_8H_{17}NO_2 \cdot HCl$ Store at 2-8°C

Aminocaproic Acid — BOC-6-Aca

Synonyms: BOC-6-Aminohexanoic Acid

Senn Chem 44016   MW 231.3

USBio B2248   MW 231.3 $C_{11}H_{21}NO_4$ ≥99%

Aminocaproic Acid — CBZ-6-Aca

Synonyms: CBZ-6-Aminohexanoic Acid

USBio C2098-25   MW 265.3 $C_{14}H_{19}NO_4$ ≥99%

Aminocaproic Acid — FMOC-6-Aca

Senn Chem 44047   MW 353.4

Aminocaproic Acid — Z-6-Aca

Senn Chem 44067   MW 265.3

Aminocaproic Acid — ε-Amino-*N*-Caproic Acid

Synonyms: 6-Amino-*N*-Hexanoic Acid; Plasmin Inhibitor

USBio A1377   MW 131.2 $C_6H_{13}NO_2$ ≥98%; white crystalline powder; specific rotation: provided with every lot analysis batch certificate.

Aminohexanoic Acid — FMOC-6-Acp

Synonyms: FMOC-εAcp; N^α-9-FMOC-ε-Aca

USBio F5300   MW 353.3 $C_{21}H_{23}NO_4$ ≥99%

USBio F5317   MW 353.3 $C_{21}H_{23}NO_4$ ≥99%

Aminoisobutyric Acid — Aib-OtBu Hydrochloride

Synonyms: α-Me-Ala-OtBu

Bachem E-3060.0001   MW 195.69 $C_8H_{17}NO_2 \cdot HCl$ Store at 2-8°C

Aminoisobutyric Acid — BOC-Aib

Synonyms: BOC-α-Aib; BOC-α-Methylalanine

Peptides International BLX-5340-PI   MW 203.24   >98% (HPLC); white crystalline powder

Aminoisobutyric Acid — DL-3-Aib

Senn Chem 44087   MW 103.1

Aminoisobutyric Acid — N-Me-Aib

Synonyms: N-Me-α-Me-Ala

Bachem F-1765.0001   MW 117.15 $C_5H_{11}NO_2$ Store at RT

Aminosuberic Acid — BOC-Asu-OBzl-Dicyclohexylamine

USBio B2284   MW 560.8 $C_{20}H_{29}NO_6 \cdot C_{12}H_{23}N$ ≥98%

Aminosuberic Acid — FMOC-Asu-OtBu

USBio F5347   MW 467.6 $C_{27}H_{33}NO_6$ ≥98%

Arginine — (R)-N₂-(DiphenylAc)-N-((n-Hydroxyphenyl)Me¹)-Arg Amide

Synonyms: BIBP 3226; Neuropeptide YY₁ Receptor Antagonist

American peptide 60-1-22   MW 473.6 $C_{27}H_{31}N_5O_3$ Pheng, LH et al, *Eur J Pharmacol*, 327: 163, 1997

Arginine — (Z-Arg)₂ MR·TFA

Synonyms: Trypsin Substrate

Kamiya   MW 1070   >99% (HPLC); Salt

Arginine — 1-Arg (U-¹⁴C) Hydrochloride

Amersham CFB63

Arginine — 4-Nitro-Z-D-Arg

Bachem C-3360.0001   MW 353.34 $C_{14}H_{19}N_5O_6$ Store at -15°C

Arginine — Ac-Arg

Bachem E-1025.0005   MW 216.24 $C_8H_{16}N_4O_3$ Store at RT

Arginine — Ac-Arg Amide

Bachem E-2800.0001   MW 215.26 $C_8H_{17}N_5O_2$ Store at 2-8°C

Arginine — Ac-Arg-OMe Hydrochloride

Synonyms: C1r Substrate

Bachem E-1030.0001   MW 266.73 $C_9H_{18}N_4O_3 \cdot HCl$ Store at RT

Peptides International SRM-3078   MW 266.73 $C_9H_{18}N_4O_3 \cdot HCl$ >96% (HPLC); amorphous powder; hydrochloride | Sim, RB in, *Proteolytic Enzymes Part C*, Methods in Enzymology, Vol 80, L Lorand, Ed, Academic Press, New York, 1981, pp 26-42

Arginine — Ac-Arg-Pmc

Bachem E-3015.0001   MW 482.6 $C_{22}H_{34}N_4O_6S$ Store at 2-8°C

Arginine — Ac-Arg-pNA Hydrochloride

Synonyms: L-AAPA

Bachem L-1025.0250   MW 372.81 $C_{14}H_{20}N_6O_4 \cdot HCl$ Store at -15°C

Arginine — Arg

Bachem E-1360.0025   MW 174.2 $C_6H_{14}N_4O_2$ Store at RT

Arginine — Arg Amide Dihydrochloride

Bachem E-1365.0001   MW 246.14 $C_6H_{15}N_5O \cdot 2HCl$ Store at RT

Arginine — Arg Polyamine Dihydrochloride

Synonyms: sFTX-3.3; FTX-3.3

Calbiochem 344595   Synthetic   MW 360.3 $C_{12}H_{29}N_7O \cdot 2HCl$ ≥95% (HPLC); lyophilized solid; soluble in H_2O; fully active as tested by the method of Dupere, et al; Harmful: LD_{50} ≤2000 mg/kg | Synthetic analog of the funnel web spider toxin (Calbiochem No. 344594) that has been shown to block neuronal voltage-gated P-type (IC_{50}=240 μM) & N-type (IC_{50}=700 μM) Ca^{2+} channels; Also acts as a potent inhibitor of low-voltage activated T-type Ca^{2+} channels (IC_{50}=10 nM); Dupere, JR et al, *Neuropharmacology*, 35: 1, 1996; Norris, TM et al, *Mol Pharmacol*, 50: 939, 1996; Scott, RH et al, *Br J Pharmacol*, 106: 199, 1992

Calbiochem 344594   *Agelenopsis aperta* (funnel web spider)   MW 419.3 $C_{12}H_{31}N_7 \cdot 4HCl$ ≥95% (HPLC); lyophilized solid; soluble in H_2O; fully active as tested by the method of Dupere, et al; Harmful: LD_{50} ≤2000 mg/kg | An arginine polyamine-type funnel web spider toxin that has been shown to block neuronal voltage-gated P-type (IC_{50}=130 μM) & N-type (IC_{50}=240 μM) Ca^{2+} channels; Also reported to block L type Ca^{2+} channels at higher concentrations; Dupere, JR et al, *Neuropharmacology*, 35: 1, 1996; Norris, TM et al, *Mol Pharmacol*, 50: 939, 1996; Brown, AM et al, *J Physiol*, 475: 197, 1994

Arginine — Arg-(NH₂) Flavianate

Synonyms: L-NAA Flavianate

Bachem E-2890.0025   MW 503.45 $C_6H_{15}N_5O_2 \cdot C_{10}H_6N_2O_8S$ Store at -15°C

Arginine — Arg-4MβNA

Bachem J-1035.0250   MW 329.4 $C_{17}H_{23}N_5O_2$ Store at -15°C

Arginine — Arg-Agarose

ICN 191283   5 atoms hydrophilic spacer arm; used for purification of plasminogen activator, prothrombin

Arginine — Arg-AMC

Synonyms: Cathepsin H Substrate

Peptides International MAR-3113-v   MW 331.37 $C_{16}H_{21}N_5O_3$ >99% (HPLC); lyophilized amorphous powder | Kanaoka, Y et al, *Chem Pharm Bull*, 25:3126, 1977

Arginine — Arg-AMC Dihydrochloride

Synonyms: Cathepsin H Substrate II; Aminopeptidase B Substrate

Bachem I-1050.0050   MW 404.3 $C_{16}H_{21}N_5O_3 \cdot 2HCl$ Store at -15°C

ICN 195958 $C_{16}H_{21}N_5O_3 \cdot 2HCl$ Substrate for quantitative determination of cathepsin H & aminopeptidase B

Arginine — Argininosuccinic Acid

Sigma A 5707 $C_{10}H_{16}N_4O_6Na_2$

Arginine — Arg-Mbs

Bachem E-2765.0001   MW 344.39 $C_{13}H_{20}N_4O_5S$ Store at -15°C

Arginine — Arg-Me Flavianate

Synonyms: L-NMMA Flavianate; L-NMA Flavianate

Bachem F-2450.0025   MW 502.46 $C_7H_{16}N_4O_2 \cdot C_{10}H_6N_2O_8S$ Store at -15°C

Arginine — Arg-Me Hydrochloride

Synonyms: L-NMMA; L-NMA

Bachem E-2770.0050   MW 224.69 $C_7H_{16}N_4O_2 \cdot HCl$ Store at -15°C

Arginine — Arg-Mtr

Bachem E-2205.0001   MW 386.47 $C_{16}H_{26}N_4O_5S$ Store at RT

Arginine — Arg-Mtr-OtBu Free Base

Bachem E-3075.0001   MW 442.58 $C_{20}H_{34}N_4O_5S$ Store at -15°C

Arginine — Arg-Nitro

Synonyms: L-NA; L-NNA; *N'*-Nitro-L-Arg

Bachem F-1870.0025   MW 219.2 $C_6H_{13}N_5O_4$ Store at RT

Peptides International XNR-2005   MW 219.20   >98% homogeneity (paper electrophoresis); colorless crystalline powder | Nitric oxide synthase inhibitor; Y Kobayashi & K Hattori, *Jpn J Pharmacol*, 52:167, 1990

USBio A3555   MW 219.2 $C_6H_{13}N_5O_4$ ≥99%

Arginine — Arg-Nitro-OBzl *p*-Tosyl

Bachem F-1880.0005   MW 481.53 $C_{13}H_{19}N_5O_4 \cdot C_7H_8O_3S$ Store at RT

Arginine — Arg-Nitro-OBzl-Tosyl

Synonyms: *N'*-Nitro-L-Arg-OBzl-Tosyl

Peptides International EBR-2048   MW 481.53   >97% (TLC); white powder

Arginine — Arg-Nitro-OMe Hydrochloride

Synonyms: L-NAME

Bachem F-1885.0005   MW 269.69 $C_7H_{15}N_5O_4 \cdot HCl$ Store at 2-8°C

Arginine — Arg-Nitro-pNA Hydrobromide

Bachem E-3380.0001   MW 420.23 $C_{12}H_{17}N_7O_5 \cdot HBr$ Store at -15°C

Arginine — Arg-OEt Dihydrochloride

Bachem E-1370.0005   MW 275.18 $C_8H_{18}N_4O_2 \cdot 2HCl$ Store at RT

Arginine — Arg-OMe Dihydrochloride

Bachem E-1375.0005   MW 261.15 $C_7H_{16}N_4O_2 \cdot 2HCl$ Store at RT

Arginine — Arg-OtBu Dihydrochloride

Bachem E-3460.0001   MW 303.23 $C_{10}H_{22}N_4O_2 \cdot 2HCl$ Store at -15°C

Arginine — Arg-Pbf

Bachem E-3110.0001   MW 426.53 $C_{19}H_{30}N_4O_5S$ Store at 2-8°C

Arginine — Arg-Pbf-OAll Hydrochloride

Bachem E-3350.0001   MW 503.07 $C_{22}H_{34}N_4O_5S \cdot HCl$ Store at -15°C

Arginine — Arg-Pbf-OMe Hydrochloride

Bachem E-3325.0001   MW 440.56 $C_{20}H_{32}N_4O_5S$ Store at -15°C

Arginine — Arg-Pmc

Bachem E-2885.0001   MW 440.56 $C_{20}H_{32}N_4O_5S$ Store at RT

Arginine — Arg-Pmc-OtBu Free Base

Bachem E-3085.0001   MW 496.68 $C_{24}H_{40}N_4O_5S$ Store at -15°C

Arginine — Arg-pNA Dihydrochloride

Bachem L-1120.0250   MW 367.24 $C_{12}H_{18}N_6O_3 \cdot 2HCl$ Store at -15°C

Arginine — Arg-Tosyl

Synonyms: *N'*-Tosyl-L-Arg

Bachem E-2460.0001   MW 328.39 $C_{13}H_{20}N_4O_4S$ Store at 2-8°C

Peptides International XTR-2101-PI   MW 328.39   >98% (TLC); white powder

Arginine — Arg-Z₂

Bachem E-1800.0001   MW 442.47 $C_{22}H_{26}N_4O_6$ Store at -15°C

Arginine — Arg-βNA Hydrochloride

Synonyms: Cathepsin H Substrate I; Aminopeptidase B Substrate

Bachem K-1080.0250   MW 335.84 $C_{16}H_{21}N_5O \cdot HCl$ Store at -15°C

ICN 195957 $C_{16}H_{21}N_5O \cdot HCl$ Substrate for quantitative determination of cathepsin H & aminopeptidase B

Arginine — BOC-Arg

Bachem A-1200.0001   MW 274.32 $C_{11}H_{22}N_4O_4$ Store at -15°C

Arginine — BOC-Arg Hydrochloride

Bachem A-1205.0005   MW 310.78 $C_{11}H_{22}N_4O_4 \cdot HCl$ Store at -15°C

Arginine — BOC-Arg Hydrochloride Hydrate

USBio B2262   MW 328.8 $C_{11}H_{22}N_4O_4HCl \cdot H_2O$ ≥99%

Arginine — BOC-Arg-BOC₂

Bachem A-2935.0001   MW 474.56 $C_{21}H_{38}N_4O_8$ Store at -15°C

Arginine — BOC-Arg-Mbs

Bachem A-2715.0001   MW 444.51 $C_{18}H_{28}N_4O_7S$ Store at -15°C

Arginine — BOC-Arg-Mtr

Bachem A-2105.0001   MW 486.59 $C_{21}H_{34}N_4O_7S$ Store at 2-8°C

USBio B2263   MW 487.6 $C_{21}H_{35}N_4O_7S$ ≥99%

Arginine — BOC-Arg-Mts

USBio B2264   MW 487.6 $C_{21}H_{35}N_4O_7S$ ≥99%

Arginine — BOC-Arg-Mts CHA

Bachem A-2110.0001   MW 555.74 $C_{20}H_{32}N_4O_6S \cdot C_6H_{13}N$ Store at 2-8°C

Arginine — BOC-Arg-Nitro

Synonyms: N^α-BOC-*N'*-Nitro-L-Arg

Bachem A-2115.0005   MW 319.32 $C_{11}H_{21}N_5O_6$ Store at -15°C

Peptides International BLR-2058   MW 319.32   >98% (HPLC); white powder; keep dry in freezer | Typically co-crystallize with ¼- 1 molecule H_2O &/or Et acetate/molecule AA

USBio B2266   MW 319.3 $C_{11}H_{21}N_5O_6$ ≥99%

Arginine — BOC-Arg-Pbf

Bachem A-3565.0001   MW 526.65 $C_{24}H_{38}N_4O_7S$ Store at -15°C

Arginine — BOC-Arg-Pmc

Bachem A-2875.0001   MW 540.69 $C_{25}H_{40}N_4O_7S$ Store at -15°C

Arginine — BOC-Arg-SBzl Hydrochloride

Bachem A-3845.0250   MW 416.97 $C_{18}H_{28}N_4O_3S \cdot HCl$ Store at -15°C

Arginine — BOC-Arg-Tosyl

Synonyms: N^α-BOC-N^r-Tosyl-L-Arg

American Peptide BLARG15   MW 428.5

Bachem A-2315.0001   MW 428.51 $C_{18}H_{28}N_4O_6S$ Store at -15°C

Peptides International BLR-2125   MW 428.51   >98% (HPLC); white crystalline powder | Typically co-crystallize with ¼-1 molecule H_2O &/or Et acetate/molecule AA

USBio B2268   MW 428.5 $C_{18}H_{28}N_4O_6S$ ≥99%

Arginine — BOC-Arg-Tosyl-® (200-400 mesh)

Bachem D-1465.0005   Store at RT

Arginine — BOC-Arg-Tosyl-O-Resin

American Peptide RMRB110   1% DVB cross-linked: 100-200 mesh | t-BOC protected AA resin; preattached resins are useful for synthesis of peptides with C-terminal carboxyl group by t-BOC chemistry

Arginine — BOC-Arg-Tosyl-PAM-® (200-400 mesh)

Bachem D-1280.0001   Store at 2-8°C

Arginine — BOC-Arg-Tosyl-PAM-Resin

American Peptide RPAM110   1% DVB cross-linked: 100-200 mesh | t-BOC protected AA resin; preattached resins are useful for synthesis of peptides with C-terminal carboxyl group by t-BOC chemistry

Arginine — BOC-Arg-Z₂

Bachem A-1565.0001   MW 542.59 $C_{27}H_{34}N_4O_8$ Store at -15°C

Arginine — BOC-Arg-Z₂-OSu

Bachem A-1570.0001   MW 639.66 $C_{31}H_{37}N_5O_{10}$ Store at -15°C

Arginine — BOC-D-Arg Hydrochloride

Bachem A-1210.0001   MW 310.78 $C_{11}H_{22}N_4O_4 \cdot HCl$ Store at -15°C

Arginine — BOC-D-Arg Hydrochloride Hydrate

Neosystem BA00201   MW 328.8

USBio B2261   MW 328.8 $C_{11}H_{22}N_4O_4HCl \cdot H_2O$ ≥99%

Arginine — BOC-D-Arg-Mtr

Bachem A-2805.0001   MW 486.59 $C_{21}H_{34}N_4O_7S$ Store at -15°C

Arginine — BOC-D-Arg-Mts CHA

Bachem A-2595.0001   MW 555.74 $C_{20}H_{32}N_4O_6S \cdot C_6H_{13}N$ Store at -15°C

Arginine — BOC-D-Arg-Nitro

Bachem A-2120.0001   MW 319.32 $C_{11}H_{21}N_5O_6$ Store at -15°C

USBio B2265   MW 319.3 $C_{11}H_{21}N_5O_6$ ≥99%

Arginine — BOC-D-Arg-Pbf

Bachem A-3750.0001   MW 526.65 $C_{24}H_{38}N_4O_7S$ Store at -15°C

Arginine — BOC-D-Arg-Pmc

Bachem A-4185.0001   MW 540.69 $C_{25}H_{40}N_4O_7S$ Store at 2-8°C

Arginine — BOC-D-Arg-Tosyl

Synonyms: N^α-BOC-N^r-Tosyl-D-Arg

American Peptide BDARG20   MW 428.5

Bachem A-2320.0001   MW 428.51 $C_{18}H_{28}N_4O_6S$ Store at -15°C

Neosystem BA00206   MW 428.5

Peptides International BDR-2609   MW 428.51   >98% (HPLC); white crystalline powder | Typically co-crystallize with ¼-1 molecule H_2O &/or Et acetate/molecule AA

USBio B2267   MW 428.5 $C_{18}H_{28}N_4O_6S$ ≥99%

Arginine — BOC-D-Arg-Z₂

Bachem A-3525.0001   MW 542.59 $C_{27}H_{34}N_4O_8$ Store at -15°C

Arginine — BOC-L-Arg Hydrochloride Hydrate

Neosystem BA00202   MW 328.8

Arginine — BOC-L-Arg-(Z₂)

Neosystem BA00204   MW 542.6

Arginine — BOC-L-Arg-Alloc₂ Dicyclohexylamine

Neosystem BA00203   MW 623.9

Arginine — BOC-L-Arg-Mtr

Neosystem BA00210   MW 486.6

Arginine — BOC-L-Arg-Nitro

Neosystem BA00205   MW 319.3

Arginine — BOC-L-Arg-Tosyl

Neosystem BA00207   MW 428.5

Arginine — BOC-L-Arg-Tosyl-MBHA Resin

Neosystem RN00207

Arginine — BOC-L-Arg-Tosyl-O-CH₂-φ-CH₂-COOH Dicyclohexylamine

Neosystem LP00202   MW 758.0

Arginine — BOC-L-Arg-Tosyl-PAM Resin

Neosystem RP00202

Arginine — BOC-N-Me-Arg-Mtr

Bachem A-3430.0001   MW 500.62 $C_{22}H_{36}N_4O_7S$ Store at -15°C

Arginine — BOC-N-Me-Arg-Tosyl

Bachem A-2955.0001   MW 442.54 $C_{19}H_{30}N_4O_6S$ Store at -15°C

Arginine — BOC-*N*-Me-D-Arg-Tosyl

Bachem A-3320.0001   MW 442.54 $C_{19}H_{30}N_4O_6S$ Store at -15°C

Arginine — Bpoc-Arg-Mtr

Bachem E-2610.0001   MW 624.76 $C_{32}H_{40}N_4O_7S$ Store at -15°C

Arginine — Bz-Arg

Bachem E-1455.0010   MW 278.31 $C_{13}H_{18}N_4O_3$ Store at RT

Arginine — Bz-Arg Amide Hydrochloride

Bachem E-1460.0005   MW 313.79 $C_{13}H_{19}N_5O_2 \cdot HCl$ Store at RT

Arginine — Bz-Arg Amide Hydrochloride Hydrate

Synonyms: Trypsin Substrate

Peptides International SRA-3002   MW 331.81 $C_{13}H_{19}N_5O_2 \cdot HCl \cdot H_2O$ >99% (HPLC); amorphous powder | Walsh, KA in *Proteolytic Enzymes*, Methods in Enzymology, Vol 19, (GE Perlmann & L Lorand, Eds), Academic Press, New York, 1970, pp 41-63

Arginine — Bz-Arg-4MβNA Hydrochloride

Bachem J-1050.0050   MW 470 $C_{24}H_{27}N_5O_3 \cdot HCl$ Store at -15°C

Arginine — Bz-Arg-AMC

Synonyms: Trypsin Substrate

Peptides International MBR-3092-v   MW 435.48 $C_{23}H_{25}N_5O_4$ >98% (HPLC); lyophilized amorphous powder | Kanaoka, Y et al, *Chem Pharm Bull*, 25:3126, 1977

Arginine — Bz-Arg-AMC Hydrochloride

Bachem I-1070.0050   MW 471.94 $C_{23}H_{25}N_5O_4 \cdot HCl$ Store at -15°C

Arginine — Bz-Arg-OEt Hydrochloride

Synonyms: L-BAEE; Trypsin Substrate

Bachem M-1130.0005   MW 342.83 $C_{15}H_{22}N_4O_3 \cdot HCl$ Store at -15°C

Peptides International SRE-3001   MW 342.82 $C_{15}H_{22}N_4O_3 \cdot HCl$ >98% (HPLC); amorphous powder | Walsh, KA in *Proteolytic Enzymes*, Methods in Enzymology, Vol 19, (GE Perlmann & L Lorand, Eds), Academic Press, New York, 1970, pp 41-63

Arginine — Bz-Arg-OMe

Synonyms: L-BAME

Bachem M-1140.0005   MW 292.34 $C_{14}H_{20}N_4O_3$ Store at -15°C

Arginine — Bz-Arg-pNA Hydrochloride

Synonyms: L-BAPA

Bachem L-1130.0250   MW 434.88 $C_{19}H_{22}N_6O_4 \cdot HCl$ Store at -15°C

Arginine — Bz-Arg-βNA Hydrochloride

Synonyms: L-BANA

Bachem K-1120.0250   MW 439.95 $C_{23}H_{25}N_5O_2 \cdot HCl$ Store at RT

Arginine — Bz-D-Arg-pNA Hydrochloride

Synonyms: D-BAPA

Bachem L-1135.0025   MW 434.88 $C_{19}H_{22}N_6O_4 \cdot HCl$ Store at -15°C

Arginine — Bz-DL-Arg Hydrochloride

Bachem F-3490.0005   MW 314.78 $C_{13}H_{18}N_4O_3 \cdot HCl$ Store at RT

Arginine — Bz-DL-Arg-AMC Hydrochloride

Bachem I-1075.0050   MW 471.9 $C_{23}H_{25}N_5O_4 \cdot HCl$ Store at -15°C

Arginine — Bz-DL-Arg-pNA Hydrochloride

Synonyms: DL-BAPA; Trypsin-Like Proteases Substrate

Bachem L-1140.0001   MW 434.88 $C_{19}H_{22}N_6O_4 \cdot HCl$ Store at -15°C

Peptides International SRN-3013   MW 434.88 $C_{19}H_{22}N_6O_4 \cdot HCl$ >98% (HPLC); amorphous powder | Thomas, KA & RA Bradshaw in *Proteolytic Enzymes Part C*, Methods in Enzymology, Vol 80, (L Lorand, Ed), Academic Press, New York, 1981, pp 609-620; Grant, GA et al in *Proteolytic Enzymes Part C*, Methods in Enzymology, Vol 80, (L Lorand, Ed), Academic Press, New York, 1981, pp 722-734

Arginine — Bz-DL-Arg-βNA Hydrochloride

Synonyms: DL-BANA

Bachem K-1125.0001   MW 439.95 $C_{23}H_{25}N_5O_2 \cdot HCl$ Store at RT

Arginine — Bz-L-Arg-AMC Hydrochloride

ICN 150444 $C_{23}H_{25}N_5O_4 \cdot HCl$ Fluorogenic substrate

Arginine — Bz-L-Arg-pNA Hydrochloride

Synonyms: Trypsin-Like Proteases Substrate; Papain Substrate

Peptides International SRN-3057   MW 434.88 $C_{19}H_{22}N_6O_4 \cdot HCl$ >99% (HPLC); amorphous powder | Walsh, KA in *Proteolytic Enzymes*, Methods in Enzymology, Vol 19, (GE Perlmann & L Lorand, Eds), Academic Press, New York, 1970, pp 41-63; Arnon, R in *Proteolytic Enzymes*, Methods in Enzymology, Vol 19, (GE Perlmann & L Lorand, Eds), Academic Press, New York, 1970, pp 226-244; Grant, GA et al in *Proteolytic Enzymes Part C*, Methods in Enzymology, Vol 80, (L Lorand, Ed), Academic Press, New York, 1981, pp 722-734

Arginine — CBZ-Arg

USBio C2098-28   MW 308.3 $C_{14}H_{20}N_4O_4$ ≥99%

Arginine — CBZ-Arg Hydrochloride

USBio C2098-30   MW 344.8 $C_{14}H_{20}N_4O_4 \cdot HCl$ ≥99%

Arginine — CBZ-Arg-4MβNA Hydrochloride

USBio C2098-32   MW 619.7 $C_{25}H_{29}N_5O_4$ ≥98%; appearance: white to lt. tan powder; specific rotation: provided with every lot analysis batch certificate

Arginine — CBZ-Arg-AMC Hydrochloride

USBio C2098-31   MW 502 $C_{24}H_{27}N_5O_5 \cdot HCl$ ≥99%

Arginine — CBZ-Arg-Arg-4MβNA Diacetate

USBio C2098-35   MW 739.8 $C_{31}H_{41}N_9O_5 \cdot 2C_2H_4O_2$ ≥99%

Arginine — CBZ-Arg-Arg-AMC

USBio C2098-34   MW 621.7 $C_{30}H_{39}N_9O_6$ ≥99%

Arginine — CBZ-Arg-Mtr-CHA

USBio C2098-38   MW 619.8 $C_{24}H_{32}N_4O_7S \cdot C_6H_{13}N$ ≥99%

Arginine — CBZ-Arg-Nitro

USBio C2098-39   MW 353.3 $C_{14}H_{19}N_5O_6$ ≥99%

Arginine — CBZ-Arg-pNA Hydrochloride

USBio C2098-33   MW 464.9 $C_{20}H_{24}N_6O_5 \cdot HCl$ ≥99%

Arginine — CBZ-Arg-Tosyl

USBio C2098-40   MW 462.5 $C_{21}H_{26}N_4O_6S$ ≥99%

Arginine — CBZ-Arg-Z$_2$

USBio C2098-36   MW 576.6 $C_{30}H_{32}N_4O_8$ ≥99%

Arginine — CBZ-Arg-Z$_2$-ONp

USBio C2098-37   MW 697.7 $C_{36}H_{35}N_5O_{10}$ ≥99%

Arginine — CBZ-D-Arg

USBio C2098-29   MW 308.3 $C_{14}H_{20}N_4O_4$ ≥99%

Arginine — CH$_3$SO$_2$-D-CHA-Abu-Arg-pNA·AcOH

Synonyms: Pefachrome®FVIIa; Pefa-5979

Pentapharm 093-20, 093-01   MW 670.8   Highly sensitive chromogenic peptide substrate for factor VIIa; sensitivity is significantly increased in the presence of TF

Arginine — Dansyl-Arg-*N*-(3-Et-1,5-Pentanediyl) Amide

Synonyms: DAPA; Thrombin Inhibitor

ICN 194937 $C_{25}H_{39}O_3N_6SCl$ Specific & potent synthetic thrombin inhibitor; Nesheim, ME etal, *Biochem*, 18:996, 1979

Arginine — Dansyl-L-Arg

ICN 100030

Arginine — D-Arg

Synonyms: (R)-2-Amino-5-Guanidinopentanoic Acid

Bachem F-1240.0005   MW 174.2 $C_6H_{14}N_4O_2$ Store at RT

Fluka 11015   MW 174.2 $C_6H_{14}N_4O_2$ ≥99% (titration); ~0.5% loss on drying; ≤0.05% residue on ignition; mp: ~240°C

Fluka 11017   MW 174.2 $C_6H_{14}N_4O_2$ ≥99% (titration); ≤5% loss on drying; ≤0.05% residue on ignition; mp: ~240°C

ICN 191238 $C_6H_{14}N_4O_2$

Neosystem AA00216   MW 174.2

Peptides International ADR-2802   MW 174.20   >99.9% optical purity by RP-HPLC; white powder

Rexim   MW 174.2 $C_6H_{14}N_4O_2$

Sigma A 2646 $C_6H_{14}N_4O_2$

USBio A3550-10   MW 174.2 $C_6H_{14}N_4O_2$ ≥99%

Arginine — D-Arg (1-^{14}C)

ARC ARC-1277   MW 174.2
$H_2NC(NH)NH(CH_2)_2CH_2CH(NH_2)COOH$ 50-60 mCi/mmol; 1.85-2.22 GBq/mmol; in EtOH:H$_2$O (1:1) | Radiochemical

Arginine — D-Arg Amide Dihydrochloride

Bachem F-3905.0001   MW 246.14 $C_6H_{15}N_5O \cdot 2HCl$ Store at 2-8°C

Arginine — D-Arg Free Base

Alexis 101-005   MW 174.2 $C_6H_{14}N_4O_2$ White solid; soluble in H$_2$O

Arginine — D-Arg Hydrochloride

Fluka 11050   MW 210.67 $C_6H_{14}N_4O_2 \cdot HCl$ ≥99%

ICN 100730 $C_6H_{14}N_4O_2 \cdot HCl$

Rexim   MW 210.7 $C_6H_{15}ClN_4O_2$

Sigma A 6757 $C_6H_{14}N_4O_2 \cdot HCl$

USBio A3570-10   MW 210.7 $C_6H_{14}N_4O_2 \cdot HCl$ ≥99%

Arginine — D-Arg-Mtr

Bachem F-1835.0001   MW 386.47 $C_{16}H_{26}N_4O_5S$ Store at RT

Arginine — D-Arg-Nitro

Bachem F-1875.0001   MW 219.2 $C_6H_{13}N_5O_4$ Store at RT

Arginine — D-Arg-Nitro-OMe Hydrochloride

Bachem F-1890.0001   MW 269.69 $C_7H_{16}N_5O_4Cl$ Store at 2-8°C

Arginine — D-Arg-OMe Dihydrochloride

Bachem F-2715.0005   MW 261.15 $C_7H_{16}N_4O_2 \cdot 2HCl$ Store at 2-8°C

Arginine — D-Arg-Pbf

Bachem F-3255.0001   MW 426.54 $C_{19}H_{30}N_4O_5S$ Store at 2-8°C

Arginine — D-Arg-Pmc

Bachem F-2645.0001   MW 440.56 $C_{20}H_{32}N_4O_5S$ Store at RT

Arginine — D-Arg-Tosyl

Bachem F-2750.0001   MW 328.39 $C_{13}H_{20}N_4O_4S$ Store at -15°C

Arginine — Dibenzoyl-L-Arg Hydrochloride

ICN 157616 $C_{20}H_{22}N_4O_4 \cdot HCl$

Arginine — Diphenylacetyl-D-Arg-4-Hydroxybenzylamide

Synonyms: Neuropeptide Y1 Receptor Antagonist; BIBP3226

Bachem E-3620.0050   MW 473.58 $C_{27}H_{31}N_5O_3$ Store at -15°C

Arginine — DL-Arg

Synonyms: (±)-2-Amino-5-Guanidinopentanoic Acid

Fluka 11020   MW 174.2 $C_6H_{14}N_4O_2$ ≥97% (titration); ~2% H$_2$O; ≤0.1% residue on ignition; mp: 228-232°C

Arginine — DL-Arg (1-^{14}C)

ARC ARC-1081   MW 174.2
$H_2NC(NH)NH(CH_2)_2CH_2CH(NH_2)COOH$ 50-60 mCi/mmol; 1.85-2.22 GBq/mmol; in 0.01 *N* HCl | Radiochemical

Arginine — DL-Arg Hydrochloride

Synonyms: Nitric Oxide Synthase Substrate

ICN 100733 $C_6H_{14}N_4O_2 \cdot HCl$

Sigma A 4881 $C_6H_{14}N_4O_2 \cdot HCl$ ≥98% (TLC) | Induces insulin release by a nitric oxide-dependent mechanism; also available as part of a kit

Arginine — FMOC-Arg

Bachem B-1040.0001   MW 396.45 $C_{21}H_{24}N_4O_4$ Store at -15°C

Peptides International FLR-1708-PI   MW 396.45   >98% (HPLC); white crystalline powder

USBio F5318   MW 396.4 $C_{21}H_{24}N_4O_4$ ≥99%

Arginine — FMOC-Arg-(Pfb)

USBio F5324   MW 648.8 $C_{34}H_{40}N_4O_7S$ ≥99%

Arginine — FMOC-Arg-BOC₂

Bachem B-2335.0001	MW 596.68	$C_{31}H_{40}N_4O_8$	Store at -15°C

Arginine — FMOC-Arg-Dimethyl

Bachem B-2745.0001	MW 424.5	$C_{23}H_{28}N_4O_4$	Store at -15°C; asymmetrical
Bachem B-3345.0001	MW 424.5	$C_{23}H_{28}N_4O_4$	Store at -15°C; symmetrical

Arginine — FMOC-Arg-Mtr

Bachem B-1385.0001	MW 608.72	$C_{31}H_{36}N_4O_7S$	Store at -15°C
USBio F5320	MW 608.7	$C_{31}H_{36}N_4O_7S$	≥99%

Arginine — FMOC-Arg-Mtr-® (200–400 mesh)

Bachem D-1125.0001	Store at RT

Arginine — FMOC-Arg-Mtr-OPfp

Bachem B-1605.0001	MW 774.77	$C_{37}H_{35}F_5N_4O_7S$	Store at -15°C
USBio F5321	MW 775	$C_{37}H_{37}N_4F_5O_7S$	≥99%

Arginine — FMOC-Arg-Mtr-SASRIN™-® (200–400 mesh)

Bachem D-1305.0001	Store at 2-8°C

Arginine — FMOC-Arg-Nitro

Bachem B-2785.0001	MW 441.45	$C_{21}H_{23}N_5O_6$	Store at 2-8°C
USBio F5322	MW 441.4	$C_{21}H_{23}N_5O_6$	≥99%

Arginine — FMOC-Arg-Pbf

Synonyms: N^α-FMOC-N^r-Pbf-L-Arg

American Peptide FARG115	MW 662.8		
Bachem B-2375.0001	MW 648.78	$C_{34}H_{40}N_4O_7S$	Store at -15°C
Peptides International FLR-1746-PI	MW 648.77		>98%
(HPLC); white crystalline powder			

Arginine — FMOC-Arg-Pbf-® (200–400 mesh)

Bachem D-1860.0001	Store at RT

Arginine — FMOC-Arg-Pbf-OPfp

Bachem B-2355.0001	MW 814.84	$C_{40}H_{39}F_5N_4O_7S$	Store at -15°C

Arginine — FMOC-Arg-Pbf-SASRIN™-® (200–400 mesh)

Bachem D-1865.0001	Store at 2-8°C

Arginine — FMOC-Arg-Pmc

Bachem B-1670.0001	MW 662.81	$C_{35}H_{42}N_4O_7S$	Store at -15°C
USBio F5326	MW 662.8	$C_{35}H_{42}N_4O_7S$	≥99%

Arginine — FMOC-Arg-Pmc-® (200–400 mesh)

Bachem D-1555.0001	Store at RT

Arginine — FMOC-Arg-Pmc-NH-® (200–400 mesh)

Bachem D-1605.0001	Store at 2-8°C

Arginine — FMOC-Arg-Pmc-OPfp

Bachem B-1695.0001	MW 828.86	$C_{41}H_{41}F_5N_4O_7S$	Store at -15°C

Arginine — FMOC-Arg-Pmc-SASRIN™-® (200–400 mesh)

Bachem D-1565.0001	Store at 2-8°C

Arginine — FMOC-Arg-Tosyl

Synonyms: N^α-FMOC-N^r-Tosyl-L-Arg

Bachem B-1435.0001	MW 550.64	$C_{28}H_{30}N_4O_6S$	Store at -15°C
Peptides International FLR-1710-PI	MW 550.63		>98%
(TLC); white powder			
USBio F5328	MW 550.6	$C_{28}H_{30}N_4O_6S$	≥99%

Arginine — FMOC-D-Arg Hydrochloride

Bachem B-3040.0001	MW 432.91	$C_{21}H_{24}N_4O_4 \cdot HCl$	Store at -15°C

Arginine — FMOC-D-Arg-(Pfb)

USBio F5323	MW 648.8	$C_{34}H_{40}N_4O_7S$	≥99%

Arginine — FMOC-D-Arg-BOC₂

Bachem B-2940.0001	MW 596.68	$C_{31}H_{40}N_4O_8$	Store at -15°C

Arginine — FMOC-D-Arg-Mtr

Bachem B-1390.0001	MW 608.71	$C_{31}H_{36}N_4O_7S$	Store at -15°C
USBio F5319	MW 608.7	$C_{31}H_{36}N_4O_7S$	≥99%

Arginine — FMOC-D-Arg-Mts

Bachem B-2370.0001	MW 578.69	$C_{30}H_{34}N_4O_6S$	Store at -15°C

Arginine — FMOC-D-Arg-Pbf

Synonyms: N^α-FMOC-N^r-Pbf-D-Arg

Bachem B-2555.0001	MW 648.78	$C_{34}H_{40}N_4O_7S$	Store at -15°C
Peptides International FDR-1801-PI	MW 648.77		>98%
(HPLC); white crystalline powder			

Arginine — FMOC-D-Arg-Pbf-® (200–400 mesh)

Bachem D-1995.0001	Store at RT

Arginine — FMOC-D-Arg-Pbf-OPfp

Bachem B-2560.0001	MW 814.84	$C_{40}H_{39}F_5N_4O_7S$	Store at -15°C

Arginine — FMOC-D-Arg-Pbf-SASRIN™-® (200–400 mesh)

Bachem D-1990.0001	Store at 2-8°C

Arginine — FMOC-D-Arg-Pmc

Bachem B-1680.0001	MW 662.81	$C_{35}H_{42}N_4O_7S$	Store at -15°C
USBio F5325	MW 662.8	$C_{35}H_{42}N_4O_7S$	≥99%

Arginine — FMOC-D-Arg-Pmc-® (200–400 mesh)

Bachem D-1560.0001	Store at RT

Arginine — FMOC-D-Arg-Pmc-OPfp

Bachem B-1820.0001   MW 828.86 $C_{41}H_{41}N_4O_7SF_5$ Store at -15°C

Arginine — FMOC-D-Arg-Pmc-SASRIN™-® (200-400 mesh)

Bachem D-1835.0001   Store at 2-8°C

Arginine — FMOC-D-Arg-Tosyl

USBio F5327   MW 550.6 $C_{28}H_{30}N_4O_6S$ ≥99%

Arginine — FMOC-L-Arg-Pbf

Neosystem FA00212   MW 648.8

Arginine — FMOC-L-Arg-Pbf-Wang Resin

American Peptide RFWAN12   Preattached Wang resins are usful for synthesis of peptides with C-terminal carboxyl groups; peptides can be cleaved from the resin with 90% TFA in the presence of scavengers

Arginine — FMOC-L-Arg-Pmc

Neosystem FA00206   MW 662.8

Arginine — FMOC-L-Arg-Pmc-Wang Resin

Neosystem RW00204

Arginine — FMOC-L-Arg-Tosyl

Neosystem FA00208   MW 550.6

Arginine — FMOC-N-Me-Arg-Mtr

Bachem B-2840.0001   MW 622.74 $C_{32}H_{38}N_4O_7S$ Store at -15°C

Arginine — L-Arg

Synonyms: NOS Substrate; (S)-2-Amino-5-Guanidinopentanoic Acid; L-α-Amino-δ-Guanidinovaleric Acid; (S)-2-Amino-5-((Aminoiminomethyl)-Amino)-Pentanoic Acid; Nitric Oxide Synthase Substrate

Amersham US11490

Fluka 11009   MW 174.2 $C_6H_{14}N_4O_2$ ≥99.5%; ≤0.05% residue on ignition; ≤0.1% loss on drying; ≤0.01% Cl, SO_4; ≤0.005% K, Na; ≤0.0005% Al, Ba, Bi, Cd, Co, Cr, Cu, Fe, Li, Mg, Mn, Mo, Ni, Pb, Sr, Zn; ≤0.001% Ca; ≤0.00001% As; ≤0.3% foreign AA | Galston, AW et al, *Plant Sci Lett*, 11: 69, 1978; *CRC Handbook of Microbiology*, Laskin, AI & Lachevalier, HA, eds, CRC Press, Cleveland, Ohio, 4: 1, 1974

Fluka 11010   MW 174.2 $C_6H_{14}N_4O_2$ ≥99.0%; ≤0.1% residue on ignition; ≤0.5% loss on drying; ≤0.005% K, Na, Cl, SO_4; ≤0.0005% Cd, Co, Cr, Cu, Fe, Mg, Mn, Mo, Ni, Pb, Zn; ≤0.001% Ca; ≤0.3% foreign AA | Galston, AW et al, *Plant Sci Lett*, 11: 69, 1978; *CRC Handbook of Microbiology*, Laskin, AI & Lachevalier, HA, eds, CRC Press, Cleveland, Ohio, 4: 1, 1974

ICN 100736 $C_6H_{14}N_4O_2$

ICN 194626 $C_6H_{14}N_4O_2$

Neosystem AA00203   MW 174.2

Rexim   MW 174.2 $C_6H_{14}N_4O_2$ White crystals or crystalline powder

Sigma A 1018   Matrix: 4% beaded agarose; activation: cyanogen bromide; attachment: amino; spacer: 1 atom; ligand immobilized: 5-10 μmoles/ml; form: suspension in 2.0 M NaCl containing 0.02% thimerosal

Sigma A 1843   FW 174.2 $C_6H_{14}N_4O_2$ USP

Sigma A 3784   FW 174.2 $C_6H_{14}N_4O_2$ Crystalline | Cell culture tested; insect cell culture tested

Sigma A 5383 $C_4H_7NO_4 \cdot C_6H_{14}N_4O_2$ Induces insulin release by a nitric oxide-dependent mechanism

Sigma A 5756 $C_{11}H_{23}N_5O_6$ Induces insulin release by a nitric oxide-dependent mechanism

Sigma A 8405   Matrix: cross-linked 4% beaded agarose; activation: epoxy; attachment: amino; spacer: 12 atoms; ligand immobilized: ≥5 μmoles/ml; form: lyophilized powder stabilized with lactose; swelling: 1 g swells to ~12 mL

USBio A3550   MW 174.2 $C_6H_{14}N_4O_2$ ≥98%; white, crystalline powder; specific rotation (C=8, 6 N HCl): +26.3° to +27.7°; pH: 10.5-12.0; heavy metals (Pb): ≤0.001%; endotoxin ≤2 EU/mg; cell culture tested: murine myeloma SP2/0-Ag14

Arginine — L-Arg (1-^{14}C)

ARC ARC-338   MW 174.2 $H_2NC(NH)NH(CH_2)_2CH_2CH(NH_2)COOH$ 45-55 mCi/mmol; 1.67-2.04 GBq/mmol; in 0.01 N HCl | Radiochemical

Sigma A 7965   40-60 Ci/mmol; radiochemical purity ≥95% (HPLC); 0.01 N HCl solution in Combi-vial | Radiochemical

Arginine — L-Arg (2,3,4,5-^{3}H) Hydrochloride

Amersham TRK698

Arginine — L-Arg (2,3-^{3}H)

ARC ART-210   MW 174.2 $H_2NC(NH)NH(CH_2)_2CH_2CH(NH_2)COOH$ 40-80 Ci/mmol; 1.48-2.96 TBq/mmol; in 0.01 N HCl | Radiochemical

Sigma A 3680   MW 174.2 $H_2NC(NH)NH(CH_2)_2CH_2CH(NH_2)COOH$ 70-85 Ci/mmol; radiochemical purity ≥98% (HPLC); aqueous solution containing 2% EtOH in serum bottle | Radiochemical

Arginine — L-Arg (4,5-^{3}H)

ICN 20069 $H_2NC(=NH)NH(CH_2)_2CH_2CH(NH_2)COOH$ Magagnin, S etal, *JBC*, 267:15384, 1992

ICN 20069E $H_2NC(=NH)NH(CH_2)_2CH_2CH(NH_2)COOH$ Magagnin, S etal, *JBC*, 267:15384, 1992

Arginine — L-Arg (Guanido-^{14}C)

ARC ARC-1214   MW 174.2 $H_2NC(NH)NH(CH_2)_2CH_2CH(NH_2)COOH$ 50-60 mCi/mmol; 1.85-2.22 GBq/mmol; in 0.1 N HCl | Radiochemical

Arginine — L-Arg (U-^{14}C)

ICN 10042 $H_2NC(=NH)NH(CH_2)_3CH(NH_2)COOH$

ICN 10042E $H_2NC(=NH)NH(CH_2)_3CH(NH_2)COOH$

ICN 10042L $H_2NC(=NH)NH(CH_2)_3CH(NH_2)COOH$

Arginine — L-Arg (U-^{14}C) Hydrochloride

Sigma A 3839   FW 174.2 $C_6H_{14}N_4O_2$ 180-360 mCi/mmol; radiochemical purity ≥98% (HPLC); aqueous solution containing 2% EtOH in Combi-vial | Radiochemical

Arginine — L-Arg Aspartate Salt

ICN 198886 $C_4H_{23}NO_4 \cdot C_6H_{14}N_4O_2$

Arginine — L-Arg Free Base

Alexis 101-004   MW 174.2 $C_6H_{14}N_4O_2$ White solid; soluble in H_2O | Physiological precursor for the formation of NO by NOS; L-Arginine enhances the release of NO & causes insulin release from pancreatic B cells; Palmer, RMJ et al, *BBRC*, 153: 1251, 1988; Palmer, RMJ et al, *Nature*, 333: 664, 1988; Schmidt, HHHW et al, *Science*, 255: 721, 1992

Calbiochem 1820   MW 174.2 $C_6H_{14}N_4O_2$ Solid; soluble in H_2O | NO precursor; substrate for NO; reverses the inhibition of NOS caused by arginine analogs; causes the release of insulin from pancreatic β cells by a NO-dependent mechanism; Schmidt, HHHW et al, *Science*, 255: 721, 1992; Garthwaite, J, *Trends Neurosci*, 14: 60, 1991; *Merck Index*, 12: 817

Arginine — L-Arg Hydrochloride

Synonyms: Argivene; Levargin; Detoxargin; (S)-2-Amino-5-((Aminoiminomethyl)-Amino)-Pentanoic Acid; Minophagen A; Nitric Oxide Synthase Substrate

Amersham US11500

Fluka 11039   MW 210.67 $C_6H_{14}N_4O_2 \cdot HCl$ ≥99.5%; ≤0.05% residue on ignition, loss on drying; ≤0.005% K, SO₄, Na; ≤0.0005% Al, Ba, Bi, Cd, Co, Cr, Cu, Fe, Li, Mg, Mn, Mo, Ni, Pb, Sr, Zn; ≤0.001% Ca; ≤0.00001% As; ≤0.3% foreign AA

Fluka 11040   MW 210.67 $C_6H_{14}N_4O_2 \cdot HCl$ ≥99.0%; ≤0.1% residue on ignition, loss on drying; ≤0.005% K, Na, SO₄; ≤0.0005% Cd, Co, Cr, Cu, Fe, Mg, Mn, Mo, Ni, Pb, Zn; ≤0.001% Ca; ≤0.3% foreign AA

ICN 100743 $C_6H_{14}N_4O_2 \cdot HCl$

ICN 194627 $C_6H_{14}N_4O_2 \cdot HCl$

Peptides International ALR-2702   MW 210.66   >99.9% optical purity by RP-HPLC; white powder

Rexim   MW 210.7 $C_6H_{15}ClN_4O_2$ White crystals or crystalline powder

Sigma A 3909   FW 210.7 $C_6H_{14}N_4O_2 \cdot HCl$ Crystalline | Cell culture tested; insect cell culture tested

Sigma A 5131 $C_6H_{14}N_4O_2 \cdot HCl$ ≥98% (TLC) | Induces insulin release by a nitric oxide-dependent mechanism; also available as part of a kit

Sigma A 5949 $C_6H_{14}N_4O_2 \cdot HCl$ Induces insulin release by a nitric oxide-dependent mechanism

USBio A3570   MW 210.7 $C_6H_{14}N_4O_2 \cdot HCl$ ≥98%; white, crystalline powder; specific rotation (C=8, 6 N HCl): +22.1° to +22.9°; pH: 4.7-6.2; heavy metals (Pb): ≤0.001%; endotoxin: ≤2 EU/mg; cell culture tested: murine myeloma SP2/0-Ag14

Arginine — L-Arg Hydroxamate Hydrochloride

Sigma A 7380 $C_6H_{15}N_5O_2 \cdot HCl$

Arginine — L-Arg α-Kg (1:1)

Synonyms: L-Arg 2-Oxopentanedioate

Rexim   MW 320.3 $C_{11}H_{20}N_4O_7$

Arginine — L-Arg α-Kg (2:1)

Synonyms: L-Arg 2-Oxopentanedioate

Rexim   MW 466.4 $C_{16}H_{26}N_4O_{12}$

Arginine — L-Arg β-Naphthylamide Hydrochloride

Sigma A 6512 $C_{16}H_{21}N_5O \cdot HCl$

Arginine — L-Arg-(U-¹⁴C)

ARC ARC-669   MW 174.2 $H_2NC(NH)NH(CH_2)_3CH(NH_2)COOH$ 200-300 mCi/mmol; 7.4-11.1 GBq/mmol; in EtOH:H₂O (2:98) | Radiochemical

Arginine — L-Arg-4-Methoxy-β-Naphthylamide Hydrochloride

Sigma A 5907 $C_{17}H_{23}N_5O_2 \cdot HCl$

Arginine — L-Arg-AMC

Synonyms: Cathepsin Substrate

ICN 150386   Fluorogenic substrate

ICN 150388   Barrett, AJ, *Biochem J*, 187:909, 1980

Sigma A 2027 $C_{16}H_{21}N_5O_3$

Arginine — L-Argininamide Dihydrochloride

Sigma A 3913 $C_6H_{15}N_5O \cdot 2HCl$

Arginine — L-Argininic Acid

Synonyms: L-α-Hydroxy-δ-Guanidinovaleric Acid

Sigma A 5146 $C_6H_{13}N_3O_3$

Arginine — L-Arg-L-2-Pyrolidone-5-Carboxylate

Synonyms: Arg Pyroglutamate

Rexim   MW 303.3 $C_{11}H_{21}N_5O_5$ White crystals or crystalline powder

Arginine — L-Arg-Nitro

Neosystem AA00217   MW 219.2

Arginine — L-Arg-Nᵍ-Amine

Synonyms: Nᵍ-Amino-L-Arg

Sigma A 1186 $C_6H_{15}N_5O_2 \cdot C_{10}H_6N_2O_8S$

Arginine — L-Arg-OEt Dihydrochloride

ICN 100744 $C_8H_{18}N_4O_2 \cdot 2HCl$

Sigma A 2883 $C_8H_{18}N_4O_2 \cdot 2HCl$

Arginine — L-Arg-OMe Dihydrochloride

Fluka 11030   MW 261.15 $C_7H_{16}N_4O_2 \cdot 2HCl$ ≥98% (titration); mp: ~190°C

ICN 100753 $C_7H_{16}N_4O_2 \cdot 2HCl$

Arginine — L-Arg-pNA Dihydrochloride

Sigma A 4566 $C_{12}H_{18}N_6O_3 \cdot 2HCl$

Arginine — Mbs-D-Arg

Bachem C-3365.0001   MW 344.35 $C_{13}H_{20}N_4O_5S$ Store at -15°C

Arginine — MTH-DL-Arg Hydrochloride

Sigma M 3756 $C_8H_{15}N_5OS \cdot HCl$

Arginine — N²-(1,3-Dicarboxypropyl)-L-Arg

Synonyms: Nopaline

Sigma N 4138 $C_{11}H_{20}N_4O_6$ Plant tumor metabolite; Nester, EW et al, *Ann Rev Plant Physiol*, 35: 387, 1984

Sigma N 6134 $C_{11}H_{20}N_4O_6$

Arginine — N-2,4-Dnp-L-Arg

Sigma D 8129 $C_{12}H_{16}N_6O_6$

Arginine — N-Ac-L-Arg

ICN 100078 $C_8H_{16}N_4O_3$

Arginine — Nᵍ-Nitro-L-Arg

USBio N2580   MW 219.2 $C_6H_{13}N_5O_4$ ≥99%

Arginine — N-O-Hydroxy-Nor-L-Arg

Synonyms: L-2-Amino-4-(2'-Hydroxyguanidino)-Butyric Acid

Bachem F-3685.0050   MW 176.18 $C_5H_{12}N_4O_3$ Store at -15°C

Arginine — N-O-Nps-L-Arg

Sigma N 4878 $C_{10}H_{11}N_3O_5S \cdot C_{12}H_{23}N$

Arginine — Nᵃ,Nᵍ,Nᵍ'-Tri-Z-L-Arg

Fluka 95935   MW 576.61 $C_{30}H_{32}N_4O_8$ ~97.0% (titration); mp: 134-138°C

Arginine — Nᵃ,Nᵍ,Nᵍ'-Tri-CBZ-L-Arg

Sigma C 1170 $C_{30}H_{32}N_4O_8$

Arginine — N^α-Ac-L-Arg

Sigma A 3133 $C_8H_{16}N_4O_3$

Arginine — N^α-BOC-L-Arg

Fluka 15374   MW 274.32 $C_{11}H_{22}N_4O_4$ ~95% (titration); mp: 117-119°C; ~5% N-BuOH

Arginine — N^α-BOC-L-Arg Hydrochloride

Fluka 15063   MW 310.78 $C_{11}H_{22}N_4O_4 \cdot$ HCl   ≥98% (TLC)

Arginine — N^α-BOC-N^δ,N^ω-Di-Z-L-Arg

Fluka 15493   MW 542.59 $C_{27}H_{34}N_4O_8$ ≥98% (TLC); mp: 140°C

Arginine — N^α-BOC-N^ω-Mtr-L-Arg

Fluka 15465   MW 486.59 $C_{21}H_{34}N_4O_7S$ Protected derivative of arginine; MTR-guanidine protection of Arg is more easily cleaved with acid than the tosyl protection; Fujino, M et al, *Chem Pharm Bull*, 29: 2825, 1981

Arginine — N^α-BOC-N^ω-Mts-L-Arg Cyclohexylamine Salt

Fluka 15545   MW 555.73 $C_{20}H_{32}N_4O_6S \cdot C_6H_{13}N$ ≥99% (TLC); mp: 130°C

Arginine — N^α-BOC-N^ω-Nitro-L-Arg

Fluka 15470   MW 319.32 $C_{11}H_{21}N_5O_6$ ~99% (titration)

Arginine — N^α-BOC-N^ω-Pbf-L-Arg

Fluka 15038   MW 526.65 $C_{24}H_{38}N_4O_7S$ ≥98% (TLC); may contain~10% solvent

Arginine — N^α-BOC-N^ω-Tosyl-D-Arg

Fluka 15184   MW 428.5 $C_{18}H_{28}N_4O_6S$ ≥98% (TLC)

Arginine — N^α-BOC-N^ω-Tosyl-L-Arg

Fluka 15506   MW 428.51 $C_{18}H_{28}N_4O_6S$ ~2% H_2O; mp: 90°C; crystallized with ~10% ethyl acetate

Arginine — N^α-Bz-D-Arg-pNA Hydrochloride

Synonyms: D-BAPNA

Sigma B 1891 $C_{19}H_{22}N_6O_4 \cdot$ HCl

Arginine — N^α-Bz-DL-Arg

Sigma B 4125 $C_{13}H_{18}N_4O_3$

Arginine — N^α-Bz-DL-Arg β-Naphthylamide Hydrochloride

Synonyms: BANA; Trypsin Substrate

Sigma B 4750 $C_{23}H_{25}N_5O_2 \cdot$ HCl   Chromogenic

Arginine — N^α-Bz-DL-Arg-2-Naphthylamide Hydrochloride

Synonyms: BANA

Fluka 12910   MW 439.95 $C_{23}H_{25}N_5O_2 \cdot$ HCl   ≥97% (titration); ≤2% H_2O; mp: 196-199°C; ≤1% DL-Arg | Glenner, GG, *Nature*, 185: 846, 1960

Arginine — N^α-Bz-DL-Arg-4-NA Hydrochloride

Synonyms: BANI; DL-BAPA; BAPNA

Fluka 12920   MW 434.89 $C_{19}H_{22}N_6O_4 \cdot$ HCl   ≥98% (titration); ≤0.2% residue on ignition; mp: ~275°C; ≤0.1% DL-Arg | Arnon, R, *Meth Enzymol*, 19: 226, 1970; Erlanger, BF et al, *Arch Biochem Biophys*, 95: 271, 1961; Englund, PT et al, *Biochemistry*, 7: 163, 1968

Arginine — N^α-Bz-DL-Arg-AMC Hydrochloride

Sigma B 7385 $C_{23}H_{25}N_5O_4 \cdot$ HCl

Arginine — N^α-Bz-DL-Arg-pNA Hydrochloride

Synonyms: DL-BAPNA; Trypsin Substrate; BAPNA

ICN 100090 $C_{19}H_{22}N_6O_4 \cdot$ HCl   Chromogenic
Sigma B 4875 $C_{19}H_{22}N_6O_4 \cdot$ HCl   Chromogenic

Arginine — N^α-Bz-L-Arg

Fluka 12870   MW 278.31 $C_{13}H_{18}N_4O_3$ ≥99% (titration); mp: ~290°C
ICN 100082 $C_{13}H_{18}N_4O_3$
Sigma B 8881 $C_{13}H_{18}N_4O_3$

Arginine — N^α-Bz-L-Arg 4-Methoxy-β-Naphthylamide Hydrochloride

Sigma B 0885 $C_{24}H_{27}N_5O_3 \cdot$ HCl

Arginine — N^α-Bz-L-Arg-4-NA Hydrochloride

Synonyms: L-BAPA

Fluka 12915   MW 434.89 $C_{19}H_{22}N_6O_4 \cdot$ HCl   ≥98% (enzyme); mp: 223-226°C | Mole, JE & Horton, HR, *Biochemistry*, 12: 816, 1973; Ota, S et al, *Biochemistry*, 3: 180, 1964; Geiger, R & Fritz, H, *Methods of Enzymatic Analysis*, Bergmeyer, HU, ed, Verlag Chemie Weinheim, vol 5: 3rd Edition, 119, 1984

Arginine — N^α-Bz-L-Arg-AMC Hydrochloride

Synonyms: Papain Substrate; Trypsin Substrate

Sigma B 7260 $C_{23}H_{25}N_5O_4 \cdot$ HCl   Sensitive fluorogenic substrate for papain & trypsin; kanaoka, y et al, *chem pharm bull*, 25: 3126, 1977

Arginine — N^α-Bz-L-Argininamide Hydrochloride Hydrate

Sigma B 4375 $C_{13}H_{19}N_5O_2 \cdot$ HCl $\cdot H_2O$

Arginine — N^α-Bz-L-Arginyl-4-Amino-Benzoic Acid Hydrochloride

Synonyms: BAPABA; N^α-Bz-L-Arg-4-Carboxyanilide

Fluka 12925   MW 433.89 $C_{20}H_{23}N_5O_4 \cdot$ HCl   ≥99% (TLC) | Zassenhaus, PH et al, *Anal Biochem*, 76: 321, 1976

Arginine — N°-Bz-L-Arg-OEt Hydrochloride

Synonyms: Papain Substrate; BAEE; Trypsin Substrate

Fluka 12880 MW 342.83 $C_{15}H_{22}N_4O_3 \cdot HCl$ ≥99%
(titration); ≤0.05% residue on ignition; mp: 128-130°C | Schwert,
GW & Takenaka, Y, *Biochem Biophys Acta*, 16: 570, 1955; Walsh,
KA & Wilcox, PE, *Meth Enzymol*, 19: 31, 1970; Fezdy, FJ et al,
Biochemistry, 4: 2302, 1965; Glazer, AN, *JBC*, 242: 433, 1967;
Smith, EL & Parker, MJ, *JBC*, 233: 1387, 1958; Arnon, R, *Meth
Enzymol*, 19: 226, 1979; Fiedler, F, *Meth Enzymol*, 45: 289, 1976;
Colman, RW & Bagdasarian, A, *Meth Enzymol*, 45: 303, 1976;
Hammond, BR & Gutfreund, H, *Biochem J*, 72: 349, 1959;
Bernhard, SA & Gutfreund, H, *Biochem J*, 63: 61, 1956; Kafatos,
FC et al, *JBC*, 242: 1477, 1967; Hruska, JF & Neurath, JH, *JBC*,
235: 99, 1960; Murachi, T, *Meth Enzymol*, 19: 273, 1979

ICN 100088 $C_{15}H_{22}N_4O_3 \cdot HCl$ Schwert, GW & Y Takenaka,
BBA, 16:570, 1955; Smith, EL & M Parker, *JBC*, 233:1387, 1958

Sigma B 4500 $C_{15}H_{22}N_4O_3 \cdot HCl$ The prototype substrate for
trypsin

Arginine — N°-Bz-L-Arg-OMe

Synonyms: BAME

Sigma B 4256 $C_{14}H_{20}N_4O_3 \cdot H_2CO_3$

Arginine — N°-Bz-L-Arg-OMe Hydrochloride

Synonyms: BAME; Kallikrein Substrate; Trypsin Substrate

ICN 100899 $C_{14}H_{20}N_4O_3 \cdot HCl$

Sigma B 1007 $C_{14}H_{20}N_4O_3 \cdot HCl$

Arginine — N°-Bz-L-Arg-pNA Hydrochloride

Synonyms: L-BAPNA; Trypsin Substrate; Protease Substrate

ICN 154829 $C_{19}H_{22}N_6O_4 \cdot HCl$

Sigma B 3133 $C_{19}H_{22}N_6O_4 \cdot HCl$ Chromogenic substrate;
Gaertner, HF & Puigserver, AJ, *Enzyme Microb Technol*, 14: 150,
1992; Gravett, PS et al, *Int J Biochem*, 23: 1085, 1991

Sigma B 3279 $C_{19}H_{22}N_6O_4 \cdot HCl$ Chromogenic substrate;
Gaertner, HF & Puigserver, AJ, *Enzyme Microb Technol*, 14: 150,
1992; Gravett, PS et al, *Int J Biochem*, 23: 1085, 1991

Arginine — N°-CBZ-L-Arg

ICN 101185 $C_{14}H_{20}N_4O_4$

Sigma C 4751 $C_{14}H_{20}N_4O_4$

Arginine — N°-CBZ-L-Arg Hydrochloride

ICN 150571 $C_{14}H_{20}N_4O_4 \cdot HCl$

Arginine — N°-CBZ-L-Arg-AMC Hydrochloride

Sigma C 8022 $C_{24}H_{27}N_5O_5 \cdot HCl$

Arginine — N°-CBZ-L-Arg-pNA Hydrochloride

Sigma C 4893 $C_{20}H_{24}N_6O_5 \cdot HCl$

Arginine — N°-CBZ-N°-Nitro-L-Arg

Sigma C 0545 $C_{14}H_{19}N_5O_6$

Arginine — N°-FMOC-L-Arg

Fluka 47589 MW 396.45 $C_{21}H_{24}N_4O_4$ ~98% (HPLC); mp:
145-160°C

ICN 158164 $C_{21}H_{24}N_4O_4$

Sigma F 8882 $C_{21}H_{24}N_4O_4$

Arginine — N°-FMOC-N°-Pbf-L-Arg

Sigma F 8916 $C_{34}H_{40}N_4O_7S$ Carpino, LA et al, *Tetrahedron
Lett*, 34: 7829, 1993

Arginine — N°-FMOC-N°-Pmc-L-Arg

Sigma F 0269 $C_{35}H_{42}N_4O_7S$ Ramage, R & Green, J,
Tetrahedron Lett, 28: 2287, 1987; Ramage R et al, *Peptides,
Chemistry & Biology*, Proc of the 10th Amer Peptide Symposium, St.
Louis, p 157, GR Marshall, ed, ESCOM, Leiden, 1988

Arginine — N°-FMOC-N°-Mtr-D-Arg

Fluka 03591 MW 608.72 $C_{31}H_{36}N_4O_7S$ ≥98% (HPLC); may
contain ~5% diisopropyl ether

Arginine — N°-FMOC-N°-Mtr-L-Arg

Fluka 47632 MW 608.72 $C_{31}H_{36}N_4O_7S$ ≥98% (TLC) |
Protected arginine derivative used in solid phase peptide synthesis;
the MTR-guanidine protection of Arg is more acid-sensitive than the
tosyl group; Fujino, M et al, *Chem Pharm Bull*, 29: 2825, 1981;
Williamson, MP et al, *Eur J Biochem*, 158: 527, 1986; Atherton, E
et al, *J Chem Soc*, Perkin I 2065, 1985

ICN 158173 $C_{31}H_{36}N_4O_7S$

Sigma F 5010 $C_{31}H_{36}N_4O_7S$

Arginine — N°-FMOC-N°-Mtr-L-Arg 4-Benzyloxybenzyl Ester Polymer Bound

Fluka 47664 0.4-0.6 mmol/g resin; carrier: polystyrene,
crosslinked with 1% DVB; 200-400 mesh particle size

Arginine — N°-FMOC-N°-Mts-L-Arg

Fluka 47573 MW 578.68 $C_{30}H_{34}N_4O_6S$ ≥98% (HPLC)

Arginine — N°-FMOC-N°-Nitro-L-Arg

Fluka 47527 MW 441.44 $C_{21}H_{23}N_5O_6$ ≥98% (HPLC); may
contain ~8% solvent

Arginine — N°-FMOC-N°-Pbf-D-Arg

Fluka 47348 MW 648.77 $C_{34}H_{40}N_4O_7S$ ≥98% (TLC)

Arginine — N°-FMOC-N°-Pbf-L-Arg

Fluka 47349 MW 648.77 $C_{34}H_{40}N_4O_7S$ ≥98% (HPLC)

Arginine — N°-FMOC-N°-Pbf-L-Arg 4-Benzyloxybenzyl Ester Polymer Bound

Fluka 47362 0.4-0.8 mmol/g resin; carrier: polystyrene,
crosslinked with 1% DVB; 100-200 mesh particle size

Arginine — N°-FMOC-N°-Pmc-L-Arg 4-Benzyloxybenzyl Ester Polymer Bound

Fluka 47665 0.4-0.6 mmol/g resin; carrier: polystyrene,
crosslinked with 1% DVB; 200-400 mesh particle size

Arginine — N°-FMOC-N°-Tosyl-L-Arg

Fluka 47534 MW 550.63 $C_{28}H_{30}N_4O_6S$ ~97% (HPLC); mp:
80°C

ICN 158176 $C_{28}H_{30}N_4O_6S$

Sigma F 5635 $C_{28}H_{30}N_4O_6S$

Arginine — N°-Lauroyl-L-Arg

Fluka 61727 MW 356.51 $C_{18}H_{36}N_4O_3$ ≥99.0% (TLC)

Arginine — N°-p-Tosyl-L-Arg-OMe Hydrochloride

Synonyms: TAME; Plasmin Substrate; Trypsin Substrate;
Thrombin Substrate; Protease Substrate

ICN 103086 MW 378.9 $C_{14}H_{22}N_4O_4S \cdot HCl$ Crystalline |
Biochem, 11:3267, 1972

Sigma T 4626 $C_{14}H_{22}N_4O_4S \cdot HCl$ Hummel, BCW, *Can J
Biochem Physiol*, 37: 1393, 1959; Castellino, FJ & Sodetz, JM, *Meth
Enzymol*, 45: 273, 1976

Arginine — N^α-t-BOC-L-Arg Hydrochloride

Sigma B 4019 $C_{11}H_{22}N_4O_4 \cdot HCl$

Arginine — N^α-t-BOC-L-Arg Hydrochloride Hydrate

ICN 150481 $C_{11}H_{22}N_4O_4 \cdot HCl \cdot H_2O$

Arginine — N^α-t-BOC-N^g-Mts-L-Arg

Sigma B 7144 $C_{20}H_{32}N_4O_6S \cdot C_6H_{13}N$

Arginine — N^α-t-BOC-N^g-CBZ-L-Arg

Sigma B 9770 $C_{19}H_{28}N_4O_6$

Arginine — N^α-t-BOC-N^g-Mtr-L-Arg

Sigma B 1642 $C_{21}H_{34}N_4O_7S$

Arginine — N^α-t-BOC-N^g-Nitro-L-Arg

ICN 105096 $C_{11}H_{21}N_5O_6$

Arginine — N^α-t-BOC-N^g-p-Tosyl-L-Arg

Sigma B 9001 $C_{18}H_{28}N_4O_6S$

Arginine — N^α-t-BOC-N^g-Tosyl-L-Arg

ICN 150506 $C_{18}H_{28}N_4O_6S$

Arginine — N^α-t-BOC-ω-Nitro-L-Arg

Sigma B 7751 $C_{11}H_{21}N_5O_6$

Arginine — N^g-Tosyl-L-Arg-OMe Hydrochloride

Synonyms: TAME; Trypsin Substrate; Thrombin Substrate

Fluka 90170   MW 378.88 $C_{14}H_{22}N_4O_4S \cdot HCl$ ≥99.0%
(titration); mp: 146-150°C; ≤0.1% residue on ignition | Not
attacked by chymotrypsin; substrate for the assay of trypsin &
thrombin; Hummel, BCW, *Can J Biochem Physiol*, 37: 1393, 1959;
Siegelman, AM et al, *Arch Biochem Biophys*, 97: 159, 1962;
Ottesen, M & Svensen, I, *Methods of Enzymatic Analysis* (HU
Bergmeyer, ed), vol 5, 3rd, 159, 1984, Verlag Chemie Weinheim;
Castellino, FJ & Sodetz, JM, *Meth Enzymol*, 45: 273, 1976; Bedi,
GS & Back, N, *Prep Biochem*, 14: 257, 1984; Matsuda, Y et al,
Chem Pharm Bull, 30: 2512, 1982; Matsuda, Y et al, *Chem Pharm
Bull*, 32: 2371, 1984

Arginine — N^α-Z-D-Arg

Fluka 95925   MW 308.34 $C_{14}H_{20}N_4O_4$ ≥98.0% (HPLC)

Arginine — N^α-Z-L-Arg

Fluka 95930   MW 308.34 $C_{14}H_{20}N_4O_4$ ~99.0% (titration);
mp: 171-174°C; 3% H_2O

Arginine — N^α-Z-L-Arg 4-Nitrobenzyl Ester Hydrochloride

Fluka 95939   MW 479.92 $C_{21}H_{25}N_5O_6 \cdot HCl$ ≥98.0% (TLC)

Arginine — N^α-Z-L-Arg Hydrobromide

Fluka 95936   MW 389.25 $C_{14}H_{20}N_4O_4 \cdot HBr$ ≥98.0% (HPLC)

Arginine — N^α-Z-L-Arg Hydrochloride

Fluka 95937   MW 344.80 $C_{14}H_{20}N_4O_4 \cdot HCl$ ≥98.0% (TLC)

Arginine — N^α-Z-L-Arg-AMC Hydrochloride

Synonyms: Z-Arg-AMC; Carbobenzoxy-L-Arg-AMC; Trypsin
Substrate; Papain Substrate

Fluka 95938   MW 501.97 $C_{24}H_{27}N_5O_5 \cdot HCl$ ~95.0%
(titration); mp: 204-205°C | Sensitive fluorogenic substrate;
Zimmermann, M et al, *Anal Biochem*, 78: 47, 1977; Kanaoka, Y et
al, *Chem Pharm Bull*, 25: 3126, 1977

Arginine — N^α-Z-N^g-Mtr-L-Arg Cyclohexylamine Salt

Fluka 96918   MW 619.78 $C_{24}H_{32}N_4O_7S \cdot C_6H_{13}N$ ≥98.0%
(TLC)

Arginine — N^α-Z-N^g-Nitro-L-Arg

Fluka 96930   MW 353.34 $C_{14}H_{19}N_5O_6$ ~99% (titration); mp:
130-132°C

Arginine — N^α-Z-N^g-Pbf-L-Arg Cyclohexylamine Salt

Fluka 96970   MW 659.84 $C_{27}H_{36}N_4O_7S \cdot C_6H_{13}N$ ~98%
(TLC)

Arginine — N^g Nitro-L-Arg (2,3,4,5-^{3}H) Hydrochloride

Amersham TRK927

Arginine — N^i,N^i-Di-CBZ-L-Arg

Sigma C 6545 $C_{22}H_{26}N_4O_6$

Arginine — N^i,N^i-Dimethyl-L-Arg Di-(p-Hydroxyazo-benzene-p'-Sulfonate) Hydrate

Synonyms: ADMA; Arg-Dimethyl

Alexis 106-005 $C_8H_{18}N_4O_2 \cdot 2C_{12}H_{10}N_2O_4S \cdot H_2O$
Asymmetrical; inhibits NO synthesis; this Salt form is more stable
than the dihydrochloride; Vallance, P et al, *Lancet*, 339: 572, 1992;
Hibbs, JB Jr et al, *J Immunol*, 138: 550, 1987

Arginine — N^i,N^i-Dimethyl-L-Arg Di(p-Hydroxyazo-benzene-p'-Sulfonate) Hydrate

Synonyms: SDMA; Arg-Dimethyl

Alexis 106-007 $C_8H_{18}N_4O_2 \cdot 2C_{12}H_{10}N_2O_4S \cdot H_2O$
Symmetrical; SDMA does not inhibits nitric oxide synthesis *in vitro*
or *in vivo*; this Salt form is more stable than the dihydrochloride;
Vallance, P et al, *Lancet*, 339: 572, 1992; Hibbs, JB Jr et al, *J
Immunol*, 138: 550, 1987

Arginine — N^i,N^i-Dimethyl-L-Arg Dihydrochloride

Synonyms: SDMA; Arg-Dimethyl

Alexis 106-008 $C_8H_{18}N_4O_2 \cdot 2HCl$ Symmetrical; does not
inhibits NO synthesis *in vitro* or *in vivo*; Vallance, P et al, *Lancet*,
339: 572, 1992

Calbiochem 311204   MW 275.2 $C_8H_{18}N_4O_2 \cdot 2HCl$ Single
component purity (TLC); solid; soluble in H_2O | Kotani, K et al, *J
Neurochem*, 58: 1127, 1992

Arginine — N^i,N^i-Dimethyl-L-Arg Dihydrochloride

Synonyms: ADMA; Arg-Dimethyl; Nitric Acid Synthase Inhibitor

Alexis 106-006 $C_8H_{18}N_4O_2 \cdot 2HCl$ Asymmetrical; inhibits NO
synthesis; Vallance, P et al, *Lancet*, 339: 572, 1992

Calbiochem 311203   MW 275.2 $C_8H_{18}N_4O_2 \cdot 2HCl$ >98%
(TLC); solid; soluble in H_2O | Reversible inhibitor of NOS *in vitro* &
in vivo; causes dose-dependent vasoconstriction & bradycardic
effects; Faraci, FM et al, *Am J Physiol*, 269: H1522, 1995;
Gardiner, SM et al, *Br J Pharmacol*, 110: 1457, 1993; Kotani, K et
al, *J Neurochem*, 58: 1127, 1992; Vallance, P et al, *Lancet*, 339:
572, 1992

Arginine — N'-Allyl-L-Arg

Synonyms: N^ω-Allyl-L-Arg

Calbiochem 128100 MW 214.3 $C_9H_{18}N_4O_2$ ≥98% (TLC); lyophilized solid; soluble in H_2O | Competitive reversible inhibitor & time-dependent inactivator of nNOS; Zhang, HQ et al, *JACS*, 119: 10888, 1997; Zhang, HQ et al, *J Med Chem*, 40: 3869, 1997; Olken, NM & Marletta, MA, *J Med Chem*, 35: 1137, 1992

Arginine — N'-Amino-L-Arg Hydrochloride

Synonyms: L-NAA; Nitric Acid Synthase Inhibitor

Alexis 106-014 MW 225.7 $C_6H_{15}N_5O_2 \cdot$ HCl ≥97% with traces of L-Orn & L-Arg; white powder; soluble in H_2O | Potent inhibitor of endothelial & inducible nitric oxide synthase (NOS III & NOS II); Fukuto, JM et al, *BBRC*, 168: 458, 1990; Gross, SS et al, *BBRC*, 170: 96, 1990; Vargas, HM et al, *J Pharmacol Exp Ther*, 257: 1208, 1991; Lambert, L et al, *Life Sci*, 48: 69, 1991

ICN 193970 $C_{11}H_{15}N_6O_2 \cdot$ HCl Potent inhibitor of endothelial & inducible NOS; Fukuto, JM etal, BBRC, 170:96, 1990

Arginine — N'-Hydroxy-L-Arg

Synonyms: L-HOArg AcOH; NOHA; Arginase Inhibitor

Alexis 106-004 $C_6H_{14}N_4O_3 \cdot CH_3COOH$ Intermediate in the biosynthesis of NO from L-arginine by NOS; Marletta, MA et al, *Biochemistry*, 27: 8706, 1988; Stuehr, DJ et al, *J Biol Chem*, 266: 6259, 1991; Zembowicz, A et al, *PNAS*, 88: 11172, 1991; Wallace, GC et al, *BBRC*, 176: 528, 1991; Wallace, GC & Fukuto, JM, *J Med Chem*, 34: 1746, 1991; Feldman, PL, *THL*, 32: 875, 1991; Hecker, M et al, *FEBS Lett*, 294: 221, 1991; Pufahl, RA et al, *Biochemistry*, 31: 6822, 1992; Daghigh, F et al, *BBRC*, 202: 174, 1994; Boucher, J-L et al, *BBRC*, 203: 1614, 1994

ICN 193661 $C_6H_{14}N_4O_3 \cdot C_2H_4O_2$ Intermediate in the conversion of L-Arg into NO & citrulline

Sigma H 7278 $C_6H_{14}N_4O_3 \cdot C_2H_4O_2$ Intermediate in the conversion of arginine to NO & citrulline by NO synthase; as a competitive inhibitor of liver & macrophage arginase, compound may be a modulator in the NO biosynthesis pathway; Klatt, P et al, *J Biol Chem*, 268: 14781, 1993; Campos, KL et al, *J Biol Chem*, 279: 1721, 1995; Pufahl, RA et al, *Biochemistry*, 34: 1930, 1995; Daghigh, P et al, *Biochem Biophys Res Commun*, 202: 174, 1994; Boucher, JL et al, *Biochem Biophys Res Commun*, 203: 1614, 1994

Arginine — N'-Hydroxy-L-Arg Monoacetate Salt

Synonyms: NOHA

Calbiochem 399250 MW 250.3 $C_6H_{14}N_4O_3 \cdot CH_3CO_2H$ >98% (TLC); solid; soluble in H_2O | Key intermediate in the biosynthesis of NO by cNOS; efficiently oxidized to NO & citrulline by cytochrome P450; potent inhibitor of liver & macrophage arginase; Boucher, JL et al, *Biochem Biophys Res Comm*, 203: 1614, 1994; Pufahl, RA & Marletta, MA, *Biochem Biophys Res Comm*, 193: 963, 1993; Stuehr, DJ et al, *J Biol Chem*, 266: 6259, 1991; Boucher, JL et al, *Biochem Biophys Res Comm*, 187: 880, 1992

Arginine — N'-Me-D-Arg

Sigma M 7034 $C_7H_{16}N_4O_2 \cdot C_2H_4O_2$ No significant effect on NOS; used as a negative control in studies of L-NMMA activity

Arginine — N'-Me-D-Arg Acetate Salt

ICN 159625 $C_7H_{16}N_4O_2 \cdot C_2H_4O_2$

Arginine — N'-Me-L-Arg

Synonyms: N^5-Methylamidino-L-Orn; Nitric Acid Synthase Inhibitor

Sigma M 7033 $C_7H_{16}N_4O_2 \cdot C_2H_4O_2$ Endothelium-derived relaxing factor inhibitor; inhibits the generation of NO from arginine; competitive inhibitor of all 3 isoforms of NOS; Sakuma, I et al, *Proc Natl Acad Sci USA*, 85: 8664, 1988

Arginine — N'-Me-L-Arg Acetate Salt

Synonyms: Endothelium-Derived Relaxing Factor Inhibitor

ICN 155470 $C_7H_{16}N_4O_2 \cdot C_2H_4O_2$ Sakuma, I etal, *PNAS*, 85:8664, 1988

Arginine — N'-Monoethyl-L-Arg Monoacetate Salt

Synonyms: L-NMEA; NMEA, AcOH; Nitric Acid Synthase Inhibitor

Alexis 106-010 MW 262.4 $C_8H_{18}N_4O_2 \cdot CH_3COOH$ White solid; soluble in H_2O

Calbiochem 475883 MW 262.3 $C_8H_{18}N_4O_2 \cdot CH_3CO_2H$ Single component purity (TLC); solid; soluble in H_2O | NOS inhibitor; Komori, Y et al, *Arch Biochem Biophys*, 315: 213, 1994; Torres, M et al, *J Neurochem*, 63: 988, 1994

Arginine — N'-Monomethyl-D-Arg Monoacetate Salt

Synonyms: D-NMMA; N^ω-Me-D-Arg; N'-Me-D-Arg AcOH

Alexis 106-002 MW 248.3 $C_7H_{16}N_4O_2 \cdot CH_3COOH$ ≥99%; white solid; soluble in H_2O

Calbiochem 475892 MW 248.3 $C_7H_{16}N_4O_2 \cdot CH_3CO_2H$ ≥99% (TLC); solid; soluble in H_2O & MeOH | Negative control for N'-monomethyl-L-arginine; used to investigate non-specific L-NMMA activity; no significant effect on NOS; Rees, DD et al, *Br J Pharmacol*, 101: 746, 1990

Arginine — N'-Monomethyl-L-Arg Acetate Salt

Synonyms: L-NMMA; Nitric Acid Synthase Inhibitor

Sigma M-125 MW 248.3 $C_7H_{16}N_4O_2 \cdot CH_3CO_2H$ >94%; white solid; mp 186-188°C; soluble in H_2O (28 mg/ml) | NOS inhibitor; Blocks formation of endothelium-derived relaxing factor (EDRF); Sakuma et al, *PNAS*, 85: 8664, 1988; Olken et al, *BBRC*, 177: 828, 1991

Arginine — N'-Monomethyl-L-Arg Di-*p*-Hydroxyazobenzene-*p'*-Sulfonate Salt

Synonyms: L-NMMA; N^ω-Me-L-Arg; N'-Me-L-Arg diHABS

Calbiochem 475902 MW 744.8 $C_7H_{16}N_4O_2 \cdot C_{24}H_{20}N_4O_8S_2$ >98% (TLC); solid; soluble in DMSO

Arginine — N'-Monomethyl-L-Arg Monoacetate Salt

Synonyms: L-NMMA; N^ω-Me-L-Arg; N'-Me-L-Arg AcOH; Nitric Acid Synthase Inhibitor

Alexis 106-001 MW 248.3 $C_7H_{16}N_4O_2 \cdot CH_3COOH$ ≥95%; white solid; soluble in H_2O | Inhibitor of NOS; suppresses the release of NO from endothelial cells; Sakuma, I et al, *PNAS*, 85: 8664, 1988; Rees, DD et al, *Br J Pharmacol*, 96: 418, 1989

Calbiochem 475886 MW 248.3 $C_7H_{16}N_4O_2 \cdot CH_3CO_2H$ ≥99% (TLC); solid; soluble in H_2O & MeOH | L-Arginine analog that acts as a competitive, inhibitor of all three isoforms of NOS; inhibits histamine- & acetylcholine-induced relaxation of intact norepinephrine-constricted guinea pig pulmonary artery; Reif, DW & McCreedy, SA, *Arch Biochem Biophys*, 320: 170, 1995; O'Kane, KP et al, *Br J Clin Pharmacol*, 38: 311, 1994; Kubes, P et al, *PNAS*, 88: 4651, 1991; Mehta, JL et al, *BBRC*, 173: 438, 1990; Sakuma, I et al, *PNAS*, 85: 8664, 1988

Arginine — N'-Monomethyl-L-Arg *p*-Hydroxyazobenzene-*p'*-Sulfonate Salt

Synonyms: L-NMMA; N^ω-Me-L-Arg HABS; Nitric Acid Synthase Inhibitor

Calbiochem 475891 MW 484.5 $C_7H_{16}N_4O_2 \cdot C_{12}H_{10}N_2O_4S$ >99% (TLC); solid; soluble in MeOH | Novel inhibitor of histamine- & acetylcholine-induced relaxation of norepinephrine-constricted pulmonary artery by endothelium derived relaxing factor dependent mechanism; inhibits rat brain NOS; Barjavel, MJ & Bhagara, HN, *Neurosci Lett*, 181: 27, 1994; Thomas, G et al, *Biochem Biophys Res Comm*, 158: 177, 1989; Sakuma, I et al, *PNAS*, 85: 8664, 1988

Arginine — N^r-Nitro-D-Arg

Alexis 105-002 MW 219.2 $C_6H_{13}N_5O_4$ White solid; soluble in 0.1 N HCl, moderately in H_2O, insoluble in MeOH

Calbiochem 483121 MW 219.2 $C_6H_{13}N_5O_4$ One major component (TLC); solid | Less active enantiomer of N^r-Nitro-L-arginine that can be used as a negative control; Khanna, JM et al, *Brain Res Bull*, 32: 43, 1993

Arginine — N^r-Nitro-D-Arg-OMe Hydrochloride

Synonyms: N^ω-Nitro-D-Arg-OMe; D-NAME

Alexis 105-004 MW 269.7 $C_7H_{15}N_5O_4 \cdot$ HCl Off-white solid; soluble in H_2O

Calbiochem 483124 MW 269.7 $C_7H_{15}N_5O_4 \cdot$ HCl ≥98% (TLC); solid; soluble in H_2O | Less active enantiomer of N^r-Nitro-L-arginine methyl ester, HCl that can be used as a negative control; Chung, BH & Chang, KC, *Gen Pharmacol*, 25: 893, 1994; Li, Q et al, *J Pharm Pharmacol*, 46: 510, 1994; Miller, J et al, *J Pharmacol Exp Ther*, 264: 11, 1993

Arginine — N^r-Nitro-L-Arg

Synonyms: L-Nitro-Arg; Nitric Acid Synthase Inhibitor; L-NNA; Nitric Acid Synthase Inhibitor

Alexis 105-001 MW 219.2 $C_6H_{13}N_5O_4$ White solid; soluble in 0.1 N HCl, moderately in H_2O, insoluble in MeOH | Moore, PK et al, *Br J Pharmacol*, 99: 408, 1990; Gibson, A et al, *Br J Pharmacol*, 99: 802, 1990; Ishii, K et al, *Eur J Pharmacol*, 176: 219, 1990; Toda, N et al, *Life Sci*, 47: 345, 1990; Brave, SR et al, *BBRC*, 179: 1017, 1991; Hecker, M et al, *FEBS Lett*, 294: 221, 1991; Dwyer, MA et al, *BBRC*, 176: 1136, 1991; Chen, X & Gillis, CN, *BBRC*, 186: 1522, 1992; Quock, RM & Nguyen, E, *Life Sci*, 51: PL255, 1992; Tam, FS-F & Hillier, K, *Life Sci*, 51: 1277, 1992; Kolesnikov, YA et al, *PNAS*, 90: 5162, 1993; Furfine, ES et al, *Biochemistry*, 32: 8512, 1993; Chang, KC et al, *BBRC*, 191: 509, 1993; mayer, B et al, *FEBS Lett*, 333: 203, 1993

Calbiochem 483120 MW 219.2 $C_6H_{13}N_5O_4$ ≥99% (TLC); solid; soluble in H_2O | Potent & reversible inhibitor of bNOS & eNOS; inhibits iNOS at much higher concentrations; causes a reduction in cardiac output & profound vasoconstriction; Modin, A et al, *Neuroscience*, 62: 189, 1994; Reif, DW & McCreedy, SA, *Arch Biochem Biophys*, 320: 170, 1995; Michel, AD et al, *Br J Pharmacol*, 109: 287, 1993; Furfine, ES et al, *Biochemistry*, 32: 8512, 1993; Dawson, VL et al, *PNAS*, 88: 6368, 1991

Arginine — N^r-Nitro-L-Arg-OMe Hydrochloride

Synonyms: N^ω-Nitro-L-Arg-OMe; L-NAME; Nitric Acid Synthase Inhibitor

Alexis 105-003 MW 269.7 $C_7H_{15}N_5O_4 \cdot$ HCl Off-white solid; soluble in H_2O | Inhibitor of NOS; more soluble than N^r-Nitro-L-Arginine; suppresses the release of NO from endothelial cells; has a different effect on superoxide generation by NOS than L-NMMA; Moore, PK et al, *Br J Pharmacol*, 99: 408, 1990; Pou, S et al, *J Biol Chem*, 267: 24173, 1992

Calbiochem 483125 MW 269.7 $C_7H_{15}N_5O_4 \cdot$ HCl ≥99% (TLC); solid | More soluble analog of arginine & a competitive, slowly reversible inhibitor of endothelial NOS; causes a prolonged inhibition of acetylcholine-induced relaxation of rat aortic rings; Moncada, S et al, *Pharmacol Review*, 43: 109, 1991; Dawson, VL et al, *PNAS*, 88: 6368, 1991; Baylis, C et al, *J Pharmacol Exp Ther*, 274: 1135, 1995; Kubes, P et al, *PNAS*, 88: 4651, 1991

Arginine — N^r-Amino-L-Arg (2,4-Dinitro)-1-Naphthol-7-Sulfonic Acid Salt

Synonyms: Nitric Oxide Synthase Inhibitor

Fluka 06804 MW 503.44 $C_6H_{15}N_5O_2 \cdot C_{10}H_6N_2O_8S$ ≥98% (TLC) | Griffith, OW & Kilbourn, RG, *Meth Enzymol*, 268: 375, 1996

Arginine — N^r-Me-L-Arg Acetate Salt

Synonyms: NMMA; N^r-Monomethyl-L-Arg Acetate

Fluka 65825 MW 248.3 $C_7H_{16}N_4O_2 \cdot C_2H_4O_2$ ≥95.0% (HPLC); ≤2% H_2O | Endothelium-derived relaxing factor inhibitor; Sakuma, I et al, *PNAS*, 85: 8664, 1988

Arginine — N^r-Mtr-L-Arg

ICN 155446 $C_{16}H_{26}N_4O_5S$

Arginine — N^r-Nitro-D-Arg-OMe Hydrochloride

Synonyms: D-NAME

Sigma N 4770 $C_7H_{15}N_5O_4 \cdot$ HCl Inactive isomer of NAME

Arginine — N^r-Nitro-L-Arg

Synonyms: L-NNA; N^r-Nitro-L-Arg; N^5-(Nitroamidino)-L-2,5-Diaminopentanoic Acid; Nitric Oxide Synthase Substrate

ICN 102503 $C_6H_{13}N_5O_4$

Sigma N 5501 $C_6H_{13}N_5O_4$ Irreversible inhibitor of nNOS; reversible inhibitor of iNOS

Arginine — N^r-Nitro-L-Arg-OBzl

Sigma N 2644 $C_{13}H_{19}N_5O_4 \cdot C_7H_8O_3S$

Arginine — N^r-Nitro-L-Arg-OBzl Di(p-Tosyl) Salt

ICN 155845 $C_{13}H_{19}N_5O_4 \cdot C_7H_8O_3S$

Arginine — N^r-Nitro-L-Arg-OMe Hydrochloride

Synonyms: L-NAME; Nitric Oxide Synthase Substrate

ICN 155846 $C_7H_{15}N_5O_4 \cdot$ HCl

Sigma N 5751 $C_7H_{15}N_5O_4 \cdot$ HCl More soluble than N^r-Nitro-L-arginine; irreversible inhibitor of nNOS; reversible inhibitor of iNOS

Arginine — N^r-Nitro-L-Arg-pNA

Sigma N 2268 $C_{12}H_{17}N_7O_5 \cdot$ HBr Actual content on label | Contains ≤20% dioxane as solvent of crystallization

Arginine — N^r-Phospho-L-Arg Sodium Salt

Synonyms: L-Arg Phosphate

Fluka 79424 MW 276.2 $C_6H_{14}N_4NaO_5P$ ≥97.0% (TLC); ≤5% H_2O; ≤0.1% P_i | Ratto, A & Christen, R, *Eur J Biochem*, 173: 667, 1988

Arginine — N^r-Propyl-L-Arg

Synonyms: N^5-(Imino(Propylamino)-Me)-L-Orn; Nitric Oxide Synthase Inhibitor

Alexis 270-203 MW 216.3 $C_9H_{20}N_4O_2$ ≥98%; crystalline solid; soluble in aqueous buffer | Potent & selective inhibitor of nNOS (NOS I) relative to iNOS (NOS II) (3158-fold) & eNOS (NOS III) (149-fold); Zhang, HQ et al, *J Med Chem*, 40: 3869, 1997

Arginine — Phospho-L-Arg

Synonyms: N^5-(Imino(Phosphonoamino)-Me)-2,5-Diaminopentanoic Acid

Sigma P 5139 $C_6H_{14}N_4O_5PNa$ High-energy phosphate reservoir in the muscle of some invertebrates

Arginine — Pth-Arg Hydrochloride

Sigma P 7158 $C_{13}H_{17}N_5OS \cdot$ HCl Useful as standards in protein sequencing by the Edman degradation

Arginine — p-Tosyl-L-Arg

ICN 104924 MW 328.4 $C_3H_{20}N_4O_4S$

Arginine — TentaGel S PHB-Arg-Mtr-FMOC

Fluka 86368 0.22 mmol protected AA/g; ~90 µm particle size

Arginine — TentaGel S PHB-Arg-Pbf-FMOC

Fluka 86381 0.22 mmol protected AA/g; ~90 µm particle size

Arginine — TentaGel S PHB-Arg-Pmc-FMOC

Fluka 86369   0.20 mmol protected AA/g; ~90 μm particle size

Arginine — TentaGel S Trt-Arg-Mtr-FMOC

Fluka 86409   0.18 mmol protected AA/g; ~90 μm particle size

Arginine — TentaGel S Trt-Arg-Pbf-FMOC

Fluka 86442   0.20 mmol protected AA/g; ~90 μm particle size

Arginine — TentaGel S Trt-Arg-Pmc-FMOC

Fluka 86411   0.18 mmol protected AA/g; ~90 μm particle size

Arginine — Tosyl-Arg

Bachem E-2465.0005   MW 328.39 $C_{13}H_{20}N_4O_4S$ Store at RT

Arginine — Tosyl-Arg-OMe Hydrochloride

Synonyms: L-TAME; Trypsin Substrate; TAME

Bachem M-1745.0005   MW 378.88 $C_{14}H_{22}N_4O_4S \cdot HCl$ Store at -15°C

Peptides International STR-3003   MW 378.88 $C_{14}H_{22}N_4O_4S \cdot HCl$ >98% (HPLC); amorphous powder; hydrochloride | Walsh, KA in *Proteolytic Enzymes*, Methods in Enzymology, Vol 19, (GE Perlmann & L Lorand, Eds), Academic Press, New York, 1970, pp 41-63

Arginine — Z-Arg

Synonyms: Benzyloxycarbonyl-L-Arg

Bachem C-1285.0025   MW 308.34 $C_{14}H_{20}N_4O_4$ Store at RT

Peptides International ZLR-2008-PI   MW 308.34   >98% (TLC); white powder

Arginine — Z-Arg Hydrochloride

Bachem C-1290.0025   MW 344.8 $C_{14}H_{20}N_4O_4 \cdot HCl$ Store at RT

Arginine — Z-Arg-(*p*ONb)

Bachem C-1300.0250   MW 443.46 $C_{21}H_{25}N_5O_6$ Store at RT

Arginine — Z-Arg-4MβNA Hydrochloride

Bachem J-1100.0250   MW 500.01 $C_{25}H_{29}N_5O_4 \cdot HCl$ Store at -15°C

Arginine — Z-Arg-AMC Hydrochloride

Bachem I-1130.0050   MW 501.97 $C_{24}H_{27}N_5O_5 \cdot HCl$ Store at -15°C

Arginine — Z-Arg-BOC₂ CHA

Bachem C-3710.0001   MW 607.75 $C_{24}H_{36}N_4O_8 \cdot C_6H_{13}N$ Store at -15°C

Arginine — Z-Arg-Mbs Dicyclohexylamine

Bachem C-3345.0001   MW 659.85 $C_{21}H_{26}N_4O_7S \cdot C_{12}H_{23}N$ Store at RT

Arginine — Z-Arg-Mtr·CHA

Bachem C-2250.0001   MW 619.78 $C_{24}H_{32}N_4O_7S \cdot C_6H_{13}N$ Store at RT

Arginine — Z-Arg-Mtr-OtBu

Bachem C-3750.0001   MW 576.71 $C_{28}H_{40}N_4O_7S$ Store at -15°C

Arginine — Z-Arg-Nitro

Synonyms: N^α-Benzyloxycarbonyl-$N^\prime$-Nitro-L-Arg

Bachem C-2270.0005   MW 353.34 $C_{14}H_{19}N_5O_6$ Store at RT

Peptides International ZLR-2009-PI   MW 353.34   >98% (TLC); white crystalline powder

Arginine — Z-Arg-OBzl-*p*-Nitro Hydrobromide

Synonyms: Carbobenzoxy-L-Arg-(*p*ONb)

Peptides International SRB-3157   MW 524.37 $C_{21}H_{25}N_5O_6 \cdot HBr$ >98% (HPLC); amorphous powder

Arginine — Z-Arg-Pbf CHA

Bachem C-3715.0001   MW 659.85 $C_{27}H_{36}N_4O_7S \cdot C_6H_{13}N$ Store at RT

Arginine — Z-Arg-Pmc CHA

Bachem C-3490.0001   MW 673.87 $C_{28}H_{38}N_4O_7S \cdot C_6H_{13}N$ Store at RT

Arginine — Z-Arg-pNA Hydrochloride

Synonyms: L-ZAPA

Bachem L-1220.0250   MW 464.91 $C_{20}H_{24}N_6O_5 \cdot HCl$ Store at -15°C

Arginine — Z-Arg-Tosyl

Bachem C-2657.0005   MW 462.53 $C_{21}H_{26}N_4O_6S$ Store at RT

Arginine — Z-Arg-Z₂

Bachem C-2975.0001   MW 576.61 $C_{30}H_{32}N_4O_8$ Store at -15°C

Arginine — Z-Arg-Z₂-OSu

Bachem C-2985.0001   MW 673.68 $C_{34}H_{35}N_5O_{10}$ Store at -15°C

Arginine — Z-Arg-βNA Hydrochloride

Bachem K-1165.0250   MW 469.97 $C_{24}H_{27}N_5O_3 \cdot HCl$ Store at -15°C

Arginine — Z-D-Arg

Bachem C-1295.0005   MW 308.34 $C_{14}H_{20}N_4O_4$ Store at RT

Neosystem ZA00201   MW 308.3

Arginine — Z-D-Arg Hydrochloride

Bachem C-3860.0005   MW 344.8 $C_{14}H_{20}N_4O_4 \cdot HCl$ Store at RT

Arginine — Z-D-Arg-Nitro

Bachem C-2275.0001   MW 353.34 $C_{14}H_{19}N_5O_6$ Store at RT

Arginine — Z-D-Arg-Pbf CHA

Bachem C-3830.0001   MW 659.85 $C_{27}H_{36}N_4O_7S \cdot C_6H_{13}N$ Store at RT

Arginine — Z-D-Arg-Pmc CHA

Bachem C-3590.0001   MW 673.87 $C_{28}H_{38}N_4O_7S \cdot C_6H_{13}N$ Store at RT

Arginine — Z-D-Arg-Z₂

Bachem C-2980.0001   MW 576.61 $C_{30}H_{32}N_4O_8$ Store at -15°C

Arginine — Z-D-Arg-Z₂-OSu

Bachem C-3235.0001   MW 673.68 $C_{34}H_{35}N_5O_{10}$ Store at -15°C

Arginine — Z-L-Arg

Neosystem ZA00202   MW 308.3

Arginine — Z-L-Arg-(Z₂)

Neosystem ZA00204   MW 576.6

Arginine — Z-L-Arg-Nitro

Neosystem ZA00218   MW 353.3

Arginine — α-Dansyl-L-Arg Hydrochloride

Sigma D 0250 $C_{18}H_{25}N_5O_4S \cdot HCl$

Asparagine — Ac-Asn-Trt

Bachem E-3020.0001   MW 416.48 $C_{25}H_{24}N_2O_4$ Store at 2-8°C

Asparagine — Ac-D-Asn

Bachem F-3090.0001   MW 174.16 $C_6H_{10}N_2O_4$ Store at RT

Asparagine — Asn Amide Hydrochloride

Bachem E-1425.0500   MW 167.6 $C_4H_9N_3O_2 \cdot HCl$ Store at RT

Asparagine — Asn Hydrate

Bachem E-1380.0025   MW 132.12 $C_4H_8N_2O_3$ Store at RT

Asparagine — Asn-(pONb) Hydrobromide

Bachem E-2760.0001   MW 348.15 $C_{11}H_{13}N_3O_5 \cdot HBr$ Store at -15°C

Asparagine — Asn-AMC

Sigma A 8046 $C_{14}H_{15}N_3O_4 \cdot C_2HF_3O_2$

Asparagine — Asn-AMC TFA

Bachem I-1435.0050   MW 403.32 $C_{14}H_{15}N_3O_4 \cdot C_2HF_3O_2$ Store at -15°C

Asparagine — Asn-Mtt

Bachem E-2995.0001   MW 388.47 $C_{24}H_{24}N_2O_3$ Store at 2-8°C

Asparagine — Asn-OMe Hydrochloride

Bachem E-3485.0001   MW 182.61 $C_5H_{10}N_2O_3 \cdot HCl$ Store at -15°C

Asparagine — Asn-OtBu

Bachem E-1385.0001   MW 188.23 $C_8H_{16}N_2O_3$ Store at -15°C

Asparagine — Asn-OtBu Hydrochloride

Bachem E-1390.0001   MW 224.69 $C_8H_{16}N_2O_3 \cdot HCl$ Store at -15°C

Asparagine — Asn-Trt

Bachem E-2810.0001   MW 374.44 $C_{23}H_{22}N_2O_3$ Store at RT

Asparagine — Asn-βNA

Bachem K-1090.0250   MW 257.3 $C_{14}H_{15}N_3O_2$ Store at -15°C

Asparagine — BOC-Asn

American Peptide BLASN10   MW 232.2

Bachem A-1215.0005   MW 232.24 $C_9H_{16}N_2O_5$ Store at RT

Peptides International BLN-2060   MW 232.24   >98% (HPLC); white crystalline powder

USBio B2270   MW 232.2 $C_9H_{16}N_2O_5$ ≥99%

Asparagine — BOC-Asn-ONp

Synonyms: BOC-L-Asn-(pONp)

Bachem A-1230.0005   MW 353.33 $C_{15}H_{19}N_3O_7$ Store at -15°C

Bachem A-3890.0005   MW 353.33 $C_{15}H_{19}N_3O_7$ Store at -15°C

Peptides International BLN-2077   MW 353.33   >98% (HPLC); white crystalline powder; keep dry in freezer

USBio B2272   MW 353.3 $C_{15}H_{19}N_3O_7$ ≥99%

Asparagine — BOC-Asn-O-Resin

American Peptide RMRB125   1% DVB cross-linked: 100-200 mesh

Asparagine — BOC-Asn-OSu

Bachem A-1225.0001   MW 329.31 $C_{13}H_{19}N_3O_7$ Store at -15°C

Asparagine — BOC-Asn-PAM-® (200-400 mesh)

Bachem D-1220.0001   Store at 2-8°C

Asparagine — BOC-Asn-PAM-Resin

American Peptide RPAM125   1% DVB cross-linked: 100-200 mesh

Asparagine — BOC-Asn-Xan

Synonyms: N^α-BOC-N^β-Xan-L-Asn

American Peptide BLASN20   MW 412.4

Bachem A-2485.0005   MW 412.44 $C_{22}H_{24}N_2O_6$ Store at -15°C

Peptides International BLN-5351-PI   MW 412.4   >98% (HPLC); white crystalline powder

USBio B2273   MW 412.4 $C_{22}H_{24}N_2O_6$ ≥99%

Asparagine — BOC-Asn-Xan-® (200-400 mesh)

Bachem D-1185.0005   Store at RT

Asparagine — BOC-Asp-Chx

American Peptide BLASP25   MW 315.4

Asparagine — BOC-Asp-OcHex-O-Resin

American Peptide RMRB135   1% DVB cross-linked: 100-200 mesh | t-BOC protected AA resin; preattached resins are useful for synthesis of peptides with C-terminal carboxyl group by t-BOC chemistry

Asparagine — BOC-Asp-OcHex-PAM-Resin

American Peptide RPAM135   1% DVB cross-linked: 100-200 mesh | t-BOC protected AA resin; preattached resins are useful for synthesis of peptides with C-terminal carboxyl group by t-BOC chemistry

Asparagine — BOC-D-Asn

Bachem A-1220.0001 MW 232.24 $C_9H_{16}N_2O_5$ Store at RT

Fluka 15064 MW 232.24 $C_9H_{16}N_2O_5$ ≥98% (TLC)

Neosystem BA00301 MW 232.2

Peptides International BDN-2626 MW 232.24 >98% (HPLC); white crystalline powder

USBio B2269 MW 232.2 $C_9H_{16}N_2O_5$ ≥99%

Asparagine — BOC-D-Asn-ONp

Bachem A-1235.0001 MW 353.33 $C_{15}H_{19}N_3O_7$ Store at -15°C

Peptides International BDN-2620 MW 353.33 >98% (HPLC); white crystalline powder; keep dry in freezer

USBio B2271 MW 353.3 $C_{15}H_{19}N_3O_7$ ≥99%

Asparagine — BOC-D-Asn-PAM-® (200-400 mesh)

Bachem D-1570.0001 Store at 2-8°C

Asparagine — BOC-D-Asn-Xan

Synonyms: N^α-BOC-N^β-Xan-D-Asn

Bachem A-4030.0001 MW 412.45 $C_{22}H_{24}N_2O_6$ Store at -15°C

Peptides International BDN-2651-PI MW 412.45 >98% (HPLC); white crystalline powder

Asparagine — BOC-L-Asn

Fluka 15381 MW 232.24 $C_9H_{16}N_2O_5$ ~99% (TLC); mp: ~180°C; ≤0.2% residue on ignition

Neosystem BA00302 MW 232.2

Asparagine — BOC-L-Asn-(4-ONp)

Fluka 15065 MW 353.33 $C_{15}H_{19}N_3O_7$ ≥98% (HPLC)

Asparagine — BOC-L-Asn-MBHA Resin

Neosystem RN00302

Asparagine — BOC-L-Asn-O-CH₂-φ-CH₂-COOH

Neosystem LP00302 MW 380.4

Asparagine — BOC-L-Asn-PAM Resin

Neosystem RP00302

Asparagine — BOC-L-Asn-Xan

Neosystem BA00305 MW 412.4

Asparagine — BOC-L-Asp 1-Amide

Synonyms: BOC-L-Isoalanine

Fluka 15037 MW 232.24 $C_9H_{16}N_2O_5$ ≥98% (TLC); mp: 150-154°C

Asparagine — Bsmoc-Asn

Peptides International BLN-5615-PI MW 354.34 >98% (HPLC); white to off-white powder | LA Carpino, et al, *JACS*, 119:9915, 1997

Asparagine — Bsmoc-Asn-Trt

Synonyms: N^α-Bsmoc-N^β-Trt-L-Asn

Peptides International BLN-5614-PI MW 596.66 >98% (HPLC); white to off-white powder | LA Carpino, et al, *JACS*, 119:9915, 1997

Asparagine — Bz-Asn-pNA

Bachem L-1645.0050 MW 356.3 $C_{17}H_{16}N_4O_5$ Store at -15°C

Asparagine — CBZ-Asn

USBio C2098-41 MW 266.3 $C_{12}H_{14}N_2O_5$ ≥99%

Asparagine — CBZ-Asn-ONp

USBio C2098-43 MW 387.3 $C_{18}H_{17}N_3O_7$ ≥99%

USBio C2098-44 MW 387.3 $C_{18}H_{17}N_3O_7$ ≥99%

Asparagine — CBZ-D-Asn

USBio C2098-42 MW 266.3 $C_{12}H_{14}N_2O_5$ ≥99%

Asparagine — Dansyl-L-Asn

ICN 100036

Sigma D 0375 $C_{16}H_{19}N_3O_5S$

Asparagine — D-Asn

Bachem F-1245.0025 MW 132.12 $C_4H_8N_2O_3$ Store at RT

Asparagine — D-Asn Anhydrous

Synonyms: (R)-2-Aminosuccinic Acid 4-Amide; (D)-Asp 4-Amide

Fluka 11155 MW 132.12 $C_4H_8N_2O_3$ ≥99% (titration); mp: ~280°C

Asparagine — D-Asn Hydrate

Synonyms: (R)-2-Aminosuccinic Acid 4-Amide; (D)-Asp 4-Amide

Fluka 11170 MW 150.14 $C_4H_8N_2O_3 \cdot H_2O$ ≥99% (titration)

ICN 100785 $C_4H_8N_2O_3 \cdot H_2O$

Neosystem AA00301 MW 150.1

Peptides International AND-2815 MW 150.14 >99.9% optical purity by RP-HPLC; white crystalline powder

Sigma A 8131 $C_4H_8N_2O_3 \cdot H_2O$

USBio A3825-10 MW 150.1 $C_4H_8N_2O_3 \cdot H_2O$ ≥99%

Asparagine — D-Asn-Mtt

Bachem F-2870.0001 MW 388.47 $C_{24}H_{24}N_2O_3$ Store at -15°C

Asparagine — D-Asn-Trt

Bachem F-2755.0001 MW 374.44 $C_{23}H_{22}N_2O_3$ Store at -15°C

Asparagine — Dinitropyridyl-L-Asn

ICN 100791 $C_9H_9N_5O_7$

Asparagine — DL-Asn Hydrate

Synonyms: (±)-2-Aminosuccinic Acid 4-Amide; DL-Asp 4-Amide

Fluka 11180 MW 150.14 $C_4H_8N_2O_3 \cdot H_2O$ ≥99.0%; ≤0.1% residue on ignition

ICN 100788 $C_4H_8N_2O_3 \cdot H_2O$

Sigma A 8256 $C_4H_8N_2O_3 \cdot H_2O$

Asparagine — Dnp-L-Asn

ICN 100789 $C_{10}H_{10}N_4O_7$

Asparagine — FMOC-Asn

Synonyms: FMOC-L-Asn

American Peptide FASN100 MW 354.4

Bachem B-1045.0005 MW 354.36 $C_{19}H_{18}N_2O_5$ Store at RT

Peptides International FLN-1704-PI MW 354.36 >98% (HPLC); white amorphous powder

USBio F5330   MW 354.4 $C_{19}H_{18}N_2O_5$ ≥99%

Asparagine — FMOC-Asn-(GlcNAcAc₃-β-D)

Synonyms: N^α-FMOC-N^γ-(2-Acetamido-2-Deoxy-3,4,6-Tri-O-Ac-β-D-Glucopyranosyl)-L-Asn

Bachem B-2480.0100   MW 683.67 $C_{33}H_{37}N_3O_{13}$ Store at -15°C

Asparagine — FMOC-Asn-DOD

Bachem B-1290.0001   MW 580.64 $C_{34}H_{32}N_2O_7$ Store at -15°C

Asparagine — FMOC-Asn-Mtt

Bachem B-2045.0001   MW 610.71 $C_{39}H_{34}N_2O_5$ Store at -15°C

Asparagine — FMOC-Asn-Mtt-® (200-400 mesh)

Bachem D-1790.0001   Store at 2-8°C

Asparagine — FMOC-Asn-Mtt-OPfp

Bachem B-2210.0001   MW 776.77 $C_{45}H_{33}N_2O_5F_5$ Store at -15°C

Asparagine — FMOC-Asn-Mtt-SASRIN™-® (200-400 mesh)

Bachem D-1820.0001   Store at 2-8°C

Asparagine — FMOC-Asn-ONp

Bachem B-1055.0001   MW 475.46 $C_{25}H_{21}N_3O_7$ Store at -15°C

Asparagine — FMOC-Asn-OPfp

Bachem B-1105.0001   MW 520.41 $C_{25}H_{17}F_5N_2O_5$ Store at -15°C

USBio F5331   MW 520.5 $C_{25}H_{17}N_2O_5F_5$ ≥99%

Asparagine — FMOC-Asn-OtBu

Synonyms: N^α-9-FMOC-β-tBu-L-Asp

USBio F5344   MW 411.5 $C_{23}H_{25}NO_6$ ≥99%

Asparagine — FMOC-Asn-OtBu-OPfp

Synonyms: N^α-9-FMOC-t-L-Asp-OPfp; N^α-9-FMOC-β-tBu-L-Asp-OPfp

USBio F5336   MW 762.8 $C_{44}H_{33}N_2O_5F_5$ ≥99%

USBio F5345   MW 577.5 $C_{29}H_{24}NO_6F_5$ ≥99%

Asparagine — FMOC-Asn-Tmob

Bachem B-1705.0001   MW 534.57 $C_{29}H_{30}N_2O_8$ Store at -15°C

Asparagine — FMOC-Asn-Trt

Synonyms: N^α-FMOC-N^β-Trt-L-Asn

American Peptide FASN115   MW 597.8

Bachem B-1785.0001   MW 596.68 $C_{38}H_{32}N_2O_5$ Store at -15°C

Peptides International FLN-1754-PI   MW 596.68   >98% (HPLC); white crystalline powder

USBio F5335   MW 596.7 $C_{38}H_{32}N_2O_5$ ≥99%

Asparagine — FMOC-Asn-Trt-® (200-400 mesh)

Bachem D-1695.0001   Store at RT

Asparagine — FMOC-Asn-Trt-OPfp

Bachem B-2200.0001   MW 762.73 $C_{44}H_{31}F_5N_2O_5$ Store at -15°C

Asparagine — FMOC-Asn-Trt-SASRIN™-® (200-400 mesh)

Bachem D-1680.0001   Store at 2-8°C

Asparagine — FMOC-Asp-tBu

American Peptide FASP105   MW 411.5

Asparagine — FMOC-D-Asn

Synonyms: N^α-FMOC-D-Asn

Bachem B-1050.0001   MW 354.36 $C_{19}H_{18}N_2O_5$ Store at RT

Neosystem FA00301   MW 354.4

Peptides International FDN-1804-PI   MW 354.36   >98% (HPLC); white powder

USBio F5329   MW 354.4 $C_{19}H_{18}N_2O_5$ ≥99%

Asparagine — FMOC-D-Asn-Mbh

USBio F5332   MW 580.6 $C_{34}H_{32}N_2O_7$ ≥99%

Asparagine — FMOC-D-Asn-Mtt

Bachem B-2055.0001   MW 610.71 $C_{39}H_{34}N_2O_5$ Store at -15°C

Asparagine — FMOC-D-Asn-OPfp

Bachem B-1795.0001   MW 520.42 $C_{25}H_{17}N_2O_5F_5$ Store at -15°C

Asparagine — FMOC-D-Asn-Trt

Synonyms: N^α-FMOC-N^β-Trt-D-Asn

Bachem B-2010.0001   MW 596.68 $C_{38}H_{32}N_2O_5$ Store at -15°C

Peptides International FDN-1805-PI   MW 596.68   >98% (HPLC); white crystalline powder

USBio F5334   MW 596.7 $C_{38}H_{32}N_2O_5$ ≥99%

Asparagine — FMOC-D-Asn-Trt-® (200-400 mesh)

Bachem D-1950.0001   Store at 2-8°C

Asparagine — FMOC-D-Asn-Trt-SASRIN™-® (200-400 mesh)

Bachem D-1780.0001   Store at 2-8°C

Asparagine — FMOC-L-Asn

Fluka 47617   MW 354.36 $C_{19}H_{18}N_2O_5$ ~98% (titration); mp: 190°C

Neosystem FA00302   MW 354.4

Asparagine — FMOC-L-Asn-Mbh

USBio F5333   MW 580.6 $C_{34}H_{32}N_2O_7$ ≥99%

Asparagine — FMOC-L-Asn-OPfp

Fluka 47443   MW 520.41 $C_{25}H_{15}F_5N_2O_5$ ≥98% (titration)

Asparagine — FMOC-L-Asn-Trt

Neosystem FA00303   MW 596.7

Asparagine — FMOC-L-Asn-Trt-Wang Resin

American Peptide RFWAN18

Asparagine — FMOC-L-Asn-Wang Resin

American Peptide RFWAN17

Asparagine — FMOC-L-Asp-tBu-Wang Resin

American Peptide RFWAN23   Preattached Wang resins are usful for synthesis of peptides with C-terminal carboxyl groups; peptides can be cleaved from the resin with 90% TFA in the presence of scavengers

Asparagine — FMOC-L-Isoalanine

Fluka 47315   MW 354.36 $C_{19}H_{18}N_2O_5$ ≥98% (HPLC)

Asparagine — L-Asn

Synonyms: (*S*)-2-Aminosuccinic Acid 4-Amide; L-Asp 4-Amide

Amersham US11590   Non-essential AA

Amersham US11595   Non-essential AA

Fluka 11150   MW 132.12 $C_4H_8N_2O_3$ ≥99.0%; ≤0.1% residue on ignition; ≤0.005% Cl, K, Na, Mg, SO$_4$; ≤0.0005% Cd, Co, Cr, Cu, Fe, Mn, Ni, Pb, Zn; ≤0.02% Ca; ≤0.00001% As; ≤0.1% foreign AA

ICN 100794 $C_4H_8N_2O_3$

ICN 194630 $C_4H_8N_2O_3$

Sigma A 0884 $C_4H_8N_2O_3$

Sigma A 1143   Matrix: cross-linked 4% beaded agarose; activation: cyanogen bromide; attachment: amino; spacer: 1 atom; ligand immobilized: 5-10 μmoles/mL; form: suspension in 2.0 M NaCl containing 0.02% thimerosal

Asparagine — L-Asn (³H(G))

ARC ART-500   MW 132.12 $NH_2COCH_2CH(NH_2)COOH$ 100-500 mCi/mmol; 3.7-18.5 GBq/mmol; in EtOH:H$_2$O (2:98) | Radiochemical

Sigma A 8090   200-800 mCi/mmol; radiochemical purity ≥95% (HPLC); aqueous solution containing 2% EtOH in Combi-vial | Radiochemical

Asparagine — L-Asn (U-¹⁴C)

ARC ARC-1244   MW 132.1 $NH_2COCH_2CH(NH_2)COOH$ 150-200 mCi/mmol; 5.55-7.4 GBq/mmol; in EtOH | Radiochemical

Sigma A 8215   120-240 mCi/mmol; radiochemical purity ≥95% (HPLC); aqueous solution containing 2% EtOH in Combi-vial | Radiochemical

Asparagine — L-Asn Anhydrous

Synonyms: (*S*)-2-Aminosuccinic Acid 4-Amide; L-Asp 4-Amide

Fluka 11149   MW 132.12 $C_4H_8N_2O_3$ ≥99.5%; ≤0.05% residue on ignition; ≤0.005% K, Ca, Mg, Na, Cl, SO$_4$; ≤0.0005% Al, Ba, Bi, Cd, Co, Cr, Cu, Fe, Li, Mn, Mo, Ni, Pb, Sr, Zn; ≤0.00001% As; ≤1% foreign AA | Lange, PW, *Anal Chem Symp Ser 5*, 187, 1980; Yagi, T et al, *Electrophoresis '83*, H. Hirai (ed), de Gruyter, Berlin, 503, 1984; *CRC Handbook of Microbiology*, Laskin, AI & Lachevalier, HA, eds, CRC Press, Cleveland, Ohio, 4: 1, 1974

Sigma A 4159   FW 132.1 $C_4H_8N_2O_3$ ≥98% (TLC) | Cell culture tested; insect cell culture tested

USBio A3820   MW 132.1 $C_4H_8N_2O_3$ ≥98.5%; white, crystalline powder; solubility (10%, 6 N HCl): clear, colorless & complete; identity (IR): conforms to reference; specific rotation (C=8, 6 N HCl): +33.5° to +36.5°; loss on drying: ≤0.5%; residue on ignition: ≤0.1%; chloride: ≤0.02%; sulfate: ≤0.02%; heavy metals (Pb): ≤0.0005%; endotoxin: ≤2 EU/mg; cell culture analysis: no inhibition of growth; murine myeloma SP2/0-Ag14

Asparagine — L-Asn Hydrate

Synonyms: (*S*)-(2,4-Diamino)-4-Oxo-Butanoic Acid Monohydrate; L-2-Amino-3-Carbamoyl-Propionic Acid Monohydrate

Fluka 11159   MW 150.14 $C_4H_8N_2O_3 \cdot H_2O$ ≥99.5%; ≤0.05% residue on ignition; ≤0.005% K, Na, Cl, SO$_4$; ≤0.0005% Ba, Bi, Cd, Co, Cr, Cu, Fe, Li, Mg, Mn, Mo, Ni, Pb, Sr, Zn; ≤0.00001% As; ≤0.001% Al; ≤0.1% foreign AA

Fluka 11160   MW 150.14 $C_4H_8N_2O_3 \cdot H_2O$ ≥99.0%; ≤0.05% residue on ignition; ≤0.005% K, Na, Cl, SO$_4$; ≤0.0005% Cd, Co, Cr, Cu, Fe, Mg, Mn, Ni, Pb, Zn; ≤0.001% Ca; ≤0.3% foreign AA

ICN 100795 $C_4H_8N_2O_3 \cdot H_2O$

ICN 194631 $C_4H_8N_2O_3 \cdot H_2O$

Neosystem AA00302   MW 150.1

Peptides International ALN-2703   MW 150.14   >99.9% optical purity by RP-HPLC; white crystalline powder

Rexim   MW 150.1 $C_4H_{10}N_2O_4$ White crystals or crystalline powder

Sigma A 4284   FW 150 $C_4H_8N_2O_3 \cdot H_2O$ Cell culture tested

Sigma A 8381 $C_4H_8N_2O_3 \cdot H_2O$

Sigma A 8824 $C_4H_8N_2O_3 \cdot H_2O$ Anhydrous

USBio A3825   MW 150.1 $C_4H_8N_2O_3 \cdot H_2O$ ≥98.5%; white, crystalline powder; specific rotation (C=8, 6 N HCl): +33.5° to +36.5°; heavy metals (Pb): ≤0.001%; endotoxin: ≤2 EU/mg; cell culture tested: murine myeloma SP2/0-Ag14

Asparagine — L-Asn β-Naphthylamide

Sigma A 3788 $C_{14}H_{15}N_3O_2$

Asparagine — L-Asn-OtBu Hydrochloride

Sigma A 0502 $C_8H_{16}N_2O_3 \cdot HCl$

Asparagine — L-Asparaginamide Hydrochloride

Synonyms: L-Asp Diamide

Sigma A 6171 $C_4H_9N_3O_2 \cdot HCl$

Asparagine — N-2,4-Dnp-L-Asn

Sigma D 8254 $C_{10}H_{10}N_4O_7$

Asparagine — N-CBZ-L-Asn

ICN 100208 $C_{12}H_{14}N_2O_5$

Asparagine — N-CBZ-L-Asn-(pONp)

ICN 101253 $C_{18}H_{17}N_3O_7$

Asparagine — N-FMOC-L-Asn (DOD)

Sigma F 9263 $C_{34}H_{32}N_2O_7$ 4,4'-Dimethoxybenzhydryl derivative

Asparagine — N-O-Nps-L-Asn

ICN 102524 $C_{10}H_{11}N_3O_5S$

Asparagine — N-t-BOC-L-Asn Pam Resin Ester

Sigma B 4643   For peptide synthesis by the Merrifield method; Mitchell, AR et al, *J Org Chem*, 43: 2845, 1978

Asparagine — Nᵅ-Ac-D-Asn

Sigma A 5250 $C_6H_{10}N_2O_4$

Asparagine — Nᵅ-Ac-L-Asn

Synonyms: Asparaginase II Substrate

Fluka 00915   MW 174.16 $C_6H_{10}N_2O_4$ ≥98% (titration); ≤0.5% H$_2$O; ≤0.05% residue on ignition; mp: 168-170°C | Substrate for hydrolysis & hydroxylaminolysis reactions of asparaginase II of *Saccharomyces cerevisiae*; Dunlop, PC et al, *JBC*, 255: 1542, 1980

Asparagine — Nᵅ-BOC-Nᵝ-9-Xanthenyl-L-Asn

Fluka 15430   MW 412.45 $C_{22}H_{24}N_2O_6$ ≥98% (TLC); mp: 176°C

Asparagine — N$^\alpha$-BOC-N$^\prime$-Trt-L-Asn

Fluka 15562   MW 474.56 $C_{28}H_{30}N_2O_5$ ≥99% (TLC); mp: 205°C

Asparagine — N$^\alpha$-CBZ-DL-Asn

Sigma C 5251 $C_{12}H_{14}N_2O_5$

Asparagine — N$^\alpha$-FMOC-D-Asn

ICN 158165 $C_{19}H_{18}N_2O_5$

Asparagine — N$^\alpha$-FMOC-L-Asn

ICN 151133 $C_{19}H_{18}N_2O_5$

Asparagine — N$^\alpha$-FMOC-L-Asn (DOD) Resin Ester

Sigma F 7636

Asparagine — N$^\alpha$-FMOC-L-Asn-(DOD) SASRIN Resin Ester

Sigma F 0514

Asparagine — N$^\alpha$-FMOC-L-Asn-pONp

ICN 158166 $C_{25}H_{21}N_3O_7$

Asparagine — N$^\alpha$-FMOC-N$^\prime$-(2,4,6-Trimethoxybenzyl)-L-Asn

Fluka 47537   MW 534.56 $C_{29}H_{30}N_2O_8$ ≥98% (TLC)

Asparagine — N$^\alpha$-FMOC-N$^\prime$-(4,4'-Dimethoxybenzhydryl)-L-Asn

Fluka 47521   MW 580.64 $C_{34}H_{32}N_2O_7$ ~98% (HPLC)

Asparagine — N$^\alpha$-FMOC-N$^\prime$-(4,4'-Dimethoxybenzhydryl)-L-Asn 4-Benzyloxybenzyl Ester Polymer Bound

Fluka 47656   0.4-0.6 mmol/g resin; carrier: polystyrene, crosslinked with 1% DVB; 200-400 mesh particle size

Asparagine — N$^\alpha$-FMOC-N$^\prime$-Trt-L-Asn

Fluka 47672   MW 596.68 $C_{38}H_{32}N_2O_5$ ≥97% (HPLC); mp: 210-214°C

Sigma F 1393 $C_{38}H_{32}N_2O_5$

Asparagine — N$^\alpha$-FMOC-N$^\prime$-Trt-L-Asn 4-Benzyloxybenzyl Ester Polymer Bound

Fluka 47388   0.4-0.8 mmol/g resin; carrier: polystyrene, crosslinked with 1% DVB; 100-200 mesh particle size

Asparagine — N$^\alpha$-t-BOC-L-Asn

ICN 150482 $C_9H_{16}N_2O_5$

Sigma B 5376 $C_9H_{16}N_2O_5$

Asparagine — N$^\alpha$-t-BOC-L-Asn-(pONp)

Sigma B 5501 $C_{15}H_{19}N_3O_7$

Asparagine — N$^\alpha$-t-BOC-N$^\prime$-Xan-L-Asn

Sigma B 2011 $C_{22}H_{24}N_2O_6$

Asparagine — Pth-Asn

Sigma P 9376 $C_{11}H_{11}N_3O_2S$

Asparagine — TentaGel S PHB-Asn-TrtFMOC

Fluka 86378   0.20 mmol protected AA/g; ~90 μm particle size

Asparagine — TentaGel S Trt-Asn-Trt-FMOC

Fluka 86412   0.20 mmol protected AA/g; ~90 μm particle size

Asparagine — Z-Asn

Synonyms: Benzyloxycarbonyl-L-Asn

Bachem C-1305.0025   MW 266.25 $C_{12}H_{14}N_2O_5$ Store at RT

Peptides International ZLN-2010-PI   MW 266.25   >98% (HPLC); white to off-white powder

Asparagine — Z-Asn-Mtt

Bachem C-3630.0001   MW 522.6 $C_{32}H_{30}N_2O_5$ Store at 2-8°C

Asparagine — Z-Asn-OMe

Bachem C-4050.0001   MW 280.28 $C_{13}H_{16}N_2O_5$ Store at -15°C

Asparagine — Z-Asn-ONp

Bachem C-1320.0005   MW 387.35 $C_{18}H_{17}N_3O_7$ Store at -15°C

Asparagine — Z-Asn-OtBu

Bachem C-1315.0001   MW 322.36 $C_{16}H_{22}N_2O_5$ Store at -15°C

Asparagine — Z-Asn-Trt

Bachem C-3370.0001   MW 508.57 $C_{31}H_{28}N_2O_5$ Store at -15°C

Asparagine — Z-D-Asn

Bachem C-3550.0001   MW 266.25 $C_{12}H_{14}N_2O_5$ Store at RT

Fluka 95941   MW 266.25 $C_{12}H_{14}N_2O_5$ ≥98.0% (TLC)

Asparagine — Z-D-Asn-Mtt

Bachem C-3620.0001   MW 522.6 $C_{32}H_{30}N_2O_5$ Store at 2-8°C

Asparagine — Z-D-Asn-Trt

Bachem C-3555.0001   MW 508.57 $C_{31}H_{28}N_2O_5$ Store at -15°C

Asparagine — Z-DL-Asn

Bachem C-1310.0025   MW 266.25 $C_{12}H_{14}N_2O_5$ Store at RT

Fluka 95945   MW 266.25 $C_{12}H_{14}N_2O_5$ ≥98.0% (TLC)

Asparagine — Z-L-Asn

Fluka 95940   MW 266.25 $C_{12}H_{14}N_2O_5$ ≥99.0% (titration); mp: 164-166°C

Neosystem ZA00301   MW 266.3

Asparagine — Z-L-Asn-(4-ONp)

Fluka 95955   MW 387.35 $C_{18}H_{17}N_3O_7$ ≥98.0% (TLC)

Asparagine — Z-L-Asn-OtBu

Fluka 95950   MW 322.36 $C_{16}H_{22}N_2O_5$ ≥98.0% (TLC)

Aspartic Acid — 1-Asp (U-^{14}C)

Amersham CFB64

Aspartic Acid — Ac-Asp Amide

Synonyms: Ac-Ias; Ac-L-Ias

Bachem E-2700.0050   MW 174.16 $C_6H_{10}N_2O_4$ Store at -15°C

   User's Guide pp. 1-2, Glossary/Indexes pp. 963-1063

Aspartic Acid — Ac-Asp-OMe
Bachem E-1035.0250   MW 189.17 $C_7H_{11}NO_5$ Store at RT

Aspartic Acid — Ac-Asp-OtBu
Bachem E-3025.0001   MW 231.25 $C_{10}H_{17}NO_5$ Store at 2-8°C

Aspartic Acid — Ac-Asp-pNA
Bachem L-1035.0050   MW 295.25 $C_{12}H_{13}N_3O_6$ Store at -15°C

Aspartic Acid — Asp
Bachem E-1395.0025   MW 133.1 $C_4H_7NO_4$ Store at RT

Aspartic Acid — Asp Amide
Synonyms: Isoasn; L-Isoasparagine
Bachem F-1260.0250   MW 132.12 $C_4H_8N_2O_3$ Store at RT

Aspartic Acid — Asp-4MβNA
Bachem J-1045.0250   MW 288.3 $C_{15}H_{16}N_2O_4$ Store at -15°C

Aspartic Acid — Asp-AMC
Bachem I-1060.0050   MW 290.28 $C_{14}H_{14}N_2O_5$ Store at -15°C
Bachem I-1775.0050   MW 290.28 $C_{14}H_{14}N_2O_5$ Store at -15°C

Aspartic Acid — Asp-OBzl
Synonyms: L-Asp-(β-OBzl)
Bachem E-1400.0001   MW 223.23 $C_{11}H_{13}NO_4$ Store at RT
Bachem E-1405.0005   MW 223.23 $C_{11}H_{13}NO_4$ Store at RT
Peptides International XBD-2093-PI   MW 223.23   >98% (TLC); white powder

Aspartic Acid — Asp-OBzl-AMC Hydrochloride
Bachem I-1780.0250   MW 416.9 $C_{21}H_{20}N_2O_5 \cdot HCl$ Store at -15°C

Aspartic Acid — Asp-OBzl-OBzl *p*-Tosyl
Bachem E-1430.0010   MW 485.56 $C_{18}H_{19}NO_4 \cdot C_7H_8O_3S$ Store at RT

Aspartic Acid — Asp-OBzl-OBzl-Tosyl
Synonyms: L-Asp (α,β-Dibenzyl Ester)-Tosyl
Peptides International EBD-2045   MW 485.55   >98% (TLC); white powder

Aspartic Acid — Asp-OMe
Bachem E-2880.0001   MW 147.13 $C_5H_9NO_4$ Store at 2-8°C

Aspartic Acid — Asp-OMe Hydrochloride
Bachem E-3115.0001   MW 183.59 $C_5H_9NO_4 \cdot HCl$

Aspartic Acid — Asp-OMe-OMe Hydrochloride
Bachem E-1440.0005   MW 197.62 $C_6H_{11}NO_4 \cdot HCl$ Store at 2-8°C

Aspartic Acid — Asp-OtBu
Bachem E-1410.0001   MW 189.21 $C_8H_{15}NO_4$ Store at 2-8°C
Bachem E-2935.0001   MW 189.21 $C_8H_{15}NO_4$ Store at -15°C

Aspartic Acid — Asp-OtBu Amide Hydrochloride
Synonyms: IsoAsn-OtBu; L-IsoAsn-OtBu
Bachem E-1415.0001   MW 224.69 $C_8H_{16}N_2O_3 \cdot HCl$ Store at 2-8°C

Aspartic Acid — Asp-OtBu-OAll Hydrochloride
Bachem E-3355.0001   MW 265.74 $C_{11}H_{19}NO_4 \cdot HCl$ Store at -15°C

Aspartic Acid — Asp-OtBu-OMe Hydrochloride
Bachem E-1420.0001   MW 239.7 $C_9H_{17}NO_4 \cdot HCl$ Store at 2-8°C

Aspartic Acid — Asp-OtBu-OtBu Hydrochloride
Bachem E-1435.0001   MW 281.78 $C_{12}H_{23}NO_4 \cdot HCl$ Store at 2-8°C

Aspartic Acid — Asp-pNA Hydrochloride
Bachem L-1525.0250   MW 289.7 $C_{10}H_{11}N_3O_5 \cdot HCl$ Store at -15°C
Bachem L-1690.0250   MW 289.7 $C_{10}H_{11}N_3O_5 \cdot HCl$ Store at -15°C

Aspartic Acid — Asp-βNA
Bachem K-1095.0250   MW 258.3 $C_{14}H_{14}N_2O_3$ Store at -15°C

Aspartic Acid — BOC-Asp
Bachem A-3615.0005   MW 233.22 $C_9H_{15}NO_6$ Store at RT
USBio B2279   MW 233.2 $C_9H_{15}NO_6$ ≥99%

Aspartic Acid — BOC-Asp Amide
Synonyms: BOC-Ias; BOC-L-Ias
Bachem A-1810.0001   MW 232.24 $C_9H_{16}N_2O_5$ Store at RT

Aspartic Acid — BOC-Asp-(O-1-Ada)
USBio B2274   MW 367.4 $C_{19}H_{29}NO_6$ ≥99%

Aspartic Acid — BOC-Asp-(O-2-Ada)
USBio B2275   MW 367.4 $C_{19}H_{29}NO_6$ ≥99%

Aspartic Acid — BOC-Asp-(PAM-®)-OFm (200-400 mesh)
Bachem D-1670.0001   Store at 2-8°C

Aspartic Acid — BOC-Asp-FMK
Synonyms: Pan-Caspase Inhibitor
Kamiya   MW 263 $C_{11}H_{18}FNO_5$

Aspartic Acid — BOC-Asp-OBzl
Synonyms: BOC-L-Asp-(β-OBzl)
Bachem A-1240.0001   MW 323.35 $C_{16}H_{21}NO_6$ Store at RT
Bachem A-1245.0005   MW 323.35 $C_{16}H_{21}NO_6$ Store at RT
Peptides International BLD-2059   MW 323.35   >98% (HPLC); white crystalline powder
USBio B2277   MW 323.3 $C_{16}H_{21}NO_6$ ≥99%
USBio B2281   MW 323.3 $C_{16}H_{21}NO_6$ ≥99%

Aspartic Acid — BOC-Asp-OBzl-CMK
Bachem N-1430.0250   MW 355.82 $C_{17}H_{22}ClNO_5$ Store at -15°C

Aspartic Acid — BOC-Asp-OBzl-ONp

Bachem A-1260.0001   MW 444.44 $C_{22}H_{24}N_2O_8$ Store at -15°C

Aspartic Acid — BOC-Asp-OBzl-OSu

Bachem A-1255.0001   MW 420.42 $C_{20}H_{24}N_2O_8$ Store at -15°C

Aspartic Acid — BOC-Asp-OcHex

Synonyms: BOC-L-Asp-(β-OcHex)

Bachem A-1285.0001   MW 315.37 $C_{15}H_{25}NO_6$ Store at RT

Peptides International BLD-2132   MW 315.37   >98% (HPLC); white crystalline powder

USBio B2282   MW 315.4 $C_{15}H_{25}NO_6$ ≥99%

Aspartic Acid — BOC-Asp-OcHex Dicyclohexylamine

Bachem A-2755.0001   MW 496.69 $C_{15}H_{25}NO_6 \cdot C_{12}H_{23}N$ Store at RT

Aspartic Acid — BOC-Asp-OcHex-® (200-400 mesh)

Bachem D-1500.0005   Store at RT

Aspartic Acid — BOC-Asp-OcHex-OSu

Bachem A-2790.0001   MW 412.44 $C_{19}H_{28}N_2O_8$ Store at -15°C

Aspartic Acid — BOC-Asp-OcHex-PAM-® (200-400 mesh)

Bachem D-1270.0001   Store at 2-8°C

Aspartic Acid — BOC-Asp-OCHX-Dicyclohexylamine

USBio B2278   MW 496.7 $C_{15}H_{25}NO_6 \cdot C_{12}H_{23}N$ ≥98%

Aspartic Acid — BOC-Asp-OcPent

Bachem A-1290.0001   MW 301.34 $C_{14}H_{23}NO_6$ Store at RT

Aspartic Acid — BOC-Asp-OFm

Bachem A-2930.0005   MW 411.46 $C_{23}H_{25}NO_6$ Store at -15°C

Bachem A-2945.0005   MW 411.46 $C_{23}H_{25}NO_6$ Store at 2-8°C

USBio B2283   MW 411.5 $C_{23}H_{25}NO_6$ ≥98%

Aspartic Acid — BOC-Asp-OFm Alpha Ester

Synonyms: BOC-L-Asp-OFm

Peptides International BLD-5571-PI   MW 411.46   >98% (HPLC); white crystalline powder

Aspartic Acid — BOC-Asp-OFm Beta Ester

Synonyms: BOC-L-Asp-OFm

Peptides International BLD-5662-PI   MW 411.46   >98% (HPLC); white crystalline powder

Aspartic Acid — BOC-Asp-OMe-CH₂F

Synonyms: Caspase Inhibitor III; Interleukin Iβ-Converting Enzyme-Like Inhibitor; Interleukin Converting Enzyme Inhibitor

Calbiochem 218745   MW 263.3 $C_{11}H_{18}FNO_5$ Single spot purity (TLC); semi-solid; soluble in DMSO | Cell-permeable, irreversible, broad-spectrum caspase inhibitor; D'Mello, SR et al, *J Neurochem*, 70: 1809, 1998

ICN 193606   Supplied as an oil | Useful in apoptosis research

Aspartic Acid — BOC-Asp-OSu-OBzl

Bachem A-3350.0001   MW 420.43 $C_{20}H_{24}N_2O_8$ Store at -15°C

Aspartic Acid — BOC-Asp-OtBu

Bachem A-2950.0001   MW 289.33 $C_{13}H_{23}NO_6$ Store at 2-8°C

Bachem A-3120.0001   MW 289.33 $C_{13}H_{23}NO_6$ Store at -15°C

Aspartic Acid — BOC-Asp-OtBu Dicyclohexylamine

Bachem A-1265.0001   MW 470.65 $C_{13}H_{23}NO_6 \cdot C_{12}H_{23}N$ Store at RT

Bachem A-2605.0001   MW 470.65 $C_{13}H_{23}NO_6 \cdot C_{12}H_{23}N$ Store at RT

Aspartic Acid — BOC-Asp-OtBu-Dicyclohexylamine

USBio B2276   MW 470.6 $C_{13}H_{23}NO_6 \cdot C_{12}H_{23}N$ ≥99%

Aspartic Acid — BOC-Asp-OtBu-ONp

Bachem A-1280.0001   MW 410.42 $C_{19}H_{26}N_2O_8$ Store at -15°C

Aspartic Acid — BOC-Asp-OtBu-OSu

Bachem A-1275.0001   MW 386.4 $C_{17}H_{26}N_2O_8$ Store at -15°C

Aspartic Acid — BOC-D-Asp

Bachem A-4190.0001   MW 233.22 $C_9H_{15}NO_6$ Store at RT

Aspartic Acid — BOC-D-Asp Amide

Synonyms: BOC-D-Ias; BOC-D-Ias

Bachem A-3630.0001   MW 232.24 $C_9H_{16}N_2O_5$ Store at RT

Aspartic Acid — BOC-D-Asp-(4-OBzl)

Fluka 15067   MW 323.35 $C_{16}H_{21}NO_6$ ≥98% (HPLC)

Aspartic Acid — BOC-D-Asp-(4-OcHex)

Fluka 15073   MW 315.37 $C_{15}H_{25}NO_6$ ≥98% (TLC)

Aspartic Acid — BOC-D-Asp-OBzl

Bachem A-1250.0001   MW 323.35 $C_{16}H_{21}NO_6$ Store at RT

Bachem A-2665.0001   MW 323.35 $C_{16}H_{21}NO_6$ Store at RT

Neosystem BA00403   MW 323.4

Peptides International BDD-2616   MW 323.35   >98% (HPLC); white crystalline powder

USBio B2280   MW 323.3 $C_{16}H_{21}NO_6$ ≥99%

Aspartic Acid — BOC-D-Asp-OcHex

Synonyms: BOC-D-Asp-(β-OcHex)

Bachem A-2625.0001   MW 315.37 $C_{15}H_{25}NO_6$ Store at RT

Neosystem BA00405   MW 315.4

Peptides International BDD-2617   MW 315.37   >98% (HPLC); white crystalline powder

Aspartic Acid — BOC-D-Asp-OFm

Bachem A-3150.0001   MW 411.46 $C_{23}H_{25}NO_6$ Store at 2-8°C

Bachem A-4195.0001   MW 411.46 $C_{23}H_{25}NO_6$ Store at 2-8°C

Peptides International BDE-2618-PI   MW 411.46   >98% (HPLC); white crystalline powder

Aspartic Acid — BOC-D-Asp-OtBu

Bachem A-3505.0001   MW 289.33 $C_{13}H_{23}NO_6$ Store at 2-8°C

Aspartic Acid — BOC-D-Asp-OtBu Dicyclohexylamine

Bachem A-1270.0001   MW 470.65 $C_{13}H_{23}NO_6 \cdot C_{12}H_{23}N$
Store at RT

Aspartic Acid — BOC-L-Asp

Fluka 15344   MW 233.22 $C_9H_{15}NO_6$ ≥99%

Neosystem BA00401   MW 233.2

Aspartic Acid — BOC-L-Asp 4-Bzl 1-(4-Nitrophenyl) Ester

Fluka 15071   MW 444.44 $C_{22}H_{24}N_2O_8$ ≥98% (TLC)

Aspartic Acid — BOC-L-Asp 4-Bzl 1-Hydroxysuccinimide Ester

Fluka 15069   MW 420.42 $C_{20}H_{24}N_2O_8$ ≥98%; mp: 98-102°C

Aspartic Acid — BOC-L-Asp 4-tBu 1-Hydroxysuccinimide Ester

Fluka 15072   MW 386.4 $C_{17}H_{26}N_2O_8$ ≥98%; mp: 98-102°C

Aspartic Acid — BOC-L-Asp-(1-OBzl)

Fluka 15066   MW 323.35 $C_{16}H_{21}NO_6$ ≥98% (TLC)

Aspartic Acid — BOC-L-Asp-(4-OBzl)

Fluka 15386   MW 323.35 $C_{16}H_{21}NO_6$ ~99% (HPLC); mp: 98-102°C

Aspartic Acid — BOC-L-Asp-(4-OBzl) N-Carboxy Anhydride

Fluka 15414   MW 349.34 $C_{17}H_{19}NO_7$ mp: 98-105°C

Aspartic Acid — BOC-L-Asp-(4-OcHex)

Fluka 15398   MW 315.37 $C_{15}H_{25}NO_6$ ≥98% (titration); mp: 93-95°C

Aspartic Acid — BOC-L-Asp-(4-OcPent)

Fluka 15074   MW 301.34 $C_{14}H_{23}NO_6$ ≥98% (TLC)

Aspartic Acid — BOC-L-Asp-(4-OMe)

Fluka 15076   MW 247.25 $C_{10}H_{17}NO_6$ ≥98% (TLC)

Aspartic Acid — BOC-L-Asp-(4-OtBu)

Fluka 15429   MW 289.33 $C_{13}H_{23}NO_6$ ≥99% (TLC); mp: 64-67°C

Aspartic Acid — BOC-L-Asp-OAll

Neosystem BA00402   MW 273.2

Aspartic Acid — BOC-L-Asp-OBzl

Neosystem BA00404   MW 323.4

Neosystem BA00408   MW 323.4

Aspartic Acid — BOC-L-Asp-OcHex

Neosystem BA00406   MW 315.4

Aspartic Acid — BOC-L-Asp-OcHex-MBHA Resin

Neosystem RN00406

Aspartic Acid — BOC-L-Asp-OcHex-O-CH₂-ϕ-CH₂-COOH

Neosystem LP00404   MW 463.6

Aspartic Acid — BOC-L-Asp-OcHex-PAM Resin

Neosystem RP00404

Aspartic Acid — BOC-L-Asp-OFm

Neosystem BA00407   MW 411.5

Aspartic Acid — Bsmoc-Asp-OtBu

Synonyms: Benzo-β-Thiophenesulfone-2-Methoxycarbonyl-L-Asp-(β-OtBu)

Peptides International BLD-5616-PI   MW 411.43   >96% (HPLC); white crystalline powder | LA Carpino, et al, *JACS*, 119:9915, 1997

Aspartic Acid — Bz-Asp-OMe

Bachem E-1465.0250   MW 251.24 $C_{12}H_{13}NO_5$ Store at RT

Aspartic Acid — Caged Asp Sodium Salt

Synonyms: N-(1-(2-Nitrophenyl)-Ethyloxycarbonyl)-Asp

Calbiochem 189110   MW 348.2 $C_{13}H_{13}N_2O_8Na$ >98% (HPLC); solid; soluble in H_2O | Photolabile derivative of aspartic acid that is biologically inactive until photolyzed; sold under license of European Patent 0233403 issued to the Medical Research Council

Aspartic Acid — Carbamoyl-Asp Magnesium Salt/Carbamoyl-Asp Dipotassium Salt (1:1)

Bachem M-2240.0001   MW 450.73 $C_{10}H_{12}K_2MgN_4O_{10}$ Store at 2-8°C

Aspartic Acid — Carbobenzoxy-Asp-CH₂-((2,6-Dichlorobenzoyl)Oxy)-Methane

Synonyms: Caspase Inhibitor

Peptides International ICE-3174-v   Synthetic   MW 454.26 $C_{20}H_{17}NO_7Cl_2$ >99% (HPLC); lyophilized amorphous powder | Inhibitor; Dolle, RE et al, *J Med Chem*, 37:563, 1994; Mashima, T et al, *BBRC*, 209:907, 1995

Aspartic Acid — Carbobenzoxy-Glu-Lys-Biotinyl-Asp-CH₂-((2,6-Dimethylbenzoyl)Oxy)-Methane

Synonyms: Caspase Affinity Ligand

Peptides International ICA-3189-v   Synthetic   MW 897.02 $C_{43}H_{56}N_6O_{13}S$ >99% (HPLC); lyophilized amorphous powder | Inhibitor; Martins, LM et al, *JBC*, 272:7421, 1997; Martins, LM et al, *Blood*, 90:4283, 1997

Aspartic Acid — CBZ-Asp

USBio C2098-45   MW 267.2 $C_{12}H_{13}NO_6$ ≥99%

Aspartic Acid — CBZ-Asp-OBzl

USBio C2098-47   MW 357.4 $C_{13}H_{13}NO_6$ ≥99%

USBio C2098-48   MW 357.4 $C_{13}H_{13}NO_6$ ≥99%

Aspartic Acid — CBZ-Asp-OBzl-ONp

USBio C2098-49   MW 478.5 $C_{25}H_{22}N_2O_8$ ≥99%

Aspartic Acid — CBZ-Asp-OtBu-Dicyclohexylamine

USBio C2098-50   MW 504.7 $C_{16}H_{21}NO_6 \cdot C_{12}H_{23}N$ ≥99%

Aspartic Acid — CBZ-D-Asp

USBio C2098-46   MW 267.2 $C_{12}H_{13}NO_6$ ≥99%

Aspartic Acid — D-(+)-*threo*-β-HydroxyAla

ICN 193662 $C_4H_7NO_5$ Neurotoxic inhibitor of L-Glu & L-Asp uptake

Aspartic Acid — Dansyl-DL-Asp

Sigma D 8381 $C_{16}H_{18}N_2O_6S \cdot (C_6H_{13}N)_2$

Aspartic Acid — Dansyl-L-Asp

ICN 100039 $C_{16}H_{18}N_2O_6S \cdot (C_6H_{13}N)_2$
Sigma D 0500 $C_{16}H_{18}N_2O_6S \cdot (C_6H_{13}N)_2$

Aspartic Acid — D-Asp

Synonyms: (R)-2-Aminosuccinic Acid; D-Asp

Alexis 101-011   MW 133.1 $C_4H_7NO_4$
Bachem F-1255.0025   MW 133.1 $C_4H_7NO_4$ Store at RT
Fluka 11200   MW 133.11 $C_4H_7NO_4$ ≥99%
ICN 150401 $C_4H_7NO_4$
Neosystem AA00415   MW 133.1
Peptides International ADD-2814   MW 133.10   >99.9% optical purity by RP-HPLC; white crystalline powder
Rexim   MW 133.1 $C_4H_7NO_4$
Sigma A 8881 $C_4H_7NO_4$ Inactive isomer of aspartic acid
USBio A3880-10   MW 133.1 $C_4H_7NO_4$ ≥99%

Aspartic Acid — D-Asp (2,3-³H)

Synonyms: D-Asp Receptor Ligand, (^{3}H)

Amersham TRK606   0.37-1.1 TBq/mmol, 10-30 Ci/mmol; 37 MBq/mL, 1 mCi/mL | Excitatory AA receptor ligand; non-selective; endogenous
ARC ART-212   MW 133.1 $HOOCCH_2CH(NH_2)COOH$ 10-25 Ci/mmol; 370-925 GBq/mmol; in 2% aqueous EtOH | Radiochemical

Aspartic Acid — D-Asp (4-¹⁴C)

ARC ARC-213   MW 133.1 $HOOCCH_2CH(NH_2)COOH$ 50-60 mCi/mmol; 1.85-2.22 GBq/mmol; in 0.01 N HCl | Radiochemical

Aspartic Acid — D-Asp Amide

Synonyms: D-Isoasn; D-Isoasparagine

Bachem F-3120.0001   MW 132.12 $C_4H_8N_2O_3$ Store at RT

Aspartic Acid — D-Asp β-Hydroxamate

Sigma A 9009 $C_4H_8N_2O_4$

Aspartic Acid — D-Asp-(α-OBzl)

Sigma A 1166 $C_{11}H_{13}NO_4$

Aspartic Acid — D-Asp-OBzl

Bachem F-1265.0001   MW 223.23 $C_{11}H_{13}NO_4$ Store at RT
Bachem F-1270.0005   MW 223.23 $C_{11}H_{13}NO_4$ Store at RT
Neosystem AA00401   MW 223.2

Aspartic Acid — D-Asp-OBzl-OBzl *p*-Tosyl

Bachem F-2605.0005   MW 485.56 $C_{25}H_{27}NO_7S$ Store at RT

Aspartic Acid — D-Asp-OMe Hydrochloride

Bachem F-3135.0001   MW 183.59 $C_5H_9NO_4 \cdot HCl$ Store at 2-8°C

Aspartic Acid — D-Asp-OMe₂ Hydrochloride

Sigma A 1421 $C_6H_{11}NO_4 \cdot HCl$

Aspartic Acid — D-Asp-OMe-OMe Hydrochloride

Bachem F-3320.0001   MW 197.62 $C_6H_{11}NO_4 \cdot HCl$ Store at 2-8°C

Aspartic Acid — D-Asp-OtBu

Bachem F-1275.0001   MW 189.21 $C_8H_{15}NO_4$ Store at -15°C

Aspartic Acid — D-Asp-OtBu Hydrochloride

Bachem F-3030.0001   MW 189.21 $C_8H_{15}NO_4$ Store at -15°C

Aspartic Acid — Dinitropyridyl-L-Asp

ICN 100801 $C_9H_8N_4O_8$

Aspartic Acid — DL-Asp

Synonyms: (±)-2-Aminosuccinic Acid

Fluka 11210   MW 133.11 $C_4H_7NO_4$ ≥99% (titration); ≤0.05% residue on ignition; mp: ~280°C
ICN 100803 $C_4H_7NO_4$
ICN 100806
ICN 194632 $C_4H_7NO_4$
Sigma A 2025 $C_4H_6NO_4K$ Principal neurotransmitter for fast synaptic excitation
Sigma A 2150 $C_4H_6NO_4 \cdot ½Mg$ Principal neurotransmitter for fast synaptic excitation
Sigma A 9006 $C_4H_7NO_4$ Principal neurotransmitter for fast synaptic excitation
USBio A3880-15   MW 133.1 $C_4H_7NO_4$ ≥99%

Aspartic Acid — DL-Asp (4-¹⁴C)

ARC ARC-1088   MW 133.1 $HOOCCH_2CH(NH_2)COOH$ 50-60 mCi/mmol; 1.85-2.22 GBq/mmol; in 0.01 N HCl | Radiochemical
ICN 10051 $HOOCCH_2CH(NH_2)COOH$

Aspartic Acid — DL-Asp Free Acid

Sigma A 4409   FW 133.1 $C_4H_7NO_4$ Crystalline | Cell culture tested

Aspartic Acid — DL-Asp Hemihydrate Potassium Salt

Fluka 11240   MW 180.2 $C_4H_6KNO_4 \cdot 5H_2O$ ~99% (titration)

Aspartic Acid — DL-Asp Magnesium Salt

Fluka 11270   MW 360.57 $C_8H_{12}MgN_2O_8 \cdot 4H_2O$ ≥98% (titration)

Aspartic Acid — DL-Asp β-Hydroxamate Hydrate

Sigma A 9756 $C_4H_8N_2O_4 \cdot H_2O$

Aspartic Acid — DL-Asp-(4-OMe) Hydrochloride

Fluka 11245   MW 183.6 $C_5H_9NO_4 \cdot HCl$ ≥97% (titration); mp: 196-200°C

Aspartic Acid — DL-Asp-(β-OMe) Hydrochloride

Sigma A 1457 $C_5H_9NO_4 \cdot HCl$

Aspartic Acid — DL-Asp-OMe

Bachem F-3210.0001   MW 147.13 $C_5H_9NO_4$ Store at 2-8°C

Aspartic Acid — DL-Asp-OMe Hydrochloride

Bachem F-3240.0001   MW 183.59 $C_5H_9NO_4 \cdot HCl$ Store at 2-8°C

Aspartic Acid — DL-Asp-OMe₂

Sigma A 9631

Aspartic Acid — DL-Asp-OMe-OMe Hydrochloride

Bachem F-1280.0005   MW 197.62 $C_6H_{11}NO_4 \cdot HCl$ Store at 2-8°C

Aspartic Acid — DL-*threo*-β-Me-Asp

Synonyms: 2-Amino-3-Methylsuccinic Acid

Sigma M 6126 $C_5H_9NO_4$

Aspartic Acid — DL-α-Me-Asp

Synonyms: 2-Amino-2-Methylsuccinic Acid; 3-Me-2-Aminosuccinic Acid

ICN 154149 $C_5H_9NO_4$ Non-natural AA

ICN 155472 $C_5H_9NO_4$ Mixture of erythro & *threo* diastereoisomers

Aspartic Acid — Dnp-L-Asp

ICN 105078 $C_{10}H_9N_3O_8$

Aspartic Acid — FMOC-Asp

Bachem B-2485.0005   MW 355.35 $C_{19}H_{17}NO_6$ Store at RT

Aspartic Acid — FMOC-Asp 2-Phenylisopropyl Ester

Bachem B-2475.0001   MW 473.53 $C_{28}H_{27}NO_6$ Store at -15°C

Aspartic Acid — FMOC-Asp-(O-1-Ada)

USBio F5338   MW 489.5 $C_{29}H_{31}NO_6$ ≥99%

Aspartic Acid — FMOC-Asp-(O-1-Ada)-OPfp

USBio F5340   MW 655.6 $C_{35}H_{30}NO_6F_5$ ≥99%

Aspartic Acid — FMOC-Asp-(O-2-Ada)

USBio F5339   MW 489.5 $C_{29}H_{31}NO_6$ ≥99%

Aspartic Acid — FMOC-Asp-(O-2-Ada)-OPfp

USBio F5341   MW 655.6 $C_{35}H_{30}NO_6F_5$ ≥99%

Aspartic Acid — FMOC-Asp-AMC

Neosystem FA00431   MW 512.51

Aspartic Acid — FMOC-Asp-OAll

Bachem B-2715.0001   MW 395.41 $C_{22}H_{21}NO_6$ Store at -15°C

Aspartic Acid — FMOC-Asp-OAll Alpha Ester

Synonyms: FMOC-L-Asp-(α-OAll)

Peptides International FLD-5536-PI   MW 395.42   >98% (HPLC); off-white crystalline powder

Aspartic Acid — FMOC-Asp-OBzl

Synonyms: FMOC-L-Asp-(β-OBzl)

Bachem B-1060.0001   MW 445.47 $C_{26}H_{23}NO_6$ Store at RT

Bachem B-2780.0001   MW 445.47 $C_{26}H_{23}NO_6$ Store at 2-8°C

Peptides International FLD-1712-PI   MW 445.47   >98% (HPLC); white powder

Aspartic Acid — FMOC-Asp-OBzl-OPfp

Bachem B-1760.0001   MW 611.53 $C_{32}H_{22}NO_6F_5$ Store at -15°C

USBio F5342   MW 445.5 $C_{26}H_{23}NO_6$ ≥99%

Aspartic Acid — FMOC-Asp-OcHex

Bachem B-2770.0001   MW 437.49 $C_{25}H_{27}NO_6$ Store at -15°C

Aspartic Acid — FMOC-Asp-ODmab

Bachem B-3240.0001   MW 666.77 $C_{39}H_{42}N_2O_8$ Store at -15°C

Aspartic Acid — FMOC-Asp-ODmb

Bachem B-2175.0001   MW 505.52 $C_{28}H_{27}NO_8$ Store at -15°C

Aspartic Acid — FMOC-Asp-OFm

Bachem B-2490.0001   MW 533.58 $C_{33}H_{27}NO_6$ Store at -15°C

Aspartic Acid — FMOC-Asp-OtBu

Synonyms: FMOC-L-Asp-(β-OtBu)

Bachem B-1065.0001   MW 411.46 $C_{23}H_{25}NO_6$ Store at RT

Bachem B-1735.0001   MW 411.46 $C_{23}H_{25}NO_6$ Store at RT

Peptides International FLD-1713-PI   MW 411.46   >98% (HPLC); white crystalline powder

USBio F5337   MW 411.5 $C_{23}H_{25}NO_6$ ≥99%

Aspartic Acid — FMOC-Asp-OtBu-® (200-400 mesh)

Bachem D-1055.0001   Store at 2-8°C

Aspartic Acid — FMOC-Asp-OtBu-OPfp

Bachem B-1110.0001   MW 577.51 $C_{29}H_{24}F_5NO_6$ Store at -15°C

Aspartic Acid — FMOC-Asp-OtBu-OSu

Bachem B-1075.0001   MW 508.53 $C_{27}H_{28}N_2O_8$ Store at -15°C

USBio F5346   MW 508.5 $C_{27}H_{28}N_2O_8$ ≥99%

Aspartic Acid — FMOC-Asp-OtBu-SASRIN™-® (200-400 mesh)

Bachem D-1315.0001   Store at 2-8°C

Aspartic Acid — FMOC-Asp-pNA

Neosystem FA00430   MW 475.45   Useful for the solid phase synthesis of peptides-4-nitroanilides on 2-chloro-Trt resins; Kaspari, A et al, *Int J Pept Prot Res*, 48:486, 1996

Aspartic Acid — FMOC-D-Asp

Bachem B-3110.0001   MW 355.35 $C_{19}H_{17}NO_6$ Store at RT

Aspartic Acid — FMOC-D-Asp 2-Phenylisopropyl Ester

Bachem B-2955.0001   MW 473.53 $C_{28}H_{27}NO_6$ Store at -15°C

Aspartic Acid — FMOC-D-Asp-OAll

Neosystem FA00414   MW 395.4

Neosystem FA00417   MW 395.4

Aspartic Acid — FMOC-D-Asp-OBzl

Bachem B-3135.0001   MW 445.47 $C_{26}H_{23}NO_6$ Store at RT

Aspartic Acid — FMOC-D-Asp-ODmb

Bachem B-2945.0001   MW 505.53 $C_{28}H_{27}NO_8$ Store at -15°C

Aspartic Acid — FMOC-D-Asp-OFm

Bachem B-2950.0001   MW 533.59 $C_{33}H_{27}NO_6$ Store at -15°C

Aspartic Acid — FMOC-D-Asp-OtBu

Synonyms: FMOC-D-Asp-(β-OtBu)

Bachem B-1070.0001	MW 411.46	$C_{23}H_{25}NO_6$	Store at RT
Bachem B-2305.0001	MW 411.46	$C_{23}H_{25}NO_6$	Store at RT
Neosystem FA00404	MW 411.5		

Peptides International FDD-1806-PI   MW 411.46   >98% (HPLC); white crystalline powder

USBio F5343   MW 411.5 $C_{23}H_{25}NO_6$ ≥99%

Aspartic Acid — FMOC-D-Asp-OtBu-OPfp

Bachem B-1845.0001   MW 577.5 $C_{29}H_{24}NO_6F_5$ Store at -15°C

Aspartic Acid — FMOC-L-Asp 4-tBu 1-(4-Benzyloxybenzyl) Ester Polymer Bound

Fluka 47646   0.4-0.6 mmol/g resin; carrier: polystyrene, crosslinked with 1% DVB; 200-400 mesh particle size

Aspartic Acid — FMOC-L-Asp 4-tBu 1-Hydroxysuccinimide Ester

Fluka 47444   MW 508.53 $C_{27}H_{28}N_2O_8$ ≥98%; mp: 128-132°C

Aspartic Acid — FMOC-L-Asp-(1-OAll)

Fluka 47578   MW 395.41 $C_{22}H_{21}NO_6$ ≥97% (HPLC); mp: 95°C

Aspartic Acid — FMOC-L-Asp-(4-OAll)

Fluka 47579   MW 395.41 $C_{22}H_{21}NO_6$ ≥98% (HPLC); mp: 111-114°C

Aspartic Acid — FMOC-L-Asp-(4-OBzl)

Fluka 47593   MW 445.47 $C_{26}H_{23}NO_6$ ≥98% (HPLC); mp: 120-130°C; ≤1% H_2O

Aspartic Acid — FMOC-L-Asp-(4-OtBu)

Fluka 47618   MW 411.46 $C_{23}H_{25}NO_6$ ≥98% (HPLC); mp: 148-150°C

Aspartic Acid — FMOC-L-Asp-OAll

Neosystem FA00401	MW 395.4
Neosystem FA00418	MW 395.4

Aspartic Acid — FMOC-L-Asp-OBzl

Neosystem FA00403   MW 445.5

Aspartic Acid — FMOC-L-Asp-OcHex

Neosystem FA00407   MW 437.5

Aspartic Acid — FMOC-L-Asp-OtBu

Neosystem FA00405   MW 411.5

Aspartic Acid — FMOC-L-Asp-OtBu-Wang Resin

Neosystem RW00402

Aspartic Acid — FMOC-N-Me-Asp-OBzl

Synonyms: FMOC-N-Me-L-Asp-(β-OBzl)

Peptides International FMD-1792-PI   MW 459.5   >95% (HPLC); white crystalline powder

Aspartic Acid — FMOC-N-Me-Asp-OtBu

Bachem B-2195.0001   MW 425.48 $C_{24}H_{27}NO_6$ Store at -15°C

Aspartic Acid — FMOC-N-Me-D-Asp-OBzl

Synonyms: FMOC-N-Me-D-Asp-(β-OBzl)

Peptides International FMD-1890-PI   MW 459.50   >97% (HPLC); white crystalline powder

Aspartic Acid — Formimino-L-Asp

Sigma F 8501 $C_5H_8N_2O_4$

Aspartic Acid — Indole-3-Ac-L-Asp

Synonyms: IAA-L-Asp

Sigma I 9387   Plant cell culture tested

Aspartic Acid — L-(-)-*threo*-β-HydroxyAla

ICN 193663 $C_4H_7NO_5$ Neurotoxic inhibitor of L-Glu & L-Asp uptake

Aspartic Acid — L-Asp

Synonyms: (S)-2-Aminosuccinic Acid; (S)-Aminobutanedioic Acid; (S)-Amino-Butanedioic Acid; L-Aminosuccinic Acid; L-Asparagic Acid; L-Asparaginic Acid

Alexis 101-010   MW 133.1 $C_4H_7NO_4$

Amersham US11620   Non-essential AA

Calbiochem 1890   MW 133.1 $C_4H_7NO_4$ >99% (TLC); solid; soluble in aqueous NaOH; may be carcinogenic/teratogenic | Found at high concentration in the brain; has powerful excitatory effects on neurons; principal neurotransmitter for fast synaptic excitation; *Merck Index*, 12: 875

Fluka 11189   MW 133.11 $C_4H_7NO_4$ ≥99.5%; ≤0.05% residue on ignition; ≤0.005% K, Cl, SO_4; ≤0.02% NH_4^+; ≤0.0005% Al, Ba, Bi, Cd, Co, Cr, Cu, Fe, Li, Mg, Mn, Mo, Ni, Pb, Sr, Zn; ≤0.1% loss on drying; ≤0.00001% As; ≤0.001% Ca; ≤0.01% Na; ≤0.3% foreign AA | Olney, JW & Price, MT, *Meth Enzymol*, 103: 379, 1983; *CRC Handbook of Microbiology*, Laskin, AI & Lachevalier, HA, eds, CRC Press, Cleveland, Ohio, 4: 1, 1974

Fluka 11190   MW 133.11 $C_4H_7NO_4$ ≥99.5%; ≤0.005% K, Cl; ≤0.05% Na, NH_4^+; ≤0.0005% Cd, Co, Cr, Cu, Fe, Mg, Mn, Ni, Pb, Zn; ≤0.1% loss on drying, residue on ignition; ≤0.001% Ca; ≤0.01% Na; ≤0.3% foreign AA | Kunec, EK & Robins, DJ, *J Chem Soc*, Perkin I: 1089, 1987; Sham, HL et al, *Chem Comm*, 683, 1987; *ibid*, 1792, 1987; Thomas, EJ & Williams, AC, *ibid*, 992, 1987; Baldwin, JE et al, *Tetrahedron Lett*, 28: 3167, 1987

ICN 100809 $C_4H_7NO_4K$

ICN 102369 $C_4H_7NO_4Na$ Hygroscopic

ICN 150406 $C_8H_{12}N_2O_8Mg$

ICN 194633 $C_4H_7NO_4$

Neosystem AA00416   MW 133.1

Peptides International ALD-2704   MW 133.10   >99.9% optical purity by RP-HPLC; white crystalline powder

Rexim   MW 133.1 $C_4H_7NO_4$ White crystals or crystalline powder

Sigma A 3394   Matrix: cross-linked 4% beaded agarose; activation: cyanogen bromide; attachment: amino; spacer: 1 atom; ligand immobilized: 5-10 µmoles/mL; form: suspension in 2.0 M NaCl containing 0.02% thimerosal

Sigma A 6558 $C_4H_6NO_4K$ Principal neurotransmitter for fast synaptic excitation

Sigma A 6683 $C_4H_6NO_4Na$ Principal neurotransmitter for fast synaptic excitation

Sigma A 8949 $C_4H_7NO_4$ Principal neurotransmitter for fast synaptic excitation

Sigma A 9256 $C_4H_7NO_4$ Principal neurotransmitter for fast synaptic excitation

Sigma A 9506 $C_4H_6NO_4 \cdot ½Mg$ Principal neurotransmitter for fast synaptic excitation

USBio A3880   MW 133.1 $C_4H_7NO_4$ ≥99%; white, crystalline powder; specific rotation (C=8, 6 N HCl): +24.8° to +25.8°; heavy metals (Pb): ≤0.001%; endotoxin: ≤2 EU/mg; cell culture tested: murine myeloma SP2/0-Ag14

Aspartic Acid — L-Asp (^{15}N)

ICN 540394 $HOOCCH_2CH(NH_2)COOH$

Aspartic Acid — L-Asp (2,3-^{3}H)

ARC ART-211   MW 133.1 $HOOCCH_2CH(NH_2)COOH$ 10-25 Ci/mmol; 370-925 GBq/mmol; in 2% aqueous EtOH | Radiochemical

ICN 20011E $HOOCCH_2CH(NH_2)COOH$

Aspartic Acid — L-Asp (2,3-^{3}H)-Asp

Amersham TRK574

Aspartic Acid — L-Asp (4-^{14}C)

ARC ARC-226   MW 133.1 $HOOCCH_2CH(NH_2)COOH$ 50-60 mCi/mmol; 1.85-2.22 GBq/mmol; in 0.01 N HCl | Radiochemical

Aspartic Acid — L-Asp (U-^{14}C)

ARC ARC-670   MW 133.1 $HOOCCH_2CH(NH_2)COOH$ 150-250 mCi/mmol; 5.55-9.25 GBq/mmol; in EtOH:H_2O (2:98) | Radiochemical

ICN 10054E $HOOCCH_2CH(NH_2)COOH$
ICN 10054L

Sigma A 7553   FW 133.1 $HOCOCH_2CH(NH_2)COOH$ 120-240 mCi/mmol; radiochemical purity ≥98% (HPLC); aqueous solution containing 2% EtOH in serum bottle | Radiochemical

Aspartic Acid — L-Asp 4-AMC

Fluka 11235   MW 290.28 $C_{14}H_{14}N_2O_5$ ≥98% (TLC)

Aspartic Acid — L-Asp Amide

Synonyms: Isoasparagine

Sigma A 1291 $C_4H_8N_2O_3$

Aspartic Acid — L-Asp Dihydrate Magnesium Salt

Synonyms: Magnesium L-Asp

Fluka 11260   MW 324.54 $C_8H_{12}MgN_2O_8 \cdot 2H_2O$ ≥98% (titration)

Aspartic Acid — L-Asp Free Acid

Sigma A 4534   FW 133.1 $C_4H_7NO_4$ ≥98% (TLC) | Cell culture tested; insect cell culture tested

Aspartic Acid — L-Asp Hemihydrate Potassium Salt

Synonyms: Potassium L-Asp

Fluka 11230   MW 180.2 $C_4H_6KNO_4 \cdot 5H_2O$ ≥99% (titration); ≤0.02% NH_4^+; ≤0.005% Cl, Ca, Mg, SO_4; ≤0.0005% Cd, Co, Cr, Cu, Fe, Mg, Mn, Ni, Pb, Zn

Fluka 11232   MW 180.2 $C_4H_6KNO_4 \cdot 5H_2O$ ≥98% (titration)

Aspartic Acid — L-Asp Hydrate Sodium Salt

USBio A3881   MW 173.1 $C_4H_5NO_4Na \cdot H_2O$ ≥99%

Aspartic Acid — L-Asp Potassium Magnesium Salt

Synonyms: Potassium Magnesium L-Asp

Fluka 11250   MW 630.94 $C_{16}H_{24}K_2MgN_4O_{16}$ ~97% (titration); ~50% potassium Salt; ~50% magnesium Salt; ~5% loss on drying

Aspartic Acid — L-Asp Receptor Ligand (^{3}H)

Amersham TRK575   Excitatory AA receptor ligand; non-selective; endogenous

Aspartic Acid — L-Asp Sodium Salt Hydrate

Synonyms: Sodium L-Asp

Fluka 11195   MW 173.11 $C_4H_6NaNO_4 \cdot H_2O$ mp: ~140°C

Aspartic Acid — L-Asp α-(β-Naphthylamide)

Sigma A 8027 $C_{14}H_{14}N_2O_3$

Aspartic Acid — L-Asp β-AMC

Sigma A 1057 $C_{14}H_{14}N_2O_5$

Aspartic Acid — L-Asp β-Hydroxamate

Synonyms: L-Asp-β-Hydroxamic Acid

ICN 104859 $C_4H_8N_2O_4$

Sigma A 6508 $C_4H_8N_2O_4$ Tranquilizer; used to calm domestic animals

Aspartic Acid — L-Asp-(4-OBzl)

Fluka 11211   MW 223.23 $C_{11}H_{13}NO_4$ ≥99% (titration); mp: ~225°C

Aspartic Acid — L-Asp-(4-OtBu)

Fluka 11214   MW 189.22 $C_8H_{15}NO_4$ ≥98% (TLC); mp: ~220°C

Aspartic Acid — L-Asp-(β-OBzl)

Synonyms: β-Bz-L-Asp

ICN 100662 $C_{11}H_{13}NO_4$

Aspartic Acid — L-Asp-(β-OMe) Hydrochloride

Sigma A 8291 $C_5H_9NO_4 \cdot HCl$

Aspartic Acid — L-Asp-OAll$_2$

Sigma A 4928 $C_{10}H_{15}NO_4 \cdot C_7H_8O_3S$

Aspartic Acid — L-Asp-OBzl

Neosystem AA00402   MW 223.2

Aspartic Acid — L-Asp-OBzl$_2$

ICN 150402 $C_{18}H_{19}NO_4 \cdot C_7H_8O_3S$
Sigma A 6275 $C_{18}H_{19}NO_4 \cdot C_7H_8O_3S$

Aspartic Acid — L-Asp-OBzl$_2$-Tosyl

Fluka 11215   MW 485.56 $C_{18}H_{19}NO_4 \cdot C_7H_8O_3S$

Aspartic Acid — L-Asp-OcHex

Neosystem AA00405   MW 215.3

Aspartic Acid — L-Asp-OEt$_2$ Hydrochloride

ICN 150404 $C_8H_{15}NO_4 \cdot HCl$

Aspartic Acid — L-Asp-OMe$_2$ Hydrochloride

ICN 150405 $C_6H_{11}NO_4 \cdot HCl$
Sigma A 7541 $C_6H_{11}NO_4 \cdot HCl$

Aspartic Acid — L-Asp-OtBu

Neosystem AA00403   MW 189.2

Aspartic Acid — L-Asp-OtBu₂ Dibenzenesulfimide Salt

Fluka 11225   MW 542.68 $C_{12}H_{23}NO_4 \cdot C_{12}H_{11}NO_4S_2$ ≥99%; mp: 150-155°C

Aspartic Acid — L-Asp-OtBu₂ Hydrochloride

ICN 150403 $C_{12}H_{23}NO_4 \cdot HCl$

Sigma A 0877 $C_{12}H_{23}NO_4 \cdot HCl$

Aspartic Acid — Mg (L-Asp)₂ Dihydrate

Rexim   MW 324.5 $C_8H_{16}MgN_2O_{10}$ White crystals or crystalline powder

Aspartic Acid — MTH-DL-Asp

Sigma M 4006 $C_6H_8N_2O_3S$

Aspartic Acid — N-(3-Nitro-2-Pyridinesulfenyl)-L-Asp-(β-OtBu)

Sigma N 4140 $C_{13}H_{17}N_3O_6S$

Aspartic Acid — N-2,4-Dnp-L-Asp

Sigma D 8379 $C_{10}H_9N_3O_8$

Aspartic Acid — N-Ac-DL-Asp

Sigma A 5625 $C_6H_9NO_5$

Aspartic Acid — N-Ac-L-Asp

Synonyms: L-2-Acetylamino-Succinic Acid

Fluka 00920   MW 175.14 $C_6H_9NO_5$ ≥99% (titration); mp: 143-144°C | Gibson, GE & Shimada, M, *Biochem Pharmacol*, 29: 167, 1980

Rexim   MW 175.1 $C_6H_9NO_5$ White crystals or crystalline powder

Sigma A 5875 $C_6H_9NO_5$

Aspartic Acid — N-Ac-L-Asp α-pNA

Sigma A 3927 $C_{12}H_{13}N_3O_6$

Aspartic Acid — N-Carbamyl-DL-Asp

Synonyms: DL-Ureidosuccinic Acid

ICN 101247 $C_5H_8N_2O_5$

Aspartic Acid — N-CBZ-DL-Asp

ICN 100210 $C_{12}H_{13}NO_6$

Aspartic Acid — N-CBZ-L-Asp

ICN 100211 $C_{12}H_{13}NO_6$

Sigma C 5751 $C_{12}H_{13}NO_6$

Aspartic Acid — N-CBZ-L-Asp α-(2,6-Dichlorobenzoyloxyethylketone)

Synonyms: Protease Inhibitor

Sigma C 2086 $C_{20}H_{17}Cl_2NO_7$ Inhibitor of ICE-like proteases

Aspartic Acid — N-CBZ-L-Asp-(α-OMe)

Sigma C 4797 $C_{13}H_{15}NO_6$

Aspartic Acid — N-CBZ-L-Asp-(β-OBzl)

Sigma C 2019 $C_{19}H_{19}NO_6$

Aspartic Acid — N-CBZ-L-Asp-(β-OMe)

Sigma C 3803 $C_{13}H_{15}NO_6$

Aspartic Acid — N-CBZ-L-Asp-(β-OtBu)

Sigma C 9906 $C_{16}H_{21}NO_6$

Aspartic Acid — N-CBZ-β-Bzl-L-Asp p–Nitrophenyl Ester

Sigma C 1894 $C_{25}H_{22}N_2O_8$

Aspartic Acid — N-CBZ-β-tBu-L-Asp N-Hydroxysuccinimide Ester

Sigma C 0920 $C_{20}H_{24}N_2O_8$

Aspartic Acid — N-FMOC-L-Asp (U-¹⁴C)

ARC ARC-1053   MW 355.3   100-200 mCi/mmol; 3.7-7.4 GBq/mmol; in EtOH | Radiochemical

Aspartic Acid — N-FMOC-L-Asp-(β-OtBu)

Sigma F 9132 $C_{23}H_{25}NO_6$

Aspartic Acid — N-FMOC-β-tBu-L-Asp N-Hydroxysuccinimide Ester

ICN 158167 $C_{27}H_{28}N_2O_8$

Aspartic Acid — N-FMOC-β-tBu-L-Asp SASRIN Resin Ester

Sigma F 0639

Aspartic Acid — N-For-L-Asp

Sigma F 9126 $C_5H_7NO_5$

Aspartic Acid — N-Lauroyl-L-Asp

Fluka 61729   MW 315.41 $C_{16}H_{29}NO_5$ ≥99.0% (TLC)

Aspartic Acid — N-Me-Asp

Bachem E-2745.0250   MW 147.13 $C_5H_9NO_4$ Store at RT

Aspartic Acid — N-Me-Asp-OtBu

Bachem E-3190.0250   MW 203.24 $C_9H_{17}NO_4$ Store at -15°C

Aspartic Acid — N-Me-D-Asp

Synonyms: N-Methyl-D-Aspartate Receptor Agonist; N-Methyl-D-Aspartate; (R)-2-Methylaminosuccinic Acid

Alexis 105-009   MW 147.1 $C_5H_9NO_4$ White crystalline powder; soluble in dilute aqueous base, slightly in H₂O or EtOH | Selective ionotropic NMDA receptor agonist; Watkins, JC & Evans, RH, *Ann Rev Pharmacol Toxicol*, 21: 165, 1981; Stone, TW & Burton, NR, *Progr Neurobiol*, 30: 333, 1988

Bachem F-2415.0250   MW 147.13 $C_5H_9NO_4$ Store at RT

Calbiochem 454575   MW 147.1 $C_5H_9NO_4$ >96% (HPLC); solid; soluble in H₂O; LD₅₀≤2000 mg/kg | Excitatory AA neurotransmitter; selective agonist of the glutamate receptor that regulates Ca²⁺ channels; important in long-term potentiation, ischemia & epilepsy; NMDA receptors are involved in the "fine tuning" of synaptic connections in the developing brain; chronic treatment with NMDA produces structural changes in synaptic morphology; over-excitation of NMDA receptors causes neuronal degeneration & cell death; Mattson, MP et al, *J Neurosci*, 13: 4575, 1993; Yen, L-H et al, *J Neurosci*, 13: 4949, 1993; Swann, JW et al, *Epilepsy Res Suppl*, 9: 115, 1992; *Merck Index*, 12: 6760

Fluka 65831   MW 147.13 $C_5H_9NO_4$ ≥98.0% (TLC) | Excitatory AA, neurotransmitter; Watkins, JC, *J Med Pharm Chem*, 5: 1187, 1962; Collingridge, GL, *Trends Pharmacol Sci*, 6: 407, 1985; Nedergaard, S et al, *Cell Mol Neurobiol*, 7: 367, 1987

ICN 153672 $C_5H_9NO_4$

Sigma M 3262 $C_5H_9NO_4$ NMDA; excitotoxic AA; prototypic agonist at the glutamate receptor that regulates ion channels; important in long-term potentiation, ischemia & epilepsy; Olney, JW et al, *Life Sci*, 25: 537, 1979; Watkins, JC et al, *Ann Rev Pharmacol Toxicol*, 21: 165, 1981

Sigma M-102   MW 147.13 $C_5H_9NO_4$ >98%; white solid; mp 189-190°C; moderately soluble in H_2O or EtOH; soluble in 0.1 *N* NaOH | Williams et al, *J Recept Res*, 8: 195, 1988; Collingridge et al, *Pharmacological Rev*, 40: 143, 1989; Hansen et al, *Med Res Rev*, 10: 55, 1990

Aspartic Acid — *N*-Me-DL-Asp

Synonyms: (±)-2-Methylaminosuccinic Acid

Sigma M 2137 $C_5H_9NO_4$

Aspartic Acid — *N*-Me-DL-Asp Hydrate

Synonyms: NMA; (±)-2-Methylaminosuccinic Acid

Fluka 65833   MW 165.15 $C_5H_9NO_4 \cdot H_2O$ ≥99.0% (titration) | Excitatory AA, to make axon-sparing lesions of hypothalamus; Nadler, JV & Evenson, DA, *Meth Enzymol*, 103: 393, 1983

Aspartic Acid — *N*-Me-L-Asp

Synonyms: NMLA; (*S*)-2-Methylaminosuccinic Acid

Alexis 105-010   MW 147.1 $C_5H_9NO_4$ ≥98%; white crystalline powder; soluble in dilute aqueous base

Fluka 65832   MW 147.13 $C_5H_9NO_4$ ≥98.0% (TLC) | Selective *N*-methylaspartate receptor agonist; Davies, J et al, *Wenner-Gren Cent Int Symp Ser*, 39: 43, 1982; Thomson, AM et al, *Neuroscience Lett*, 54: 21, 1985

ICN 198769 $C_5H_9NO_4$ Inactive isomer of *N*-methylaspartic acid; Watkins, JC, *J Med Pharm Chem*, 5:1187, 1962

Sigma M 3387 $C_5H_9NO_4$ Inactive isomer of *N*-methylaspartic acid; Watkins, JC, *J Med Pharm Chem*, 5: 1187, 1962

Aspartic Acid — *N*-O-Nps-L-Asp Dicyclohexylammonium Salt

ICN 102525 $C_{10}H_{10}N_2O_6S \cdot C_{12}H_{23}N$

Aspartic Acid — *N*-t-BOC-D-Asp-(β-OBzl)

Sigma B 4644 $C_{16}H_{21}NO_6$

Aspartic Acid — *N*-t-BOC-L-Asp-(α-OBzl)

Sigma B 7257 $C_{16}H_{21}NO_6$
Sigma B 5626 $C_{16}H_{21}NO_6$

Aspartic Acid — *N*-t-BOC-L-Asp-(β-OcHex)

Sigma B 0641 $C_{15}H_{25}NO_6$

Aspartic Acid — *N*-t-BOC-L-Asp-(β-OMe)

Sigma B 7274 $C_{10}H_{17}NO_6$

Aspartic Acid — *N*-t-BOC-L-Asp-(β-t-OBzl)

Sigma B 4504 $C_{12}H_{23}NO_6 \cdot C_{12}H_{23}N$

Aspartic Acid — *N*-t-BOC-β-Bzl-L-Asp Pam Resin Ester

Sigma B 4768   For peptide synthesis by the Merrifield method; Mitchell, AR et al, *J Org Chem*, 43: 2845, 1978

Aspartic Acid — *N*-t-BOC-β-Bzl-L-Asp Resin Ester

Sigma B 9646   For peptide synthesis by the Merrifield method

Aspartic Acid — *N*-t-BOC-β-tBu-L-Asp *N*-Hydroxysuccinimide Ester

Sigma B 1517 $C_{17}H_{26}N_2O_8$

Aspartic Acid — *N*ᵉ-FMOC-L-Asp

Synonyms: Citrulline

Fluka 47518   MW 397.43 $C_{21}H_{23}N_3O_5$ ≥98% (HPLC)

Aspartic Acid — *N*ᵉ-FMOC-L-Asp-(β-OtBu)

ICN 151134 $C_{23}H_{25}NO_6$

Aspartic Acid — *N*ᵉ-t-BOC-L-Asp

ICN 150483 $C_9H_{15}NO_6$

Aspartic Acid — *N*ᵉ-t-BOC-L-Asp-(α-OBzl)

ICN 150484 $C_{16}H_{21}NO_6$
ICN 101067 $C_{16}H_{21}NO_6$

Aspartic Acid — *N*ᵉ-t-BOC-L-Asp-(β-OtBu)

ICN 101056 $C_{13}H_{23}NO_6 \cdot C_{12}H_{23}N$

Aspartic Acid — Pth-Asp

Sigma P 9501 $C_{11}H_{10}N_2O_3S$

Aspartic Acid — TentaGel S PHB-Asp-tBu-FMOC

Fluka 86379   0.22 mmol protected AA/g; ~90 μm particle size | FMOC-L-Asp-(4-OtBu) 1-(4-(Poly(Ethylenoxy))-OBzl) Polymer Bound

Aspartic Acid — TentaGel S Trt-Asp-tBu-FMOC

Fluka 86413   0.18 mmol protected AA/g; ~90 μM particle size

Aspartic Acid — Z-Asp

Synonyms: Benzyloxycarbonyl-L-Asp

Bachem C-1325.0025   MW 267.24 $C_{12}H_{13}NO_6$ Store at RT

Peptides International ZLD-2011-PI   MW 267.24   >98% (HPLC); white powder

Aspartic Acid — Z-Asp-(2,6-Dichlorobenzoyloxymethylketone)

Synonyms: Caspase I Inhibitor III

Alexis 260-029   MW 454.3 $C_{20}H_{17}NO_7Cl_2$ ≥97%; white solid; soluble in DMSO or MeOH | Novel aspartate-based caspase-1 inhibitor which prevents apoptosis caused by various types of cytotoxic agents; Dolle, RE et al, *J Med Chem*, 37: 563, 1994; Mashima, T et al, *BBRC*, 209: 907, 1995

Aspartic Acid — Z-Asp-CH₂-DCB

Synonyms: Z-Asp-(2,6-Dichlorobenzoyloxymethylketone); Caspase I Inhibitor V; Interleukin Converting Enzyme Inhibitor VI

Bachem N-1445.0100   MW 454.26 $C_{20}H_{17}Cl_2NO_7$ Store at -15°C

Calbiochem 400019   MW 454.3 $C_{20}H_{17}Cl_2NO_7$ ≥95% (HPLC); solid; soluble in DMSO | Potent inhibitor of caspase-1-like proteases; blocks apoptotic cell death in human myeloid leukemia U937 cells; Mashima, T et al, *Biochem Biophys Res Comm*, 209: 907, 1995; Dolle, RE et al, *J Med Chem*, 37: 563, 1994

Aspartic Acid — Z-Asp-CH₂OC(O)₂-6-Dichlorobenzene

Synonyms: Caspase I Inhibitor VI

American Peptide 81-7-05   MW 454.2 $C_{20}H_{17}NO_7Cl_2$ Selective inhibitor of ICE/ced₃ family proteases; Shimizu, T et al, *Gan To Kagaku Ryoho*, 24(2): 211, 1997; Mashima, T et al, *Biochem Biophys Res Comm*, 209: 907, 1995

Aspartic Acid — Z-Asp-OBzl

Bachem C-1340.0001   MW 357.36 $C_{19}H_{19}NO_6$ Store at RT
Bachem C-1350.0001   MW 357.36 $C_{19}H_{19}NO_6$ Store at RT

Aspartic Acid — Z-Asp-OBzl-OSu
Bachem C-3635.0001   MW 454.44 $C_{23}H_{22}N_2O_8$ Store at -15°C

Aspartic Acid — Z-Asp-OMe
Bachem C-3475.0001   MW 281.27 $C_{13}H_{15}NO_6$ Store at -15°C

Bachem C-3780.0001   MW 281.27 $C_{13}H_{15}NO_6$ Store at 2-8°C

Aspartic Acid — Z-Asp-ONp-OBzl
Bachem C-1345.0001   MW 478.46 $C_{25}H_{22}N_2O_8$ Store at -15°C

Aspartic Acid — Z-Asp-OSu-OBzl
Bachem C-3640.0001   MW 454.44 $C_{23}H_{22}N_2O_8$ Store at -15°C

Aspartic Acid — Z-Asp-OSu-OMe
Bachem C-3470.0001   MW 378.34 $C_{17}H_{18}N_2O_8$ Store at -15°C

Aspartic Acid — Z-Asp-OtBu
Bachem C-1355.0001   MW 323.35 $C_{16}H_{21}NO_6$ Store at 2-8°C

Aspartic Acid — Z-Asp-OtBu Dicyclohexylamine
Bachem C-3385.0001   MW 504.67 $C_{16}H_{21}NO_6 \cdot C_{12}H_{23}N$ Store at 2-8°C

Aspartic Acid — Z-Asp-OtBu-bromomethylketone
Bachem N-1425.0250   MW 400.27 $C_{17}H_{22}BrNO_5$ Store at -15°C

Aspartic Acid — Z-Asp-OtBu-ONp
Bachem C-1365.0001   MW 444.44 $C_{22}H_{24}N_2O_8$ Store at -15°C

Aspartic Acid — Z-Asp-OtBu-OSu
Bachem C-1370.0001   MW 420.42 $C_{20}H_{24}N_2O_8$ Store at -15°C

Aspartic Acid — Z-D-Asp
Bachem C-1330.0005   MW 267.24 $C_{12}H_{13}NO_6$ Store at RT

Fluka 95965   MW 267.24 $C_{12}H_{13}NO_6$ ≥98.0% (TLC)

Aspartic Acid — Z-D-Asp-OBzl
Bachem C-3165.0001   MW 357.36 $C_{19}H_{19}NO_6$ Store at RT

Aspartic Acid — Z-D-Asp-OtBu
Bachem C-1360.0001   MW 323.35 $C_{16}H_{21}NO_6$ Store at 2-8°C

Aspartic Acid — Z-DL-Asp
Bachem C-1335.0005   MW 267.24 $C_{12}H_{13}NO_6$ Store at RT

Fluka 95975   MW 267.24 $C_{12}H_{13}NO_6$ ≥98.0% (TLC)

Aspartic Acid — Z-DL-Asp-OBzl
Bachem C-3810.0001   MW 357.36 $C_{19}H_{19}NO_6$ Store at RT

Aspartic Acid — Z-L-Asp
Fluka 95970   MW 267.24 $C_{12}H_{13}NO_6$ ≥99.0% (titration); mp: 115-118°C

Neosystem ZA00401   MW 267.2

Aspartic Acid — Z-L-Asp 1-OMe
Fluka 95998   MW 281.27 $C_{13}H_{15}NO_6$ ≥98.0% (TLC)

Aspartic Acid — Z-L-Asp 4-Bzl 1-(4-Nitrophenyl) Ester
Fluka 95985   MW 478.46 $C_{25}H_{22}N_2O_8$ ≥98.0% (HPLC)

Aspartic Acid — Z-L-Asp 4-tBu 1-(N-Succinimidyl) Ester
Fluka 95991   MW 420.42 $C_{20}H_{24}N_2O_8$ ~99.0%; mp: 150-152°C

Aspartic Acid — Z-L-Asp-(4-OBzl)
Fluka 95980   MW 357.36 $C_{19}H_{19}NO_6$ ≥98.0% (TLC)

Aspartic Acid — Z-L-Asp-(4-OMe)
Fluka 96002   MW 281.27 $C_{13}H_{15}NO_6$ ≥98.0% (TLC)

Aspartic Acid — Z-L-Asp-(4-OtBu)
Fluka 02378   MW 323.35 $C_{16}H_{21}NO_6$ ≥98.0% (TLC)

Aspartic Acid — Z-L-Asp-(4-OtBu) 1-Bromomethyl Ketone
Fluka 95990   MW 400.27 $C_{17}H_{22}BrNO_5$ ≥98.0%

Aspartic Acid — Z-L-Asp-OBzl₂
Fluka 95995   MW 447.49 $C_{26}H_{25}NO_6$ ≥98.0% (TLC)

Aspartic Acid — Z-L-Asp-OtBu
Neosystem ZA00402   MW 323.3

Aspartic Acid — Z-N-Me-Asp-OtBu Dicyclohexylamine
Bachem C-3800.0001   MW 518.69 $C_{17}H_{23}NO_6 \cdot C_{12}H_{23}N$ Store at -15°C

Aspartic Acid — α-Me-DL-Asp
Synonyms: 2-Amino-2-Methylsuccinic Acid
Sigma M 6001 $C_5H_9NO_4$

Aspartic Acid — β-Bzl-L-Aspartate
Synonyms: L-Asp-(4-OBzl)
Sigma B 2129 $C_{11}H_{13}NO_4$

Aspartic Acid — β-D-Asp-Aminomethyl-Phosphonic Acid
Sigma A 3421 $C_5H_{11}N_2O_6P$ NMDA antagonist; Davies, J et al, *Comp Biochem Physiol*, 72°C: 211, 1982

Citrulline — BOC-Cit
Synonyms: BOC-Orn-Carbamoyl
Bachem A-3165.0001   MW 275.31 $C_{11}H_{21}N_3O_5$ Store at -15°C

USBio B2288   MW 275.3 $C_{11}H_{21}N_3O_5$ ≥98%

Citrulline — BOC-Cit-ONp
Synonyms: BOC-Orn-Carbamoyl-ONp
Bachem A-1530.0001   MW 396.4 $C_{17}H_{24}N_4O_7$ Store at -15°C

Citrulline — BOC-D-Cit
Synonyms: BOC-D-Orn-Carbamoyl

Bachem A-2845.0001   MW 275.31 $C_{11}H_{21}N_3O_5$ Store at 2-8°C

Citrulline — BOC-L-Thiocitrulline-OtBu
Bachem A-3700.0001   MW 347.48 $C_{15}H_{29}N_3O_4S$ Store at -15°C

Citrulline — Bz-Cit-OMe
Synonyms: Bz-Orn-Carbamoyl-OMe

Bachem F-2525.0001   MW 293.32 $C_{14}H_{19}N_3O_4$ Store at 2-8°C

Citrulline — Cit-AMC Hydrobromide
Synonyms: Orn-Carbamoyl-AMC HBr

Bachem I-1460.0025   MW 413.27 $C_{16}H_{20}N_4O_4 \cdot HBr$ Store at -15°C

Citrulline — Cit-AMC TFA
Synonyms: Orn-Carbamoyl-AMC TFA

Bachem I-1170.0050   MW 446.38 $C_{16}H_{20}N_4O_4 \cdot C_2HF_3O_2$ Store at -15°C

Citrulline — Cit-βNA Hydrobromide
Synonyms: Orn-Carbamoyl-βNA HBr

Bachem K-1665.0050   MW 381.27 $C_{16}H_{20}N_4O_2 \cdot HBr$ Store at -15°C

Citrulline — D-Cit
Synonyms: D-Orn-Carbamoyl

Bachem F-2435.0001   MW 175.19 $C_6H_{13}N_3O_3$ Store at 2-8°C

Citrulline — FMOC-Cit
Synonyms: FMOC-Orn-Carbamoyl; N^α-9-FMOC-L-2-Amino-5-Ureido-N-Valeric Acid

Bachem B-2090.0001   MW 397.43 $C_{21}H_{23}N_3O_5$ Store at RT
USBio F5352   MW 397.4 $C_{21}H_{23}N_3O_5$ ≥99%

Citrulline — FMOC-Cit-® (200-400 mesh)
Synonyms: FMOC-Orn-Carbamoyl-®

Bachem D-1840.0001   Store at RT

Citrulline — FMOC-D-Cit
Synonyms: FMOC-D-Orn-Carbamoyl

Bachem B-2075.0001   MW 397.43 $C_{21}H_{23}N_3O_5$ Store at RT

Citrulline — FMOC-L-Cit
Neosystem FA10601   MW 397.44

Citrulline — L-Cit
Synonyms: N^5-(Aminocarbonyl)-L-Orn; α-Amino-5-Ureidovaleric Acid; δ-Ureidonorvaline; N^5-Carbamoyl-L-Orn

Rexim   MW 175.2 $C_6H_{13}N_3O_3$ White crystals or crystalline powder

Citrulline — S-Me-L-Thiocitrulline
Bachem F-3180.0050   MW 205.28 $C_7H_{15}N_3O_2S$ Store at -15°C

Cysteine — (+)-S-Trt-L-Cys
Fluka 93450   MW 363.48 $C_{22}H_{21}NO_2S$ mp: 185°C

Cysteine — (Ac-Cys-OMe)₂ Disulfide Bond
Bachem E-1770.0001   MW 352.43 $C_{12}H_{20}N_2O_6S_2$ Store at -15°C

Cysteine — (BOC-Cys)₂ Disulfide Bond
Bachem A-1005.0005   MW 440.54 $C_{16}H_{28}N_2O_8S_2$ Store at RT

Cysteine — (Cys Amide)₂ Dihydrochloride Disulfide Bond
Bachem E-2380.0250   MW 311.26 $C_6H_{14}N_4O_2S_2 \cdot 2HCl$ Store at RT

Cysteine — (Cys)₂ Disulfide Bond
Bachem E-1760.0025   MW 240.3 $C_6H_{12}N_2O_4S_2$ Store at RT

Cysteine — (Cys-4MβNA)₂ Disulfide Bond
Bachem J-1175.0250   MW 550.7 $C_{28}H_{30}N_4O_4S_2$ Store at -15°C

Cysteine — (Cys-OAll)₂ 2 p-Tosyl Disulfide Bond
Bachem E-3465.0005   MW 664.84 $C_{12}H_{20}N_2O_4S_2 \cdot {}_2 C_7H_8O_3S$ Store at -15°C

Cysteine — (Cys-OEt)₂ Dihydrochloride Disulfide Bond
Bachem E-2860.0005   MW 369.33 $C_{10}H_{20}N_2O_4S_2 \cdot 2HCl$ Store at -15°C

Cysteine — (Cys-OMe)₂ Dihydrochloride Disulfide Bond
Bachem E-1765.0005   MW 341.28 $C_8H_{16}N_2O_4S_2 \cdot 2HCl$ Store at RT

Cysteine — (Cys-OtBu)₂ Dihydrochloride Disulfide Bond
Bachem E-2855.0005   MW 425.44 $C_{14}H_{28}N_2O_4S_2 \cdot 2HCl$ Store at -15°C

Cysteine — (Cys-pNA)₂ Disulfide Bond
Bachem L-1255.0250   MW 480.53 $C_{18}H_{20}N_6O_6S_2$ Store at -15°C

Cysteine — (Cys-βNA)₂ Dihydrochloride Disulfide Bond
Bachem K-1235.0001   MW 563.57 $C_{26}H_{26}N_4O_2S_2 \cdot 2HCl$ Store at RT

Cysteine — (Cys-βNA)₂ Disulfide Bond
Bachem K-1225.0001   MW 490.65 $C_{26}H_{26}N_4O_2S_2$ Store at RT

Cysteine — (D-Cys-OMe)₂ Dihydrochloride Disulfide Bond
Bachem F-3345.0250   MW 341.28 $C_8H_{16}N_2O_4S_2 \cdot 2HCl$ Store at RT

Cysteine — (FMOC-Cys)₂ Disulfide Bond
Bachem B-1615.0001   MW 684.79 $C_{36}H_{32}N_2O_8S_2$ Store at RT

Cysteine — (FMOC-Cys-OSu)₂ Disulfide Bond
Bachem B-1555.0001   MW 878.94 $C_{44}H_{38}N_4O_{12}S_2$ Store at -15°C

Cysteine — (Z-Cys)₂ Disulfide Bond

Bachem C-3175.0005 MW 508.57 $C_{22}H_{24}N_2O_8S_2$ Store at RT

Cysteine — 3,3-Dimethyl-D-Cys Free Base

Synonyms: D-Penicillamine

ICN 151805 MW 149.2 $C_5H_{11}NO_2S$ 99%; crystalline | Physiological chelating agent for heavy metals

Cysteine — 3,3-Dimethyl-D-Cys Hydrochloride

Synonyms: D-Penicillamine

ICN 151806 MW 185.7 $C_5H_{11}NO_2S \cdot HCl$ ~99%; crystalline

Cysteine — 3,3-Dimethyl-DL-Cys

Synonyms: DL-Penicillamine; β-Mercapto-DL-Val

ICN 102590 MW 149.2 $C_5H_{11}NO_2S$ Crystalline | Typical degradation product of penicillin antibiotics; a heavy metal chelating agent

Cysteine — 3,3-Dimethyl-L-(+)-Cys

Synonyms: L-(+)-Penicillamine

ICN 151807 MW 149.2 $C_5H_{11}NO_2S$ ≥99%; crystalline | Metal chelating agent

Cysteine — 5-Me-L-Cys

ICN 102316 MW 135.2 $C_4H_9NO_2S$ Crystalline

Cysteine — Ac-Cys-Dodecyl-CMK

Bachem N-1785.0250 MW 363.99 $C_{18}H_{34}ClNO_2S$ Store at -15°C

Cysteine — Ac-Cys-Farnesyl

Synonyms: AFC

Bachem M-1935.0100 MW 367.55 $C_{20}H_{33}NO_3S$ Store at -15°C

Cysteine — Ac-Cys-Farnesyl-OMe

Synonyms: AFCME

Bachem F-2930.0100 MW 381.58 $C_{21}H_{35}NO_3S$ Store at -15°C

Calbiochem 121740 MW 381.6 $C_{21}H_{35}NO_3S$ ≥95% (TLC); light yellow oil; soluble in MeOH; packaged under inert gas | Substrate for studying the reversibility of carboxyl methylation in signal transducing G-proteins; an enzyme present in bovine retinal rod outer segment (ROS) membranes can demethylate the methylated transducin γ-subunit, small G-proteins & AFCME; Perez-Sala, D et al, *PNAS*, 88: 3043, 1991; Tan, EW & Rando, RR, *Biochemistry*, 31: 5572, 1992; Maltese, WA, *FASEB J*, 4: 3319, 1990

Cysteine — Ac-Cys-Trt

Bachem E-3030.0001 MW 405.52 $C_{24}H_{23}NO_3S$ Store at 2-8°C

Cysteine — Bis-δ-(L-α-Aminoadipyl)-L-Cystine

Synonyms: Bis-AC

Bachem G-4250.0050 Store at -15°C

Cysteine — Bis-δ-(L-α-Aminoadipyl)-L-Cystinyl-Bis-D-α-Aminobutyric Acid

Synonyms: Bis-ACAbu

Bachem H-1516.0025 Store at -15°C

Cysteine — BOC₂-Cystine

USBio B2294 MW 440.6 $C_{16}H_{28}N_2O_8S_2$ ≥99%

Cysteine — BOC-Cys

USBio B2293 MW 221.3 $C_8H_{15}NO_4S$ ≥99%

Cysteine — BOC-Cys-(3,4-Dimethylbenzyl)

Bachem A-3965.0005 MW 339.46 $C_{17}H_{25}NO_4S$ Store at -15°C

Cysteine — BOC-Cys-(pMeOBzl)

USBio B2292 MW 341.4 $C_{16}H_{23}NO_5S$ ≥99%

Cysteine — BOC-Cys-(SEt) Dicyclohexylamine

Bachem A-1600.0001 MW 462.72 $C_{10}H_{19}NO_4S_2 \cdot C_{12}H_{23}N$ Store at RT

Cysteine — BOC-Cys-4-Methylbenzyl

Synonyms: BOC-S-4- Methylbenzyl -L-Cys

Peptides International BLC-2129 MW 325.43 >98% (HPLC); white crystalline powder

Cysteine — BOC-Cys-Acm

Synonyms: BOC-S-Acm-L-Cys

American Peptide BLCYS10 MW 292.4

Bachem A-1045.0005 MW 292.36 $C_{11}H_{20}N_2O_5S$ Store at -15°C

Peptides International BLC-2121 MW 292.36 >98% (HPLC); white crystalline powder

USBio B2289 MW 292.3 $C_{11}H_{20}N_2O_5S$ ≥99%

Cysteine — BOC-Cys-Acm-ONp

Bachem A-3530.0001 MW 413.45 $C_{17}H_{23}N_3O_7S$ Store at -15°C

Cysteine — BOC-Cys-Acm-O-Resin

American Peptide RMRB145 1% DVB cross-linked: 100-200 mesh

Cysteine — BOC-Cys-Acm-OSu

Bachem A-3710.0001 MW 389.43 $C_{15}H_{23}N_3O_7S$ Store at -15°C

Cysteine — BOC-Cys-Bzl

Synonyms: BOC-S-Bzl-L-Cys

Bachem A-1300.0005 MW 311.4 $C_{15}H_{21}NO_4S$ Store at RT

Peptides International BLC-2061 MW 311.40 >98% (HPLC); white crystalline powder

USBio B2290 MW 311.4 $C_{15}H_{21}NO_4S$ ≥99%

Cysteine — BOC-Cys-Bzl-ONp

Bachem A-1310.0001 MW 432.5 $C_{21}H_{24}N_2O_6S$ Store at -15°C

Cysteine — BOC-Cys-Bzl-OSu

Bachem A-1305.0001 MW 408.48 $C_{19}H_{24}N_2O_6S$ Store at -15°C

Cysteine — BOC-Cys-Dpm

Bachem A-2550.0001 MW 387.5 $C_{21}H_{25}NO_4S$ Store at -15°C

Cysteine — BOC-Cys-Et

Bachem A-4060.0001 MW 249.33 $C_{10}H_{19}NO_4S$ Store at -15°C

Cysteine — BOC-Cys-Fm

Bachem A-3435.0001   MW 399.51 $C_{22}H_{25}NO_4S$ Store at -15°C

Cysteine — BOC-Cys-Me

Bachem A-3455.0005   MW 235.3 $C_9H_{17}NO_4S$ Store at -15°C

Cysteine — BOC-Cys-Methylbenzyl

Synonyms: BOC-*S*-*p*-Methylbenzyl-L-Cys

American Peptide BLCYS20   MW 325.4

Bachem A-2030.0005   MW 325.43 $C_{16}H_{23}NO_4S$ Store at RT

Peptides International BLC-2078   MW 341.43   >98% (HPLC); white crystalline powder

Cysteine — BOC-Cys-Methylbenzyl-OSu

Bachem A-3395.0001   MW 422.5 $C_{20}H_{26}N_2O_6S$ Store at -15°C

Cysteine — BOC-Cys-Methylbenzyl-PAM-® (200-400 mesh)

Bachem D-1535.0001   Store at 2-8°C

Cysteine — BOC-Cys-Mob

Bachem A-1990.0001   MW 341.43 $C_{16}H_{23}NO_5S$ Store at RT

Cysteine — BOC-Cys-Mob-OSu

Bachem A-2000.0001   MW 438.5 $C_{20}H_{26}N_2O_7S$ Store at -15°C

Cysteine — BOC-Cys-Mob-PAM-® (200-400 mesh)

Bachem D-1260.0001   Store at 2-8°C

Cysteine — BOC-Cys-Npys

Bachem A-2825.0001   MW 375.43 $C_{13}H_{17}N_3O_6S_2$ Store at -15°C

Cysteine — BOC-Cys-*p*-Methylbenzyl

USBio B2291   MW 325.4 $C_{16}H_{23}NO_4S$ ≥99%

Cysteine — BOC-Cys-*p*-Methylbenzyl-O-Resin

American Peptide RMRB150   1% DVB cross-linked: 100-200 mesh

Cysteine — BOC-Cys-*p*-Methylbenzyl-PAM-Resin

American Peptide RPAM150   1% DVB cross-linked: 100-200 mesh

Cysteine — BOC-Cys-Sulfo Disodium Salt

Bachem A-3625.0001   MW 345.3 $C_8H_{13}NNa_2O_7S_2$ Store at -15°C

Cysteine — BOC-Cys-tBu

Synonyms: BOC-*S*-tBu-L-Cys

Peptides International BLC-2130   MW 277.37   >98% (HPLC); white crystalline powder

Cysteine — BOC-Cys-tBu Dicyclohexylamine

Bachem A-4005.0005   MW 458.71 $C_{12}H_{23}NO_4S \cdot C_{12}H_{23}N$ Store at -15°C

Cysteine — BOC-Cys-Trt

Bachem A-2355.0001   MW 463.6 $C_{27}H_{29}NO_4S$ Store at -15°C

USBio B2295   MW 463.6 $C_{27}H_{29}NO_4S$ ≥99%

Cysteine — BOC-Cys-Z-Aminoethyl

Bachem A-3535.0001   MW 398.48 $C_{18}H_{26}N_2O_6S$ Store at -15°C

Cysteine — BOC-D-Cys-4-Methylbenzyl

Synonyms: BOC-*S*-4-Methylbenzyl-D-Cys

Neosystem BA00507   MW 325.4

Peptides International BDC-2611   MW 325.43   >98% (HPLC); white crystalline powder

Cysteine — BOC-D-Cys-Acm

Bachem A-3930.0001   MW 292.36 $C_{11}H_{20}N_2O_5S$ Store at RT

Cysteine — BOC-D-Cys-Bzl

Bachem A-3905.0001   MW 311.4 $C_{15}H_{21}NO_4S$ Store at RT

Cysteine — BOC-D-Cys-Methylbenzyl

Bachem A-3545.0001   MW 325.43 $C_{16}H_{23}NO_4S$ Store at RT

Cysteine — BOC-D-Cys-Methylbenzyl-OSu

Bachem A-3635.0001   MW 422.5 $C_{20}H_{26}N_2O_6S$ Store at -15°C

Cysteine — BOC-D-Cys-Mob

Bachem A-1995.0001   MW 341.43 $C_{16}H_{23}NO_5S$ Store at RT

Cysteine — BOC-D-Cys-Npys

Bachem A-3470.0001   MW 375.42 $C_{13}H_{17}N_3O_6S_2$ Store at -15°C

Cysteine — BOC-D-Cys-*p*-Methylbenzyl-O-CH₂-♦-CH₂-COOH

Neosystem LP00503   MW 473.6

Cysteine — BOC-D-Cys-Trt

Bachem A-3265.0001   MW 463.6 $C_{27}H_{29}NO_4S$ Store at -15°C

Cysteine — BOC-L-Cys

Fluka 15411   MW 221.27 $C_8H_{15}NO_4S$ ≥99.5% (HPLC); mp: 76-79°C

Cysteine — BOC-L-Cys-4-Methoxybenzyl

Neosystem BA00506   MW 341.4

Cysteine — BOC-L-Cys-4-Methylbenzyl

Neosystem BA00508   MW 325.4

Cysteine — BOC-L-Cys-Acm

Neosystem BA00501   MW 292.4

Cysteine — BOC-L-Cys-Acm-MBHA Resin

Neosystem RN00501

Cysteine — BOC-L-Cys-Acm-PAM Resin

Neosystem RP00502

Cysteine — BOC-L-Cys-Allocam Dicyclohexylamine

Neosystem BA00502   MW 516.7

Cysteine — BOC-L-Cys-Bzl

Neosystem BA00503   MW 311.4

Cysteine — BOC-L-Cys-p-Methylbenzyl-MBHA Resin

Neosystem RN00508

Cysteine — BOC-L-Cys-p-Methylbenzyl-O-CH₂-φ-CH₂-COOH

Neosystem LP00504   MW 473.6

Cysteine — BOC-L-Cys-p-Methylbenzyl-PAM Resin

Neosystem RP00504

Cysteine — BOC-S-4-Methylbenzyl-L-Cys

Fluka 15439	MW 341.42	$C_{16}H_{23}NO_5S$	~98% (HPLC)
Fluka 15444	MW 325.42	$C_{16}H_{23}NO_4S$	~98% (HPLC)

Cysteine — BOC-S-4-Methylbenzyl-L-Cys Dicyclohexylamine

Fluka 15463   MW 522.77 $C_{16}H_{23}NO_5S \cdot C_{12}H_{23}N$ mp: 127-130°C

Cysteine — BOC-S-4-Methylbenzyl-L-Cys N-Carboxy Anhydride

Fluka 15468   MW 367.42 $C_{17}H_{21}NO_6S$ ~99%; mp: 85-88°C

Cysteine — BOC-S-Acm-L-Cys

Fluka 15376   MW 292.36 $C_{11}H_{20}N_2O_5S$ ~98% (titration); mp: 111-114°C

Cysteine — BOC-S-Bzl-L-Cys

Fluka 15383   MW 311.4 $C_{15}H_{21}NO_4S$ ≥99% (titration); mp: 86-88°C

Cysteine — BOC-S-t-Butylmercapto-L-Cys

Fluka 15375   MW 309.45 $C_{12}H_{23}NO_4S$ ≥99%; mp: 117-119°C; ≤0.05% residue on ignition

Cysteine — Bsmoc-Cys-Trt

Synonyms: Bsmoc-S-Trt-L-Cys

Peptides International BLC-5618-PI   MW 585.70   >98% (HPLC); white to off-white powder | LA Carpino, et al, *JACS*, 119:9915, 1997

Cysteine — CBZ-Cys-(pMeOBzl)

USBio C2098-53   MW 375.3 $C_{19}H_{21}NO_5S$ ≥99%

Cysteine — CBZ-Cys-Bzl

USBio C2098-51   MW 345.4 $C_{18}H_{19}NO_4OS$ ≥99%

Cysteine — CBZ-Cys-Bzl-ONp

USBio C2098-52   MW 466.5 $C_{24}H_{22}N_2O_6S$ ≥99%

Cysteine — Cys Hydrochloride

Bachem E-1755.0025   MW 157.62 $C_3H_7NO_2S \cdot HCl$ Store at RT

Cysteine — Cys Hydrochloride Hydrate

Fluka 30122	MW 175.64	$C_3H_7NO_2S \cdot HCl \cdot H_2O$	pH Eu
Sigma C 3357	FW 175.6	$C_3H_7NO_2S \cdot HCl \cdot H_2O$	USP Grade

Cysteine — Cys-Acm Hydrochloride

Bachem E-1000.0001   MW 228.7 $C_6H_{12}N_2O_3S \cdot HCl$ Store at 2-8°C

Cysteine — Cys-Aminoethyl Hydrochloride

Bachem E-1355.0001   MW 200.69 $C_5H_{12}N_2O_2S \cdot HCl$ Store at 2-8°C

Cysteine — Cys-Bzl

Bachem E-1540.0025   MW 211.29 $C_{10}H_{13}NO_2S$ Store at RT

Cysteine — Cys-Bzl-AMC

Bachem I-1090.0050   MW 368.5 $C_{20}H_{20}N_2O_3S$ Store at -15°C

Cysteine — Cys-Bzl-OBzl p-Tosyl

Bachem E-2925.0005   MW 473.62 $C_{24}H_{27}NO_5S_2$ Store at RT

Cysteine — Cys-Bzl-OEt Hydrochloride

Bachem E-2720.0005   MW 275.8 $C_{12}H_{17}NO_2S \cdot HCl$ Store at 2-8°C

Cysteine — Cys-Bzl-OMe Hydrochloride

Bachem E-1545.0005   MW 261.77 $C_{11}H_{15}NO_2S \cdot HCl$ Store at RT

Cysteine — Cys-Bzl-pNA

Bachem L-1145.0001   MW 331.4 $C_{16}H_{17}N_3O_3S$ Store at -15°C

Cysteine — Cys-Bzl-βNA

Bachem K-1150.0250   MW 336.46 $C_{20}H_{20}N_2OS$ Store at RT

Cysteine — Cys-Carbamoyl

Bachem E-3440.0005   MW 173.2 $C_4H_8N_2O_3S$ Store at 2-8°C

Cysteine — Cys-Dpm

Bachem E-2945.0001   MW 287.38 $C_{16}H_{17}NO_2S$ Store at RT

Cysteine — Cys-Farnesyl

ARC ARCD-117   MW 325.3   In EtOH:H₂O (1:1)

Cysteine — Cys-Farnesyl-OMe

ARC ARCD-140   MW 376.3   Neat liquid

Cysteine — Cys-Fm Hydrochloride

Bachem E-3070.0001   MW 335.85 $C_{17}H_{17}NO_2S \cdot HCl$ Store at 2-8°C

Cysteine — Cys-Mob

Bachem E-2120.0005   MW 241.31 $C_{11}H_{15}NO_3S$ Store at RT

Cysteine — Cys-Phenyl

Bachem E-3400.0001   MW 197.26 $C_9H_{11}NO_2S$ Store at -15°C

Cysteine — Cys-StBu

Bachem E-2710.0005   MW 209.33 $C_7H_{15}NO_2S_2$ Store at 2-8°C

Cysteine — Cys-Sulfo Sodium Salt

Bachem E-3130.0005   MW 223.21 $C_3H_6NNaO_5S_2$ Store at -15°C

Cysteine — Cys-tBu Hydrochloride

Bachem E-1650.0001   MW 213.73 $C_7H_{15}NO_2S \cdot HCl$ Store at 2-8°C

Cysteine — Cys-Trt

Bachem E-2495.0005   MW 363.48 $C_{22}H_{21}NO_2S$ Store at RT

Cysteine — Cys-Trt Amide

Bachem E-2930.0001   MW 362.5 $C_{22}H_{22}N_2OS$ Store at -15°C

Cysteine — Cys-Trt-2-Chloro-Trt Resin

American Peptide RGEN167   Enantiomerization free

Cysteine — D-Cys

Fluka 30095   MW 121.16 $C_3H_7NO_2S$ ≥99% (titration); mp: 230°C

Cysteine — D-Cys Free Base

Sigma C 8882   FW 121.2 $C_3H_7NO_2S$ ≥98% (TLC) | Inactive isomer

Cysteine — D-Cys Hydrochloride

Bachem F-2200.0001   MW 157.62 $C_3H_7NO_2S \cdot HCl$ Store at RT

ICN 101438   MW 157.62 $C_3H_7NO_2S \cdot HCl$ Crystalline

Cysteine — D-Cys Hydrochloride Anhydrous

Fluka 30110   MW 157.62 $C_3H_7NO_2S \cdot HCl$ ~99% (titration); mp: 185°C

Cysteine — D-Cys Hydrochloride Hydrate

Fluka 30140   MW 175.64 $C_3H_7NO_2S \cdot HCl \cdot H_2O$ ~99% (titration

Peptides International ADC-2817   MW 175.64   >99.9% optical purity by RP-HPLC; white crystalline powder

Sigma C 8005   FW 175.6 $C_3H_7NO_2S \cdot HCl \cdot H_2O$ ≥98% (TLC) | Inactive isomer

USBio C9010-10   MW 175.6 $C_3H_7NO_2S \cdot HCl \cdot H_2O$ ≥99%

Cysteine — D-Cys-Acm Hydrochloride

Bachem F-2470.0001   MW 228.7 $C_6H_{13}N_2O_3SCl$ Store at 2-8°C

Cysteine — D-Cys-Bzl

Bachem F-2195.0001   MW 211.29 $C_{10}H_{13}NO_2S$ Store at RT

Cysteine — D-Cys-Fm

Bachem E-3550.0001   MW 299.39 $C_{17}H_{17}NO_2S$ Store at -15°C

Cysteine — D-Cys-Methylbenzyl

Bachem F-3060.0001   MW 225.31 $C_{11}H_{15}NO_2S$ Store at 2-8°C

Cysteine — D-Cys-Mob

Bachem F-2885.0001   MW 241.31 $C_{11}H_{15}NO_3S$ Store at 2-8°C

Cysteine — D-Cys-tBu Hydrochloride

Bachem F-2355.0001   MW 213.73 $C_7H_{15}NO_2S \cdot HCl$ Store at -15°C

Cysteine — D-Cystine

Fluka 30210   MW 240.3 $C_6H_{12}N_2O_4S_2$ ~99% (titration); mp: 250°C

Sigma C 8505   FW 240.3 $C_6H_{12}N_2O_4S_2$ Crystalline | Also available as part of a kit

USBio C9050-10   MW 240.3 $C_6H_{12}N_2O_5S_2$ ≥99%

Cysteine — D-Cys-Trt

Bachem F-2770.0001   MW 363.48 $C_{22}H_{21}NO_2S$ Store at -15°C

Cysteine — D-Glucose-L-Cys

ICN 157198   MW 283.3 $C_9H_{17}NO_7S$ Crystalline

Cysteine — D-Homocystine

Synonyms: D-4,4'-Dithio-Bis-2-Aminobutanoic Acid

Sigma H 5134   FW 268.3 $C_8H_{16}N_2O_4S_2$

Cysteine — Di-Z-L-Cystine

Fluka 96080   MW 508.57 $C_{22}H_{24}N_2O_8S_2$ ~99% (titration); mp: 118-122°C

Cysteine — DL-Cys

Fluka 30097   MW 121.16 $C_3H_7NO_2S$ ≥99% (titration); mp: 225°C

Cysteine — DL-Cys Free Base

Synonyms: NMDA agonist

Sigma C 4022   FW 121.2 $C_3H_7NO_2S$ ≥98% (TLC) | NMDA agonist at quisqualate receptors at high concentration

Cysteine — DL-Cys Hydrochloride Anhydrous

Synonyms: NMDA Agonist

Fluka 30150   MW 157.62 $C_3H_7NO_2S \cdot HCl$ ≥97% (titration)

Sigma C 9768   FW 157.6 $C_3H_7NO_2S \cdot HCl$ ≥98% (TLC) | NMDA agonist at quisqualate receptors at high concentration

Cysteine — DL-Cys Hydrochloride Hydrate

Synonyms: NMDA Agonist

Fluka 30152   MW 175.64 $C_3H_7NO_2S \cdot HCl \cdot H_2O$ ≥97% (titration)

Sigma C 8256   FW 157.6 $C_3H_7NO_2S \cdot HCl$ ≥98% (TLC) | NMDA agonist at quisqualate receptors at high concentration

Cysteine — DL-Cystine

Fluka 30220   MW 240.3 $C_6H_{12}N_2O_4S_2$ ≥99% (titration); meso free

Sigma C 8630   FW 240.3 $C_6H_{12}N_2O_4S_2$ Crystalline | Also available as part of a kit

Cysteine — DL-Cystine Dihydrochloride

Sigma C 5632   FW 313.2 $C_6H_{12}N_2O_4S_2 \cdot 2HCl$ Crystalline

Cysteine — DL-Homocystine

Synonyms: meso-4,4'-Dithio-Bis-2-Aminobutanoic Acid

Fluka 53550   MW 268.36 $C_8H_{16}N_2O_4S_2$ ≥99.0% (titration); mp: >300°C; ≤0.05% residue on ignition

Sigma H 0501   FW 268.3 $C_8H_{16}N_2O_4S_2$

Cysteine — FA-Cys-Farnesyl

Synonyms: 3-(2-Furyl)-Acryloyl-Cys-Farnesyl; Methyltransferase Substrate

Bachem M-1925.0100   MW 445.62 $C_{25}H_{35}NO_4S$ Store at -15°C

Calbiochem 341250   MW 445.6 $C_{25}H_{35}NO_4S$ ≥95% (TLC); light yellow powder; soluble in MeOH; packaged under inert gas | Potential substrate for the methyltransferase activity in studies involving methylation of C-terminal S-farnesylcysteine residue

Cysteine — FA-Cys-Farnesyl-OMe

Bachem M-1920.0100   MW 459.65 $C_{26}H_{37}NO_4S$ Store at -15°C

Cysteine — FMOC-Cys-(pMeOBzl)

Synonyms: FMOC-S-p-Methylbenzyl-L-Cys

Peptides International FLC-1717-PI   MW 463.51   >98% (HPLC); white crystalline powder

USBio F5360   MW 463.6 $C_{26}H_{25}NO_5S$ ≥99%

Cysteine — FMOC-Cys-(t-Butylcarboxymethyl)

Bachem B-3395.0001   MW 457.55 $C_{24}H_{27}NO_6S$ Store at 2-8°C

Cysteine — FMOC-Cys-2-Hydroxyethyl

Bachem B-2630.0001   MW 387.46 $C_{20}H_{21}NO_5S$ Store at -15°C

Cysteine — FMOC-Cys-3-(BOC-Amino)-Propyl

Bachem B-3120.0001   MW 500.62 $C_{26}H_{32}N_2O_6S$ Store at -15°C

Cysteine — FMOC-Cys-4-Methoxytrityl

Bachem B-2540.0001   MW 615.76 $C_{38}H_{33}NO_5S$ Store at -15°C

Cysteine — FMOC-Cys-Acm

Synonyms: FMOC-S-Acm-L-Cys

Bachem B-1005.0001   MW 414.48 $C_{21}H_{22}N_2O_5S$ Store at RT

Peptides International FLC-1714-PI   MW 414.48   >98% (TLC); off-white crystalline powder

USBio F5353   MW 414.5 $C_{21}H_{22}N_2O_5S$ ≥99%

Cysteine — FMOC-Cys-Acm-® (200-400 mesh)

Bachem D-1040.0001   Store at RT

Cysteine — FMOC-Cys-Acm-OPfp

Bachem B-1115.0001   MW 580.53 $C_{27}H_{21}F_5N_2O_5S$ Store at -15°C

USBio F5354   MW 580.6 $C_{27}H_{21}F_2N_5O_5S$ ≥99%

Cysteine — FMOC-Cys-Acm-SASRIN™-® (200-400 mesh)

Bachem D-1320.0001   Store at 2-8°C

Cysteine — FMOC-Cys-Bzl

Bachem B-1880.0001   MW 433.53 $C_{25}H_{23}NO_4S$ Store at RT

USBio F5358   MW 433.5 $C_{25}H_{23}NO_4S$ ≥99%

Cysteine — FMOC-Cys-Bzl-OPfp

Bachem B-1955.0001   MW 599.58 $C_{31}H_{22}F_5NO_4S$ Store at -15°C

Cysteine — FMOC-Cys-Et

Bachem B-1970.0001   MW 371.46 $C_{20}H_{21}NO_4S$ Store at 2-8°C

Cysteine — FMOC-Cys-Et-® (200-400 mesh)

Bachem D-1715.0001   Store at 2-8°C

Cysteine — FMOC-Cys-Et-SASRIN™-® (200-400 mesh)

Bachem D-1720.0001   Store at 2-8°C

Cysteine — FMOC-Cys-Me

Bachem B-2510.0001   MW 357.43 $C_{19}H_{19}NO_4S$ Store at 2-8°C

Cysteine — FMOC-Cys-Methylbenzyl

Bachem B-2775.0001   MW 447.56 $C_{26}H_{25}NO_4S$ Store at -15°C

Cysteine — FMOC-Cys-Mob

Bachem B-1375.0001   MW 463.55 $C_{26}H_{25}NO_5S$ Store at RT

Cysteine — FMOC-Cys-Mob-OPfp

Bachem B-2225.0001   MW 629.61 $C_{32}H_{24}NO_5F_5S$ Store at -15°C

Cysteine — FMOC-Cys-Mtt

Bachem B-3340.0001   MW 599.75 $C_{38}H_{33}NO_4S$ Store at -15°C

Cysteine — FMOC-Cys-p-Methylbenzyl

USBio F5359   MW 477.6 $C_{26}H_{25}NO_4S$ ≥99%

Cysteine — FMOC-Cys-StBu

Bachem B-1530.0001   MW 431.58 $C_{22}H_{25}NO_4S_2$ Store at -15°C

USBio F5357   MW 431.6 $C_{22}H_{25}NO_4S_2$ ≥99%

Cysteine — FMOC-Cys-StBu-® (200-400 mesh)

Bachem D-1415.0001   Store at 2-8°C

Cysteine — FMOC-Cys-Sulfo Disodium Salt

Bachem B-2415.0001   MW 467.42 $C_{18}H_{15}NNa_2O_7S_2$ Store at -15°C

Cysteine — FMOC-Cys-tBu

Synonyms: FMOC-S-tBu-L-Cys

Bachem B-1220.0001   MW 399.51 $C_{22}H_{25}NO_4S$ Store at 2-8°C

Peptides International FLC-1741-PI   MW 399.51   >98% (TLC); white powder

USBio F5355   MW 399.5 $C_{22}H_{25}NO_4S$ ≥99%

Cysteine — FMOC-Cys-tBu-® (200-400 mesh)

Bachem D-1065.0001   Store at 2-8°C

Cysteine — FMOC-Cys-tBu-OPfp

Bachem B-1590.0001	MW 565.56	$C_{28}H_{24}F_5NO_4S$	Store at -15°C
USBio F5356	MW 553.5	$C_{27}H_{24}F_5NO_4S$	≥99%

Cysteine — FMOC-Cys-tBu-SASRIN™-® (200-400 mesh)

Bachem D-1325.0001	Store at 2-8°C

Cysteine — FMOC-Cys-Trt

Synonyms: FMOC-S-Trt-L-Cys

American Peptide FCYS145	MW 585.7		
Bachem B-1440.0001	MW 585.72	$C_{37}H_{31}NO_4S$	Store at -15°C
Peptides International FLC-1718-PI	MW 585.72	>98% (HPLC); white crystalline powder	
USBio F5362	MW 585.7	$C_{37}H_{31}NO_4S$	≥99%

Cysteine — FMOC-Cys-Trt-® (200-400 mesh)

Bachem D-1700.0001	Store at RT

Cysteine — FMOC-Cys-Trt-OPfp

Bachem B-1120.0001	MW 751.77	$C_{43}H_{30}F_5NO_4S$	Store at -15°C
USBio F5363	MW 751.8	$C_{43}H_{30}F_5NO_4S$	≥99%

Cysteine — FMOC-Cys-Trt-SASRIN™-® (200-400 mesh)

Bachem D-1330.0001	Store at 2-8°C

Cysteine — FMOC-Cys-Xan

Synonyms: FMOC-S-Xan-L-Cys

Peptides International FLC-5358-PI	MW 523.61	>98% (HPLC); white crystalline powder

Cysteine — FMOC-Cys-Xan-OPfp

Synonyms: FMOC-S-Xan-L-Cys-OPfp

Peptides International FLC-5374-PI	MW 689.66	>98% (TLC); white crystalline powder

Cysteine — FMOC-D-Cys-Acm

Bachem B-1890.0001	MW 414.48	$C_{21}H_{22}N_2O_5S$	Store at RT

Cysteine — FMOC-D-Cys-Bzl

Bachem B-3390.0001	MW 433.53	$C_{25}H_{23}NO_4S$	Store at RT

Cysteine — FMOC-D-Cys-Methylbenzyl

Bachem B-2330.0001	MW 447.56	$C_{26}H_{25}NO_4S$	Store at RT

Cysteine — FMOC-D-Cys-Mob

Bachem B-2170.0001	MW 463.55	$C_{26}H_{25}NO_5S$	Store at RT

Cysteine — FMOC-D-Cys-StBu

Bachem B-2875.0001	MW 431.57	$C_{22}H_{25}NO_4S_2$	Store at -15°C

Cysteine — FMOC-D-Cys-tBu

Bachem B-1645.0001	MW 399.51	$C_{22}H_{25}NO_4S$	Store at 2-8°C

Cysteine — FMOC-D-Cys-Trt

Synonyms: FMOC-S-Trt-D-Cys

Bachem B-2030.0001	MW 585.72	$C_{37}H_{31}NO_4S$	Store at -15°C
Neosystem FA00509	MW 585.7		
Peptides International FDC-1807-PI	MW 585.72	>98% (HPLC); white crystalline powder	
USBio F5361	MW 585.7	$C_{37}H_{31}NO_4S$	≥99%

Cysteine — FMOC-D-Cys-Trt-OPfp

Bachem B-2025.0001	MW 751.79	$C_{43}H_{30}NO_4SF_5$	Store at -15°C

Cysteine — FMOC-D-Cys-Trt-SASRIN™-® (200-400 mesh)

Bachem D-1985.0001	Store at -15°C

Cysteine — FMOC-L-Cys-4-Methylbenzyl

Neosystem FA00508	MW 447.6

Cysteine — FMOC-L-Cys-Acm

Neosystem FA00502	MW 414.5

Cysteine — FMOC-L-Cys-Allocam

Neosystem FA00503	MW 457.5

Cysteine — FMOC-L-Cys-StBu

Neosystem FA00507	MW 431.6

Cysteine — FMOC-L-Cys-tBu

Neosystem FA00506	MW 399.5

Cysteine — FMOC-L-Cys-Trt

Neosystem FA00510	MW 585.7

Cysteine — FMOC-L-Cys-Trt-Wang Resin

American Peptide RFWAN29
Neosystem RW00504

Cysteine — FMOC-S-4-Methylbenzyl-L-Cys

Fluka 47575	MW 447.55	$C_{26}H_{25}NO_4S$ ≥99% (HPLC); mp: 150-158°C

Cysteine — FMOC-S-Bzl-L-Cys-OPfp

Fluka 47445	MW 599.57	$C_{31}H_{22}F_5NO_4S$ ≥98% (HPLC)

Cysteine — Guanyl-Cys

Bachem F-3360.0001	MW 163.2	$C_4H_9N_3O_2S$	Store at -15°C

Cysteine — L-Cys

Fluka 30089 MW 121.16 $C_3H_7NO_2S$ ≥99.5% (titration); ≤0.5% foreign AA; ≤0.005% K, Na, Cl, SO$_4$; ≤0.0005% Al, Ba, Bi, Cd, Co, Cr, Cu, Fe, Li, Mg, Mn, Mo, Ni, Pb, Sr, Zn; ≤0.05% loss on drying, residue on ignition; ≤0.00001% As; ≤0.001% Ca; ≤0.05% NH_4^+

Fluka 30090 MW 121.16 $C_3H_7NO_2S$ ≥99% (titration); ≤0.1% residue on ignition, loss on drying; ≤0.005% K, Na; ≤0.0005% Cd, Co, Cr, Cu, Fe, Mg, Mn, Ni, Pb, Zn; ≤0.05% NH_4^+; ≤0.01% Cl, SO$_4$; ≤0.001% Ca; ≤0.3% foreign AA

Sigma C 4424 FW 121.2 Free Base $C_3H_7NO_2S$ Crystalline | Plant cell culture tested

Sigma C 7895 Matrix: 4% beaded agarose; activation: cyanogen bromide; attachment: amino; spacer: 1 atom; ligand immobilized: 0.5-1.0 μmoles/mL; form: suspension in 0.5 M NaCl, 0.01 M citrate, pH 4.5, containing 0.01% thimerosal

Sigma C 8152 FW 121.2 Free Base $C_3H_7NO_2S$ ≥98% (TLC); crystalline | Cell culture tested; insect cell culture tested

USBio C9000 MW 121.2 $C_3H_7NO_2S$ ≥98%; white, crystalline powder; specific rotation (C=8, 1 N HCl): +7° to +9.5°; pH: 4.5-5.5; heavy metals (Pb): ≤0.001%; endotoxin: ≤2 EU/mg; cell culture tested: murine myeloma SP2/0-Ag14

Cysteine — L-Cys (1-¹⁴C)

ARC ARC-841 MW 121.2 $HSCH_2CH(NH_2)COOH$ 25-50 mCi/mmol; 0.925-1.85 GBq/mmol; in 0.01 N HCl

Cysteine — L-Cys (3-¹⁴C)

ARC ARC-842 MW 121.2 $HSCH_2CH(NH_2)COOH$ 25-50 mCi/mmol; 0.925-1.85 GBq/mmol; in 0.01 N HCl

Cysteine — L-Cys (³⁵S)

Amersham SJ15232 Aqueous, containing 20 mM potassium acetate, 5 mM DTT, stabilized with 15 mM pyridine-3,4-dicarboxylic acid; >37 TBq/mmol, >1000 Ci/mmol; 370 MBq/mL, 10 mCi/mL

Amersham SJ232 $HSCH_2CH(NH_2)CO_2H$ Aqueous, containing 20 mM potassium acetate, 5 mM DTT; >37 TBq/mmol, >1000 Ci/mmol; 370 MBq/mL, 10 mCi/mL

ARC ARS-101 MW 121.2 $HSCH_2CH(NH_2)COOH$ 500-1k Ci/mmol; 18.5-37 TBq/mmol; in 10 mM BME in H_2O | BME

ICN 51002 MW 121.2 $HSCH_2CH(NH_2)COOH$ SA >800 Ci/mmol, >29.6 TBq/mmol; aqueous solution stabilized with 50 mM L-Lys, pH 7.4, 10 mM BME; ~10 mCi/mL; ~370 MBq/mL

Cysteine — L-Cys (³⁵S) Hydrochloride

Amersham SJ141 Aqueous solution; 0.74-5.5 GBq/mmol, 20-150 mCi/mmol

Cysteine — L-Cys Free Base

Synonyms: β-Mercapto-L-Ala; N-Methyl-D-Aspartate Agonist

ICN 101444 MW 121.2 $C_3H_7NO_2S$ Crystalline

ICN 194646 MW 121.2 $C_3H_7NO_2S$ Crystalline | Cell culture reagent

Sigma C 7755 FW 121.2 $C_3H_7NO_2S$ ≥98% (TLC) | NMDA agonist at quisqualate receptors at high concentration; also available as part of a kit

Cysteine — L-Cys Hydrate

ICN 194647 MW 175.6 $C_3H_7NO_2S \cdot HCl \cdot H_2O$ Cell culture reagent

Cysteine — L-Cys Hydrochloride

Alexis 101-012 MW 157.7 $C_3H_7NO_2S$ White solid; soluble in H_2O | Hygroscopic

Amersham US14035 98.5-101.5% (dry basis, monohydrate)

Calbiochem 2430 MW 157.6 $C_3H_7NO_2S \cdot HCl$ >98% (titration); solid; soluble in H_2O

Cysteine — L-Cys Hydrochloride Anhydrous

Fluka 30120 MW 157.62 $C_3H_7NO_2S \cdot HCl$ ≥99% (titration); ≤0.3% foreign AA, loss on drying; ≤0.005% K, Na; ≤0.0005% Cd, Co, Cr, Cu, Fe, Mg, Mn, Ni, Pb, Zn; ≤0.1% residue on ignition; ≤0.05% NH_4^+; ≤0.001% Ca; ≤0.01% SO_4

Sigma C 1276 FW 157.6 $C_3H_7NO_2S \cdot HCl$ ≥98% (TLC) | NMDA agonist that is a quisqualate receptor agonist at high concentration

Sigma C 2529 FW 157.6 $C_3H_7NO_2S \cdot HCl$ Crystalline | Cell culture tested

USBio C9005 MW 157.6 $C_3H_7NO_2S \cdot HCl$ ≥98%; white, crystalline powder; specific rotation (C=8, 1 N HCl): +6.7° to +7.3°; heavy metals (Pb): ≤0.001%; endotoxin: ≤2 EU/mg; cell culture tested: murine myeloma SP2/0-Ag14

Cysteine — L-Cys Hydrochloride Hydrate

Synonyms: NMDA Agonist

Fluka 30119 MW 157.62 $C_3H_7NO_2S \cdot HCl$ ≥99.5% (titration); ≤0.5% foreign AA, loss on drying; ≤0.005% K, Na; ≤0.0005% Al, Ba, Bi, Cd, Co, Cr, Cu, Fe, Li, Mg, Mn, Mo, Ni, Pb, Sr, Zn; ≤0.05% residue on ignition, NH_4^+; ≤0.001% Ca; ≤0.01% SO_4

Fluka 30129 MW 175.64 $C_3H_7NO_2S \cdot HCl$ ≥99% (titration); ≤0.5% foreign AA; ≤0.005% K, Na; ≤0.0005% Al, Ba, Bi, Cd, Co, Cr, Cu, Fe, Li, Mg, Mn, Mo, Ni, Pb, Sr, Zn; ≤0.05% residue on ignition, NH_4^+; ≤0.00005% As; ≤0.001% Ca; ≤0.01% SO

Fluka 30130 MW 175.64 $C_3H_7NO_2S \cdot HCl \cdot H_2O$ ≥99% (titration); ≤1% foreign AA; ≤0.005% K, Na; ≤0.0005% Cd, Co, Cr, Cu, Fe, Mg, Mn, Ni, Pb, Zn; ≤0.1% residue on ignition; ≤0.05% NH_4^+; ≤0.001% Ca; ≤0.01% SO

ICN 101446 MW 175.6 $C_3H_7NO_2S \cdot HCl \cdot H_2O$ Crystalline

Neosystem AA00501 MW 175.6

Peptides International ALC-2705 MW 175.64 >99.9% optical purity by RP-HPLC; white crystalline powder

Sigma C 4820 FW 175.6 $C_3H_7NO_2S \cdot HCl \cdot H_2O$ SigmaUltra >99% (TLC); residue on ignition: <0.4%; solubility (1 M in H_2O, 20°C): complete, colorless; insoluble matter: <0.1%; Al, Cu: <0.0005%; Ca, K: 0.005%; Fe: <0.003%; Mg: <0.001% | NMDA agonist at quisqualate receptors at high concentration

Sigma C 7880 FW 175.6 $C_3H_7NO_2S \cdot HCl \cdot H_2O$ ≥98% (TLC | NMDA agonist at quisqualate receptors at high concentration; non-profit & government laboratories may request 1X5 g gratis

Sigma C 8277 FW 175.6 $C_3H_7NO_2S \cdot HCl \cdot H_2O$ Crystalline | Cell culture tested

USBio C9010 MW 175.6 $C_3H_7NO_2S \cdot HCl \cdot H_2O$ ≥98%; white, crystalline powder; specific rotation (C=8, 1 N HCl): +5.5° to +7.0°; pH: 1.0-2.0; heavy metals (Pb): ≤0.001%; endotoxin: ≤2 EU/mg; cell culture tested: murine myeloma SP2/0-Ag14

Cysteine — L-Cys S-Sulfate

Synonyms: S-Sulfo-L-Cys; NMDA Receptor Ligand

Fluka 30190 MW 201.21 $C_3H_7NO_5S_2$ ≥99% (TLC); ≤8% H_2O

Sigma C 2196 FW 201.2 $C_3H_7NO_5S_2$ ≥98% (TLC) | Olverman, HJ et al, *Neuroscience*, 26: 17, 1988; Mewett, KN et al, in *CNS Receptors: From Molecular Pharmacology to Behavior*, Mandel, P & DeFeudis, FV, eds, Raven Press, New York, 1983, p. 163

Cysteine — L-Cys Sulfinic Acid

ICN 101448 MW 153.2 $C_3H_7NO_4S$ Crystalline

Sigma C 4418 FW 153.2 $C_3H_7NO_4S$ Putative excitatory AA neurotransmitter; Griffiths, R, *Prog Neurobiol*, 35: 313, 1990

Cysteine — L-Cys-Bisally Ester Bis-Tosyl

Fluka 30230 MW 664.84 $C_{12}H_{20}N_2O_4S_2 \cdot 2C_7H_8O_3S$ ~97% (TLC); mp: 175-180°C

Cysteine — L-Cys-Bzl

Neosystem AA00502 MW 211.3

Cysteine — L-Cys-OEt Hydrochloride

Fluka 30100 MW 185.68 $C_5H_{11}NO_2S \cdot HCl$ ≥99% (titration); mp: 126-128°C

ICN 101443 MW 185.7 $C_5H_{11}NO_2S \cdot HCl$ Crystalline; ~98%

ICN 194648 MW 185.7 $C_5H_{11}NO_2S \cdot HCl$ ~98% | Cell culture reagent

Sigma C 2757 FW 185.7 $C_5H_{11}NO_2S \cdot HCl$ ~98% (TLC)

Cysteine — L-Cys-OMe Hydrochloride

Fluka 30160	MW 171.65	$C_4H_9NO_2S \cdot HCl$	≥99% (titration);
mp: 140°C			
ICN 150750	MW 171.6	$C_4H_9NO_2S \cdot HCl$	Crystalline
Sigma C 8255	FW 171.6	$C_4H_9NO_2S \cdot HCl$	

Cysteine — L-Cystine

Amersham US14050 98.0-102.0%; ≤0.004% heavy metals (Pb)

Fluka 30199 MW 240.3 $C_6H_{12}N_2O_4S_2$ ≥99.5% (titration); ≤0.3% foreign AA; ≤0.005% K, Na; ≤0.0005% Al, Ba, Bi, Cd, Co, Cr, Cu, Fe, Li, Mg, Mn, Mo, Ni, Pb, Sr, Zn; ≤0.05% residue on ignition, loss on drying, Cl; ≤0.001% Ca; ≤0.00001% As; ≤0.01% SO_4

Fluka 30200 MW 240.3 $C_6H_{12}N_2O_4S_2$ ≥99% (titration); ≤0.3% foreign AA; ≤0.005% K, Na, Ca; ≤0.0005% Cd, Co, Cr, Fe, Mg, Mn, Ni, Pb, Zn; ≤0.1% residue on ignition, loss on drying; ≤0.05% Cl; ≤0.0006% Cu; ≤0.01% SO_4

Peptides International ALC-2706 MW 240.30 >99.9% optical purity by RP-HPLC; white crystalline powder

Sigma C 6195 FW 240.3 $C_6H_{12}N_2O_4S_2$ SigmaUltra >99% (TLC); residue on ignition: <0.1%; solubility (0.5 M in 1.0 M HCl, 20°C): complete, colorless; insoluble matter: <0.1%; Al, Cu, Ca, Fe, Mg: <0.0005%; K: 0.005%%

Sigma C 8755 FW 240.3 $C_6H_{12}N_2O_4S_2$ Crystalline; Sigma Grade | Also available as part of a kit

Sigma C 8786 FW 240.3 $C_6H_{12}N_2O_4S_2$ Crystalline | Cell culture tested

USBio C9050 MW 240.3 $C_6H_{12}N_2O_4S_2$ ≥98.5%; white, crystalline powder; specific rotation (C=2, 1 N HCl): -215° to -225°; pH: 5.0-6.5; heavy metals (Pb): ≤0.001%; endotoxin: ≤2 EU/mg; cell culture tested: murine myeloma SP2/0-Ag14

Cysteine — L-Cystine Dihydrochloride

Sigma C 2526	FW 313.2	$C_6H_{12}N_2O_4S_2 \cdot 2HCl$	
Sigma C 7777	FW 313.2	$C_6H_{12}N_2O_4S_2 \cdot 2HCl$	Crystalline |
Cell culture tested; insect cell culture tested			

USBio C9055 MW 313.2 $C_6H_{12}N_2O_4S_2 \cdot 2HCl$ ≥98%; identification by IR absorption spectrum; specific rotation: -165° to -177°; solubility: clear, colorless & complete; residue after ignition: ≤0.1%; heavy metals (as Pb): ≤0.002%; iron: ≤0.003%; chloride: ≤23.5%

Cysteine — L-Cystine Disodium Hydrate

USBio C9057 MW 302.3 $C_6H_{10}N_2O_4S_2 \cdot 2Na \cdot H_2O$ ≥98%; appearance: white to off-white crystalline powder; solubility (1%): clear, colorless & complete; specific rotation (C=2, 1 N HCl): -168.5 to -178.9°C; loss on drying: ≤6.0%; pH (1%): 9.5-10.5; heavy metals (Pb): ≤0.002%; cell culture analysis: murine myeloma SP2/0-Ag14

Cysteine — L-Cystine-Di-2-Naphthylamide

Fluka 30260	MW 490.65	$C_{26}H_{26}N_4O_2S_2$	≥98% (HPLC)

Cysteine — L-Cystine-Di-β-Naphthylamide Dihydrochloride

Sigma C 9130	FW 563.6	$C_{26}H_{26}N_4O_2S_2 \cdot 2HCl$	Possibly
carcinogenic			

Cysteine — L-Cystine-Di-β-Naphthylamide Free Base Dihydrochloride

Sigma C 9005	FW 490.6	$C_{26}H_{26}N_4O_2S_2 \cdot 2HCl$	Possibly
carcinogenic			

Cysteine — L-Cystine-OEt₂ Dihydrochloride

Sigma C 4772	FW 369.3	$C_{10}H_{20}N_2O_4S_2 \cdot 2HCl$

Cysteine — L-Cystine-OMe₂ Dihydrochloride

Fluka 30250	MW 341.28	$C_8H_{16}N_2O_4S_2 \cdot 2HCl$	~97%
(HPLC); mp: 177°C			
Sigma C 8880	FW 341.3	$C_8H_{16}N_2O_4S_2 \cdot 2HCl$	

Cysteine — L-Homocystine

Synonyms: L-4,4'-Dithio-Bis-2-Aminobutanoic Acid

Fluka 53540	MW 268.36	$C_8H_{16}N_2O_4S_2$	≥99.0% (titration);
mp: 281-284°C			
Sigma G 4137	FW 268.3	$C_8H_{16}N_2O_4S$	Cell culture tested
Sigma H 6010	FW 268.3	$C_8H_{16}N_2O_4S_2$	

Cysteine — N-(3-(2-Furyl)-Acryloyl)-S-Farnesyl-L-Cys

Sigma F 3792	FW 445.6	$C_{25}H_{35}NO_4S$	≥90% (TLC)

Cysteine — N-(S-Bzl-L-Cys)-2-Aminoacridone

Fluka 13345	MW 403.5	$C_{23}H_{21}N_3O_2S$	≥98% (TLC); mp:
243°C; soluble in DMF, DMSO			

Cysteine — N,N'-bist-BOC-L-Cys

Sigma B 5148	$C_{16}H_{28}N_2O_8S$

Cysteine — N,N-Di(2,4-Dnp)-L-Cystine

Sigma D 8879	FW 572.5	$C_{18}H_{16}N_6O_{12}S_2$

Cysteine — N,N'-Dibenzoyl-L-Cystine

Fluka 33562	MW 448.52	$C_{20}H_{20}N_2O_6S_2$	mp: 196-200°C

Cysteine — N,N'-Didansyl-L-Cystine

Sigma D 0625	FW 706.9	$C_{30}H_{34}N_4O_8S_4$	~95%

Cysteine — N,S-Di(2,4-Dnp)-L-Cys

Sigma D 8754	FW 453.3	$C_{15}H_{11}N_5O_{10}S$	~95%

Cysteine — N,S-Di-CBZ-L-Cys

Sigma C 6626	FW 389.4	$C_{19}H_{19}NO_6S$

Cysteine — N,S-Di-Dnp-L-Cys

ICN 101439	MW 453.3	$C_{15}H_{11}N_5O_{10}S$	~95%; crystalline

Cysteine — N,S-Di-Z-L-Cys

Fluka 96070	MW 389.42	$C_{19}H_{19}NO_6S$	≥98.0% (HPLC)

Cysteine — N-Ac-L-Cys

Alexis 151-005 MW 163.2 $C_5H_9NO_3S$ ≥99%; white solid; soluble in H_2O or alcohol

Amersham US10120 98-102% | Mucolytic agent; induces apoptosis in human & rat smooth muscle cells

Calbiochem 106425 MW 163.2 $C_5H_9NO_3S$ ≥98% (titration); solid; soluble in H_2O

Fluka 01039 MW 163.2 $C_5H_9NO_3S$ ≥99% (titration); mp: 110-112°C

ICN 100098 MW 163.2 $C_5H_9NO_3S$ Crystalline; >96% | A mucolytic agent for isolation of mycobacteria from sputum

ICN 194603 MW 163.2 $C_5H_9NO_3S$ Crystalline; >96% | A mucolytic agent for isolation of mycobacteria from sputum; cell culture reagent

Sigma A 7250 $C_5H_9NO_3S$ Facilitates NO generation by platelets; increases cellular pools of free radical scavengers; prevents apoptotic death of neuronal cells but induces apoptosis in human & rat smooth muscle cells; mucolytic agent for the isolation of mycobacteria from sputum; inhibitor of HIV replication in cell lines & peripheral blood mononuclear cells; Kubica, GP et al, *Am Resp Dis*, 87: 775, 1963; Mulcahy, JD et al, *J Conf State & Prov Pub Health Lab Dir*, 23: 5, 1990; Roederer, M et al, *PNAS USA*, 87: 4884, 1990

Sigma A 8199 $C_5H_9NO_3S$ Facilitates NO generation by platelets; increases cellular pools of free radical scavengers; prevents apoptotic death of neuronal cells but induces apoptosis in human & rat smooth muscle cells; mucolytic agent for the isolation of mycobacteria from sputum; inhibitor of HIV replication in cell lines & peripheral blood mononuclear cells; Kubica, GP et al, *Am Resp Dis*, 87: 775, 1963; Mulcahy, JD et al, *J Conf State & Prov Pub Health Lab Dir*, 23: 5, 1990; Roederer, M et al, *PNAS USA*, 87: 4884, 1990

Sigma A 9165 FW 163.2 $C_5H_9NO_3S$ Cell culture tested

USBio A0526-50 MW 163.2 $C_5H_9NO_3S$

Cysteine — *N*-Ac-L-Cys (1-^{14}C)

ARC ARC-838 MW 163.2 $HSCH_2CH(NHCOCH_3)COOH$ 25-50 mCi/mmol; 0.925-1.85 GBq/mmol; in EtOH

ICN 16010 MW 163.2 $HSCH_2CH(NHCOCH_3)COOH$ EtOH solution; 25-50 mCi/mmol, 0.925-1.85 GBq/mmol

Cysteine — *N*-Ac-L-Cys-OMe

Fluka 01042 MW 177.22 $C_6H_{11}NO_3S$ ~97% (HPLC); mp: 77-81°C

Cysteine — *N*-Ac-L-Cystine-S-(2R,3S,4R)-3-Hydroxy-2-((1S)-1-Hydroxy-2-Methylpropyl)-4-Me-5-Oxo-2-Pyrolidinecarbonyl

Synonyms: Lactacystin; Proteasome Inhibitor

Peptides International ILC-4368-v Microbial MW 376.43 $C_{15}H_{24}N_2O_7S$ Lyophilized amorphous powder; integrity assessed by activity | Omura, S et al, *J Antibiotics*, 44:113, 1991; Omura, S et al, *J Antibiotics*, 44:117, 1991; Fenteany, G et al, *Science*, 268:726, 1995; Imajoh-Ohmi, S et al, *BBRC*, 217:1070, 1995

Cysteine — *N*-Ac-S-Farnesyl-L-Cys

Synonyms: S-Farnesyl Cysteine Methyltransferase Inhibitor; S-Farnesyl Cysteine Methyltransferase Substrate

Alexis 290-001 MW 367.6 $C_{20}H_{33}NO_3S$ ≥98%; clear oil; soluble in DMSO, MeOH or EtOH

ARC ARCD-120 MW 367.5 Solid

Calbiochem 110110 MW 367.6 $C_{20}H_{33}NO_3S$ ≥98% (TLC); clear oil; soluble in DMSO & EtOH

ICN 159030 MW 367.5 98% | Specific inhibitor; also prevents carboxyl methylation of p21^{ras} platelet RAP 1 & the transduction of γ subunit; Volker, C et al, *JBC*, 266:21515, 1991; Huzoor-Akabar etal, ibid, 266:4387, 1991; Perez-Sala, D etal, *PNAS*, 88: 3043, 1991

Sigma A 9201 $C_{20}H_{33}NO_3S$ High affinity for S-farnesylcysteine methyl transferase, thereby inhibiting carboxyterminal methylation of proteins like p21^{ras} & Rap 1; Volker, C et al, *JBC*, 266: 21515, 1991; Huzoor-Akbar, et al, *JBC*, 266: 4387, 1991

Cysteine — *N*-Ac-S-Farnesyl-L-Cys (1-^{3}H)

ARC ART-468 MW 343.48 10-20 Ci/mmol; 370-740 GBq/mmol; in EtOH

ICN 26020 MW 367.5 10-20 Ci/mmol, 370-740 MBq/mmol; EtOH

Cysteine — *N*-Ac-S-Farnesyl-L-Cys-OMe

Alexis 290-010 MW 381.6 $C_{21}H_{35}NO_3S$ ≥96%; slightly yellow solid; soluble in MeOH

Sigma A 5332 $C_{21}H_{35}NO_3S$ Substrate for studying reversible methyl esterification of signal-transducing G proteins; Tan, EW et al, *JBC*, 266: 10719, 1991

Cysteine — *N*-Ac-S-Geranylgeranyl-L-Cys

Synonyms: AGGC

Alexis 290-008 MW 435.7 $C_{25}H_{41}O_3S$ ≥98%; viscous yellow oil; soluble in DMSO or EtOH

Calbiochem 110115 MW 435.7 $C_{25}H_{41}NO_3S$ ≥98% (TLC); yellow viscous oil; soluble in DMSO & EtOH

ICN 159845 MW 435.7 98% | Specifically inhibits methyl esterification of geranylgeranylated proteins; also blocks signal transduction in human neutrophils that are receptor mediated; Phillips, MR etal, *Science*, 259:977, 1993; Volker, C etal, *FEBS Lett*, 295:189, 1991

Cysteine — *N*-Ac-S-Geranyl-L-Cys

Synonyms: AGC

Alexis 290-007 MW 299.4 $C_{15}H_{25}O_3S$ ≥98%; clear viscous oil; soluble in DMSO or EtOH

Calbiochem 110113 MW 299.4 $C_{15}H_{25}NO_3S$ ≥98% (TLC); clear viscous oil; soluble in DMSO & EtOH

ICN 195681 MW 299.4 98% | A biologically inactive close structural analog of AFC & AGGC; used as a negative control

Cysteine — *N*-BOC-S-Et-Carbamoyl-L-Cys

American Peptide BLCYS35 MW 292.3

Cysteine — *N*-Bzl-L-Cys-4-NA

Fluka 13340 MW 331.39 $C_{16}H_{17}N_3O_3S$ ≥98% (titration); mp: 98-100°C

Cysteine — *N*-CBZ-S-Bzl-L-Cys

Sigma C 5876 FW 345.4 $C_{18}H_{19}NO_4S$

Cysteine — *N*-FMOC-S-Acm-L-Cys

Sigma F 9268 FW 414.5 $C_{21}H_{22}N_2O_5S$

Cysteine — *N*-FMOC-S-tBu-L-Cys

Sigma F 1258 FW 399.5 $C_{22}H_{25}NO_4S$

Cysteine — *N*-FMOC-S-tBu-L-Cys Resin Ester

Sigma F 1136 For peptide synthesis; Wang, SS, *J Am Chem Soc*, 95: 1328, 1973; Lu, G et al, *J Org Chem*, 46: 3433, 1981

Cysteine — *N*-FMOC-S-Trt-L-Cys

ICN 158177 MW 585.7 $C_{37}H_{31}NO_4$

Sigma F 6510 FW 585.7 $C_{37}H_{31}NO_4S$

Cysteine — *N*-Isobutyryl-D-Cys

Fluka 58689 MW 191.25 $C_7H_{13}NO_3S$ ≥97%; mp: 97-101°C; enantiomer ratio: ≥99.5:0.5

Cysteine — *N*-Isobutyryl-L-Cys

Fluka 58698 MW 191.25 $C_7H_{13}NO_3S$ ≥97%; mp: 97-101°C; enantiomer ratio: ≥99.5:0.5

Cysteine — *N*-O-Nps-L-Cys Hydrate

ICN 102526

Cysteine — *N*-t-BOC-S-Acm-L-Cys

Sigma B 8891 FW 292.3 $C_{11}H_{20}N_2O_5S$

Cysteine — *N*-t-BOC-*S*-Acm-L-Cys Resin Ester
Sigma B 2772   For peptide synthesis by the Merrifield method

Cysteine — *N*-t-BOC-*S*-Bzl-L-Cys
Sigma B 5751 $C_{15}H_{21}NO_4S$

Cysteine — *N*-t-BOC-*S*-*p*-Methylbenzyl-L-Cys
Sigma B 0265 $C_{16}H_{23}NO_5S$
Sigma B 9017   FW 325.4 $C_{16}H_{23}NO_4S$

Cysteine — *N*-t-BOC-*S*-*p*-Methylbenzyl-L-Cys Pam Resin Ester
Sigma B 3143   For peptide synthesis by the Merrifield method; Mitchell, AR et al, *J Org Chem*, 43: 2845, 1978

Cysteine — *N*-t-BOC-*S*-Trt-L-Cys
Sigma B 7006 $C_{27}H_{29}NO_4$

Cysteine — *N*-Trt-Cys
ICN 157142   MW 317.4 $C_{21}H_{19}NO_2$

Cysteine — *N*ᵃ,*N*ᵃ'-Di-BOC-L-Cystine
Fluka 15387   MW 440.54 $C_{16}H_{28}N_2O_8S_2$ mp: 140-145°C

Cysteine — *N*ᵃ-BOC-*S*-3-Nitro-2-Pyridylthio-L-Cys
Fluka 15554   MW 375.41 $C_{13}H_{17}N_3O_6S_2$ ~99% (TLC); mp: 160°C

Cysteine — *N*ᵃ-BOC-*S*-Trt-L-Cys
Fluka 15511   MW 463.59 $C_{27}H_{29}NO_4S$ ~99% (TLC)

Cysteine — *N*ᵃ-FMOC-*S*-4-Methylbenzyl-L-Cys
Fluka 47525   MW 463.55 $C_{26}H_{25}NO_5S$ ≥98% (HPLC); mp: 140°C

Cysteine — *N*ᵃ-FMOC-*S*-Acm-L-Cys
Fluka 47603   MW 414.48 $C_{21}H_{22}N_2O_5S$ ~97% (HPLC); mp: 147-150°C

Cysteine — *N*ᵃ-FMOC-*S*-Acm-L-Cys 4-Benzyloxybenzyl Ester Polymer Bound
Fluka 47613   0.4-0.6 g/resin; carrier: polystyrene, crosslinked with 1% DVB; 200-400 mesh particle size

Cysteine — *N*ᵃ-FMOC-*S*-Acm-L-Cys-OPfp
Fluka 47439   MW 580.53 $C_{27}H_{21}F_5N_2O_5S$ ≥97% (titration)

Cysteine — *N*ᵃ-FMOC-*S*-Butyl-L-Cys-OPfp
Fluka 47448   MW 565.56 $C_{28}H_{24}F_5NO_4S$ ≥98% (TLC)

Cysteine — *N*ᵃ-FMOC-*S*-StBu-L-Cys
Fluka 47631   MW 431.57 $C_{22}H_{25}NO_4S_2$ ~97% (titration); mp: 73-77°C

Cysteine — *N*ᵃ-FMOC-*S*-StBu-L-Cys 4-Benzyloxybenzyl Ester Polymer Bound
Fluka 47651   0.4-0.6 mmol/g resin; carrier: polystyrene, crosslinked with 1% DVB; 200-400 mesh particle size

Cysteine — *N*ᵃ-FMOC-*S*-tBu-L-Cys
Fluka 47516   MW 399.51 $C_{22}H_{25}NO_4S$ ≥98% (TLC)
ICN 151136   MW 399.5 $C_{22}H_{25}NO_4S$

Cysteine — *N*ᵃ-FMOC-*S*-Trt-L-Cys
Fluka 47695   MW 585.72 $C_{37}H_{31}NO_4S$ ~97% (HPLC); mp: 171-174°C

Cysteine — *N*ᵃ-FMOC-*S*-Trt-L-Cys-OPfp
Fluka 47476   MW 751.77 $C_{43}H_{30}F_5NO_4S$ ≥98% (HPLC)

Cysteine — *N*ᵃ-t-BOC-EthylMercapto-L-Cys Dicyclohexylammonium Salt
Sigma B 9266   FW 462.7 $C_{10}H_{19}NO_4S \cdot C_{12}H_{23}N$

Cysteine — *N*ᵃ-t-BOC-*S*-Bzl-L-Cys
ICN 101065   MW 311.4 $C_{15}H_{21}NO_4S$ Crystalline

Cysteine — *N*ᵃ-t-BOC-*S*-*p*-Methylbenzyl-L-Cys
ICN 150504   MW 341.4 $C_{16}H_{23}NO_5S$

Cysteine — *N*ᵃ-Z-*S*-Bzl-L-Cys
Fluka 96012   MW 345.41 $C_{18}H_{19}NO_4S$ ≥98.0% (TLC)

Cysteine — *N*ᵃ-Z-*S*-Bzl-L-Cys-(4-ONp)
Fluka 96014   MW 466.51 $C_{24}H_{22}N_2O_6S$ ≥98.0% (HPLC)

Cysteine — Palmitoyl-Cys-((*RS*)-2,3-Di(Palmitoyloxy)-Propyl)
Synonyms: Pam³-Cys
Bachem F-2630.0250   MW 910.48 $C_{54}H_{103}NO_7S$ Store at -15°C

Cysteine — Pth-*S*-Carboxymethyl-Cys
Sigma P 9141   FW 296.4 $C_{12}H_{12}N_2O_3S_2$ ~99% | Useful as standards in protein sequencing by the Edman degradation

Cysteine — Pth-*S*-Phenylthiocarbamyl-Cys
Sigma P 9266   FW 373.5 $C_{17}H_{15}N_3OS_3$ ~99% | Useful as standards in protein sequencing by the Edman degradation

Cysteine — *S*-2-(4-Pyridyl)Et-L-Cys
Fluka 82910   MW 226.30 $C_{10}H_{14}N_2O_2S$ ≥99.0% (titration); mp: 212-214°C

Cysteine — *S*-2-Aminoethyl-L-Cys Hydrochloride
Synonyms: Thialsine
ICN 100616   MW 200.7 $C_5H_{12}N_2O_3S \cdot HCl$ Crystalline | Can be used with trypsin to break peptides at –SH groups; Work, E, *BBA*, 62:173, 1962
Sigma A 2636 $C_5H_{12}N_2O_2S \cdot HCl$

Cysteine — *S*-Acm-L-Cys Hydrochloride
ICN 154674   MW 228.7 $C_6H_{12}N_2O_3S \cdot HCl$

Cysteine — *S*-Adenosyl-L-Cys
Sigma A 7772 $C_{13}H_{18}N_6O_5S$

Cysteine — *S*-Allyl-L-Cys
Synonyms: L-Deoxyalliin
ICN 193749   MW 161.2 $C_6H_{11}NO_2$

Cysteine — *S*-Bzl-L-Cys
Fluka 13310   MW 211.29 $C_{10}H_{13}NO_2S$ ≥98% (titration); mp: ~220°C
ICN 100911   MW 211.3 $C_{10}H_{13}NO_2S$ Crystalline

Cysteine — S-Bzl-L-Cys-OEt Hydrochloride

ICN 154841	MW 275.8	$C_{12}H_{17}NO_2S \cdot HCl$	Crystalline
Sigma B 7875	$C_{12}H_{17}NO_2S \cdot HCl$		

Cysteine — S-Bzl-L-Cys-OMe Hydrochloride

ICN 154842	MW 261.8	$C_{11}H_{17}NO_2S \cdot HCl$	Crystalline
Sigma B 8000	FW 261.8	$C_{11}H_{15}NO_2S \cdot HCl$	

Cysteine — S-Bzl-L-Cys-pNA

ICN 100915	MW 331.4	$C_{16}H_{17}N_3O_3S$	Crystalline

Cysteine — S-Carboxyethyl-L-Cys

Sigma C 9544	FW 193.2	$C_6H_{11}NO_4S$

Cysteine — S-Carboxymethyl-L-Cys

Fluka 21905	MW 179.2	$C_5H_9NO_4S$	≥98% (titration); mp: 210-213°C
Sigma C 7757	FW 179.2	$C_5H_9NO_4$	Crystalline

Cysteine — S-CBZ-L-Cys

Sigma C 6501	FW 255.3	$C_{11}H_{13}NO_4S$

Cysteine — Seleno-DL-Cystine

Sigma S 1650	FW 334.1	$C_6H_{12}N_2O_4Se_2$	Yellow-brown powder | Very toxic

Cysteine — S-Et-L-Cys

ICN 101670	MW 125.1	$C_5H_{11}NO_2S$
Sigma E 1878	FW 149.2	$C_5H_{11}NO_2S$

Cysteine — S-Me-L-Cys

Fluka 66560	MW 135.19	$C_4H_9NO_2S$	≥99.0% (titration); mp: 240°C; ≤0.5% residue on ignition
Sigma M 6626	FW 135.2	$C_4H_9NO_2S$	

Cysteine — S-N-Benzylthio-Carbamoyl-L-Cys

Synonyms: Glutathione S-Transferase Inducer

ICN 159071	MW 270.4	$C_{11}H_{14}N_2O_2S_2$	Crystalline; >98% | Zheng, G etal, *J Med Chem*, 35:185, 1992

Cysteine — S-tBu-L-Cys Hydrochloride

ICN 154909	MW 213.7	$C_7H_{15}NO_2S \cdot HCl$

Cysteine — S-t-Butylmercapto-L-Cys

Synonyms: S-StBu-L-Cys; 3-t-Butyldithiol-L-Ala

Fluka 20235	MW 209.33	$C_7H_{15}NO_2S_2$	~97% (titration); ≤0.05% residue on ignition; mp: 190°C; ≤0.5% free Cys
Sigma B 5890	$C_7H_{15}NO_2S_2$		

Cysteine — S-Trt-L-Cys

ICN 103137	MW 363.5	$C_{22}H_{21}NO_2S$	Crystalline

Cysteine — S-β-4-Pyridylethyl-L-Cys

Sigma P 0139	FW 226.3	$C_{10}H_{14}N_2O_2S$	Internal standard for AA analysis; Cavins, JF & Friedman, M, *Anal Biochem*, 35: 489, 1970; Fullmer, CS, *Anal Biochem*, 142: 336, 1984

Cysteine — TentaGel S PHB-Cys-Acm-FMOC

Fluka 86383	0.22 mmol protected AA/g; ~90 μm particle size

Cysteine — TentaGel S PHB-Cys-StBu-FMOC

Synonyms: N^α-FMOC-S-Tert-Butylthio-L-Cys 4-(Poly(Ethylenoxy))-OBzl Polymer Bound

Fluka 86385	0.20 mmol protected AA/g; ~90 μm particle size

Cysteine — TentaGel S PHB-Cys-tBu-FMOC

Synonyms: N^α-FMOC-S-tBu-L-Cys 4-(Poly(Ethylenoxy))-OBzl Polymer Bound

Fluka 86384	0.22 mmol protected AA/g; ~90 μm particle size
Fluka 86415	0.20 mmol protected AA/g; ~90 μm particle size

Cysteine — TentaGel S PHB-Cys-Trt-FMOC

Fluka 86386	0.20 mmol protected AA/g; ~90 μm particle size

Cysteine — TentaGel S Trt-Cys-Acm-FMOC

Fluka 86414	0.20 mmol protected AA/g; ~90 μm particle size

Cysteine — TentaGel S Trt-Cys-StBu-FMOC

Fluka 86416	0.20 mmol protected AA/g; ~90 μm particle size

Cysteine — TentaGel S Trt-Cys-Trt-FMOC

Fluka 86417	0.18 mmol protected AA/g; ~90 μm particle size

Cysteine — Trt-Cys-Trt

Bachem E-2940.0001	MW 605.8	$C_{41}H_{35}NO_2S$	Store at -15°C

Cysteine — Trt-Cys-Trt DEA

Bachem E-1815.0001	MW 678.94	$C_{41}H_{35}NO_2S \cdot C_4H_{11}N$	Store at -15°C

Cysteine — Z-Cys-Bzl

Synonyms: Benzyloxycarbonyl-S-Bzl-L-Cys

Bachem C-1380.0005	MW 345.42	$C_{18}H_{19}NO_4S$	Store at RT
Peptides International ZLC-2014-PI	MW 345.41	>98% (HPLC); white powder	

Cysteine — Z-Cys-Bzl-ONp

Bachem C-1385.0005	MW 466.52	$C_{24}H_{22}N_2O_6S$	Store at -15°C

Cysteine — Z-Cys-Z

Bachem C-2910.0005	MW 389.43	$C_{19}H_{19}NO_6S$	Store at 2-8°C

Cysteine — Z-L-Cys-Bzl

Neosystem ZA00501	MW 345.4

Cysteine — α,α-Dimethyl-D-Cys-Trt

Synonyms: D-Pen-Trt

Bachem F-3070.0001	MW 391.53	$C_{24}H_{25}NO_2S$	Store at 2-8°C

Cystine — L-Cys (^{35}S)

Amersham SJ126	$(HO_2CCH(NH_2)CH_2S-)_2$	Solid; 1.5-9.3 GBq/mmol, 40-250 mCi/mmol

Dab — BOC-Dab

Bachem A-3215.0001	MW 218.25	$C_9H_{18}N_2O_4$	Store at RT

Dab — BOC-Dab-Aloc

Bachem A-4125.0001   MW 302.33 $C_{13}H_{22}N_2O_6$ Store at 2-8°C

Dab — BOC-Dab-BOC Dicyclohexylamine

Bachem A-3480.0001   MW 499.69 $C_{14}H_{26}N_2O_6 \cdot C_{12}H_{23}N$ Store at RT

Dab — BOC-Dab-FMOC

Bachem A-3520.0001   MW 440.5 $C_{24}H_{28}N_2O_6$ Store at RT

Dab — BOC-Dab-OtBu Hydrochloride

Bachem A-4415.0001   MW 310.82 $C_{13}H_{26}N_2O_4 \cdot HCl$ Store at 2-8°C

Dab — BOC-Dab-Z Dicyclohexylamine

Bachem A-2905.0001   MW 533.71 $C_{17}H_{24}N_2O_6 \cdot C_{12}H_{23}N$ Store at RT

Dab — BOC-D-Dab

Bachem A-4215.0001   MW 218.25 $C_9H_{18}N_2O_4$ Store at RT

Dab — BOC-D-Dab-FMOC

Bachem A-4230.0001   MW 440.5 $C_{24}H_{28}N_2O_6$ Store at 2-8°C

Dab — BOC-D-Dab-Z Dicyclohexylamine

Bachem A-4260.0001   MW 533.71 $C_{17}H_{24}N_2O_6 \cdot C_{12}H_{23}N$ Store at RT

Dab — Dab Dihydrochloride

Bachem F-3050.0001   MW 191.06 $C_4H_{10}N_2O_2 \cdot 2HCl$ Store at 2-8°C

Dab — Dab-BOC

Bachem A-3305.0001   MW 218.25 $C_9H_{18}N_2O_4$ Store at RT

Dab — Dab-BOC-OMe Hydrochloride

Bachem E-3360.0001   MW 268.74 $C_{10}H_{20}N_2O_4 \cdot HCl$ Store at -15°C

Dab — D-Dab Dihydrochloride

Bachem F-3055.0001   MW 191.06 $C_4H_{10}N_2O_2 \cdot 2HCl$ Store at 2-8°C

Dab — FMOC-Dab

Bachem B-2300.0001   MW 340.38 $C_{19}H_{20}N_2O_4$ Store at 2-8°C

Dab — FMOC-Dab-Adpoc

Bachem B-2860.0001   MW 560.69 $C_{33}H_{40}N_2O_6$ Store at -15°C

Dab — FMOC-DabBOC

Bachem B-1800.0001   MW 440.5 $C_{24}H_{28}N_2O_6$ Store at 2-8°C

Dab — FMOC-Dab-FMOC

Bachem B-2270.0001   MW 562.62 $C_{34}H_{30}N_2O_6$ Store at 2-8°C

Dab — FMOC-Dab-Z

Bachem B-3250.0001   MW 474.51 $C_{27}H_{26}N_2O_6$ Store at 2-8°C

Dab — FMOC-D-Dab

Bachem B-2365.0001   MW 340.38 $C_{19}H_{20}N_2O_4$ Store at 2-8°C

Dab — FMOC-D-DabBOC

Bachem B-2960.0001   MW 440.5 $C_{24}H_{28}N_2O_6$ Store at 2-8°C

Dab — Z-Dab

Bachem C-3705.0001   MW 252.27 $C_{12}H_{16}N_2O_4$ Store at RT

Dab — Z-Dab-BOC Dicyclohexylamine

Bachem C-3510.0001   MW 533.71 $C_{17}H_{24}N_2O_6 \cdot C_{12}H_{23}N$ Store at RT

Dab — Z-Dab-Z

Bachem C-3690.0001   MW 386.41 $C_{20}H_{22}N_2O_6$ Store at RT

Dab — Z-D-Dab

Bachem C-3770.0001   MW 252.27 $C_{12}H_{16}N_2O_4$ Store at RT

Dab — Z-D-Dab-BOC Dicyclohexylamine

Bachem C-3765.0001   MW 533.71 $C_{17}H_{24}N_2O_6 \cdot C_{12}H_{23}N$ Store at RT

Dap — BOC-Dap

Bachem A-3220.0001   MW 204.23 $C_8H_{16}N_2O_4$ Store at RT

Dap — BOC-Dap-Aloc

Bachem A-4115.0001   MW 288.3 $C_{12}H_{20}N_2O_6$ Store at 2-8°C

Dap — BOC-Dap-BOC Dicyclohexylamine

Bachem A-3475.0001   MW 485.67 $C_{13}H_{24}N_2O_6 \cdot C_{12}H_{23}N$ Store at RT

Dap — BOC-Dap-Bromoacetyl

Bachem A-4130.0005   MW 325.16 $C_{10}H_{17}BrN_2O_5$ Store at 2-8°C

Dap — BOC-Dap-Dnp

Bachem A-4290.0001   MW 370.32 $C_{14}H_{18}N_4O_8$ Store at 2-8°C

Dap — BOC-Dap-Dnp-OSu

Bachem A-4295.0001   MW 467.39 $C_{18}H_{21}N_5O_{10}$ Store at 2-8°C

Dap — BOC-Dap-FMOC

Bachem A-3580.0001   MW 426.47 $C_{23}H_{26}N_2O_6$ Store at RT

Dap — BOC-Dap-Z Dicyclohexylamine

Bachem A-3000.0001   MW 519.68 $C_{16}H_{22}N_2O_6 \cdot C_{12}H_{23}N$ Store at RT

Dap — BOC-D-Dap

Bachem A-3590.0001   MW 204.23 $C_8H_{16}N_2O_4$ Store at RT

Dap — BOC-D-Dap-FMOC

Bachem A-4235.0001   MW 426.47 $C_{23}H_{26}N_2O_6$ Store at RT

Dap — BOC-D-Dap-Z Dicyclohexylamine

Bachem A-4265.0001 MW 519.68 $C_{16}H_{22}N_2O_6 \cdot C_{12}H_{23}N$
Store at RT

Dap — Dap Hydrochloride

Bachem F-3040.0001 MW 140.57 $C_3H_8N_2O_2 \cdot HCl$ Store at 2-8°C

Dap — Dap-BOC-OMe Hydrochloride

Bachem F-3420.0001 MW 254.71 $C_9H_{18}N_2O_4 \cdot HCl$ Store at 2-8°C

Dap — D-Dap Hydrochloride

Bachem F-3045.0001 MW 140.57 $C_3H_8N_2O_2 \cdot HCl$ Store at 2-8°C

Dap — FMOC-Dab-Aloc

Bachem B-2850.0001 MW 424.45 $C_{23}H_{24}N_2O_6$ Store at 2-8°C

Dap — FMOC-Dap

Bachem B-2385.0001 MW 326.35 $C_{18}H_{18}N_2O_4$ Store at 2-8°C

Dap — FMOC-Dap-Adpoc

Bachem B-2865.0001 MW 547.67 $C_{32}H_{39}N_2O_6$ Store at -15°C

Dap — FMOC-Dap-Aloc

Bachem B-2845.0001 MW 410.43 $C_{22}H_{22}N_2O_6$ Store at 2-8°C

Dap — FMOC-Dap-BOC

Bachem B-2380.0001 MW 426.47 $C_{23}H_{26}N_2O_6$ Store at 2-8°C

Dap — FMOC-Dap-Dnp

Bachem B-2995.0001 MW 492.45 $C_{24}H_{20}N_4O_8$ Store at -15°C

Dap — FMOC-Dap-FMOC

Bachem B-2265.0001 MW 548.6 $C_{33}H_{28}N_2O_6$ Store at 2-8°C

Dap — FMOC-D-Dap

Bachem B-3055.0001 MW 326.35 $C_{18}H_{18}N_2O_4$ Store at 2-8°C

Dap — FMOC-D-Dap-BOC

Bachem B-2965.0001 MW 426.47 $C_{23}H_{26}N_2O_6$ Store at 2-8°C

Dap — n-Butyloxycarbonyl-Dap

Bachem F-3250.0001 MW 204.23 $C_8H_{16}N_2O_4$ Store at 2-8°C

Dap — Z-Dap

Bachem C-3315.0001 MW 238.24 $C_{11}H_{14}N_2O_4$ Store at RT

Dap — Z-Dap-BOC

Bachem C-3685.0001 MW 338.36 $C_{16}H_{22}N_2O_6$ Store at RT

Dap — Z-Dap-Z

Bachem C-3695.0001 MW 372.38 $C_{19}H_{20}N_2O_6$ Store at RT

Dap — Z-D-Dap

Bachem C-3755.0001 MW 238.24 $C_{11}H_{14}N_2O_4$ Store at RT

Dap — Z-D-Dap-BOC

Bachem C-3760.0001 MW 338.36 $C_{16}H_{22}N_2O_6$ Store at RT

Glutamic Acid — (2R,4R)-4-Me-Glu Hydrochloride

Alexis 550-326 MW 197.7 $C_6H_{11}NO_4 \cdot HCl$ Off-white solid; soluble in H_2O

Glutamic Acid — (2S)-α-Et-Glu

Synonyms: Group II mGluR Antagonist; EGLU

Alexis 550-324 MW 175.2 $C_7H_{13}NO_4$ ≥98%; white solid; soluble in dilute aqueous base | Presumed group II mGluR antagonist; Thomas, NK et al, *Br J Pharmacol*, 117: 70P, 1996; Jane, DE, *Neuropharmacology*, 35: 1029, 1996

ICN 198735 MW 175.2 $C_7H_{13}O_4N$

Glutamic Acid — (2S,4R)-4-Me-Glu

Alexis 550-186 MW 161.2 $C_6H_{11}NO_4$ White powder; soluble in dilute aqueous acid

Glutamic Acid — (2S,4R)-γ-Hyglu

Bachem F-3335.0250 MW 163.13 $C_5H_9NO_5$ Store at -15°C

Glutamic Acid — (2S,4S)-4-Me-Glu Hydrochloride

Alexis 550-327 MW 197.7 $C_6H_{11}NO_4 \cdot HCl$ Off-white solid; soluble in H_2O

Glutamic Acid — (2S,4S)-γ-Hyglu

Bachem F-3330.0250 MW 163.13 $C_5H_9NO_5$ Store at -15°C

Glutamic Acid — 4-Amino-N^{10}-Methylpteroyl-L-Glu

Fluka 06563 MW 454.45 $C_{20}H_{22}N_8O_5$ ≥99% (HPLC)
Fluka 06564 MW 454.45 $C_{20}H_{22}N_8O_5$ ≥85%

Glutamic Acid — 4-Aminopteroyl-Glu

Synonyms: 4-Aminofolic Acid; Aminopterin; Folic Acid Antagonist; Dihydrofolate Reductase Inhibitor

Alexis 440-041 MW 440.4 $C_{19}H_{20}N_8O_5$ ≥98%; yellow solid; soluble in dilute aqueous base or DMSO; slightly soluble in H_2O or EtOH

ICN 100623 MW 440.4 $C_{19}H_{20}N_8O_5$ ~98%
ICN 194622 MW 440.4 $C_{19}H_{20}N_8O_5$ ~98% | Cell culture reagent
Sigma A 1784 FW 440.4 $C_{19}H_{20}N_8O_5$ ~98% | Potent inhibitor of dihydrofolate reductase
Sigma A 3411 FW 440.4 $C_{19}H_{20}N_8O_5$ ~98%; | Cell culture tested

Glutamic Acid — 4-Aminopteroyl-L-Glu Dihydrate

Fluka 09328 MW 476.45 $C_{19}H_{20}N_8O_5 \cdot 2H_2O$ ~98% (UV); mp: 230-235°C

Glutamic Acid — Ac-Glu Amide

Synonyms: Ac-Igln; Ac-L-Igln

Bachem E-2695.0250 MW 188.18 $C_7H_{12}N_2O_4$ Store at -15°C

Glutamic Acid — Ac-Glu-OMe

Bachem E-1050.0250 MW 203.2 $C_8H_{13}NO_5$ Store at 2-8°C

Glutamic Acid — Ac-Glu-OSu-OBzl

Bachem E-1045.0001 MW 376.37 $C_{18}H_{20}N_2O_7$ Store at 2-8°C

Glutamic Acid — Ac-Glu-OtBu
Bachem E-3040.0001　　MW 245.28　$C_{11}H_{19}NO_5$　Store at 2-8°C

Glutamic Acid — Ac-Glu-pNA
Bachem L-1535.0050　　MW 309.28　$C_{13}H_{15}N_3O_6$　Store at -15°C

Glutamic Acid — BOC-D-Glu
Bachem A-4240.0001　　MW 247.25　$C_{10}H_{17}NO_6$　Store at RT

Glutamic Acid — BOC-D-Glu Amide
Synonyms: BOC-D-Igln
Bachem A-2635.0001　　MW 246.26　$C_{10}H_{18}N_2O_5$　Store at RT

Glutamic Acid — BOC-D-Glu-(1-OBzl)
Fluka 15106　　MW 337.37　$C_{17}H_{23}NO_6$　≥98% (TLC)

Glutamic Acid — BOC-D-Glu-(5-OtBu)
Fluka 15113　　MW 303.36　$C_{14}H_{25}NO_6$　≥98% (TLC)

Glutamic Acid — BOC-D-Glu-OBzl
Synonyms: BOC-D-Glu-(γ-OBzl)
Bachem A-1635.0001　　MW 337.37　$C_{17}H_{23}NO_6$　Store at RT
Bachem A-1645.0001　　MW 337.37　$C_{17}H_{23}NO_6$　Store at RT
Neosystem BA00603　　MW 337.4
Peptides International BDE-2625　　MW 337.37　>98% (HPLC); white crystalline powder
USBio B2305　　MW 337.4　$C_{17}H_{23}NO_6$　≥99%
USBio B2310　　MW 337.4　$C_{17}H_{23}NO_6$　≥99%

Glutamic Acid — BOC-D-Glu-OcHex
Bachem A-3425.0001　　MW 329.39　$C_{16}H_{27}NO_6$　Store at RT
Neosystem BA00605　　MW 329.4

Glutamic Acid — BOC-D-Glu-OFm
Bachem A-3270.0001　　MW 425.48　$C_{24}H_{27}NO_6$　Store at 2-8°C
Bachem A-4245.0001　　MW 425.48　$C_{24}H_{27}NO_6$　Store at 2-8°C

Glutamic Acid — BOC-D-Glu-OFm Alpha Ester
Synonyms: BOC-D-Glu-OFm
Peptides International BDE-5595-PI　　MW 425.49　>98% (HPLC); white powder

Glutamic Acid — BOC-D-Glu-OFm Gamma Ester
Synonyms: BOC-D-Glu-(γ-9-OFm)
Peptides International BDE-5687-PI　　MW 425.49　>98% (HPLC); white to off-white powder

Glutamic Acid — BOC-D-Glu-OMe
Bachem A-3880.0001　　MW 261.28　$C_{11}H_{19}NO_6$　Store at 2-8°C

Glutamic Acid — BOC-D-Glu-OtBu
Bachem A-1680.0001　　MW 303.36　$C_{14}H_{25}NO_6$　Store at 2-8°C
Bachem A-3415.0001　　MW 303.36　$C_{14}H_{25}NO_6$　Store at 2-8°C

Glutamic Acid — BOC-D-Glu-OtBu-ONp
Bachem A-2520.0001　　MW 424.45　$C_{20}H_{28}N_2O_8$　Store at -15°C

Glutamic Acid — BOC-DL-Glu-OBzl
Bachem A-4120.0001　　MW 337.37　$C_{17}H_{23}NO_6$　Store at RT

Glutamic Acid — BOC-Glu
Bachem A-3620.0005　　MW 247.25　$C_{10}H_{17}NO_6$　Store at RT

Glutamic Acid — BOC-Glu Amide
Synonyms: BOC-L-Igln
Bachem A-1820.0001　　MW 246.26　$C_{10}H_{18}N_2O_5$　Store at RT

Glutamic Acid — BOC-Glu-Chx
American Peptide BLGLU35　　MW 329.3

Glutamic Acid — BOC-Glu-OBzl
Synonyms: BOC-L-Glu-(γ-OBzl)
Bachem A-1630.0001　　MW 337.37　$C_{17}H_{23}NO_6$　Store at RT
Bachem A-1640.0005　　MW 337.37　$C_{17}H_{23}NO_6$　Store at RT
Peptides International BLE-2103　　MW 337.37　>98% (HPLC); white crystalline powder
USBio B2306　　MW 337.4　$C_{17}H_{23}NO_6$　≥99%
USBio B2311　　MW 337.4　$C_{17}H_{23}NO_6$　≥99%

Glutamic Acid — BOC-Glu-OBzl Dicyclohexylamine
Bachem A-1650.0005　　MW 518.69　$C_{17}H_{23}NO_6 \cdot C_{12}H_{23}N$　Store at RT

Glutamic Acid — BOC-Glu-OBzl-ONp
Bachem A-1665.0001　　MW 458.47　$C_{23}H_{26}N_2O_8$　Store at -15°C

Glutamic Acid — BOC-Glu-OBzl-OSu
Bachem A-1655.0001　　MW 434.45　$C_{21}H_{26}N_2O_8$　Store at -15°C

Glutamic Acid — BOC-Glu-OBzl-PAM-® (200-400 mesh)
Bachem D-1285.0001　　Store at 2-8°C

Glutamic Acid — BOC-Glu-OcHex
Synonyms: BOC-L-Glu-(γ-OcHex)
Bachem A-2770.0001　　MW 329.39　$C_{16}H_{27}NO_6$　Store at RT
Bachem A-4085.0001　　MW 329.4　$C_{16}H_{27}NO_6$　Store at 2-8°C
Peptides International BLE-2134　　MW 329.39　>98% (HPLC); white crystalline powder
USBio B2307　　MW 329.4　$C_{16}H_{27}NO_6$　≥98%
USBio B2312　　MW 329.4　$C_{16}H_{27}NO_6$　≥99%

Glutamic Acid — BOC-Glu-OcHex-® (200-400 mesh)
Bachem D-1505.0005　　Store at RT

Glutamic Acid — BOC-Glu-OcHex-O-Resin
American Peptide RMRB165　　1% DVB cross-linked: 100-200 mesh

Glutamic Acid — BOC-Glu-OcHex-PAM-Resin
American Peptide RPAM165　　1% DVB cross-linked: 100-200 mesh

Glutamic Acid — BOC-Glu-OFm
Bachem A-2940.0005　　MW 425.48　$C_{24}H_{27}NO_6$　Store at 2-8°C
Bachem A-3460.0001　　MW 425.48　$C_{24}H_{27}NO_6$　Store at -15°C
USBio B2313　　MW 425.5　$C_{24}H_{27}NO_6$　≥98%

Glutamic Acid — BOC-Glu-OFm Alpha Ester

Synonyms: BOC-L-Glu-OFm

Peptides International BLE-5591-PI   MW 425.49   >98% (HPLC); white crystalline powder

Glutamic Acid — BOC-Glu-OFm Gamma Ester

Synonyms: BOC-L-Glu-(γ-9-OFm)

Peptides International BLE-5683-PI   MW 425.49   >98% (HPLC); white crystalline powder

Glutamic Acid — BOC-Glu-OMe

Bachem A-3440.0001   MW 261.28 $C_{11}H_{19}NO_6$ Store at 2-8°C

USBio B2308   MW 261.4 $C_{11}H_{19}NO_6$ ≥98%

Glutamic Acid — BOC-Glu-OMe Dicyclohexylamine

Bachem A-1695.0005   MW 442.6 $C_{11}H_{19}NO_6 \cdot C_{12}H_{23}N$ Store at -15°C

Glutamic Acid — BOC-Glu-OMe-ONp

Bachem A-2525.0001   MW 382.36 $C_{17}H_{22}N_2O_8$ Store at -15°C

Glutamic Acid — BOC-Glu-OPh

Bachem A-4540.0001   MW 332.35 $C_{16}H_{21}NO_6$ Store at 2-8°C

Glutamic Acid — BOC-Glu-OSu-OBzl

Bachem A-2530.0001   MW 434.45 $C_{21}H_{26}N_2O_8$ Store at -15°C

Glutamic Acid — BOC-Glu-OSu-OtBu

Bachem A-2910.0001   MW 400.43 $C_{18}H_{28}N_2O_8$ Store at -15°C

Glutamic Acid — BOC-Glu-OtBu

Bachem A-1670.0001   MW 303.36 $C_{14}H_{25}NO_6$ Store at 2-8°C

Bachem A-1675.0001   MW 303.36 $C_{14}H_{25}NO_6$ Store at 2-8°C

USBio B2309   MW 303.4 $C_{14}H_{25}NO_6$ ≥98%

Glutamic Acid — BOC-Glu-OtBu-ONp

Bachem A-1685.0001   MW 424.45 $C_{20}H_{28}N_2O_8$ Store at -15°C

Glutamic Acid — BOC-Glu-OtBu-OSu

Bachem A-2675.0001   MW 400.43 $C_{18}H_{28}N_2O_8$ Store at -15°C

Glutamic Acid — BOC-L-Glu

Fluka 15345   MW 247.25 $C_{10}H_{17}NO_6$ ≥98% (titration); mp: 110°C

Neosystem BA00601   MW 247.2

Glutamic Acid — BOC-L-Glu-(1-OPh)

Fluka 15115   MW 323.35 $C_{16}H_{21}NO_6$ ≥98% (TLC)

Glutamic Acid — BOC-L-Glu-(5-OBzl)

Fluka 15418   MW 337.38 $C_{17}H_{23}NO_6$ ≥98% (titration); mp: 69-71°C

Glutamic Acid — BOC-L-Glu-(5-OBzl) Dicyclohexylamine

Fluka 15419   MW 518.7 $C_{17}H_{23}NO_6 \cdot C_{12}H_{23}N$ ≥99% (titration); mp: 140-142°C

Glutamic Acid — BOC-L-Glu-(5-OBzl) N-Carboxy Anhydride

Fluka 15425   MW 363.37 $C_{18}H_{21}NO_7$ mp: 74-78°C

Glutamic Acid — BOC-L-Glu-(5-OcHex)

Fluka 15437   MW 329.39 $C_{16}H_{27}NO_6$ ~99% (TLC); mp: 54-57°C

Glutamic Acid — BOC-L-Glu-(5-OtBu)

Fluka 15436   MW 303.36 $C_{14}H_{25}NO_6$ ~99% (TLC); mp: 102-105°C

Glutamic Acid — BOC-L-Glu-5-Bzl 1-(4-Nitrophenyl) Ester

Fluka 15108   MW 458.47 $C_{23}H_{26}N_2O_8$ ~98% (TLC)

Glutamic Acid — BOC-L-Glu-Aminol

Bachem A-3300.0001   MW 232.28 $C_{10}H_{20}N_2O_4$ Store at -15°C

Glutamic Acid — BOC-L-Glu-OAll

Neosystem BA00602   MW 287.3

Glutamic Acid — BOC-L-Glu-OBzl

Neosystem BA00604   MW 337.4

Neosystem BA00608   MW 337.4

Glutamic Acid — BOC-L-Glu-OcHex

Neosystem BA00606   MW 329.4

Glutamic Acid — BOC-L-Glu-OcHex-MBHA Resin

Neosystem RN00606

Glutamic Acid — BOC-L-Glu-OcHex-O-CH₂-φ-CH₂-COOH

Neosystem LP00604   MW 477.6

Glutamic Acid — BOC-L-Glu-OcHex-PAM Resin

Neosystem RP00604

Glutamic Acid — BOC-L-β-Glu-(5-OBzl)

Fluka 03691   MW 337.37 $C_{17}H_{23}NO_6$ ≥98% (HPLC)

Glutamic Acid — BOC-L-β-HGlu-(6-OBzl)

Fluka 14977   MW 351.4 $C_{18}H_{25}NO_6$ ≥98% (TLC)

Glutamic Acid — BOC-N-Me-Glu-OBzl

Bachem A-2050.0001   MW 351.4 $C_{18}H_{25}NO_6$ Store at RT

Glutamic Acid — Bsmoc-Glu-OtBu

Synonyms: Bsmoc-L-Glu-(γ-OtBu)

Peptides International BLE-5624-PI   MW 425.46   >97% (HPLC); white crystalline powder | LA Carpino, et al, *JACS*, 119:9915, 1997

Glutamic Acid — Caged Glu Sodium Salt

Calbiochem 351015   MW 362.3 $C_{14}H_{15}N_2O_8Na$ >98% (HPLC); solid; soluble in H_2O

Glutamic Acid — Calcium L-Glu Dihydrate
Rexim   MW 368.4 $C_{10}H_{20}CaN_2O_{10}$

Glutamic Acid — Calcium L-Glu Tetrahydrate
Fluka 21151   MW 404.39 $C_{10}H_{16}CaN_2O_8 \cdot 4H_2O$ ≥98%

Glutamic Acid — CBZ-D-Glp
USBio C2099-57   MW 263.2 $C_{13}H_{13}NO_5$ ≥99%

Glutamic Acid — CBZ-D-Glu
USBio C2098-55   MW 281.3 $C_{13}H_{15}NO_6$ ≥99%

Glutamic Acid — CBZ-D-Glu-OBzl
USBio C2098-57   MW 271.4 $C_{20}H_{21}NO_6$ ≥99%

Glutamic Acid — CBZ-D-Glu-OtBu
USBio C2098-62   MW 337.4 $C_{17}H_{23}NO_6$ ≥99%

Glutamic Acid — CBZ-Glp
USBio C2099-56   MW 263.2 $C_{13}H_{13}NO_5$ ≥99%

Glutamic Acid — CBZ-Glp-ONp
USBio C2099-58   MW 384.3 $C_{19}H_{16}NO_7$ ≥99%

Glutamic Acid — CBZ-Glu
USBio C2098-54   MW 281.3 $C_{13}H_{15}NO_6$ ≥99%

Glutamic Acid — CBZ-Glu-OBzl
USBio C2098-56   MW 271.4 $C_{20}H_{21}NO_6$ ≥99%
USBio C2098-58   MW 371.4 $C_{20}H_{21}NO_6$ ≥99%

Glutamic Acid — CBZ-Glu-OBzl-ONp
USBio C2098-59   MW 492.8 $C_{26}H_{24}N_2O_8$ ≥99%

Glutamic Acid — CBZ-Glu-OtBu
USBio C2098-60   MW 337.4 $C_{17}H_{23}NO_6$ ≥99%
USBio C2098-63   MW 337.4 $C_{17}H_{23}NO_6$ ≥99%

Glutamic Acid — CBZ-Glu-OtBu-Dicyclohexylamine
USBio C2098-61   MW 518.7 $C_{24}H_{46}N_2O_6$ ≥99%

Glutamic Acid — cis-2,4-Methano-Glu
Synonyms: cis-ACBD; 1-Aminocyclobutane-cis-1,3-Dicarboxylic Acid; trans-1-Amino-1,3-Dicarboxycyclobutane; NMDA Agonist
Sigma A 4066   FW 159.1 $C_6H_9NO_4$ ~95% | The most potent & selective NMDA agonist known

Glutamic Acid — D-(γ-Carboxy)-Glu
Sigma C 4272   FW 191.1 $C_6H_9NO_6$

Glutamic Acid — Dansyl-DL-Glu Dicyclohexylammonium Salt
Sigma D 8756   FW 578.8 $C_{17}H_{20}N_2O_6S \cdot (C_6H_{13}N)_2$ ~99%

Glutamic Acid — Dansyl-L-Glu Dicyclohexylammonium Salt
ICN 100051   MW 578.8 $C_{17}H_{20}N_2O_6S \cdot (C_6H_{13}N)_2$
Sigma D 0750   FW 578.8 $C_{17}H_{20}N_2O_6S \cdot (C_6H_{13}N)_2$ ~99%

Glutamic Acid — D-Glp
Synonyms: D-5-Oxo-2-Pyrrolidinecarboxylic Acid
Fluka 83165   MW 129.12 $C_5H_7NO_3$ ≥99.0% (titration); ≤0.05% residue on ignition; mp: 155-158°C; enantiomer ratio: D:L ≥99:1
Fluka 83167   MW 129.12 $C_5H_7NO_3$ ~97% (titration); mp: 155-158°C; enantiomer ratio: D:L ≥96:4
ICN 156461   MW 129.1 $C_5H_7NO_3$
Sigma P 2426   FW 129.1 $C_5H_7NO_3$

Glutamic Acid — D-Glp-OEt
Synonyms: D-5-Oxo-2-Pyrrolidinecarboxylic Acid-OEt
Fluka 83173   MW 157.17 $C_7H_{11}NO_3$ ≥99.0% (HPLC); mp: 53-55°C; enantiomer ratio: D:L ≥99:1
ICN 156462   MW 157.2 $C_7H_{11}NO_3$
Sigma P 9916   FW 157.2 $C_7H_{11}NO_3$

Glutamic Acid — D-Glu
Synonyms: D-2-Aminopentanedioic Acid; D-α-Aminoglutaric Acid; (2R)-2-Aminopentanedioic Acid; (R)-2-Aminopentanedioic Acid
Alexis 101-019   MW 147.1 $C_5H_9NO_4$ White solid; soluble in H_2O
Bachem F-1550.0025   MW 147.13 $C_5H_9NO_4$ Store at RT
Fluka 49460   MW 147.13 $C_5H_9NO_4$ ≥99% (titration); mp: 200°C
ICN 101796   MW 147.1 $C_5H_9NO_4$ ≥98%; crystalline
Neosystem AA00622   MW 147.1
Peptides International ADE-2804   MW 147.13   >99.9% optical purity by RP-HPLC; white crystalline powder
Rexim   MW 147.1 $C_5H_9NO_4$
Sigma G 1001   FW 147.1 $C_5H_9NO_4$ ≥99% (TLC) | Also available as part of a kit
USBio G7115-10   MW 147.1 $C_5H_9NO_4$ ≥99%

Glutamic Acid — D-Glu (1-^{14}C)
ARC ARC-314   MW 147.1 $HO_2C(CH_2)_2CH(NH_2)COOH$ 50-60 mCi/mmol; 1.85-2.22 GBq/mmol; in 0.01 N HCl

Glutamic Acid — D-Glu (2,3,4-^{3}H)
ARC ART-232   MW 147.1 $HO_2C(CH_2)_2CH(NH_2)COOH$ 30-60 Ci/mmol; 1.11-2.22 TBq/mmol; in 0.01 N HCl

Glutamic Acid — D-Glu (5-^{14}C)
ARC ARC-315   MW 147.1 $HO_2C(CH_2)_2CH(NH_2)COOH$ 50-60 mCi/mmol; 1.85-2.22 GBq/mmol; in 0.01 N HCl

Glutamic Acid — D-Glu Amide
Synonyms: D-Isogln; D-Isoglutamine
Bachem F-2255.0001   MW 146.15 $C_5H_{10}N_2O_3$ Store at RT

Glutamic Acid — D-Glu-(γ-OBzl)
Synonyms: γ-Bzl-D-Glutamate
ICN 101759   MW 237.3 $C_{12}H_{15}NO_4$ Crystalline

Glutamic Acid — D-Glu-AMC
Bachem I-1190.0050   MW 304.3 $C_{15}H_{16}N_2O_5$ Store at -15°C

Glutamic Acid — D-Glu-OBzl
Bachem F-1555.0001   MW 237.26 $C_{12}H_{15}NO_4$ Store at RT
Bachem F-1560.0001   MW 237.26 $C_{12}H_{15}NO_4$ Store at RT
Neosystem AA00601   MW 237.3

Glutamic Acid — D-Glu-OBzl-OBzl *p*-Tosyl

Bachem F-1570.0005   MW 499.59 $C_{19}H_{21}NO_4 \cdot C_7H_8O_3S$
Store at RT

Glutamic Acid — D-Glu-OMe

Bachem F-3220.0001   MW 161.16 $C_6H_{11}NO_4$ Store at 2-8°C

Glutamic Acid — D-Glu-OMe-OMe Hydrochloride

Bachem F-1575.0005   MW 211.65 $C_7H_{13}NO_4 \cdot HCl$ Store at 2-8°C

Glutamic Acid — D-Glu-OtBu

Bachem F-1565.0001   MW 203.24 $C_9H_{17}NO_4$ Store at -15°C

Glutamic Acid — D-Glu-OtBu Hydrochloride

Bachem F-3000.0001   MW 239.7 $C_9H_{17}NO_4 \cdot HCl$ Store at -15°C

Bachem F-3165.0001   MW 295.81 $C_{13}H_{25}NO_4 \cdot HCl$ Store at -15°C

Glutamic Acid — D-Glu-OtBu-OAll Hydrochloride

Bachem F-3935.0001   MW 279.76 $C_{12}H_{21}NO_4 \cdot HCl$ Store at -15°C

Glutamic Acid — D-Glu-pNA

Bachem L-1840.0050   MW 267.24 $C_{11}H_{13}N_3O_5$ Store at -15°C

Glutamic Acid — DL-(γ-Carboxy)-Glu

Sigma C 3767   FW 191.1 $C_6H_9NO_6$

Glutamic Acid — DL-Glp

Synonyms: DL-5-Oxo-2-Pyrrolidinecarboxylic Acid

Fluka 83170   MW 129.12 $C_5H_7NO_3$ ≥99.0% (titration); mp: 182-185°C

ICN 102780   MW 129.1 $C_5H_7NO_3$ Crystalline

Sigma P 0506   FW 129.1 $C_5H_7NO_3$

Glutamic Acid — DL-Glu

Synonyms: DL-Glu; (±)-2-Aminopentanedioic Acid; Kainate Receptor Agonist; NMDA Receptor Agonist; Quisqualate Receptor Agonist

Sigma G 1126   FW 147.1 $C_5H_9NO_4$ ≥98% (TLC) | Neurotransmitter at fast synapses; also available as part of a kit

Sigma G 5513   FW 147.1 $C_5H_9NO_4$ Crystalline | Cell culture tested

Glutamic Acid — DL-Glu (1-^{14}C)

ARC ARC-583   MW 147.1 $HO_2C(CH_2)_2CH(NH_2)COOH$ 50-60 mCi/mmol; 1.85-2.22 GBq/mmol; in 0.01 *N* HCl

Glutamic Acid — DL-Glu Hydrate

Synonyms: DL-2-Aminopentanedioic Acid

Fluka 49480   MW 165.15 $C_5H_9NO_4 \cdot H_2O$ ≥99.0% (titration); mp: 180-185°C

ICN 101791   MW 165.1 $C_5H_9NO_4 \cdot H_2O$ ≥99%; crystalline

ICN 194675   MW 147.1 $C_5H_9NO_4 \cdot H_2O$ ≥99% | Cell culture reagent

USBio G7115-15   MW 165.2 $C_5H_9NO_4 \cdot H_2O$ ≥99%

Glutamic Acid — DL-Glu γ-Anilide

Sigma G 1628   FW 222.2 $C_{11}H_{14}N_2O_3$

Glutamic Acid — DL-Glu-Anilide

Bachem F-1585.0001   MW 238.25 $C_{11}H_{14}N_2O_4$ Store at RT

Glutamic Acid — Dnp-DL-Glu

ICN 101789   MW 313.2 $C_{11}H_{11}N_3O_8$ Crystalline

Glutamic Acid — Dnp-L-Glu

ICN 101794   MW 313.2 $C_{11}H_{11}N_3O_8$ Crystalline; 99%

Glutamic Acid — FMOC-D-Glu-OAll

Neosystem FA00612   MW 409.4

Neosystem FA00625   MW 409.4

Glutamic Acid — FMOC-D-Glu-OBzl

Bachem B-3140.0001   MW 459.5 $C_{27}H_{25}NO_6$ Store at RT

Glutamic Acid — FMOC-D-Glu-OcHex

Neosystem FA00606   MW 451.5

Glutamic Acid — FMOC-D-Glu-OFm

Bachem B-3370.0001   MW 547.61 $C_{34}H_{29}NO_6$ Store at -15°C

Glutamic Acid — FMOC-D-Glu-OtBu

Synonyms: FMOC-D-Glu-(γ-OtBu)

Bachem B-1320.0001   MW 425.48 $C_{24}H_{27}NO_6$ Store at RT

Bachem B-2245.0001   MW 425.49 $C_{24}H_{27}NO_6$ Store at RT

Neosystem FA00604   MW 425.5

Peptides International FDE-1810-PI   MW 425.48   >98% (HPLC); white crystalline powder

USBio F5376   MW 425.5 $C_{24}H_{27}NO_6$ ≥99%

Glutamic Acid — FMOC-D-Glu-OtBu-OPfp

Bachem B-1840.0001   MW 591.5 $C_{30}H_{26}NO_6F_5$ Store at -15°C

Glutamic Acid — FMOC-D-Glu-OtBu-SASRIN™-® (200-400 mesh)

Bachem D-1775.0001   Store at 2-8°C

Glutamic Acid — FMOC-Gla-OtBu₂

USBio F5365   MW 525.6 $C_{29}H_{35}NO_8$ ≥98%

Glutamic Acid — FMOC-Glu

Bachem B-2445.0005   MW 369.37 $C_{20}H_{19}NO_6$ Store at RT

Glutamic Acid — FMOC-Glu 2-Phenylisopropyl Ester

Bachem B-2500.0001   MW 487.55 $C_{29}H_{29}NO_6$ Store at -15°C

Glutamic Acid — FMOC-Glu-OAll

Bachem B-2720.0001   MW 409.44 $C_{23}H_{23}NO_6$ Store at -15°C

Bachem B-3255.0001   MW 409.44 $C_{23}H_{23}NO_6$ Store at -15°C

Glutamic Acid — FMOC-Glu-OAll Alpha Ester

Synonyms: FMOC-L-Glu-(α-OAll)

Peptides International FLE-5538-PI   MW 409.44   >98% (HPLC); white crystalline powder

Glutamic Acid — FMOC-Glu-OBzl

Synonyms: FMOC-L-Glu-(γ-OBzl)

Bachem B-1310.0001	MW 459.5	$C_{27}H_{25}NO_6$	Store at RT

Peptides International FLE-1719-PI   MW 459.50   >98% (HPLC); white crystalline powder

USBio F5379	MW 459.5	$C_{27}H_{25}NO_6$ ≥99%

Glutamic Acid — FMOC-Glu-ODmab

Bachem B-3005.0001	MW 680.81	$C_{40}H_{44}N_2O_8$	Store at -15°C
Bachem B-3010.0001	MW 680.81	$C_{40}H_{44}N_2O_8$	Store at -15°C

Glutamic Acid — FMOC-Glu-OFm

Bachem B-2495.0001	MW 547.61	$C_{34}H_{29}NO_6$	Store at -15°C

Glutamic Acid — FMOC-Glu-OSu-OtBu

Bachem B-2310.0001	MW 522.55	$C_{28}H_{30}N_2O_6$	Store at -15°C

Glutamic Acid — FMOC-Glu-OtBu

Synonyms: FMOC-L-Glu-(γ-OtBu)

Bachem B-1315.0001	MW 425.48	$C_{24}H_{27}NO_6$	Store at RT
Bachem B-1595.0001	MW 425.48	$C_{24}H_{27}NO_6$	Store at RT

Peptides International FLE-1720-PI   MW 425.48   >98% (HPLC); off-white crystalline powder

USBio F5377	MW 425.5	$C_{24}H_{27}NO_6$ ≥99%

Glutamic Acid — FMOC-Glu-OtBu-® (200-400 mesh)

Bachem D-1100.0001	Store at 2-8°C

Glutamic Acid — FMOC-Glu-OtBu-OPfp

Bachem B-1130.0001	MW 591.53	$C_{30}H_{26}F_5NO_6$	Store at -15°C
USBio F5378	MW 591.5	$C_{30}H_{26}F_5NO_6$	≥99%

Glutamic Acid — FMOC-Glu-OtBu-OSu

Bachem B-1325.0001	MW 522.56	$C_{28}H_{30}N_2O_6$	Store at -15°C

Glutamic Acid — FMOC-Glu-OtBu-SASRIN™-® (200-400 mesh)

Bachem D-1340.0001	Store at 2-8°C

Glutamic Acid — FMOC-Glu-tBu

American Peptide FGLU105	MW 425.5

Glutamic Acid — FMOC-L-Glu-(1-OAll)

Fluka 47702	MW 409.44	$C_{23}H_{23}NO_6$	≥99% (HPLC); mp: 118-122°C

Glutamic Acid — FMOC-L-Glu-(5-OAll)

Fluka 47703	MW 409.44	$C_{23}H_{23}NO_6$	~98% (HPLC)

Glutamic Acid — FMOC-L-Glu-(5-OBzl)

Fluka 47571	MW 459.5	$C_{27}H_{25}NO_6$	≥98% (HPLC); mp: 70°C; ≤1% H_2O

Glutamic Acid — FMOC-L-Glu-(5-OtBu)

Fluka 47625	MW 425.49	$C_{24}H_{27}NO_6$	~97% (HPLC); mp: 83-87°C; ~3% H_2O

Glutamic Acid — FMOC-L-Glu-(5-OtBu) N-Carboxy Anhydride

Fluka 47684	MW 451.48	$C_{25}H_{25}NO_7$	≥97%; mp: 123-126°C

Glutamic Acid — FMOC-L-Glu-(EDANS)

Neosystem FA00618	MW 617.68

Glutamic Acid — FMOC-L-Glu-5-tBu 1-(4-Benzyloxybenzyl) Ester Polymer Bound

Fluka 47658   0.4-0.6 mmol/g resin; carrier: polystyrene, crosslinked with 1% DVB; 200-400 mesh particle size

Glutamic Acid — FMOC-L-Glu-OAll

Neosystem FA00601	MW 409.4
Neosystem FA00626	MW 409.4

Glutamic Acid — FMOC-L-Glu-OBzl

Neosystem FA00603	MW 459.5

Glutamic Acid — FMOC-L-Glu-OcHex

Neosystem FA00607	MW 451.5

Glutamic Acid — FMOC-L-Glu-OtBu Hydrate

Neosystem FA00605	MW 443.5

Glutamic Acid — FMOC-L-Glu-OtBu-Wang Resin

Neosystem RW00602

Glutamic Acid — FMOC-L-Glu-tBu-Wang Resin

American Peptide RFWAN33

Glutamic Acid — FMOC-L-β-Glu-(5-OtBu)

Fluka 03689	MW 459.5	$C_{27}H_{25}NO_6$	≥98% (HPLC)

Glutamic Acid — FMOC-L-β-HGlu-(6-OtBu)

Fluka 47837	MW 439.51	$C_{25}H_{29}NO_6$	≥98% (HPLC)

Glutamic Acid — FMOC-N-Me-Glu-OBzl

Synonyms: FMOC-N-Me-L-Glu-(γ-OBzl)

Peptides International FME-1782-PI   MW 473.53   >95% (HPLC); white crystalline powder

Glutamic Acid — FMOC-N-Me-Glu-OtBu

Bachem B-2395.0001	MW 439.51	$C_{25}H_{29}NO_6$	Store at -15°C

Glutamic Acid — FMOC-γ-Carboxy-D-Glu-OtBu₂

Synonyms: FMOC-D-Gla-OtBu₂

Bachem B-2760.0250	MW 525.6	$C_{29}H_{35}NO_8$	Store at -15°C

Glutamic Acid — FMOC-γ-Carboxy-Glu-OtBu₂

Synonyms: FMOC-Gla-OtBu₂

Bachem B-1265.0001	MW 525.6	$C_{29}H_{35}NO_8$	Store at -15°C

Glutamic Acid — Formimino-L-Glu Hemibarium Salt

Synonyms: FIGLU

Sigma F 8626	FW 241.8	$C_6H_9N_2O_4 \cdot ½Ba$	~90%

Glutamic Acid — Glp

Bachem E-2340.0025	MW 129.12	$C_5H_7NO_3$	Store at RT

Glutamic Acid — Glp-4MβNA
Bachem J-1295.0250   MW 284.32 $C_{16}H_{16}N_2O_3$ Store at -15°C

Glutamic Acid — Glp-Opcp
Bachem E-2350.0001   MW 377.44 $C_{11}H_6Cl_5NO_3$ Store at -15°C

Glutamic Acid — Glp-pNA
Bachem L-1375.0250   MW 249.23 $C_{11}H_{11}N_3O_4$ Store at -15°C

Sigma P 2664   FW 249.2 $C_{11}H_{11}N_3O_4$

Glutamic Acid — Glp-βNA
Bachem K-1480.0250   MW 254.28 $C_{15}H_{14}N_2O_2$ Store at RT

Glutamic Acid — Glu
Bachem E-1860.0100   MW 147.13 $C_5H_9NO_4$ Store at RT

Glutamic Acid — Glu 5-(3-Carboxy-4-NA) Ammonium Salt
Fluka 49525   MW 328.3 $C_{12}H_{16}N_4O_7$ ≥99.0% (TLC); ≤6% H_2O

Glutamic Acid — Glu Amide
Synonyms: Isogln; L-Isoglutamine
Bachem F-1680.0001   MW 146.15 $C_5H_{10}N_2O_3$ Store at RT

Glutamic Acid — Glu-4MβNA
Bachem J-1180.0050   MW 302.33 $C_{16}H_{18}N_2O_4$ Store at -15°C

Bachem J-1185.0050   MW 302.3 $C_{16}H_{18}N_2O_4$ Store at -15°C

Glutamic Acid — Glu-AMC
Bachem I-1180.0050   MW 304.3 $C_{15}H_{16}N_2O_5$ Store at -15°C

Bachem I-1185.0050   MW 304.3 $C_{15}H_{16}N_2O_5$ Store at -15°C

Glutamic Acid — Glu-Anilide
Bachem E-1935.0001   MW 238.25 $C_{11}H_{14}N_2O_4$ Store at RT

Glutamic Acid — Glu-OBzl
Synonyms: L-Glu-(γ-OBzl)
Bachem E-1865.0001   MW 237.26 $C_{12}H_{15}NO_4$ Store at 2-8°C

Bachem E-1870.0005   MW 237.26 $C_{12}H_{15}NO_4$ Store at RT

Peptides International XBE-2003   MW 237.26   >98% (TLC); white powder

Glutamic Acid — Glu-OBzl₂-Tosyl
Synonyms: L-Glu α,γ-Dibenzyl Ester-Tosyl
Peptides International EBE-2046   MW 499.58   >98% (TLC); white powder

Glutamic Acid — Glu-OBzl-OBzl p-Tosyl
Bachem E-1895.0005   MW 499.59 $C_{19}H_{21}NO_4 \cdot C_7H_8O_3S$ Store at RT

Glutamic Acid — Glu-OBzl-OtBu Hydrochloride
Bachem E-3535.0001   MW 329.82 $C_{16}H_{23}NO_4 \cdot HCl$ Store at 2-8°C

Glutamic Acid — Glu-O-Chloroanilide
Bachem E-1940.0001   MW 272.69 $C_{11}H_{13}N_2O_4Cl$ Store at RT

Glutamic Acid — Glu-OEt
Bachem E-2780.0005   MW 175.19 $C_7H_{13}NO_4$ Store at 2-8°C

Glutamic Acid — Glu-OEt-OEt Hydrochloride
Bachem E-2670.0005   MW 239.7 $C_9H_{17}NO_4 \cdot HCl$ Store at 2-8°C

Glutamic Acid — Glu-OMe
Synonyms: L-Glu-(γ-OMe)
Bachem E-1910.0005   MW 161.16 $C_6H_{11}NO_4$ Store at 2-8°C

Bachem E-2875.0001   MW 161.16 $C_6H_{11}NO_4$ Store at 2-8°C

Peptides International XME-2001-PI   MW 161.16   >98% (HPLC); white to off-white powder

Glutamic Acid — Glu-OMe Amide Hydrochloride
Synonyms: Igln-OMe; L-Igln-OMe
Bachem E-3540.0001   MW 196.63 $C_6H_{12}N_2O_3 \cdot HCl$ Store at 2-8°C

Glutamic Acid — Glu-OMe-OMe Hydrochloride
Bachem E-1905.0005   MW 211.65 $C_7H_{13}NO_4 \cdot HCl$ Store at 2-8°C

Glutamic Acid — Glu-OMe-OtBu Hydrochloride
Bachem E-3515.0001   MW 217.27 $C_{10}H_{19}NO_4$ Store at -15°C

Glutamic Acid — Glu-OtBu
Bachem E-1875.0001   MW 203.24 $C_9H_{17}NO_4$ Store at 2-8°C

Glutamic Acid — Glu-OtBu Amide Hydrochloride
Synonyms: Igln-OtBu; L-Igln-OtBu
Bachem E-1880.0001   MW 238.71 $C_9H_{18}N_2O_3 \cdot HCl$ Store at 2-8°C

Glutamic Acid — Glu-OtBu Hydrochloride
Bachem E-3605.0001   MW 239.7 $C_9H_{17}NO_4 \cdot HCl$ Store at -15°C

Glutamic Acid — Glu-OtBu-OAll Hydrochloride
Bachem E-3585.0001   MW 279.76 $C_{12}H_{21}NO_4 \cdot HCl$ Store at -15°C

Glutamic Acid — Glu-OtBu-OMe Hydrochloride
Bachem E-1885.0001   MW 253.73 $C_{10}H_{19}NO_4 \cdot HCl$ Store at 2-8°C

Glutamic Acid — Glu-OtBu-OtBu Hydrochloride
Bachem E-1900.0001   MW 295.81 $C_{13}H_{25}NO_4 \cdot HCl$ Store at -15°C

Glutamic Acid — Glu-pNA
Bachem L-1265.0001   MW 267.3 $C_{11}H_{13}N_3O_5$ Store at -15°C

Bachem L-1540.0250   MW 267.24 $C_{11}H_{13}N_3O_5$ Store at -15°C

Glutamic Acid — Glu-pNA Hydrate

Synonyms: γ-Glutamyl Transpeptidase Substrate

Peptides International SEN-3066   MW 285.26 $C_{11}H_{13}N_3O_5 \cdot$ H_2O >98% (HPLC); amorphous powder

Glutamic Acid — Glu-αNA

Bachem M-1445.0250   MW 272.3 $C_{15}H_{16}N_2O_3$ Store at RT
Bachem K-1265.0250   MW 272.3 $C_{15}H_{16}N_2O_3$ Store at RT
Bachem K-1270.0250   MW 272.3 $C_{15}H_{16}N_2O_3$ Store at RT

Glutamic Acid — L-(γ-Carboxy)-Glu

Sigma C 4147   FW 191.1 $C_6H_9NO_6$

Glutamic Acid — L-Glu (1,5-^{14}C)

ARC ARC-770   MW 147.1 $HO_2C(CH_2)_2CH(NH_2)COOH$ 50-60 mCi/mmol; 1.85-2.22 GBq/mmol; in EtOH: H_2O (2:98)

Glutamic Acid — L-Glp

Synonyms: 5-Oxoproline; L-Pyrrolidone Carboxylic Acid; L-5-Oxo-2-Pyrrolidinecarboxylic Acid; L-5-Oxo-2-Pyrrolidine Carboxylic Acid

Fluka 83160   MW 129.12 $C_5H_7NO_3$ ≥99.0% (titration); ≤0.05% residue on ignition; mp: 155-158°C; enantiomer ratio: L:D ≥99:1

ICN 102781   MW 129.1 $C_5H_7NO_3$ Crystalline

Neosystem AA06301   MW 129.1

Peptides International ALU-2724-PI   MW 129.12   >98% optical purity by TLC; white crystalline powder

Sigma P 3634   FW 129.1 $C_5H_7NO_3$ Crystalline

Sigma P 5960   FW 129.1 $C_5H_7NO_3$ SigmaUltra; residue on ignition: <0.1%; solubility (1 *M* in H_2O, 20°C): complete, colorless; insoluble matter: <0.1%; NH_4^+, Cl, SO_4: <0.05%; K, Na: <0.005%; Al, Ca, Cu, Fe, Mg, P, Zn: <0.0005%; Pb: <0.001%

USBio P9540   MW 129.1 $C_5H_7NO_3$ ≥99%

Glutamic Acid — L-Glp 2-Naphthylamide

Fluka 83176   MW 254.30 $C_{15}H_{14}N_2O_2$ ≥99.0% (TLC)

Glutamic Acid — L-Glp Pentachlorophenyl Ester

Sigma P 9644   FW 377.4 $C_{11}H_6Cl_5NO_3$

Glutamic Acid — L-Glp β-Naphthylamide

Synonyms: L-Pyrrolidonyl-β-Naphthylamide; Pyroglutamate Aminopeptidase Substrate

ICN 156463   MW 254.3 $C_{15}H_{14}N_2O_2$ Szewczuk, A & M Mulcyzk, Eur *J Biochem*, 8:63, 1969

Sigma P 5891   FW 254.3 $C_{15}H_{14}N_2O_2$ Possibly carcinogenic; Szewczuk, A & Mulczyk, M, *Eur J Biochem*, 8: 63, 1969

Glutamic Acid — L-Glp-AMC

Synonyms: Pyroglutamyl Peptidase Substrate

Fluka 83174   MW 286.29 $C_{15}H_{14}N_2O_4$ ≥99.0% (HPLC); ≤4% H_2O

ICN 151995   MW 286.3 $C_{15}H_{14}N_2O_4$ Fujiwara, K etal, *J Biochem*, 83:1145, 1978

Sigma P 3149   FW 286.3 $C_{15}H_{14}N_2O_4$ Similar to Sigma P 4079 but prepared for Sigma

Sigma P 4079   FW 286.3 $C_{15}H_{14}N_2O_4$ ≥98%

Glutamic Acid — L-Glp-OEt

Fluka 83175   MW 157.17 $C_7H_{11}NO_3$ ≥99.0% (HPLC); mp: 54-56°C; enantiomer ratio: L:D ≥99:1

Glutamic Acid — L-Glp-pNA

ICN 156464   MW 249.2 $C_{11}H_{11}N_3O_4$

Glutamic Acid — L-Glu

Synonyms: (*S*)-2-Aminoglutaric Acid; (2*S*)-2-Aminopentanedioic Acid; α-Aminoglutaric Acid; 1-Aminopropane-1,3-Dicarboxylic Acid; (*S*)-2-Aminopentanedioic Acid; Kainate Receptor Agonist; *N*-Methyl-D-Aspartate Receptor Agonist; Quisqualate Receptor Agonist; L-α-Aminoglutaric Acid

Alexis 101-018   MW 147.1 $C_5H_9NO_4$ White solid; soluble in H_2O

Amersham US16215   98.5-101.5% dsb

Calbiochem 3510   MW 147.1 $C_5H_9NO_4$ Solid; soluble in H_2O

Fluka 09581   MW 147.13 $C_5H_9NO_4$ European Pharmacopoeia

Fluka 49449   MW 147.13 $C_5H_9NO_4$ ≥99.5% (titration); ≤0.05% residue on ignition, loss on drying; ≤0.005% K; ≤0.02% Na; ≤0.0005% Al, Ba, Bi, Cd, Co, Cr, Cu, Fe, Li, Mg, Mn, Mo, Ni, Pb, Sr, Zn; ≤0.001% Ca; ≤0.01% NH_4^+, Cl, SO_4; ≤0.3% foreign AA | Lesion-sparing agent; has excitotoxic properties; may function as precipitating agent in protein crystallization; Olney, JW etal, *Meth Enzymol*, 101:379, 1983; Gilliland, GL, etal *Meth Enzymol*, 104:370, 1984

Fluka 49450   MW 147.13 $C_5H_9NO_4$ ≥99.0% (titration); ≤0.1% residue on ignition, loss on drying; ≤0.005% K; ≤0.0005% Cd, Co, Cr, Cu, Fe, Mg, Mn, Ni, Pb, Zn; ≤0.02% Ca, Na; ≤0.01% NH_4^+, Cl, SO_4; ≤0.3% foreign AA

Neosystem AA00623   MW 147.1

Peptides International ALE-2708   MW 147.13   >99.9% optical purity by RP-HPLC; white crystalline powder

Rexim   MW 147.1 $C_5H_9NO_4$ White crystals or crystalline powder

Sigma G 1376   FW 164.2 $C_5H_8NO_4 \cdot NH_4$ ≥99% (TLC) | Excitatory AA neurotransmitter

Sigma G 2759   Matrix: cross-linked 4% beaded agarose; activation: cyanogen bromide; attachment: amino; spacer: 1 atom; ligand immobilized: 5-10 μmoles/mL; form: suspension in 2.0 *M* NaCl containing 0.02% thimerosal

Sigma G 5638   FW 147.1 (free acid) $C_5H_9NO_4$ ≥99%; crystalline | Cell culture tested; insect cell culture tested

USBio G7115   MW 147.1 $C_5H_9NO_4$ ≥98.5%; identification by FCC tests; specific rotation (@ 20°C): 31.5° to 32.2°; loss on drying: ≤0.1%; heavy metals (as Pb): ≤0.002%; lead: ≤0.001%; chloride: ≤0.2%; arsenic: ≤0.0003%; store @ 15° to 35°C

Glutamic Acid — L-Glu (1-^{14}C)

Amersham CFA531 $HO_2C(CH_2)_2CH(NH_2)CO_2H$ Aqueous, 2% EtOH, sterilized; 1.85-2.29 GBq/mmol, 50-62 mCi/mmol; 1.85 MBq/mL, 50 μCi/mL

ARC ARC-240   MW 147.1 $HO_2C(CH_2)_2CH(NH_2)COOH$ 50-60 mCi/mmol; 1.85-2.22 GBq/mmol; in 0.01 *N* HCl

Glutamic Acid — L-Glu (^{15}N)

ICN 540139   MW 148.1 $HOOC(CH_2)_2CH(NH_2)COOH$ 99% ^{15}N atomic purity

Glutamic Acid — L-Glu (2,3,4-^{3}H)

ARC ART-132   MW 147.1 $HO_2C(CH_2)_2CH(NH_2)COOH$ 30-60 Ci/mmol; 1.11-2.22 TBq/mmol; in 2% aqueous EtOH

Glutamic Acid — L-Glu (3,4-^{3}H)

ICN 20019   MW 147.1 $HOOCCH_2CH_2CH(NH)_2COOH$ 30-50 Ci/mmol, 1.11-1.85 TBq/mmol, 0.01 *N* HCl

ICN 20019E   MW 147.1 $HOOCCH_2CH_2CH(NH)_2COOH$ 30-50 Ci/mmol, 1.11-1.85 TBq/mmol, sterile EtOH:H_2O (2:98)

Sigma G 5787   FW 147.1 $C_5H_9NO_4$ 30-80 Ci/mmol; radiochemical purity ≥95% (HPLC); EtOH: H_2O (2:98) solution in serum bottle

Glutamic Acid — L-Glu (5-^{14}C)

ARC ARC-241   MW 147.1 $HO_2C(CH_2)_2CH(NH_2)COOH$ 50-60 mCi/mmol; 1.85-2.22 GBq/mmol; in 0.01 *N* HCl

Glutamic Acid — L-Glu (G-³H)

Amersham TRK445 $HO_2C(CH_2)_2CH(NH_2)CO_2H$ Aqueous, 2% EtOH, sterilized; 0.74-2.2 TBq/mmol, 20-60 Ci/mmol; 37 MBq/mL, 1 mCi/mL

ARC ART-130 MW 147.1 $HO_2C(CH_2)_2CH(NH_2)COOH$ 15-30 Ci/mmol; 0.55-1.11 TBq/mmol; in 0.01 N HCl

Glutamic Acid — L-Glu (U-¹⁴C)

Amersham CFB65 $HO_2C(CH_2)_2CH(NH_2)CO_2H$ Aqueous, 2% EtOH, sterilized; >9.25 GBq/mmol, >250 mCi/mmol; 1.85 MBq/mL, 50 μCi/mL

ARC ARC-165 MW 147.1 $HO_2C(CH_2)_2CH(NH_2)COOH$ >200 mCi/mmol; >7.4 GBq/mmol; in 0.01 N HCl

ARC ARC-165A MW 147.1 $HO_2C(CH_2)_2CH(NH_2)COOH$ >200 mCi/mmol; >7.4 GBq/mmol; in EtOH: H_2O (2:98)

ICN 10065 MW 147.1 $HOOC(CH_2)_2CH(NH_2)COOH$ 225-275 mCi/mmol, 8.33-10.18 GBq/mmol, 0.01 N HCl

ICN 10065E MW 147.1 $HOOC(CH_2)_2CH(NH_2)COOH$ 225-275 mCi/mmol, 8.33-10.18 GBq/mmol, sterile EtOH:H_2O (2:98)

Glutamic Acid — L-Glu (U-¹⁴C) Hydrochloride

Sigma G 5398 FW 147.1 $C_5H_9NO_4$ 200-300 mCi/mmol; radiochemical purity ≥98% (HPLC); aqueous solution containing 2% EtOH in Combi-vial

Glutamic Acid — L-Glu 1-(4-NA)

Fluka 49622 MW 267.24 $C_{11}H_{13}N_3O_5$ ≥98.0% (titration); mp: 184-186°C

Glutamic Acid — L-Glu 5-(4-NA) Hydrate

Fluka 49623 MW 285.25 $C_{11}H_{13}N_3O_5 \cdot H_2O$ ≥99.0% (titration); mp: 186-188°C

Glutamic Acid — L-Glu Amide

Synonyms: L-Igln

ICN 157213 MW 146.1 $C_5H_{10}N_2O_3$

Sigma G 3521 FW 146.1 $C_5H_{10}N_2O_3$

Glutamic Acid — L-Glu Calcium Salt

ICN 101795

Glutamic Acid — L-Glu Free Acid

Synonyms: L-2-Aminopentanedioic Acid; (S)-2-Aminopentanedioic Acid; Kainate Receptor Agonist; N-Methyl-D-Aspartate Receptor Agonist; Quisqualate Receptor Agonist

ICN 101793 MW 147.1 $C_5H_9NO_4$ 99-100%; crystalline

ICN 194676 MW 147.1 $C_5H_9NO_4$ 99-100%; crystalline | Cell culture reagent

Sigma G 1251 FW 147.1 $C_5H_9NO_4$ ≥99% (TLC) | Excitatory neurotransmitter; also available as part of a kit

Sigma G 6904 FW 147.1 $C_5H_9NO_4$ SigmaUltra; ≥99% (TLC); residue on ignition <0.1%; solubility (1 M in 1.0 M HCl, 20°C): complete, colorless; insoluble matter <0.1%; Cl, SO_4: <0.05%; Al, Cu, Fe: <0.0005%; Ca, K: <0.005% | Excitatory AA neurotransmitter

Glutamic Acid — L-Glu Hemi-Magnesium Salt

Fluka 49605 MW 388.62 $C_{10}H_{16}MgN_2O_8 \cdot 4H_2O$ ≥98% (titration); mp: 130-135°C; tetrahydrate

Glutamic Acid — L-Glu Hydrate

Synonyms: Kaglutam; Kainic Acid; 2-Carboxy-3-Carboxymethyl-4-Isopropenylpyrrolidine

Rexim MW 203.2 $C_5H_{10}KNO_5$ White crystals or crystalline powder

Sigma I 0250 *Digenea simplex* (seaweed) FW 213.2 $C_{10}H_{15}NO_4$ Conformationally restricted analog of L-Glu

Glutamic Acid — L-Glu Hydrate Monopotassium Salt

Fluka 49601 MW 203.24 $C_5H_8KNO_4 \cdot H_2O$ ≥99.0% (titration); ≤0.05% Na; ≤0.0005% Cd, Co, Cr, Cu, Fe, Mg, Mn, Ni, Pb, Zn; ≤0.02% Ca, Na; ≤0.02% NH_4^+; ≤0.01% Cl, SO_4; ≤0.5% foreign AA

Glutamic Acid — L-Glu Hydrate Monosodium Salt

Fluka 49621 MW 187.14 $C_5H_8NaNO_4 \cdot H_2O$ ≥99.0% (titration); mp: 165°C

Glutamic Acid — L-Glu Hydrochloride

Synonyms: Acidalin; Acidogen; Acidoride; Acidothyn; Acidulen; Acidulin; Aciglumin; Aclor; Acridogen; Acridoride; Antalka; Gastuloric; Glusatin; Kainic Acid-OMe_2

Fluka 49569 MW 183.60 $C_5H_9NO_4 \cdot HCl$ ≥99.5% (titration); ≤0.05% residue on ignition, loss on drying; ≤0.00001% As; ≤0.0005% Al, Ba, Bi, Cd, Co, Cr, Cu, Fe, Li, Mg, Mn, Mo, Ni, Pb, Sr, Zn; ≤0.001% Ca; ≤0.01% NH_4^+; ≤0.005% K, Na, SO_4; ≤0.5% foreign AA

Fluka 49570 MW 183.60 $C_5H_9NO_4 \cdot HCl$ ≥99.0% (titration); ≤0.05% residue on ignition; ≤0.0005% Cd, Co, Cr, Cu, Fe, Mg, Mn, Ni, Pb, Zn; ≤0.001% Ca; ≤0.01% NH_4^+; ≤0.005% K, Na, SO_4; ≤0.5% foreign AA

ICN 101799 MW 183.6 $C_5H_9NO_4 \cdot HCl$ 99%; crystalline

Rexim MW 183.6 $C_5H_{10}ClNO_4$ White crystals or crystalline powder

Sigma I 4877 FW 277.7 $C_{12}H_{19}NO_4 \cdot HCl$

Glutamic Acid — L-Glu Hydrochloride Free Acid

Synonyms: (S)-2-Aminopentanedioic Acid; Kainate Receptor Agonist; NMDA Receptor Agonist; Quisqualate Receptor Agonist

Sigma G 2128 FW 183.6 $C_5H_9NO_4 \cdot HCl$ ≥98% (TLC) | Excitatory AA neurotransmitter

Glutamic Acid — L-Glu L-Arginine Salt

ICN 100738 $C_{11}H_{23}N_5O_6$ The arginine salt of Glu—not a dipeptide

Glutamic Acid — L-Glu Monopotassium Salt

Synonyms: (S)-2-Aminopentanedioic Acid; Kainate Receptor Agonist; NMDA Receptor Agonist; Quisqualate Receptor Agonist

Sigma G 1149 FW 185.2 $C_5H_8NO_4K$ ≥99%; crystalline | Cell culture tested; insect cell culture tested

Sigma G 1501 FW 185.2 $C_5H_8NO_4K$ ≥99% (TLC) | Excitatory AA neurotransmitter

Glutamic Acid — L-Glu Monosodium Salt

Synonyms: L-α-Aminoglutaric Acid; Monosodium Glutamate; (S)-2-Aminopentanedioic Acid; Kainate Receptor Agonist; N-Methyl-D-Aspartate Receptor Agonist; Quisqualate Receptor Agonist

Amersham US16245 ≥99.0%

ICN 101800 MW 169.1 $C_5H_9NO_4Na$ ≥99%; crystalline

ICN 194677 MW 169.1 $C_5H_9NO_4Na$ ≥99% | Cell culture reagent

Sigma G 1626 FW 169.1 $C_5H_8NO_4Na$ ≥99% (TLC) | Excitatory AA neurotransmitter

Sigma G 5889 FW 169.1 $C_5H_8NO_4Na$ ≥99%; crystalline | Cell culture tested; insect cell culture tested

Glutamic Acid — L-Glu Potassium Salt

USBio G7115-05 MW 185.2 $C_5H_9NO_4K$

Glutamic Acid — L-Glu Sodium Salt

USBio G7115-07 MW 169.1 $C_5H_9NO_4Na$

Glutamic Acid — L-Glu γ-(2,2,2-Trichloroethyl) Ester

Sigma G 5529 FW 278.5 $C_7H_{10}Cl_3NO_4$

Glutamic Acid — L-Glu γ-(3-Carboxy-4-Hydroxyanilide)

Sigma G 0512	FW 282.3	$C_{12}H_{14}N_2O_6$

Glutamic Acid — L-Glu γ-(3-Carboxy-4-NA) Ammonium Salt

Sigma G 5008	FW 328.3	$C_{12}H_{12}N_3O_7NH_4$

Glutamic Acid — L-Glu γ-(4-Methoxy-β-Naphthylamide)

Sigma G 0141	FW 302.3	$C_{16}H_{18}N_2O_4$

Glutamic Acid — L-Glu γ-(α-Naphthylamide)

Sigma G 7754	FW 272.3	$C_{15}H_{16}N_2O_3$
Sigma G 3626	FW 272.3	$C_{15}H_{16}N_2O_3$

Glutamic Acid — L-Glu γ-AMC

Sigma G 7261	FW 304.3	$C_{15}H_{16}N_2O_5$

Glutamic Acid — L-Glu γ-Hydrazide

Sigma G 7257	FW 161.2	$C_5H_{11}N_3O_3$

Glutamic Acid — L-Glu γ-Monohydroxamate

Sigma G 2253	FW 162.1	$C_5H_{10}N_2O_4$	>90%

Glutamic Acid — L-Glu γ-pNA Free Base

Synonyms: L-γ-Glu-pNA; γ-Glutamyl Transpeptidase Substrate

Sigma G 1135	FW 267.2	$C_{11}H_{13}N_3O_5$

Glutamic Acid — L-Glu γ-pNA Hydrochloride

Synonyms: γ-Glutamyl Transpeptidase Substrate

Sigma G 6133	FW 303.7	$C_{11}H_{13}N_3O_5 \cdot HCl$

Glutamic Acid — L-Glu-(5-OBzl)

Fluka 49510	MW 237.26	$C_{12}H_{15}NO_4$	≥99.0% (titration); mp: 172-180°C

Glutamic Acid — L-Glu-(5-OEt)

Fluka 49490	MW 175.19	$C_7H_{13}NO_4$	≥99.0% (titration); mp: 179°C

Glutamic Acid — L-Glu-(5-OMe)

Fluka 49610	MW 161.16	$C_6H_{11}NO_4$	≥99.0% (titration); mp: 185°C

Glutamic Acid — L-Glu-(Bis-OAll) Toluene 4-Sulfonate

Fluka 49515	MW 399.47	$C_{11}H_{17}NO_4 \cdot C_7H_8O_3S$	≥98.0%; mp: 90-93°C

Glutamic Acid — L-Glu-(γ-OBzl)

Synonyms: γ-Bzl-L-Glu

ICN 101756	MW 237.3	$C_{12}H_{15}NO_4$	Crystalline
Sigma G 8653	FW 237.3	$C_{12}H_{15}NO_4$	

Glutamic Acid — L-Glu-(γ-OEt)

Synonyms: γ-Et-L-Glu

ICN 101797	MW 175.2	$C_7H_{13}NO_4$	Crystalline
Sigma G 1876	FW 175.2	$C_7H_{13}NO_4$	

Glutamic Acid — L-Glu-(γ-OMe)

Synonyms: γ-Me-L-Glu

ICN 101798	MW 161.2	$C_6H_{11}NO_4$	Crystalline

Sigma G 1751	FW 161.2	$C_6H_{11}NO_4$

Glutamic Acid — L-Glu-AMC

ICN 157214	MW 304.3	$C_{15}H_{16}N_2O_5$	Crystalline

Glutamic Acid — L-Glu-OAll₂ p-Tosyl

Sigma G 0648	FW 399.5	$C_{11}H_{17}NO_4 \cdot C_7H_8O_3S$

Glutamic Acid — L-Glu-OBzl

Neosystem AA00602	MW 237.3

Glutamic Acid — L-Glu-OcHex

Neosystem AA00606	MW 229.3

Glutamic Acid — L-Glu-OEt₂ Hydrochloride

Synonyms: Diethyl-L-Glu

Fluka 49550	MW 239.7	$C_9H_{17}NO_4 \cdot HCl$	≥99.0% (titration); mp: 113-115°C
ICN 157216	MW 239.7	$C_9H_{17}NO_4 \cdot HCl$	Crystalline
Sigma G 9378	FW 239.7	$C_9H_{17}NO_4 \cdot HCl$	Non-selective agonist

Glutamic Acid — L-Glu-OMe₂ Hydrochloride

Synonyms: Dimethyl-L-Glu

Fluka 49560	MW 211.65	$C_7H_{13}NO_4 \cdot HCl$	≥99.0% (titration); 2% H_2O
ICN 157217	MW 211.6	$C_7H_{13}NO_4 \cdot HCl$	Crystalline
Sigma G 9253	FW 211.6	$C_7H_{13}NO_4 \cdot HCl$	

Glutamic Acid — L-Glu-OtBu₂ Dibenzenesulfimide Salt

Fluka 49555	MW 556.7	$C_{13}H_{25}NO_4 \cdot C_{12}H_{11}NO_4S_2$	≥99.0%; mp: 139-141°C

Glutamic Acid — L-Glu-OtBu₂ Hydrochloride

Synonyms: Di-tBu-L-Glu

ICN 157215	MW 295.8	$C_{13}H_{25}NO_4 \cdot HCl$	Crystalline
Sigma G 7501	FW 295.8	$C_{13}H_{25}NO_4 \cdot HCl$	

Glutamic Acid — Methotrexate L-Glu (1-¹⁴C) Sodium Salt

ARC ARC-1266	MW 454.5	50-60 mCi/mmol; 1.85-2.22 TBq/mmol; in EtOH: H_2O (1:1)

Glutamic Acid — Mg (L-Glu)₂ Hydrate

Rexim	MW 388.6	$C_{10}H_{24}MgN_2O_{12} \cdot 4H_2O$	White crystals or crystalline powder

Glutamic Acid — N-(p-(((2-Amino-4-Hydroxy-6-Pteridinyl)-Methyl)Amino)Benzoyl))Glutamic Acid

Synonyms: Folic Acid

USBio F5800-05 ≥98% purity by HPLC; lyophilized | Suitable for antigenic applications in immunological protocols

Glutamic Acid — N-2,4-Dnp-DL-Glu

Sigma D 9129	FW 313.2	$C_{11}H_{11}N_3O_8$

Glutamic Acid — N-2,4-Dnp-DL-Glu Dicyclohexylammonium Salt

Sigma D 9254	FW 511.6	$C_{11}H_{11}N_3O_8 \cdot 2C_6H_{13}N$

Glutamic Acid — N-3-Nitro-2-Pyridinesulfenyl-L-Glu-(γ-OtBu)

Sigma N 4390	FW 357.4	$C_{14}H_{19}N_3O_6S$

Glutamic Acid — *N*-4-Aminobenzoyl-L-Glu

Fluka 07080   MW 266.26 $C_{12}H_{14}N_2O_5$ ~99% (titration); mp: ~175°C

Glutamic Acid — N^5-For-(5,6,7,8-Tetrahydropteroyl)-L-Glu Calcium Salt

Synonyms: Citrovorum Factor; Leucovorin; Folinic Acid

Sigma F 7878   FW 511.5 $C_{20}H_{21}N_7O_7Ca$ 90-95%

Glutamic Acid — *N*-Ac-DL-Glu

Sigma A 8875   FW 189.2 $C_7H_{11}NO_5$

Glutamic Acid — *N*-Ac-L-Glu

Synonyms: L-2-Acetylamino-Glutaric Acid

Fluka 01160   MW 189.17 $C_7H_{11}NO_5$ ≥99% (titration); mp: 194-196°C

ICN 100077   MW 189.2 $C_7H_{11}NO_5$ Crystalline

Rexim   MW 189.2 $C_7H_{11}NO_5$ White crystals or crystalline powder

Glutamic Acid — *N*-Carbamyl-DL-Glu

ICN 101248   MW 190.2 $C_6H_{10}N_2O_5$ Crystalline; 99%

Glutamic Acid — *N*-Carbamyl-L-Glu

Sigma C 4375   FW 190.2 $C_6H_{10}N_2O_5$

Glutamic Acid — *N*-CBZ-D-Glu-(α-OBzl)

Sigma C 3663   FW 371.4 $C_{20}H_{21}NO_6$

Glutamic Acid — *N*-CBZ-L-Glp

Sigma C 5761   FW 263.2 $C_{13}H_{13}NO_5$

Glutamic Acid — *N*-CBZ-L-Glu

ICN 100215   MW 281.3 $C_{13}H_{15}NO_6$ Crystalline

Sigma C 6876   FW 281.3 $C_{13}H_{15}NO_6$

Glutamic Acid — *N*-CBZ-L-Glu-(α-OMe)

Sigma C 4922   FW 295.3 $C_{14}H_{17}NO_6$

Glutamic Acid — *N*-CBZ-L-Glu-(γ-OtBu)

Sigma C 0532   FW 337.4 $C_{17}H_{23}NO_6$

Glutamic Acid — *N*-CBZ-γ-tBu-L-Glu *N*-Hydroxysuccinimide Ester

Sigma C 0795   FW 434.4 $C_{21}H_{26}N_2O_8$

Glutamic Acid — *N*-FMOC-L-Glu-(γ-OtBu)

Sigma F 1758   FW 425.5 $C_{24}H_{27}NO_6$

Glutamic Acid — *N*-FMOC-γ-tBu-L-Glu SASRIN Resin Ester

Sigma F 1389   200-400 mesh; substitution range: 0.3-0.8 mmole FMOC AA/g resin | For peptide synthesis; *N*-FMOC AA acyl ester of 3-methoxy-4-oxymethyl-phenoxymethylated 1% divinylbenzene cross-linked polystyrene; peptides can be cleaved from the resin with 0.5-1% CF_3CO_2H in CH_2Cl_2; Mergler, M et al, *in Peptides: Chemistry & Biology* (Proc 10th Am Peptide Symp), Marhall, GR, ed, 259, 1988

Glutamic Acid — *N*-FMOC-γ-tBu-L-Glu-OPfp

Sigma F 9514   FW 591.5 $C_{30}H_{26}F_5NO_6$ ≥90% (TLC)

Glutamic Acid — *N*-Me-DL-Glu

ICN 155554   MW 161.2 $C_6H_{11}NO_4$ Crystalline

Sigma M 5263   FW 161.2 $C_6H_{11}NO_4$

Glutamic Acid — *N*-Me-Glu

Bachem E-2130.0001   MW 161.16 $C_6H_{11}NO_4$ Store at RT

Glutamic Acid — *N*-Me-L-Glu

ICN 155555   MW 161.2 $C_6H_{11}NO_4$

Sigma M 4017   FW 161.2 $C_6H_{11}NO_4$

Glutamic Acid — *N*-*p*-Aminobenzoyl-L-Glu

ICN 100606   MW 266.3 $C_{12}H_{14}N_2O_5$ Crystalline

Sigma A 0879   FW 266.3 $C_{12}H_{14}N_2O_5$

Glutamic Acid — *N*-Phthaloyl-L-Glu

Synonyms: PhGA

Fluka 79840   MW 277.24 $C_{13}H_{11}NO_6$ ≥98.0% (titration); mp: 160-162°C

Sigma P 1801   FW 277.2 $C_{13}H_{11}NO_6$

Glutamic Acid — *N*-*p*-Nitrobenzoyl-L-Glu

Synonyms: N-4-Nitrobenzoyl-L-Glu

ICN 155862   MW 296.3 $C_{12}H_{12}N_2O_7$

Glutamic Acid — *N*-t-BOC-D-Glu-(α-OBzl)

Sigma B 3280   FW 337.4 $C_{17}H_{23}NO_6$

Glutamic Acid — *N*-t-BOC-L-Glu-(α-OBzl)

Sigma B 9003   FW 337.4 $C_{17}H_{23}NO_6$ Substrate for Vitamin K-dependent carboxylation; Kappel, WK & Romiti, S, *J Biochem Biophys Meth*, 11: 59, 1985

Glutamic Acid — *N*-t-BOC-L-Glu-(α-OPh)

Synonyms: Staphylococcal Protease Substrate

Sigma B 3016   FW 323.3 $C_{16}H_{21}NO_6$ Chromogenic substrate; Houmard, J, *int J Peptide Protein Res*, 8: 199, 1976

Glutamic Acid — *N*-t-BOC-L-Glu-(α-OtBu)

Sigma B 2522   FW 303.4 $C_{14}H_{25}NO_6$

Glutamic Acid — *N*-t-BOC-L-Glu-(γ-OBzl) Free Acid

Sigma B 6876   FW 337.4 $C_{17}H_{23}NO_6$

Glutamic Acid — *N*-t-BOC-L-Glu-(γ-OcHex) Free Acid

Sigma B 1900   FW 329.4 $C_{16}H_{27}NO_6$

Glutamic Acid — *N*-t-BOC-L-Glu-(γ-OtBu)

Sigma B 7504   FW 303.4 $C_{14}H_{25}NO_6$

Glutamic Acid — *N*-t-BOC-γ-Bzl-L-Glu Resin Ester

Sigma B 8521   *N*-t-BOC AA ester of methylated 1% divinylbenzene cross-linked polystyrene; 200-400 mesh; substitution range: 0.2-0.6 mmole t-BOC-AA/g resin | For peptide synthesis by the Merrifield method

Glutamic Acid — *N*-TFA-L-Glu-(α-OMe) Dicyclohexylammonium Salt

Sigma T 6401   FW 438.5 $C_8H_{10}F_3NO_5 \cdot C_{12}H_{23}N$

Glutamic Acid — *N*-Tosyl-L-Glu-OMe

Synonyms: *N*-*p*-Toluenesulfonyl-L-Glu

Sigma T 3630   FW 301.3 $C_{12}H_{15}NO_6S$ May contain ~10% isomer

Glutamic Acid — *N*-α-FMOC-L-Glu-(γ-OtBu)

ICN 151142   MW 425.5 $C_{24}H_{27}NO_6$

Glutamic Acid — *N*-α-t-BOC-L-Glu

ICN 150491   MW 247.2 $C_{10}H_{17}NO_6$ Crystalline

Glutamic Acid — *N*-α-t-BOC-L-Glu-(α-OBzl)

ICN 150492   MW 337.4 $C_{17}H_{23}NO_6$ Crystalline

Glutamic Acid — *N*-α-t-BOC-L-Glu-(γ-OBzl) Free Acid

ICN 101066   MW 337.4 $C_{17}H_{23}NO_6$ Crystalline

Glutamic Acid — *N*-α-t-BOC-L-Glu-(γ-OtBu)

ICN 150493   MW 303.4 $C_{14}H_{25}NO_6$ Crystalline

Glutamic Acid — *N*-α-t-BOC-L-Glu-(γ-OtBu) Dicyclohexylammonium Salt

ICN 105098   MW 484.4 $C_{14}H_{25}NO_6 \cdot C_{12}H_{23}N$

Glutamic Acid — Pteroyl-L-Glu

Fluka 47620   MW 441.41 $C_{19}H_{19}N_7O_6$ ≥97% (HPLC); ≤0.1% residue on ignition; ≤0.005% K; ≤0.0005% Cd, Co, Cr, Cu, Fe, Mg, Mn, Ni, Pb, Zn; ≤0.05% Na; ≤0.001% Ca; 8% H_2O

Glutamic Acid — Pth-Glu

Sigma P 0127   FW 264.3 $C_{12}H_{12}N_2O_3S$ ~99% | Useful as standards in protein sequencing by the Edman degradation

Glutamic Acid — TentaGel S PHB-Glu-tBu-FMOC

Fluka 86389   0.20 mmol protected AA/g; ~90 µm particle size

Glutamic Acid — TentaGel S Trt-Glu-tBu-FMOC

Fluka 86421   0.18 mmol protected AA/g; ~90 µm particle size

Glutamic Acid — *trans*-2,4-Methanoglutamic Acid

Synonyms: *trans*-ACBD; 1-Aminocyclobutane-*trans*-1,3-Dicarboxylic Acid; *cis*-1-Amino-1,3-Dicarboxycyclobutane

Sigma A 4816   FW 159.1 $C_6H_9NO_4$ ~95% | Potent inhibitor of glutamate uptake

Glutamic Acid — Z-(γ-Carboxy)-D-Glu-OtBu₂

Synonyms: Z-D-Gla-OtBu₂

Bachem C-1515.0250   MW 437.49 $C_{22}H_{31}NO_8$ Store at -15°C

Glutamic Acid — Z-(γ-Carboxy)-DL-Glu-OtBu₂

Synonyms: Z-DL-Gla-OtBu₂

Bachem C-1520.0250   MW 437.49 $C_{22}H_{31}NO_8$ Store at -15°C

Glutamic Acid — Z-(γ-Carboxy)-Glu-OtBu₂

Synonyms: Z-Gla-OtBu₂

Bachem C-1510.0250   MW 437.49 $C_{22}H_{31}NO_8$ Store at -15°C

Glutamic Acid — Z-D-Glp

Bachem C-2560.0001   MW 263.25 $C_{13}H_{13}NO_5$ Store at RT

Peptides International ZDU-2613   MW 263.25   >98% (HPLC); white crystalline powder

Glutamic Acid — Z-D-Glu

Bachem C-1565.0005   MW 281.27 $C_{13}H_{15}NO_6$ Store at RT
Neosystem ZA00624   MW 281.3

Glutamic Acid — Z-D-Glu-OBzl

Bachem C-3120.0001   MW 371.39 $C_{20}H_{21}NO_6$ Store at RT
Bachem C-3675.0001   MW 371.39 $C_{20}H_{21}NO_6$ Store at RT

Glutamic Acid — Z-D-Glu-OMe

Bachem C-3820.0001   MW 295.29 $C_{14}H_{17}NO_6$ Store at -15°C
Bachem C-3930.0001   MW 295.29 $C_{14}H_{17}NO_6$ Store at -15°C

Glutamic Acid — Z-D-Glu-OtBu

Bachem C-1600.0001   MW 337.37 $C_{17}H_{23}NO_6$ Store at 2-8°C

Glutamic Acid — Z-Glp

Synonyms: Benzyloxycarbonyl-L-Pyroglutamic Acid; *N*-Benzyloxycarbonyl-L-Pyrrolidone Carboxylic Acid

Bachem C-2555.0005   MW 263.25 $C_{13}H_{13}NO_5$ Store at RT
Peptides International ZLU-2117   MW 263.25   >98% (HPLC); white crystalline powder

Glutamic Acid — Z-Glp-ONp

Bachem C-2565.0001   MW 384.34 $C_{19}H_{16}N_2O_7$ Store at -15°C

Glutamic Acid — Z-Glp-OSu

Bachem C-3335.0001   MW 360.33 $C_{17}H_{16}N_2O_7$ Store at -15°C

Glutamic Acid — Z-Glu

Bachem C-1560.0025   MW 281.27 $C_{13}H_{15}NO_6$ Store at RT
Peptides International ZLE-2015-PI   MW 281.27   >98% (HPLC); white crystalline powder

Glutamic Acid — Z-Glu Amide

Synonyms: Z-L-Isoglutamic Acid; Z-Isoglutamic Acid

Bachem C-3735.0001   MW 280.28 $C_{13}H_{16}N_2O_5$ Store at RT

Glutamic Acid — Z-Glu-OBzl

Bachem C-1570.0001   MW 371.39 $C_{20}H_{21}NO_6$ Store at RT
Bachem C-1575.0001   MW 371.39 $C_{20}H_{21}NO_6$ Store at RT

Glutamic Acid — Z-Glu-OBzl-ONp

Bachem C-1585.0001   MW 492.49 $C_{26}H_{24}N_2O_8$ Store at -15°C

Glutamic Acid — Z-Glu-OBzl-OSu

Bachem C-3980.0001   MW 468.47 $C_{24}H_{24}N_2O_8$ Store at -15°C

Glutamic Acid — Z-Glu-OMe

Bachem C-3295.0005   MW 295.29 $C_{14}H_{17}NO_6$ Store at -15°C
Bachem C-3465.0001   MW 295.29 $C_{14}H_{17}NO_6$ Store at -15°C

Glutamic Acid — Z-Glu-OMe-OSu

Bachem C-3945.0001   MW 392.37 $C_{18}H_{20}N_2O_8$ Store at -15°C

Glutamic Acid — Z-Glu-OSu-OBzl

Bachem C-1580.0001   MW 468.46 $C_{24}H_{24}N_2O_8$ Store at -15°C

Glutamic Acid — Z-Glu-OtBu

Bachem C-1595.0005   MW 337.37 $C_{17}H_{23}NO_6$ Store at 2-8°C

Bachem C-3905.0001   MW 337.37 $C_{17}H_{23}NO_6$ Store at 2-8°C

Glutamic Acid — Z-Glu-OtBu Dicyclohexylamine

Bachem C-1590.0001   MW 518.69 $C_{17}H_{23}NO_6 \cdot C_{12}H_{23}N$ Store at RT

Glutamic Acid — Z-Glu-OtBu-ONp

Bachem C-1610.0001   MW 458.47 $C_{23}H_{26}N_2O_8$ Store at -15°C

Glutamic Acid — Z-Glu-OtBu-OSu

Bachem C-1605.0001   MW 434.45 $C_{21}H_{26}N_2O_8$ Store at -15°C

Glutamic Acid — Z-L-Glu

Fluka 96120   MW 281.27 $C_{13}H_{15}NO_6$ ≥99.0% (titration); mp: 117-119°C

Neosystem ZA00601   MW 281.3

Glutamic Acid — Z-L-Glu 5-tBu-1-(N-Succinimidyl) Ester

Fluka 96131   MW 434.45 $C_{21}H_{26}N_2O_8$ ≥99.0%; mp: 105-107°C

Glutamic Acid — Z-L-Glu-(1-OBzl)

Fluka 96125   MW 371.39 $C_{20}H_{21}NO_6$ ≥98.0% (TLC)

Glutamic Acid — Z-L-Glu-(1-OMe)

Fluka 96140   MW 295.29 $C_{14}H_{17}NO_6$ ≥99.0% (TLC); mp: 64-67°C

Glutamic Acid — Z-L-Glu-(5-OtBu)

Fluka 96129   MW 337.38 $C_{17}H_{23}NO_6$ ≥99.0% (titration); mp: 85-87°C

Glutamic Acid — Z-L-Glu-OtBu

Neosystem ZA00602   MW 337.4

Glutamic Acid — Z-N-Me-Glu-OtBu

Bachem C-3805.0001   MW 351.4 $C_{18}H_{25}NO_6$ Store at -15°C

Glutamic Acid — α-Me-DL-Glu Hydrate

ICN 100497   MW 161.2 $C_6H_{11}NO_4$ 99%; crystalline | MW is anhydrous

Sigma M 9626   FW 161.2 $C_6H_{11}NO_4$

Glutamic Acid — β-Glu

Synonyms: 3-Aminopentanedioic Acid

Sigma G 1763   FW 147.1 $C_5H_9NO_4$

Glutamic Acid — β-Glu Hydrochloride

Fluka 03688   MW 183.59 $C_5H_9NO_4 \cdot HCl$ ≥98.0% (TLC)

Glutamic Acid — γ-Carboxy-D-Glu

Synonyms: D-Gla

Bachem F-1385.0010   MW 191.14 $C_6H_9NO_6$ Store at -15°C

Glutamic Acid — γ-Carboxy-DL-Glu

Synonyms: DL-Gla

Bachem F-1390.0010   MW 191.14 $C_6H_9NO_6$ Store at -15°C

Fluka 21885   MW 191.14 $C_6H_9NO_6$

Glutamic Acid — γ-Carboxy-Glu

Synonyms: Gla

Bachem F-1380.0010   MW 191.14 $C_6H_9NO_6$ Store at -15°C

Glutamic Acid — γ-Carboxy-L-Glu

Fluka 21880   MW 191.14 $C_6H_9NO_6$ ≥98% (TLC)

Glutamic Acid — γ-D-Glu-Aminomethylphosphonic Acid

Synonyms: NMDA Antagonist

ICN 157219   MW 240.2 $C_6H_{13}N_2O_6P$ Davies, M etal, *Comp Biochem Physiol*, 72:211, 1982

Sigma G 4019   FW 240.2 $C_6H_{13}N_2O_6P$ Davies, J et al, *Comp Biochem Physiol*, 72: 211, 1982

Glutamic Acid — γ-D-Glu-Aminomethylsulfonic Acid

Synonyms: GAMS; Kainate Receptor Agonist; Quisqualate Receptor Agonist

ICN 157220   MW 240.2 $C_6H_{12}N_2O_6S$ Davies, M etal, *Comp Biochem Physiol*, 72:211, 1982

Sigma G 3894   FW 240.2 $C_6H_{12}N_2O_6S$ Similar to Sigma G 2897, but prepared for Sigma; Kainate/quisqualate selective antagonist; anticonvulsant; Davies, J et al, *Comp Biochem Physiol*, 72: 211, 1982

Glutamic Acid — γ-D-Glutamylaminomethylsulfonic Acid

Synonyms: GAMS; Kainate Receptor Agonist; Quisqualate Receptor Agonist

Sigma G 2897   FW 240.2 $C_6H_{12}N_2O_6S$ ~95%, balance primarily ammonium acetate buffer | Kainate/quisqualate selective antagonist; anticonvulsant; Davies, J et al, *Comp Biochem Physiol*, 72: 211, 1982

Glutamic Acid — γ-D-Glu-Taurine

Synonyms: γ-D-Glutamylaminoethylsulfonic Acid; Kainite Antagonist; Quisqualate Antagonist; Kainate Receptor Agonist; Quisqualate Receptor Agonist

ICN 193656   MW 254.3 $C_7H_{14}N_2O_6S$

Sigma G 4777   FW 254.3 $C_7H_{14}N_2O_6S$ ~98% | Kainate/quisqualate selective antagonist; Davies, J et al, *Comp Biochem Physiol*, 72: 211, 1982

Glutamic Acid — γ-L-Glu Hydrazide

Synonyms: L-Glu-γ-Hydrazide; Glutamine Antagonist

ICN 101764   MW 161.2 $C_5H_{11}N_3O_3$

Glutamic Acid — γ-L-Glu-pNA Hydrate

Synonyms: γ-Glutamyl Transpeptidase Substrate

ICN 100663   MW 267.2 $C_{11}H_{13}N_3O_5$ Crystalline; white to cream colored; <0.1% free p-nitroaniline

Glutamic Acid — γ-L-Glu-pNA Hydrochloride

Synonyms: γ-Glutamyl Transpeptidase Substrate

ICN 151495   MW 303.7 $C_{11}H_{13}N_3O_5 \cdot HCl$ Crystalline | A water soluble form of ICN 100663

Glutamic Acid — γ-Methylene-DL-Glu

Bachem F-1815.0050	MW 159.14	$C_6H_9NO_4$	Store at RT	
ICN 155533	MW 159.1	$C_6H_9NO_4$	Crystalline	
Sigma M 5388	FW 159.1	$C_6H_9NO_4$	Crystalline	

Glutamine — Ac-Gln Amide

Bachem E-2865.0250   MW 187.2 $C_7H_{13}N_3O_3$ Store at -15°C

Glutamine — Ac-Gln-Trt

Bachem E-3035.0001   MW 430.5 $C_{26}H_{26}N_2O_4$ Store at 2-8°C

Glutamine — BOC-D-Gln

Synonyms: BOC-D-Gln

Bachem A-1710.0001	MW 246.26	$C_{10}H_{18}N_2O_5$	Store at RT
Fluka 15098	MW 246.26	$C_{10}H_{18}N_2O_5$	≥98% (TLC)
Neosystem BA00701	MW 246.3		
Peptides International BDQ-2623	MW 246.26	>98% (HPLC); white crystalline powder	
USBio B2301	MW 246.3	$C_{10}H_{18}N_2O_5$	≥99%

Glutamine — BOC-D-Gln-(4-ONp)

Fluka 15103   MW 367.36 $C_{16}H_{21}N_3O_7$ ≥98% (TLC)

Glutamine — BOC-D-Gln-ONp

Synonyms: BOC-D-Gln-(pONp)

Bachem A-1725.0001   MW 367.36 $C_{16}H_{21}N_3O_7$ Store at -15°C

Peptides International BDQ-2621   MW 367.36   >98% (HPLC); white crystalline powder; keep dry in freezer

Glutamine — BOC-D-Gln-Xan

Bachem A-3330.0001   MW 426.47 $C_{23}H_{26}N_2O_6$ Store at -15°C

Glutamine — BOC-Gln

Synonyms: BOC-L-Gln

American Peptide BLGLN10	MW 246.3		
Bachem A-1705.0005	MW 246.26	$C_{10}H_{18}N_2O_5$	Store at RT
Peptides International BLQ-2062	MW 246.26	>98% (HPLC); white crystalline powder	
USBio B2302	MW 246.3	$C_{10}H_{18}N_2O_5$	≥99%

Glutamine — BOC-Gln-ONp

Synonyms: BOC-L-Gln-(pONp)

Bachem A-1720.0005   MW 367.36 $C_{16}H_{21}N_3O_7$ Store at -15°C

Peptides International BLQ-2079   MW 367.36   >98% (HPLC); white crystalline powder; keep dry in freezer

USBio B2303   MW 367.4 $C_{16}H_{21}N_3O_7$ ≥99%

Glutamine — BOC-Gln-O-Resin

American Peptide RMRB175   1% DVB cross-linked: 100-200 mesh

Glutamine — BOC-Gln-OSu

Bachem A-1715.0001   MW 343.34 $C_{14}H_{21}N_3O_7$ Store at -15°C

Glutamine — BOC-Gln-PAM-® (200-400 mesh)

Bachem D-1290.0001   Store at 2-8°C

Glutamine — BOC-Gln-PAM-Resin

American Peptide RPAM175   1% DVB cross-linked: 100-200 mesh

Glutamine — BOC-Gln-Xan

Synonyms: $N^α$-BOC-$N^γ$-Xan-L-Gln

American Peptide BLGLN15	MW 426.5		
Bachem A-2490.0005	MW 426.47	$C_{23}H_{26}N_2O_6$	Store at -15°C
Peptides International BLQ-5349-PI	MW 426.47	>98% (HPLC); white crystalline powder	
USBio B2304	MW 426.5	$C_{23}H_{26}N_2O_6$	≥99%

Glutamine — BOC-Glp

Bachem A-3850.0001   MW 229.23 $C_{10}H_{15}NO_5$ Store at RT

Glutamine — BOC-Glp Dicyclohexylamine

Bachem A-3925.0005   MW 410.55 $C_{10}H_{15}NO_5 \cdot C_{12}H_{23}N$ Store at RT

Glutamine — BOC-L-Gln

Fluka 15412	MW 246.26	$C_{10}H_{18}N_2O_5$	≥99% (titration)
Neosystem BA00702	MW 246.3		

Glutamine — BOC-L-Gln Hydroxysuccinimide Ester

Fluka 15101   MW 343.34 $C_{14}H_{21}N_3O_7$ ≥98%; mp: 121-124°C

Glutamine — BOC-L-Gln-(4-ONp)

Fluka 15105   MW 367.36 $C_{16}H_{21}N_3O_7$ ≥98% (HPLC)

Glutamine — BOC-L-Gln-MBHA Resin

Neosystem RN00702

Glutamine — BOC-L-Gln-O-CH₂-φ-CH₂-COOH

Neosystem LP00702   MW 394.5

Glutamine — BOC-L-Gln-PAM Resin

Neosystem RP00702

Glutamine — Bsmoc-Gln-Trt

Synonyms: $N^α$-Bsmoc-$N^γ$-Trt-L-Gln

Peptides International BLQ-5622-PI   MW 610.69   >98% (HPLC); white to off-white powder | LA Carpino, et al, *JACS*, 119:9915, 1997

Glutamine — CBZ-D-Gln

USBio C2098-67   MW 280.3 $C_{13}H_{16}N_2O_5$ ≥99%

Glutamine — CBZ-Gln

USBio C2098-66   MW 280.3 $C_{13}H_{16}N_2O_5$ ≥99%

Glutamine — CBZ-Gln-ONp

USBio C2098-68   MW 401.4 $C_{19}H_{19}N_3O_7$ ≥99%

Glutamine — Dansyl-L-Gln

ICN 100066

Glutamine — Dansyl-L-Gln Free Acid

Sigma D 0631   FW 379.4 $C_{17}H_{21}N_3O_5S$ ~99%

Glutamine — D-Gln

Synonyms: D-2-Aminoglutaramic Acid

Alexis 101-017   MW 146.2 $C_5H_{10}N_2O_3$ White solid; soluble in H_2O or dilute aqueous acid

Bachem F-1580.0001   MW 146.15 $C_5H_{10}N_2O_3$ Store at -15°C

Fluka 49410   MW 146.15 $C_5H_{10}N_2O_3$ ≥99.0% (titration); ≤0.05% residue on ignition, loss on drying

ICN 101801   MW 146.1 $C_5H_{10}N_2O_3$ Crystalline

Neosystem AA00701   MW 146.2

Sigma G 9003   FW 146.1 $C_5H_{10}N_2O_3$ ≥98% (TLC) | Inactive isomer of glutamine

USBio G7120-10   MW 146.2 $C_5H_{10}N_2O_3$ ≥99%

Glutamine — D-Gln Amide Hydrochloride

Bachem F-3245.0001   MW 181.62 $C_5H_{11}N_3O_2 \cdot HCl$ Store at 2-8°C

Glutamine — D-Gln-Mtt

Bachem F-2865.0001   MW 402.51 $C_{25}H_{26}N_2O_3$ Store at -15°C

Glutamine — D-Gln-Trt

Bachem F-2760.0001   MW 388.47 $C_{24}H_{24}N_2O_3$ Store at -15°C

Glutamine — Dnp-L-Gln

ICN 101804   MW 313.2 $C_{11}H_{12}N_4O_7$ Crystalline

Glutamine — FMOC-D-Gln

Synonyms: N^α-FMOC-D-Gln

Bachem B-1585.0001   MW 368.39 $C_{20}H_{20}N_2O_5$ Store at RT

Fluka 47459   MW 368.39 $C_{20}H_{20}N_2O_5$ ≥98% (HPLC)

Neosystem FA00701   MW 368.4

Peptides International FDQ-1808-PI   MW 368.39   >98% (HPLC); off-white crystalline powder

USBio F5366   MW 368.4 $C_{20}H_{20}N_2O_5$ ≥99%

Glutamine — FMOC-D-Gln-Mbh

USBio F5370   MW 594.7 $C_{35}H_{34}N_2O_7$ ≥99%

Glutamine — FMOC-D-Gln-Mtt

Bachem B-2060.0001   MW 624.74 $C_{40}H_{36}N_2O_5$ Store at -15°C

Glutamine — FMOC-D-Gln-Mtt-® (200-400 mesh)

Bachem D-1925.0001   Store at 2-8°C

Glutamine — FMOC-D-Gln-Mtt-SASRIN™-® (200-400 mesh)

Bachem D-1920.0001   Store at 2-8°C

Glutamine — FMOC-D-Gln-OPfp

Bachem B-1850.0001   MW 534.4 $C_{26}H_{19}N_2O_5F_5$ Store at -15°C

Glutamine — FMOC-D-Gln-Trt

Synonyms: N^α-FMOC-N^γ-Trt-D-Gln

Bachem B-2015.0001   MW 610.72 $C_{39}H_{34}N_2O_5$ Store at -15°C

Peptides International FDQ-1809-PI   MW 610.71   >98% (HPLC); off-white crystalline powder

USBio F5372   MW 610.7 $C_{39}H_{34}N_2O_5$ ≥99%

Glutamine — FMOC-D-Gln-Trt-® (200-400 mesh)

Bachem D-1945.0001   Store at 2-8°C

Glutamine — FMOC-D-Gln-Trt-SASRIN™-® (200-400 mesh)

Bachem D-1940.0001   Store at 2-8°C

Glutamine — FMOC-Gln

Synonyms: FMOC-L-Gln

American Peptide FGLN100   MW 368.4

Bachem B-1300.0005   MW 368.39 $C_{20}H_{20}N_2O_5$ Store at RT

Peptides International FLQ-1721-PI   MW 368.39   >98% (HPLC); white powder

USBio F5367   MW 368.4 $C_{20}H_{20}N_2O_5$ ≥99%

Glutamine — FMOC-Gln-1-Adamantyl

Bachem B-2435.0001   MW 502.61 $C_{30}H_{34}N_2O_5$ Store at RT

Glutamine — FMOC-Gln-1-Adamantyl-® (200-400 mesh)

Bachem D-1955.0001   Store at 2-8°C

Glutamine — FMOC-Gln-DOD

Bachem B-1295.0001   MW 594.66 $C_{35}H_{34}N_2O_7$ Store at -15°C

Glutamine — FMOC-Gln-DOD-® (200-400 mesh)

Bachem D-1095.0001   Store at 2-8°C

Glutamine — FMOC-Gln-DOD-SASRIN™-® (200-400 mesh)

Bachem D-1335.0001   Store at 2-8°C

Glutamine — FMOC-Gln-Mbh

USBio F5371   MW 594.7 $C_{35}H_{34}N_2O_7$ ≥99%

Glutamine — FMOC-Gln-Mtt

Bachem B-2050.0001   MW 624.74 $C_{40}H_{36}N_2O_5$ Store at -15°C

Glutamine — FMOC-Gln-Mtt-® (200-400 mesh)

Bachem D-1815.0001   Store at 2-8°C

Glutamine — FMOC-Gln-Mtt-OPfp

Bachem B-2215.0001   MW 790.79 $C_{46}H_{35}N_2O_5F_5$ Store at -15°C

Glutamine — FMOC-Gln-Mtt-SASRIN™-® (200-400 mesh)

Bachem D-1810.0001   Store at 2-8°C

Glutamine — FMOC-Gln-Mtt-Ser-(Psi(Me,Me)-Pro)

Synonyms: (4S)-3-(FMOC-Gln-Mtt)-(2,2-Dimethyl)-Oxazolidine-4-Carboxylic Acid

Bachem B-3450.0001   MW 751.88 $C_{46}H_{45}N_3O_7$ Store at -15°C

Glutamine — FMOC-Gln-ONp

Bachem B-1305.0001   MW 489.49 $C_{26}H_{23}N_3O_7$ Store at -15°C

USBio F5368   MW 489.5 $C_{26}H_{23}N_3O_7$ ≥99%

Glutamine — FMOC-Gln-OPfp

Bachem B-1125.0001   MW 534.44 $C_{26}H_{19}F_5N_2O_5$ Store at -15°C

USBio F5369   MW 534.4 $C_{26}H_{19}N_2O_5F_5$ ≥99%

Glutamine — FMOC-Gln-OtBu

USBio F5375   MW 425.5 $C_{24}H_{27}NO_6$ ≥99%

Glutamine — FMOC-Gln-Tmob

Bachem B-1700.0001   MW 548.59 $C_{30}H_{32}N_2O_8$ Store at -15°C

Glutamine — FMOC-Gln-Trt

Synonyms: N^α-FMOC-N^γ-Trt-L-Gln

American Peptide FGLN115   MW 611.8

Bachem B-1790.0001   MW 610.71 $C_{39}H_{34}N_2O_5$ Store at -15°C

Peptides International FLQ-1755-PI   MW 610.71   >98% (HPLC); white crystalline powder

USBio F5373   MW 610.7 $C_{39}H_{34}N_2O_5$ ≥99%

Glutamine — FMOC-Gln-Trt-® (200-400 mesh)

Bachem D-1690.0001   Store at RT

Glutamine — FMOC-Gln-Trt-OPfp

Bachem B-2205.0001   MW 776.76 $C_{45}H_{33}F_5N_2O_5$ Store at -15°C

USBio F5374   MW 776.8 $C_{45}H_{35}N_2O_5F_5$ ≥99%

Glutamine — FMOC-Gln-Trt-SASRIN™-® (200-400 mesh)

Bachem D-1685.0001   Store at 2-8°C

Glutamine — FMOC-L-Gln

Fluka 47626   MW 368.39 $C_{20}H_{20}N_2O_5$ ~97% (titration); mp: 224-226°C

Neosystem FA00702   MW 368.4

Glutamine — FMOC-L-Gln-(4-ONp)

Fluka 47461   MW 475.46 $C_{25}H_{21}N_3O_7$ ~98% (HPLC)

Glutamine — FMOC-L-Gln-Trt

Neosystem FA00703   MW 610.7

Glutamine — FMOC-L-Gln-Trt-Wang Resin

American Peptide RFWAN38

Neosystem RW00704

Glutamine — FMOC-L-Gln-Wang Resin

American Peptide RFWAN37

Glutamine — Gln

Bachem E-1915.0025   MW 146.15 $C_5H_{10}N_2O_3$ Store at RT

Glutamine — Gln Amide Hydrochloride

Bachem E-1890.0001   MW 181.62 $C_5H_{11}N_3O_2 \cdot HCl$ Store at 2-8°C

Glutamine — Gln-(p-ONb) Hydrobromide

Bachem E-1925.0005   MW 362.18 $C_{12}H_{15}N_3O_5 \cdot HBr$ Store at -15°C

Glutamine — Gln-AMC Hydrobromide

Bachem I-1175.0050   MW 384.23 $C_{15}H_{17}N_3O_4 \cdot HBr$ Store at -15°C

Glutamine — Gln-DOD-OMe Hydrochloride

Bachem E-2625.0001   MW 422.91 $C_{21}H_{26}N_2O_5 \cdot HCl$ Store at -15°C

Glutamine — Gln-Isopropyl

Bachem E-3285.0001   MW 188.23 $C_8H_{16}N_2O_3$ Store at -15°C

Glutamine — Gln-Mtt

Bachem E-2990.0001   MW 402.51 $C_{25}H_{26}N_2O_3$ Store at 2-8°C

Glutamine — Gln-OtBu Hydrochloride

Bachem E-1920.0001   MW 238.71 $C_9H_{18}N_2O_3 \cdot HCl$ Store at -15°C

Glutamine — Gln-Trt

Bachem E-2815.0001   MW 388.47 $C_{24}H_{24}N_2O_3$ Store at RT

Glutamine — Gln-βNA Hydrochloride

Bachem K-1245.0250   MW 307.8 $C_{15}H_{17}N_3O_2 \cdot HCl$ Store at -15°C

Glutamine — Hydroxy-3-Me-Gln

Synonyms: Coenzyme A DL-3-Glu (3-^{14}C)

ARC ARC-1000   MW 911.7   50-60 mCi/mmol; 1.85-2.22 GBq/mmol; in aqueous solution, pH 5.0

Glutamine — L-Gln

Synonyms: L-α-Aminoglutamic Acid; L-2-Aminoglutaramic Acid; 2-Aminoglutaramic Acid; (S)-(2,5-Diamino)-5-Oxopentanoic Acid; Glu Amide; L-2-Aminoglutaramidic Acid; Levoglutamide Pentanoic Acid

Alexis 101-016   MW 146.2 $C_5H_{10}N_2O_3$ White solid; soluble in H_2O or dilute aqueous acid

Amersham US16285   ≥98.5%

Fluka 49419   MW 146.15 $C_5H_{10}N_2O_3$ ≥99.5% (titration); ≤0.1% residue on ignition, loss on drying; ≤0.00001% As; ≤0.0005% Al, Ba, Bi, Cd, Co, Cr, Cu, Fe, Li, Mg, Mn, Mo, Ni, Pb, Sr, Zn; ≤0.001% Ca; ≤0.005% K, Na, Cl, SO₄; ≤0.3% foreign AA

Fluka 49420   MW 146.15 $C_5H_{10}N_2O_3$ ≥99.0% (titration); ≤0.1% residue on ignition, loss on drying; ≤0.0005% Cd, Co, Cr, Cu, Fe, Mg, Mn, Ni, Pb, Zn; ≤0.001% Ca; ≤0.005% K, Na, Cl, SO₄; ≤0.3% foreign AA

ICN 101806   MW 146.1 $C_5H_{10}N_2O_3$ Crystalline; 99-100%; completely homogeneous by TLC showing one spot

ICN 1580113 ICN 1580115 ICN 1580116 ICN 1580117   Powder

ICN 1680146 ICN 1680149   200 mM solution

ICN 1680249   Gold standard; 200 mM solution, 100X, 0.9% saline; ≤0.03 endotoxin U/mL; sterile filtered

ICN 194678   MW 146.1 $C_5H_{10}N_2O_3$ Crystalline; 99-100% | Cell culture reagent

Neosystem AA00702   MW 146.2

Peptides International ALQ-2707   MW 146.15   >99.9% optical purity by RP-HPLC; white crystalline powder

Rexim   MW 146.1 $C_5H_{10}N_2O_3$ White crystals or crystalline powder

Sigma G 2884   Matrix: 4% beaded agarose; activation: cyanogen bromide; attachment: amino; spacer: 1 atom; ligand immobilized: 5-10 μmoles/mL; form: suspension in 2.0 M NaCl containing 0.02% thimerosal

Sigma G 3126   FW 146.1 $C_5H_{10}N_2O_3$ ≥99% (TLC); may contain ~0.5% Glu & <0.2% free ammonia | Glutamic acid precursor & metabolite; excitatory AA

Sigma G 5763   FW 146.1 $C_5H_{10}N_2O_3$ ≥99% (TLC) | Cell culture tested; insect cell culture tested

Sigma G 6392   0.292 g L-glutamine in 50 mL amber serum vial with rubber stopper; prepares a 200 mM solution when reconstituted to 10 mL with sterile H_2O; sterilized by γ-irradiation | Endotoxin tested; cell culture tested

Sigma G 7029   FW 146.1 $C_5H_{10}N_2O_3$ SigmaUltra; ≥99% (TLC); residue on ignition <0.1%; solubility (1 M in 1.0 M HCl, 20°C): complete, colorless; insoluble matter <0.1%; Cl, SO_4: <0.05%; Al, Cu, Fe, Ca: <0.0005%; K: <0.005% | Glutamic acid precursor & metabolite; excitatory AA

Sigma G 7513   200 mM; prepared in tissue culture grade H_2O; sterile-filtered | Endotoxin tested; cell culture tested

Sigma G 9273   FW 146.1 $C_5H_{10}N_2O_3$ ≥99% (TLC); crystalline | Plant cell culture tested

Sigma G 9644   Matrix: cross-linked 4% beaded agarose; activation: epoxy; attachment: amino; spacer: 12 atoms; ligand immobilized: 0.5-1.0 μmoles/mL; form: suspension in 2.0 M NaCl containing 0.02% thimerosal

USBio G7120   MW 146.2 $C_5H_{10}N_2O_3$ ≥98%; white, crystalline powder; specific rotation: C=4, H_2O +6.3° to +7.3°; pH: 4.0-6.0; heavy metals (Pb): ≤0.001%; endotoxin: ≤2 EU/mg; cell culture tested: murine myeloma SP2/0-Ag14

Glutamine — L-Gln (2,3,4-³H)

ARC ART-149   MW 146.2 $H_2NCO(CH_2)_2CH(NH_2)COOH$ 30-60 Ci/mmol; 1.11-2.22 TBq/mmol; in EtOH: H_2O (1:99)

ARC ART-149A   MW 146.2 $H_2NCO(CH_2)_2CH(NH_2)COOH$ 30-60 Ci/mmol; 1.11-2.22 TBq/mmol; in sterile H_2O

Sigma G 3914   20-60 Ci/mmol; radiochemical purity ≥95% (HPLC); EtOH: H_2O (2:98) solution in Combi-vial

Glutamine — L-Gln (G-³H)

Amersham TRK459 $H_2NCO(CH_2)_2CHCNH_2CO_2H$ Aqueous, 2% EtOH, sterilized; 0.74-1.85 TBq/mmol, 20-50 Ci/mmol; 37 MBq/mL, 1 mCi/mL

Glutamine — L-Gln (U-¹⁴C)

Amersham CFB81 $H_2NCO(CH_2)_2CHCNH_2CO_2H$ Aqueous, 2% EtOH, sterilized; 9.25 GBq/mmol, >250 mCi/mmol; 1.85 MBq/mL, 50 μCi/mL

ARC ARC-196   MW 146.2 $H_2NCO(CH_2)_2CH(NH_2)COOH$ >200 mCi/mmol; >7.4 GBq/mmol; in sterile H_2O

Glutamine — L-Gln Hybri-Max®

Sigma G 1517   FW 146.1 $C_5H_{10}N_2O_3$ ≥99% (TLC) | Endotoxin tested; hybridoma tested

Sigma G 2150   200 mM solution; prepared in tissue culture grade H_2O; sterilized by γ-irradiation | Endotoxin tested; Hybridoma tested

Glutamine — L-Gln-OtBu Hydrochloride

Sigma G 9375   FW 238.7 $C_9H_{18}N_2O_3 \cdot HCl$ Crystalline

Glutamine — L-HGln

Synonyms: 5-Aminohexanedioic Acid; ε-Oxolysine

Sigma H 4017   FW 160.2 $C_6H_{12}N_2O_3$

Glutamine — L-β-Gln Hydrochloride

Fluka 03653   MW 198.61 $C_5H_{10}N_2O_4 \cdot HCl$ ≥98.0% (TLC)

Glutamine — L-β-HGln Hydrochloride

Fluka 03663   MW 196.63 $C_6H_{12}N_2O_3 \cdot HCl$ ≥98.0% (TLC)

Glutamine — N-2,4-Dnp-L-Gln

Sigma D 9379   FW 312.2 $C_{11}H_{12}N_4O_7$

Glutamine — N-Ac-L-Gln

ICN 100099   MW 188.2 $C_7H_{12}N_2O_4$ Crystalline

Sigma A 9125   FW 188.2 $C_7H_{12}N_2O_4$

Glutamine — N-Bz-(N',N''-Dipropyl)-DL-Isoglutamic Acid

Synonyms: Proglumide; CCK Antagonist

Sigma P 4160   FW 334.4 $C_{18}H_{26}N_2O_4$ Selectively blocks CNS effects

Glutamine — N-CBZ-L-Gln

ICN 100213   MW 280.3 $C_{13}H_{16}N_2O_5$ Crystalline

Sigma C 7126   FW 280.3 $C_{13}H_{16}N_2O_5$

Glutamine — N-O-Nps-L-Gln

ICN 102527   MW 299.3 $C_{11}H_{13}N_3O_5S$ Yellow crystals

Glutamine — N-O-Nps-L-Gln Dicyclohexylammonium Salt

Sigma N 5253   FW 480.6 $C_{11}H_{13}N_3O_5S \cdot C_{12}H_{23}N$ Yellow crystals

Glutamine — N-t-BOC-L-Gln Resin Ester

Sigma B 0272   N-t-BOC AA ester of methylated 1% divinylbenzene cross-linked polystyrene; 200-400 mesh; substitution range: 0.2-0.6 mmole t-BOC-AA/g resin | For peptide synthesis by the Merrifield method

Glutamine — Nᵉ-(1-D-Mannityl)-L-Gln

Synonyms: Mannopine

Sigma M 5020   FW 310.3 $C_{11}H_{22}N_2O_8$

Glutamine — Nᵉ-Ac-L-Gln

Fluka 01150   MW 188.18 $C_7H_{12}N_2O_4$ ≥99% (titration); ≤0.05% residue on ignition; mp: 206-208°C

Glutamine — Nᵅ-BOC-Nᵟ-(9-Xanthenyl)-L-Gln

Fluka 15438   MW 426.47 $C_{23}H_{26}N_2O_6$ ≥97% (TLC); mp: 150°C; ~3% Et acetate

Glutamine — Nᵅ-BOC-Nᵟ-Trt-L-Gln

Fluka 15563   MW 488.58 $C_{29}H_{32}N_2O_5$ ≥98% (TLC); crystallized with ~1 mol diisopropyl ether

Glutamine — Nᵅ-CBZ-L-Gln-(pONp)

Sigma C 0420   FW 401.4 $C_{19}H_{19}N_3O_7$

Glutamine — Nᵅ-FMOC-D-Gln

ICN 158171   MW 368.4 $C_{20}H_{20}N_2O_5$

Glutamine — Nᵅ-FMOC-L-Gln

ICN 151143   MW 368.4 $C_{20}H_{20}N_2O_5$

Glutamine — Nᵅ-FMOC-L-Gln (DOD)

Sigma F 9388   FW 594.7 $C_{35}H_{34}N_2O_7$ 4,4'-Dimethoxybenzhydryl derivative

Glutamine — Nᵅ-FMOC-L-Gln (DOD) Resin Ester

Sigma F 6260   For peptide synthesis; Wang, SS, *J Am Chem Soc*, 95: 1328, 1973; Lu, G et al, *J Org Chem*, 46: 3433, 1981

Glutamine — N$^\alpha$-FMOC-L-Gln (DOD) SASRIN Resin Ester

Sigma F 1514   200-400 mesh; substitution range: 0.3-0.8 mmole FMOC AA/g resin | For peptide synthesis; *N*-FMOC AA acyl ester of 3-methoxy-4-oxymethyl-phenoxymethylated 1% divinylbenzene cross-linked polystyrene; peptides can be cleaved from the resin with 0.5-1% CF$_3$CO$_2$H in CH$_2$Cl$_2$; Mergler, M et al, *in Peptides: Chemistry & Biology* (Proc 10th Am Peptide Symp), Marhall, GR, ed, 259, 1988

Glutamine — N$^\alpha$-FMOC-L-Gln-(pONp)

ICN 158172   MW 489.5   C$_{26}$H$_{23}$N$_3$O$_7$

Glutamine — N$^\alpha$-FMOC-N$^\delta$-(2,4,6-Trimethoxybenzyl)-L-Gln

Fluka 47538   MW 548.59   C$_{30}$H$_{32}$N$_2$O$_8$ ≥98% (TLC)

Glutamine — N$^\alpha$-FMOC-N$^\delta$-(4,4'-Dimethoxybenzhydryl)-L-Gln 4-Benzyloxybenzyl Ester Polymer Bound

Fluka 47657   0.4-0.6 mmol/g resin; carrier: polystyrene, crosslinked with 1% DVB; 200-400 mesh particle size

Glutamine — N$^\alpha$-FMOC-N$^\delta$-Trt-L-Gln

Fluka 47674   MW 610.71   C$_{39}$H$_{34}$N$_2$O$_5$ ~98% (HPLC); mp: 170-173°C

Sigma F 6768   FW 610.7   C$_{39}$H$_{34}$N$_2$O$_5$ ~95%

Glutamine — N$^\alpha$-FMOC-N$^\delta$-Trt-L-Gln 4-Benzyloxybenzyl Ester Polymer Bound

Fluka 47402   0.4-0.8 mmol/g resin; carrier: polystyrene, crosslinked with 1% DVB; 100-200 mesh particle size

Glutamine — N$^\alpha$-Lauroyl-L-Gln

Fluka 61732   MW 328.45   C$_{17}$H$_{32}$N$_2$O$_4$ ≥99.0% (TLC)

Glutamine — N$^\alpha$-t-BOC-L-Gln

ICN 101061   MW 427.3   C$_{10}$H$_{18}$N$_2$O$_5$ · C$_{12}$H$_{23}$N

Sigma B 6501   FW 246.3   C$_{10}$H$_{18}$N$_2$O$_5$

Glutamine — N$^\alpha$-t-BOC-L-Gln Dicyclohexylammonium Salt

ICN 101069   MW 246.3   C$_{10}$H$_{18}$N$_2$O$_5$ Crystalline

Glutamine — N$^\alpha$-t-BOC-L-Gln-(pONp)

ICN 101070   MW 367.4   C$_{16}$H$_{21}$N$_3$O$_7$ Crystalline

Glutamine — N$^\alpha$-t-BOC-N$^\delta$-Xan-L-Gln

Sigma B 2136   FW 426.5   C$_{23}$H$_{26}$N$_2$O$_6$

Glutamine — N$^\alpha$-BOC-L-β-Gln

Fluka 03651   MW 246.26   C$_{10}$H$_{18}$N$_2$O$_5$ ≥98% (TLC)

Glutamine — N$^\alpha$-BOC-L-β-Hgln

Fluka 03667   MW 260.29   C$_{11}$H$_{20}$N$_2$O$_5$ ≥98% (TLC)

Glutamine — N$^\alpha$-FMOC-L-β-Gln

Fluka 03652   MW 368.39   C$_{20}$H$_{20}$N$_2$O$_5$ ≥98% (HPLC)

Glutamine — N$^\alpha$-FMOC-L-β-Hgln

Fluka 03666   MW 382.42   C$_{21}$H$_{22}$N$_2$O$_5$ ≥98% (HPLC)

Glutamine — N$^\gamma$-Et-L-Gln

Synonyms: L-Glu γ-(Ethylamide); L-Theanine

Sigma E 4393   FW 175.2   C$_7$H$_{15}$N$_2$O$_3$ AA derivative found in tea

Glutamine — Pth-Gln

Sigma P 0377   FW 263.3   C$_{12}$H$_{13}$N$_3$O$_2$S   ~99% | Useful as standards in protein sequencing by the Edman degradation

Glutamine — TentaGel S PHB-Gln-TrtFMOC

Fluka 86388   0.20 mmol protected AA/g; ~90 μm particle size

Glutamine — TentaGel S Trt-Gln-Trt-FMOC

Fluka 86419   0.18 mmol protected AA/g; ~90 μm particle size

Glutamine — Z-D-Gln

Bachem C-1620.0001   MW 280.28   C$_{13}$H$_{16}$N$_2$O$_5$ Store at RT

Fluka 96098   MW 280.28   C$_{13}$H$_{16}$N$_2$O$_5$ ≥98.0% (HPLC)

Glutamine — Z-D-Gln-Mtt

Bachem C-3615.0001   MW 536.64   C$_{33}$H$_{32}$N$_2$O$_5$ Store at 2-8°C

Glutamine — Z-D-Gln-Trt

Bachem C-3560.0001   MW 522.61   C$_{32}$H$_{30}$N$_2$O$_5$ Store at -15°C

Glutamine — Z-Gln

Synonyms: Benzyloxycarbonyl-L-Gln

Bachem C-1615.0005   MW 280.28   C$_{13}$H$_{16}$N$_2$O$_5$ Store at RT

Peptides International ZLQ-2017-PI   MW 280.28   >98% (HPLC); white powder

Glutamine — Z-Gln-DOD

Bachem C-1545.0001   MW 506.56   C$_{28}$H$_{30}$N$_2$O$_7$ Store at -15°C

Glutamine — Z-Gln-Mtt

Bachem C-3625.0001   MW 536.63   C$_{33}$H$_{32}$N$_2$O$_5$ Store at 2-8°C

Glutamine — Z-Gln-OMe

Bachem C-3055.0001   MW 294.31   C$_{14}$H$_{18}$N$_2$O$_5$ Store at -15°C

Glutamine — Z-Gln-ONp

Bachem C-1630.0005   MW 401.38   C$_{19}$H$_{19}$N$_3$O$_7$ Store at -15°C

Glutamine — Z-Gln-OSu

Bachem C-1625.0005   MW 377.35   C$_{17}$H$_{19}$N$_3$O$_7$ Store at -15°C

Glutamine — Z-Gln-Trt

Bachem C-3375.0001   MW 522.6   C$_{32}$H$_{30}$N$_2$O$_5$ Store at -15°C

Glutamine — Z-L-Gln

Fluka 96100   MW 280.28   C$_{13}$H$_{16}$N$_2$O$_5$ ~99% (titration); mp: 134-138°C

Neosystem ZA00701   MW 280.3

Glycine — (±)-(3-Carboxyphenyl)-Gly

ICN 159740   MW 195.2 $C_9H_9NO_4$

Glycine — (±)-N-3-Mercapto-2-Benzylpropionyl-Gly

Fluka 89035   MW 253.3 $C_{12}H_{15}NO_3S$ ≥99.0% (TLC); mp: 138-140°C

Glycine — (±)-N-CBZ-α-Phosphono-Gly-OMe₃

Sigma C 9794   FW 331.3 $C_{13}H_{18}NO_7P$

Glycine — (±)-Z-α-Phosphono-Gly-OMe₃

Fluka 97040   MW 331.27 $C_{13}H_{18}NO_7P$ ≥97.0% (HPLC); mp: 78-80°C

Glycine — (±)-α-Me-4-Carboxyphenyl-Gly

Synonyms: (±)-MCPG; Metabotropic Glutamate Receptors Antagonist; L-AP4-Like Presynaptic Glutamate Autoreceptors Antagonist

Sigma M 4796   FW 209.2 $C_{10}H_{11}NO_4$ Competitive antagonist at PI-linked metabotropic glutamate receptors & L-AP4-like presynaptic glutamate autoreceptors; Eaton, SA et al, *Eur J Pharmacol*, 244: 195, 1993

Glycine — (±)-α-Me-4-Phosphonophenyl-Gly

Synonyms: MPPG; L-AP4-Sensitive Receptor Antagonist

Sigma M 4921   FW 245.2 $C_9H_{12}NO_5P$ Jane, DE et al, *Neuropharmacology*, 34: 851, 1995

Glycine — (±)-α-Me-4-Sulfonophenyl-Gly

Synonyms: MSPG

Sigma M 5046   FW 245.2 $C_9H_{11}NO_5S$ Blocks A4-induced monosynaptic excitationof rat motoneurons; Jane, DE et al, *Neuropharmacology*, 34: 85, 1995

Glycine — ((RS)-2-Bzl-3-Mercaptopropionyl)-Gly

Synonyms: DL-Thiorphan; (DL-3-Mercapto-2-Benzylpropanoyl)-Gly

Bachem N-1195.0025   MW 253.32 $C_{12}H_{15}NO_3S$ Store at -15°C

Glycine — (-)-Bzl (R)-Glycidyl Ether

Fluka 13425   MW 164.2 $C_{10}H_{12}O_2$ ≥99% (GC)

Glycine — (+)-Bzl (S)-Glycidyl Ether

Fluka 13423   MW 164.2 $C_{10}H_{12}O_2$ ≥99% (GC)

Glycine — (2R,1S,2'R,3'S)-2-(2'-Carboxy-3'-Phenylcyclopropyl)-Gly

Alexis 550-349

Glycine — (2S,1'R,2'S)-2-(2-Carboxycyclopropyl)-Gly

Synonyms: L-CCG-IV; NMDA Receptor Agonist

Sigma C 1585   FW 159.1 $C_6H_9NO_4$ Potent NMDA receptor agonist; Yamashita, H et al, *Eur J Pharmacol*, 289: 387, 1995

Glycine — (2S,1'S,2'S)-2-(2-Carboxycyclopropyl)-Gly

Synonyms: Metabotropic Glutamate Receptors Agonist

Sigma C 1710   FW 159.1 $C_6H_9NO_4$ Yamashita, H et al, *Eur J Pharmacol*, 289: 387, 1995

Glycine — (2S,1'S,2'S)-2-(Carboxycyclopropyl)-Gly

Calbiochem 217370   MW 159.1 $C_6H_9NO_4$ ≥97% (HPLC); solid; soluble in dilute aqueous base

Glycine — (2S,1'S,2'S,3'R)-2-(2'-Carboxy-3'-Phenylcyclopropyl)-Gly

Alexis 550-311

Glycine — (2S,2'R,3'R)-2-(2',3'-Dicarboxycyclopropyl)-Gly

Alexis 550-334

Glycine — (2S,3R,4S)-α-(Carboxycyclopropyl)-Gly

Synonyms: NMDA Receptor Agonist

ICN 193623   MW 159.1 $C_6H_9NO_4$ A powerful NMDA receptor agonist; Kawai etal, *Eur J Pharmacol*, 211:195, 1992

Glycine — (2S,3S,4S)-2-Me-2-(Carboxycyclopropyl)-Gly

Alexis 550-333   MW 173.2 $C_7H_{11}NO_4$ Soluble in dilute aqueous base

Glycine — (Nᵅ-Bz)-Gly-L-α-Hydroxy-δ-Guanidinovaleric Acid Hydrochloride

Synonyms: Nᵅ-Hippuryl-L-Argininic Acid

Bachem M-1765.0050   MW 372.81 $C_{15}H_{20}N_4O_5 \cdot HCl$ Store at -15°C

Glycine — (R)-4-Carboxy-3-Hydroxyphenyl-Gly

Synonyms: (R)-4C3H-PG; NMDA Receptor Antagonist

Alexis 550-199   MW 211.2 $C_9H_9NO_5$ White powder; soluble in dilute aqueous base | Potent NMDA receptor antagonist; weak AMPA/kainate receptor antagonist; Watkins, J & Collingridge, G, *TIPS*, 15: 333, 1994; Ornstein, PL et al, *Current Pharm Design*, 1: 355, 1995; Birse, EF et al, *Neuroscience*, 52: 481, 1993

Glycine — (RS)-3-Carboxy-4-Hydroxyphenyl-Gly

Synonyms: (RS)-3C4H-PG

Alexis 550-196   MW 211.2 $C_9H_9NO_5$ White powder; soluble in dilute aqueous base | Watkins, J & Collingridge, G, *TIPS*, 15: 333, 1994; Ornstein, PL et al, *Current Pharm Design*, 1: 355, 1995

Glycine — (RS)-4-Carboxy-3-Hydroxyphenyl-Gly

Synonyms: (RS)-4C3H-PG

Alexis 550-198   MW 211.2 $C_9H_9NO_5$ White powder; soluble in dilute aqueous base | Watkins, J & Collingridge, G, *TIPS*, 15: 333, 1994; Ornstein, PL et al, *Current Pharm Design*, 1: 355, 1995

Glycine — (RS)-α-Cyclopropyl-4-Phosphonophenyl-Gly

Synonyms: CPPG; Group II/III mGlu Receptor Antagonist

ICN 198734   MW 271.2 $C_{11}H_{14}NO_5P$ Potent group II/III mGlu receptor antagonist

Glycine — (RS)-α-Me-3-Carboxy-4-Hydroxyphenyl-Gly

Alexis 550-156   MW 225.2 $C_{10}H_{11}NO_5$ Soluble in dilute aqueous base

Glycine — (RS)-α-Me-3-Carboxymethylphenyl-Gly

Alexis 550-157   MW 223.2 $C_{11}H_{13}NO_4$ White solid; soluble in dilute aqueous base

Glycine — (RS)-α-Me-4-Carboxyphenyl-Gly

Alexis 550-054   MW 209.2 $C_{10}H_{11}NO_4$ White solid; soluble in dilute aqueous base; insoluble in H_2O, EtOH, DMSO

Calbiochem 472500   MW 209.2 $C_{10}H_{11}NO_4$ ≥96% (HPLC); solid; soluble in dilute aqueous base

Glycine — (*RS*)-α-Me-4-Phosphonophenyl-Gly

Alexis 550-122   MW 245.2 $C_9H_{12}NO_5P$ White solid; soluble in dilute aqueous base

Glycine — (*RS*)-α-Me-4-Sulphonophenyl-Gly

Alexis 550-123   MW 245.3 $C_9H_{12}NO_5S$ White solid; soluble in dilute aqueous base

Glycine — (*RS*)-α-Me-4-Tetrazoylphenyl-Gly

Alexis 550-124   MW 234.2 $C_{10}H_{12}N_5O_2$ White solid; soluble in dilute aqueous base

Glycine — (*S*)-(3,4-Dihydroxyphenyl)-Gly

Synonyms: Group I mGlu Receptor Agonist

ICN 198732   MW 183.2 $C_8H_9NO_4$ H_2O soluble | Selective Group I mGlu receptor agonist

Glycine — (*S*)-3-Carboxy-4-Hydroxyphenyl-Gly

Synonyms: (*S*)-3C4H-PG

Alexis 550-197   MW 211.2 $C_9H_9NO_5$ White powder; soluble in dilute aqueous base | Watkins, J & Collingridge, G, *TIPS*, 15: 333, 1994; Ornstein, PL et al, *Current Pharm Design*, 1: 355, 1995; Sekiyama, N et al, *Br J Pharmacol*, 117: 1493, 1996

Glycine — (*S*)-4-Carboxy-3-Hydroxyphenyl-Gly

Synonyms: (*S*)-4C3H-PG; Group II mGluR Agonist; Group I mGluR1 Antagonist

Alexis 550-119   MW 211.2 $C_9H_9NO_5$ ≥98% *S*-isomer; white powder; soluble in dilute aqueous base | Agonist for group II mGluRs; competitive antagonist for group I mGluR1; mixed effects at mGluR5; Watkins, J & Collingridge, G, *TIPS*, 15: 333, 1994; Ornstein, PL et al, *Current Pharm Design*, 1: 355, 1995; Birse, EF et al, *Neuroscience*, 52: 481, 1993; Sekiyama, N et al, *Br J Pharmacol*, 117: 1493, 1996; Thomsen, C et al, *Eur J Pharmacol*, 267: 77, 1994; Opitz, T et al, *Neuropharmacology*, 33: 715, 1994; Orlando, LR et al, *Neurosci Lett*, 202: 109, 1995; Kingston, AE et al, *Neuropharmacology*, 34: 887, 1995

Glycine — (*S*)-4-Carboxyphenyl-Gly

Synonyms: (*S*)-4C-PG; Group I mGluR Antagonist

Alexis 550-120   MW 195.2 $C_9H_9NO_4$ White solid; soluble in dilute aqueous base | Competitive antagonist for group I mGluR; antagonist with selectivity for mGluR1 over mGluR5

Glycine — (*S*)-α-Me-4-Carboxyphenyl-Gly

Alexis 550-075   MW 209.2 $C_{10}H_{11}NO_4$ White solid; soluble in dilute aqueous base

Glycine — (*S*)-α-Me-4-Phosphonophenyl-Gly

Alexis 550-312   MW 245.2 $C_9H_{12}NO_5P$ White solid; soluble in dilute aqueous base

Glycine — 2-Fluoro-DL-α-Phenyl-Gly

Fluka 47353   MW 169.13 $C_8H_8FNO_2$ ≥98% (titration); mp: >300°C

Glycine — 4-Fluoro-DL-α-Phenyl-Gly

Fluka 47358   MW 169.13 $C_8H_8FNO_2$ ~99% (titration); mp: >300°C

Glycine — 4-Fluoro-D-α-Phenyl-Gly

Fluka 47355   MW 169.13 $C_8H_8FNO_2$ ≥99% (titration); mp: >300°C

Glycine — 4-Fluoro-L-α-Phenyl-Gly

Fluka 47352   MW 169.13 $C_8H_8FNO_2$ ≥99% (titration); mp: >300°C

Glycine — 4-Hydroxy-D-Phenyl-Gly

Alexis 106-035   MW 167.2 $C_8H_9NO_3$ ≥99%; white powder; soluble in alkali; Irritant

Fluka 56155   MW 167.17 $C_8H_9NO_3$ ≥98.0%; mp: 240°C

Glycine — 4-Hydroxy-L-Phenyl-Gly

Fluka 56160   MW 167.17 $C_8H_9NO_3$ ≥99.0% (titration); mp: 240°C

Glycine — Ac-DL-Phg

Bachem F-2310.0025   MW 193.2 $C_{10}H_{11}NO_3$ Store at RT

Glycine — Ac-Gly

Bachem E-1055.0025   MW 117.11 $C_4H_7NO_3$ Store at RT

Glycine — Ac-Gly-NHMe

Bachem E-1060.0001   MW 130.15 $C_5H_{10}N_2O_2$ Store at RT

Glycine — Ac-Gly-ONp

Bachem E-1065.0001   MW 238.2 $C_{10}H_{10}N_2O_5$ Store at -15°C

Glycine — Ac-*N*-Gly (1-^{14}C)

ARC ARC-881   MW 117.1 $CH_3CONHCH_2COOH$ 50-60 mCi/mmol; 1.85-2.22 GBq/mmol; in EtOH

Glycine — Ac-Phg-OMe

Bachem F-1070.0250   MW 207.23 $C_{11}H_{13}NO_3$ Store at 2-8°C

Glycine — Ac-Propargyl-DL-Gly-OEt

Synonyms: Ac-DL-Pra-OEt

Bachem F-1080.0001   MW 183.21 $C_9H_{13}NO_3$ Store at -15°C

Glycine — Betaine-Gly (1-^{14}C)

ARC ARC-748   MW 117.15 $(CH_3)_3NCH_2COOH$ 50-60 mCi/mmol; 1.85-2.22 GBq/mmol; in EtOH: H_2O (9:1)

Glycine — Bisphenol F Diglycidyl Ether

Fluka 15144   MW 312.37 $C_{19}H_{20}O_4$ ≥97% (GC); mixture of 3 isomers: ortho-ortho, ortho-para, para-para

Glycine — BOC *N*-Me-Gly

Synonyms: BOC-Sar

USBio B2389   MW 189.2 $C_8H_{15}NO_4$ ≥99%

Glycine — BOC-(^{15}N)-Gly

Bachem A-3790.0100   MW 176.18 $C_7H_{13}^{15}NO_4$ Store at -15°C

Glycine — BOC-Cyclohexyl-D-Gly

Bachem A-4470.0001   MW 257.33 $C_{13}H_{23}NO_4$ Store at 2-8°C

Glycine — BOC-Cyclohexyl-Gly

Bachem A-4465.0001   MW 257.33 $C_{13}H_{23}NO_4$ Store at 2-8°C

Glycine — BOC-D-(Allyl)-Gly Dicyclohexylamine

Fluka 09808   MW 396.57 $C_{10}H_{17}NO_4 \cdot C_{12}H_{23}N$ ≥98% (TLC)

Glycine — BOC-D-(Allyl)-Gly Dicyclohexylammonium Salt

Neosystem BA14901   MW 396.57

Glycine — BOC-D-2-Propargyl-Gly Dicyclohexylamine

Fluka 09797   MW 394.55 $C_{10}H_{15}NO_4 \cdot C_{12}H_{23}N$ ≥98% (TLC)

Glycine — BOC-D-Phg

Synonyms: BOC-D-Phg

Bachem A-3720.0001   MW 251.28 $C_{13}H_{17}NO_4$ Store at RT

Neosystem BA04201   MW 251.3

Peptides International BDP-2600-PI   MW 251.28   >98% (HPLC); white to off-white powder

USBio B2385   MW 251.3 $C_{13}H_{17}NO_4$ ≥99%

Glycine — BOC-D-α-Cyclohexyl-Gly

Fluka 15089   MW 257.33 $C_{13}H_{23}NO_4$ ≥98% (TLC)

Glycine — BOC-D-α-Phg

Fluka 15487   MW 251.29 $C_{13}H_{17}NO_4$ ≥99% (titration); mp: 88-91°C

Glycine — BOC-D-α-tBu-Gly

Synonyms: BOC-D-t-Leu

USBio B2285   MW 231.3 $C_{11}H_{21}NO_4$ ≥99%

Glycine — BOC-Gly

American Peptide BLGLY10   MW 175.2

Bachem A-1730.0005   MW 175.19 $C_7H_{13}NO_4$ Store at -15°C

Fluka 15420   MW 175.19 $C_7H_{13}NO_4$ ~98% (titration); ≤0.05% residue on ignition; mp: 86-89°C

Neosystem BA00801   MW 175.2

Peptides International BLG-2054   MW 175.18   >98% (HPLC); white crystalline powder

USBio B2314   MW 175.2 $C_7H_{13}NO_4$ ≥99%

Glycine — BOC-Gly Hycram™ Resin

Fluka 15422   0.66 mmol Gly/g resin; 100-200 mesh particle size; carrier: polystyrene, crosslinked with 1% DVB

Glycine — BOC-Gly Hydroxysuccinimide Ester

Fluka 15423   MW 272.26 $C_{11}H_{16}N_2O_6$ ≥99%; ≤0.05% residue on ignition; mp: 165-167°C

Glycine — BOC-Gly *N*-Carboxy Anhydride

Fluka 15428   MW 201.18 $C_8H_{11}NO_5$ ≥97%; mp: 136°C

Glycine — BOC-Gly-(4-ONp)

Fluka 15117   MW 296.28 $C_{13}H_{16}N_2O_6$ ≥98% (HPLC)

Glycine — BOC-Gly-® (200-400 mesh)

Bachem D-1195.0005   Store at RT

Glycine — BOC-Gly-MBHA Resin

Neosystem RN00801

Glycine — BOC-Gly-O-CH₂-φ-CH₂-COOH

Neosystem LP00801   MW 323.4

Glycine — BOC-Gly-ONp

Bachem A-1740.0005   MW 296.28 $C_{13}H_{16}N_2O_6$ Store at -15°C

Glycine — BOC-Gly-O-Resin

American Peptide RMRB185   1% DVB cross-linked: 100-200 mesh

Glycine — BOC-Gly-OSu

Bachem A-1735.0005   MW 272.26 $C_{11}H_{16}N_2O_6$ Store at -15°C

Glycine — BOC-Gly-OtBu

Fluka 15116   MW 231.29 $C_{11}H_{21}NO_4$ ≥98% (GC)

Glycine — BOC-Gly-PAM Resin

Neosystem RP00801

Glycine — BOC-Gly-PAM-® (200-400 mesh)

Bachem D-1545.0001   Store at 2-8°C

Glycine — BOC-Gly-PAM-Resin

American Peptide RPAM185   1% DVB cross-linked: 100-200 mesh

Glycine — BOC-L-(Allyl)-Gly Dicyclohexylammonium Salt

Neosystem BA14902   MW 396.57

Glycine — BOC-L-2-Allyl-Gly Dicyclohexylamine

Fluka 09809   MW 396.57 $C_{10}H_{17}NO_4 \cdot C_{12}H_{23}N$ ≥98% (TLC)

Glycine — BOC-L-2-Propargyl-Gly Dicyclohexylamine

Fluka 09796   MW 394.55 $C_{10}H_{15}NO_4 \cdot C_{12}H_{23}N$ ≥98% (TLC)

Glycine — BOC-L-Phg

Synonyms: BOC-L-Phg

Neosystem BA04202   MW 251.3

Glycine — BOC-L-α-Cyclohexyl-Gly

Fluka 15091   MW 257.33 $C_{13}H_{23}NO_4$ ≥98% (TLC)

Glycine — BOC-L-α-Phenyl-Gly

Fluka 15488   MW 251.28 $C_{13}H_{17}NO_4$ ≥99% (titration); mp: 88-91°C

Glycine — BOC-L-α-tBu-Gly

Synonyms: BOC-L-t-Leu

Neosystem BA04102   MW 231.3

Glycine — BOC-*N*-Me-Gly

Synonyms: BOC-Sar

Neosystem BA00802   MW 189.2

USBio B2359   MW 189.2 $C_8H_{15}NO_4$ ≥99%

Glycine — BOC-*N*-Me-Phg

Bachem A-2080.0001   MW 265.31 $C_{14}H_{19}NO_4$ Store at RT

Glycine — BOC-Phg

Synonyms: BOC-L-Phg

Bachem A-2225.0001   MW 251.28 $C_{13}H_{17}NO_4$ Store at RT

Peptides International BFG-5323-PI   MW 251.28   >98% (HPLC); white crystalline powder

USBio B2386   MW 251.3 $C_{13}H_{17}NO_4$ ≥99%

Glycine — BOC-Phg-OSu

Bachem A-2230.0001 MW 348.36 $C_{17}H_{20}N_2O_6$ Store at -15°C

Glycine — BOC-tBu-DL-Gly

Synonyms: BOC-DL-Tle

Bachem A-1425.0001 MW 231.29 $C_{11}H_{21}NO_4$ Store at RT

Glycine — BOC-tBu-Gly

Synonyms: BOC-Tle

Bachem A-3910.0001 MW 231.29 $C_{11}H_{21}NO_4$ Store at RT

Glycine — BOC-α-(FMOC-Amino)-DL-Gly

Synonyms: BOC-α-Amino-DL-Gly-FMOC; FMOC-α-(BOC-Amino)-DL-Gly; FMOC-α-Amino-DL-Gly-BOC

Bachem A-4320.0001 MW 412.44 $C_{22}H_{24}N_2O_6$ Store at -15°C

Glycine — BOC-α-tBu-Gly

Synonyms: BOC-D-t-Leu

USBio B2286 MW 231.3 $C_{11}H_{21}NO_4$ ≥99%

Glycine — Boro Gly-(N-Ethylamide)

Synonyms: Ammonia-N-EthylCarbamoylborane

ICN 154630 MW 101.9 $BC_3H_{11}N_2O$ >97%

Glycine — Boro Gly-OMe

Synonyms: Ammonia-Carbomethoxyborane

ICN 154627 MW 88.9 $BC_2H_8NO_2$ >97%

Glycine — Boro-Gly

Synonyms: Ammonia-Carboxyborane; β-Hydroxy-β-Methylglutaryl-CoA Reductase Inhibitor

ICN 154608 MW 74.9 BCH_6NO_2 >97% | Potent hypocholesterolemic agent; Hall, IH etal, *J Pharm Sci*, 70:339, 1981

Glycine — BorotrimethylGly

Synonyms: Borobetaine;Trimethylamine-Carboxyborane; Fatty Acid Synthetase Inhibitor

ICN 154611 MW 117 $BC_4H_{12}NO_2$ >97%; LD_{50}: 1800 mg/kg in mice | Hall, IH etal, *J Pharm Sci*, 70:339, 1981

Glycine — Bpoc-Gly Dicyclohexylamine

Bachem E-3145.0001 MW 494.68 $C_{18}H_{19}NO_4 \cdot C_{12}H_{23}N$ Store at -15°C

Glycine — Bsmoc-Gly

Synonyms: Bsmoc-Gly

Peptides International BLG-5625-PI MW 297.29 >98% (HPLC); off-white crystalline powder | LA Carpino, et al, *JACS*, 119:9915, 1997

Glycine — Buffer Solutions

Sigma 105-2 0.2 *M*, pH 10.4 at 25°C
Sigma 292-6 0.7 *M*, pH 9.4 at 25°C

Glycine — Bzl-Gly Hydrochloride

Bachem E-1550.0001 MW 201.65 $C_9H_{11}NO_2 \cdot HCl$ Store at RT

Glycine — Caged Gly Sodium Salt

Calbiochem 357125 MW 290.2 $C_{11}H_{11}N_2O_6Na$ >98% (HPLC); powder; soluble in H_2O

Glycine — CBZ-Gly

USBio C2098-70 MW 209.2 $C_{10}H_{11}NO_4$ ≥99%

Glycine — CBZ-Gly-ONp

USBio C2098-82 MW 330.3 $C_{16}H_{14}N_2O_6$ ≥99%

Glycine — CBZ-Hydrazido-Gly TFA Salt

Synonyms: Gly-NHNH-CBZ

Sigma C 5537 FW 337.3 $C_{10}H_{13}N_3O_3 \cdot C_2HF_3O_2$

Glycine — CBZ-Me-Gly

Synonyms: CBZ-Sar

USBio C2099-59 MW 223.3 $C_{11}H_{13}NO_4$ ≥99%

Glycine — CBZ-Phg

USBio C2099-48 MW 285.3 $C_{16}H_{15}NO_4$ ≥99%

Glycine — Chloroacetyl-Gly

Bachem E-1730.0005 MW 151.55 $C_4H_6ClNO_3$ Store at RT

Glycine — Crotyl-Gly

Synonyms: Methionine Antagonist

ICN 101427

Glycine — Cyclohexyl-D-Gly

Bachem F-3765.0001 MW 157.21 $C_8H_{15}NO_2$ Store at 2-8°C

Glycine — Cyclohexyl-Gly

Bachem F-3760.0001 MW 157.21 $C_8H_{15}NO_2$ Store at 2-8°C

Glycine — D-(-)-(p-Hydroxyphenyl)-Gly

ICN 151305 MW 167.2 $C_8H_9NO_3$ Intermediate for synthesis of penicillin, amoxicillin & other antibiotics

Glycine — D-(-)-α-Phenyl-Gly

Synonyms: D-(-)-α-Aminophenylacetic Acid

Fluka 78570 MW 151.17 $C_8H_9NO_2$ ≥99.0% (titration); mp: >300°C; ≤0.05% residue on ignition
ICN 151832 MW 151.2 $C_8H_9NO_2$ Crystalline | Pharmaceutical intermediate

Glycine — D-(-)-α-Phenyl-Gly Chloride Hydrochloride

Fluka 78583 MW 206.07 $C_8H_8ClNO \cdot HCl$ ~97.0% (titration); mp: 177°C

Glycine — D-(c-Allyl)-Gly

Fluka 05957 MW 115.13 $C_5H_9NO_2$ ≥99% (titration); ≤2% H_2O

Glycine — D-2-tBu-Gly

Synonyms: D-t-Leu

ICN 154911 MW 131.2 $C_6H_{13}NO_2$
USBio B9000-10 MW 131.2 $C_6H_{13}NO_2$ ≥99%

Glycine — Dansyl Gly Free Acid

Sigma D 0875 FW 308.4 $C_{14}H_{16}N_2O_4S$ ~99%

Glycine — Diglycine Hydrochloride

Synonyms: Gly Hemihydrochloride

Sigma D 6882 FW 186.6 $(C_2H_5NO_2)_2 \cdot HCl$ Compound containing 2 moles Gly & 1 mole HCl | Do not confuse with iminodiacetic acid HCl or glycylglycine HCl

Glycine — Dihydroxyethyl-Gly

ICN 151490 White crystals or powder; ~98% | Chelating agent

Glycine — Dinitropyridyl-L-Gly

ICN 101835 MW 242.1 $C_7H_6N_4O_6$

Glycine — DL-(Allyl)-Gly

Synonyms: 2-Amino-4-Pentenoic Acid

ICN 100413 MW 115.1 $C_5H_9NO_2$ 99%, crystalline

Glycine — DL-(c-Allyl)-Gly

Fluka 05960 MW 115.13 $C_5H_9NO_2$ ≥99% (titration); mp: ~265°C

Glycine — DL-(Propargyl)-Gly

Synonyms: 2-Amino-4-pentynoic Acid

Sigma P 7888 FW 113.1 $C_5H_7NO_2$

Glycine — DL-(Tetrazol-5-yl)-Gly

ICN 193721 MW 143.1 $C_3H_5N_5O_2$

Glycine — DL-(Vinyl)-Gly

ICN 158296 MW 101.1 $C_4H_7NO_2$

Glycine — DL-(Vinyl)-Gly Free base

Synonyms: 2-Amino-3-Butenoic Acid

Sigma V 7252 FW 101.1 $C_4H_7NO_2$

Glycine — DL-(Vinyl)-Gly TFA Salt

Synonyms: 2-Amino-3-Butenoic Acid

Sigma V 7131 FW 215.1 $C_4H_7NO_2 \cdot C_2HF_3O_2$

Glycine — DL-2-(2-Thienyl)-Gly

Fluka 09487 MW 157.19 $C_6H_7NO_2S$ ~98% (titration); mp: 208-212°C

Glycine — DL-3-Mercapto-2-Benzylpropanoyl-Gly

Synonyms: Thiorphan, DL-; Enkephalinase inhibitor

Sigma T 6031 FW 253.3 $C_{12}H_{15}NO_3S$ ≥97% (TLC) | Antinociceptive activity; Roques, BP et al, *Nature*, 288: 286, 1980

Glycine — DL-3-Mercapto-2-Bzl-Propanoyl-Gly

ICN 152841 MW 253.3 $C_{12}H_{15}NO_3S$

Glycine — DL-Phg (1-^{14}C)

ARC ARC-1004 MW 151.17 $C_6H_5CH(NH_2)\mathbf{C}OOH$ 50-60 mCi/mmol; 1.85-2.22 GBq/mmol; in EtOH

Glycine — DL-α-2-Thienyl-Gly

ICN 156853 MW 157.2 $C_6H_7NO_2S$
Sigma T 8153 FW 157.2 $C_6H_7NO_2S$

Glycine — DL-α-3-Thienyl-Gly

ICN 156856 MW 157.2 $C_6H_7NO_2S$
Sigma T 8528 FW 157.2 $C_6H_7NO_2S$

Glycine — DL-α-Neopentyl-Gly

Fluka 72138 MW 145.2 $C_7H_{15}NO_2$ mp: 272-274°C

Glycine — DL-α-Phenyl-Gly

Synonyms: DL-α-Aminophenylacetic Acid

ICN 151833 MW 151.2 $C_8H_9NO_2$ Crystalline

Glycine — DL-α-Phg

Fluka 78580 MW 151.17 $C_8H_9NO_2$ ≥99.0% (titration); mp: >300°C

Glycine — Dnp-Gly

ICN 101833 MW 241.2 $C_8H_7N_3O_6$ Crystalline

Glycine — D-Phg

Bachem F-2015.0025 MW 151.17 $C_8H_9NO_2$ Store at RT

Glycine — D-Phg (1-^{14}C)

ARC ARC-1002 MW 151.17 $C_6H_5CH(NH_2)\mathbf{C}OOH$ 50-60 mCi/mmol; 1.85-2.22 GBq/mmol; in EtOH

Glycine — D-Phg Amide Hydrochloride

Bachem F-3895.0001 MW 186.64 $C_8H_{10}N_2O \cdot HCl$ Store at -15°C

Glycine — D-Phg-OEt Hydrochloride

Bachem F-3865.0001 MW 215.68 $C_{10}H_{13}NO_2 \cdot HCl$ Store at 2-8°C

Glycine — D-Phg-OMe (1-^{14}C) Hydrochloride

ARC ARC-1245 MW 201.66 $C_6H_5CH(NH_2)\mathbf{C}OOCH_3 \cdot HCl$ 50-60 mCi/mmol; 1.85-2.22 GBq/mmol; in EtOH

Glycine — D-Phg-OMe Hydrochloride

Bachem F-2570.0005 MW 201.65 $C_9H_{11}NO_2 \cdot HCl$ Store at 2-8°C

Glycine — D-Phg-OtBu Hydrochloride

Bachem F-3995.0001 MW 243.73 $C_{12}H_{17}NO_2 \cdot HCl$ Store at -15°C

Glycine — D-α-2-Thienyl-Gly

ICN 156852 MW 157.2 $C_6H_7NO_2S$
Sigma T 8028 FW 157.2 $C_6H_7NO_2S$

Glycine — D-α-3-Thienyl-Gly

ICN 156855 MW 157.2 $C_6H_7NO_2S$
Sigma T 8403 FW 157.2 $C_6H_7NO_2S$

Glycine — Et$_2$-Gly

Synonyms: Deg

Peptides International ALX-5041-PI MW 131.17 >98% optical purity by RP-HPLC; white crystals

Glycine — Ethylene Glycol Diglycidyl Ether

Fluka 03800 MW 174.2 $C_8H_{14}O_4$ ~50% (GC); bp: 112°C

Glycine — Et-*N*-Diphenylmethylene-Glycinate

USBio E8605 MW 267.3 $C_{17}H_{17}NO_2$ ≥99%

Glycine — FA-Gly

Bachem M-1360.0001 MW 195.18 $C_9H_9NO_4$ Store at RT

Glycine — FMOC-(^{15}N)-Gly

Bachem B-2650.0100 MW 298.3 $C_{17}H_{15}{}^{15}NO_4$ Store at -15°C

Glycine — FMOC-Cyclohexyl-D-Gly
Bachem B-3275.0001   MW 379.46 $C_{23}H_{25}NO_4$ Store at -15°C

Glycine — FMOC-Cyclohexyl-Gly
Bachem B-3270.0001   MW 379.46 $C_{23}H_{25}NO_4$ Store at -15°C

USBio F5351   MW 379.5 $C_{23}H_{25}NO_3$ ≥98%

Glycine — FMOC-D-(Allyl)-Gly
Neosystem FA14901   MW 337.38

Glycine — FMOC-D-2-Allyl-Gly
Fluka 09807   MW 337.38 $C_{20}H_{19}NO_4$ ~98% (HPLC)

Glycine — FMOC-D-2-Propargyl-Gly
Fluka 09798   MW 335.36 $C_{20}H_{17}NO_4$ ≥98% (HPLC)

Glycine — FMOC-D-Phg
Synonyms: FMOC-D-Phg
Neosystem FA04201   MW 373.4

Peptides International FDX-1862-PI   MW 373.41   >98% (HPLC); white crystalline powder

Glycine — FMOC-D-α-Phg
Fluka 00211   MW 373.41 $C_{23}H_{19}NO_4$ ≥98% (HPLC)

Glycine — FMOC-D-α-tBu-Gly
USBio F5348   MW 353.5 $C_{21}H_{23}NO_4$ ≥99%

Glycine — FMOC-Et₂-Gly
Peptides International FLX-1799-PI   MW 353.42   >98% (HPLC); off-white crystalline powder

Glycine — FMOC-Gly
Synonyms: FMOC-Gly
American Peptide FGLY100   MW 297.3
Bachem B-1330.0005   MW 297.31 $C_{17}H_{15}NO_4$ Store at RT
Fluka 47627   MW 297.32 $C_{17}H_{15}NO_4$ ≥98% (titration); mp: 174-175°C
Neosystem FA00801   MW 297.3
Peptides International FLG-1701-PI   MW 297.31   >98% (HPLC); white crystalline powder
USBio F5380   MW 297.3 $C_{17}H_{15}NO_4$ ≥99%

Glycine — FMOC-Gly 2-Chlorotrityl Ester Polymer Bound
Fluka 47621   0.4-0.6 mmol/g resin; carrier: polystyrene, crosslinked with 1% DVB

Glycine — FMOC-Gly 4-Benzyloxybenzyl Ester Polymer Bound
Fluka 47659   0.4-0.6 mmol/g resin; carrier: polystyrene, crosslinked with 1% DVB; 200-400 mesh particle size

Glycine — FMOC-Gly N-Carboxy Anhydride
Fluka 47685   MW 323.31 $C_{18}H_{13}NO_5$ ≥97%; mp: 145-160°C

Glycine — FMOC-Gly-® (100-200 mesh)
Bachem D-1745.0001   Store at RT

Glycine — FMOC-Gly-® (200-400 mesh)
Bachem D-1105.0001   Store at RT

Glycine — FMOC-Gly-HMPB 4-Methylbenzhydrylamine Ester Polymer Bound
Fluka 47558   ~0.7 mmol/g resin; carrier: polystyrene, crosslinked with 1% DVB; 200-400 mesh particle size

Glycine — FMOC-Gly-OPfp
Bachem B-1135.0001   MW 463.36 $C_{23}H_{14}F_5NO_4$ Store at -15°C

USBio F5381   MW 463.4 $C_{23}H_{14}F_5NO_4$ ≥99%

Glycine — FMOC-Gly-OSu
Bachem B-1335.0001   MW 394.38 $C_{21}H_{18}N_2O_6$ Store at -15°C
USBio F5382   MW 394.4 $C_{21}H_{18}N_2O_6$ ≥99%

Glycine — FMOC-Gly-SASRIN™-® (200-400 mesh)
Bachem D-1345.0001   Store at 2-8°C

Glycine — FMOC-Gly-Wang Resin
American Peptide RFWAN40
Neosystem RW00801

Glycine — FMOC-L-(Allyl)-Gly
Neosystem FA14902   MW 337.38

Glycine — FMOC-L-2-Allyl-Gly
Fluka 00347   MW 337.38 $C_{20}H_{19}NO_4$ ≥98% (HPLC)

Glycine — FMOC-L-2-Propargyl-Gly
Fluka 00397   MW 335.36 $C_{20}H_{17}NO_4$ ≥98% (HPLC)

Glycine — FMOC-L-Phg
Synonyms: FMOC-L-Phg
Neosystem FA04202   MW 373.4

Glycine — FMOC-L-α-Phg
Fluka 47531   MW 373.41 $C_{23}H_{19}NO_4$ ≥98% (HPLC); mp: 176-180°C

Glycine — FMOC-L-α-tBu-Gly
Synonyms: FMOC-L-t-Leu
Neosystem FA04102   MW 353.4

Glycine — FMOC-Me-Gly
Synonyms: N^α-9-FMOC-Sar; FMOC-Sar
USBio F5419   MW 311.3 $C_{18}H_{17}NO_4$ ≥98%
USBio F5455   MW 311.3 $C_{18}H_{17}NO_4$ ≥99%

Glycine — FMOC-N-(N'-BOC-Aminoethyl)-Gly
Bachem B-3285.0001   MW 440.5 $C_{24}H_{28}N_2O_6$ Store at -15°C

Glycine — FMOC-N-2-Hydroxy-4-Methoxybenzyl-Gly
Bachem B-3490.0001   MW 433.46 $C_{25}H_{23}NO_6$ Store at 2-8°C

Glycine — FMOC-N-Cyclohexyl-Gly
Synonyms: FMOC-N-Chg
Neosystem FA13501   MW 379.47

Glycine — FMOC-*N*-Me-Gly

Synonyms: FMOC-Sar

Neosystem FA00802   MW 311.3

Peptides International FLX-1751-PI   MW 311.33   >95% (HPLC); white crystalline powder

Glycine — FMOC-Phg

Synonyms: FMOC-L-Phg

Bachem B-2980.0001   MW 373.41 $C_{23}H_{19}NO_4$ Store at 2-8°C

Peptides International FFG-1789-PI   MW 373.41   >98% (HPLC); white crystalline powder

Glycine — FMOC-tBu-D-Gly

Synonyms: FMOC-D-Tle

Bachem B-2970.0001   MW 353.42 $C_{21}H_{23}NO_4$ Store at -15°C

Glycine — FMOC-tBu-Gly

Synonyms: FMOC-Tle

Bachem B-2110.0001   MW 353.42 $C_{21}H_{23}NO_4$ Store at RT

Glycine — FMOC-α-Allyl-DL-Gly

Synonyms: FMOC-DL-2-Amino-4-Pentenoic Acid

Bachem B-3280.0001   MW 337.38 $C_{20}H_{19}NO_4$ Store at -15°C

Glycine — FMOC-α-tBu-Gly

USBio F5349   MW 353.5 $C_{21}H_{23}NO_4$ ≥99%

Glycine — For-D-Phg

Bachem F-3945.0001   MW 179.18 $C_9H_9NO_3$ Store at RT

Glycine — Gly

Synonyms: Aminoacetic Acid; Aminoethanoic Acid; Glicoamin; Glycocoll; Glycolixir; Glycosthene; Padil; Sodium glycinate; *N*-Methyl-D-Aspartate Receptor Regulator

Bachem E-3000.0100   MW 75.07 $C_2H_5NO_2$ Store at RT

Calbiochem 3570   MW 75.1 $C_2H_5NO_2$ >99%; solid; soluble in H_2O; may be carcinogenic/teratogenic

Fluka 15527   MW 75.07 $C_2H_5NO_2$ 99-101%; ≤0.1% sulfated ash; ≤0.2% loss on drying; ≤0.005% SO_4, Cl; ≤0.001% heavy metals | DAB, European Pharmacopoeia, BP, USP, Pharmacopee Francaise

Fluka 33226   MW 75.07 $C_2H_5NO_2$ 99.7-101%; ≤0.05% sulfated ash; ≤0.1% H_2O; ≤0.02% NH_4^+; ≤0.002% SO_4; ≤0.003% Cl; ≤0.001% heavy metals; ≤0.0005% Fe | Analytical Reagent

Fluka 50046   MW 75.07 $C_2H_5NO_2$ ≥99.0% (titration); ≤0.05% residue on ignition, loss on drying; ≤0.00001% As; ≤0.0005% Al, Ba, Bi, Cd, Co, Cr, Cu, Fe, Li, Mg, Mn, Mo, Ni, Pb, Sr, Zn; ≤0.001% Ca; ≤0.02% NH_4^+; ≤0.005% K, Na, Cl, SO_4; no detectable DNases, RNases, proteases, phosphatases | For Molecular Biology

Fluka 50049   MW 75.07 $C_2H_5NO_2$ ≥99.0% (titration); ≤0.05% residue on ignition, loss on drying; ≤0.00001% As; ≤0.0005% Al, Ba, Bi, Cd, Co, Cr, Cu, Fe, Li, Mg, Mn, Mo, Ni, Pb, Sr, Zn; ≤0.001% Ca; ≤0.02% NH_4^+; ≤0.005% K, Na, Cl, SO_4

Fluka 50050   MW 75.07 $C_2H_5NO_2$ ≥99.0% (titration); ≤0.05% residue on ignition, loss on drying; ≤0.3% foreign AA; ≤0.0005% Cd, Co, Cr, Cu, Fe, Mg, Mn, Ni, Pb, Zn; ≤0.001% Ca; ≤0.02% NH_4^+; ≤0.005% K, Na, Cl, SO_4; mp: 250°C

Fluka 50052   MW 75.07 $C_2H_5NO_2$ ≥98.5% (titration); ≤0.02% loss on drying; ≤0.1% residue on ignition

Fluka 50056   MW 75.07 $C_2H_5NO_2$ ≥99.0% (titration); ≤0.05% residue on ignition, loss on drying; ≤0.0005% Al, Ba, Bi, Cd, Co, Cr, Cu, Fe, Li, Mg, Mn, Mo, Ni, Pb, Sr, Zn; ≤0.001% Ca; ≤0.02% NH_4^+; ≤0.005% K, Na | For Luminescence

Fluka 50058   MW 75.07 $C_2H_5NO_2$ European Pharmacopoeia

ICN 194825   MW 75.1 $C_2H_5NO_2$ ≥99% | Ideal for all molecular biology applications & buffer preparations

ICN 808822 ICN 808822   MW 75.1 $C_2H_5NO_2$ >99.5%; <10 ppm heavy metals (as Pb) | Electrophoresis grade; used for preparation of Tris-glycine buffers

Neosystem AA00801   MW 75.1

Peptides International ALG-2709   MW 75.07   >99.9% optical purity by RP-HPLC; white crystalline powder

Rexim   MW 75.1 $C_2H_5NO_2$ White crystals or crystalline powder

Sigma G 2414   FW 75.07 $C_2H_5NO_2$ USP Grade

Sigma G 3009   Matrix: 4% beaded agarose; activation: cyanogen bromide; attachment: amino; spacer: 1 atom; ligand immobilized: 2-10 µmoles/mL; form: suspension in 2.0 *M* NaCl containing 0.02% thimerosal

Sigma G 6143   FW 75.07 $C_2H_5NO_2$ Crystalline; essentially ammonia-free | Plant cell culture tested

Sigma G 6761   FW 97.05 $C_2H_4NO_2Na$ ≥99% (TLC) | Inhibitory neurotransmitter in spinal cord; allosteric regulator of NMDA receptors

Glycine — Gly (1,2-^{13}C)

ICN 530092   MW 77 NH_2CH_2COOH 99% ^{13}C atom

Glycine — Gly (1-^{13}C)

ICN 530090   MW 76 NH_2CH_2COOH 99% ^{13}C atom

Glycine — Gly (1-^{14}C)

Amersham CFA30 $H_2NCH_2CO_2H$ Aqueous, 2% EtOH, sterilized; 1.85-2.29 GBq/mmol, 50-62 mCi/mmol; 7.4 MBq/mL, 200 µCi/mL

ARC ARC-166   MW 75.1 H_2NCH_2COOH 40-60 mCi/mmol; 1.48-2.22 GBq/mmol; in 0.01 *N* HCl

ICN 10074   MW 75.1 NH_2CH_2COOH >50 mCi/mmol, >1.85 GBq/mmol; 0.01 *N* HCl

ICN 10074E   MW 75.1 NH_2CH_2COOH >50 mCi/mmol, >1.85 GBq/mmol; EtOH:H_2O (2:98), sterile

Glycine — Gly (^{15}N)

Bachem E-3230.0100   MW 76.07 $C_2H_5{}^{15}NO_2$ Store at -15°C

Fluka 50055   MW 76.06 NH_2CH_2COOH ≥98% (atom % ^{15}N)

ICN 540132   MW 76 NH_2CH_2COOH 99% ^{15}N atom

Glycine — Gly (2-^{13}C)

ICN 530091   MW 76 NH_2CH_2COOH 99% ^{13}C atom

Glycine — Gly (2-^{14}C)

ARC ARC-317   MW 75.1 H_2NCH_2COOH 50-60 mCi/mmol; 1.85-2.22 GBq/mmol; in 0.01 *N* HCl

ICN 10077   MW 75.1 NH_2CH_2COOH >40 mCi/mmol, >1.48 GBq/mmol; 0.01 *N* HCl

ICN 10077E   MW 75.1 NH_2CH_2COOH >40 mCi/mmol, >1.48 GBq/mmol; EtOH:H_2O (2:98), sterile

Sigma G 9535   FW 75.07 $C_2H_5NO_2$ 40-60 mCi/mmol; radiochemical purity ≥98% (HPLC); aqueous solution containing 2% EtOH in serum bottle

Glycine — Gly (2-^{3}H)

ARC ART-161   MW 75.1 H_2NCH_2COOH 30-60 Ci/mmol; 1.11-2.22 TBq/mmol; in 0.01 *N* HCl

ARC ART-161A   MW 75.1 H_2NCH_2COOH 30-60 Ci/mmol; 1.11-2.22 TBq/mmol; in EtOH: H_2O (2:98)

ICN 20020   MW 75.1 NH_2CH_2COOH 10-40 mCi/mmol, 0.37-1.48 TBq/mmol; 0.01 *N* HCl | Schneider, CI & S Urwyler, *Biochem Pharmacol*, 43:1693, 1992

ICN 20020E   MW 75.1 NH_2CH_2COOH 10-40 mCi/mmol, 0.37-1.48 TBq/mmol; EtOH:H_2O (2:98), sterile

Sigma G 3649   FW 75.07 $C_2H_5NO_2$; NH_2CH_2COOH 30-60 Ci/mmol; aqueous solution containing 2% EtOH in serum bottle

Glycine — Gly (^3H)
Amersham TRK71 $H_2NCH_2CO_2H$ Aqueous, 2% EtOH, sterilized; 0.37-1.1 TBq/mmol, 10-30 Ci/mmol; 37 MBq/mL, 1 mCi/mL

Glycine — Gly (^5D)
ICN 521260   MW 80.1 ND_2CD_2COOD 98% D atom

Glycine — Gly (Carboxy-^{14}C) Hydrochloride
Sigma G 0398   FW 111.5 $C_2H_5NO_2 \cdot HCl$ 10-55 mCi/mmol; aqueous solution containing 10% EtOH in serum bottle

Glycine — Gly (U-^{14}C)
Amersham CFB66 $H_2NCH_2CO_2H$ Aqueous, 2% EtOH, sterilized; >3.7 GBq/mmol, >100 mCi/mmol; 1.85 MBq/mL, 50 µCi/mL

ARC ARC-292   MW 75.1 H_2NCH_2COOH 50-120 mCi/mmol; 1.85-4.44 GBq/mmol; in 0.01 N HCl

ICN 10071   MW 75.1 NH_2CH_2COOH >90 mCi/mmol, >3.33 GBq/mmol; 0.01 N HCl

ICN 10071E   MW 75.1 NH_2CH_2COOH >90 mCi/mmol, >3.33 GBq/mmol; EtOH:H_2O (2:98), sterile

ICN 10071L   MW 75.1 NH_2CH_2COOH 65-90 mCi/mmol, 2.4-3.3 GBq/mmol; >95%; EtOH:H_2O (2:98), sterile

Glycine — Gly (U-^{14}C) Hydrochloride
Sigma G 5523   FW 75.07 $C_2H_5NO_2$ 80-120 mCi/mmol; radiochemical purity ≥98% (HPLC); aqueous solution containing 2% EtOH in Combi-vial

Glycine — Gly Amide Acetate
Bachem E-1945.0001   MW 134.14 $C_2H_6N_2O \cdot C_2H_4O_2$ Store at RT

Glycine — Gly Amide Hydrochloride
Bachem E-1950.0005   MW 110.54 $C_2H_6N_2O \cdot HCl$ Store at RT

Neosystem AA00805   MW 110.5

Glycine — Gly Anhydride
Synonyms: 2,5-Peperazinedione; 2,5-Diketopiperazine
Fluka 50080   MW 114.1 $C_4H_6N_2O_2$ ~99%; mp: 315°C
ICN 101824   MW 114.1 $C_4H_6N_2O_2$
Sigma G 7251   FW 114.1 $C_4H_6N_2O_2$

Glycine — Gly CC
Synonyms: Aminoacetic Acid; Glycocoll; Aminoethanoic Acid
USBio G8155   MW 75.07 $C_2H_5NO_2$ Glycine — Special grade of Gly used specifically for cell culture applications; ≥99%; pH: 5.9-6.5; heavy metals (Pb) ≤0.002%; endotoxin: ≤2 EU/mg; cell culture tested: murine myeloma SP2/0-Ag14

Glycine — Gly Cresol Red
Synonyms: 3,3'-Di-(N-Carboxymethylamino-methyl)-O-Cresolsulfonphthalein
Fluka 50100   MW 578.58 $C_{27}H_{27}N_2NaO_9S$
ICN 151199   MW 556.6 $C_{27}H_{28}N_2O_9S$ Red-brown hygroscopic powder | Photometric reagent for heavy metals

Glycine — Gly Cresol Red Sodium Salt
Synonyms: 3',3''-Bis-(N-(Carboxymethyl)-Aminomethyl)-O-Cresolsulfonphthalein
Sigma G 8282   FW 578.6 $C_{27}H_{27}N_2O_9SNa$

Glycine — Gly Free Acid
Synonyms: Aminoacetic Acid
ICN 100570   MW 75.1 $C_2H_5NO_2$ Crystalline
ICN 194049   MW 75.1 $C_2H_5NO_2$ ≥98.5%; <10 ppm heavy metals (as Pb) | ACS reagent grade
ICN 194681   MW 75.1 $C_2H_5NO_2$ Crystalline | Cell culture reagent

Glycine — Gly Free Base
Synonyms: Aminoacetic Acid; Aminoethanoic Acid; Glycocoll; N-Methyl-D-Aspartate Receptor Regulator
Amersham US16407   ≥99.0%; nuclease free
Sigma G 6388   FW 75.07 $C_2H_5NO_2$ ≥99% (TLC); crystalline; essentially ammonia-free | Cell culture tested; insect cell culture tested
Sigma G 7032   FW 75.07 $C_2H_5NO_2$ ACS reagent; ≥98.5%; residue on ignition <0.1%; heavy metals (as Pb): ≤0.002%; SO_4, Cl, NH_4: ≤0.005%; substances darkened by sulfuric acid: passes test; hydrolyzable substances: passes test | Inhibitory neurotransmitter in spinal cord; allosteric regulator of NMDA receptors
Sigma G 7126   FW 75.07 $C_2H_5NO_2$ ≥99% (TLC); <0.1% ammonia | Inhibitory neurotransmitter in spinal cord; allosteric regulator of NMDA receptors
Sigma G 7403   FW 75.07 $C_2H_5NO_2$ SigmaUltra; >99% (titration); loss on drying (HV, 20°C): <0.05%; residue on ignition <0.05%; solubility (0.1 M in 1 M HCl, 20°C): complete, colorless; insoluble matter passes filter test; SO_4, Cl: <0.005%; Ca: <0.001%; Al, Ba, Bi, Cd, Co, Cr, Cu, Fe: <0.0005%; As: <0.00001%; A_{260}<0.01; A_{280}<0.01 (1 M in 1 M HCl) | Inhibitory neurotransmitter in spinal cord; allosteric regulator of NMDA receptors

Glycine — Gly Hydrochloride
Synonyms: NMDA Receptor Regulator
Sigma G 2879   FW 111.5 $C_2H_5NO_2 \cdot HCl$ ≥99% (TLC) | Inhibitory neurotransmitter in spinal cord; allosteric regulator of NMDA receptors

Glycine — Gly Hydroxamate
Sigma G 2753   FW 90.08 $C_2H_6N_2O_2$ >90%

Glycine — Gly Sodium Hydroxide
Fluka 82617   20 mM Glycine-NaOH buffer solution; conductance: ~0.8 mS/cm | Buffer solution, pH 11.0; for HPCE; prepared under cleanroom conditions & filtered through a 0.2 µm membrane filter

Glycine — Gly Sodium Hydroxide Solution Sodium Chloride
Synonyms: FIXANAL
Fluka 82576   ~0.051 M Gly, ~0.049 M NaOH, ~0.051 M NaCl | Buffer Solution, pH 11.0; limited shelf life
Fluka 82666   Ampoule; for 500 mL buffer solution | Buffer Concentrate, pH 13.00
Fluka/RdH 33552   Solution ready for use | Buffer Solution, pH 13.00
Fluka/RdH 38752   For 500 mL buffer solution | Buffer Concentrate, pH 13.00

Glycine — Gly Thymol Blue
Synonyms: 3,3'-Di(N-Carboxymethylamino-Methyl)-Thymolsulfonphthalein; 3',3"-Bis-(N-(Carboxymethyl)-Aminomethyl)-Thymolsulfonphthalein
Fluka 33928   MW 662.73 $C_{33}H_{39}N_2NaO_9S$
ICN 151200   MW 642.7 $C_{33}H_{39}N_2O_9S$ Dark orange-brown powder; hygroscopic | Photometric reagent for Cu, Cd Pd, & rare earth metals; Cheng, KL etal, "Handbook of Organic Analytical Reagents," CRC Press, 1982

Sigma G 1004   FW 662.7 $C_{33}H_{39}N_2O_9SNa$ Dye content: ~70%; brown crystals

Glycine — Gly β-Naphthylamide Free Base
Sigma G 8388   FW 200.2 $C_{12}H_{12}N_2O$

Glycine — Gly β-Naphthylamide Sodium Salt Hydrochloride
Sigma G 5761   FW 236.7 $C_{12}H_{12}N_2O \cdot HCl$

Glycine — Gly-4MβNA Hydrochloride
Bachem J-1195.0250   MW 266.7 $C_{13}H_{14}N_2O_2 \cdot HCl$ Store at -15°C

Glycine — Gly-AMC Hydrobromide
Bachem I-1210.0050   MW 313.15 $C_{12}H_{12}N_2O_3 \cdot HBr$ Store at -15°C
ICN 157243

Glycine — Gly-AMC Hydrochloride
Sigma G 4894   FW 313.2 $C_{12}H_{12}N_2O_3 \cdot HBr$

Glycine — Glycinamide Hydrochloride
Sigma G 7378   FW 110.5 $C_2H_6N_2O \cdot HCl$ pKa=8.0 at 25°C; useful pH range 7.4-8.8

Glycine — Gly-DL-α-Amino-n-Butyric Acid
ICN 101846   MW 160.2 $C_6H_{12}N_2O_3$ Crystalline

Glycine — Gly-N-Carboxyanhydride
Bachem F-2675.0001   MW 101.06 $C_3H_3NO_3$ Store at -15°C

Glycine — Gly-NHMe Hydrochloride
Bachem E-1995.0001   MW 124.57 $C_3H_8N_2O \cdot HCl$ Store at 2-8°C

Glycine — Gly-NHNZ
Bachem E-1990.0001   MW 337.25 $C_{12}H_{14}N_3O_5F_3$ Store at RT

Glycine — Gly-NMe₂ Acetate
Bachem E-1985.0001   MW 162.19 $C_4H_{10}N_2O \cdot C_2H_4O_2$ Store at 2-8°C

Glycine — Gly-OBg TFA
Neosystem AA00810   MW 412.4

Glycine — Gly-OBzl Hydrochloride
Synonyms: Bzl Glycinate
Bachem E-1970.0005   MW 201.65 $C_9H_{11}NO_2 \cdot HCl$ Store at RT
ICN 157244   MW 201.7 $C_9H_{11}NO_2 \cdot HCl$
Sigma G 3267   FW 201.7 $C_9H_{11}NO_2 \cdot HCl$

Glycine — Gly-OBzl p-Tosyl
Synonyms: Bzl Glycinate
Bachem E-1975.0025   MW 337.4 $C_9H_{11}NO_2 \cdot C_7H_8O_3S$ Store at RT
Sigma G 7376   FW 337.4 $C_9H_{11}NO_2 \cdot C_7H_8O_3S$

Glycine — Gly-OBzl-Tosyl
Synonyms: Gly-OBzl-Tosyl
Fluka 50051   MW 337.4 $C_9H_{11}NO_2 \cdot C_7H_8O_3S$ ~99% (titration); mp: 132-134°C
Peptides International EBG-2047   MW 337.39   >98% (TLC); white powder

Glycine — Gly-OEt (1-¹⁴C) Hydrochloride
ARC ARC-244   MW 139.6 $H_2NCH_2COOC_2H_5 \cdot HCl$ 50-60 mCi/mmol; 1.85-2.22 GBq/mmol; in EtOH

Glycine — Gly-OEt (2-¹⁴C) Hydrochloride
ARC ARC-712   MW 139.6 $H_2NCH_2COOC_2H_5 \cdot HCl$ 50-60 mCi/mmol; 1.85-2.22 GBq/mmol; in EtOH
Sigma G 8037   FW 139.6 $C_4H_9NO_2 \cdot HCl$ 40-60 mCi/mmol; radiochemical purity ≥95% (HPLC); solid in screw cap bottle

Glycine — Gly-OEt Hydrochloride
Synonyms: Gly-OEt; Et Glycinate
Bachem E-2735.0025   MW 139.58 $C_4H_9NO_2 \cdot HCl$ Store at RT
Fluka 50060   MW 139.58 $C_4H_9NO_2 \cdot HCl$ ≥99% (titration); ≤0.05% residue on ignition; ≤0.001% Fe; mp: 145-148°C
Fluka 50061   MW 139.58 $C_4H_9NO_2 \cdot HCl$ ≥98% (titration); mp: 144-148°C
ICN 101829   MW 139.6 $C_4H_9NO_2 \cdot HCl$ ~99%; crystalline
Neosystem AA00803   MW 139.6
Peptides International EEG-2037-PI   MW 139.58
Sigma G 8001   FW 139.6 $C_4H_9NO_2 \cdot HCl$ ~99%

Glycine — Gly-OEt Phosphate Salt
Synonyms: Et glycinate
Sigma G 3629   FW 201.1 $C_4H_9NO_2 \cdot H_3PO_4$

Glycine — Gly-OMe Hydrochloride
Synonyms: Methyl Glycinate
Bachem E-2000.0025   MW 125.56 $C_3H_7NO_2 \cdot HCl$ Store at RT
Fluka 50110   MW 125.56 $C_3H_7NO_2 \cdot HCl$ ≥99% (titration); ≤0.05% residue on ignition; mp: 175-176°C
ICN 157245   MW 125.6 $C_3H_7NO_2 \cdot HCl$ Crystalline
Neosystem AA00804   MW 125.6
Sigma G 8126   FW 125.6 $C_3H_7NO_2 \cdot HCl$ May contain ~3% free Gly

Glycine — Gly-ONp Hydrochloride
Bachem E-2005.0001   MW 232.62 $C_8H_8N_2O_4 \cdot HCl$ Store at -15°C

Glycine — Gly-OtBu (1-¹⁴C)
ARC ARC-1108   MW 131.14 $H_2NCH_2COOC(CH_3)_3$ 50-60 mCi/mmol; 1.85-2.22 GBq/mmol; in DMF in sealed ampoule

Glycine — Gly-OtBu Dibenzenesulfimide Salt
Fluka 50085   MW 428.53 $C_6H_{13}NO_2 \cdot C_{12}H_{11}NO_4S_2$ ≥99%; mp: 154-157°C

Glycine — Gly-OtBu Hydrochloride
Synonyms: tBu-Glycinate
Bachem E-1980.0001   MW 167.64 $C_6H_{13}NO_2 \cdot HCl$ Store at 2-8°C
ICN 101826   MW 167.7 $C_6H_{13}NO_2 \cdot HCl$ Crystalline
Sigma G 4128   FW 167.6 $C_6H_{13}NO_2 \cdot HCl$

Glycine — Glyphosate-Gly (2-¹⁴C)

ARC ARC-1312 MW 339.2 $HO_2CCH_2NHCH_2PO_3H_2$ 10-30 mCi/mmol; 370-1110 MBq/mmol; in sterile H_2O

Sigma G 1661 10-30 mCi/mmol; aqueous solution

Glycine — Glyphosate-Gly (2-¹⁴C) Sodium Salt

Sigma G 9035 FW 405.2 $C_6H_{14}N_2Na_3O_{10}P_2$ 10-30 mCi/mmol; aqueous solution

Glycine — Gly-pNA

Synonyms: Serum Oxytocinase Substrate

Bachem L-1280.0001 MW 195.18 $C_8H_9N_3O_3$ Store at -15°C

ICN 101836 MW 195.2 $C_8H_9N_3O_3$ Crystalline

Glycine — Gly-pNA Hydrobromide

Sigma G 4254 FW 195.2 $C_8H_9N_3O_3$

Glycine — Gly-βNA

Bachem K-1550.0250 MW 200.24 $C_{12}H_{12}N_2O$ Store at RT

Glycine — Gly-βNA Hydrochloride

Bachem K-1295.0250 MW 236.7 $C_{12}H_{12}N_2O \cdot HCl$ Store at RT

Glycine — Gly-β-Naphthylamide Hydrochloride

ICN 101827

Glycine — Gly-γ–Aminobutyric Acid

Sigma G 3004 FW 160.2 $C_6H_{12}N_2O_3$

Glycine — Indole-3-Ac-Gly

Sigma I 9512 FW 232.2 $C_{12}H_{12}N_2O_3$ Plant cell culture tested

Glycine — L-(+)-Phenyl-Gly

Synonyms: L-α-Aminophenylacetic Acid; L-Phg

USBio P4065 MW 151.2 $C_8H_9NO_2$ ≥99%

Glycine — L-(+)-α-Phg

Fluka 78565 MW 151.17 $C_8H_9NO_2$ ≥99.0% (titration)

ICN 151834 MW 151.2 $C_8H_9NO_2$ Crystalline

Glycine — L-(c-Allyl)-Gly

Fluka 05958 MW 115.13 $C_5H_9NO_2$ ≥99% (titration); ≤2% H_2O; mp: ~283°C

Glycine — L-(Vinyl)-Gly

Synonyms: (S)-2-Amino-3-Butenoic Acid

Fluka 94997 MW 101.11 $C_4H_7NO_2$ ≥99.0% (TLC)

ICN 158297 MW 101.1 $C_4H_7NO_2$

Sigma V 4255 FW 101.1 $C_4H_7NO_2$

Glycine — L-2-Cyclohexyl-Gly Hydrochloride

USBio C8800 MW 193.7 $C_8H_{15}NO_2 \cdot HCl$ ≥99%

Glycine — L-2-tBu-Gly

Synonyms: L-t-Leu

ICN 154912 MW 131.2 $C_6H_{13}NO_2$

USBio B9000 MW 131.2 $C_6H_{13}NO_2$ ≥99%

Glycine — L-Gly-(C-Propargyl)-Gly

Fluka 81838 MW 113.12 $C_5H_7O_2$ ≥99.0% (TLC)

Glycine — L-Phg (1-¹⁴C)

ARC ARC-1003 MW 151.17 $C_6H_5CH(NH_2)COOH$ 50-60 mCi/mmol; 1.85-2.22 GBq/mmol; in EtOH

Glycine — L-α-(2-Aminoethoxyvinyl) Gly Hydrochloride

Synonyms: AVG; (S)-trans-2-Amino-4-(2-Aminoethoxy)-3-Butenoic Acid

Sigma A 1284 FW 196.6 $C_6H_{12}N_2O_3 \cdot HCl$

Glycine — L-α-2-Thienyl-Gly

ICN 156854 MW 157.2 $C_6H_7NO_2S$

Sigma T 8278 FW 157.2 $C_6H_7NO_2S$

Glycine — L-α-3-Thienyl-Gly

ICN 156857 MW 157.2 $C_6H_7NO_2S$

Sigma T 8653 FW 157.2 $C_6H_7NO_2S$

Glycine — L-α-4-Carboxyphenyl-Gly

Synonyms: (S)-4C-PG; (S)-α-Amino-4-Carboxyphenylacetic Acid); Metabotropic Glutamate Receptors Agonist

Sigma C 1835 FW 195.2 $C_9H_9NO_4$ Competitive antagonist at PI-linked metabotropic glutamate receptors; Eaton, SA et al, *Eur J Pharmacol*, 244: 195, 1993

Glycine — MTH-Gly Hydrochloride

Sigma M 4381 FW 130.2 $C_4H_6N_2OS \cdot HCl$ ~95%

Glycine — Myr-Gly

Bachem F-2550.0050 MW 285.43 $C_{16}H_{31}NO_3$ Store at RT

Glycine — N-(3,5-Dinitrobenzoyl)-D-α-Phg

Fluka 42032 MW 345.27 $C_{15}H_{11}N_3O_7$ ≥99% (HPLC); mp: 217-218°C

Glycine — N-(3,5-Dinitrobenzoyl)-L-α-Phg

Fluka 42033 MW 345.27 $C_{15}H_{11}N_3O_7$ ≥98% (HPLC); mp: 215-217°C

Glycine — N-(Imino(Phosphonoamino)Me)-N-Me-Gly Disodium Salt

Synonyms: Phosphocreatine

Sigma P 6502 Synthetic FW 255.1 $C_4H_8N_3O_5PNa_2$ ~98% | High-energy phosphate reservoir in vertebrate & some invertebrate muscle; provides phosphate for ADP-ATP conversion

Glycine — N-(Imino(Phosphonoamino)Me)-N-Me-Gly Di-Tris Salt

Synonyms: Phosphocreatine

Sigma P 1937 FW 453.4 $C_4H_{10}N_3O_5P \cdot 2C_4H_{11}NO_3$ ~98%; enzymatic; sodium content: <0.1%; <0.02% free creatine | High-energy phosphate reservoir in vertebrate & some invertebrate muscle; provides phosphate for ADP-ATP conversion; for use in systems where alkali metal ions are undesirable

Sigma P 4635 Synthetic FW 453.4 $C_4H_{10}N_3O_5P \cdot 2C_4H_{11}NO_3$ ~98% | High-energy phosphate reservoir in vertebrate & some invertebrate muscle; provides phosphate for ADP-ATP conversion; for use in systems where alkali metal ions are undesirable

Glycine — N-(Imino(Phosphonoamino)Me)-N-Me-Gly Hydrate Disodium Salt

Synonyms: Phosphocreatine

Sigma P 6915 FW 255.1 $C_4H_8N_3O_5PNa_2$ Enzymatic: ~98%; <0.02% free creatine | High-energy phosphate reservoir in vertebrate & some invertebrate muscle; provides phosphate for ADP-ATP conversion; this type not always available; to avoid delay will substitute Sigma P 7936 unless specifically asked not to do so

Sigma P 7936 FW 255.1 $C_4H_8N_3O_5PNa_2$ Enzymatic: ~98%; <0.02% free creatine | High-energy phosphate reservoir in vertebrate & some invertebrate muscle; provides phosphate for ADP-ATP conversion; the same as Sigma P 6915 but prepared using a different enzymatic procedure

Sigma 520-6 Synthetic FW 255.1 $C_4H_8N_3O_5PNa_2$ ~98%; preweighed vial of 45 mg (Sigma P 6502) | High-energy phosphate reservoir in vertebrate & some invertebrate muscle; provides phosphate for ADP-ATP conversion

Glycine — N-(N-β-BOC-Aminoethyl)-Gly-OEt

Synonyms: PNA Building Block; Peptide Nucleic Acid Building Block

Bachem A-3705.0001 MW 246.31 $C_{11}H_{22}N_2O_4$ Store at -15°C

Glycine — N-(p-Aminobenzoyl)-Gly

Synonyms: p-Aminohippuric Acid

ICN 102570 MW 194.2 $C_9H_{10}N_2O_3$ Crystalline | Useful in kidney function test

Glycine — N-(p-Aminobenzoyl)-Gly (C-^{14}C)

Synonyms: p-Aminohippuric Acid

ICN 17048 MW 194.2 $H_2NC_6H_4CONHCH_2COOH$ 2-20 mCi/mmol, 74-740 MBq/mmol; aqueous, sterile

Glycine — N-(p-Aminobenzoyl)-Gly Sodium Salt

Synonyms: p-Aminohippuric Acid

ICN 154746 MW 216.2 $C_9H_{10}N_2O_3Na$ Crystalline

Glycine — N-(Phthaloyl)-Gly

Fluka 79850 MW 205.17 $C_{10}H_7NO_4$ ≥99.0% (titration); ≤0.05% residue on ignition; mp: 193-199°C

Glycine — N,N-Bis-(N'-FMOC-3-Aminopropyl)-Glycine Potassium Hemisulfate

Neosystem FA09301 MW 769.94

Glycine — N,N-Bis-2-Hydroxyethyl-Gly

Synonyms: Bicine

Fluka 14871 MW 163.18 $C_6H_{13}NO_4$ ≥99.5% (titration); ≤0.1% residue on ignition; ≤0.005% K, Na, Cl, SO$_4$; ≤0.0005% Al, Ba, Bi, Cd, Co, Cr, Cu, Fe, Li, Mg, Mn, Mo, Ni, Pb, Sr, Zn; ≤0.1% loss on drying; ≤0.00001% As; ≤0.001% Ca

Fluka 14872 MW 163.18 $C_6H_{13}NO_4$ ≥99% (titration); ≤0.1% residue on ignition; ≤0.005% K, Na, SO$_4$; ≤0.0005% Cd, Co, Cr, Cu, Fe, Mg, Mn, Ni, Pb, Zn; ≤0.5% loss on drying; ≤0.05% Cl; ≤0.001% Ca

ICN 101005 MW 163.2 $C_6H_{13}NO_4$ pKa 8.3, 25°C; useful pH 7.6-9.0

Glycine — N,N-BisPhosphonomethyl-Gly

Synonyms: Glyphosine

Fluka 15149 MW 263.08 $C_4H_{11}NO_8P_2$ ≥98% (titration)

ICN 156238 MW 263.1 $C_4H_{11}NO_8P_2$ ≥85%; crystalline

Glycine — N,N-Dimethyl-Gly

Fluka 40360 MW 103.12 $C_4H_9NO_2$ ≥99% (titration); mp: 178-182°C

Glycine — N,N-Dimethyl-Gly Free Base

Synonyms: N,N-Dimethylaminoacetic Acid

ICN 157819 MW 103.1 $C_4H_9NO_2$ 99%

Sigma D 1156 FW 103.1 $C_4H_9NO_2$ ~99%

Glycine — N,N-Dimethyl-Gly Hydrochloride

Synonyms: N,N-Dimethylaminoacetic Acid

Fluka 40380 MW 139.58 $C_4H_9NO_2 \cdot HCl$ ≥99% (titration); mp: 190-193°C

ICN 101597 MW 139.6 $C_4H_9NO_2 \cdot HCl$ Crystalline

Sigma D 6382 FW 139.6 $C_4H_9NO_2 \cdot HCl$ Crystalline

Glycine — N,N-Dimethyl-Gly-OEt

Fluka 40370 MW 131.18 $C_6H_{13}NO_2$ ≥98% (GC)

ICN 157820 MW 131.2 $C_6H_{13}NO_2$ ~0.93 g/mL

Glycine — N-2,4-Dnp-Gly

Sigma D 9504 FW 241.2 $C_8H_7N_3O_6$

Glycine — N-2-Furoyl-Gly

Sigma F 5402 FW 169.1 $C_7H_7NO_4$ ~98% (TLC)

Glycine — N-2-Mercaptopropionyl-Gly

Fluka 63794 MW 163.2 $C_5H_9NO_3S$ ≥98.0% (titration); mp: 93-96°C

ICN 151606 MW 163.2 $C_5H_9NO_3S$ >98%; crystalline | Stimulates NADPH-dependent lipid peroxidation by rat liver microsomes; minimizes the effect of the carcinogen dimethylbenzanthracene (DMBA) in regenerating mouse liver; Harata, J etal, *Biochem intern*, 8:49, 1984; Dixit, A & AR Rao, *Biochem intern*, 7:695, 1983

Sigma M 6635 FW 163.2 $C_5H_9NO_3S$

Glycine — N-4-Aminobenzoyl-Gly

Fluka 08088 MW 194.19 $C_9H_{10}N_2O_3$ ≥99% (titration); mp: 200-202°C

Fluka 08090 MW 194.19 $C_9H_{10}N_2O_3$ ≥98% (titration); mp: 198-202°C

Glycine — N-Ac-DL-(Allyl)-Gly

Sigma A 7637 FW 157.2 $C_7H_{11}NO_3$

Glycine — N-Ac-DL-Phg

Sigma A 9027 FW 193.2 $C_{10}H_{11}NO_3$

Glycine — N-Ac-Gly

Synonyms: Aceturic Acid

Fluka 01180 MW 117.11 $C_4H_7NO_3$ ≥99% (titration); mp: 207-209°C

ICN 100085 MW 117.1 $C_4H_7NO_3$ Crystalline

Glycine — N-Bz-Gly Ammonium Salt

Synonyms: Hippuric Acid; Benzoylaminoacetic Acid

Sigma H 8255 FW 196.2 $C_9H_9NO_3 \cdot NH_3$ ~99%

Glycine — N-Bz-Gly Free Acid

Synonyms: Hippuric Acid; Benzoylaminoacetic Acid

Sigma H 6375 FW 179.2 $C_9H_9NO_3$ ~99%

Glycine — *N*-Bz-Gly Sodium Salt

Synonyms: Hippuric Acid; Benzoylaminoacetic Acid

Sigma H 6529   FW 201.2 $C_9H_8NO_3Na$ SigmaUltra; ~99%; solubility (0.5 *M* in H_2O, 20°C): complete, colorless; insoluble matter: <0.1%; K: <0.005%; NH_4^+, SO_4, Cl: <0.05%; Al, Mg, P, Zn, Ca, Cu, Fe: <0.0005%; Pb; <0.001%

Sigma H 9380   FW 201.2 $C_9H_8NO_3Na$ ~99%

Glycine — *N*-Bz-Gly-OEt

Fluka 13430   MW 193.25 $C_{11}H_{15}NO_2$ ≥97% (GC)

Glycine — *N*-CBZ-Gly

ICN 100216   MW 209.2 $C_{10}H_{11}NO_4$ Crystalline

Sigma C 7376   FW 209.2 $C_{10}H_{11}NO_4$

Glycine — *N*-CBZ-Gly-(*p*ONp)

ICN 101250   MW 330.3 $C_{16}H_{14}N_2O_6$ Crystalline

Sigma C 7626   FW 330.3 $C_{16}H_{14}N_2O_6$

Glycine — *N*-CBZ-Glycinamide

Sigma C 7501   FW 208.2 $C_{10}H_{12}N_2O_3$

Glycine — *N*-Chloroacetyl-Gly

Sigma C 0878   FW 151.5 $C_4H_6NO_3Cl$

Glycine — *N*-Diphenylmethylene-Gly-OEt

Fluka 43121   MW 267.33 $C_{17}H_{17}NO_2$ ~98% (GC); mp: 51-53°C; ≤5% benzophenone

Glycine — *N*-FMOC-Gly

Sigma F 2008   FW 297.3 $C_{17}H_{15}NO_4$

Glycine — *N*-FMOC-Gly (1-¹⁴C)

ARC ARC-495   MW 297.3   50-60 mCi/mmol; 1.85-2.22 GBq/mmol; in EtOH

Glycine — *N*-FMOC-Gly (2-³H)

ARC ART-416   MW 297.3   10-20 Ci/mmol; 370-740 GBq/mmol; in EtOH

Glycine — *N*-FMOC-Gly Resin Ester

Sigma F 1386   For peptide synthesis; Wang, SS, *J Am Chem Soc*, 95: 1328, 1973; Lu, G et al, *J Org Chem*, 46: 3433, 1981

Glycine — *N*-FMOC-Gly SASRIN Resin Ester

Sigma F 1639   200-400 mesh; substitution range: 0.3-0.8 mmole FMOC AA/g resin | For peptide synthesis; *N*-FMOC AA acyl ester of 3-methoxy-4-oxymethyl-phenoxymethylated 1% divinylbenzene cross-linked polystyrene; peptides can be cleaved from the resin with 0.5-1% CF_3CO_2H in CH_2Cl_2; Mergler, M et al, *in Peptides: Chemistry & Biology* (Proc 10th Am Peptide Symp), Marhall, GR, ed, 259, 1988

Glycine — *N*-FMOC-Gly-OPfp

Sigma F 9764   FW 463.4 $C_{23}H_{14}F_5NO_4$ ≥95% (HPLC)

Glycine — *N*-For-Gly

Fluka 47723   MW 103.08 $C_3H_5NO_3$ ≥98% (titration); mp: 149-151°C

ICN 151164   MW 103.1 $C_3H_5NO_3$ Crystalline

Sigma F 3127   FW 103.1 $C_3H_5NO_3$

Glycine — *N*-For-Gly-OEt

Fluka 47719   MW 131.13 $C_5H_9NO_3$ ≥98% (GC); bp: 267-269°C

Glycine — *N*-For-Gly-OMe

Fluka 47724   MW 117.11 $C_3H_5NO_3$ ≥97% (GC); mp: 29-31°C

Glycine — *N*-Glycylmuscimol Hydrobromide

Sigma G 0393   FW 252.1 $C_6H_9N_3O_3 \cdot HBr$

Glycine — *N*-Guanyl-*N*-Me-Gly

Fluka 27890   MW 131.14 $C_4H_9N_3O_2$ ≥99% (titration); ≤0.5% H_2O; mp: 295°C

Glycine — *N*-Guanyl-*N*-Me-Gly Hydrate

Fluka 27900   MW 149.15 $C_4H_9N_3O_2 \cdot H_2O$ ≥99% (titration); ≤0.05% residue on ignition; mp: 295°C

Glycine — *N*-Me-DL-Phg

Bachem F-3595.0001   MW 165.19 $C_9H_{11}NO_2$ Store at RT

Glycine — *N*-Me-Gly

Synonyms: Sarcosine

Neosystem AA00811   MW 89.1

Glycine — *N*-Me-Phg

Bachem E-2175.0001   MW 165.19 $C_9H_{11}NO_2$ Store at RT

Glycine — *N*-MSOC-Gly Dicyclohexylammonium Salt

ICN 155737   MW 407.6 $C_6H_{11}NO_6S \cdot C_{12}H_{23}N$ Crystalline

Glycine — *N*-O-Nps-Gly Dicyclohexylammonium Salt

ICN 102528   MW 409.5 $C_8H_8N_2O_4S \cdot C_{12}H_{23}N$ Yellow crystals

Sigma N 5378   FW 409.5 $C_8H_8N_2O_4S \cdot C_{12}H_{23}N$ Yellow crystals

Glycine — *N*-Phg

Synonyms: *N*-Phenylaminoacetic Acid

Fluka 78560   MW 151.17 $C_8H_9NO_2$ ≥97.0% (titration); mp: 121-123°C

ICN 151835   MW 151.2 $C_8H_9NO_2$ Crystalline

Sigma P 7001   FW 151.2 $C_8H_9NO_2$ Yellow crystals

Glycine — *N*-Phosphonomethyl-Gly

Fluka 79533   MW 169.07 $C_3H_8NO_5P$ ≥97.0% (titration)

ICN 156237   MW 169.1 $C_3H_8NO_5P$ 95%

Sigma P 5671   FW 169.1 $C_3H_8NO_5P$ ~95%

Glycine — Npys-Gly

Bachem E-2910.0001   MW 215.21 $C_7H_7N_2O_4S$ Store at -15°C

Glycine — *N*-t-BOC Gly (1-¹⁴C)

ARC ARC-1182   MW 175.2   50-60 mCi/mmol; 1.85-2.22 GBq/mmol; in EtOH

Glycine — *N*-t-BOC Gly (2-³H)

ARC ART-644   MW 175.2   30-60 Ci/mmol; 1.11-2.22 TBq/mmol; in EtOH

Glycine — *N*-t-BOC-DL-2-tBu-Gly

Sigma B 1392   FW 231.3 $C_{11}H_{21}NO_4$

Glycine — *N*-t-BOC-D-α-Cyclohexyl-Gly

Sigma B 6905   FW 257.3 $C_{13}H_{23}NO_4$

Glycine — N-t-BOC-Gly

Sigma B 1268   FW 175.2 $C_7H_{13}NO_4$

Glycine — N-t-BOC-Gly N-Hydroxysuccinimide Ester

Sigma B 0391   FW 272.3 $C_{11}H_{16}N_2O_6$

Glycine — N-t-BOC-Gly Pam Resin Ester

Sigma B 5268   N-t-BOC AA ester of 4-(Oxymethyl)-Phenyl-Acm 1% divinylbenzene cross-linked polystyrene; 200-400 mesh; substitution range: 0.2-0.6 mmole t-BOC-AA/g resin | For peptide synthesis by the Merrifield method; Mitchell, AR et al, *J Org Chem*, 43: 2845, 1978

Glycine — N-t-BOC-Gly Resin Ester

Sigma B 5896   N-t-BOC AA ester of methylated 1% divinylbenzene cross-linked polystyrene; 200-400 mesh; substitution range: 0.2-0.6 mmole t-BOC-AA/g resin | For peptide synthesis by the Merrifield method

Glycine — N-t-BOC-Gly-(pONp)

Sigma B 3508   FW 296.3 $C_{13}H_{16}N_2O_6$

Glycine — N-t-BOC-L-α-Cyclohexyl-Gly

Sigma B 7030   FW 257.3 $C_{13}H_{23}NO_4$

Glycine — N-TFA-Gly-OMe

Sigma T 4006   FW 185.1 $C_5H_6F_3NO_3$ ~98%

Glycine — N-Tris-Hydroxymethyl-Me-Gly

Amersham US22561   ≥98.5 dsb; ≤10ppm lead

Glycine — N-Z-D-α-Cyclohexyl-Gly

Fluka 96065   MW 291.35 $C_{16}H_{21}NO_4$ ≥98.0% (TLC)

Glycine — N$^\epsilon$-FMOC-Gly

ICN 151144   MW 297.3 $C_{17}H_{15}NO_4$

Glycine — N$^\epsilon$-t-BOC-Gly

ICN 101071   MW 175.2 $C_7H_{13}NO_4$

Glycine — N$^\epsilon$-Aminoethyl-Gly

Bachem E-3175.0001   MW 118.14 $C_4H_{10}N_2O_2$ Store at -15°C

Glycine — N$^\epsilon$-Aminoethyl-Gly-OEt Dihydrochloride

Bachem E-3210.0001   MW 219.11 $C_6H_{14}N_2O_2 \cdot 2HCl$ Store at -15°C

Glycine — OBz-Gly-DL-β-Phenyllactic Acid Sodium Salt

Synonyms: O-Hippuryl-DL-β-Phenyllactic Acid

Bachem M-1500.0001   MW 349.32 $C_{18}H_{16}NNaO_5$ Store at 2-8°C

Glycine — OBz-Gly-L-β-Phenyllactic Acid Sodium Salt

Synonyms: O-Hippuryl-L-β-Phenyllactic Acid

Bachem M-1495.0001   MW 349.32 $C_{18}H_{16}NNaO_5$ Store at 2-8°C

Glycine — Phenylac-Gly

Bachem F-2625.0001   MW 193.2 $C_{10}H_{11}NO_3$ Store at -15°C

Glycine — Phg

Bachem F-2010.0005   MW 151.17 $C_8H_9NO_2$ Store at RT

Glycine — Phg Amide Hydrochloride

Bachem F-3835.0001   MW 186.64 $C_8H_{10}N_2O \cdot HCl$ Store at 2-8°C

Glycine — Phg-OEt Hydrochloride

Bachem F-3860.0001   MW 215.68 $C_{10}H_{13}NO_2 \cdot HCl$ Store at 2-8°C

Glycine — Phg-OMe Hydrochloride

Bachem F-2025.0001   MW 201.65 $C_9H_{11}NO_2 \cdot HCl$ Store at 2-8°C

Glycine — Phg-OtBu

Bachem E-2640.0001   MW 207.27 $C_{12}H_{17}NO_2$ Store at -15°C

Glycine — Phg-OtBu Hydrochloride

Bachem F-2020.0001   MW 243.73 $C_{12}H_{17}NO_2 \cdot HCl$ Store at -15°C

Glycine — Propargyl-D-Gly

Synonyms: D-Pra

Bachem F-2900.0001   MW 113.12 $C_5H_7NO_2$ Store at RT

Glycine — Propargyl-DL-Gly

Synonyms: DL-Pra

Bachem F-2890.0001   MW 113.12 $C_5H_7NO_2$ Store at -15°C

Glycine — Pth-Gly

Sigma P 0627   FW 192.2 $C_9H_8N_2OS$ ~99% | Useful as standards in protein sequencing by the Edman degradation

Glycine — p-Tosyl-Gly-OBzl

ICN 101828

Glycine — S-(+)-2-(3'-Carboxybicyclo(1.1.1)Pentyl)-Gly

Synonyms: S-(+)-CBPG; UPF 596; Group I mGluR Antagonist

Alexis 550-323   MW 185.2 $C_8H_{11}NO_4$ ≥98%; white solid; soluble in DMSO or H_2O | Potent & selective, structurally new group I mGluR antagonist; Pellicciari, R et al, *J Med Chem*, 39: 2874, 1996; Pellicciari, R et al, *Neuropharmacology*, 35: A23Abs90, 1996; Moroniet, F et al, *Eur J Pharmacol*, 347: 189, 1998

Glycine — Suc-Gly Amide

Bachem E-2400.0001   MW 174.16 $C_6H_{10}N_2O_4$ Store at 2-8°C

Glycine — tBu-D-Gly

Synonyms: D-Tle

Bachem F-1340.0001   MW 131.18 $C_6H_{13}NO_2$ Store at RT

Glycine — tBu-DL-Gly

Synonyms: DL-Tle

Bachem F-1345.0001   MW 131.18 $C_6H_{13}NO_2$ Store at RT

Glycine — tBu-Gly

Synonyms: Tle

Bachem F-1335.0001   MW 131.18 $C_6H_{13}NO_2$ Store at RT

Glycine — tBu-Gly-OtBu Hydrochloride

Synonyms: Tle-OtBu

Bachem F-1350.0001   MW 223.74 $C_{10}H_{21}NO_2 \cdot HCl$ Store at -15°C

Glycine — t-Butyloxycarbonyl-Diethylglycine

Synonyms: BOC-Deg

Peptides International BLX-5341-PI   MW 231.29   >98% (HPLC); white crystalline powder

Glycine — TentaGel S PHB-Gly-FMOC

Fluka 86391   0.22 mmol protected AA/g; ~90 μm particle size

Glycine — TentaGel S Trt-Gly-FMOC

Fluka 86422   0.20 mmol protected AA/g; ~90 μm particle size

Glycine — TFA-Gly

Bachem E-2415.0001   MW 171.08 $C_4H_4F_3NO_3$ Store at 2-8°C

Glycine — Tosyl-Gly-OSu

Bachem E-2480.0005   MW 326.33 $C_{13}H_{14}N_2O_6S$ Store at -15°C

Glycine — Tris-Glycine Buffer

Sigma T 4904   10X concentration; sterile-filtered; 1 L 0.25 mole Tris base & 1.92 moles Gly; no mEtOH; upon dilution pH ~8.3; protease: none detected

Glycine — Tris-Glycine-SDS Buffer

Sigma T 7777   10X concentration; sterile-filtered; working solution 0.025 *M* Tris, 0.192 *M* Gly & 0.1% SDS, pH ~8.6

Glycine — Trt-Gly

Bachem E-2835.0005   MW 317.39 $C_{21}H_{19}NO_2$ Store at -15°C

Glycine — Z-D-Phg

Bachem C-3825.0001   MW 285.3 $C_{16}H_{15}NO_4$ Store at RT

Glycine — Z-Gly

Synonyms: Benzyloxycarbonyl-Gly

Bachem C-1655.0025   MW 209.2 $C_{10}H_{11}NO_4$ Store at RT

Fluka 96160   MW 209.20 $C_{10}H_{11}NO_4$ ~99% (titration); ≤0.05% residue on ignition; mp: 118-122°C

Neosystem ZA00801   MW 209.2

Peptides International ZLG-2018-PI   MW 209.20   >98% (HPLC); white powder

Glycine — Z-Gly Amide

Bachem C-1660.0005   MW 208.22 $C_{10}H_{12}N_2O_3$ Store at RT

Glycine — Z-Gly *N*-Succinimidyl Ester

Fluka 96185   MW 306.27 $C_{14}H_{14}N_2O_6$ ~97%; mp: 111-115°C

Glycine — Z-Gly-(4-ONp)

Fluka 96178   MW 330.30 $C_{16}H_{14}N_2O_6$ ~98% (TLC)

Glycine — Z-Gly-ONp

Bachem M-1250.0005   MW 330.3 $C_{16}H_{14}N_2O_6$ Store at -15°C

Glycine — Z-Gly-OSu

Bachem C-1665.0005   MW 306.28 $C_{14}H_{14}N_2O_6$ Store at -15°C

Glycine — Z-L-Vinyl-Gly-OMe

Fluka 97335   MW 249.27 $C_{13}H_{15}NO_4$ ≥99.0% (HPLC); mp: 34-37°C

Glycine — Z-*N*-(*N*ᵋ-BOC-Aminoethyl)-Gly

Bachem C-3865.0001   MW 352.39 $C_{17}H_{24}N_2O_6$ Store at -15°C

Glycine — Z-Phg

Bachem C-2450.0001   MW 285.3 $C_{16}H_{15}NO_4$ Store at RT

Glycine — Z-Phg-OSu

Bachem C-2455.0001   MW 382.37 $C_{20}H_{18}N_2O_6$ Store at -15°C

Glycine — Z-tBu-Gly Dicyclohexylamine

Synonyms: Z-Tle ·DCHA

Bachem C-4065.0001   MW 446.63 $C_{14}H_{19}NO_4 \cdot C_{12}H_{23}N$ Store at RT

Histidine — 1-Me-L-His

Synonyms: (*S*)-1-Methylimidazole-5-Ala; π-Me-L-His

ICN 155569   MW 169.2 $C_7H_{11}N_3O_2$ Crystalline

Sigma M 9005   FW 169.2 $C_7H_{11}N_3O_2$ This structure corresponds to IUPAC 3-Me-L-His

Histidine — 2,5-Diiodo-His Hydrochloride

Bachem F-3460.0001   MW 443.41 $C_6H_7I_2N_3O_2 \cdot$ HCl   Store at -15°C

Histidine — 2-Mercapto-His

Bachem F-3615.0001   MW 187.22 $C_6H_9N_3O_2S$ Store at -15°C

Histidine — 3-Me-L-His

Synonyms: (*S*)-1-Methylimidazole-4-Ala; π-Me-L-His

ICN 155570   MW 169.2 $C_7H_{11}N_3O_2$ Crystalline

Sigma M 3879   FW 169.2 $C_7H_{11}N_3O_2$ This structure corresponds to IUPAC 1-Me-L-His

Histidine — Ac-His-1-Trt

Synonyms: Ac-His-(τ-Trt)

Bachem E-3045.0001   MW 439.52 $C_{27}H_{25}N_3O_3$ Store at 2-8°C

Histidine — Ac-His-NHMe

Bachem E-1070.0250   MW 210.24 $C_9H_{14}N_4O_2$ Store at RT

Histidine — Ac-His-OMe

Bachem E-1075.0250   MW 211.22 $C_9H_{13}N_3O_3$ Store at RT

Histidine — BOC-1-Me-His

USBio B2350   MW 269.3 $C_{12}H_{19}N_3O_4$ ≥98%

Histidine — BOC-3-Me-His

USBio B2351   MW 269.3 $C_{12}H_{19}N_3O_4$ ≥98%

Histidine — BOC-D-His

Bachem A-1770.0001   MW 255.27 $C_{11}H_{17}N_3O_4$ Store at RT

Histidine — BOC-D-His-1-Me

Synonyms: BOC-D-His-(τ-Me)

Bachem A-3015.0250   MW 269.3 $C_{12}H_{19}N_3O_4$ Store at RT

Histidine — BOC-D-His-3-Bom

Synonyms: BOC-D-His-(π-Bom)

Bachem A-1375.0001 MW 375.43 $C_{19}H_{25}N_3O_5$ Store at -15°C

Histidine — BOC-D-His-3-Me

Synonyms: BOC-D-His-(π-Me)

Bachem A-3020.0250 MW 269.3 $C_{12}H_{19}N_3O_4$ Store at RT

Histidine — BOC-D-His-BOC

American Peptide BDHIS40 MW 355.3

Histidine — BOC-D-His-BOC Benzene

Bachem A-3495.0001 MW 433.51 $C_{16}H_{25}N_3O_6 \cdot C_6H_6$ Store at -15°C

Histidine — BOC-D-His-Bzl

Bachem A-1325.0001 MW 345.4 $C_{18}H_{23}N_3O_4$ Store at RT

Histidine — BOC-D-His-Dnp IpOH

Bachem A-3955.0001 MW 481.46 $C_{17}H_{19}N_5O_8 \cdot C_3H_8O$ Store at -15°C

Neosystem BA00907 MW 481.5

Histidine — BOC-D-His-Tosyl

Bachem A-2330.0001 MW 409.46 $C_{18}H_{23}N_3O_6S$ Store at -15°C

Histidine — BOC-D-His-Tosyl Dicyclohexylamine

Synonyms: N^α-BOC- N^{im}-Tosyl-D-His

Peptides International BDH-2605 MW 590.78 >98% (HPLC); white crystalline powder

Histidine — BOC-His

Bachem A-1765.0001 MW 255.27 $C_{11}H_{17}N_3O_4$ Store at RT

USBio B2315 MW 255.3 $C_{11}H_{17}N_3O_4$ ≥99%

Histidine — BOC-His Hydrazide

Bachem A-1780.0001 MW 269.3 $C_{11}H_{19}N_5O_3$ Store at -15°C

Histidine — BOC-His-(Bom)

Synonyms: N^α-BOC-N^π-Bom-L-His

American Peptide BLHIS30 MW 375.0

Peptides International BLH-2138 MW 375.42 >98% (HPLC); white crystalline powder

USBio B2317 MW 375.4 $C_{19}H_{25}N_3O_5$ ≥99%

Histidine — BOC-His-(Bom)-O-Resin

American Peptide RMRB190 1% DVB cross-linked: 100-200 mesh

Histidine — BOC-His-1-Me

Synonyms: BOC-His-(τ-Me)

Bachem A-2560.0250 MW 269.3 $C_{12}H_{19}N_3O_4$ Store at -15°C

Histidine — BOC-His-3-Bom

Synonyms: BOC-His-(π-Bom)

Bachem A-1370.0001 MW 375.43 $C_{19}H_{25}N_3O_5$ Store at -15°C

Histidine — BOC-His-3-Bom-® (200-400 mesh)

Synonyms: BOC-His-(π-Bom)-®

Bachem D-1470.0001 Store at 2-8°C

Histidine — BOC-His-3-Bom-OMe Hydrochloride

Synonyms: BOC-His-(π-Bom)-OMe

Bachem A-1380.0001 MW 425.91 $C_{20}H_{27}N_3O_5 \cdot HCl$ Store at -15°C

Histidine — BOC-His-3-Bom-OSu

Synonyms: BOC-His-(π-Bom)-OSu

Bachem A-2775.0001 MW 427.5 $C_{23}H_{28}N_4O_7$ Store at -15°C

Histidine — BOC-His-3-Me

Synonyms: BOC-His-(π-Me)

Bachem A-2565.0250 MW 269.3 $C_{12}H_{19}N_3O_4$ Store at -15°C

Histidine — BOC-His-BOC

American Peptide BLHIS20 MW 355.3

USBio B2316 MW 355.4 $C_{16}H_{25}N_3O_6$ ≥99%

Histidine — BOC-His-BOC benzene

Bachem A-1010.0001 MW 433.51 $C_{16}H_{25}N_3O_6 \cdot C_6H_6$ Store at -15°C

Histidine — BOC-His-BOC-OSu

Bachem A-1015.0001 MW 452.46 $C_{20}H_{28}N_4O_8$ Store at -15°C

Histidine — BOC-His-Bzl

Bachem A-1320.0001 MW 345.4 $C_{18}H_{23}N_3O_4$ Store at RT

Histidine — BOC-His-Dnp IpOH

Bachem A-1595.0005 MW 481.46 $C_{17}H_{19}N_5O_8 \cdot C_3H_8O$ Store at -15°C

Histidine — BOC-His-Dnp-IpOH

USBio B2318 MW 481.5 $C_{17}H_{19}N_3O_8 \cdot C_3H_8O$ ≥99%

Histidine — BOC-His-Mts Dicyclohexylamine

Bachem A-2510.0001 MW 618.84 $C_{20}H_{27}N_3O_6S \cdot C_{12}H_{23}N$ Store at -15°C

Histidine — BOC-His-OMe

Bachem A-1775.0005 MW 269.3 $C_{12}H_{19}N_3O_4$ Store at -15°C

Histidine — BOC-His-Tosyl

Synonyms: N^α-BOC-N^τ-Tosyl-L-His

American Peptide BLHIS35 MW 409.5

Bachem A-2325.0001 MW 409.46 $C_{18}H_{23}N_3O_6S$ Store at -15°C

Peptides International BLH-2109 MW 409.46 >98% (HPLC); white crystalline powder; keep dry in freezer

USBio B2319 MW 409.5 $C_{18}H_{23}N_3O_6S$ ≥99%

Histidine — BOC-His-Tosyl Dicyclohexylamine

Bachem A-2335.0001 MW 590.78 $C_{18}H_{23}N_3O_6S \cdot C_{12}H_{23}N$ Store at 2-8°C

Histidine — BOC-His-Tosyl-PAM-® (200-400 mesh)

Bachem D-1255.0001 Store at 2-8°C

Histidine — BOC-His-Z

Bachem A-1460.0005 MW 389.41 $C_{19}H_{23}N_3O_6$ Store at -15°C

Histidine — BOC-L-His

Neosystem BA00901 MW 255.3

Histidine — BOC-L-His-BOC

Neosystem BA00906 MW 355.3

Histidine — BOC-L-His-Dnp IpOH

Neosystem BA00908 MW 481.5

Histidine — BOC-L-His-Tosyl Dicyclohexylamine

Neosystem BA00911 MW 590.8

Histidine — BOC-L-His-Tosyl-MBHA Resin

Neosystem RN00911

Histidine — BOC-Me-His

USBio B2349 MW 269.3 $C_{12}H_{19}N_3O_4$ ≥98%

Histidine — Bsmoc-His-Trt

Synonyms: N^α-Bsmoc-N^{im}-Trt-L-His

Peptides International BLH-5626-PI MW 619.70 >98% (HPLC); white to off-white powder | LA Carpino, et al, *JACS*, 119:9915, 1997

Histidine — Bz-DL-His

ICN 100134 MW 235.3 $C_{13}H_{17}NO_3$ Crystalline

Histidine — Bz-His

Bachem E-1470.0005 MW 259.27 $C_{13}H_{13}N_3O_3$ Store at RT

Histidine — Bz-His-OMe

Bachem E-1475.0005 MW 273.29 $C_{14}H_{15}N_3O_3$ Store at RT

Histidine — Bzl-His

Bachem E-1555.0001 MW 245.28 $C_{13}H_{15}N_3O_2$ Store at RT

Histidine — Bzl-His-Bzl

Bachem E-1535.0001 MW 335.41 $C_{20}H_{21}N_3O_2$ Store at RT

Histidine — Bzl-His-OMe Dihydrochloride

Bachem E-3200.0001 MW 332.23 $C_{14}H_{17}N_3O_2 \cdot 2HCl$ Store at 2-8°C

Histidine — CBZ-His

USBio C2098-86 MW 289.3 $C_{14}H_{15}N_3O_4$ ≥99%

Histidine — CBZ-His Hydrazide

USBio C2098-88 MW 303.3 $C_{14}H_{17}N_5O_3$ ≥99%

Histidine — CBZ-His-Bzl

USBio C2098-87 MW 379.4 $C_{21}H_{21}N_3O_4$ ≥99%

Histidine — CBZ-His-Z

USBio C2098-89 MW 423.4 $C_{22}H_{20}N_3O_6$ ≥99%

Histidine — D-(-)-(α-Hydrazino)-His Hydrochloride

ICN 154153 MW 206.6 $C_6H_{10}N_4O_2 \cdot HCl$ An inhibitor of histidine decarboxylase

Histidine — Deamino-His

Synonyms: 3-(Imidazol-4-yl)-Propionic Acid; Dihydrourocanic Acid

Bachem F-3185.0001 MW 140.14 $C_6H_8N_2O_2$ Store at -15°C

Histidine — D-His

Synonyms: (R)-2-Amino-3-(4-Imidazolyl)-Propionic Acid

Bachem F-2330.0005 MW 155.16 $C_6H_9N_3O_2$ Store at RT

Fluka 53321 MW 155.16 $C_6H_9N_3O_2$ ~99.0% (titration); mp: 280-290°C

Peptides International ADH-2805 MW 155.16 >99.9% optical purity by RP-HPLC; white crystalline powder

Rexim MW 155.2 $C_6H_9N_3O_2$

USBio H5100-10 MW 155.2 $C_6H_9N_3O_2$ ≥99%

Histidine — D-His Free Base

Synonyms: D-α-Amino-β-Imidazolepropionic Acid

ICN 151263 MW 155.2 $C_6H_9N_3O_2$

Sigma H 3751 FW 155.2 $C_6H_9N_3O_2$ ≥98% (TLC) | Inactive isomer

Histidine — D-His Hydrochloride Hydrate

Synonyms: D-α-Amino-β-Imidazolepropionic Acid

Alexis 101-058 MW 209.7 $C_6H_9N_3O_2 \cdot HCl \cdot H_2O$ White solid; soluble in H_2O

Fluka 53380 MW 209.63 $C_6H_9N_3O_2 \cdot HCl \cdot H_2O$ ≥99.0% (titration); mp: 180°C

ICN 101949 MW 209.6 $C_6H_9N_3O_2 \cdot HCl \cdot H_2O$ Crystalline

Sigma H 7625 FW 209.6 $C_6H_9N_3O_2 \cdot HCl \cdot H_2O$ ≥98% (TLC) | Inactive isomer

USBio H5105-10 MW 209.6 $C_6H_9N_3O_2 \cdot HCl \cdot H_2O$ ≥99%

Histidine — D-His-1-Me

Synonyms: D-His-(τ-Me)

Bachem F-2595.0250 MW 169.18 $C_7H_{11}N_3O_2$ Store at RT

Histidine — D-His-1-Trt

Synonyms: D-His-(τ-Trt)

Bachem F-2860.0001 MW 397.48 $C_{25}H_{23}N_3O_2$ Store at -15°C

Histidine — D-His-1-Trt-OMe Hydrochloride

Synonyms: D-His-(τ-Trt)-OMe

Bachem F-3020.0001 MW 447.97 $C_{26}H_{25}N_3O_2 \cdot HCl$ Store at 2-8°C

Histidine — D-His-3-Me

Synonyms: D-His-(π-Me)

Bachem F-2600.0250 MW 169.18 $C_7H_{11}N_3O_2$ Store at RT

Histidine — D-His-Bzl

Bachem F-2695.0001 MW 245.28 $C_{13}H_{15}N_3O_2$ Store at RT

Histidine — D-His-OMe Dihydrochloride

Bachem F-1600.0001 MW 242.11 $C_7H_{11}N_3O_2 \cdot 2HCl$ Store at 2-8°C

Histidine — Dinitropyridyl-L-His

ICN 101956 MW 322.2 $C_{11}H_{10}N_6O_6$

Histidine — DL-His

Synonyms: DL-α-Amino-β-imidazole Propionic Acid

USBio H5100-15   MW 155.2 $C_6H_9N_3O_2$ ≥99%

Histidine — DL-His Free Base

Synonyms: DL-α-Amino-β-Imidazolepropionic Acid

ICN 101950   MW 155.2 $C_6H_9N_3O_2$ ≥99%; crystalline

Sigma H 7750   FW 155.2 $C_6H_9N_3O_2$ ≥99% (TLC)

Histidine — DL-His Hydrochloride

Synonyms: (±)-2-Amino-3-(4-Imidazolyl)-Propionic Acid

Fluka 53330   MW 155.16 $C_6H_9N_3O_2$ ≥99.0% (titration); mp: 280-285°C; ~1% DL-His

Histidine — DL-His Hydrochloride Hydrate

Synonyms: DL-α-Amino-β-Imidazolepropionic Acid

Fluka 53390   MW 209.63 $C_6H_9N_3O_2 \cdot HCl \cdot H_2O$ ≥99.0% (titration)

ICN 101951   MW 209.6 $C_6H_9N_3O_2 \cdot HCl \cdot H_2O$ 99%; crystalline

Sigma H 7875   FW 209.6 $C_6H_9N_3O_2 \cdot HCl \cdot H_2O$ ≥99% (TLC) | Also available as part of a kit

Histidine — FMOC-D-His

Bachem B-2295.0001   MW 377.4 $C_{21}H_{19}N_3O_4$ Store at 2-8°C

Histidine — FMOC-D-His-1-Trt

Synonyms: FMOC-D-His-(τ-Trt)

Bachem B-1710.0001   MW 619.72 $C_{40}H_{33}N_3O_4$ Store at -15°C

Histidine — FMOC-D-His-1-Trt-SASRIN™-® (200-400 mesh)

Synonyms: FMOC-D-His-(τ-Trt)-SASRIN™-® (200-400 mesh)

Bachem D-1980.0001   Store at 2-8°C

Histidine — FMOC-D-His-BOC

Bachem B-1825.0001   MW 477.5 $C_{26}H_{27}N_3O_6$ Store at -15°C

USBio F5383   MW 477.5 $C_{26}H_{21}N_3O_6$ ≥99%

Histidine — FMOC-D-His-FMOC

American Peptide FHIS120   MW 599.6

Bachem B-1575.0001   MW 599.64 $C_{36}H_{29}N_3O_6$ Store at -15°C

Histidine — FMOC-D-His-Mtt

Bachem B-2585.0001   MW 633.75 $C_{41}H_{35}N_3O_4$ Store at -15°C

Histidine — FMOC-D-His-Trt

Synonyms: N^{α}-FMOC- N^{im}-Trt-D-His

Neosystem FA00903   MW 619.7

Peptides International FDH-1812-PI   MW 619.72   >98% (HPLC); white crystalline powder

USBio F5388   MW 619.7 $C_{40}H_{33}N_3O_4$ ≥99%

Histidine — FMOC-His

Bachem B-2600.0001   MW 377.4 $C_{21}H_{19}N_3O_4$ Store at 2-8°C

Histidine — FMOC-His-(Bum)

USBio F5385   MW 463.5 $C_{26}H_{20}N_3O_5$ ≥99%

Histidine — FMOC-His-(Bz)

USBio F5386   MW 467.5 $C_{28}H_{25}N_3O_4$ ≥99%

Histidine — FMOC-His-1-Me

Bachem B-3375.0250   MW 391.43 $C_{22}H_{21}N_3O_4$ Store at 2-8°C

Histidine — FMOC-His-1-Trt

Synonyms: FMOC-His-(τ-Trt)

Bachem B-1570.0001   MW 619.72 $C_{40}H_{33}N_3O_4$ Store at -15°C

Histidine — FMOC-His-1-Trt-® (200-400 mesh)

Synonyms: FMOC-His-(τ-Trt)-® (200-400 mesh)

Bachem D-1705.0001   Store at RT

Histidine — FMOC-His-1-Trt-OPfp

Synonyms: FMOC-His-(τ-Trt)-OPfp

Bachem B-1650.0001   MW 785.77 $C_{46}H_{32}F_5N_3O_4$ Store at -15°C

Histidine — FMOC-His-1-Trt-SASRIN™-® (200-400 mesh)

Synonyms: FMOC-His-(τ-Trt)-SASRIN™-® (200-400 mesh)

Bachem D-1530.0001   Store at 2-8°C

Histidine — FMOC-His-3-Bom

Synonyms: FMOC-His-(π-Bom)

Bachem B-2805.0001   MW 497.55 $C_{29}H_{27}N_3O_5$ Store at -15°C

Histidine — FMOC-His-3-Me

Bachem B-3365.0250   MW 391.43 $C_{22}H_{21}N_3O_4$ Store at -15°C

Histidine — FMOC-His-BOC

Bachem B-1830.0001   MW 477.5 $C_{26}H_{27}N_3O_6$ Store at -15°C

Histidine — FMOC-His-BOC-OPfp

Bachem B-2000.0001   MW 659.57 $C_{32}H_{26}N_3O_7F_5$ Store at -15°C

USBio F5384   MW 643 $C_{32}H_{26}N_3O_6F_5$ ≥99%

Histidine — FMOC-His-Bzl

Synonyms: N^{α}-Fluorenylmethoxycarbonyl-N^{im}-Bzl-L-His

Bachem B-1540.0001   MW 467.5 $C_{28}H_{25}N_3O_4$ Store at RT

Peptides International FLH-1742-PI   MW 467.53   >98% (TLC); white powder

Histidine — FMOC-His-FMOC

Synonyms: N^{α},N^{im}-Di-FMOC-L-His

American Peptide FHIS115   MW 599.6

Bachem B-1000.0001   MW 599.64 $C_{36}H_{29}N_3O_6$ Store at -15°C

Peptides International FLH-1702-PI   MW 599.64   >98% (HPLC); white crystalline powder

USBio F5387   MW 599.6 $C_{36}H_{29}N_3O_6$ ≥99%

Histidine — FMOC-His-FMOC-® (200-400 mesh)

Bachem D-1085.0001   Store at 2-8°C

Histidine — FMOC-His-FMOC-OPfp

Bachem B-1140.0001   MW 765.7 $C_{42}H_{28}N_3O_6F_5$ Store at -15°C

Histidine — FMOC-His-Mtt

Bachem B-2190.0001   MW 633.75 $C_{41}H_{35}N_3O_4$ Store at -15°C

Histidine — FMOC-His-Trt

Synonyms: N^α-FMOC-N^{im}-Trt-L-His

American Peptide FHIS125   MW 619.7

Peptides International FLH-1740-PI   MW 619.72   >98% (HPLC); white crystalline powder

USBio F5389   MW 619.7 $C_{40}H_{33}N_3O_4$ ≥99%

Histidine — FMOC-His-Trt-OPfp

USBio F5390   MW 786 $C_{46}H_{33}N_3O_4F_5$ ≥99%

Histidine — FMOC-L-His-BOC-Wang Resin

American Peptide RFWAN45

Histidine — FMOC-L-His-Trt

Neosystem FA00904   MW 619.7

Histidine — FMOC-L-His-Trt-Wang Resin

Neosystem RW00905

Histidine — His

Bachem E-2010.0025   MW 155.16 $C_6H_9N_3O_2$ Store at RT

Histidine — His Amide Dihydrochloride

Bachem E-2015.0001   MW 227.09 $C_6H_{10}N_4O \cdot 2HCl$ Store at RT

Histidine — His-1-Me Hydrochloride

Synonyms: His-(τ-Me)

Bachem F-2480.0001   MW 205.64 $C_7H_{11}N_3O_2 \cdot HCl$ Store at RT

Histidine — His-1-Me-OMe Hydrochloride

Synonyms: His-(τ-Me)-OMe

Bachem E-2795.0001   MW 219.67 $C_8H_{13}N_3O_2 \cdot HCl$ Store at 2-8°C

Histidine — His-1-Trt

Synonyms: His-(τ-Trt)

Bachem E-2980.0001   MW 397.48 $C_{25}H_{23}N_3O_2$ Store at 2-8°C

Histidine — His-1-Trt-OMe Hydrochloride

Synonyms: His-(τ-Trt)-OMe

Bachem E-2985.0001   MW 447.97 $C_{26}H_{25}N_3O_2 \cdot HCl$ Store at -15°C

Histidine — His-1-Trt-OtBu

Synonyms: His-(τ-Trt)-OtBu

Bachem E-3600.0001   MW 453.58 $C_{29}H_{31}N_3O_2$ Store at -15°C

Histidine — His-3-Me

Synonyms: His-(π-Me)

Bachem E-2845.0250   MW 169.18 $C_7H_{11}N_3O_2$ Store at RT

Histidine — His-AMC

Bachem I-1485.0050   MW 312.33 $C_{16}H_{16}N_4O_3$ Store at -15°C

Histidine — His-Bzl

Bachem E-2435.0001   MW 245.28 $C_{13}H_{15}N_3O_2$ Store at RT

Histidine — His-Bzl-OMe Dihydrochloride

Bachem E-1560.0001   MW 332.23 $C_{14}H_{17}N_3O_2 \cdot 2HCl$ Store at RT

Histidine — His-Dnp-OMe Dihydrochloride Ethylacetate

Bachem E-1820.0001   MW 452.25 $C_{13}H_{13}N_5O_6 \cdot \frac{1}{2}C_4H_8O_2 \cdot$ 2HCl   Store at -15°C

Histidine — His-Mtt

Bachem E-3005.0001   MW 411.5 $C_{26}H_{25}N_3O_2$ Store at -15°C

Histidine — His-OEt Dihydrochloride

Bachem E-2630.0025   MW 256.13 $C_8H_{13}N_3O_2 \cdot 2HCl$ Store at -15°C

Histidine — His-OMe Dihydrochloride

Synonyms: L-His-OMe

Bachem E-2020.0025   MW 242.11 $C_7H_{11}N_3O_2 \cdot 2HCl$ Store at RT

Peptides International EMH-2038-PI   MW 249.10   >98% (TLC); white crystalline powder

Histidine — His-pNA

Bachem L-1785.0250   MW 275.27 $C_{12}H_{13}N_5O_3$ Store at -15°C

Histidine — His-βNA

Bachem K-1350.0250   MW 280.33 $C_{16}H_{16}N_4O$ Store at RT

Histidine — im-Bzl-L-His

Sigma B 8125   FW 245.3 $C_{13}H_{15}N_3O_2$

Histidine — L-His

Synonyms: (S)-2-Amino-3-(4-Imidazolyl)-Propionic Acid; (S)-α-Amino-1H-Imidazole-4-Propanoic Acid; L-2-Amino-3-(4'-Imidazolyl)-Propionic Acid; (S)-4-(2-Amino-2-Carboxyethyl)-Imidazole

Fluka 53319   MW 155.16 $C_6H_9N_3O_2$ ≥99.5% (titration); ≤0.1% residue on ignition, loss on drying; ≤0.3% foreign AA; ≤0.00001% As; ≤0.0005% Al, Ba, Bi, Cd, Co, Cr, Cu, Fe, Li, Mg, Mn, Mo, Ni, Pb, Sr, Zn; ≤0.001% Ca; ≤0.005% Na, K; ≤0.05% NH_4^+; ≤0.01% Cl, SO_4; mp: 285°C | Nguyen, NY & Chrambach, A, *Anal Biochem*, 94: 202, 1979; Brawn, K & Fridovich, I, *Acta Physiol Scand Suppl*, 492: 9, 1980; *CRC Handbook of Microbiology*, Laskin, AI & Lachevalier, HA, eds, 4: 1, 1974, CRC Press, Cleveland, Ohio

Fluka 53320   MW 155.16 $C_6H_9N_3O_2$ ≥99.0% (titration); ≤0.1% residue on ignition; ≤0.5% loss on drying; ≤0.3% foreign AA; ≤0.0005% Cd, Co, Cr, Cu, Fe, Mg, Mn, Ni, Pb, Zn; ≤0.001% Ca; ≤0.005% Na, K; ≤0.05% NH_4^+; ≤0.01% Cl, SO_4; mp: 285°C

Neosystem AA00901   MW 155.2

Rexim   MW 155.2 $C_6H_9N_3O_2$ White crystals or crystalline powder

Sigma H 0767   Matrix: cross-linked 4% beaded agarose; activation: epoxy; attachment: amino; spacer: 12 atoms; ligand immobilized: 1-2 μmoles/mL; form: suspension in 2.0 M NaCl containing 0.02% thimerosal

Sigma H 3257   Matrix: cross-linked 4% beaded agarose; activation: cyanogen bromide; attachment: amino; spacer: 1 atom; ligand immobilized: 2-10 μmoles/mL; form: suspension in 2.0 M NaCl containing 0.02% thimerosal

USBio H5100   MW 155.2 $C_9H_9N_3O_2$ ≥98.5%; white, crystalline powder; specific rotation(C=11, 6 N HCl): +12.0° to +12.8°; pH: 7.0-8.5; heavy metals (Pb): ≤0.001%; endotoxin: ≤2 EU/mg; cell culture tested: murine myeloma SP2/0-Ag14

Histidine — L-His (2,5-³H)

Amersham TRK199   Aqueous, 2% EtOH, sterilized; 1.48-2.59 TBq/mmol, 40-70 Ci/mmol; 37 MBq/mL, 1 mCi/mL

ICN 20071E   MW 155.2 $C_3H_3N_2CH_2CH(NH_2)COOH$ 30-60 Ci/mmol, 1.11-2.22 TBq/mmol; sterile EtOH:H_2O (2:98); purity analysis by HPLC

Sigma H 8144   FW 155.2 $C_6H_9N_3O_2$ 30-60 Ci/mmol; radiochemical purity ≥95% (HPLC); EtOH solution containing 2% EtOH in Combi-vial

Histidine — L-His (Carboxyl-¹⁴C)

ARC ARC-503   MW 155.2 $C_3H_3N_2CH_2CH(NH_2)$**COOH**   50-60 mCi/mmol; 1.85-2.22 GBq/mmol; in 0.01 N HCl

Histidine — L-His (Ring 2,5-³H)

ARC ART-234   MW 155.2 $C_3H_3N_2CH_2CH(NH_2)COOH$ 40-60 Ci/mmol; 1.48-2.22 TBq/mmol; in EtOH: H_2O (1:1)

Histidine — L-His (Ring 2-¹⁴C) Hydrochloride

Sigma H 4529   FW 191.6 $C_6H_9N_3O_2 \cdot HCl$ 40-60 mCi/mmol; aqueous solution containing 2% EtOH in serum bottle

Histidine — L-His (U-¹⁴C)

Amersham CFB140   Aqueous, 2% EtOH, sterilized; >11 GBq/mmol, >300 mCi/mmol; 1.85 MBq/mL, 50 μCi/mL

ARC ARC-671   MW 155.2 $C_3H_3N_2CH_2CH(NH_2)$**COOH**   200-300 mCi/mmol; 7.4-11.1 GBq/mmol; in EtOH: H_2O (2:98)

ICN 11081E   MW 155.2   270-330 mCi/mmol, 10-12.2 GBq/mmol; sterile EtOH:H_2O (2:98); purity analysis by HPLC

Histidine — L-His Dihydrochloride

Fluka 53340   MW 228.08 $C_6H_9N_3O_2 \cdot 2HCl$ ≥99.0% (titration); ≤0.1% residue on ignition; ≤0.00001% As; ≤0.0005% Cd, Co, Cu, Fe, Ni, Pb, Zn; ≤0.001% Ca; ≤0.05% NH_4^+; ≤0.005% SO_4; mp: 240-245°C

ICN 101953   MW 228.1 $C_6H_9N_3O_2 \cdot 2HCl$ Crystalline

Histidine — L-His Free Base

Synonyms: L-α-Amino-β-Imidazolepropionic Acid; Histidine Decarboxylase Substrate

Amersham US17070   98.5-101.5% dsb ≤0.0015% heavy metals (Pb)

ICN 101954   MW 155.2 $C_6H_9N_3O_2$ ≥99%; crystalline

ICN 194684   MW 155.2 $C_6H_9N_3O_2$ ≥99% | Cell culture reagent

Sigma H 8000   FW 155.2 $C_6H_9N_3O_2$ ≥99% (TLC) | Precursor of histamine by histidine decarboxylase

Sigma H 8776   FW 155.2 $C_6H_9N_3O_2$ SigmaUltra; >99% (TLC); solubility (1 M in 0.5 M HCl, 20°C): complete, colorless; insoluble matter: <0.1%; NH_4^+, SO_4, Cl: <0.05%; Mg, Al, Ca, Cu, P, Zn, Fe: <0.0005%; Na, K: <0.005%; Pb: <0.001% | Precursor of histamine by histidine decarboxylase

Sigma H 9386   FW 155.2 $C_6H_9N_3O_2$ Crystalline | Cell culture tested; insect cell culture tested

Histidine — L-His Hydrate Monochloride

Amersham US17080   ≥98.0%; ≤10ppm heavy metals (Pb)

Histidine — L-His Hydrochloride Hydrate

Synonyms: DL-α-Amino-β-Imidazolepropionic Acid; Histidine Decarboxylase Substrate

Alexis 101-057   MW 209.7 $C_6H_9N_3O_2 \cdot HCl \cdot H_2O$ White solid; soluble in H_2O

Fluka 53369   MW 209.63 $C_6H_9N_3O_2 \cdot HCl \cdot H_2O$ ≥99.5% (titration); ≤0.05% residue on ignition; ≤0.3% foreign AA; ≤0.00001% As; ≤0.0005% Al, Ba, Bi, Cd, Co, Cr, Cu, Fe, Li, Mg, Mn, Mo, Ni, Pb, Sr, Zn; ≤0.001% Ca; ≤0.005% Na, K; ≤0.05% NH_4^+; ≤0.01% SO_4

Fluka 53370   MW 209.63 $C_6H_9N_3O_2 \cdot HCl \cdot H_2O$ ≥99.0% (titration); ≤0.1% residue on ignition; ≤0.3% foreign AA; ≤0.0005% Cd, Co, Cr, Cu, Fe, Mg, Mn, Ni, Pb, Zn; ≤0.001% Ca; ≤0.05% NH_4^+; ≤0.005% K, Na, SO_4

ICN 101957   MW 209.6 $C_6H_9N_3O_2 \cdot HCl \cdot H_2O$ Crystalline

ICN 194685   MW 209.6 $C_6H_9N_3O_2 \cdot HCl \cdot H_2O$ Crystalline | Cell culture reagent

Peptides International ALH-2710   MW 209.64   >99.9% optical purity by RP-HPLC; white crystalline powder

Rexim   MW 209.6 $C_6H_{12}ClN_3O_3$ White crystals or crystalline powder

Sigma H 8125   FW 209.6 $C_6H_9N_3O_2 \cdot HCl \cdot H_2O$ ≥98% (TLC)) | Precursor of histamine by histidine decarboxylase

Sigma H 9511   FW 209.6 $C_6H_9N_3O_2 \cdot HCl \cdot H_2O$ Crystalline | Cell culture tested; insect cell culture tested

USBio H5105   MW 209.6 $C_6H_9N_3O_2 \cdot HCl \cdot H_2O$ ≥98.5%; white, crystalline powder; specific rotation(C=11, 6 N HCl): +8.9° to +9.6°; pH: 3.5-4.5; heavy metals (Pb): ≤0.001%; endotoxin: ≤2 EU/mg; cell culture tested: murine myeloma SP2/0-Ag14

Histidine — L-His Hydroxamate

Sigma H 8500   FW 170.2 $C_6H_{10}N_4O_2$ >90%

Histidine — L-His Monochloride

Amersham US17085   98.5-101.0%; ≤10ppm heavy metals (Pb); histamine-free

Histidine — L-His β-Naphthylamide

Sigma H 6759   FW 280.3 $C_{16}H_{16}N_4O$ White to faint yellow powder

Histidine — L-His-AMC

Sigma H 9906   FW 312.3 $C_{16}H_{16}N_4O_3$

Histidine — L-His-OBzl p-Tosyl

Sigma H 8140   FW 589.7 $C_{13}H_{15}N_3O_2 \cdot 2C_7H_8O_3S$ ≥95%; | Jones, et al, *Syn Comm,* 16: 1515, 1986; Akabori, et al, *Bull Chem Soc Jap,* 31: 784, 1958

Histidine — L-His-OMe Dihydrochloride

Fluka 53360   MW 242.11 $C_7H_{11}N_3O_2 \cdot 2HCl$ ≥99.0% (titration); mp: 200°C

ICN 101960   MW 242.1 $C_7H_{11}N_3O_2 \cdot 2HCl$ ≥99%; crystalline

Neosystem AA00903   MW 242.2

Sigma H 8625   FW 242.1 $C_7H_{11}N_3O_2 \cdot 2HCl$

Histidine — L-Histidinamide Dihydrochloride

Sigma H 5016   FW 227.1 $C_6H_{10}N_4O \cdot 2HCl$

Histidine — L-Histidinol Dihydrochloride

Bachem F-1605.0001   MW 214.09 $C_6H_{11}N_3O \cdot 2HCl$ Store at 2-8°C

Histidine — L-His-Trt

Neosystem AA00904   MW 397.5

Histidine — L-ThiolHis
ICN 103040   MW 187.2 $C_6H_9N_3O_2S$

Histidine — N,N'-Di(2,4-Dnp)-L-His
Sigma D 9629   FW 487.3 $C_{18}H_{13}N_7O_{102}$

Histidine — N-1-Trt-Deamino-His
Synonyms: 3-(N-1-Trt-Imidazol-4-yl)-Propionic Acid; N-1-Trt-Dihydrourocanic Acid
Bachem F-3190.0001   MW 382.46 $C_{25}H_{22}N_2O_2$ Store at -15°C

Histidine — N-Ac-L-His
ICN 100092   MW 197.2 $C_8H_{11}N_3O_3$ Crystalline

Histidine — N-Bzl-(im)-Bzl-L-His
Synonyms: N,1-Dibenzyl-L-His
ICN 154840   MW 335.4 $C_{20}H_{21}N_3O_2$ Crystalline

Histidine — N-Bz-L-His
ICN 100130   MW 259.3 $C_{13}H_{13}N_3O_3$ Crystalline

Histidine — N-Carbamyl-L-His Hydrochloride
Sigma C 4625   FW 234.6 $C_7H_{10}N_4O_3 \cdot HCl$ ~80% (TLC)

Histidine — N-CBZ-DL-His
ICN 100218   MW 289.3 $C_{14}H_{15}N_3O_4$ Crystalline

Histidine — N-CBZ-L-His
ICN 100219   MW 289.3 $C_{14}H_{15}N_3O_4$ Crystalline
Sigma C 9895   FW 289.3 $C_{14}H_{15}N_3O_4$

Histidine — N-For-L-His
ICN 151165

Histidine — N-Me-D-His Hydrochloride
Bachem F-2260.0250   MW 205.64 $C_7H_{11}N_3O_2 \cdot HCl$ Store at RT

Histidine — N-Me-His Hydrochloride
Bachem E-2135.0250   MW 205.64 $C_7H_{11}N_3O_2 \cdot HCl$ Store at RT

Histidine — N-Me-His-OMe Hydrochloride
Bachem E-3300.0001   MW 219.67 $C_8H_{13}N_3O_2 \cdot HCl$ Store at -15°C

Histidine — N-t-BOC-(im)-Tosyl-L-His Free Acid
Sigma B 6505   FW 409.5 $C_{18}H_{23}N_3O_6S$

Histidine — N-t-BOC-L-His Hydrazide
Sigma B 7754   FW 369.3 $C_{11}H_{19}N_5O_3$

Histidine — N$^\alpha$-(3-Nitro-2-Pyridinesulfenyl)-N^{im}-Tosyl-L-His
Sigma N 5640   FW 463.5 $C_{18}H_{17}N_5O_6S_2$

Histidine — N$^\alpha$,N^{im}-Di-BOC-L-His Dicyclohexylamine
Fluka 15427   MW 536.72 $C_{16}H_{25}N_3O_6 \cdot C_{12}H_{23}N$ ≥99% (titration); mp: 160-165°C

Histidine — N$^\alpha$,N^{im}-Di-FMOC-L-His
Fluka 47592   MW 599.64 $C_{36}H_{29}N_3O_6$ ≥97% (TLC); mp: 160°C

Histidine — N$^\alpha$-Ac-L-His
Sigma A 8133   FW 197.2 $C_8H_{11}N_3O_3$

Histidine — N$^\alpha$-BOC-D-His
Fluka 15119   MW 255.27 $C_{11}H_{17}N_3O_4$ ≥98% (TLC)

Histidine — N$^\alpha$-BOC-L-His
Fluka 15426   MW 255.27 $C_{11}H_{17}N_3O_4$ ≥99% (titration); mp: 195°C

Histidine — N$^\alpha$-BOC-N^{im}-Benzyloxymethoxy-L-His Hydrate
Fluka 15391   MW 393.45 $C_{19}H_{25}N_3O_5 \cdot H_2O$ ≥98% (HPLC); mp: 122-125°C | Ideally protected derivative of L-His which overcomes the problems hitherto associated with His in peptide synthesis, such as racemization, side reactions & solubility; Brown, T et al, *J Chem Soc*, Perkin I 1553, 1982

Histidine — N$^\alpha$-BOC-N^{im}-Bom-D-His
Fluka 15077   MW 375.42 $C_{19}H_{25}N_3O_5$ ≥98% (TLC)

Histidine — N$^\alpha$-BOC-N^{im}-Bom-L-His
Fluka 14968   MW 375.42 $C_{19}H_{25}N_3O_5$ ≥98% (TLC)

Histidine — N$^\alpha$-BOC-N^{im}-Bzl-L-His
Fluka 15534   MW 345.4 $C_{18}H_{23}N_3O_4$ ≥98% (TLC); mp: 181-184°C

Histidine — N$^\alpha$-BOC-N^{im}-Dnp-L-His
Fluka 15287   MW 481.46 $C_{17}H_{19}N_5O_8 \cdot C_3H_8O$ mp: 98-100°C; ~10% IpOH

Histidine — N$^\alpha$-BOC-N^{im}-Mts-L-His Dicyclohexylamine
Fluka 15546   MW 618.83 $C_{20}H_{27}N_3O_6S \cdot C_{12}H_{23}N$ ≥98% (TLC)

Histidine — N$^\alpha$-BOC-N^{im}-Tosyl-L-His
Fluka 15559   MW 409.46 $C_{18}H_{23}N_3O_6S$ ≥98% (TLC); mp: 125°C

Histidine — N$^\alpha$-BOC-N^{im}-Tosyl-L-His Dicyclohexylamine
Synonyms: (R)-2-(BOC-Amino)-5-Phenyl-4-Pentenoic Acid
Fluka 15507   MW 590.78 $C_{18}H_{23}N_3O_6S \cdot C_{12}H_{23}N$ mp: 165°C

Histidine — N$^\alpha$-BOC-N^{im}-Trt-L-His
Fluka 15449   MW 497.59 $C_{30}H_{31}N_3O_4$ ≥98% (TLC); mp: 130°C

Histidine — N$^\alpha$-Bz-L-His
Sigma B 2254   FW 259.3 $C_{13}H_{13}N_3O_3$

Histidine — N$^\alpha$-Bz-L-His-OMe Dihydrochloride
Sigma B 1519   FW 332.2 $C_{14}H_{17}N_3O_2 \cdot 2HCl$

Histidine — N$^\alpha$-CBZ-N^{im}-Bzl-L-His
Sigma C 9783   FW 379.4 $C_{21}H_{21}N_3O_4$

Histidine — N$^\alpha$-FMOC-(im)-Bzl-L-His

ICN 151135	MW 467.5	C$_{28}$H$_{25}$N$_3$O$_4$
Sigma F 9257	FW 467.5	C$_{28}$H$_{25}$N$_3$O$_4$

Histidine — N$^\alpha$-FMOC-(im)-Trt-L-His

Sigma F 1518	FW 619.7	C$_{40}$H$_{33}$N$_3$O$_4$

Histidine — N$^\alpha$-FMOC-N^{im}-(t-Butoxymethyl)-L-His

Fluka 47566   MW 463.53   C$_{26}$H$_{29}$N$_3$O$_5$ ~95% (HPLC); mp: 180-190°C | Protected His resistant to hydrogenolysis; Colombo, R et al, *Chem Comm*, 292: 1994

Histidine — N$^\alpha$-FMOC-N^{im}-4-Methyltrityl-L-His

Fluka 47526   MW 633.75   C$_{41}$H$_{35}$N$_3$O$_4$ ≥98% (HPLC)

Histidine — N$^\alpha$-FMOC-N^{im}-BOC-L-His Cyclohexylamine Salt

Synonyms: N^{im}-BOC-N$^\alpha$-FMOC-L-His

Fluka 47515   MW 576.7   C$_{26}$H$_{27}$N$_3$O$_6$ · C$_6$H$_{13}$N   ≥98% (TLC)

Histidine — N$^\alpha$-FMOC-N^{im}-Bom-L-His

Fluka 47446   MW 497.55   C$_{29}$H$_{27}$N$_3$O$_5$ ≥98% (TLC)

Histidine — N$^\alpha$-FMOC-N^{im}-Bzl-L-His

Fluka 47591   MW 467.52   C$_{28}$H$_{25}$N$_3$O$_4$ ≥98% (HPLC); mp: 167-171°C

Histidine — N$^\alpha$-FMOC-N^{im}-Tosyl-L-His

Fluka 47535   MW 531.58   C$_{28}$H$_{25}$N$_3$O$_6$S   ≥98%; mp: 170-174°C

Histidine — N$^\alpha$-FMOC-N^{im}-Trt-L-His

Fluka 47639   MW 619.73   C$_{40}$H$_{33}$N$_3$O$_4$ ≥98% (HPLC); mp: 145°C

Histidine — N$^\alpha$-FMOC-N^{im}-Trt-L-His-OPfp

Fluka 47477   MW 785.77   C$_{46}$H$_{32}$F$_5$N$_3$O$_4$ ~80% (TLC); ~1 mol pentafluorophenol & ~1 mol cyclohexane

Sigma F 0898   FW 1054.0   1 mole each pentafluorophenol & cyclohexane from crystallization

Histidine — N$^\alpha$-FMOC-π-Bom-L-His

Sigma F 9393   FW 497.5   C$_{29}$H$_{27}$N$_3$O$_5$

Histidine — N$^\alpha$-N^{im}-Di-FMOC-L-His Resin Ester

Sigma F 1761   For peptide synthesis; Wang, SS, *J Am Chem Soc*, 95: 1328, 1973; Lu, G et al, *J Org Chem*, 46: 3433, 1981

Histidine — N$^\alpha$-N^{im}-Di-FMOC-L-His-OPfp

Fluka 47464   MW 765.69   C$_{42}$H$_{28}$F$_5$N$_3$O$_6$ ≥97% (titration)

Histidine — N$^\alpha$-t-BOC-(im)-Bzl-L-His

Sigma B 5876   FW 345.4   C$_{18}$H$_{23}$N$_3$O$_4$

Histidine — N$^\alpha$-t-BOC-L-His

ICN 150496	MW 255.3	C$_{11}$H$_{17}$N$_3$O$_4$ Crystalline
Sigma B 1895	FW 255.3	C$_{11}$H$_{17}$N$_3$O$_4$

Histidine — N$^\alpha$-t-BOC-N^{im}-Bzl-L-His

ICN 101068   MW 345.4   C$_{18}$H$_{23}$N$_3$O$_4$ Crystalline

Histidine — N$^\alpha$-t-BOC-N^{im}-Dnp-L-His

Sigma B 9774   FW 421.4   C$_{17}$H$_{19}$N$_5$O$_8$

Histidine — N$^\alpha$-t-BOC-N^{im}-Mts-L-His Dicyclohexylammonium Salt

Sigma B 7394   FW 618.8   C$_{20}$H$_{27}$N$_3$O$_6$S · C$_{12}$H$_{23}$N

Histidine — N$^\alpha$-t-BOC-N^{im}-Tosyl-L-His Dicyclohexylammonium Salt

ICN 150507   MW 590.8   C$_{18}$H$_{23}$N$_3$O$_6$S · C$_{12}$H$_{23}$N

Histidine — N$^\alpha$-Z-L-His

Fluka 96290   MW 289.29   C$_{14}$H$_{15}$N$_3$O$_4$ ≥99.0% (TLC); mp: 160°C

Histidine — N$^\alpha$-Z-N^{im}-Bzl-L-His

Fluka 96016   MW 379.42   C$_{21}$H$_{21}$N$_3$O$_4$ ≥98.0% (TLC)

Histidine — N$^\alpha$-Me-L-His

Synonyms: 3-Me-L-His; 3-(1-Methylimidazol-4-yl)-L-Ala

Fluka 67520   MW 169.18   C$_7$H$_{11}$N$_3$O$_2$ ~99.0%; mp: 240°C; ≤1% H$_2$O | Natural but non-proteinogenic AA; employed as index of muscle protein breakdown; Young, VR et al, *JBC*, 247: 293, 1972; Ballard, FJ & Tomas, FM, *Clin Sci*, 65: 209, 1983; Betto, P et al, *J Chromatogr*, 584: 256, 1992

Histidine — Pth-His

Sigma P 0752   FW 272.3   C$_{13}$H$_{12}$N$_4$OS   ~99% | Useful as standards in protein sequencing by the Edman degradation

Histidine — TentaGel S PHB-His-Trt-FMOC

Fluka 86392   0.20 mmol protected AA/g; ~90 μm particle size

Histidine — TentaGel S Trt-His-Trt-FMOC

Fluka 86423   0.18 mmol protected AA/g; ~90 μm particle size

Histidine — Z-D-His

Bachem C-1950.0001   MW 289.29   C$_{14}$H$_{15}$N$_3$O$_4$ Store at RT

Histidine — Z-His

Bachem C-1945.0005   MW 289.29   C$_{14}$H$_{15}$N$_3$O$_4$ Store at RT

Histidine — Z-His Amide

Bachem C-1955.0001   MW 288.31   C$_{14}$H$_{16}$N$_4$O$_3$ Store at RT

Histidine — Z-His Hydrazide

Bachem C-1960.0005   MW 303.32   C$_{14}$H$_{17}$N$_5$O$_3$ Store at -15°C

Histidine — Z-His-Bzl

Bachem C-1390.0001   MW 379.42   C$_{21}$H$_{21}$N$_3$O$_4$ Store at RT

Histidine — Z-His-OBzl

Bachem C-4140.0005   MW 379.42   C$_{21}$H$_{21}$N$_3$O$_4$ Store at -15°C

Histidine — Z-His-Z EtOH

Bachem C-2915.0005   MW 469.49   C$_{24}$H$_{27}$N$_3$O$_7$ Store at -15°C

Histidine — Z-His-Z-ONp

Bachem C-2920.0005   MW 544.52   C$_{28}$H$_{24}$N$_4$O$_8$ Store at -15°C

Histidine — α-Me-DL-His Dihydrochloride

Bachem F-1795.0250   MW 242.11 $C_7H_{11}N_3O_2 \cdot 2HCl$ Store at RT

ICN 100512   MW 242.1 $C_7H_{11}N_3O_2 \cdot 2HCl$ Histidine decarboxylase inhibitor; Robinson & Stepherd, *JCS*, 5038, 1961

Sigma M 8628   FW 242.1 $C_7H_{11}N_3O_2 \cdot 2HCl$

Histidine — γ-Aminobutyryl-L-His Free Base

Synonyms: L-Homocarnosine

Sigma H 4885   FW 240.3 $C_{10}H_{16}N_4O_3$

Histidine — γ-Aminobutyryl-L-His Sulfate Salt

Synonyms: L-Homocarnosine

Sigma H 0251   FW 338.3 $C_{10}H_{16}N_4O_3 \cdot H_2SO4$

Homoarginine — BOC-D-HarEt₂

Bachem A-3780.0050   MW 344.46 $C_{16}H_{32}N_4O_4$ Store at -15°C

Homoarginine — BOC-Har Hydrochloride

Bachem A-4040.0001   MW 324.81 $C_{12}H_{24}N_4O_4 \cdot HCl$ Store at -15°C

Homoarginine — BOC-HarEt₂

Bachem A-3775.0050   MW 344.46 $C_{16}H_{32}N_4O_4$ Store at -15°C

Homoarginine — BOC-Har-Nitro

Bachem A-3935.0001   MW 333.35 $C_{12}H_{23}N_5O_6$ Store at -15°C

Homoarginine — FMOC-Har-Pmc

Bachem B-3130.0001   MW 676.83 $C_{36}H_{44}N_4O_7S$ Store at -15°C

Homoarginine — Har

Synonyms: L-α-Amino-ε-Guanidinohexanoic Acid

Bachem F-2780.0005   MW 188.23 $C_7H_{16}N_4O_2$ Store at RT

Homoarginine — L-Har Hydrochloride

Synonyms: (S)-2-Amino-6-Guanidinohexanoic Acid; 2-Amino-6-Guanidinohexanoic Acid

Fluka 53460   MW 224.7 $C_7H_{16}N_4O_2 \cdot HCl$ ~99.0% (TLC)

ICN 101967 $C_7H_{16}N_4O_2 \cdot HCl$

Sigma H 1007 $C_7H_{16}N_4O_2 \cdot HCl$

Homoarginine — L-β-Har Hydrochloride

Synonyms: (S)-3-Amino-6-Guanidinohexanoic Acid

Fluka 03672   MW 224.7 $C_7H_{16}N_4O_2 \cdot HCl$ ≥98.0% (TLC)

Homoarginine — N'-Monomethyl-L-Har Monoacetate Salt

Synonyms: L-NMMHA; NMMHA

Alexis 106-011   MW 262.4 $C_8H_{18}N_4O_2 \cdot CH_3COOH$ White solid; soluble in H_2O

Calbiochem 475890   MW 262.3 $C_8H_{18}N_4O_2 \cdot CH_3CO_2H$ Single component purity (TLC); solid; soluble in H_2O | Novel modulator of NOS

Homocitrulline — BOC-D-Hci

Synonyms: BOC-D-Lys-Carbamoyl

Bachem A-2870.0001   MW 289.33 $C_{12}H_{23}N_3O_5$ Store at 2-8°C

Homocitrulline — BOC-Hci

Synonyms: BOC-Lys-Carbamoyl

Bachem A-3465.0001   MW 289.33 $C_{12}H_{23}N_3O_5$ Store at -15°C

Homocitrulline — D-Hci

Synonyms: D-Lys-Carbamoyl

Bachem F-2735.0001   MW 189.22 $C_7H_{15}N_3O_3$ Store at -15°C

Homocitrulline — FMOC-D-Hci

Synonyms: FMOC-D-Lys-Carbamoyl

Bachem B-2390.0001   MW 411.46 $C_{22}H_{25}N_3O_5$ Store at RT

Homocitrulline — FMOC-Hci

Synonyms: FMOC-Lys-Carbamoyl

Bachem B-2250.0001   MW 411.46 $C_{22}H_{25}N_3O_5$ Store at RT

Homocitrulline — Hci

Synonyms: Lys-Carbamoyl

Bachem F-2995.0001   MW 189.22 $C_7H_{15}N_3O_3$ Store at 2-8°C

Homocitrulline — Z-D-Hci

Synonyms: Z-D-Lys-Carbamoyl

Bachem C-3965.0001   MW 323.35 $C_{15}H_{21}N_3O_5$ Store at 2-8°C

Homocysteine — (R)-S-(2-Amino-2-Carboxyethyl)-L-Hcy

Synonyms: L-(+)-Cystathionine

Sigma C 7505   FW 222.3 $C_7H_{14}N_2O_4S$ ~90%; crystalline

Homocysteine — BOC-D-Hcy-Methylbenzyl

Synonyms: (R)-2-(BOC-Amino)-4-(4-Methylbenzylsulfanyl)-Butyric Acid

Bachem A-4255.0001   MW 339.46 $C_{17}H_{25}NO_4S$ Store at -15°C

Homocysteine — BOC-Hcy-Bzl

USBio B2320   MW 325.4 $C_{16}H_{23}NO_4S$ ≥98%

Homocysteine — BOC-Hcy-Methylbenzyl

Synonyms: (S)-2-(BOC-Amino)-4-(4- Methylbenzylsulfanyl)-Butyric Acid

Bachem A-3420.0001   MW 339.46 $C_{17}H_{25}NO_4S$ Store at -15°C

USBio B2321   MW 339.4 $C_{17}H_{23}NO_4S$ ≥98%

Homocysteine — BOC-Hcy-Trt

Synonyms: (S)-2-(BOC-Amino)-4-Tritylsulfanyl-Butyric Acid

Bachem A-3610.0001   MW 477.62 $C_{28}H_{31}NO_4S$ Store at -15°C

Homocysteine — D-Hcy Sulfinic Acid

Synonyms: NMDA Agonist

Sigma H 3899   FW 167.2 $C_4H_9NO_4S$ ≥98% | Potent, fast-acting NMDA agonist

Homocysteine — D-Hcy Thiolactone Hydrochloride

Fluka 53525   MW 153.63 $C_4H_7NOS \cdot HCl$ ≥99.0%; mp: 185-187°C

Homocysteine — DL-Hcy

Synonyms: 2-Amino-4-Mercaptobutyric Acid

Fluka 53510   MW 135.19 $C_4H_9NO_2S$ ≥95.0% (titration); mp: 232-233°C

Sigma H 4628   FW 135.2 $C_4H_9NO_2S$ ≥95%

Homocysteine — DL-Hcy Thiolactone

Synonyms: DL-2-Amino-4-Mercaptobutyric Acid 1,4-Thiolactone

Sigma H 8762   FW 117.2 C_4H_7NOS

Homocysteine — DL-Hcy Thiolactone Hydrochloride

Synonyms: DL-2-Amino-4-Mercaptobutyric Acid 1,4-Thiolactone

Fluka 53530   MW 153.63 $C_4H_7NOS \cdot HCl$ ≥99.0% (titration); mp: 203°C; ≤0.5% loss on drying

ICN 101976   MW 153.6 $C_4H_7NOS \cdot HCl$ Crystalline

Sigma H 0376   FW 153.6 $C_4H_7NOS \cdot HCl$

Homocysteine — FMOC-Hcy-Trt

Synonyms: (S)-2-(FMOC-Amino)-4-Tritylsulfanyl-Butyric Acid

Bachem B-2405.0001   MW 599.75 $C_{38}H_{33}NO_4S$ Store at -15°C

Homocysteine — L-Hcy

Synonyms: 2-Amino-4-Mercaptobutyric Acid

ICN 101975   MW 135.2 $C_4H_9NO_2S$ >97%; crystalline

USBio H7000   MW 268.3 $C_8H_{16}N_2O_4S_2$ ≥99%

Homocysteine — L-Hcy Sulfinic Acid

Sigma H 4024   FW 167.2 $C_4H_9NO_4S$ ≥98% | Neurotransmitter candidate

Homocysteine — L-Hcy Thiolactone Hydrochloride

Synonyms: D-2-Amino-4-Mercaptobutyric Acid 1,4-Thiolactone; L-2-Amino-4-Mercaptobutyric Acid 1,4-Thiolactone

Fluka 53527   MW 153.63 $C_4H_7NOS \cdot HCl$ ≥99.0%; mp: 185-187°C

Sigma H 2767   FW 153.6 $C_4H_7NOS \cdot HCl$

Sigma H 6503   FW 153.6 $C_4H_7NOS \cdot HCl$

Homocysteine — L-β-Hcy Hydrochloride

Fluka 03655   MW 171.64 $C_4H_9NO_2S \cdot HCl$ ≥98.0% (TLC)

Homocysteine — N,S'-Di-Dnp-DL-Hcy

ICN 101977   MW 313.2 $C_{15}H_{11}N_3O_8$

Homocysteine — N-Ac-DL-Hcy Thiolactone

Synonyms: Citiolone

Fluka 01190   MW 159.21 $C_6H_9NO_2S$ ≥99%; ≤0.05% residue on ignition; mp: 109-111°C

ICN 100097   MW 159.2 $C_6H_9NO_2S$ Crystalline | Reagent for insolubilizing antibodies

Sigma A 9375   FW 159.2 $C_6H_9NO_2S$ Citiolone

Homocysteine — S-2-Amino-2-Carboxyethyl-Hcy

Synonyms: Cystathionine

Sigma C 3633   FW 222.3 $C_7H_{14}N_2O_4S$ ≥90%; mixture of 4 stereoisomers: L-, D-, L-allo-, D-allo-cystathionine | Stekol, JA, *JBC*, 173: 153, 1948

Homocysteine — S-5'-Adenosyl-L-Hcy

Fluka 02090   MW 384.41 $C_{14}H_{20}N_6O_5S$ ≥99% (HPLC)

Homocysteine — S-Adenosyl-L-Hcy

Synonyms: 5'-Deoxy-S-Adenosyl-D-Hcy

Sigma A 7389 $C_{14}H_{20}N_6O_5S$

Sigma A 9384 $C_{14}H_{20}N_6O_5S$

Homocysteine — S-Bzl-DL-Hcy Hydrochloride

Sigma B 8250   FW 225.3 $C_{11}H_{15}NO_2S \cdot HCl$ Cream colored crystals

Homoglutamic Acid — BOC-L-Hglu

Fluka 15054   MW 261.27 $C_{11}H_{19}NO_6$ ≥98% (TLC)

Homoglutamic Acid — D-Hglu

Fluka 06654   MW 161.16 $C_6H_{11}NO_4$ ≥98% (titration); mp: ~205°C

Homoglutamic Acid — L-Hglu

Fluka 06653   MW 161.16 $C_6H_{11}NO_4$ ≥98% (titration); mp: 203-205°C

Homoglutamic Acid — L-β-Hglu Hydrochloride

Fluka 03765   MW 197.62 $C_6H_{11}NO_4 \cdot HCl$ ≥98.0% (TLC)

Homohistidine — L-β-Hhi Hydrochloride

Synonyms: (S)-3-Amino-4-(4-Imidazolyl)-Butyric Acid

Fluka 03684   MW 205.64 $C_7H_{11}N_3O_2 \cdot HCl$ ≥98.0% (HPLC)

Homohistidine — N^6-BOC-N^{im}-Dnp-L-β-Hhi

Fluka 03687   MW 435.39 $C_{18}H_{21}N_5O_8$ ≥98%

Homohistidine — N^6-FMOC-N^{im}-Trt-L-β-Hhi

Fluka 03686   MW 633.75 $C_{41}H_{35}N_3O_4$ ≥98% (HPLC)

Homoisoleucine — FMOC-L-β-Hil

Synonyms: (3R,4S)-3-(FMOC-Amino)-4-Methylhexanoic Acid

Fluka 03671   MW 367.45 $C_{22}H_{25}NO_5$ ≥98% (HPLC)

Homoisoleucine — L-β-Hil Hydrochloride

Synonyms: (3R,4S)-3-Amino-4-Methylhexanoic Acid

Fluka 03669   MW 181.66 $C_7H_{15}NO_2 \cdot HCl$ ≥98.0% (TLC)

Homoleucine — BOC-L-β-Hle

Synonyms: (S)-3-(BOC-Amino)-5-Methylhexanoic Acid

Fluka 14975   MW 245.32 $C_{12}H_{23}NO_4$ ≥98% (TLC) | Building blocks for the synthesis of "carba" peptides; Rodriguez, M et al, *Tetrahedron Lett*, 31: 7319, 1990; Seebach, D et al, *Helv Chim Acta*, 79: 913, 1996; *ibid*, 79: 2043, 1996

Neosystem BA04402   MW 245.3

Homoleucine — FMOC-D-Hle

Neosystem FA11901   MW 367.45

Homoleucine — FMOC-L-Hle

Neosystem FA11902   MW 367.45

Homoleucine — FMOC-L-β-Hle

Synonyms: (S)-3-(FMOC-Amino)-5-Methylhexanoic Acid

Fluka 47946   MW 367.45 $C_{22}H_{25}NO_4$ ~98% (HPLC)

Homoleucine — L-β-Hle Hydrochloride

Synonyms: (S)-3-Amino-5-Methylhexanoic Acid

Fluka 03764   MW 181.66 $C_7H_{15}NO_2 \cdot HCl$ ≥98.0% (TLC)

Homolysine — L-β-Hly Hydrochloride
Synonyms: (S)-3,7-Diaminoheptanoic Acid

Fluka 03759 MW 196.68 $C_7H_{16}N_2O_2 \cdot$ HCl ≥98.0% (TLC)

Homolysine — Nᵋ-BOC-Nᵅ-Z-L-Hly
Synonyms: (S)-3-(BOC-Amino)-7-Z-AminoHeptanoic Acid

Fluka 14978 MW 394.47 $C_{20}H_{30}N_2O_6$ ≥98% (TLC)

Homolysine — Nᵋ-FMOC-Nᵅ-BOC-L-β-Hly
Synonyms: (S)-7-(BOC-Amino)-3-(FMOC-Amino)-Heptanoic Acid

Fluka 47874 MW 482.58 $C_{27}H_{34}N_2O_6$ ~97% (TLC)

Homomethionine — BOC-L-β-Hme
Synonyms: (R)-3-(BOC-Amino)-5-(Methylthio-Pentanoic Acid

Fluka 03661 MW 263.35 $C_{11}H_{21}NO_4S$ ≥98% (TLC)

Homomethionine — FMOC-L-β-Hme
Synonyms: (R)-3-(FMOC-Amino)-5-(Methylthio-Pentanoic Acid

Fluka 03658 MW 385.48 $C_{21}H_{23}NO_4S$ ≥98% (HPLC)

Homomethionine — L-β-Hme Hydrochloride
Synonyms: (R)-3-Amino-5-(Methylthio-Pentanoic Acid

Fluka 03681 MW 199.7 $C_6H_{13}NO_2S \cdot$ HCl ≥98.0% (TLC)

Homophenylalanine — BOC-D-Hph
Synonyms: (R)-2-(BOC-Amino)-4-Phenylbutyric Acid

American Peptide BDPHE35 MW 280.4
Bachem A-1195.0001 MW 279.34 $C_{15}H_{21}NO_4$ Store at RT
Fluka 15043 MW 279.34 $C_{15}H_{21}NO_4$ ≥98% (TLC)

Homophenylalanine — BOC-Hph
Synonyms: (S)-2-(BOC-Amino)-4-Phenylbutyric Acid

Bachem A-1190.0001 MW 279.34 $C_{15}H_{21}NO_4$ Store at RT
Peptides International BLF-5371-PI MW 279.34 >98% (HPLC); white to off-white powder

Homophenylalanine — BOC-L-Hph
American Peptide BLPHE35 MW 280.4
Fluka 15469 MW 279.34 $C_{15}H_{21}NO_4$ ≥98% (TLC); mp: 76-80°C
Neosystem BA04502 MW 279.3
USBio B2322 MW 279.3 $C_{15}H_{21}NO_4$ ≥99%

Homophenylalanine — BOC-L-β-Hph
Synonyms: (S)-3-(BOC-Amino)-4-Phenylbutyric Acid

Fluka 14979 MW 279.34 $C_{15}H_{21}NO_4$ ≥98% (TLC); mp: 100-104°C

Homophenylalanine — D-Hph
Synonyms: (R)-2-Amino-4-Phenylbutyric Acid

Bachem F-1615.0001 MW 179.22 $C_{10}H_{13}NO_2$ Store at RT
Fluka 53565 MW 179.22 $C_{10}H_{13}NO_2$ ≥97.0%; mp: >300°C

Homophenylalanine — DL-Hph
Synonyms: (RS)-2-Amino-4-Phenylbutyric Acid; (±)-2-Amino-4-Phenylbutyric Acid

Bachem F-1620.0001 MW 179.22 $C_{10}H_{13}NO_2$ Store at RT
Sigma H 0762 FW 179.2 $C_{10}H_{13}NO_2$

Homophenylalanine — FMOC-D-Hph
Synonyms: (R)-2-(FMOC-Amino)-4-Phenylbutyric Acid

American Peptide FPHE147 MW 402.5

Bachem B-2810.0001 MW 401.46 $C_{25}H_{23}NO_4$ Store at RT
Fluka 47429 MW 401.46 $C_{25}H_{23}NO_4$ ≥98% (HPLC)

Homophenylalanine — FMOC-Hph
Synonyms: (S)-2-(FMOC-Amino)-4-Phenylbutyric Acid

Bachem B-1535.0001 MW 401.46 $C_{25}H_{23}NO_4$ Store at RT

Homophenylalanine — FMOC-Hph-® (200-400 mesh)
Synonyms: FMOC-(+)-α-Amino-4-Phenylbutyric Acid-®

Bachem D-2005.0001 Store at RT

Homophenylalanine — FMOC-L-Hph
Synonyms: (S)-2-(FMOC-Amino)-4-Phenylbutyric Acid

American Peptide FPHE146 MW 402.5
Fluka 47430 MW 401.46 $C_{25}H_{23}NO_4$ ≥98% (HPLC)

Homophenylalanine — FMOC-L-β-Hph
Synonyms: (S)-3-(FMOC-Amino)-4-Phenylbutyric Acid

Fluka 47878 MW 401.46 $C_{25}H_{23}NO_4$ ~97% (HPLC)

Homophenylalanine — Hph
Synonyms: (S)-2-Amino-4-Phenylbutyric Acid

Bachem F-1610.0001 MW 179.22 $C_{10}H_{13}NO_2$ Store at RT

Homophenylalanine — L-Hph
Synonyms: (S)-2-Amino-4-Phenylbutyric Acid

Fluka 53570 MW 179.22 $C_{10}H_{13}NO_2$ ≥97.0%; mp: >300°C

Homophenylalanine — L-β-Hph Hydrochloride
Synonyms: (S)-3-Amino-4-Phenylbutyric Acid

Fluka 03769 MW 215.68 $C_{10}H_{13}NO_2 \cdot$ HCl ≥98.0% (TLC)

Homophenylalanine — L-β-HphOAll Hydrochloride
Synonyms: Allyl (S)-3-Amino-4-Phenylbutyrate

Fluka 53575 MW 255.74 $C_{13}H_{17}NO_2 \cdot$ HCl ≥98.0% (TLC); mp: 56-59°C

Homophenylalanine — N-FMOC-D-Hph
Synonyms: (R)-N-FMOC-2-Amino-4-Phenylbutyric Acid

Sigma F 7170 FW 401.5 $C_{25}H_{23}NO_4$

Homophenylalanine — N-t-BOC-L-Hph
Sigma B 7155 FW 279.3 $C_{15}H_{21}NO_4$

Homophenylalanine — Z-D-Hph
Synonyms: (R)-2-Z-Amino-4-Phenylbutyric Acid

Bachem C-1280.0001 MW 313.35 $C_{18}H_{19}NO_4$ Store at RT

Homophenylalanine — Z-Hph
Synonyms: (S)-2-Z-Amino-4-Phenylbutyric Acid

Bachem C-1275.0001 MW 313.35 $C_{18}H_{19}NO_4$ Store at RT

Homophenylalanine — Z-L-Hph
Synonyms: (S)-2-Z-Amino-4-Phenylbutyric Acid

Fluka 09968 MW 313.35 $C_{18}H_{19}NO_4$ ≥98.0% (TLC)

Homoproline — BOC-D-Hpr
Synonyms: BOC-D-Pipecolic Acid

Bachem A-3125.0001 MW 229.28 $C_{11}H_{19}NO_4$ Store at RT

Homoproline — BOC-Hpr

Synonyms: BOC-L-Pipecolic Acid

Bachem A-2830.0001   MW 229.28 $C_{11}H_{19}NO_4$ Store at RT

Homoproline — BOC-L-β-HomoPro

Synonyms: (S)-2-(1-BOC-2-Pyrrolidinyl) Acetic Acid

Fluka 14982   MW 229.28 $C_{11}H_{19}NO_4$ ≥98% (TLC) | Building block for a pyrrolizidine alkaloid; Knight, DW et al, *J Chem Soc, Perkin I* 1615, 1991

Homoproline — D-Hpr

Synonyms: D-Pipecolic Acid

Bachem F-1630.0001   MW 129.16 $C_6H_{11}NO_2$ Store at RT

Homoproline — D-Hpr-OMe Hydrochloride

Synonyms: D-Pipecolic Acid-OMe

Bachem F-3125.0001   MW 179.65 $C_7H_{13}NO_2 \cdot HCl$ Store at 2-8°C

Homoproline — DL-Hpr

Synonyms: DL-Pipecolic Acid

Bachem F-2915.0025   MW 129.16 $C_6H_{11}NO_2$ Store at RT

Homoproline — FMOC-D-Hpr

Synonyms: FMOC-D-Pipecolic Acid

Bachem B-2290.0001   MW 351.4 $C_{21}H_{21}NO_4$ Store at RT

Homoproline — FMOC-Hpr

Synonyms: FMOC-L-Pipecolic Acid

Bachem B-2285.0001   MW 351.4 $C_{21}H_{21}NO_4$ Store at RT

Homoproline — FMOC-L-β-Hpr

Synonyms: (S)-2-(1-FMOC-2-Pyrrolidinyl) Acetic Acid

Fluka 47912   MW 351.4 $C_{21}H_{21}NO_4$ ≥98% (HPLC)

Homoproline — Hpr

Synonyms: L-Pipecolic Acid

Bachem F-1625.0001   MW 129.16 $C_6H_{11}NO_2$ Store at RT

Homoproline — Hpr-OMe Hydrochloride

Synonyms: L-Pipecolic Acid-OMe

Bachem F-2465.0001   MW 179.65 $C_7H_{13}NO_2 \cdot HCl$ Store at 2-8°C

Homoserine — BOC-O-Bzl-L-β-Hse

Fluka 03697   MW 309.36 $C_{16}H_{23}NO_5$ ≥98% (TLC)

Homoserine — D-Hse

Synonyms: (R)-2-Amino-4-Hydroxybutyric Acid; 2-Amino-4-hydroxy-D-Butyric Acid

Fluka 53595   MW 119.12 $C_4H_9NO_3$ ≥99.0%; mp: 205°C

Rexim   MW 119.1 $C_4H_9NO_3$

Sigma H 4021   FW 119.1 $C_4H_9NO_3$

Homoserine — DL-Hse

Synonyms: DL-2-Amino-4-Hydroxybutyric Acid; (±)-2-Amino-4-Hydroxybutyric Acid

ICN 101982   MW 119.1 $C_4H_9NO_3$ Crystalline

Sigma H 1001   FW 119.1 $C_4H_9NO_3$

USBio H7015-15   MW 119.1 $C_4H_9NO_3$ ≥99%

Homoserine — DL-Hse (1-^{14}C)

ARC ARC-1187   MW 119.12 $HOCH_2CH_2CH(NH_2)COOH$ 50-60 mCi/mmol; 1.85-2.22 GBq/mmol; in 0.01 *N* HCl | Radiochemical

Homoserine — DL-Hse Lactone Hydrobromide

Synonyms: 2-Amino-4-Butyrolactone

Fluka 07253   MW 182.02 $C_4H_7NO_2 \cdot HBr$ ≥97% (titration); mp: 220-225°C

ICN 157389   MW 182 $C_4H_7NO_2 \cdot HBr$

Homoserine — FMOC-O-tBu-L-β-Hse

Fluka 03696   MW 397.47 $C_{23}H_{27}NO_5$ ≥98% (HPLC)

Homoserine — L-Hse

Synonyms: Homoserine Dehydrogenase Substrate; L-2-Amino-4-Hydroxy-Butyric Acid; (S)-2-Amino-4-Hydroxybutyric Acid

Fluka 53600   MW 119.12 $C_4H_9NO_3$ ~99.0% (titration); mp: 205°C; ≤0.5% H_2O | Assay substrate; Datta, P & Gest, H, *Meth Enzymol*, 17A: 703, 1970

ICN 101983   MW 119.1 $C_4H_9NO_3$ ≥99%

Rexim   MW 119.1 $C_4H_9NO_3$

Sigma H 6515   FW 119.1 $C_4H_9NO_3$

USBio H7015   MW 119.1 $C_4H_9NO_3$ ≥98%

Homoserine — L-Hse Lactone Hydrobromide

Synonyms: (S)-2-Amino-4-Butyrolactone

Fluka 07252   MW 182.02 $C_4H_7NO_2 \cdot HBr$ ≥98% (titration); mp: ~245°C | Dekhane, M et al, *Tetrahedron Lett*, 37: 1883, 1996; Koch, T & Buchardt, O, *Synthesis*, 1065, 1993; Krief, A & Trabelsi, M, *Synth Comm*, 19: 1203, 1989; Silks, LA et al, *ibid*, 20: 1555, 1990; Boyle, PH et al, *Chem Comm*, 1875, 1994; *Tetrahedron: Asymmetry*, 6: 2819, 1995

Homoserine — L-Hse Lactone Hydrochloride

Synonyms: (S)-2-Amino-4-Butyrolactone

Fluka 07254   MW 137.57 $C_4H_7NO_2 \cdot HCl$ ≥99% (TLC); mp: 210-220°C | Bajgrowicz, JA et al, *Tetrahedron*, 41: 1833, 1985; Son, JK et al, *Synthesis 240*, 1988; Kemp, DS & McNamara, PE, *J Org Chem*, 49: 2286, 1984

Sigma H 7890   FW 137.6 $C_4H_7NO_2 \cdot HCl$

Homoserine — L-β-Hse Hydrochloride

Synonyms: (R)-3-Amino-4-Hydroxybutyric Acid

Fluka 03694   MW 155.58 $C_4H_9NO_3 \cdot HCl$ ≥98.0% (TLC)

Homoserine — N-Octanoyl-DL-Hse Lactone

Synonyms: N-Capryloyl-D-L-Hse Lactone

Fluka 10940   MW 227.3 $C_{12}H_{21}NO_3$ ≥97.0% (HPLC) | Induces violacein expression in a *Chromobacterium violaceum* mutant usually not able to produce homoserinlactones; McClean, KH et al, *Microbiology*, 143: 3703, 1997

Homoserine — N-t-BOC-L-Hse

Sigma B 4645   FW 219.2 $C_9H_{17}NO_5$

Homoserine — N-Tetradecanoyl-DL-Hse Lactone

Synonyms: N-Myr-DL-Hse Lactone

Fluka 10937   MW 311.46 $C_{18}H_{33}NO_3$ ≥97.0% (HPLC) | Induces violacein expression in a *Chromobacterium violaceum* mutant usually not able to produce homoserinlactones; McClean, KH et al, *Microbiology*, 143: 3703, 1997

Homoserine — N-Z-L-Hse Lactone

Synonyms: (S)-α-Z-Amino-γ-Butyrolactone

Fluka 96304   MW 235.24 $C_{12}H_{13}NO_4$ ≥98.0% (HPLC); mp: 127-132°C

Homoserine — o-Succinyl-L-Hse

Sigma S 7129   FW 219.2 $C_8H_{13}NO_6$

Homoserine — Trt-Hse

USBio T8660   MW 361.4 $C_{13}H_{24}NO_3$ ≥98%

Homoserine — Z-L-Hse-OBzl

Synonyms: (S)-2-Z-Amino-4-Hydroxybutyric Acid-OBzl

Fluka 96302   MW 343.38 $C_{19}H_{21}NO_5$ ~97% (HPLC); mp: 51-55°C

Homothreonine — BOC-O-Bzl-L-β-Hth

Synonyms: (3R,4R)-4-Benzyloxy-3-(BOC-Amino)-Pentanoic Acid

Fluka 14976   MW 323.39 $C_{17}H_{25}NO_5$ ≥98% (TLC)

Homothreonine — FMOC-O-tBu-L-β-Hth

Synonyms: (3R,4R)-4-tBu-3-(FMOC-Amino)-Pentanoic Acid

Fluka 47911   MW 411.5 $C_{24}H_{29}NO_5$ ~98% (TLC)

Homothreonine — L-β-Hth Hydrochloride

Synonyms: (3R,4S)-3-Amino-4-Hydroxypentanoic Acid

Fluka 03767   MW 169.61 $C_5H_{11}NO_3 \cdot HCl$ ≥98.0% (TLC)

Homotryptophan — L-β-Htr Hydrochloride

Synonyms: (S)-3-Amino-4-3-Indolylbutyric Acid

Fluka 03790   MW 254.72 $C_{12}H_{14}N_2O_2 \cdot HCl$ ≥98.0% (HPLC)

Homotryptophan — N^i-BOC-L-β-Htr

Synonyms: (S)-3-(BOC-Amino)-4-3-Indolylbutyric Acid

Fluka 14981   MW 318.37 $C_{17}H_{22}N_2O_4$ ≥98% (TLC)

Homotryptophan — N^i-BOC-L-β-Htr-OMe

Fluka 17008   MW 332.4 $C_{18}H_{24}N_2O_4$ ≥98% (TLC)

Homotryptophan — N^i-FMOC-L-β-Htr

Synonyms: (S)-3-(FMOC-Amino)-4-3-Indolylbutyric Acid

Fluka 47901   MW 440.5 $C_{27}H_{24}N_2O_4$ ~98% (HPLC)

Homotyrosine — BOC-O-Bzl-L-β-Hty

Fluka 03693   MW 385.46 $C_{22}H_{27}NO_5$ ≥98% (HPLC)

Homotyrosine — FMOC-OtBu-L-β-Hty

Fluka 03692   MW 473.57 $C_{29}H_{31}NO_5$ ≥98% (HPLC)

Homotyrosine — L-β-Hty Hydrochloride

Synonyms: (S)-3-Amino-4-(4-Hydroxyphenyl)-Butyric Acid

Fluka 03758   MW 231.68 $C_{10}H_{13}NO_3 \cdot HCl$ ≥98.0% (HPLC)

Hydroxyaspartic Acid — DL-*threo*-β-Hyasp

Synonyms: threo-2-Amino-3-Hydroxysuccinic Acid

Calbiochem 390185   MW 149.1 $C_4H_7NO_5$ >99% (TLC); solid; soluble in H_2O | An AA derivative that blocks glutamate uptake; causes neuronal degeneration, particularly of cholinergic & GABAergic neurons; constituent of the polypeptide antibiotic cinnamycin; Velasco, I et al, *J Neurosci Res,* 44: 551, 1996; Marini, AM et al, *J Neurosci*, 9: 3665, 1989; Kaletta, C et al, *Eur J Biochem*, 199: 411, 1991; McBean, GJ & Roberts, PJ, *J Neurochem*, 44: 247, 1985

ICN 157414 $C_4H_7NO_5$

Sigma H 2775 $C_4H_7NO_5$ Glutamate uptake inhibitor; Bender, AS et al, *Neurochem Res*, 14: 641, 1989; Marini, A & Novelli, A, *Eur J Pharmacol*, 194: 131, 1991

Hydroxylysine — L-Hyl Dihydrochloride

Synonyms: 5-Hydroxy-L-Lys

Rexim   MW 235.1 $C_6H_{16}Cl_2N_2O_3$

Hydroxylysine — δ-DL-allo-Hyl Hydrochloride

ICN 101480   MW 198.6 $C_6H_{14}N_2O_3 \cdot HCl$ Crystalline

Hydroxylysine — δ-Hyl Hydrochloride

Synonyms: 2,6-Diamino-5-Hydroxyhexanoic Acid

Sigma H 0377   FW 198.6 $C_6H_{14}N_2O_3 \cdot HCl$ Mixed DL & DL-allo

Hydroxymethionine — DL-α-Hym Calcium Salt

Synonyms: 2-Hydroxy-4-Methylthiobutanoic Acid; 2-Hydroxy-4-Methylthiobutyric Acid; DL-α-Hydroxy-γ-Methylthio-Butyrate

Rexim   MW 338.5 $C_{10}H_{18}CaO_6S_2$ White to very slightly yellowish crystals or crystalline powder

Hydroxynorvaline — DL-3-Hyn

Synonyms: 2-Amino-3-Hydroxypentanoic Acid; DL-β-Hyn

Fluka 56058   MW 133.15 $C_5H_{11}NO_3$ ≥98.0% (titration); ≤1% Nva; ≤0.5% Gly | Schwartz, TW, *JBC*, 263: 11504, 1988

Hydroxynorvaline — DL-β-Hyn

Synonyms: 3-Hydroxy-2-Aminopentanoic Acid; α-Amino-β-Hydroxyvaleric Acid

Sigma H 4002   FW 133.1 $C_5H_{11}NO_3$ A specific threonine antagonist; Shiio, I et al, *J Biochem*, 68: 859, 1970

Hydroxyphenylalanine — DL-3,4-Di-Hyph (Ala-1-^{14}C)

ICN 16032   MW 197.2 $3,4-(OH)_2C_6H_3CH_2CH(NH_2)COOH$ EtOH solution; SA 50-60 mCi/mmol, 1.85-2.22 GBq/mmol

Hydroxyphenylalanine — DL-β-3,4-Di-Hyph

Synonyms: DL-DOPA

ICN 101577   MW 197.2 $C_9H_{11}NO_4$ Crystalline

Hydroxyphenylalanine — L-3,4 Di-Hyph (Ring 2,5,6-^{3}H)

ARC ART-235   MW 197.2 $3,4-(OH)_2C_6H_3CH_2CH(NH_2)COOH$ 20-60 Ci/mmol; 0.74-2.22 TBq/mmol; in 0.2 N HOAc: EtOH (9:1) | Radiochemical; L-DOPA

Hydroxyphenylalanine — L-3,4-Di-Hyph (Ring 2,5,6-^{3}H)

ICN 26030   MW 197.2 $3,4-(OH)_2C_6H_3CH_2CH(NH_2)COOH$ 0.01 N KH_2PO_4 solution; SA 20-60 Ci/mmol, 0.74-2.22 TBq/mmol

Hydroxyphenylalanine — L-β-3,4-Di-Hyph

Synonyms: L-DOPA; L-3-Hyty; Tyrosine Aminotransferase Inhibitor

ICN 101578   MW 197.2 $C_9H_{11}NO_4$ Crystalline; 99% | Useful for treatment of Parkinsonism

Hydroxyphenylalanine — L-β-3,4-Di-Hyph Hydroxamic Acid

Synonyms: L-DOPA Hydroxamic Acid

ICN 101579

Hydroxyphenylalanine — α-Me-DL-β-3,4-Di-Hyph

Synonyms: DL-α-Me-DOPA

ICN 155517   MW 211.2 $C_{10}H_{13}NO_4$

Hydroxyphenylalanine — α-Me-L-β-3,4-Di-Hyph

Synonyms: L-α-Me-DOPA

ICN 155519   MW 211.2 $C_{10}H_{13}NO_4$

Hydroxyproline — 4-Hydroxy-L-Pro

Synonyms: L-4-Hydroxy-2-Pyrrolidine-Carboxylic Acid

ICN 102015   MW 131.1 $C_5H_9NO_3$ *trans* isomer; crystalline

Hydroxyproline — 4-Hydroxy-L-Pro *Trans* Isomer

Synonyms: L-4-Hydroxy-2-Pyrrolidine-Carboxylic Acid

ICN 102015   MW 131.1 $C_5H_9NO_3$ Crystalline | Cell culture reagent

Hydroxyproline — Ac-Hyp

Bachem F-1040.0005   MW 173.17 $C_7H_{11}NO_4$ Store at RT

Hydroxyproline — allo-Hyp-L-4(G-³H)

ARC ART-458   MW 131.11 $HNCH_2CHOHCH_2CHCOOH$ 15-30 Ci/mmol; 0.55-1.11 TBq/mmol; in 0.01 *N* HCl | Radiochemical

Hydroxyproline — BOC-cis-Hyp-OMe

Bachem A-3115.0001   MW 245.28 $C_{11}H_{19}NO_5$ Store at -15°C

Hydroxyproline — BOC-Hyp

Bachem A-1790.0005   MW 231.25 $C_{10}H_{17}NO_5$ Store at RT

Hydroxyproline — BOC-HypBzl

Synonyms: BOC-O-Bzl-L-Hyp

Bachem A-2820.0001   MW 321.37 $C_{17}H_{23}NO_5$ Store at RT

Peptides International BLX-2116   MW 321.37   >98% (HPLC); white crystalline powder

Hydroxyproline — BOC-Hyp-OBzl

Bachem A-4535.0001   MW 321.37 $C_{17}H_{23}NO_5$ Store at 2-8°C

Hydroxyproline — BOC-Hyp-OMe

Bachem A-1795.0001   MW 245.28 $C_{11}H_{19}NO_5$ Store at 2-8°C

Hydroxyproline — BOC-Hyp-OtBu

Bachem A-4405.0001   MW 287.36 $C_{14}H_{25}NO_5$ Store at -15°C

Hydroxyproline — BOC-L-Hyp

Fluka 15544   MW 231.25 $C_{10}H_{17}NO_5$ ≥98% (TLC); mp: 123-127°C

Neosystem BA04802   MW 231.2

Hydroxyproline — BOC-O-Bzl-L-Hyp

Fluka 15535   MW 321.37 $C_{17}H_{23}NO_5$ ≥98% (TLC)

Hydroxyproline — BOC-O-Bzl-L-β-Homohydroxyproline Dicyclohexylamine

Fluka 03683   MW 516.71 $C_{18}H_{25}NO_5 \cdot C_{12}H_{23}N$ ≥98% (TLC)

Hydroxyproline — cis-3-Hydroxy-DL-Pro

Synonyms: (±)-*cis*-3-Hydroxypyrrolidine-2-Carboxylic Acid

Fluka 56245   MW 131.14 $C_5H_9NO_3$ ≥97.0% (TLC); mp: 232°C

Sigma H 6152   FW 131.1 $C_5H_9NO_3$

Hydroxyproline — cis-4-Hydroxy-D-Pro

Synonyms: (2R,4R)-4-Hydroxypyrrolidine-2-Carboxylic Acid; allo-D-Hyp; (2R,4S)-4-Hydroxy-2-Pyrrolidinecarboxylic Acid

Fluka 56246   MW 131.13 $C_5H_9NO_3$ ≥99.0% (titration); ≤0.3% foreign AA; ≤0.5% loss on drying

Sigma H 5877   FW 131.1 $C_5H_9NO_3$

Hydroxyproline — cis-4-Hydroxy-L-Pro

Synonyms: (2S,4S)-4-Hydroxypyrrolidine-2-Carboxylic Acid; Proline Antagonist; L-allo-Hyp; (2S,4S)-4-Hydroxy-2-Pyrrolidinecarboxylic Acid

Fluka 56248   MW 131.13 $C_5H_9NO_3$ ≥99.0% (titration); mp: 256-257°C | Pro antagonist in collagen synthesis; inhibits human skin fibroblast growth & collagen production in culture; specifically inhibits cultured tumor cells that synthesize basement-membrane collagen; inhibits secretion of procollagen; Tan, EML et al, *J Invest Dermatol*, 80: 261, 1983; Klohs, WD et al, *J Natl Cancer Inst*, 75: 253, 1985; Lewko, WM et al, *Cancer Res*, 41: 2855, 1981; Schein, J et al, *Arch Biochem Biophys*, 183: 416, 1977

Sigma H 1637   FW 131.1 $C_5H_9NO_3$ Constituent of *Amanita* species peptide toxins but not of animal proteins like collagen

Hydroxyproline — cis-Hyp

Bachem F-1025.0001   MW 131.13 $C_5H_9NO_3$ Store at RT

Hydroxyproline — cis-Hyp-OMe Hydrochloride

Bachem F-2680.0001   MW 181.62 $C_6H_{11}NO_3 \cdot HCl$ Store at 2-8°C

Hydroxyproline — Dansyl-Hydroxy-L-Pro Cyclohexylammonium Salt

ICN 100079

Hydroxyproline — D-cis-Hyp

Synonyms: (4R)-4-Hydroxy-D-Pro

Bachem F-1395.0001   MW 131.13 $C_5H_9NO_3$ Store at RT

Rexim   MW 131.1 $C_5H_9NO_3$

Hydroxyproline — D-Hyp

Bachem F-2980.0001   MW 131.13 $C_5H_9NO_3$ Store at RT

Hydroxyproline — Dnp-Hydroxy-L-Pro

ICN 102724   MW 297.2 $C_{11}H_{11}N_3O_7$ Crystalline

Hydroxyproline — FMOC-D-cis-Hyp

Bachem B-3065.0001   MW 353.38 $C_{20}H_{19}NO_5$ Store at 2-8°C

Hydroxyproline — FMOC-D-HyptBu

Bachem B-2975.0001   MW 409.48 $C_{24}H_{27}NO_5$ Store at 2-8°C

Hydroxyproline — FMOC-Hyp

Bachem B-2800.0005   MW 353.38 $C_{20}H_{19}NO_5$ Store at -15°C

Hydroxyproline — FMOC-HypBzl

Bachem B-2450.0001   MW 443.5 $C_{27}H_{25}NO_5$ Store at 2-8°C

Hydroxyproline — FMOC-HyptBu

Synonyms: FMOC-O-tBu-L-Hyp

Bachem B-1525.0001 MW 409.48 $C_{24}H_{27}NO_5$ Store at -15°C

Peptides International FLX-1771-PI MW 457.53 >98% (HPLC); white powder

Hydroxyproline — FMOC-L-Hyp

Neosystem FA04802 MW 353.4

Hydroxyproline — FMOC-L-Pro 4-Hyp

Fluka 47686 MW 353.37 $C_{20}H_{19}NO_5$ ≥98% (HPLC); mp: 189-193°C

Hydroxyproline — FMOC-OtBu-L-Hyp

Fluka 47517 MW 409.48 $C_{24}H_{27}NO_5$ ≥98% (HPLC); mp: 63°C

Hydroxyproline — FMOC-O-tBu-L-Hyp

Synonyms: FMOC-L-Hyp-tBu

Neosystem FA04804 MW 409.5

Hydroxyproline — FMOC-O-tBu-L-β-Homohydroxyproline

Fluka 03751 MW 423.51 $C_{25}H_{29}NO_5$ ≥98% (HPLC)

Hydroxyproline — Hyp

Bachem F-1655.0025 MW 131.13 $C_5H_9NO_3$ Store at RT

Hydroxyproline — Hyp-AMC

Bachem I-1230.0050 MW 288.3 $C_{15}H_{16}N_2O_4$ Store at -15°C

Hydroxyproline — HypBzl-OMe Hydrochloride

Bachem E-3315.0001 MW 271.74 $C_{13}H_{17}NO_3 \cdot HCl$ Store at -15°C

Hydroxyproline — Hyp-OBzl Hydrochloride

Bachem F-2775.0005 MW 257.72 $C_{12}H_{15}NO_3 \cdot HCl$ Store at -15°C

Hydroxyproline — Hyp-OMe Hydrochloride

Bachem F-1660.0005 MW 181.62 $C_6H_{11}NO_3 \cdot HCl$ Store at 2-8°C

Hydroxyproline — HyptBu

Bachem F-2205.0001 MW 187.24 $C_9H_{17}NO_3$ Store at RT

Hydroxyproline — Hyp-βNA

Bachem K-1370.0250 MW 256.3 $C_{15}H_{16}N_2O_2$ Store at RT

Hydroxyproline — Hyp-βNA Hydrochloride

Bachem K-1545.0250 MW 292.77 $C_{15}H_{16}N_2O_2 \cdot HCl$ Store at RT

Hydroxyproline — L-4-Hyp

Synonyms: (2S,4R)-4-Hydroxypyrrolidine-2-Carboxylic Acid

Fluka 56250 MW 131.13 $C_5H_9NO_3$ ≥99.0% (titration); ≤0.1% loss on drying, residue on ignition; ≤0.3% foreign AA; ≤0.0005% Cd, Co, Cr, Cu, Fe, Mg, Mn, Ni, Pb, Zn; ≤0.001% Ca; ≤0.005% Na, K; ≤0.05% NH₄⁺; ≤0.01% Cl, NH₄⁺, SO₄ | Adams, E, *Meth Enzymol*, 17B: 266, 1971

Hydroxyproline — L-Hpr

Peptides International ALX-2711 MW 131.13 >99.9% optical purity by RP-HPLC; white crystalline powder

Hydroxyproline — L-Hyp

Synonyms: (4R)-4-Hydroxy-L-Pro; (-)-4-Hydroxy-2-Pyrrolidine Carboxylic Acid

Neosystem AA04802 MW 131.1

Rexim MW 131.1 $C_5H_9NO_3$ White crystals or crystalline powder

Hydroxyproline — L-Hyp 4(G-³H)

ARC ART-231 MW 131.11 $HNCH_2CHOHCH_2CHCOOH$ 15-30 Ci/mmol; 0.55-1.11 TBq/mmol; in 0.01 N HCl | Radiochemical

Hydroxyproline — L-β-Homohydroxyproline Hydrochloride

Synonyms: (2S,4R)-4-Hydroxy-2-Pyrrolidinylacetic Acid

Fluka 03698 MW 181.62 $C_6H_{11}NO_3 \cdot HCl$ ≥98.0% (TLC)

Hydroxyproline — MTH-DL-Hyp

Sigma M 4631 FW 186.2 $C_7H_{10}N_2O_2S$ ~95%

Hydroxyproline — N-Ac-Hydroxy-Pro L-4(G-³H)

ARC ART-554 MW 173.2 15-30 Ci/mmol; 0.55-1.11 TBq/mmol; in EtOH | Radiochemical

Hydroxyproline — N-Ac-L-Hyp

Synonyms: N-Ac-L-Hyp; 1-Ac-4-Hyp

Fluka 01192 MW 173.17 $C_7H_{11}NO_4$ ≥99% (titration); mp: 132-133°C

Rexim MW 173.2 $C_7H_{11}NO_4$ White crystals or crystalline powder

Hydroxyproline — N-BOC-O-Bzl-L-β-Homohydroxyproline-OMe

Fluka 17007 MW 349.43 $C_{19}H_{27}NO_5$ ≥98% (TLC)

Hydroxyproline — N-CBZ-Hydroxy-L-Pro

ICN 100220 MW 265.3 $C_{13}H_{15}NO_5$ Crystalline

Sigma C 1127 FW 265.3 $C_{13}H_{15}NO_5$ ≥98% (TLC)

Hydroxyproline — N-CBZ-Hydroxy-L-Pro-OMe

Sigma C 1009 FW 279.3 $C_{14}H_{17}NO_5$ Viscous liquid

Hydroxyproline — N-Decyl-Hyp

Bachem E-2545.0001 MW 271.4 $C_{15}H_{29}NO_3$ Store at RT

Hydroxyproline — N-Decyl-L-Hyp

Sigma D 1394 FW 271.4 $C_{15}H_{29}NO_3$

Hydroxyproline — N-Heptyl-Hyp

Bachem F-1590.0001 MW 229.32 $C_{12}H_{23}NO_3$ Store at RT

Hydroxyproline — N-Heptyl-L-Hyp

ICN 157334 MW 229.3 $C_{12}H_{23}NO_3$ Useful HPLC stationary phase for resolution of racemic α-AA; Devankov, VA etal, *Chromatographia*, 13:677, 1980

Sigma H 1010 FW 229.3 $C_{12}H_{23}NO_3$ HPLC stationary phase for resolution of α-AA; Devankov, VA et al, *Chromatographia*, 13: 677, 1980

Hydroxyproline — N-Hexadecyl-Hyp

Bachem F-1595.0001 MW 355.56 $C_{21}H_{41}NO_3$ Store at RT

Hydroxyproline — *N*-Hexadecyl-L-Hyp

ICN 157353 MW 355.6 $C_{21}H_{41}NO_3$ Useful HPLC stationary phase for resolution of racemic α-AA; Devankov, VA etal, *Chromatographia*, 13:677, 1980

Hydroxyproline — *N*-Me-*cis*-Hyp

Bachem F-2370.0250 MW 145.16 $C_6H_{11}NO_3$ Store at RT

Hydroxyproline — *N*-O-Nps-L-Hyp Dicyclohexylammonium Salt

ICN 102529 MW 465.6 $C_{11}H_{12}N_2O_5S \cdot C_{12}H_{23}N$

Sigma N 5503 FW 465.6 $C_{11}H_{12}N_2O_5S \cdot C_{12}H_{23}N$

Hydroxyproline — *N*-t-BOC-L-Hyp

Sigma B 5923 FW 231.2 $C_{10}H_{17}NO_5$

Hydroxyproline — *N*ᵃ-t-BOC-L-Hyp Dicyclohexylammonium Salt

ICN 150497 MW 412.6 $C_{10}H_{17}NO_5 \cdot C_{12}H_{23}N$

Hydroxyproline — Pth-4-Hyp

Sigma P 0877 FW 248.3 $C_{12}H_{12}N_2O_2S$ ~99% | Useful as standards in protein sequencing by the Edman degradation

Hydroxyproline — *trans*-3-Hydroxy-L-Pro

Synonyms: (2S,3S)-3-Hydroxypyrrolidine-2-Carboxylic Acid

Fluka 56244 MW 131.14 $C_5H_9NO_3$ ≥98.0% (titration)

Hydroxyproline — *trans*-4-Hydroxy-L-Pro

Synonyms: L-Hyp; (2S,4R)-4-Hydroxy-2-Pyrrolidinecarboxylic Acid; Hydroxy-L-Pro; L-4-Hydroxy-2-Pyrrolidinecarboxylic Acid

Sigma H 6002 FW 131.1 $C_5H_9NO_3$ Crystalline | Natural constituent of animal proteins like collagen & elastin; also available as part of a kit

Sigma H 7279 FW 131.1 $C_5H_9NO_3$ SigmaUltra; residue on ignition: <0.1%; solubility (0.5 *M* in H_2O, 20°C): complete, colorless; insoluble matter: <0.1%; NH_4^+, Cl, SO_4: <0.05%; K, Na: <0.005%; Al, Ca, Cu, Fe, Mg, P, Zn: <0.0005%; Pb: <0.001% | Natural constituent of animal proteins like collagen & elastin

Sigma H 9261 Mammalian FW 131.1 $C_5H_9NO_3$ Cell culture tested; insect cell culture tested

Hydroxyproline — *trans*-4-Hydroxy-L-Pro-AMC Hydrochloride

Sigma H 0157 FW 324.8 $C_{15}H_{16}N_2O_4 \cdot HCl$

Hydroxyproline — Z-*cis*-Hyp

Bachem C-3200.0001 MW 265.27 $C_{13}H_{15}NO_5$ Store at RT

Hydroxyproline — Z-Hyp

Bachem C-2015.0005 MW 265.27 $C_{13}H_{15}NO_5$ Store at RT

Hydroxyproline — Z-Hyp-OMe

Bachem C-2020.0001 MW 279.29 $C_{14}H_{17}NO_5$ Store at -15°C

Hydroxyproline — Z-HyptBu

Bachem C-3150.0001 MW 321.37 $C_{17}H_{23}NO_5$ Store at RT

Hydroxyproline — Z-L-4-Hyp

Fluka 96310 MW 265.27 $C_{13}H_{15}NO_5$ ≥99.0% (titration); mp: 105-107°C

Hydroxytryptophan — 5-Hydroxy-DL-Trp

Synonyms: DL-2-Amino-3-(5-Hydroxyindolyl)-Propionic Acid; DL-5-HTP; Aromatic Amino Acid L-Decarboxylase Substrate

Fluka 56590 MW 220.23 $C_{11}H_{12}N_2O_3$ ~99.0% (titration); mp: 295°C; ≤0.05% residue on ignition

ICN 100749 MW 220.2 $C_{11}H_{12}N_2O_3$ ~99%, crystalline | Precursor of serotonin

Sigma H 8127 FW 220.2 $C_{11}H_{12}N_2O_3$ Sigma grade; ~99% | Immediate precursor of serotonin

Hydroxytryptophan — 5-Hydroxy-DL-Trp-OEt Hydrochloride

ICN 195228 MW 284.7 $C_{13}H_{16}N_2O_3 \cdot HCl$ Crystalline

Sigma H 8377 FW 284.7 $C_{13}H_{16}N_2O_3 \cdot HCl$

Hydroxytryptophan — 5-Hydroxy-D-Trp

Synonyms: D-2-Amino-3-(5-Hydroxyindolyl)-Propionic Acid; D-5-Hydroxytryptophan

ICN 157510 MW 220.2 $C_{11}H_{12}N_2O_3$ Crystalline

Sigma H 8002 FW 220.2 $C_{11}H_{12}N_2O_3$ Inactive isomer

Hydroxytryptophan — 5-Hydroxy-L-Trp

Synonyms: L-5- Hydroxytryptophan; Aromatic L-Amino Acid Decarboxylase Substrate

Alexis 106-023 MW 220.2 $C_{11}H_{12}N_2O_3$ White solid; soluble in H_2O

Calbiochem 399698 MW 220.2 $C_{11}H_{12}N_2O_3$ ≥95% (HPLC); solid; soluble in H_2O & acidic solutions; LD_{50}≤2000 mg/kg | Serotonin precursor; *Merck Index*, 12: 4895

Fluka 56570 MW 220.23 $C_{11}H_{12}N_2O_3$ ~99.0% (titration); mp: 275°C; ≤0.05% residue on ignition; ≤2% loss on drying | Substrate of choice for the assay of aromatic L-AA decarboxylase; Lovenberg, W, *Meth Enzymol*, 17A: 652, 1970

ICN 100751 MW 220.2 $C_{11}H_{12}N_2O_3$ 99%, crystalline | Precursor of serotonin

Hydroxytryptophan — 5-Hydroxy-L-Trp Hydrate

Synonyms: L-2-Amino-3-(5-Hydroxyindolyl)-Propionic Acid; L-5-HTP

Sigma H 9772 FW 220.2 $C_{11}H_{12}N_2O_3$ Immediate precursor of serotonin, L-aromatic AA decarboxylase substrate

Hydroxytryptophan — BOC-L-5-Hytr

American Peptide BLTRP20 MW 320.4

Isoglutamine — *N*ᵃ-t-BOC-L-Igln

Sigma B 6764 FW 246.3 $C_{10}H_{18}N_2O_5$

Isoleucine — 5-L-Ile (Tyrosyl-3,5-³H(N))

Synonyms: Angiotensin II

ARC ART-700 MW 1046.1 40-80 Ci/mmol; 1.48-2.96 TBq/mmol; in 0.05 *M* acetic acid in silanized vial | Radiochemical

Isoleucine — Ac-Ile

Bachem E-1080.0001 MW 173.21 $C_8H_{15}NO_3$ Store at RT

Isoleucine — Ac-Ile Amide

Bachem E-1085.0001 MW 172.23 $C_8H_{16}N_2O_2$ Store at RT

Isoleucine — Ac-Ile-NHMe

Bachem E-1090.0250 MW 186.25 $C_9H_{18}N_2O_2$ Store at RT

Isoleucine — Ac-Ile-OMe

Bachem E-1095.0001 MW 187.24 $C_9H_{17}NO_3$ Store at RT

Isoleucine — allo-Ile

Bachem F-1160.0250 MW 131.18 $C_6H_{13}NO_2$ Store at RT

Isoleucine — BOC-allo-Ile

Synonyms: BOC-allo-Ile

Bachem A-3345.0001 MW 231.29 $C_{11}H_{21}NO_4$ Store at RT

Peptides International BLX-5342-PI MW 231.29 >98% (TLC); white powder

Isoleucine — BOC-D-allo-Ile

Bachem A-3735.0001 MW 231.29 $C_{11}H_{21}NO_4$ Store at RT

Isoleucine — BOC-D-Ile

Bachem A-1830.0001 MW 231.29 $C_{11}H_{21}NO_4$ Store at RT

Fluka 15127 MW 231.29 $C_{11}H_{21}NO_4$ ≥98% (TLC)

Isoleucine — BOC-D-Ile Hemihydrate

Synonyms: BOC-D-Ile Hemihydrate

Neosystem BA01001 MW 240.3

Peptides International BDI-5318-PI MW 240.3 >98% (HPLC); white crystalline powder

Isoleucine — BOC-DL-Ile

Bachem A-2495.0001 MW 231.29 $C_{11}H_{21}NO_4$ Store at RT

Isoleucine — BOC-Ile

American Peptide BLILE10 MW 240.3

Bachem A-1825.0005 MW 231.29 $C_{11}H_{21}NO_4$ Store at RT

Isoleucine — BOC-Ile Hemihydrate

Synonyms: BOC-L-Ile Hemihydrate

Bachem A-3450.0005 MW 240.3 $C_{11}H_{21}NO_4 \cdot \frac{1}{2}H_2O$ Store at RT

Peptides International BLI-2065 MW 240.30 >98% (HPLC); white crystalline powder

USBio B2325 MW 240.3 $C_{11}H_{21}NO_4 \cdot \frac{1}{2}H_2O$ ≥99%

Isoleucine — BOC-Ile-® (200-400 mesh)

Bachem D-1025.0005 Store at RT

Isoleucine — BOC-Ile-ONp

Bachem A-1840.0001 MW 352.39 $C_{17}H_{24}N_2O_6$ Store at -15°C

Isoleucine — BOC-Ile-O-Resin

American Peptide RMRB200 1% DVB cross-linked: 100-200 mesh | t-BOC protected AA resin; preattached resins are useful for synthesis of peptides with C-terminal carboxyl group by t-BOC chemistry

Isoleucine — BOC-Ile-OSu

Bachem A-1835.0001 MW 328.37 $C_{15}H_{24}N_2O_6$ Store at -15°C

Isoleucine — BOC-Ile-PAM-® (200-400 mesh)

Bachem D-1225.0001 Store at 2-8°C

Isoleucine — BOC-Ile-PAM-Resin

American Peptide RPAM200 1% DVB cross-linked: 100-200 mesh | t-BOC protected AA resin; preattached resins are useful for synthesis of peptides with C-terminal carboxyl group by t-BOC chemistry

Isoleucine — BOC-L-allo-Ile

Fluka 15532 MW 231.29 $C_{11}H_{21}NO_4$ ≥99% (TLC); mp: 60-64°C

Isoleucine — BOC-L-Ile

Fluka 15440 MW 231.29 $C_{11}H_{21}NO_4$ ~98% (titration); mp: 66-69°C

Isoleucine — BOC-L-Ile Hemihydrate

Neosystem BA01002 MW 240.3

Isoleucine — BOC-L-Ile Hydroxysuccinimide Ester

Fluka 15442 MW 328.37 $C_{15}H_{24}N_2O_6$ ≥98% (HPLC); mp: 103-108°C

Isoleucine — BOC-L-Ile N-Carboxy Anhydride

Fluka 15441 MW 257.29 $C_{12}H_{19}NO_5$ ≥97% (titration); mp: 109-112°C | Produced by Propeptide under an exclusive license from Bioresearch

Isoleucine — BOC-L-Ile-(4-ONp)

Fluka 15128 MW 352.39 $C_{17}H_{24}N_2O_6$ ≥98% (TLC)

Isoleucine — BOC-L-Ile-MBHA Resin

Neosystem RN01002

Isoleucine — BOC-L-Ile-O-CH₂-φ-CH₂-COOH

Neosystem LP01002 MW 379.5

Isoleucine — BOC-L-Ile-PAM Resin

Neosystem RP01002

Isoleucine — BOC-Me-Ile

USBio B2352 MW 245.3 $C_{12}H_{23}NO_4$ ≥98%

Isoleucine — BOC-N-Me-allo-Ile

Bachem A-2025.0001 MW 245.32 $C_{12}H_{23}NO_4$ Store at RT

Isoleucine — BOC-N-Me-D-allo-Ile

Bachem A-3730.0001 MW 245.32 $C_{12}H_{23}NO_4$ Store at RT

Isoleucine — BOC-N-Me-D-Ile

Bachem A-3725.0001 MW 245.32 $C_{12}H_{23}NO_4$ Store at RT

Isoleucine — BOC-N-Me-Ile

Bachem A-2055.0001 MW 245.32 $C_{12}H_{23}NO_4$ Store at RT

Isoleucine — BOC-N-Me-L-Ile

Neosystem BA01003 MW 245.3

Isoleucine — BOC-N-Me-L-Ile Dicyclohexylamine

Fluka 15445 MW 426.64 $C_{12}H_{23}NO_4 \cdot C_{12}H_{23}N$ ~99% (TLC); mp: 116-119°C

Isoleucine — Bsmoc-Ile

Synonyms: Bsmoc-L-Ile

Peptides International BLI-5628-PI MW 353.40 >98% (HPLC); white to off-white powder | LA Carpino, et al, *JACS*, 119:9915, 1997

Isoleucine — Bzl-Ile

Bachem E-1565.0001 MW 221.3 $C_{13}H_{19}NO_2$ Store at RT

Isoleucine — CBZ-Ile Oil
USBio C2098-93   MW 265.4 $C_{14}H_{19}NO_4$ ≥99%

Isoleucine — CBZ-Ile-ONp
USBio C2098-94   MW 386.4 $C_{20}H_{22}N_2O_6$ ≥99%

Isoleucine — CBZ-Ile-OSu
USBio C2098-95   MW 362.4 $C_{18}H_{22}N_2O_6$ ≥99%

Isoleucine — CBZ-Me-Ile
USBio C2099-23   MW 279.3 $C_{15}H_{21}NO_4$ ≥99%

Isoleucine — Chloroacetyl-Ile
Bachem E-1840.0005   MW 207.66 $C_8H_{14}ClNO_3$ Store at RT

Isoleucine — D-allo-Ile
Synonyms: D-allo-Ile; D-allo-2-Amino-3-Methylpentanoic Acid; (2R,3S)-2-Amino-3-Methylpentanoic Acid

Bachem F-1165.0250   MW 131.18 $C_6H_{13}NO_2$ Store at RT
Fluka 05705   MW 131.18 $C_6H_{13}NO_2$ ≥99% (titration); ~1% D-Ile, L-allo-Ile; mp: ~285°C
ICN 155095   MW 131.2 $C_6H_{13}NO_2$ Crystalline
Sigma I 0380   FW 131.2 $C_6H_{13}NO_2$

Isoleucine — Dansyl-L-Ile Cyclohexylammonium Salt
ICN 100084   MW 463.6 $C_{18}H_{24}N_2O_4S \cdot C_6H_{13}N$
Sigma D 1125   FW 463.6 $C_{18}H_{24}N_2O_4S \cdot C_6H_{13}N$ ~99%

Isoleucine — D-Ile
Synonyms: (-)-2-Amino-3-Methylpentanoic Acid; (2R,3R)-2-Amino-3-Methylpentanoic Acid

Bachem F-1685.0001   MW 131.18 $C_6H_{13}NO_2$ Store at RT
ICN 155094   MW 131.2 $C_6H_{13}NO_2$ Crystalline; ≤10% allo-isomer
Peptides International ADI-2819   MW 131.17   >99.9% optical purity by RP-HPLC; white crystalline powder
Sigma I 7634   FW 131.2 $C_6H_{13}NO_2$ ≥98% (TLC); may contain ≤10% allo-isomer | Also available as part of a kit
USBio I8940-10   MW 131.2 $C_6H_{13}NO_2$ ≥99%

Isoleucine — D-Ile-OBzl p-Tosyl
Bachem F-3575.0001   MW 393.5 $C_{13}H_{19}NO_2 \cdot C_7H_8O_3S$ Store at RT

Isoleucine — Dinitropyridyl-L-Ile Monocyclohexylammonium Salt
ICN 102080   MW 397.4 $C_{11}H_{14}N_4O_6 \cdot C_6H_{13}N$

Isoleucine — DL-4-ThiaIle
Synonyms: 2-Amino-3-Methylthiobutyric Acid

Sigma T 6259   FW 149.2 $C_5H_{11}NO_2S$

Isoleucine — DL-allo-Ile
Synonyms: (2RS,3SR)-2-Amino-3-Methylpentanoic Acid; DL-allo-Ile

Bachem F-1170.0001   MW 131.18 $C_6H_{13}NO_2$ Store at RT
Fluka 05706   MW 131.18 $C_6H_{13}NO_2$ ≥99% (titration); ≤0.6% DL-Ile; mp: ~285°C

Isoleucine — DL-Ile
Synonyms: N-2-Methylpropionyl-D-Cys; 2-Amino-3-Methylpentanoic Acid; DL-2-Amino-3-Me-Pentanoic Acid

Fluka 58883   MW 131.18 $C_6H_{13}NO_2$ Mixture of 50% DL-Ile & 50% DL-allo-Ile; mp: 280°C

Fluka 58884   MW 131.18 $C_6H_{13}NO_2$ ≥99% (titration); ≤1% DL-allo-Ile
ICN 102082   MW 131.2 $C_6H_{13}NO_2$ ≥99%; crystalline; essentially free of Leu
Sigma I 5393   FW 131.2 $C_6H_{13}NO_2$ ≥98% (TLC); essentially free of Leu; a mixture of 4 stereoisomers; specific rotation: 0±2° | Also available as part of a kit
Sigma I 6268   FW 131.2 $C_6H_{13}NO_2$ Essentially free of Leu | A mixture of 4 stereoisomers; cell culture tested

Isoleucine — Dnp-L-Ile
ICN 102079   MW 297.3 $C_{12}H_{15}N_3O_6$ Crystalline

Isoleucine — FMOC-allo-Ile
Synonyms: FMOC-L-allo-Ile

Bachem B-2880.0001   MW 353.42 $C_{21}H_{23}NO_4$ Store at RT
Peptides International FLI-1757-PI   MW 353.42   >98% (HPLC); white crystalline powder

Isoleucine — FMOC-D-allo-Ile
Bachem B-2230.0001   MW 353.42 $C_{21}H_{23}NO_4$ Store at RT

Isoleucine — FMOC-D-Ile
Bachem B-1630.0001   MW 353.42 $C_{21}H_{23}NO_4$ Store at RT
Neosystem FA01001   MW 353.4
USBio F5395   MW 353.4 $C_{21}H_{23}NO_4$ ≥99%

Isoleucine — FMOC-D-Ile-SASRIN™-® (200-400 mesh)
Bachem D-1760.0001   Store at 2-8°C

Isoleucine — FMOC-D-Ile
Synonyms: FMOC-D-Ile

Peptides International FDI-1856-PI   MW 353.42   >98% (HPLC); white powder

Isoleucine — FMOC-Ile
Synonyms: FMOC-L-Ile

American Peptide FILE110   MW 353.4
Bachem B-1340.0005   MW 353.42 $C_{21}H_{23}NO_4$ Store at RT
Peptides International FLI-1724-PI   MW 353.42   >98% (HPLC); white crystalline powder
USBio F5396   MW 353.4 $C_{21}H_{23}NO_4$ ≥99%

Isoleucine — FMOC-Ile-® (200-400 mesh)
Bachem D-1110.0001   Store at RT

Isoleucine — FMOC-Ile-ol
Synonyms: FMOC-L-isoleucinol

Senn Chem 44060   MW 339.4

Isoleucine — FMOC-Ile-OPfp
Bachem B-1145.0001   MW 519.47 $C_{27}H_{22}F_5NO_4$ Store at -15°C
USBio F5397   MW 519.5 $C_{27}H_{22}F_5NO_4$ ≥99%

Isoleucine — FMOC-Ile-SASRIN™-® (200-400 mesh)
Bachem D-1355.0001   Store at 2-8°C

Isoleucine — FMOC-L-Ile
Fluka 47628   MW 353.43 $C_{21}H_{23}NO_4$ ≥98% (titration); mp: 146-147°C
Neosystem FA01002   MW 353.4

826

Isoleucine — FMOC-L-Ile 4-Benzyloxybenzyl Ester Polymer Bound

Fluka 47661 0.4-0.6 mmol/g resin; carrier: polystyrene, crosslinked with 1% DVB; 200-400 mesh particle size

Isoleucine — FMOC-L-Ile *N*-Carboxy Anhydride

Fluka 47687 MW 379.41 $C_{22}H_{21}NO_5$ ≥97% (GC); mp: 115-118°C | Produced by Propeptide under an exclusive license from Bioresearch

Isoleucine — FMOC-L-Ile-OPfp

Fluka 47466 MW 519.47 $C_{27}H_{22}F_5NO_4$ ≥98% (TLC)

Isoleucine — FMOC-L-Ile-Wang Resin

American Peptide RFWAN50 Preattached Wang resins are usful for synthesis of peptides with C-terminal carboxyl groups; peptides can be cleaved from the resin with 90% TFA in the presence of scavengers

Neosystem RW01002

Isoleucine — FMOC-Me-Ile

USBio F5420 MW 367.4 $C_{22}H_{25}NO_4$ ≥99%

Isoleucine — FMOC-*N*-Me-Ile

Synonyms: FMOC-*N*-Me-L-Ile

Peptides International FMI-1793-PI MW 367.45 >97% (HPLC); white crystals

Isoleucine — FMOC-*N*-Me-L-Ile

Fluka 47596 MW 367.45 $C_{22}H_{25}NO_4$ ≥98% (HPLC); mp: 181-184°C; crystallized with 1 mol diethyl ether

Isoleucine — Ile

Bachem E-2025.0025 MW 131.18 $C_6H_{13}NO_2$ Store at RT

Isoleucine — Ile Amide Hydrochloride

Bachem E-2030.0001 MW 166.65 $C_6H_{14}N_2O \cdot HCl$ Store at RT

Isoleucine — Ile-AMC

Bachem I-1420.0250 MW 288.4 $C_{16}H_{20}N_2O_3$ Store at -15°C

Isoleucine — Ile-AMC TFA

Bachem I-1235.0050 MW 402.4 $C_{16}H_{20}N_2O_3 \cdot C_2HF_3O_2$ Store at -15°C

Isoleucine — Ile-NHOH

Bachem E-2805.0001 MW 178.19 $C_6H_{14}N_2O_2$ Store at -15°C

Isoleucine — Ile-OBzl *p*-Tosyl

Bachem E-2035.0005 MW 393.5 $C_{13}H_{19}NO_2 \cdot C_7H_8O_3S$ Store at RT

Isoleucine — Ile-OBzl-Tosyl

Synonyms: L-Ile-OBzl-Tosyl

Peptides International EBI-2111 MW 393.50 >98% (TLC); white powder

Isoleucine — Ile-ol

Synonyms: L-Isoleucinol

Senn Chem 44097 MW 117.2

Isoleucine — Ile-OMe Hydrochloride

Synonyms: L-Ile-OMe

Bachem E-2045.0005 MW 181.66 $C_7H_{15}NO_2 \cdot HCl$ Store at RT

Peptides International EMI-2039-PI MW 181.66 >98% (TLC); white powder

Isoleucine — Ile-OtBu Hydrochloride

Bachem E-2040.0001 MW 223.74 $C_{10}H_{21}NO_2 \cdot HCl$ Store at -15°C

Isoleucine — Ile-pNA

Bachem L-1815.0001 MW 251.29 $C_{12}H_{17}N_3O_3$ Store at -15°C

Isoleucine — Ile-βNA

Bachem K-1375.0250 MW 256.35 $C_{16}H_{20}N_2O$ Store at RT

Isoleucine — L-allo-Ile

Synonyms: L-allo-Ile; L-allo-2-Amino-3-Methylpentanoic Acid; (2S,3R)-2-Amino-3-Methylpentanoic Acid

Fluka 05703 MW 131.18 $C_6H_{13}NO_2$ ≥99% (titration); mp: ~285°C

ICN 155096 MW 131.2 $C_6H_{13}NO_2$ Crystalline

Sigma I 8754 FW 131.2 $C_6H_{13}NO_2$

Isoleucine — L-Ile

Synonyms: 2-Amino-3-Methylvaleric Acid; (2S,3S)-2-Amino-3-Methylpentanoic Acid; D-allo-Ile

Amersham US17825 98.5-101.5% dsb; ≤0.0015% heavy metals (Pb)

Fluka 58879 MW 131.18 $C_6H_{13}NO_2$ ≥99.5% (titration); free of allo-Ile; ≤0.05% residue on ignition, loss on drying; ≤0.3% foreign AA; ≤0.00001% As; ≤0.0005% Al, Ba, Bi, Cd, Co, Cr, Cu, Fe, Li, Mg, Mn, Mo, Ni, Pb, Sr, Zn; ≤0.001% Ca; ≤0.005% Na, K; ≤0.01% Cl, NH₄⁺, SO₄ | *CRC Handbook of Microbiology*, Laskin, AI & Lachevalier, HA, eds, 4: 1, 1974, CRC Press, Cleveland, Ohio

Fluka 58880 MW 131.18 $C_6H_{13}NO_2$ ≥99.0% (titration); free of allo-Ile; ≤0.1% residue on ignition, loss on drying; ≤0.3% foreign AA; ≤0.0005% Cd, Co, Cr, Cu, Fe, Mg, Mn, Ni, Pb, Zn; ≤0.001% Ca; ≤0.005% Na, K; ≤0.01% Cl, NH₄⁺, SO₄

ICN 102092 MW 131.2 $C_6H_{13}NO_2$ 99%; crystalline

ICN 194689 MW 131.2 $C_6H_{13}NO_2$ 99%; crystalline | Cell culture reagent

Neosystem AA01002 MW 131.2

Peptides International ALI-2712 MW 131.17 >99.9% optical purity by RP-HPLC; white crystalline powder

Rexim MW 131.2 $C_6H_{13}NO_2$ White crystals or crystalline powder

Sigma 38,648-0 FW 132.2 $C_6H_{13}{}^{15}NO_2$ 98% atom % ^{15}N; screw cap bottle

Sigma I 2752 FW 131.2 $C_6H_{13}NO_2$ ≥98% (TLC) | Also available as part of a kit

Sigma I 2877 FW 131.2 $C_6H_{13}NO_2$ ~1:1 mixture of diastereomers | Often erroneously called DL-Ile; see separate listing of DL-Ile, which is a mixture of 4 stereoisomers

Sigma I 4505 Matrix: 4% beaded agarose; activation: cyanogen bromide; attachment: amino; spacer: 1 atom; ligand immobilized: 3-7 µmoles/mL; form: suspension in 2.0 *M* NaCl containing 0.02% thimerosal | Deutsch, DG & Mertz, ET, *Proc Fed Amer Soc Exp Biol*, 29: 647, 1970; Deutsch, DG & Mertz, ET, *Science*, 170: 1095, 1970; Vuento, M & Vaheri, A, *Biochem J*, 183: 331, 1979

Sigma I 7268 FW 131.2 $C_6H_{13}NO_2$ SigmaUltra; ≥98% (TLC); residue on ignition: <0.3%; solubility (0.5 *M* in 1 *M* NH₄OH): complete, colorless; insoluble matter: <0.1%; NH₄⁺: <0.2%; Cl, SO₄: <0.05%; K, Na: <0.005%; Al, Ca, Cu, Fe, Mg, Zn: <0.0005%; Pb, P: <0.001%

Sigma I 7383 FW 131.2 $C_6H_{13}NO_2$ Crystalline | Cell culture tested; insect cell culture tested

Sigma I 8905	FW 131.2	$C_6H_{13}NO_2$	USP Grade

USBio I8940   MW 131.2 $C_6H_{13}NO_2$ ≥98.5%; appearance: white crystalline powder; pH: 5.5-6.5; specific rotation: C=4, 6 *N* HCl +39.5° to +41.5°; heavy metals (Pb): ≤0.001%; endotoxin: ≤2 EU/mg; cell culture tested: murine myeloma SP2/0-Ag14

Isoleucine — L-Ile (1-^{14}C)

ARC ARC-326   MW 131.2 $CH_3CH_2CH(CH_3)CH(NH_2)COOH$ 50-60 mCi/mmol; 1.85-2.22 GBq/mmol; in EtOH: H_2O (2:98) | Radiochemical

Isoleucine — L-Ile (4,5-^{3}H(N))

ARC ART-233   MW 131.2 $CH_3CH_2CH(CH_3)CH(NH_2)COOH$ 30-60 Ci/mmol; 1.11-2.22 TBq/mmol; in 0.01 *N* HCl | Radiochemical

Isoleucine — L-Ile (4,5-^{3}H)

Amersham TRK585   Aqueous, 2% EtOH, sterilized; 3-4.4 TBq/mmol, 80-120 Ci/mmol; 37 MBq/mL, 1 mCi/mL

ICN 20031E   MW 131.2 $CH_3CH_2CH(CH_3)CH(NH_2)COOH$ 30-60 Ci/mmol, 1.11-2.22 TBq/mmol; sterile EtOH:H_2O (2:98)

Isoleucine — L-Ile (U-^{14}C)

Amersham CFB68   Aqueous, 2% EtOH, sterilized; >11 GBq/mmol, >300 mCi/mmol; 1.85 MBq/mL, 50 µCi/mL

ARC ARC-672   MW 131.2 $CH_3CH_2CH(CH_3)CH(NH_2)COOH$ 200-300 mCi/mmol; 7.4-11.1 GBq/mmol; in EtOH: H_2O (2:98) | Radiochemical

ICN 10083E   MW 131.2 $CH_3CH_2CH(CH_3)CH(NH_2)COOH$ >270 mCi/mmol, >12.2 GBq/mmol; sterile EtOH:H_2O (2:98)

ICN 10083L   MW 131.2 $CH_3CH_2CH(CH_3)CH(NH_2)COOH$ 200-270 mCi/mmol, 9.0-12.2 GBq/mmol; >90%; sterile EtOH:H_2O (2:98)

Isoleucine — L-Ile (U-^{14}C) Hydrochloride

Sigma I 3388 $CH_3CH_2CH(CH_3)CH-(NH_2)COOH$ 240-360 mCi/mmol; radiochemical purity ≥98% (HPLC); aqueous solution containing 2% EtOH in Combi-vial | Radiochemical;

Isoleucine — L-Ile Amide Hydrochloride

Neosystem AA01003   MW 166.7

Isoleucine — L-Ile Hydroxamate

Sigma I 4142	FW 146.2	$C_6H_{14}N_2O_2$

Isoleucine — L-Ile β-Naphthlyamide

ICN 155100   MW 256.3 $C_{16}H_{20}NO_2$ Crystalline | Possible carcinogen

Isoleucine — L-Ile β-Naphthylamide

Sigma I 4879	FW 256.3	$C_{16}H_{20}N_2O$

Isoleucine — L-Ile-AMC TFA Salt

Sigma I 5025	FW 402.4	$C_{16}H_{20}N_2O_3 \cdot C_2HF_3O_2$

Isoleucine — L-Ile-OAll p-Tosyl

ICN 155097	MW 343.4	$C_9H_{17}NO_2 \cdot C_7H_8O_3S$	
Sigma I 5013	FW 343.4	$C_9H_{17}NO_2 \cdot C_7H_8O_3S$	

Isoleucine — L-Ile-OMe Hydrochloride

Fluka 58920   MW 181.66 $C_7H_{15}NO_2 \cdot HCl$ ≥98%; mp: 98-100°C

ICN 155099	MW 181.7	$C_7H_{15}NO_2 \cdot HCl$	Crystalline
Neosystem AA01004	MW 181.7		
Sigma I 6252	FW 181.7	$C_7H_{15}NO_2 \cdot HCl$	

Isoleucine — L-Ile-OtBu Dibenzenesulfimide Salt

Fluka 58905   MW 484.64 $C_{10}H_{21}NO_2 \cdot C_{12}H_{11}NO_4S_2$ ≥99%; mp: 134-135°C

Isoleucine — L-Ile-OtBu Hydrochloride

ICN 155098	MW 223.7	$C_{10}H_{21}NO_2 \cdot HCl$	Crystalline
Sigma I 2000	FW 223.7	$C_{10}H_{21}NO_2 \cdot HCl$	

Isoleucine — L-Isoleucinamide Hydrochloride

Sigma I 5886	FW 166.7	$C_6H_{14}N_2O \cdot HCl$

Isoleucine — L-Isoleucinol Hydrochloride

Bachem F-1690.0250   MW 153.65 $C_6H_{15}NO \cdot HCl$ Store at -15°C

Isoleucine — N-2,4-Dnp-L-Ile

Sigma D 9879	FW 297.3	$C_{12}H_{15}N_3O_6$

Isoleucine — N-Ac-D-allo-Ile

ICN 100053	MW 173.2	$C_8H_{15}NO_3$	Crystalline

Isoleucine — N-Bz-L-Ile

Sigma B 1394	FW 221.3	$C_{13}H_{19}NO_2$

Isoleucine — N-CBZ-Ile-(pONp)

ICN 101265	MW 386.4	$C_{20}H_{22}N_2O_6$	Crystalline

Isoleucine — N-CBZ-L-Ile

ICN 100221	MW 265.3	$C_{14}H_{19}NO_4$	Viscous oil

Isoleucine — N-CBZ-L-Ile N-Hydroxysuccinimide Ester

Sigma C 1045	FW 362.4	$C_{18}H_{22}N_2O_6$

Isoleucine — N-FMOC-L-Ile

Sigma F 0134	FW 353.4	$C_{21}H_{23}NO_4$

Isoleucine — N-FMOC-L-Ile Resin Ester

Sigma F 1511   For peptide synthesis; Wang, SS, *J Am Chem Soc*, 95: 1328, 1973; Lu, G et al, *J Org Chem*, 46: 3433, 1981

Isoleucine — N-Me-allo-Ile-OBzl p-Tosyl

Bachem F-1760.0250   MW 407.53 $C_{14}H_{21}NO_2 \cdot C_7H_8O_3S$ Store at RT

Isoleucine — N-Me-Ile

Bachem E-2140.0001   MW 145.2 $C_7H_{15}NO_2$ Store at RT

Isoleucine — N-Me-Ile-OMe Hydrochloride

Bachem F-3915.0250   MW 195.69 $C_8H_{17}NO_2 \cdot HCl$ Store at RT

Isoleucine — N-MSOC-L-Ile

ICN 155738	MW 281.3	$C_{10}H_{19}NO_6S$	Crystalline

Isoleucine — N-O-Nps-L-Ile Dicyclohexylammonium Salt

ICN 102530   MW 465.6 $C_{12}H_{16}N_2O_4S \cdot C_{12}H_{23}N$ Yellow crystals

Sigma N 5628   FW 465.6 $C_{12}H_{16}N_2O_4S \cdot C_{12}H_{23}N$ Yellow crystals

Isoleucine — *N*-t-BOC-L-Ile Resin Ester

Sigma B 8145   *N*-t-BOC AA ester of methylated 1% divinylbenzene cross-linked polystyrene; 200-400 mesh; substitution range: 0.2-0.6 mmole t-BOC-AA/g resin | For peptide synthesis by the Merrifield method

Isoleucine — *N*$^\alpha$-FMOC-L-Ile

ICN 151145   MW 353.4 $C_{21}H_{23}NO_4$

Isoleucine — *N*$^\alpha$-t-BOC-D-Ile

Sigma B 8147   FW 231.3 $C_{11}H_{21}NO_4$

Isoleucine — *N*$^\alpha$-t-BOC-L-Ile

Sigma B 7126   FW 231.3 $C_{11}H_{21}NO_4$

Isoleucine — *N*$^\alpha$-t-BOC-L-Ile Hemihydrate

ICN 104917   MW 240.3 $C_{11}H_{21}NO_4 \cdot \frac{1}{2}H_2O$ Crystalline

Isoleucine — Pth-Ile

Sigma P 1002   FW 248.3 $C_{13}H_{16}N_2OS$ ~99% | Useful as standards in protein sequencing by the Edman degradation

Isoleucine — TentaGel S PHB-Ile-FMOC

Synonyms: FMOC-L-Ile 4-Poly(Ethylenoxy)-OBzl Polymer Bound

Fluka 86393   0.18 mmol protected AA/g; ~90 μm particle size

Isoleucine — TentaGel S Trt-Ile-FMOC

Fluka 86424   0.18 mmol protected AA/g; ~90 μm particle size

Isoleucine — Z-D-allo-Ile Dicyclohexylamine

Bachem C-4075.0001   MW 446.63 $C_{14}H_{19}NO_4 \cdot C_{12}H_{23}N$ Store at -15°C

Isoleucine — Z-Ile

Synonyms: Benzyloxycarbonyl-L-Ile

Bachem C-2025.0025   MW 265.31 $C_{14}H_{19}NO_4$ Store at -15°C

Peptides International ZLI-2028-PI   MW 265.31   98% (TLC); syrup

Isoleucine — Z-Ile Amide

Bachem C-3045.0001   MW 264.32 $C_{14}H_{20}N_2O_3$ Store at RT

Isoleucine — Z-Ile-ONp

Bachem C-2040.0005   MW 386.41 $C_{20}H_{22}N_2O_6$ Store at -15°C

Isoleucine — Z-Ile-OSu

Bachem C-2035.0005   MW 362.38 $C_{18}H_{22}N_2O_6$ Store at -15°C

Isoleucine — Z-L-Ile

Synonyms: (*S*)-α-Z-Amino-γ-Butyrolactone

Fluka 96618   MW 265.31 $C_{14}H_{19}NO_4$ ~97% (HPLC)

Neosystem ZA01001   MW 265.3

Isoleucine — Z-L-Ile Dicyclohexylamine

Fluka 96620   MW 446.64 $C_{14}H_{19}NO_4 \cdot C_{12}H_{23}N$ mp: 159-160°C

Isoleucine — Z-L-Ile Hydroxysuccinimide Ester

Fluka 96621   MW 362.38 $C_{18}H_{22}N_2O_6$ mp: 111-112°C

Isoleucine — Z-L-Ile-(4-ONp)

Fluka 96640   MW 386.40 $C_{20}H_{22}N_2O_6$ ≥98.0% (TLC)

Isoleucine — Z-*N*-Me-Ile

Bachem C-3775.0001   MW 279.34 $C_{15}H_{21}NO_4$ Store at RT

Isoleucine — Z-*N*-Me-L-Ile

Fluka 00419   MW 279.34 $C_{15}H_{21}NO_4$ ≥98.0% (HPLC)

Isoleucine — α-Keto-Ile Calcium Salt

Synonyms: Calcium-3-Me-2-Oxo-Valerate; (*S*)-3-Me-2-Oxopentanoic Acid

Rexim   MW 298.4 $C_{12}H_{18}CaO_6$ White crystals or crystalline powder

Isoserine — (2*R*,3*S*)-3-Phenyl-Ise

Synonyms: (2*R*,3*S*)-3-Amino-2-Hydroxy-3-Phenylpropanoic Acid

Bachem F-3890.0250   MW 181.19 $C_9H_{11}NO_3$ Store at 2-8°C

Isoserine — (2*R*,3*S*)-BOC-3-Phenyl-Ise

Synonyms: (2*R*,3*S*)-3-(BOC-Amino)-2-Hydroxy-3-Phenylpropanoic Acid

Bachem A-4530.0250   MW 281.31 $C_{14}H_{19}NO_5$ Store at -15°C

Isoserine — (2*R*,3*S*)-FMOC-3-Phenyl-Ise

Synonyms: (2*R*,3*S*)-3-(FMOC-Amino)-2-Hydroxy-3-Phenylpropanoic Acid

Bachem B-3405.0250   MW 403.43 $C_{24}H_{21}NO_5$ Store at 2-8°C

Isoserine — DL-Ile

Synonyms: DL-2-Hydroxy-3-Aminopropionic Acid

ICN 155130   MW 105.1 $C_3H_7NO_3$ Crystalline

Sigma I 2503   FW 105.1 $C_3H_7NO_3$

Isoserine — DL-Ise

Synonyms: DL-α-Amino-α-Hydroxypropionic Acid; (±)-3-Amino-2-Hydroxypropionic Acid; 3-Amino-2-Hydroxypropanoic Acid; 2-Hydroxy-β-Ala; 3-Aminolactic Acid

Bachem F-1695.0001   MW 105.09 $C_3H_7NO_3$ Store at -15°C

Fluka 59805   MW 105.09 $C_3H_7NO_3$ ≥99% (titration); mp: 235°C

Rexim   MW 105.1 $C_3H_7NO_3$

Isoserine — FMOC-DL-Ise

Bachem B-2590.0001   MW 327.34 $C_{18}H_{17}NO_5$ Store at RT

Isoserine — L-Ise

Synonyms: (*S*)-3-Amino-2-Hydroxypropanoic Acid

Rexim   MW 105.1 $C_3H_7NO_3$

Isovaline — DL-Val

Synonyms: DL-2-Amino-2-Methylbutyric Acid; α-Amino-α-Methylbutyric Acid

ICN 105143   MW 117.2 $C_5H_{11}NO_2$

Leucine — ((2*R*,3*R*)-3-Amino-2-Hydroxy-4-Phenylbutanoyl)-L-Leu Hydrochloride

Synonyms: Epibestatin; Metallo-Protease Inhibitor

Sigma E 0381   FW 344.8 $C_{16}H_{24}N_2O_4 \cdot HCl$ ≥97% (HPLC) | Selectively inhibits aminopeptidases; bioactive peptide

Leucine — ((2S,3R)-3-Amino-2-Hydroxy-4-Phenylbutanoyl)-Leu Free Acid

ICN 152844   MW 308.4 $C_{16}H_{24}N_2O_4$ Bestatin; inhibitor for aminopeptidase B & leucine aminopeptidase; Umezawa, H et al, *J Antibiot*, 29:97, 1976

Leucine — ((2S,3R)-3-Amino-2-Hydroxy-4-Phenylbutanoyl)-L-Leu

Synonyms: Ubenimex; Bestatin; Aminopeptidase B Inhibitor; Leu Aminopeptidase Inhibitor

Peptides International IBS-4093-PI   Microbial   MW 308.38 $C_{16}H_{24}N_2O_4$ 98%; white crystalline powder; bulk | Inhibitor; Umezawa, H et al, *J Antibiotics*, 29:97, 1976

Leucine — ((2S,3R)-3-Amino-2-Hydroxy-4-Phenylbutyryl)-L-Leu

Synonyms: Bestatin

Bachem F-1215.0005   MW 308.38 $C_{16}H_{24}N_2O_4$ Store at 2-8°C

Leucine — ((RS)-2-Carboxy-3-Phenylpropionyl)-Leu

Bachem N-1185.0025   MW 307.35 $C_{16}H_{21}NO_5$ Store at -15°C

Leucine — (2S,3R)-3-Amino-2-Hydroxy-4-(4-Nitrophenyl)-Butanoyl-D-Leu Hydrochloride

ICN 195665   MW 389.8 $C_{16}H_{23}N_3O_6 \cdot HCl$ Bestatin analog with inhibitory activity 5 times greater

Leucine — (2S,3R)-3-Amino-2-Hydroxy-4-(4-Nitrophenyl)-Butanoyl-L-Leu Hydrochloride

Synonyms: Bestatin Analog; Metallo-Protease Inhibitor

Sigma A 5921   FW 389.8 $C_{16}H_{23}N_3O_6 \cdot HCl$ ≥97% (HPLC) | Bioactive; Enzyme inhibitor; bestatin analog with 5X the inhibitory activity; Nishizawa, R et al, *J Med Chem*, 20: 510, 1977

Leucine — (2S,3R)-3-Amino-2-hydroxy-4-Phenylbutanoyl)-L-Leu Hydrochloride

Synonyms: Bestatin Analog; Bestatin

Sigma B 8385   FW 344.8 $C_{16}H_{24}N_2O_4 \cdot HCl$ ≥98% (HPLC) | Enzyme inhibitor; bioactive peptide

Alexis 260-012   Synthetic   MW 344.9 $C_{16}H_{21}N_2O_4 \cdot HCl$ ≥98%; White lyophilized powder; soluble in MeOH or acetic acid | Aminopeptidase inhibitor; activates macrophages & T lymphocytes; antitumor properties; Umezawa, H et al, *J Antibiot*, 23: 549, 1981; Umezawa, H, *Ann Rev Microbiol*, 36: 75, 1982

Leucine — (2S,3S)-Ethyl-trans-Epoxysuccinyl-L-Leucylamido-3-Methylbutane

Synonyms: E-64d; Cathepsin B Inhibitor; Cathepsin L Inhibitor; Calpain Inhibitor

Neosystem SC878   MW 342.44   A membrane permeable ester derivative of E-64c; McGowan, EB et al, *BBRC*, 158:432-435, 1989

Leucine — (2S,3S)-L-trans-Epoxysuccinyl-L-Leucylamido-3-Methylbutane

Synonyms: E-64c; Cathepsin B Inhibitor; Cathepsin L Inhibitor

Neosystem SC344   MW 314.38   Barrett, AJ et al, *Biochem J*, 201:189-198, 1982

Leucine — (2S,3S)-trans-Epoxysuccinyl-L-Leucylamido-3-Methylbutane

Synonyms: E-64c; Thiol Protease Inhibitor

ICN 195985   MW 314.4 $C_{15}H_{26}N_2O_5$

Leucine — (2S,3S)-trans-Epoxysuccinyl-L-Leucylamido-3-Methylbutane-OEt

Synonyms: E-64d; EST; Calpain Inhibitor; Cysteine Protease Inhibitor

ICN 195986   MW 342.4 $C_{17}H_{30}N_2O_5$ membrane-permeable lysosomal cysteine protease inhibitor, similar to E-64 without charged groups

Leucine — (L-3-trans-Carboxyoxiran-2-Carbonyl)-L-Leu-Agmatin

Synonyms: E-64

Alexis 260-007   Synthetic   MW 366.4 $C_{15}H_{27}N_5O_5 \cdot \frac{1}{2}H_2O$ Specific inhibitor of thiol proteases; Hanada, K et al, *Agric Biol Chem*, 42: 523 & 529, 1978

Leucine — (L-3-trans-Carboxyoxiran-2-Carbonyl)-L-Leu-Agmatine

Synonyms: E-64; Thiol Protease Inhibitor

Peptides International IES-4096-v   Synthetic   MW 357.41 $C_{15}H_{27}N_5O_5$ >99% (HPLC); lyophilized amorphous powder

Leucine — (L-3-trans-Carboxyoxiran-2-Carbonyl)-L-Leu-Agmatine Hemihydrate

Synonyms: E-64; Thiol Protease Inhibitor

Peptides International IES-4096   Synthetic   MW 366.42 $C_{15}H_{27}N_5O_5 \cdot \frac{1}{2}H_2O$ >99% (HPLC); amorphous powder; bulk | Inhibitor; Hanada, K et al, *Agric Biol Chem*, 42:523, 1978; Hanada, K et al, *Agric Biol Chem*, 42:529, 1978

Leucine — (L-3-trans-Carboxyoxirane-2-Carbonyl)-L-Leu (3-Methylbutyl) Amide

Synonyms: E-64-c; Thiol Protease Inhibitor; Cathepsin B/H/L Inhibitor; Calpain Inhibitor

Peptides International IEC-4320-v   Synthetic   MW 314.38 $C_{15}H_{26}N_2O_5$ >99% (HPLC); lyophilized amorphous powder | Inhibitor; Hashida, S et al, *J Biochem*, 88:1805, 1980; Tamai, M et al, *J Biochem*, 90:255, 1981; Barrett, AJ et al, *Biochem J*, 201:189, 1982

Leucine — (L-3-trans-Ethoxycarbonyloxirane-2-Carbonyl)-L-Leu (3-Methylbutyl) Amide

Synonyms: E-64-d; Thiol Protease Inhibitor; Cathepsin B/H/L Inhibitor; Calpain Inhibitor

Peptides International IED-4321-v   Synthetic   MW 342.44 $C_{17}H_{30}N_2O_5$ >99% (HPLC); lyophilized amorphous powder | Inhibitor; membrane permeable analog of E-64-c; Tamai, M et al, *J Pharmacobio-Dyn*, 9:672, 1986; Tamai, M et al, *Chem Pharm Bull*, 35:1098, 1987

Leucine — 4,5-Dehydro-Leu

Bachem F-2985.0001   MW 129.16 $C_6H_{11}NO_2$ Store at -15°C

Leucine — 4-Aza-DL-Leu Dihydrochloride

ICN 100837   MW 205.1 $C_5H_{12}N_2O_2 \cdot 2HCl$ Crystalline | Leucine antagonist

Leucine — Ac-4,5-Dehydro-Leu

Bachem F-3175.0001   MW 171.2 $C_8H_{13}NO_3$ Store at -15°C

Leucine — Ac-D-Leu

Bachem F-1045.0001   MW 173.21 $C_8H_{15}NO_3$ Store at RT

Leucine — Ac-Leu

Bachem E-1100.0005   MW 173.21 $C_8H_{15}NO_3$ Store at RT

Leucine — Ac-Leu Amide

Bachem E-1105.0001   MW 172.23 $C_8H_{16}N_2O_2$ Store at RT

Leucine — Ac-Leu-NHMe

Bachem E-1110.0250   MW 186.25 $C_9H_{18}N_2O_2$ Store at RT

Leucine — Ac-Leu-OMe

Bachem E-1115.0001   MW 187.24 $C_9H_{17}NO_3$ Store at RT

Leucine — Ac-Leu-pNA

Bachem L-1040.0001   MW 293.32 $C_{14}H_{19}N_3O_4$ Store at -15°C

Leucine — BOC-4,5-Dehydro-Leu Dicyclohexylamine

Bachem A-3485.0001   MW 410.6 $C_{11}H_{19}NO_4 \cdot C_{12}H_{23}N$ Store at -15°C

Leucine — BOC-CyclohomoLeu

Synonyms: 1-(BOC-Amino)-Cyclohexanecarboxylic Acid

Fluka 03582   MW 243.3 $C_{12}H_{21}NO_4$ ≥98% (TLC); mp: 176-178°C

Leucine — BOC-Cycloleucine

Synonyms: 1-(BOC-Amino)-Cyclopentanecarboxylic Acid

Fluka 03583   MW 229.28 $C_{11}H_{19}NO_4$ ≥98% (TLC)

Leucine — BOC-D-Leu

American Peptide BDLEU10   MW 240.3

Fluka 15129   MW 231.29 $C_{11}H_{21}NO_4$ ≥98% (TLC)

Leucine — BOC-D-Leu Hydrate

Synonyms: BOC-D-Leu Hydrate

Bachem A-1855.0001   MW 249.31 $C_{11}H_{21}NO_4 \cdot H_2O$ Store at RT

Neosystem BA01101   MW 249.3

Peptides International BDL-2603   MW 249.31   >98% (HPLC); white crystalline powder

USBio B2329   MW 249.3 $C_{11}H_{21}NO_4 \cdot H_2O$ ≥99%

Leucine — BOC-D-Leu-O-CH₂-φ-CH₂-COOH

Neosystem LP01101   MW 379.5

Leucine — BOC-D-Leu-OSu

Bachem A-1865.0001   MW 328.37 $C_{15}H_{24}N_2O_6$ Store at -15°C

Leucine — BOC-D-Leu-PAM Resin

Neosystem RP01101

Leucine — BOC-D-Leu-PAM-® (200-400 mesh)

Bachem D-1590.0001   Store at 2-8°C

Leucine — BOC-DL-Leu Hydrate

Bachem A-2505.0005   MW 249.31 $C_{11}H_{21}NO_4 \cdot H_2O$ Store at RT

Leucine — BOC-D-Me-Leu

USBio B2353   MW 245.3 $C_{12}H_{23}NO_4$ ≥98%

Leucine — BOC-D-t-Leu

Synonyms: N^α-BOC-D-α-tBu-Gly

USBio B2327   MW 231.3 $C_{11}H_{21}NO_4$ ≥99%

Leucine — BOC-Leu

American Peptide BLLEU10   MW 240.3

Leucine — BOC-Leu (^{15}N)

Bachem A-3795.0100   MW 232.3 $C_{11}H_{23}^{15}NO_5$ Store at -15°C

Leucine — BOC-Leu Hydrate

Synonyms: BOC-L-Leu Hydrate

Bachem A-1850.0005   MW 249.31 $C_{11}H_{21}NO_4 \cdot H_2O$ Store at RT

Peptides International BLL-2055   MW 249.31   >98% (HPLC); white crystalline powder

USBio B2330 $C_{11}H_{21}NO_4 \cdot H_2O$ ≥99%

Leucine — BOC-Leu-CMK

Bachem N-1440.0001   MW 263.76 $C_{12}H_{22}ClNO_3$ Store at -15°C

Leucine — BOC-Leu-OEt

Synonyms: BOC-L-Leu-OEt

Peptides International BEL-5540-PI   MW 259.35   >98% (TLC); colorless oil

Leucine — BOC-Leu-ol

Synonyms: BOC-L-Leucinol

Senn Chem 44034   MW 217.3

Leucine — BOC-Leu-ONp

Bachem A-1870.0001   MW 352.39 $C_{17}H_{24}N_2O_6$ Store at -15°C

Leucine — BOC-Leu-O-Resin

American Peptide RMRB205   1% DVB cross-linked: 100-200 mesh | t-BOC protected AA resin; preattached resins are useful for synthesis of peptides with C-terminal carboxyl group by t-BOC chemistry

Leucine — BOC-Leu-OSu

Bachem A-1860.0001   MW 328.37 $C_{15}H_{24}N_2O_6$ Store at -15°C

Leucine — BOC-Leu-PAM-® (200-400 mesh)

Bachem D-1230.0001   Store at 2-8°C

Leucine — BOC-Leu-PAM-Resin

American Peptide RPAM205   1% DVB cross-linked: 100-200 mesh | t-BOC protected AA resin; preattached resins are useful for synthesis of peptides with C-terminal carboxyl group by t-BOC chemistry

Leucine — BOC-L-Leu Hydrate

Fluka 15450   MW 249.31 $C_{11}H_{21}NO_4 \cdot H_2O$ ~99% (titration); ≤0.5% residue on ignition; mp: 85-87°C

Neosystem BA01102   MW 249.3

Leucine — BOC-L-Leu Hydroxysuccinimide Ester

Fluka 15454   MW 328.37 $C_{15}H_{24}N_2O_6$ ≥99%; mp: 111-113°C | Used for the quantitative determination of D- & L-AA; Mitchell, A et al, Anal Chem, 50: 637, 1978

Leucine — BOC-L-Leu N-Carboxy Anhydride

Fluka 15452   MW 257.29 $C_{12}H_{19}NO_5$ mp: 55-58°C | Produced by Propeptide under an exclusive license from Bioresearch

Leucine — BOC-L-Leu-(4-ONp)

Fluka 15130   MW 352.39 $C_{17}H_{24}N_2O_6$ ~98% (TLC)

Leucine — BOC-L-Leu-MBHA Resin

Neosystem RN01102

Leucine — BOC-L-Leu-O-CH₂-φ-CH₂-COOH

Neosystem LP01102   MW 379.5

Leucine — BOC-L-Leu-ol

Bachem A-3190.0001   MW 217.31 $C_{11}H_{23}NO_3$ Store at -15°C

Leucine — BOC-L-Leu-PAM Resin

Neosystem RP01102

Leucine — BOC-L-t-Leu

Synonyms: (S)-N-BOC-2-Amino-3,3-Dimethylbutyric Acid

Fluka 15451   MW 231.29 $C_{11}H_{12}NO_4$ ≥99% (titration); mp: 118-121°C

Leucine — BOC-L-β-Leu

Synonyms: (R)-3-(BOC-Amino)-4-Methylpentanoic Acid; BOC-L-β-Hva

Fluka 03678   MW 231.29 $C_{11}H_{21}NO_4$ ≥98% (TLC)

Leucine — BOC-Me-Leu

USBio B2354   MW 245.3 $C_{12}H_{23}NO_4$ ≥98%

Leucine — BOC-Me-Nle

USBio B2355   MW 245.3 $C_{12}H_{23}NO_4$ ≥98%

Leucine — BOC-N-Et-Leu Dicyclohexylamine

Bachem A-3355.0001   MW 440.67 $C_{13}H_{25}NO_4 \cdot C_{12}H_{23}N$ Store at RT

Leucine — BOC-Nle

USBio B2369   MW 231.3 $C_{11}H_{21}NO_4$ ≥99%

Leucine — BOC-N-Me-D-Leu

Bachem A-4050.0001   MW 245.32 $C_{12}H_{23}NO_4$ Store at RT
Fluka 02677   MW 245.32 $C_{12}H_{23}NO_4$ ≥98% (HPLC)

Leucine — BOC-N-Me-Leu

Synonyms: BOC-N-Me-L-Leu

Bachem A-2060.0001   MW 245.32 $C_{12}H_{23}NO_4$ Store at RT
Peptides International BML-5313-PI   MW 245.32   >98% (HPLC); white to off-white powder

Leucine — BOC-N-Me-L-Ile

Fluka 02678   MW 245.32 $C_{12}H_{23}NO_4$ ≥98% (HPLC)
Fluka 15446   MW 245.32 $C_{12}H_{23}NO_4$ ~99% (TLC); mp: 54-57°C
Neosystem BA01103   MW 245.3

Leucine — BOC-Thiono-Leu-1-(6-Nitro)-Benzotriazolide

Synonyms: (S)-2-(BOC-Amino)-4-Me-Pentanethioic-O-Acid-1-(6-Nitro)-Benzotriazolide

Bachem A-4345.0001   MW 393.47 $C_{17}H_{23}N_5O_4S$ Store at -15°C

Leucine — BOC-t-Leu

Synonyms: BOC-α-tBu-Gly

USBio B2328   MW 231.3 $C_{11}H_{21}NO_4$ ≥99%

Leucine — Bsmoc-Leu

Synonyms: Bsmoc-L-Leu

Peptides International BLL-5629-PI   MW 353.40   >98% (HPLC); white to off-white powder | LA Carpino, et al, *JACS*, 119:9915, 1997

Leucine — Carbamoyl-Leu

Synonyms: L-4-Me-2-Ureidopentanoic Acid

Bachem F-2430.0001   MW 174.2 $C_7H_{14}N_2O_3$ Store at RT

Leucine — CBZ-D-Leu Oil

USBio C2098-97   MW 265.3 $C_{14}H_{19}NO_4$ ≥99%

Leucine — CBZ-Leu Oil

USBio C2098-96   MW 265.3 $C_{14}H_{19}NO_4$ ≥99%

Leucine — CBZ-Leu-ONp

USBio C2098-98   MW 386.4 $C_{20}H_{22}N_2O_6$ ≥99%

Leucine — CBZ-Leu-OSu

USBio C2098-99   MW 362.4 $C_{18}H_{22}N_2O_6$ ≥99%

Leucine — CBZ-Me-Leu

USBio C2099-24   MW 279.3 $C_{15}H_{21}NO_4$ ≥99%

Leucine — Cycloleucine

Synonyms: 1-Aminocyclopentane-1-Carboxylic Acid

Sigma A 4,810-5   FW 129.2 $C_6H_{11}NO_2$ NMDA antagonist acting at the glycine site; formerly Sigma C 6630

Leucine — Dansyl-DL-Leu Cyclohexylammonium Salt

Sigma D 8881   FW. 463.6 $C_{18}H_{24}N_2O_4S \cdot C_6H_{13}N$ ~99%

Leucine — Dansyl-L-Leu

ICN 100091   MW 463.6 $C_{18}H_{24}N_2O_4S \cdot C_6H_{13}N$

Leucine — Dansyl-L-Leu Cyclohexylammonium Salt

Sigma D 1375   FW. 463.6 $C_{18}H_{24}N_2O_4S \cdot C_6H_{13}N$ ~99%

Leucine — D-Leu

Synonyms: (R)-2-Amino-4-Methylpentanoic Acid; D-2-Amino-4-Methylpentanoic Acid

Bachem F-1700.0005   MW 131.18 $C_6H_{13}NO_2$ Store at -15°C
Fluka 61830   MW 131.18 $C_6H_{13}NO_2$ ≥99.0% (titration)
ICN 102152   MW 131.2 $C_6H_{13}NO_2$ Crystalline
Neosystem AA01101   MW 131.2
Peptides International ADL-2806   MW 131.17   >99.9% optical purity by RP-HPLC; white crystalline powder
Sigma L 7750   FW 131.2 $C_6H_{13}NO_2$ ≥98% (TLC) | Also available as part of a kit
USBio L2020-10   MW 131.2 $C_6H_{13}NO_2$ ≥99%

Leucine — D-Leu (1-¹⁴C)

ARC ARC-562   MW 131.2 $(CH_3)_2CH_2CH(NH_2)COOH$ 50-60 mCi/mmol; 1.85-2.22 GBq/mmol; in 0.01 N HCl | Radiochemical

Leucine — D-Leu Amide

Bachem F-1705.0001   MW 130.19 $C_6H_{14}N_2O$ Store at -15°C

Leucine — D-Leu Amide Hydrochloride

Bachem F-1710.0001   MW 166.65 $C_6H_{14}N_2O \cdot HCl$ Store at -15°C

Leucine — D-Leu-OBzl p-Tosyl

Bachem F-2720.0005   MW 393.5 $C_{13}H_{19}NO_2 \cdot C_7H_8O_3S$ Store at RT

Leucine — D-Leu-OBzl PTSA

Neosystem AA01109   MW 393.5

Leucine — D-Leu-ol

Senn Chem 44099   MW 117.2

Leucine — D-Leu-OMe Hydrochloride

Bachem F-1720.0001   MW 181.66 $C_7H_{15}NO_2 \cdot HCl$ Store at 2-8°C

Leucine — D-Leu-OtBu Hydrochloride

Bachem F-1715.0001   MW 223.74 $C_{10}H_{21}NO_2 \cdot HCl$ Store at -15°C

Leucine — D-Leu-pNA

Bachem L-1310.0250   MW 251.29 $C_{12}H_{17}N_3O_3$ Store at -15°C

Leucine — D-Leu-βNA Hydrochloride

Bachem K-1390.0250   MW 292.81 $C_{16}H_{20}N_2O \cdot HCl$ Store at RT

Leucine — DL-Leu

Synonyms: (±)-Amino-4-Methylpentanoic Acid; DL-2-Amino-4-Me-Pentanoic Acid

Bachem F-2905.0025   MW 131.18 $C_6H_{13}NO_2$ Store at RT

Fluka 61840   MW 131.18 $C_6H_{13}NO_2$ ≥99.0% (titration); ≤0.05% residue on ignition; mp: 295°C

ICN 102154   MW 131.2 $C_6H_{13}NO_2$ ≥99%; crystalline

ICN 194693   MW 131.2 $C_6H_{13}NO_2$ ≥99%; crystalline | Cell culture reagent

Sigma L 1387   FW 131.2 $C_6H_{13}NO_2$ Crystalline | Cell culture tested

Sigma L 7875   FW 131.2 $C_6H_{13}NO_2$ ≥98% (TLC) | Also available as part of a kit

USBio L2020-15   MW 131.2 $C_6H_{13}NO_2$ ≥99%

Leucine — DL-Leu (4,5-³H)

Amersham TRK75   Aqueous, 2% EtOH, sterilized; 0.92-1.85 TBq/mmol, 25-50 Ci/mmol; 37 MBq/mL, 1 mCi/mL

Leucine — DL-Leu Amide Hydrochloride

Bachem F-2360.0005   MW 166.65 $C_6H_{14}N_2O \cdot HCl$ Store at RT

Leucine — DL-Leucinamide Hydrochloride

Sigma L 8250   FW 166.7 $C_6H_{14}N_2O \cdot HCl$

Leucine — DL-Leu-OMe Hydrochloride

Bachem F-2925.0025   MW 181.66 $C_7H_{15}NO_2 \cdot HCl$ Store at 2-8°C

Leucine — DL-Nle

USBio N5000-15   MW 131.2 $C_6H_{13}NO_2$ ≥99%

Leucine — DL-t-Leu

Synonyms: (±)-2-Amino-3,3-Dimethylbutyric Acid; DL-α-tBu-Gly

Fluka 61837   MW 131.18 $C_6H_{13}NO_2$ ≥99.0% (titration); 0.05% residue on ignition; mp: >300°C

Leucine — DL-γ-Me-Leu

Sigma M 4018   FW 145.2 $C_7H_{15}NO_2$

Leucine — D-Nle

USBio N5000-10   MW 131.2 $C_6H_{13}NO_2$ ≥99%

Leucine — Dnp-L-Leu

ICN 102155   MW 297.3 $C_{12}H_{15}N_3O_6$ Crystalline

Leucine — D-t-Leu

Synonyms: (R)-2-Amino-3,3-Dimethylbutyric Acid; D-α-tBu-Gly

Fluka 61835   MW 131.18 $C_6H_{13}NO_2$ ≥99.0% (titration); ≤0.5% H_2O; mp: >300°C | Unnatural, "fat" AA; used as chiral auxiliary, e.g. in the synthesis of aldehydes by double alkylation of unsaturated aldimines of t-Leu esters; Kogen, H et al, *Tetrahedron*, 37: 3951, 1981; Whittaker, M et al, *Can J Chem*, 63: 2844, 1985; Tomioka, K et al, *Tetrahedron*, 45: 643, 1989; Denmark, SE et al, *Tetrahedron Lett*, 30: 2469, 1989; Schoellkopf, U & Neubauer, HJ, *Synthesis*, 861: 1982; Hayashi, T et al, *J Org Chem*, 48: 2195, 1983

Leucine — D-γ-Me-Leu

Sigma M 3768   FW 145.2 $C_7H_{15}NO_2$

Leucine — FMOC-4,5-Dehydro-Leu

Bachem B-2255.0001   MW 351.4 $C_{21}H_{21}NO_4$ Store at -15°C

Leucine — FMOC-Cycloleucine

Synonyms: 1-(FMOC-Amino)-Cyclopentanecarboxylic Acid

Fluka 47512   MW 351.4 $C_{21}H_{21}NO_4$ ≥98% (TLC)

Leucine — FMOC-D-Leu

Synonyms: FMOC-D-Leu

Bachem B-1350.0001   MW 353.42 $C_{21}H_{23}NO_4$ Store at RT

Fluka 47316   MW 353.42 $C_{21}H_{23}NO_4$ ~98% (TLC)

Neosystem FA01101   MW 353.4

Peptides International FDL-1814-PI   MW 353.42   >98% (HPLC); white crystalline powder

USBio F5400   MW 353.5 $C_{21}H_{23}NO_4$ ≥99%

Leucine — FMOC-D-Leu-SASRIN™-® (200-400 mesh)

Bachem D-1765.0001   Store at 2-8°C

Leucine — FMOC-D-Me-Leu

USBio F5421   MW 367.4 $C_{22}H_{25}NO_4$ ≥98%

Leucine — FMOC-D-t-Leu

Synonyms: FMOC-D-α-tBu-Gly

USBio F5398   MW 353.5 $C_{21}H_{23}NO_4$ ≥99%

Leucine — FMOC-Homocycloleucine

Synonyms: 1-(FMOC-Amino)-Cyclohexanecarboxylic Acid

Fluka 04061   MW 365.4 $C_{22}H_{23}NO_4$ ≥98% (titration); mp: 185-187°C

Leucine — FMOC-Leu

Synonyms: FMOC-L-Leu

American Peptide FLEU100   MW 353.4

Bachem B-1345.0005   MW 353.42 $C_{21}H_{23}NO_4$ Store at RT

Peptides International FLL-1725-PI   MW 353.42   >98%
(HPLC); white crystalline powder

USBio F5401   MW 353.5 $C_{21}H_{23}NO_4$ ≥99%

Leucine — FMOC-Leu (^{15}N)

Bachem B-2655.0100   MW 353.42 $C_{21}H_{23}NO_4$ Store at
-15°C

Leucine — FMOC-Leu-® (200-400 mesh)

Bachem D-1115.0001   Store at RT

Leucine — FMOC-Leu-ol

Synonyms: FMOC-L-Leucinol

Senn Chem 44062   MW 339.4

Leucine — FMOC-Leu-ONp

Bachem B-1355.0001   MW 474.51 $C_{27}H_{26}N_2O_6$ Store at
-15°C

Leucine — FMOC-Leu-OPfp

Bachem B-1150.0001   MW 519.47 $C_{27}H_{22}F_5NO_4$ Store at
-15°C

USBio F5402   MW 519.5 $C_{27}H_{22}F_5NO_4$ ≥99%

Leucine — FMOC-Leu-OSu

USBio F5403   MW 450.5 $C_{25}H_{26}N_2O_6$ ≥99%

Leucine — FMOC-Leu-SASRIN™-® (200-400 mesh)

Bachem D-1360.0001   Store at 2-8°C

Leucine — FMOC-L-Leu

Fluka 47633   MW 353.42 $C_{21}H_{23}NO_4$ ≥97% (titration); mp:
152-154°C

Neosystem FA01102   MW 353.4

Leucine — FMOC-L-Leu 4-Benzyloxybenzyl Ester Polymer Bound

Fluka 47662   0.4-0.6 mmol/g resin; carrier: polystyrene,
crosslinked with 1% DVB; 200-400 mesh particle size

Leucine — FMOC-L-Leu N-Carboxy Anhydride

Fluka 47688   MW 379.41 $C_{22}H_{21}NO_5$ ≥97% (titration); mp:
119-122°C | Produced by Propeptide under an exclusive license
from Bioresearch

Leucine — FMOC-L-Leu-OPfp

Fluka 47468   MW 519.47 $C_{27}H_{22}F_5NO_4$ ≥98% (TLC)

Leucine — FMOC-L-Leu-Wang Resin

American Peptide RFWAN52   Preattached Wang resins are
usful for synthesis of peptides with C-terminal carboxyl groups;
peptides can be cleaved from the resin with 90% TFA in the
presence of scavengers

Neosystem RW01102

Leucine — FMOC-L-t-Leu

Fluka 47524   MW 353.42 $C_{21}H_{23}NO_4$ ≥98% (HPLC); mp:
124-127°C

Leucine — FMOC-L-β-Leu

Synonyms: (R)-3-(FMOC-Amino)-4-Methylpentanoic Acid; FMOC-L-
β-Hva

Fluka 03676   MW 353.42 $C_{21}H_{23}NO_4$ ≥98% (HPLC)

Leucine — FMOC-Me-Leu

USBio F5422   MW 367.4 $C_{22}H_{25}NO_4$ ≥98%

Leucine — FMOC-N-Me-D-Leu

Synonyms: FMOC-N-Me-D-Leu

Fluka 02451   MW 367.45 $C_{22}H_{25}NO_4$ ≥98% (TLC)

Peptides International FML-1784-PI   MW 367.45   >98%
(HPLC); white crystalline powder

Leucine — FMOC-N-Me-Leu

Synonyms: FMOC-N-Me-L-Leu

Bachem B-2035.0001   MW 367.45 $C_{22}H_{25}NO_4$ Store at RT

Peptides International FML-1794-PI   MW 367.45   >97%
(HPLC); white crystalline powder

Leucine — FMOC-N-Me-L-Leu

Fluka 47597   MW 367.45 $C_{22}H_{25}NO_4$ ≥99% (HPLC); mp:
113-116°C

Leucine — FMOC-t-Leu

Synonyms: FMOC-α-tBu-Gly

USBio F5399   MW 353.5 $C_{21}H_{23}NO_4$ ≥99%

Leucine — Glutaryl-Leu 2 Dicyclohexylamine

Bachem E-2870.0005   MW 607.92 $C_{11}H_{19}NO_5 \cdot 2C_{12}H_{23}N$
Store at RT

Leucine — Leu

Bachem E-2050.0025   MW 131.18 $C_6H_{13}NO_2$ Store at RT

Leucine — Leu (1-^{14}C)

ICN 10085E   MW 131.2 $(CH_3)_2CHCH_2CH(NH_2)COOH$ >50
mCi/mmol, >1.85 GBq/mmol; sterile EtOH:H2O (2:98)

Leucine — Leu (^{15}N)

Bachem E-3235.0100   MW 132.18 $C_6H_{13}{}^{15}NO_2$ Store at
-15°C

Leucine — Leu Amide

Bachem E-2650.0005   MW 130.19 $C_6H_{14}N_2O$ Store at RT

Leucine — Leu Amide Hydrochloride

Synonyms: Aminopeptidase Substrate

Bachem E-2055.0005   MW 166.65 $C_6H_{14}N_2O \cdot HCl$ Store
at RT

Peptides International SLN-3027   MW 166.65 $C_6H_{14}N_2O \cdot$
HCl   >99% (HPLC); amorphous powder

Leucine — Leu-4MβNA Hydrochloride

Bachem J-1235.0250   MW 322.84 $C_{17}H_{22}N_2O_2 \cdot HCl$ Store
at -15°C

Leucine — Leu-AMC

Synonyms: Aminopeptidase Substrate

Bachem I-1240.0050   MW 288.35 $C_{16}H_{20}N_2O_3$ Store at
-15°C

Peptides International MLM-3091-v   MW 288.35
$C_{16}H_{20}N_2O_3$ >99% (HPLC); lyophilized amorphous powder |
Saifuku, K et al, *Clin Chim Acta*, 84:85, 1978

Leucine — Leu-AMC Hydrochloride

Synonyms: Aminopeptidase Substrate I

Bachem I-1245.0050   MW 324.81 $C_{16}H_{20}N_2O_3$ Store at
-15°C

ICN 195885   MW 324.8 $C_{16}H_{20}N_2O_3 \cdot HCl$ ≥99% |
Aminopeptidase fluorogenic substrate

Leucine — Leu-CMK Hydrochloride

Bachem N-1105.0250   MW 200.11 $C_7H_{14}ClNO \cdot HCl$ Store
at -15°C

Leucine — Leu-NHNH-Z TFA

Bachem E-2075.0001   MW 393.36 $C_{14}H_{21}N_3O_3 \cdot C_2HF_3O_2$
Store at RT

Leucine — Leu-NHOH Acetate

Bachem E-3455.0001   MW 206.24 $C_6H_{14}N_2O_2 \cdot C_2H_4O_2$
Store at -15°C

Leucine — Leu-OAll p-Tosyl

Bachem E-3220.0001   MW 344.45 $C_9H_{18}NO_2 \cdot C_7H_8O_3S$
Store at 2-8°C

Leucine — Leu-OBzl p-Tosyl

Bachem E-2060.0005   MW 393.5 $C_{13}H_{19}NO_2 \cdot C_7H_8O_3S$
Store at RT

Leucine — Leu-OBzl-Tosyl

Synonyms: L-Leu-OBzl-Tosyl

Peptides International EBL-2112   MW 393.50   >98% (TLC);
white powder

Leucine — Leu-OEt Hydrochloride

Synonyms: L-Leu-OEt

Bachem E-2070.0025   MW 195.69 $C_8H_{17}NO_2 \cdot HCl$ Store
at RT

Peptides International EEL-2040-PI   MW 195.69   >98%
(TLC); white powder

Leucine — Leu-ol

Synonyms: L-Leucinol

Senn Chem 44098   MW 117.2   Liquid | Reversible inhibitor
of protein synthesis

Leucine — Leu-OMe Hydrochloride

Bachem E-2080.0025   MW 181.66 $C_7H_{15}NO_2 \cdot HCl$ Store
at RT

Leucine — Leu-OtBu Hydrochloride

Bachem E-2065.0001   MW 223.74 $C_{10}H_{21}NO_2 \cdot HCl$ Store
at -15°C

Leucine — Leu-pNA

Synonyms: Aminopeptidase Substrate

Bachem L-1305.0001   MW 251.29 $C_{12}H_{17}N_3O_3$ Store at
-15°C

Peptides International SLN-3014   MW 251.29 $C_{12}H_{17}N_3O_3$
>99% (HPLC); amorphous powder | Peleiderer, G in *Proteolytic
Enzymes*, Methods in Enzymology, Vol 19, (GE Perlmann & L
Lorand, Eds), Academic Press, New York, 1970, pp 514-521

Leucine — Leu-αNA

Bachem M-1520.0250   MW 256.35 $C_{15}H_{20}N_2O$ Store at RT
Bachem K-1380.0001   MW 256.35 $C_{16}H_{20}N_2O$ Store at RT

Leucine — Leu-βNA Hydrochloride

Bachem K-1385.0001   MW 292.81 $C_{16}H_{20}N_2O \cdot HCl$ Store
at RT

Leucine — L-Leu

Synonyms: (S)-2-Amino-4-Methylpentanoic Acid; L-2-Amino-4-
Methylpentanoic Acid; L-α-Aminoisocaproic Acid

Amersham US18280   ≥98.5%; ≤30ppm heavy metals (Pb)

Amersham US18285   ≥98.5%; ≤30ppm heavy metals (Pb);
Met-free

Fluka 61819   MW 131.18 $C_6H_{13}NO_2$ ≥99.5% (titration);
≤0.1% residue on ignition, loss on drying; ≤2% foreign AA;
≤0.00001% As; ≤0.0005% Al, Ba, Bi, Cd, Co, Cr, Cu, Fe, Li, Mg,
Mn, Mo, Ni, Pb, Sr, Zn; ≤0.001% Ca; ≤0.005% K; ≤0.01% Cl, Na,
NH_4^+, SO_4 | Spacer in isotachophoresis of serum proteins; Yagi, T
et al, *Elctrophoresis '83*, ed H Hirai, 503, 1984, de Gruyter, Berlin;
Hine, T, *ibid*, 541; *CRC Handbook of Microbiology*, Laskin, AI &
Lachevalier, HA, eds, 4: 1, 1974, CRC Press, Cleveland, Ohio

Fluka 61820   MW 131.18 $C_6H_{13}NO_2$ ≥99.5% (titration);
≤0.1% residue on ignition, loss on drying; ≤0.3% foreign AA;
≤0.0005% Cd, Co, Cr, Cu, Fe, Li, Mg, Mn, Mo, Ni, Pb, Sr, Zn;
≤0.001% Ca; ≤0.005% K; ≤0.01% Cl, Na, NH_4^+, SO_4

ICN 102158   MW 131.2 $C_6H_{13}NO_2$ Crystalline; substantially
free of Ile, Met

ICN 194694   MW 131.2 $C_6H_{13}NO_2$ Crystalline; substantially
free of Ile, Met | Cell culture reagent

Neosystem AA01102   MW 131.2

Peptides International ALL-2713   MW 131.17   >99.9%
optical purity by RP-HPLC; white crystalline powder

Rexim   MW 131.2 $C_6H_{13}NO_2$ White crystals or crystalline
powder

Sigma L 0389   Solution; 79 *mM* (10.4 mg/mL); L-Leu in tissue
culture grade H_2O; sterile-filtered | Endotoxin tested; cell culture
tested

Sigma L 1512   FW 131.2 $C_6H_{13}NO_2$ Substantially free of Ile
& Met | Cell culture tested; insect cell culture tested

Sigma L 5506   Matrix: 4% beaded agarose; activation:
cyanogen bromide; attachment: amino; spacer: 1 atom; ligand
immobilized: 2-10 µmoles/mL; form: suspension in 2.0 *M* NaCl |
Deutsch, DG & Mertz, ET, *Proc Fed Amer Soc Exp Biol*, 29: 647,
1970; Deutsch, DG & Mertz, ET, *Science*, 170: 1095, 1970; Vuento,
M & Vaheri, A, *Biochem J*, 183: 331, 1979

Sigma L 5652   FW 131.2 $C_6H_{13}NO_2$ SigmaUltra; >99%;
residue on ignition: <0.1%; solubility (0.1 *M* in 1.0 *M* HCl, 20°C):
complete, colorless; insoluble matter: <0.1%; NH_4^+: <0.2%; Na,
KCl, SO_4: <0.005%; Al, Cu, Zn, Ca, Fe, P: <0.0005%; Pb:
<0.001%; A_{260}<0.1; A_{280}<0.1 (1 *M* in 1.0 *M* HCl)

Sigma L 6911   FW 131.2 $C_6H_{13}NO_2$ USP Grade

Sigma L 8000   FW 131.2 $C_6H_{13}NO_2$ ≥98% (TLC);
substantially free of Ile & Met | Also available as part of a kit

Sigma L 8125   FW 131.2 $C_6H_{13}NO_2$ Purified grade; may
contain small amounts of AA impurities (Met, L-Val, L-Ile)

USBio L2020   MW 131.2 $C_6H_{13}NO_2$ ≥98.5%; appearance:
white crystalline powder; pH: 5.5-7.0; specific rotation: C=4, 6 *N*
HCl +14.9° to +17.3°; heavy metals (Pb): ≤0.001%; endotoxin:
≤2 EU/mg; cell culture tested: murine myeloma SP2/0-Ag14

Leucine — L-Leu (1-^{14}C)

ARC ARC-156   MW 131.2 $(CH_3)_2CH_2CH(NH_2)$**COOH**   50-60
mCi/mmol; 1.85-2.22 GBq/mmol; in 0.01 *N* HCl | Radiochemical

ARC ARC-156A   MW 131.2 $(CH_3)_2CH_2CH(NH_2)$**COOH**   50-60
mCi/mmol; 1.85-2.22 GBq/mmol; in EtOH: H_2O (2:98) |
Radiochemical

ARC ARC-156C   MW 131.2 $(CH_3)_2CH_2CH(NH_2)$**COOH**   50-60
mCi/mmol; 1.85-2.22 GBq/mmol; in sterile saline | Radiochemical

ICN 10088E   MW 131.2 $(CH_3)_2CHCH_2CH(NH_2)$**COOH**   >50
mCi/mmol, >1.85 GBq/mmol; sterile EtOH:H_2O (2:98); HPLC
analyzed for purity

Leucine — L-Leu (^{15}N)

ICN 540146   MW 131.2 $(CH_3)_2CHCH_2CH(NH_2)COOH$ 99%
^{15}N atomic purity

Leucine — L-Leu (2,3,4,5-^{3}H)

ICN 20032   MW 131.2 $(CH_3)_2CHCH_2CH(NH_2)COOH$ >110 Ci/mmol, >4.1 TBq/mmol; 0.01 N HCl; HPLC analyzed for purity | Vento, R etal, *Exp Eye Res*, 59:221, 1994; Novak, DA etal, *Biochem J*, 301:671, 1994; Magagnin, S etal, *JBC*, 267:15384, 1992

ICN 20032E   MW 131.2 $(CH_3)_2CHCH_2CH(NH_2)COOH$ >110 Ci/mmol, >4.1 TBq/mmol; sterile EtOH:H$_2$O (2:98); HPLC analyzed for purity

Leucine — L-Leu (3,4,5-^{3}H(N))

Sigma L 7772   FW 131.2 $(CH)_2CHCH_2CH(NH_2)COOH$ >140 Ci/mmol; aqueous solution containing 2% EtOH in serum bottle | Radiochemical

Leucine — L-Leu (3,4,5-^{3}H)

ARC ART-470   MW 131.2 $(CH)_2CHCH_2CH(NH_2)COOH$ 100-150 Ci/mmol; 3.7-5.55 TBq/mmol; in EtOH: H$_2$O (2:98) | Radiochemical

Leucine — L-Leu (4,5-^{3}H(N))

ARC ART-140   MW 131.2 $(CH)_2CHCH_2CH(NH_2)COOH$ 40-60 Ci/mmol; 1.48-2.22 TBq/mmol; in 0.01 N HCl | Radiochemical

ARC ART-140A   MW 131.2 $(CH)_2CHCH_2CH(NH_2)COOH$ 40-60 Ci/mmol; 1.48-2.22 TBq/mmol; in EtOH: H$_2$O (2:98) | Radiochemical

Sigma L 5897   FW 131.2 $(CH_3)_2CHCH_2CH(NH_2)COOH$ 40-60 Ci/mmol; aqueous solution containing 2% EtOH in serum bottle | Radiochemical

Leucine — L-Leu (4,5-^{3}H)

Amersham TRK170   Aqueous, 2% EtOH, sterilized; 1.7-3.1 TBq/mmol, 45-85 Ci/mmol; 37 MBq/mL, 1 mCi/mL

Amersham TRK510   Aqueous, 2% EtOH, sterilized; 4.4-7.0 TBq/mmol, 120-190 Ci/mmol; 37 MBq/mL, 1 mCi/mL

Amersham TRK636   Aqueous, sterilized; 4.4-7.0 TBq/mmol, 120-190 Ci/mmol; 37 MBq/mL, 1 mCi/mL

Amersham TRK683   Aqueous, sterilized; 4.4-7.0 TBq/mmol, 120-190 Ci/mmol; 185 MBq/mL, 5 mCi/mL

Amersham TRK754   Aqueous, sterilized; 1.7-3.1 TBq/mmol, 45-85 Ci/mmol; 185 MBq/mL, 5 mCi/mL

ICN 20036   MW 131.2 $(CH_3)_2CHCH_2CH(NH_2)COOH$ 40-60 Ci/mmol, 1.48-2.22 TBq/mmol; 0.01 N HCl; HPLC analyzed for purity | Hoorn, CM & RA Roth, *Am J Physiol*, 262:L740, 1992; Guest, I & DR Varma, *J Toxicol Environ Health*, 36:27, 1992

ICN 20036E   MW 131.2 $(CH_3)_2CHCH_2CH(NH_2)COOH$ 40-60 Ci/mmol, 1.48-2.22 TBq/mmol; sterile EtOH:H$_2$O (2:98); HPLC analyzed for purity

Leucine — L-Leu (U-^{14}C)

Amersham CFA273   Aqueous, 2% EtOH, sterilized; 1.85-2.29 GBq/mmol, 50-62 mCi/mmol; 1.85 MBq/mL, 50 μCi/mL

Amersham CFB183   Aqueous, sterilized; >11 GBq/mmol, >300 mCi/mmol; 1.85 MBq/mL, 50 μCi/mL

Amersham CFB67   Aqueous, 2% EtOH, sterilized; >11 GBq/mmol, >300 mCi/mmol; 1.85 MBq/mL, 50 μCi/mL

ARC ARC-656A   MW 131.2 $(CH_3)_2CH_2CH(NH_2)COOH$ 200-300 mCi/mmol; 7.4-11.1 GBq/mmol; in EtOH: H$_2$O (2:98) | Radiochemical

ICN 10089   MW 131.2 $(CH_3)_2CHCH_2CH(NH_2)COOH$ 270-330 mCi/mmol, 10-12.2 GBq/mmol; 0.01 N HCl; HPLC analyzed for purity

ICN 10089E   MW 131.2 $(CH_3)_2CHCH_2CH(NH_2)COOH$ 270-330 mCi/mmol, 10-12.2 GBq/mmol; sterile EtOH:H$_2$O (2:98); HPLC analyzed for purity

Sigma L 5770   FW 131.2 $(CH_3)_2CHCH_2CH(NH_2)COOH$ 240-360 mCi/mmol; radiochemical purity ≥98% (HPLC); aqueous solution containing 2% EtOH in Combi-vial | Radiochemical

Leucine — L-Leu 3-Carboxy-4-Hydroxyanilide Hydrochloride

ICN 155192   MW 302.8 $C_{13}H_{18}N_2O_4 \cdot HCl$

Sigma L 9384   FW 302.8 $C_{13}H_{18}N_2O_4 \cdot HCl$

Leucine — L-Leu 4-Methoxy-β-Naphthylamide Free Base

Sigma L 1136   FW 286.4 $C_{17}H_{22}N_2O_2$

Leucine — L-Leu 4-Methoxy-β-Naphthylamide Hydrochloride

Sigma L 3387   FW 322.8 $C_{17}H_{22}N_2O_2 \cdot HCl$

Leucine — L-Leu Amide Hydrochloride

Neosystem AA01103   MW 166.7

Leucine — L-Leu CMK Hydrochloride

ICN 155193   MW 200.1 $C7H14ClNO \cdot HCl$

Sigma L 1636   FW 200.1 $C_7H_{14}ClNO \cdot HCl$

Leucine — L-Leu Hydroxamate

Sigma L 0252   FW 146.2 $C_6H_{14}N_2O_2$

Leucine — L-Leu U- (^{14}C)

ARC ARC-656   MW 131.2 $(CH_3)_2CH_2CH(NH_2)COOH$ 200-300 mCi/mmol; 7.4-11.1 GBq/mmol; in 0.01 N HCl | Radiochemical

Leucine — L-Leu β-Naphthylamide Free Base

Sigma L 1635   FW 256.3 $C_{16}H_{20}N_2O$ Very low free β-naphthylamine; not H$_2$O soluble | Caution: Upon hydrolysis this chemical produces β-naphthylamine which is believed to be a carcinogen that causes bladder tumors; substrate for aminopeptidase M

Leucine — L-Leu β-Naphthylamide Hydrochloride

Sigma L 0376   FW 292.8 $C_{16}H_{20}N_2O \cdot HCl$ Very low free β-naphthylamine; soluble in H$_2$O | Caution: Upon hydrolysis this chemical produces β-naphthylamine which is believed to be a carcinogen that causes bladder tumors; suitable substrate for leucine aminopeptidase determination in colorimetric & histochemical procedures; Niinobe, H & Fujii, S, *J Biochem* (*Tokyo*), 87: 195, 1980

Leucine — L-Leu-2-Naphthylamide Hydrochloride

Synonyms: Leucine Aminopeptidase Substrate

Fluka 61900   MW 292.81 $C_{16}H_{20}N_2O \cdot HCl$ ≥97.0% (titration); mp: 255-260°C; 2-4% H$_2$O | Substrate for determination of leucine aminopeptidase; Green, MN et al, *Arch Biochem Biophys*, 57: 458, 1955; Bodansky, O, *Meth Enzymol*, 17B: 875, 1971; Prescott, JM & Wilkes, SH, *Meth Enzymol*, 45: 530, 1976; Coburn, JT et al, *Anal Chem*, 57: 1669, 1985

Leucine — L-Leu-4-NA

Synonyms: Aminopeptidase *Aeromonas* Chromogenic Substrate

Fluka 61910   MW 251.29 $C_{12}H_{17}N_3O_3$ ≥99.0% (titration); mp: 88-90°C; ≤0.1% H$_2$O | Chromogenic substrate for the spectrophotometric assay of aminopeptidase from *Aeromonas*; Prescott, JM & Wilkes, SH, *Meth Enzymol*, 45: 530, 1976; Roncari, G et al, *Meth Enzymol*, 45: 522, 1976; Pfleiderer, G, *Meth Enzymol*, 19: 514, 1970; Achstetter, T et al, *Arch Biochem Biophys*, 226: 292, 1983

Leucine — L-Leu-AMC Hydrochloride

Synonyms: Leu-AMC; L-Leu 7-Amido-4-Methyl Coumarin; L-Leu-4-Methylum-Belliferylamide; Leucine Aminopeptidase Fluorogenic Substrate

Fluka 61888   MW 324.81 $C_{16}H_{20}N_2O_3 \cdot HCl$ Kanaoka, Y et al, *Chem Pharm Bull*, 25: 362, 1977

Sigma L 2145   FW 324.8 $C_{16}H_{20}N_2O_3 \cdot HCl$

Leucine — L-Leu-AMC-Tosyl

ICN 151545   Substrate for leucine aminopeptidase

Leucine — L-Leucinamide Hydrochloride

Sigma L 8375   FW 166.7 $C_6H_{14}N_2O \cdot HCl$

Leucine — L-Leucinol Hydrochloride

Bachem F-1725.0250   MW 153.65 $C_6H_{15}NO \cdot HCl$ Store at -15°C

Leucine — L-Leucylthiol Dihydrochloride Oxidized

Synonyms: Dithio-Bis-2-Amino-4-Methylpentane; Aminopeptidase M Inhibitor; Aminopeptidase B Inhibitor; Metallo-Protease Inhibitor

Sigma L 8397   FW 337.4 $C_{12}H_{28}N_2S_2 \cdot 2HCl$ More powerful inhibitor of aminopeptidases M & B than Bestatin; Chan, WW-C, *Biochem Biophys Res Commun*, 116: 297, 1983; Ocain, TD & Rich, DH, *ibid*, 145: 1038, 1987; Ocain, TD & Rich, DH, *J Med Chem*, 31: 2193, 1988

Leucine — L-Leu-OAll Hydrochloride

Neosystem AA01108   MW 207.7

Leucine — L-Leu-OAll p-Tosyl

ICN 155190   MW 343.4 $C_9H_{17}NO_2 \cdot C_7H_8O_3S$

Sigma L 6271   FW 343.4 $C_9H_{17}NO_2 \cdot C_7H_8O_3S$

Leucine — L-Leu-OBzl p-Tosyl

Sigma L 9759   FW 393.5 $C_{20}H_{27}NO_5S$

Leucine — L-Leu-OBzl PTSA

Neosystem AA01107   MW 393.5

Leucine — L-Leu-OBzl-p-Tosyl

ICN 102161   MW 393.5 $C_{20}H_{27}NO_5S$ Crystalline

Leucine — L-Leu-OBzl-Tosyl

Fluka 61872   MW 393.51 $C_{13}H_{19}NO_2 \cdot C_7H_8O_3S$ mp: 151-155°C

Leucine — L-Leu-OEt Hydrochloride

Fluka 61850   MW 195.69 $C_8H_{17}NO_2 \cdot HCl$ ≥99.0% (titration); mp: 134-136°C

ICN 102163   MW 195.7 $C_8H_{17}NO_2 \cdot HCl$

Leucine — L-Leu-OMe Hydrochloride

Fluka 61890   MW 181.66 $C_7H_{15}NO_2 \cdot HCl$ ≥99.0% (titration); mp: 151-153°C

Neosystem AA01105   MW 181.7

Sigma L 9000   FW 181.7 $C_7H_{15}NO_2 \cdot HCl$

Leucine — L-Leu-OtBu Hydrochloride

ICN 155191   MW 223.7 $C_{10}H_{21}NO_2 \cdot HCl$ Crystalline

Sigma L 2125   FW 223.7 $C_{10}H_{21}NO_2 \cdot HCl$

Leucine — L-Leu-pNA

ICN 100792   MW 251.3 $C_{12}H_{17}N_3O_3$ Colorimetric substrate for leucine aminopeptidase

Leucine — L-Leu-pNA Free Base

Synonyms: Leucine Aminopeptidase Substrate

Sigma L 9125   FW 251.3 $C_{12}H_{17}N_3O_3$ Suitable for the colorimetric determination of leucine aminopeptidase

Leucine — L-Leu-pNA Hydrochloride

Sigma L 2158   FW 287.7 $C_{12}H_{17}N_3O_3 \cdot HCl$ Suitable for the colorimetric determination of leucine aminopeptidase

Leucine — L-t-Leu

Synonyms: (S)-2-Amino-3,3-Dimethylbutyric Acid; L-α-tBu-Gly; 3-Me-L-Val; L-2-Amino-3,3-Dimethylbutyric Acid; L-Pseudoleucine; (S)-2-Amino-3,3-Dimethylbutanoic Acid

Fluka 61825   MW 131.18 $C_6H_{13}NO_2$ ≥99.0% (titration); 0.5% H_2O; mp: >300°C

Rexim   MW 131.2 $C_6H_{13}NO_2$ White crystals or crystalline powder

Leucine — L-t-Leu-OMe Hydrochloride

Fluka 61891   MW 181.66 $C_7H_{15}NO_2 \cdot HCl$ ≥99.0% (titration); mp: 183-186°C

Leucine — L-trans-Epoxysuccinyl-Leu-3-Methylbutylamide

Synonyms: E-64c

Bachem N-1655.0001   MW 314.38 $C_{15}H_{26}N_2O_5$ Store at -15°C

Leucine — L-trans-Epoxysuccinyl-Leu-3-Methylbutylamide-OEt

Synonyms: E-64d; L-trans-Epoxysuccinyl-OEt-Leu-3-Methylbutylamide

Bachem N-1650.0001   Store at -15°C

Leucine — L-β-Leu Hydrochloride

Synonyms: (R)-3-Amino-4-Methylpentanoic Acid; L-β-Hva

Fluka 03675   MW 167.64 $C_6H_{13}NO_2 \cdot HCl$ ≥98.0% (TLC)

Leucine — L-γ-Me-Leu

Sigma M 3893   FW 145.2 $C_7H_{15}NO_2$

Leucine — N-((2S,3R)-3-Amino-2-Hydroxy-4-Phenylbutyryl)-L-Leu Hydrochloride

Synonyms: Bestatin; Aminopeptidase B Inhibitor; Leucine Aminopeptidase Inhibitor

Fluka 08170   MW 344.84 $C_{16}H_{24}N_2O_4 \cdot HCl$ ≥99% (TLC); mp: 216-218°C; ≤0.5% epibestatin; ≤0.2% Gly; ≤0.1% Leu, Phe | Specific inhibitor; Miura, K et al, *J Antibiotics*, 39: 734, 1986; Harbeson, SL & Rich, DH et al, *JMC*, 32: 1378, 1989; Umezawa, H, *Ann Rev Microbiol*, 36: 75, 1982

Leucine — N-((R,S)-2-Carboxy-3-Phenylpropionyl)-L-Leu

ICN 152832   MW 307.3 $C_{16}H_{21}NO_5$ A potent & highly-specific Enkephalinase inhibitor; Fournie-Zaluski, MC etal, *J Med Chem*, 26:60, 1983

Leucine — n-2,4-Dnp-L-Leu

Sigma D 0130   FW 297.3 $C_{12}H_{15}N_3O_6$

Leucine — N-3-Nitro-2-Pyridinesulfenyl-L-Leu

Sigma N 4765   FW 285.3 $C_{11}H_{15}N_3O_4S$

Leucine — *N*-Ac-D-Leu
Sigma A 0876 FW 173.2 $C_8H_{15}NO_3$ Sigma Grade

Leucine — *N*-Ac-DL-Leu
ICN 100103 MW 173.2 $C_8H_{15}NO_3$ Crystalline
Sigma A 1001 FW 173.2 $C_8H_{15}NO_3$

Leucine — *N*-Ac-L-Leu
Fluka 01260 MW 173.21 $C_8H_{15}NO_3$ ~99% (titration); mp: ~185°C
ICN 100107 MW 173.2 $C_8H_{15}NO_3$ Crystalline
Sigma A 1894 FW 173.2 $C_8H_{15}NO_3$

Leucine — *N*-Ac-L-Leu-pNA
Sigma A 0292 FW 293.3 $C_{14}H_{19}N_3O_4$

Leucine — *N*-Bz-DL-Leu
Sigma B 1504 FW 235.3 $C_{13}H_{17}NO_3$

Leucine — *N*-Bz-DL-Leu β-Naphthylamide
Sigma B 3628 FW 360.5 $C_{23}H_{24}N_2O_2$

Leucine — *N*-CBZ-D-Leu
Sigma C 1283 FW 265.3 $C_{14}H_{19}NO_4$

Leucine — *N*-CBZ-Leu-(*p*ONp)
ICN 101251 MW 386.4 $C_{20}H_{22}N_2O_6$ Crystalline

Leucine — *N*-CBZ-L-Leu
ICN 150573 MW 265.3 $C_{14}H_{19}NO_4$ Viscous oil
Sigma C 2502 FW 265.3 $C_{14}H_{19}NO_4$ Viscous liquid

Leucine — *N*-CBZ-L-Leu (4,5-³H)
ARC ART-625 MW 265.3 40-60 Ci/mmol; 1.48-2.22 TBq/mmol; in EtOH | Radiochemical

Leucine — *N*-CBZ-L-Leu *N*-Hydroxysuccinimide Ester
Sigma C 0670 FW 362.4 $C_{18}H_{22}N_2O_6$

Leucine — *N*-CBZ-L-Leu β-Naphthylamide
Sigma C 7148 FW 390.5 $C_{24}H_{26}N_2O_3$

Leucine — *N*-CBZ-*N*-Me-L-Leu
Sigma C 1158 FW 279.3 $C_{15}H_{21}NO_4$

Leucine — *N*-Chloroacetyl-L-Leu
Sigma C 1128 FW 207.7 $C_8H_{14}ClNO_3$

Leucine — *N*-Chloroacetyl-L-Leu Cyclohexylammonium Salt
ICN 101339 MW 207.7 $C_8H_{14}ClNO_3$ Crystalline

Leucine — *N*-FMOC-D-Leu (1-¹⁴C)
ARC ARC-1094 MW 353.4 50-60 mCi/mmol; 1.85-2.22 GBq/mmol; in EtOH | Radiochemical

Leucine — *N*-FMOC-L-Leu
Sigma F 0259 FW 353.4 $C_{21}H_{23}NO_4$

Leucine — *N*-FMOC-L-Leu (1-¹⁴C)
ARC ARC-662 MW 353.4 50-60 mCi/mmol; 1.85-2.22 GBq/mmol; in EtOH | Radiochemical

Leucine — *N*-FMOC-L-Leu (4,5-³H)
ARC ART-415 MW 353.4 10-20 Ci/mmol; 370-740 GBq/mmol; in EtOH | Radiochemical

Leucine — *N*-FMOC-L-Leu Resin Ester
Sigma F 1636 For peptide synthesis; Wang, SS, *J Am Chem Soc*, 95: 1328, 1973; Lu, G et al, *J Org Chem*, 46: 3433, 1981

Leucine — *N*-FMOC-L-Leu SASRIN Resin Ester
Sigma F 2014 200-400 mesh; substitution range: 0.3-0.8 mmole FMOC AA/g resin | For peptide synthesis; *N*-FMOC AA acyl ester of 3-methoxy-4-oxymethyl-phenoxymethylated 1% divinylbenzene cross-linked polystyrene; peptides can be cleaved from the resin with 0.5-1% CF_3CO_2H in CH_2Cl_2; Mergler, M et al, *in Peptides: Chemistry & Biology* (Proc 10th Am Peptide Symp), Marhall, GR, ed, 259, 1988

Leucine — *N*-For-L-Leu
ICN 151167 MW 159.2 $C_7H_{13}NO_3$ Crystalline
Sigma F 9251 FW 159.2 $C_7H_{13}NO_3$ White powder

Leucine — *N*-L-Leu-2-Aminoacridone
Synonyms: Leucine Aminopeptidase Fluorogenic Substrate
Fluka 61965 MW 323.39 $C_{19}H_{21}N_3O_2$ ≥98.0% (TLC); mp: 270°C | Fluorogenic substrate for the determination of leucine aminopeptidase; Moser, R, *Appl Fluoresc Tech*, 1: 15, 1989

Leucine — *N*-Me-Cycloleucine
Synonyms: N-Me-1-Aminocyclopentanecarboxylic Acid
Alexis 550-053 MW 143.2 $C_7H_{13}NO_2$ ≥97%; white powder; soluble in H_2O or dilute aqueous base

Leucine — *N*-Me-DL-Leu
Bachem F-1775.0001 MW 145.2 $C_7H_{15}NO_2$ Store at RT
ICN 155596 MW 145.2 $C_7H_{15}NO_2$
Sigma M 3502 FW 145.2 $C_7H_{15}NO_2$

Leucine — *N*-Me-Leu
Bachem E-2145.0001 MW 145.2 $C_7H_{15}NO_2$ Store at RT

Leucine — *N*-Me-Leu-OBzl *p*-Tosyl
Bachem E-2150.0001 MW 407.53 $C_{14}H_{21}NO_2 \cdot C_7H_8O_3S$ Store at RT

Leucine — *N*-Me-L-Leu
ICN 155597 MW 145.2 $C_7H_{15}NO_2$
Sigma M 2382 FW 145.2 $C_7H_{15}NO_2$ Crystalline

Leucine — *N*-Me-L-Leu Hydrochloride
Fluka 02674 MW 181.66 $C_7H_{15}NO_2 \cdot HCl$ ≥98.0% (TLC)

Leucine — *N*-MSOC-L-Leu Dicyclohexylammonium Salt
ICN 155739 $C_{10}H_{19}NO_6S \cdot C_{12}H_{23}N$ Crystalline

Leucine — *N*-O-Nps-L-Leu cyclohexylammonium Salt
ICN 103324 MW 465.6 $C_{12}H_{16}N_2O_4S \cdot C_{12}H_{23}N$ Yellow crystals

Leucine — N-Phthaloyl-L-Leu

Sigma P 4503   FW 261.3 $C_{14}H_{15}NO_4$ May contain 1-4% free AA

Leucine — N-t-BOC-D-Leu

Sigma B 8513   FW 231.3 $C_{11}H_{21}NO_4$

Leucine — N-t-BOC-L-Leu

Sigma B 0643   FW 231.3 $C_{11}H_{21}NO_4$

Leucine — N-t-BOC-L-Leu N-Hydroxysuccinimide Ester

Sigma B 2005   FW 328.4 $C_{15}H_{24}N_2O_6$ ~95% (TLC)

Leucine — N-t-BOC-N-Me-L-Leu

Sigma B 2662   FW 245.3 $C_{12}H_{23}NO_4$

Leucine — N-TFA-L-Leu-OMe

Sigma T 4506   FW 241.2 $C_9H_{14}F_3NO_3$ ~98%

Leucine — Nᵅ-FMOC-L-Leu

ICN 151146   MW 353.4 $C_{21}H_{23}NO_4$

Leucine — Nᵅ-t-BOC-L-Leu Hydrate

ICN 101073   MW 249.3 $C_{11}H_{21}NO_4 \cdot H_2O$ Crystalline

Leucine — Phthaloyl-Leu

Bachem E-2285.0005   MW 261.28 $C_{14}H_{15}NO_4$ Store at RT

Leucine — Pth-Leu

Sigma P 1252   FW 248.3 $C_{13}H_{16}N_2OS$ ~99% | Useful as standards in protein sequencing by the Edman degradation

Leucine — TentaGel S PHB-Leu-FMOC

Synonyms: FMOC-L-Leu 4-(Poly(Ethylenoxy))-OBzl Polymer Bound

Fluka 86394   0.20 mmol protected AA/g; ~90 μm particle size

Leucine — TentaGel S Trt-Leu-FMOC

Fluka 86425   0.20 mmol protected AA/g; ~90 μm particle size

Leucine — trans-Epoxysuccinyl-Leucylamido-4-Guanidino Butane

Synonyms: E-64; (L-3-trans-Carboxy-Oxiran-2-Carbonyl)-Leu-Agmatin; Thiol Protease Inhibitor

ICN 152846   MW 357.4 $C_{15}H_{27}N_5O_5$ Hanada, A etal, *Agric Biol Chem*, 42:529, 1978; Barrett, AJ etal, *Biochem J*, 201:189, 1982

Leucine — Z-D-Leu

Fluka 96705   MW 265.31 $C_{14}H_{19}NO_4$ ≥98.0% (TLC)

Leucine — Z-D-Leu Oil

Bachem C-2105.0001   MW 265.31 $C_{14}H_{19}NO_4$ Store at 2-8°C

Leucine — Z-D-Leu-ONp

Bachem C-2125.0001   MW 386.41 $C_{20}H_{22}N_2O_6$ Store at -15°C

Leucine — Z-D-Leu-OSu

Bachem C-2115.0001   MW 362.38 $C_{18}H_{22}N_2O_6$ Store at -15°C

Leucine — Z-DL-Leu

Bachem C-3210.0025   MW 265.31 $C_{14}H_{19}NO_4$ Store at 2-8°C

Fluka 96715   MW 265.31 $C_{14}H_{19}NO_4$ ≥98.0% (TLC)

Leucine — Z-Leu

Synonyms: Benzyloxycarbonyl-L-Leu

Peptides International ZLL-2029-PI   MW 265.31   98% (TLC); syrup

Leucine — Z-Leu Oil

Bachem C-2100.0025   MW 265.31 $C_{14}H_{19}NO_4$ Store at 2-8°C

Leucine — Z-Leu-CMK

Bachem N-1260.0001   MW 297.78 $C_{15}H_{20}ClNO_3$ Store at -15°C

Leucine — Z-Leu-ONp

Bachem C-2120.0005   MW 386.41 $C_{20}H_{22}N_2O_6$ Store at -15°C

Leucine — Z-Leu-OSu

Bachem C-2110.0005   MW 362.38 $C_{18}H_{22}N_2O_6$ Store at -15°C

Leucine — Z-L-Leu

Fluka 96710   MW 265.31 $C_{14}H_{19}NO_4$ ~98% (titration)

Neosystem ZA01101   MW 265.3

Leucine — Z-L-Leu Dicyclohexylamine

Neosystem ZA01102   MW 446.6

Leucine — Z-L-Leu Hydroxysuccinimide Ester

Fluka 96631   MW 362.38 $C_{18}H_{22}N_2O_6$ mp: 118-120°C

Leucine — Z-L-Leu-(4-ONp)

Fluka 96742   MW 386.40 $C_{20}H_{22}N_2O_6$ ≥98.0% (TLC)

Leucine — Z-L-t-Leu Dicyclohexylamine

Fluka 96720   MW 446.63 $C_{14}H_{19}NO_4 \cdot C_{12}H_{23}N$ ≥99.0% (HPLC); mp: 163-167°C

Leucine — Z-N-Me-Leu

Bachem C-2240.0001   MW 279.34 $C_{15}H_{21}NO_4$ Store at -15°C

Leucine — Z-N-Me-L-Leu

Fluka 96925   MW 279.34 $C_{15}H_{21}NO_4$ ≥98.0% (HPLC)

Leucine — α-Keto-Leu Calcium Salt

Synonyms: 4-Me-2-Oxopentanoic Acid; α-Ketoisocaproic Acid; Calcium α-Oxoisocaproate

Rexim   MW 298.4 $C_{12}H_{18}CaO_6$ White crystals or crystalline powder

Leucine — α-Me-DL-Leu

Bachem F-1800.0001   MW 145.2 $C_7H_{15}NO_2$ Store at RT

ICN 155595   MW 145.2 $C_7H_{15}NO_2$

Sigma M 4142   FW 145.2 $C_7H_{15}NO_2$

Leucine — γ-Aminobutyryl-L-Leu Hydrobromide

ICN 101757   MW 297 $C_{10}H_{20}N_2O_3 \cdot HBr$

Leucine — γ-Me-D-Leu
ICN 155592   MW 145.2 $C_7H_{15}NO_2$

Leucine — γ-Me-DL-Leu
ICN 155593   MW 145.2 $C_7H_{15}NO_2$

Leucine — γ-Me-L-Leu
ICN 155594   MW 145.2 $C_7H_{15}NO_2$

Lysine — (5R)-5-Hydroxy-L-Lys Dihydrochloride Hydrate
Synonyms: (2S,5R)-2,6-Diamino-5-Hydroxycaproic Acid

Fluka 55501   MW 253.17 $C_6H_{14}N_2O_3 \cdot 2HCl \cdot H_2O$ ≥99.0% (titration

Lysine — Ac-DL-Lys-Ac
Bachem F-3205.0001   MW 230.26 $C_{10}H_{18}N_2O_4$ Store at -15°C

Lysine — Ac-L-Lys
Synonyms: L-2,6-Diaminocaproic Acid

Rexim   MW 206.2 $C_8H_{18}N_2O_4$ White crystals or crystalline powder

Lysine — Ac-Lys
Bachem E-1120.0001   MW 188.23 $C_8H_{16}N_2O_3$ Store at RT

Lysine — Ac-Lys Amide Hydrochloride
Bachem E-1125.0050   MW 223.7 $C_8H_{17}N_3O_2 \cdot HCl$ Store at RT

Lysine — Ac-Lys-Ac
Bachem E-1775.0001   MW 230.26 $C_{10}H_{18}N_2O_4$ Store at -15°C

Lysine — Ac-Lys-AMC Acetate
Bachem I-1010.0050   MW 405.5 $C_{18}H_{23}N_3O_4 \cdot C_2H_4O_2$ Store at -15°C

Lysine — Ac-Lys-BOC
Bachem E-1040.0250   MW 288.34 $C_{13}H_{24}N_2O_5$ Store at 2-8°C

Lysine — Ac-Lys-NHMe
Bachem E-1130.0250   MW 201.27 $C_9H_{19}N_3O_2$ Store at 2-8°C

Lysine — Ac-Lys-OMe Hydrochloride
Synonyms: L-ALME

Bachem M-1045.0001   MW 238.71 $C_9H_{18}N_2O_3 \cdot HCl$ Store at -15°C

Lysine — Ac-Lys-pNA Hydrochloride
Bachem L-1045.0050   MW 344.8 $C_{14}H_{20}N_4O_4 \cdot HCl$ Store at -15°C

Lysine — Aloc-Lys-BOC Dicyclohexylamine
Bachem E-3635.0001   MW 511.7 $C_{15}H_{26}N_2O_6 \cdot C_{12}H_{23}N$ Store at -15°C

Lysine — BOC-D-Lys
American Peptide BDLYS10   MW 246.3

Bachem A-2705.0001   MW 246.31 $C_{11}H_{22}N_2O_4$ Store at 2-8°C

Lysine — BOC-D-Lys-2-Chloro-Z
Bachem A-1520.0001   MW 414.89 $C_{19}H_{27}ClN_2O_6$ Store at RT

Neosystem BA01208   MW 414.9

USBio B2339   MW 414.9 $C_{19}H_{27}ClN_2O_6$ ≥99%

Lysine — BOC-D-Lys-2-Chloro-Z-PAM Resin
Neosystem RP01203

Lysine — BOC-D-Lys-BOC
American Peptide BDLYS35   MW 346.4

Lysine — BOC-D-Lys-BOC Dicyclohexylamine
Bachem A-4285.0005   MW 527.75 $C_{16}H_{30}N_2O_6 \cdot C_{12}H_{23}N$ Store at -15°C

Neosystem BA01205   MW 527.7

Lysine — BOC-D-Lys-Chloro-Z
Synonyms: N^α-BOC-N^ϵ-2-Chloro-Benzyloxycarbonyl-D-Lys

Peptides International BDK-2628   MW 488.03   >98% (HPLC); white crystalline powder

Lysine — BOC-D-Lys-FMOC
Bachem A-3985.0001   MW 468.55 $C_{26}H_{32}N_2O_6$ Store at RT

Lysine — BOC-D-Lys-Nicotinoyl
Bachem A-3130.0001   MW 351.4 $C_{17}H_{25}N_3O_5$ Store at 2-8°C

USBio B2344   MW 351.4 $C_{17}H_{25}N_3O_6$ ≥98%

Lysine — BOC-D-Lys-TFA
Bachem A-2700.0001   MW 342.32 $C_{13}H_{21}F_3N_2O_5$ Store at -15°C

Lysine — BOC-D-Lys-Z
Bachem A-1470.0001   MW 380.44 $C_{19}H_{28}N_2O_6$ Store at RT

Neosystem BA01202   MW 380.4

USBio B2346   MW 380.5 $C_{19}H_{28}N_2O_6$ ≥99%

Lysine — BOC-L-Lys-2-Chloro-Z
Neosystem BA01209   MW 414.9

Lysine — BOC-L-Lys-2-Chloro-Z-MBHA Resin
Neosystem RN01209

Lysine — BOC-L-Lys-2-Chloro-Z-O-CH₂-φ-CH₂-COOH
Neosystem LP01204   MW 563.1

Lysine — BOC-L-Lys-2-Chloro-Z-PAM Resin
Neosystem RP01204

Lysine — BOC-L-Lys-Alloc Dicyclohexylamine
Neosystem BA01201   MW 511.7

Lysine — BOC-L-Lys-BOC Dicyclohexylamine
Neosystem BA01207   MW 527.7

Lysine — BOC-L-Lys-FMOC
Neosystem BA01210   MW 468.6

Lysine — BOC-L-Lys-TFA
Neosystem BA01224   MW 342.3

Lysine — BOC-L-Lys-Z

Neosystem BA01203   MW 380.4

Lysine — BOC-Lys

American Peptide BLLYS10   MW 246.3

Bachem A-1920.0001   MW 246.31 $C_{11}H_{22}N_2O_4$ Store at 2-8°C

USBio B2331   MW 246.3 $C_{11}H_{22}N_2O_4$ ≥98%

Lysine — BOC-Lys-(Isopropyl,Z) Oil

Bachem A-2890.0001   MW 422.52 $C_{22}H_{34}N_2O_6$ Store at -15°C

Lysine — BOC-Lys-(Isopropyl,Z)-Dicyclohexylamine

USBio B2343   MW 603.8 $C_{22}H_{34}N_2O_6 \cdot C_{12}H_{23}N$ ≥98%

Lysine — BOC-Lys-2-Bromo-Z

Bachem A-1415.0001   MW 459.34 $C_{19}H_{27}BrN_2O_6$ Store at RT

Lysine — BOC-Lys-2-Chloro-Z

American Peptide BLLYS40   MW 414.9

Bachem A-1510.0005   MW 414.89 $C_{19}H_{27}ClN_2O_6$ Store at RT

USBio B2340   MW 414.9 $C_{19}H_{27}ClN_2O_6$ ≥99%

Lysine — BOC-Lys-2-Chloro-Z t-Butylamine

Bachem A-1515.0001   MW 488.02 $C_{19}H_{27}ClN_2O_6 \cdot C_4H_{11}N$ Store at -15°C

Lysine — BOC-Lys-2-Chloro-Z-® (200-400 mesh)

Bachem D-1200.0005   Store at RT

Lysine — BOC-Lys-2-Chloro-Z-O-Resin

American Peptide RMRB215   1% DVB cross-linked: 100-200 mesh | t-BOC protected AA resin; preattached resins are useful for synthesis of peptides with C-terminal carboxyl group by t-BOC chemistry

Lysine — BOC-Lys-2-Chloro-Z-OSu

Bachem A-3515.0001   MW 511.96 $C_{23}H_{30}ClN_3O_8$ Store at -15°C

Lysine — BOC-Lys-2-Chloro-Z-PAM-® (200-400 mesh)

Bachem D-1235.0001   Store at 2-8°C

Lysine — BOC-Lys-2-Chloro-Z-PAM-Resin

American Peptide RPAM215   1% DVB cross-linked: 100-200 mesh | t-BOC protected AA resin; preattached resins are useful for synthesis of peptides with C-terminal carboxyl group by t-BOC chemistry

Lysine — BOC-Lys-4-Nitro-Z

Bachem A-4545.0005   MW 425.44 $C_{19}H_{27}N_3O_8$ Store at -15°C

Lysine — BOC-Lys-Ac

Bachem A-1050.0001   MW 288.34 $C_{13}H_{24}N_2O_5$ Store at RT

Lysine — BOC-Lys-Ac-AMC

Bachem I-1875.0250   MW 445.52 $C_{23}H_{31}N_3O_6$ Store at -15°C

Lysine — BOC-Lys-AMC

Bachem I-1880.0250   MW 403.48 $C_{21}H_{29}N_3O_5$ Store at -15°C

Lysine — BOC-Lys-Biotin

American Peptide BLLYS25   MW 472.6

Lysine — BOC-Lys-Biotinyl

Bachem A-3770.0250   MW 472.61 $C_{21}H_{36}N_4O_6S$ Store at -15°C

USBio B2333   MW 489.6 $C_{21}H_{36}N_4O_7S$ ≥98%

Lysine — BOC-Lys-BOC

American Peptide BLLYS30   MW 346.4

USBio B2334   MW 346.4 $C_{16}H_{30}N_2O_6$ ≥99%

Lysine — BOC-Lys-BOC Dicyclohexylamine

Synonyms: N^α,N^ϵ-Di-BOC-L-Lys

Bachem A-1020.0005   MW 527.75 $C_{16}H_{30}N_2O_6 \cdot C_{12}H_{23}N$ Store at 2-8°C

Peptides International BLK-5328-PI   MW 527.75   >98% (HPLC); white crystalline powder

Lysine — BOC-Lys-BOC-Dicyclohexylamine

USBio B2335   MW 527.7 $C_{16}H_{30}N_2O_6 \cdot C_{12}H_{23}N$ ≥99%

Lysine — BOC-Lys-BOC-ONp

Bachem A-1030.0005   MW 467.52 $C_{22}H_{33}N_3O_8$ Store at -15°C

Lysine — BOC-Lys-BOC-OSu

Bachem A-1025.0001   MW 443.5 $C_{20}H_{33}N_3O_8$ Store at -15°C

Lysine — BOC-Lys-Chloro-Z

Synonyms: N^α-BOC-N^ϵ-2-Chlorobenzyloxycarbonyl-L-Lys

Peptides International BLK-2135   MW 414.89   >98% (HPLC); white crystalline powder

Lysine — BOC-Lys-Dimethyl

Bachem A-3885.0001   MW 274.36 $C_{13}H_{26}N_2O_4$ Store at -15°C

USBio B2342   MW 274.4 $C_{13}H_{26}N_2O_4$ ≥98%

Lysine — BOC-Lys-FMOC

American Peptide BLLYS50   MW 468.6

Bachem A-1610.0001   MW 468.55 $C_{26}H_{32}N_2O_6$ Store at RT

USBio B2341   MW 468.6 $C_{26}H_{32}N_2O_6$ ≥99%

Lysine — BOC-Lys-FMOC-OMe

Bachem A-1615.0001   MW 482.58 $C_{27}H_{34}N_2O_6$ Store at 2-8°C

Lysine — BOC-Lys-For

Bachem A-1620.0001   MW 274.32 $C_{12}H_{22}N_2O_5$ Store at 2-8°C

Lysine — BOC-Lys-Nicotinoyl

Bachem A-3135.0001   MW 351.4 $C_{17}H_{25}N_3O_5$ Store at 2-8°C

USBio B2345   MW 351.4 $C_{17}H_{25}N_3O_6$ ≥98%

Lysine — BOC-Lys-OMe

Bachem A-1925.0001   MW 260.33 $C_{12}H_{24}N_2O_4$ Store at -15°C

Lysine — BOC-Lys-TFA

Bachem A-2345.0001   MW 342.32 $C_{13}H_{21}F_3N_2O_5$ Store at -15°C

Lysine — BOC-Lys-TFA-OSu

Bachem A-2350.0001   MW 439.39 $C_{17}H_{24}F_3N_3O_7$ Store at -15°C

Lysine — BOC-Lys-Tosyl Dicyclohexylamine

Bachem A-2540.0001   MW 581.82 $C_{18}H_{28}N_2O_6S \cdot C_{12}H_{23}N$
Store at 2-8°C

Lysine — BOC-Lys-Tosyl-ONp

Bachem A-3680.0001   MW 521.59 $C_{24}H_{31}N_3O_8S$ Store at -15°C

Lysine — BOC-Lys-Z

Synonyms: N^α-BOC-N^ϵ-Benzyloxycarbonyl-L-Lys

Bachem A-1465.0005   MW 380.44 $C_{19}H_{28}N_2O_6$ Store at RT

Peptides International BLK-2108   MW 380.44   >98% (HPLC); white crystalline powder

USBio B2347   MW 380.5 $C_{19}H_{28}N_2O_6$ ≥99%

Lysine — BOC-Lys-Z Dicyclohexylamine

Bachem A-3600.0005   MW 561.76 $C_{19}H_{28}N_2O_6 \cdot C_{12}H_{23}N$
Store at RT

Lysine — BOC-Lys-Z-ONp

Bachem A-1480.0001   MW 501.34 $C_{25}H_{31}N_3O_8$ Store at -15°C

Lysine — BOC-Lys-Z-OSu

Bachem A-1475.0001   MW 477.52 $C_{23}H_{31}N_3O_8$ Store at -15°C

Lysine — BOC-Lys-Z-pNA

Bachem L-1510.0001   MW 500.6 $C_{25}H_{32}N_4O_7$ Store at -15°C

Lysine — BOC-N-Me-Lys-Z Dicyclohexylamine

Bachem A-3690.0001   MW 575.79 $C_{20}H_{30}N_2O_6 \cdot C_{12}H_{23}N$
Store at 2-8°C

Lysine — Bsmoc-Lys-BOC

Synonyms: N^α-Bsmoc-N^ϵ-BOC-L-Lys

Peptides International BLK-5630-PI   MW 468.53   >98% (HPLC); white to off-white powder | LA Carpino, et al, *JACS*, 119:9915, 1997

Lysine — Bz-Lys

Bachem E-1480.0001   MW 250.3 $C_{13}H_{18}N_2O_3$ Store at RT

Lysine — Bz-Lys-OMe Hydrochloride

Bachem E-1485.0250   MW 300.79 $C_{14}H_{20}N_2O_3 \cdot HCl$ Store at RT

Lysine — CBZ-D-Lys

USBio C2099-04   MW 280.3 $C_{14}H_{20}N_2O_4$ ≥99%

Lysine — CBZ-D-Lys-BOC

USBio C2099-15   MW 380.4 $C_{19}H_{28}N_2O_6$ ≥99%

Lysine — CBZ-D-Lys-OMe Hydrochloride

USBio C2099-08   MW 330.8 $C_{15}H_{22}N_2O_4 \cdot HCl$ ≥99%

Lysine — CBZ-Lys

USBio C2099-03   MW 280.3 $C_{14}H_{20}N_2O_4$ ≥99%
USBio C2099-06   MW 280.3 $C_{14}H_{20}N_2O_4$ ≥99%

Lysine — CBZ-Lys Thiobenzyl Ester Hydrochloride

USBio C2099-09   MW 423 $C_{21}H_{26}N_2O_3S \cdot HCl$ ≥99%

Lysine — CBZ-Lys-BOC

USBio C2099-16   MW 380.4 $C_{19}H_{28}N_2O_6$ ≥99%

Lysine — CBZ-Lys-BOC-Dicyclohexylamine

USBio C2099-17   MW 561.7 $C_{19}H_{28}N_2O_6 \cdot C_{12}H_{23}N$ ≥99%

Lysine — CBZ-Lys-BOC-OSu

USBio C2099-18   MW 477.4 $C_{23}H_{31}N_3O_8$ ≥99%

Lysine — CBZ-Lys-OBzl-Benzenesulfonate

USBio C2099-14   MW 528.5 $C_{27}H_{32}N_2O_7S$ ≥99%

Lysine — CBZ-Lys-OMe Hydrochloride

USBio C2099-11   MW 330.9 $C_{15}H_{22}N_2O_4 \cdot HCl$ ≥99%
USBio C2099-12   MW 330.8 $C_{15}H_{22}N_2O_4 \cdot HCl$ ≥99%

Lysine — CBZ-Lys-ONp Hydrochloride

USBio C2099-13   MW 437.9 $C_{20}H_{23}N_3O_6 \cdot HCl$ ≥99%

Lysine — CBZ-Lys-OtBu Hydrochloride

USBio C2099-10   MW 372.9 $C_{18}H_{28}N_2O_4 \cdot HCl$ ≥99%

Lysine — CBZ-Lys-Z

USBio C2099-19   MW 414.5 $C_{22}H_{26}N_2O_6$ ≥99%

Lysine — Dde-Lys-FMOC

Bachem E-3385.0001   MW 532.64 $C_{31}H_{36}N_2O_6$ Store at -15°C

Lysine — Didansyl-L-Lys

Sigma D 1500   FW 612.8 $C_{30}H_{36}N_4O_6S_2$ ~95%; crystalline

Lysine — Di-N-TFA-L-Lys-OMe

Sigma T 4631   FW 352.2 $C_{11}H_{14}F_6N_2O_4$ ~98%

Lysine — DL-5-Hydroxy-L-Lys Hydrochloride

Synonyms: 2,6-Diamino-5-Hydroxycaproic Acid

Fluka 55500   MW 198.65 $C_6H_{14}N_2O_3 \cdot HCl$ ≥97.0% (titration); mp: 230°C; mixture of DL- & DL-allo-form)

Lysine — DL-Lys

Synonyms: (±)-2,6-Diaminocaproic Acid

Fluka 62860   MW 146.19 $C_6H_{14}N_2O_2$ ~97.0% (titration); mp: 175°C; ≤2% H_2O

Lysine — DL-Lys Dihydrochloride

Fluka 62910   MW 219.11 $C_6H_{14}N_2O_2 \cdot 2HCl$ ≥99.0% (titration); mp: 190-195°C; ≤0.05% residue on ignition

ICN 102213   MW 219.1 $C_6H_{14}N_2O_2 \cdot 2HCl$ Crystalline

Lysine — DL-Lys Free Base

Synonyms: DL-2,6-Diaminohexanoic Acid

Sigma L 2513 FW 146.2 $C_6H_{14}N_2O_2$ ≥98% (TLC)

Lysine — DL-Lys Hydrochloride

Synonyms: DL-2,6-Diaminohexanoic Acid

Amersham US18580 ≥98.0%

Bachem F-2210.0025 MW 182.65 $C_6H_{14}N_2O_2 \cdot$ HCl Store at RT

Fluka 62960 MW 182.65 $C_6H_{14}N_2O_2 \cdot$ HCl ≥99.0% (titration); mp: 265-270°C; ≤0.05% residue on ignition

ICN 102214 MW 182.6 $C_6H_{14}N_2O_2 \cdot$ HCl Crystalline

Sigma L 6001 FW 182.6 $C_6H_{14}N_2O_2 \cdot$ HCl ≥98% (TLC) | Also available as part of a kit

USBio L9070-20 MW 182.6 $C_6H_{14}N_2O_2 \cdot$ HCl ≥99%

Lysine — DL-Lys Solution

Synonyms: (±)-2,6-Diaminocaproic Acid

Fluka 62870 MW 146.19 $C_6H_{14}N_2O_2$ 50% in H_2O

Lysine — D-Lys

Synonyms: (R)-2,6-Diaminocaproic Acid

Fluka 62830 MW 146.19 $C_6H_{14}N_2O_2$ ~98.0% (titration); mp: 221-225°C; ≤5% H_2O

Lysine — D-Lys (1-^{14}C)

ARC ARC-600 MW 146.2 $H_2N(CH_2)_4CH(NH_2)COOH$ 50-60 mCi/mmol; 1.85-2.22 GBq/mmol; in 0.01 N HCl | Radiochemical

Lysine — D-Lys (4,5-^{3}H)

ARC ART-797 MW 146.2 $H_2NCH_2CH_2CH_2CH_2CH(NH_2)COOH$ 40-60 Ci/mmol; 1.48-2.22 TBq/mmol; in 0.01 N HCl | Radiochemical

Lysine — D-Lys Amide Dihydrochloride

Bachem E-3390.0001 MW 218.13 $C_6H_{15}N_3O \cdot$ 2HCl Store at -15°C

Lysine — D-Lys Free Base

Synonyms: D-2,6-Diaminohexanoic Acid

Sigma L 8021 FW 146.2 $C_6H_{14}N_2O_2$ ≥98% (TLC)

Lysine — D-Lys Hydrochloride

Synonyms: D-2,6-Diaminohexanoic Acid

Bachem F-1730.0001 MW 182.65 $C_6H_{14}N_2O_2 \cdot$ HCl Store at RT

Fluka 62950 MW 182.65 $C_6H_{14}N_2O_2 \cdot$ HCl ≥99.0% (titration); mp: 260°C; ≤0.05% residue on ignition

ICN 102212 MW 182.6 $C_6H_{14}N_2O_2 \cdot$ HCl ≥99%; crystalline

ICN 151572 MW 182.6 $C_6H_{14}N_2O_2 \cdot$ HCl ~99%; crystalline

Neosystem AA01201 MW 182.6

Peptides International ADK-2813 MW 182.65 >99.9% optical purity by RP-HPLC; white crystalline powder

Rexim MW 182.7 $C_6H_{15}ClN_2O_2$

Sigma L 5876 FW 182.6 $C_6H_{14}N_2O_2 \cdot$ HCl ≥98% (TLC)

USBio L9070-15 MW 182.7 $C_6H_{14}N_2O_2 \cdot$ HCl ≥99%

Lysine — D-Lys-2-Chloro-Z

Bachem F-2880.0001 MW 314.77 $C_{14}H_{19}ClN_2O_4$ Store at RT

Lysine — D-Lys-Ac

Bachem F-3110.0001 MW 188.23 $C_8H_{16}N_2O_3$ Store at RT

Lysine — D-Lys-Aloc

Bachem F-3690.0001 MW 230.27 $C_{10}H_{18}N_2O_4$ Store at -15°C

Lysine — D-Lys-BOC

Bachem F-1302.0001 MW 246.31 $C_{11}H_{22}N_2O_4$ Store at RT

USBio B2337 MW 246.3 $C_{11}H_{22}N_2O_4$ ≥99%

Lysine — D-Lys-BOC-OAll Hydrochloride

Bachem F-3520.0001 MW 322.84 $C_{14}H_{26}N_2O_4 \cdot$ HCl Store at -15°C

Lysine — D-Lys-BOC-OtBu Hydrochloride

Bachem F-2455.0001 MW 338.88 $C_{15}H_{30}N_2O_4 \cdot$ HCl Store at -15°C

Lysine — D-Lys-Z

Bachem F-1400.0005 MW 280.32 $C_{14}H_{20}N_2O_4$ Store at RT

Lysine — D-Lys-Z-OBzl Hydrochloride

Bachem F-2475.0001 MW 406.91 $C_{21}H_{26}N_2O_4 \cdot$ HCl Store at RT

Lysine — D-Lys-Z-OMe Hydrochloride

Bachem F-2230.0001 MW 330.81 $C_{15}H_{22}N_2O_4 \cdot$ HCl Store at 2-8°C

Lysine — DL-δ-Hydroxy-DL-Lys Hydrochloride

Bachem F-1650.0001 MW 198.65 $C_6H_{14}N_2O_3 \cdot$ HCl Store at 2-8°C

Lysine — DL-δ-Hydroxy-DL-Lys-BOC

Bachem F-2335.0250 MW 262.31 $C_{11}H_{22}N_2O_5$ Store at -15°C

Lysine — FMOC-D-Lys Hydrochloride

Bachem B-2020.0001 MW 404.9 $C_{21}H_{25}N_2O_4Cl$ Store at RT

Lysine — FMOC-D-Lys-Aloc

Bachem B-2900.0001 MW 452.51 $C_{25}H_{28}N_2O_6$ Store at -15°C

Lysine — FMOC-D-Lys-Biotin

American Peptide FLYS115 MW 594.8

Lysine — FMOC-D-Lys-BOC

Synonyms: $N^α$-FMOC-$N^ε$-BOC-D-Lys

Bachem B-1085.0001 MW 468.55 $C_{26}H_{32}N_2O_6$ Store at -15°C

Neosystem FA01206 MW 468.6

Peptides International FDK-1815-PI MW 468.55 >98% (HPLC); white crystalline powder

USBio F5406 MW 468.6 $C_{26}H_{32}N_2O_6$ ≥99%

Lysine — FMOC-D-Lys-BOC-OPfp

Bachem B-1835.0001 MW 634.6 $C_{32}H_{31}F_5N_2O_6$ Store at -15°C

Lysine — FMOC-D-Lys-BOC-SASRIN™-® (200-400 mesh)

Bachem D-1785.0001 Store at 2-8°C

Lysine — FMOC-D-Lys-Dnp

Bachem B-3145.0250	MW 534.53	$C_{27}H_{26}N_4O_8$ Store at -15°C

Lysine — FMOC-D-Lys-FMOC

American Peptide FLYS130	MW 594.8
Bachem B-3050.0001	MW 590.68 $C_{36}H_{34}N_2O_6$ Store at 2-8°C

Lysine — FMOC-D-Lys-Mtt

Synonyms: N^α-FMOC-N^ε-4-Methyltrityl-D-Lys

Bachem B-2620.0001	MW 624.78	$C_{41}H_{40}N_2O_4$ Store at -15°C
Peptides International FDK-1898-PI	MW 624.79	>98% (HPLC); white crystalline powder

Lysine — FMOC-D-Lys-Nicotinoyl

Bachem B-3300.0001	MW 473.53	$C_{27}H_{27}N_3O_5$ Store at -15°C
USBio F5412	MW 473.5	$C_{27}H_{27}N_3O_5$ ≥98%

Lysine — FMOC-D-Lys-Z

Bachem B-3000.0001	MW 502.57	$C_{29}H_{30}N_2O_6$ Store at 2-8°C
USBio F5414	MW 502.6	$C_2H_{30}N_2O_6$ ≥99%

Lysine — FMOC-L-Lys-2-Chloro-Z

Neosystem FA01209	MW 537.0

Lysine — FMOC-L-Lys-Alloc

Neosystem FA01203	MW 452.5

Lysine — FMOC-L-Lys-Biotin

Neosystem FA01220	MW 594.7
Neosystem FA01220	MW 594.7

Lysine — FMOC-L-Lys-BOC

Neosystem FA01207	MW 468.6

Lysine — FMOC-L-Lys-BOC-Wang Resin

American Peptide RFWAN55 Preattached Wang resins are usful for synthesis of peptides with C-terminal carboxyl groups; peptides can be cleaved from the resin with 90% TFA in the presence of scavengers

Neosystem RW01202

Lysine — FMOC-L-Lys-DABCYL

Neosystem FA01221	MW 619.7

Lysine — FMOC-L-Lys-Dnp

Neosystem FA01210	MW 534.5

Lysine — FMOC-L-Lys-FMOC

Neosystem FA01212	MW 590.7

Lysine — FMOC-L-Lys-FMOC-Wang Resin

American Peptide RFWAN56 Preattached Wang resins are usful for synthesis of peptides with C-terminal carboxyl groups; peptides can be cleaved from the resin with 90% TFA in the presence of scavengers

Lysine — FMOC-L-Lys-Z

Neosystem FA01205	MW 502.6

Lysine — FMOC-Lys

American Peptide FLYS100	MW 368.5	
USBio F5404	MW 368.4	$C_{21}H_{24}N_2O_4$ ≥99%

Lysine — FMOC-Lys Hydrochloride

Bachem B-1870.0001	MW 404.89	$C_{21}H_{24}N_2O_4 \cdot HCl$ Store at RT

Lysine — FMOC-Lys-(Biotinyl-ε-Aminocaproyl)

Bachem B-2580.0250	MW 707.9	$C_{37}H_{49}N_5O_7S$ Store at -15°C

Lysine — FMOC-Lys-(Extended-Biotin)

American Peptide FLYS111	MW 707.8

Lysine — FMOC-Lys-(Isopropyl-BOC)

USBio F5411	MW 510.6	$C_{29}H_{38}N_2O_6$ ≥98%

Lysine — FMOC-Lys-2-Chloro-Z

Bachem B-2790.0001	MW 537.01	$C_{29}H_{29}ClN_2O_6$ Store at -15°C
USBio F5409	MW 538	$C_{29}H_{30}N_2O_6Cl$ ≥99%

Lysine — FMOC-Lys-4-Methoxytrityl

Bachem B-3215.0001	MW 639.77	$C_{41}H_{39}N_2O_5$ Store at -15°C

Lysine — FMOC-Lys-Ac

Bachem B-3520.0001	MW 410.47	$C_{23}H_{26}N_2O_5$ Store at -15°C
USBio F5405	MW 410.5	$C_{23}H_{26}N_2O_5$ ≥99%

Lysine — FMOC-Lys-Adpoc

Bachem B-2520.0001	MW 588.75	$C_{35}H_{44}N_2O_6$ Store at -15°C

Lysine — FMOC-Lys-Aloc

Bachem B-2240.0001	MW 452.51	$C_{25}H_{28}N_2O_6$ Store at -15°C

Lysine — FMOC-Lys-Biotin

American Peptide FLYS110	MW 594.8

Lysine — FMOC-Lys-Biotinyl

Bachem B-2640.0250	MW 594.74	$C_{31}H_{38}N_4O_6S$ Store at -15°C

Lysine — FMOC-Lys-BOC

Synonyms: N^α-FMOC-N^ε-BOC-L-Lys

American Peptide FLYS135	MW 468.6	
Bachem B-1080.0001	MW 468.55	$C_{26}H_{32}N_2O_6$ Store at -15°C
Peptides International FLK-1726-PI	MW 468.55	>98% (HPLC); white crystalline powder
USBio F5407	MW 468.6	$C_{26}H_{32}N_2O_6$ ≥99%

Lysine — FMOC-Lys-BOC-® (200-400 mesh)

Bachem D-1060.0001	Store at 2-8°C

Lysine — FMOC-Lys-BOC-Isopropyl

Bachem B-2455.0001	MW 510.64	$C_{29}H_{38}N_2O_6$ Store at -15°C

Lysine — FMOC-Lys-BOC-OPfp

Bachem B-1155.0001 MW 634.6 $C_{32}H_{31}F_5N_2O_6$ Store at -15°C

USBio F5408 MW 634.6 $C_{32}H_{31}F_5N_2O_6$ ≥99%

Lysine — FMOC-Lys-BOC-SASRIN™-® (200-400 mesh)

Bachem D-1365.0001 Store at 2-8°C

Lysine — FMOC-Lys-Dansyl

Bachem B-2820.0250 MW 601.73 $C_{33}H_{35}N_3O_6S$ Store at -15°C

Lysine — FMOC-Lys-Dde

Bachem B-3015.0001 MW 532.64 $C_{31}H_{36}N_2O_6$ Store at -15°C

Lysine — FMOC-Lys-Dimethyl Hydrochloride

Bachem B-3290.0250 MW 432.95 $C_{23}H_{28}N_2O_4 \cdot HCl$ Store at -15°C

Lysine — FMOC-Lys-Dnp

Bachem B-2885.0250 MW 534.53 $C_{27}H_{26}N_4O_8$ Store at -15°C

Lysine — FMOC-Lys-FMOC

Synonyms: N^α, N^ϵ-Di-FMOC-L-Lys

American Peptide FLYS125 MW 590.8

Bachem B-1610.0001 MW 590.68 $C_{36}H_{34}N_2O_6$ Store at 2-8°C

Peptides International FLK-1750-PI MW 590.67 >98% (HPLC); white crystalline powder

USBio F5410 MW 590.8 $C_{36}H_{34}N_2O_6$ ≥99%

Lysine — FMOC-Lys-FMOC-OPfp

Bachem B-1675.0001 MW 756.73 $C_{42}H_{33}F_5N_2O_6$ Store at -15°C

Lysine — FMOC-Lys-For

Bachem B-1865.0001 MW 396.5 $C_{22}H_{24}N_2O_5$ Store at 2-8°C

Lysine — FMOC-Lys-ivDde

Bachem B-3515.0001 MW 574.72 $C_{34}H_{42}N_2O_6$ Store at 2-8°C

Lysine — FMOC-Lys-Mtt

Synonyms: N^α-FMOC-N^ϵ-4-Methyltrityl-L-Lys

Bachem B-2535.0001 MW 624.78 $C_{41}H_{40}N_2O_4$ Store at -15°C

Peptides International FLK-1798-PI MW 624.79 >98% (HPLC); white powder

Lysine — FMOC-Lys-Nde

Bachem B-3380.0001 MW 583.6 $C_{32}H_{29}N_3O_8$ Store at -15°C

Lysine — FMOC-Lys-Nicotinoyl

Bachem B-3295.0001 MW 473.53 $C_{27}H_{27}N_3O_5$ Store at -15°C

USBio F5413 MW 473.5 $C_{27}H_{27}N_3O_5$ ≥98%

Lysine — FMOC-Lys-Palmitoyl

Bachem B-2530.0001 MW 606.85 $C_{37}H_{54}N_2O_5$ Store at -15°C

Lysine — FMOC-Lys-Tnm

Synonyms: FMOC-Lys-(1,5-Dioxaspiro(5·5)Undecane-3-Nitro-3-Methoxycarbonyl)

Bachem B-2740.0001 MW 625.68 $C_{32}H_{39}N_3O_{10}$ Store at -15°C

Lysine — FMOC-Lys-Trimethyl chloride

Bachem B-2685.0250 MW 446.97 $C_{24}H_{31}ClN_2O_4$ Store at -15°C

Lysine — FMOC-Lys-Z

Synonyms: N^α-Fluorenylmethoxycarbonyl-N^ϵ-Benzyloxycarbonyl-L-Lys

Bachem B-1270.0001 MW 502.57 $C_{29}H_{30}N_2O_6$ Store at 2-8°C

Peptides International FLK-1727-PI MW 502.57 >98% (HPLC); white crystalline powder

USBio F5415 MW 502.6 $C_2H_{30}N_2O_6$ ≥99%

Lysine — L-Lys

Synonyms: (S)-2,6-Diaminocaproic Acid; Aminutrin; Lys-oh

Fluka 62840 MW 146.19 $C_6H_{14}N_2O_2$ ≥98.0% (titration); crystallized; mp: 215°C; ≤1% H_2O

Rexim MW 146.2 $C_6H_{14}N_2O_2$ Yellowish, aqueous solution; 50%

Sigma L 1014 L-Lys HCl solution; 80 *mM* (14.6 mg/mL); L-Lys HCl in tissue culture grade H_2O; sterile-filtered

Sigma L 5631 Matrix: 4% beaded agarose; activation: cyanogen bromide; attachment: amino; spacer: 1 atom; ligand immobilized: 5-15 μmoles/mL; form: suspension in 2.0 *M* NaCl containing 0.02% thimerosal | Deutsch, DG & Mertz, ET, *Proc Fed Amer Soc Exp Biol*, 29: 647, 1970; Deutsch, DG & Mertz, ET, *Science*, 170: 1095, 1970; Vuento, M & Vaheri, A, *Biochem J*, 183: 331, 1979

Sigma L 6132 Matrix: sepharose 4b; activation: cyanogen bromide; attachment: α-amino; spacer: 1 atom; ligand immobilized: ~4 μmoles/mL; form: lyophilized powder stabilized with lactose & dextran | Deutsch, DG & Mertz, ET, *Proc Fed Amer Soc Exp Biol*, 29: 647, 1970; Deutsch, DG & Mertz, ET, *Science*, 170: 1095, 1970; Vuento, M & Vaheri, A, *Biochem J*, 183: 331, 1979

Lysine — L-Lys (1-^{14}C)

ARC ARC-459 MW 146.2 $H_2N(CH_2)_4CH(NH_2)COOH$ 50-60 mCi/mmol; 1.85-2.22 GBq/mmol; in 0.01 *N* HCl | Radiochemical

Lysine — L-Lys (4,5-^{3}H(N))

ARC ART-243 MW 146.2 $H_2NCH_2CH_2CH_2CH_2CH(NH_2)COOH$ 40-60 Ci/mmol; 1.48-2.22 TBq/mmol; in 0.01 *N* HCl | Radiochemical

Sigma L 6022 FW 146.2 $H_2NCH_2CH_2CH_2CH_2CH(NH_2)COOH$ 40-110 Ci/mmol; radiochemical purity ≥95% (HPLC); aqueous solution containing 2% EtOH in Combi-vial | Radiochemical

Lysine — L-Lys (4,5-^{3}H)

ICN 20037 MW 146.2 $H_2NCH_2CH_2CH_2CH_2CH(NH_2)COOH$ 40-60 Ci/mmol, 1.48-2.22 TBq/mmol; 0.01 *N* HCl solution; HPLC analyzed for purity

ICN 20037E MW 146.2 $H_2NCH_2CH_2CH_2CH_2CH(NH_2)COOH$ 40-60 Ci/mmol, 1.48-2.22 TBq/mmol; sterile EtOH:H_2O (2:98); HPLC analyzed for purity

Lysine — L-Lys (4,5-^{3}H) Hydrochloride

Amersham TRK520 Aqueous, sterilized; 2.7-3.7 TBq/mmol, 75-100 Ci/mmol; 37 MBq/mL, 1 mCi/mL

Amersham TRK752 Aqueous, sterilized; 2.7-3.7 TBq/mmol, 75-100 Ci/mmol; 37185 MBq/mL, 5 mCi/mL

Lysine — L-Lys (6-^{14}C)

ARC ARC-1075   MW 146.2   H$_2$NCH$_2$CH$_2$CH$_2$CH$_2$CH(NH$_2$)COOH
50-60 mCi/mmol; 1.85-2.22 GBq/mmol; in 0.01 N HCl |
Radiochemical

Lysine — L-Lys (6-^{3}H)

ARC ART-597   MW 146.2   H$_2$NCH$_2$CH$_2$CH$_2$CH$_2$CH(NH$_2$)COOH
40-60 Ci/mmol; 1.48-2.22 TBq/mmol; in 0.01 N HCl |
Radiochemical

Lysine — L-Lys (U-^{14}C)

ARC ARC-673   MW 146.2   H$_2$N(CH$_2$)$_4$CH(NH$_2$)COOH   200-
300 mCi/mmol; 7.4-11.1 GBq/mmol; in EtOH: H$_2$O (2:98) |
Radiochemical

Lysine — L-Lys (U-^{14}C) Hydrochloride

Amersham CFB69   Aqueous, 2% EtOH, sterilized; >11
GBq/mmol, >300 mCi/mmol; 1.85 MBq/mL, 50 µCi/mL

ICN 10092E   MW 182.7   NH$_2$CH$_2$(CH$_2$)$_3$CH(NH$_2$)COOH · HCl
270-330 mCi/mmol, 10-12.2 GBq/mmol; sterile EtOH:H$_2$O (2:98);
HPLC analyzed for purity

Sigma L 3273   FW 146.2   H$_2$N(CH$_2$)$_4$CH(NH$_2$)COOH   240-
360 mCi/mmol; radiochemical purity ≥98% (HPLC); aqueous
solution containing 2% EtOH in Combi-vial | Radiochemical

Lysine — L-Lys Acetate Salt

Synonyms: L-2,6-Diaminohexanoic Acid

Sigma L 1884   FW 206.2   C$_6$H$_{14}$N$_2$O$_2$ · C$_2$H$_4$O$_2$ ≥98% (TLC)

Lysine — L-Lys Dihydrochloride

Synonyms: L-2,6-Diaminohexanoic Acid

Fluka 62900   MW 219.11   C$_6$H$_{14}$N$_2$O$_2$ · 2HCl   ≥99.0%
(titration); ≤0.05% residue on ignition, loss on drying; ≤0.3%
foreign AA; ≤0.0005% Cd, Co, Cr, Cu, Fe, Mg, Mn, Ni, Pb, Zn;
≤0.001% Ca; ≤0.005% K, Na, SO$_4$; ≤0.01% NH$_4$$^+$

ICN 102220   MW 219.1   C$_6$H$_{14}$N$_2$O$_2$ · 2HCl   Crystalline

Sigma L 5751   FW 219.1   C$_6$H$_{14}$N$_2$O$_2$ · 2HCl   ≥98% (TLC)

Lysine — L-Lys Free Base

Synonyms: L-2,6-Diaminohexanoic Acid

ICN 190224   MW 146.2   C$_6$H$_{14}$N$_2$O$_2$ Crystalline

ICN 194696   MW 146.2   C$_6$H$_{14}$N$_2$O$_2$ Crystalline | Cell culture
reagent

Sigma L 1137   FW 146.2   C$_6$H$_{14}$N$_2$O$_2$ Crystalline

Sigma L 5501   FW 146.2   C$_6$H$_{14}$N$_2$O$_2$ ≥98% (TLC)

Lysine — L-Lys Hydrate

Synonyms: (S)-2,6-Diaminohexanoic Acid; L-2,6-Diaminocaproic
Acid; L-2,6-Di-Aca

Fluka 62855   MW 164.21   C$_6$H$_{14}$N$_2$O$_2$ · H$_2$O   ≥98.0%
(titration)

Rexim   MW 164.2   C$_6$H$_{16}$N$_2$O$_3$ White crystals or crystalline
powder

USBio L9060   MW 164.2   C$_6$H$_{14}$N$_2$O$_2$ · H$_2$O   ≥98.5%; specific
rotation: 25.5º to 27.0º; loss on drying 10.0-11.5%; residue after
ignition: ≤0.1%; arsenic: ≤0.0001%; sulfate: ≤0.02%; iron:
≤0.001%; heavy metals (as Pb): ≤0.001%; Nitrogen Content:
16.5-17.5%; cell culture analysis: no inhibition of growth with
murine myeloma SP2/0-Ag14

Lysine — L-Lys Hydrochloride

Synonyms: Darvyl; Lysion; (S)-2,6-Diaminohexanoic Acid; L-2,6-
Diaminocaproic Acid; L-2,6-Diaminohexanoic Acid

Amersham US18585   ≥98.5%; ≤10ppm heavy metals (Pb)

Fluka 62929   MW 182.65   C$_6$H$_{14}$N$_2$O$_2$ · HCl   ≥99.5%
(titration); ≤0.1% residue on ignition; ≤0.5% loss on drying; ≤0.3%
foreign AA; ≤0.00001% As; ≤0.0005% Al, Ba, Bi, Cd, Co, Cr, Cu,
Fe, Li, Mg, Mn, Mo, Ni, Pb, Sr, Zn; ≤0.001% Ca; ≤0.005% K, Na;
≤0.01% NH$_4$$^+$, SO$_4$

Fluka 62930   MW 182.65   C$_6$H$_{14}$N$_2$O$_2$ · HCl   ≥99.0%
(titration); ≤0.1% residue on ignition; ≤0.5% loss on drying; ≤0.3%
foreign AA; ≤0.0005% Cd, Co, Cr, Cu, Fe, Mg, Mn, Ni, Pb, Zn;
≤0.001% Ca; ≤0.005% Ca, K, Na; ≤0.01% NH$_4$$^+$, SO$_4$

ICN 102218   MW 182.6   C$_6$H$_{14}$N$_2$O$_2$ · HCl   ≥99%; crystalline

ICN 194697   MW 182.6   C$_6$H$_{14}$N$_2$O$_2$ · HCl   ≥99%; crystalline
| Cell culture reagent

Neosystem AA01202   MW 182.6

Peptides International ALK-2714   MW 182.65   >99.9%
optical purity by RP-HPLC; white crystalline powder

Rexim   MW 182.7   C$_6$H$_{15}$ClN$_2$O$_2$ White crystals or crystalline
powder

Sigma L 1262   FW 182.6   C$_6$H$_{14}$N$_2$O$_2$ · HCl   Crystalline |
Cell culture tested; insect cell culture tested

Sigma L 5626   FW 182.6   C$_6$H$_{14}$N$_2$O$_2$ · HCl   ≥98% (TLC) |
Also available as part of a kit

USBio L9070   MW 182.7   C$_6$H$_{14}$N$_2$O$_2$ · HCl   ≥98%;
appearance: white crystalline powder; specific rotation (C=8, 6 N
HCl): +20.4° to +21.4°; arsenic: ≤0.00015%; heavy metals (Pb):
≤0.0015%; endotoxin: ≤2 EU/mg; cell culture tested: murine
myeloma SP2/0-Ag14

USBio L9070-10   MW 182.7   C$_6$H$_{14}$N$_2$O$_2$ · HCl   ≥98%;
peptide grade; appearance: white crystalline powder; pH: 5.0-6.0;
specific rotation (C=8, 6 N HCl): +20.4° to +21.4°; heavy metals
(Pb): ≤0.001%

Lysine — L-Lys Hydroxamate Hydrochloride

Sigma L 9001   FW 197.7   C$_6$H$_{15}$N$_3$O$_2$ · HCl

Lysine — L-Lys HyperD

Sigma L 7661   Matrix: Ceramic HyperD F Hydrogel composite;
binding capacity: >20 mg BSA/mL; average particle size: 50 µm;
form: suspension in 1 M NaCl with 20% EtOH; pH from 2-12 |
HyperD are highly porous rigid ceramic beads filled with a
derivatized hydrogel; characteristics are tolerance of very high flow
rates without bed compression & high exchange capacity; stable to
most common solvents; Deutsch, DG & Mertz, ET, *Proc Fed Amer
Soc Exp Biol*, 29: 647, 1970; Deutsch, DG & Mertz, ET, *Science*,
170: 1095, 1970; Vuento, M & Vaheri, A, *Biochem J*, 183: 331,
1979

Lysine — L-Lys Lactam

Synonyms: L-(-)-α-Amino-ε-Caprolactam; (S)-3-Amino-
Hexahydro-2-Azepinone

Fluka 07257   MW 128.17   C$_6$H$_{12}$N$_2$O   ~97% (titration)

Lysine — L-Lys L-Aspartate

Synonyms: L-2,6-Diaminohexanoic Acid

ICN 155288   MW 279.3   C$_6$H$_{14}$N$_2$O$_2$ · C$_4$H$_7$NO$_4$ Aspartic acid
Salt of L-Lys; not a dipeptide

Sigma L 6752   FW 279.3   C$_6$H$_{14}$N$_2$O$_2$ · C$_4$H$_7$NO$_4$

Lysine — L-Lys L-Glutamate Salt

Synonyms: L-2,6-Diaminohexanoic Acid

ICN 155289   MW 293.3   C$_6$H$_{14}$N$_2$O$_2$ · C$_5$H$_9$NO$_4$ Crystalline |
Glutamic acid Salt of L-Lys; not a dipeptide

Sigma L 8876   FW 293.3   C$_6$H$_{14}$N$_2$O$_2$ · C$_5$H$_9$NO$_4$

Lysine — L-Lys L-Malate

Rexim   MW 280.3   C$_{10}$H$_{20}$N$_2$O$_7$

Lysine — L-Lys Orotate

ICN 155292   MW 302.3   C$_6$H$_{14}$N$_2$O$_2$ · C$_5$H$_4$N$_2$O$_4$ Crystalline

Lysine — ʟ-Lys Solution

Synonyms: (S)-2,6-Diaminocaproic Acid

Fluka 62850   MW 146.19 $C_6H_{14}N_2O_2$ 50% in H_2O

Lysine — ʟ-Lys β-Naphthylamide Carbonate Salt

ICN 155290   MW 333.4 $C_{16}H_{22}N_3O \cdot HCO_3$

Lysine — ʟ-Lysinamide Dihydrochloride

Sigma L 0141   FW 218.1 $C_6H_{15}N_3O \cdot 2HCl$

Lysine — ʟ-Lys-OEt Dihydrochloride

Fluka 62880   MW 247.17 $C_8H_{18}N_2O_2 \cdot 2HCl$ ≥99.0% (titration); mp: 150°C; ≤0.05% residue on ignition

ICN 102221   MW 247.2 $C_8H_{18}N_2O_2 \cdot 2HCl$ Crystalline

Sigma L 5754   FW 247.2 $C_8H_{18}N_2O_2 \cdot 2HCl$ ~95%

Lysine — ʟ-Lys-OMe Dihydrochloride

ICN 102222   MW 233.1 $C_7H_{16}N_2O_2 \cdot 2HCl$ Crystalline

Sigma L 0645   FW 233.1 $C_7H_{16}N_2O_2 \cdot 2HCl$

Lysine — ʟ-Lys-pNA Dihydrochloride

ICN 155291   MW 428.1 $C_{12}H_{18}N_4O_3 \cdot 2HBr$ Crystalline

Lysine — ʟ-N^6-1-Iminoethyl-Lys Dihydrochloride

Synonyms: L-NIL

Alexis 270-010   MW 260.2 $C_8H_{17}N_3O_2 \cdot 2HCl$ ≥98%; white solid; soluble in H_2O; very hygroscopic | Selective inhibitor of inducible nitric oxide synthase (iNOS, NOS II); Moore, WM et al, *J Med Chem*, 37: 3886, 1994

Lysine — ʟ-N^6-1-Iminoethyl-Lys Hydrochloride

Synonyms: L-NIL; L-Lys ω-Acetamidine; iNOS Inhibitor; Nitric Oxide Synthase (Inducible) Inhibitor

Sigma I 8021   FW 223.7 $C_8H_{17}N_3O_2 \cdot HCl$ Selective inhibitor of inducible nitric oxide synthase; Moore, WM et al, *J Med Chem*, 37: 3886, 1994

Lysine — Lys Amide Dihydrochloride

Bachem E-2090.0001   MW 218.13 $C_6H_{15}N_3O \cdot 2HCl$ Store at 2-8°C

Lysine — Lys Hydrochloride

Bachem E-2085.0025   MW 182.65 $C_6H_{14}N_2O_2 \cdot HCl$ Store at RT

Lysine — Lys Iron Agar

Fluka 62915   Composition: 5 g/L meat peptone, 3 g/L yeast extract, 10 g/L ʟ-Lys monohydrochloride, 1 g/L glucose, 0.5 g/L ferric ammonium citrate, 0.04 g/L sodium thiosulfate, 0.02 g/L bromocresol purple, 12.5 g/L agar | Test agar for the simultaneous detection of lysine decarboxylase & formation of hydrogen sulfide in the identification of *Enterobacteriaceae*, in particular *Salmonella* & Arizona according to Edwards & Fife, *Appl Microbiol*, 9: 478, 1961; Timms, L *Med Lab Tech*, 28: 150, 1971

Lysine — Lys β-Naphthylamide Carbonate Salt

Sigma L 6259   FW 333.4 $C_{16}H_{22}N_3O \cdot HCO_3$

Lysine — Lys-(2,4-Dichloro-Z)-OBzl Hydrochloride

Bachem E-3305.0005   MW 475.8 $C_{21}H_{24}Cl_2N_2O_4 \cdot HCl$ Store at -15°C

Lysine — Lys-(ᴅʟ-2-Amino-2-Carboxyethyl) Dihydrochloride

Synonyms: Lysinoalanine

Bachem F-1195.0050   MW 306.19 $C_9H_{19}N_3O_4 \cdot 2HCl$ Diastereomeric mixture: LL + LD; store at -15°C

Lysine — Lys-2-Chloro-Z

Bachem E-2725.0005   MW 314.77 $C_{14}H_{19}ClN_2O_4$ Store at RT

Lysine — Lys-2-Chloro-Z-OBzl Hydrochloride

Bachem E-3215.0005   MW 441.37 $C_{21}H_{25}ClN_2O_4 \cdot HCl$ Store at RT

Lysine — Lys-4-Nitro-Z

Bachem E-2960.0005   MW 325.32 $C_{14}H_{19}N_3O_6$ Store at RT

Lysine — Lys-4-Nitro-Z-pyrrolidide

Bachem N-1710.0001   MW 378.43 $C_{18}H_{26}N_4O_5$ Store at -15°C

Lysine — Lys-Ac

Bachem E-2705.0001   MW 188.23 $C_8H_{16}N_2O_3$ Store at RT

Lysine — Lys-Acetimidoyl

Synonyms: Lys-1-Iminoethyl; *N*-(5-Amino-5-Carboxypentyl)-Acetamidine

Bachem F-3325.0001   MW 187.24 $C_8H_{17}N_3O_2$ Store at -15°C

Lysine — Lys-Agarose

ICN 191281   5 atoms hydrophilic spacer arm; 10-18 μmoles of ʟ-Lys/mL gel; in distilled H_2O, 0.02% NaN_3 | Useful for purification of plasminogen activator, rRNA, ds DNA

Lysine — Lys-Aloc

Bachem E-3065.0001   MW 230.27 $C_{10}H_{18}N_2O_4$ Store at 2-8°C

Lysine — Lys-AMC

Synonyms: Aminopeptidase Substrate

Peptides International MKM-3132-v   MW 303.36 $C_{16}H_{21}N_3O_3$ >98% (HPLC); lyophilized amorphous powder

Lysine — Lys-AMC Acetate

Bachem I-1255.0050   MW 423.5 $C_{16}H_{21}N_3O_3 \cdot C_2H_4O_2$ Store at -15°C

Lysine — Lys-Biotinyl

Synonyms: Biocytin

Bachem E-3435.0250   MW 372.49 $C_{16}H_{28}N_4O_4S$ Store at -15°C

Lysine — Lys-Biotinyl Amide

Synonyms: Biocytin

Bachem E-3430.0250   MW 371.51 $C_{16}H_{29}N_5O_3S$ Store at -15°C

Lysine — Lys-BOC

Bachem E-1610.0001   MW 246.31 $C_{11}H_{22}N_2O_4$ Store at RT

Lysine — Lys-BOC Amide Hydrochloride

Bachem A-3335.0001   MW 281.78 $C_{11}H_{23}N_3O_3 \cdot HCl$ Store at -15°C

Lysine — Lys-BOC-OAll Hydrochloride

Bachem E-3345.0001 MW 322.84 $C_{14}H_{26}N_2O_4 \cdot HCl$ Store at -15°C

Lysine — Lys-BOC-OMe Hydrochloride

Bachem E-1620.0001 MW 296.8 $C_{12}H_{24}N_2O_4 \cdot HCl$ Store at 2-8°C

USBio B2338 MW 296.8 $C_{12}H_{24}N_2O_4 \cdot HCl$ ≥99%

Lysine — Lys-BOC-OtBu Hydrochloride

Bachem E-1615.0001 MW 338.88 $C_{15}H_{30}N_2O_4 \cdot HCl$ Store at -15°C

USBio B2336 MW 338.9 $C_{15}H_{30}N_2O_4$ ≥99%

Lysine — Lys-BOC-pNA

Bachem L-1575.0001 MW 366.42 $C_{17}H_{26}N_4O_5$ Store at -15°C

Lysine — Lys-Dimethyl Hydrochloride

Bachem E-1810.0250 MW 210.7 $C_8H_{18}N_2O_2 \cdot HCl$ Store at 2-8°C

Lysine — Lys-FMOC

Bachem B-1875.0001 MW 368.44 $C_{21}H_{24}N_2O_4$ Store at RT

Lysine — Lys-FMOC-OMe Hydrochloride

Bachem E-2665.0001 MW 418.92 $C_{22}H_{26}N_2O_4 \cdot HCl$ Store at -15°C

Lysine — Lys-Isopropyl

Bachem E-3395.0001 MW 188.27 $C_9H_{20}N_2O_2$ Store at -15°C

Lysine — Lys-Me Hydrochloride

Bachem E-2155.0250 MW 196.68 $C_7H_{16}N_2O_2 \cdot HCl$ Store at 2-8°C

Lysine — Lys-Nicotinoyl Hydrochloride

Bachem E-3590.0001 MW 287.75 $C_{12}H_{17}N_3O_3 \cdot HCl$ Store at -15°C

Lysine — Lys-OEt Dihydrochloride

Bachem E-2785.0005 MW 247.16 $C_8H_{18}N_2O_2 \cdot 2HCl$ Store at RT

Lysine — Lys-OMe Dihydrochloride

Bachem E-2095.0005 MW 233.14 $C_7H_{16}N_2O_2 \cdot 2HCl$ Store at RT

Lysine — Lys-pNA Dihydrobromide

Bachem L-1315.0001 MW 428.1 $C_{12}H_{18}N_4O_3 \cdot 2HBr$ Store at -15°C

Sigma L 7002 FW 428.1 $C_{12}H_{18}N_4O_3 \cdot 2HBr$ Crystalline

Lysine — Lys-Sepharose 4B

Fluka 62961 Lys is attached through the α-amino group to Sepharose 4B; the carboxy & ε-amino groups are free | For affinity chromatography; for the separation of the 4 molecular forms of human plasminogen; Nieuwenhuizen, W & Traas, DW, *Thromb Haemostasis*, 61: 208, 1989; De Munk, GAW et al, *Biochemistry*, 28: 7318, 1989

Lysine — Lys-TFA

Bachem E-2615.0001 MW 242.2 $C_8H_{13}F_3N_2O_3$ Store at -15°C

Lysine — Lys-Tosyl

Bachem E-2645.0005 MW 300.38 $C_{13}H_{20}N_2O_4S$ Store at -15°C

Lysine — Lys-Trimethyl Chloride

Bachem F-2665.0050 MW 224.73 $C_9H_{21}ClN_2O_2$ Store at -15°C

Lysine — Lys-Z

Synonyms: N^ϵ-Carbobenzoxy-L-Lys

Bachem E-1702.0025 MW 280.32 $C_{14}H_{20}N_2O_4$ Store at RT

Peptides International XZK-2006-PI MW 280.32 >98% (TLC); white powder

Lysine — Lys-Z-AMC Hydrochloride

Bachem I-1155.0250 MW 473.9 $C_{24}H_{27}N_3O_5 \cdot HCl$ Store at -15°C

Lysine — Lys-Z-OBzl Hydrochloride

Bachem E-1705.0005 MW 406.91 $C_{21}H_{26}N_2O_4 \cdot HCl$ Store at RT

Lysine — Lys-Z-OMe Hydrochloride

Bachem E-1715.0005 MW 330.81 $C_{15}H_{22}N_2O_4 \cdot HCl$ Store at RT

Lysine — Lys-Z-OtBu Hydrochloride

Bachem E-1710.0001 MW 372.89 $C_{18}H_{28}N_2O_4 \cdot HCl$ Store at -15°C

Lysine — Lys-βNA

Bachem K-1415.0001 MW 271.36 $C_{16}H_{21}N_3O$ Store at RT

Lysine — Methoxycarbonyl-Lys-Z Dicyclohexylamine

Bachem E-3445.0001 MW 519.69 $C_{16}H_{22}N_2O_6 \cdot C_{12}H_{23}N$ Store at 2-8°C

Lysine — N,N'-Di(2,4-Dnp)-L-Lys

Sigma D 0255 FW 478.4 $C_{18}H_{18}N_6O_{10}$

Lysine — N,N'-Di-CBZ-L-Lys Free Acid

Sigma C 3127 FW 414.5 $C_{22}H_{26}N_2O_6$

Lysine — N^6-1-Iminoethyl-L-Lys Hydrochloride

Synonyms: Nitric Oxide Synthase Inhibitor

ICN 195056 MW 260.2 $C_8H_{17}N_3O_2 \cdot HCl$ ≥98% | Selective inhibitor

Lysine — N-FMOC-L-Lys-FMOC-N-(4,5-³H)

ARC ART-581 MW 590.7 10-20 Ci/mmol; 370-740 GBq/mmol; in EtOH | Radiochemical

Lysine — N-Me-Lys Hydrochloride

Bachem E-3180.0250 MW 196.68 $C_7H_{16}N_2O_2 \cdot HCl$ Store at RT

Lysine — N-Me-Lys-Ac-OMe Hydrochloride

Bachem F-3920.0250 MW 252.74 $C_{10}H_{20}N_2O_3 \cdot HCl$ Store at -15°C

Lysine — N-Me-Lys-Z

Bachem E-3185.0250 MW 294.35 $C_{15}H_{22}N_2O_4$ Store at RT

Lysine — Nps-Lys-BOC Dicyclohexylamine
Bachem E-2790.0001   MW 580.79 $C_{17}H_{25}N_3O_6S \cdot C_{12}H_{23}N$
Store at 2-8°C

Lysine — N^α,N^ϵ-Bis-Carboxymethyl-L-Lys Hydrate
Synonyms: (S)-N-(5-Amino-1-Carbozypentyl)-Iminodiacetic Acid
Fluka 14580   MW 262.26 $C_{10}H_{18}N_2O_6$ aq; ≥97% (HPLC); mp: ~75°C; ~2 mol H_2O & ~10% inorganic Salts | Tricarboxylic acid used for preparing chelating resins employed for the chromatography of proteins; Hochuli, E et al, *J Chromatography*, 411: 177, 1987

Lysine — N^α,N^ϵ-Bis-Carboxymethyl-L-Lys TFA Salt
Synonyms: N-(5-Amino-1-Carboxypentyl)-Iminodiacetic Acid
Sigma C 3205   FW 376.3 $C_{10}H_{18}N_2O_6 \cdot C_2F_3O_2H$ ~95% (TLC) | Hochuli, E et al, *J Chrom*, 411: 177, 1987

Lysine — N^α,N^ϵ-Di-CBZ-L-Lys N-Hydroxysuccinimide Ester
Sigma C 3034   FW 511.5 $C_{26}H_{29}N_3O_8$

Lysine — N^α,N^ϵ-Di-FMOC-L-Lys
Fluka 47317   MW 590.67 $C_{36}H_{34}N_2O_6$ ≥98% (HPLC)

Lysine — N^α,N^ϵ-Di-t-BOC-L-Lys Dicyclohexylammonium Salt
Sigma B 9255   FW 527.7 $C_{16}H_{30}N_2O_6 \cdot C_{12}H_{23}N$

Lysine — N^α,N^ϵ-Di-t-BOC-L-Lys N-Hydroxysuccinimide Ester
Sigma B 7019   FW 443.5 $C_{20}H_{33}N_3O_8$

Lysine — N^α,N^ϵ-Di-Z-DL-Lys
Fluka 96837   MW 414.46 $C_{22}H_{26}N_2O_6$ ≥98.0% (TLC)

Lysine — N^α,N^ϵ-Di-Z-L-Lys
Fluka 96835   MW 414.46 $C_{22}H_{26}N_2O_6$ ≥98.0% (TLC)

Lysine — N^α,N^ϵ-Di-Z-L-Lys Hydroxysuccinimide Ester
Fluka 96885   MW 511.53 $C_{26}H_{29}N_3O_8$ ~98% (HPLC)

Lysine — N^α,N^ϵ-Di-Z-L-Lys-(4-ONp)
Fluka 96893   MW 535.55 $C_{28}H_{29}N_3O_8$ ≥98.0% (TLC)

Lysine — N^α-Ac-L-Lys
ICN 150234   MW 188.2 $C_8H_{16}N_2O_3$ Crystalline, 98%
Sigma A 2010   FW 188.2 $C_8H_{16}N_2O_3$

Lysine — N^α-Ac-L-Lys N-Methylamide Free Base
Sigma A 7012   FW 201.3 $C_9H_{19}N_3O_2$

Lysine — N^α-Ac-L-Lys-OMe Hydrochloride
Sigma A 9885   FW 238.7 $C_9H_{18}N_2O_3 \cdot HCl$

Lysine — N^α-BOC-L-Lys
Fluka 15456   MW 246.31 $C_{11}H_{22}N_2O_4$ ≥99% (titration); mp: 205°C

Lysine — N^α-BOC-N^ϵ-2-Bromo-Z-L-Lys
Fluka 15536   MW 459.34 $C_{19}H_{27}BrN_2O_6$ ≥98% (TLC); mp: 72-76°C

Lysine — N^α-BOC-N^ϵ-2-Chloro-Z-D-Lys
Fluka 15084   MW 414.88 $C_{19}H_{27}ClN_2O_6$ ≥98% (TLC)

Lysine — N^α-BOC-N^ϵ-2-Chloro-Z-L-Lys
Fluka 15399   MW 414.89 $C_{19}H_{27}ClN_2O_6$ ≥98% (titration); mp: 70-73°C | Protected Lys especially suited for solid phase peptide synthesis; 2-CZ protection is 50 times more stable than the Z-group; Erickson, BW & Merrifield, RB, *JACS*, 95: 3757, 1973

Lysine — N^α-BOC-N^ϵ-FMOC-L-Lys
Fluka 15435   MW 468.55 $C_{26}H_{32}N_2O_6$ ~99% (TLC); mp: 93-96°C

Lysine — N^α-BOC-N^ϵ-For-L-Lys
Fluka 15542   MW 274.32 $C_{12}H_{22}N_2O_5$ ≥99% (TLC); mp: 125-128°C

Lysine — N^α-BOC-N^ϵ-TFA-L-Lys
Fluka 15503   MW 342.31 $C_{13}H_{21}F_3N_2O_5$ ≥97% (HPLC); mp: 102-105°C

Lysine — N^α-BOC-N^ϵ-Z-L-Lys Dicyclohexylamine
Synonyms: N^ϵ-Z-N^α-BOC-L-Lys
Fluka 15540   MW 561.77 $C_{19}H_{28}N_2O_6 \cdot C_{12}H_{23}N$ ≥99% (titration); mp: 114-116°C

Lysine — N^α-BOC-N^ϵ-Z-L-Lys Hydroxysuccinimide Ester
Synonyms: N^ϵ-Z-N^α-BOC-L-Lys Hydroxysuccinimide Ester
Fluka 15541   MW 477.52 $C_{23}H_{31}N_3O_8$ mp: 108-114°C

Lysine — N^ϵ-CBZ-D-Lys
Sigma C 1408   FW 280.3 $C_{14}H_{20}N_2O_4$

Lysine — N^ϵ-CBZ-L-Lys
ICN 150574   MW 280.3 $C_{14}H_{20}N_2O_4$
Sigma C 7130   FW 280.3 $C_{14}H_{20}N_2O_4$

Lysine — N^ϵ-CBZ-L-Lys Thiobenzyl Ester Hydrochloride
Sigma C 3647   FW 423.0 $C_{21}H_{26}N_2O_3S \cdot HCl$

Lysine — N^ϵ-CBZ-L-Lys-(pONp) Hydrochloride
Sigma C 3637   FW 437.9 $C_{20}H_{23}N_3O_6 \cdot HCl$

Lysine — N^ϵ-CBZ-N^α-t-BOC-D-Lys
Sigma C 0533   FW 380.4 $C_{19}H_{28}N_2O_6$

Lysine — N^ϵ-FMOC-L-Lys
ICN 151147   MW 368.4 $C_{21}H_{24}N_2O_4$

Lysine — N^α-FMOC-N^ϵ-(1-(4,4-Dimethyl-2,6-Dioxocyclohexylidene)Et)-D-Lys
Synonyms: N^α-FMOC-N^ϵ-Dde-D-Lys
Fluka 09765   MW 532.63 $C_{31}H_{36}N_2O_6$ ~98% (HPLC)

Lysine — N^α-FMOC-N^ω-(1-(4,4-Dimethyl-2,6-Dioxocyclohexylidene)Et)-L-Lys

Synonyms: N^α-FMOC-N^ω-Mtr-L-Lys

Fluka 47562   MW 532.63 $C_{31}H_{36}N_2O_6$ ≥97% (HPLC) |
Protected arginine derivative used in solid phase peptide synthesis; the MTR-guanidine protection of arg is more acid-sensitive than the tosyl group; Fujino, M et al, *Chem Pharm Bull*, 29: 2825, 1981; Williamson, MP et al, *Eur J Biochem*, 158: 527, 1986; Atherton, E et al, *J Chem Soc*, Perkin I 2065, 1985

Lysine — N^α-FMOC-N^ε-2-Chloro-Z-L-Lys

Fluka 47567   MW 537.01 $C_{29}H_{29}ClN_2O_6$ ≥97% (HPLC); mp: 135-140°C

Lysine — N^α-FMOC-N^ε-Alloc-L-Lys

Fluka 47583   MW 452.51 $C_{25}H_{25}N_2O_6$ ~97% (HPLC); mp: 87-91°C

Lysine — N^α-FMOC-N^ε-BOC-L-Lys

Fluka 47624   MW 468.55 $C_{26}H_{32}N_2O_6$ ≥98% (titration); mp: 132-133°C

Lysine — N^α-FMOC-N^ε-BOC-L-Lys 4-Benzyloxybenzyl Ester Polymer Bound

Fluka 47647   0.4-0.6 mmol/g resin; carrier: polystyrene, crosslinked with 1% DVB; 200-400 mesh particle size

Lysine — N^α-FMOC-N^ε-BOC-L-Lys Resin Ester

Sigma F 2011   For peptide synthesis; Wang, SS, *J Am Chem Soc*, 95: 1328, 1973; Lu, G et al, *J Org Chem*, 46: 3433, 1981

Lysine — N^α-FMOC-N^ε-BOC-L-Lys-OPfp

Synonyms: N^ε-BOC-N^α-FMOC-L-Lys-OPfp

Fluka 47447   MW 634.6 $C_{32}H_{31}F_5N_2O_6$ ~98% (TLC)

Lysine — N^α-FMOC-N^ε-CBZ-L-Lys

ICN 158169   MW 502.6 $C_{29}H_{30}N_2O_6$

Sigma F 0885   FW 502.6 $C_{29}H_{30}N_2O_6$

Lysine — N^α-FMOC-N^ε-t-BOC-L SASRIN Resin Ester

Sigma F 0764   200-400 mesh; substitution range: 0.3-0.8 mmole FMOC AA/g resin | For peptide synthesis; N-FMOC AA acyl ester of 3-methoxy-4-oxymethyl-phenoxymethylated 1% divinylbenzene cross-linked polystyrene; peptides can be cleaved from the resin with 0.5-1% CF_3CO_2H in CH_2Cl_2; Mergler, M et al, *in Peptides: Chemistry & Biology* (Proc 10th Am Peptide Symp), Marhall, GR, ed, 259, 1988

Lysine — N^α-FMOC-N^ε-t-BOC-L-Lys

Sigma F 1133   FW 468.5 $C_{26}H_{32}N_2O_6$

Lysine — N^α-FMOC-N^ε-Z-L-Lys

Fluka 47577   MW 502.57 $C_{29}H_{30}N_2O_6$ ≥98% (HPLC); mp: 105-115°C

Lysine — N^α-FMOC-ε-t-BOC-L-Lys

ICN 151140   MW 468.6 $C_{26}H_{32}N_2O_6$

Lysine — N^ε-Lauroyl-L-Lys

Fluka 61734   MW 328.49 $C_{18}H_{36}N_2O_3$ ≥99.0% (TLC)

Lysine — N^{α}-N^ε-Di-BOC-L-Lys Hydroxysuccinimide Ester

Fluka 15131   MW 443.5 $C_{20}H_{33}N_3O_8$ ≥98%; mp: 184-188°C

Lysine — N^ε-p-Tosyl-L-Lys CMK Hydrochloride

Synonyms: TLCK; Papain Inhibitor; Trypsin Inhibitor; 1-Chloro-3-Tosylamido-7-Amino-L-2-Heptanone; Trypsin-Like Serine Protease Inhibitor

ICN 152152   MW 369.3 $C_{14}H_{21}ClN_2O_3S \cdot HCl$ Crystalline |
Specific reagent for essential His site in trypsin

Sigma T 7254   FW 369.3 $C_{14}H_{21}ClN_2O_3S \cdot HCl$ White to pink powder | Selective, irreversible inhibitor of trypsin-like serine proteases; blocks iNOS production by interfering with gene transcription; Griscavage, JM et al, *Biochem Biophys Res Commun*, 215: 721, 1995

Lysine — N^ε-p-Tosyl-L-Lys-OMe Hydrochloride

ICN 156950   MW 350.9 $C_{14}H_{21}N_2O_4S \cdot HCl$ Crystalline

Sigma T 5012   FW 350.9 $C_{14}H_{22}N_2O_4S \cdot HCl$

Lysine — N^ε-t-BOC-L-Lys

ICN 150502   MW 246.3 $C_{11}H_{22}N_2O_4$

ICN 150503   MW 246.3 $C_{11}H_{22}N_2O_4$ Crystalline

Sigma B 5011   FW 246.3 $C_{11}H_{22}N_2O_4$

Lysine — N^ε-t-BOC-N^α-CBZ-L-Lys Dicyclohexylammonium Salt

ICN 101057   MW 561.8 $C_{19}H_{28}N_2O_6 \cdot C_{12}H_{23}N$

Lysine — N^ε-t-BOC-N^α-2-Chloro-CBZ-D-Lys

Sigma B 4269   FW 414.9 $C_{12}H_{27}ClN_2O_6$

Lysine — N^ε-t-BOC-N^α-2-Chloro-CBZ-L-Lys

ICN 150490   MW 414.9 $C_{19}H_{27}ClN_2O_6$ Crystalline

Sigma B 8389   FW 414.9 $C_{12}H_{27}ClN_2O_6$

Lysine — N^ε-t-BOC-N^α-2-Chloro-CBZ-L-Lys Pam Resin Ester

Sigma B 4018   N-t-BOC AA ester of 4-(Oxymethyl)-Pheny-lAcm 1% divinylbenzene cross-linked polystyrene; 200-400 mesh; substitution range: 0.2-0.6 mmole t-BOC-AA/g resin | For peptide synthesis by the Merrifield method; Mitchell, AR et al, *J Org Chem*, 43: 2845, 1978

Lysine — N^ε-t-BOC-N^α-2-Chloro-CBZ-L-Lys Resin Ester

Sigma B 9521   N-t-BOC AA ester of methylated 1% divinylbenzene cross-linked polystyrene; 200-400 mesh; substitution range: 0.2-0.6 mmole t-BOC-AA/g resin | For peptide synthesis by the Merrifield method

Lysine — N^ε-t-BOC-N^α-Ac-L-Lys

Sigma B 7643   FW 288.3 $C_{13}H_{24}N_2O_5$

Lysine — N^ε-t-BOC-N^α-CBZ-L-Lys (4,5-^{3}H)

ARC ART-686   MW 380.4   100-150 mCi/mmol; 3.7-5.55 GBq/mmol; solid | Radiochemical

Lysine — N^ε-t-BOC-N^α-CBZ-L-Lys Dicyclohexylammonium Salt

ICN 150489   MW 561.8 $C_{19}H_{28}N_2O_6 \cdot C_{12}H_{23}N$

Sigma B 8379   FW 561.8 $C_{19}H_{28}N_2O_6 \cdot C_{12}H_{23}N$

Lysine — N^ε-t-BOC-N^α-CBZ-L-Lys Free Acid

ICN 150488   MW 380.4 $C_{19}H_{28}N_2O_6$

Sigma B 8254   FW 380.4 $C_{19}H_{28}N_2O_6$

Lysine — N^α-t-BOC-N^ϵ-CBZ-L-Lys N-Hydoxysuccinimide Ester

Sigma B 3256 FW 477.5 $C_{23}H_{31}N_3O_8$ ~80%

Lysine — N^α-t-BOC-N^ϵ-For-L-Lys

ICN 101058 MW 274.3 $C_{12}H_{22}N_2O_6$ Crystalline
Sigma B 8004 FW 274.3 $C_{12}H_{22}N_2O_5$

Lysine — N^α-t-BOC-N^ϵ-p-Tosyl-L-Lys

Sigma B 1258 FW 400.5 $C_{18}H_{28}N_2O_6S$

Lysine — N^ϵ-Tosyl-L-Lys CMK Hydrochloride

Synonyms: TLCK; (3S)-7-Amino-1-Chloro-3-Tosylamino-2-Heptanone; (3S)-1-Chloro-3-Tosylamido-7-Amino-2-Heptanone; Tosyl-L-Lysyl-Chloromethane

Fluka 90182 MW 369.31 $C_{14}H_{21}ClN_2O_3S \cdot HCl$ ≥99.0% (titration); mp: 165°C | An alkylating reagent that reacts with His or Cys residues at the active site of enzymes; inactivates proteolytic enzymes such as trypsin; Shaw, E et al, *Biochemistry*, 4: 2219, 1965; Porter, WH et al, *JBC*, 246: 7675, 1971; Shaw, E & Glover, G, *Arch Biochem Biophys*, 139: 298, 1970; Glover, G & Shaw, E, *JBC*, 246: 4594, 1971; Whitaker, JR & Perez-Villasenor, J, *Arch Biochem Biophys*, 124: 70, 1968; Matsuda, Y et al, *Chem Pharm Bull*, 30: 2512, 1982; Kupfer, A et al, *PNAS*, 76: 3073, 1979; Polakoski, KL & McRorie, RA, *JBC*, 248: 8183, 1973; Kolb, WP, *Biochemistry*, 21: 294, 1982; Solomon, DH et al, *FEBS Lett*, 190: 342, 1985; Urban, MK et al, *Biochemistry*, 18: 3952, 1979

Lysine — N^α-Tosyl-Lys CMK Hydrochloride

Synonyms: TLCK; Trypsin-Like Serine Proteinase Inhibitor; Trypsin Inactivator

Calbiochem 616382 MW 369.3 $C_{14}H_{21}ClN_2O_3S \cdot HCl$ ≥95% (EA); solid; soluble in aqueous solutions; LD$_{50}$≤2000 mg/kg | Inactivates trypsin irreversibly without affecting chymotrypsin; prevents nitric oxide production by activated macrophages by interfering with transcription of the iNOS gene; blocks cell-cell adhesion & binding of HIV-1 virus to the target cells; in macrophages, blocks nitric oxide synthase induced by interferon-γ & lipopolysaccharides; prevents endonucleolysis accompanying apoptotic death of HL-60 leukemia cells & normal thymocytes; Griscavage, JM et al, *Biochem Biophys Res Comm*, 215: 721, 1995; Kim, H et al, *J Immunol*, 154: 4741, 1995; Bourinbaiar, AS & Nagorny, R, *Cell Immunol*, 155: 230, 1994; Bruno, S et al, *Leukemia*, 6: 1113, 1992; Lee, SF, *Infect Immunol*, 60: 4032, 1992; Schmidt, HH et al, *Mol Pharmacol*, 41: 615, 1992

Oncogene 616382 MW 369.3 $C_{14}H_{21}ClN_2O_3S \cdot HCl$ Solid; ≥95% (EA); may cause irritation | Inactivates trypsin irreversibly without affecting chymotrypsin; prevents nitric oxide production by activated macrophages by interfering with transcription of the iNOS gene; blocks cell-cell adhesion & binding of HIV-1 virus to the target cells; prevents endonucleolysis accompanying apoptotic death of HL-60 leukemia cells & normal thymocytes; Grascavage, JM et al, *BBRC*, 215: 721, 1995; Kim, H et al, *J Immunol*, 154: 4741, 1995; Bourinbaiar, AS & Nagorny, R, *Cell Immunol*, 155: 230, 1994; Bruno, S et al, *Leukemia*, 6: 1113, 1992; Lee, SF, *Infect Immunol*, 60: 4032, 1992; Schmidt, HH et al, *Mol Pharmacol*, 41: 615, 1992

Lysine — N^ϵ-Z-D-Lys

Fluka 96825 MW 280.32 $C_{14}H_{20}N_2O_4$ ≥98.0% (TLC)

Lysine — N^ϵ-Z-L-Lys

Fluka 96830 MW 280.33 $C_{14}H_{20}N_2O_4$ ~98% (titration); mp: 226-230°C | Scott, AI & Wilkonson, TJ, *Synth Comm*, 10: 127, 1980

Lysine — N^ϵ-Z-L-Lys-(4-ONp) Hydrochloride

Fluka 96895 MW 437.88 $C_{20}H_{23}N_3O_6 \cdot HCl$ ≥98.0% (titration); mp: 149-152°C

Lysine — N^α-Z-N^ϵ-BOC-D-Lys

Fluka 96019 MW 380.44 $C_{19}H_{28}N_2O_6$ ≥98.0% (TLC)

Lysine — N^α-Z-N^ϵ-BOC-L-Lys-(4-ONp)

Fluka 96021 MW 501.54 $C_{25}H_{31}N_3O_8$ ≥98.0% (TLC)

Lysine — N^α,N^ϵ-(CBZ)$_2$-L-Lys

ICN 101547 MW 414.5 $C_{22}H_{26}N_2O_6$

Lysine — N^ϵ-(+)-Biotinyl-L-Lys

Synonyms: Bct; Biocytin

Fluka 14409 MW 372.49 $C_{16}H_{28}N_4O_4S$ ≥98% (HPLC); mp: ~245°C | Ebrahim, H et al, *Anal Biochem*, 154: 282, 1986; Cuatrecasa, P & Wilchek, M, *BBRC*, 33: 235, 1968

Lysine — N^ϵ-(+)-Biotinyl-L-Lys Hydrazide

Synonyms: Biocytin

Fluka 14413 MW 386.51 $C_{16}H_{30}N_6O_3S$ ~95% (HPLC)

Lysine — N^ϵ-(+)-Biotinyl-L-Lys Hydrazide Hydrochloride

Synonyms: Biocytin

Fluka 14415 MW 422.97 $C_{16}H_{30}N_6O_3S \cdot HCl$ ~90% (HPLC)

Lysine — $N^\alpha,N^\epsilon,N^\epsilon$-Trimethyl-Lys

Sigma T 1660 FW 188.3 $C_9H_{20}N_2O_2$ ≥97% (TLC); AA content ~75%; balance Salts & H_2O

Lysine — N^ϵ-2,4-Dnp-D-Lys Hydrochloride

Sigma D 2285 FW 348.7 $C_{12}H_{16}N_4O_6 \cdot HCl$

Lysine — N^ϵ-2,4-Dnp-L-Lys Hydrochloride

Sigma D 0380 FW 348.7 $C_{12}H_{16}N_4O_6 \cdot HCl$

Lysine — N^ϵ-Ac-L-Lys

ICN 150235 MW 188.2 $C_8H_{16}N_2O_3$ Crystalline, 99%
Sigma A 4021 FW 188.2 $C_8H_{16}N_2O_3$

Lysine — N^ϵ-Biotinyl-N^α-BOC-L-Lys

Synonyms: N^α-BOC-Biocytin

Fluka 15083 MW 472.6 $C_{21}H_{36}N_4O_6S$ ≥98% (TLC)

Lysine — N^ϵ-BOC-L-Lys

Fluka 15453 MW 246.31 $C_{11}H_{22}N_2O_4$ ~97%; mp: 250°C

Lysine — N^ϵ-BOC-Lys

American Peptide BLLYS15 MW 246.3
USBio B2332 MW 246.3 $C_{11}H_{22}N_2O_4$ ≥99%

Lysine — N^ϵ-BOC-N^α-Z-L-Lys Dicyclohexylamine

Synonyms: N^α-Z-N^ϵ-BOC-L-Lys

Fluka 15550 MW 561.77 $C_{19}H_{28}N_2O_6 \cdot C_{12}H_{23}N$ ≥98% (HPLC); mp: 154-156°C

Lysine — N^ϵ-CBZ-D-Lys

Sigma C 1787 FW 280.3 $C_{14}H_{20}N_2O_4$

Lysine — N^ϵ-CBZ-L-Lys

ICN 101263 MW 280.3 $C_{14}H_{20}N_2O_4$ 98%
Sigma C 7573 FW 280.3 $C_{14}H_{20}N_2O_4$

Lysine — *N*ᵉ-CBZ-ʟ-Lys-AMC Hydrochloride

Sigma C 7283   FW 474.0 $C_{24}H_{27}N_3O_5 \cdot HCl$

Lysine — *N*ᵉ-CBZ-ʟ-Lys-OBzl Hydrochloride

Sigma C 3252   FW 406.9 $C_{21}H_{26}N_2O_4 \cdot HCl$

Lysine — *N*ᵉ-CBZ-ʟ-Lys-OMe Hydrochloride

Sigma C 3377   FW 330.8 $C_{15}H_{22}N_2O_4 \cdot HCl$

Lysine — *N*ᵉ-Dansyl-ʟ-Lys Free Acid

Sigma D 9006   FW. 379.5 $C_{18}H_{25}N_3O_4S$

Lysine — *N*ᵉ-FMOC-ʟ-Lys

ICN 151148   MW 368.4 $C_{21}H_{24}N_2O_4$

Lysine — *N*ᵉ-FMOC-α-t-BOC-ʟ-Lys

ICN 151141   MW 468.6 $C_{26}H_{32}N_2O_6$

Lysine — *N*ᵉ-For-ʟ-Lys

ICN 151168   MW 174.2 $C_7H_{14}N_2O_3$

Sigma F 9376   FW 174.2 $C_7H_{14}N_2O_3$

Lysine — *N*ᵉ-Me-ʟ-Lys Hydrochloride

Sigma M 6004   FW 196.7 $C_7H_{16}N_2O_2 \cdot HCl$

Lysine — *N*ᵉ-Phthaloyl-ʟ-Lys Hydrochloride

Sigma P 3058   FW 312.8 $C_{14}H_{16}N_2O_4 \cdot HCl$

Lysine — *N*ᵉ-p-Tosyl-ʟ-Lys-OEt Hydrochloride

Sigma T 1411   FW 364.9 $C_{15}H_{24}N_2O_4S \cdot HCl$

Lysine — *N*ᵉ-t-BOC-ʟ-Lys

Sigma B 7376   FW 246.3 $C_{11}H_{22}N_2O_4$

Lysine — *N*ᵉ-t-BOC-*N*ᵉ-CBZ-ʟ-Lys Dicyclohexylammonium Salt

Sigma B 7882   FW 561.8 $C_{19}H_{28}N_2O_6 \cdot C_{12}H_{23}N$

Lysine — *N*ᵉ-TFA-ʟ-Lys

Fluka 91696   MW 242.2 $C_8H_{13}F_3N_2O_3$ ≥99.0% (TLC); ≤1% H_2O

Lysine — *N*ᵉ-Tosyl-ʟ-Lys

Sigma T 1286   FW 300.4 $C_{13}H_{20}N_2O_4S$

Lysine — *N*ᵉ-Z-ʟ-Lys

Fluka 96840   MW 280.32 $C_{14}H_{20}N_2O_4$ ~99% (titration); mp: 250°C; ≤0.05% residue on ignition

Lysine — *N*ᵉ-Z-ʟ-Lys-OBzl Hydrochloride

Fluka 96880   MW 406.91 $C_{21}H_{26}N_2O_4 \cdot HCl$ ≥99.0% (titration); mp: 138-140°C

Lysine — *N*ᵉ-Z-ʟ-Lys-OMe Hydrochloride

Fluka 96890   MW 330.81 $C_{15}H_{22}N_2O_4 \cdot HCl$ ~99% (titration); mp: 115-118°C

Lysine — Pth-ε-Phenylthiocarbamyl-Lys

Sigma P 1502   FW 398.5 $C_{20}H_{22}N_4O_2S$ ~99% | Useful as standards in protein sequencing by the Edman degradation

Lysine — TentaGel S PHB-Lys-BOC-FMOC

Synonyms: *N*ᵅ-FMOC-*N*ᵉ-BOC-Lys 4-(Poly(Ethylenoxy))-OBzl Polymer Bound

Fluka 86395   0.20 mmol protected AA/g; ~90 µm particle size

Lysine — TentaGel S Trt-Lys-BOC-FMOC

Fluka 86426   0.20 mmol protected AA/g; ~90 µm particle size

Lysine — Tosyl-Lys-BOC

Bachem E-3120.0001   MW 400.5 $C_{18}H_{28}N_2O_6S$ Store at RT

Lysine — Tosyl-Lys-CMK Hydrochloride

Synonyms: TLCK

Bachem N-1155.0100   MW 369.31 $C_{14}H_{21}ClN_2O_3S \cdot HCl$ Store at -15°C

Lysine — Tosyl-Lys-OMe Hydrochloride

Synonyms: L-TLME; Trypsin Substrate

Bachem E-2485.0001   MW 350.87 $C_{14}H_{22}N_2O_4S \cdot HCl$ Store at RT

Peptides International STK-3054   MW 350.87 $C_{14}H_{22}N_2O_4S$ · HCl   >98% (HPLC); amorphous powder; hydrochloride | Widmer, F & JT Johansen, *Carlsberg Res Commun*, 44:37, 1979; Widmer, F et al, *Proc 16th European Peptide Symposium* (K Brunfeldt, Ed), Scriptor, Copenhagen 1981, pp 46-55

Lysine — Tosyl-Lys-Z

Bachem E-2475.0001   MW 434.51 $C_{21}H_{26}N_2O_6S$ Store at 2-8°C

Lysine — *trans*-4,5-Dehydro-ᴅʟ-Lys

Synonyms: ᴅʟ-*trans*-2,6-Diamino-4-Hexenoic Acid

Bachem F-2970.0250   MW 144.17 $C_6H_{12}N_2O_2$ Store at -15°C

Lysine — Z-ᴅʟ-Lys-Z

Bachem C-3040.0005   MW 414.46 $C_{22}H_{26}N_2O_6$ Store at RT

Lysine — Z-ᴅ-Lys

Bachem C-2205.0001   MW 280.32 $C_{14}H_{20}N_2O_4$ Store at RT

Lysine — Z-ᴅ-Lys-BOC

Bachem C-1430.0001   MW 380.44 $C_{19}H_{28}N_2O_6$ Store at 2-8°C

Lysine — Z-ᴅ-Lys-BOC-OSu

Bachem C-3285.0001   MW 477.52 $C_{23}H_{31}N_3O_8$ Store at -15°C

Lysine — Z-ᴅ-Lys-OBzl Benzenesulfonate

Bachem C-3875.0001   MW 528.62 $C_{21}H_{26}N_2O_4 \cdot C_6H_6O_3S$ Store at 2-8°C

Lysine — Z-ᴅ-Lys-Z

Bachem C-3035.0001   MW 414.46 $C_{22}H_{26}N_2O_6$ Store at RT

Lysine — Z-Lys

Bachem C-2200.0005   MW 280.32 $C_{14}H_{20}N_2O_4$ Store at RT

Lysine — Z-Lys Amide Hydrochloride

Bachem C-3600.0001   MW 315.8 $C_{14}H_{21}N_3O_3 \cdot HCl$ Store at 2-8°C

Lysine — Z-Lys-Ac Amide
Bachem C-3955.0001   MW 321.38 $C_{16}H_{23}N_3O_4$ Store at RT

Lysine — Z-Lys-BOC
Bachem C-1420.0001   MW 380.44 $C_{19}H_{28}N_2O_6$ Store at 2-8°C

Lysine — Z-Lys-BOC Dicyclohexylamine
Bachem C-1425.0005   MW 561.76 $C_{19}H_{28}N_2O_6 \cdot C_{12}H_{23}N$
Store at RT

Lysine — Z-Lys-BOC-OMe
Bachem C-3480.0005   MW 394.47 $C_{20}H_{30}N_2O_6$ Store at -15°C

Lysine — Z-Lys-BOC-ONp
Bachem C-1440.0001   MW 501.54 $C_{25}H_{31}N_3O_8$ Store at -15°C

Lysine — Z-Lys-BOC-OSu
Bachem C-1435.0001   MW 477.52 $C_{23}H_{31}N_3O_8$ Store at -15°C

Lysine — Z-Lys-FMOC
Bachem C-4125.0001   MW 502.57 $C_{29}H_{30}N_2O_6$ Store at RT

Lysine — Z-Lys-Isopropyl
Bachem C-3960.0001   MW 322.4 $C_{17}H_{26}N_2O_4$ Store at -15°C

Lysine — Z-Lys-OBzl Benzenesulfonate
Bachem C-2210.0005   MW 528.62 $C_{21}H_{26}N_2O_4 \cdot C_6H_6O_3S$
Store at RT

Lysine — Z-Lys-OMe Hydrochloride
Bachem M-1290.0250   MW 330.81 $C_{15}H_{22}N_2O_4 \cdot HCl$ Store at -15°C

Lysine — Z-Lys-ONp Hydrochloride
Bachem M-1295.0250   MW 437.88 $C_{20}H_{23}N_3O_6 \cdot HCl$ Store at -15°C

Lysine — Z-Lys-SBzl Hydrochloride
Synonyms: Enterokinase Substrate; Trypsin-Like Enzyme Substrate

Bachem M-1300.0050   MW 422.98 $C_{21}H_{26}N_2O_3S \cdot HCl$
Store at -15°C

ICN 195982   MW 423 $C_{21}H_{26}N_2O_3S \cdot HCl$ Excellent substrate

Lysine — Z-Lys-Tosyl-ONp
Bachem C-3795.0001   MW 555.61 $C_{27}H_{29}N_3O_8S$ Store at -15°C

Lysine — Z-Lys-Z
Synonyms: N^α,N^ϵ-Di-Benzyloxycarbonyl-L-Lys

Bachem C-2925.0005   MW 414.46 $C_{22}H_{26}N_2O_6$ Store at RT

Peptides International ZLK-2019-PI   MW 414.46   >98% (HPLC); white powder

Lysine — Z-Lys-Z-ONp
Bachem C-2935.0005   MW 535.55 $C_{28}H_{29}N_3O_8$ Store at -15°C

Lysine — Z-Lys-Z-OSu
Bachem C-2930.0005   MW 511.53 $C_{26}H_{29}N_3O_8$ Store at -15°C

Lysine — ε-Dnp-L-Lys
ICN 102215   MW 312.3 $C_{12}H_{16}N_4O_6$

Lysine — ε-Dnp-L-Lys Hydrochloride
ICN 102216   MW 348.8 $C_{12}H_{16}N_4O_6 \cdot HCl$ Crystalline

Lysine — ε-N-Me-L-Lys Hydrochloride
ICN 101647   MW 196.7 $C_7H_{16}N_2O_2 \cdot HCl$

Methionine — Ac-DL-Met
Bachem F-3300.0005   MW 191.25 $C_7H_{13}NO_3S$ Store at RT

Methionine — Ac-D-Met
Bachem F-2300.0005   MW 191.25 $C_7H_{13}NO_3S$ Store at RT

Methionine — Ac-Met
Bachem E-1135.0005   MW 191.25 $C_7H_{13}NO_3S$ Store at RT

Methionine — Ac-Met Amide
Bachem E-1140.0001   MW 190.27 $C_7H_{14}N_2O_2S$ Store at RT

Methionine — Ac-Met α-Naphthyl Ester
Bachem M-1050.0100   MW 317.41 $C_{17}H_{19}NO_3S$ Store at -15°C

Methionine — Ac-Met-AMC
Bachem I-1800.0050   MW 348.4 $C_{17}H_{20}N_2O_4S$ Store at -15°C

Methionine — Ac-Met-o
Bachem F-2940.0001   MW 207.25 $C_7H_{13}NO_4S$ Store at 2-8°C

Methionine — Ac-Met-OMe
Bachem E-1145.0001   MW 205.28 $C_8H_{15}NO_3S$ Store at -15°C

Methionine — Adenosyl-L-Met S-(Carboxyl-¹⁴C)
ARC ARC-343   MW 398.5   40-60 mCi/mmol; 1.48-2.22 GBq/mmol; in sulfuric acid, pH 2.0: EtOH (9:1) | Radiochemical

Methionine — Adenosyl-L-Met S-(Me-¹⁴C)
ARC ARC-344   MW 398.5   40-60 mCi/mmol; 1.48-2.22 GBq/mmol; in sulfuric acid, pH 2.0: EtOH (9:1) | Radiochemical

Methionine — Adenosyl-L-Met S-(Me-³H)
ARC ART-288   MW 398.5   15-30 Ci/mmol; 0.55-1.11 TBq/mmol; in sulfuric acid, pH 2.0: EtOH (9:1) | Radiochemical

Methionine — BOC-DL-Met
Bachem A-1945.0005   MW 249.33 $C_{10}H_{19}NO_4S$ Store at RT

Methionine — BOC-D-Met
American Peptide BDMET10   MW 249.3

Bachem A-1940.0001   MW 249.33 $C_{10}H_{19}NO_4S$ Store at RT

Fluka 15132   MW 249.33 $C_{10}H_{19}NO_4S$ ≥98% (TLC)

Neosystem BA01301   MW 249.3

Peptides International BDM-2608   MW 249.33   >98% (HPLC); white crystalline powder

USBio B2362   MW 249.3 $C_{10}H_{19}NO_4S$ ≥99%

Methionine — BOC-D-Met Dicyclohexylamine

Fluka 15133   MW 430.65 $C_{10}H_{19}NO_4S \cdot C_{12}H_{23}N$ ≥98%

Methionine — BOC-D-Met-OSu

Bachem A-1955.0001   MW 346.41 $C_{14}H_{22}N_2O_6S$ Store at -15°C

Methionine — BOC-L-Met

Fluka 15462   MW 249.33 $C_{10}H_{19}NO_4S$ ≥98% (TLC); mp: 47-50°C

Neosystem BA01302   MW 249.3

Neosystem BA01304   MW 265.3

Methionine — BOC-L-Met Hydroxysuccinimide Ester

Fluka 15461   MW 346.4 $C_{14}H_{22}N_2O_6S$ ≤0.05% residue on ignition; mp: 120-126°C

Methionine — BOC-L-Met N-Carboxy Anhydride

Fluka 15464   MW 275.32 $C_{11}H_{17}NO_5S$ ≥97%; mp: 96-99°C

Methionine — BOC-L-Met-(4-ONp)

Fluka 15134   MW 370.42 $C_{16}H_{22}N_2O_6S$ ≥98% (HPLC)

Methionine — BOC-L-Met(O)-o-CH₂-φ-CH₂-COOH Dicyclohexylamine

Neosystem LP01304   MW 594.8

Methionine — BOC-L-Met-MBHA Resin

Neosystem RN01302

Methionine — BOC-L-Met-O-CH₂-φ-CH₂-COOH

Neosystem LP01302   MW 397.5

Methionine — BOC-L-Met-O-MBHA Resin

Neosystem RN01304

Methionine — BOC-L-Met-O-PAM Resin

Neosystem RP01304

Methionine — BOC-L-Met-PAM Resin

Neosystem RP01302

Methionine — BOC-Met

Synonyms: BOC-L-Met

American Peptide BLMET10   MW 249.3

Bachem A-1935.0005   MW 249.33 $C_{10}H_{19}NO_4S$ Store at RT

Peptides International BLM-2104   MW 249.33   >98% (HPLC); white crystalline powder

USBio B2363   MW 249.3 $C_{10}H_{19}NO_4S$ ≥99%

Methionine — BOC-Met-(O₂)

Bachem A-2885.0001   MW 281.33 $C_{10}H_{19}NO_6S$ Store at RT

USBio B2365   MW 281.3 $C_{10}H_{19}NO_6S$ ≥99%

Methionine — BOC-Met-O

Bachem A-1965.0001   MW 265.33 $C_{10}H_{19}NO_5S$ Store at RT

USBio B2364   MW 265.3 $C_{10}H_{19}NO_5S$ ≥99%

Methionine — BOC-Met-ol

Senn Chem 44035   MW 235.3

Methionine — BOC-Met-ONp

Bachem A-1960.0001   MW 370.43 $C_{16}H_{22}N_2O_6S$ Store at -15°C

Methionine — BOC-Met-O-Resin

American Peptide RMRB225   1% DVB cross-linked: 100-200 mesh | t-BOC protected AA resin; preattached resins are useful for synthesis of peptides with C-terminal carboxyl group by t-BOC chemistry

---

Methionine — BOC-Met-OSu

Bachem A-1950.0001   MW 346.41 $C_{14}H_{22}N_2O_6S$ Store at -15°C

Methionine — BOC-Met-PAM-® (200-400 mesh)

Bachem D-1595.0001   Store at 2-8°C

Methionine — BOC-Met-PAM-Resin

American Peptide RPAM225   1% DVB cross-linked: 100-200 mesh | t-BOC protected AA resin; preattached resins are useful for synthesis of peptides with C-terminal carboxyl group by t-BOC chemistry

---

Methionine — Bpoc-Met Dicyclohexylamine

Bachem E-1645.0001   MW 568.82 $C_{21}H_{25}NO_4S \cdot C_{12}H_{23}N$ Store at -15°C

Methionine — Bsmoc-Met

Synonyms: Bsmoc-L-Met

Peptides International BLM-5632-PI   MW 371.43   >95% (HPLC); white to off-white powder | LA Carpino, et al, *JACS*, 119:9915, 1997

Methionine — Bz-DL-Met

Bachem F-1285.0005   MW 253.32 $C_{12}H_{15}NO_3S$ Store at RT

Methionine — Bz-Met

Bachem E-1490.0005   MW 253.32 $C_{12}H_{15}NO_3S$ Store at RT

Methionine — Bz-Met Amide

Bachem E-1495.0001   MW 252.34 $C_{12}H_{16}N_2O_2S$ Store at RT

Methionine — Caproyl-Met

Bachem E-3495.0250   MW 247.36 $C_{11}H_{21}NO_3S$ Store at -15°C

Methionine — Carbonyl Bis-L-Met-(pONp)

ICN 154935   MW 566.4 $C_{23}H_{26}N_4O_9S_2$ Cross-linking agent; Busse, WD & FH Carpenter, *Biochem*, 15:1649, 1976

Methionine — CBZ-D-Met

USBio C2099-21   MW 283.4 $C_{13}H_{17}NO_4S$ ≥99%

Methionine — CBZ-Met

USBio C2099-20	MW 283.4	$C_{13}H_{17}NO_4S$	≥99%

Methionine — Dansyl-DL-Met Cyclohexylammonium Salt

Sigma D 9131	FW. 481.7	$C_{17}H_{22}N_2O_4S_2 \cdot C_6H_{13}N$

Methionine — Dansyl-L-Met Cyclohexylammonium Salt

ICN 100112	MW 481.7	$C_{17}H_{22}N_2O_4S_2 \cdot C_6H_{13}N$
Sigma D 1625	FW. 481.7	$C_{17}H_{22}N_2O_4S_2 \cdot C_6H_{13}N$

Methionine — Dinitropyridyl-DL-Met

ICN 102280	MW 316.3	$C_{10}H_{12}N_4O_6S$

Methionine — DL-Met

Synonyms: (±)-2-Amino-4-Methylmercaptobutyric Acid; DL-2-Amino-4-Methylthiobutanoic Acid; α-Amino-γ-Methylmercaptobutyric Acid; Acimetion; Banthionine; Cynaron; Lobamine; Meonine; Methilanin; Metione; Neston

Amersham US18910	≥99.0% dsb; ≤20ppm heavy metals

Bachem F-2910.0025	MW 149.21	$C_5H_{11}NO_2S$	Store at RT

Fluka 64340 MW 149.21 $C_5H_{11}NO_2S$ ≥99.0% (titration); ≤0.05% residue on ignition; mp: 280°C

ICN 102281	MW 149.2	$C_5H_{11}NO_2S$	≥99%; crystalline
ICN 190955	99% | Feed grade		

ICN 194706 MW 149.2 $C_5H_{11}NO_2S$ ≥99%; crystalline | Cell culture reagent

Rexim MW 149.2 $C_5H_{11}NO_2S$ White crystals or crystalline powder

Sigma M 2768 FW 149.2 $C_5H_{11}NO_2S$ ~99.5%; crystalline | Cell culture tested; insect cell culture tested

Sigma M 9500 FW 149.2 $C_5H_{11}NO_2S$ ≥99% (TLC) | Also available as part of a kit

USBio M3015 MW 149.2 $C_5H_{11}NO_2S$ ≥98.5%; identification: IR conforms to structure; specific rotation: 22.4° to 24.7°; loss on drying ≤0.3%; residue after ignition: ≤0.4%; pH (1%): 5.6 to 6.1; heavy metals (as Pb): ≤0.0015%; chloride: ≤0.05%; sulfate: ≤0.03%; iron: ≤0.003%; organic volatile impurities: passes

Methionine — DL-Met DL-Sulfoximine

Synonyms: DL-(3-Amino-3-Carboxypropyl)-Me Sulfoximine

Sigma M 9503	FW 180.2	$C_5H_{12}N_2O_3S$

Methionine — DL-Met Hemi-Calcium

Synonyms: DL-2-Amino-4-Methylthiobutanoic Acid

ICN 102013 MW 169.2 $C_5H_9O_3S \cdot \frac{1}{2}Ca$ ~90%; tan powder | Hydroxy analog

Methionine — DL-Met Hydroxamate

Sigma M 4253	FW 164.2	$C_5H_{12}N_2O_2S$

Methionine — DL-Met Methylsulfonium Chloride

Synonyms: MMS; (3-Amino-3-Carboxypropyl)-Dimethyl Sulfonium Chloride; Me-Met Sulphonium Chloride; Vitamin U

Fluka 64382 MW 199.2 $C_6H_{14}ClNO_2S$ ≥99.0% (titration); ≤0.1% residue on ignition; ≤0.5% loss on drying; mp: 139-140°C; ≤0.002% heavy metals

Methionine — DL-Met Sulfone

Synonyms: DL-2-Amino-4-Methylsulfonyl-Butanoic Acid

Fluka 64410 MW 181.21 $C_5H_{11}NO_4S$ ≥99.0% (titration); mp: 250°C

ICN 102285	MW 181.2	$C_5H_{11}NO_4S$	Crystalline
Sigma M 0751	FW 181.2	$C_5H_{11}NO_4S$	

Methionine — DL-Met Sulfoxide

Synonyms: DL-2-Amino-4-Methylsulfinyl-Butanoic Acid

Fluka 64430 MW 165.21 $C_5H_{11}NO_3S$ ~99.0% (titration); ≤0.05% residue on ignition; mp: 240°C

ICN 102287	MW 165.2	$C_5H_{11}NO_3S$	Crystalline

Methionine — DL-Met Sulfoximine

ICN 195312	MW 180.2	$C_5H_{12}N_2O_3S$	Crystalline

Methionine — DL-Met β-Naphthylamide Hydrochloride

ICN 155399 MW 310.8 $C_{15}H_{18}N_2OS \cdot HCl$ Crystalline | Possibly carcinogenic

Sigma M 0626 FW 310.8 $C_{15}H_{18}N_2OS \cdot HCl$ Possibly carcinogenic

Methionine — DL-Met-OEt Hydrochloride

Bachem F-1740.0005 MW 213.73 $C_7H_{15}NO_2S \cdot HCl$ Store at 2-8°C

Methionine — DL-Met-ol

Senn Chem 44101	MW 135.2	Liquid

Methionine — DL-Met-OMe Hydrochloride

Bachem F-2535.0005 MW 199.7 $C_6H_{13}NO_2S \cdot HCl$ Store at 2-8°C

ICN 155394	MW 199.7	$C_6H_{13}NO_2S \cdot HCl$	>95%; crystalline
Sigma M 0251	FW 199.7	$C_6H_{13}NO_2S \cdot HCl$	95-97%

Methionine — DL-Met-S-Methylsulfonium Chloride

Synonyms: Racemic Vitamin U; DL-Met S-Methochloride

ICN 155397	MW 199.7	$C_6H_{14}ClNO_2S$	Crystalline
Sigma M 0501	FW 199.7	$C_6H_{14}ClNO_2S$	

Methionine — DL-Met-βNA Hydrochloride

Bachem K-1435.0250 MW 310.85 $C_{15}H_{18}N_2OS \cdot HCl$ Store at RT

Methionine — D-Met

Synonyms: (R)-2-Amino-4-Methylmercaptobutyric Acid; D-2-Amino-4-Methylthiobutanoic Acid

Bachem F-1735.0005	MW 149.21	$C_5H_{11}NO_2S$	Store at RT
Fluka 64330	MW 149.21	$C_5H_{11}NO_2S$	≥99.0% (titration)
ICN 102278	MW 149.2	$C_5H_{11}NO_2S$	~99%; crystalline
Neosystem AA01301	MW 149.2		

Peptides International ADM-2807 MW 149.21 >99.9% optical purity by RP-HPLC; white crystalline powder

Rexim	MW 149.2	$C_5H_{11}NO_2S$	

Sigma M 9375 FW 149.2 $C_5H_{11}NO_2S$ ≥98% (TLC) | Also available as part of a kit

USBio M3020-10	MW 149.2	$C_5H_{11}NO_2S$	≥99%

Methionine — D-Met-OMe Hydrochloride

ICN 155393	MW 199.7	$C_6H_{13}NO_2S \cdot HCl$	Crystalline
Sigma M 3629	FW 199.7	$C_6H_{13}NO_2S \cdot HCl$	

Methionine — Dnp-DL-Met Sulfone

ICN 102292	MW 347.3	$C_{11}H_{13}N_3O_8S$	Crystalline

Methionine — FA-Met

Bachem M-2400.0001 MW 269.32 $C_{12}H_{15}NO_4S$ Store at -15°C

Methionine — Farnesyl-Met-OMe

Bachem F-3505.0250 MW 367.6 $C_{21}H_{37}NO_2S$ Store at -15°C

Methionine — FMOC-D-Met

Synonyms: FMOC-D-Met

Bachem B-1365.0001	MW 371.46	$C_{20}H_{21}NO_4S$	Store at RT
Fluka 47329	MW 371.45	$C_{20}H_{21}NO_4S$	≥98% (HPLC)
Neosystem FA01301	MW 371.5		
Peptides International FDM-1816-PI	MW 371.45	>97% (TLC); white crystalline powder	
USBio F5431	MW 371.5	$C_{20}H_{21}NO_4S$	≥99%

Methionine — FMOC-L-Met

Fluka 47634	MW 371.45	$C_{20}H_{21}NO_4S$	≥97% (titration); mp: 121-123°C
Neosystem FA01302	MW 371.5		

Methionine — FMOC-L-Met 4-Benzyloxybenzyl Ester Polymer Bound

Fluka 47663 0.4-0.6 mmol/g resin; carrier: polystyrene, crosslinked with 1% DVB; 200-400 mesh particle size

Methionine — FMOC-L-Met-OPfp

Fluka 47469	MW 537.5	$C_{26}H_{20}F_5NO_4S$	~98% (HPLC)

Methionine — FMOC-L-Met-Wang Resin

American Peptide RFWAN60 Preattached Wang resins are usful for synthesis of peptides with C-terminal carboxyl groups; peptides can be cleaved from the resin with 90% TFA in the presence of scavengers

Neosystem RW01302

Methionine — FMOC-Met

Synonyms: FMOC-L-Met

American Peptide FMET100	MW 371.5		
Bachem B-1360.0005	MW 371.46	$C_{20}H_{21}NO_4S$	Store at RT
Peptides International FLM-1703-PI	MW 371.45	>97% (TLC); off-white crystalline powder	
USBio F5432	MW 371.5	$C_{20}H_{21}NO_4S$	≥99%

Methionine — FMOC-Met-(O₂)

Bachem B-1905.0001	MW 403.46	$C_{20}H_{21}NO_6S$	Store at RT

Methionine — FMOC-Met-® (200-400 mesh)

Bachem D-1120.0001 Store at RT

Methionine — FMOC-Met-o

Bachem B-2130.0001	MW 387.46	$C_{20}H_{21}NO_5S$	Store at RT

Methionine — FMOC-Met-ONp

Bachem B-1370.0001	MW 492.55	$C_{26}H_{24}N_2O_6S$	Store at -15°C

Methionine — FMOC-Met-OPfp

Bachem B-1160.0001	MW 537.51	$C_{26}H_{20}F_5NO_4S$	Store at -15°C
USBio F5433	MW 537.5	$C_{26}H_{20}F_5NO_4S$	≥99%

Methionine — FMOC-Met-SASRIN™-® (200-400 mesh)

Bachem D-1370.0001 Store at 2-8°C

Methionine — For-DL-Met

Bachem F-2365.0025	MW 177.22	$C_6H_{11}NO_3S$	Store at RT

Methionine — For-Met

Bachem E-1850.0005	MW 177.22	$C_6H_{11}NO_3S$	Store at RT

Methionine — For-Met-βNA

Bachem K-1240.0050	MW 302.4	$C_{16}H_{18}N_2O_2S$	Store at RT

Methionine — L-Met

Synonyms: (S)-2-Amino-4-Methylthiobutanoic Acid; 2-Amino-4-Methylthiobutyric Acid; L-2-Amino-4-Methylthiobutanoic Acid; (S)-2-Amino-4-Methylmercaptobutyric Acid

Amersham US18915 ≥98.5-101.5% dsb; ≤15ppm heavy metals (Pb)

Fluka 64319 MW 149.21 $C_5H_{11}NO_2S$ ≥99.5% (titration); ≤0.1% residue on ignition, loss on drying; ≤0.5% foreign AA; ≤0.00001% As; ≤0.0005% Al, Ba, Bi, Cd, Co, Cr, Cu, Fe, Li, Mg, Mn, Mo, Ni, Pb, Sr, Zn; ≤0.001% Ca; ≤0.005% K, Cl, Na; ≤0.01% NH₄⁺, SO₄ | Dollinger, G et al, *Meth Enzymol*, 127: 649, 1986; Hine, T, *Electrophoresis '83*, ed H Hirai, 541, 1984, de Gruyter, Berlin; Kreis, W, *Cancer Treat Rep*, 63: 1069, 1979; Kreis, W et al, *Cancer Res*, 40: 634, 1980; *CRC Handbook of Microbiology*, Laskin, AI & Lachevalier, HA, eds, 4: 1, 1974, CRC Press, Cleveland, Ohio

Fluka 64320 MW 149.21 $C_5H_{11}NO_2S$ ≥99.0% (titration); ≤0.1% residue on ignition, loss on drying; ≤0.3% foreign AA; ≤0.0005% Cd, Co, Cr, Cu, Fe, Mg, Mn, Ni, Pb, Zn; ≤0.001% Ca; ≤0.005% K, Cl, SO₄; ≤0.01% Na, NH₄⁺

ICN 102291	MW 149.2	$C_5H_{11}NO_2S$	≥99%; crystalline
ICN 194707	MW 149.2	$C_5H_{11}NO_2S$	≥99%; crystalline | Cell culture reagent
Neosystem AA01302	MW 149.2		

Peptides International ALM-2715 MW 149.21 >99.9% optical purity by RP-HPLC; white crystalline powder

Rexim MW 149.2 $C_5H_{11}NO_2S$ White crystals or crystalline powder

Sigma M 2893 FW 149.2 $C_5H_{11}NO_2S$ Crystalline | Cell culture tested; insect cell culture tested

Sigma M 5010 Matrix: cross-linked 4% beaded agarose; activation: cyanogen bromide; attachment: amino; spacer: 1 atom; ligand immobilized: 2-10 μmoles/mL; form: suspension in 2.0 *M* NaCl | Deutsch, DG & Mertz, ET, *Proc Fed Amer Soc Exp Biol*, 29: 647, 1970; Deutsch, DG & Mertz, ET, *Science*, 170: 1095, 1970; Vuento, M & Vaheri, A, *Biochem J*, 183: 331, 1979

Sigma M 6039 FW 149.2 $C_5H_{11}NO_2S$ SigmaUltra; >99% (TLC); residue on ignition: <0.1%; solubility (1 *M* in 1.0 *M* HCl, 20°C): complete, colorless; insoluble matter: <0.1%; NH₄⁺,: <0.05%; K, Na: <0.005%; Al, Cu, Fe, Mg, Zn: <0.0005%; Pb, Ca: <0.001%

Sigma M 7520 20 m*M*; 3.0 mg/mL in tissue culture grade H_2O; sterile-filtered | Endotoxin tested; cell culture tested

Sigma M 9625 FW 149.2 $C_5H_{11}NO_2S$ ≥98% (TLC) | Also available as part of a kit

Sigma M 9679 FW 149.2 $C_5H_{11}NO_2S$ USP Grade

USBio M3020 MW 149.2 $C_5H_{11}NO_2S$ ≥98.5%; appearance: white crystalline powder; pH: 5.6-6.1; specific rotation: C=2, 6 *N* HCl +21.9° to +24.1°; heavy metals (Pb): ≤0.0015%; endotoxin: ≤2 EU/mg; cell culture tested: murine myeloma SP2/0-Ag14; loss on drying: ≤0.5%; residue on ignition: ≤0.1%; arsenic: ≤0.00015%

Methionine — L-Met (1-¹⁴C)

ARC ARC-271 MW 149.2 $CH_3S(CH_2)_2CH(NH_2)COOH$ 50-60 mCi/mmol; 1.85-2.22 GBq/mmol; in EtOH: H_2O (7:3) | Radiochemical

ARC ARC-271A MW 149.2 $CH_3S(CH_2)_2CH(NH_2)COOH$ 50-60 mCi/mmol; 1.85-2.22 GBq/mmol; in sterile H_2O | Radiochemical

Sigma M 2678 40-60 mCi/mmol; radiochemical purity ≥95% (HPLC); EtOH: H_2O (7:3) solution in Combi-vial | Radiochemical

Methionine — L-Met (^{35}S)

Amersham SJ1015 CH$_3$S(CH$_2$)$_2$CH(NH$_2$)CO$_2$H   Aqueous, 0.1% BME, 15 mM pyridine 3,4-dicarboxylic acid; >37 TBq/mmol, >1000 Ci/mmol; 370 MBq/mL, 10 mCi/mL | Cell labeling grade

Amersham SJ123 CH$_3$S(CH$_2$)$_2$CH(NH$_2$)CO$_2$H   Aqueous; 1.5-18.5 GBq/mmol, 40-500 mCi/mmol; 370 MBq/mL, 10 mCi/mL

Amersham SJ1515 CH$_3$S(CH$_2$)$_2$CH(NH$_2$)CO$_2$H   Aqueous, 0.1% BME, 15 mM pyridine 3,4-dicarboxylic acid; >37 TBq/mmol, >1000 Ci/mmol; 555 MBq/mL, 15 mCi/mL

Amersham SJ204 CH$_3$S(CH$_2$)$_2$CH(NH$_2$)CO$_2$H   20 mM KOAc, 0.1% BME; >37 TBq/mmol, >1000 Ci/mmol; 555 MBq/mL, 15 mCi/mL

Amersham SJ235 CH$_3$S(CH$_2$)$_2$CH(NH$_2$)CO$_2$H   Aqueous, 0.1% BME; >37 TBq/mmol, >1000 Ci/mmol; 555 MBq/mL, 15 mCi/mL

Amersham SJ5050 CH$_3$S(CH$_2$)$_2$CH(NH$_2$)CO$_2$H   Aqueous, 0.1% BME, 15 mM pyridine 3,4-dicarboxylic acid; >37 TBq/mmol, >1000 Ci/mmol; 1.85 GBq/mL, 50 mCi/mL

ARC ARS-104 MW 149.2   CH$_3$S(CH$_2$)$_2$CH(NH$_2$)COOH   >400 Ci/mmol; >14.8 TBq/mmol; 0.1-2% BME solution; in sterile H$_2$O under nitrogen | Radiochemical translation grade

ARC ARS-104A MW 149.2   CH$_3$S(CH$_2$)$_2$CH(NH$_2$)COOH >800 Ci/mmol; >29.6 TBq/mmol; 0.1-2% BME solution; in sterile H$_2$O under nitrogen | Radiochemical translation grade

ARC ARS-119 MW 149.2   CH$_3$S(CH$_2$)$_2$CH(NH$_2$)COOH   >1000 Ci/mmol; >37 TBq/mmol; in sterile H$_2$O under nitrogen with Blue Dye; Stabilized for 4°C storage | Radiochemical translation grade

ICN 51001 MW 149.2   CH$_3$S(CH$_2$)$_2$CH(NH$_2$)COOH   *in vitro* translation grade;>400 Ci/mmol, >14.8 TBq/mmol; aq solution, 50 mM L-Lys, pH 7.4, 10 mM BME; ~10 mCi/mL, ~370 MBq/mL; packaged under N$_2$ in the Versatainer system

ICN 51001H MW 149.2   CH$_3$S(CH$_2$)$_2$CH(NH$_2$)COOH   *in vitro* translation grade;>1000 Ci/mmol, >37 TBq/mmol; aq solution, 50 mM L-Lys, pH 7.4, 10 mM BME; ~10 mCi/mL, ~370 MBq/mL; packaged under N$_2$ in the Versatainer system

ICN 51004 MW 149.2   CH$_3$SCH$_2$CH$_2$CH(NH$_2$)COOH   >400 Ci/mmol, >14.8 TBq/mmol; aqueous solution, 50 mM L-Lys, pH 7.4, 10 mM BME; ~10 mCi/mL, ~370 MBq/mL; packaged under N$_2$ in Versatainer | No-thaw version of ICN 51001

Methionine — L-Met (Me-^{13}C)

ICN 530799 MW 150.2   CC$_4$H$_{11}$NO$_2$S   98% ^{13}C purity; crystalline

Methionine — L-Met (Me-^{14}C)

Amersham CFA152 Lyophilized under N$_2$; 1.85-2.29 GBq/mmol, 50-62 mCi/mmol

ARC ARC-345 MW 149.2   CH$_3$S(CH$_2$)$_2$CH(NH$_2$)COOH   40-55 mCi/mmol; 1.48-2.04 GBq/mmol; in EtOH: H$_2$O (7:3) | Radiochemical

ICN 10102 MW 149.2   CH$_3$SCH$_2$CH$_2$CH(NH$_2$)COOH   30-60 mCi/mmol, 1.11-2.22 GBq/mmol; EtOH:H$_2$O (3:7); 10 mM BME

Sigma M 3053 40-60 mCi/mmol; radiochemical purity ≥95% (HPLC); EtOH: H$_2$O (7:3) solution in Combi-vial | Radiochemical

Methionine — L-Met (Me-^{3}D)

ICN 520875 MW 150.2   CD$_3$SCH$_2$CH$_2$CH(NH$_2$)COOH   98% D atom purity; crystalline

Methionine — L-Met (Me-^{3}H)

Amersham TRK583 Aqueous, 0.2% BME, sterilized; 2.6-3.1 TBq/mmol, 70-85 Ci/mmol; 37 MBq/mL, 1 mCi/mL

Amersham TRK705 Aqueous, 0.2% BME, sterilized; 2.6-3.1 TBq/mmol, 70-85 Ci/mmol; 185 MBq/mL, 5 mCi/mL

ARC ART-169 MW 149.2   CH$_3$S(CH$_2$)$_2$CH(NH$_2$)COOH   30-80 Ci/mmol; 1.11-2.96 TBq/mmol; in EtOH: H$_2$O (7:3) | Radiochemical

ICN 20039 MW 149.2   CH$_3$SCH$_2$CH$_2$CH(NH$_2$)COOH   1.5-1 Ci/mmol, 18.5-37 GBq/mmol; >94%; EtOH:H$_2$O (3:7); 10 mM BME

ICN 20039H MW 149.2   CH$_3$SCH$_2$CH$_2$CH(NH$_2$)COOH   5-15 Ci/mmol, 185-555 GBq/mmol; >94%; EtOH:H$_2$O (3:7); 10 mM BME

Methionine — L-Met Amide Hydrochloride

Neosystem AA01304 MW 184.7

Methionine — L-Met Sulfone

Synonyms: L-2-Amino-4-Methylsulfonylbutanoic Acid

Fluka 64400 MW 181.21   C$_5$H$_{11}$NO$_4$S   ≥99.0% (titration); mp: 275°C

ICN 102295 MW 181.2   C$_5$H$_{11}$NO$_4$S   99%; crystalline

Sigma M 0876 FW 181.2   C$_5$H$_{11}$NO$_4$S

Methionine — L-Met Sulfoxide

Synonyms: L-2-Amino-4-Methylsulfinylbutanoic Acid

Fluka 64420 MW 165.21   C$_5$H$_{11}$NO$_3$S   ≥99.0% (titration); mp: 255°C

ICN 102296 MW 165.2   C$_5$H$_{11}$NO$_3$S   ≥99%; crystalline

Sigma M 1126 FW 165.2   C$_5$H$_{11}$NO$_3$S

Methionine — L-Met Sulfoximine

Synonyms: Glutamine Synthetase Inhibitor; L-S-(3-Amino-3-Carboxypropyl)-S-Methylsulfoximine; Glutamine Transferase Inhibitor

Fluka 64435 MW 180.22   C$_5$H$_{12}$N$_2$O$_3$S   ≥99.0% (titration); mp: 240-245°C | Weisbrod, RE & Meister, A, *JBC*, 248: 3997, 1973

ICN 151619 MW 180.2   C$_5$H$_{12}$N$_2$O$_3$S   Crystalline | Substrate for microbiological determination of L-Met

Sigma M 5379 FW 180.2   C$_5$H$_{12}$N$_2$O$_3$S

Methionine — L-Met-AMC TFA

Synonyms: L-Met 4-Methylumbelliferylamide TFA; Calpain Substrate

Fluka 64367 MW 420.41   C$_{15}$H$_{18}$N$_2$O$_3$S · C$_2$HF$_3$O$_2$ ≥99.0% (HPLC) | Sasaki, T et al, *JBC*, 259: 12489, 1984

Methionine — L-Methioninamide Hydrochloride

Sigma M 3504 FW 184.7   C$_5$H$_{12}$N$_2$OS · HCl

Methionine — L-Met-OAll-Tosyl

Fluka 64360 MW 361.48   C$_8$H$_{15}$NO$_2$S · C$_7$H$_8$O$_3$S   ≥99.0% (titration); mp: 114-116°C

Methionine — L-Met-OEt Hydrochloride

Fluka 64350 MW 213.73   C$_7$H$_{15}$NO$_2$S · HCl   ~99.0% (titration); mp: 84-87°C

ICN 102293 MW 199.7   C$_7$H$_{15}$NO$_2$S · HCl Chromatographically pure; crystalline

Methionine — L-Met-ol

Bachem F-1745.0001 MW 135.23   C$_5$H$_{13}$NOS   Store at 2-8°C

Methionine — L-Met-OMe (1-^{14}C) Hydrochloride

ARC ARC-1150 MW 199.7   CH$_3$S(CH$_2$)$_2$CH(NH$_2$)**C**O$_2$CH$_3$ · HCl 50-60 mCi/mmol; 1.85-2.22 GBq/mmol; in MeOH | Radiochemical

Methionine — L-Met-OMe (Me-^{3}H) Hydrochloride

ARC ART-721 MW 199.7   CH$_3$S(CH$_2$)$_2$CH(NH$_2$)CO$_2$CH$_3$ · HCl 30-80 Ci/mmol; 1.11-2.96 TBq/mmol; in MeOH | Radiochemical

Methionine — L-Met-OMe Hydrochloride

Fluka 64370 MW 199.7   C$_6$H$_{13}$NO$_2$S · HCl   ≥99.0% (titration); mp: 150°C

ICN 102294 MW 199.7   C$_6$H$_{13}$NO$_2$S · HCl   Crystalline

Sigma M 0376 FW 199.7   C$_6$H$_{13}$NO$_2$S · HCl

Methionine — L-Met-pNA

ICN 155400	MW 269.3	$C_{11}H_{15}N_3O_3S$
Sigma M 3529	FW 269.3	$C_{11}H_{15}N_3O_3S$

Methionine — L-Met-S-Methylsulfonium Bromide

ICN 155396	MW 244.1	$C_6H_{14}O_2NSBr$	95%; crystalline

Methionine — L-Met-S-Methylsulfonium Iodide

Synonyms: L-Met S-Methoiodide

ICN 155398	MW 291.1	$C_6H_{14}NIO_2S$	Crystalline
Sigma M 1881	FW 291.1	$C_6H_{14}INO_2S$	

Methionine — Met

Bachem E-2100.0025	MW 149.21	$C_5H_{11}NO_2S$	Store at RT

Methionine — Met Amide Hydrochloride

Bachem E-2105.0001	MW 184.69	$C_5H_{12}N_2OS \cdot HCl$	Store at RT

Methionine — Met-(O₂)

Bachem F-2895.0001	MW 181.21	$C_5H_{11}NO_4S$	Store at 2-8°C

Methionine — Met-AMC

Synonyms: Aminopeptidase Substrate

Peptides International MMM-3149-v   MW 306.38
$C_{15}H_{18}N_2O_3S$ >96% (HPLC); lyophilized amorphous powder; Tosyl form

Methionine — Met-AMC Acetate

Bachem I-1265.0050	MW 366.4	$C_{15}H_{18}N_2O_3S \cdot C_2H_4O_2$	Store at -15°C

Methionine — Met-o

Bachem F-2945.0005	MW 165.21	$C_5H_{11}NO_3S$	Store at 2-8°C

Methionine — Met-OEt Hydrochloride

Bachem E-2110.0005	MW 213.73	$C_7H_{15}NO_2S \cdot HCl$	Store at RT

Methionine — Met-ol

Synonyms: L-Methioninol

Senn Chem 44100	MW 135.2	Crystalline

Methionine — Met-OMe Hydrochloride

Synonyms: L-Met-OMe

Bachem E-2115.0005	MW 199.7	$C_6H_{13}NO_2S \cdot HCl$	Store at RT

Peptides International EMM-2036-PI   MW 199.70   >97% (HPLC); white to off-white powder

Methionine — Met-OtBu Hydrochloride

Bachem E-2730.0001	MW 241.78	$C_9H_{19}NO_2S \cdot HCl$	Store at -15°C

Methionine — Met-pNA

Bachem L-1320.0250	MW 269.33	$C_{11}H_{15}N_3O_3S$	Store at -15°C

Methionine — Met-βNA

Bachem K-1430.0250	MW 274.39	$C_{15}H_{18}N_2OS$	Store at RT

Methionine — MTH-DL-Met

Sigma M 5131	FW 204.3	$C_7H_{12}N_2OS_2$	~95%

Methionine — N-2,4-Dnp-DL-Met

Sigma D 0505	FW 315.3	$C_{11}H_{13}N_3O_6S$

Methionine — N-2,4-Dnp-DL-Met Sulfone

Sigma D 0630	FW 347.3	$C_{11}H_{13}N_3O_8S$

Methionine — N-2,4-Dnp-DL-Met Sulfoxide

Sigma D 0755	FW 331.3	$C_{11}H_{13}N_3O_7S$	~95%

Methionine — N-Ac-DL-Met

Synonyms: DL-2-Acetylamino-4-Methylthiobutyric Acid

ICN 100113	MW 191.2	$C_7H_{13}NO_3S$	Crystalline
Rexim	MW 191.3	$C_7H_{13}NO_3S$	White crystals or crystalline powder

Methionine — N-Ac-DL-Met Magnesium Salt

Sigma A 1876	FW 404.8	$C_{14}H_{24}N_2O_6S_2Mg$

Methionine — N-Ac-D-Met

Sigma A 1501	FW 191.2	$C_7H_{13}NO_3S$	~99%

Methionine — N-Ac-L-Met

Synonyms: Thiomedon; L-2-Acetylamino-4-Methylthiobutyric Acid; Methionamine

Fluka 01310	MW 191.25	$C_7H_{13}NO_3S$	≥99% (titration); ≤0.05% residue on ignition; mp: 103-106°C
ICN 154705	MW 191.2	$C_7H_{13}NO_3S$	Crystalline
Rexim	MW 191.3	$C_7H_{13}NO_3S$	White crystals or crystalline powder
Sigma A 3258	FW 191.2	$C_7H_{13}NO_3S$	

Methionine — N-Bz-DL-Met

Sigma B 1754	FW 253.3	$C_{12}H_{15}NO_3S$

Methionine — N-CBZ-DL-Met

ICN 100224	MW 283.3	$C_{13}H_{17}NO_4S$	Crystalline

Methionine — N-CBZ-L-Met

ICN 100225	MW 283.3	$C_{13}H_{17}NO_4S$	Crystalline
Sigma C 3752	FW 283.3	$C_{13}H_{17}NO_4S$	

Methionine — N-FMOC-L-Met Resin Ester

Sigma F 1886   For peptide synthesis; Wang, SS, *J Am Chem Soc*, 95: 1328, 1973; Lu, G et al, *J Org Chem*, 46: 3433, 1981

Methionine — N-For-DL-Met

ICN 158182	MW 177.2	$C_6H_{11}NO_3S$	Crystalline

Methionine — N-For-L-Met

ICN 151169	MW 177.2	$C_6H_{11}NO_3S$	Crystalline
Sigma F 3377	FW 177.2	$C_6H_{11}NO_3S$	

Methionine — N-For-L-Met β-Naphthylamide

Sigma F 2260	FW 302.4	$C_{16}H_{18}N_2O_2S$

Methionine — N-For-L-Met-pNA

ICN 158183

Methionine — *N*-O-Nps-L-Met Cyclohexylammonium Salt

ICN 102531   MW 483.7 $C_{11}H_{14}N_2O_4S_2 \cdot C_{12}H_{23}N$ Yellow crystals,

Methionine — *N*-O-Nps-L-Met Dicyclohexylammonium Salt

Sigma N 5878   FW 483.7 $C_{11}H_{14}N_2O_4S_2 \cdot C_{12}H_{23}N$ Yellow crystals

Methionine — *N*-Phthaloyl-DL-Met

Sigma P 4628   FW 279.3 $C_{13}H_{13}NO_4S$ May contain 1-4% free AA

Methionine — Nps-Met DCHA

Bachem E-2225.0001   MW 483.7 $C_{11}H_{14}N_2O_4S_2 \cdot C_{12}H_{23}N$ Store at -15°C

Methionine — *N*-t-BOC-D-Met

Sigma B 4138   FW 249.3 $C_{10}H_{19}NO_4S$

Methionine — *N*-t-BOC-L-Met Free Acid

Sigma B 9004   FW 249.3 $C_{10}H_{19}NO_4S$

Methionine — *N*-TFA-L-Met-OMe

Sigma T 4011   FW 259.2 $C_8H_{12}F_3NO_3S$ ~98%

Methionine — *N*ᵅ-FMOC-L-Met

ICN 151149   MW 371.5 $C_{20}H_{21}NO_4$

Methionine — *N*ᵅ-t-BOC-L-Met Dicyclohexylammonium Salt

ICN 101059   MW 430.6 $C_{10}H_{19}NO_4S \cdot C_{12}H_{23}N$

Methionine — *N*ᵅ-t-BOC-L-Met Free Acid

ICN 104916   MW 249.3 $C_{10}H_{19}NO_4S$ Crystalline

Methionine — Palmitoyl-Met

Bachem E-3365.0250   MW 387.63 $C_{21}H_{41}NO_3S$ Store at -15°C

Methionine — Pth-Met

Sigma P 1627   FW 266.4 $C_{12}H_{14}N_2OS_2$ ~99% | Useful as standards in protein sequencing by the Edman degradation

Methionine — Pth-Met Sulfone

Sigma P 1752   FW 298.4 $C_{12}H_{14}N_2O_3S_2$ ~99% | Useful as standards in protein sequencing by the Edman degradation

Methionine — Redivue L-Met (³⁵S)

Amersham AG1094 $CH_3S(CH_2)_2CH(NH_2)CO_2H$ Redivue stabilized & colored, aqueous; >37 TBq/mmol, >1000 Ci/mmol; 370 MBq/mL, 10 mCi/mL

Amersham AG1594 $CH_3S(CH_2)_2CH(NH_2)CO_2H$ Redivue stabilized & colored, aqueous; >37 TBq/mmol, >1000 Ci/mmol; 555 MBq/mL, 15 mCi/mL

Methionine — *S*-Adenosyl-L-Met

Synonyms: SAM; Active Methionine; *S*-(5'-Adenosyl)-L-Met

Fluka 02110   MW 570.63 $C_{22}H_{30}N_6O_8S_2$ ≥85% (HPLC); ≤5% H_2O; Tosyl

Methionine — *S*-Adenosyl-L-Met (Carboxyl-¹⁴C)

Synonyms: SAM

Amersham CFA477   Solution in H_2SO_4; pH 2.5-3.5; 1.85-2.29 GBq/mmol, 50-62 mCi/mmol; 925 kBq/ml, 25 μCi/ml

Methionine — *S*-Adenosyl-L-Met (Me-¹⁴C)

Synonyms: SAM

Amersham CFA360   Solution in H_2SO_4; pH 2.5-3.5; 1.85-2.29 GBq/mmol, 50-62 mCi/mmol; 925 kBq/ml, 25 μCi/ml

ICN 14007   MW 398.5   40-60 mCi/mmol, 1.48-2.22 GBq/mmol; sulfuric acid, pH 2:EtOH (9:1)

Sigma A 8300   40-60 Ci/mmol; radiochemical purity ≥95% (HPLC); Sulfuric acid solution, pH 1-3, containing 10% EtOH in Combi-vial | Radiochemical

Methionine — *S*-Adenosyl-L-Met (Me-³H)

Synonyms: SAM

Amersham TRA236   Solution in HCl; pH 2.0-2.5; 18.5 GBq/mmol, 500 Ci/mmol; 18.5 MBq/ml, 500 μCi/ml

Amersham TRK236   Solution in HCl; pH 2.0-2.5; 555 GBq/mmol, 15 Ci/mmol; 37 MBq/ml, 1 mCi/ml

Amersham TRK581   Solution in HCl; pH 2.0-2.5; 2.2-3.1 TBq/mmol, 60-85 Ci/mmol; 37 MBq/ml, 1 mCi/ml

Amersham TRK614   Solution in HCl:EtOH (9:1); pH 2.0-2.5; 550 GBq/mmol, 15 Ci/mmol; 37 MBq/ml, 1 mCi/ml

Amersham TRK865   Solution in HCl:EtOH (9:1); pH 2.0-2.5; 2.2-3.1 TBq/mmol, 60-85 Ci/mmol; 37 MBq/ml, 1 mCi/ml

ICN 24051   MW 398.5   5-15 Ci/mmol, 185-555 GBq/mmol; sulfuric acid, pH 2:EtOH (9:1)

ICN 24051H   MW 398.5   >50 Ci/mmol, >1.85 TBq/mmol; sulfuric acid, pH 2:EtOH (9:1)

Sigma A 4553   FW 399.4 $C_{15}H_{23}N_6O_5S$ 10-20 Ci/mmol; Sulfuric acid solution, pH 1-3, containing 10% EtOH in serum bottle | Radiochemical

Methionine — *S*-Adenosyl-L-Met Chloride

Synonyms: SAM; Active Methionine; *S*-(5'-Adenosyl)-L-Met

Fluka 02095   MW 434.9 $C_{15}H_{23}ClN_6O_5S$ ≥80% (HPLC); ≤6% solvent; ≤8% H_2O | Prevents ischemic neuronal death in rats; Matsui, Y, *Eur J Pharmacol*, 144: 211, 1987

Methionine — *S*-Adenosyl-L-Met Chloride Salt

Synonyms: SAM

Sigma A 7007   FW 434.9 $C_{15}H_{23}N_6O_5SCl$ ~70%; ~80-90% when prepared, but very unstable; as much as 10% loss per day at 25°C has been noted; SAM isolated from yeast ~18% unnatural *S*(+)-isomer; purity based on UV & HPLC | Prepared from L-Met-enriched yeast; methyl donor; cofactor for COMT, PNMT, & other enzyme-catalyzed methylations; Stolowitz, ML & Minch, MJ, *J Am Chem Soc*, 103: 6015, 1981; Shapiro, SK & Ehninger, DJ, *J Anal Biochem*, 15: 323, 1966

Methionine — *S*-Adenosyl-L-Met Hydrogen Sulfate Salt

Synonyms: SAM

Amersham US10605   Methyl donor for use with methylases

Methionine — *S*-Adenosyl-L-Met Iodide

Synonyms: SAM; Active Methionine; *S*-(5'-Adenosyl)-L-Met

Fluka 02100   MW 526.35 $C_{15}H_{23}IN_6O_5S$ ≥80% (HPLC) | Cantoni, GL et al, *Ann Rev Biochem*, 44: 435, 1975; *Biochem J*, 213: 1, 1983

Methionine — *S*-Adenosyl-L-Met Iodide Salt

Synonyms: SAM

ICN 190690   MW 526.3 $C_{15}H_{23}N_6O_5SI$ 85-90% (enzymatic analysis & column chromatography)

Sigma A 4377   FW 526.3 $C_{15}H_{23}N_6O_5SI$ ~80%; very unstable at room temperature; SAM isolated from yeast ~18% unnatural $S(+)$-isomer; purity based on UV & HPLC | Prepared from L-Met-enriched yeast; methyl donor; cofactor for COMT, PNMT, & other enzyme-catalyzed methylations; Stolowitz, ML & Minch, MJ, *J Am Chem Soc*, 103: 6015, 1981; Shapiro, SK & Ehninger, DJ, *J Anal Biochem*, 15: 323, 1966

Methionine — S-Adenosyl-L-Met p-Tosyl
Synonyms: SAM

ICN 190526   ~90%

Sigma A 2408   FW 399.4 $C_{15}H_{23}N_6O_5S$ ~90%; very unstable at room temperature; SAM isolated from yeast ~18% unnatural $S(+)$-isomer; purity based on UV & HPLC | Prepared from L-Met-enriched yeast; MW is free base; methyl donor; cofactor for COMT, PNMT, & other enzyme-catalyzed methylations; Stolowitz, ML & Minch, MJ, *J Am Chem Soc*, 103: 6015, 1981; Shapiro, SK & Ehninger, DJ, *J Anal Biochem*, 15: 323, 1966

Methionine — Seleno-DL-Met
Synonyms: S-Adenosylmethionine Synthetase Substrate

Fluka 84925   MW 196.10 $C_5H_{11}NO_2Se$ ≥99.0% (TLC) | Reacts more rapidly than Met in the S-adenosylmethionine synthetase reaction; Markham, GD et al, *JBC*, 255: 9082, 1980

Sigma S 3875   FW 196.1 $C_5H_{11}NO_2Se$ Very toxic

Methionine — Seleno-L-Met
Synonyms: (S)-2-Amino-4-Methylselenobutyric Acid

Fluka 09975   MW 196.11 $C_5H_{11}NO_2Se$ ≥98.0% (TLC)

Sigma S 3132   FW 196.1 $C_5H_{11}NO_2Se$ Very toxic

Methionine — S-Me-DL-Met-Sulfonium Chloride
Synonyms: MMS; 3-Amino-3-Carboxypropyl-Dimethylsulfonium

Rexim   MW 199.7 $C_6H_{14}ClNO_2S$ Hygroscopic white crystals or crystalline powder

Methionine — TentaGel S PHB-Met-FMOC
Synonyms: FMOC-L-Met-4-Poly(Ethylenoxy)-OBzl Polymer Bound

Fluka 86396   0.20 mmol protected AA/g; ~90 µm particle size

Methionine — TentaGel S Trt-Met-FMOC
Fluka 86427   0.20 mmol protected AA/g; ~90 µm particle size

Methionine — Trt-Met DEA
Bachem E-3525.0001   MW 464.67 $C_{24}H_{25}NO_2S \cdot C_4H_{11}N$ Store at -15°C

Methionine — Z-DL-Met
Fluka 96912   MW 283.34 $C_{13}H_{17}NO_4S$ ≥98.0% (TLC)

Methionine — Z-D-Met
Bachem C-3395.0005   MW 283.35 $C_{13}H_{17}NO_4S$ Store at RT

Methionine — Z-L-Met
Fluka 96910   MW 283.35 $C_{13}H_{17}NO_4S$ ~99% (titration); mp: 67-68°C

Neosystem ZA01301   MW 283.4

Methionine — Z-L-Met-OMe
Fluka 96915   MW 297.37 $C_{14}H_{19}NO_4S$ ≥99.0% (HPLC); mp: 41-43°C | Intermediate for the synthesis of derivatives of L-vinylglycine ((S)-2-Amino-3-Butenoic Acid); Afzali-Arkadani, A & Rapoport, H, *J Org Chem*, 45: 4817, 1980; Meffre, P et al, *Synth Comm*, 19: 3457, 1989; Tashiro, T et al, *Chem Pharm Bull*, 36: 893, 1988

Methionine — Z-Met
Synonyms: Benzyloxycarbonyl-L-Met

Bachem C-2215.0025   MW 283.35 $C_{13}H_{17}NO_4S$ Store at RT

Peptides International ZLM-2020-PI   MW 283.34   >95% (HPLC); white powder

Methionine — α-Me-DL-Met
Synonyms: Methionine Antagonist

Sigma M 4252   FW 163.2 $C_6H_{13}NO_2S$

Norleucine — 6-Diazo-5-oxo-D-Nle
Synonyms: 6-Diazo-5-Oxo-D-2-Aminohexanoic Acid; D-DON

Bachem F-2185.0025   MW 171.16 $C_6H_9N_3O_3$ Store at -15°C

ICN 157612   MW 171.2 $C_6H_9N_3O_3$

Sigma D 0778   FW 171.2 $C_6H_9N_3O_3$

Norleucine — 6-Diazo-5-Oxo-L-Nle
Synonyms: DON; (S)-2-Amino-6-Diazo-5-Oxocaproic Acid

Fluka 33515   MW 171.16 $C_6H_9N_3O_3$ ≥98% (UV); mp: 145°C | Inhibits purine synthesis; an analog of glutamine, inhibiting glutamine-requiring enzymes

ICN 150845   MW 171.2 $C_6H_9N_3O_3$ Light yellow crystals | Exhibits teratogenic effects, anti-microbial action, enzymatic inhibition; King, GL etal, *JBC*, 253:3933, 1978; Morgan, PR & RM Pratt, *Teratology*, 15:281, 1977

Sigma D 2141   FW 171.2 $C_6H_9N_3O_3$ Light yellow crystals

Norleucine — 6-Diazo-5-oxo-Nle
Synonyms: 6-Diazo-5-Oxo-L-2-Aminohexanoic Acid; L-DON

Bachem F-1510.0025   MW 171.16 $C_6H_9N_3O_3$ Store at -15°C

Norleucine — BOC-D-Nle
Synonyms: BOC-D-2-Aminohexanoic Acid; BOC-D-2-Aminocaproic Acid

Bachem A-2145.0001   MW 231.29 $C_{11}H_{21}NO_4$ Store at -15°C

Fluka 15176   MW 231.29 $C_{11}H_{21}NO_4$ ≥98% (TLC)

Neosystem BA05101   MW 231.3

Norleucine — BOC-L-Nle
Synonyms: BOC-L-2-Aminocaproic Acid

Fluka 15555   MW 231.29 $C_{11}H_{21}NO_4$ ≥99% (TLC)

Neosystem BA05102   MW 231.3

Norleucine — BOC-Nle
Synonyms: BOC-L-Nle

American Peptide BLNLE10   MW 231.2   Syrup

Peptides International BLX-5325-PI   MW 231.29   >98% (HPLC); syrup

Norleucine — BOC-Nle Dicyclohexylamine
Synonyms: BOC-L-2-Aminohexanoic Acid

Bachem A-2140.0001   MW 412.61 $C_{11}H_{21}NO_4 \cdot C_{12}H_{23}N$ Store at RT

Norleucine — BOC-Nle Syrup
Synonyms: BOC-L-2-Aminohexanoic Acid Syrup

Bachem A-2135.0001   MW 231.29 $C_{11}H_{21}NO_4$ Store at -15°C

Norleucine — BOC-Nle-OSu

Synonyms: BOC-L-2-Aminohexanoic Acid N-Hydroxysuccinimide Ester)

Bachem A-2150.0001   MW 328.37 $C_{15}H_{24}N_2O_6$ Store at -15°C

Norleucine — BOC-N-Me-L-Nle

Fluka 00435   MW 245.32 $C_{12}H_{23}NO_4$ ≥98% (TLC)

Norleucine — BOC-N-Me-Nle

Synonyms: BOC-N-Me-L-2-Aminohexanoic Acid

Bachem A-4440.0001   MW 245.32 $C_{12}H_{23}NO_4$ Store at RT

Norleucine — Bzl-Nle-OMe Hydrochloride

Synonyms: Bzl-L-2-Aminohexanoic Acid-OMe

Bachem F-3910.0001   MW 271.79 $C_{14}H_{21}NO_2 \cdot HCl$ Store at RT

Norleucine — CBZ-Nle

USBio C2099-27   MW 265.3 $C_{14}H_{19}NO_4$ ≥99%

Norleucine — CBZ-Nle-ONp

USBio C2099-28   MW 386.3 $C_{20}H_{22}O_6N$ ≥99%

Norleucine — Dansyl-DL-Nle Cyclohexylammonium Salt

ICN 100095   MW 463.6 $C_{18}H_{24}N_2O_4S \cdot C_6H_{13}N$

Sigma D 0756   FW. 463.6 $C_{18}H_{24}N_2O_4S \cdot C_6H_{13}N$ ~99%

Norleucine — Diazoacetyl-DL-Nle-OMe

Synonyms: Diazoacetyl-DL-2-Aminohexanoic Acid-OMe

Bachem F-2220.0250   MW 213.24 $C_9H_{15}N_3O_3$ Store at -15°C

Sigma D 4512   FW 213.2 $C_9H_{15}N_3O_3$ Yellow crystals

Norleucine — DL-Nle

Synonyms: DL-2-Aminohexanoic Acid; DL-α-Amino-N-Caproic Acid; DL-2-Amino-N-Hexanoic Acid

Bachem F-1925.0025   MW 131.18 $C_6H_{13}NO_2$ Store at RT

Fluka 74580   MW 131.18 $C_6H_{13}NO_2$ ~99.0% (titration); mp: 325°C | (±)-2-Aminocaproic Acid

ICN 102482   MW 131.2 $C_6H_{13}NO_2$ 98-99%; crystalline

Sigma N 1398   FW 131.2 $C_6H_{13}NO_2$ Also available as part of a kit

Norleucine — DL-Nle Hydroxamate

Sigma N 3752   FW 146.2 $C_6H_{14}N_2O_2$

Norleucine — DL-Nle-OMe Hydrochloride

Synonyms: DL-2-Aminohexanoic Acid-OMe

Bachem F-1935.0005   MW 181.66 $C_7H_{15}NO_2 \cdot HCl$ Store at 2-8°C

ICN 155953   MW 181.7 $C_7H_{15}NO_2 \cdot HCl$ Crystalline

Sigma N 7002   FW 181.7 $C_7H_{15}NO_2 \cdot HCl$

Norleucine — D-Nle

Synonyms: D-2-Aminohexanoic Acid; D-2-Aminocaproic Acid; D-α-Amino-N-Caproic Acid

Bachem F-1920.0001   MW 131.18 $C_6H_{13}NO_2$ Store at RT

Fluka 74570   MW 131.18 $C_6H_{13}NO_2$ ≥99.0% (titration); mp: >300°C; enantiomer ratio: D-:L-≥99.5:0.5 | (R)-2-Aminocaproic Acid

ICN 102480   MW 131.2 $C_6H_{13}NO_2$ ≥99%; crystalline

Neosystem AA05101   MW 131.2

Sigma N 6627   FW 131.2 $C_6H_{13}NO_2$ Also available as part of a kit

Norleucine — Dnp-DL-Nle

ICN 102486   MW 297.3 $C_{12}H_{15}N_3O_6$ Crystalline

Norleucine — FMOC-DL-Nle

Synonyms: FMOC-DL-2-Aminohexanoic Acid

Bachem B-2410.0001   MW 353.42 $C_{21}H_{23}NO_4$ Store at RT

Norleucine — FMOC-D-Nle

Synonyms: FMOC-D-2-Aminohexanoic Acid; FMOC-D-2-Aminocaproic Acid

Bachem B-2925.0001   MW 353.42 $C_{21}H_{23}NO_4$ Store at RT

Fluka 47781   MW 353.42 $C_{21}H_{23}NO_4$ ≥98% (HPLC)

Neosystem FA05101   MW 353.4

Norleucine — FMOC-L-Nle

Synonyms: FMOC-L-Nle; FMOC-L-2-Aminocaproic Acid

Fluka 47692   MW 353.42 $C_{21}H_{23}NO_4$ ≥98% (HPLC); mp: 141-144°C

Neosystem FA05102   MW 353.4

Norleucine — FMOC-L-Nle-OPfp

Fluka 47473   MW 519.47 $C_{27}H_{22}F_5NO_4$ ~98% (HPLC)

Norleucine — FMOC-Me-Nle

USBio F5423   MW 367.4 $C_{22}H_{25}NO_4$ ≥98%

Norleucine — FMOC-Nle

Synonyms: FMOC-L-2-Aminohexanoic Acid; FMOC-L-Nle

American Peptide FNLE100   MW 353.4

Bachem B-1400.0001   MW 353.42 $C_{21}H_{23}NO_4$ Store at RT

Peptides International FLU-1728-PI   MW 353.42   >98% (HPLC); white crystalline powder

USBio F5434   MW 353.4 $C_{21}H_{23}NO_4$ ≥99%

Norleucine — FMOC-Nle-® (200-400 mesh)

Synonyms: FMOC-L-2-Aminohexanoic Acid-®

Bachem D-1795.0001   Store at RT

Norleucine — FMOC-Nle-OPfp

Synonyms: FMOC-L-2-Aminohexanoic Acid-OPfp

Bachem B-1165.0001   MW 519.47 $C_{27}H_{22}F_5NO_4$ Store at -15°C

USBio F5435   MW 519.5 $C_{27}H_{22}F_5NO_4$ ≥99%

Norleucine — FMOC-Nle-OSu

Synonyms: FMOC-L-2-Aminohexanoic Acid N-Hydroxysuccinimide Ester

Bachem B-2460.0001   MW 450.49 $C_{25}H_{26}N_2O_6$ Store at -15°C

Norleucine — FMOC-Nle-SASRIN™-® (200-400 mesh)

Synonyms: FMOC-L-2-Aminohexanoic Acid-SASRIN™-®)

Bachem D-1730.0001   Store at 2-8°C

Norleucine — FMOC-N-Me-L-Nle

Fluka 02450   MW 367.45 $C_{22}H_{25}NO_4$ ~98% (HPLC)

Norleucine — FMOC-N-Me-Nle

Synonyms: FMOC-N-Me-L-2-Aminohexanoic Acid

Bachem B-3230.0001   MW 367.45 $C_{22}H_{25}NO_4$ Store at RT

Norleucine — For-Nle

Synonyms: For-L-2-Aminohexanoic Acid

Bachem E-2905.0001 MW 159.19 $C_7H_{13}NO_3$ Store at RT

Norleucine — L-Nle

Synonyms: (S)-2-Aminocaproic Acid; L-2-Aminohexanoic Acid; L-2-Aminocaproic Acid; L-α-Amino-N-Caproic Acid

Fluka 74560 MW 131.18 $C_6H_{13}NO_2$ ≥99.0% (titration); mp: >300°C; enantiomer ratio: L-:D-≥99.5:0.5

ICN 151775 MW 131.2 $C_6H_{13}NO_2$ 99%; crystalline

Neosystem AA05102 MW 131.2

Sigma N 6877 FW 131.2 $C_6H_{13}NO_2$

Sigma N 8513 Internal standard used for AA analysis employing ninhydrin, o-phthaldialdehyde & phenyl isothiocyanate as detection reagents; highly purified internal standard

USBio N5000 MW 131.2 $C_6H_{13}NO_2$ ≥98%; appearance: white crystalline powder; pH: 5.6-6.1; specific rotation: C=2, 6 N HCl +23.0° to +25.0°; heavy metals (Pb): ≤0.001%; endotoxin: ≤2 EU/mg; cell culture tested: murine myeloma SP2/0-Ag14

Norleucine — L-Nor

Peptides International ALU-2728-PI MW 131.17 >98% optical purity by TLC; white crystalline powder

Norleucine — N-2,4-Dnp-DL-Nle

Sigma D 0880 FW 297.3 $C_{12}H_{15}N_3O_6$

Norleucine — N-CBZ-L-Nle

ICN 150575 MW 265.3 $C_{14}H_{19}NO_4$

Norleucine — N-FMOC-L-Nle

Sigma F 2917 FW 353.4 $C_{21}H_{23}NO_4$

Norleucine — Nle

Synonyms: L-2-Aminohexanoic Acid

Bachem F-1915.0001 MW 131.18 $C_6H_{13}NO_2$ Store at RT

Norleucine — Nle-OMe Hydrochloride

Synonyms: L-2-Aminohexanoic Acid-OMe

Bachem F-1930.0001 MW 181.66 $C_7H_{15}NO_2 \cdot HCl$ Store at 2-8°C

Norleucine — N-Me-Nle

Synonyms: N-Me-L-2-Aminohexanoic Acid

Bachem F-3750.0001 MW 145.2 $C_7H_{15}NO_2$ Store at -15°C

Norleucine — N-O-Nps-DL-Nle Cyclohexylammonium Salt

ICN 102532 MW 465.6 $C_{12}H_{16}N_2O_4S \cdot C_{12}H_{23}N$ Yellow crystals

Norleucine — N-t-BOC-L-Nle Dicyclohexylammonium Salt

Sigma B 0131 FW 412.6 $C_{23}H_{44}N_2O_4$

Norleucine — N'-t-BOC-L-Nle Free Acid

ICN 101088 MW 231.3 $C_{11}H_{21}NO_4$

Norleucine — Pth-Nle

Sigma P 1877 FW 248.3 $C_{13}H_{16}N_2OS$ ~99% | Useful as standards in protein sequencing by the Edman degradation

Norleucine — TentaGel S PHB-Nle-FMOC

Fluka 86397 0.20 mmol protected AA/g; ~90 μm particle size

Norleucine — TentaGel S Trt-Nle-FMOC

Fluka 86428 0.20 mmol protected AA/g; ~90 μm particle size

Norleucine — Z-DL-Nle

Fluka 96945 MW 265.31 $C_{14}H_{19}NO_4$ ≥98.0% (TLC)

Norleucine — Z-D-Nle

Synonyms: Z-D-2-Aminohexanoic Acid

Bachem C-2285.0001 MW 265.31 $C_{14}H_{19}NO_4$ Store at RT

Fluka 96940 MW 265.31 $C_{14}H_{19}NO_4$ ≥98.0% (HPLC)

Norleucine — Z-D-Nle-(4-ONp)

Fluka 96950 MW 386.40 $C_{20}H_{22}N_2O_6$ ≥98.0%; mp: 63-67°C

Norleucine — Z-D-Nle-ONp

Synonyms: Z-D-2-Aminohexanoic Acid-(pONp)

Bachem C-2295.0001 MW 386.41 $C_{20}H_{22}N_2O_6$ Store at -15°C

Norleucine — Z-Nle

Synonyms: Z-L-2-Aminohexanoic Acid

Bachem C-2280.0001 MW 265.31 $C_{14}H_{19}NO_4$ Store at RT

Norleucine — Z-Nle-ONp

Synonyms: Z-L-2-Aminohexanoic Acid-(pONp)

Bachem C-2290.0001 MW 386.41 $C_{20}H_{22}N_2O_6$ Store at -15°C

Norleucine — Z-Nle-OSu

Synonyms: Z-L-2-Aminohexanoic Acid N-Hydroxysuccinimide Ester

Bachem C-3725.0001 MW 362.38 $C_{18}H_{22}N_2O_6$ Store at -15°C

Norleucine — Z-N-Me-Nle

Synonyms: Z-N-Me-L-2-Aminohexanoic Acid

Bachem C-4020.0001 MW 279.34 $C_{15}H_{21}NO_4$ Store at RT

Norvaline — BOC-D-Nva

Synonyms: BOC-D-Nva; BOC-D-2-Aminovaleric Acid

Bachem A-4180.0001 MW 217.27 $C_{10}H_{19}NO_4$ Store at RT

Neosystem BA05201 MW 217.3

Norvaline — BOC-L-Nva

Synonyms: BOC-L-Nva; BOC-L-2-Aminovaleric Acid

Fluka 15556 MW 217.27 $C_{10}H_{19}NO_4$ ≥98% (TLC)

Neosystem BA05202 MW 217.3

Norvaline — BOC-N-Me-Nva

Synonyms: BOC-N-Me-L-2-Aminovaleric Acid; BOC-L-2-Aminovaleric Acid; BOC-L-Nva

Bachem A-4435.0001 MW 231.29 $C_{11}H_{21}NO_4$ Store at RT

Bachem A-2155.0001 MW 217.27 $C_{10}H_{19}NO_4$ Store at RT

Peptides International BLX-5320-PI MW 217.27 >98% (HPLC); syrup

USBio B2370 MW 217.3 $C_{10}H_{19}NO_4$ ≥99%

Norvaline — Dansyl-DL-Nva Piperidinium Salt

Sigma D 9256 FW. 435.6 $C_{17}H_{22}N_2O_4S \cdot C_5H_{11}N$ Dansyl = 5-dimethylaminonaphthalene-1-sulfonyl

Norvaline — Dansyl-L-Nva Cyclohexylammonium Salt

Sigma D 9381 FW. 449.6 $C_{17}H_{22}N_2O_4S \cdot C_6H_{13}N$

Norvaline — DL-Nva

Synonyms: (±)-2-Aminopentanoic Acid; DL-2-Aminopentanoic Acid

Fluka 74670 MW 117.15 $C_5H_{11}NO_2$ ~99.0% (titration); mp: >300°C

ICN 102497 MW 117.1 $C_5H_{11}NO_2$ Crystalline

Sigma N 7502 FW 117.1 $C_5H_{11}NO_2$ Sigma grade | Also available as part of a kit

Norvaline — DL-Nva Hydroxamate

Sigma N 1253 FW 132.2 $C_5H_{12}N_2O_2$

Norvaline — Dnp-DL-Nva

ICN 102498 MW 283.2 $C_{11}H_{13}N_3O_6$ Crystalline

Norvaline — D-Nva

Synonyms: D-2-Aminovaleric Acid; (R)-2-Aminopentanoic Acid; D-2-Aminopentanoic Acid

Bachem F-1945.0005 MW 117.15 $C_5H_{11}NO_2$ Store at RT

Fluka 74660 MW 117.15 $C_5H_{11}NO_2$ ≥99.0% (titration); mp: >300°C; ≤0.05% residue on ignition

ICN 102494 MW 117.1 $C_5H_{11}NO_2$ Crystalline

Neosystem AA05201 MW 117.1

Sigma N 7377 FW 117.1 $C_5H_{11}NO_2$ Sigma grade

USBio N5304 MW 117.1 $C_5H_{11}NO_2$ ≥99%

Norvaline — D-Nva-OBzl *p*-Tosyl

Synonyms: D-2-Aminovaleric Acid-OBzl

Bachem F-3580.0001 MW 379.48 $C_{12}H_{17}NO_2 \cdot C_7H_8O_3S$ Store at -15°C

Norvaline — FMOC-D-Nva

Synonyms: FMOC-D-2-Aminovaleric Acid

Bachem B-2935.0001 MW 339.39 $C_{20}H_{21}NO_4$ Store at RT

Fluka 47790 MW 339.39 $C_{20}H_{21}NO_4$ ~98% (HPLC)

Neosystem FA05201 MW 339.4

Norvaline — FMOC-L-Nva

Synonyms: FMOC-L-Nva; FMOC-L-2-Aminovaleric Acid

Fluka 47529 MW 339.39 $C_{20}H_{21}NO_4$ ≥98% (HPLC); mp: 151-153°C

Neosystem FA05202 MW 339.4

Norvaline — FMOC-*N*-Me-Nva

Synonyms: FMOC-*N*-Me-L-2-Aminovaleric Acid

Bachem B-3225.0001 MW 353.42 $C_{21}H_{23}NO_4$ Store at RT

Norvaline — FMOC-Nva

Synonyms: FMOC-L-2-Aminovaleric Acid; FMOC-L-Nva

Bachem B-1945.0001 MW 339.39 $C_{20}H_{21}NO_4$ Store at RT

Peptides International FLX-1785-PI MW 339.39 >98% (HPLC); white crystalline powder

USBio F5436 MW 339.4 $C_{20}H_{21}NO_4$ ≥99%

Norvaline — L-Nva

Synonyms: (S)-2-Aminopentanoic Acid; Ornithine Carbamyltransferase Inhibitor; L-2-Aminopentanoic Acid; L-2-Aminovaleric Acid

Fluka 74650 MW 117.15 $C_5H_{11}NO_2$ ≥99.0% (titration); mp: >300°C; ≤0.05% residue on ignition | Competitive inhibitor of ornithine carbamyltransferase; Nakamura, M & Jones, ME, *Meth Enzymol*, 17A: 286, 1970; Marshall, M & Cohen, PP, *JBC*, 247: 1654, 1972

ICN 102495 MW 117.1 $C_5H_{11}NO_2$ Crystalline

Neosystem AA05202 MW 117.1

Sigma N 7627 FW 117.1 $C_5H_{11}NO_2$ Sigma grade

USBio N5305 MW 117.1 $C_5H_{11}NO_2$ ≥99%

Norvaline — *N*-2,4-Dnp-DL-Nva

Sigma D 1005 FW 283.2 $C_{11}H_{13}N_3O_6$

Norvaline — *N*-Carbamyl-DL-Nva

Sigma C 5375 FW 160.2 $C_6H_{12}N_2O_3$

Norvaline — *N*-CBZ-DL-Nva

ICN 100227 MW 251.3 $C_{13}H_{17}NO_4$ Crystalline

Norvaline — *N*-CBZ-L-Nva

ICN 150576 MW 251.3 $C_{13}H_{17}NO_4$

Norvaline — *N*-Me-Nva

Synonyms: *N*-Me-L-2-Aminovaleric Acid

Bachem F-3745.0001 MW 131.18 $C_6H_{13}NO_2$ Store at 2-8°C

Norvaline — *N*-O-Nps-DL-Nva Cyclohexylammonium Salt

ICN 102533 MW 451.6 $C_{11}H_{14}N_2O_4S \cdot C_{12}H_{23}N$ Yellow crystals

Norvaline — *N*-t-BOC-L-Nva Dicyclohexylammonium Salt

Sigma B 3763 FW 398.6 $C_{10}H_{19}NO_4 \cdot C_{12}H_{23}N$

Norvaline — Nva

Synonyms: L-2-Aminovaleric Acid

Bachem F-1940.0005 MW 117.15 $C_5H_{11}NO_2$ Store at RT

Norvaline — Nva Amide Hydrochloride

Synonyms: L-2-Aminovaleric Acid Amide

Bachem F-2445.0250 MW 152.62 $C_5H_{12}N_2O \cdot HCl$ Store at 2-8°C

Norvaline — Nva-OMe Hydrochloride

Synonyms: L-2-Aminovaleric Acid-OMe

Bachem F-3495.0001 MW 167.64 $C_6H_{13}NO_2 \cdot HCl$ Store at 2-8°C

Norvaline — *N*ª-t-BOC-L-Nva Free Acid

ICN 101089 MW 217.3 $C_{10}H_{19}NO_4$

Norvaline — Pth-Nva

Sigma P 2002 FW 234.3 $C_{12}H_{14}N_2OS$ ~99% | Useful as standards in protein sequencing by the Edman degradation

Norvaline — Z-DL-Nva

Fluka 96955 MW 251.28 $C_{13}H_{17}NO_4$ ≥98.0% (TLC)

Norvaline — Z-*N*-Me-Nva

Synonyms: Z-*N*-Me-L-2-Aminovaleric Acid

Bachem C-4015.0001 MW 265.31 $C_{14}H_{19}NO_4$ Store at RT

Norvaline — Z-Nva

Synonyms: Z-L-α-Aminovaleric Acid

Bachem C-2300.0001 MW 251.28 $C_{13}H_{17}NO_4$ Store at RT

Norvaline — Z-Nva-OSu

Synonyms: Z-L-α-Aminovaleric Acid *N*-Hydroxysuccinimide Ester

Bachem C-3740.0001   MW 348.36 $C_{17}H_{20}N_2O_6$ Store at -15°C

Octahydroindolyl Carboxylic Acid — BOC-Oic

Synonyms: N^α-t-BOC-L-Octahydroindole-2-Carboxylic Acid

USBio B2371   MW 269.3 $C_{14}H_{23}NO_4$ ≥98%

Octahydroindolyl Carboxylic Acid — FMOC-Oic

Synonyms: N^α-9-FMOC-L-Octahydroindole-2-Carboxylic Acid

USBio F5437   MW 391.5 $C_{24}H_{25}NO_4$ ≥98%

Ornithine — Ac-Orn

Bachem E-1150.0001   MW 174.2 $C_7H_{14}N_2O_3$ Store at RT

Ornithine — Aloc-DL-Orn-BOC Dicyclohexylamine

Bachem F-4015.0001   MW 497.68 $C_{14}H_{24}N_2O_6 \cdot C_{12}H_{23}N$ Store at -15°C

Ornithine — Aloc-D-Orn-BOC Dicyclohexylamine

Bachem F-4030.0001   MW 497.68 $C_{14}H_{24}N_2O_6 \cdot C_{12}H_{23}N$ Store at -15°C

Ornithine — Aloc-Orn-BOC Dicyclohexylamine

Bachem E-3640.0001   MW 497.68 $C_{14}H_{24}N_2O_6 \cdot C_{12}H_{23}N$ Store at -15°C

Ornithine — BOC-D-Orn

Bachem A-3380.0001   MW 232.28 $C_{10}H_{20}N_2O_4$ Store at RT

Ornithine — BOC-D-Orn-BOC

Bachem A-3385.0001   MW 332.4 $C_{15}H_{28}N_2O_6$ Store at 2-8°C

Ornithine — BOC-D-Orn-FMOC

Synonyms: N^α-BOC-N^δ-FMOC-D-Orn

Bachem A-3375.0001   MW 454.52 $C_{25}H_{30}N_2O_6$ Store at RT

Peptides International BDX-5355-PI   MW 454.52   >98% (HPLC); white to off-white powder

Ornithine — BOC-D-Orn-Z

Bachem A-1495.0001   MW 366.41 $C_{18}H_{26}N_2O_6$ Store at RT

Ornithine — BOC-Orn

Synonyms: N^α-BOC-L-Orn

American Peptide BLORT10   MW 232.3

Bachem A-2865.0001   MW 232.28 $C_{10}H_{20}N_2O_4$ Store at RT

Peptides International BLO-5321-PI   MW 232.28   >98% (TLC); white powder

USBio B2372   MW 232.3 $C_{10}H_{20}N_2O_4$ ≥99%

Ornithine — BOC-Orn-2-Chloro-Z

Bachem A-4020.0001   MW 400.86 $C_{18}H_{25}ClN_2O_6$ Store at -15°C

USBio B2373   MW 400.9 $C_{18}H_{25}N_2O_6Cl$ ≥99%

Ornithine — BOC-Orn-BOC

Synonyms: N^α,N^δ-Di-BOC-L-Orn

Bachem A-2645.0005   MW 332.4 $C_{15}H_{28}N_2O_6$ Store at 2-8°C

Peptides International BLO-5361-PI   MW 332.4   >96% (HPLC); white powder

Ornithine — BOC-Orn-Chloro-Z

Synonyms: N^α-BOC-N^δ-2-Chlorobenzyloxycarbonyl-L-Orn

Peptides International BLO-5326-PI   MW 400.86   >96% (HPLC); white crystalline powder

Ornithine — BOC-Orn-Dnp

Bachem A-4430.0250   MW 398.37 $C_{16}H_{22}N_4O_8$ Store at -15°C

Ornithine — BOC-Orn-FMOC

Synonyms: N^α-BOC-N^δ-FMOC-L-Orn

Bachem A-3325.0001   MW 454.52 $C_{25}H_{30}N_2O_6$ Store at RT

Peptides International BLX-5368-PI   MW 454.52   >98% (HPLC); white to off-white powder

USBio B2374   MW 454.5 $C_{25}H_{30}N_2O_7$ ≥98%

Ornithine — BOC-Orn-OtBu Hydrochloride

Bachem A-3755.0001   MW 324.85 $C_{14}H_{28}N_2O_4 \cdot HCl$ Store at -15°C

Ornithine — BOC-Orn-Pyrazinylcarbonyl

Bachem A-3865.0001   MW 338.36 $C_{15}H_{22}N_4O_5$ Store at -15°C

Ornithine — BOC-Orn-Z

American Peptide BLORT20   MW 366.4

Bachem A-1490.0001   MW 366.41 $C_{18}H_{26}N_2O_6$ Store at RT

USBio B2375   MW 366.4 $C_{18}H_{26}N_2O_6$ ≥99%

Ornithine — BOC-Orn-Z-OSu

Bachem A-3510.0001   MW 463.49 $C_{22}H_{29}N_3O_8$ Store at -15°C

Ornithine — Bz-Orn

Bachem E-1500.0001   MW 236.27 $C_{12}H_{16}N_2O_3$ Store at RT

Ornithine — CBH-Orn

USBio C2099-31   MW 266.3 $C_{13}H_{18}N_2O_4$ ≥99%

Ornithine — CBZ-Orn

USBio C2099-30   MW 266.3 $C_{13}H_{18}N_2O_4$ ≥99%

Ornithine — CBZ-Orn-BOC

USBio C2099-32   MW 266.4 $C_{18}H_{26}N_2O_6$ ≥99%

Ornithine — CBZ-Orn-Z

USBio C2099-34   MW 400.4 $C_{21}H_{24}N_2O_6$ ≥99%

Ornithine — D L-Orn (1-^{14}C) Hydrochloride

Amersham CFA423   Aqueous, 2% EtOH, sterilized; 1.85-2.29 GBq/mmol, 50-62 mCi/mmol; 185 MBq/mL, 50 μCi/mL

Ornithine — DL-Orn (1-^{14}C)

ARC ARC-200   MW 132.1 $H_2NCH_2(CH_2)_2CH(NH_2)COOH$ 50-60 mCi/mmol; 1.85-2.22 GBq/mmol; in 0.01 *N* HCl | Radiochemical; Ernestus, R & Rohn, G et al, *J Neuro-Oncology*, 29: 167, 1996

Ornithine — DL-Orn Hydrochloride

Synonyms: (±)-2,5-Diaminopentanoic Acid Hydrochloride; DL-2,5-Diaminopentanoic Acid

Fluka 75490   MW 168.62 $C_5H_{12}N_2O_2 \cdot HCl$ ~99.0% (titration); mp: 233°C

ICN 102516   MW 168.6 $C_5H_{12}N_2O_2 \cdot HCl$ ~99%; crystalline

Sigma O 2250   FW 168.6 $C_5H_{12}N_2O_2 \cdot$ HCl   ~99%; essentially ammonia-free | Product of arginine degradation by arginase

Ornithine — DL-γ-Orn Dihydrochloride

Synonyms: 4,5-Diaminopentanoic Acid; 4,5-Diaminovaleric Acid

Sigma O 2753   FW 205.1 $C_5H_{12}N_2O_2 \cdot$ 2HCl   ≥95% (TLC)

Ornithine — D-Orn (1-^{14}C)

ARC ARC-625   MW 132.1 $H_2NCH_2(CH_2)_2CH(NH_2)COOH$ 50-60 mCi/mmol; 1.85-2.22 GBq/mmol; in 0.01 *N* HCl | Radiochemical

Ornithine — D-Orn Hydrochloride

Synonyms: (*R*)-2,5-Diaminopentanoic Acid Hydrochloride; D-2,5-Diaminopentanoic Acid

Alexis 101-037   MW 168.6 $C_5H_{12}N_2O_2 \cdot$ HCl

Bachem F-1970.0005   MW 168.62 $C_5H_{12}N_2O_2 \cdot$ HCl   Store at RT

Fluka 75480   MW 168.62 $C_5H_{12}N_2O_2 \cdot$ HCl   ≥99.0% (titration)

ICN 102514   MW 168.6 $C_5H_{12}N_2O_2 \cdot$ HCl   ~99%; crystalline

Rexim   MW 168.6 $C_5H_{13}ClN_2O_2$

Sigma O 5250   FW 168.6 $C_5H_{12}N_2O_2 \cdot$ HCl   ~98%

USBio O8031-10   MW 168.6 $C_5H_{12}N_2O_2 \cdot$ HCl   ≥99%

Ornithine — D-Orn-BOC

Bachem F-2845.0001   MW 232.28 $C_{10}H_{20}N_2O_4$ Store at RT

Ornithine — D-Orn-Z

Bachem F-1405.0001   MW 266.3 $C_{13}H_{18}N_2O_4$ Store at RT

Ornithine — FMOC-D-Orn-Aloc

Bachem B-2895.0001   MW 438.49 $C_{24}H_{26}N_2O_6$ Store at -15°C

Ornithine — FMOC-D-Orn-BOC

Synonyms: N^α-FMOC-N^δ-BOC-D-Orn

Bachem B-1095.0001   MW 454.52 $C_{25}H_{30}N_2O_6$ Store at RT

Peptides International FDX-1818-PI   MW 454.52   >98% (HPLC); white crystalline powder

Peptides International FDX-1818-PI   MW 454.52   >98% (HPLC); white crystalline powder

Ornithine — FMOC-Orn Hydrochloride

Bachem B-2625.0001   MW 390.87 $C_{20}H_{22}N_2O_4 \cdot$ HCl   Store at RT

Ornithine — FMOC-Orn-Aloc

Bachem B-2890.0001   MW 438.49 $C_{24}H_{26}N_2O_6$ Store at -15°C

Ornithine — FMOC-Orn-BOC

Synonyms: N^α-FMOC-N^δ-BOC-L-Orn; N^α-9-FMOC-N^δ-t-BOC-L-Orn

Bachem B-1090.0001   MW 454.52 $C_{25}H_{30}N_2O_6$ Store at RT

Peptides International FLO-1729-PI   MW 454.52   >98% (HPLC); white to off-white powder

USBio F5438   MW 454.5 $C_{25}H_{30}N_2O_6$ ≥99%

Ornithine — FMOC-Orn-BOC-® (200-400 mesh)

Bachem D-1800.0001   Store at 2-8°C

Ornithine — FMOC-Orn-BOC-OPfp

Bachem B-2155.0001   MW 620.57 $C_{31}H_{29}F_5N_2O_6$ Store at -15°C

Ornithine — FMOC-Orn-Dde

Bachem B-3185.0001   MW 518.62 $C_{30}H_{34}N_2O_6$ Store at -15°C

Ornithine — FMOC-Orn-Dnp

Bachem B-3220.0250   MW 520.51 $C_{26}H_{24}N_4O_8$ Store at -15°C

Ornithine — FMOC-Orn-FMOC

Synonyms: N^α,N^δ-Di-FMOC-L-Orn

Bachem B-2260.0001   MW 576.66 $C_{35}H_{32}N_2O_6$ Store at 2-8°C

Peptides International FLO-1788-PI   MW 576.65   >98% (HPLC); white to off-white powder

Ornithine — FMOC-Orn-Mtt

Synonyms: N^α-FMOC-N^δ-4-Methyltrityl-L-Orn

Peptides International FLO-1747-PI   MW 610.76   >98% (HPLC); white crystalline powder

Ornithine — FMOC-Orn-Pyrazinylcarbonyl

Bachem B-2710.0001   MW 460.49 $C_{25}H_{24}N_4O_5$ Store at -15°C

Ornithine — L-N^5-1-Iminoethyl-Orn Dihydrochloride

Synonyms: L-NIO; Nitric Oxide Synthase Inhibitor

Alexis 270-002   MW 246.2 $C_7H_{15}N_3O_2 \cdot$ 2HCl   ≥97%; off-white solid; soluble in MeOH, slightly in H_2O | Scannell, JP et al, *J Antibiot*, 25: 179, 1972; Palacios, M et al, *BBRC*, 165: 802, 1989; Rees, DD et al, *Br J Pharmacol*, 101: 746, 1990; McCall, TB et al, *Br J Pharmacol*, 102: 234, 1991; Mulligan, MS et al, *Br J Pharmacol*, 107: 1159, 1992

Calbiochem 400600   MW 246.1 $C_7H_{15}N_3O_2 \cdot$ 2HCl   ≥98% (TLC); solid; soluble in H_2O | ~5 times more potent as an inhibitor of endothelial cell nitric oxide synthase than other arginine analogs such as L-NAME & L-NMMA; inhibits acetylcholine induced relaxation of rat aorta rings & causes dose-dependent increase in mean arterial blood pressure in the rat; Moore, WM et al, *J Med Chem*, 37: 3886, 1994; Rees, DD et al, *Br J Pharmacol*, 101: 746, 1990

Ornithine — L-N^5-1-Iminoethyl-Orn Hydrochloride

Synonyms: L-NIO; L-Orn ψ-Acetamidine

Sigma I 8768   FW 209.7 $C_7H_{15}N_3O_2 \cdot$ HCl   Potent irreversible inhibitor of nitric oxide synthase; McCall, TB et al, *Br J Pharmacol*, 102: 234, 1991

Ornithine — L-Orn (1-^{14}C)

ARC ARC-199   MW 132.1 $H_2NCH_2(CH_2)_2CH(NH_2)COOH$ 50-60 mCi/mmol; 1.85-2.22 GBq/mmol; in 0.01 *N* HCl | Radiochemical

ICN 10107   MW 132.1 $NH_2CH_2(CH_2)_2CH(NH_2)COOH$ 50-60 mCi/mmol, 1.85-2.22 GBq/mmol, 0.01 *N* HCl; HPLC analyzed for purity

Ornithine — L-Orn (1-^{14}C) Hydrochloride

Amersham CFA491   Aqueous, 2% EtOH, sterilized; 1.85-2.29 GBq/mmol, 50-62 mCi/mmol; 185 MBq/mL, 50 µCi/mL

Ornithine — L-Orn (2,3-^{3}H)

ARC ART-226   MW 132.1 $H_2NCH_2CH_2CH_2CH(NH_2)COOH$ 30-60 Ci/mmol; 1.11-2.22 TBq/mmol; in 0.01 *N* HCl | Radiochemical

Sigma O 0256   FW 132.2 $H_2NCH_2CH_2CH_2CH(NH_2)COOH$ 40-70 Ci/mmol; aqueous solution containing 2% EtOH in serum bottle | Radiochemical

Ornithine — L-Orn (5-^{14}C)

ARC ARC-239 MW 132.1 $H_2NCH_2(CH_2)_2CH(NH_2)COOH$ 50-60 mCi/mmol; 1.85-2.22 GBq/mmol; in 0.01 N HCl | Radiochemical

Ornithine — L-Orn (Carboxy-^{14}C) Hydrochloride

Sigma O 2381 FW 168.6 $H_2N(CH_2)_3CH(NH_2)CO_2H$ 5-20 mCi/mmol; aqueous solution containing 10% EtOH in serum bottle | Radiochemical

Ornithine — L-Orn (U-^{14}C)

ARC ARC-1027 MW 132.1 $H_2NCH_2(CH_2)_2CH(NH_2)COOH$ >250 mCi/mmol; >9.25 GBq/mmol; in 0.01 N HCl | Radiochemical

ICN 16012 MW 132.1 $NH_2CH_2(CH_2)_2CH(NH_2)COOH$ >250 mCi/mmol, >9.25 GBq/mmol, 0.01 N HCl

Ornithine — L-Orn (U-^{14}C) Hydrochloride

Amersham CFB180 Aqueous, 2% EtOH, sterilized; >9.25 GBq/mmol, >250 mCi/mmol; 185 MBq/mL, 50 μCi/mL

Ornithine — L-Orn 2-Oxopentanedioate

Rexim	MW 314.3	$C_{10}H_{22}N_2O_9$	L-Orn, α-Kg (1:1) · 2H$_2$O
Rexim	MW 446.5	$C_{15}H_{34}N_4O_{11}$	L-Orn, α-Kg (2:1) · 2H$_2$O

Ornithine — L-Orn Aspartic Salt

Synonyms: L-2,5-Diaminopentanoic Acid

Sigma O 7125 FW 265.3 $C_9H_{19}N_3O_6$ White powder | Product of arginine degradation by arginase

Ornithine — L-Orn Dihydrochloride

Synonyms: (S)-2,5-Diaminopentanoic Acid; L-2,5-Diaminopentanoic Acid

Fluka 75440 MW 205.09 $C_5H_{12}N_2O_2$ · 2HCl ≥99.0% (titration); ≤0.1% residue on ignition; ≤0.00001% As; ≤0.01% SO$_4$; ≤0.0005% Cd, Co, Cu, Fe, Ni, Pb, Zn; ≤0.01% NH$_4^+$; mp: 197-199°C

ICN 100419 MW 205.1 $C_5H_{12}N_2O_2$ · 2HCl 98-99%; crystalline

Ornithine — L-Orn Hydrochloride

Synonyms: (S)-2,5-Diaminopentanoic Acid; L-2,5-Diaminopentanoic Acid; L-2,5-Diaminovaleric Acid

Alexis 101-036 MW 168.6 $C_5H_{12}N_2O_2$ · HCl

Amersham US19830 ≥98.5%

Fluka 75469 MW 168.62 $C_5H_{12}N_2O_2$ · HCl ≥99.5% (titration); ≤0.1% residue on ignition, loss on drying; ≤0.3% foreign AA; ≤0.00001% As; ≤0.005% K, Na, SO$_4$; ≤0.0005% Al, Ba, Bi, Cd, Co, Cr, Cu, Fe, Li, Mg, Mn, Mo, Ni, Pb, Sr, Zn; ≤0.001% Ca; ≤0.002% NH$_4^+$

Fluka 75470 MW 168.62 $C_5H_{12}N_2O_2$ · HCl ≥99.0% (titration); ≤0.1% residue on ignition; ≤0.3% foreign AA; ≤0.5% loss on drying; ≤0.0005% Cd, Co, Cu, Fe, Ni, Pb, Zn; ≤0.01% NH$_4^+$, SO$_4$

ICN 100421 MW 168.6 $C_5H_{12}N_2O_2$ · HCl ~99%; crystalline; essentially free of citrulline & ammonia, hydrochloride

ICN 194718 MW 168.6 $C_5H_{12}N_2O_2$ · HCl Cell culture reagent;~99%; crystalline; essentially free of citrulline & ammonia, hydrochloride

Neosystem AA08602 MW 168.6

Peptides International ALO-2716 MW 168.62 >99.9% optical purity by RP-HPLC; white crystalline powder

Rexim MW 168.6 $C_5H_{13}ClN_2O_2$ White crystals or crystalline powder

Sigma O 2375 FW 168.6 $C_5H_{12}N_2O_2$ · HCl ~99%; essentially ammonia- & citrulline-free | Product of arginine degradation by arginase

Sigma O 4386 FW 168.6 $C_5H_{12}N_2O_2$ · HCl SigmaUltra; ~99%; residue on ignition: <0.1%; solubility (1 M in H$_2$O, 20°C): complete, colorless; insoluble matter: <0.1%; SO$_4$, NH$_4^+$,: <0.05%; K, Na: <0.005%; Al, Cu, Fe, Mg, Zn, Ca: <0.0005%; P: <0.002%; Pb: <0.001%; essentially ammonia- & citrulline-free | Product of arginine degradation by arginase

Sigma O 6503 FW 168.6 $C_5H_{12}N_2O_2$ · HCl ~99%; essentially ammonia & citrulline free; Crystalline; cell culture tested

USBio O8030 MW 168.6 $C_5H_{12}N_2O_2$ · HCl ≥98%; appearance: white crystalline powder; specific rotation (C=4, 6 N HCl): +22.5° to +24.5°; pH: 5.0-6.0; heavy metals: ≤0.002%; endotoxin: ≤2 EU/mg; cell culture tested: murine myeloma SP2/0-Ag14

Ornithine — N^5-1-Imino-3-Butenyl-L-Orn

Synonyms: Vinyl-L-NIO

Alexis 270-216 MW 199.3 $C_9H_{17}N_3O_2$ ≥95%; soluble in H$_2$O | Very selective & potent nNOS (NOS I) inhibitor; Babu, BR & Griffith, OW, *J Biol Chem*, 273: 8882, 1998

Ornithine — N^5-1-Iminoethyl-L-Orn Dihydrochloride

Synonyms: L-NIO; Nitric Oxide Synthase Inhibitor

ICN 193668 MW 246 $C_7H_{15}N_3O_2$ · 2HCl Powerful inhibitor of endothelial nitric oxide synthase

Ornithine — N-Me-Orn Hydrochloride

Bachem E-3630.0250 MW 182.65 $C_6H_{14}N_2O_2$ · HCl Store at RT

Ornithine — N^α-Ac-L-Orn

Sigma A 3626 FW 174.2 $C_7H_{14}N_2O_3$

Ornithine — N^α-BOC-N^δ-2-Chloro-Z-L-Orn

Fluka 15537 MW 400.86 $C_{18}H_{25}ClN_2O_6$ ≥98% (TLC); mp: 114-118°C

Ornithine — N^α-BOC-N^δ-Alloc-L-Orn

Synonyms: BOC-L-Orn-Alloc

Neosystem BA08603 MW 316.4

Ornithine — N^α-BOC-N^δ-FMOC-L-Orn

Synonyms: N^δ-FMOC-N^α-BOC-L-Orn

Fluka 15539 MW 454.52 $C_{25}H_{30}N_2O_6$ ≥99% (TLC); mp: 150-154°C

Ornithine — N^α-BOC-N^δ-Z-D-Orn

Synonyms: N^δ-Z-N^α-BOC-D-Orn

Fluka 15193 MW 366.41 $C_{18}H_{26}N_2O_6$ ≥98% (TLC)

Ornithine — N^α-BOC-N^δ-Z-L-Orn

Synonyms: N^δ-Z-N^α-BOC-L-Orn

Fluka 15565 MW 366.41 $C_{18}H_{26}N_2O_6$ ≥98% (TLC)

Ornithine — N^α-CBZ-L-Orn

ICN 150577 MW 266.3 $C_{13}H_{18}N_2O_4$

Ornithine — N^α-FMOC-N^δ-(1-(4,4-Dimethyl-2,6-Dioxocyclohexylidene)Et)-L-Orn

Synonyms: N^α-FMOC-N^δ-Dde-L-Orn

Fluka 09766 MW 518.61 $C_{30}H_{34}N_2O_6$ ≥98% (HPLC)

Ornithine — N^α-FMOC-N^δ-Alloc-L-Orn

Synonyms: FMOC-L-Orn-Alloc

Neosystem FA08603 MW 438.5

Ornithine — Nᵅ-FMOC-Nᵟ-BOC-L-Orn

Synonyms: Nᵟ-BOC-Nᵅ-FMOC-L-Orn; FMOC-L-Orn-BOC

Fluka 47560   MW 454.52 $C_{25}H_{30}N_2O_6$ ~97% (HPLC); mp: 111-115°C

Neosystem FA05304   MW 454.5

Ornithine — Nᵅ-FMOC-Nᵟ-BOC-L-Orn-OPfp

Synonyms: Nᵟ-BOC-Nᵅ-FMOC-L-Orn-OPfp

Fluka 00232   MW 620.57 $C_{31}H_{29}F_5N_2O_6$ ~98% (HPLC)

Ornithine — Nᵅ-t-BOC-L-Orn Hydrate

ICN 152037   MW 250.3 $C_{10}H_{20}N_2O_4 \cdot H_2O$ Crystalline

Ornithine — Nᵅ-t-BOC-Nᵟ CBZ-L-Orn

Sigma B 5273   FW 366.4 $C_{18}H_{26}N_2O_6$

Ornithine — Nᵟ-2,4-Dnp-L-Orn Hydrochloride

Sigma D 1130   FW 334.7 $C_{11}H_{14}N_4O_6 \cdot HCl$

Ornithine — Nᵟ-CBZ-D-Orn

Sigma C 7783   FW 266.3 $C_{13}H_{18}N_2O_4$

Ornithine — Nᵟ-CBZ-L-Orn

ICN 150578   MW 266.3 $C_{13}H_{18}N_2O_4$ Crystalline

Sigma C 6779   FW 266.3 $C_{13}H_{18}N_2O_4$

Ornithine — Nᵟ-t-BOC-L-Orn

Sigma B 8126   FW 232.3 $C_{10}H_{20}N_2O_4$

Ornithine — Orn Hydrochloride

Bachem F-1965.0025   MW 168.62 $C_5H_{12}N_2O_2 \cdot HCl$ Store at RT

Ornithine — Orn-AMC Dihydrochloride

Bachem I-1280.0050   MW 362.3 $C_{15}H_{19}N_3O_3 \cdot 2HCl$ Store at -15°C

Ornithine — Orn-BOC

Bachem E-1625.0001   MW 232.28 $C_{10}H_{20}N_2O_4$ Store at RT

Ornithine — Orn-BOC-OMe Hydrochloride

Bachem A-3490.0001   MW 282.77 $C_{11}H_{22}N_2O_4 \cdot HCl$ Store at -15°C

Ornithine — Orn-FMOC

Bachem E-2740.0001   MW 354.41 $C_{20}H_{22}N_2O_4$ Store at RT

Ornithine — Orn-OMe Dihydrochloride

Bachem F-2785.0005   MW 219.12 $C_6H_{14}N_2O_2 \cdot 2HCl$ Store at 2-8°C

Ornithine — Orn-Pyrazinylcarbonyl

Bachem E-3280.0001   MW 238.25 $C_{10}H_{14}N_4O_3$ Store at -15°C

Ornithine — Orn-Z

Bachem E-1720.0005   MW 266.3 $C_{13}H_{18}N_2O_4$ Store at RT

Ornithine — Orn-Z-OBzl Hydrochloride

Bachem E-1725.0005   MW 392.88 $C_{20}H_{24}N_2O_4 \cdot HCl$ Store at RT

Ornithine — Orn-Z-OMe Hydrochloride

Bachem E-3125.0005   MW 316.79 $C_{14}H_{20}N_2O_4 \cdot HCl$ Store at 2-8°C

Ornithine — Orn-Z-OtBu Hydrochloride

Bachem E-3090.0001   MW 358.87 $C_{17}H_{27}N_2O_4Cl$ Store at -15°C

Ornithine — Orn-βNA

Bachem K-1440.0001   MW 257.34 $C_{15}H_{19}N_3O$ Store at -15°C

Ornithine — TentaGel S PHB-Orn-BOC-FMOC

Fluka 86398   0.18 mmol protected AA/g; ~90 μm particle size

Ornithine — Z-D-Orn

Bachem C-3605.0001   MW 266.3 $C_{13}H_{18}N_2O_4$ Store at RT

Ornithine — Z-D-Orn-BOC

Bachem C-3070.0001   MW 366.41 $C_{18}H_{26}N_2O_6$ Store at RT

Ornithine — Z-D-Orn-Z

Bachem C-3610.0001   MW 400.43 $C_{21}H_{24}N_2O_6$ Store at RT

Ornithine — Z-Orn

Bachem C-2305.0001   MW 266.3 $C_{13}H_{18}N_2O_4$ Store at RT

Ornithine — Z-Orn-BOC

Bachem C-1450.0001   MW 366.41 $C_{18}H_{26}N_2O_6$ Store at RT

Ornithine — Z-Orn-FMOC

Bachem C-3325.0001   MW 488.55 $C_{28}H_{28}N_2O_6$ Store at RT

Ornithine — Z-Orn-Z

Bachem C-2965.0005   MW 400.43 $C_{21}H_{24}N_2O_6$ Store at RT

Ornithine — Z-Orn-Z-OSu

Bachem C-2970.0005   MW 497.5 $C_{25}H_{27}N_3O_8$ Store at -15°C

Ornithine — α-Difluoro-Me-DL-Orn

Synonyms: DFMO; Eflornithine

Bachem F-2395.0050   MW 182.17 $C_6H_{12}F_2N_2O_2$ Store at -15°C

Ornithine — α-Hydrazino-Orn Hydrochloride

ICN 157396   MW 183.6 $C_5H_{13}N_3O_2 \cdot HCl$ ~90%

Ornithine — α-Me-Orn Hydrochloride

Synonyms: Ornithine Decarboxylase Inhibitor

ICN 155615   MW 182.6 $C_6H_{14}N_2O_2 \cdot HCl$

Sigma M 3511   FW 182.6 $C_6H_{14}N_2O_2 \cdot HCl$ Induces reversible cytostasis in 9L rat brain tumor cells

Penicillamine — BOC-D-Pen-(pMeOBzl)

USBio B2378   MW 369.2 $C_{18}H_{27}NO_5S$ ≥99%

Penicillamine — BOC-D-Pen-Acm

Synonyms: BOC-β,β-Dimethyl-D-Cys-Acm

Bachem A-2970.0001   MW 320.41 $C_{13}H_{24}N_2O_5S$ Store at 2-8°C

Penicillamine — BOC-D-Pen-Meb Dicyclohexylamine

Synonyms: BOC-*S*-D-Methylbenzyl-D-Penicillamine

Peptides International BDX-5220-PI MW 534.80 >98%
(TLC); white crystals

Penicillamine — BOC-D-Pen-MeBzl Dicyclohexylamine

Synonyms: BOC-β,β-Dimethyl-D-Cys-Methylbenzyl

Bachem A-3665.0001 MW 534.8 $C_{18}H_{27}NO_4S \cdot C_{12}H_{23}N$
Store at 2-8°C

Penicillamine — BOC-D-Pen-Mob

Synonyms: BOC-β,β-Dimethyl-D-Cys-Mob; BOC-*S*-p-Methylbenzyl-D-Penicillamine

Bachem A-3990.0001 MW 369.48 $C_{18}H_{27}NO_5S$ Store at 2-8°C

Peptides International BDX-2615-PI MW 369.48 >98%
(HPLC); white crystalline powder

Penicillamine — BOC-D-Pen-Npys

Synonyms: BOC-β,β-Dimethyl-D-Cys-Npys

Bachem A-3655.0001 MW 403.48 $C_{15}H_{21}N_3O_6S_2$ Store at -15°C

Penicillamine — BOC-D-Pen-p-Methylbenzyl

USBio B2377 MW 353.2 $C_{18}H_{27}NO_4S$ ≥99%

Penicillamine — BOC-D-Pen-p-Methylbenzyl-Dicyclohexylamine

USBio B2376 MW 534.7 $C_{18}H_{27}NO_4S \cdot C_{12}H_{23}N$ ≥99%

Penicillamine — BOC-D-Pen-Trt

Synonyms: BOC-β,β-Dimethyl-D-Cys-Trt

Bachem A-3555.0001 MW 491.65 $C_{29}H_{33}NO_4S$ Store at -15°C

Penicillamine — BOC-Pen-Acm

Synonyms: BOC-β,β-Dimethyl-Cys-Acm

Bachem A-2965.0001 MW 320.41 $C_{13}H_{24}N_2O_5S$ Store at 2-8°C

Penicillamine — BOC-Pen-Meb Dicyclohexylamine

Synonyms: BOC-*S*-p-Methylbenzyl-L-Penicillamine

Peptides International BLX-5221-PI MW 534.80 >98%
(HPLC); white crystalline powder

Penicillamine — BOC-Pen-MeBzl Dicyclohexylamine

Synonyms: BOC-β,β-Dimethyl-Cys-Methylbenzyl

Bachem A-3660.0001 MW 534.8 $C_{18}H_{27}NO_4S \cdot C_{12}H_{23}N$
Store at 2-8°C

Penicillamine — BOC-Pen-Mob

Synonyms: BOC-β,β-Dimethyl-Cys-Mob; BOC-*S*-p-Methylbenzyl-L-Penicillamine

Bachem A-2900.0001 MW 369.48 $C_{18}H_{27}NO_5S$ Store at -15°C

Peptides International BLX-5828-PI MW 369.48 >98%
(HPLC); white to off-white powder

Penicillamine — BOC-Pen-Npys

Synonyms: BOC-β,β-Dimethyl-Cys-Npys

Bachem A-3650.0001 MW 403.48 $C_{15}H_{21}N_3O_6S_2$ Store at -15°C

Penicillamine — BOC-Pen-Trt

Synonyms: BOC-β,β-Dimethyl-Cys-Trt

Bachem A-3550.0001 MW 491.65 $C_{29}H_{33}NO_4S$ Store at -15°C

Penicillamine — BOC-S-4-Methylbenzyl-D-Pen

Fluka 15135 MW 369.48 $C_{18}H_{27}NO_5S$ ≥98% (TLC)

Penicillamine — BOC-S-4-Methylbenzyl-L-Pen

Fluka 15158 MW 369.48 $C_{18}H_{27}NO_5S$ ≥98% (TLC)

Penicillamine — D-(-)-Pen Free Base

Synonyms: D-(-)-2-Amino-3-Mercapto-3-Methylbutanoic Acid; β,β-Dimethyl-D-Cys

Sigma P 4875 FW 149.2 $C_5H_{11}NO_2S$ ~99%

Penicillamine — D-(-)-Pen Hydrochloride

Synonyms: D-(-)-2-Amino-3-Mercapto-3-Methylbutanoic Acid; β,β-Dimethyl-D-Cys

Sigma P 5000 FW 185.7 $C_5H_{11}NO_2S \cdot HCl$ ~99%

Penicillamine — DL-Pen

Synonyms: DL-2-Amino-3-Mercapto-3-Methylbutanoic Acid; DL-β-Mercaptovaline; β,β-Dimethyl-DL-Cys

Sigma P 5125 FW 149.2 $C_5H_{11}NO_2S$

Penicillamine — D-Pen

Synonyms: D-(-)-2-Amino-3-Mercapto-3-Methylbutanoic AAcid

Peptides International AXP-5020-PI MW 149.21 >98%
(TLC); off white crystalline powder

USBio P3150-10 MW 149.2 $C_5H_{11}NO_2S$ ≥99%

Penicillamine — D-Pen Disulfide

Synonyms: 3,3'-Dithio-Bis-2-Amino-3-Methylbutanoic Acid

Sigma P 1271 FW 296.4 $C_{10}H_{20}N_2O_4S_2$

Penicillamine — D-Pen-p-Methylbenzyl

Synonyms: D-Penicillamine; (*S*-p-Methylbenzyl-L-Pen)

Peptides International XDM-5120-PI MW 253.37 >98%
(TLC); white powder

Penicillamine — FMOC-D-Pen-Acm

Synonyms: FMOC-β,β-Dimethyl-D-Cys-Acm; FMOC-*S*-Acm-D-Penicillamine; D-N^α-9-FMOC-*S*-Acm-β,β-Dimethyl-D-Cys

Bachem B-1915.0001 MW 442.54 $C_{23}H_{26}N_2O_5S$ Store at RT

Peptides International FDX-1827-PI MW 442.54 >98%
(HPLC); white crystalline powder

USBio F5439 MW 442.5 $C_{23}H_{26}N_2O_5S$ ≥99%

Penicillamine — FMOC-D-Pen-Acm-® (200-400 mesh)

Synonyms: FMOC-β,β-Dimethyl-D-Cys-Acm-®

Bachem D-1875.0001 Store at 2-8°C

Penicillamine — FMOC-D-Pen-Acm-SASRIN™-® (200-400 mesh)

Synonyms: FMOC-β,β-Dimethyl-D-Cys-Acm-SASRIN™-®

Bachem D-1845.0001 Store at 2-8°C

Penicillamine — FMOC-D-Pen-Bzl

Synonyms: FMOC-β,β-Dimethyl-D-Cys-Bzl

Bachem B-1545.0001 MW 461.58 $C_{27}H_{27}NO_4S$ Store at RT

Penicillamine — FMOC-D-Pen-Meb

Synonyms: FMOC-*S*-*p*-Methylbenzyl-D-Penicillamine

Peptides International FDX-1861-PI   MW 475.64   >98% (HPLC); white crystalline powder

Penicillamine — FMOC-D-Pen-Trt

Synonyms: FMOC-β,β-Dimethyl-D-Cys-Trt

Bachem B-2320.0001   MW 613.78 $C_{39}H_{35}NO_4S$ Store at -15°C

Penicillamine — FMOC-D-Pen-Trt-® (200-400 mesh)

Synonyms: FMOC-β,β-Dimethyl-D-Cys-Trt-®

Bachem D-1870.0001   Store at 2-8°C

Penicillamine — FMOC-D-Pen-Trt-SASRIN™-® (200-400 mesh)

Synonyms: FMOC-β,β-Dimethyl-D-Cys-Trt-SASRIN™-®

Bachem D-1855.0001   Store at 2-8°C

Penicillamine — FMOC-L-Pen-Acm

Synonyms: N^α-9-FMOC-*S*-Acm-β,β-Dimethyl-L-Cys; L- N^α-9-FMOC-*S*-Acm-β,β-Dimethyl-D-Cys

USBio F5440   MW 442.5 $C_{23}H_{26}N_2O_5S$ ≥99%

Penicillamine — FMOC-Pen-Acm

Synonyms: FMOC-β,β-Dimethyl-Cys-Acm; FMOC-*S*-Acm-L-Penicillamine

Bachem B-1885.0001   MW 442.54 $C_{23}H_{26}N_2O_5S$ Store at RT

Peptides International FLX-1711-PI   MW 442.54   >98% (HPLC); white crystalline powder

Penicillamine — FMOC-Pen-Acm-® (200-400 mesh)

Synonyms: FMOC-β,β-Dimethyl-Cys-Acm-®

Bachem D-1890.0001   Store at 2-8°C

Penicillamine — FMOC-Pen-Acm-SASRIN™-® (200-400 mesh)

Synonyms: FMOC-β,β-Dimethyl-Cys-Acm-SASRIN™-®

Bachem D-1885.0001   Store at 2-8°C

Penicillamine — FMOC-Pen-Meb

Synonyms: FMOC-*S*-*p*-Methylbenzyl-L-Penicillamine

Peptides International FLX-1760-PI   MW 475.64   >98% (TLC); white powder

Penicillamine — FMOC-Pen-Trt

Synonyms: FMOC-β,β-Dimethyl-Cys-Trt

Bachem B-2315.0001   MW 613.78 $C_{39}H_{35}NO_4S$ Store at -15°C

Penicillamine — FMOC-Pen-Trt-® (200-400 mesh)

Synonyms: FMOC-β,β-Dimethyl-Cys-Trt-®

Bachem D-1880.0001   Store at 2-8°C

Penicillamine — FMOC-Pen-Trt-SASRIN™-® (200-400 mesh)

Synonyms: FMOC-β,β-Dimethyl-Cys-Trt-SASRIN™-®

Bachem D-1850.0001   Store at 2-8°C

Penicillamine — L-(+)-Pen

Synonyms: L-(+)-2-Amino-3-Mercapto-3-Methylbutanoic Acid; β,β-Dimethyl-L-Cys

Sigma P 1771   FW 149.2 $C_5H_{11}NO_2S$

Penicillamine — L-Pen

Synonyms: L-(+)-2-Amino-3-Mercapto-3-Methylbutanoic Acid

Peptides International AXL-5021-PI   MW 149.21   >98% (HPLC); white powder

USBio P3150   MW 149.2 $C_5H_{11}NO_2S$ ≥99%

Penicillamine — L-Pen Acetone Adduct Hydrochloride

Synonyms: *N*,*S*-Isopropylidene-D-Pen

Sigma P 2051   FW 225.7 $C_8H_{15}NO_2S \cdot HCl$ ≥95%

Penicillamine — L-Pen-*p*-Methylbenzyl

Synonyms: L-Penicillamine (*S*-*p*-Methylbenzyl-L-Pen)

Peptides International XLM-5121-PI   MW 253.37   >98% (TLC); white powder

Penicillamine — *N*-t-BOC-*S*-Me-L-Pen Dicyclohexylammonium Salt

Sigma B 1772   FW 444.7 $C_{11}H_{21}NO_4S \cdot C_{12}H_{23}N$ ≥99%

Penicillamine — *N*-t-BOC-*S*-*p*-Methylbenzyl-L-Pen

Sigma B 7525   FW 369.5 $C_{18}H_{27}NO_5S$

Penicillamine — Pen

Synonyms: β,β-Dimethyl-Cys

Bachem F-2515.0001   MW 149.21 $C_5H_{11}NO_2S$ Store at RT

Penicillamine — Pen-Trt

Synonyms: β,β-Dimethyl-Cys-Trt

Bachem F-3065.0001   MW 391.53 $C_{24}H_{25}NO_2S$ Store at -15°C

Phenylalanine — (±)-3-CarboxyPhe

ICN 159739   MW 209.2 $C_{10}H_{11}NO_4$

Phenylalanine — (4-(bis)₂-Chloroethyl)Amino-L-Phe

ICN 155345   MW 305.2 $C_{13}H_{18}Cl_2N_2O_2$ 95%; light yellow powder

Phenylalanine — 1,4-Bis-((N^α-2-Naphthylsulfonyl)-3-Amidino-(D,L)-Phe)-Piperazide Dihydrochloride

Synonym Pefabloc®Tryp; Pefa-2887s:

Pentapharm 390-10   Synthetic   MW 917.9 $C_{44}H_{44}O_6N_8S_2 \cdot 2HCl$ Low MW inhibitor for mast cell tryptase

Phenylalanine — 2,4,5-Trihydroxy-DL-Phe

Synonyms: 2,5-Dihydroxy-DL-Tyr; 6-Hydroxy-DOPA

Fluka 91960   MW 213.19 $C_9H_{11}NO_5$ mp: 255-260°C

Phenylalanine — 2-Chloro-L-Phe

Fluka 25926   MW 199.64 $C_9H_{10}ClNO_2$ ≥98% (titration); ≤2% H_2O

Phenylalanine — 2-Fluoro-DL-Phe

Fluka 47300   MW 183.18 $C_9H_{10}FNO_2$ ≥98% (titration); ≤0.05% residue on ignition; mp: 250°C

Phenylalanine — 2-Fluoro-D-Phe

Fluka 47298   MW 183.18 $C_9H_{10}FNO_2$ ≥99% (HPLC); D-:L- ≥99.5:0.5

Phenylalanine — 2-Fluoro-L-Phe

Fluka 47296   MW 183.18 $C_9H_{10}FNO_2$ ≥99% (HPLC); L-:D- ≥99.5:0.5

USBio F5101-01   MW 183.2 $C_9H_{10}FNO_2$ ≥98%

Phenylalanine — 3,4-D-Dichloro-Phe
USBio D7850-10   MW 234.2 $C_9H_9N_2O_2Cl_2$ ≥99%

Phenylalanine — 3,4-Dichloro-D-Phe
Bachem F-3400.0001   MW 234.08 $C_9H_9Cl_2NO_2$ Store at RT

Phenylalanine — 3,4-Dichloro-Phe
Bachem F-3395.0001   MW 234.08 $C_9H_9Cl_2NO_2$ Store at RT
USBio D7850   MW 234.2 $C_9H_9N_2O_2Cl_2$ ≥99%

Phenylalanine — 3-Carboxy-DL-Phe
Sigma C 3336   FW 209.2 $C_{10}H_{11}NO_4$ ≥98%

Phenylalanine — 3-Fluoro-DL-Phe
Fluka 47310   MW 183.18 $C_9H_{10}FNO_2$ ≥97% (titration); ≤0.05% residue on ignition; mp: 240°C

Phenylalanine — 3-Fluoro-D-Phe
Fluka 47308   MW 183.18 $C_9H_{10}FNO_2$ ≥99% (HPLC)

Phenylalanine — 3-Fluoro-L-Phe
Fluka 47306   MW 183.18 $C_9H_{10}FNO_2$ ≥99% (HPLC)
USBio F5101-02   MW 183.2 $C_9H_{10}FNO_2$ ≥98%

Phenylalanine — 4-(1-Pyrenyl)butyryl-Phe
Bachem E-3560.0250   MW 435.52 $C_{29}H_{25}NO_3$ Store at 2-8°C

Phenylalanine — 4-(Bis-(2-Chloroethyl)Amino)-L-Phe
Synonyms: L-PAM; Melphalan; Phenylalanine Mustard
Fluka 63648   MW 305.2 $C_{13}H_{18}Cl_2N_2O_2$ ≥90.0% (TLC); mp: 180°C; ≤10% H_2O | Kohn, KW et al, *Nucl Acids Res*, 15: 10531, 1987; Feyns, LV, *Anal Profiles of Drug Subst*, 13: 265, 1984
Sigma M 2011   FW 305.2 $C_{13}H_{18}Cl_2N_2O_2$ ≥95% | Antineoplastic agent; forms DNA intrastrand crosslinks by bifunctional alkylation in 5'-GGC sequences; Nawata, S et al, *J Thorac Cardiovasc Surg*, 112: 1542, 1996; Orlandi, L et al, *Br J Cancer*, 74: 1924, 1996; Bauer, GB & Povirk, LF, *Nucleic Acids Res*, 25: 1211, 1997

Phenylalanine — 4-Amino-(3,5-Diiodo)-L-Phe
Sigma A 4038   FW 432.0 $C_9H_{10}I_2N_2O_2$

Phenylalanine — 4-Amino-(3,5-Diiodo)-Phe
Bachem F-1200.0001   MW 432 $C_9H_{10}I_2N_2O_2$ Store at 2-8°C

Phenylalanine — 4-Bromo-L-Phe
Fluka 18055   MW 244.09 $C_9H_{10}BrNO_2$ ≥98% (titration); mp: 265°C

Phenylalanine — 4-Chloro-L-Phe
Fluka 25920   MW 199.64 $C_9H_{10}ClNO_2$ ~98% (titration); mp: 260°C

Phenylalanine — 4-Cyano-D-Phe
Bachem F-3980.0001   MW 190.2 $C_{10}H_{10}N_2O_2$ Store at -15°C

Phenylalanine — 4-Cyano-Phe
Bachem F-3610.0001   MW 190.2 $C_{10}H_{10}N_2O_2$ Store at -15°C

Phenylalanine — 4-Fluoro-DL-Phe
Fluka 47320   MW 183.18 $C_9H_{10}FNO_2$ ≥99% (titration); ≤0.1% residue on ignition; mp: 270°C

Phenylalanine — 4-Fluoro-D-Phe
Fluka 47318   MW 183.18 $C_9H_{10}FNO_2$ ≥99% (HPLC); D-:L- ≥99:1; mp: 245°C

Phenylalanine — 4-Fluoro-L-Phe
Fluka 47290   MW 183.18 $C_9H_{10}FNO_2$ ≥99% (HPLC); L-:D- ≥99:1; mp: 255°C

Phenylalanine — 4-Methoxyphenylazoformyl-Phe Potassium Salt
Synonyms: Carboxypeptidase A Substrate; AAFP
Peptides International SAA-3197-v   MW 327.34 $C_{17}H_{17}N_3O_4$ >96% (HPLC); lyophilized amorphous powder | Mock, WL et al, *Anal Biochem*, 239:218, 1996

Phenylalanine — 4-Phenyl-D-Phe
Synonyms: H-α-4-Biphenylyl-D-Ala; H-D-Bip
Bachem F-3650.0001   MW 241.29 $C_{15}H_{15}NO_2$ Store at RT

Phenylalanine — 4-Phenyl-Phe
Synonyms: H-α-4-Biphenylyl-Ala; H-Bip
Bachem F-3645.0001   MW 241.29 $C_{15}H_{15}NO_2$ Store at RT

Phenylalanine — Abu-Phe
Bachem G-1455.0250   MW 250.3 $C_{13}H_{18}N_2O_3$ Store at -15°C

Phenylalanine — Ac-DL-Phe
Bachem F-1055.0005   MW 207.23 $C_{11}H_{13}NO_3$ Store at RT

Phenylalanine — Ac-DL-Phe β-Naphthyl Ester
Bachem M-1055.0001   MW 333.39 $C_{21}H_{19}NO_3$ Store at RT

Phenylalanine — Ac-DL-Phe-ONp
Bachem F-1060.0001   MW 328.32 $C_{17}H_{16}N_2O_5$ Store at -15°C

Phenylalanine — Ac-DL-Phe-pNA
Bachem L-1055.0001   MW 327.34 $C_{17}H_{17}N_3O_4$ Store at -15°C

Phenylalanine — Ac-D-Phe
Bachem F-1050.0001   MW 207.23 $C_{11}H_{13}NO_3$ Store at RT

Phenylalanine — Ac-D-Phe-OMe
Bachem F-2975.0001   MW 221.26 $C_{12}H_{15}NO_3$ Store at 2-8°C

Phenylalanine — Ac-p-Amino-Phe-OMe
Bachem F-1015.0001   MW 236.27 $C_{12}H_{16}N_2O_3$ Store at -15°C

Phenylalanine — Ac-p-Bromo-DL-Phe
Bachem F-2275.0001   MW 286.13 $C_{11}H_{12}BrNO_3$ Store at RT

Phenylalanine — Ac-p-Bz-D-Phe
Synonyms: Ac-D-Bpa
Bachem F-3265.0001   MW 311.34 $C_{18}H_{17}NO_4$ Store at RT

Phenylalanine — Ac-Phe
Bachem E-1155.0005   MW 207.23 $C_{11}H_{13}NO_3$ Store at RT

Phenylalanine — Ac-Phe Amide
Bachem E-1160.0001   MW 206.24 $C_{11}H_{14}N_2O_2$ Store at RT

Phenylalanine — Ac-Phe-NHMe
Bachem E-1170.0250   MW 220.27 $C_{12}H_{16}N_2O_2$ Store at RT

Phenylalanine — Ac-Phe-OEt
Bachem E-1165.0005   MW 235.28 $C_{13}H_{17}NO_3$ Store at RT
Peptides International SFE-3006   MW 235.28 $C_{13}H_{17}NO_3$
>99% (HPLC); amorphous powder

Phenylalanine — Ac-Phe-OMe
Bachem E-1175.0005   MW 221.26 $C_{12}H_{15}NO_3$ Store at RT

Phenylalanine — Ac-Phe-pNA
Bachem L-1050.0001   MW 327.34 $C_{17}H_{17}N_3O_4$ Store at -15°C

Phenylalanine — Ac-p-Iodo-D-Phe
Bachem F-3015.0001   MW 333.13 $C_{11}H_{12}INO_3$ Store at 2-8°C

Phenylalanine — Aloc-D-Phe Dicyclohexylamine
Bachem F-4010.0001   MW 430.59 $C_{13}H_{15}NO_4 \cdot C_{12}H_{23}N$
Store at -15°C

Phenylalanine — BOC-(^{15}N)-Phe
Bachem A-3800.0100   MW 266.31 $C_{14}H_{19}{}^{15}NO_4$ Store at -15°C

Phenylalanine — BOC-(2,4-Dichloro)-D-Phe
Fluka 14991   MW 334.2 $C_{14}H_{17}Cl_2NO_4$ ≥98% (TLC); mp: 132-136°C

Phenylalanine — BOC-(2,4-Dichloro)-L-Phe
Fluka 14992   MW 334.2 $C_{14}H_{17}Cl_2NO_4$ ≥98% (TLC); mp: 132-136°C

Phenylalanine — BOC-(3,4-Dichloro)-D-Phe
Synonyms: BOC-(3,4-Dichloro)-D-Phe
Bachem A-4045.0001   MW 334.2 $C_{14}H_{17}Cl_2NO_4$ Store at -15°C
Fluka 15041   MW 334.2 $C_{14}H_{17}Cl_2NO_4$ ≥98% (TLC)
Neosystem BA05701   MW 334.2
USBio B2298   MW 334.2 $C_{14}H_{17}Cl_2NO_4$ ≥98%

Phenylalanine — BOC-(3,4-Dichloro)-L-Phe
Synonyms: BOC-(3,4-Dichloro)-L-Phe
Fluka 15042   MW 334.2 $C_{14}H_{17}Cl_2NO_4$ ≥98% (TLC)
Neosystem BA05702   MW 334.2

Phenylalanine — BOC-(3,4-Dichloro)-Phe
USBio B2299   MW 334.2 $C_{14}H_{17}Cl_2NO_4$ ≥98%

Phenylalanine — BOC-(3,4-Difluoro)-D-Phe
Fluka 14993   MW 301.29 $C_{14}H_{17}F_2NO_4$ ≥98% (TLC)

Phenylalanine — BOC-(3,4-Difluoro)-L-Phe
Fluka 14994   MW 301.29 $C_{14}H_{17}F_2NO_4$ ≥98% (TLC)

Phenylalanine — BOC-2-Chloro-D-Phe
Fluka 15018   MW 299.75 $C_{14}H_{18}ClNO_4$ ≥98% (TLC); mp: 94-98°C

Phenylalanine — BOC-2-Chloro-L-Phe
Fluka 15021   MW 299.75 $C_{14}H_{18}ClNO_4$ ≥98% (TLC); mp: 94-98°C

Phenylalanine — BOC-2-Cyano-D-Phe
Fluka 14983   MW 290.32 $C_{15}H_{18}N_2O_4$ ≥98% (TLC)

Phenylalanine — BOC-2-Cyano-L-Phe
Fluka 14984   MW 290.32 $C_{15}H_{18}N_2O_4$ ≥98% (HPLC)

Phenylalanine — BOC-2-Fluoro-D-Phe
Fluka 15023   MW 283.3 $C_{14}H_{18}FNO_4$ ≥98% (TLC); mp: 94-98°C

Phenylalanine — BOC-2-Fluoro-L-Phe
Fluka 15024   MW 283.3 $C_{14}H_{18}FNO_4$ ≥98% (TLC); mp: 92-96°C

Phenylalanine — BOC-2-Me-D-Phe
Fluka 14997   MW 279.34 $C_{15}H_{21}NO_4$ ≥98% (TLC)

Phenylalanine — BOC-2-Me-L-Phe
Fluka 14998   MW 279.34 $C_{15}H_{21}NO_4$ ≥98% (TLC)

Phenylalanine — BOC-2-Trifluorormethyl-D-Phe
Fluka 15009   MW 333.31 $C_{15}H_{18}F_3NO_4$ ≥98% (TLC)

Phenylalanine — BOC-2-Trifluorormethyl-L-Phe
Fluka 15011   MW 333.31 $C_{15}H_{18}F_3NO_4$ ≥98% (TLC)

Phenylalanine — BOC-3-(1-Naphthyl)-L-Phe
Fluka 15347   MW 315.37 $C_{18}H_{21}NO_4$ ≥97% (HPLC)

Phenylalanine — BOC-3-(2-Naphthyl)-D-Phe
Fluka 15478   MW 315.37 $C_{18}H_{21}NO_4$ ≥97% (HPLC); mp: 90°C

Phenylalanine — BOC-3-Cyano-D-Phe
Fluka 14985   MW 290.32 $C_{15}H_{18}N_2O_4$ ≥98% (TLC)

Phenylalanine — BOC-3-Cyano-L-Phe
Fluka 14986   MW 290.32 $C_{15}H_{18}N_2O_4$ ≥98% (TLC)

Phenylalanine — BOC-3-Fluoro-D-Phe
Fluka 14995   MW 283.3 $C_{14}H_{18}FNO_4$ ≥98% (TLC); mp: 75-78°C

Phenylalanine — BOC-3-Fluoro-L-Phe
Fluka 14996   MW 283.3 $C_{14}H_{18}FNO_4$ ≥98% (TLC); mp: 75-78°C

Phenylalanine — BOC-3-Me-D-Phe
Fluka 14999   MW 279.34 $C_{15}H_{21}NO_4$ ≥98% (TLC)

Phenylalanine — BOC-3-Me-L-Phe
Fluka 15002   MW 279.34 $C_{15}H_{21}NO_4$ ≥98% (TLC)

Phenylalanine — BOC-3-Trifluorormethyl-D-Phe
Fluka 15012   MW 333.31 $C_{15}H_{18}F_3NO_4$ ≥98% (TLC); mp: 135-138°C

Phenylalanine — BOC-3-Trifluorormethyl-L-Phe
Fluka 15013   MW 333.31 $C_{15}H_{18}F_3NO_4$ ≥98% (TLC); mp: 135-138°C

Phenylalanine — BOC-4-Amino-(3,5-Diiodo)-Phe
Bachem A-1160.0001   MW 532.12 $C_{14}H_{18}I_2N_2O_4$ Store at -15°C

Phenylalanine — BOC-4-Bz-D-Phe
Fluka 09776   MW 369.42 $C_{21}H_{23}NO_5$ ~98% (HPLC)

Phenylalanine — BOC-4-Bz-L-Phe
Fluka 09775   MW 369.42 $C_{21}H_{23}NO_5$ ~98% (HPLC)

Phenylalanine — BOC-4-Chloro-D-Phe
Fluka 15471   MW 299.75 $C_{14}H_{18}ClNO_4$ ~98% (TLC); mp: 110°C

Phenylalanine — BOC-4-Chloro-L-Phe
Synonyms: BOC-*p*-Chloro-L-Phe
Fluka 15472   MW 299.75 $C_{14}H_{18}ClNO_4$ ~98% (TLC); mp: 110°C
Neosystem BA05602   MW 299.7

Phenylalanine — BOC-4-Cyano-D-Phe
Bachem A-4575.0001   MW 290.32 $C_{15}H_{18}N_2O_4$ Store at -15°C
Fluka 14987   MW 290.32 $C_{15}H_{18}N_2O_4$ ≥97% (TLC)

Phenylalanine — BOC-4-Cyano-L-Phe
Fluka 14988   MW 290.32 $C_{15}H_{18}N_2O_4$ ≥98% (TLC)

Phenylalanine — BOC-4-Cyano-Phe
Bachem A-4375.0001   MW 290.32 $C_{15}H_{18}N_2O_4$ Store at -15°C

Phenylalanine — BOC-4-Fluoro-D-Phe
Fluka 15351   MW 283.3 $C_{14}H_{18}FNO_4$ ≥99% (TLC); mp: 83-86°C

Phenylalanine — BOC-4-Fluoro-L-Phe
Fluka 15352   MW 283.3 $C_{14}H_{18}FNO_4$ ≥99% (TLC); mp: 83-86°C

Phenylalanine — BOC-4-Iodo-D-Phe
Fluka 15044   MW 391.21 $C_{14}H_{18}INO_4$ ≥98% (TLC)

Phenylalanine — BOC-4-Iodo-L-Phe
Fluka 15346   MW 391.21 $C_{14}H_{18}INO_4$ ≥99% (TLC); mp: 150°C

Phenylalanine — BOC-4-Me-D-Phe
Fluka 15003   MW 279.34 $C_{15}H_{21}NO_4$ ≥98% (TLC); mp: 84-88°C

Phenylalanine — BOC-4-Me-L-Phe
Fluka 15006   MW 279.34 $C_{15}H_{21}NO_4$ ≥98% (TLC); mp: 84-88°C

Phenylalanine — BOC-4-Nitro-D-Phe
Synonyms: BOC-*p*-Nitro-D-Phe
Fluka 15174   MW 310.31 $C_{14}H_{18}N_2O_6$ ≥98% (TLC)
Neosystem BA06001   MW 310.3

Phenylalanine — BOC-4-Nitro-L-Phe
Synonyms: BOC-*p*-Nitro-L-Phe
Fluka 15348   MW 310.31 $C_{14}H_{18}N_2O_6$ ~99% (titration); mp: 120-123°C
Neosystem BA06002   MW 310.3

Phenylalanine — BOC-4-Phenyl-D-Phe
Synonyms: BOC-β-4-Biphenylyl-D-Ala; BOC-D-Bip
Bachem A-4390.0001   MW 341.41 $C_{20}H_{23}NO_4$ Store at RT

Phenylalanine — BOC-4-Phenyl-Phe
Synonyms: BOC-β-4-Biphenylyl-Ala; BOC-Bip
Bachem A-4385.0001   MW 341.41 $C_{20}H_{23}NO_4$ Store at RT

Phenylalanine — BOC-4-Trifluorormethyl-D-Phe
Fluka 15016   MW 333.31 $C_{15}H_{18}F_3NO_4$ ≥98% (TLC)

Phenylalanine — BOC-4-Trifluorormethyl-L-Phe
Fluka 15017   MW 333.31 $C_{15}H_{18}F_3NO_4$ ≥98% (TLC); mp: 131-135°C

Phenylalanine — BOC-D-(3,4-Dichloro)-Phe
American Peptide BDPHE45   MW 336.2

Phenylalanine — BOC-D-2-Chloro-Phe
American Peptide BDPHE15   MW 300.8

Phenylalanine — BOC-D-2-Fluoro-Phe
American Peptide BDPHE25   MW 284.3

Phenylalanine — BOC-D-4-Chloro-Phe
American Peptide BDPHE20   MW 300.8

Phenylalanine — BOC-D-4-Fluoro-Phe
American Peptide BDPHE30   MW 284.3

Phenylalanine — BOC-D-4-Nitro-Phe
American Peptide BDPHE40   MW 311.3

Phenylalanine — BOC-D-Dip
Synonyms: BOC-D-3,3-Diphenylalanine
Peptides International BDF-5530-PI   MW 341.41   >98% (TLC); off-white powder

Phenylalanine — BOC-D-Homophenylalanine
Synonyms: BOC-D-Hph
Neosystem BA04501   MW 279.3
Peptides International BDF-5373-PI   MW 279.34   >98% (HPLC); white crystalline powder

Phenylalanine — BOC-Dip
Synonyms: BOC-L-3,3-Diphenylalanine
Peptides International BDF-5528-PI   MW 341.41   >98% (HPLC); white crystalline powder

Phenylalanine — BOC-DL-Phe
Bachem A-2170.0005   MW 265.31 $C_{14}H_{19}NO_4$ Store at RT

Phenylalanine — BOC-D-Pentafluoro-Phe

Synonyms: BOC-Pentafluoro-D-Phe

Peptides International BDF-5385-PI   MW 355.26   >98% (HPLC); white to off-white powder

Phenylalanine — BOC-D-Phe

American Peptide BDPHE10   MW 265.3

Bachem A-2165.0001   MW 265.31 $C_{14}H_{19}NO_4$ Store at RT

Fluka 15484   MW 265.31 $C_{14}H_{19}NO_4$ ≥99% (TLC); mp: 77-80°C

Neosystem BA01401   MW 265.3

Peptides International BDF-2604   MW 265.31   >98% (HPLC); white crystalline powder

USBio B2379   MW 265.3 $C_{14}H_{19}NO_4$ ≥99%

Phenylalanine — BOC-D-Phe Hydroxysuccinimide Ester

Fluka 15178   MW 362.38 $C_{18}H_{22}N_2O_6$ ~98%; mp: 149-153°C

Phenylalanine — BOC-D-Phe-(4-ONp)

Fluka 15181   MW 386.41 $C_{20}H_{22}N_2O_6$ ≥98% (HPLC)

Phenylalanine — BOC-D-Phe-Nitro

Synonyms: BOC-p-Nitro-D-Phe

Peptides International BDF-5222-PI   MW 310.31   >98% (HPLC); white crystalline powder

Phenylalanine — BOC-D-Phe-O-CH₂-♦-CH₂-COOH

Neosystem LP01401   MW 413.5

Phenylalanine — BOC-D-Phe-ONp

Bachem A-2190.0001   MW 386.41 $C_{20}H_{22}N_2O_6$ Store at -15°C

Phenylalanine — BOC-D-Phe-OSu

Bachem A-2180.0001   MW 362.38 $C_{18}H_{22}N_2O_6$ Store at -15°C

Phenylalanine — BOC-D-Phe-PAM Resin

Neosystem RP01401

Phenylalanine — BOC-D-Phe-PAM-® (200-400 mesh)

Bachem D-1580.0001   Store at 2-8°C

Phenylalanine — BOC-D-Phe-p-Chloro

USBio B2381   MW 299.8 $C_{14}H_{18}ClNO_4$ ≥99%

Phenylalanine — BOC-D-Phe-p-Fluoro

USBio B2383   MW 283.2 $C_{14}H_{18}FNO_4$ ≥99%

Phenylalanine — BOC-D-p-Iodo Phe

American Peptide BDPHE50   MW 392.2

Phenylalanine — BOC-L-(3,4-Dichloro)-Phe

American Peptide BLPHE45   MW 336.2

Phenylalanine — BOC-L-2-Chloro-Phe

American Peptide BLPHE15   MW 300.8

Phenylalanine — BOC-L-2-Fluoro-Phe

American Peptide BLPHE25   MW 284.3

Phenylalanine — BOC-L-4-Chloro-Phe

American Peptide BLPHE20   MW 300.8

Phenylalanine — BOC-L-4-Fluoro-Phe

American Peptide BLPHE30   MW 284.3

Phenylalanine — BOC-L-4-Nitro-Phe

American Peptide BLPHE40   MW 311.3

Phenylalanine — BOC-L-Phe

Fluka 15480   MW 265.31 $C_{14}H_{19}NO_4$ ≥99% (TLC); mp: 85-87°C

Neosystem BA01402   MW 265.3

Phenylalanine — BOC-L-Phe Hydroxysuccinimide Ester

Fluka 15481   MW 362.38 $C_{18}H_{22}N_2O_6$ ≥98% (TLC); mp: 150-152°C

Phenylalanine — BOC-L-Phe N-Carboxy Anhydride

Fluka 15479   MW 291.3 $C_{15}H_{17}NO_5$ mp: 102-105°C | Produced by Propeptide under an exclusive license from Bioresearch

Phenylalanine — BOC-L-Phe-(4-ONp)

Fluka 15482   MW 386.41 $C_{20}H_{22}N_2O_6$ mp: 127-128°C

Phenylalanine — BOC-L-Phe-MBHA Resin

Neosystem RN01402

Phenylalanine — BOC-L-Phenylalaninol

Bachem A-3970.0001   MW 251.33 $C_{14}H_{21}NO_3$ Store at -15°C

Phenylalanine — BOC-L-Phe-O-CH₂-♦-CH₂-COOH

Neosystem LP01402   MW 413.5

Phenylalanine — BOC-L-Phe-OMe

Fluka 15179   MW 279.34 $C_{15}H_{21}NO_4$ ≥98% (TLC); mp: 38-41°C

Phenylalanine — BOC-L-Phe-PAM Resin

Neosystem RP01402

Phenylalanine — BOC-L-p-Iodo-Phe

American Peptide BLPHE50   MW 392.2

Phenylalanine — BOC-Me-Phe

USBio B2356   MW 279.3 $C_{15}H_{21}NO_4$ ≥98%

Phenylalanine — BOC-Me-Phe-Dicyclohexylamine

USBio B2357   MW 460.7 $C_{15}H_{21}NO_4 \cdot C_{12}H_{23}N$ ≥98%

Phenylalanine — BOC-N-Me-D-Phe

Bachem A-3900.0001   MW 279.34 $C_{15}H_{21}NO_4$ Store at RT

Phenylalanine — BOC-N-Me-L-Phe

Neosystem BA01403   MW 279.3

Phenylalanine — BOC-N-Me-L-Phe Dicyclohexylamine

Fluka 15447   MW 460.66 $C_{15}H_{21}NO_4 \cdot C_{12}H_{23}N$ ≥99%; mp: 174-177°C

Phenylalanine — BOC-*N*-Me-*p*-Chloro-D-Phe

| Bachem A-2880.0001 | MW 313.78 | C$_{15}$H$_{20}$ClNO$_4$ | Store at RT |

Phenylalanine — BOC-*N*-Me-Phe

Synonyms: BOC-*N*-Me-L-Phe

| Bachem A-2075.0001 | MW 279.34 | C$_{15}$H$_{21}$NO$_4$ | Store at RT |
| Peptides International BXF-5322-PI | | MW 279.34 | >98% |

(HPLC); white crystalline powder

Phenylalanine — BOC-*N*-Me-*p*-Nitro-Phe Dicyclohexylamine

| Bachem A-2070.0001 | MW 505.66 | C$_{15}$H$_{20}$N$_2$O$_6$ · C$_{12}$H$_{23}$N |

Store at RT

Phenylalanine — BOC-*p*-Amino-D-Phe

| Bachem A-2980.0001 | MW 280.32 | C$_{14}$H$_{20}$N$_2$O$_4$ | Store at -15°C |

Phenylalanine — BOC-*p*-Amino-D-Phe-FMOC

| Bachem A-4065.0001 | MW 502.57 | C$_{29}$H$_{30}$N$_2$O$_6$ | Store at -15°C |

Phenylalanine — BOC-*p*-Amino-Phe

| Bachem A-1185.0001 | MW 280.32 | C$_{14}$H$_{20}$N$_2$O$_4$ | Store at -15°C |

Phenylalanine — BOC-*p*-Amino-Phe-FMOC

| Bachem A-3975.0001 | MW 502.57 | C$_{29}$H$_{30}$N$_2$O$_6$ | Store at -15°C |

Phenylalanine — BOC-*p*-Aminophenylalanine

| USBio B2260 | MW 454.5 | C$_{25}$H$_{30}$N$_2$O$_7$ | ≥98% |

Phenylalanine — BOC-*p*-Amino-Phe-Z

| Bachem A-1455.0001 | MW 414.46 | C$_{22}$H$_{26}$N$_2$O$_6$ | Store at 2-8°C |

Phenylalanine — BOC-*p*-Azido-D-Phe

| Bachem A-4200.0001 | MW 306.32 | C$_{14}$H$_{18}$N$_4$O$_4$ | Store at -15°C |

Phenylalanine — BOC-*p*-Azido-Phe

| Bachem A-3570.0001 | MW 306.32 | C$_{14}$H$_{18}$N$_4$O$_4$ | Store at -15°C |

Phenylalanine — BOC-*p*-Bromo-D-Phe

| Bachem A-4205.0001 | MW 344.21 | C$_{14}$H$_{18}$BrNO$_4$ | Store at RT |

Phenylalanine — BOC-*p*-Bromo-Phe

| Bachem A-3695.0001 | MW 344.21 | C$_{14}$H$_{18}$BrNO$_4$ | Store at RT |

Phenylalanine — BOC-*p*-Bz-D-Phe

Synonyms: BOC-D-Bpa

| Bachem A-3560.0001 | MW 369.42 | C$_{21}$H$_{23}$NO$_5$ | Store at RT |

Phenylalanine — BOC-*p*-Bz-Phe

Synonyms: BOC-Bpa

| Bachem A-3295.0001 | MW 369.42 | C$_{21}$H$_{23}$NO$_5$ | Store at RT |

Phenylalanine — BOC-*p*-Carboxy-Phe-OtBu Dicyclohexylamine

| Bachem A-4325.0001 | MW 546.75 | C$_{19}$H$_{27}$NO$_6$ · C$_{12}$H$_{23}$N |

Store at -15°C

Phenylalanine — BOC-*p*-Chloro-D-Phe

Synonyms: BOC-*p*-Chloro-D-Phe

| Bachem A-2655.0001 | MW 299.75 | C$_{14}$H$_{18}$ClNO$_4$ | Store at RT |
| Peptides International BDF-5316-PI | | MW 299.76 | >98% |

(HPLC); white to off-white powder

Phenylalanine — BOC-*p*-Chloro-Phe

Synonyms: BOC-*p*-Chloro-L-Phe

| Bachem A-1525.0001 | MW 299.75 | C$_{14}$H$_{18}$ClNO$_4$ | Store at RT |
| Peptides International BLF-5315-PI | | MW 299.76 | >98% |

(HPLC); white crystalline powder

Phenylalanine — BOC-Pentafluoro-D-Phe

| Bachem A-3960.0001 | MW 355.26 | C$_{14}$H$_{14}$F$_5$NO$_4$ | Store at -15°C |

Phenylalanine — BOC-Pentafluoro-Phe

Synonyms: BOC-Pentafluoro-L-Phe

| Bachem A-3915.0001 | MW 355.26 | C$_{14}$H$_{14}$F$_5$NO$_4$ | Store at -15°C |
| Peptides International BLF-5383-PI | | MW 355.26 | >98% |

(HPLC); off-white crystalline powder

Phenylalanine — BOC-*p*-Fluoro-DL-Phe

| Bachem A-1605.0001 | MW 283.3 | C$_{14}$H$_{18}$FNO$_4$ | Store at RT |

Phenylalanine — BOC-*p*-Fluoro-D-Phe

| Bachem A-2835.0001 | MW 283.3 | C$_{14}$H$_{18}$FNO$_4$ | Store at RT |
| Peptides International BDF-5335-PI | | MW 283.31 | >98% |

(HPLC); off-white crystalline powder

Phenylalanine — BOC-*p*-Fluoro-Phe

Synonyms: BOC-*p*-Fluoro-L-Phe

| Bachem A-3065.0001 | MW 283.3 | C$_{14}$H$_{18}$FNO$_4$ | Store at RT |
| Peptides International BLF-5329-PI | | MW 283.31 | >98% |

(HPLC); white crystalline powder

Phenylalanine — BOC-Phe

Synonyms: BOC-L-Phe

American Peptide BLPHE10		MW 265.3	
Bachem A-2160.0005	MW 265.31	C$_{14}$H$_{19}$NO$_4$	Store at RT
Peptides International BLF-2068		MW 265.31	>98%

(HPLC); white crystalline powder

| USBio B2380 | MW 265.3 | C$_{14}$H$_{19}$NO$_4$ | ≥99% |

Phenylalanine — BOC-Phe-(NH$_2$)

Synonyms: BOC-*p*-Amino-L-Phe

| Peptides International BLF-5337-PI | | MW 280.33 | >98% |

(TLC); white powder

Phenylalanine — BOC-Phe-(NH-Z)

Synonyms: N^α-BOC-L-Phe (*p*-Amino-Carbobenzoxy)

| Peptides International BLX-5230-PI | | MW 414.46 | >98% |

(TLC); white powder

Phenylalanine — BOC-Phe-® (200-400 mesh)

| Bachem D-1440.0005 | Store at RT |

Phenylalanine — BOC-Phe-Nitro

Synonyms: BOC-*p*-Nitro-L-Phe

| Peptides International BLF-5333-PI | | MW 309.30 | >98% |

(HPLC); white crystalline powder

Phenylalanine — BOC-Phe-OBzl
Bachem A-2555.0001   MW 355.43 $C_{21}H_{25}NO_4$ Store at RT

Phenylalanine — BOC-Phe-OEt
Synonyms: BOC-L-Phe-OEt

Peptides International BEF-5841-PI   MW 293.36   >98% (HPLC); white crystalline powder

Phenylalanine — BOC-Phe-ol
Synonyms: BOC-L-Phenylalaninol

Senn Chem 44036   MW 251.3

Phenylalanine — BOC-Phe-OMe
Bachem A-3995.0001   MW 279.34 $C_{15}H_{21}NO_4$ Store at 2-8°C

Phenylalanine — BOC-Phe-ONp
Bachem A-2185.0001   MW 386.41 $C_{20}H_{22}N_2O_6$ Store at -15°C

Phenylalanine — BOC-Phe-O-Resin
American Peptide RMRB235   1% DVB cross-linked: 100-200 mesh | t-BOC protected AA resin; preattached resins are useful for synthesis of peptides with C-terminal carboxyl group by t-BOC chemistry

Phenylalanine — BOC-Phe-OSu
Bachem A-2175.0001   MW 362.38 $C_{18}H_{22}N_2O_6$ Store at -15°C

Phenylalanine — BOC-Phe-PAM-® (200-400 mesh)
Bachem D-1520.0001   Store at 2-8°C

Phenylalanine — BOC-Phe-PAM-Resin
American Peptide RPAM235   1% DVB cross-linked: 100-200 mesh | t-BOC protected AA resin; preattached resins are useful for synthesis of peptides with C-terminal carboxyl group by t-BOC chemistry

Phenylalanine — BOC-Phe-p-Chloro
USBio B2382   MW 299.8 $C_{14}H_{18}ClNO_4$ ≥99%

Phenylalanine — BOC-Phe-p-Fluoro
USBio B2384   MW 283.2 $C_{14}H_{18}FNO_4$ ≥99%

Phenylalanine — BOC-Phe-psi(CH2NH)-Phe
Neosystem BB01401   MW 398.49

Phenylalanine — BOC-p-Iodo-DL-Phe
Bachem A-1805.0001   MW 391.21 $C_{14}H_{18}INO_4$ Store at RT

Phenylalanine — BOC-p-Iodo-D-Phe
Bachem A-3640.0001   MW 391.21 $C_{14}H_{18}INO_4$ Store at RT

Phenylalanine — BOC-p-Iodo-Phe
Bachem A-1800.0001   MW 391.21 $C_{14}H_{18}INO_4$ Store at RT

Phenylalanine — BOC-p-Me-D-Phe
Bachem A-4500.0001   MW 279.34 $C_{15}H_{21}NO_4$ Store at RT

Phenylalanine — BOC-p-Me-Phe
Bachem A-4495.0001   MW 279.34 $C_{15}H_{21}NO_4$ Store at RT

Phenylalanine — BOC-p-Nitro-D-Phe
Bachem A-2130.0001   MW 310.31 $C_{14}H_{18}N_2O_6$ Store at RT

Phenylalanine — BOC-p-Nitro-Phe
Bachem A-2125.0001   MW 310.31 $C_{14}H_{18}N_2O_6$ Store at RT

USBio B2368   MW 310.3 $C_{14}H_{18}N_2O_6$ ≥98%

Phenylalanine — BOC-p-tBu-D-Phe
Bachem A-4485.0001   MW 321.42 $C_{18}H_{27}NO_4$ Store at -15°C

Phenylalanine — BOC-p-tBu-Phe
Bachem A-4490.0001   MW 321.42 $C_{18}H_{27}NO_4$ Store at -15°C

Phenylalanine — BOC-Thiono-Phe-1-(6-Nitro)-Benzotriazolide
Synonyms: (S)-2-(BOC-Amino)-3-Phenyl-Propanethioic-O-Acid-1-(6-Nitro)-Benzotriazolide

Bachem A-4355.0001   MW 427.48 $C_{20}H_{21}N_5O_4S$ Store at -15°C

Phenylalanine — Bsmoc-Phe
Synonyms: Bsmoc-L-Phe

Peptides International BLF-5633-PI   MW 387.41   >98% (HPLC); white to off-white powder | LA Carpino, et al, *JACS*, 119:9915, 1997

Phenylalanine — Bz-DL-Phe
Bachem F-1290.0005   MW 269.3 $C_{16}H_{15}NO_3$ Store at RT

ICN 100922   MW 269.3 $C_{16}H_{15}NO_3$ Crystalline

Phenylalanine — Bz-DL-Phe β-Naphthyl Ester
Bachem M-1150.0001   MW 395.46 $C_{26}H_{21}NO_3$ Store at RT

Phenylalanine — Bz-DL-Phe-βNA
Bachem K-1140.0001   MW 394.47 $C_{26}H_{22}N_2O_2$ Store at RT

Phenylalanine — Bz-Phe
Bachem E-1505.0005   MW 269.3 $C_{16}H_{15}NO_3$ Store at RT

Phenylalanine — Bz-Phe Amide
Bachem E-1510.0001   MW 268.32 $C_{16}H_{16}N_2O_2$ Store at RT

Phenylalanine — CBZ-D-Hph
Synonyms: Z-(-)-α-Amino-4-Phenylbutyric Acid

USBio C2098-90   MW 313.4 $C_{18}H_{19}NO_4$ ≥99%

Phenylalanine — CBZ-D-Phe
USBio C2099-36   MW 299.3 $C_{17}H_{17}NO_4$ ≥99%

Phenylalanine — CBZ-L-Hph
Synonyms: Z-(+)-α-Amino-4-Phenylbutyric Acid; N^α-Benzyloxycarbonyl-(+)-α-Amino-4-Phenylbutyric Acid

USBio C2098-91   MW 313.4 $C_{18}H_{19}NO_4$ ≥99%

Phenylalanine — CBZ-Me-Phe-Dicyclohexylamine
USBio C2099-25   MW 495.5 $C_{17}H_{17}NO_4 \cdot C_{12}H_{23}N$ ≥99%

Phenylalanine — CBZ-Phe
USBio C2099-35   MW 299.3 $C_{17}H_{17}NO_4$ ≥99%

Phenylalanine — CBZ-Phe Amide
USBio C2099-37   MW 298.3 $C_{17}H_{18}N_2O_3$ ≥99%

Phenylalanine — CBZ-Phe-CMK
USBio C2099-38   MW 331.8 $C_{18}H_{18}ClNO_3$ ≥99%

Phenylalanine — CBZ-Phe-ONp
USBio C2099-39   MW 420.4 $C_{23}H_{20}N_2O_6$ ≥99%

Phenylalanine — Chloroacetyl-DL-Phe
Bachem F-1415.0005   MW 241.67 $C_{11}H_{12}ClNO_3$ Store at RT

Phenylalanine — Chloroacetyl-D-Phe
Bachem F-1410.0001   MW 241.67 $C_{11}H_{12}ClNO_3$ Store at RT

Phenylalanine — Cinnamoyl-*trans*-Phe
Bachem E-1750.0250   MW 295.34 $C_{18}H_{17}NO_3$ Store at -15°C

Phenylalanine — Cyclo(-D-α-Hydroxyisovaleryl-L-*N*-Methylphenylalanyl)₃ Hydrate
Synonyms: Beauvericin

ICN 153064   MW 784 $C_{45}H_{57}N_3O_9$ Crystalline | Hamill, R etal, *Tetrahed Lett*, 49:4255, 1969; Suzuki, A etal, *Tetrahed Lett*, 25:2167, 1977

Phenylalanine — Cyclo(-D-α-Hydroxyisovaleryl-L-*N*-Methylphenylalanyl)₃
Synonyms: Beauvericin

Sigma B 7510   FW 784.0 $C_{45}H_{57}N_3O_9$ ≥97% (HPLC) | Bioactive peptide; Suzuki, A et al, *Tetrahedron Lett*, 2167, 1997; Hamill, RL et al, *Tetrahedron Lett*, 4255, 1969

Phenylalanine — Cyclo(-D-α-Hydroxyisovaleryl-*N*-Me-Phe)₃
Synonyms: Beauvericin

Bachem H-2135.0005   Store at -15°C

Phenylalanine — D-3,4-Dihydroxyphenylalanine
Synonyms: D-DOPA; D-3-Hyty

Sigma D 9378 $C_9H_{11}NO_4$ Immediate precursor of dopamine; inactive isomer

Phenylalanine — D-3,4-Dihydroxyphenylalanine (Ala-1-¹⁴C)
Synonyms: D-DOPA; DL-DOPA; Dihydroxyphenylalanine

ARC ARC-1276   MW 197.2 $3,4\text{-}(OH)_2C_6H_3CH_2CH(NH_2)COOH$ 50-60 mCi/mmol; 1.85-2.22 GBq/mmol; in 0.2 N HOAc: EtOH (9:1) | Radiochemical

ARC ARC-252   MW 197.2 $3,4\text{-}(OH)_2C_6H_3CH_2CH(NH_2)COOH$ 50-60 mCi/mmol; 1.85-2.22 GBq/mmol; in 0.2 N HOAc: EtOH (9:1) under N₂ | Radiochemical

Phenylalanine — Dansyl-DL-Phe Cyclohexylammonium Salt
Sigma D 9506   FW 497.7 $C_{21}H_{22}N_2O_4S \cdot C_6H_{13}N$

Phenylalanine — Dansyl-D-Phe Free Acid
Sigma D 2514   FW 398.5 $C_{21}H_{22}N_2O_4S$

Phenylalanine — Dansyl-L-Phe
ICN 100116   MW 398.5 $C_{21}H_{22}N_2O_4S$

Phenylalanine — Dansyl-L-Phe Free Acid
Sigma D 1750   FW 398.5 $C_{21}H_{22}N_2O_4S$

Phenylalanine — D-Hph
Synonyms: (-)-2-Amino-4-Phenylbutyric Acid

USBio H7010-10   MW 179.2 $C_{10}H_{13}NO_2$ ≥99%

Phenylalanine — Dinitropyridyl-DL-Phe
ICN 102621   MW 332.3 $C_{14}H_{12}N_4O_6$

Phenylalanine — DL-3,4-Dihydroxyphenylalanine
Synonyms: DL-DOPA; DL-3-Hyty

Sigma D 9503 $C_9H_{11}NO_4$ Also available as part of a kit | Immediate precursor of dopamine; product of tyrosine hydroxylase

Phenylalanine — DL-*p*-ChloroPhe
Synonyms: PCP; PCPA; Tryptophan Hydroxylase Inhibitor

Sigma C 6506   FW 199.6 $C_9H_{10}ClNO_2$ Crosses blood-brain barrier

Phenylalanine — DL-*p*-Chloro-Phe
Synonyms: DL-Phe-*p*-Chloro

USBio C4500-15   MW 199.6 $C_9H_{10}ClNO_2$ ≥99%

Phenylalanine — DL-*p*-Chloro-Phe-OEt Hydrochloride
Synonyms: Tryptophan Hydroxylase Inhibitor; Phenylalanine 4-Hydroxylase Inhibitor

Sigma 15,678-7   FW 264.2 $C_{11}H_{14}ClNO_2 \cdot HCl$ Aldrich Brand; formerly Sigma C 9144; crosses blood-brain barrier

Phenylalanine — DL-*p*-Chloro-Phe-OMe Hydrochloride
Synonyms: Tryptophan Hydroxylase Inhibitor

Sigma C 3635   FW 250.1 $C_{10}H_{12}ClNO_2 \cdot HCl$ Crosses blood-brain barrier

Phenylalanine — DL-Phe
Synonyms: (±)-2-Amino-3-Phenylpropionic Acid; DL-2-Amino-3-Phenylpropanoic Acid

Bachem F-1980.0025   MW 165.19 $C_9H_{11}NO_2$ Store at RT

Fluka 78040   MW 165.19 $C_9H_{11}NO_2$ ≥99.0% (titration); ≤0.05% residue on ignition; ≤0.1% loss on drying; mp: 271-273°C

ICN 102618   MW 165.2 $C_9H_{11}NO_2$ 99%; crystalline

ICN 194722   MW 165.2 $C_9H_{11}NO_2$ Cell culture reagent; 99%

Sigma P 1876   FW 165.2 $C_9H_{11}NO_2$ ≥98% (TLC) | Also available as part of a kit

Sigma P 4905   FW 165.2 $C_9H_{11}NO_2$ Crystalline | Cell culture tested; insect cell culture tested

USBio P4060-15   MW 165.2 $C_9H_{11}NO_2$ ≥99%

Phenylalanine — DL-Phe (1-¹⁴C)
ARC ARC-1123   MW 165.2 $C_6H_5CH_2CH(NH_2)COOH$ 50-60 mCi/mmol; 1.85-2.22 GBq/mmol; in 0.01 N HCl | Radiochemical

Phenylalanine — DL-Phe Amide Hydrochloride
Bachem F-1990.0001   MW 200.67 $C_9H_{12}N_2O \cdot HCl$ Store at -15°C

Phenylalanine — DL-Phe Hydroxamate
Sigma P 3881   FW 180.2 $C_9H_{12}N_2O_2$

Phenylalanine — DL-Phenylalaninamide Hydrochloride
Sigma P 2251   FW 200.7 $C_9H_{12}N_2O \cdot HCl$

Phenylalanine — DL-Phe-OBzl p-Tosyl

Bachem F-2340.0005   MW 427.52 $C_{16}H_{17}NO_2 \cdot C_7H_8O_3S$
Store at RT

Phenylalanine — DL-Phe-OMe Hydrochloride

Bachem F-2005.0005   MW 215.68 $C_{10}H_{13}NO_2 \cdot HCl$ Store at -15°C

Sigma P 0416   FW 215.7 $C_{10}H_{13}NO_2 \cdot HCl$

Phenylalanine — Dnp-L-Phe

ICN 102617   MW 331.3 $C_{15}H_{13}N_3O_6$ Crystalline

Phenylalanine — D-p-Chloro-Phe

Synonyms: D-PCP; D-PCPA

Sigma C 9419   FW 199.6 $C_9H_{10}ClNO_2$ Inactive isomer of PCP

USBio C4500-10   MW 199.6 $C_9H_{10}ClNO_2$ ≥99%

Phenylalanine — D-p-Fluoro-Phe

USBio F5200-10   MW 183.2 $C_9H_{10}FNO_2$ ≥99%

Phenylalanine — D-Phe

Synonyms: (R)-2-Amino-3-Phenylpropionic Acid; D-α-Amino-β-Phenylpropanoic Acid; D-2-Amino-3-Phenylpropanoic Acid

Bachem F-1975.0005   MW 165.19 $C_9H_{11}NO_2$ Store at RT

Fluka 78030   MW 165.19 $C_9H_{11}NO_2$ ≥99.0% (titration); ≤0.05% residue on ignition; mp: 275°C

ICN 102616   MW 165.2 $C_9H_{11}NO_2$ 99%; crystalline

Neosystem AA01401   MW 165.2

Peptides International ADF-2808   MW 165.19   >99.9% optical purity by RP-HPLC; white crystalline powder

Rexim   MW 165.2 $C_9H_{11}NO_2$

Sigma P 1751   FW 165.2 $C_9H_{11}NO_2$ ≥98% (TLC) | Also available as part of a kit

USBio P4060-10   MW 165.2 $C_9H_{11}NO_2$ ≥99%

Phenylalanine — D-Phe (1-^{14}C)

ARC ARC-1116   MW 165.2 $C_6H_5CH_2CH(NH_2)COOH$ 50-60 mCi/mmol; 1.85-2.22 GBq/mmol; in 0.01 N HCl | Radiochemical

Phenylalanine — D-Phe Amide

Bachem F-1985.0001   MW 164.21 $C_9H_{12}N_2O$ Store at -15°C

Phenylalanine — D-Phe-OBzl p-Tosyl

Bachem F-1995.0005   MW 427.52 $C_{16}H_{17}NO_2 \cdot C_7H_8O_3S$
Store at RT

Phenylalanine — D-Phe-ol

Senn Chem 44103   MW 151.2

Phenylalanine — D-Phe-OMe Hydrochloride

Bachem F-2000.0005   MW 215.68 $C_{10}H_{13}NO_2 \cdot HCl$ Store at -15°C

Sigma P 8040   FW 215.7 $C_{10}H_{13}NO_2 \cdot HCl$

Phenylalanine — D-Phe-pNA

Bachem L-1360.0001   MW 285.3 $C_{15}H_{15}N_3O_3$ Store at -15°C

Phenylalanine — FA-Phe

Bachem M-1390.0250   MW 285.3 $C_{16}H_{15}NO_4$ Store at -15°C

Phenylalanine — FA-Phe-OMe

Bachem M-1395.0250   MW 299.33 $C_{17}H_{17}NO_4$ Store at -15°C

Phenylalanine — Farnesyl-Phe-OMe

Bachem F-3510.0250   MW 338.57 $C_{25}H_{37}NO_2$ Store at -15°C

Phenylalanine — FMOC-(^{15}N)-Phe

Bachem B-2660.0100   MW 388.43 $C_{24}H_{21}^{15}NO_4$ Store at -15°C

Phenylalanine — FMOC-(2,4-Dichloro)-D-Phe

Fluka 47808   MW 456.33 $C_{24}H_{19}Cl_2NO_4$ ≥98% (HPLC)

Phenylalanine — FMOC-(2,4-Dichloro)-L-Phe

Fluka 47809   MW 456.33 $C_{24}H_{19}Cl_2NO_4$ ≥98% (HPLC)

Phenylalanine — FMOC-(3,3-Diphenyl)-Ala

Synonyms: FMOC-Dip

Fluka 09895   MW 463.53 $C_{30}H_{25}NO_4$ ≥98% (HPLC)

Phenylalanine — FMOC-(3,4-Dichloro)-D-Phe

Synonyms: FMOC-(3,4-Dichloro)-D-Phe

Fluka 47425   MW 456.33 $C_{24}H_{19}Cl_2NO_4$ ~98% (HPLC)

Neosystem FA05701   MW 456.3

Phenylalanine — FMOC-(3,4-Dichloro)-L-Phe

Synonyms: FMOC-(3,4-Dichloro)-L-Phe

Fluka 47426   MW 456.33 $C_{24}H_{19}Cl_2NO_4$ ~98% (HPLC)

Neosystem FA05702   MW 456.3

Phenylalanine — FMOC-(3,4-Difluoro)-D-Phe

Fluka 47812   MW 423.42 $C_{24}H_{19}F_2NO_4$ ≥98% (HPLC)

Phenylalanine — FMOC-(3,4-Difluoro)-L-Phe

Fluka 47813   MW 423.42 $C_{24}H_{19}F_2NO_4$ ≥98% (HPLC)

Phenylalanine — FMOC-2-Fluoro-D-Phe

Fluka 47767   MW 405.43 $C_{24}H_{20}FNO_4$ ≥98% (TLC)

Phenylalanine — FMOC-2-Fluoro-L-Phe

Fluka 47769   MW 405.43 $C_{24}H_{20}FNO_4$ ~98% (HPLC)

Phenylalanine — FMOC-2-Me-D-Phe

Fluka 47816   MW 401.46 $C_{25}H_{23}NO_4$ ≥98% (HPLC)

Phenylalanine — FMOC-2-Me-L-Phe

Fluka 47817   MW 401.46 $C_{25}H_{23}NO_4$ ≥98% (HPLC)

Phenylalanine — FMOC-2-Trifluorormethyl-D-Phe

Fluka 47824   MW 455.43 $C_{25}H_{20}F_3NO_4$ ≥98% (HPLC)

Phenylalanine — FMOC-2-Trifluorormethyl-L-Phe

Fluka 47826   MW 455.43 $C_{25}H_{20}F_3NO_4$ ≥98% (HPLC)

Phenylalanine — FMOC-3-Cyano-D-Phe

Fluka 47804   MW 412.44 $C_{25}H_{20}N_2O_4$ ≥98% (HPLC); ~15% 2-propanol

Phenylalanine — FMOC-3-Cyano-L-Phe

Fluka 47805   MW 412.44 $C_{25}H_{20}N_2O_4$ ≥98% (HPLC); ~15% 2-propanol

Phenylalanine — FMOC-3-Fluoro-D-Phe

Fluka 47814   MW 405.43 $C_{24}H_{20}FNO_4$ ~98% (HPLC)

Phenylalanine — FMOC-3-Fluoro-L-Phe

Fluka 47815   MW 405.43 $C_{24}H_{20}FNO_4$ ≥98% (HPLC)

Phenylalanine — FMOC-3-Me-D-Phe

Fluka 47818   MW 401.46 $C_{25}H_{23}NO_4$ ≥98% (HPLC); ~10% IpOH

Phenylalanine — FMOC-3-Me-L-Phe

Fluka 47819   MW 401.46 $C_{25}H_{23}NO_4$ ≥98% (HPLC); ~10% IpOH

Phenylalanine — FMOC-3-Trifluorormethyl-D-Phe

Fluka 47832   MW 455.43 $C_{25}H_{20}F_3NO_4$ ≥98% (HPLC); may contain ~10% solvent

Phenylalanine — FMOC-3-Trifluorormethyl-L-Phe

Fluka 47833   MW 455.43 $C_{25}H_{20}F_3NO_4$ ≥98% (HPLC); may contain ~10% solvent

Phenylalanine — FMOC-4-Bromo-D-Phe

Synonyms: FMOC-*p*-Bromo-D-Phe

Neosystem FA01406   MW 466.34

Phenylalanine — FMOC-4-Bromo-L-Phe

Synonyms: FMOC-*p*-Bromo-L-Phe

Neosystem FA01407   MW 466.34

Phenylalanine — FMOC-4-Bz-D-Phe

Fluka 09773   MW 491.54 $C_{31}H_{25}NO_5$ ≥98% (HPLC)

Phenylalanine — FMOC-4-Bz-L-Phe

Fluka 09774   MW 491.54 $C_{31}H_{25}NO_5$ ≥98% (HPLC)

Phenylalanine — FMOC-4-Chloro-D-Phe

Synonyms: FMOC-*p*-Chloro-D-Phe

Fluka 47420   MW 421.88 $C_{24}H_{20}ClNO_4$ ≥98% (HPLC)

Neosystem FA05601   MW 421.90

Phenylalanine — FMOC-4-Chloro-L-Phe

Synonyms: FMOC-*p*-Chloro-L-Phe

Fluka 47424   MW 421.88 $C_{24}H_{20}ClNO_4$ ≥98% (TLC)

Neosystem FA05602   MW 421.9

Phenylalanine — FMOC-4-Cyano-D-Phe

Bachem B-3480.0001   MW 412.45 $C_{25}H_{20}N_2O_4$ Store at -15°C

Fluka 47806   MW 412.44 $C_{25}H_{20}N_2O_4$ ≥98% (HPLC)

Phenylalanine — FMOC-4-Cyano-L-Phe

Fluka 47807   MW 412.44 $C_{25}H_{20}N_2O_4$ ≥98% (HPLC)

Phenylalanine — FMOC-4-Cyano-Phe

Bachem B-3125.0001   MW 412.45 $C_{25}H_{20}N_2O_4$ Store at -15°C

Phenylalanine — FMOC-4-Fluoro-D-Phe

Synonyms: FMOC-*p*-Fluoro-D-Phe

Fluka 47427   MW 405.43 $C_{24}H_{20}FNO_4$ ≥98% (TLC)

Neosystem FA05801   MW 405.44

Phenylalanine — FMOC-4-Fluoro-L-Phe

Synonyms: FMOC-*p*-Fluoro-L-Phe

Fluka 47428   MW 405.43 $C_{24}H_{20}FNO_4$ ≥98% (HPLC)

Neosystem FA05802   MW 405.44

Phenylalanine — FMOC-4-Iodo-L-Phe

Fluka 47431   MW 513.33 $C_{24}H_{20}INO_4$ ≥98% (TLC)

Phenylalanine — FMOC-4-Me-D-Phe

Synonyms: FMOC-*p*-Me-D-Phe

Fluka 47821   MW 401.46 $C_{25}H_{23}NO_4$ ≥98% (HPLC)

Neosystem FA01410   MW 401.46

Phenylalanine — FMOC-4-Me-L-Phe

Synonyms: FMOC-*p*-Me-L-Phe

Fluka 47823   MW 401.46 $C_{25}H_{23}NO_4$ ≥98% (HPLC)

Neosystem FA01411   MW 401.46

Phenylalanine — FMOC-4-Nitro-D-Phe

Synonyms: FMOC-*p*-Nitro-D-Phe

Fluka 47434   MW 432.43 $C_{24}H_{20}N_2O_6$ ≥98% (TLC)

Neosystem FA06001   MW 432.4

Phenylalanine — FMOC-4-Nitro-L-Phe

Synonyms: FMOC-*p*-Nitro-L-Phe

Fluka 47472   MW 432.43 $C_{24}H_{20}N_2O_6$ ~98% (HPLC)

Neosystem FA06002   MW 432.4

Phenylalanine — FMOC-4-Phenyl-D-Phe

Synonyms: FMOC-β-4-Biphenylyl-D-Ala; FMOC-D-Bip

Bachem B-3160.0001   MW 463.43 $C_{30}H_{25}NO_4$ Store at 2-8°C

Phenylalanine — FMOC-4-Phenyl-Phe

Synonyms: FMOC-Bip; FMOC-β-4-Biphenylyl-Ala

Bachem B-3155.0001   MW 463.53 $C_{30}H_{25}NO_4$ Store at 2-8°C

Phenylalanine — FMOC-4-Phosphonomethyl-L-Phe

Fluka 09768   MW 481.44 $C_{25}H_{24}NO_7P$ ≥98% (TLC)

Phenylalanine — FMOC-4-Trifluorormethyl-D-Phe

Fluka 47834   MW 455.43 $C_{25}H_{20}F_3NO_4$ ≥98% (HPLC)

Phenylalanine — FMOC-4-Trifluorormethyl-L-Phe

Fluka 47835   MW 455.43 $C_{25}H_{20}F_3NO_4$ ≥98% (HPLC)

Phenylalanine — FMOC-D-(3,4-Dichloro)-Phe

American Peptide FPHE151   MW 458.4

Phenylalanine — FMOC-D-2-Chloro-Phe

American Peptide FPHE115   MW 422.9

Phenylalanine — FMOC-D-2-Fluoro-Phe

American Peptide FPHE135   MW 406.4

Phenylalanine — FMOC-D-3,4-dimethoxyphenylalanine

Neosystem FA15201   MW 447.49

Phenylalanine — FMOC-D-4-Chloro-Phe

American Peptide FPHE125   MW 422.9

Phenylalanine — FMOC-D-4-Fluoro-Phe

American Peptide FPHE145   MW 406.4

Phenylalanine — FMOC-D-4-Nitro-Phe

American Peptide FPHE149   MW 433.4

Phenylalanine — FMOC-D-Hph

USBio F5391   MW 401.5 $C_{25}H_{23}NO_4$ ≥98%

Phenylalanine — FMOC-D-Me-Phe

USBio F5424   MW 401.5 $C_{25}H_{23}NO_4$ ≥98%

Phenylalanine — FMOC-D-p-Chloro-Phe

Synonyms: FMOC-p-Chloro-D-Phe

Peptides International FDF-1863-PI   MW 421.88   >98% (HPLC); off-white powder

Phenylalanine — FMOC-D-p-Fluoro-Phe

Synonyms: FMOC-p-Fluoro-D-Phe

Peptides International FDF-1774-PI   MW 405.43   >98% (HPLC); white crystalline powder

Phenylalanine — FMOC-D-Phe

Bachem B-1410.0001   MW 387.44 $C_{24}H_{21}NO_4$ Store at RT

Fluka 47378   MW 387.44 $C_{24}H_{21}NO_4$ ≥98% (TLC)

Neosystem FA01401   MW 387.4

Peptides International FDF-1830-PI   MW 387.44   >98% (HPLC); white crystalline powder

USBio F5441   MW 387.4 $C_{24}H_{21}NO_4$ ≥99%

Phenylalanine — FMOC-D-Phe-(3,4-Dichloro)

USBio F5447   MW 456.4 $C_{24}H_{19}Cl_2NO_4$ ≥98%

Phenylalanine — FMOC-D-Phe-Nitro

Synonyms: FMOC-p-Nitro-D-Phe

Peptides International FDF-1783-PI   MW 432.44   >98% (HPLC); white crystalline powder

Phenylalanine — FMOC-D-Phe-p-Chloro

USBio F5445   MW 421.9 $C_{24}H_{20}ClNO_4$ ≥98%

Phenylalanine — FMOC-D-Phe-p-Fluoro

USBio F5448   MW 405.4 $C_{24}H_{20}FNO_4$ ≥98%

Phenylalanine — FMOC-D-Phe-SASRIN™-® (200-400 mesh)

Bachem D-1770.0001   Store at 2-8°C

Phenylalanine — FMOC-D-p-Iodo-Phe

American Peptide FPHE153   MW 514.3

Phenylalanine — FMOC-D-β-Phe

Synonyms: (S)-3-(FMOC-Amino)-3-Phenylpropionic Acid

Fluka 00396   MW 387.44 $C_{24}H_{21}NO_4$ ≥98% (HPLC)

Phenylalanine — FMOC-Hph

USBio F5392   MW 401.5 $C_{25}H_{23}NO_4$ ≥98%

Phenylalanine — FMOC-L-(3,4-Dichloro)-Phe

American Peptide FPHE150   MW 458.4

Phenylalanine — FMOC-L-2-Chloro-Phe

American Peptide FPHE110   MW 422.9

Phenylalanine — FMOC-L-2-Fluoro-Phe

American Peptide FPHE130   MW 406.4

Phenylalanine — FMOC-L-3,4-dimethoxyphenylalanine

Neosystem FA15202   MW 447.49

Phenylalanine — FMOC-L-4-Chloro-Phe

American Peptide FPHE120   MW 422.9

Phenylalanine — FMOC-L-4-Fluoro-Phe

American Peptide FPHE140   MW 406.4

Phenylalanine — FMOC-L-4-Nitro-Phe

American Peptide FPHE148   MW 433.4

Phenylalanine — FMOC-L-Phe

Fluka 47635   MW 387.44 $C_{24}H_{21}NO_4$ ≥98% (titration); mp: 181-182°C

Neosystem FA01402   MW 387.4

Phenylalanine — FMOC-L-Phe 4-Benzyloxybenzyl Ester Polymer Bound

Fluka 47666   0.4-0.8 mmol/g resin; carrier: polystyrene, crosslinked with 1% DVB; 200-400 mesh particle size

Phenylalanine — FMOC-L-Phe N-Carboxy Anhydride

Fluka 47693   MW 413.43 $C_{25}H_{19}NO_5$ ~90% (NMR); mp: 66-72°C; ~5% solvent constituents (alkanes, methylene chloride); ~5% FMOC-L-Phe | Produced by Propeptide under an exclusive license from Bioresearch

Phenylalanine — FMOC-L-Phe-OPfp

Fluka 47474   MW 553.49 $C_{30}H_{20}F_5NO_4$ ~98% (HPLC)

Phenylalanine — FMOC-L-Phe-Wang Resin

American Peptide RFWAN70   Preattached Wang resins are usful for synthesis of peptides with C-terminal carboxyl groups; peptides can be cleaved from the resin with 90% TFA in the presence of scavengers

Neosystem RW01402

Phenylalanine — FMOC-L-p-Iodo-Phe

American Peptide FPHE152   MW 514.3

Phenylalanine — FMOC-L-β-Phe

Synonyms: (R)-3-(FMOC-Amino)-3-Phenylpropionic Acid

Fluka 09795   MW 387.44 $C_{24}H_{21}NO_4$ ≥98% (HPLC)

Phenylalanine — FMOC-Me-Phe

USBio F5425   MW 401.5 $C_{25}H_{23}NO_4$ ≥99%

Phenylalanine — FMOC-*m*-Fluoro-Phe
Bachem B-2595.0001   MW 405.43 $C_{24}H_{20}FNO_4$ Store at 2-8°C

Phenylalanine — FMOC-*N*-Me-D-Phe
Fluka 02399   MW 401.46 $C_{25}H_{23}NO_4$ ≥98% (HPLC)

Peptides International FMF-1896-PI   MW 401.47   >98% (HPLC); off-white crystalline powder

Phenylalanine — FMOC-*N*-Me-L-Phe
Fluka 47598   MW 401.46 $C_{25}H_{23}NO_4$ ≥99% (HPLC); mp: 143-146°C

Phenylalanine — FMOC-*N*-Me-Phe
Synonyms: FMOC-*N*-Me-L-Phe

Bachem B-1725.0001   MW 401.46 $C_{25}H_{23}NO_4$ Store at RT

Peptides International FMF-1795-PI   MW 401.47   >98% (HPLC); off-white crystalline powder

Phenylalanine — FMOC-*p*(CH₂-PO₃Et₂)-D-Phe
Neosystem FA05301   MW 537.56   Garbay-Jaureguiberry, C et al, *Int J Pept Prot Res*, 39:523, 1992

Phenylalanine — FMOC-*p*(CH₂-PO₃Et₂)-L-Phe
Neosystem FA05302   MW 537.56   Garbay-Jaureguiberry, C et al, *Int J Pept Prot Res*, 39:523, 1992

Phenylalanine — FMOC-*p*-Amino-D-Phe-BOC
Bachem B-2930.0001   MW 502.57 $C_{29}H_{30}N_2O_6$ Store at -15°C

Phenylalanine — FMOC-*p*-Amino-Phe
Bachem B-2070.0001   MW 402.45 $C_{24}H_{22}N_2O_4$ Store at -15°C

Phenylalanine — FMOC-*p*-Amino-Phe-BOC
Bachem B-1995.0001   MW 502.57 $C_{29}H_{30}N_2O_6$ Store at -15°C

Phenylalanine — FMOC-*p*-Azido-Phe
Bachem B-2360.0001   MW 428.45 $C_{24}H_{20}N_4O_4$ Store at -15°C

Phenylalanine — FMOC-*p*-Bz-D-Phe
Synonyms: FMOC-D-Bpa

Bachem B-2340.0001   MW 491.54 $C_{31}H_{25}NO_5$ Store at RT

Phenylalanine — FMOC-*p*-Bz-Phe
Synonyms: FMOC-Bpa

Bachem B-2220.0001   MW 491.54 $C_{31}H_{25}NO_5$ Store at RT

Phenylalanine — FMOC-*p*-Carboxy-Phe-OtBu
Bachem B-3070.0001   MW 487.55 $C_{29}H_{29}NO_6$ Store at -15°C

Phenylalanine — FMOC-*p*-Chloro-D-Phe
Bachem B-1900.0001   MW 421.88 $C_{24}H_{20}ClNO_4$ Store at RT

Phenylalanine — FMOC-*p*-Chloro-Phe
Synonyms: FMOC-L-*p*-Chloro-L-Phe

Bachem B-2115.0001   MW 421.88 $C_{24}H_{20}ClNO_4$ Store at RT

Peptides International FLF-1762-PI   MW 421.88   >98% (HPLC); white crystalline powder

Phenylalanine — FMOC-*p*-Fluoro-DL-Phe
Bachem B-1550.0001   MW 405.43 $C_{24}H_{20}FNO_4$ Store at RT

Phenylalanine — FMOC-*p*-Fluoro-D-Phe
Bachem B-3210.0001   MW 405.43 $C_{24}H_{20}FNO_4$ Store at RT

Phenylalanine — FMOC-*p*-Fluoro-Phe
Synonyms: FMOC-*p*-Fluoro-L-Phe

Bachem B-2835.0001   MW 405.43 $C_{24}H_{20}FNO_4$ Store at RT

Peptides International FLF-1770-PI   MW 405.43   >98% (HPLC); white crystalline powder

Phenylalanine — FMOC-Phe
Synonyms: FMOC-L-Phe

American Peptide FPHE100   MW 387.4

Bachem B-1405.0005   MW 387.44 $C_{24}H_{21}NO_4$ Store at RT

Peptides International FLF-1730-PI   MW 387.44   >98% (HPLC); white crystalline powder

USBio F5442   MW 387.4 $C_{24}H_{21}NO_4$ ≥99%

Phenylalanine — FMOC-Phe-® (200-400 mesh)
Bachem D-1140.0001   Store at 2-8°C

Phenylalanine — FMOC-Phe-Nitro
Synonyms: FMOC-*p*-Nitro-L-Phe

Peptides International FLF-1781-PI   MW 432.44   >98% (HPLC); white crystalline powder

USBio F5450   MW 432.4 $C_{24}H_{20}N_2O_6$ ≥99%

Phenylalanine — FMOC-Phe-ol
Synonyms: FMOC-L-phenylalaninol

Senn Chem 44063   MW 373.5

Phenylalanine — FMOC-Phe-OPfp
Bachem B-1170.0001   MW 553.49 $C_{30}H_{20}F_5NO_4$ Store at -15°C

USBio F5443   MW 553.5 $C_{30}H_{20}F_5NO_4$ ≥99%

Phenylalanine — FMOC-Phe-OSu
Bachem B-1415.0001   MW 484.51 $C_{28}H_{24}N_2O_6$ Store at -15°C

USBio F5444   MW 484.5 $C_{28}H_{24}N_2O_6$ ≥99%

Phenylalanine — FMOC-Phe-*p*-Chloro
USBio F5446   MW 421.9 $C_{24}H_{20}ClNO_4$ ≥98%

Phenylalanine — FMOC-Phe-*p*-Fluoro
USBio F5449   MW 405.4 $C_{24}H_{20}FNO_4$ ≥98%

Phenylalanine — FMOC-Phe-SASRIN™-® (200-400 mesh)
Bachem D-1375.0001   Store at 2-8°C

Phenylalanine — FMOC-Phe-Ser-tBu
Bachem B-1505.0001   MW 530.62 $C_{31}H_{34}N_2O_6$ Store at -15°C

Phenylalanine — FMOC-*p*-Iodo-Phe
Bachem B-2750.0001   MW 513.33 $C_{24}H_{20}INO_4$ Store at 2-8°C

Phenylalanine — FMOC-*p*-Me-D-Phe

Bachem B-3330.0001   MW 401.46 $C_{25}H_{23}NO_4$ Store at 2-8°C

Phenylalanine — FMOC-*p*-Me-Phe

Bachem B-3335.0001   MW 401.46 $C_{25}H_{23}NO_4$ Store at 2-8°C

Phenylalanine — FMOC-*p*-Nitro-D-Phe

Bachem B-2350.0001   MW 432.43 $C_{24}H_{20}N_2O_6$ Store at RT

Phenylalanine — FMOC-*p*-Nitro-Phe

Bachem B-1395.0001   MW 432.43 $C_{24}H_{20}N_2O_6$ Store at RT

Phenylalanine — FMOC-*p*-Nitro-Phe-® (200-400 mesh)

Bachem D-1135.0001   Store at RT

Phenylalanine — FMOC-*p*-tBu-D-Phe

Bachem B-3325.0001   MW 443.54 $C_{28}H_{29}NO_4$ Store at 2-8°C

Phenylalanine — FMOC-*p*-tBu-Phe

Bachem B-3320.0001   MW 443.54 $C_{28}H_{29}NO_4$ Store at 2-8°C

Phenylalanine — FMOC-α-Me-D-Phe

Bachem B-3355.0001   MW 401.46 $C_{25}H_{23}NO_4$ Store at -15°C

Phenylalanine — For-Phe

Bachem E-2690.0005   MW 193.2 $C_{10}H_{11}NO_3$ Store at RT

Phenylalanine — For-Phe-OMe

Bachem E-2825.0001   MW 207.23 $C_{11}H_{13}NO_3$ Store at 2-8°C

Phenylalanine — Glutamic Acid — CBZ-Glu-Phe

USBio C2098-64   MW 428.4 $C_{22}H_{24}N_2O_7$ ≥99%

Phenylalanine — Glutaryl-L-Phe-4-NA

Synonyms: Chymotrypsin Substrate

Fluka 49738   MW 399.4 $C_{20}H_{21}N_3O_6$ ≥98.0%; mp: 200°C | Erlanger, BF et al, *Biochemistry*, 3: 1880, 1964

Phenylalanine — Glutaryl-L-Phe-AMC

Synonyms: Glutaryl-Phe-AMC; *N*-Glutaryl-L-Phe-AMC; Chymotrypsin Substrate

Fluka 49737   MW 436.47 $C_{24}H_{24}N_2O_6$ ≥98.0% (HPLC) | Sensitive fluorogenic substrate; Zimmermann, M et al, *Anal Biochem*, 70: 258, 1976; *ibid*, 78: 47, 1977

Phenylalanine — Glutaryl-Phe-AMC

Bachem I-1205.0050   MW 436.46 $C_{24}H_{24}N_2O_6$ Store at -15°C

Phenylalanine — Glutaryl-Phe-pNA

Bachem L-1275.0250   MW 399.4 $C_{20}H_{21}N_3O_6$ Store at -15°C

Phenylalanine — Glutaryl-Phe-βNA

Bachem K-1290.0250   MW 404.47 $C_{24}H_{24}N_2O_4$ Store at RT

Phenylalanine — Indole-3-Ac-L-Phe

Synonyms: IAA-L-Phe

Sigma I 9637   FW 322.4 $C_{19}H_{18}N_2O_3$ White to pink powder | Plant cell culture tested

Phenylalanine — L-(3,4-Dihydroxy)-Phe

Synonyms: L-DOPA; L-3-Hyty

Sigma D 9628 $C_9H_{11}NO_4$ Natural isomer of the immediate precursor of dopamine; product of tyrosine hydroxylase

Phenylalanine — L-(3,4-Dihydroxy)-Phenyl(3-¹⁴C)-Alanine

Amersham CFA439

Phenylalanine — L-(3,4-Dihydroxy)-Phe-OMe Hydrochloride

Synonyms: Methyl L-DOPA

Sigma D 1507 $C_{10}H_{13}NO_4 \cdot HCl$ Dopamine precursor; anti-Parkinsonian

Phenylalanine — L-Hph

Synonyms: (+)-2-Amino-4-Phenylbutyric Acid

USBio H7010   MW 179.2 $C_{10}H_{13}NO_2$ ≥99%

Phenylalanine — L-O-ChloroPhe

Sigma C 9294   FW 199.6 $C_9H_{10}ClNO_2$

Phenylalanine — L-*p*-ChloroPhe

Synonyms: L-PCP; L-PCPA; Tryptophan Hydroxylase Inhibitor

Sigma C 8655   FW 199.6 $C_9H_{10}ClNO_2$ Crosses blood-brain barrier

Phenylalanine — L-*p*-Chlorophenylalanine

USBio C4500   MW 199.6 $C_9H_{10}ClNO_2$ ≥99%

Phenylalanine — L-Phe

Synonyms: (*S*)-2-Amino-3-Phenylpropionic Acid; L-2-Amino-3-Phenylpropanoic Acid; (*S*)-α-Amino-β-Phenylpropionic Acid; (*S*)-α-Aminobenzenepropanoic Acid; (*S*)-α-Aminohydrocinnamic Acid

Amersham US20105   98.5-101.5% dsb; ≤0.0015% heavy metals (Pb)

Fluka 78019   MW 165.19 $C_9H_{11}NO_2$ ≥99.0% (titration); ≤0.1% residue on ignition, loss on drying; ≤0.3% foreign AA; ≤0.00001% As; ≤0.0005% Al, Ba, Bi, Cd, Co, Cr, Cu, Fe, Li, Mg, Mn, Mo, Ni, Pb, Sr, Zn; ≤0.001% Ca; ≤0.005% K, Cl, Na, SO₄; ≤0.01% NH₄⁺ | *CRC Handbook of Microbiology*, Laskin, AI & Lachevalier, HA, eds, 4: 1, 1974, CRC Press, Cleveland, Ohio

Fluka 78020   MW 165.19 $C_9H_{11}NO_2$ ≥99.0% (titration); ≤0.1% residue on ignition; ≤0.2% loss on drying; ≤0.3% foreign AAs; ≤0.0005% Cd, Co, Cr, Cu, Fe, Mg, Mn, Ni, Pb, Zn; ≤0.001% Ca; ≤0.005% K, Cl, Na, SO₄; ≤0.01% NH₄⁺

ICN 102623   MW 165.2 $C_9H_{11}NO_2$ >99%; crystalline

ICN 194723   MW 165.2 $C_9H_{11}NO_2$ Cell culture reagent; >99%

Neosystem AA01402   MW 165.2

Peptides International ALF-2717   MW 165.19   >99.9% optical purity by RP-HPLC; white crystalline powder

Rexim   MW 165.2 $C_9H_{11}NO_2$ White crystals or crystalline powder

Sigma P 2126   FW 165.2 $C_9H_{11}NO_2$ SigmaUltra; ≥98% (TLC) | Also available as part of a kit

Sigma P 2168   FW 165.2 $C_9H_{11}NO_2$ Crystalline | Plant cell culture tested

Sigma P 3018   Matrix: cross-linked 4% beaded agarose; activation: cyanogen bromide; attachment: amino; spacer: 1 atom; ligand immobilized: 2-10 μmoles/mL; form: suspension in 2.0 *M* NaCl containing 0.02% thimerosal | Deutsch, DG & Mertz, ET, *Proc Fed Amer Soc Exp Biol*, 29: 647, 1970; Deutsch, DG & Mertz, ET, *Science*, 170: 1095, 1970; Vuento, M & Vaheri, A, *Biochem J*, 183: 331, 1979

Sigma P 5030   FW 165.2 $C_9H_{11}NO_2$ Crystalline | Cell culture tested; insect cell culture tested

Sigma P 8324   FW 165.2 $C_9H_{11}NO_2$ SigmaUltra; >99% (TLC); pH (0.1M in H_2O, 20°C): 5.0-7.0; residue on ignition: <0.1%; solubility (1 *M* in BME, 20°C): complete, colorless; insoluble matter: <0.1%; SO_4, Cl, NH_4^+,: <0.05%; K, Na: <0.005%; Al, Cu, Fe, Mg, Zn: <0.0005%; P: <0.01%; Pb, Ca: <0.001%

Sigma P 9596   FW 165.2 $C_9H_{11}NO_2$ USP Grade

USBio P4060   MW 165.2 $C_9H_{11}NO_2$ ≥98.5%; appearance: white crystalline powder; pH: 5.4-6.0; specific rotation (C=2, H_2O): -32.7° to -34.7°; heavy metals (Pb): ≤0.0015%; endotoxin: ≤2 EU/mg; cell culture tested: murine myeloma SP2/0-Ag14

Phenylalanine — L-Phe (2,3,4,5,6-³H)

Amersham TRK648   Aqueous, 2% EtOH, sterilized; 4.07-5.18 TBq/mmol, 100-140 Ci/mmol; 37 MBq/mL, 1 mCi/mL

Phenylalanine — L-Phe (2,6-³H)

Amersham TRK552   Aqueous, 2% EtOH, sterilized; 1.5-2.2 TBq/mmol, 40-60 Ci/mmol; 37 MBq/mL, 1 mCi/mL

Phenylalanine — L-Phe (3,4,5-³H)

ARC ART-614   MW 165.2 $C_6H_5CH_2CH(NH_2)COOH$ 80-100 Ci/mmol; 2.96-3.7 TBq/mmol; in 0.01 *N* HCl | Radiochemical

Phenylalanine — L-Phe (4-³H)

Amersham TRK204   Aqueous, 2% EtOH, sterilized; 1.55-1.1 TBq/mmol, 15-30 Ci/mmol; 37 MBq/mL, 1 mCi/mL

ARC ART-411   MW 165.2 $C_6H_5CH_2CH(NH_2)COOH$ 15-30 Ci/mmol; 0.55-1.11 TBq/mmol; in 0.01 *N* HCl | Radiochemical

Phenylalanine — L-Phe (Ala-1-¹⁴C)

ARC ARC-212   MW 165.2 $C_6H_5CH_2CH(NH_2)$**C**OOH   50-60 mCi/mmol; 1.85-2.22 GBq/mmol; in 0.01 *N* HCl | Radiochemical

Phenylalanine — L-Phe (Ala-¹⁴C)-OEt Hydrochloride

ARC ARC-751   MW 165.2 $C_6H_5CH_2CH(NH_2)$**C**OOC_2H_5 · HCl   50-60 mCi/mmol; 1.85-2.22 GBq/mmol; in EtOH | Radiochemical

Phenylalanine — L-Phe (Ring 2,6-³H(N))

Sigma P 6053   FW 165.2 $2,6-H_2-C_6H_3CH_2CH(NH_2)COOH$ 40-60 Ci/mmol; aqueous solution containing 2% EtOH in serum bottle | Radiochemical

Phenylalanine — L-Phe (Side chain ³H)

ICN 20044   MW 165.2 $C_6H_5CH_2CH(NH_2)COOH$ 15-25 Ci/mmol, 555-925 GBq/mmol; 0.01 *N* HCl; HPLC analyzed for purity

ICN 20044E   MW 165.2 $C_6H_5CH_2CH(NH_2)COOH$ 15-25 Ci/mmol, 555-925 GBq/mmol; sterile EtOH:H_2O (2:98); HPLC analyzed for purity

Phenylalanine — L-Phe (U-¹⁴C)

Amersham CFB70   Aqueous, 2% EtOH, sterilized; >16.6 GBq/mmol, >450 mCi/mmol; 1.85 MBq/mL, 50 μCi/mL

ARC ARC-675   MW 165.2 $C_6H_5CH_2CH(NH_2)COOH$ 250-450 mCi/mmol; 9.3-15.7 GBq/mmol; in EtOH: H_2O (2:98) | Radiochemical

ICN 10113E   MW 165.2 $C_6H_5CH_2CH(NH_2)$**C**OOH   405-495 mCi/mmol, 15-18.3 GBq/mmol; sterile EtOH:H_2O (2:98); HPLC analyzed for purity

Phenylalanine — L-Phe (U-¹⁴C) Hydrochloride

Sigma P 9055   FW 165.2 $C_6H_5CH_2CH(NH_2)COOH$ 360-540 mCi/mmol; radiochemical purity ≥98% (HPLC); aqueous solution containing 2% EtOH in Combi-vial | Radiochemical

Phenylalanine — L-Phe Amide

Neosystem AA01403   MW 164.2

Phenylalanine — L-Phe Bzl Hydrochloride

Fluka 78060   MW 291.78 $C_{16}H_{17}NO_2$ · HCl   ≥99.0% (titration); mp: 197-200°C

Phenylalanine — L-Phe CMK Hydrobromide

ICN 156153   MW 278.6 $C_{10}H_{12}ClNO$ · HBr

Phenylalanine — L-Phe β-Naphthylamide

Sigma P 3762   FW 290.4 $C_{19}H_{18}N_2O$ Possibly carcinogenic

Phenylalanine — L-Phe-AMC

ICN 151823   MW 322.4 $C_{19}H_{18}N_2O_3$ Fluorogenic ENZYME SUBSTRATE; Zimmerman, M, *Anal Biochem*, 78:47, 1977

Phenylalanine — L-Phe-AMC TFA

Synonyms: L-Phe 4-Methylumbelliferylamide TFA

Fluka 78089   MW 436.39 $C_{19}H_{18}N_2O_3$ · $C_2HF_3O_2$ ≥99.0% (HPLC); mp: 200°C | Ishiura, S et al, *J Biochem*, 102: 1023, 1987

Phenylalanine — L-Phe-AMC TFA Salt

Sigma P 7023   FW 436.4 $C_{19}H_{18}N_2O_3$ · $C_2HF_3O_2$

Phenylalanine — L-Phe-Cyclohexylamide

Synonyms: N-Cyclohexyl-L-Phenylalaninamide

Fluka 78082   MW 246.35 $C_{15}H_{22}N_2O$ ≥98.0% (TLC); mp: 96-99°C

Phenylalanine — L-Phenylalaninamide Free Base

Sigma P 1883   FW 164.2 $C_9H_{12}N_2O$

Phenylalanine — L-Phenylalaninamide Hydrochloride

Sigma P 2376   FW 200.7 $C_9H_{12}N_2O$ · HCl

Phenylalanine — L-Phe-OAll Hydrochloride

Neosystem AA01412   MW 241.7

Phenylalanine — L-Phe-OAll Toluene 4-Sulfonate

Fluka 78055   MW 377.46 $C_{12}H_{15}NO_2$ · $C_7H_8O_3S$ mp: 155-158°C | Useful in peptide & *N*-glycopeptide synthesis; the ester is selectively deblocked with $(Ph_3P)_3RhCl$; Waldmann, H & Kunz, H, *Liebigs Ann Chem*, 1712, 1983

Phenylalanine — L-Phe-OBzl *p*-Tosyl

Sigma P 1764   FW 427.5 $C_{16}H_{17}NO_2$ · $C_7H_8O_3S$

Phenylalanine — L-Phe-OBzl PTSA

Neosystem AA01413   MW 427.5

Phenylalanine — L-Phe-OEt Hydrochloride

Fluka 78050   MW 229.71 $C_{11}H_{15}NO_2$ · HCl   ≥99.0% (titration); mp: 153-155°C; ≤0.05% residue on ignition

ICN 102624   MW 229.7 $C_{11}H_{15}NO_2$ · HCl   ~99%; crystalline

Sigma P 2751   FW 229.7 $C_{11}H_{15}NO_2$ · HCl

Phenylalanine — L-Phe-ol

Synonyms: Phenylalaninol

Bachem F-2950.0001   MW 151.21 $C_9H_{13}NO$ Store at -15°C

Phenylalanine — L-Phe-OMe Hydrochloride

Fluka 78090   MW 215.69 $C_{10}H_{13}NO_2 \cdot HCl$ ≥99.0% (titration); mp: 158-160°C

ICN 102625   MW 215.7 $C_{10}H_{13}NO_2 \cdot HCl$ Crystalline

Neosystem AA01404   MW 215.7

Sigma P 3126   FW 215.7 $C_{10}H_{13}NO_2 \cdot HCl$

Phenylalanine — L-Phe-OtBu Hydrochloride

Fluka 78080   MW 257.76 $C_{13}H_{19}NO_2 \cdot HCl$ ≥99.0% (HPLC); mp: 240°C

ICN 156152   MW 257.8 $C_{13}H_{19}NO_2 \cdot HCl$ Crystalline

Sigma P 0881   FW 257.8 $C_{13}H_{19}NO_2 \cdot HCl$

Phenylalanine — L-Phe-pNA

Sigma P 4673   FW 285.3 $C_{15}H_{15}N_3O_3$

Phenylalanine — Malonate-Phe Broth

Fluka 63286   Composition: 2 g/L ammonium sulfate, 1 g/L yeast extract, 0.6 g/L dipotassium hydrogen phosphate, 0.4 g/L potassium dihydrogen phosphate, 2 g/L NaCl, 3 g/L sodium malonate, 1 g/L L-Phe, 0.025 g/L bromothymol blue | Test medium to detect the utilization of malonate & Phe deamination; Shaw & Clarke, *J Gen Microbiol*, 13: 155, 1955

Phenylalanine — m-Fluoro-DL-Phe

Synonyms: Protein Synthesis Inhibitor; Phenylalanine Antagonist

Bachem F-2135.0001   MW 183.18 $C_9H_{10}FNO_2$ Store at RT

ICN 101706   MW 183.2 $C_9H_{10}FNO_2$ Crystalline

Sigma F 5126   FW 183.2 $C_9H_{10}FNO_2$

Phenylalanine — m-Fluoro-D-Phe

Bachem F-3290.0001   MW 183.18 $C_9H_{10}FNO_2$ Store at RT

Phenylalanine — m-Fluoro-D-Phe-OMe Hydrochloride

Bachem F-3295.0001   MW 233.67 $C_{10}H_{12}FNO_2 \cdot HCl$ Store at 2-8°C

Phenylalanine — m-Fluoro-Phe

Bachem F-3285.0001   MW 183.18 $C_9H_{10}FNO_2$ Store at RT

Phenylalanine — N^α-(2,4,6-Triisopropyl-Phenylsulfonyl)-3-Amidino-L-Phe-4-Ethoxycarbonylpiperazide Hydrochloride

Synonyms: Pefabloc®uPA; Pefa-0888

Pentapharm 382-12, 382-03   Synthetic   MW 650.3 $C_{32}H_{47}O_5N_5S \cdot HCl$ Low MW inhibitor for urokinase (uPA); Stürzebecher J et al, *Bioorg Med Letters*, 93147, 1999

Phenylalanine — N^α-(2-Naphthylsulfonyl)-3-Amidino-D,L-Phe-4-Cyclobutanoylpiperazide Hydrochloride

Synonyms: Pefabloc®Try1420; Pefa-1420

Pentapharm 384-02   Synthetic   MW 584.1 $C_{29}H_{33}O_4N_5S \cdot HCl$ Low MW inhibitor selective for trypsin; Stürzebecher J et al, *J Med Chem*, 40:3091, 1997

Phenylalanine — N^α-(2-Naphthylsulfonyl)-3-Amidino-D,L-Phe-4-Dimethylcarbamoylppiperazide Hydrochloride

Synonyms: Pefabloc®TH1158; Pefa-1158

Pentapharm 381-02   MW 573.1 $C_{27}H_{32}O_4N_6S \cdot HCl$ Solubility: 100 mg/mL in H_2O (175 *mM*) | Low MW competitive inhibitor selective for thrombin with high solubility in H_2O; used to exclude undesired thrombin activity; Stürzebecher J et al, *J Med Chem*, 40:3091, 1997

Phenylalanine — N^α-(2-Naphthylsulfonyl)-3-Amidino-D,L-Phe-Nipecotic Acid-Thiophene-2-Methylamide Hydrochloride

Synonyms: Pefabloc®PL; Pefa-2532

Pentapharm 383-12, 383-03   Synthetic   MW 640.2 $C_{31}H_{33}O_4N_5S_2 \cdot HCl$ Low MW inhibitor for plasmin

Phenylalanine — N-(3-(2-Furyl)-Acryloyl)-L-Phe

Sigma F 6010   FW 285.3 $C_{16}H_{15}NO_4$

Phenylalanine — N-(4-Methoxyphenylazoformyl)-Phe Potassium Salt

Bachem M-2245.0100   Store at 2-8°C

Phenylalanine — N-(N-Glutaryl-L-Phe)-2-Aminoacridone

Synonyms: Chymotrypsin α-Substrate

Fluka 49742   MW 471.51 $C_{27}H_{25}N_3O_5$ For continuous fluorescence assay at 450/570 nm; Baustert, JH et al, *Anal Biochem*, 171: 393, 1988

Phenylalanine — N-(N-Suc-L-Phe)-2-Aminoacridone

Synonyms: Chymotrypsin α-Substrate

Fluka 85992   MW 457.49 $C_{26}H_{23}N_3O_5$ ≥98.0% (TLC); mp: 249-253°C; soluble in DMF, DMSO, water | Fluorogenic substrate; Baustert, JH et al, *Anal Biochem*, 171: 393, 1988

Phenylalanine — N-(N-Tosyl-L-Phe)-2-Aminoacridone

Synonyms: Chymotrypsin α-Substrate

Fluka 90185   MW 511.6 $C_{29}H_{25}N_3O_4S$ ≥98.0% (TLC); mp: 227°C; soluble in DMF, DMSO | Fluorogenic substrate

Phenylalanine — N,N-Dimethyl-L-Phe

Fluka 41389   MW 193.25 $C_{11}H_{15}NO_2$ ≥99% (HPLC); mp: 225-228°C

Phenylalanine — N-2,4-Dnp-L-Phe

Sigma D 1380   FW 331.3 $C_{15}H_{13}N_3O_6$

Phenylalanine — N-Ac-DL-Phe

ICN 150240   MW 207.2 $C_{11}H_{13}NO_3$ Crystalline

Sigma A 4001   FW 207.2 $C_{11}H_{13}NO_3$ Crystalline

Phenylalanine — N-Ac-DL-Phe β-Naphthyl Ester

Sigma A 7512   FW 333.4 $C_{21}H_{19}NO_3$ Crystalline

Phenylalanine — N-Ac-DL-Phe-(pONp)

Sigma A 4376   FW 328.3 $C_{17}H_{16}N_2O_5$ Crystalline

Phenylalanine — N-Ac-DL-Phe-pNA

Synonyms: Chymotrypsin Substrate

ICN 100121   MW 327.3 $C_{17}H_{17}N_3O_4$ Crystalline

Sigma A 9133   FW 327.3 $C_{17}H_{17}N_3O_4$ Crystalline

Phenylalanine — *N*-Ac-D-Phe

ICN 100128	MW 207.2	$C_{11}H_{13}NO_3$	Crystalline
Sigma A 3876	FW 207.2	$C_{11}H_{13}NO_3$	Sigma Grade

Phenylalanine — *N*-Ac-L-Phe

ICN 150241	MW 207.2	$C_{11}H_{13}NO_3$	Crystalline
Sigma A 4126	FW 207.2	$C_{11}H_{13}NO_3$	Sigma Grade

Phenylalanine — *N*-Ac-L-Phe-OEt

Sigma A 4251	FW 235.3	$C_{13}H_{17}NO_3$	Sigma Grade

Phenylalanine — *N*-Ac-m-Fluoro-DL-Phe

ICN 100110	MW 225.2	$C_{11}H_{12}FNO_3$	Crystalline

Phenylalanine — *N*-Ac-O-Fluoro-DL-Phe

ICN 150230	MW 225.2	$C_{11}H_{12}FNO_3$	Crystalline

Phenylalanine — *N*-Ac-p-Fluoro-DL-Phe

ICN 102579	MW 225.2	$C_{11}H_{12}FNO_3$	Crystalline

Phenylalanine — *N*-Bz-DL-Phe

Sigma B 1879	FW 269.3	$C_{16}H_{15}NO_3$

Phenylalanine — *N*-Bz-DL-Phe β-Naphthol As-Bi Ester

Sigma B 4506	FW 623.5	$C_{34}H_{27}BrN_2O_5$

Phenylalanine — *N*-Bz-DL-Phe β-Naphthy Ester

Sigma B 2379	FW 395.5	$C_{26}H_{21}NO_3$

Phenylalanine — *N*-Bz-DL-Phe β-Naphthyl Ester

Synonyms: BPANE

ICN 100918	MW 395.5	$C_{26}H_{21}NO_3$	Crystalline

Phenylalanine — *N*-Bz-L-Phe

ICN 150447	MW 269.3	$C_{16}H_{15}NO_3$	Crystalline
Sigma B 9382	FW 269.3	$C_{16}H_{15}NO_3$	

Phenylalanine — *N*-CBZ-DL-Phe

ICN 100228	MW 299.3	$C_{17}H_{17}NO_4$	Crystalline

Phenylalanine — *N*-CBZ-D-Phe

Sigma C 4127	FW 299.3	$C_{17}H_{17}NO_4$

Phenylalanine — *N*-CBZ-L-Phe

ICN 101191	MW 299.3	$C_{17}H_{17}NO_4$	Crystalline
Sigma C 4377	FW 299.3	$C_{17}H_{17}NO_4$	

Phenylalanine — *N*-CBZ-L-Phe CMK

Synonyms: ZPCK; Chymotripsin Aγ Inhibitor

Sigma C 9511   FW 331.8 $C_{18}H_{18}ClNO_3$ Useful as a site-specific inhibitor for the study of bovine chymotrypsin Aγ; Segal, DM et al, *Biochemistry*, 10: 3728, 1971

Phenylalanine — *N*-CBZ-L-Phe-(*p*ONp)

ICN 101257	MW 420.4	$C_{23}H_{20}N_2O_6$	Crystalline

Phenylalanine — *N*-CBZ-L-Phe-Bromomethylketone

Synonyms: ZPBK

ICN 103310	MW 376.3	$C_{18}H_{18}BrNO_3$	99%

Phenylalanine — *N*-CBZ-L-Phe-CMK

Synonyms: ZPCK

ICN 103309	MW 331.8	$C_{18}H_{18}ClNO_3$

Phenylalanine — *N*-CBZ-L-Phe-pNA

Synonyms: Chymotripsin Substrate

ICN 101274	MW 419.4	$C_{23}H_{21}N_3O_5$	Crystalline
Sigma C 0271	FW 419.4	$C_{23}H_{21}N_3O_5$	

Phenylalanine — *N*-Chloroacetyl-DL-Phe

Sigma C 1378	FW 241.7	$C_{11}H_{12}ClNO_3$

Phenylalanine — *N*-Chloroacetyl-L-Phe

Sigma C 1503	FW 241.7	$C_{11}H_{12}ClNO_3$

Phenylalanine — *N*-FMOC-4-Chloro-L-Phe

Sigma F 6545	FW 421.9	$C_{24}H_{20}ClNO_4$

Phenylalanine — *N*-FMOC-D-Phe

Sigma F 3042	FW 387.4	$C_{24}H_{21}NO_4$

Phenylalanine — *N*-FMOC-D-Phe ($1\text{-}^{14}\text{C}$)

ARC ARC-1109   MW 387.4   50-60 mCi/mmol; 1.85-2.22 GBq/mmol; in EtOH | Radiochemical

Phenylalanine — *N*-FMOC-L-Phe

Sigma F 0509	FW 387.4	$C_{24}H_{21}NO_4$

Phenylalanine — *N*-FMOC-L-Phe ($1\text{-}^{14}\text{C}$)

ARC ARC-1033   MW 387.4   50-60 mCi/mmol; 1.85-2.22 GBq/mmol; in EtOH | Radiochemical

Phenylalanine — *N*-FMOC-L-Phe ($4\text{-}^3\text{H}$)

ARC ART-410   MW 387.4   20-30 Ci/mmol; 0.74-1.11 TBq/mmol; in EtOH | Radiochemical

Phenylalanine — *N*-FMOC-L-Phe Resin Ester

Sigma F 2261   For peptide synthesis; Wang, SS, *J Am Chem Soc*, 95: 1328, 1973; Lu, G et al, *J Org Chem*, 46: 3433, 1981

Phenylalanine — *N*-FMOC-L-Phe SASRIN Resin Ester

Sigma F 2389   200-400 mesh; substitution range: 0.3-0.8 mmole FMOC AA/g resin | For peptide synthesis; *N*-FMOC AA acyl ester of 3-methoxy-4-oxymethyl-phenoxymethylated 1% divinylbenzene cross-linked polystyrene; peptides can be cleaved from the resin with 0.5-1% CF_3CO_2H in CH_2Cl_2; Mergler, M et al, *in Peptides: Chemistry & Biology* (Proc 10th Am Peptide Symp), Marhall, GR, ed, 259, 1988

Phenylalanine — *N*-FMOC-p-Fluoro-D-Phe

Sigma F 6920	FW 405.4	$C_{24}H_{20}FNO_4$

Phenylalanine — *N*-FMOC-p-Nitro-L-Phe

ICN 158174	MW 432.4	$C_{24}H_{20}N_2O_6$	
Sigma F 8261	FW 432.4	$C_{24}H_{20}N_2O_6$	≥98% (TLC)

Phenylalanine — *N*-For-DL-Phe

Sigma F 3387	FW 193.2	$C_{10}H_{11}NO_3$

Phenylalanine — *N*-For-L-Phe

ICN 151171	MW 193.2	$C_{10}H_{11}NO_3$
Sigma F 3889	FW 193.2	$C_{10}H_{11}NO_3$

Phenylalanine — *N*-Glutaryl-L-Phe-AMC

ICN 157224	MW 436.5	$C_{24}H_{24}N_2O_6$
Sigma G 6758	FW 436.5	$C_{24}H_{24}N_2O_6$

Phenylalanine — *N*-Glutaryl-L-Phe-pNA

ICN 101813	MW 399.4	$C_{20}H_{21}N_3O_6$
Sigma G 2505	FW 399.4	$C_{20}H_{21}N_3O_6$

Phenylalanine — *N*-Glutaryl-L-Phe-β-Naphthylamide

ICN 101812

Phenylalanine — *N*-Me-D-Phe

Bachem F-1785.0001	MW 179.22	$C_{10}H_{13}NO_2$	Store at RT
Sigma M 5516	FW 179.2	$C_{10}H_{13}NO_2$	

Phenylalanine — *N*-Me-L-Phe

Sigma M 9515	FW 179.2	$C_{10}H_{13}NO_2$

Phenylalanine — *N*-Me-L-Phe Hydrochloride

Fluka 02452   MW 215.68 $C_{10}H_{13}NO_2 \cdot HCl$ ≥98.0% (TLC); mp: 293-294°C

Phenylalanine — *N*-Me-Phe

Bachem E-2165.0001   MW 179.22 $C_{10}H_{13}NO_2$ Store at RT

Phenylalanine — *N*-Me-Phe-OBzl *p*-Tosyl

Bachem E-2755.0001   MW 441.55 $C_{17}H_{19}NO_2 \cdot C_7H_8O_3S$ Store at RT

Phenylalanine — *N*-Me-Phe-OMe Hydrochloride

Bachem E-2170.0250   MW 229.71 $C_{11}H_{15}NO_2 \cdot HCl$ Store at RT

Phenylalanine — *N*-Me-*p*-Nitro-Phe

Bachem F-1780.0001   MW 224.22 $C_{10}H_{12}N_2O_4$ Store at RT

Phenylalanine — *N*-MSOC-L-Phe

ICN 155740   MW 315.3 $C_{13}H_{17}NO_6S$ Crystalline

Phenylalanine — *N*-O-Nps-L-Phe Dicyclohexylammonium Salt

ICN 102534	MW 499.7	$C_{15}H_{14}N_2O_4S \cdot C_{12}H_{23}N$	Yellow crystals
Sigma N 6253	FW 499.7	$C_{15}H_{14}N_2O_4S \cdot C_{12}H_{23}N$	Yellow crystals

Phenylalanine — *N*-Phenylacetyl-L-Phe

ICN 156151	MW 283.3	$C_{17}H_{17}NO_3$
Sigma P 4164	FW 283.3	$C_{17}H_{17}NO_3$

Phenylalanine — *N*-Phthaloyl-L-Phe

Fluka 79950	MW 295.3	$C_{17}H_{13}NO_4$	≥99.0% (TLC); mp: 181-185°C
Sigma P 4753	FW 295.3	$C_{17}H_{13}NO_4$	

Phenylalanine — Nps-Phe DCHA

Bachem F-1960.0001   MW 499.68 $C_{15}H_{14}N_2O_4S \cdot C_{12}H_{23}N$ Store at 2-8°C

Phenylalanine — *N*-Succinyl-L-Phe-pNA

Synonyms: Chymotrypsin Substrate

ICN 102974   Crystalline

Sigma S 2628   FW 385.4 $C_{19}H_{19}N_3O_6$ Nagel, W et al, *Hoppe-Seyler's Z Physiol Chem*, 330: 1, 1965

Phenylalanine — *N*-t-BOC-D-Phe

Sigma B 5269   FW 265.3 $C_{14}H_{19}NO_4$

Phenylalanine — *N*-t-BOC-D-Phe *N*-Hydroxysuccinimide Ester

Sigma B 4394   FW 362.4 $C_{18}H_{22}N_2O_6$

Phenylalanine — *N*-t-BOC-L-(1,2,3,4-Tetrahydroisoquinoline)-3-Carboxylic Acid

Synonyms: BOC-L-Phe

Sigma B 1671   FW 277.3 $C_{15}H_{19}NO_4$ ≥95% | Methylene-bridged derivative of BOC-L-Phe

Phenylalanine — *N*-t-BOC-L-(3,4-Dihydroxy)-Phe Dicyclohexylammonium Salt

Synonyms: BOC-L-DOPA

Sigma B 5398   FW 478.6 $C_{14}H_{19}NO_6 \cdot C_{12}H_{23}N$

Phenylalanine — *N*-t-BOC-L-Phe (2,6-³H)

ARC ART-505   MW 265.3   30-60 Ci/mmol; 1.11-2.22 TBq/mmol; in EtOH | Radiochemical

Phenylalanine — *N*-t-BOC-L-Phe (3,4,5-³H)

ARC ART-615   60-90 Ci/mmol; 2.22-3.33 TBq/mmol; in EtOH | Radiochemical

Phenylalanine — *N*-t-BOC-L-Phe Free Acid

Sigma B 5394   FW 265.3 $C_{14}H_{19}NO_4$

Phenylalanine — *N*-t-BOC-L-Phe *N*-Hydroxysuccinimide Ester

Sigma B 1880   FW 362.4 $C_{18}H_{22}N_2O_6$

Phenylalanine — *N*-t-BOC-L-Phe Resin Ester

Sigma B 2896   *N*-t-BOC AA ester of methylated 1% divinylbenzene cross-linked polystyrene; 200-400 mesh; substitution range: 0.2-0.6 mmole t-BOC-AA/g resin | For peptide synthesis by the Merrifield method

Phenylalanine — *N*-t-BOC-L-Phe-(*p*ONp)

Sigma B 7135   FW 386.4 $C_{20}H_{22}N_2O_6$

Phenylalanine — *N*-t-BOC-L-Phe-Pam Resin Ester

Sigma B 5393   *N*-t-BOC AA ester of 4-(Oxymethyl)-Phenyl-Acm 1% divinylbenzene cross-linked polystyrene; 200-400 mesh; substitution range: 0.2-0.6 mmole t-BOC-AA/g resin | For peptide synthesis by the Merrifield method; Mitchell, AR et al, *J Org Chem*, 43: 2845, 1978

Phenylalanine — *N*-t-BOC-*N*-Me-D-Phe Dicyclohexylammonium Salt

Sigma B 3513   FW 460.7 $C_{15}H_{21}NO_4 \cdot C_{12}H_{23}N$

Phenylalanine — *N*-t-BOC-*N*-Me-L-Phe Dicyclohexylammonium Salt

Sigma B 3638   FW 460.7 $C_{15}H_{21}NO_4 \cdot C_{12}H_{23}N$

Phenylalanine — *N*-t-BOC-*p*-Chloro-D-Phe

Sigma B 9649   FW 299.8 $C_{14}H_{18}ClNO_4$

Phenylalanine — N-t-BOC-p-Chloro-L-Phe

Sigma B 8268   FW 299.8 $C_{14}H_{18}ClNO_4$

Phenylalanine — N-t-BOC-p-Fluoro-D-Phe

Sigma B 0296   FW 283.3 $C_{14}H_{18}FNO_4$

Phenylalanine — N-t-BOC-p-Fluoro-L-Phe

Sigma B 0421   FW 283.3 $C_{14}H_{18}FNO_4$

Phenylalanine — N-t-BOC-p-Iodo-D-Phe

Sigma B 0671   FW 391.2 $C_{14}H_{18}INO_4$

Phenylalanine — N-t-BOC-p-Iodo-L-Phe

Sigma B 0796   FW 391.2 $C_{14}H_{18}INO_4$

Phenylalanine — N-t-BOC-p-Nitro-L-Phe Dicyclohexylammonium Salt

Sigma B 2513   FW 491.6 $C_{14}H_{18}N_2O_6 \cdot C_{12}H_{23}N$

Phenylalanine — N-t-BOC-p-Nitro-L-Phe Free Acid

Sigma B 8390   FW 310.3 $C_{14}H_{18}N_2O_6$

Phenylalanine — N-TFA-L-Phe-OMe

Sigma T 5006   FW 275.2 $C_{12}H_{12}F_3NO_3$ ~98%

Phenylalanine — N-α-FMOC-L-Phe

ICN 151150   MW 387.4 $C_{24}H_{21}NO_4$

Phenylalanine — N$^\alpha$-t-BOC-L-Phe Free Acid

ICN 101090   MW 265.3 $C_{14}H_{19}NO_4$ Crystalline

Phenylalanine — N$^\alpha$-t-BOC-L-Phe-(pONp)

ICN 101091   MW 386.4 $C_{20}H_{22}N_2O_6$ Crystalline

Phenylalanine — N$^\alpha$-Tosyl-D,L-HPH-4-Amidinoanilide Hydrochloride

Synonyms: Pefabloc®PK; Pefa-2094

Pentapharm 380-03   MW 487.0 $C_{24}H_{26}N_4O_3S \cdot HCl$ Readily soluble in organic solvents (MeOH, EtOH, DMSO), soluble in H_2O (4mM) & aqueous buffers | Inhibits plasma kallikrein (PK) competitively; used in chromogenic assays for determination of factor XIIa activity in plasma with Pefachrome®FXIIa; suppresses undesired cleavage of substrate by plasma kallikrein; Stürzebecher J et al, *Thromb Res*, 55:709, 1989; Vieweg H et al, *Pharmazie*, 38818, 1983

Phenylalanine — N$^\alpha$-Tosyl-Phe CMK

Synonyms: TPCK; Chymotrypsin Inhibitor

Calbiochem 616387   MW 351.5 $C_{17}H_{18}ClNO_3S$ ≥99% (TLC); solid; soluble in EtOH or MeOH; LD_{50}≤2000 mg/kg | Irreversible inhibitor of chymotrypsin; inhibits chymotrypsin activity in trypsin preparations; inhibits apoptosis in thymocytes; blocks the induction of nitric oxide synthase by γ-interferon & lipopolysaccharides; Griscavage, JM et al, *Biochem Biophys Res Comm*, 215: 721, 1995; Fearnhead, HO et al, *FEBS Lett*, 357: 242, 1995; Weaver, VM et al, *Biochem Cell Biol*, 71: 488, 1993; Wilson, WE et al, *J Biol Chem*, 264: 17777, 1989

Oncogene 616387   MW 351.5 $C_{17}H_{18}ClNO_3S$ Solid; ≥99% (TLC); may cause irritation | Irreversible inhibitor of chymotrypsin; useful for inhibiting chymotrypsin activity in trypsin preparations; inhibits apoptosis in thymocytes; blocks the induction of nitric oxide synthase by γ-interferon & lipopolysaccharides; Grascavage, JM et al, *BBRC*, 215: 721, 1995; Fearnhead, HO et al, *FEBS Lett*, 357: 242, 1995; Weaver, VM et al, *Biochem Cell Biol*, 71: 488, 1993; Wilson, WE et al, *JBC*, 264: 17777, 1989

Phenylalanine — o-Chloro-DL-Phe

ICN 101355   MW 199.6 $C_9H_{10}ClNO_2$ Crystalline

Phenylalanine — o-Fluoro-DL-Phe

Sigma F 7263   FW 183.2 $C_9H_{10}FNO_2$

Phenylalanine — o-Fluoro-DL-Phe Hydrate

ICN 102518   MW 201.2 $C_9H_{10}FNO_2 \cdot H_2O$ Crystalline

Phenylalanine — p-Amino-DL-Phe

Bachem F-1230.0001   MW 180.21 $C_9H_{12}N_2O_2$ Store at -15°C

ICN 150359   MW 180.2 $C_9H_{12}N_2O_2$ Crystalline

Sigma A 1505   FW 180.2 $C_9H_{12}N_2O_2$

Phenylalanine — p-Amino-D-Phe

ICN 150358   MW 180.2 $C_9H_{12}N_2O_2$

Phenylalanine — p-Amino-D-Phe Hydrochloride

Bachem F-2855.0001   MW 216.67 $C_9H_{12}N_2O_2 \cdot HCl$ Store at -15°C

Sigma A 5338   FW 216.7 $C_9H_{12}N_2O_2 \cdot HCl$

Phenylalanine — p-Amino-L-Phe

Sigma A 8154   FW 180.2 $C_9H_{12}N_2O_2$

USBio A1379   MW 180.2 $C_9H_{12}N_2O_2$ ≥99%

Phenylalanine — p-Amino-L-Phe Hydrochloride

ICN 150360   MW 225.7 $C_9H_{12}N_2O_2 \cdot HCl \cdot \frac{1}{2}H_2O$

Phenylalanine — p-Amino-Phe Hydrochloride

Bachem F-1225.0001   MW 216.67 $C_9H_{12}N_2O_2 \cdot HCl$ Store at -15°C

Phenylalanine — p-Azido-Phe

Bachem F-3075.0001   MW 206.2 $C_9H_{10}N_4O_2$ Store at -15°C

Phenylalanine — p-Bromo-DL-Phe

Bachem F-1310.0001   MW 244.09 $C_9H_{10}BrNO_2$ Store at RT

ICN 101151   MW 244.1 $C_9H_{10}BrNO_2$ Crystalline

Sigma B 0880   FW 244.1 $C_9H_{10}BrNO_2$ Crystalline

Phenylalanine — p-Bromo-D-Phe

Bachem F-3700.0001   MW 244.09 $C_9H_{10}BrNO_2$ Store at RT

Phenylalanine — p-Bromo-Phe

Bachem F-1305.0001   MW 244.09 $C_9H_{10}BrNO_2$ Store at RT

Phenylalanine — p-Bz-D-Phe

Synonyms: D-Bpa

Bachem F-2810.0001   MW 269.3 $C_{16}H_{15}NO_3$ Store at RT

Phenylalanine — p-Bz-Phe

Synonyms: Bpa

Bachem F-2800.0001   MW 269.3 $C_{16}H_{15}NO_3$ Store at -15°C

Phenylalanine — p-Carboxy-Phe

Bachem F-3590.0001   MW 209.2 $C_{10}H_{11}NO_4$ Store at RT

Phenylalanine — p-Carboxy-Phe-OtBu

Bachem F-3585.0001   MW 265.31 $C_{14}H_{19}NO_4$ Store at -15°C

Phenylalanine — *p*-Chloro-DL-Phe

Alexis 106-030   MW 199.6 $C_9H_{10}ClNO_2$ ≥99%; white hygroscopic solid; soluble in dilute aqueous base

Bachem F-1450.0005   MW 199.64 $C_9H_{10}ClNO_2$ Store at RT

ICN 101365   MW 199.6 $C_9H_{10}ClNO_2$ Crystalline

Phenylalanine — *p*-Chloro-DL-Phe-OEt Hydrochloride

Synonyms: Phenylalanine-4-Hydrolase Inhibitor; Tryptophan-5-Hydrolase Inhibitor

ICN 150639   MW 264.2 $C_{11}H_{14}ClNO_2 \cdot HCl$

Phenylalanine — *p*-Chloro-DL-Phe-OMe Hydrochloride

Synonyms: Serotonin Biosynthesis Inhibitor

Bachem F-1455.0001   MW 250.12 $C_{10}H_{12}ClNO_2 \cdot HCl$ Store at 2-8°C

ICN 101366   MW 250.1 $C_{10}H_{12}ClNO_2 \cdot HCl$ Crystalline | Selective inhibitor

Phenylalanine — *p*-Chloro-D-Phe

Bachem F-2520.0001   MW 199.64 $C_9H_{10}ClNO_2$ Store at RT

Phenylalanine — *p*-Chloro-D-Phe-OMe Hydrochloride

Bachem F-2690.0001   MW 250.12 $C_{10}H_{12}ClNO_2 \cdot HCl$ Store at 2-8°C

Phenylalanine — *p*-Chloro-L-Phe

Alexis 106-031   MW 199.6 $C_9H_{10}ClN_2O_2$ ≥99%; white solid; soluble in acetic acid

Phenylalanine — *p*-Chloro-Phe

Bachem F-1445.0001   MW 199.64 $C_9H_{10}ClNO_2$ Store at RT

Phenylalanine — *p*-Fluoro-DL-Phe

Synonyms: Tryptophan Hydroxylase Inhibitor; Tyrosine Aminotransferase Inhibitor

Bachem F-1535.0001   MW 183.18 $C_9H_{10}FNO_2$ Store at RT

Sigma F 5251   FW 183.2 $C_9H_{10}FNO_2$ Less potent than *p*-chloroPhe in depleting brain serotonin; inhibits mitosis & reversibly arrests HeLa cells in G2

Phenylalanine — *p*-Fluoro-DL-Phe-OMe Hydrochloride

Bachem F-1540.0001   MW 233.67 $C_{10}H_{12}FNO_2 \cdot HCl$ Store at 2-8°C

Phenylalanine — *p*-Fluoro-D-Phe

Bachem F-2320.0001   MW 183.18 $C_9H_{10}FNO_2$ Store at RT

ICN 101708   MW 183.2 $C_9H_{10}FNO_2$ 99%; crystalline

ICN 158141   MW 183.2 $C_9H_{10}FNO_2$

Sigma F 4391   FW 183.2 $C_9H_{10}FNO_2$ Inactive isomer of *p*-fluorophenylalanine

Phenylalanine — *p*-Fluoro-D-Phe-OMe Hydrochloride

Bachem F-3850.0001   MW 233.67 $C_{10}H_{12}FNO_2 \cdot HCl$ Store at 2-8°C

Phenylalanine — *p*-Fluoro-L-Phe

Synonyms: Tryptophan Hydroxylase Inhibitor; Tyrosine Aminotransferase Inhibitor

Sigma F 4646   FW 183.2 $C_9H_{10}FNO_2$ Less potent than *p*-chloroPhe in depleting brain serotonin; inhibits mitosis & reversibly arrests HeLa cells in G2

USBio F5200   MW 183.2 $C_9H_{10}FNO_2$ ≥98%

Phenylalanine — *p*-Fluoro-Phe

Bachem F-1530.0001   MW 183.18 $C_9H_{10}FNO_2$ Store at RT

Phenylalanine — *p*-Fluoro-Phe-OEt Hydrochloride

Bachem F-3820.0001   MW 247.7 $C_{11}H_{14}FNO_2 \cdot HCl$ Store at -15°C

Phenylalanine — Phe

Bachem E-2230.0025   MW 165.19 $C_9H_{11}NO_2$ Store at RT

Phenylalanine — Phe (^{15}N)

Bachem E-3240.0100   MW 166.19 $C_9H_{11}NO_2$ Store at -15°C

Phenylalanine — Phe (Ring 2,4-^{3}H)

ICN 20053   MW 165.2 $C_6H_5CH_2CH(NH_2)COOH$ 40-60 Ci/mmol, 1.48-2.22 TBq/mmol; sterile $EtOH:H_2O$ (2:98); HPLC analyzed for purity

Phenylalanine — Phe Agar

Fluka 78052   Composition: 3 g/L yeast extract, 2 g/L DL-Phe, 1 g/L disodium hydrogen phosphate, 5 g/L NaCl, 12 g/L agar | Test medium for detecting phenylalanine deaminase for the identification of proteus & *providencia* alongside other *Enterobacteriaceae* according to Ewing et al, 1957; *Schweizerisches Lebensmittelbuch*, 5th ed, 56A

Phenylalanine — Phe Amide

Bachem E-2235.0001   MW 164.21 $C_9H_{12}N_2O$ Store at -15°C

Phenylalanine — Phe Amide Hydrochloride

Bachem E-2240.0001   MW 200.67 $C_9H_{12}N_2O \cdot HCl$ Store at RT

Phenylalanine — Phe-4MβNA

Bachem J-1270.0250   MW 320.4 $C_{20}H_{20}N_2O_2$ Store at -15°C

Phenylalanine — Phe-AMC

Synonyms: Aminopeptidase Substrate

Peptides International MFM-3148-v   MW 322.36 $C_{19}H_{18}N_2O_3$ >99% (HPLC); lyophilized amorphous powder; Tosyl form

Phenylalanine — Phe-AMC TFA

Bachem I-1285.0050   MW 436.39 $C_{19}H_{18}N_2O_3 \cdot C_2HF_3O_2$ Store at -15°C

Phenylalanine — Phe-CMK Hydrochloride

Bachem *N*-1060.0250   MW 234.13 $C_{10}H_{12}ClNO \cdot HCl$ Store at -15°C

Phenylalanine — Phe-Cyclohexylamide

Bachem E-3270.0001   MW 246.36 $C_{15}H_{22}N_2O$ Store at 2-8°C

Phenylalanine — Phe-NHNH-Z TFA

Bachem E-2265.0001   MW 427.38 $C_{17}H_{19}N_3O_3 \cdot C_2HF_3O_2$ Store at RT

Phenylalanine — Phenylacetyl-Phe

Bachem E-2675.0005   MW 283.33 $C_{17}H_{17}NO_3$ Store at RT

Phenylalanine — Phe-OBzl Hydrochloride

Bachem E-3080.0005   MW 291.78 $C_{16}H_{17}NO_2 \cdot HCl$ Store at RT

Phenylalanine — Phe-OBzl *p*-Tosyl

Bachem E-2245.0005   MW 427.52 $C_{16}H_{17}NO_2 \cdot C_7H_8O_3S$
Store at RT

ICN 102619   MW 427.5 $C_{16}H_{17}NO_3 \cdot C_7H_8O_3S$ Crystalline

Phenylalanine — Phe-OBzl-Tosyl

Synonyms: L-Phe-OBzl-Tosyl

Peptides International EBF-2110   MW 427.52   >98%
(TLC); white powder

Phenylalanine — Phe-OEt Hydrochloride

Synonyms: L-Phe-OEt

Bachem E-2260.0025   MW 229.71 $C_{11}H_{15}NO_2 \cdot HCl$ Store
at RT

Peptides International EEF-2041-PI   MW 229.71   >98%
(TLC); white powder

Phenylalanine — Phe-ol

Synonyms: L-Phenylalaninol

Senn Chem 44102   MW 151.2   Inhibits binding of Phe to Phe
t-RNA-synthetase

Phenylalanine — Phe-OMe Hydrochloride

Bachem E-2270.0005   MW 215.68 $C_{10}H_{13}NO_2 \cdot HCl$ Store
at RT

Phenylalanine — Phe-OtBu Hydrochloride

Bachem E-2255.0001   MW 257.76 $C_{13}H_{19}NO_2 \cdot HCl$ Store
at 2-8°C

Phenylalanine — Phe-pNA

Bachem L-1355.0001   MW 285.3 $C_{15}H_{15}N_3O_3$ Store at -15°C

Phenylalanine — Phe-Pyrrolidide

Bachem E-3275.0001   MW 218.3 $C_{13}H_{18}N_2O$ Store at 2-8°C

Phenylalanine — Phe-SBzl Free Base

Bachem E-2850.0001   MW 271.38 $C_{16}H_{17}NOS$ Store at
-15°C

Phenylalanine — Phe-βNA

Bachem K-1445.0001   MW 290.37 $C_{19}H_{18}N_2O$ Store at RT

Phenylalanine — *p*-Iodo-DL-Phe

Synonyms: 2-Amino-3-(4-Iodophenyl)-Propanoic Acid

Bachem F-1675.0001   MW 291.09 $C_9H_{10}INO_2$ Store at RT

ICN 155064   MW 291.1 $C_9H_{10}NO_2I$ Crystalline

Sigma I 4628   FW 291.1 $C_9H_{10}INO_2$

Phenylalanine — *p*-Iodo-D-Phe

Synonyms: 2-Amino-3-(4-Iodophenyl)-Propanoic Acid

Bachem F-1670.0250   MW 291.09 $C_9H_{10}INO_2$ Store at RT

ICN 155063   MW 291.1 $C_9H_{10}NO_2I$

Sigma I 7509   FW 291.1 $C_9H_{10}INO_2$

Phenylalanine — *p*-Iodo-L-Phe

Synonyms: 2-Amino-3-(4-Iodophenyl)-Propanoic Acid

ICN 155065   MW 291.1 $C_9H_{10}NO_2I$

Sigma I 8757   FW 291.1 $C_9H_{10}INO_2$

Phenylalanine — *p*-Iodo-Phe

Bachem F-1665.0001   MW 291.09 $C_9H_{10}INO_2$ Store at RT

Phenylalanine — *p*-Me-D-Phe

Bachem F-3785.0001   MW 179.22 $C_{10}H_{13}NO_2$ Store at RT

Phenylalanine — *p*-Me-Phe

Bachem F-3780.0001   MW 179.22 $C_{10}H_{13}NO_2$ Store at RT

Phenylalanine — *p*-Nitro-DL-Phe

Synonyms: Phenylalanine Antagonist

Bachem F-1905.0005   MW 210.19 $C_9H_{10}N_2O_4$ Store at RT

ICN 102468   MW 210.2 $C_9H_{10}N_2O_4$ Light yellow crystals

Sigma N 7014   FW 210.2 $C_9H_{10}N_2O_4$

Phenylalanine — *p*-Nitro-D-Phe

Bachem F-1900.0001   MW 210.19 $C_9H_{10}N_2O_4$ Store at RT

Phenylalanine — *p*-Nitro-D-Phe Hydrate

Sigma N 0759   FW 228.2 $C_9H_{10}N_2O_4 \cdot H_2O$

Phenylalanine — *p*-Nitro-L-Phe Hydrate

ICN 151757   MW 228.2 $C_9H_{10}N_2O_4 \cdot H_2O$ Crystalline

Sigma N 0884   FW 228.2 $C_9H_{10}N_2O_4 \cdot H_2O$

Phenylalanine — *p*-Nitro-Phe

Bachem F-1895.0005   MW 210.19 $C_9H_{10}N_2O_4$ Store at RT

Phenylalanine — *p*-Nitro-Phe-OMe Hydrochloride

Bachem F-1910.0001   MW 260.68 $C_{10}H_{14}N_2O_4 \cdot HCl$ Store
at 2-8°C

Phenylalanine — *p*-tBu-D-Phe

Bachem F-3795.0001   MW 221.3 $C_{13}H_{19}NO_2$ Store at RT

Phenylalanine — *p*-tBu-Phe

Bachem F-3790.0001   MW 221.3 $C_{13}H_{19}NO_2$ Store at RT

Phenylalanine — Pth-Phe

Sigma P 2127   FW 282.4 $C_{16}H_{14}N_2OS$ ~99% | Useful as
standards in protein sequencing by the Edman degradation

Phenylalanine — Suc-Phe-OMe

Bachem E-2820.0001   MW 279.29 $C_{14}H_{17}NO_5$ Store at 2-8°C

Phenylalanine — Suc-Phe-pNA

Bachem L-1420.0001   MW 385.38 $C_{19}H_{19}N_3O_6$ Store at
-15°C

Phenylalanine — TentaGel S PHB-Phe-FMOC

Synonyms: FMOC-L-Phe 4-(Poly(Ethylenoxy))-OBzl Polymer Bound

Fluka 86399   0.22 mmol protected AA/g; ~90 μm particle size

Phenylalanine — TentaGel S Trt-Phe-FMOC

Fluka 86431   0.20 mmol protected AA/g; ~90 μm particle size

Phenylalanine — TFA-Phe-OMe

Bachem E-3625.0001   MW 275.23 $C_{12}H_{12}F_3NO_3$ Store at
-15°C

Phenylalanine — Tosyl-L-Phe CMK

Synonyms: TPCK; (S)-1-Chloro-4-Phenyl-3-Tosylamino-2-Butanone; (S)-1-Chloro-3-Tosylamido-4-Phenyl-2-Butanone; Tosyl-L-Phe-Chloromethane; Chymotrypsin Inhibitor; Protease Inhibitor

Fluka 90184 MW 351.85 $C_{17}H_{18}ClNO_3S$ ~98.0% (HPLC); mp: 101-106°C | Schoellmann, G & Shaw, E, *Biochemistry*, 2: 252, 1963; Goldberg, AL et al, *Meth Enzymo*, 80: 680, 1981; Bender, ML & Brubacher, LJ, *Am Chem Soc*, 88: 5880, 1966; Wolthers, BC, *FEBS Lett*, 2: 143, 1969; Urban, MK et al, *Biochemistry*, 18: 3952, 1979

Phenylalanine — Tosyl-Phe

Bachem E-2490.0010 MW 319.38 $C_{16}H_{17}NO_4S$ Store at RT

Phenylalanine — Tosyl-Phe-CMK

Synonyms: TPCK

Bachem N-1160.0500 MW 351.85 $C_{17}H_{18}ClNO_3S$ Store at -15°C

Phenylalanine — Trt-D-Phe DEA

Bachem F-2590.0001 MW 480.65 $C_{28}H_{25}NO_2 \cdot C_4H_{11}N$ Store at -15°C

Phenylalanine — Trt-Phe DEA

Bachem E-2830.0001 MW 480.65 $C_{28}H_{25}NO_2 \cdot C_4H_{11}N$ Store at -15°C

Phenylalanine — Z-4-Phenyl-D-Phe

Synonyms: Z-D-Bip; Z-β-4-Biphenylyl-D-Ala

Bachem C-4035.0001 MW 375.42 $C_{23}H_{21}NO_4$ Store at RT

Phenylalanine — Z-4-Phenyl-Phe

Synonyms: Z-β-4-Biphenylyl-Ala; Z-Bip

Bachem C-4030.0001 MW 375.42 $C_{23}H_{21}NO_4$ Store at RT

Phenylalanine — Z-DL-Phe

Bachem C-2320.0025 MW 299.33 $C_{17}H_{17}NO_4$ Store at RT

Fluka 97005 MW 299.33 $C_{17}H_{17}NO_4$ ≥98.0% (TLC)

Phenylalanine — Z-D-Phe

Bachem C-2315.0001 MW 299.33 $C_{17}H_{17}NO_4$ Store at RT

Neosystem ZA01401 MW 299.3

Phenylalanine — Z-D-Phe-(4-ONp)

Fluka 97018 MW 420.42 $C_{23}H_{20}N_2O_6$ ≥98.0% (HPLC)

Phenylalanine — Z-D-Phe-ONp

Bachem C-2350.0001 MW 420.42 $C_{23}H_{20}N_2O_6$ Store at -15°C

Phenylalanine — Z-D-Phe-OSu

Bachem C-3925.0001 MW 396.4 $C_{21}H_{20}N_2O_6$ Store at -15°C

Phenylalanine — Z-L-Phe

Synonyms: Thermolysin Inhibitor

Fluka 97000 MW 299.33 $C_{17}H_{17}NO_4$ ≥99.0% (titration); ≤0.05% residue on ignition; mp: 85-87°C | Shieh, TL & Byrn, SR, *JMC*, 25: 403, 1982

Neosystem ZA01402 MW 299.3

Phenylalanine — Z-L-Phe 2-Naphthyl Ester

Fluka 97015 MW 425.48 $C_{27}H_{23}NO_4$ ≥98.0% (TLC)

Phenylalanine — Z-L-Phe-(4-ONp)

Fluka 97022 MW 420.42 $C_{23}H_{20}N_2O_6$ ≥98.0% (HPLC)

Phenylalanine — Z-N-Me-L-Phe

Fluka 96927 MW 313.35 $C_{18}H_{19}NO_4$ ≥98.0% (TLC)

Phenylalanine — Z-N-Me-Phe

Bachem C-2245.0001 MW 313.35 $C_{18}H_{19}NO_4$ Store at RT

Phenylalanine — Z-p-Carboxy-Phe-OtBu

Bachem C-3975.0001 MW 399.44 $C_{22}H_{25}NO_6$ Store at -15°C

Phenylalanine — Z-p-Fluoro-Phe

Bachem C-3525.0001 MW 317.32 $C_{17}H_{16}FNO_4$ Store at RT

Phenylalanine — Z-p-Fluoro-Phe-CMK

Bachem N-1265.0001 MW 349.79 $C_{18}H_{17}ClFNO_3$ Store at -15°C

Phenylalanine — Z-Phe

Synonyms: Benzyloxycarbonyl-L-Phe

Bachem C-2310.0005 MW 299.33 $C_{17}H_{17}NO_4$ Store at RT

Peptides International ZLF-2021-PI MW 299.33 >98% (TLC); white powder

Phenylalanine — Z-Phe Amide

Bachem C-2325.0001 MW 298.34 $C_{17}H_{18}N_2O_3$ Store at RT

Phenylalanine — Z-Phe-CMK

Synonyms: ZPCK; CBZ-PCK; N^α-Benzyloxycarbonyl-L-Phe-CMK

Bachem N-1035.0250 MW 331.8 $C_{18}H_{18}ClNO_3$ Store at -15°C

USBio C2099-47 MW 331.8 $C_{18}H_{18}ClNO_3$ ≥99%

Phenylalanine — Z-Phe-OBzl

Bachem C-2330.0001 MW 389.45 $C_{24}H_{23}NO_4$ Store at RT

Phenylalanine — Z-Phe-ol

Synonyms: Z-L-Phenylalaninol

Senn Chem 44076 MW 285.3

Phenylalanine — Z-Phe-OMe

Bachem C-2340.0001 MW 313.35 $C_{18}H_{19}NO_4$ Store at -15°C

Phenylalanine — Z-Phe-ONp

Bachem C-2345.0005 MW 420.42 $C_{23}H_{20}N_2O_6$ Store at -15°C

Phenylalanine — Z-Phe-OSu

Bachem C-2335.0005 MW 396.4 $C_{21}H_{20}N_2O_6$ Store at -15°C

Phenylalanine — Z-Phe-pNA

Bachem L-1520.0001 MW 419.44 $C_{23}H_{21}N_3O_5$ Store at -15°C

Phenylalanine — α-Keto-Phe Calcium Salt

Synonyms: α-Oxo-Benzenepropanoic Acid; Phenylpyroracemic Acid; Phenylpyruvate; 2-Oxo-3-Phenylpropionate,

Rexim MW 366.4 $C_{18}H_{14}CaO_6$ White crystals or crystalline powder

Phenylalanine — α-Me-DL-Phe

Synonyms: (±)-2-Amino-2-Me-3-Phenylpropionic Acid

Bachem F-1805.0001   MW 179.22 $C_{10}H_{13}NO_2$ Store at RT

Fluka 68627   MW 179.22 $C_{10}H_{13}NO_2$ ≥98.0%; mp: 293-294°C

ICN 155626   MW 179.2 $C_{10}H_{13}NO_2$

Sigma M 3635   FW 179.2 $C_{10}H_{13}NO_2$ Crystalline

Phenylalanine — α-Me-DL-Phe-OMe Hydrochloride

Synonyms: (±)-Me-2-Amino-2-Me-3-Phenylpropionate

Bachem F-2805.0001   MW 229.71 $C_{11}H_{15}NO_2 \cdot HCl$ Store at 2-8°C

Fluka 68628   MW 229.7 $C_{11}H_{15}NO_2 \cdot HCl$ ≥98.0% (titration); mp: 140-143°C

Phenylalanine — α-Me-D-Phe

Bachem F-3115.0001   MW 179.22 $C_{10}H_{13}NO_2$ Store at RT

Phenylalanine — α-Me-*m*-Methoxy-DL-Phe

ICN 155600   MW 209.2 $C_{11}H_{15}NO_3$

Sigma M 4377   FW 209.2 $C_{11}H_{15}NO_3$

Phenylalanine — α-Me-Phe

Bachem E-3150.0001   MW 179.22 $C_{10}H_{13}NO_2$ Store at RT

Pra — Pra

Synonyms: Propargyl-Gly

Bachem F-2040.0001   MW 113.12 $C_5H_7NO_2$ Store at -15°C

Pra — Pra-OMe Hydrochloride

Synonyms: Propargyl-Gly-OMe

Bachem F-2075.0250   MW 163.6 $C_6H_9NO_2 \cdot HCl$ Store at 2-8°C

Proline — (2S)-1-(3-Mercapto-2-Me-Propionyl)-L-Pro

Synonyms: Captopril

ICN 154928   MW 217.3 $C_9H_{15}NO_3S$

Proline — 3,4-Dehydro-DL-Pro

Synonyms: 3,4-Didehydro-DL-Pro; (±)-3-Pyrroline-2-Carboxylic Acid; (±)-2,5-Dihydro-1H-Pyrrole-2-Carboxylic Acid; L-Pro Antagonist

Bachem F-2705.0250   MW 113.12 $C_5H_7NO_2$ Store at -15°C

Fluka 30900   MW 113.12 $C_5H_7NO_2$ ≥99% (titration); mp: 245-250°C | Mauger, AB & Witkop, B, *Chem Rev*, 66: 47, 1966

ICN 190318   MW 113.1 $C_5H_7NO_2$ Specific & powerful

Sigma D 0265   FW 113.1 $C_5H_7NO_2$

Proline — 3,4-Dehydro-L-Pro

Synonyms: 3,4-Didehydro-L-Pro; (S)-3-Pyrroline-2-Carboxylic Acid

Fluka 30890   MW 113.12 $C_5H_7NO_2$ ≥99% (titration); mp: 239-241°C | Mauger, AB & Witkop, B, *Chem Rev*, 66: 47, 1966

ICN 157537   MW 113.1 $C_5H_7NO_2$ Crystalline

Sigma D 4893   FW 113.1 $C_5H_7NO_2$

USBio D2000   MW 113.1 $C_5H_7NO_2$ ≥99%

Proline — 3,4-Dehydro-L-Pro-OMe Hydrochloride

ICN 157538   MW 163.6 $C_6H_9NO_2 \cdot HCl$ Crystalline

Sigma D 5018   FW 163.6 $C_6H_9NO_2 \cdot HCl$

Proline — 3,4-Dehydro-Pro

Bachem F-1490.0050   MW 113.12 $C_5H_7NO_2$ Store at -15°C

Proline — 3,4-Dehydro-Pro Amide Hydrochloride

Bachem F-1495.0050   MW 148.59 $C_5H_8N_2O \cdot HCl$ Store at -15°C

Proline — 3,4-Dehydro-Pro-OMe Hydrochloride

Bachem F-1500.0250   MW 163.6 $C_6H_9NO_2 \cdot HCl$ Store at -15°C

Proline — Ac-DL-Pro

Bachem F-2380.0005   MW 157.17 $C_7H_{11}NO_3$ Store at RT

Proline — Ac-D-Pro

Bachem F-1075.0001   MW 157.17 $C_7H_{11}NO_3$ Store at RT

Proline — Ac-Pro

Bachem E-1180.0005   MW 157.17 $C_7H_{11}NO_3$ Store at RT

Proline — Ac-Pro Amide

Bachem E-1185.0001   MW 156.19 $C_7H_{12}N_2O_2$ Store at RT

Proline — Ac-Pro-OMe

Bachem E-1195.0001   MW 171.2 $C_8H_{13}NO_3$ Store at RT

Proline — BOC-3,4-Dehydroproline

American Peptide BLPRO15   MW 213.3

Bachem A-1550.0001   MW 213.23 $C_{10}H_{15}NO_4$ Store at -15°C

USBio B2296   MW 213.2 $C_{10}H_{15}NO_4$ ≥99%

Proline — BOC-4-Piperidyl-L-Pro

Neosystem BA09602   MW 298.38

Proline — BOC-Dehydroproline

Synonyms: BOC-3,4-Dehydro-L-Pro

Peptides International BLX-5324-PI   MW 197.23   >98% (HPLC); white to off-white powder

Proline — BOC-D-Pro

American Peptide BDPRO10   MW 215.3

Bachem A-2240.0001   MW 215.25 $C_{10}H_{17}NO_4$ Store at RT

Neosystem BA01501   MW 215.3

Peptides International BDP-2610   MW 215.25   >98% (HPLC); white crystalline powder

USBio B2387   MW 215.3 $C_{10}H_{17}NO_4$ ≥99%

Proline — BOC-D-Pro-ol

Senn Chem 44042   MW 201.3   Liquid

Proline — BOC-D-Pro-OSu

Bachem A-2250.0001   MW 312.32 $C_{14}H_{20}N_2O_6$ Store at -15°C

Proline — BOC-HypBzl

USBio B2324   MW 321.4 $C_{17}H_{23}NO_5$ ≥99%

Proline — BOC-Hyp-Dicyclohexylamine

USBio B2323   MW 412.6 $C_{10}H_{17}NO_5 \cdot C_{12}H_{23}N$ ≥99%

Proline — BOC-L-Pro

Fluka 15490   MW 215.25 $C_{10}H_{17}NO_4$ ≥99% (titration); ≤0.05% residue on ignition; mp: 133-135°C

Neosystem BA01502   MW 215.3

Proline — BOC-L-Pro Hycram™-Resin

Fluka 15492 0.74 mmol Pro/g resin; 100-200 mesh particle size; carrier: polystyrene, crosslinked with 1% DVB | BOC-L-Pro esterified with Hycram™-resin; useful for synthesis of peptides; Kunz, H & Dombo, B, *Angew Chem*, 100: 732, 1988

Proline — BOC-L-Pro Hydroxysuccinimide Ester

Fluka 15491 MW 346.4 $C_{14}H_{20}N_2O_6$ ~99%; ≤0.05% residue on ignition; mp: 133-135°C

Proline — BOC-L-Pro-MBHA Resin

Neosystem RN01502

Proline — BOC-L-Pro-O-CH₂-◆-CH₂-COOH

Neosystem LP01502 MW 363.5

Proline — BOC-L-Pro-PAM Resin

Neosystem RP01502

Proline — BOC-L-Thioproline

Synonyms: BOC-L-Thiazolidine-4-Carboxylic Acid

Bachem A-3945.0005 MW 233.29 $C_9H_{15}NO_4S$ Store at 2-8°C

Proline — BOC-Pro

Synonyms: BOC-L-Pro

American Peptide BLPRO10 MW 215.3

Bachem A-2235.0005 MW 215.25 $C_{10}H_{17}NO_4$ Store at RT

Peptides International BLP-2056 MW 215.25 >98% (HPLC); white crystalline powder

USBio B2388 MW 215.3 $C_{10}H_{17}NO_4$ ≥99%

Proline — BOC-Pro-® (200-400 mesh)

Bachem D-1445.0005 Store at RT

Proline — BOC-Pro-ol

Synonyms: BOC-L-Prolinol

Senn Chem 44041 MW 201.3 Liquid

Proline — BOC-Pro-ONp

Bachem A-2255.0005 MW 336.35 $C_{16}H_{20}N_2O_6$ Store at -15°C

Proline — BOC-Pro-O-Resin

American Peptide RMRB245 1% DVB cross-linked: 100-200 mesh | t-BOC protected AA resin; preattached resins are useful for synthesis of peptides with C-terminal carboxyl group by t-BOC chemistry

Proline — BOC-Pro-OSu

Bachem A-2245.0001 MW 312.32 $C_{14}H_{20}N_2O_6$ Store at -15°C

Proline — BOC-Pro-PAM-® (200-400 mesh)

Bachem D-1540.0001 Store at 2-8°C

Proline — BOC-Pro-PAM-Resin

American Peptide RPAM245 1% DVB cross-linked: 100-200 mesh | t-BOC protected AA resin; preattached resins are useful for synthesis of peptides with C-terminal carboxyl group by t-BOC chemistry

Proline — BOC-Pro-psi(CH₂NH)-Gly

Neosystem BB01501 MW 258.31

Proline — Bsmoc-Pro

Synonyms: Bsmoc-L-Pro

Peptides International BLP-5634-PI MW 337.35 >98% (HPLC); white to off-white powder | LA Carpino, et al, *JACS*, 119:9915, 1997

Proline — CBZ-D-Pro

USBio C2099-50 MW 249.3 $C_{13}H_{15}NO_4$ ≥99%

Proline — CBZ-Hyp

USBio C2098-92 MW 265.3 $C_{13}H_{15}NO_5$ ≥99%

Proline — CBZ-Pro

USBio C2099-49 MW 249.3 $C_{13}H_{15}NO_4$ ≥99%

Proline — CBZ-Pro-ONp

USBio C2099-54 MW 370.4 $C_{19}H_{18}N_2O_6$ ≥99%

Proline — cis-4-Fluoro-Pro

Synonyms: (2S,4S)-4-Fluoro-Pyrrolidine-2-Carboxylic Acid

Bachem F-3970.0250 MW 133.12 $C_5H_8FNO_2$ Store at RT

Proline — cis-4-Hydroxy-L-Pro

Peptides International ALX-5330-PI MW 131.13 >98% (TLC); off white powder

Proline — Dansyl-L-Pro Cyclohexylammonium Salt

ICN 100125

Proline — Dansyl-L-Pro Free Acid

Sigma D 1875 FW. 348.4 $C_{17}H_{20}N_2O_4S$

Proline — Dinitropyridyl-L-Pro Monocyclohexylammonium Salt

ICN 102727 MW 381.4 $C_{10}H_{10}N_4O_6 \cdot C_6H_{13}N$

Proline — DL-Pro

Synonyms: (±)-Pyrrolidine-2-Carboxylic Acid

Bachem F-2050.0005 MW 115.13 $C_5H_9NO_2$ Store at RT

Fluka 81720 MW 115.13 $C_5H_9NO_2$ ~99.0% (titration); mp: 210°C

ICN 102725 MW 115.1 $C_5H_9NO_2$ Crystalline

Sigma P 0255 FW 115.1 $C_5H_9NO_2$ ≥99% (TLC) | Also available as part of a kit

USBio P9010-15 MW 115.1 $C_5H_9NO_2$ ≥99%

Proline — DL-Pro Amide Hydrochloride

Bachem F-2385.0001 MW 150.61 $C_5H_{10}N_2O \cdot HCl$ Store at -15°C

Proline — DL-Pro-OMe Hydrochloride

Bachem F-2070.0005 MW 165.62 $C_6H_{11}NO_2 \cdot HCl$ Store at 2-8°C

Proline — Dnp-L-Pro

ICN 102723 MW 281.2 $C_{11}H_{11}N_3O_6$ Crystalline

Proline — Dnp-Pro

Bachem E-1835.0250 MW 281.23 $C_{11}H_{11}N_3O_6$ Store at -15°C

Proline — D-Pro

Synonyms: (R)-Pyrrolidine-2-Carboxylic Acid

Bachem F-2045.0001 MW 115.13 $C_5H_9NO_2$ Store at RT

Fluka 81705 MW 115.13 $C_5H_9NO_2$ ≥99.0% (titration); mp: 220-223°C | Asato, AE et al, *Tetrahedron Lett*, 33: 3105, 1992

ICN 195456 MW 115.1 $C_5H_9NO_2$ ≥99%; crystalline

Neosystem AA01501 MW 115.1

Peptides International ADP-2816 MW 115.13 >99.9% optical purity by RP-HPLC; white crystalline powder

Rexim MW 115.1 $C_5H_9NO_2$

Sigma 85,891-9 FW 115.1 $C_5H_9NO_2$ ≥99% (TLC) | Aldrich Brand, formerly Sigma P 4266

USBio P9010-10 MW 115.1 $C_5H_9NO_2$ ≥99%

Proline — D-Pro Amide

Bachem F-2960.0001 MW 114.15 $C_5H_{10}N_2O$ Store at 2-8°C

Proline — D-Pro Amide Hydrochloride

Bachem F-2965.0001 MW 150.61 $C_5H_{10}N_2O \cdot HCl$ Store at 2-8°C

Proline — D-Pro-OBzl Hydrochloride

Bachem F-2055.0001 MW 241.72 $C_{12}H_{15}NO_2 \cdot HCl$ Store at 2-8°C

Proline — D-Pro-ol

Synonyms: R(-)-2-Hydroxymethylpyrrolidine

Senn Chem 44105 MW 101.2 Liquid

Proline — D-Pro-OMe Hydrochloride

Bachem F-2065.0001 MW 165.62 $C_6H_{11}NO_2 \cdot HCl$ Store at 2-8°C

Proline — D-Pro-OtBu Syrup

Bachem F-2060.0001 MW 171.24 $C_9H_{17}NO_2$ Store at -15°C

Proline — FA-Pro

Bachem M-1415.0250 MW 235.24 $C_{12}H_{13}NO_4$ Store at -15°C

Proline — FMOC-3,4-Dehydro-L-Pro

USBio F5364 MW 335.4 $C_{20}H_{17}NO_4$ ≥99%

Proline — FMOC-3,4-Dehydroproline

Bachem B-1660.0001 MW 335.36 $C_{20}H_{17}NO_4$ Store at -15°C

Proline — FMOC-4-Piperidyl-L-Pro

Neosystem FA09602 MW 420.51

Proline — FMOC-D-Pro

Bachem B-1625.0001 MW 337.38 $C_{20}H_{19}NO_4$ Store at RT

Fluka 47532 MW 337.38 $C_{20}H_{19}NO_4$ ≥98% (TLC)

Neosystem FA01501 MW 337.4

Peptides International FDP-1826-PI MW 337.38 >98% (HPLC); white crystals

USBio F5451 MW 337.4 $C_{20}H_{19}NO_4$ ≥99%

Proline — FMOC-Hyp

USBio F5393 MW 353.4 $C_{20}H_{19}NO_5$ ≥99%

Proline — FMOC-HyptBu

USBio F5394 MW 409 $C_{28}H_{27}NO_5$ ≥99%

Proline — FMOC-L-Pro

Fluka 47636 MW 337.38 $C_{20}H_{19}NO_4$ ≥99% (HPLC); mp: 113-116°C

Neosystem FA01502 MW 337.4

Proline — FMOC-L-Pro 2-Chlorotrityl Ester Polymer Bound

Fluka 47643 0.4-0.6 mmol/g resin; carrier: polystyrene, crosslinked with 1% DVB | Barlos, K et al, *Tetrahedron Lett*, 30: 3947, 1989

Proline — FMOC-L-Pro 4-Benzyloxybenzyl Ester Polymer Bound

Fluka 47667 0.4-0.6 mmol/g resin; carrier: polystyrene, crosslinked with 1% DVB; 200-400 mesh particle size

Proline — FMOC-L-Pro-Wang Resin

American Peptide RFWAN75 Preattached Wang resins are usful for synthesis of peptides with C-terminal carboxyl groups; peptides can be cleaved from the resin with 90% TFA in the presence of scavengers

Neosystem RW01502

Proline — FMOC-Pro

Synonyms: FMOC-L-Pro

American Peptide FPRO100 MW 337.4

Bachem B-1420.0005 MW 337.38 $C_{20}H_{19}NO_4$ Store at RT

Peptides International FLP-1731-PI MW 337.38 >98% (HPLC); white crystalline powder

USBio F5452 MW 337.4 $C_{20}H_{19}NO_4$ ≥99%

Proline — FMOC-Pro-® (200-400 mesh)

Bachem D-1145.0001 Store at RT

Proline — FMOC-Pro-DHPP-® (200-400 mesh)

Bachem D-1830.0001 Store at 2-8°C

Proline — FMOC-Pro-ol

Senn Chem 44064 MW 323.5

Proline — FMOC-Pro-ONp

Bachem B-1425.0001 MW 458.47 $C_{26}H_{22}N_2O_6$ Store at -15°C

Proline — FMOC-Pro-OPfp

Bachem B-1175.0001 MW 503.43 $C_{26}H_{18}F_5NO_4$ Store at -15°C

USBio F5453 MW 503.4 $C_{26}H_{18}F_5NO_5$ ≥99%

Proline — FMOC-Pro-OSu

USBio F5454 MW 434.5 $C_{28}H_{24}N_2O_6$ ≥99%

Proline — FMOC-Pro-SASRIN™-® (200-400 mesh)

Bachem D-1380.0001 Store at 2-8°C

Proline — FMOC-*trans*-4-Phenyl-L-Pro

Synonyms: (2S,4S)-1-FMOC-4-Phenylpyrrolidine-2-Carboxylic Acid

Fluka 00417 MW 413.47 $C_{26}H_{23}NO_4$ ≥98% (HPLC)

Proline — Hydroxy-L-Pro

Synonyms: L-4-Hydroxy-Pyrrolidine Carboxylic Acid

USBio H9110-10   MW 131.1 $C_5H_9NO_3$ ≥98%; peptide grade; appearance: white to off-white crystalline powder; specific rotation (C=4, H_2O): -74.0° to -77.0°; heavy metals (Pb): ≤0.001%

Proline — L-Pro

Synonyms: (*S*)-Pyrrolidine-2-Carboxylic Acid; (-)-2-Pyrrolidinecarboxylic Acid

Amersham US20730   98.5-101.5% dsb; ≤0.0015% heavy metals (Pb)

Fluka 81709   MW 115.13 $C_5H_9NO_2$ ≥99.5% (titration); ≤0.1% residue on ignition, loss on drying; ≤0.3% foreign AA; ≤0.00001% As; ≤0.0005% Al, Ba, Bi, Cd, Co, Cr, Cu, Fe, Li, Mg, Mn, Mo, Ni, Pb, Sr, Zn; ≤0.001% Ca; ≤0.005% Na, K; ≤0.01% Cl, SO_4; ≤0.002% NH_4^+ | Paleg, LG et al, *Aust J Plant Physiol*, 8: 107, 1981; *CRC Handbook of Microbiology*, Laskin, AI & Lachevalier, HA, eds, 4: 1, 1974, CRC Press, Cleveland, Ohio

Fluka 81710   MW 115.13 $C_5H_9NO_2$ ≥99.0% (titration); ≤1.5% D-Pro; ≤0.1% residue on ignition; ≤0.2% loss on drying; ≤0.3% foreign AA; ≤0.0005% Cd, Co, Cr, Cu, Fe, Mg, Mn, Ni, Pb, Zn; ≤0.001% Ca; ≤0.005% Na, K; ≤0.01% Cl, SO_4; ≤0.002% NH_4^+ | Agami, C et al, *J Chem Soc Chem Comm*, 418: 1981; Asato, AE et al, *Tetrahedron Lett*, 33: 3105, 1992

ICN 102730   MW 115.1 $C_5H_9NO_2$ ≥99%; crystalline; hydroxy-L-Pro free

ICN 194728   MW 115.1 $C_5H_9NO_2$ Cell culture reagent;≥99%; crystalline; hydroxy-L-Pro free

Neosystem AA01502   MW 115.1

Peptides International ALP-2718   MW 115.13   >99.9% optical purity by RP-HPLC; white crystalline powder

Rexim   MW 115.1 $C_5H_9NO_2$ White crystals or crystalline powder

Sigma P 0380   FW 115.1 $C_5H_9NO_2$ ≥99% (TLC); Hydroxy-L-Pro free | Also available as part of a kit

Sigma P 1428   FW 115.1 $C_5H_9NO_2$ Crystalline; Hydroxy-L-Pro free | Plant cell culture tested

Sigma P 3268   Matrix: 4% beaded agarose; activation: cyanogen bromide; attachment: amino; spacer: 1 atom; ligand immobilized: 2-10 µmoles/mL; form: suspension in 2.0 *M* NaCl containing 0.02% thimerosal | Deutsch, DG & Mertz, ET, *Proc Fed Amer Soc Exp Biol*, 29: 647, 1970; Deutsch, DG & Mertz, ET, *Science*, 170: 1095, 1970; Vuento, M & Vaheri, A, *Biochem J*, 183: 331, 1979

Sigma P 4097   FW 115.1 $C_5H_9NO_2$ USP Grade

Sigma P 4655   FW 115.1 $C_5H_9NO_2$ Crystalline; hydroxy-L-Pro free | Cell culture tested; insect cell culture tested

Sigma P 8449   FW 115.1 $C_5H_9NO_2$ SigmaUltra; >99% (TLC); pH (1 *M* in H_2O, 20°C): 6.0-7.0; residue on ignition: <0.1%; solubility (1 *M* in H_2O, 20°C): complete, colorless; insoluble matter: <0.1%; SO_4, Cl, NH_4^+,: <0.05%; K, Na: <0.005%; Al, Ca, Cu, Fe, Mg, Zn, P: <0.0005%; Pb: <0.001%; A_{260}<0.05; A_{280}<0.05 (1 *M* in H_2O)

USBio P9010   MW 115.1 $C_5H_9NO_2$ ≥98%; appearance: white crystalline powder; pH: 5.6-6.9; specific rotation: C=4, H_2O -84.5° to -86.0°; heavy metals (Pb): ≤0.001%; endotoxin tested; cell culture tested: murine myeloma SP2/0-Ag14

Proline — L-Pro (2,3,4,5-³H)

Amersham TRK534   Aqueous, 2% EtOH, sterilized; 3.7-4.8 TBq/mmol, 85-130 Ci/mmol; 37 MBq/mL, 1 mCi/mL

Amersham TRK750   Aqueous, sterilized; 3.15-4.81 TBq/mmol, 85-130 Ci/mmol; 185 MBq/mL, 5 mCi/mL

ARC ART-475   MW 115.1   90-120 Ci/mmol; 3.33-4.44 TBq/mmol; in 0.01 *N* HCl | Radiochemical

ICN 20073   MW 115.1   60-100 Ci/mmol, 2.22-3.7 TBq/mmol; 0.01 *N* HCl solution | Schwartz, Z etal, *Endocrinology*, 136:402, 1995

ICN 20073E   MW 115.1   60-100 Ci/mmol, 2.22-3.7 TBq/mmol; sterile EtOH:H_2O solution (2:98)

Proline — L-Pro (2,3-³H)

Amersham TRK638   Aqueous, 2% EtOH, sterilized; 1.1-2.2 TBq/mmol, 30-60 Ci/mmol; 37 MBq/mL, 1 mCi/mL

ARC ART-190   MW 115.1   20-40 Ci/mmol; 0.74-1.48 TBq/mmol; in 0.01 *N* HCl | Radiochemical

Sigma P 6178   FW 115.1 $C_5H_9NO_2$ 20-40 Ci/mmol; aqueous solution containing 2% EtOH in serum bottle | Radiochemical

Proline — L-Pro (3,4-³H(N))

ARC ART-157   MW 115.1   40-60 Ci/mmol; 1.48-2.22 TBq/mmol; in 0.01 *N* HCl | Radiochemical

Proline — L-Pro (3,4-³H)

ICN 20060E   MW 115.1   40-60 Ci/mmol, 1.48-2022 TBq/mmol; sterile EtOH:H_2O solution (2:98)

Proline — L-Pro (4-³H(N))

Sigma P 6553   FW 115.1 $C_5H_9NO_2$ 15-30 Ci/mmol; aqueous solution containing 2% EtOH in serum bottle | Radiochemical

Proline — L-Pro (5-³H)

Amersham TRK323   Aqueous, sterilized; 0.55-1.5 TBq/mmol, 15-40 Ci/mmol; 37 MBq/mL, 1 mCi/mL

ARC ART-191   MW 115.1   5-15 Ci/mmol; 185-555 GBq/mmol; in 0.01 *N* HCl | Radiochemical

ICN 20034   MW 115.1   15-30 Ci/mmol, 0.55-1.11 TBq/mmol; 0.01 *N* HCl solution | Schwartz, Z etal, *Endocrinology*, 136:402, 1995

Proline — L-Pro (U-¹⁴C)

Amersham CFB71   Aqueous, 2% EtOH, sterilized; >9.25 GBq/mmol, >250 mCi/mmol; 1.85 MBq/mL, 50 µCi/mL

ARC ARC-654   MW 115.1 $HNCH_2CH_2CH_2CHCOOH$ 200-300 mCi/mmol; 7.4-11.1 GBq/mmol; in 0.01 *N* HCl | Radiochemical

ARC ARC-654A   MW 115.1 $HNCH_2CH_2CH_2CHCOOH$ 200-300 mCi/mmol; 7.4-11.1 GBq/mmol; in EtOH: H_2O (2:98) | Radiochemical

ICN 10117   MW 115.1   >225 mCi/mmol, >8.33 GBq/mmol; 0.01 *N* HCl solution

ICN 10117E   MW 115.1   >225 mCi/mmol, >8.33 GBq/mmol; sterile EtOH:H_2O solution (2:98)

Proline — L-Pro (U-¹⁴C) Hydrochloride

Sigma P 4801   FW 115.1 $C_5H_9NO_2$ 200-300 mCi/mmol; radiochemical purity ≥98% (HPLC); aqueous solution containing 2% EtOH in Combi-vial | Radiochemical

Proline — L-Pro 4-Methoxy-β-Naphthylamide Hydrochloride

Sigma P 9655   FW 306.8 $C_{16}H_{18}N_2O_2 \cdot HCl$

Proline — L-Pro Amide

Neosystem AA01503   MW 114.1

Proline — L-Pro β-Naphthylamide Hydrochloride

Sigma P 1380   FW 276.8 $C_{15}H_{16}N_2O \cdot HCl$ ~95%

Proline — L-Pro-2-Naphthylamide Hydrochloride

Synonyms: Pro β-Naphthylamidase

Fluka 81742   MW 276.8 $C_{15}H_{16}N_2O \cdot HCl$ ≥99.0% (TLC) | Purification & characterization of Pro β-naphthylamidase; Takahashi, T et al, *JBC*, 264: 11565, 1989

Proline — L-Pro-AMC Hydrobromide

ICN 156379

ICN 198702   MW 353.2 $C_{14}H_{16}N_2O_3 \cdot HBr$

Sigma P 5898   FW 353.2 $C_{15}H_{16}N_2O_3 \cdot$ HBr

Proline — L-Prolinamide Free Base

Sigma P 7517   FW 114.1 $C_5H_{10}N_2O$

Proline — L-Prolinamide Hydrochloride

Sigma P 5010   FW 150.6 $C_5H_{10}N_2O \cdot$ HCl

Proline — L-Pro-OBg TFA

Neosystem AA01507   MW 452.4

Proline — L-Pro-OBzl Hydrochloride

Fluka 81723   MW 241.81 $C_{12}H_{15}NO_2 \cdot$ HCl   ≥99.0% (titration); mp: 146-148°C | Building block for peptide synthesis; useful as chiral auxiliary in DA-cycloaddition of the *N*-acryloyl derivative; Waldmann, H, *J Org Chem*, 53: 6133, 1988

ICN 102731   MW 241.7 $C_{12}H_{15}NO_2 \cdot$ HCl   Crystalline

Sigma P 9007   FW 241.7 $C_{12}H_{15}NO_2 \cdot$ HCl

Proline — L-Pro-OMe Hydrochloride

Fluka 81740   MW 165.62 $C_6H_{11}NO_2 \cdot$ HCl   ≥99.0% (titration); mp: 68-71°C

ICN 156380   MW 165.6 $C_6H_{11}NO_2 \cdot$ HCl   Crystalline

Neosystem AA01504   MW 165.6

Sigma P 3135   FW 165.6 $C_6H_{11}NO_2 \cdot$ HCl

Proline — L-Pro-OtBu Dibenzenesulfimide Salt

Fluka 81725   MW 241.81 $C_9H_{17}NO_2 \cdot C_{12}H_{11}NO_4S_2$ ≥99.0% (TLC); mp: 160-162°C

Proline — L-Pro-OtBu Dibenzenesulfonimide Salt

Sigma P 8282   FW 468.6 $C_9H_{17}NO_2 \cdot C_{12}H_{11}NO_4S_2$

Proline — L-Pro-OtBu Free Base

ICN 151952   MW 171.2 $C_9H_{17}NO_2$

Sigma P 7764   FW 171.2 $C_9H_{17}NO_2$

Proline — L-Pro-pNA TFA Salt

ICN 156381   MW 349.3 $C_{11}H_{15}N_3O_3 \cdot C_2HF_3O_2$

Sigma P 5267   FW 349.3 $C_{11}H_{13}N_3O_3 \cdot C_2HF_3O_2$

Proline — L-Pro-β-Naphthylamide Hydrobromide

ICN 102732   Crystalline

Proline — MTH-DL-Pro

Sigma M 5631   FW 170.2 $C_7H_{10}N_2OS$ ~95%

Proline — N-2,4-Dnp-L-Pro

Sigma D 1505   FW 281.3 $C_{11}H_{11}N_3O_6$

Proline — N-Ac-DL-Pro

ICN 100133   MW 157.2 $C_7H_{11}NO_3$

Sigma A 4876   FW 157.2 $C_7H_{11}NO_3$ Sigma Grade

Proline — N-Acetylhydroxy-L-Pro

ICN 154700   MW 173.2 $C_7H_{11}NO_4$ Crystalline

Proline — N-Ac-L-Pro

Synonyms: *N*-Ac-L-Pyrrolidine-2-Carboxylic Acid

ICN 150242   MW 157.2 $C_7H_{11}NO_3$

Rexim   MW 157.2 $C_7H_{11}NO_3$ White crystals or crystalline powder

---

Sigma A 0783   FW 157.2 $C_7H_{11}NO_3$

Proline — N-Bzl-D-Pro-OEt

Fluka 13836   MW 233.31 $C_{14}H_{19}NO_2$ ≥97% (GC)

Proline — N-Bzl-L-Prolinol

Synonyms: S(−)-Benzylpyrrolidine-2-Me

Fluka 13839   MW 191.27 $C_{12}H_{17}NO$ ≥98% (GC) | Chiral auxiliary; Itsuno, S et al, *J Chem Soc*, Perkin I, 2887, 1984

Proline — N-Bzl-L-Pro-OEt

Fluka 13838   MW 233.31 $C_{14}H_{19}NO_2$ ≥97% (GC) | Starting material for the synthesis of various chiral auxiliaries derived from L-Pro; Enders, D et al, *Bull Soc Chim Belg*, 97: 691, 1988

Proline — N-CBZ-L-Pro

ICN 100229   MW 249.3 $C_{13}H_{15}NO_4$ Crystalline

Sigma C 4752   FW 249.3 $C_{13}H_{15}NO_4$

Proline — N-CBZ-L-Pro N-Hydroxysuccinimide Ester

Sigma C 0295   FW 346.3 $C_{17}H_{18}N_2O_6$

Proline — N-CBZ-L-Pro-(pONp)

ICN 104982   MW 370.4 $C_{19}H_{18}N_2O_6$ Crystalline

Sigma C 4877   FW 370.4 $C_{19}H_{18}N_2O_6$

Proline — N-Dansyl-trans-4-Hydroxy-L-Pro Cyclohexylammonium Salt

Sigma D 1250   FW. 463.6 $C_{17}H_{20}N_2O_5S \cdot C_6H_{13}N$ ~99%

Proline — N-FMOC-L-Pro

Sigma F 0634   FW 337.4 $C_{20}H_{19}NO_4$

Proline — N-FMOC-L-Pro (3,4-³H)

ARC ART-501   MW 337.4   30-60 Ci/mmol; 1.11-2.22 TBq/mmol; in EtOH | Radiochemical

Proline — N-FMOC-L-Pro SASRIN Resin Ester

Sigma F 2514   200-400 mesh; substitution range: 0.3-0.8 mmole FMOC AA/g resin | For peptide synthesis; *N*-FMOC AA acyl ester of 3-methoxy-4-oxymethyl-phenoxymethylated 1% divinylbenzene cross-linked polystyrene; peptides can be cleaved from the resin with 0.5-1% CF_3CO_2H in CH_2Cl_2; Mergler, M et al, *in Peptides: Chemistry & Biology* (Proc 10th Am Peptide Symp), Marhall, GR, ed, 259, 1988

Proline — N-Me-L-Pro

ICN 155645   MW 129.2 $C_6H_{11}NO_2$

Sigma M 8021   FW 129.2 $C_6H_{11}NO_2$

Proline — N-Me-Pro

Bachem E-2185.0001   MW 129.16 $C_6H_{11}NO_2$ Store at RT

Proline — N-MSOC-L-Pro Dicyclohexylammonium Salt

ICN 155741   MW 446.6 $C_9H_{15}NO_6S \cdot C_{12}H_{23}N$ Crystalline

Proline — N-O-Nps-L-Pro Dicyclohexylammonium Salt

ICN 102535   MW 449.6 $C_{11}H_{12}N_2O_4S \cdot C_{12}H_{23}N$ Yellow crystals

Sigma N 6378   FW 449.6 $C_{11}H_{12}N_2O_4S \cdot C_{12}H_{23}N$ Yellow crystals

Proline — *N*-Succinyl-L-Pro

Synonyms: Angiotensin Converting Enzyme Inhibitor

ICN 156700   Ondetti, MA etal, *Science*, 196:441, 1977

Sigma S 6633   FW 215.2 $C_9H_{13}NO_5$ Ondetti, MA et al, *Science*, 196: 441, 1977

Proline — *N*-t-BOC-3,4-Dehydro-L-Pro

Sigma B 2511   FW 213.2 $C_{10}H_{15}NO_4$

Proline — *N*-t-BOC-4-Thio-L-Pro

Synonyms: *N*-t-BOC-L-Thiazolidine-4-Carboxylic Acid

Sigma B 9513   FW 233.3 $C_9H_{15}NO_4S$

Proline — *N*-t-BOC-D-Pro

Sigma B 4769   FW 215.2 $C_{10}H_{17}NO_4$

Proline — *N*-t-BOC-D-Pro *N*-Hydroxysuccinimide Ester

Sigma B 1142   FW 312.3 $C_{14}H_{20}N_2O_6$

Proline — *N*-t-BOC-L-Pro

Sigma B 0894   FW 215.2 $C_{10}H_{17}NO_4$

Proline — *N*-t-BOC-L-Pro *N*-Hydroxysuccinimide Ester

Sigma B 1755   FW 312.3 $C_{14}H_{20}N_2O_6$

Proline — *N*-t-BOC-L-Pro Resin Ester

Sigma B 3021   *N*-t-BOC AA ester of methylated 1% divinylbenzene cross-linked polystyrene; 200-400 mesh; substitution range: 0.2-0.6 mmole t-BOC-AA/g resin | For peptide synthesis by the Merrifield method

Proline — *N*-t-BOC-L-Pro-Pam Resin Ester

Sigma B 3768   *N*-t-BOC AA ester of 4-(Oxymethyl)-Phenyl-Acm 1% divinylbenzene cross-linked polystyrene; 200-400 mesh; substitution range: 0.2-0.6 mmole t-BOC-AA/g resin | For peptide synthesis by the Merrifield method; Mitchell, AR et al, *J Org Chem*, 43: 2845, 1978

Proline — *N*ᵅ-BOC-D-Thioproline

Synonyms: BOC-D-Thz

USBio B2406   MW 233.3 $C_9H_{15}NO_4S$ ≥98%

Proline — *N*ᵅ-BOC-L-Thioproline

Synonyms: BOC-Thz

USBio B2407   MW 233.3 $C_9H_{15}NO_4S$ ≥98%

Proline — *N*ᵅ-FMOC-L-Pro

ICN 151151   MW 337.4 $C_{20}H_{19}NO_4$

Proline — *N*ᵅ-t-BOC-L-Pro

ICN 101092   MW 215.2 $C_{10}H_{17}NO_4$ Crystalline

Proline — *o*-Acetylhydroxy-L-Pro

Sigma A 9637   FW 173.2 $C_7H_{11}NO_4$

Proline — Pro

Bachem E-2290.0025   MW 115.13 $C_5H_9NO_2$ Store at RT

Proline — Pro Amide

Bachem E-2295.0001   MW 114.15 $C_5H_{10}N_2O$ Store at -15°C

Proline — Pro Amide Hydrochloride

Bachem E-2300.0001   MW 150.61 $C_5H_{10}N_2O \cdot HCl$ Store at -15°C

Proline — Pro-(*p*ONb) Hydrobromide

Bachem E-2330.0001   MW 331.17 $C_{12}H_{14}N_2O_4 \cdot HBr$ Store at -15°C

Proline — Pro-2-Chloro Trt Resin

American Peptide RGEN168   Enantiomerization free

Proline — Pro-2-Chlorotrityl-® (200-400 mesh)

Bachem D-2000.0001   Store at -15°C

Proline — Pro-4MβNA Hydrochloride

Bachem J-1280.0250   MW 306.79 $C_{18}H_{18}N_2O_2 \cdot HCl$ Store at -15°C

Proline — Pro-AMC Hydrobromide

Bachem I-1290.0050   MW 353.22 $C_{15}H_{16}N_2O_3 \cdot HBr$ Store at -15°C

Proline — Pro-NHEt Hydrochloride

Bachem E-3140.0001   MW 178.66 $C_7H_{14}N_2O \cdot HCl$ Store at -15°C

Proline — Pro-NMe₂

Bachem E-2320.0001   MW 142.2 $C_7H_{14}N_2O$ Store at -15°C

Proline — Pro-OBzl Hydrochloride

Synonyms: L-Pro-OBzl

Bachem E-2305.0001   MW 241.72 $C_{12}H_{15}NO_2 \cdot HCl$ Store at RT

Peptides International EBP-2049   MW 241.72   >98% (TLC); white powder

Proline — Pro-ol

Synonyms: S(+)-2-Hydroxymethylpyrrolidine

Senn Chem 44104   MW 101.2

Proline — Pro-OMe Hydrochloride

Bachem E-2325.0005   MW 165.62 $C_6H_{11}NO_2 \cdot HCl$ Store at -15°C

Proline — Pro-OtBu Dibenzenesulfonimide

Bachem E-2315.0001   MW 468.6 $C_9H_{17}NO_2 \cdot C_{12}H_{11}NO_4S_2$ Store at -15°C

Proline — Pro-OtBu Syrup

Bachem E-2310.0001   MW 171.24 $C_9H_{17}NO_2$ Store at -15°C

Proline — Pro-pNA

Bachem L-1370.0250   MW 235.24 $C_{11}H_{13}N_3O_3$ Store at -15°C

Proline — Pro-pNA Hydrobromide

Bachem L-1735.0250   MW 316.16 $C_{11}H_{13}N_3O_3 \cdot HBr$ Store at -15°C

Proline — Pro-βNA Hydrochloride

Bachem K-1465.0001   MW 276.77 $C_{15}H_{16}N_2O \cdot HCl$ Store at RT

Proline — Pth-DL-Pro

Sigma P 2252 FW 232.3 $C_{12}H_{12}N_2OS$ ~99% | Useful as standards in protein sequencing by the Edman degradation

Proline — Pth-L-Pro

Sigma P 2377 FW 232.3 $C_{12}H_{12}N_2OS$ ~99% | Useful as standards in protein sequencing by the Edman degradation

Proline — Pz-Pro

Bachem E-2275.0001 Store at 2-8°C

Proline — Pz-Pro-OSu

Bachem E-2280.0001 Store at 2-8°C

Proline — Suc-Pro

Bachem E-2410.0001 MW 215.21 $C_9H_{13}NO_5$ Store at 2-8°C

Proline — TentaGel S PHB-Pro-FMOC

Fluka 86401 0.20 mmol protected AA/g; ~90 µm particle size | Immobilized FMOC-Pro useful for the SPPS of peptides with C-terminal Pro; Zhang, L et al, *Solid Phase Synthesis*, 3rd Intl Symp (R Epton, ed), 717, 1997, Mayflower, Birmingham

Proline — TentaGel S Trt-Pro-FMOC

Fluka 86433 0.18 mmol protected AA/g; ~90 µm particle size

Proline — Thioproline

Synonyms: L-Thz

USBio T5007 MW 133.2 $C_4H_7NO_2S$ ≥98%

Proline — Tosyl-D-Pro

Bachem F-2745.0001 MW 269.32 $C_{12}H_{15}NO_4S$ Store at RT

Proline — Tosyl-Pro

Bachem E-2915.0005 MW 269.32 $C_{12}H_{15}NO_4S$ Store at RT

Proline — *trans*-4-Fluoro-Pro

Synonyms: (2S,4R)-4-Fluoro-Pyrrolidine-2-Carboxylic Acid

Bachem F-3975.0250 MW 133.12 $C_5H_8FNO_2$ Store at RT

Proline — *trans*-4-Hydroxy-L-Pro

Synonyms: L-4-Hydroxy-Pyrrolidine Carboxylic Acid; Hyp

USBio H9110 MW 131.1 $C_5H_9NO_3$ ≥99%; appearance: white to off-white crystalline powder; pH: 5.0 to 6.5; specific rotation (C=4, H_2O): -74.0° to -77.0°; loss on drying: ≤1%; residue on ignition: ≤1%; ammonium: ≤0.02%; chloride: ≤0.02%; sulfate: ≤0.02%; arsenic: ≤0.001%; heavy metals (Pb): ≤0.0025%; endotoxin: ≤2 EU/mg; cell culture analysis: murine myeloma SP2/0-Ag14

Proline — *trans*-4-Hydroxy-L-Pro β-Naphthylamide

Sigma H 7259 FW 256.3 $C_{15}H_{16}N_2O_2$

Proline — Z-D-Pro

Bachem C-2465.0001 MW 249.27 $C_{13}H_{15}NO_4$ Store at RT

Neosystem ZA01501 MW 249.3

Proline — Z-D-Pro Amide

Bachem C-3030.0001 MW 248.28 $C_{13}H_{16}N_2O_3$ Store at RT

Proline — Z-D-Pro-OSu

Bachem C-3660.0001 MW 346.34 $C_{17}H_{18}N_2O_6$ Store at -15°C

Proline — Z-D-Pro-OtBu

Bachem C-3515.0001 MW 305.37 $C_{17}H_{23}NO_4$ Store at -15°C

Proline — Z-L-Pro

Fluka 97090 MW 249.27 $C_{13}H_{15}NO_4$ ≥99.0% (titration); mp: 75-77°C

Neosystem ZA01502 MW 249.3

Proline — Z-L-Pro Hydroxysuccinimide Ester

Fluka 97101 MW 346.34 $C_{17}H_{28}N_2O_6$ mp: 88-90°C

Proline — Z-Pro

Synonyms: Benzyloxycarbonyl-L-Pro

Bachem C-2460.0005 MW 249.27 $C_{13}H_{15}NO_4$ Store at RT

Peptides International ZLP-2022-PI MW 249.27 >98% (TLC); white powder

Proline — Z-Pro Amide

Bachem C-2470.0005 MW 248.28 $C_{13}H_{16}N_2O_3$ Store at RT

Proline — Z-Pro-ONp

Bachem C-2480.0005 MW 370.36 $C_{19}H_{18}N_2O_6$ Store at -15°C

Proline — Z-Pro-OSu

Bachem C-2475.0005 MW 346.34 $C_{17}H_{18}N_2O_6$ Store at -15°C

Proline — Z-Pro-OtBu

Bachem C-3450.0005 MW 305.37 $C_{17}H_{23}NO_4$ Store at -15°C

Proline — Z-Pro-βNA

Bachem K-1215.0001 MW 374.44 $C_{23}H_{22}N_2O_3$ Store at RT

Proline — Z-Thioprolyl-Thiazolidine

American Peptide 95-0-12 MW 338.4 $C_{16}H_{20}N_2O_2S_2$

Proline — α-Me-Pro

Bachem F-3440.0001 MW 129.16 $C_6H_{11}NO_2$ Store at -15°C

Pyroglutamic Acid — Glp-AMC

Bachem I-1300.0050 MW 286.29 $C_{15}H_{14}N_2O_4$ Store at -15°C

Sarcosine — BOC-Sar

Synonyms: BOC-*N*-Methylglycine

American Peptide BLSAR10 MW 189.2

Bachem A-2265.0005 MW 189.21 $C_8H_{15}NO_4$ Store at RT

Fluka 15495 MW 189.21 $C_8H_{15}NO_4$ ≥99% (titration); mp: 89-90°C

Neosystem BA00802 MW 189.2

Peptides International BSR-2120 MW 189.21 >98% (HPLC); white crystalline powder

Sarcosine — BOC-Sar-OSu

Synonyms: BOC-*N*-Me-Gly-OSu

Bachem A-2270.0005   MW 286.29 $C_{12}H_{18}N_2O_6$ Store at -15°C

Sarcosine — BoroSar

Synonyms: Boromethylglycine, Methylamine-Carboxyborane; Dihydrofolate Reductase Inhibitor; DNA Polymerase Inhibitor

ICN 154609   MW 88.9 $BC_2H_8NO_2$ >97% | Hall, IH etal, *J Pharm Sci*, 74:755, 1985

Sarcosine — Boro-Sar *N*-Ethylamide

Synonyms: Methylamine-*N*-Ethylcarbamoylborane

ICN 154631   MW 115.97 $BC_4H_{13}N_2O$ >97%

Sarcosine — Boro-Sar-OMe

Synonyms: Methylamine-Carbomethoxyborane

ICN 154628   MW 102.9 $BC_3H_{10}NO_2$ >97%

Sarcosine — Dansyl-Sar Piperidinium Salt

Sigma D 9631   FW. 407.5 $C_{15}H_{18}N_2O_4S \cdot C_5H_{11}N$

Sarcosine — FMOC-Sar

Synonyms: FMOC-*N*-Me-Gly

Bachem B-1720.0005   MW 311.34 $C_{18}H_{17}NO_4$ Store at RT

Sarcosine — FMOC-Sar Hydrate

Synonyms: FMOC-*N*-Methylglycine

Fluka 47595   MW 329.36 $C_{18}H_{17}NO_4 \cdot H_2O$ ≥98% (HPLC); mp: 115°C

Sarcosine — FMOC-Sar-® (200-400 mesh)

Synonyms: FMOC-*N*-Me-Gly-®

Bachem D-1805.0001   Store at RT

Sarcosine — FMOC-sarcosine

Synonyms: FMOC-*N*-Me-Gly

Neosystem FA00802   MW 311.3

Sarcosine — FMOC-SarOPfp

Synonyms: FMOC-*N*-Me-Gly-OPfp

Bachem B-1730.0001   MW 477.39 $C_{24}H_{16}F_5NO_4$ Store at -15°C

Fluka 00233   MW 477.39 $C_{24}H_{16}F_5NO_4$ ~98%

Sarcosine — FMOC-Sar-SASRIN™-® (200-400 mesh)

Synonyms: FMOC-*N*-Me-Gly-SASRIN™-®

Bachem D-1645.0001   Store at 2-8°C

Sarcosine — *N*-Amidinosarcosine Disodium Salt Tetrahydrate

Synonyms: Creatine Phosphate

Fluka 27920   MW 327.14 $C_4H_9N_3Na_2O_5P \cdot 4H_2O$ ≥98% (titration); ≤0.6% free creatine | Substrate for the assay of creatine kinase; Gerhardt, W et al, *Methods of Enzymatic Analysis*, HU Bergmeyer, ed, vol 3, 3rd ed: 510, 1983, Verlag Chemie Weinheim

Sarcosine — *N*-Lauroyl-Sar

Synonyms: *N*-Dodecanoyl-*N*-Methylglycine; Sarkosyl L

Fluka 61739   MW 271.4 $C_{15}H_{29}NO_3$ ~99.0% (GC); ≤0.1% residue on ignition; ≤0.0005% Cd, Co, Cr, Cu, Fe, Mg, Mn, Ni, Pb; ≤0.002% Ca, Zn; ≤0.005% K, Cl, ≤0.02% Na; ≤0.01% SO₄; ≤0.2% H₂O; mp: 44.5-47.5°C

Sarcosine — *N*-Lauroyl-Sar Free Acid

ICN 190110   MW 271.4 $C_{15}H_{29}NO_3$ ~95%

Sigma L 5000   FW 271.4 $C_{15}H_{29}NO_3$ ≥95% | May be liquid at room temperature

Sarcosine — *N*-Lauroyl-Sar Sodium Salt

Synonyms: Hexokinase Inhibitor

ICN 190289   MW 293.4 $C_{15}H_{28}NO_3Na$ ≥97%; crystalline | Useful for solubilizing membrane proteins

ICN 194008   MW 293.4 $C_{15}H_{28}NO_3Na$ Molecular biology reagent; ≥97% | Useful in concentrated Salt solutions useful in the cell lysis step of RNA purification

ICN 194009   MW 293.4 $C_{15}H_{28}NO_3Na$ Ultra pure grade; ≥97% | Useful for solubilizing membrane proteins

Sigma L 5125   FW 293.4 $C_{15}H_{28}NO_3Na$ ≥94%

Sigma L 5777   FW 293.4 $C_{15}H_{28}NO_3Na$ SigmaUltra; ≥97%; solubility (0.1 *M* in H₂O, 20°C): complete, very faint yellow; insoluble matter: <0.1%; NH₄⁺: <0.05%; Fe, Ca, Mg: <0.005%; Al, Cu, Zn: <0.0005%; Pb: <0.01%; P: <0.1%; K: ≤100 ppm

Sarcosine — *N*-t-BOC-Sar

Sigma B 9637   FW 189.2 $C_8H_{15}NO_4$

Sarcosine — Phenylsulfonyl-Sar

Synonyms: Phenylsulfonyl-*N*-Me-Gly

Bachem F-3140.0001   MW 229.26 $C_9H_{11}NO_4S$ Store at 2-8°C

Sarcosine — Polysarcosine

Synonyms: `

Sigma P 2379   Fessler, JH & Ogston, AG, *trans Faraday Soc*, 47: 667, 1951

Sarcosine — Sar

Synonyms: *N*-Methylaminoacetic Acid; *N*-Methylglycine

Fluka 84529   MW 89.1 $C_3H_7NO_2$ ≥99.0% (titration); ≤0.05% residue on ignition, loss on drying; ≤0.00001% As; ≤0.0005% Al, Ba, Bi, Cd, Co, Cr, Cu, Fe, Li, Mg, Mn, Mo, Ni, Pb, Sr, Zn; ≤0.001% Ca; ≤0.005% K, Cl, Na, SO₄

Fluka 84530   MW 89.1 $C_3H_7NO_2$ ≥99.0% (titration); ≤0.1% residue on ignition, loss on drying; ≤0.00001% As; ≤0.0005% Cd, Co, Cr, Cu, Fe, Ni, Pb, Zn; ≤0.001% Ca; ≤0.005% Cl, SO₄ | Riccio, P et al, *Dev Bioenerg Biomembr*, 6: 361, 1983

Sigma S 7672   FW 89.09 $C_3H_7NO_2$ SigmaUltra; residue on ignition: <0.1%; solubility (1 *M* in H₂O, 20°C): complete, colorless; insoluble matter: <0.1%; Cl, SO₄, Na, NH₄⁺,: <0.05%; K: <0.005%; Al, Ca, Cu, Fe, Mg, Zn, P: <0.0005%; Pb: <0.001%

Sarcosine — Sar Amide Hydrochloride

Synonyms: *N*-Me-Gly Amide

Bachem E-2355.0001   MW 124.57 $C_3H_8N_2O \cdot HCl$ Store at RT

Sarcosine — Sar Anhydride

Synonyms: 1,4-Dimethyl-2,5-Piperazinedione; 1,4-Dimethyl-2,5-Dioxopiperazine

Fluka 84550   MW 142.16 $C_6H_{10}N_2O_2$ ≥99.0%; mp: 144-147°C

Sigma S 1375   FW 142.2 $C_6H_{10}N_2O_2$ ≥99%; Colorless to light yellow crystals

Sarcosine — Sar Free Base

Synonyms: *N*-Methylaminoacetic Acid; *N*-Methylglycine

Sigma S 9881   FW 89.09 $C_3H_7NO_2$

Sarcosine — Sar Hydrochloride

Synonyms: *N*-Methylaminoacetic Acid; *N*-MethylGly

Sigma S 1500   FW 125.6 $C_3H_7NO_2 \cdot HCl$

Sarcosine — Sarcosinamide Hydrochloride

Sigma S 4763   FW 124.6 $C_3H_8N_2O \cdot HCl$

Sarcosine — Sar-NMe₂

Synonyms: *N*-Me-Gly-NMe₂

Bachem E-2365.0001   MW 116.16 $C_5H_{12}N_2O$ Store at -15°C

Sarcosine — Sar-OBzl *p*-Tosyl

Synonyms: *N*-Me-Gly-OBzl

Bachem E-1580.0005   MW 351.42 $C_{10}H_{13}NO_2 \cdot C_7H_8O_3S$ Store at RT

Sarcosine — Sar-OEt Hydrochloride

Fluka 84539   MW 153.61 $C_5H_{11}NO_2 \cdot HCl$ ≥99.0% (titration); ≤0.05% residue on ignition; mp: 125°C

Sigma S 5887   FW 153.6 $C_5H_{11}NO_2 \cdot HCl$

Sarcosine — Sar-OMe Hydrochloride

Synonyms: *N*-Me-Gly-OMe

Bachem E-2370.0005   MW 139.58 $C_4H_9NO_2 \cdot HCl$ Store at RT

Fluka 84570   MW 139.58 $C_4H_9NO_2 \cdot HCl$ ≥97.0% (titration); mp: 117-119°C

Sarcosine — Sar-OtBu Hydrochloride

Synonyms: *N*-Me-Gly-OtBu

Bachem E-2360.0001   MW 181.66 $C_7H_{15}NO_2 \cdot HCl$ Store at -15°C

Sarcosine — Sodium *N*-Lauroylsarcosinate

Synonyms: *N*-Lauroylsarcosine; Sarcosyl; Sarkosyl NL; Sodium *N*-Dodecanoyl-*N*-Methylglycinate

Fluka 61743   MW 293.39 $C_{15}H_{28}NNaO_3$ ≥99.0% (HPLC); ≤1% residue on ignition; DNases, RNases, proteases, phosphatases: none detected; ≤0.01% peroxides; ≤0.005% K, Cl, SO₄; ≤0.0005% Al, Ba, Bi, Cd, Co, Cr, Cu, Fe, Li, Mg, Mn, Mo, Ni, Pb, Sr, Zn; ≤0.001% Ca | Useful for rupturing eukaryote cells in transcription studies; improves purity of the initial RNA precipitate & avoids excessive foaming in the guanidine HCl method for the isolation of RNA; MacDonald, RJ, *Meth Enzymol*, 152: 219, 1987; Gariglio, P et al, *FEBS Lett*, 44: 330, 1974; Schmookler, RJ et al, *Virology*, 57: 122, 1974; Wroblewski, W et al, *Biochimie*, 60: 389, 1978; Mahoney, DE et al, *Appl Environ Microbiol*, 51: 521, 1986

Fluka 61744   MW 293.39 $C_{15}H_{28}NNaO_3$ ≥99.0% (HPLC); ≤1% residue on ignition; ≤0.01% peroxides; ≤0.005% K, Cl, SO₄; ≤0.0005% Al, Ba, Bi, Cd, Co, Cr, Cu, Fe, Li, Mg, Mn, Mo, Ni, Pb, Sr, Zn; ≤0.001% Ca | Davies, RL et al, *J Immunol Meth*, 134: 215, 1990; Frangioni, JV & Neel, BG, *Anal Biochem*, 210: 179, 1993; Finel, M et al, *Eur J Biochem*, 226: 237, 1994; Gould, W et al, *Appl Environ Microbiol*, 49: 28, 1985; Iyer, VN et al, *J Ind Chem Soc*, 62: 507, 1985

Fluka 61745   MW 293.39 $C_{15}H_{28}NNaO_3$ ≥97.0% (HPLC); ≤7% H_2O

Sarcosine — Sodium *N*-Lauroylsarcosinate Solution

Fluka 61747   MW 293.39 $C_{15}H_{28}NNaO_3$ ≥97.0% (HPLC); 30% aqueous solution; ≤1% residue on ignition; ≤3% free fatty acid; ≤0.001% K, Fe, Mg; ≤0.0005% Cd, Co, Cr, Cu, Mn, Ni, Pb, Zn; ≤0.002% Ca | Useful for rupturing eukaryote cells in transcription studies; improves purity of the initial RNA precipitate & avoids excessive foaming in the guanidine HCl method for the isolation of RNA; MacDonald, RJ, *Meth Enzymol*, 152: 219, 1987; Gariglio, P et al, *FEBS Lett*, 44: 330, 1974; Schmookler, RJ et al, *Virology*, 57: 122, 1974; Wroblewski, W et al, *Biochimie*, 60: 389, 1978; Mahoney, DE et al, *Appl Environ Microbiol*, 51: 521, 1986

Sarcosine — Z-Sar

Synonyms: Z-*N*-Me-Gly

Bachem C-2570.0005   MW 223.23 $C_{11}H_{13}NO_4$ Store at RT

Sarcosine — Z-Sar-OSu

Synonyms: Z-*N*-Me-Gly-OSu

Bachem C-2575.0001   MW 320.3 $C_{15}H_{16}N_2O_6$ Store at -15°C

Serine — (±)-*threo*-Dihydroxyphenylserine

Synonyms: (±)-*threo*-DOPS

ICN 153736   MW 213.2 $C_9H_{11}NO_5$ Noradrenaline precursor which elevates brain noradrenaline concentrations

Serine — (*RS*)-α-Me-Ser-O-Phosphate

Synonyms: (*RS*)-MSOP; Group III mGluR Antagonist

Alexis 550-331   MW 199.1 $C_4H_{10}NO_6P$ White solid; soluble in NaOH | Selective antagonist; Jane, DE et al, *Phosphorous, Sulphur & Silicon*, 109-110: 313, 1996; Thomas, NK et al, *Neuropharmacology*, 35: 637, 1996

Serine — 1,2-Dimyristoyl-*sn*-Glycero-3-Phospho-L-Ser Sodium Salt

Synonyms: DMPS

Alexis 300-021   MW 701.9 $C_{34}H_{65}NNaO_{10}P$

Serine — 1,2-Dioleoyl-*sn*-Glycero-3-Phospho-L-Ser Sodium Salt

Synonyms: DOPS

Alexis 300-102   MW 810.0 $C_{42}H_{77}NNaO_{10}P$ ≥99%; white solid | Shows similar membrane properties to the naturally occurring phosphoserine from bovine brain; stable & therefore can be handled more easily

Serine — 1,2-Dipalmitoyl-*sn*-Glycero-3-Phospho-L-Ser Sodium Salt

Synonyms: DPPS

Alexis 300-022   MW 758.0 $C_{38}H_{73}NNaO_{10}P$

Serine — 1,2-Distearoyl-*sn*-Glycero-3-Phospho-L-Ser Sodium Salt

Synonyms: DSPS

Alexis 300-023   MW 814.1 $C_{42}H_{81}NNaO_{10}P$

Serine — 1-Palmitoyl-*sn*-Glycero-3-Phospho-L-Ser Sodium Salt

Alexis 300-024   Synthetic   MW 519.5 $C_{22}H_{43}NNaO_9P$ ≥99%

Serine — 3-Hydroxymethylserine (1-¹⁴C)

ARC ARC-645   MW 135.1 $(HOCH_2)_2C(NH_2)COOH$ 50-60 mCi/mmol; 1.85-2.22 GBq/mmol; in 0.01 *N* HCl | Radiochemical

Serine — 3-*sn*-Lysophosphatidyl-L-Ser

Fluka 62968   Bovine brain   ≥98.0% (TLC); fatty acid composition: ~75% 18:0 ~3% 18:1, ~2% 20:4

Serine — 3-*sn*-Phosphatidyl-L-Ser

Fluka 79406   Bovine brain   Powder; ≥99.0% (TLC); ≤1% lyso-PS; fatty acid composition: 1% 16:0, 41% 18:0, 30% 18:1, 4% 18:3, 1% 18:4, 1% 20:4, 2% 20:5, 14% 22:6

Serine — 3-*sn*-Phosphatidyl-L-Ser Solution

Synonyms: PS; Phosphatidylserine; L-α-Phosphatidyl-L-Ser

Fluka 79405   Bovine brain   Solution; 10 mg/mL chloroform:MeOH 95:5; ≥98.0% (TLC); ≤1% lyso-PS; fatty acid composition: 140% 18:0, 28% 18:1, 4% 18:3, 1% 18:4, 1% 20:4, 2% 20:5, 9% 22:6; powder

Serine — Ac-Ser-OMe

Bachem E-1205.0250   MW 161.16 $C_6H_{11}NO_4$ Store at RT

Serine — Ac-Ser-tBu

Bachem E-3010.0001   MW 203.24 $C_9H_{17}NO_4$ Store at -15°C

Serine — Aloc-Ser-OMe

Bachem E-3265.0001   MW 203.2 $C_8H_{13}NO_5$ Store at -15°C

Serine — BOC-D-Ser

Bachem A-2280.0001   MW 205.21 $C_8H_{15}NO_5$ Store at RT

Fluka 15182   MW 205.21 $C_8H_{15}NO_5$ ≥98% (TLC)

USBio B2390   MW 205.2 $C_8H_{15}NO_5$ ≥99%

Serine — BOC-D-Ser Dicyclohexylamine

Neosystem BA01601   MW 386.5

Serine — BOC-D-Ser-Bzl

Synonyms: BOC-O-Bzl-D-Ser

Bachem A-1335.0001   MW 295.34 $C_{15}H_{21}NO_5$ Store at RT

Neosystem BA01604   MW 295.3

Peptides International BDS-2627   MW 295.34   >98% (HPLC); white crystalline powder

USBio B2393   MW 295.3 $C_{15}H_{21}NO_5$ ≥99%

Serine — BOC-D-Ser-Bzl-OSu

Bachem A-1345.0001   MW 392.41 $C_{19}H_{24}N_2O_7$ Store at -15°C

Serine — BOC-D-Ser-tBu Dicyclohexylamine

Bachem A-1435.0001   MW 442.64 $C_{12}H_{23}NO_5 \cdot C_{12}H_{23}N$ Store at RT

Serine — BOC-L-Ser

Fluka 15500   MW 205.21 $C_8H_{15}NO_5$ ~99% (titration) | Building block in peptide synthesis; Pansare, SV et al, *Org Synth*, 70: 10, 1992

Neosystem BA01602   MW 205.2

Serine — BOC-L-Ser Hydrate

Neosystem BA01611   MW 223.2

Serine — BOC-L-Ser-Alloc Dicyclohexylamine

Neosystem BA01603   MW 470.7

Serine — BOC-L-Ser-Bzl

Neosystem BA01605   MW 295.3

Serine — BOC-L-Ser-Bzl-MBHA Resin

Neosystem RN01602

Serine — BOC-L-Ser-Bzl-O-CH₂-φ-CH₂-COOH Dicyclohexylamine

Neosystem LP01602   MW 443.5

Serine — BOC-L-Ser-Bzl-PAM Resin

Neosystem RP01602

Serine — BOC-L-Ser-OMe

Fluka 15183   MW 219.24 $C_9H_{17}NO_5$ ≥98% (TLC)

Serine — BOC-N-Me-L-Ser

Fluka 15552   MW 219.24 $C_9H_{17}NO_5$ ≥98% (TLC); mp: 83-87°C

Serine — BOC-N-Me-Ser

Synonyms: BOC-N-Me-L-Ser

Bachem A-2085.0001   MW 219.24 $C_9H_{17}NO_5$ Store at RT

Peptides International BMS-5310-PI   MW 219.24   >98% (HPLC); white powder

Serine — BOC-O-Bzl-D-Ser

Fluka 15078   MW 295.34 $C_{15}H_{21}NO_5$ ≥98% (HPLC)

Serine — BOC-O-Bzl-L-Ser

Fluka 15390   MW 295.34 $C_{15}H_{21}NO_5$ ≥99% (titration); mp: 58-61°C; ≤0.05% residue on ignition

Serine — BOC-O-Bzl-L-Ser Hydroxysuccinimide Ester

Fluka 15079   MW 392.41 $C_{19}H_{24}N_2O_7$ ≥98% (TLC)

Serine — BOC-O-Bzl-L-Ser N-Carboxy Anhydride

Fluka 15415   MW 321.33 $C_{16}H_{19}NO_6$ ≥97% (titration); mp: 99-102°C | Produced by Propeptide under an exclusive license from Bioresearch

Serine — BOC-O-Dimethylphosphous-L-Ser

American Peptide BLSER30   MW 313.2

Serine — BOC-O-tBu-L-Ser Dicyclohexylamine

Fluka 15432   MW 442.64 $C_{12}H_{23}NO_5 \cdot C_{12}H_{23}N$ ~99%; mp: 161-164°C

Serine — BOC-Ser

American Peptide BLSER10   MW 205.2

Bachem A-2275.0005   MW 205.21 $C_8H_{15}NO_5$ Store at RT

USBio B2391   MW 205.2 $C_8H_{15}NO_5$ ≥99%

Serine — BOC-Ser Bzl

American Peptide BLSER20   MW 295.3

Serine — BOC-Ser Bzl-O-Resin

American Peptide RMRB255   1% DVB cross-linked: 100-200 mesh | t-BOC protected AA resin; preattached resins are useful for synthesis of peptides with C-terminal carboxyl group by t-BOC chemistry

Serine — BOC-Ser Bzl-PAM-Resin

American Peptide RPAM255   1% DVB cross-linked: 100-200 mesh | t-BOC protected AA resin; preattached resins are useful for synthesis of peptides with C-terminal carboxyl group by t-BOC chemistry

Serine — BOC-Ser-(PO₃Bzl₂)

USBio B2395   MW 465.3 $C_{22}H_{28}NO_8P$ ≥98%

Serine — BOC-Ser-Ac Dicyclohexylamine

Bachem A-2610.0001   MW 428.54 $C_{10}H_{17}NO_6 \cdot C_{12}H_{23}N$ Store at 2-8°C

Serine — BOC-Ser-Bzl

Synonyms: BOC-O-Bzl-L-Ser

Bachem A-1330.0005   MW 295.34 $C_{15}H_{21}NO_5$ Store at RT

Peptides International BLS-2102   MW 295.34   >98% (HPLC); white crystalline powder

USBio B2394   MW 295.3 $C_{15}H_{21}NO_5$ ≥99%

Serine — BOC-Ser-Bzl-® (200-400 mesh)

Bachem D-1170.0005   Store at RT

Serine — BOC-Ser-Bzl-ONp

Bachem A-1350.0001   MW 416.43 $C_{21}H_{24}N_2O_7$ Store at -15°C

Serine — BOC-Ser-Bzl-OSu

Bachem A-1340.0001   MW 392.41 $C_{19}H_{24}N_2O_7$ Store at -15°C

Serine — BOC-Ser-Bzl-PAM-® (200-400 mesh)

Bachem D-1275.0001   Store at 2-8°C

Serine — BOC-Ser-Dimethylphospho

USBio B2396   MW 313.1 $C_{10}H_{20}NO_8P$ ≥98%

Serine — BOC-Ser-Diphenylphospho

USBio B2397   MW 437.3 $C_{20}H_{24}NO_8P$ ≥98%

Serine — BOC-Ser-Me Dicyclohexylamine

Bachem A-3830.0001   MW 400.56 $C_9H_{17}NO_5 \cdot C_{12}H_{23}N$ Store at -15°C

Serine — BOC-Ser-OBzl

Bachem A-2285.0001   MW 295.34 $C_{15}H_{21}NO_5$ Store at -15°C

Serine — BOC-Ser-OMe

Bachem A-3835.0001   MW 219.24 $C_9H_{17}NO_5$ Store at -15°C

Serine — BOC-Ser-OtBu

Bachem A-3025.0001   MW 261.32 $C_{12}H_{23}NO_5$ Store at -15°C

Serine — BOC-Ser-p-Chlorobenzyl Dicyclohexylamine

Bachem A-1500.0001   MW 511.1 $C_{15}H_{20}ClNO_5 \cdot C_{12}H_{23}N$ Store at RT

Serine — BOC-Ser-tBu Dicyclohexylamine

Bachem A-1430.0001   MW 442.64 $C_{12}H_{23}NO_5 \cdot C_{12}H_{23}N$ Store at RT

Serine — BOC-Ser-tBu Syrup

Bachem A-2915.0001   MW 261.32 $C_{12}H_{23}NO_5$ Store at -15°C

Serine — BOC-Ser-tBu-Dicyclohexylamine

USBio B2392   MW 442.6 $C_{12}H_{23}NO_5 \cdot C_{12}H_{23}N$ ≥99%

Serine — BOC-ThionoSer-Bzl-1-(6-Nitro)-Benzotriazolide

Synonyms: (S)-2-(BOC-Amino)-3-Benzyloxy-Propanethioic-O-Acid-1-(6-Nitro)-Benzotriazolide

Bachem A-4365.0001   MW 457.41 $C_{21}H_{23}N_5O_5S$ Store at -15°C

Serine — Bpoc-Ser-tBu CHA

Bachem E-2970.0001   MW 498.66 $C_{23}H_{29}NO_5 \cdot C_6H_{13}N$ Store at -15°C

Serine — Bsmoc-Ser-tBu

Synonyms: Bsmoc-O-tBu-L-Ser

Peptides International BLS-5635-PI   MW 383.42   >98% (HPLC); white to off-white powder | LA Carpino, et al, *JACS*, 119:9915, 1997

Serine — Bzl-N-Me-Ser

Bachem E-1570.0001   MW 209.25 $C_{11}H_{15}NO_3$ Store at RT

Serine — Bzl-Ser

Bachem E-2620.0001   MW 195.22 $C_{10}H_{13}NO_3$ Store at 2-8°C

Serine — Bzl-Ser-Bzl

Bachem E-1795.0005   MW 285.34 $C_{17}H_{19}NO_3$ Store at RT

Serine — CBZ-D-Ser

USBio C2099-61   MW 239.2 $C_{11}H_{13}NO_5$ ≥99%

Serine — CBZ-D-Ser-tBu

USBio C2099-64   MW 295.3 $C_{15}H_{21}NO_5$ ≥99%

Serine — CBZ-Ser

USBio C2099-60   MW 239.2 $C_{11}H_{13}NO_5$ ≥99%

Serine — CBZ-Ser-Bzl

USBio C2099-63   MW 329.4 $C_{18}H_{19}NO_5$ ≥99%

Serine — CBZ-Ser-OBzl

USBio C2099-62   MW 329.4 $C_{18}H_{19}NO_5$ ≥99%

Serine — CBZ-Ser-OMe

USBio C2099-66   MW 253.2 $C_{12}H_{15}NO_5$ ≥99%

Serine — CBZ-Ser-tBu

USBio C2099-65   MW 295.3 $C_{15}H_{21}NO_5$ ≥99%

Serine — Dansyl-L-Ser Cyclohexylammonium Salt

ICN 100131   MW 437.6 $C_{15}H_{18}N_2O_5S \cdot C_6H_{13}N$

Serine — D-Cycloserine

Synonyms: (R)-4-Amino-3-Isoxazolidone; D-4-Amino-3-Isoxazolidinone; Antibiotic Bactericidal

ICN 194788   MW 102.1 $C_3H_6N_2O_2$ γ-irradiated | Molecular biology reagent; inhibits cell wall synthesis

Fluka 30020   Microbial   MW 102.09 $C_3H_6N_2O_2$ ≥98% (titration); mp: 150°C | Inhibitor; Jain, MK, *Handbook of Enzyme Inhibitors*, 112: 1982, John Wiley & Sons New York

Sigma C 6880   Microbial   FW 102.1 $C_3H_6N_2O_2$ Partial agonist at the glycine modulatory site of MNDA receptors; blocks amnesia

ICN 100535   Synthetic   MW 102.1 $C_3H_6N_2O_2$ Crystalline

Sigma C 3909   Synthetic   FW 102.1 $C_3H_6N_2O_2$ Partial agonist at the glycine modulatory site of MNDA receptors; blocks amnesia

Serine — Dinitropyridyl-DL-Ser

ICN 102866   MW 272.2 $C_8H_8N_4O_7$

Serine — DL-Cycloserine

Synonyms: DL-4-Amino-3-Isoxazolidinone; Alanine Aminotransferase Inhibitor; Ketosphinganine Synthetase Inhibitor

Bachem F-1485.0250   MW 102.09 $C_3H_6N_2O_2$ Store at -15°C

ICN 190315   MW 102.1 $C_3H_6N_2O_2$ Crystalline

Sigma C 1034   FW 102.1 $C_3H_6N_2O_2$ Similar to C 7005, but produced by Sigma | Blocks sphingosine biosynthesis by inhibition of ketosphinganine synthetase; Edmondson, *J Biochem Biophys Res Commun*, 76: 751, 1977

Sigma C 7005   FW 102.1 $C_3H_6N_2O_2$ Crystalline | Blocks sphingosine biosynthesis by inhibition of ketosphinganine synthetase; Edmondson, *J Biochem Biophys Res Commun*, 76: 751, 1977

Serine — DL-Ser

Synonyms: (±)-2-Amino-3-Hydroxypropionic Acid; DL-2-Amino-3-Hydroxypropionic Acid; 2-Amino-3-Hydroxypropanoic Acid; *N*-Methyl-D-Aspartate Agonist

Bachem F-2920.0025   MW 105.09 $C_3H_7NO_3$ Store at RT

Fluka 84980   MW 105.09 $C_3H_7NO_3$ ≥99.0% (titration); mp: 230-240°C

ICN 102868   MW 105.1 $C_3H_7NO_3$ Crystalline

Rexim   MW 105.1 $C_3H_7NO_3$ White crystals or crystalline powder

Sigma S 4375   FW 105.1 $C_3H_7NO_3$ ≥98% (TLC) | NMDA agonist acting at the glycine site; precursor of glycine by serine hydroxymethyltransferase; also available as part of a kit

Sigma S 5386   FW 105.1 $C_3H_7NO_3$ ≥98% (TLC); crystalline; | Cell culture tested; insect cell culture tested

USBio S1000-15   MW 105.1 $C_3H_7NO_3$ ≥99%

Serine — DL-Ser (1-¹⁴C)-OMe Hydrochloride

ARC ARC-602   MW 155.6 $HOCH_2CH(NH_2)COOCH_3 \cdot HCl$ 50-60 mCi/mmol; 1.85-2.22 GBq/mmol; in EtOH | Radiochemical

Serine — DL-Ser Hydroxamate

Sigma S 4503   FW 120.1 $C_3H_8N_2O_3$

Serine — DL-Ser-OMe Hydrochloride

Bachem F-2610.0025   MW 155.58 $C_4H_9NO_3 \cdot HCl$ Store at 2-8°C

ICN 102871   MW 155.6 $C_4H_9NO_3 \cdot HCl$ Crystalline

Sigma S 5000   FW 155.6 $C_4H_9NO_3 \cdot HCl$

Serine — DL-*threo*-β-(3,4-Dihydroxyphenyl)-Ser

Synonyms: DL-*threo*-DOPS

Sigma D 2384   FW 213.2 $C_9H_{11}NO_5$ *Droxidopa* precursor of norepinephrine, elevates brain norepinephrine content

Serine — DL-*threo*-β-Phenyl-Ser

Synonyms: 2-Amino-3-Hydroxy-3-Phenylpropanoic Acid

ICN 101000   MW 181.2 $C_9H_{11}NO_3$ Crystalline

Sigma P 8376   FW 181.2 $C_9H_{11}NO_3$ Essentially free of the erythro form

Serine — DL-α-(2-Thienyl)-DL-Ser

Synonyms: ((2RS,3RS)-2-Amino-3-hydroxy-3-(2-Thienyl)-Propionic Acid

Bachem F-2120.0001   MW 187.22 $C_7H_9NO_3S$ Store at -15°C

Serine — DL-α-(3,4-Dihydroxyphenyl)-DL-Ser

Synonyms: ((2RS,3RS)-2-Amino-3-(3,4-Dihydroxy-Phenyl)-3-Hydroxy-Propionic Acid

Bachem F-1525.0050   MW 213.19 $C_9H_{11}NO_5$ Store at -15°C

Serine — DL-α-Phosphatidyl-L-Ser Dipalmitoyl (C16:0)

Synonyms: 1,2-Dihexadecanoyl-rac-Glycero-3-Phospho-L-Ser

Sigma P 1902   FW 736.0 $C_{38}H_{74}NO_{10}P$ ~98%

Serine — Dnp-L-Ser

ICN 102874   MW 271.2 $C_9H_9N_3O_7$ Crystalline

Serine — D-Ser

Synonyms: (R)-2-Amino-3-Hydroxypropionic Acid; D-2-Amino-3-Hydroxypropionic Acid; 2-Amino-3-Hydroxypropanoic Acid; β-Hydroxyalanine

Bachem F-2090.0005   MW 105.09 $C_3H_7NO_3$ Store at RT

Fluka 84970   MW 105.09 $C_3H_7NO_3$ ≥99.0% (titration); mp: 215-225°C

ICN 102865   MW 105.1 $C_3H_7NO_3$ Crystalline

Neosystem AA01601   MW 105.1

Peptides International ADS-2818   MW 105.09   >99.9% optical purity by RP-HPLC; white powder

Rexim   MW 105.1 $C_3H_7NO_3$

Sigma S 4250   FW 105.1 $C_3H_7NO_3$ ≥98% (TLC) | Also available as part of a kit

USBio S1000-10   MW 105.1 $C_3H_7NO_3$ ≥99%

Serine — D-Ser (1-¹⁴C)

ARC ARC-377   MW 105.1 $HOCH_2CH(NH_2)COOH$ 50-60 mCi/mmol; 1.85-2.22 GBq/mmol; in 0.01 *N* HCl | Radiochemical

Serine — D-Ser (G-³H)

ARC ART-728   MW 105.1 $HOCH_2CH(NH_2)COOH$ 15-40 Ci/mmol; 555 GBq-1.48 TBq/mmol; in EtOH: H_2O (2:98) | Radiochemical

Serine — D-Ser-(PO₃H₂)

Bachem F-2035.0500   MW 185.07 $C_3H_8NO_6P$ Store at -15°C

Serine — D-Ser-Ac Hydrochloride

Bachem F-1085.0001   MW 147.13 $C_5H_9NO_4$ Store at RT

Serine — D-Ser-Bzl

Bachem F-1297.0001   MW 195.22 $C_{10}H_{13}NO_3$ Store at RT

Serine — D-Ser-OMe Hydrochloride

Bachem F-2105.0001   MW 155.58 $C_4H_9NO_3$ HCl   Store at -15°C

Serine — D-Ser-Sulfo

Bachem F-3370.0001   MW 185.16 $C_3H_7NO_6S$ Store at -15°C

Serine — D-Ser-tBu

Bachem F-1355.0001   MW 161.2 $C_7H_{15}NO_3$ Store at 2-8°C

Serine — D-Ser-tBu-OMe Hydrochloride

Bachem F-1360.0001   MW 211.69 $C_8H_{17}NO_3 \cdot HCl$ Store at 2-8°C

Serine — FMOC-D-Ser

American Peptide FSER140   MW 383.4

Bachem B-1980.0001   MW 327.34 $C_{18}H_{17}NO_5$ Store at RT

Fluka 47533   MW 327.34 $C_{18}H_{17}NO_5$ ≥98% (HPLC); mp: 108-112°C

Serine — FMOC-D-Ser-BSi

Bachem B-1230.0001   MW 441.6 $C_{24}H_{31}NO_5Si$ Store at -15°C

Serine — FMOC-D-Ser-Bzl

Bachem B-1205.0001   MW 417.46 $C_{25}H_{23}NO_5$ Store at RT

Serine — FMOC-D-Ser-tBu

Synonyms: FMOC-O-tBu-D-Ser

Bachem B-1240.0001   MW 383.44 $C_{22}H_{25}NO_5$ Store at RT

Neosystem FA01605   MW 383.4

Peptides International FDS-1831-PI   MW 383.45   >98% (HPLC); white crystalline powder

USBio F5457   MW 383.4 $C_{22}H_{25}NO_5$ ≥99%

Serine — FMOC-D-Ser-tBu-ODhbt

Bachem B-1855.0001   MW 528.57 $C_{29}H_{28}N_4O_6$ Store at -15°C

Serine — FMOC-D-Ser-tBu-SASRIN™-® (200-400 mesh)

Bachem D-1430.0001   Store at 2-8°C

Serine — FMOC-L-Ser

Fluka 47601   MW 327.34 $C_{18}H_{17}NO_5$ ≥98% (HPLC)

Neosystem FA01601   MW 327.3

Serine — FMOC-L-Ser tBu-Wang Resin

American Peptide RFWAN80   Preattached Wang resins are usful for synthesis of peptides with C-terminal carboxyl groups; peptides can be cleaved from the resin with 90% TFA in the presence of scavengers

Serine — FMOC-L-Ser-Bzl

Neosystem FA01604   MW 417.5

Serine — FMOC-L-Ser-tBu

Neosystem FA01606   MW 383.4

Serine — FMOC-L-Ser-tBu-Wang Resin

Neosystem RW01602

Serine — FMOC-Me-Ser-Bzl

USBio F5426   MW 431.5 $C_{26}H_{25}NO_5$ ≥99%

Serine — FMOC-N-Me-Ser-tBu

Bachem B-3400.0001   MW 397.47 $C_{23}H_{27}NO_5$ Store at -15°C

Serine — FMOC-O-Benzylphospho-L-Ser

Fluka 09769   MW 497.44 $C_{25}H_{24}NO_8P$ ≥97% (HPLC)

Serine — FMOC-O-Bzl-L-Ser

Fluka 47678   MW 417.46 $C_{25}H_{23}NO_5$ ≥98% (HPLC)

Serine — FMOC-O-Bzl-N-Me-L-Ser

Fluka 00548   MW 431.49 $C_{26}H_{25}NO_5$ ≥98% (HPLC)

Serine — FMOC-O-tBu-D-Ser

Fluka 47311   MW 383.44 $C_{22}H_{25}NO_5$ ≥98% (TLC)

Serine — FMOC-O-tBu-L-Ser

Fluka 47619   MW 383.44 $C_{22}H_{25}NO_5$ ~98% (titration); mp: 130°C

Serine — FMOC-O-tBu-L-Ser 4-Benzyloxybenzyl Ester Polymer Bound

Fluka 47648   0.4-0.6 mmol/g resin; carrier: polystyrene, crosslinked with 1% DVB; 200-400 mesh particle size

Serine — FMOC-O-tBu-L-Ser N-Carboxy Anhydride

Fluka 47679   MW 409.44 $C_{23}H_{23}NO_6$ ≥97%; mp: 97-100°C | Produced by Propeptide under an exclusive license from Bioresearch

Serine — FMOC-O-tBu-L-Ser-OPfp

Fluka 00231   MW 549.49 $C_{28}H_{24}F_5NO_5$ ~98%

Serine — FMOC-O-Trt-L-Ser

Fluka 47563   MW 569.66 $C_{37}H_{31}NO_5$ ≥98% (HPLC); mp: 200-210°C

Serine — FMOC-O-Trt-L-Ser 4-Benzyloxybenzyl Ester Polymer Bound

Fluka 47403   0.4-0.8 mmol/g resin; carrier: polystyrene, crosslinked with 1% DVB; 100-200 mesh particle size

Serine — FMOC-Ser

American Peptide FSER130   MW 383.4

Bachem B-1430.0001   MW 327.34 $C_{18}H_{17}NO_5$ Store at RT

USBio F5456   MW 327.3 $C_{18}H_{17}NO_5$ ≥99%

Serine — FMOC-Ser tBu

American Peptide FSER110   MW 383.4

Serine — FMOC-Ser-(GalNAcAc₃-α-D)

Synonyms: N^α-FMOC-O-β-(2-Acetamido-2-Deoxy-3,4,6-Tri-O-Ac-α-D-Galactopyranosyl)-L-Ser

Bachem B-2565.0025   MW 656.64 $C_{32}H_{36}N_2O_{14}$ Store at -15°C

Serine — FMOC-Ser-(PO₃H₂)

Bachem B-2505.0001   MW 407.32 $C_{18}H_{18}NO_8P$ Store at -15°C

Serine — FMOC-Ser-(PO-OBzlOH)

Bachem B-3455.0001   MW 497.44 $C_{25}H_{24}NO_8P$ Store at -15°C

Serine — FMOC-Ser-Ac

Bachem B-1010.0001   MW 369.37 $C_{20}H_{19}NO_6$ Store at 2-8°C

Serine — FMOC-Ser-BSi

Bachem B-1225.0001   MW 441.6 $C_{24}H_{31}NO_5Si$ Store at -15°C

Serine — FMOC-Ser-Bzl

Synonyms: FMOC-O-Bzl-L-Ser

Bachem B-1200.0001   MW 417.46 $C_{25}H_{23}NO_5$ Store at RT

Peptides International FLS-1732-PI   MW 417.46   >98% (HPLC); white crystalline powder

USBio F5459   MW 418.5 $C_{25}H_{24}NO_4$ ≥99%

Serine — FMOC-Ser-Dibutylphospho

USBio F5460   MW 520.5 $C_{26}H_{35}NO_8P$ ≥99%

Serine — FMOC-Ser-OMe

Bachem B-2695.0001   MW 341.37 $C_{19}H_{19}NO_5$ Store at -15°C

Serine — FMOC-Ser-tBu

Synonyms: FMOC-O-tBu-L-Ser

Bachem B-1235.0001   MW 383.44 $C_{22}H_{25}NO_5$ Store at RT

Peptides International FLS-1733-PI   MW 383.45   >98% (HPLC); white crystalline powder

USBio F5458   MW 383.4 $C_{22}H_{25}NO_5$ ≥99%

Serine — FMOC-Ser-tBu-® (200-400 mesh)

Bachem D-1070.0001   Store at 2-8°C

Serine — FMOC-Ser-tBu-ODhbt

Bachem B-1640.0001   MW 528.57 $C_{29}H_{28}N_4O_6$ Store at -15°C

Serine — FMOC-Ser-tBu-OPfp

Bachem B-1180.0001   MW 549.49 $C_{28}H_{24}F_5NO_5$ Store at -15°C

Serine — FMOC-Ser-tBu-SASRIN™-® (200-400 mesh)

Bachem D-1385.0001   Store at 2-8°C

Serine — FMOC-Ser-Trt

Bachem B-2550.0001   MW 569.66 $C_{37}H_{31}NO_5$ Store at -15°C

Serine — FMOC-Ser-β-Lactone

Bachem B-3525.0001   MW 309.32 $C_{18}H_{15}NO_4$ Store at -15°C

Serine — L-

Sigma S 1054   FW 105.1 $C_3H_7NO_3$ USP Grade

Serine — L-3-Phosphatidyl-L-Ser (3-^{14}C) 1,2-Dioleoyl

Amersham CFA757   Toluene:EtOH (1:1); 1.85-2.29 GBq/mmol, 50-62 mCi/mmol; 370 kBq/mL, 10 µCi/mL

Serine — L-Cycloserine

Synonyms: (S)-4-Amino-3-Isoxazolidone; Aspartate Aminotransferase Inhibitor; Vitamin B6 Antagonist; L-4-Amino-3-Isoxazolidinone; Transaminase Inhibitor; Ketosphinganine Synthetase Inhibitor

Bachem F-1475.0250   MW 102.09 $C_3H_6N_2O_2$ Store at -15°C

Fluka 30018   MW 102.09 $C_3H_6N_2O_2$ ≥98% (TLC); mp: 145-150°C; ≤3% H_2O | A vitamin B6 antagonist of natural origin; inhibits mitochondrial & cytosolic aspartate aminotransferase; Klosterman, HJ, *Meth Enzymol*, 62: 483, 1979; Janski, AM & Cornell, NW, *Biochem J*, 194: 1027, 1981

ICN 190316   MW 102.1 $C_3H_6N_2O_2$ Crystalline

Sigma C 1159   FW 102.1 $C_3H_6N_2O_2$ Blocks sphingosine biosynthesis by inhibition of ketosphinganine synthetase; Edmondson, *J Biochem Biophys Res Commun*, 76: 751, 1977

Serine — L-Ser

Synonyms: 1-2-Amino-3-Hydroxypropionic Acid; (S)-2-Amino-3-Hydroxypropionic Acid; L-2-Amino-3-Hydroxypropionic Acid

Amersham US21505   98.5-101.5%; ≤0.0015% heavy metals (Pb)

Fluka 84959   MW 105.09 $C_3H_7NO_3$ ≥99.5% (titration); ≤0.1% residue on ignition, loss on drying; ≤0.3% foreign AA; ≤0.00001% As; ≤0.0005% Al, Ba, Bi, Cd, Co, Cr, Cu, Fe, Li, Mg, Mn, Mo, Ni, Pb, Sr, Zn; ≤0.005% Na, K, Ca, Cl, SO_4; ≤0.01% NH_4^+ | Dollinger, G et al, *Meth Enzymol*, 127: 649, 1986; Arakawa, T & Timasheff, SN, *Meth Enzymol*, 114: 49, 1985; Nguyen, NY & Chrambach, A, *Anal Biochem*, 94: 202, 1979; *CRC Handbook of Microbiology*, Laskin, AI & Lachevalier, HA, eds, 4: 1, 1974, CRC Press, Cleveland, Ohio

Fluka 84960   MW 105.09 $C_3H_7NO_3$ ≥99.0% (titration); ≤0.1% residue on ignition; ≤0.2% loss on drying; ≤0.3% foreign AA; ≤0.0005% Cd, Co, Cr, Cu, Fe, Mg, Mn, Ni, Pb, Zn; ≤0.001% Ca; ≤0.005% Na, K, Cl, SO_4; ≤0.01% NH_4^+

ICN 102873   MW 105.1 $C_3H_7NO_3$ ≥99%; crystalline

ICN 194737   MW 105.1 $C_3H_7NO_3$ ≥99%; crystalline | Cell culture reagent

Neosystem AA01602   MW 105.1

Peptides International ALS-2719   MW 105.09   >98% optical purity; white powder; 0.5-1.5% D-Ser

Rexim   MW 105.1 $C_3H_7NO_3$ White crystals or crystalline powder

Sigma S 3881   Matrix: 4% beaded agarose; activation: cyanogen bromide; attachment: amino; spacer: 1 atom; ligand immobilized: 2-10 µmoles/mL; form: suspension in 2.0 *M* NaCl containing 0.02% thimerosal | Deutsch, DG & Mertz, ET, *Proc Fed Amer Soc Exp Biol*, 29: 647, 1970; Deutsch, DG & Mertz, ET, *Science*, 170: 1095, 1970; Vuento, M & Vaheri, A, *Biochem J*, 183: 331, 1979

Sigma S 4500   FW 105.1 $C_3H_7NO_3$ ≥99% (TLC) | Also available as part of a kit

Sigma S 5511   FW 105.1 $C_3H_7NO_3$ Cell culture tested; insect cell culture tested

Sigma S 8407   FW 105.1 $C_3H_7NO_3$ SigmaUltra; >99% (TLC); pH (1 *M* in H_2O, 20°C): 5.0-7.0; residue on ignition: <0.1%; solubility (1 *M* in H_2O, 20°C): complete, colorless; insoluble matter: <0.1%; SO_4, Cl, NH_4^+,: <0.05%; K, Na: <0.005%; Al, Ca, Cu, Fe, Mg, Zn, P: <0.0005%; Pb: <0.001%; A_{260}<0.05; A_{280}<0.05 (1 *M* in H_2O) | Precursor of glycine by serine hydroxymethyltransferase

USBio S1000   MW 105.1 $C_3H_7NO_3$ ≥99%; appearance: white crystalline powder; pH: 5.5-6.5; specific rotation @ C=10, 2 *N* HCl: +14.4° to +15.5°; heavy metals (Pb): ≤0.001%; endotoxin: ≤2 EU/mg; cell culture tested: murine myeloma SP2/0-Ag14

Serine — L-Ser (1-^{14}C)

ARC ARC-374   MW 105.1 $HOCH_2CH(NH_2)COOH$ 50-60 mCi/mmol; 1.85-2.22 GBq/mmol; in 0.01 *N* HCl | Radiochemical

Serine — L-Ser (3-^{14}C)

Amersham CFA151 $HOCH_2CH(NH_2)CO_2H$ Aqueous, 2% EtOH, sterilized; 1.85-2.29 GBq/mmol, 50-62 mCi/mmol; 1.85 MBq/mL, 50 µCi/mL

ARC ARC-603   MW 105.1 $HOCH_2CH(NH_2)COOH$ 50-60 mCi/mmol; 1.85-2.22 GBq/mmol; in EtOH: H_2O (2:98) | Radiochemical

ICN 10123   MW 105.1 $HOCH_2CH(NH_2)COOH$ 40-60 mCi/mmol, 1.48-2.22 GBq/mmol; 0.01 *N* HCl

---

Serine — L-Ser (3-³H)

Amersham TRK308 HOCH₂CH(NH₂)CO₂H Aqueous, sterilized; 0.74-1.48 TBq/mmol, 20-40 Ci/mmol; 37 MBq/mL, 1 mCi/mL

Serine — L-Ser (G-³H)

ARC ART-246 MW 105.1 HOCH₂CH(NH₂)COOH 5-25 Ci/mmol; 185-925 GBq/mmol; in 0.01 N HCl | Radiochemical

ICN 20050E MW 105.1 HOCH₂CH(NH₂)COOH 5-25 Ci/mmol, 185-925 GBq/mmol; sterile EtOH:H₂O (2:98)

Sigma S 9522 FW 105.1 HOCH₂CH(NH₂)COOH 15-40 Ci/mmol; aqueous solution containing 2% EtOH in serum bottle | Radiochemical

Serine — L-Ser (U-¹⁴C)

Amersham CFB72 HOCH₂CH(NH₂)CO₂H Aqueous, 2% EtOH, sterilized; >5.5 GBq/mmol, >150 mCi/mmol; 1.85 MBq/mL, 50 μCi/mL

ARC ARC-676 MW 105.1 HOCH₂CH(NH₂)COOH 100-150 mCi/mmol; 3.7-5.5 GBq/mmol; in EtOH: H₂O (2:98) | Radiochemical

ICN 10122E MW 105.1 HOCH₂CH(NH₂)COOH 135-165 mCi/mmol, 5-6.1 GBq/mmol; sterile EtOH:H₂O (2:98)

Serine — L-Ser (U-¹⁴C) Hydrochloride

Sigma S 9771 FW 105.1 HOCH₂CH(NH₂)COOH 120-180 mCi/mmol; radiochemical purity ≥98% (HPLC); aqueous solution containing 2% EtOH in Combi-vial | Radiochemical

Serine — L-Ser β-Naphthylamide

ICN 156612 MW 230.3 C₁₃H₁₄N₂O₂ Crystalline

Sigma S 4630 FW 230.3 C₁₃H₁₄N₂O₂

Serine — L-Ser-AMC Hydrochloride

ICN 156611 MW 298.7 C₁₃H₁₄N₂O₄ · HCl

Sigma S 6393 FW 298.7 C₁₃H₁₄N₂O₄ · HCl

Serine — L-Ser-AMC Hydrochloride Hydrate

ICN 198703 MW 298.7 C₁₃H₁₄N₂O₄ · HCl

Serine — L-Ser-Bzl

Neosystem AA01603 MW 195.2

Serine — L-Serinamide Hydrochloride

Sigma S 4750 FW 140.6 C₃H₈N₂O₂ · HCl

Serine — L-Ser-OBzl Hydrochloride

ICN 152054 MW 231.7 C₁₀H₁₃NO₃ · HCl

Sigma S 5633 FW 231.7 C₁₀H₁₃NO₃ · HCl

Serine — L-Ser-OEt Hydrochloride

ICN 152055 MW 169.6 C₅H₁₁NO₃ · HCl

Serine — L-Ser-OMe Hydrochloride

Fluka 85000 MW 155.58 C₄H₉NO₃ · HCl ≥99.0% (titration); mp: 160-165°C; enantiomer ratio: L:D ≥99.5:0.5

ICN 102875 MW 155.6 C₄H₉NO₃ · HCl Crystalline

Neosystem AA01610 MW 155.6

Sigma S 5125 FW 155.6 C₄H₉NO₃ · HCl

Serine — L-Ser-O-Sulfate Potassium Salt

Sigma S 5006 FW 223.2 C₃H₆NO₆SK

Serine — L-α-Glycerophosphorylserine

Sigma G 4257 FW 259.2 C₆H₁₄NO₈P ~98%; 5 mg/mL in MeOH: H₂O

Serine — L-α-Lysophosphatidyl-L-Ser Palmitoyl (C16:0) Sodium Salt

Synonyms: 1-Palmitoyl-sn-Glycero-3-Phospho-L-Ser

Sigma L 3401 FW 519.5 C₂₂H₄₃NO₉PNa

Serine — L-α-Lysophosphatidyl-L-Ser Sodium Salt

Synonyms: Monoacyl-sn-Glycero-3-Phospho-L-Ser

Sigma L 5772 ~99%; primarily stearic acid | Degradation product of phosphatidylserine by phospholipase A

Serine — L-α-Phosphatidyl-L-Ser

Synonyms: 1,2-Diacyl-sn-Glycerol-3-Phospho-(1-D-myo-Inositol 4-Phosphate); Diphosphoinositide; PIP

Sigma P 6641 Bovine brain ~98%; chloroform: MeOH (95:5) solution; actual concentration given on label | A slowly metabolized structural phospholipid found mainly in gray matter; solutions exposed to room temperature to decompose ~0.5%/day

Sigma P 7769 Bovine brain ~98%; amorphous powder | A slowly metabolized structural phospholipid found mainly in gray matter; solutions exposed to room temperature will decompose ~0.5%/day

Sigma P 0474 Soybean ~98% | A slowly metabolized structural phospholipid found mainly in gray matter; solutions exposed to room temperature will decompose ~0.5%/day

Serine — L-α-Phosphatidyl-L-Ser Dioleoyl Sodium Salt

Synonyms: 1,2-Di-(cis-9-Octadecenoyl)-sn-Glycero-3-Phospho-L-Ser

Sigma P 1060 FW 810.0 C₄₂H₇₇NO₁₀PNa ~95%

Serine — L-α-Phosphatidyl-L-Ser Dipalmitoyl (C16:0) Sodium Salt

Synonyms: 1,2-Dihexadecanoyl-sn-Glycero-3-Phospho-L-Ser

Sigma P 1185 FW 758.0 C₃₈H₇₃NO₁₀PNa ≥99%

Serine — L-α-Phosphatidyl-L-Ser Sodium Salt

Synonyms: 1,2-Diacyl-sn-Glycerol-3-Phospho-(1-D-myo-Inositol 4-Phosphate); Diphosphoinositide; PIP

Sigma P 5660 Bovine brain ~98; washed with EDTA & converted to a sodium Salt | A slowly metabolized structural phospholipid found mainly in gray matter; solutions exposed to room temperature will decompose ~0.5%/day

Serine — N-(3-Nitro-2-Pyridinesulfenyl)-O-tBu-L-Ser Dicyclohexylammonium Salt

Sigma N 5390 FW 496.7 C₁₂H₁₇N₃O₅S · C₁₂H₂₃N

Serine — N-2,4-Dnp-L-Ser

Sigma D 1755 FW 271.2 C₉H₉N₃O₇

Serine — N-Ac L-Ser (1-¹⁴C)

ARC ARC-837 MW 147.1 HOCH₂CH(NHCOCH₃)COOH 25-50 mCi/mmol; 0.925-1.85 GBq/mmol; in EtOH | Radiochemical

Serine — N-Ac-DL-Ser

Sigma A 2638 FW 147.1 C₅H₉NO₄

Serine — N-Carbamyl-DL-Ser Potassium Salt

Sigma C 9883 FW 186.2 C₄H₇N₂O₄K

Serine — N-CBZ-DL-Ser

Sigma C 5127	FW 239.2	$C_{11}H_{13}NO_5$

Serine — N-CBZ-L-Ser

ICN 104865	MW 239.2	$C_{11}H_{13}NO_5$	Crystalline
Sigma C 5252	FW 239.2	$C_{11}H_{13}NO_5$	

Serine — N-CBZ-O-tBu-L-Ser

Sigma C 0157	FW 295.3	$C_{15}H_{21}NO_5$

Serine — N-Dansyl-DL-Ser Cyclohexylammonium Salt

Sigma D 9756	FW. 437.6	$C_{15}H_{18}N_2O_5S \cdot C_6H_{13}N$

Serine — N-Dansyl-L-Ser Cyclohexylammonium Salt

Sigma D 2000	FW. 437.6	$C_{15}H_{18}N_2O_5S \cdot C_6H_{13}N$

Serine — N-Dichloroacetyl-DL-Ser Sodium Salt

Synonyms: Blevidon

ICN 101552

Serine — N-Dichloroacetyl-L-Ser Sodium Salt

ICN 101550

Serine — N-FMOC-O-tBu-L-Ser

Sigma F 1383	FW 383.4	$C_{22}H_{25}NO_5$

Serine — N-FMOC-O-tBu-L-Ser 3,4-Dihydro-4-Oxo-1,2,3-Benzotriazine Ester

Sigma F 0148	FW 528.6	$C_{29}H_{28}N_4O_5$	~10% solvents

Serine — N-Me-Ser

Bachem E-2190.0001	MW 119.12	$C_4H_9NO_3$	Store at RT

Serine — N-Me-Ser-tBu

Bachem E-3565.0001	MW 175.23	$C_8H_{17}NO_3$	Store at 2-8°C

Serine — N-O-Nps-L-Ser Dicyclohexylammonium Salt

Sigma N 1390	FW 439.6	$C_9H_{10}N_2O_5S \cdot C_{12}H_{23}N$

Serine — Nps-Ser-tBu Dicyclohexylamine

Bachem E-2220.0001	MW 495.68	$C_{13}H_{18}N_2O_5S \cdot C_{12}H_{23}N$
Store at 2-8°C		

Serine — N-t-BOC-D-Ser

Sigma B 9766	FW 205.2	$C_8H_{15}NO_5$

Serine — N-t-BOC-D-Ser (1-^{14}C)

ARC ARC-1291	MW 205.2	50-60 mCi/mmol; 1.85-2.22 GBq/mmol; in EtOH	Radiochemical

Serine — N-t-BOC-L-Ser

Sigma B 8751	FW 205.2	$C_8H_{15}NO_5$

Serine — N-t-BOC-L-Ser (1-^{14}C)

ARC ARC-1292	MW 205.2	50-60 mCi/mmol; 1.85-2.22 GBq/mmol; in EtOH	Radiochemical

Serine — N-t-BOC-L-Ser-OMe

Sigma B 41,048-9	FW 219.2	$C_9H_{17}NO_5$	≥95% (GC); Aldrich brand	Formerly Sigma B 8786

Serine — N-t-BOC-O-Bzl-D-Ser

Sigma B 9141	FW 295.3	$C_{15}H_{21}NO_5$

Serine — N-t-BOC-O-Bzl-L-Ser

Sigma B 6001	FW 295.3	$C_{15}H_{21}NO_5$

Serine — N-t-BOC-O-Bzl-L-Ser Resin Ester

Sigma B 8646   N-t-BOC AA ester of methylated 1% divinylbenzene cross-linked polystyrene; 200-400 mesh; substitution range: 0.2-0.6 mmole t-BOC-AA/g resin | For peptide synthesis by the Merrifield method

Serine — N-Trt-L-Ser Lactone

Synonyms: (S)-3-Tritylamino-2-Oxetanone

Fluka 93470   MW 329.4 $C_{22}H_{19}NO_2$ ≥98.0% (TLC); mp: 191-196°C | Chiral building block for preparing unnatural AA with nucleophiles; Arnold, LD et al, JACS, 110: 2237, 1988

Serine — N-Z-O-tBu-L-Ser

Fluka 96028	MW 295.34	$C_{15}H_{21}NO_5$	≥98.0% (TLC)

Serine — N-Z-O-tBu-L-Ser Dicyclohexylamine

Fluka 96030	MW 476.66	$C_{15}H_{21}NO_5 \cdot C_{12}H_{23}N$	≥99.0%
(titration)			

Serine — N'-BOC-O-Dibenzylphospho-L-Ser

Fluka 03581	MW 465.44	$C_{22}H_{28}NO_8P$	~98% (TLC)

Serine — N'-BOC-O-Dimethylphospho-L-Ser

Fluka 03571	MW 313.24	$C_{10}H_{20}PNO_8$	≥98% (TLC)

Serine — N'-BOC-O-Diphenylphospho-L-Ser

Fluka 00082	MW 437.39	$C_{20}H_{24}NO_8P$	~98% (HPLC)

Serine — N'-FMOC-O-tBu-L-Ser

ICN 151137	MW 399.5	$C_{22}H_{25}NO_4S$

Serine — N'-t-BOC-L-Ser

ICN 101093	MW 205.2	$C_8H_{15}NO_5$	Crystalline

Serine — N'-t-BOC-O-Bzl-L-Ser

ICN 150486	MW 295.3	$C_{15}H_{21}NO_5$	Crystalline

Serine — N'-t-BOC-O-Bzl-L-Ser Dicyclohexylammonium Salt

ICN 105097	MW 476.9	$C_{27}H_{44}N_2O_5$

Serine — o-Ac-L-Ser

ICN 105006	MW 147.1	$C_5H_9NO_4$	Crystalline

Serine — o-Ac-L-Ser Hydrochloride

Sigma A 6262	FW 183.6	$C_5H_9NO_4 \cdot HCl$

Serine — o-Bzl-DL-Ser

Synonyms: (±)-2-Amino-3-Benzyloxypropionic Acid

Fluka 13920	MW 195.22	$C_{10}H_{13}NO_3$	≥99% (titration); mp: ~227°C
Sigma B 2876	FW 195.2	$C_{10}H_{13}NO_3$	

Serine — o-Bzl-D-Ser

Synonyms: (R)-2-Amino-3-Benzyloxypropionic Acid

Fluka 13910	MW 195.22	$C_{10}H_{13}NO_3$	≥99% (titration); mp: ~227°C
Sigma B 5401	FW 195.2	$C_{10}H_{13}NO_3$	

Serine — o-Bzl-L-Ser

Synonyms: (*S*)-2-Amino-3-Benzyloxypropionic Acid

Fluka 13900	MW 195.22	$C_{10}H_{13}NO_3$	≥99% (titration); mp: ~227°C

Sigma B 3005	FW 195.2	$C_{10}H_{13}NO_3$	Crystalline

Serine — o-Diazoacetyl-L-Ser

Synonyms: Azaserine

Fluka 11430	MW 173.1	$C_5H_7N_3O_4$	≥99% (TLC) | Segel, GB et al, *JBC*, 264: 16399, 1989; Honda, A et al, *Biochem J*, 261: 627, 1989

ICN 154805	MW 173.1	$C_5H_7N_3O_4$	Crystalline

Sigma A 4142	FW 173.1	$C_5H_7N_3O_4$

Serine — o-Diazoacetyl-L-Ser Hybri-Max®

Synonyms: Azaserine; Hybri-Max®

Sigma A 1164   50X; lyophilized; when reconstituted to 10 mL, each vial contains 2.85×10^{-5} M azaserine; sterilized by γ-irradiation \| Endotoxin tested; hybridoma tested

Serine — o-Diazoacetyl-L-Ser-Hypoxanthine Hybri-Max®

Synonyms: Azaserine-Hypoxanthine; Hybri-Max®

Sigma A 9666   50X; lyophilized; when reconstituted to 10 mL, each vial contains 2.85×10^{-5} M azaserine & 5×10^{-3} M hypoxanthine; sterilized by γ-irradiation \| Endotoxin tested; hybridoma tested

Serine — o-Me-DL-Ser

ICN 155659	MW 119.1	$C_4H_9NO_3$	Crystalline

Sigma M 7378	FW 119.1	$C_4H_9NO_2$

Serine — o-Phospho-DL-Ser

Synonyms: DL-Ser Monophosphoric Acid; DL-2-Amino-3-Hydroxypropanoic Acid 3-Phosphate; DL-SOP; Metabotropic AP4 Receptor GluR₄ Agonist

Fluka 79710	MW 185.07	$C_3H_8NO_6P$	≥98.0% (titration); ≤1% H_2O; ≤0.5% free Ser

Sigma P 0753	FW 185.1	$C_3H_8NO_6P$

Serine — o-Phospho-DL-Ser Barium Salt

ICN 102647   Inorganic phosphorus free

Serine — o-Phospho-D-Ser

Synonyms: D-2-Amino-3-Hydroxypropanoic Acid 3-Phosphate; D-SOP

ICN 156241	MW 185.1	$C_3H_8NO_6P$	Crystalline

Sigma P 5506	FW 185.1	$C_3H_8NO_6P$	Inactive isomer

Serine — o-Phospho-L-Ser

Synonyms: L-Phosphoserine; L-AP4 Agonist; Competitive *N*-Methyl-D-Aspartate Antagonist; L-Serine-O-Phosphate; L-2-Amino-3-Hydroxypropanoic Acid 3-Phosphate; Metabotropic AP4 Receptor GluR₄ Agonist

Alexis 106-024   MW 185.1 $C_3H_8NO_6P$ White solid; soluble in H_2O \| Affinity for type IV, but not type I, metabotropic glutamate receptor; Klunk, WE et al, *J Neurochem*, 56: 1997, 1991; Thomsen, C & Suzdak, PD, *NeuroReport*, 4: 1099, 1993; Ornstein, PL et al, *Current Pharm Design*, 1: 355, 1995

Fluka 79715	MW 185.07	$C_3H_8NO_6P$	≥95.0% (TLC); ≤2% H_2O; ≤0.5% P_i

ICN 105189	MW 185.1	$C_3H_8NO_6P$	Crystalline

Sigma P 0878	FW 185.1	$C_3H_8NO_6P$

Serine — o-tBu-L-Ser

Fluka 20587	MW 161.2	$C_7H_{15}NO_3$	≥97% (titration); mp: 220°C

Sigma B 6278	FW 161.2	$C_7H_{15}NO_3$

Serine — o-tBu-L-Ser-OtBu Hydrochloride

Fluka 20589   MW 253.77 $C_{11}H_{23}NO_3 \cdot HCl$ ~99% (titration); mp: 177°C \| Protected L-Ser useful in peptide synthesis; Paulsen, H & Adermann, K, *Liebigs Ann Chem*, 751, 1989; *ibid*, 771, 1989; Tsuda, Y et al, *Chem Pharm Bull*, 39: 607, 1991; Turcotte, JG et al, *Chem Phys Lipids*, 58: 81, 1991

Serine — Phosphatidyl-L-Ser

ICN 100969   ~98%

Serine — Pth-Ser

Sigma P 2502   FW 222.3 $C_{10}H_{10}N_2O_2S$ ~99% \| Useful as standards in protein sequencing by the Edman degradation

Serine — rac-1,2-Dipalmitoylglycero-3-Phospho-L-Ser

Synonyms: rac-Phosphatidyl-L-Ser, 1,2-Dipalmitoyl

Fluka 42548   MW 736 $C_{38}H_{74}NO_{10}P$ ≤2% H_2O \| Nikolelis, DP et al, *Analyst*, 116: 1221, 1991

Serine — Ser

Bachem E-2375.0025	MW 105.09	$C_3H_7NO_3$	Store at RT

Serine — Ser Amide Hydrochloride

Bachem E-2440.0001	MW 140.57	$C_3H_8N_2O_2 \cdot HCl$	Store at RT

Serine — Ser-(PO₃H₂)

Bachem F-2030.0001	MW 185.07	$C_3H_8NO_6P$	Store at -15°C

Serine — Ser-4MβNA

Bachem J-1300.0250	MW 260.29	$C_{14}H_{16}N_2O_3$	Store at -15°C

Serine — Ser-Ac Hydrochloride

Bachem E-1200.0001	MW 183.59	$C_5H_9NO_4 \cdot HCl$	Store at 2-8°C

Serine — Ser-AMC Hydrochloride

Bachem I-1305.0050	MW 298.73	$C_{13}H_{14}N_2O_4 \cdot HCl$	Store at -15°C

Serine — Ser-Bzl

Bachem E-1575.0001	MW 195.22	$C_{10}H_{13}NO_3$	Store at RT

Serine — Ser-OBzl Hydrochloride

Bachem E-2950.0005	MW 231.68	$C_{10}H_{13}NO_3 \cdot HCl$	Store at RT

Serine — Ser-OEt Hydrochloride

Bachem E-2680.0005	MW 169.61	$C_5H_{11}NO_3 \cdot HCl$	Store at RT

Serine — Ser-OMe Hydrochloride

Synonyms: L-Ser-OMe

Bachem E-2395.0005	MW 155.58	$C_4H_9NO_3 \cdot HCl$	Store at RT

Peptides International EMS-2042-PI   MW 155.58   >98% (TLC); white powder

Serine — Ser-Sulfo

Bachem F-3365.0001   MW 185.16 $C_3H_7NO_6S$ Store at -15°C

Serine — Ser-tBu

Bachem E-1655.0001   MW 161.2 $C_7H_{15}NO_3$ Store at 2-8°C

Serine — Ser-tBu-AMC Hydrochloride

Bachem I-1510.0250   MW 354.8 $C_{17}H_{22}N_2O_4 \cdot HCl$ Store at -15°C

Serine — Ser-tBu-OMe Hydrochloride

Bachem E-1665.0001   MW 211.69 $C_8H_{17}NO_3 \cdot HCl$ Store at 2-8°C

Serine — Ser-tBu-OtBu

Bachem E-1660.0001   MW 217.31 $C_{11}H_{23}NO_3$ Store at -15°C

Serine — Ser-βNA

Bachem K-1495.0001   MW 230.27 $C_{13}H_{14}N_2O_2$ Store at RT

Serine — TentaGel S PHB-Ser-tBu-FMOC

Synonyms: FMOCO-tBu-L-Ser 4-(Poly(Ethylenoxy))-OBzl Polymer Bound

Fluka 86402   0.22 mmol protected AA/g; ~90 μm particle size

Serine — TentaGel S Trt-Ser-tBu-FMOC

Fluka 86434   0.20 mmol protected AA/g; ~90 μm particle size

Serine — Z-DL-Ser

Bachem C-2595.0005   MW 239.23 $C_{11}H_{13}NO_5$ Store at RT

Fluka 97195   MW 239.23 $C_{11}H_{13}NO_5$ ≥98.0% (TLC)

Serine — Z-D-Ser

Bachem C-3060.0001   MW 239.23 $C_{11}H_{13}NO_5$ Store at RT

Serine — Z-D-Ser-OBzl

Bachem C-3645.0001   MW 329.35 $C_{18}H_{19}NO_5$ Store at RT

Serine — Z-D-Ser-tBu

Bachem C-1460.0001   MW 295.34 $C_{15}H_{21}NO_5$ Store at -15°C

Serine — Z-D-Ser-tBu-OMe

Bachem C-3575.0001   MW 309.36 $C_{16}H_{23}NO_5$ Store at 2-8°C

Serine — Z-L-Ser

Fluka 97190   MW 239.23 $C_{11}H_{13}NO_5$ ~99% (titration); mp: 116-119°C | Building block in peptide synthesis; starting material for the synthesis of various α-AA via the β-lactone; Pansare, SV et al, *Org Synth*, 70: 1, 1992

Neosystem ZA01601   MW 239.2

Serine — Z-L-Ser-tBu

Neosystem ZA01606   MW 295.3

Serine — Z-Ser

Synonyms: Benzyloxycarbonyl-L-Ser

Bachem C-2590.0005   MW 239.23 $C_{11}H_{13}NO_5$ Store at RT

Peptides International ZLS-2023-PI   MW 239.23   >98% (TLC); white powder

Serine — Z-Ser Hydrazide

Bachem C-2610.0005   MW 235.26 $C_{11}H_{15}N_3O_4$ Store at RT

Serine — Z-Ser-Bzl

Bachem C-1400.0001   MW 329.35 $C_{18}H_{19}NO_5$ Store at RT

Serine — Z-Ser-Bzl-OSu

Bachem C-3520.0001   MW 426.43 $C_{22}H_{22}N_2O_7$ Store at -15°C

Serine — Z-Ser-OBzl

Bachem C-2600.0005   MW 329.35 $C_{18}H_{19}NO_5$ Store at RT

Serine — Z-Ser-OMe

Bachem C-2605.0005   MW 253.26 $C_{12}H_{15}NO_5$ Store at -15°C

Serine — Z-Ser-tBu

Bachem C-1455.0001   MW 295.34 $C_{15}H_{21}NO_5$ Store at -15°C

Serine — Z-Ser-tBu Hydrazide

Bachem C-1465.0001   MW 309.37 $C_{15}H_{23}N_3O_4$ Store at RT

Serine — Z-Ser-tBu-OSu

Bachem C-1470.0001   MW 392.41 $C_{19}H_{24}N_2O_7$ Store at -15°C

Serine — Z-Ser-Tosyl-OMe

Bachem C-3460.0005   MW 407.44 $C_{19}H_{21}NO_7S$ Store at -15°C

Serine — Z-Ser-Trt

Bachem C-4085.0001   MW 481.55 $C_{30}H_{27}NO_5$ Store at -15°C

Serine — α-Me-DL-Ser

ICN 100516   MW 119.1 $C_4H_9NO_3$ Crystalline

Sigma M 6877   FW 119.1 $C_4H_9NO_3$

Serine — β-2-Thienyl-DL-Ser

ICN 100483   MW 187.2 $C_7H_9NO_3S$ Crystalline

Sigma T 5000   FW 187.2 $C_7H_9NO_3S$

Statine — BOC-Cyclohexylstatine

Synonyms: (3S,4S)-4-(BOC-Amino)-5-Cyclohexyl-3-Hydroxy-Pentanoic Acid; BOC-ACHPA

Bachem A-3340.0250   MW 315.41 $C_{16}H_{29}NO_5$ Store at 2-8°C

Statine — BOC-epi-Statine

Synonyms: (3R,4S)-4-(BOC-Amino)-3-Hydroxy-6-Me-Heptanoic Acid

Bachem F-3990.0250   MW 275.35 $C_{13}H_{25}NO_5$ Store at 2-8°C

Statine — BOC-Phenylstatine

Synonyms: (3S,4S)-4-(BOC-Amino)-3-Hydroxy-5-Phenyl-Pentanoic Acid

Bachem A-4100.0500   MW 309.36 $C_{16}H_{23}NO_5$ Store at 2-8°C

Statine — BOC-Sta

Synonyms: N^{α}-t-(BOC-3S,4S-4-Amino)-3-Hydroxy-6-Methylheptacoic Acid; (3S,4S)-4-(BOC-Amino)-3-Hydroxy-6-Methylheptanoic Acid; (3S,4S)-Statine

American Peptide BXSTA10   MW 275.3

Bachem A-1180.0250   MW 275.35 $C_{13}H_{25}NO_5$ Store at 2-8°C

Fluka 15397   MW 275.35 $C_{13}H_{25}NO_5$ ~95% (HPLC); mp: 120-122°C | A novel AA of pepstatin, a low molecular weight inhibitor of acid protease; Umezawa, H et al, *J Antibiot*, 23: 259, 1970; Rich, DH et al, *BBRC*, 74: 762, 1977

Neosystem BA08901   MW 275.34

USBio B2398   MW 275.4 $C_{13}H_{25}NO_5$ ≥99%

Statine — FMOC-Sta

Synonyms: N^{α}-9-FMOC-3S,4S-4-Amino-3-Hydroxy-6-Methylheptanoic Acid

Neosystem FA08901   MW 397.48

USBio F5461   MW 397.5 $C_{23}H_{27}NO_5$ ≥99%

Statine — N-t-BOC-5-Cyclohexylstatine

Synonyms: N-t-BOC-(3S,4S)-4-Amino-5-Cyclohexyl-3-Hydroxypentanoic Acid

Sigma B 7032   FW 315.4 $C_{16}H_{29}NO_5$ ≥98% (TLC)

Statine — N-t-BOC-Sta

Synonyms: N-t-BOC-(3S,4S)-4-Amino-3-Hydroxy-6-Methylheptanoic Acid

Sigma B 8885   FW 275.3 $C_{13}H_{25}NO_5$

Statine — Sta

Synonyms: (3S,4S)-4-Amino-3-Hydroxy-6-Methylheptanoic Acid

Bachem F-1220.0050   MW 175.23 $C_8H_{17}NO_3$ Store at 2-8°C

Sigma S 5508   FW 175.2 $C_8H_{17}NO_3$ Novel AA found in microbial pepsin inhibitors pepsinostreptin & pepstatin

USBio S7970   MW 175.2 $C_8H_{17}NO_3$ ≥98%

Tetrahydroisoquinolone Carboxylic Acid — BOC-D-Tic

American Peptide BDXTIC5   MW 277.3

Tetrahydroisoquinolone Carboxylic Acid — BOC-L-Tic

American Peptide BLXTIC5   MW 277.3

Tetrahydroisoquinolone Carboxylic Acid — FMOC-D-Tic

American Peptide FXTIC15   MW 399.5

Tetrahydroisoquinolone Carboxylic Acid — FMOC-L-Tic

American Peptide FXTIC10   MW 399.5

Threonine — (MEBMT-β-³H)Cyclosporin-A (Me-Butenyl-Me-Thr-β-³H)

Amersham TRK904   EtOH solution; 185-740 GBq/mmol, 5-20 Ci/mmol; 37 MBq/mL, 1 mCi/mL

Threonine — Ac-Thr

Bachem E-1210.0250   MW 161.16 $C_6H_{11}NO_4$ Store at RT

Threonine — Ac-Thr-Ac

Bachem E-1780.0250   MW 203.19 $C_8H_{13}NO_5$ Store at -15°C

Threonine — Ac-Thr-OMe

Bachem E-1215.0250   MW 175.19 $C_7H_{13}NO_4$ Store at RT

Threonine — Ac-Thr-tBu

Bachem E-3050.0001   MW 217.27 $C_{10}H_{19}NO_4$ Store at 2-8°C

Threonine — allo-Thr

Bachem F-1175.0250   MW 119.12 $C_4H_9NO_3$ Store at RT

Threonine — allo-Thr-OMe Hydrochloride

Bachem F-2545.0001   MW 169.61 $C_5H_{11}NO_3 \cdot HCl$ Store at 2-8°C

Threonine — allo-Thr-tBu

Bachem F-2540.0001   MW 175.23 $C_8H_{17}NO_3$ Store at 2-8°C

Threonine — BOC-Dimethylphosphous-L-Thr

American Peptide BLTHR30   MW 327.2

Threonine — BOC-D-Thr

Bachem A-3390.0001   MW 219.24 $C_9H_{17}NO_5$ Store at RT

Neosystem BA01701   MW 219.2

Threonine — BOC-D-Thr-Bzl

Synonyms: BOC-O-Bzl-D-Thr

Bachem A-1360.0001   MW 309.36 $C_{16}H_{23}NO_5$ Store at RT

Neosystem BA01704   MW 309.4

Peptides International BDT-2624   MW 309.36   >98% (HPLC); white crystalline powder

USBio B2401   MW 309.4 $C_{16}H_{23}NO_5$ ≥99%

Threonine — BOC-D-Thr-Bzl-O-CH₂-φ-CH₂-COOH Dicyclohexylamine

Neosystem LP01701   MW 457.6

Threonine — BOC-D-Thr-Bzl-PAM Resin

Neosystem RP01701

Threonine — BOC-D-Thr-tBu

Bachem A-1445.0001   MW 275.35 $C_{13}H_{25}NO_5$ Store at 2-8°C

Threonine — BOC-L-Thr

Fluka 15505   MW 219.24 $C_9H_{17}NO_5$ ~99% (titration); ≤0.2% residue on ignition

Neosystem BA01702   MW 219.2

Threonine — BOC-L-Thr-Alloc Dicyclohexylamine

Neosystem BA01703   MW 484.7

Threonine — BOC-L-Thr-Bzl

Neosystem BA01705   MW 309.4

Threonine — BOC-L-Thr-Bzl-MBHA Resin

Neosystem RN01705

Threonine — BOC-L-Thr-Bzl-O-CH₂-φ-CH₂-COOH Dicyclohexylamine

Neosystem LP01702   MW 457.6

Threonine — BOC-L-Thr-Bzl-PAM Resin

Neosystem RP01702

Threonine — BOC-Me-Thr-Bzl

USBio B2358	MW 323.4	$C_{17}H_{25}NO_5$	≥98%

Threonine — BOC-N-Me-Thr

Bachem A-4445.0001	MW 233.27	$C_{10}H_{19}NO_5$	Store at RT

Threonine — BOC-N-Me-Thr-Bzl CHA

Bachem A-4105.0001   MW 422.57 $C_{17}H_{25}NO_5 \cdot C_6H_{13}N$
Store at RT

Threonine — BOC-O-Bzl-D-Thr

Fluka 15081	MW 309.36	$C_{16}H_{23}NO_5$	≥98% (TLC)

Threonine — BOC-O-Bzl-L-Thr

Fluka 15405   MW 309.36 $C_{16}H_{23}NO_5$ ≥99% (titration); mp:
112-114°C

Threonine — BOC-O-Bzl-L-Thr Hydroxysuccinimide Ester

Fluka 15082	MW 406.43	$C_{20}H_{26}N_2O_7$	≥98% (TLC)

Threonine — BOC-O-Bzl-L-Thr N-Carboxy Anhydride

Fluka 15416   MW 335.36 $C_{17}H_{21}NO_6$ ≥97% (titration); mp:
77-80°C | Produced by Propeptide under an exclusive license from
Bioresearch

Threonine — BOC-O-Bzl-N-Me-L-Thr

Fluka 00044	MW 323.39	$C_{17}H_{25}NO_5$	≥98% (HPLC)

Threonine — BOC-O-tBu-L-Thr

Fluka 15433   MW 275.35 $C_{13}H_{25}NO_5$ ~99% (TLC; mp: 95-
98°C

Threonine — BOC-Thr

American Peptide BLTHR10	MW 219.2		
Bachem A-2305.0005	MW 219.24	$C_9H_{17}NO_5$	Store at RT

Threonine — BOC-Thr Bzl

American Peptide BLTHR15	MW 309.4

Threonine — BOC-Thr Bzl-O-Resin

American Peptide RMRB265   1% DVB cross-linked: 100-200
mesh | t-BOC protected AA resin; preattached resins are useful
for synthesis of peptides with C-terminal carboxyl group by t-BOC
chemistry

Threonine — BOC-Thr Bzl-PAM-Resin

American Peptide RPAM265   1% DVB cross-linked: 100-200
mesh | t-BOC protected AA resin; preattached resins are useful
for synthesis of peptides with C-terminal carboxyl group by t-BOC
chemistry

Threonine — BOC-Thr-Bzl

Synonyms: BOC-O-Bzl-L-Thr

Bachem A-1355.0005	MW 309.36	$C_{16}H_{23}NO_5$	Store at RT
Peptides International BLT-2070	MW 309.36	>98%	

(HPLC); white crystalline powder

USBio B2402	MW 309.4	$C_{16}H_{23}NO_5$	≥99%

Threonine — BOC-Thr-Bzl-® (200-400 mesh)

Bachem D-1015.0005   Store at RT

Threonine — BOC-Thr-Bzl-OSu

Bachem A-1365.0001   MW 406.44 $C_{20}H_{26}N_2O_7$ Store at
-15°C

Threonine — BOC-Thr-Bzl-PAM-® (200-400 mesh)

Bachem D-1265.0001   Store at 2-8°C

Threonine — BOC-Thr-Dimethylphospho

USBio B2404	MW 327.2	$C_{11}H_{22}NO_8$	≥99%

Threonine — BOC-Thr-Diphenylphospho

USBio B2405	MW 451.4	$C_{21}H_{26}NO_8$	≥99%

Threonine — BOC-Thr-OSu

Bachem A-2310.0001   MW 316.31 $C_{13}H_{20}N_2O_7$ Store at
-15°C

Threonine — BOC-Thr-p-Chlorobenzyl Dicyclohexylamine

Bachem A-1505.0001   MW 525.13 $C_{16}H_{22}ClNO_5 \cdot C_{12}H_{23}N$
Store at RT

Threonine — BOC-Thr-tBu

Bachem A-1440.0001	MW 275.35	$C_{13}H_{25}NO_5$	Store at 2-8°C
USBio B2403	MW 275.3	$C_{13}H_{25}NO_5$	≥99%

Threonine — Bpoc-Thr-tBu CHA

Bachem E-1635.0001   MW 512.69 $C_{24}H_{31}NO_5 \cdot C_6H_{13}N$
Store at -15°C

Threonine — Bpoc-Thr-tBu-OSu

Bachem E-1640.0001   MW 510.59 $C_{28}H_{34}N_2O_7$ Store at
-15°C

Threonine — Bsmoc-Thr-tBu

Synonyms: Bsmoc-O-tBu-L-Thr

Peptides International BLT-5636-PI   MW 397.45   >98%
(HPLC); white to off-white powder | LA Carpino, et al, *JACS*,
119:9915, 1997

Threonine — Bz-D-Thr-OMe

Bachem F-1295.0001   MW 237.26 $C_{12}H_{15}NO_4$ Store at 2-
8°C

Threonine — Bz-Thr-OMe

Bachem E-1515.0001	MW 237.26	$C_{12}H_{15}NO_4$	Store at RT

Threonine — CBZ-D-Thr

USBio C2099-68	MW 253.3	$C_{12}H_{15}NO_5$	≥99%

Threonine — CBZ-Thr

USBio C2099-67	MW 253.3	$C_{12}H_{15}NO_5$	≥99%

Threonine — CBZ-Thr-Bzl

USBio C2099-70	MW 343.4	$C_{19}H_{21}NO_5$	≥99%

Threonine — CBZ-Thr-OBzl

USBio C2099-69	MW 343.4	$C_{19}H_{21}NO_5$	≥99%

Threonine — CBZ-Thr-tBu-Dicyclohexylamine

USBio C2099-71	MW 490.7	$C_{16}H_{23}NO_5 \cdot C_{12}H_{23}N$	≥99%

Threonine — D-allo-Thr

Synonyms: (2R,3R)-2-Amino-3-Hydroxybutyric Acid; D-allo-Thr

Bachem F-1180.0250	MW 119.12	$C_4H_9NO_3$	Store at RT
Fluka 05756	MW 119.12	$C_4H_9NO_3$	≥99%; mp: ~278°C
ICN 156905	MW 119.1	$C_4H_9NO_3$	Crystalline
Sigma T 3013	FW 119.1	$C_4H_9NO_3$	

Threonine — Dansyl-DL-Thr Cyclohexylammonium Salt

ICN 100135	MW 451.6	$C_{16}H_{20}N_2O_4S \cdot C_6H_{13}N$

Threonine — DL-allo-Thr

Bachem F-2635.0250	MW 119.12	$C_4H_9NO_3$	Store at RT
Sigma T 9643	FW 119.1	$C_4H_9NO_3$	

Threonine — DL-Thr

Synonyms: 2-Amino-3-Hydroxybutyric Acid; DL-2-Amino-3-Hydroxybutyric Acid; DL-α-Amino-β-Hydroxybutyric Acid

Fluka 89200	MW 119.12	$C_4H_9NO_3$	≥99.0% (titration); free of allo-Thr; mp: 245°C
ICN 103050	MW 119.1	$C_4H_9NO_3$	99%; crystalline
ICN 194752	MW 119.1	$C_4H_9NO_3$	Cell culture reagent; 99%; crystalline
Sigma T 1520	FW 119.1	$C_4H_9NO_3$	Crystalline; allo-free | Cell culture tested; insect cell culture tested
Sigma T 8375	FW 119.1	$C_4H_9NO_3$	≥98% (TLC); allo-free | Also available as part of a kit
USBio T5020-15	MW 119.1	$C_4H_9NO_3$	≥99%

Threonine — DL-Thr Hydroxamate

Sigma T 6629	FW 134.1	$C_4H_{10}N_2O_3$

Threonine — DL-Thr-OMe Hydrochloride

Sigma T 8750	FW 169.6	$C_5H_{11}NO_3 \cdot HCl$

Threonine — Dnp-L-Thr

ICN 103049	MW 285.2	$C_{10}H_{11}N_3O_7$	Crystalline

Threonine — D-Thr

Synonyms: (2R,3S)-2-Amino-3-Hydroxybutyric Acid; D-2-Amino-3-Hydroxybutyric Acid; D-α-Amino-β-Hydroxybutyric Acid

Bachem F-2125.0005	MW 119.12	$C_4H_9NO_3$	Store at RT
Fluka 89190	MW 119.12	$C_4H_9NO_3$	≥99.0% (titration); free of allo-Thr
ICN 103048	MW 119.1	$C_4H_9NO_3$	99%; crystalline
Neosystem AA01701	MW 119.1		
Peptides International ADT-2809	MW 119.12	>99.9% optical purity by RP-HPLC; white crystalline powder	
Sigma T 8250	FW 119.1	$C_4H_9NO_3$	≥98% (TLC); allo-free | Also available as part of a kit

Threonine — D-Thr allo-free

USBio T5020-10	MW 119.1	$C_4H_9NO_3$	≥99%

Threonine — D-Thr-Bzl Amide Hydrochloride

Bachem F-2495.0001	MW 244.72	$C_{11}H_{17}N_2O_2Cl$	Store at RT

Threonine — D-Thr-Bzl-OBzl Oxalate (1:1)

Bachem F-2215.0005	MW 389.41	$C_{18}H_{21}NO_3 \cdot C_2H_2O_4$ Store at RT

Threonine — D-Thr-OBzl Oxalate (1:1)

Bachem F-2130.0005	MW 299.28	$C_{11}H_{15}NO_3 \cdot C_2H_2O_4$ Store at RT

Threonine — D-Thr-ol

Senn Chem 44108	MW 105.1

Threonine — D-Thr-OMe Hydrochloride

Bachem F-2350.0005	MW 169.61	$C_5H_{11}NO_3 \cdot HCl$	Store at 2-8°C

Threonine — D-Thr-tBu

Bachem F-1365.0001	MW 175.23	$C_8H_{17}NO_3$	Store at 2-8°C

Threonine — D-Thr-tBu-OMe Hydrochloride

Bachem F-2100.0001	MW 225.72	$C_9H_{19}NO_3 \cdot HCl$	Store at 2-8°C

Threonine — FMOC-allo-Thr

Bachem B-3100.0001	MW 341.36	$C_{19}H_{19}NO_5$	Store at RT

Threonine — FMOC-allo-Thr-tBu

Bachem B-1815.0001	MW 397.47	$C_{23}H_{27}NO_5$	Store at 2-8°C

Threonine — FMOC-allo-Thr-tBu-ODhbt

Bachem B-1810.0001	MW 542.59	$C_{30}H_{30}N_4O_6$	Store at -15°C

Threonine — FMOC-D-allo-Thr

Bachem B-3090.0001	MW 341.36	$C_{19}H_{19}NO_5$	Store at RT

Threonine — FMOC-D-Thr

American Peptide FTHR140	MW 397.5		
Bachem B-3095.0001	MW 341.36	$C_{19}H_{19}NO_5$	Store at RT

Threonine — FMOC-D-Thr-Bzl

Bachem B-1805.0001	MW 431.49	$C_{26}H_{25}NO_5$	Store at RT

Threonine — FMOC-D-Threoninol

Bachem B-3085.0001	MW 327.38	$C_{19}H_{21}NO_4$	Store at -15°C

Threonine — FMOC-D-Thr-tBu

Synonyms: FMOC-O-tBu-D-Thr

Bachem B-1250.0001	MW 397.47	$C_{23}H_{27}NO_5$	Store at 2-8°C
Peptides International FDT-1828-PI	MW 397.47	>98% (HPLC); white crystalline powder	
USBio F5465	MW 397.5	$C_{23}H_{27}NO_5$	≥99%

Threonine — FMOC-L-Thr

Neosystem FA01701	MW 341.4

Threonine — FMOC-L-Thr Hydrate

Fluka 47602	MW 359.38	$C_{19}H_{19}NO_5 \cdot H_2O$	≥98% (HPLC); mp: 115°C

Threonine — FMOC-L-Thr tBu-Wang Resin

American Peptide RFWAN83 Preattached Wang resins are usful for synthesis of peptides with C-terminal carboxyl groups; peptides can be cleaved from the resin with 90% TFA in the presence of scavengers

Threonine — FMOC-L-Thr-Bzl

Neosystem FA01703   MW 431.5

Threonine — FMOC-L-Thr-tBu

Neosystem FA01705   MW 397.5

Threonine — FMOC-L-Thr-tBu-Wang Resin

Neosystem RW01702

Threonine — FMOC-Me-Thr-Bzl

USBio F5427   MW 445.5 $C_{27}H_{27}NO_5$ ≥99%

Threonine — FMOC-*N*-Me-Thr

Bachem B-3235.0001   MW 355.39 $C_{20}H_{21}NO_5$ Store at RT

Threonine — FMOC-*N*-Me-Thr-tBu

Bachem B-3420.0001   MW 411.5 $C_{24}H_{29}NO_5$ Store at -15°C

Threonine — FMOC-O-Benzylphospho-L-Thr

Fluka 09771   MW 511.47 $C_{26}H_{26}NO_8P$ ≥97% (HPLC)

Threonine — FMOC-O-Bzl-L-Thr

Fluka 47513   MW 431.49 $C_{26}H_{25}NO_5$ ≥98% (TLC)

Threonine — FMOC-O-Bzl-*N*-Me-L-Thr

Fluka 00547   MW 445.52 $C_{27}H_{27}NO_5$ ≥98% (HPLC)

Threonine — FMOC-O-tBu-D-Thr

Fluka 47312   MW 397.47 $C_{23}H_{27}NO_5$ ≥98% (TLC); mp: 130°C

Threonine — FMOC-O-tBu-L-Thr

Fluka 47622   MW 397.47 $C_{23}H_{27}NO_5$ ≥98% (HPLC); mp: 131-132°C

Threonine — FMOC-O-tBu-L-Thr 4-Benzyloxybenzyl Ester Polymer Bound

Fluka 47652   0.4-0.6 mmol/g resin; carrier: polystyrene, crosslinked with 1% DVB; 200-400 mesh particle size

Threonine — FMOC-O-tBu-L-Thr-OPfp

Fluka 00224   MW 563.52 $C_{29}H_{26}F_5NO_5$ ~98%

Threonine — FMOC-Thr

American Peptide FTHR130   MW 397.5
Bachem B-1745.0005   MW 341.36 $C_{19}H_{19}NO_5$ Store at RT
USBio F5464   MW 341.4 $C_{19}H_{19}NO_5$ ≥99%

Threonine — FMOC-Thr-(GalNAcAc₃-α-D)

Synonyms: N^α-FMOC-O-β-(2-Acetamido-2-Deoxy-3,4,6-Tri-O-Ac-α-D-Galactopyranosyl)-L-Thr
Bachem B-2570.0025   MW 670.67 $C_{33}H_{38}N_2O_{14}$ Store at -15°C

Threonine — FMOC-Thr-(PO₃H₂)

Bachem B-3425.0001   MW 421.34 $C_{19}H_{20}NO_8P$ Store at -15°C

Threonine — FMOC-Thr-(PO-OBzl-OH)

Bachem B-3460.0001   MW 511.47 $C_{26}H_{26}NO_8P$ Store at -15°C

Threonine — FMOC-Thr-Bzl

Synonyms: FMOC-O-Bzl-L-Thr
Bachem B-1210.0001   MW 431.49 $C_{26}H_{25}NO_5$ Store at RT
Peptides International FLT-1734-PI   MW 431.49   >98% (TLC); white powder
USBio F5467   MW 431.5 $C_{26}H_{25}NO_5$ ≥99%

Threonine — FMOC-Thr-Dibutylphospho

USBio F5468   MW 534.6 $C_{27}H_{36}NO_8P$ ≥99%

Threonine — FMOC-Thr-ol

Synonyms: FMOC-L-Threoninol
Senn Chem 44065   MW 327.4

Threonine — FMOC-Thr-tBu

Synonyms: FMOC-O-tBu-L-Thr
Bachem B-1245.0001   MW 397.47 $C_{23}H_{27}NO_5$ Store at 2-8°C
Peptides International FLT-1700-PI   MW 397.47   >98% (HPLC); white crystalline powder
USBio F5466   MW 397.5 $C_{23}H_{27}NO_5$ ≥99%

Threonine — FMOC-Thr-tBu-® (200-400 mesh)

Bachem D-1075.0001   Store at 2-8°C

Threonine — FMOC-Thr-tBu-ODhbt

Bachem B-1635.0001   MW 542.59 $C_{30}H_{30}N_4O_6$ Store at -15°C

Threonine — FMOC-Thr-tBu-OPfp

Bachem B-1185.0001   MW 563.52 $C_{29}H_{26}F_5NO_5$ Store at -15°C

Threonine — FMOC-Thr-tBu-SASRIN™-® (200-400 mesh)

Bachem D-1390.0001   Store at 2-8°C

Threonine — L-allo-Thr

Synonyms: (2S,3S)-2-Amino-3-Hydroxybutyric Acid; L-allo-Thr
Fluka 05753   MW 119.12 $C_4H_9NO_3$ ≥99% (TLC); mp: ~278°C
ICN 156906   MW 119.1 $C_4H_9NO_3$
Sigma 21,026-9   FW 119.1 $C_4H_9NO_3$ Aldrich brand; formerly Sigma T 2888

Threonine — L-Thr

Synonyms: (2S,3R)-2-Amino-3-Hydroxybutyric Acid; 2-Amino-3-Hydroxybutyric Acid; (R-(R,S))-2-Amino-3-Hydroxybutanoic Acid; L-α-Amino-β-Hydroxybutyric Acid

Amersham US22290   ≥98.5%; ≤15 ppm heavy metals (Pb)
Fluka 89179   MW 119.12 $C_4H_9NO_3$ ≥99.5% (titration); ≤0.1% residue on ignition, loss on drying; ≤0.00001% As; ≤0.005% K, Na, Cl, SO₄; ≤0.0005% Al, Ba, Bi, Cd, Co, Cr, Cu, Fe, Li, Mg, Mn, Mo, Ni, Pb, Sr, Zn; ≤0.001% Ca; ≤0.05% NH₄⁺; ≤0.3% foreign AA | Kitano, T et al, *Meth Enzymol*, 124: 349, 1986; Nguyen, NY & Chrambach, A, *Anal Biochem*, 94: 202, 1979; Hine, T, *Electrophoresis '83*, (H. Hirai ed), 541, 1984, de Gruyter, Berlin; *CRC Handbook of Microbiology*, (AI Laskin & HA Lechevalier, eds), 4: 1, 1974
Fluka 89180   MW 119.12 $C_4H_9NO_3$ ≥99.0% (titration); ≤0.1% residue on ignition, loss on drying; ≤0.005% K, Ca, Na, Cl, SO₄; ≤0.0005% Cd, Co, Cr, Cu, Fe, Mg, Mn, Ni, Pb, Zn; ≤0.05% NH₄⁺; ≤0.3% foreign AA; free of allo-Thr
ICN 103053   MW 119.1 $C_4H_9NO_3$ 99%; crystalline
ICN 194753   MW 119.1 $C_4H_9NO_3$ Cell culture reagent; 99%; crystalline
Neosystem AA01702   MW 119.1

Peptides International ALT-2720   MW 119.12   >99.9%
optical purity by RP-HPLC; white crystalline powder

Rexim   MW 119.1 $C_4H_9NO_3$ White crystals or crystalline powder

Sigma T 0387   Matrix: cross-linked 4% beaded agarose; activation: cyanogen bromide; attachment: amino; spacer: 1 atom; ligand immobilized: 2-10 μmoles/mL; form: suspension in 2.0 M NaCl containing 0.02% thimerosal | Deutsch, DG & Mertz, ET, *Proc Fed Amer Soc Exp Biol*, 29: 647, 1970; Deutsch, DG & Mertz, ET, *Science*, 170: 1095, 1970; Vuento, M & Vaheri, A, *Biochem J*, 183: 331, 1979

Sigma T 1645   FW 119.1 $C_4H_9NO_3$ Crystalline | Cell culture tested; insect cell culture tested

Sigma T 5062   FW 119.1 $C_4H_9NO_3$ USP Grade

Sigma T 8534   FW 119.1 $C_4H_9NO_3$ SigmaUltra; >99% (TLC); pH (0.5 M in H_2O, 20°C): 5.0-7.0; residue on ignition: <0.1%; solubility (0.5 M in H_2O, 20°C): complete, colorless; insoluble matter: <0.1%; SO_4, Cl, NH_4^+,: <0.05%; K, Na: <0.005%; Al, Ca, Cu, Fe, Mg, P, Zn: <0.0005%; Pb: <0.001%; A_{260}<0.05; A_{280}<0.05 (0.5 M in H_2O)

Sigma T 8625   FW 119.1 $C_4H_9NO_3$ ≥98% (TLC) | Also available as part of a kit

USBio T5020   MW 119.1 $C_4H_9NO_3$ ≥99%; appearance: white crystalline powder; pH: 5.0-6.5; specific rotation @ C=6, H_2O: -27.6° to -29.0°; heavy metals (Pb): ≤0.001%; endotoxin: ≤2 EU/mg; cell culture tested: murine myeloma SP2/0-Ag14

Threonine — L-Thr (1-^{14}C)

ARC ARC-506   MW 119.1 $CH_3CH(OH)CH(NH_2)COOH$ 50-60 mCi/mmol; 1.85-2.22 GBq/mmol; in 0.01 N HCl | Radiochemical

Threonine — L-Thr (3-^{3}H)

Amersham TRK716   Aqueous, 2% EtOH, sterilized; 185-740 GBq/mmol, 5-20 Ci/mmol; 37 MBq/mL, 1 mCi/mL

Threonine — L-Thr (G-^{3}H)

ARC ART-330   MW 119.1 $CH_3CHOHCH(NH_2)COOH$ 5-20 Ci/mmol; 185-740 GBq/mmol; in EtOH: H_2O (2:98) | Radiochemical

ICN 20052E   MW 119.1 $CH_3CHOHCH(NH_2)COOH$ 15-25 Ci/mmol, 555-925 GBq/mmol; sterile EtOH:H_2O (2:98)

Threonine — L-Thr (U-^{14}C)

ARC ARC-677   MW 119.1 $CH_3CH(OH)CH(NH_2)COOH$ 150-200 mCi/mmol; 5.55-7.4 GBq/mmol; in EtOH: H_2O (2:98) | Radiochemical

ICN 10126E   MW 119.1 $CH_3CHOHCH(NH_2)COOH$ 180-220 mCi/mmol, 6.7-8.1 GBq/mmol; sterile EtOH:H_2O (2:98)

ICN 10126L   MW 119.1 $CH_3CHOHCH(NH_2)COOH$ 110-220 mCi/mmol, 4.1-6.7 GBq/mmol; sterile EtOH:H_2O (2:98); >95%

Threonine — L-Thr (U-^{14}C) Hydrochloride

Sigma T 3907   FW 119.1 $CH_3CHOHCH(NH_2)COOH$ 160-240 mCi/mmol; radiochemical purity ≥98% (HPLC); aqueous solution containing 2% EtOH in Combi-vial | Radiochemical

Threonine — L-Thr-AMC Hydrochloride

Sigma T 8934   FW 312.8 $C_{14}H_{16}N_2O_4 \cdot$ HCl

Threonine — L-Thr-Bzl

Neosystem AA01706   MW 209.2

Threonine — L-Threoninamide Hydrochloride

Sigma T 3887   FW 154.6 $C_4H_{10}N_2O_2 \cdot$ HCl

Threonine — L-Thr-ol

Synonyms: L-Threoninol

Bachem F-3200.0001   MW 105.14 $C_4H_{11}NO_2$ Store at -15°C

Threonine — L-Thr-OMe Hydrochloride

Fluka 89210   MW 169.61 $C_5H_{11}NO_3 \cdot$ HCl   ≥97.0% (titration); mp: 64-67°C

ICN 156907   MW 169.6 $C_5H_{11}NO_3 \cdot$ HCl

Neosystem AA01704   MW 169.6

Sigma T 5898   FW 169.6 $C_5H_{11}NO_3 \cdot$ HCl

Threonine — L-Thr-OtBu Hemioxalate

Fluka 89205   MW 440.49 $(C_8H_{17}NO_3)_2 \cdot C_2H_2O_4$ ≥99.0% (titration); mp: 150°C

Threonine — N-2,4-Dnp-L-Thr

Sigma D 1880   FW 285.2 $C_{10}H_{11}N_3O_7$

Threonine — N-Bz-D-Thr-OMe

Sigma B 7518   FW 237.3 $C_{12}H_{15}NO_4$

Threonine — N-CBZ-L-Thr

ICN 100209   MW 253.3 $C_{12}H_{15}NO_5$ Crystalline

Sigma C 1760   FW 253.3 $C_{12}H_{15}NO_5$

Threonine — N-Dansyl-DL-Thr Cyclohexylammonium Salt

Sigma D 0881   FW. 451.6 $C_{16}H_{20}N_2O_5S \cdot C_6H_{13}N$

Threonine — N-Dansyl-L-Thr Cyclohexylammonium Salt

Sigma D 2125   FW. 451.6 $C_{16}H_{20}N_2O_5S \cdot C_6H_{13}N$

Threonine — N-FMOC-O-tBu-L-Thr

Sigma F 1508   FW 397.5 $C_{23}H_{27}NO_5$

Threonine — N-FMOC-O-tBu-L-Thr Resin Ester

Sigma F 2636   For peptide synthesis; Wang, SS, *J Am Chem Soc*, 95: 1328, 1973; Lu, G et al, *J Org Chem*, 46: 3433, 1981

Threonine — N-Me-Thr

Bachem E-3480.0001   MW 133.15 $C_5H_{11}NO_3$ Store at 2-8°C

Threonine — N-Me-Thr-Bzl Hydrochloride

Bachem E-3545.0001   MW 259.73 $C_{12}H_{17}NO_3 \cdot$ HCl   Store at 2-8°C

Threonine — N-Me-Thr-tBu

Bachem E-3570.0250   MW 189.26 $C_9H_{19}NO_3$ Store at 2-8°C

Threonine — N-O-Nps-L-Thr Dicyclohexylammonium Salt

ICN 102536   MW 453.6 $C_{10}H_{12}N_2O_5S \cdot C_{12}H_{23}N$ Yellow crystals

Sigma N 6503   FW 453.6 $C_{10}H_{12}N_2O_5S \cdot C_{12}H_{23}N$ Yellow crystals

Threonine — N-t-BOC-L-Thr

Sigma B 9129   FW 219.2 $C_9H_{17}NO_5$

Threonine — N-t-BOC-O-Bzl-L-Thr

Sigma B 4379   FW 309.4 $C_{16}H_{23}NO_5$

Threonine — *N*-t-BOC-O-Bzl-L-Thr Resin Ester

Sigma B 9771   *N*-t-BOC AA ester of methylated 1% divinylbenzene cross-linked polystyrene; 200-400 mesh; substitution range: 0.2-0.6 mmole t-BOC-AA/g resin | For peptide synthesis by the Merrifield method

Threonine — *N*-t-BOC-O-tBu-L-Thr

Sigma B 1769   FW 275.3 $C_{13}H_{25}NO_5$

Threonine — *N*-TFA-DL-Thr-OMe

Sigma T 5631   FW 229.2 $C_7H_{10}F_3NO_4$ ~98%

Threonine — *N*-Z-O-tBu-L-Thr Dicyclohexylamine

Fluka 96040   MW 490.70 $C_{16}H_{23}NO_5 \cdot C_{12}H_{23}N$ ~99.0%

Threonine — *N*ᵃ-FMOC-O-tBu-L-Thr

ICN 151138   MW 397.5 $C_{23}H_{27}NO_5$

Threonine — *N*ᵃ-t-BOC-L-Thr

ICN 101094   MW 219.2 $C_9H_{17}NO_5$ Crystalline

Threonine — *N*ᵃ-t-BOC-O-Bzl-L-Thr

ICN 150487   MW 309.4 $C_{16}H_{23}NO_3$ Crystalline

Threonine — o-Bzl-L-Thr Hydrochloride

Sigma B 0402   FW 245.7 $C_{11}H_{15}NO_3 \cdot HCl$

Threonine — o-Me-L-Thr

Synonyms: (±)-2-Amino-2-Me-3-Phenylpropionic Acid; L-Thr Methyl Ether

Fluka 69405   MW 133.15 $C_5H_{11}NO_3$ ≥98.0% (TLC)

Sigma M 1630   FW 133.1 $C_5H_{11}NO_3$

Threonine — o-Phospho-DL-Thr

Synonyms: DL-2-Amino-3-Hydroxybutanoic Acid 3-Phosphate

ICN 102646   MW 199.1 $C_4H_{10}NO_6P$ Crystalline; inorganic phosphorus free

Sigma P 1003   FW 199.1 $C_4H_{10}NO_6P$

Threonine — o-Phospho-L-Thr

Synonyms: L-Thr-O-Phosphate; L-2-Amino-3-Hydroxybutanoic Acid 3-Phosphate

Fluka 79717   MW 199.1 $C_4H_{10}NO_6P$ ≥98.0% (TLC); ≤2% H_2O

Sigma P 1053   FW 199.1 $C_4H_{10}NO_6P$

Threonine — o-Phospho-L-Thr Free Acid

Synonyms: D-2-Amino-3-Hydroxybutanoic Acid 3-Phosphate

ICN 156242   MW 199.1 $C_4H_{10}NO_6P$

Threonine — o-tBu-L-Thr

Fluka 20644   MW 175.23 $C_8H_{17}NO_3$ ≥98% (titration); mp: 244-247°C

Sigma B 0277   FW 175.2 $C_8H_{17}NO_3$

Threonine — o-tBu-L-Thr-OtBu Acetate

Fluka 20646   MW 291.39 $C_{12}H_{25}NO_3 \cdot C_2H_4O_2$ ~97%; mp: 59-61°C | Protected threonine useful in peptide synthesis, e.g. of glycopeptides; Forrow, NJ & Batchelor, MJ, *Tetrahedron Lett*, 31: 3493, 1990; Chernyak, AY et al, *Carbo Res*, 216: 381, 1991

Threonine — TentaGel S PHB-Thr-tBu-FMOC

Synonyms: FMOCO-tBu-L-Thr 4-(Poly(Ethylenoxy))-OBzl Polymer Bound

Fluka 86403   0.22 mmol protected AA/g; ~90 µm particle size

Threonine — TentaGel S Trt-Thr-tBu-FMOC

Fluka 86435   0.20 mmol protected AA/g; ~90 µm particle size

Threonine — Thr

Bachem E-2420.0025   MW 119.12 $C_4H_9NO_3$ Store at RT

Threonine — Thr Amide Hydrochloride

Bachem E-2425.0001   MW 154.6 $C_4H_{10}N_2O_2 \cdot HCl$ Store at RT

Threonine — Thr tBu-2-Chloro-Trt Resin

American Peptide RGEN169

Threonine — Thr-AMC

Bachem I-1360.0050   MW 276.29 $C_{14}H_{16}N_2O_4$ Store at -15°C

Threonine — Thr-Bzl Hydrochloride

Bachem E-1585.0001   MW 245.71 $C_{11}H_{15}NO_3 \cdot HCl$ Store at RT

Threonine — Thr-Bzl-OBzl Oxalate (1:1)

Bachem E-1590.0005   MW 389.41 $C_{18}H_{21}NO_3 \cdot C_2H_2O_4$ Store at RT

Threonine — Thr-NHMe

Bachem E-2450.0001   MW 132.16 $C_5H_{12}N_2O_2$ Store at -15°C

Threonine — Thr-OBzl Oxalate (1:1)

Bachem E-2430.0005   MW 299.28 $C_{11}H_{15}NO_3 \cdot C_2H_2O_4$ Store at RT

Threonine — Thr-ol

Synonyms: L-Threoninol

Senn Chem 44107   MW 105.1

Threonine — Thr-OMe Hydrochloride

Bachem E-2455.0005   MW 169.61 $C_5H_{11}NO_3 \cdot HCl$ Store at 2-8°C

Threonine — Thr-OtBu Hydrochloride

Bachem E-3290.0001   MW 211.69 $C_8H_{17}NO_3 \cdot HCl$ Store at -15°C

Threonine — Thr-tBu

Bachem E-1670.0001   MW 175.23 $C_8H_{17}NO_3$ Store at RT

Threonine — Thr-tBu-AMC Hydrochloride

Bachem I-1505.0250   MW 368.86 $C_{18}H_{24}N_2O_4$ Store at -15°C

Threonine — Thr-tBu-OAll Hydrochloride

Bachem E-3340.0001   MW 251.75 $C_{11}H_{21}NO_3 \cdot HCl$ Store at -15°C

Threonine — Thr-tBu-OMe Hydrochloride

Bachem E-1680.0001   MW 225.72 $C_9H_{19}NO_3 \cdot HCl$ Store at 2-8°C

Threonine — Thr-tBu-OtBu

Bachem E-1675.0001   MW 231.34 $C_{12}H_{25}NO_3$ Store at -15°C

Threonine — Thr-tBu-pNA

Bachem L-1870.0001   MW 295.3 $C_{14}H_{21}N_3O_4$ Store at -15°C

Threonine — Thr-βNA

Bachem K-1520.0001   MW 244.29 $C_{14}H_{16}N_2O_2$ Store at RT

Threonine — Z-allo-Thr-tBu Dicyclohexylamine

Bachem C-3390.0001   MW 490.68 $C_{16}H_{23}NO_5 \cdot C_{12}H_{23}N$ Store at 2-8°C

Threonine — Z-DL-Thr

Fluka 97332   MW 251.28 $C_{13}H_{17}NO_4$ ≥98.0% (TLC)

Threonine — Z-D-Thr

Bachem C-2635.0001   MW 253.26 $C_{12}H_{15}NO_5$ Store at RT

Fluka 97225   MW 253.26 $C_{12}H_{15}NO_5$ ≥98.0% (TLC)

Threonine — Z-D-Thr-Bzl Dicyclohexylamine

Bachem C-3075.0001   MW 524.7 $C_{19}H_{21}NO_5 \cdot C_{12}H_{23}N$ Store at RT

Threonine — Z-D-Thr-tBu Dicyclohexylamine

Bachem C-1480.0001   MW 490.68 $C_{16}H_{23}NO_5 \cdot C_{12}H_{23}N$ Store at RT

Threonine — Z-L-Thr

Fluka 97230   MW 253.26 $C_{12}H_{15}NO_5$ ≥99.0% (titration); mp: 101-103°C

Neosystem ZA01701   MW 253.3

Threonine — Z-L-Thr-OMe

Fluka 97234   MW 267.28 $C_{13}H_{17}NO_5$ ≥98.0% (TLC)

Threonine — Z-L-Thr-tBu Dicyclohexylamine

Neosystem ZA01705   MW 490.7

Threonine — Z-N-Me-Thr

Bachem C-4025.0001   MW 267.28 $C_{13}H_{17}NO_5$ Store at RT

Threonine — Z-Thr

Synonyms: Benzyloxycarbonyl-L-Thr

Bachem C-2630.0005   MW 253.26 $C_{12}H_{15}NO_5$ Store at RT

Peptides International ZLT-2024-PI   MW 253.26   >98% (TLC); white powder

Threonine — Z-Thr Amide

Bachem C-4055.0001   MW 252.27 $C_{12}H_{16}N_2O_4$ Store at RT

Threonine — Z-Thr Hydrazide

Bachem C-2645.0005   MW 267.29 $C_{12}H_{17}N_3O_4$ Store at -15°C

Threonine — Z-Thr-Bzl Dicyclohexylamine

Bachem C-1405.0001   MW 524.7 $C_{19}H_{21}NO_5 \cdot C_{12}H_{23}N$ Store at RT

Threonine — Z-Thr-OBzl

Bachem C-2640.0005   MW 343.38 $C_{19}H_{21}NO_5$ Store at RT

Threonine — Z-Thr-ol

Synonyms: Z-L-Threoninol

Senn Chem 44077   MW 243.3

Threonine — Z-Thr-OMe

Bachem C-2655.0005   MW 267.28 $C_{13}H_{17}NO_5$ Store at -15°C

Threonine — Z-Thr-OSu

Bachem C-2650.0005   MW 350.33 $C_{16}H_{18}N_2O_7$ Store at -15°C

Threonine — Z-Thr-tBu Dicyclohexylamine

Bachem C-1475.0001   MW 490.68 $C_{16}H_{23}NO_5 \cdot C_{12}H_{23}N$ Store at RT

Threonine — Z-Thr-tBu-OSu

Bachem C-1490.0001   MW 406.44 $C_{20}H_{26}N_2O_7$ Store at -15°C

Tryptophan — 1',1-Ethylidene-Bis-L-Trp

Sigma E 0266   FW 434.5 $C_{24}H_{26}N_4O_4$ ≥90% (HPLC) | Tryptophan contaminant linked to eosinophilia-myalgia syndrome; Belongia EA et al, *N Eng J Med*, 323: 357, 1990; mayeno, AN et al, *Science*, 250: 1707, 1990

Tryptophan — 1-Me-DL-Trp

ICN 155680   MW 218.3 $C_{12}H_{14}N_2O_2$ 95%; off-white to tan powder

Tryptophan — 3,5-Dibromo-D-Tyr

Bachem F-3825.0001   MW 338.98 $C_9H_9Br_2NO_3$ Store at RT

Tryptophan — 3,5-Dibromo-Tyr

Bachem F-1520.0005   MW 338.98 $C_9H_9Br_2NO_3$ Store at RT

Tryptophan — 3,5-Diiodo-D-Tyr

Bachem F-3005.0005   MW 432.98 $C_9H_9I_2NO_3$ Store at RT

Tryptophan — 4-Fluoro-DL-Trp

ICN 158155   MW 222.2 $C_{11}H_{11}FN_2O_2$ Off-white crystals

Tryptophan — 4-Me-DL-Trp

Bachem F-1820.0250   MW 218.26 $C_{12}H_{14}N_2O_2$ Store at 2-8°C

ICN 155681   MW 218.3 $C_{12}H_{14}N_2O_2$

Sigma M 8502   FW 218.3 $C_{12}H_{14}N_2O_2$ CNS stimulant

Tryptophan — 5-Benzyloxy-DL-Trp

ICN 150453   MW 310.4 $C_{18}H_{18}N_2O_3$ Crystalline

Sigma B 2251   FW 310.4 $C_{18}H_{18}N_2O_3$ Light yellow crystals

Tryptophan — 5-Fluoro-DL-Trp

Fluka 47570   MW 222.22 $C_{11}H_{11}FN_2O_2$ ~99% (titration); ≤0.05% residue on ignition; mp: 265°C

ICN 158156   MW 222.2 $C_{11}H_{11}FN_2O_2$

Sigma F 0896   FW 222.2 $C_{11}H_{11}FN_2O_2$ White powder

Tryptophan — 5-Fluoro-L-Trp

Fluka 47568   MW 222.22 $C_{11}H_{11}FN_2O_2$ ≥98% (HPLC); L-:D- ≥99.5:0.5 | Since fluorine-19 is a very useful reporter group, 5-Fluoro-L-Trp can be useful as substrate analog to study enzyme mechanisms by NMR spectroscopy; Miles, EW et al, *Biochemistry*, 25: 4240, 1986; Boschelli, F et al, *JBC*, 256: 11595, 1981; Rule, GS et al, *Biochemistry*, 26: 549, 1987; Gerig, JT, *Biol Magn Reson*, 1: 139, 1978; Ho, C et al, *Curr Top Bioenerg*, 14: 53,1985

Tryptophan — 5-Me-DL-Trp

Fluka 69560   MW 218.26 $C_{12}H_{14}N_2O_2$ ~99.0% (titration); mp: 275°C

ICN 195320   MW 218.3 $C_{12}H_{14}N_2O_2$ Crystalline

Sigma M 0534   FW 218.3 $C_{12}H_{14}N_2O_2$

Tryptophan — 5-Methoxy-DL-Trp

Fluka 65380   MW 234.26 $C_{12}H_{14}N_2O_3$ ≥97.0% (titration); mp: 265°C

ICN 155448   MW 234.3 $C_{12}H_{14}N_2O_3$ Light yellow crystals

Sigma M 4001   FW 234.3 $C_{12}H_{14}N_2O_3$ Light yellow crystals

Tryptophan — 6-Fluoro-DL-Trp

ICN 158157   MW 222.2 $C_{11}H_{11}FN_2O_2$ White to light brown crystals

Sigma F 7626   FW 222.2 $C_{11}H_{11}FN_2O_2$ White to light brown crystals

Tryptophan — 6-Me-DL-Trp

Fluka 69570   MW 218.25 $C_{12}H_{14}N_2O_2$ ≥99.0%; mp: 290°C

ICN 155683   MW 218.3 $C_{12}H_{14}N_2O_2$ Light yellow to orange crystals

Sigma M 8752   FW 218.3 $C_{12}H_{14}N_2O_2$ Light yellow to orange crystals

Tryptophan — 7-Benzyloxy-DL-Trp

ICN 154848   MW 310.4 $C_{18}H_{18}N_2O_3$ Crystalline

Tryptophan — 7-Me-DL-Trp

ICN 155684   MW 218.3 $C_{12}H_{14}N_2O_2$ Off-white crystals

Sigma M 8379   FW 218.3 $C_{12}H_{14}N_2O_2$ Off-white crystals

Tryptophan — Ac-5-Me-DL-Trp

Bachem F-2305.0001   MW 260.29 $C_{14}H_{16}N_2O_3$ Store at RT

Tryptophan — Ac-D-Trp

Bachem F-1090.0001   MW 246.27 $C_{13}H_{14}N_2O_3$ Store at RT

Tryptophan — Ac-Trp

Bachem E-1220.0005   MW 246.27 $C_{13}H_{14}N_2O_3$ Store at RT

Tryptophan — Ac-Trp Amide

Bachem E-1225.0001   MW 245.28 $C_{13}H_{15}N_3O_2$ Store at RT

Tryptophan — Ac-Trp-3,5-Bis-Trifluorormethyl-OBzl

Bachem E-3135.0250   MW 472.39 $C_{22}H_{18}F_6N_2O_3$ Store at -15°C

Tryptophan — Ac-Trp-OEt

Bachem M-1075.0001   MW 274.32 $C_{15}H_{18}N_2O_3$ Store at -15°C

Peptides International SWE-3034   MW 274.32 $C_{15}H_{18}N_2O_3$ >99% (HPLC); amorphous powder

Tryptophan — Ac-Trp-OMe

Bachem E-1230.0001   MW 260.29 $C_{14}H_{16}N_2O_3$ Store at RT

Tryptophan — Ac-Trp-ONp

Bachem E-1235.0001   MW 367.36 $C_{19}H_{17}N_3O_5$ Store at -15°C

Tryptophan — Adpoc-Trp

Bachem E-2250.0001   MW 424.55 $C_{25}H_{32}N_2O_4$ Store at -15°C

Tryptophan — BOC-D-Trp

American Peptide BDTRP10   MW 304.4

Bachem A-2365.0001   MW 304.35 $C_{16}H_{20}N_2O_4$ Store at RT

Neosystem BA01801   MW 304.4

Peptides International BDW-2602   MW 304.35   >98% (HPLC); white crystalline powder

USBio B2411   MW 304.3 $C_{16}H_{20}N_2O_4$ ≥99%

Tryptophan — BOC-D-Trp-CHO

Synonyms: N^{α}-BOC- N^{in}-For-D-Trp

Peptides International BDW-2614   MW 332.36   >98% (HPLC); white crystalline powder

Tryptophan — BOC-D-Trp-For

Bachem A-2500.0001   MW 332.36 $C_{17}H_{20}N_2O_5$ Store at 2-8°C

USBio B2413   MW 332.4 $C_{17}H_{20}N_2O_5$ ≥99%

Tryptophan — BOC-D-Trp-For solvent

Neosystem BA01812   MW 332.4.n.solvent

Tryptophan — BOC-L-Trp

Neosystem BA01802   MW 304.4

Tryptophan — BOC-L-Trp-For

Neosystem BA01807   MW 332.4

Tryptophan — BOC-L-Trp-For solvent

Neosystem BA01813   MW 332.4   Solvent

Tryptophan — BOC-L-Trp-For-MBHA Resin

Neosystem RN01807

Tryptophan — BOC-L-Trp-For-O-CH₂-⦙-CH₂-COOH

Neosystem LP01804   MW 480.6

Tryptophan — BOC-L-Trp-For-PAM Resin

Neosystem RP01804

Tryptophan — BOC-L-Trp-Mts-MBHA Resin

Neosystem RN01808

Tryptophan — BOC-L-Trp-Mts-PAM Resin

Neosystem RP01806

Tryptophan — BOC-Trp

Synonyms: BOC-L-Trp

American Peptide BLTRP10   MW 304.4

Bachem A-2360.0005   MW 304.35 $C_{16}H_{20}N_2O_4$ Store at RT

Peptides International BLW-2057   MW 304.35   >98% (HPLC); white crystalline powder

USBio B2412   MW 304.3 $C_{16}H_{20}N_2O_4$ ≥99%

Tryptophan — BOC-Trp (CHO)
American Peptide BLTRP15 MW 332.4

Tryptophan — BOC-Trp-® (200-400 mesh)
Bachem D-1450.0005 Store at RT

Tryptophan — BOC-Trp-CHO
Synonyms: N^α-BOC-N^{in}-For-L-Trp

Peptides International BLW-2115 MW 332.36 >98%
(HPLC); white crystalline powder

Tryptophan — BOC-Trp-For
Bachem A-1625.0001 MW 332.36 $C_{17}H_{20}N_2O_5$ Store at 2-8°C

USBio B2414 MW 332.4 $C_{17}H_{20}N_2O_5$ ≥99%

Tryptophan — BOC-Trp-For-® (200-400 mesh)
Bachem D-1510.0005 Store at 2-8°C

Tryptophan — BOC-Trp-For-O-Resin
American Peptide RMRB285 1% DVB cross-linked: 100-200 mesh | t-BOC protected AA resin; preattached resins are useful for synthesis of peptides with C-terminal carboxyl group by t-BOC chemistry

Tryptophan — BOC-Trp-For-OSu
Bachem A-3410.0001 MW 429.43 $C_{21}H_{23}N_3O_7$ Store at -15°C

Tryptophan — BOC-Trp-For-PAM-Resin
American Peptide RPAM285 1% DVB cross-linked: 100-200 mesh | t-BOC protected AA resin; preattached resins are useful for synthesis of peptides with C-terminal carboxyl group by t-BOC chemistry

Tryptophan — BOC-Trp-Hoc
Synonyms: N^α-BOC-N^{in}-Cyclohexyloxycarbonyl-L-Trp

Peptides International BLW-2139 MW 430.50 >98%
(HPLC); white crystalline powder | Y Nishiuchi, et al, *Tetrahedron Lett*, 37:7529, 1996

Tryptophan — BOC-Trp-Mts Dicyclohexylamine
Bachem A-2515.0001 MW 667.91 $C_{25}H_{30}N_2O_6S \cdot C_{12}H_{23}N$ Store at -15°C

Tryptophan — BOC-Trp-Mts-Dicyclohexylamine
USBio B2415 MW 667.9 $C_{37}H_{53}N_3O_6S$ ≥99%

Tryptophan — BOC-Trp-OMe
Bachem A-4510.0001 MW 318.37 $C_{17}H_{22}N_2O_4$ Store at -15°C

Tryptophan — BOC-Trp-ONp
Bachem A-2375.0001 MW 425.44 $C_{22}H_{23}N_3O_6$ Store at -15°C

Tryptophan — BOC-Trp-O-Resin
American Peptide RMRB275 1% DVB cross-linked: 100-200 mesh | t-BOC protected AA resin; preattached resins are useful for synthesis of peptides with C-terminal carboxyl group by t-BOC chemistry

Tryptophan — BOC-Trp-OSu
Bachem A-2370.0001 MW 401.42 $C_{20}H_{23}N_3O_6$ Store at -15°C

Tryptophan — BOC-Trp-PAM-® (200-400 mesh)
Bachem D-1660.0001 Store at 2-8°C

Tryptophan — BOC-Trp-PAM-Resin
American Peptide RPAM275 1% DVB cross-linked: 100-200 mesh | t-BOC protected AA resin; preattached resins are useful for synthesis of peptides with C-terminal carboxyl group by t-BOC chemistry

Tryptophan — Bsmoc-Trp-BOC
Synonyms: N^α-Bsmoc-N^{in}-t-Butyloxycarbonyl-L-Trp

Peptides International BLW-5638-PI MW 526.57 >98%
(HPLC); white to off-white powder | LA Carpino, et al, *JACS*, 119:9915, 1997

Tryptophan — Bz-Trp
Bachem E-1520.0005 MW 308.34 $C_{18}H_{16}N_2O_3$ Store at RT

Tryptophan — CBZ-D-Trp
USBio C2099-73 MW 338.4 $C_{19}H_{18}N_2O_4$ ≥99%

Tryptophan — CBZ-Trp
USBio C2099-72 MW 338.4 $C_{19}H_{18}N_2O_4$ ≥99%

Tryptophan — CBZ-Trp-ONp
USBio C2099-74 MW 459.5 $C_{25}H_{21}N_3O_6$ ≥99%

Tryptophan — Chloroacetyl-Trp
Bachem E-1735.0005 MW 280.71 $C_{13}H_{13}ClN_2O_3$ Store at RT

Tryptophan — Dansyl-DL-Trp Cyclohexylammonium Salt
ICN 100138 MW 536.7 $C_{23}H_{23}N_3O_4S \cdot C_6H_{13}N$

Tryptophan — DL-4-FluoroTrp
Sigma F 7376 FW 222.2 $C_{11}H_{11}FN_2O_2$ Off-white crystals

Tryptophan — DL-7-Azatryptophan Hydrate
ICN 100844 MW 223.2 $C_{10}H_{11}N_3O_2 \cdot H_2O$ Crystalline

Tryptophan — DL-Trp
Synonyms: (±)-2-Amino-3-3-Indolylpropionic Acid; DL-2-Amino-3-Indolepropionic Acid; (±)-α-Amino-3-Indolepropionic Acid; 3β-Indolylalanine; DL-α-Amino-3-Indolepropionic Acid; DL-3β-Indolylalanine

Bachem F-2150.0025 MW 204.23 $C_{11}H_{12}N_2O_2$ Store at RT

Fluka 93680 MW 204.23 $C_{11}H_{12}N_2O_2$ ≥99.0% (titration); mp: 295°C

ICN 103147 MW 204.2 $C_{11}H_{12}N_2O_2$ 99%; crystalline

ICN 194757 MW 204.2 $C_{11}H_{12}N_2O_2$ 99%; crystalline | Cell culture reagent

Sigma T 3300 FW 204.2 $C_{11}H_{12}N_2O_2$ Prepared from L-Trp produced by fermentation | AA precursor of serotonin & melatonin

Sigma T 7425 FW 204.2 $C_{11}H_{12}N_2O_2$ Prepared from L-Trp by fermentation | Cell culture tested

USBio T9015 MW 204.2 $C_{11}H_{12}N_2O_2$ ≥98%

Tryptophan — DL-Trp Amide Hydrochloride
Bachem F-2155.0001 MW 239.7 $C_{11}H_{13}N_3O \cdot HCl$ Store at RT

Sigma T 0504 FW 239.7 $C_{11}H_{13}N_3O \cdot HCl$

916

Tryptophan — DL-Trp Octyl Ester Hydrochloride
Sigma T 1379 FW 352.9 $C_{19}H_{28}N_2O_2 \cdot HCl$

Tryptophan — DL-Trp-OBu Hydrochloride
Sigma T 9510 FW 296.8 $C_{15}H_{20}N_2O_2 \cdot HCl$

Tryptophan — DL-Trp-OEt Hydrochloride
Bachem F-2160.0005 MW 268.74 $C_{13}H_{17}N_2O_2Cl$ Store at 2-8°C

ICN 103149 MW 268.7 $C_{13}H_{16}N_2O_2 \cdot HCl$ ≥99%; crystalline

Sigma T 1129 FW 268.7 $C_{13}H_{16}N_2O_2 \cdot HCl$

Tryptophan — DL-Trp-OMe Hydrochloride
Bachem F-3435.0005 MW 254.72 $C_{12}H_{14}N_2O_2 \cdot HCl$ Store at 2-8°C

ICN 157157 MW 254.7 $C_{12}H_{14}N_2O_2 \cdot HCl$

Sigma T 1254 FW 254.7 $C_{12}H_{14}N_2O_2 \cdot HCl$

Tryptophan — Dnp-L-Trp
ICN 103153 MW 370.3 $C_{17}H_{14}N_4O_6$ Crystalline

Tryptophan — D-Trp
Synonyms: (R)-2-Amino-3-3-Indolylpropionic Acid; D-2-Amino-3-Indolepropionic Acid; D-α-Amino-3-Indolepropionic Acid

Bachem F-2145.0005 MW 204.23 $C_{11}H_{12}N_2O_2$ Store at RT

Fluka 93670 MW 204.23 $C_{11}H_{12}N_2O_2$ ≥99.0% (titration); ≤0.1% residue on ignition

ICN 103144 MW 204.2 $C_{11}H_{12}N_2O_2$ 99%; crystalline

Neosystem AA01801 MW 204.3

Peptides International ADW-2810 MW 204.23 >99.9% optical purity by RP-HPLC; white powder

Sigma T 3762 Matrix: 4% beaded agarose; activation: cyanogen bromide; attachment: amino; spacer: 1 atom; ligand immobilized: 1-3 μmoles/mL; form: suspension in 2.0 M NaCl containing 0.02% thimerosal | Deutsch, DG & Mertz, ET, *Proc Fed Amer Soc Exp Biol*, 29: 647, 1970; Deutsch, DG & Mertz, ET, *Science*, 170: 1095, 1970; Vuento, M & Vaheri, A, *Biochem J*, 183: 331, 1979

Sigma T 9753 FW 204.2 $C_{11}H_{12}N_2O_2$ ≥98% (TLC) | Inactive isomer of Trp; also available as part of a kit

USBio T9020-10 MW 204.2 $C_{11}H_{12}N_2O_2$ ≥99%

Tryptophan — D-Trp-BOC
Bachem F-2840.0001 MW 304.35 $C_{16}H_{20}N_2O_4$ Store at -15°C

Tryptophan — D-Trp-OAll Hydrochloride
Bachem F-3530.0001 MW 266.73 $C_{13}H_{14}N_2O_2 \cdot HCl$ Store at -15°C

Tryptophan — D-Trp-OBzl Hydrochloride
Bachem F-2850.0001 MW 330.81 $C_{18}H_{18}N_2O_2 \cdot HCl$ Store at RT

Tryptophan — D-Trp-OMe Hydrochloride
Bachem F-2280.0005 MW 254.72 $C_{12}H_{14}N_2O_2 \cdot HCl$ Store at 2-8°C

ICN 157156 MW 254.7 $C_{12}H_{14}N_2O_2 \cdot HCl$

Sigma T 0756 FW 254.7 $C_{12}H_{14}N_2O_2 \cdot HCl$

Tryptophan — FMOC-D-Trp
American Peptide FTRP105 MW 426.5

Bachem B-1450.0001 MW 426.47 $C_{26}H_{22}N_2O_4$ Store at RT

Fluka 47478 MW 426.47 $C_{26}H_{22}N_2O_4$ ≥98% (HPLC)

Neosystem FA01801 MW 426.5

Peptides International FDW-1832-PI MW 426.47 >98% (HPLC); off-white powder

USBio F5472 MW 426.5 $C_{20}H_{22}N_2O_4$ ≥99%

Tryptophan — FMOC-D-Trp 4-Benzyloxybenzyl Ester Polymer Bound
Fluka 47404 0.4-0.8 mmol/g resin; carrier: polystyrene, crosslinked with 1% DVB; 100-200 mesh particle size

Tryptophan — FMOC-D-Trp-BOC
Synonyms: N^α-FMOC-N^{in}-BOC-D-Trp

Bachem B-2125.0001 MW 526.6 $C_{31}H_{30}N_2O_6$ Store at -15°C

Neosystem FA01805 MW 526.6

Peptides International FDW-1829-PI MW 526.59 >97% (HPLC); off-white powder

USBio F5475 MW 526.6 $C_{31}H_{30}N_2O_6$ ≥99%

Tryptophan — FMOC-D-Trp-BOC-® (200-400 mesh)
Bachem D-1905.0001 Store at 2-8°C

Tryptophan — FMOC-D-Trp-BOC-SASRIN™-® (200-400 mesh)
Bachem D-1915.0001 Store at 2-8°C

Tryptophan — FMOC-D-Trp-OPfp
Bachem B-2105.0001 MW 592.52 $C_{32}H_{21}F_5N_2O_4$ Store at -15°C

Tryptophan — FMOC-L-Trp
Fluka 47637 MW 426.47 $C_{26}H_{22}N_2O_4$ ≥97% (titration); mp: 167-168°C

Neosystem FA01802 MW 426.5

Tryptophan — FMOC-L-Trp 4-Benzyloxybenzyl Ester Polymer Bound
Fluka 47668 0.4-0.6 mmol/g resin; carrier: polystyrene, crosslinked with 1% DVB; 200-400 mesh particle size

Tryptophan — FMOC-L-Trp-BOC
Neosystem FA01804 MW 526.6

Tryptophan — FMOC-L-Trp-OPfp
Fluka 47479 MW 608.52 $C_{32}H_{21}F_5N_2O_5$ ≥98% (HPLC)

Tryptophan — FMOC-L-Trp-Wang Resin
American Peptide RFWAN85 Preattached Wang resins are usful for synthesis of peptides with C-terminal carboxyl groups; peptides can be cleaved from the resin with 90% TFA in the presence of scavengers

Neosystem RW01802

Tryptophan — FMOC-N-Me-Trp
Bachem B-3430.0250 MW 440.5 $C_{27}H_{24}N_2O_4$ Store at RT

Tryptophan — FMOC-Trp
Synonyms: FMOC-L-Trp

American Peptide FTRP100 MW 426.5

Bachem B-1445.0005 MW 426.47 $C_{26}H_{22}N_2O_4$ Store at RT

Peptides International FLW-1735-PI MW 426.47 >98% (HPLC); off-white crystalline powder

USBio F5473 MW 426.5 $C_{20}H_{22}N_2O_4$ ≥99%

Tryptophan — FMOC-Trp BOC

American Peptide FTRP115   MW 426.5

Tryptophan — FMOC-Trp-® (200-400 mesh)

Bachem D-1150.0001   Store at RT

Tryptophan — FMOC-Trp-BOC

Synonyms: N^α-FMOC-N^{in}-BOC-L-Trp

Bachem B-2065.0005   MW 526.59 $C_{31}H_{30}N_2O_6$ Store at -15°C

Peptides International FLW-1769-PI   MW 526.59   >98% (HPLC); white crystalline powder

USBio F5476   MW 526.6 $C_{31}H_{30}N_2O_6$ ≥99%

Tryptophan — FMOC-Trp-BOC-® (200-400 mesh)

Bachem D-1900.0001   Store at 2-8°C

Tryptophan — FMOC-Trp-BOC-SASRIN™-® (200-400 mesh)

Bachem D-1910.0001   Store at 2-8°C

Tryptophan — FMOC-Trp-OPfp

Bachem B-1190.0001   MW 592.52 $C_{32}H_{21}F_5N_2O_4$ Store at -15°C

USBio F5474   MW 592.5 $C_{32}H_{21}F_5N_2O_4$ ≥99%

Tryptophan — FMOC-Trp-SASRIN™-® (200-400 mesh)

Bachem D-1395.0001   Store at 2-8°C

Tryptophan — L-Trp

Synonyms: (S)-2-Amino-3-3-Indolylpropionic Acid; L-2-Amino-3-Indolepropionic Acid; α-Amino-3-Indolepropionic Acid; L-α-Amino-3-Indolepropionic Acid

Alexis 101-051   MW 204.2 $C_{11}H_{12}N_2O_2$ Soluble in H_2O or dilute aqueous acid

Amersham US22765   98.5-101.5% dsb; ≤0.0015% heavy metals (Pb)

Calbiochem 6540   MW 204.2 $C_{11}H_{12}N_2O_2$ ≥98% (titration); solid; soluble in H_2O; may be carcinogenic/teratogenic | *Merck Index*, 12: 9929

Fluka 93659   MW 204.23 $C_{11}H_{12}N_2O_2$ ≥99.5% (titration); ≤0.1% residue on ignition, loss on drying; ≤0.3% foreign AA; ≤0.00001% As; ≤0.0005% Al, Ba, Bi, Cd, Co, Cr, Cu, Fe, Li, Mg, Mn, Mo, Ni, Pb, Sr, Zn; ≤0.001% Ca; ≤0.005% Na, K, Cl, SO₄ | Porath, J, *J Chromatogr*, 177: 201, 1979; Hine, T, *Electophoresis '83*, H Hirai, 541, 1984, de Gruyter, Berlin; *CRC Handbook of Microbiology*, Laskin, AI & Lachevalier, HA, eds, 4: 1, 1974, CRC Press, Cleveland, Ohio

Fluka 93660   MW 204.23 $C_{11}H_{12}N_2O_2$ ≥99.0% (titration); ≤0.1% residue on ignition, loss on drying; ≤0.3% foreign AA; ≤0.0005% Cd, Co, Cr, Cu, Fe, Mg, Mn, Ni, Pb, Zn; ≤0.001% Ca; ≤0.005% K, Cl; ≤0.01% Na, SO₄

ICN 103151   MW 204.2 $C_{11}H_{12}N_2O_2$ ≥99%; crystalline

ICN 194758   MW 204.2 $C_{11}H_{12}N_2O_2$ ≥99%; crystalline | Cell culture reagent

Neosystem AA01802   MW 204.3

Peptides International ALW-2721   MW 204.23   >99.9% optical purity by RP-HPLC; white powder

Sigma T 0254   FW 204.2 $C_{11}H_{12}N_2O_2$ ≥98% (TLC) | AA precursor of serotonin & melatonin; also available as part of a kit

Sigma T 0271   FW 204.2 $C_{11}H_{12}N_2O_2$ Cell culture tested; insect cell culture tested

Sigma T 0655   FW 204.2 $C_{11}H_{12}N_2O_2$ Sigma Grade; crystalline | Plant cell culture tested

Sigma T 6812   FW 204.2 $C_{11}H_{12}N_2O_2$ USP Grade

Sigma T 8659   FW 204.2 $C_{11}H_{12}N_2O_2$ ≥98% (TLC); SigmaUltra; residue on ignition: <0.1%; solubility (0.1 *M* in 0.5 *M* HCl, 20°C): complete, colorless; insoluble matter: <0.1%; NH₄⁺, Cl, SO₄: <0.05%; K, Na, P: <0.005%; Al, Ca, Cu, Fe, Mg, Zn: <0.0005%; Pb: <0.001% | AA precursor of serotonin & melatonin

Tryptophan — L-Trp

USBio T9020   MW 204.2 $C_{11}H_{12}N_2O_2$ ≥99%; appearance: white to slight yellow crystalline powder; pH: 5.5-6.5; specific rotation@ C=1, H_2O: -30.5° to -32.5°; heavy metals (Pb): ≤0.001%; endotoxin: ≤2 EU/mg; cell culture tested: murine myeloma SP2/0-Ag14

Tryptophan — L-Trp (5-³H)

Amersham TRK460   EtOH:H_2O (1:1); 0.74-1.1 TBq/mmol, 20-30 Ci/mmol; 37 MBq/mL, 1 mCi/mL

ARC ART-244   MW 204.2   20-30 Ci/mmol; 0.74-1.11 TBq/mmol; in EtOH: H_2O (1:1) | Radiochemical

Tryptophan — L-Trp (Side chain 1-¹⁴C)

ARC ARC-255   MW 204.2 $C_8H_6NCH_2CH(NH_2)COOH$ 50-60 mCi/mmol; 1.85-2.22 GBq/mmol; in EtOH: H_2O (2:98) | Radiochemical

Tryptophan — L-Trp (Side chain 3-¹⁴C)

ARC ARC-254   MW 204.2 $C_8H_6NCH_2CH(NH_2)COOH$ 50-60 mCi/mmol; 1.85-2.22 GBq/mmol; in EtOH: H_2O (2:98) | Radiochemical

Tryptophan — L-Trp Amide Hydrochloride

Neosystem AA01803   MW 239.7

Sigma T 0629   FW 239.7 $C_{11}H_{13}N_3O \cdot HCl$

Tryptophan — L-Trp Hydroxamate

Sigma T 1255   FW 219.2 $C_{11}H_{13}N_3O_2$

Tryptophan — L-Trp β-Naphthylamide

Sigma T 2016   FW 329.4 $C_{21}H_{19}N_3O$

Tryptophan — L-Trp-OAll Hydrochloride

Neosystem AA01811   MW 280.8

Tryptophan — L-Trp-OBzl Hydrochloride

Sigma T 0879   FW 330.8 $C_{18}H_{18}N_2O_2 \cdot HCl$

Tryptophan — L-Trp-OEt Hydrochloride

Fluka 93690   MW 268.74 $C_{13}H_{16}N_2O_2 \cdot HCl$ ≥99.0% (titration); mp: 220-225°C

ICN 103152   MW 268.7 $C_{13}H_{16}N_2O_2 \cdot HCl$ 99%; crystalline

Sigma T 8755   FW 268.7 $C_{13}H_{16}N_2O_2 \cdot HCl$

Tryptophan — L-Trp-OMe Hydrochloride

Fluka 93730   MW 254.72 $C_{12}H_{14}N_2O_2 \cdot HCl$ ≥99.0% (titration); mp: 220°C

ICN 103155   MW 254.7 $C_{12}H_{14}N_2O_2 \cdot HCl$ Crystalline

Sigma T 5505   FW 254.7 $C_{12}H_{14}N_2O_2 \cdot HCl$

Tryptophan — m-Fluoro-DL-Trp

ICN 158158   MW 199.2 $C_9H_{10}FNO_3$

Tryptophan — MTH-DL-Trp

Sigma M 6006   FW 259.3 $C_{13}H_{13}N_3OS$ ~95%

Tryptophan — N-(3-(2-Furyl)-Acryloyl)-L-Trp-OMe

Sigma F 3005   FW 338.4 $C_{19}H_{18}N_2O_4$

Tryptophan — N-(3-(2-Furyl)-Acryloyl)-Trp Dicyclohexylammonium Salt

ICN 158215	MW 505.7	$C_{18}H_{16}N_2O_4 \cdot C_{12}H_{23}N$

Tryptophan — N-(3-(2-Furyl)-Acryloyl)-Trp-OMe

ICN 158216	MW 338.4	$C_{19}H_{18}N_2O_4$	Crystalline

Tryptophan — N-2,4-Dnp-L-Trp

Sigma D 2005	FW 370.3	$C_{17}H_{14}N_4O_6$

Tryptophan — N-Ac-DL-Trp

Fluka 01520	MW 246.27	$C_{13}H_{14}N_2O_3$	≥99% (titration); ≤0.5% free Trp; ≤0.1% residue on ignition; mp: 205-207°C
ICN 100142	MW 246.3	$C_{13}H_{14}N_2O_3$	99%, crystalline
Sigma A 6251	FW 246.3	$C_{13}H_{14}N_2O_3$	

Tryptophan — N-Ac-D-Trp

ICN 100140	MW 246.3	$C_{13}H_{14}N_2O_3$	Crystalline
Sigma A 6001	FW 246.3	$C_{13}H_{14}N_2O_3$	

Tryptophan — N-Ac-L-Trp

ICN 100147	MW 246.3	$C_{13}H_{14}N_2O_3$	Crystalline
Sigma A 6376	FW 246.3	$C_{13}H_{14}N_2O_3$	

Tryptophan — N-Ac-L-Trp 3,5-Bis-Trifluorormethyl-OBzl

Sigma A 5330	FW 472.4	$C_{22}H_{18}F_6N_2O_3$

Tryptophan — N-Ac-L-Trp-OEt

Sigma A 2648	FW 274.3	$C_{15}H_{18}N_2O_3$

Tryptophan — N-Ac-L-Tryptophanamide

Sigma A 6501	FW 245.3	$C_{13}H_{15}N_3O_2$

Tryptophan — N-Carbamyl-L-Trp

Sigma C 1401	FW 247.3	$C_{12}H_{13}N_3O_3$

Tryptophan — N-CBZ-DL-Trp

ICN 100231	MW 338.4	$C_{19}H_{18}N_2O_4$
Sigma C 9640	FW 338.4	$C_{19}H_{18}N_2O_4$

Tryptophan — N-CBZ-L-Trp

ICN 100232	MW 338.4	$C_{19}H_{18}N_2O_4$	Crystalline
Sigma C 5377	FW 338.4	$C_{19}H_{18}N_2O_4$	

Tryptophan — N-CBZ-L-Trp-(pONp)

Sigma C 5502	FW 459.5	$C_{25}H_{21}N_3O_6$

Tryptophan — N-Chloroacetyl-L-Trp

ICN 101343	MW 280.7	$C_{13}H_{13}N_2O_3Cl$	Crystalline
Sigma C 1753	FW 280.7	$C_{13}H_{13}ClN_2O_3$	

Tryptophan — N-FMOC-L-Trp Resin Ester

Sigma F 2761	For peptide synthesis; Wang, SS, *J Am Chem Soc*, 95: 1328, 1973; Lu, G et al, *J Org Chem*, 46: 3433, 1981

Tryptophan — N-For-DL-Trp

ICN 158189	MW 232.2	$C_{12}H_{12}N_2O_3$	
Sigma F 9626	FW 232.2	$C_{12}H_{12}N_2O_3$	White powder

Tryptophan — N-Me-Trp

Synonyms: L-Abrine

Bachem E-3255.0250	MW 218.26	$C_{12}H_{14}N_2O_2$	Store at 2-8°C

Tryptophan — N-O-Nps-L-Trp Dicyclohexylammonium Salt

ICN 102537	MW 515.7	$C_{15}H_{14}N_2O_5S \cdot C_{12}H_{23}N$	Yellow crystals
Sigma N 6753	FW 538.7	$C_{17}H_{15}N_3O_4S \cdot C_{12}H_{23}N$	Yellow crystals

Tryptophan — N-t-BOC-L-Trp

Sigma B 9126	FW 304.3	$C_{16}H_{20}N_2O_4$

Tryptophan — N^α-BOC-D-Trp

Fluka 15185	MW 304.35	$C_{16}H_{20}N_2O_4$	≥98% (TLC)

Tryptophan — N^α-BOC-L-Trp

Fluka 15512	MW 304.35	$C_{16}H_{20}N_2O_4$	≥99% (TLC); mp: 138°C

Tryptophan — N^α-BOC-L-Trp Hydroxysuccinimide Ester

Fluka 15516	MW 401.42	$C_{20}H_{23}N_3O_6$	mp: 146-148°C

Tryptophan — N^α-BOC-L-Trp N^α-Carboxy Anhydride

Fluka 15514	MW 330.34	$C_{17}H_{18}N_2O_5$	≥98%; mp: 135°C

Tryptophan — N^α-BOC-L-Trp-(4-ONp)

Fluka 15186	MW 425.44	$C_{22}H_{23}N_3O_6$	≥98% (TLC)

Tryptophan — N^α-BOC-N^{in}-(Mesitylene-2-Sulfonyl)-L-Trp Dicyclohexylamine

Fluka 15547	MW 667.9	$C_{25}H_{30}N_2O_6S \cdot C_{12}H_{23}N$	≥98% (TLC); mp: 174-178°C

Tryptophan — N^α-BOC-N^{in}-L-Trp

Fluka 15349	MW 332.36	$C_{17}H_{20}N_2O_5$	≥97% (HPLC); mp: 100°C

Tryptophan — N^α-Dansyl-DL-Trp Cyclohexylammonium Salt

Sigma D 1006	FW. 536.7	$C_{23}H_{23}N_3O_4S \cdot C_6H_{13}N$

Tryptophan — N^α-Dansyl-L-Trp Cyclohexylammonium Salt

Sigma D 2250	FW. 536.7	$C_{23}H_{23}N_3O_4S \cdot C_6H_{13}N$

Tryptophan — N^α-FMOC-L-Trp

ICN 151152	MW 426.5	$C_{26}H_{22}N_2O_4$

Tryptophan — N^α-FMOC-N^{in}-BOC-D-Trp

Fluka 47309	MW 526.59	$C_{31}H_{30}N_2O_6$	≥90% (TLC)

Tryptophan — N^α-FMOC-N^{in}-BOC-L-Trp

Synonyms: N^{in}-BOC-N^α-FMOC-L-Trp

Fluka 47561	MW 526.59	$C_{31}H_{30}N_2O_6$	≥97% (HPLC); mp: 80°C; ≤10% IpOH

Tryptophan — N^α-Me-L-Trp

Synonyms: L-Abrine

Fluka 69555	MW 218.26	$C_{12}H_{14}N_2O_2$	≥97.0%; mp: >300°C

ICN 154667   MW 218.3 $C_{12}H_{14}N_2O_2$

Tryptophan — N^α-t-BOC-L-Trp
ICN 101095   MW 304.3 $C_{16}H_{20}N_2O$ Crystalline

Tryptophan — N^α-t-BOC-N^{in}-For-L-Trp
Sigma B 9641   FW 332.4 $C_{17}H_{20}N_2O_5$

Tryptophan — N^α-t-BOC-N^{in}-Mts-L-Trp Dicyclohexylammonium Salt
Sigma B 7269   FW 667.9 $C_{25}H_{30}N_2O_6S \cdot C_{12}H_{23}N$

Tryptophan — N^α-Z-N^{in}-BOC-L-Trp Dicyclohexylamine
Fluka 96023   MW 619.80 $C_{24}H_{26}N_2O_6 \cdot C_{12}H_{23}N$ ≥98.0% (TLC)

Tryptophan — Pth-Trp
Sigma P 2752   FW 321.4 $C_{18}H_{15}N_3OS$ ~99% | Useful as standards in protein sequencing by the Edman degradation

Tryptophan — TentaGel S PHB-Trp-BOC-FMOC
Fluka 86418   0.22 mmol protected AA/g; ~90 µm particle size

Tryptophan — TentaGel S PHB-Trp-FMOC
Synonyms: FMOC-L-Trp 4-(Poly(Ethylenoxy))-OBzl Polymer Bound
Fluka 86404   0.22 mmol protected AA/g; ~90 µm particle size

Tryptophan — TentaGel S Trt-Trp-BOC-FMOC
Fluka 86443   0.20 mmol protected AA/g; ~90 µm particle size

Tryptophan — TentaGel S Trt-Trp-FMOC
Fluka 86436   0.20 mmol protected AA/g; ~90 µm particle size

Tryptophan — Trp
Bachem E-2500.0025   MW 204.23 $C_{11}H_{12}N_2O_2$ Store at RT

Tryptophan — Trp Amide Hydrochloride
Bachem E-2505.0001   MW 239.7 $C_{11}H_{13}N_3O \cdot HCl$ Store at 2-8°C

Tryptophan — Trp-AMC
Bachem I-1670.0250   MW 361.4 $C_{21}H_{19}N_3O_3$ Store at -15°C

Tryptophan — Trp-BOC
Bachem E-2975.0005   MW 304.35 $C_{16}H_{20}N_2O_4$ Store at -15°C

Tryptophan — Trp-BOC-OtBu Hydrochloride
Bachem E-3370.0001   MW 396.91 $C_{20}H_{28}N_2O_4 \cdot HCl$ Store at -15°C

Tryptophan — Trp-OBzl Hydrochloride
Bachem E-2390.0005   MW 330.81 $C_{18}H_{18}N_2O_2 \cdot HCl$ Store at 2-8°C

Tryptophan — Trp-OEt Hydrochloride
Bachem E-2510.0005   MW 268.74 $C_{13}H_{16}N_2O_2 \cdot HCl$ Store at RT

Tryptophan — Trp-OMe Hydrochloride
Bachem N-1165.0005   MW 254.72 $C_{12}H_{14}N_2O_2 \cdot HCl$ Store at -15°C

Tryptophan — Trp-OtBu Hydrochloride
Bachem E-3105.0005   MW 296.8 $C_{15}H_{20}N_2O_2 \cdot HCl$ Store at -15°C

Tryptophan — Trp-βNA
Bachem K-1525.0250   MW 329.4 $C_{21}H_{19}N_3O$ Store at RT

Tryptophan — Z-DL-Trp
Bachem C-2671.0005   MW 338.36 $C_{19}H_{18}N_2O_4$ Store at RT

Tryptophan — Z-D-Trp
Bachem C-2670.0005   MW 338.36 $C_{19}H_{18}N_2O_4$ Store at RT
Neosystem ZA01801   MW 338.4

Tryptophan — Z-L-Trp
Fluka 97240   MW 338.37 $C_{19}H_{18}N_2O_4$ ≥99.0% (titration); mp: 125-127°C; ≤0.05% residue on ignition
Neosystem ZA01802   MW 338.4

Tryptophan — Z-L-Trp-(4-ONp)
Fluka 97260   MW 459.46 $C_{25}H_{21}N_3O_6$ ~99%; mp: 102-104°C

Tryptophan — Z-Trp
Synonyms: Benzyloxycarbonyl-L-Trp
Bachem C-2665.0005   MW 338.36 $C_{19}H_{18}N_2O_4$ Store at RT
Peptides International ZLW-2025-PI   MW 338.36   >98% (TLC); white powder

Tryptophan — Z-Trp Amide
Bachem C-2675.0001   MW 337.38 $C_{19}H_{19}N_3O_3$ Store at RT

Tryptophan — Z-Trp-BOC Dicyclohexylamine
Bachem C-3995.0005   MW 619.81 $C_{24}H_{26}N_2O_6 \cdot C_{12}H_{23}N$ Store at 2-8°C

Tryptophan — Z-Trp-ONp
Bachem C-2685.0005   MW 459.46 $C_{25}H_{21}N_3O_6$ Store at -15°C

Tryptophan — Z-Trp-OSu
Bachem C-2680.0005   MW 435.44 $C_{23}H_{21}N_3O_6$ Store at -15°C

Tryptophan — α-Me-DL-Trp
Bachem F-1810.0001   MW 218.26 $C_{12}H_{14}N_2O_2$ Store at RT
ICN 155679   MW 218.3 $C_{12}H_{14}N_2O_2$ Light yellow crystals
Sigma M 8377   FW 218.3 $C_{12}H_{14}N_2O_2$ Light yellow crystals

Tryptophan — α-Me-DL-Trp-OMe
Bachem F-2240.0001   MW 232.28 $C_{13}H_{16}N_2O_2$ Store at 2-8°C

Tryptophan — α-Me-Trp L-(Me-^{14}C)
ARC ARC-1107   MW 218.3   50-60 mCi/mmol; 1.85-2.22 GBq/mmol; in EtOH | Radiochemical

Tryptophan — α-Me-Trp L-(Me-^{3}H)
ARC ART-622   MW 218.3   70-87 Ci/mmol; 2.59-3.21 TBq/mmol; in EtOH | Radiochemical

Tyrosine — (3,5-Diiodo)-Tyr
Bachem F-2225.0005   MW 432.98 $C_9H_9I_2NO_3$ Store at RT

Tyrosine — (3,5-Diiodo)-Tyr-OMe Hydrochloride

Bachem E-2385.0001   MW 483.47 $C_{10}H_{11}I_2NO_3 \cdot HCl$ Store at -15°C

Tyrosine — (R)-α-Hydroxymethyl-Tyr

Synonyms: α-Hydroxymethyl-L-Tyr; (+)-α-Hydroxymethyl Tyr

Sigma H 2772   FW 211.2 $C_{10}H_{13}NO_4$

Tyrosine — (S)-α-Hydroxymethyl-Tyr

Synonyms: α-Hydroxymethyl-D-Tyr; (-)-α-Hydroxymethyl-Tyr

Sigma H 2897   FW 211.2 $C_{10}H_{13}NO_4$

Tyrosine — 1-(N,O-Bis-5-Isoquinolinesulfonyl-N-Me-L-Tyr)-4-Phenyl-Piperazine

Synonyms: KN-62

ICN 158944   MW 721.8 $C_{38}H_{35}N_5O_6S_2$ 98% | Potently & specifically inhibits Ca^{2+}/Calmodulin kinase II; Tokumitsu, H et al, *JBC*, 265:4315, 1990

Tyrosine — 1-(N,O-Bis-5-Isoquinolinesulfonyl-N-Me-L-Tyrosyl)-4-Phenylpiperazine

Synonyms: KN-62; Ca^{2+}/Calmodulin-Dependent Protein Kinase II Inhibitor

Alexis 430-024   MW 721.9 $C_{38}H_{35}N_5O_6S_2$ ≥95%; off-white powder; soluble in MeOH or DMSO | Selective inhibitor of rat brain Ca^{2+}/calmodulin-dependent protein kinase II; inhibits growth of K562 cells in a dose-dependent manner; Tokumitsu, H et al, *J Biol Chem*, 265: 4315, 1990; Ishii, A et al, *BBRC*, 176: 1051, 1991; Ito, I et al, *Neurosci Lett*, 121: 119, 1991; Tohda, M et al, *Neurosci Lett*, 129: 47, 1992; Wenham, RM et al, *BBRC*, 189: 128, 1992; Minami, H et al, *BBRC*, 199: 241, 1994

Tyrosine — 3,5-Dibromo-L-Tyr

ICN 101544   MW 339 $C_9H_9Br_2NO_3$ Antithyroid agent
Sigma D 5632   FW 339.0 $C_9H_9BrNO_3$

Tyrosine — 3,5-Dibromo-Tyr

USBio D3500   MW 339 $C_9H_9Br_2NO_3$ ≥99%

Tyrosine — 3,5-Dihydroxy-Bz-Tyr

Bachem E-1805.0250   MW 317.3 $C_{16}H_{15}NO_6$ Store at RT

Tyrosine — 3,5-Diiodo-D-Tyr

Sigma D 1553   FW 433.0 $C_9H_9I_2NO_3$ ≥98%

Tyrosine — 3,5-Diiodo-L-Tyr

Synonyms: 3,5-Diiodo-4-Hydroxy-β-Phenylalnine

ICN 150914   MW 433 $C_9H_9I_2NO_4$ White crystals
Sigma D 0754   FW 433.0 $C_9H_9I_2NO_3$ White crystals
Sigma D 8392   FW 433.0 $C_9H_9I_2NO_3$ Tan crystals

Tyrosine — 3,5-Diiodo-L-Tyr Dihydrate

Synonyms: Iodogorgoic Acid; Aminotransferase (Halogenated Tyrosine) Substrate; Aminotransferase (Thyroid Hormone) Substrate

Fluka 38130   MW 469.02 $C_9H_9I_2NO_3 \cdot 2H_2O$ ≥98% (titration); mp: 199-201°C | Substrate for the assay of halogenated tyrosine & thyroid hormone aminotransferase; Nakano, M, *JBC*, 242: 73, 1967; Nakano, M, *Meth Enzymol*, 17A: 660, 1970

Tyrosine — 3,5-Dinitro-L-Tyr

ICN 101609   MW 271.2 $C_9H_9N_3O_7$ Crystalline

Tyrosine — 3-Amino-L-Tyr Dihydrochloride

Sigma A 9383   FW 269.1 $C_9H_{12}N_2O_3 \cdot 2HCl$

Tyrosine — 3-Amino-Tyr Dihydrochloride

Bachem F-3675.0001   MW 269.13 $C_9H_{12}N_2O_3 \cdot 2HCl$ Store at -15°C

Tyrosine — 3-Chloro-L-Tyr Hydrochloride

ICN 150661   MW 252.1 $C_9H_{10}ClNO_3 \cdot HCl$ ~97-99% | Metabolite of p-chlorophenylalanine
Sigma C 5897   FW 252.1 $C_9H_{10}ClNO_3 \cdot HCl$ 95-98%

Tyrosine — 3-Fluoro-L-Tyr

Fluka 47545   MW 199.18 $C_9H_{10}FNO_3$ ≥98% (TLC)

Tyrosine — 3-Iodo-L-Tyr

Synonyms: Tyrosine Hydroxylase Inhibitor; S(-)-3-Iodo-4-Hydroxy-Phenylamine; 3-Monoiodo-L-Tyr; 3-Monoiodo-Tyr

Fluka 58120   MW 304.09 $C_9H_{10}INO_3$ ≥99% (HPLC); mp: 204-208°C | Shiman, R & Kaufman, S, *Meth Enzymol*, 17A: 609, 1970; Jain, MK, *Handbook of Enzyme Inhibitors*, (1965-1977), 362, 1982, John Wiley & Sons, New York
ICN 102367   MW 307.1 $C_9H_{10}INO_3$ ~95%; crystalline; ≤5% Tyr
ICN 153684   MW 307.1 $C_9H_{10}INO_3$ 99%
ICN 155069   MW 307.1 $C_9H_{10}INO_3$ ~97%; crystalline; ≤3% Tyr
Sigma I 8250   FW 307.1 $C_9H_{10}INO_3$ ~5% Tyr
USBio I8448   MW 307.1 $C_9H_{10}INO_3$ ≥99%

Tyrosine — 3-Iodo-Tyr

Bachem F-3350.0001   MW 307.09 $C_9H_{10}INO_3$ Store at RT

Tyrosine — 3-Methoxy-DL-Tyr

Synonyms: DL-4-Hydroxy-3-Methoxyphenylalanine; DL-3-OMe-DOPA

ICN 155450   MW 211.2 $C_{10}H_{13}NO_4$ Crystalline
Sigma M 1132   FW 211.2 $C_{10}H_{13}NO_4$ Crystalline

Tyrosine — 3-Methoxy-L-Tyr Hydrate

Synonyms: 4-Hydroxy-3-Methoxy-L-Phe; L-3-OMe-DOPA

ICN 155451   MW 229.2 $C_{10}H_{13}NO_4 \cdot H_2O$ Off-white to yellow powder
Sigma M 4255   FW 211.2 $C_{10}H_{13}NO_4$ Off-white to yellow powder

Tyrosine — 3-Nitro-L-Tyr

Synonyms: Nitrotyrosine

Alexis 106-020   MW 226.2 $C_9H_{10}N_2O_5$ ≥99%; yellow-green crystals; soluble in acidic aqueous solution | Major product from the spontaneous reaction of peroxynitrite with Tyr; formation of nitrotyrosine *in vivo* can indicate the formation of peroxynitrite by a nitric oxide-dependent oxidative damage; Ishiropoulos, H et al, *Arch Biochem Biophys*, 298: 431, 1992; Beckman, JS et al, *Biol Chem Hoppe-Seyler*, 375: 81, 1994; Ischiropoulos, H & Al-Mehdi, AB, *FEBS Lett*, 364: 279, 1995; Crow, JP et al, *Meth Enzymol*, 269: 185, 1996; Ye, YZ et al, *Meth Enzymol*, 269: 201, 1996
Calbiochem 487926   MW 226.2 $C_9H_{10}N_2O_5$ ≥98% (TLC); solid; soluble in aqueous acids | A marker for peroxynitrite, a powerful oxidant & cytotoxic agent; the major product obtained by the nitration of Tyr by peroxynitrite; Royall, JA et al, *J Leukoc Biol*, 56: 759, 1994; Kooy, NW et al, *Am J Respir Crit Care Med*, 151: 1250, 1995; Ichiropoulos, H et al, *Am J Physiol*, 269: L158, 1995; Ichiropoulos, H & al-Mehdi, AB, *FEBS Lett*, 364: 279, 1995
Fluka 74090   MW 226.19 $C_9H_{10}N_2O_5$ ≥98.0% (titration); mp: 230°C
ICN 102460   MW 226.2 $C_9H_{10}N_2O_5$ ~99%; yellowish-green crystals

Sigma N 7389   FW 226.2 $C_9H_{10}N_2O_5$ Yellow-green crystals | Oxidant & cytotoxic agent; marker for peroxynitrite

Tyrosine — 3-Nitro-L-Tyr-OEt Hydrochloride

Fluka 74095   MW 290.7 $C_{11}H_{14}N_2O_5 \cdot HCl$ ≥99.0% (titration); mp: 185°C | Modifies proteins via carboxylic group; Desvages, G, *Eur J Biochem*, 105: 259, 1980

ICN 158219   MW 290.7 $C_{11}H_{14}N_2O_5 \cdot HCl$ Useful as a protein modifier

Sigma N 1018   FW 290.7 $C_{11}H_{14}N_2O_5 \cdot HCl$ ≥99% (titration)

Tyrosine — 3-Nitro-Tyr

Synonyms: 4-Hydroxy-3-Nitro-Phe

Bachem F-3340.0005   MW 226.19 $C_9H_{10}N_2O_5$ Store at -15°C

Tyrosine — Ac-(3,5-Dinitro)-Tyr-OEt

Bachem F-1030.0001   MW 341.28 $C_{13}H_{15}N_3O_8$ Store at -15°C

Tyrosine — Ac-D-Tyr

Bachem F-2620.0001   MW 223.23 $C_{11}H_{13}NO_4$ Store at RT

Tyrosine — Ac-Tyr

Bachem E-1240.0005   MW 223.23 $C_{11}H_{13}NO_4$ Store at RT

Tyrosine — Ac-Tyr Amide

Bachem E-1245.0001   MW 222.24 $C_{11}H_{14}N_2O_3$ Store at RT

Peptides International SYA-3009   MW 222.24 $C_{11}H_{14}N_2O_3$ >98% (HPLC); amorphous powder

Tyrosine — Ac-Tyr Hydrazide

Bachem E-1250.0001   MW 237.26 $C_{11}H_{15}N_3O_3$ Store at RT

Tyrosine — Ac-Tyr-Me-OMe

Synonyms: Ac-4-Methoxy-Phe-OMe

Bachem E-2605.0001   MW 251.28 $C_{13}H_{17}NO_4$ Store at 2-8°C

Tyrosine — Ac-Tyr-NHMe

Bachem E-1255.0250   MW 236.27 $C_{12}H_{16}N_2O_3$ Store at RT

Tyrosine — Ac-Tyr-OEt

Bachem M-1080.0005   MW 251.28 $C_{13}H_{17}NO_4$ Store at -15°C

Tyrosine — Ac-Tyr-OEt Hydrate

Synonyms: Chymotrypsin Substrate; C1s Substrate

Peptides International SYE-3008   MW 269.3 $C_{13}H_{17}NO_4 \cdot H_2O$ >99% (HPLC); amorphous powder | Wilcox, PE in *Proteolytic Enzymes*, Methods in Enzymology, Vol 19, GE Perlmann & L Lorand, Eds, Academic Press, New York, 1970, pp 64-108; Sim, RB in *Proteolytic Enzymes Part C*, Methods in Enzymology, Vol 80, L Lorand, Ed, Academic Press, New York, 1981, pp 26-42

Tyrosine — Ac-Tyr-OMe

Bachem E-1260.0005   MW 237.26 $C_{12}H_{15}NO_4$ Store at RT

Tyrosine — Ac-Tyr-tBu

Bachem E-3055.0001   MW 279.34 $C_{15}H_{21}NO_4$ Store at 2-8°C

Tyrosine — BOC-(^{15}N)-Tyr

Bachem A-3805.0100   MW 282.31 $C_{14}H_{19}{}^{15}NO_5$ Store at -15°C

Tyrosine — BOC-(3,5-Dibromo)-D-Tyr

Bachem A-4220.0001   MW 439.1 $C_{14}H_{17}Br_2NO_5$ Store at -15°C

Tyrosine — BOC-(3,5-Dibromo)-Tyr

Bachem A-1555.0001   MW 439.1 $C_{14}H_{17}Br_2NO_5$ Store at RT

Tyrosine — BOC-(3,5-Dibromo)-Tyr-Bzl

USBio B2297   MW 529.2 $C_{21}H_{23}NO_5$ ≥98%

Tyrosine — BOC-(3,5-Diiodo)-D-Tyr

Bachem A-4225.0001   MW 533.1 $C_{14}H_{17}I_2NO_5$ Store at -15°C

Tyrosine — BOC-(3,5-Diiodo)-L-Tyr

Fluka 15092   MW 533.1 $C_{14}H_{17}I_2NO_5$ ≥98% (TLC)

Tyrosine — BOC-(3,5-Diiodo)-L-Tyr Hydroxysuccinimide Ester

Fluka 15093   MW 630.17 $C_{18}H_{20}I_2N_2O_7$ ≥97%; mp: 144-148°C

Tyrosine — BOC-(3,5-Diiodo)-Tyr

Bachem A-1580.0001   MW 533.1 $C_{14}H_{17}I_2NO_5$ Store at RT

Tyrosine — BOC-(3,5-Diiodo)-Tyr-(2',6'-Dichlorobenzyl)

Bachem A-2570.0001   MW 692.12 $C_{21}H_{21}Cl_2I_2NO_5$ Store at RT

Tyrosine — BOC-(3,5-Diiodo)-Tyr-(3'-Bromobenzyl)

Bachem A-1410.0001   MW 702.12 $C_{21}H_{22}BrI_2NO_5$ Store at RT

Tyrosine — BOC-(3,5-Diiodo)-Tyr-3-Bromobenzyl

USBio B2300   MW 702.1 $C_{21}H_{22}BrI_2NO_4$ ≥98%

Tyrosine — BOC-(3,5-Diiodo)-Tyr-OMe

Bachem A-1590.0001   MW 547.13 $C_{15}H_{19}I_2NO_5$ Store at -15°C

Tyrosine — BOC-(3,5-Diiodo)-Tyr-OSu

Bachem A-1585.0001   MW 630.18 $C_{18}H_{20}I_2N_2O_7$ Store at -15°C

Tyrosine — BOC-3-Iodo-Tyr-3'-Bromobenzyl

USBio B2326   MW 576.2 $C_{21}H_{23}BrINO_5$ ≥99%

Tyrosine — BOC-DL-Tyr-(2,6-Dichlorobenzyl)

Bachem A-3675.0001   MW 440.32 $C_{21}H_{23}Cl_2NO_5$ Store at RT

Tyrosine — BOC-DL-Tyr-Me

Synonyms: BOC-4-Methoxy-DL-Phe

Bachem A-2005.0001   MW 295.34 $C_{15}H_{21}NO_5$ Store at RT

Tyrosine — BOC-D-Tyr

American Peptide BDTYR30   MW 281.3

Bachem A-2410.0001   MW 281.31 $C_{14}H_{19}NO_5$ Store at RT

922

Fluka 15187	MW 281.31	$C_{14}H_{19}NO_5$	≥98% (HPLC)	
Neosystem BA01901	MW 281.3			
USBio B2416	MW 281.3	$C_{14}H_{19}NO_5$	≥99%	

Tyrosine — BOC-D-Tyr-(2,6-Dichlorobenzyl)

Bachem A-4025.0001	MW 440.32	$C_{21}H_{23}Cl_2NO_5$	Store at -15°C	
Neosystem BA01908	MW 440.3			
USBio B2424	MW 440.3	$C_{21}H_{23}Cl_2NO_5$	≥99%	

Tyrosine — BOC-D-Tyr-2-Bromo-Z

American Peptide BDTYR20	MW 494.4			
Bachem A-2710.0001	MW 494.34	$C_{22}H_{24}BrNO_7$	Store at -15°C	
Neosystem BA01906	MW 494.4			
USBio B2421	MW 494.4	$C_{22}H_{24}BrNO_7$	≥99%	

Tyrosine — BOC-D-Tyr-Bromo-Z

Synonyms: BOC-O-2-Bromobenzyloxycarbonyl-D-Tyr

Peptides International BDY-2612	MW 494.34	>98%

(HPLC); white crystalline powder

Tyrosine — BOC-D-Tyr-Bzl

Bachem A-1395.0001	MW 371.43	$C_{21}H_{25}NO_5$	Store at RT
USBio B2418	MW 371.4	$C_{21}H_{25}NO_5$	≥99%

Tyrosine — BOC-D-Tyr-Dichlorobenzyl

Synonyms: BOC-O-(2,6-Dichlorobenzyl)-D-Tyr

Peptides International BDY-2622	MW 440.32	>98%

(HPLC); white crystalline powder

Tyrosine — BOC-D-Tyr-Et

Synonyms: BOC-O-Et-D-Tyr; BOC-pEtO-D-Phe

Bachem A-2765.0001	MW 309.36	$C_{16}H_{23}NO_5$	Store at RT
Peptides International BDY-5338-PI	MW 309.36	>98%	
USBio B2426	MW 309.4	$C_{16}H_{23}NO_5$	≥99%

(HPLC); white crystalline powder

Tyrosine — BOC-D-Tyr-Me

Synonyms: BOC-4-Methoxy-D-Phe; BOC-OMe-D-Tyr; BOC-p-Methoxy-D-Phe

Bachem A-3670.0001	MW 295.34	$C_{15}H_{21}NO_5$	Store at RT
Peptides International BDY-5336-PI	MW 295.34	>98%	

(HPLC); white crystalline powder

Tyrosine — BOC-D-Tyr-OMe

Bachem A-2425.0001	MW 295.34	$C_{15}H_{21}NO_5$	Store at -15°C
Fluka 15188	MW 295.34	$C_{15}H_{21}NO_5$	≥98% (TLC)

Tyrosine — BOC-L-Tyr

Fluka 15520	MW 281.31	$C_{14}H_{19}NO_5$	~99% (titration); mp: 136-138°C
Neosystem BA01902	MW 281.3		

Tyrosine — BOC-L-Tyr 2,4,5-Trichlorophenyl Ester

Fluka 15189	MW 460.74	$C_{20}H_{20}Cl_3NO_5$	≥98% (TLC)

Tyrosine — BOC-L-Tyr Hydroxysuccinimide Ester

Fluka 15525	MW 378.38	$C_{18}H_{22}N_2O_7$	≥99%; mp: 190°C

Reagent used in a rapid method for iodotyrosylation of peptides; Assoian, RK et al, *Anal Biochem*, 103: 70, 1980

Tyrosine — BOC-L-Tyr-(2,6-Dichlorobenzyl)

Neosystem BA01909	MW 440.3

Tyrosine — BOC-L-Tyr-(All)

Neosystem BA01904	MW 321.3

Tyrosine — BOC-L-Tyr-2-Bromo-Z

Neosystem BA01907	MW 494.4

Tyrosine — BOC-L-Tyr-2-Bromo-Z-MBHA Resin

Neosystem RN01907

Tyrosine — BOC-L-Tyr-2-Bromo-Z-O-CH₂-φ-CH₂-COOH

Neosystem LP01902	MW 642.6

Tyrosine — BOC-L-Tyr-2-Bromo-Z-PAM Resin

Neosystem RP01906

Tyrosine — BOC-L-Tyr-Bzl

Neosystem BA01905	MW 371.4

Tyrosine — BOC-Me-Tyr-Bzl

USBio B2360	MW 385.5	$C_{22}H_{27}NO_5$	≥98%

Tyrosine — BOC-Me-Tyr-Dichlorobenzyl

USBio B2361	MW 454.3	$C_{22}H_{25}Cl_2NO_5$	≥98%

Tyrosine — BOC-N-Me-D-Tyr Dicyclohexylamine

Bachem A-4420.0001	MW 476.66	$C_{15}H_{21}NO_5 \cdot C_{12}H_{23}N$

Store at RT

Tyrosine — BOC-N-Me-D-Tyr-Bzl

Bachem A-4425.0001	MW 385.46	$C_{22}H_{27}NO_5$	Store at RT

Tyrosine — BOC-N-Me-Tyr Dicyclohexylamine

Synonyms: BOC-N-Me-L-Tyr

Bachem A-2620.0001	MW 476.66	$C_{15}H_{21}NO_5 \cdot C_{12}H_{23}N$

Store at RT

Peptides International BMY-5311-PI	MW 476.66	>98%

(HPLC); white to off-white powder

Tyrosine — BOC-N-Me-Tyr-(2,6-Dichlorobenzyl)

Bachem A-2045.0001	MW 454.35	$C_{22}H_{25}Cl_2NO_5$	Store at RT

Tyrosine — BOC-N-Me-Tyr-Ac Dicyclohexylamine

Bachem A-2020.0001	MW 518.69	$C_{17}H_{23}NO_6 \cdot C_{12}H_{23}N$

Store at 2-8°C

Tyrosine — BOC-N-Me-Tyr-Bzl

American Peptide BLTYR35	MW 323.4		
Bachem A-2040.0001	MW 385.46	$C_{22}H_{27}NO_5$	Store at RT

Tyrosine — BOC-N-Me-Tyr-Me Dicyclohexylamine

Synonyms: BOC-N-Me-4-Methoxy-Phe

Bachem A-2065.0001	MW 490.68	$C_{16}H_{23}NO_6 \cdot C_{12}H_{23}N$

Store at RT

Tyrosine — BOC-O-(2,6-Dichlorobenzyl)-L-Tyr

Fluka 15393	MW 440.33	$C_{21}H_{23}Cl_2NO_5$	≥97% (TLC); mp: 105°C

Tyrosine — BOC-O-2-Bromo-Z-L-Tyr

Fluka 15431   MW 494.34 $C_{22}H_{24}BrNO_7$ ~99% (TLC); mp: 112-115°C

Tyrosine — BOC-O-Bzl-L-Tyr

Fluka 15410   MW 371.44 $C_{21}H_{25}NO_5$ mp: 110-112°C

Tyrosine — BOC-O-Bzl-L-Tyr Hydroxysuccinimide Ester

Fluka 15377   MW 468.51 $C_{25}H_{28}N_2O_7$ ~99%; mp: 149-150°C

Tyrosine — BOC-O-Bzl-L-Tyr *N*-Carboxy Anhydride

Fluka 15417   MW 397.43 $C_{22}H_{23}NO_6$ mp: 118-121°C | Produced by Propeptide under an exclusive license from Bioresearch

Tyrosine — BOC-O-Dimethylphosphous-L-Tyr

American Peptide BLTYR30   MW 389.2

Tyrosine — BOC-O-Et-D-Tyr

Fluka 15095   MW 309.36 $C_{16}H_{23}NO_5$ ≥98% (TLC)

Tyrosine — BOC-O-Et-L-Tyr

Fluka 15096   MW 309.36 $C_{16}H_{23}NO_5$ ≥98% (TLC)

Tyrosine — BOC-OMe-L-Tyr

Fluka 15161   MW 295.34 $C_{15}H_{21}NO_5$ ≥98% (TLC)

Tyrosine — BOC-OtBu-L-Tyr

Fluka 15434   MW 337.42 $C_{18}H_{27}NO_5$ ~99% (TLC); mp: 113-116°C

Tyrosine — BOC-Tyr

American Peptide BLTYR10   MW 281.3

Bachem A-2405.0005   MW 281.31 $C_{14}H_{19}NO_5$ Store at RT

USBio B2417   MW 281.3 $C_{14}H_{19}NO_5$ ≥99%

Tyrosine — BOC-Tyr 2-Bromo-Z

American Peptide BLTYR15   MW 494.4

Tyrosine — BOC-Tyr 2-Bromo-Z-O-Resin

American Peptide RMRB290   1% DVB cross-linked: 100-200 mesh | t-BOC protected AA resin; preattached resins are useful for synthesis of peptides with C-terminal carboxyl group by t-BOC chemistry

Tyrosine — BOC-Tyr 2-Bromo-Z-PAM-Resin

American Peptide RPAM290   1% DVB cross-linked: 100-200 mesh | t-BOC protected AA resin; preattached resins are useful for synthesis of peptides with C-terminal carboxyl group by t-BOC chemistry

Tyrosine — BOC-Tyr-(2,6-Dichlorobenzyl)

Bachem A-1575.0001   MW 440.32 $C_{21}H_{23}Cl_2NO_5$ Store at RT

USBio B2425   MW 440.3 $C_{21}H_{23}Cl_2NO_5$ ≥99%

Tyrosine — BOC-Tyr-(PO₃Bzl₂)

USBio B2429   MW 541.4 $C_{28}H_{32}NO_8P$ ≥99%

Tyrosine — BOC-Tyr-2-Bromo-Z

Bachem A-1420.0001   MW 494.34 $C_{22}H_{24}BrNO_7$ Store at -15°C

USBio B2422   MW 494.4 $C_{22}H_{24}BrNO_7$ ≥99%

Tyrosine — BOC-Tyr-2-Bromo-Z-® (200-400 mesh)

Bachem D-1515.0005   Store at 2-8°C

Tyrosine — BOC-Tyr-2-Bromo-Z-PAM-® (200-400 mesh)

Bachem D-1240.0001   Store at 2-8°C

Tyrosine — BOC-Tyr-Ac

Bachem A-1055.0001   MW 323.35 $C_{16}H_{21}NO_6$ Store at RT

Tyrosine — BOC-Tyr-BOC

Bachem A-1040.0005   MW 381.43 $C_{19}H_{27}NO_7$ Store at -15°C

Tyrosine — BOC-Tyr-Bromo-Z

Synonyms: BOC-O-2-Bromobenzyloxycarbonyl-L-Tyr

Peptides International BLY-2114   MW 494.34   >98% (HPLC); white crystalline powder

Tyrosine — BOC-Tyr-Bzl

Synonyms: BOC-O-Bzl-L-Tyr

Bachem A-1390.0005   MW 371.43 $C_{21}H_{25}NO_5$ Store at RT

Peptides International BLY-2071   MW 371.43   >98% (HPLC); white crystalline powder

USBio B2419   MW 371.4 $C_{21}H_{25}NO_5$ ≥99%

Tyrosine — BOC-Tyr-Bzl-aldehyde

Bachem A-4595.0001   MW 355.43 $C_{21}H_{25}NO_4$ Store at -15°C

Tyrosine — BOC-Tyr-Bzl-ol

Bachem B-3500.0001   MW 357.45 $C_{21}H_{27}NO_4$ Store at 2-8°C

Tyrosine — BOC-Tyr-Bzl-OSu

Bachem A-1400.0001   MW 468.51 $C_{25}H_{28}N_2O_7$ Store at -15°C

USBio B2420   MW 468.5 $C_{25}H_{28}N_2O_7$ ≥99%

Tyrosine — BOC-Tyr-Dichlorobenzyl

Synonyms: BOC-O-(2,6-Dichlorobenzyl)-L-Tyr

Peptides International BLY-2119   MW 440.32   >98% (HPLC); white crystalline powder

Tyrosine — BOC-Tyr-Dimethylphospho

USBio B2430   MW 389.2 $C_{16}H_{24}NO_8P$ ≥99%

Tyrosine — BOC-Tyr-Diphenylphospho

USBio B2431   MW 513.4 $C_{26}H_{28}NO_8P$ ≥99%

Tyrosine — BOC-Tyr-Et

Synonyms: BOC-O-Et-L-Tyr; BOC-*p*OEt-Phe

Bachem A-2925.0001   MW 309.36 $C_{16}H_{23}NO_5$ Store at RT

Peptides International BLY-5334-PI   MW 309.36   >98% (HPLC); white powder

USBio B2427   MW 309.4 $C_{16}H_{23}NO_5$ ≥99%

Tyrosine — BOC-Tyr-Me

Synonyms: BOC-4-Methoxy-Phe; BOC-OMe-L-Tyr; BOC-*p*-Methoxy-L-Phe; BOC-*p*OMe-Phe

Bachem A-2090.0001   MW 295.34 $C_{15}H_{21}NO_5$ Store at RT

Tyrosine — BOC-Tyr-OBzl (continued)

Peptides International BLY-5332-PI   MW 295.34   >98% (HPLC); white crystalline powder

USBio B2428   MW 295.3 $C_{15}H_{21}NO_5$ ≥99%

Tyrosine — BOC-Tyr-OBzl

Bachem A-3585.0001   MW 371.43 $C_{21}H_{25}NO_5$ Store at RT

Tyrosine — BOC-Tyr-ol

Senn Chem 44002   MW 267.3

Tyrosine — BOC-Tyr-OMe

Bachem A-2420.0005   MW 295.34 $C_{15}H_{21}NO_5$ Store at -15°C

Tyrosine — BOC-Tyr-ONp

Bachem M-1180.0001   MW 402.4 $C_{20}H_{22}N_2O_7$ Store at -15°C

Tyrosine — BOC-Tyr-OSu

Bachem A-2415.0001   MW 378.38 $C_{18}H_{22}N_2O_7$ Store at -15°C

Tyrosine — BOC-Tyr-tBu

Bachem A-1450.0001   MW 337.42 $C_{18}H_{27}NO_5$ Store at RT

USBio B2423   MW 337.4 $C_{18}H_{27}NO_5$ ≥99%

Tyrosine — Bsmoc-Tyr-tBu

Synonyms: Bsmoc-O-tBu-L-Tyr

Peptides International BLY-5640-PI   MW 459.52   >98% (HPLC); white to off-white powder | LA Carpino, et al, *JACS*, 119:9915, 1997

Tyrosine — Bz-Tyr Amide

Bachem E-1525.0005   MW 284.32 $C_{16}H_{16}N_2O_3$ Store at RT

Tyrosine — Bz-Tyr-OEt

Synonyms: Chymotrypsin Substrate

Bachem E-1530.0005   MW 313.35 $C_{18}H_{19}NO_4$ Store at RT

Peptides International SYE-3010   MW 313.35 $C_{18}H_{19}NO_4$ >99% (HPLC); amorphous powder

Tyrosine — Bz-Tyr-pNA

Synonyms: Chymotrypsin Substrate

Bachem L-1155.0500   MW 405.41 $C_{22}H_{19}N_3O_5$ Store at -15°C

Peptides International SYN-3015 $C_{22}H_{19}N_3O_5$ >98% (HPLC); amorphous powder

Tyrosine — Bz-Tyr-βNA

Bachem K-1145.0250   MW 410.5 $C_{26}H_{22}N_2O_3$ Store at RT

Tyrosine — CBZ-D-Tyr

USBio C2099-76   MW 315.3 $C_{17}H_{17}NO_5$ ≥99%

Tyrosine — CBZ-D-Tyr-tBu-Dicyclohexylamine

USBio C2099-78   MW 552.8 $C_{21}H_{25}NO_5 \cdot C_{12}H_{23}N$ ≥99%

Tyrosine — CBZ-Tyr

USBio C2099-75   MW 315.3 $C_{17}H_{17}NO_5$ ≥99%

Tyrosine — CBZ-Tyr-Bzl

USBio C2099-77   MW 405.4 $C_{24}H_{23}NO_5$ ≥99%

Tyrosine — CBZ-Tyr-ONp

USBio C2099-80   MW 436.4 $C_{23}H_{20}NO_7$ ≥99%

Tyrosine — CBZ-Tyr-tBu-Dicyclohexylamine

USBio C2099-79   MW 552.8 $C_{21}H_{25}NO_5 \cdot C_{12}H_{23}N$ ≥99%

Tyrosine — Chloroacetyl-Tyr

Bachem E-1740.0005   MW 257.67 $C_{11}H_{12}ClNO_4$ Store at RT

Tyrosine — Dansyl-L-Tyr

ICN 100141   MW 414.5 $C_{21}H_{22}N_2O_5S$

Tyrosine — DL-m-Tyr

Synonyms: (±)-2-Amino-3-(3-Hydroxyphenyl)-Propionic Acid; DL-3-(3-Hydroxyphenyl)-Ala

Fluka 93852   MW 181.19 $C_9H_{11}NO_3$ ≥98.0% (titration); mp: 285°C

Fluka 93853   MW 181.19 $C_9H_{11}NO_3$ ≥98.0% (titration); mp: 285°C

ICN 103179   MW 181.2 $C_9H_{11}NO_3$ White to off-white crystals

Sigma T 3629   FW 181.2 $C_9H_{11}NO_3$ White to off-white crystals

Tyrosine — DL-O-Tyr

Synonyms: (±)-2-Amino-3-(2-Hydroxyphenyl)-Propionic Acid; DL-3-(2-Hydroxyphenyl)-Ala

Fluka 93851   MW 181.19 $C_9H_{11}NO_3$ ≥98.0% (titration); mp: 263°C

ICN 103181   MW 181.2 $C_9H_{11}NO_3$ ~98%; crystalline

Sigma T 3504   FW 181.2 $C_9H_{11}NO_3$ ~98%

Tyrosine — DL-Tyr

Synonyms: (±)-2-Amino-3-(4-Hydroxyphenyl)-Propionic Acid; DL-3-(4-Hydroxyphenyl)-Ala

Fluka 93850   MW 181.19 $C_9H_{11}NO_3$ ~99.0% (titration); mp: >300°C

ICN 103176   MW 181.2 $C_9H_{11}NO_3$ 99%; crystalline

Sigma T 3379   FW 181.2 $C_9H_{11}NO_3$ ≥98% (TLC) | AA precursor of dopamine & other catecholamines; also available as part of a kit

USBio T9230-15   MW 181.2 $C_9H_{11}NO_3$ ≥99%

Tyrosine — DL-Tyr-Me

Synonyms: 4-Methoxy-DL-Phe

Bachem F-1750.0001   MW 195.22 $C_{10}H_{13}NO_3$ Store at RT

Tyrosine — DL-Tyr-OMe Hydrochloride

Sigma T 9130   FW 231.7 $C_{10}H_{13}NO_3 \cdot HCl$

Tyrosine — DL-α-Me-m-Tyr

ICN 100521   MW 195.2 $C_{10}H_{13}NO_3$ Crystalline

Tyrosine — DL-α-Me-p-Tyr

ICN 100513   MW 195.2 $C_{10}H_{13}NO_3$ Crystalline

Tyrosine — D-Tyr

Synonyms: (R)-2-Amino-3-(4-Hydroxyphenyl)-Propionic Acid; D-3-(4-Hydroxyphenyl)-Ala

Bachem F-2165.0005   MW 181.19 $C_9H_{11}NO_3$ Store at RT

Fluka 93840   MW 181.19 $C_9H_{11}NO_3$ ≥99.0% (titration); mp: >300°C

ICN 103175   MW 181.2 $C_9H_{11}NO_3$ ≥98.5%; crystalline

Neosystem AA01901   MW 181.2

Peptides International ADY-2812   MW 181.19   >99.9% optical purity by RP-HPLC; white powder

Rexim   MW 181.2 $C_9H_{11}NO_3$

Sigma T 3254   FW 181.2 $C_9H_{11}NO_3$ ≥98% (TLC) | AA precursor of dopamine & other catecholamines; inactive isomer

USBio T9230-10   MW 181.2 $C_9H_{11}NO_3$ ≥99%

Tyrosine — D-Tyr (1-^{14}C)

ARC ARC-1050   MW 181.2 p-HOC$_6$H$_4$CH$_2$CH(NH$_2$)COOH   50-60 mCi/mmol; 1.85-2.22 GBq/mmol; in 0.01 N HCl | Radiochemical

Tyrosine — D-Tyr-Bzl

Bachem F-1300.0001   MW 271.32 $C_{16}H_{17}NO_3$ Store at RT

Tyrosine — D-Tyr-Me

Synonyms: 4-Methoxy-D-Phe

Bachem F-3145.0001   MW 195.22 $C_{10}H_{13}NO_3$ Store at RT

Tyrosine — D-Tyr-OBzl p-Tosyl

Bachem F-3010.0005   MW 443.52 $C_{16}H_{17}NO_3$ · $C_7H_8O_3S$ Store at RT

Tyrosine — D-Tyr-OMe Hydrochloride

Bachem F-2175.0005   MW 231.68 $C_{10}H_{13}NO_3$ · HCl   Store at 2-8°C

Tyrosine — D-Tyr-OtBu

Bachem F-3100.0001   MW 237.3 $C_{13}H_{19}NO_3$ Store at 2-8°C

Tyrosine — D-Tyr-tBu

Bachem F-2170.0001   MW 237.3 $C_{13}H_{19}NO_3$ Store at 2-8°C

Tyrosine — D-Tyr-tBu-OAll Hydrochloride

Bachem F-3525.0001   MW 313.82 $C_{16}H_{23}NO_3$ · HCl   Store at -15°C

Tyrosine — FMOC-(3,5-Dibromo)-Tyr

Bachem B-1275.0001   MW 561.23 $C_{24}H_{19}Br_2NO_5$ Store at 2-8°C

Tyrosine — FMOC-(3,5-Diiodo)-L-Tyr

Fluka 47457   MW 655.23 $C_{24}H_{19}I_2NO_5$ ≥98% (HPLC)

Tyrosine — FMOC-(3,5-Diiodo)-Tyr

Bachem B-1285.0001   MW 655.23 $C_{24}H_{19}I_2NO_5$ Store at 2-8°C

Tyrosine — FMOC-(3,5-Dinitro)-Tyr

Bachem B-3265.0001   MW 493.43 $C_{24}H_{19}N_3O_9$ Store at -15°C

Tyrosine — FMOC-3-Iodo-Tyr

Bachem B-1740.0001   MW 529.33 $C_{24}H_{20}INO_5$ Store at 2-8°C

Tyrosine — FMOC-3-Nitro-L-Tyr

Synonyms: FMOC-3-Nitro-L-Tyr

Fluka 47780   MW 448.43 $C_{24}H_{20}N_2O_7$ ≥98% (HPLC)

Neosystem FA07102   MW 448.4

Tyrosine — FMOC-3-Nitro-Tyr

Synonyms: FMOC-4-Hydroxy-3-Nitro-Phe

Bachem B-2690.0001   MW 448.43 $C_{24}H_{20}N_2O_7$ Store at -15°C

Tyrosine — FMOC-D-Tyr

American Peptide FTYR105   MW 403.5

Bachem B-2085.0001   MW 403.43 $C_{24}H_{21}NO_5$ Store at RT

USBio F5477   MW 403.4 $C_{24}H_{21}NO_5$ ≥99%

Tyrosine — FMOC-D-Tyr-(2,6-Dichlorobenzyl)

USBio F5484   MW 562.5 $C_{31}H_{25}NO_5Cl_2$ ≥99%

Tyrosine — FMOC-D-Tyr-(All)

Neosystem FA01919   MW 443.5

Tyrosine — FMOC-D-Tyr-Dimethylphospho

Bachem B-2275.0001   MW 511.47 $C_{26}H_{26}NO_8P$ Store at -15°C

Tyrosine — FMOC-D-Tyr-Et

Bachem B-1775.0001   MW 431.49 $C_{26}H_{25}NO_5$ Store at RT

Tyrosine — FMOC-D-Tyr-Me

Synonyms: FMOC-4-Methoxy-D-Phe

Bachem B-2430.0001   MW 417.46 $C_{25}H_{23}NO_5$ Store at RT

Tyrosine — FMOC-D-Tyr-tBu

Synonyms: FMOC-O-tBu-D-Tyr

Bachem B-1260.0001   MW 459.54 $C_{28}H_{29}NO_5$ Store at RT

Peptides International FDY-1833-PI   MW 459.54   >98% (HPLC); white crystalline powder

USBio F5481   MW 459.6 $C_{28}H_{29}NO_5$ ≥99%

Tyrosine — FMOC-L-Tyr

Neosystem FA01902   MW 403.4

Tyrosine — FMOC-L-Tyr tBu-Wang Resin

American Peptide RFWAN90   Preattached Wang resins are usful for synthesis of peptides with C-terminal carboxyl groups; peptides can be cleaved from the resin with 90% TFA in the presence of scavengers

Tyrosine — FMOC-L-Tyr-(2,6-Dichlorobenzyl)

Neosystem FA01907   MW 562.5

Tyrosine — FMOC-L-Tyr-(All)

Neosystem FA01903   MW 443.5

Tyrosine — FMOC-L-Tyr-2-Bromo-Z

Neosystem FA01906   MW 616.5

Tyrosine — FMOC-L-Tyr-tBu

Neosystem FA01905   MW 459.5

Tyrosine — FMOC-L-Tyr-tBu-Wang Resin

Neosystem RW01902

Tyrosine — FMOC-Me-Tyr-Bzl

USBio F5428   MW 507.6 $C_{32}H_{29}NO_5$ ≥99%

Tyrosine — FMOC-O-Bzl-L-Tyr

Fluka 47565 MW 493.56 $C_{31}H_{27}NO_5$ ≥98% (HPLC); mp: 157-161°C

Tyrosine — FMOC-O-Dimethylphospho-L-Tyr

Fluka 47523 MW 511.47 $C_{26}H_{26}NO_8P$ ≥98% (HPLC); mp: 100-102°C

Tyrosine — FMOC-OtBu-D-Tyr

Fluka 47319 MW 459.55 $C_{28}H_{29}NO_5$ ≥98% (HPLC)

Tyrosine — FMOC-OtBu-L-Tyr

Fluka 47623 MW 459.55 $C_{28}H_{29}NO_5$ ~98% (titration); mp: 150-151°C

Tyrosine — FMOC-OtBu-L-Tyr 4-Benzyloxybenzyl Ester Polymer Bound

Fluka 47654 0.4-0.6 mmol/g resin; carrier: polystyrene, crosslinked with 1% DVB; 200-400 mesh particle size

Tyrosine — FMOC-OtBu-L-Tyr-OPfp

Fluka 47456 MW 625.59 $C_{34}H_{28}F_5NO_5$ ~98% (HPLC)

Tyrosine — FMOC-Tyr

American Peptide FTYR100 MW 403.5

Bachem B-1580.0001 MW 403.43 $C_{24}H_{21}NO_5$ Store at RT

USBio F5478 MW 403.4 $C_{24}H_{21}NO_5$ ≥99%

Tyrosine — FMOC-Tyr (^{15}N)

Bachem B-2665.0100 MW 404.43 $C_{24}H_{21}^{15}NO_5$ Store at -15°C

Tyrosine — FMOC-Tyr tBu

American Peptide FTYR130 MW 459.5

Tyrosine — FMOC-Tyr-(2,6-Dichlorobenzyl)

Bachem B-1280.0001 MW 562.45 $C_{31}H_{25}Cl_2NO_5$ Store at RT

USBio F5485 MW 562.5 $C_{31}H_{25}NO_5Cl_2$ ≥99%

Tyrosine — FMOC-Tyr-(3,5-Diiodo)

USBio F5486 MW 655.2 $C_{24}H_{19}I_2NO_5$ ≥99%

Tyrosine — FMOC-Tyr-(Malonyl-Di-OtBu)

Bachem B-2825.0100 MW 617.7 $C_{35}H_{39}NO_9$ Store at -15°C

Tyrosine — FMOC-Tyr-(PO₃(MDPSE)₂)

Bachem B-2910.0001 MW 932.17 $C_{54}H_{54}NO_8PSi_2$ Store at -15°C

Tyrosine — FMOC-Tyr-(PO₃H₂)

Bachem B-2470.0001 MW 483.41 $C_{24}H_{22}NO_8P$ Store at -15°C

Tyrosine — FMOC-Tyr-2-Bromo-Z

Bachem B-2795.0001 MW 616.48 $C_{32}H_{26}NO_7Br$ Store at 2-8°C

USBio F5480 MW 616.5 $C_{32}H_{26}NO_7Br$ ≥99%

Tyrosine — FMOC-Tyr-Bzl

Synonyms: FMOC-O-Bzl-L-Tyr

Bachem B-1215.0001 MW 493.56 $C_{31}H_{27}NO_5$ Store at RT

Peptides International FLY-1736-PI MW 493.56 >98% (HPLC); white powder

USBio F5479 MW 493.6 $C_{31}H_{27}NO_5$ ≥99%

Tyrosine — FMOC-Tyr-Bzl-SASRIN™-® (200-400 mesh)

Bachem D-1620.0001 Store at 2-8°C

Tyrosine — FMOC-Tyr-Dibutylphospho

USBio F5487 MW 596.6 $C_{32}H_{38}NO_8P$ ≥99%

Tyrosine — FMOC-Tyr-Dichlorobenzyl

Synonyms: FMOC-O-(2,6-Dichlorobenzyl)-L-Tyr

Peptides International FLY-1738-PI MW 562.45 >98% (TLC); white powder

Tyrosine — FMOC-Tyr-Dimethylphospho

Bachem B-1990.0001 MW 511.47 $C_{26}H_{26}NO_8P$ Store at -15°C

Tyrosine — FMOC-Tyr-Sulfo

Bachem B-1600.0001 MW 483.5 $C_{24}H_{19}NO_8S$ Store at -15°C

Tyrosine — FMOC-Tyr-tBu

Synonyms: FMOC-O-tBu-L-Tyr

Bachem B-1255.0001 MW 459.54 $C_{28}H_{29}NO_5$ Store at 2-8°C

Peptides International FLY-1737-PI MW 459.54 >98% (HPLC); white crystalline powder

USBio F5482 MW 459.6 $C_{28}H_{29}NO_5$ ≥99%

Tyrosine — FMOC-Tyr-tBu-® (200-400 mesh)

Bachem D-1080.0001 Store at 2-8°C

Tyrosine — FMOC-Tyr-tBu-OPfp

Bachem B-1195.0001 MW 625.59 $C_{34}H_{28}F_5NO_5$ Store at -15°C

USBio F5483 MW 625.6 $C_{34}H_{28}F_5NO_5$ ≥99%

Tyrosine — FMOC-Tyr-tBu-SASRIN™-® (200-400 mesh)

Bachem D-1400.0001 Store at 2-8°C

Tyrosine — L-Tyr

Synonyms: (S)-2-Amino-3-(4-Hydroxyphenyl)-Propionic Acid; (-)-α-Amino-p-Hydroxyhydrocinnamic Acid; (S)-α-Amino-4-Hydroxybenzenepropanoic Acid; 4-Hydroxy-L-Phe

Amersham US22910 ≥98.5% dsb; ≤15ppm heavy metals (Pb)

Fluka 93829 MW 181.19 $C_9H_{11}NO_3$ ≥99.0% (titration); ≤0.1% residue on ignition; ≤0.2% loss on drying; ≤0.3% foreign AA; ≤0.00001% As; ≤0.01% SO₄; ≤0.02% Cl, Na, NH₄⁺; ≤0.0005% Al, Ba, Bi, Cd, Co, Cr, Cu, Fe, Li, Mg, Mn, Mo, Ni, Pb, Sr, Zn; ≤0.001% Ca; ≤0.005% K | *CRC Handbook of Microbiology*, Laskin, AI & Lachevalier, HA, eds, 4: 1, 1974, CRC Press, Cleveland, Ohio

Fluka 93830 MW 181.19 $C_9H_{11}NO_3$ ≥99.0% (titration); ≤0.1% residue on ignition; ≤0.5% loss on drying; ≤0.3% foreign AA; ≤0.01% Ca, Mg, SO₄; ≤0.02% Cl, Na, NH₄⁺; ≤0.0005% Cd, Co, Cr, Cu, Fe, Mn, Ni, Pb, Zn; ≤0.005% K

Neosystem AA01902 MW 181.2

Peptides International ALY-2722 MW 181.19 >99.9% optical purity by RP-HPLC; white powder

Rexim MW 181.2 $C_9H_{11}NO_3$ White crystals or crystalline powder

Sigma T 7187 FW 181.2 $C_9H_{11}NO_3$ USP Grade

Sigma T 8016 Matrix: 4% beaded agarose; activation: epoxy; attachment: amino; spacer: 12 atoms; ligand immobilized: 15-25 μmoles/mL; form: suspension in 0.15 M sodium phosphate, pH 6.8, containing 0.02% sodium azide | Deutsch, DG & Mertz, ET, *Proc Fed Amer Soc Exp Biol*, 29: 647, 1970; Deutsch, DG & Mertz, ET, *Science*, 170: 1095, 1970; Vuento, M & Vaheri, A, *Biochem J*, 183: 331, 1979

USBio T9230 MW 181.2 $C_9H_{11}NO_3$ ≥98%; appearance: white crystalline powder; specific rotation (C=5, 1 N HCl): -11.3° to -12.1°; chloride:≤0.02%; ammonium:≤0.02%; sulfate:≤0.02%; iron:≤0.001%; heavy metals (as Pb):≤0.001%; arsenic:≤0.0001%; Loss on drying:≤0.2%; sulfated:≤0.1%; pH:5.0 to 6.5; pyrogen: nonpyrogenic; endotoxin: ≤2 EU/mg; cell culture analysis: murine myeloma SP2/0-Ag14

Tyrosine — L-Tyr (1-^{14}C)

ARC ARC-507 MW 181.2 p-HOC$_6$H$_4$CH$_2$CH(NH$_2$)COOH 50-60 mCi/mmol; 1.85-2.22 GBq/mmol; in 0.01 N HCl | Radiochemical

Tyrosine — L-Tyr (^{15}N)

ICN 540150 MW 182.2 HOC$_6$H$_4$CH$_2$CH(NH$_2$)COOH 95% ^{15}N atomic purity

Tyrosine — L-Tyr (2,3,5,6-^{3}H)

Amersham TRK530 Aqueous, 2% EtOH, sterilized; 2.59-4.26 TBq/mmol, 70-115 Ci/mmol; 37 MBq/mL, 1 mCi/mL

Tyrosine — L-Tyr (3,5-^{3}H)

Amersham TRK200 Aqueous, 2% EtOH, sterilized; 1.5-2.2 TBq/mmol, 40-60 Ci/mmol; 37 MBq/mL, 1 mCi/mL

ICN 20055E MW 181.2 25-50 Ci/mmol, 0.92-1.85 TBq/mmol; sterile EtOH:H$_2$O (2:98)

Tyrosine — L-Tyr (Carboxyl-^{14}C)

Amersham CFA565 Aqueous, 2% EtOH, sterilized; 1.85-2.29 GBq/mmol, 50-62 mCi/mmol; 1.85 MBq/mL, 50 μCi/mL

Tyrosine — L-Tyr (Ring 3,5-^{3}H)

ARC ART-195 MW 181.2 HCCHC(OH)CHCHCCH$_2$CH(NH$_2$)COOH 40-60 mCi/mmol; 1.48-2.22 TBq/mmol; in 0.01 N HCl | Radiochemical

ARC ART-195A MW 181.2 HCCHC(OH)CHCHCCH$_2$CH(NH$_2$)COOH 40-60 mCi/mmol; 1.48-2.22 TBq/mmol; in EtOH: H$_2$O (2:98) | Radiochemical

Sigma T 3533 FW 181.2 HCCHC(OH)CHCHCCH$_2$CH(NH$_2$)COOH 40-60 Ci/mmol; aqueous solution containing 2% EtOH in serum bottle | Radiochemical

Tyrosine — L-Tyr (Side chain 2,3-^{3}H)

ARC ART-399 MW 181.2 p-HOC$_6$H$_4$CH$_2$CH$_2$(NH$_2$)COOH 30-50 Ci/mmol; 1.11-1.85 TBq/mmol; in 0.01 N HCl | Radiochemical

Tyrosine — L-Tyr (U-^{14}C)

Amersham CFB74 Aqueous, 2% EtOH, sterilized; >16.6 GBq/mmol, >450 mCi/mmol; 1.85 MBq/mL, 50 μCi/mL

ARC ARC-655 MW 181.2 p-HOC$_6$H$_4$CH$_2$CH(NH$_2$)COOH 400-500 mCi/mmol; 14.8-18.5 GBq/mmol; in EtOH: H$_2$O (2:98) | Radiochemical

ICN 10135 MW 181.2 p-HOC$_6$H$_4$CH$_2$CH(NH$_2$)COOH 405-495 mCi/mmol; 15-18.3 GBq/mmol; 0.01 N HCl solution

ICN 10135E MW 181.2 p-HOC$_6$H$_4$CH$_2$CH(NH$_2$)COOH 405-495 mCi/mmol; 15-18.3 GBq/mmol; sterile EtOH:H$_2$O (2:98)

Tyrosine — L-Tyr (U-^{14}C) Hydrochloride

Sigma T 4032 FW 181.2 HOC$_6$H$_4$CH$_2$CH(NH$_2$)COOH 360-540 mCi/mmol; radiochemical purity ≥98% (HPLC); aqueous solution containing 2% EtOH in Combi-vial | Radiochemical

Tyrosine — L-Tyr Amide Hydrobromide

ICN 103188 ≥99%; hygroscopic

Tyrosine — L-Tyr Disodium Salt

Synonyms: L-3-(4-Hydroxyphenyl)-Ala

Amersham US22927 ≥98.5% dsb

Sigma T 2269 FW 225.2 $C_9H_9NO_3Na_2$ ≥98% (TLC; off-white to tan powder | AA precursor of dopamine & other catecholamines; inactive isomer; also available as part of a kit

Tyrosine — L-Tyr Disodium Salt Hydrate

USBio T9235 MW 261.2 $C_9H_9NO_3Na_2 \cdot H_2O$ ≥98%; appearance: off-white crystalline powder; solubility (10%): clear, pale yellow & complete; specific rotation: -12.6° to -14.4°; pH (1%): 11.5± 0.5; heavy metals (as Pb): ≤0.0015%; endotoxin: ≤2 EU/mg; cell culture tested: no inhibition of growth with murine myeloma SP2/0-Ag14

Tyrosine — L-Tyr Free Base

Synonyms: β-p-Hyph; L-3-(4-Hydroxyphenyl)-Ala

Calbiochem 6570 MW 181.2 $C_9H_{11}NO_3$ ≥98% (titration); crystalline solid; soluble in H$_2$O | *Merck Index*, 12: 9970

ICN 103183 MW 181.2 $C_9H_{11}NO_3$ Crystalline

ICN 194759 MW 181.2 $C_9H_{11}NO_3$ Crystalline | Cell culture reagent

Sigma T 1020 FW 181.2 $C_9H_{11}NO_3$ Crystalline | Cell culture tested; insect cell culture tested

Sigma T 1655 FW 181.2 $C_9H_{11}NO_3$ Crystalline | Plant cell culture tested

Sigma T 3754 FW 181.2 $C_9H_{11}NO_3$ ≥98% (TLC) | AA precursor of dopamine & other catecholamines; inactive isomer; also available as part of a kit

Sigma T 8909 FW 181.2 $C_9H_{11}NO_3$ >99% (TLC; SigmaUltra; residue on ignition: <0.1%; solubility (0.5 M in 1.0 M HCl, 20°C): complete, colorless to faint yellow; insoluble matter: <0.1%; NH$_4^+$, SO$_4$: <0.05%; K: <0.005%; Al, Cu, Fe, Zn: <0.0005%; Pb, P, Ca: <0.001%; Na: <0.01% | AA precursor of dopamine & other catecholamines; inactive isomer

Tyrosine — L-Tyr Hydrazide

Fluka 93915 MW 192.22 $C_9H_{13}N_3O_2$ ≥97.0% (titration); mp: 196-197°C | Chiral base used for the resolution of AA derivatives; Vogler, K & Lanz, P, *Helv Chim Acta*, 49: 1348, 1966; Grzonka, Z & Liberek, B, *Tetrahedron*, 27: 1783, 1971; Boggs, NT et al, *J Org Chem*, 44: 2262, 1979

ICN 157171 MW 195.2 $C_9H_{13}N_3O_2$ Crystalline

Tyrosine — L-Tyr Hydrochloride

Synonyms: L-3-(4-Hydroxyphenyl)-Ala; 3-(4-Hydroxyphenyl)-L-Ala

Amersham US22935 ≥98.5% dsb; ≤15ppm heavy metals (Pb)

ICN 103192 MW 217.7 $C_9H_{11}NO_3 \cdot HCl$ Crystalline

ICN 194760 MW 217.7 $C_9H_{11}NO_3 \cdot HCl$ Crystalline | Cell culture reagent

Sigma T 2006 FW 217.7 $C_9H_{11}NO_3 \cdot HCl$ ≥98% (TLC) | AA precursor of dopamine & other catecholamines; inactive isomer

Sigma T 2025 FW 217.7 $C_9H_{11}NO_3 \cdot HCl$ Crystalline | Cell culture tested

USBio T9236 MW 217.7 $C_9H_{11}NO_3 \cdot HCl$ ≥98%; appearance: white crystalline powder; solubility: clear, colorless & complete; specific rotation (C=4, 1 N HCl): -7° to -9°; chloride: 16.5 ± 0.25%; heavy metals (as Pb): ≤0.002%; endotoxin: ≤2 EU/mg; cell culture tested: murine myeloma SP2/0-Ag14

Tyrosine — L-Tyr Hydroxamate

Sigma T 1380 FW 196.2 $C_9H_{12}N_2O_3$

Tyrosine — L-Tyr Sodium Salt

ICN 105573 MW 225.2 $C_9H_9NO_3Na_2$ Crystalline

Sigma T 1145	FW 225.2	$C_9H_9NO_3Na_2$	Crystalline |

Extremely hygroscopic

Tyrosine — L-Tyr β-Naphthylamide

Sigma T 3384	FW 306.4	$C_{19}H_{18}N_2O_2$

Tyrosine — L-Tyr-AMC

ICN 157168	MW 338.4	$C_{19}H_{18}N_2O_4$	
Sigma T 2141	FW 338.4	$C_{19}H_{18}N_2O_4$	

Tyrosine — L-Tyr-OAll p-Tosyl

ICN 157167	MW 393.5	$C_{12}H_{15}NO_3 \cdot C_7H_8O_3S$	
Sigma T 2657	FW 393.5	$C_{12}H_{15}NO_3 \cdot C_7H_8O_3S$	

Tyrosine — L-Tyr-OBzl p-Tosyl

ICN 103186	MW 443.5	$C_{16}H_{17}NO_3 \cdot C_7H_8O_3S$	Crystalline
Sigma T 9505	FW 443.5	$C_{16}H_{17}NO_3 \cdot C_7H_8O_3S$	

Tyrosine — L-Tyr-OEt Free Base

ICN 157170	MW 209.2	$C_{11}H_{15}NO_3$	Crystalline
Sigma T 4754	FW 209.2	$C_{11}H_{15}NO_3$	

Tyrosine — L-Tyr-OEt Hydrochloride

Synonyms: TEE

Fluka 93890	MW 245.71	$C_{11}H_{15}NO_3 \cdot HCl$	≥99.0%

(titration); mp: 165-167°C

ICN 103189	MW 245.7	$C_{11}H_{15}NO_3 \cdot HCl$	Crystalline
Sigma T 4879	FW 245.7	$C_{11}H_{15}NO_3 \cdot HCl$	

Tyrosine — L-Tyr-OMe

Fluka 93920	MW 195.22	$C_{10}H_{13}NO_3$	≥99.0% (titration); mp:

134-136°C

Tyrosine — L-Tyr-OMe Free Base

ICN 157172	MW 195.2	$C_{10}H_{13}NO_3$	Crystalline

Tyrosine — L-Tyr-OMe Hydrochloride

Fluka 93930	MW 231.68	$C_{10}H_{13}NO_3 \cdot HCl$	≥99.0%

(titration); ≤0.05% residue on ignition; mp: 190°C

ICN 103193	MW 231.7	$C_{10}H_{13}NO_3 \cdot HCl$	98%; crystalline
Neosystem AA01905	MW 231.7		
Sigma T 5004	FW 231.7	$C_{10}H_{13}NO_3 \cdot HCl$	

Tyrosine — L-Tyr-OtBu

Fluka 93902	MW 237.30	$C_{13}H_{19}NO_3$	≥99.0% (titration); mp:

141-143°C

ICN 157169	MW 237.3	$C_{13}H_{19}NO_3$	Crystalline
Sigma T 1128	FW 237.3	$C_{13}H_{19}NO_3$	

Tyrosine — L-β-Phenyllactoyl-Tyr

Bachem F-2615.0050	MW 329.4	$C_{18}H_{19}NO_5$	Store at -15°C

Tyrosine — m-Fluoro-DL-Tyr

Sigma F 4505	FW 199.2	$C_9H_{10}FNO_3$

Tyrosine — MTH-DL-Tyr

Sigma M 6131	FW 236.3	$C_{11}H_{12}N_2O_2S$	~95%

Tyrosine — N-(3-Nitro-2-Pyridinesulfenyl)-O-tBu-L-Tyr

Sigma N 5890	FW 391.4	$C_{18}H_{21}N_3O_5S$

Tyrosine — N,O-Diacetyl-L-Tyr-OEt

ICN 157573	MW 293.3	$C_{15}H_{19}NO_5$	Crystalline

Tyrosine — N,O-Didansyl-L-Tyr Monocyclohexylammonium Salt

Sigma D 2375	FW 746.9	$C_{33}H_{33}N_3O_7S_2 \cdot C_6H_{13}N$	~95%

Tyrosine — N-Ac-(3,5-Diiodo)-L-Tyr

ICN 100063	MW 475.9	$C_{11}H_{11}I_2NO_4$	Crystalline
Sigma A 7375	FW 475.0	$C_{11}H_{11}I_2NO_4$	

Tyrosine — N-Ac-(3,5-Dinitro)-L-Tyr

ICN 154693	MW 313.2	$C_{11}H_{11}N_3O_8$	Crystalline
Sigma A 7625	FW 313.2	$C_{11}H_{11}N_3O_8$	

Tyrosine — N-Ac-(3,5-Dinitro)-L-Tyr-OEt

Sigma A 7750	FW 341.3	$C_{13}H_{15}N_3O_8$

Tyrosine — N-Ac-L-3-Nitro-Tyr-OEt

Sigma A 2635	FW 296.3	$C_{13}H_{16}N_2O_6$

Tyrosine — N-Ac-L-Tyr

Synonyms: L-2-Acetylamino-3-p-Hydroxyphenyl-Propionic Acid

Fluka 01527	MW 223.23	$C_{11}H_{13}NO_4$	≥98% (titration);

≤0.2% loss on drying; mp: 149-152°C

ICN 100150	MW 223.2	$C_{11}H_{13}NO_4$	Crystalline
Rexim	MW 223.2	$C_{11}H_{13}NO_4$	White crystals or crystalline

powder

Sigma A 2513	FW 223.2	$C_{11}H_{13}NO_4$

Tyrosine — N-Ac-L-Tyr Hydrazide

ICN 100148	MW 237.3	$C_{11}H_{15}N_3O_3$	Crystalline

Tyrosine — N-Ac-L-Tyr-OEt

Synonyms: ATEE; Chymotrypsin Substrate

ICN 100156	MW 251.3	$C_{13}H_{17}NO_4$	Crystalline
Sigma A 6751	FW 251.3	$C_{13}H_{17}NO_4$	Sigma Grade

Tyrosine — N-Ac-L-Tyr-OEt Hydrate

Synonyms: ATEE; Chymotrypsin Esterase Substrate

Fluka 01530	MW 269.30	$C_{13}H_{17}NO_4 \cdot H_2O$	≥99% (HPLC);

mp: 78-81°C \| Widely used substrate for the determination of the esterase activity of chymotrypsin; Schwert, GW & Takenake, Y, *Biochem Biophys Acta*, 16: 570, 1955; Sumar, S & Hein, GE, *Anal Biochem*, 30: 203, 1969; Garrell, J & Cuchillo, CM, *FEBS Lett*, 190: 329, 1985; Laskowski, M, *Meth Enzymol*, 2: 8, 1955; Wilcox, PE, *Meth Enzymol*, 19: 64, 1970

Tyrosine — N-Ac-L-Tyrosinamide

Sigma A 6626	FW 222.2	$C_{11}H_{14}N_2O_3$

Tyrosine — N-Ac-OMe-L-Tyr-OMe

Sigma A 2510	FW 251.3	$C_{13}H_{17}NO_4$

Tyrosine — N-Bz-L-Tyr-4-Aminobenzoic Acid Sodium Salt Hydrate

Synonyms: Chymotrypsin Substrate

Fluka 13135	MW 444.42	$C_{23}H_{19}N_2NaO_5 \cdot H_2O$	≥99%

(HPLC); mp: ~210°C \| Substrate for the assay of chymotrypsin in crude biological materials; Imondi, AR et al, *Anal Biochem*, 54: 199, 1973; Pemberton, PW & Lobley, RW, *Biochem Soc trans*, 13: 175, 1985

Tyrosine — N-Bz-L-Tyr-OEt

Synonyms: BTEE; Chymotrypsin Substrate; Trypsin-Resistant Substrate

Fluka 13110   MW 313.36 $C_{18}H_{19}NO_4$ ≥98% (titration); mp: 118-122°C | Hummel, BCW, *Can J Biochem Physiol*, 37: 1393, 1959; Walsh, KA & Wilcox, PE, *Meth Enzymol*, 19: 31, 1979; Rao, KN & Lombardi, B, *Anal Biochem*, 65: 548, 1975

ICN 100917   MW 313.4 $C_{18}H_{19}NO_4$ Crystalline

Sigma B 6125   FW 313.4 $C_{18}H_{19}NO_4$

Tyrosine — N-Bz-L-Tyrosinamide

Sigma B 6000   FW 284.3 $C_{16}H_{16}N_2O_3$

Tyrosine — N-Bz-L-Tyr-p-Aminobenzoic Acid Sodium Salt

Synonyms: Bentiromide

Sigma B 0262   FW 426.4 $C_{23}H_{19}N_2O_5Na$

Tyrosine — N-Bz-L-Tyr-pNA

Sigma B 6760   FW 405.4 $C_{22}H_{19}N_3O_5$

Tyrosine — N-CBZ-D-Tyr

Sigma C 3158   FW 315.3 $C_{17}H_{17}NO_5$

Tyrosine — N-CBZ-L-Tyr

ICN 100203   MW 315.3 $C_{17}H_{17}NO_5$ Crystalline

Sigma C 5627   FW 315.3 $C_{17}H_{17}NO_5$

Tyrosine — N-CBZ-L-Tyr-(pONp)

ICN 101259   MW 436.4 $C_{23}H_{20}N_2O_7$ 98%; crystalline

Sigma C 6002   FW 436.4 $C_{23}H_{20}N_2O_7$ ~95%

Tyrosine — N-CBZ-O-tBu-L-Tyr Dicyclohexylammonium Salt

Sigma C 0407   FW 552.8 $C_{21}H_{25}NO_5 \cdot C_{12}H_{23}N$

Tyrosine — N-Chloroacetyl-L-Tyr

Sigma C 1878   FW 257.7 $C_{11}H_{12}ClNO_4$

Tyrosine — N-FMOC-(3,5-Diiodo)-L-Tyr

ICN 158170   MW 655.2 $C_{24}H_{19}I_2NO_6$

Tyrosine — N-FMOC-L-Tyr (3,5-³H)

ARC ART-498   MW 403.4   30-60 Ci/mmol; 1.11-2.22 TBq/mmol; in EtOH | Radiochemical

Tyrosine — N-FMOC-O-tBu-L-Tyr

Sigma F 1633   FW 459.5 $C_{28}H_{29}NO_5$

Tyrosine — N-FMOC-o-Dimethylphospho-L-Tyr

Sigma F 3898   FW 511.5 $C_{26}H_{26}NO_8P$ Valerio, RM et al, *Int J Pept Prot Res*, 33: 428, 1989

Tyrosine — N-For-L-Tyr

ICN 151173   MW 209.2 $C_{10}H_{11}NO_4$

Sigma F 9751   FW 209.2 $C_{10}H_{11}NO_4$ White powder

Tyrosine — N-Me-D-Tyr

Bachem F-3715.0250   MW 195.22 $C_{10}H_{13}NO_3$ Store at RT

Tyrosine — N-Me-Tyr

Bachem F-2235.0250   MW 195.22 $C_{10}H_{13}NO_3$ Store at RT

Tyrosine — N-Me-Tyr-Me

Synonyms: N-Me-4-Methoxy-Phe

Bachem E-2160.0250   MW 209.25 $C_{11}H_{15}NO_3$ Store at RT

Tyrosine — N-O-Nps-L-Tyr Dicyclohexylammonium Salt

Sigma N 6628   FW 515.7 $C_{15}H_{14}N_2O_5S \cdot C_{12}H_{23}N$ Yellow crystals

Tyrosine — N-t-BOC-(3,5-Diiodo)-L-Tyr

Sigma B 5771   FW 533.1 $C_{14}H_{17}I_2Cl_2NO_5$

Tyrosine — N-t-BOC-D-Tyr-OMe

Sigma B 2019   FW 295.3 $C_{15}H_{21}NO_5$

Tyrosine — N-t-BOC-L-Tyr

Sigma B 0518   FW 281.3 $C_{14}H_{19}NO_5$

Tyrosine — N-t-BOC-L-Tyr 2,4,5-Trichlorophenyl Ester

Sigma B 8759   FW 460.7 $C_{20}H_{20}Cl_3NO_5$

Tyrosine — N-t-BOC-L-Tyr N-Hydroxysuccinimide Ester

Sigma B 0142   FW 378.4 $C_{18}H_{22}N_2O_7$

Tyrosine — N-t-BOC-O-(2,6-Dichlorobenzyl)-L-Tyr

Sigma B 1761   FW 440.3 $C_{21}H_{23}Cl_2NO_5$

Tyrosine — N-t-BOC-O-(2,6-Dichlorobenzyl)-L-Tyr Pam Resin Ester

Sigma B 3518   N-t-BOC AA ester of 4-(Oxymethyl)-Phenyl-Acm 1% divinylbenzene cross-linked polystyrene; 200-400 mesh; substitution range: 0.2-0.6 mmole t-BOC-AA/g resin | For peptide synthesis by the Merrifield method; Mitchell, AR et al, *J Org Chem*, 43: 2845, 1978

Tyrosine — N-t-BOC-O-(2,6-Dichlorobenzyl)-L-Tyr Resin Ester

Sigma B 5646   N-t-BOC AA ester of methylated 1% divinylbenzene cross-linked polystyrene; 200-400 mesh; substitution range: 0.2-0.6 mmole t-BOC-AA/g resin | For peptide synthesis by the Merrifield method

Tyrosine — N-t-BOC-O-Ac-L-Tyr

Sigma B 2644   FW 323.3 $C_{16}H_{21}NO_6$

Tyrosine — N-t-BOC-O-Bzl-L-Tyr

Sigma B 6126   FW 371.4 $C_{21}H_{25}NO_5$

Tyrosine — N-t-BOC-OMe-L-Tyr

Sigma B 4648   FW 295.3 $C_{15}H_{21}NO_5$

Tyrosine — N-TFA-L-Tyr-OMe

Sigma T 6131   FW 291.2 $C_{12}H_{12}F_3NO_4$ ~98%

Tyrosine — N-Z-OtBu-L-Tyr Dicyclohexylamine

Fluka 96048   MW 552.75 $C_{21}H_{25}NO_5 \cdot C_{12}H_{23}N$ ≥98.0% (TLC)

Fluka 96050   MW 552.75 $C_{21}H_{25}NO_5 \cdot C_{12}H_{23}N$ ≥99.0% (titration)

Tyrosine — *N*,*O*-Di(2,4-Dnp)-L-Tyr

Sigma D 2130   FW 513.4 $C_{21}H_{15}N_5O_{11}$

Tyrosine — *N*ª-Ac-L-Tyr Amide

Fluka 01540   MW 222.25 $C_{11}H_{14}N_2O_3$ ≥99% (TLC)

Tyrosine — *N*ª-BOC-O-Bzl-*N*ª-Me-L-Tyr

Fluka 00043   MW 385.46 $C_{22}H_{27}NO_5$ ≥98% (HPLC)

Tyrosine — *N*ª-BOC-O-Dibenzylphospho-L-Tyr

Fluka 03574   MW 541.54 $C_{28}H_{32}NO_8P$ ~97% (TLC)

Tyrosine — *N*ª-FMOC-L-Tyr

Fluka 47751   MW 403.43 $C_{24}H_{21}NO_5$ ≥97% (HPLC); mp: 182-187°C

Tyrosine — *N*ª-FMOC-O-Bis-Benzyloxyphosphoryl-L-Tyr

Fluka 47564   MW 663.66 $C_{38}H_{34}NO_8P$ ~95% (HPLC); amorphous powder | Building block for the synthesis of peptides with *O*-phospho-L-Tyr; Kitas, EA et al, *Helv Chim Acta*, 74: 1314, 1991

Tyrosine — *N*ª-FMOC-O-Phospho-L-Tyr

Fluka 00147   MW 483.41 $C_{24}H_{22}NO_8P$ ~98% (HPLC)

Tyrosine — *N*ª-FMOC-O-tBu-L-Tyr

ICN 151139   MW 459.5 $C_{28}H_{29}NO_5$

Tyrosine — *N*ª-t-BOC-L-Tyr

ICN 101054   MW 281.3 $C_{14}H_{19}NO_5$ Crystalline

Tyrosine — *N*ª-t-BOC-O-Bzl-L-Tyr

ICN 101060   MW 371.4 $C_{21}H_{25}NO_5$ Crystalline

Tyrosine — *o*-(4-Hydroxy-3-Iodophenyl)-(3,5-Diiodo)-L-Tyr Free Acid

Synonyms: 3,3',5-Triiodo-L-Thyronine; Liothyronine; T₃
Sigma T 2877   FW 651.0 $C_{15}H_{12}I_3NO_4$ 95-98%

Tyrosine — *o*-(4-Hydroxy-3-Iodophenyl)-(3,5-Diiodo)-L-Tyr Sodium Salt

Synonyms: 3,3',5-Triiodo-L-Thyronine; Liothyronine; T₃
Sigma T 2752   FW 673.0 $C_{15}H_{11}I_3NO_4Na$ ~98%

Tyrosine — *o*-Bzl-L-Tyr

Fluka 14010   MW 271.32 $C_{16}H_{17}NO_3$ ≥99% (titration)

Tyrosine — *o*-CBZ-L-Tyr

Sigma C 6761   FW 315.3 $C_{17}H_{17}NO_5$

Tyrosine — *o*-Dansyl-L-Tyr Free Acid

Sigma D 0386   FW. 414.5 $C_{21}H_{22}N_2O_5S$

Tyrosine — *o*-Me-DL-Tyr

Synonyms: 4-Methoxy-DL-Phe
ICN 155685   MW 195.2 $C_{10}H_{13}NO_3$ White powder
Sigma M 8504   FW 195.2 $C_{10}H_{13}NO_3$ White powder

Tyrosine — *o*-Me-D-Tyr

Fluka 69575   MW 195.22 $C_{10}H_{13}NO_3$ ≥99.0% (titration)

Tyrosine — *o*-Me-L-Tyr

Fluka 69576   MW 195.22 $C_{10}H_{13}NO_3$ ≥98.0% (titration); ≤1% H_2O; mp: 244°C

Tyrosine — *o*-Me-L-Tyr Free Base

Synonyms: 4-Methoxy-L-Phe
Sigma M 4775   FW 195.2 $C_{10}H_{13}NO_3$

Tyrosine — *o*-Me-L-Tyr Hydrochloride

Synonyms: 4-Methoxy-L-Phe
Sigma M 5149   FW 231.7 $C_{10}H_{13}NO_3 \cdot HCl$

Tyrosine — *o*-Mono-2,4-Dnp-L-Tyr

Sigma D 2255   FW 347.3 $C_{15}H_{13}N_3O_7$

Tyrosine — *o*-Phospho-DL-Tyr

Synonyms: DL-3-(4-Hydroxyphenyl)-Ala 4'-Phosphate
Sigma P 5024   FW 261.2 $C_9H_{12}NO_6P$

Tyrosine — *o*-Phospho-D-Tyr

Synonyms: D-3-(4-Hydroxyphenyl)-Ala 4'-Phosphate
Sigma P 2532   FW 261.2 $C_9H_{12}NO_6P$

Tyrosine — *o*-Phospho-L-Tyr

Synonyms: L-Tyr-O-Phosphate; L-3-(4-Hydroxyphenyl)-Ala 4'-Phosphate
Fluka 79720   MW 261.17 $C_9H_{12}NO_6P$ ~95.0% (HPLC) | Cooper, JA et al, *Meth Enzymol*, 99: 387, 1983
Sigma P 9405   FW 261.2 $C_9H_{12}NO_6P$

Tyrosine — *o*-tBu-L-Tyr-OtBu Hydrochloride

Sigma B 2519   FW 329.9 $C_{17}H_{27}NO_3 \cdot HCl$

Tyrosine — Pth-Tyr

Sigma P 3002   FW 298.4 $C_{16}H_{14}N_2O_2S$ ~99% | Useful as standards in protein sequencing by the Edman degradation

Tyrosine — TentaGel S PHB-Tyr-tBu-FMOC

Synonyms: FMOC-O-tBu-L-Tyr 4-(Poly(Ethylenoxy))-OBzl Polymer Bound
Fluka 86405   0.22 mmol protected AA/g; ~90 μm particle size

Tyrosine — TentaGel S Trt-Tyr-tBu-FMOC

Fluka 86437   0.20 mmol protected AA/g; ~90 μm particle size

Tyrosine — Tyr

Bachem E-2515.0025   MW 181.19 $C_9H_{11}NO_3$ Store at RT

Tyrosine — Tyr (¹⁵N)

Bachem E-3245.0100   MW 182.19 $C_9H_{11}{}^{15}NO_3$ Store at -15°C

Tyrosine — Tyr Amide

Bachem E-2520.0001   MW 180.21 $C_9H_{12}N_2O_2$ Store at -15°C

Tyrosine — Tyr Amide Hydrochloride

Bachem E-2525.0001   MW 216.67 $C_9H_{12}N_2O_2 \cdot HCl$ Store at RT

Tyrosine — Tyr-AMC

Bachem I-1375.0050   MW 338.36 $C_{19}H_{18}N_2O_4$ Store at -15°C

Tyrosine — Tyr-AMC TFA
Bachem I-1665.0050   MW 452.39 $C_{19}H_{18}N_2O_4 \cdot C_2HF_3O_2$
Store at -15°C

Tyrosine — Tyr-Bzl
Bachem E-1595.0005   MW 271.32 $C_{16}H_{17}NO_3$ Store at RT

Tyrosine — Tyr-Bzl-OBzl p-Tosyl
Bachem E-1600.0005   MW 533.65 $C_{23}H_{23}NO_3 \cdot C_7H_8O_3S$
Store at RT

Tyrosine — Tyr-Bzl-OMe Hydrochloride
Bachem E-1605.0005   MW 321.8 $C_{17}H_{19}NO_3 \cdot HCl$ Store at RT

Tyrosine — Tyr-D-Tic Amide
Synonyms: Tyr-D-Tic Amide
Bachem G-4350.0050   Store at -15°C

Tyrosine — Tyr-L-Tic Amide
Synonyms: Tyr-Tic Amide
Bachem G-4345.0050   Store at -15°C

Tyrosine — Tyr-Me
Synonyms: 4-Methoxy-Phe
Bachem E-3260.0001   MW 195.22 $C_{10}H_{13}NO_3$ Store at RT

Tyrosine — Tyr-OBzl
Bachem E-2955.0005   MW 271.32 $C_{16}H_{17}NO_3$ Store at RT

Tyrosine — Tyr-OBzl p-Tosyl
Bachem E-2530.0005   MW 443.52 $C_{16}H_{17}NO_3 \cdot C_7H_8O_3S$
Store at RT

Tyrosine — Tyr-OEt Free Base
Bachem E-2900.0025   MW 209.25 $C_{11}H_{15}NO_3$ Store at -15°C

Tyrosine — Tyr-OEt Hydrochloride
Synonyms: L-Tyr-OEt
Bachem E-2550.0025   MW 245.71 $C_{11}H_{15}NO_3 \cdot HCl$ Store at RT

Peptides International EEY-2043-PI   MW 245.71   >98% (TLC); white powder

Tyrosine — Tyr-OMe
Bachem E-2750.0025   MW 195.22 $C_{10}H_{13}NO_3$ Store at -15°C

Tyrosine — Tyr-OMe Hydrochloride
Bachem E-2555.0025   MW 231.68 $C_{10}H_{13}NO_3 \cdot HCl$ Store at RT

Tyrosine — Tyr-OtBu
Bachem E-2535.0001   MW 237.3 $C_{13}H_{19}NO_3$ Store at -15°C

Tyrosine — Tyr-Sulfo Sodium Salt
Bachem E-3645.0001   MW 251.17 $C_9H_{10}NNaO_6$ Store at -15°C

Tyrosine — Tyr-tBu
Bachem E-1685.0001   MW 237.3 $C_{13}H_{19}NO_3$ Store at RT

Tyrosine — Tyr-tBu-OAll Hydrochloride
Bachem E-3335.0001   MW 313.82 $C_{16}H_{23}NO_3 \cdot HCl$ Store at -15°C

Tyrosine — Tyr-tBu-OMe Hydrochloride
Bachem E-1695.0001   MW 287.79 $C_{14}H_{21}NO_3 \cdot HCl$ Store at 2-8°C

Tyrosine — Tyr-tBu-OtBu Hydrochloride
Bachem E-1690.0001   MW 329.87 $C_{17}H_{27}NO_3 \cdot HCl$ Store at -15°C

Tyrosine — Tyr-βNA
Bachem K-1530.0250   MW 306.36 $C_{19}H_{18}N_2O_2$ Store at RT

Tyrosine — Z-D-Tyr
Bachem C-2735.0005   MW 315.33 $C_{17}H_{17}NO_5$ Store at RT
Fluka 97285   MW 315.33 $C_{17}H_{17}NO_5$ ≥98.0% (TLC)

Tyrosine — Z-D-Tyr-Bzl
Bachem C-1415.0001   MW 405.45 $C_{24}H_{23}NO_5$ Store at RT

Tyrosine — Z-D-Tyr-tBu Dicyclohexylamine
Bachem C-1500.0001   MW 552.76 $C_{21}H_{25}NO_5 \cdot C_{12}H_{23}N$ Store at RT

Tyrosine — Z-L-Tyr Dihydrate
Neosystem ZA01902   MW 351.4

Tyrosine — Z-L-Tyr Hydrate
Fluka 97290   MW 315.33 $C_{12}H_{15}NO_5$ ≥98.0%; mp: 57-60°C; ≤10% H_2O

Tyrosine — Z-L-Tyr-(4-ONp)
Fluka 97300   MW 436.42 $C_{23}H_{20}N_2O_7$ ≥99.0% (HPLC); mp: 156-157°C

Tyrosine — Z-L-Tyr-tBu Dicyclohexylamine
Neosystem ZA01904   MW 552.8

Tyrosine — Z-O-Bzl-L-Tyr
Fluka 96018   MW 405.45 $C_{24}H_{23}NO_5$ ≥98.0% (TLC)

Tyrosine — Z-Tyr
Synonyms: Benzyloxycarbonyl-L-Tyr
Bachem C-2730.0005   MW 315.33 $C_{17}H_{17}NO_5$ Store at RT
Peptides International ZLY-2026-PI   MW 315.33   >98% (HPLC); off-white crystalline powder

Tyrosine — Z-Tyr Amide
Bachem C-2740.0001   MW 314.34 $C_{17}H_{18}N_2O_4$ Store at RT

Tyrosine — Z-Tyr Hydrazide
Bachem C-2745.0001   MW 329.36 $C_{17}H_{19}N_3O_4$ Store at -15°C

Tyrosine — Z-Tyr-4MβNA
Bachem J-1155.0250   MW 470.5 $C_{28}H_{26}N_2O_5$ Store at -15°C

Tyrosine — Z-Tyr-Bzl
Bachem C-1410.0005   MW 405.45 $C_{24}H_{23}NO_5$ Store at RT

Tyrosine — Z-Tyrex
Bachem C-3215.0250   MW 399.49 $C_{23}H_{29}NO_5$ Store at RT

Tyrosine — Z-Tyr-OMe
Bachem C-4040.0001   MW 329.35 $C_{18}H_{19}NO_5$ Store at -15°C

Tyrosine — Z-Tyr-ONp
Bachem M-1315.0001   MW 436.42 $C_{23}H_{20}N_2O_7$ Store at -15°C

Peptides International SYN-3016   MW 436.42 $C_{23}H_{20}N_2O_7$ >96% (HPLC); amorphous powder

Tyrosine — Z-Tyr-tBu Dicyclohexylamine
Bachem C-1495.0001   MW 552.76 $C_{21}H_{25}NO_5 \cdot C_{12}H_{23}N$ Store at RT

Tyrosine — Z-Tyr-tBu-OMe
Bachem C-3065.0005   MW 385.46 $C_{22}H_{27}NO_5$ Store at -15°C

Tyrosine — Z-Tyr-tBu-OSu
Bachem C-1505.0001   MW 468.51 $C_{25}H_{28}N_2O_7$ Store at -15°C

Tyrosine — α-Me-DL-m-Tyr
Synonyms: AMMT; DL-2-Me-3-(3-Hydroxyphenyl)-Ala; α-MMT; L-Amino Acid Decarboxylase Inhibitor
Sigma M 8877   FW 195.2 $C_{10}H_{13}NO_3$

Tyrosine — α-Me-DL-m-Tyr-OMe Hydrate
Synonyms: DL-2-Me-3-(3-Hydroxyphenyl)-Ala-OMe
ICN 155686   MW 263.7 $C_{11}H_{15}NO_3 \cdot HCl \cdot H_2O$ Crystalline

Tyrosine — α-Me-DL-m-Tyr-OMe Hydrochloride Hydrate
Synonyms: DL-2-Me-3-(3-Hydroxyphenyl)-Ala-OMe
Sigma M 9006   FW 263.7 $C_{11}H_{15}NO_3 \cdot HCl \cdot H_2O$

Tyrosine — α-Me-DL-p-Tyr
Synonyms: AMPT; DL-2-Me-3-(4-Hydroxyphenyl)-Ala; Tyrosine Hydroxylase Inhibitor
Sigma M 7628   FW 195.2 $C_{10}H_{13}NO_3$

Tyrosine — α-Me-DL-p-Tyr-OMe Hydrochloride
Synonyms: DL-2-Me-3-(4-hydroxyphenyl)-Ala-OMe; DL-2-Me-3-(4-hydroxyphenyl)-Ala Methyl Ester; Tyrosine Hydroxylase Inhibitor
ICN 155687   MW 245.7 $C_{11}H_{15}NO_3 \cdot HCl$ 98%; crystalline
Sigma M 3281   FW 245.7 $C_{11}H_{15}NO_3 \cdot HCl$ ~98% (TLC)

Tyrosine — α-Me-DL-Tyr
Synonyms: Tyrosine Hydroxylase Inhibitor
Fluka 69577   MW 195.22 $C_{10}H_{13}NO_3$ ≥95.0% (titration); ≤0.05% residue on ignition; mp: >300°C | Moore, KE & Dominic, JA, Federation Proc, 30: 859, 1971

Tyrosine — α-Me-DL-Tyr-OMe Hydrochloride
Synonyms: DL-2-Me-4-Hyph-OMe; AMPT; Catecholamine Depletor; Tyrosine Hydroxylase Inhibitor
Fluka 69578   MW 245.7 $C_{11}H_{15}NO_3 \cdot HCl$ ≥99.0% (TLC); ≤2% H_2O | Mackay, AVP, Brit J Pharm, 51: 509, 1974; Weinberger, J et al, Stroke, 16: 864, 1985

Tyrosine — α-Me-L-p-Tyr
Synonyms: L-AMPT; L-2-Me-3-(4-Hydroxyphenyl)-Ala; Tyrosine Hydroxylase Inhibitor
Sigma M 8131   FW 195.2 $C_{10}H_{13}NO_3$ Active isomer

Tyrosine — α-Me-L-Tyr
Fluka 69574   MW 195.22 $C_{10}H_{13}NO_3$ ≥99.0% (titration)

Valine — (4,4,4,5,5,5-Hexafluoro)-DL-Val
Fluka 52518 $C_5H_5F_6NO_2$ ~97% (titration); mp: 210-212°C

Valine — 3,4-Dehydro-D-Val
Bachem F-4005.0001   MW 115.13 $C_5H_9NO_2$ Store at -15°C

Valine — 3,4-Dehydro-Val
Bachem F-4000.0001   MW 115.13 $C_5H_9NO_2$ Store at -15°C

Valine — 3-Fluoro-DL-Val
Fluka 47581   MW 135.14 $C_5H_{10}FNO_2$ ≥99% (titration); mp: 210°C

Valine — Ac-D-Val
Bachem F-1095.0001   MW 159.19 $C_7H_{13}NO_3$ Store at RT

Valine — Ac-Val
Bachem E-1265.0005   MW 159.12 $C_7H_{13}NO_3$ Store at RT

Valine — Ac-Val Amide
Bachem E-1270.0001   MW 158.2 $C_7H_{14}N_2O_2$ Store at RT

Valine — Ac-Val-NHMe
Bachem E-1275.0250   MW 172.23 $C_8H_{16}N_2O_2$ Store at RT

Valine — Ac-Val-OMe
Bachem E-1280.0001   MW 173.21 $C_8H_{15}NO_3$ Store at RT

Valine — BOC-3,4-dehydro-D-Val
Bachem A-4585.0001   MW 215.25 $C_{10}H_{17}NO_4$ Store at -15°C

Valine — BOC-3,4-dehydro-Val
Bachem A-4580.0001   MW 215.25 $C_{10}H_{17}NO_4$ Store at -15°C

Valine — BOC-DL-Val
Bachem A-2465.0005   MW 217.27 $C_{10}H_{19}NO_4$ Store at RT
Fluka 17096   MW 217.27 $C_{10}H_{19}NO_4$ ≥98% (TLC)

Valine — BOC-D-Val
American Peptide BDVAL10   MW 217.3
Bachem A-2460.0001   MW 217.27 $C_{10}H_{19}NO_4$ Store at RT
Fluka 15191   MW 217.27 $C_{10}H_{19}NO_4$ ≥98% (TLC)
Neosystem BA02001   MW 217.3
Peptides International BDV-2619   MW 217.27   >98% (HPLC); white crystalline powder
USBio B2432   MW 217.3 $C_{10}H_{19}NO_4$ ≥99%

Valine — BOC-D-Val-O-CH₂-φ-CH₂-COOH
Neosystem LP02001   MW 365.5

Valine — BOC-D-Val-ol

Senn Chem 44044   MW 203.3

Valine — BOC-D-Val-PAM Resin

Neosystem RP02001

Valine — BOC-D-Val-PAM-® (200-400 mesh)

Bachem D-1575.0001   Store at 2-8°C

Valine — BOC-L-Val

Fluka 15528   MW 217.27 $C_{10}H_{19}NO_4$ ≥99% (titration); ≤0.05% residue on ignition; mp: 77-80°C

Neosystem BA02002   MW 217.3

Valine — BOC-L-Val Hydroxysuccinimide Ester

Fluka 15531   MW 314.34 $C_{14}H_{22}N_2O_6$ ≥99%; ≤0.05% residue on ignition; mp: 126-128°C

Valine — BOC-L-Val N-Carboxy Anhydride

Fluka 15529   MW 243.26 $C_{11}H_{17}NO_5$ mp: 120°C | Produced by Propeptide under an exclusive license from Bioresearch

Valine — BOC-L-Val-MBHA Resin

Neosystem RN02002

Valine — BOC-L-Val-O-CH₂-♦CH₂-COOH Dicyclohexylamine

Neosystem LP02002   MW 546.8

Valine — BOC-L-Val-PAM Resin

Neosystem RP02002

Valine — BOC-Me-Val

USBio B2366   MW 231.3 $C_{11}H_{21}NO_4$ ≥99%

Valine — BOC-Me-Val-Dicyclohexylamine

USBio B2367   MW 412.6 $C_{11}H_{21}NO_4 \cdot C_{12}H_{23}N$ ≥99%

Valine — BOC-N-Me-DL-Val

Bachem A-3500.0001   MW 231.29 $C_{11}H_{21}NO_4$ Store at RT

Valine — BOC-N-Me-D-Val

Bachem A-3980.0001   MW 231.29 $C_{11}H_{21}NO_4$ Store at RT

Valine — BOC-N-Me-L-Val

Fluka 15538   MW 231.29 $C_{11}H_{21}NO_4$ ≥99% (TLC)
Neosystem BA02003   MW 231.3

Valine — BOC-N-Me-L-Val Dicyclohexylamine

Fluka 15448   MW 412.61 $C_{11}H_{21}NO_4 \cdot C_{12}H_{23}N$ ~99%; mp: 110-113°C

Valine — BOC-N-Me-Val

Synonyms: BOC-N-Me-L-Val

Bachem A-2100.0001   MW 231.29 $C_{11}H_{21}NO_4$ Store at RT
Peptides International BMV-5312-PI   MW 231.30   >98% (HPLC); white crystalline powder

Valine — BOC-Thiono-Val-1-(6-Nitro)-Benzotriazolide

Synonyms: (S)-2-(BOC-Amino)-3-Methylbutanethioic-O-Acid-1-(6-Nitro)-Benzotriazolide

Bachem A-4350.0001   MW 379.44 $C_{16}H_{21}N_5O_4S$ Store at -15°C

Valine — BOC-Val

Synonyms: BOC-L-Val

American Peptide BLVAL10   MW 217.3
Bachem A-2455.0005   MW 217.27 $C_{10}H_{19}NO_4$ Store at RT
Peptides International BLV-2105   MW 217.27   >98% (HPLC); white crystalline powder
USBio B2433   MW 217.3 $C_{10}H_{19}NO_4$ ≥99%

Valine — BOC-Val (¹⁵N)

Bachem A-3810.0100   MW 218.27 $C_{10}H_{19}{}^{15}NO_4$ Store at -15°C

Valine — BOC-Val-ol

Senn Chem 44043   MW 203.3

Valine — BOC-Val-ONp

Bachem A-2475.0001   MW 338.36 $C_{16}H_{22}N_2O_6$ Store at -15°C

Valine — BOC-Val-O-Resin

American Peptide RMRB300   1% DVB cross-linked: 100-200 mesh | t-BOC protected AA resin; preattached resins are useful for synthesis of peptides with C-terminal carboxyl group by t-BOC chemistry

Valine — BOC-Val-OSu

Bachem A-2470.0001   MW 314.34 $C_{14}H_{22}N_2O_6$ Store at -15°C

Valine — BOC-Val-PAM-® (200-400 mesh)

Bachem D-1550.0001   Store at 2-8°C

Valine — BOC-Val-PAM-Resin

American Peptide RPAM300   1% DVB cross-linked: 100-200 mesh | t-BOC protected AA resin; preattached resins are useful for synthesis of peptides with C-terminal carboxyl group by t-BOC chemistry

Valine — BOC-α-Me-DL-Val

Bachem A-4145.0001   MW 231.29 $C_{11}H_{21}NO_4$ Store at RT

Valine — Bsmoc-Val

Synonyms: Bsmoc-L-Val

Peptides International BLV-5642-PI   MW 339.37   >98% (HPLC); white to off-white powder | LA Carpino, et al, JACS, 119:9915, 1997

Valine — Bz-Val

Bachem E-2540.0005   MW 221.26 $C_{12}H_{15}NO_3$ Store at RT

Valine — CBZ-D-Val

USBio C2099-82   MW 251.3 $C_{13}H_{17}NO_4$ ≥99%

Valine — CBZ-Me-Val

USBio C2099-26   MW 265.3 $C_{14}H_{19}NO_4$ ≥99%

Valine — CBZ-Nva

USBio C2099-29   MW 251.3 $C_{13}H_{17}NO_4$ ≥99%

Valine — CBZ-Val

USBio C2099-81	MW 251.3	$C_{13}H_{17}NO_4$	≥99%

Valine — CBZ-Val-ONp

USBio C2099-83	MW 372.4	$C_{19}H_{20}N_2O_6$	≥99%

Valine — Dansyl-DL-Val Cyclohexylammonium Salt

ICN 100145	MW 449.6	$C_{17}H_{22}N_2O_4S \cdot C_6H_{13}N$
Sigma D 1131	FW. 449.6	$C_{17}H_{22}N_2O_4S \cdot C_6H_{13}N$

Valine — Dansyl-L-Val Cyclohexylammonium Salt

Sigma D 2500	FW. 449.6	$C_{17}H_{22}N_2O_4S \cdot C_6H_{13}N$

Valine — Dinitropyridyl-L-Val Monocyclohexylammonium Salt

ICN 103240	MW 383.4	$C_{10}H_{12}N_4O_6 \cdot C_6H_{13}N$

Valine — DL-Val

Synonyms: (±)-α-Aminoisovaleric Acid; DL-2-Amino-3-Methylbutanoic Acid; DL-α-Aminoisovaleric Acid; 2-Amino-3-Methylbutanoic Acid

Bachem F-3025.0025	MW 117.15	$C_5H_{11}NO_2$	Store at RT
Fluka 94640	MW 117.15	$C_5H_{11}NO_2$	≥99.0% (titration); mp: 295°C
ICN 103229	MW 117.1	$C_5H_{11}NO_2$	99%; crystalline
ICN 194768	MW 117.1	$C_5H_{11}NO_2$	99%; crystalline | Cell culture reagent
Rexim	MW 117.1	$C_5H_{11}NO_2$	White crystals or crystalline powder
Sigma V 0375	FW 117.1	$C_5H_{11}NO_2$	≥98% (TLC) | Also available as part of a kit
Sigma V 6379	FW 117.1	$C_5H_{11}NO_2$	Crystalline | Cell culture tested; insect cell culture tested
USBio V1020-15	MW 117.1	$C_5H_{11}NO_2$	≥99%

Valine — DL-Val Hydroxamate

Sigma V 3251	FW 132.2	$C_5H_{12}N_2O_2$

Valine — DL-Val-OMe Hydrochloride

Sigma V 0875	FW 167.6	$C_6H_{13}NO_2 \cdot HCl$

Valine — Dnp-L-Val

ICN 103235	MW 283.2	$C_{11}H_{13}N_3O_6$	Crystalline

Valine — D-Val

Synonyms: (R)-α-Aminoisovaleric Acid; D-2-Amino-3-Methylbutanoic Acid

Bachem F-2180.0005	MW 117.15	$C_5H_{11}NO_2$	Store at RT
Fluka 94630	MW 117.15	$C_5H_{11}NO_2$	~99.0% (titration)
ICN 103226	MW 117.1	$C_5H_{11}NO_2$	99%; crystalline
ICN 194767	MW 117.1	$C_5H_{11}NO_2$	99%; crystalline | Cell culture reagent
Neosystem AA02001	MW 117.2		
Peptides International ADV-2811	MW 117.15	>99.9% optical purity by RP-HPLC; white crystalline powder	
Sigma V 0250	FW 117.1	$C_5H_{11}NO_2$	≥98% (TLC) | Also available as part of a kit
Sigma V 1255	FW 117.1	$C_5H_{11}NO_2$	Crystalline | Cell culture tested
USBio V1020-10	MW 117.1	$C_5H_{11}NO_2$	≥99%

Valine — D-Val (1-^{14}C)

ARC ARC-434	MW 117.2	$(CH_3)_2CHCH(NH_2)COOH$	40-60 mCi/mmol; 1.48-2.22 GBq/mmol; in 0.01 N HCl | Radiochemical

Valine — D-Val-OBzl p-Tosyl

Bachem F-3500.0001	MW 379.48	$C_{12}H_{17}NO_2 \cdot C_7H_8O_3S$
Store at RT		

Valine — D-Val-ol

Senn Chem 44110	MW 103.2	Liquid | Skin irritant; light sensitive

Valine — D-Val-OMe Hydrochloride

Bachem F-3160.0001	MW 167.64	$C_6H_{13}NO_2 \cdot HCl$	Store at 2-8°C
Fluka 94665	MW 167.64	$C_6H_{13}NO_2 \cdot HCl$	≥99.0% (titration); mp: 170°C

Valine — D-Val-OtBu Hydrochloride

Bachem F-3170.0001	MW 209.72	$C_9H_{19}NO_2 \cdot HCl$	Store at -15°C

Valine — Farnesyl-Val-OBzl

Bachem F-3515.0250	MW 411.63	$C_{27}H_{41}NO_2$	Store at 2-8°C

Valine — FMOC-3,4-Dehydro-D-Val

Bachem B-3510.0001	MW 337.38	$C_{20}H_{19}NO_4$	Store at -15°C

Valine — FMOC-3,4-dehydro-Val

Bachem B-3505.0001	MW 337.38	$C_{20}H_{19}NO_4$	Store at -15°C

Valine — FMOC-D-Me-Val

USBio F5429	MW 353.4	$C_{21}H_{23}NO_4$	≥99%

Valine — FMOC-D-Val

Bachem B-1460.0001	MW 339.39	$C_{20}H_{21}NO_4$	Store at RT
Fluka 47481	MW 339.39	$C_{20}H_{21}NO_4$	≥98% (HPLC)
Neosystem FA02001	MW 339.4		
Peptides International FDV-1834-PI	MW 339.39	>98% (HPLC); white powder	
USBio F5488	MW 339.4	$C_{20}H_{21}NO_4$	≥99%

Valine — FMOC-D-Val 4-Benzyloxybenzyl Ester Polymer Bound

Fluka 47407	0.4-0.8 mmol/g resin; carrier: polystyrene, crosslinked with 1% DVB; 100-200 mesh particle size

Valine — FMOC-L-Val

Fluka 47638	MW 339.39	$C_{20}H_{21}NO_4$	≥98% (titration); mp: 143-145°C
Neosystem FA02002	MW 339.4		

Valine — FMOC-L-Val 4-Benzyloxybenzyl Ester Polymer Bound

Fluka 47669	0.4-0.6 mmol/g resin; carrier: polystyrene, crosslinked with 1% DVB; 200-400 mesh particle size

Valine — FMOC-L-Val N-Carboxy Anhydride

Fluka 47699	MW 365.39	$C_{21}H_{19}NO_5$	≥97%; mp: 88-91°C | Produced by Propeptide under an exclusive license from Bioresearch

Valine — FMOC-L-Val-OPfp

Fluka 47507	MW 505.44	$C_{26}H_{20}F_5NO_4$	≥98% (HPLC)

Valine — FMOC-L-Val-Wang Resin

American Peptide RFWAN93   Preattached Wang resins are usful for synthesis of peptides with C-terminal carboxyl groups; peptides can be cleaved from the resin with 90% TFA in the presence of scavengers

Neosystem RW02002

Valine — FMOC-Me-Val

USBio F5430   MW 353.4 $C_{21}H_{23}NO_4$ ≥99%

Valine — FMOC-N-Me-D-Val

Fluka 47347   MW 353.42 $C_{21}H_{23}NO_4$ ≥98% (TLC)

Peptides International FMV-1787-PI   MW 353.42   >98% (HPLC); white crystalline powder

Valine — FMOC-N-Me-L-Val

Fluka 47599   MW 353.42 $C_{21}H_{23}NO_4$ ≥98% (HPLC); mp: 187-190°C

Valine — FMOC-N-Me-Val

Synonyms: FMOC-N-Me-L-Val

Bachem B-1380.0001   MW 353.42 $C_{21}H_{23}NO_4$ Store at RT

Peptides International FMV-1796-PI   MW 353.42   >97% (HPLC); white powder

Valine — FMOC-Val

Synonyms: FMOC-L-Val

American Peptide FVAL100   MW 339.4

Bachem B-1455.0005   MW 339.39 $C_{20}H_{21}NO_4$ Store at RT

Peptides International FLV-1739-PI   MW 339.39   >98% (HPLC); white crystalline powder

USBio F5489   MW 339.4 $C_{20}H_{21}NO_4$ ≥99%

Valine — FMOC-Val (^{15}N)

Bachem B-2670.0100   MW 340.38 $C_{20}H_{21}{}^{15}NO_4$ Store at -15°C

Valine — FMOC-Val-® (200-400 mesh)

Bachem D-1155.0001   Store at RT

Valine — FMOC-Val-ol

Synonyms: FMOC-L-Valinol

Senn Chem 44066   MW 325.3

Valine — FMOC-Val-OPfp

Bachem B-1565.0001   MW 505.44 $C_{26}H_{20}F_5NO_4$ Store at -15°C

USBio F5490   MW 505.4 $C_{26}H_{20}F_5NO_5$

Valine — FMOC-Val-OSu

Bachem B-1465.0001   MW 436.46 $C_{24}H_{24}N_2O_6$ Store at -15°C

Valine — FMOC-Val-SASRIN™-® (200-400 mesh)

Bachem D-1405.0001   Store at 2-8°C

Valine — For-Val

Bachem E-1855.0005   MW 145.16 $C_6H_{11}NO_3$ Store at RT

Valine — L-Val

Synonyms: (S)-α-Aminoisovaleric Acid; L-2-Amino-3-Methylbutanoic Acid; (S)-α-Amino-β-Methylbutyric Acid

Amersham US23205   98.5-101.5% dsb; ≤0.0015% heavy metals (Pb)

Fluka 94619   MW 117.15 $C_5H_{11}NO_2$ ≥99.5% (titration); ≤0.05% residue on ignition, loss on drying; ≤0.3% foreign AA; ≤0.00001% As; ≤0.005% Cl, Ca, K, Na, SO₄; ≤0.01% NH₄⁺; ≤0.0005% Al, Ba, Bi, Cd, Co, Cr, Cu, Fe, Li, Mg, Mn, Mo, Ni, Pb, Sr, Zn | Protects enzymes against heat inactivation; Paleg, LG et al, *Aust J Plant Physiol*, 8: 107, 1981; Yagi, T et al, *Electrophoresis '83*, ed H Hirai, 503, 1984, de Gruyter, Berlin; Hine, T, *ibid*, 541; *CRC Handbook of Microbiology*, Laskin, AI & Lachevalier, HA, eds, 4: 1, 1974, CRC Press, Cleveland, Ohio

Fluka 94620   MW 117.15 $C_5H_{11}NO_2$ ≥99.0% (titration); ≤0.1% residue on ignition, loss on drying; ≤0.3% foreign AA; ≤0.005% Cl, Ca, K, Na, SO₄; ≤0.01% NH₄⁺; ≤0.0005% Cd, Co, Cr, Cu, Fe, Mg, Mn, Ni, Pb, Zn

ICN 104760   MW 117.1 $C_5H_{11}NO_2$ 99%; crystalline

ICN 194769   MW 117.1 $C_5H_{11}NO_2$ 99%; crystalline | Cell culture reagent

Neosystem AA02002   MW 117.2

Peptides International ALV-2723   MW 117.15   >99.9% optical purity by RP-HPLC; white crystalline powder

Rexim   MW 117.1 $C_5H_{11}NO_2$ White crystals or powder

Sigma V 0258   FW 117.1 $C_5H_{11}NO_2$ ≥98% (TLC); SigmaUltra; residue on ignition: <0.1%; solubility (1 *M* in 1.0 *M* HCl, 20°C): complete, colorless; insoluble matter: <0.1%; SO₄, Cl, NH₄⁺,: <0.05%; K, Na: <0.005%; Al, Cu, Ca, Fe, Mg, Zn, P: <0.0005%; Pb: <0.001%; A_{260}<0.1; A_{280}<0.1 (1 *M* in 1.0 *M* HCl)

Sigma V 0500   FW 117.1 $C_5H_{11}NO_2$ ≥98% (TLC) | Also available as part of a kit

Sigma V 6504   FW 117.1 $C_5H_{11}NO_2$ Crystalline | Cell culture tested; insect cell culture tested

Sigma V 7888   FW 117.2 $C_5H_{11}NO_2$ USP Grade

USBio V1020   MW 117.2 $C_5H_{11}NO_2$ ≥98.5%; appearance: white crystalline powder; pH: 5.5-7.0; specific rotation (C=8, 6 *N* HCl): +27.6° to +29.0°; heavy metals (Pb): ≤0.001%; endotoxin: ≤EU/mg; cell culture tested: murine myeloma SP2/0-Ag14

Valine — L-Val (1-^{14}C)

ARC ARC-277   MW 117.2 $(CH_3)_2CHCH(NH_2)COOH$ 40-60 mCi/mmol; 1.48-2.22 GBq/mmol; in 0.01 *N* HCl | Radiochemical

Valine — L-Val (2,3,4-^3H)

ARC ART-466   MW 117.2 $(CH_3)_2CHCH(NH_2)COOH$ 60-90 Ci/mmol; 2.22-3.33 TBq/mmol; in 0.01 *N* HCl | Radiochemical

Valine — L-Val (2,3-^3H)

ARC ART-245   MW 117.2 $(CH_3)_2CHCH(NH_2)COOH$ 30-60 Ci/mmol; 1.11-2.22 TBq/mmol; in 0.01 *N* HCl | Radiochemical

ICN 20057E   MW 117.2 $(CH_3)_2CHCH(NH_2)COOH$ 10-25 Ci/mmol, 370-925 GBq/mmol; sterile EtOH:H₂O (2:98) | Beckett, PR etal, *Br J Nutr*, 68:139, 1992

Valine — L-Val (3,4(n)-^3H)

Amersham TRK533   Aqueous, 2% EtOH, sterilized; 0.93-1.85 TBq/mmol, 25-50 Ci/mmol; 37 MBq/mL, 1 mCi/mL

Valine — L-Val (3,4-^3H)

ARC ART-230   MW 117.2 $(CH_3)_2CHCH(NH_2)COOH$ 45-65 Ci/mmol; 1.67-2.4 TBq/mmol; in 0.01 *N* HCl | Radiochemical

Sigma V 1383 $(CH_3)_2CHCH(NH_2)COOH$ 25-60 Ci/mmol; 0.01 *N* HCl solution | Radiochemical

Valine — L-Val (U-^{14}C)

Amersham CFB75   Aqueous, 2% EtOH, sterilized; >9.25 GBq/mmol; >250 mCi/mmol; 1.85 MBq/mL, 50 µCi/mL

ARC ARC-678   MW 117.2 $(CH_3)_2CHCH(NH_2)COOH$ 150-250 mCi/mmol; 5.5-9.2 GBq/mmol; in EtOH: H₂O (2:98) | Radiochemical

ICN 10139E   MW 117.2 $(CH_3)_2CHCH(NH_2)COOH$ 225-275 mCi/mmol, 8.3-10.2 GBq/mmol; sterile EtOH:H₂O (2:98)

ICN 10139L MW 117.2 (CH$_3$)$_2$CHCH(NH$_2$)COOH 170-225 mCi/mmol, 6.3-8.3 GBq/mmol; sterile EtOH:H$_2$O (2:98); >90%

Valine — L-Val (U-^{14}C) Hydrochloride

Sigma V 8004 FW 117.1 (CH$_3$)$_2$CHCH(NH$_2$)COOH 200-300 mCi/mmol; radiochemical purity ≥98% (HPLC); aqueous solution containing 2% EtOH in Combi-vial | Radiochemical

Valine — L-Val Amide Hydrobromide

ICN 103228 MW 197 C$_5$H$_{11}$N$_2$O · HBr Crystalline

Valine — L-Val Amide Hydrochloride

ICN 152185 MW 152.6 C$_5$H$_{11}$N$_2$O · HCl Crystalline

Neosystem AA02003 MW 152.6

Valine — L-Val p-Nitrobenzyl Ester Hydrobromide

ICN 158275 MW 333.2 C$_{12}$H$_{16}$N$_2$O$_4$ · HBr Crystalline

Valine — L-Val β-Naphthylamide

ICN 158274 MW 242.3 C$_{15}$H$_{18}$N$_2$O Crystalline

Valine — L-Val β-Naphthylamide Hydrochloride

Sigma V 6001 FW 242.3 C$_{15}$H$_{18}$N$_2$O

Valine — L-Val-AMC TFA Salt

Sigma V 3262 FW 389.4 C$_{15}$H$_{19}$N$_3$O$_3$ · C$_2$HF$_3$O$_2$

Valine — L-Valinamide Hydrochloride

Sigma V 0625 FW 152.6 C$_5$H$_{12}$N$_2$O · HCl

Valine — L-Val-OAll-Tosyl

Fluka 94645 MW 329.42 C$_8$H$_{15}$NO$_2$ · C$_7$H$_8$O$_3$S ≥99.0% (titration); mp: 117-120°C | Used in peptide & N-glycopeptide synthesis; Waldmann, H & Kunz, H, *Liebigs Ann Chem*, 1712, 1983

Valine — L-Val-OBzl Hydrochloride

ICN 158272 MW 243.7 C$_{12}$H$_{17}$NO$_2$ · HCl Crystalline

Sigma V 0750 FW 243.7 C$_{12}$H$_{17}$NO$_2$ · HCl

Valine — L-Val-OBzl p-Tosyl

Sigma V 2627 FW 379.5 C$_{12}$H$_{17}$NO$_2$ · C$_7$H$_8$O$_3$S

Valine — L-Val-OBzl-p-Tosyl

ICN 103242 MW 379.5 C$_{12}$H$_{17}$NO$_2$ · C$_7$H$_8$O$_3$S Crystalline

Valine — L-Val-OBzl-Tosyl

Fluka 94651 MW 379.48 C$_{12}$H$_{17}$NO$_2$ · C$_7$H$_8$O$_3$S ~99.0%; mp: 160-162°C

Valine — L-Val-OEt Hydrochloride

ICN 152186 MW 181.7 C$_7$H$_{15}$NO$_2$ · HCl Crystalline

Valine — L-Val-OMe Hydrochloride

Fluka 94670 MW 167.64 C$_6$H$_{13}$NO$_2$ · HCl ≥99.0% (titration); mp: 165-170°C

ICN 103234 ICN 103245 MW 167.6 C$_6$H$_{13}$NO$_2$ · HCl Crystalline

Sigma V 1000 FW 167.6 C$_6$H$_{13}$NO$_2$ · HCl

Valine — L-Val-OtBu Hydrochloride

Fluka 94660 MW 209.72 C$_9$H$_{19}$NO$_2$ · HCl ≥99.0% (titration)

ICN 158273 MW 209.7 C$_9$H$_{19}$NO$_2$ · HCl

Sigma V 1125 FW 209.7 C$_9$H$_{19}$NO$_2$ · HCl

Valine — L-Val-pNA Hydrochloride

Sigma V 6005 FW 273.7 C$_{11}$H$_{15}$N$_3$O$_3$ · HCl

Valine — N,N'-Carbonyl Bis-L-Val Diisopropyl Ester

Synonyms: L-Val Ureide Isopropyl Ester

ICN 154936 MW 344.5 C$_{17}$H$_{32}$N$_2$O$_5$ Chiral stationary phase for gas chromatographic resolution of optical isomers; Suzuki, S etal, *Bunseki Kogaku*, 30:479, 1981; Lochmueller, CH & JV Hinshaw Jr, *J Chromatogr*, 178:411, 1979

Valine — N-2,4-Dnp-L-Val

Sigma D 2380 FW 283.2 C$_{11}$H$_{13}$N$_3$O$_6$

Valine — N-Ac-3-Mercapto-DL-Val

Synonyms: N-Ac-DL-Pen

Fluka 01425 MW 191.25 (CH$_3$)$_2$C(SH)CH(NHCOCH$_3$)CO$_2$H ≥98% (titration); mp: 186-189°C

ICN 104825 MW 191.2 C$_7$H$_{13}$NO$_3$S Crystalline

Valine — N-Ac-3-Mercapto-D-Val

Synonyms: N-Ac-D-Pen

Fluka 01423 MW 191.25 C$_7$H$_{13}$NO$_3$S ≥99% (titration); mp: 185-190°C | Chiral reagent for the precolumn derivatization of AA or amino alcohols; the diastereoisomers formed can be efficiently resolved by HPLC on conventional reversed-phased columns; Buck, RH & Krummen, K; *J Chromatography*, 387: 255, 1987

ICN 154707 MW 191.2 C$_7$H$_{13}$NO$_3$S Chiral reagent for precolumn derivatization of enantiomeric AA or amino alcohols; Buck, RH & KJ Krummen, *Chromatogr*, 387:255, 1987

Valine — N-Ac-DL-Val

Fluka 01550 MW 159.19 C$_7$H$_{13}$NO$_3$ ~99% (titration); mp: 148°C

ICN 100162 MW 159.2 C$_7$H$_{13}$NO$_3$ Crystalline

Sigma A 7001 FW 159.2 C$_7$H$_{13}$NO$_3$

Valine — N-Ac-D-Val

Sigma A 6876 FW 159.2 C$_7$H$_{13}$NO$_3$ Sigma Grade

Valine — N-Ac-L-Val

ICN 150249 MW 159.2 C$_7$H$_{13}$NO$_3$ ≥99%, crystalline

Sigma A 6894 FW 159.2 C$_7$H$_{13}$NO$_3$

Valine — N-Bz-DL-Val

ICN 100151 MW 221.3 C$_{12}$H$_{15}$NO$_3$

Sigma B 6500 FW 221.3 C$_{12}$H$_{15}$NO$_3$

Valine — N-Carbamyl-DL-Val

Sigma C 6250 FW 160.2 C$_6$H$_{12}$N$_2$O$_3$

Valine — N-CBZ-DL-Val

ICN 100205 MW 251.3 C$_{13}$H$_{17}$NO$_4$ Crystalline

Valine — N-CBZ-L-Val

ICN 100207 MW 251.3 C$_{13}$H$_{17}$NO$_4$ Crystalline

Sigma C 6252 FW 251.3 C$_{13}$H$_{17}$NO$_4$

Valine — N-Chloroacetyl-DL-Val

ICN 101344 MW 193.6 C$_7$H$_{12}$ClNO$_3$ Crystalline

Sigma C 2128 FW 193.6 C$_7$H$_{12}$ClNO$_3$

Valine — *N*-Chloroacetyl-L-Val

ICN 101345	MW 193.6	$C_7H_{12}ClNO_3$	Crystalline
Sigma C 2253	FW 193.6	$C_7H_{12}ClNO_3$	Sigma Grade

Valine — *N*-FMOC-D-Val

ICN 158178	MW 339.4	$C_{20}H_{21}NO_4$

Valine — *N*-FMOC-D-Val (1-¹⁴C)

ARC ARC-1095   MW 339.4   50-60 mCi/mmol; 1.85-2.22 GBq/mmol; in EtOH | Radiochemical

Valine — *N*-FMOC-L-Val

Sigma F 4508	FW 339.4	$C_{20}H_{21}NO_4$

Valine — *N*-FMOC-L-Val (1-¹⁴C)

ARC ARC-661   MW 339.4   50-60 mCi/mmol; 1.85-2.22 GBq/mmol; in EtOH | Radiochemical

Valine — *N*-FMOC-L-Val (3,4-³H)

ARC ART-527   MW 337.4   30-60 Ci/mmol; 1.11-2.22 TBq/mmol; in EtOH | Radiochemical

Valine — *N*-FMOC-L-Val (U-¹⁴C)

Sigma F 0422   40-60 mCi/mmol; Purity ≥98% (HPLC); EtOH solution in Combi-vial | Radiochemical

Valine — *N*-FMOC-L-Val Resin Ester

Sigma F 3011   For peptide synthesis; Wang, SS, *J Am Chem Soc*, 95: 1328, 1973; Lu, G et al, *J Org Chem*, 46: 3433, 1981

Valine — *N*-For-DL-Val

ICN 158190	MW 145.2	$C_6H_{11}NO_3$	White powder

Valine — *N*-Me-DL-Val

Bachem F-1790.0001	MW 131.18	$C_6H_{13}NO_2$	Store at RT
ICN 155703	MW 131.2	$C_6H_{13}NO_2$	Crystalline
Sigma M 9377	FW 131.2	$C_6H_{13}NO_2$	

Valine — *N*-Me-Val

Bachem E-2195.0001	MW 131.18	$C_6H_{13}NO_2$	Store at RT

Valine — *N*-Me-Val-OBzl *p*-Tosyl

Bachem E-2200.0001   MW 393.5 $C_{13}H_{19}NO_2 \cdot C_7H_8O_3S$ Store at RT

Valine — *N*-Me-Val-OMe Hydrochloride

Bachem E-1830.0001   MW 181.66 $C_7H_{15}NO_2 \cdot HCl$ Store at RT

Valine — *N*-O-Nps-L-Val Dicyclohexylammonium Salt

ICN 102539	MW 451.6	$C_{11}H_{14}N_2O_4S \cdot C_{12}H_{23}N$	Yellow crystals
Sigma N 6878	FW 451.6	$C_{11}H_{14}N_2O_4S \cdot C_{12}H_{23}N$	Yellow crystals

Valine — Nps-Val DCHA

Bachem F-2190.0001   MW 451.63 $C_{11}H_{14}N_2O_4S \cdot C_{12}H_{23}N$ Store at 2-8°C

Valine — *N*-t-BOC-DL-Val

Sigma B 8518	FW 217.3	$C_{10}H_{19}NO_4$

Valine — *N*-t-BOC-D-Val

Sigma B 8768	FW 217.3	$C_{10}H_{19}NO_4$

Valine — *N*-t-BOC-L-Val (2,3,4-³H)

ARC ART-671   MW 217.3   60-90 Ci/mmol; 2.22-3.33 TBq/mmol; in EtOH | Radiochemical

Valine — *N*-t-BOC-L-Val Free Acid

Sigma B 4151	FW 217.3	$C_{10}H_{19}NO_4$	Similar to B 9501 but prepared by Sigma
Sigma B 9501	FW 217.3	$C_{10}H_{19}NO_4$	

Valine — *N*-t-BOC-L-Val Resin Ester

Sigma B 6271   *N*-t-BOC AA ester of methylated 1% divinylbenzene cross-linked polystyrene; 200-400 mesh; substitution range: 0.2-0.6 mmole t-BOC-AA/g resin | For peptide synthesis by the Merrifield method

Valine — *N*-TFA-L-Val-OMe

Sigma T 6256	FW 227.2	$C_8H_{12}F_3NO_3$	~98%

Valine — *N*ᵅ-FMOC-L-Val

ICN 151153	MW 339.4	$C_{20}H_{21}NO_4$

Valine — *N*ᵅ-t-BOC-L-Val

ICN 150508	MW 217.3	$C_{10}H_{19}NO_4$	Crystalline

Valine — *N*ᵅ-t-BOC-L-Val Dicyclohexylammonium Salt

ICN 101055	MW 398.3	$C_{10}H_{19}NO_4 \cdot C_{12}H_{23}N$

Valine — *p*-Nitro-CBZ-L-Val *N*-Hydroxysuccinimide Ester

Sigma N 5530	FW 393.4	$C_{17}H_{19}N_3O_8$

Valine — Pth-Val

Sigma P 3127   FW 234.3 $C_{12}H_{14}N_2OS$ ~99% | Useful as standards in protein sequencing by the Edman degradation

Valine — TentaGel S PHB-Val-FMOC

Synonyms: FMOC-L-Val 4-(Poly(Ethylenoxy))-OBzl Polymer Bound

Fluka 86406   0.22 mmol protected AA/g; ~90 μm particle size

Valine — TentaGel S Trt-Val-FMOC

Fluka 86438   0.20 mmol protected AA/g; ~90 μm particle size

Valine — Val

Bachem E-2560.0025	MW 117.15	$C_5H_{11}NO_2$	Store at RT

Valine — Val (¹⁵N)

Bachem E-3250.0100   MW 118.15 $C_5H_{11}{}^{15}NO_2$ Store at -15°C

Valine — Val Amide Hydrobromide

Bachem E-2565.0005   MW 197.08 $C_5H_{12}N_2O \cdot HBr$ Store at 2-8°C

Valine — Val Amide Hydrochloride

Bachem E-2570.0005   MW 152.62 $C_5H_{12}N_2O \cdot HCl$ Store at RT

Valine — Val-(*p*ONb) Hydrobromide

Bachem E-2600.0001   MW 333.18 $C_{12}H_{16}N_2O_4 \cdot HBr$ Store at -15°C

Valine — Val-4MβNA Hydrochloride

Bachem J-1315.0250   MW 308.8 $C_{16}H_{20}N_2O_2 \cdot HCl$ Store at -15°C

Valine — Val-AMC

Bachem I-1380.0050   MW 274.32 $C_{15}H_{18}N_2O_3$ Store at -15°C

Valine — Val-AMC TFA

Bachem I-1385.0050   MW 388.34 $C_{15}H_{18}N_2O_3 \cdot C_2HF_3O_2$ Store at -15°C

Valine — Val-NHtBu Hydrochloride

Bachem E-2585.0001   MW 208.73 $C_9H_{20}N_2O \cdot HCl$ Store at -15°C

Valine — Val-OAll p-Tosyl

Bachem E-3510.0001   MW 329.42 $C_8H_{15}NO_2 \cdot C_7H_8O_3S$ Store at 2-8°C

Valine — Val-OBzl Hydrochloride

Bachem E-2575.0005   MW 243.73 $C_{12}H_{17}NO_2 \cdot HCl$ Store at RT

Valine — Val-OBzl p-Tosyl

Bachem E-2580.0005   MW 379.48 $C_{12}H_{17}NO_2 \cdot C_7H_8O_3S$ Store at RT

Valine — Val-OEt Hydrochloride

Bachem E-1825.0005   MW 181.66 $C_7H_{15}NO_2 \cdot HCl$ Store at RT

Valine — Val-ol

Synonyms: L-Valinol

Senn Chem 44109   MW 103.2   Inhibits protein synthesis; skin irritant; light sensitive

Valine — Val-OMe Hydrochloride

Synonyms: L-Val-OMe

Bachem E-2595.0005   MW 167.64 $C_6H_{13}NO_2 \cdot HCl$ Store at RT

Peptides International EMV-2044-PI   MW 167.63   >98% (TLC); white crystalline powder

Valine — Val-OtBu Hydrochloride

Bachem E-2590.0001   MW 209.72 $C_9H_{19}NO_2 \cdot HCl$ Store at 2-8°C

Valine — Val-pNA

Bachem L-1440.0250   MW 237.26 $C_{11}H_{15}N_3O_3$ Store at -15°C

Valine — Val-pNA Hydrochloride

Bachem L-1855.0250   MW 273.72 $C_{11}H_{15}N_3O_3 \cdot HCl$ Store at -15°C

Valine — Val-βNA

Bachem K-1535.0250   MW 242.32 $C_{15}H_{18}N_2O$ Store at RT

Valine — Z-D-Val

Bachem C-2810.0001   MW 251.28 $C_{13}H_{17}NO_4$ Store at RT

Valine — Z-D-Val-OSu

Bachem C-2825.0001   MW 348.36 $C_{17}H_{20}N_2O_6$ Store at -15°C

Valine — Z-L-Val

Fluka 97330   MW 251.28 $C_{13}H_{17}NO_4$ ~99% (titration); mp: 58-61°C

Neosystem ZA02002   MW 251.3

Valine — Z-L-Val-(4-ONp)

Fluka 97333   MW 372.38 $C_{19}H_{20}N_2O_6$ ≥98.0% (TLC)

Valine — Z-N-Me-L-Val

Fluka 00913   MW 265.31 $C_{14}H_{19}NO_4$ ≥98.0% (HPLC)

Valine — Z-N-Me-Val

Bachem C-3700.0001   MW 265.31 $C_{14}H_{19}NO_4$ Store at RT

Valine — Z-Val

Synonyms: Benzyloxycarbonyl-L-Val

Bachem C-2805.0005   MW 251.28 $C_{13}H_{17}NO_4$ Store at RT

Peptides International ZLV-2027-PI   MW 251.28   >98% (HPLC); white crystalline powder

Valine — Z-Val-ONp

Bachem C-2830.0005   MW 372.38 $C_{19}H_{20}N_2O_6$ Store at -15°C

Valine — Z-Val-OSu

Bachem C-2820.0005   MW 348.36 $C_{17}H_{20}N_2O_6$ Store at -15°C

Valine — α-Me-DL-Val

Bachem F-3355.0001   MW 131.18 $C_6H_{13}NO_2$ Store at RT

Valine — α-Me-D-Val

Bachem F-3540.0001   MW 131.18 $C_6H_{13}NO_2$ Store at RT

Valine — α-Me-Val

Bachem F-3535.0001   MW 131.18 $C_6H_{13}NO_2$ Store at RT

Part 6. Appendices

Appendix A. Reagent Suppliers

Alexis

USA: Alexis Corporation
San Diego, CA 92121-4727
Phone: 619-658-0065/800-900-0065
Fax: 619-658-9224/800-900-9224
E-mail: alexis-usa@alexis-corp.com
Web: www.alexi-corp.com

Switzerland: Alexis Corporation
CH-4448 Läufelfingen
Phone: 41-62-299-29-20
Fax: 41-62-299-24-80
E-mail: alexis-ch@alexis-corp.com

Germany: Alexis Deutschland GMBH
D-35305 Grünberg
Phone: 49-6401-90077/800-253-9472
Fax: 49-6401-90078
E-mail: alexis-d@alexis-corp.com

UK: Alexis Corporation (UK) Ltd.
Bingham, Nottingham NG13 8QG
Phone: 44-1949-83611
Fax: 44-1949-836222
E-mail: alexis-uk@alexis-corp.com

USA: Fisher Scientific
Pittsburgh, PA
Phone: 800-766-7000
Fax: 800-926-1166

American Peptide

USA: American Peptide Company, Inc.
Sunnyvale, CA 94086
Phone: 408-733-7604/800-926-8272
Fax: 408-733-7603/888-670-0070
Web: www.americanpeptide.com

Japan: Itoham Foods, Inc.
Kitasouma, Ibaraki 302-01
Phone: 0297-45-6311
Fax: 0297-45-6353
E-mail: itoham2@fureai.or.jp

Amersham

USA: Amersham Pharmacia Biotech
Piscataway, NJ 08855-1327
Phone: 800-526-3593
Fax: 877-295-8102
E-mail: apbcsus@am.apbiotech.com
Web: www.apbiotech.com
UK: Amersham Pharmacia Biotech UK Ltd.
Little Chalfont, Buckinghamshire HP7 9NA

Sweden: Amersham Pharmacia Biotech AB
SE-751 84 Uppsala

ARC

USA: American Radiolabeled Chemicals, Inc.
St. Louis, MO 63146
Phone: 314-991-4545/800-331-6661
Fax: 314-991-4692/800-999-9925
E-mail: arcinc@arc-inc.com
Web: www.arc-inc.com

Australia: Bio-Scientific Pty. Ltd.
Gymea, N.S.W. 2227
Phone: 61-2-9521-2177/800-251-437
Fax: 61-2-9542-3100
E-mail: TechServ@biosci.com.au
Web: www.biosci.com.au

Austria: Humos Diagnostika GmbH
A-5026 Salzburg
Phone: 43-662-620560
Fax: 43-662-620560

Belgium: Isobio
BE-6224 Fleurus
Phone: 32-71-81-41-45
Fax: 32-71-81-05-53
E-mail: isobio@skynet.be

Chile: Biocronogen Limitada
Santiago
Phone: 56-2-287-3802
Fax: 56-2-287-3802
E-mail: advanced@asianet.net.hk

Hong Kong: Advanced Technology & Industrial Co., Ltd.
Tai Kok Tsui
Phone: 852-23902293
Fax: 852-27898314
E-mail: sales@advtechind.com

Germany: Biotrend Chemikalien GmbH
D-50876 Köln
Phone: 49-221-9498320
Fax: 49-221-9498325
E-mail: jaeger@biotrend.com
Web: www.biotrend.com

Greece: Labo-Chem
GR-11527 Athens
Phone: 30-1-770-9474
Fax: 30-1-775-6090
E-mail: kakavoulis@ath.forthnet.gr

Israel: Ornat Biochemicals & Laboratory Equipment Ltd.
Rehovot 76702
Phone: 972-8-9477077
Fax: 972-8-9363034
E-mail: Ornatbio@ornat.co.il

Japan: Muromachi Yakuhin Kaisha Ltd.
Tokyo 103-0022
Phone: 81-03-3242-1601
Fax: 81-03-3242-1601

South Korea: New Korea Industrial Co., Ltd.
Seoul
Phone: 82-2-552-2531
Fax: 82-2-557-0763

Mexico: Accesorios para Laboratorios SA de CV
Mexico D.F. 11590
Phone: 52-50-08-05
Fax: 52-55-55-20
E-mail: accesolab@accesolab.com
Web: www.accesolab.com

New Zealand: Nuclear Supplies Ltd.
Auckland 6
Phone: 64-9-535-6285
Fax: 64-9-535-6287

Sweden: Bio Nuclear AB
S-161 26 Bromma
Phone: 46-8-26-61-65
Fax: 46-8-26-61-47
E-mail: kari.eriksson@bio-nuclear.se
Web: www.bio-nuclear.se

Spain: ITISA Biomedica
Madrid
Phone: 34-91-6-57-23-93/4
Fax: 34-91-6-62-17-69
E-mail: itisabiomedica@itisabiomedica.es

Switzerland: ANAWA Trading SA
CH-8602 Wangen
Phone: 41-1-805-76-81
Fax: 41-1-805-76-75
E-mail: hassler@anawa.ch
Web: www.anawa.ch

Taiwan: Feng Jih Biomedical & Instruments Co., Ltd.
Taipei Hsien
Phone: 886-2-2695-9990
Fax: 886-2-2695-9963; 886-2-2692-3410
E-mail: fengjih@ms3.hinet.net

UK: Tocris Cookson Ltd.
Bristol BS11 8TA
Phone: 44-0117-9826551
Fax: 44-0117-9826552
E-mail: customerservice@tocris.co.uk
Web: www.tocris.com

Bachem

Switzerland: Bachem AG
CH-4416 Bubendorf
Phone: +41-61-931-23-33
Fax: +41-61-931-25-49
E-mail: sales.ch@bachem.com
Web: www.bachem.com

France: Bachem Biochimie SARL
F-78961 Voisins-le-Bretonneux
Phone: 1/30-12-15-95
Fax: 1/30-57-38-82
E-mail: vente@fr.bachem.com

Germany: Bachem Biochemica GmbH
D-69126 Heidelberg
Phone: 06221/3305-0
Fax: 06221/3305-99
E-mail: verkauf.de@bachem.com

UK: Bachem (UK) Ltd.
Merseyside WA9 3AJ, England
Phone: +44-(0)1744-61-21-08
Fax: +44-(0)1744-73-00-64
E-mail: sales.uk@bachem.com

USA: Bachem Bioscience Inc.
King of Prussia, PA 19406
Phone: 800-634-3183
Fax: 610-239-0800
E-mail: sales@us.bachem.com
Web: www.bachem.com

USA: Bachem California Inc.
Torrance, CA 90505
Phone: 310-539-4171
Fax: 310-530-1571
E-mail: sales@us.bachem.com
Web: www.bachem.com

Biodesign

USA: Biodesign International
Kennebunk, Maine 04043
Phone: 207-985-1944/888-530-0140
Fax: 207-985-6322
E-mail: info@biodesign.com
Web: www.biodesign.com

Australia: Jomar Diagnostics
Magill 5072
Phone: 61-8-8364-0021
Fax: 61-8-8364-2061
E-mail: jomar@kern.com.au

France: Interchim
F-03100 Montlucon
Phone: 33-4-7003-8855
Fax: 33-4-7003-8260
E-mail: intrchim@calva.net

Germany: Dunn Labortechnik GmbH
D-53567 Asbach
Phone: 49-2-6834-3094
Fax: 49-2-6834-2776
E-mail: dunnlab@t-online.de

Japan: Wako Pure Chemical Ind., Ltd.
Chuo-Ku Osaka 541
Phone: 81-6-203-3741
Fax: 81-6-201-5964
E-mail: KYM02031@niftyserve.or.jp

Japan: Sunfco, Ltd.
Tokyo 103
Phone: 81-6-314-3003
Fax: 81-6-364-1658

Korea: Fine Life Sciences Co., Ltd.
Seoul 011-350
Phone: 82-2-744-7859
Fax: 82-2-744-5281

Sweden: AMS Biotechnology A.B.
183 62 Taby
Phone: 46-8-630-0232
Fax: 46-8-756-9490
E-mail: ams.bio@mbox200.swipnet.se

The Netherlands: Campro Scientific
3900AH Veenendaal
Phone: 31-3-1852-9437
Fax: 31-3-1854-2181
E-mail: campro@pi.net

UK: AMS Biotechnology Ltd.
Witney Oxon OX8 7GE
Phone: 44-1993-706-500
Fax: 44-1933-706-006
E-mail: ams.biotech@dial.pipex.com

Biogenesis

UK: Biogenesis Ltd
Poole, BH17 7DA, England
Phone: (0)-1202-660006
Fax: (0)-1202-660020
E-mail:
sales@biogenesis.co.uk/technical@biogenesis.co.uk
Web: www.biogenesis.co.uk

USA: Biogenesis Inc.
Kingston, NH 03848
Phone: 603-642-8302
Fax: 603-642-8322
E-mail: biogenesis@sprintmail.com

BioSource International

USA: Biosource International
Camarillo, CA 93012
Phone: 800-242-0607
Web: www.biosource.com

Biotrend

Germany: Biotrend Chemikalien GmbH
D-50933 Köln
Phone: 49-221-9-49-83-20
Fax: 49-221-949-83-25
E-mail: jaeger@biotrend.com
Web: www.biotrend.com

Calbiochem

USA: Calbiochem-Novabiochem Corporation
La Jolla, CA 92039-2087
Phone: 619-453-3552/800-854-3417 (Calbiochem)/800-228-9622 (Novabiochem)
Fax: 619-453-3552/800-776-0999
E-mail: orders@calbiochem.com;
technical@calbiochem.com
Web: www.calbiochem.com

Argentina: Biodynamic Srl
Phone: 541-383-3000; 541-381-7962
Fax: 541-384-7316
E-mail: info@biodynamics.com.ar
Web: www.biodynamics.com.ar

Australia: Calbiochem-Novabiochem Pty.
Alexandria, NSW 2015
Phone: 02-9318-0322; 0416-250-739/800-023-956
Fax: 02-9319-2440
E-mail: calbiochem_novabiochem@compuserve.com

Austria: Bio-Trade
A-1230 Wien
Phone: 0043-1-889-18-19
Fax: 0043-1-889-18-19-20
E-mail: bio-trade@telecom.at

Belgium: Euro Biochem Scrl
B-1301 Bierges
Phone: 32-10-43-70-60; 32-10-43-70-61
Fax: 32-10-43-70-69

Brazil: Imprint CB-Tech Do Brasil
Campinas SP
Phone: 019-230-5961
Fax: 019-230-7465
E-mail: cbtech@correionet.com.br
Web: www.imprint-corp.com

Brazil: Promicro Comércio Representaçáo Ltda.
05331030-Sâo Paulo-SP
Phone: 55-11-869-0699
Fax: 55-11-869-0699
E-mail: promicro@originet.com.br

Canada: Cedarlane Laboratories Ltd.
Hornby, Ontario LOP 1EO
Phone: 800-268-5058
Fax: 905-878-7800
E-mail: info@cedarlanelabs.com
Web: www.cedarlanelabs.com

Canada: InterSciences Inc.
Markham, Ontario L3R 1A9
Phone: 905-940-1831/877 SCIENCE (724-3623)
Fax: 905-940-1832/888-673-3148
E-mail: marketing@interscience.com
Web: www.interscience.com

Chile: Fermelo S.A.
Santiago
Phone: 562-247-2976
Fax: 562-247-2977

China: Fudan Biotechnology, Inc.
Shanghai 200438
Phone: 86-21-655-72386
Fax: 86-21-655-72384
E-mail: dkfdbiot@uninet.com.cn

Denmark: Bie & Berntsen A-S
DK-2610 Roedovre
Phone: 45-44-94-88-22
Fax: 45-44-94-27-09

Finland: Ya-Kemia Oy
Fin-00700 Helsinki
Phone: 09-350-9250
Fax: 09-350-92555
E-mail: info@ya-kemia.fi

France: France Biochem
92190 Meudon
Phone: 01-46-26-78-70
Fax: 01-45-34-25-20
E-mail: biochem@pratique.fr

Germany: Calbiochem-Novabiochem GmbH
65824 Schwalbach
Phone: 0-6196-63955/800-6931-000
Fax: 0-6196-62361/800-62361-00
E-mail: customer.service@calbiochem-novabiochem.de

Greece: Biodata Hellas Ltd.
102 10 Athens
Phone: 30-1-8840613-5
Fax: 30-1-8840614-5
E-mail: biodata@ath.forthnet.gr
Web: www.mednet.gr/biodata; www.atlantis.gr/biodata

Hong Kong: Onwon Trading Limited
Kowloon
Phone: 852-27577569; 852-27966985
Fax: 852-27577211
E-mail: mlliu@ibm.net

Hong Kong: Gene Company, Ltd.
Chai Wan
Phone: 852-2896-6283
Fax: 852-2515-9371; 852-2557-1283
E-mail: genehk@hkstar.com

India: Biobusiness Development Agency
New Delhi 110 018
Phone: 91-11-559-68-20
Fax: 91-11-559-68-20; 91-11-514-43-16
E-mail: bio.multani@axcess.net.in

Israel: Megapharm, Ltd.
Hod Ha Sharon 45105
Phone: 09-760-4596
Fax: 09-760-4514

Italy: Inalco S.P.A.
20139 Milano
Phone: 02-55213005
Fax: 02-5694518
E-mail: inalco@inalco.it
Web: www.Inalco.it

Japan: Calbiochem-Novabiochem Japan Ltd.
Tokyo 108-0014
Phone: 03-5443-0281
Fax: 03-5443-0271
E-mail: cnjapan@po.globe.or.jp

Japan: Cosmo Bio Co. Ltd.
Tokyo 135-0016
Phone: 03-5632-9610
Fax: 03-5632-9619
Web: www.globe.or.jp/cosmobio/

Japan: Iwai Chemicals Company Ltd.
Tokyo 103-0023
Phone: 03-3241-0376
Fax: 03-3255-2825
E-mail: infocent@iwai.dp.u-netsurf.or.jp
Web: www.u-netsurf.or.jp/~iwai

Japan: Nacalai Tesque, Inc.
Kyoto 604-0855
Phone: 06-6381-8121
Fax: 06-6381-8137
E-mail: info.intl@nacalai.co.jp

Japan: Shigematsu & Co., Ltd.
Phone: 06-6231-6146
Fax: 06-6231-6149

Japan: Wako Pure Chemical Ind., Ltd.
Osaka 540-8605
Phone: 0120-052099
Fax: 0120-052806
E-mail: labchem-tec@wako-chem.co.jp (Osaka); labchem-tect@wako-chem.co.jp (Tokyo)
Web: www.wako-chem.co.jp

Korea: Koram Biotech Corp.
Seoul 135-080
Phone: 822-556-0311
Fax: 822-556-0828
E-mail: korambio@nuri.net

Mexico: Control Tecnico Y Representaciones
Monterrey, N.L.C.P. 64320
Phone: 528-371-6050; 528-370-1571
Fax: 528-373-2891
E-mail: controltec@infosel.net.mx
Mexico: Control Tecnico Y Representaciones
Mexico City
Phone: 525-399-2840
Fax: 525-399-2870

Norway: Bio Test A/S
1254 Oslo
Phone: 47-22-619-660
Fax: 47-22-619-661
E-mail: jbiotest@online.no ; wold@online.no

Portugal: Biocontec Lda.
2777 Carcavelos Codex
Phone: 351-1-361-36-20
Fax: 351-1-362-56-15
E-mail: biocontec@mail.eunet.pt

Singapore: Scimed (Asia) Pte. Ltd.
128384
Phone: 65-779-3388
Fax: 65-266-3086; 65-799-0100
E-mail: scimed@singnet.com.sg

South Africa: South African Scientific Products (PTY) Ltd.
Randburg (Johannesburg) 2125
Phone: 27-11-886-4710
Fax: 27-11-787-9598

Spain: Bionova Cientifica S.L.
28007 Madrid
Phone: 34-91-5515403
Fax: 34-91-4334545
E-mail: bionova@logiccontrol.es

Sweden: Labkemi AB
135 70 Stockholm
Phone: 46-8-7424200
Fax: 46-8-7424243
E-mail: Labkemi@stockholm.mail.telia.com

Sweden: Göteborg
Gothenburg
Phone: 46-31-892980
Fax: 46-31-477890

Sweden: Lund/Malmö
226 60 Lund
Phone: 46-46-155140
Fax: 46-46-189066

Switzerland: Juro Supply AG
6000 Lucerne 4
Phone: 41-41-226-20-60
Fax: 41-41-226-20-66
E-mail: juro_lucerne@mail.tic.ch
Web: www.tic.ch/juro

Taiwan: Cashmere Scientific Co.
Chai-Yi 625
Phone: 886-5-3451214; 886-2-25416188/080-222095
Fax: 886-5-3453548; 886-2-25627894
E-mail: cashmere@tcts.seed.net.tw

The Netherlands: Omnilabo International B.V.
4817 Bl Breda
Phone: 31-0-76-5795-795/800-099-7775
Fax: 31-0-76-5876-236
E-mail: omnilabo@wxs.nl
Web: www.tip.nl/users/omnilabo

Turkey: Biomar
41470 Gebze/Kocaeli
Phone: 90-262-646-9811
Fax: 90-262-642-4759

UK: Calbiochem-Novabiochem (UK) Ltd.
Beeston, Nottingham NG9 2JR
Phone: 0115-943-0840/800-622-935
Fax: 0115-943-0951
E-mail: customer.service@cnuk.uk
Web: www.cnuk.co.uk

Venezuela: Vargas Scientific C.A.
Caracas 1060
Phone: 582-263-2202; 582-265-0891
Fax: 582-263-0924
E-mail: vss@ven.net

Chemicon

USA: Chemicon International, Inc.
Temecula, CA 92590
Phone: 909-676-8080/800-437-7500
Fax: 909-676-9209
E-mail: custserv@chemicon.com
Web: www.chemicon.com

UK: Chemicon International, Ltd.
Harrow HA3 5UT
Phone: 44-20-8863-0415
Fax: 44-20-8863-0416

Germany: Chemicon International, Ltd.
D-65719 Hofheim
Phone: 49-6192-2959-00
Fax: 49-6192-2959-55

Australia: Silenus
Victoria 3155
Phone: 61-3-9839-2000
Fax: 61-3-9887-3912

Cortex

USA: Cortex Biochem, Inc.
San Leandro, CA 94577
Phone: 510-568-2228/800-888-7713
Fax: 510-568-2467
E-mail: cortbioc@ix.netcom.com
Web: www.cortex-biochem.com

Fitzgerald

USA: Fitzgerald Industries International, Inc.
Concord, MA 01742-3049
Phone: 978-371-6446/800-370-2222
Fax: 978-371-2266
E-mail: antibodies@fitzgerald-fii.com
Web: www.fitzgerald-fii.com

Fluka

Switzerland: Fluka Chemie AG
CH-9471 Buchs
Phone: 41-81-755-25-11
Fax: 41-81-756-54-49
E-mail: fluka@sial.com
Web: www.sigma-aldrich.com

Germany: RdH Laborchemikalien GmbH & Co. KG
D-30926 Seelze
Phone: 49-5137-82-38-0
Fax: 49-5137-82-38-120
E-mail: riedel@sial.com
Web: www.sigma-aldrich.com

Argentina: Sigma-Aldrich de Argentina S.A.
Phone: 54-11-4807-0321
Fax: 54-11-4807-0346

Australia: Sigma-Aldrich Pty., Ltd.
Sydney
Phone: 61-2-9841-0555/800-800-097
Fax: 61-2-9841-0500/800-800-096

Austria: Sigma-Aldrich Handels GmbH
Phone: 01-605-81-10
Fax: 01-605-81-20

Belgium: Sigma-Aldrich NV/SA
Phone: 03-899-13-01/0800-14-747
Fax: 03-899-13-11/0800-14-745

Brazil: Sigma-Aldrich Química Brasil Ltda.
Phone: 011-231-1866
Fax: 011-257-9079

Canada: Sigma-Aldrich Canada., Ltd.
Phone: 905-829-9500/800-565-1400
Fax: 905-829-9292/800-265-3858

Czech Republic: Sigma-Aldrich s.r.o.
Phone: 02-2176-1300
Fax: 02-2176-3300

Denmark: Sigma-Aldrich Denmark A-S
Phone: 43-56-59-10
Fax: 43-56-59-05

Finland: Sigma-Aldrich Finland/YA-Kemia Oy
Phone: 09-350-9250
Fax: 09-350-92555

France: Sigma-Aldrich Chimie S.A.R.L
Phone: 04-74-82-28-40/0800-21-14-08
Fax: 04-74-95-68-08/0800-03-10-52

Germany: Sigma-Aldrich Chemie GmbH
Phone: 089-6513-0/0800-5155-000
Fax: 089-6513-1160/0800-6490-000

Greece: Sigma-Aldrich Ltd.
Phone: 01-9948031
Fax: 01-9943831

Hungary: Sigma-Aldrich Kft
Phone: 06-1-269-6474/06-80-355-355
Fax: 06-1-235-9050/06-80-344-344

India: Sigma-Aldrich
Bangalore
Phone: 08-851-8797; 08-851-8942
Fax: 08-851-8358
New Delhi
Phone: 011-689-9826; 011-689-7830
Fax: 011-689-9827

Ireland: Sigma-Aldrich Ireland Ltd.
Phone: 01-404-1900/800-200-888
Fax: 01-404-1910/800-600-222

Israel: Sigma Israel Chemicals Ltd.
Phone: 08-9484-555/800-70-2222
Fax: 08-9484-200

Italy: Sigma-Aldrich s.r.l.
Phone: 02-33417-310/800-827018
Fax: 02-38010737

Japan: Sigma-Aldrich Japan K.K.
Tokyo
Phone: 03-5821-3171
Fax: 03-5821-3170

Korea: Sigma-Aldrich Korea Ltd.
Phone: 02-783-5211/080-023-7111
Fax: 02-783-5011/080-023-8111

Mexico: Sigma-Aldrich Química S.A. de C.V.
Phone: 01-800-007-5300
Fax: 01-800-712-9920

Norway: Sigma-Aldrich Norway AS
Phone: 22-09-15-70
Fax: 22-09-15-55

Poland: Sigma-Aldrich Sp. z.o.o.
Phone: 061-823-24-84
Fax: 061-823-27-81

Portugal: Sigma-Aldrich Química, S.A.
Phone: 351-1-9242555/0800-20-21-82
Fax: 351-1-9242610/0800-20-21-78

Russia: Sigma-Aldrich Russia/TechCare Systems, Inc.
Phone: 7-095-975-3321
Fax: 7-095-975-4792

Singapore: Sigma-Aldrich Pte. Ltd.
Phone: 65-271-10-89
Fax: 65-271-15-71

South Africa: Sigma-Aldrich (Pty) Ltd.
Phone: 011-397-88-86 JHB/0800-11-00-75 (SA)
Fax: 011-397-88-59 JHB/0800-11-00-79 (SA)

Spain: Sigma-Aldrich Química S.A.
Phone: 34-91-661-99-77/900-10-13-76
Fax: 34-91-661-96-42/900-1020-28

Sweden: Sigma-Aldrich Sweden AB
Phone: 020-35-05-11/46-8-742-42-00
Fax: 020-35-25-22/46-8-742-72-99

Switzerland: Fluka Chemie AG
Phone: 081-755-28-28/0800-80-00-80
Fax: 081-755-28-15

The Netherlands: Sigma-Aldrich Chemie BV
Phone: 078-620-54-11/0800-022-90-88
Fax: 078-620-54-21/0800-022-90-89

UK: Sigma-Aldrich Co. Ltd.
Phone: 01747-82-22-11/0800-71-71-81
Fax: 01747-82-37-79/0800-37-85-38

USA: Fluka Chemical Corp.
Phone: 414-273-3850/800-558-9160
Fax: 414-273-4979/800-962-9591
24-Hour Emergency Phone: 414-273-3850
E-mail: Fluka-RdH@sial.com
Web: www.sigma-aldrich.com

Harlan

USA: Harlan Bioproducts for Science, Inc.
Indianapolis, IN 46229-0176
Phone: 317-894-7536/800-9-SCIENCE
Fax: 317-894-9458
E-mail: hbps@harlan.com

France: Harlan France SARL
03800 Gannat
Phone: 04-70-90-02-20
Fax: 04-70-90-68-08

Germany: Harlan Winkelmann GmbH
D-33176 Borchen
Phone: 0-5251-131510
Fax: 0-5251-388962

Israel: Harlan Biotech Israel Ltd./Harlan Laboratories Limited
Jerusalem 91120
Phone: 02-6439398
Fax: 02-6439403

Italy: Harlan Italy
20050 Correzzana (Milan)
Phone: 039-6064621
Fax: 039-6981665

Mexico: Harlan Mexico, S.A. de C.V.
06760 Mexico D.F.
Phone: 525-264-86-57
Fax: 525-574-32-25

Spain: Harlan Interfauna Ibérica S.A.
Barcelona
Phone: 93-866-12-61
Fax: 93-866-03-73

The Netherlands: Harlan Nederland
NL-5960 Horst
Phone: 0478-578300
Fax: 0478-571117

UK: Harlan Interfauna Limited
Wyton, Huntingdon, Cambs PE17 2DT
Phone: 01480-455335
Fax: 01480-455337

UK: Harlan Isotec/UK Limited/Harlan Teklad
Blackthorn, Bicester, Oxon OX6 0TP
Phone: 01869-243241
Fax: 01869-246759

UK: Harlan-Sera-Lab Limited
Loughborough, Leicestershire LE12 9TE
Phone: 01530-222794
Fax: 01530-222807

USA: Harlan Sprague Dawley, Inc.
Indianapolis, IN 46229-0176
Phone: 317-894-7521
Fax: 317-894-1840

USA: Harlan Teklad
Madison, WI 53744-4220
Phone: 608-277-2070
Fax: 608-277-2066

IBT

Germany: Immunological and Biochemical Testsystems
GmbH
D-72770 Reutlingen
Phone: +49-7121-51463-4
Fax: +49-7121-51463-5
E-mail: service@anaspec.com
Web: www.anaspec.com

ICN

USA: ICN Biomedicals, Inc.
Costa Mesa, CA 92626
Phone: 800-854-0530
Fax: 800-334-6999
E-mail: biomark@icnbiomed.com
Web: www.icnbiomed.com

Argentina: ICN Argentina
1417 Capital Federal
Phone: 54-1-502-7788
Fax: 54-1-502-7788

Australia: ICN Biomedicals Australasia Pty. Ltd.
Seven Hills NSW 2147
Phone: 61-2-9-838-7422
Fax: 61-2-9-838-7390

Belgium: ICN Biomedicals, NV/SA
B-1731 Asse-Relegem
Phone: 32-2-466-0000
Fax: 32-2-466-2642

Bolivia: Diagnotest
La Paz
Phone: 59-1-236-6390
Fax: 59-1-239-1136

Brazil: Instrucom
Sao Paulo
Phone: 55-11-560-7833
Fax: 55-11-530-0895

Brazil: Interlab
Sao Paulo 04330-130
Phone: 55-11-5564-9500
Fax: 55-11-276-6942

Brazil: Activa
01050-904 Sao Paulo
Phone: 55-11-255-9500
Fax: 55-11-255-9986

Chile: Immunotech
Providencia
Phone: 56-2-204-6469

Colombia: Quimiolab
Bogota
Phone: 571-540-1213; 571-418-3763
Fax: 571-413-2194

Czech Republic: Starlab s.r.o.
108 03 Praha 10-Malesice
Phone: 420-2-6702-1391
Fax: 420-2-6702-1393

Denmark: Life Science Danmark
DK-3480 Fredensborg
Phone: 45-48-47-50-51
Fax: 45-48-47-50-61

Dominican Republic: Bionuclear
Santo Dominog
Phone: 809-567-8172

Egypt: Medico
12655 Cairo
Phone: 202-336-1918
Fax: 202-336-1918

Egypt: MPT Medicopharmatrade
Dokki
Phone: 202-349-8311
Fax: 202-348-1468

Egypt: Omyk Medical Comp. Lab. Med & Dental Equipment
& Supply
Nasr City 11762 Cairo
Fax: 202-291-8059

El Salvador: D & R Co.
San Salvador
Phone: 503-279-3075
Fax: 503-279-3582

Finland: Tamro Corporation
FIN-01641 Vantaa
Phone: 358-204-454700
Fax: 358-204-454717

France: ICN Pharmaceuticals France S.A.
91893 Orsay Cedex
Phone: 33-1-60-19-37-37/800-130-373
Fax: 33-1-60-19-34-60

Germany: ICN Biomedicals GmbH
D-37269 Eschwege
Phone: 00800-426-67337
Fax: 00800-329-67337

Greece: Labo-Chem
GR-11527 Athens
Phone: 30-1-770-94-74
Fax: 30-1-775-60-90

Hong Kong: Advanced Technology & Industrial Co.
Tai Kok Tsui, Kln
Phone: 852-239-02293; 852-239-45546
Fax: 852-2789-8314

Hong Kong: Gold Pacific Enterprises
283 Queens Rd.
Fax: 852-2526-7053

Hungary: ICN Alkaloida Hungary
H-4440 Tiszavasvari PF1
Phone: 36-42-372-511/087
Fax: 36-42-372-137

Hungary: Izinta Trading Co., Ltd.
H-1121 Budapest
Phone: 36-1-395-9273
Fax: 36-1-395-9267

Iceland: Icelanding Pharmaceuticals Ltd.
IC-458 Reykjavik
Phone: 354-540-8000
Fax: 354-540-8001

India: Span Diagnostics, Ltd.
Udhna 394 210, Surat
Phone: 91-261-677143
Fax: 91-261-679319
Telex: 188-284-Spanin

Ireland: Medical Supply Co., Ltd.
Dublin 9
Phone: 353-184-26644

Israel: Enco Scientific Services Ltd.
Petaca-Tiqva 49127
Phone: 972-3-934-9922
Fax: 972-3-934-9876
Telex: 25615 Jamil il

Italy: ICN Biomedicals, S.r.l.
20090 Opera (MI)
Phone: 39-2-57601041
Fax: 39-2-57601610

Japan: ICN Pharmaceuticals, K.K.
Chuo-ku, Tokyo 104
Phone: 81-3-3275-8020
Fax: 81-3-3275-4055

Japan: Cosmo Bio Co. Ltd.
Tokyo 135
Phone: 81-3-5632-9630
Fax: 81-3-5632-9623

Japan: Dainippon Pharmaceutical K.K. Laboratory Group
Osaka 564
Phone: 81-6-386-2164
Fax: 81-6-337-1606

Japan: Iwai Kagaku Yakuhin K.K.
Tokyo 103
Phone: 81-3-3255-2781
Fax: 81-3-3255-2825

Japan: Kanto Chemical K.K.
Tokyo 103
Phone: 81-3-3663-7631
Fax: 81-3-3667-8277

Japan: Nacalai Tesque, Inc.
Kyoto 603
Phone: 81-75-251-1723
Fax: 81-75-251-1762

Japan: Seikagaku Corp.
Tokyo 135
Phone: 81-3-3245-1951
Fax: 81-3-3242-5335

Japan: Toho Biochemical Co. Ltd.
Tokyo 162
Phone: 81-33235-0218
Fax: 81-33235-0304

Jordan: Arab Supplies & General Trade Co.
Amman 11118
Phone: 962-6-661187; 962-6-661188; 962-6-661189
Fax: 962-6-601345
Telex: 21554 Arabco Jo

Korea: Koram Biotech Corp.
Seoul 135-080
Phone: 822-556-0311
Fax: 822-556-0828

Korea: Merck
Seoul
Phone: 82-2-557-1440
Fax: 82-2-555-5964

Korea: Sugsan D & P Inc.
Seoul
Phone: 82-2-706-1230
Fax: 82-2-702-3835

Kuwait: Al Sedan Trading & Contracting Co.
Safat 13021
Phone: 965-245-2763
Fax: 965-240-0111

Lebanon: Medek S.A.R.L.
Jdeidet El Metn
Phone: 961-1-884-114
Fax: 961-1-881-042

Malaysia: FC-Bios SDN BHD
Selangor Darul, Ehsan
Phone: 60-3-735-1559
Fax: 60-3-753-1306

Mexico: ICN Farmaceutica S.A. de C.V.
09080 Mexico-DF
Phone: 52-5-670-0739
Fax: 52-5-581-4938

New Zealand: Nuclear Supplies
Aukland
Phone: 64-9-535-6285

New Zealand: Scientific Supplies
Aukland
Phone: 64-9-274-7579
Fax: 64-9-274-0268

Norway: Nerliens Kemisk-Tekniske AS
N-0654 Oslo
Phone: 47-22-68-50-70
Fax: 47-22-7-65-06

Pakistan: Pakistan Hospital & Industrial Co.
Lahore
Phone: 92-42-732-4276
Fax: 92-42-723-3013
Telex: 44679-47528

Panama: Casa Del Medico
Colinia Bellavista 16891
Phone: 507-227-5311

Panama: Mac International
Panama City
Phone: 507-213-0627

Peru: Ares Mercantile, S.R.L.
Lima 17
Phone: 51-461-9767
Fax: 51-1-470-1855

Poland: Abo
80-255 Gdansk
Phone: 48-058-341-2143
Fax: 48-058-341-2143

Poland: Arcid
60-461 Poznan
Phone: 48-61-8557-265
Fax: 48-61-8557-266

Poland: Hand-Prod SP. z.o.o. ul.
PL-01-113 Warsaw
Phone: 48-22-37-42-35
Fax: 48-22-668-4303

Poland: ICN Polfa Rzeszow
Phone: 48-22-646-51-78
Fax: 48-22-651-20-35

Qatar: Al-Haidous Trading Establishment
Doha
Phone: 974-428716
Fax: 974-431126
Telex: 4336-AHTE DH

Qatar: SciTech Arabia
Doha
Phone: 974-411-605
Fax: 974-611-606

Russia: ICN Pharmaceuticals, Inc.
Moscow 119048
Phone: 7-095-234-3183; 7-095-232-6600
Fax: 7-095-234-3183; 7-095-234-3187

Saudi Arabia: Ebrahim M. Almana & Bros. Co.
Riyadh 11564
Phone: 966-1-463-2897
Fax: 966-3-864-3603
Telex: 406598 Najeeb Sj

Singapore: SciMed (Asia) Pte. Ltd.
128384
Phone: 65-779-3388
Fax: 65-266-3086

South Africa: Laboratory Specialtiess
Randburg 2125
Phone: 27-11-792-2190
Fax: 27-11-793-1064

South Africa: Separations/Biotechniques
Randburg 2125
Phone: 27-11-792-3428
Fax: 27-11-792-1043

Spain: ICN Iberica, S.A.
Barcelona
Phone: 34-93-688-28-34
Fax: 34-93-688-04-01

Sri Lanka: Aristons (PVT) Ltd.
Kirullapone, Colombo 5
Phone: 94-1-825-111; 94-1-825-112
Fax: 94-1-822-986
Telex: 21302 Ruwani Ce

Sweden: Chemicon AB
S-20039 Malmo
Phone: 46-40-21-21-00
Fax: 46-40-22-22-81

Taiwan: Chinese Tycoon Enterprise Co. Ltd.
Taipei
Phone: 886-2-366-1681
Fax: 886-2-366-1538

Taiwan: General Biologicals, Corp.
Hsin-Chu
Phone: 886-35-779-221
Fax: 886-35-578-4065

Taiwan: Level Biotechnology
Taipei
Fax: 886-2-695-0403

Thailand: Bang Trading 1992 Co. Ltd.
Suanluang Bangkok 10250
Phone: 66-2-718-333, ext. 460
Fax: 66-2-718-3588; 66-2-718-3577

Ukraine: ICN Pharmaceuticals
252004, Kiev
Phone: 380-44-2284712
Fax: 380-44-2284712

United Arab Emirates: Al Thabat Trading Est.
Dubai
Phone: 971-4-630113
Fax: 971-4-631-361

United Arab Emirates: Horizon Medical Supplies
Deira, Dubai
Phone: 971-4-698-666
Fax: 971-4-698-667

UK: ICN Pharmaceuticals, Ltd.
Basingstoke, Hampshire RG 24 8WG
Phone: 44-1-256-374-620/0800-282-474
Fax: 44-1-256-374-621/0800-614735

Uruguay: Buro Ltda.
11300-Montevideo
Phone: 598-707-4318

Venezuela: Log-On Corp.
La Urbina, Caracas
Phone: 582-241-4080
Fax: 582-2241-4834

Venezuela: Scientec
Guacara, Carabobo 2015
Phone: 5845-64-7881

Yugoslavia: ICN Yugoslavia
Belgrade, Zemun
Phone: 381-11-619-333
Fax: 381-11-199-424

Kamiya

USA: Kamiya Biomedical Company
Seattle, WA 98188
Phone: 206-575-8068
Fax: 206-575-8094
E-mail: kassays@kamiyabiomedical.com
Web: www.kamiyabiomedical.com

Neosystem

France: Neosystem Groupe SNPE
67100 Strasbourg, France
Phone: +33-3-88-79-08-79
Fax: +33-3-88-79-18-56
E-mail: neo@neosystem.worldnet.fr
Web: www.neosystem.fr

Italy: Chem Progress srl
20098 Sesto Ulteriano (MI)
Phone: (02)-98-80-735/6/7
Fax: (02)-98-81-106

Japan: SNPE Japan
Tokyo T 105
Phone: (3)-34-33-28-80
Fax: (3)-34-33-66-52

UK: SNPE England
Croydon, Surrey Cro I PE
Phone: (0181)-649-74-74
Fax: (0181)-760-04-20

USA: Multiple Peptide Systems
San Diego, CA 92121
Phone: 858-455-3710/800-338-4965
Fax: 858-455-3713/800-654-5592
E-mail: mps@mps-sd.com
Web: www.mps-sd.com

USA: SNPE North America LLC
Princeton, NJ 08540
Phone: 609-987-9424
Fax: 609-987-2767

Oncogene

USA: Oncogene Research Products
Cambridge, MA 02142
Phone: 617-577-1639/800-662-2616
Fax: 617-577-8015/800-828-4871
E-mail: customer.service@oncresprod.com
Web: www.oncresprod.com; www.apoptosis.com;
www.neuroproducts.com

Australia: Calbiochem-Novabiochem Pty. Ltd.
Alexandria, NSW 1435
Phone: 02-9318-0322/800-023-956
Fax: 02-9319-2440
E-mail: calbiochem_novabiochem@compuserve.com

Australia: Amrad Biotech
Boronia, Vic 3155
Phone: 03-9839-2000/800-252-265
Fax: 03-9887-3912
E-mail: apbcs@amrad.com.au
Web: www.amrad.com.au

Austria: Bio-Trade
1230 Wien
Phone: 0043-1-889-18-19
Fax: 0043-1-889-18-19-20
E-mail: bio-trade@telecom.at

Belgium: Euro Biochem Scrl
B-1301 Bierges
Phone: 32-10-43-70-60; 32-10-43-70-61
Fax: 32-10-43-70-69

Canada: Cedarlane Laboratories, Ltd.
Hornby, Ontario LOP 1EO
Phone: 905-878-8891/800-268-5058
Fax: 905-878-7800
E-mail: info@cedarlanelabs.com
Web: www.cedarlanelabs.com

Croatia: HEBE D.O.O.
21000 Split
Phone: 385-21-364-701
Fax: 385-21-364-701

Denmark: Bie & Berntsen A-S
Roedovre DK-2650
Phone: 45-44-94-88-22
Fax: 45-44-94-27-09

Egypt: Medico Pharma Trade
Giza
Phone: 20-2-349-3734
Fax: 20-2-348-1468

Finland: Ya-Kemia Oy
Fin-00700 Helsinki
Phone: 358-9-350-9250
Fax: 358-9-350-92555
E-mail: info@ya-kemia.fi
Web: www.ya-kemia.fi

France: France Biochem
92190 Meudon
Phone: 01-46-26-78-70
Fax: 01-45-34-25-20
E-mail: biochem@pratique.fr

Germany: Calbiochem-Novabiochem GmbH
65824 Schwalbach
Phone: 49-6196-63955; 49-6196-63956/800-6931000
Fax: 49-6196-62361
E-mail: customer.service@calbiochem-novabiochem.de

Germany: Dianova GmbH
20354 Hamburg
Phone: 040-45-06-70
Fax: 040-45-06-74-90
E-mail: info@dianova.de
Web: www.dianova.de

Greece: Biodata Hellas Ltd.
102 10 Athens
Phone: 30-1-8840613-5
Fax: 30-1-8840614-5
E-mail: biodata@ath.forthnet.gr
Web: www.mednet.gr/biodata; www.atlantis.gr/biodata

Greece: Biodynamics S.A.
Athens 11471
Phone: 30-1-644-9421
Fax: 30-1-644-2266
E-mail: biodynamics@ath.forthnet.gr

Hong Kong: Gene Company, Ltd.
Chai Wan
Phone: 852-2896-6283
Fax: 852-2515-9371; 852-2557-1283
E-mail: genehk@hkstar.com

India: Biobusiness Development Agency
110018 New Delhi
Phone: 91-11-559-68-20
Fax: 91-11-559-68-20; 91-11-514-43-16
E-mail: bio.multani@axcess.net.in

India: OSB Agencies
Delhi 110031
Phone: 91-11-224-9973
Fax: 91-11-221-6736
E-mail: bhatia.osb@axcess.net.in

Israel: Megapharm, Ltd.
Hod Ha Sharon 45105
Phone: 09-760-4596
Fax: 09-760-4514

Italy: Inalco S.p.A.
20139 Milano
Phone: 39-02-55213005
Fax: 39-02-5694518
E-mail: inalco@inalco.it
Web: www.Inalco.it

Japan: Calbiochem-Novabiochem Japan
Tokyo 108-0014
Phone: 81-3-5443-0281
Fax: 81-3-5443-0271

Japan: Cosmo Bio Co. Ltd.
Tokyo 135-0016
Phone: 81-3-5632-9610
Fax: 81-3-5632-9619
Web: www.globe.or.jp/cosmobio/

Japan: Iwai Chemicals Company
Tokyo 103-0023
Phone: 81-3-3279-6363
Fax: 81-3-3270-2462

Japan: Nacalai Tesque, Inc.
Kyoto 604-0855
Phone: 81-75-251-1723/0120-489-552
Fax: 81-75-251-1762

Japan: Wako Pure Chemical Ind., Ltd.
Osaka 540-8605
Phone: 06-203-1788/0120-052099
Fax: 06-201-5965/0120-052806
E-mail: labchem-tec@wako-chem.co.jp (Osaka); labchem-tect@wako-chem.co.jp (Tokyo)
Web: www.wako-chem.co.jp

Korea: Kormed Corp.
Seoul 135-010
Phone: 82-2-540-4663
Fax: 82-2-544-4539

Norway: Bio Test A/S
N 1254 Oslo
Phone: 47-22-619-660
Fax: 47-22-619-661
E-mail: jbiotest@online.no ; wold@online.no

Singapore: Scimed (Asia) Pte. Ltd.
Phone: 65-779-3388
Fax: 65-266-3086
E-mail: scimed@singnet.com.sg

South Africa: Aec-Amersham
Kelvin 2054
Phone: 27-11-444-4330
Fax: 27-11-444-5457
E-mail: aecamjhb@lafrica.com

Spain: ITISA Biomedicina S.A.
Madrid
Phone: 34-91-657-23-93/4
Fax: 34-91-662-17-69

Sweden: Labkemi AB
S-135 70 Stockholm
Phone: 46-08-742-42-00/020-35-05-10
Fax: 46-08-742-42-43/020-35-25-22
E-mail: Labkemi@stockholm.mail.telia.com

Switzerland: Juro Supply AG
CH-6000 Lucerne 4
Phone: 41-41-226-20-60
Fax: 41-41-226-20-66
E-mail: juro_lucerne@mail.tic.ch
Web: www.tic.ch/juro

Taiwan: Cashmere Scientific Co.
Chai-Yi 625
Phone: 886-5-3451214/886-2-25416188
Fax: 886-5-3453548/886-2-25627894
E-mail: cashmere@tcts.seed.net.tw

The Netherlands: Omnilabo International B.V.
4817 Bl Breda
Phone: 31-0-76-5795-795/800-099-7775
Fax: 31-0-76-5876-236
E-mail: omnilabo@wxs.nl
Web: www.tip.nl/users/omnilabo

Turkey: Bio-Kem, Ltd.
Findikzade-Istanbul
Phone: 90-212-534-01-03
Fax: 90-212-631-20-61
E-mail: biokem@aidata.com.tr

Turkey: Biomar
41470 Gebze/Kocaeli
Phone: 90-262-646-9811

Fax: 90-262-642-4759
UK: Calbiochem-Novabiochem (UK) Ltd.
Beeston, Nottingham NG9 2JR
Phone: 44-0115-943-0840/800-622-935
Fax: 44-0115-943-0951
E-mail: customer.service@cnuk.uk
Web: www.cnuk.co.uk

Venezuela: Vargas Scientific C.A.
Caracas 1060
Phone: 582-263-2202; 582-265-0891
Fax: 582-263-0924
E-mail: vss@ven.net

Pentapharm
Switzerland: Pentapharm Ltd.
CH-4002 Basel
Phone: +41-61-706-4848
Fax: +41-61-319-9619
Web: www.pentapharm.com

Peptides International
USA: Peptides International
Louisville, KY 40224
Phone: 502-266-8787/800-777-4779
Fax: 502-267-1FAX
E-mail: peptides@pepnet.com
Web: www.pepnet.com

Austria: ict Chemietechnik GmbH
1060 Wien
Phone: +43/1-585-77-00
Fax: +43/1-585-77-01
E-mail: office@ict-int.net

Germany: ict Handel-GmbH
61352 Bad Homburg
Phone: +49-6172/4063-0
Fax: +49-6172/4063-79
E-mail: office@ict-int.net

Japan: Peptide Institute, Inc.
Osaka 562
Phone: (0727) 29-4121
Fax: (0727) 29-4124
Web: www.peptide.co.jp

PeproTech
USA: Rocky Hill, NJ 08553-0275
Phone: 609-497-0253/800-436-9910
Fax: 609-497-0321
E-mail: info@peprotech.com
Web: www.peprotech.com

UK: PeproTech EC Ltd.
London SW1Y 4JH
Phone: 44-0-171-603-8288; 44-171-839-8688
Fax: 44-0-171-603-8233
E-mail: info@peprotechec.com
Web: www.peprotechec.com

France: TEBU
78610 Le Perray en Yvelines
Phone: 33-1-30-46-39-00
Fax: 33-1-30-46-39-11

Israel: Cytolab Ltd.
Rehovot 76124
Phone: 972-0-8-940-3302
Fax: 972-0-8-940-3736

Promega
USA: Promega Corporation
Madison, WI
Phone: 608-274-4330/800-356-9526
Fax: 608-277-2516/800-356-1970
E-mail: custserv@promega.com
Web: www.promega.com

Argentina: Biodynamics, Srl.
Buenos Aires
Phone: 11-4383-3000
Fax: 11-4384-7316
E-mail: info@biodynamics.com.ar

Australia: Promega Corporation
Sydney, Melbourne
Phone: 02-9565-1100/800-225-123
Fax: 02-9550-4454/800-626-017
E-mail: aus_custserv@promega.com

Brazil: bioBras
Belo Horizonte
Phone: 31-291-9877
Fax: 31-291-5369
E-mail: biomol@biobras.com.br

Brazil: Promicro
Sâo Paulo
Phone: 11-869-0699
Fax: 11-869-0699
E-mail: promicro@originet.com.br

Canada: Fisher Scientific, Ltd.
Nepean, Ontario
Phone: 613-226-8874/800-267-7424
Fax: 613-226-8639

Canada: VWR CanLab
Mississauga, Ontario
Phone: 905-813-7377
Fax: 905-813-5245

Chile: Fermelo, S.A.
Santiago
Phone: 2-247-2976
Fax: 2-247-2977
E-mail: fermelo@entelchile.net

China: Promega Beijing
Beijing
Phone: 10-6849-8287
Fax: 10-6849-8390
E-mail: prombs@bj.com.cn

China: Shanghai Promega Biological Products, Ltd.
Shanghai
Phone: 21-6483-5136
Fax: 21-6470-0176
E-mail: spromega@srcb.ac.cn

Colombia: Biologia Molecular, Ltda.
Bogota
Phone: 1-611-2287
Fax: 1-611-2131
E-mail: biomol@hotmail.com

Czech Republic: East Port Scientific
Prague
Phone: 02-301-8177
Fax: 02-333-12428
E-mail: eastport@login.cz

Denmark: Bie & Berntsen A-S
Roedovre
Phone: 44-94-88-22
Fax: 44-94-27-09
E-mail: bie-bern@post4.tele.dk

Ecuador: Electromedics
Quito
Phone: 3-258-2483
Fax: 3-295-9363
E-mail: gleoro@interactive.net.ec

Egypt: Lab Technology
Cairo
Phone: 2-245-1785
Fax: 2-242-8366
E-mail: sahar@link.com.eg

Finland: Biofellows OY
Helsinki
Phone: 358-9-755-2550
Fax: 358-9-755-25555
E-mail: info@biofellows.com

France: Promega France
Lyon
Phone: 04-37-22-50-00/0800-48-79
Fax: 04-37-22-50-10
E-mail: fr_custserv@fr.promega.com

Germany: Promega GmbH
Mannheim
Phone: 49-0-621-8501-0/00800-77663422
Fax: 49-0-621-8501-220/00800-77663423
E-mail: de_custserv@de.promega.com

Greece: BioAnalytica, S.A.
Athens
Phone: 1-6436138
Fax: 1-6462748
E-mail: bioanalyt@hol.gr

Hong Kong: Bio-Gene Technology
Phone: 2646-6101
Fax: 2686-8806
E-mail: biogene@speednet.net

Hungary: Bio-Science Kft.
Budapest
Phone: 361-463-5077
Fax: 361-463-5261
E-mail: bio-sci@bio-science.hu

India: Hysel India Pvt. Ltd.
New Delhi
Phone: 11-684-6265; 11-631-8659
Fax: 11-691-5917; 11-644-7032
E-mail: hysel@del2.vsnl.net.in

Indonesia: PT Diastika Biotekindo
Jakarta Selatan
Phone: 21-489-1718
Fax: 21-475-4707
E-mail: diastika@dnet.net.id

Ireland: Medical Supply Co., Ltd.
Dublin
Phone: 01-8224222
Fax: 01-8224100
E-mail: msc@internet-ireland.ie

Israel: Ornat Biochemicals & Laboratory Equipment
Rehovot
Phone: 89477077
Fax: 89363034
E-mail: ornatbio@ornat.co.il

Italy: Promega Italia s.r.l.
Milan
Phone: 02-290-6651/800-69-1818
Fax: 02-2901-7365
E-mail: it_custserv@it.promega.com

Japan: Promega KK
Tokyo, Osaka
Phone: 03-3669-7981
Fax: 03-3669-7982
E-mail: prometec@jp.promega.com

Jordan: Masoud Est. for Medical and Scientific Supplies
Amman
Phone: 64648094
Fax: 64648093
E-mail: masoud.est@firstnet.com.jo

Korea: Seoulin Scientific Co., Ltd.
Seoul
Phone: 02-478-5911
Fax: 02-478-5572
E-mail: sltrade@seoulin.co.kr

Lebanon: Atom Medical Company, sarl
Beirut
Phone: 1-249836
Fax: 1-249838
E-mail: atommed@cyberia.net.lb

Malaysia: Research Instruments
Petaling Jaya Selangor
Phone: 3-704-8600
Fax: 3-704-8599
E-mail: resinst@po.jaring.my

Mexico: Uniparts, S.A.
Mexico City
Phone: 5-281-4718
Fax: 5-281-4722
E-mail: uniparts@uniparts.com.mx

Morocco: Genome Biotechnologies sarl
Casablanca
Phone: 02-343376
Fax: 02-344135
E-mail: genome@lemel.fr

New Zealand: Dade Behring Diagnostics
Auckland
Phone: 9-366-4784
Fax: 9-379-8308

Norway: AH Diagnostics AS
Oslo
Phone: 23-233-260
Fax: 23-233-270
E-mail: ahdiag@ahdiag.no

Panama: Servi-Lab
Panama City
Phone: 507-229-1233
Fax: 507-229-3934
E-mail: servilab@sinfo.net

Poland: Symbios Sp. z.o.o.
Gdansk
Phone: 58-344-1980
Fax: 58-341-4726
E-mail: symbios@gd.onet.pl

Portugal: Biocontec Biotecnologia e Ambiente, Ltda.
Carcavelos
Phone: 1-361-3620
Fax: 1-362-5615
E-mail: biocontec@mail.eunet.pt

Romania: Dexter Com S.R.L.
Bucharest
Phone: 401-231-1629
Fax: 401-231-1629
E-mail: dexter@itcnet.ro

Russia: Bion
Moscow
Phone: 095-135-4206
Fax: 095-135-4206
E-mail: bion@glas.apc.org

Singapore: Promega-Representative Office
Beijing
Phone: 65-452-4230
Fax: 65-452-4518
E-mail: nng@promega.com

Singapore: Research Instruments
Phone: 775-7284
Fax: 775-9228
E-mail: resinst@singnet.com.sg

Slovak Republic: Lambda Life a.s.
Bratislava
Phone: 07-44-880-160
Fax: 07-44-880-165
E-mail: lambda@netlab.sk

Slovenia: Kemomed d.o.o.
Kranj
Phone: 0-64-351-510
Fax: 0-64-351-511
E-mail: kemomed@siol.net

South Africa: Whitehead Scientific
Cape Town
Phone: 21-981-1560
Fax: 21-981-5789
E-mail: whitesci@iafrica.com

Spain: Innogenetics Diagnostica y Tecnologia
Barcelona
Phone: 93-404-52-14/902-500010
Fax: 93-404-54-85/902-500011
E-mail: molecular@innogenetics.es

Sweden: SDS Scandinavian Diagnostic

Falkenberg
Phone: 0346-83050
Fax: 0346-84840
E-mail: info@sdsscan.se

Switzerland: Catalys AG
Wallisellen
Phone: 01-830-70-37
Fax: 01-830-55-78
E-mail: catalys_custserv@catalys.promega.com

Taiwan: Genelabs Life Science
Taipei
Phone: 02-238-25378
Fax: 02-231-18524
E-mail: genelabs@ms1.hinet.net

Thailand: Bio-Active Co., Ltd.
Bangkok
Phone: 66-2-683-01114
Fax: 66-2-295-4805
E-mail: active@infonews.co.th

The Netherlands: Promega Benelux b.v.
Leiden
Phone: 31-0-71-5324244/0800-18098 (Belgium)/0800-0221910 (NL)
Fax: 31-0-71-5324907/0800-16971 (Belgium)/0800-0226545 (NL)
E-mail: bnl_custserv@nl.promega.com

Turkey: Ermanak Miskciyan
Istanbul
Phone: 216-385-8321
Fax: 216-385-4649
E-mail: miskciyan@hotmail.com

UK: Promega UK
Southampton
Phone: 0800-378994
Fax: 0800-181037
E-mail: ukcustserve@uk.promega.com

USA: Promega Biosciences
San Luis Obispo, CA
Phone: 805-544-8524/888-234-7437
Fax: 805-543-1531/800-463-2996

Venezuela: Biochrom Laboratorio, C.A.
Caracas
Phone: 2-265-0891
Fax: 2-263-0924
E-mail: vss@ven.net

R&D Systems

USA: R&D Systems, Inc.
Minneapolis, MN 55413
Phone: 612-379-2956/800-343-7475
Fax: 612-379-6580
E-mail: info@rndsystems.com

UK: R&D Systems Europe Ltd.
Abingdon OX14 3YS
Phone: 44-0-1235-529449/800-30-20-30-20
Fax: 44-0-1235-533420
E-mail: infor@rndsystems.co.uk

Germany: R&D Systems GmbH
65205 Wiesbaden
Phone: 49-0-6122-90980/800-909-4455
Fax: 49-0-6122-909819
E-mail: infogmbh@rndsystems.co.uk

Argentina
Phone: 54-114942-3654

Australia
Phone: 61-2-9521-2177

Austria
Phone: 43-01-291-0754

Belgium
Phone: 800-10-468

Brazil
Phone: 55-21-592-6642

Chile
Phone: 56-2-2641576

Denmark
Phone: 80-01-85-92

Egypt
Phone: 20-02-525-7212

Finland
Phone: 358-3-682-2758/800-30-20-30-20

France
Phone: 800-90-72-49

Greece
Phone: 30-031-32-25-25

Hong Kong
Phone: 852-2649-9988

Israel
Phone: 972-3-6459649

Italy
Phone: 39-02-2575377

Japan
Phone: 80-3-5684-1622

Korea
Phone: 82-02-569-0781

Mexico
Phone: 52-5-612-0085

New Zealand
Phone: 64-9-377-3336

Norway
Phone: 800-11033

Poland
Phone: 48-022-720-44-54

Portugal
Phone: 351-21-862-0550

Singapore
Phone: 65-873-5997
Spain
Phone: 34-91-535-39-60

Switzerland
Phone: 800-55-2482

Taiwan
Phone: 886-2-2368-3600

Thailand
Phone: 662-246-7243

The Netherlands
Phone: 31-029-75-688-93/88-0225-607

Turkey
Phone: 90-0216-347-49-50

Venezuela
Phone: 58-2-238-6237

Rexim

France: Rexim SA
F-92400 Courbevoie
Phone: +33-1-4188-1515
Fax: +33-1-4188-1500

Germany: Rexim SA/ Degussa AG
D-60287 Frankfurt am Main
Phone: +49-69-218-2385
Fax: +49-69-218-3292

Japan: Degussa Japan Co., Ltd.
Tokyo 163-09
Phone: +81-3-5323-7312
Fax: +81-3-5323-7397

USA: Degussa Corporation Chemical Group
Ridgefield Park, NJ 07660
Phone: 201-807-3254
Fax: 201-807-3111

Sigma

USA: Sigma
St. Louis, MO 63178
Phone: 314-771-5750/800-325-3010; 800-521-8956
Fax: 314-771-5757/800-325-5052
Web: www.sigma-aldrich.com

Argentina: Sigma-Aldrich de Argentina S.A.
Phone: 54-1-807-0321
Fax: 54-1-807-0346

Australia: Sigma-Aldrich Pty., Ltd.
Sydney
Phone: 61-2-9841-0555/800-800-097
Fax: 61-2-9841-0500/800-800-096

Austria: Sigma-Aldrich Handels GmbH
Phone: 01-605-81-10
Fax: 01-605-81-20

Belgium: Sigma-Aldrich NV/SA
Phone: 03-899-13-01/0800-14747
Fax: 03-899-13-11/0800-14745

Brazil: Sigma-Aldrich Química Brasil Ltda.
Phone: 011-231-1866
Fax: 011-257-9079

Canada: Sigma-Aldrich Canada., Ltd.
Phone: 905-829-9500/800-565-1400
Fax: 905-829-9292/800-265-3858

Czech Republic: Sigma-Aldrich s.r.o.
Phone: 02-2176-361
Fax: 02-2176-356

Denmark: Sigma-Aldrich Denmark A-S
Phone: 43-56-59-00
Fax: 43-56-59-05

Finland: Sigma-Aldrich Finland/YA-Kemia Oy
Phone: 09-350-9250
Fax: 09-350-92555

France: Sigma-Aldrich Chimie S.à.r.l.
Phone: 04-74-82-28-20/0800-21-14-08
Fax: 04-74-95-68-08/0800-03-10-52

Germany: Sigma-Aldrich Chemie GmbH
Phone: 089-65-13-0/0800-5155-000
Fax: 089-65-13-1169/0800-6490-000

Greece: Sigma-Aldrich (O.M.) Ltd.
Phone: 01-9948041
Fax: 01-9943831

Hungary: Sigma-Aldrich Kft
Phone: 06-1-269-6474/06-80-355-355
Fax: 06-1-235-9050/06-80-344-344

India: Sigma-Aldrich Corp.
Bangalore
Phone: 08-851-8797
Fax: 08-851-8358
New Delhi
Phone: 011-689-9826; 011-689-7830
Fax: 011-689-9827

Ireland: Sigma-Aldrich Ireland Ltd.
Phone: 01-404-1900/800-200-888
Fax: 01-404-1910/800-600-222

Israel: Sigma Israel Chemicals Ltd.
Phone: 08-9484-222/800-70-2222
Fax: 08-9484-200

Italy: Sigma-Aldrich S.r.l.
Phone: 02-33417-310/167-827018
Fax: 02-38010737

Japan: Sigma-Aldrich Japan K.K.
Tokyo
Phone: 03-640-8885
Fax: 03-5640-8857

Korea: Sigma-Aldrich Korea Ltd.
Phone: 02-783-5211/080-023-7111
Fax: 02-783-5011/080-023-8111

Mexico: Sigma-Aldrich Química S.A. de C.V.
Mexico City
Phone: 01-5-631-3671/01-800-007-5300
Fax: 01-5-631-3780

Norway: Sigma-Aldrich Norway AS
Phone: 22-09-15-00
Fax: 22-09-15-10

Poland: Sigma-Aldrich Sp. z.o.o.
Phone: 061-823-2484
Fax: 061-823-2781

Portugal: Sigma-Aldrich Química, S.A.
Phone: 351-1-9242555/0800-20-21-80
Fax: 351-1-9242610/0800-20-21-78

Russia: Sigma-Aldrich Russia/TechCare Systems, Inc.
Phone: 7-095-975-3321
Fax: 7-095-975-4792

Singapore: Sigma-Aldrich Pte. Ltd.
Phone: 65-271-1089
Fax: 65-271-1571

South Africa: Sigma-Aldrich (Pty) Ltd.
Phone: 011-397-88-86
Fax: 011-397-88-59

Spain: Sigma-Aldrich Química S.A.
Phone: 34-91-661-9977/900-10-1376
Fax: 34-91-661-9642/900-10-2028

Sweden: Sigma-Aldrich Sweden AB
Phone: 020-35-05-10/08-7424200
Fax: 020-35-25-22/08-7427299

Switzerland: Sigma Chemie
Phone: 081-755-27-21/0800-80-00-80
Fax: 081-755-28-40

The Netherlands: Sigma-Aldrich Chemie BV
Phone: 078-620-54-11/0800-022-90-88
Fax: 078-620-54-21/0800-022-90-89

UK: Sigma-Aldrich Co. Ltd.
Phone: 01202-733114/0800-717181
Fax: 01202-715460/0800-378785

Scipac
UK: Scipac Ltd
Kent ME9 8AQ
Phone: +44-(0)1795-423077
Fax: +44-(0)1795-426942
E-mail: mail@scipac.com
Web: www.scipac.com

Senn Chem
USA: Senn Chemicals
La Jolla, CA 92037
Phone: 858-450-6091/858-450-6096
E-mail: usa@sennchem.com

France: Senn Chemicals International
F-94250 Gentilly
Phone: +33-1-49-69-10-60
Fax: +33-1-49-69-10-64
E-mail: france@sennchem.com

Switzerland: Senn Chemicals AG
CH-8157 Dielsdorf
Phone: +41-1-854-90-54
Fax: +41-1-854-90-55
E-mail: swiss@sennchem.com

Upstate
USA: Upstate Biotechnology, Inc.
Waltham, MA 02451
Phone: 781-890-8845/800-233-3991
Fax: 781-890-7738
E-mail: info@upstatebiotech.com
Web: www.upstatebiotech.com

Australia/New Zealand: Auspep Ltd. Pty.
Parkville, Victoria, Australia 3052
Phone: 613-9-328-1211/800-805-393
Fax: 613-9-326-8810
E-mail: auspep@auspep.com.au
Web: www.auspep.com.au

Austria: Biomedica GmbH
A-1210 Wien
Phone: +43-1-291-07-54
Fax: +43-1-291-07-71
E-mail: sales.biomedica@bmgrp.co.at
Web: www.biomedica.co.at

Belgium/The Netherlands: Campro Scientific
3900 AH Veenendaal, The Netherlands
Phone: +31-318-529-437
Fax: +31-318-542-181
E-mail: campro@pi.net

Brazil: Sellex, Inc
05130-001 Sao Paolo SP
Phone: +3-872-2015/USA: 202-686-3044
Fax: +3-872-1024/USA: 202-686-3177
E-mail: vendas@sellex.com/USA: washington@sellex.com
Web: www.sellex.com

Chile: Bios Chile Ingenieria Genetica SA
Santiago
Phone: +562-238-1878
Fax: +562-239-4250
E-mail: ventas@bioschile.cl
Web: www.bioschile.cl

Czech Republic/Hungary/Poland/Germany: Biomol
Feinchemikalien GmbH
D-22769 Hamburg, Germany
Phone: +49-(0)40-85-32-600/Germany: 0130-112-793
Fax: +49-(0)40-85-32-6022
E-mail: info@biomol.de
Web: www.biomol.de

Denmark: Trichem APS
DK 3600 Frederikssund
Phone: +45-4738-6628
Fax: +45-4738-6627
E-mail: jlh@trichem.dk

Finland: YA Kemia OY
FIN-00700 Helsinki
Phone: +358-(0)9-350-9250
Fax: +358-(0)9-350-92555
E-mail: info@ya-kemia.fi
Web: www.ya-kemia.fi

France: Euromedex
F-67458 Mundolsheim
Phone: +33-(0)3-88-18-07-20
Fax: +33-(0)3-88-18-07-25
E-mail: info@euromedex.com
Web: www.euromedex.com

Germany: Biozol Diagnostica Vertrieb GmbH
D-85386 Eching
Phone: +49-(0)89-37-99-666-6/0800-024-69-65/08000-BIOZOL
Fax: +49-(0)89-37-99-666-99
E-mail: info@biozol.de
Web: www.biozol.de

Greece: Chemilab Ltd
Athens
Phone: 00-30-1-67-75-363
Fax: 00-30-1-67-75-455

Hong Kong/China: Onwon Trading Ltd.
Kowloon, Hong Kong, PRC
Phone: +852-275-77569
Fax: +852-275-77211
E-mail: info@onwon.com.hk

Israel: Ornat
Rehovot 76120
Phone: +972-(0)8-947-7077
Fax: +972-(0)8-936-3034
E-mail: ornatbio@ornat.co.il
Web: www.ornat.co.il

Italy: DBA
20090 Segrate (MI)
Phone: +39-(0)2-2692-2300
Fax: +39-(0)2-2692-6058/+39-(0)2-2692-3535
E-mail: dba@interbusiness.it

Japan: Cosmo Bio Co.
Tokyo 135
Phone: +81-(0)3-5632-9630
Fax: +81-(0)3-5632-9965
E-mail: mail@cosmobio.co.jp
Web: www.cosmobio.co.jp

Japan: Wako Pure Chemicals
Osaka 541
Phone: +81-(0)66-203-3741/Tokyo: +81-(0)3-3270-8121
Fax: +81-(0)66-201-5965/+81-(0)3-3242-6501
E-mail: labchem-tec@wako-chem.co.jp
Web: www.wako-chem.co.jp

Japan: Iwai Chemicals Company
Tokyo 103-0023
Phone: +81-3-3255-2781
Fax: +81-3-3255-2825
E-mail: infocent@iwai-chem.co.jp
Web: www.iwai-chem.co.jp

Korea: Koma Biotech Inc.
Seoul 135-270
Phone: +82-(0)2-579-8787
Fax: +82-(0)2-578-7042
E-mail: komainfo@komabiotech.com
Web: www.komabiotech.com

Norway/Iceland: Medprobe AS
N–0131 Oslo, Norway
Phone: +47-2332-7380
Fax: +47-2332-7390
E-mail: medprobe@medprobe.com
Web: www.medprobe.com

Singapore: Scimed (Asia) Pte Ltd.
Singapore 128384
Phone: +65-779-3388
Fax: +65-266-3086/+65-779-0100
E-mail: scimed@singnet.com.sg

Spain: Reactiva SA
08004 Barcelona
Phone: +34-3-329-2595
Fax: +34-3-443-0668
E-mail: reactiva@intercom.es

Sweden: Kelab
S–414 74 Goteborg
Phone: +46-(0)31-125-160
Fax: +46-(0)31-148-260
E-mail: goteborg@kelab-biochem.com
Web: www.kelab-biochem.com

Switzerland: Lucerna Chem AG
CH-6000 Lucerne 10
Phone: +41-(0)41-420-9636
Fax: +41-(0)41-420-9656
E-mail: lucerna-chem@lucerna-chem.ch
Web: www.lucerna-chem.ch

Taiwan: Level Biotechnology
Taipei Hsien
Phone: +886-2-2695-9935
Fax: +886-2-2695-0403
E-mail: info@level.com.tw
Web: www.level.com.tw

UK/Ireland: TCS Biologicals
Buckingham MK 182LR, England
Phone: +44-(0)1296-714-071
Fax: +44-(0)1296-715-753
E-mail: sales@tcsbiologicals.co.uk
Web: www.tcsbiologicals.co.uk

USBio

USA: United States Biological
Swampscott, MA 01907
Phone: 800-520-3011
Fax: 781-639-1768
E-mail: chemicals@usbio.net
Web: www.usbio.net

Argentina: ETC Internacional SA
Buenos Aires
Phone: +541-639-3488/+541-639-4136
Fax: +541-639-6771
E-mail: etcint@vianetworks.net.ar

Australia: Jomar Diagnostics
Stepney, South Australia
Phone: 61-8-83626766
Fax: 61-8-83626388
E-mail: jomar@adelaide.on.net

Belgium/The Netherlands: ImmunoSource
Zoersel-Halle
Phone: +32-3385-3685
Fax: +32-3384-3818
E-mail: info@immunosource.com
Web: www.immunosource.com

Chile: Fermelo SA
Santiago
Phone: (56)-2-247-2976
Fax: (56)-2-247-2977
E-mail: fermelo@entelchile.net

Hong Kong: Advanced Technology & Industrial Company
Phone: +852-23902293
Fax: +852-27898314
E-mail: sales@advtechind.com; advanced@asianet.net.hk

Taiwan: Rainbow Biotechnology Co., Ltd.
E-mail: rainbbio@tpts5.seed.net.tw

Denmark: Saveen Biotech ApS
Phone: +45-86-20-16-16
Fax: +45-86-20-15-11
E-mail: mail@saveen-bio.se

Estonia: Naxo Ltd.
Tartu
Phone: +372-7-428001
Fax: +372-7-477131
E-mail: indrek@mail.sxpress.com
Web: www.naxolab.com

Finland: Bio-Mediator Ky
Vantaa
Phone: +358-9-852-4898
Fax: +358-9-852-4884
E-mail: info@bio-mediator.com
Web: www.bio-mediator.com

France: Euromedex
Souffelweyersheim
Phone: +33-3-88-18-07-22
Fax: +33-3-88-18-07-25
E-mail: info@euromedex.com

Germany: DPC Bierham GmbH
Herford NRW
Phone: +49-5221-349817
Fax: +49-5221-349566
E-mail: HJSoll.acris@t-online.de

Greece: Diachel, Ltd. (Diagnostics Chemical Instrumentation)
Athens
Phone: 0030-1-72-39-306/72-39-307-7
Fax: 0030-1-72-19-874
E-mail: diachel@hol.gr

India: DNA Technologies, Pvt. Ltd.
Gorakhpur
Phone: 011-91-55-133-7355

India: Gaurav Enterprise
D-29 Kamla Nagar Agra-282 005 (U.P.)
Phone: 562-383724
Fax: 562-381414
E-mail: akshayco@nde.vsnl.net.in

India: PMK International
Indianapolis, IN, USA
Phone: 317-872-4957
Fax: 317-875-9150
E-mail: pmkintl@iquest.com

India: Tropica Diagnostics
Delhi
Phone: +91-11-5122368
E-mail: ramneek@cbs.dtu.dk

Italy: Diagnostic Brokers Associated (DBA)
Milano
Phone: +02-26922300
Fax: +02-26923535
E-mail: dba@sente.it

Japan: Dainippon Pharmaceutical Co., Ltd.
Osaka
Phone: +81-6-6386-2164
Fax: +81-6-6337-1606
E-mail: shinzo-nishimura@dainippon-pharm.co.jp
Web: www.dainippon-pharm.co.jp

Korea: Dongil Biotech
Seoul
Phone: +82-2-571-3790
Fax: +82-2-571-3792
E-mail: dibio@chollian.net

Malaysia: BioSynTech Sdn Bhd
Selangor DE
Phone: +603-736-9198
Fax: +603-736-9242
E-mail: info@bst-asia.com

Mexico: Control Tecnico y Representaciones (CTR
Scientific)
Nuevo Leon
Phone: 52-8-370-1771
Fax: 52-8-373-2891
E-mail: controltec@infosel.net.mx

Mexico/Columbia/Peru: Y Corporation of America, Inc
Coral Gables, FL
Phone: 305-629-8808
Fax: 305-629-8809
E-mail: pyunis@aol.com

Norway: Saveen Biotech AS
Phone: +47-22-22-87-87
Fax: +47-35-53-07-99
E-mail: mail@saveen-bio.se

Poland: Pointe Scientific
Warsaw
Phone: +48-22-6426779/+48-22-8428049
Fax: +48-22-6426779
E-mail: mkawcz@friko4.onet.pl

Saudi Arabia: Al-Habib Company Ltd.
Riyadh
Phone: 00-966-1-460-2188
Fax: 00-966-1-460-2216

Spain: Cultek
Madrid
Phone: +34-91-729-0333
Fax: +34-91-3580761
E-mail: sgarcia@cultek.com; mmatas@cultek.com

Sweden: Saveen Biotech AB
Malmo
Phone: +46-40-51-00-00
Fax: +46-40-16-45-00
E-mail: mail@saveen-bio.se

Switzerland: Lucerna Chem AG
CH-6000 Lucerne
Phone: +41-(0)41-420-9636
Fax: +41-(0)41-420-9656
E-mail: lucerna-chem@lucerna-chem.ch
Web: www.lucerna-chem.ch

Taiwan: Bio-Genesis Technologies Inc
Taipei
Phone: 886-2-23825378
Fax: 886-2-23716316
E-mail: f4903180@ms13.hinet.net

Thailand: Pacific Science Co., Ltd.
Bangkok
Phone: 66-2-880-8750/880-9473
Fax: 66-2-880-8751
E-mail: pacscien@ksc.th.com

UK: Europa Bioproducts, Ltd.
Cambridge
Phone: +1-353-721118
Fax: +1-353-624589
E-mail: info@europa-bioproduts.com

Appendix B. Glossary of Abbreviations & Acronyms

2

2-Br-Z	2-Bromobenzyloxycarbonyl
2-Cl-Z	2-Chlorobenzyloxycarbonyl

4

4MeBzl	4-Methylbenzyl

A

A	Alanine
AA	Amino acid(s)
Aaa	Adamantaneacetyl
AAA	Agglutinin, *Anguilla anguilla*
AAP	Antiarrhythmic Peptide
ABA	Agglutinin, *Agaricus bisporus*
Abu	Aminobutyric acid
Abz	Aminobenzoic acid
Ac	Acetyl
AC	Adenylate Cyclase
Aca	Aminocaproic acid
ACAP	Adenylate Cyclase Activating Peptide
ACE	Angiotensin Converting Enzyme
Acm	Acetamidomethyl
ACP	Acyl Carrier Protein
Acrylodan	6-Acryloyl-2-dimethylaminonaphthalene
ACT	Antichymotrypsin
ACTH	Adrenocorticotropic Hormone
ADNF	Neurotrophic Factor, Activity-Dependent
AFC	7-Amido-4-trifluoromethylcoumarin
AFP	Fetoprotein, α-
AFPr	Fetoprotein Receptor, α-
AG	Acid Glycoprotein
Aga	Agatoxin
AGP	Acid Glycoprotein
AGRP	Agouti Related Peptide
AgTx	Agitoxin
AHMHA	(3*S*,4*S*)-4-Amino-3-hydroxy-6-methyl-heptanoic acid
Ahx	6-Aminohexanoic acid
Aib	Aminoisobutyric acid
AIF	Adipogenesis Inhibitory Factor
AIGF	Androgen Induced Growth Factor
AIP	Autocamitide II Related Inhibitory Peptide
AKAP	A-Kinase Anchoring Protein
AKH	Adipokinetic Hormone
Ala	Alanine
Alloc	Allyloxycarbonyl
Allocam	Allyloxycarbonylaminomethyl
ALS	Acid labile subunit

AM	Adrenomedullin
AMC	7-Amido-4-methylcoumarin (*See*, MCA)
ANP	Atrial (A-Type) Natriuretic Peptide
ANT	Atrial Natriuretic Factor
Aoc	*t*-Amyloxycarbonyl
AP	Anthopleurin
APA	Agglutinin, *Abrus precatorius*
APC	Activated Protein C
APC	Allophycocyanin
APMA	4-Aminophenylmercuric acetate
APO	Apolipoprotein
APP	Amyloid Protein Precursor
AR	Amphiregulin
AR	Androgen Receptor Protein N-Terminal Peptide
Arg	Arginine
ASIF	Aldosterone Secretin Inhibiting Factor
Asn	Asparagines
Asp	Aspartic acid
Asu	L-α-Aminosuberic acid
AT	Antithrombin
AVP	Vasopressin, Arginine
AZM	Macroglobulin, αII

B

BAF	B-Cell Activating Factor
BAM	Bovine Adrenal Medulla
BCA	B-Cell Attracting Chemokine
BD	Defensin, β-
BDF	B-Cell Differentiation Factor
BDNF	Neurotrophic Factor, Brain Derived
BGP	Bone Gla Protein
BHA	Benzhydrylamine
Bhg	5H-Dibenzo(a,d)cycloheptene-5-acetic acid
BINP	Brain Injury-Derived Neurotrophic Peptide
BIP	Immunoglobulin Heavy Chain Binding Protein
BLP	Bombesin-Like Peptide
BME	β-Mercaptoethanol
BMP	Bone Morphogenic Protein
BNP	Natriuretic Peptide, Brain
Boc/BOC/ *t*-Boc	*t*-Butyloxycarbonyl
Bom	Benzyloxymethyl
BOP	Benzotriazolyl-*N*-oyl-tris (dimethylamino)phosphonium hexafluorophosphate (Castro's reagent)
BPA	Agglutinin, *Bauhinia purpuria*
BPE	Bovine Pituitary Extract
Bpoc	2-*p*-Biphenylisopropyloxycarbonyl

BPP	Bradykinin Potentiating Pentapeptide
BS	Agglutinin, *Bandeiraea simplicifolia*
BSA	Albumin, Bovine Serum
BSF	B-Cell Stimulatory Factor
Bsmoc	Benzo[6]thiophenesulfone-2-methoxycarbonyl
BSP	Bone Sialoprotein
BTC	Betacellulin
BTD	(3*S*,6*S*,9*R*)-2-Oxo-3-amino-7-thia-1-aza-bicyclo(430)nonane-9-carboxylic acid
BTG	Thromboglobulin, β-
Bu	Butyl
BuOH	Butanol
BuTx	Bungarotoxin
Bz	Benzoyl
Bzl	Benzyl

C

C	Cysteine
CA	Cancer Antigen
CAA	Agglutinin, *Caragana aborescens*
CAA	Cancer Associated Antigen
CaC	Calcicludine
CALP	Calcium-Like Peptide
Calpain	Calcium Dependent Cysteine Protease
CaM	Calmodulin
CaMK	Calmodulin-Dependent Protein Kinase
CaM-PDE	Calmodulin Dependent Phosphodiesterase
CaN	Calcineurin
CAP	C-Terminal Adjacent Peptide
CART	Cocaine & Amphetamine Regulated Transcript
CaS	Calciseptine
CBG	Corticosteroid Binding Globulin
CBG	Cortisol Binding Globulin
Cbz, CBZ, Z	Benzyloxycarbonyl; carbobenzoxy
CCAP	Crustacean Cardioactive Peptide
CCK	Cholecystokinin
CCK-8	Cholecystokinin Octapeptide
CCR	C-C Chemokine Receptor
CDK	Cell-Cycle Dependent Kinase
CDK	Cyclin-Dependent Kinase
CEA	Carcinoembryonic Antigen
CFTR	Cystic Fibrosis Transmembrane Conductance Regulator Protein
CFU-SA	Colony Forming U-Stimulating Activity
CG	Chorionic Gonadotropin
CGRP	Calcitonin Gene Regulated Peptide
CgTx	Conotoxin
Cha	Cyclohexylalanine; β-Cyclohexyl-L-alanine

CHO	Aldehyde
CHO	Chinese hamster ovary
ChTx	Charybdotoxin
Chx	Cyclohexyl
CIF	Colony Inhibition Factor
CINC	Cytokine-Induced Neutrophil Chemoattractant
CIP	Corticotropin Inhibiting Peptide
Cit	Citrulline
CK	Casein Kinase
CLIP	Corticotropin-Like Intermediate Peptide
CLMF	Cytotoxic Lymphocyte Maturation Factor
CMK	Chloromethylketone
CMV	Cytomegalovirus
CNP	C-Type Natriuretic Peptide
CNS	Central nervous system
CNTF	Neurotrophic Factor, Ciliary
ConA	Concanavalin A; Agglutinin, *Conavalia ensiformis*
Cpd	Capreomycidine; (*S*,*S*)-α-(2-iminohexahydro-4-pyrimidyl)glycine
CRF	Corticotropin Releasing Factor
CRIF	Corticotropin Release-Inhibiting Factor
CRP	C-Reactive Protein
CS	Calpastatin
CSF	Colony Stimulating Factor
CSIF	Cytokine Synthesis Inhibitory Factor
CSL	Agglutinin, *Cytisus scoparius*
CST	Cortistatin
CT	Calcitonin
CT	Cardiotrophin
CT	Crotoxin
C-TAC	Cutaneous T-Cell Attracting Chemokine
Cys	Cysteine

D

Dabcyl, DABCYL	4-(4-Dimethylaminophenylazo)benzoyl *N*-(7-dimethylamino-4-methylcoumarinyl)maleimide
Dansyl	5-Dimethylaminonaphthalene-1-sulfonyl
DAP	Diabetes Associated Peptide; Dipeptidyl Aminopeptidase
Dapa	Diaminopropionic acid
DBA	Agglutinin, *Dolichos biflorus*
DBI	Diazepam Binding Inhibitor
DCB	Dichlorobenzoyl
DCC	Dicyclohexylcarbodiimide
DCHA	Dicyclohexylammonium salt
DDAVP	Desamino-(D-Arg)-Vasopressin
Deg, DEG	Diethylglycine
DFF	DNA Fragmentation Factor

DGAP	Dermorphin Gene Associated Peptide
DIA	Differentiation Inhibition Activity
DiBr	Dibromo
Dip	3,3-Diphenylalanine
Dmob	2,4-Dimethyloxybenzyl
Dmpc	cis-2,6-Dimethylpiperidinocarbonyl
Dnp	Dinitrophenyl; 2,4-Dinitrophenyl
DSIP	Sleep Inducing Peptide, Delta
DTx	Dendrotoxin
Dyn	Dynamin

E

E	Glutamic acid
EAE	Experimental Allergic Encephalitogenic; Experimental Allergic Encephalomyelitis
EBV	Epstein Barr Virus
ECA	Agglutinin, *Erythrina christagalli*
ECE	Endothelin Converting Enzyme
ECGF	Endothelial Cell Growth Factor
ECGS	Endothelial Cell growth Supplement
EcorA	Agglutinin, *Erythrina corallodendron*
EDANS	5-((2-Aminoethyl)amino)-naphthalene-1-sulfonic acid
EDF	Eosinophil Differentiating Factor
EEA	Agglutinin, *Euonymus europaeus*
EGF	Epidermal Growth Factor
EGF-R	Epidermal Growth Factor Receptor
ELH	Egg Laying Hormone
EMAP	Endothelial-Monocyte Activating Polypeptide
EMP	Erythropoietin Mimetic Peptide
ENA	Epithelial Neutrophil Activating Peptide
EO-CSF	Eosinophil Colony Stimulating Factor
EOT	Eotaxin
EP	Endogenous Pyrogen
EPO	Erythropoietin
ER	Estrogen Receptor
ERCC	Excision Repair Cross Complement
ES	Embryonic Stem Cells
ES	Endostatin
ET	Endothelin
Et	Ethyl
ETH	Ecdysis-Triggering Hormone
EtOH	Ethanol

F

F	Phenylalanine
FA	3-(2-Furyl)acryloyl
FABP	Fatty Acid Binding Protein
FAOM	Fluoroacyloxymethylketone

FBP	Folate Binding Protein
FDP	Fibrin Degradation Product
FeLV	Leukemia Virus, Feline
FGF	Fibroblast Growth Factor
FITC	Fluorescein isothiocyanate
FMK	Fluoromethylketone
Fmoc, FMOC	9-Fluorenylmethoxycarbonyl
For	Formyl
FPA	Fibrinopeptide A
FSH	Follicle Stimulating Hormone
Ftase	Farnesyltransferase
FTN	Fibroblast Fibronectin
FTS	Serum Thymic Factor

G

G	Glycine
gA	Gramicidin A
GABA	γ-Aminobutyric acid
GAD	Glutamic Acid Decarboxylase
GAPDH	Glyceraldehyde-3-Phosphate Dehydrogenase
GCE-γ	Glutathione Monoethyl Ester
G-CSF	Granulocyte Colony Stimulating Factor
GDNF	Neurotrophic Factor, Glial Derived
GFAP	Glial Fibrillary Acidic Protein
GH	Growth Hormone
GHRH	Growth Hormone Releasing Hormone (Factor)
GIF	Growth Hormone Release Inhibiting Factor
GIP	Gastric Inhibitory Peptide
GIP	Glucose Dependent Insulinotropic Hormone
Gla	γ-Carboxyglutamic acid
Gln	Glutamine
GLP	Glucagon-Like Peptide
Glp	Pyroglutamic acid
Glt	Glutaryl
Glu	Glutamic acid
GluR	Glutamate Receptor
GLUT	Glucose Transporter
Gly	Glycine
GMAP	Galanin Message Associated Peptide
GM-CSF	Granulocyte Macrophage Colony Stimulating Factor
GNL	Agglutinin, *Galanthus nivalis*
GnRH	Gonadotropin Releasing Hormone
GP	Glycogen Phosphorylase
GP	G-Protein
GPH	Glycoprotein Hormone
GR	Glucocorticoid Receptor Peptide
GRF	Growth Hormone Releasing Factor

Gro/GRO	Growth-Releated Oncogene
GRP	Gastrin Releasing Peptide
GRP	Gonadotropin Releasing Factor
GS	Glycogen Synthase
GSH	Glutathione, Reduced
GSMEE	Glutathione Monoethyl Ester
GSSG	Glutathione, Oxidized
GTx	Geographutoxin

H

H	Histidine
HA	Hemagglutinin
Har	Homoarginine
HARP	Heparin Affin Regulatory Peptide
HAV	Hepatitis A Virus
Hb	Hemoglobin
HBBM	Heparin Binding Brain Mitogen
HB-EGF	Heparin-Binding Epidermal Growth Factor-Like Growth Factor
HB-GAM	Heparin Binding Growth Associated Molecule
HBGF	Heparin Binding Growth Factor
HBTU	2-(1H-Benzotriazol-1-yl)-1,1,3,3-tetramethyluronium hexafluorophosphate
HBV	Hepatitis B Virus
HCC	Hemofiltrate CC Chemokine
hCG	Corionic Gonadotropin, Human
HCG	Human Chorionic Gonadotropin
Hci	Homocitrulline
HCNP	Hippocampal Cholinergic Neurostimulating Peptide
HCV	Hepatitis C Virus
Hcy	Homocysteine
HDL	High Density Lipoprotein
HDV	Hepatitis D Virus
HEPP	Human IgE Pentapeptide
hF-GRP	Human Follicular Gonadotropin Releasing Peptide
HGF	Growth Factor, Hepatocyte
HGF	Hepatocyte Growth Factor
HGF	Hyperglycemic-Glycogenolytic Factor
hGH	Growth Hormone, Human
Hgln	Homoglutamine
Hglu	Homoglutamic acid
HHA	Agglutinin, Helix aspersa
Hhi	Homohistidine
HILDA	Human Interleukin for DA Cells
HIP	Hemagglutination Inhibiting Peptide
His	Histidine
HIV	Human Immunodeficiency Virus
Hle	Homoleucine

Hly	Homolysine
Hme	Homomethionine
HMM	Meromyosin, Heavy
hNP	Neutrophil Peptide, Human
Hobt	1-Hydroxybenzotriazole
Hof	Homophenylalanine
Hol	Homoleucine
Hosu	N-Hydroxysuccinimide
HPA	Agglutinin, Helix pomatia
Hph	Homophenylalanine
HPL	Pituitary Lactogen, Human
hPL	Placental Lactogen, Human
Hpr	Homoproline
HRG	Histidine Rich Glycoprotein
HS	Hymenistatin
HSA	Albumin, Human Serum
Hse	Homoserine
HSF	Heat Shock Factor
HSF	Hepatocyte Stimulating Factor
HSP	Heat Shock Protein
HSPG	Heparin Sulfate Proteoglycan
HSPI	Human Seminal Plasma Inhibin
HSV	Herpes Simplex Virus
Hth	Homothreonine
HTLV	Human T-Cell Lymphotropic Virus
HTP	hydroxytryptophan
Htr	Homotryptophan
Hty	Homotyrosine
Hva	Homovaline
Hyal	Hydroxyalanine
Hyasp	Hydroxyaspartic acid
Hyl	Hydroxylysine
Hylva	Hydroxyisovaleric acid
Hym	Hydroxymethionine
Hyn	Hydroxynorvaline
Hyp	Hydroxyproline
Hyph	Hydroxyphenylalanine
Hytr	Hydroxytryptophan
Hyty	Hydroxytyrosine

I

I	Isoleucine
IAPP	Insulinoma or Islet Amyloid Polypeptide
Ias	Isoasparagine
IbTx	Iberiotoxin
ICAM	Intracellular Adhesion Molecule
ICE	Interleukin Converting Enzyme

IFN	Interferon
IGF	Insulin-Like Growth Factor
IGFBP	Insulin-Like Growth Factor Binding Protein
IGIF	Interferon-γ Inducing Factor
IgIn	Isoglutamine
IgIu	Isoglutamic acid
IL	Interleukin
Ile	Isoleucine
IN	Integrase
IP	Interferon-γ Inducible Protein
IPA	Isopropanol
IPase	Isopeptidase
IpOH	Isopropanol
Ipr	Isopropyl
IpTxa	Imperatoxin A
IRRP	Interferon-α Receptor Recognition Peptide
IRS	Insulin Receptor Substrate
Ise	Isoserine
I-TAC	Interferon-Inducible T-Cell α-Chemoattractant
IU	International unit

K

K	Lysine
KAF	Keratinocyte Autocrine Factor
KGF	Keratinocyte Growth Factor
KKI	Kallikrein Inhibitor
KL	c-Kit Ligand
KLH	Hemocyanin, Keyhole Limpet
KSPG	Keratan Sulfate Proteoglycan
KTx	Kaliotoxin

L

L	Leucine
L	Leukokinin
LAA	Agglutinin, *Laburnum alpinum*
Lac	Lactic acid
LAF	Lymphocyte Activating Factor
LAMP	Lysosome-Associated Membrane Protein
LAP	Latency Associated Peptide
LBA	Agglutinin, *Phaseolus limensis*
LCF	Lymphocyte Chemoattractant Factor
LCH	Agglutinin, *Lens culinaris*
LCMV	Lymphocytic Choriomeningitis Virus Peptide
LCRF	Luminal Cholecystokinin Releasing Factor
LDL	Low Density Lipoprotein
LEA	Agglutinin, *Lycopersicon esculentum*
LEC	Liver Expressed Chemokine
Leu	Leucine

LFA	Agglutinin, *Limax flavus*
LH	Luteinizing Hormone
LHRH	Luteinizing Hormone Releasing Hormone
LIF	Leukemia Inhibitory Factor
LIFR	Leukemia Inhibitory Factor Soluble Receptor
LIX	Lipopolysaccharide Induced cxc Chemokine
LMS	Leukomyosuppressin
LOA	Agglutinin, *Lathyrus odoratus*
LORF	Leech Osmoregulatory Factor
LP	Lipoprotein
LPA	Agglutinin, *Limulus polyphemus*
LPK	Leukopyrokinin
LTA4	Leukotriene A4
LTH	Leuteotropic Hormone
LTIF	Lethal Toxin Inhibiting Factor
LTx	Latrotoxin
LVP	Vasopressin, Lysine
Lys	Lysine

M

M	Macroglobulin
M	Methionine
MAA	Agglutinin, *Maackia amurensis*
MAGE	Melanoma Associated Antigen
MAGP	Microfibril-Associated Glycoprotein
MAP	Microtubule Associating Protein
MAP	Mitogen-Activated Protein
MAP	Multiple Antigenic Peptide
MARCKS	Myristoylated Alanine Rich C-Kinase Substrate
Mas	Mastoparan
MBHA	Methylbenzhydrylamine
MBNA	Methoxy-β-napthylamide
MBP	Mannan Binding Protein
MBP	Myelin Basic Protein
Mbzl, MeBzl	4-Methoxybenzyl
MCA	7-Methoxycoumarin-4-yl acetyl; α-(4-Methyl-coumaryl-7-amide); 7-Amido-4-methylcoumarin (See, AMC)
MCA	Agglutinin, *Momordica charantia*
MCA	Mammary Carcinoma
MCAF	Macrophage/Monocyte Chemotactic & Activating Factor
MCC	Moth Cytochrome C
MCD	Mast Cell Degranlating
MCGF	Mast Cell Growth Factor
MCH	Melanin Concentrating Hormone
MCP	Macrophage/Monocyte Chemoattractant Protein
MCP	Macrophage/Monocyte Chemotactic Protein

968

M-CSF	Colony Stimulating Factor, Macrophage
MDC	Macrophage Derived Chemokine
MDP	Muramyl Dipeptide
Me	Methyl
Me-Gly	Methylglycine
MeOH	methanol
Met	Methionine
MGF	Mast Cell Growth Factor
MGOP	Melanin Concentrating Hormone-Gene-Overprinted Polypeptide
MGSA	Melanoma Growth Stimulating Activity
MgTx	Margatoxin
MIF	Macrophage Migration Inhibitory Factor
MIF	Melanocyte Stimulating Hormone Release Inhibiting Factor
MIG	Monokine, Interferon-γ Induced
MIP	Macrophage Inflammatory Peptide
MK	Midkine
MMA	Metamorphosin A
MMP	Matrix Metalloproteinase
MOC	7-Methoxycoumarin
MOG	Myelin Oligodendrocyte Glycoprotein
Mox	Methoxinine
MP	Mitogenic Protein
MP	Myelopeptide
MPA	Agglutinin, *Maclura pomifera*
MPF	Melanotropin Potentiating Factor
Mpr	3-(4-Methylbenzyl)-thiopropionic acid
MSH	Melanocyte Stimulating Hormone
MT	Melanotan
MT	Muscarinic Toxin
MTH	Methylthiohydantoin
Mtr	4-Methoxy-2,3,6-trimethylbenzenesulfonyl
Mts	Mesitylene-2-sulfonyl
Mtt	3-(4,5-Dimethylthiazo(-2-yl)-2,5-diphenyltetrazolium bromide
MTx	Muscarinic Toxin
Mu	Morpholinoureidyl
Myr	Myristoyl

N

N	Asparagine
NA	Nitroanilide
NAC	Non-β-Amyloid
NAIP	Neuronal Apoptosis Inhibitory Protein
Nal	2-Naphthylalanine
NAP	Neuroprotective Factor, Activity-Dependent
NAP	Neutrophil Activating Protein
NaPB	Sodium phosphate buffer

NCA	Non-Specific Cross Reacting Antigen
NEI	Neuropeptide EI
N-g	Guanidine nitrogen
NGE	Neuropeptide GE
NGF	Nerve Growth Factor
NH₄HCO₃	Ammonium bicarbonate
NHEt	*N*-Ethylamide
NHNH₂	Hydrazide
Nic	*N*-α-Nicotinyl
N-im	Imidazole nitrogen
N-in	Indole nitrogen
NIPP	Anti-Nuclear Inhibitor of Protein Phosphatase
NIPP	Nuclear Inhibitor of Protein Phosphatase
NK	Neurokinin
NKSF	Natural Killer Cell Stimulatory Factor
NKSF	Natural Killer Cell Stimulatory Factor
Nle	Norleucine
NMa	*N*-Methylanthranoyl
NMDA	*N*-Methyl-D-Aspartate
NMe	*N*-Methylamide
NMe₂	*N,N*-Dimethylamide
NMU	Neuromedin U
NO₂	Nitro
NOS	Nitric oxide synthase
NOS	Nitric Oxide Synthase
NPA	Agglutinin, *Narcissus pseudonarcissus*
Nps	2-Nitrophenylsulfonyl
NPTx	Nephilatoxin
NPY	Neuropeptide Y
Npys, NPYS	3-Nitro-2-pyridinesulfonyl
NT	Neurotrophin
NTB	Naltriben
NTN	Neurturin
Nva	Norvaline

O

OAll	Allyl ester
OB	Obese Protein
OBg	*N*-Benzhydryl-glycoamide ester
OBzl	Benzyl ester
OChx, OcHex	Cyclohexyl ester
OCIF	Osteoclastogenesis Inhibitory Factor
OCN	Octadecaneuropeptide
OcPent	Cyclopentyl ester
ODN	Octadecane Neuropeptide
OEt	Ethyl ester
OFm	α-9-Fluorenylmethyl ester

OGP	Osteogenic Growth Peptide
Oic	(3aS,7aS)-Octahydroindolyl-2-carboxylic acid
OMe	Methyl ester
ONb	Nitrobenzyl ester
ONp	Nitrophenyl ester
OPfp	Pentafluorophenyl ester
OPG	Osteoprotegerin
OPh	Phenyl ester
OR	Orexin Receptor
ORL1	Opioid Receptor-Like 1
Orn	Ornithine
OSF	Osteoblast Specific Factor
OSM	Oncostatin M
OT	Oxytocin
OtBu	*t*-Butyl ester
OVA	Ovalbumin

P

P	Proline
PA	Agglutinin, *Pseudomonas auerginosa*
PACAP	Pituitary Adenylate Cyclase Activating Polypeptide
PAI	Plasminogen Activator Inhibitor
Pal	Pyridylalanine
PAM	Phenylacetamidomethyl
PAMP	Pro-Adrenalmedullin
PAPP-A	Protein A Pregnancy Associated Plasma
PAR	Proteinase Activated Receptor
PARC	Pulmonary & Activation Regulated Chemokine
Pbf	2,2,4,6,7-Pentamethyl-dihydrobenzofuran-5-sulfonyl
PBR	Peripheral-Type Benzodiazepine Receptor
PBSF	Pre-B-Cell Growth Stimulating Factor
PBSF	Pre-B-Cell Stimulation Factor
PCA	Agglutinin, *Phaseolus coccineus*
PCT-GF	Plasmacytoma Growth Factor
PD-ECGF	Platelet Derived Endothelial Cell Growth Factor
PDGF	Platelet Derived Growth Factor
PEG	Polyethylene glycol
Pen	Penicillamine
PEX	Pituitary Extract
PF	Platelet Factor
Pfp	Pentafluorophenyl ester
PG	Proteoglycan
pGlu	Pyroglutamic acid (*See* Glp)
PGP	Protein Gene Product
PHA	Phytohemagglutinin
PHA-L	Leukoagglutinin

PHAS	Phosphorylated Heat- & Acid-Stable Protein
Phe	Phenylalanine
Phg	Phenylglycine
PHI	Peptide Histidine Isoleucine
PHM	Peptide Histidine Methionine
PI	Phosphatidylinositol
PI 3-K	Phosphatidylinositol 3-Kinase
pIL	Interleukin Precursor
Pip	Pipecolyl
PKA	Protein Kinase A
PKC	Protein Kinase C
PKCi	Protein Kinase C Inhibitor
PKG	Protein Kinase G
PKI	Protein Kinase Inhibitor
PKI-tide	Protein Kinase Inhibitor Peptide
PL	Lactogen, Placental
PL	Placental Lactogen
PLA2	Phospholipase A2
PLAP	Phospholipase A_2 Activating Peptide
PLGF	Placenta Growth Factor
PLGF	Placental Growth Factor
PLP	Proteolipid Protein
Pmc	2,2,5,7,8-Pentamethylchroman-6-sulfonyl
p-MeBzl	4-Methylbenzyl
PMEL	Melanocyte Protein
PMN	Peptone Peptonize Milk Nutrient
PMP	Peroxisomal Membrane Protein
Pmp	β-Mercapto-b, β-Cyclopentamethylene propionic acid
PMSG	Gonadotropin, Pregnant Mare Serum
PMT	*Pasteurella multocida* Toxin
PNA	Agglutinin, Peanut
pNA	*p*-Nitroanilide; 4-Nitroanilide
PP	Placental Protein
PP	Protein Phosphatase
PPA	Agglutinin, *Ptilota plumosa*
PPI	Protein Phosphatase Inhibitor
PPIase	Peptidyl Prolyl Isomerase
Pra	Propargyl
PRL	Prolactin
Pro	Proline
PrP	Prion Protein
PrRP	Prolactin Releasing Peptide
PSA	Agglutinin, *Pisum sativum*
PSA	Prostate Specific Antigen
PSI	Proteosome Inibitor
PSP	Persephin
PSP	Prostate Specific Protein

PSTAIR, PSTAIRE	p34cdc2 Peptide		SGLT	Sodium Glucose Transporter
PTH	Parathyroid Hormone		SHA	Agglutinin, *Salvia horminum*
Pth	Phenylthiohydantoin		SHBG	Globulin, Sex Hormone Binding
PTHRP	Parathyroid Hormone Related Peptide		SHBG	Sex Hormone Binding Globulin
PTH-rP	Parathyroid Hormone Related Peptide		ShK	Neurotoxin, *Stichodactyla helianthus*
PTN	Pleiotrophin		ShNA	Neurotoxin, *Stichodactyla helianthus*
PTN	Pleiotrophin		SIMP	
PTP	Protein Tyrosine Phosphatase		SJA	Agglutinin, *Sophora japonica*
PVP	Polyvinylpyrrolidone		SLPI	Secretory Leukocyte Protease Inhibitor
PWM	Pokeweed Mitogen		SNA	Agglutinin, *Sambucus nigra*
Pya	Pyridylalanine		SOD	Super oxide dismutase
Pyr	Pyroglutamyl (*See* Glp)		SOD	Super Oxide Dismutase
PYY	Peptide YY		SOP	Serine-O-Phosphate
			SP	Glycoprotein, Specific

Q

Q	Glutamine		SpA	Protein A, *Staphylococcus aureus*
			SPAI	Sodium Potassium ATPase Inhibitor
			SPARC	

R

R	Arginine		SPDP	3-(2-Pyridyldithio)propionic acid
RANTES	Regulated on activation, normal T-cell expressed and secreted		SPF	Seminal Plasmin Fragment
RAR	Retinoic Acid Receptor		SRIF	Somatotropin Release Inhibiting Factor
RBP	Retinol Binding Protein		SrTx	Sarafotosin
RBP	Riboflavin Binding Protein		S-S	Sulfhydryl (disulfide) bond
RCA	Agglutinin, *Ricinus communis*		SSA	Agglutinin, *Salvia sclarea*
RPA	Agglutinin, *Robinia pseudoacacia*		SSB	Single Strand Binding Protein
R-PE	Phycoerythrin		SSB	Single Stranded DNA Binding Protein
RSA	Albumin, Rat Serum		St	Stearated
Rsk	Ribosomal S6 Kinase		STA	Agglutinin, *Solanum tuberosum*
RSV	Respiratory Snycytial Virus		Sta	Statine; (3*S*,4*S*)-4-Amino-3-hydroxy-6-methylheptanoic acid
			STAT	Signal Transducer & Activator of Transcription
			StBu	*t*-Butylthio
			STF	Serum Thymic Factor
			Su, Suc	Succinyl

S

S	Serine			
SA	Specific activity			
SAA	Serum Amyloid A			

T

SAP	Serum Amyloid P		T	Threonine
SAP	Sperm Activating Protein		T4	Thyroxine
Sar	Sarcosine		TARC	Thymus and Activation-Regulated Chemokine
SBA	Agglutinin, *Soybean*; Agglutinin, *Glycine max*		TBG	Globulin, Thyroxine Binding
SCF	Stem Cell Factor		TBG	Thyroxine Binding Globulin
SCGF	Stem Cell Growth Factor		TBTU	2-(1H-Benzotrialzole-1-yl) 1,1,3,3-tetramethyluronium tetrafluoroborate
SCPa	Cardioactive Peptide A, Small		tBu	*t*-Butyl
SDF	Stromal Cell Derived Factor		TCA	T-Cell Activation Gene
SDF	Stromal Cell Derived Factor		TCGF	T-Cell Growth Factor
SDS	Sodium dodecyl sulfate		TCx	Taicatoxin
SEB	Staphylococcal Enterotoxin B		TeBG	Testosterone-Estradiol Binding Globulin
Ser	Serine		TeBG	Testosterone-Estradiol Binding Globulin
SGI	Secretogranin		TECK	Thymus-Expressed Chemokine

TF	Tissue Factor
Tfa, tfa, TFA	Trifluoroacetyl; Trifluoroacetate
TG	Globulin, Thyro-
Tg	Thyroglobulin
Tg	Thyroglobulin
TGF	Transforming Growth Factor
THA	threo-Hydroxyaspartic Acid
THG	Glycoprotein, Tamm-Horsfall
Thi	β-(2-Thienyl) alanine
Thr	Threonine
Thz	Thioproline; Thiazolidine-4-carbozylic acid
Tic	1,2,3,4-Tetrahydroisoquinoline-3-carboxylic acid
TIMP	Tissue Inhibitor of Metalloproteinase
TKA	Agglutinin, *Trichosanthes kinlowii*
Tle	*t*-Leucine
TMA	Thyroid Microsomal Antigen
TMA	Thyroid Microsomal Antigen
Tn	Troponin
TNBS	2,4,6-Trinitrobenzenesulfonic acid
TNF	Tumor Necrosis Factor
Tosyl, Tos	4-Toluenesulfonyl tosylate
TP	Thymopoietin
TPA	Tissue Polypeptide Antigen
t-PA/TPA	Tissue Plasminogen Activator
TPO	Thrombopoietin
TRAIL	Tissue Necrosis Factor Related Apoptosis Inducing Ligand
TRAP	Thrombin Receptor Activator Peptide
TRF	T-Cell Replacing Factor
TRH	Thyrotropin Releasing Hormone
TRITC	Tetramethylrhodamine isothiocyanate
Trp	Tryptophan
Trt	Trityl
TSF	Toluenesulfonylfluoride
TSH	Thyroid Stimulating Hormone
TSST	Toxic Shock Syndrome Toxin
TSST	Toxic Shock Syndrome Toxin
TsTx	Tityustoxin
TTC	Tetanus Toxin C Fragment
TTN	Tissue Necrosis Factor
Tyr	Tyrosine
Tyr(Me)	O-Methyl-L-tyrosine

U

U	Unit
Ub	Ubiquitin
UEA	Agglutinin, *Ulex europaeus*
UP	Urinary Protein
USP	*United States Pharmacopoeia*
UTI	Urinary Trypsin Inhibitor

V

V	Valine
VAA	Agglutinin, *Viscum album*
Val	Valine
VCA	Viral Capsid Antigen
VEGF	Vascular Endothelial Cell Growth Factor
VEGF	Vascular Endothelial Growth Factor
VFA	Agglutinin, *Vicia faba*
VIC	Vasoactive Intestinal Contractor (Constrictor)
VIP	Vasoactive Intestinal Peptide
VLDL	Very Low Density Lipoprotein
VNP	Vasonatrin Peptide
VP	Vasopressin
VPF	Vasculotropin
VQY	Valosin Peptide
VSA	Agglutinin, *Vicia sativa*
VSV	Vesicular Stomatitis Virus
VVA	Agglutinin, *Vicia villosa*

W

W	Tryptophan
WBA	Agglutinin, Wax Bean
WFA	Agglutinin, *Wisteria floribunda*
WGA	Agglutinin, Wheat Germ; Agglutinin, *Triticum vulgaris*

X

X	Linker
Xan	Xanthyl
XIAP	X-Linked Inhibitor of Apoptosis Protein
XIP	Exchange Inhibitory Peptide
XP	Xenopsin-Related Peptide

Y

Y	Tyrosine

Z

Z, Cbz	Benzyloxycarbonyl; Carbobenzoxy

β

βAPP	Amyloid β Precursor Protein
βNA	β-Naphthylamide

Part 7. Indexes

Index A. Peptide Sequences

H

I

M

User's Guide pp. 1-2, Glossary/Indexes pp. 963-1063

Index B. General Index

Compound Names, Synonyms & Derivatives

User's Guide pp. 1-2, Glossary/Indexes pp. 963-1063

Amino Acid Mixtures · 733, 734
Amino Acid Oxidase
    Apo-D- · 43
    D- · 43
Aminobenzoic Acid · 734, 735
Aminobutyric Acid · 735
Aminocaproic Acid · 735
Aminoenkephalinase Inhibitor · 283
Aminohexanoic Acid · 735
Aminoisobutyric Acid · 735, 736
Aminooctanoic Acid · 612
Aminopeptidase A Inhibitor · 668
Aminopeptidase N Inhibitor · 663
Aminopeptidase Substrate · 216, 460, 717, 726, 727, 785, 834, 835, 836, 837, 847, 858, 887
    II · 216
Aminopeptidases Inhibitor · 668
Aminosuberic Acid · 736
Aminotrifluoromethylcoumarin Tetrapeptide Conjugate · 647
AMP-Activated Protein Kinase Substrate · 433
Amphibian Osmoregulatory Peptide · 341
Amphiregulin · 43, 964
Amphiregulin Epidermal Growth Factor · 43
Amylin · 209, 249, 433, 481, 482, 559, 584, 609, 683
Amyloglucosidase · 43, 147
Amyloid · 43, 229, 236, 239, 261, 262, 285, 289, 291, 292, 293, 294, 295, 328, 347, 349, 364, 370, 372, 375, 381, 386, 393, 413, 419, 420, 430, 431, 440, 446, 447, 448, 470, 475, 481, 485, 488, 494, 502, 512, 521, 529, 565, 613, 614, 653, 659, 662, 663, 668, 672, 673, 674, 964, 969, 971, 972
    A · 43, 229, 440, 662, 663
    A, Apo-Serum · 43
    A4 Precursor Protein · 239, 262, 289
    A4 Splitting Enzyme · 229
    A4-Generating Enzyme Substrate · 440, 662, 663
    Bri Protein · 328, 349, 370
    Bri Protein Precursor 277 · 328, 349, 370
    P · 43, 291, 292, 293, 294, 372, 375, 381, 413, 420, 431, 481, 529, 613, 659, 673, 674, 964, 971
    Peptide · 291, 292, 293, 294, 420, 659, 673, 674
    Protein · 372, 375, 431, 481, 964
    Protein Precursor · 431, 481, 964
    Protein Precursor 770 · 565
    Protein, Non-Aβ Component · 375
    α/A4 Precursor Protein 770 · 239, 262, 431, 494
    α-Protein · 236, 261, 285, 291, 292, 293, 372, 386, 393, 420, 470
    β Precursor Protein · 674, 972
    β/A4 Generating Enzyme Substrate · 662
    β/A4 Precursor Protein · 239, 262, 289, 448, 521, 565, 653, 674
    β/A4 Precursor Protein 770 · 289, 448, 521, 565, 653
    β-Protein · 239, 291, 292, 294, 295, 347, 364, 419, 420, 446, 447, 485, 488, 502, 512, 613, 614, 659, 668
Anantin · 319, 413
Anaphylatoxic Peptide C3a · 432
Anaphylatoxin C3a · 248, 432
Androgen Induced Growth Factor · 85, 964
Androgen Receptor Protein N-Terminal Peptide · 500, 964
Angiogenesis Inhibitor · 317, 550
Angiogenin · 43, 371, 664
Angiogenin Complementary Sequence · 664
Angiogenin C-Terminal Peptide · 371
Angiostatin · 43
Angiotensin · 44, 210, 244, 254, 271, 284, 285, 286, 295, 296, 297, 298, 299, 312, 353, 368, 378, 386, 395, 396, 398, 403, 414, 426, 431, 446, 452, 491, 510, 513, 517,

520, 525, 527, 555, 556, 557, 559, 610, 626, 628, 667, 674, 675, 676, 677, 686, 824, 895, 964
    Antagonist · 296
    AT2 Receptor Agonist · 628
    I · 210, 244, 254, 271, 284, 285, 286, 295, 296, 297, 298, 299, 312, 353, 378, 386, 396, 398, 403, 426, 431, 446, 510, 513, 525, 555, 556, 557, 559, 610, 626, 628, 667, 674, 675, 676, 677, 824
    I Converting Enzyme Inactivator · 525
    I Converting Enzyme Inhibitor · 254, 271, 353, 510, 676, 677
    I Converting Enzyme Substrate · 386, 396, 398, 403, 431, 513
    I/II · 296, 446, 555, 626, 667
    II · 244, 284, 285, 286, 296, 297, 298, 299, 312, 378, 426, 446, 555, 556, 557, 559, 610, 626, 628, 667, 674, 824
    II Agonist · 244, 296
    II Antagonist · 555, 556, 610
    II Antipeptide · 378
    II Heptapeptide · 284
    II Inhibitor · 555, 556, 610
    II Receptor Ligand · 628, 674
    II Selective Antagonist · 555, 556
    III · 284, 285, 426, 557, 667
    III Antagonist · 557
    III Antipeptide · 426
    III Selective Antagonist · 284
    IV · 667, 674
    Receptor Antagonist · 378
    Receptor Ligand · 675
Angiotensin Converting Enzyme · 44, 368, 395, 403, 414, 431, 452, 491, 510, 517, 520, 525, 527, 686, 895, 964
    Inactivator · 368, 525
    Inhibiting Peptide · 452
    Inhibitor · 368, 491, 510, 520, 686, 895
    Substrate · 395, 403, 431, 517, 527
Angiotensinogen · 44, 210, 298, 660, 675
Ankyrin · 343, 344, 555
Annexin · 44, 45
    I · 44
    II · 44
    III · 44
    IV · 44
    V · 44, 45
    VI · 45
Anorectic Peptide, New · 450
Anorexigenic Peptide · 355
Anserine, L- · 235
Antagonist · 240, 276, 282, 394, 464, 492, 524, 544, 740, 778, 790, 794, 796, 797, 799, 808, 822, 860, 883, 888, 890, 898, 903, 906
    D · 276
    G · 282
Anthopleura Toxin A · 45, 426
Anthopleurin · 426, 964
    A · 426
Antho-RF Amide · 352, 353
    Neuropeptide · 353
Antho-RN Amide · 455
Antho-RP Amide
    I · 349, 470
    II · 349
Antho-RW Amide
    I · 353, 368
    II · 353
Anthranilyl-HIV Protease Substrate · 208, 209, 557
    III · 557
    IV · 209
    V · 208, 209

O · 598, 620
Sulfoxide · 619
-Thr · 612, 642, 643
Enkephalinamide · 234, 401, 598, 603, 604, 605, 606,
615, 620, 629, 643, 644
Enkephalinase
Inhibitor · 401, 472, 520
Substrate · 217, 234, 520
eNOS Blocking Peptide · 554
Entero-Hylambatin · 312
Entero-Kassinin · 302
Enterokinase Substrate · 391, 853
Enterostatin · 244, 664
Enterotoxin B · 76
Eosinophil Chemotactic Factor Of Anaphylaxis · 659
Eosinophil Colony Stimulating Factor · 124, 966
Eosinophil Differentiating Factor · 124, 966
Eosinophilotactic Peptide · 234, 659
Eosinophilotactic Tetrapeptide · 658, 659
Eotaxin · 76, 142, 416, 684, 966
C-C Chemokine · 76
II C-C Chemokine · 76
III C-C Chemokine · 76
Receptor Ligand · 684
Epiamastatin · 668
Epidermal Growth Factor · 76, 77, 144, 267, 291, 316,
332, 343, 479, 480, 653, 684, 685, 703, 966, 967
Long · 77, 685
nm- · 684
-R Fragment · 685, 703
rh- · 684
α- · 685
Epidermal Growth Factor Receptor · 77, 144, 267, 291,
316, 479, 480, 653, 685, 966
Extracellular Domain · 77
Kinase Inhibitor · 653
Kinase Substrate · 144
Peptide · 316, 685
Thr669 Protein Kinase Substrate · 479
Epidermal Mitosis Inhibiting Pentapeptide · 352, 376
Epidermal Receptor Factor
Receptor Tyrosine Kinase Inhibitor · 653
Epinephrine · 77
Epithelial Neutrophil Activating Peptide 78 · 77, 685
Epoxomicin · 447
Epstein Barr Virus Antigen · 77
Epstein Barr Virus Early Antigen · 78
D · 78
R Complex p17 · 78
Epstein Barr Virus Membrane Protein gp350/250 · 78
Epstein Barr Virus Nuclear Antigen · 78
I · 78
Epstein Barr Virus Viral Capsid Antigen · 78
125 · 78
160 · 78
Erabutoxin
A · 78
B · 78
ERK Kinase Substrate · 244
ERT Protein Kinase Substrate · 479
Erythroagglutinin · 23, 24
Erythromycin Resistance Peptide · 497
Erythrophore Concentrating Hormone · 455
Erythropoietin · 78, 246, 641, 685, 966
Mimetic Peptide Sequence 20 · 641
Natural · 78
Soluble Receptor · 78
Eskine · 186
Esterase Inhibitor · 675

Estrogen Receptor · 79, 229, 287, 325, 338, 384, 506,
602, 966
β · 384, 602
β Immunizing Peptide · 602
α N-Terminal Peptide · 287, 506
α Peptide · 384
β C-Terminal Peptide · 325, 338
β N-Terminal Peptide · 229
ETA-Selective Antagonist · 472
ETB Receptor Agonist · 338, 581
ETB Selective Antagonist · 473
Exchange Inhibitory Peptide · 257, 972
Excision Repair Cross Complement Group
III · 79
VI · 79
Excitatory AA
Antagonist · 377
Receptor Ligand · 685
Excitatory Peptide · 393
Exendin · 307, 378, 430, 438, 588
III · 438
IV · 378, 430
Exodus · 6, 140
II · 6, 140
III · 140
III/ELC · 140
Exorphin C · 635
Experimental Allergenic Encephalitogenic Peptide · 685
Experimental Allergic Encephalitogenic Peptide · 528
Experimental Allergic Encephalogenic Peptide · 528
Experimental Allergic Encephalomyelitis Inducing Peptide ·
439, 500, 625, 659, 660
Experimental Autoimmune Encephalomyelitis
Complementary Peptide · 663
Eye Derived Growth Factor I · 86
EZMix™ · 200, 692, 693

F

F7 · 587
F-8-F · 521, 522
Factor
B · 64, 79
D · 64, 79
H · 79
I · 79, 80, 136, 175, 747
II · 79, 80, 136, 175
IX · 79
P · 79
V · 79, 174, 175, 205, 471, 517, 591
VII · 79, 175, 205, 471, 517
VIIa Substrate · 517
VIIa Tf Substrate · 471
X · 79, 80, 175, 288, 346, 347, 352, 369, 411, 444, 675
X Gla-Domainless · 80
Xa Inhibitor · 411
Xa Substrate · 352, 444
XI · 80, 288, 346, 347, 369
XIa Substrate · 369
XII · 80, 288, 346, 347
XIIa Substrate · 346, 347
XIII · 80, 288
XIIIa · 288
α- · 595
FADD · 80
FAGLA · 406
FAGPA · 411
FALGPA · 460
FAPGG · 517

N

T